Who's Who in Frontier Science and Technology

Biographical Reference Works
Published by Marquis Who's Who

Who's Who in America

Who's Who in America supplements:

Who's Who in America Index by Geographic Location and Professional Area

Who's Who in America College Alumni Directory

Who's Who in America Birthdate Index

Who's Who in the World

Who Was Who in America

Historical Volume (1607-1896)

Volume I (1897-1942)

Volume II (1943-1950)

Volume III (1951-1960)

Volume IV (1961-1968)

Volume V (1969-1973)

Volume VI (1974-1976)

Volume VII (1977-1981)

Who Was Who in American History—Arts and Letters

Who Was Who in American History—The Military

Who Was Who in American History—Science and Technology

Who's Who in the Midwest

Who's Who in the East

Who's Who in the South and Southwest

Who's Who in the West

Who's Who in Frontier Science and Technology

Who's Who of American Women

Who's Who in Finance and Industry

Who's Who in American Law

World Who's Who in Science

Directory of Women in Marquis Who's Who Publications

Marquis Who's Who Publications/Index to All Books 1984

Volume 1: Alphabetic

Volume 2: Geographic

Professional Publications
from Marquis Who's Who

Directory of Medical Specialists

Directory of Professionals and Resources in Cancer

Directory of Professionals and Resources in Rehabilitation

Biographical Directory of the Computer Graphics Industry

Biographical Directory of Online Professionals

Who's Who in Frontier Science and Technology®

1st edition

1984-1985

MARQUIS

Who's Who

Marquis Who's Who, Inc.
200 East Ohio Street
Chicago, Illinois 60611 U.S.A.

 For information, address Marquis Who's Who Incorporated, 200 East Ohio Street, Chicago, Illinois 60611.

Library of Congress Catalog Card Number 82-82015
International Standard Book Number 0-8379-5701-X
Product Code Number 030290

Manufactured in the United States of America

Table of Contents

Preface

The first edition of *Who's Who in Frontier Science and Technology* meets the need for biographical information on individuals engaged in the forefront fields of science. The volume includes approximately 16,500 scientists and technologists who are currently working in North America in the frontier areas of their respective specialties.

The biographical sketches include the content elements found in *Who's Who in America,* including vital statistics, education, career history, awards, publications, and memberships. In addition, using a computer-coded list of frontier fields and subspecialties, each biographee has provided his or her subspecialties of science and a description of current research or work interest. This narrative statement of current work enables the biographee to describe narrower subspecialties than those covered in the list. The information on specialties is also found in the Index of Fields and Subspecialties, in which biographees are listed under approximately 350 subspecialties arranged within 24 major fields.

The challenge of defining frontier areas of science was met by going to the scientific community for current descriptions of fields at the cutting edge of science and technology. Research by Marquis staff resulted in a two-fold general description of frontier science that included both new directions in traditional fields of research, such as genetic engineering and gravitational biology, and work using the most advanced technology, like lasers and computers. Based on these descriptions, Marquis researchers developed a working list of frontier topics covering fields from agriculture to zoology. This working list was sent to 1,500 scientists, including editors of scholarly scientific journals, for additions, deletions, and other changes and comments. Response from the scientists was enthusiastic and strong, indicating an awareness of the need for the book as a new reference tool. The scientists' comments were embodied in the expanded final list of frontier subspecialties, which became the basis for definition of frontiers of science in each field.

The list was sent to each biographee, who was asked to select one or two subspecialties that reflected personal research or work interest. Thus, the comprehensive Index and the sketches contain two subspecialty entries for many listees in *Who's Who in Frontier Science and Technology.* If a subspecialty was found in more than one field of science, the name of the field was appended to the subspecialty description in the sketch, as in Biomass (agriculture) and Biomass (energy science and technology). Biographees with subspecialties not encompassed in the list are found in the Index at the end of each major field under the category "Other."

The Index serves two important reference functions. It provides a current listing of the frontier fields of science and technology, as defined by those within the fields. It also allows additional access to sketch information beyond biographee name; the user can find practitioners within specific new fields of science and technology.

The distinguished Consultants and members of the Board of Advisors played an important role in the development of *Who's Who in Frontier Science and Technology,* especially in their contributions to the list of fields and subspecialties and in their nomination of individuals for inclusion in the volume. Consultants Dean Bruno A. Boley and Dr. Lawrence D. Grouse worked closely with the Marquis editor in the critical early stages of book development. Advisors provided valuable feedback from colleagues in science during the course of book production.

As in all Marquis Who's Who biographical volumes, the principle of current reference value determined selection of biographees. Reference interest is based on either position of responsibility or noteworthy achievement. Those in high positions related to frontiers of science include many eminent scientists. The noteworthy achievers are not all familiar names, since much advanced scientific research is being conducted by younger practitioners who are not yet known outside their own fields.

In most instances, biographees furnished their own data and reviewed the sketches to assure accurate, current information. In some cases where individuals failed to supply needed information, Marquis staff members compiled the data through independent research.

Marquis Who's Who editors exercise diligent care in the preparation of each biographical sketch for publication. Despite precautions, errors do occasionally occur. Users of this directory are asked to report such errors to the publisher for correction in a later edition.

The editors would welcome comments and suggestions from scientists and other users on topics encompassed in the frontiers of science. Such comments will assist the editors in describing the fields of frontier science in future editions.

Adele Hast, Ph.D.
Editor-in-Chief

Board of Advisors

Marquis Who's Who gratefully acknowledges the contribution of the following distinguished individuals to the first edition of *Who's Who in Frontier Science and Technology.* Members of the Board of Advisors have nominated outstanding scientists and technologists for inclusion in the book and have made themselves available for review and evaluation of contents. However, the Board of Advisors, either collectively or individually, is in no way responsible for the selection of names or for the accuracy of the information in this volume.

Consultants

Bruno A. Boley, Sc.D.
Dean, Technological Institute
Northwestern University

Lawrence D. Grouse, M.D., Ph.D.
Senior Medical Editor
Lifetime Television Network

Advisors

George N. Aagaard, M.D.
Professor, Medicine and Pharmacology
University of Washington

Robert E. Boyer, Ph.D.
Dean, College of Natural Sciences
University of Texas at Austin

Joseph F. Bunnett, Ph.D.
Editor, *Accounts of Chemical Research*
University of California, Santa Cruz

Theodore Cooper, M.D., Ph.D.
Executive Vice President
The Upjohn Company

James F. Danielli, Ph.D., F.R.S.
Chief Editor
Journal of Theoretical Biology

Charles H. Davis, Ph.D.
Dean, Graduate School of Library and Information Science
University of Illinois at Urbana-Champaign

Dorothy E. Denning, Ph.D.
SRI International
Menlo Park, California

Jay M. Enoch, Ph.D.
Dean, School of Optometry
University of California, Berkeley

Herman Feshbach, Ph.D.
Head, Department of Physics
Massachusetts Institute of Technology

Walter Freiberger, Ph.D.
Managing Editor
Quarterly of Applied Mathematics
Brown University

Yuan-Cheng B. Fung, Ph.D.
Department of Bioengineering & Applied Mechanics
University of California, San Diego

Pierre M. Galletti, M.D., Ph.D.
Vice President, Division of Biology and Medicine
Brown University

William R. Kuhn, Ph.D.
Chairman, Department of Atmospheric & Oceanic Science
The University of Michigan

Lee N. Miller, Ph.D.
Managing Editor
Ecological Society of America
Cornell University

Jeremiah P. Ostriker, Ph.D.
Director, Princeton University Observatory

Robert G. Petersdorf, M.D.
Vice Chancellor, Health Sciences
University of California San Diego

Richard F. Thompson, Ph.D.
Professor, Department of Psychology
Stanford University

Charles A. Wert, Ph.D.
Head, Department of Metallurgy and Mining Engineering
University of Illinois at Urbana-Champaign

Sylvan H. Wittwer, Ph.D.
Director, Agricultural Experiment Station
Michigan State University

Standards of Admission

Selection of biographees for *Who's Who in Frontier Science and Technology* is determined by reference interest, based on involvement in research or other work at the frontiers of science. Reference interest derives from either or both of two factors: (1) incumbency in a defined position or (2) attainment of a significant level of achievement.

Admission based on position includes the following examples:

Heads of selected university research institutes and programs

Directors of selected independent research institutes

Directors of selected governmental research centers

Heads of selected industrial research laboratories

Principals of selected consulting organizations

Deans of selected academic divisions

Selected members of honorary organizations in science, such as the National Academy of Sciences, the National Academy of Engineering, and the Institute of Medicine

Recipients of selected national and international awards in science

Admission by the factor of significant achievement is based on objective criteria for measuring accomplishments within the scientific profession.

Key to Information in this Directory

❶ **DAVIES, STEPHEN FRANCIS,** ❷ physics educator; ❸ b. Evanston, Ill., Oct. 8, 1930; ❹ s. Paul Harwell and Mary Louise (Ryan) D.; ❺ m. Elizabeth C. Swan, June 10, 1956; ❻ children: Robert Dwight, Mary Adele, William Fremont. ❼ B.S., Princeton U., 1953; Ph.D., MIT, 1959. ❽ Cert. safety profl., Mass. ❾ Research asst. radiation lab. physics dept. MIT, Cambridge, 1959-61; assoc. physicist Battelle Meml. Inst., Columbus, Ohio, 1961-64; asst. prof. physics Columbia U., N.Y.C., 1965-69, assoc. prof., 1969-75, prof. radiation physics, 1975—; vis. scientist Am. Inst. Physics, 1978; vis. prof. Rensselaer Poly. Inst., Troy, N.Y., 1981-82; chmn. commn. radiation physics Internat. Union Pure and Applied Physics, 1979-81; mem. Nat. Sci. Bd., 1982—. ❿ Editor: Radiation Testing Standards, 1976; contbr. articles to profl. jours. ⓫ Trustee Cornell U.; bd. dirs. Fairfield County chpt. Am. Cancer Soc., Greenwich Civic Orch. ⓬ Served with USN, 1948-49. ⓭ Recipient Research prize N.Y. Acad. Scis., 1980; fellow Sloan Found., 1974-76, Guggenheim Found., 1981. ⓮ Mem. AAAS, Am. Phys. Soc., Nat. Acad. Scis., Am. Nuclear Soc., Sigma Xi. ⓯ Republican. ⓰ Episcopalian. ⓱ Clubs: Down Town (N.Y.C.); Rolling Meadows Country. ⓲ Lodge: Masons. ⓳ Subspecialties: Radiation physics; Nuclear physics. ⓴ Current work: Research in nondestructive testing. ㉑ Home: 411 Wolf Pit Rd Greenwich CT 06830 ㉒ Office: Dept Physics Columbia U New York NY 10002

KEY

❶ Name
❷ Occupation
❸ Vital Statistics
❹ Parents
❺ Marriage
❻ Children
❼ Education
❽ Professional certifications
❾ Career
❿ Writings and special achievements
⓫ Civic and political activities
⓬ Military record
⓭ Awards and fellowships
⓮ Professional and association memberships
⓯ Political affiliation
⓰ Religion
⓱ Clubs
⓲ Lodges
⓳ Subspecialties
⓴ Current work
㉑ Home address
㉒ Office address

The biographical listings in *Who's Who in Frontier Science and Technology* are arranged in alphabetical order according to the first letter of the last name of the biographee. Each sketch is presented in a uniform order as in the sample sketch above. The many abbreviations used in the sketches are explained in the Table of Abbreviations.

Table of Abbreviations

The following abbreviations and symbols are frequently used in this Directory.

A.A. Associate in Arts
AAAL American Academy of Arts and Letters
AAAS American Association for the Advancement of Science
AAHPER Alliance for Health, Physical Education and Recreation
AAU Amateur Athletic Union
AAUP American Association of University Professors
AAUW American Association of University Women
A.B. Arts, Bachelor of
AB Alberta
ABA American Bar Association
ABC American Broadcasting Company
AC Air Corps
acad. academy, academic
acct. accountant
acctg. accounting
ACDA Arms Control and Disarmament Agency
ACLU American Civil Liberties Union
ACP American College of Physicians
ACS American College of Surgeons
ADA American Dental Association
a.d.c. aide-de-camp
adj. adjunct, adjutant
adj. gen. adjutant general
adm. admiral
adminstr. administrator
adminstrn. administration
adminstrv. administrative
ADP automatic data processing
adv. advocate, advisory
advt. advertising
A.E. Agricultural Engineer (for degrees only)
A.E. and P. Ambassador Extraordinary and Plenipotentiary
AEC Atomic Energy Commission
aero. aeronautical, aeronautic
aerodyn. aerodynamic
AFB Air Force Base
AFL-CIO American Federation of Labor and Congress of Industrial Organizations
AFTRA American Federation TV and Radio Artists
agr. agriculture
agrl. agricultural
agt. agent
AGVA American Guild of Variety Artists
agy. agency
A&I Agricultural and Industrial
AIA American Institute of Architects
AIAA American Institute of Aeronautics and Astronautics
AID Agency for International Development
AIEE American Institute of Electrical Engineers
AIM American Institute of Management
AIME American Institute of Mining, Metallurgy, and Petroleum Engineers
AK Alaska
AL Alabama
ALA American Library Association
Ala. Alabama
alt. alternate
Alta. Alberta
A&M Agricultural and Mechanical
A.M. Arts, Master of
Am. American, America
AMA American Medical Association
A.M.E. African Methodist Episcopal
Amtrak National Railroad Passenger Corporation
AMVETS American Veterans of World War II, Korea, Vietnam
anat. anatomical
ann. annual
ANTA American National Theatre and Academy
anthrop. anthropological
AP Associated Press
APO Army Post Office
Apr. April
apptd. appointed
apt. apartment
AR Arkansas
ARC American Red Cross
archeol. archeological
archtl. architectural
Ariz. Arizona
Ark. Arkansas
ArtsD. Arts, Doctors of
arty. artillery
ASCAP American Society of Composers, Authors and Publishers
ASCE American Society of Civil Engineers
ASHRAE American Society of Heating, Refrigeration, and Air Conditioning Engineers
ASME American Society of Mechanical Engineers

assn. association
assoc. associate
asst. assistant
ASTM American Society for Testing and Materials
astron. astronomical
astrophys. astrophysical
ATSC Air Technical Service Command
AT&T American Telephone & Telegraph Company
atty. attorney
AUS Army of the United States
Aug. August
aux. auxiliary
Ave. Avenue
AVMA American Veterinary Medical Association
AZ Arizona

B. Bachelor
b. born
B.A. Bachelor of Arts
B.Agr. Bachelor of Agriculture
Balt. Baltimore
Bapt. Baptist
B. Arch. Bachelor of Architecture
B.A.S. Bachelor of Agricultural Science
B.B.A. Bachelor of Business Administration
BBC British Broadcasting Corporation
B.C., BC British Columbia
B.C.E. Bachelor of Civil Engineering
B.Chir. Bachelor of Surgery
B.C.L. Bachelor of Civil Law
B.C.S. Bachelor of Commercial Science
B.D. Bachelor of Divinity
bd. board
B.E. Bachelor of Education
B.E.E. Bachelor of Electrical Engineering
B.F.A. Bachelor of Fine Arts
bibl. biblical
bibliog. bibliographical
biog. biographical
biol. biological
B.J. Bachelor of Journalism
Bklyn. Brooklyn
B.L. Bachelor of Letters
bldg. building
B.L.S. Bachelor of Library Science
Blvd. Boulevard
bn. battalion
B.&O.R.R. Baltimore & Ohio Railroad
bot. botanical
B.P.E. Bachelor of Physical Education
br. branch
B.R.E. Bachelor of Religious Education
brig. gen. brigadier general
Brit. British, Brittanica
Bros. Brothers
B.S. Bachelor of Science
B.S.A. Bachelor of Agricultural Science
B.S.D. Bachelor of Didactic Science
B.S.T. Bachelor of Sacred Theology
B.Th. Bachelor of Theology
bull. bulletin
bur. bureau
bus. business
B.W.I. British West Indies

CA California

CAA Civil Aeronautics Administration
CAB Civil Aeronautics Board
Calif. California
C.Am. Central America
Can. Canada, Canadian
CAP Civil Air Patrol
capt. captain
CARE Cooperative American Relief Everywhere
Cath. Catholic
cav. cavalry
CBC Canadian Broadcasting Company
CBI China, Burma, India Theatre of Operations
CBS Columbia Broadcasting System
CCC Commodity Credit Corporation
CCNY City College of New York
CCU Cardiac Care Unit
CD Civil Defense
C.E. Corps of Engineers, Civil Engineer (in firm's name only or for degree)
cen. central (To be used for court system only)
CENTO Central Treaty Organization
CERN European Organization of Nuclear Research
cert. certificate, certification, certified
CETA Comprehensive Employment Training Act
CFL Canadian Football League
ch. church
Ch.D. Doctor of Chemistry
chem. chemical
Chem.E. Chemical Engineer
Chgo. Chicago
chirurg. chirurgical
chmn. chairman
chpt. chapter
CIA Central Intelligence Agency
CIC Counter Intelligence Corps
Cin. Cincinnati
cir. circuit
Cleve. Cleveland
climatol. climatological
clin. clinical
clk. clerk
C.L.U. Chartered Life Underwriter
C.M. Master in Surgery
C.&N.W.Ry. Chicago & Northwestern Railway
CO Colorado
Co. Company
COF Catholic Order of Foresters
C. of C. Chamber of Commerce
col. colonel.
coll. college
Colo. Colorado
com. committee
comd. commanded
comdg. commanding
comdr. commander
comdt. commandant
commd. commissioned
comml. commercial
commn. commission
commr. commissioner
condr. conductor
Conf. Conference
Congl. Congregational, Congressional
Conglist. Congregationalist
Conn. Connecticut
cons. consultant, consulting
consol. consolidated
constl. constitutional
constn. constitution
constrn. construction
contbd. contributed
contbg. contributing
contbn. contribution
contbr. contributor
Conv. Convention
coop. cooperative
CORDS Civil Operations and Revolutionary Development Support
CORE Congress of Racial Equality
corp. corporation, corporate
corr. correspondent, corresponding, correspondence
C.&O.Ry. Chesapeake & Ohio Railway
C.P.A. Certified Public Accountant
C.P.C.U. Chartered property and casualty underwriter
C.P.H. Certificate of Public Health
cpl. corporal
CPR Cardio-Pulmonary Resuscitation
C.P.Ry. Canadian Pacific Railway
C.S. Christian Science
C.S.B. Bachelor of Christian Science
CSC Civil Service Commission
C.S.D. Doctor of Christian Science
CT Connecticut
ct. court
ctr. center
CWS Chemical Warfare Service
C.Z. Canal Zone

d. daughter
D. Doctor
D.Agr. Doctor of Agriculture
DAR Daughters of the American Revolution
dau. daughter
DAV Disabled American Veterans
D.C., DC District of Columbia
D.C.L. Doctor of Civil Law
D.C.S. Doctor of Commercial Science
D.D. Doctor of Divinity
D.D.S. Doctor of Dental Surgery
DE Delaware
dec. deceased
Dec. December
def. defense
Del. Delaware
del. delegate, delegation
Dem. Democrat, Democratic
D.Eng. Doctor of Engineering
denom. denomination, denominational
dep. deputy
dept. department
dermatol. dermatological
desc. descendant
devel. development, developmental
D.F.A. Doctor of Fine Arts
D.F.C. Distinguished Flying Cross
D.H.L. Doctor of Hebrew Literature
dir. director
dist. district
distbg. distributing
distbn. distribution
distbr. distributor
disting. distinguished
div. division, divinity, divorce
D.Litt. Doctor of Literature
D.M.D. Doctor of Medical Dentistry
D.M.S. Doctor of Medical Science
D.O. Doctor of Osteopathy
D.P.H. Diploma in Public Health
D.R. Daughters of the Revolution
Dr. Drive, Doctor
D.R.E. Doctor of Religious Education
Dr.P.H. Doctor of Public Health, Doctor of Public Hygiene
D.S.C. Distinguished Service Cross
D.Sc. Doctor of Science
D.S.M. Distinguished Service Medal
D.S.T. Doctor of Sacred Theology
D.T.M. Doctor of Tropical Medicine
D.V.M. Doctor of Veterinary Medicine
D.V.S. Doctor of Veterinary Surgery

E. East
ea. eastern (use for court system only)
E. and P. Extraordinary and Plenipotentiary
Eccles. Ecclesiastical
ecol. ecological
econ. economic
ECOSOC Economic and Social Council (of the UN)
E.D. Doctor of Engineering
ed. educated
Ed.B. Bachelor of Education
Ed.D. Doctor of Education
edit. edition
Ed.M. Master of Education
edn. education
ednl. educational
EDP electronic data processing
Ed.S. Specialist in Education
E.E. Electrical Engineer (degree only)
E.E. and M.P. Envoy Extraordinary and Minister Plenipotentiary
EEC European Economic Community
EEG electroencephalogram
EEO Equal Employment Opportunity

EEOC Equal Employment Opportunity Commission
EKG electrocardiogram
E.Ger. German Democratic Republic
elec. electrical
electrochem. electrochemical
electrophys. electrophysical
elem. elementary
E.M. Engineer of Mines
ency. encyclopedia
Eng. England
engr. engineer
engring. engineering
entomol. entomological
environ. environmental
EPA Environmental Protection Agency
epidemiol. epidemiological
Episc. Episcopalian
ERA Equal Rights Amendment
ERDA Energy Research and Development Administration
ESEA Elementary and Secondary Education Act
ESL English as Second Language
ESSA Environmental Science Services Administration
ethnol. ethnological
ETO European Theatre of Operations
Evang. Evangelical
exam. examination, examining
exec. executive
exhbn. exhibition
expdn. expedition
expn. exposition
expt. experiment
exptl. experimental

F.A. Field Artillery
FAA Federal Aviation Administration
FAO Food and Agriculture Organization (of the UN)
FBI Federal Bureau of Investigation
FCA Farm Credit Administration
FCC Federal Communication Commission
FCDA Federal Civil Defense Administration
FDA Food and Drug Administration
FDIA Federal Deposit Insurance Administration
FDIC Federal Deposit Insurance Corporation
F.E. Forest Engineer
FEA Federal Energy Administration
Feb. February
fed. federal
fedn. federation
FERC Federal Energy Regulatory Commission
fgn. foreign
FHA Federal Housing Administration
fin. financial, finance
FL Florida
Fla. Florida
FMC Federal Maritime Commission
FOA Foreign Operations Administration
found. foundation
FPC Federal Power Commission
FPO Fleet Post Office
frat. fraternity
FRS Federal Reserve System
FSA Federal Security Agency
Ft. Fort
FTC Federal Trade Commission

G-1 (or other number) Division of General Staff
Ga., GA Georgia
GAO General Accounting Office
gastroent. gastroenterological
GATT General Agreement of Tariff and Trades
gen. general
geneal. genealogical
geod. geodetic
geog. geographic, geographical
geol. geological
geophys. geophysical
gerontol. gerontological
G.H.Q. General Headquarters
G.N. Ry. Great Northern Railway
gov. governor
govt. government
govtl. governmental
GPO Government Printing Office
grad. graduate, graduated
GSA General Services Administration
Gt. Great
GU Guam
gynecol. gynecological

hdqrs. headquarters
HEW Department of Health, Education and Welfare
H.H.D. Doctor of Humanities
HHFA Housing and Home Finance Agency
HHS Department of Health and Human Services
HI Hawaii
hist. historical, historic
H.M. Master of Humanics
homeo. homeopathic
hon. honorary, honorable
Ho. of Dels. House of Delegates
Ho. of Reps. House of Representatives
hort. horticultural
hosp. hospital
HUD Department of Housing and Urban Development
Hwy. Highway
hydrog. hydrographic

IA Iowa
IAEA International Atomic Energy Agency
IBM International Business Machines Corporation
IBRD International Bank for Reconstruction and Development
ICA International Cooperation Administration
ICC Interstate Commerce Commission
ICU Intensive Care Unit
ID Idaho
IEEE Institute of Electrical and Electronics Engineers
IFC International Finance Corporation
IGY International Geophysical Year
IL Illinois
Ill. Illinois
illus. illustrated
ILO International Labor Organization
IMF International Monetary Fund
IN Indiana
Inc. Incorporated
ind. independent
Ind. Indiana
Indpls. Indianapolis
indsl. industrial
inf. infantry
info. information
ins. insurance
insp. inspector
insp. gen. inspector general
inst. institute
instl. institutional
instn. institution
instr. instructor
instrn. instruction
intern. international
intro. introduction
IRE Institute of Radio Engineers
IRS Internal Revenue Service
ITT International Telephone & Telegraph Corporation

JAG Judge Advocate General
JAGC Judge Advocate General Corps
Jan. January
Jaycees Junior Chamber of Commerce
J.B. Jurum Baccolaureus
J.C.B. Juris Canoni Baccalaureus
J.C.D. Juris Canonici Doctor, Juris Civilis Doctor
J.C.L. Juris Canonici Licentiatus
J.D. Juris Doctor
j.g. junior grade
jour. journal
jr. junior
J.S.D. Juris Scientiae Doctor
J.U.D. Juris Utriusque Doctor
jud. judicial

Kans. Kansas
K.C. Knights of Columbus
K.P. Knights of Pythias
KS Kansas

K.T. Knight Templar
Ky., KY Kentucky

La., LA Louisiana
lab. laboratory
lang. language
laryngol. laryngological
LB Labrador
lectr. lecturer
legis. legislation, legislative
L.H.D. Doctor of Humane Letters
L.I. Long Island
lic. licensed, license
L.I.R.R. Long Island Railroad
lit. literary, literature
Litt.B. Bachelor of Letters
Litt.D. Doctor of Letters
LL.B. Bachelor of Laws
LL.D. Doctor of Laws
LL.M. Master of Laws
Ln. Lane
L.&N.R.R. Louisville & Nashville Railroad
L.S. Library Science (in degree)
lt. lieutenant
Ltd. Limited
Luth. Lutheran
LWV League of Women Voters

m. married
M. Master
M.A. Master of Arts
MA Massachusetts
mag. magazine
M.Agr. Master of Agriculture
maj. major
Man. Manitoba
Mar. March
M.Arch. Master in Architecture
Mass. Massachusetts
math. mathematics, mathematical
MATS Military Air Transport Service
M.B. Bachelor of Medicine
MB Manitoba
M.B.A. Master of Business Administration
MBS Mutual Broadcasting System
M.C. Medical Corps
M.C.E. Master of Civil Engineering
mcht. merchant
mcpl. municipal
M.C.S. Master of Commercial Science
M.D. Doctor of Medicine
Md, MD Maryland
M.Dip. Master in Diplomacy
mdse. merchandise
M.D.V. Doctor of Veterinary Medicine
M.E. Mechanical Engineer (degree only)
ME Maine
M.E.Ch. Methodist Episcopal Church
mech. mechanical
M.Ed. Master of Education
med. medical
M.E.E. Master of Electrical Engineering
mem. member
meml. memorial
merc. mercantile
met. metropolitan
metall. metallurgical
Met.E. Metallurgical Engineer
meteorol. meteorological
Meth. Methodist
Mex. Mexico
M.F. Master of Forestry
M.F.A. Master of Fine Arts
mfg. manufacturing
mfr. manufacturer
mgmt. management
mgr. manager
M.H.A. Master of Hospital Administration
M.I. Military Intelligence
MI Michigan
Mich. Michigan
micros. microscopic, microscopical
mid. middle (use for Court System only)
mil. military
Milw. Milwaukee
mineral. mineralogical
Minn. Minnesota
Miss. Mississippi
MIT Massachusetts Institute of Technology
mktg. marketing
M.L. Master of Laws
MLA Modern Language Association
M.L.D. Magister Legnum Diplomatic
M.Litt. Master of Literature
M.L.S. Master of Library Science
M.M.E. Master of Mechanical Engineering
MN Minnesota
mng. managing
Mo., MO Missouri
moblzn. mobilization
Mont. Montana
M.P. Member of Parliament
M.P.E. Master of Physical Education
M.P.H. Master of Public Health
M.P.L. Master of Patent Law
Mpls. Minneapolis
M.R.E. Master of Religious Education
M.S. Master of Science
MS, Ms. Mississippi
M.Sc. Master of Science
M.S.F. Master of Science of Forestry
M.S.T. Master of Sacred Theology
M.S.W. Master of Social Work
MT Montana
Mt. Mount
MTO Mediterranean Theatre of Operations
mus. museum, musical
Mus.B. Bachelor of Music
Mus.D. Doctor of Music
Mus.M. Master of Music
mut. mutual
mycol. mycological

N. North
NAACP National Association for the Advancement of Colored People
NACA National Advisory Committee for Aeronautics
NAD National Academy of Design
N.Am. North America
NAM National Association of Manufacturers
NAPA National Association of Performing Artists
NAREB National Association of Real Estate Boards
NARS National Archives and Record Service
NASA National Aeronautics and Space Administration
nat. national
NATO North Atlantic Treaty Organization
NATOUSA North African Theatre of Operations
nav. navigation
N.B., NB New Brunswick
NBC National Broadcasting Company
N.C., NC North Carolina
NCCJ National Conference of Christians and Jews
N.D., ND North Dakota
NDEA National Defense Education Act
NE Nebraska
NE Northeast
NEA National Education Association
Nebr. Nebraska
NEH National Endowment for Humanities
neurol. neurological
Nev. Nevada
NF Newfoundland
NFL National Football League
Nfld. Newfoundland
N.G. National Guard
N.H. NH New Hampshire
NHL National Hockey League
NIH National Institutes of Health
NIMH National Institute of Mental Health
N.J., NJ New Jersey
NLRB National Labor Relations Board
NM New Mexico
N.Mex. New Mexico
No. Northern
NOAA National Oceanographic and Atmospheric Administration
NORAD North America Air Defense
NOW National Organization for Women
Nov. November
N.P.Ry. Northern Pacific Railway
nr. near
NRC National Research Council
N.S., NS Nova Scotia
NSC National Security Council

NSF National Science Foundation
N.T. New Testament
NT Northwest Territories
numis. numismatic
NV Nevada
NW Northwest
N.W.T. Northwest Territories
N.Y., NY New York
N.Y.C. New York City
NYU New York University
N.Z. New Zealand

OAS Organization of American States
ob-gyn obstetrics-gynecology
obs. observatory
obstet. obstetrical
O.D. Doctor of Optometry
OECD Organization of European Cooperation and Development
OEEC Organization of European Economic Cooperation
OEO Office of Economic Opportunity
ofcl. official
OH Ohio
OK Oklahoma
Okla. Oklahoma
ON Ontario
Ont. Ontario
ophthal. ophthalmological
ops. operations
OR Oregon
orch. orchestra
Oreg. Oregon
orgn. organization
ornithol. ornithological
OSHA Occupational Safety and Health Administration
OSRD Office of Scientific Research and Development
OSS Office of Strategic Services
osteo. osteopathic
otol. otological
otolaryn. otolaryngological

Pa., PA Pennsylvania
P.A. Professional Association
paleontol. paleontological
path. pathological
P.C. Professional Corporation
PE Prince Edward Island
P.E.I. Prince Edward Island (text only)
PEN Poets, Playwrights, Editors, Essayists and Novelists (international association)
penol. penological
P.E.O. women's organization (full name not disclosed)
pfc. private first class
PHA Public Housing Administration
pharm. pharmaceutical
Pharm.D. Doctor of Pharmacy
Pharm. M. Master of Pharmacy
Ph.B. Bachelor of Philosophy
Ph.D. Doctor of Philosophy
Phila. Philadelphia
philharm. philharmonic
philol. philological
philos. philosophical
photog. photographic
phys. physical
physiol. physiological
Pitts. Pittsburgh
Pkwy. Parkway
Pl. Place
P.&L.E.R.R. Pittsburgh & Lake Erie Railroad
P.O. Post Office
PO Box Post Office Box
polit. political
poly. polytechnic, polytechnical
PQ Province of Quebec
P.R., PR Puerto Rico
prep. preparatory
pres. president
Presbyn. Presbyterian
presdl. presidential
prin. principal
proc. proceedings
prod. produced (play production)
prodn. production
prof. professor
profl. professional
prog. progressive
propr. proprietor
pros. atty. prosecuting attorney
pro tem pro tempore
PSRO Professional Services Review Organization
psychiat. psychiatric
psychol. psychological
PTA Parent-Teachers Association
ptnr. partner
PTO Pacific Theatre of Operations, Parent Teacher Organization
pub. publisher, publishing, published
pub. public
publ. publication
pvt. private

quar. quarterly
q.m. quartermaster
Q.M.C. Quartermaster Corps.
Que. Quebec

radiol. radiological
RAF Royal Air Force
RCA Radio Corporation of America
RCAF Royal Canadian Air Force
RD Rural Delivery
Rd. Road
REA Rural Electrification Administration
rec. recording
ref. reformed
regt. regiment
regtl. regimental
rehab. rehabilitation
rep. representative
Rep. Republican
Res. Reserve
ret. retired
rev. review, revised
RFC Reconstruction Finance Corporation
RFD Rural Free Delivery
rhinol. rhinological
R.I., RI Rhode Island
R.N. Registered Nurse
roentgenol. roentgenological
ROTC Reserve Officers Training Corps
R.R. Railroad
Ry. Railway

s. son
S. South
SAC Strategic Air Command
SALT Strategic Arms Limitation Talks
S.Am. South America
san. sanitary
SAR Sons of the American Revolution
Sask. Saskatchewan
savs. savings
S.B. Bachelor of Science
SBA Small Business Administration
S.C., SC South Carolina
SCAP Supreme Command Allies Pacific
Sc.B. Bachelor of Science
S.C.D. Doctor of Commercial Science
Sc.D. Doctor of Science
sch. school
sci. science, scientific
SCLC Southern Christian Leadership Conference
SCV Sons of Confederate Veterans
S.D., SD South Dakota
SE Southeast
SEATO Southeast Asia Treaty Organization
sec. secretary
SEC Securities and Exchange Commission
sect. section
seismol. seismological
sem. seminary
s.g. senior grade
sgt. sergeant
SHAEF Supreme Headquarters Allied Expeditionary Forces
SHAPE Supreme Headquarters Allied Powers in Europe
S.I. Staten Island
S.J. Society of Jesus (Jesuit)
S.J.D. Scientiae Juridicae Doctor
SK Saskatchewan
S.M. Master of Science
So. Southern
soc. society
sociol. sociological
S.P. Co. Southern Pacific Company
spl. special

splty. specialty
Sq. Square
sr. senior
S.R. Sons of the Revolution
SS Steamship
SSS Selective Service System
St. Saint, Street
sta. station
stats. statistics
statis. statistical
S.T.B. Bachelor of Sacred Theology
stblzn. stabilization
S.T.D. Doctor of Sacred Theology
subs. subsidiary
SUNY State University of New York
supr. supervisor
supt. superintendent
surg. surgical
SW Southwest

TAPPI Technical Association of Pulp and Paper Industry
Tb Tuberculosis
tchr. teacher
tech. technical, technology
technol. technological
Tel.&Tel. Telephone & Telegraph
temp. temporary
Tenn. Tennessee
Ter. Territory
Terr. Terrace
Tex. Texas
Th.D. Doctor of Theology
theol. theological
Th.M. Master of Theology
TN Tennessee
tng. training
topog. topographical
trans. transaction, transferred
transl. translation, translated
transp. transportation
treas. treasurer
TV television
TVA Tennessee Valley Authority
twp. township
TX Texas
typog. typographical

U. University
UAW United Auto Workers
UCLA University of California at Los Angeles
UDC United Daughters of the Confederacy
U.K. United Kingdom
UN United Nations
UNESCO United Nations Educational, Scientific and Cultural Organization
UNICEF United Nations International Children's Emergency Fund
univ. university
UNRRA United Nations Relief and Rehabilitation Administration
UPI United Press International
U.P.R.R. United Pacific Railroad
urol. urological
U.S. United States
U.S.A. United States of America
USAAF United States Army Air Force
USAF United States Air Force
USAFR United States Air Force Reserve
USAR United States Army Reserve
USCG United States Coast Guard
USCGR United States Coast Guard Reserve
USES United States Employment Service
USIA United States Information Agency
USMC United States Marine Corps
USMCR United States Marine Corps Reserve
USN United States Navy
USNG United States National Guard
USNR United States Naval Reserve
USO United Service Organizations
USPHS United States Public Health Service
USS United States Ship
USSR Union of the Soviet Socialist Republics
USV United States Volunteers
UT Utah

VA Veterans' Administration
Va., VA Virginia
vet. veteran, veterinary
VFW Veterans of Foreign Wars
V.I., VI Virgin Islands
vice pres. vice president
vis. visiting
VISTA Volunteers in Service to America
VITA Volunteers in Technical Service
vocat. vocational
vol. volunteer, volume
v.p. vice president
vs. versus
Vt., VT Vermont

W. West
WA Washington (state)
WAC Women's Army Corps
Wash. Washington (state)
WAVES Women's Reserve, U.S. Naval Reserve
WCTU Women's Christian Temperance Union
we. Western (use for court system only)
W. Ger. Germany, Federal Republic of
WHO World Health Organization
WI, Wis. Wisconsin
W.I. West Indies
WSB Wage Stabilization Board
WV West Virginia
W.Va. West Virginia
WY Wyoming
Wyo. Wyoming
YK Yukon Territory (for address)
YMCA Young Men's Christian Association
YMHA Young Men's Hebrew Association
YM & YWHA Young Men's and Young Women's Hebrew Association
Y.T. Yukon Territory
YWCA Young Women's Christian Association
yr. year

zool. zoological

Alphabetical Practices

Names are arranged alphabetically according to the surnames, and under identical surnames according to the first given name. If both surname and first given name are identical, names are arranged alphabetically according to the second given name. Where full names are identical, they are arranged in order of age—with the elder listed first.

Surnames, beginning with De, Des, Du, however capitalized or spaced, are recorded with the prefix preceding the surname and arranged alphabetically, under the letter D.

Surnames beginning with Mac and Mc are arranged alphabetically under M.

Surnames beginning with Saint or St. appear after names that begin Sains, and are arranged according to the second part of the name, e.g. St. Clair before Saint Dennis.

Surnames beginning with Van, Von or von are arranged alphabetically under letter V.

Compound hyphenated surnames are arranged according to the first member of the compound. Compound unhyphenated surnames are treated as hyphenated names.

Parentheses used in connection with a name indicate which part of the full name is usually deleted in common usage. Hence Abbott, W(illiam) Lewis indicates that the usual form of the given name is W. Lewis. In such a case, the parentheses are ignored in alphabetizing. However if the name is recorded Abbott, (William) Lewis, signifying that the entire name William is not commonly used, the alphabetizing would be arranged as though the name were Abbott, Lewis.

Who'sWho in Frontier Science and Technology

AADLAND, DONALD INGVALD, electrical and mechanical engineer; b. Britton, S.D., Apr. 20, 1936; s. Ingvald Martin and Mabel Luverne (Hickock) A.; m. Georgia Doris Miller, Jan. 1957; children: Elizabeth, Donald, Kirsten, Danon, Darren. B.S. in Engring. Physics, S.D. State U., 1959. Registered profl. engr., Ariz., Colo., Tex., N.Mex. Sr. electronic engr. Govt. Electronics div. Motorola, Scottsdale, Ariz., 1962-70; profl. engr. in pvt. practice, Scottsdale, 1971—. Mem. adv. bd. Maricopa Tech. Community Coll., Phoenix; mem. Plumbing and Mech. Adv. Bd., Scottsdale. Served to capt. USAF, 1959-62. Decorated Air Force Commendation medal. Mem. ASHRAE, AIAA, ASME. Jehovah's Witness. Subspecialties: Solar energy; Electrical engineering. Current work: Solar thermal, and biomass electrical power generation systems, solar air conditioning systems, acoustics, ultrasonics, electronics. Design and development of electrical, mechanical and electronic systems; scientific analysis as expert witness. Office: PO Box 340 Scottsdale AZ 85252

AAGAARD, GEORGE NELSON, medical educator; b. Mpls., Aug. 16, 1913; s. George N. and Lucy T. (Nelson) A.; m. Lorna D. Docken, Aug. 26, 1939; children: Diane Louise, George Nelson, Richard Nelson, David Nelson, Steven Nelson. B.S., U. Minn., 1934, M.B., 1936, M.D., 1937. Intern Mpls. Gen. Hosp., 1936-37; successively fellow, instr., asst. prof. internal medicine U. Minn. Med. Sch., 1941-47, assoc. prof., dir. continuing med. edn., 1948-51; prof. medicine, dean Southwestern Med. Sch., U. Tex., 1952-54; dean U. Washington Sch. Medicine, 1954-64, prof. medicine, 1954-78, disting. prof. medicine and pharmacology, 1978—, head div. clin. pharmacology, 1964-79; mem. Nat. Adv. Council for Health Research Facilities USPHS, 1954-58; mem. nat. adv. heart council NIH, 1961-65; mem. spl. med. adv. group VA, 1970-74; chmn. bd. trustees Network for Continuing Med. Edn., 1966-78. Bd. dirs., editorial bd.: Western Jour. Medicine. Mem. Am. Heart Assn. (trustee), Assn. Am. Med. Colls. (pres. 1960-61), AMA (dir., chmn. com. continuing profl. edn. programs 1972), Pharm. Mfrs. Assn. Found. (mem. sci. adv. com. 1967-74), Am. Soc. Clin. Pharmacology and Therapeutics (pres. 1977, Flexner award 1983), N.Y. Acad. Scis., A.A.A.S., Washington, King County med. socs., Alpha Omega Alpha. Subspecialties: Internal medicine; Pharmacology. Home: 3810 49th Ave NE Seattle WA 98105

AAROE, WILLIAM HENRY, state official, industrial hygienist; b. Perth Amboy, N.J., June 18, 1920; s. Christian P. and Hilda (Petersen) A.; m. Mildred Isabella Ramsay, Sept. 21, 1946; children: Wendy Lou, William Christian. B.S. in San. Engring, Rutgers U., 1942; M.P.H., U. Mich., 1960. Engr./draftsman Raritan Copper Works, Perth Amboy, 1946-48; field engr. Pitometer Co., Inc., N.Y.C., 1948-49; pub. health engr. N.J. Dept. Health, Trenton, 1950-58, prin. pub. health engr. and radiation physicist, 1958-59, radiation physicist, 1960-62, chief radiol. health program, 1962-70; dep. dir. office radiation control N.Y.C. Dept. Pub. Health, 1970-74; acting chief environ. engring. sect. Phila. Dept. Pub. Health, 1975-76, chief occupational and radiol. health sect., 1974-76; dir. indls. hygiene div. W.Va. Dept. Health, South Charleston, 1977—; guest lectr. Rutgers U. Extension, 1963-70, Manhattan Coll., 1972; adj. asst. prof. environ. medicine NYU Med. Ctr., 1973-76; chmn. Conf. Radiol. Health, 1969. Contbr. articles to profl. jours. Served to lt. comdr. USNR, 1943-62; to capt. USPHSR, 1962—. Mem. AAAS, Am. Water Works Assn., Health Physics Soc., Conf. Radiation Control Program Dirs., Am. Conf. Govtl. Indsl. Hygienists (chmn. waste disposal com.), Commd. Officers Assn. USPHS, Am. Indsl. Hygiene Assn., Health Physics Soc., Am. Nuclear Soc. Subspecialties: Nuclear physics; Toxicology (medicine). Home: 5308 Pamela Circle Cross Lanes WV 25313 Office: 151 11th Ave South Charleston WV 25303

AARONSON, MARC, astronomer; b. Los Angeles, Aug. 24, 1950; s. Simon and Rena (Silverstein) A.; m. Marianne Gabrielle Kun, Aug. 20, 1972; children: Laura, Jamie. B.S., Calif. Inst. Tech., 1972; M.A., Harvard U., 1974, Ph.D., 1977. Teaching fellow Harvard U., 1973-75, research asst., 1974-76; research assoc. Steward Obs., U. Ariz., Tucson, 1977-80, asst. astronomer, 1980-82, assoc. astronomer, 1982-83, assoc. prof., 1983—. Contbr. articles to profl. jours. Recipient Fulbright award U. Ariz., 1965, Bart J. Bok prize Harvard U., 1983. Mem. Am. Astron. Soc., Astron. Soc. Pacific, Sigma Xi. Subspecialties: Cosmology; Infrared optical astronomy. Current work: Infrared stellar and extragalactic photometry, observational cosmology. Home: 2210 Springmill Rd Fort Wayne IN 46825 Office: Steward Obs U Ariz Tucson AZ 85721

AARONSON, STUART ALAN, cancer researcher; b. Mt. Clemens, Mich., Feb. 28, 1942; s. Michael and Frances (Leviant) A.; m. Gayle R. Ziff, Aug. 21, 1971; children: Lauren, David, Daniel. B.S., U. Calif.-Berkeley, 1962, M.S., 1965; M.D., U. Calif.-San Francisco, 1966. Staff assoc. viral carcinogenesis br. Nat. Cancer Inst., Bethesda, Md., 1967-69, sr. staff fellow, 1969-70, head molecular biology sect., 1970-77, chief lab. celluar and molecular biology, 1977—. Contbr. over 200 articles to profl. jours. Served with USPHS, 1967-70, 75—. Recipient Meritorious Service medal USPHS, 1982. Mem. Am. Assn. Cancer Research (Rhoads Meml. award 1982), Am. Soc. Microbiology, AAAS, Fedn. Am. Socs. Exptl. Biology. Subspecialties: Cell study oncology; Virology (medicine). Office: NCI Bldg 37 Room 1A07 Bethesda MD 20205

AASLESTAD, HALVOR GUNERIUS, government science administrator; b. Birmingham, Ala., Sept. 6, 1937; s. Knut and Geraldine W. (Dobson) A.; m. Barbara N. Wohn, July 30, 1960; children: Katherine, Karen, Peter, Lauren. B.S., La. State U., 1960, Ph.D., 1965; M.S., Pa. State U., 1961. Research fellow U.S. Army Biol. Labs., Frederick, Md., 1965-68; asst. prof. U. Ga., 1968-70; research investigator Wistar Inst., Phila., 1970-73; sr. scientist Frederick Cancer Research Ctr., 1973-76; exec. sec. mammalian genetics study sect., div. research grants NIH, Bethesda, Md., 1976-81, asst. br. chief div. research grants, 1981—. Contbr. numerous articles to tech. jours. Recipient Quality Increase award NIH, 1980; NIH fellow, 1963-65; NRC-NSF fellow, 1965-67; Nat. Inst. Allergy and Infectious Disease grantee, 1969-70. Mem. Am. Soc. Microbiology, AAUP, Sigma XI, Phi Kappa Phi. Club: Chesapeake Bay Yacht Racing Assn. Subspecialties: Science Administration; Molecular biology. Current work: Science administration/review; molecular biology, genetics, cell biology, virology. Home: 4498 Willowtree Dr Middletown MD 21769 Office: Westwood Bldg Room 334 Bethesda MD 20205

ABBOTT, FRED, engring. adminstr.; b. Burnley, Eng., July 5, 1941; s. Jack and Eva A.; m. Christine J. Sugden, Feb. 15, 1963; children: Mark, Robert Heather. B.Sc. with honors in Physics, U. Leicester, Eng., 1963. Head optical devel. Rank Taylor Hobson, Leicester, 1963-66; supr. advanced lens testing lab. Itek Corp., Lexington, Mass., 1966-69; v.p. devel. Hudson Precision Optical Co., Hudson, N.H., 1969-72; mgr. internat. engring. tech. 3M Co., St. Paul, 1972—; mem. adv. bd. Optical Pub. Co.; gen. chmn. CLEO Conf., 1981. Author 1 book; contbr. articles to profl. jours. Fellow Inst. Physics (London); mem. Optical Soc. Am. (pres. Minn. sect. 1981-82). Lodge: Lions. Subspecialty: Optical engineering. Office: 3M Co PO Box 33800 St Paul MN 55133

ABBOTT, MITCHEL THEODORE, biochemistry educator; b. Los Angeles, June 6, 1930; s. Chester and Judith (Wasserman) A.; m. Florine Lutze, Oct. 27, 1962; children: Valerie Michele, Mark Nelson, Chester Bruce. B.S., UCLA, 1957, Ph.D., 1962. Postdoctoral fellow biochemistry dept. NYU Med. Sch., 1963-64; vis. scientist Roche Inst. Molecular Biology, Nutley, N.J., 1972-73; prof. chemistry San Diego State U., 1964—. Contbr. articles on biochemistry to profl. jours. and books. Served with U.S. Army, 1951-53. NIH grantee, 1965-83. Mem. Am. Soc. Biol. Chemists, AAAS. Subspecialties: Biochemistry (biology); Biochemistry (medicine). Current work: Enzymology and the regulation of metabolism, especially as these pertain to oxygenases and pyrimidine metabolism. Office: Chemistry Dept San Diego State U San Diego CA 92182

ABBOTT, RICHARD NEWTON, JR., geology educator; b. Newton, Mass., Jan. 12, 1949; s. Richard Newton and Myrtle Louise (vonHagen) A.; m. Lesley MacVane, June 26, 1971; children: Andrew MacVane, Tyson William. B.A., Bowdoin Coll., 1971; M.Sc., U. Maine, 1973; Ph.D., Harvard U., 1977. Postdoctoral fellow Dalhousie U., Halifax, N.S., Can., 1977-79; asst. prof. geology Appalachian State U., 1979—. Mem. Mineral Assn. Can. (Hawley award 1979), Mineral. Soc. Am. (Kraus-Golss award 1982), Geol. Soc. Maine, Phi Beta Kappa, Sigma Xi, Phi Kappa Phi. Subspecialties: Crystallography; Igneous and Metamorphic petrology. Current work: Petrogenesis of granites; phase relationships of mineral assemblages in igneous and metamorphic rocks. Home: Route 4 Box 475 Boone NC 28607 Office: Dept Geology Appalachian Stae U Boone NC 28608

ABBOTT, ROBERT DEAN, psychology educator; b. Twin Falls, Idaho, Dec. 19, 1946; s. Charles Dean and Billie June (Moore) A.; m. Sylvia Patricia Keim, Dec. 16, 1967; children: Danielle, Matthew. B.A., Calif. Western U., 1967; M.S., U. Wash., 1968, Ph.D., 1970. Asst. then assoc. prof. psychology Calif. State U., Fullerton, 1970-75; prof. edn1. psychology U. Wash., Seattle, 1975—; cons. Edmonds Sch. Dist., 1979-82, Orting Sch. Dist., 1978-80, Seattle Sch. Dist., 1982—. Author: Multivariate Analysis, 1983; contbr. articles to profl. jours. Calif. State scholar, 1964-67; fellow, 1967. Mem. Am. Psychol. Assn., Psychometric Soc., Am. Statis. Assn. (chpt. v.p. 1975—), Am. Ednl. Research Assn., Nat. Soc. for Study of Edn. Methodist. Club: Porsche of Am. Subspecialty: Quantitative psychology. Current work: Construct validity of measures of individual differences; multivariate statistical analysis; cognitive processes and writing. Home: 7357 57 St NE Seattle WA 98115 Office: Univ Wash DQ12 Seattle WA 98195

ABBOTT, ROBINSON SHEWELL, biology educator, researcher; b. Phila., June 23, 1926; s. Charles Shewell and Margaret Elizabeth (Robinson) A.; m. RoseMarie Savelkoul, June 9, 1956; children: Rodman Peyton, Jane Paxson, Bartram Shewell, Braden Keim. B.S., Bucknell U., 1949; Ph.D., Cornell U., 1956. Teaching asst. Bucknell U., 1947-49; substitute tchr. Hempstead and Garden City (L.I.) high schs., 1949-51; teaching asst. Cornell U., 1951-52, research asst., 1952-56; instr. in biology Smith Coll., 1956-61; asst. prof. biology U. Minn.-Morris, 1961-64, assoc. prof., 1964-71, prof., 1971—; researcher Radioecology Inst., Oak Ridge Inst. Nuclear Studies, Eagle Lake Assn., Willmar, Minn., Padre Island Nat. Seashore, Corpus Christi, Tex. Research publs. in field. Mem. AAUP, Bot. Soc. Am., Am. Inst. Biol. Scis., Ecol. Soc. Am., Phycological Soc. Am., Internat. Phycological Soc., Minn. Acad. Scis., Nature Conservancy, Sigma xi. Mem. Democratic-Farm Labor Party. Subspecialties: Ecosystems analysis; Resource management. Current work: Zonation of aquatic organisms; overwinter of algae; off-road vehicle impact on salt marshes in South Texas. Home: 205 E 6th St Morris MN 56267 Office: Dept Biology U Minn Morris MN 56267

ABBRECHT, PETER HERMAN, physiologist, educator; b. Toledo, Nov. 27, 1930; s. Hermann Richard and Paula Katherine (Schwenk) A.; m. Anne Patterson Lampman, Feb. 16, 1957; children—Elaine, Brian. B.S., Purdue U., 1952; M.S., U. Mich., 1953, Ph.D. in Chem. Engring, 1957, M.D., 1962. Diplomate: Am. Bd. Internal Medicine. Sr. chem. engr. Minn. Mining & Mfg. Co., Detroit, 1956-58; intern U. Calif. Hosp., Los Angeles, 1962-63; mem. faculty U. Mich. Med. Sch., Ann Arbor, 1963-80, prof. physiology, 1972-80, chmn. bioengring. program, 1972-77, prof. internal medicine, 1976-80; prof. internal medicine and physiology Uniformed Services U. of Health Scis., Bethesda, Md.; and cons. physician Walter Reed Army Med. Center, 1980—; guest scientist Naval Med. Research Inst., 1980—; resident in internal medicine U. Mich. Hosp., 1971-72, fellow in pulmonary disease, 1974-75; vis. prof. bioengring. U. Calif., San Diego, 1973; dir. physiology and biomed. engring. program NIGMS-NIH, 1977-78; cons. VA, NASA, Air Force Office Sci. Research, NIH, NSF; mem. nat. research resources advisory council, 1975-78. Editor in chief: Internat. Jour. Biomed. Engring, 1972-74; Editorial bd.: Jour. Biomechanics; editor-in-chief: Jour. Bioengring, 1979—; Contbr. articles to profl. jours. Recipient outstanding research award Mich. Heart Assn., 1960; research career devel. award NIH, 1969-73. Fellow ACP; mem. Biomed. Engring. Soc. (dir. 1970-72); Mem. Am. Physiol. Soc., Am. Thoracic Soc., Soc. Exptl. Biology and Medicine. Subspecialties: Physiology (medicine); Biomedical engineering. Current work: Control of respiration; high frequency ventilators. Home: 2806 Spencer Rd Chevy Chase MD 20015

ABDEL-LATIF, ATA ABDEL-HAFEZ, biochemist, educator; b. Palestine, Jan. 22, 1933; s. Abdel-hafez M. and Ayshah A. (Abour) Abdel-L.; m. Iris K. Graham, Sept. 10, 1957; children: Rhonda, David, Joseph, Rhadi. B.S. in Chemistry, DePaul U., 1955, M.S., 1958; Ph.D. in Biochemistry, Ill. Inst. Tech., 1963. Postdoctoral fellow U. Ill., Chgo., 1963-65; med. research assoc. U. Ill.-State of Ill. Pediatric Inst., Chgo., 1965-67; assoc. prof. biochemistry Med. Coll. Ga., 1967-74, prof. cell and molecular biology, 1974—. Contbr. articles on biochemistry of nerve and muscle to profl. publs. NIH grantee, 1963—. Mem. Am. Soc. Biol. Chemistsy, Am. Physiol. Soc., Internat. Soc. Neurochemistry, others. Subspecialty: Biochemistry (medicine). Current work: Effects of neurotransmitters on phospholipid metabolism in nerve and muscle. Release of arachidonate from membrance phospholipids for prostaglandin synthesis in nerve and muscle. Home: 123 Avondale Dr Augusta GA 30907 Office: Dept Cell and Molecular Biology Med Coll GA Augusta GA 30912

ABDOU, MOHAMED AZIZ, nuclear engineer; b. Egypt, July 10, 1945; came to U.S., 1969, naturalized, 1980; s. Abdelaziz A. Abdou and Ensaf M. Basha; m. Zahira M. El-Derini, Sept. 17, 1969; 1 child, Shareef. B.S., U. Alexandria, Egypt, 1967; M.S., U. Wis., 1971, Ph.D., 1973. Nuclear engr. Argonne (Ill.) Nat. Lab., 1974-78, assoc. dir. fusion power program, 1979—; head fusion sect., applied physics div.; assoc. prof. Ga. Inst. Tech., Atlanta, 1978-79; mgr. nuclear engring. and materials br. Fusion Engring. Design Center, Oak Ridge Nat. Lab., 1982—. Mem. Am. Nuclear Soc. Subspecialties: Nuclear fusion; Nuclear fission. Current work: Development of fusion engineering and technology; special interest in nuclear components (first wall, blanket, shield, impurity control) of fusion reactors and overall fusion reactor system analysis and design. Home: 733 Chateaugay Dr Naperville IL 60540 Office: Argonne National Laboratory 9700 S Cass Ave Bldg 205 Argonne IL 60439

ABDOU, NABIH I., physician, educator; b. Cairo, Egypt, Oct. 11, 1934; came to U.S., 1962, naturalized, 1972; s. Ibrahim and Galila (Azer) A.; m. Nancy L. Layle, Aug. 26, 1939; children: Mark L., Marie L. M.D., Cairo U., 1959; Ph.D., McGill U., 1969. Prof. medicine U. Kans. Med. Center, Kansas City, 1978—. Fulbright scholar, 1962-65. Fellow A.C.P., Am. Acad. Allergy; mem. Am. Assn. Immunologists, Am. Rheumatism Assn., Central Soc. Clin. Research. Subspecialties: Immunology (medicine); Allergy. Current work: Cellular immunology, autoimmunity. Office: U Kans Med Center 4035B Kansas City KS 66103

ABDULHAY, GAZI, obstetrician, genecologist, educator; b. Antakya-Hatay, Turkey, Sept. 14, 1948; s. Ali and Mediha (Bekfilavi) A.; m. Suzanne Moorman, Aug. 18, 1979; 1 child, Leyla. M.D. (Turkish Sci. and Tech. Research Assn. scholar), U. Istanbul, 1972. Diplomate: Am. Bd. Ob-Gyn. Extern Soldier's Home, Chelsea, Mass., 1972-73; intern Rush-Presbyn.-St. Luke's Med. Ctr., Chgo., 1973-74, resident in ob-gyn, 1974-77; clin. fellow in gynecologic oncology Georgetown U., Washington, 1977-78, U. Calif.-Irvine, 1978-80; asst. prof. ob-gyn Oreg. Health Scis. U., Portland, 1980—, chief gynecologic oncology, 1980—. Contbr. numerous sci. articles to profl. jours. Mem. Med Assn. Arab Ams., Am. Coll. Ob-Gyn, AMA, Soc. Gynecologic Oncologists, ACS, Western Assn. Gynecologic Oncologists. Subspecialties: Oncology; Gynecological oncology. Current work: Cervical cancer interferon in gynecological malignancies; clinical and basic science research relatied to gynecologic oncology. Home: 4268 Albert Circle Lake Oswego OR 97034 Office: 3181 SE Sam Jackson Park Rd Portland OR 97201

ABDULLA, ABDULLA MOHAMMED, cardiologist, medical educator; b. Bombay, India, June 20, 1942; came to U.S., 1971; s. Mohammed Abedeen and Zubaida (Siddick) A.; m. Shahien Shahzada, Dec. 20, 1970 (div. 1978); children: Aaliya, Mikaal; m. Sue Ann Swan, Apr. 6, 1979; 1 child, Siara. B.Sc. (honors), U. Karachi, Pakistan, 1965, M.B., B.S., 1970. Diplomate: Am. Bd. Internal Medicine, Am. Bd. Cardiology. Intern Dow. Med. Coll., U. Karachi, Pakistan, 1970-71; intern Mt. Carmel Mercy Med. Center, Detroit, 1971-72, resident in internal medicine, 1972-74; fellow in cardiology Med. Coll. Ga., Augusta, 1974-76, attending physician in cardiology, 1976—, asst. chief cardiology, 1983—, assoc. prof. medicine, 1980—; cons. computer aided instructions to med. schs., 1980—, lectr. in field. Author: Cardiovascular Physical Diagnosis, 1983; contbr. chpts. to books; reviewer: Jour. AMA, 1982—, Jour. Am. Acad. Dermatology, 1982—; Contbr. to profl. jours. Fellow Am. Coll. Cardiology, ACP; mem. So. Soc. Clin. Investigations, AMA, So. Med. Assn. (vice chmn. chest diseases sect. So. chpt. 1982), Am. Heart Assn. (teaching scholar 1980-83), Ga. Heart Assn. (dir. 1982—). Republican. Subspecialties: Cardiology. Current work: Cardiac diseases; computer-assisted learning; medical education. Developer multi-media computer-aided learning in med. edn., (Gov.s award of Excellence Am. Coll. Cardiology 1982), 1980. Home: 2333 Kings Way Augusta GA 30904 Office: Med Coll Ga 15th St Suite 6125 Augusta GA 30912

ABELES, TOM PETER, renewable energy consultant, researcher; b. Louisville, Feb. 25, 1941; s. Gerd Hans and Ilse Rachael (Lowenstein) A.; m. Ruth Elaine Brink, 1980; children: Katrina Victoria Brink, Christopher Mathew. B.S. cum laude (Nat. Merit Scholar), Wilmington (Ohio) Coll., 1963; Ph.D. (NDEA scholar), U. Louisville, 1969. Asso. prof. U. Wis., Green Bay, 1970-76; pres., chmn. bd. i.e. assocs., inc., Mpls., 1976—. Contbr. articles to profl. jours. Fellow, v.p. Minn. chpt. Partners of the Ams. Mem. Soc. Gen. Systems Research, World Future Soc., others. Subspecialties: Biomass (energy science and technology); Solar energy. Current work: Research and development of renewable energy, financial and economic development, energy systems, agricultural energy management, overseas energy development, policy analysis. Home: 3704 11th Ave S Minneapolis MN 55407 Office: i e associates 3702 E Lake St Minneapolis MN 55406

ABELOFF, MARTIN DAVID, physician, oncologist; b. Shenandoah, Pa., Apr. 4, 1942; s. Aaron and Cele (Freid) A.; m. Diane Kaufman, Jan. 7, 1967; children: Elisa, Jennifer. A.B., Johns Hopkins U., 1963, M.D., 1966. Diplomate: Am. Bd. Internal Medicine. Intern U. Chgo., 1966-67; clin. assoc. Balt. Cancer Research Ctr., 1967-69; sr. asst. resident Beth Israel Hosp., Boston, 1969-70; fellow clin. hematology New England Med. Ctr., Boston, 1970-71; fellow clin. oncology Johns Hopkins U., 1971-72, instr. medicine, 1972-75, asst. prof. oncology, 1974-79, asst. prof. medicine, 1975-80, dir. oncology outpatient dept., 1976—, assoc. prof. oncology, 1979—, assoc. prof. medicine, 1980—, chief med. oncology, 1983—. Editor: Complications of Cancer, 1979. Mem. Am. Soc. Clin. Oncology, Am. Assn. Cancer Research. Subspecialties: Oncology; Chemotherapy. Current work: Clinical research in breast and lung cancer. Home: 2213 Sulgrave Ave Baltimore MD 21209 Office: Johns Hopkins Oncology Center 600 N Wolfe St Baltimore MD 21205

ABERG, GUNNAR A.K., pharmacologist; b. Falkenberg, Sweden, Mar. 16, 1936; came to U.S., 1978, naturalized, 1980; s. Thorsten Gustaf and Birgit (Lindahl) A.; m. Britt-Marie Wiman, Sept. 15, 1964 (div. Apr. 1974); 1 dau., Maria; m. Laurie Jane Seagondollar, May 27, 1978; 1 son, Peter. Ph.D. in Physiology, U. Gothenberg, Sweden, 1965, U. Linkoping, Sweden, 1972. Sect. head Bofors Nobel-Pharma, Molndal, Sweden, 1964-68, dir. pharmacology, 1968-73; group leader Haessle, Gothenburg, 1973-78; dir. cardiovascular pharmacology Astra, Worcester, Mass., 1978-80, Ciba-Geigy, Summit, N.J., 1980-82; dir. pharmacology Squibb Corp., Princeton, N.J., 1982—; assoc. prof. U. Linkoping, 1974—. Contbr. articles to profl. jours. Served with Swedish Army, 1956-57. Lutheran. Subspecialties: Pharmacology; Membrane biology. Current work: Cardiovascular drug research. Patentee in field. Office: Squibb Inst for Med Research PO Box 4000 Princeton NJ 08540 Home: 519 Bergen St Lawrenceville NJ 08648

ABERNATHY, BILL J., energy engineer, educator, consultant; b. Cape Girardeau, Mo., Oct. 19, 1922; s. Leonard W. and Anna E. (Austin) A.; m. June E. Abernathy, June 19, 1923; children: Janice Abernathy Ririe, Barbara Abernathy Rummel, Steven. Vocat. class A credential, UCLA. Mem. faculty Orange Coast Coll., 1961—, successively asst. prof. energy engring., assoc. prof., now prof., also chmn. tech. div., cons. energy; solar seminar dir.; energy mgmt. trainer. Served with USN. Decorated Presdl. citation. Current work: Energy conservation in residential and commercial structures; passive solar energy. Patentee portable recompression chamber. Office: 2701 Fairview Rd Costa Mesa CA 92626

ABERNATHY, RICHARD PAUL, nutrition educator; b. McCaysville, Ga., Mar. 22, 1932; m., 1957; 3 children. B.S., Berry Coll., 1952; M.S., U. Ga., 1957; Ph.D., Cornell U., 1960. Asst. prof., nutritionist Ga. Expt. Sta., 1960-66; assoc. prof., prof. nutrition Va. Poly. Inst. and State U., 1966-74; head dept. foods and nutrition Purdue U., West Lafayette, Ind., 1974—. Mem. Am. Inst. Nutrition, Inst. Food Tech., Soc. Nutritional Edn. Subspecialty: Nutrition (biology). Office: Dept Foods and Nutrition Purdue U West Lafayette IN 47907

ABID, SYED HASAN, virologist; b. Delhi, India, Apr. 27, 1945; came to U.S., 1968, naturalized, 1976; s. Syed Mohammad and Sahar Zainab (Bilgrami) Askari; m. Barbara Bozena, Dec. 1, 1973; children: Sahra Sheerin, Anisa Farah. M.Sc., U. Karachi, Pakistan, 1966; M.S., No. Ill. U., DeKalb, 1970; Ph.D., U. Ill., 1975. Lectr. dept. biol. scis. U. Ill., Chgo., 1970-71; teaching asst. dept. microbiology U. Ill. Med. Ctr., 1971-74, research assoc. dept. biochemistry, 1975; virologist research dept. Met. San. Dist. Greater Chgo., 1975-83; dir. virology program Pub. Health Labs., Westchester County, Valhalla, N.Y., 1983—; cons. dept. microbiology and botany U. Kuwait, 1981—; cons. United Nations Devel. Programme, 1983—. Recipient Summer Study award Internat. Inst. Edn., 1969; Am. Cancer Soc. instl. grantee, 1973-74. Mem. Am. Soc. Microbiology. Islam. Subspecialties: Animal virology; Cell and tissue culture. Current work: Development of new methods for virus detection in environmental and clinical specimen, development of methods and detection of mutagens and carcinogens using short term mutagenic/carcinogenic tests in environmental samples; assessment of health risks resulting due to microbial and chemical burden in the environment.

ABILDSKOV, J.A., physician; b. Salem, Utah, Sept. 22, 1923; s. John and Annie Marie (Peterson) A.; m. Mary Helen McKell, Dec. 14, 1944; children: Becky Ann, Alan, Mary Karen, Marilyn. B.A., U. Utah, 1944, M.D., 1946. Diplomate: Am. Bd. Internal Medicine. Instr. medicine Tulane U., New Orleans, 1947-54; from asst. prof. to prof. SUNY, Syracuse, 1956-68; prof. medicine U. Utah, Salt Lake City, 1968—. Contbr. articles to profl. jours. Served to capt. U.S. Army, 1954-56. Recipient Disting. Research award U. Utah, 1977. Fellow Am. Coll. Cardiology; mem. Assn. Am. Physicians, Am. Soc. for Clin. Investigation (emeritus), Assn. Univ. Cardiologists. Republican. Mormon. Subspecialties: Cardiology; Psychophysiology. Current work: Research in cardiac electrophysiology. Office: U Utah Nora E Harrison Cardiovascular Research and Tng Inst Bldg 100 Salt Lake City UT 84112

ABINANTI, FRANCIS RALPH, veterinary microbiology educator; b. San Francisco, Mar. 2, 1916; s. Frank and Florence Christina (Lawson) A.; m. Elizabeth McDougle, Aug. 1943; m. Margery Jean Maggs, June 12, 1951; children: Lawson, Andrea, Michael. B.S., D.V.M., Wash. State U., 1941; Ph.D., Cambridge U., 1958. Veterinarian U.S. Dept. Agr., 1942-43; practice vet. medicine, Sacramento, 1943-45; mem. Calif. Poultry Improvement Adv. Bd., 1945-49; sr. asst. veterinarian viral and rickettsial diseases lab. USPHS, Berkeley, Calif., 1949-53; veterinarian div. biologic standards NIH, Bethesda, Md., 1955-57; sr. vet. officer lab. infectious diseases Nat. Inst. Allergy and Infectious Diseases, NIH, Bethesda, 1957-64, vet. office dir., 1964-67, assoc. dir. extramural programs, 1967-70, research veterinarian, Hamilton, Mont., 1973; vis. scientist Scripps Clinic and Research Found., La Jolla, Calif., 1970-73; vet. pathologist livestock and poultry disease diagnostic lab. Calif. State Dept. Agr., Petaluma, 1974; vet. virologist dept. vet. sci. Coll. Agr., U. Ariz., Tucson, 1974-76; vet. microbiologist Coll. Vet. Medicine, Wash. State U., Pullman, 1976—. Mem. AVMA, Conf. Research Workers in Animal Diseases, Am. Pub. Health Assn., San Diego Vet. Med. Assn., Am. Coll. Vet. Microbiologists (diplomate), Wash. State Pub. Health Assn. Subspecialties: Virology (veterinary medicine); Preventive medicine (veterinary medicine). Current work: Teaching of preventive veterinary medicine and food hygiene. Office: Dept Microbiology Coll Vet Medicine Wash State U Pullman WA 99164

ABLASHI, DHARAM VIR, virologist, immunologist, lecturer, researcher; b. Lahore, India, Oct. 8, 1931; came to U.S., 1959, naturalized, 1970; s. Maharaj Krishan and Lajwanti Devi D.; m. Kristine Louise Upper, Aug. 13, 1966; children: Kerin Louise, Davinder Kumar, Sameer Kumar. Diploma radio engring., Poly. Tech. Inst., Delhi, India, 1950; D.V.M., Panjab Veterinary Coll., India, 1955; M.S., U. R.I., 1962. Sr. research assoc. Indian Veterinary Research Inst., Izatnagar, India, 1955-59; research assoc. Animal Pathology dept. U. R.I., Kingston, 1960-66; research veterinarian Cobb Breeding Corp., Concord, Mass., 1967-69; research viorologist Flow Labs., Inc., Rockville, Md., 1967-69; research microbiologist Nat. Cancer Inst. NIH, Bethesda, Md., 1969-73, head primate virus sect., 1973-79, microbiologist, 1979—, mem., 1972-77; mem. internat. faculty WHO/ Internat. Agy. Research on Cancer, Lyon, France, 1974-76, 78; cons. in field. Editor: (with others) Epstein Barr Virus Production, Concentration and Purification, 1975, Nasopharyngeal Carcinoma, 1980, Nasopharyngeal Carcinoma-Current Concepts and Prospects, 1983; contbr.: numerous articles to profl. jours. Nasopharyngeal Carcinoma-Current Concepts and Prospects. Chmn. Environ. com. Chincoteague Bay Trails END Assn. Inc., 1977—, dir., 1979—. NIH grantee, 1963-65. Mem. Am. Soc. Microbiology, Am. Assn. Cancer Research, Internat. Assn. Comparative Research on Leukemia and Related Diseases, Soc. Exptl. Biology and Medicine, Veterinary Cancer Soc. Republican. Hindu. Subspecialties: Cancer research (medicine); Virology (medicine). Current work: Etiologic role, mechanism of action and control of DNA viruses of primates with oncogenesis with particular emphasis on herpesviruses; study in nasopharyngeal carcinoma in relation to virus, chemicals, etc.

ABLE, KENNETH PAUL, biology educator, animal behaviorist; b. Louisville, Feb. 5, 1944; s. William Morris and Viola (Bridwell) A.; m. Mary Frances Allen, Jan. 28, 1967; 1 son, Joshua. B.A., U. Louisville, 1966, M.S., 1968; Ph.D, U. Ga., 1971. Asst. prof. dept. biology SUNY, Albany, 1971-77, assoc prof., 1977—; dir. Cranberry Lake (N.Y.) Biol. Sta., 1973. Contbr. chpt. to book. NSF grantee, 1974, 76, 80, 83. Fellow Am. Ornithologists Union (treas. 1981—); mem. Am. Soc. Naturalists, Animal Behavior Soc., Ecol. Soc. Am., Cooper Ornithological Soc. Subspecialty: Ethology. Current work: Mechanisms of animal orientation and navigation, especially birds; behavioral ecology especially as it relates to migratory behavior in birds. Home: Irish Hill Rd Berne NY 12023 Office: Department Biology State University New York 1400 Washington Ave Albany NY 12222

ABLIN, RICHARD JOEL, immunologist; b. Chgo., May 15, 1940; s. Robert B. and Minnie E. (Gordon) A.; m. Linda L. Lutwack, Jan. 2, 1964; 1 son, Michael D. A.B., Lake Forest Coll., 1962; Ph.D., SUNY, Buffalo, 1967. Diplomate: Am. Bd. Clin. Immunology and Allergy; cert. Nat. Registry Microbiologists. Instr., research asst. Rosary Hill Coll., Buffalo, 1965-66; USPHS postdoctoral fellow in microbiology SUNY Sch. Medicine-Buffalo, 1966-68; AID research cons. Paraguay Program in Med. Edn., 1968; dir. div. immunology Millard Fillmore Hosp., Buffalo, 1968-70; head immunology sect. renal unit and cons. med. staff Meml. Hosp. of Springfield, Ill., 1970-73; asst. prof. medicine So. Ill. U. Sch. Medicine, Springfield, 1971-73; cons. med. staff dept. pediatrics St. Johns Hosp., Springfield, 1971-73; dir. immuno-biology sect. div. urology dept. surgery Cook County Hosp. and Hektoen Inst. Med. Research, Chgo., 1973-75, sr. sci. officer div. immunology, 1976—; sr. sci. staff, clin. immunologist Cook County Hosp., 1973-75; assoc. prof. microbiology Univ. Health Scis./Chgo. Med. Sch., 1973-74. Editor or contbg. editor numerous sci. jours. and books; contbr. articles to sci. publs. Fellow Am. Assn. Clin. Immunology and Allergy, Am. Coll. Cryosurgery, Nat. Acad. Clin. Biochemistry; mem. AAAS, Am. Assn. Cancer Research, Am. Assn. Immunologists, Am. Fedn. Clin. Research, Am. Soc. Immunology of Reprodn., Internat. Soc. Immunology of Reprodn, Assn. Surg. Oncology (fgn.), Chgo. Assn. Immunologists, Comite Internacional de Andrologia, Internat. Soc. Chronobiology, Internat. Soc. Cryosurgery (U.S. del. 1980-83, pres. 1977-80, hon. life pres.), Japan Soc. Low Temperature Medicine, Nat. Acad. Clin. Biochemistry, N.Y. Acad. Scis., Reticuloendothelial Soc., Soc. Cryobiology, Soc. Exptl. Biology and Medicine, Soc. Protozoologists, Soc. Study Reprodn., Transplanation Soc., Sigma Xi. Subspecialties: Immunology (medicine); Cancer research (medicine). Current work: Aberrations of immunologic responsiveness; immuno-logic consequences of experimentally and clinically induced hypo- and hyperthermia; immunoparasitic relationships; transplantation and tumor antigens. Office: 1835 W Harrison St Chicago IL 60612

ABLOW, CLARENCE MAURICE, mathematician; b. N.Y.C., Nov. 6, 1919; s. Abraham and Bertha (Wopfers) A.; m. Stella Jane Kahn, Aug. 6, 1945. B.A., UCLA, 1940, M.A., 1942; Ph.D., Brown U., 1951. Mathematician Boeing Airplane Co., Seattle, 1951-55, SRI Internat., Menlo Park, Calif., 1955—. Served to lt. USNR, 1942-46. Mem. Soc. Indsl. and Applied Math. (pres. No. Calif. Sect. 1970-79), Am. Math. Soc. Democrat. Subspecialties: Applied mathematics; Mathematical software. Current work: Mathematical modeling of physical phenomena; solid mechanics, reactive flow, catalytic chemistry. Office: SRI Internat 333 Ravenswood Ave Menlo Park CA 94025

ABLOWITZ, MARK JAY, mathematics and computer science educator; b. Bronx, N.Y., June 5, 1945; s. Ben and Mae (Markoff) A.; m. Enid Bate, June 9, 1968. B.A., U. Rochester, 1967; Ph.D., MIT, 1971. Teaching asst. MIT, Cambridge, Mass., 1967-71, instr., 1970; asst. prof. math. Clarkson Coll. Tech., Potsdam, N.Y., 1971-75, assoc. prof. math., 1975-76, prof. math. and computer sci., 1977—, chmn. dept. math. and computer sci., 1979—. Author: (with H. Segur) Solitons and the Inverse Scattering Transform, 1981. Recipient Clarkson Graham Research award Clarkson Coll. Tech., 1976; Sloan fellow, 1975-77; Guggenheim fellow, 1984. Mem. Soc. Indsl. and Applied Math., Math. Assn. Am., Am. Math. Soc., Sigma Xi, Tau Beta Pi. Subspecialties: Applied mathematics. Current work: Solutions of nonlinear evolution equations of physical interest. Office: Dept Math and Computer Sci Clarkson Coll Tech Potsdam NY 13676

ABOLINS, MARIS ARVIDS, physics educator, researcher; b. Liepaja, Latvia, Feb. 5, 1938; came to U.S., 1949, naturalized, 1956; s. Arvids Gustavs and Olga Elizabete (Grintals) A.; m. Frances Junia Delano, Dec. 19, 1959; children: Mark, Krista. B.S., U. Wash., 1960; M.S., U. Calif.-San Diego, 1962, Ph.D., 1965. Research assoc. U. Calif., San Diego, 1960-65; physicist Lawrence Berkeley Lab., Berkeley, Calif., 1965-68; assoc. prof. Mich State U., East Lansing, 1968-72, prof. physics, 1972—; vis. researcher CERN, Geneva, 1976-77, CEN, Saclay, France, 1977. Contbr. articles to profl. jours. Chmn. Fermilab Users Exec. Com., Batavia, Ill., 1982-83. Gen Motors scholar U. Wash., 1956-60. Mem. Am. Phys. Soc., AAAS; mem. Phi Beta Kappa; Mem. Patria-Latvian Frat. Subspecialty: Particle physics. Current work: Studying the electro-weak force by means of high energy beams of neutrinos, using neutrinos to investigate the structure of nucleons. Home: 220 Loree Dr East Lansing MI 48823 Office: Physics Dept Mich State U East Lansing MI 48824

ABORN, MURRAY, science administrator; b. N.Y.C., July 15, 1919; s. Maurice and Rose (Kaufman) A.; m. Barbara Anne Soltow, Dec. 22, 1960; children: Shoshanah, David Asher. B.S.S., CCNY, 1942; M.A., Columbia U., 1947, Ph.D., 1950. Research psychologist Air Research and Devel. Command, U.S. Air Force, 1953-57; ednl. specialist Indsl. Coll. Armed Forces, Washington, 1957-58; exec. sec. behavioral scis. NIH, Bethesda, Md., 1958-62, Nat. Inst. Gen. Med. Scis., Bethesda, 1962-63; head social measurement and analysis sect. NSF, Washington, 1963—. Served with USAAF, 1942-45; PTO. Recipient Meritorious Service award NSF, 1973. Fellow Am. Psychol. Assn., N.Y. Acad. Sci.; mem. Psychonomic Soc., AAAS, Sigma Xi. Democrat. Jewish. Subspecialties: Behavioral psychology; Information systems (information science). Current work: Quantitative methodology in social science; applications of cognitive science and information science to social science. Address: NSF 1800 G St NW Washington DC 20550

ABOU-DONIA, MARTHA MAY, neurochemist; b. Cedar Creek, Ark., June 6, 1944; d. James S. and Helen E. (Hughes) Davis; m. Mohamed Bahie-el-dien Abou-Donia, Feb. 1, 1967; children: Tarek, Sheref, Suzanne. B.S., U. Calif.-Davis, 1965; M.S., U. Calif.-Berkeley, 1967; Ph.D., Duke U., 1981. Sr. research technician depts. biochemistry and animal sci. Tex. A&M U., 1967-70; sr. research technician dept. pharmacology Duke U., Durham, N.C., 1973-74; research scientist I dept. microbiology Burroughs Wellcome Co., 1974-77, research scientist II dept. medicinal biochemistry, 1977-82, research scientist III, 1982—. Contbr. articles to profl. jours. Mem. Soc. Neurosci. Subspecialties: Neurochemistry; Neuropharmacology. Current work: Regulatory role of pteridine cofactor in peripheral and central nervous system. Home: 106 Catawba Ct Chapel Hill NC 27514 Office: 3030 Cornwallis Rd Research Triangle Park NC 27709

ABOU-DONIA, MOHAMED BAHIE, pharmacologist; b. Domiat, Egypt, Nov. 3, 1939; s. Ahmed Awad and Fathia Abdo (Abou-hindia) A-D.; m. Martha Davis, Feb. 1, 1967; children: Ricky, Steve, Suzanne. B.S., Alexandria (Egypt) U., 1960; Ph.D., U. Calif.-Berkeley, 1967. Diplomate: Am. Bd. Toxicology, 1981. Research assoc. Tex. A&M U., College Station, 1967-70; asst. prof. Alexandria U., 1970-73; research assoc. Duke U. Med. Ctr., Durham, N.C., 1973-74, asst. prof. pharmacology, 1978-79, assoc. prof., 1979—, dep. dir. toxicology program, 1981—; cons. EPA, U.S. Dept. Labor, Nat. Cottonseed Assn., Monsanto Agrl. Products, UN. Contbr. articles to profl. jours. Fellow Am. Inst. Chemistry; mem. Soc. Toxicology (pres. neurotoxicology specialty sect. 1983—, counselor N.C. chpt. 1982—), AAAS, Am. Assn. Pathologists, Am. Chem. Soc., Am. Coll. Toxicology, Am. Soc. Neurochemistry, Am. Soc. Pharmacology and Exptl. Therapeutics, Entomol. Soc. Am., N.Y. Acad. Scis., Soc. Environ. Toxicology and Chemistry, Soc. Neurosci., Sigma Xi. Subspecialties: Environmental toxicology; Neurochemistry. Current work: neurotoxicity, pharmacokinetics.

ABOULELA, MOHAMED, biology educator; b. Alexandria, Egypt, Mar. 10, 1918; s. Mohamed and Eisha (Shihata) A.; m. Jean Wygle, Jan. 1, 1951; children: Shareen, Noreen, Nasreen, Fareed, Amir, Hosam. B.S., Cairo U., 1940; M.S., Iowa State U., 1947, Ph.D., 1950. Tchr. Sch.Agr., Damanhoor, Egypt, 1940-42; instr. U. Alexandria, Egypt, 1942-50, asst. prof., 1950, prof., to 1963; research scientist Tex. A&M U., College Station, 1962-65; assoc. prof. biology Tex. Woman's U., Denton, 1965—; cons. Agrarian Reform Authority, Cairo, Egypt, 1956-62. Co-author: (in Arabic): Introduction to Experimental Design, 1965. Mem. AAAS, Am. Inst. Biol. Sci., Tex. Acad. Scis., Sigma Xi. Democrat. Subspecialty: Plant physiology (biology). Current work: Effect of radiation on enzymatic activity of growing bean seedlings. Home: Rural Route 1 Box 409 H Denton TX 76201 Office: Dept Tex Woman's U Denton TX 76204

ABRAHAM, GEORGE, research physicist, engineer; b. N.Y.C., July 15, 1918; s. Herbert and Dorothy (Jacoby) A.; m. Hilda Mary Wenz, Aug. 26, 1944; children: Edward H., Dorothy J., Anne H., Alice J. Sc.B., Brown U., 1940; S.M., Harvard U., 1942; Ph.D., U. Md., 1972; postgrad., MIT, George Washington U. Registered profl. engr., D.C. Chmn. bd., pres. bd. Intercollegiate Broadcasting System, N.Y., 1941—; radio engr. RCA, Camden, N.J., 1941; with Naval Research Lab., Washington, 1942—, head sci. edn., head exptl. devices and microelectronics sects., 1945-69, head systems applications Office of Dir. Research, 1969-75, research physicist Office Research and Tech. Applications, cons., 1975—; lectr. U. Md., 1945-52, George Washington U., 1952-67, Am. U., 1979; indsl. cons.; mem. Bd. Registration Profl. Engrs. Contbr. chpts. to books, articles to profl. jours. Chmn. bd. Canterbury Sch., Accokeek, Md.; mem. schs. and scholarships com. Harvard U.; active PTA, Boy Scouts Am. Served to capt. USNR, World War II. Recipient Group Achievement award Fleet Ballistic Missile Program U.S. Navy, 1963; Edison award Naval Research Lab., 1971; Research Publ. award, 1974; Patent awards, 1959-75; D.C. Sci. citation, 1982; others. Fellow IEEE (Harry Diamond award 1981), Washington Acad. Scis. (pres. 1974-75), N.Y. Acad. Scis., AAAS; Mem. Am. Phys. Soc., Am. Assn. Physics Tchrs., Am. Soc. Naval Engrs., Washington Soc. Engrs. (pres. 1974, award 1981), Philos. Soc. Washington, AAUP, Sierra Club, Sigma Xi, Sigma Pi Sigma, Tau Beta Pi, Sigma Tau, Eta Kappa Nu, Iota Beta Sigma. Clubs: Cosmos, Harvard (Washington); Appalachian Mountain (Boston); Sierra (San Francisco). Subspecialties: Computer engineering; Integrated circuits. Current work: Solid state elecyronics, microelectronics, multivalued logic implementation, solid state physics, electrical engineering. Patentee in field. Home: 3107 Westover Dr SE Washington DC 20020 Office: Naval Research Lab Washington DC 20375

ABRAHAMS, CLARK RICHARD, operations research manager, consultant; b. San Francisco, Mar. 7, 1951; s. Harold Jay and Ida (Krasne) A.; m. Judy Carolyn Whitesides, Oct. 24, 1980; 1 son, Robert Bradford. A.B., U. Calif.-Berkeley, 1973; M.S., Stanford U., 1974. Project mgr. Standard Oil Co. of Ind., Chgo., 1974-76; project mgr. Fair, Isaac & Co., Inc., San Rafael, Calif., 1976-78; pres. StatComp, San Francisco, 1978-80; asst. v.p. Bank of Calif., San Francisco, 1980-81; corp. mgr. ops. research Barclays Am. Corp., Charlotte, N.C., 1981—; pres. StatComp, Charlotte, N.C., 1983—. Inventor: computerized process Genosys, 1978, Credit Management Tool RE&PF Systems, 1979. Mem. Am. Statis. Assn., Ops. Research Soc. Am. Democrat. Presbyterian. Club: Mt. Diablo Jaycees. Subspecialties: Operations research (mathematics); Statistics. Current work: Discrete multivariate statistics; design implementation of statistical computing software; automated generation-selection of optimal statistical decision models; stochastic modeling of consumer behavior; financial modeling. Home: 216 Altondale Ave Charlotte NC 28207 Office: Barclays Am. Corp 201 S Tryon St Charlotte NC 28231

ABRAHAMSON, HARMON BRUCE, chemistry educator, researcher; b. Cokato, Minn., Aug. 26, 1952; s. Harvey B. and Ruth A. (Werner) A.; m. Julie K. Conrad, May 22, 1982; m. Cathy L. Vandenheuvel, May 24, 1974 (div. Apr. 1979). B.Chem., U. Minn., 1974; Ph.D., MIT, 1978. Undergrad. research asst. U. Minn., Mpls., 1971-74; grad. asst. MIT, Cambridge, 1974-78; asst. prof. chemistry U. Okla., Norman, 1978—. Contbr. articles to profl. jours. Nat. merit scholar U. Minn., 1970-74; NSF grad. fellow MIT, 1975-78; recipient Freshman Chemistry award Alpha Chi Sigma, 1971; M. Cannon Sneed award dept. chemistry U. Minn., 1974. Mem. Am. Chem. Soc., Interam. Photochem. Soc., Sigma Xi, Tau Beta Pi, Phi Lambda Upsilon. Democrat. Lutheran. Subspecialties: Inorganic chemistry; Photochemistry. Current work: Spectroscopy and photochemistry of metal-alkyl complexes, photochemical reactions of metal-metal bonds, e.g. with organic disulfides, synthesis of metal complexes of sulfur monoxide. Office: Dept Chemistry U Okla 620 Parrington Oval Norman OK 73019

ABRAHAMSON, WARREN GENE, II, biologist, educator; b. Ludington, Mich., Mar. 26, 1947; s. Warren Gene and Alice Enid (Johnson) A.; m. Christy Raye Abrahamson, Aug. 16, 1969; 1 dau., Jill Raye. B.S. in Botany, U. Mich., 1969; M.S. in Biology, Harvard U., 1971, Ph.D., 1973. Asst. prof. biology Bucknell U., Lewisburg, Pa., 1973-79, assoc. prof., 1979-83, David Burpee prof. plant genetics, 1983—; vis. asst. prof. Mich. State U., East Lansing, 1976; research fellow Archbold Biol. Sta., Lake Placid, Fla., 1980-81, research assoc., 1976—. Contbr. articles to sci. jours. Recipient Bradley-Moore-Davis award U. Mich., 1969; Lindback award, 1975-76; Class of 1956 Lectureship, 1982-83; NSF grantee, 1974—. Mem. Audubon Soc. (chpt. program chmn. 1978-80, chpt. pres. 1982—), Bot. Soc. Am., Ecol. Soc. Am., Fla. Acad. Sci., Internat. Soc. Plant Population Biologists, Soc. Study of Evolution, Torrey Bot. Club, Sigma Xi, Phi Sigma. Democrat. Methodist. Subspecialties: Evolutionary biology; Population biology. Current work: Plant-insect interactions, goldenrod-stem gall insect-parasitoid guild interaction, plant life histories, fire ecology of Florida. Home: 153 Mountain View Rd Lewisburg PA 17837 Office: Bucknell U Lewisburg PA 17837

ABRAIRA, CARLOS, physician, hosptial section administrator, researcher; b. Buenos Aires, Mar. 25, 1936; U.S., 1963, naturalized, 1976; s. Jose B. and Maria (Cela) A.; m. Rosa M.E. Saffier, July 11, 1963; children: Daniel, Irene. M.D., U. Buenos Aires, 1962, B.S., 1953. Diplomate: Am. Bd. Internal Medicine. Intern U. Buenos Aires, 1962-63, Mercy Hosp., Chgo., 1963-64; resident Mt. Sinai Hosp., Chgo., 1964-66, Michael Reese Hosp., 1967-69; fellow Evanston (Ill.) Hosp., 1966-67; asst. prof. U. Ill., 1970-78, assoc. prof., 1978—; asst. chief endocrinology Hines (Ill.) VA Hosp., 1970-72, chief, 1972—, chief diabetes research lab., 1971—. Editor: (with others) Learning to be Your Manger, 1981; contbr. reviews and articles to profl. jours. Allstate Found. grantee, 1966-67; VA grantee, 1976—; The Sugar Found. grantee, 1983—. Fellow A.C.P.; mem. Am. Soc. Clin. Nutrition, Am. Inst. Nutrition, Am. Diabetes Assn., Chgo. Med. Soc. (coordinator diatbetes program 1980—). Clubs: Chgo. Endocrine, Sugar. Subspecialties: Endocrinology; Nutrition (medicine). Current work: Glucose metabolism; diabetes, nutrtion and lipidmetabolism; vascular disease in diabetes. Office: Hines VA Hospital Medical Service 111C Hines IL 60141

ABRAMS, HERBERT LEROY, radiologist, educator; b. N.Y.C., Aug. 16, 1920; s. Morris and Freda (Sugarman) A.; m. Marilyn Spitz, Mar. 23, 1943; children—Nancy, John. B.A., Cornell U., 1941; M.D., State U. Medicine, N.Y., 1946. Diplomate: Am. Bd. Radiology. Intern L.I. Coll. Hosp., 1946-47; resident in internal medicine Montefiore Hosp., Bronx, N.Y., 1947-48; resident in radiology Stanford U. Hosp., 1948-51; practice medicine specializing in radiology Stanford, Calif., 1951-67; faculty Sch. Medicine Stanford, 1951-67, dir. div. diagnostic roentgenology, 1961-67, prof. radiology, 1962-67; Philip H. Cook prof. radiology Harvard U., 1967-80, chmn. dept. radiology, 1967-80; radiologist-in-chief Peter Bent Brigham Hosp., Boston, 1967-80; chmn. dept. radiology Brigham and Women's Hosp., Boston, 1981—; radiologist-in-chief Sidney Farber Cancer Inst., Boston, 1974—; R.H. Nimmo vis. prof. U. Adelaide, Australia; mem. radiation study sect. NIH, 1962-66; cons. to hosps., profl. socs. Author: (with others) Congenital Heart Disease, 1965; Coronary Arteriography: A Practical Approach, 1983; Editor: Angiography, 1961, 3d edit., 1983, Investigative Radiology; editor-in-chief: Cardiovascular and Interventional Radiology Postgrad. Radiology; Contbr. (with others) articles to profl. jours. Nat. Cancer Inst. fellow, 1950; Spl. Research fellow Nat. Heart Inst., 1960, 73-74; David M. Gould Meml. lectr. Johns Hopkins, 1964; William R. Whitman Meml. lectr., 1968; Leo G. Rigler lectr. Tel-Aviv U., 1969; Holmes lectr. New Eng. Roentgen Ray Soc., Boston, 1970; Ross Golden lectr. N.Y. Roentgen Ray Soc., N.Y.C., 1971; Stauffer Meml. lectr. Phila. Roentgen Ray Soc., 1971; J.M.T. Finney Fund lectr. Md. Radiol. Soc., Ocean City, 1972; Aubrey Hampton lectr. Mass. Gen. Hosp., Boston, 1974; Kirklin-Weber lectr. Mayo Clinic, 1974; Crookshank lectr. Royal Coll. Radiology, 1980; W.H. Herbert lectr. U. Calif.; Caldwell lectr. Am. Roentgen Ray Soc., 1982; Percy lectr. McMaster Med. Sch., 1983; Henry J. Kaiser sr. fellow Center for Advanced Study in Behavioral Sci., 1980-81. Hon. fellow Royal Coll. Radiology; mem. Assn. Univ. Radiologists, Inst. Medicine, Am. Coll. Radiology, Am. Heart Assn., Am. Soc. Nephrology, Radiol. Soc. N.Am., N.Am. Soc. Cardiac Radiology (pres. 1979-80), Soc. Cardiovascular Radiology, Mass. Radiologic Soc., Soc. Chmn. Acad. Radiology Depts. (pres. 1970-71), Phi Beta Kappa, Alpha Omega Alpha. Subspecialty: Radiology. Home: 433 Walnut St Brookline MA 02146 Office: Harvard Med Sch Boston MA 02115

ABRAMS, RICHARD FRANCIS, engineering, systems development manager; b. Cambridge, Mass., Nov. 14, 1948; s. Abraham and Dorothea (Bannen) A.; m. Jean Parker, June 10, 1972; children: Keith, Julie. B.S. in Chem. Engring, Worcester Polytech. Inst., 1970. Pilot plant supr. Artisan Industries Inc., Waltham, Mass., 1971-74; process engr. CTI Cyrogenics, Waltham, 1974-76; supr. process engring. Helix Tech, Westboro, Mass., 1976-78, task force leader, 1978-80; mgr. engring./devel. Koch Process Systems (formerly Helix/CTI), 1980—. Men. Am. Inst. Chem. Engrs., Am. Nuclear Soc. Subspecialties: Nuclear engineering; Gas cleaning systems. Current work: Develop and design radioactive waste treatment systems for nuclear facilities, gaseous and solid wastes. Patentee improved control system, 1980, radioactive waste process, 1982. Home: 9 Jasper St Westboro MA 01581

ABRAMS, ROBERT JAY, physics educator; b. Chgo., June 16, 1938; s. Philip and Esther (Marmelstein) A.; m. Dorothy E. Manaster, Jan. 30, 1966; children: Elizabeth, Sherri, Susan. B.S., Ill. Inst. Tech., 1959; M.S., U. Ill., 1961, Ph.D. in Physics, 1966. Research assoc. Brookhaven Nat. Lab., Upton, N.Y., 1966-68, asst. physicist, 1968-70; asst. prof. physics U. Ill.-Chgo., 1970-74, assoc. prof., 1974—; cons. Electri-Graphics, Inc., 1980—. Contbr. articles to profl. jours. NSF grantee, 1974—. Mem. Am. Phys. Soc. Subspecialties: Particle physics; Graphics, image processing, and pattern recognition. Current work: High transverse momentum collisions, production of dimuons, collisions at 2 Tev CM energy, industrial particle accelerator systems. Home: 1444 Dartmouth Ln Deerfield IL 60015 Office: Dept Physics U Ill Box 4348 Chicago IL 60680

ABRAMS, ROBERT MARLOW, physiologist, researcher; b. Rome, N.Y., June 8, 1931; s. Marlow Edward and Gladys Olivia (Brown) A.; m. Patricia Theobalnd, June 29, 1956; children: William, Thomas, Julie, Sarah. A.B., Hamilton Coll., Clinton, N.Y., 1953; M.S., U. Pa., 1960, D.D.S., 1960, Ph.D., 1963. Asst. fellow John B. Pierce Found., New Haven, Conn., 1963-68; asst. prof. epidemiology and public health Yale U., 1968-69; asst. prof. U. Fla. (dept. obstetrics and gynecology), 1969-74, assoc. prof., 1974—. Contbr. numerous articles in field to profl. jours. Served with U.S. Army, 1953-55. Fogarty Sr. Internat. fellow NIH, 1977-78; recipient Career Devel. award NIH, 1954-59. Mem. Am. Physiology Soc., Soc. Gynecologic Investigation. Subspecialties: Reproductive biology (medicine); Physiology (biology). Current work: Reproductive physiology; fetal brain energy metabolism, sound environment of fetus. Office: University of Florida Kealth Center Box J-294 Gainesville FL 32610

ABRAMSON, EDWARD E(RIC), psychology educator, clinical psychologist; b. Bklyn., July 7, 1944; s. Morris B. and Helen (Landau) A.; m. Alina M. Rosette, Dec. 21, 1968; children: Anne, Jeremy. B.A., SUNY-Stony Brook, 1965; Ph.D., Cath. U. Am., 1971. Lic. psychologist, Calif. Clin. psychol. intern VA Hosp., Martinsburg, W.Va., 1969-70; asst. prof., assoc. prof. Calif. State U.-Chico, 1970-79, prof. psychology, 1979—; clin. psychologist York Clinic, Guy's Hosp., London, Eng., 1978, Glenn Gen. Hosp., Willows, Calif., 1980—; oral examiner Bd. Med. Quality Assurance, Calif. Psychology Examining Com., Sacramento, 1979—; mem. Butte County Mental Health Adv. Bd., Choco, 1974-77. Author: Behavioral Approaches to Weight Control, 1977; contbr. chpt. to book in field. Mem. Am. Psychol. Assn., Assn. for Advancement of Behavior Therapy, Assn. for Media Psychology (membership com. 1982—), Western Psychol. Assn. Subspecialty: Behavioral psychology. Current work: Behavioral treatment of obesity, emotional and social variables in eating disorders, health psychology. Home: Rt 2 Box 200 A Chico CA 95926 Office: Dept Psychology Calif State Univ Chico CA 95929

ABRAMSON, HYMAN NORMAN, engring. and sci. research exec.; b. San Antonio, Mar. 4, 1926; s. Nathan and Pearl (Westerman) A.; m. Idelle Rebecca Ringel, Apr. 20, 1947; children— Phillip David, Mark Donald. B.S.M.E., Stanford U., 1950, M.S. in Engring. Mechanics, 1951; Ph.D. in Engring. Mechanics (So. Fellowship Fund fellow), U. Tex., Austin, 1956. Engr. U.S. Naval Air Missile Test Center, Point Mugu, Calif., 1947-48; project engr. Chance Vought Aircraft Co., Dallas, 1951-52; asso. prof. aero. engring. Tex. A&M U., 1952-55; sect. mgr., dept. dir. S.W. Research Inst., San Antonio, 1956-72, v.p. div. engring. scis., 1972—; mem. research adv. com. USCG; adv. panel engring. mechanics NSF; com. U.S. Dept. Commerce. Author: An Introduction to the Dynamics of Airplanes, 1958, reprinted, 1971; contbr. numerous articles to profl. publs.; editor: (with others) Applied Mechanics Surveys, 1966, The Dynamic Behavior of Liquids in Moving Containers, 1966; asso. editor: Applied Mechanics Revs, 1954—; editorial adv. bd.: Jour. Computers and Structures, 1970—, Aeros. and Astronautics, 1975—. Mem. Greater San Antonio C. of C., and City of San Antonio Market Sq. Adv. Com., 1973-77; mem. U.S. Bicentennial Com. of San Antonio, 1975-76. Served with USN, 1943-45. Mem. AIAA (Disting. Service award 1973, dir.), ASME (v.p.; hon.), Nat. Acad. Engring., Soc. Naval Architects and Marine Engrs., AAAS, Sigma Xi. Republican. Jewish. Subspecialties: Theoretical and applied mechanics; Fluid mechanics. Current work: Fluid-structure interaction. Home: 1511 Spanish Oaks Dr San Antonio TX 78213 Office: 6220 Culebra Rd San Antonio TX 78284

ABRASS, ITAMAR B., physician, educator; b. Chedera, Israel, Nov. 15, 1941; came to U.S., 1950, naturalized, 1955; m. Christine Kreger, Nov. 23, 1974; children: David L., Rachel B. Student, U. Calif., Berkeley, 1959-62, B.S., 1963, M.D., 1966. Diplomate: Am. Bd. Internal Medicine. Intern Columbia-Presbyn. Hosp., N.Y.C., 1966-67, resident in internal medicine, 1967-68; clin. assoc. NIH, Bethesda, Md., 1968-70; fellow in endocrinology U. Calif. Sch. Medicine, San Diego, 1970-71, asst. prof. medicine, 1972-76, UCLA Sch. Medicine, 1976-80, assoc. prof., 1980—; clin. dir. Geriatric Research , Edn. and Clin. Center, VA Med. Center, Sepulveda, Calif., 1976—, co-dir., 1982—. Author 2 books, numerous sci. articles. Fellow A.C.P.; mem. Endocrine Soc., Am. Fedn. for Clin. Research, Gerontol. Soc. Am., AAAS. Subspecialties: Endocrinology; Gerontology. Current work: Mechanisms of hormone action in aging. Home: 2117 Banyan Dr Los Angeles CA 90049 Office: VA Med Center 16111 Plummer St Sepulveda CA 91343

ABRUTYN, ELIAS, physician, educator; b. Jersey City, Apr. 15, 1940; s. Samuel Bruce and Eva (Honigberg) A.; m. Leslye Silver, Dec. 23, 1975; 1 son, Alex. B.A., U. Pa., 1960; postgrad., U. Basel, Switzerland, 1960-63; M.D., U. Pitts., 1966. Diplomate: Am. Bd. Internal Medicine. Intern U. Pa. Sch. Medicine, 1966-67, resident, 1967-68, chief resident, 1970-71, asst. prof. medicine, 1970-78; acting asst. chief immunizations br., field services div. Epidemic Intelligence Service Ctrs. Disease Control, Atlanta, 1968-70; assoc. prof. Med. Coll. Pa., 1979-82, prof., 1982—, asst. chief med. services, 1978—. Contbr. articles, chpts. to profl. publs.; assoc. editor: Annals of Internat. Medicine, 1978—. Served to surgeon USPHS, 1968-70. Fellow Infectious Diseases Soc. Am., ACP; mem. Hosp. Infection Soc., Am. Fedn. Clin. Research, soc. Hosp. Epidemiologists Am., AMA, Alpha Omega Alpha. Subspecialties: Internal medicine; Infectious diseases. Current work: Antibiotic pharmacology; hospital epidemiology. Home: 209 Rhyl Ln Bala-Cynwyd PA 19004 Office: VA Hosp Philadelphia PA 19104

ABSOLOM, DARRYL ROBIN, immunologist; b. Germiston, Transvaal, S. Africa, Dec. 15, 1954; s. George Dalziel and Margaret (Marlowe) A. B.Sc., U. Cape Town, S. Africa, 1973, 1974, Ph.D., 1977. Research asst. prof. SUNY-Buffalo, 1978-82, research assoc. prof., 1983—; cancer scientist Roswell Park Meml. Inst., Buffalo, 1978-79; scientist Hosp. for Sick Children, Toronto, Ont., Can., 1982—; asst. prof. U. Toronto, 1982-83, assoc. prof., 1983—. Editor: Molecular Immunology, 1983; Editorial advisor: Immunological Communication Jour, 1980—, Preparative Biochemistry Jour, 1980—, Biol. Dispersions and Surfaces Jour, 1983—. Rotary Internat. Scholarship awardee, 1977; Ont. Heart Found. Fellowship awardee, 1978-82; Can. Heart Found. Sr. Fellowship awardee, 1983—. Mem. Internat. Surface Sci. Group, Am. Soc. Artificial Internal Organs, N.Y. Acad. Sci., Am. Chem. Soc. Subspecialties: Biomaterials; Biophysics (biology). Current work: Biomaterials, cell adhesion, protein adsorption, phagocytosis, cell-cell interactions, applied surface thermodynamics, antigen-antibody interactions. Office: Hosp for Sick Children Research Inst 555 University Ave Toronto ON Canada M5G 1X8

ABUZZAHAB, FARUK SAID, SR., psychiatrist, educator; b. Beirut, Lebanon, Oct. 12, 1932; came to U.S., 1959, naturalized, 1978; s. Said Salim and Nehmat (Muezzin) A.; m. Kathryn Buoen, Oct. 1982; children: Nada, Jennifer, Faruk, Jeffrey, Mark, Brita. M.D., Am. U. Beirut, 1959; Ph.D., U. Minn., 1968. Diplomate: Am. Bd. Psychiatry and Neurology. Rotating intern Am. U. Hosp., Beirut, 1958-59; jr., then sr. asst. resident Johns Hopkins Hosp., Balt., 1959-62; vis. resident psychiatry Balt. City Hosp., 1960-62; fellow dept. psychiatry Johns Hopkins U. Sch. Medicine, 1959-62; research psychiatrist Henry Phipps Psychiat. Clinic, 1959-62; research fellow depts. psychiatry and pharmacology U. Minn., Mpls., 1962-66, instr., 1966-67, asst. prof., 1967-71, clin. asst. prof., 1971-73, clin. asso. prof., 1973-79, clin. prof., 1979—; clin. dir. Hastings (Minn.) State Hosp., 1963-78; practice medicine specializing in adult psychiatry and clin. psycho-pharmacology, Mpls., 1965-73; staff psychiatrist Dunn County Health Care Center, Menomonie, Wis., 1971—; active med. staff St. Mary's Hosp., Mpls., 1971—; asso. in psychiatry Fairview Hosp., Mpls., 1971—; also cons.; pres. Clin. Psychopharmacology Cons., P.A., Mpls., 1973—. Editorial adv. bd.: Jour. Practical Therapeutics, 1979—, Am. Jour. Psychiatry, 1980—; Contbr. articles to sci. publs. Adminstr. Pharmacopsychiatry Fund, Minn. Med. Found., Mpls., 1972—. Fellow Am. Psychiat. Assn., Am. Coll. Clin. Pharmacology; mem. Johns Hopkins Med. and Surg. Soc., Med. Alumni Assn. of Am. U. Beirut (life), AMA, Minn. Psychiat. Soc., Minn. State Med. Soc., Hennepin County Med. Soc., Hennepin County Psychiat. Soc. (pres. 1976-78), Am. Soc. Clin. Pharmacology and Therapeutics, Soc. Biol. Psychiatry, Minn. Soc. Neurol. Sci., Collegium Internationale Neuropsychopharmacologicum, Am. Soc. Pharmacology and Exptl. Therapeutics, Am. Assn. Geriatric Psychiatry, Sigma Xi, Alpha Omega Alpha. Subspecialties: Psychiatry; Psychopharmacology. Current work: Psychopharmacology, neuropsychopharmacology research, teaching and clinical practice. Home: 2601 E Lake of Isles Blvd Minneapolis MN 55408 Office: 606 24th Ave S Suite 818 Minneapolis MN 55454

ACHENBACH, JAN DREWES, engineering scientist, educator; b. Leeuwarden, Netherlands, Aug. 20, 1935; came to U.S., 1959, naturalized, 1978; s. Johannes and Elizabeth (Schipper) A.; m. Marcia Graham Fee, July 15, 1961. Kand. Ir., Delft U. Tech., 1959; Ph.D., Stanford U., 1962. Preceptor Columbia U., 1962-63; asst. prof. Northwestern U., Evanston, Ill., 1963, assoc. prof., 1966-69, prof. dept. civil engring., 1969—, Walter P. Murphy prof. civil engring. and applied math., 1981—; vis. assoc. prof. U. Calif., San Diego, 1969; vis. prof. Tech. U. Delft, 1970-71; cons. Huazhong Inst. Sci. and Tech., 1981, prof., 1981; mem. at large U.S. Nat. Com. Theoretical and Applied Mechanics, 1972-78. Author: Wave Propagation in Elastic Solids, 1973, A Theory of Elasticity with Microstructure for Directionally Reinforced Composites, 1975, (with A.K. Gankson and H. McMaken) Ray Methods for Waves in Electric Solids, 1982; editor: (with J. Miklowitz) Modern Problems in Elastic Wave Propagation, 1978; editor-in-chief: Wave Motion, 1979—. Recipient award C. Gelderman Found., 1970, C.W. McGraw Research award Am. Soc. Engring. Edn., 1975. Fellow Am. Acad. Mechanics (pres. 1978-79), ASME; mem. Am. Geophys. Union, AIAA., U.S. Nat. Acad. Engring. Subspecialties: Theoretical and applied mechanics; Applied mathematics. Current work: The propagation of mechanical disturbances in solids; with applications to scattering of ultrasonic waves by cracks for nondestructive evalutation of materials and acoustic radiation by material damage regions; fracture mechanics. Home: 574 Ingleside Park Evanston IL 60201 Office: Dept Civil Engring Northwestern U Evanston IL 60201

ACHENBACH, THOMAS M., psychologist; b. Stamford, Conn., Dec. 29, 1940; s. Hans and Mary (Hill) A.; m. Susan Young, June 18, 1966; children: Gretchen, Christopher. B.A., Yale U., 1962; postgrad., Heidelberg (W.Ger.) U., 1962-63; Ph.D., U. Minn., 1966. Lic. psychologist, Vt. Postdoctoral fellow, research staff Yale U., New Haven, 1966-68, from asst. prof. to assoc. prof. psychology, 1968-75; research psychologist NIMH, Bethesda, Md., 1978-80; prof. psychiatry U. Vt., Burlington, 1980—; cons. NIMH, NIH, others. Author: Developmental Psychopathology, 2d edit, 1982, Research in Developmental Psychiatry, 1978, Manual for the Child Behavior Checklist, 1983. Social Sci. Research Council fellow, 1971-72; Am. Psychol. Found. grantee, 1982—; Spencer Found. grantee, 1982—; March of Dimes Found. grantee, 1982—. Fellow Am. Psychol. Assn.; mem. Soc. for Research in Child Devel., AAAS, Internat. Soc. Study Behavioral Devel. Subspecialty: Developmental psychology. Current work: Assessment of children's behavioral problems and competencies; epidemiology; outcome of treatment. Home: Marble Island Rd Colchester VT 05446 Office: Child Psychiatry U Vt 1 S Prospect St Burlington VT 05401

ACHESON, WILLARD P(HILLIPS), applied physicist, researcher; s. Joseph Willard and Alice (Phillips) A.; m. Patricia L. Marshall, Aug. 26, 1949; children: Louise S., Ann L., Amy J. B.S., Westminster Coll., 1948; M.S., Pa. State U., 1950; Ph.D., U. Pitts., 1961. Fellow Mellon Inst., 1953-60; asst. prof. Muskingum Coll., 1960-62; sr. research assoc. Gulf Research & Devel. Co., Pitts., 1962—. Mem. Am. Phys. Soc., Am. Assn. Physics Tchrs., Soc. Indsl. and Applied Math., Soc. Petroleum Engrs. Subspecialties: Petroleum engineering; Fuels and sources. Current work: Applications of fluid mechanics, heat transfer, and thermodynamics by means of experiments and numerical simulation to the enhanced recovery of petroleum and related resources. Patentee in field. Office: Gulf Research & Devel Co PO Box 2038 Pittsburgh PA 15230

ACHORD, JAMES LEE, medical educator; b. Dayton, Ohio, Sept. 24, 1931; s. Lonnie M. and Ethel (Collins) A.; m. Pat Moore, Dec. 18, 1954; children: Michael, Ann, Andrew. M.D., Emory U., 1956. From instr. to asst. prof. Emory Med. Sch., Atlanta, 1962-71; med. dir. Med. Center Central Ga., Macon, 1971-75; prof., assoc. dean East Tenn. State U. Sch. Medicine, Johnson City, 1975-76; prof. medicine, dir. div. digestive diseases U. Miss. Sch. Medicine, Jackson, 1976—; chief of staff Univ. Med. Center, Jackson, 1981-82. Editor: Chronic Inflammatory Bowel Disease, 1974; author articles. Mem. adminstrv. bd. Meth. Ch., Jackson, 1983. Served to capt. M.C. U.S. Army, 1957-59. Fellow Am. Coll. Gastroenterology (pres. 1984), A.C.P., Am. Gastroenterological Assn., Am. Assn. for Study of Liver Disease, Fedn. Digestive Disease Socs. (vice chmn. 1982-83), Ga. Gastroenterological Soc. (founding pres. 1971-73). Subspecialties: Gastroenterology; Internal medicine. Current work: Liver disease, peptic ulcer disease, others. Home: 202 Wexford Ct Jackson MS 39208 Office: U Miss Sch Medicine 2500 N State St Jackson MS 39216

ACHTARIDES, THEODOROS ANTONIOU, naval architect, ocean engineering educator, consultant; b. Thessaloniki, Greece, July 30, 1948; came to U.S., 1981; s. Antonios Athanasiou and Vasilia Athanasiou (Gatzaki) A. B.A. summa cum laude, Augustana Coll., Sioux Falls, S.D., 1968; S.M., MIT, 1972, O.E., 1973; Ph.D., U. Newcastle Upon Tyne, Eng., 1979. Registered profl. engr., USA, U.K.; profl. lic., Greece. Teaching asst. M.I.T., 1968-69, 70-73, research fellow, 1969-70, research asst., 1972; research engr. Am. Bur. Shipping, N.Y.C., 1973; demonstrator U. Newcastle Upon Tyne, 1975-78, Ridley research fellow, 1974-77; asst. research engr. Stevens Inst. Tech., 1981—, asst. prof. dept. ocean engring., 1981—; cons. Hydromechanics, Inc., Plainview, N.Y., 1982. Contbr. articles to profl. jours. Served with Hellenic Army Signal Corps., 1980. Recipient 1st prize Naval Architects and Marine Engrs., 1973, Naval Survey 1st prize, 1975. Fellow Hellenic Inst. Marine Technology; mem. Royal Inst. Naval Architects, Inst. Marine Engrs., Marine Tech. Soc., Soc. Underwater Technology, Underwater Assn., Assn. Technique Maritime Et Aeronautique, Soc. Naval Architects and Marine Engrs., Insts. Engrs. and Shipbuilders in Scotland, Sigma Xi. Current work: Wave excited vibration; ship frequencies prediction; ultimate strength of ships; elasto-plastic analysis and design of plates. Office: Department Ocean Engineering Stevens Institute of Technology Castle PointStation Hoboken NJ 07030

ACKERMAN, ALLAN DOUGLAS, energy conservation cons., architect; b. London, June 13, 1947; s. John B. and Anne Faith (Donaldson) A.; m. Mary Abigail King, Mar. 14, 1970; children: Molly, Samuel. B.A., Dartmouth Coll., 1968; M.Arch., Harvard U., 1974. Project mgr. ABT Assocs., Cambridge, Mass., 1974-75; v.p. Technology & Econs., Cambridge, 1975-77; pres. Energyworks, Inc., West Newton, Mass., 1977—. Bd. overseers Shady Hill Sch. Served to 1st lt. C.E. U.S. Army, 1969-71; Vietnam. Decorated Bronze Star with one oak leaf cluster.; Disting. Mil. Grad. U.S. Army Engr. Officer Candidate Sch., 1969. Subspecialties: energy conservation; Information systems, storage, and retrieval (computer science). Current work: Energy efficiency problems in buildings: approach combines energy auditing and service with information products, publications and software. Home: 54 Irving St Arlington MA 02174 Office: 45 Border St West Newton MA 02165

ACKERMAN, RALPH EMIL, psychology educator; b. N.Y.C., Aug. 21, 1921; s. Meyer H. and Frieda (Broadman) A.; m. Eileen E. Dotta, Mar. 13, 1945. B.S.Ed., Ohio U., 1948, M.Ed., 1949; postgrad., Columbia U., 1952; Ph.D., U. Conn., 1960. Lic. psychologist, Pa. Tchr. biology Bobyns-Bennett High Sch., Kingsport, Tenn., 1949-54; tchr. math. and sci. Glastonbury, (Conn.) High Sch., 1954-61; vis. prof. Marshall U., Huntington, W.Va., summer 1961; prof. psychology Edinboro (Pa.) U., 1961—; cons. Police Merit Bd., Erie, Pa., 1970-75. Contbr. articles to profl. jour. Served with U.S. Army, 1942-46; PTO. Named Outstanding Educator Edinboro U., 1969. Fellow Pa. Psychol. Assn. (pres. 1973-74); mem. Am. Psychol. Assn., Am. Ednl. Research Assn., NEA, Pa. Edn. Assn., Phi Delta Kappa. Republican. Lodges: Masons; Shriners. Subspecialties: Learning; Developmental psychology. Current work: Research on sex-role development and aging process. Home: 124 Sunset Dr PO Box 29 Edinboro PA 16412 Office: Edinboro Univ Edinboro PA 16444

ACKERMAN, ROBERT HAROLD, neuroscientist, neurologist/ neuroradiologist; b. N.Y.C., June 1, 1935; s. Myron and Leona (Auerbach) Ackerman. B.A. in History, Brown U., 1957; postgrad., Columbia U., 1959-60; M.D., U. Rochester, 1964; M.P.H., Harvard U., Boston, 1972. Diplomate: Am. Bd. Psychiatry and Neurology, Am. Bd. Radiology. Rotating intern Bassett Hosp., Cooperstown, N.Y., 1964-65; resident in medicine U. N.C.-Chapel Hill, 1965-66, resident in neurology, 1966-68; fellow in neuropathology Mass. Gen. Hosp., Boston, 1968-69, resident in neurology, 1969-70, fellow in neuroradiology, 1972-75, dir. Carotid Evaluation lab. and Lab. Cerebral Blood Flow and Metabolism, 1975—, assoc. neurologist, assoc. radiologist, 1980—; cerebrovascular research fellow Nat. Hosp. Nervous Diseases, London, 1970-71; instr. in neurology Harvard U., 1972-75, asst. prof. radiology, 1975-80, assoc. prof., 1980—; cons. to dept. pharm. research New Eng. Nuclear Corp., 1982. Contbr.

numerous articles to profl. jours.; editorial bd.: Archives Neurology, 1980—, Stroke, 1980—. Served to pfc. USAR, 1958-64. Dalton scholar, 1970; Nat. Insts. Neurol. and Communicative Disorders and Stroke spl. fellow, 1972-75; grantee, 1979, 81. Fellow Am. Acad. Neurology, Am. Heart Assn. Stroke Council (exec. com. 1979-81); mem. Am. Assn. Neuropathologists, Am. Coll. Radiology, AMA, Am. Neurol. Assn., Am. Soc. Neuroradiology, Assn. Univ. Radiologists, Soc. Neurosci., Soc. Nuclear Medicine. Clubs: Harvard of N.Y., Eastern Point Yacht. Subspecialties: Neurology; Diagnostic radiology. Current work: Development and application of new technology for diagnosis and treatment cerebrovascular and other neurological diseases; measurement of cerebro vascular and brain physiology. Home: Edgemoor Rd Gloucester MA 01930 Office: Mass Gen Hosp Boston MA 02114

ACKERMAN, ROY ALAN, biochemical engineer, research corporation executive; b. Bklyn., Sept. 9, 1951; s. Jack and Estelle (Kuchlik) A.; m.; 1 child, Shanna Avrah. B.S. in Chem. Engring., Poly. Inst., Bklyn., 1972; M.S. in Chem. Engring. and Biochem. Engring., MIT, 1974, Ph.D., U. Va., 1984. Registered profl. engr. Chem. engr. Tri-Flo Research Labs., Ltd., Queens, N.Y., 1972-74; sr. project engr. Thetford Corp., Ann Arbor, Mich., 1975; dir. research and devel. Astre Corporate Group, Charlottesville, Va., 1976-79, tech. dir., 1979—, also dir.; adj. prof. biomed. tech. George Washington U., Washington, 1977, 80, NYU, N.Y.C., 1979-80, Poly. Inst. N.Y., Bklyn., 1980-81; dir. Biofiltration Technols., Ltd., Manhasset, N.Y., 1980-82, Indsl. Microgenics, Ltd., Charlottesville and Roanoke, Va., 1980—, Edlaw Preparations, Inc., Farmingdale, N.Y., 1982—. Holder 10 U.S. patents; contbr. over 30 tech. articles to profl. publs. Bd. dirs. Jaycees, Charlottesville, 1979, adminstrv. v.p., 1980; mem. Albemarle County (Va.) Citizens Zoning Rev. Com., 1980-81, Charlottesville-Albemarle Airport Bd., 1983—. Samuel Ruben scholar, 1968-72. Mem. Am. Inst. Chemists, Am. Inst. Chem. Engrs., Am. Chem. Soc., Am. Soc. Microbiology, Am. Soc. Artificial Internal Organs, Soc. Indsl. Microbiology, Internat. Soc. Artificial Organs, Water Pollution Control Fedn. Subspecialties: Enzyme technology; Artificial organs. Current work: Medical device design and development; genetic engineering; hemodialysis; genetic recombination; respiratory therapy; hazardous waste treatment; water reuse and recycling. Office: AstreCorporate Group 1130 E Market St PO Box 5072 Charlottesville VA 22905

ACKERMANN, NORBERT JOSEPH, JR., energy technology company executive; b. Chattanooga, July 3, 1942; s. Norbert Joseph and Lusella (Smith) A.; m. (div.); children: Dori, Nancy, Andy, Jill. B.S., U. Tenn., 1965, M.S., 1967, Ph.D., 1971. Research instr. U. Tenn., Knoxville, 1967-68; nuclear engr. Union Carbide, Oak Ridge, 1968-71, head fast reactor measurement methods, 1971, head reactor controls devel. sect., 1971-76; pres. Tech. for Energy Corp., Knoxville, 1975—; dir. spl. instrumentation group Three Mile Island Recovery Operation, 1979; mem. adv. bd. dirs. United Am. Bank; bd. dirs. Energy Opportunities Consortium's Innovation Ctr. Contbr. articles in field of power plant surveillance and diagnostics and specialized measurement systems to profl. jours. Bd. dirs. Jr. Achievement. Herman Hickman fellow, 1966-67; AEC fellow, 1966-68; recipient Outstanding Engring. Alumunus award U. Tenn. Coll. Engring., 1974. Mem. Am. Nuclear Soc., IEEE, Am. Mgmt. Assn., AAAS, Energy Conservation Soc., Southeastern Conf. Football Ofcls. Assn. Current work: Technical services and instrumentation and computer systems business serving the electric power industry. Home: 12220 Bluff Shore Dr Knoxville TN 37922 Office: 1 Energy Ctr Pellissippi Pkwy Knoxville TN 37922

ACOSTA, DANIEL, pharmacologist, educator; b. El Paso, Tex., Mar. 25, 1945; s. Dan and Maria Luisa (Hernandez) A.; m. Patricia C. Clune, Aug. 18, 1973; children: Anna, Elise, Dani. B.S. in Pharmacy, U. Tex., Austin, 1968; Ph.D., U. Kans., 1974. Registered pharmacist, Tex., Ga. Research asst. U. Kans., Lawrence, 1970-74; asst. prof. pharmacology U. Tex., Austin, 1974-79, asso. prof., 1979—. Contbr. articles on toxicology, pharmacology and cell culture to sci. jours. Del. Travis County Democratic Conv., 1976, 80, 82. Served with U.S. Army, 1968-70. Recipient award Nat. Chicano Council on Higher Edn., 1979; Ford Found. fellow, 1978-79. Mem. Tissue Culture Assn., Am. Heart Assn., Soc. Toxicology, Am. Soc. Pharmacology and Exptl. Therapeutics, Phi Delta Chi (Alumunus of Distinction award Lambda chpt. 1981). Roman Catholic. Subspecialties: Toxicology (medicine); Cellular pharmacology. Current work: Cellular toxicology; drug metabolism; hepatic toxicology; cardiac injury. Home: 6210 Turkey Hollow Austin TX 78750 Office: Coll Pharmacy U Tex Austin TX 78712

ACOSTA, GUSTAVO, physician, clin. researcher; b. Cardenas, Cuba, Sept. 13, 1933; came to U.S., 1962, naturalized, 1969; s. Gustavo and Daria R. (Rua) A; m. Josefina A. Acosta, Dec. 27, 1958; children: Gustavo, Marie. B.S., Trinitarian Coll., Cardenas, 1951; student medicine, U. Havana, 1961; M.D., U. Salamanca, Spain, 1963. Intern U. Salamanca, 1963; staff physician Huntington (W.VA.) State Hosp., 1964-66; asst. sect. chief Osawatomie (Kans.) State Hosp., 1966-69; regional med. dir. Latin Am. region Am. Cyanamid Internat. (Lederle Labs), 1970-71; asso. med. dir. Internat. div. Abbott Labs., 1971-76; asso. dir. corp. med. affairs Marion Labs., 1976-79; asso. dir. clin. research dept. Stuart Pharms. div. ICI Ams., Inc., Wilmington, Del., 1979—. Author: The Role of the Physician in the Pharmaceutical Industry, 1980; editorial council: Medico Intermaericano. Recipient Presdl. awards Abbott Labs., 1974, 75. Mem. Am. Soc. Clin. Pharmacology and Therapeutics, Neurosci. Soc., Am. Burn Soc., Internat. Soc. Burn Injuries, Internat. Anesthesia Research Soc. Republican. Roman Catholic. Subspecialties: Anesthesiology; Psychiatry. Current work: Extensive clin. research in C.N.S. and anesthesiology, as well as infectious diseases.

ACRIVOS, ANDREAS, chem. engr., educator; b. Athens, Greece, June 13, 1928; came to U.S., 1947, naturalized, 1962; s. Athanasios and Anna (Besi) A.; m. Juana Vivo, Sept. 1, 1956. B.S. in Chem. Engring, Syracuse U., 1950; M.S., U. Minn., 1951, Ph.D., 1954. Instr., asst. prof., asso. prof. U. Calif., Berkeley, 1954-62; prof. chem. engring. Stanford U., 1962—. Contbr. tech. articles to profl. jours. Guggenheim fellow, 1959, 76. Mem. Am. Inst. Chem. Engrs. (Profl. Progress award), Am. Chem. Soc., Am. Phys. Soc., Soc. Rheology, Nat. Acad. Engring. Subspecialty: Chemical engineering. Office: Dept Chem Engring Stanford U Stanford CA 94305

ACRIVOS, JUANA VIVO, chemist, educator; b. Habana, Cuba; came to U.S., 1951, naturalized, 1962; d. Adolfo Vivo and Lilia Azpeitia; m. Andreas Acrivos, Apr. 1, 1956. Ph.D. in Phys. Chemistry and Math, U. Minn., 1956. Postdoctoral fellow Stanford U., 1956-59, U. Calif., Berkeley, 1959-62; lectr. chemistry U. Minn., Mpls., 1961; asst. prof. chemistry San Jose (Calif.) State U., 1963-67, assoc. prof., 1967-72, prof., 1972—. Contbr. articles to profl. jours. Roman Catholic. Subspecialties: Physical chemistry; Solid state chemistry. Current work: Development of magnetic resonance (semiquinones, MgO, and metal ammonia systems) and in interpretation of the metal insulator transitions (Mott-Anderson Theory and layer compounds) produced chemically. Office: Dept Chemistry San Jose State U 125 S 7th St San Jose CA 95192

ADAIR, GERALD MICHAEL, somatic cell geneticist, educator; b. St. Louis, June 8, 1949; s. Raymond S. and Audrey C. (Boelling) A.; m. Martha P. McNulty, Sept. 30, 1972; 1 dau., Jennifer E. B.A. cum laude in Biology, Washington U., St. Louis, 1971, Ph.D. in Biology, 1975. Research assoc. Lawrence Livermore (Calif.) Lab., Calif., 1975-78; asst. biologist Sci. Park Research Div., U. Tex. System Cancer Ctr., Smithville, 1978-82, assoc. biologist, 1982—, asst. prof. biology, 1982—. Mem. Am. Soc. Cell Biology, Environ. Mutagenesis Soc., Genetics Soc. Am. Democrat. Baptist. Subspecialties: Gene actions; Genome organization. Current work: Mammalian cell mutagenesis; somatic cell genetics; DNBA repair; gene mapping. Office: U Tex System Cancer Ctr Sci Park Smithville TX 78957

ADAIR, ROBERT KEMP, physicist, educator; b. Ft. Wayne, Ind., Aug. 14, 1924; s. Robert Clel and Margaret (Wiegman) A.; m. Eleanor Reed, June 21, 1952; children—Douglas McVeigh, Margaret Guthrie, James Cleland. Ph.B., U. Wis., 1947, Ph.D., 1951. Instr. U. Wis., 1950-53; physicist Brookhaven Nat. Lab., 1953-58; mem. faculty Yale, 1958—, prof. physics, 1961-72, Eugene Higgins prof. physics, 1972—, chmn. dept., 1967-70, dir. div. phys. scis., 1977—. Author: (with Earle C. Fowler) Strange Particles, 1963, Concepts in Physics, 1969; Asso. editor: Phys. Rev, 1963-66; asso. editor: Phys. Rev. Letters, 1974-76; editor, 1978—. Served with inf. AUS, 1943-46. Guggenheim fellow, 1954; Ford Found. fellow, 1962-63; Sloane Found. fellow, 1962-63. Fellow Am. Phys. Soc. (chmn. div. particles and fields 1972-73); mem. Nat. Acad. Scis. Subspecialties: Nuclear physics; Particle physics. Home: 50 Deepwood Dr Hamden CT 06517 Office: JW Gibbs Lab Yale Univ New Haven CT 06520

ADAMANTIADES, ACHILLES G., nuclear safety specialist, researcher; b. Volos, Greece, Apr. 22, 1934; came to U.S., 1960; s. Grammenos and Penelope (Theodorou) A.; m. Masha Michaelovna Kolosha, July 14, 1963; children: Grammenos, Penelope, Mikhail. Diploma, Nat. Metsovion Poly., Athens, Greece, 1957; Ph.D., M.I.T., 1965. Plant engr. Pub. Power Corp. Greece, Ptolemais, 1960; asst. prof. Iowa State U., Ames, 1966-71; vis. prof. U. Patras, Greece, 1971-73; cons. Fed. Power Commn., Washington, 1973-74; profl. lectr. George Washington U., Washington, 1974-78; project mgr. Electric Power Research Inst., Palo Alto, Calif., 1974—; cons. World Bank, Washington, 1974-78, Sci. Research and Tech. Agy., Athens, 1979-80; cons. assoc. prof. Stanford U., Palo Alto, 1981-82. Co-author: Handbook for Energy Technology and Economics, 1983, Guidebook to Nuclear Power Technology, 1984; contbr. articles to profl. jours. Council mem. Greek Orthodox Ch., Bethesda Md.; bd. dirs. Greek Community Sch. Served with Greek Navy, 1957-60. Fulbright grantee, 1960. Mem. Am. Nuclear Soc., AAAS, Tech. Chamber Greece, Democritos Soc. (pres. 1981-82), Sigma Xi. Club: Helenic Am. Profl. Assn. Subspecialties: Nuclear fission; Nuclear engineering. Current work: Nuclear safety, research and development project selection and management, fuel transient behavior, systems planning. Office: Electric Power Research Inst 3412 Hillview Ave Palo Alto CA 94303 Home: 1143 Hopkins Ave Palo Alto CA 94301

ADAMCZAK, ROBERT LEONARD, chemist; b. Buffalo, Aug. 16, 1927; s. Leo Stanley and Lottie Dorothy (Pawlak) A.; m. Patricia Chupinski, Aug. 23, 1958; children: Michael R., Bruce W., Karen M. B.A., U. Buffalo, 1951, M.A., 1954, Ph.D. in Phys. Chemistry, 1956. Chemist Olin-Mathieson Corp., Niagara Falls, N.Y., 1951-53; research chemist Esso Research & Engring. Co., Linden, N.J., 1957-59; inorganic polymer chemist USAF Material Lab., Dayton, Ohio, 1959-60, chief tribology/lubrication br., 1960-68, sci. adminstr., 1968-75; chief of plans Wright Aero Labs., Wright-Patterson AFB, Ohio, 1975—. Contbr. articles to profl. jours. Served with U.S. Army, 1947-48. Union Carbide fellow, 1954-55. Fellow Am. Inst. Chemists, ASME; mem. Am. Chem. Soc., Ohio Acad. Sci., N.Y. Acad. Sci., Am. Soc. Lubrication Engrs., Sigma Xi. Republican. Roman Catholic. Subspecialties: Aerospace engineering and technology; Aeronautical engineering. Current work: Planning for future aerospace technology capabilities. Home: 7556 Beldale Ave Huber Heights OH 45424 Office: Wright Aero Labs AFWAL/XR Wright-Patterson AFB OH 45433

ADAMS, CHARLES HENRY, animal science educator; b. Burdick, Kans., Nov. 7, 1918; s. Henry Lory and Bertha Frances (Westbrook) A.; m. Eula Mae Peters, Apr. 23, 1943. B.S., Kans. State U., 1941, M.S., 1942; Ph.D., Mich. State U., 1964. Instr. Kans. State U., Manhattan, 1946-47; asst. prof. animal sci. U. Nebr.-Lincoln, 1947-64, assoc. prof., 1964-70, prof., 1970—; asst. dean Coll. Agr., Inst. Agrl. and Natural Resources, 1973—. Served to lt. U.S. Army, 1943-46. Fellow Am. Soc. Animal Sci. (Disting. Teaching award 1972); mem. Am. Meat Sci. Assn. (Disting. Teaching award 1969), Inst. Food Tech., AAAS, Lincoln C. of C., Sigma Xi, Alpha Zeta. Republican. Mem. Disciples of Christ Ch. Lodge: Rotary. Current work: Teaching muscle biology, meats in animal science. Home: 7101 Colby St Lincoln NE 68505 Office: 103 Agr Hall U Nebr Lincoln NE 68586

ADAMS, DAVID BACHRACH, psychology educator; b. Mo., 1939; s. Robert M. and Elizabeth (Bachrach) A. B.A., Columbia Coll., 1962; Ph.D., Yale U., 1967. Fellow Yale U., 1968-70; prof. dept. psychology Wesleyan U., 1970—; mem. Nat. Acad. Sci. Exchange to, Soviet Union, 1976; Fulbright exchange lectr. in, Soviet Union, 1981. Contbr. articles in field to profl. jours. Pres. Conn. Assn. Am.-Soviet Friendship. Mem. Internat. Soc. Research on Aggression, Behavior Genetics Assn., Neurosci. Soc., Animal Behavior Soc., Am. psychol. Assn., AAAS. Subspecialties: Physiological psychology; Comparative neurobiology. Current work: Human and animal social behavior; neural and cultural factors; this includes brain mechanisms of aggression, ethological studies and cultural studies of human sex and warfare. Office: Dept Psychology Wesleyan U Middletown CT 06457

ADAMS, DONALD FREDERICK, mech. engr., educator, cons.; b. Streator, Ill., Sept. 25, 1935; s. Fred Mathew and Margaret Ann (Doerr) A.; m. Roberta Ann, June 22, 1957; children: David Allan, Daniel Scott, Douglas John, Jayne Lynn. B.S., U. Ill., 1957, Ph.D. in Applied Mechanics, 1963; M.S., U. So. Calif., 1960. Registered profl. engr., Wyo. Engr. Northrop Aircraft, Hawthorne, Calif., 1957-60; instr. U. Ill., 1960-63; supr. Aeronutronic div. Ford Motor Co., Newport Beach, Calif., 1963-67; mem. staff Rand Corp., Santa Monica, Calif., 1967-72; assoc. prof. mech. engring. U. Wyo., 1972-75, prof., 1975—; legal and indsl. cons.; internat. travel for UN Indsl. Devel. Orgn. Contbr. numerous articles to profl. publs. Named Outstanding Faculty Mem. U. Wyo., 1973. Mem. AIAA, ASME, ASTM, Soc. Automotive Engrs. (Ralph-Teetor award 1978), Am. Acad. Mechanics, Wyo. Engring. Soc., Nat. Soc. Profl. Engrs., Wyo. Soc. Profl. Engrs., Am. Soc. Metals, Soc. Exptl. Stress Analysis, Am. Soc. Engring. Edn. Roman Catholic. Subspecialties: Composite materials; Solid mechanics. Current work: Fiber reinforced composite materials. Home: 421 S 19th St Laramie WY 82070 Office: Mech Engring Dept U Wyo Laramie WY 82071

ADAMS, DONALD ROBERT, anatomist, educator; b. Red Bluff, Calif., Aug. 3, 1937; s. Robert William and Nellie May (Longcor) A.; m. Carol Gene Gribler, July 19, 1974; children: Shawn Ryan, Robert James. A.B. in Zoology, U. Calif., Davis, 1960, Ph.D. in Anatomy, 1970; M.A. in Biology, Chico (Calif.) State Coll., 1967. Tchr. Malangali Govt. Secondary Sch., Tanganyika, East Africa, 1961-63, C.K. McClatchy Sr. High Sch., Sacramento, 1964-65; asst. prof. anatomy Mich. State U., 1970-74; asst. prof. vet. anatomy Iowa State U., Ames, 1974-76, assoc. prof., 1976—. Named Norden Outstanding Tchr. Coll. Vet. Medicine, 1982. Mem. AAAS, Am. Assn. Anatomists, Am. Assn. Vet. Anatomists, World Assn. Vet. Anatomists, Iowa Microbeam Soc. Republican. Subspecialties: Anatomy and embryology; Pulmonary medicine. Current work: Morphological adaption of vertebrates to environ. factors with emphasis on upper respiratory system. Home: 812 Onyx Circle Ames IA 50010 Office: 1034 Coll of Veterinary Medicine Ames IA 50011

ADAMS, GAIL D., radiation physicist, educator; b. Cleve., Jan. 27, 1918; s. Gail D. and Edna M. (Baker) A.; m. Lucile A. Pyne, May 26, 1942; children: Kenneth, Frances, Paul, Susan, Anne; m. Helen L. Newman, Oct., 1966; m. Reba I. Smith, Jan. 17, 1976. B.S. magna cum laude in Physics, Case Inst. Tech., 1940; M.S., U. Ill., 1942, Ph.D., 1943. Research asst. prof. U. Ill., Urbaba, 1943-50; assoc. prof. U. Calif., San Francisco, 1951-58, prof., 1958-64; prof., vice head dept. radiol. scis. U. Okla., Oklahoma City, 1965—; pres. Radiation Physics, Inc., 1957—. Contbr. articles to profl. jours. Neighborhood commr. Boy Scouts Am., 1952-62; mem. budget rev. com. United Appeal, Oklahoma City, 1974-80. Fellow Am. Phys. Soc., Am. Coll. Radiology; mem. Am. Assn. Physicists in Medicine (Coolidge award 1982), Health Physics Soc., Radiation Research Soc., AAAS, Radiol. Soc. N.Am., N.Y. Acad. Scis., Okla. Acad. Scis., Phi Beta Kappa, Sigma Xi, Tau Beta Pi. Republican. Unitarian. Subspecialties: Medical physics; Atomic and molecular physics. Current work: Iodine kinetics in the human; renal kinetics; radiation safety officer. Patentee in field. Home: 1009 East Dr Edmond OK 73034 Office: Radiol Scis Box 26901 Oklahoma City OK 73190

ADAMS, GEORGE G., mechanical engineering educator; b. Bklyn., Sept. 12, 1948; s. George Gabriel and Sally Saydah A. B.S.M.E., Cooper Union, 1969; M.S., U. Calif.-Berkeley, 1972, Ph.D., 1975. Assoc. engr. Curtiss-Wright Co., Woodridge, N.J., 1969-70; asst. prof. Clarkson Coll. Tech., Potsdam, N.Y., 1975-78; vis. scientist IBM Research Lab., San Jose, Calif., 1978-79; vis. scholar U. Calif.-Berkeley, 1979; asst. prof. Northeastern U., Boston, 1979-82, assoc. prof. mech. engring., 1982—. Contbr. articles in field to profl. jours. NSF fellow, 1970-73; NSF initiation grantee, 1976-78; NSF engring. grantee, 1979-80. Mem. Am. Soc. Engring. Edn., Soc. Engring. Sci., Tau Beta Pi, Pi Tau Sigma. Democrat. Subspecialties: Solid mechanics; Theoretical and applied mechanics. Current work: Study of the response of elastic bodies to moving loads, study of the stress distribution at the interface of bonded elastic materials. Home: 135 Pleasant St Apt 306 Brookline MA 02146 Office: Northeastern U Mech Engring 458HO Boston MA 02115

ADAMS, HENRY B(ETHUNE), clinical psychologist, consultant; b. Charlotte, N.C., Aug. 26, 1925; s. Hal B. and Mabel (Cooper) A. A.B., U. N.C.-Chapel Hill, 1949; M.A., Duke U., 1953; Ph.d., Purdue U., 1956. Lic. psychologist, D.C. Instr. Nebr. U. Med. Sch., Omaha, 1956-59; research psychologist VA Hosp., Richmond, Va., 1959-63; staff psychologist Nat. Tng. Sch., Washington, 1963-64; chief psycology service Fed. Reformatory, Alderson, W.Va., 1964-67; psychology coordinator Area C Mental Health Center, WAshington, 1967-80; clin. psychologist St. Elizabeth's HOsp., WAshington, 1980—; research cons. George Mason U., 1977—; project cons. Ciga Rest Center, Lakewood, N.J., 1980—. Contbr. chpts. to books. Mem. D.C. Polit. Action Com., 1982—. Mem. Am. Psychol. Assn., D.C. Psychol. Assn. (treas. 1970-73), Soc. Personality Assessment, Soc. for Psychologists in Substance Abuse, Soc. Advancement Environ. Therapies (chmn. steering com. 1980—), Phi Beta Kappa, Sigma Xi, Alpha Kappa Delta. Methodist. Subspecialties: Sensory processes; Alocholism and substance abuse. Current work: Therapeutic uses of reduced environmental stimulation or sensory deprivation; research and treatment methods for alcoholism and substance abuse. Home: 3001 Branch Ave SE Apt 722 Temple Hills MD 20748 Office: O'Malley Alcoholism Program St Elizabeth's Hosp M Bldg Washington DC 20032

ADAMS, JAMES HALL, JR., research physicist; b. Statesville, N.C., Aug. 7, 1943; s. James Hall and Mary Elizabeth (Powell) A.; m. Rebekah Ellis, June 4, 1968; children: Mary Amanda, David Alexander. B.S., N.C. State U., 1966, M.S., 1968, Ph.D., 1972. Research asst. Johnson Space Center, Houston, 1968-72; research asso. NRC, Washington, 1972-74; research physicist U.S. Naval Research Lab., Washington, 1974—. Mem. AAAS, IEEE, Am. Phys. Soc., Am. Geophys. Union, Am. Astron. Soc. Subspecialty: Cosmic ray high energy astrophysics. Current work: Cosmic ray composition, low energy cosmic rays origin, cosmic ray physics, magnetospheric physics, cosmic ray effects on microelectronics. Office: US Naval Research Lab Code 4022 Washington DC 20375

ADAMS, JOHN DAVID, consultant, lecturer; b. Wooster, Ohio, Sept. 13, 1942; s. Lyman Harry and Catherine Ruth (Whittlesey) A.; m. Sarah Jane Dalbey, Aug. 15, 1964; children: Samantha Lee, Gillian Lindsey. A.B. in Math, Wittenberg U., 1963; B.S. in Mgmt. Sci, Case Inst. Tech., 1965; Ph.D. in Orgnl. Behavior, CAse-Western Res. U., 1965. Cert. Consultants Internat. Vis. lectr. U. Leeds, Yorkshire, U.K., 1969-71; mgr. Nat. Tng. Labs. Inst., Washington, 1971-74, div. dir., 1974-75; consultant, Arlington, Va., 1975—. Author: Transition, 1976, Understanding and Managing Stress (3 vols.), 1980; editor: New Technology in Organizational Development, Vol. 2, 1975, Organizational Development in Health Care Organizations, 1982. Bd. trustees Potomac Rugby Union, 1975—. NDEA fellow, 1965-69. Mem. Am. Psychol. Assn., Assn. Humanistic Psychology, Orgn. Devel. Network (dir. 1971-74), Blue Key, Tau Beta Pi. Subspecialties: Social psychology; Preventive medicine. Current work: Health protection, risk reduction, stress management, belief systems, personal peak performance, personal transformation, expanded consciousness, cultural hypnosis, paradigm shift, holistic. Home and Office: 2914 27th St N Arlington VA 22207

ADAMS, JOHN LESTER, JR., agrl. educator, writer, researcher; b. Manitou, Okla., Oct. 9, 1921; s. John Lester and Melinda Cleveland (Manis) A.; m. Dorothy Fern Hessel, Dec. 4, 1942; children: John Michael, David Patrick. B.S., Okla. A&M Coll., 1943; M.S., U. Wis., 1947, Ph.D., 1951. Asst. prof. poultry sci. U. Wis., Madison, 1951-55, assoc. prof., 1955-57; prof., chmn. poultry sci. dept. U. Nebr., Lincoln, 1957-64, dir. coop. extension service, 1964-75, prof. agrl. communication, extension sci. writer, 1975—; sec. Nebr. Council on Pub. Relations for Agr., 1976—. Contbr. articles to tech. jours. and mags. Mem. Am. Legion. Republican. Presbyterian. Subspecialties: Animal husbandry; Communications research. Current work: Endocrinology, communications. Home: 4340 Normal Blvd Lincoln NE 68506 Office: East Campus U Nebr 108 Agrl Communications Bldg Lincoln NE 68506

ADAMS, KARYL ANN, defense aeronautical laboratory electronics enginner, computer systems analysist; b. Urbana, Ohio, Jan. 8, 1952; d. Robert Vincent Adams and Mildred Marie (Neff) Richardson. B.S., Ohio No. U., 1975; M.S., Air Force Inst. Tech., 1983. Elec. engr. Air Force Aerospace Med. Research Lab., Wright Patterson AFB, Ohio, 1975, Air Force Flight Dynamics Lab., 1976-79; electronics engr. Air Force Wright Aero. Lab., 1980—, guest lectr., 1976—, computer graphics researcher, 1980—. Activity organizer Campus Chest-United Way, Ada, Ohio, 1971-75. Olin fellow Washington U., St. Louis, 1975;

recipient Sustained Superior Performance award U.S. Air Force, 1977, Spl. Act of Service award, 1981; named Jr. Scientist and Engr. of Yr. Air Force Flight Dynamics Lab., 1978. Mem. Assn. Computing Machinery, IEEE, Mortarboard, Alpha Omicron Pi. Methodist. Subspecialties: Aerospace engineering and technology; Computer engineering. Current work: Managing a technical team evaluating the performance of the forward swept wing technology demonstrator; developing an interactive graphics environment for the design of cockpit displays for advanced aircraft systems. Home: 7578 Mount Hood Huber Heights OH 45424 Office: Air Force Wright Aero Labs Wright Patterson AFB OH 45341

ADAMS, LUDWIG, civil engineer, consultant; b. N.Y.C., Aug. 12, 1916; m. Alberta Anne Howard, June 28, 1941; children: Alberta Anne, Ludwig Howard. B.S. cum laude in Civil Engring., NYU, 1936; M.C.E., Rensselaer Poly. Inst., 1937. Registered profl. engr., Pa., 9 other states. With Pitts. Des Moines Steel Co., 1937-42; mem. NASA, 1942; with Mellon Inst. Indsl. Research, 1942-46, U. Pitts., 1943-44, Steel Structures Painting Council, 1952; PDM Steel Co., 1946-57, Blaw Knox Co., 1957-60, Va. Erection Co., 1960-63, Rust Painting Co., 1962-63; asst. chief engr. PDM Corp., Pitts., 1961-81, mgr. engring. services, 1981-82; owner, mgr. AHALA Cons., Pitts., 1982—. Contbr. articles to profl. jours. Mem. Ross Twp. Planning Commn., 1974—, Northland Library Writers Workshop. Mem. ASCE, ASME, ASTM, ASHRAE, Am. Welding Soc., Tau Beta Pi, Sigma Xi, Iota Alpha, Sigma Pi Sigma, Sigma Beta Alpha, Sigma Beta Sigma. Republican. Methodist. Club: Shannopin Country. Lodges: Elks; Shriners. Subspecialties: Cryogenics; Aerospace engineering and technology. Current work: Thermal Conductivity of thermal insulations and gaseous fuels. Home: 205 Thompson Dr Pittsburgh PA 15229

ADAMS, MAX D., pharmacologist, researcher; b. St. Marys, W.Va., July 25, 1941; s. Guy and Alfarata (Johnson) A.; m. Mergie Ann Wilson, July 5, 1963; children: Julie, Christy. B.S., W. Va. U., 1965; M.S., Purdue U., 1968, Ph.D. in Pharmacology, 1970. Diplomate: Am. Bd. Toxicology. Instr. Med. Coll. Va., 1972-76, asst. prof., 1972-76; sr. research pharmacolgist Mallinckrodt, Inc., St. Louis, 1977-81, research mgr. pharmacology and toxicology, 1981—. Mem. Am. Soc. Pharmacology and Exptl. Therapeutics. Methodist. Subspecialties: Pharmacology; Toxicology (medicine). Current work: Pharmacology and toxicology of diagnostic pharmaceuticals.

ADAMS, PERRIE MILTON, psychology educator, researcher; b. Miami, Fla., Jan. 5, 1943; s. Perrie M. and Vera Marie (Perkins) A.; m. Carolyn Anne Haynes, Aug. 27, 1966; 1 son: Richard Todd. B.A., Fla. State U., 1966, M.S., 1968, Ph.D., 1970. Psychologist, Tex. Research asst. Fla. State U., 1966-68, teaching asst., 1968-70; asst. prof. psychology U. Tex. Med. Br., 1970-74, assoc. prof., 1974-82, prof., 1982—. Contbr. articles to profl. jours., chpts. to books. Bd. dirs. Galveston Community Theater, 1973-75. NIH grantee, 1973-75, 82—. Mem. Am. Psychol. Assn. (chmn. com. on animal research and experimentation 1979-81), Neurosci. Soc., Teratology Soc., AAAS, Behavioral Teratology Soc. Subspecialties: Psychobiology; Neuropharmacology. Current work: Reseach directed toward the study of the effects of parental drug exposure on the behavioral development of the offspring using animal models. Home: 180 San Marino St Galveston TX 77550 Office: U Tex Med Br Dept Psychiatry and Behavioral Sci Galveston TX 77550

ADAMS, RALPH MELVIN, radiological physicist; b. College Place, Wash., Apr. 16, 1921; s. Walter Lee and Minnie Florence A.; m. Ruth Ona Berg, Sept. 26, 1943; children: Shirley, Marilyn. B.A., La Sierra Coll. Diplomate: Am. Bd. Radiology. Radiol. physicist White Meml. Med. Center, Los Angeles, 1951-62, Cedars of Lebanon Hosp., 1962-68; physicist, nuclear radiology Loma Linda (Calif.) U. Med. Center, 1970—; IAEA tech. assistance expert, Iraq, 1961, Israel, 1971, Pakistan, 1974, Thailand, 1982. Contbr. articles to profl. jours. Served with U.S. Army, 1943-46. Mem. Soc. Nuclear Medicine, Radiol. Soc. N.Am., Am. Coll. Radiology, Am. Assn. Physicists in Medicine. Seventh-Day Adventist. Subspecialty: Nuclear medicine. Current work: Development of color-coded display systems in nuclear medicine. Development of methods to measure temporal resolution and spatial distortion of scintillation cameras. Development of computer software for image processing and display. Patentee in field. Office: Nuclear Radiology Loma Linda U Loma Linda CA 92350

ADAMS, RICHARD L., computer scientist; b. Fayetteville, Ark., May 12, 1943; s. John L. and Helen L. A. Div. B.S. in Math, Calif. State Poly. Coll., 1967; M.E.E. in Computer Sci, U. Calif.-Santa Barbara, 1978. Mathematician Navy Civil Service, Point Mugu, Calif., 1969-78; computer scientist Computer Scis. Corp., Oxnard, Calif., 1979—. Served to 1st lt. Signal Corps U.S. Army, 1967-69. Mem. Assn. Computing Machinery, AIAA. Unitarian Universalist. Subspecialties: Software engineering; Artificial intelligence. Current work: Integrated, automated aids to software development. Currently developing an integrated, automated program development station to include system specification, design, development, documentation, testing and maintenance.

ADAMS, ROBERT EDWARD, federal government research program executive; b. Memphis, Nov. 4, 1929; s. Jesse Bryant and Bessie (Parks) A.; m. Betty Jean Hickman, Oct. 28, 1949; 1 dau., Emily Jean. B.S., Memphis State U., 1954; M.S., U. Miss., 1956. Research staff mem. Oak Ridge Nat. Lab., 1956-70, research mgr., 1975—; pres. Air-Tec, Inc., Oak Ridge, 1970-75. Served with USAF, 1951-52. Mem. Am. Chem. Soc., Am. Nuclear Soc. Baptist. Current work: Study of behavior of aerosols within nuclear reactor containment, engineering safeguards for aerosol control. Home: 112 Amanda Dr Oak Ridge TN 37830 Office: Bldg 9108 MS2 Oak Ridge Nat Lab PO Box Y Oak Ridge TN 37830

ADAMS, STEVEN PAUL, research chemist; b. Oceanside, Calif., Dec. 31, 1952; s. Lawrence Carter and Liane (Bertoch) A.; m. Lynette Knowles, Dec. 27, 1974; children: Alisa, Ammon Steven, Jared Paul. B.A. magna cum laude, U. Utah, 1976; Ph.D., U. Wis., 1980. Research asst. U. Utah, 1975-76; teaching asst. U. Wis., 1976-78, research asst., 1978-80; sr. research scientist Monsanto Co., St. Louis, 1980—, project leader chemistry sect. molecular biology program, 1980—. Mem. Am. Chem. Soc., Sigma Xi. Mormon. Subspecialties: Organic chemistry; Molecular biology. Current work: Nucleic acid chemistry, DNA/RNA synthesis, organic chemical applications to molecular biology, peptide/DNA interactions. Patentee in field.

ADAMS, WILLIAM EUGENE, chemical corporation executive, mechanical engineer; b. Mt. Vernon, Ohio, Oct. 18, 1930; s. Elmer William and Crystine Merle (Reichert) A.; m. Helen Virginia Drollinger, Aug. 29, 1953; children; Eric, Barbara. B.M.E., Ohio State U., 1955. Registered profl. engr., Ohio. With Ethyl Corp., Richmond, Va., 1955—, research adviser air cleanliness div., Detroit, 1970-74, dir. auto research, 1975-76, mgr., 1976—. Contbr. numerous sci. articles to profl. publs. Mem. Soc. Automotive Engrs. (Horning Meml. award 1964), Engring. Soc. Detroit. Presbyterian. Club: Recess (Detroit). Subspecialties: Fuels; Mechanical engineering. Current work: Additives for fuels and lubricants; manage research laboratories and direct research activities in fuel and lubricant areas. Home: 4306 Arlington Dr Royal Oak MI 48072 Office: Detroit Research Labs 1600 W Eight Mile Rd Ferndale MI 48220

ADAMS, WILLIAM HENRY, chemist, educator; b. Balt., Dec. 21, 1933; s. William Henry and Mary Ellen (Verlander) A.; m. Violeta Lourdes Flores, Dec. 20, 1958 (dec.); 1 dau., Maryna Noelle. A.B., Johns Hopkins U., 1955; S.M., U. Chgo., 1956, Ph.D., 1960. Asst. prof. Pa. State U., 1962-66; asst., then assoc. prof. Rutgers U., New Brunswick, N.J., 1966-75, prof., 1975—. NSF postdoctoral fellow, 1960-62. Mem. Am. Phys. Soc. Subspecialty: Theoretical chemistry. Current work: Quantum theory of intermolecular interactions. Home: 3 Meadowbrook Ln Piscataway NJ 08854 Office: Dept Chemistry Rutgers U New Brunswick NJ 08903

ADAMSON, ARTHUR WILSON, chemistry educator; b. Shanghai, China, Aug. 15, 1919; s. Arthur Quintin and Ethel (Rhoda) A.; m. Virginia Louise Dillman, Mar. 24, 1942; children: Carol Ann, Janet Louise, Jean Elizabeth. B.S. with honors, U. Calif.-Berkeley, 1940; Ph.D. in Phys. Chemistry, U. Chgo., 1944. Research asst. plutonium project Oak Ridge Nat. Lab., 1942-44, research assoc., 1944-46; mem. faculty U. So. Calif., Los Angeles, 1946—, prof. chemistry, 1953—. Author: Concepts of Inorganic Photochemistry, 1975, Understanding Physical Chemistry, 3rd edit, 1980, Textbook of Physical Chemistry, 2d edit, 1979, Physical Chemistry of Surfaces, 4th edit, 1982; contbr. over 200 sci. articles to profl. publs. Mem. Am. Chem. Soc. (So. Calif. Tolman medal 1967, Kendall award in colloid or surface chemistry 1976, Inorganic Chemistry Disting. Service award 1983, Chem. Edn. award 1984), Am. Inst. Chemists (Western div. Honor Scroll award 1976), AAAS, Sigma Xi. Republican. Club: Palos Verdes Tennis. Subspecialties: Inorganic chemistry; Surface chemistry. Current work: Photochemistry of coordination compounds; physical absorption of vapors; contact angle and wetting. Office: Chemistry Dept Univ So Calif Los Angeles CA 90089

ADAMSON, THOMAS CHARLES, JR., aerospace engineering educator, consultant; b. Cicero, Ill., Mar. 24, 1924; s. Thomas Charles and Helen Emily (Koubek) A.; m. Susan Elizabeth Huncilman, Sept. 16, 1949; children: Thomas Charles, William Andros, Laura Eliabeth. B.S., Purdue U., 1949; M.S., Calif. Inst. Tech., 1950, Ph.D., 1954. Research engr. Jet Propulsion Lab., Pasadena, Calif., 1952-54; assoc. research engr. U. Mich., Ann Arbor, 1954-56, asst. prof., 1956-57, assoc. prof., 1957-61, prof., 1961—, chmn. dept. aerospace engring., 1983—. Editor: (with M.F. Platzer) Transonic Flow Problems in Turbo Machinery, 1977; contbr. articles to profl. jours. Served with U.S. Army, 1943-46; ETO. Guggenheim fellow, 1950-51. Fellow AIAA; mem. Combustion Inst., Sigma Xi. Episcopalian. Subspecialties: Aeronautical engineering; Aerospace engineering and technology. Current work: Fluid mechanics; steady and unsteady transonic flow fields; propulsion. Home: 667 Worthington Pl Ann Arbor MI 48103 Office: University of Michigan Department of Aerospace Engineering 2508 Patterson Pl Ann Arbor MI 48109

ADDICOTT, WARREN OLIVER, geologist; b. Fresno, Calif., Feb. 17, 1930; s. Irwin Oliver and Astrid (Jensen) A.; m. Suzanne Aubin, Oct. 2, 1976; m. Susanne Smith, Aug. 20, 1956 (div. 1974); children: Eric Oliver, Carol. B.A., Pomona Coll., 1951; M.A., Stanford U., 1952; Ph.D., U. Califf.-Berkeley, 1956. Teaching asst. U. Calif.-Berkeley, 1952-54; geologist Mobil Oil Co., Los Angeles, 1954-62; research geologist U.S. Geol. Survey, Menlo Park, Calif., 1962—; cons. prof. Stanford (Calif.) U., 1971—; dep. chmn. Circum-Pacific Map Project, Menlo Park, Calif., 1979-82, gen. chmn., 1982—. Fellow AAAS, Geol. Soc. Am., Calif. Acad. Scis.; mem. Paleontol. Soc. (pres. 1979-80), Am. Assn. Petroleum Geologists. Congregationalist. Subspecialties: Paleontology; Geology. Current work: Coordinating geological, geophysical and resource mapping program of Circum-Pacific region involving 35 countries and representatives from more than 135 organizations. Home: 957 Los Robles Ave Palo Alto CA 94306 Office: US Geol Survey 345 Middlefield Rd Menlo Park CA 94025

ADDY, JOHN KEITH, chemist, educator; b. Sheffield, Eng., June 30, 1937; came to U.S., 1966, naturalized, 1976; s. John Arthur and Gladys (Powell) A.; m. Jean Elizabeth Powell, July 23, 1963; children: John David, June Elizabeth. B.Sc. with spl. honors, Kings Coll., U. London, 1958; Ph.D., U. Southampton, Eng., 1962. Postdoctoral research assoc. U. Oreg., 1961-62; Lectr. chemistry N.E. Essex Coll., Colchester, Eng., 1962-63, John Dalton Coll. Tech., Manchester, Eng., 1963-66; asst. prof. chemistry Wagner Coll., Staten Island, N.Y., 1966-69, assoc. prof., 1969-74, prof., 1974—. Contbr. articles to chem. jours. Fellow Royal Soc. Chemistry (treas. U.S. sec. 1981—); mem. Am. Chem. Soc. (nat. councillor 1975-81). Mem. Evangelical Free Ch. Am. Lodge: Rotary (dir. 1976—). Subspecialties: Physical chemistry; Polymer chemistry. Current work: Crystal structure of polymers. Home: 166 Chelsea St Staten Island NY 10307 Office: Dept Chemistry Wagner Coll Staten Island NY 10301

ADELBERG, EDWARD ALLEN, genetics educator, researcher; b. Cedarhurst, N.Y., Dec. 6, 1920; s. Max and Janet (Ehrlich) A.; m. Marion Sanders, Nov. 28, 1942; children: Michael G., David E., Arthur W. B.S., Yale U., 1942, M.S., 1947, Ph.D., 1949. From instr. to prof. U. Calif.-Berkeley, 1949-61, chmn. dept. bacteriology, 1957-61; prof. microbiology Yale U., New Haven, 1961-74, chmn. dept. microbiology, 1961-64, 1970-72, prof. human genetics, 1972—; cons. in field. Author: (with Jawetz and J. Jemlnick) Review of Medical Microbiology, (with R. Y. Stanier and M. Doudoroff) The Microbial World; contbr. articles on genetics to profl. jours. Served to maj. USAAF, 1942-46. Guggenheim fellow, 1956-57, 1965-66. Fellow Am. Acad. Arts and Scis.; mem. Nat. Acad. Scis., Conn. Acad. Sci. and Engring. (council 1978-81). Subspecialties: Membrane biology; Gene actions. Current work: Using genetic and biochemical techniques to study structure-function relationships of mammalian membrane transport systems. Home: 204 Prospect St New Haven CT 06511 Office: Dept Human Genetics Yale U Med Sch New Haven CT 06510

ADELMAN, GEORGE, editor, librarian, consultant; b. Boston, Sept. 17, 1926; s. Morris and Anna (Cohen) A.; m. Sondra Cohen, July 15, 1957; children: Merle, Marjorie. B.A., Dartmouth Coll., 1947; M.A. in Psychology, Boston U., 1950; M.S. in Library Sci., Simmons Coll., 1950. Gen. mgr. Mass. Chem. Co., Boston, 1947-55; reference librarian Boston Public Library, 1955-56; tech. info. officer U.S. Office of Naval Research, Boston, 1955-64; dir. publs., editor, librarian Neuroscis. Research Program M.I.T., Boston, 1964-82, editorial cons., 1983—; cons. in field. Editor: Neuroscis. Research Program Bull., 1965-82; co-editor: The Organization of the Cerebral Cortex, 1981, The Neurosciences: Paths of Discovery, 1977, The Neurosciences: Study Program I-IV, 1967-77, Neurosciences Symposium Summaries, 1968-80; editor-in-chief: Ency. Neurosci., 1983—; contbr. articles and revs. on neuroscis. to profl. jours. Active Boston Mus. Fine Arts. Served with USN, 1943-46. Recipient Meritorious Service award Office Naval Research, 1963. Mem. Soc. Neurosci., Council Biology Editors. Democrat. Jewish. Club: Print and Drawing. Subspecialty: Interdisciplinary neuroscience. Current work: editorial and program consultant. Home: 1904 Beacon St Brookline MA 02146

ADELMAN, SAUL JOSEPH, physics educator; b. Atlantic City, N.J., Nov. 18, 1944; s. Benjamin and Kitty (Sandler) A.; m. Carol Jeanne Sugarman, Mar. 28, 1970; children: Aaron, Barry, David. B.S. in Physics, U. Md., 1966; Ph.D., Calif. Inst. Tech., 1971. NAS/NRC postdoctoral resident research assoc. NASA Goddard Space Flight Center, Greenbelt, Md., 1972-74; asst. prof. astronomy Boston U., 1974-78; asst. prof. physics The Citadel, Charleston, S.C., 1978—. Author: Bound for the Stars, 1981, (with Benjamin Adelman); contbr. articles to profl. jours. NASA grantee, 1975-76, 79-81, 82—; NSF grantee, 1976-79, 81-82. Fellow Royal Astron. Soc., British Interplanetary Soc.; mem. Internat. Astron. Union, Am. Astron. Soc., Optical Soc. Am., Astron. Soc. Pacific, Phi Beta Kappa, Phi Kappa Phi, Sigma Xi, Sigma Pi Sigma. Subspecialties: space science; Optical astronomy. Current work: Research peculiar B and A type stars; main sequence B, A and F type stars, planetary engineering; photoelectric spectrophotometry. Office: Dept Physics The Citadel Charleston SC 29409 Home: 1434 Fairfield Ave Charleston SC 29407

ADELMAN, STEVEN A., chemistry educator, research scientist; b. Chgo., July 4, 1945; s. Hyman and Sarah A.; m. Barbara Stolberg, May 13, 1974. Ph.D. (NSF fellow), Harvard U., 1972. Postdoctoral fellow M.I.T., Cambridge, 1972-73; U. Chgo., 1973-74; asst. prof. chemistry Purdue U., West Lafayette, Ind., 1975-77, assoc. prof., 1977-82, prof., 1982—; cons. Exxon Research Co. Contbr. articles on chemistry to profl. jours. Alfred P. Sloan fellow, 1976-80; Guggenheim fellow, 1982-83; NSF grantee, 1976. Mem. Am. Chem. Soc., Am. Phys. Soc. Subspecialties: Theoretical chemistry; Statistical mechanics. Current work: Applications of statistical mechanics to problems in chemistry, theories for chemical reaction dynamics in liquid solution. Home: 2827 Wilshire St West Lafayette IN 47906 Office: Dept Chemistry Purdue U West Lafayette IN 47906

ADELSTEIN, ROBERT SIMON, research physician, cardiologist; b. N.Y.C., Jan. 15, 1934; s. George and Belle (Schild) A.; m. Miriam Appleman, June 25, 1961; children: Ben, Sandra, Michael. A.B., Princeton U., 1955; M.D., Harvard U., 1959. Commd. med. officer USPHS, 1961—; research assoc. Nat. Heart, Lung, and Blood Inst., Bethesda, Md., 1961-64, sr. scientist, 1965-70, head sect. on molecular cardiology, 1970-81, chief lab. molecular cardiology, 1981—. Contbr. articles to sci. jours. Co-chmn. Commn. Concerned Scientists, N.Y.C., 1976-78. Recipient commendation medal USPHS, 1980. Mem. Am. Soc. Biol. Chemistry, Am. Soc. Clin. Investigation. Subspecialties: Biochemistry (medicine); Cardiology. Current work: Investigation of regulation of contractile proteins in muscle and nonmuscle cells. Home: 611 Crocus Dr Rockville MD 20850 Office: Nat Heart Lung Blood Inst Bldg 10 Room 7B-09 Bethesda MD 20205

ADKINSON, N. FRANKLIN, JR., physician, educator; b. Forest City, N.C., May 18, 1943; s. N. Frank and Estelle (Stembridge) A.; m. Judy Faye Hyder, Aug. 20, 1966; children: Anna Estelle, Carter Franklin. B.A. with highest honors, U. N.C., Chapel Hill, 1965; M.D., Johns Hopkins U., 1969. Intern, resident Osler Med. Service, Johns Hopkins Hosp., Balt., 1969-71; clin. assoc. Lab. Immunology Nat. Cancer Inst., Bethesda, Md., 1971-73; mem. faculty Johns Hopkins U. Sch. Medicine, Balt., 1973—, assoc. prof. medicine, 1980—. Served to surgeon USPHS, 1971-73. NIH grantee. Mem. Am. Acad. Allergy, AAAS; mem. Am. Assn. Clin. Investigation; Mem. Am. Fedn. Clin. Research. Subspecialties: Allergy; Immunology (medicine). Current work: Drug hypersensitivity states; IgE methodology; arachidonate metabolism in inflammation. Office: Johns Hopkins U at Good Samaritan Hosp 5601 Loch Raven Blvd Baltimore MD 21239

ADKISSON, PERRY LEE, university official; b. Hickman, Ark., Mar. 11, 1929; s. Robert Louise and Imogene (Perry) A.; m. Frances Rozelle, Dec. 29, 1956; 1 dau., Jean Amanda. B.S., U. Ark., 1950, M.S., 1954; Ph.D. in Entomology, Kans. State U., 1956. Asst. prof. entomology U. Mo., 1956-58; asso. prof. Tex. A&M U., 1958-63, prof., 1963—; Disting. prof. entomology, 1979—, head dept. entomology, 1967—, v.p. for agr. and renewable resources, 1978-80, dep. chancellor for agr., 1980-83, dep. chancellor, 1983—; cons. Internat. AEC, Vienna, 1969-74; Chmn. sci. adv. panel Gov. Tex. on Agrl. Chems., 1970-72; chmn. Tex. Pesticide Adv. Com., 1971—; mem. panel experts on integrated pest control UN/FAO, Rome, 1971—; mem. Structural Pest Control Bd., Tex., 1971-78, NRC World Food and Nutrition Study Team, 1977; chmn. com. biology pest species NRC, 1974; mem. environ. studies bd., study group problems pest control Nat. Acad. Sci.-NRC, 1972—; mem. vis. group Internat. Center Insect Physiology and Ecology, Nairobi, Kenya, 1974-75; mem. U.S. directorate UNESCO Man and the Biosphere Program, 1975—. Mem.: editorial com. Ann. Rev. Entomology, 1973—; contr. articles to profl. jours. Served with M.C. AUS, 1951-53. Recipient Alexander Von Humboldt award, 1980; Disting. Alumnus award Kans. State U., 1980; USPHS postdoctoral fellow Harvard U., 1963-64. Fellow AAAS; mem. Entomol. Soc. Am. (governing bd. 1971-79, pres. 1974, Bussart Meml. award 1967), Kans. Entomol. Soc., Internat. Orgn. Biol. Control, Am. Registry Profl. Entomologists (governing council 1976-78, pres. 1977), Nat. Acad. Scis, Phi Kappa Phi, Sigma Xi. Subspecialty: Integrated pest management. Current work: Research in insect physiology, photoperiodism, ecology and integrated control of crop pests.

ADKISSON, WILLIAM MILTON, electrical engineer; b. Okmulgee, Okla., Oct. 21, 1921; s. Robert Walker and Radie (Ward) A.; m. Marijean Feik, June 26, 1954; children: Robert Lewis, Richard Ward, James William. B.S.E.E., Tex. A&M U., 1943; postgrad., U. Minn., 1950-53. With Honeywell Inc., 1946—, mgr. dept. applied research, electronic data processing div., Waltham, Mass., 1968-72; sr. staff engr. Med. Systems Ctr., Mpls., 1972-77, sr. staff engr. test instruments div., Denver, 1977—. Guest editor: Honeywell Computer Jour, spring 1960. Mem. Am. Radio Relay League. Served to capt. Signal Corps U.S. Army, 1943-46. Mem. IEEE, Assn. Computing Machinery, Optical Soc. Am., Soc. Photographic Scientists and Engrs. Republican. Methodist. Subspecialties: Systems engineering; Biomedical engineering. Current work: Color graphics, integrated circuit wafer-mask inspection, replacement for film-screen in med. X-ray systems. Patentee pulse type receiver circuits, course control apparatus, scanning microscope system with automatic cell find and autofocus, programmed sample pyrolysis for mass spectrometer. Home: 7215 S Xanthia St Englewood CO 80112 Office: 4800 E Dry Creek Rd Denver CO 80217

ADLER, ALAN DAVID, chemistry educator; b. Nyack, N.Y., Oct. 5, 1931; s. Edward Jay and Martha (Margulies) A.; m. Jean Evamay Tremble, June 14, 1954; children: Lucas Steven, Dana Ann, Chris Paula. B.S., U. Rochester, 1953; Ph.D., U. Pa., 1960. Asst. prof. molecular biology U. Pa., Phila., 1960-67; assoc. prof. biophys. chemistry New Eng. Inst., Ridgefield, Conn., 1967-74; prof. chemistry Western Conn. State U., Danbury, 1974—; cons. in field. Editor monographs. Active Boy Scouts Am., 1970—, Mark Twain Library Fair Assn., Redding, Conn., 1972—. Recipient various grants. Mem. Am. Chem. Soc., Am. Phys. Soc., Am. Soc. Photobiology, AAAS, Am. Assn. Clin. Chemistry, Sigma Xi. Subspecialties: Physical chemistry; Biophysical chemistry. Current work: Biophysical chemical studies of porphyrins, chlorins, hemeproteins and related structures; from solid state studies to clinical applications. Home: 11 Long Ridge Rd West Redding CT 06896 Office: Western Conn State U 181 White St Danbury CT 06810

ADLER, IRVING, mathematics educator, author, researcher in mathematical biology; b. N.Y.C., Apr. 27, 1913; s. Marcus and Celia (Kress) A.; m. Ruth Relis, June 2, 1935 (dec. 1968); children: Stephen Louis, Peggy Adler Robohm; m. Joyce Theresa Lifshutz, Sept. 16, 1968. B.S., CCNY, 1931; M.A., Columbia U., 1938, Ph.D., 1961. Math. tchr. pub. high schs., N.Y.C., 1932-46; chmn. math. dept. Textile High Sch., N.Y.C., 1946-52; nat. dir. Nat. Council Arts, Scis.

and Professions, N.Y.C., 1953-54; instr. Columbia U., N.Y.C., 1957-60, Bennington (Vt.) Coll., 1961; free lance author, 1954—; cons. ednl. policies commn. NEA, Washington, 1940-41; keynote speaker Conf. State Suprs. of Math., U.S. Office Edn., 1961. Author: The New Mathematics; author 54 other books; co-author: The Reason Why Series, 1960-77; contbr. articles to profl. publs. Pres. Vt. in Miss. Corp., Vt. and Miss., 1965-67; mem. Sch. Bd. Shaftsbury, Vt., 1976-82, chmn., 1979-80; mem. sch. bd. Mt. Anthony High Sch. Dist., Bennington, Vt., 1981—; sec. county com. Democratic party, Bennington, 1972-75. Recipient awards for sci. books for children Nat. Sci. Tchrs. Assn. and Children's Book Council, 1972, 75, 80, N.Y. State Assn. Supervision and Curriculum Devel., 1961. Fellow AAAS; mem. Vt. Acad. Arts and Scis. (pres. 1978-81), Am. Math. Soc., Math. Assn. Am., Soc. Indsl. and Applied Math., Nat. Council Tchrs. of Math., Phi Beta Kappa, Kappa Delta Pi. Jewish. Subspecialties: Morphology; Applied mathematics. Current work: Models of phyllotaxis; models of pre-biotic chemical evolution. Home: RD 2 North Bennington VT 05257

ADLER, JULIUS, biologist, biochemist, educator; b. Edelfingen, Germany, Apr. 30, 1930; came to U.S., 1938, naturalized, 1943; s. Adolf and Irma (Stern) A.; m. Hildegard Wohl, Oct. 15, 1963; children: David Paul, Jean Susan. A.B., Harvard U., 1952; M.S., U. Wis., 1954, Ph.D., 1957; Postdoctoral fellow, Washington U., St. Louis, 1957-59, Stanford U., 1959-60. Asst. prof. biochemistry and genetics U. Wis., Madison, 1960-63, asso. prof., 1963-66, prof., 1966—, Edwin Bret Hart prof. biochemistry and genetics, 1972—, Steenbock prof. microbiol. scis., 1982—. Research, publs. in field. Mem. Am. Acad. Arts and Scis., Nat. Acad. Scis. Jewish. Subspecialties: Biochemistry (biology); Molecular biology. Current work: Behavior of simple organisms, especially bacteria. Home: 1234 Wellesley Rd Madison WI 53705 Office: Dept Biochemistry U Wis Madison WI 53706

ADLER, KRAIG (KERR), biology educator; b. Lima, Ohio, Dec. 6, 1940; s. William Charles and Jennie Belle (Noonan) A.; m. Dolores Rose Pochocki, Mar. 25, 1967; 1 son, Todd David. B.A., Ohio Wesleyan U., 1962; M.S., U. Mich., 1965, Ph.D., 1968. Asst. prof. biology U. Notre Dame, 1968-72; assoc. prof. biology Cornell U., 1972-80, prof., 1980—, chmn. sect. neurobiology and behavior, 1976-79; Baer Meml. lectr. Milw. Pub. Mus., 1977; Hefner lectr. Miami U., Oxford, Ohio, 1980; Anderson Meml. lectr. Rutgers U., 1982; Am. del. 16th Internat. Ethological Congress, Vancouver, 1979, 1st Herpetological Congress of Socialist Countries, Budapest, 1981; sec. gen. I World Congress Herpetology Internat., 1982—. Editor: Herpetological Studies in Eastern United States, 1978, Herpetological Explorations in the Great American West, 1978; contbr. numerous articles to profl. jours. NSF grantee, 1971-83; NIH grantee, 1983—. Fellow Acad. Zoology, AAAS; mem. Soc. Study Amphibians and Reptiles (pres. 1982), Animal Behavior Soc., Am. Soc. Naturalists, Soc. Study Evolution. Subspecialties: Evolutionary biology; Behavioral ecology. Current work: Animal orientation and navigation, especially sensory aspects; behavior and evolution of amphibians and reptiles. Home: 12 Eagles Head Rd Ithaca NY 14850 Office: Section of Neurobiology and Behavior Cornell University Seeley G Mudd Hall Ithaca NY 14853

ADLER, LEONORE LOEB, psychologist; b. Karlsruhe, Ger., May 2, 1921; came to U.S., 1938, naturalized, 1944; d. Leo and Elsie (Laemle) Loeb; m. Helmut Ernest Adler, May 22, 1943; children: Barry Peter, Beverly Sharmaine, Evelyn Renee. B.A., Queen's Coll., CUNY, 1968; Ph.D., Adelphi U., 1972. Research asst. dept. mammalogy Am. Mus. Natural History, N.Y.C., 1956—; adj. asst. prof. dept. psychology Coll. S.I., CUNY, 1974-80; research assoc. Mystic (Conn.) Marine Life Aquarium, 1976—; adj. asst. prof. dept. psychology York Coll., CUNY, 1978-80; dir. Inst. Cross-Cultural and Cross-Ethnic Studies, Molloy Coll., Rockville Centre, N.Y., 1980—. Author, editor: Cross Cultural Research at Issue, 1982; author, translator: This is the Dachshund, 1966, 75; editor, co-author: Issues in Cross-Cultural Research, 1976; mng. editor: Internat. Jour. Group Tensions, 1978—. Fellow N.Y. Acad. Scis. (sect. psychology and women in sci.); mem. Am. Psychol. Assn., N.Y. State Psychol. Assn. (pres. div. social psychology 1978-79, 80-82, pres. div. acad. psychology 1982-83, chair com. women's issues 1982—), Eastern Psychol. Assn., Internat. Council Psychologists (chair publs. policy com. 1981—, treas. 1983—), Internat. Assn. Cross-Cultural Psychology, Soc. Cross-Cultural Research, Assn. Women in Sci., Animal Behavior Soc., Internat. Orgn. Study Group Tensions, Internat. Soc. History Behavioral and Social Scis. Subspecialties: Social psychology; Comparative psychology. Current work: Cross-cultural research with children's drawings, social distances, attitudes, sea mammals. Home: 162-14 86th Ave Jamaica NY 11432 Office: Inst Cross-Cultural & Cross-Ethnic Studies Molloy Coll 1000 Hempstead Ave Rockville Centre NY 11570

ADLER, MARTIN W., pharmacologist, researcher, educator; b. Phila., Oct. 30, 1929; s. Jack and Sonia (Coopersmith) A.; m. Toby Wisotsky, June 28, 1953; children: Charles H., Eve R. B.A., N.Y. U., 1949; B.S. cum laude, Bklyn. Coll. Pharmacy, 1953; M.S., Columbia U., 1957; Ph.D., Albert Einstein Coll. Medicine, 1960. Instr. pharmacology Temple U. Sch. Medicine, Phila., 1960-62, asst. prof., 1962-66, assoc. prof., 1966-73, prof., 1973—; chmn. biomed. research rev. com. Nat. Inst. Drug Abuse, 1980-82; mem. exec. com. Commn. on Problems of Drug Dependence, 1980—. Contbr. articles to profl. jours.; mem. editorial bd.: Jour. Pharmacology and Exptl. Therapeutics, Substance and Alcohol Actions/Misuse. Served with U.S. Army, 1953-55. NIMH grantee, 1960-70; Nat. Inst. Drug Abuse grantee, 1970—; NATO research grantee, 1973-74. Fellow AAAS, Am. Coll. Neuropsychopharmacology; mem. Am. Soc. Pharmacology and Exptl. Therapeutics. Jewish. Subspecialties: Neuropharmacology; Pharmacology. Current work: Neuropsychopharmacology, drugs of abuse, mechanisms of opiate actions, interactions of drugs of abuse. Home: 3247 W Bruce Dr Dresher PA 19025 Office: 3420 N Broad St Philadelphia PA 19140

ADLER, PAUL NEIL, geneticist, educator; b. Bronx, Jan. 6, 1948; s. Aaron and Elsie (Broner) A.; m. Ann Louise Beyer, Aug. 20, 1978. Ph.D., M.I.T., 1975. Jane Coffin Childs Found. postdoctoral fellow U. Calif.-Irvine, 1975-77; asst. prof. biology U. Va., Charlottesville, 1977-82, assoc. prof., 1982—. NIH Research Career Devel. awardee, 1980—. Mem. Genetics Soc. Am., Soc. Developmental Biology, AAAS. Subspecialties: Developmental biology; Gene actions. Current work: Pattern formation and regulation in drosophila - genetic control of same. Office: U Va Dept Biology Gilmer Hall Charlottesville VA 22901

ADLER, ROBERT, electronics engineer; b. Vienna, Austria, Dec. 4, 1913; came to U.S., 1940, naturalized, 1945; s. Max and Jenny (Herzmark) A.; m. Mary F. Buehl, 1946. Ph.D. in Physics, U. Vienna, 1937. Asst. to patent atty., Vienna, 1937-38; lab. Sci. Acoustics, Ltd., London, Eng., 1939-40, Asso. Research, Inc., Chgo., 1940-41; research group Zenith Radio Corp., Chgo., 1941-52, asso. dir. research, 1952-63, v.p., 1959-77, dir. research, 1963-77, EXTEL Corp., Northbrook, Ill., 1978-79, v.p. research, 1979-82; tech. cons. Zenith Corp., 1982—. Contbr. numerous articles profl. publs. Fellow IEEE (Edison medal 1980); mem. Nat. Acad. Engring. Subspecialties: Electronics; Acoustics. Current work: Improvement of cathode ray tubes; applications of optics in information display devices; surface acoustic wave resonators and filters; ultrasonic testing of structures. Developed various electron beam tubes for frequency modulation transmitters, for TV receivers, electron beam parametric amplifier; pioneer ultrasonic remote control for TV, ultrasonic light deflection for laser projection TV. Home: 327 Latrobe Ave Northfield IL 60093 Office: Zenith Ctr 1000 Milwaukee Ave Glenview IL 60025

ADLER, STEPHEN LOUIS, physicist; b. N.Y.C., Nov. 30, 1939; s. Irving and Ruth (Relis) A.; m. Judith Ann Curtis, Oct. 27, 1962; children: Jessica Wendy, Victoria Stephanie, Anthony Curtis. A.B. summa cum laude, Harvard U., 1961; Ph.D., Princeton U., 1964. Jr. fellow Soc. of Fellows Harvard U., 1964-66; research asso. Calif. Inst. Tech., 1966; mem. nst. for Advanced Study, Princeton U., N.J., 1966-69; prof. Sch. Natural Scis., Inst. for Advanced Study, 1969-79, N.J. Albert Einstein prof., 1979—; vis. lectr. dept. physics Princeton U., 1969—; cons. in field. Author: (with R.F. Dashen) Current Algebras, 1968; contbr. articles to profl. jours. Fellow Am. Acad. Arts and Scis., Am. Phys. Soc.; mem. Nat. Acad. Scis., Phi Beta Kappa, Sigma Xi. Subspecialties: Particle physics; Theoretical physics. Current work: Quantum gravitation; effective action models for quark confinement. Home: 9 Veblen Circle Princeton NJ 08540 Office: Inst for Advanced Study Princeton NJ 08540

ADLER, WILLIAM FRED, materials scientist; b. Chgo., Aug., 19, 1937; s. Fred William and Margaret Ann (Haak) A.; m. Dorothy Joanne Weinmann, June 8, 1958; 1 dau., Elizabeth Anne. B.S. in Civil Engring, Ill. Inst. Tech., 1958, M.S. in Mechanics, 1961; Ph.D. in Engring. Mechanics, Columbia U., 1965. Sr. research engr. Martin-Marietta Corp., Denver, 1961-62; sr. scientist Battelle Meml. Inst., Columbus, Ohio, 1965-71; prin. scientist Bell Aerospace Corp., Buffalo, 1971-76; mgr. materials group Effects Tech., Inc., Santa Barbara, Calif., 1976-82; mgr. materials sci. sect. Gen. Research Corp., Santa Barbara, 1982—; chmn. nat. materials adv. bd. com. on conservation of materials in energy systems through the reduction of erosion Nat. Acad. Scis., 1976-77. Editor: Inelastic Behavior of Solids, 1970, Erosion: Prevention and Useful Applications, 1979; Contbr. articles to profl. jours. Mem. Am. Ceramic Soc., ASTM (chmn. subcom. on erosion by solid particle impingement 1974-80), Soc. Engring. Sci., Sigma Xi, Tau Beta Pi, Chi Epsilon. Subspecialties: Materials (engineering); Fracture mechanics. Current work: Liquid and solid particle erosion; dynamical response of materials and structures; fracture mechanics of metals, ceramics, polymers, and composites; computer and analytical modeling of material behavior in hostile environments; life predictions; management of personnel, marketing, technical report preparation and review for contract and development activities.

ADNEY, JOSEPH ELLIOTT, JR., mathematics educator; b. DeLand, Fla., Aug. 20, 1923; m., 1952; 2 children. B.S., Stetson U., 1944; M.A., Ohio State U., 1949, Ph.D., 1954. Assoc. Research Found., Ohio State U., 1952-54, instr. math., 1954-55; asst. prof. Purdue U., 1955-64; assoc. prof. Mich. State U., East Lansing, 1964-69, prof. math., 1969—, chmn. dept., 1974—; cons. USAF, 1957, 59. Mem. Am. Math. Soc. Subspecialty: Abstract algebra. Office: Dept Math Mich State U East Lansing MI 48823

ADRION, WILLIAM RICHARD, government executive, consultant; b. Alexandria, La., Nov. 2, 1943; s. Vernon Richards and Mary Leone (Carlock) A.; m. Jacqueline Cotner, July 3, 1971; children: Carrie Buchanan, Emily Richards. B.S., Cornell U., 1966, M.E., 1967; Ph.D., U. Tex.-Austin, 1971. Computer engr. Honeywell EDP, Waltham, Mass., 1969-70; asst. prof. U. Tex.-Austin, 1971-72; area chmn., asst. prof. Oreg. State U., Corvallis, 1972-78; program dir. NSF, Washington, 1976-78, 80—; group mgr. Nat. Bur. Standards, 1978-80; cons. Radio Free Europe/Radio Liberty, Munich, West Germany, 1981-82, Applied Theory Assocs., Corvallis, Oreg., 1973-78, Tektronic, Portland, Oreg., 1974-76, Am. U., Washington, 1976-77; lectr. George Washington U., Washington, 1978. Contbr. articles in field to profl. jours. Named Outstanding Young Faculty Am. Soc. Engring. Edn., 1973. Mem. Assn. Computing Machinery (vice-chmn. SIGSOFT 1981—), IEEE, AAAS, Soc. Indsl. and Applied Math., N.Y. Acad. Scis., Sigma Xi. Subspecialties: Software engineering. Current work: Research in software tools, techniques and support systems, structure of and support for experimental computer science. Home: 5007 N 34th Rd Arlington VA 22207 Office: NSF 1800 G St NW Washington DC 20550

ADUSS, HOWARD, orthodontist, educator; b. N.Y.C., Mar. 31, 1932; s. Irving and Yvette (Rothwax) A.; m. Marcia I. Katz, Dec. 27, 1953; children: Kathy, Laura, Robert, Deborah. B.S., Purdue U., 1954; D.D.S., Northwestern U.-Chgo., 1957; M.S., U. Rochester, 1962. Diplomate: Am. Bd. Orthodontists. Prof. orthodontics U. Ill.-Chgo., 1962—; also prof. Rush Med. Coll., Chgo., 1974—; sec.-treas. Craniofacial Biology Group, Chgo., 1969-81. Contbr. articles profl. jours. Pres. Eastman Orthodontic Alumni Assn., 1964-66. Served to lt. USNR, 1957-59. Office Naval Research award, 1962; fellow USPHS, 1960-62. Fellow Am. Coll. Dentists; mem. Am. Assn. Orthodontists, Am. Assn. Dental Research, Am. Soc. Human Genetics, Teratology Soc., Am. Clef Palate Assn. (pres. 1974-75), Sigma Xi. Jewish. Club: Chgo. Yacht. Subspecialties: Orthodontics; Developmental biology. Current work: Craniofacial malformations. Home: 237 Lakeside Pl Highland Park IL 60035 Office: University of Illinois Medical Center PO Box 6998 Chicago IL 60680

AFRICANO, JOHN LOUIS, astronomer; b. St. Louis, Feb. 8, 1951; s. John L. and Dorothy (McDonald) A.; m. Linda A., Feb. 12, 1972; children: James, Brian, Monica. M.S., Vanderbilt U., 1974. Support scientist McDonald Obs., U. Tex., Ft. Davis, 1974-78; mgr. Cloudcroft (N.Mex.) Obs., AURA, 1978-81; mgr. telescope support group Kitt Peak Nat. Obs., Tucson, 1981—. Contbr. articles to profl. jours. Mem. Am. Astronom. Soc. Subspecialty: Optical astronomy. Current work: Lunar occultations, solar-stellar connection photoelectric photometry of RS CVn, BY Dra, FK Comae variable stars. Home: 9511 E 42nd St Tucson AZ 85710 Office: Kitt Peak Obs PO Box 26732 Tucson AZ 85726

AFSHAR, SIROOS K., software engineer; b. Qom, Tehran, Iran, July 3, 1949; came to U.S., 1975, naturalized, 1980; s. Bagher and Mastooreh (Manii) Afshar-K.; m. Simin Manii, Aug. 11, 1974; children: Pedram, Afsheen. B.S.E.E., Sharif U., Tehran, Iran, 1972; M.S.E.E., U. Mich., 1976, Ph.D., 1979. Design engr. Nat. Iranian Oil Co., Tehran, 1972-75; research asst. U. Mich., Ann Arbor, 1976-79; asst. prof. La. State U., Baton Rouge, 1979-82; mem. tech. staff AT&T Info. Systems, Lincroft, N.J., 1982—; Predoctoral fellow U. Mich., 1978. Recipient Outstanding Achievement award U. Mich., 1979. Mem. IEEE, Assn. Computing Machinery, Soc. Indsl. and Applied Math., Phi Kappa Phi, Tau Beta Pi, Eta Kappa Nu. Subspecialties: Software engineering; Theoretical computer science. Current work: Research on, experiments with, and implementation of tools, design languages, and methodologies for the development process of large computer systems. Home: 10 Constitution Ct Manalapan NJ 07726 Office: AT&T Info Systems 307 Middletown-Lincroft Rd Lincroft NJ 07738

AGARD, EUGENE THEODORE, medical physicist; b. Barbados, W.I., Aug. 15, 1932; came to U.S., 1959; s. Samuel and Pearl Doris (Best) A.; m. Joyce Elaine Phillips, June 8, 1960; children: Noel A., Ian C., Wendy T., Linda G. B.Sc. with 1st class honors, U. W.I., Jamaica, 1954-56; M.Sc., U. London, 1958; Ph.D., U. Toronto, Ont., Can., 1970. Cert. Am. Bd. Radiology. Asst. physisist Mass. Gen. Hosp., Boston, 1959-63; lectr., then asst. prof. U. W.I., Trinidad, 1963-72; dir med. physics program P.R. Nuclear Center, 1972-76; asso. clin. prof. Wright State U., Dayton, 1977-80; med. physicist, radiation safety officer Kettering Med. Center, Dayton, Ohio, 1976—; cons., researcher. Contbr. articles to profl. jours. Mem. Am. Assn. Physicists in Medicine (mem. internat. relations com. 1974-77), Health Physics Soc. (chmn. affirmative action com. 1980-81), Miami Valley Alt. Energy Assn. (dir. 1980-81), Radiol. Soc. N. Am., Am. Coll. Radiology. Christian. Subspecialties: Biophysics (physics); Imaging technology. Current work: Radiation dosimetry and biological effects of radiation. Office: 3535 Southern Blvd Dayton OH 45429

AGEE, LANCE JAMES, nuclear engineer; b. Bridgeport, Calif., Sept. 18, 1942; s. Robert James and Thelma V. (Volkman) A.; m. Olga Maria Fuentes, July 25, 1964; children: Sonia Maria, Robert Kenneth. B.S. in Engring. Sci, U. Nev., 1965, M.S. in Nuclear Engring, 1966; postgrad., U. Wash., 1968-70. Registered profl. engr., Calif. Engr. Douglas United Nuclear Inc., Richland, Wash., 1966-68, sr. engr., 1970-71, Combustion Engring. Co., Windsor, Conn., 1971-75, lead engr., 1974-75; project mgr. Elec. Power Research Inst., Palo Alto, Calif., 1975-80, program mgr., 1980—; cons. advanced code rev. group NRC, Washington, 1975-81; U.S. rep. Com. Safety Nuclear Installations, Paris, 1978. Editor: Conf. Proc. Internat. Retran Conf, 1980, 82. AEC fellow, 69-70. Mem. Am. Nuclear Soc., Sigma Tau. Subspecialties: Nuclear engineering; Nuclear fission. Current work: Develop nuclear reactor simulators that predict normal and accident conditions in nuclear plants; RETRAN and RASP are industry tools. Home: 19294 Dehavilland Dr Saratoga CA 95070 Office: Electric Power Research Inst 3420 Hillview Ave Palo Alto CA 94304

AGHAJANIAN, GEORGE KEVORK, psychiatry and pharmacology educator; b. Beirut, Apr. 14, 1932; U.S., 1933; s. Ghevont M. and Araxi (Movsessian) A.; m. Anne Elaine Hammond, Ja. 10, 1959; children: Michael, Andrew, Carol, Laura. A.B., Cornell U., 1954; M.D., Yale U., 1958. Intern Jackson Meml. Hosp., Miami, Fla., 1958-59; resident in psychiatry Yale U., 1959-63, asst. prof. psychiatry, 1965-68, assoc. prof. psychiatry, 1968-70, assoc. prof. psychiatry and pharmacology, 1970-74, prof. psychiatry and pharmacology, 1974—. Contbr. numerous articles to sci. publs.; mem. editorial bd.: Brain Research, 1968—, Ann. Rev. Neurosci, 1981—, Jour. Neurosci, 1980—, European Jour. Pharmacology, 1983—, Neuropharmacology, 1972—. Served to capt. M.C. U.S. Army, 1963-67. Recipient Scheele medal Swedish Acad. Pharm. Scis., 1981; co-recipient Founds. Fund research prize in psychiatry Am. Psychiat. Assn., 1981. Mem. Am. Soc. Pharmacology and Exptl. Therapeutics, Neurochem. Soc., Soc. Neurosci., Am. Coll. Neuropsychopharmacology (Efron award 1975), Internat. Brain Research Orgn. Subspecialty: Neuropharmacology. Current work: Neurotransmitter systems in the brain and the mechanism of action of psychotropic drugs. Office: 34 Park St New Haven CT 06508

AGNELLO, VINCENT, physician, researcher; b. Bklyn., Aug. 1, 1938; m. Carole Fry; 4 children. B.S., Rensselaer Poly. Inst., 1960; M.D., U. Rochester, 1964. Diplomate: Am. Bd. Internal Medicine. Intern Vanderbilt Hosp., Nashville, 1964-65; postdoctoral fellow, Rockefeller U., N.Y.C., 1967-71; resident N.Y. Hosp., 1972-73; asst. physician Rockefeller Univ. Hosp., 1967-71, assoc. physician, 1971-73; asst. prof. Rockefeller U., 1971-73; assoc. prof. Tufts U. Sch. Medicine, Boston, 1973—; chief div. rheumatology/clin. immunology New Eng. Med. Ctr. Hosp., Boston, 1973-81; dir. clin. immunology lab. Tufts-New Eng. Med. Ctr., Boston, 1981—; mem. staff Edith Nourse Rogers VA Hosp., 1981—; mem. adv. panel Am. Bd. Med. Lab. Immunology. Editorial bd.: Diagnostic Immunology. Bd. dirs. Lupus Found.; trustee Mass. chpt. Arthritis Found. Served to lt., M.C. USNR, 1965-67. NIH awardee, 1974-79. Subspecialty: Immunology (medicine). Current work: Human immune complex disease; complement deficiency states; idiotypic studies of autoantibodies in man. Home: 11 French Rd Weston MA 02193 Office: 171 Harrison Ave Boston MA 02111

AGNEW, DOUGLAS CRAIG, environmental consultant; b. East Meadow, N.Y., June 6, 1957; s. Edwin Lee and Mary Margaret (Reardon) A. B.S. in Environ. Biology, Stockton State Coll., 1979. Aquatic biologist Lakes and Waterways Mgmt. Inc., Pompano Beach, Fla., 1979-80, div. mgr., Tampa, Fla., 1980-82, v.p., 1982; environ. cons., pres. Aqua Sci., Inc., Dunedin, Fla., 1982—. Active Greater Tampa C. of C., Environ. Quality Council. Served with USCGR, 1976-82. Mem. Fla. Aquatic Plant Mgmt. Soc., Marine Tech. Soc., N. Am. Lake Mgmt. Soc. Subspecialties: Ecosystems analysis; Resource management. Home: 129 Edgewater Terr Dunedin FL 33528 Office: Aqua Sci Inc 129 Edgewater Terr Dunedin FL 33528

AGNEW, HAROLD MELVIN, physicist; b. Denver, Mar. 28, 1921; s. Sam E. and Augusta (Jacobs) A.; m. Beverly Jackson, May 2, 1942; children: Nancy E. Agnew Owens, John S. A.B., U. Denver, 1942; M.S., U. Chgo., 1948, Ph.D., 1949. With Los Alamos Sci. Lab., 1943-46, alt. div. leader, 1949-61, leader weapons div., 1964-70, dir., 1970-79; pres. GA Techs. Inc., San Diego, 1979—, Blaws Corp., 1967-72; sci. adviser Supreme Allied Comdr. in Europe, Paris, France, 1961-64; Chmn. Army Sci. Adv. Panel, 1965-70, mem., 1970-74; mem. aircraft panel Pres.'s Sci. Adv. Com., 1965-73; mem. USAF Sci. Adv. Bd., 1957-69, Def. Sci. Bd., 1965-70, Gov. N.Mex. Radiation Adv. Council, 1959-61; sec. N.Mex. Health and Social Services, 1971-73; chmn. gen. adv. com. ACDA, 1974-77, mem., 1977-81; mem. aerospace safety adv. panel NASA, 1968-74; mem. U.S. Army Sci. Bd., 1978-80, White House Sci. Council, 1982—. Mem. council engring. NRC, 1978—; Mem. Los Alamos Bd. Ednl. Trustees, 1950-55, pres., 1955; mem. Woodrow Wilson Nat. Fellowship Found., 1973—; Mem. N.Mex. Senate, 1955-61; sec. N.Mex. Legis. Council, 1957-61; chmn. N.Mex. Senate Corp. Commn., 1957-61; bd. dirs. Fedn. Rocky Mountain States, Inc., 975-77. Recipient Ernest Orlando Lawrence award AEC, 1966; Enrico Fermi award Dept. Energy, 1978. Fellow Am. Phys. Soc., AAAS; mem. Nat. Acad. Scis., Nat. Acad. Engring. (Assembly of Engring.), Council on Fgn. Relations, Phi Beta Kappa, Sigma Xi, Omicron, Delta Kappa. Subspecialty: Nuclear physics. Home: 322 Punta Baja Solana Beach CA 92075 Office: PO Box 81608 San Diego CA 92138

AGNEW, WILLIAM GEORGE, mech. engr., engring. lab. adminstr.; b. Oak Park, Ill., Jan. 12, 1926; s. Dupre L. and Marion S. (Roberts) A.; m. Norma Jean Light, Mar. 9, 1957; children—Brian R., Daniel D., Dalen W. B.S. in Mech. Engring, Purdue U., 1948, M.S., 1950, Ph.D., 1952. Research engr. Project Squid, U.S. Navy, Purdue U., 1948-50; with General Motors Research Labs., Warren, Mich., 1952—, dept. head fuels and lubricants, 1967-70, dept. head emissions research, 1970-71, tech. dir., 1971—. Contbr. articles to profl. jours. Served with Manhattan Dist. U.S. Army, 1944-46. Mem. Soc. Automotive Engrs. (Horning Meml. award 1960), ASME, Combustion Inst. (bd. dirs. 1960-76), Nat. Acad. Engring., AAAS, Engring. Soc. Detroit, Sigma Xi. Subspecialties: Mechanical engineering; Combustion processes. Current work: Combustion, engines, vehicle structures, aerodynamics, fules, lubricants, energy resources, power transmissions, air pollution, atmospheric chemistry, auto safety.

Home: 3450 31 Mile Rd Romeo MI 48065 Office: Research Laboratories General Motors Technical Center Warren MI 48090

AGNIHOTRI, KRISHNA VENKTESH, high technology company executive; b. Bombay, India, Jan. 6, 1946; came to U.S., 1969, naturalized, 1977; s. Venktesh V and Kusumavati (Navalgund) A.; m. Varsha K. Kurtkoti, Aug. 5, 1973; 1 son, Vikram. B.S., Indian Inst. Tech., Bombay, 1969; M.S., Stevens Inst. Tech., 1972, now postgrad. Grad. teaching asst. Stevens Inst. Tech., Hoboken, N.J., 1970-71; mfg. engr. Nat. Beryllia Corp., Haskell, N.J., 1972-74, spl. project scientist, 1974-79, v.p., dir. tech., 1979—. Author: High Thermal Package Development, 1980; inventor auto/reverse plating, 1975. Recipient cert. Nat. Beryllia Corp., 1975; Indian Inst. Tech. Scholar, 1964-69. Subspecialties: Metallurgical engineering; Ceramic engineering. Current work: Advanced engineering materials/processes, resolution of unique metallurgical parameters in microelectronics. Home: 228 Perry St Dover NJ 07801 Office: Nat Beryllia Corp Greenwood Ave Haskell NJ 07420

AGOSTA, WILLIAM CARLETON, chemist, educator; b. Dallas, Jan. 1, 1933; s. Angelo N. and Helen Carleton (Jones) A.; m. Karin Solveig Engstrom, July 2, 1958; children—Jennifer Ellen, Christopher William. B.A., Rice Inst., 1954; A.M., Harvard U., 1955, Ph.D., 1957. NRC postdoctoral fellow Oxford (Eng.) U., 1957-58; Pfizer postdoctoral fellow U. Ill., Urbana, 1958-59; asst. prof. U. Calif., Berkeley, 1959-61; liaison scientist U.S. Navy, Frankfurt, Germany, 1961-63; asst. prof. Rockefeller U., 1963-67, asso. prof., 1967-74, prof., 1974—; cons. in field; officer, dir. Chiron Press, Inc. Contbr. articles to profl. jours. John Angus Erskine fellow, U. Canterbury (N.Z.), 1981—. Mem. Am. Chem. Soc., Chem. Soc. London, Interam. Photochem. Soc., Am. Soc. Photobiology, Phi Beta Kappa, Sigma Xi. Subspecialties: Organic chemistry; Photochemistry. Current work: Organic photochemistry, conformational analysis, mammalian pheromones. Home: 32 Washington Sq New York NY 10011 Office: Rockefeller U 1230 York Ave New York NY 10021

AGRAWAL, HARISH CHANDRA, neurobiologist, researcher; b. Allahabad, Uttar Pradesh, India; came to U.S., 1970, naturalized, 1982; s. Shambhu and Rajmani Devi A.; m. Daya Kumari Bhushan, Feb. 6, 1960; children: Sanjay, Sanjeev. B.Sc., Allahabad U., 1957, M.Sc., 1959, Ph.D., 1964. Med. research assoc. Thudichum Psychiat. Lab., Galesburg, Ill., 1964-68; lectr. dept. biochemistry Charing Cross Hosp., London, 1968-70; prof. pediatrics Wash. U. Sch. Medicine, St. Louis, 1970—; mem. neurology study sect. NIH, 1979-82. Author: Handbook of Neurochemistry, 1969, Developmental Neurobiology, 1970, Biochemistry of Developing Brain, 1971, Membranes and Receptors, 1974, Proteins of the Nervous System, 1980, Biochemistry of Brain, 1980, Handbook of Neurochemistry, 1983. Jr. research fellow Council Sci. and Indsl. Research, New Delhi, 1960-62; sr. research fellow, 1963-64; recipient Research Career Devel. award Nat. Inst. Neurol. and Communicative Disorders, 1974-79. Mem. Internat. Soc. Neurochemistry, Internat. Brain Research Orng., Am. Soc. Neurochemistry, Am. Soc. Biol. Chemists, Am. Soc. Physiologists Soc., Soc. Neurosci. Subspecialties: Neurochemistry; Neurobiology. Current work: Acylation, glycosylation and phosphorylation of proteins of myelin, myelin or glial surface antigens responsible for demyelination of neural tissues. Home: 18 Chafford Woods Saint Louis MO 63144 Office: Dept Pediatrics Washington U 500 S Kingshighway Blvd Saint Louis MO 63110

AGRAWAL, KRISHNA CHANDRA, pharmacology, educator, consultant, researcher; b. Calcutta, India, Mar. 15, 1937; came to U.S., 1962; s. Prasadi Lal and Asarfi Devi (Agrawal) A.; m. Mani Agrawal, Dec. 2, 1960; children: Sunil, Lina, Nira. B.S. in Pharmacy, Andhra U., Waltair, India, 1959, M.S., 1960; Ph.D., U. Fla., 1965; cert. in pharm. chemistry. Research asso. dept. pharmacology Yale U. Sch. Medicine, New Haven, 1966-69, instr., 1969-70, asst. prof., 1970-76, asso. prof., 1976; assoc. prof. dept. pharmacology Tulane U. Sch. Medicine, New Orleans, 1976-81, prof., 1981—; cons. mem. Southeastern Cancer Study Group, 1980—; mem. adv. com. on instnl. grants Am. Cancer Soc., 1980—. Contbr. numerous articles on cancer chemotherapy and devel. of radiosensitizing agts. for use in radiotherapy to sci. jours. Nat. Cancer Inst. grantee, 1976—; WHO grantee, 1979-82; La. Bd. Regents grantee, 1981-82. Fellow Am. Inst. Chemists; mem. Am. Chem. Soc., Am. Assn. for Cancer Research, Radiation Research Soc., Am. Soc. Pharmacology and Exptl. Therapeutics, Sigma Xi. Subspecialties: Molecular pharmacology; Cancer research (medicine). Current work: Development of anticancer agents; development of radiation sensitizing agents for selective sensitization of hypoxic tumor cells to radiotherapy; studies related to molecular mechanism of action of hyperthermia and cancer chemotherapeutic agents. Patentee radiosensitizers for hypoxic tumor cells and compositions thereof. Home: 6760 Bamberry Dr New Orleans LA 70126 Office: Dept Pharmacology Tulane U Sch Medicine New Orleans LA 70112

AGRAWAL, RAM KUMAR, mechanical engineer, consultant; b. Chhattar, Haryana, India, Aug. 15, 1938; came to U.S., 1968, naturalized, 1977; s. Lakhi Ram and Chandroli Devi A.; m. Bimla Kumari, Apr. 4, 1945; children: Vipin K., Vini K., Vina K., Binu K. B.S. in M.E, G.N. Engring. Coll., Ludhiana, India, 1964, M.S., U. Houston, 1970; Ph.D., Tex. A&MU., 1973. Registered profl. engr., Tex. Lectr. mech. engring. Bits, Pilani, India, 1965-68; research asst., instr. U. Houston, 1968-70; instr., cons. Tex. A&MU., 1970-73; sr. and prin. mech. engr. Heat Research Corp., Houston, 1973-78; pvt. practice cons. engring., Houston, 1978-80; sr. mech. engr. Fluor Engrs. Inc., Houston, 1980—. Mem. ASME. Subspecialties: Mechanical engineering. Current work: Specific interests in designing and engineering of direct fired equipments such as pyralysis furnaces, reformer furnaces, fired heaters and boilers.

AGRIOS, GEORGE N., plant pathologist, educator; b. Galarinos, Halkidiki, Greece, Jan. 16, 1936; came to U.S., 1963, naturalized, 1966; s. Nicholas G. and Olga (Kotsioudis) A.; m. Annette E. Braynard, Nov. 11, 1962; children: Nicholas, Anthony, Alexander. B.S., U. Thessaloniki, Greece, 1957; Ph.D., Iowa State U., Ames, 1960. Asst. prof. dept. plant pathology U. Mass., Amherst, 1963-69, assoc. prof., 1969-75, prof., 1975—. Author: Plant Pathology, 1969, 2d edit, 1978. Served with C.E. AUS, 1960-61. Mem. Am. Phytopath. Soc., Can. Soc. Plant Pathology, AAAS, N.Y. Acad. Scis. Greek Orthodox. Subspecialties: Plant virology; Plant cell and tissue culture. Current work: Transmission of viruses through tissue culture techniques. Detection, identification and manipulation of genes for resistance to viruses through tissue culture techniques; effect of antiviral compounds. Home: 20 Valley View Circle Amherst MA 01002 Office: Dept Plant Pathology U Mass Amherst MA 01003

AGUIAR, ADAM MARTIN, chemistry educator, cons., researcher; b. Newark, Aug. 11, 1929; s. Joaquim and Emilea (Nunes) A.; m. Harriet Joan Greenberg, Dec. 23, 1940; m. Laura Estelle Brand, July 10, 1957; children: Justine Diane, David Laurence. B S., Fairleigh Dickinson U., 1955; M.A., Columbia U., 1957, Ph.D. (Union Carbide fellow), 1960. Chemist Otto B. May, Newark, 1948-55; NIH Postdoctoral fellow Columbia, U., N.Y.C., 1959-60; prof. chemistry Tulane U., New Orleans, 1963-72; research fellow Roche Inst. Molecular Biology, 1982, Tulane U., 1969-70; dean grad. and research programs William Paterson Coll., Wayne, N.J., 1972-73; prof. chemistry Fairleigh Dickinson U., Madison, N.J., 1973—; pres. Seltox Corp., N.J.; vice chmn. Seltox Internat. Corp., Nev.; pres. A & B Assos.; cons. chemists; hon. research prof. Birkbeck Coll., U. London, 1970, Rutgers U., 1973-74. Contbr. articles on chemistry to profl. jours. Numerous research grants, 1972-77. Mem. Am. Chem. Soc., AAAS, N.Y. Acad. Scis., Oral Health Research Center (Fairleigh Dickinson U. Dental Sch. Hackensack, N.J.), Sigma Xi. Subspecialties: Organic chemistry; Medicinal chemistry. Current work: The synthesis and development of novel organophosphorus medicinals for third world and other protozoal and parasitic diseases. Home: 530 Valley Rd 5F Upper Montclair NJ 07043 Office: Madison Ave 11-C Madison NJ 07940

AGUS, ZALMAN S., physician; b. Chgo., Apr. 3, 1941; m., 1963; 3 children. B.A., Johns Hopkins U., 1961; M.D., U. Md., 1965. Intern Sch. Medicine, U. Md., 1965-66, resident, 1966-68; research fellow Sch. Medicine U. Pa., 1968-71; attending physician nephrology USAF, Lackland AFB, Tex., 1971-73; asst. prof. Sch. Medicine, U. Pa., Phila., 1973-78, assoc. prof. med. nephrology, 1978—, chief renal-electrolyte sect., 1979—; NIH fellow, 1969-71, clin. investigator, VA, 1973-76. Recipient Research Career Devel. award NIH, 1977. Fellow ACP; mem. Am. Fedn. Clin. Research, AAAS, Am. Soc. Nephrology, Internat. Soc. Nephrology, N.Y. Acad. Sci. Subspecialties: Nephrology; Physiology (medicine). Office: 860 Gates Bldg 3400 Spruce St Philadelphia PA 10104

AHEARNE, DANIEL PAUL, physicist; b. New Britain, Conn., Apr. 23, 1931; s. Daniel Paul and Balbena Marion (Baloski) A.; m. Germaine Marie Sirois, Feb. 4, 1954; children: Michael Jude, Douglas James. B.S. in Applied Physics, Calif. State U., 1962; postgrad, UCLA, 1963-65, U. N.Mex., 1966-67, N.Mex. Inst. Tech., 1967-69. Jr. engr. Gen. Dynamics/Electronics, San Diego, 1961-62; physicist N. Am. Rockwell, Los Angeles, 1963-65; weapons system research staff TRW Systems, Redondo Beach, Calif., 1965-66; br. chief U.S. Air Force Weapon Lab., Kirtland AFB, N. Mex., 1967-69; chief scientist Safeguard Systems Comm. Agy., Washington, 1969-70; pres. Mesa Cons., 1971-73; tech. dir. ITT Grinnell, Washington, 1981—; cons. in field; asso. adminstr. FEA, 1974-75; pres. Team Inc., 1976-81; tech. dir. ITT Grinnell Energy Products Group, 1981-82; participant USAF Communications Master Plan Formulation, others. Contbr. articles to profl. jours. Exec. dir. Dem. Party., N.M., 1970-72. Served with U.S. Army, 1948-57. Decorated D.S.C., Silver Star, Bronze Star medals (2), Purple Heart (3).; Nuclear Sci. fellow, 1962; N. Am Rockwell Corp. fellow, 1964; Colo. U. research assistantship, 1965. Mem. Am Phys. Soc., N.Y. Acad. Sci., Va. Acad. Sci., Ala. Acad. Sci. Roman Catholic. Subspecialties: Plasma physics; Laser fusion. Current work: Work relating to energy systems including laser, nuclear, and solar energy systems; weapons system and thermonuclear design and effects analysis/design. Patentee in field. Home: 7221 Briarcliff Dr Springfield VA 22153

AHLBRANDT, CALVIN DALE, educator; b. Scotts Bluff, Nebr., Aug. 13, 1940; s. Herman and Catherine (Ehrlich) A.; m. Evelyn Elaine Keller, Dec. 31, 1961; children: Robert, William, Michael. B.S., U. Wyo., 1962; M.A., U. Okla., 1965, Ph.D., 1968. Phys. sci. aid U.S. Bur. Mines, Laramie, Wyo., 1960-62; grad. asst. in math. U. Okla., 1962-66, research asst. in math., 1966-68, spl. instr. math., 1968; asst. prof. math. U. Mo., Columbia, 1968-72, assoc. prof., 1972-82, prof. math., 1982—; lectr. in field. Contbr. articles to profl. jours. Mem. Am. Math. Soc., Soc. for Indsl. and Applied Math., Sigma Xi, Sigma Pi Sigma. Current work: Boundary value problems for differential systems. Home: 2236 Walcox Rt 3 Columbia MO 65201 Office: Dept Math Univ Mo Columbia MO 65211 Home: 2236 Walcox Rt 3 Columbia MO 65201

AHLFORS, LARS VALERIAN, mathematician, educator; b. Helsingfors, Finland, Apr. 18, 1907; s. Karl Axel and Sieva (Helander) A.; m. Erna Lehnert, June 22, 1933; children—Cynthia, Vanessa, Caroline. Ph.D., LL.D., Boston Coll., 1951; Sc.D., London U., 1978. Adj. math. U. Helsingfors, 1932-35, prof., 1938-44; asst. prof. math. Harvard, 1935-38; prof. U. Zurich, 1944-46; asso. prof. Harvard, 1946, prof. math., 1946—, named W.C. Graustein prof., 1964, chmn. math. dept., 1948-50. Author: Complex Analysis, 1953; Contbr. papers on conformal mapping, Riemann surfaces, other brs. Theory of Function of a Complex Variable to profl. lit. Recipient Field's medal for math. research Internat. Congress of Mathematicians, Oslo, 1936, Wolf prize for Math., 1981. Mem. Am. Math. Soc., Am. Math. Assn., Societas Scientiarum Fennica, Academia Scientarum Fennica, Swedish Royal, Nat. acads. sci. Club; Faculty (Harvard). Subspecialty: Complex analysis. Home: 160 Commonwealth Ave Boston MA 02116 Office: Harvard Cambridge MA 02138

AHLGREN, CLIFFORD ELMER, forester; b. Toimi, Minn., Apr. 22, 1922; s. Herman and Olga Marie (Kopponen) A.; m. Isabel F. Fulton, Apr. 7, 1954; children: Clifford L., Olga Marie. B.S., U. Minn., 1948, M.S., 1952; D.Sc. (hon.), Cornell U., 1976. Forester Iron Range Resources and Rehab. Com., 1948; dir. research Wilderness Research Found., Duluth, Minn., 1948—; research asso. U. Minn., 1960—. Contbr. articles to profl. jours. Served with USCG, 1941-42. Fellow Soc. Am. Foresters; mem. Ecol. Soc. Am., Am. Forestry Assn., Phytopath Soc. Am., Finnish Forestry Assn. (corr.). Presbyterian. Lodges: Masons; Shriners. Subspecialties: Ecology; Genetics and genetic engineering (agriculture). Current work: Anthropogenic alteration of wilderness, forest succession. Home and Office: 215 W Oxford St Duluth MN 55803

AHLUWALIA, BHAGWAT DATTA, radiol. physicist, educator; b. Pindibhatian, India, Apr. 28, 1939; came to U.S., 1965, naturalized, 1978; s. Shri Kundan Lal and Gauran A.; m. Ekta Walia, Mar. 11, 1972; children: Ana, Atul, Sumit. B.Sc., Hons Sch., 1963; M.Sc., Hons Sch., Panjab U., 1964; Ph.D., Boston U., 1972. Diplomate: Am. Bd. Sci. in Nuclear Medicine. Asst. physicist dept. radiology Mass. Gen. Hosp., Boston, 1973-77; assoc. Harvard U. Med. Sch., dept. radiology, 1973-77; vis. scientist dept. radiation therapy Postgrad. Inst. Medicine and Research, Chandigarh, India, 1979; asst. prof., radiol. physicist U. Okla. Health Sci. Center, Oklahoma City, 1977—. Recipient Medal Panjab U., 1963, Young Investigator award NIH, 1975; Ford Found. grantee, 1965. Mem. Am. Assn. Physicists in Medicine, Soc. Nuclear Medicine, Health Physics Assn., IEEE. Subspecialty: Medical Physics. Current work: Med. physics, imaging and therapy.

AHLUWALIA, HARJIT SINGH, educator; b. Bombay, India, May 13, 1934; s. Sewa Singh and Jaswant (Kaur) A.; m. Manjit Kaur, Nov. 29, 1964; children: Suvinder Singh, Davinder Singh. B.Sc., Panjab U., 1953; Ph.D., Gujarat U., 1960. Sr. research fellow Phys. Research Lab., Ahmedabad, India, 1954-62; tech. assistance expert UNESCO, Paris, 1962-63; research assoc. Southwest Ctr. for Advanced Studies, Dallas, 1963-64; vis. prof. IAEA, Vienna, 1965-67; sci. dir. Lab. de Fisica Cosmica, U. Mayor de San Andres, LaPaz, Bolivia, 1965-67; vis. prof. Pam Am U., Washington, 1967-68; prof. U. N.Mex., 1968—; nat. rep. of Bolivia Internat. Union Pure and Applied Physics, 1966-69, Com. on Space Research, 1966-67, Space and Radio Monitoring Orgn., 1966-67; high energy group sec. to Internat. Council Sci. Unions, 1974—; mem. Cosmic Ray Commn. Internat. Union Pure and Applied Physics, 1966-69. Contbr. articles to profl. jours. Research grantee USAF, 1962-68, NSF, 1964-67, 69-71, 76-77, 78-81, NASA, 1972-73, 73-77, Sandia Nat. Lab., 1969-71. Mem. Am. Geophys. Union, Am. Phys. Soc., IEEE, Am. Astron. Soc., AAUP, Sigma Xi. Democrat. Sikh. Subspecialties: Cosmic ray high energy astrophysics. Current work: Study of cosmic ray intensity variations and anistropies. Home: 13000 Cedar Brook NE Albuquerque NM 87111 Office: 800 Yale Blvd NE Albuquerque NM 87131

AHMADI, GOODARZ, engineering educator, researcher; b. Tehran, Iran, July 23, 1943; s. Mahmod and Parvindokht (Ahmadi) A.; m. Behnaz Kafi, July 28, 1972; 1 dau. Anahita. B.S., Tehran U., 1965; M.S., Purdue U., 1968; Ph.D., 1970. Prof. engring. Shiraz (Iran) U., 1970-81, dean engring., 1979-80; vis. prof. U. Sask., 1974-75; vis. scholar Princeton U., 1975; vis. prof. U. Calgary, Alta., 1981-82; prof. engring. Clarkson Coll., Potsdam, N.Y., 1982—; cons. Atomic Energy Orgn. Iran, 1976-80; mem. research council Ministry of Sci. and Higher Edn., Iran, 1977-79. Author: (with others) Mathematical Methods in Engineering and Science, 2 vols, 1977; Assoc. editor: Iranian Jour. Sci. and Tech, 193, 75-78; editor, 1973-74, 78-80. Recipient Disting Scientist award Alborz Found., Iran, 1976; Research medal Minsitry of Sci. and Higher Edn., 1978. Mem. Soc. Engring. Sci., Internat. Solar Energy Soc., Soc. Indsl and Applied Mat., Earthquake Engring. Research Inst., Internat. Assn. for Structural Mechanics in Reactor Tech. Subspecialties: Theoretical and applied mechanics; Fluid mechanics. Patentee aeroelastic wind energy converter, 1979. Home: J1 Meadow East Apts Potsdam NY 13676 Office: Clarkson Coll Dept Mech and Indsl Engring Potsdam NY 13676

AHMANN, DAVID LAWRENCE, physician, oncologist, educator; b. St. Cloud, Minn., May 21, 1933; s. Norbert T. and Clotilda (Hall) A.; m. Rosemary Morrissey, Dec. 29, 1956; children: David, Mary, Mark, Carla, Gregory, Christopher. M.S., Marquett U., 1954, M.D., 1958. Diplomate: Am. Bd. Internal Medicine, 1969. Inst. Mayo Med. Sch., Rochester, Minn., 1968-72, asst. prof., 1972-74, assoc. prof., 1975-77, prof., 1977—; chmn. div. med. oncology dept. oncology Mayo Clinic, 1972—. Served with U.S. Army, 1959-62. Mem. Am. Assn. Cancer Research, Am. Soc. Clin. Oncology (sec.-treas. 1982), AMA, Minn. State Med. Soc., Zumbro Valley Med. Soc., Sigma Xi., Alpha Omega Alpha, Alpha Kappa Kappa (pres. 1957). Subspecialties: Internal medicine; Oncology. Office: Div Med Oncology Mayo Clinic 200 1st St SW Rochester MN 55905 Home: 521 SW 14th Ave Rochester MN 55901

AHMED, FARID EL MAMOUN, research exec.; b. Cairo, Egypt., Oct. 26, 1946; s. Mohamed Ahmed and Zeirab Hassan (Gheita) Soleiman; m. Leila F. Carolyn Rogers, May 30, 1949; 1 son, Khaled. B.S., Ain Shams U., Cairo, 1965, M.S., 1970; M.S., Ohio State U., 1974, Ph.D., 1975. Instr. Ain Shams U., 1965-69; chief med. technologist Scarborough (Ont., Can.) Gen. Hosp., 1971; teaching, research asso. Ohio State U., 1971-75; asst. scientist Brookhaven Nat. Lab., Upton, N.Y., 1975-79; dir. toxicology ops. Pharmacopathic Research Labs., Inc., Laurel, Md., 1979—. Contbr. numerous articles to profl. jours. Mem. AAAS, Am. Assn. Cancer Research, Am. Chem. Soc., Soc. Toxicology, Environ. Mutagen Soc., Genetic Toxicology Assn. Subspecialties: Toxicology (agriculture); Genetics and genetic engineering (biology). Current work: Genetic and biochemical toxicology, biotechnology, biohazard evaluation. Office: 9705 Washington Blvd Laurel MD 20707

AHMED, NAHED K., biochemist, educator; b. Cairo, Egypt, Sept. 29, 1945; s. Ahmed I. and Safia I. Khalil; m. Mahmoud S. Ahmed, Feb. 17, 1968; children: Tamer S., Sonya S. B.S., Cairo U., 1966, M.S., 1970; Ph.D., U. Tenn., 1975. Vis. fellow Balt. Cancer Research Ctr., Nat. Cancer Inst., Balt., 1975-77; research assoc. dept. biochem. and clin. pharmacology St. Jude Children's Research Hosp., Memphis, 1977-79, asst. mem. div. biochem. and clin. pharmacology, 1979—; asst. prof. U. Tenn., 1981—. Contbr. articles to profl. jours. Am. Cancer Soc. grantee, 1973; NSF grantee, 1981-84. Mem. Am. Assn. Cancer Research. Subspecialties: Cancer research (medicine); Chemotherapy. Current work: Cancer research, pyrimidine and anthraeycline metabolism, chemotherapy, enzymology, drug metabolism and biochemical pharmacology. Office: 332 N Lauderdale Memphis TN 38101

AHRENS, RICHARD AUGUST, nutrition educator, researcher; b. Manitowoc, Wis., Sept. 18, 1936; s. Richard William and Gladys LaVerne (Bierman) A.; m. Joan Ellen Morley, Aug. 19, 1961; children: Deborah Joan, Jill LaVerne, David Richard. B.S., U. Wis., 1958; Ph.D., U. Calif.-Davis, 1963. Registered dietitian. Research physiologist U.S. Dept. Agr., Beltsville, Md., 1963-66; assoc. prof. U. Md., College Park, 1966 75, prof. food and nutrition, 1975—; lectr. U, London, 1973. Contbr.: articles to profl. jours. including Jour. Animal Sci. Treas. Boys and Girls Club, Berwyn Heights, Md., 1979-80; election judge Precinct 21-7 Prince Georges County, 1970-81, chief election judge, 1982—. Research grantee Nutrition Found. U.S. Dept. Agr., 1966—. Mem. Am. Inst. Nutrition, Am. Home Econs. Assn. (Md. state pres. 1982-83), Am. Dietetic Assn., Soc. for Nutrition Edn. Republican. Lutheran. Subspecialty: Nutrition (medicine). Current work: Dietary sucrose and blood pressure elevation. Home: 6216 Seminole Pl Berwyn Heights MD 20740 Office: FNIA Dep Univ Maryland College Park MD 20742

AHRENS, THOMAS J., geophysics educator; b. Wichita Falls, Tex., Apr. 25, 1936; m., 1956; 3 children. B.S., MIT, 1957; M.S., Calif. Inst. Tech., 1958; postgrad., Rensselaer Poly. Inst., 1962. Intermediate exploration geophysicist Pan Am. Petroleum Corp., 1958-59; asst. geophysics Rensselaer Poly. Inst., 1962; geophysicist Poulter Research Labs. Stanford Research Inst., 1962-66, head geophysics group, 1966-67; assoc. prof. Calif. Inst. Tech., Pasadena, 1967-76, prof. geophysics, 1976—; mem. earth sci. adv. panel NSF, 1972-75. Assoc. editor: Rev. Sci. Instruments and Jour. Geophys. Research, 1971-74; editor: Jour. Geophys. Research, 1979—. Mem. AAAS, Am. Geophys. Union, Soc. Exploration Geophysicists, Am. Phys. Soc., Royal Astron. Soc. Subspecialty: Geophysics. Office: Div Geology and Planetary Sci Calif Inst Tech Pasadena CA 91125

AIKMAN, GEORGE CHRISTOPHER LAWRENCE, astronomer; b. Ottawa, Ont., Can., Nov. 11, 1943; s. Cecil Howard and Gwendolen Ellery (Read) A.; m. Beverly Mildred Boylan, May 27, 1972 (div.); children: Michael Donovan, Tabatha Karen.; m. Judith Anne Rackharm, Nov. 19, 1983. B.Sc. with honors, Bishop's U., Lennoxville, Que., 1965; M.Sc., U. Toronto, 1968. Research officer Dominion Astrophys. Obs., Victoria, B.C., Can., 1968—, assoc. research officer, 1979—. Contbr. articles in field to profl. jours. Mem. Am. Astronom. Soc., Canadian Astron. Soc., Royal Astron. Soc. Can., Astron. Soc. Pacific. Subspecialty: Optical astronomy. Current work: Origin of chem. peculiarities in stars. Office: Dominion Astrophys Obs 5071 W Saanich Rd Rural Route 5 Victoria BC Canada V8X 4M6

AISENBERG, ALAN C., research physician; b. N.Y.C., Dec. 7, 1926; s. Jacob and Celia (Able) A.; m. Nadya L. Margulies, Oct. 2, 1952; children: James, Margaret. S.B., Harvard U., 1945, M.D., 1950; Ph.D., U. Wis., 1956. Diplomate: Am. Bd. Internal Medicine. Intern and resident in medicine Presbyn. Hosp., N.Y.C., 1950-53; Am. Cancer Soc. fellow U. Wis., 1954-56; asst. in medicine Mass. Gen. Hosp., 1957-61; instr. medicine Harvard Med. Sch., 1957-61; asst. physician Mass. Gen. Hosp., 1961-69, physician and head lymphoma clinic, 1969—; asst. prof. medicine Harvard Med. Sch., 1961-69, assoc. prof.,

1969—. Author: Glycolysis and Respiration of Tumors, 1961, also over 100 articles. Guggenheim fellow, 1964. Mem. Am. Assn. Immunologists, Am. Assn. Cancer Research, Am. Assn. Clin. Oncology, Am. Fedn. Clin. Research, ACP. Subspecialties: Cancer research (medicine); Oncology. Current work: Cellular immunology, malignant lymphoma. Home: 124 Chestnut St Boston MA 02108 Office: Mass Gen Hosp Boston MA 02114

AISNER, JOSEPH, physician, oncologist, cancer researcher; b. Munich, Germany, Jan. 5, 1944; came to U.S., 1948, naturalized, 1954; s. Philip and Faye A.; m. Seena C. Feldman, June 26, 1948; children: Dara L., Leon A. B.S. in Chemistry, Wayne State U., 1965, M.D., 1970. Diplomate: Am. Bd. Internal Medicine, Sub-Bd. Oncology. Intern Sinai Hosp., Detroit, 1970-71; resident in internal medicine Georgetown U. Hosp., Washington, 1971-72; clin. assoc. Balt. Cancer Research Ctr., Nat. Cancer Inst., 1972-74, sr. investigator, 1974-76, chief med. oncology, 1976-81; head div. med. oncology U. Md. Cancer Ctr., Balt., 1981—, prof. medicine and oncology, 1982—, dep. dir. clin. affairs, 1982—; cons. in field. Contbr. numerous articles and abstracts to sci. jours., also chpts. to books. Served with USPHS, 1972-81. Mem. ACP, Am. Soc. Clin. Oncology, Am. Soc. Hematology, Am. Fedn. Clin. Research, AAAS, Cancer Leukemia Group B, Am. Assn. Cancer Research. Subspecialties: Cancer research (medicine); Oncology. Current work: Cancer research. Home: 1404 Berwick Rd Ruxton MD 21204 Office: 22 S Greene St Baltimore MD 21201

AIST, JAMES ROBERT, plant pathologist; b. Cheverly, Md., Feb. 20, 1945; s. Arthur Stewart and Carolyn Roberta (Thornberg) A.; m. Sheila Jean Beckwith, Jan. 21, 1967; children: Beverly, Gregory, Liesel. B.S., U. Ark., 1966, M.S., 1968; Ph.D., U. Wis., 1971. NATO postdoctoral fellow Swiss Fed. Inst. Tech., Zurich, 1971-72; asst. prof. Cornell U., Ithaca, N.Y., 1972-78, assoc. prof., 1978—; vis. assoc. research biologist U. Calif., Irvine, 1979-80. Asso. editor: Phythopathology, 1979-82, Exptl. Mycology, 1978—; contbr. articles to profl. jours. NSF grantee, 1976—; Dep. Def. grantee, 1978-80; USDA grantee, 1979—. Mem. AAAS, Am. Phytopathological Soc., Am. Soc. Cell Biology. Roman Catholic. Subspecialties: Plant pathology; Cell biology. Current work: Plant disease resistance, mitosis, analytical microscopy research. Home: 414 Snyder Hill Rd Ithaca NY 14850 Office: Dept Plant Pathology Cornell U Ithaca NY 14853

AKASOFU, SYUN-ICHI, geophysicist; b. Nagano-Ken, Japan, Dec. 4, 1930; came to U.S., 1958; s. Shigenori and Kumiko (Koike) A.; m. Emiko Endo, Sept. 25, 1961; children: Ken-Ichi, Keiko. B.S., Tohoku U., 1953, M.S., 1957; Ph.D., U. Alaska, 1961. Sr. research asst. Nagasaki U., 1953-55; research asst. Geophys. Inst., U. Alaska, 1958-61, mem. faculty, 1961—, prof. geophysics, 1964—. Author: Polar and Magnetospheric Substorms (Russian edit. 1971), 1968, The Aurora: A Discharge Phenomenon Surrounding the Earth (in Japanese), 1975, Physics of Magnetospheric Substorms, 1977, Aurora Borealis: The Amazing Northern Lights (Japanese edit. 1981), 1979; co-author: Sydney Chapman, Eighty, 1968, Solar-Terrestrial Physics (Russian edit. 1974); editor: Dynamics of the Magnetosphere, 1979; co-editor: Physics of Auroral Arc Formation, 1980—; editorial bd.: Planet and Earth Sci; co-editor: Space Sci. Revs. Recipient Chapman medal Royal Astron. Soc., 1976, award Japan Acad., 1977; named Disting. Alumnus U. Alaska, 1980. Fellow Am. Geophys. Union (John Adam Fleming medal 1979); mem. AAAS, Sigma Xi. Subspecialties: Aeronomy; Plasma physics. Current work: Auroral physics, magnetospheric physics. As a researcher of earth sciences, I feel that an artist and a scientist have something very much in common. Both watch carefully a natural object such as the aurora, a glacier, migrating birds, the Arctic Ocean, etc., and abstract whatever they feel the most essential part from the object. Then, an artist paints his abstraction on a canvas, while a scientist puts his abstraction into the form of equations.

AKBAR, HUZOOR, pharmacologist, educator, researcher; b. Karachi, Pakistan, Dec. 2, 1948; came to U.S., 1977; s. Hasan and Taeed Fatima (Rehbar) A.; m. Ildiko St. George, Apr. 9, 1977; children: Vazeer Daniel, Imran Shaan. B.S. with honors, Karachi U., Pakistan, 1971; M.S., 1972; Ph.D., Australian Nat. U., Canberra, 1978. Research asoc. in pathology Vanderbilt U., Nashville, 1977-78; in medicine Boston U., 1978-79; in pharmacology Ohio State, Columbus, 1979-81; asst. prof. pharmacology and biomed. sci. Ohio U., Athens, 1981—. Contbr. articles on pharmacology to profl. jours. Am. Heart Assn. grantee, 1982-84. Mem. Am. Physiol. Soc., Internat. Soc. on Thrombosis and Haemostatis, Am. Soc. for Pharmacology and Exptl. Therapeutics. Subspecialties: Pharmacology; Hematology. Current work: Mechanisms of blood platelet aggregation and secretion as well as the mechanisms of the actions of drugs which inhibit aggregation and secretion from platelets.

AKERA, TAI, pharmacologist, educator; b. Wakayama, Japan, July 13, 1932; came to U.S., 1971; s. Jibusuke and Ayako (Omata) A.; m. Chiseko Masuda; children: Atsushi, Yuka, Chika. M.D., Keio U., Tokyo, 1958, Ph.D., 1965. Diplomate: med. diplomate Japanese Ministry Welfare. Intern Keio U. Hosp., 1958-59; postdoctoral fellow U. Mich., Ann Arbor, 1962-64; instr. Keio U., 1964-66, asst. prof., 1966-71; vis. asst. prof. dept. pharmacology and toxicology Mich. State U., East Lansing, 1967-70, asso. prof., 1971-74, prof., 1974—; mem. drug abuse biomed. research rev. com. Nat. Inst. on Drug Abuse, 1980—. Contbr. over 120 articles to profl. jours., also rev. papers. NIH grantee, 1972—; Mich. Heart Assn. grantee, 1972-82. Mem. Japanese Med. Assn., Japanese Pharmacology Soc., Am. Soc. for Pharmacology and Exptl. Therapeutics. Subspecialties: Pharmacology; Toxicology (medicine). Current work: Cardiotonic drugs; ion transport across the cell membranes; basic research on mechanisms of drug actions and reactions. Home: 1873 Ridgewood Dr East Lansing MI 48823 Office: Dept Pharmacology and Toxicology Mich State U East Lansing MI 48824

AKERS, FRANCIS IRVING, optical company executive; b. Rural Retreat, Va., Mar. 10, 1947; s. Irving and Frances Lynn (Neff) A.; m.; children by previous marriage: Kimberly Dawn, James Christopher. B.S. in Chemistry, Va. Poly. Inst., 1969, M.S., 1974. Tchr. Roanoke (Va.) Pub. Schs., 1969-77; research asst. Va. Poly. Inst., Blacksburg, 1973-74; adj. lectr. chemistry Va. Western Community Coll., Roanoke, 1974-77; group mgr. Optical Fiber Engring., Roanoke, 1977—; cons. and lectr. in field. Contbr. articles to profl. jours. and lectr. to profl. confs. Mem. Am. Chem. Soc. (tchr. of yr. 1972-73, editor newsletter Blue Ridge sect.), Va. Acad. Sci. (com. sci. edn.), Internat. Mgmt. Council, Sigma Xi (grad. research award 1974), Phi Kappa Phi, Phi Lambda Upsilon, Phi Delta Kappa. Subspecialty: Fiber optics. Current work: Process and product engineering for all type optical fibers; radiation curing; fiber optic cable; fiber optic systems; fiber optic sensors; chemical vapor deposition, single mode fiber. Patentee in optical wave guide, optical fiber. Office: Box 7065 Roanoke VA 24019

AKERS, STUART WILLIAM, plant-stress physiologist; b. Sentinel, Okla., July 6, 1945; s. William Frederick and Nettie Rachel (Holt) A.; m. Carolyn Sue Pruett, Dec. 20, 1970; children: Jonathon S., Susan C. B.S., U. Okla., 1967; M.S., N.C. State U., 1969, Ph.D., 1975. Asst. prof. S.W. Okla. State U., Weatherford, 1975-76; research assoc U. Ky., Lexington, 1976-77, Purdue U., West Lafayette, Ind., 1977-81; asst. prof. Okla. State U., Stillwater, 1981—. Served as 1st lt. U.S. Army, 1969-71. Decorated Bronze Star. Mem. Am. Soc. Plant Physiologist, Am. Soc. Horticulture Sci., Plant Growth Regulator Soc. Am. Democrat. Methodist. Subspecialties: Plant physiology (agriculture); Space agriculture. Current work: Investigations on plant response to environmental stresses that limit horticultural crop productivity and propagation effectiveness. Home: 1017 E Ridgecrest Ave Stillwater OK 74074

AKI, KEIITI, seismologist, educator; b. Tokyo, Japan, Mar. 3, 1930; came to U.S., 1966, naturalized, 1976; s. Koichi and Humiko (Kojima) A.; m. Haruko Uyeda, Mar. 25, 1956; children: Shota, Zenta. B.S., U. Tokyo, 1952, Ph.D., 1958. Research fellow Calif. Inst. Tech., 1958-60, vis. prof., 1963; instr. Internat. Inst. Seismology and Earthquake Engring., 1961-62; asso. prof. U. Tokyo, 1964-66; prof. geophysics MIT, Cambridge, 1966—, R.R. Shrock prof. earth and planetary scis., 1982—; vis. prof. U. Chile, 1970, 72, U. Paris, 1983; WAE geophysicist U.S. Geol. Survey, 1967-75; vis. scientist Royal Norwegian Council for Sci. and Indsl. Research, 1974; cons. Sandia Corp., 1976—; vis. scientist Los Alamos Sci. Labs., U. Calif., 1977—; cons. Del Mar Assos., 1977-78, Nuclear Regulatory Commn., 1978, UN, 1979, Time-Life, Inc., 1980-81, NSF, 1981—; vis. scientist Japan Soc. Promotion of Sci., 1978; chmn. com. on seismology Nat. Acad. Sci., 1978-79; Disting. vis. prof. U. Alaska, 1981; Mem. Nat. Council for Earthquake Prediction Evaluation, 1980—. Author: Stochastic Phenomena in Physics, 1956, Quantitative Seismology: Theory and Methods, Vols. I and II, 1980; editor-in-chief: Pure and Applied Geophysics; Mem. editorial com.: Tectonophysics, 1974—; assoc. editor: Geophys. Research Letters, 1977-82; adv. editor: Jour. Physics of Earth and Planetary Interiors. Fulbright postdoctoral fellow, 1958-60. Fellow Am. Acad. Arts and Scis.; mem. Nat. Acad. Scis., Am. Geophys. Union (mem. com. fellows 1975-76, pres. seismology sect. 1980), Seismological Socs. Am. (dir. 1971-74, v.p. 1978, pres. 1979), Japan), Royal Astron. Soc. Subspecialties: Geophysics; Tectonics. Current work: Earthquake, volcano, earth's interior, lithosphere, asthenosphere, seismology, seismic waves, fault zone, strong motion, hazard. Home: 56 Park Ln Newton MA 02159 Office: 77 Massachusetts Ave Cambridge MA 02139

ALAPOUR, ADEL, nuclear engineer; b. Tehran, Iran, June 17, 1948; came to U.S., 1974; s. Mostafa and Mehry (Latifi) A; m. Cynthia Ann Jackson, Feb. 20, 1980; children: Kavon A., Vida N. B.S.M.E., Arya-Mehr U. Tech.-Iran, 1970; M.S.N.E., Ga. Inst. Tech., 1975, Ph.D., 1980. Operation coordinator Iran Electric Power Generation & Transmission Co., Tehran, 1970-72, plant analyst, 1972-74; grad. research asst. dept. nuclear engring. Ga. Inst. Tech., Atlanta., 1974-78; nuclear engr. Brookhaven Nat. Lab., Upton, N.Y., 1978-80, So. Co. Services Inc., Birmingham, Ala., 1980-81, sr. engr., 1981—. Mem. Am. Nuclear Soc. (Birmingham sect. chmn. 1982—), N.Y. Acad. Scis., So. Co. Services Leadership Devel. Assn., Sigma Xi. Subspecialties: Nuclear engineering; Nuclear fission. Current work: Steady state and time dependent analysis of Nuclear reactor systems. Home: 5193 Selkirk Cir Birmingham AL 35243 Office: So Co Services Inc PO Box 2625 Birmingham AL 35202

ALAVI, ABASS, physician; b. Tabriz, Iran, Mar. 15, 1938; came to U.S., 1966, naturalized, 1977; s. Mohsen and Fatemeh A.; m. Jane Bradley, Jan. 2, 1971. M.D., U. Tehran, 1964. Diplomate: Am. Bd. Internal Medicine, Am. Bd. Nuclear Medicine. Intern, resident internal medicine Albert Einstein Med. Ctr., Phila. VA Hosp., 1966-70; fellow nuclear medicine U. Pa., 1971-73, instr. radiology, 1973-74, asst. prof., 1974-77, assoc. prof., 1977-82, assoc. prof. neurology, 1979-82, acting chief div. nuclear medicine, 1978-79, prof. radiology and neurology, 1982—, chief div. nuclear medicine, 1979—; assoc. dir. Positron Emission Tomography Ctr., 1979—. Editor: (with P. Arger, Grune and Stratton) Abdomen: Nuclear Medicine, CT and Ultrasound, 1980; contbr. articles to profl. jours. Served with Iranian Health Corps, 1964-66. Subspecialties: Internal medicine; Nuclear medicine. Current work: Evaluation of the positron emission tomographic technique in central nervous system physiology and disorders, research in nuclear magnetic resonance and monoclonal antibodies. Home: 939 Remington Rd Wynnewood PA 19096 Office: Nuclear Medicine Hosp U Pa Philadelphia PA 19104

ALAVIAN, FARID, computer scientist; b. Iran, May 31, 1951; came to U.S., 1974; s. Paymon and Farideh A.; m. Michelle M. Mortensen, Jan. 1, 1981. B.S., U. Tehran, Iran, 1974; M.S., UCLA, 1976, Ph.D., 1981. Research engr. UCLA, Los Angeles, 1975-79; computer scientist Computer Scis. Corp., El Segundo, Calif., 1979-82, United Techs., Westlake Village, Calif., 1982—. Subspecialties: Database systems; Distributed systems and networks. Current work: Database management systems, distributed systems, information systems, software engineering, real-time applications. Office: United Technologies Lexar 31829 La Tienda Dr Westlake Village CA 91362

ALBACH, RICHARD ALLEN, microbiology educator; b. Chgo., Mar. 31, 1930; s. Maurice and Martha (Silverman) A.; m. Janice Elaine Boewe, Jan. 23, 1962; children: Michael, Karren, Kimala, David, Brian, Julie, Barry. B.S., U. Ill., 1956, M.S., 1958; Ph.D., Northwestern U., 1963. Asst. prof. U. Health Scis., Chgo. Med. Sch., 1968-69, North Chicago, Ill., assoc. prof., 1969-73, prof., 1973—, vice chmn., 1975-82, acting chmn., 1982—; editorial cons. Yearbook Med. Pubs., Chgo., 1975-81. Contbr. articles to profl. jours. Served with U.S. Army, 1953-55. NIH grantee, 1965-78; LAbbott Found. fellow, 1961; Trustees Research award Chgo. Med. Sch., 1968; Teaching Prof. of the Yr., 1976, 78, 82. Fellow Am. Acad. Microbiology; mem. Am. Soc. Microbiology, Soc. Protozoologists (bus. adv. com. 1977-79), Am. Soc. Parasitologists, Ill. Soc. Microbiology (membership chmn. 1969-70). Subspecialties: Microbiology (medicine); Parasitology. Current work: Molecular biology of parasitic protozoa. Address: Univ Health Sci Chicago Med Sch 3333 Green Bay Rd North Chicago IL 60064

AL-BAGDADI, FAKHRI ABDUL KAREEM, veterinary medicine educator; b. Baghdad, Iraq, Mar. 2, 1940; s. Kareem A. Hadi and Ghena (Jawad) Al-B; m. Lone Ingrid Jensen, Nov. 14, 1965; 1 child, Talal. B.V.M.S., Coll. Vet. Medicine, Baghdad; M.S., Royal Vet. and Agr. Coll., Denmark, Iowa State U.; Ph.D., U. Ill. Instr. Coll. Vet. Medicine, Baghdad, 1963-68, asst. dean students, 1968-69; postdoctoral fellow in gerontology Iowa State U., Ames, 1970-71; instr. biol. structure dept. Coll. Vet. Medicine U. Ill., Champaign-Urbana, 1971-74; asst. prof. human anatomy St. Louis U., 1974-75; asst. prof. vet. anatomy La. State U., Baton Rouge, 1975-79, assoc. prof., 1979—. Contbr. chpts. to textbooks, articles to profl. jours. Danish Govt. Ministry of Agr. grantee, 1964-65; Gulbankian scholar, Portugal, 1969-70; FAO fellow; grantee. Mem. Am. Assn. Anatomists, Electron Microscopy Assn., AVMA, World Assn. Vet. Anatomists, Am. Assn. Vet. Anatomists, Phi Zeta. Muslim. Subspecialties: Gerontology; Cytology and histology. Current work: Changes in tissues caused by pollution and carcinogens in environment; hemal lymph nodes changes caused by leukemia; melanocytes and mast cells response to carcinogens; cellular gerontology; integument and hair cycle. Office: 2522 Vet Med Sch Baton Rouge LA 70803

ALBANO, JOANNE EDVIGE, neurobiologist; b. N.Y.C., Dec. 2, 1948; d. Dominick and Camille (DiScenza) A. B.A. in Psychology, Calif. State U., Los Angeles, 1971; postgrad., 1971-72; Ph.D., Duke U., 1977. Research asst. dept. ophthalmology Duke U., 1972-73, NIMH predoctoral trainee, 1973-74, research asso., 1976-77; Nat. Research Service award postdoctoral fellow Lab. Neurobiology NIMH, 1977-79; staff fellow Lab. Sensorimotor Research, Nat. Eye Inst., NIH, 1979—. Contbr. articles to profl. jours. Mem. Assn. Research in Vision and Ophthalmology, Assn. Women in Soc., NIH Bicycle Commuters Assn. (co-pres.), NOW, Nature Conservancy, Smithsonian Assocs., Potomac Appalachian Trail Club, Potomac Peddler Touring Assn., Washington Area Bicyclists Assn., NIH Sailing Club. Subspecialties: Neurophysiology; Neuropsychology. Current work: Neural mechanisms of vision and ocular motility. Office: Lab Sensorimotor Research Nat Eye Inst Bldg 9 Room B1E14 Bethesda MD 20205

ALBANO, WILLIAM A., surg. oncologist; b. Summit, N.J., Aug. 6, 1945; s. Alexander and Marguerite (Pedecine) A.; m. Marjo Friese, Aug. 20, 1968; children: William, Alexander, Andrew, Michelle. B.S., Seton Hall U., 1967; M.D., Creighton U., 1971. Diplomate: Am. Bd. Surgery. Resident in surgery Creighton U., Omaha, 1971-75; fellow in surg. oncology City of Hope, Duarte, Calif., 1975-76; asst. prof. surgery, chief div. surg.oncology Creighton Cancer Center, Omaha, 1976—. Contbr. articles to profl. jours. Fellow Soc. Surg. Oncology, A.C.S., Southwestern Surg. Soc.; mem. North Central Cancer Group, Nebr. Cancer Soc. (pres. 1981-82). Subspecialties: Surgery; Cancer research (medicine). Current work: Hereditary cancer. Office: Creighton Cancer Center 601 N 30th St Omaha NE 68131

AL-BAZZAZ, FAIQ JABER, physician, educator, researcher; b. Baghdad, Iraq, July 1, 1939; came to U.S., 1966; s. Jaber Mehdi and Fadelah Hassoun (Ismail) Al-B.; m. Paulette Dodds, Nov. 15, 1969; children: Nesreen, Basheer, Senan. M.B., Ch.B., U. Baghdad, 1962. Diplomate: Am. Bd. Internal Medicine, Am. Bd. Pulmonary Diseases. Gen. practice Medicine, 1965; resident Mosul Med. Coll., 1965-66, U. Miss. Med. Center, 1966-68, U. Minn. and Mpls. VA. Hosp., 1968-69; clin. and research fellow Mass. Gen. Hosp. and Harvard U., Boston, 1969-71; asst. prof. medicine U. Ill., Chgo., 1971-78, assoc. prof. clin. medicine, 1978-81, assoc. prof. medicine, 1981—; chief respiratory and critical care sect. West Side VA Med. Ctr., Chgo., 1977—. Contbr. articles to profl. jours. Chgo. Lung Assn. grantee, 1972, 73, 77; VA grantee, 1976—; Am. Heart Assn. grantee, 1979-81. Fellow ACP, Am. Coll. Chest Physicians, Royal Coll. Physicians Can.; mem. Am. Physiol. Soc., Central Soc. Clin. research, Am. Thoracic Soc., Am. Fedn. Clin. Research. Moslem. Subspecialties: Pulmonary medicine; Membrane biology. Current work: Elucidation of mechanisms of ion and fluid transport across respiratory epithelia; role of cyclic nucleotides, cytosolic calcium and prostaglandins in modulation of membrane and paracellular conductance; role of neuropeptides in regulation of ion transport. Home: 343 Jefferson Woodstock IL 60098

ALBERS, HENRY ELLIOTT, biomedical researcher; b. Ames, Iowa, Apr. 7, 1953; s. Henry H. and Marjorie (Klein) A. B.A., U. Nebr., Lincoln, 1974; M.S., Tulane U., 1978, Ph.D., 1979. Research fellow in physiology Harvard Med. Sch., Boston, 1979-82; research assoc. in endocrinology Worcester Found. for Exptl. Biology, Shrewsbury, Mass., 1981-82; sr. research assoc. Worcester Found. Exptl. Biology, 1982—. Contbr. sci. articles to profl. jours. USPHS grantee, 1982—. Mem. Soc. Neurosci., Internat. Soc. Chronobiology. Subspecialties: Neuropsychology; Neuroendocrinology. Current work: Studies of the neural and hormonal control of circadian behavior. Home: 104 Lakeside Dr Shrewsbury MA 01545 Office: Worcester Found Exptl Biology 222 Maple Ave Shrewsbury MA 01545

ALBERS, MARK ALAN, research physicist; b. Denver, Dec. 11, 1952; s. Vernon Leo and Lea (Fletcher) A. B.S., Colo. Sch. Mines, 1975, M.S., 1978. Research fellow CSM, Golden, Colo., 1975-78; research physicist Manville Corp., Denver, 1978—. Mem. Am. Phys. Soc. Republican. Lutheran. Club: 6502 Microprocessor Group. Subspecialties: Aerosoles and Particulates; Heat Transfer. Current work: Research in various modes of thermal transport through materials and insulations; radiative transfer; infrared backscattering, absorption, extinction. Home: 1194 W Stanford Pl Littleton CO 80127 Office: Manville Corp PO Box 5108 Denver CO 80217

ALBERT, MARY DAY, biology laboratory adminstrator, researcher; b. Manchester, N.H., Mar. 2, 1926; d. Charles Howard and Sara (Forbes) Day; m.; children: Eric, Ross. B.S., U. N.H., 1948; M.A., Bryn Mawr Coll., 1950; Ph.D., Brown U., 1955. Research fellow in medicine Harvard U., 1955-61; lectr. Northeastern U., Boston, 1961-62, Wellesley Coll., 1963-64; asst. prof. biology Newton (Mass.) Coll., 1964-75; dir. biol. labs. Boston Coll., Chestnut Hill, Mass., 1975—. Chmn. Com. on the Handicapped, Newton Centre, Mass. Subspecialties: Reproductive biology; Endocrinology. Current work: The effects of THC (marijuana) on the uterus, vagina, testes, ventral prostate, adrenals of male and female rats. Home: 56 Chapin Rd Newton Centre MA 02159 Office: Dept Biology Boston Coll Chestnut Hill MA 02167

ALBERTE, RANDALL SHELDON, plant cell biology educator, researcher; b. Newark, June 7, 1947; s. Frank and Josephine I. (Kline) A. B.A., Gettysburg Coll., 1969; Ph.D., Duke U., 1974. Teaching asst. Duke U., 1969-74; NSF fellow, research assoc. UCLA, 1974-75, NIH fellow, research assoc., 1975-77, acting asst. prof. biology, 1975-75; asst. prof. U. Chgo., 1977-80, assoc. prof., 1981—; cons. Ency. Brit., 1981-82. Contbr. articles to profl. jours. Mellon Found. fellow, 1979. Mem. Am. Soc. Plant Physiology, AAAS, Am. Inst. Biol. Sci., Soc. Exptl. Biology, Bot. Soc., Scandinavian Soc. Plant Physiology. Subspecialties: Cell biology; Photosynthesis. Current work: Functional organization of photosynthetic apparatus in oxygen evolving plants; the adaptive physiology of photosynthesis and respiration in terrestrial and marine plants; the control of chloroplast development, development, structure, and function of pigment proteins. Office: Barnes Lab 5630 S Ingleside Ave Chicago IL 60637

ALBERTS, BRUCE MICHAEL, molecular biologist; b. Chgo., Apr. 14, 1938; s. Harry C. and Lillian (Surasky) A.; m. Betty Neary, June 14, 1960; children—Beth, Jonathan, Michael. A.B., Harvard, 1960, Ph.D., 1965. Postdoctoral fellow Inst. de Biologie Moleculaire, Geneva, 1965-66; mem. faculty Princeton, 1966-76, Damon Pfeiffer prof. life scis., 1973-76, acting chmn. dept. biochem. scis., 1973-74; prof. biochemistry U. Calif., San Francisco, 1976—; bd. sci. advisers Jane Coffin Childs Meml. Fund for Med. Research, 1978—. Editorial bd.: Jour. Biol. Chemistry; Contbr. numerous articles, papers profl. jours. Recipient award molecular biology U.S. Steel Found., 1975; grantee NIH, 1966—; Am. Cancer Soc. Lifetime Research prof., 1980. Fellow Am. Acad. Arts and Scis.; mem. Nat. Acad. Scis., Am. Chem. Soc. (Eli Lilly award 1972), Am. Soc. Biol. Chemists, Phi Beta Kappa, Sigma Xi. Subspecialty: Molecular biology. Address: Dept Biochemistry and Biophysics Univ Calif San Francisco CA 94143

ALBERTS, WALTER WATSON, health sci. adminstr.; b. Los Angeles, Dec. 31, 1929; s. Hugo William and Ruth Lucia (Watson) A.; m. Marilyn West, Mar. 22, 1959; children: Allison C., Allan W. A.B., U. Calif. Berkeley, 1951, Ph.D., 1956. Research physiologist U. Calif. Med. Ctr., San Francisco, 1955-56; biophysicist Mt. Zion Hosp. and Med. Ctr., San Francisco, 1956-72; grants assoc. NIH, Bethesda, Md., 1972-73; spl. asst. to assoc. dir. collaborative and field research, 1973-74, head research contracts sect., 1974-75; asst. dir. contract research programs, extramural activity program Nat. Inst. Neurol. and

Communicative Disorders and Stroke, NIH, 1975-77; adminstrv. dir. Smith Kettlewell Inst., San Francisco, 1977-78; dep. dir. fundamental neuroscis. program Nat. Inst. Neurol. and Communicative Disorders and Stroke, NIH, Bethesda, Md., 1979—. Contbr. articles to profl. jours. Nat. Inst. Neurol. Diseases and Blindness research career program awardee, 1963-68. Fellow AAAS; mem. Am. Physiol. Soc., Biophys. Soc., IEEE, Soc. Neurosci., Sierra Club, Phi Beta Kappa, Sigma Xi. Congregationalist. Subspecialties: Neurophysiology; Biophysics (physics). Current work: Neurophysiology, biological and medical physics, particularly the central nervous system of man. Home: 9205 Friars Rd Bethesda MD 20817 Office: NIH Bethesda MD 20205

ALBERTY, ROBERT ARNOLD, educator; b. Winfield, Kans., June 21, 1921; s. Luman Harvey and Mattie (Arnold) A.; m. Lillian Jane Wind, May 22, 1944; children—Nancy Lou, Steven Charles, Catherine Ann. B.S., U. Nebr., 1943, M.S., 1944, D.Sc., 1967; Ph.D., U. Wis., 1947; D.Sc., Lawrence U., 1967. Engaged in research blood plasma fractionation for U.S. Govt., 1944-46; mem. faculty U. Wis., 1947-67, prof. chemistry, 1955-67, assoc. dean letters and sci., 1961-63, dean, 1963-67, Sch. Sci., Mass. Inst. Tech., 1967-72; cons. NSF, 1958-83, NIH, 1962-72; chmn. commn. on human resources NRC, 1974-77; dir. Colt Industries, 1978—, Inst. for Def. Analysis, 1980—. Co-Author: Physical Chemistry, 6th edit, 1983, Experimental Physical Chemistry, 3d edit., 1970. Guggenheim fellow Calif. Inst. Tech., 1950-51; recipient Eli Lilly award biol. chemistry, 1955. Fellow AAAS; mem. Am. Chem. Soc. (chmn. com. on chemistry and public affairs 1978-80), Am. Soc. Biol. Chemists, Nat. Acad. Sci., Inst. Medicine, Am. Acad. Arts and Scis., Phi Beta Kappa, Sigma Xi. Subspecialties: Physical chemistry; Thermodynamics. Current work: Thermodynamics and kinetics of reactions in complex organic systems. Home: 7 Old Dee Rd Cambridge MA 02138

ALBINI, BORIS, immunopathologist, researcher; b. Zagreb, Croatia, Yugoslavia, Mar. 7, 1943; came to U.S., 1974, naturalized, 1977; s. Julius and Maja (Skojic) A.; m. Christine Helen Seymanski, Dec. 16, 1978; children: Thomas, Paul. M.A., Acad. Music, Vienna, Austria, 1963; M.D., U. Vienna, 1969. Intern surgery Floridsdorf Hosp., Vienna, Austria, 1969-70; clin. asst. prof. exptl. pathology U. Vienna, 1970-73; research asst. prof. microbiology SUNY-Buffalo, 1974-76, asst. prof., 1976-78, assoc. prof., 1978-83, prof., 1983—; cons. dept. medicine, 1981—. Author: (with others) The Immunopathology of the Kidney, 1979; Co-editor: Immunopathology, 1979; assoc. editor: Clin. Exptl. Immunology, 1980, Immunological Communications, 1981, Internat. Archives Allergy and Applied Immunology, 1981. Vice-pres. Friends of Vienna, Buffalo, 1976—. Mem. Assn. Immunology, Am. Assn. Immunology, Assn. Immunology of Austria. Subspecialties: Immunology (medicine); Nephrology. Current work: Immunopathology of kidney diseases, immune response in immune complex-mediated diseases, immunopathology and autoimmunity in spontaneous autoimmune thyroiditis of obese strain chickens, immunopathology of the gastrointestinal tract. Home: 31 Kim Circle Williamsville NY 14221 Office: SUNY-Buffalo 240B Cary Hall Buffalo NY 14214

ALBISSER, ANTHONY MICHAEL, biomedical researcher; b. Johannesburg, South Africa, Sept. 5, 1941; s. Albert and Anna (Benode) A.; m. Marianne Sperlich, Sept. 7, 1964; children: Brian, David, Jeremy, Gregory. B.Eng., McGill U., 1964; M.A.Sc., U. Toronto, 1966, Ph.D., 1968. Registered profl. engr., Ont. Dir. dept. med. engring. Hosp. for Sick Children, Toronto, Ont., Can., 1971-76, sr. scientist dept. surgery div. biomed. research, 1978—, dir. div. biomed. research, 1978—; assoc. prof. dept. medicine U. Toronto, 1977—, spl. lectr. elec. engring., 1968-72; cons. life sci. instruments Miles Labs., Elkhart, Ind., 1974-80; mem. exec. com. Banting and Best Diabetes Ctr. Contbr. chpts. to books, articles to profl. jours. Recipient Career Achievement award Becton-Dickinson, 1981; David Rumbough award Juvenile Diabetes Found., 1981. Mem. Am. Diabetes Assn., Can. Soc. for Clin. Investigation, Assn. for Advancement Med. Instrumentation, Can. Med. and Biol. Engring. Soc., Profl. Engrs. Ont., European Assn. for Study of Diabetes, Nat. Diabetes Research Interchange. Subspecialties: Biomedical engineering; Computer engineering. Current work: Research into the pathophysiology of diabetes mellitus and the control of the disease using exogenous insulin delivery devices. Patentee in field. Office: 555 University Ave Toronto ON Canada M5G 1X8

ALBRECHT, THOMAS BLAIR, microbiology educator; b. Phila., July 31, 1943; s. Blair Robson and Deborah Hawley (Smedley) A.; m. Isis de Alencar, July 28, 1967; children: Christine, Thomas Edward, Alan Wayne. B.S., Brigham Young U., 1967, M.S., 1969; Ph.D., Pa. State U., 1973. Research fellow Harvard U., 1974-75; asst. prof. microbiology U. Tex., 1976-80; assoc. prof., 1980—. Contbr. articles in field to profl. jours. Grantee NIH, 1976-79, Moody Found., 1979—, Cooper Labs., Inc., 1979—. Mem. Am. Soc. Microbiology., Am. Soc. Cell Biology, Soc. Exptl. Biology and Medicine, Sigma Xi. Methodist. Subspecialties: Microbiology (medicine); Cell biology (medicine). Current work: Cell responses to herpes virus, particularly human cytomegalo-viruses. Home: 1905 Back Bay Dr Galveston TX 77551 Office: Microbiology U Texas Med Br Galveston TX 77550

ALBRIGHT, JOHN GROVER, chemistry educator, researcher; b. Winfield, Kans., June 29, 1934; s. Penrose Strong and Mary (Lucas) A.; m. Sharon Rae Rudd, June 11, 1960; children: Mary Kathrine, David Louis. B.A. in Chemistry and Physics, Wichita U., 1956; Ph.D. in Phys. Chemistry, U. Wis., 1962. Postdoctoral fellow Australian Nat. U., 1963-65; research fellow Inst. Enzyme Research, Madison, Wis., 1965-66; prof. chemistry Tex. Christian U., 1966—; researcher; cons. Lawrence Livermore Nat. Lab. Contbr. articles profl.jours. NSF grantee, 1963-65; Robert A. Welch Found. grantee, 1967-73; NSF grantee, 1981-83. Mem. Am. Chem. Soc. Congregationalist. Lodge: S.W. Lions. Subspecialties: Physical chemistry; Thermodynamics. Current work: Relation of viscosity to diffusion processes in multicomponent liquid systems, isotope effects in liquid diffusion processes. Home: 4332 Lanark St Fort Worth TX 76109 Office: Dept Chemistry Christian U Fort Worth TX 76129

ALBRIGHT, JOSEPH FINLEY, immunology educator, researcher; b. New Tazewell, Tenn., Mar. 9, 1927; s. Philip N. and Louise (Harris) A.; m. Julia Wan, Nov. 29, 1975; children by previous marriage: Emily Christine, Kendra Suzanne. B.S., Southwestern U., 1949; Ph.D., Ind. U., 1956. Postdoctoral fellow Oak Ridge Nat. Lab., 1956-58, biologist, 1960-70; asst. prof. Med. Coll. Va., Richmond, 1958-60; sr. immunologist Smith, Kline Labs., Phila., 1970-72; prof. dept. life scis. Ind. State U., Terre Haute, 1972—; program dir. NSF, Washington, 1982-83; cons. Argonne (Ill.) Nat. Lab., 1976-78, Nat. Cancer Inst., Bethesda, Md., 1974-79, Bio-Response, Inc., San Francisco, 1978-82. Contbr. chpts. to books, articles to profl. publs. Pres. Friends of Responsible Edn., Oak Ridge, 1968-70. NIH predoctoral fellow, 1954-56; recipient First Faculty Research-Creativity award Ind. State U., 1981; various research grants NSF, NIH, Eagles' Cancer Fund, 1970-83. Mem. AAAS, Soc. Devel. Biology, Am. Assn. ZImmunologists, Fedn. Am. Soc. Exptl. Biology. Subspecialties: Immunology (medicine); Infectious diseases. Current work: Cellular interactions in immune responses; interaction of parasites and host's immune system; natural resistance to parasites; aging and immunity. Office: Dept Life Scis Ind State Univ Terre Haute IN 47809

ALBRIGHT, JULIA WAN, immunologist, researcher, educator; b. Peking, China, Apr. 29, 1940; came to U.S., 1956, naturalized, 1967; d. Kao-Ping and Kwei-Jean (Wong) Wan; m. Joseph F. Albright, Nov. 29, 1975. B.S., East Tenn. U., 1962; M.S., U. Akron, 1972; Ph.D., Ind. State U., 1978. Research biologist Oak Ridge Nat. Lab., 1965-68; research asst. U. Rochester, N.Y., 1968-70; microbiologist Aging Inst. NIH, Balt., 1972-76; asst. prof. immunology Ind. State U., Terre Haute, 1978-81, research assoc. prof., 1981—, dir. immunol. and cellular aging studies, 1982—; research assoc. prof. Med. Sch. George Washington U., 1981-82. Contbr. articles to profl. jours. Bd. dirs. Am. Family Services Assn., Terre Haute, 1981—. NIH grantee, 1982-85; NSF grantee, 1983—. Mem. Gerontol. Soc. Am., Tissue Culture Assn., Am. Assn. Immunologists (recipient Travel awards), Am. Soc. Microbiology, Sigma Xi. Subspecialties: Immunology (medicine); Gerontology. Current work: Immunological resistance to and interactions with parasites; aging of the immune system and the role of the immune system in aging. Office: Department of Life Sciences Indiana State University Terre Haute IN 47809

ALBRIGHT, LOUIS DEMONT, agrl. engr., educator; b. Ithaca, N.Y., Dec. 31, 1940; s. Richard and Catherine A.; m.; children: Adam, Amy. B.S.A.E., Cornell U., 1963, M.S., 1965, Ph.D., 1972. Asst. prof. dept. agrl. engring. U. Calif., Davis, 1972-73; mem. faculty dept. agrl. engring. Cornell U., Ithaca, N.Y., 1974—, assoc. prof., 1978—. Contbr. articles to profl. jours. Served to capt. U.S. Army, 1965-68. Mem. Am. Soc. Agrl. Engrs. (Young Researcher of Yr. award 1979), ASHRAE, Internat. Solar Energy Soc., Internat. Hort. Soc. Subspecialty: Agricultural engineering. Current work: Environmental control and energy management in buildings. Office: 206 Riley-Robb Hall Cornell U Ithaca NY 14853

ALCENA, VALIERE, internist, educator; b. Haiti, West Indies, Aug. 24, 1934; came to U.S., 1960, naturalized, 1965; s. Lamartine and Florisane (Lacoste) A. B.A., Queens Coll., Flushing, N.Y., 1970; M.D., Albert Einstein Coll., Bronx, N.Y., 1973. Diplomate: Am. Bd. Internal Medicine. Intern Montefiore Hosp. and Med. Center, Bronx, N.Y., 1973-74, resident in medicine, 1974-74, chief resident, 1975-76, fellow in hematology, 1976-77, fellow in oncology, 1977-78; clin. instr. medicine Albert Einstein Coll., Bronx, N.Y., 1978—; clin. instr. medicine in psychiatry Cornell U. Med. Sch., N.Y.C., 1978—; adj. attending physician in medicine Montefiore Hosp., Bronx, 1978—, asst. attending physician in oncology, 1980—; assoc. attending physician in medicine and hematology White Plains (N.Y.) Hosp., 1978—; assoc. attending physician in medicine St. Agnes Hosp., White Plains, 1978—; clin. affiliate, cons. medicine N.Y. Hosp., White Plains, 1978—. Contbr. sci. articles to profl. publs. Chmn. public end. com. Am. Cancer Soc., Central Westchester, N.Y., 1979-81, v.p., 1982; mem. exec. com. Westchester chpt., 1983—, mem. bd. dirs., 1982—; pres. Central Westchester unit, Am. Cancer Soc., 1983. King-Kennedy scholar, 1968-69. Mem. AMA, Am. Soc. Clin. Oncology, Am. Soc. Hematology, ACP, Nat. Soc. Internal Medicine, N.Y. Soc. Internal Medicine, Westchester Soc. Internal Medicine. Democrat. Roman Catholic. Subspecialties: Chemotherapy; Hematology. Current work: Academic solo practitioner; clinical teaching. Home: 137 Trails End New City NY 10956 Office: 170 Maple Ave White Plains NY 10601

ALCORN, STANLEY MARCUS, plant pathologist; b. Modesto, Calif., June 18, 1926; s. Timothy Marshal and Marian (Boehne) A.; m. Esther Eastvold, June 19, 1949; children: Steven, Joseph, Eric, Mark. A.A., Modesto Jr. Coll., 1946; B.S., U. Calif., 1948, Ph.D., 1953. Postdoctoral fellow U. Calif., Berkeley, 1953-55; plant pathologist U.S. Dept. Agr., Tucson, Ariz., 1955-63; assoc. prof. U. Ariz., Tucson, 1963-65, prof. plant pathology, 1965—. Served with USAAF, 1944-45. Fellow AAAS; mem. Am. Phytopath. Soc., Am. Soc. Microbiology, Am. Soc. Hort. Sci., Am. Farm Bur., Sigma Xi. Republican. Methodist. Subspecialty: Plant pathology. Current work: Diseases of new crops, of cacti, phytobacteriaology, soil-borne fungi, etiology, epidemiology, control, guayule, jojoba, guar, cacti, bio-mass plants. Office: Dept. Plant Pathology U Ariz Tucson AZ 85721

ALCOUFFE, RAYMOND EDMOND, nuclear engr.; b. Calistoga, Calif., Dec. 15, 1940; s. Paul Claude and Roberta (Salmon) A.; m. Joan Frances Davenport, June 27, 1964; children: Michael, Matthew, John, Douglas, Carole. B.S. in Physics, St. Mary's Coll., 1962; Ph.D. in Nuclear Engring, U. Wash., 1968. Physicist Phillips Petroleum, Idaho Falls, Idaho, 1962-64; mem. staff Los Alamos Lab., 1968—; cons. Nuclear Regulatory Commn., Bethesda, Md., 1975-77. Mem. Am. Nuclear Soc. Subspecialties: Nuclear engineering; Nuclear fission. Current work: Numerical modeling of neutron transport. Home: 711 Kris Ct Los Alamos NM 87544 Office: Los Alamos Nat Lab PO Box 1663 Los Alamos NM 87545

ALDER, BERNI JULIAN, physicist; b. Duisburg, Ger., Sept. 9, 1925; came to U.S., 1941, naturalized, 1944; s. Ludwig and Ottilie (Gottschalk) A.; m. Esther Berger, Dec. 28, 1956; children—Kenneth, Daniel, Janet. B.S., U. Calif., Berkeley, 1947, M.S., 1948; Ph.D., Calif. Inst. Tech., 1951. Instr. chemistry U. Calif., Berkeley, 1951-54; theoretical physicist Lawrence Livermore Lab., Livermore, Calif., 1955—; van der Waals prof. U. Amsterdam, Netherlands, 1971; prof. associé U. Paris, 1972. Author: Methods of Computational Physics, 1963; editor: Jour. Computational Physics, 1966—. Served with USN, 1944-46. Guggenheim fellow, 1954-55; NSF sr. postdoctoral fellow, 1963-64. Fellow Am. Phys. Soc.; mem. Nat. Acad. Scis., Am. Chem. Soc. Republican. Jewish. Subspecialties: Statistical physics; Theoretical chemistry. Current work: Computer simulation of classical and quantum mechasnical many-bodied systems. Office: PO Box 808 Lawrence Livermore Lab Livermore CA 94550

ALDER, GUY MICHAEL, biotech. co. exec.; b. Denver, May 7, 1943; s. Guy Davis and Afton Lenora (Worthen) A.; m. Marian Farnsworth, July 26, 1968; children: Christopher Guy, Andrew Michael, Lara Ann, Mathew Noble, Jonathan Kimball. B.S. in Botany, U. Utah, 1968, M.S. in Biology, 1970. Range technician Utah Dept. Agr. Forest Service, 1966-70; suprv. growing div. Engh Floral Corp., Salt Lake City, 1970-71; salesmen Porter Walton Co., Salt Lake City, 1971-72; v.p. Interpretive Planning Concept Design Assocs., Salt Lake City, 1972-74; founder, pres., chief ops. Native Plants, Inc. and Plant Resource Inst., Salt Lake City, 1974—. Dist. committeeman Gt. Salt Lake council Boy Scouts Am. Served with USAR, 1961-69. Recipient Ralph Sargent Indsl. award Rocky Mountain Ctr. on Environ., 1977. Mem. Soc. for Soil and Water Conservation, Internat. Erosion Control. Assn., Utah Native Plant Soc. (dir.). Republican. Mormon. Club: Salt Lake Men's Garden. Subspecialties: Ecosystems analysis; Biomass (agriculture). Current work: Research in harsh site plant establishment techniques and materials, new product development in applied biotechnology. Co-inventor Tubepak-R- plant growing system, 1977. Office: 360 Wakara Way Salt Lake City UT 84108

ALDERMAN, EDWIN L., cardiologist, researcher; b. Rochester, N.Y., Feb. 1, 1938; s. Allen M. and Catherine (Cramer) A.; m. Sylvia V. Rich, June 13, 1967; children: David, Judith, Joel. B.A., U. Rochester, 1959; M.D., Johns Hopkins U., 1963. Intern, then resident Bronx Mcpl. Hosp., N.Y.C., 1963-68; resident in cardiology Montefiore Hosp., N.Y.C., 1968-69; asst. prof. medicine Stanford U., 1971-78, assoc. prof. medicine, 1978—; dir. Stanford Med. Ctr. Catheterization/Angio Lab, 1973—. Served with USN, 1969-71. Jewish. Subspecialty: Cardiology. Current work: Cardiol hemodynamics, coronary artery disease, clinical therapeutic trials. Office: Stanford U Med Ctr Cardiology Stanford CA 94305

ALDERMAN, MICHAEL HARRIS, public health educator; b. New Haven, Mar. 26, 1936; s. Julius and Ann (Vener) A.; m. Betsy Feinstein, July 21, 1968; children: John F., Peter B. B.A. magna cum laude, Harvard Coll., 1958; M.D., Yale U., 1962; postgrad., Johns Hopkins Sch. Pub. Health, 1965-66. Vis. physician, assoc. dir. Albert Einstein Coll. Medicine, Bronx, N.Y., 1962-64, 68-70; prof. pub. health and medicine Cornell U. Med. Ctr., N.Y.C., 1964-68, 70—; vis. lectr. U. West Indies, Jamaica, 1970-71; chmn. bd. African Med. Research Found., N.Y.C., 1981—; cons. UN Med. Service, N.Y.C., 1981; mem. exec. com. Nat. Council on Internat. Health, Washington. Editor: Clinical Medicine for the Occupational Physician, 1982, Hypertension: The Nurse's Role in Ambulatory Care, 1977. Served with USPHS, 1964-66. Glorney Raisbeck fellow N.Y. Acad. Medicine, 1967-68; traveling fellow WHO, Switzerland, 1973, 77. Fellow Am. Acad. Physicians; mem. Am. Fedn. Clin. Research, Internat. Assn. Occupational Health, Soc. Epidemiological Research, N.Y. Heart Assn. (chmn. council on community programs 1975), Am. Soc. Tropical Medicine and Hygiene, N.Y. Acad. Medicine, Am. Pub. Health Assn., Council on Fgn. Relations, Explorers Club, N.Y. State Civil Rights Commn. Club: Harvard (N.Y.C.). Subspecialties: Internal medicine; Epidemiology. Current work: The epidemiology of high blood pressure and its treatment in the community as well as health in the developing world. Office: Cornell U Med Coll 1300 York Ave New York NY 10021

ALDRICH, JEFFREY RICHARD, research entomologist; b. Columbus, Ohio, July 14, 1949; s. Richard John and June Ellen (Ellison) A.; m. Barbara Dee, June 28, 1980. B.S. cum laude in Biochemistry, U. MO., 1971, M.S. in Entomology, 1974, Ph.D., U. Ga., 1977. Postdoctoral research assoc. dept. entomology U. Ga., Athens, 1977-78; postdoctoral research assoc. dept. entomology N.Y. Agr. Exptl. Sta., Geneva, 1978-80; research entomologist Agr. Research Sta., U.S. Dept. Agr., Beltsville, Md., 1980—. Contrbr. articles to profl. jours. Mem. Entomol. Soc. Am., Entomol. Soc. Washington, Am. Soc. Zologists. Subspecialties: Entomology; Organic chemistry. Current work: Chemistry of the endocrine and exocrine secretions of insects in the order Hemiptera suborder Heteroptera. Patentee synthetic pheromone. Office: US Dept Agr Bldg 467 BARC-East Beltsville MD 20705

ALDRICH, RICHARD JOHN, research agronomist, agronomy educator; b. Fairgrove, Mich., Apr. 16, 1925; s. George and Eva Ann (Misner) A.; m. June Ellen Ellison, Apr. 5, 1943; children: Judith, Sharon, Jeffrey. B.S., Mich. State U., 1948; Ph.D., Ohio State U., 1950. Agronomist, research specialist U.S. Dept. Agr., Rutgers U., New Brunswick, N.J., 1950-57; asst. dir. Agrl. Expt. Sta., Mich. State U., East Lansing, 1957-64; assoc. dir., assoc. dean U. Mo., Columbia, 1964-76, prof. agronomy, 1978-81, research agronomist, prof., 1981—; adminstr. Coop. State Research Service, U.S. Dept. Agr., Washington, 1976-78; contractee/cons. Office Tech. Assessment, U.S. Congress, Washington, 1978-79. Author: Weed-Crop Ecological Principles in Weed Management, 1984. Served to 1st lt. USAAF, 1943-46. Mem. Am. Soc. Agronomy (sect. chmn. 1952), Weed Sci. Soc. Am., Agrl. Research Inst. (pres. 1974-75), AAAS (sect. nominating com.), Sigma Xi, Phi Kappa Phi, Gamma Sigma Delta. Subspecialties: Integrated pest management; Ecology. Current work: Integrated systems of weed management, weed prevention, antimicrobial properties of weed seeds. Office: US Dept Agr U Mo 206A Waters Hall Columbia MO 65211 Home: 1715 Woodrail Ave Columbia MO 65201

ALDRIDGE, MELVIN DAYNE, elec. engr., educator, cons.; b. Crab Orchard, W.Va., July 20, 1941; s. William Bert and Gladys Revel (Deck) A.; m. Nancy Dickinson, June 6, 1963; children: Kenrick Lee, Randal Jay. B.S.E.E., W.Va. U., 1963; M.E.E., U. Va., 1965, D.Sc., 1968. Registered profl. engr., W.Va. Electronic engr. NASA, 1963-68; asst. prof. dept. elec. engring. W.Va. U., Morgantown, 1968-72, assoc. prof., 1972-76, prof., 1976—; dir. Energy Research Ctr., 1978—; cons. in field. Contbr. articles on electronic monitoring and control in coal mines to profl. jours. Mem. IEEE, Am. Soc. for Engring. Edn., Soc. Mining Engrs., W.Va. Soc. Profl. Engrs. (Outstanding Young Engr. award 1978), Sigma Xi. Baptist. Lodge: Rotary. Subspecialties: Electrical engineering; Coal. Current work: Application of electronics to coal mining and related activities; research administrator. Home: 353 Rotary St Morgantown WV 26505 Office: Energy Research Center WVa U Morgantown WV 26506

ALEKMAN, STANLEY LAWRENCE, chemical company executive; b. N.Y.C., Mar. 21, 1938; s. Harry and Evelyn A.; m. Alice Finkelstein, June 25, 1961; children: Rachel, Eric, Elizabeth. B.A., CCNY, 1960; Ph.D., U. Del., 1968. Research chemist E.I. duPont de Nemours & Co., 1968, research supr., 1970-72, mfg. supr., 1972-74, merchandising rep., 1974-75, mfg. mgr., 1976-78, research and devel. mgr., 1978-79; tech. dir. Houdry div. Air Products and Chems., Inc., Linwood, Pa., 1979-81, tech. dir. performance chems. div., 1981-83; tech. dir. NUDDEX Inc., Piscataway, N.J., 1983—. Served with U.S. Army, 1962-64. Mem. Am. Chem. Soc., AAAS, N.Y. Acad. Scis., Sigma Xi. Subspecialties: Organic chemistry; Kinetics. Current work: Surfactants, coatings, colorants, lubricants, anti-microbials, organometallics, catalysis. Home: 505 Windsor Dr Newark DE 19711 Office: Turner Pl PO Box 365 Piscataway NJ 08854

ALESHIRE, MERLE J., aerospace company executive, researcher; b. Carthage, Ill., Dec. 30, 1933; s. Howard and Eula Juanita (Payne) A.; m. Dorothy Ann Ikerd, May 2, 1958; children: Benton, Brett, Barron. B.S., Fresno State Coll., 1960; M.S., U.C.L.A., 1971; Ph.D., U.S. Internat. U., 1979. Chief engr. systems analysis Gen. Dynamics Corp., San Diego, 1974-77, mgr. ops. research, 1977-79, mgr. advanced programs, 1979-80, dir. research and bidding, 1980-82, dir. program devel., 1982—; cons. Nat. Acad. Sci., 1977—, Nat. Security Indsl. Assn., 1975—, Am. Def. Preparedness Assn., 1976—, Naval Reseach Adv. Com., 1980—. Pres. Escondido (Calif.) Union Sch. Dist. Bd. Edn., 1974-75, 78-79; pres. bd. dirs. Boys & Girls Club, Escondido, Calif., 1975-76; vice-chmn. Palomar council Boy Scouts Am., San Diego, 1970-71; bd. dirs. Greater San Luis Rey Planning Council, Escondido, Calif., 1980—. Mem. numerous profl. orgns., including Navy League, Air Force Assn. Clubs: Meadow Lake Country, Flying. Lodge: Elks. Subspecialties: Aerospace engineering and technology; Operations research (engineering). Current work: Director company sponsored research and development. Home: 925 Mills St Escondido CA 92027 Office: General Dynamics PO Box 80847 San Diego CA

ALESSANDRO, DANIEL, med. physicist; b. Senglea, Malta, Feb. 16, 1933; s. Joseph and Elsie A.; m.; children: Paul J., Deborah A. B.S., Bklyn. Coll., 1960; M.S., N.Y. U., 1964, postgrad., 1967-70. Physicist Columbia-Presbyn. Med. Center, N.Y.C., 1960-62, N.Y. U. Med. Center, 1962-70, Overlook Hosp., Summit, N.J., 1971-74; physicist St. Elizabeth Hosp., Elizabeth, N.J., 1977-82, cons. physicist, 1969—; physicist Good Samaritan Hosp., Suffern, N.Y., 1969—; cons. physicist, 1969—. Served with USN, 1951-55. Mem. Am. Assn. Pysicists in Medicine, N.J. Med. Physics Soc. (co-founder 1974, pres. 1977—78). Roman Catholic. Subspecialty: Medical physics. Current work: Physics research. Home Cottage Pl: Westfield NJ 07090

ALEVY, YAEL GRIS, immunologist; b. Krakow, Poland, May 13, 1948; d. Leon Arie and Deborah (Friedman) G.; m. Mitchell Alan Alevy, July 15, 1973. B.Sc., Bar-Ilan U., Ramat Gan, Israel, 1971; Ph.D., Albert Einstein Coll. Medicine, Bronx, N.Y., 1975. Postdoctoral fellow dept. microbiology and immunology Washington U. Sch. Medicine, St. Louis, 1975-77; research assoc. St. Louis U. Sch. Medicine, 1977-79, asst. research prof., 1979-83, assoc. prof. dept. medicine, 1983—. Contbr. articles to profl. jours. NIH grantee, 1980-86. Mem. Am. Assn. Immunologists, N.Y. Acad. Scis., AAAS. Subspecialty: Immunobiology and immunology. Current work: Immune response and immunoregulation in uremia, experimental animal model; immune response and immunoregulation in patients with chronic uremia maintained on hemodialysis; immunoregulation of IgE synthesis in vitro. Home: 11916 Kendon Dr St Louis MO 63131 Office: 1402 S Grand Blvd 208 Doisy Hall St Louis MO 63104

ALEXANDER, CHARLES KENNETH, JR., engineering educator, consultant; b. Amherst, Ohio, Aug. 4, 1943; s. Charles Kenneth and June Elaine (Carter) A.; m. Suann Carol Miller, Aug. 31, 1963; children: Christina, Tamara, Jennifer. B.S.E.E., Ohio No. U., 1965; M.S.E.E., Ohio U., 1967, Ph.D., 1971. Registered profl. engr., Ohio. Instr. elec. engring Ohio U., 1967-71, asst. prof., 1971-72; assoc. prof. elec. engring. Youngstown State U., 1972-80, Disting. prof., 1977; NASA/Am. Soc. Engring. Edn. summer faculty fellow Auburn U. and Marshall Space Flight Ctr., summer 1973; prof., chmn. dept. elec. engring. Tenn. Tech. U., 1980—; also cons. Contbr. articles to profl. jours. Recipient award City of Youngstown, 1980; named Outstanding Prof. Wm. Rayen Sch. Engring., 1978; grantee. Mem. Am. Soc. Engring. Edn., IEEE, Ohio Solar Energy Assn., Sigma Xi, Tau Beta Pi, Eta Kappa Nu. Subspecialties: Computer engineering; Solar energy. Current work: Digital system design using Register Transfer Languages, solar energy system modeling and simulation. Home: 1475 Pilot Dr Cookeville TN 38501 Office: Tenn Technol U Box 5004 Elec Engring Cookeville TN 38501

ALEXANDER, DENNIS JAY, engineering company administrator; b. Washington, July 20, 1944; s. James Wessley and Judie O'Del (Hamilton) A.; m. Brenda Joyce Allison, July 4, 1970; children: Brian Keith, James Michael. B.S. in Indsl. Engring, U. Okla., Norman, 1967; postgrad., U.S. Naval Nuclear Power Sch., Mare Island, Calif., 1967. Lead power engr. Stone & Webster Engring. Corp., Boston, 1972-75, mgr. power div., Cherry Hill, N.J., 1975-79, asst. mgr. engring., 1979-80, project mgr., 1980—. Served to lt. USN, 1967-72. Mem. Am. Nuclear Soc., Frankford Radio Club (pres. awards chmn.). Lodge: K.C. Subspecialties: Nuclear engineering; Mechanical engineering. Current work: Managing and directing the technical and administrative aspects of large and small projects. Most of these projects are in the electrical power industry. Home: RD 3 Butterworth Bogs Rd Vincentown NJ 08088 Office: Stone & Webster Engring Corp PO Box 5200 Cherry Hill NJ 08034

ALEXANDER, EDWARD RUSSELL, disease research administrator; b. Chgo., June 15, 1928; s. Russell Green and Ethelyn Satterlee (Abel) A. Ph.B., U. Chgo., 1948, B.S., 1950, M.D., 1953. Intern Cin. Gen. Hosp.; chief surveillance sect. Communicable Disease Center, Atlanta, 1955-57, 59-60; resident, instr. dept. pediatrics U. Chgo., 1954-55, 57-59; asst. prof. dept. preventive medicine and dept. pediatrics U. Wash., Seattle, 1961-65, asso. prof., 1965-69, prof., 1969-79, chmn. dept. epidemiology, 1970-75; prof. dept. pediatrics U. Ariz., Tucson, 1979-83; dir. research br., venereal diseases control div. Centers for Disease Control, Atlanta, 1983—. Contbr. articles to profl. jours. Markle scholar, 1962-67. Mem. Am. Acad. Pediatrics, Am. Pediatric Soc., Am. Pub. Health Assn., Assn. Tchrs. Preventive Medicine, Am. Epidemiol. Soc., Soc. Epidemiol. Research, Internat. Epidemiol. Soc., King County Med. Soc. Subspecialties: Epidemiology; Microbiology (medicine). Current work: Infectious disease epidemiology with particular concern for sexually transmitted disease and its maternal-neonatal consequences. Office: Research Br. Venereal Disease Control Div Centers for Disease Control Atlanta GA 30333

ALEXANDER, HAROLD, biomedical engineering, educator, consultant, researcher; b. N.Y.C.; s. Jack and Freda (Koltun) A.; m. Sheila E., Dec. 20, 1964; children: Robin, Andrea. B.S., N.Y.U., 1962, M.S., 1963, Ph.D., 1967. Head Lab. for Balloon Tech., 1968-73, asst. prof., 1968-71, assoc. prof., 1971-77, co-dir., 1973-77; assoc. prof. surgery U. Medicine and Dentistry N.J.-N.J. Med. Sch., Newark, 1977-81, prof., 1981—, dir., 1977—; cons. med. device industry; officer CAS, Inc. Contbr. articles to profl. jours. Recipient Founder's Day award N.Y.U., 1967, Outstanding Research Paper award Internat. Research Soc. for Orthopaedics and Traumatology, 1981. Mem. Orthopaedic Research Soc., ASME, ASTM, Sigma Xi, Sigma Gamma Tau. Subspecialties: Biomedical engineering; Physiology (medicine). Current work: Research and development of orthopaedic implants; orthopaedic biomechanics educator; manufacturing and sales of medical devices. Co-patentee bio-adsorbable composite tissue scaffold. Home: 47 Elmwood Pl Short Hills NJ 07078 Office: Sect Orthopaedic Surgery U Medicine and Dentistry NJ-NJ Med Sch 100 Bergen St Newark NJ 07103

ALEXANDER, JAMES L., behavioral ecologist, educator, research; b. San Francisco, June 5, 1945; s. William Preston and Lenore E. (Funk) A.; m. Elizabeth Pascoe, Sept. 3, 1977; children: Alan James, Lauren Elizabeth. B.B.A., U. Tex.-Austin, 1967; M.A., U. Houston, 1977, Ph.D., 1978. Research asst. in psychiatry Baylor Coll. Medicine, Houston, 1972-73, asst. prof. rehab., 1979—; research asst. in behavorial ecology Inst. for Rehab. and Research, Houston, 1973-78, assoc. dir. behavorial ecology, 1978-79, dir. behavorial ecology, 1979—; clin. asst. profl. psychology U. Houston, 1978—, project dir., 1982—; v.p. Planning Assocs., Houston, 1982—; vis. architecture critic Rice U., Houston, 1976. Developer: measurement tool self-observation and report technique, 1982. Served to lt. U.S. Army, 1968-71; Vietnam. Decorated Bronze Star; recipient Outstanding Community Service award Pres. Carter, 1979. Mem. Am. Congress Rehab. Medicine, Am. Psychol. Assn., AAAS, Sigma Xi. Subspecialties: Behavioral ecology; Physical medicine and rehabilitation. Current work: Current research interests involve applying the principles of behavorial ecology: (1) to establish better systems of outpatient and follow-up care for physically disabled persons, and (2) to study the processes of adjustment to a physical disability. Further interests include applying the perspective of behavorial ecology to solve problems in business, health care, government, and educational organizations. Office: Inst for Rehab and Research 1333 Moursund Houston TX 77030

ALEXANDER, JOHN CHARLES, pharmaceutical executive, physician, researcher; b. Perth Amboy, N.J., Dec. 28, 1943; s. Charles John and Agnes (Maloney) A.; m. Margaret Ann Kohler, July 19, 1969; children: Laurel, Jennifer, Anna. B.S., St. Francis Coll., Loretto, Pa., 1965; M.D., St. Louis U., 1970; M.P.H., Johns Hopkins U., 1972. Intern Barnes Hosp., St. Louis, 1970-71; asst. dir. Charlottesville (Va.) Health Dept., 1974-76; asst. to assoc. clin. dir. Squibb Inst. Med. Research, Princeton, N.J., 1976-79, dir. clin. research, 1979-82, v.p. cardiovascular clin. research, 1982—. Served to lt. comdr. U.S. Navy, 1972-74. Recipient Beaumont prize for med. research St. Louis U., 1970. Mem. Alpha Omega Alpha. Subspecialties: Pharmacology; Preventive medicine. Current work: Cardiovascular clinical research, supervision of cardiovascular drug development. Home: 179 Longview Dr Princeton NJ 08540 Office: E R Squibb & Sons PO Box 4000 Princeton NJ 08540

ALEXANDER, JONATHAN, cardiologist, consultant; b. N.Y.C., Nov. 29, 1947; s. Josef and Hannah (Margolis) A.; m. Karen Deborah Einhorn, Aug. 8, 1971; children: Jessica, Beth, Daniel Lewis. B.A., Harvard U., 1968; M.D., Albert Einstein Coll. Medicine, 1973. Intern, resident Yale-New Haven Hosp., 1973-76; fellow dept. cardiology Yale U. Sch. Medicine, New Haven, 1976-78, asst. clin. prof. medicine, 1978-83, assoc. clin. prof. medicine, 1983—; attending physician Danbury (Conn.) Hosp., 1978—, West Haven (Conn.) Vets. Hosp., 1978—, New Milford (Conn.) Hosp., 1980—; dir. cardiac rehab. unit and nuclear cardiology Danbury (Conn.) Hosp., 1978—. Bd. dirs. Ives Festival Artists, Danbury, Conn., 1979-82; sec. bd. dirs. Western Conn. Symphony Orch., Danbury, 1981-83. Recipient Samuel Kushlan award Yale-New Haven Hosp., 1974, Revlon award 11th Internat. Congress Chemotherapy, 1983. Fellow ACP, Am. Coll. Cardiology, Am. Coll. Chest Physicians, Am. Heart Assn.; mem. Soc. Nuclear Medicine, N.Y. Acad. Scis., Alpha Omega Alpha. Jewish. Subspecialties: Cardiology; Internal medicine. Current work: Hospital-based cardiologist directing cardiac rehabilitation and nuclear cardiology programs. Office: Cardiology Dept Danbury Hosp Danbury CT 06877

ALEXANDER, LECKIE FREDERICK, mechanical engineering, educator, consultant; b. Dundee, Scotland, Mar. 26, 1929; came to U.S., 1978; s. Frederick and Mary Baxter (Barclay) L.; m. Alison Elizabeth Wheelwright, Mar. 30, 1957; children: Gavin F., Gregor W., Sean C. B.S.C., St. Andrews U., 1949; M.S., Stanford U., 1955, Ph.D., 1958. Cons. civil engr. Mott, Hay & Anderson, London, 1949-51; forschungs assistent Technische Hochschule, Hanover, W.Ger., 1957-58; univ. lectr. Cambridge (Eng.) U., 1958-68, fellow and dir. studies, 1959-68; prof. engring. U. Leicester, Eng., 1968-78; prof. mech. engring. and theoretical and applied mechanics U. Ill., Urbana, 1978—; cons. solid mechanics. Author: (with E. C. Pestel) Matrix Methods in Elaso-mechanics, 1963; contbr. numerous articles to profl. jours.; editorial bd. various jours. Bd. govs. Bedales Sch., Petersfield, Eng., 1975-78. Served to flight lt. RAF, 1951-54. Recipient Halliburton award U. Ill., 1982. Fellow Am. Acad. Mechanics; mem. ASME. Subspecialties: Theoretical and applied mechanics; High-temperature materials. Current work: Strength of materials, computer-aided design. Home: 50 Lake Park Champaign IL 61820 Office: 1206 W Green St Urbana IL 61801

ALEXANDER, MARY LOUISE, biology educator; b. Ennis, Tex., Jan. 15, 1926; d. Emmett F. and Florence (Hill) A. B.A., U.Tex., 1947, M.A., 1949, Ph.D., 1951. Instr., research asst. Genetics Found., U. Tex., Austin, 1944-51, research asst., 1962-67; postdoctoral fellow biology div. AEC, Oak Ridge, 1951-52; postdoctoral research fellow dept. zoology U.Tex., Austin, 1952-55; research assoc. M.D. Anderson Hosp. and Tumor Inst., Houston, 1956-58, asst. biologist, 1959-62; research cons. Brookhaven Nat. Lab., Upton, N.Y., 1955; researcher Oak Ridge Inst. Nuclear Studies, 1951-77; assoc. prof. biology S.W. Tex State U., San. Marcos, 1966-69, prof., 1970—. Nat. Cancer Inst. fellow Inst. Animal Genetics, Edinburgh, Scotland, 1960-61. Mem. Genetics Soc. Am., Radiation Research Soc., Am. Soc. Human Genetics, Sigma Xi, Gamma Phi Beta, Phi Sigma, Alpha Epsilon Delta. Subspecialties: Gene actions; Animal genetics. Current work: Mutagensis. Home: Hunter's Glen Route 2 Box 119 San Marcos TX 78666 Office: Dept Biology SW Tex State U San Marcos TX 78666

ALEXANDER, MILLARD HENRY, chemist, educator; b. Boston, Feb. 17, 1943; s. Benjamin and Marie M. (Mayer) A.; m. Francoise Bore, Dec. 23, 1966 (div.); 1 dau., Stephanie Andrea. B.A. magna cum laude, Harvard Coll., 1964; Ph.D., U. Paris-South, 1967. Research fellow Harvard U., 1968-71, lectr. summer sch., 1971; asst. prof. Inst. Molecular Physics, U. Md., College Park, 1971-72, asst. prof. dept. chemistry, 1972-75, assoc. prof., 1975-79, prof., 1979—. Contbr. numerous articles to profl. jours. Recipient Outstanding Young Tchr. award D.C. Inst. Chemists, 1977. Mem. Am. Phys. Soc., Am. Chem. Soc. Subspecialties: Theoretical chemistry; Atomic and molecular physics. Current work: Theory of inelastic and reactive collisions involving atoms and molecules. Office: Dept Chemistry U Md College Park MD 20742

ALEXANDER, NANCY JEAN, molecular geneticist; b. Ithaca, N.Y., Jan. 14, 1947; d. Ralph William and Gladys (Robin) A. B.S., SUNY, Oswego, 1968; M.A., Duke U., 1971, Ph.D., 1977. Research assoc. Ohio State U., Columbus, 1977-80; molecular geneticist No. Regional Research Lab., U.S. Dept. Agr., Peoria, Ill., 1980—. Mem. Genetics Soc. Am., Am. Soc. Cell Biology, Am. Soc. Microbiology, Sigma Xi. Subspecialties: Genetics and genetic engineering (agriculture); Biomass (agriculture). Current work: Conversion of biomass to useful chemicals using genetic engineering techniques. Address: 1815 N University St Peoria IL 61604

ALEXANDER, PETER, energy technology company executive, physicist, geophysicist; b. N.Y.C., Feb. 14, 1935; s. John and Beatrice (Strauss) A.; m. Iris Rubin, June 7, 1956; children: Susan, Robin, Scott. B.S. in Physics, M.I.T., 1956; Ph.D. in Physics (X-R fellow), Purdue U., 1961; research fellow, Calif. Inst. Tech., 1961-63. Mgr. physics dept. Teledyne Inc., Westwood, N.J., 1965-72; head adv. tech. br. Office Naval Intelligence, Washington, 1972-74; dir. adv. tech. div. Bendix Corp., Grand Junction, Colo., 1974-77; mgr. energy tech. TRW Inc., Denver, 1977—; presents numerous symposia in, U.S. and Europe. Author of numerous technical papers, patents and tech. reports. Recipient NASA patent award. Mem. Am. Nuclear Soc. (chmn. Nuclear Standards commn. 1973-74), IEEE, Soc. for Control & Instrumentation Energy Processes (dir. 1980), Soc. Exploration Geophysicists, Am. Physical Soc. Subspecialties: Oil shale; Geophysics. Current work: Exploitation of energy resources through insitu extraction technology. Development of technology for disposal of high and low level nuclear wastes. Office: TRW 200 Union Blvd Lakewood CO 80228

ALEXANDER, RICHARD DALE, educator, zoologist; b. White Heath, Ill., Nov. 18, 1929; s. Archie Dale and Elizabeth (Heath) A.; m. Lorraine Kearnes, Aug. 19, 1950; children—Susan Dale, Nancy Lorraine. B.S., Ill. State Normal U., 1950; M.S., Ohio State U., 1951, Ph.D., 1956. Grad. research fellow Ohio State U., 1954-56; research asso. Rockefeller Found., 1956-57; mem. faculty U. Mich., Ann Arbor, 1957—, prof. zoology, 1966—. Author: Darwinism and Human Affairs; Contbr. articles to profl. jours.; Bd. editors: Animal Behaviour, 1962—, Evolution, 1961-63, Ethology and Sociobiology, 1979—. Served with AUS, 1951-53. Guggenheim fellow, 1968-69; recipient Elliot medal Nat. Acad. Sci., 1971. Fellow Ohio Acad. Sci., AAAS. (Newcomb Cleveland prize 1962); mem. Soc. Study Evolution (council 1965-67), Animal Behavior Soc., Entom. Soc. Am., Nat. Acad. Sci., Sigma Xi. Subspecialty: Evolutionary biology. Current work: Insect systematics, acoustical communication; evolution of behavior. Home: 10731 Bethel Church Rd Manchester MI 48158

ALEXANDER, SAMUEL ADAM, plant pathologist; b. Tallulah, La., May 15, 1941; s. Lloyd Clevel and Ruby Bell (Osborne) A.; m. Connie Hall, May 30, 1949; children: Adam Hall, Scott Lloyd. B.S., La. Tech. U., 1964, M.S., 1970; Ph.D., Va. Poly. Inst. and State U., 1973. Grad. research asst. dept. botany and bacteriology La. Tech. U., 1968-70; grad. research asst. dept. plant pathology and physiology Va. Poly. Inst. and State U., 1970-73; research assoc. dept. plant pathology Pa. State U., 1974-76; asst. prof. plant pathology Va. Poly. Inst. and State U., Blacksburg, 1976-82, assoc. prof., 1982—. Contbr. articles to profl. jours. Served with U.S. Army, 1965-68. Mem. Am. Phytopathological Soc., Soc. Am. Foresters, Sigma Xi, Phi Sigma, Gamma Sigma Delta. Subspecialties: Plant pathology; Integrated pest management. Current work: Root diseases of forest trees, development of predictive models, and application of integrated forest pest management strategies. Office: Dept Plant Pathology and Physiology Va Poly Inst and State U Blacksburg VA 24061

ALEXANDER, SAMUEL CRAIGHEAD, JR., anesthesiologist; b. Upper Darby, Pa., May 3, 1930; m., 1951; 3 children. B.S., Davidson Coll., 1951; M.D., U. Pa., 1955. Intern Phila. Gen. Hosp., 1955-56; sr. asst. surgeon USPHS, 1956-58; instr. pharmacology U. Pa., 1958-60, instr. anesthesiology, 1960-63, assoc., 1963-65, asst. prof., 1965-70; prof., head dept. U. Conn., 1970-72; prof. anesthesiology Med. Sch. U. Wis., Madison, 1972—, chmn. dept., 1972-77; Pharm. Mfrs. fellow clin. pharmacology, 1957-58; cons. Phila. VA Hosp., 1964-68, Madison VA Hosp., 1972—; vis. scientist clin. physiology Bispebjerg Hosp., Copenhagen, Denmark, 1968-69. Recipient Career Devel. award USPHS, 1965-70. Mem. Am. Soc. Pharmacology and Exptl. Therapeutics, Am. Soc. Anesthesiology, Assn. Univ. Anesthetists. Subspecialty: Anesthesiology. Office: Dept Anesthesiology U Wis Med Sch Madison WI 53706

ALEXANDER, WILLIAM NEBEL, dentist, former army officer, educator; b. Pitts., May 22, 1929; s. William Harrison and Ida Margaret (Nebel) A.; m. Lorrain Michaela Berg, Nov. 29, 1958; children: Kathleen, Gregory, Christopher, Jeffrey, Steven. Student, St. Vincent Coll., 1947-49; D.D.S., U. Pitts., 1953; cert., U. Pa., 1961; M.S. in Edn, Jackson State U., 1982. Diplomate: Am. Bd. Oral Medicine. Commd. 2d lt. U.S. Army, 1952, advanced through grades to col., 1972; chief oral medicine service (Letterman Army Med. Ctr.), San Francisco, 1961-66, chief clinician, Kaiserslautern, W.Ger., 1966-69, comdg. officer, Vietnam, 1969-70, dir. gen. dental residency, Ft. Hood, Tex., 1970-74, dir. and chmn. dental edn., Tacoma, 1974-78, ret., 1978; prof., dir. patient admissions U. Miss. Sch. Dentistry, Jackson, 1978—; cons. oral medicine; dental cons. Miss. State Dept. Edn., Jackson, 1980—. Co-editor: Problem Oriented Dental Record, 1980-81; contbg. author: Oral Health of the Elderly, 1983; author manual/clin. lab. medicine for dentists, 1980-81; video tape TMJ/MFPD Syndrome Diagnosis, 1977. Tchr. piloting U.S. Power Squadrons, San Francisco, Tacoma, 1965-78; co-chmn. parish council Christ the King Parish, Belton, Tex., 1971-83; usher-greeter St. Paul's Parish, Brandon, Miss., 1978—. Decorated Bronze Star, Legion of Merit. Fellow Am. Coll. Dentists, Am. Acad. Oral Medicine; mem. Internat. Assn. Dental Research, AAAS, Omicron Kappa Upsilon (pres. chpt. 1983), Sigma Xi, Phi Kappa Phi. Republican. Roman Catholic. Clubs: Presidio Yacht (San Francisco); Loyal Order of Boar. Subspecialties: Oral Pathology; Oral Medicine. Current work: Diagnosis and treatment of temporomandibular joint disturbances, research and treatment oral manifestation of systemic disease, geriatric oral diseases. Home: 300 Forest Point Dr Brandon MS 39042 Office: U Miss Med Center 2500 MS State St Jackson MS 39216

ALFANO, MICHAEL CHARLES, drug company executive, periodontist; b. Newark, Aug. 8, 1947; s. Michael Ferdin and Anne Marie (Barrington) A.; m. Jo-Ann Mary Coletta, Mar. 30, 1969; children: Michael Anthony, Kristin Lynn. D.M.D., Coll. Medicine and Dentistry N.J., Newark, 1971; Ph.D., MIT, 1975; postgrad., Harvard U., 1971-73. Cert. in periodontics. Asst. prof. periodontics Fairleigh Dickinson U., 1974-77, assoc. prof., 1977-80, prof., 1980-82, dir., 1977-82, asst. dean, 1981-82; v.p. dental research Block Drug Co., Jersey City, 1982—; cons. NIH, 1975—, Internat. Life Scis. Inst., 1980-82, 15 maj. food and pharm. cos., 1975-82; chmn. 6th Conf. Foods, Nutrition and Dental Health, Am. Dental Assn., 1982; Mem. Sci. Adv. Council, Office of the Gov., N.J., 1981-84; spl. grants rev. com. Nat. Inst. Dental Research, 1978-81; vis. prof. Nat. Dairy Council, 1981. Editor: Changing Perspectives in Nutrition and Caries, 1979; contbr. articles profl. jours. Postdoctoral fellow NIH, 1971; research grantee, 1974-82; recipient citation for leadership YMCA, Newark, 1966. Fellow Am. Coll. Dentists; mem. Am. Inst. Nutrition, Student Clinicians of Am. Dental Assn. (Nat. Achievement award 1978), Am. Dental Assn. Roman Catholic. Subspecialties: Periodontics; Nutrition (biology). Current work: Innovative dental materials, new diagnostic aids for dentists and public at large. Home: 818 High Mountain Rd Franklin Lakes NJ 07417 Office: Block Drug Co 257 Cornelison Ave Jersey City NJ 07302

ALFANO, ROBERT RICHARD, physicist, cons., educator; b. N.Y.C., May 7, 1941. B.S., Fairleigh Dickinson U., 1963, M.S. in Physics, 1964, Ph.D., N.Y. U., 1972. Mem. research staff GTE, 1964-72; prof. physics CCNY, 1972—; dir. Ultrafast Spectroscopy and Laser Lab., 1982—. Contbr. numerous articles to profl. jours. A.P. Sloan fellow, 1974-78. Fellow Am. Phys. Soc.; mem. Am. Biophys. Soc. Subspecialties: Condensed matter physics; Biophysics (physics). Current work: Picosecond and femtosecond lasers and techniques, ultrafast spectroscopy in materials. Patentee in field. Office: City University of New York Convent Ave and 138th St New York NY 10031

ALFELD, PETER WILHELM, research mathematician; b. Schwarzenberg, Sachsen, Germany, Feb. 20, 1950; came to U.S., 1977; s. Wilhelm and Edeltraud (Rosel) A.; m. Michelle Ann Poulin, Dec. 27, 1976; children: Christopher, Anna. B.S., Hamburg U., 1974; M.Sc., U. Dundee, 1975, Ph.D., 1977. Instr. U. Utah, Salt Lake City, 1977-79, asst. prof., 1979-82, assoc. prof. math., 1982—. Served with German Air Force, 1969-70. Recipient Scholarship German Nat. Scholar Found., 1972-77; Utah Consortium award Dept. Energy, 1980; Recognition for Superior Teaching Ability Ut Utah, 1982. Mem. Soc. Indsl. and Applied Math., Am. Math. Soc., Assn. Computing Machinery, Spl. Interest Group for Numerical Analysis. Subspecialties: Numerical analysis; Mathematical software. Current work: Multivariate interpolation and approximation. Office: U Utah Dept Math Salt Lake City UT 84112

ALFIDI, RALPH JOSEPH, radiologist, educator; b. Rome, Italy, Apr. 20, 1932; s. Lucas and Angeline (Panella) A.; m. Rose Ester Senesac, Sept. 3, 1956; children—Sue, Lisa, Christine, Catherine, Mary, John. A.B., Ripon (Wis.) Coll., 1955; M.D., Marquette U., Milw., 1959. Intern Oakwood Hosp., Dearborn, Mich., 1959-60; resident, chief resident, A.C.S. fellow U. Va., 1960-63; practice medicine, specializing in radiology, Cleve., 1965—; staff mem. Cleve. Clinic, 1965-78, head dept. hosp. radiology, 1968-78; dir. dept. radiology Univ. Hosps., Cleve.; cons. VA Hosp., Hillcrest Hosp., Cleve.; chmn. dept. radiology Case Western Res. U. Sch. Medicine, 1978—; Chmn. staff Cleve. Clinic Found., 1975-76. Author: Complications and Legal Implications of Special Procedures, 1972, Computed Tomography of the Human Body: An Atlas of Normal Anatomy, 1977; Editor: Whole Body Computed Tomography, 1977; Contbr. articles to radiology jours. Served to capt., M.C. U.S. Army Res., 1963-65. Picker Found. grantee, 1969-70; NRC grantee, 1969-70. Fellow Am. Coll. Radiology; mem. AMA, Radiol. Soc. N. Am., Am. Roentgen Ray Soc., Am. Heart Assn., Soc. Cardiovascular Radiology, Soc. Gastrointestinal Radiology, Soc. Computed Body Tomography (pres. 1977-78), Eastern Radiol. Soc., Ohio Radiol. Soc., Cleve. Radiol. Soc. (pres. 1976-77). Roman Catholic. Clubs: Hillbrook, Chagrin Valley Racquet. Subspecialties: Diagnostic radiology; Imaging technology. Current work: Research in NMR imaging grating of vascular and other systems. Home: 742 Coy Ln Chagrin Falls OH 44022 Office: Case Western Res U Dept Radiology 2074 Abington Rd Cleveland OH 44106

ALFORD, GEARY SIMMONS, experimental clinical psychologist, educator; b. McComb, Miss., Apr. 11, 1945; s. Percy Knapp and Murrell (Dodds) A.; m. Catherine Elizabeth Alford; children: Alexander Geary, Zeb Burton. Diploma, Goethe-Inst., 1965; B.A., Millsaps Coll., 1968; M.A., U. Ariz., 1971, Ph.D., 1972. Intern, then resident in psychology U. Miss. Med. Center, Jackson, 1971-72, clin. asst. prof., 1972-73, asst. prof., 1973-79, assoc. prof. psychiatry-psychology, 1980—, asst. prof. pharmacology-toxicology, 1979—, clin. asst. prof. family medicine, 1975—; cons. in field; dir. Protective Service Life Ins. Co., Jackson, 1976—; mem. adv. bd. Drug Research and Edn. Assn. Miss. Contbr. articles, chpts. to profl. publs.; editorial bd.: Behavior Modification, 1982—; editorial reviewer for various nat., internat. profl. jours. Chmn. Diocesan Comm. on Alcohol and Drug Dependency, Episcopal Diocese Miss., 1980—. Sigma Xi traveling scholar, 1971; Nat. Inst. Drug Abuse postdoctoral fellow, 1975; Nat. Inst. Alcoholism and Alcohol Abuse postdoctoral fellow, 1976. Fellow Behavior Therapy and Research Soc. (clin.); mem. Am. Psychol. Assn., Miss. Psychol. Assn. (pres. 1977-78), Assn. Advancement Behavior Therapy, Sigma Xi. Subspecialties: Behavioral psychology; Neuropharmacology. Current work: Development and evaluation of behavioral therapeutic procedures derived from basic experimental psychology; clinical neuropharmacology, addictive disorders and their treatment. Office: Dept Psychiatry and Human Behavior U Miss Med Center 2500 N State St Jackson MS 39216 Home: 20 Sheffield Crescent Jackson MS 39211

ALFVEN, HANNES OLOF GOSTA, physicist; b. May 30, 1908. Ph.D., U. Uppsala, Sweden, 1934. Prof. theory of electricity Royal Inst. Tech., Stockholm, 1940-45, prof. electronics, 1945-63, prof. plasma physics, 1963-73; prof. applied physics and info. sci. U. Calif., San Diego, 1967—; past mem. Swedish Sci. Adv. Council, Swedish Atomic Energy Commn.; past bd. govs. Swedish Def. Research Inst., Swedish Atomic Energy Co.; past sci. adv. Swedish Govt.; past pres. Pugwash Confs. Sci. and World Affairs; mem. panel comets and astroids NASA. Author: Cosmical Electrodynamics, 1950, On the Origin of the Solar System, 1954, Cosmical Electrodynamics: Fundamental Principles, 1963, Worlds-Antiworlds, 1966, The Tale of the Big Computer, 1968, Atom, Man and the Universe, 1969, Living on the Third Planet, 1972, Evolution of the Solar System, 1976, Cosmic Plasma, 1981. Recipient Nobel prize for physics, 1970; Lomonsov Gold medal USSR Acad. Scis., 1971; Franklin medal, 1971. Mem. Swedish Acad. Scis., Akademia NAUK (USSR), U.S. Acad. Scis. (fgn. asso.), Royal Soc. (fgn. mem.), numerous others. Subspecialty: Plasma physics. Office: Dept Electrical Engineering and Computer Science U Calif La Jolla CA 92093

ALIKHAN, MAHMOOD, physician, consultant; b. Rampur, U.P., Brit. Indies, Nov. 15, 1945; came to U.S., 1970; s. Anwar and Nadir (Jehan) A.; m. Linda Dorothy Linn, Oct. 16, 1971; 1 dau., Leah. M.B., B.S., Dow Med. Coll., Karachi, Pakistan, 1968; M.P.H., John Hopkins U., 1981. Diplomate: Am. Bd. Internal Medicine. House officer cardiology and surgery Jinnah and Civil Hosp., Karachi, Pakistan, 1969-70; intern and resident in medicine Union Meml., Balt., 1970-73; NIH fellow in cardiology U. Vt., Burlington, 1973-75; asst chief clin. investigations USPHS Hosp., Balt., 1975-81; assoc. chief and dir. cardiac catheterization lab. USPHS, Balt., 1977-81; dir. cardiology floor and rehab. Greater Balt. Med. Ctr., 1982—; dir. Greater Balt. Med. Ctr. Med. Assoc., Towson, Md., 1982—; v.p. and dir. Clin. Assocs. P.A., Towson, 1982—. Fellow Am. Coll. Cardiology, Am. Coll. Chest Physicians, Am. Coll. Angiology; mem. ACP, Am. Fedn. Clin. Research. Subspecialties: Cardiology; Gerontology. Current work: Cardiovascular physiology; cardiac catheterization and angiography; cardiovascular epidemiology; cardiovascular diseases in the elderly. Home: 1602 Pot Spring Rd Timonium MD 21093 Office: Clin Assocs P A 660 Kenilworth Dr Towson MD 21204

ALKANA, RONALD LEE, neuroscientist, pharmacologist, educator; b. Los Angeles, Oct. 17, 1945; s. Sam A. and Madelyn J. Davis; m. Linda Kelly, Sept. 12, 1970; 1 son, Alexander Philippe Kelly. Lic. pharmacist, Calif., Nev. Zoology student UCLA, 1966; Pharm.D. U. So. Calif., 1970; Ph.D. U. Calif.-Irvine, 1975; Postdoctoral fellow Nat. Inst. on Alcohol Abuse and Alcoholism U. Calif.-Irvine, 1974-76, resident assisting dir., div. neurochemistry, dept. psychiatry, 1976; asst. prof. pharmacology and toxicology Sch. Pharmacy U. So. Calif., Los Angeles, 1976-82, assoc. prof., 1982—; cons. Nat. Alcohol Research Center U. Calif. - Irvine. Contbr. articles on pharmacology and behavior to profl. jours. Nat. Inst. on Alcohol Abuse and Alcoholism grantee, 1980-85. Mem. Am. Soc. for Pharmacology and Exptl. Therapeutics, Research Soc. on Alcoholism, Am. Coll. Clin. Pharmacology, Internat. Soc. Biomed. Research on Alcoholism, Soc. for Neurosci., Sigma Xi. Subspecialties: Neuropharmacology; Psychobiology. Current work: Effects of drugs on brain and behavior, biological basis of behavior, hyperbaric induced antagonism of ethanol's behavioral effects, ethanol's effects on learning and memory, temperature dependence of ethanol's effect. Office: 1985 Zonal Ave Los Angeles CA 90033

ALLARA, DAVID LAWRENCE, chemist; b. Vallejo, Calif., Nov. 3, 1937; s. Lawrence Phillip and Lillian (Johansen) A.; m. Judith Havir, June 9, 1968; children: Michael Joseph, Amy Elizabeth. B.S., U. Calif.-Berkeley, 1959; Ph.D., UCLA, 1964. NSF postdoctoral fellow Oxford (Eng.) U., 1964-65; postdoctoral fellow, mem. staff Stanford Research Inst., Menlo Park, Calif., 1965-67; assoc. prof. San Francisco State U., 1967-69; mem. tech. staff Bell Labs., Murray Hill, N.J., 1969—. Editor: (with others) Fundamental Development in Stabilization and Degradation of Polymers, 1978; Bd. editors: Surface and Interface Analysis. Mem. Am. Chem. Soc. (bd. editors book series), Am. Phys. Soc., AAAS. Subspecialties: Physical chemistry; Surface chemistry. Current work: Structure and reaction meachanisms of organic interfaces and surfaces; vibrational surface spectroscopy (IR and Raman); kinetics and thermochemistry; computational modeling of reaction mechanisms. Office: Bell Labs 7F-212 Murray Hill NJ 07974

ALLARD, GILLES OLIVIER, geology educator, consultant; b. Rhougemont, Que., Can., Dec. 12, 1927; came to U.S., 1958; s. Alcide and Jeanne (Favreau) A.; m. Bernadette C. Martineau, Sept. 27, 1952; children: Claude B., Martine S., Michel L. B.A., U. Montreal, 1948, B.S., 1951; M.A., Queen's U., 1953; Ph.D., Johns Hopkins U., 1956. Exploration mgr. Chibougamau Mining and Smelting, Que., 1955-58; asst. prof. geology U. Va., Charlottesville, 1958-59; prof. geology Centro de Aperfeicoamento e Pesquisas Petrobras, Salvador, Brazil, 1959-64; vis. lectr. geoology U. Calif.-Riverside, 1964-65; prof. geology U. Ga., Athens, 1965—; cons. Quebec Ministry Energy and Resources, 1966—. Recipient Retty prize econ. geology U. Montreal, 1951; Johns Hopkins Scholarship, 1954-55; faculty research award Sigma Xi, U. Ga., 1970; Sandy Beaver Teaching Prof. chair U. Ga., 1978-82. Fellow Geol. Soc. Am., Mineral. Soc. Am.; mem. Can. Inst. Mining and Metallurgy, Soc. Econ. Geologists, Mineral. Assn. Can., Ga. Geol. Soc. (pres. 1968-69). Roman Catholic. Club: Torch. Subspecialties: Geology; Petrology. Current work: Volcanogenic theory of ore deposits—applications to field mapping especially in high grade metamorphic terranes; layered complexes-petrology-metamorphism-associated ore deposits. Home: 225 Hampton Ct Athens GA 30605 Office: Dept Geology U Ga Athens GA 30602

ALLARD, ROBERT WAYNE, geneticist, educator; b. Los Angeles, Sept. 3, 1919; s. Glenn A. and Alma A. (Roose) A.; m. Ann Catherine Wilson, June 16, 1944; children: Susan, Thomas, Jane, Gillian, Stacie. B.S., U. Calif. at Davis, 1941; Ph.D., U. Wis., 1946. From asst. to asso. prof. U. Calif. at Davis, 1946—, prof. genetics, 1955—. Author books; contbr. articles to profl. jours. Served to lt. USNR. Recipient Crop Sci. award Am. Soc. Agronomy, 1964, DeKalb Disting. Career award Crop Sci. Soc. Am., 1983; Guggenheim fellow, 1954, 60; Fulbright fellow, 1955. Mem. Nat. Acad. Scis., Am. Soc. Naturalists (pres. 1974-75), Genetics Soc. Am. (pres. 1983-84), Phi Beta Kappa, Sigma Xi, Alpha Gamma Rho, Alpha Zeta. Democrat. Unitarian. Subspecialties: Genetics and genetic engineering (biology); Animal genetics. Current work: Population genetic, evolutionary genetics, plant breeding. Home: 622 Fillmore St Davis CA 95616

ALLAUDEEN, HAMEEDSULTHAN SHEIK, molecular biologist; b. Punganur, India, Apr. 12, 1943; s. Hameed Sulthan and Jannath Beevi; m. Sadhika; children: Jerina, Nazima. Ph.D., Indian Inst. Sci., Bangalore, India, 1971. Research staff scientist molecular biophysics and biochemistry dept. Yale U., New Haven, 1971-73, lectr. pharmacology dept., 1975-77, asst. prof., 1977-; asst. div. natural product pharmacology Smith Kline & French Labs., Phila., 1982—. Contbr. articles to profl. jours. Leukemia Soc. Am. fellow, 1973-75. Mem. Am. Assn. Cancer Research, Am. Soc. Microbiology. Subspecialties: Cancer research (medicine); Molecular pharmacology. Current work: Molecular virology; regulation of DNA replication in tumor cells. Home: 738 Champlain Dr King of Prussia PA 19403 Office: Smith Kline & French Labs Philadelphia PA 19101

ALLEN, CLARENCE RODERIC, geologist, educator; b. Palo Alto, Cal., Feb. 15, 1925; s. Hollis Partridge and Delight (Wright) A. B.A., Reed Coll., 1949; M.S., Cal. Inst Tech., 1951, Ph.D., 1954. Asst. prof. geology U. Minn., 1954-55; mem. faculty Cal. Inst. Tech., 1955—, prof. geology and geophysics, 1964—; interim dir. Seismological Lab., 1965-67, acting chmn. division of geological scis., 1967-68; Chmn. cons. bd. earthquake analysis Cal. Dept. Water Resources, 1965-74; chmn. geol. hazards adv. com. for program Cal. Resources Agy., 1965-66; mem. earth scis. adv. panel NSF, 1965-68, chmn., 1967-68, mem. adv. com. environmental scis., 1970-72; mem. U.S. Geol. Survey adv. panel to Nat. Center Earthquake Research, 1966-75, Cal. Mining and Geology Bd., 1969-75, chmn., 1975; mem. task force on earthquake hazard reduction Office Sci. and Tech., 1970-71. Served to 1st lt. USAAF, 1943-46. Recipient G.K. Gilbert award seismic geology Carnegie Instn., 1960. Fellow Am. Geophys. Union, Geol. Soc. Am. (counselor 1968-70, pres. 1973-74), Am. Acad. Arts Scis.; mem. Am. Assn. Petroleum Geologists, Nat. Acad. Scis., Earthquake Engring. Research Inst., Seismological Soc. Am. (dir. 1970—, pres. 1975-76), Assn. Engring. Geologists, Nat. Acad. Engring., Soc. Exploration Geophysicists, Phi Beta Kappa. Subspecialties: Geophysics; Tectonics. Current work: Seismotectonics; Evaluation of seismic hazard. Home: 700 S Lake Ave Apt 322 Pasadena CA 91106

ALLEN, DELMAS JAMES, neuroanatomy educator, researcher; b. Hartsville, S.C., Aug. 13, 1937; s. James Paul and Sara Francis (Segars) A.; m. Sarah Virginia Bahous, July 5, 1958; children: Carolyn, James, Susan. B.S., Am. U. Beirut, 1965, M.S. in Biology, 1967; postgrad. in human anatomy, Med. Coll. Ga., 1968; cert. radiation sci., Colo. State U., 1969; Ph.D. in Anatomy, U. N.D., 1974. Teaching fellow dept. biology Am. U. Beirut, 1965-67; instr. Clarke Coll., Dubuque, Iowa, 1968-69, asst. prof. chmn. dept. biology, coordinator med. tech. and nursing programs, 1969-72; grad. teaching fellow and research asst., dept. anatomy U. N.D., Grand Forks, 1972-74; asst. prof. U. South Ala., Mobile, 1974-75; asst. prof. dept. anatomy Med. Coll. Ohio, Toledo, 1975-77, assoc. prof., 1981—, prof., 1981—, asst. dean Grad. Sch., 1979—; vis. prof. U. Uberlandia, Brazil, 1980, Riyad U. Coll. Medicine, Saudi Arabia, 1981; del. 5th Internat. Symosium on Morphol. Scis., 1982. Co-author 2 books; co-editor 1 book. Served to capt. USAR, 1960-63. NSF-AEC fellow, 1969; recipient Outstanding Teaching award Clarke Coll., 1972; A. Rodger Denison award N.D. Acad. Sci., 1973; Merit award Sigma Xi, 1974; Golden Apple award for Teaching Med. Coll. Ohio, 1977, 78, 79, 80, 82; Segment I Teaching award, 1979, 80, 81; award Brazilian Acad. Medicine, 1980; Research award Electron Microscopy Soc. N.W. Ohio, 1980; NSF grantee, 1970-72; others; grantee Ala. Heart assn., Am. Cancer Soc., NIH, Nat. Spinal Cord Injury Found. Fellow Ohio Acad. Sci. (sec. v.p. 1978-79); mem. AAAS, Midwest Anatomists Assn., Electron Microscopy Soc. Am., S.E. Electron Microscopy Soc., So Soc. Anatomists, Am. Assn. Anatomists, Soc. Neurosci., Brit. Brain Research Assn. (hon.), European Brain Research Assn. (hon.), Am. Heart Assn., Am. Soc. Cell Biology, Pan Am. Assn. Anatomy, N.Y. Acad. Scis., Sigma Xi (chpt. sec.-treas. 1974-75). Subspecialties: Anatomy and embryology; Cytology and histology. Home: 3341 W Bancroft Toledo OH 43606 Office: Med Coll Ohio Toledo OH 43609

ALLEN, DONALD ORRIE, pharmacologist, educator; b. Belding, Mich., Jan. 11, 1939; s. Orrie Burt and Bessie E. (Elmendorf) A.; m. Arlene E. Stalberger, Dec. 30, 1961; children: Michael, Mark, Kelley, Paul. B.S., Ferris Inst., Big Rapids, Mich., 1962; Ph.D., Marquette U., Milw., 1967. Asst. prof. Ind. U. Sch. Medicine, 1967-72, asso. prof., 1972-75; prof. pharmacology U. S.C. Sch. Medicine, Columbia, 1975—, chmn. dept., 1977—. Contbr. articles to profl. jours. Grantee NIH, Am. Diabetes Assn. Mem. Am. Soc. Pharmacology and Exptl. Therapeutics, Internat. Soc. Biochem. Pharmacology, Southeastern Pharmacology Soc. Roman Catholic. Subspecialties: Cellular pharmacology; Biochemistry (medicine). Current work: Control of metabolism, cyclic nucleotides, hormone action, adipose tissue, lipolysis. Office: Dept Pharmacology Sch Medicine Univ SC Columbia SC 29208

ALLEN, GARY IRVING, neurophysiologist; b. Lockport, N.Y., Apr. 7, 1942; s. Ralph Willard and Lois Marie (Chamberlin) A.; m. Elaine Irene Main, June 13, 1964; children: Michelle Irene, Elisa Joy, Scott Jeremy. B.S. in E.E, Cornell U., 1965; Ph.D. in Physiology, SUNY, Buffalo, 1969. Asst. prof. Lab. Neurobiology, dept. physiology Sch. Medicine, SUNY, Buffalo, 1971-75, asst. prof., dir., 1975-76; lectr., vis. scholar dept. physiology and anatomy U. Calif., Berkeley, 1976-79; dir. Christian Embassy at UN, 1979—; adj. asst. prof. dept. physiology N.Y. Med. Coll., Valhalla, N.Y., 1981—. Contbr. articles to profl. jours. Mem. Am. Physiol. Soc., Soc. for Neurosci., Internat. Brain Research Orgn., Asia Soc., Japan Soc., Am.-Nepal Soc. Subspecialty: Neurophysiology. Current work: Cerebro-cerebellar loops in initiation and control of movement. Home: 965 Knollwood Rd White Plains NY 10603

ALLEN, GIL C., chiropractor, neurophysiologist; b. Queens, N.Y., July 2, 1945; s. Harrison and Ethel (Garby) A.; m. Rise Hepner, May 30, 1968; children: Jonathan, Douglas. B.A., Queens Coll., 1969, M.A., 1972; M.S., U. Bridgeport, 1980; Ph.D., CUNY, 1978; D.C., Nat. Coll. Chiropractic, 1966. Practice chiropractics, nutrition, applied kinesiologist, neurophysiologist, acupuncturist, Flushing, N.Y., 1968—. Mem. Am. Biomagnetic Assn., Am. Soc. Clin. Hypnosis, N.Y. Acad. Scis., Council on Nutrition, Council on Roentgenology, Tridology Soc. Am., Am. Chiropractic Assn., Soc. Neurosci., Found. Chiropractic Edn. and Research, N.Y. State Chiropractic Assn. (pres. Queens dist.). Current work: Developing new treatment procedures to better handle chronic ailments as well as the physical, emotional, biochemical, biomechanical and psychological concomitants of such ailments; all encompassing holistic health care concepts. Office: 142-01 37th Ave Flushing NY 11354

ALLEN, JERRY MICHAEL, aerospace technologist; b. Forest City, N.C., Jan. 19, 1940; s. William Howard and Lois (Henderson) A.; m. Carolyn Harris, Aug. 19, 1962; children: Michael, Scott, Kellie. B.S. in Aerospace Engring, N.C. State U., 1962, M.S., U. Va., 1967. Research scientist Langley Research Center, Hampton, Va., 1962—. Contbr. tech. articles to profl. publs. and confs. Recipient NASA Apollo Achievement Award, 1970. Assoc. fellow AIAA. Baptist. Clubs: Glendale Recreation, Centre Court Racquet. Subspecialties: Aeronautical engineering; Aerospace engineering and technology. Current work: Analytical and experimental research in supersonic missile aerodynamics. Home: 117 Huxley Pl Newport News VA 23606 Office: Langley Research Center Langley Field Hampton VA 23665

ALLEN, JOE HASKELL, geophysicist, national geophysical data center executive; b. Oklahoma City, June 6, 1939; s. William Haskell and Gertrude Lilly (Arrington) A.; m. Charlotte Anne Moody, July 28, 1939; children: Melissa Anne, Susan Michelle, Melinda Kaye. B.S. in Physics, U. Okla., 1961; M.S. in Engring. Geoscis, U. Calif.-Berkeley, 1966. Tchr. Casady Sch., The Village, Okla., 1961-63; NOAA, with U.S. Dept. Commerce, Washington, 1963—; head Central Info. Exchange Office for Internat. Magnetospheric Study, Boulder, Colo., 1976-79, chief solar-terrestrial physics div., Boulder, 1981—; dir. World Data Ctr.-A for STP, Boulder, 1981—; Mem. various panels Nat. Acad. Scis., 1976—. Contbr. over 35 sci. articles to profl. publs. Deacon East Boulder Baptist Ch., 1978-80; mem. exec. bd. Colo. State Sci. Fair., Denver, 1968—. Recipient Silver medal U.S. Dept. Commerce, 1978. Mem. Research Soc. Am., Am. Geophys. Union, Internat. Assn. Geomagnetism and Aeronomy. Democrat. Subspecialties: Geophysics; Solar-Terrestrial Physics. Current work: Transfer of non-radient energy from Sun to near-Earth space; administration, planning and research. Home: 2880 Colby Dr Boulder CO 80303 Office: NOAA/D63 325 Broadway St Boulder CO 80303

ALLEN, JOHN CHRISTOPHER, JR., geology educator, consultant, researcher; b. Dallas, Sept. 17, 1935; s. John Christopher and Dorothy (Holloway) A.; m. Joan Audrey Stout, Apr. 1, 1961; children: Julia Lee, John Christopher III. B.S., So. Methodist U., 1958; M.A., Princeton U., 1960, Ph.D., 1961. Instr. Wesleyan U., 1961, asst. prof., 1961-63; asst. prof. geology Bucknell U., 1963-67, assoc. prof., 1967-74, prof., 1974—, chmn. dept. geology, 1982—. Mem. Lewisburg (Pa.) Planning Com. on Floodplain Use, 1975-78. Pirtle scholar, 1957. Fellow Mineral. Soc. Am.; mem. Geol. Soc. Am., AAAS. Subspecialties: Petrology; Mineralogy. Current work: Stability of amphiboles in basalt and andesites at high pressures and temperatures. Office: Dept Geology Bucknell U Lewisburg PA 17837 Home: 26 S 3d St PO Box 207 Lewisburg PA 17837

ALLEN, JOHN MALONE, research engineer; b. Cleve., Apr. 15, 1921; s. Horace J. and Eleanor Malone A.; m. Anne L. Stevens, Mar. 29, 1952; children: Timothy, Anne, Katherine. B.S., Haverford Coll., 1943. Research engr. NACA, NASA, Cleve., 1943-47, Battelle Meml. Inst., Columbus, Ohio, 1948—. Mem. Air Pollution Control Assn., AIAA. Unitarian. Club: Leatherships Yacht. Subspecialties: Combustion processes; Environmental engineering. Current work: Control of emissions from burning wood and other fuels in industrial and residential applications and hazardous waste destruction. Home: 2690 W Granville Rd Worthington OH 43085 Office: Battelle Meml Inst 505 King Ave Columbus OH 43201

ALLEN, JOHN PAUL, neuroendocrinologist; b. Portland, Oreg., Aug. 17, 1942; s. Richard Taft and Jean Ester (Smith) A.; m. Catherine Jean Fowler (div.); 1 son, Matthew; m. Rebecca Francis Lyons; 1 dau., Sarah. B.S., Portland State Coll., 1965; M.D., U. Oreg., 1967. Asst. prof. U. Tex. Health Sci. Center, San Antonio, 1975-78; assoc. prof.neuroscis. U. Ill. Coll. Medicine, Peoria, 1978—; cons. VA, Peoria, 1980—. Advisor Med. Explorer Post, Boy Scouts Am., Peoria, 1980—. Served to maj. M.C. USAF, 1973-75. Mem. Soc. for Neurosci., Endocrine Soc., N.Y. Acad. Sci., Am. Acad. Neurology, Internat. Soc. Psychoneruoendocrinology, Sigma Xi. Democrat. Unitarian. Subspecialties: Neuroendocrinology; Neurology. Current work: Neural control of ACTH secretion. Opiate regulation on hormone secretion. Home: 2719 W Greenbrier Peoria IL 61614 Office: U Ill Coll Medicine 1 Illini Dr Peoria IL 61656

ALLEN, JONATHAN, electrical engineering educator; b. Hanover, N.H., June 4, 1934; m., 1960; 2 children. A.B., Dartmouth Coll., 1956, M.S., 1957; Ph.D. in Elec. Engring, MIT, 1968. Supr. human factors engring. Bell Telephone Labs., 1961-67; asst. prof. elec. engring. MIT, Cambridge, 1968-76, assoc. prof. elec. engring., assoc. dir. research lab. electronics, 1976—; cons. Lincoln Lab. and Sperry Rand Research Corp., 1968—. Mem. IEEE. Subspecialty: Electronics. Office: Dept Elec Engring MIT Cambridge MA 02139

ALLEN, LEW, JR., former air force officer, jet propulsion laboratory executive; b. Miami, Fla., Sept. 30, 1925; s. Lew and Zella (Holman) A.; m. Barbara Frink, Aug. 19, 1949; children: Barbara Allen Miller, Lew, Marjorie Allen Dauster, Christie, James. B.S., U.S. Mil. Acad., 1946; M.S., U. Ill., 1952, Ph.D., 1954. Commd. 2d lt. USAAF, 1946; advanced through grades to gen. USAF, 1977; physicist AEC, Los Alamos, 1954-57, Air Force Weapons Lab., 1957-61; mem. staff Office Sec. Def., 1961-65; with Air Force Space Program, 1965-72; dir. Nat. Security Agy., Ft. Meade, Md., 1973-77; comdr. Air Force Systems Command, 1977; vice chief staff USAF, 1978, chief staff, 1978-82; ret., 1982; dir. Jet Propulsion Lab., Pasadena, Calif., 1982—. Decorated D.S.M. (5), Legion of Merit with 2 oak leaf clusters, Joint Service Commendation medal. Mem. Am. Phys. Soc., Am. Geophys. Union, Nat. Acad. Engring., Sigma Xi. Subspecialty: Space science research management.. 1040 S Arroyo Blvd Pasadena CA 91105 Office: Jet Propulsion Lab California Inst Tech 4800 Oak Grove Dr Pasadena CA 91125

ALLEN, MARK ANDREW, astrochemist, research scientist; b. N.Y.C., Sept. 29, 1949; s. Israel and Lucille (Goldberg) A.; m. Emily Anne Bergman, June 20, 1982. B.A., Columbia Coll., N.Y.C., 1971; Ph.D., Calif. Inst. Tech., 1976. NRC resident research assoc. NASA Goddard Inst. Space Studies, N.Y.C., 1976-78; research fellow div. geol. and planetary scis. Calif. Inst. Tech., 1978-81, sr. scientist, 1981—. Contbr. articles to profl. publs. Mem. Am. Astron. Soc., Am. Chem. Soc., Am. Geophys. Union, AAAS, Phi Beta Kappa, Sigma Xi, Phi Lambda Upsilon. Subspecialties: Planetary atmospheres; Space chemistry. Current work: Chemical models and remote sensing of atmospheres of earth and other planets and moons, of comets, and of interstellar clouds. Home: 1101 N Geneva St Glendale CA 91207 Office: MS 183-601 Jet Propulsion Lab 4800 Oak Grove Dr Pasadena CA 91109

ALLEN, MARTIN, chemistry educator; b. N.Y.C., Mar. 26, 1918; s. Isidor and Frances (Gudowitz) A.; m. Sophie Parker, June 13, 1942;

children: Susan, Scott, Barbara, Robert. B.A., Bklyn. Coll., 1938; M.S., U. Minn., 1941, Ph.D., 1944. Instr. U. Minn.-Mpls., 1943-45; research assoc. Alleghany Ballistics Lab., Cumberland, Md., 1945; mem. sr. tech. staff B.F. Goodrich Co., Akron, Ohio, 1947-47; assoc. prof. chemistry Butler U., Indpls., 1947-56; mem. faculty Coll. of St. Thomas, St. Paul, 1956—, prof. chemistry, 1957—, chmn. chemistry dept., 1975—, dir. scis. and math. div., 1977-83, dir. insts., 1958-70; vis. prof. chem. engring. dept. U. Minn-Mpls, 1981-83. Contbr. over 600 chem. abstracts to profl. publs. Mem. citizens com. Pub. Edn., Mpls., 1965-70. Recipient Gold medal Bklyn. Coll., 1938. Fellow AAAS; mem. Am. Chem. Soc. (chmn. Minn. sect. 1968), Midwestern Assn. Chemistry Tchers. in Liberal Arts Colls. (pres. 1965-66), Fedn. Am. Scientists. Subspecialties: Physical chemistry; Surface chemistry. Current work: Thermodynamics of solutions; chemistry of surfactants. Home: 4620 Bassett Creek Ln Golden Valley MN 55422 Office: 2215 Summit Ave Saint Paul MN 55105

ALLEN, PATTON TOLBERT, virologist, farmer; b. Lockhart, Tex., Oct. 7, 1939; s. Wilson Tolbert and Hester (Patton) A.; m. Jacqueline Diane Cockerell, Nov. 25, 1961; children: Scot Tolbert, Sharon Elaine. B.S. in Biology, S.W. Tex. State U., 1962; Ph.D. in Microbiology, U. Tex.-Austin, 1966. Asst. prof. U. Tex. M.D. Anderson Hosp. and Tumor Inst., Houston, 1971-74; vis. scientist U. Leuven (Belgium) Rega Inst., 1974-75; staff scientist Nat. Cancer Inst., Frederick, Md., 1975—. Contbr. articles to profl. jours. Served to capt. USAF, 1966-71. Mem. Am. Soc. Microbiology, N.Y. Acad. Sci., Sigma Xi. Subspecialties: Cell study oncology; Virology (medicine). Current work: Research in viral and chemical oncogenesis; regulaton of cell growth as influenced by growth factors, interferons, or other biomodifiers. Office: Nat Cancer Inst Frederick Cancer Research Facility Room 111 Bldg 538 Frederick MD 21701

ALLEN, ROBERT ARTHUR, clinical psychologist; b. Stockton, Calif., July 24, 1952; s. Arthur Hagy and Peggy Irmgard (Oetjen) A.; m. Joann May Meluskey, Aug. 3. B.A., DePauw U., 1975; M.A., U. Ottawa, Ont., Can., 1977; Ph.D., Calif. Sch. Profl. Psychology, 1980. Lic. clin. psychologist, Calif. Postdoctoral intern/research asst. Stanford U. Med. Center, 1977-81; psychologist No. Calif. Psychol. Services, San Francisco, 1979-82; pvt. practice clin. psychology, Cupertino, Calif. Contbr. articles to profl. jours. Mem. Am. Psychol. Assn., Calif. Psychol. Assn., Biofeedback Soc. Am., Am. Assn. Study Headache. Subspecialties: Neuropsychology; Psychobiology. Current work: Research in psychobiology and neuropsychology. Office: 20333 Stevens Creek Blvd Cupertino CA 95014

ALLEN, ROBERT CARTER, pathology educator; b. Natick, Mass., Feb. 5, 1930; s. Roy Henry and Helen Louise (Carter) A.; m. Carol Lillian Chase, Oct. 26, 1956; children: Robert Carter, Roger, Chase. B.S., U. Vt., 1952, M.S., 1955; Ph.D., Va. Poly. Inst., 1959. Postdoctoral fellow Jackson Meml. Lab., Bar Harbor, Maine, 1960-62; sr. scientist Oak Ridge (Tenn.) Nat. Lab., 1962-68, Oak Ridge Tech. Enterprises Corp., 1968-73; assoc. prof. pathology Med. U. S.C., Charleston, 1973-76, prof. pathology, 1981—, prof., chmn. lab. animal medicine, 1976—; cons. Corning Med., Sullivan Park, N.Y., 1979, Upjohn, Kalamazoo, Mich, 1980, E.I. duPont de Nemours & Co., Wilmington, Del., 1982. Author, sr. editor: Electrophoresis, 1981, Marker Proteins of Immflamation, 1982, Electrophoresis and Isoelectric Focusing, 1974; assoc. editor: Electrophoresis, 1980—. Mem. Electrophoresis Soc. (pres. 1980-82, sec.-treas. 1980—). Clubs: Sertoma (Charleston, S.C.); Hobcaw Yacht (Mount Pleasant, S.C.) (mem. bd. 1975-78). Subspecialties: Genetics and genetic engineering (medicine); Pathology (medicine). Current work: Development of methodology for the study of genetic polymorphisms and molecular expressions of disease. Home: 501 Palm Blvd Isle of Palms SC 29451 Office: Med U SC 171 Ashley Ave Charleston SC 29425

ALLEN, ROBERT CHARLES, physician, biochemist; b. Pueblo, Colo., Aug. 12, 1945; s. Noel Charles and Gladys Louise (Puig) A.; m. Joan Marie Lindsay, June 26, 1976; 1 son, Robert Lindsay. B.S., Southeastern La. U., 1967; Ph.D., Tulane U., 1973, M.D., 1977, Asso. in biochemistry, 1973-76. Diplomate: La. Bd. Med. Examiners. Commd. capt. M.C. U.S. Army, 1977, advanced through grades to Maj., 1980; intern Brooke Army Med. Center, Ft. Sam Houston, Tex., 1977-78; infectious disease officer U.S. Army Inst. Surg. Research, 1978—. Contbr. articles to profl. jours. Served with U.S. Army, 1968-70. Recipient Clin. Pathology award La. Pathology Soc., 1977; Leah Seidman award Tulane U., 1977. Mem. Am. Assn. Immunology, Am. Chem. Soc., AMA, Am. Soc. Biol. Chemistry, Am. Soc. Photobiology, Am. Soc. Microbiology, Biophys. Soc., Surg. Infection Soc., Reticuloendothelial Soc., Sigma Xi. Subspecialties: Infectious diseases; Biochemistry (medicine). Current work: Excited state and radical chemistry; chemiluminescence and single photon counting; biological redox mechanisms; information aspects of humoral immune systems; phagocyte microbicidal mechanisms. Discovered the phenomenon of native granulocytic leukocyte luminescence.

ALLEN, THERESA OHOTNICKY, research psychologist; b. Torrington, Conn., Apr. 27, 1948; d. Frank Richard and Helen Theresa (Drozdenko) Ohotnicky; m. Thomas Atherton Allen, Aug. 12, 1972; children: Melanie Atherton, Abigail Baldwin. B.A., U. Conn., 1970; M.S., Villanova U., 1975; Ph.D., Duke U., 1978. Fellow in neurol. sci. U. Pa., Phila., 1978-80, fellow in psychology, 1980-81, research assoc. dept. psychology, 1981—; research specialist dept. elec. and computer engring. Drexel U., 1983—. Contbr. articles in field to profl. jours. Mem. AAAS, Animal Behavior Soc., Soc. Study of Reproduction, Soc. for Neurosci., Sigma Xi. Subspecialties: Neurobiology; Ethology. Current work: Researcher in devel. computer methods to analyse neurobiol. data. Office: University of Pennsylvania Dept Psychology 3815 Walnut St Philadelphia PA 19104

ALLEN, THOMAS CORT, JR., plant pathologist; b. Madison, Wis., Oct. 28, 1931; s. Thomas Cort and Esther Elsa (Liening) A.; m. Donna Jeanne Hillebrand, June 27, 1953; 1 dau., Kathleen Deanne Gillispie. B.S., U. Wis., Madison, 1953; Ph.D., U. Calif., Davis, 1956. Head microbiology sect. Agrl. Research Lab., Stauffer Chem. Co., Mountain View, Calif., 1958-62; asst. prof. dept. botany and plant pathology Oreg. State U., Corvallis, 1962-66, assoc. prof., 1966-76, prof., 1976—; community edn. art instr. Linn-Benton Community Coll., 1972—. Pres. Corvallis Arts Center, 1963-64. Served as 1st lt. U.S. Army, 1956-58. Allied Chem. Co. fellow, 1957; NSF fellow, 1964; NATO sr. fellow, 1970. Fellow Royal Hort. Soc. London; mem. Am. Phytopath. Soc., Electron Microscopy Soc. Am., Oreg. Electron Microscopists (pres.), Watercolor Soc. Oreg. (pres.), Sigma Xi (chpt. pres., regional lectr.). Republican. Congregationalist. Club: Soc. Preservation Barbershop Quartet Singing in Am. Lodge: Rotary. Subspecialties: Plant virology; Plant pathology. Current work: Creation of virus-free plants in tissue culture, electronmicroscopy, ultrastructural pathology, ornamentals and potatoes. Patentee in field. Home: 3989 SW Fairhaven Ct Corvallis OR 97333 Office: Dept Botany Oreg State U Corvallis OR 97331

ALLEN, VERNON L(ESLIE), psychology educator; b. Lineville, Ala., June 6, 1933; s. Harvey N. and Hassie S. A.; m. Patricia S. Shumcke, Dec. 31, 1956; children: Derek R., Craig R. B.A., U. Ala., 1955; M.S., Tufts U., 1958; Ph.D., U. Calif.-Berkeley, 1962. Asst. prof. U. Wis., Madison, 1963-66, assoc. prof., 1966-69, prof. psychology, 1969—. Editor: Psychological Factors in Poverty, 1970, Children as Teachers, 1976, Cognative Learning in Children, 1976, Role Transitions, 1983. Served to 1st lt. U.S. Army, 1956-58. Postdoctoral fellow W.F.M. Ltd., Stanford U., 1962-63; Fulbright fellow to, Eng., 1969-70; fellow Netherlands Instn. for Advanced Study in Humanities and Social Sci., Wassenaar, 1979-80. Fellow Am. Psychol. Assn.; mem. Am. Sociol. Assn., Soc. for Psychol. Study of Social Issues, Soc. for Exptl. Social Psychology, Brit. Psychol. Soc. Subspecialty: Social psychology. Current work: Small groups, social roles, applied social psychology. Office: Univ Wisconsin Psychology Bldg Madison WI 53706

ALLEN, WILLIAM HAND, electrical engineer; b. Sharon, Pa., Apr. 2, 1952; s. William Henry and Barbara Frances (Yazvac) A.; m. Deana J. Bradstock, Oct. 14, 1977; children: William Yazvac, Daniel Burker. B.E.E., Ga. Inst. Tech., 1975; M.A. in Bus. Adminstrn, Nat. U., San Diego, 1981; postgrad., Johns Hopkins U., 1981—. Registered profl. engr., Md., Calif. Nuclear engr. Balt. Gas & Electric Co., 1981-82; sr. elec engr. Mueller Assocs. Inc., Balt., 1982-83; prin. elec. engr. H.A. Schlenger & Assocs., Balt., 1983—. Mem. Cromwell Valley Community Assn., Towson, Md., 1982—. Served to lt. U.S. Navy, 1975-81; now officer USNR. Mem. IEEE, Illuminating Engring. Soc. Republican. Unitarian Universalist. Subspecialties: Electrical engineering; Solar energy. Current work: Power distribution, lighting, fire detection and alarm, security, solar and energy management systems for commercial, industrial, and institutional facilities. Home: 924 Beaverbank Circle Towson MD 21204 Office: H A Schlenger & Assocs Baltimore MD 21201

ALLER, JOHN EARL, mechanical engineer; b. East St. Louis, Ill., Dec. 1, 1946; s. Harold L. and Ruth E. (Sheppard) A.; m. Janis B. Wehrle, June 3, 1967; children: Thomas, Bethany. B.S. in Mech. Engring. with honors, U. Ill.-Urbana, 1969, M.S., 1971. Registered profl. engr. Engr. Nat. Distillers & Chem. Co., Tuscola, Ill., 1971-77, engring. mgr., Cin., 1977—. Contbr. articles in field to profl. publs. Nat. Merit scholar, 1964-69. Mem. ASME, AM. Soc. Metals. Mem. Ch. of Christ. Subspecialties: Mechanical engineering; Materials (engineering). Current work: Design and applicatons of high pressure equipment; application and technology of materials in engineering design; failure analysis. Office: 11499 Chester Rd Room 502 Cincinnati OH 45246

ALLER, LAWRENCE HUGH, astronomer; b. Tacoma, Sept. 24, 1913; s. Leslie E. and Lenabelle (Davis) A.; m. Rosalind Duncan Hall, Apr. 24, 1941; children—Hugh Duncan, Raymond Donald, Gwendolyn Jean. A.B., U. Calif., 1936; M.A., Harvard, 1938, Ph.D., 1942. Jr. fellow Soc. Fellows, Harvard U., 1939-42, instr. physics, 1942-43; research physicist U. Calif. at Berkeley, 1943-45; asst. prof. astronomy Ind. U., 1945-48; asso. prof. U. Mich., 1948-54, prof. astronomy, 1954-62; vis. prof. Australian Nat. U., Canberra, 1960-61, U. Toronto, 1961-62; prof. astronomy U. Calif. at Los Angeles, 1962—; Guest investigator Dominion Astrophys. Obs., 1951, Mt. Wilson Obs.; vis. prof. U. Sydney, U. Tasmania, 1968-69, U. Queensland, 1977-78; guest investigator C.S.I.R.O. Australia, 1968-69, 71, 77-78, Anglo-Australian Obs., 1978. Author: Atoms, Stars and Nebulae, 2d edit, 1971, Astrophysics, 1954, Gaseous Nebulae, 1956, Abundances of the Elements, 1961, Atmospheres of the Sun and Stars, 1963. Mem. Internat. Astron. Union, Royal Astron. Soc., Am. Acad. Arts and Scis., Nat. Acad. Scis. Subspecialties: Optical astronomy. Current work: Determination of chemical compositions of nebulae stars; physical progress in attenuated nebulae plasmas. Home: 18118 W Kingsport Dr Malibu CA 90265 All actions must be judged by their ultimate effects and one must weigh the implications of any deed against its possible consequences. Above all, one must remember that destruction stalks the unwary and the gullible. The tiger is always waiting in the shadows. Experience is truth. Be not deceived by the preachments of self-annointed experts. Investigate for yourself. Gullibility is the trademark of the fool; conformity is often only the emblem of mediocrity.

ALLEY, KEITH EDWARD, anatomist; b. Palm Springs, Calif., June 27, 1943; s. William and Kathryn (Carmody) A.; m. Jean Lane Doyle, June 18, 1966; 1 son, Colin. D.D.S., U. Ill., 1968, Ph.D., 1972. Postdoctoral trainee U. Ill. Coll. Medicine, Chgo., 1968-72; spl. fellow in physiology U. Iowa, Iowa City, 1972-74; asst. prof. anatomy Case Western Res. U., Cleve., 1974-80, assoc. prof. oral biology, 1980—. Contbr. articles to profl. jours., chpts. to books. NIH grantee, 1977-82. Mem. Soc. Neurosci., AAAS, Internat. Assn. Dental Research. Subspecialties: Comparative neurobiology; Oral biology. Current work: Development of neuronal systems responsible for motor activity. Home: 2763 Derbyshire Rd Cleveland Heights OH 44106 Office: 2123 Abington Rd Cleveland OH 44106

ALLEY, THOMAS LEROY, mechanical engineer; b. Winfield, Kans., Dec. 5, 1936; s. Glenn and Reba Gertrude (Notestine) A.; m. Mary Anne Richard, Apr. 17, 1971; children: Margaret Theresa, Joseph Glenn. B.S., Kans. State U., 1958; M.S., Ohio State U., 1960, Ph.D., 1964; M.S., U. So. Calif., 1970, 1976. Registered profl. engr., Ohio. Mem. tech. staff Aerospace Corp., El Segundo, Calif., 1964—. Mem. ASME. Subspecialty: Mechanical engineering. Home: 6304 Sattes Dr Rancho Palos Verdes CA 90274 Office: 2350 E El Segundo Blvd El Segundo CA 90245

ALLGOWER, EUGENE LEO, mathematics educator; b. Chgo., Aug. 11, 1935; s. Eugene and Martha (Kettner) A.; m. Solveig Odland, Aug. 30, 1958; 1 son, Chris Eugene. B.S., Ill. Inst. Tech., 1957, M.S., 1959, Ph.D., 1964. Instr. math. Ill. Inst. Tech., Chgo., 1960-62, U. Ariz., Tucson, 1962-64; asst. prof. math. Sacramento State Coll., 1964-65; assoc. prof. math. U. Tex.-El Paso, 1965-66; asst. prof. math. Colo. State U., Ft. Collins, 1966-71, assoc. prof., 1971-75, prof., 1975—; vis. prof. ETH, Zurich, Switzerland, 1972-73, U. Bonn, W.Ger., 1977-78, 80-81, U. Stuttgart, 1978, U. Hamburg, 1983; numerous invited lecures. Mem. Am. Math. Soc., Soc. Indsl. and Applied Math., Assn for Computing Machinery (sec.-trea. Rocky Mountain chpt. 1980—). Subspecialties: Numerical analysis; Applied mathematics. Current work: Continuation methods and simplicial algorithms. Office: Math Dept Colo State U Fort Collins CO 80523

ALLISON, DAVID COULTER, surgeon; b. Detroit, June 2, 1942; s. William David and Dorothy (Watson) A.; m. Elizabeth Ward, Sept. 19, 1970; children: Tiffany, Elizabeth, Mathew. M.D., U. Mich., 1967; Ph.D., U. Chgo., 1976. Diplomate: Am. Bd. Surgery. Intern U. Chgo., 1967-68; resident, 1974-78, practice medicine, specializing in surgery, Albuquerque, 1978—; faculty dept. surgery U. N.Mex., Albuquerque, 1978—; asst. chief surgery VA Med. Center, Albuquerque, 1978—. Contbr. articles to profl. jours. Served to capt. U.S. Army, 1968-70. VA grantee, 1981, 82. Fellow ACS; mem. Histochemistry Soc., Soc. Cell Kinetics. Subspecialties: Surgery; Cell biology (medicine). Current work: Cytometry of cancer and TCDD exposed cells to identify factors associated with tumor progression and cell damage. Office: Univ NM Med Sch 2211 Lomas Blvd NE Dept Surgery Albuquerque NM 87131

ALLMAN, JOHN MORGAN, biology educator; b. Columbus, Ohio, May 17, 1943; s. John Morgan and Edna Blanche (Danford) A.; m. EveLynn McGuinness, June 28, 1977. B.A., U. Va., 1965; Ph.D., U. Chgo., 1970. Postdoctoral fellow U. Wis., Madison, 1970-73; research asst. prof. Vanderbilt U., Nashville, 1973-74; asst. prof. dept. biology Calif. Inst. Tech., Pasadena, 1974-77, assoc. prof., 1977—. Sloan Found. fellow, 1974-76; USPHS Career devel. awardee, 1976-81; NSF grantee, 1977-80; NIH grantee, 1974—. Mem. Soc. Neurosci., Internat. Brain Research Orgn., Assn. Research in Vision and Ophthalmology, Internat. Primatological Soc., Am. Soc. Primatologists. Subspecialties: Neurophysiology; Comparative neurobiology. Current work: Functional organization of visual cortex in primates; primate brain evolution. Office: 216-76 Calif Inst Tech Pasadena CA 91125

ALLMAN, NORRIS C., nuclear engineer; b. Bklyn., May 30, 1952; s. Bernard and and Florence (Boxill) A.; m. Migail Igus, Apr. 24, 1976; 1 son, Ryan. B.S. in Physics, Poly. Inst. Bklyn., 1975. Engr. Pub. Service Co., Newark, 1976-77, lead engr., 1977-79, sr. engr., 1979-81, prin. staff radiation analyst, 1981—. Mem. Am. Nuclear Soc., ASME, Health Physics Soc. Subspecialty: Nuclear fission. Current work: Provide health physics calculation to estimate environmental radiation levels outside nuclear power plants. Home: 536 Pierson St Westfield NJ 07090 Office: 80 Park Plaza Suite T-16D Newark NJ 07101

ALLMANN, DAVID WILLIAM, biochemist, educator; b. Peru, Ind., May 20, 1935; s. Frederick Carl and Eunice (Vermillion) A.; m. Mary Ann Van Der Weele, Feb. 11, 1956; children: Victoria Lynn, Judith Ann. B.S., Ind. U.-Bloomington, 1958, Ph.D., 1964. Postdoctoral researcher Inst. Enzyme Research, Madison, Wis., 1964-66, asst. prof., 1966-70; assoc. prof. biochemistry Ind. U. Sch. Medicine, Indpls., 1970-80, prof., 1980—. Contbr. numerous articles, abstracts to profl. jours. Nat. Inst. Dental Research grantee, 1977-83. Mem. Am. Assn. Dental Research (sec./treas. Pharmacology Toxicology and Therapeutics Group 1981—), Am. Assn. Dental Schs. (pres. Nutrition Sect. 1980), Am. Inst. Nutrition, Am. Assn. Biol. Chemists. Clubs: Red Devil Rooters (pres. 1982), Band Booster (Indpls.) (pres. 1976). Subspecialty: Biochemistry (medicine). Current work: Determination of metabolic effects of Na F in vivo and in vitro. Office: Dept Biochemistry Ind U Sch Medicine 635 Barnhill Dr Indianpolis IN 46223 Home: 4101 Melbourne Rd Indianapolis IN 46208

ALLOCCA, JOHN ANTHONY, medical research scientist; b. Bklyn., Aug. 27, 1948; s. Frank and Dorothy (Aulicino) A.; m.; children: Jennifer, Jerry. A.A.S., SUNY-Farmingdale, 1972; B.A., SUNY-Old Westbury, 1975; M.S., Poly. Inst. N.Y., 1979; D.Sc., Pacific Western U., 1981. Adminstr. Hofstra U., 1967-71; psychotherapist Creedmore State Hosp., 1975-76; biomed. engr. Doll Research Inc., 1971-77; research scientist Albert Einstein Coll. Medicine, 1977-78; research cons. L.I. Coll. Hosp., 1979-80; research scientist, tech. dir. pulmonary labs. Mt. Sinai Med. Center, 1980-82; research scientist Langer Biomech. Group, Inc., 1983—. Mem. IEEE, Assn. Advancement Med. Instrumentation, AAAS, Am. Assn. Physicists in Medicine, N.Y. Acad. Scis. Subspecialties: Neurology; Biophysics (physics). Current work: Biomedical research, biophysics. Home: 234-05 133d Ave Rosedale NY 11422

ALM, ALVIN ARTHUR, forestry educator; b. Albert Lea., Minn., July 30, 1935; m. Martha Ann Peterson; children: Bonnie, Susan. B.S., U. Minn., 1961, M.S., 1965, Ph.D., 1971. Assoc. forester Forestry Cons. Services, Jackson, Mich., 1961-62; property appraiser Bur. Pub. Rds., Washington, 1965-66; research fellow U. Minn. Coll. Forestry, Cloquet, 1966—, prof. forestry, 1980—. Contbr. articles to profl. publs. Served with U.S. Army, 1954-56. N.W. Paper Found. fellow, 1970. Mem. Soc. Am. Foresters, Sigma Xi, Gamma Sigma Delta, Xi Sigma Pi. Lutheran. Subspecialty: Forestry. Current work: Silviculture; forest regeneration; site preparation. Home: 402 Dalewood Ave Cloquet MN 55720 Office: 175 University Rd Cloquet MN 55720

ALMODOVAR, LUIS RAUL, marine scientist, educator; b. San German, P.R., Jan. 19, 1930; s. Jose Pablo and Elisa A. Ph.D., Fla. State U., 1958. Teaching asst. Fla. State U., 1955-57; asst. prof. marine scis. U. P.R., Mayaguez, 1960-64, assoc. prof., 1964-67, prof., 1967—; cons. in field. Contbr. numrous articles to profl. jours. Served to sgt. USMC, 1951-53. Recipient Diploma De Reconocimiento U. Autonoma, Sto. Domingo, 1978; Guggenheim Found. fellow, 1969-70. Mem. Phycological Soc. Am., Brit. Phycological Soc., Internat. Phycological Soc., Sigma Xi, Phi Sigma, Tri Beta. Subspecialty: Ecology. Current work: Marine algae of tropical West Atlantic. Office: Dept Marine Scis U PR Mayaguez PR 00708

ALPEN, EDWARD LEWIS, biophysicist, educator; b. San Francisco, May 14, 1922; s. Edward Lawrence and Margaret Lilly (Shipley) A.; m. Wynella June Dosh, Jan. 6, 1945; children: Angela Marie, Jeannette Elise. B.S., U. Calif., Berkeley, 1946, Ph.D., 1950. Br. chief, then dir. biol. and med. scis. Naval Radiol. Def. Lab., San Francisco, 1952-68; mgr. environ. and life scis. Battelle Meml. Inst., Richland, Wash., 1968-69, asso. dir., then dir. Pacific N.W. div., 1969-75; dir. Donner Lab., U. Calif., Berkeley; also asso. dir. Lawrence Berkeley Lab., 1975—; prof. biophysics U. Calif., Berkeley, 1975—, prof. radiology, San Francisco, 1976—; mem. Nat. Council Radiol. Protection, 1969—; Mem. Gov. Wash. Council Econ. Devel., 1973-75; bd. dirs. Wash. Bd. Trade, 1973-76. Author papers, abstracts in field. Served to capt. USNR, 1942-46, 50-51. Recipient Navy Sci. medal, 1962, Disting. Service medal Dept. Def., 1963, Sustaining Members medal Assn. Mil. Surgeons, 1971; fellow Guggenheim Found., 1960-61; sr. fellow NSF, 1958-59. Fellow Calif. Acad. Scis.; mem. Bioelectomagnetics Soc. (pres. 1979-80), Am. Physiol. Soc., Radiation Research Soc., Soc. Exptl. Biology and Medicine, Biophys. Soc., Am. Philatelic Soc., Sigma Xi. Episcopalian. Subspecialty: Biophysics (physics). Current work: Radiation biology. Home: 1182 Miller Ave Berkeley CA 94708 Office: 466 Donner U Calif Berkeley CA 94720

ALPER, ALLEN MYRON, precision materials co. exec.; b. N.Y.C., Oct. 23, 1932; s. Joseph and Pauline (Frohlich) A.; m. Barbara Marshall, Dec. 20, 1959; children—Allen Myron, Andrew Marshall. B.S., Bklyn. Coll., 1954; Ph.D. (Univ. fellow, Dyckman Inst. scholar, Univ. Pres's. scholar), Columbia U., 1957. Sr. mineralogist Corning Glass Works, N.Y., 1957-59, research mineralogist, 1959-62, mgr. ceramic research, sr. research asso., 1962-69; with GTE Sylvania Inc. div. Sylvania Inc., Towanda, Pa., 1969—, chief engr., 1971-72, dir. research and engring., 1972-78, mgr. ops., from 1978; now pres. GTE Walmet, Royal Oak, Mich.; mem. Pa. Gov's Adv. Panel on Materials, 1971—; chmn. adv. com. Materials Research Lab., Pa. State U. Editor: Phase Diagrams: Materials Science and Technology, 1970, High Temperature Oxides, 1970-71; editorial bd.: High Temperature Sci. jour, 1969—, High Temperature Chemistry, 1973—, Materials Handbook, 1974—; editor: Materials Sci. and Tech. Series, 1972—; contbr. articles to profl. jours. Mem. exec. bd. Gen. Sullivan council Boy Scouts Am. N.Mex. Bur. Mines. grantee, 1954-57; also fellow. Fellow Am. Ceramic Soc., Geol. Soc. Am., Am. Inst. Chemists; mem. Brit. Ceramic Soc., Am. Soc. Metals, Am. Chem. Soc., Sigma Xi. Presbyterian. Club: Towanda Country. Subspecialties: Composite materials; High-temperature materials. Current work: Ceramic, carbides, nitrides, phase equilibria, new compounds, composites, new materials, high temperature materials (research, development and engineering. Patentee in field. Home: 880 Great Oaks Blvd Rochester MI 48063 Office: GTE Walmet Royal Oak MI 48068

ALPER, HOWARD, chemistry educator; b. Montreal, Que, Can., Oct. 17, 1941; s. Max and Frema (Weinstein) A.; m. Anne Elizabeth, June 4, 1966; children: Ruth, Lara. B.Sc. in Chemistry with honors, Sir George Williams U., Can., 1963; Ph.D., McGill U., 1967. NATO postdoctoral fellow, 1967-68; Asst. prof. SUNY-Binghamton, 1968-71, assoc. prof., 1971-74; assoc. prof. chemistry U. Ottawa, Ont., Can., 1975-77, prof., 1978—, chmn. dept., 1982—. Contbr. 155 articles to sci. jours. Recipient E.W.R. Steacie award Natural Scis. and Engring. Research Council Can., 1980. Mem. Am. Chem. Soc., Chem. Inst. Can. (award inorganic chemistry 1980), Chem. Soc. (London). Subspecialties: Organic chemistry; Inorganic chemistry. Current work: Metal complexes as catalysts, reagents, and intermediates in synthesis; phase transfer catalysis. Patentee in field (5). Office: 365 Nicholas St Ottawa ON Canada KIN 9B4

ALPERT, JOEL JACOBS, physician, educator; b. New Haven, May 9, 1930; s. Herman H. and Alice (Jacobs) A.; m. Barbara E. Wasserstrom, July 13, 1957; children—Norman, Mark, Deborah. A.B. with high honors, Yale U., 1952; M.D., Harvard U., 1956. Diplomate: Am. Bd. Pediatrics. Intern Children's Hosp. Med. Center, Boston, 1956-57, jr. asst. resident in medicine, 1957-58; sr. registrar St. Mary's Hosp. Med. Sch., London, 1958-59; chief resident for ambulatory services Children's Hosp. Med. Center, Boston, 1961-62, also fellow in medicine, 1961-62, asst. dir. family health care program, 1962-63, asst. in medicine, 1963-64, asso. in medicine, 1964-66; chief Family and Child Health Div., 1964-72, sr. asso. in medicine, 1966-73; cons. in child and family medicine, 1972—; research fellow in pediatrics Harvard U. Med. Sch., Boston, 1962-63; practice medicine specializing in pediatrics, Boston, 1964—; asst. dir. Family Health Care Programs Harvard U. Med. Sch., 1962-64; instr. in pediatrics, 1963-66, asso. in pediatrics, 1966-68, med. dir. Family Health Care Program, 1963-72, asst. prof. pediatrics, 1968-69, asso. prof., 1969-72, lectr. in pediatrics, 1972—; asso. in medicine Beth Israel Hosp., Boston, 1964-73; asso. pediatrician Boston Hosp. for Women, 1966—; lectr. in medicine Simmons Coll., 1964-72; prof., chmn. dept. pediatrics Boston U., 1972—, prof. dept. socio-med. sci., 1976—; dir. pediatric service Boston City Hosp., 1972—; vis. lectr. health services Harvard U. Sch. Public Health, 1976-79; cons. pediatrics Joseph P. Kennedy Meml. Hosp., 1972—, Carney Hosp., 1972—; mem. health services research study sect. HEW, 1968-72; cons. Bur. Health Services, 1972—, Bur. Health Hanpower, 1976—; vis. research asso. dept. econs. Northeastern U., 1975—; chmn. exec. com. Boston Poison Info. Center, 1976-78; chmn. adv. com. Mass. Poison Control System, 1979—. Author: (with F. H. Lovejoy, Jr.) A Handbook for Acute Childhood Poisoning, 1971, (with E. Charney) The Education of Physicians for Primary Care, 1973, (with others) Towards Changing the Medical Care System, 1974; contbr. numerous articles on poisoning, pediatrics, primary care and med. edn. to med. jours. Mem. adv. com. spl. edn. City of Winchester, Mass., 1969-79, chmn., 1978-79; mem. exec. com. Mass. Commn. on Children and Youth, 1972—. Served as capt. M.C. AUS, 1959-61. Harvard Sch. Public Health spl. fellow, 1972. Mem. Am. Pediatric Soc., Soc. Pediatric Research, Inst. Medicine, Nat. Acad. Scis., New Eng. Pediatric Soc., Am. Acad. Pediatrics, AAAS, Mass. Med. Soc., Ambulatory Pediatric Assn. (pres. 1969), Am. Assn. Poison Control Centers (dir. 1965-68). Clubs: Aesculapian, Yale of Boston, Harvard of Boston. Subspecialty: Pediatrics. Current work: Health care for low income families; primary care pediatrics. Home: 193 Marlborough St Boston MA 02116 Office: 818 Harrison Ave Boston MA 02118

ALPHIN, REEVIS STANCIL, pharmacologist, drug. co. adminstr.; b. Mt. Olive, N.C., Apr. 21, 1929; s. Fred and Carla (Dail) A.; m. Barbara Gilliam, Sept. 3, 1955; children: Robert Stancil, Carla Gilliam. B.A., U. N.C., 1951; M.A., Duke U., 1955; Ph.D., Med. Coll. Va., 1966. Pharmacologist Eli Lilly Co., Indpls., 1956-60; group mgr. pharmacology A.H. Robins Co., Richmond, Va., 1960-71, asso. dir. pharmacology, 1971—. Contbr. numerous articles to profl. publs. Served to capt. USAF, 1951-53. Mem. AAAS, N.Y. Acad. Scis., Am. Physiol. Soc., Am. Pharmacology Soc. Presbyterian. Lodge: Masons. Subspecialties: Molecular pharmacology; Gastroenterology. Current work: Research in area of gastrointestinal pharmacology. Patentee in field.

AL SAADI, ABDUL AMIR, clinical geneticist, cytogeneticist; b. Baqubah, Iraq, Oct. 20, 1935; s. Zahra J. and Majeed S. A.S.; m. Karen Eileen Svendsen, Nov. 20, 1961; children: Neda, Yasmin, Sami, Laith. B.Sc. magna cum laude, U. Baghdad, 1955; M.A., U. Kans., 1959; Ph.D., U. Mich., 1963. Research asso. U. Mich., 1963-65, asst. prof., 1965-70; chief genetics William Beaumont Hosp., Royal Oak, Mich., 1970—; dir. Sch. Histotech., 1976—; clin. prof. Wayne State U., 1973—, Oakland U., 1977—. Contbr. numerous articles to profl. jours., chpts. to books. Mem. Am. Soc. Human Genetics, Am. Soc. Cell Biology, Am. Assn. Cancer Research, Tissue Culture Assn., AAAS, Am. Thyroid Assn., Am. Fedn. Clin. Research. Subspecialties: Genetics and genetic engineering (medicine); Cell biology (medicine). Current work: Clinical genetics, role of chromosomal abnormalities in cancer. Home: 2325 Adare St Ann Arbor MI 48108 Office: William Beaumont Hosp Royal Oak MI 48072

AL-SARRAF, MUHYI, oncologist, educator, researcher; b. Baghdad, Iraq, Sept. 15, 1938; came to U.S., 1963, naturalized, 1971; s. Abdul Hussien K. and Neisemah J. (Jawad) Al-S.; m. Ellen Grace Connors, Aug. 29, 1976; children: Renee, Ramsey. M.B., Ch.B., Baghdad U., 1961. Diplomate: Am. Bd. Internal Medicine.; Lic. physician, Mich., Ill. Intern Providence Hosp., Washington, 1963-64; resident in internal medicine Grace Hosp., Detroit, 1964-67; fellow in oncology Wayne State U., Detroit, 1967-68; Detroit Inst. Cancer Research, 1967-68; instr. oncology Wayne State U., 1968-69, asst. prof., 1970-73, assoc. prof., 1973-81, prof., 1981—; mem. Met. Detroit Cancer Control Program, Comprehensive Cancer Center Met. Detroit; vis. prof. Med. City Teaching Hosp. and Med. Coll., Baghdad U., 1975, U. Okla. Health Scis. Center and Okla. Med. Research Found., 1975, Cancer Control Agy. B.C., Vancouver, 1975; chief oncology div., acting chief hematology div. King Faisal Specialist Hosp. and Research Center, Riyadh, Saudi Arabia, 1975-76; mem. diagnostic research adv. com. div. cancer biology and diagnosis Nat. Cancer Inst., NIH, HEW, 1980—; mem. staff Grace, Detroit Gen., Harper, Hutzel hosps.; cons. VA. Author: Immunosuppressive Therapy, 1971; contbr. articles and abstracts to profl. jours., chpts. to books. Fellow Royal Coll. Physicians Can., ACP.; mem. Wayne County Med. Soc., Mich. Med. Soc., Am. Soc. Clin. Oncology, Am. Assn. Cancer Research, Detroit Physiol. Soc., Am. Assn. Cancer Edn., Am. Soc. Preventive Oncology (founding mem.), S.W. Oncology Group, Radiation Therapy Oncology Group. Subspecialties: Cancer research (medicine); Chemotherapy. Current work: Teaching and tng. med. oncologists, clin. trials, tumor markers, . Home: 2952 Westview Ct N Bloomfield Hills MI 48013 Office: 3990 John R St Detroit MI 48201

ALSPAUGH, MARGARET ANN, immunologist, educator, researcher; b. Detroit, Apr. 7, 1938; d. Robert Eugene and Frances (Zemanek) A. B.A. in Biology, U. Hawaii, 1965; Ph.D. in Immunology (NIH fellow), Wash. State U., 1972. Asst. in biology Eastern Mich. U., 1958-59, asst. in chemistry, 1959-61; fellow Scripps Clinic and Research Found., 1972-77; asst. prof. medicine and microbiology La. State U., 1977-79, asso. prof, 1979-80, dir. Clin. Rheumatology Lab., 1977-80; asso. prof. medicine and pathology, asso. scientist Cancer Research Center, U. Mo., Columbia, 1980—, co-dir., 1980—; cons. for diagnostic Rheumatology Labs. Contbr. chpt., articles to profl. publs. NIH postdoctoral fellow, 1972-75; Am. Assn. Immunologists travel fellow, 1974; Arthritis Found. fellow, 1975-78. Mem. Am. Rheumatism Assn., Am. Assn. Immunologists, Am.Fedn. Clin. Research. Subspecialties: Immunology (medicine); Rheumatology. Current work: Research on role of discovered antigen-antibody systems in pathogenesis of forms of arthritis. Discovered antigen-antibody systems.

ALTAN, TAYLAN, mechanical engineer; b. Trabzon, Turkey, Feb. 13, 1938; came to U.S., 1962, naturalized, 1971; s. Seref and Sadife (Baysal) A.; m. Susan Barbara Borah, July 18, 1964; children: Peri, Aylin. Diploma Engr., Tech. U. Hannover, 1962; Ph.D., U. Calif., Berkeley, 1966. Research scientist E.I. duPont de Nemours & Co., Wilmington, Del., 1966-68; sr. research leader Battelle Columbus Lab., Columbus, Ohio, 1968—; adj. prof. Ohio State U., 1977—. Co-author: Forging Equipment, Materials and Practices, 1973; sr. author: Metalforming: Fundamentals and Applications, 1983. Mem. Internat. Prodn. Engring. Research Inst., Am. Soc. Metals (chmn. forging com. 1978—), Soc. Mfg. Engrs., ASME (chmn. prodn. div. 1976). Subspecialties: Materials processing; Solid mechanics. Current work: Forging, extrusion, rolling, sheet metal forming, computer aided design and manufacturing, materials processing, behavior of materials. Home: 1380 Sherbrook Pl Columbus OH 43209 Office: 505 King Ave Columbus OH 43201

ALTER, H. WARD, business executive; b. Taxila, Punjab, India, Dec. 26, 1923; s. Joseph C. and Marjorie (Ward) A.; m. Mary Elnora Newton, Sept. 27, 1944 (div. Dec. 1979); childen: Nancy, Dale, Ralph; m. Mary Crystal, Dec. 26, 1979. A.B., U. Calif.-Berkeley, 1943, Ph.D. in Chemistry, 1948. Research assoc. Gen. Electric Co., Schenectady, 1948-54, mgr. nuclear chemistry, 1954-57, various positions, Pleasanton, Calif., 1957-68, mgr. nucleonics lab., 1968-71; mgr. NTAO, 1971-74; pres. Terradex Corp., Walnut Creek, Calif., 1974—. Served in U.S. Army, 1943-45. Recipient G.E.Coffin award Gen. Electric Co., 1950. Fellow Am. Nuclear Soc.; mem. Am. Chem. Soc. (councillor 1950), Geochem. Soc., Phi Beta Kappa, Sigma Xi. Republican. Subspecialties: Nuclear fission; Geochemistry. Current work: Inorganic solution chemistry, mass spectrometry, nuclear physics, thermodynamics. Patentee in field (15). Home: 410 N Civic St #308 Walnut Creek CA 94596 Office: 460 N Wiget Ln Walnut Creek CA 94598

ALTIERO, NICHOLAS JAMES, engineering mechanics educator, researcher, consultant; b. Youngstown, Ohio, Sept. 22, 1947; s. Nicholas James and Fanny E. (Parise) A.; m. Amy Jean Johnson, Nov. 4, 1978; 1 dau., Elizabeth Claire. B.S. in Aero. Engring, U. Notre Dame, 1969, M.S., U. Mich., 1970, M.A. in Math, 1971, Ph.D. in Aero. Engring, 1974. Research asst. U. Notre Dame, 1968-69; research asst. U. Mich., 1969-74; asst. prof. Mich. State U., 1975-79, assoc. prof., 1979—; cons. U.S. Bur. MInes. Contbr. articles in field. Fulbright Hays fellow, 1981; Alexander von Humboldt fellow, 1982. Mem. Am. Acad. Mechanics, ASME, Internat. Soc. Computational Methods, Soc. Engring. Sci., Sigma Xi, Tau Beta Pi. Subspecialties: Solid mechanics; Numerical analysis. Current work: Computational methods in solid mechanics, fracture mechanics, structural analysis. Home: 2696 Lake Lansing Rd East Lansing MI 48823 Office: Michigan State University 359 Engineering Bldg East Lansing MI 48824

ALTMAN, LEONARD, allergist, educator; b. Fresno, Calif., Sept. 1, 1944; s. Martin and Ida (Sharnoff) A.; m. Gaylene Bouska, Dec. 26, 1970; children: Jonathan D., Matthew C. B.A., U. Pa., 1965; M.D. cum laude, Harvard U., 1969. Diplomate: Am. Bd. Internal Medicine, 1975, Am. Bd. Allergy and Immunology, 1980. Intern U. Wash. Affiliated Hosps., Seattle, 1969-70, resident in medicine, 1970-71; chief med. resident and instr. medicine Harborview Med. Ctr., 1974-75, asst. prof. medicine div. allergy and infectious diseases dept. medicine, 1975-79, assoc. prof. medicine, 1979—, chief allergy div., 1979—; research assoc. immunology sect. lab. Microbiology and Immunology Nat. Inst. Dental Research, NIH, Bethesda, Md., 1971-73, sr. research assoc., 1973-74. Contbr. articles to profl. jours. Served to lt. comdr. USPHS, 1971-74. Fellow Am. Acad. Allergy, ACP; mem. Am. Assn. Immunologists, Western Soc. Clin. Research, AAAS, Am. Fedn. Clin. Research, Reticuloendothelial Soc., Puget Sound Allergy Soc. (pres. 1977), Infectious Disease Soc. Am., King County Med. Soc., Wash. State Med. Assn. Subspecialties: Immunology (medicine); Allergy. Current work: Phagocyte function, lung injury, leukocyte chemotaxis. Home: 1015 Belmont Pl E Seattle WA 98102 Office: Dept Medicine U Wash Seattle WA 98195

ALTSCHULER, BRUCE ROBERT, dentist, air force officer; b. Bklyn., Feb. 17, 1947; s. Frank Philip and Sarah Gertrude (Cloder) A.; m. Ruth Phyllis Gass, Oct. 27, 1974; children: Joan Ellen, Wendy Karen. B.A., Bklyn. Coll., 1967; D.D.S., Temple U., 1971. Commd. capt. U.S. Air Force, 1971, advanced through grades to lt. col., 1980; chief dental laser holography Dental Investigations Service (USAF Sch. Aerospace Medicine), Brooks AFB, Tex., 1976-80, chief dental computer-laser tech., 1980-82; clin. asst. prof. dental diagnostic sci. U. Tex. Health Sci. Center Dental Sch., San Antonio, 1976-82; chief advanced systems research Info. Processing Tech. Br., Avionics Lab., Wright Patterson AFB, Ohio, 1982—; Mem. dental x-ray equipment subcom. Am. Nat. Standards Inst., Bethesda, Md., 1980—. Bd. dirs. Am. Cancer Soc., San Antonio, 1980-82; spl. awards judge Alamo Regional Sci. Fair, San Antonio, 1980-82. Recipient Research and Devel. award USAF, 1981, Clin. award Tex. Dental Assn., 1982. Mem. ADA, Am. Assn. Dental Research, N.Y. Acad. Scis., Soc. Photo-Optical Instrumentation Engrs., Am. Soc. Photogrammetry, Sigma Xi. Jewish. Subspecialties: Laser dentistry; Optical signal processing. Current work: Real-time 3-D topographic mapping and 3-D machine perception using electro-optics, unconventional computer architectures and signal processing for space physiology, machine intelligence, dentistry and avionics. Patentee dental x-ray alignment system, 1977, optical surface topography mapping system, 1978, topographic comparator, 1981. Home: 121 Meehan Dr Dayton OH 45431 Office: Advanced Systems Research Group AFWAL/AAAT-3 Wright-Patterson AFB OH 45433

ALTSCHULER, MARTIN DAVID, med. physicist, educator; b. Bklyn., Feb. 25, 1940; s. Frank Philip and Sarah Gertrude (Cloder) A.; m. Susan Jane, Sept. 2, 1962; children: Steven Jeffrey, Daniel Lewis, Rachel Lyra. B.S. summa cum laude, Poly. Inst. Bklyn., 1960; M.S., Yale U., 1961, Ph.D., 1964. Scientist High Altitude Obs., Nat. Center Atmospheric Research, Boulder, Colo., 1965-76; assoc. prof. adj. dept. astrogeophysics U. Colo., Boulder, 1966-76; assoc. prof. computer sci. SUNY-Buffalo, 1976-81; assoc. prof. radiation therapy Sch. Medicine, U. Pa., 1981—; fellow, mem. summer faculty Air Force Office Sci. Research, 1981. Contbr. numerous articles to profl. jours. Recipient Career Devel. award NIH, 1977-82. Mem. Am. Phys. Soc., Am. Astron. Soc., Internat. Astron. Union, Am. Assn. Physicists in Medicine, Soc. Photo-Optical Instrumentation Engrs., IEEE. Subspecialties: Mathematical software; Biomedical engineering. Current work: Three-dimensional mapping of human body: interior: CAT scanning and x-rays for cardiology, brachytherapy and treatment planning; exterior: laser mapping for anthropometry. Patentee topographic comparator, brachytherapy algorithms. Office: Hosp U Pa Dept Radiation Therapy 3400 Spruce Philadelphia PA 19104

ALTURA, BURTON MYRON, physiologist, educator; b. N.Y.C., Apr. 9, 1936; s. Barney and Frances (Dorfman) A.; m. Bella Tabak, Dec. 27, 1961; 1 dau., Rachel Allison. B.A., Hofstra U., 1957; M.S., N.Y. U., 1961, Ph.D. (USPHS fellow), 1964. Teaching fellow in biology N.Y. U., N.Y.C., 1960-61; instr. exptl. anesthesiology Sch. Medicine, 1964-65, asst. prof., 1965-66; asst. prof. physiology and anesthesiology Albert Einstein Coll. Medicine, N.Y.C., 1967-70, asso. prof., 1970-74, vis. prof., 1974-76; prof. physiology SUNY Downstate Med. Center, Bklyn., 1974—; research fellow Bronx Mcpl. Hosp. Center, 1967-76; mem. spl. study sect. on toxicology Nat. Inst. Environ. Health Scis., 1977—; mem. Alcohol Biomed. Research Rev. Com., Nat. Inst. Alcohol Abuse and Alcoholism, 1978—; cons. NSF, Nat. Heart, Lung and Blood Inst., CUNY, Miles Inst., Upjohn Co., Bayer AG, Zyma SA. Author: Microcirculation, 3 vols., 1977-80, Vascular Endothelium and Basement Membranes, 1980, Ionic Regulation of the Microcirculation, 1980; editor-in-chief: Physiology and Patho-physiology Series, 1976—, Microcirculation, 1980—; mem. editorial bd.: Jour. Circulatory Shock, 1973—, Advances in Microcirculation, 1976—, Jour. Cardiovascular Pharmacology, 1977—, Prostaglandins and Medicine, 1978—, Substance and Alcohol Actions/Misuse, 1979—; asso. editor: Jour. of Artery, 1974—, Microvascular Research, 1978—, Agents and Actions, 1981—; contbr. over 450 articles to profl. jours. Recipient Research Career Devel. award USPHS, 1968-72; travel awards NIH, 1968, Am. Soc. Pharm. and Exptl. Therapeutics, 1969; NIH grantee, 1968—; NIMH grantee, 1974—; Nat. Inst. Drug Abuse grantee, 1979—. Fellow Am. Heart Assn. (mem. council on stroke 1973—, council basic sci. 1969—, council on thrombosis 1971—, council on circulation 1978—, council on high blood pressure 1978—, cardiovascular A study sect. 1978—), Am. Coll. Nutrition; mem. Microcirculatory Soc. (past mem. exec. council, mem. nominating com. 1973-74), Am. Physiol. Soc. (mem. circulation group 1971—, public info. com. 1980—), Soc. Exptl. Biology and Medicine (editorial bd. 1976—), AAUP, Am. Public Health Assn., Am. Chem. Soc. (div. medicinal chemistry), Am. Soc. Pharm. and Exptl. Therapeutics, Endocrine Soc., Harvey Soc., Am. Coll. Toxicology, Research Soc. on Alcoholism, Am. Thoracic Soc., Soc. for Neurosci., Shock Soc. (founding), Am. Fedn. Clin. Research, AAAS, European Conf. Microcirculation, Internat. Anesthesia Research Soc., Internat. Soc. Thrombosis and Haemostasis, Internat. Soc. Biorheology, Soc. Environ. Geochemistry and Health, Soc. Neurosci., Reticuloendothelial Soc., Gerontol. Soc., Internat. Platform Assn., Am. Inst. Biol. Sci., Assn. Gnotobiotics, Am. Microscopical Soc., Am. Soc. Zoologists, Am. Soc. Cell Biology, N.Y. Acad. Scis., Am. Public Health Assn., N.Y. Heart Assn., Sigma Xi. Subspecialties: Physiology (medicine); Pharmacology. Current work: Excitation contraction coupling of vascular smooth muscle, microcirculation; role of magnesium in cardiovascular patho-physiology; cardiovascular research. Office: 450 Clarkson Ave Brooklyn NY 11203

ALVARES, ALVITO PETER, pharmacologist, educator, science researcher; b. Bombay, India, Dec. 25, 1935; came to U.S., 1958, naturalized, 1971; s. Amancio and Diva A.; m. Joy Ann Schmidt, Aug. 31, 1969; children: Christopher, Kevin. B.Sc., U. Bombay, 1955, 1957; M.S. in Biochemistry, U. Detroit, 1961; Ph.D. in Pharmacology, U. Chgo., 1966. Sr. research biochemist Burroughs Wellcome & Co., 1967-70; research asso. Rockefeller U., N.Y.C., 1970-71, asst. prof., 1972-75, asso. prof., 1975-77; asso. prof. dept. pharmacology Uniformed Services U. Health Scis., Bethesda, Md., 1977-78, prof., 1978—. Mem. editorial bd.: Pharmacology, Clin. Pharmacology and Therapeutics; contbr. numerous articles to sci. jours., chpts. in books. Recipient career devel. award NIH, 1975-77; Irma T. Hirschl scholar, 1975-77. Mem. Am. Soc. Pharmacology and Exptl. Therapeutics, Am. Soc. Biol. Chemists, Am. Soc. for Clin. Pharmacology and Therapeutics, Soc. Toxicology, N.Y. Acad. Scis. Subspecialties: Molecular pharmacology; Toxicology (medicine). Current work: Professor of pharmacology; research in toxicology; teach medical and graduate students; environmental toxicological research. Office: Dept Pharmacology Uniformed Services U 4301 Jones Bridge Rd Bethesda MD 20814

ALVARES, OLAV FILOMENO, dentist, dental educator; b. Bombay, Maharashtra, c4India, Nov. 1, 1939; U.S., 1961, naturalized, 1980; s. Amancio Bernadino and Diva Noemia (Vaz) A.; m. Dorthea Anne Johnson, July 19, 1976; children: Bryan Olav, Stacy Marie. D.D.S., U. Bombay, India, 1960; M.S., U. Detroit, 1963; Ph.D., U. Ill.-Chgo., 1971. Instr. U. Ill., Chgo., 1965-68, asst. prof. oral pathology, 1972-74; research assoc. prof. oral biology U. Wash., Seattle, 1974-81; assoc. prof. periodontics U. Tex., San Antonio, 1981—; vis. prof. dentistry U. Benin, Nigeria, 1981. Contbr. numerous sci. articles to profl. publs. Mem. spl. grants review com. NIH, Bethesda, Md., 1981-85; sec. oral biology sect. Am. Assn. Dental Schs., Washington, 1981-1983. Recipient Teaching award U. Wash. Dental Sch., 1975, 78, 79, 81; NIH grantee, 1969; NIH research career devel. awardee, 1976-81. Fellow Am. Acad. Oral Pathology; mem. Internat. Assn. Dental Research, Sigma Xi. Roman Catholic. Subspecialties: Pathology (medicine); Periodontics. Current work: Teaching undergraduate and postgraduate students; research in area of nutrition as it relates to oral health and diseases. Office: Univ Tex 7703 Floyd Curl Dr San Antonio TX 78284

ALVAREZ, LUIS W., physicist; b. San Francisco, June 13, 1911; s. Walter C. and Harriet S. (Smyth) A.; m. Geraldine Smithwick, 1936; children—Walter, Jean; m. Janet L. Landis, 1958; children—Donald and Helen. B.S., U. Chgo., 1932, M.S., 1934, Ph.D., 1936, Sc.D., 1967; Sc.D., Carnegie-Mellon U., 1968, Kenyon Coll., 1969, Notre Dame U., 1976, Ain Shams U., Cairo, 1979. Research asso., instr., asst. prof., asso. prof. U. Calif., 1936-45, prof. physics, 1945-78, prof. emeritus, 1978—; asso. dir. Lawrence Radiation Lab., 1954-59, 75-78; radar research and devel. Mass. Inst. Tech., 1940-43, Los Alamos, 1944-45; dir. Hewlett Packard Co. Recipient Collier Trophy, 1946; Medal for Merit, 1948; John Scott medal, 1953; Einstein medal, 1961; Nat. Medal of Sci., 1964; Michelson award, 1965; Nobel prize in physics, 1968; Wright prize, 1981; named Calif. Scientist of Year, 1960; named to Nat. Inventors Hall of Fame, 1978. Fellow Am. Phys. Soc. (pres. 1969); mem. Nat. Acad. Scis., Nat. Acad. Engring., Am. Philos. Soc., Am. Acad. Arts and Scis., Phi Beta Kappa, Sigma Xi; asso. mem. Institut D'Egypte. Current work: Experimental and theoretical aspects of the major biological paleo-extinctions; stabilized optical devices; particle physics.

ALVAREZ, WALTER, geology educator; b. Berkeley, Calif., Oct. 3, 1940; s. Luis W. and Geraldine (Smithwick) A.; m. Mildred M Millner, May 8, 1965. B.A., Carleton Coll., 1962; Ph.D., Princeton U., 1967. Sr. geologist Am. Overseas Petroleum Ltd., The Hague, Netherlands, 1967-70, Tripoli, Libya, 1967-70; research asso. Lamont-Doherty Geol. Obs., Columbia U., Palisades, N.Y., 1971-77; NATO fellow British Sch. Archaeology, Rome, 1970-71; asst. prof. geology Dept. Geology & Geophysic U. Calif.-Berkeley, 1977-79, assoc. prof., 1979-81, prof., 1981—. Fellow Geol. Soc. Am.; mem. Am. Geophys. Union, AAAS, Societa Geologica Italiana, Sigma Xi. Subspecialties: Tectonics; Geophysics. Current work: Mass extinctions; paleomagnetism; plate tectonics; apennine structure, stratigraphy, tectonics; mediterranean tectonics. Address: Dept Geology and Geophysics Univ Calif Berkeley CA 94720

ALVES, LEO MANUEL, plant physiologist; b. Phila., May 21, 1945; s. Manuel Louis and Rose Catherine (Fargoniere) A. B.S., St. Norbert Coll., 1968; Ph.D., U. Chgo., 1975. Asst. prof. Lincoln U., Pa., 1975-76; research plant pathologist Eastern Regional Research Center, Agrl. Research Service, U.S. Dept. Agr., Phila., 1976-78, collaborator, 1978—; research asso., asst. prof. Manhattan Coll., Bronx, N.Y., 1978. Contbr. articles in field to profl. jours. Mem. Am. Phytopath. Soc.,

Am. Soc. Hort. Sci., Am. Soc. Plant Physiologists, Japanese Soc. Plant Physiologists, Sigma Xi. Club: Mensa. Subspecialties: Plant physiology (biology); Plant pathology. Current work: Researcher in exploring the role of sesquiterpenoid stress metabolites in the mediation of expression of disease resistance in the plants of the Solanaceae. Office: Manhattan College Plant Morphogenesis Laboratory Bronx NY 10471

AMADOR, JOSE MANUEL, plant pathologist, consultant; b. Cuba, Mar. 3, 1938; s. Louis Felipe and Blanca Rosa (Muniz) A.; m. Silvia G. Garcia, Nov. 25, 1965; children: Silvia, Marian, Daniel. B.S., La. State U., [illegible], M.S., [illegible], Ph.D., [illegible]. Extension plant pathologist Tex. Agrl. Extension Service, Tex. A&M U., 1964—; cons. internat. crops. Recipient Tex. award for superior service, 1980. Mem. Am. Phytopath. Soc. Roman Catholic. Subspecialties: Plant pathology; Integrated pest management. Current work: Extension plant pathology. Home: 1400 Yucca St McAllen TX 78501 Office: 2401 E Hwy 83 Weslaco TX 78501

AMAR, AMAR-DEV, productions/operations systems educator; b. Bhakkar, Punjab, India; came to U.S., 1972, naturalized, 1981; s. Prem Datt Shakir and Kaushlaya (Khirbat) Shakir; m. Sneh Lata Chopra, Mar. 16, 1975; children: Harpriye Amar Juneja, Januj Amar Juneja. B.S., Panjab U., 1969; M.S., Mont. State U., 1973; M.B.A., Baruch Coll., 1980; Ph.D., CUNY, 1980, M.Phil., 1980. Engr.-in-charge Orisun Machine Tools, Chandigarh, India, 1966-67; asst. engr. Teledyne Pacific Indsl. Controls, Oakland, Calif., 1972; indsl. design engr. Vornado/Store Decorating Co., Garfield, N.J., 1973-76; adj. asst. prof. Baruch Coll., N.Y.C., 1978-81; project assoc. Research Found. CUNY, N.Y.C., 1980-82; asst. prof. prodns./ops. systems Montclair State Coll., Upper Montclair, N.J., 1975-83, dir., 1981-83; assoc. prof. prodn./ops. mgmt. and indsl. relations Seton Hall U., South Orange, N.J., 1983—; cons. in field. Mem. Inst. Indsl. Engrs. (sr.), Inst. Mgmt. Scis., Ops. Research Soc. Am. Subspecialties: Distributed systems and networks; Industrial engineering. Current work: Scheduling large systems, ordering, combinatorics; planning outputs, establishing sequence, workcenter planning, scheduling activities and events in a large system. Home: 567 Rutgers Ln Parsippany NJ 07054 Office: Dept Mgmt and Indsl Relations Seton Hall U South Orange NJ 07079

AMATRUDA, JOHN MICHAEL, physician, educator; b. New Haven, Oct. 31, 1944; s. Andrew A. and Concetta (Fusco) A.; m. Mary-Jo Cipriano, Aug. 10, 1968; children: Matthew Steven, J. Kristen. B.A., Yale U., 1966; M.D., Med. Coll. Wis., 1970. Intern, resident in medicine Johns Hopkins U. Sch. Medicine, Balt., 1970-72, fellow endocrinology, 1972-75; head endocrinology sect. Naval Med. Research Inst., Bethesda, Md., 1975-77; asst. prof. medicine U. Rochester (N.Y.) Sch. Medicine, 1977-81, assoc prof. medicine, 1981—. Contbr. chpts. to sci. vols., articles to sci. jours. Bd. dirs. Rochester Regional Diabetes Assn. Served to lt. comdr. USN, 1975-77. NIH career devel. research grantee, 1978-83; Mellon Found. fellow, ctl2-84. Mem. Am. Fedn. Clin. Research, Endocrine Soc., Am. Diabetes Assn., Am. Soc. Biol. Chemists, Am. Soc. Clin. Investigation. Subspecialties: Endocrinology; Biochemistry (medicine). Current work: factors which affect the ability of the body to respond to insulin; insulin receptors.

AMBLER, ERNEST, govt. ofcl.; b. Bradford, Eng., Nov. 20, 1923; came to U.S., 1953, naturalized, 1958; s. William and Sarah Alice (Binns) A.; m. Alice Virginia Seiler, Nov. 19, 1955; children—Christopher William, Jonathan Ernest. B.A., New Coll., Oxford U., 1945, M.A., 1949, Ph.D., 1953. With Armstrong Siddeley Motors, Ltd., Coventry, Eng., 1944-48; Nuffield Research fellow Oxford U., 1953; with Nat. Bur. Standards, Commerce Dept., 1953—, div. chief inorganic materials div., Washington, 1965-68; dir. Inst. for Basic Standards, Washington, 1968-73; dep. dir. Nat. Bur. Standards, Washington, 1973, acting dir., 1975-78, dir. bur., 1978—; Liaison rep. to div. phys. scis. Nat. Acad. Sci.-NRC, 1968-69; Sponsor's del. Nat. Conf. Standards Lab., 1968; U.S. rep. Internat. Com. on Weights and Measures, 1972—. Recipient Arthur S. Flemming award Washington Jr. C. of C., 1961; John Simon Guggenheim Meml. Found. fellow, 1963; recipient William A. Wildmack award in metrology, 1976, Pres.'s award for Distinguished Fed. Civilian Service, 1977. Mem. Am. Phys. Soc. (editor Rev. Modern Physics 1966-69), Washington Acad. Scis., AAAS. Subspecialties: Low temperature physics; Condensed matter physics. Patentee low temperature refrigeration apparatus. Home: 6920 Blaisdell Rd Bethesda MD 20817 Office: Nat Bur Standards Washington DC 20234

AMBRE, JOHN JOSEPH, physician, toxicologist, pharmacologist, educator; b. Aurora, Ill., Sept. 14, 1937; s. Frederick Mathias and Cecelia Angela (Petit) A.; m. Anita Marie Sievert, Nov. 3, 1962; children: Susan, Peter, Denise, Matthew. B.S., Notre Dame U., 1959; M.D., Loyola U.-Chgo., 1963; M.S., U. Iowa, 1970, Ph.D., 1972. Med. fellow Mayo Clinic, Rochester, Minn., 1966-68; asst. to assoc. prof. U. Iowa Coll. Medicine, Iowa City, 1972-78; med. dir. CBT Labs., Highland Park, Ill., 1978—; assoc. prof. Northwestern U. Med. Sch., Chgo., 1980—; cons. Abbott Labs., North Chicago, Ill.; toxicology cons. Metpath Labs., Teterboro, N.J.; cons. Hyland Diagnostics, Round Lake, Ill., 1981. Author: Drug Assay, 1983; contbr. articles to sci. publs. Served to capt. AUS, 1964-66. U. Iowa VA Clin. Investigator, 1973. Mem. Am. Fedn. Clin. Research, Am. Soc. Clin. Pharmacology and Therapeutics, Am. Soc. Pharmacology and Exptl. Therapeutics, Am. Assn. Clin. Chemistry, AAAS. Subspecialties: Pharmacology; Toxicology (medicine). Current work: Clinical pharmacology; clinical toxicology; drug analysis in biological fluids; therapeutic drug monitoring; drug metabolism; drug metabolite identification; metabolism of drugs and chemicals causing poisoning especially drugs of abuse. Drug patentee, 1972. Home: 1210 Walden Ln Deerfield IL 60015 Office: Northwestern U Med School 303 E Superior Chicago IL 60611

AMBROSE, AUDREY BELSON, endocrinologist, educator; b. N.Y.C., Aug. 23, 1941; d. Robert S. and Betty L. (Trent) Claycomb; m. Darrell T. Ambrose, Mar. 10, 1971; children: Nicholas, Suzanne. B.A., NYU, 1962; M.D., Northwestern U., 1966. Diplomate: Am. Bd. Internal Medicine. Resident Northwestern Meml. Hosp., Chgo., 1967-69; fellow in diabetes and endocrinology, 1969-71; asst. clin. prof. medicine U. Pa. Sch. Medicine, Phila., 1972-75, prof., 1975—. Grantee Am. Diabetes Assn., 1978-80. Mem. ACP, Am. Fedn. Clin. Research, Am. Diabetes Assn., N. Am. Assn. Study of Obesity. Subspecialties: Endocrinology; Internal medicine. Office: Werik Bldg Lower Level Suite 100 9600 Rosevelt Blvd Philadelphia PA 19115

AMDAHL, GENE MYRON, computer company executive; b. Flandreau, S.D., Nov. 16, 1922; s. Anton E. and Inga (Brendsel) A.; m. Marian Quissell, June 23, 1946; children: Carlton Gene, Beth Delaine, Andrea Leigh. B.S.E.E., S.D. State U., 1948, D.Eng. (hon.), 1974; Ph.D., U. Wis., 1952, D.Sc. (hon.), 1979, Luther Coll., 1980. Project mgr. IBM Corp., Poughkeepsie, N.Y., 1952-55; group head Ramo-Wooldridge Corp., Los Angeles, 1956; mgr. systems design Aeronutronics, Los Angeles, 1956-60; mgr. systems design advanced data processing systems IBM Corp., N.Y.C., Los Gatos, Calif., Menlo Park, Calif., 1960-70; founder Amdahl Corp.; founder, chmn. Trilogy Ltd. Served with USN, 1942-44. Recipient Disting. Alumnus award S.D. State U., 1973, Data Processing Man of Yr. award Data Processing Mgmt. Assn., 1976, Disting. Service citation U. Wis., 1976, Michelson-Morley award Case-Western Res. U., 1977, Harry Goode Meml. award for outstanding contbns. to design and manufacture of large, high-performance computers Am. Fedn. Info. Processing Socs., 1983; IBM fellow, 1965; IEEE fellow, 1969. Fellow Brit. Computer Soc.; mem. Nat. Acad. Engring., IEEE (W.W. McDowell award 1976), Quadrato della Radio, Pontecchio Marcon. Lutheran. Club: La Rinconada Country (Saratoga, Calif.). Subspecialties: Computer architecture; Computer engineering. Current work: Advanced large scale integrated circuitry and high performance computer organization. Patentee in field. Home: 165 Patricia Dr Atherton CA 94025 Office: 10500 Ridgeview Ct Cupertino CA [illegible]

AMEN, RALPH DUWAYNE, biology educator; b. Cheyenne, Wyo., Feb. 26, 1928; s. Adolph and Amilia (Sitzman) A.; m. Shirley Diane George, Aug. 1, 1952; children: Christine, duWayne, Katherine, Eric. B.A., U. No. Colo., 1952, M.A., 1954; M.B.S., U. Colo., 1959, Ph.D., 1962. Asst. prof. biology Wake Forest Coll., Winston-Salem, 1962-67, assoc. prof., 1967-80, prof., 1980—, chmn. dept. biology, 1967-71. Contbr. articles to profl. jours. Served with USAF, 1951-53. Recipient award for paper Sci. Citation Index, 1980. Mem. Sigma Xi. Subspecialties: Plant physiology (biology); Developmental biology. Current work: Hormonal regulation of development; embryogenesis. Home: 100 Friendship Cir Winston-Salem NC 27106 Office: Wake Forest Univ 7325 Reynolda Station Winston-Salen NC 27109

AMER, AHMAD (EL SAYED), engring. and fabrication co. exec., metall. engr., cons.; b. Cairo, Egypt, June 17, 1940; came to U.S., 1968, naturalized, 1973. B.S. in Metall. Engring, Cairo U., 1964. Registered profl. engr., Calif. Process metallurgist Gen. Orgn. for Industrialization, Cairo, 1964-1968; plant metallurgist Phoenix Steel Corp., Claymont, Del., 1968-71; mgr. quality control Pipeco Steel Co., Wilmington, Del., 1971-72; chief metallurgist Cann & Saul Co., Royersford, Pa., 1972-74; sr. metall. engr. Bechtel Power Corp., Gaithersburg, Md., 1974-77; pres. Amer Indsl. Techs., Inc., Wilmington, 1977—. Mem. ASME, Am. Welding Soc., Egyptian Welding Soc., Am. Soc. Metals. Club: Rodney Sq. Lodges: Masons; Shriners. Subspecialties: Materials processing; Nuclear engineering. Current work: Fabrication of pressure vessels and piping, weldments for nuclear and chem. processing industries. Developed process for continuous casting of high strength low alloy steel, process for rolling, heat treating and testing explosive-bonded titanium cladding. Home: 1515 Forsythia Ave Wilmington DE 19810 Office: 1000 S Madison St Wilmington DE 19801

AMER, M. SAMIR, pharm. co. exec., researcher; b. Egypt, Sept. 2, 1935; s. M. Mohamed and Zeinab H. (Saad) A.; m. Laila E. El-Fatatry, June 12, 1958; children: Amre S., Nancy S., Mona S., Suzanne S. Ph.D., U. Ill., 1962; M.B.A., Columbia U., 1980. Dir. cardiovascular research Mead Johnson & Co., Evansville, Ind., 1969-77; dir. biol. research Bristol Myers Internat., N.Y.C., 1977-82, dir. strategic mktg., 1982—. Contbr. articles to profl. jours. Mem. Am. Soc. Pharmacology and Exptl. Therapeutics, Am. Soc. Biol. Chemists. Subspecialties: Molecular biology; Pharmacology. Current work: Mechanisms of hormone action, blood pressure control, strategic marketing, new trends in pharmacology and medicine. Home: 155 North St Greenwich CT 06830 Office: Bristol Myers Internat 345 Park Ave New York NY 10154

AMES, BRUCE N(ATHAN), biochemist, geneticist; b. N.Y.C., Dec. 16, 1928; s. Maurice U. and Dorothy (Andres) A.; m. Giovanna Ferro-Luzzi, Aug. 26, 1960; children—Sofia, Matteo. B.A., Cornell U., 1950; Ph.D., Calif. Inst. Tech., 1953. Chief sect. microbial genetics NIH, Bethesda, Md., 1953-68; prof. biochemistry U. Calif., Berkeley, 1968—. Mem. Nat. Cancer Adv. Bd. Recipient Eli Lilly award Am. Chem. Soc., 1964, Flemming award, 1966; Rosenstiel award, 1976; FASEB award, 1976; Environ. Mutagen Soc. award, 1977; Felix Wankel award, 1978; John Scott medal, 1979; New Brunswick award, 1980; Corson medal, 1980. Mem. Am. Soc. Biol. Chemists, Am. Soc. Microbiology, Environ. Mutagen Soc., Genetics Soc., Am. Assn. Cancer Research, Soc. of Toxicology, Am. Chem. Soc., Am. Acad. Arts and Scis., Nat. Acad. Scis. Subspecialties: Biochemistry (biology); Genetics and genetic engineering (biology). Current work: Mutagens; biochemical genetics. Research, publs. on bacterial molecular biology, histidine biosynthesis and its control; RNA and regulation; mutagenesis; detection of environmental mutagens and carcinogens, genetic toxicology. Home: 1324 Spruce St Berkeley CA 94709

AMES, GIOVANNA FERRO-LUZZI, biochemistry educator; b. Rome, Italy, Jan. 20, 1936; d. Giovanni and Sofia (Saltzman) Ferro-Luzzi; m. Bruce Nathan Ames, Aug. 26, 1960; children: Sofia, Matteo. D. Biology, U. Rome, 1958. Research scientist NIH, 1967; prof. biochemistry U. Calif.-Berkeley, 1967—; consultant. Contbr. articles profl. jours. Recipient Agnes Fay Morgan triennial award Iota Sigma Pi, 1975. Mem. Am. Soc. Microbiology, Am. Soc. Biol. Chemists. Democrat. Subspecialties: Biochemistry (biology); Gene actions. Office: Biochemistry Dept U Calif Berkeley CA 94720

AMES, IRA H., biologist; b. Bklyn., Apr. 27, 1937; s. Lawrence and Blanche (Tannenbaum) A.; m. Joyce T. Surnamer, June 26, 1958; children: Michael, Sarah. B.A., Bklyn. Coll., 1959; M.S., NYU, 1962, Ph.D., 1966. Instr. biology Bklyn. Coll., 1960-63; research assoc. Brookhaven Nat. Lab., 1966-68; asst. prof. biology SUNY-Upstate Med. Ctr., Syracuse, 1968-73, assoc. prof., 1973—. Contbr. chpt. to book. Mem. Fayetteville Residents Assn., 1981-82. NSF grad. fellow., 1964-66; recipient Founders Day award NYU, 1967. Mem. AAAS, Am. Soc. Cell Biology, Am. Assn. Anatomists, Phi Beta Kappa, Sigma Xi. Democrat. Jewish. Subspecialties: Oncology; Cell study oncology. Current work: Morphological characteristics of normal and abnormal mouse mammary tissue. Home: 105 Woodmancy Ln Fayetteville NY 13066 Office: Anatomy Dept SUNY Upstate Med Ctr Syracuse NY 13210

AMES, LYNFORD LENHART, chemist, educator, researcher; b. Fresno, Ohio, May 20, 1938; s. Robert Jonathan and Magalene Elaine (Miller) A.; m. Judith Elaine Kilbourne, Mar. 23, 1963; children: Graham, Constance. B.S., Muskingum Coll., 1960; Ph.D., Ohio State U., 1965. NSF postdoctoral fellow Oxford (Eng.) U., 1965-66; asst. prof. N. Mex. State U., 1966-70, assoc. prof., 1970-78, prof., 1978—, head dept. chemistry, 1976—. Mem. Am. Chem. Soc., AAAS, Sigma Xi. Republican. Presbyterian. Lodge: Rotary. Subspecialties: High temperature chemistry; Physical chemistry. Current work: High temperature chemistry; spectroscopy. Home: 685 Farney Rd Las Cruces NM 88005 Office: Box 3C N Mex State U Las Cruces NM 88005

AMES, MATTHEW MARTIN, pharmacologist, research scientist, educator; b. Richland, Wash., Dec. 17, 1947; s. Milo E. and Betty J. (Hill) A.; m. Sharon Jeter, Mar. 19, 1977. B.A., Whitman Coll., 1970; Ph.D. in Pharm. Chemistry, U. Calif.-San Francisco, 1976. Pharmacology research assoc. NIH, Bethesda, Md., 1975-77; research assoc., instr. pharmacology Mayo Clnic, Rochester, Minn., 1977-78, assoc. cons., asst. prof., 1978-79, assoc. prof., 1982—, cons. oncology, 1980—. Contbr. articles to prfl. jours. Recipient Patent Fund award U. Calif., 1974-75; NIH trainee, 1975-77; Cancer Research fellow Ladies Aux. VFW, 1977-78; Research Career Devel. awardee Nat. Cancer Inst., 1981-86. Mem. Am. Chem. Soc., Am. Soc. Pharmacology and Exptl. Therapeutics, Am. Assn. Cancer Research. Subspecialties: Cancer research (medicine); Pharmacology. Current work: Metabolism and mechanism of action of antitumor agents. Office: Mayo Clnic 442 Guggenheim Rochester MN 55905

AMEY, RALPH LEONARD, chemistry educator, researcher; b. Huntington Park, Calif., June 5, 1937; s. Leonard Garwin and Dorothy Jessie (Gorn) A.; m. Ruth Ann Fortune, May, 1964 (div.); children: Stephen, Mark. A.B., Pomona Coll., 1959; Ph.D., Brown U., 1964. Materials research engr. Space Environ. Lab., Douglas Aircraft Co., Santa Monica, Calif., 1963-65; asst. prof. chemistry Occidental Coll., 1965-74, assoc. prof., 1974—. Mem. Am. Chem. Soc., Calif. Assn. Chemistry Tchrs. Subspecialties: Biophysical chemistry; Physical chemistry. Current work: Molecular interactions in liquids, solutions and solids by spectroscopic and thermodynamic methods. Office: 1600 Campus Rd Los Angeles CA 90041

AMHERD, NOEL A., project manager, technology scientist; b. San Francisco, Dec. 12, 1940; s. Francis and Laurence A.; m. Charlene; children: Alicia, Keven. B.S., Ariz. State U., 1963; M.S., U. Colo., 1969; Ph.D., U. Wash., 1973. Tech. staff Hughes Aircraft Co., El Segundo, Calif.; asst. prof. Princeton U., N.J.; now sr. project mgr. Electric Power Research Inst., Palo Alto, Calif. Bd. dirs.: Jour. Fusion Energy. Masters fellow Hughes Aircraft Co.; NSF research grantee; recipient NSF grad. tng. Mem. Am. Phys. Soc., Am. Nuclear Soc. (exec. com. fusion energy), Sigma Xi. Subspecialties: Nuclear fusion; Aerospace engineering and technology. Current work: Engineering of fusion energy systems, resolution of societal problems using advanced technology, and long range planning. Office: Electric Power Research Inst 3412 Hillview Ave Palo Alto CA 94303

AMIR, JACOB, physician; b. Anvers, Belgium, Jan. 17. 1932; came to U.S., 1974, naturalized, 1982; s. Leon and Aliza (Gatenio) Manoah; m. Aviva Kehati, June 11, 1957; children: Ariel, Yoram. M.D., Hebrew U., Jerusalem, 1960. Intern Hadassah U. Hosp., Jerusalem, Israel, 1960, resident, 1961, Beilinson Hosp., Tel Aviv, Israel, 1965-70, chief physician medicine, 1969-75; sr. lectr. Tel Aviv U., 1970-75; dir. Little Rock Diagnostic Clinic, 1975—, sr. physician, 1977—; asst. chief oncology VA Hosp., Cleve., 1975-77; asst. prof. medicine Case Western Res. U., 1975-77; asst. clin. prof. U. Ark., Little Rock, 1978—. Dep. mayor Govt. Kyriat-Ono, Israel, 1968-70. Served to capt. Israeli Def. Army. Fellow ACP; mem. Am. Soc. Clin. Oncology. Jewish. Subspecialties: Chemotherapy; Hematology. Office: Little Rock Diagnostic Clinic 10001 Lile Dr Little Rock AR 72205

AMIRIKIAN, ARSHAM, engineering company executive; b. Armenia, May 17, 1899; U.S., 1919, naturalized, 1927; s. Paravon and Pearl (Delbarian) A.; m. Philomena Elizabeth Boardman, Aug. 8, 1925; children: Richard Armen, Joyce Eleanor (Mrs. Robert A. Harrison). B.S., Ecole superieure des Ponts et Chaussees, Constantinople, 1919; C.E., Cornell U., 1923; D.Tech.Sc., Technische Hochschule, Vienna, 1960. Steel fabricator draftsman and designer 1923-28; various engring. positions to chief engring. cons. Naval Facilities Engring. Command, U.S. Navy Dept., Washington, 1928-71; pres. Amirikian Engring. Co., 1971—; cons. engr. shore and floating structures, harbor and docking facilities; adj. prof. engring. George Washington U., 1965-66. Author: Analysis of Rigid Frames, 1942; Contbr. articles tech. periodicals. Recipient Fuertes Grad. gold medal Cornell U., 1943, Lincoln gold medal Am. Welding Soc., 1949, A.E. Lindau award Am. Concrete Inst., 1958, Distinguished Service award Dept. of Navy, 1966, Def. Dept., 1969, Civilian Career Achievement award Dept. Navy, 1971, Goethals medal Soc. Am. Mil. Engrs., 1971. Fellow Am. Concrete Inst., Soc. Am. Mil. Engrs.; mem. Nat. Acad. Engring., ASCE (hon., E.E. Howard award 1978), Am. Welding Soc. (hon.), Soc. Naval Architects and Marine Engrs., Internat. Inst. Welding (hon.), Sigma Xi. Subspecialties: Ocean engineering; Offshore technology. Current work: Harbor and drydocking facilities; special shore and floating structures. Inventor of Ammi lift dock and transfer system, biserrated rib framing, split-beam prestressing, thin-shell hollow-rib and cellular precast concrete framing systems. Home: 6526 Western Ave Chevy Chase MD 20815 Office: 35 Wisconsin Circle Chevy Chase MD 20815

AMIRKHANIAN, JOHN DAVID, geneticist; b. Esfahan, Iran, Nov. 10, 1927; came to U.S., 1979, naturalized, 1979; s. Gregor D. and Astghik H. (Alexanderian) A.; m. Romelia Grigorian, Jan 30, 1957; children: Varoujan, Areg, Aspet. B.Sc., Tehran U., 1973; Ph.D., U. London, 1977. Ctlt. community coll. tchr. biol. scis., Calif. Instr., vis. prof. U. So. Calif, Los Angeles, 1979-81; research assoc. Natural History Mus., Los Angeles, 1980—; asst. prof. Sch. Pub. Health, Tehran U., 1977-79; sr. research assoc. Charles Drew Postgrad. Med. Sch., Los Angeles, 1982—. Contbr. articles to profl. jours. WHO grantee, 1969; recipient Sci. Research prize Sci. Research Council of Tehran U., 1973. Mem. Genetics Soc. Am., Genetical Soc. Eng., Inst. Biology London. Mem. Apostolic Ch. of Armenia. Subspecialties: Neuropharmacology; Environmental toxicology. Current work: Genetic repair and recombination in eukaryotes. Office: Lab C Charles R Drew Postgrad Med Sch 1621 E 120th St Los Angeles C 90059

AMIRTHARAJAH, APPIAH, environmental engineering educator; b. Colombo, Sri Lanka, Apr. 4, 1940; came to U.S., 1976; s. Arumugam and Sinnamma Appiah; m. Uma Hymavati, July 13, 1968; children: Rajeevan, Mohana. B.S. with honors in Engring, U. Ceylon, 1963; M.S., Iowa State U., 1970, Ph.D., 1971. Registered profl. engr., Mont. Project design engr. dept. water supply, Ratmalana, Sri Lanka, 1963-74, vis. prof. univs., Sri Lanka, 1973-76; chief engr. Nat. Water Bd., Ratmalana, 1975-76; asst. prof. dept. civil engring. Mont. State U., Bozeman, 1976-77, assoc. prof., 1977-81, prof., 1981—, coordinator environ. programs, 1979—. Contbr. chpts. to books and articles in field to profl. jours.; reviewer, NSF, 1979—. Fulbright scholar, 1968-69; NSF research grantee, 1980-83. Mem. Am. Water Works Assn. (coagulation com. 1980—, reviewer jour. 1976—), ASCE (publ. com. acting editor 1982—), Water Pollution Control Fedn., Internat. Assn. Water Pollution Research, Pub. Service Engrs. Assn. (v.p. 1974), Sigma Xi, Phi Kappa Phi. Subspecialties: Water supply and wastewater treatment; Civil engineering. Current work: Research and teaching in environmental engineering; granular media filtration coagulation and water supply in developing nations. Office: Mont State U Dept Civil Engring Bozeman MT 59717 Home: 1104 W Koch Bozeman MT 59715

AMJAD, HASSAN, physician; b. Jhang, Punjab, Pakistan, Nov. 27, 1947; came to U.S., 1971, naturalized, 1980; s. Jafar and Anwer (Fatima) H.; m. Lolita Quezon, Oct. 27, 1973; children: Urooj, Quartel Ayne, Shabnaum. M.D., King Edward Med. Coll., Pakistan, 1970. Diplomate: Am. Bd. Internal Medicine, Am. Bd. Hematology. Intern Riverside Hosp., Toledo, Ohio, 1971-72, SUNY-Buffalo, 1972-73, William Beaumont Hosp., Royal Oak, Mich., 1972-73; resident Wayne State U. Sch. Medicine-Harper Grace Hosp., Detroit, 1974-76; asst. clin. prof. medicine Marshall U. Med. Ctr., Huntington, W.Va., 1977-82; chief med. services VA, Beckley, W.Va., 1980-82; chief medicine Plateau Med. Ctr., Oak Hill, W.Va., 1982—, dir. intensive care unit, 1982—; cons. in medicine and cancer diseases Appalachian Regional Hosp., 1977—; also Raleigh Gen. Hosp. Contbr. articles to profl. jours. Bd. dirs. Am. Cancer Soc., Raleigh County, W.Va., 1981—, Raleigh County Heart Assn., 1981—. Recipient Hands and Heart award VA, 1981. Fellow ACP, Internat. Soc. Internal Medicine, Internat. Soc. Hematology; mem. Am. Soc. Hematology, Am. Soc.

Clin. Oncology, AAAS, Am. Fedn. Clin. Research. Muslim. Subspecialties: Chemotherapy; Hematology. Current work: Biomedical use of lasers, recombinant DNA, health care in underdeveloped. Address: 32 Hummingbird Ln Beckley WV 25801

AMMANN, EUGENE OTTO, laser researcher; b. Portland, Oreg., June 26, 1935; s. Eugene Otto and Frances Elizabeth (Bowker) A.; m. Christina Aparicio, June 20, 1970; 1 dau., Alicia. B.S. in Gen. Engring, U. Portland, 1957; M.S.E.E., Stanford U., 1959, Ph.D. in Elec. Engring, 1963. Research assoc. Stanford (Calif.) U., 1963; laser researcher GTE Sylvania, Mountain View, Calif., 1963—, now sr. engring. specialist.; Soccer coach, 1982. Contbr. articles to sci. jours., chpts. to books. NSF fellow, 1958-60. Mem. IEEE, Optical Soc. Am., Am. Phys. Soc. Subspecialties: Laser research; Electro-optics. Current work: Lasers; electro-optics; nonlinear optics; stimulated Raman scattering; optical birefringent filters. Patentee in field (6). Office: 100 Ferguson Dr PO Box 7188 Mail Stop 4G07 Montain View CA 94039

AMOLS, HOWARD IRA, med. physicist, educator, researcher; b. N.Y.C., Feb. 11, 1949; s. Nathan and Esther Ruth (Rauchwarger) A.; m. Koren Lynn, Sept. 1, 1970; children: Amy, Rachel. B.S. summa cum laude in Physics N.Y. State Regents scholar), Cooper Union, 1970; M.S., Brown U., 1973; Ph.D. in Physics (NDEA Title IV grad. fellow, 1970-73), 1974. Nat. Cancer Inst. postdoctoral fellow Los Alamos (N.Mex.) Nat. Lab., 1974-76, staff mem., 1974-79; asst. prof. radiology U. N.Mex., Albuquerque, 1979-81; assoc. physicist R.I. Hosp., Providence, 1981—; asst. prof. radiation medicine Brown U., Providence, 1981—; vis. scientist Karlsruhe Nuclear Research Center, W. Ger., 1977-78. Contbr. articles to profl. jours. Prin. investigator, research grantee Nat. Cancer Inst., 1981—. Mem. Radiation Research Soc., Am. Assn. Physicists in Medicine, Soc. Physics Students, Sigma Xi. Subspecialties: Oncology; Radiology. Current work: Dosimetry, microdosimetry, radiobiology. Home: 22 Sutton Pl Cranston RI 02910 Office: RI Hosp Dept Radiation Oncology Providence RI 02902

AMOS, DENNIS BERNARD, physician; b. Bromley, Eng., Apr. 16, 1923; m., 1949; 5 children. M.B., B.S., U. London, 1951; M.D., 1963. Intern Guy's Hosp., London, 1951; from assoc. cancer research scientist to prin. cancer research scientist Roswell Park Meml. Inst., 1956-62; James B. Duke prof. immunology and exptl. surgery Med. Center, Duke U., Durham, N.C., 1962—; research fellow Guy's Hosp., London, 1952-55; sr. research fellow Roswell Park Meml. Inst., 1955-56; chmn. human lymphocyte-antigen standards com. AACHT and NIH/Nat. Inst. Allergy and Infectious Diseases; mem. nomenclature com. luekocyte antigens WHO, Internat. Union Immunol. Socs. Mem. Am. Assn. Immunologists, Am. Assn. Cancer Research, Transplantation Soc., Am. Assn. Clin. Histocomptabilitiy Testing, AAAS, Nat. Acad. Scis. Subspecialties: Immunology (medicine); Transplantation. Office: Duke U Med Center Box 3010 Durham NC 27710

AMOS, DEWEY HAROLD, geology educator; b. Harrisville, W.Va., Feb. 27, 1925; s. Worthy and Lona Virginia (Maxwell) A.; m. Dora Cornelia Ames, Nov. 26, 1948; children: Susan Lynne, Alan Scott, Melinda Kaye. B.S., Marietta Coll., 1948; M.A., U. Ill.-Urbana, 1950, Ph.D., 1957. Geologist U.S. Geol. Survey, Knoxville, Tenn., 1952-55; asst. prof. So. Ill. U., Carbondale, 1955-65; assoc. prof. Eastern Ill. U., Charleston, 1965-68, prof. geology, 1968—; geologist U.S. Geol. Survey, Denver, 1955—. Served to 1st lt. U.S. Army, 1943-46, 51-52; ETO. Fellow Geol. Soc. Am.; mem. Soc. Econ. Geologists, Phi Beta Kappa, Sigma Xi. Subspecialties: Tectonics; Petrology. Current work: Tectonics and structure of Ozark region. Home: 2003 University Dr Charleston IL 61921 Office: Geology Dept Eastern Ill U Charleston IL 61920

AMOSS, MAX ST. CLAIR, JR., neuroendocrinology educator, researcher; b. Balt., May 9, 1937; s. Max and Mary G. (Myers) A.; m. Helen Elizabeth Clark, June 28, 1955; children: Bridget D. Amoss Michael, Max St. Clair III, Michael C. B.S., Pa. State U., 1962; M.S., Tex. A&M U., 1965; Ph.D., Baylor U. Coll. Medicine, 1969, postgrad., 1969-70. Asst. prof. Baylor Coll. Medicine, Houston, 1970; asst. research prof. Salk Inst. La Jolla, Calif., 1970-75; assoc. prof. Tex. A&M U., College Station, 1975—; cons. Hoffman La Roche, Nutley, N.J., 1976—. Mem. Am. Physiol. Soc., Soc. Study of Reprodn. (editorial bd. 1978-83), Endocrine Soc., Am. Soc. Animal Sci., Internat. Soc. Neuroendocrinology. Republican. Episcopalian. Subspecialties: Neuroendocrinology; Reproductive biology. Current work: Neuroendocrine control of anterior pituitary function; co-inventor analogs of LHRH. Co-inventor analogs of CHRH. Office: Tex A&M U Dept Vet Physiology College Station TX 77843

AMSBURY, WAYNE, computer science educator, writer; b. Topeka, Nov. 11, 1935; s. Leonard L. and Annette (Walker) A.; m. Carlene Cox, Dec. 14, 1963; children: Burl Wayne, Kimberly Noel. B.A. in Physics, Rice U., 1957; M.A. in Math, U. Tenn.-Knoxville, 1968, Ph.D., 1972. Engr.-programmer AEC Facilities, Oak Ridge, Tenn., 1960-68; faculty, math. and computer sci. Northwest Mo. State U., Maryville, 1972-78, assoc. prof. computer sci., 1978—. Author: Structured BASIC, 1980. Mem. Math. Assn. Am., Am. Math. Soc., Soc. Indsl. and Applied Math., IEEE. Subspecialties: Computer architecture; Algorithms. Current work: Technical writing. Home: 709 S Vine Maryville MO 64468 Office: Northwest Mo State Univ Maryville MO 64468

AMUNDSEN, KEITH BYRON, electrical engineer; b. Freeport, N.Y., July 1, 1954; s. Lloyd Alfred and Joyce Lynn (Cassidy) Amundsen G. B.S. in Elec. Engring, MIT, 1976; M.S. in Computer Architecture, U. Calif.-Berkeley, 1976-77. Lab. asst. MIT Digital Systems Lab., Cambridge, 1973; engr. Long Island (N.Y.) Lighting Co., 1974-76; teaching asst. digital lab., instr. analog lab. U. Calif.-Berkeley, 1976-77; prin. engr. Digital Equipment Corp., Maynard, Mass., 1977—; pres. Solar Wind Assocs., Cambridge, 1978—; cons. in field. Editor: Digital Equipment Co. Unibus Handbook, 1979; puslisher: Massbus Interface Standard, 1979. Recipient Bausch & Lombe award in physics, 1972. Mem. IEEE, Assn. Computing Machinery, Am. Nat. Standards Inst. (U.S. del. to Internat. Standards Orgn.), Sigma Xi, Pi Lambda Phi. Republican. Lutheran. Club: Unqua Corinthian Yacht. Subspecialties: Distributed systems and networks; Computer architecture. Current work: Communication architecture; VLSI architecture; multiprocessors. Home: 44 Dorothy Rd Arlington MA 02174 Office: 146 Main St ML 3-5/U26 Maynard MA 01754

ANAGNOSTAKIS, SANDRA LEE, agricultural scientist; b. Coffeyville, Kans., May 14, 1939; d. Donald Clinton and Zella Blanche (McGowen) Fowler; m. Christopher Anagnostakis, June 10, 1969; 1 dau.: Kathryn M. B.A., U. Calif.-Riverside, 1961; M.A., U. Tex.-Austin, 1966. With Conn. Agrl. Expt. Sta., New Haven, 1966—, mem. staff genetics dept., 1966-78, assoc. agrl. scientist plant pathology and botany, 1978—. Contbr. numerous articles to profl. jours. Mem. Mycol. Soc. Am., AAAS, Am. Phytopath. Soc., Conn. Acad. Arts and Sci. Episcopalian. Subspecialties: Genetics and genetic engineering (agriculture); Plant pathology. Current work: Biological control of fungi that cause plant diseases, using genetic techniques to study mycological and plant pathological problems. Office: Conn Agrl Expt Sta Box 1106 New Haven CT 06504

ANAND, SUBHASH C., civil engring. educator; b. Lyallpur, India, July 27, 1933; came to U.S., 1964, naturalized, 1974; s. Bhagat Singh and Vidya (Khera) A.; m. Vera Barata, Aug. 28, 1965; children: Mina Louise, Indu Stacey. B.Sc., Banaras Hindu U., India, 1955; M.S., Northwestern U., 1965, Ph.D., 1968. Registered profl. engr., S.C. Design engr. Stahlbau Humboldt, Cologne, W.Ger., 1960-64; asst. prof. civil engring. Calif. State U., Sacramento, 1968-70; assoc. prof. civil engring. Ill. Inst. Tech., Chgo., 1970-72, Clemson (S.C.) U., 1972-76, prof., 1976—; cons. in field. Treas. chpt. Am. Field Service, Clemson, 1978—. Recipient McQueen Quattlebaum Faculty Achievement award Clemson U., 1979; Fulbright lectr., 1980; NATO lectr., 1981; NSF grantee, 1982-83. Mem. ASCE (sec. 1981—, chmn. inelastic behavior 1979-81, editorial bd. 1979—), Am. Soc. Engring. Edn. (chmn. civil engring. com. 1981-82), Am. Acad. mechanics, Sigma Xi, Chi Epsilon. Subspecialties: Civil engineering; Theoretical and applied mechanics. Current work: Developing finite element models to understand the behavior of masonry. Also developing finite element methodology to predict the flow of contaminants in saturated groundwater flow. Developing models of risk analysis for earthquake damage in eastern U.S. Office: Dept Civil Engring Clemson U Clemson SC 29631

ANANTHASWAMY, HONNAVARA NARASIMHAMURTHY, microbiologist, researcher; b. India, Aug. 12, 1938; s. Honnavara Srikantia and Gouramma (Ramanna) Narasimhamurthy; m. Regina Loffland, Apr. 26, 1975. B.S., U. Mysore, India, 1961; M.S., U. Bombay, 1970; Ph.D., U. Mo., Columbia, 1975. Sci. officer Bhaba Atomic Research Center, Bombay, 1964-70; grad. teaching asst. U. Mo., Columbia, 1970-74, postdoctoral fellow, 1974-76, U. Calif., Berkeley, 1976-78; scientist Frederick (Md.) Cancer Research Facility, 1978—; researcher. Contbr. writings to profl. publs. in field. Recipient Nat. Research Service award NIH, 1976-78. Mem. AAAS, Am. Soc. Photobiology, Sigma Xi. Subspecialties: Cell biology; Cancer research (medicine). Current work: Ultraviolet radiation-induced carcinogenesis, DNA damage and repair, gene transfer in mammalian cells. Office: Nat Cancer Ins Frederick Cancer Research Facility Frederick MD 21701

ANCKER-JOHNSON, BETSY, physicist, automotive company executive; b. St. Louis, Apr. 29, 1927; d. Clinton James and Fern (Lalan) Ancker; m. Harold Hunt Johnson, Mar. 15, 1958; children: Ruth P. Johnson, David H. Johnson, Paul A. Johnson, Martha H. Johnson. B.A. in Physics with high honors (Pendleton scholar), Wellesley Coll., 1949; Ph.D. magna cum laude, U. Tuebingen, Germany, 1953; D.Sc. (hon.), Poly. Inst. N.Y., 1979, LL.D., Bates Coll., 1980. Instr., jr. research physist U. Calif., 1953-54; physicist Sylvania Microwave Physics Lab., 1956-58; mem. tech. staff RCA Labs., 1958-61; research specialist Boeing Co., 1961-70, exec., 1970-73; asst. sec. commerce for sci. and tech., 1973-77; dir. phys. research Argonne Nat. Lab., Ill., 1977-79; v.p. environ. activities staff Gen. Motors Tech. Center, Warren, Mich., 1979—; affiliate prof. elec. engring. U. Wash., 1964-73; dir. Gen. Mills; mem. Energy Research Adv. Bd. Dept. Energy. Author. Mem. staff Inter-Varsity Christian Fellowship, 1954-56; Trustee Wellesley Coll., 1972-77. AAUW fellow, 1950-51; Horton Hollowell fellow, 1951-52; NSF grantee, 1967-72. Fellow Am. Phys. Soc. (councillor-at-large 1973-76), IEEE; mem. Nat. Acad. Engring., Phi Beta Kappa, Sigma Xi. Patentee in field. Office: Environmental Activities Staff GM Technical Center Warren MI 48090

ANCONA, ANTONIO, nuclear physicist, consultant; b. Bari, Puglia, Italy, Mar. 9, 1945; came to U.S., 1961; s. Francesco and Maria (Scotella) A.; m. Carole Louise Prindle, June 30, 1978. B.S.E.E., U. Hartford, 1968; M.S.N.S., Rensselaer Poly. Inst., 1970, Ph.D. in Nuclear Engring, 1977. Electronic engr. Chandler Evans Co., West Hartford, Conn., 1968-70; physicist Combustion Engring. Co., Windsor, Conn., 1970-75; research asst. Rensselaer Poly. Inst., 1975-77; nuclear physicist Nuclear Assocs. Internat., Rockville, Md., 1977—; prin. Ancona & Assocs., Annapolis, Md., 1979—. Contbr.: articles to Nuclear Sci. and Engring, Transactions Am. Nuclear Soc. Recipient Outstanding Achievement award Control Data Corp., Mpls., 1979. Mem. Am. Nuclear Soc., Nat. Soc. Profl. Engrs. Subspecialties: Nuclear physics; Numerical analysis. Current work: Research in advanced iteration schemes for diffusion and transport theory solutions of nuclear reactor neutron flux. Patentee digital function generator, digital controlled oscillator. Home: 2616 Quiet Water Cove Annapolis MD 21401 Office: Nucleare Assocs Internat 6003 Executive Blvd Rockville MD 20852

ANDERS, EDWARD, educator, chemist; b. Libau, Latvia, June 21, 1926; came to U.S., 1949, naturalized, 1955; s. Adolph and Erica (Leventals) Alperovitch; m. Joan Elizabeth Fleming, Nov. 12, 1955; children: George Charles, Nanci Elizabeth. Student, U. Munich, Germany, 1949; A.M., Columbia U., 1951, Ph.D., 1954. Instr. U. Ill., 1954-55; mem. faculty U. Chgo., 1955—, prof. chemistry, 1962-73, Horace B. Horton prof. chemistry, 1973—; vis. prof. Calif. Inst. Tech., 1960, U. Berne, Switzerland, 1963-64, 70, 78, 80, 83; research asso. Field Mus. Natural History, Chgo., 1968—; resident research asso. NASA, 1961; cons., 1961-69, mem. lunar sample analysis planning team, 1967-69. Asso. editor: Geochimica et Cosmochimica Acta, 1966-73, Icarus, 1970—, The Moon and the Planets, 1974—; contbr. articles to profl. jours. Recipient Univ. medal for excellence Columbia U., 1966; J. Lawrence Smith medal Nat. Acad. Scis., 1971; Quantrell award for excellence in undergrad. teaching U. Chgo., 1973; NASA medal for exceptional sci. achievement, 1973; Guggenheim fellow, 1973-74. Fellow AAAS (Newcomb Cleveland prize 1959), Meteoritical Soc. (v.p. 1968-72, Leonard medal 1974), Am. Acad. Arts and Sci., Am. Geophys. Union; asso. Royal Astron. Soc.; mem. Nat. Acad. Sci., Am. Astron. Soc. (chmn. div. planetary scis. 1971-72), Internat. Astron. Union (pres. com. on moon 1976-79), Am. Chem. Soc., Geochem. Soc., Sigma Xi. Subspecialties: Planetary science; Space chemistry. Current work: Origin of solar system; chemical and isotopic studies of meteorites and lunar rocks. Spl. research origin, age, composition of meteorites and lunar rocks, origin moon and planets. Office: Enrico Fermi Inst 5630 S Ellis Ave Chicago IL 60637

ANDERSEN, BARBARA LEE, psychology educator; b. Elgin, Ill., May 2, 1951; d. Edgar Alfred and Gladys Viola (Jensen) A.; m. John T. Cacioppo, May 17, 1981. B.S., U. Ill., 1973, M.A., 1978, Ph.D., 1980. Lic. psychologist, Iowa. Vis. instr. U. Ill., Urbana, 1978-79; UAF psychology fellow UCLA Neuropsychiat. Inst., 1979-80; asst. prof. psychology U. Iowa, Iowa City, 1980—. Author: Behavior Modificiation, 1979; contbr. articles to profl. jours. Edmund J. James scholar, 1971-73. Mem. Am. Psychol. Assn. (program com.), Soc. Behavioral Medicine, Midwestern Psychol. Assn., Sigma Xi, Psi Chi, Phi Kappa Phi. Subspecialty: Behavioral psychology. Current work: Conducting longitudinal behavioral medicine research on psychosocial adjustment of cancer patients following diagnosis and treatment and psychological reaction to cancer treatments. Office: Dept Psychology U Iowa Iowa City IA 52242

ANDERSEN, BURTON ROBERT, physician, educator; b. Chgo., Aug. 27, 1932; s. Burton R. and Alice C. (Mara) A.; m. Louise R. Gross, July 23, 1960; children: Ellen C., Julia A., Brian E. B.S., U. Ill.-Urbana, 1955; M.S., U. Ill. at Chgo., 1957, M.D., 1957. Lic. physician, Ill. Clin. assoc. NIH, 1961-64; asst. prof. U. Rochester, 1964-67; assoc. prof. Northwestern U., 1967-70; prof. medicine and microbiology U. Ill.-Chgo., 1973—; chief infectious diseases West Side VA Hosp., Chgo., 1970—. Contbr. numerous articles to profl. jours. Pres. Civic Arts Council, Oak Park, Ill., 1981-82. Served with USPHS, 1961-64. Mem. ACP, Am. Soc. Clin. Investigation. Subspecialty: Infectious diseases. Current work: Physiology of phagocytic cells. Office: 820 S Damen Ave Chicago IL 60612

ANDERSON, ARTHUR OSMUND, anatomic pathologist, immunobiologist, electron microscopist; b. Staten Island, N.Y., Mar. 12, 1945; s. Arthur Edmond and Florence Ranveig (Osmundsen) A.; m. Julane Pynn, Oct. 4, 1969. B.S. in Biology, Wagner Coll., 1966; M.D., U. Md., 1970. Diplomate: Am. Bd. Pathology. Intern in pathology Johns Hopkins Hosp., 1970-71, resident in pathology, 1971-74; fellow in exptl. pathology Johns Hopkins U. Sch. Medicine, 1970-74; pathologist, prin. investigator in immunopathology U.S. Army Med. Research Inst. Infectious Diseases, Ft. Detrick, Md., 1974-80; asst. prof. pathology and biology U. Pa., 1980—; vis. fellow NIH/Nat. Inst. Chronic Diseases Gerontology Research Center, Balt., 1972-74; guest lectr. Johns Hopkins Immunology Council, Found. Advanced Edn. in Scis., NIH, Bethesda, Md.; lectr.; seminar speaker; bd. dirs. Frederick (Md.) Am. Cancer Soc., 1975-70. Contbr. articles, chpts., abstracts to profl. publs.; reviewer profl. jours. Bd. dirs. Frederick County (Md.) Group Homes Inc., 1975-80, Univ. City Hist. Soc., Phila., 1980—. Served to lt. Col. M.C. U.S. Army, 1974-80. Decorated Meritorious Service medal; recipient Alumni Achievement award Wagner Coll., 1979. Mem. Am. Assn. Pathologists, Am. Assn. Immunologists, Internat. Acad. Pathology, N.Y. Acad. Scis., AAAS. Republican. Episcopalian. Subspecialties: Pathology (medicine); Immunology (medicine). Current work: Immunobiological mechanisms controlling chronic inflammatory diseases (like arthritis), lymphocytes, antigen presenting cells, lymphatic tissues, vascular endothelium, chemotaxis, locomotion of cells. Cons. hist. restoration; restored hist. properties, Frederick, Phila.

ANDERSON, CARL JOHN, physicist; b. Mpls., May 28, 1952; s. Laurel Ethan and Dorothy Mildred (Fobes) A.; m. Brynn Jane Anderson, Oct. 13, 1951; children: Kirsten, Byron, Erin. B.S., U. Mo.-Columbia, 1974; M.S., U. Wis.-Madison, 1976, Ph.D., 1979. Research staff IBM Research, Yorktown Heights, N.Y., 1979—. Mem. Am. Phys. Soc., Sigma Xi. Presbyterian. Subspecialties: Superconductors; Atomic and molecular physics. Current work: Design of GaAs digital devices and circuits.

ANDERSON, CARL WILLIAM, molecular biologist, geneticist; b. Washington, May 19, 1944; s. Carl Elmore and Laverne Ann-Marie (Larsen) A.; m. Mary Elizabeth Daniell, Apr. 20, 1968; children: Carl E., Christine E. B.A., Harvard Coll., 1966; Ph.D. in Microbiology, Washington U., St. Louis, 1970. With Cold Spring Harbor Lab., N.Y., 1970-75; genetecist biology dept. Brookhaven Nat. Lab., Upton, N.Y., 1975—; assoc. prof. microbiology SUNY, Stony Brook. Mem. AAAS, Am. Soc. Microbiology, Am. Soc. Virology. Subspecialties: Molecular biology; Virology (biology). Current work: Regulation of gene expression, molecular biology of DNA tumor viruses. Home: 23 Shelbourne Ln Stony Brook NY 11790 Office: Biology Dept Brookhaven Nat Lab Upton NY 11973

ANDERSON, CHARLES EDWARD, animal science educator; b. Rolla, Mo., Jan. 24, 1953; s. Drue Edward and Edna May (Coffman) A.; m. Cheryl Lynn. B.S. in Agr, U. Mo., 1975; M.S. in Animal Nutrition, Tex. A&M U. Grad. teaching asst. Tex. A&M U., 1975-78; instr. horse sci. and animal sci. Western Ky. U., Bowling Green, 1978-81, asst. prof., 1981—. Subspecialty: Animal nutrition. Current work: Horse nutrition and exercise physiology. Office: Dept Agr Western Ky U Bowling Green KY 42101

ANDERSON, CHARLES THOMAS, chemistry educator; b. Fairmont, W.Va., Feb. 26, 1921; s. Charles Thomas and Versa Esther (Hickman) A.; m. Mary Virginia Hendrick, June 10, 1950; children: Katherine Sue, David Murray. A.B., Fairmont State Coll., 1942; Ph.D., Ohio State U., 1955. Instr. Ohio U., Athens, 1946-51; asst. prof. Eastern Mich. U., Ypsilanti, 1955-58, assoc. prof., 1958-62, prof. chemistry, 1962—. Mem. Am. Chem. Soc., AAUP, Mich. Coll. Chemistry Tchrs. Assn. Methodist. Lodge: Kiwanis. Subspecialties: Analytical chemistry; Inorganic chemistry. Home: 720 Kewanee Ave Ypsilanti MI 48197 Office: Eastern Mich U Ypsilanti MI 48197

ANDERSON, CONRAD VICTOR, mech. engr.; b. Mpls., Jan. 30, 1948; s. Arthur John and Pearl Inez (Albrecht) A.; m. Anita Marguerite, Mar. 19, 1971. B.M.E., U. Minn., 1970. Design engr. Standard Convenor Co., St. Paul., 1973-74; sr. design engr. Pako Corp., Mpls., 1974-78; mgr. research and devel. engring. Zero Max div. Barry Wright Corp., Mpls., 1978—. Mem. ASME. Subspecialty: Mechanical engineering. Current work: Products for mechanical power tranmission. Patentee leader belt stabilizer, demand driven clutch, transport system for processor of photosensitive web material.

ANDERSON, DALE ARDEN, aerospace engineering educator, researcher; b. Alta, Iowa, Aug. 11, 1936; s. Everett and Inez (Burwell) A.; m. Marleen Marie Ankerson, June 15, 1958; children: Gregory, Lisa. B.S. in Aero. Engring, Parks Coll., St. Louis U., EAst St. Louis, Ill., 1957; M.S. in Aerospace Engring, Iowa State U., 1959; Ph.D. in 4616spa space and Elec. Engring, Iowa State U., 1964. Registered profl. engr., Iowa. Asst. prof. aerospace engring. Iowa State U., 1961-64, assoc. prof., 1965-74, prof., 1974—, dir., 1980—; mem. tech. staff Aerospace Corp., San Bernardino, Calif., 1964-65; cons. in field. Author: Computational Fluid Mechanics and Heat Transfer, 1983; contbr. numerous articles to profl. jours. NASA grantee, 1968—. Mem. AIAA, Sigma Xi. Presbyterian. Subspecialties: Aerospace engineering and technology; Numerical analysis. Current work: Numerical analysis; hyperbolic partial differential equations; computational fluid dynamics; numerical methods. Office: Iowa State U 498 Town Engring Bldg Ames IA 50010

ANDERSON, DANIEL R(AYMOND), psychology educator, consultant; b. South Milwaukee, Wis., Sept. 5, 1944; s. Raymond M. and Lillian M. (Nurmi) A. B.S. with honors, U. Wis.-Madison, 1966; Am.M., Brown U., 1968, Ph.D., 1971. Asst. prof. psychology U. Mass., 1970-76, assoc. prof., 1972-82, prof., 1982—; cons. in field. Contbr. chpt., articles on TV viewing to profl. jours.; editor: Children's Understanding of TV, 1983. NSF grantee, 1973—; NIMH grantee, 1978—; Markle Found. grantee, 1982. Mem. Am. Psychol. Assn., Soc. Research in Child Devel., Internat. Communication Assn., Psychonomic Soc., Sigma Xi. Subspecialties: Developmental psychology; Cognition. Current work: Research on cognitive processing of television. Office: Dept Psychology U Mass Amherst MA 01003

ANDERSON, DAVID CARLETON, mechanical engineer, educator; b. Pontiac, Mich., July 14, 1948; s. Norwood Keith and Shirley Ellen (Williams) A.; m. Vicki Ann Niewoehner, Aug. 29, 1970; children: Christopher David, Jonathan Robert. B.S. in Mech. Engring. with highest distinction, Purdue U., 1970, M.S., 1971, Ph.D., 1974. Mem. faculty Purdue U., West Lafayette, Ind., 1975—; assoc. prof. mech. engring. and computer sci. Sch. Mech. Engring., 1979—; dir. Computer Aided Design and Graphics Lab., Inst. for Interdisciplinary Engring. Studies, A. Potter Engring. Ctr., 1981—; cons. Kodak Co.,

Rochester, N.Y., 1980, Whirlpool Corp., Benton Harbor, Mich., 1981, Control Data Corp., La Jolla, Calif., 1978—. Contbr. articles to profl. publs. Purdue U.-NASA grantee, 1973-76, 77-78, 79-82; Siegesmund Fund grantee, 1975-77; V.P. Reilly Fund grantee, 1975-77; Structural Dynamics Research Corp. grantee, 1978-79; Control Data Corp. grantee, 1979-83. Mem. ASME, Assn. Computing Machinery. Subspecialties: Mechanical engineering; Graphics, image processing, and pattern recognition. Current work: Computer aided design; computer graphics; mechanical engineering design. Office: Purdue Univ Mech Engring West Lafayette IN 47907

ANDERSON, DAVID WALTER, physicist, educator; b. Heron Lake, Minn., June 18, 1937; s. Walter O. and Martha G. (Anderson); m. Jane L. Friedlund, Dec. 17, 1960; children: Bonnie Jean, Brian David. B.S., Hamline U., 1959; Ph.D., Iowa State U., 1965. Diplomate: in radiol. physics Am. Bd. Radiology. Prof. radiation physics U. Okla. Health Scis. Ctr., Oklahoma City, 1975-82; dir. radiol. physics, prof. radiology City of Faith Med. and Research Ctr., Tulsa, 1982—; cons. in radiol. physics, 1972—. Contbr. numerous articles to profl. jours. AEC postdoctoral research fellow, 1965-66. Mem. Am. Assn. Physicists in Medicine, Am. Phys. Soc., Am. Coll. Radiology. Democrat. Methodist. Subspecialties: Biophysics (physics); Biomedical engineering. Current work: Radiological physics, radiation dosimetry, photonuclear reactions. Office: Radiology Dept 8181 S Lewis Ave Tulsa OK 74136

ANDERSON, DON LYNN, educator, geophysicist; b. Frederick, Md., Mar. 5, 1933; s. Richard Andrew and Minola (Phares) A.; m. Nancy Lois Ruth, Sept. 15, 1956; children: Lynn Ellen, Lee Weston. B.S., Rensselaer Poly. Inst., 1955; M.S., Calif. Inst. Tech., 1959, Ph.D., 1962. With Chevron Oil Co., Mont., Wyo., Calif., 1955-56; with Air Force Cambridge Research Center, Boston, 1956-58, Arctic Inst. N.Am., 1958; mem. faculty Calif. Inst. Tech., Pasadena, 1962—, asso. prof. geophysics, 1964-68, prof., 1968—, dir. seismol. lab., 1967—; Prin. investigator King Mars Seismic Expt.; mem. various coms. NASA, Nat. Acad. Scis.; chmn. seismology com. Nat. Acad. Sci., 1975, mem. Acad., 1982—. Asso. editor: Jour. Geophys. Research, 1965-67, Tectonophysics, 1974-77; editor: Physics of the Earth and Planetary Interiors. Recipient Exceptional Sci. Achievement award NASA, 1977; Sloan Found. fellow, 1965-67. Fellow Am. Geophys. Union (James B. Macelwane award, 1966, pres. tectonophysics sect. 1971-72, chmn. Macelwane award com. 1975), Geol. Soc. Am. (asso. editor bull. 1971—); mem. Am. Acad. Arts and Scis., AAAS, Royal Astron. Soc., Seismol. Soc. Am., Sigma Xi. Subspecialty: Geophysics. Current work: Composition of earth, seismology. Home: 669 E Alameda St Altadena CA 91001 Office: 1201 E California Blvd Pasadena CA 91109

ANDERSON, DONALD KEITH, chem. engr.; b. Iron Mountain, Mich., July 15, 1931; s. Milton Eugene and Edna Olive (Van Court) A.; m. Gina Dale Garrett, July 12, 1957; children—Shannon Elizabeth, Amanda Juliet. B.S., U. Ill., 1956; M.S., U. Wash., 1958, Ph.D., 1960. Asst. prof. chem. engring. Mich. State U., 1960-64, asso. prof., 1964-69, prof. chem. engring. and physiology, 1970—, chmn. dept. chem engring., 1977—; cons. to industry. Contbr. numerous articles to profl. jours. Recipient Disting. Faculty award Mich State U., 1973. Mem. Am. Inst. Chem. Engrs., Am. Chem. Soc., Am. Soc. Engring. Edn., Sigma Xi, Tau Beta Pi, Sigma Tau, Phi Lambda Upsilon, Omega Chi Epsilon. Club: Lions. Subspecialties: Chemical engineering; Physiology (biology). Current work: Mass transfer, diffusion. Office: Dept Chem Engring Mich State U East Lansing MI 48824

ANDERSON, DUWAYNE MARLO, earth and polar scientist, university dean; b. Lehi, Utah, Sept. 9, 1927; s. Duwayne LeRoy and Fern Francell (Fagan) A.; m. June B. Hodgin, Apr. 2, 1980; children by previous marriage: Lynna Nadine, Christopher Kent, Lesleigh Leigh. B.S., Brigham Young U., 1954; Ph.D. (Purdue Research Found. fellow), Purdue U., 1958. Prof. soil physics U. Ariz., Tucson, 1958-63; research scientist, chief earth scis. br. (Cold Regions Research and Engring. Lab.), Hanover, N.H., 1963-76; chief scientist, div. polar programs NSF, Washington, 1976-78, mem. Viking sci. team, 1969-76; dean faculty natural scis. and math. SUNY, Buffalo, 1978—, Pegrum lectr., 1980; cons. NASA, 1964—, NSF, 1979—; sr. U.S. rep., Antarctica, 1976, 77, vis. prof., lectr. numerous univs. Editor: (with O.B. Andersland) Geotechnical Engineering for Cold Regions, 1978; Cons. editor: Soil Sci, 1965—, (with O.B. Andersland) Cold Regions Sci. and Tech, 1978-82; Contbr. numerous sci. and tech. articles to profl. jours. Bd. dirs. Ford K. Sayre Meml. Ski Council, Hanover, 1969-71; bd. dirs. Grafton County Fish and Game Assn., 1965—, pres., 1968-70; bd. dirs. Hanover Conservation Council, 1970-76, v.p., 1970-73; bd. dirs. Buffalo Mus. Sci., 1980—, v.p., 1982—. Served in USAF, 1946-49. Recipient Sci. Achievement award Cold Regions Research and Engring. Lab., 1968; Sec. of Army Research fellow, 1966. Fellow Am. Soc. Agronomy; mem. Internat. Glaciological Soc., Am. Polar Soc., Am. Geophys. Union, AAAS, Soil Sci. Soc. Am., Niagara Frontier Assn. Research and Devel. Dirs. (pres. 1983), Sigma Xi, Sigma Gamma Epsilon. Republican. Subspecialties: Organic geochemistry; Geochemistry. Current work: Behavior of earth materials at low temperatures; surface chemistry of clay minerals; planetary geology. Home: 188 Koster Row Amherst NY 14226 Office: Clemens Hall SUNY Buffalo NY 14260

ANDERSON, EDWARD EVERETT, university dean, researcher, author; b. Algonia, Iowa, Jan. 9, 1941; s. Everett Joseph and Juen Arlene (Rasmussen) A.; m. Sharon Ann Sanders, Apr. 13, 1963; children: David Edward, Julie Ann, Jill Leigh. B.S. in Mech. Engring, Iowa State U., 1964, M.S., 1966; Ph.D., Purdue U., 1972. Registered profl. engr., Iowa. Vis. asst. prof. Iowa State U., Ames, 1971-72; asst. prof. U. Southwestern La., Lafayette, 1972-74, S.D. Sch. Mines & Tech., Rapid City, 1974-76; prof. U. Nebr., Lincoln, 1976—, asst. dean engring. and tech., 1980—. Author: Fundamentals of Solar Energy Conversion, 1983; also articles. Mem. ASME, AIAA. Lutheran. Subspecialties: Mechanical engineering; Fluid mechanics. Current work: Alternate energy sources, heat transfer, academic administration, engineering research, engineering education. Home: 5208 Cameron Ct Lincoln NE 68512 Office: U Nebr W181 Nebr Hall Lincoln NE 68588

ANDERSON, ERIC EDWARD, health psychology educator, consultant; b. Mpls., Jan. 24, 1951; s. Charles Eric and Elizabeth B. (Engstr) A.; m. Florence Kaye, June 18, 1978; 1 dau., Cara Elizabeth. B.A. in Psychology summa cum laude, U. Minn.-Mpls., 1973; M.A. in Theology Fuller Theol. Sem, 1977; Ph.D. in Clin. Psychology, 1978. Lic. cons. psychologist, Minn. Research asst. Fuller Theol. Sem., 1973-76; instr. in psychology Pasadena City Coll., 1976-78; intern in psychology Long Beach (Calif.) VA Hosp., 1977-78; postdoctoral intern U. Minn.,-Mpls., 1978-79, asst. prof. health psychology, coordinator tng. in aging, 1979—; dir. profl. services Kiel Clinic. Mpls., St. Paul, 1980—; cons. in field; expert witness First Examiner Hennepin County (Minn.) 4th Jud. Dist., 1982—. Contbr. chpt., articles to profl. publs.; editorial cons.: Jour. Gerontology, 1981—. Recipient Outstanding Achievement award Am. Acad. Achievement, 1969. Mem. Am. Psychol. Assn., Minn. Psychol. Assn., Gerontol. Soc. Am., Minn. Pub. Health Assn., Phi Beta Kappa. Subspecialties: Developmental psychology; Behavioral psychology. Current work: Psychology of aging, promotion of mental health, life span development, health psychology. Office: Program in Health Psychology Sch Public Health U Minn Box 717 Mayo Minneapolis MN 55455

ANDERSON, FRANZ ELMER, sedimentologist; b. Cleve., July 23, 1938; s. Elmer C. and Flo (Jordan) A.; m. Harmony Wilson, Apr. 22, 1965 (div. Juen 1973); children: Toren, Tristan; m. Ann Laundon, July 31, 1976; 1 dau., Brynanne. B.A., Ohio Wesleyan U., 1960; M.S., Northwestern U., Evanston, Ill., 1962; Ph.D., U. Wash.-Seattle, 1967. Research asst. U. Wash.-Seattle, 1965-67; asst. prof. U. N.H.-Durham, 1967-71, assoc. prof., 1971-82, prof. dept. earth scis., 1982—. Author: Introduction to Oceanography Lab Manual, 1978. Fulbright grantee, Izmir, Turkey, 1973; Nat. Environ. Research Council sr. fellow, Scotland, 1980-81. Fellow Geol. Soc. Am; mem. Soc. Econ. Paleontologists/Mineralogists, Internat. Assn. Sedimentologists, Marine Edn. Assn., New Eng. Estuarine Research Soc. Subspecialties: Sedimentology; Oceanography. Current work: Estuarine sedimentation, especially processes which control intertidal fine-grained sediment erosion and deposition. Home: 74 Shore Rd Cape Neddick ME 03902 Office: Dept Earth Scis U NH Durham NH 03824

ANDERSON, GARY BRUCE, animal scientist, physiologist, educator; b. Kent City, Mich., Apr. 21, 1947; s. Russell William and Frances (Johnson) A.; m. Dianne Lynn Anderson, Aug. 6, 1977; 1 dau., Ann Marie. B.S., Mich. State U., 1969; Ph.D., Cornell U., 1973. Asst. prof. animal sci. U. Calif.-Davis, 1974-78, assoc. prof., 1978-83, prof., 1983—. Mem. Soc. Study of Reproduction, Am. Soc. Animal Sci., Internat. Embryo Transfer Soc. Subspecialties: Animal breeding and embryo transplants; Genetics and genetic engineering (biology). Current work: Preimplantation embryo development; embryo culture; embryo transfer in laboratory and farm animal species. Office: Dept Animal Science U Calif Davis CA 95616

ANDERSON, GREGORY JOSEPH, biologist, educator; b. Chgo., Nov. 26, 1944. B.S. (Kiwanis Club of Pipestone, Minn., fellow), St. Cloud State Coll., 1966; Ph.D. (Floyd Meml. fellow), Ind. U., 1971. Asst. prof. U. Nebr., Lincoln, 1971-73; assoc. prof. biology U. Conn., 1973—; vis. scientist Research Inst., U. Adelaide, South Australia, Australia, 1979-80; Sigma Xi grantee, U. Nebr. Research Council grantee, S. Am., 1972; mem. organizing com. 2d Internat. Symposium on Solanaceae, 1981—; speaker profl. seminars. Contbr. articles, chpts., revs. to profl. publs. U. Conn. Research Found. grantee, 1973, 74, 75, 79, 80, 81; NSF grantee, 1978, 79, 81. Mem. AAUP, Am. Soc. Naturalists, Am. Soc. Plant Taxonomists (George R. Cooley award 1981—), Bot. Soc. Am., Internat. Assn. Plant Taxonomy, Soc. Econ. Botany (governing council 1978-81, sec. 1981—), Soc. Study Evolution, Sigma Xi. Club: Joshuas Tract Land Trust. (vice chmn.). Subspecialties: Evolutionary biology; Systematics. Current work: Plant systematics and evolution; economic botany; reproductive and pollination biology. Office: U Conn Biol Sci Group Box U-43 Storrs CT 06268

ANDERSON, JAMES ALFRED, psychologist, neuroscientist, researcher, educator; b. Detroit, July 31, 1940; s. Courtney Alfred and Catherine Plummer (Bullock) A.; m. Diana M. De Vincenzi, Nov. 1, 1969; 1 son, Eric David. S.B., M.I.T., 1962, Ph.D., 1967. Postdoctoral fellow depts. anatomy, physiology Brain Research Inst., Space Biology Lab., UCLA, 1967-71; research asso., postdoctoral fellow Rockefeller U., N.Y.C., 1971-73; asst. prof. divs. applied math. and biol. and med. scis. Brown U., Providence, 1973-77, asst. prof. psychology and neurosci., 1977-78, assoc. prof. dept. psychology and Center for Neural Sci., 1978—. Editor: (with Geoffrey Hinton) Parallel Models of Associative Memory, 1981; contbr. articles to profl. jours. NSF grantee, 1977-78, 80—. Mem. Soc. for Neurosci., Soc. for Math. Psychology, Psychonomic Soc., Cognitive Sci. Soc., Sigma Xi. Subspecialties: Cognition; Neurophysiology. Current work: Mathematical models of cognition; models of cognitive function inspired by the structure and properties of the nervous system. Home: 22 Blaisdell Ave Pawtucket RI 02860 Office: Dept Psychology Brown U Providence RI 02912

ANDERSON, JAMES ARTHUR, mining company executive; b. Aurelia, Iowa, Mar. 25, 1935; s. Vernon and Agnes (Weil) A.; m. Ann Charlene Sutherland, Sept. 9, 1956. B.S. in Geol. Engring, U. Utah, 1957; M.S. in Mining Geology, Harvard U., 1960; Ph.D. in Enon. Geology, Harvard U., 1965; M.B.A., Stanford U., 1978. Registered geologist, Calif. Mem. geol. staff Kennecott Copper Corp., Salt Lake City, 1960-68; v.p., mgr. U.S. metal exploration Occidental Minerals Corp., Denver, 1968-75; v.p. gen. mgr. Homestake Mining Co., San Francisco, 1975-81, exec. v.p. exploration and bus. devel., 1981—, also dir. Contbr. chpt. to book and article in field to profl. jour. Mem. bd. Calif. State Mining & Geology Bd., 1978—. Mem. AIME(mining engrs.), Geol. Soc. Am., Calif. Mining & Mettallurgical Soc. Am., World Affairs Council, Calif., Colo. Mining Assn. Republican. Methodist. Clubs: Commonwealth of Calif., Bankers of San Francisco. Subspecialties: Geology; Economic geology; mining geology. Current work: Management of all aspects of metals exploration and development, business development and government relations in United States and foreign countries. Office: Homestake Mining Co 650 California St 9th floor San Francisco CA 94108

ANDERSON, JAMES BRYAN, mathematician, engineer; b. Sandusky, Ohio, Aug. 23, 1950; s. Rodney Jack and Ann Marie (Glenn) A. B.S., U. Toledo, 1977, M.S., 1980, M.S. in Engring. Sci, 1982. Systems analyst (project leader) Teledyne Continental Aircraft Co., Toledo, 1980-82; sr. engr. FMC Corp., Mpls., 1982—. Contbr. articles to profl. jours. Mem. Am. Soc. Computing Machinery, Soc. Indsl. and Applied Math., AIAA, Pi Mu Epsilon, Eta Kappa Nu. Subspecialties: Numerical analysis; Mathematical software. Current work: Laser welding simulation, finite difference schemes for partial differential equations, two point boundary value problems associated with optimal process control; system reliability simulation. Home: 7862 Yates Ave N Brooklyn Park MN 55443 Office: FMC Corp No Ordnance 4800 East River Rd Minneapolis MN 55421

ANDERSON, JAMES HENRY, ednl. adminstr.; b. Odum, Ga., Jan. 11, 1926; s. James Tillman and Mamie (Aspinwall) A.; m. Dorothy Allen, Dec. 29, 1951; children—Alicia Carol, Laurie Beth, James Hampton, Sue Ellen, John Allen. B.S., U. Ga., 1949; M.S., N.C. State U., 1955; Ph.D., Iowa State U., 1957. Bar: Registered profl. engr., Miss. Prof., head agrl. engring. dept U. Tenn., Knoxville, 1960-61; prof., head agrl. engring. dept. Miss. State U., Starkville, 1961-68; dean resident instrn. Coll. Agr., 1967-68; dir. Miss. Agrl. and Forestry Exptl. Sta., Starkville, 1969-77; dean Coll. Agrl. and Natural Resources, Mich. State U., Lansing, 1977—. Contbr. articles to various publs. Mem. exec. com S. Bapt. Conv., 1973-77. Served with U.S. Army, 1944-46. Fellow Am. Soc. Agrl. Engrs. Baptist. Club: Rotary. Subspecialty: Agricultural engineering. Current work: Administration of broad and comprehensive agriculture research program. Home: 3850 Roxbury Ave Okemos MI 48864 Office: Michigan State U 104 Agriculture Hall East Lansing MI 48824

ANDERSON, JAMES HILBERT, mechanical engineer; b. Lewisburg, Pa., Oct. 20, 1908; s. John and Annie Mahalia (Wolford) A.; m. Emilie Frederica Aschbach, Oct. 6, 1937; children: Gretchen, James Hilbert. B.S. in M.E, Pa. State U., 1930, M.S., 1933. Registered profl. engr., Pa., Nev., Calif. Jr. engr. Ingersoll Rand Co., Phillipsburg, N.J., 1930-32; marine engr. Grace Lines, N.Y.C., 1933-34; mgr. research Ingersoll Rand Co., 1934-52; chief engr. York div. Borg-Warner Co., York, Pa., 1952-63; pres. Sea Solar Power, Inc., 1973—, J. Hilbert Anderson, Inc., 1974—; cons. engr., York, 1963—. Contbr. articles to profl. jours. Fellow ASME; mem. ASHRAE, Pa. Soc. Profl. Engrs. Lutheran. Lodge: Rotary. Subspecialty: Mechanical engineering. Patentee in field. Office: 2422 S Queen St York PA 17402

ANDERSON, JAMES OTTO, research plant scientist, consultant, educator; b. Eighty Four, Pa., Aug. 3, 1945; s. Otto Frank and Ethel Marie (Irey) A.; m. Martha Alene Cain, Sept. 9, 1966; 1 dau., Heather Lynn. B.S. in Edn, Calif. (Pa.) State Coll., 1967; Ph.D., W.Va. U., 1972. NDEA fellow in plant sci. W.Va. U., Morgantown, 1967-71, Inst. Biol. Sci. fellow, 1971-72; vis. scientist Inst. Low Temperature Sci., Hokkaido U., Sapporo, Japan, 1972-73; research assoc. U.S. Dept. Agr. Pasture Lab., Pa. State U., University Park, 1973-75; postdoctoral research scientist W. Alton Jones Cell Sci. Ctr., Lake Placid, N.Y., 1975-76; adj. asst. prof. plant sci. U. Ariz., Tucson, 1976—; pres. Vari-Ident Labs., Tucson, 1982—. Author computer based inventory program. Japan Soc. Sci. fellow, 1972. Mem. Soc. Cryobiology, Am. Soc. Plant Physiologists, Internat. Soc. Plant Tissue Culture, N.Y. Acad. Scis., Sigma Xi. Clubs: Tucson Rod and Gun, W.Va. Wildwater. Subspecialties: Genetics and genetic engineering (agriculture); Plant cell and tissue culture. Current work: Plant biotechnology (plant tissue culture, protoplast fusion), plant germplasm conservation and biochemical identification of varieties, plant physiology (crop productivity and stress). Office: Dept Plant Scis Univ Ariz Tucson AZ 85721 Home: 5075 E Cooper St Tucson AZ 85711

ANDERSON, JAY OSCAR, poultry nutritionist; b. Brigham, Utah, Dec. 5, 1921; s. Oscar Carl and Ada Ruth (Petersen) A.; m. Nelda Elaine Huber, Jan. 12, 1944; children: Wayne J., Raymond C., Russell K., Elaine Anderson Young, Dale H., Keith J. M.S., U. Md., 1948, Ph.D., 1950. Chemist, Merck & Co., Rahway, N.J., 1950-51; asst. prof. animal sci. Utah State U., Logan, 1951-54, asso. prof., 1954-59, prof., 1959—. Contbr. articles to profl. jours. Served to 1st lt. U.S. Army, 1943-46. Mem. Poultry Sci. Assn., Am. Inst. Nutrition, AAAS, World Poultry Sci. Assn., Sigma Xi, Phi Kappa Phi. Mormon. Subspecialties: Animal nutrition; Nutrition (biology). Current work: Protein and amino acid requirements of poultry.

ANDERSON, KINSEY A., physicist, educator; b. Preston, Minn., Sept. 18, 1926; s. Malvin R. and Allene (Michener) A.; m. Lilica Athena Vassiliades, May 29, 1954; children—Danae, Sindri. B.A., Carleton Coll., 1949; Ph.D., U. Minn., 1955. Research assoc., faculty U. Iowa, Iowa City, 1955-59; Guggenheim fellow Royal Inst. Tech., Stockholm, Sweden, 1959-60; faculty U. Calif. at Berkeley, 1960—, prof. physics, 1966—; dir. Space Sci. Lab., 1970-79; Cons. NSF, NASA. Contbr. numerous articles to profl. jours. Mem. Nat. Acad. Scis., Am. Geophys. Union, Am. Phys. Soc., Am. Astron. Soc., Internat. Astron. Union, Phi Beta Kappa, Sigma Xi. Subspecialty: Solar physics. Research in space plasma physics, magnetospheric particles, magnetic and electric fields using balloon, rocket and satellite instruments. Home: 8321 Buckingham Dr El Cerrito CA 94530 Office: Space Sci Lab U Calif Berkeley CA 94720

ANDERSON, KURT STEVEN JARL, astronomer, educator; b. Rockford, Ill., Oct. 8, 1941; s. Edwin A. and Karin (Jarl) A.; m. Marcia Beth Prunty, Sept. 7, 1968; children: Sara M., Joshua C. B.A. in Astronomy, Calif. Inst. Tech., 1963, Ph.D. in Astronomy and Astrophysics, 1969. Postgrad. research astronomer Lick Obs., U. Calif.-Santa Cruz, 1968-70; research fellow Harvard Coll. Obs., 1970-72; vis. astronomer Kitt Peak Nat. Obs., 1972-73; asst. prof. physics U. Houston, 1973-74; asst., then assoc. prof. astronomy N.Mex. State U.-Las Cruces, 1974—, head dept., 1981—; bd. dirs. Southwest Regional Conf. Astronomy and Astrophysics. Contbr. articles to profl. jours. Mem. Am. Astron. Soc., Astron. Soc. of Pacific, N.Mex. Acad. Sci., Planetary Soc. Subspecialties: Optical astronomy; Theoretical astrophysics. Current work: Nuclei of active galaxies, instrumentation, atmospheric optics. Academic department head, graduate and undergraduate teaching; optical astronomical observations, photometry and spectroscopy, modeling of active galactic nuclei. Home: 204 W Ethel St Las Cruces NM 88005 Office: Dept Astronomy N Mex State U Las Cruces NM 88003

ANDERSON, LLOYD LEE, animal science educator; b. Nevada, Iowa, Nov. 18, 1933; s. Clarence and Carrie G. A.; m. JaNelle R. Sanny, June 15, 1970; children: Marc C., James R. B.S., Iowa State U., 1957, Ph.D., 1961. NIH fellow Iowa State U., Ames, 1961-62, asst. prof. animal sci., 1961-65, assoc. prof. animal sci., 1965-71, prof., 1971—. Contbr. numerous articles on animal sci. to profl. jours. Served with U.S. Army, 1953-55. Lalor Found. fellow Inst. Nat. Research Agronomy, Jouy-en-Josas, France, 1963-64; NIH fellow, 1962. Mem. Endocrine Soc., Am. Physiol. Soc., Am. Assn. Anatomists, Soc. for Exptl. Biology and Medicine (council 1980-83), Am. Soc. Animal Sci., Soc. for Study of Reproduction, Sigma Xi, Gamma Sigma Delta. Methodist. Subspecialties: Animal physiology; Neuroendocrinology. Current work: Hypothalamic regulation of pituitary hormone secretion in beef cattle and pigs, neurosurgery, radioimmunoassay of protein and steroid hormones, relaxin, growth and reproduction. Home: 1703 Maxwell Ave Ames IA 50010 Office: Dept Animal Sci Iowa State U 11 Kildee Hall Ames IA 50011

ANDERSON, LOUISE ELEANOR, biochemistry educator; b. Cleve., May 18, 1934; d. Bertil Gottfrid and Lorraine Dorothy (Ossian) A. A.B., Augustana Coll., Rock Island, Ill., 1956; Ph.D., Cornell U., 1961. Research assoc. Washington U., St. Louis, 1960-62, Dartmouth Med. Sch., Hanover, N.H., 1962-64, 66-67; Kettering Found. internat. fellow Sydney (Australia) U., 1964-65; research assoc. U. Tenn.- Oak Ridge Grad. Sch., 1967-68; asst. prof., prof. biology dept. U. Ill.-Chgo., 1968—; panel mem. NSF, 1979-82; mem. nat. com. for internat. union biol. scis. Nat. Acad. Scis., 1980—. Contbr. chpt. to book, articles to profl. jours. Katzir-Katchalsky fellow Weizmann Inst., Rehobot, Israel, 1974-75; Fogarty internat. fellow NIH, 1981-82; research grantee NSF, Dept. Energy, U.S. Dept. Agr.; recipient Outstanding Achievement award Augustana Coll. Alumni Assn., 1979. Mem. Am. Soc. Biol. Chemists, Am. Soc. Plant Physiologists. Lutheran. Subspecialties: Photosynthesis; Biochemistry (biology). Current work: Control of carbon metabolism in green plants; SO2 effects on green plants. Home: 1130 S Michigan Apt 1103 Chicago IL 60605 Office: Biol Scis Univ Ill Box 4348 Chicago IL 60680

ANDERSON, MAURITZ GUNNAR, biology educator, consultant, researcher; b. Chgo., Aug. 11, 1918; s. Carl Gunnar and Signe (Holme) A.; m. Jeannette Stacey, Dec. 12, 1942; children: William Stacey, John Gunnar. A.B., U. Mich., 1942; postgrad., Harvard U., 1946-47; M.S. in Botany, Ind. U., 1962, Northwestern U. Med. Ch., 1953-54; Ph.D. in Mycology, Va. Poly. Inst. and State U., 1976. Supr. histology and microscopy div. Swift and Co., Chgo., 1949-56, research histologist and microscopist, 1958-63; chief histologist Norwich Pharm. Co., N.Y., 1957; instr. Towson (Md.) State U., 1963-64, asst. prof., 1964-69, assoc. prof., 1969-82, prof., 1982—; cons. McCrone Assocs., Chgo., McCormick Spices, Balt., Best Foods Products, Union, N.J., T.J. Lipton Co., Englewood Cliffs, N.J. Bd. dirs. YMCA, Wheaton, Ill., 1959,60, Y Men's chmn., 1953-54; organizer, chmn. Y Men's Club; organizer and coach swim team YMCA, 1953; swim team coach City of Wheaton, 1964; pres. PTA, Wheaton, 1960. Recipient Spl. Service award YMCA, Wheaton, 1962; Towson State U. yearbook dedicated in honor, 1968. Mem. Am. Bot. Soc., Mycological Soc. Am., Inst. Food Technologists (nat. and local chpts.), Am. Assn. Feed Microscopists, Am. Inst. Biol. Scis., Internat. Soc. Aerobiology, AAUP, Sigma Xi, Beta Beta Beta, Phi Gamma Delta. Clubs: Ind. U. Alumni, U. Mich., Va. Poly. Inst. and State U. Alumni., M of U. Mich. Subspecialties: Microbiology; Microscopy. Current work: Nutrition of aquatic molds and microscopy of industrial products. Discoverer method of puffing bacon rinds, 1958, method of destroying Trichinella spiralis in meat infested therewith, 1967. Home: 18 Maryland Ave Towson MD 21204 Office: Dept Biol Scis Towson MD 21204

ANDERSON, NEIL ALBERT, educator; b. Mpls., Oct. 21, 1928; s. Seth Albert and Hedvig Mary (Johnson) A.; m. Barbara Ann Anderson, Aug. 6, 1960; children: Mary, Elizabeth. B.S., U. Minn., 1951, M.S., 1957, Ph.D., 1960. Asst. prof. plant pathology U. Minn., St. Paul, 1960-64, asso. prof., 1964-70, prof., 1970—. Served with U.S. Army, 1952-54. Mem. Am. Phytopath. Soc., Mycological Soc. Am., AAAS, Sigma Xi. Subspecialties: Plant pathology; Gene actions. Current work: Researcher in genetics of parasitism. Office: University of Minnesota 204 Stakman Hall Dept Plant Pathology St Paul MN 55108

ANDERSON, PER HOLME, astrophysicist; b. Herning, Denmark, Aug. 31, 1945; emigrated to Can., 1954, naturalized, 1968; s. Carl and Vera Johanne (Petersen) A. B.Sc. with honors in Astronomy, U. Victoria, B.C., 1968; M.Sc. in Astronomy, U. Western Ont., 1970, Ph.C., U. Wash.-Seattle, 1973. Lectr. dept. physics and astronomy Brandon (Man.) U., 1979-80; asst. prof. dept. physics Brock U., St. Catharines, Ont., 1980-81; asst. prof. dept. math. and astronomy U. Man.-Winnipeg, 1981—, NRC of Can. scholar, 1968-70. Contbr. articles in field to profl. jours. Mem. Am. Astron. Soc., Canadian Astron. Soc. Subspecialties: Optical astronomy; Theoretical astrophysics. Current work: Theory and observation of stellar atmospheres, mass loss, spectral line formation, novae. Office: U Manitoba 347 University Coll Winnipeg MB Canada R3T 2N2

ANDERSON, PHILIP WARREN, physicist; b. Indpls., Dec. 13, 1923; s. Harry W. and Elsie (Osborne) A.; m. Joyce Gothwaite, July 31, 1947; 1 dau., Susan Osborne. B.S., Harvard U., 1943, M.A., 1947, Ph.D., 1949; D.Sc. (hon.), U. Ill., 1979. Mem. staff Naval Research Lab., 1943-45; mem. tech. staff Bell Telephone Labs., Murray Hill, N.J., 1949—, chmn. theoretical physics dept., 1959-60, asst. dir. phys. research lab., 1974-76, cons. dir., 1976—; Fulbright lectr. U. Tokyo, 1953-54; Loeb lectr. Harvard U., 1964; prof. theoretical physics Cambridge (Eng.) U., 1967-75; prof. physics Princeton U., 1975—; Overseas fellow Churchill Coll., Cambridge U., 1961-62; fellow Jesus Coll., 1969-75, hon. fellow, 1978—. Author: Concepts in Solids, 1963. Recipient Oliver E. Buckley prize Am. Physical Soc., 1964; Dannie Heinemann prize Göttingen (Ger.) Acad. Scis., 1975; Nobel prize in physics, 1977; Guthrie medal Inst. of Physics, 1978; Nat. Medal Sci., 1982. Fellow Am. Phys. Soc., Am. Acad. Arts and Scis., AAAS; mem. Nat. Acad. Scis., Royal Soc. (fgn.), Phys. Soc. Japan, European Phys. Soc. Subspecialty: Theoretical physics. Research in quantum theory, especially theoretical physics of solids, spectral line broadening, magnetism, superconductivity. Address: Bell Telephone Labs Murray Hill NJ 07974

ANDERSON, RALPH ROBERT, animal science educator; b. Fords, N.J., Nov. 1, 1932; s. Harry W. and Johanna K. (Damgaard) A.; m. LaVeta Ann Phillips, Jan. 28, 1961; children: Richard, Laura. B.S., Rutgers U., 1953, M.S., 1958; Ph.D., U. Mo.-Columbia, 1961. Cert. animal scientist. Instr. U. Mo., Columbia, 1961-62, asst. prof., 1965-68, assoc. prof., 1968-76, prof. animal sci., 1976—; asst. prof. Iowa State U., Ames, 1962-64; trainee in endocrinology U. Wis., Madison, 1964-65. Editor: Relaxin, 1982; co-editor: The Endocrine Pancreas, 1979, Hormones and Energy Metabolism, 1979, The Renin-Angiotension System, 1980. Served with U.S. Army, 1954-56. NIH predoctoral fellow U. Mo., 1960-61, U. Wis., 1964-65; Fulbright fellow, N.Z., 1973-74; recipient Grad. Superior Teaching award Gamma Sigma Delta, U. Mo., 1982. Mem. Endocrine Soc., Am. Physiol. Soc., Am. Dairy Sci. Assn. (bd. editors), Sigma Xi. Club: Track. Lodge: Kiwanis. Subspecialties: Animal physiology; Endocrinology. Current work: Hormones effecting mammary growth and lactation; present emphasis is on effects of relaxin, placental lactogen, estrogens and progesterone in vivo and in vitro. Home: 2517 Shepard Blvd Columbia MO 65201 Office: Dept Dairy Sci Univ Mo 162 Animal Sci Research Center Columbia MO 65211

ANDERSON, REBECCA J., pharmacologist, educator; b. Ft. Madison, Iowa, Aug. 21, 1949; s. Charley R. and Anna L. (Herrmuth) A. B.A. cum laude, Coe Coll., Cedar Rapids, Iowa, 1971; Ph.D. with distinction, Georgetown U., 1975. Adj. assoc. prof. toxicology U. Mich., 1983—; Med. Research Council Can. fellow U. Toronto, 1975-77; asst. prof. pharmacology George Washington U., 1977-80, assoc. prof., 1980-83; group leader neuro. diseases Warner-Lambert Co., Ann Arbor, Mich., 1983—; cons. Bur. Med. Devices, FDA, Life Systems, Inc. Contbr. articles to profl. jours. Mem. Soc. Neurosci., Am. Soc. Pharmacology and Exptl. Therapeutics, Am. Eplepsy Soc., Am. Chem. Soc., AAAS, Biometric Soc., Sigma Xi, Phi Kappa Phi. Subspecialties: Neuropharmacology; Toxicology (medicine). Current work: Neuropharmacology of motor systems, environmental and industrial neurotoxicology. Office: Dept Pharmacology Warner-Lambert Co 2800 Plymouth Rd Ann Arbor MI 48105

ANDERSON, RICHARD COOPER, consultant, electronics engineer; b. Berkeley, Calif., Jan. 30, 1935; s. Harold Rudolph and Elsie Carolyn (Karsten) A.; m. Linda Euler, July 21, 1962; children: Kristin Ruth, Karen Lynn. M.S.E.E., Stanford U., 1957; postgrad., U. Md., 1958-60. Engr. Fairchild Semicondr., Palo Alto, Calif., 1960-66; mgr. computer systems engring. Varian Assocs., Palo Alto, 1966-72; program dir. Am. Videonetics, Sunnyvale, Calif., 1973-75; pres. R.C. Anderson, Inc., Los Altos, Calif., 1975—. Served to lt. comdr. U.S. Navy, 1957-60. Mem. IEEE, Profl. and Tech. Cons. Assn. Subspecialties: Integrated circuits; Computer engineering. Current work: Development of custom integrated circuits; custom integrated circuits, VLSI; silicon foundry; IC design and tooling standards. Patentee in field. Office: 900 N San Antonio Rd Suite 204 Los Altos CA 94022

ANDERSON, ROBERT CLARK, metallurgist consultant; b. Galesburg, Ill., July 18, 1926; s. Clark Leonard and Hildure Josephine (Lofgren) A.; m. Aileen English, Dec. 31. 1954; children: Stephen, Randall. B.S. in Metall. Engring, U. Ill., 1950. Registered profl. engr., Tex., La.; registered corrosion specialist; cert. mfg. engr. Metall. engr. Sheffield div. Armco Steel Corp., Houston, 1950-52, W-K-M div. ACF Industries, 1952-57; pres. Anderson & Assocs., Inc., Houston, 1957—, Metall. supply Co., Andreco Corp., Anderson & Assocs., D.F.W., Inc.; pres. Brazoria Broadcasting Co. Author: Inspection of Metals, Visual Examination, Vol. 1, 1983; contbr. articles to profl. jours. Served with USN, 1944-46. Fellow Am. Soc. for Metals (past chmn. Houston chpt.), Tex. Soc. Profl. Engrs., Nat. Soc. Profl. Engrs., ASTM, Am. Welding Soc. (past chmn. Houston chpt.), ASME, AIME, Nat. Assn. Corrosion Engrs., Houston Engring. and Sci. Soc., Internat. Metallographic Soc., SME, AM. Soc. for Non-Destructive Testing, Am. Council Ind. Labs. Subspecialties: Corrosion. Current work: Failure analysis of metals and associated litigation. Office: 919 F M Rd 1959 Houston TX 77034

ANDERSON, ROBERT SIMPERS, immunologist; b. Bryn Mawr, Pa., Jan. 4, 1939; s. Paul Alexander Anderson and Ella (Trew) Simpers; m. Lucy Anne Macdonald, Aug. 29, 1964; children: Robert Simpers, Donald Paul. B.S., Drexel U., 1961; M.S., Hahnemann Med. U., 1968; Ph.D., U. Del., 1971. Postdoctoral fellow U. Minn., 1970-73; asst. mem. Sloan-Kettering Inst. Cancer Research, 1973-82, lab. head, 1974-82; asst. prof. Cornell U., 1975-82; immunologist Aberdeen Proving Ground, Md., 1982—. Contbr. articles to profl. jours; mem. editorial bd.: Jour. Invertebrate Pathology, 1977, Jour. Comparative and Developmental Immunology, 1977—. Mem. Marshland Conservancy, Rye, N.Y., 1978—. Whitehall Found. grantee, 1974-81; NSF grantee, 1977-83; EPA grantee, 1980-83; Griffis Found. grantee, 1981-83. Mem. Am. Assn. Immunologists, Am. Entomol. Soc., Am. Soc. Zoologists, Internat. Soc. Developmental and Comparative Immunology, N.Y. Acad. Scis., Phila. Physiol. Soc., Soc. Invertebrate Pathology, Sigma Xi. Republican. Episcopalian. Subspecialties: Immunology (agriculture); Environmental toxicology. Current work: The effects of environmental chemicals on immune response. Office: Chem Research and Devel Ctr Toxicology Br Aberdeen Proving Ground MD 21010

ANDERSON, ROGER CLARK, biologist, educator; b. Wausau, Wis., Oct. 30, 1941; s. Jerome Alfred and Virginia S. (Anderson); m. Mary Rebecca Blocher, Aug. 5, 1967; children: John Allen, Nancy Lynn. B.S., Wis. State Coll., 1963; M.S., U. Wis., Madison, 1965, Ph.D., 1968. Asst. prof. botany So. Ill. U., Carbondale, 1968-70; asst. prof. U. Wis., Madison, 1970-73; assoc. prof. biology Central State U., Edmond, Okla., 1973-76; assoc. prof. Ill. State U., Normal, 1976-79, prof., 1980—; dir. U. Wis.-Madison Arboretum, 1973-76. Contbr. articles on biology to profl. jours. Treas. Miller Park Zoo Bd., 1982; pres. So. Ill. Audubon Soc., 1970. NSF grantee, 1981-83; Dept. Agr. grantee, 1982; Ill. Dept. Conservation grantee, 1981-83; Met. San. Dist. Greater Chgo. grantee, 1978-79. Mem. Am. Bot. Soc., Ecol. Soc., Am., Ill. Acad. Sci., Sigma Xi. Subspecialty: Ecology. Current work: Community ecology of Grasslands. Home: 14 McCormick Blvd Normal IL 61761 Office: Dept Biology Ill State U Normal IL 61761

ANDERSON, STEPHEN CLARK, metallurgist; b. Houston, Nov. 13, 1953; s. Robert Clark and Addie Aileen (English) A.; m. Victoria Ellen Kasparian, Nov. 3, 1979. B.S. in Engring. Tech, U. Houston, 1979. With Anderson & Assocs., Inc., Houston, 1970—, v.p., 1978—; pres. Andreco Corp., 1976—, Anderson and Assocs. D.F.W., Inc., 1981—. Mem. Am. Soc. Metals (Young Mem. of Yr. 1982), Am. Welding Soc., Microbeam Analysis Soc., Electron Microscopy Soc. Am. Subspecialties: Metallurgical engineering; Metallurgy. Current work: Metal failure analysis. Office: 919 FM1959 Houston TX 77034

ANDERSON, THOMAS FOXEN, educator, scientist; b. Manitowoc, Wis., Feb. 7, 1911; s. Anton Oliver and Mabel (Foxen) A.; m. Wilma Fay Ecton, Dec. 28, 1937; children—Thomas Foxen, Jessie Dale. B.S., Calif. Inst. Tech., 1932, Ph.D., 1936; student, U. Munich, Germany, 1932-33. Instr. chemistry U. Chgo., 1936-37; investigator botany U. Wis., 1937-39, instr. phys. chemistry, 1939-40; RCA fellow of NRC, 1940-42; asso. Johnson Found., U. Pa., 1942-46; mem. faculty U. Pa., 1946—, prof. biophysics, 1958—; sr. mem. Inst. Cancer Research, 1958—. Fulbright and Guggenheim fellow Inst. Pasteur, Paris, France, 1955-57; recipient Silver medal, 1957. Mem. Nat. Acad. Scis. (chmn. U.S. nat. com. pure and applied biophysics), Electron Microscope Soc. Am. (pres. 1955, Disting. award 1978), Internat. Fedn. Electron Misscroscope Socs. (pres. 1960-64), Biophys. Soc. (pres. 1965), Am. Soc. Naturalists, AAAS, Soc. Gen. Physiologists, Deutsche Gesellschaft für Elektronenmikroskopie, Soc. Francaise de Microscopie Electronique (hon. mem.), Am. Soc. for Microbiology, Sigma Xi. Subspecialties: Molecular biology; Virology (biology). Current work: Electron microscopy of viruses and bacteria; bacterial gentics; structrue and function of bacteriophages and bacteria. Spl. research raman spectra, molecular structure, surface chemistry, biol. applications electron microscopy, genetics and structure viruses and bacteria. Home: 326 Zane Ave Philadelphia PA 19111

ANDERSON, THOMAS RICHARD, cell biologist, researcher; b. Chgo., Nov. 29, 1933; s. William Matthew and Katherine Helen (Weckart) A.; m. Frances Bullock, May 4, 1969; children: Eric, Fleda. B.S., Calif. State U.-Long Beach, 1960, M.A., 1964; Ph.D., U. Ill.-Urbana, 1969. Postdoctoral tng. fellow in neuroscis. Duke u., 1969-71, research assoc. in pharmacology, 1971-75; research analyst in biochemistry of hard tissues Dental Research Ctr. U. N.C.-Chapel Hill, 1975-80, research assoc. prof., 1980—. Contbr. articles to sci. jours. Served to cpl. U.S. Army, 1951-54. Mem. Fedn. Am. Socs. Exptl. Biology, N.Y. Acad. Scis., Sigma Xi. Subspecialties: Biochemistry (biology); Cell biology. Current work: Function of acid phosphatases in iron metabolism. Home: 3040 Rosebriar Dr Durham NC 27705 Office: Dental Research Ctr U NC Chapel Hill NC 27514

ANDERSON, TOM, med. oncologist, educator; b. Ekalaka, Mont., Sept. 4, 1942; s. Walter and Morine (Renshaw) A.; m. Rita Lynn Jurica, Sept 12, 1964; children: Matthew Scott, Gregory Stewart. B.S. cum laude, U. Mont., 1964; M.D., Stanford U., 1969. Diplomate: Nat. Bd. Med. Examiners, Am. Bd. Internal Medicine. Intern Strong Meml. Hosp., Rochester, N.Y., 1969-70, asst. resident, 1970-71; clin. assoc. Solid Tumor Service, Medicine Br., NIH, Behesda, Md., 1971-74; fellow in oncology U. Rochester-Strong Meml. Hosp., 1974-75, chief resident in medicine, 1974-75; sr. investigator, attending physician Med. Br., Nat. Cancer Int., NIH, Bethesda, 1975-79, head sect. clin. pharmacology, 1977-79; assoc. prof., chief sect. hematology-oncology, dept. medicine Med. Coll. Wis., Milw., 1979—. Contbr. articles to profl. jours. Served with USPHS, 1972-74, 75-79. Mem. Am. Assn. for Cancer Research, Am. Soc. Clin. Oncology, Am. Soc. Hematology, Phi Kappa Phi, Alpha Omega Alpha. Subspecialties: Chemotherapy; Hematology. Current work: Chemotherapy of malignant disease. Structure-activity relationships of nitrosoureas, and platinum coordination compounds. Metabolic abnormalities of tumors and/or therapy. Home: 6132 Washington Circle Wauwatosa WI 53213 Office: Milwaukee County Med Complex Box 170 8700 W Wisconsin Ave Milwaukee WI 53226

ANDERSON, VERNON L., animal scientist; b. Watertown, S.D., Dec. 29, 1947; s. George C. and Effie N. (Lundquist) A.; m. Janis K. Bonzer, June 6, 1970; children: Gretchen, Pehr. B.S. in Agrl. Edn, S.D. State U., 1970, M.S. in Animal Sci, 1979. Grad. asst. S.D. State U., 1976-79; asst. animal husbandman Carrington Irrigation Br. Sta. N.D. State U., 1979—. Leader 4-H Club, Foster County, N.D. Served to capt. U.S. Army, 1970-76. Mem. Am. Soc. Animal Sci., Council of Agrl. Sci. and Tech. Republican. Lutheran. Subspecialty: Animal nutrition. Current work: Beef cattle production in confinement; irrigated forage; drylot cow/calf production; forage quality; crop residue utilization. Office: PO Box 219 Carrington ND 58421

ANDERSON, W. FRENCH, research scientist; b. Tulsa, Dec. 31, 1936; s. Daniel French and LaVere Frances (Schoenfelt) A.; m. Kathryn Dorothy Duncan, June 24, 1961. A.B. magna cum laude, Harvard U., 1958, M.D., 1963; M.A. with honors, Cambridge (Eng.) U., 1963. Diplomate: Nat. Bd. Med. Examiners. Intern Children's Hosp., Boston, 1963-64; Head sect. human biochemistry Lab. Clin. Biochemistry, Nat. Heart Intst., NIH, 1968-71, head sect. molecular hematology, 1971-73, chief molecular hematology br., 1973-77, chief, 1977—; adj. prof. grad. genetics program George Washington U.; mem. faculty dept. medicine and physiology NIH Grad. Program. Author 3 books.; contbr. numerous articles to profl. jours. Served as commd. officer USPHS, 1965-67. Recipient ann. award for sci. achievement biol. scis. Washington Acad. Scis., 1971, Superior Service award HEW, 1975, Thomas B. Cooley award for sci. achievement Cooley's Anemia Blood and Research Found. for Children, 1977, Outstanding Performance award HHS, 1982. Mem. Am. Chem. Soc., Am. Fedn. Clin. Research, Am. Soc. Cell Biology, Am. Soc. Biol. Chemists, Am. Soc. Clin. Investigation, Am. Soc. Hematology, Am. Soc. Human Genetics, Assn. Am. Physicians, Peripatetic Club, Hastings Center. Subspecialties: Genetics and genetic engineering (medicine); Molecular biology. Current work: Gene therapy; genetic engineering; molecular genetics. Home: 6820 Melody Ln Bethesda MD 20817 Office: National Institutes of Health Bldg 10 Room 7D 20 Bethesda MD 20205

ANDERSON, WAYNE KEITH, medicinal chemistry educator; b. Pine Falls, Man., Can., Apr. 4, 1941; came to U.S., 1964; s. Sigward Emmanuel and Verna Madelaine (Sorbo) A.; m. Ellen Lorraine Robertson, Aug. 31, 1962; children: Brian Ross, Laura Elizabeth, Shari Lynn. B.Sc. in Pharmacy, U. Man., 1962; M.Sc., 1965; Ph.D. in Pharm. Chemistry, U. Wis., 1968. Instr. U. Wis.-Madison, 1967-68; asst. prof. medicinal chemistry SUNY-Buffalo, 1968-72, assoc. prof., 1972-81, prof., 1981—. Contbr. articles to profl. jours. NIH grantee; Nat. Cancer Inst. grantee. Mem. Am. Chem. Soc., The Chem. Soc., Sigma Xi, Rho Chi. Subspecialties: Medicinal chemistry; Cancer research (medicine). Current work: The design and synthesis of new cancer chemotherapeutic agents; the design and synthesis of prodrugs; synthetic organic chemistry—heterocyclic, aromatic, alicyclic, and aliphatic. Patentee in field. Home: 505 Sprucewood Terr Williamsville NY 14221 Office: SUNYBuffalo Sch Pharmacy Dept Medicinal Chemistry Buffalo NY 14260

ANDERSON, WILLIAM JUDSON, aerospace engring. educator, cons.; b. Yale, Iowa, Nov. 18, 1935; s. Frank Albert and Agnes Delores A.; m. Elizabeth Maurine Anderson, June 21, 1957; children: Anne Christine, Ellen Carol, Glen Richard. B.S., Iowa State U., 1957, M.S., 1958; Ph.D., Calif. Inst. Tech., 1963. Registered profl. engr., Mich. Technician Chance Vought Aircraft, Dallas, 1954, jr. engr., 1955, 56; engr. Bendix Aerospace Systems, Ann Arbor, Mich., 1962; asst. prof. aerospace engring. U. Mich., Ann Arbor, 1965-68, assoc. prof., 1968-74, prof., 1974—; cons. to various corps. Contbr. articles to profl. jours. Served to 1st lt. USAF, 1962-65. Recipient Standard Oil award for excellence in teaching, 1966. Mem. AIAA. Presbyterian. Club: U. Mich. Volleyball. Subspecialties: Aerospace engineering and technology; Numerical analysis. Current work: Research in finite element methods, structural dynamics, teaching and research in finite element methods, consultant on stress, heat conduction and vibration problems. Office: Aerospace Engineering Dept U Michigan Ann Arbor MI 48109

ANDERSON, WYATT WHEATON, genetics educator; b. New Orleans, Mar. 27, 1939; s. William Wyatt and Lottie (Johnson) A.; m. Margaret Shugart, July 28, 1962; children: James, Elizabeth, Karen. B.S., U. Ga., 1960, M.S., 1962; Ph.D., Rockefeller U., N.Y.C., 1967. Lectr. in biology Yale U., New Haven, 1966-67, asst. prof. biology, 1967-71, assoc. prof., 1971-72; assoc. prof. zoology U. Ga., Athens, 1972-75; prof., 1975—, head dept. molecular and population genetics, 1980—; mem. NSF Panel on Population Biology and Physiol. Ecology, 1977-80. Assoc. editor: Genetics, 1977-82. Served to capt. U.S. Army, 1967-69. NSF grantee, 1979, 1973; NIH Grantee, 1970; U.S. Dept. Commerce-NOAA grantee, 1974. Mem. AAAS, Am. Soc. Human Genetics, Am. Soc. Naturalists, Ecol. Soc. Am., Soc. for Study of Evolution (v.p. 1981), Genetics Soc. Am. Subspecialties: Evolutionary biology; Population biology. Current work: Population genetics and population biology, particularly connections between genetics and ecology of populations. Office: Dept Genetics Univ Georgia Athens GA 30602

ANDRE, MICHAEL, medical physicist; b. Des Moines, Apr. 25, 1951; s. Paul Leo and Pauline (Vermie) A. B.A., Central U. Iowa, 1972; postgrad., U. Ariz., 1972-73; M.S., UCLA, 1975, Ph.D. in Med. Physics, 1979. Research assoc. Inst. Atmospheric Physics, Tucson, 1972-73; mem. tech. staff Hughes Aircraft Co., Los Angeles, 1973-74; researcher UCLA, 1974-77; med. radiation physicist Los Angeles County Olive View Med. Center, Los Angeles, 1977-81; asst. prof. radiol. scis. UCLA Sch. Medicine, 1980-81; asst. prof. radiology U. Calif. and VA Med. Ctr., San Diego, 1981—; sr. radiation physicist Cedars-Sinai Med. Ctr., Los Angeles, 1980—. Contbr. articles to profl. jours. Coordinator Ariz. Public Interest Research Group, 1972-73. Recipient Cum Laude award Radiol. Soc. N.Am., 1979; Louis B. Silverman award Health Physics Soc., 1979. Mem. Am. Assn. Physicists in Medicine, Soc. Photo-optical Instrumentation Engrs., Soc. Nuclear Medicine, American Alpine Sierra Club. Current work: Digital image processing of CT, ultrasound, x-ray data. Home: 536 Glenmont Dr Solana Beach CA 92075 Office: Dept Radiology U Calif V114 3350 La Jolla Village Dr La Jolla CA 92161

ANDREASSI, JOHN L(AWRENCE), psychology educator, researcher; b. N.Y.C., Oct. 23, 1934; s. Croce and Agnes M. (Aiello) A.; m. Gina Bearzatto, Mar. 29, 1969; children: John L., Jeanine, Cristina. B.A., CCNY, 1956; M.A., Fordham U., 1959; Ph.D., Case Western Res. U., 1964. Lic. psychologist, N.Y. Psychologist Dunlop & Assos., Stanford, Conn., 1958-61; USPHS fellow Western Res. U., Cleve., 1961-64; psychologist researcher U.S. Naval Tng. Device Center, Sands Point, N.Y., 1964-67; assoc. prof. N.Y. U., N.Y.C., 1967-73; prof. psychology Baruch Coll., CUNY, N.Y.C., 1973—; cons. ITT, Paramus, N.J., 1960-61, Naval Applied Scis. Lab., Bklyn., 1969-72. Author: Psychophysiology, 1980; contbr. articles to profl. jours., sect. to ency. Research awardee Office Naval Research, 1969-73, 73-79, Air Force Office Sci. Research, 1979—. Fellow Internat. Orgn. Psychophysiology (bd. dirs. 1981—, editorial bd.); mem. Am. Psychol. Assn. Subspecialties: Physiological psychology; Sensory processes. Current work: Research on event-related brain potentials and related sensory-perceptual processes, especially with respect to brain hemispheric differences in performance and physiological response. Home: 38 Gainsborough Rd Scarsdale NY 10583 Office: Baruch Coll Box 512 17 Lexington Ave New York NY 10010

ANDREEFF, MICHAEL, physician, researcher, educator; b. Berlin, Mar 11, 1943; U.S., 1977; s. Ljuben A. and Ursula H. (Nitzsche) A. Student, U. Muenster, 1962-64; M.D. magna cum laude, U. Heidelberg, 1968, Ph.D., 1976. Intern Mcpl. Hosp., Ludwigshafen, Germany, 1968-70; resident in medicine, sci. asst. U. Heidelberg, Germany, 1970-75; vis. scientist Karolinska Inst., Stockholm, 1973; staff clinician U. Mainz, Germany, 1976-77; mem. med. faculty U. Heidelberg, 1976—; vis. investigator Sloan-Kettering Inst. for Cancer Research, N.Y.C., 1977-79, assoc., 1979—; assoc. prof. medicine Cornell U., 1982—; asst. attending hematopathologist, asst. attending physician Leukemia/Lymphoma Service, Meml. Hosp., N.Y.C., 1981—. Contbr. numerous articles to profl. jours. Grantee German Research Council, 1977, Nat. Cancer Inst., 1980, Am. Cancer Soc., 1980. Mem. German Cancer Soc., German Soc. Hematology and Oncology, European Study Group for Cell Proliferation, European Assn. Cancer Research, Am. Assn. Cancer Research, Am. Soc. Clin. Oncology, Cell Kinetics Soc., N.Y. Acad. Sci. Subspecialties: Hematology; Cell study oncology. Current work: Hematology, laser in medicine, flow cytometry, leukemia cell biology, chemotherapy of cancer. Office: Meml Sloan-Kettering Cancer Center 1275 York Ave New York NY 10021

ANDRES, RONALD PAUL, chem. engr., educator; b. Chgo., Jan. 9, 1938; s. Harold William and Amanda Ann (Breuhaus) A.; m. Jean Mills Elwood, July 15, 1961; children—Douglas, Jennifer, Mark. B.S., Northwestern U., 1959; Ph.D., Princeton U., 1962. Asst. prof. Princeton U., 1962-68, asso. prof., 1968-76, prof. chem. engring., 1976-81, Purdue U., West Lafayette, Ind., 1981—, head, 1981—. Mem. Am. Chem. Soc., Am. Inst. Chem. Engrs., AAAS, Sigma Xi, Tau Beta Pi, Pi Mu Epsilon, Phi Lambda Upsilon, Phi Eta Sigma. Subspecialties: Chemical engineering; Surface chemistry. Current work: Physics and chemistry of small molecular clusters; process control; nucleation; atmospheric chemistry. Office: Sch Chem Engring Purdue U West Lafayette IN 47907

ANDRESEN, BRIAN DEAN, pharmacology educator; b. Reed City, Mich., Jan. 20, 1947; s. Arthur and Clarice A. B.S., Fla. State U., 1969; M.S., MIT, 1971, 72, Ph.D., 1974. Asst. prof. Coll. Pharmacy, U. Fla., Gainesville, 1974-79; also asst. dir. analytical toxicology lab.; assoc. prof. depts. pharmacology and ob-gyn Coll. Medicine, Ohio State U., Columbus, 1979-83; sr. scientist Lawrence Livermore Nat. Lab., Livermore, Calif., 1983—. Appearances on: radio shows on drug abuse Reye's Syndrome; Contbr. articles to sci. jours. Active Central Ohio Heary Chpt., Inc. Grantee NIH, Am. Diabetes Assn., Central Ohio Heart Assn., Nat. Inst. Environ. Scis., Nat. Reye's Syndrome Found.; others; NIH trainee, 1967-75. Mem. Am. Chem. Soc., Nat Oceanographic Soc., Am. Soc. Mass Spectrometry, Am.Soc. Clin. Pharmacology. Lutheran. Subspecialties: Mass spectrometry; Obstetrics and gynecology. Current work: Reye's Syndrome; food mutagens and toxin; mass spectrometry; amniotic fluid assay. Office: L-453/LLNL PO Box 5505 Livermore CA 94550

ANDRESEN, MICHAEL CHRISTIAN, neurophysiologist; b. Lynwood, Calif., Dec. 1, 1949; s. John Christian and Yvonne Elsie (Meyer) A.; m. Pamela Sue Rice, Nov. 23, 1974. B.Sc., U. Calif.-Irvine, 1971; M.S., San Diego State U., 1973; Ph.D., U. Tex.-Galveston, 1978. Grad. teaching and research asst. U. Tex. Med. Br., Galveston, 1975-77, research assoc., 1977-78, NIH postdoctoral fellow, 1978-79, research scientist, 1979-80; asst. prof. dept. physiology and biophysics, 1981—; vis. scientist Baker Med. Research Inst., Melbourne, Australia, 1980-81; mem. central research rev. com. Tex. affiliate Am. Heart Assn., Austin, 1983—. Contbr. articles to profl. jours. NIH postdoctoral research fellow, 1978; Am. Heart Assn. research grantee, 1981-84; NIH research grantee, 1983-85. Mem. Biophys. Soc., Am. Physiol. Soc., AAAS. Subspecialties: Cardiology; Neurophysiology. Current work: Neural control of the circulation and cardiovascular system using electrophysiological approaches, especially baroreceptors and barorefiexes. Office: U Tex Med Br Dept Physiology and Biophysics Galveston TX 77550

ANDREW, CLIFFORD GEORGE, neuroscientist, neurologist; b. St. Louis, Sept. 10, 1946; s. Eugene Ashton and Anna Louise (Hanish) A.; m. Louise Collier Briggs, June 13, 1970; children: Galen Michael, Amalie Linnea. A.B. (Coll. scholar), Columbia U., 1968; Ph.D. in Biochemistry, Duke U., 1974; M.D. (Duke Avelor scholar), Duke U., 1975. Diplomate: Am. Bd. Psychiatry and Neurology. Intern Duke U. Hosp., 1975-1976; resident in neurology Johns Hopkins U., 1976-79, neuromuscular fellow, 1979-80, asst. prof. neurology, 1980—. Contbr. articles to profl. jours. Recipient Tchr. Investigator Devel. award Nat. Inst. Neurologic Communicative Disease and Stroke, 1980—; Muscular Dystrophy Assn. basic research grantee, 1981—. Mem. Columbia Chemists, AAAS, Am. Soc. Neurochemistry, Neuroscis. Soc., Am. Acad. Neurology, Physicians for Social Responsibility, Chesapeake Bay Found., Severn River Assn., Pointfield Landing Community Assn. Independent Democrat. Presbyterian. Subspecialties: Neurobiology; Membrane biology. Current work: Interested in role of biol membranes and their constituent proteins and glycoproteins in neuromuscular devel., function, and disease; investigating membrane factors responsible for developmentally regulated muscle susceptibility to Coxsackie A2 Virus. Patentee Intern bd. game, microcomputer game. Home: 474 Old Orchard Cirle Point Field Landing on the Severn Millersville MD 21108 Office: Dept Neurology 5134 Meyer Neuroscis Center Johns Hopkins U Sch Medicine Baltimore MD 21205

ANDREWS, FRANK CLINTON, chemistry educator; b. Manhattan, Kans., May 29, 1932; s. A. Clinton and Jessie A. (Yahn) A.; m. Jean M. Langford, June 21, 1964; children: Karen, Elizabeth. B.S. with highest honors in Chemistry, Kans. State U., 1954; A.M. in Phys. Chemistry, Harvard U., 1959; Ph.D. in Chem. Physics, Harvard U., 1960. Postdoctoral fellow U. Calif.-Berkeley, 1960-61; asst. prof. chemistry U. Wis.-Madison, 1961-66; assoc. prof. chemistry U. Calif.-Santa Cruz, 1967-73, prof., 1973—; vis. asst. prof. dept. theoretical physics Oxford U., 1966-67, Harvard U., 1966; vis. prof. Dartmouth Coll., 1979, 80. Author: Equilibrium Statistical Mechanics, 1963, 2d edit., 1975, Thermodynamics - Principles and Applications, 1976; contbr. articles to profl. jours. Served with U.S. Army, 1955-57. Recipient Disting. Chemistry Alumnus award Kans. State U., 1969; Gustav Ohaus award for innovations in coll. si. teaching Nat. Sci. Tchrs. Assn., 1975; Fulbright scholar, 1954-55; Alfred P. Sloan fellow, 1963-67. Mem. Am. Phys. Soc., Assn. Humanistic Psychology, Educators for Social Responsibility. Subspecialties: Statistical mechanics; Thermodynamics. Current work: Equilibrium Statistical mechanics of fluids; general problem solving and psychological unblocking; science and human values; quality of life. Home: 1025 Laurent St Santa Cruz CA 95060 Office: Dept Chemistry U Calif Santa Cruz NS II Santa Cruz CA 95064

ANDREWS, GEORGE HAROLD, educator; b. Syracuse, N.Y., July 31, 1932; s. George Harold and Marion Louise (Downing) A.; m. Marlene Gertrude Erickson, Dec. 31, 1955; children: Erik, Geoffrey, Elise, Christopher. A.B., Oberlin Coll., 1954; M.A., U. Mich., 1955, Ph.D., 1963. Actuarial trainee Conn. Gen. Life Ins. Co., Hartford, Conn., 1955-56; teaching asst. U. Mich., Ann Arbor, 1958-62, vis. prof., 1975-76; prof. math. Oberlin (Ohio) Coll., 1962—; vis. scholar U. Mich., 1982-83; cons. in field. Author: (with Ben Noble) Book and Videotaped Course in numerical analysis, 1972. Served with U.S. Army, 1956-58. NSF fellow, 1969. Mem. Soc. Actuaries, Math. Assn. Am., Soc. Indsl. and Applied Math., Sigma Xi. Democrat. United Ch. of Christ. Subspecialties: Numerical analysis. Current work: Population projections, particularly as they are developed and used in estimating social security costs. Home: 174 E College St Oberlin OH 44074 Office: Oberlin Coll Lorain St Oberlin OH 44074

ANDREWS, HENRY NATHANIEL, JR., scientist, educator; b. Melrose, Mass., June 15, 1910; s. Henry Nathaniel and Florence Clara (Hollings) A.; m. Elisabeth Claude Ham, Jan. 12, 1939; children—Hollings T., Henry III, Nancy R. B.S., M.I.T., 1934; M.S., Washington U., St. Louis, 1937, Ph. D., 1939; student, Cambridge (Eng.) U., 1937-38. Mem. staff Washington U., 1939-64; prof. Bot. Garden, St. Louis, 1941-64, asst. to dir., 1944-47; part time employee U.S. Geol. Survey; chmn. botany dept. U. Conn., Storrs, 1965-67, head systematic and environ. biol. sect., 1967-70, prof. emeritus, 1975—. NSF postdoctoral fellow Swedish Mus. Natural History, Stockholm, 1964-65; Fulbright teaching fellow Poona Univ., India, 1960-61; vis. prof. Aarhus (Denmark) U., 1976. Mem. Bot. Soc. Am. (Merit award 1966), Nat. Acad. Scis., Paleontol. Soc., Torrey Bot. Club, Phi Beta Kappa, Sigma Xi. Subspecialties: Paleobiology; Evolutionary biology. Current work: Early land plants. Research work deals primarily with fossil plants of central coal fields, Devonian age Fossils, Arctic paleobotany. Home: RFD 1 Box 146 Laconia NH 03246

ANDREWS, HUGH ROBERT, nuclear physicist, researcher; b. Frederickon, N.B., Can., Apr. 29, 1940; s. John V. and Bertha T. (Smith) A.; m. Josephine D. Kennedy, May 22, 1971; children: J. Matthew, E. Louise. B.Sc. (Lord Beaverbrook scholar), U. N.B., 1962; A.M., Harvard U., 1963, Ph.D., 1970. Research officer Chalk River Nuclear Labs., Atomic Energy of Can., 1971—. Contbr. articles to sci. publs. Recipient prize Can. Math. Congress, 1958, Rutherford prize Royal Soc. Can., 1968; Frank Knox fellow, 1962-64. Mem. Can. Assn. Physicists, Am. Phys. Soc., Sigma Xi. Anglican. Subspecialties: Nuclear physics; Condensed matter physics. Current work: Experimental nuclear structure, particularly high spin states; gamma spectroscopy; hyperfine interactions; energy loss phenomena; ultra-sensentive mass spectrometry. Home: 66 Rutherford Ave PO Box 1356 Deep River ON Canada KOJ 1PO Office: Chalk River Nuclear Labs Station 49 Chalk River ON Canada K0J 1J0

ANDREWS, JAMES BARCLAY, II, nuclear engineer; b. Roanoke, Va., Nov. 25, 1939; s. James Barclay and Louise (Kirkwood) A.; m. Helen Louise Munsey, June 8, 1962; children: JamesBarclay III, Donna Louise, David Edward. B.S.E.E., Va. Poly. Inst., 1961, M.S., 1965; Ph.D., MIT, 1967. Sr. engr. Babcock & Wilcox, Lynchburg, Va., 1969-71, unit mgr., 1971-82, sect. mgr., 1982—. Served to capt. U.S. Army, 1967-69. Mem. Am. Nuclear Soc. Presbyterian. Subspecialties: Nuclear engineering; Nuclear fission. Current work: Nuclear reactor fuel. Home: 2301 Hawthorne Rd Lynchburg VA 24503 Office: Babcock & Wilcox PO Box 1260 Lynchburg VA 24505

ANDREWS, JOHN THOMAS, quaternary geologist; b. Millom, Cumberland, Eng., Nov. 8, 1937; came to U.S., 1968, naturalized, 1976; s. George and Dorothy (Black) A.; m. Martha Lee Tuthill, Dec. 16, 1961; children—Melissa Margaret, Thomas George. B.A., U. Nottingham, 1959, Ph.D., 1965, D.Sc., 1978; M.Sc., McGill U., 1961. Research scientist Govt. Can., Ottawa, 1961-67; prof. geol. scis. Inst. Arctic and Alpine Research, U. Colo., Boulder, 1968—. Author: Glacial Isosasty, 1974, Glacial Systems, 1975. Fellow Geol. Soc. Am. (Kirk Bryan medal 1973), Arctic Inst. N.Am.; mem. Internat. Glaciol. Soc., Am. Geophys. Union, Canadian Geol. Assn., Colo. Sci. Soc. (Past President's award 1969). Subspecialty: Geology. Current work: Glaciology. Home: 1407 Kennedy Ct Boulder CO 80303 Office: Inst Arctic and Alpine Research Univ Colo Boulder CO 80309

ANDREWS, LUTHER DAVID, animal science educator; b. Rogers, Ark., Oct. 24, 1923; s. Mark William and Dovie A.; m. Betty Jo Roper, May 31, 1965; children: Rebecca Diane, Michael David. B.A., U. Ark., 1952, M.S.A., 1953; Ph.D., U. Mo., 1966. Mgr. pedigree operation Ark. Farmers Poultry Breeding Farm, Fayetteville, 1953-57; instr. animal sci. U. Ark., Fayetteville, 1957-61, asst. prof., 1961-68, assoc. prof., 1968-75, prof., 1975—. Contbr. articles to prfl. jours. Served with USAF, 1946-49. Mem. Poultry Sci. Assn., Sigma Xi, Gamma Sigma Delta, Phi Eta Sigma, Alpha Zeta. Baptist. Subspecialties: Animal genetics; Poultry management. Current work: Caged broilers, test broiler strains, broiler breeders. Home: 2025 Wedington Dr Fayetteville AR 72701 Office: Dept Animal Sci U Ark Fayetteville AR 72701

ANDREWS, MARY LOU, systems engineer, researcher; b. Hillsdale, Mich., Oct. 17, 1935; d. Earl R. and Carolyn Constance (Drust) A.; m. William Leonard Harkness, Dec. 26, 1956 (div.); children: Judy Lynn Harkness Edighoffer, Tammy Lee, William Andrew. B.S., Mich. State U., 1956; M.A., Pa. State U., 1965. Cert. secondary edn., Mich., Pa. Instr. Mich. State U., East Lansing, 1957-59, Pa. State U., University Park, 1960-66; asst. prof. statistics Calif. State U., Hayward, 1966-67; instr. Pa. State U., 1967-77; asst. prof. bus. adminstrn. and computation Hillsdale Coll., Mich., 1980-81; sr. engr. HRB Singer, Inc., State College, Pa., 1978-80, 81—; cons. Desmatics, State College, 1981—. Author: Study Guide for Statistical 200, 1980, Statistics Laborator Manual, 1965. Mem. Soc. Indsl. and Applied Math., Am. Stats. Assn., Assn. Computing Machinery, Am. Math. Soc., Math. Assn. Am. Democrat. Roman Catholic. Subspecialties: Systems engineering; Statistics. Current work: As a consultant, I statistically analyze results of studies carried out by organizations; as a systems engineer, I develop concepts and algorithms for future software systems and analyze current algorithms in systems. Office: HRB Singer Inc Box 60 Science Park State College PA 16801

ANDREWS, PETER WALTER, developmental biologist, educator, researcher; b. London, June 5, 1950; s. Walter and Sheila A.; m. Dale Joan Andrews, Sept. 26, 1981; 1 dau., Whitney Elizabeth. B.Sc. with 1st class honors in Biochemistry, U. Leeds, Eng., 1971; D.Phil., Oxford (Eng.) U., 1975. Research fellow Inst. Pasteur, Paris, 1974-75, Sloan-Kettering Inst., N.Y.C., 1976-78; research investigator Wistar Inst., Phila., 1978-80, research assoc., 1980-82, asst. prof., 1983—. Contbr. articles to sci. jours. Mem. Genetics Soc. Am., Soc. Developmental Biology. Subspecialties: Developmental biology; Cell and tissue culture. Current work: Analysis of cellular differentiation, especially using human and mouse teratocarcima cell lines and employing immunogenetic approaches that include the use of monoclonal antibody-defined differentiation antigens. Office: Wistar Inst 36th and Spruce Sts Philadelphia PA 19104

ANDREWS, RICHARD WAYNE, statistics educator, consultant; b. Dayton, Ohio, Dec. 11, 1940; s. Gus W. and Mildred F. (Milton) A.; m. Elizabeth Ann Cervinski, May 20, 1967; children: David, Daniel, Martha. B.S., U.S. Nval Acad., 1964; M.S., Mich. State U., 1970; Ph.D., Va. Poly. Inst. and State U., 1973. Commd. 2nd lt. U.S. Marine Corps, 1964, advanced through grades to capt., 1966; served in, Vietnam, 1965-66, resigned in, 1968; asst. prof. U. North Fla., Jacksonville, 1972-74; assoc. prof. stats. U. Mich., Ann Arbor, 1974—; statistician Naval Undersea Ctr., San Diego, 1972; vis. prof. U. Southampton, Eng., 1980-81; audit sampling cont. leader Ernst & Whinney, Cleve., 1976-82. Grantee Peat, Marwick, Mitchell & Co., 1979, Ernst & Whinney Found., 1980, Office Naval Resrch, 1981. Mem. Am. Statis. Assn., Ops. Research Soc. Am. Subspecialties: Statistics; Operations research (mathematics). Current work: Audit sampling using Bayesian methodology; computer simulation output analysis. Home: 1615 South Blvd Ann Arbor MI 48104 Office: U Mich Ann Arbor MI 48109

ANDROULAKIS, JOHN GEORGE, mechanical engineer; b. Archanes, Crete, Greece, May 15, 1924; came to U.S., 1953, naturalized, 1962; s. George John and Maria Stylianos (Terjakis) A.; m. Vassiliki John Androulakis, Jan. 3, 1959; children: George John. B.Engring., Ecole Nationale Superieure du Genie Maritime, France, 1953; M.S.M.E., Poly. Inst. Bklyn., 1955. Design engr. H.N. Whittelsey, Inc., N.Y.C., 1954-58; devel. engr. Brookhaven Nat. Lab., Upton, N.Y., 1958-64; project engr. Grumman Aerospace Corp., Bethpage, N.Y., 1964-75; tech. dir. Intersystem Design & Tech. Corp., Lauderdale, Fla., 1975-80; prin. mem. tech. staff RCA/AStro Electronics, Princeton, N.J., 1980—. Contbr. articles to profl. jours. Served with Greek Air Force, 1942-46. Decorated Brit. Mil. Medal. Greek Orthodox. Subspecialties: Aerospace engineering and technology; Satellite studies. Current work: Meteorol. and communication satellite design. Patentee in field.

ANDRUS, JAN FREDERICK, mathematics educator; b. Washington, Sept. 17, 1932; s. Charles Frederick and Margaret (Grow) A.; m. Joanne Louise Newton, June 16, 1955 (div. 1960); m. Nancy Rose Swing, July 15, 1961; children: Lynn, Leslie. B.S., Coll. Charleston, 1954; M.A., Emory U., 1955; Ph.D., U. Fla., 1958. Math. specialist Lockheed Aircraft Corp., Marietta, Ga., 1958-62; math. cons. Gen. Electric Co., Huntsville, Ala., 1962-66; mem. sr. tech. staff Northrop Corp., Huntsville, 1966-73; assoc. prof. math. U. New Orleans, 1973-79, prof., 1979—. USAF research grantee, 1978. Mem. Soc. Indsl. and Applied Math., Am. Math. Soc., Phi Kappa Phi, Sigma Pi Sigma, Sigma Alpha Phi. Democrat. Subspecialty: Applied mathematics. Current work: Research in numerical problems associated with systems of ordinary differential equations. Home: 237 Dorrington Blvd Metairie LA 70005 Office: Math Dept U New Orleans New Orleans LA 70148

ANDRUSHKIW, ROMAN IHOR, mathematics educator, consultant; b. Lviw, Ukraine, May 3, 1937; came to U.S., 1949; s. Joseph Wasyl and Sofia A.; m. Svitlana Maria Lutzky, Nov. 22, 1975; 1 son, Pavlo Marian. B.E., Stevens Inst. Tech., 1959, Ph.D. in Math., 1973; M.S. E.E., Newark Coll. Engring., 1964; M.S.in Math, U. Chgo., 1967. Elec. engr. Weston Instruments and Electronics, Newark, 1959-64; instr. Newark Coll. Engring., 1964-66, 68-70; research asst. U. Chgo., 1966-68; asst. prof. N.J. Inst. Tech., Newark, 1970-78, assoc. prof. math., 1978—; cons., instr. Bendix Corp., Teterboro, N.J., 1981-82; cons. E.T. Killam Assos., Millburn, N.J., 1982; reviewer Am. Math. Soc., Ann Arbor, Mich., 1977—. Contbr. research papers to publs. Assoc. dir. Newark chpt. Plast, Inc., 1980-82. NSF grantee, 1966. Mem. Soc. Computer Simulation, Soc. Indsl. and Applied Math., Internat. Assn. Math. and Computers in Simulation, Am. Math. Soc., Shevchenko Sci. Soc. (exec. council 1974—), Ukranian Engrs. Soc. Am. Subspecialties: Applied mathematics; Numerical analysis. Current work: Finite difference and finite element analysis of initial-boundary value problems in fluid dynamics, heat and mass transfer, wave propagation, Stefan-type problems; spectral theory and eigenvalue approximation of operator-valued functions. Co-inventor, pantentee. Office: NJ Inst Tech 323 High St Newark NJ 07102

ANDRYSCO, ROBERT MICHAEL, specialist in human, animal relations, researcher; b. Cleve., Nov. 11, 1954; s. Robert and Bernice Beatrice A.; m. Paula Ann Hohenbrink, Oct. 17, 1982. B.S. in Biology and Chemistry, Ashland Coll., 1976; M.S. in Vet. Physiology and Pharmacology, Ohio State U., 1978; Ph.D. in Vet. Medicine and Psychology, Ohio State U., 1982. Surg. asst. Brookville Animal Hosp., Parma, Ohio, 1975; research asso. Ohio Agrl. Research and Devel. Center, Wooster, 1976; grad. research assoc., teaching asso. vet. physiology Ohio State U., Columbus, 1976-78, grad. research asso., 1978-81; specialist in human and animal relations, Columbus, Ohio, 1980—; cons. Capital Area Humane Soc., Columbus; 14 Columbus area vet. clinics Ohio State U. Vet. Clinic, others. Contbr. articles on human-animal relations to profl. jours. Named Outstanding Young Man of Am. U.S. Jaycees, 1979. Mem. Soc. for Neurosci., European Brain and Behavior Soc., Brit. Brain Research Assn., Delta Soc. Roman Catholic. Subspecialty: Human/animal bond. Current work: Pet-facilitated therapy: companion animal behavior modification; establishment of Campanion Animal Services, Inc.

ANFINSEN, CHRISTIAN BOEHMER, biochemist; b. Monessen, Pa., Mar. 26, 1916; s. Christian Boehmer and Sophie (Rasmussen) A.; m. Florence Bernice Kenenger, Nov. 29, 1941; children: Carol Bernice, Margot Sophie, Christian Boehmer; m. Libby Shulman Ely, 1979. B.A., Swarthmore Coll., 1937, D.Sc., 1965; M.S., U. Pa., 1939, D.Sc., 1967; Ph.D., Harvard, 1943; D.Sc. (hon.), Georgetown U., 1967, Swarthmore Coll., 1965, N.Y. Med. Coll., 1969, Gustav Adolphus U., 1975, Brandeis U., 1977, Providence Coll., 1978. Am.-Scandinavian Found. fellow Carlsberg Lab., Copenhagen, 1939; sr. cancer research fellow Nobel Inst., Stockholm, 1947; Markle scholar, 1948—; asst. prof. biol. chemistry Harvard U. Med. Sch.; prof. biochemistry Harvard Med. Sch., 1962-63; Guggenheim fellow Weizmann Inst., Rehovot, Israel, 1958; chief lab. cellular physiology and metabolism Nat. Heart Inst., Bethesda, Md., 1950-62; chief lab. chem. biology Nat. Inst. Arthritis and Metabolic Diseases, Bethesda, 1963-82; mem. faculty dept. biology Johns Hopkins U., Balt., 1982—; Bd. govs. Weizmann Inst. Sci., Rehovot, Israel. Author: The Molecular Basis of Evolution, 1959; Contbr. to sci. publs. Recipient Rockefeller Public Service award, 1954-55; Nobel prize in chemistry, 1972; Myrtle Wreath Hadassah, 1977. Mem. Am. Soc. Biol. Chemists (pres. 1971-72), Am. Acad. Arts and Scis., Am. Philos. Soc., Nat. Acad. Scis., Washington Acad. Scis., Fedn. Am. Scientists (treas. 1958-59, vice chmn. 1959-60, 73-76). Subspecialty: Biochemistry (biology). Current work: Proteins. Home: 1740 Vineyard Trail Epping Forest Annapolis MD Office: Dept Biology Johns Hopkins U. 34th and Charles St Baltimore MD 21218

ANGEL, JAMES ROGER PRIOR, astronomer; b. St. Helens, Eng., Feb. 7, 1941; came to U.S., 1967; s. James Lee and Joan (Prior) A.; m. Ellinor M. Goonan, Aug. 21, 1965; children—Jennifer, James. B.A., Oxford (Eng.) U., 1963, D.Phil., 1967; M.S., Calif. Inst. Tech., 1966. From research asso. to asso. prof. physics Columbia U., 1967-74; vis. asso. prof. astronomy U. Tex., Austin, 1974; mem. faculty U. Ariz., Tucson, 1974—, prof. astronomy, 1975—. Sloan fellow, 1970-74. Mem. Am. Astron. Soc. (Pierce prize 1976). Research on white dwarf stars, quasars, astron. instruments. Office: Steward Obs Univ Ariz Tucson AZ 85721

ANGEL, THOMAS MICHAEL, oil company executive; b. Long Run, Ohio, Apr. 16, 1939; s. Joseph Vincent and Mary Lucille (Stock) A.; m. Bette Jean Miller, Sept. 21, 1964; children: Nicole Lee, Tommie Jean, Martine Renee. Hon. lifetime faculty mem. Fla. Inst. Tech., 1977; Mgr. Sanford Bros. Divers, Morgan City, La., 1963-66; v.p. Sanford Marine Service, Morgan City, La., 1966-70; mgr. Fluor Ocean Services, Houma, La., 1970-73; Santa Fe Engring. and Constrn. Co., Nouma, 1973-79; v.p., mgr. Santa Fe Underwater Services Inc., Houma, 1979—; adv. bd. Fla. Inst. Tech., 1977—; tech. com. Am. Bur. Shipping, N.Y., 1980—; adv. bd. Santa Barbara City (Calif.) Coll., 1974—. Contbr. articles to profl. jours. Active Houma C. of C. Served with USN, 1957-59. Mem. Am. Mgmt. Assn., Inst. Diving, Am. Welding Soc., Assn. Diving Contractors (pres. 1973-74), Am. Petroleum Inst., Marine Tech. Soc., Under Seas Med. Soc. Republican. Roman Catholic. Subspecialties: Robotics; Offshore technology. Current work: Professional deep sea oil field related commercial diving manned/unmanned remote controlled underwater vehicles. Home: 25 Mary Hughes Ct Houma LA 70360 Office: Santa Fe Under Water Services Inc PO Box 3518 Houma LA 70361

ANGINO, ERNEST EDWARD, geology educator, researcher, administrator; b. Winsted, Conn., Feb. 16, 1932; s. Alfred and F. Mabel (Serluco) A.; m. Margaret Lachat, June 26, 1954; children: Cheryl Ann, Kimberly Ann. B.S. in Mining Engring, Lehigh U., 1954; M.S. in Geology, U. Kans., 1958, Ph.D., 1961. Instr. geochemistry U. Kans., Lawrence, 1961-62; asst. prof. oceanography Tex. A&M U., College Station, 1962-65; chief geochemistry Kans. Geol. Survey, Lawrence, 1965-70, assoc. dir. and assoc. state geologist, 1970-72;

prof./ chmn. Dept. Geology, U. Kans., Lawrence, 1972—; cons. numerous orgns. Author: (with G.K. Billing) Atomic Absorption Spectrometry in Geology, 1972; author-editor: (with D.T. Long) Geochemistry of Bismuth, 1979; Editor: (with G.K. Billing) Geochemistry Subsurface Brines, 1969, (with R. Hardy) Precedure 3rd Forum-Geology Industrial Minerals, 1967. Active Police Community Relations Com., 1969-71. Served with U.S. Army, 1955-57. NSF fellow, 1963; Angino Buttress named in honor, 1967; recipient Antarctic Service medal Dept. Defense, 1969. Fellow Geol. Soc. Am.; mem. Soc. Environ. Geochemistry and Health (pres. 1978-79), Geochem. Soc. (sec. 1970-76), Internat. Assn. Geochemists and Cosmochemistry (treas. 1980—), AAAS, Am. Assn. Petroleum Geologists. Roman Catholic. Subspecialties: Geochemistry; Geology. Current work: Trace elements in natural rock systems, chemistry of water, geochemistry of natural systems. Home: 1215 W 27th St Lawrence KS 66044 Office: Dept Geology U Kans Lawrence KS 66045

ANGUS, JOHN COTTON, chemical engineering educator; b. Grand Haven, Mich., Feb. 22, 1934; s. Francis Clark and Margaret (Cotton) A.; m. Caroline Helen Gezon, June 25, 1960; children—Lorraine Margaret, Charles Thomas. B.S. in Chem. Engring, U. Mich., 1956, M.S., 1958, Ph.D. in Engring, 1960. Registered profl. engr., Ohio, Mich. Research engr. Minn. Mining & Mfg. Co., St. Paul, 1960-63; prof. Case Inst. Tech., Cleve., 1963-67, prof. chem. engring., 1967—, chmn. dept., 1974-80; vis. lectr. U. Edinburgh, Scotland, 1972-73; vis. prof. Northwestern U., 1980-81. Vice pres. ARC, Inc.; trustee Ohio Coal Research Lab.; chmn. bd. trustees Ohio Scottish Games. NSF fellow, 1956-57; NATO sr. fellow, 1972-73. Mem. Am. Inst. Chem. Engrs., Am. Chem. Soc., Electrochem. Soc., Sigma Xi, Tau Beta Pi, Phi Lambda Upsilon. Subspecialties: Chemical engineering; Electrochemistry. Research in fields of crystal growth, laser applications, coal gasification, sulfur removal processes, electrochemical devices, thermodynamics. Home: 2716 Colchester Rd Cleveland OH 44106 Office: Dept Chem Engring Case Western Res U Cleveland OH 44106

ANISMAN, HYMIE, psychologist, educator, researcher; b. Munich, Germany, Apr. 3, 1948; emigrated to Can., 1950; s. Simon M. and Helen (Pulver) A.; m. Maida S. Silverstone, Aug. 30, 1969; children: Simon, Rebecca, Jessica. B.A., Sir George Williams U., Montreal, Que., 1969; M.A., Meml. U. Nfld., St. John's, 1970; Ph.D., U. Waterloo, Ont., 1972. Asst. prof. U. Waterloo, 1972-73, Wilfred Laurier U., Waterloo, 1973-74; asst. prof. Carleton U., Ottawa, Ont., 1974-76, assoc. prof., 1976-82, prof., 1982—. Editor: Psychopharmacology of Aversive Motivated Behavior, 1978; contbr. articles to profl. jours. Fellow Can. Psychol. Assn.; mem. Am. Psychol. Assn., AAAS, Soc. for Neurosci. Subspecialties: Neuropharmacology; Neuropsychology. Current work: Stress and its relation to neurochemical change. Implications for physical and pscyhological pathology. Office: Carleton Univ Ottawa ON Canada K1S 5B6

ANNIS, LAWRENCE VINCENT, JR., clinical psychologist, mental health administrator; b. Augusta, Ga., Dec. 28, 1946; s. Lawrence Vincent and Betty (Allen) A.; m. Kathy Ann Kirkwood, June 12, 1971 (div. Aug. 1973); m. Christy Adele Baker, Aug. 22, 1982. B.A., Augusta Coll., 1968; M.A., Western Carolina U., Cullowhee, N.C., 1974; Ph.D., U. Miss., 1981. Staff psychologist Alpine Psychoedn. Center, Gainesville, Ga., 1974-75; psychologist N. Miss. Retardation Center, Oxford, 1978-79; psychology intern Devereux Found., Devon, Pa., 1979-80; sr. behavior specialist Gracewood State Sch., Ga., 1980-81; clin. psychologist Fla. State Hosp., Chattahoochee, 1981-82, 83—; clin. dir. S.E. Region Mental Health Center, Lumberton, N.C., 1982-83. Served with USN, 1969-72. Mem. Internat. Council Psychologists, Am. Psychol. Assn., Southeastern Psychol. Assn. Unitarian. Subspecialty: Behavioral psychology. Current work: Forensic psychology, juvenile delinquency, sex offender rehabilitation. Home: 1600 Pullen Rd #2-H Tallahassee FL 32303 Office: Forensic Service Fla State Hosp Chattahoochee FL 32324

ANSARI, AFTAB A., immunoligist; b. Allahabad, India, July 1, 1950; s. Abdul Aziz and Amina (Khatoon) A.; m. Hashima Hasan, Sept. 19, 1980. M.Sc., Aligarh (India) Muslim U., 1968, M.Phil., 1970, Ph.D., 1971. Lectr. Aligarh Muslim U., 1970-73; vis. fellow NIH, Bethesda, Md., 1974-76; vis. assoc. Nat. Inst. Environ. Health Scis., Research Triangle Park, N.C., 1977-78; clin. asst. prof. U. N.C., Chapel Hill, 1979-80, clin. assoc. prof., 1980—; vis. scientist, head immunology Nat. Inst. Environ. Health Scis., Research Triangle Park, 1979-81; sr. scientist Northrop Services, Inc., Research Triangle Park, 1981—. Contbr. articles to profl. jours. Mem. Genetics Soc. Am., Environ. Mutagen Soc., Am. Assn. Immunologists, Fedn. Am. Soc. Exptl. Biologists, AAAS, Tissue Culture Assn., N.Y. Acad. Sci., Sigma Xi. Subspecialties: Immunobiology and immunology; Environmental toxicology. Current work: Monoclonal antibodies, gene cloning, mutation monitoring. Home: 3200-8 Stonesthrow Ln Durham NC 27713 Office: 5-A Triangle Dr Research Triangle Park NC 27709

ANSELL, GEORGE STEPHEN, metallurgist, educator; b. Akron, Ohio, Apr. 1, 1934; s. Frederick Jesse and Fanny (Soletsky) A.; m. Marjorie Boris, Dec. 18, 1960; children: Frederick Stuart, Laura Ruth, Benjamin Jesse. B.Met.E., Rensselaer Poly. Inst., 1954; M.Met.E., 1955; Ph.D., 1960. Registered profl. engr., N.Y. State. Phys. metallurgist U.S. Naval Research Lab., Washington, 1957-58; faculty Rensselaer Poly. Inst., Troy, N.Y., 1960—, Robert W. Hunt prof. metall. engring., 1965—, chmn. materials div., 1969-74, dean engring., 1974—; Cons. to pvt. cos. Recipient Curtis W. McGraw award Am. Soc. for Engring. Edn., 1971. Fellow Am. Soc. for Metals (Alfred H. Geisler award Eastern N.Y. chpt. 1964, Bradley Stoughton award 1968), Metall. Soc.; mem. Am. Inst. Mining, Metall. and Petroleum Engrs. (Hardy Gold medal 1961, chmn. Inst. Metals div. 1974), Nat. Soc. Profl. Engrs., Sci. Research Soc. Am., Sigma Xi, Tau Beta Pi, Phi Lambda Upsilon. Subspecialties: Metallurgy; High-temperature materials. Research, publs. on theoretical and exptl. analysis of relationships between defect structure and properties of crystaline solids. Home: 6 Colonial Green Loudonville NY 12211 Office: Sch of Engring Rensselaer Poly Inst Troy NY 12181

ANSELL, JULIAN SAMUEL, urologist; b. Portland, Maine, June 30, 1922; m., 1951; 5 children. B.A., Bowdoin Coll., 1947; M.D., Tufts U., 1951; Ph.D., U. Minn., 1959. Instr. urology, mem. staff Univ. Hosp. U. Minn., Mpls., 1956-59; asst. prof., assoc. prof. Sch. Medicine, U. Wash. and Affiliated Teaching Hosps., Seattle, 1959-65, head div. urology, 1959—, prof., chmn. dept., 1965—; chief urology VA Hosp., Mpls., 1956-59; urologist in chief King County Hosp; cons. VA Hosp., Seattle, 1959. Mem. AAAS, AMA, ACS, Am. Urol. Assn. Subspecialty: Urology. Office: Dept Urology U Wash Sch Medicine Seattle WA 98195

ANSUINI, FRANK JOSEPH, consulting corrosion engineer; b. Providence, Mar. 29, 1942; s. Frank and Iola Norma (Lombardi) A.; m. Sandra Carrington, Nov. 20, 1965; children: F. Douglas, Karen L. B.Sc., MIT, 1963; M.Sc., N.Y. U., 1969. Registered profl. engr., Calif. Research metallurgist INCO, Suffern, N.Y., 1963-70, applications engr., N.Y.C., 1971-72; sr. applications engr. Kennecott Copper Co., Lexington, Mass., 1973-80; cons. engr., Wayland, Mass., 1980—. Contbr. articles to profl. jours. Recipient merit award Materials Engring. Mag., 1981. Mem. Nat. Assn. Corrosion Engrs. (sect. chmn. 1976-77), Marine Tech. Soc., Am. Soc. Metals, Soc. Naval arch. and Marine Engrs., Soc. Marine Cons. Subspecialties: Corrosion; Alloys. Current work: Creative application of materials in corrosive environments. Patentee alloys. Office: 4 Juniper Ln Wayland MA 01778

ANTAL, MICHAEL JERRY, JR., energy specialist and consultant; b. Monroe, Mich., May 18, 1947; s. Michael Jerry and Carolyn Sarah (McAdam) A.; m. Ann Gorsuch Slaughter, July 14, 1949; children: Dickinson James, Rachel Caroline. A.B., Dartmouth Coll., 1969; M.S., Harvard U., 1970, Ph.D., 1973. Mem. staff Los Alamos Sci. Lab., N.Mex., 1973-75; asst. prof. mech. and aerospace engring. Princeton U., 1975-81; Coral Industries prof. renewable energy resources U. Hawaii, Honolulu, 1981—; cons. Council on Environ. Quality, Office Tech. Assessment, Exxon Corp. Contbr. articles to sci. jours. Recipient numerous grants fed. and state agys. Mem. Am. Chem. Soc., Am. Phys. Soc., Soc. Indsl. and Applied Math., Combustion Inst., AAAS. Christian Scientist. Subspecialties: Biomass (energy science and technology); Solar energy. Current work: Pyrolysis chemistry, thermochemical biomass conversion, high temperature solar thermal energy. Home: 705 Hoopii St Honolulu HI 96825 Office: U Hawaii 2540 Dole St Honolulu HI 96822

ANTCZAK, DOUGLAS FRANCIS, veterinary immunologists, researcher; b. Waterbury, Conn., Dec. 19, 1947; s. Francis A. and Harriet H. (Kolakowski) A.; m. Wendy Susan Robinson, Jan. 4, 1978; 1 dau., Elizabeth Catherine. B.A., Cornell U., 1969; V.M.D. (Thomas scholar), U. Pa., 1973; Ph.D., U. Cambridge (Eng.), 1978. Asst. prof. immunology James Boker Inst. Animal Health Cornell U., Ithaca, N.Y. Contbr. articles to sci. jours. Mem. Am. Assn. Vet. Immunologists, Am. Assn. Immunology, others. Democrat. Roman Catholic. Clubs: Pitt (Cambridge); Cornell (N.Y.C.). Subspecialties: Immunology (agriculture); Animal genetics. Current work: Immunogenetics of domestic animals; histocompatibility; monoclonal antibody technology.

ANTMAN, ELLIOTT MARSHALL, cardiologist, hospital coronary care unit director, educator; b. N.Y.C., May 9, 1950; s. Charles Harold and Mimi (Elbaum) A.; m. Karen Hamm, Aug. 15, 1976; children: Amy, David. M.D., Columbia U., 1974. Med. intern. Presbyn. Hosp., N.Y.C., 1974-75, resident in medicine, 1975-77; fellow in cardiovascular medicine Peter Bent Brigham Hosp.-Harvard U., Boston, 1977-80; dir. CCU Brigham and Women's Hosp., Boston, 1980—; asst. prof. medicine Harvard U. Med. Sch., 1980—. Subspecialties: Cardiology; Internal medicine. Current work: Cardiology. Home: 200 Hartman Rd Newton Center MA 02159 Office: Cardiovascular Div Brigham and Women's Hosp 75 Francis St Boston MA 02115

ANTONACCIO, MICHAEL JOHN, pharmaceutical company executive; b. Yonkers, N.Y., Mar. 6, 1943; s. Mario and Frances (Renda) A.; m. Jeanne Borman, June 20, 1970; m. Patricia Ann McDevitt, July 8, 1978; 1 son, Nicholas. B.S., Duquesne U., 1966; Ph.D., U. Mich., 1970. Sr. scientist I Geigy Pharms., 1970-72; sr. scientist II CIBA-Geigy Pharms., 1972-73, sr. staff scientist, 1973-75, mgr. cardiovascular pharmacology, 1975-77; dir. pharmacology Squibb Inst. Med. Research, Princeton, N.J., 1977-81; v.p. new drug discovery Schering Corp., Bloomfield, N.J., 1981-82; v.p. cardiovascular research and devel. Bristol-Myers Co., Evansville, Ind., 1982—. Contbr. articles to profl. jours. Recipient Bernard Schiller award for excellence in the humanities, 1966, Harry Goldblatt award in cardiovascular research, 1981. Fellow Council High Blood Pressure, Council Circulation Am. Heart Assn.; mem. Am. Pharm. Assn., AAAS, N.Y. Acad. Sci., Am. Soc. Pharmacology and Exptl. Therapeutics, Phila. Physiol. Soc., Soc. for Neurosci., Internat. Soc. Hypertension, Internat. Soc. Heart Research, Internat. Soc. Immunopharmacology, Rho Chi. Subspecialties: Pharmacology; Neuropharmacology. Current work: Cardiovascular research; new drug discovery. Home: 1732 S Sassafras Dr Evansville IN 44712 Office: 2204 Pennsylvania Ave Evansville IN 47721

ANTONUCCI, TONI CLAUDETTE, research psychologist, educator; b. Bklyn., Sept. 9, 1948; d. Santino and Dorothy (Ritch) A.; m. James S. Jackson, Dec. 1, 1979; 1 dau., Ariana. B.A., Hunter Coll., 1969; M.A., Wayne State U., 1972, Ph.D., 1973. Asst. prof. psychology Syracuse U., 1973-79; lectr. in psychology U. Mich., 1979, asst. prof. family practice, 1979—, postdoctoral fellow, 1977-79, asst. research scientist, 1979-82, assoc. research scientist, 1982—; cons. Nat. Acad. Sci. Author: Life-span Development and Behavior, 1980. Wayne State U./U. Mich. summer fellow, 1971; Wayne State U. Grad. Profl. Sch. fellow, 1971-72; U. Mich. Postdoctoral fellow and research trainee, 1977; Nat. INst. Aging grantee, 1979-81. Mem. Midwestern Psychol. Assn., Soc. Research in Child Devel., Gerontol. Soc. Am., Am. Psychol. Assn., Soc. Tchrs. in Family Medicine. Subspecialties: Family practice; Gerontology. Current work: Attachment and social support across the life span; relationships between social supports. Home: 517 Fairview Ypsilanti MI 48197 Office: Inst Social Research/Survey Research Center PO Box 1248 Ann Arbor MI 48106

ANTUNES, JOÃO LOBO, neurosurgeon, researcher; b. Lisbon, Portugal, June 4, 1944; s. João Lobo and Margarida (Lima) A.; m.; children: Margarida, Maria-João, Paula. M.D., U. Lisbon, 1968. Diplomate: Am. Bd. Neurosurgery, 1978. Research fellow Parkinson's Disease Found., Neurol. Inst. N.Y., 1971; asst. prof. neurol. surgery Columbia U. Coll. Physicians and Surgeons, N.Y.C., 1978-81, assoc. prof., 1981—. Contbr. articles to med. jours. Fulbright-Hayes grantee, 1971; Matheson Found. grantee, 1975-77. Mem. ACS, Am. Assn. Neurol. Surgeons, Congress Neurol. Surgeons, Endocrine Soc., Soc. Neurosci. Subspecialty: Neuroendocrinology. Current work: Hypothalamus-pituitary axis; neural control of reproduction; neural regeneration. Office: care R Ricardo Espirito Santo 2 Cascais 2750 Portugal 10032

ANUSAVICE, KENNETH JOHN, dental educator, researcher; b. Worcester, Mass., Aug. 6, 1940; s. Frank Anthony and Jennie (Angies) A.; m. Sandra Lee Hatch; children: Jennifer Leigh, Joel Scott. B.S., Worcester Poly. Inst., 1962; Ph.D., U. Fla., 1970; D.M.D., Med. Coll. Ga., 1977. Design engr. AVCO Corp., Wilmington, Mass., 1962; grad. research asst. U. Fla., Gainesville, 1963-69; metall. engr. U.S. AEC, Aiken, S.C., 1969-71, tech. analyst, 1971-73; asst. prof. Med. Coll. Ga., Augusta, 1973-77, assoc. prof., 1977-83; prof., chmn. dept. dental biomaterials U. Fla., Gainesville, 1983—; mem., cons. oral biology and medicine ad-hoc study sect. Nat. Inst. Dental Research, Bethesda, Md., 1981-83; chmn. porcelain-metal systems subcom. Am. Nat. Standards Com., Chgo., 1980-83. Contbr. chpts. to publs. in field. Lector St. Mary's on the Hill Ch., Augusta, 1982-83; vol. dentist Children and Youth Clinic, Augusta, 1977-82. Research career devel. awardee NIH/Nat. Inst. Dental Research, 1980. Mem. ADA, Internat. Assn. Dental Research, Am. Assn. Dental Research, Internat. Assn. Dental Research (chmn. constitution and by-laws com. Dental Materials Group 1982—), Pi Tau Sigma, Alpha Sigma Mu, Omicron Kappa Upsilon. Roman Catholic. Clubs: Atomic Energy (Aiken) (pres. 1972-73); Am. Soc. for Metals (chmn. Savannah River Chpt. 1972-73). Subspecialties: Biomaterials; Prosthodontics. Current work: Research in dental materials including metallic, ceramic, metal-ceramic restorations. Specific areas of research include fractography, finite element stress analysis, metal-ceramic adherence, thermally-induced stresses. Home: 1000 NW 112th Terr Gainesville FL 32601 Office: Coll Dentistry U Fla Gainesville FL 32610

APEL, JOHN RALPH, physicist, govt. ofcl.; b. Absecon, N.J., June 14, 1930; s. Ezio A. and Grace A. (Rose) Baltera; m. Martha Eleise Davis, Sept. 8, 1956; children—Denise Alison, Jacqueline Jeanne. B.S., U. Md., 1957, M.S., 1961; Ph.D. (William S. Parsons fellow), Johns Hopkins, 1970. With Applied Physics Lab., Johns Hopkins, Silver Spring, Md., 1957-70, sr. physicist, 1961-70, asst. group supr., 1966-70; dir. Ocean Remote Sensing Lab., Atlantic Oceanographic and Meteorol. Labs., Miami, Fla., 1970-75, Pacific Marine Environ. Lab., NOAA, Seattle, 1976—; Adj. prof. physics U. Miami, 1970-76; affiliate prof. atmospheric scis. and oceanography U. Wash., 1976—; cons. NASA, Dept. Def., 1971—, UNESCO, Intergovtl. Oceanographic Commn., 1975—; chmn. ocean dynamics adv. sub com. NASA, 1973-76; mem. Internat. Union of Radio Scis., Commn. F, 1974—, Inter-Union Commn. Radio Meteorology, 1975-82; Sr. fellow Joint Inst. for Study of Atmosphere and Ocean, 1977—; chmn. Sea Use Council Sci. and Tech. Bd., 1977-81; sr. fellow Joint Inst. for Marine and Atmospheric Research, 1978—; chmn. aerospace and remote sensing Internat. Council for Exploration of Sea, 1979—; trustee Pacific Sci. Center, 1977—; mem. Sci. Commn. Ocean Research, 1980—. Contbg. author: Ballistic Missile and Space Technology, Vol. 4, 1961, Advances in Geophysics, Vol. 9, 1962, Advances in Astronautical Sciences, Vol. 30, 1974, Remote Sensing: Energy-Related Studies, 1975, Annual Review of Earth and Planetary Sciences, vol. 8, 1980; guest co-editor: Boundary Layer Meteorology, 1977; Contbr. articles to profl. jours. Served with USNR, 1951-52. Recipient Disting. Authorship award Nat. Oceanic and Atmospheric Adminstrn., 1976; Gold medal for meritorious service Dept. Commerce, 1974. Mem. Am. Phys. Soc., Am. Meteorol. Soc., N.Y. Acad. Scis., A.A.A.S., Am. Geophys. Union, Sigma Xi, Sigma Pi Sigma, Phi Eta Sigma, Phi Delta Theta. Clubs: Cosmos (Washington); Explorers (N.Y.C.). Subspecialty: Physics laboratory administration. Current work: Physical oceanography; remote sensing. Home: 5754 63d Ave NE Seattle WA 98105 Office: 3711 15th Ave NE Seattle WA 98105

APGAR, BARBARA JEAN, research chemist; b. Tyler, Tex., Mar. 4, 1936; d. Albert Edward and Mary Agnes Francis; m.; children: Katherine, Michael, John. Ph.D., Cornell U., 1964. Research chemist U.S. Plant, Soil and Nutrition Lab., Ithaca, N.Y., 1959—; courtesy appointment asst. prof. dept. animal sci., div. nutritional sci. Cornell U., Ithaca, 1971—. Mem. Am. Soc. Animal Sci., Am. Inst. Nutrition. Subspecialty: Animal nutrition. Current work: Zinc nutrition. Office: US Plant Soil and Nutrition Lab Ithaca NY 14853

APICELLA, MICHAEL ALLEN, med. researcher; b. Bklyn., Apr. 4, 1938; s. Anthony and Fay (Kahn) A.; m. Agnes Maria, Dec. 16, 1961; children: Michael P., Christopher A., Peter N. B.A., Coll. Holy Cross, 1959; M.D., SUNY Downstate Med. Ctr., 1963. Diplomate: Am. Bd. Internal Medicine. Intern, then resident Ohio State U., Columbus, 1963-66; fellow in medicine Johns Hopkins Hosp., Balt., 1966-68, practice medicine, San Antonio, 1968-70, Buffalo, 1970-78, 81—, Reno, 1979-81; prof. medicine and microbiology Med. Sch. SUNY-Buffalo, 1970—. Contbr. articles to profl. jours. Served to maj. USAF, 1968-70. Recipient USPHS career devel. award, 1974-79. Mem. Infectious Disease Soc., Central Soc. Clin. Research, Western Soc. Clin. Research, Am. Soc. Microbiology. Roman Catholic. Subspecialties: Internal medicine; Infectious diseases. Current work: bacterial antigens and their relationship to human disease. Office: Med Sch SUNY Buffalo NY 14215

APIRION, DAVID, geneticist; b. Petah-Tikva, Israel, July 17, 1935; came to U.S., 1963, naturalized, 1970; s. Shlomo Zalman and Zehava Golda (Shevkes) A.; m. Mary Riddle McKinley, Sept. 10, 1963; children: Jonathan, Michael, Alison. M.Sc., Hebrew U., Jerusalem, 1960; Ph.D., Glasgow U., 1963. Lectr. Glasgow U., 1963; research fellow Harvard U., Cambridge, Mass., 1963-65; asst. prof. Washington U., St. Louis, 1965-70, assoc. prof., 1970-78, prof. microbiology/ immunology, 1978—. Contbr. over 140 articles to profl. jours. Active Big Bros. of St. Louis. Served with Israeli Army, 1953-55. Jewish. Subspecialties: Biochemistry (biology); Genetics and genetic engineering (biology). Current work: RNA processing, differentiation. Home: 40 Hillvale St Clayton MO 63105 Office: 660 S Euclid St PO Box 8093 Saint Louis MO 63110

APLEY, WALTER JULIUS, JR., nuclear engineer; b. Salem, Oreg., July 9, 1948; s. Walter Julius and Kathryn (Mallory) A.; m. Gail Anna Lillie, June 12, 1971. B.S. in Aero. Engring, Stanford U., 1971, M.S., 1971; M.S. in Nuclear Engring, U. Wash., 1977. Registered profl. engr., Wash. Sr. researcher Battelle Northwest Richland, Wash., 1977—, tech.leader nuclear systems, 1979—, NRC operator license examiner, 1981—. Democratic precinct committeeman, 1978—. Served to lt. USN, 1971-76. Mem. Am. Nuclear Soc., ASME, Nat. Soc. Profl. Engrs., Am. Soc. Naval Engrs. Episcopalian. Subspecialties: Nuclear engineering; Nuclear fission. Current work: Energy systems research with emphasis on nuclear systems control, specifically procedure and test program optimization. Office: PO Box 999 Richland WA 99352 Home: 1220 Rd 64 Pasco WA 99301

APOSTAL, MICHAEL CHRISTOPHER, engring. mechanics research and devel. co. exec., energy-related tech. cons. co. exec., educator; b. Monterey, Calif., July 21, 1944; s. Emanuel and Iris (Kalman) A.; m. Rose Maryann Parente, June 7, 1980; 1 dau., Nikki. B.S. in Civil Engring., U. R.I., 1967; M.S. in Civil Engring, U. Conn., 1974, Ph.D., SUNY, Buffalo, 1974. Sr. structures engr. Bell Aerospace Corp., Buffalo, 1969-74; sr. structures engr. Marc Analysis Research Corp., Providence, 1974-76; pres. Jordan, Apostal, Ritter Assos., Inc., Davisville, R.I., 1977—; v.p. Drilling Resources Devel. Corp. subs. Jordan, Apostal, Ritter Assos., Inc., Tulsa, 1980—; adj. asso. prof. civil and environ. engring. U. R.I.; Mem. North Kingstown (R.I.) C. of C. Contbr. articles to profl. jours. Mem. ASME, Soc. Petroleum Engrs., Soc. Computer Simulation, Petroleum Club R.I. Greek Orthordox. Clubs: Warwick (R.I.); Sportsman's Assn. Subspecialties: Fracture mechanics; Theoretical and applied mechanics. Current work: Numerically modeling (using the finite element method) three-dimensional dynamic behavior characteristics of a rotating, drilling, bottom hole assembly. Office: Jordan Apostal Ritter Assos Inc Adminstrn Bldg 7 Davisville RI 02854

APOSTOLAKOS, DIANE, molecular biologist; b. Duluth, Minn., Apr. 4, 1953; d. Peter Constantine and Ruth Millie (Ziegler) A. B.A. in Chemistry and B.A. in Microbiology, Calif. State U.-Chico, 1975; M.S. in Microbiology, Ariz. State U., 1978. Asst. scientist chem. process research Pfizer Inc., Groton, Conn., 1979-80, assoc. scientist, 1980-81, assoc. scientist molecular genetics, 1981—. U. Mich. Biol. Sta. grantee, 1975. Mem. Am. Soc. for Microbiology, Soc. for Indsl. Microbiology. Current work: bacterial genetics, process research on penicillins, recombinant DNA technology for chemical processes. Office: Pfizer Inc Central Research Bldg 118 Groton CT 06340

APOSTOLOU, SPYRIDON F., med. devices mfg. co. exec.; b. Athens, June 9, 1944; U.S., 1965, naturalized, 1970; s. Photios K. and Maria A. (Gikontes) A.; m. Christina O. Nerrie, Dec. 12, 1981. B.S. in Physics, Bradley U., 1968; M.S., Rensselaer Poly. Inst., 1970; Ph.D. in Solid State Physics, Washington U., St. Louis, 1972; Sc.D. in Mech. Engring. and Material Engring, 1976. Postdoctoral researcher Sch.

Engring., Washington U., 1976-77; project engr. Becton-Dickinson, Rutherford, N.J., 1977-78, mgr. materials, 1979-80, asst. dir. research and devel., 1980-81, dir. sci. and tech., 1981—. Contbr. articles to profl. jours. Mem. Soc. Plastics Engrs. (chmn. mktg. div.), Am. Mgmt. Assn., N.Y. Acad. Sci. Greek Orthodox. Subspecialties: Biomaterials; Polymer physics. Current work: Metallurgy technology, polymeric and radiation technology, needle technology, coating and lubrication technology, new processes, new materials technology. Home: 79 Alexander Ave Upper Montclair NJ 07043 Office: Stanley St East Rutherford NJ 07070

APPEL, ANTOINETTE RUTH, neuropsychologist; b. N.Y.C., Mar. 31, 1943; d. Leon S. and Augusta (Marienberg) A. B.A., U. Vt., Burlington, 1964; M.A., Mt. Holyoke Coll., 1965; Ph.D., CUNY, 1972. Lic. psychologist, Conn, Mass., R.I.; Diplomate Am. Bd. Profl. Neuropsychology. Fellow, instr. Mt. Sinai Sch. Medicine, N.Y.C., 1971-74; asst. prof. So. Ill. U. Sch. Medicine, Carbondale, 1974-76; USPHS intern Conn. Valley Hosp., Middletown, 1976-77; asst. prof. U. Conn. Health Ctr., Farmington, 1977-79, Brown U., Providence, 1979-83; cons., Providence, 1981-83; neuropsychologist Butler Hosp., Providence, 1979-81; dir. neuropsychol. assessment and treatment Center Neurol. Scis., Fort Lauderdale, Fl, 1983—; cons. Fuller Meml. Hosp., Attleboro, Mass., 1979-81, VA Hosp., Providence, 1979-81, Dept. Social Rehab. Services, 1981-83. Contbr. articles to profl. jours. Bd. dirs. Sojourner House, Providence, 1979-81, Hartford (Conn.) Interval House, 1978-79, Combined Hosp. Alcoholism Program, Hartford, 1978. Served with WAC, 1963-64. Recipient cert. of recognition Psi Chi, 1974; Hartford Salute award Greater Hartford Conv. Bur., 1979. Mem. Am. Psychol. Assn., Eastern Psychol. Assn., N.Y. Acad. Scis., Sigma Xi, Psi Chi. Democrat. Subspecialties: Neuropsychology; Health services research. Current work: Methods of identifying physical diseases which masquerade as psychiatric syndromes; mechanisms of dysfunction in closed head injuries. Inventor eye movement monitor. Home: 6622 Racquet Club Dr Lauderhill FL 33319 Office: Center Neurol Services 5601 N Dixie Hwy Suite 407 Fort Lauderdale FL 33334

APPELBE, WILLIAM FREDERICK, computer systems educator, consultant; b. Bristol, Eng., Sept. 26, 1942; s. Frederick James and Barbara Jean (Tucker) A. B.S. with honors, Monash U., Australia, 1973; M.S. in Computer Sci, U. B.C., 1975; Ph.D. in Computer Sci. and Elec. Engring, U. B.C., 1978. Asst. prof. computer sci. So. Meth. U., 1977-78; asst. prof. elec. engring. and computer scis. U. Calif.-San Diego, 1978—; vis. prof. Copenhagen U., 1980; cons. to industry and govt.; program chmn. UNICOM Conf., 1983. Can. Commonwealth scholar, 1973. Mem. IEEE, Assn. Computing Machinery, Sigma Xi. Subspecialties: Distributed systems and networks; Programming languages. Current work: Design and implementation of system software. Home: 8332 Regents St B2 San Diego CA 92122 Office: U Calif San Diego C 014 EECS Dept LaJolla CA 92093

APPELL, GERALD FRANCIS, government scientific project office, mechanical engineer; b. N.Y.C., June 9, 1942; s. Francis Joseph and Evelyn Ann (Brosch) A.; m. Marika Jombach, Oct. 31, 1964; children: Stephen, Michael. B.S. in Mech. Engring, Bklyn. Poly. Inst., 1964. Mech. engr. U.S. Naval Oceanographic Office, Washington, 1964-70, Nat. Oceanographic Instrumentation Center, 1970; project leader NOAA, Rockville, Md., 1970—. Editor, contbg. author articles, report to profl. publs. Recipient spl. achievement award Dept. Commerce, NOAA, 1974, research publ. award Naval Ocean Research and Devel. Activity, 1980. Mem. Marine Tech. Soc., IEEE (vice chmn. current measurement tech. com. 1980—). Republican. Lutheran. Subspecialty: Ocean engineering. Current work: Performance evaluation of oceanographic measurement systems, measurement data quality assurance programs, ocean current measurement technology specialist. Inventor high pressure lighting system. Home: 11805 Earnshaw Ct Brandywine MD 20613 Office: Dept Commerce NOAA 6010 Executive Blvd Rockville MD 20852

APPELMAN, HENRY D., pathologist, medical educator; b. Chgo., Dec. 16, 1935; s. Morris J. and H. Esther (Rosenstein) A. M.D., U. Mich., 1961. Diplomate: Am. Bd. Pathology. Intern Wayne County Gen. Hosp., Eloise, Mich., 1961-62; resident in pathology U. Mich. Hosp., Ann Arbor, 1962-66; asst. prof. pathology Hahnemann Med. Coll., Phila., 1968-69, U. Mich., Ann Arbor, 1969-70, assoc. prof. pathology, 1970-75, prof. pathology, 1975—. Served as capt. USAR, 1966-68. Mem. Gastrointestinal Pathology Club (pres. 1980-81), Am. Gastroenterologic Assn., Internat. Acad. Pathology, Am. Assn. Pathology, Am. Assn. Study Liver Disease. Subspecialty: Pathology (medicine). Current work: Clinical-pathologic correlative studies of diseases of the gastrointestinal tract and liver. Office: Dept Pathology U Mich Med Ctr PO Box 045 Ann Arbor MI 48109

APPLE, MARTIN ALLEN, computer-biotech. co. exec.; b. Duluth, Minn., Sept. 17, 1939; s. Samuel Ben and Sylvia (Mintz) A.; m. Grace Ann Canfield, 1960; children: Deborah Dawn, Pamela Ruth, Nathan Herschel, Rebeccah Lynn. A.B., U. Minn., 1959, M.Sc., 1963; Ph.D., U. Calif., 1968. Asst. prof. U. Calif., 1971-78; asst. prof. Sch. Pharmacy and Cancer Research Inst., U. Calif., 1971-78; pres., chief exec. officer Internat. Plant Research Inst., Inc., San Carlos, Calif., 1978-82; adj. prof. U. Calif., San Francisco, 1982; chmn., chief exec. officer Ean-Tech, Daly City, Calif., 1982—; dir. Internat. Plant Research Inst., San Carolos, 1978-82, Conation Techs., San Francisco, 1982—. Contbr. articles in field to profl. jours. Chmn. Zone 7 Sch. Council, San Francisco, 1970-72. Recipient awards Am. Cancer Soc., 1970, 71, 73, 75, Grantee NIH, 1973-78, U. Calif., 1975-78. Fellow Am Inst. Chemists, Am. Coll. Clin. Pharmacologists; mem. AAAS, Am. Inst. Chemists, Am. Coll. Clin. Pharmacology, Am Soc. Pharmacology and Exptl. Therapeutics, Phi Beta Kappa. Subspecialties: Genetic Engineering and biotech.; Computer intelligence/graphics. Current work: Biomolecular computer components, artificial intelligence, bioactive molecule design exec., computer software-genetic engring. product design and mktg. Patentee in field. Office: 699-A Serramonte Blvd Daly City CA 94015

APPLEBY, JOHN FREDERICK, physicist; b. Houston, Aug. 22, 1948; s. Walter Goode and Helen Marie (Norris) A.; m. Jeannine Poma, Dec. 28, 1977; 1 dau, Danielle Jeanne-Helen. B.A., Oberlin Coll., 1970; M.S., U. Mass.-Amherst, 1972; Ph.D., SUNY-Stony Brook, 1980. Research scientist Planetary Sci. Inst., Pasadena, 1979-81; sr. scientist Jet Propulsion Lab., Pasadena, 1981—. Contbr. articles to profl. jours. Mem. Am. Astron. Soc. (dir. planetary sci.), Sigma Xi. Subspecialties: Planetary atmospheres; Planetary science. Current work: Atmospheric structures of the Jovian Planets and Titan; remote retrieval of atmospheric parameters; non-local thermodynamic equilibrium processes; Jupiter's clouds and NH3 ice scattering. Office: 4800 Oak Grove Dr MS 183 301 Pasadena CA 91109

APT, JEROME, III, planetary physicist; b. Springfield, Mass., Apr. 28, 1949; s. Jerome and Joan (Frank) A. A.B., Harvard U., 1971; Ph.D., M.I.T., 1976. Research fellow in planetary physics Harvard U., 1976-79, research assoc., 1979-80, asst. dir. div. applied sci., 1978-80; mem. tech. staff Jet Propulsion Lab., Calif. Inst. Tech., Pasadena, Calif., 1980—, dir., 1981—. Contbr. articles to sci. jours. Mem. Am. Phys. Soc., Am. Astron. Soc., Am. Geophys. Union, Nat. Assn. Rocketry (trustee). Subspecialties: Planetary science; Remote sensing (geoscience). Current work: Composition of planetary surfaces and atmospheres; utilize earth-orbital, deep-space, and groundbased remote sensing techniques in the visible and near-infrared.

AQUADRO, CHARLES FREDERICK, geneticist; b. Wilmington, Del., July 16, 1953; s. Lawrence Charles and Anne (Hallowell Klutey) A.; m. Gwen Sholl, June 14, 1975; 1 dau., Christine Anne. B.S. in Biology, St. Lawrence U., 1975; M.S. in Zoology, U. Vt., 1978; Ph.D. in Genetics, U. Ga., 1981. Staff fellow Lab. Genetics, Nat. Inst. Environ. Health Scis., NIH, Research Triangle Park, N.C., 1981—. Contbr. articles to profl. jours. Recipient Theodore Roosevelt Meml. award Am. Mus. Natural History 1979; NIH trainee, 1978-81. Mem. Genetics Soc. Am., Soc. for Study of Evolution, Am. Soc. Mammalogists, Omicron Delta Kappa, Beta Beta Beta. Subspecialties: Evolutionary biology; Gene actions. Current work: Molecular evolution and evolutionary genetics; drosophila, gene regulation, alcohol dehydrogenase, humans, mitochondrial DNA, DNA sequence variation. Home: 812 Chalice St Durham NC 27705 Office: Lab Genetics Nat Inst Environ Health Scis PO Box 12233 Research Triangle Park NC 27709

ARAVE, CLIVE WENDELL, animal scientist, educator; b. Idaho Falls, Idaho, May 12, 1931; s. Joseph Clarence and Rhoda Elvera (Peterson) A.; m. Carley McMurtrey, Oct. 10, 1950; children: Wendy, Stephanie, Joe, Christine, Lorraine, James. B.S., Utah State U., 1956, M.S., 1957; Ph.D., U. Calif.-Davis, 1963. Asst. mgr. Lavacre Farms, Modesto, Calif., 1957-59; asst. prof. agr. Calif. State U.-Chico, 1963-65; asst. prof. to assoc. prof. animal, dairy and vet. scis. Utah State U.-Logan, 1965—. Contbr. articles to profl. jours. Served to sgt. AUS, 1951-53. U.S.-N.Z. Coop. Sci. Program grantee NSF, 1980-81. Mem. Am. Dairy Sci. Assn., Animal Behavior Soc., Sigma Xi, Phi Kappa Phi. Republican. Mormon. Lodge: North Logan Lions. Subspecialties: Animal genetics; Behaviorism. Current work: Genetic polymorphisms, animal sensory perception, early experience, animal behavior, sensory perception, operant conditioning, social behavior, early experience. Home: 1460 E 2100 N Logan UT 84321 Office: Utah State U UMC 48 Logan UT 84322

ARCHARD, HOWELL OSBORNE, dental educator, oral pathology consultant; b. Yonkers, N.Y., Mar. 25, 1929; s. Howell Osborne and Rachel (Richmond) A.; m. Nellie Joan Lamkie, Dec. 14, 1961; 1 dau., Karen Rae. B.Sc., Rutgers U., 1951; D.D.S., Columbia U., 1955. Diplomate: Am. Bd. Oral Pathology. Clin. assoc. Columbia U., 1958-60; resident in oral pathology Nat. Inst. Dental Research and Armed Forces Inst. Pathology, Bethesda, Md., and Washington, 1960-64; chief diagnostic pathology Nat. Inst. Dental Research, Bethesda, 1968-79; assoc. prof. dentistry SUNY-Stony Brook, 1979—; vis. scientist Inst. Dental Research of U. Ala.-Birmingham, 1973-78; cons. com. dental edn. Am. Dental Assn., 1977—. Contbr. articles sci. jours. Served with USPHS, 1955-57, 60-79. Recipient cert. for disting. performance U.S. Air Force and Armed Forces Inst. Pathology, 1962. Fellow Am. Acad. Oral Pathology; mem. Internat. Assn. Oral Pathologists (charter), Am. Dental Assn., Internat. Assn. Dental Research, Soc. Investigative Dermatology. Republican. Clubs: Rutgers of Washington (sec. 1968-72), Columbia Dental Alumni.). Subspecialty: Oral pathology. Current work: Clinico-pathologic studies of human oral mucosal diseases, studies of inherited and acquired metabolic diseases affecting the dentition. Home: 9 Stern Dr Port Jefferson NY 11777 Office: School of Dental Medicine State University of New York Stony Brook NY 11794

ARCHER, JUANITA ALMETTA HINNANT, physician, educator; b. Washington, Nov. 3, 1934; d. Roy E. and Anna O. (Blakeasey) Hinnant; m. Frederick I. Archer, June 8, 1958. B.S., Howard U., 1956, M.S., 1958, M.D., 1965. Intern Freedmen's Hosp., Washington, 1965-66; resident Howard U., Washington, 1966-68, fellow, 1970-71, instr., 1971-75, asst. prof., 1975-78, assoc. prof. medicine, 1978—; attending staff D.C. Gen. Hosp.; attending staff, dir. endocrine metabolic lab. Howard U. Hosp., Washington, 1975; cons. NIH, 1979-82, Gen. Clin. Research Ctr., 1979—, Nat. Inst. Allergy and Metabolic Disease, 1979. Am. Diabetes Assn. grantee, 1976; NIH grantee, 1979; Howard U. Women's Aux. grantee, 1981; Josiah Macy Found. faculty fellow, 1974-77. Baptist. Subspecialties: Receptors; Internal medicine. Current work: Insulin action and insulin receptor activity in human erthrocytes. Home: 4305 Ranger Ave Temple Hills MD 20748 Office: Howard U Hosp 2041 Georgia Ave NW Washington DC 20060

ARCHER, MICHAEL CHRISTOPHER, cancer researcher, educator; b. Stoke-on-Trent, Eng., May 9, 1943; s. William Thomas and Mabel (Bailey) A.; m. Carolyn D. Hoskins, Oct. 10, 1970; children: Emily Winpenny, Jason Warwick. M.A., Cambridge (Eng.) U., 1965; M.Sc., U. Warwick, Eng., 1967; Ph.D., U. Toronto, Ont., Can., 1970. Research assoc. dept. nutrition and food sci. M.I.T., 1970-72, asst. prof., 1972-76, assoc. prof., 1976-79; assoc. prof. dept. med. sci. U. Toronto, 1979-82; sr. scientist Ont. Cancer Inst., Toronto, 1979-82, prof., sr. scientist, 1982—; cons. in field. Contbr. articles to profl. jours. Recipient Future Leader award Nutrition Found., 1974-76; NIH grantee, 1977—. Mem. Royal Soc. Chemistry (fellow), Am. Chem. Soc., AAAS, Am. Soc. Biol. Chemists, Soc. Toxicology, Am. Assn. Cancer Research. Subspecialties: Cancer research (medicine); Toxicology (medicine). Current work: Chemical carcinogenesis. Office: Ont Cancer Inst 500 Sherbourne St Toronto Canada M4X 1K9

ARCHER, RICHARD EARL, product designer and alternative energy design cons.; b. Springfield, Ill., Aug. 24, 1945; s. Earl Wiley and Era Marie (Fentress) A.; m. Elizabeth Lou Lutz, Aug. 9, 1969; children: Jeremy Richard, William Earl. B.A. in Design, So. Ill. U., Carbondale, 1970; M.S., Gov.'s State U., 1979. Instr. design So. Ill. U., 1971-79, coordinator design program, 1979-80, asst. prof. comprehensive planning and design, 1980—; dir. Applied Alternatives.; Mem. Nat. Alcohol Fuels Commn., 1980; chmn. Carbondle Energy Futures Task Force, 1980-81; mem. Ill. Legislature Alternative Energy Commn., 1981—; mem. adv. panel U.S. Congl. Office Tech. Assessment, 1982. Editor: Ill. Solar Resource Adv. Council Grants Newsletter, 1979-81; contbr. articles to profl. jours. Recipient Outstanding Tchr. Year award Coll. Human Resources, So. Ill. U., 1979; U.S. Dept. Energy grantee, 1979-81; U.S. Dept. Labor grantee, 1978-79; Ill. Dept. Energy grantee, 1980-81. Mem. Internat. Solar Energy Soc., Am. Wind Energy Assn., Internat. Biomass Inst., Nat. Alcohol Fuels Producers Assn., Solar Lobby (dir.). Subspecialties: Solar energy; Biomass (energy science and technology). Current work: Solar energy, ethanol prodn., energy and chem. feedstocks from biomass, community econs. of conservation and alternative energy. Home: Box 168 Rural Route 1 DeSoto IL 62924 Office: Design Program Bldg 0720 So Ill U Carbondale IL 62901

ARCHER, RICHARD LLOYD, psychology educator; b. Wichita, Kans., Dec. 28, 1948; s. George Louis and Mary Lee (Brower) A.; m. Cheryl Ann Brandner, Mar. 29, 1971. B.A. in Psychology, U. Kans., 1971; Ph.D. in Social Psychology, Duke U., 1976. Asst. prof. U. Tex.-Austin, 1975-82; asst. prof. Southwest Tex. State U., San Marcos, 1982—; cons. in field; reviewer NSF, 1980—. Contbr. articles on psychology to profl. jours. Beulah Morrison grantee, 1971; NSF fellow, 1971-74. Mem. Soc. of Exptl. Social Psychology, Am. Psychol. Assn., Southwestern Psychol. Assn., Phi Beta Kappa, Omicron Delta Kappa. Democrat. Subspecialty: Social psychology. Current work: Attraction and relationships, especially self-disclosure altruism and helping, especially empathy. Home: 4706 Fieldstone Dr Austin TX 78735 Office: Dept Psychology Southwest Tex State U San Marcos TX 78666

ARCHIBALD, PATRICIA A., biologist, educator; b. Olney, Ill., July 18, 1934; d. Stanley Ray and Mabel Ellen (Seed) A. B.S., Ball State U., 1953, M.A. in Biol. Sci, 1961; Ph.D. in Botany, U. Tex.-Austin, 1969. Mem. faculty Slippery Rock State U. (Pa.), 1969—, prof. biology, 1979—; sci. rev. adminstr. competitive grants program EPA, 1980-82; exchange scientist Czechoslovak Acad. Sci., 1977, USSR Acad. Sci., 1978. Contbr. articles to sci. jours. Mem. Am. Phycol. Soc., Brit. Phycol. Soc., Internat. Phycol. Soc., Am. Assn. Systematics, AAAS. Subspecialties: Physiology ([illegible]); Morphology. Office: Slippery Rock State Univ Slippery Roc PA 16057 Home: PO Box 429 Grove City PA 16127

ARDEN, DANIEL DOUGLAS, oil company geologist, researcher; b. Bainbridge, Ga., Sept. 24, 1922; s. Daniel Douglas and Caroline (Battle) A.; m. Mary Moore, Oct. 23, 1943; children: Dana, Daniel D. V., Nancy, Laurie. A.B., Emory U., 1948, M.S., 1949; Ph.D., U. Calif.-Berkeley, 1961. Registered profl. geologist, Ga., Calif. Geologist Calif. Exploration Co., San Francisco, 194-56, Sohio Petroleum Co., Oklahoma City, 1956-65; exploration adv. Signal OIl & Gas Co., Los Angeles, 1965-70; prof. Ga. Southwestern Coll., 1970-82, emeritus, 1982—, head dept., 1975-82; staff geologist La. Land & Explroation Co., New Orleans, 1982—; geol. cons. Geophys. Service Inc., Dallas, 1971-82. Editor: Southeastern Geology Geol. Symposium, 1982. Served to capt. USAF, 1951-53; PTO, Korea. Am. Chem. Soc. grantee, 1973; NASA grantee, 1974-76. Fellow Geol. Soc. Am.; mem. Am. Geophys. Union, Am. Inst. Profl. Geologists. Club: New Orleans PEtroleum. Subspecialties: Tectonics; Geophysics. Current work: Tectonics and structural geology as related to petroleum exploration. Home: 4141 Loire Dr Kenner LA 70062 Office: La Land & Exploration Co PO Box 60350 New Orleans LA 70160

AREF, HASSAN, engineering educator; b. Alexandria, Egypt, Sept. 28, 1950; came to U.S., 1975; s. Moustafa and Jytte (Adolphsen) A.; m. Susanne Eriksen, Aug. 3, 1974; children: Michael, Thomas. Cand. Scient., U. Copenhagen, 1975; Ph.D., Cornell U., 1980. Asst. prof. engring. Brown U., Providence, 1980—. NATO fellow, 1975; fellow Cornell U., 1976-79; summer fellow Woods Hole (Mass.) Oceanog. Instn., 1980. Mem. Am. Phys. Soc., Soc. Indsl. and Applied Math., N.Y. Acad. Scis., Am. Acad. Mechanics. Subspecialties: Fluid mechanics; Theoretical physics. Current work: Vortex dynamics, computational fluid mechanics, coherent structures in turbulent flows, theory of dynamical systems. Office: Div Engring Brown Univ Providence RI 02912

ARENBERG, DAVID LEE, research psychologist; b. Balt., Nov. 14, 1927; s. Morris and Dorothy (Silver) A. A.B., Johns Hopkins U., Balt., 1951; Ph.D., Duke U., 1960. Chief learning and problem solving sect. Nat. Inst. on Aging, NIH, Balt., 1960—; vis. scholar Inst. for Child Study, U. Md., 1978-79. Co-editor, author: New Directions in Memory and Aging, 1980; Assoc. editor: Jour. Gerontology, 1970-72; cons. editor, mem. editorial bd., reviewer numerous jours. in gerontology, psychology and sci.; contbr. chpts. to books, articles to profl. jours. Served with AUS, 1946-47. Fellow Am. Psychol. Assn. (pres. div. on adult devel. and aging 1975-76, numerous coms.), Gerontol. Soc. Am. (mem. exec. com. 1971-73). Subspecialty: Cognition. Current work: Research on adult age differences and changes in memory, problem solving, and information processing. Home: 2215 South Rd Baltimore MD 21209 Office: Gerontology Research Center NIH Baltimore City Hosps Baltimore MD 21224

ARIANO, MARJORIE A., research scientist, educator; b. Tokyo, Japan, Feb. 13, 1951; d. Richard A. and Marjorie W. (Farr) A.; m. George J. Rederich, Dec. 21, 1974. B.S., UCLA, 1972, Ph.D., 1977. Postdoctoral fellow in molecular biology U. So. Calif., 1977-80; asst. prof. anatomy and neurobiology U. Vt., Burlington, 1980—; Mem. sci. research bd. Vt. Heart Assn.; sci. reviewer Hereditary Disease Found., NSF. Contbr. articles to profl. publs. Grantee NIH, Vt. Heart Assn., NSF, 1972—. Mem Soc. Neurosci. Subspecialties: Neurochemistry; Immunology (medicine). Current work: Role of cyclic Nucleotides in mediation of neural transmission within the basal ganglia. Office: Dept Anatomy and Neurobiology Given Med Bldg Burlington VT 05405

ARIS, RUTHERFORD, applied mathematician, educator; b. Bournemouth, Eng., Sept. 15, 1929; came to U.S., 1955, naturalized, 1962; s. Algernon Pollock and Janet (Elford) A.; m. Claire Mercedes Holman, Jan. 1, 1958. B.Sc. (spl.) with 1st class honours in Math, London (Eng.) U., 1948, Ph.D., 1960, D.Sc., 1964; student, Edinburgh (Scotland) U., 1948-50. Tech. officer Billingham div. I.C.I. Ltd., 1950-55; research fellow U. Minn., 1955-56; lectr. tech. math. Edinburgh U., 1956-58; mem. faculty U. Minn., 1958—, prof. chem. engring., 1963—, Regents' prof., 1978—; O.A. Hougen vis. prof. U. Wis., 1979; Sherman Fairchild Disting. Scholar Calif. Inst. Tech., 1980-81; cons. to industry, lectr., 1961—. Author: Optimal Design of Chemical Reactors, 1961, Vectors, Tensors and the Basic Equations of Fluid Mechanics, 1962, Discrete Dynamic Programming, 1964, Introduction to the Analysis of Chemical Reactors, 1965, Elementary Chemical Reactor Analysis, 1969, (with N.R. Amundson) First-Order Partial Differential Equations with Applications, 1973, (with W. Strieder) Variational Methods Applied to Problems of Diffusion and Reaction, 1973, The Mathematical Theory of Diffusion and Reaction in Permeable Catalysts, 1975, Mathematical Modelling Techniques, 1978, Chemical Engineering in the University Context, 1982, Springs of Scientific Creativity, 1982; co-editor: An Index of Scripts for E.A. Lowe's Codices Latini Antiquiores, 1982. Sr. research fellow NSF, 1964-65; Guggenheim fellow, 1971-72; Recipient E. Harris Harbison award for distinguished teaching, 1969; Alpha Chi Sigma award Am. Inst. Chem. Engrs., 1969; Chem. Engring. lectr. award Am. Soc. Engring. Edn., 1973. Fellow Inst. Math. and Applications; mem. Nat. Acad. Engring., Soc. Nat. Philosophy, Soc. for Math. Biology, Soc. Indsl. and Applied Math., Am. Chem. Soc., Am. Inst. Chem. Engrs. (R.H. Wilhelm award 1975), Mediaeval Acad. Am., Soc. Scribes and Illuminators, Soc. Textual Scholarship, Internat. Soc. Math. Modeling. Lutheran. Subspecialties: Applied mathematics; Chemical engineering. Current work: Mathematical modelling of chemical reactors; Latin palacography. Office: Dept of Chemical Engineering Univ of Minnesota Minneapolis MN 55455

ARKILIC, GALIP MEHMET, applied scientist, educator; b. Sivas, Turkey, Mar. 10, 1920; came to U.S., 1943, naturalized, 1960; s. Sabir Mehmet and Zehra Fatima (Hocazade) A.; m. Ann A. Bryan, Mar. 31, 1956; children: Victor, Dennis, Layla, Errol. B.M.E., Cornell U., 1946; M.S., Ill. Inst. Tech., 1948; Ph.D., Northwestern U., 1954. Mech. engr. Miehle Printing Press and Mfg. Co., Chg., 1948-49, analyst, 1954-56; research and devel. engr. Mech. and Chem. Industries, Turkey, 1949-52; asst. prof. Pa. State U., University Park, 1956-58; asso. prof. dept. civil engring. George Washington U., Washington, 1958-63, prof. applied sci., 1963—, chmn. dept. engring. mechanics, 1966-69, asst. dean, 1969-74. Contbr. articles to sci. jours. Vice pres. Courtland Civic Assn., Arlington, Va., 1965-66; pres. Am. Turkish Assn., Washington, 1967-71. Served to 2d lt. Turkish Army, 1939-41. Recipient Disting. Leadership award Am. Turkish Assn., 1972; Recognition of Service award Sch. Engring. and Applied Sci., George Washington U., 1976; Air Force Office of Sci. Research grantee, 1963-69. Mem. ASME, AAUP, Am. Acad. Mechanics, Sigma Xi. Club: George Washington U. (Washington). Subspecialties: Theoretical and applied mechanics;

Applied mathematics. Current work: Analysis of plates and shell structures, applied ordinary differential equationss. Home: 8403 Camden St Alexandria VA 22308 Office: George Washington U Washington DC 20052

ARLIAN, LARRY G., biological sciences educator; b. Aspen, Colo., Aug. 5, 1944; s. Briece E. and Della J. (Stringer) A.; m. Nancy E. Nelson, June 4, 1966; children: Heidi J., Heather Renee. B.S., Colo. State U., 1964, M.S., 1966; Ph.D., Ohio State U., 1972. Asst. prof. dept. biol. scis. Wright State U., Dayton, 1972-76, assoc. prof., 1976-82, prof., 1982—. Contbr. articles to profl. jours. NIH grantee, 1978—. Mem. Am. Soc. Zoologists, Entomol. Soc. Am., Am. Soc. Parasitologists, Acarological Soc. Am., Sigma Xi. Subspecialty: Parasitology. Current work: Osmoregulation and water balance physiology, house dust mite allergy and scabies infestations. Home: 470 Beacon Dr Fairborn OH 45324 Office: Dept Biol Scis Wright State U Colonel Glenn Hwy Dayton OH 45435

ARMBRUSTER, CHARLES WILLIAM, chemistry educator; b. St. Louis, Mar. 24, 1937; s. Charles Edward and Grace Doyne (Williams) A. B.S., U. Notre Dame, 1958; Ph.D. in Organic Chemistry, Washington U., St. Louis, 1966. Asst. prof. U. Mo., St. Louis, 1962-66, assoc. prof., 1966—, chmn. div. sci., 1963-67, chmn. dept. chemistry, 1967-75. Mem. Am. Chem. Soc. Subspecialties: Organic chemistry; Chemical education. Home: 20 Sunswept Creve Coeur MO 63141 Office: U Mo Saint Louis MO 63121

ARMENDAREZ, PETER XAVIER, chemistry educator; b. San Pedro, Calif., Sept. 7, 1930; s. Pedro Macias and Carmen (Miranda) A.; m. Charlene Towery, Oct. 23, 1954; children: Peter X., Lawrence P., William J., Jennifer L. B.S., Loyola U.-Los Angeles, 1952; M.A., Washington U., 1954; Ph.D., U. Ariz., 1963; M.B.A., Murray State U., 1981. Registered profl. engr., Ky. Instr. chemistry Odessa (Tex.) Coll., 1958-59; teaching asst. U. Ariz., Tucson, 1959-63; asst. prof. chemistry U. Tenn.-Martin, 1963-65; mem. faculty dept. chemistry Brescia Coll., Owensboro, ky., 1965—, prof., 1968—; indsl. supr. Nat. Southwire Aluminum, Hawesville, Ky., 1973-74. Served to capt. USAF, 1954-57. Fellow Am. Inst. Chemists; mem. Am. Chem. Soc., Am. Crystallographic Assn. Roman Catholic. Subspecialties: Physical chemistry; Inorganic chemistry. Current work: Infra-red, ultraviolet and visible spectroscopy and structure of inorganic complexes. Home: 1224 Parrish Ave Owensboro KY 42301 Office: Brescia Coll Owensboro KY 42301

ARMISTEAD, WILLIS WILLIAM, univ. adminstr.; b. Detroit, Oct. 28, 1916; s. Eber Merrill and Josephine Brunell (Kindred) A.; m. Martha Sidney Clark, Sept. 17, 1938 (dec. 1964); children—Willis William, Jack Murray, Sidney Merrill; m. Mary Wallace Nelson, 1967. D.V.M., Tex. A. and M. Coll., 1938; M.Sc., Ohio State U., 1950; Ph.D., U. Minn., 1955. Diplomate: hon. diplomate, 1975. Pvt. practice veterinary medicine, 1938-40; instr. Sch. Veterinary Medicine, Tex. A. and M. Coll., 1940-42, asst. prof. to prof., 1946-53; dean Sch. Veterinary Medicine, 1953-57, Coll. Veterinary Medicine, Mich. State U., East Lansing, 1957-74, Coll. Veterinary Medicine, U. Tenn., Knoxville, 1974-79; v.p. agr. U. Tenn. System, 1979—; collaborator animal diseases and parasite research div. Dept. Agr., 1954-65; cons., adviser commn. veterinary edn. of South So. Regional Edn. Bd., 1953-56; mem. gov.'s sci. adv. bd., 1958-60; nat. cons. to Air Force Surgeon Gen., 1960-62; mem. adv. council Inst. Lab. Animal Resources, NRC, 1962-66; pres. Assn. Am. Veterinary Med. Colls., 1964-65, 73-74; veterinary med. resident investigators selection com. U.S. VA, 1967-70; veterinary medicine rev. com. Bur. Health Professions Edn. and Manpower Tng., HEW, 1967-71; mem. Nat. Bd. Veterinary Med. Examiners, 1970-74; mem. adv. panel for veterinary medicine Inst. Medicine, Nat. Acad. Scis., 1972-74; mem. bd. agr. and renewable resources NRC, 1976-77; 1st Allam lectr. Am. Coll. Veterinary Surgeons, 1972. Contbg. author: Canine Surgery, rev. edit, 1957, Canine Medicine, rev. edit, 1959; editor: The N.Am. Veterinarian, 1950-56, Jour. Veterinary Med. Edn, 1974-80; asso. editor: Jour. Am. Animal Hosp. Assn, 1964-70; contbr. tech. articles to profl. jours. Bd. dirs. Tenn. Farm Bur. Fedn., 1979—. Served from 1st lt. to maj. Veterinary Corps AUS, 1942-46. Recipient Meritorious Service award Selective Service System, 1972; hon. alumnus Mich. State U., 1972, Disting. Alumnus award Coll. Vet. Medicine, Tex. A. and M. U., 1980. Mem. AAAS, U.S. Animal Health Assn., Am. Veterinary Med. Assn. (pres. 1957-58, award 1977), Tex Veterinary Med. Assn. (pres. 1947-48), Mich. Veterinary Med. Assn. (trustee Edn. and Sci. Trust 1970-74), Tenn. Veterinary Med. Assn., Inst. Medicine of Nat. Acad. Scis., N.Y. Acad. Scis., Sigma Xi, Phi Kappa Phi, Alpha Zeta, Phi Zeta, Omega Tau Sigma (nat. Gamma Award Ohio State U. 1962). Episcopalian. Club: Rotary. Subspecialties: Administration, agricultural research; Administration, veterinary medical research. Home: 1101 Cherokee Blvd Knoxville TN 37919 Office: Box 1071 Knoxville TN 37901

ARMITAGE, JAMES OLEN, physician; b. Los Angeles, Dec. 19, 1946; s. Bernard O. and Thelma A. (Young) A.; m. Nancy Elaine Roker, Aug. 12, 1967; children: Amy Jolane, Gregory Olen, Anne Marie, Joel Donald. B.S., U. Nebr.-Lincoln, 1969; M.D., U. Nebr.-Omaha, 1973. Diplomate: Am. Bd. Internal Medicine with subsplty. in med. oncology. Intern, resident U. Nebr. Med. Center, Omaha, 1973-75; fellow U. Iowa Hosp. and Clinics, Iowa City, 1975-77; practice medicine specializing in oncology, hematology, Omaha, 1977-79; asst. prof. medicine, dir. bone marrow transplanation unit U. Iowa, 1979-82; asso. prof. medicine, vice chmn. dept. internal medicine U. Nebr.-Omaha, 1982—. Contbr. articles to profl. jours. Mem. Central Soc. Clin. Research, Am. Soc. Hematology, Am. Assn. Cancer Research, Am Soc. Clin. Oncology, Internat. Soc. Exptl. Hematology, ACP, Phi Beta Kappa, Alpha Omega Alpha. Subspecialties: Oncology; Marrow transplant. Current work: Bone marrow transplantation; mgmt. leukemia and lymphona. Home: 3716 S 94th Circle Omaha NE 68124 Office: 42d and Dewey Ave Omaha NE 68105

ARMSTRONG, ANDREW THURMAN, chemical consultant, educator; b. Haslet, Tex., May 26, 1935; s. Andrew Thurman and Ila (Kitchen) A.; m. Kay Frances Masters, Sept. 7, 1938; children: Michael Andrew, Marion Kay, Benjamin Neil. B.S., North Tex. State U., 1958, M.S., 1959; Ph.D., La. State U., 1967. Asst. prof. chemistry U. Tex., Arlington, 1969-74, assoc. prof. chemistry, 1974-84; pres., owner Armstrong Forensic Lab., Arlington, 1981. Contbr. articles to profl. jours. Mem. Tarrant County Adv. Council on Arson. Recipient White Helmet award Arlington Fire Dept., 1976; named Outstanding Tchr. Sch. Sci., U. Tex., Arlington, 1975-76. Fellow Am. Inst. Chemists; mem. Am. Chem. Soc. (W.T. Doherty award Dallas-Fort Worth Chpt. 1982), Am. Assn. Forensic Scientists, Am. Assn. Indsl. Hygiene. Baptist. Subspecialty: Analytical chemistry. Current work: Forensic chemistry of fire utilizing. Office: 3008 W Division Suite C Arlington TX 76012

ARMSTRONG, DAVID MICHAEL, biologist, educator; b. Louisville, July 31, 1944; s. John David and Elizabeth Ann (Horine) A.; m. Ann Beddoes, June 11, 1966; children: John David, Laura Christine. B.S., Colo. State U., 1966; M.A.T., Harvard U., 1967; Ph.D., U. Kans., 1971. Asst. prof. biol. sci. U. Colo., Boulder, 1971-77, assoc. prof., 1977—, mus. assoc. curator, 1971—; dir. Ctr. for Interdisciplinary Studies, 1981—; sci. collaborator Nat. Park Service, Canyonlands Nat. Park, Moab, Utah, 1971-77; cons. ecologist, 1971—. Author: Distribution of Mammals in Colorado, 1972, Rocky Mountain Mammals, 1975, Mammals of the Canyon Country, 1982; editor: Southwestern Naturalist, 1978-81, Jour. Mammalogy, 1982—. Mem. non-game adv. council Colo. Div. Wildlife, 1973-76; mem., chmn. Natural Areas Council, State of Colo., 1978-81. Recipient Teaching award Colo. Alumni Assn., 1975. Mem. Southwestern Assn. Naturalists, Am. Soc. Mammalogists, Ecol. Soc. Am., Colo.-Wyo. Acad. Scis. Subspecialties: Species interaction; Systematics. Current work: Ecology, systematics, biogeography of mammals of western North America; management of native ecosystems; history of evolutionary biology; science teaching for non-scientists. Home: 5653 Baseline Rd Boulder CO 80303 Office: U Colo Ctr Interdisciplinary Studies Campus Box 331 Boulder CO 80309

ARMSTRONG, DONALD, educator; b. Hamilton, Ont., Can., July 20, 1933; s. Alfred George and Dorothy Emma (Burden) A.; m. Christine Marie Medieros, June 13, 1954; children: Donald, David, Dennis, Sandra, Kenneth, Elizabeth. B.S., San Diego State U., 1957; M.S., U. Colo., 1969; Ed.D., U. Tulsa, 1974; Ph.D., U. Oslo, 1980. Instr. San Diego State U., 1960-62, U. Oreg. Med. Sch., Portland, 1963-67, U. Colo. Med. Ctr., Denver, 1967-70; clin. chemist, dir. research ctr. Hillcrest Med. Ctr., Tulsa, 1970-74; asst. prof. U. Colo. Med. Ctr., 1974-81; assoc. prof. U. Fla. Med. Sch., Gainesville, 1981—; dir., clin. chemist Allied Vet. Med. Lab., Tulsa, 1970-74; research cons. Hillcrest Med. Ctr., Tulsa, 1974-76; cons., clin. chem. toxicologist Ft. Lyons VA Hosp., Colo., 1975-81; cons. Sigma Chem. Co., St. Louis, 1979—. Editor: Ceroid-Lipofuscinosis (Batten's Disease), 1982, Free Radicals in Molecular Biology and Aging, 1984. Bd. dirs. N. Fla. Sight Found.; fields events Official U. Fla. Track Office. Served with U.S. Army, 1953-55. Children's Brain Disease grantee, 1973-83; March of Dimes grantee, 1974; Hillcrest Med. Ctr. grantee, 1974-78; Nat. Retinitis Pigmentosa Found. grantee, 1975-76. Mem. Am. Aging Assn. (bd. dirs. 1975-83, symposium chmn. 1982-83), Assn. Research in Vision and Ophthalmology, Research to Prevent Blindness, Ophthalmic Assn. Democrat. Roman Catholic. Club: Lions (Gainesville, Fla.). Subspecialties: Biochemistry (biology); Ophthalmology. Current work: Autoxidation of membrane lipids induced by free radicals and the subsequent formation of fluorescent materials. Home: 2036 NW 20th Ln Gainesville FL 32605 Office: U Fla Box J-284 Gainesville FL 32610

ARMSTRONG, GREGORY DAVENPORT, horticulturist; b. La Crosse, Wis., May 30, 1943; s. Miles T. and Beth (Davenport) A.; m. Elizabeth Marchant; children: Miles, Marjorie. B.S., U. Wis., 1967; M.A., Smith Coll., 1980; diploma, Royal Bot. Gardens, Kew Eng., 1970. Dir. Bot. Garden, lect. biol. scis. Smith Coll., Northampton, Mass., 1971—; v.p., dir. Child Park Inc. Mem. mus. com. Hist. Deerfield. Mem. Am. Assn. Bot. Gardens and Arboretums. Congregationalist. Home: Kosior Dr Hadley MA 01035 Office: Smith Coll Northampton MA 01063

ARMSTRONG, JOHN ALLAN, physicist, research and development executive; b. Schenectady, July 1, 1934; s. Orlo Lucius and Mary Kathryn (Moffitt) A.; m. Elizabeth Saunders, Sept. 20, 1958; children: Sarah Richardson, Jennifer Mary. A.B., Harvard U., 1956, Ph.D., 1961. Postdoctoral fellow Harvard U., 1961-63; mem. research staff IBM, Yorktown Heights, N.Y., 1963-74, dir. phys. scis., 1975-80, mem. corp. tech. com., 1980-81, mgr. materials and tech. devel., East Fishkill, N.Y., 1981-83, v.p. logic and memory research div., Yorktown Heights, 1983—. Contbr. articles to profl. jours. Fellow Am. Phys. Soc., Optical Soc. Am.; mem. AAAS. Subspecialties: Microchip technology (engineering); Spectroscopy. Current work: Director of semiconductor materials, device and circuit research; semiconductor packaging. Office: IBM Research Box 218 Yorktown Heights NY 10598

ARMSTRONG, NEAL EARL, civil engineering educator; b. Dallas, Jan. 29, 1941; m.; 3 children. B.A., U. Tex., 1962, M.A., 1965, Ph.D., 1968. Research engr. Engring. Sci., Inc., 1967-68; asst. office mgr., cons. san. engring., 1968-70; mgr. Washington Research and Devel. Lab., 1970-71; assoc. prof. U. Tex.-Austin, 1971-79, prof. civil engring., 1979—. Mem. Ecol. Soc. Am., Am. Soc. Limnology and Oceanography, Water Pollution Control Fedn., Internat. Assn. Water Pollution Research, Estuarine Research Fedn. (v.p. 1975-77). Subspecialties: Environmental engineering; Ecology. Current work: Pollution ecology, effects of waste discharges on receiving area ecology, chiefly estuarine ecology, quantitative correlations between waste discharges and ecosystem response. Office: Dept Civil Engring U Tex Austin TX 78712

ARMSTRONG, NEIL A., computer systems company executive, former astronaut; b. Wapakoneta, Ohio, Aug. 5, 1930; s. Stephen A.; m. Janet Shearon; children: Eric, Mark. B.S. In Aero. Engring, Purdue U., 1955; M.S. in Aero. Engring, U. So. Calif. With Lewis Flight Propulsion Lab., NACA, 1955; then aero. research pilot for NACA (later NASA, High Speed Flight Sta.), Edwards, Calif.; astronaut Manned Spacecraft Center, NASA, Houston, 1962-70; command pilot Gemini 8, Mar. 1966; comdr. Apollo II; dep. asso. adminstr. for aeros. Office Advanced Research and Tech., Hdqrs. NASA, Washington, 1970-71; prof. aerospace engring. U. Cin., 1971-79; chmn. bd. Cardwell Internat., Ltd., Lebanon, Ohio, 1980-82; chmn. CTA, Inc., 1982—; dir. Gates Learjet Corp., Cin. Gas & Electric Co., Eaton Corp., Taft Broadcasting Co., Cin. Milacron, UAL, Inc. Chmn. bd. trustees Cin. Mus. Natural History. Served as naval aviator USN, 1949-52; Korea. Recipient numerous awards, including Octave Chanute award Inst. Aero. Scis., 1962, Presdl. Medal for Freedom, 1969, Exceptional Service medal NASA, Hubbard Gold medal Nat. Geog. Soc., 1970, Kitty Hawk Meml. award, 1969, Pere Marquette medal, 1969, Arthur S. Fleming award, 1970, Congl. Space Medal of Honor, Explorers Club medal. Fellow AIAA (hon., Astronautics award 1966), Internat. Astronautical Fedn. (hon.), Soc. Exptl. Test Pilots; mem. Nat. Acad. Engring. Subspecialties: Astronautics. 1st man to walk on moon, 1969. Office: 31 N Broadway Lebanon OH 45036

ARNAS, OZER ALI, mechanical engineering educator; b. Izmir, Turkey, Sept. 1, 1936; s. Hulusi Ahmet and Saadet Hayriye (Maro) A.; m. Ozden Oncu, Dec. 23, 1960; children: Neyla, Hale, Burak. B.S. in Mech. Engring, Robert Coll., Istanbul, Turkey, 1958, M.S., Duke U., 1961; Ph.D., N.C. State U.-Raleigh, 1965. Registered profl. mech. engr., La. Mem. faculty La. State U., Baton Rouge, prof. mech. engring., 1970—; vis. prof. and research fellow U. Liege, Belgium, 1972-73, 79-80; vis. prof. Eindoven Inst. Tech., Netherlands, 1979-80; cons. prof. Cath. U. Am., Washington, 1967-72, U. Costa Rica, San Jose, 1967-68; research analyst Wright Patterson AFB, Dayton, Ohio, 1967-68. Contbr. numerous articles in field to profl. publs. Served to lt. Turkish Army, 1968-70. Recipient Halliburton Excellence in Teaching award Halliburton Ednl. Found., 1967; Western Electric Fund award Am. Soc. Engring. Edn., 1979. Mem. ASME (vice chmn. Baton Rouge sect. 1982-83, mem. exec. com. advanced energy systems 1982—), Sigma Xi. Subspecialties: Mechanical engineering; Solar energy. Current work: Applications of thermodynamics to the analyses of engineering systems; heat transfer in enclosures and multi-phase systems; technology transfer to developing countries; heat transfer in conduits of arbitrary shape for compact heat exchanger design. Home: 5513 Valley Forge Ave Baton Rouge LA 70808 Office: Mech Engring La State Univ CEBA 2505 Baton Rouge LA 70803-6413

ARNASON, BARRY GILBERT WYATT, neurologist; b. Winnipeg, Man., Can., Aug. 30, 1933; s. Ingolfur Gilbert and Elsie (Wyatt) A.; m. Joan Frances Morton, Dec. 27, 1961; children: Stephen, Jon, Eva. M.D., U. Man., 1957. Intern Winnipeg Gen. Hosp., 1956-57; resident in neurology Mass. Gen. Hosp., Boston, 1958-62; asst. prof. neurology Harvard Med. Sch., 1965-71, assoc. prof., 1971-76; prof., chmn. dept. neurology U. Chgo. Pritzker Sch. Medicine, 1976—; mem. med. adv. bd. Nat. Multiple Sclerosis Soc., 1977-83, Amyotrophic Lateral Sclerosis Soc. Am., 1976-79, Hereditary Disease Found., 1977—. Contbr. articles to med. jours. Nat. Multiple Sclerosis Soc. research fellow, 1959-61, 62-64; grantee NSF, other founds. Mem. Am. Soc. Clin. Investigation, Am. Assn. Immunology, Am. Neurol. Assn., Am. Acad. Neurology, Am. Assn. Neuropathology, Soc. Neurosci. Subspecialties: Neurology; Neuroimmunology. Current work: Immunologic aspects of neurologic disease. Home: 4832 S Ellis Ave Chicago IL 60615 Office: 950 E 59th St Chicago IL 60637

ARNASON, BARRY GILBERT WYATT, neurologist; b. Winnipeg, Man., Can., Aug. 30, 1933; s. Ingolfur Gilbert and Elsie (Wyatt) A.; m. Joan Frances Morton, Dec. 27, 1961; children—Stephen, Jon. M.D., U. Man., 1957. Intern Winnipeg (Man., Can.) Gen. Hosp., 1956-57; resident in neurology Mass. Gen. Hosp., Boston, 1958-62; asst. prof. neurology Harvard Med. Sch., 1965-71, asso. prof., 1971-76; prof., chmn. dept. neurology U. Chgo. Pritzker Sch. Medicine, 1976—. Contbr. articles to med. jours., chpts. in med. books. Mem. med. adv. bd. Nat. Multiple Sclerosis Soc., 1977—, Amyotrophic Lateral Sclerosis Soc. Am., 1976-79, Hereditary Disease Found., 1977—. Nat. Multiple Sclerosis Soc. research fellow, 1959-61, 62-64; recipient numerous research grants NSF, founds. Mem. Am. Soc. Clin. Investigation, Am. Assn. Immunology, Am. Neurol. Assn., Am. Acad. Neurology, Am. Assn. Neuropathology, Soc. Neuroscience. Subspecialty: Neuroimmunology. Home: 4832 S Ellis Ave Chicago IL 60615 Office: 950 E 59th St Chicago IL 60637

ARNOLD, BEN ALLEN, med. physicist; b. Columbia, Ky., Jan. 28, 1945; s. Winfrey Lee and Estell Cora (Overstreet) A.; m. Margaret Salce, Apr. 8, 1971; children: Matthew, Rebecca, Elizabeth. B.A., Centre Coll., 1967; M.S., Yale U., 1969; Ph.D., Harvard U., 1977. Physicist Lawrence Radiation Lab., Livermore, Calif., 1969; physicist Harvard Med. Sch., Peter Bent Brigham Hosp., Boston, 1970-73; research fellow in radiology Mass. Gen. Hosp., Boston, 1976; asst. prof. U. Calif.-Irvine, 1977-79; dir. research and physics South Bay Hosp., Redondo Beach, Calif., 1979-82; founder Health Care Affiliates, Inc., Laguna Hills, Calif.; adj. asst. prof. UCLA Sch. Medicine, 1979—. Contbr. articles to profl. jours. Mem. Am. Assn. Physicists in Medicine, Soc. Photo-Optical Instrumentation Engrs. Democrat. Baptist. Subspecialties: Imaging technology; Medical physics. Current work: Research in medical x-ray imaging, research and development in digital radiography; development of computerized imaging facilities. Home: 35 Redhawk Irvine CA 92714 Office: 514 N Prospect Ave Redondo Beach CA 90277

ARNOLD, JAMES RICHARD, chemist, educator; b. New Brunswick, N.J., May 5, 1923; s. Abraham Samuel and Julia (Jacobs) A.; m. Louise Clark, Oct. 11, 1952; children: Robert C., Theodore J., Kenneth C. A.B., Princeton U., 1943, M.A., 1945, Ph.D., 1946. Postdoctoral fellow Inst. Nuclear Studies, U. Chgo., 1946-47, mem. faculty, 1948-55; NRC fellow Harvard U., 1947-48; mem. faculty chemistry Princeton U., 1955-58; asso. prof. chemistry U. Calif., San Diego, 1958-60, prof., 1960—, Harold C. Urey prof., 1983—, chmn. dept. chemistry, 1960-63; asso. Manhattan Project, 1943-46; dir. Calif. Space Inst., 1980—; mem. various bds. NASA, 1959—; mem. space sci. bd. Nat. Acad. Sci., 1970-74, mem. com. on sci. and public policy, 1973-77. Mem. editorial bd.: Ann. Rev. Nuclear Chemistry, 1972; asso. editor: Revs. Geophysics and Space Physics, 1972-75, Moon, 1972—; contbr. articles to profl. jours. Pres. Torrey Pines Elem. Sch. PTA, 1964-65; pres. La Jolla Democratic Club, 1965-66; mem. nat. council World Federalists-U.S.A., 1970-72. Recipient E.O. Lawrence medal AEC, 1968; Leonard medal Meteoritical Soc., 1976; asteroid 2143 named Jimarnold in his honor, 1980; Guggenheim fellow, India, 1972-73. Mem. Nat. Acad. Sci., Am. Acad. Arts and Scis., Am. Chem. Soc., AAAS, Fedn. Am. Scientists, World Federalist Assn., Internat. Acad. Astronautics (corr.).

ARNOLD, JONATHAN, statistical geneticist; b. N.Y.C., Nov. 27, 1953; s. Christopher John and Eleanor Marlin (Williams) A.; m. Julia Mae Phillips, Apr. 26, 1980. B.S., Yale U., 1975, M.Phil., 1975, Ph.D., 1982. Adj. instr. dept. stats. Rutgers U., 1981-82; asst. prof. dept. genetics and dept. stats. U. Ga., Athens, 1982—. Contbr. articles to profl. jours. Mem. Inst. Math. Stats., Am. Statis. Assn., Soc. for Study of Evolution, Genetics Soc. Am. Subspecialties: Statistics; Population biology. Current work: Data analysis, population genetics. Office: Dept Genetics U Ga Athens GA 30602

ARNOLD, LESLIE KINGSLAND, systems analyst, mathematician; b. Larned, Kans., Oct. 18, 1938; s. Kingsland and Mary Ellen (Bentley) A.; m. Beth Randall, Aug. 17, 1963; children: Sarah, John. B.A., Rice U., 1961; Ph.D., Brown U., 1966. Assoc. Daniel H. Wagner Assocs., Paoli, Pa., 1966-74, sr. assoc., 1974-83; systems analyst Gen. Electric Co., King of Prussia, Pa., 1983—. Mem. Am. Math. Soc., Math. Assn. Am., London Math. Soc., Soc. Indsl. and Applied Math, Sigma Xi. Subspecialties: Applied mathematics; Operations research (mathematics). Current work: Ergodic theory; search theory; optimization, interactive software; military operations analyst. Home: 540 Col Dewees Rd Wayne PA 19087 Office: Valley Forge Space Center King of Prussia PA 19406

ARNOLD, ORVILLE EDWARD, cons. engr.; b. Sparta, Wis., Sept. 30, 1933; s. Donald E. and Lenice K. (Reilly) A.; m.; children: Donald, David, Beth, Sandra. B.S.C.E., U. Wis.-Madison, 1955. Registered profl. engr., Wis. Design engr. Inland Steel Co., East Chicago, Ind., 1955-56; chief structural engr. Flad & Assocs. (Architects), Madison, Wis., 1956-63; pres. Arnold & O'Sheridan Cons (Engrs.), Madison, 1964—. Bd. dirs. Madison Opportunity Ctr., Hospice Care.; pres. Mendota/Monoma Lake Problems Assn., 1974. Recipient Engr. of Yr. in Pvt. Practice award Wis. Soc. Profl. Engrs., 1978; Disting. Service citation U. Wis. Coll. Engring., 1982. Mem. Nat. Soc. Profl. Engrs., Wis. Soc. Profl. Engrs., ASCE (pres. Wis. sect. 1972), Am. Concrete Inst. Roman Catholic. Subspecialty: Civil engineering. Current work: Design of structural, mechanical and electrical engineering systems for buildings. Home: 1521 Edgehill Dr Madison WI 53705 Office: 815 Forward Dr Madison WI 53711

ARNOLD, WILLIAM HOWARD, energy company executive; b. Jefferson Barracks, Mo., May 13, 1931; s. William Howard and Elizabeth Welsh (Mullen) A.; m. Josephine Inman Routheau, June 13, 1952; children: William, Frances, Edward, David, Thomas. A.B., Cornell U., 1951; A.M., Princeton U., 1953, Ph.D., 1955. Instr., research assoc. Princeton U., 1955; sr. engr., mgr. reactor physics Westinghouse Atomic Power Div., Pitts., 1955-61; dir. nuclear fuel mgmt. NUS Corp., Washington, 1961-62; various mgmt. positions to gen. mgr. PWR Systems, Pitts., 1962-79; pres. Westinghouse Nuclear Internat., Pitts., 1979-80, gen. mgr. Advanced Reactor div., 1981—, gen. mgr. Advanced Energy Systems div., 1983—; tchr. nuclear tech.

U. Pitts., 1957, U. Ala., 1963. Contbr. articles to profl. jours. Fellow Am. Nuclear Soc., AAAS; mem. Nat. Acad. Engring., Am. Phys. Soc., Sigma Xi. Clubs: Longue Vue Country, Chevy Chase. Subspecialty: Nuclear engineering. Office: Box 158 Madison PA 15663

ARNON, DANIEL I(SRAEL), biochemist, educator; b. Poland, Nov. 14, 1910; s. Leon and Rachel (Chodes) A.; m. Lucile Jane Soule, Feb. 24, 1940; children: Anne Arnon Hodge, Ruth Arnon Hanham, Stephen, Nancy, Dennis. B.S., U. Calif., 1932; Ph.D., 1936; Ph.D. hon. doctorate, U. Bordeaux, France, 1975. Instr. U. Calif. at Berkeley, 1936-41, asst. prof., assoc. prof., 1941-50, prof. plant physiology, 1950-60, prof. cell physiology, 1960—, research biochemist, 1978—; founding chmn. dept. cell physiology, 1961-78; biochemist Calif. Agrl. Expt. Sta., 1958-78; Guggenheim fellow, Cambridge U., Eng., 1947-48; lectr. Belgian Am. Found., U. Liège, Belgium, 1948; Fulbright research scholar Max-Planck Inst., Berlin-Dahlem, Germany, i955-56. Author sci. articles. Served from lt. to maj. AUS, 1943-46. Recipient Gold medal U. Pisa, 1958; Charles F. Kettering award photosynthesis research; Nat. medal of Sci.; Guggenheim fellow, 1962-63. Fellow Am. Acad. Arts and Scis., AAAS (recipient Newcomb Cleveland prize 1940); mem. Nat. Acad. Scis., Royal Swedish Acad. Scis., Acad. d'Agriculture de France, Deutsche Akademie der Naturforscher Leopoldina, Am. Chem. Soc., Am. Soc. Biol. Chemists, Biochem. Soc. (London), Am. Soc. Photobiology, Am. Soc. Plant Physiologists (pres. 1952-53, Stephen Hales prize, Charles Reid Barnes life membership award), Scandinavian Soc. Plant Physiologists, Spanish Biochem. Soc. (hon.). Subspecialties: Photosynthesis; Biochemistry (biology). Current work: Mechanisms for the conversion of solar energy into biologically useful chemical energy as manifested by photosynthetic electron transport and photophosphorylation. Home: 28 Norwood Ave Berkeley CA 94707

ARNOTT, STRUTHER, univ. ofcl., molecular biologist; b. Larkhall, Scotland, Sept. 25, 1934. B.Sc., Glasgow (Scotland) U., 1956, Ph.D., 1960, D.Sc., 1978. Scientist biophysics unit Med. Research Council Kings Coll., London, 1960-70; prof. biology Purdue U., West Lafayette, Ind., 1970—, chmn. dept. biol. scis., 1975-80, v.p. research, dep. provost, dean grad. sch., 1980—. Contbr. numerous articles on molecular structures of nucleic acids, fibrous proteins and polysaccharides to profl. jours. Fellow Royal Soc. Chemistry. Subspecialties: Biochemistry (biology); Biophysics (biology). Current work: Automic details of the molecular structures of fibrous biopolymers like nucleic acids, polysaccharides and proteins. Home: 421 Robinson St West Lafayette IN 47906 Office: Grad Sch Purdue U West Lafayette IN 47907

ARNOULT, MALCOLM DOUGLAS, psychology educator; b. New Orleans, Feb. 15, 1923; s. Albert Eager and Phoebe (Voss) A.; m. Billye Keith, Jan. 19, 1952; children: Sharon Louise, Douglas Edwards. B.A., Tulane U., 1943, M.S., 1948; Ph.D., U. Tex., 1951. Cert. psychologist, Tex. Research psychologist USAF Personnel and Tng. Research Center, San Antonio, 1950-57; vis. prof. U. Tex., Austin, 1957; assoc. prof. psychology U. Miss., Oxford, 1957-59, Tex. Christian U., Ft. Worth, 1959-62, prof., 1962—; cons. Life Scis., Inc., Ft. Worth, 1963—, NASA, Langley, Va., 1979, others. Author: Fundamentals of Scientific Method in Psychology, 1972. Served to sgt. U.S. Army, 1943-46. Fellow Am. Psychol. Assn.; mem. Psychonomic Soc., Southwestern Psychol. Assn. (pres. 1967), So. Soc. for Philosophy and Psychology (pres. 1971), Sigma Xi. Subspecialties: Sensory processes; Cognition. Current work: Determinants of form perception; problems in perceptual learning; effects of noise on cognitive processes. Office: Tex Christian U Box 32878 Fort Worth TX 76129

ARNSDORF, MORTON FRANK, cardiologist; b. Chgo., Aug. 7, 1940; s. Selmar N. and Irmgard C. (Steinmann) A.; m. Mary Hunter Tower, Dec. 26, 1963 (div. 1982). B.A. magna cum laude, Harvard U., 1962; M.D., Columbia U., 1966. Diplomate: Am. Bd. Internal Medicine. Housestaff officer U. Chgo., 1966-69; fellow cardiology Columbia-Presbyn. Med. Ctr., N.Y.C., 1969-71; asst. prof. medicine U. Chgo., 1973-79, assoc. prof., 1979-83, prof., 1983—, chief sect. cardiology, 1981—; mem. pharmacology study sect. NIH, 1981—. Contbr. articles to profl. jours. Served to maj. USAF, 1971-73. Recipient Research Career Devel. award. NIH, 1976-81; research grantee Chgo. Heart Assn., 1976-78, NIH, 1977—. Fellow Am. Coll. Cardiology; mem. Am. Heart Assn. (dir. 1981-83, chmn. exec. com. basic sci. council 1981-83), Chgo. Heart Assn. (bd. govs., chmn. research council), AMA, Am. Fedn. Clin. Research, Assn. U. Cardiologists, Central Soc. Clin. Research (counsellor), Chgo. Med. Soc., Ill. Med. Soc. Club: Quadrangle. Subspecialties: Cardiology. Current work: Cellular electrical activity assessed by microelectrode techniques in mammalian heart tissue; mechanism of arrhythmogeneis and the manner in which antiarrhythmic drugs work. Office: Chief Sect Cardiology Univ Chicago Hosps and Clinics Box 423 950 E 59th St Chicago IL 60637

ARNTZEN, CHARLES JOEL, biochemistry educator; b. Granite Falls, Minn., July 20, 1941; m.; 1 child. B.S., U. Minn., 1965, M.S., 1967; Ph.D., Purdue U., 1970. NSF fellow photosynthesis C.F. Kettering Research Lab., 1969-70; from asst. prof. to prof. botany U. Ill., Urbana, 1970-80; plant physiologist Sci. Edn. Adminstrn., U.S. Dept. Agr., 1976-80; prof. biochemistry, dir. dept. energy, plant research lab. Mich. State U., East Lansing, 1980—. Mem. Am. Soc. Plant Physiologists (Charles Albert Shull award 1980), Am. Soc. Agronomy, Am. Soc. Cell Biology, Am. Biophys. Soc., Weed Sci. Soc. Am. Subspecialties: Photosynthesis; Plant physiology (agriculture). Office: Dept Energy Plant Research La Mich State U East Lansing MI 48824

ARNY, DEANE CEDRIC, plant pathology educator, researcher; b. St. Paul, May 22, 1917; s. Albert Cedric and Mary Katherine (Hummel) A.; m. Edith Boardman, Jan. 11, 1947; children: Margaret, Barbara, Michael, Carol, Philip. B.S., U. Minn., 1939; Ph.D., U. Wis.-Madison, 1943. Mem. faculty U. Wis.-Madison, 1943—, prof. plant pathology, 1956—; prof. U. Ife, Nigeria, 1966-68. Mem. Am. Soc. Agronomy, Crop Sci. Soc., Am. Phytopath Soc. Subspecialty: Plant pathology. Current work: Diseases of field crops. Patentee method for reducing temperature at which plants freeze. Home: 5401 Whitcomb Dr Madison WI 53711 Office: 1630 Linden Dr Madison WI 53706

ARONIN, NEIL, endocrinologist, educator; b. Washington, Mar. 16, 1948; s. David and Thelma (Borenstein) A.; m. Marion DiFiglia, Jan. 1, 1980; children: Elizabeth H., Leah M. B.A. magna cum laude, Duke U., 1970; M.D., U. Pa., 1974. Diplomate: Am. Bd. Internal Medicine, subsplty. bd. endocrinology and metabolism. Intern in medicine Duke U. Med. Center, Durham, N.C., 1974-75, resident, 1975-77; clin. fellow in endocrinology and metabolism Mt. Sinai Med. Center, N.Y.C., 1977-79, instr. dept. medicine, 1979-80; NIH research fellow in neuroendocrinology, dept. neurology Mass. Gen. Hosp.-Harvard Med. Sch., Boston, 1979-81; asst. prof. medicine and physiology U. Mass. Med. Center, Worcester, 1981—; clin. assoc. in neurendocrinology Mass. Gen Hosp., 1981—. NIH grantee, 1982—. Mem. Soc. Neurosci., Endocrine Soc., N.Y. Acad. Scis., Am. Fedn. Clin. Research. Democrat. Jewish. Subspecialty: Neuroendocrinology. Current work: Anatomy and physiology of peptide systems in brain; regulation of hypothalamic-pituitary axis.

ARONOW, LEWIS, pharmacologist, educator; b. N.Y.C., Mar. 28, 1927; s. Max and Pearl (Lewiton) A.; m. Gladys Levy, June 7, 1953; children: Bruce, Joyce, Ellen, Amy. B.S., CCNY, 1950; M.S., Georgetown U., 1952; Ph.D., Harvard U., 1956. Mem. faculty dept. pharmacology Stanford (Calif.) U. Sch. Medicine, 1956-76, prof., 1970-76, acting chmn. dept., 1974-76; prof. dept pharmacology Uniformed Services U. Health Scis., Bethesda, Md., 1976—, chmn, dept., 1976—. Author numerous books and articles on cellular and molecular pharmacology. Recipient Premio Martin Vegas Sociedad Venezolana de Dermatologia, Venereologia y Leprologia, 1964, Commendation medal Uniformed Services U. Health Scis., 1981; Eleanor Roosevelt Internat. Cancer fellow, 1970. Mem. Am. Soc. for Pharmacology and Exptl. Therapeutics (sec.-treas. 1981), Assn. for Med. Sch. Pharmacology (councilor 1981). Subspecialties: Receptors; Molecular pharmacology. Current work: Glucocorticoid receptors. Office: 4301 Jones Bridge Rd Bethesda MD 20814

ARONS, JONATHAN, astrophysicist, educator; b. Phila, Aug. 16, 1943; s. David and Gladys A. (Jaffe) A.; m. Claire Ellen Max, Dec. 22, 1974; 1 son: Samuel Max. B.A. in Physics, Williams Coll., 1965; A.M., Harvard U., 1969, Ph.D. in Astronomy, 1970. Research assoc. Princeton U. Obs., 1970-71; mem. Inst. Advanced Study, 1971-72; asst. prof. astronomy U. Calif.-Berkeley, 1972-76, assoc. prof. astronomy, 1976-80, assoc. prof. astronomy and physics, 1980-82, prof., 1982—. Danforth Found. grad. fellow, 1965-70; Guggenheim Meml. Found. fellow, 1980-81. Mem. Am. Astron. Soc., Am. Geophys. Union, Am. Phys. Soc. (vice chmn. astrophys. div. 1983-84, mem., chmn. 1984-85); Mem. Internat. Astron. Union. Subspecialties: Theoretical astrophysics; Plasma physics. Current work: Theoretical studies of dynamics and collective emission phenomena in high energy astrophysical systems. Office: 601 Campbell Hall U Calif Berkeley CA 94720

ARONSON, CARL EDWARD, pharmacologist, toxicologist, educator; b. Providence, Mar. 14, 1936; s. Carl Ivar and Ruth (Workman) A.; m. Marjorie B. Aronson, Dec. 17, 1960; children: Linda, Kristen. A.B., Brown U., 1958; Ph.D., U. Vt., 1966; M.A. (hon.), U. Pa., 1973. Research technician Worcester Found. Exptl. Biology, Shrewsbury, Mass., 1959-60; postdoctoral fellow U. Pa., 1965-67, lectr., 1965-74, instr., 1967-70, assoc. dept. pharmacology, 1970-71, asst. prof., 1971-73, head labs. pharmacology and toxicology dept. animal biology, 1972—, assoc. prof., 1973—, acting assoc. dean student affairs, 1974-75; cons. in field. Contbr. numerous articles to profl. jours. Active numerous civic orgns. Served with USAFR, 1958-65. Recipient Norden Disting. Tchr. award U. Pa., 1982. Mem. AAAS, AAUP, Am. Acad. Vet. Pharmacology and Therapeutics (pres. 1983), Am. Coll. Vet. Toxicologists, Am. Soc. Pharmacology and Exptl. Therapeutics, John Morgan Soc., Physiol. Soc. Phila., Sigma Xi. Subspecialties: Pharmacology; Toxicology (medicine). Current work: Biomedical pharmacology and toxicology with emphasis on in vitro methods of assessment of cardiotoxicity of drugs and chemicals. Office: Labs Pharmacology and Toxicology U Pa Sch Vet Medicine 3800 Spruce St Philadelphia PA 19104

ARONSON, MIRIAM KLAUSNER, gerontologist, consultant, research, educator; b. N.Y.C., July 12, 1940; d. Joseph and Martha (Sklower) Klausner; m. Sidney R. Aronson, July 3, 1962 (div. Apr. 1977); children: Eric, Andrew, Elliott. A.B., Barnard Coll., N.Y.C., 1961; M.Ed., Tchrs. Coll. Columbia U., 1970, ED.D., 1980. Cons., researcher geriatric facilities, N.Y. and N.J., 1969-75; dir. geriatric program Soundview-Throgs Neck Community Mental Health Ctr., Bronx, N.Y., 1975-78; chief services to elderly Bronx-Lebanon Hosp. Community Mental Health Ctr., Bronx, 1978-79; dir. longterm Care Gerontol. Ctr. Albert Einstein Coll. Medicine, Bronx, 1979—, asst. prof. neurology and psychiatry, 1980—. Author, dir.: film series Teaching series on Alzheimers Disease, 1980; author (with R. Bennett and B. Gurland); editor: The Acting Out Elderly, 1983. Mem. Hillsdale (N.J.) Bd. Health, 1975-82, pres., 1977-81; mem. planning and policy com. Outreach Health Service Program, Bergen County, N.J., 1976. N.Y. State Regents scholar, 1957-61; Adminstrn. Aging grantee, 1968-70, 74-75; Nat. Council Community Mental Health Ctrs. Best Outreach Program award, 1976; Alzheimers Disease Soc. Greater N.Y. award, 1982. Fellow Gerontol. Soc. Am. (dir. task force on long team care 1981—), Am. Orthopsychiat. Soc.; mem. Am. Pub. Health Assn., Am. Geriatrics Soc., Nat. Alzheimers Disease & Related Disorders Assn. (dir. edn. and pub. awareness com. 1979—). Club: Altrusa. Subspecialty: Gerontology. Current work: Current research, practice and education interests relate to risk factors for and clinical course of Alzheimer's Disease and other dementing illnesses and their impact on the family. Home: 305 Ell Rd Hillsdale NJ 07642 Office: Albert Einstein College of Medicine Resnick Gerontology Center 1300 Morris Park Ave Belfer Bldg 802 Bronx NY 10461

ARONSON, STANLEY MAYNARD, educator, physician; b. N.Y.C., May 28, 1922; s. Eliuh and Lena (Hassner) A.; m. Betty Ellis, June 3, 1947; children—Susan, Lisa, Sarah. B.S., CCNY, 1943; M.D., N.Y. U., 1947; M.A., Brown U., 1971; M.P.H., Harvard U. Sch. Public Health, 1981. Diplomate: Am. Bd. Pathology. Residence tng. Bellevue Hosp., Meml. Center for Cancer, Mt. Sinai Hosp., all N.Y.C., 1947-53; faculty Columbia Coll. Physicians and Surgeons, 1951-54; prof. pathology, asst. dean SUNY, Bklyn., 1954-70; prof. med. sci., dean medicine Brown U., 1970-81, Univ. prof. med. sci., 1981—; dir. labs. Kings County Hosp. Center, Bklyn., 1965-70; pathologist-in-chief Miriam Hosp., Providence, 1970-75; cons. physician neuropathology Jewish Chronic Disease Hosp., Bklyn., 1951—, NIH, 1962—, VA, N.Y.C. and; Providence, 1965—; cons. physician R.I. Hosp., Roger Williams Hosp., Meml. Hosp., Providence VA Hosp., Butler Hosp., Providence, Luth. Med. Center, N.Y.C.; lectr. Yale Sch. Medicine, 1964-65; lectr. pathology Tufts U. Sch. Medicine, 1978—. Author: (with B.W. Volk) Cerebral Sphingolipidoses, 1962, Inborn Disorders of Sphingolipid Metabolism, 1966, Sphingolipids, Sphingolipidoses and Allied Disorders, 1972, (with A. Sahs and E Hartman) Guidelines for Stroke Care, 1976; also numerous articles; mem. editorial bd.: Jour. Submicroscopic Cytology. Mem. Nat. Adv. Commn. on Multiple Sclerosis, 1973-74; commr. U.S. Commn. Control of Huntington's Disease, 1976—; mem. NIH Perinatal Research Commn., Joint Commn. on Stroke Facilities; mem. med. adv. bd. Nat. Multiple Sclerosis Soc., Dysautonomia Found., Nat. Tay-Sachs Assn., Nat. Fund for Med. Edn., Hospice R.I., Interfaith Health Care Ministries. Served with U.S. Army, 1942-46. Mem. AMA, Am. Neurol. Assn., Am. Assn. Neuropathology (pres. 1971-72), N.Y. Acad. Medicine, Am. Acad. Neurology, Am. Assn. Pathologists and Bacteriologists, Internat. Soc. Neuropathology, Assn. Am. Med. Colls., N.Y. Neurol. Soc. Subspecialties: Pathology (medicine); Preventive medicine. Research on genetics epidemiology, pathology and diagnostic features of cerebral degenerative diseases, population dynamics, pathology and epidemiology of cerebral vascular disease and stroke. Home: 26 Elm St Rehoboth MA 02769 Office: Office Med Affairs Brown U Providence RI 02912

ARONSTAM, ROBERT STEVEN, pharmacologist, cons.; b. S.I., N.Y., Aug. 5, 1950; s. Robert Harwood and Margaret Louise (Fairbanks) A.; m. Joan Lincoln, Sept. 8, 1948; children: Emily Lincoln, Robert Andrew. B.A., Columbia U., 1972; Ph.D. (NIMH fellow), U. Rochester, 1978. NIMH postdoctoral fellow dept. pharmacology and exptl. therapeutics U. Md. Sch. Medicine, 1977-79; asst. prof. pharmacology Med. Coll. Ga., 1980—; cons. in field. Contbr. numerous articles, abstracts, chpts. to profl. publs. Nat. Inst. Drug Abuse grantee, 1981-83; Nat. Inst. Neurol. Communicative Diseases and Strokes grantee, 1981-84. Mem. Soc. Neurosci., Soc. Exptl. Biology and Medicine, Soc. Neurochemistry, N.Y. Acad. Scis., Southeastern Pharmacology Soc., Sigma Xi. Subspecialties: Neuropharmacology; Neurochemistry. Current work: Biochem. and pharmacol. characterization of cholinergic synaptic transmission. Office: Dept Pharmacology Med Coll Ga 30912

ARSENEAU, JAMES CHARLES, medical oncologist, researcher; b. Syracuse, N.Y., Aug. 29, 1942; s. James Howard and Glenna (Wurth) A.; m. Jane Ellen Macy, July 2, 1966; children: Marc, David. A.B., Syracuse U., 1964; M.D., Albany (N.Y.) Med. Coll., 1968. Cert. Am. Bd. Internal Medicine. Intern Strong Meml. Hosp., Rochester, N.Y., 1968-69, resident, 1969-70; instr. trainee U. Rochester, 1968-70; clin. assoc. Nat. Cancer Inst., Bethesda, 1970-73; fellow med. oncology U. Rochester, 1973-74, asst. prof. oncology, 1974-80, assoc. prof., 1980—; chmn. new drug com. Gynecologic Oncology Group, Phila., 1980—, mem. exec. com., 1982—; co-chmn. Melanoma group Eastern Coop. Oncology Group, Madison, Wis., 1980—. Cons. editor: Investigational New Drugs, 1982—; contbr. articles to profl. jours. Bd. dirs. United Cancer Council, Rochester, 1977-80, Make Today Count, Rochester, 1977—. Served with USPHS, 1970-73. Wilson fellow, 1973-74; Am. Cancer Soc. fellow, 1974-77. Mem. AMA, N.Y. State Med. Soc., Rochester Acad. Medicine, Am. Soc. Clin. Oncology, N.Y. Acad. Scis. Republican. Subspecialties: Chemotherapy; Cancer research (medicine). Current work: Clincial research of new anti-neoplastic drugs, biologicals, anti-cancer agents clinical studies of new anti-cancer therapies. Office: Rochester Gen Hosp 1425 Portland Ave Rochester NY 14621 Home: 24 Vincent Dr Pittsford NY 14534

ARTHUR, LARRY OTTIS, immunologist, researcher; b. Many, La., Mar. 14, 1944; s. Ottis G. and Velma (Lee) A.; m. Alice Rose LeBlanc, Sept. 5, 1964; children: John, David, Matthew. B.S., Northwestern State U., Natchitoches, La., 1966, M.S., 1968; Ph.D., La. State U., 1970. Postdoctoral fellow No. Regional Lab., Peoria, Ill., 1970-72; asst. prof. Grand Valley State Coll., Allendale, Mich., 1972-73; with Nat. Cancer Inst.-Frederick Cancer Research Facility, Frederick, Md., 1973—, sr. research scientist, 1979—, head biol. prodn. and monoclonal antibody labs., 1982—; assoc. prof. immunology Hood Coll., 1976—, charter mem. adv. com. for biomed. scis. grad. program, 1976—. Contbr. numerous articles, abstracts to profl. pubs. Mem. AAAS, Am. Soc. Microbiology, Tissue Culture Assn., Assn. Schs. Allied Health Professions, Am. Assn. Cancer Research, Sigma Xi, Phi Kappa Phi. Subspecialties: Immunobiology and immunology; Virology (biology). Office: Nat Cancer Inst-Frederick Cancer Research Facility Frederick MD 21701

ARTHUR, RANSOM JAMES, psychiatrist, educator; b. N.Y.C., Dec. 5, 1925; s. Ransom James and Barbara Remick A.; m. Frances Nickolls, Dec. 18, 1954; children: Jane, Shelley. A.B with honors, U. Calif., Berkeley, 1947; M.D. cum laude, Harvard U., 1951. Intern Mass. Gen. Hosp., Boston, 1951-52; teaching fellow Harvard Med. Sch., 1951-54; resident in pediatrics Children's Med. Center, Boston, 1952-54; resident in psychiatry Queens Hosp., Honolulu, 1954-55; commd. lt U.S. Navy, 1958, advanced through grades to capt., 1968; resident U.S. Naval Hosp., Bethesda, Md., 1957-60; comdg. officer U.S. Naval Med. Neuropsychiat. Res. Unit, San Diego, 1963-74, ret., 1974; prof. psychiatry, assoc. dean Sch. Medicine (Sch. Medicine); dir. Neuropsychiat. Inst. Hosp. and Clinics, UCLA, 1974-79; dean Sch. Medicine, U. Oreg., Portland, 1979-83; prof. UCLA, 1983—; chief of staff Brentwood VA Hosp., Los Angeles, 1983—; cons. Founder Masters' Swimming, 1970, nat. chmn., 1970-72, nat. chmn. goals and objectives, 1972—. Author: An Introduction to Social Psychiatry, 1971; contbr. articles to profl. jours.; mem. editorial bd.: Mil. Medicine, 1972—, Am. Jour. Psychiatry, 1975-79. Served with USMCR, 1943-46. Decorated Legion of Merit (2). Mem. Am. Coll. Psychiatrists, Am. Psychiat. Assn., Royal Soc. Medicine, AMA (Physicians Recognition award), Assn. Am. Med. Colls., Assn. Mil. Surgeons, Phi Beta Kappa, Sigma Xi. Subspecialty: Psychiatry. Current work: Research in psychosocial stress in relationship to psychiatric and other illnesses. Home: 2666 SW Buckingham Ave Portland OR 97201

ARUNKUMAR, KOOVAPPADI ANANTHASUBRAMONY, physicist, educator; b. Tiruchirapally, India, June 19, 1949; came to U.S., 1979, naturalized, 1982; s. Koovappadi Subramani and Jayalakshmi (Arunachalam) Ananthasubramony; m. Radha Arunkumar, Aug. 20, 1976; 1 child, Amiethab. B.Sc., U. Kerala, India, 1969, M.Sc., 1971; Ph.D., Indian Inst. Tech., Madras, 1975, U. Hull, Eng., 1979. Vis. asst. prof. U. Ky., Lexington, 1979-80, sr. research assoc., 1980—. Contbr. articles to profl. jours. Indian Inst. Tech. research fellow, 1971-72; Council Sci. and Indsl. Research jr. research fellow, 1972-75; U.K. Commonwealth scholar, 1975-79. Mem. Am. Phys. Soc., Smithsonian Instn., Sigma Xi. Hindu. Current work: Laser systems, free electron lasers, laser plasma interaction, surface enhanced Raman scattering. Laser Raman spectroscopy as applied to study of catalysis in connection with methanation. Home: 151 Todds Rd Apt 298 Lexington KY 40509 Office: Dept Elec Engring U Ky Lexington KY 40506

ARVIDSON, RAYMOND ERNST, earth and planetary science educator; b. Bklyn., Jan. 22, 1948; m., 1969; 2 children. B.A., Temple U., 1969; M.S., Brown U., 1971, Ph.D., 1974. Research assoc. planetary geology Brown U., 1973; asst. prof. earth and planetary sci. Washington U. St. Louis, 1974-78, assoc. prof., 1978—; fellow McDonnel Ctr. Space Sci., Washington U., 1976—; dir. NASA Planetary Image Facility, 1978—; team leader Viking Lander Imaging Team, 1978—; mem. Space Sci. Bd., chmn. Com. Data Mgmt. and Computers, Nat. Acad. Sci., 1981—; mem. NASA Planetary Rev. Panel, 1981—. Mem. Am. Soc. Photogrammetry, Am. Geophys. Union, AIAA. Subspecialties: Remote sensing (geoscience); Planetary science. Office: Dept Earth and Planetary Sci Washington U Saint Louis MO 63130

ARVIND, computer science educator and researcher, consultant; b. Lucknow, India, May 15, 1947; s. Devki Nandan and Ruckmani (Mithal); m. Gita Singh Mithal, Dec. 4, 1973; children: Divakar Singh, Prabhakar Singh. B.Tech. in Elec. Engring, Indian Inst. Tech., Kanpur, 1969; M.S. in Computer Sci, U. Minn., 1972, Ph.D., 1973. Asst. prof. info. and computer sci. U. Calif.-Irvine, 1974-78; assoc. prof. elec. engring. and computer sci. MIT, 1979—; cons. IBM, TRW Array Processors, 1980-81. Contbr. articles to profl. jours. Mem. Assn. Computing Machinery, IEEE Computer Soc. Subspecialties: Computer architecture; Programming languages. Current work: Dataflow architectures and functional programming languages; data flow; functional languages; parallel computing. Home: 34 Lombard Rd Arlington MA 02174 Office: MIT Lab for Computer Sci 545 Technology Square Cambridge MA 02139

ARZBAECHER, ROBERT, electrical and computer engineering educator; b. Chgo., Oct. 28, 1931; m., 1956; 5 children. B.S., Fournier Inst. Tech., 1953; M.S., U. Ill., 1958, Ph.D., 1960. Elec. engr. Argonne Nat. Lab., 1954-60; from asst. prof. to prof. elec. engring. Christian Bros. Coll., Tenn., 1960-67; from assoc. prof. to prof. U. Ill.-Chgo., 1967-76; prof., chmn. dept. elec. and computer engring. U. Iowa, Iowa City, 1976—, prof. internal medicine, 1978—; mem. com.

electrocardiography Am. Heart Assn., 1971—. Subspecialties: Electrical engineering; Computers in medicine. Office: Dept Elec and Computer Engineering U Iowa Iowa City IA 52242

ASAKURA, TOSHIO, medical educator and researcher; b. Osaka, Japan, Aug. 21, 1935; came to U.S., 1967; s. Shigeji and Sachiko (Arita) A.; m. Sumiko Kimura, Oct. 30, 1967; children: Emi, Kenji. B.S. in Biology, Kyotofuritsu U., 1956; M.D., Kyoto Med. Coll., 1960; Ph.D. in Biochemistry, Tokyo U. Sch. Medicine, 1965; hon. degree, U. Pa., 1967. Intern Tokyo (Japan) Univ. Hosp., 1960-61; asst. prof. physiochemistry U. Tokyo, 1965-67; research assoc. dept. biophysics U. Pa., Phila., 1967-69, asst. prof. phys. biochemistry, 1969-73, asst. prof. phys. biochemistry and pediatrics, 1973-74; assoc. prof. pediatrics and biochemistry U. Pa. and Children's Hosp. Phila., 1974-76, prof. pediatrics, biochemistry biophysics, 1976—. Author: Pentose Metabolism, 1969, Metabolism of Sugar Alcohols, 1974. Vice pres. Japan Assn. Phila., 1979. Recipient Career Devel. award NIH, 1972, citation City of Phila., 1981. Mem. Am. Soc. Biol. Chemists, Am. Chem. Soc., Biophys. Soc., Soc. Pediatric Research, John Morgan Soc. Subspecialties: Biochemistry (biology); Hematology. Current work: Hemoglobin, sickle cell disease, blood storage, red blood cells, enzymes, clinical testings, artificial blood and protein polymerization. Home: 28 Mary Watersford Rd Bala Cynwyd PA 19004 Office: Children's Hosp of Phila 34th and Civic Center Blvd Philadelphia PA 19104

ASANUMA, HIROSHI, physiologist; b. Kobe, Japan, Aug. 17, 1926; came to U.S., 1965, naturalized, 1966; s. Kisaburo and Yukiko (Takahashi) A.; m. Reiko Shimasu, Dec. 18, 1953; children—Chisato, Mari. M.D., Keio U., Tokyo, 1952; D.Med.Sci., Kobe (Japan) Med. Coll., 1959. Intern Saseikai Hosp., Kobe, 1952-53; instr. dept. physiology Kobe Med. Coll., 1953-59; asst. prof. dept. physiology Osaka City U. Med. Sch., 1959-65; asso. prof. dept. physiology N.Y. Med. Coll., N.Y.C., 1965-71, prof., 1971-72, Rockefeller U., N.Y.C., 1972—; guest investigator Rockefeller Inst., 1961-63. Contbr. articles to profl. jours. Rockefeller Found. Internat. fellow, 1961-62; City of N.Y. Career Scientist awardee, 1965-72. Mem. Am. Physiol. Soc., Japanese Physiol. Soc., Harvey Soc., Soc. for Neurosci. Subspecialties: Neurophysiology; Neurobiology. Current work: To study the functional role of sensory input to the motor cortex during willed movements. Home: 505 E 79th St New York NY 10021 Office: Rockefeller U 1230 York Ave New York NY 10021

ASATO, YUKIO, biologist; b. Waipahu, Hawaii, Jan. 19, 1934; s. Kame and Kama (Tamashiro) A.; m. Sue Akemi, Nov. 12, 1937; children: April A., Lynn K. B.A., U. Hawaii, 1957, M.S., 1966, Ph.D., 1969. Postdoctoral research scientist NASA, Moffett Field, Calif., 1969-71; asst. prof. Southereastern Mass. U., North Dartmouth, 1971-75, assoc. prof., 1975-80, prof. biology, 1980—. Contbr. articles to profl. jours. Served with AUS, 1957-60. Mem. Am. Soc. Microbiology, AAAS, Genetic Soc. Am. Subspecialties: Genome organization; Molecular biology. Current work: Research in molecular biology of the blue-green bacteria; regulation of cell cycle events in cyanobacteria. Office: Dept Biology Southeastern Mass Univ North Dartmouth MA 02747

ASAWA, GEORGE NOBUO, dentist; b. Norwalk, Calif., Apr. 29, 1925; s. Zensuke and Moto (Murata) A.; m. Masako Sadao, July 7, 1957; 1 dau., Elizabeth. B.A., U. Calif.-Berkeley, 1950, B.S., 1953; D.D.S., Washington U., St. Louis, 1963. Clin. asst. prof. U. So. Calif. Sch. Dentistry, Los Angeles, 1979—. Fellow Acad. Gen. Dentistry; mem. Internat. Assn. Dental Research, Am. Assn. Dental Research, Fedn. Internat. Dentaires, Am. Dental Assn. Democrat. Quaker. Subspecialty: Prosthodontics. Current work: Teaching prosthodontics at University Southern California School of Dentistry. Home: 930 S Knott Ave Anaheim CA 92804

ASBURY, ARTHUR KNIGHT, neurologist, educator; b. Cin., Nov. 22, 1928; s. Eslie and Mary (Knight) A.; m. Carolyn Holstein, May 17, 1980; children by previous marriage: Dana, Patricia Knight, William Francis. Grad., Phillips Acad., Andover, Mass., 1946; student, Stanford, 1947-48; B.S., U. Ky., 1951; M.D., U. Cin., 1958; M.A. (hon.), U. Pa., 1974. Intern in medicine Mass. Gen. Hosp., Boston, 1958-59, resident, 1959-63, fellow, 1963-65, staff neurologist, 1965-69; chief neurology San Francisco VA Hosp., 1969-74; prof. dept. neurology U. Pa., Phila., 1974—, chmn. dept. neurology, 1974-82; teaching fellow Harvard Med. Sch., 1958-65, instr., 1965-68, assoc., 1968-69; assoc. prof. neurology U. Calif. at San Francisco, 1969-73, prof., vice-chmn., 1969-74. Sr. editor: Internat. Med. Rev. Series-Neurology, Butterworth & Co., London, 1980—; Asso. editor: Archives of Neurology, 1975-76; Assoc. editor: Annals of Neurology, 1976-81; mem. editorial bd.: Muscle and Nerve, 1977—, Neurology, 1981—, Jour. Neuropathology and Exptl. Neurology, 1981-83; contbr. chpts. to med. textbooks, articles to med. jours. Vice-pres., bd. dirs. Forest Retreat Farms Inc., Carlisle, Ky., 1970—. Served with AUS, 1951-53. USPHS grantee, 1967—; Muscular Dystrophy Assn. grantee, 1974—. Fellow Am. Acad. Neurology (v.p. 1977-79); mem. Am. Neurol. Assn. (councillor 1976-81, pres. 1982-83), Am. Assn. Neuropathologists (v.p. 1983-84), Soc. Neurosci., Assn. Univ. Profs. Neurology (pres. 1980-82). Episcopalian (vestryman). Subspecialty: Neurology. Home: 2409 Naudain St Philadelphia PA 19146 Office: Dept Neurology Hospital Univ Pa Philadelphia PA 19104

ASCENSÃO, JOÃO L., physician, educator; b. Maputo, Mozambique, July 6, 1948; came to U.S., 1974; s. João F. and Maria (Almeida) A.; m. Virginia F. Baresch, Aug. 7, 1976 (div. 1981). M.D., U. Lisbon, Portugal, 1972. Diplomate: Am. Bd. Internal Medicine. Resident in medicine U. Lisbon Hosp., 1972-74; fellow in immunology Meml. Sloan Kettering Center/Sloan Kettering Inst., N.Y.C., 1974-76; resident in medicine U. Minn. Hosps., Mpls., 1977-79, fellow in hematology/oncology, 1979-81; instr. medicine U. Minn., Mpls., 1981-82, asst. prof., 1982—. Mem. editorial bd.: Exptl. Hematology, 1980-82. Portuguese Sci. Found. fellow, 1974; Ministry of Edn. fellow, 1975; J.M. Found. fellow, 1975. Fellow A.C.P.; mem. Am. Fedn. for Clin. Research, Am. Soc. Hematology, Internat. Soc. Exptl. Hematology, Portuguese Soc. Immunology. Subspecialties: Hematology; Oncology. Current work: Study of the blood cell precursors and th regulatory mechanisms of blood cell production. Home: 2604 37th Ave S Minneapolis MN 55406 Office: VA Med Center 48th St and 54th Ave S Minneapolis MN 55417

ASCHE, DAVID ROBERT, medical radiation physicist; b. Oakland, Calif., May 22, 1951; s. Robert Franklin and Margaret (Chamberlain) A.; m. Constance Joanne Fries, July 28, 1973; children: Richard, Ryan, Jonathan. B.S., U. Calif.-Davis, 1973; M.S., Calif. State U.-Sacramento, 1976. Acting chief physicist Radiation Oncology Center, Sacramento, 1981-82, asst. radiation physicist, 1976—, radiation safety officer, 1981—; guest lectr. biomed. engring. Calif. State U.-Sacramento, 1976—. Mem. Am. Assn. Physicists in Medicine, Am. Assn. Physicists in Medicine (Bay Area chpt.). Mem. Evangel. Free Ch. of Am. Subspecialty: Therapeutic radiological physics. Current work: Applying the principals of physics and engineering to the problems arising in the environment of a clinical therapeutic radiology center. Office: Radiation Oncology Center 5271 F St Sacramento CA 95819

ASCHER, MICHAEL CHARLES, engineering company executive; b. N.Y.C., Jan. 8, 1944; s. Benjamin Paul and Miriam (Schechter) A.; m. Alexa Fern Rubin, June 12, 1966; children: Stacey Nicole, Jennifer Sharyn. B.M.E., CCNY, 1966; M.S. in Nuclear Engring, L.I. U., 1971. Registered profl. engr., N.Y., N.J., Tenn. Mech. engr. L.I. Lighting Co., Hicksville, N.Y., 1966-72; sr. engr. and supr. nuclear engr. Burns and Roe, Inc., Hempstead, N.Y., 1972-76, project mgr., Woodbury, N.Y., 1976-78, project engring. mgr., Oradell, N.J., 1978-79, project mgr. breeder reactor div., Oradell, 1979-83, dep. dir., 1983—. Contbr. tech. articles to profl. publs. Tech. del. L.I. (N.Y.) Space Shuttle Task Force, 1972. Mem. ASME (chmn. L.I. exec. com. 1971-72, Valued Service cert. 1972, 75), Am. Nuclear Soc., Nat. Soc. Profl. Engrs., Assn. L.I. Engrs. and Scientists (vice chmn. 1972-73). Republican. Lodge: B'nai B'rith (treas. Oradell chpt. 1979-80). Subspecialty: Nuclear fission. Current work: Responsible for development of breeder reactor energy technology through the design of the nation's first large scale liquid metal fast breeder reactor demonstration project. Office: Burns and Roe Inc 700 Kinderkamack Rd Oradell NJ 07649

ASCOLI, FRANK ANTHONY, radiol. physicist; b. Providence, Jan. 2, 1950; s. Joseph and Jennie (Durante) S.; m. Paulette Brousseau, Apr. 28, 1973; children: Amy, Luke. B.S. in Elec. Engring, Northeastern U., 1972; M.S., U. Cin., 1976. Diplomate: Am. Bd. Radiology. Physicist Boston City Hosp. and Univ. Hosp., 1972-75; health physicist Yale U.-New Haven Hosp., 1976-77; med. physicist St. Vincent Hosp., Worcester, Mass., 1977—, lectr., 1977—. Mem. Am. Assn. Physicists in Medicine, New Eng. Soc. Radiation Oncology. Subspecialties: Radiology; Imaging technology. Current work: Radiation therapy and radiology research. Home: RFD #1 Dresser Hill Rd Southbridge MA 01550 Office: St Vincent Hosp Worcester MA 01604

ASH, SIDNEY ROY, geologist, educator; b. Albuquerque, Nov. 25, 1928; s. Oliver Knox and Ellen Rosena (Tavernier) A.; m. Shirley Martha Arviso, Jan. 23, 1962; children: Kathleen Ellen, Randolph Henry. B.A., Midland Lutheran Coll., 1951, U. N.Mex., 1957, M.A., 1961; Ph.D., U. Reading, Eng., 1966. Geologist, U.S. Geol. Survey, Albuquerque, 1958-64; asst. prof. earth sci. Midland Luth. Coll., 1966-69; asst. prof. geology Ft. Hays Kans. State Coll., 1969-70; asso. prof. geology and geography Weber State Coll., 1970-76, prof., 1976—, chmn. dept. geology/geography, 1977-83. Contbr. numerous articles to profl. jours. Served to lt. USNR, 1952-55. NSF grantee, 1971-73, 73-75, 79-82; NSF-Australian Dept. Sci. Coop. Sci. Program grantee, 1976-77. Mem. Paleontol. Soc., Internat. Organ. Paleobotany, Bot. Soc. Am., Paleontol. Assn., Sigma Xi. Lutheran. Subspecialties: Geology; Paleontology. Current work: Plant fossils of Mesozoic and their evolution and biostratigraphy. Home: 1341 Henderson Dr Ogden UT 84404 Office: Dept Geology/Geography Weber State Coll Ogden UT 84408

ASHCROFT, NEIL WILLIAM, physics educator; b. London, Nov. 27, 1938; m., 1961; 2 children. B.Sc., U. N.Z., 1958, M.Sc. with honors, 1960; Ph.D., U. Cambridge, 1964. Sci. research council sr. fellow Cavendish Lab. U. Cambridge, 1973-74, vis. fellow Clare Hall, 1973-74; assoc. theoretical physics Cornell U., Ithaca, N.Y., 1965-66, from asst. prof. to assoc. prof., 1966-75, prof. physics, 1975—, dir., 1979—; cons. Los Alamos Nat. Lab. Fellow Am. Phys. Soc.; mem. Am. Inst. Physics, AAAS. Subspecialties: Condensed matter physics; Theoretical physics. Office: Clark Hall Cornell U Ithaca NY 14853

ASHER, SANFORD ABRAHAM, chemistry educator; b. Landberg, Germany, June 18, 1947; s. Leo Dow and Pearl (Lon) A.; m.; children: David, Dianne; m. Nancy Lee Day, June 29, 1976; 1 dau., Rachel Marie. B.A., U. Mo.-St. Louis, 1971; M.S., U. Calif.-Berkeley, 1973, Ph.D., 1977. Research asst. Petrolite Corp., St. Louis, 1966-71; research fellow Harvard U., Cambridge, Mass., 1977-80; asst. prof. chemistry U. Pitts., 1980—. Contbr. articles to profl. jours. Grantee NIH, 1982-84, NSF, 1982, Petroleum Research Fund, 1981-83, Research Corp., 1982-83. Mem. Am. Chem. Soc., AAAS, Biophys. Soc., Soc. Analytical Chemists Pitts., Soc. Applied Spectroscopy. Jewish. Subspecialties: Analytical chemistry; Biophysical chemistry. Current work: Development of new spectroscopic techniques for studying complex biomolecules and polymers, laser spectroscopy, instrumentation. Office: Dept Chemistry U Pitts Pittsburgh PA 15260 Home: 6425 Bartlett St Pittsburgh PA 15217

ASHER, STEVEN ROBERT, educational psychology educator; b. Newark, Feb. 19, 1945; s. Emil and Martha (Stone) A.; m. Nancy Weinberg, Nov. 30, 1969; children: Matthew, David. B.A., Rutgers U., 1966; M.A., U. Wis.-Madison, 1968, Ph.D., 1971. Ednl. evaluation specialist U. Wis.-Madision, 1969-71; asst. prof. ednl. psychology U. Ill., Champaign, 1971-76, assoc.prof., 1976-78, assoc. prof. ednl. psychology and psychology, 1978-82, prof., 1982—; dir. Bur. Ednl. Research, 1980—. Editorial bd.: Child Devel, 1976—; editorial bd.: Merrill-Palmer Quar, 1979—; editorial cons. numerous profl. jours.; editor: The Development of Children's Friendships, 1981; cons.: Champaign and Urbana Schs, 1971—. Grantee Nat. Inst. Edn., 1973-75, Nat. Inst. Child Health and Human Devel., 1973-77; fellow Center Advanced Study, U. Ill., 1972-73. Fellow Am. Psychol. Assn.; mem. Am. Ednl. Research Assn., Soc. Research in Child Devel. Democrat. Jewish. Subspecialty: Developmental psychology. Current work: Children's social development, peer relationships, and social skills. Home: 1012 W Daniel Champaign IL 61821 Office: 1310 S 6th St Champaign IL 61820

ASHLEY, HOLT, aero. scientist, educator; b. San Francisco, Jan. 10, 1923; s. Harold Harrison and Anne (Oates) A.; m. Frances M. Day, Feb. 1, 1947. Student, Calif. Inst. Tech., 1940-43; B.S., U. Chgo., 1944; S.M., Mass. Inst. Tech., 1948, Sc.D., 1951. Faculty Mass. Inst. Tech., 1946-67, prof. aero., 1960-67; prof. aeros. and astronautics Stanford U., 1967-; spl. research aeroelasticity, aerodynamics; cons. govt. agys., research orgns., indsl. corps.; dir. office of exploratory research and problem assessment and div. advanced tech. applications NSF, 1972-74; mem. sci. adv. bd. USAF, 1958-80; research adv. com. structural dynamics NASA, 1952-60, research adv. com. on aircraft structures, 1962-70, chmn. research adv. com. on materials and structures, 1974-77; mem. Kanpur Indo-American program Indian Inst. Tech., 1964-65; AIAA Wright Bros. lectr., 1981. Co-author: Aeroelasticity, 1955, Principles of Aeroelasticity, 1962, Aerodynamics of Wings and Bodies, 1969, Engineering Analysis of Flight Vehicles, 1974. Mem. Greater Boston coordinating council Boy Scouts Am., also mem.-at-large and adviser air explorer squadron. Recipient Goodwin medal M.I.T., 1952; Exceptional Civilian Service award U.S. Air Force, 1972, 80; Public Service award NASA, 1981; named one of 10 outstanding young men of year Boston Jr. C. of C., 1956. Fellow Am. Acad. Arts and Scis.; hon. fellow AIAA (asso. editor jour., v.p. tech. 1971, pres. 1973, Structures, Structural Dynamics and Materials award 1969); mem. Am. Meterol. Soc. (profl., recipient 50th Anniversary medal 1971), AAAS, Nat. Acad. Engring. (aeros. and space engring. bd. 1977—), Phi Beta Kappa, Sigma Xi, Tau Beta Pi. Subspecialties: Aeronautical engineering; Aerospace engineering and technology. Current work: Aeroelasticity; unsteady aerodynamics; theory of wind. Address: 475 Woodside Dr Woodside CA 94062

ASHLOCK, PETER DUNNING, entomology educator; b. San Francisco, Aug. 22, 1929; s. Charles Louis Dunning A. and Josephine (Dunning) Myer; m. Virginia Jane Harris, June 26, 1956; children: Daniel, Joseph. B.S., U. Calif.-Berkeley, 1952, Ph.D., 1966; M.S., U. Conn., 1956. Mem. faculty dept. entomology U. Kans., Lawrence, 1968—, prof., 1981—. Served with U.S. Army, 1952-54. Mem. Soc. Systematic Zoology, Soc. for Study of Evolution, Entomol. Soc. Am. Subspecialty: Systematics. Current work: Systematics of hemipterous insects, especially Lygaeidae; biogeography; systematic theory. Home: 800 Indiana St Lawrence KS 66044 Office: Dept Entomology U Kans Lawrence KS 66045

ASHWORTH, CLINTON PAUL, mechanical engineer; b. Salt Lake City, Oct. 25, 1928; s. Paul P. and Jane D. (Ferrin) A.; m. Rachel Calder, Oct. 9, 195o; children: Jeff C., Lucy Ashworth Pugh, James F., Alan S., Paul L., Sally, Jane. B.S.E.E., U. Utah, 1950. Registered profl. engr., Calif. Engr. Gen. Electric Co., Schenectady and San Jose, Calif., 1950-57; mech. engr. Pacific Gas & Electric Co., San Francisco, 1957—, supr., 1971—. Mem. Am. Nuclear Soc., ASME. Republican. Mem. Ch. Jesus Christ of Latter-day Saints. Subspecialties: Fusion; Nuclear fission. Home: 3187 Cafeto Dr Walnut Creek CA 94598 Home: 3187 Cafeto Dr Walnut Creek CA 94598 Office: 77 Beale St Rm 2547 San Francisco CA 94106

ASKEW, ELDON WAYNE, army officer, biochemist, researcher; b. Pontiac, Ill, Aug. 23, 1942; s. Robert Eldon and Doria Elizabeth (Carter) A.; m. Sharon Lee, Feb. 13, 1982; children: REbecca C., Jennifer J. B.S., U. Ill., 1964, M.S., 1966; Ph.D., Mich. State U.-East Lansing, 1969. Commd. 2d lt. U.S. Army, 1964, advanced through grades to maj., 1978; chief lipid br. (U.S. Army Med. Research and Nutrition Lab.), Denver, 1969-74, chief energy metabolism br. dept. nutrition, San Francisco, 1974-76, chief radiation services group, 1977-82, asst. chief dept. clin. investigation, Honolulu, 1982—; affiliate prof. Colo. State U., Ft. Collins, 1977-78; mem. speakers bur. Ross Labs., Columbus, Ohio, 1977—. Contbr. articles to profl. jours. Tchr. Sunday sch. Mililani Presbyterian Ch., Miliani Town, Hawaii, 1982. Decorated Meritorius Service medal. Mem. Am. Inst. Nutrition, Fedn. Am. Soc. Exptl. Biology, Soc. Exptl. Biology and Medicine, Am. Coll. Sports Medicine, Nutrtion Today Soc., Sigma Xi, Alpha Zeta. Club: Mililani Sportsman. Subspecialties: Nutrition (medicine); Animal nutrition. Current work: Nutrition and physical performance; adipose tissue energy metabolism. Home: 94-074 Kaweloahii Pl Mililani Town HI 96789 Office: Department of Clinical Investigation Tripler Army Medical Center Honolulu HI 96859

ASKINS, ROBERT ARTHUR, zoology educator; b. Waltham, Mass., Aug. 24, 1947; s. Arthur Alexander and Frankie D. (McCormick) A.; m. Karen Lichtenwanger, May 3, 1969; children: Stephen Alexander, Michael Andrew. B.S., U. Mich., 1970; M.S., U. Minn., 1977, Ph.D., 1981. Teaching asst. dept. ecology and behavioral biology U. Minn., Mpls., 1974-78; project asst. Bell Mus. Natural History, Mpls., 1980-81; asst. prof. zoology Conn. Coll., New London, 1981—. Contbr. chpt. to book and articles in field to profl. jours. Mem. bd. trustees Thames Sci. Ctr., New London, 1983. Francis E. Andrews Mpls. Found. fellow U. Minn., 1977; U. Minn. fellow, 1978. Mem. AAAS, Am. Ornithologist Union, Animal Behavior Soc., Ecol. Soc. Am., Wilson Ornithological Soc. Subspecialties: Ecology; Ethology. Current work: Effect of the size and isolation of forest patches on the composition of bird communities; communication in woodpeckers. Home: 3 River Ridge Rd New London CT 06320 Office: Dept Zoology Conn Coll New London CT 06320

ASLESON, GARY LEE, chemistry educator, researcher; b. St. Peter, Minn., June 14, 1948; s. Gordon A. and Adeline (Bieraugel) A.; m. Meredith A. Knutson, Aug. 21, 1971; 1 dau., Kristin. B.A., Gustavus Adolphus Coll., 1970; Ph.D., U. Iowa, 1975. Teaching asst. U. Iowa, Iowa City, 1970-73, research asst., 1973-75; asst. prof. chemistry Coll. Charleston, S.C., 1975-80, assoc. prof., 1980—. Mem. Am. Chem. Soc., Sigma Xi., Alpha Chi Sigma. Lutheran. Club: James Island Yacht (Charleston, S.C.). Subspecialties: Analytical chemistry; Nuclear magnetic resonance (chemistry). Current work: High performance liquid chromatography, chromatographic optimization, C-13 nuclear magnetic resonance of tetracyclines. Home: 1015 Birchdale Dr Charleston SC 29412 Office: Dept Chemistry Coll Charleston Charleston SC 29424

ASPLUND, JOHN MALCOLM, animal scientist, educator, cons., researcher; b. Raymond, Alta., Can., Mar. 19, 1930; came to U.S., 1967, naturalized, 1974; s. Charles Owen and Julia Ellen (Russell) A.; m. Patricia Jean Havens, May 1, 1956; children: Curtis, Virginia, Leila, Liisa, Kirsti, John, Nicholl, Renee. B.S. in Agr, U. Alta.-Edmonton, 1951, M.S. in Animal Sci, 1957; Ph.D. in Nutrition, U. Wis. Madison, 1960. Asst. prof. U. Alta., 1959-64; research assoc. Utah State U., 1967-69; assoc. prof. animal sci. U. Mo.-Columbia, 1969-74, prof., 1974—; cons. in field. Contbr. articles to profl. jours. Recipient Bd. Govs. prize in agr., 1951; Wis. Alumni Research Found. scholar, 1955-57; grantee Nat. Arthritis Metabolism and Digestive Diseases Council, NIH, Dept. Agr., Amax Corp. Mem. Am. Soc. Animal Sci., Am. Inst. Nutrition, Agr. Inst. Can., Alta. Inst. Agrologists, Can. Soc. Animal Prodn., Internat. Soc. Parenteral Nutrition, Utah Acad. Sci. Arts and Letters, Sigma Xi, Gamma Delta Sigma. Mormon. Subspecialties: Animal nutrition; Nutrition (medicine). Current work: Ruminant nutrition, nitrogen utilization, parenteral nutrition.

ASPREY, MARGARET WILLIAMS, nuclear engineer, reactor safety analyst; b. Chicago Heights, Ill., Dec. 21, 1921; d. Russell John and Dorothy (Atkinson) Williams; m. Larned Brown Asprey, May 3, 1944; children: Peter L., Barbara A., Elizabeth A., Robert R., Margaret S., Thomas A., William J. Student, U. Chgo., 1941-43; B.S., Coll. Santa Fe, 1966; M.S., U. N.Mex., 1978. Technician Manhattan Project, Chgo., 1943-45; tchr. math. Espanola (N.Mex.) Pub. Schs., 1966-67; mem. staff Los Alamos Nat. Lab., 1967—; vis. scientist Kernforschungszentrum, Karlsruhe, W.Ger., 1981-82. Author manual. Mem. Am. Nuclear Soc., AAAS. Democrat. Subspecialties: Nuclear engineering; Nuclear fission. Current work: Calculations for safety analysis of Clinch River Breeder Reactor. Office: Los Alamos Nat Lab Lab Group X-7 MS-B257 Los Alamos NM 87545 Home: 720 46th St Los Alamos NM 87544

ASTERITA, MARY FRANCES, educator; b. N.Y.C., Dec. 30, 1937; d. Martin and Mary Lucy (diPalma) A. B.A., Marymount Coll., 1961; M.S., NYU, 1969; Ph.D., Cornell U., 1973; postgrad., Ind. U. N.W., 1983. Postdoctoral fellow Yale U. Sch. Medicine, New Haven, 1973-75; Faculty Ind. U. Sch. Medicine, Gary, 1975—, now prof. physiology and pharmacology; postdoctoral fellow Yale U. Sch. Medicine, New Haven, 1973-75. Contbr. articles to profl. jours.; author: Physiology of Stress, 1984. NIH fellow, 1969-73, 74-75; Conn. Heart Assn. fellow, 1973-74. Mem. Am. Physiol. Soc., Biofeedback Soc. Am., Biofeedback Soc. Ill. (pres.), Ind. Acad. Scis. (chmn. 1979), Am. Heart Assn. Subspecialties: Physiology (biology); Biofeedback. Current work: Cellular transport; physiology of stress; biofeedback; stress management. Office: Indiana Univ Sch Medicine 3400 Broadway Gary IN 46408

ASTLEY, EUGENE ROY, seamless tube manufacturing executive; b. Alameda, Calif., Dec. 5, 1926; s. Frank Robert and Mary (Barr) A.; m. Peggy Lund, June 20, 1948; children: Clifford Andrew, Michael J., William Lawrence. B.S. in Physics, U. Oreg., 1948, M.S., Oreg. State U., 1950. Engr. Gen. Electric, Schenectady, 1950-54; physicist Gen.

Electric-Hanford, Richland, Wash., 1955-59, mgr.-maintenance engr., 1960-64; dir. FFTF project Battelle Northwest, Richland, 1965-71; v.p. Exxon Nuclear Co., Inc., Bellevue, Wash., 1971-83, dir., 1979-83; pres. Sandvik Spl. Metals, 1983—; dir. Jersey Nuclear Avco Isotopes, Inc., Exxon Nuclear Idaho Co., Idaho Falls, United Nuclear Industries; assoc. prof. U. Wash., Seattle, 1958. Contbr. articles to profl. jours. Pres. Lower Columbia YMCA, Richland, 1963-64; mgr. Colt League team, 1964-65; bd. dirs. United Way, Richland, 1979—. Mem. Am. Nuclear Soc. (chmn. local chpt. 1961); mem. Atomic Indsl. Forum, Tri-City C. of C. (bd. dirs.), Sigma Pi Sigma, Sigma Xi. Republican. Unitarian. Subspecialties: Fuels; Nuclear engineering. Current work: Development and manufacture of nuclear fuels for light water cooled reactors. Patentee nuclear instrumentation (6). Office: Sandvik Spl Metals Co PO Box 6027 Kennewick WA 99336

ASTON, ROY, pharmacology educator; b. Windsor, Ont., Can., Dec. 31, 1929; came to U.S.; s. Harold and Lillian Isbelle (Brydges) A.; m. Claudia Louise Macomber, May 5, 1973; children: Roderick, Gregory, Victoria, Christine. B.A., U. Windsor, 1950; M.Sc., Wayne State U., 1954; Ph.D., U. Toronto, 1958. Asst. prof. pharmacology Wayne State U. Sch. Medicine, 1963-70, assoc. prof., 1970-78; prof. pharmacology U. Detroit Dental Sch., 1978—. Contbr. articles to sci. jours., chpts. to books. Bd. dirs. Wayne-Westland Fed. Credit Union. Mem. Am. Soc. Pharmacology and Exptl. Therapeutics, Soc. Toxicology, Soc. Exptl. Biology and Medicine. Episcopalian. Subspecialties: Pharmacology; Neuropharmacology. Current work: Research related to clinical aspects of benzodiazepine use; research in intolerance to depressants. Home: 1867 Treadwell St Westland MI 48185 Office: U Detroit Dental Sch 2985 E Jefferson St Detroit MI 48207

ATCHLEY, WILLIAM REID, genetics educator; b. Stilwell, Okla., Sept. 6, 1942; s. Reid Kenneth and Velma Alice (Mays) A.; m. Wilinda Landon, Sept. 4, 1964; children: Erika Leigh, Kevin Landon. B.S., Eastern N.Mex. U., 1964; M.A., U. Kans., 1966, Ph.D., 1969; postgrad., U. Melbourne, Australia, 1969-70, U. Wis., 1976-77. Asst. prof. entomology U. Kans., Lawrence, 1970-71; asst. prof. biology and stats. Tex. Tech. U., Lubbock, 1971-74, assoc. prof., 1974-77; assoc. prof. entomology U. Wis., Madison, 1977-80, prof. entomology and genetics, 1980—; vis. research fellow in genetics U. Melbourne, 1974; cons. statistician Eastern N.Mex. U., 1974-77; vis. research fellow in population biology Australian Nat. U., Canberra, 1980-81. Editor: Evolution and Speciation, 1981, Multivariate Statistical Methods, 2 vols, 1975. NSF fellow, 1966-69; Fulbright fellow, 1969-70; NIH fellow, 1977; recipient Outstanding Alumni award Eastern N.Mex. U., 1982. Mem. Soc. Study of Evolution (councillor 1980-82), Soc. Systematic Zoology (councillor 1981-83), AAAS, Am. Soc. Naturalists. Subspecialties: Gene actions; Evolutionary biology. Current work: Quantitative genetics, genetics of growth and development, biostatistics, evolutionary biology, population biology. Office: Dept Entomology and Genetics Univ Wis Madison WI 53706

ATEMA, JELLE, biologist, educator; b. Deventer, Netherlands, Dec. 9, 1940; s. Ate and Cornelia M.J. (Lindenburg) A.; m. Hilda Mary Maingay, June 26, 1963; children: Ate, Jurgen, Sven; m. Ilyse Jill Rosenthal, Apr. 3, 1974; children: Annemieke, Adriana. Cand. degree, U. Utrecht, Netherlands, 1962, Doctorandus, 1966; Ph.D., U. Mich., 1969. Research asst. U. Utrecht, 1960-66; research assoc. U. Mich., 1966-69; asst. scientist Woods Hole Oceanographic Instn., 1970-74; asso. prof. Boston U., 1974—; vis. prof. U. Regensburg, W.Ger., 1979-80. Editorial bd.: Sensory Processes; guest editor: Oceanus Mag, fall 1980. Mem. NSF Rev. Panel Psychobiology NRC Panel.; Bd. dirs. Falmouth Music Assn., pres., 1973, 82—. Recipient 1st prize Cape Cod Symphony Music Competition, 1974; research grantee NSF, EPA, AEC, Dept. Energy, Culpepper Found., Whitehall Found., others, 1970—. Fellow AAAS; mem. Am. Soc. Zoologists, Animal Behavior Soc., European Chemoreception Research Orgn., Assn. Chemoreception Sci., Soc. Neurosci., Crustacean Soc. Subspecialties: Behavioral ecology; Neurophysiology. Current work: Physiological and behavioral function of chemoreceptors in lobsters and catfish. Behavioral ecology of aquatic animals. Evolution of chemoreceptor organs. Office: Boston U Marine Program Marine Biological Lab Woods Hole MA 02543

ATKINS, CHARLES GILMORE, medical school administrator; b. Stambaugh, Mich., July 4, 1939; s. Howard Burgl and Bernice Mary (Gilmore) A.; m. Kay Roberta Atkins, Dec. 28, 1958; children: Robert Howard, Karla Marie, James Charles. B.A., Albion Coll., 1961; postgrad., U. Mich. Med. Sch., 1960-62; M.S., Eastern Mich. U., 1963; Ph.D., N.C. State U., 1969. Instr. Coe Coll., 1963-66; lectr. genetics Cornell (Iowa) Coll., 1964-65; NIH trainee N.C. State U., 1966-69; asst prof. microbiology Ohio U., 1969-74, dir., 1972-74, assoc. prof. microbiology and biomed. scis., 1974—, dir. willed body program, 1976-77, assoc. dean for basic scis., 1976—. Contbr. articles to profl. jours. Scoutmaster Boy Scouts Am., 1972-82, dist. and council commr., 1978—. Served to maj. USAR, 1981—. Recipient Silver Wreath award Nat. Eagle Scout Assn., 1981. Mem. Genetics Soc. Am., Am. Soc. Microbiology, AAAS, Assn. Am. Med. Colls. Presbyterian (elder). Lodge: Rotary Internat. Subspecialties: Gene actions; Microbiology. Current work: Genetic and biochemical analysis of the enteric bacteria (E. coli and several speciesd of salmonella) and of hybrids of these species; recombination studies with T4 bacteriophage; phage as models in invertebrate systems. Home: 6 Riverview Dr Athens OH 45701 Office: Ohio U Coll Osteo Medicine 226 Irvine Hall Athens OH 45701

ATKINSON, STEVEN ALBERT, engineering administrator, heat transfer analyst; b. Powell, Wyo., Sept. 15, 1940; s. Albert Nye and Minnie Bell (Watson) A.; m. Patsy Marie Grimes, June 26, 1962; children: Aaron, John, Mark. B.S. in M.E, Colo. State U., 1962, M.S., U. Idaho, Moscow, 1969. Registered profl. engr., Idaho. Reactor operator Atomic Energy Div., Phillips Petroleum Co., Idaho Falls, 1962-64, engr., 1964-66; sr. engr. Idaho Nuclear Corp., Idaho Falls, 1966-71; assoc. project engr. Aerojet Nuclear Corp., Idaho Falls, 1971-76; project engr., engring. supr. EG & G Idaho, Inc., Idaho Falls, 1976-80, sr. engring. specialist, 1980—. Contbr. articles to profl. jours. Chmn. Bonneville County Young Republicans, Idaho Falls, 1967; elder, chmn. commn. First Presbyn. Ch., Idaho Falls, 1975-77; chmn. membership Boy Scouts Am., Idaho Falls, 1980-83, leader webelos den, 1982—. Mem. ASME (treas. Idaho sect. 1976-77, sec. 1977-80), Am. Nuclear Soc. Club: Toastmasters. Subspecialties: Nuclear engineering; Mechanical engineering. Current work: Application of statistical and probabilistic methods to prediction of nuclear core heat transfer conditions during accidents in nuclear reactors. Home: 1034 Mojave St Idaho Falls ID 83401 Office: EG & G Idaho Inc PO Box 1625 Idaho Falls ID 83415

ATLAS, DAVID, research scientist; b. Bklyn., May 25, 1924; s. Isadore and Rose (Jaffee) A.; m. Lucille Rosen, Sept. 26, 1948; children: Joan Linda, Robert Fred. B.Sc., NYU, 1946; M.Sc., MIT, 1951, D.Sc. in Meteorology, 1955. Chief weather radar br. Air Force Cambridge Research Labs., Bedford, Mass., 1948-66; prof. meterology U. Chgo., 1966-72; dir. atmospheric tech. div. Nat. Center for Atmospheric Research, Boulder, Colo., 1972-73, dir. nat. hail research expt., 1974-75; dir. lab. for atmospheric sci. NASA, Goddard Space Flight Center, Greenbelt, Md., 1977—; Chmn. Nat. Acad. Scis. Panel Remote Atmospheric Probing, also mem. com. on atmospheric scis., 1975-82. Served as 1st lt. USAAF, 1943-46. Recipient Loeser award Air Force Cambridge Research Labs., 1957, O'Day award, 1964; Robert M. Losey award AIAA, 1966; NASA Outstanding Leadership medal, 1982; NSF sr. postdoctoral fellow Imperial Coll., London, Eng., 1959-60. Fellow Am. Meteorol. Soc. (councilor 1961-64, 72-74, Meisinger award 1957, asso. editor publs. 1957-74, pres. 1975—, Cleveland Abbé award 1983), Am. Geophys. Union, Am. Astronautical Soc., Royal Meteorol. Soc., AAAS; mem. Internat. Radio Sci. Union (pres. inter-union commn. on radio meteorology 1969-72). Subspecialties: Meteorology; Remote sensing (atmospheric science). Current work: Meterology and climatology; satellite meterology, radar meterology, remote sensing, research management. Inventor weather radar devices. Home: 7420 Westlake Terr Bethesda MD 20034

ATLEE, JOHN LIGHT, III, anesthesiologist, researcher; b. Lancaster, Pa., Feb. 22, 1941; s. John Light and Ann Conyham (Stevens) A.; m. Barbara Sheaffer, June 20, 1964 (div. dec. 1967); m. Barbara Sanford, Feb. 3, 1968; children: Sarah Sanford, John Light. B.A., Franklin & Marshall Coll., 1963; M.D., Temple U., 1967, M.S. in Pharmacology, 1971. Diplomate: Am. Bd. Anesthesiology.; Cert. Nat. Bd. Med. Examiners, Am. Coll. Anesthesiologists. Intern Germantown Dispensary and Hosp., Phila., 1967-68; resident Temple U. Hosp., 1968-71; asst. prof. anesthesiology U. Wis.-Madison, 1973-78, assoc. prof., 1978—; vis. prof. U. Pa., 1976, U. Man., Can., 1977, Dalhousie U., Halifax, N.S., Can., 1978, U. Ill. Med. Ctr., Chgo., 1978, U. Va., Charlottesville, 1978. Contbr. chpts. to books and articles in field to profl. jours. Served with USN, 1971-73. Grantee in field. Fellow Am. Coll. Anesthesiologists, Phila. Coll. Physicians; mem. AMA, Am. Soc. Anesthesiologists, Am. Soc. Pharmacology and Exptl. Therapeutics, Soc. Research on Circulatory Effects of Anesthetics, Soc. Cardiovascular Anesthesiologists, Dane County Med. Soc., Internat. Anesthesia Research Soc., Wis. Soc. Anesthesiologists, Wis. State Med. Soc., Sigma Xi. Republican. Episcopalian. Club: Social Register Assn. Subspecialty: Anesthesiology. Current work: Cardiac electrophysiology, electropharmacology; drug interactions. Patentee neuromuscular function monitoring device. Home: 3873 State Hwy 19 DeForest WI 53532 Office: B6/386 Clin Sci Ctr 600 Highland Ave Madison WI 53792

ATTALLAH, ABDELFATTAH, immunobiologist; b. Dakahlia, Egypt, Feb. 2, 1944; s. Mohamed M. A. B.S., Alexandria (Egypt) U., 1967; D.E.A., Paris U., 1971; Ph.D., George Washington U., 1974. Research scientist Research Found. Children's Nat. Med. Ctr., Washington, 1972-76; research scientist Naval Med. Research Inst., Bethesda, Md., 1976-78; chief immunology sect. Nat. Ctr. Drugs and Biologics, NIH, Bethesda, 1978—; vis. scientist Naval Med. Research Inst., 1978-82; adj. prof. Georgetown U., 1982—, George Washington U. Contbr. chpts. to books, articles to profl. jours. Mem. Am. Assn. Immunologists, AAAS, Am. Soc. Microbiology, N.Y. Acad. Scis., Assn. Egyptian-Am. Scholars, Sigma Xi. Subspecialties: Cellular engineering; Immunopharmacology. Current work: Cell biology, cell and tissue culture, immunobiology, infectious diseases, genetics and genetic engring. and cancer research. Home: 5919 Beech Ave Bethesda MD 20817 Office: FDA 8000 Rockville Pike Bethesda MD 20205

ATTAYA, HOSNY M(OUSTAFA), nuclear engineer, consultant; b. Alexandria, Egypt, Nov. 29, 1945; came to U.S., 1976, naturalized, 1983; s. Moustafa K. Attaya and Zakia (Ali) Ali; m. Hala T. Abou-El-Nasr, Ot. 6, 1977; children: Shariff, Nabil. B.S. in Nuclear Engring, U. Alexandria, 1967, M.S., 1974, M.S., U. Wis.-Madison, 1977, Ph.D., 1981. Research assoc. Atomic Energy Authority of Egypt, Cairo, 1967-76; research asst. U. Wis.-Madison, 1976-81, project assoc., 1981—; cons. Fusion Power Assocs., GAithersburg, Md., 1982-83. Mem. Am. Nuclear Soc., Am. Soc. Metals, Metall. Soc. Subspecialties: Nuclear engineering; Nuclear fusion. Current work: Materials particle transport in material, magneto-statico; magnet design in fusion reactors; interactions of ferro-magnetic materials in magnetic fusion reactors. Home: 4801 Sheboygan Ave Apt 702 Madison WI 53705 Office: Dept Nuclear Engring U Wis-Madison 1500 Johnson Dr Madison WI 53706

ATTINGER, ERNST OTTO, biochemical engineering education administrator; b. Zurich, Switzerland, Dec. 27, 1922; came to U.S., 1952, naturalized, 1965; s. Ernst and Martha (Padrutt) A.; m. Francoise Marie L. Daubige, Feb. 4, 1947; children: Christopher, Nathalene, Joelle. B.A., Winterthur, Switzerland, 1971; M.D., Zurich U., 1948; M.S., Drexel Inst. Tech., 1961; Ph.D. in Biomed Engring., U. Pa., 1965. Research fellow Am. Heart Assn. Tufts U., Boston, 1956-59; sr. investigator Presbyn. Hosp., Phila, 1961-62; research dir. Presbyn. Hosp. U. Pa. Med. Center, Phila., 1962-67; chmn. div. biomed. engring. U. Va., Charlottesville, 1967—. Author: Pulsatile Blood Flow, 1964, Global Systems Dynamics, 1970, Biomedical Engineering in Dentistry, 1977. Recipient Pres.'s and Visitor's Research prize U. Va., 1976. Fellow IEEE; mem. Biomed. Engring. Soc., Am. Physiol. Soc., N.Y. Acad. Scis., Sigma Xi. Subspecialties: Biomedical engineering; Systems engineering. Current work: Analysis of biological systems, systems analysis of control hierarchies, assessment of health care technology. Office: Med Center Dept Biomed Engring U Va Box 377 Charlottesville VA 22908

ATTIX, FRANK HERBERT, medical physics educator, researcher; b. Portland, Oreg., Apr. 2, 1925; s. Ulysses Sheldon and Alma Katherine (Michelsen) A.; m. Evelyn Louise Van Scoy, Apr. 19, 1946; m. Shirley Adeline Lohr, Jan. 24, 1959; children: Shelley Anne, Richard Haven. A.B. in Physics with honors, U. Calif., Berkeley, 1949, M.S., U. Md., College Park, 1953. With NIH, Bethesda, Md., 1949-50, Nat. Bur. Standards, Washington, 1950-57, Am. Car & Foundry Co., 1957-58, Naval Research Lab., 1958-76; prof. med. physics U. Wis., Madison, 1976—. Editor in chief, contbg. author: Radiation Dosimetry, 2d edit. vols. I-III, 1966-69; contbr. other books, sci. papers to publs. Served to lt. USN, 1943-46. Recipient ann. award for applied sci. Research Soc. Am.-Naval Research Lab., 1969. Mem. Am. Phys. Soc., Am. Assn. Physicists in Medicine (bd. dirs. 1978-80), Health Physics Soc. (bd. 1968-71), Phi Beta Kappa, Sigma Xi, Sigma Pi Sigma. Club: August Derleth Soc. (Sauk City, Wis.) (bd.). Subspecialties: Medical physics; Radiation Dosimetry. Current work: Radiol. physics and dosimetry. Patentee. Home: 3333 Westview Ln Madison WI 53713 Office: U Wis 1300 University Ave Room 1530 Madison WI 53706

ATWELL, CONSTANCE WOODRUFF, research grant administrator; b. Phila., Jan. 27; d. Marston True and Viola Estelle (Habel) Woodruff; m. Michael Harry Kovar, Aug. 31, 1963 (div. 1972); m. Robert Herron Atwell, Sept. 4, 1972; children: Catherine, Cynthia. A.B., Mt. Holyoke Coll., South Hadley, Mass., 1963; M.A., UCLA, 1965, Ph.D., 1968. Asst. prof. to prof. Pitzer Coll. and Claremont (Calif.) Grad. Sch., 1967-78; research assoc., lectr. Univ. Coll., Nairobi, Kenya, 1968-69; grants assoc. NIH, Bethesda, Md., 1978-79; chief strabismus, amblyopia and visual processing br. Nat. Eye Inst., Bethesda, 1979—. Editor: Nutrition, Pharmacology and Vision, 1983; Contbr. articles to profl. jours. Reader Rec. for the Blind, 1973-78; trustee Claremont Collegiate Sch., 1975-77; pres. Cabin John Jr. High Sch. PTA, Potomac, Md., 1981-82, bd. dirs., 1980—; bd. dirs. Winston Churchill High Sch., Potomac, 1980—. USPHS trainee UCLA, 1964-67; NIH grantee, 1975-78; Grant Found. grantee, 1975. Mem. AAAS, Am. Psychol. Assn., Assn. for Research in Vision and Ophthalmology, Assn. for Women in Sci., Women in Eye Research, Phi Beta Kappa. Subspecialty: Physiological psychology. Current work: Management of branch that supports research on the prevention, treatment and rehabilitation of disorders of central visual processing. Research fields typically involved include neuroanatomy, neurophysiology, neurochemistry, visual psychophysics, physiological optics, and pediatric ophthalmology. Home: 8608 Timber Hill Ln Potomac MD 20854 Office: Nat Eye Inst Bldg 31/6A49 Bethesda MD 20205

ATWOOD, CHARLES LEROY, computer services executive; b. Decatur, Ill., Dec. 24, 1951; s. Basil Leroy and Ruth Caroline (Peters) A.; m. Carla Sue Barber, Feb. 19, 1952; children: Eric Charles, Alison Suzanne. B.S., U. Ill.-Champaign-Urbana, 1974, M.Arch., 1976. Cert. in data processing. Assoc., asst. dir. computer service Skidmore, Owings & Merrill, Chgo., 1976-81; v.p., dir. computer service Hellmuth, Obata & Kassabaum, St. Louis, 1981—. Recipient James M. White award U. Ill., 1974; Plym fellow, 1975-76. Mem. AIA, Assn. Computing Machinery, Digital Equipment Computer Users Soc., Data Processing Mgmt. Assn., Soc. Computer Applications in Engring. Planning and Architecture. Subspecialties: Computer applications in engineering; Architectural computer-aided design. Current work: Development of HOK systems for computer-aided design and drafting to serve architectural/engineering/interior applications. Office: Hellmuth, Obata & Kassabaum 100 N Broadway Saint Louis MO 63102

ATWOOD, DONALD JESSE, JR., automotive executive; b. Haverhill, Mass., May 25, 1924; s. Donald Jesse and Doris Albertine (French) A.; m. Curina Harian, Sept. 8, 1946; children: Susan Albertine, Donald Jesse. B.S.E.E., MIT, 1948, M.S.E.E., 1950. With AC Electronics div. Gen. Motors Corp., 1961-70, dir. ops., Milw., 1968-70; mgr. Indpls. ops. Detroit Diesel Allison div. (Gen. Motors Corp.), 1970-73, 1st gen. mgr. Transp. Systems div., 1973-74, gen. mgr. Delco Electronics div., Kokomo, Ind., 1974-78, gen. mgr., Detroit, 1978-80; v.p., group exec. Gen. Motors Corp., 1981—; dir. Charles Stark Draper Lab. Corp. Bd. dirs. Automotive Hall of Fame, Inc., Western Hwy. Inst. Served with AUS, 1943-46. Mem. Am. Helicopter Assn., Soc. Automotive Engrs., Motor Vehicle Mfrs. Assn. (policy com.), Nat. Acad. Engring., AIAA, Air Force Assn., Assn. U.S. Army, Navy League. Subspecialties: Electrical engineering; Electronics. Home: Franklin MI 48025 Office: 31 Judson St Pontiac MI 48058

ATWOOD, JIM D., chemistry educator; b. Springfield, Mo., June 3, 1950; s. Harvey D. and F. Louise (Young) A.; m. Mary L. Nicholson; children: Angela, Amanda. B.S., S.W. Mo. State U., Springfield, 1971; Ph.D., U. Ill., 1975. Mem. U.S.-USSR Program Cooperation in Catalysis, Cornell U., 1975-76; asst. prof. chemistry SUNY-Buffalo, 1977-81, assoc. prof., 1981—. Sloan fellow, 1983; Theron S. Piper fellow, 1975. Mem. Am. Chem. Soc. Subspecialty: Inorganic chemistry. Current work: Primary interest is in mechanistic organometallic chemistry; including studies of stoichiometric and homogeneously catalyzed reactions. Home: 80 Greentree Tonawanda NY 14150 Office: Dept Chemistry State Univ New York Buffalo NY 14214

ATWOOD, JOAN DOLORES, psychology educator, psychotherapist, researcher, counselor; b. Queens, N.Y., Sept. 15, 1943; d. Louis and Helen (Juarez) Armagno; m. William R. Atwood, Mar. 21, 1981; children: Debby, Barbara, Lisa, Janine, Brian. B.A. in Psychology magna cum laude, SUNY-Stony Brook, 1973, M.A., 1975, M.A. in Sociology, 1977, Ph.D., 1981. Research assoc. dept. psychology SUNY-Stony Brook, 1973-75, dept. psychiatry, 1975-80; asst. prof. psychology SUNY-Farmingdale, 1980—; cons. Biofeedback Services, 1975—; pvt. practice counseling, 1975—, workshop/group coordinator, 1975—. Author: Making Contact with Human Sexuality, 1981. Mem. Citizens Alliance, Nassau, N.Y., 1980—; mem. task force NOW, Nassau, 1981—. Fellow Internat. Council Sex Edn. and Parenthood; mem. Am. Psychol. Assn., Am. Assn. Sex Educators, Counselors, and Therapists, Sex Info. and Edn. Council U.S. (assoc.), Am. Assn. Marriage and Family Therapists (clin.), Environ. Control Assn., Phi Alpha Sigma. Subspecialties: Social psychology; Developmental psychology. Current work: Marriage, family and sex therapy; sexual behavior; social relationships. Home: 542 Lakeview Ave Rockville Centre NY 11570 Office: SUNY Route 110 Farmingdale NY 11375

AUBERT, EUGENE JAMES, environ. lab. exec.; b. North Bergen, N.J., Mar. 6, 1921; s. Charles Fernan and Agnes Catherine (Robertson) A.; m. Dorothy M. Stephens, Nov. 7, 1942; children—Donald E., Patricia M., Allan C., Richard S. Student, Montclair State Coll., 1938-41; B.S., N.Y. U., 1946, M.S., 1947; Ph.D., M.I.T., 1957. Chief technique applications Meteorology Lab., Geophys. Research Directorate, Bedford, Mass., 1951-60; v.p., dir. atmospheric and oceanographic dept. Travelers Research Center, Hartford, Conn., 1960-70; dir. Internat. Field Year for Gt. Lakes, 1971-79, Gt. Lakes Environ. Research Lab., Ann Arbor, Mich., 1974—; dir. Travelers Research Corp.; mem. Gt. Lakes Basin Commn. Author: Study of the Feasibility of National Data Buoy Systems, 1967, Environmental Prediction, 1968, The International Field Year for the Great Lakes-U.S. Viewpoint, 1972, Great Lakes, 1976, The Relevance of IFYGL, 1978. Served with USN, 1942-46. Recipient Gold medal Dept. Commerce, 1973. Methodist. Club: North Cape Yacht. Subspecialty: Environmental research management. Office: NOAA/Gt Lakes Environ Research Lab 2300 Washtenaw Ave Ann Arbor MI 48104

AUBIN, WILLIAM M., aerospace company executive; b. Detroit, Dec. 17, 1929; s. Hector M. and Alice C. (Nittinger) A.; m. Joyce N. Aubin, Apr. 26, 1952; children: Mark, Julie, Denise, Brian, Bruce, Allison, Elaine. B.Aerospace Engring., U. Detroit, 1953; M.S., Adelphi Coll., 1956. With Grumman Aerospace Corp., Bethpage, N.Y., 1953—, dir. materials and structural mechanics, until 1982, asst. to gen. mgr. product devel. ctr., 1982—. Vice-pres. West Islip (N.Y.) Library Bd., 1958-62. Mem. AIAA. Republican. Roman Catholic. Subspecialties: Aerospace engineering and technology; Database systems. Current work: Computer aided design and manufacturing; establishing a quick response development center for one-of-a-kind hardware. Office: Grumman Aerospace Corp Mail Stop C44/005 Bethpage NY 11714

AUBREY, ROGER FREDERICK, psychology educator; b. Waterloo, Iowa, Nov. 1, 1929; s. Earl Folsom and Ruth Marguerite (Schminke) A.; m. Dixie Lee Cook, Mar. 15, 1963; children: Joshua, David, Christopher. A.B., U. Miami, 1954; M.A., U. Chgo., 1964; Ed.D., Boston U., 1975. Tchr. social studies East Moline (Ill.) High Sch., 1959-62; psychologist Riverdale (Ill.) Sch. Dist., 1963-64; dir. counseling U. Chgo., 1964-69; dir. guidance and health Brookline (Mass.) Pub. Schs., 1969-77; prof. Vanderbilt U., 1977—; cons. numerous pub., pvt. sch. systems, colls., univs., test pubs., hosps., clinics. Author: Experimenting with Living: Pros and Cons, 1973, The School Counselor and Drug Abuse, 1973, Career and Occupational Development of Thirteen-Year-Olds, 1978; contbr. numerous articles to profl. jours., chpts. to books; co-editor: The Practice of Guidance, 1972, Guidance: Strategies and Techniques, 1975. Served with U.S. Army, 1950-51. Mem. Am. Personnel and Guidance Assn. (senator 1974), Assn. Counselor Edn. and Supervision (pres. 1973-74), Am. Psychol. Assn., Am. Ednl. Research Assn. Democrat. Unitarian. Subspecialty: Developmental psychology. Current work: Life-span developmental psychology; social issues of the future; historical analysis of counseling, mental fields, developmental education. Home:

6304 Torrington Rd Nashville TN 37205 Office: Peabody Coll Vanderbilt U Box 52 Nashville TN 37203

AUDET, JOHN JAMES, JR., physical scientist, oceanographer, marine data official; b. Ridgway, Pa., July 2, 1937; s. John James and Doris Ellen (Goldy) A.; m. Mary Ellen Daly, Oct. 20, 1962; children: Daniel James, Sharon Elaine, Gregory John. Student, Pa. State U., 1957-58; A.B. in Phys. Sci, Lycoming Coll., 1960; postgrad., George Washington U., 1962-72; M.S. in Sci. and Tech, Am. U., 1975. Geodesist Naval Hydrographic, Suitland, Md., 1961-62; oceanographer Naval Oceanographic, Suitland, 1962-72; staff oceanographer Office Naval Research, Washington, 1972-75; phys. scientist Nat. Oceanographic Data Ctr./NOAA, Washington, 1975—; mgr. Central Coordination and Referral Office, Ocean Pollution Data and Info. Network, Washington, 1981—. Contbr. writings in field to profl. publs. PTA pres. Kettering Elem. Sch., Largo, Md., 1978; coach Kettering Boys/Girls Club, Largo, 1974-77; sci. fair judge Prince Georges High Schs., 1977—; council mem. Holy Family Parish, Mitchellville, Md., 1981-82. Served with U.S. Army, 1960-61. Recipient silver medal Dept. Commerce, 1979, alumni outstanding achievement award Lycoming Coll., 1981. Mem. Marine Tech. Soc., Coastal Soc., Nat. Wildlife Fedn., Nat. Resources Def. Council, Internat. Oceanographic Found. Democrat. Roman Catholic. Lodge: K.C. Subspecialties: Information systems (information science); Distributed systems and networks. Current work: Provide coordination of marine pollution data and information activities for involved federal agency and non-federal users of the network. Office: Nat. Oceanographic Data Center/NOAA/Dept Commerce Page Bldg 1 2001 Wisconsin Ave Washington DC 20235

AUDETTE, LOUIS GIRARD, II, energy co. exec.; b. Orange, N.J., Sept. 24, 1939; s. Charles LaPointe and Mary Ford (Haggart) A.; m. Anna Brita Held, Aug. 15, 1964; children: Jessie, Alexis. B.A., Yale U., 1962, B.F.A., 1963, M.S., 1965. Instr. anatomy and dir. med. TV Yale U., New Haven, 1964-69; dir. dept. biomed. communications U. Conn. Health Ctr., Farmington, 1969-78; regional library cons. for New Eng. Nat. Library of Medicine, Bethesda, Md., 1977-79; dir. learning resources ctr. Project HOPE, Alexandria, Egypt, 1977-78; founder, pres. New Eng. Alt. Fuels, Inc., Brattleboro, Vt., 1979—; adv. Vt. Energy Office, N.H. Gov.'s Council on Energy. Chmn. Brattleboro Alternative Energy Com., 1980; mem. Monadnock Energy Project. Dept. of Energy appropriate tech. grantee, 1980-81. Subspecialties: Fuels and sources; Combustion processes. Current work: Development of small scale modular power plants using non-traditional fuels such as cellulosic waste, landfill gas, wood chips. Office: 67 Main St Brattleboro VT 05301

AUE, WALTER ALOIS, chemistry educator; b. Vienna, Austria, Jan. 20, 1935. Ph.D., U. Vienna, 1963. Research investigator organic chemistry Western Res. U., 1963-65; asst. prof. analytical chemistry U. Mo.-Columbia, 1965-69, assoc. prof. chemistry, research assoc., 1969-73; prof. chemistry Dalhousie U., Halifax, N.S., Can., 1973—. U.S. Dept. Agr. grantee, 1967-70; USPHS grantee, 1967-70; NSF grantee, 1970-72; EPA grantee, 1972-75; NRC Can. grantee, 1973—; Can. Def. Research Bd. grantee, 1974-76; Agr. Can. grantee, 1974-79; Environ. Can. grantee, 1978-80; recipient Fisher award, Can., 1980. Subspecialty: Analytical chemistry. Office: Dept Chemistr Dalhousie U Halifax NS Canada B3H 3J5

AUER, MARTIN TUCKER, engineering educator; b. Syracuse, N.Y., Sept. 14, 1948; s. Martin Swift and Wilma (Tucker) A.; m. Nancy Ann Arnold, Dec. 29, 1973. B.S., SUNY-Syracuse, 1972; M.S., Syracuse U., 1973; Ph.D., U. Mich., 1979. Research engr. U. Mich., Ann Arbor, 1979-81; asst. prof. environ. engring. Mich. Tech. U., Houghton, 1981—; cons. LTI, Inc., 1979—, Environ. Engring., Syracuse, 1982. Spl. editor: Jour. Great Lakes Research, 1982; contbr. articles to profl. jours.; author: Limnology of Grand, 1976. Leader Boy Scouts Am., Syracuse, 1968-72; asst. coach Youth Hockey Program, Ann Arbor, 1981. ; EPA grantee, 1979, 82. Mem. Water Pollution Control Fedn., ASCE, Internat. Assn. Great Lakes Research, Assn. Environ. Engring. Profs., Am. Soc. Limnology and Oceanography, Internat. Limnological Soc. Subspecialties: Environmental engineering; Ecosystems analysis. Current work: Math. modeling and exptl. study of impacts of anthropogenic perturbation on aquatic systems; algal ecology; limnology. Home: 723 Green Acres Rd Houghton MI 49931 Office: Mich Tech Univ Dept Civil Engring Houghton MI 49931

AUER, PETER LOUIS, plasma physicist, educator; b. Budapest, Hungary, Jan. 12, 1928; came to U.S., 1937, naturalized, 1942; s. Laszlo and Irma (Morgenstern) A.; m. Rheta E. Siegel, Aug. 27, 1952; children—Deborah, Douglas, Andrea, Matthew. A.B., Cornell U., 1947; Ph.D., Calif. Inst. Tech., 1951. Physicist Gen. Electric Research Lab., Schenectady, 1954-61; head plasma physics Sperry Rand Research Center, Sudbury, Mass., 1961-64; dir. Ballistic Missile Def., Office Sec. Def., Washington, 1964-66; prof. aerospace engring. Cornell U., Ithaca, N.Y., 1966—, dir. lab. plasma studies, 1967-74; cons. Office Sec. Def., Gen. Electric Co., Electric Power Research Inst., AEC, NRC, Nat. Acad. Scis., Dept. of Energy, Inst. for Energy Analysis; vis. scientist Frascati, Italy, 1960-61; vis. prof. Oxford U., 1972-73. Editor: Plasma Physics, 1970; asso. editor: Energy, 1976; Contbr. articles to profl. jours. Guggenheim fellow, 1960-61. Fellow Am. Phys. Soc. Subspecialties: Plasma physics; Fusion. Current work: Plasma physics and fusion reactor studies. Pantentee in field. Home: 220 Devon Rd Ithaca NY 14850

AUERBACH, ISAAC LEVIN, computer scientist, publisher, cons.; b. Phila., Oct. 9, 1921; s. Philip and Rose (Levin) A.; m.; 1 son, Philip B. B.S. in Elec. Engring, Drexel U., 1943; M.S. in Applied Physics, Harvard U., 1947. Research engr. Sperry Univac, 1947-48; dir. spl. products div. Burroughs Corp., 1949-57; chmn. bd., pres. Auerbach Corp. for Sci. and Tech., Phila., 1957—; pres. Auerbach Assocs., Inc., 1957-76, Auerbach Cons. (cons.), 1976—; chmn. Auerbach Pubs., Inc., 1960—; pub. Auerbach Computer Tech Reports; treas., dir. The Baupost Group Inc., 1982—; dir. The Software Group Inc., 1983—; U.S. cons. on info. processing and automation UNESCO, 1957-60; chmn. U.S. com. 1st Internat. Conf. Info. Processing, 1959; founder, 1st pres. Internat. Fedn. for Info. Processing, 1960-65, hon. life mem., 1969; mem. Nat. Planning Assn. Editor: The Auerbach Annual-Best Computer Papers, 1971-74, 79-80. Trustee Fedn. Jewish Agys.; internat. bd. govs. Boy's Town Jerusalem; bd. dirs. Jewish Publ. Soc., 1966-73; assoc. trustee, bd. govs. Drexel U. Served as lt. (j.g.) USNR, 1943-46. Recipient Grand medal City of Paris, 1959, alumni citation Drexel U., 1961, Tower of David award State of Israel, 1969. Fellow IEEE (Phila. sect. award 1961), AAAS; distinguished fellow Brit. Computer Soc.; mem. Nat. Acad. Engring. (publs. com. 1983), Franklin Inst., Japan Computer Soc. (hon. fellow), Nat. Acad. Sci. (com. inter sci. and tech. info. programs), Am. Friends Hebrew U., Sigma Xi, Eta Kappa Nu.; Mem. B'nai B'rith. Current work: Computer, communications and management consultant, covering all phases of computer science, data processing and information systems. Pioneered devel., design and use computers and digital communications systems; dir. devel. 1st ICBM guidance computer. Home: 900 Centennial Rd Narberth PA 19072 Office: 455 Righters Mill Rd Narberth PA 19072

AUERBACH, OSCAR, pathologist, educator; b. N.Y.C., Jan. 2, 1905; s. Max and Jennie (Geller) A.; m. Dora Herman, Mar. 20, 1932; children: Richard C., Bruce E. B.S., N.Y.U., 1925; M.D., N.Y. Med. Coll., 1929. Diplomate: Am. Bd. Pathology. Intern Morrisania Hosp., N.Y.C., 1929-31; research fellow U. Vienna, 1931-32; pathologist Sea View Hosp., S.I., N.Y., 1932-47; chief lab. VA Hosp., Staten Island, 1947-51, East Orange, N.J., 1952-59, sr. med. investigator, 1960-78; disting. physician VA Med. Center, East Orange, 1978-83; prof. pathology U. Medicine and Dentistry, Newark, N.J., 1966-69, 71—, asst. dean student affairs, 1980-83; vis. instr. Washington U. Med. Sch., St. Louis, 1944; assoc. prof. pathology N.Y. Med. Coll., 1949-61, prof., 1962-71; cons. pathologist Richmond Gen. Hosp., S.I., 1938-47, VA Hosp., Castle Point, N.Y., 1946-47, East Orange Gen. Hosp., 1962-68, St. Vincent's Med. Center, S.I., 1965-78, St. Barnabas Med. Center, Livingston, N.J., 1973—. Contbr. articles in field to profl. jours. Served to lt. M.C. USN, 1944-46. Named Sr. Med. Investigator VA, 1960; recipient Career Service award Nat. Civil Service League, 1963, cert. of Commendation Adminstrn. Vets. Affairs, 1963, Selman Waksman award Am. Coll. Chest Physicians, 1968, Miriam Goldberg Levin Meml. award, 1970, Edward J. Ill award Acad. Medicine N.J., 1971, Golden Apple award N.J. Med. Sch. Student AMA, 1973, Alumni Sci. medal N.Y. Med. Coll., 1973, Spl. award Coll. Medicine and Dentistry N.J., N.J. Med. Sch., 1978, Trudeau medal Am. Lung Assn., 1979, Dennis J. Sullivan award of merit N.J. Pub. Health Assn., 1979, Exceptional Merit award Pres. Coll. Medicine and Dentistry N.J., 1981, Disting. Scientist award Grad. Student Assn. Coll. Medicine and Dentistry N.J., 1981, Golden Apple award N.J. Med. Sch. Student Council, 1982. Fellow ACP; mem. Internat. Assn. Study Lung Cancer, AMA, Am. Assn. Pathologists and Bacteriologists, Am. Trudeau Soc., Assn. Am. Med. Colls., Am. Assn. Thoracic Surgeons, Mexican Soc. Study Tb and Chest Diseases (hon.), Soc. Pathologists and Chilean Soc. Study Chest Diseases (hon.), N.Y. Acad. Medicine, Am. Lung Assn. (hon., life), Cuban Soc. Anatomical Pathologists (hon.), Sigma Xi. Subspecialties: Pathology (medicine); Cancer research (medicine).

AUERBACH, VICTOR HUGO, biochemist; b. N.Y.C., Oct. 2, 1928; s. Leo and Goldie (Ratner) A.; m. Helen Matalas, June 1, 1956. A.B., Columbia U., 1951, postgrad., 1951-52; A.M., Harvard U., 1955, Ph.D., 1957. Instr. U. Wis. Med. Sch., Madison, 1957-58; asst. prof. Temple U. Sch. Medicine, Phila., 1958-64, assoc. prof., 1964-68, research prof. in pediatrics (biochemistry), 1968—; dir. enzyme lab. St. Christopher's Hosp. for Children, Phila., 1958—, dir. clin. labs., 1976—. Contbr. chpts. to books, articles to profl. jours. Eli Lilly & Co. fellow, 1952-53; Arthur Lehman fellow, 1953-54; Andelot fellow, 1954-55; Arthritis and Rheumatism Found. fellow, 1954-56; NIH grantee, 1958-80. Mem. Biochem. Soc. (London), Am. Assn. Cancer Research, Am. Soc. Biol. Chemists, Am. Soc. Human Genetics, Soc. Pediatric Research, Am. Soc. Clin. Nutrition, Am. Assn. Clin. Chemists, Nat. Acad. Clin. Biochemistry, others. Subspecialties: Clinical chemistry; Biochemistry (medicine). Current work: Biochemistry of various genetic diseases of man as they occur in children and infants. Home: 1244 Hoffman Rd Ambler PA 19002 Office: St Christopher's Hosp for Children 2600 N Lawrence St Philadelphia PA 19133

AUGENSEN, HARRY JOHN, astronomer, physicist, educator; b. Chgo., July 18, 1951; s. Harry Clarence and Margaret Barbara (Schramm) A.; m. Anna DiEgidio, Oct. 17 1981. B.A., Elmhurst Coll., 1973; M.S. in Astronomy, Northwestern U., 1974, Ph.D., 1978. Lectr. in physics and astronomy Northwestern U., 1978-80; lectr. in astronomy Swarthmore Coll., 1980-82, Widener U., 1981, asst. prof. physics and astronomy, 1982—; coordinator Astro-Sci. Workshop, Chgo., 1979-80; lectr. Harlow Shapley Vis. Lectureship Series at Colls. and Univs.; summer faculty fellow NASA-Goddard Space Flight Center, 1982; vis. observer Cerro Tololo Obs., 1976, 77, 81. Contbr. articles to profl. jours. Recipient Small Grant award Am. Astron. Soc., 1981. Mem. Am. Astron. Soc., Royal Astron. Soc., Astron. Soc. Pacific, Sigma Xi. Presbyterian. Subspecialties: Optical astronomy; Ultraviolet high energy astrophysics. Current work: Astronomy edn.; planetary nebulae; high-velocity stars; Internat. Ultraviolet Explorer satellite observations of central stars of planetary nebulae; optical spectroscopy of central stars. Home: 217 Plush Mill Rd Wallingford PA 19086 Office: Kirkbridge Hall Widener U Chester PA 19013

AUGUST, GILBERT PAUL, pediatric endocrinologist; b. Jersey City, Sept. 18, 1936; m. Bernice Ide, July 3, 1960; children: Sharon, Lauren. B.S., CCNY, 1958; M.D., NYU, 1962. Diplomate: Am. Bd. Pediatrics. Intern Bellevue Hosp., N.Y.C., 1962-63, resident in pediatrics, 1963-65; endocrinologist Children's Hosp. of D.C., Washington, 1969—; mem. cert. subcom. for pediatric endocrinology Am. Bd. Pediatrics, 1977-81. Author: Pediatric Endocrinology, 1983. Served as surgeon USPHS, 1965-67. Mem. Endocrine Soc., Lawson Wilkins Pediatric Endocrine Soc. Subspecialties: Pediatrics; Endocrinology. Current work: Somatomedin and MSA receptors. Office: Children's Hospital National Medical Center 111 Michigan Ave NW Washington DC 20010

AUKRUST, EGIL, metallurgist; b. Lom, Norway, June 16, 1933; s. Paal and Torlang (Borresen) A.; m. Rose, Mar. 23, 1962. Diploma Engr., Tech. U. Norway, 1957, Dr. Ing., 1960. Research assoc. Pa. State U., University Park, 1960-62; research engr. Jones & Laughlin Steel Corp., Pitts., 1962-65, research supr., 1965-69, asst. dir. process metallurgy, 1969-71, dir. process metallurgy, 1971-74, gen. mgr. research, 1974-81, sr. dir. tech., 1981—. Contbr. articles to profl. jours. Served with Norwegian Army, 1952-53. Mem. Am. Iron and Steel Inst., Indsl. Research Inst., Metals Soc. (U.K.), AIME, Assn. Iron and Steel Engrs., Am. Soc. Metals, Engrs. Soc. Western Pa. Subspecialty: Metallurgical engineering. Office: J & L Steel Corp Research Lab 900 Agnew Rd Pittsburgh PA 15227

AULENBACH, DONALD BRUCE, environmental engineering educator, environmental engineering consultant; b. Berwick, Pa., Mar. 7, 1928; s. Henry Israel and Mildred Clara (Schlasman) A.; m. Marie Pauline Wertz, Aug. 16, 1952; children: Louise M. Trakimas, Bruce D., Nancy J. Baker, Brent T. B.S., Franklin and Marshall Coll., 1950; M.S., Rutgers U., 1952, Ph.D., 1954. Registered profl. engr., N.Y. Chemist-bacteriologist Del. Water Pollution Commn., Dover, 1954-60; asst. prof. Rensselaer Poly. Inst., 1960-65, assoc. prof. environ. engring., 1965-73, prof., 1973—; cons. Gen. Electric Co., 1965—, Kinderhook Lake Assn., Niverville, N.Y., 1982, Travelers Ins. Co., Albany, 1981—; project mgr. Environ. Technol. Group, Waterford, N.Y., 1981—. Author: Register of Undergraduate Programs in Environmental Engineering, 1977. Pres. Sand Lake Council Chs., 1965-66, West Sand Lake PTA, 1965-66. Mem. Water Pollution Control Fedn. (dir. 1977-79), Am. Chem. Soc., Am. Water Works Assn., Assn. Environ. Engring. Profls., Nat. Soc. Profl. Engrs., Sigma Xi. Republican. Mem. Reformed Ch. (elder, 1973—). Subspecialties: Environmental engineering; Water supply and wastewater treatment. Current work: Removal of heavy metals in wastewaters using sodium aluminate; nutrient budgets in lakes; treatment of sewage by rapid infiltration into sand. Home: 24 Valencia Ln Clifton Park NY 12065 Office: Dept Environmental Engring Rensselaer Poly Inst Troy NY 12181

AULICK, CHARLES MARK, mathematics and computer science educator; b. Vero Beach, Fla., July 25, 1952; s. Donald Loraine Garrard. and Frances Louise (Smith) A. Student, Fla. Technol. U., 1970-72; B.S., Stetson U., 1975; M.S., Fla. State U., 1977; Ph.D., Duke U., 1981. Teaching asst. Fla. State U., 1976-77; instr. Stetson U., 1977-78; grad. asst. Duke U., 1978-81; asst. prof. math. and computer sci. La. State U.-Shreveport, 1981—. Fla. State U. Math. Dept. univ. fellow, 1975. Mem. Soc. Indsl. and Applied Math., Assn. Computing Machinery, Omicron Delta Kappa. Club: Baptist Student Union (faculty adv. 1977-78, 82—). Subspecialty: Numerical analysis. Current work: Ill-posed problems; error analysis for ill-conditioned linear systems. Home: PO Box 5861 Shreveport LA 71135 Office: Dept Math and Computer Sci 8515 Youree Dr Shreveport LA 71115

AULL, JOHN LOUIS, biochemist, educator, researcher; b. Newberry, S.C., May 7, 1939; s. Louis Eugene and Helen (Lomineck) A.; m. Judy Capps, June 1, 1963; children: Amber Kristina, Ashley Caroline. A.B. in Chemistry, U. N.C., Chapel Hill, 1964; Ph.D. in Biochemistry, N.C. State U., 1972. Postdoctoral research asso. U. S.C., 1973; asst. prof. chemistry Auburn U., 1974-80, assoc. prof., 1980—. Contbr. articles to profl. jours. Served to lt. USN, 1964-67. Decorated Nat. Service medal.; NIH grantee, 1979—. Mem. Am. Soc. Biol. Chemists, Am. Chem. Soc. (chmn. local sect.). Episcopalian. Subspecialties: Biochemistry (biology); Nuclear magnetic resonance (biotechnology). Current work: Biochemistry-enzymology; enzyme inhibition; nuclear magnetic resonance studies of enzymes. Home: 1029 Cumberland Dr Auburn AL 36830 Office: 305 Saunders Hall Dept Chemistry Auburn U Auburn AL 36849

AULT, ADDISON, chemist, educator; b. Boston, July 3, 1933; s. Warren Ortman and Myrtle Lavina (Wilcock) A.; m. Janet Ruth Meade, Aug. 23, 1958; children: Margaret Ruth, Warren James, Addison David, Peter Harwell, Emily Elizabeth. B.A., Amherst Coll., 1955; Ph.D., Harvard U., 1960. Asst. prof. Grinnell Coll., Iowa, 1959-61; research asso. Argonne (Ill.) Nat. Lab., 1961-62; prof. chemistry Cornell Coll., Mt. Vernon, Iowa, 1962—. Author: Problems in Organic Structure Determination, 1967, Techniques and Experiments for Organic Chemistry, 1973, 2d edit., 1976, 3d edit., 1980, 4th edit., 1983, (with G.O. Dudek) NMR: An Introduction to Proton Nuclear Magnetic Resonance Spectroscopy, 1976, (with Margaret Ault) A Handy and Systematic Catalog of NMR Spectra, 1980. NSF Sci. Faculty fellow, 1967. Mem. Am. Chem. Soc., Chem. Soc. (London), AAAS, AAUP, Phi Beta Kappa, Sigma Xi. Subspecialty: Organic chemistry. Current work: Stereochemistry; nuclear magnetic resonance; enzyme kinetics. Home: 519 N 2d St W Mount Vernon IA 52314 Office: Cornell Coll Mount Vernon IA 52314

AULT, JEFFREY GEORGE, cell biologist; b. Fargo, N.D., June 18, 1953; s. Aldon G. and Delores L. (Dittus) A. B.A. in Zoology, N.D. State U., 1975, M.S. in Agronomy, 1977; Ph.D. in Zoology, Ariz. State U., 1981. Research fellow dept. zoology Duke U., Durham, N.C., 1981—. Contbr. articles to profl. jours. Charles W. Hargitt fellow, 1982—. Mem. Am. Soc. for Cell Biology, Ariz.-Nev. Acad. Sci., Ariz. Soc. for Electron Microscopy and Microbeam Analysis, Electron Microscopy Soc. Am., Genetics Soc. Am., Sigma Xi. Subspecialty: Cell biology. Current work: Ultrastructural mechanism of chromosome movement utilizing the combined techniques of cine analysis and serial section electron microscopy. Home: 2752 Middleton St 31 N Holly Hills Durham NC 27705 Office: Dept Zoology Duke U Durham NC 27706

AUNE, THOMAS MARTIN, immunologist, educator; b. Winona, Minn., Apr. 23, 1951; s. Henrik Joakim and Edna Mae (Crisp) A.; m. Patricia Ann Hurt, Dec. 9, 1978. B.S., Southwestern U. of Memphis, 1973; Ph.D., U. Tenn., 1976. Postdoctoral fellow Stanford U., 1977-78; postdoctoral fellow Washington U., St. Louis, 1979-80, instr., 1981-82, asst. prof. dept. pathology, 1982—. Recipient Jr. Faculty Research award Am. Cancer Soc., 1982-85; NSF grantee, 1982-85; Council Tobacco Research grantee, 1982-84. Mem. Am. Assn. Immunologists, Am. Assn. Pathologists. Subspecialties: Immunology (medicine); Biochemistry (medicine). Current work: Regulation of cell proliferation and immune responses by non-antibody lumphocyte products. Home: 650E Adams Kirkwood MO 63122 Office: Dept Pathology Jewish Hosp 216 S Kingshighway Saint Louis MO 63110

AUPPERLE, ERIC MAX, computer network management executive, computer engineering educator; b. Batavia, N.Y., Apr. 14, 1935; s. Max Karl and Hedwig Elise Helen (Haas) A.; m. Nancy Ann Jach, June 21, 1958; children: Bryan, Lisa. B.S.E.E., U. Mich., Ann Arbor, 1957, 1957, M.S.E. in Nuclear Engring, 1958, Instm.E., 1964. Registered profl. engr., Mich. With U. Mich., Ann Arbor, 1957—, asst. research engr., 1963-67, research engr., 1967-69, 1969-74, research scientist, 1974—, project leader, 1969-73, assoc. dir., 1973-74, dir., 1974—, assoc. dir. communications, 1981—; lectr. Wayne State U., 1975, U. Mich.-Dearborn, 1972; cons. to industry; panelist Instructional Sci. Equipment Program NSF, 1978; program com. Nat. Electronics Conf., 1962-65, 70-73. Sr. mem. IEEE (dir. Southeastern Mich. sect. 1963-64, 65-67), Acacia Fraternity. Subspecialties: Computer engineering; Distributed systems and networks. Current work: Engineering packet switching computer networks. Patentee in field. Office: U. Mich 5115 IST Bldg Ann Arbor MI 48109

AUPPERLE, KENNETH ROBERT, computer architecture specialist; b. Bklyn., Aug. 27, 1957; s. Robert Wolfgang and Margarete Elisabeth (Fischer) A.; m. Laura Jean Marie Welch, Aug. 8, 1981. B.S. in Elec. Engring, Poly. Inst. N.Y., 1979, M.S. in Computer Sci, 1979. Regional computer architecture specialist Intel Corp., Hauppauge, N.Y., 1979—; adj. lectr. Poly. Inst. N.Y., Farmingdale, 1979—. Mem. IEEE, Assn. Computing Machinery, Tau Beta Pi. Lutheran. Subspecialties: Computer engineering; Computer architecture. Current work: Software and system architecture considerations in microprocessor systems. Office: 300 Vanderbilt Motor Pkwy Hauppauge NY 11787

AURAND, LEONARD WILLIAM, food chemist, researcher, educator; b. Shamokin Dam, Pa., Feb. 5, 1920; s. James Wilson and Esther Matilda (Weissinger) A.; m. Eleanor May Nichols, Feb. 22, 1943; children: Rebecca Louise Aurand Newton, Thomas James, Sarah Jane Aurand Anderson. B.S., Pa. State U., 1941, Ph.D., 1949; M.S., U. N.H., 1947. Asst. prof. N.C. State U., 1949-55, assoc. prof., 1955-60, prof. food sci. and biochemistry, 1960—. Author: Food Composition and Analysis, 1963, Food Chemistry, 1973, Laboratory Manual in Food Chemistry, 1977; contbr. numerous articles to profl. jours. Cert. hunger interpreter Bd. Global Ministeries, United Methodist Ch., N.Y.C., and, N.C. Conf. United methodist Ch. Served with USNR, 1942-46. Mem. Am. Inst. Nutrition, Inst. Food Technologists. Democrat. Lodge: Masons. Subspecialties: Food science and technology; Nutrition (biology). Current work: Lipid oxidation in foods and problems of hunger in U.S. and Third World nations. Home: 921 Trailwood Dr Raleigh NC 27606 Office: 236 Schaub Hall NC State U Raleigh NC 27650

AUSICH, WILLIAM IRL, geology educator; b. Kewanee, Ill., Feb. 2, 1952; s. Anton and Sarah Margaret (Tubbs) A.; m. Regina Sharlene Dolk, Aug. 4, 1973; children: Elizabeth Claire, Arlene Elejandria. B.S., U. Ill., 1974; A.M., Ind. U., 1976, Ph.D., 1978. Assoc. instr. Ind.-U.-Bloomington, 1974-78, instr., summer 1978; asst. prof. geology Wright State U., Dayton, Ohio, 1978-82, assoc. prof., 1982—. Contbr. numerous articles to profl. jours., chpts. to books. Mem. AAAS, Paleontol. Soc., Internat. Paleontol Assn., Geol. Soc. Am., Sigma Xi. Subspecialties: Paleobiology; Paleoecology. Current work: Paleoecology, evolution, taxonomy crinoids; community paleoecology. Office: Dept Geol Scis Wright State U Dayton OH 45435

AUST, STEVEN DOUGLAS, biochemistry and toxicology educator, researcher; b. South Bend, Wash., Mar. 11, 1938; s. Emil and Helen Mae (Crawford) A.; m. Nancy Lee Haworth, June 5, 1960 (dec.); children: Teresa, Brian; m. Ann Elizabeth Lacy, Feb. 4, 1972. B.S. in Agr., Wash. State U., 1960, M.S. in Nutrition, 1962; Ph.D. in Dairy Sci., U. Ill., 1965. Postdoctoral fellow dept. toxicology Karolinska Inst., Stockholm, 1966; N.Z. Facial Eczema sr. postdoctoral fellow Ruakura Agrl. Research Center, Hamilton, 1975-76; mem. faculty dept. biochemistry Mich. State U., East Lansing, 1967—, prof., 1977—, assoc. dir. Ctr. for Environ. Toxicology, 1980—; mem. toxicology study sect. NIH, 1979-83; mem. environ. measurements com., sci. adv. bd. EPA, 1980-83; mem. toxicology data bank, peer rev. com. Nat. Library of Medicine, 1983—; mem. Mich. Toxic Substance Control Commn., 1979-82, chmn., 1981-82. Contbr. numerous articles to sci. jours. Recipient Nat. Research Service award, 1966. Mem. Am. Soc. Biol. Chemists, Am. Soc. Pharmacology and Exptl. Therapeutics, Soc. Toxicology, Am. Soc. Photobiology. Subspecialties: Biochemistry (medicine); Toxicology (medicine). Current work: Toxicology of polyhalogenated aromatic hydrocarbons; mechanisms of lipid peroxidation. Office: Dept Biochemistry Mich State U East Lansing MI 48824

AUSTEN, K(ARL) FRANK, physician; b. Akron, Ohio, Mar. 14, 1928; s. Karl and Bertle (Jehle) A.; m. Joycelyn Chapman, Apr. 11, 1959; children: Leslie Marie, Karla Ann, Timothy Frank, Jonathan Arthur. A.B., Amherst Coll., 1950; M.D., Harvard U., 1954. Intern in medicine Mass. Gen. Hosp., 1954-55, asst. resident, 1955-56, sr. resident, 1958-59, chief resident, 1961, asst. in medicine, 1962-63, asst. physician, 1963-66, chief pulmonary unit, 1964-66, also cons. in medicine; practice medicine, specializing in internal medicine, allergy and immunology, Boston, 1962—; USPHS postdoctoral research fellow Nat. Inst. Med. Research, Mill Hill, London, 1959-61; asst. in medicine Harvard Med. Sch., 1960-1961, instr., 1961-62, asso. in medicine, 1962-64, asst. prof., 1965-66, asso. prof., 1966-68, prof., 1969-72, Theodore B. Bayles prof., 1972—; physician-in-chief Robert B. Brigham Hosp., 1966—; physician Peter Bent Brigham Hosp., 1966—; chmn. dept. rheumatology and immunology Brigham and Women's Hosp., 1980—; mem. fellowship subcom. Arthritis Found., 1968-71, chmn., 1971; mem. council Infectious Disease Soc. Am., 1969-71; mem. arthritis tng. grants com. Nat. Inst. Arthritis and Metabolic Diseases, NIH, 1970-73; mem. directing group, task force on immunology and disease Nat. Inst. Allergy and Infectious Diseases, 1972-73; bd. dirs. Arthritis Found., 1972—, mem. manpower study com., 1972-73, chmn. research com., 1972—, Med. Found., Inc., 1972—; mem. Am. Bd. Allergy and Immunology, 1973-78, Nat. Commn. on Arthritis and Related Musculoskeletal Diseases, 1975-76, Allergy and Immunology Research Commn., 1975—, chmn., 1976-79. Mem. editorial bd.: Arthritis and Rheumatism, 1968—, Proc. of Transplantation Soc, 1968—, Jour. Infectious Diseases, Jour. Exptl. Medicine, 1971—, Immunol. Communications, 1972—, Clin. Immunology and Immunopathology, 1972—, Proc. of Nat. Acad. Scis, 1978—, Clin. and Exptl. Immunology, 1978—, Immunopharmacology, 1979—, Receptors and Recognition, 1980—, Rheumatology Internat, 1980—, Clin. Immunology Revs, 1981—; contbr. articles to profl. jours. Trustee Amherst Coll., 1981—. Served to capt., M.C. U.S. Army, 1956-58. Mem. Nat. Acad. Scis. (chmn. sect. on med. microbiology and immunology 1983—), Am. Soc. Pharm. and Exptl. Therapeutics, Am. Soc. Exptl. Pathology, Am. Assn. Immunologists (pres. 1977-78), Brit. Soc. Immunology, Am. Soc. Clin. Investigation, Am. Rheumatism Assn., A.C.P., Transplantation Soc., Am. Acad. Arts and Scis., Assn. Am. Physicians (recorder 1978—), Am. Acad. Allergy (exec. com. 1970-72, sec. 1977-80, pres. 1981), Interurban Clin. Club, Fedn. Am. Soc. Exptl. Biology (dir. 1977—). Subspecialties: Immunobiology and immunology; Internal medicine. Current work: Molecular and cell biology, immunopharmacology of mediators of inflammation. Home: 34 Bradford Rd Wellesley Hills MA 02181 Office: Brigham and Women's Hosp 75 Francis St Boston MA 02115 also 250 Longwood Ave Boston MA 02115

AUSTIN, EDWARD MARVIN, mech. engr.; b. Rome, Ga., Nov. 15, 1933; s. Marvin Hart and Sarah Katherine (Youngblood) A.; m. Elizabeth Maria Geisz, Dec. 17, 1955; children: Jean, Diane, Judy. B.M.E., Ga. Inst. Tech., 1955, M.S. M.E., 1957. Registered profl. engr., N.C. Assoc. aircraft engr. Lockheed Aircraft Corp., Marietta, Ga., 1955; sr. engr. Safeguard System, Western Electric Co., Greensboro, N.C., 1968-71; project engr. Sandia Nat. Labs., Albuquerque, 1957—. Pres. Heights br. YMCA, Albuquerque, 1974-75. Mem. ASME, Nat. Soc. Profl. Engrs., Sigma Xi, Phi Eta Sigma, Pi Tau Sigma, Kappa Kappa Sigma, Tau Beta Pi, Phi Kappa Phi. Subspecialties: Systems engineering; Nuclear fission. Current work: Project engineer for nuclear warhead development. Home: 3017 Matador Dr NE Albuquerque NM 87111 Office: Sandia Nat Labs Box 5800 Albuquerque NM 87185

AUSTIN, GEOFFREY LEONARD, atmospheric physicist; b. London, Dec. 26, 1943; emigrated to Can., 1969, naturalized, 1978; s. Leonard Arthur and Louisa Maud (Read) A.; m. Lydia Huckett, Sept. 18, 1965; children: Barney, Kim, Nicola, Jeremy. B.A., Cambridge (Eng.) U., 1965; M.Sc., Canterbury U., N.Z., 1967, Ph.D., 1969. Lextr. Canterbury U., N.Z., 1968-69; asst. prof. physics McGill U., Montreal, Que., Can., 1969-73, assoc. prof., 1973-83, prof., 1983—; dir. McGill Weather Radar Obs., 1978—. Contbr. numerous sci. articles to profl. publs. Fellow Royal Meteorol. Soc.; mem. Can. Meteorol. Soc. Subspecialty: Meteorology. Home: PO Box 11 1452 Chemin du Fieuve Les Cedres PQ Canada JOP 1L0 Office: McGill Weather Radar Ob MacDonald Coll Ste Anne-de-Bellevu PQ Canada H9X 3M1

AUSTIN, SAM M., physicist, educator; b. Columbus, Wis., June 6, 1933; s. A. Wright and Mildred G. (Reinhard) A.; m. Mary E. Herb, Aug. 15, 1959; children: Laura Gail, Sara Kay. B.S., U. Wis., 1955, M.S., 1957, Ph.D., 1960. Research assoc. U. Wis., 1960; NSF postdoctoral fellow Oxford (Eng.) U., 1960-61; asst. prof. Stanford U., 1961-65; assoc. prof. physics Mich. State U., East Lansing, 1965-69, prof. physics, 1969—; assoc. dir. (Cyclotron Lab.), 1976-79, chmn. dept. physics and astronomy, 1980—; cons. NSF, Dept. Energy, Argonne Nat. Lab. Contbr. articles to profl. jours. Alfred P. Sloan fellow, 1963-66; recipient Sr. Research award Sigma Xi, 1977. Fellow Am. Phys. Soc. (vice chmn. div. nuclear physics 1981-82, Chmn. 1982-83); mem. Am. Assn. Physics Tchrs., AAAS. Subspecialties: Nuclear physics; Nuclear Astrophysics. Current work: Research (experimental) in nuclear structure and nuclear reactions; in nuclear astrophysics and nitrogen fixation. Office: Dept Physics and Astronomy Mich State U East Lansing MI 48824

AUSTIN, WILLIAM GEORGE, psychologist; b. Flint, Mich., Aug. 19, 1947; s. George Edward and Dorothy Grace (Ferguson) A.; m. Diane S. Hoyle, Jan. 9, 1977; children: Jason, Jennie, Benjamin. B.A., Western Mich. U., 1969; M.S., U. Wis.-Madison, 1972, Ph.D., 1974; postgrad., U. Va. Sch. Law, Charlottesville, 1977-79, cert., 1982. Asst. prof. psychology U. Va., Charlottesville, 1974-81; resident clin. psychology U. Va. Med. Ctr., Charlottesville, 1981-82; dir. psychol. services Southeastern Mental Health Ctr., Wilmington, N.C., 1982—. Editor: (with S. Worchel) The Social Psychology of Intergroup Relations, 1979; editorial bd.: Social Psychology Quarterly, 1979-83; contbr. articles in field to profl. jours. Bd. dirs. Child Advocacy Commn., Wilmington, N.C., 1982. NIMH grantee, 1975-76; Law Enforcement Assistance Adminstrn. and Inst. Justice grantee, 1976-78; Russell Sage Found. grantee, 1977-79; others. Mem. Am. Psychol. Assn., Am. Sociol. Assn., Am. Psychology and Law Soc., Soc. Exptl. Social Psychology. Democrat. Baptist. Subspecialties: Social psychology; Clinical and forensic psychology. Current work: Clinical and foresic psychology, forensic psychology, social psychology. Office: Southeastern Mental Health Center 2023 S 17th St Wilmington NC 28401

AUSTRIAN, ROBERT, physician; b. Balt., Apr. 12, 1916; s. Charles Robert and Florence (Hochschild) A.; m. Babette Friedmann, Dec. 29, 1963; stepchildren—Jill Bernstein, Toni Bernstein. A.B., Johns Hopkins U., 1937, M.D., 1941; D.Sc. honoris causa, Hahnemann Med. Coll., 1980, Phila. Coll. Pharmacy and Sci., 1981. Diplomate: Am. Bd. Internal Medicine. House officer Johns Hopkins Hosp., 1941-50, asst. dir. med. out-patient dept., 1951-52; assoc. prof. medicine, then prof. medicine State U. N.Y. Coll. Medicine, 1952-62; John Herr Musser prof., chmn. dept. research medicine U. Pa. Sch. Medicine, 1962—; attending physician Hosp. U. Pa.; Tyndale vis. lectr. and prof. Coll. Medicine U. Utah, 1964; spl. research infectious diseases, bacterial genetics; mem. com. Meningococcal Infractions Commn., 1964-72; spl. research infectious diseases, bacterial genetics; mem. commn. Acute Respiratory Disease, 1965-72; spl. research infectious diseases, bacterial genetics; mem. com. Commn. Streptoccal and Staphylococcal Diseases, 1970-72, Armed Forces Epidemiol. Bd.; cons. surg. gen. U.S. Army Research and Devel. Command, 1966-69; subcom. streptococcus and pneumococcus Internat. Com. Bacteriol. Nomenclature; mem. allergy and immunology study sect. Nat. Inst. Allergy and Infectious Diseases, 1965-69, mem. bd. sci. counselors, 1967-70, chmn., 1969-70. Mem. editorial bd.: Jour. Bacteriology, 1964-69, Am. Rev. Respiratory Diseases, 1963-66, Bacteriol. Rev, 1967-71, Jour. Infectious Diseases, 1969-74, Antimicrobial Agents and Chemotherapy, 1972—, Infection and Immunity, 1973-81, Revs. of Infectious Diseases, 1979—. Trustee Johns Hopkins U., 1963-69. Served to capt. M.C. AUS, 1943-45. Recipient U.S. Typhus Commn. medal, 1947; Albert Lasker Clin. Med. Research award, 1978; Phila. award, 1979; Willard O. Thompson award Am. Geriatric Soc., 1981, others. Fellow ACP (master, James D. Bruce Meml. award 1979), N.Y. Acad. Scis., Am. Acad. Microbiology, AAAS (chmn. sect. on med. scis. 1975); mem. Assn. Am. Physicians, Am. Soc. Clin. Investigation, Am. Clin. and Climatol. Assn., Am. Soc. Microbiology (v.p. N.Y. br. 1961-62), Nat. Acad. Scis., Soc. Exptl. Biology and Medicine, Harvey Soc., Am. Fedn. Clin. Research, Balt. Med. Soc., Am. Assn. Immunologists, N.Y. Acad. Medicine (sec. sect. microbiology 1961-62), Phila. County Med. Soc. (Strittmatter award 1979), Coll. Physicians Phila. (award of Meritorious Service 1980), Interurban Clin. Club (pres. 1970), Infectious Disease Soc. Am. (pres. 1971, Maxwell Finland lecture award 1974), Johns Hopkins Soc. Scholars, Phi Beta Kappa, Sigma Xi, Alpha Omega Alpha, Omicron Delta Kappa. Club: 14 W. Hamilton Street (Balt.). Subspecialties: Infectious diseases; Microbiology (medicine). Current work: Studies of the epidemiology, bacteriology and prevention by vaccination of pneumococcal infection. Address: Dept Research Medicine U Pa Sch Medicine Philadelphia PA 19104

AUVIL, PAUL R., physics and astronomy educator; b. Charleston, W.Va., Aug. 4, 1937; s. Paul R. and Florence M. (Malseed) A.; m. Carole Ann Koehler, June 18, 1960; children: Pamela D., Paul R. B.A., Dartmouth Coll., 1959; Ph.D., Stanford U., 1963. NSF postdoctoral fellow Imperial Coll., London, 1963-65; asst. prof. physics Northwestern U., Evanston, Ill., 1966-70, assoc. prof., 1970—, dir. integrated sci. program, 1979—; vis. scientist CERN, Geneva, 1969-70. Contbr. articles to profl. jours. Mem. Am. Phys. Soc., AAAS, N.Y. Acad. Scis., Phi Beta Kappa, Sigma Xi. Club: Michigan Shores (Wilmette, Ill.). Subspecialties: Particle physics; Theoretical physics. Home: 1231 Ashland Wilmette IL 60091 Office: Dept Physics and Astronomy Northwestern U Evanston IL 60201

AVADIAN, JOHN MARK, computer scientist, distributed system architectures consultant; b. La Reole, Gironde, France, Nov. 29, 1943; came to U.S., 1970; s. Simon and Linda (Ruffini) A.; m. Jacqueline Verdier, Aug. 10, 1968; 1 dau., Cecile. Math. cert., U. Bordeaux, France, 1963; M.B.A., Pepperdine U., 1979; M.S. in Computer Sci, UCLA, 1982. Systems programmer IBM, Poughkeepsie, N.Y., 1968-71; data systems specialist Computer Sci. Corp., N.Y.C., 1971-73, sr. tech. cons., El Segundo, Calif., 1974-79; research mgr. Systems Devel. Corp., Santa Monica, Calif., 1973-74; chief architect Transaction Tech.Inc., Los Angeles, 1979-81; prin. scientist InnovaTech Corp., Los Angeles, 1981—, also dir.; dir. Calsoft Mgmt. Cons., Los Angeles, 1981—, Credifax, Inc., 1982—; tech. cons. Internat. Cons., El Toro, Calif., 1982—. Served to 1st lt. French Air Force, 1966-68. Mem. Assn. Computing Machinery. Subspecialties: Distributed systems and networks; Software engineering. Current work: Research in architectural design of distributed cognitive systems involving multiple interacting, non-cooperating systems; specializing in models of distributed heterogeneous systems with emphasis on conflicts, security problems, logistics issues, etc., regarding the distribution of algorithms; postulating that algorithm ownership and distribution control, which is often ignored, is as important as data management. Home: 2119 Holmby Ave Los Angeles CA 90025 Office: InnovaTech Corp 12304 Santa Monica Blvd Los Angeles CA 90025

AVAMPATO, JAMES ERWIN, biologist; b. Torrington, Conn., Feb. 15, 1924; s. Carl Charles and Mildred Ava A.; m. Joan Ming Lieu, Sept. 21, 1957; children: Lynn, Gail Steve. B.S., U. Conn., 1949, M.S., 1952. Research technologist in animal nutrition U. Conn., 1949-52, in animal diseases, 1953-56; research biologist Lederle Labs., Pearl River, N.Y., 1956-70; staff virologist Franklin Labs., Amarillo, Tex., 1970-74; prin. Ada Sold Mine Realtors, Amarillo, 1975-76; research assoc. Tex. A&M U., 1977—. Contbr. articles to profl. publs. Served with U.S. Army, 1943-46. Decorated Bronze Star. Mem. Am. Soc. Microbiology, Tissue culture Assn., Pandandle Amateur Wrestling Assn. Democrat. Unitarian. Club: Aide (Amarillo). Subspecialties: Virology (veterinary medicine); Tissue culture. Current work: Viruses diseases of cattle; seeking causes and cures of shipping fever complex in cattle; production and effects of bovine interferon on cattle performance. Home: 3711 Langtry Dr Amarillo TX 79109 Office: 6500 Amarillo Blvd W Amarillo TX 79106

AVASTHI, PRATAP SHANKER, nephrologist, educator, researcher; b. Lucknow, India, Jan. 15, 1936; came to U.S., 1965; m. Pushpa Upadhyaya, Dec. 12, 1964; children: Surabbi, Smita, Swati. B.S., King George Med. Coll., Lucknow, 1958, M.D., 1962. Diplomate: Am. Bd. Internal Medicine. Intern Meml. Hosp., Worcester, Mass., 1965; resident U. N.Mex. Hosp., 1966-69; pool officer All India Inst. Med. Scis., New Delhi, 1970-71, asst. prof., 1971-72; asst. prof. medicine U. N.Mex. Sch. Medicine, Albuquerque, 1972-79, assoc. prof., 1979—; staff physician VA Med. Ctr., Albuquerque, 1972—, chief renal sect., 1977—; con. scientist, vascular physiologist research div. Lovelace Med. Found., 1983. Contbr. articles to med. jours. Mem. med. adv. bd. N.Mex. Kidney Found., Albuquerque, 1977; facility rep. VA Med. Ctr., Albuquerque, 1981. Fellow ACP; mem. Am. Fedn. Clin. Research, Am. Heart Assn., Am. Soc. Nephrology, Internat. Soc. Nephrology, Sigma Xi, Alpha Omega Alpha. Subspecialty: Nephrology. Current work: Doppler blood flow measurement and noninvasive diagnosis of renal vascular lesions. Office: VA Med Center 2100 Ridgecrest Dr SE Albuquerque NM 87108

AVENI, ANTHONY FRANCIS, astronomer, archaeoastronomer, educator; b. New Haven, Mar. 5, 1938; s. Anthony Mark and Frances Elsie (Cremonie) A.; m. Lorraine Reiner, Oct. 18, 1941; children: Patricia, Anthony. A.B. in Physics, Boston U., 1960; Ph.D. in Astronomy, U. Ariz., 1963. Mem. faculty Colgate U., Hamilton, N.Y., 1963—, prof. astronomy, 1976-81, C.A. Pana prof. astronomy and anthropology, 1982—. Author: Skywatchers of Ancient Mexico, 1980; editor: Archaeoastronomy in Pre Columbian America, 1975, Native American Astronomy, 1977, New World Archaeoastronomy, 1982. Recipient Prof. of Yr. award Council for Advancement and Support of Edn., 1982. Fellow AAAS; mem. N.Y. Astron. Soc., N.Y. Acad. Sci. Subspecialties: Archaeoastronomy; History of Astronomy. Current work: Maya calendar; astronomical orientations; development of ancient science; work in Mexico, Peru. Home: RD 2 Hamilton NY 13346 Office: Colgate Hamilton NY 13346

AVERBUCH, MARTIN PHILIP, heat recovery equipment manufacturing executive; b. Bklyn., Mar. 8, 1952; s. Jack Arnold and Arlene Joan A. B.S. in Econs, U. Pa., 1974; J.D./M.B.A., U. Chgo., 1977. Bar: Bar: Ill 1977. Founder, exec. v.p. Energy Saving Devices, Inc., Lynbrook, N.Y., 1977—; spl. prof. fin. Hofstra U., 1977-78. Mem. ABA Ill. Bar Assn., Beta Gamma Sigma. Subspecialty: Heat Recovery. Current work: Heat recovery from exhaust gases; condensing heat recovery. Patentee waste heat recovery equipment. Home: 250 Dune Rd West Hampton Beach NY 11978 Office: 27 Wilbur St Lynbrook NY 11563

AVERY, JAMES KNUCKEY, dental educator; b. Holly, Colo., Aug. 6, 1921; s. Willard Smith and Bertha (Knuckey) A.; m. Dorothy Jane Thuerk, Aug. 26, 1950; children—Nancy Jane, David Lloyd, Robert Hugh. B.A., U. Rochester, 1948, Ph.D., 1952; D.D.S., U. Kansas City, 1945. Instr. anatomy U. Rochester Dental Sch., 1952-54; mem. faculty U. Mich. Med. and Dental Sch., 1954—; prof. oral biology Sch. Dentistry, 1963—, prof. anatomy, 1970—, chmn. dept. oral biology, 1977—; dir. Dental Research Inst., 1975—; mem. dental tng. com. NIH, 1964-68; research cons. VA, Ann Arbor, 1964—. Editorial bd.: Jour. Dental Research, 1968-72. Served to lt. (j.g.) USNR, 1945-47. Recipient award Acad. Dental Medicine. Fellow Am. Coll. Dentists, AAAS (chmn. dentistry sect. 1976); mem. ADA (cons. sci. session 1960-70), Internat. Assn. Dental Research (pres. 1974-75), Am. Assn. Anatomists, Electron Microscopic Soc. Am., Teratology Soc., Sigma Xi (hon.), Omicron Kappa Upsilon o3(hon.). Subspecialty: Oral biology. Home: 2465 Adare St Ann Arbor MI 48104

AVERY, ROBERT TOLMAN, electro-mech. engr.; b. San Luis Obispo, Calif., Feb. 7, 1926; s. Harold Tolman and Elizabeth Ella (Murphey) A.; m. Beverly Gail Beckman, July 11, 1948; children: Scott Murphey, Leslie Ann. B.S., U. Minn., 1946; M.S., Stanford U., 1948; D.Eng., U. Calif.-Berkeley, 1974. Registered profl. engr., Calif. Bridge engr. State of Calif., San Francisco, 1948-50; design engr. U. Calif. Radiation Lab., Berkeley, 1950-54; research engr. Chromatic TV Labs., Oakland, Calif., 1954-55; sr. engr. Varian Assos., Palo Alto, Calif., 1955-61; project mgr. Brobeck Assocs. (Cons. Engrs.), Berkeley, 1961-66; sr. staff scientist Lawrence Berkeley Lab., U. Calif., 1966—; cons. in field. Contbr. articles to profl. jours. Mem. Nat. Soc. Profl. Engrs., IEEE, ASME, Internat. Soc. Rock Mechanics. Subspecialties: Mechanical engineering; Applied magnetics. Current work: Engineering design of novel particle accelerators, electromagnets, vacuum systems and related equipment. Patentee in field. Home: 1408 Camino Peral Moraga CA 94556 Office: 1 Cyclotron Rd Berkeley CA 94720

AVIV, ABRAHAM, pediatric nephrologist; b. Tel-Aviv, May 9, 1944; U.S., 1965; s. Chaim and Rachel (Dickstein) Cwaigrach; m.; 1 son, Daniel. M.D., SUNY Upstate Med. Ctr., Syracuse, N.Y., 1972. Intern N.Y. Hosp. Cornell Med. Ctr. (Pediatrics), N.Y.C., 1972-73, resident, 1973-74; fellow in pediatric nephrology Albert Einstein Med. Ctr., Bronx, N.Y., 1974-77; pediatric nephrologist Columbia-Presbyn. Babies Hosp., N.Y.C., 1977-79, N.J. Med. Sch., Newark, 1979—, physiologist, 1979—. Contbr. numerous sci. articles to profl. publs. NIH grantee and fellow, 1982-85; Am. Heart Assn. grantee, 1982-85; Am. Kidney Found. grantee, 1982. Mem. Am. Fedn. Clin. Research, Am. Soc. Nephrology, Am. Soc. Pediatric Nephrology, Internat. Soc. Nephrology. Subspecialties: Nephrology; Cell biology (medicine). Current work: Pathophysiology and hypertension; the Na-K-ATpase in vascular tissue, trace element and the Na-K-ATpase; sodium homeostasis during development. Home: 22 Deer Trail Rd North Caldwell NJ 07006 Office: NJ Med Sch 100 Bergen St Newark NJ 07103

AVNER, ELLIS DAVID, pediatric nephrologist; b. Pitts., June 12, 1948; s. Elkan Avrom and Cecelia (Zalud) A.; m. Jane Adashko, June 12, 1970; 1 son, Benjamin Saul. A.B., Princeton U., 1970; M.D., U. Pa., 1975. Pediatric nephrologist Children's Hosp. of Pitts., 1980—; mem. med. adv. bd. Kidney Found. Western Pa., 1980—. March of Dimes grantee, 1982; Kidney Found. Western Pa. grantee, 1981. Mem. Am. Soc. Nephrology, Internat. Soc. Nephrology, Am. Soc. Pediatric Nephrology, Am. Fedn. Clin. Research, Alpha Omega Alpha. Subspecialties: Nephrology; Cell and tissue culture. Current work: Studies of normal and abnormal kidney development in tissue and organ culture. Office: Children's Hosp of Pitts 125 DeSoto St Pittsburgh PA 15213

AWAD, WILLIAM MICHEL, JR., physician, biochemist, researcher, educator; b. Shanghai, Kiangsu, China, Nov. 5, 1927; s. William Michel and Lily Teresa (Affounso) A.; m. Marilyn Suzanne Wells, June 1, 1957; children: Lily, William Michel. B.S., Manhattan Coll., 1950; M.D., SUNY Downstate Med. Ctr., 1954; Ph.D., U. Wash., 1965. Intern SUNY-Kings County Hosp., N.Y.C., 1954-55; asst. resident in pathology St. Vincent's Hosp., N.Y.C., 1955-56; asst. resident Medicine Albert Einstein Coll. Medicine, N.Y.C., 1956-57; fellow dept. Medicine U. Minn. Hosps., Mpls., 1957-58; fellow cardiology N.Y. Hosp., N.Y.C., 1958-59; asst. prof. medicine U. Miami, 1965-73, assoc. prof., 1973-78, prof., 1978—, dir., 1973—. Contbr. articles to profl. jours. Served in U.S. Army, 1946-47. Damon Runyon Found. fellow, 1960-63; Nat. Cancer Inst. spl. fellow, 1963-65. Mem. Am. Soc. Biol. Chemists, Am. Chem. Soc., Am. Soc. Hematology, So. Soc. Clin. Investigation. Roman Catholic. Subspecialties: Hematology; Oncology. Current work: Protein chemistry, biochemical methylation reactions, macrophage function, alcoholism. Home: 6309 Castaneda St Coral Gables FL 33146 Office: Dept Medicine R-123 U Miami Sch Medicine PO Box 16960 Miami FL 33101

AXEL, LEON, radiologist; b. Lakewood, N.J., Nov. 1, 1947; s. Milton and Alice (Terry) A. B.S., Syracuse U., 1967; Ph.D., Princeton U., 1971; M.D., U. Calif., San Francisco., 1976. Diplomate: Am. Bd. Radiology. Clin. instr. U. Calif., San Francisco, 1980-81; asst. prof. radiology U. Pa., Phila., 1981—. Mem. AAAS, Radiol. Soc. N.Am., Am. Roentgen Ray Soc., Am. Assn. Physicists in Medicine, Soc. Magnetic Resonance in Medicine, Am. Inst. Ultrasound in Medicine. Subspecialties: Diagnostic radiology; Imaging technology. Current work: Computed tomography, ultrasound, nuclear magnetic resonance, noninvasive measurement of blood flow. Office: Dept Radiology 3400 Spruce St Philadelphia PA 19104

AXEL, RICHARD, pathologist; b. Bklyn., July 2, 1946; m., 1975. A.B., Columbia U., 1967; M.D., Johns Hopkins U., 1970. Intern pathology Columbia U. Coll. Physicians and Surgeons, 1970-71; fellow pathology and oncology Inst. Cancer Research, 1971-72; joint appointment vis. fellow dept. pathology Columbia U., 1971-72; research assoc. NIH, 1972-74; asst. prof. dept. pathology and inst., 1974-78; prof. dept. pathology and biochemistry and Inst. Cancer Research, Columbia U., N.Y.C., 1978—; asst. attending physician Presbyn. Hosp., N.Y.C., 1974—. Recipient Young Scientist award Passano Found., 1979; Waterman award, 1982. Subspecialty: Genetics and genetic engineering (medicine). Office: Inst Cancer Research 701 W 168th St New York NY 10032

AXELROD, NORMAN NATHAN, optical engineering company executive; b. N.Y.C., Aug. 26, 1934; s. Louis E. and Said (Katz) A.; m. Victoria Ann Grant, Mar. 21, 1975; children: Lauren, Brian. A.B., Cornell U., 1954; Ph.D., U. Rochester, 1960. Postdoctoral fellow U. London, 1960-61, NASA Goddard Space Flight tr., 1959-60; asst. prof. U. Del., 1961-65; mem. tech. staff Bell Telephone Labs., Murray Hill, N.J., 1965-72; prin., pres. Norman N. Axelrod Assocs., N.Y.C., 1972—. Mem. Optical Soc. Am., IEEE. Subspecialties: Optical engineering; Laser data storage and reproduction. Current work: Plan, design and development of optical and laser systems; materials characterization, sensing, data storage/retrieval/reproduction/display, industrial automation. Home: 445 E 86th St New York NY 10028 Office: 56 W 45th St New York NY 10036

AXELROOD, HELEN BLAU, psychologist; b. Chgo., Feb. 13; d. Morris and Goldie (Bookstien) Blau; m. Jack Axelrood, June 27, 1948; children: Lisa, Barney, Larry Michael. B.A., Roosevelt U., 1951, M.A., 1977; Ph.D., Marquette U., 1982. Tchr., art dir., teaching supr. Chgo. schs., 1951-75; pvt. practice psychotherapy, Evanston, Ill., 1977—; intern in drug and alcohol rehab. Luth. Gen. Hosp., Park Ridge, Ill., 1979; dir. Weight Care Inst. and Counseling Clinic, Evanston, 1981—; lectr. on eating disorders to various Chgo. area groups and orgns. TV and state actress and writer, Chgo., 1951-74. Recipient Dedicated Service award Beth Emet Temple, Evanston, 1960; Outstanding Service award Temple Beth El, Chgo., 1970. Mem. Am. Psychol. Assn., Am. Personnel and Guidance Assn., Assn. Specialists in Group Work, Ill. Psychol. Assn., Ill. Personnel and Guidance Assn. Subspecialty: Behavioral psychology. Current work: Research on personality differences with weight changes. Home: 2022 Hawthorne Ln Evanston IL 60201 Office: Weight Care Institute & Counseling Clinic 1601 Sherman St Suite 402 Evanston IL 60201

AXEN, KENNETH, biomedical engineer, researcher, educator; b. N.Y.C., Mar. 22, 1943; s. John and Edith (Turkia) A.; m. Geraldine Miscione, July 6, 1966 (div. Sept. 1980); m. Kathleen Vermitsky, Nov. 25, 1980; 1 dau., Laurel. B.E.E., CCNY, 1966; M.S.E.E., N.Y. U., 1967, Ph.D., 1972. Research scientist N.Y. U. Med. Center, 1972-77, instr., 1977-79, asst. prof., 1979—. Author: (with others) Pulmonary Therapy and Rehabilitation: Principles and Practice, 1979; contbr. articles to profl. jours. Mem. Am. Physiol. Soc. Subspecialties: Biomedical engineering; Physiology (biology). Current work: Investigation of mechanisms regulating ventilation in humans. Office: Inst Rehab Medicine NY U 400 E 34th St New York NY 10016

AYALA, FRANCISCO JOSE, geneticist, educator; b. Madrid, Mar. 12, 1934; U.S., 1961, naturalized, 1971; s. Francisco and Soledad (Pereda) A.; children—Francisco Jose, Carlos Alberto. B.S., U. Madrid, 1954; M.A., Columbia U., 1963, Ph.D., 1964; D. honoris causa, Universidad de León (Spain), 1982. Research asso. Rockefeller U., 1964-65; asst. prof. Providence Coll., 1965-67, Rockefeller U., 1967-71; asso. prof. to prof. genetics U. Calif., Davis, 1971—; nat. adv. council Nat. Inst. Gen. Med. Scis.; exec. com. EPA. Author: Population and Evolutionary Genetics, 1982, Modern Genetics, 1980, Evolving: the Theory and Processes of Organic Evolution, 1979, Evolution, 1977, Molecular Evolution, 1976, Studies in the Philosophy of Biology, 1974. Recipient medal College de France, 1979. Fellow AAAS; mem. Soc. Study of Evolution (pres. 1979-80), Am. Soc. Naturalists (sec. 1973-76), Genetics Soc. Am., Am. Genetic Assn., Ecology Soc. Am., Nat. Acad. Scis., Am. Acad. Arts and Scis. Subspecialties: Evolutionary biology; Genetics and genetic engineering (biology). Current work: Genetics of the evolutionary process; the process of speciation; molecular evolution; ecology; philosophy of science. Home: 747 Plum Ln Davis CA 95611 Office: Dept of Genetics University of California Davis CA 95616

AYASLI, SERPIL, physicist; b. Ankara, Turkey, May 22, 1951; came to U.S., 1979; d. Nazmi and Ayse (Torun) Evrensel; m. Yalcin Ayasli, Oct. 23, 1973; 1 dau., Ceylan. B.S.E.E., Middle East Tech. U., Ankara, Turkey, 1973, M.S., 1975, Ph.D., 1978. Teaching asst. physics dept. Middle East Tech. U., 1974-78, instr., 1978-79; postdoctoral researcher M.I.T., Cambridge, 1979-82; mem. staff Lincoln Lab., Lexington, 1982—. Contbr. articles in field to profl. jours. Mem. Am. Astron. Soc. Subspecialties: Theoretical astrophysics; 1-ray high energy astrophysics. Current work: Research and data analysis on radars, electromagnetic wave propagation. Office: MIT Lincoln Lab PO Box 73 Room C-484 Lexington MA 02173

AYDELOTTE, LEE C., computer scientist; b. Inglewood, Calif., Sept. 24, 1956; s. Charles W. and Frances B. (Nyberg) A.; m. Elizabeth A. Peterson, Mar. 29, 1981; 1 dau., Laura E. B.S. in Math, Calif Inst. Tech., 1978, M.S. in Computer Sci. Computer scientist ALS Electronics, Anaheim, Calif., 1978-81; spl. systems analyst Genisco Computer Corp., Costa Mesa, Calif., 1981—. Mem. Assn. Computing Machinery, IEEE Computer Soc. Subspecialties: Graphics, image processing, and pattern recognition; Software engineering. Current work: Development of three dimensional graphics display system, development of high level local interactive graphics software for standard graphics displays. Office: 3545 Cadillac Ave Costa Mesa CA 92626

AYERS, ARNOLD LESLIE, SR., chemical engineer; b. Dayton, Ohio, July 30, 1917; s. Leslie Arnold and Marie Antoinette (Ritter) A.; m. Doris Evelyn Lotspeich, Mar. 19, 1938; children: Arnold Leslie, Kenneth Duanne, Vicki Lynn Ayers Kershaw. B.S. in Chem. Engring, Iowa State U., 1938. Research chemist Magic City Printing Co., Omaha, 1938-40; sr. chem. engr. Phillips Petroleum Co., Bartlesville, Okla., 1940-51, supt. ops., Idaho Falls, Idaho, 1951-67; tech. dir. Allied Chem. Corp., Morristown, N.J., 1967-71, Allied-Gen. Nuclear Service, Barnwell, S.C., 1971-79; mgr. TWTF project EG&G-Idaho, Inc., Idaho Falls, 1979-82, cons., 1982—; cons. Gorleban Hearings, Lower Saxony-State, Hanover, W.Ger., 1979. Served to maj. U.S. Army, 1942-45. Recipient Merit award Chem. Engring., 1972; prof. Achievement citation Iowa State U.-Ames, 1975. Fellow Am. Inst. Chem. Engrs.; mem. Am. Nuclear Soc., Am. Chem. Soc. Subspecialties: Chemical engineering; Nuclear engineering. Current work: Design of nuclear reprocessing plants and nuclear waste treatment facilities. Patentee electro pulse separater, method of dissolving spent nuclear fuel. Home: 887 Linden Pl Idaho Falls ID 83401 Office: EG&G Idaho PO Box 1625 Idaho Falls ID 83415

AYLES, G. BURTON, genetcist; b. Prince Albert, Sask., Can., Nov. 28, 1945; s. W. George and Leonis W. (Farnsworth) A. B.Sc., U. B.C., 1967, M.Sc., 1969; Ph.D., U. Toronto, 1972. Research scientist Freshwater Inst., Winnipeg, Man., Can., 1972-80, regional planning officer, 1980-82, dir. research, 1982—. Contbr. articles to profl. jours. Subspecialties: Genetics and genetic engineering (biology); Animal genetics. Current work: Aquaculture, genetics of fish. Office: 501 University Crescent Winnipeg MB Canada R3T 2N6

AYOUB, GEORGE TANIOS, engineering company executive; b. Lebanon, Jan. 24, 1951; s. Tanios Tannoos and Hellen (Ishaya) A. E.E., U. St. Joseph, Lebanon, 1974; Ph.D., U. Nebr.-Lincoln, 1976, M.B.A., 1978. Project mgr., sr. engr. Marathon Oil Co., 1978-80, research scientist, Denver, 1980-82; sr. staff exec. AREC, Abu Dhabi, United Arab Emirates, 1982—. Author: Optical Characterization of Thin Films on Water by Ellipsometry, 1978, New Applications of Coherency in Seismic Exploration, 1983. U. Nebr. grantee, 1974; U. Nebr. postdoctoral fellow, 1976. Mem. Soc. Exploration Geophysicists, Optical Soc. Am., Sigma Xi. Subspecialties: Petroleum engineering; Geophysics. Current work: Exploration techniques - computer science applied to exploration and production of petroleum, signal processing and fiber optics. Office: PO Box 7658 Abu Dhabi UAE

AZARNOFF, DANIEL LESTER, pharm. co. exec.; b. Bklyn., Aug. 4, 1926; s. Samuel J. and Kate (Asarnow) A.; m. Joanne Stokes, Dec. 26, 1951; children: Rachel, Richard, Martin. B.S., Rutgers U., 1947, M.S., 1948; M.D., U. Kans., 1955. Instr. anatomy U. Kans., 1949-50, research fellow, 1950-52, intern, 1955-56; Nat. Heart Inst. resident research fellow, 1956-58, asst. prof. medicine, 1962-64, assoc. prof., 1964-68, dir. clin. pharmacology study unit, 1964-68, assoc. prof. pharmacology, 1965-68, prof. medicine and pharmacology, 1968, dir., 1967-68, Disting. prof., 1973-78; also prof. medicine; Nat. Inst. Neurol. Diseases and Blindness pl. trainee Washington U., St. Louis, 1958-60; asst. prof. medicine St. Louis U., 1960-62; vis. scientist, Fulbright scholar Karolinska Inst., Stockholmm, 1968; sr. v.p. worldwide research and devel. G.D. Searle & Co., Chgo., 1978; pres. Searle Research and Devel., Skokie, Ill., 1979—; prof. pathology, clin. prof. pharmacology Northwestern U., 1978—; William N. Creasy vis. prof. clin. pharmacology Med. Coll. Va., 1975; Bruce Hall Meml. lectr. St. Vincents Hosp., Sydney, 1976; 7th Sir Henry Hallett Dale lectr. Johns Hopkins U., 1978; professorial lectr. U. Chgo., 1979; dir. 2d Workshop on Prins. Drug Evaluation in Man, 1970; chmn. com. on problems of drug safety NRC-Nat. Acad. Sci., 1972-76; cons. numerous govtl. agencies. Editorial bd. Jours., AMA, many others; editor: Rev. of Drug Interactions, 1974-77, Yearbook of Drug Therapy, 1977-79; series editor: Monographs in Clin. Pharmacology, 1977-. Served with AUS, 1945-46. Recipient Ginsburg award in phys. diagnosis U. Kans. Med. Center, 1953, Outstanding Intern award, 1956, Ciba award for gernotol. research, 1958, Rectors medal U. Helsinki, 1968; John and Mary R. Markle scholar, 1962; Burroughs Wellcome scholar, 1964. Fellow ACP, N.Y. Acad. Scis.; mem. Am. Soc. Clin. Nutrition, Am. Nutrition Inst., Am. Soc. Pharmacology and Exptl. Theapeutics (chmn. clin. pharmacology div. 1969-71, exec. com. 1966-73, 78—, del. 1975-78, bd. publ. trustees), Am. Soc. Clin. Pharmacology and Therapeutics, Am. Fedn. Clin. Research, Brit. Pharmacol. Soc., AMA (vice chmn. council on drugs 1971-72), Central Soc. Clin. Research, Royal Soc. Promotion of Health, Inst. Medicine Nat. Acad. Scis., Soc. Exptl. Biology and Medicine (Councillor), Internat. Union Pharmacologists (sec. clin. pharmacology sect.), Sigma Xi. Subspecialties: Pharmacology; Internal medicine. Current work: Integrate development of new drugs, including those produced by bioengineering. Home: 1030 Lake Shore Blvd Evanston IL 60202 Office: 4901 Searle Pkwy Skokie IL 60077

AZAROFF, LEONID VLADIMIROVITCH, physics educator; b. Moscow, June 19, 1926; U.S., 1939, naturalized, 1945; s. Vladimir Ivanovitch and Maria Yulievna (Odlen) A.; m. Carmen Wade, Mar. 9, 1946 (div. July 1968); m. Beth Sulzer, Mar. 4, 1972; children: David, Richard, Lenore. B.S. cum laude, Tufts Coll., 1948; Ph.D., M.I.T., 1954. Research physicist Armour Research Found., Chgo., 1953-54, sr. scientist, 1954-57; asso. prof. metall. engring. Ill. Inst. Tech., 1957-61, prof., 1961-66; prof. physics, dir. Inst. Material Sci., U. Conn., 1966—; guest physicist Brookhaven Nat. Lab., 1961, 62, 64; vis. prof. U. Mass., 1978-79; cons. Owens-Ill., Philips Electronics, Hilger-Watts, Inc.; U.S. del. Internat. Union Crystallography, teaching commn., 1963-69; dir. Conn. Product Devel. Corp., Rogers Corp. Asso., Conn. Devel. Corp., 1977—. Author: 7 books, including X-Ray Diffraction and X-Ray Spectroscopy, 1973; also articles. Served with AUS, 1944-46. Fellow Am. Phys. Soc. (cons. editor), Mineral. Soc. Am.; mem. AAAS (dir.), IEEE (sr.), Am. Soc. Engring. Edn., Conn. Acad. Sci. and Engring. (pres. 1976-82), Am. Crystallographic Assn., Am. Inst. Mining Engrs., Internat. Union Physics, Internat. Union Crystal Growth, Sigma Xi, Phi Kappa Phi, Sigma Pi Sigma. Subspecialties: Alloys, Crystallography. Current work: Electronic structure of alloys; liquid crystal structures. Home: PO Box 103 Storrs CT 06268 I have always adhered to the principle that anything worth doing at all is worth doing as well as possible. Therefore, I select very carefully the tasks to undertake.

AZMITIA, EFRAIN CHARLES, medical researcher, biology educator; b. Tampa, Fla., Sept. 18, 1946; s. Efrain Charles and Angela Alicia (Gutierrez) A.; children: Bianka, Anthea. B.A. cum laude, Washington U., St. Louis, 1968; Ph.D., Rockefeller U., 1973; M.A., Cambridge (Eng.) U., 1976. Vis scientist lab. neuropharmacology NIMH, Washington, 1973-75; lab. neurobiology Nat. Inst. Med. Research, London, 1975-76; research fellow dept. anatomy Darwin Coll., Cambridge U., 1976-78; dir. lab. neurohistology Ecole Pratique des Hautes Etudes, La Salpetriere, Paris, 1981-82; assoc. prof. anatomy Mt. Sinai Med. Center, N.Y.C., 1978-83; prof. biology NYU, N.Y.C., 1983—. Contbr. articles to profl. jours.; Guest editor: Jour. Histochemistry and Cytochemistry. Active PTA. Grantee NSF, 1979—, NIH, 1981—; NIH fellow, 1976-78; recipient Career Scientist award I. Hirschl Trust, 1979—; others. Mem. Am. Assn. Anatomists, Soc. Neurosci., N.Y. Acad. Medicine, Royal Philos. Soc., Internat. Soc. Developmental Neurosci., Harvey Soc. Roman Catholic. Club: Cajal. Subspecialties: Anatomy and embryology; Neurochemistry. Current work: Multidisciplinary analysis of Serotonin neurons; brain plasticity after experimental intracerebral microinjections of chemicals and fetal neurons (brain transplants). Home: 7 Baylis Pl Syosset NY 11791 Office: Dept Biology 1009 Main Bldg NYU Washington Sq New York NY 10003

AZZIZ, NESTOR JALIL, physics educator; b. Durazno, Uruguay, Dec. 11, 1932; s. Mateo and Saturnina (Jozami) A.; m. Juana Maria Baumgartner, Mar. 30, 1957; children: Ricardo, Rodolfo, Cecilia, Eduardo. B.Engring., U. Uruguay, 1958; M.S. in Nuclear Tech, U. P.R., 1960; Ph.D., Pa. State U., 1963. Pres. Azziz Heating, Montevideo, Uruguay, 1951-59; fellow physicist Westinghouse Atomic Power Div., Pitts., 1963-70; prof. dept. physics U. P.R., Mayaguez, 1971—; cons. and lectr. in field. Contbr. articles to profl. jours. Dept. Energy grantee, 1976-78. Mem. N.Y. Acad. Sci., Am. Phys. Soc. Subspecialties: Nuclear physics; Atomic and molecular physics. Current work: Pure physics, theoretical physics and applied physics. Patentee in field. Office: Dept Physics University of Puerto Rico Mayaguez PR 00708

BABB, ALBERT LESLIE, nuclear engineering educator; b. Vancouver, C., Can., Nov. 7, 1925; came to U.S., 1948, naturalized, 1954; s. Clarence Stanley and Mildred (Gutterdige) B.; m. Marion A. McDougall; children: Eugene Matthew, Philip Leslie, Christine Louise. B.A. Sc., U. B.C., 1948; M.S., U. Ill., 1949, Ph.D., 1951; student, Internat. Sch. Nuclear Sci. and Engring., Argonne Nat. Lab., 1956-57. Chem. engr. Nat. Research Council Can., 1948; research engr. Rayonier Inc., 1951-52; mem. faculty U. Wash., Seattle, 1952—, chmn. nuclear engring. group, 1957-65, prof. chem. engring., 1960—, dir. nuclear reactor labs., 1962-72, prof. nuclear engring., 1965—, chmn. dept. nuclear engring., 1965-81; del. Japan-U.S. Seminar on Nuclear Engring. Edn., 1974; lectr. hemodialysis engring. USSR Ministry of Health, Moscow, 1976; lectr. biomed. engring. Norwegian Nephrological Soc., Oslo, 1980; lectr. hemodialysis engring. Kuratorium für Hemodialyse, Munster, Germany, 1980, Clinique Iser, Munich, W. Germany, 1980, Meml. Hosp., Hvidovre, Denmark, 1980, State Hosp., Copenhagen, 1980; cons. in field. Contbr. chpts. to books., articles to profl. jours. Active local services to Children's Orthopaedic Hosp., Northwest Artificial Kidney Ctr.; trustee Pacific Sci. Ctr. Found., mem. exec. com., 1973-80. Recipient citation Wash. Joint Legis. Com. Nuclear Energy, 1968; named Engr. of Yr. Wash. State Profl. Engrs. Assn., 1969; award for excellence Sigma Xi. Fellow Am. Inst. Chem. Engrs., Am. Inst. Chemists; mem. Am. Nuclear Soc. (dir. 1976—), Am. Inst. Chem. Engrs. (Engr. of Distinction), Engrs. Joint Council, Nat. Acad. Engring. (membership com.), Am. Soc. Engring. Edn. (chmn. nuclear engring. div. 1965-66), Am. Nephrology Soc., Am. Soc. Artificial Internal Organs, European Dialysis and Transplantation Assn., Sigma Xi, Tau Beta Pi, Pi Mu Epsilon, Alpha Chi Sigma. Presbyterian. Subspecialties: Biomedical engineering; Nuclear fission. Current work: Mathematical modelling of human respiratory systems; synthetic fuel producing nuclear energy systems, dialysis engineering. Co-inventor of artificial kidney systems, techniques for early diagnosis of cystic fibrosis in children using nuclear reactors, formulated dialysis index for hemodialysis patients; co-inventor system for treatment of sickle cell anemia, computerized insulin pump for diabetics; patentee of structured permafrost, artificial kidneys and artificial pancreas. Home: 3237 Lakewood Ave S Seattle WA 98144 Office: Dept Nuclear Engring U Wash Seattle WA 98195

BABB, HAROLD, psychology educator; b. Mosheim, Tenn., Sept. 4, 1926; s. Ray Edward and Mary Louise (Brown) B.; m. Marjorie Craig Leask, Sept. 27, 1947; children: Patricia Craig, Barbara Lou, David Edward. B.A., Wayne State U., 1950; M.A., Ohio State U., 1951, Ph.D., 1953. Cert. psychologist, N.Y. Asst. to assoc. prof., chmn Coe Coll., Cedar Rapids, Iowa, 1953-58; prof., chmn. Hobart and William Smith Coll., Geneva, N.Y., 1958-63; exec. sec., grants specialist NIH/NIMH, Bethesda, Md., 1963-64; prof., chmn. U. Mont., Missoula, 1964-71; prof. psychology SUNY-Binghamton, 1971—. Contbr. articles to profl. jours. NSF grantee, 1968-69; NIMH tng. grantee, 1967-71; research grantee, 1965-68; SUNY/NIMH research grantee, 1981—. Mem. AAUP (chpt. pres. 1981-83). Subspecialties: Behavioral psychology; Learning. Current work: research on aversive conditioning with emphasis on self-punitive or vicious circle behavior. Home: 2309 Hemlock Ln Vestal NY 13850 Office: SUNY Binghamton Dept Psychology Binghamton NY 13901

BABCOCK, HOPE MADELINE, lawyer; b. N.Y.C., Feb. 13, 1941; d. Edgar M. and Barbara (Schiff) Sinauer. B.A. magna cum laude, Smith Coll., 1963; LL.B., Yale U., 1966. Asst. regional counsel N.E. regional office OEO, 1966-68; asso. firm LeBeouf, Lamb, Leiby & McRae, 1971-77; dep. asst. sec. energy and minerals Dept. Interior, 1977-79; partner firm Blum & Nash, Washington, 1979—. Bd. dirs., treas. Environ. Policy Center, Environ. Policy Inst. Mem. D.C. Bar Assn. Subspecialty: Optical astronomy. Office: 1015 18th St NW Washington DC 20036

BABERO, BERT BELL, parasitologist, educator; b. St. Louis, Oct. 9, 1918; s. Andras and Bertha (Bell) B.; m. Harriett King, Feb. 14, 1950; children: Bert Bell, Andras Fanfiero. B.S., U. Ill.-Champaign-Urbana, 1949, M.S., 1950, Ph.D., 1957. Parasitologist Arctic Health Research Ctr., USPHS, Anchorage, 1950-54; research asst. U. Ill., 1954-57; prof. zoology Ft. Valley State Coll., 1957-59; prof. So. U., Baton Rouge, 1959-60; lectr. Fed. Emergency Sci., Lagos, Nigeria, West Africa, 1960-62; parasitologist U. Baghdad, Iraq, 1962-65; prof. zoology U. Nev., Las Vegas, 1965—. Contbr. numerous articles to profl. jours. Commr. Nev.State Equal Rights Commn., Las Vegas, 1968. Served with U.S. Army, 1943-46; PTO. La. State U. fellow, 1969. Mem. Am. Soc. Parasitologists, Helminthol. Soc. Wash., Rocky Mountain Conf. Parasitologists (pres. 1975-76), So. Calif. Parasitologists, Am. Soc. Protozoologists, NAACP. Subspecialties: Parasitology; Taxonomy. Current work: Tropical diseases in Africa, Middle East and Central America; zoonosis of animal parasites. Home: 2202 Golden Arrow Dr Las Vegas NV 89109 Office: U Nev 4505 Maryland Pkwy Las Vegas NV 98154

BABINGTON, RONALD GLENN, pharm. researcher; b. Dayton, Ohio, Oct. 18, 1937; s. William Oscar and Christine (Townsend) B.; m. Eileen Marie Wilson, Nov. 11, 1972. Student, Miami U., Oxford, Ohio, 1955-57; B.S. in Pharmacy, Ohio State U., 1960; M.S., Purdue U., 1963; Ph.D. in Pharmacology, Purdue U., 1965. Sr. research investigator Squibb Inst. for Med. Research, Princeton, N.J., 1965-75; head central nervous system sect. Sandoz Pharms. Inc., East Hanover, N.J., 1975-80, assoc. dir. clin. research, 1980—. Contbr. articles to profl. jours., chpts. to books. Mem. Am. Soc. Pharmacology and Exptl. Therapeutics, AAAS, Soc. for Neurosci., N.Y. Acad. Scis., Sigma Xi, Rho Chi. Subspecialty: Neuropharmacology. Current work: Neuropsychiatric clinical research. Patentee in field. Office: Sandoz Pharms Inc East Hanover NJ 07936

BABRAKZAI, NOORULLAH, biologist, educator, researcher; b. Nadershah Koat, Afghanistan, May 10, 1945; came to U.S., 1970; s. Said Akbar and Amila Bibi B.; m. Sianoosh Samsam, July 18, 1973 (div. 1981). B.Sc., U. Peshawar, Abbottabad, Pakistan, 1965, M.Sc., 1967; Ph.D., U. Ariz., 1975. Lectr. Kabul (Afghanistan) U., 1968-70; grad. asst. U. Ariz., 1971-75, research asst. prof., 1977-81; teaching assoc. Pima Community Coll., Tucson, 1980-81; asst. prof. dept. biology Central Mo. State U., 1981—. Recipient Gold medal in zoology U. Peshawar, 1967; Meritorious Teaching Asst. award U. Ariz. Found., 1975. Mem. AAAS, Am. Malacological Union, Am. Soc. Zoologists, Mo. Acad. Sci., Sigma Xi. Subspecialty: Cytology and histology. Current work: Cytology, cytotaxonomy, cytogenetics of pulmonate land snails. Home: 610 Christopher Warrensburg MO 64093 Office: Dept Biology Central Mo State Univ Warrensburg MO 64093

BACCANARI, DAVID PATRICK, research microbiologist; b. Wilkes-Barre, Pa., Apr. 15, 1947; s. Samuel M. and Helen A. Sokolski B.; m. Frances M. Gluc, June 22, 1968; children: David P., Nina A. B.S. in Chemistry, Wilkes Coll., 1968; Ph.D. in Biology and Med. Sci, Brown U., 1972. Research assoc. Brown U., Providence, 1972-73; research microbiologist Wellcome Research Labs., Research Triangle Park, N.C., 1973—; mem. spl. study sect. NIH, Bethesda, Md., 1981—. Contbr. articles on microbiology to profl. jours. Mem. Am. Soc. for Microbiology, Am. Soc. Biol. Chemists. Subspecialties: Biochemistry (medicine); Microbiology (medicine). Current work: Enzymology-isolation, characterization and inhibition of enzymes of chemotherapeutic importance. Office: Wellcome Research Labs 3030 Cornwallis Rd Research Triangle Park NC 27709

BACH, DEBORAH, psychologist; b. Springfield, Mass., Dec. 7, 1937; d. Kenneth William and Ethel (Martin) Donaldson; m. Edward

August Bach, Mar. 26, 1958 (div. 1980); children: Cynthia, Hillary, Melanie, Edward, Andrea. B.S. in Edn. and B.A. in Psychology, Am. Internat. Coll., 1975, M.S., 1977, M.A., 1980. Cert. social worker, Mass. Tchr. Hampden (Mass.) Schs., 1963-74; instr. Springfield (Mass.) Tech. Community Coll., 1977; psychologist Providence Hosp., Holyoke, Mass., 1976—; cons. drug edn. Springfield schs., 1976—, Concerned Parents, Region I, Mass., 1976—; psychol. evaluator Springfield Police Cadets, 1982. Mem. Republican Town Com., Wilbraham, Mass., 1970-71; v.p. PTO, Wilbraham, 1972-73; del. White House Conf. on Families, 1980. Mem. Am. Psychol. Assn., LWV. Republican. Roman Catholic. Subspecialty: Clinical psychology. Current work: Personality assessment/research of drug-dependent individuals. Office: 210 Elm St Holyoke MA 01040

BACH, MARILYN LEE, immunology educator, education council science researcher; b. Lynn, Mass., Apr. 24, 1937; d. Samuel and Ida (Callum) Brenner; m. Fritz Heinz Bach, June 18, 1958 (div. Sept. 1980); children: David, Peter, Wendy. B.S., Simmons Coll., 1958; postgrad., MIT, 1958-60; Ph.D., N.Y. U., 1960. Fellow in oncology U. Wis.-Madison, 1966-67, project assoc. dept. genetics, 1967-68, research assoc. dept. med. genetics, 1968-70, asst. prof., 1970-75, assoc. prof. dept. genetics, 1975-78; assoc. prof. depts. of lab. of medicine/pathology Med. Sch., Sch. Pub. Health-Health Services Research Ctr., U. Minn., Mpls., 1978—; vis. prof. U. Leiden, Netherlands, 1973; invited lectr. U. Hosp. Bloodbank, 1974; sci. liaison Office for Med. Applications of Research, NIH, Bethesda, Md., 1979-80; spl. asst. to dir. for program devel. Nat. Inst. Allergy and Infectious Disease, 1980-81; fellow in sci. and pub. policy Brookings Instn., Washington, 1981-82; vis. fellow Am. Council on Edn., Washington, 1982—. Contbr. sects. to books, numerous articles in field to profl. jours.; patentee primed lymphocyte typing, use of lymphocyte blastoid cell lines for primed lymphocyte typing. Mem. nat. adv. council Nat. Inst. Allergy and Infectious Disease, 1974-77; mem. Basil O'Connor starter research grant panel Nat. Found., N.Y.C., 1975-78. Recipient Founders Day award N.Y. U., 1967, Faculty Research award Am. Cancer Soc., 1971. Mem. Am. Assn. Immunologists (chmn. nominating com. 1978-79, com. mem. on status of women 1982—), AAAS, Transplantation Soc. (editorial bd. Transplantation 1974-77, symposium lectr. internat. congress 1976). Democrat. Jewish. Club: MIT (Washington). Subspecialties: Immunogenetics; Transplantation. Current work: Research directed at genetic and cellular immunology mechanisms of transplantation; concurrent interest in science policy, particularly university/industry relationships. Home: 5412 Lambeth Rd Bethesda MD 20814 Office: Am Council on Edn Suite 824 1 Dupont Circle Washington DC 20036

BACH, MICHAEL KLAUS, research immunopharmacologist; b. Stuttgart, West Germany, Oct. 2, 1931; came to U.S., naturalized, 1947; s. Rudolph and Ruth (Meyer) B.; m. Shirley Rosenberg, June 20, 1955; children: Mark Allen, David Scott. M.S., U. Wis., 1955, Ph.D., 1957. Research scientist Union Carbide Chem. Co., South Charleston, W.Va., 1957-60; research scientist The Upjohn Co., Kalamazoo, Mich., 1960-71, sr. research scientist, 1971-83, sr. scientist, 1983—, disting. scientist, 1983—; mem. study sect. on allergy and immunology NIH, 1982—. Editor: Immediate Mypersensitivity, 1978; assoc. editor: Jour. Immunology, 1979-83, Internat. Jour. Immunopharmacology, 1981—. Mem. Am. Assn. Immunologists, Am. Soc. Biol. Chemistry, Collegium Internationale Allergologicum. Subspecialties: Immunopharmacology; Biochemistry (medicine). Current work: Biochemistry and pharmacology of prostaglandins, leukotrienes, other eicosanoids and other mediators of anaphylaxis; membrane biochemistry, especially receptors for immunoglobulins. Office: The Upjohn Company Kalamazoo MI 49001

BACHMAN, RICHART T., oceanographer; b. Los Angeles, May 10, 1943; s. John Randolph and Elizabeth (Lapsley) B.; m. Nancy Merideth Hines, July 8, 1967; children: Richard Randolph, Sarah Catherine. B.A., San Diego State U., 1966. Oceanographer Naval Ocean System Ctr., San Diego, 1967—. Fellow Geol. Soc. Am. Subspecialties: Sedimentology; Geophysics. Current work: Acoustic properties of marine sediments. Home: 9162 Brier Rd La Mesa CA 92041 Office: Naval Ocean Systems Ctr San Diego CA 92152

BACHMAN, WALTER CRAWFORD, ship designer, marine cons.; b. Pitts., Dec. 24, 1911; s. Clarence E. and Mary Elizabeth (Crawford) B.; m. Helen Elizabeth Van Cleaf, Mar. 25, 1938; children—Van Cleaf, Elizabeth Crawford Bachman Ramjoué. B.S. in Indsl. Engring, Lehigh U., 1933, M.S., 1935. Tchr. mech. engring. Lehigh U., 1935-36; marine engr. Fed. Shipbldg. and Dry Dock Co., 1936, Gibbs & Cox, Inc., N.Y.C., 1936-70, chief engr., 1958-63, v.p., chief engr., 1963-70; marine cons., Short Hills, N.J., 1970—. Fellow ASME; mem. Nat. Acad. Engring. (mem. marine bd. 1967-75), Soc. Naval Architects and Marine Engrs., Am. Soc. Naval Engrs., N.Y. Acad. Scis. Club: Yacht (Beaulieu-St. Jean). Subspecialties: Mechanical engineering; Ocean engineering. Address: 21 Wayside Short Hills NJ 07078

BACHRACH, HOWARD L., biochemist; b. Faribault, Minn., May 21, 1920; s. Harry and Elizabeth (Panovitz) B.; m. Shirley F. Lichterman, June 13, 1943; children: Eve E., Harrison J. B.A. in Chemistry, U. Minn., 1942, Ph.D. in Biochemistry, 1949. Research asst. explosives research lab. Nat. Def. Research Com. project Carnegie Inst. Tech., Pitts., 1942-45; research asst. U. Minn., Mpls., 1945-49; biochemist foot-and-mouth disease mission U.S. Dept. Agr., Denmark, 1949-50; research biochemist biochem. and virus lab. U. Calif. at Berkeley, 1950-53; chief scientist, head biochem. and phys. investigation Plum Island Animal Disease Center, Greenport, N.Y., 1953-80, research chemist, advisor to dir., 1981—; sr. exec. U.S. Dept. Agr., 1979; mem. viral and rickettsial grants subcom. Walter Reed Army Inst. Research, 1982—. Recipient Naval Ordnance Devel. award, 1945; Certificate of Merit U.S. Dept. Agr., 1960; Disting. Service award U.S. Dept. Agr., 1982; Presdl. citation, 1965, U.S. Sr. Exec. Service award, 1980; Newcomb Cleveland prize AAAS, 1982; Nat. Award for Agrl. Excellence, 1983; Alexander von Humboldt award, 1983. Fellow N.Y. Acad. Scis.; affiliate Am. Coll. Veterinary Microbiologists; mem. Am. Chem. Soc. (Kenneth A. Smith award 1983), Am. Soc. Microbiology, Soc. Exptl. Biology and Medicine, Nat. Acad. Scis., Am. Soc. Virology, Sigma Xi., Gamma Alpha, Phi Lambda Upsilon. Subspecialties: Animal virology; Genetics and genetic engineering. Current work: Structure and function of viruses; vironome strategies; cloned subunit protein vaccines; mapping and synthesis of antigenic sites; direction of research. Home: Dayton Rd Southold NY 11971

BACIOCCO, ALBERT JOSEPH, JR., naval officer; b. San Francisco, Mar. 4, 1931; s. Albert Joseph and Florence Beatrice (Wiegner) B.; m. Mary Jane Rivera, June 25, 1955; children—David Anthony, Delora Ann, Andrew Joseph, Mary Susan. B.S., U.S. Naval Acad., 1953. Commd. ensign U.S. Navy, 1953; advanced through grades to rear adm.; comdr. in U.S.S. Gato, 1965-69; with Submarine Div. 42, 1969-71, Submarine Squadron 4, Charleston, S.C., 1974-76; former mem. chief naval ops. staff, Washington, chief naval research, chief naval devel., dep. chief naval material, Arlington, Va. Decorated Legion of Merit (3), Meritorious Service medal, Navy Commendation medal. Mem. U.S. Naval Inst., Am. Def. Preparedness Assn., Am. Soc. Naval Engrs., U.S. Naval Acad. Alumni Assn. Roman Catholic. Subspecialty: Research administration.

BACKER, DONALD CHARLES, astronomer; b. Plainfield, N.J., Nov. 9, 1943. B.Engring., Cornell U., 1966, Ph.D., 1971; M.Sc., U. Manchester, Eng., 1968. Research assoc. Nat. Radio Astronomy Obs., Charlottesville, Va., 1971-73; NRC fellow Goddard Space Flight Ctr., 1973-75; research astronomer, lectr. U. Calif., Berkeley, 1975—. Mem. Am. Astron. Soc., Internat. Radio Sci. Union, Internat. Astron. Union, AAAS. Subspecialty: Radio and microwave astronomy. Current work: Compact radio sources in our galaxy; pulsars; radio interferometry. Office: 601 Campbell Hall U Calif Berkeley CA 94720

BACKHUS, DEWAYNE ALLAN, geoscience educator, energy researcher, consultant; b. Hope, Kans., Aug. 1, 1944; s. William F. and Hilda S. (Ottensmeier) B.; m. Martha L. Burden, Aug. 14, 1966. B.S. in Math. and Phys. Scis, Emporia State U., 1966; M.A. in Planetary Sics, Harvard U., 1967. Asst. prof. to assoc. prof. geoscis. Emporia State U., 1967-81, 3d Roe R. Cross Disting. prof., 1981—; cons. to public utilities, ednl. instns., pvt. corps. Contbr. articles to profl. jours. NSF grantee Tenn. Energy Authority, 1979; Harvard U. Prize fellow. Mem. Am. Solar Energy Soc., Nat. Sci. Tchrs. Assn., Kans. Acad. Sci., Kans. Assn. Tchrs. Sci., Sigma Pi Sigma, Kappa Mu Epsilon. Subspecialties: Fuels; Solar energy. Current work: Energy conservation and passive solar applications; teaching in astronomy, beginning geology, and energy-related topics.

BACKUS, CHARLES EDWARD, elec. engr., cons., researcher, educator; b. Wadestown, W.Va., Sept. 17, 1937; s. Clyde Harvey and Opal Daisy (Strader) B.; m. Judith Ann, Sept. 1, 1957; children: David, Elizabeth, Amy. B.S.M.E., Ohio U., 1959; M.S., U. Ariz., 1961, Ph.D., 1965. Supr. system dynamic analysis Westinghouse Astronuclear Lab., Pitts., 1965-68; asst. prof. engring Ariz. State U., 1968-71, assoc. prof., 1971-76, prof., 1976—, asst. dean, 1979—, dir., 1981—; with Los Alamos Sci. Lab., summers 1969-72, Lawrence Livermore (Calif.) Lab., summer 1973; cons. in field. Author 60 papers on photovoltaics and advanced energy conversion. Recipient Faculty Achievement award Ariz. State U. Alumni Assn., 1976. Fellow IEEE; mem. Am. Nuclear Soc., AAAS, Am. Soc. Engring. Edn., ASME, IEEE, Internat. Solar Energy Soc., Mesa C. of C., Sigma Xi, Pi Mu Epsilon, Tau Beta Pi. Methodist. Subspecialty: Solar energy. Current work: Research on photovoltaic concentration.

BACKUS, GEORGE EDWARD, geophysicist, educator; b. Chgo., May 24, 1930; s. Milo Morlan and Dora Etta (DAre) B.; m. Elizabeth Evelyn Allen, Nov. 25, 1961; m. Varda Esther Peller, Jan. 8. 1977; children: Benjamin, Brian, Emily. Ph.B., U. Chgo., 1947, S.B., 1948, S.M., 1950, Ph.D., 1956. Asst. examiner U. Chgo., 1949-50; jr. mathematician Inst. Air Weapons Research, Chgo., 1951-53; physicist Project Matterhorn, Princeton, N.J., 1957-58; asst. prof. math. M.I.T., 1958-60; assoc. prof. geophysics U. Calif.-San Diego, 1960-62, prof., 1962—. Research publs. in theoretical seismology, geomagnetism, plate tectonics, sci. inference, differential geometry and numerical analysis. Guggenheim fellow, 1963, 70; AEC grantee, 1954-57; NSF grantee, 1963-74, 83—. Fellow Am. Geophys. Union; mem. U.S. Nat. Acad. Scis., Am. Acad. Arts and Scis., Royal Astron. Soc., N.Y. Acad. Scis., Am. Math. Soc., Am. Phys. Soc. Subspecialties: Geophysics; Planetary science. Current work: Anisotrpic elasticity; electrical conductivity of earth's mantle; motion of earth's core; geomagnetic dynamo theory. Office: 1GPP U Calif-San Diego A-025 La Jolla CA 92093

BACKUS, JOHN, computer scientist; b. Phila., Dec. 3, 1924; s. Cecil Franklin and Elizabeth (Edsall) B.; m. Una Stannard, 1968; children: Karen, Paula. B.S., Columbia U., 1949, A.M., 1950. Programmer IBM, N.Y.C., 1950-53, mgr. programming research, 1954-59; staff mem. IBM T.J. Watson Research Center, Yorktown Heights, N.Y., 1959-63; IBM fellow IBM Research, Yorktown Heights and San Jose, Calif., 1963—. Editorial bd.: Internat. Jour. Computer and Info. Scis. Served with AUS, 1943-46. Recipient W. Wallace McDowell award IEEE, 1967; Nat. medal of Sci., 1975; Harold Pender award Moore Sch. Elec. Engring., U. Pa., 1983; Achievement award Indsl. Research Inst., Inc., 1983. Mem. Nat. Acad. Engring., Nat. Acad. Scis., Assn. Computing Machinery (Turing award 1977), Am. Math. Soc., European Assn. Theoretical Computer Sci. Subspecialties: Programming languages; Foundations of computer science. Current work: Function level programming; algebra of programs; algebraic transformation and optimization; functional equations, combinatory foundations for programming. System designer IBM 704, Fortran programming lang., Backus-Naur Form Lang., functional level programming; mem. design group ALGOL 60 lang. Home: 91 St Germain Ave San Francisco CA 94114 Office: IBM Research Lab 5600 Cottle Rd San Jose CA 95193

BACON, JONATHAN PETER, neurobiologist, educator; b. Sheffield, Eng., Aug. 26, 1950; s. Norwood Peter and Margaret Joan (Barker) B.; m. Marion Margarete Nicolai Hedwig, Oct. 26, 1956. B.A., Cambridge U., 1972; M.Sc., Manchester (Eng.) U., 1975, Ph.D., 1977. Postdoctoral research fellow Max-Planck Institut fur Verhaltensphysiologie, Seewiesen, W. Ger., 1977-80; research assoc. SUNY, Albany, 1980—, tchr. neurobiology, 1980—. Mem. Am. Soc. for Neuroscis., U.K. Soc. for Exptl. Biology, Deutsche Zoologische Gesellschaft. Subspecialties: Neurobiology; Developmental biology. Current work: Physiology, anatomy and development of simple invertebrate nervous systems. Office: Dept Biology SUNY 1400 Washington Ave Albany NY 12222

BACON, VINTON WALKER, civil engr., educator; b. Estelline, S.D., Dec. 21, 1916; s. Ernest Vinton and Emma Omar (Edwards) B.; m. Margaret Ann Pratt, May 29, 1940; children—Robert Vinton, Kathryn Ann, Don Edwards, Vinton Walker. A.A., Los Angeles Jr. Coll., 1937; B.S. in Civil Engring, U. Calif., Berkeley, 1940. Diplomate: Am. Acad. Environ. Engrs. Engrs. East Bay City Sewage Disposal Survey, Berkeley, 1940-41; designer Los Angeles County Sanitation Dists., 1941-43, 46; office engr. Orange County Sewerage Survey, Santa Ana, Calif., 1946-49; exec. officer Calif. State Water Pollution Control Bd., Sacramento, 1950-56; exec. sec. N.W. Paper and Pulp Assn., Tacoma, 1956-62; gen. supt. Met. San. Dist. Greater Chgo., 1962-70; prof. civil engring. U. Wis., Milw., 1970—; cons. in field. Contbr. numerous papers to profl. jours. Chmn. Wis. Gov.'s Solid Waste Recycling Task Force, 1971-76; mem. Wis. Solid Waste Recycling Authority, 1974-76. Served with USPHS, 1943-46. Named Constrn. Man of Year Engring. News-Record, 1967; Recipient Silver Beaver award Boy Scouts Am., 1962. Fellow ASCE (awards 1956, 67), Inst. Water Resources (hon.), Am. Pub. Health Assn.; mem. Water Pollution Control Fedn., Am. Water Works Assn., Nat. Acad. Engrs. Subspecialties: Civil engineering; Water supply and wastewater treatment. Current work: Wastewater reclamation and reuse, sewage sludge application to enhance crop production, toxicology aspects of reuse of waters and sludge, socio- economic aspects of points non-point sources of wastes. Originator deep tunnel project and land utilization of sewage sludge project Met. San. Dist. Greater Chgo., 1965. Home: 4634 N Wilshire Rd Milwaukee WI 53211 Office: EMS Bldg University of Wisconsin-Milwaukee 3200 N Cramer St Milwaukee WI 53201

BACON-PRUE, ANSLEY, psychology educator; b. Bay City, Mich., Aug. 31, 1951; d. Harry Lentz and Barbara Ann (Kennedy) Bacon; m. Donald Michael Prue, June 13, 1979. B.A., Western Mich. U., 1973, M.A., 1974; Ph.D., W.Va. U., 1977. Lic. psychologist, Miss. Dir. psychology dept. Hudspeth Ctr., Whitfield, Miss., 1977-78; dir. Univ. Affiliated Program, Jackson, Miss., 1978—; asst. prof. psychology U. So. Miss., Hattiesburg, 1979; adj. asst. prof. psychology U. Miss. Med. Ctr., Jackson, 1979—. Editor: Proc. Ky. Autism Conf, 1982; contbr. chpts. to books, articles to profl. jours. Bd. dirs. Supervised Apt. Project, Jackson, Willowood Devel. Ctr., Jackson; adv. bd. Resource Access Project Head Start; mem. Devel. Disabilities Planning Council. Dept. HHS grantee, 1978, 79, 80, 81, 82. Mem. Am. Psychol. Assn., Miss. Psychol. Assn., Assn. Behavior Analysis, Am. Assn. Mental Deficiency (chpt. pres.-elect 1982, Miss. chpt. pres. 1980-81). Subspecialty: Behavioral psychology. Current work: Research interests: manpower development and program development in the area of developmental disabilities, use of technology in mental retardation, teaching vocational and independent living skills to adults with developmental disabilities. Office: Univ Affiliated Program 1102 Robert E Lee Bldg Jackson MS 39201 Home: 146 Westlake Dr Brandon MS 39042

BACOPOULOS, NICHOLAS G., neurobiol research scientist, educator; b. Athens, Greece, Mar. 13, 1949; s. George and Antigoni (Vithoulka) B.; m. Calypso Gounti, Aug. 28, 1980. Ph.D. in Pharmacology, U. Iowa, 1976. Postdoctoral fellow Yale U., 1976-79; asst. prof. pharmacology and psychiatry Dartmouth Med. Sch., Hanover, N.H., 1979-83, adj. asso. prof., 1983—; head CNS sect. Pfizer Central Research, Groton, Conn., 1983—. Contbr. articles to profl. jours., chpts. to books. Mem. Am. Soc. Pharmacology and Exptl. Therapeutics, Soc. Neurosci. N.Y. Acad. Scis. Subspecialties: Pharmacology; Neurochemistry. Current work: Pharmacology of central dopamine receptors. Home: 28 Twin Lakes Dr Waterford CT Office: Pfizer Central Research Eastern Point Rd Groton CT 06340

BADDER, ELLIOTT, physician, surgeon; b. Phila., Feb. 7, 1943; s. Bernard and Charlotte (Brownstein) B.; m. Susan Stevenson, Jan. 9, 1971; 1 son, Nathaniel. B.A., U. Pa., 1963; M.D., Thomas Jefferson U., 1967. Resident in surgery Columbia-Presbyn. Hosp., N.Y.C., 1967-72; asst. prof. surgery Pa. State U., Hershey, 1975-77, U. Md., Balt., 1977-81, assoc. prof., 1981—; acting chief surgery Mercy Hosp., Balt., 1983. Served to maj. USAF, 1973-75. Fellow Soc. Univ. Surgeons, Endocrine Soc., ACS, Balt. Acad. Surgery, Am. Assn. Endocrine Surgeons. Subspecialty: Surgery. Current work: Determinants of sympathetic activity. Office: 301 Saint Paul Pl Baltimore MD 21202

BADGLEY, FRANKLIN ILSLEY, atmospheric science educator; b. Mansfield, Ohio, Dec. 20, 1914; m., 1943; 2 children. B.S., U. Chgo., 1935; M.S., NYU, 1949, Ph.D., 1951. Chemist Swift & Co., 1936-42; meteorologist Trans World Airlines, Inc., 1947; from instr. to assoc. prof. U. Wash., Seattle, 1950-67, prof. meteorology, 1967—, prof. atmospheric sci., 1973—, dept. chmn., 1977—; assoc. dir. Quarternary Research Ctr., U. Wash., 1973-76. Mem. AAAS, Am. Meteorol. Soc., Am. Geophys. Union. Subspecialty: Meteorology. Office: Dept Atmospheric Sci U Wash Seattle WA 98195

BAER, ERIC, educator; b. Nieder-Weisel, Germany, July 18, 1932; came to U.S., 1947, naturalized, 1952; s. Arthur and Erna (Kraemer) B.; m. Ana Golender, Aug. 5, 1956; children—Lisa, Michelle. M.A., Johns Hopkins, 1953, D.Engring., 1957. Research engr., polychems. dept. E.I. du Pont de Nemours & Co., Inc., 1957-60; asst. prof. chemistry and chem. engring. U. Ill., 1960-62; asso. prof. engring. Case Inst. Tech., 1962- 66; prof., head dept. polymer sci. Case Western Res. U., 1966-78; dean Case Inst. Tech., 1978—; cons. to industry, 1961—. Author articles in field.; Editor: Engineering Design for Plastics, 1963, Polymer Engineering and Science, 1967—. Recipient Curtis W. McGraw award ASEE, 1968. Mem. Am. Chem. Soc. (Borden award 1981), Am. Phys. Soc., Am. Inst. Chem. Engring., Soc. Plastics Engring. (internat. award 1980), Plastics Inst. Am. (trustee). Subspecialties: Polymer chemistry; Physical chemistry. Home: 2 Mornington Ln Cleveland Heights OH 44106 Office: Case Western Res Univ: Cleveland OH 44106

BAER, FERDINAND, meteorologist, educator; b. Dinkelsbuhl, Germany, Aug. 30, 1929; s. Julius F. and Leonie (Rothschild) B.; m. Karen K. Klein, July 25, 1977; children: Darius S., Robin A., Jason F., Gavin L. B.A., U. Chgo., 1950, M.S., 1954, Ph.D., 1961. Asst. prof. Colo. State U., 1961-65, assoc. prof., 1965-71; prof. U. Mich., Ann Arbor, 1972-77; prof. meteorology U. Md., College Park, 1977—, dir. meteorology program, 1977-79, chmn. dept. meteorology, 1979—; cons. McGraw-Hill, 1968-79; U. Md. rep. to Univ. Corp. for Atmospheric Research, 1977-79; mem. bd. atmospheric scis. and climate NRC-Nat. Acad. Scis., 1982—; mem. adv. com. Nat. Ctr. Atmospheric Research, 1982—. Contbr. articles to profl. jours. Bd. dirs. Washington Recorder Soc., 1979—; active Am. Recorder Soc. Research grantee NSF, Dept. Def., NOAA, NASA, Dept. of Energy. Fellow Am. Meteorol. Soc., Royal Meteorol. Soc.; mem. Am. Geophys. Union, Japanese Meteorol. Soc., AAAS, Can. Meteorol. and Oceanic Soc., Sigma Xi. Subspecialties: Meteorology; Atmospheric modeling. Current work: Numerical weather prediction, atmospheric waves, atmospheric dynamics, research into atmospheric flow. Office: U Maryland 2207 Space Scis Bldg College Park MD 20742

BAER, ROBERT LLOYD, nuclear engineering manager; b. N.Y.C., Jan. 29, 1931; s. Leonard J. and Rose (Ritz) B.; m. Marlene Furst, Dec. 19, 1954 (dec. Aug. 29, 1974); children: James, Pamela; m. Edith Muller, Apr. 4, 1976; 1 son, Leonard. B.M.E., Clarkson Coll., 1952; M.S.M.E., Rensselaer Poly, Inst., 1958. Registered profl. engr., N.Y., Md. Engr. Pratt & Whitney, Middletown, Con., 1954-57, sr. engr., 1957-59; project engr. Martin Co., Balt., 1959-63; dept. mgr. Hittman Corp., Columbia, Md., 1963-68, v.p., 1968-70; sr. project mgr. U.S. NRC, Bethesda, Md., 1971-77, br. chief, 1977—; lectr. dept. nuclear engring. Cath. U. Am., 1963-76. Served to 1st lt. U.S. Army, 1952-54; Korea. Mem. ASME (chmn. honors and awards Balt. sect. 1968-71, sect. bd. dirs. 1970-71), Am. Nuclear Soc., Tau Beta Pi, Pi Tau Sigma. Democrat. Jewish. Subspecialties: Nuclear fission; Mechanical engineering. Current work: Safety evaluations and probablistic risk analysis of commercial nuclear power plants. Home: 2844 Blue Spruce Ln Wheaton MD 20906 Office: US NRC Washington DC 20555

BAER, THOMAS MICHAEL, physicist; b. Baraboo, Wis., July 28, 1952; s. Joseph and Mary Rosina (Lazar) B.; m. Barbara Jo Weesen, July 7, 1979. B.A., Lawrence U., 1974; M.S., U. Chgo., 1976, Ph.D., 1979. Teaching/research asst. U. Chgo., 1974-79; research asst. Argonne (Ill.) Nat. Lab., 1976; postdoctoral research asso. Joint Inst. for Lab. Astrophysics, Boulder, Colo., 1979-81; sr. scientist Research and Advanced Devel. Group Spectra-Physics, Mountain View, Calif., 1981—. Contbr. articles to profl. jours. Mem. Optical Soc. Am., Am. Phys. Soc., Nat. Geographic Soc., Amateur Radio Relay League. Subspecialty: Atomic and molecular physics. Current work: High resolution atomic/molecular spectroscopy, laser and laser applications. Office: 1250 W Middlefield Rd 2 45 Mountain View CA 94042

BAER, THOMAS STRICKLAND, health and environmental safety executive; b. Huntington, W.Va., May 24, 1942; s. Peter Harrison and Virginia (Strickl) B.; m. Margaret Thresa Durkin, Nov. 21, 1964; children: Kathleen Nancy, Thomas Holman. B.S., U.S. Naval Acad., 1964; M.S., U. Cin., 1971, Ph.D., 1973. Sta. engr. Met. Edison Co., Middletown, Tenn., 1973-74; v.p., gen. mgr. Protective Packaging Inc., Louisville, 1974-76; v.p. Nuclear Engring. Co., Louisville, 1976-81; health physics and environ. safety mgr. Bechtel Nat. Inc., Oak Ridge, 1981—. Cub scout leader Old Ky. Home council, Louisville, 1975-78, commr., 1978-81; scoutmaster Great Smokey Mountain council Boy Scouts Am., Knoxville, Tenn., 1981—. Served to lt. USN, 1964-69; Vietnam. Mem. Am. Nuclear Soc., Health Physics Soc., Scientists and Engrs. for Secure Energy. Republican. Roman Catholic. Subspecialties: Nuclear engineering; Environmental engineering. Current work: Development of methods and technique to safely dispose of hazardous materials. Home: 1004 High Springs Rd Knoxville TN 37922 Office: Bechtel Nat Inc Oak Ridge TN 37830

BAERTSCH, RICHARD DUDLEY, electrical engineer; b. Mpls., Mar. 14, 1936; s. Dudley and Sylvia Pearl (Kuglin) B.; m. Diane Caroll Maltby, July 2, 1960; children: Robert, Suzanne, Laura. B.S.E.E., U. Minn., 1959, M.S.E.E., 1961, Ph.D., 1964. Mem. tech. staff Gen. Electric Research and Devel. Center, Schenectady, 1964—. Contbr. articles to profl. jours. Mem. IEEE (sr.). Subspecialties: Integrated circuits; Microchip technology (engineering). Current work: Integrated circuits. Patentee in field. Office: Gen Electric Corp Research and Devel PO Box 8 Schenectady NY 12301

BAGCHI, NANDALAL, endocrinologist, medical educator; b. Kushtia, Bangladesh, Aug. 1, 1936; came to U.S., 1963; s. Aswini Kumar and Joydurga (Sanyal) B.; m. Arleta Jane Flansburg, Jan. 6, 1968; children: Ranjan, Monika. B.Sc. with honors, U. Calcutta, 1954, M.B.B.S., 1960; Ph.D., U. Alta., Can., 1972. Lic. physician, Mich., diplomate: Am. Bd. Internal Medicine (subcert. in endocrinology and metabolism). Asst. prof. medicine Wayne State U., Detroit, 1973-78, assoc. prof., 1978—; chief sect. of endocrinology and metabolism Hutzel Hosp., Detroit, 1979—. Author: Special Topics in Endocrinology and Metabolism, 1982; contbr. articles to med. jours. Research grantee Skillman Found., 1974-77, Mich. Heart Found., 1975-77, Mich. Kidney Found., 1980-81. Fellow Royal Coll. Physicians and Surgeons (Can.), ACP; mem. Endocrine Soc., Am. Thyroid Assn. Subspecialties: Endocrinology; Physiology (medicine). Current work: Control of thyroid gland secretion; basic mechanisms of thyroid biochemistry. Home: 2833 Palmerston Troy MI 48084 Office: Hutzel Hosp Wayne State U 47-7 Antoine Detroit MI 48201

BAGGETT, BILLY, biochemist, consultant; b. Oxford, Miss., Oct. 23, 1928; s. Lee and Estelle (Brown) B.; m. Harriette Brady Lane, Nov. 23, 1949 (div.); children: Sallie, William, Mary, Teresa, H. Lane. B.A., U. Miss., 1947; Ph.D., St. Louis U., 1952. Asst. biochemist Mass. Gen. Hosp., Boston, 1952-57; instr. biol. chemistry Harvard U., 1952-57; asst. prof.pharmacology and biochemistry U. N.C., Chapel Hill, 1957-59, asso. prof., 1959-69, prof., 1969; prof. biochemistry Med. U. S.C., Charleston, 1969—, chmn. dept., 1969-78, prof. ob-gyn, 1980—; biochemist Nat. Inst. Child Health and Human Devel., 1978-80; cons. Contbr. many articles to profl. jours. ; NIH sr. research fellow, 1957-62; recipient numerous grants NIH, Population Council, Am. Heart Assn. Mem. AAAS, Am. Chem. Soc., Am. Soc. Biol. Chemists, Endocrine Soc., Am. Assn. Cancer Research, Soc. Study Reprodn. Subspecialties: Biochemistry (biology); Reproductive biology. Current work: Biochemistry of sex hormones and of implantation, research on biochemistry of implantation and of androgen action; consultation on projects related to side effects of oral contraceptives. Home: PO Box 623 Folly Beach SC 29439 Office: Dept Ob-Gyn Med S.C. Charleston SC 29425

BAGLEY, BRIAN G., materials science researcher; b. Racine, Wis., Nov. 20, 1934; s. Wesley John and Ethel Sophie (Rasmussen) B.; m. Dorothy Elizabeth Olson, Nov. 20, 1959; children: Brian John, James David, Kristin Marie. B.S., U. Wis., 1958, M.S., 1959; A.M., Harvard U., 1964, Ph.D., 1968. Mem. tech. staff Bell Labs., Murray Hill, N.J., 1967—. Contbr. articles in field to profl. jours.; patentee in field. Served to lt. U.S. Army, 1960-61. Xerox predoctoral fellow Harvard U., 1964-66; R.J. Painter fellow, 1966-67. Mem. Am. Phys. Soc., Materials Research Soc. Lutheran. Subspecialties: Electronic materials; Fiber optics. Current work: Basic research in the physics and chemistry of amorphous materials, including preparation, characterization and applications in electronic and optical devices. Home: 467 Ridge Rd Watchung NJ 07060 Office: Bell Labs Rm 1D-445 600 Mountain Ave Murray Hill NJ 07974

BAGNALL, LARRY OWEN, engineering educator; b. Enumclaw, Wash., Dec. 19, 1935; s. Richard Henry and Juanita May (Wheeler) B.; m. Shirley Anne VonAhn, July 26, 1958; children: Juanita Jean, Mark Owen, Edith Anne. B.S.A.E., Wash. State U.-Pullman, 1957; Ph.D., Cornell U., 1967. Registered profl. engr., Fla. Engring. scientist Allis Chalmers, West Allis, Wis., 1957-66; grad. asst. Cornell U., Ithaca, N.Y., 1962-66; research engr. Allis-Chalmers, West Allis, 1966-68; asst. prof. U. Fla., Gainesville, 1969-74, assoc. prof., 1974-81, prof. engring., 1981—; panelist Nat. Acad. Scis., Washington, 1973, 75; instr. AID, Washington, 1977. Scoutmaster Boy Scouts Am., Gainesville, Fla., 1975-83. Mem. Am. Soc. Agrl. Engrs., Aquatic Plant Mgmt. Soc. Lutheran. Subspecialties: Agricultural engineering; Biomass (energy science and technology). Current work: Management, harvesting, processing and utilization of aquatic plants, including determination of properties, machine and system development and production of feed, soil ammendments and biogas. Inventor radial expansion compensating support, waterhyacinth harvester-processor. Home: 2128 NW 29th Pl Gainesville FL 32605 Office: Agrl Engring Dept U Fla Gainesville FL 32611

BAGWELL, JOYCE MARIE BURRIS, geology-chemistry educator, researcher; b. Charleston, S.C., July 6, 1932; d. Chalmers Eugene and Jennie Louise (Hall) Burris; m. W. Howard Bagwell, June 12, 1954; children: W. Howard Jr., John Frederick. B.A., Furman U., Greenville, S.C., 1954; M.S., Clemson U., S.C., 1962. Tchr. Pendleton (S.C.) High Sch., 1954-55; tchr. sci. S.W. Daniel High Sch., Clemson, 1955-57, 58-60, McCants Jr. High Sch., Anderson, S.C., 1960-65; tchr. biology Anderson Jr. Coll., 1962-63; tchr. sci. 1st Baptist High Sch., Charleston, 1966-67; asst. prof. geology and chemistry Baptist Coll., Charleston, 1967—; prin. investigator lower S.C. mini-network U.S. Geol. Survey, 1976—. Contbr. articles to profl. jours. Mem. Eastern Seismological Soc. Am., Nat. Tchrs. Assn., S.C. Tchrs. Assn., Alpha Delta Kappa. Republican. Baptist. Subspecialties: Geology; Seismology. Current work: Isoseismal studies of felt earthquakes in lower South Carolina; monitoring eleven station, thirty one channel seismic network in lower South Carolina; teaching chemistry and geology. Home: 127 Sycamore Dr Summerville SC 29483 Office: Dept Chemistry/Geology Baptist Coll PO Box 10087 Charleston SC 29411

BAHCALL, JOHN NORRIS, astrophysicist; b. Shreveport, La., Dec. 30, 1934; s. Malcolm and Mildred (Lazarus) B.; m. Neta Assaf, Sept. 21, 1966; children—Ron Assaf, Dan Ophir, Orli Gilat. B.A., U. Calif., 1956; M.Sc., U. Chgo., 1957; Ph.D., Harvard U., 1961. Research asso. Ind. U., 1961-62; theoretical physicist, research asso., asst. prof., then asso. prof. theoretical physics Calif. Inst. Tech., 1962-70; mem. Inst. Advanced Study, Princeton, N.J., 1968-70, prof. natural scis., 1971—; mem. physics advisory panel NSF; mem.-at-large, large space telescope mgmt. and ops. working group NASA; mem. com., div. high energy astrophysics; mem. x-ray astron com. URA. Author: (with Field and Arp) The Redshift Controversy, 1973. Sloan Found. fellow, 1968-71. Fellow Am. Phys. Soc.; mem. Nat. Acad. Scis., Nat. Acad. Arts and Scis., Am. Astron. Soc. (Helen B. Warner prize 1969), Internat. Astron. Union. Jewish. Subspecialty: 1-ray high energy astrophysics. Address: Inst Advanced Study Princeton NJ 08540

BAHCALL, NETA ASSAF, astrophysicist, educator, consultant, lecturer; b. Israel, Dec. 16, 1942; d. Yehezkel Oscar and Gita (Zilberstein) Assaf; m. John Norris Bahcall, Mar. 21, 1966; children: Ron Assaf, Dan Ophir, Orli Gilat. B.S., Hebrew U., Jerusalem, Israel, 1963; M.S., Weizmann Inst. Sci., Israel, 1965; Ph.D., Tel Aviv U., Israel, 1970. Research fellow Calif. Inst. Tech., 1970-71; staff mem. Princeton (N.J.) U., 1971-75, research astronomer, 1975-79, sr. research astronomer, 1979—. Contbr. articles to profl. jours. Mem. Am. Astron. Soc. Subspecialties: Optical astronomy; Cosmology. Current work: Research in galactic and extragalactic astronomy, including the study of the nature and properties of galaxies, clusters of galaxies, superclusters and the large-scale structure of the universe. Office: Peyton Hall Princeton U Princeton NJ 08540

BAHE, LOWELL WARREN, chemistry educator, researcher; b. Sycamore, Ill., Jan. 30, 1927; s. Herman William and Myra May (Snow) B.; m. Virginia Ruth Peterson, June 12, 1954 (div.); children: Margaret, Laurel, Ellen. B.S., Purdue U., 1949; M.A., Princeton U., 1951, Ph.D., 1953. Research chemist Allis Chalmers Mfg. Co., West Allis, Wis., 1953-57; asst. prof., then assoc. prof., prof. chemistry U. Wis., Milw., 1957—. Contbr. articles on chemistry to profl. jours. Served with USN, 1945-46. Mem. Am. Chem. Soc., Wis., Acad. Arts, Letters and Sci., Sigma Xi. Unitarian. Subspecialties: Physical chemistry; Thermodynamics. Current work: Electrolyte solutions, dielectric constants, research on behavior of ions in media with high dielectric constants. Home: 2106 E Newton Blvd Shorewood WI 53211 Office: Chemistry Dept U Wis Milwaukee WI 53201

BAHN, ARTHUR NATHANIEL, microbiologist, immunologist, researcher; b. Boston, Jan. 5, 1926; s. Benjamin Edward and Mollie Sarah (Abramson) B.; m. Marna Jean Katz, Dec. 7, 1952; children: Lisa M., Janice L., Bruce R. A.B., Boston U., 1949; M.A., U. Kans.-Lawrence, 1952; Ph.D., U. Wis.-Madison, 1956. Instr. microbiology, immunology U. Ill., Chgo., 1956-58; asst. prof. Northwestern U., 1958-60, assoc. prof., 1960-70; prof., chmn. So. Ill. U., 1971—; panel mem. oral cavity drug products FDA, Washington, 1974-80; cons. Am. Dental Assn., U.S. Navy. Mem. Am. Assn. Dental Research, Internat. Assn. Dental Research, AAAS. Jewish. Subspecialties: Oral biology; Infectious diseases. Current work: Immunization against dental caries with bacterial enzymes, bacterial neuraminidases in adherence of bacteria to oral glycoprotein. Home: 112 Beau Jardin Ct Saint Louis MO 63141 Office: So Ill U 2800 College Ave Alton IL 62002

BAHNER, CARL TABB, chemist, educator; b. Conway, Ark., July 14, 1908; s. Gustavus Lonsford and Augusta Thomas (Moore) B.; m. Mary Catharine Garrott, Sept. 17, 1931; children: Thomas Maxfield, Mary Catharine, Frances Jane. A.B., Hendrix Coll., 1927; M.S., U. Chgo., 1928; Th.M., So. Bapt. Theol. Sem., 1931; postgrad., Yale U., 1931-32; Ph.D., Columbia U., 1937. Head physics dept. Union U., 1936-37; head chemistry dept., prof. Carson-Newman Coll., Jefferson City, Tenn., 1937-67, research coordinator, 1967-73; assoc. prof. Walters State Community Coll., Morristown, Tenn., 1973-78; prof. chemistry Bluefield (Va.) Coll., 1979—; cons. TVA, 1941-45; med. div. Oak Ridge Inst. Nuclear Studies, 1950-56, Oak Ridge Nat. Labs., 1948-80; head natural sci unit, test devel. sect. U.S. Civil Service, 1945; sr. research chemist Roswell Park Meml. Hosp., 1956; research chemist Chester Beatty Research Inst., 1957. Contbr. articles to profl. jours. Chmn. Charter Revision Com., City of Jefferson City, 1978-79. Recipient Fla. award Fla. sect. Am. Chem. Soc., 1964; award for excellence in chemistry teaching Mfg. Chemists Assn., 1967; Hendrix Coll. Disting. Alumnus award, 1969; Algernon Sydney award, 1969; Disting. Service award Walters State Community Coll., 1978. Fellow AAAS, Am. Inst. Chemists, Tenn. Acad. Sci. (pres. 1951); mem. Am. Assn. Cancer Research, AAUP, Am. Chem. Soc. (chmn. E. Tenn. sect. 1951), Tenn. Inst. Chemists (pres. 1972-73), W.Va. Acad. Sci., Sigma Xi, Alpha Chi, Phi Lambda Upsilon, Sigma Pi Sigma. Subspecialties: Organic chemistry; Chemotherapy. Current work: Cancer chemotherapy, carcinogenesis, structure activity relations; radiation effects on organic compounds. Patentee in field. Home: PO Box 549 Jefferson City TN 37760 Office: Bluefield Coll Bluefield VA 24605

BAHNG, JOHN DEUCK RYONG, astronomy educator; b. Bookchung, Korea, Mar. 21, 1927; m., 1961; 3 children. B.S., St. Norbert Coll., 1950; M.S., U. Wis., 1954, Ph.D., 1957. Research assoc. astronomy Princeton U., 1957-62; asst. prof. Northwestern U., Evanston, Ill., 1962-66, chmn. dept., 1975-77, assoc. prof. astronomy, 1966—, dir., 1975—. Mem. Am. Astron. Soc., Royal Astron. Soc., Internat. Astron. Union. Subspecialty: Optical astronomy. Office: Dearborn Obs Northwestern U Evanston IL 60201

BAI, TAEIL, astrophysicist; b. Jeonnam, Korea, July 16, 1945; came to U.S., 1972; s. Jong-hun and Soo-bong (Suh) B.; m. Suin Kim; children: Samuel, Jean, Helen. B.S., Kyung Hee U., Seoul, Korea, 1967; Ph.D., U. Md., 1977. Research assoc. U. Md., 1977-78; postgrad. research physicist U. Calif.-San Diego, 1978-80; asst. research physicist U. San Diego, 1980-82; sr. research assoc. Stanford U., Calif., 1982—. Contbr. articles to profl. jours. Served with Korean Army, 1967-69. Recipient Donald E. Billings award in astrogeophysics U. Colo., Boulder, 1978. Mem. Am. Phys. Soc., Am. Astron. Soc. Subspecialty: Solar physics. Current work: High energy phenomena in solar flares; x-ray astronomy. Office: Inst Plasma Research Stanford U Stanford CA 94305

BAILEY, FREDERICK EUGENE, JR., polymer scientist; b. Bklyn., Oct. 8, 1927; s. Frederick Eugene and Florence (Berkeley) B.; m. Mary Catherine Lowder, May 7, 1979. B.A., Amherst Coll., 1948; M.S., Yale U., 1950; Ph.D., 1952. Sr. chemist Union Carbide Research Devel., 1952-59, group leader, 1959-62, asst. dir., 1962-69, mgr. mktg. research, N.Y.C., 1969-71, sr. research scientist, South Charleston, W.V.A., 1971—; adj. prof. chemistry Marshall U., Huntington, W.Va., 1975—; adj. prof. chem. engring. W.Va. Coll. Grad. Studies, 1981—; lectr. polymer chemistry U. Charleston, Morris Harvey Coll., 1962-63, 65; mem. grad. faculty W.Va. U., 1959-61; chmn. Gordon Research Conf. on Polymers, 1972, N.H. chmn., 1984; mem. Gordon Research Conf. Council. Author: Polyethylene Oxide, 1976; editor: Initiation of Polymerization, 1983, (with K.N. Edwards) Urethane Chemistry and Applications, 1981. Addison Brown scholar Amherst Coll., 1948; Forrest Jewett Moore fellow, 1949. Fellow AAAS, Am. Inst. Chemists (cert. chemist), N.Y. Acad. Scis.; mem. Am. Chem. Soc. (chmn. divisional officer caucus 1980—, chmn. div. polymer chemistry 1976, councilor div. 1978-84, com. sci. 1978, 82-83, gen. sec. Macromolecular secretariat 1978, Top DOG award 1983). Republican. Episcopalian. Clubs: Williams (N.Y.C.); Tennis (Charleston, W.Va.). Subspecialties: Polymer chemistry; Polymer physics. Current work: Synthesis and characterization of high polymer systems and market applications of such systems, current emphasis on polyurethane materials. Patentee in field. Home: 848 Beaumont Rd Charleston WV 25314 Office: Union Carbide Corp Tech Center South Charleston WV 25303

BAILEY, GORDON BURGESS, biochemistry educator, researcher; b. Worcester, Mass., Feb. 13, 1934; s. John M. and Emma B. B.; m. Dorothy C. Cummings, July 5, 1958; children: Andrea L., Amy B., Joshua J. B.A., Brown U., 1956; M.A., U. Mass., 1961; Ph.D., U.Fla., 1966. Mem. field staff Rockefeller Found., 1966-73; asst. prof. St. Louis U. Med. Sch. (overseas), Thailand, 1973-76; cons. WHO, Thailand, 1975-76; prof. bio-chemistry Morehouse Sch. Medicine, 1976—. Contbr. articles on biochemistry and cell biology to profl. jours. Served to lt. comdr. USN, 1956-59. Mem. AAAS, Am. Soc. Microbiology, Am. Soc. Parasitology/Protozoology. Subspecialties: Biochemistry (medicine); Cell biology (medicine). Current work: Differentiation, chemotropic and pathogentic mechanisms of Entamoeba. Home: 2133 Eldorado Dr Atlanta GA 30345 Office: Dept Biochemistry Morehouse Sch Medicine Atlanta GA 30310

BAILEY, IAN LAURENCE, optometry educator; b. Melbourne, Australia, Dec. 22, 1940; came to U.S., 1976; s. Laurence Frank and Lucie Lillian (Downward) B.; m. Valerie May, Feb. 16, 1963; children: Adina Michele, David Laurence. M.S., Ind. U., 1970; B.Applied Sci., U. Melbourne, Australia, 1962; postgrad., City U., Eng., 1965-66. Diplomate: in low vision Am. Acad. Optometry. Asst. prof. Sch. Optometry, U. Calif.-Berkeley, 1976-80, assoc. prof., 1980—. Life gov. Assn. for Blind, Melbourne. Fellow Am. Acad. Optometry (chmn. low vision diploma program 1982—), Brit. Coll. Optometrists, Brit. Optical Assn. Subspecialties: Optometry; Psychophysics. Current work: Low vision, ophthalmic optics, physiological optics, occupational vision. Office: Sch Optometry U Calif Berkeley CA 94720

BAILEY, JAKE S., elec. engr.; b. Middlesboro, Ky., Dec. 29, 1927; s. Charles Wise and Mary Elizabeth (Nice) B.; m. Barbara Jean McClelland, Sept. 11, 1947; children: Linda Heguy, Mimi McDonough, Alan. B.S., U. Ala., 1949; postgrad., U. Minn., 1954. Registered profl. engr., N.Y., N.J., Pa., Del., Md., Fla., Kans. Substa. design engr. Memphis Light Gas and Water, 1949-52; project electronics engr. Boeing Corp., Wichita, Kans., 1952-54; autopilot design engr. Honeywell Aero., Mpls., 1954-58; electronics design supr. Link/Singer, Binghamton, N.Y., 1958-60; Mgr. exptl. methods Gen. Electric Aerospace, Phila., 1960-69; pres. B&G Corp., Valley Forge, Pa., 1969-74; elec. engring. cons., Phoenixville, Pa., 1974-81; mgr. elec. engring. ADCI, Milan, Italy, 1981; chief elec. engr. Haines, Lundbere Waehler, N.Y.C., 1981—. Served with USNR. Sr. mem. IEEE; Mem. Illumination Engrs. Soc., Aircraft Owners and Pilots Assn. Subspecialties: Electrical engineering; Electronics. Current work: Electrical design, control design, energy studies. Office: 2 Park Ave New York NY 10016

BAILEY, JOHN ALBERT, mech. engr., cons., researcher, educator; b. Liverpool, Eng., June 8, 1937; came to U.S., 1963, naturalized, 1972; s. John Albert and Phylis Cathrine (Monk) B.; m. Anne, May 7, 1963; children: Michelle Allison, Sharon Denise. B.Sc. in Metallurgy, Univ. Coll., Swansea, U.K., 1960, Ph.D., 1963. Asst. prof. mech. engring. Ga. Inst. Tech., 1963-65, assoc. prof., 1965-67; assoc. prof. mech. engring. N.C. State U., 1967-71, prof., 1971—. Contbr. numerous articles to profl. jours. Recipient numerous fed. contracts for research. Fellow Instn. Metallurgists; mem. ASME, Soc. Mfg. Engrs., Sigma Xi, Phi Kappa Phi. Subspecialties: Materials processing; Metallurgy. Current work: Metal machining, surface integrity, mechanical properties of materials, analyses of metal working operations. Home: 1214 Gray Owl Garth Cary NC 27511 Office: Dept Mech Engring NC State U Raleigh NC 27650

BAILEY, ROY ALDEN, geologist; b. Providence, July 28, 1929; m., 1958; 3 children. A.B., Brown U., 1951; M.Sc., Cornell U., 1954; Ph.D., Johns Hopkins U., 1978. Geologist U.S. Geol. Survey, Reston, Va., 1953-55, 57—, now coordinator volcanic hazards program. Mem. Geol. Soc. Am., Mineral. Soc. Am., Am. Geophys. Union. Subspecialty: Volcanology. Office: US Geol Survey Nature Center 951 Reston VA 22092

BAILEY, STUART LOHR, project engineer; b. Harper, Kans., Apr. 1, 1927; s. Vernon A. and Esther Rhea (Goodnight) B.; m. Anita Jo Bean, Feb. 22, 1927; children: Michael, Patti. Student, Okla. State U., 1946-48. Draftsman O.E.M. Mfg. Co., Houston, 1948-56, supr., 1956-65, design engr., 1965-67, sr. engr., 1967-79, sr. test. engr., 1979-81, project engr., 1981—. Mem ASME. Republican. Subspecialties: Mechanical engineering; Materials (engineering). Current work: Development of oil field drilling equipment; create, develop, test, prodn. specifications, quality assurance, manufacturing trouble shooting, performance analysis. Patentee valves; single acting piston. Home: 15215 Moss Way San Antonio TX 78232 Office: 6110 Rittiman Rd San Antonio TX 78218

BAILEY, WILLIAM F., chemist, educator, researcher; b. N.J., Dec. 8, 1946. B.S., St. Peters Coll., 1968; Ph.D., U. Notre Dame, 1973. Postdoctoral assoc. U. N.C., 1973, Yale U., 1974-75; asst. prof. chemistry U. Conn., 1975-80, assoc. prof., 1980—; cons. in field. Contbr. numerous articles, abstracts to profl. publs. Mem. Am. Chem. Soc., Internat. Assn. Dental Research. Subspecialties: Organic chemistry; Nuclear magnetic resonance (chemistry). Current work: Molecular structure and energetics; Carbon-13 nuclear magnetic resonance; conformational analysis; chemistry of group I organometalics; wear behavior of dental restoratives.

BAILEY, WILLIAM JAMES, public health educator, drug abuse prevention consultant; b. Gary, Ind., Nov. 5, 1947; s. William Arthur and Alice Jeanne (Tittle) B. B.S., Ind. U., 1976, M.S., 1977, M.P.H., 1979, postgrad., 1979-83. Drug abuse cons., Bloomington, Ind., 1974-80, coordinator risk reduction project, lectr. pub. health, 1980—. Author: Drug Use In American Society, 1981; editor: periodical Bibliog. Index of Health Edn. Periodicals, 1981—. Pub. edn. chmn. Am. Cancer Soc.-Monore County, Bloomington, 1981—. Mem. Am. Pub. Health Assn., Assn. for Ednl. Communication and Tech., Eta Sigma Gamma. Republican. Roman Catholic. Subspecialties: Preventive medicine; Information systems (information science). Current work: Research in prevention of drug abuse particularly in university students and adolescents, including multivariate causation models of drug problems (tobacco, alcohol, illicit drugs); development of automated information retrieval systems in health education. Home: Apt 2-T 800 N Smith Rd Bloomington IN 47401 Office: Ind U Dept Health Edn HPER Bldg Room 116 Bloomington IN 47405

BAILIN, GARY, biochemistry educator; b. Bklyn., Apr. 2, 1936; s. Morris and Sadie (Rudolph) B.; m. Ellen Sue Seckler, Sept. 16, 1946; children: Alison, Leslie. B.S. in Chemistry, CCNY, 1958; Ph.D. in Biochemistry, Adelphi U., 1965. Assoc. mem. Inst. Muscle. Disease, N.Y.C., 1964-74; research asst. prof. Mt.Sinai Sch. Medicine, N.Y.C., 1974-77; assoc. prof. U. Med. and Dentistry N.J.-N.J. Sch. Osteo. Medicine, Piscataway, 1977—. Mem. Am. Soc. Biol. Chemists. Subspecialties: Biochemistry (biology); Biochemistry (medicine). Current work: Protein and enzyme chemistry, regulation of muscle contraction. Office: PO Box 55 Piscataway NJ 08854

BAILLIEUL, JOHN BROUARD, mathematician; b. Boise, Idaho, May 13, 1945; s. Paul Brouard and Geneva (Gillam) B.; m. Patricia Pfeiffer, June 15, 1968; children: Emily, Charlotte. B.A., U. Mass., 1967; M.Math., U. Waterloo, Ont., Can., 1969; M.S., Harvard U., 1973, Ph.D., 1975. Sr. programmer, analyst/engr., fellow medicine Mass. Gen. Hosp., Boston, 1969-74; asst. prof. math. Georgetown U., Washington, 1975-79; sr. mathematician Sci. Systems, Inc., Cambridge, Mass., 1979—, mgr. control and systems engring. div., 1981—; vis. scientist Harvard U., 1983-84. Assoc. editor: IEEE Transactions Automatic Control. Mem. Math. Assn. Am., Am. Math. Soc., Soc. Indsl. and Applied Math. Subspecialties: Applied mathematics; Systems engineering. Current work: Control and estimation of nonlinear systems, geometric and topological methods for dynamical systems. Home: 181 Common St Belmont MA 02178 Office: Sci Systems Inc 54 Rindge Ave Extension Cambridge MA 02140

BAINBRIDGE, KENNETH TOMPKINS, physicist, educator; b. Cooperstown, N.Y., July 27, 1904; s. William Warin and Mae (Tompkins) B.; m. Margaret Pitkin, Sept. 8, 1931 (dec. 1967); children—Martin Keeler, Joan, Margaret Tompkins; m. Helen Brinkley King, Oct. 11, 1969. S.B., Mass. Inst. Tech., 1926, S.M., 1926; M.A., Princeton, 1927, Ph.D., 1929; M.A. (hon.), Harvard, 1942. Physicist, 1928-29; Nat. Research Council fellow Bartol Research Found., 1929-31, Bartol Research Found. fellow, 1931-33; Guggenheim Meml. Found. fellow at Cavendish Lab., Cambridge, Eng., 1933-34; asst. prof. physics Harvard, 1934-38, asso. prof., 1938-46, prof., 1946—, chmn. dept. physics, 1953-55, George Vasmer Leverett prof. physics, 1961-75, emeritus, 1975—; now design cons. linear direct current motors; tech. cons. Nat. Def. Research Council, 1940-44, M.I.T. Radiation Lab., 1940-43, Los Alamos Lab., 1943-45; dir. Alamogordo Atomic Bomb Test, Feb.-Sept. 1945. Contbr.: tech. articles to Jour. of Franklin Inst. Trustee Asso. Univs., Inc., 1957-59; Mem. 7th Solvay Chemistry Congress, 1947. Awarded Louis Edward Levy medal Franklin Inst., 1933; Presdl. certificate of merit for work on radar, 1948. Mem. Am. Phys. Soc., Nat. Acad. Scis., Am. Acad. Arts and Scis., Alpha Tau Omega, Tau Beta Pi. Subspecialties: Nuclear physics; Applied magnetics. Current work: Linear D.C. motors. Holder of patents on photo electric cells, electronic multiplier and electro magnetic pumps. Address: 5 Nobscot Rd Weston MA 02193

BAINE, WILLIAM BRENNAN, internal medicine and microbiology educator, physician; b. Washington, Aug. 10, 1945; s. John Raymond and Alice (Brennan) B.; m. Martha Scott, Aug. 30, 1969; 1 son, Britton Alexander. A.B., Princeton U., 1966, cert. proficiency in East Asian studies, 1966; M.D., Vanderbilt U., 1970. Diplomate: Am. Bd. Internal Medicine. Intern and asst. resident in medicine Cleve. Met. Gen. Hosp., 1970-72; with Epidemic Intelligence Service, Ctr. Disease Control, Atlanta, 1972-74; resident in internal medicine Parkland Meml. Hosp., Dallas, 1974-75; fellow in infectious diseases U. Tex. Health Sci Center-Dallas, 1975-77, asst. prof. internal medicine and microbiology, 1981—; med. epidemiologist Ctr. Disease Control, Atlanta, 1977-79, Rome, 1979-81. Served to sr. surgeon USPHS, 1972, 75-81. Recipient cert. recognition and appreciation Ctr. Disease Control, Atlanta, 1978. Fellow Am. Coll. Epidemiology, ACP, Infectious Diseases Soc. Am.; mem. AAAS, Am. Fedn. Clin. Research, Am. Soc. Microbiology, Soc. Epidemiologic Research, Tex. Infectious Disease Soc. Episcopalian. Club: Colonial of Princeton U. Subspecialties: Microbiology (medicine); Epidemiology. Current work: pathogenesis and pathophysiology of legionellosis; epidemiology of acute diseases. Office: Dept Internal Medicine U Tex Health Sci Center 323 Harry Hines Blvd Dallas TX 75235 Home: 4914 Abbott Ave Dallas TX 75205

BAINTON, CEDRIC ROLAND, anesthesiologist, educator; b. New Haven, Feb. 10, 1931; s. Roland Herbert and Ruth (Woodruff) B.; m. Dorothy Dee Ford, Nov. 28, 1959; children: Roland, Bruce, James. B.A., Oberlin Coll., 1953; M.D., U. Rochester, 1958. Intern Strong Meml. Hosp., Rochester, N.Y., 1958-60; resident in anesthesia U. Calif.-San Francisco, 1962-66, mem. faculty, 1966—, prof. anesthesiology, 1980—. Contbr. articles to profl. publs. Served with USPHS, 1960-62. Recipient Merit Rev. award VA, San Francisco, 1976; NJH grantee, 1968-81. Mem. Am. Physiol. Soc., Assn. Univ. Anesthetists, Am. Soc. Anesthesiology. Democrat. Congregationalist. Subspecialty: Anesthesiology. Current work: Control of breathing. Home: 50 Ventura St San Francisco CA 94116 Office: San Franciso VA Med Center 3150 Clement St San Francisco CA 94121

BAINTON, DOROTHY DEE FORD, research physician, educator; b. Magnolia, Miss., June 18, 1933; d. Aubrey Ratcliff and Leta (Brumfield) Ford; m. Cedric Roland Bainton, Nov. 28, 1959; children: Roland J., Bruce G., James H. B.S., Millsaps Coll., Jackson, Miss., 1955; M.D., Tulane U., 1958. Intern U. Rochester Sch. Medicine, Rochester, N.Y., 1958, resident, to 1960, U. Wash., Seattle, 1960-62; prof. pathology U. Calif., San Francisco, 1962—. Recipient Research Career Devel. award NIH, 1970-75. Mem. Am. Soc. Cell Biology, Am. Assn. Pathologists, Am. Assn. Hematologists. Democrat. Quaker. Subspecialties: Hematology; Cell biology (medicine). Current work: Studies on differentiation and function of hematopoetic cells.

BAIR, SCOTT SLAYBAUGH, III, research mechanical engineer; b. Balt., Apr. 17, 1950; s. Scott Slaybaugh and Joan Patricia (Matthews) B.; m. Lynn Broom, Jan. 23, 1971; children: Christianne, Carribeth, Brennan. B.S. in Mech. Engring, Ga. Tech. Inst., 1972, M.S., 1974. Research engr. Ga. Inst. Tech., Atlanta, 1974—. Contbr. articles to profl. jours. Mem. ASME. Subspecialties: Mechanical engineering; Fluid mechanics. Current work: Tribology, high pressure rheology. Patentee in field. Home: 1603 Trentwood Pl Atlanta GA 30319 Office: Dept Mech Engring Ga Inst Tech Atlanta GA 30332

BAIRD, HENRY WELLES, III, pediatric neurologist; b. Fort Leavenworth, Kans., Oct. 10, 1922; s. Henry Welles and Elizabeth (Tower) B.; m. Eleanora C. Gordon, Apr. 21, 1950; children—Henry Welles IV, Douglas G., Bruce C., Matthew C. B.S., Yale U., 1945, M.D., 1949. Fellow, resident neurology and pediatrics Temple U. Sch. Medicine, Phila., 1950-53, faculty pediatrics, 1953—, asso. prof., 1963-68, prof. pediatrics, 1968—; practice medicine specializing in pediatric neurology, Phila., 1953—; attending pediatrician St. Christopher's Hosp. for Children, Phila., 1966—. Author: The Child with Convulsions, 1972, Neurologic Evaluation of Infants and Children, 1983; Mem. editorial bd.: Devel. Medicine and Child Neurology, 1971—; editor, 1977—; contbr. articles to profl. jours. Served to capt. M.C. AUS, 1950-56. Mem. Soc. Pediatric Research, Am. Acad. Pediatrics, Am. Acad. Cerebral Palsy. Subspecialties: Pediatrics; Neurophysiology. Current work: Research in pediatric neurology, convulsive disorders, electro-encephalograpy and evoked cortical responses. Home: 263 Kent Rd Wynnewood PA 19096 Office: 2600 N Lawrence St Philadelphia PA 19133 I have gotten into more difficulty by believing facts that proved to be incorrect than in any other way.

BAIRD, JOHN JEFFERS, zoology educator; b. North English, Iowa, Jan. 1, 1921; s. William Simon and Ruth Carolyn (Jeffers) B.; m. Grace Geraldine Garner, Oct. 13, 1945; 1 dau., Stephanie Lynn. B.A., U. No. Iowa, 1948; M.S., U. Iowa, 1953, Ph.D., 1957. Chief pilot M&T Aerial Spray Co., Cedar Falls, Iowa, 1948; tchr. Muscatine (Iowa) Sch. Dist., 1948-54; asst. prof. zoology Calif. State U.-Long Beach, 1956-59, assoc. prof., 1959-65, prof., 1965—, dep. dean programs, 1967-78. Trustee Savanna Elem. Sch. Dist., Anaheim, Calif., 1965-81, pres., 1975-78. Served to capt. USAF, 1942-46; ETO. Mem. Am. Soc. Zoologists, AAAS, Western Soc. Naturalists, Soc. Calif. Acad. Scis. (pres. 1975-77, treas. 1980-82), Sigma Xi. Subspecialties: Developmental biology; Comparative neurobiology. Current work: Peripheral control of developing motor cells and spinal ganglia-amphibia. Home: 3239 W Ravenswood Dr Anaheim CA 92804 Office: Calif State U 1250 Bellflower Blvd Long Beach CA 90840

BAIRD, SAMUEL DEMPSEY, oceanic sensor design researcher, consultant underwater sensing; b. Greenock, Scotland, Jan. 12, 1951; emigrated to Canada, 1971; s. Samuel Dempsey and Beatrice (Lucas) B.; m. Anne Dorrian, June 19, 1971 (div. 1982); children: Sean-Paul, Nicole. Elec. Engr., Watt Meml. Inst., Greenock, 1971. Systems designer Govt. Can., Burlington, Ont., Can., 1971—. Contbr. numerous articles to sci. jours. Recipient Govt. Can. Sci. Merit award, 1982. Mem. Marine Tech. Soc., IEEE, Ont. Assn. Cert. Engring. Technologists. Subspecialties: Systems engineering; Sensor research and design. Current work: Research and design of sensors and systems to measure physical, biological and chemical oceanographic parameters within the Arctic Ocean below the ice pack. Patentee in field. Home: 801-421 Maple Ave Burlington ON Canada L7S 1L9 Office: Govt Can 867 Lakeshore Rd Burlington ON Canada L7R 4A6

BAIZER, JOAN SUSAN, physiologist; b. N.Y.C., Sept. 12, 1946; d. Manuel Mannheim and Mary Martha (Meshkov) B. B.A., Bryn Mawr Coll., 1968; M.S., Brown U., 1970, Ph.D., 1973. Postdoctoral fellow Lab. Neurobiology, NIMH, Bethesda, Md., 1973-76; asst prof. SUNY-Buffalo, 1976—. Mem. NOW, Assn. Women in Sci., Women in Neurosci., Soc. Neurosci, AAAS, Assn. Research in Vision and Ophthalmology. Subspecialty: Neurobiology. Current work: Organization and function of extrastriate visual cortex. Office: 4234 Ridge Lea Rd Amherst NY 14226 Home: 84 Russell Bell NY 14214

BAK, DAVID ARTHUR, chemist; b. Yankton, S.D., Feb. 6, 1939; s. Arthur E. and Helen E. (Munkvold) B.; m. Rita Mae Bak, Nov. 29, 1964 (div.); children: John David, Mikkle DeWayne. B..A., Augustana Coll., 1961; Ph.D., Kans. State U., 1965. Teaching Asst. Kans. State U., Manhattan, 1961-62, research asst., 1962-65; research assoc. Mich. State U., 1965-66; vis. prof. chemistry N. Tex. State U., Denton, 1977-78; asst. prof. chemistry Hartwick Coll., Oneonta, N.Y., 1966-71, assoc. prof., 1971-77, prof., 1977—, chmn. dept., 1974—, chmn. div. phys. and life scis., 1976-77. Commr., merit badge counselor Boy Scouts Am., 1976—; trustee Hartwick Coll., 1972, 74, 82. Served with U.S. Army, 1957-60. Mem. Am. Chem. Soc., AAAS, AAUP, Sigma Xi, Phi Lambda Upsilon. Subspecialty: Organic chemistry. Current work: Organic synthetic electrochemistry, intramolecular nonconjugated pi electron interactions. Home: 109 Clinton St A Oneonta NY 13820 Office: Hartwick Coll Oneonta NY 13820

BAK, MARTIN JOSEPH, electronics engr.; b. Washington, Apr. 15, 1947; s. Anthony F. and Irene L. (Hutton) B.; m. Tina Mauree Bak, Apr. 11, 1949; children: Mauree, E., Brian H. B.S.E.E., U. Md. Electronics test technician Electronics for Life Scis. Inc., Rockville, Md., 1964-69; electronics engr. NIH (Nat. Inst. Neurol. and Communicative Disorders and Stroke, Lab. Neural Control), Bethesda, MD., 1970—; also cons. Contbr. articles to profl. jours. Served with USAR, 1969-75. Mem. Soc. Neurosci., IEEE, AAAS. Subspecialties: Electronics; Biomedical engineering. Current work: Design of instrumentation for acquisition and processing of neurological signals. Design of chronically implantable electrodes for long-term recording and stimulation of neural tissue. Office: Bldg 36 Room 5A-29 Bethesda MD 20205

BAKER, BRENDA S., computer scientist; b. Oakland, Calif., Dec. 19, 1948; d. James G. and Elizabeth K. (Breitenstein) B.; m. Eric H. Grosse, June 27, 1982. B.A., Radcliffe Coll., 1969; M.S., Harvard U., 1970, Ph.D., 1973. Research fellow Harvard U., Cambridge, Mass., 1973-74; mem. tech staff Bell Labs., Murray Hill, N.J., 1974-77, 79—; vis. lectr. U. Calif.-Berkeley, 1977-78; asst. prof. U. Mich., Ann Arbor, 1978-79. Mem. Assn. Computing Machinery, Spl. Interest Group Automata and Computability Theory (treas. 1979-81). Subspecialty: Theoretical computer science. Current work: Complexity of algorithms, vlsi, programming languages and compilers. Office: Bell Labs 600 Mountain Ave Murray Hill NJ 07974

BAKER, CHARLES CLAYTON, nuclear engr., adminstr.; b. Racine, Wis., Mar. 13, 1943; s. Edwin Lee and Mildred Marie (Andersen) B.; m. Susan June Beyer, Jan. 23, 1965; children: Michael Charles, Andrew Mark. B.S., U. Wis.-Madison, 1966, M.S., 1967, Ph.D., 1972. Dep. mgr. Gen. Atomic Co. fusion div., San Diego, 1972-77; fusion program dir. Argonne (Ill.) Nat. Lab., 1977—. Contbr. articles on nuclear engring. to profl. jours. Chmn. com. Boy Scouts Am., Naperville, Ill., 1978—. Recipient Cert. of Appreciation Dept. Energy, 1982. Mem. Am. Nuclear Soc. (chmn. fusion energy div. 1981-82), Phi Eta Sigma. Subspecialties: Fusion; Nuclear engineering. Current work: Fusion technology and systems analysis, plasma engineering. Home: 6S 272 Concord Rd Naperville IL 60540 Office: Argonne Nat Lab 9700 S Cass Ave Argonne IL 60439

BAKER, CON JACYN, research plant pathologist; b. Chgo., July 23, 1948; s. Con James and Ethel Marie (Sckowbo) B.; m. Rose Yvette MacKenzie, Aug. 10, 1968 (div.); 1 dau., Constance Lawrane; m. Jane Krantz Wren, Sept. 18, 1982; 1 son, Frank Wren. B.S., U. Wis., Madison, 1969; Ph.D., Cornell U., 1978. Postdoctoral fellow Cornell U., Ithaca, N.Y., 1978-79, U. Wis.-Madison, 1979-81, U. Del., Beltsville, Md., 1981-82; research plant pathologist U.S. Dept. Agr., Beltsville, Md., 1982—. Contbr. articles in field to profl. jours. Served with USMC, 1970-72. NIH trainee, 1979-80. Mem. Am. Phytopath. Soc., Sigma Xi, Phi Kappa Phi. Subspecialties: Plant pathology; Plant physiology (agriculture). Current work: Plant cuticle and cell wall degradation, induced resistance, biocontrol. Home: 10773 Cordage Walk Columbia MD 21044 Office: US Dept Agr BARC-W Room 18 Bldg 011 Beltsville MD 20705

BAKER, DANIEL CLIFTON, surgery educator; b. N.Y.C., Dec. 11, 1942; s. Daniel Clifton and Geraldine Elizabeth (Dieck) B.; m. Mary Bradwell Conley, Feb. 9, 1974; children: Daniel Clifton IV, John Conley. A.B., Columbia U., 1964, M.D., 1968. Asst. prof. plastic surgery NYU Med. Sch., 1979—; cons. in field. Editor: Facial Paralysis, 1979, Symposium on Rhinoplasty, 1983. Served to maj. U.S. Army, 1970-72; Vietnam. Recipient alumni achievement award Farleigh Dickinson U., 1980. Mem. N.Y. County Med. Soc., Am. Soc. Plastic and Reconstructive Surgeons, Internat. Soc. Reconstructive Microsurgery, Am. Soc. Aesthetic Plastic Surgery (best. sci. presentation 1979). Clubs: N.Y. Athletic; Meadow (Southampton, N.Y.). Subspecialties: Surgery; Microsurgery. Current work: Primary interest and specialization in plastic and reconstructive surgery of the face and neck, reconstructive surgery and microsurgery following trauma and cancer. Home: 630 Park Ave New York NY 10021 Office: Inst Reconstructive Plastic Surgery 550 1st Ave New York NY 10016

BAKER, DONALD JAMES, JR., oceanographer, educator; b. Long Beach, Calif., Mar. 23, 1937; s. Donald James and Lillian Mae (Pund) B.; m. Emily Lind Delman, Sept. 7, 1968. B.S., Stanford U., 1958; Ph.D., Cornell U., 1962. Postdoctoral fellow Grad. Sch. Oceanography, U. R.I., Kingston, 1962-63; NIH fellow in chem. biodynamics Lawrence Radiation Lab., U. Calif., Berkeley, 1963-64; research fellow, asst. prof., asso. prof. phys. oceanography Harvard U., 1964-73; research prof. dept. oceanography, sr. oceanographer, applied physics lab. U. Wash., Seattle, 1973-77; sr. fellow Joint Inst. for Study Atmosphere and Ocean, 1977—, prof., chmn. dept. oceanography, 1979—; dean Coll. Ocean and Fishery Scis., 1981—; group leader deep-sea physics group Pacific Marine Environ. Lab., NOAA, Seattle, 1977-79; co-chmn. exec. com. Internat. So. Ocean Studies (NSF project), 1974—; chmn. bd. govs. Joint Oceanographic Instns., Inc., 1980—; mem. ocean scis. bd. Nat. Acad. Scis., 1979-82, mem. com. on atmospheric scis., 1978-81, vice-chmn. joint panel, global weather experiment, 1976—. Co-editor in chief: Dynamics of Atmospheres and Oceans Jour, 1975-79; Contbr. sci. articles to profl. jours. Fellow Explorers Club; mem. Am. Geophys. Union, Am. Meteorol. Soc., Sigma Xi. Subspecialties: Oceanography; Remote sensing (geoscience). Current work: General ocean circulation, air-sea interaction in polar regions, ocean instrumentation, ocean measurements from satellites. Patentee deep-sea pressure gauge. Office: Dept Oceanography WB-10 U Wash Seattle WA 98195

BAKER, DONALD ROY, geology educator; b. Norfolk, Va., May 8, 1927; m., 1948; 2 children. B.S., Calif. Inst. Tech., 1950; Ph.D., Princeton U., 1955. Instr. petroleum and chem. geology Northwestern U., 1954-56; sr. research geologist Denver Research Ctr., Marathon Oil Co., 1956-66; assoc. prof. Rice U., Houston, 1966-72, chmn. dept., 1977-80, prof. geology, 1972—. Fellow Geol. Soc. Am.; mem. Am. Assn. Petroleum Geologists, Soc. Econ. Paleontologists and Mineralogists, Geochem. Soc., Mineral. Soc. N.Am. Subspecialties: Petrology; Geochemistry. Office: Dept Geology Rice U Houston TX 77001

BAKER, HERBERT GEORGE, botany educator; b. Brighton, Eng., Feb. 23, 1920; came to U.S., 1957; s. Herbert Reginald and Alice (Bambridge) B.; m. Irene Williams, Apr. 4, 1945; 1 dau., Ruth Elaine. B.S., U. London, 1941, Ph.D., 1945. Research chemist, asst. plant physiologist Hosa Research Labs., Sunbury-on-Thames, Eng., 1940-45; lectr. botany U. Leeds, Eng., 1945-54; research fellow Carnegie Instn., Washington, 1948-49; prof. botany U. Coll. Ghana, 1954-57; faculty U. Calif., Berkeley, 1957—, assoc. prof. botany, 1957-60, prof., 1960—, dir. bot. garden, 1957-69. Author: Plants and Civilization, 1965, 70, 78; Editor: (with G.L. Stebbins) Genetics of Colonizing Species, 1965; series editor: Bot. Monographs, 1971—; Contbr. articles to sci. jours. Fellow AAAS (pres. Pacific div.), Assn. Tropical Biology; mem. Am. Inst. Biol. Sci., Internat. Assn. Botanic Gardens (past v.p.), Internat. Orgn. Plant Biosystematists, Ecol. Soc. Am., Soc. for Study Evolution (past pres.), Bot. Soc. Am. (past pres.), Sigma Xi. Subspecialties: Evolutionary biology; Ecology. Current work: Reproductive biology of plants (pollination biology, breeding systems, population structure) especially of tropical species. Evolution and ecology of plants under human influence, particulary weeds. Chemistry of nectar and pollen. Home: 635 Creston Rd Berkeley CA 94708

BAKER, HOWARD CRITTENDEN, physics educator; b. Clay City, Ky., Sept. 4, 1943; s. Arlie Howard and Ruby (Mays) B.; m. Dorothy Kay Kvarve. B.A., Berea Coll., 1965; Ph.D., Washington U., St. Louis, 1972. Prof. physics Berea (Ky.) Coll., 1972—; vis. prof. U. Conn., Storrs, 1974-75, U. Ky., Lexington, 1982-83; vis. scientist Oak Ridge Nat. Lab. and Johns Hopkins U., 1979-80. Contbr. articles to profl. jours. Mem. Am. Phys. Soc. Subspecialty: Theoretical physics. Current work: Laser-atom interactions; field theory multiphoton interactions with atoms; atomic laser-induced collisional phenomena; coherence phenomena. Home: Route 3 Dogwood Heights Berea KY 40403 Office: Dept Physics Berea Coll Berea KY 40404

BAKER, IRENE, research botanist; b. Tredegar, Gwent, Wales, Feb. 22, 1918; d. William John and Mary Jane (Jones) Williams; m. Herbert George Baker, Apr. 4, 1945; 1 dau., Ruth Elaine. B.Sc. with honors, U. Wales, 1940, diploma in edn, 1941. Lectr. in biology Leeds Tech. Coll., Eng., 1945; research asst. zoology U. Leeds, 1945-48; research assoc. Govt. of Tsetse Control, Ghana, 1954-57; asst., then lectr. biology Mills Coll., 1958-69; research assoc. botany U. Calif.-Berkeley, 1974—. Contbr. articles sci. jours., chpts. in books. Mem. Bot. Soc. Am., Soc. Woman Geographers, Biosystematists, Sigma Xi. Subspecialty: Reproductive biology. Current work: Pollination biology and chemistry of flower products. Home: 635 Creston Rd Berkeley CA 94708 Office: Botany Dept U Calif Berkeley CA 94720

BAKER, JOHN PATTON, JR., physiologist, researcher, educator; b. Wichita Falls, Tex., Aug. 30, 1947; s. John P. and Virginia (Gilbert) B.; m. Evalyn Torppa, June 8, 1968. B.S. E.E., U. Idaho, 1969; M.S. E.E., U. Ky., 1971, Ph.D. in Physiology, 1976. Fellow dept. physiology U. Tex. Med. Br., Galveston, 1977-80, research asst. prof., dept. physiology, 1980—. Contbr. chpt. to books, articles to publs. in field. New investigator grantee NIH, 1981—; postdoctoral fellow, 1978-80. Mem. Am. Physiol. Soc., Soc. for Neurosci. Subspecialties: Physiology (medicine); Neurophysiology. Current work: Extracellular and intracellular recording of respiratory neurons (located in the medulla and pons) involved in respiratory phase transitions and effects of afferent inputs on them. Office: Dept Physiology and Biophysics U Tex Med Br Galveston TX 77550

BAKER, KENNETH FRANK, plant pathologist, educator; b. Ashton, S.D., June 3, 1908; s. Frank and Laura May (Boyer) B.; m. Katharine Cummings, June 17, 1944. B.S., Wash. State U., 1930, Ph.D., 1934; postgrad., U. Wis., 1934-35, U. Adelaide, Australia, 1961-62. With Pineapple Expt. Sta., Honolulu, 1936-39; mem. faculty dept. plant pathology UCLA, 1939-61; prof. plant pathology U. Calif.-Berkeley, 1961-75, prof. emeritus, 1975—; collaborator Hort. Crops Research Lab., U.S. Dept. Agr., Corvallis, 1976—; prof. dept. botany and plant pathology Oreg. State U., Corvallis, 1976—. Author: U.C. System for Producing Healthy Container-grown Plants, 1957, Ecology of Soil-borne Plant Pathogens, 1965, Biological Control of Plant Pathogens, 1974, Nature and Practice of Biological Control of Plant Pathogens, 1983. Recipient awards Calif. State Florists Assn., 1956, Am. Assn. Nurserymen, 1959, Calif. Assn. Nurserymen, 1966, Fedn. Australian Nurserymen, 1969, Soc. Am. Florists, 1976. Fellow AAAS, Am. Phytopath. Soc.; Mem. Brit. Mycological Soc., Mycological Soc. Am., Netherlands Plant Path. Soc., Assn. Applied Biologists, Australian Plant Path. Soc., Brit. Soc. Plant Pathology. Subspecialties: Plant pathology; Microbiology. Current work: Biological control of plant pathogens, seed pathology; thermotherapy. Home: 6952 NW Cardinal Dr Corvallis OR 97330 Office: 3420 NW Orchard Ave Corvallis OR 97330

BAKER, KILE BARTON, space plasma physicist; b. Bozeman, Mont., Mar. 27, 1950; s. Graeme Levo and Hazel Arvilla (Swenson) B. B.S. in Physics, Mont. State U., 1971, Ph.D., Stanford U., 1979. Postdoctoral researcher Inst. for Plasma Research, Stanford (Calif.) U., 1978, Boston U., 1979-80, asst. prof. astronomy dept., 1980-81; postdoctoral researcher Applied Physics Lab., Johns Hopkins U., Laurel, Md., 1981-83, staff scientist, 1983. Contbr.: papers to sci. jours. including Jour. Geophys. Research. Active ACLU, Md. Civil Liberties Union. NSF fellow, 1971. Mem. Am. Geophys. Union, Am. Astron. Soc. Subspecialties: Satellite studies; Solar physics. Current work: Planetary magnetospheres; plasma physics in space; space sci. research. Office: Johns Hopkins Applied Physics Lab Johns Hopkins Rd Laurel MD 20707

BAKER, LAURENCE HOWARD, osteopathic internist, educator; b. Bklyn., Jan. 14, 1943; s. Jacob and Sylvia (Tannenbaum) B.; m. Maxine V. Friedman, July 25, 1964; children: Mindy, Jennifer. B.A., Bklyn. Coll., 1962; D.O., U. Osteopathic Medicine and Surgery-Des Moines, 1966. Intern Flint (Mich.) Osteo. Hosp., 1966-67; resident in internal Medicine Detroit Osteo. Hosp., 1967-68; mem. faculty Wayne State U., Detroit, 1972—, prof. internal medicine, 1979—, chief oncology div. Med. Center, 1982—; coordinator cancer program, 1981—; dep. dir. Comprehensive Cancer Center, Detroit, 1981—. Editor: New Agents and Pharmacology Soft Tissue Sarcomas, 1983.

Assoc. chmn. S.W. Oncology Group, San Antonio, 1981—; chmn. Intergroup Sarcoma Study Group, Detroit, 1982—. Served to maj. U.S. Army, 1968-70; Viet Nam War. Mem. Am. Soc. Cancer Research, Am. Soc. Clin. Oncology, Am. Soc. Clin. Pharmacology and Therapeutics, Am. Soc. Clin. Research, Am. Soc. Cancer Edn. Subspecialties: Cancer research (medicine); Chemotherapy. Current work: Anticancer drug development. Office: Wayne State Univ Univ Health Center 4201 Saint Antoine St 7C Detroit MI 48201

BAKER, LAWRENCE JOHN, mathematician, geophysicist; b. Rome, N.Y., Nov. 21, 1950; s. Abraham Harris and Ruth (Flanagan) B.; m. Dorthy J. Zayatz, May 26, 1973; children: Elizabeth Eve, Daniel Abraham. B.S., MIT, 1972; M.A., U. Mich., 1974; Ph.D., U. Md., 1978. Cryptanalytic mathematician Nat. Security Agy., Ft. Meade, Md., 1974-76; sr. research specialist Exxon Prodn. Research Co., Houston, 1979—. Mem. Am. Math. Soc., Soc. Exploration Geophysicists, Acoustical Soc. Am., Soc. Indsl. and Applied Math. Republican. Subspecialties: Applied mathematics; Geophysics. Current work: Modelling acoustic logging and seismic recording; seismic data processing. Home: 13923 Pinerock Houston TX 77079 Office: Exxon Prodn Research Co 3120 Buffalo Speedway PO Box 2189 Houston TX 77001

BAKER, R. RALPH, phytopathologist, educator; b. Houston, Aug. 31, 1924; m. Eleanor Jo Damer, June 19, 1965; children: Jock Maxwell, Kit Edmund, Jennifer Katherine, Sean Randal, Dawn Michelle, Nicole Diane. B.A., Colo. State U., 1948, M.S., 1950; Ph.D., U. Calif.-Berkeley, 1954. Prof. plant pathology Colo. State U., Ft. Collins, 1954—; vis. prof. U. Calif.-Berkeley, 1963-64, Cambridge (Eng.) U., 1969. Contbr. articles to profl. jours. Served to lt. (j.g.) USN, 1943-46; PTO. Recipient Group Achievement award NASA, 1974, 76, 78. Fellow Am. Phytopath. Soc. (pres. Pacific div. 1981), AAAS; mem. Sigma Xi. Baptist. Subspecialties: Plant pathology; Gravitational biology. Current work: Biological control of soilborne plant pathogens, space biology. Home: 1216 Southridge Dr Fort Collins CO 80521 Office: Dept Botany and Plant Pathology Colo State U Fort Collins CO 80523

BAKER, ROBERT FRANK, molecular biologist, educator; b. Weiser, Idaho, Apr. 9, 1936; s. Robert C. and Beulah B.; m. Mary Margaret Murphy, May 29, 1965; children: Allison Leslie, Steven Mark. B.S., Stanford U., 1959; Ph.D., Brown U., 1966. Postdoctoral fellow Stanford U., 1966-68; asst. prof. U. So. Calif., 1968-72; vis. assoc. prof. Harvard U. Med. Sch., 1975-76; assoc. prof. molecular biology U. So. Calif., 1972-82, prof., 1982—, dir. molecular biology div., 1978-80. Contbr. articles to profl. jours. USPHS grantee, 1968, 72,75,78; NSF grantee, 1983—. Subspecialties: Molecular biology; Genetics and genetic engineering (medicine). Current work: Genetic regulation in differentiating cells; genetic engineering in culture cells. Home: 607 Almar Ave Pacific Palisades CA 90272 Office: Molecular Biology U So Calif ACBR406 Los Angeles CA 90089

BAKER, ROBERT MAURICE, manufacturing company executive; b. Odessa, Mo., Feb. 11, 1928; s. William F. and Lillian L. B.; m. Marilyn D. Strode, Feb. 2, 1979; children—Deborah E., James E., Richard M., Ross A., Michelle L. B.S. in Bus. Adminstrn, Central Mo. State Coll., 1951. C.P.A., Mo. Audit mgr. Arthur Andersen & Co., C.P.A.s, St. Louis, 1951-63; treas. Angelica Corp., St. Louis, 1963—. Active local Boy Scouts Am., 1966-72; chmn. com. Sing-Out Florissant Valley, 1971-75. Served with AUS, 1946-47. Mem. Tax Execs. Inst. (officer St. Louis chpt. 1983—). Methodist. Home: 242 Pennington Ln Chesterfield MO 63017 Office: 10176 Corporate Square Dr St Louis MO 63132

BAKER, THOMAS, pharmacology educator, research consultant; b. Mineola, N.Y, Sept. 19, 1933; s. Raymond Ira and Agatha (Carroll) B.; m. Marion Whitaker, Feb. 7, 1933; children: Patricia Anne, Susan, Thomas, Peter, David, Marian. B.A., Hunter Coll., CUNY, 1968; M.S., Cornell U., 1971. Assoc. research prof. Cornell U. Med. Coll., N.Y.C., 1962-83; dir. research dept. anesthesiology St. Joseph's Hosp. and Med. Ctr., Paterson, N.J., 1983—, research cons. to anesthesiology dept., Paterson, N.J. Contbr. articles on pharmacology to profl. jours. Mem. Ramapo Voluntary Ambulance Corps. NIH grantee, 1971-82. Mem. AAAS, Am. Soc. Pharmacology and Exptl. Therapeutics, N.Y. Acad. Scis., Soc. Exptl. Biology and Medicine, Soc. Neurosci., Soc. Toxicology. Subspecialties: Neuropharmacology; Toxicology (medicine). Current work: Effects of drugs and toxic agents on the nervous system, neuropharmacology, neurotoxicology. Office: 703 Main St Paterson NJ 07503

BAKER, THOMAS IRVING, microbiologist, researcher, educator; b. LaRue, Ohio, Sept. 28, 1931; s. John Benjamin and Irene (Allyn) B.; m. Donna Gail Lewis, Sept. 11, 1975; children: Debra, Karen, Thomas, James. B.S., Kent State U., 1954; M.S., Ohio State U., 1956; Ph.D., Western Res. U. Sch. Medicine, 1965. NIH postdoctoral fellow genetics dept. U. Wash., Seattle, 1965-67; asst. prof. microbiology U. N.Mex. Sch. Medicine, Albuquerque, 1976-74, assoc. prof., 1974—. Author: The Genetics Learning System, 1980, The Microbial Genetics Learning Series, 1981. Mem. Am. Soc. Microbiology, Genetics Soc. Am., AAAS. Subspecialties: Microbiology; Molecular biology. Current work: Mutation and DNA repair. Office: Dept Microbiology Sch Medicine U New Mexico Albuquerque NM 87131

BAKER, WILLIAM KAUFMAN, geneticist; b. Portland, Ind., Dec. 2, 1919; s. Frank K. and Jennie (Schaeffer) B.; m. Margaret Stewart, Mar. 4, 1944; children—Bruce, Ann, Brian. B.A., Coll. of Wooster, 1941; M.A., U. Tex., 1943, Ph.D. (NRC fellow), 1948. Asst. prof. U. Tenn., 1948-51; sr. biologist Oak Ridge Nat. Lab., 1951-55; asso. prof. U. Chgo., 1955-59, prof. zoology, 1959-77, chmn. dept. biology, 1968-72, U. Utah, 1977-80, prof., 1977—. Co-editor: Am. Naturalist, 1965-70; Author articles on genetics. Served to 1st lt. USAAF, 1943-46. NSF sr. postdoctoral fellow U. Rome, 1963-64; NIH spl. postdoctoral fellow, Madrid, 1972-73. Mem. Am. Soc. Naturalists (sec., v.p. 1976), Genetics Soc. Am. (pres. 1980). Subspecialty: Genetics and genetic engineering (biology). Home: 4499 Gilead Way Salt Lake City UT 84117

BAKER, WILLIAM OLIVER, research chemist; b. Chestertown, Md., July 15 1915; s. Harold May and Helen (Stokes) B.; m. Frances Burrill, Nov. 15, 1941; 1 son, Joseph Burrill. B.S., Washington Coll., 1935, Sc.D., 1957; Ph.D., Princeton, 1938; Sc.D., Georgetown U., 1962, U. Pitts., 1963, Seton Hall U., 1965, U. Akron, 1968, U. Mich., 1970, St. Peter's Coll., 1972, Poly. Inst. N.Y., 1973, Trinity Coll., Dublin, Ireland, 1975, Northwestern U., 1976, U. Notre Dame, 1978, Tufts U., 1981, N.J. Coll. Medicine and Dentistry, 1981; D.Eng., Stevens Inst. Tech., 1962, N.J. Inst. Tech., 1978; LL.D., U. Glasgow, 1965, U. Pa., 1974, Kean Coll., N.J., 1976, Lehigh U., 1980, Drew U., 1981; L.H.D., Monmouth Coll., 1973, Clarkson Coll. Tech., 1974. With Bell Telephone Labs., 1939-80, in charge polymer research and devel., 1948-51, asst. dir. chem. and metall. research, 1951-54, dir. research, phys. scis., 1954-55, v.p. research, 1955-73, pres., 1973-79, chmn. bd., 1979-80; dir. Ann. Revs., Inc., Summit and Elizabeth Trust Co.; vis. lectr. Northwestern U., Princeton U., Duke; Schmitt lectr. U. Notre Dame, 1968; Harrelson lectr. N.C. State U., 1971; Herbert Spencer lectr. U. Pa., 1974; Charles M. Schwab Meml. lectr. Am. Iron and Steel Inst., 1976; NIH lectr., 1958, Metall. Soc. Am. Inst. Mining Engrs./Am. Soc. Metals disting. lectr., 1976; Miles Conrad Meml. lectr. Nat. Fedn. Abstracting and Indexing Services, 1977; Wulff lectr. M.I.T., 1979; other lectureships; cons. Office Sci. and Tech., 1977—; Mem. Princeton Grad. Council, 1956-64; bd. visitors Tulane U., 1963—; mem. commn. sociotech. systems NRC, 1974-78, also chmn. adv. bd. on mil. personnel supplies, 1964-78; mem. com. on phys. chemistry of div. chemistry and chem. tech., 1963-70; also steering com. Pres.'s Food and Nutrition Study Commn. Internat. Relations Nat. Acad. Scis.-NRC, 1975; mem. panel on phys. chemistry Office Naval Research, 1948-51; past mem. Pres.'s Sci. Adv. Com., 1957-60; nat. sci. bd. NSF, 1960-66; past chmn. Nat. Sci. Info. Council, 1959-61; mem. sci. adv. bd. Nat. Security Agy., 1959-76, cons., 1976—, Dept. Def., 1958-71, 1963-73, ; Panel of Ops. Evaluation Group, USN, 1960-62; mem. N.J. Bd. Higher Edn., 1967—, exec. com., 1970—, vice chmn., 1970-72; mem. liaison com. for sci. and tech. Library of Congress, 1963-73; mem. Pres.'s Fgn. Intelligence Adv. Bd., 1959-77; chmn. Pres.'s Adv. Group Anticipated Advances in Sci. and Tech., 1975-76; vice chmn. Pres.'s Com. Sci. and Tech., 1976-77; bd. regents Nat. Library Medicine, 1969-73; bd. visitors Air Force Systems Command, 1962-73; mem. mgmt. adv. council Oak Ridge Nat. Lab., 1970—; mem. Nat. Commn. on Libraries and Info. Scis., 1971-75, Commn. on Critical Choices for Ams., 1973-75, Nat. Cancer Adv. Bd., 1974-80; mem. panel adv. Inst. Materials Research, Nat. Bur. Standards, 1966-69; mem. Council Trends and Perspectives, U.S. C. of C., 1966-74; chmn. tech. panels adv. to Nat. Bur. Standards, Nat. Acad. Scis.-NRC, 1969-78; mem. Nat. Council Ednl. Research, 1973-75; mem. energy research and devel. adv. council Energy Policy Office, 1973-75; mem. Project Independence adv. com. Fed. Energy Adminstrn., 1974-75, Gov.'s Com. to Evaluate Capital Needs N.J., 1974-75; mem. governing bd. Nat. Enquiry into Scholarly Communication, 1975-79; adv. council N.J. Regional Med. Library, 1975—, Fed. Emergency Mgmt. Adv. Bd., 1980—, Gas Research Inst. Adv. Bd., 1978—; mem. adv. bd. N.J. Sci./Tech. Center, 1980—; Mem. sci. adv. bd. Robert A. Welch Found., 1968—; vis. com. for chemistry Harvard, 1959-72; mem. council Marconi Fellowships, 1978—; vis. com., div. chemistry and chem. engring. Calif. Inst. Tech., 1969-72; vis. com. on scis. and math. Drew U., 1969—; asso. in univ. seminar on tech. and social change Columbia, 1969—; vis. com., dept. materials sci. and engring. M.I.T., 1973-76; bd. overseers Coll. Engring. and Applied Sci. U. Pa., 1975—; bd. dirs. Council on Library Resources, 1970—, Health Effects Inst., 1980—; Clin. Scholar Program Robert Wood Johnson Found., 1973-76, Third Century Corp., 1973-76. Contbr.: High Polymers, 1945, Symposium on Basic Research, AAAS, 1959, Rheology, Vol. III, 1960, Technology and Social Change, 1964, Science: The Achievement and the Promise, 1968, Ann. Rev. Materials Sci, 1976, various other books.; Mem. editorial adv. bd.: Jour. Info. Sci; past mem. adv. editorial bd.: Chem. and Engring. News; hon. editorial adv. bd.: Carbon; Contbr. numerous articles to tech. jours. Trustee Urban Studies, Inc., 1960-78, Aerospace Corp., 1961-76, Carnegie-Mellon U., 1967—, Princeton, 1964—, Fund N.J., 1974—, Harry Frank Guggenheim Found., 1976—, Gen. Motors Cancer Research Found., 1978—, Charles Babbage Inst., 1978—, Newark Mus., 1979—, Rockefeller U., 1960—; chmn. Rockefeller U., 1980—; trustee Andrew W. Mellon Found., 1965—, chmn., 1975—. Named 1 of 10 top scientists in U.S. industry, 1954; recipient Perkin medal, 1963; Honor scroll Am. Inst. Chemist, 1962; award to execs. ASTM, 1967; Edgar Marburg award, 1967; Indsl. Research Inst. medal, 1970; Frederik Philips award IEEE, 1972; Indsl. Research Man of Year award, 1973; Procter prize Sigma Xi, 1973; James Madison medal Princeton, 1975; Mellon Inst. award, 1975; Soc. Research Adminstrs. award for distinguished contbns., 1976; von Hippel award Materials Research Soc., 1978; Fahrney medal Franklin Inst., 1977; N.J. Sci/Tech. medal, 1980; Harvard fellow, 1937-38; Procter fellow, 1938-39; Jefferson medal N.J. Patent Law Assn., 1981; David Sarnoff prize AFCEA, 1981. Fellow Am. Phys. Soc., Am. Inst. Chemists (Gold medal 1975), Franklin Inst., Am. Acad. Arts and Scis.; mem. Dirs. of Indsl. Research, Am. Chem. Soc. (past mem. com. nat. def., cons., past mem. com. chemistry and pub. affairs, Priestley medal 1966, Parsons award 1976, Willard Gibbs award 1978, Madison Marshall award 1980), Am. Philos. Soc., Nat. Acad. Scis. (council 1969-72, com. sci. and pub. policy 1966-69), Nat. Acad. Engring., Inst. Medicine (council 1973-75), Indsl. Research Inst. (dir. 1960-63), Sigma Xi, Phi Lambda Upsilon, Omicron Delta Kappa. Clubs: Chemists of N.Y. (hon.), Cosmos, Princeton of Northwestern N.J. Subspecialties: Solid state chemistry; Materials (engineering). Current work: Formation and structure of polymers; viscoelasticity of biopolymers; ultrathin films for information storage and integrated circuit patterns; photonics for communication systems; organization of research and development. Holder 13 patents. Home: Spring Valley Rd Morristown NJ 07960 Office: 600 Mountain Ave Murray Hill NJ 07974

BAKSHI, PRADIP M., physics educator, researcher; b. Baroda, India, Aug. 21, 1936; came to U.S., 1956; s. M.M. and R. M. Baxi; m. Hansika Parekh, Sept. 14, 1967; children: Vaishali, Ashesh. B.Sc., Bombay U., 1955; M.A., Harvard U., 1957, Ph.D., 1962. Postdoctoral fellow in physics Harvard U., 1963; sr. research physicist Air Force Cambridge Research Labs., 1963-66; sr. research assoc. Physics Dept., Brandeis U., 1966-68, vis. assoc. prof., 1968-70; research assoc. prof. physics dept. Boston Coll., 1970-75, research prof., 1975—. Contbr. articles to sci. jours. Subspecialties: Theoretical physics; Plasma physics. Current work: Mathematical physics, non-perturbative techniques, theoretical plasma physics. Home: 122 Florence Rd Waltham MA 02154 Office: Physics Dept Boston Coll Chestnut Hill MA 02167

BAKSI, SAMARENDRA NATH, endocrinologist; b. Rajshahi, India, Dec. 28, 1940; s. Rishi and Haribhabini (Majumdar) B.; m. Shila Baksi, Aug. 16, 1971; children: Samudra Neal, Subir Kumar. B.V.Sc., Bihar U., India, 1960; M.S., U. Mo., Columbia, 1967, Ph.D., 1971. Fellow in reproductive physiology Worcster Found. Exptl. Biology, Shrewsbury, Mass., 1971-72; vis. fellow Nat. Inst. Environ. Health Sci., NIH, Research Triangle Park, N.C., 1972-74; research assoc. in pharmacology U. Tex. Med. Br., Galveston, 1975, Tex. Tech. U. Health Scis. Center, Lubbock, 1976-79, research assoc. in physiology, 1980—. Contbr. articles to sci. jours. Mem. Endocrine Soc., Am. Physiol. Soc., Am. Soc. Pharmacology and Exptl. Therapeutics, Am. Soc. Bone and Mineral Research. Subspecialties: Neuroendocrinology; Molecular pharmacology. Current work: Neurotransmitter physiology; central catecholamine metabolism alteration by estrogen, lead and calcium. Regulation of vitamin D endocrine system. Cardiac histamine receptor physiology; toxicology. Home: 2804 53rd St Lubbock TX 79413 Office: Tex Tech U Health Scis Center Lubbock TX 79430

BALANIS, GEORGE NICK, electrical engineer; b. Athens, Greece, Oct. 7, 1944; came to U.S., 1963, naturalized, 1977; s. Nicholas G. and Mary (Traganoudaki) B.; m. Toula Koutis, Nov. 15, 1971; 1 son, Nikolas. B.S. with honors, Calif. Inst. Tech., 1967, M.S., 1968, Ph.D., 1972. Research asst. Calif. Inst. Tech., Pasadena, 1968-71; staff scientist Applied Theory, Inc., Los Angeles, 1971-77, Arete Assocs., 1977-80; sr. engr. Garrett Airesearch, Torrance, Calif., 1980—. Contbr. articles to profl. jours. Mem. IEEE, ASME, Soc. Indsl. and Applied Math., Am. Math. Soc., Am. Acad. Mechanics, Tau Beta Pi, Sigma Chi. Greek Orthodox. Subspecialties: Applied mathematics; Electrical engineering. Current work: Inverse scattering, computer aided engineering. Home: 2349 Hill St Santa Monica CA 90405

BALASCIO, JOSEPH FRANCIS, mfg. co. exec.; b. Wilkes-Barre, Pa., Aug. 14, 1941; s. Joseph Francis and Anna Marie (Fristic) B.; m. Irma Marie Wagner, Sept. 10, 1966; children: John, Johanna, Catherine. B.S., King's Coll., Wilkes-Barre, 1963; M.S., Pa. State U., 1966, Ph.D. in Solid State Sci, 1972. Sr. scientist ISOMET Corp., Oakland, N.J., 1972-74, mgr. crystal dept., Springfield, Va., 1975-76; sr. staff engr. Motorola, Inc., Carlisle, Pa., 1976-82, prin. staff engr., 1982—; assoc. mem. sci. adv. bd. assn., 1981—. Contbr. articles to profl. jours. Mem. Am. Chem. Soc., Am. Ceramic Soc., Optical Soc. Am., Am. Assn. Crystal Growth. Subspecialties: Solid state chemistry; Crystallography. Current work: Research and development on the growth of alpha quartz single crystals so that this material may be routinely fabricated employing semiconductor processing technology and used in satellite applications. Home: 734 Sherwood Dr Carlisle PA 17013 Office: 2510 Ritner Hwy Carlisle PA 17013

BALASUBRAHMANYAN, VRIDDHACHALAM KRISHNASWAMY, astrophysicist; b. Tiruchendoor, Madras, India, Nov. 11, 1926; came to U.S., 1962; s. Vriddhachalam and Mangalam (Ammal) Krishnaswamy; m. Saroja Balasubrahmanyan, May 25, 1954; children: Ravishankar, Raghu. M.Sc., Calcutta U., 1949; Ph.D., Bombay U., 1961. Research asst. Tatainstitute of Fundamental Research, Bombay, India, 1959, fellow, 1959-62; resident research assoc. Nat. Acad. Sci., Washington, 1962-65; astrophysicst Goddard Space Flight Center, Greenbelt, Md., 1965—. Contbr. articles to sci. jours. Mem. Am. Phys. Soc., Am. Geophys. Union, Indian Physics Assn. Subspecialty: Cosmic ray high energy astrophysics. Current work: Research on cosmic ray composition at high energies; propogation and origin of cosmic rays. Home: 9333 Wellington St Seabrook MD 20706 Office: Goddard Space Flight Center Greenbelt Rd Greenbelt MD 20771

BALAZS, TIBOR, research toxicologist, educator; b. Sarbogard, Hungary, Mar. 1, 1922; came to U.S., 1963, naturalized, 1968; s. Armin and Bella (Stern) B.; m. Eva Bokor, Dec. 18, 1949; 1 dau., Anna. Diplomate: Acad. Toxicol. Scis. Mag. Pharm. Peter Pazmany U., Budapest, Hungary, 1948; D.V.M. Vet. U. Budapest, 1949; Toxicologist Food and Drug Directoarate, Ottawa, Ont., Can., 1959-63; research group leader Lederle Lab., Pearl River, N.Y., 1963-69; asst. dir. Smith Kline Co., Phila., 1969-71; chief drug toxicology br., div. drug biology Nat. Ctr. Drugs and Biologics, FDA, Washington, 1971—; vis. prof. pharmacology Howard U. Coll. Medicine. Research numerous publs. in drug toxicology; editor: Cardiac Toxicology, 1981. Recipient Commendable Service award FDA, 1978, Merit award, 1982. Fellow Am. Coll. Vet. Toxicology; mem. Am. Soc. Pharmacology and Exptl. Therapeutics, Soc. Toxicology, AVMA. Subspecialties: Pharmacology; Toxicology (medicine). Current work: Direct and conduct drug toxicology research. Office: 200 C St SW Washington DC 20204

BALBINDER, ELIAS, biologist; b. Warsaw, Poland, Jan. 22, 1926; s. Aaron Lejba and Chaja Pessa (Kratka) B.; m. Evelyn Weissman, May 10, 1955 (dec.); children: Rachel Naomi, Sara Elizabeth; m. Glory Hirshfeld, May 17, 1980. B.S., U. Mich., 1949; Ph. D., Ind. U., 1957. Research assoc. Carnegie Inst. Wash., Cold Spring Harbor, N.Y., 1957-60; Am. Cancer Soc. postdoctoral research fellow U. Calif., San Diego, 1960-62; asst. prof. genetics Syracuse U., 1962-67, assoc. prof., 1967-71, prof., 1971-76; dir. genetics and carcinogenesis Am. Med. Center-Cancer Research Center, Lakewood, Colo., 1976-82; adj. prof. biochemistry, biophysics and genetics U. Colo. Health Scis. Ctr., Denver, 1982—; mem. genetic biology panel NSF, 1977-79. Contbr. articles to profl. jours. NSF, NIH grantee, 1963-78; Fulbright Hays awardee, Argentina, 1963, Colombia, 1982; NIH spl. fellow Osaka (Japan) U., 1970. Fellow AAAS; mem. Am. Soc. Microbiology, Genetics Soc. Am., Environ. Mutagen Soc., Sigma Xi. Democrat. Jewish. Subspecialties: Gene actions; Cancer research (medicine). Current work: Research in role of DNA repaid and mutagens/carcinogens in causing rearrangements. Home: 2160 E Columbia Pl Denver CO 80210 Office: 4200 E 9th Ave Denver CO 80262

BALCERZAK, STANLEY PAUL, physician; b. Pitts., Apr. 27, 1930; s. Stanley P. and Margaretta R. (Giel) B.; m. Mary E. Kicher, Aug. 29, 1953; children: Paul, William, Margaret, John, James, Thomas, Eric. B.S., U. Pitts., 1953; M.D., U. Md., 1955. Diplomate: Am. Bd. Internal Medicine. Intern U. Pitts., 1955-56; asst. resident medicine U. Chgo., 1956-59, chief resident, 1959-60, instr. medicine, 1959-60; instr. medicine, asst. prof. U. Pitts., 1962-67; asso. prof. medicine Ohio State U., Columbus, 1967-71, prof., 1971—, dir. div. hematology and oncology, 1969—; mem. med. adv. com. ARC, 1975—; mem. clin. fellowship review com. Am. Cancer Soc., 1976-82. Contbr. articles in field to profl. jours. Served with U.S. Army, 1960-62. Fellow ACP; mem. Am. Soc. Clin. Oncology, Am. Assn. Cancer Research, Am. Fedn. Clin. Research, Am. Soc. Hematology, Central Soc. Clin. Research (councillor 1980—), S.W. Oncology Group (chmn. sarcoma com. 1981—), Polycythemia Vera Study Group (standardization com., publication com., quality control com. 1969—, myeloproliferative com. 1973—), Phi Beta Kappa, Alpha Omega Alpha. Subspecialty: Hematology. Office: 410 W 10th Ave N-1035 Columbus OH 43210

BALDESCHWIELER, JOHN DICKSON, chemist, educator; b. Elizabeth, N.J., Nov. 14, 1933; s. Emile L. and Isobel (Dickson) B.; m. Marcia Ewing, June 20, 1959; children—John Eric, Karen Anne, David Russell. B. Chem. Engring., Cornell U., 1956; Ph.D., U. Calif. at Berkeley, 1959. From instr. to asso. prof. chemistry Harvard U., 1960-65; faculty Stanford (Calif.) U., 1965-71, prof. chemistry, 1967-71; chmn. adv. bd. Synchrotron Radiation Project, 1972-75; vis. scientist Synchrotron Radiation Lab., 1977; dep. dir. Office Sci. and Tech., Exec. Office Pres., Washington, 1971-73; prof. chemistry Calif. Inst. Tech., Pasadena, 1973—, chmn. div. chemistry and chem. engring., 1973-78; OAS vis. lectr. U. Chile, 1969; spl. lectr. in chemistry U. London, Queen Mary Coll., 1970; vis. scientist Bell Labs., 1978; Mem. Pres.'s Sci. Adv. Com., 1969—, vice chmn., 1970-71; mem. Def. Sci. Bd., 1973-80, vice chmn., 1974-76; mem. carcinogenesis adv. panel Nat. Cancer Inst., 1973—; mem. com. planning and instl. affairs NSF, 1973-77; adv. com. Arms Control and Disarmament Agy., 1974-76; mem. Nat. Acad. Sci. Bd. Sci. and Tech. for Internat. Devel., 1974-76, ad hoc com. on fed. sci. policy, 1979, task force on synfuels, 1979; mem. Pres.'s Com. on Nat. Medal, 1974-76, Pres.'s Adv. Group on Sci. and Tech., 1975-76; mem. governing bd. Reza Shah Kabir U., 1975—; mem. Sloan Commn. on Govt. and Higher Edn., 1977-79, U.S.-USSR Joint Commn. on Sci. and Tech. Cooperation, 1977-79; vice chmn. del. on pure and applied chemistry to People's Republic of China, 1978, mem. com. on scholarly communication with, 1978—; mem. research adv. council Ford Motor Co., 1979—; mem. Chem. and Engring. Adv. Bd., 1981—. Mem. editorial adv. bd.: Chem. Physics Letters, 1979—. Served to 1st lt. AUS, 1959-60. Sloan Found. fellow, 1962-64, 64-65; recipient Fresenius award Phi Lambda Upsilon, 1968. Mem. Nat. Acad. Scis., Am. Chem. Soc. (award in pure chemistry 1967), Council on Sci. and Tech. for Devel., Am. Acad. Arts and Scis., Am. Philos. Soc. Subspecialties: Physical chemistry; Nuclear magnetic resonance (chemistry). Current work: Drug delivery via encapsulation using phospholipid vesicles. Molecular specrtoscopy and structure determination using EXAFS. Home: 619 S Hill Ave Pasadena CA 91106 Office: PO Box 5886 Pasadena CA 91107

BALDINO, FRANK, JR., neuropharmacologist, consultant; b. Passaic, N.J., May 13, 1953; s. Frank and Sally (Cannizzaro) B.; m. Judith A. Jones, Sept. 4, 1976; 1 son, Jeffrey Paul. B.S., Muhlenberg Coll., 1975; Ph.D. in Pharmacology, Temple U., 1979. Postdoctoral fellow Rutgers Med. Sch., 1979-81, instr., 1981-82; prin. scientist research and devel. neurobiology E.I. DuPont de Nemours and Co., Glenolden, Pa., 1982—. NIH fellow, 1979. Mem. AAAS, Soc. Neuroscience. Subspecialties: Neuropharmacology; Neurobiology. Current work: In vitro-electrophysiology; single-unit studies in organotypic tissue cultures. Office: E I Du Pont de Nemours Glenolden Lab 500 S Ridgeway Ave Glenolden PA 19036

BALDO, GEORGE JESSE, research physiologist, physiology educator; b. Herkimer, N.Y., Aug. 14, 1952; s. Sullivan George and Dellalouise (Crane) B.; m. Linda Kathryn Brown, Aug. 22, 1981. B.S. with honors, Union Coll., 1974; Ph.D., SUNY-Stony Brook, 1982. Tchr. asst. Bd. Coop. Ednl. Services II, Patchogue, N.Y., 1974-76; tchr. asst. SUNY-Stony Brook, 1977-81, research assoc., 1981-83. Contbr. articles to profl. jours. Mem. Consumer Services Com., Stony Brook, Civic Assn., South Setauket Park, 1981-83. Mem. AAAS, Am. Physiol. Soc., Sigma Xi. Republican. Roman Catholic. Club: Nat. Corvette Owners Assn. (N.Y.C.). Subspecialty: Neurophysiology. Current work: Synaptic transmission at the neuromuscular junction; factors controlling the release of neurotransmitter from motor nerve terminals. Office: Dept Physiology and Biophysics Health Scis Center SUNY-Stony Brook NY 11794

BALDONADO, ARDELINA-ERIKA ALBANO, nursing educator; b. Ilocos, Norte, Philippines, May 18, 1936; came to U.S., 1960; d. Rosalino and Jovita Crisencia (Acosta) Albano; m. Alfredo Pulido Baldonado, Feb 2, 1963; children: Rozelda Fredelyn, Bradshaw Mark, Erika Gina. B.S.N., Santo Thomas U., Manila, Philippines, 1959; M.S., DePaul U., 1965; Ph.D., Loyola U., 1982. Instr. nursing edn. dept. Northwestern U., Chgo., 1966-70; instr. U. Ill. Coll. Nursing, Chgo., 1973-76; asst. prof. nursing Loyola U., Chgo., 1976—; acting chmn. St. Francis Hosp. Sch. Nursing, 1972-73; research coordinator Hospice of the Northshore, Wilmette, Ill., 1982—. Author: Cancer Nursing: A Holistic Multidisciplinary Approach, 1978; contbr. articles to profl. jours. Mem. Am. Edn. Research Assn., Am. Nurses Assn., Council Nurse Researchers, Ill. Nurses Assn., Philippine Nurses Assn. Chgo. (v.p. 1966-67, bd. dirs. 1967-70), Ill. League Nursing, Am. Assn. Critical Care Nurses, Nat. League Nursing. Roman Catholic. Subspecialties: Research methodology; Behavioral psychology. Current work: Meta-analysis of research; cancer research; pain and symptom control; correlation of personality traits and job characteristics and professional success; application of nursing theoretical formulations by graduate students during clinical practicum. Home: 1808 Dobson St Evanston IL 60202 Office: Loyola U Chgo Niehoff Sch Nursing 6525 N Sheridan Rd Chicago IL 60626

BALDWIN, GARY DALE, laser physicist; b. Harrisburg, Pa., Apr. 3, 1938; s. Oscar Bankus and Evelyn Louisa (Moyer) B.; m. Lucille Elizabeth Hillpot, Mar. 26, 1960; children: Katherin Lynn, Lora Jean. B.S. in Physics, Drexel Inst. Tech., 1961. From asst. engr. to engr. in advanced devel. engring. Surface div. Westinghouse Electric Corp., Balt., 1961-67, sr. engr. on laser systems, 1967-78, fellow engr., 1978—. Contbr. articles to profl. jours. Organist Christ Meml. Presbyterian Ch., Columbia, Md., 1969—. Mem. Optical Soc. Am. Subspecialty: Laser Research. Current work: Laser materials, system, optics and beam propagation; research and development in military lasers. Patentee in field. Office: Westinghouse Electric Corp Box 746 MS 360 Baltimore MD 21203

BALDWIN, ROBERT LESH, educator, biochemist; b. Madison, Wis., Sept. 30, 1927; s. Ira Lawrence and Mary (Lesh) B.; m. Anne Theodora Norris, Aug. 28, 1965; children—David Norris, Eric Lawrence. B.A., U. Wis., 1950; D.Phil. (Rhodes scholar), Oxford (Eng.) U., 1954. Asst. prof., then asso. prof. biochemistry U. Wis., 1955-59; mem. faculty Stanford, 1959—, prof. biochemistry, 1964—; vis. prof. Collège de France, Paris, 1972. Asso. editor: Jour. Molecular Biology, 1964-68, 75-79; mem. editorial bd.: Trends Biochem. Sci, 1977—. Served with AUS, 1946-47. Guggenheim fellow, 1958-59. Mem. Nat. Acad. Scis., Am. Soc. Biol. Chemists, Am. Chem. Soc., Am. Biophysics Soc. (council 1977-81), Am. Acad. Arts and Scis. Subspecialties: Biophysical chemistry; Molecular biology. Current work: Mechanism of protein folding. Home: 1243 Los Trancos Rd Portola Valley CA 94025 Office: Dept Biochemistry Stanford Med Sch Stanford CA 94305

BALFOUR, HENRY H., JR., medical educator; b. Jersey City, Feb. 9, 1940; m., 1967; 3 children. B.A., Princeton U., 1962; M.D., Columbia U., 1966. Attending pediatrician, Wright-Patterson AFB, Ohio, 1968-70; asst. prof. U. Minn., Mpls., 1972-75, assoc. prof., 1975-79, prof. lab. medicine and pathology and pediatrics, 1979—, dir. div. clin. microbiology, 1974—; cons. clin. virology VA Hosp., Mpls., 1973—; prin. investigator NIH grant, 1976—. Center Disease Control grantee, 1977-79; Wellcome Found. grantee, 1978-82; NIH grantee, 1977-79. Mem. Am. Soc. Microbiology, Soc. Exptl. Biology and Medicine, Soc. Pediatric Research, Infectious Disease Soc. Am. Subspecialty: Virology (medicine). Office: Health Sci Center Box 437 240 Delaware St SE Minneapolis MN 55455

BALIAN, GARY, researcher, biochemistry educator; b. Cairo, Oct. 23, 1942; U.S., 1975; s. Assadour and Veronica (Semsarian) B.; m. Joan Sheila Nelson, Aug. 12, 1967; children: Jeremy, Sarah. B.Sc., U. London, 1965, Ph.D., 1969. Sr. fellow U. Wash., Seattle, 1969-72, research asst. prof., 1975-79; research assoc. U. Bristol, Eng., 1972-75; assoc. prof. U. Va., Charlottesville, 1979—. Mem. Biochem. Soc., Am. Soc. Biol. Chemistry, Am. Soc. Cell Biology, Orthopaedic Research Soc. Subspecialties: Biochemistry (medicine); Cell biology (medicine). Current work: Instruction of biochemistry, research in connective tissue. Home: 1609 Yorktown Dr Charlottesville VA 22901 Office: U Va Box 374 Charlottesville VA 22908

BALICK, MICHAEL JEFFREY, biologist, adminstr., cons.; b. Phila., July 21, 1952; s. Jacob and Lillian (Rosen) B.; m. Daphne Allon, Aug. 19, 1980. Student, Tel Aviv U., 1972-73; B.S., U. Del., 1975; A.M., Harvard U., 1976, Ph.D. in Biology, 1980. Research asst. Bot. Mus. of Harvard U., Cambridge, Mass., 1979-80, research assoc. in plant domestication, 1980—; exec. asst. to pres., asst. curator N.Y. Bot Garden, Bronx, 1980—; adj. asst. prof. CUNY, 1982—; agribus. cons. Contbr. articles on biology to profl. jours. Recipient G. H. M. Lawrence Meml. award, 1979; Charles A. Lindbergh grantee, 1980. Fellow Linnean Soc.; mem. Internat. Assn. Plant Taxonomy, Soc. for Econ. Botany, Palm Soc., Am. Assn. Bot. Gardens and Arboreta, Sigma Xi. Subspecialties: Taxonomy; Economic botany. Current work: Exploration for and identification of little known useful plants, especially palms, of New World Tropics, follow-up agronomic domestication studies. Office: New York Botanical Garden Bronx NY 10458

BALINSKY, DORIS, research biochemist, educator; b. Frankfurt, Ger., Dec. 3, 1934; came to U.S., 1975; d. Robert Emil and Else Leonore (Machol) Goldschmidt; m. John Boris Balinsky, Mar. 29, 1958; children: Andrew Paul, Martin George. Research biochemist South African Inst. Med. Research, Johannesburg, 1960-76, head enzyme research unit, 1969-76; research assoc. Columbia U., 1975-76; adj. assoc. prof. dept. biochemistry and biophysics Iowa State U., 1978-80, adj. prof. biochemistry, 1980—. Contbr. articles in field to profl. jours. Witwatersrand Council Edn. fellow, 1957-59; AAUW fellow, 1968-69; recipient Rebecca Lurie Brown prize, 1955. Mem. Am. Soc. Biol. Chemists, Am. Assn. Cancer Research, Sigma Xi, Iota Sigma Pi. Subspecialties: Biochemistry (medicine); Cancer research (medicine). Current work: Biochemical, especially isozymic changes in cancer and in aging. Office: Department of Biochemistry Iowa State University Ames IA 50011

BALL, GEORGE WILLIAM, computer sci. educator; b. Warren, Pa., Apr. 3, 1941; s. William Lincoln and Dorothy Frances (Levey) B.; m. Sandra Shirk, June 6, 1963 (dec. Feb. 1975); 1 son, Christopher; m. Becca Flemming Barnett, Jan. 5, 1978. B.S., Union Coll., 1963; M.S., Syracuse U., 1965, Ph.D., 1972. Prof. math. Alfred (N.Y.) U., 1968—. Treas. St. Peter's Ch., Dansville, N.Y., 1981—; treas. CDH Triparish, Hornell, N.Y., 1982-83. Mem. Math. Assn. Am. Episcopalian. Subspecialties: Graphics, image processing, and pattern recognition. Current work: Compiler construction, Ada. Home: PO Box 56 North Cohocton NY 14868 Office: Alfred U Alfred NY 14802

BALL, JOHN ALLEN, radio astronomer; b. Ravenna, Nebr., Dec. 8, 1935; s. Wallace Ray and Christine (Brehm) B.; m. Audrey Ann Roth, Aug. 21, 1955; children: Fifine Diane, Desiree Belinda, Laurie Gabriella, Kevin Dana. B.S.E., U. Nebr., 1957; Ph.D., Harvard U., 1969. Staff mem. M.I.T. Lincoln Lab., Lexington, Mass., 1961-69; research fellow, then research asso, then dir. radio astronomy facilities and lectr. Harvard Coll. Obs., Cambridge, 1969—. Contbr. articles to profl. jours.; author: Algorithms for RPN Calculators, 1978. Served to lt. USAF, 1957-60. Recipient Shared Rumford award Am. Acad. Arts and Scis., 1971. Mem. IEEE, Am. Astron. Soc., AAAS, Union Radio Scientifique Internat., Audio Engring. Soc., Internat. Astron. Union, Am. Assn. Physics Tchrs. Unitarian Universalist. Subspecialties: Radio and microwave astronomy; Electronics. Co-discoverer several new molecules in interstellar space. Home: Oak Hill Rd Harvard Ma 01451 Office: Harvard Coll Obs Harvard MA 01451

BALL, LAURENCE ANDREW, biochemist, educator, researcher; b. York, Yorkshire, Eng., July 9, 1944; s. Laurence Elinger and Christine Mary (Howe) B.; m. Ann Marguerite, July 20, 1968; children: Jennifer Susan, Katherine Sarah. B.A., Oxford (Eng.) U., 1966, Ph.D., 1969. Postdoctoral fellow U. Wis., 1969-71, assoc. prof., Madison, 1979-82, prof. biochemistry and biophysics, 1982—; sci. staff mem. Nat. Inst. Med. Research, Mill Hill, London, Eng., 1972-74; asst. prof. in residence U. Conn., Storrs, 1974-78, assoc. prof., 1978-79; active NIH Study Sect. Virology, 1980—. Editorial bds.: Virology, Jour. Interferon Research; contbr. writings to profl. publs. in field. Recipient NIH Research Career Devel. Award, 1978—; NIH research grantee, 1979—, 81—. Mem. Am. Soc. Microbiology, Am. Soc. Virology, Am. Soc. Biol. Chemists. Subspecialties: Biochemistry (biology); Virology (biology). Current work: Control of gene expression of animal viruses; mechanism of action of interferon. Office: 1525 Linden Dr Biophysics Lab Madison WI 53706

BALL, WILFRED RANDOLPH, biology educator; b. Chgo., Jan. 3, 1932; s. Wilfred R. and Mary (Saunders) B.; m. Jane Lee, Apr. 1, 1958; children: Janet, Carol, Wendy, Cris. B.S., Morehouse Coll., Atlanta, 1952; M.S., Atlanta U., 1955; Ph.D., Ohio State U., 1965. Instr. biology So. U., 1955-60; research assoc. Ohio State U., 1965-66; assoc. prof. biology Alcorn Coll., Lorman, Miss., 1968-69, Knoxville Coll., 1969-70; research assoc. Dept. Agr., Beltsville, Md., summer 1970; prof. biology Wilberforce (Ohio) U., 1972—. NSF fellow, 1960-62. Mem. Am. Soc. Tropical Medicine and Hygiene, Beta Kappa Chi, Beta Beta Beta. Subspecialties: Cell biology; Physiology (biology). Current work: Effects of chicken malaria on erythrocye survival, aspects of hermatology; histochemistry of rat liver. Home: 1395 Corry St Yellow Springs OH 45387 Office: Wilberforce University Wilberforce OH 45384

BALL, WILLIAM ERNEST, computer science educator, consultant; b. Emporia, Kans., May 26, 1930; s. Russell I. and Helen I. (Whipkey) B.; m. Eva S. Ross, May 17, 1953; children: Susan H. Ball Tiffany, Steven E. B.S. in Chem. Engring, Washington U., St. Louis, 1952, D.Sc., 1958. Profl. engr. Mo. Group leader engring. Monsanto, St. Louis, 1958-62; prof. computer sci. Washington U., St. Louis, 1962—; bd. dirs. Auto Comp, Inc., St. Louis, 1974—; limited ptnr. Metme Communications, St. Louis, 1982—; cons. Gen. Dynamics, St. Louis, 1980. Served with U.S. Army, 1952-54. Mem. Assn. for Computing Machinery, AAUP, Soc. Indsl. and Applied Math., IEEE (affiliate). Subspecialties: Database systems; Artificial intelligence. Current work: Design and implementation of intelligent knowledge databases. Use of logic programming for query analysis and response. Home: 14161 Trailtop Dr Chesterfield MO 63017 Office: Washington U Box 1045 Saint Louis MO 63130

BALLAL, RAGHU VEER (FORMERLY N. RAGHUVEERA BALLAL), molecular biologist, physician; b. Udupi, India, Apr. 7, 1941; came to U.S., 1967, naturalized, 1981; s. Nidambur Ramdas and Kalyani (Rao) B.; m. Kanakadurga R., Oct. 27, 1966; children: Sunita, Aarthi. B.Sc., U. Mysore, India, 1961, M.Sc. (Univ. Grants Commn. postgrad. merit scholar), 1963; Ph.D. in Chemistry, U. Poona, India, 1968; M.D., Universidad Autonoma, Ciudad Juarez, Mexico, 1982. Postdoctoral fellow UCLA, 1967-71; asst. prof. Birla Inst. Sci. and Tech., Pilani, India, 1971-72; postdoctoral fellow Baylor Coll. Medicine, Houston, 1973-74, instr., 1974-75, asst. prof. pharmacology, 1975-80; intern St. Joseph's Hosp. and Med. Ctr., Paterson, N.J., 1982-83. Contbr. chpts., numberous articles to profl. publs. Am. Cancer Soc. instnl. grantee, 1974-75; Nat. Cancer Inst. program project grantee, 1977-82. Mem. Am. Assn. Cancer Research. Hindu. Subspecialties: Molecular biology; Cancer research (medicine). Current work: Mechanism and specificity of eukaryotic gene expression; chromatin structure and function; histones, conjugated histones and nonhistone chromosomal proteins; biochemical and immunologic differences between normal and malignant cells; understanding mechanism of malignant growth. Home: 78A Primrose Ln Paramus NJ 07652 Office: 703 Main St Paterson NJ 07503

BALLARD, CHARLES HENRY, utility company executive; b. Detroit, June 3, 1946; s. Ray Eveleth and Violet May (Richards) B.; m. Violet Jean Bahr, May 4, 1968; children: Jennifer Elizabeth, Charles Douglas. B.S. in Nuclear Engring, U. Mich., 1968; M.S., U. Calif.-Berkeley, 1974. Registered profl. engr., Calif. Sr. ops. engr. Gen. Electric Co., San Jose, Calif., 1968-82; sr. project engr. Pa. Power & Light Co., Allentown, 1982—; Research jr. assoc. NSF, Smithsonian Instn., 1969. Mem. Am. Nuclear Soc., Am. Radio Relay League (life.). Subspecialties: Nuclear engineering. Current work: Application of computer technology to data acquisition, control and display in nuclear power plants. Patentee irradiated component scanner. Office: Pa Power & Light Co PO Box 4498 Allentown PA 18105

BALLARD, ROBERT DUANE, scientist; b. Wichita, Kans., June 30, 1942; s. Chester Patrick and Harriet Nell (May) B.; m. Marjorie C. Jacobsen, July 1, 1966; children: Todd, Doug. B.S., U. Calif.-Santa Barbara, 1965; postgrad. in Oceanography, U. Hawaii, 1965-66, in Geology and Geophysics, U. So. Calif., 1966-67; Ph.D. in Marine Geology and Geophysics, U. R.I., 1974. Asst. scientist Woods Hole (Mass.) Oceanographic Instn., 1974-76, assoc. scientist, 1976-83, sr. scientist, 1983—; pvt. cons. Benthos, Inc., North Falmouth, Mass., 1982-83; cons. dep. chief naval ops. for submarine warfare, 1983—; vis. scholar Stanford U., 1979-80, cons. prof., 1980-81, dir. Deep Submergence Lab., 1983—; pres. Deep Ocean Search and Survey, 1983—. Author: Exploring Our Living Planet, 1983. Served with U.S. Army, 1965-67; with USN, 1967-70. Recipient Sci. award Underwater Soc. Am., 1976; Newcomb Cleveland prize AAAS, 1981; Cutty Sark Sci. award, 1982. Mem. Geol. Soc. Am., Marine Tech. Soc. (Compass Disting. Achievement award 1977), Am. Geophys. Union, Explorers Club. Subspecialties: Sea floor spreading; Ocean engineering. Office: Woods Hole Oceanographic Instn Woods Hole MA 02543 Home: 538 Hatchville Rd Hatchville MA 02536

BALLHAUS, WILLIAM FRANCIS, engineering executive; b. San Francisco, Aug. 15, 1918; s. William Frederick and Eva Rose Callero (O'Connor) B.; m. Edna Dooley, Feb. 13, 1944; children: William Francis, Katherine Louise, Martin Dennis, Mary Susan. B.S., Stanford U., 1940, M.E. (Switzer research fellow, Rosenberg research fellow 1940-42), 1942; Ph.D., Calif. Inst. Tech., 1947. Registered profl. engr. Mem. tech. adv. panel on aeros. Office Sec. Def., 1954-60; mem. NACA, 1954-57; chief engr. Northrop Aircraft, 1953-57, v.p. engring., 1957; v.p., gen. mgr. Nortronics, 1957-61; exec. v.p., dir. Northrop Corp., Beverly Hills, Calif., 1961-64; now dir.; pres. Beckman Instruments, Inc., Fullerton, Calif., 1965-82, chief exec. officer, 1983-84; dir., Ames Research Center NASA, Moffett Field, Ca., 1984—; dir. Amerace Corp., Northrop Corp., Union Oil Co. Calif.; Cons. Office of Critical Tables, Nat. Acad. Scis., 1958-65. Trustee Northrop U., Harvey Mudd Coll.; fellow Claremont U. Center; adv. council Sch. Engring., Stanford. Fellow AIAA; mem. Nat. Acad. Engring., Assn. U.S. Army (pres. Greater Los Angeles chpt. 1963-65, council of trustees 1965-69). Subspecialty: Aeronautical engineering. Office: Ames Research Center NASA Moffett Field CA 94305

BALLUFFI, ROBERT WEIERTER, physical metallurgist; b. Bayshore, N.Y., Apr. 18, 1924; s. Frank William and Louise (Weierter) B.; m. Ruth S. Nickse, June 1, 1973; children by previous marriage: Andrew W., Barbara W., Frank C. S.B. in Metallurgy, Mass. Inst. Tech., 1947, Sc.D., 1950. Sr. engr. Sylvania Elec. Co., 1950-54; asst. prof. to prof. U. Ill., 1954-64; prof. to Francis N. Bard prof. metallurgy, dir. dept. Cornell U., Ithaca, N.Y., 1964-78; prof. metallurgy M.I.T., 1978—; cons. Oak Ridge Nat. Lab., Argonne Nat. Lab., Nat. Acad. Sci., AEC, Dept. Energy. Contbr. articles to profl. jours., chpts. in books. Served with AUS, 1943-46. Decorated Bronze Star. Fellow Am. Phys. Soc., AIME; mem. Am. Acad. Scis., Am. Acad. Arts and Scis., Inst. Metals, Sigma Xi. Subspecialties: Materials; Metallurgy. Current work: Research on defects in crystals, diffusion and kinetic processes, teaching in area of materials science and physical metallurgy. Research in phys. metallurgy, including crystal defects, diffusion, radiation damage. Home: 58 Monmouth St Brookline MA 02146 Office: 13-5078 Mass Inst Tech Cambridge MA 02139

BALLY, JOHN, astrophysicist; b. Szombathely, Hungary, Jan. 11, 1950; s. Istvan and Livia (Bally) Pogacsas; m. Kim Ruhland, June 12, 1982. M.A., U. Calif., Berkeley, 1972; M.S., U. Mass., Amherst, 1977, Ph.D., 1981. Research asst. Five Coll. Radio Astronomy Obs., U. Mass., Amherst, 1977-81; staff Bell Labs., Holmdel, N.J., 1981—. Mem. Am. Astron. Soc. Subspecialty: Radio and microwave astronomy. Current work: Investigation of physics of star formation; radio astron. observations of energetic outflows in star forming clouds, molecular cloud structure instrumentation; superconducting (SIS) receivers, microwave and mm-wave electronics. Office: HOH L-245 Bell Labs Holmdel NJ 07733

BALSLEY, LAWRENCE EDWARD, mech. engr.; b. Chgo., Dec. 10, 1945; s. Lorne Neal and Marie (Bernds) B. B.S. in M.E, U. Ill., 1973. Project engr. Crane Packing Co., Morton Grove, Ill., 1973-74; design engr. Ilg Industries, Chgo., 1974-78; project engr. Joy Mfg., New Philadelphia, Ohio, 1978-81, Lawrence Pump and Engine Co., Mass., 1981-82; design analyst, marine engr Ingalls Shipbldg., Pascagoula, Miss., 1982—. Served with U.S. Army, 1968-71. Mem. ASME, Soc. Automotive Engrs. Club: VFW. Lodges: Elks; Moose. Subspecialties: Mechanical engineering; Solid mechanics. Current work: Vibration and design of axial and centrifugal pumps and fans; computer programming of engineering design problems. Home: 3500 Chico Rd B206 Pascagoula MS 39567 Office: PO Box 149 Pascagoula MS 39567

BALSTER, FREDERICK WERDEN, mechanical engineer; b. Monmouth, Iowa, July 16, 1929; s. Ernst Leo and Stella Ann (Werden) B.; m. Ruth Ann Montgomery, Sept. 18, 1955; children: Martha, Brian, David. B.S., Iowa State U., 1957. With Wagner Electric Co., 1957-62; packaging engr. Apollo, Collins Radio Corp., 1962-70; designer, sr. project engr. Xonics Med. Corp., Des Plaines, Ill., 1970—. Served with U.S. Army, 1951-53; Korea. Mem. ASME. Methodist. Subspecialties: Mechanical engineering; Biomedical engineering. Current work: Design of medical x-ray equipment. Patentee car and truck brakes. Home: 1361 Hassell Dr Hoffman Estates IL 60195 Office: 515 E Touhy Ave Des Plaines IL 60018 Home: 1361 Hassell Dr Hoffman Estates IL 60195

BALSTER, ROBERT LOUIS, psychopharmacology educator; b. St. Cloud, Minn., Oct. 12, 1944; s. Louis and Marion M. (Vandergon) B.; m. Sandra K. Herwig, June 25, 1966; 1 dau., Sarah E. B.A., U. Minn., 1966; Ph.D., U. Houston, 1970. Postdoctoral fellow in psychiatry and pharmacology U. Chgo., 1970-72; assoc. in med. psychology Duke U. Med. Sch., 1972-73, asst. prof. med. psychology, 1973; asst. prof. pharmacology Med. Coll. Va., Va. Commonwealth U., Richmond, 1973-78, assoc. prof., 1978—, dir. grad. studies in pharmacology, 1980—; govt. cons. Contbr. articles, book chpts, papers and abstracts to profl. lit. Fellow Am. Psychol. Assn.; mem. Am. Soc. Pharmacology and Exptl. Therapeutics, Behavioral Pharmacology Soc. Subspecialties: Psychopharmacology; Physiological psychology. Current work: Laboratory research in area of drug abuse. Office: Pharmacology Dept Virginia Commonwealth U Richmond VA 23298

BALTIMORE, DAVID, microbiologist, educator; b. N.Y.C., Mar. 7, 1938; s. Richard I. and Gertrude (Lipschitz) B.; m. Alice S. Huang, Oct. 5, 1968; 1 dau., Teak. B.A. with high honors in Chemistry, Swarthmore Coll., 1960; postgrad., Mass. Inst. Tech., 1960-61; Ph.D., Rockefeller U., 1964. Research assoc. Salk Inst. Biol. Studies, LaJolla, Calif., 1965-68; assoc. prof. microbiology Mass. Inst. Tech., Cambridge, 1968-71, prof. biology, 1972—, Am. Cancer Soc. prof. microbiology, 1973—, dir. Whitehead Inst. Biomed. Research, 1982—. Editorial bd.: Jour. Virology. Recipient Gustav Stern award in virology, 1971; Warren Triennial prize Mass. Gen. Hosp., 1971; Eli Lilly and Co. award in microbiology and immunology, 1971; U.S. Steel Found. award in molecular biology, 1974; Gairdner Found. ann. award, 1974; Nobel prize in physiology or medicine, 1975. Fellow AAAS; mem. Nat. Acad. Scis., Am. Acad. Arts and Scis., Pontifical Acad. Scis. Subspecialties: Molecular biology; Immunobiology and immunology. Current work: Abelson murine leukemia virus, poliovirus replication, gene expression, immunoglobulin transcription. Home: 28 Donnell St Cambridge MA 02138 Office: Mass Inst Tech Cambridge MA 02139

BALWANZ, WILLIAM WALTER, corporation executive, consultant; b. Glencoe, Ohio, Mar. 9, 1913; s. William A. and Ardella (Gibbons) B.; m. Margaret M. Connelly, Mar. 4, 1936; children: William W., Mary F., Barbara A. B.E.E., George Washington U., 1941; M.S.E.E., U. Md., 1949. Registered profl. engr., Va., D.C. Research engr. Naval Research Lab., Washington, 1946-80; also radio engr. Naval Air Sta., Patuxent River, Md., 1942-46; pres. Mattox, Inc., Oak Grove, Va., 1980—; instr. George Washington U., 1950-60, U. Md., 1960-62. Contbr. numerous articles profl. jours. Mem. IEEE, Combustion Inst., AIAA, Profl. Group on Antennas and Propagation (chmn. D.C. chpt. 1965-66). Lodges: Eagles; Masons. Subspecialties: Plasma engineering; Electrical engineering. Current work: Physical scientific research, electrical, chemical and mechanical phases of propulsion systems. Patentee in field. Home: Box 1973 Oak Grove VA 22443 Office: Mattox Inc Oak Grove VA 22443

BALZHISER, RICHARD EARL, energy researcher; b. Elmhurst, Ill., May 27, 1932; m., 1951; 4 children. B.S.E., U. Mich., 1955, M.S.E., 1956, Ph.D., 1961. Prof. chem. engring. U. Mich., 1961-71, chmn., 1970-71; asst. dir. Office Sci. and Tech., 1971-73; dir. Fossil Fuel and Advanced Systems, 1973-79; v.p. research and devel. Elec. Power Research Inst., Palo Alto, Calif., 1979—; cons. Allegany Ballistics Lab., 1961-62, E. I. duPont de Nemours & Co., Inc., 1967-70; White House fellow, 1967-68; cons. prof. Stanford U., 1974-79; mem. adv. bd. U. Calif.-Berkeley, 1974-77, Argonne Nat. Lab., 1974-80, Oak Ridge Nat. Lab., 1976-79, Calif. Tech. Energy Com., 1976—, Gas Research Inst., 1979—, Inst. Energy Analysis, 1980—. Mem. AAAS, Am. Inst. Chem. Engrs., AIME. Subspecialty: Energy research management. Office: Elec Power Research Inst PO Box 10412 Palo Alto CA 94303

BAMBARA, ROBERT ANTHONY, biochemistry educator, microbiology educator; b. Chgo., Jan. 3, 1949; s. Philip Henry and Phyllis (Zullo) B.; m. Bonnie Messersmith, Sept. 15, 1974. B.A., Northwestern U., 1970; Ph.D., Cornell U., 1974. Postdoctoral fellow Stanford (Calif.) U., 1975-76; asst. prof. U. Rochester, N.Y., 1977-81, assoc. prof. biochemistry and microbiology, 1981—. Contbr.: articles to sci. jours. including Biochemistry. Research fellow Jane Coffin Childs Fund, 1975-76; recipient Faculty Research award Am. Cancer Soc., 1980—. Mem. Am. Assn. Biol. Chemists, Sigma Xi. Subspecialties: Biochemistry (biology); Microbiology. Current work: DNA replication in bacteria and mammalian cells; DAN polymerases and their mechanism of action. Home: 105 Hunters Ln Rochester NY 14618 Office: Dept Biochemistry Univ Rochester 601 Elmwood Ave Rochester NY 14642

BAMBURG, JAMES ROBERT, biochemistry educator, researcher; b. Chgo., Aug. 20, 1943; s. Leslie H. and Rose M. (Abrahams) B.; m. Alma Y. Vigo, June 7, 1970; children: Eric G., Leslie A. B.S. in Chemistry, U. Ill., 1965; Ph.D. in Biochemistry, U. Wis., 1969; postgrad., Stanford U., 1969-71. Asst. prof. biochemistry Colo. State U., 1971-76, assoc. prof., 1976-81, prof., 1981—, interim chmn., 1982-83; vis. prof. Cambridge (Eng.) U. Med. Sch., 1978-79; mem. Biomedl Sci. Study sect. NIH, 1982—. Contbr. articles in field to profl. jours. W.H. Peterson fellow, 1968; Nat. Multiple Sclerosis Soc. fellow, 1969-71; Guggenheim fellow, 1978-79. Mem. Am. Soc. Biol. Chemists, Am. Soc. Cell Biology, Am. Chem. Soc., Internat. Soc. Neurochemistry. Subspecialties: Biochemistry (biology); Cell biology. Current work: Role of cytoskeletal proteins in cell motility and nerve growth. Office: Department of Biochemistry Colorado State University Fort Collins CO 80523

BAME, SAMUEL JARVIS, JR., research scientist; b. Lexington, N.C., Jan. 12, 1924; s. Samuel Jarvis and Stella Blanche (Davis) B.; m. Joyce Carleton Fancher, June 21, 1956; children: Karen, Dorthe, Barbara. B.S., U. N.C., 1947; Ph.D., Rice U., 1951. Staff mem. Los Alamos (N.Mex.) Nat. Lab., 1951-81, fellow, 1981—; mem. numerous NASA adv. coms. Contbr. articles to profl. jours. Served with AUS, 1943-46. Recipient Disting. Performance award Los Alamos Nat. Lab., 1980. Fellow Am. Phys. Soc., AAAS; mem. Am. Geophys. Union, Am. Astron. Soc. Subspecialties: Space plasma physics; Solar physics. Current work: Study of interplanetary solar wind and magnetospheric plasmas; design and implementation of space instrumentation for scientific studies; analysis and interpretation of data received from space. Home: 164 Dos Brazos Los Alamos NM 87544 Office: MS D438 Los Alamos Nat Lab Los Alamos NM 87545

BAMFORTH, STUART SHOOSMITH, biology educator; b. White Plains, N.Y.; s. Arthur and Eva Madeleine (Shoosmith) B.; m. Olivia Mary Birdsell, July 8, 1952; children: John, Marjory. B.A., Temple U., 1951; M.S., U. Pa., 1954, Ph.D., 1957. Mem. faculty dept. biology Tulane U., New Orleans, 1957—, prof., 1972—; cons. Acad. Natural Scis., 1956-73. Contbr. chpts. to books. Fellow London Zool. Soc.; mem. Soc. Protozoologists (pres. 1978-79), Am. Micros. Soc. (pres. 1977-78), Ecol. Soc. Am., Am. Soc. Zoologists. Episcopalian. Subspecialty: Ecology. Current work: Distribution of soil protozoa and population dynamics of protozoa on artificial substrates in waters with applications to pollution studies. Home: 2512 Pine St New Orleans LA 70125 Office: Dept Biology Newcomb Coll Tulane U New Orleans LA 70118

BAN, STEPHEN DENNIS, mechanical engineer; b. Gary, Ind., Dec. 16, 1940; m., 1966; 3 children. B.S.M.E., Rose Poly. Inst., 1962; M.S.M.E., Case Inst. Tech., 1964, Ph.D., 1967. Research engr. Battelle Meml. Inst., 1967-68, sr. research engr., 1968-69, sr. project leader, 1969-70, assoc. chief fluid and gas dynamics div., 1970-71, chief, 1971-77; dir. R&D devel. & testing lab. Bituminous Mat Co., 1977-80; mem. staff Gas Research Inst., Chgo., 1980—; adj. prof. mech. engring. Rose Hulman Inst. Tech., 1980—; v.p. Gas Research Inst., Chgo. Mem. AIAA, Sigma Xi. Subspecialty: Mechanical engineering. Office: Gas Research Inst 8600 W Bryn Mawr Chicago IL 60631

BANAS, PAUL ANTHONY, organizational psychologist, human resources consultant; b. Norwich, Conn., June 12, 1929; s. Anthony Paul and Anna (Lucas) B.; m. Charlotte Shilling, Mar. 30, 1963; children: Charles, Darnell, Paul. B.A., U. Conn., 1951; M.A., U. Minn., 1959, Ph.D., 1963. Research psychologist U.S. Army Personnel Research Office, Washington, 1963-64; project mgr. Human Scis. Research, McLean, Va., 1964-66; personnel research cons. Ford Motor Co., Dearborn, Mich., 1966, mgr. research, 1966—; adj. prof. Oakland U., 1972—; advisor SUNY-Buffalo, 1982—. Trustee Ball Found., Wheaton, Ill. Served with USNR, 1952-56. Mem. Am. Soc. Personnel Adminstrn., Acad. Mgmt., Personnel Human Resources (pres. 1971-72), Am. Psychol. Assn., Indsl. Relations Research Assn. Club: Fairlane (Dearborn, Mich.). Subspecialties: Organizational; Social psychology. Current work: Organizational change; human resources management organization development; productivity improvement; human resource systems design, employee involvement systems; selection systems. Home: 5140 Driftwood St Milford MI 48042 Office: Ford Motor Co Room 431 WHQ American Rd Dearborn MI 48121

BANAVAR, JAYANTH RAMA RAO, physicist, researcher; b. Bangalore, India, Oct. 8, 1953; d. K. Rama and Rathna Rama Rao; m. Suchitra Banavar, June 4, 1982. Ph.D., U. Pitts., 1978. Research assoc. U. Chgo., 1979-80; mem. tech. staff Bell Labs., Murray Hill, N.J., 1981-83; mem. profl. staff Schlumberger-Doll Research, Ridgefield, Conn., 1983—. Contbr. articles to sci. jours. Mem. Am. Phys. Soc. Subspecialties: Condensed matter physics; Statistical physics. Current work: Physics of disordered systems. Patentee semicondr. devices. Office: PO Box 307 Ridgefield CT 06877

BANCROFT, GEORGE MICHAEL, chemical physicist, educator; b. Saskatoon, Sask., Can., Apr. 3, 1942; s. Fred and Florence Jean B.; m. Joan Marion MacFarlane, Sept. 16, 1967; children: David Kenneth, Catherine Jean. B.Sc., U. Man., 1963; M.Sc., 1964; Ph.D., Cambridge (Eng.) U., 1967, M.A., 1970, Sc.D. (E.W. Staecie fellow), 1979. Univ. demonstrator Cambridge U.; then teaching fellow Christ Coll.; mem. faculty U. Western Ont., London; now prof. chem. physics, also dir. Centre Chem. Physics. Author: Mössbauer Spectroscopy, 1973; also articles in photoelectron spectroscopy; revs. Mössbauer Spectroscopy. Recipient Harrison Meml. prize, 1972, Meldola medal, 1972, Rutherford Meml. medal, 1980; Guggenheim fellow, 1982-83. Fellow Royal Soc. Can.; mem. Royal Soc. Chemistry, Can. Chem. Soc., Can. Geol. Soc., Can. Physics Soc. Mem. United Ch. Can. Clubs: Curling, Tennis (London). Subspecialty: Physical chemistry. Office: Chemistry Dept U Western Ont London ON N6A 5B7 Canada

BAND, HENRETTA TRENT, geneticist; b. Danville, VA., June 28, 1932; d. Oscar Edward and Lucile Hughes (Allen) Trent; m. Rudolph Neal Band, July 15, 1955; 1 dau., Elizabeth Lee. B.S. (Exeter exchange scholar), Coll. William and Mary, 1954; Ph.D., U. Calif., Berkeley, 1959. NIH predoctoral fellow U. Calif., Berkeley, 1957-58; NIH Postdoctoral fellow Amherst (Mass.) Coll., 1958-59; NIH spl. fellow in genetics Cambridge (Eng.) U., 1965-66; research fellow U. Calif., Davis, 1972-73; lectr. zoology U. B.C., 1959-60, 61-62, research assoc., 1960-63, Mich. State U., 1963-65, asst. prof. temporary, 1965; ind. investigator, East Lansing, 1965—. Contbr. articles to profl. jours. Mem. Am. Soc. Naturalists, Genetics Soc. Am., Soc. Study Evolution, Sigma Xi, Phi Beta Kappa. Republician. Subspecialties: Population biology; Developmental biology. Current work: Research in host plant utilization, overwintering and hemolymph proteins in endemic and geographically widespread drosophilid species, especially Chymomyza amoena, Drosophilia melanogaster and Drosophilia simulans.

BAND, JEFFREY DAVID, med. educator, researcher; b. Detroit, Feb. 24, 1948; s. Herman Allen and Dorothy (Green) Cooperman; m. Meredith E. Weston, May 31, 1973; children: Joshua, Marissa. B.S., U. Mich., 1969, M.D., 1973. Diplomate: Nat. Bd. Med. Examiners, Am. Bd. Internal Medicine. Intern U. Mo. Med. Ctr., Columbia, 1973-74, resident, 1974-76; fellow U. Wis.-Madison, 1976-78; epidemic intelligence service officer Ctrs. for Disease Control, Atlanta, 1978-80, chief spl. pathogens br., 1980-81; assoc. prof. Wayne State U., Detroit, 1981—; cons. epidemiologist Pan Am. Health Orgn., Washington, 1978-81. Contbr. numerous articles to med. jours. Served with USPHS, 1978-81. Recipient Commendation medal UPSHS, 1981. Fellow ACP; mem. Infectious Disease Soc., Soc. Hosp. Epidemiologists Am., Am. Soc. Microbiology, AMA. Subspecialties: Infectious diseases; Epidemiology. Current work: Nosocomial infection prevention, prevention and control of bacterial meningitis, toxic-shock syndrome, Legionnaires Disease. Office: Hutzel Hosp Detroit MI 48201

BANDT, CARL LEE, periodontist, educator; b. Wisconsin Rapids, Wis., Mar. 22, 1938; s. Lawrence Edward and Ethel Marie (Schultz) B.; m. Mary Virginia Rice, June 22, 1963; children—Laura Marie, Mary Louise, Daniel Michael, Matthew Phillip. Student, U. Wis., 1956-57; B.S., U. Minn., 1960, D.D.S., 1962, M.S.D., 1966, M.S., 1968. Instr. Sch. Dentistry, U. Minn., 1966-68, asso. prof., dir. clin. periodontology, 1968-70, asso. prof., dir. clin. systems, 1970-74, prof., 1974—, asst. dean clin. affairs, 1974-78, chmn. dept. periodontology, 1978—; cons. in field. Contbr. articles to profl. jours. Mem. ADA, Am. Acad. Periodontology, Internat. Assn. Dental Research, Am. Assn. Dental Schs., Minn. State Dental Soc., Mpls. Dist. Dental Soc., Phi Eta Sigma, Omicron Kappa Upsilon. Subspecialty: Periodontics. Current work: Clinical studies of periodontal disease emphasizing disease preventing risk assessment, and mointoring state of health and disease. Office: Sch Dentistry Univ of Minn Minneapolis MN 55455

BANDYOPADHYAY, PROMODE RANJAN, researcher, educator; b. Raniswar, Bihar, India, Apr. 3, 1948; came to U.S., 1981; s. Bhupati Nath and Kanak Lata B.; m. Utpala Mukherji, Nov. 24, 1980. B.E., U. North Bengal, 1968; M.E., U. Calcutta, 1970; Ph.D., Indian Inst. Tech., Madras, 1974, U. Cambridge, Eng., 1978. Aero. engr. Hindustan Aero. Ltd., Lucknow, India, 1974-75; research asst. Cambridge U., 1978-79; sr. design engr. Hindustan Aero. Ltd., Bangalore, India, 1979-81; research scientist dept. mech. engring. U. Houston, 1981—. Contbr. in field. Nat. scholar in mech. engring. Ministry of Edn. Govt. India, New Delhi, 1974, fellow Wolfson Coll., Cambridge, 1978. Mem. AIAA, Am. Phys. Soc., ASME. Club: Univ. Oaks Civic (Houston). Subspecialties: Fluid mechanics; Aerospace engineering and technology. Current work: Basic aspects of turbulent boundary layer, transition and other shear flows. Office: Dept Mechanical Engineering University of Houston 4800 Calhoun St Houston TX 77004 Home: 4362 Varsity St Houston TX 77004

BANERJEE, AMIYA KUMAR, molecular virologist, educator; b. Rangoon, Burma, May 3, 1936; came to U.S., 1965, naturalized, 1978; s. Phanindra Nath and Bibhati (Ghosal) B.; m. Sipra Datta, Jan. 23, 1965; children: Antara, Arjun. M.A., Calcutta U., 1958, Ph.D., 1964, D.Sc., 1970. Research assoc. Albert Einstein Coll. Medicine, Yeshiva U., Bronx, N.Y., 1966-69; asst. mem. Roche Inst. Molecular Biology, Nutley, N.J., 1970-73, assoc. mem., 1974-79, full mem., 1980—; adj. prof. dept. cell biology N.Y.U. Med. Sch., N.Y.C., 1980—; Mem. virology study sect. NIH. Editorial bd.: Jour. Virology, Virology; Contbr. articles and abstracts to sci. jours. Recipient Phoebe Weinstein award NIH, 1977. Mem. AAAS, Am. Soc. Biol. Chemists, Am. Soc. Microbiology, Harvey Soc., N.Y. Acad. Scis. Club: Tagore Soc. (N.Y.C.). Subspecialties: Virology (veterinary medicine); Molecular biology. Current work: I am primarily interested in the molecular biology of vesicular stomatitis virus gene expression. Areas studied include: mode of genome transcription and replication in vitro, structure and function of the genetic material, biochemical functions of purified viral proteins, mechanisms of defective virus production and the interference phenomenon. Home: 113 Fairway Ave Verona NJ 07044 Office: Roche Institute of Molecular Biology Nutley NJ 07110

BANERJEE, KRISHNADAS, radiol. physicist; b. Calcutta, India, Jan. 3, 1934; came to U.S., 1961, naturalized, 1977; s. Ramkanai and Hemlata (Mukherjee) B.; m. Rama Dhar, June 19, 1968; 1 dau., Rini. M.S., Calcutta U., India, 1956; Ph.D., U. Pitts., 1965. Cert. radio. physicist Am. Bd. Radiology. Research fellow Saha Inst. Nuclear Physics, Calcutta, 1967-69; postdoctoral fellow U. Pitts., 1969-70; dir. radiol. physics and radiation biology St. Francis Gen. Hosp., Pitts., 1970—; adj. prof. radiol. physics U. Pitts. Grad. Sch. Public Health, Pitts., 1978—. Fulbright scholar, 1961. Mem. Am. Assn. Physicists in Medicine, Health Physics Soc., Radiol. Soc. N.Am. Hindu. Subspecialties: Radiological physics; Biophysics (physics). Current work: Radiological physics, nuclear medicine, radiation therapy, medical physics, nuclear magnetic resonance. Home: 641 Ravencrest Rd Pittsburgh PA 15215 Office: St Francis Gen Hosp Pittsburgh PA 15201

BANERJEE, MUKUL RANJAN, physiology educator, researcher; b. Dacca, India, Jan. 3, 1937; came to U.S., 1960; s. Chunilal and Snehalata (Mukherjee) B.; m. Gita Ganguly, Dec. 4, 1968; children: Amit, Sumita. B.V.Sc., Calcutta U., 1957; Ph.D., La. State U.-Baton Rouge, 1964. Research assoc. Ind. U., 1965-67, asst. prof., 1968-72; research fellow U. Fla., Gainesville, 1973-74; assoc. prof. Tenn. State U., 1975-76; assoc. prof. physiology Meharry Med. Coll., Nashville, 1977-79, prof., 1980—. Mem. Am. Physiol. Soc., Internat. Soc. Biometeor, Sigma Xi. Subspecialty: Physiology (medicine). Current work: Respiratory physiology, environmental physiology. Home: 999 Windrowe Dr Nashville TN 37205 Office: Dept Physiology Meharry Med Coll 1005 18th Ave N Nashville TN 37208

BANERJEE, RANJIT, geneticist; b. Calcutta, India, June 6, 1948; s. Ajit Kumar and Maya (Mukherjee) B. M.S., U. Calcutta, 1971, NYU, 1979, Ph.D., 1981. Research assoc. dept. pathology Temple U. Med. Center, Phila., 1981-83; staff assoc. Inst. Cancer Research, Columbia U., N.Y.C., 1983—. Contbr. articles to profl. jours. Mem. Am. Soc. Cell Biology, Genetics Soc. Am., AAAS, N.Y. Acad. Sci., Am. Soc. Microbiology. Subspecialties: Genetics and genetic engineering (biology); Molecular biology. Current work: Researcher in genetics and cytogenetics of cancer, gene regulation in mammalian cells and recombinant DNA. Office: Inst Cancer Research Columbia U New York NY 10032

BANERJEE, SIPRA, biochemist, educator; b. Calcutta, India, Feb. 20, 1939; came to U.S., 1965, naturalized, 1978; d. Sanmatha Nath and Sudharani Datta; m. Amiya Banerjee, Jan. 23, 1965; children: Antara, Arjun. Ph.D. in Biochemistry, Calcutta U., 1965. Postdoctoral fellow dept. biochemistry Albert Einstein Coll. Medicine, Bronx, N.Y., 1966-69; research biochemist, dept. pharmacology N.Y. Med. Coll., N.Y.C., 1969-70; fellow dept. biochemistry and drug metabolism Hoffman-LaRoche Inc., Nutley, N.J., 1972-73; assoc. research scientist, dept. environ. medicine N.Y.U. Med. Ctr., N.Y.C., 1972-78; asst. prof. Inst. Environ. Medicine, 1978—. Contbr. articles to profl. jours. Mem. Am. Soc. Biol. Chemists, Am. Assn. for Cancer Research, N.Y. Acad. Scis. Clubs: Milani (Bay Shore, N.Y.); Kallol (N.J.). Subspecialties: Biochemistry (biology); Cancer research (medicine). Current work: Biochemical studies of mulecular mechanisms of chemical carcinogenesis. Home: 113 Fairway Ave Verona NJ 07044 Office: 550 1st Ave MSB Room 207 New York NY 10016

BANERJEE, UTPAL, computer scientist; b. Calcutta, India, Aug. 4, 1942; came to U.S., 1965, naturalized, 1970; s. Santosh Kumar and Santi Rani (Mukherjee) B.; m. Aloka Mukherjee, June 11, 1969; 1 dau., Sanchita. B.Sc., Calcutta U., 1961, M.Sc., 1963, M.S., Carnegie-Mellon U., Pitts., 1967, Ph.D., 1970; M.S., U. Ill., Urbana-Champaign, 1976, Ph.D., 1979. Asst. prof. math. U. Cin., 1969-75; prin. analyst Honeywell, Phoenix, 1979-81; mem. research staff Fairchild Labs., Palo Alto, Calif., 1981-82; cons. Control Data Corp., Sunnyvale, Calif., 1982—. Contbr. articles to profl. jours. Mem. Am. Math. Soc., Assn. Computing Machinery, IEEE, Sigma Xi. Subspecialty: Computer architecture. Current work: Parallel processing for supercomputers. Office: MS SVL 144F Control Data Corp Box 3492 Sunnyvale CA 94088

BANIK, NARENDRA LAL, neuroscientist; b. Ganganagar, Bengal, India, Jan. 2, 1938; s. Surendra Lal and Radharani B.; m. Meena Banik, Mar. 7, 1968; children: Manendu, Nandini. B.Sc., U. Calcutta, 1959; M.Sc., U. London, 1966, Ph.D. in biochemistry, 1970. Research asst. Inst. Psychiatry, London, 1962-65; research asst. Charing Cross Hosp. Med. Sch., London, 1966-70; lectr. Inst. Neurology, London, 1970-74; research asso. Stanford U., 1974-76; asst. prof. Med. U. S.C., Charleston, 1977-80; assoc. prof. U. S.C., 1980—. Contbr. articles to profl. jours. NATO grantee, 1971; Welcome Research Found., grantee, 1974. Mem. Biochem. Soc. London, Internat. Soc. Neurochemistry, Am. Soc. Neurochemistry, AAAS, Soc. Neurosci., Biochem. Soc. London, Am. Soc. Biol. Chemists. Subspecialties: Neurochemistry; Biochemistry (biology). Current work: Membrane biochemistry, myelination and mechanism of demyelination, spinal cord injury research. Home: 750 Olney Rd Charleston SC 29407 Office: 171 Ashley Ave Neurology Charleston SC 29425

BANJAVIC, RICHARD ALAN, educator, consultant; b. Chgo., June 13, 1947; s. John and Dorothy (Kostadin) B.; m. Judith Caplan Banjavic, Jan. 29, 1982. B.A. in Physics with honors, Johns Hopkins U., 1969, M.S., U. Ill., 1971, U. Wis., Madison, 1973, Ph.D., 1978. Research assoc. Materials Sci. Lab., U. Ill., 1970-72; research asst. nuclear medicine U. Wis., 1972-74; research asst. radiotherapy physics and ultrasound, 1974-78, postdoctoral research assoc. ultrasound, 1978-79; research assoc. ultrasound U. Colo. Health Sci. Center, Denver, 1979-80, instr., 1980-81, asst. prof., 1981-83; ultrasound physicist Rocky Mountain Med. Physics, Inc., Denver, 1983—; cons in field. Contbr. articles in field to profl. jours. Served with USNR, 1971-80. Woodrow Wilson fellow, 1969; NSF fellow, 1970. Mem. Am. Inst. Ultrasound in Medicine, Am. Assn. Physicists in Medicine, Assn. Advancement Physics, Acoustical Soc. Am., Soc. Magnetic Resource Imaging, IEEE. Democrat. Roman Catholic. Subspecialties: Imaging technology; Acoustics. Current work: Study of diagnostic ultrasonic imaging itself as a separate modality and in conjunction with the other types of diagnostic modalities presently available.

BANKA, VIDYA SAGAR, cardiologist; b. Lyallpur, India, Mar. 5, 1941; came to U.S., 1970; s. Ram Ch and Hukam (Devi) B.; m. Reena Sagar Banka, Jan. 26, 1972; children: Sahil, Sarovar. M.B.B.S., Med. Coll. Mraitsar, India, 1964; M.D., Panjab U., India, 1967. Diplomate: Am. Bd. Internal Medicine. Registrat incardiology Postgrad. Inst. Med. Edn. and Research, Chandigarh, India, 1967-69; dir. cardiac catheterization labs. Presbyn. U. Pa. Med. Center, Phila., 1974-83; dir. interventional cardiovascular medicine Episcopal Hosp., Phila., 1983—; assoc. prof. medicine U. Pa. Sch. Medicine, Phila., 1979-82, assoc. prof. clin. medicine, 1982—. Author: A Clinical and Angiographic Approach to Coronary Heart Disease, 1978. Fellow ACP, Am. Coll. Cardiology, Am. Coll. Chest Physicians, Am. Heart Assn. (council clin. cardiology). Subspecialty: Cardiology. Current work: Coronary heart disease; coronary angioplasty and use of laser during angioplasty. Patentee in field. Home: 237 Staley Rd Narberth PA 19072 Office: Presbyn UPa Med Center 51 N 39th St Philadelphia PA 19104

BANKER, GILBERT STEPHEN, pharmacy educator; b. Tuxedo Park, N.Y., Sept. 12, 1931; s. Gilbert Miller and Mary Edna (Gladstone) B.; m. Gwenivere May Hughes, Mar. 31, 1956; children: Stephen, Susan, David, William. B.S., Union U., Albany, N.Y., 1953; M.S., Purdue U., 1955, Ph.D., 1957. Registered pharmacist, N.Y., Ind. Mem. faculty Purdue U., West Lafayette, Ind., 1955—, prof. indsl. pharmacy Sch. Pharmacy, 1964—, head indsl. and phys. pharmacy dept., 1967—; cons. Richardson-Vicks, Mt. Vernon, N.Y., 1975—, G.D. Searle Co., Skokie, Ill., 1978—, FMC Corp., Phila., 1979—, Menley James Labs., 1981—; mem. revision com. U.S. Pharmacopeia, 1975—. Editor: Modern Pharmaceutics, 1970; editor: Pharmaceutics and Pharmacy Practice, 1980; contbr. over 100 articles to sci. publs. Recipient service recognition award Kappa Psi, 1970, Outstanding Alumnus award Union U., 1977. Fellow Acad. Pharm. Scis. (v.p. 1971-72, chmn. indsl. pharm. tech. sect. 1968-69); mem. Am. Pharm. Assn. (Indsl. Pharmacy award 1971, mem. ho. of dels. 1977-80), Sigma Xi (pres. Purdue U. chpt. 1971-72), Rho Chi (exec. council 1964-66). Republican. Presbyterian. Lodges: Rotary; Elks. Subspecialty: Medicinal chemistry. Current work: Design of new drug delivery systemss employing polymers and colloids, including development of bioadhesive and transdermal systems. Home: 1210 Western Dr West Lafayette IN 47906 Office: Sch Pharmacy and Pharmacol Scis Purdue U West Lafayette IN 47907

BANKERT, RICHARD BURTON, immunologist; b. Phila., Apr. 22, 1940; children: Lauren, Darin. B.A., Gettysburg Coll., 1962; V.M.D., U. Pa., 1968, Ph.D., 1973. Postdoctoral fellow U. Pa., 1968-70, postdoctoral research fellow in immunology, 1970-72; assoc. veterinarian Atlantic Vet. Clinic, Pomona, N.J., 1968-69; owner, mgr. Spruce Hill Vet. Clinic, Pa., 1968-73; assoc. chief molecular immunology Roswell Park Meml. Inst., Buffalo, 1973—; prof. cell and molecular biology SUNY-Buffalo. Served with AUS. Nat. Cancer Inst. grantee, 1978—; Am. Cancer Soc. grantee, 1979—; Allergy and Immunology grantee, 1983—. Mem. AVMA, AAAS, Am. Assn. Immunologists, N.Y. Acad. Scis., Sigma Xi. Subspecialties: Immunology (agriculture); Cancer research (veterinary medicine). Current work: Immunology, microbiology.

BANKHURST, ARTHUR DALE, physician, educator; b. Cleve., July 21, 1937; s. John William and Daisy (Howard) B.; m. Lois Hull, Feb. 20, 1969; children: Anne, Claire, Benjamin, Noah. B.S., M.I.T., 1958; M.D., Case Western Res. U., 1962. Diplomate: Am. Bd. Internal Medicine., subsplty. bd. rheumatology. Intern in medicine Univ. Hosps. of Clev., 1962-63, resident, 1965-69; research fellow Walter and Eliz Hall Inst., Melbourne, Australia, 1969-71, WHO, Geneva, Switzerland, 1971-73; asst. prof. medicine U. N.Mex., Albuquerque, 1973-77, assoc. prof., 1977-81; prof., 1981—, chief clin. immunology and rheumatic diseases, 1979—; cons. N.Mex. Profl. Standard Rev. Orgn.; chief of rheumatology Albuquerque VA Med. Center. Contbr. articles to numerous profl. jours. Served to lt. comdr. U.S. Navy, 1963-65. Cleveland fellow Walter and Eliza Hall Inst., 1969-71; Nat. Arthritis Found. sr. investigator, 1974-79. Mem. AAAS, Am. Assn. Immunologists, Am. Fedn. Clin. Research, Am. Rheumatic Assn., Brit. Soc. Immunology, Western Soc. Clin. Investigation. Subspecialties: Immunology (medicine); Immunopharmacology. Office: Dept Medicine 2211 Lomas Blvd NE Albuquerque NM 87131

BANKS, EDWIN MELVIN, biological sciences educator; b. Chgo., Mar. 21, 1926; s. David Louis and Eleanor (Johnson) B.; m. Hilda Markoff, June 20, 1950; children—Daniel, Ronald, Ellen. Ph.B., U. Chgo., 1948, B.S., 1949, M.S., 1950; Ph.D., U. Fla., 1955. Asst. prof. biology U. Ill., 1955-60, asso. prof., 1960-63, prof. ethology, head dept., 1965—; asso. prof. zoology U. Toronto, 1963-65. Author: Vertebrate Social Organization, 1977, Animal Behavior, 1977; author articles in field. Served with USNR, 1943-46. H.F. Guggenheim Found. grantee, 1977-78. Fellow AAAS, Animal Behavior Soc. (pres. 1972); mem. Am. Soc. Zoologists, Am. Soc. Mammalogists, Ecol. Soc. Am., Internat. Ethological Soc., Soc. Study of Aggression, Sigma Xi. Jewish. Club: Rotary. Subspecialty: Ethology. Office: 515 Morrill Hall University of Illinois 505 Goodwin Ave Urbana IL 61801

BANKS, PETER MORGAN, educator; b. San Diego, May 21, 1937; s. George Willard and Mary Margaret (Morgan) B.; m. Paulett M. Behanna, May 21, 1983; children by previous marriage: Kevin, Michael, Steven, David. M.S. in E.E, Stanford U., 1960; Ph.D. in Physics, Pa. State U., 1965. Postdoctoral fellow Institut d'Aeronomie Spatiale de Belgique, Brussels, Belgium, 1965-66; prof. applied physics U. Calif., San Diego, 1966-76; prof. physics Utah State U., 1976-81, head dept. physics, 1976-81; prof. elec. engring. Stanford U., 1981—, vis. asso. prof., 1972-73; vis. scientist Max Planck Inst. for Aeronomie, Ger., 1975; pres. La Jolla Scis. Inc., 1973-77, Upper Atmosphere Research Corp., 1978-82. Author: (with G. Kockarts) Aeronomy, 1973, (with J.R. Doupnik) Introduction to Computer Science, 1976; Assoc. editor Jour.: Geophys. Research, 1974-77, Planetary and Space Sci, 1977—; regional editor, 1983—; Contbr. (with J.R. Doupnik) numerous articles in field to profl. jours. Mem. space sci. adv. council NASA, 1976-80. Served with U.S. Navy, 1960-63. Recipient Appleton prize Royal Soc. London, 1978; Space Sci. award Am. Inst. Aeros. and Astronautics, 1981. Mem. Am. Geophys. Union, Internat. Union Radio Sci. Episcopalian. Clubs: Am. Youth Soccer Assn., Cosmos. Subspecialty: Space physics. Home: 23 Peter Coults Circle Stanford CA 94305 Office: Elec Engring Dept Stanford U Stanford CA 94305

BANKS, WILLIAM LOUIS, JR., biochemistry and surgery educator, cancer center administr.; b. Paterson, N.J., Mar. 26, 1936; s. William Louis and Martha (Roughgarden) B.; m. Sharon R. Hazelton, Aug. 1, 1965; 1 dau., Heather Michelle. B.S., Rutgers U., 1958; M.S., Bucknell U., 1961; Ph.D., Rutgers U., 1963. Teaching asst. Bucknell U., 1958-60; research asst. Rutgers U., 1960-62; lectr. chemistry St. Mary's U., 1963-65; asst. prof. biochemistry Med. Coll. Va., 1965-69, assoc. prof., 1969-74, prof., 1974—; co-dir. MCV/VCU Cancer Center, 1974—. Contbr. in field. Bd. dirs. Epilepsy Found. Va., 1974-76. Served to capt. USAF, 1963-65. Johnson & Johnson Research fellow, 1962-63; Alfred P. Sloan Found. Faculty scholar, 1975-76; grantee in field. Mem. AAAS, Am. Assn. Cancer Edn., Am. Assn. Cancer Research, AAUP, Am. Chem. Soc., Am. Council on Sci. and Health, Am. Inst. Nutrition, N.Y. Acad. Sci., Nutrition Today Soc., Soc. Exptl. Biology and Medicine, Sigma Xi, Rho Chi, Sigma Zeta. Subspecialties: Biochemistry (medicine); Oncology. Current work: Nutrition and cancer. Office: Medical College Virginia PO Box 37 Richmond VA 23298

BANNON, PETER RICHARD, meteorology educator, dynamic meteorology researcher; b. N.Y.C., Apr. 30, 1949; s. John V. and Helen (Muller) B.; m. Helen Robb Whitehouse, Jan. 8, 1972; 1 son, James. B.S., Cornell U., 1971; M.S., U. Colo.-Boulder, 1973, Ph.D., 1979. Phys. scientist Geophys. Fluid Dynamics Lab., Princeton, N.J., 1974-76; postdoctoral fellow Nat. Ctr. Atmospheric Research, Boulder, 1978-80; asst. prof. meteorology U. Chgo., 1980—. Mem. Am. Meteorol. Soc. (Max A. Eaton award 1979, v.p. Chgo. chpt. 1981-82). Subspecialty: Meteorology. Current work: Dynamic meteorology, monsoon dynamics, mountain meteorology. Office: U Chicago 5734 S Ellis Ave Chicago IL 60637

BANSCHBACH, MARTIN WAYNE, biochemistry educator, researcher; b. Glen Cove, N.Y., July 5, 1946; s. Martin Luther and Dorothy Elizabeth (Bach) B.; m. Susan Virginia Karl, Sept. 7, 1968; children: Kimberly Sue, Eric Brian. A.B., Susquehanna U., 1968; Ph.D., Va. Tech. U., 1972. NIH fellow U. Wis. Med. Sch., Madison, 1972-75; asst. prof. biochemistry La. State U. Med. Sch., Shreveport, 1975-80; assoc. prof. biochemistry Okla. Coll. Osteo. Medicine, Tulsa, 1980—, dir. research, 1980—; cons. Boots Pharms., Shreveport, 1979-80. Contbr. articles to profl. jours. Pres. Lutheran Ch. Council, Shreveport, 1980; judge pub. sch. sci. fair, Shreveport, 1979. NIH grantee, 1980; March of Dimes grantee, 1980; NIMH grantee, 1977; Boots Pharms. grantee, 1980. Mem. Am. Chem. Soc., AAAS, Assn. Med. Depts. Biochemistry, Soc. Exptl. Biology and Medicine, Sigma Xi. Republican. Subspecialties: Biochemistry (medicine); Membrane biology. Current work: Medical research (cystic fibrosis), basic research cell membrane structure and function. Home: 5902 S 92d E Ave Tulsa OK 74145 Office: Okla Coll Osteo Medicine PO Box 2280 Tulsa OK 74101

BANTLE, JOHN ALBERT, II, zoology educator; b. Detroit, Jan. 20, 1946; s. John Albert and Corinne Helen (Schulte) B.; m. Donetta Louise Lee, Feb. 1, 1969; children: Kelly Ann, John Albert. A.B., Eastern Mich. U., 1968, M.S., 1970; Ph.D., Ohio State U., 1973. Grad. teaching asst. Eastern Mich. U., Ypsilanti, 1968-70; grad. teaching assoc. Ohio State U., Columbus, 1970-73; research assoc. U. Colo. Med. Sch., Denver, 1973-76; asst. prof. zoology Okla. State U., Stillwater, 1976-81, assoc. prof., 1981—. Contbr. articles to profl. jours. NSF grantee, 1977; NIH grantee, 1982; March of Dimes Found. grantee, 1982; Okla. Water Resources and Research Inst. grantee, 1982. Mem. Am. Soc. Zoologists, Am. Soc. Cell Biology, Okla. Acad. Scis., Soc. Envrion. Toxicology and Chemistry. Democrat. Roman Catholic. Subspecialties: Developmental biology; Cell and tissue culture. Current work: Exploring the effect of genotoxic agents on embryonic development using recombinant DNA in a genotoxicity assay; investigating effect of chronic alcohol treatment on polyadenylation of RNA. Home: 1417 Surrey Dr Stillwater OK 74074 Office: Dept Zoology Okla State U Stillwater OK 74078

BANWART, GEORGE J., microbiology educator; b. Algona, Iowa; s. George W. and Leah R. (Schneider) B.; m. Sally F. Foss, Mar. 18, 1955; children: Deborah J., Geoffrey D. B.S., Iowa State U., 1950, Ph.D., 1955. Research asst. Iowa State U., 1950-55; asst. prof. U. Ga., Athens, 1955-57; head egg products U.S. Dept. Agr., Chgo. and Washington, 1957-62; assoc. prof. Purdue U., West Lafayette, Ind., 1962-65; research microbiology Agrl. Research Service, U.S. Dept. Agr., Beltsville, Md., 1965-69; prof. microbiology Ohio State U., Columbus, 1969—; cons. in field. Author: Basic Food Mircobiology, 1979. Mem. Prince George Pub. Sch. Bd., Bowie, Md., 1967-68. NIH grantee, 1963-65; Ohio State U. grantee. Mem. Inst. Food Tech. (chmn. scholarship com. 1975, exec. com. food microbiology div. 1981-83), Am. Soc. Microbiology, Soc. Applied Bacteriology, Phi Tau Sigma. Subspecialties: Microbiology; Food science and technology. Current work: Methods to detect salmonella enterotoxin; bacillus cereus diarrheagenic toxin; methods to detect campylobacter. Office: Dept Microbiology Ohio State U 484 W 12th Ave Columbus OH 43210

BAPNA, MAHENDRA SINGH, biomaterials educator, researcher; b. Udaipur, Rajasthan, India, Sept. 8, 1939; came to U.S., 1962; s. Both Lal and Chagan Bai (Nahar) B.; m. Prabha Bhandari, May 31, 1966; children: Manish, Mitali. B.Sc., Rajasthan U., Udaipur, 1957; B.S. in Metallurgy, Banaras (Utar Pradesh, India) U., 1961; M.S. in Materials Sci, Marquette U., 1964; Ph.D., Northwestern U., 1969. Asst. prof. Loyola U. Chgo., Maywood, Ill., 1970-73; postdoctoral fellow Northwestern U., 1973-74; asst. prof. biomaterials Coll. Dentistry, U. Ill.-Chgo., 1974-77, assoc. prof., dir. biomaterials, 1977-83, prof., dir. biomaterials, 1983—. Contbr. articles to profl. jours. Univ. scholar, 1955-57; Murphy fellow, 1966-68; NIH grantee, 1972; univ. grantee, 1977. Mem. Am. Soc. Metals (hon., ednl. com.), Internat. Assn. Dental Research, Am. Assn. Dental Research, Sigma Xi. Subspecialties: Biomaterials; Materials. Current work: Understanding the existing and developing new restorative dental materials; effects of oral environment on mechanical and physical properties of restorative dental materials; examining the elastic interaction among defects in anisotropic metals. Office: U Ill Coll Dentistry 801 S Paulina St Chicago IL 60612 Home: 711 N Jefferson St Hinsdale IL 60521

BARANKIN, EDWARD WILLIAM, statistics educator; b. Phila., Dec. 18, 1920; s. Myer and Esther (Grossman) B.; m. Claire Chertcoff, June 22, 1941 (div. 1973); children: Joseph Paul, Barry Alexander. A.B., Princeton U., 1941; M.A., U. Calif.-Berkeley., 1942, Ph.D., 1946. Instr. dept. math. U. Calif.-Berkeley, 1947-48, asst. prof. math., 1948-52, assoc. prof. math., then assoc. prof. stats., 1952-59, prof. stats., 1959—; cons. in field; vis. prof. U. Paris, 1956-57, Kyoto U., 1962-63, Nat. U. Mex., 1962, 64, 73, 74, 76, 81, Inst. Statis. Math., Tokyo, 1970, 74, 75, 82, Tamkang Coll., Taiwan, 1970, 74, U. Veracruz, Mex., 1981; lectr. internat. confs. Contbr. articles to profl. jours. Bd. dirs. Consumers Coop., Berkeley, 1961-68. Guggenheim fellow, 1956-57; Fulbright-Hays fellow, 1962-63, 64; Japan Found. fellow, 1974. Fellow Inst. Math. Stats.; mem. Am. Math. Soc., Bernoulli Soc., Am. Statis. Assn. (1st chmn. com. sci. freedom and human rights 1979-81), Internat. Assn. Math. Modeling, Econometric Soc., Inst. Mgmt. Scis. (charter, editorial bd. 1955-69), N.Y. Acad. Sci., Japan Assn. Philosophy of Sci. (fgn.), Albert Einstein Soc., Assn. Mems. of Inst. for Advanced Study. Office: Dept Stats U Calif Berkeley CA 94720

BARANOV, ANDREY I(PPOLITOVICH), botanist; b. Harbin, China, Oct. 17, 1917; s. Ippolit G. and Varvara M. B.; m. Nina M., June 14, 1946; 1 dau., Elena. First class diploma (equivalent of LL.B), Harbin Law Sch., 1938; post grad., U. Wash., 1960-61; M.S. in Biology, Northeastern U., 1973. Research fellow Harbin Regional Museum, 1946-50, Academia Sinica inst. Forestry and Soil Sci., Shenyang, China, 1950-58; herbarium asst. Arnold Arboretum, Harvard U., 1963-67; bibliography researcher World Life Research Inst., Colton, Calif., 1967-68; mem. bd. advs. Inst. for Traditional Medicine and Preventive Health Care, 1979—. Author: Basic Latin for Plant Taxonomists, 1971, Studies in Begoniaceae, 1981; contbr. numerous articles to profl. jours.; author monographs. Mem. New Eng. Bot. Club, Am. Fern Soc., Bot. Soc. Am., Internat. Assn. Plant Taxonomy, Mus. Russian Culture, Sigma Xi (assoc.). Russian Orthodox. Subspecialties: Taxonomy; Ethnobotany. Current work: Plant taxonomy; ethnopharmacology of Far Eastern nations, primarily study of ginseng. Office: PO Box 131 Cambridge MA 02140

BARANOWSKI, TOM, health psychology researcher, educator; b. Bklyn., Dec. 3, 1946; s. Joseph Anthony and Vera Mashkoff (Willard) B.; m. Jimmie Dolores Miller, June 8, 1968; children: Tanya Elise, Todd Michael. A.B., Princeton U., 1968; M.A., U. Kans., Lawrence, 1970, Ph.D., 1974. Evaluator, acting dir. planning and evaluation Kans. Regional Med. Program, Kansas City, 1971-73; research scientist Battelle Meml. Inst., Seattle, 1974-76; asst. prof. community medicine W.Va. U. Med. Center, Charleston, 1976-80; asst. prof. pediatrics U. Tex. Med. Br., Galveston, 1980—; cons. N.D. State U., Fargo, 1980—; proposal reviewer NSF, 1982—; cons. Nat. Inst. Occupational Safety and Health, Cin., 1980. Author: Health Promotion Disease Prevention, 1983; Contbr. chpts. to books, articles to profl. jours. Bd. dirs. W.Va. affiliate Am. Diabetes Assn., 1979-80; chmn. W.Va. Hypertension Control Com., Charleston, 1978-80; linesman, asst. coach Bay Area Soccer League, Galveston, 1980—. Recipient citation W.Va. affiliate Am. Diabetes Assn., 1980; USPHS fellow, 1969-71; Regent's scholar N.Y. Bd. Regents, 1964. Mem. Am. Psychol. Assn., Am. Pub. Health Assn. (task force on reimbursement 1980-82), Ambulatory Pediatric Assn., Am. Heart Assn., Evaluation Network. Current work: Family intervention for cardiovascular risk reduction; social learning theory and social support aspects of behavior change; social epidemiology of salt consumption and of breast feeding. Home: 1706 Church St Galveston TX 77550 Office: Dept Pediatrics Univ Tex Med Br Shearn Moody Plaza Suite 7020 Galveston TX 77550

BARATTA, ANTHONY JOHN, nuclear engring. educator, researcher; b. Bayonne, N.J., Dec. 24, 1945; s. Anthony John and Mary Elizabeth (Palmore) B.; m. Barbara Kay Hill, Feb. 3, 1968; children: Anthony John, Jaime Lee. B.A., B.S. Columbia U, 1968; M.S., Brown U., 1970, Ph.D., 1978. Project engr. (nuclear) div. naval reactors AEC, Washington, 1974-76, project officer, 1976-78; asst. prof. nuclear engring. Pa. State U., University Park, 1978—; cons. Naval Research Lab., Washington, 1978—, EG&G, Idaho Falls, Idaho, 1979-80. Pres. Oakview Recreation Corp., Silver Spring, Md., 1976-78; mem. USCG Aux., Solomons, Md., 1975—. Served to lt. USN, 1969-74. U.S. Navy/Am. Soc. Engring. Edn. summer fellow, 1981. Mem. Am. Phys. Soc., Am. Nuclear Soc. (program com., edn. div. 1980-82). Roman Catholic. Subspecialties: Nuclear engineering; Condensed matter physics. Current work: Fission/fusion reactor safety, design, instrumentation, condensed matter physics, electronic properties and radiation interaction. Inventor waterlevel gauge for nuclear reactors, radio iodine monitor. Home: RD 5 B151 Tyrone PA 16686 Office: Pa State U 231 Sackett University Park PA 16802

BARBACH, LONNIE, psychologist, writer; b. Newark, Oct. 6, 1946; d. Marvin and Terry (Sokolow) B.; m. Franklin Garfield, June 19, 1967 (div. Sept. 1973); m. Alberto Villoldo, Sept. 3, 1979. B.S., Simmons Coll., 1968; M.A., Wright Inst., Berkeley, Calif., 1972, Ph.D., 1974. Co-dir. clin. tng. U. Calif.-San Francisco, 1973-76, asst. clin. prof., 1978—; assoc. prof. psychology Antioch Coll. West, San Francisco, 1977-78; psychologist Nexus, San Francisco, 1976-81. Author: For Yourself, 1975 (Am. Med. Writers Assn. hon. mention 1978), Women Discover Orgasn, 1980, (with Linda Levine) Shared Intimacies, 1980, For Each Other, 1982. Mem. Am. Assn. Sex Educators, Counselors and Therapists (3d Ann. Regional award 1978), Am. Psychol. Assn., Soc. Sci. Study of Sex. Subspecialties: Behavioral psychology; Clinical psychology. Current work: Research and writing pertaining to intimacy, relationships, and sexuality. Office: Nexus 1968 Green St San Francisco CA 94123

BARBAN, STANLEY, government science administrator, biochemist, virologist, researcher; b. N.Y.C., Mar. 16, 1921; s. Isidore and Pauline (Wagner) B.; m. Barbara Lois Rosenberg, June 18, 1950; children: Beth, Lisa. B.S., CCNY, 1943; M.S., U. Mich., 1949; Ph.D., Washington U., St.Louis, 1953. Research asst. Columbia U. Med. Coll., N.Y.C., 1946-47; bacteriologist Syracuse (N.Y.) Dept. Health, 1950-51; instr. microbiology SUNY Upstate Med. Ctr., 1952-53; research biochemist NIH, Bethesda, Md., 1954-78, scientist adminstr., 1978—. Contbr. articles to profl. jours. Served to sgt. M.C. U.S. Army, 1943-46; Italy. USPHS postdoctoral fellow, 1953-54. Mem. Am. Soc. Biol. Chemists, Am. Soc. Microbiology, Sigma Xi, Phi Sigma. Subspecialties: Genetics and genetic engineering (biology); Virology (biology). Current work: Recombinant DNA technology, human genetics, animal virology tumor viruses. Office: Office Recombinant DNA Activities NIH Bethesda MD 20205

BARBAS, HELEN, neuroscientist; b. Kyrenia, Cyprus, Dec. 16, 1949; d. Theophanis L. and Evdokia (Serghiou) B. B.A., Kean Coll., 1972; M.S., Kans. State U., 1974; Ph.D., McGill U, Montreal, Que., Can., 1979. Research asst. Clin. Neurophysiol. Lab., Allan Meml. Inst., McGill U, Montreal, 1974-76, research asst., 1974-78; research fellow in neurology Harvard Med. Sch, Beth Israel Hosp., Boston, 1979-81; asst. prof. health scis. Boston U., 1981—. Contbr. numerous sci. articles to profl. publs. Fulbright scholar, 1968; Med. Research Council scholar McGill U., 1975-77; Univ. fellow, 1977-8; Faculty of Medicine fellow, 1978; NIH Individual fellow, 1981. Mem. Soc. Neurosci., Montreal Physiol. Soc. Subspecialties: Neurophysiology; Anatomy. Current work: Behavorial, neurophysiological and anatomical approaches in the study of afferent sensory and limbic input affecting complex motor behavior in the frontal cortex of cats and animals. Office: Boston U 36 Cummington St B-11 Boston MA 02215

BARBEHENN, CRAIG E(DWIN), nuclear engineer; b. Plainfield, N.J., May 24, 1953; s. Edwin William and Sheila (Maginess) B. B.S., U. Va., 1975; Engr.'s degree, M.I.T., 1977, M.Sc., 1977; M.B.A., U. Conn., 1982. Reactor physicist Combustion Engring., Windsor, Conn., 1977-82, test engr., San Onofre, Calif., 1982—. Mem. Am. Nuclear Soc., Tau Beta Pi, Phi Beta Gamma. Subspecialties: Nuclear engineering; Nuclear fission. Current work: Nuclear reactor technology, specifically reactor physics, fuel management, reactor engineering and start-up testing. Home: 317 DelMar #B San Clemente CA 92672 Office: San Onofre Nuclear Generation Sta PO Box 4019 San Clemente CA 92672

BARBIERI, RICHARD CHARLES, quality assurance engr., cons., educator; b. Bell, Calif., Nov. 9, 1941; s. Richard Stanley and Charlotte Francis (Nearman) Sieler; m. Joyce Ann Barbieri, June 14, 1965; children—Richard Charles, Kimberly, Susan, Darran; m. Marie Barbieri, Nov. 4, 1977. A.A.S., Wentworth Inst., Boston, 1967; B.S.I.T., Northeastern U., 1972. Test Engr. Digital Equipment Corp., Maynard, Mass., 1967-69; asst. quality assurance mgr. Environ. Equipment div. EG&T, Waltham, Mass., 1969-71; quality assurance mgr. Iotron Corp., Bedford, Mass., 1971-75; quality assurance mgr., chief test engr. Spacetac, Bedford, 1975-78; quality assurance mgr Nypro, Clinton, Mass., 1978-81; group quality assurance mgr. Dennison Mfg. Co., Framingham, Mass., 1981—82; dir. quality Electronic Designs, Inc., Hopkinton, Mass., 1982—; quality systems cons.; tchr. quality-tech.-mgmt. North Shore Community Coll., Beverly, 1982—. Scoutmaster, mem. council exec. bd. Boy Scouts Am. Served with USNR, 1961-64. Mem. Am. Soc. Quality Control (cert. quality engr., chmn. sect. seminars), Internat. Assn. Quality Circles. Subspecialties: Information systems (information science); Statistics. Current work: Modernising quality-productivity systems using imaging technology, computer systems and organizational culture dynamics for cost effective non-stress results. Home: Route 119 W State Rd Ashby MA 01431 Office: Framingham MA 01701

BARBORAK, JAMES CARL, chemist, educator, researcher; b. Moulton, Tex., Sept. 8, 1941; s. Victor John and Clothilda Rose (Motal) B.; m. Shirley Ann Heczko, June 7, 1969. B.S., U. Tex., Austin, 1963, Ph.D., 1968. Research scientist Uniroyal, Inc., 1968-69; postdoctoral fellow Princeton U., 1969-70, instr., 1970-72; asst. prof. chemistry U. N.C., Greensboro, 1972-76, assoc. prof., 1976-79, prof., 1979—, dir. research in organic and organometallic chemistry, 1972—. Contbr. articles to profl. jours. Am. Chem. Soc.-Petroleum Research Fund grantee, 1978. Mem. Am. Chem. Soc., Sigma Xi. Roman Catholic. Subspecialties: Organic chemistry; Oraganometallic chemistry. Current work: Organic chem. synthesis, especially of strained ring systems; mechanistic organometallic chemistry; organic synthesis employing organometallic intermediates. Home: 205 Kingsdale Ct Jamestown NC 27282 Office: Dept Chemistry U NC Greensboro NC 27412

BARBOSA-SALDIVAR, JOSE LUIS, internist, endocrinologist, clinical nutritionist, educator; b. Asuncion, Paraguay, Dec. 24, 1946; came to U.S., 1972. B.Sc., Goethe Coll., Asuncion, 1964; M.D. summa cum laude, Asuncion U., 1970; postgrad, Balliol Coll., Oxford U., 1971-72; M.P.H., Harvard U., 1973. Diplomate: Am. Coll. Nutrition. Intern Bklyn. Hosp., SUNY-Downstate Med. Ctr., N.Y.C., 1974-75; resident Mt. Sinai Hosp., CUNY, N.Y.C., 1975-77; fellow in endocrinology, metabolism and nutrition Columbia U. Coll. Physicians and Surgeons, St. Luke's Hosp. and Inst. Human Nutrition; asst. attending physician St. Luke's-Roosevelt Med. Ctr., N.Y.C., 1979—; assoc. in clin. medicine Columbia U., 1979—; mem. adv. bd. Council on Mcpl. Performance, N.Y.C., 1974—. Contbr. articles to profl. jours. Sec. Oxford and Cambridge Soc., N.Y.C., 1981—. Recipient Univ. Gold medal Asuncion U., 1970, Gold Medal Goethe Coll., 1964; Hoescht prize, 1970. Fellow ACP, N.Y. Acad. Medicine, Royal Soc. Health; mem. Am. Inst. Nutrition, Am. Heart Assn., N.Y. Heart Assn., Am. Diabetes Assn., N.Y. Diabetes Assn. Roman Catholic. Clubs: Harvard (N.Y.C.); Union. Subspecialties:

Internal medicine; Endocrinology. Current work: Composition of diets and metabolic balance in the treatment of obesity; composition of intravenous nutrition and its metabolic effects. Office: 30 Central Park S New York NY 10028

BARBOUR, MICHAEL GEORGE, botany educator; b. Jackson, Mich., Feb. 24, 1942; s. George Jerome and Mae (Dater) B.; m. Norma Jean Yourist, Sept. 26, 1963; children: Julie Ann, Alan Benjamin. B.Sc., Mich. State U., 1963; Ph.D., Duke U., 1967. Asst. prof. dept. botany U. Calif.-Davis, 1967-71, assoc. prof., 1971-76, prof., 1976—, chmn. dept. botany, 1982—. Editor: (with Major) Terrestrial Vegatation of California, 1977; author: (with Burk and Pitts) Terrestrial Plant Ecology, 1980, (with Weier, Stocking, Rost) Botany, 6th edit, 1982. Fulbright fellow, Australia, 1964; Guggenheim fellow, 1978; NSF grantee, 1969-74; U. Calif. Sea Grant awardee, 1974-77. Mem. Brit. Ecol. Soc., Ecol. Soc. Am., Bot. Soc. Am., Am. Inst. Biol. Scis., Torrey Bot. Club. Democrat. Jewish. Subspecialties: Ecology; Species interaction. Current work: Ecology of dominant plant species in stressed habitats, such as warm desert, salt marsh, coastal dune, upper montane forest. Home: 2453 Creekhollow St Davis CA 95616 Office: Dept Botany U Calif Davis CA 95616

BARCHAS, JACK DAVID, psychiatrist, educator; b. Los Angeles, Nov. 2, 1935; s. Samuel Isaac and Cecile Margaret (Pasarow) B.; m. Patricia Ruth Corbitt, Feb. 9, 1957; 1 son, Isaac Doherty. B.A., Pomona Coll., 1956; M.D., Yale U., 1961. Intern Pritzker Sch. Medicine, U. Chgo., 1961-62; postdoctoral fellow in biochemistry and pharmacology NIH, 1962-64; resident in psychiatry Stanford Med. Sch., 1964-67, instr., 1966-67, asst. prof., 1967-71, asso. prof., 1971-76, prof., 1976—, Nancy Friend Pritzker prof. psychiatry and behavioral scis., 1976—; dir. Nancy Pritzker Lab. of Behavioral Neurochemistry, 1976—. Editor; author: Serotonin and Behavior, 1973, Neuroregulators and Psychiatric Disorders, 1977, Psychopharmacology from Theory to Practice, 1977, Catecholamines - Basic and Clinical Frontiers, 1979; contbr. articles to profl. jours. Served with USPHS, 1962-64. Recipient Psychopharmacology award Am. Psychol. Assn., 1970. Mem. Am. Psychiat. Assn., Soc. Neurosci., Am. Coll. Neuropsychopharmacology (Daniel Efron award 1978), Am. Soc. Pharmacology and Exptl. Therapeutics, Am. Physiol. Soc., Am. Soc. Neurochemistry, Am. Chem. Soc., Am. Psychosomatic Soc., Psychiat. Research Soc., Soc. Biol. Psychiatry (A.E. Bennett award 1968), Am. Psychopathol. Assn., Inst. Medicine Nat. Acad. Scis. Subspecialty: Psychopharmacology. Home: 669 Mirada Ave Stanford CA 94305 Office: Dept Psychiatry Stanford Med Sch 300 Pasteur Dr Stanford CA 94305

BARCHI, ROBERT LAWRENCE, neuroscientist, clinical neurologist, educator; b. Phila., Nov. 23, 1946; s. Henry John and Elizabeth (Pesci) B.; m. Joan E. Mollman, Sept. 25, 1976. B.S., Georgetown U., 1968, M.S., 1968; Ph.D. in Biochemistry, U. Pa., 1972, M.D., 1973. Diplomate: Am. Bd. Med. Examiners. Resident in neurology U. Pa., 1973-74, asst. prof. biochemistry, 1974-75, asst. prof. neurology and biochemistry, 1975-78, assoc. prof., 1978-81, prof., 1981—, chmn. Neurosci. Grad. Group, 1983—, dir. Inst. for Neurosci., 1983—; dir. Lab. Membrane Biophysics Henry Watts Research Center, 1978—; attending neurologist Hosp. of U. Pa., 1975—; cons. neurologist Phila. Veterans Hosp. and Grad. Hosp. Author: (with R.P. Lisak) Modern Problems in Neurology, 1982; editorial bd.: Muscle & Nerve, 1980—, Jour. Neurochemistry, 1982—; contbr. numerous articles to profl. jours. Grantee in field; Recipient NIH research career devel. award, 1977-82. Fellow Am. Acad. Neurology (cert.); Mem. Am. Neurol. Assn., Soc. Neurosci, Biophysl Soc., AAAS, Phi Beta Kappa, Alpha Omega Alpha. Subspecialties: Neurochemistry; Neurology. Current work: Molecular basis of excitable membrane function and pathogenesis of neuromuscular diseases; research on sodium channels, muscular dystrophy, med. sch. teaching and practice clin. neurology. Office: Department of Neurology 3400 Spruce St Philadelphia PA 19104

BARCUS, ROBERT A., clinical psychologist, educator; b. Cleve.; s. Sanford R. and Evelyn June (Aronoff) B. B.A., Ohio State U., 1969, M.A., 1973, Ph.D., 1978. Counselor Beechwold Clinic, Columbus, Ohio, 1972; lectr. psychology U. Buffalo, 1975-76; therapist Psychiat. Clinic, Inc., Buffalo, 1973-75, Gahanna (Ohio) Counseling Services, 1977-79; lectr. psychology Franklin U., Columbus, 1977-79, U. Md., European Div., Heidelberg, W.Ger., 1979—. Chmn. Com. Against War in Vietnam, Columbus, 1969. USPHS fellow, 1969; Ohio State U. grantee, 1978. Mem. Am. Psychol. Assn., Internat. Assn. Psychologists, Community Services Inc., Nat. Audubon Soc., Green peace. Democrat. Jewish. Subspecialty: Clinical psychology. Current work: Relationship between diet and behavior. Home: 2536 Claver Rd Cleveland OH 44118 Office: University of Marylan European Div Im Bosseldorn 30 Heidelber Federal Republic of Germany 6900

BARD, ALLEN JOSEPH, educator, chemist; b. N.Y.C., Dec. 18, 1933; s. John J. and Dora (Rosenberg) B.; m. Frances Joan Segal, June 15, 1957; children: Edward David, Sara Lynn. B.S. summa cum laude, CCNY, 1955; A.M., Harvard U., 1956, Ph.D. (NSF fellow), 1958. Research chemist Gen. Chem. Co., Morristown, N.J., 1955; faculty U. Tex., Austin, 1958—, prof. chemistry, 1967—, Jack S. Josey prof., 1980-82, Norman Hackerman prof., 1982—; cons. E.I. duPont de Nemours & Co., Wilmington, Del., Tex. Instruments, Dallas; vis. prof. Mich. State U., UCLA, U. N.C.; vis. scholar U. Ga., 1969; Fulbright prof. U. Paris, 1973; Sherman Mills Fairchild scholar Calif. Inst. Tech., 1977. Author: Chemical Equilibrium, 1966; co-author: Electrochemical Methods, 1980; Editor: Electroanalytical Chemistry-A Series of Monographs on Recent Advances, 1966—, Ency. of Electrochemistry of the Elements, 1973—, Jour. Am. Chem. Soc, 1982—; Contbr. articles to profl. jours. Recipient Ward medal, 1955; named Analyst of Year Dallas Soc. Analytical Chemists, 1976. Mem. Am. Chem. Soc. (Harrison Howe award Rochester sect. 1980), Electrochem. Soc. (Carl Wagner Meml. award 1981), AAAS, Nat. Acad. Sci., Internat. Soc. Electrochemistry, Sigma Xi. Subspecialties: Analytical chemistry; Physical chemistry. Research on electrogenerated chemiluminescence; semicondr. electrodes for solar energy conversion; co-discoverer magnetic field effects on solution spectroscopic and chemiluminescent processes; discoverer solar photosynthesis of amino acids on semicondr. powders. Home: 6202 Mountainclimb Dr Austin TX 78731

BARDACH, JOAN L(UCILE), clinical psychologist, educator; b. Albany, N.Y., Oct. 3, 1919; d. Monroe Lederer and Lucile May (Lowenberg) B. A.B., Cornell U., 1940; A.M., N.Y. U., 1951, Ph.D., 1957, postgrad., 1960. Lic. psychologist, N.Y. Supr. clin. psychology N.Y. U. Inst. Rehab. Medicine, 1959-61, dir. psychol. services, 1965-82; supr. postdoctoral program N.Y. U., 1978—, prof. clin. rehab. medicine, 1976—; research psychologist, mem. faculty N.Y. Med. Coll., 1961-62; pvt. practice clin. psychology, N.Y.C., 1967—; mem. adv. bd. Coalition on Sexuality and Disability, N.Y.C.; mem. med. adv. bd. Planned Parenthood, N.Y.C.; film cons. Time/Life, N.Y.C. Contbr. articles, chps. on psychology to profl. publs.; project dir. film: Choices: In Sexuality with Physical Disability, 1982 (Silver Hugo award Chgo. Internat. Film Festival 1982, 1st pl. Internat. Rehab. Film Festival 1982). Mem. Am. Psychol. Assn., Eastern Psychol. Assn., N.Y. Soc. Clin. Psychologists, N.Y. State Psychol. Assn., Am. Congress Rehab. Medicine; fellow Am. Orthopsychiat. Assn. Subspecialty: Clinical psychology. Current work: Human sexuality; experiential learning; criminology; psychoanalytic therapy; somatopsychology; psychology and physical disability; psychotherapy. Home: 50 E 10th St New York NY 10003

BARDANA, EMIL JOHN, JR., physician, educator, researcher; b. N.Y.C., May 21, 1935; s. Emilio Dominico and Florencia Fellicita (Perotti) B.; m. Norma Jean Olson, June 11, 1959; children: Anthony John, Anne Michelle. B.S., Georgetown U., 1957; M.D., McGill U., 1961. Diplomate: Am. Bd. Internal Medicine, Am. Bd. Allergy and Immunology. Med. intern U. Calif. Med. Ctr., San Francisco, 1961-62; resident U. Oreg. Med. Sch., Portland, 1965-68, fellow in immunology and allergy, 1969, prof. medicine, head allergy sect., 1972—; fellow in immunology Nat. Jewish Hosp., Denver, 1969-71. Contbr. articles to profl. jours., chpts. to books. Served with USN, 1962-65. Fellow ACP (gov. Oreg. 1981—), Am. Acad. Allergy and Immunology, Am. Coll. Chest Physicians; mem. N.Y. Acad. Scis., Alpha Omega Alpha. Roman Catholic. Club: Oswego Lake Country (Lake Oswego, Oreg.). Lodge: Sons of Italy (Portland). Subspecialties: Allergy; Immunology (medicine). Home: 12389 SW Clara Ln Lake Oswego OR 97034 Office: 3181 SW Sam Jakson Park Rd Portland OR 97201

BARDEEN, JOHN, physicist, emeritus educator; b. Madison, Wis., May 23, 1908; s. Charles Russell and Althea (Harmer) B.; m. Jane Maxwell, July 18, 1938; children—James Maxwell, William Allen, Elizabeth Ann Bardeen Greytak. B.S., U. Wis., 1928, M.S., 1929; Ph.D., Princeton, 1936. Geophysicist Gulf Research & Devel. Corp., Pitts., 1930-33; mem. Soc. Fellows Harvard, 1935-38; asst. prof. physics U. Minn., 1938-41; with Naval Ordnance Lab., Washington, 1941-45; research physicist Bell Telephone Labs., Murray Hill, N.J., 1945-51; prof. physics, elec. engring. U. Ill., 1951-75, emeritus, 1975—; mem. Pres.'s Sci. Adv. Com., 1959-62. Recipient Ballantine medal Franklin Inst., 1952; John Scott medal, Phila., 1955; Fritz London award, 1962; Vincent Bendix award, 1964; Nat. Medal Sci., 1966; Morley award, 1968; medal of honor IEEE, 1971; Franklin medal, 1975; co-recipient Nobel prizes in physics, 1956, 72; Presdl. medal of Freedom, 1977. Fellow Am. Phys. Soc. (Buckley prize 1954, pres. 1968-69); mem. Am. Acad. Arts and Sci., IEEE (hon.), Am. Philos. Soc., Royal Soc. Gt. Britain (fgn. mem.), Acad. Sci. USSR (fgn. mem.), Indian Nat. Sci. Acad. (fgn.), Japan Acad. (fgn.). Subspecialties: Condensed matter physics; Low temperature physics. Current work: Transport in quasi-one-dimensional metals. Home: 55 Greencroft Champaign IL 61820 Office: Dept Physics Univ Ill Urbana IL 61801

BARDEN, ROLAND EUGENE, biochemist, educator; b. Powers Lake, N.D., Sept.11, 1942; s. Harry S. and Sena Barden; m. Carolyn J. Eliason, Nov. 25, 1967; children: Carl, Janine, Ann. B.S., U. N.D., 1964; M.S., U. Wis., 1966, Ph.D., 1969. Postdoctoral fellow Case Western Res. U., Cleve., 1968-71; asst. prof., then assoc. prof. U. Wyo., Laramie, 1971-82, prof. chemistry, 1982—, chmn. dept., 1980—; Mem. biomed. sci. study sect. NIH; chmn. research com. Wyo. affiliate Am. Heart Assn.; mem. com. on research and edn. Assoc. Western Univs. Contbr. articles to sci. jours. NIH postdoctoral fellow, 1968-70; career devel. award, 1976-80. Mem. Am. Soc. Biol. Chemists, Am. Chem. Soc. Lutheran. Subspecialties: Biochemistry (biology); Organic geochemistry. Current work: Enzymes, affinity labels, fossil biopolymers, academic administration, teaching and research. Office: Dept Chemistry U Wyo Laramie WY 82071

BARDIN, CLYDE WAYNE, research association executive, biomedical research; b. McCamey, Tex., Sept. 18, 1934; s. James A. and Nora Irene B.; m. Dorothy T. Krieger, Aug. 11, 1978; children: Charlotte Elaine, Stephanie Faye. B.A., Rice U., 1957; M.S., Baylor U. Coll. Medicine, 1962, M.D., 1962. Intern N.Y. Hosp., N.Y.C., 1962-63; resident, investigator Nat. Cancer Inst., Bethesda, Md., 1964-70; prof. endocrinology Milton S. Hershey Med. Ctr., Hershey, Pa., 1970-78; v.p., dir. biomed. research Population Council, N.Y.C., 1978—. Editor: books, including Cell Biology of the Testes, 1982, Current Therapy in Endocrinology, 1983; contbr. numerous articles, revs. to profl. publs. Served to comdr. USPHS, 1964-67. Mem. Am. Fedn. Clin. Research, Am. Soc. Andrology, Endocrine Soc., Soc. Study Reprodn., Am. Assn. Physicians, Am. Soc. Clin. Investigation. Subspecialties: Endocrinology; Receptors. Current work: Broad interest in endocrinology with particular focus on mechanism of action of androgens; direct development program for contraceptives with particular emphasis on finding methods that are suitable for use in males. Office: Population Council 1230 York Ave New York NY 10021 Home: 1148 Fifth Ave New York NY 10028

BARE, CHARLES EDGAR, plant physiologist; b. Allison Twp., Ill., Nov. 6, 1939; s. Kenneth Wilford and Florence Juanita (Waters) B., m. Patricia Houle, Aug. 12, 1961; children: Charles, Theresa, Andrew, David, Ursula, Michael, Rebeca. B.S., So. Ill. U., 1966, M.S., 1969. Cert. comml. pesticide applicator. Instr. natural sci. Franconia (N.H.) Coll., 1969-72; plant physiologist, chemonarcocide unit Weed Sci. Lab., Agr. Environ. Quality Inst., Agrl. Research Service, U.D. Dept. Agr., Beltsville, Md., 1972-76, plant physiologist, 1976—. Contbr. articles to profl jours. Active Boy Scounts Am., Annapolis High Band Parents. Served with U.S. Navy, 1957-60. NSF grantee, 1971-72. Mem. Council Agrl. Sci. and Tech., Am. Soc. Agronomy, Crop Sci. Soc. Am., Soil Sci. Soc. Am., Assn. for Tropical Biology, Phi Sigma Soc. Subspecialties: Plant physiology (agriculture); Plant cell and tissue culture. Current work: Crop production, pest control, mode of action of herbicides. Membrane and cell biology. Office: USDA-ARS-BARC-W B001 R35 Beltsville Md 20705

BAREISS, LYLE EUGENE, aerospace company executive; b. Rawlins, Wyo., Nov. 4, 1945; s. Godfrey Matthew and Vera Edith (Squires) B.; m. Chris Elizabeth Bartlett, Aug. 23, 1969. B.S. in Mech. Engring, Wyo. U., 1969; postgrad., Colo. State U., 1970. Skylab systems engr. Martin Marietta Aerospace Co., Denver, 1969-73, sr. staff engr., shuttle/spacelab contamination, 1974-79, mgr. tech. unit, contamination and laser effects, 1980—. Recipient Skylab Achievement award NASA, 1973. Mem. AIAA, Sigma Alpha Epsilon. Presbyterian. Lodge: Odd Fellows. Subspecialties: Aerospace engineering and technology; Satellite studies. Current work: Architect of computer model to predict contamination of U.S. satellite systems; basic research of effects of laser irradiation on satellite/booster system and materials oxidation of satellites in low-earth orbit. Home: 8031 E Phillips Circle Englewood CO 80112 Office: Martin Marietta Aerospace Co Box 179 Denver CO 80201

BARENBERG, SUMNER A(RNOLD), scientist, consultant; b. Boston, May 11, 1945; s. Bernard Samuel and Ruth Edith (Volk) B.; m. Sally Franc Viden, Nov. 7, 1970. B.S., Case Tech., 1967; M.S., Case Western Res. U., 1970, Ph.D., 1976. Ops. dir. Cambridge (Mass.) Chem., 1970-74; fellow Cleve. Clinic Found., 1975-76; postdoctoral fellow Case Western Res. U., Cleve., 1976-77; research faculty U. Mich., Ann Arbor, 1977-80; scientist E.I. duPont de Nemours and Co., Wilmington, Del., 1980—; chmn. Gordon Research Conf. on Biomaterials, 1985; cons. U. Mich., 1980-81, B.F. Goodrich, Cleve., 1974-77, Gould Inc., Chgo., 1978-80, Diamond Shamrock, Cleve., 1978-80; cons. scientist Cleve. Clinic, 1976—. Editorial bd: Jour. Biomedical Materials Research; contbr. articles to profl. publs. NIH postdoctoral fellow, 1976-77; grantee, 1977-80. Mem. Am. Chem. Soc., N.Y. Acad. Scis., AAAS, Am. Phys. Soc., Am. Assn. Aritificial Organs, Soc. Biomaterials, Sigma Xi. Jewish. Subspecialties: Biomaterials; Polymers. Current work: Biomedical meterials, implant retrieval, high polymer physics, artificial organs, engineering polymers. Patentee molding material. Home: RD 4 PO Box 250E Hockessin DE 19707 Office: E I duPont Exptl Station 323 Wilmington DE 19898

BARES, JAN, physicist; b. Strakonice, Czechoslovakia, Mar. 4, 1938; came to U.S., 1970; s. Antonin and Emilie B.; m. Vera Folcova; children: Tom, Julie. M.S., Charles U., Prague, Czechoslovakia, 1961, Ph.D., 1965. Postdoctoral fellow Soviet Acad. Scis., Leningrad, USSR, 1965; scientist Czechoslovak Acad. Scis., Prague, 1965-68; postdoctoral fellow Poly. Milan, Italy, 1968-70, Rensselaer Poly. Inst., 1970-72; scientist Xerox Corp., Webster, N.Y., 1972-80, sr. scientist, 1980—. Author: (with others) Experiments in Polymer Science, 1973; contbr. articles on polymer sci., tech. to profl. jours. Mem. Am. Phys. Soc., Am. Chem. Soc., N.Y. Acad. Scis., Soc. Photographic Scientists and Engrs., Sigma Xi. Subspecialties: Polymer physics; Composite materials. Current work: Electrophotography, technology development. Patentee electrophotography. Office: Xerox Corp 800 Phillips Rd Webster NY 14580

BARFIELD, WALTER DAVID, physicist; b. Gainesville, Fla., Nov. 25, 1928; s. Walter H. and Myrtle Marie (May) B. B.S. in Physics and Math. cum laude, U. Fla., 1950, M.S., 1951; postgrad., U. N.Mex., Los Alamos, 1952-56, 61, 72-73; Ph.D. in Physics, Rice U., 1961. Research asst. theoretical div. Los Alamos Nat. Lab., 1951-52, mem. staff, 1952-57, 60-63, 1968-70, 1971-79, staff theoretical physics div., 1980—; tech. staff Research and Engring. Support div. Inst. Def. Analyses, Arlington, Va., 1963-67. Contbr. numerous articles to profl. jours. Mem. Am. Phys. Soc., Phi Beta Kappa, Sigma Xi. Subspecialties: Theoretical and applied mechanics; Radiative transfer. Current work: Numerical modeling of high explosives. Home: 4647 Ridgeway Dr Los Alamos NM 87544 Office: PO Box 1663 Los Alamos NM 87545

BARGER, A. CLIFFORD, physiologist; b. Greenfield, Mass., Feb. 1, 1917; s. Paul and Rose (Solomon) B.; m. Claire Basch, June 6, 1943; children—Craig, Shael, Curtis. A.B., Harvard U., 1939, M.D., 1943; D.Sc. (hon.), U. Cin., 1977. Research asst. Harvard Fatigue Lab., 1938-41; intern Peter Bent Brigham Hosp., Boston, 1943-44, med. house officer, 1945, cons. medicine, 1959—; mem. faculty Harvard Med. Sch., 1946—, prof. physiology, 1961—, Robert Henry Pfeiffer prof. physiology, 1963—, chmn. dept. physiology, 1974-76; cons. physiology Children's Med. Center, 1966—. Editor: Am. Heart Assn, 1962-66. Bd. sci. counselors Nat. Cancer Inst., 1969-73, chmn., 1972-73; Pres. Elbanobscot Found., 1964, bd. dirs., 1965-80; pres. Harvard Apparatus Found., 1970—. Served with AUS, 1944-45. Commonwealth Fund travelling fellow, 1959-60. Fellow Am. Acad. Arts and Scis.; mem. Am. Physiol. Soc. (chmn. publs. com. 1962-63, 66-69, councilor 1968-72, pres. 1970-71, co-chmn. Porter devel. com. 1971—), Inst. Medicine. Assocs., Mass. Assn. Med. Research (pres. 1956-70, 83—). Subspecialty: Animal physiology. Current work: Cardiovascular and renal physiology. Home: 14 Orchard Rd Brookline MA 02146 Office: 25 Shattuck St Boston MA 02115

BARGER, JAMES DANIEL, pathologist; b. Bismarck, N.D., May 17, 1917; s. Michael Thomas and Mary Margaret (Donohue) B.; m. Susie B. Helm, Nov. 20, 1945 (dec. 1951); children: James Daniel, Mary Susan; m. Josephine Steiner, May 30, 1952 (dec. 1971); children: Michael Thomas, Mary Elizabeth; m. Jane H. Ray Regan, Apr. 20, 1980. Student, St. Mary's Coll., Winona, Minn., 1934-35; A.B., B.S., U. N.D., 1939; M.D., U. Pa., 1941; M.S. in Pathology, U. Minn., 1949. Diplomate: Am. Bd. Pathology; registered profl. engr., Calif. Pathologist Pima County Hosp., Tucson, 1949-50, Maricopa County Hosp., Phoenix, 1950-51; chmn. dept. pathology Good Samaritan Hosp., Phoenix, 1951-63; chief clin. pathology Sunrise Hosp., Las Vegas., 1964-68, chmn. dept. lab. medicine, 1968-81, sr. staff pathologist, 1981—. Contbr. articles to profl. jours. Served to maj. U.S. Army, 1942-45. Fellow Coll. Am. Pathologists (pres. 1981—, exec. com. 1967—), Am. Soc. Clin. Pathologists, AAAS; mem. AMA, Am. Assn. History of Medicine, Internat. Acad. Pathology, Am. Assn. Pathology and Bacteriology, N.Y. Acad. Scis., Ariz. Soc. Pathologists (pres. 1955-58), Calif. Soc. Pathologists, Am. Assn. Blood Banks, Phoenix Soc. Pathologists, Phoenix Med. History Soc., Nev. State Med. Assn., Nev. Soc. Pathologists, Am. Pub. Health Assn., Am. Acad. Forensic Scis., Am. Assn. Pathologists, Assn. Advancement of Med. Instrumentation, Am. Mgmt. Assns., Am. Soc. for Quality Control, Am. Assn. Clin. Chemists, Soc. Advancement of Mgmt., Internat. Assn. Quality Circles, Sigma Xi. Democrat. Roman Catholic. Club: Rotary. Subspecialties: Pathology (medicine); Hematology. Current work: Quality circle applications in medicine. Home: 1307 Canosa Las Vegas NV 89104 Office: Sunrise Hosp PO Box 14157 Las Vegas NV 89104

BARGER, JAMES EDWIN, physicist; b. Manhattan, Kans., Dec. 28, 1934; s. Edgar Lee and Carolyn Marie (Grantham) B.; m. Mary Elizabeth Rupp, Aug. 24, 1957; children—Elaine Marie, Carolyn Ruth, James Rupp, Corinne Elizabeth. B.S., U. Mich., 1957; M.S., U. Conn., 1960; Ph.D., Harvard U., 1964. Teaching asst. Harvard U., Cambridge, 1961-64; v.p. Bolt Beranek & Newman, Inc., Cambridge, Mass., 1965-75, chief scientist, 1975—. Mem. Methods and Procedures Com., Town of Winchester, 1967-71; trustee Winchester Hosp., 1972—; corp. mem. Mt. Vernon House, 1979—. Served with USNR, 1957-63. NSF fellowship, 1960-64. Fellow Acoustical Soc. Am.; mem. Marine Tech. Soc., AAAS, Tau Beta Pi, Pi Tau Sigma. Conglist (deacon). Club: Winchester Country. Subspecialty: Acoustics. Home: 3 Lakeview Rd Winchester MA 01890 Office: 50 Moulton St Cambridge MA 02138

BARISAS, BERNARD GEORGE, JR., chemistry educator, immunology researcher; b. Shreveport, La., July 16, 1945; s. Bernard George and Edith (Bailey) B.; m. Judith Kathleen O'Rear, May 19, 1973 (div. Sept. 1978); m. Deborah Anne Roess, Aug. 6, 1981. B.A., U. Kans., 1965, Oxford U., 1967; M. Phil., Yale U., 1969, Ph.D., 1971. NIH postdoctoral trainee Yale U., 1971-72, research assoc., 1972; NIH postdoctoral fellow U. Colo., Boulder, 1973-75; asst. prof. biochemistry St. Louis U., 1975-80, assoc. prof., 1980-81; assoc. prof. chemistry and microbiology Colo. State U., 1981—. Contbr. articles to tech. jours. Sec. Mo. Rhodes Scholarship Selection Com, 1976-81; mem. Colo. Com., 1982. Rhodes scholar, 1965; Woodrow Wilson fellow, 1965; recipient Research Career Devel. award NIH, 1978. Mem. Am. Soc. Biol. Chemists, Am. Assn. Immunologists, Biophys. Soc., Am. Chem. Soc., AAAS, N.Y. Acad. Scis., Soc. Applied Spectroscopy, Phi Beta Kappa, Sigma Xi, Omicron Delta Kappa, Pi Mu Epsilon, Phi Lambda Upsilon, Delta Phi Alpha. Clubs: Am. Alpine (N.Y.C.); Colo. Mountain (Denver); St. Louis Mountain (pres. 1976-77). Subspecialties: Spectroscopy; Immunobiology and immunology. Current work: Application of laser optical and image processing techniques in bioinstrumentation, particularly applied to immunology and membrane biology. Home: 1718 Valley Forge St Fort Collins CO 80526 Office: Dept of Chemistry Colorado State University Fort Collins CO 80523

BARKAN, PHILIP, mech. engr.; b. Boston, Mar. 29, 1925; s. Philip and Blanche (Seifert) B.; m. Hinda Brody, Sept. 5, 1948 (dec. Aug. 1979); children—Ruth, David. B.S.M.E., Tufts U., 1946; M.S.M.E., U. Mich., 1948; Ph.D. in Mech. Engring, Pa. State U., 1953. Asst. prof. engring. research Pa. State U., 1948-51; sect. mgr. applied physics and mech. engring. Gen. Electric Co., Phila., 1953-77; prof. mech. engring. Stanford U., 1977—; vis. prof. Israel Inst. Tech., Haifa, 1971-72; cons.

to electric power industry, 1977—. Contbr. numerous articles to profl. publs. Pres. bd. trustees Middletown (Pa.) Free Library, 1959-61; chmn. bd. trustees Sch. in Rose Valley, 1967-68; Democratic candidate for Middletown Twp. Supr., 1959, 61, 63; pres. Middletown Dem. Club, 1960. Served with USN, 1943-46. Recipient 1st Charles P. Steinmetz medal and award Gen. Electric Co., 1973; Electric Power Research Inst. grantee, 1979. Fellow IEEE; mem. ASME, Nat. Acad. Engring., Sigma Xi. Subspecialty: Mechanical engineering. Patentee in field. Office: Design Div Dept Mech Engring Stanford U Stanford CA 94305

BARKER, CLYDE FREDERICK, surgeon, medical educator; b. Salt Lake City, Aug. 16, 1932; m., 1956; 4 children. B.A., Cornell U., 1954, M.D., 1958. Intern U. Pa. Med. Sch. Hosp., Phila., 1958-59, from asst. resident to chief resident in surgery, 1959-64, assoc., 1964-68, from asst. prof. to assoc. prof., 1968-73, prof. surgery, 1973—, Guthrie prof. surgery, 1982—, assoc. med. genetics, 1966—, fellow, 1959-64, Hartford fellow vascular surgery, 1964-65, attending surgeon assoc. chief sect. vascular surgery, 1966-82, chief sect. renal transplantation, 1969—, chief sect. vascular surgery, 1982—; Am. Cancer Soc. fellow, 1961-62, USPHS fellow med. genetics, 1965-66; Markle Found. scholar Acad. Medicine, 1968. Mem. Transplantation Soc., Soc. Univ. Surgeons, Soc. Vascular Surgery, Am. Surg. Assn., Soc. Clin. Surgery. Subspecialties: Surgery; Transplant surgery. Office: Depts Surgery and Immunolog U Pa Sch Medicine Hosp 3400 Spruce St Philadelphia PA 19104

BARKER, COLIN G., geosciences educator; b. Plymouth, Eng., Aug. 3, 1939; came to U.S., 1965; s. George H. and Hilda May (Finch) B.; m. Yvonne I. Meredith, Apr. 28, 1962; 1 son, Conan N. B.A., Oxford (Eng.) U., 1962, D.Phil., 1965. Research fellow U. Tex.-Austin, 1965-67; sr. research chemist Exxon Prodn. Research Co., Houston, 1967-69; prof. dept. geoscis. U. Tulsa, 1969—; cons. Oil & Gas Cons. Inc., Tulsa, 1980—; cons. to internat. and domestic oil cos., 1967—. Author: Organic Geochemistry in Petroleum Exploration, 1979; contbr. articles to profl. jours. Mem. Am. Assn. Petroleum Geologists (Matson award 1978, 82, assoc. editor bull. 1980—), Geochem. Soc. (assoc. editor jour. 1978—). Subspecialty: Organic geochemistry. Current work: Petroleum geology and ultradeep gas. Source rocks of petroleum. Regional patterns for oil and gas distribution. Office: Dept Geoscis U Tulsa Tulsa OK 74114 Home: 2527 E 26th Place Tulsa OK 74114

BARKER, DANIEL STEPHEN, geology educator; b. Waltham, Mass., Feb. 27, 1934; s. Kenneth Watson and Sadie Webb (Brown) B.; m. Barbara Catherine Mackin, Aug. 14, 1964; children: Molly, Amy. B.S., Yale U., 1956; M.S., Calif. Inst. Tech., 1958; Ph.D., Princeton U., 1961. Research assoc. Cornell U., Ithaca, N.Y., 1961-62; postdoctoral fellow Yale U., New Haven, Conn., 1962-63; asst. prof. geology U. Tex., Austin, 1963-67, assoc. prof., 1967-72, prof., 1972-82, F. M. Bullard prof., 1982—. Author: Igneous Rocks, 1983; contbr. articles to profl. jours. Recipient Fulbright-Hays sr. research awards, Denmark, 1974; NSF research grantee, 1966-78. Fellow Geol. Soc. Am., Mineral Soc. Am.; mem. Mineral. Assn. Can., Am. Geophys. Union. Subspecialties: Petrology; Geochemistry. Current work: Research and teaching concerning igneous rocks and associated mineral deposits and geothermal energy resources. Office: Dept Geol Scis U Tex Austin TX 78712

BARKER, HAROLD GRANT, surgeon; b. Salt Lake City, June 10, 1917; s. Frederick George and Jennetta (Stephens) B.; m. Kathleen Butler, July 29, 1949; children—Janet Stephens, Douglas Reid. A.B., U. Utah, 1939, postgrad., 1939-41; M.D., U. Pa., 1943. Diplomate: Am. Bd. Surgery. Intern. Hosp. U. Pa., 1943-44, asst. resident in surgery, 1947-51, sr. resident in surgery, 1951-52, asst. attending surgeon, 1952-53; also asst. instr., research fellow U. Pa., 1946-51, instr., research fellow, 1951-52, assoc. in surgery, 1952-53; asst. prof. surgery, Columbia, 1953-57, assoc. prof., 1957-68, prof., 1968—; asst. attending surgeon Presbyn. Hosp., 1953-57, assoc. attending surgeon, 1957-69, attending surgeon, 1969—, dir. med. affairs, 1974-82; practice medicine specializing in surgery, Phila., 1952-53, N.Y.C., 1953—. Contbr. articles med. jours. Served from 1st lt. to capt., M.C. AUS, 1944-46; ETO. Fellow A.C.S.; mem. Soc. U. Surgeons, N.Y. Surg. Soc., Am. Physiol. Soc., Soc. Exptl. Biology and Medicine, A.M.A., Halsted Soc., N.Y. State (chmn. surg. sect. 1961-62), N.Y. County med. socs., Am. Surg. Assn., N.Y. Gastroent. Assn., Sociètè Internationale de Chirurgie, Soc. Surgery Alimentary Tract, Allen O. Whipple Surg. Soc., Am. Assn. History Medicine, Collegium Internationale Chirurgiae Digestivae. Republican. Presbyn. Clubs: Century Assn.; Manursing Island (Rye, N.Y.); Am. Yacht. Subspecialties: Surgery; Physiology (medicine). Current work: Renal physiology, gastrointestinal physiology.liver physiology, cirrhosis of liver, fluid physiology. Home: 1 Forest Ave Rye NY 10580 Office: 161 Ft Washington Ave New York NY 10032

BARKER, JEFFERY LANGE, neurobiologist; b. N.Y.C., Jan. 29, 1943; s. LeBaron Russell and Eileen A. (Lange) B.; m.; children: Alexandra, Olivia, Pamela. B.A., Harvard U., 1964; M.D., Boston U., 1968. Intern in surgery Univ. Hosp., Boston, 1968; research assoc. Nat. Inst. Neurol. and Communicative Disorders and Stroke, NIH, Bethesda, Md., 1969-72, spl fellow, 1973, research scientist intramural research program, 1974-76, research scientist, 1976—, chief lab. neurophysiology, 1981—. Contbr. 140 articles to profl. jours.; editor 5 books. Served with USPHS, 1969. Recipient Pub. Health Service Commendation medal, 1981. Democrat. Subspecialty: Neurobiology. Office: NINCDS-NIH Bldg 36 Rm 2C02 Bethesda MD 20205

BARKER, LEWELLYS FRANKLIN, health care exec.; b. Balt., Sept. 9, 1933; s. William Halsey and Mary Lee (Randol) B.; m. Eileen Frances Sweeney, June 6, 1964; children: Robin Lee, Lillian Halsey, Colin MacLeod. A.B., Princeton U., 1955; M.D., Johns Hopkins U., 1959. Intern Johns Hopkins U. Hosp., Balt., 1959-60; resident in medicine Bellevue Hosp., N.Y.C., 1960-62; commd. med. officer USPHS, 1962-78; med. officer div. biologics standards NIH, Bethesda, MD., 1962-72; dep. dir. div. virology Bur. Biologics FDA, Rockville, Md., 1972-73, dir. div. blood and blood products, 1973-78; v.p. blood services ARC, Washington, 1978-81, v.p. health services, 1981—; bd. dirs. Am. Blood Commn., 1978—, Nat. Health Council, 1981—; mem. expert com. on viral diseases WHO. Recipient Meritorious service medal USPHS., 1973. Mem. AMA, AAAS, Am. Soc. Microbiology, Am. Assn. Immunologists, N.Y.Acad. Scis., Am. Epidemiological Soc.; Am. Assn. Blood Banks, Internat. Soc. Blood Transmission (councilor 1976-82), Am. Clin. Climatol. Assn., Am. Public Health Assn., Phi Beta kappa. Subspecialty: Preventive medicine. Office: ARC 17th and D Sts NW Washington DC 20006

BARKER, TIMOTHY, astronomer, educator; b. Balt., Oct. 21, 1946; s. Roy Chester and Beatrice McClellan (Steinmetz) B.; m. Gloria Frances Landon, June 23, 1973; 1 son, Eric Timothy. A.B., Swarthmore Coll., 1969; Ph.D. (NDEA fellow), U. Calif., Santa Cruz, 1974. Lectr. astronomyU. Calif., Santa Cruz, 1972-73; research asst. U. Calif., 1973-74; assoc. prof. dept. physics and astronomy Wheaton Coll., Norton, Mass., 1974—, asst. dean, 1980-83; asst. prof. Bridgewater (Mass.) State Coll., 1974—. Contbr. articles to profl. jours. NASA grantee, 1980-83; Am. Astron. Soc. grantee, 1981, 83. Mem. Am. Astron. Soc. Subspecialties: Optical astronomy; Theoretical astrophysics. Current work: Spectrophotometry of planetary nebulae and H II regions to determine physical conditions and chemical abundances. Home: 14 Taunton Ave Norton MA 02766 Office: Wheaton Coll Norton MA 02766

BARKER, WILLIAM HAMBLIN, mathematician; b. Albuquerque, Aug. 10, 1948; s. William and LaCharles (Fracaroli) B.; m. Mary L. Crcamean, Sept. 1, 1979. B.S., U. Calif.-Santa Barbara, 1970; M.S., Stanford U., 1971, Ph.D., 1975. Assoc. Daniel H. Wagner Assocs., Paoli, Pa., 1975-79, sr. assoc., 1979-83; mgr. systems analysis Tiburon Systems, San Jose, Calif., 1983—. Mem. Am. Math. Soc., Soc. Indsl. and Applied Math. Subspecialties: Operations research (mathematics); Mathematical software. Current work: Target tracking algorithms, computer assisted search. Home: 555 W Middlefield Rd #L202 Mountainview CA 95134 Office: care Tiburon Systems 611 River Oaks Pkwy San Jose CA 94535

BARKLEY, LINDA DOROTHY, reliability engineer; b. San Diego, Dec. 12, 1951; d. James Falls and Helen Patricia (Yoo) B. B.A., U. San Diego, 1974; M.S., Loyola Marymount U., 1980. Project engr. Hughes Aircraft Co., Los Angeles, 1978—. Mem. Women and Math. (So. Calif. coordinator 1979—), Am. Math. Soc., Assn. for Women in Math., Soc Indsl. and Appied Math, Soc. Women Engrs., Math./Sci. Interchange Los Angeles (v.p. 1980—), Pi Mu Epsilon. Subspecialty: Satellite studies. Current work: Reliability studies for communication satellites. Office: Hughes Aircraft Co PO Box 92919 S32/C314 Los Angeles CA 90009

BARKMEIER, WAYNE WALTER, dental research scientist; b. Friend, Nebr., Mar. 29, 1944; s. Walter Henry and Virginia Rosebud (Thompson) B.; m. Carolyn Ann Johnsen, June 27, 1964; children: Kimberly, Jennifer, Wayne Walter. Student, U. Nebr.-Lincoln, 1962-65, D.D.S., 1969; M.S., U. Tex.-Houston, 1975. Commd. 1st lt. U.S. Air Force, 1969, advanced through grades to col., 1982, dental intern, Keesler AFB, Miss., 1969-70, gen. dentist, Karamursel (Turkey) Air Sta., 1970-72, Whiteman AFB, Mo., 1972-73, chief of operative, Chanute AFB, Ill., 1975-78, Res., 1978; asst. prof. Creighton U., Omaha, 1978-82; dental research scientist L.D. Caulk Co., Milford, Del., 1982—. Articles review cons.: Jour. ADA, 1980-81, Jour. Dental Edn, 1980—; editorial rev. bd.: Jour. Nebr. Dental Assn, 1981-82; contbr. articles to profl. publs.; lectr., profl. groups. Disting. grad. U. Nebr., 1969; recipient article award Quintessence Internat., 1977. Mem. ADA, Acad. Operative Dentistry, Am. Assn. Dental Research, Internat. Assn. Dental Research, Am. Assn. Dental Schs. (del. Council of Faculties 1980-82), Omicron Kappa Upsilon. Roman Catholic. Subspecialties: Dental materials; Restorative dentistry. Current work: Research and development of dental products for dental practitioners, primarily in area of dental materials. Home: 2 Crossley Dr Dover DE 19901 Office: L D Caulk Co Div Dentsply Internat PO Box 359 Milford DE 19663

BARLOW, GEORGE WEBBER, zoology educator; b. Long Beach, Calif., June 15, 1929; s. Fred and Jessie (Kenny) B.; m. Gerta Marianna Offczarczyk, Nov. 5, 1955; children: Linda A., Bicka A., Nora G. A.B., UCLA, 1951, M.A., 1955, Ph.D., 1958. Postdoctoral fellow NIMH, Seewiesen, Germany, 1958-60; asst. prof. U. Ill., Champaign, 1960-63, assoc. prof., 1963-66; assoc. prof. dept. zoology U. Calif., Berkeley, 1966-70, prof., 1970—; mem. psychobiology panel NSF, Washington, 1965-68; Am. rep. Internat. Ethological Com., 1968-75. Author, editor: Sociobiology: Beyond Nature/Nurture?, 1980, Behavioral Development, 1981. Served to lt. (j.g.) USCG, 1951-53; PTO. NIMH postdoctoral fellow, Seewiesen, Germany, 1958-60; Miller prof. Miller Found., Berkeley, Calif., 1970-71; Winegard prof. Guelph U., 1982; G. C. Wheeler disting. lectr. U. N.D., 1982. Fellow Am. Behavior Soc. (pres. 1979), AAAS, Calif. Acad. Soc.; mem. Am. Soc. Zoologists (chmn. ecology div. 1976, chmn. animal behavior div. 1978). Subspecialties: Ethology; Behavioral ecology. Current work: Evolution and development of the mechanisms of social behavior. Home: 1460 Grizzly Peak Blvd Berkeley CA 94708 Office: Dept Zoolog U Calif Berkeley CA 94720

BARLOW, ROBERT BROWN, JR., neuroscientist; b. Trenton, July 31, 1939; s. Robert Brown and Mary Frances (Jones) B.; m. Patricia Ann Dreyer, June 17, 1961; children: Jill, Kim, Jack. B.A., Bowdoin Coll., 1961; Ph.D., Rockefeller U., 1967. Research assoc. Am. Optical Co., Southbridge, Mass., 1960; investigator Marine Biol. Labs., Woods Hole, Mass., summers 1964, 71-73, 76—; asst. prof. Syracuse U., 1967-71; assoc. prof., 1971-77, prof., 1977—. Contbr. articles to profl. jours. NIH grantee, 1971—; NSF grantee, 1978—; recipient Faculty Research award Sigma Xi, 1979. Mem. AAAS, Optical Soc. Am., Assn. Research in Vision and Opthalmology, Soc. Neurosci, Sigma Xi. Subspecialties: Neurobiology; Neurophysiology. Current work: Visual function especially modulation of visual processing via efferent input, circadian clocks and other sense modulations. Home: 4009 Gates Rd N Jamesville NY 13078 Office: Inst Sensory Research Syracuse Univ Syracuse NY 13210

BARMACK, NEAL HERBERT, neurobiologist; b. N.Y.C., Aug. 23, 1942; s. Joseph Ephraim and Therese (Mayer) B.; m. Judith Ellen Avrin, Aug. 25, 1964; children: Matthew Aaron, Erik Seth. B.S., U. Mich., 1963; Ph.D., U. Rochester, 1970. Postdoctoral fellow Inst. Neurophysiology U. Oslo, Norway, 1969-70; research assoc. Neurol. Scis. Inst., Good Samaritan Hosp. and Med. Ctr., Portland, Oreg., 1970-72, sr. research assoc., 1972-75, assoc. scientist, 1975-79, sr. scientist, 1979-81; sr. scientist dept. ophthalmology Neurol. Scis. Inst., 1982—; assoc. prof. physiology U. Conn., 1981-82. Contbr. chpts. to books, articles to profl. jours. Mem. Assn. Research in Vision and Ophthalmology, Soc. Neurosci., Am. Physiol. Soc., Internat. Brain Research Orgn. Jewish. Subspecialty: Neurophysiology. Current work: Functions of the central nervous system. Home: 2610 NW Overton Portland OR 97210 Office: Neurol Scis Inst 1120 NW 20th St Port OR 97209

BARNA, ARPAD ALEX, electronic engineer, consultant; b. Budapest, Hungary, Apr. 3, 1933; came to U.S., 1957; s. Sandor and Elizabeth (Markus) B. B.S. in Elec. Engring, Tech. U. Budapest, 1956; M.S. in Engring, Stanford U., 1966, Ph.D., 1968. Researcher Hungarian Acad. Scis., Budapest, 1956; electronic engr. Calif. Inst. Tech., Pasadena, 1957-61; sr. electronics engr. U. Chgo., 1961-63; tech. staff Stanford Linear Accelerator Center, 1963-69; assoc. prof. elec. engring. U. Hawaii, Honolulu, 1969-72; tech. staff Hewlett-Packard Lab., Palo Alto, Calif., 1972-83; cons. Stanford U. Author: High-Speed Pulse Circuits, 1970, Operational Amplifiers, 1971, (with D.I. Porat) Integrated Circuits in Digital Electronics, 1973, Introduction to Microcomputers and Microprocessors, 1976, Introduction to Digital Techniques, 1979, High Speed Pulse and Digital Techniques, 1980, Very High Speed Integrated Circuits Technologies and Tradeoffs, 1981; contbr. articles to profl. jours.; transl.: Ten Poems (Villon), 1979. Sr. mem. IEEE. Subspecialty: Electronics. Home: 4500 Soquel Dr Soquel CA 95073 Office: Hewlett-Packard Lab 1501 Page Mill Rd Palo Alto CA 94304

BARNARD, ROY JAMES, kinesiology educator, research cardiologist; b. Canton, Ohio, July 4, 1937; s. Fred Henry and Eloise Clark (Denton) B.; m. Kathleen A. Kerringan, Mar. 19, 1978; 1 dau., Kara. B.A., Kent State U., 1959, M.A., 1962; Ph.D., U. Iowa-Iowa City, 1968. Tchr. Hawthorne (Calif.) Sch. Dist., 1959-61; teaching and research asst. Kent State U., 1951-64; research asst. U. Iowa, 1964-66; fellow dept medicine UCLA, 1968-71, assoc. prof. dept. kinesiology, 1973-79, prof., 1979—, asst. research cardiologist dept. surgery, 1971-73, assoc. research cardiologist, 1973-79, research cardiologist, 1979—; dir. research Pritikin Research Found., Santa Monica, Calif., 1979—. NDEA fellow, 1966-68; Muscular Dystrophy Assn. fellow, 1970-71; recipient USPHS Research Career Devel. award, 1975-80. Fellow Am. Coll. Sports Medicine (v.p. 1975); mem. Am. Physiol. Soc., Am. Heart Assn. Council on Circulation. Subspecialties: Physiology (medicine); Physical medicine and rehabilitation. Current work: Rules of exercise and diet in the treatment and prevention of degenerative diseases. Home: 15333 DePauw St Pacific Palisades CA 90272 Office: Dept Surgery UCLA Medical Center 405 Hilgard St Los Angeles CA 90024

BARNES, AARON, physicist; b. Shenandoah, Iowa, May 9, 1939; s. Charles Raymond and Avis Irene (Ross) B.; m. Barbara JoAnne Dean, Sept. 17, 1962; children: Christopher, Stephen. S.B. with honors, U. Chgo., 1961, S.M., 1962, Ph.D., 1966. NRC resident research assoc. NASA-Ames Research Center, Moffett Field, Calif., 1966-67, research scientist, theoretical and planetary studies br., 1967—; guest scientist Max Planck Institut fur Physik und Astrophysik, Garching, W.Ger., 1973. Contbr. articles to profl. jours. Hon. Woodrow Wilson fellow, 1961; interdisciplinary scientist Internat. Solar Polar Mission, 1978—; NASA grantee; recipient Outstanding Sci. Achievement award NASA, 1982. Mem. Am. Phys. Soc., Am. Astron. Soc., Am. Geophys. Union, Internat. Astron. Union, AAAS, Phi Beta Kappa, Sigma Xi. Subspecialties: Theoretical astrophysics; Solar physics. Current work: Theory of space and astrophysical plasmas, solar wind, hydromagnetic waves and turbulence. Office: NASA-Ames Research Center 245-3 Moffett Field CA 94035

BARNES, CHARLES DEC, physiologist, researcher; b. Carroll, Iowa, Aug. 17, 1935; s. Jack Y. and Gladys R. (Beckwith) B.; m. Leona G. Wohler, Sept. 8, 1957; children: Tara L., Teagen Y., Kalee M., Kyler A. B.S., Mont. State U., 1958; M.S., U Wash., 1961; Ph.D., U. Iowa, 1962. Postdoctoral fellow U. Calif., San Francisco, 1962-64; asst. prof. Ind. U, Bloomington, 1964-68, assoc. prof., 1968-71; vis. scientist Inst. of Human Physiology, Pisa, Italy, 1968-69; prof. Ind. State U. and Ind. U. Sch. Medicine, Terre Haute, 1971-75; prof., chmn, physiology Tex. Tech U. Health Scis. Center, Sch. Medicine, Lubbock, 1975-73; prof., chmn. dept. Wash. State U., Pullman, 1983—. Contbr. numerous articles on physiology to profl. jours. Mem. Am. Physiol. Soc., Internat. Brain Research Orgn., Soc. for Neuroscis., Am. Soc. for Pharmacology and Exptl. Therapeutics, Radiation Research Soc., Western Pharmacology Soc. Subspecialties: Neurophysiology; Neuropharmacology. Current work: Brainstem-spinal cord interaction, sensory evoked potentials. Home: SE 890 Edge Knoll Pullman WA 99163 Office: Dept VCAPP Wash State U Pullman WA 99164

BARNES, DAVID EDWARD, neurobiology researcher; b. New Albany, Ind., Dec. 29, 1946; s. Cyrus Crosier and Mary Margarite (Cunningham) B. B.S., Purdue U., 1970, M.S., 1973, Ph.D., 1975. Instr. biology Purdue U., West Lafayette, Ind., 1975-77; postdoctoral researcher in neurosci. U. Fla., Gainesville, 1977-79, research scientist, 1979—; cons. VA Med. Ctr., Gainesville, 1980—. Contbr. articles on neurobiology to profl. jours. NIMH fellow, 1971-75; NIAAA fellow, 1977-79; grantee, 1979. Mem. Soc. Neurosci., Am. Assn. Anatomists, Electron Microscopy Soc. Am., Soc. Devel. Psychobiology. Subspecialties: Neurobiology; Microscopy. Current work: Mental retardation, genetic, environmental influences, fetal alcohol syndrome, quantitative microscopy, autoradiography, electron microscopy. Home: PO Box 709 Melrose FL 32666 Office: Dept Neurosci U Fla Coll Medicine Gainesville FL 32610

BARNES, FRANK STEPHENSON, electrical engineer, educator; b. Pasadena, Cal., July 31, 1932; s. Donald Porter and Thedia (Schellenberg) B.; m. Gay Dirstine, Dec. 17, 1955; children—Stephen, Amy. B.S., Princeton, 1954; M.S., Stanford, 1955, Ph.D., 1958. Fulbright prof. Coll. Engring., Baghdad, Iraq, 1957-58; research asso. Colo. Research Corp., Broomfield, 1958-59; prof. dept. elec. engring. U. Colo., Boulder, 1959—, chmn. dept., 1964-81, acting dean, 1980-81; mem. G-Ed Adcom, IEEE, 1970-77; pres. IEEE Device Soc., 1974-75; Faculty Research lectr. U. Colo., 1965. Regional editor: Electronics Letters of Brit. Inst. Elec. Engrs, 1970-75. Bd. dirs. ABET, 1980-82. Recipient Curtis W. McGraw Research award, 1965; Robert L. Stearns award, 1980. Fellow AAAS, IEEE (editor Student Jour. 1967-70, v.p. for publ. activities 1974-75, ednl. activities bd. 1976-82), Engrs. Council Profl. Devel. (dir. 1976-82, chmn. com. advanced level accreditation 1976-78). Subspecialties: 3emiconductors; Laser data storage and reproduction. Current work: The effects of electromagnetic waves on biological materials; millimeter waves devices; lasers. Home: 225 Continental View Dr Boulder CO 80303 There are always more interesting problems to solve than time to solve them. The trick is to find important problems which can be solved with an effort which is small compared to the value of the results and where one can have a good time learning new ideas at the same time.

BARNES, GEORGE LEWIS, plant pathologist; b. Detroit, Aug. 21, 1920; s. Harold Bernard and Christina Sinclair (White) B.; m. Phyllis June Dollarhite, June 14, 1947; children: William, Jeffrey, Gregory, Susan. B.S., Mich. State U., 1948, M.S., 1950; Ph.D., Oreg. State U., 1953. With Olin Mathieson Chem. Corp., Columbus, Ohio, 1953-55, Port Jefferson Sta., N.Y., 1955-58; with U.S. Dept. Agr., Stillwater, Okla., 1958-61; plant pathologist Okla. State U., 1958-80, extension plant pathologist, 1980—; cons. Allergy Labs., Inc., Oklahoma City. Contbr. articles to profl. jours. Served with USAAF, 1942-45. Mem. Am. Phytopath. Soc., Okla. Acad. Sci., Sigma Xi, Phi Kappa Phi. Subspecialties: Integrated pest management; Plant pathology. Current work: Extension, research, teaching. Home: 424 N Donaldson Dr Stillwater OK 74075 Office: Dept Plant Pathology Okla State U Stillwater OK 74078

BARNES, HUBERT LLOYD, geochemistry educator, consultant; b. Chelsea, Mass., July 20, 1928; s. George Lloyd and Mary Ellen (MacPherson) B.; m. Mary Talbot Westergaard; children: Roy Malcolm, Catherine Patricia. B.S., M.I.T., 1950; Ph.D., Columbia U., 1958. Resident geologist Peru Mining Co., Hanover, N.Mex., 1950-52; lectr. geology Columbia U., N.Y.C., 1952-54; postdoctoral fellow Geophys. Lab., Carnegie Inst., Washington, 1956-60; prof. Pa. State U., University Park, 1960—; dir. Ore Deposits Research, University Park, Sci. Systems, Inc., State College, Pa.; vis. prof. Mineralogy-Petrology Inst., Heidelberg, 1974; exchange scientist Nat. Acad. Sci., Moscow, 1974; Crosby lectr. M.I.T., Cambridge, 1983; cons. numerous corps. Author: Uranium Prospecting, 1956; editor: Geochemistry of Hydro. Ores, 1967, 79. Guggenheim fellow, 1966-67; N.L. Britton scholar, 1955-56; C.F. Davidson lectr. U. St. Andrews, Scotland, 1971; Thayer Lindsley lectr. Soc. Econ. Geologists, 1980-81. Fellow Mineral Soc. Am.; mem. Geochem. Soc. (councilor, v.p. 1982-83), Soc. Econ. Geologists (councilor 1981-83), Am. Geologic Inst. (governing bd. 1981-83), U.S. Nat. Com. Geochemistry (chmn. 1976-78). Subspecialties: Geochemistry; High temperature chemistry. Current work: Chemistry and geology of matural hydrothermal processes especially in geothermal and ore-forming systems. Patentee (2). Home: 213 E Mitchell Ave State College PA 16801 Office: Pa State U 235 Deike Bldg University Park PA 16802

BARNES, JAMES ALFORD, chemist; b. Charlotte, N.C., Aug. 20, 1944; s. James Crowell and Margaret Ruth (Alford) B.; m. Helen Scroggins, June 11, 1966; children: Mary, Curtis. B.S., Davidson (N.C.) Coll., 1966; Ph.D. (NDEA fellow), U. N.C., Chapel Hill, 1970. Chemist U. Southampton, Eng., 1970-71; NIH fellow U. S.C., 1971-72; asst. prof. Western Md. Coll., 1972-73; assoc. prof. Austin Coll., Sherman, Tex., 1973—; research chemist Naval Research Lab., Washington, 1981-82. Contbr. articles to sci. publs. NSF fellows, 1976, 81; NSF grantee, 1977. Mem. Am. Chem. Soc., AAAS, Royal Chem. Soc., AAUP. Subspecialties: Inorganic chemistry; Physical chemistry. Current work: Lithium batteries, basic chemistry, manufacturing technology, use, safety. Mem. Navy's Lead Lab. for battery safety studies. Office: Naval Surface Weapons Center Code R33 Silver Spring MD 20910

BARNES, KAREN LOUISE, neurophysiologist, educator; b. Cleve., Feb. 24, 1942; d. Bentley Tiffany and Margaret Evelyn (Rowlands) B; m. William E. Garapick, Sept. 1, 1973. A.B., Mt. Holyoke coll., 1963; A.M., U. Mich., 1965; Ph.D., Case Western Res. U., 1970. Asst. prof. psychology John Carroll U., 1969-73; research psychologist Cleve. VA Hosp., 1973-75; joint asst. prof. neurology and biomed. engring. Case Western Res. U., 1976-83, adj. assoc. prof. neurology and biomed. engring., 1983—; head sect. research dept. neurology and staff, cardiovascular research Cleve. Clinic Found., 1976—; spl. reviewer cardiovascular and pulmonary study sect. NIH, 1980. Contbr. numerous articles, abstracts to profl. jours. Mem. Soc. Neurosci., Internat. Assn. for Study Pain, Am. Acad. Neurology, Am. Physiol. Soc., Assn. Research in Nervous and Mental Disease, Phi Beta Kappa, Sigma Xi. Clubs: Edgewater Yacht, Cleve. Ski (Cleve.). Subspecialties: Neurophysiology; Neurobiology. Current work: Cardiovascular neurobiology, somatosensory systems; neurophysiol. and neuroanat. research; quantitative evaluation of cutaneous sensory function in neurol. disorders. Office: 9500 Euclid Ave Cleveland OH 44106

BARNES, LARRY DEAN, biochemist, educator; b. Red Oak, Iowa, Aug. 20, 1944; s. Stanley H. and Sarah P. (Windom) B.; m. Sandra L. Bachman, Sept. 13, 1968; children: Jaime Danielle, Brian Kyle. B.A., Rice U., 1966; Ph.D., UCLA, 1970. NIH postdoctoral fellow U. Iowa, Iowa City, 1971-73; research fellow Mayo Clinic, Rochester, Minn., 1973-76; asst. prof. U. Tex. Health Science Ctr., San Antonio, 1976-82, assoc. prof., 1982—. Mem. Am. Soc. Biol. Chemistry, Am. Soc. Cellular Biology, AAAS, Sigma Xi, Phi Lambda Upsilon. Subspecialties: Biochemistry (biology); Cell biology. Current work: Nucleotides in cellular processes; tubulin-ligand interactions; angiotension II in renal physiology. Home: 10407 Northampton San Antonio TX 78230 Office: Dept Biochemistry U Tex Heath Sci Ctr San Antonio TX 78284

BARNETT, ALLEN, research pharmacologist; b. Newark, May 5, 1937; s. Samuel and Lillian (Bloomberg) B.; m. Mary Lou Victoria Selva, June 6, 1965; children: Carole, David. B.S. in Pharmacy, Rutgers U., 1959; Ph.D. in Pharmacology, SUNY-Buffalo, 1965. Registered pharmacist, N.J., N.Y. Group leader pharmacology Roche Labs., 1965-66; group leader pharmacology Schering Corp., Bloomfield, N.J., 1966-69, sect. leader, 1969-73, mgr. pharmacology, 1974-75, assoc. dir. biol. research, 1976-79, sr. assoc. dir. biol. research, 1980—; adj. assoc. prof. Fairleigh Dickinson U. Contbr. articles to profl. jours. Served with U.S. Army, 1960-61. Mem. Am. Soc.Pharmacology and Exptl. Therapeutics, Am. Chem. Soc., Acad. Pharm. Sci., N.Y. Acad. Scis., AAAS, Am. Pain Soc., Soc. Neurosci. Subspecialties: Pharmacology; Neuropharmacology. Current work: Analgesia and endogenous pain suppression systems, dopamine receptors and behavior. Office: Schering Corp 60 Orange St Bloomfield NJ 07003

BARNETT, JOHN BRIAN, immunologist; b. Cardston, Alta., Can., Apr. 17, 1945; s. Durwood L. and Yvonne D. B.; m. Cecilia Anne Barnett, Dec. 27, 1966; children: John, Cynthia S. B.S., Mont. State U., 1967, M.S., 1969; Ph.D., U. Louisville Sch. Medicine, 1973. Postdoctoral fellow U. Tenn., Knoxville, 1973-75; asst. prof. dept. microbiology and immunology U. Ark. Med. Scis., Little Rock, 1975-82, assoc. prof., vice chmn. dept., 1982—. Contbr. articles to profl. jours. Am. Lung Assn. grantee, 1978-80; NIH grantee, 1980—. Mem. Am. Assn. Immunologists, Am. Soc. Microbiology, Sigma Xi. Presbyterian. Subspecialty: Immunology (medicine). Current work: Effect of vitamin A on immune response; effect of chlordane on immune response. Office: Dept Microbiology and Immunology U Ark Med Scis 4301 W Markham Little Rock AR 72205

BARNETT, STOCKTON GORDON, research/development director; b. East Orange, N.J., July 18, 1939; s. Stockton and Ethel (Osborn) B.; m. Lucy Estelle Gockel, Aug. 20, 1966; 1 dau., Elizabeth Anne. B.A., Dartmouth Coll., 1961; M.S., U. Iowa, 1963; Ph.D., Ohio State U., 1966. Faculty geology and earth sci. SUNY-Plattsburgh, 1966-81, prof., 1976-81; dir. research/devel. Condar Co., Hiram, Ohio, 1981—. Contbr. articles to profl. jours. Exec. com. mem Lake Champlain Commn., 1975-82. Recipient EPA award, 1979. Mem. AAAS, Soc. Econ. Paleontologists and Mineralogists, Sigma Xi. Subspecialties: Paleontology; Biomass (energy science and technology). Current work: developed first smokeless woodstove and combustion controls; developed method of measuring particulate emissions from woodstoves. Patentee in field. Home: 11782 Mills Rd Garrettsville OH 44231 Office: PO Box 6 Hiram OH 44234

BAROCAS, HARVEY A(ARON), clinical psychologist, educator; b. N.Y.C., June 5, 1942; s. Isaac and Sue (Azriel) B.; m. Carol Birnbaum, Sept. 8, 1942; children: Briana, Solon. B.B.A., CCNY, 1964, M.A., 1966; Ph.D., CUNY, 1970; postgrad. diploma, Postgrad. Ctr. for Mental Health, 1974. Lic. psychologist, N.Y. Lectr. Baruch Coll., N.Y.C., 1966-70, asst. prof., 1970-74, assoc. prof., 1974-81; prof. CUNY, N.Y.C., 1981—; pvt. practice clin. psychology, N.Y.C., 1975—; cons. B.F.S. Psychol. Assos., N.Y.C., 1972-74, N.Y.C. Police Dept., 1971-72, N.J. Chief of Police, 1970-72. Author: Personal Adjustment and Growth: A Lifespan Approach, 1982; contbr. articles to profl. jours. NIMH postdoctoral fellow, 1972-73; recipient Meml. award Postgrad. Ctr. for Mental Health, 1973, Outstanding Service award, 1980. Mem. Am. Psychol. Assn., Internat. Fedn. Psychoanalysis, N.Y. Soc. Clin. Psychologists. Subspecialties: Developmental psychology; Clinical-community psychology. Current work: Personality adjustment, stress, crisis intervention. Office: Baruch Coll CUNY 17 Lexington Ave New York NY 10010

BAROLET, RALPH YVON, dental educator; b. Montreal, Quebec, Can., Oct. 4, 1929; s. Yvon Ralph and Rose Ann (Dumas) B.; m. Lorraine Claire Raby, Nov. 26, 1955; 1 dau., Frances. B.S., Loyola Coll., Montreal, 1952, McGill U., 1954; D.D.S., U. Montreal, 1970; M.Sc., Ind. U., 1972. Plant engr. No. Electric Co., Montreal, Can., 1954-56; design engr. Union Carbide Can, Ltd., Montreal, 1956-66; prof. dentistry Laval U., Quebec City, Can., 1970—. Fellow Internat. Coll. Dentists; mem. Assn. Can. Faculties of Dentistry (pres. 1980-82), Que. Dental Soc. (pres. 1981-82), Can. Dental Assn. (council chmn. 1977-82), Assn. Internat. Dental Studies, ADA (SCADA achievement award 1980). Subspecialties: Biomaterials; Prosthodontics. Current work: Clinical investigation of esthetic dental restorative materials. Office: Laval U Sch Dental Medicine Quebec City Canada G1K 7P4

BARON, JEFFREY, educator, researcher; b. Bklyn., July 10, 1942; s. Harry Leo and Terry (Goldstein) B.; m. Judith Carol, June 27, 1965; children: Stephanie Ann, Leslie Beth, Melissa Leigh. B.S. in Pharmacy, U. Conn., 1965; Ph.D. in Pharmacology, U. Mich., 1969. Research fellow in biochemistry U. Tex. Southwestern Med. Sch., Dallas, 1969-71, research asst. prof. biochemistry and pharmacology, 1971-72; asst. prof. U. Iowa, 1972-75, assoc. prof., 1975-80, prof. pharmacology, 1980—. Contbr. articles to profl. jours. USPHS fellow U. Mich., 1965-69; recipient Research Career Devel. award NIAMDD, 1975-80. Mem. Am. Soc. Pharmacology and Exptl. Therapeutics, Am. Soc. Biol. Chemists, AAAS, Am. Assn. Cancer Research, N.Y. Acad. Scis., Soc. Exptl. Biology and Medicine. Subspecialties: Cancer research (medicine); Toxicology (medicine). Current work: Immunohistochemistry, immunocytochemistry, chemical carcinogenesis, drug metabolism; toxicology, cellular pharmacology. Home: 302 Shrader Rd Iowa City IA 52240 Office: 2-270 Bowan Sci Bld Iowa City IA 52242

BARON, MELVIN LEON, consulting engineer; b. Bklyn., Feb. 27, 1927; s. Frank and Esther (Hirskowitz) B.; m. Muriel Wicker, Dec. 24, 1950; children—Jaclyn Adele, Susan Gail. B.C.E., CCNY, 1948; M.S., Columbia U., 1949, Ph.D., 1953. Lic. profl. engr., N.Y., Mass. Structural designer Corbett-Tinghir Co., N.Y.C., 1949-50; research assoc. civil engring. Columbia U., 1951-53, asst. prof., 1953-57, adj. assoc. prof., 1958-61, adj. prof., 1961—; chief engr. Paul Weidlinger, N.Y.C., 1957-60, assoc., 1960-64, ptnr., dir. research, 1964—; also v.p. Advanced Computer Techniques Corp., N.Y.C., 1962-66; Formerly chmn. adv. panel engring. mechanics program NSF. Author: (with M.G. Salvadori) Numerical Methods in Engineering, 1952; Editor: Jour. Engring. Mechs. Div. ASCE, 1970—; Contbr. articles profl. jours. Recipient Spirit of St. Louis Jr. award Am. Soc. M.E., 1958; J. James R. Croes medal ASCE, 1963; Walter L. Huber Research prize, 1966; Arthur M. Wellington prize, 1969; Nathan M. Newmark medal, 1977. Fellow ACSE (exec. com. engring. mechanics div. 1966-69, 72-76, mem. mgmt. group C engring. mechanics div. 1972-76, mem. tech. activities com. 1974), ASME; mem. N.Y. Acad. Scis., U.S. Nat. Acad. Engring. (chmn. com. computational mechanics), U.S. Nat. Com. on Theoretical and Applied Mechanics, AAAS, Sigma Xi. Subspecialties: Theoretical and applied mechanics; Civil engineering. Current work: Dynamic response of structures to explosive loadings, structure-media interaction procedures; materials engineering. Home: 3801 Hudson Manor Terr Riverdale Bronx NY 10463 Office: 110 E 59th St New York NY 10022

BARON, MIRON, psychiatrist, educator; b. Israel, June 14, 1947; s. Ichiel and Carmela (Muzikansky) B.; m. Carmela Tal, June 5, 1974. M.D., Sackler Med. Sch., Tel Aviv,U., 1973. Diplomate: Am. Bd. Psychiatry and Neurology, Nat. Bd. Med. Examiners. Resident in psychiatry Albert Einstein Coll. Medicine, N.Y.C., 1974-77; research psychiatrist N.Y. State Psychiat. Inst., N.Y.C., 1977-80, dir. div. psychogenetics, 1980—; asst. prof. psychiatry Columbia U. Coll. Physicians and Surgeons, N.Y.C., 1977-82, assoc. prof. psychiatry, 1983—. Contbr. articles to profl. jours. Recipient A.E. Bennett research award Soc. Biol. Psychiatry, 1976, Roche Labs. award, 1976, Mead Johnson Labs. award, 1976; NIMH research scientist devel. awardee, 1978-83. Mem. Am. Psychiat. Assn., Soc. Neurosci., Am. Soc. Human Genetics, AAAS, N.Y. Acad. Sci. Subspecialties: Psychiatry; Genetics and genetic engineering (medicine). Current work: Biological psychiatry, psychiatric genetics, psychopharmacology, neuroscience. Office: 722 W 168th St New York NY 10032

BARON, ROBERT ALAN, psychology educator; b. N.Y.C., June 7, 1943; s. Bernard Paul and Ruth (Schlossberg) B.; m. Sandra Faye Lawton, June 21, 1975. B.S., Bklyn. Coll., 1964; M.A., U. Iowa, 1967, Ph.D., 1968. Asst. prof. psychology U. S.C., Columbia, 1968-71; assoc. prof. psychology Purdue U., West Lafayette, Ind., 1971-75, prof., 1975—; vis. prof. Princeton U., 1977-78; program dir. NSF, Washington, 1979-81; fellow Oxford (Eng.) U., 1982. Author: Human Aggression, 1977, Behavior in Organizations, 1983, (with D. Byrne) Social Psychology, 4th edit., 1984; assoc. editor: Jour. Personality and Social Psychology, 1977-79. NSF grantee, 1970, 72. Fellow Am. Psychol. Assn.; mem. Acad. Mgmt., Soc. Exptl. Social Psychology, Internat. Soc. Research on Aggression. Subspecialty: Social psychology. Current work: Human aggression; behavior in organizational settings. Home: 690 Cardinal Dr Lafayette IN 47905 Office: Purdue U West Lafayette IN 47907

BARON, SEYMOUR, engineering company executive; b. N.Y.C., Apr. 5, 1923; s. Benjamin and Tillie (Schuster) B.; m. Florence Chill, Aug. 27, 1950; children: Richard Mark, Paul Lawrence. B.S. Engring., Johns Hopkins U., 1944, M.S., 1947; Ph.D., Columbia U., 1950. Lab. researcher U.S. Indsl. Chem. Co., 1944-47; research asst. Columbia U., N.Y.C., 1947-50; chief engr. Burns and Roe, Inc., Oradell, N.J., 1950-64, v.p., 1964-75, sr. v.p., 1975-76, sr. corporate v.p., 1976—, dir., 1967—; bd. dirs. Argonne Asso. Univs.; mem. exec. com., spl. com. for reactor devel., reactor devel. and safety div. Argonne Univs. Assn., 1976—; mem. adv. com., engring. tech. div. Oak Ridge Nat. Lab.; mem. N.J. Commn. on Radiation Protection. Fellow ASME, Am. Nuclear Soc.; mem. Am. Inst. Chem. Engrs., Nat. Acad. Engring. Club: Lions (Oradell). Subspecialties: Mechanical engineering; Nuclear engineering. Current work: Executive management of energy projects. Office: Burns and Roe Inc 700 Kinderkamack Rd Oradell NJ 07649

BARONE, FRANK CARMEN, pharmacologist; b. Syracuse, N.Y., July 5, 1949; s. Frank and Sophie (Kurzepa) B.; m. Diane M. Osborne, June 24, 1972; children: Adam, Amy. A.A., Onondaga Community Coll., Syracuse, 1971; B.A., Syracuse U., 1973, Ph.D., 1978. Research asst. prof. biopsychology Syracuse U., 1978-82; postdoctoral scientist Smith Kline & French Labs., Phila., 1982—; editorial cons. Ankho Internat., Syracuse, 1978-82. Mng. editor: Neurobehavioral Toxicology and Teratology, 1979-81. Mem. AAAS, Soc. Neurosci., Am. Physiol. Soc., N.Y. Acad. Sci. Subspecialties: Neuropharmacology; Gastroenterology. Current work: Central neural control of gastrointestinal function; drug development for the treatment of gastrointestinal disease. Office: Smith Kline & French Labs 1500 Spring Garden St Philadelphia PA 19101

BARONE, MILO CARMINE, biology educator; b. Throop, Pa., Dec. 4, 1941; s. Emilio L. and Mary (Muto) B.; m. Marilyn M. Smith, Oct. 9, 1982. B.S., U. Scranton, 1963; M.S., John Carroll U., 1965; Ph.D., St. Bonaventure U., Olean, N.Y., 1968. Asst. prof. biology Fairfield (Conn.) U., 1968—. Mem. AAAS, N.Y. Acad. Scis., N.Y. Zoool. Soc., AAUP. Roman Catholic. Subspecialty: Animal physiology. Home: 1242 Old Academy Rd Fairfield CT 06430 Office: Biology Dept Fairfield University Fairfield CT 06430

BARONE, ROBERT MICHAEL, physician/surgeon; b. Buffalo, Apr. 2, 1941; s. Michael Horace and Antoinette (Bugcaglia) B.; m. Mary Margaret Wallin, Mar. 11, 1967; children: Susanne, Julie, Robert. B.S., Georgetown U., 1962; M.D., SUNY-Buffalo, 1966; M.S., U. Ill., 1970. Resident U. Ill., Chgo., 1966-74; staff surgeon San Diego Tumor Inst., 1980-82; asso. clin. prof. surgery U. Calif.-San Diego, 1976—; oncology staff Oncology Assos. of San Diego, 1982—; mem. U. Calif. San Diego Cancer Center, 1978—. Contbr. articles to profl. jours. Served to comdr. USNR, 1966-74. Recipient Chi. Surg. Research award Chi. Surg. Soc., 1971; NCI grantee, 1979—; Mead Johnson Research award, 1971. Fellow ACS, Soc. Surg. Oncology, Soc. Head Neck Surgeons, Sigma Xi; mem. Soc. Acad. Surgery, Soc. Gen. Surgeons San Diego, Calif. Med. Soc., San Diego County Med. Soc. Republican. Roman Catholic. Subspecialty: Psychiatry. Current work: Devel. of implantable infusion system for regional delivery chemotherapy to liver; devel. implantable devices for vascular access. Office: 3930 4th Ave San Diego CA 92103

BAROODY, WILLIAM JOSEPH, JR., research institute executive; b. Manchester, N.H., Nov. 5, 1937; s. William Joseph and Nabeeha Marion (Ashooh) B.; m. Mary Margaret Cullen, Apr. 23, 1960; children: William Joseph, Mary Nabeeha, David, Jo Ellen, Christopher, Andrew, Thomas, Philip, Paul. A.B. in English, Holy Cross Coll., 1959; postgrad. in polit. sci., Georgetown U., 1961-64; LL.D. (hon.), Seattle U., 1976; LL.D. (hon.), Marist Coll., 1976, Assumption Coll., 1981; Litt.D. (hon.), St. Mary of the Woods Coll., 1976. Legis. asst. and press sec. to Congressman Melvin R. Laird, 1961-68; research dir. House Republican Conf., Washington, 1968-69; asst. to sec and dep. sec. Dept. Def., Washington, 1969-73; spl. asst. to Pres. U.S., White House, 1973-74; asst. to Pres. U.S., White House, 1974-76; exec. v.p. Am. Enterprise Inst. for Pub. Policy Research, Washington, 1977-78; pres., 1978—. Publisher: Pub. Opinion mag., 1977—, Regulation mag., 1977—, Fgn. Policy and Def. Rev., 1977—, The AEI Economist, 1978—. Chmn. bd. Woodrow Wilson Internat. Ctr. for Scholars, 1982—; bd. dirs. St. Anselm Coll., Ctr. for Study of the Presidency, Wolftrap Found., John Carroll Soc.; mem. Pres.'s Task Force on Pvt. Sector Initiatives, 1982. Served in USN, 1959-61. Recipient Disting. Civilian Pub. Service award Dept. Def., 1973. Mem. Am. Polit. Sci. Assn. Republican. Melkite Catholic. Office: 1150 17th St NW Washington DC 20036

BAROSS, JOHN ALLEN, marine microbiologist, educator; b. San Francisco, Aug. 27, 1941. B.S., San Francisco State U., 1963, M.A., 1965; Ph.D. in Marine Microbiology, U. Wash., 1972. Research asst. Coll. Fisheries, U. Wash., Seattle, 1966-70; research assoc. Oreg. State U., Corvallis, 1970-77, asst. prof., sr. researcher, 1977—, assoc., prof. Mem. Am. Soc. Microbiology, AAAS, Soc. Indsl. Microbiology, Audobon Soc., Sigma Xi. Subspecialty: Microbiology. Office: Dept Microbiology Oreg State U Corvallis OR 97331

BARR, RICHARD ARTHUR, biology educator; b. Southport, N.Y., Mar. 12, 1925; s. Harold Arthur and Emma Marie (Ferguson) B.; m. Violet Marie Keens, Oct. 7, 1961; children: Robert Adrian, Elisa Marie. B.S., U. Vt., 1950, M.S., 1955; Ph.D., Cornell U., 1963. Asst. prof. Cornell U., Ithaca, N.Y., 1963-66, U. Mo, St. Louis, 1966-68; assoc. prof. biology Shippensburg (Pa.) State Coll., 1968-72, prof., 1972-83, Shippensburg U. Pa. (formerly Shippensburg State Coll.), 1983—. Served with U.S. Navy, 1943-46; ETO. Pa. Sci. and Engring. Found. grantee, 1973, 74, 76; NSF grantee, 1982—. Mem. Am. Inst. Biol. Sci., AAAS, Bot. Soc. Am., Sigma Xi. Republican. Presbyterian. Subspecialties: Plant physiology (agriculture); Plant cell and tissue culture. Current work: Plant cell culture, growth and differentiation; protoplast fusion. Home: Star Route 2 Box 63B Shippensburg PA 17257 Office: Shippensburg U Pa Dept Biology F 201 Shippensburg PA 17257

BARRAT, JOSEPH GEORGE, plant pathologist; b. New Haven, May 30, 1922; s. Joseph George and Ida May (Davis) B.; m. Ann Eldred, Sept. 6, 1948; children: Robert Eldred, John Davis, James Rodman. B.S. in Agr, R.I. State Coll., 1948; M.S. in Botany-Plant Pathology, U. R.I., 1951; Ph.D., U. N.H., 1958. Plant pathologist Wash. State Dept. Agr., Prosser, 1951-55; extension specialist in plant pathology Univ. Farm, W.Va. U., Kearneysville, 1958-70, plant pathologist and supt., 1971-79, plant pathologist and extension specialist, 1980—. Author articles. Served with USN, 1943-46. Mem. Am. Phytopath. Soc., Sigma Xi. Episcopalian. Club: Men's (Shepherdstown, W.Va.). Subspecialties: Plant pathology; Plant virology. Current work: Research on virus, fungal and bacterial diseases of decicuous tree fruits. Plant pathology, virus diseases of apples and peaches, fungus and bacterial diseases of apples and peaches, control recommendations. Patentee in field. Office: PO Box 303 Kearneysville WV 25430

BARRATT, ERNEST STOELTING, psychologist, educator, researcher; b. North Charleroi, Pa., Mar. 31, 1925; s. Robert Duff and Marie Agnes (Stoelting) B.; m. Bobbye Lee Rheinlander, July 27, 1947 (div. 1967); 1 dau., Robin Rhein; m. Karen Marie Creel, Dec. 18, 1968; 1 son, Christopher Robert. B.A., Tex. Christian U., Fort Worth, 1947, M.A., 1949; Ph.D., U. Tex.-Austin, 1952. Lic. psychologist, Tex. Instr. to assoc. prof. psychology U. Del., Newark, 1951-57, prof. Tex. Christian U., 1957-62; assoc. prof. to prof. U. Tex. Med. Br., Galveston, 1962—; NIH fellow UCLA Brain Research Inst., 1961-62. Contbr. chpts. to books, articles to profl. jours. Trustee Galveston Ind. Sch. Dist., 1970—. Served with USN, 1943-46. Fellow Am. Psychol. Assn.; fellow Am. EEG Soc.; mem. Soc. Biol. Psychiatry, Am. EEG Soc., Soc. Psychophysiol. Research, Neural Sci. Soc. Episcopalian. Subspecialties: Psychophysiology; Neuropsychology. Current work: Biological bases of timing and rhythm behavior, biological bases of personality. Home: 2641 Gerol Dr Galveston TX 77551 Office: Dept Psychiatry and Behavior Scis Univ Texas Medical Branch Galveston TX 77550

BARRETT, CHARLES SANBORN, emeritus metallurgy educator; b. Vermillion, S.D., Sept. 28, 1902; s. Charles H. and Laura (Dunham) B.; m. Dorothy A. Adams, Aug. 2, 1928; 1 dau., Marjorie A. B.S., U. S.D., 1925; fellow, U. Chgo., 1927-28, Ph.D., 1928. With metallurgy dept. Naval Research Lab., 1928-32; metals research lab., dept. metall. engring. Carnegie Inst. Tech., 1932-46; prof. James Franck Inst., U. Chgo., 1946-71, emeritus, 1971—; prof., sr. research engr., adj. prof. physics U. Denver, 1970—; exchange prof. U. Birmingham, Eng., 1951-52; vis. prof. U. Denver, 1961, Stanford, 1963, U. Va., 1968, 70, Ga. Inst. Tech., 1973; Eastman prof. Oxford U., Eng., 1965-66; Mem. nat. com. on crystallography, 1950-54. Author: Structure of Metals, 1943, rev. edits., 1952, (with T.B. Massalski) Structure of Metals, 1966; Co-editor vols.: Advances in X-ray Analysis; Author tech. papers, phys. metallurgy, crystallography. Recipient Mathewson medal Am. Inst. Mining and Metall. Engrs., 1934, 44, 51, Hume-Rothery award, 1975; Howe medal Am. Soc. Metals, 1939; Clamer medal Franklin Inst., 1950; Heyn medal Deutsches Gesellschaft fûr Metallkunde, 1966; Sauveur medal Am. Soc. Metals, 1966; Gold medal Japan Inst. Metals, 1976; Acta Metallurgica Gold medal, 1982. Fellow Am. Phys. Soc., Am. Soc. Metals (hon. mem., Gold medal 1976), Am. Inst. Mining and Metall. Engrs. (chmn. Inst. Metals div. 1956, hon. mem. 1980); mem. Am. Crystallographic Assn., Nat. Acad. Scis., Inst. Metals (London), Internat. Union Crystallography (editor metals sect. Structure Reports 1949-51), Phi Beta Kappa, Sigma Xi, Delta Tau Delta, Sigma Pi Sigma, Alpha Sigma Mu. Subspecialty: Materials. Current work: Applied x-ray diffraction. Office: Metallurgy Materials Sci Div Denver Research Inst U Denver Denver CO 80208

BARRETT, EDWARD JOSEPH, chemistry educator, molecular design company executive; b. N.Y.C., July 4, 1931; s. Martin A. and Mary (Gallagher) B.; m. Joanne Gallitano, Oct. 2, 1976; children: Elizabeth, William. B.A., Fordham U., 1953; M.A., Columbia U., 1961, Ph.D., 1962. Faculty Hunter Coll., CUNY, N.Y.C., 1961—, prof. chemistry, 1971—; pres. Molecular Design Inc., N.Y.C., 1978—. Contbr. research papers to profl. lit. Grantee NIH, 1958-62, Merck &

Co., 1972, CUNY Research Found., 1968-72. Mem. Am. Chem. Soc., AAAS, N.Y. Acad. Scis., Am. Inst. Chemists, Royal Soc. Chemistry, Sigma Xi, Phi Lambda Upsilon. Subspecialties: Organic chemistry; Biochemistry (biology). Current work: Molecular models. Patentee in field. Office: Dept Chemistry Hunter Coll CUNY 695 Park Ave New York NY 10021

BARRETT, JAMES EDWARD, medical psychologist, educator, researcher; b. Camden, N.J., Aug. 9, 1942; s. Thomas T. and Ruth E. (Taylor) B.; m. Maura Dean, June 10, 1962; children: Jennifer, Andrea, Stephanie. B.S., U. Md.; Ph.D. (NIMH fellow), Pa. State U., 1971. With Lab. Comparative Psychobiology, NIMH, 1966-68; postdoctoral fellow Worcester Found. Exptl. Biology, 1971-72; asst. prof. U. Md., College Park, 1972-75, assoc. prof., 1975-79, prof., 1979; assoc. prof. psychiatry and med. psychology Uniformed Services U. Health Scis., Bethesda, Md., 1979-83, prof. psychiatry, pharmacology and med. psychology, 1983—; adj. prof. psychology U. Md., 1979—; adj. prof. pharmacology and toxicology U. Md. Med. Sch., Balt., 1977—. Contbr. articles to sci. and tech. publs., chpts. to books on behavior and behavioral pharmacology. NIMH postdoctoral fellow, 1971-72; Nat. Inst. Alcohol Abuse and Alcoholism grantee, 1973-79; Nat. Inst. on Drug Abuse grantee, 1976—. Mem. Am. Soc. Pharmacology and Exptl. Therapeutics, Soc. for Neurosci., AAAS, Behavioral Pharmacology Soc., Acad. Behavioral Medicine. Subspecialties: Psychopharmacology; Psychobiology. Current work: Behavioral pharmacology, drug abuse, neuroactive peptides, neuropharmacology. Home: 1507 Live Oak Dr Silver Spring MD 20910 Office: 4301 Jones Bridge Rd Bethesda MD 20814

BARRETT, JAMES THOMAS, immunologist; b. Centerville, Iowa, May 20, 1927; s. Alfred Wesley and Mary Marjorie (Taylor) B.; m. Barbro Anna-Lill Nilsson, July 31, 1967; children: Sara, Robert, Annika, Nina. B.A., State U. Iowa, 1950, M.S., 1951, Ph.D., 1953. Asst. prof. bacteriology and parasitology U. Ark. Sch. Medicine, Little Rock, 1953-57; asst. prof. microbiology U. Mo.-Columbia Sch. Medicine, 1957-59, assoc. prof., 1959-67, prof., 1967—. Author: Textbook of Immunology, 4th edit, 1983, Basic Immunology and Its Medical Application, 2d edit, 1980. Served with USN, 1944-45. NIH fellow, 1963-64; NIH Fogarty sr. fellow, 1977-78. Mem. Am. Assn. Immunology, Am. Soc. Microbiology, Sigma Xi. Subspecialties: Immunobiology and immunology; Microbiology. Current work: Immunology of enzyme-proenzyme pairs, toxicity of Loxosceles reclusa spider venom, anaerobic bacteria-chemotaxis, phagocytosis, platelet activation by toxins and bacteria. Home: 901 Westport Dr Columbia MO 65201 Office: Sch Medicin U Mo Columbia MO 65212

BARRETT, RICHARD JOHN, nuclear engineer, energy analyst; b. West Pittston, Pa., Mar. 28, 1945; s. Robert Joseph and Margaret (O'Malley) B.; m. Margaret McNevin, June 21, 1969; children: Robert, Kathleen. B.S. in Physics, U. Scranton, Pa., 1967, Ph.D., U. Va., 1972. Postdoctoral fellow Case Western Res. U., Cleve., 1972-75; staff scientist Los Alamos (N.Mex.) Nat Lab., 1975-82; nuclear engr. U.S. Nuclear Regulatory Commn., Bethesda, Md., 1982—; energy program analyst U.S. Dept. Energy, Washington, 1980-82. Mem. Am. Nuclear Soc. Subspecialties: Nuclear fission; Fuels and sources. Current work: Determining the safe limits of operation for nuclear power reactors; analyzing the energy needs for the nation's future and the role of energy tecnologies. Home: 7501 Mill Run Dr Derwood MD 20855 Office: Nuclear Regulatory Commn 7920 Norfolk St Bethesda MD 20555

BARRETT, TERENCE WILLIAM, biophysicist, researcher; b. London, Apr. 22, 1939; s. Albert Edward and Norah Kathleen (Frost) B.; m. Constance McClintock, Sept. 7, 1968; children: Josepine, James, Nathaniel. M.A. summa cum laude, U. Edinburgh, Scotland, 1964; Ph.D., Stanford U., 1967. Research fellow Carnegie-Mellon U., Pitts., 1969-70; asst. prof. U. Tenn. Center Health Scis., Memphis, 1971-81; research physicist Naval Research Lab., Washington, 1981—. Contbr. over 100 articles to sci. jours. Served with Brit. Army, 1958-60. NATO student, 1964-67. Mem. Biophys. Soc., Am. Phys. Soc., Am. Chem. Soc., Acoustical Soc. Am., Am. Physiol. Soc., Soc. Math. Biology, Sigma Xi. Episcopalian. Subspecialties: Biophysics (physics); Biophysical chemistry. Current work: Material science, polymer science, biopolymer science, organic semiconduction; physics of macromolecular complexes; spectroscopy. Office: Naval Research Lab Chemistry Div Surface Chemistry Br Code 6170 Washington DC 20375

BARRIGA, OMAR OSCAR, Immunologist, parasitologist, educator; b. Santiago, Chile, Mar. 1, 1938; s. Simon S. and Elvira E. (Val) B.; m. Ines Quirland Rojas, Dec. 31, 1960; children: Omar Alexander, Alvaro Gonzalo. B.A., U. Chile, 1958, D.V.M., 1963; M.S., U. Ill., 1971, Ph.D., 1973. Asst. prof. to assoc. prof. parasitology and immunology U. Chile, 1964-68; instr. parasitology U. Ill., 1969-72; asst. prof. parasitology U. Pa., 1973-78; assoc. prof. parasitology Ohio State U., Columbus, 1979—; vis. prof. parasitology and immunology U. Fed. Rio Grande do Sul, Brazil, 1977-78; ad hoc cons. NIH and NSF. Author: The Immunology of Parasitic Infections, 1981; contbr. articles to sci. jours., chpts. in books. Recipient Disting. Tchr. award Norden Labs., 1975. Mem. Am. Soc. Parasitologists, Am. Soc. Immunologists, Am. Soc. Tropical Medicine and Hygiene, Phi Zeta, Phi Sigma. Subspecialties: Immunology (medicine); Parasitology. Current work: Immunomodulation by parasitic infections, immunization for control of parasitic infections. Office: 1900 Coffey Rd Columbus OH 43210

BARRON, RANDALL FRANKLIN, mech. engr., educator, cons; b. Many, La., May 16, 1936; s. Benjamin Franklin and Inez (Norseworthy) B.; m. Shirley McDuffie, Mar. 14, 1958; children: Randy, Donna Carol Barron Ellard, Steven Dale, Brian Richard. B.S., La. Poly. Inst., 1958; M.S., Ohio State U., 1961; Ph.D., 1965. Registered profl. engr., La. Instr. dept. mech.engring. Ohio State U., 1958-64, asst. prof., 1965; assoc. prof. mech. engring. La. Technol. U., 1965-70, prof., 1970—, Alumni prof., 1979, dir. div. engring. research, 1975—; cons. engring., including to Riley-Beaird, 1966—. Author: Cryogenic Systems, 1966; contbr. articles to nat., internat. profl. jours. Player agt. Dixie Baseball, Ruston, La., 1976—. Recipient Gold Medal award Pi Tau Sigma, 1968, award of merit La. Engring. Soc., 1969, Outstanding Research award Sigma Xi, 1971, Achievement award Engring./Sci. Council, 1981, Outstanding Teaching award Tau Beta Pi, 1981, La. Tech. Alumni Found. Prof. award, 1979. Mem. ASME, Am. Soc. Engring. Edn., Cryogenic Soc. Am., AAAS. Baptist. Lodge: Masons. Subspecialties: Mechanical engineering; Cryogenics. Current work: Heat transfer at cryogenic temperatures. Home: 2202 Greenbriar Dr Ruston LA 71270 Office: Dept Mech Engring La Technol U Ruston LA 71272

BARRON, SARAH KATHRYN BRASWELL, zoologist, educator, researcher; b. McKinney, Tex., Nov. 6, 1941; d. Albert Dalton and Gloria Belle (Staton) Braswell; m. John Calvin Barron, June 3, 1961; 1 dau., Lucille Ann. B.S. in Math, Trinity U., 1962; Ph.D. in Zoology, U. Tex. Austin, 1982. Cert. tchr. math./English, biol./composite sci. Tchr. math. Northside Ind. Sch. Dist., San Antonio, 1963-65, Corpus Christi Ind. Sch. Dist., Tex., 1965-66; instr. math. Christopher Jr. Coll., Corpus Christi, 1966; tchr. math./English Del Valle (Tex.) Ind. Sch. Dist., 1968, Incarnate Word Acad., Corpus Christi, 1966-68; tchr. math., sci., English Holy Cross High Sch., Austin, Tex., 1969-70; teaching asst. U. Tex. Austin, 1970, 72, 1973-75, NSF genetics tng. grantee, 1975-76, research asst., 1977, 78-79, RASSL tutor, 1980-81, specialist dept. computer scis., 1981, research asst. petroleum engring. dept., 1981-82, grad. research asst. III dept. zoology, summer, 1982, postdoctoral assoc. dept. zoology, 1982—; tchr. sci. Town-Country Sch., Austin, 1976. Contbr. reports in field. Active Walnut Creek Neighborhood Assn., Austin. Nat. Inst. Aging grantee, 1982; Ken-ichi Kojima Genetics Meml. travel fellow, 1981. Mem. AAAS, Assn. for Computing Machinery, Entomol. Soc. Am., Genetics Soc. Am., Soc. for Study of Evolution, Sigma Xi (grantee 1980). Presbyterian. (officer bd. edn., diaconate council for cultural missions). Subspecialties: Evolutionary biology; Gene actions. Current work: Genetics of life histories of drosophila, longevity and aging in drosophila, morphometrics, life histories, temperature and humidity effects, micro-evolution. Home: 510 E Braker Ln Austin TX 78753 Office: Dept Zoolog U Tex Austin TX 78712

BARRON, SAUL, chemist, educator, cons.; b. N.Y.C., Feb. 24, 1917; s. Max and Sadie (Levitt) B.; m. Phyllis Levin, Sept. 6, 1941. B.S. in Chem. Engring, Lafayette Coll., Easton, Pa., 1941, M.S., Ohio State U., 1948, Ph.D. (fellow), 1954. Registered profl. engr., Ohio, Md. Jr. naval architect Phila. Navy Yard, 1941-42; mech. engr. Wright-Patterson AFB, Ohio, 1946-51; staff scientist Martin Co., 1954-56; sr. scientist Avco, Lawrence, Mass., 1956-57; dir. research Thiokol Co., 1958-60, Bell Aerosystems Co., Wheatfield, N.Y., 1960-64; prof. chemistry SUNY, Buffalo, 1964—; pres. Space Scis. Co., Buffalo, 1964—; vis. prof. Hebrew U. Jerusalem, 1982; lectr. in field; cons. in field. Contbr. articles to tech. jours. Served to capt. USAAF, 1943-46; with USAF, 1951-52; Korea. Mem. Am. Chem. Soc. (chmn. Western N.Y. chpt.), Am. Ceramic Soc., Instruments and Control Soc., N.Y. Acad. Scis., AAUP, Sigma Xi, Tau Beta Pi, Phi Lambda Upsilon. Lodge: Masons. Subspecialties: Thermodynamics; Chemical engineering. Current work: Thermal behavior of polymers; temperature studies of mixtures; developing cooperative program in chemistry; research on heat transfer-cooling systems in supersonic aircraft, re-entry cooling of ballistc nose cones, low temperature testing of aircraft power plants, oscillographic data processing, material and environ. effects upon transport properties of propellants and their behavior in space, thermodynamics and combustion studies of high energy propellants, analytical procedures in chem. instrumentation, solid state physics. Home: 249 Troy Del Way Buffalo NY 14221 Office: Buffalo State College 1300 Elmwood Ave Buffalo NY 14222

BARRON, WILLIAM LORING, III, psychology educator; b. Houston, Feb. 28, 1954; s. William Loring and Cherry Joyce (Jones) B. B.A., Rice U., 1976; M.A., U. Tex.-Austin, 1978, Ph.D., 1980. Grad. research asst. U. Tex.-Austin, 1978-80; asst. prof. psychology William Penn Coll., Oskaloosa, Iowa, 1980—; cons. Human Devel. Assocs., Oskaloosa, 1981—. Contbr. articles to profl. jours. Bd. dirs., tchr. Mahaska County Hospice, Oskaloosa, 1981—; chmn. bd. Jack & Jill Nursery Sch., Oskaloosa, 1981—. Mem. Am. Psychol. Assn., Phi Kappa Phi, Pi Lambda Theta. Republican. Baptist. Subspecialties: Social psychology; Psychology of humor. Current work: Research in humor, psychology of music, population psychology, social psychology, law enforcement selection. Home: 1321 N Market St Oskaloosa IA 52577 Office: William Penn Coll Dept Psychology Oskaloosa IA 52577

BARROWS, EDWARD MYRON, biology educator; b. Detroit, Aug. 8, 1946; s. Sigmund E. and Meroslava (Lewandowsky) B.; m. Julie Norberg, Mar. 22, 1979. B.S., U. Mich., 1968; Ph.D., U. Kans., 1975. Asst prof. dept. biology Georgetown U., Washington, 1975-81, assoc. prof., 1981—; cons. U.S. Army, 1979, Dynamic, Rockville, Md., 1980, MacMillan Pub. Co., 1981, Nat. Wildlife Fedn., 1981. Contbr. articles to profl. jours. Served to capt. USAR. NSF fellow, 1967; Smithsonian fellow, 1975; Georgetown U. grantee, 1980; Washington Field Biology Club grantee, 1982. Mem. AAAS, Animal Behavior Soc., Ecol. Soc. Am., Am. Inst. Biol. Scis., Sigma Xi. Subspecialties: Ethology; Ecology. Current work: Behavior, ecology, evolution of insects; ecology of highly disturbed environments; human behavioral responses toward insects. Home: 21 Wellesley Cir Glen Echo MD 20812 Office: Dept Biology Georgetown U Washington DC 20057

BARRY, JOHN REAGAN, psychology educator; b. Lyndonville, N.Y., July 2, 1921; s. Stanley R. and Alice (Reagan) B.; m. Marian C. Combs, Dec. 21, 1946; children: Judith Ann, David John, Elizabeth Jeanne. A.B., Hamilton Coll., 1942; M.A., Syracuse U., 1943; Ph.D., Ohio State U., 1949. Clin. psychol. intern Ohio State U., Columbus, 1946-49; asst. prof. psychology U. Ill., Chgo., 1949-51; asst. dir. psychology Sch. Aviation Medicine, Randolph Field, Tex., 1951-55; assoc. prof. psychology U. Pitts., 1955-61; dir. Rehab. Research Inst., U. Fla., Gainesville, 1962-66; prof. psychology U. Ga., Athens, 1966—, exec. bd., 1975—; dir. Human Interaction Research Inst., Los Angeles, 1975—; cons. to dir. Ga. Dept. Rehab., Atlanta, 1966-78. Contbr. articles to profl. jours.; cons. editor various jours. Served to 2d lt. USAF, 1943-46. Recipient cert. of Merit Ga. Psychol. Assn., 1972, 82; cert. of Appreciation Pub. Offender Counseling Assn., 1978. Fellow Am. Psychol. Assn. (pres. div. 1964-65, 72-73), AAAS; mem. Am. Personnel and Guidance Assn., Gerontol. Soc. Am., Ga. Psychol. Assn. (v.p. 1969-75), Pa. Psychol. Assn. (sec. 1958-60), Pitts. Psychol. Assn. (pres. 1958-59), Sigma Phi Omega, Sigma Xi, Phi Delta Kappa. Club: Torch. Subspecialties: Developmental psychology; Social psychology. Current work: Community psychology and social welfare research; program evaluation; psychology of aging; personality theory. Home: 189 Spruce Valley Athens GA 30605 Office: Dept Psychology Univ Ga Athens GA 30602 Home: 189 Spruce Valley Athens GA 30605

BARRY, ROGER GRAHAM, climatologist, educator; b. Sheffield, Eng., Nov. 13, 1935; came to U.S., 1968; s. Graham Charles and Winifred (Watson) B.; m. Valerie Tompkin, Oct. 3, 1959; children—Rachel Elena, Jane Christina. B.A. with honors, U. Liverpool, Eng., 1957; M.Sc., McGill U., Montreal, Que., Can., 1959; Ph.D., Southampton (Eng.) U., 1965. Leverhulme Research fellow U. Liverpool, 1959; lectr. U. Southampton, 1960-66, 67-68; research scientist dept. energy, mines, resources, Ottawa, Ont., Can., 1966-67; asso. prof. geography U. Colo., 1968-71, prof., 1971—; dir. World Data Center-A for Glaciology, 1976; vis. fellow Australian Nat. U., Canberra, 1975. Co-author: Atmosphere, Weather and Climate, 1968, rev. edit., 1976, Synoptic Climatology, 1973; Co-editor: Arctic and Alpine Environments, 1975; Contbr. articles to profl. jours. NSF grantee; NOAA grantee. Mem. Am. Meteorol. Soc., AAUP, Royal Meteorol. Soc., Assn. Am. Geography, Inst. Brit. Geographers, Am. Quaternary Assn. Subspecialties: Climatology. Current work: Climatic change, snow/ice-climate interactions, mountain climates, polar environments, synoptic climatology. Office: World Data Center-A Glaciology Box 449 U of Colo Boulder CO 80309

BARSCHALL, HENRY HERMAN, physics educator; b. Berlin, Germany, Apr. 29, 1915; m. Eleanor A. Folsom; two children. A.M., Princeton U., 1939, Ph.D., 1940; Dr. rev. nat. h.c., U. Marburg (W. Ger.), 1982. Instr. Princeton U., 1940-41, U. Kans., 1941-43; mem. staff Los Alamos Sci. Lab., 1943-46, asst. div. leader, 1951-52; mem. faculty U. Wis., 1946—, prof. physics, 1950—, Bascom prof., 1973—, chmn. dept., 1951, 54, 56-57, 63-64; assoc. div. leader Lawrence Livermore Labs., 1971-73; Vis. prof. U. Calif. at Davis, 1972-73. Assoc. editor: Revs. Modern Physics, 1951-53; Asso. editor: Nuclear Physics, 1959-72; editor: Phys. Rev. C, 1972—; mem. editorial bd.: Jour. Phys. and Chem. Reference Data, 1979-84. Fellow Am. Phys. Soc. (chmn. div. nuclear physics 1968-69, mem. council 1983-86, Bonner prize 1965); mem. Nat. Acad. Sci. (chmn. physics sect. 1980-83), Am. Inst. Physics (chmn. publ. bd. 1980-82, governing bd. 1983-86), NRC (assembly math. and phys. scis. 1980-83). Subspecialties: Nuclear physics; Nuclear fission. Current work: Pure and applied neutron physics. Home: 1110 Tumalo Trail Madison WI 53711

BARSKY, ARNOLD M(ILTON), nuclear engineer; b. Chgo., June 21, 1953; s. Murray H. and Doris (Stein) B.; m. Dawn Ann Terry, Jan. 15, 1978; children: Rebecca Marie, Adam Matthew. B.S. in Physics, U. Ill., 1975, 1975; M.S. in Nuclear Engring, U. Wis., 1977. Exchange student Energieonderzoek Centrum Nederland, Holland, 1975; nuclear ops. engr. Gen. Electric Knolls Atomic Power Lab., Schenectady, 1978-81, nuclear refueling engr., Windsor, Conn., 1981—. Mem. Am. Nuclear Soc. Jewish. Subspecialties: Nuclear engineering; Nuclear fission. Current work: Naval nuclear propulsion systems, refueling and pressure vessel reuse. Home: RD3 Charlton Rd Ballston Lake NY 12019 Office: General Electric Knolls Atomic Power Lab PO Box 545 Windsor CT 06095

BARSKY, BRIAN ANDREW, computer scientist; b. Montreal, Sept. 17, 1954; s. Arthur Harold and Audrey Barbara (Epstein) B. D.C.S., McGill U., 1973, B.Sc., 1976; M.S., Cornell U., 1978; Ph.D., U. Utah, 1981. Vis. researcher Sentralinstitutt, Oslo, 1979; instr. U. Calif., Santa Cruz, 1982—, asst. prof. computer sci., Berkeley, 1981—; adj. asst. prof. U. Waterloo, Ont., Can., 1982—. Contbr. articles to profl. jours. U. Utah fellow; Natural Scis. and Engring. Research Council scholar; U. Calif. Berkeley Regents Jr. Faculty fellow; NSF grantee, 1982—. Mem. Spl. Interest Group on Graphics, Assn. Computing Machinery, Nat. Computer Graphics Assn., IEEE Computer Soc., Can. Man-Computer Communications Soc., Soc. Indsl. and Applied Math. Subspecialty: Graphics, image processing, and pattern recognition. Current work: Interactive three-dimensional computer graphics and computer aided geometric design and modeling. Office: Computer Sci Div Univ Calif Berkeley CA 94720

BARSTOW, DAVID ROBBINS, computer scientist; b. Middletown, Conn., Aug. 5, 1947; s. Robbins Wolcott and Margaret (Vanderbeek) B.; m. Linda Gail Francis, Dec. 27, 1970; 1 son, Geoffrey Francis. B.A. in Math, Carleton Coll., Northfield, Minn., 1969; M.S. Stanford U, 1970; Ph.D. in Computer Sci. with distinction, 1978. J.W. Gibbs instr. dept. computer sci. Yale U., New Haven, 1977-79, asst. prof., 1979-80; program leader for software research Schlumberger-Doll Research, Ridgefield, Conn., 1980—. Author: Knowledge-based Program Construction, 1979; editor: (with others) Interactive Programming Environments, 1984; Assoc. editor: Computing Surveys; Contbr. articles to profl. jours. Mem. Assn. for Computing Machinery, IEEE, Am. Assn. for Artificial Intelligence. Subspecialties: Artificial intelligence; Software engineering. Current work: Automatic programming, programming environments, industrial applications of artificial intelligence. Office: Schlumberger-Doll Research Old Quarry Rd Ridgefield CT 06877

BARTA, OTA, immunologist, educator; b. Ostrava, Czechoslovakia, Aug. 18, 1931; s. Otakar and Ludmila (Schirmerova) B.; m. Vera D. Dadakova, 1956; children: Marketa, Tomas. MVDr., Vet. Faculty, Agrl. U., Brno, Czechoslovakia, 1955, C.Sc., 1963; Ph.D., U. Guelph, Ont., Can., 1969; cert., Am. Coll. Vet. Microbiologists. Asst. prof. Vet. Faculty Agrl. U., Brno, 1955-61; scientist Central Research Inst. Animal Husbandry, Prague-Uhrineves, 1961-63, Inst. Vet. Med. Research, Brno, 1963-67; research asso. U. Guelph, 1967-69; asst. prof. Okla. State U., Stillwater, 1969-73, asso. prof., 1973-75, La. State U., 1975-78, prof., 1978—. Contbr. articles to profl. jours. Recipient Chaire Francqui Internationale, Brussels, 1979; Silver medal U. Liege, Belgium, 1979; Fulbright sr. lectr., 1981. Mem. Am. Soc. Microbiology, Am. Assn. Immunologists, World Assn. Vet. Microbiologists, Immunologists and Specialists in Infectious Disease, AVMA, Am. Assn. Vet. Immunologists. Subspecialties: Immunology (agriculture); Clinical Immunology. Current work: Clinical immunology, serum regulation of lymphocyte functions. Office: Dept Vet Microbiology and Parasitology La State U Baton Rouge LA 70803

BARTEL, NORBERT HARALD, astronomer; b. Wettendorf, W. Ger., Feb. 24, 1950; came to U.S., 1980; s. Heinrich J. and Irmgard E. (Kall) B.; m. Joan C. Bartel, Jan. 11, 1951; children: Hanna Siglinde, Robert H. Diplom in physics, Rheinische Friedrich-Wilhelms U., Bonn. W.Ger., 1976; Dr. rer. nat., Rheinische Friedric-Wilhelms U., Bonn. W.Ger., 1978. Research assoc. Max-Planck-Inst. fur Radioastronomie, Bonn, W. Ger., 1978-80; vis. scientist MIT, 1980-82; research assoc. Ctr. Astrophysics, Harvard U., 1983—. Contbr. articles in field to profl. jours. Recipient Otto-Hahn medal Max-Planck-Gesellschaft, 1980. Mem. Am. Astron. Soc., Astronomische Gesellschaft. Subspecialties: Infrared optical astronomy; Radio and microwave astronomy. Current work: Compact objects, pulsars, supernovae, galactic nuclei, quasars. Home: 46 Lombard Terr Arlington MA 02174 Office: 60 Garden St Cambridge MA 02138

BARTELS, RICHARD HAROLD, computer science educator; b. Ann Arbor, Mich., Jan. 10, 1939; s. Robert Christian Frank and Virginia Francis (Terwilliger) B.; m. Renate Wessendorf, June 28, 1968. B.S., U. Mich., 1961, M.S., 1963; Ph.D., Stanford U., 1968. Asst. prof. U. Tex.-Austin, 1968-69, 71-74; wissenschaftlicher mitarbeiter Tech U. Munich, W.Ger., 1970-71; asst. prof. Johns Hopkins U., Balt., 1974-78, assoc. prof., 1978-79; assoc. prof. computer sci. U. Waterloo, Ont., Can., 1979—; cons. Nat. Bur. Standards, 1978-79. Assoc. editor: Math. Programming, 1978-81. Mem. Assn. Computing Machinery (dir. SIGNUM 1979-82, chmn. SIGNUM bd. 1983-85), Soc. Indsl. and Applied Math. Subspecialties: Numerical analysis; Graphics, image processing, and pattern recognition. Current work: Nonlinear and linear programming, with application to approximation and data-fitting; splines and their use in curve and surface representation in computer graphics. Office: Dept Computer Sci U Waterloo Waterloo ON Canada N2L 3G1

BARTH, CHARLES ADOLPH, astro-geophysicist, atmospheric and space physicist; b. Phila., July 12, 1930; m., 1954; 4 children. B.S., Lehigh U., 1951; M.A., UCLA, 1955, Ph.D., 1958. Research geophysicist Inst. Geophysics, UCLA, 1957-58; research physicist Jet Propulsion Lab., Calif. Inst. Tech., 1958-65; assoc. prof. U. Colo., Boulder, 1965-67, prof. astro-geophysics, 1967—, dir. atmospheric and space physics lab., 1965—. NSF fellow, Bonn, W.Ger., 1958-59. Mem. AAAS, Am. Astron. Soc., Am. Geophys. Union. Subspecialties: Planetary atmospheres; Space physics. Office: Atmospheric and Space Physics Lab U Colo Boulder CO 80309

BARTH, DANIEL STEPHEN, neuroscientist, researcher; b. Balt., Jan. 21, 1954; s. John Simmons and Anne (Strickl) B. B.A. in Psychology summa cum laude with distinction, Boston U., 1977, M.A., UCLA, 1979, Ph.D. candidate, 1979—. Instr. physiol. psychology UCLA. Contbr. articles to profl. jours. UCLA Univ. fellow, 1979-80; NIMH grantee, 1980-83; NSF grantee, 1982; NATO grantee, 1982. Mem. Soc. for Neurosci., Soc. Psychophysiol. Research, Phi Beta Kappa. Subspecialties: Neurophysiology; Bioinstrumentation. Current work: Normal and pathological brain function, epilepsy, neuroscience, magnetoencephalography, electroencephalography, neurophysiology, bioinstrumentation. Office: Dept Psychology UCLA 405 Hilgard Ave Los Angeles CA 90024

BARTH, ROLF FREDERICK, pathologist, educator; b. N.Y.C., Apr. 4, 1937; s. Rolf L. and Josephine B.; m. Christine Ferguson, Oct. 30, 1965; children: Suzie, Alison, Rolf, Christofer. A.B., Cornell U., 1959; M.D., Columbia U., 1964. Diplomate: Am. Bd. Pathology. Prof. dept. pathology and oncology U. Kans. Med. Ctr., Kansas City, 1970-77; clin. prof. dept. pathology Med. Coll. Wis. and U. Wis.-Madison, 1977-79; prof. dept. pathology Ohio State U., Columbus, 1979—; cons. div. cancer research resources and ctrs. Nat. Cancer Inst. Contbr. articles to profl. jours. Served to sr. asst. surgeon USPHS, 1966-70. NIH grantee; Am. Cancer Soc. grantee; Dept. Energy grantee. Mem. Am. Assn. Pathologists, Am. Assn. Immunologists, Am. Assn. Cancer Research, Soc. for Nuclear Medicine. Subspecialties: Pathology (medicine); Immunology (medicine). Current work: Tumor and transplantation immunology, monoclonal antibodies, tumor antigens, neutron capture therapy. Home: 2670 Crafton Park Columbus OH 43221

BARTHOLOMEW, MERVIN JEROME, geologist, state official; b. Altoona, Pa., Nov. 22, 1942; s. Mervin Wilbur and Catherine Clara (Morris) B.; m. Dinah Jill Heberling, Sept. 14, 1963 (div. 1974); m. Sharon Elizabeth Lewis, June 26, 1975. B.S., Pa. State U., 1964; M.S., U. So. Calif., 1968; Ph.D., Va. Poly. Inst. and State U., 1971. Asst. prof. geoscis. N.C. State U., Raleigh, 1972-75; contract geologist N.C. Div. Land Resources, Raleigh, 1975-78, Va. Div. Mineral Resources, Charlottesville, 1976-79, geologist in charge regional office, 1979-82; chief geology and mineral resources div. Mont. Bur. Mines and Geology, Butte, 1983—. Fellow Geol. Soc. Am. (prin. editor spl. paper 1984); mem. Carolina Geol. Soc., Soc. Pa. Archaeology, Sigma Xi. Subspecialties: Geology; Tectonics. Current work: Grenville massifs of Appalachians; structural analysis of Blue Ridge and Valley and Ridge of Appalachians. Office: Mont Bur Mines and Geology Mont Coll Mineral Sci and Tech Butte MT 59701

BARTINE, DAVID ELLIOTT, engineering physics laboratory manager; b. Phila., Dec. 6, 1936; s. David Fenton and Elda Josephine (McClain) B.; m. Dorothy Judith Shankle, Dec. 19, 1959; children: David, Benjamin, Rebecca, Mac. B.S. in Chemistry, Eastern Bapt. Coll., 1959; M.S. in Sci. Edn, U. Pa., 1961, U. Mo.-Rolla, 1966, Ph.D., 1971. Instr. physics and chemistry Beaver Coll., Glenside, Pa., 1960-62; instr. chemistry Montgomery Jr. Coll., Takoma Park, Md., 1962-65; research staff mem. Union Carbide, Oak Ridge Nat. Lab., 1969-78, group leader, 1978-81, sect. head reactor analysis and shielding, engrng. physics div., 1981—; mem. Com. to Develop Ctr. for Advancement Radiation Edn. and Research Johns Hopkins Med. Inst., Balt., 1982—; mem. nuclear engrng. adv. com. U. Mo.-Rolla, 1982—; advisor Def. Nuclear Agy. on Radiation Transport Issues, Washington, 1975—; tech. coordinator Dept. Energy Fgn. Exchange, Washington, 1976—. Author: Radiation Transport Cross Section Sensitivity Analysis - A General Approach Illustrated, 1974; contbr. articles to profl. jours. Recipient citation for Three Mile Island assistance Dept. Energy. Mem. Am. Nuclear Soc. (treas. radiation protection and shielding div. 1978-80, vice chmn. 1981-82, chmn. 1982-83, outstanding service award 1978), AAAS, Sigma Xi. Methodist. Subspecialties: Nuclear engineering; Nuclear fission. Current work: Radiation transport methods development and application, shielding integral experiments and analysis, reactor core analysis, system conceptual design studies. Office: Oak Ridge Nat Lab PO Box X Oak Ridge TN 37830

BARTKE, ANDRZEJ, endocrinologist, educator; b. Krakow, Poland, May 23, 1939; s. Gustaw and Jadwiga (Dabrowska) B.; m. Rose Schwarz, Apr. 15, 1966. M.Sc. in Biology, Jagiellonian U., Krakow, 1962; Ph.D. in Zoology, U. Kans., 1965. Asst. prof. genetics Jagiellonian U., 1965-67; tng. fellow in reproductive physiology Worcester Found. Exptl. Biology, Shrewsbury, Mass., 1967-69; staff scientist, 1969-72, sr. scientist, 1972-78; assoc. prof. dept. ob-gyn U. Tex. Health Sci. Center, San Antonio, 1978-82; prof. U. Tex. Health Sci. Ctr., San Antonio, 1982—. Author research papers, book chpts.; Editor: Jour. Andrology, 1979-83. Recipient research career devel. award NIH, 1972-77; research grantee NSF, NIH. Mem. Endocrine Soc., Genetics Soc. Am., Soc. Endocrinology (U.K.), Soc. Study Reprodn., AAAS, Fedn. Am. Scientists, Am. Soc. Andrology, Soc. Study Fertility (U.K.), Sigma Xi. Roman Catholic. Subspecialties: Reproductive biology; Endocrinology. Current work: Male reproductive endocrinology; role of prolactin; effects of season and photoperiod; hereditary variations; marijuana and alcohol effects. Home: Route 1 Box 1311 Boerne TX 78006 Office: Dept Ob/Gy U Tex Health Sci Center San Antonio TX 78284

BARTLETT, ALAN CLAYMORE, geneticist; b. Price, Utah, June 17, 1934; s. Rulon Ashley and Emily Bertha (Hunter) B.; m. Vanice Rae Baker, Mar 16, 1956; children: Ravae Edith Johnson, Denice Alene Hardman, Lelsle Emily Jacobson, Trace Alan. A.S., Carbon Coll., 1954; B.A., U. Utah, 1956; Ph.D., Purdue U., 1962. Instr. Carbon Coll., Price, Utah, 1957-58, Purdue U., Lafayette, Ind., 1958-62; geneticist U.S. Dept. Agr., State College, Miss., 1962-67, Tucson, 1967-69, Phoenix, 1969—; adj. prof. Miss. State U., 1963-67; adj. faculty Ariz. State U., 1975-79. Contbr. articles to profl. jours. Pres. Tempe Assn. for Gifted, 1975-76; bd. dirs. Ariz. Assn. Gifted, 1977-78. Mem. AAAS, Ariz.-Nev. Acad. Sci., Genetics Soc. Am., Am. Genetic Assn., Entomology Soc. Am., Am. Inst. Biol. Scis. Mormon. Subspecialties: Genetics and genetic engineering (agriculture); Genetics and genetic engineering (biology). Current work: Development of genetic techniques for insect control. Office: 4135 E Broadway Phoenix AZ 85040

BARTLETT, DAVID FARNHAM, physicist, educator; b. N.Y.C., Dec. 13, 1938; s. Frederic Pearson and Margaret Mary (Boulton) B.; m. Roxana Ellen Stoessel, Nov. 19, 1960; children: Andrew, Susannah, Christopher, Jennifer. A.B., Harvard U., 1959; A.M., Columbia U., 1961, Ph.D., 1965. Instr. Princeton U., 1964-67, asst. prof., 1967-71; assoc. prof. physics U. Colo., Boulder, 1971-82, prof., 1982—. Editor: The Metric Debate; Contbr. articles to profl. jours. Recipient various grants. Fellow Am. Phys. Soc.; mem. Am. Assn. Physics Tchrs. Democrat. Subspecialties: Relativity and gravitation; Low temperature physics. Current work: Use of low-temperature techniques for experimental tests of general relativity. Home: 954 Lincoln Pl Boulder CO 80302 Office: U Colo Dept Physics Campus Box 390 Boulder CO 80309

BARTLETT, JANETH MARIE, nuclear pharmacy educator; b. Cooperstown, N.Y., Sept. 10, 1946; d. Harold C. and Emily (Walker) B. B.S. in Pharmacy, Temple U., 1969, M.S., 1970; Ph.D., Rutgers U., 1981. Registered pharmacist, Pa. Research assoc. E.R. Squibb & Sons, Inc., New Brunswick, N.J., 1970-74; asst. research investigator, 1974-81; asst. prof. nuclear pharmacy Purdue U., West Lafayette, Ind., 1981—. Mem. Soc. Nuclear Medicine, Am. Pharm. Assn., AAAS, Sigma Xi, Rho Chi. Current work: Radiopharmaceutical research and development, drug delivery systems, clinical nuclear pharmacy, pharmacokinetics, diagnostic imaging. Patentee in field. Home: 3304 Peppermill Dr 1B West Lafayette IN 47906 Office: Dept Medicinal Chemistry Purdue U West Lafayette IN 47907

BARTLETT, NEIL, chemist, educator; b. Newcastle-upon-Tyne, Eng., Sept. 15, 1932; s. Norman and Ann Willins (Voak) B.; m. Christina Isabel Cross, Dec. 26, 1957; children—Jeremy John, Jane Anne, Christopher, Robin. B.Sc., Kings Coll., U. Durham, Eng., 1954, Ph.D. in Inorganic Chemistry, 1957; D.Sc. (hon.), U. Waterloo, Can., 1968, Colby Coll., 1972, U. Newcastle-upon-Tyne, Eng., 1981, U. Bordeaux, France, 1976. Lectr. chemistry U. B.C., 1958-63, prof., 1963-66; prof. chemistry Princeton U., 1966-69, U. Calif.-Berkeley, 1969—; dir. Inorganic Syntheses Corp.; mem. adv. bd. on inorganic reactions and methods Verlag Chemie, 1978—; mem. adv. panel Nat. Measurement Lab., Nat. Bur. Standards, 1974-80; E.W.R. Steacie Meml. fellow NRC, Can., 1964-66; Miller vis. prof. U. Calif.-Berkeley, 1967-68; 20th G.N. Lewis Meml. lectr. Ohio State U., 1966, William Lloyd Evans Meml. lectr., 1966; A.D. Little lectr. Northeastern U., 1969; Phi Beta Upsilon lectr. U. Nebr., 1975; Henry Werner lectr. U. Kans., 1977; Jeremy Musher Meml. lectr., Israel, 1980; Brotherton vis. prof. U. Leeds (Eng.), 1981; Erskine vis. lectr. U. Canterbury (Eng.), 1983; vis. fellow All Souls Coll., Oxford U., 1984. Bd. editors: Inorganic Chemistry, 1967-78, Jour. Fluorine Chemistry, 1971-80, Synthetic Metals, Revue Chimie Minérale, Noveau Jour. de Chimie; adv. bd.: McGraw-Hill Ency. Sci. and Tech. Recipient Research Corp. award, E.W.R. Steacie prize, 1965, Elliott Cresson medal Franklin Inst., 1968; Kirkwood medal Yale U. and New Haven sect. Am. Chem. Soc., 1969; award in inorganic chemistry Am. Chem. Soc., 1970; Dannie-Heinemann prize The Göttingen Acad., 1971; Robert A. Welch award in chemistry, 1976; Alexander von Humboldt Found. award, 1977; medal Jozef Stefan Inst., Yugoslavia, 1980; Alfred P. Sloan fellow, 1964-66. Fellow Chem. Soc. London (Corday Morgan medal 1962), The Royal Soc., Am. Acad. Arts and Scis., Royal Inst. Chemistry, Chem. Inst. Can. (1st Noranda lectr. 1963); mem. Nat. Acad. Scis. U.S.A. (fgn. asso.), Leopoldina Acad. (Halle, Saale), Akademie der Wissenschaften in Göttingen, Am. Chem. Soc. (chmn. divs. fluorine chemistry 1972, inorganic chemistry 1977, W.H. Nichols award N.Y. sect. 1983), Phi Lambda Upsilon (hon.). Subspecialties: Inorganic chemistry; Solid state chemistry. Current work: Fluorine chemistry, graphite chemistry, noble-gas chemistry, electrode materials, fast-ion sollid conductors, electrochemical oxidation. Spl. research synthesis 1st true compound of xenon. Home: 6 Oak Dr Orinda CA 94563 Office: Latimer Hall Univ of Calif Berkeley CA 94720

BARTLETT, PAUL DOUGHTY, chemist; b. Ann Arbor, Mich., Aug. 14, 1907; s. George Miller and Mary Louise (Doughty) B.; m. Mary Lula Court, June 24, 1931; children—Joanna Court (Mrs. Stephen D. Kennedy), Geoffrey McSwain, Sarah Webster (Mrs. Edson H. Rafferty). A.B., Amherst Coll., 1928, Sc.D. (hon.), 1953; M.A., Harvard, 1929, Ph.D., 1931; Sc.D. (hon.), U. Chgo., 1954, U. Montpellier, 1967, U. Paris, 1968, U. Munich, 1977. NRC fellow Rockefeller Inst., 1931-32; instr. chemistry U. Minn., 1932-34; mem. faculty Harvard, 1934-75, prof. chemistry, 1946-75, Erving prof. chemistry, 1948-75, Erving prof. emeritus, 1975—, chmn. dept., 1950-53; Robert A. Welch prof. chemistry Tex. Christian U., 1974—; George Fisher Baker lectr. Cornell U., spring 1949; vis. prof. U. Calif. at Los Angeles, 1950; Walker-Ames lectr. U. Wash., 1952; guest lectr. U. Munich, Germany, 1957; speaker 15th Internat. Congress Pure and Applied Chemistry, Paris, France, 1957; Karl Folkers lectr. U. Ill., 1960; Spl. Univ. lectr. U. London, Eng., 1961; lectr. Japan Soc. for Promotion of Sci., 1978; mem. div. com. math., phys. and engrng. scis. NSF, 1957-61. Author: Nonclassical Ions, 1965, also chpts. in textbooks, research papers.; Mem. editorial bd.: Jour. Am. Chem. Soc, 1945-55, Jour. Organic Chemistry, 1954-57, Tetrahedron. Recipient award in pure chemistry Am. Chem. Soc., 1938; August Wilhelm von Hofmann gold medal German Chem. Soc., 1962; Roger Adams award organic chemistry, 1963; Willard Gibbs medal, 1963; Theodore William Richards medal, 1966; Nat. Medal of Sci., 1968; James Flack Norris award in phys. organic chemistry Am. Chem. Soc., 1969; John Price Wetherill medal, 1970; Linus Pauling award, 1976; Nichols medal, 1976; James Flack Norris award in teaching chemistry, 1978; Alexander von Humboldt sr. scientist award U. Freiburg, Germany, 1976, U. Munich, 1977; Wilfred T. Doherty award, 1980; Max Tishler award Harvard U., 1981; Robert A. Welch award, 1981; Guggenheim and Fulbright fellow, spring 1957. Hon. fellow Chem. Soc. (London; Centenary lectr. 1969, Ingold lectr. 1975); mem. Swiss Chem Soc. (hon.), Chem. Soc. Japan (hon.), Nat., N.Y. acads. scis., Am. Acad. Arts and Scis., Am. Philos. Soc., Franklin Inst. (hon.), Am. Chem. Soc. (chmn. Northeastern sect. 1953-54), Internat. Union Pure and Applied Chemistry (pres. organic div. 1967-69, program chmn. 23d internat. congress 1971), Deutsche Akademie der Naturforscher Leopoldina, Phi Beta Kappa, Sigma Xi, Phi Lambda Upsilon. Subspecialties: Organic chemistry; Reaction mechanisms. Current work: Study of the steps by which organic chemical reactions occur, as determined by molecular structure, arrangement of atoms in space, catalysts, light, etc. Research kinetics and mechanism organic reactions. Office: Dept Chemistry Texas Christian Univ Fort Worth TX 76129

BARTLETT, PETER GREENOUGH, engrng. co. exec.; b. Manchester, N.H., Apr. 22, 1930; s. Richard Cilley and Dorothy (Pillsbury) B.; m. Jeanne Eddes, July 8, 1954; children: Peter G., Marta, Lauren, Karla, Richard E. Ph.B., Northwestern U., 1955. Engr. Westinghouse Electric Co., Balt., 1955-58; mgr. mil. communications Motorola, Inc., Chgo., 1958-60; pres. Bartlett Labs., Inc., Indpls., 1960-63; assoc. prof. elec. engrng. U. S.C., Columbia, 1963-64; dir. research Eagle Signal Co., Davenport, Iowa, 1964-67; div. mgr. Struthers-Dunn, Inc., Bettendorf, Iowa, 1967-74; pres. Automation Systems, Inc., Eldridge, Iowa, 1974—. Served with U.S. Navy, 1952-54. Mem. IEEE. Republican. Presbyterian. Subspecialties: Electronics; Computer architecture. Over 50 patents in field. Office: 208 N 12th Ave Eldridge IA 52748

BARTON, DONALD WILBER, plant breeder; b. Fresno, Calif., June 12, 1921; m., 1944; 4 children. B.S., U. Calif., 1946, Ph.D., 1949. Asst. in genetics U. Calif., 1946-49; asst. prof. U. Mo., 1950-51; assoc. prof. Cornell U., Ithaca, N.Y., 1951-59, prof. vegetable crops, 1959—, head dept., 1959-60; dir. N.Y. State Agr. Expt. Sta., Geneva, 1960—; assoc. dir. research Coll. Agr., Geneva, 1960—; vis. prof. Oreg. State Coll., 1959-60. AEC fellow U. Mo., 1949-50; research grantee genetics, 1950-51; research grantee horticulture Northwest Canners and Freezers Assn., 1959—. Mem. Genetics Soc. Am., Am. Soc. Hort. Sci. Subspecialty: Plant genetics. Office: Dept Vegetable Crops NY State Agr Expt Sta Geneva NY 14456

BARTON, EDWARD JAMES, psychologist, consultant; b. Monroe, Mich., June 24, 1950; s. Aubrey and Evelyn (Masserant) B. B.A., Western Mich. U., 1972, M.A., 1973; Ph.D., Utah State U., 1978. Psychologist North Central Mich. Mental Health Center, Cadillac, 1973-75; behavior cons. Behavioral Mgmt. Co., Logan, Utah, 1975-76; assoc. prof. psychology No. Mich. U., 1977-82; psychologist Drake Beam Morin, Birmingham, Mich., 1982—; research and devel. of new approach to social skills tng., 1976. Author: Direct Observation, 1983; contbr. articles to profl. jours.; editorial bd.: Behavior Modification, 1982—; editorial cons. for books, 1979—. Recipient Research Council award Utah State U, 1976; Faculty merit award No. Mich. U., 1978, 79, 81; Faculty Research award, 1979 (2). Mem. Assn. Behavior Analysis (area resource person 1978—), Internat. Soc. Individual Instruction (dir. 1982—), Assn. Advancement Behavior Therapy, Am. Psychol. Assn. (Teaching of Psychology award 1981). Roman Catholic. Subspecialties: Behavioral psychology; Industrial/organizational psychology. Current work: Remediation of business problems, maximizing profitability, and outplacement through assessment, guidance, and management of human resources. Office: Drake Beam Morin 30300 Telegraph Rd Suite 176 Birmingham MI 48010

BARTON, RONALD WINSTON, architect-engineering company official; b. Middletown, N.Y., Dec. 31, 1942; s. Arthur Preston and Inez Lucy (Romanelli) B.; m.; stepson Richard Kipp Cox; m. Judith Ann Pickard, July 17, 1965; children: Elizabeth Pickard, Jennifer Lynn. B.M.E., Villanova U., 1964; M.S. in Mech. Engring, Rensselaer Poly. Inst., 1968. Registered profl. engr., Pa., N.Y., Conn., R.I., Maine. Advanced systems engr. United Techs. Corp., Windsor Locks, Conn., 1964-69; sr. staff engr. Combustion Engrng. Co., Windsor, Conn., 1969-73; staff engrng. supr. United Engrs. & Constructors, Inc., Phila., 1973-75, project engrng. supr., 1975-80, project engrng. mgr., 1980-82, project mgr., 1982—. Mem. Am. Nuclear Soc., ASME, Am. Mgmt. Assns., Am. Nat. Standards Inst. (com. B16H). Roman Catholic. Subspecialties: Nuclear fission; Nuclear engineering. Current work: Overall project management for engineering/design/consulting services provided to operating nuclear power plants for improvements to safety/availability/performance. Home: 715 W Prospect Ave North Wales PA 19454 Office: United Engrs & Constructors Inc 30 S 17th St Philadelphia PA 19101

BARTTER, FREDERIC CROSBY, physician, med. research adminstr.; b. Philippine Islands, Sept. 10, 1914; U.S., 1927, naturalized, 1935; s. George Charles and Frances Crosby (Buffington) B.; m. Jane H. Lillard, May 25, 1946; children—Frederic Crosby, Thaddeus, Pamela Anne. A.B., Harvard Coll., 1935, M.D., 1940. Diplomate: Am. Bd. Internal Medicine. Intern Roosevelt Hosp., N.Y.C., 1941-42; research and clin. fellow in medicine Harvard Med. Sch. and Mass. Gen. Hosp., Boston, 1946-48; clin. investigator, specializing in endocrinology, 1946—; commd. med. officer USPHS, HEW, 1942; dir. labs. USPHS Hosp., Sheepshead Bay, N.Y., 1942-44; med. officer in charge onchocerciasis investigation Pan Am. San. Bur., 1944-45; mem. staff Lab. Tropical Diseases, NIH, 1945-46; tutor biochem. scis. Harvard U., 1946-51; chief endocrinology br. Nat. Heart, Lung and Blood Inst., NIH, Bethesda, Md., 1951-73, clin. dir., 1970-76, chief hypertension-endocrine br., 1973-78; asso. prof. pediatrics Howard U., Washington, 1958-64, prof. pediatrics, 1965-78; clin. prof. medicine Georgetown U., Washington, 1965-78; asso. chief of staff for research Audie L. Murphy Vets. Hosp., 1978—; prof. medicine U. Tex. Health Sci. Center, San Antonio, 1978—. Contbr. over 300 articles on endocrinology and physiology to sci. jours. Recipient Meritorious Service medal NIH, 1970, Modern Medicine's Disting. Achievement award, 1977; Churchill fellow, Cambridge, Eng., 1968-69. Fellow (hon.) Am. Coll. Cardiology, Royal Coll. Physicians; mem. Endocrine Soc. (Fred C. Koch award 1978), Am. Physiol. Soc., Assn. Am. Physicians, Am. Soc. Clin. Investigation, Salt and Water Club, Royal Soc. Medicine, Nat. Acad. Sci., Peripatetic Club, N. Am. Mycological Assn. Anglican. Subspecialty: Endocrinology. Home: 227 Primrose Pl San Antonio TX 78209 Office: 7400 Merton Minter Blvd San Antonio TX 78284

BARTUS, RAYMOND T., neuroscientist, writer, lectr.; b. Chgo., May 19, 1947; s. Frank A. and Katherine (Bogus) B.; m. Cheryl Gyure, Feb. 11, 1967; children—Raymond T., Kristin Marie. B.A. in Psychology with honors, California (Pa.) State Coll., 1968; M.S. in Exptl. Psychology (NIH predoctoral asst.), N.C. State U., 1970; Ph.D. in Physiol. Psychology (NASA pre-doctoral fellow), N.C. State U., 1972. Postdoctoral asso. Naval Med. Research Lab., Groton, Conn., 1972-73; scientist Warner-Lambert/Parke-Davis Research Labs., Ann Arbor, Mich., 1973-77, sr. scientist, 1977-78; sr. scientist CNS research Med. Research div. Am. Cyanamid Co., Pearl River, N.Y., 1978, dir. geriatric discovery program, 1979—, group leader behavioral neurosci., 1980—; adj. prof. psychiatry N.Y. U., 1980—; research affiliate Tulane U. and Delta Regional Primate Research Center, Covington, La., 1978—; adj. asst. prof. Conn. Coll., 1973; referee, program advisor, spl. sci. cons. NSF, NIMH, Nat. Inst. Aging, Fedn. Am. Socs. Exptl. Biology; research cons.; speaker profl. cons. Editor in chief: Neurobiology of Aging; Experimental and Clinical Research, 1981; editorial adv. bd., referee: Jour. Gerontology, 1981—, Exptl. Aging Research, 1981-82, Pharmacology, Biochemistry and Behavior, 1981-82; referee: jours. Behavioral and Neurol Biology: An Interdisciplinary Jour; sci. adv. bd.: Anti-Aging News, 1980—; contbr. articles to profl. jours. Nat. Research Council postdoctoral fellow, 1972-73; recipient Exceptional Sr. Research award Calif. State Coll., 1968. Mem. Soc. Neurosci., Gerontol. Soc., Am. Aging Assn. (dir.), Am. Psychol. Assn., Sigma Xi, Psi Chi. Subspecialties: Neuropharmacology; Gerontology. Current work: Neurophysiology, understanding changes in brain with age and how to correct, psychopharmacology, cognition, learning, neurochemistry, psychiatry, neuropsychology, psychobiology, physiol. psychology. Office: American Cyanamid Co Medical Research Div Pearl River NY 10965

BARZILAI, JONATHAN, mathematician; b. Tel-Aviv, June 2, 1945; s. Zechariah and Ruhama (Zairi) B.; m. Rachel Taasse, Aug. 10, 1972; 1 son, Dan. B.Sc., Technion-IIT, Haifa, 1967, M.Sc., 1978, D.Sc., 1980. Instr. math. U. Tex.-Austin, 1981-82; asst. prof. Faculty Adminstrv. Studies, York U., Downsview, Ont., Can., 1982-83, Sch. Bus. Adminstrn. and dept. math. Dalhousie U., Halifax, N.S., Can., 1983—. Served with Israel Def. Forces, 1967-75. Mem. Ops. Research Soc. Am. Subspecialties: Algorithms; Numerical analysis. Current work: Analysis and design of algorithms for unconstrained minimization and solution of systems of equations. Office: Dalhousie U. Sch Bus 6152 Coburg Rd Halifax NS Canada B3H 1Z5

BASAVATIA, RAM A., med. radiation physicist; b. Burma, July 8, 1941; came to U.S., 1971, naturalized, 1976; s. Govind Ram and Rukmani (Devi) B.; m. Nirmala Ladsaria, Jan. 19, 1964; children: Nilam, Amar, Indira. M.S. in Physics, Western Ill. U., 1972. Cert. radiol. physicist Am. Bd. Radiology. Lectr. physics Rangoon U., Burma, 1962-71; grad. asst. Western Ill. U., 1971-72; radiation physicist VA Hines (Ill.) Hosp., 1972-73, St. Francis Hosp., Evanston, Ill., 1973-75; radiation physicist, radiation safety officer St. Joseph Hosp., Joliet, Ill., 1975—; cons. physicist hosp., radiation therapy clinics. Mem. Am. Coll. Radiology, Am. Assn. Physicists in Medicine. Subspecialty: Medical physics. Current work: Dosimetry, high energy medical linear accelerator, radiological physics services for diagnostic and therapeutic radiology fields. Home: 2405 Ingalls St Apt 13 Joliet Il 60435 Office: 333 N Madison St Joliet Il 60435

BASCOM, WILLARD NEWELL, scientist, engineer; b. N.Y.C., Nov. 7, 1916; s. Willard Newell and Pearle (Boyd) B.; m. Rhoda Nergaard, Apr. 15, 1946; children: Willard, Anitra. Grad., Colo. Sch. Mines, 1942. Registered profl. engr., Fla., D.C. Research engr. U. Calif.-Berkeley, 1945-50, Scripps Inst. Oceanography, 1950-54; exec. sec. Nat. Acad. Scis., Washington, 1954-62; pres. Ocean Sci. & Engring., Inc., Washington, 1962-72; dir. Coastal Water Research Project, Long Beach, Calif., 1973—; cons. to govt. and industry. Author: Waves and Beaches, 1964, A Hole in the Bottom of the Sea, 1961, Deep Water, Ancient Ships, 1976, over 100 articles. Recipient Disting. Achievement medal Colo. Sch. Mines, 1979; Compass Disting. Achievement award Marine Tech. Soc., 1970. Clubs: Explorers (Explorers medal 1980), Adventurers. Subspecialties: Oceanography; Archaeology. Current work: Coastal ecology, archaeology. Patentee deep ocean search/recovery system.

BASEHORE, KERRY LEE, nuclear engineer, consultant; b. Hershey, Pa., Aug. 24, 1953; s. Kenneth Leroy and Jeannette Marie (Tietsworth) B.; m. Betsy Ann Chamberlain, June 14, 1975; children: Kenneth Lawrence, Carolyn Joyce, Laura Elizabeth. B.S., Pa. State U., 1975; M.S., M.I.T., 1977, Nuclear Engr., 1977. Registered profl. engr., Va. Devel. engr. Battelle N.W. Labs., Richland, Wash., 1977-80; sr. engr. Va. Electric & Power Co., Richmond, 1980—; cons. Westinghouse-Hanford Co., Richland, Wash., 1982. Tchr. Chester United Meth. Ch., Chester, Va., 1981-82. Mem. Am. Nuclear Soc. (thermal hydraulic program com. 1981—), Nat. Soc. Profl. Engrs., Sigma Xi. Republican. Methodist. Club: Engineers (Richmond). Subspecialties: Nuclear engineering; Numerical analysis. Current work: Devel. and application of core thermal hydraulic methods, both in light water reactor and liquid metal reactor areas, reload safety analysis, LOCA analysis. Home: 10118 Remora Dr Richmond VA 23237 Office: Va Electric & Power Co PO Box 26666 Richmond VA 23261

BASERGA, RENATO LUIGI, pathology educator; b. Meda, Milan, Italy, Apr. 11, 1925; came to U.S., 1949; s. Alessandro and Giuseppina (Annoni) B.; m. Beverly Lange, Oct. 12, 1945 (div.); m. Jane Conrad, Dec. 23, 1954 (div.); children: Susan Jane, Janice Rene. M.D., U. Milan, Italy, 1949; postgrad., Northwestern U., 1956-58. Diplomate: Am. Bd. Pathology. Resident U. Milan, 1949-51; intern Columbus Hosp., Chgo., 1952, 53; resident St. Luke's Hosp. Dept. Pathology, Chgo., 1955-58; assoc. oncology Chgo. Med. Sch., 1953-54; instr. pathology Northwestern U. Med. Sch., Chgo., 1958-60; cons. Argonne (Ill.) Nat. Lab., 1959-65; asst. prof. pathology Northwestern U. Med. Sch., Chgo., 1960-64, assoc. prof., 1964-65; sr. investigator Fels Research Inst., Temple U., Phila., 1965—; prof. pathology Temple U. Sch. Medicine, Phila., 1965—, prof., chmn., 1968-73, 80—. Author: Autoradiography: Techniques and Applications, 1969, Multiplication and Division in Mammalian Cells, 1976; editor: The Cell Cycle and Cancer, 1971. Served with Vol. Forces, 1943-45; Italy. Recipient Maria Antonietta Della scholarship, Milan, 1951; sr. research fellowship USPHS, 1958-60; Research Career Devel. award, 1960-65; Louis Gross Meml. lectr., 1974; Searle lectr. Brit. Soc. Cell Biology, 1976. Mem. Am. Soc. Biol. Chemists. Subspecialties: Cell and tissue culture; Molecular biology. Current work: Genetics of cell proliferation, the identification genes and gene products that control the proliferation of animal cells. Office: Dept Pathology Temple U Med Sch 3400 N Broad St Philadelphia PA 19140

BASFORD, ROBERT EUGENE, biochemistry educator, researcher; b. Montpelier, N.D., Aug. 21, 1923; s. Eugene M. and Bertha (Cudworth) B.; m. Carol Kaufman Phebus, Dec. 23, 1965; 1 son, Lee A. Phebus. B.S., U. Wash., 1951, Ph.D., 1954. Postdoctoral fellow U. Wis.-Madison, 1954-58; asst. prof. U. Pitts., 1958-63, assoc. prof., 1963-70, prof., 1970—; cons. Mine Safety Appliance Co., Pitts., 1966-69; mem. neurol. scis. study sect. NIH, Washington, 1977-80. Mem. Am. Soc. Biol. Chemists, Reticuloendothelial Soc., Am. Soc. for Neurochemistry. Subspecialties: Biochemistry (medicine); Cell biology (medicine). Current work: Biochemical basis of phagocytosis and microbicidal activity, bioenergetics of the central nervous system. Office: U Pitts Sch Medicine Biochemistry Dept Pittsburgh PA 15261

BASH, FRANK NESS, astronomer, educator; b. Medford, Oreg., May 3, 1937; s. Frank Cozad and Kathleen Jane (Ness) B.; m. Susan Martin Fay, Sept. 10, 1960; children: Kathryn Fay, Francis Lee. B.A., Willamette U., 1959; M.A. in Astronomy, Harvard U., 1962, Ph.D., U. Va., 1967. Staff scientist Lincoln Lab., M.I.T., 1962; asso. astronomer Nat. Radio Astronomy Obs., Green Bank, W.Va., 1962-64; research asst. U. Va., 1965-67; postdoctoral faculty asso. U.Tex., Austin., 1967-69; asst. prof. astronomy, 1969-73, asso. prof., 1973-81, prof., 1981—, chmn. dept. astronomy, 1982—. Author: (with Daniel Schiller and Pilip Balamore) Astronomy, 1977; contbr. articles to profl. jours. NSF grantee, 1967—; Netherlands's NSF grantee, 1979. Mem. Am. Astron. Soc., Internat. Astron. Union, Internat. Sci. Radio Union, Tex. Assn. Coll. Tchrs. (pres. U. Tex. chpt. 1980-82). Club: Town and Gown (Austin). Subspecialties: Radio and microwave astronomy; Theoretical astrophysics. Current work: Research on process of star formation in spiral galaxies. Office: Dept Astronomy U Tex Austin TX 78712

BASHEY, REZA ISMAIL, med. biochemist, researcher; b. Bombay, India, Aug. 28, 1933; came to U.S., 1955, naturalized, 1967; s. Ismail H. and Rokya (Busheri) B.; m. Hildegarde Erika Moesaner, Apr. 15, 1961; children: Parveen, Ali, Farrah. B.S., U. Bombay, 1952, M.S., 1954; Ph.D., Rutgers U., 1958. Postdoctoral fellow U. So. Calif., Los Angeles, 1958-59, U. Miami, Fla., 1960-62; sci. biol. officer Indian Council Med. Research, New Delhi, 1962-64; asst. prof. Albert Einstein Coll. Medicine, Bronx, N.Y., 1964-70; asst. prof. assoc. prof. Hahnemann Med. Coll., Phila., 1970-75; asst. prof., assoc. prof. U. Pa., Phila., 1975—. Contbr. articles on biochemistry to profl. jours. NIH investigator, 1971-74, 1975-78. Mem. AAAS, Am. Chem. Soc., Am. Assn. Pathologists, Am. Soc. Biol. Chemists. Subspecialty: Biochemistry (medicine). Current work: Biochemistry of normal and diseased connective tissues, especially collagen in rheumatic diseases such as scleroderma andarthritis. Home: 4055 Center Ave Lafayette Hill PA 19444 Office: U Pa 573 Maloney Bldg Philadelphia PA 19104

BASIC, JOHN NICHOLAS, SR., engring. co. exec., mech. engr.; b. Chgo., Dec. 16, 1923; s. Marin and Mary (Lucin) B.; m. Marijo C. Coleman, May 20, 1950; children: Cathe Ostrowski, John Nicholas, Sarah Ann, Margaret Mary, Kerry Eileen, Laura Rene. B.S.M.E., Ill. Inst. Tech., 1947, postgrad., 1948-51; student law, Loyola U., Chgo., 1956, 59. Registered profl. engr., Ill. Jr. engr. U. Chgo. Cyclotron Project, 1948; jr. engr to mgr. maintenance Joanna Western Mills Co., Chgo., 1949-56, dir. engring., 1959-69; chief engr. Mt. Hope Machinery Co., Taunton, Mass., 1956-59; pres., owner Basic Environ. Engring., Inc., Glen Ellyn, Ill., 1970—. Served with USNR., 1944-46; PTO. Grantee Pollution Engring. Mag., 1977, 78, 79. Mem. ASME, We. Soc. Engrs., Air Pollution Control Assn. Republican. Roman Catholic. Club: Glen Ellyn Country. Subspecialties: Combustion processes; Biomass (energy science and technology). Current work: Design and build systems that burn wide spectrum of wastes with same equipment; construct solid waste water wall combustion systems to burn varied wastes, extract higher energy efficiencies and with environ. acceptable standards. Patentee in field. Home: 41 W 202 Whitney Rd Saint Charles IL 60174 Office: 21 W 161 HIll Ave Glen Ellyn IL 60137

BASILICO, CLAUDIO, genetist, educator; b. Milan, Italy, Feb. 7, 1936; came to U.S., 1967; s. Vittorio and Enrica (Belloni) B.; m. Mariapia Casartelli, Oct. 7, 1961; children: Stefano, Francesca, Enrica. M.D., U. Milan, 1960. Vis. research fellow div. biology Calif. Inst. Tech., 1962; staff Internat. Lab. Genetics and Biophysics, Naples, Italy, 1963-66; research asso. dept. cell biology Albert Einstein Coll. Medicine, 1966-67; vis. research scientist dept. pathology N.Y. U. Sch. Medicine, N.Y.C., 1967-69, research asst. prof., 1969, asso. prof., 1970-75, prof. dept. pathology, 1975—; mem. Am. Cancer Soc. adv. com. on cellular and developmental biology, 1979-82. Contbr. articles to profl. jours. Trustee Cold Spring Harbor Lab., 1981—. Mem. Am. Soc. Microbiology, Am. Soc. Cell Biology, Am. Assn. for Cancer Research, Am. Soc. Virology. Subspecialties: Genome organization; Virology (biology). Current work: Viral oncology, cell genetics, mechanisms gene expression. Office: 550 First Ave New York NY 10016

BASINSKA-LEWIN, EWA MARIA, astrophysicist; b. Warsaw, Poland, Apr. 22, 1947; came to U.S., 1979; d. Joseph and Wiktoria Maria (Malinowska) Basinski; m. Pawel Grzesik, July 12, 1972; m. Walter Hendrik Gustav Lewin, Jan. 2, 1981. M.S. in Physics summa cum laude, Copernicus U., Torun, Poland, 1970, Ph.D., 1977. Research asst. N. Copernicus Astronom. Center, Torun, Poland, 1970-77, research asso., 1978-79; vis. scientist Mass. Inst. Tech., Cambridge, 1979-81; research asso. U. Amsterdam, The Netherlands, 1981-82; staff scientist M.I.T., 1982; research physicist Regis Coll., Weston, Mass., 1982—. Contbr. articles to profl. jours. Recipient N. Copernicus U. award for best student, 1969; Polish Acad. Scis. award for work in galactic astronomy, 1979. Mem. Am. Astronom. Soc., Polish Astronom. Soc. Subspecialties: 1-ray high energy astrophysics; High energy astrophysics. Current work: Analysis of x-ray data obtained with x-ray satellites: Small Astronom. Satellite No. 3 and High Energy Astrophysical Observatory No. 1; analysis of plasma density in the ionosphere.

BASKIN, DAVID, psychologist, researcher; b. N.Y.C., May 2, 1947; s. Aharon and Rachel B. B.A., Yeshiva U., N.Y.C., 1968; M.A., New Sch. for Social Research, N.Y.C., 1971; Ph.D., CUNY, 1977. Lic. psychologist, N.Y. Lectr. Bklyn. Coll., 1974-76; dir. research and evaluation Raritan Bay Community Mental Health Ctr., Perth Amboy, N.J., 1977-78; dir. program evaluation Bronx-Lebanon Hosp., 1978-82; asst. prof. Albert Einstein Coll Medicine, Bronx, 1979—; assoc. dir. Soundview/Throgs Neck Community Mental Health Ctr., 1982—. Contbr. articles to profl. jours. Bd. dirs., v.p. Whitehall Commons, Ltd., Bronxville, N.Y., 1981-82. Mem. Am. Psychol. Assn., Am. Pub. Health Assn., N.Y. State Program Evaluators Assn. (pres. 1982-83), N.Y.C. Program Evaluators Orgn. (pres. 1980-82). Subspecialties: Psychological evaluation and measurement; Epidemiology. Current work: Epidemiology of mental illness; cross-cultural percetions and treatment of mental illness; program evaluation in mental health programs. Office: Albert Einstein Coll Medicine 2527 Glebe Ave Bronx NY 10461

BASKIN, STEVEN IVAN, cardiovascular toxicologist, pharmacologist, scientist; b. Los Angeles, Nov. 14, 1942; s. Louis A. and Rose (Wasserman) B.; m. Eileen A. Cooper, Aug. 15, 1965; children: Marcy, Lloyd. Pharm. D., U. So. Calif., 1966; Ph.D., Ohio State U., 1971. Registered pharmacist, Calif., Nev., Pa. Pharmacy aide Brentwood N.P. Hosp. U.S. VA Hosp., Los Angeles, 1960-62; pharmacy intern Los Angeles County Gen. Hosp., 1965-66; cons. chemist Theodore Hamm's Brewing Co., Los Angeles, 1966; teaching asst. natural products chemistry U. So. Calif., 1966; teaching asst. Ohio State U., 1966-67; research assoc. pharmacology Mich. State U., 1971-73; demonstrator Med. Cardiovascular Lab., 1973; asst. research prof Med. Coll. Pa., 1973-74, asst. prof., 1976-79, assoc. prof., 1979-82; team leader cardiac pathophysiology U.S. Army Med. Research Inst., Aberdeen Proving Grounds, Md., 1982—; reviewer grants NSF; cons. in field. Author: (with others) The Effects of Taurine on Excitable Tissues, 1981; editor: Neural Regulator Mechanisms During Aging, 1980; contbr. numerous articles to profl. jours. NSF fellow, 1967-66; NIH fellow, 1967-71; grantee, 1978—; Del Heart Assn. grantee, 1981-82; recipient Pharm. Mfrs. Assn. award, 1975-76, 75-77. Fellow Am. Coll. Clin. Pharmacology, Am. Coll. Cardiology; mem. Internat. Study Group for Research in Cardiac Metabolism, Am. Heart Assn., AAAS, Am. Chem. Soc., Am. Fedn. Clin. Research, Phila. Physiol. Soc., Gerontology Soc., Am. Soc. Pharmacology and Exptl. Therapeutics, Toxicology Soc., Alpha Epsilon Delta, Sigma Xi, Rho Chi. Subspecialties: Pharmacology; Toxicology (agriculture). Current work: Toxicology of cyanide, development of new cyanide antidotes; studies on the mechanism of cyanide antidotes. Patentee lyophilic waste disposal. Home: 608B Harborside Dr Joppatowne MD 21085 Office: SRRD-UV-RY U.S. Army Medical Research Institute Aberdeen Proving Grounds MD 21010

BASKIN, STEVEN MARC, clinical psychologist; b. N.Y.C., Oct. 19, 1949; s. Nathan and Vivian Francine (Kaplan) B. B.A., U. Va.-Charlottesville, 1971, M.Ed., 1973; Ph.D., North Tex. State U., 1979. Psychology intern Ohio State U. Med. Ctr., Columbus, 1976-77; dir. mental health, instr. psychology Pa. State U., Erie, 1977-79; dir. behavioral medicine New Eng. Ctr. for Headache, Cos Cob, Conn., 1979—; pvt. practice clin. psychology, Cos Cob, 1980—; cons. Greenwich (Conn.) Sch. System, 1982-83. Author: Manual of Behavioral Medicine, 1983; contbr. articles in field to profl. jorus. Pa. State U. scholarly activities grantee, 1978-79. Mem. Am. Psychol. Assn., Biofeedback Soc. Am., Am. Assn. for Study Headache. Jewish. Subspecialties: Biofeedback; Clinical psychology. Current work: Psychological treatment of primary headache disorders, family psychotherapy, clinical psychophysiology. Home: 159 E Elm St Greenwich CT 06830 Office: New Eng Ctr for Headache 40 E Putnam Ave Cos Cob CT 06807

BASOLO, FRED, chemistry educator; b. Coello, Ill., Feb. 11, 1920; s. John and Catherine (Marino) B.; m. Mary P. Nutley, June 14, 1947; children: Mary Catherine, Freddie, Margaret Ann, Elizabeth Rose. B.E., So. Ill. U., 1940; M.S., U. Ill., 1942, Ph.D. in Inorganic Chemistry, 1943. Research chemist Rohm & Haas Chem. Co., Phila., 1943-46; mem. faculty Northwestern U., Evanston, Ill., 1946—, prof. chemistry, 1958—, chmn. dept. chemistry, 1969-72, Morrison prof., 1980—; cons. and lectr. in field. Co-author: (with R.G. Pearson) Mechanisms of Inorganic Reactions, 1958, (with R.C. Johnson) Coordination Chemistry, 1964; assoc. editor: Inorganica Chemica Acta, 1967—, Inorganic Chemica Acta Letters, 1977—; mem. editorial bds. of numerous profl. jours.; contbr. articles to profl. jours. Mem. adv. bd. Who's Who in America. Recipient Bailar Medal award, 1972; So. Ill. U. Alumni achievement award, 1974; Dwyer Medal award, 1976; James Flack Norris award outstanding achievement in teaching chemistry, 1981; Oesper Meml. award, 1983; Guggenheim fellow U. Copenhagen, 1954-55; Sr. NSF fellow U. Rome, 1962; NATO disting. prof. Tech. U. Munich, 1969; Japanese Soc. Promotion of Sci. fellow, 1979; NATO Sr. Scientist fellow, Italy, 1981. Fellow AAAS; mem. Nat. Acad. Scis., Am. Chem. Soc. (bd. dirs. 1982—, pres. 1983, inorganic chemistry research award 1964, citation for excellence 1971, Disting. Service award 1975), Chem. Soc. (London), Italian Chem. Soc. (hon.), Sigma Xi, Alpha Chi Sigma, Kappa Delta Phi, Phi Lambda Upsilon (hon.). Subspecialties: Inorganic chemistry; Nuclear chemistry. Office: Dept Chemistry Northwestern U Evanston IL 60201

BASS, EUGENE LAWRENCE, physiologist, zoology educator; b. Bklyn., Sept. 27, 1942; s. Morris S. and Minnie (Ackerman) B.; m. Linda Jean Epstein, June 16, 1968; 1 dau., Elena Lisa. B.S., CUNY, 1964; Ph.D., U. Mass., 1970. Asst. prof. biology Salisbury (Md.) State Coll., 1969-72; asst. prof. biology U. Md., Princess Anne, 1972—. N.Y. State regents scholar, 1960-64; U.S. Dept. Agr. grantee, 1973-78; NIH grantee, 1981-84. Mem. Am. Soc. Zoologists, Am. Inst. Biol. Scis., Atlantic Estuarine Research Soc., Sigma Xi. Subspecialties: Comparative physiology; Toxicology (agriculture). Current work: Toxicology of pollutants and natural products especially as they affect the nervous and muscular systems of vertebrate and invertebrate species. Office: U Md Princess Anne MD 21853

BASS, HYMAN, mathematician, educator; b. Houston, Oct. 5, 1932; s. Isador and Fanny (Weiss) B.; m. Mary Ellen Popkin, June 9, 1957 (div. 1978); children: Anne Ruth, Ivan Philip; m. Dorothea Henriette Goldys, Nov. 1, 1979; 1 dau., Gabriella Sierra. B.A., Princeton U., 1955; M.S., U. Chgo., 1956, Ph.D. (NSF grad. fellow), 1959. Ritt instr. math. Columbia U., 1959-62, asst. prof., 1963-64, chmn. dept. math., 1975—; asso. prof., chmn. at Barnard Coll., 1964-65, prof., 1965—; Vis. mem. Inst. Advanced Study, Princeton, 1964, 65-66, Inst. de Hautes Etudes Scientifiques, Paris, 1968-69; vis. prof. Universidad Nacional Autónoma de Mex., 1965, Tata Inst. Fundamental Research, Bombay, 1965-66, 69, 76, 80, U. Paris, 1968, 73, 81, Cambridge U., 1973, Instituto de Matematica Pura e Applicada, Rio de Janeiro, 1977, Bar Ilan U., Israel, 1980; chmn. adv. com. pure mathematics NRC, 1970-71; adv. panel, div. math. NSF, 1973-75. Editorial bd.: Jour. Indian Math. Soc, 1968—, Cambridge Tracts in Pure and Applied Mathematics; 968: Jour. Pure and Applied Algebra, 1970—, Am. Jour. Mathematics, 1971—, North-Holland Math. Library, 1971—, Acad. Press Series in Pure and Applied Math, 1974—. NSF fellow Coll. de France, 1962-63; Sloan fellow, 1964-66; Guggenheim fellow, 1968-69; recipient Van Amridge book prize Columbia, 1969, Cole prize Am. Math. Soc., 1975. Mem. Am. Math Soc. (editorial bd. 1969—, council 1969-72), London Math. Socs., Societé Mathématique de France, Soc. Collaborateurs N. Bourbaki, Math. Assn. Am., AAAS., Am. Acad. Arts and Scis., Nat. Acad. Scis. Subspecialty: Algebra. Home: 435 Riverside Dr New York NY 10025

BASS, MICHAEL, electrical engineer, educator, administrator, consultant; b. N.Y.C., Oct. 24, 1939; s. Reuben Herman and Mary (Oboler) B.; m. Judith Harriette Rubin, June 30, 1962; children: George, Meredith. B.S. in Physics, Carnegie Mellon U, 1960; M.S., Ph.D., U. Mich., 1964. Acting asst. prof. U. Calif.-Berkeley, 1964-66; sr. research scientist Raytheon Research div., Waltham, Mass., 1966-73; assoc. dir. Ctr. for Laser Studies, U. So. Calif., Los Angeles, 1973-77, dir., 1977—, prof. elec. engring., 1981—; cons. in field. Contbr. numerous articles on physics and lasers to profl. jours. Named disting. researcher U. Soc. Calif. Sch. Engring., 1979. Fellow IEEE, Optical Soc. Am.; mem. Laser Inst. Am. (bd. dirs.). Subspecialties: Spectroscopy; Laser medicine. Current work: Studies of the optical properties of matter. Office: Ctr for Laser Studies U So Calif Los Angeles CA 90089-1112

BASSFORD, THOMAS HARVEY, mettalurgist; b. Huntington, W. Va., Apr. 25, 1927; s. Clifford Dell and Ethel Virginia (Knootz) B.; m. Joan Turley, June 1952; children: Timothy, Andrew, Janet, Colleen. B.S.M.E., Ga. Inst. Tech., 1952; M.S., U. Cin., 1976. Registered profl. engr., W.Va. Designer A.J. Boynton Cons., Chgo., 1952; metallurgist Huntington Alloys, Inc., 1952-63, sr. metallurgist, 1963-73, chief mech. test engr., 1973—. Contbr. articles to profl. jours. Served with USAF, 1945-46. Mem. ASME, Am. Soc. Metals, Soc. Exptl. Stress Analyst. Republican. Methodist. Club: Esquire Golf (Huntington). Subspecialties: High-temperature materials; Alloys. Current work: Alloy development. Patentee in field. Home: 779 Eastwood Dr Huntington WV 25705 Office: Huntington Alloys Inc PO Box 1958 Huntington WV 25721

BASSHAM, JAMES ALAN, biochemist; b. Sacramento, Nov. 26, 1922; m., 1956; 5 children. B.S., U. Calif., 1945, Ph.D., 1949. Dept. head chem. biodynamics div. Lawrence-Berkeley Lab., U. Calif.-Berkeley, 1976-81, chemist bio-organic chem. group, 1949-77, sr. staff chemist, 1978—; lectr. chemistry U., 1957-59, adj. prof. biochemistry, 1972-80; vis. prof. dept. biochemistry, biophysics U. Hawaii, 1968-69. NSF sr. fellow Oxford U., 1956-57. Mem. Am. Chem. Soc., AAAS, Am. Soc. Biol. Chemists, Am. Soc. Plant Physiologists. Subspecialties: Genetics and genetic engineering (biology); Cell and tissue culture. Office: Bldg 3 Lawrence Berkeley Lab U Calif-Berkeley 1 Cyclotron Rd Berkeley CA 94720

BASSI, SUKH DEV, microbiologist; b. Kericho, Kenya, Feb. 11, 1941; came to U.S., 1963, naturalized, 1975; s. Telu R. and Vidya B. (Gug) B.; m. Jane Gempler, Aug. 21, 1971; children: Neal, Nathah, Sean. B.A., Knox Coll., 1965; M.S., St. Louis U., 1968, Ph.D., 1970. Prof. biology Benedictine Coll., Atchison, Kans., 1971-81; tech. dir. Midwest Solvents Co., Atchison, 1981—; cons. Clark Coll., Atlanta, 1971—. Chmn. Sunflower dist. Boy Scouts Am., Atchison, 1980. Paul Harris fellow, 1982. Mem. Am. Soc. Microbiologists, Am. Chem. Soc., Assn. Am. Cereal Chemists, AAAS, C. of C. Atchison (dir. 1981, v.p. 1983), Sigma Xi. Lodges: Rotary (pres. 1979-80) (Atchison); Elks. Subspecialties: Biomass (energy science and technology); Enzyme technology. Current work: Biomass conversion to chemicals; genetic engineering; simultaneous production of glucose and alcohol from cellulose. Home: Rural Route 3 Box 159B Atchison KS 66002 Office: Midwest Solvents Co Inc 1300 Main St Atchison KS 66002

BAST, ROBERT CLINTON, JR., physician; b. Washington, Dec. 8, 1943; s. Robert Clinton and Ann Christine (Borl) E.; m. Blanche Amy Simpson, Oct. 21, 1972; 1 dau.: Elizabeth Simpson. B.A. cum laude, Wesleyan U., Middletown, Conn., 1965; M.D. magna cum laude, Harvard U., 1971. Intern dept. medicine Johns Hopkins Hosp., Balt., 1971-72; research assoc., research scientist biology br. Nat. Cancer Inst., NIH, Bethesda, Md., 1972-75; jr. asst. resident Peter Bent Brigham Hosp., Boston, 1975-76; clin. fellow to asst. physician Sidney Farber Cancer Inst., Boston, 1976—; jr. asso. to asso. physician Brigham and Women's Hosp., Boston, 1977—; asst. prof. medicine Harvard Med. Sch., Boston, 1977—; cons. Boston Hosp. for Women, 1979—, Nat. Cancer Inst., 1976—. Mem. editorial bd.: Internat. Jour. Immunopharmacology, 1979—, Jour. Biol. Response Modifiers, 1982—; contbr. articles to profl. jours. Served with USPHS, 1972-75. Recipient Henry Asbury Christian award Harvard Med. Sch., 1971; USPHS fellow and Eli Lilly student research fellow, 1967-69; Leukemia Soc. Am. scholar, 1978-83. Mem. Am. Assn. Cancer Research, Am. Assn. Immunologists, Am. Fedn. Clin. Research, Am. Soc. Clin. Oncology, Am. Soc. Microbiology, Internat. Assn. Immunopharmacology, Reticuloendothelial Soc., Sigma Xi, Alpha Omega Alpha. Subspecialties: Cancer research (medicine); Immunopharmacology. Current work: Tumor immunology, immunopharmacology, application of monoclonal antibodies to immunodiagnosis and immunotherapy of cancer. Home: 105 Arena Terr Concord MA 01742 Office: Sidney Farber Cancer Inst 44 Binney St Boston MA 02115

BASTAWI, ALY ELOUI, dental educator, consultant, researcher; b. Cairo, Oct. 13, 1928; U.S., 1961, naturalized, 1980; s. Abdel Ghafour and Amina Abdel Ghafour Mohammed; m. Khadiga Hamza, June 26, 1969; children: Akrum Eloui, Bassel Eloui. P.N.S., Cairo U., 1952, B.D.S., 1957; M.S.D., Ind. U.-Indpls., 1961; D.M.D., U. Louisville, 1975. Lectr. pedodontics Cairo U., 1965-69; asst. prof. Howard U., 1969-73, dir. postgrad. pedodontics, 1969-73; assoc. prof. U. Louisville, 1973-76, chmn. dept. pedodontics, 1976-79, prof., 1979—; dir. research, 1973-76; prof., cons. dental edn. Garyounis U., Benghazi, Libya, 1979-81, Cairo U., 1981-82. Contbr. articles to profl. jours. Internat. Coll. Dentists fellow, 1962. Mem. Am. Dental Assn., Am. Assn. Dental Schs., Am. Acad. Pedodontics, Internat. Assn. Dental Research, Egyptian Dental Assn., Omicron Kappa Upsilon. Subspecialties: Pedodontics; Psychophysiology. Current work: Dental education, consultant in dental education, research in the field of psychophysiology to study anxiety in dental patients through their autonomic responses. Home: PO Box 22496 Louisville KY 40222 Office: University of Louisville School of Dentistry Preston St and Muhammed Ali Blvd Louisville KY

BASTEDO, WILLIAM GARDNER, government executive; b. Asbury Park, N.J., Nov. 2, 1929; s. William O. and Kathleen R. (Gardner) B.; m. Mary Lou Martinez, June 7, 1958; children: Mary Ann, William Gardner Jr., Robert G., Margaret F. B.S. in Mil. Sci., U. Md., 1952; M.S. in Pub. Adminstrn., George Washington U., 1964. Capsule communicator NASA Johnson Space Ctr., Houston, 1965-68; chief instrumentation ships div. Western Test Range, Calif., 1968-70; dep. dir. plans USAF Space and Missile Test Orgn., 1970-72; officer in charge outer space affairs U.S. Dept. State, Washington, 1972-74; tech. intelligence Def. Intelligency Agy., Washington, 1974-75; dir. internat. program support Office Internat. Affairs NASA Hdqrs., 1975-81; program mgr. Space Tracking and Data Network, 1981—; instr. Golden Gate U., part-time, 1969-72, 73. Served to col. USAF, 1952-75. Decorated Air Force Commendation medal with cluster, Air Force Meritorious Service Medal, Legion of Merit, Joint Services Commendation medal; Recipient Exceptional Service medals NASA, 1979, 81; Outstanding Service award Skylab Reentry Interagy. Task Force, 1979. Roman Catholic. Lodge: K.C. Subspecialty: Aerospace engineering and technology. Office: Code TN NASA Headquarters Washington DC 20546

BASTIAN, JAMES WINSLOW, pharmaceutical company executive; b. Indpls., Apr. 17, 1926; s. Richmond Ellison and Helen Louise (Hackleman) B.; m. Helen Marie Hawkins, June 27, 1950; children: Thomas W., Keith W., Karen M. B.S., Purdue U., 1949, M.S., 1951, Ph.D., 1954. With Armour Pharm. Co., Kankakee, Ill., 1953—, dir. quality control, 1979, mgr. sect. quality control, 1979-82, mgr. sect. ops. services, 1982—. Contbr. numerous articles to profl. jours. Served with USNR, 1944-45. NSF fellow. Mem. Am. Soc. Pharmacology and Exptl. Therapeutics. Subspecialties: Molecular pharmacology; Endocrinology. Current work: Calcitonin: pharmaceutical, medical use aspects; drug screening. Patentee, inventor in field.

BATAY-CSORBA, PETER ANDREW, nuclear physicist; b. Budapest, Hungary, July 10, 1945; came to U.S., 1964; s. Alexander and Catherine (Rona) B.-C.; m. Judy Radvanyi, June 27, 1970; 1 dau., Suzy. B.S., M.I.T., 1968; Ph.D., Calif. Inst. Tech., 1975. Research asst. in nuclear physics Kellogg Lab., Calif. Inst. Tech., Pasadena, 1968-75; research assoc. in nuclear physics U. Colo., Boulder, 1975-79; research nuclear physicist Schlumberger-Doll Research Ctr., Ridgefield, Conn., 1979—. Contbr. articles to profl. Jours. Ont. scholar, 1964; recipient Ann. Citizenship award York Mills Coll., 1961. Mem. Am. Phys. Soc., Am. Nuclear Soc. Subspecialties: Nuclear physics; Fuels. Current work: Experimental and theoretical research to develop down-hole nuclear physics measurements for oil-well logging. Office: Schlumberger-Doll Research Center PO Box 307 Ridgefield CT 06877

BATEMAN, ALFRED CHANDLER, mech. engr.; b. Berkeley, Calif., July 8, 1937; s. Roy Valleau and Emilie Rusch (Perry) B.; m. Judith Pruett, May 25, 1963; children: Andrew Christopher, Jeffrey Alan, Elizabeth Emilie. B.S. in M.E, Calif. Poly. State U., 1960. Registered profl. engr., Calif., N.Mex. Staff mem. Los Alamos Nat. Lab., 1960-63, 64—; design engr. Albuquerque div. ACF Industries, 1963-64. Active PTA. Mem. ASME. Democrat. Methodist and Swedenborgian. Subspecialty: Mechanical engineering. Current work: Mechanical, electromechanical and pyrotechnic-mechanical development. Development and testing of explosively actuated multifunctional valves. Office: PO Box 1663 Mail Stop G780 Los Alamos NM 87545

BATEMAN, DURWARD FRANKLIN, plant pathologist, educator; b. Tyner, N.C., May 28, 1934; s. Benny Franklin and Grace (Cale) B.; m. Shirley Eugenia Byrum, June 23, 1953; children: Cynthia Anne, Brenda Sue, Diane Mia. B.S., N.C. State Coll., 1956; M.S., Cornell U., 1958, Ph.D., 1960. Asst. prof. dept. plant pathology Cornell U., Ithaca, N.Y., 1960-65, assoc. prof., 1965-69, prof., 1969-70, prof., chmn. dept., 1970-79, tchr. grad. course in area of disease and pathogen physiology, 1963-79, field rep. dept. plant pathology, 1966-69; cons. NIH, 1968; vis. prof. N.C. State U., Raleigh, 1975, assoc. dean; also dir. N.C. Agrl. Research Service, 1979—; vice chmn. So. Agrl. Expt. Sta. Dirs. Assn., 1983; chmn. legis. subcom. of com. on policy So. Agrl. Expt. Sta., 1983; mem. biotech. com. Nat. Assn. State Univs. and Land Grant Colls., 1982—. Contbr. articles to profl. jours. Spl. NIH fellow U. Calif., Davis, 1967. Fellow Am. Phytopath. Soc. (councilor-at-large 1973, sr. councilor-at-large 1974, v.p. 1975, pres. 1977-78); mem. AAAS, Intersoc. Consortium for Plant Protection (exec. com. 1977-80), Internat. Soc. Plant Pathology, Sigma Xi, Phi Kappa Phi, Kappa Phi Kappa, Gamma Sigma Delta. Subspecialties: Plant pathology; Plant physiology (agriculture). Current work: Research administration and enzymology of plant tissue decomposition by fungi and bacteria. Home: 4026 Glen Laurel Ln Raleigh NC 27612

BATES, MARGARET WESTBROOK, nutrition educator; b. Boston, Oct. 5, 1926; s. oscar Kenneth and Frances Guest (Westbrook) B. B.A., Wellesley Coll., 1948; M.S., Cornell U., 1950; D.Sc., Harvard U., 1954. Research fellow Harvard U., 1954-55; research assoc. U. Pitts., 1957-59, asst. prof. nutrition, 1959-72, assoc. prof., 1972—. NIH grantee, 1959-81. Mem. AAAS, Am. Inst. Nutrition, Am. Chem. Soc., Biochem. Soc., Sigma Xi. Subspecialties: Nutrition (biology); Biochemistry (biology). Current work: Fatty acid metabolism, director of clinical nutrition core laboratory. Home: 118 Mayflower Dr Pittsburgh PA 15238 Office: U Pitts DeSoto St Pittsburgh PA 15261

BATES, STEPHEN ROGER DENIS, physician, educator; b. London, Aug. 17, 1944; U.S., 1971; s. Denis W. and Kathleen Mary (Lea) B.; m. Jeannie Almind, Dec. 28, 1968; children: Stephen, Thomas, Kristina. Student, Haileybury Coll.-Eng., 1958-62; student, U. London, 1963-68; M.B.B.S., St. George's Hosp. Med. Sch.-U. London, 1968. Diplomate: Royal Coll. Obstetricians & Gynecologists. Intern, London, 1969-70, resident in pediatrics and neurology, Cin., 1971-74; chief resident in neurology U. Cin., 1974-75, instr. pediatrics, 1975-76; asst. prof. child neurology Children's Hosp. Med. Center, Cin., 1976-81, assoc. prof., 1981—, asst. chief of staff, 1975—, dir. EEG Lab., 1976—, dir. epilepsy children's program, 1980—. Contbr. articles to profl. jours. Mem. Brit. Med. Assn., Am. Acad. Neurology, Cin. Soc. Neurology and Neurosurgery, Cin. Pediatric Soc., Child Neurology Soc., Am. Epilepsy Soc., Am. Med. Electroencephalographic Assn., AAAS. Club: Cin. Raquet. Subspecialties: Neurology; Pediatrics. Current work: Study of hepatic toxicity; carbamazepine and dyslipoproteinemia; migraine in childhood and relationship to dyslipoproteinemia; nitrazepam in the treatment of childhood epilepsy. Home: 7054 Mt Vernon Cincinnati OH 45227 Office: Children's Hosp Med Center Elland & Bethesda Aves Cincinnati OH 45229

BATRA, ROMESH C(HANDER), engineering educator, consultant; b. Village Dherowal, Panjab, India, Aug. 16, 1947; came to U.S., 1969; s. Amir Ch and Dewki Bai (Dhamija); m. Manju Dhamija, June 26, 1972; children: Monica, Meenakshi. B.S., Mahendra Coll., India, 1964; B.E. Mech. Engring, Thapar Coll. Engring., India, 1968; M.A.Sc., U. Waterloo, Can., 1969; Ph.D., Johns Hopkins U., 1972. Postdoctoral assoc. Johns Hopkins U., Balt., 1972-73; research assoc. McMaster U., Hamilton, Ont., Can., 1973-74; asst. prof. U. Mon.-Rolla, 1974-76, assoc. prof., 1977-81, prof. dept. engring. mechanics, 1981—; asst. prof. U. Ala., Tuscaloosa, 1976-77; cons. Xerox, Rochester, N.Y., 1978-80, Athena Engring. Co., Tuscaloosa, 1976-77, various lawyers. Editor profl. proceedings, 1980, 82; contbr. articles to profl. jours. and publs. Govt. of India scholar, 1964-68; fellow Johns Hopkins U., 1969-72; NSF research grantee, 1981-82; recipient excellence in teaching award U. Ala., 1977. Mem. ASME, Soc. Engring. Sci., Soc. Rheology, Am. Soc. Engring. Edn., Soc. for Natural Philosophy. Hindu. Subspecialties: Solid mechanics; Theoretical and applied mechanics. Current work: Application of finite element method to solve nonlinear thermomechanical problems especially those involving rubberlike materials. Home: 211 Steeplechase Rd Rolla MO 65401 Office: Dept Engring Mechanics Univ Missouri Rolla Rolla MO 65401

BATSAKIS, JOHN GEORGE, pathologist, educator; b. Petoskey, Mich., Aug. 14, 1929; s. George John and Stella (Vlankis) B.; m. Mary Janet Savage, Dec. 27, 1957; children: Laura, Sharon, George. M.D., U. Mich., 1954. Diplomate: Am. Bd. Pathology. Intern George Washington U. Hosp., Washington, 1954-55; resident U. Mich. Hosp., Ann Arbor, 1955-59; prof. pathology U. Mich. Med. Sch., Ann Arbor, 1961-79, U. Vt. Med. Sch., Burlington, 1980-81; chmn. pathology Maine Med. Ctr., Portland, 1979-82, M.D. Anderson Hosp., Houston, 1982—; prof. pathology U. Tex. Med. Sch., Houston, 1982—. Author 5 books, sci. articles in field; editor, mem. editorial bds. profl. jours. Served to capt. U.S. Army, 1959-61. Fellow ACP, Coll. Am. Pathologist (chmn. commn. on sci. resources 1981—), Am. Soc. Clin. Pathologists. Republican. Episcopalian. Subspecialty: Pathology (medicine). Current work: Head and neck oncology, quantitative aspects of morphology. Home: 1607 Sanford Rd Houston TX 77096 Office: Dept Patholog MD Anderson Hosp Houston TX 77030

BATSON, ALAN PERCY, computer scientist; b. Birmingham, Eng., Sept. 18, 1932; m., 1957, ; 2d, 1968; 4 children. B.Sc., U. Birmingham, 1953, Ph.D., 1956. Fellow U. Birmingham, 1956-58; from instr. to assoc. prof. physics U. Va., Charlottesville, 1958-67, prof. computer sci., 1967—, dir., 1962-72. Mem. IEEE, Assn. Computer Machinery. Subspecialty: Computer systems. Office: Dept Applied Math and Computer Sci U Va Charlottesville VA 22903

BATTIN, RICHARD HORACE, aeronautical engineer; b. Atlantic City, Mar. 3, 1925; s. Horace Leslie and Martha Esther (Scheu) B.; m. Margery Katheryn Milne, Aug. 25, 1947; children: Thomas, Pamela, Jeffrey. B.S., M.I.T., 1945, Ph.D., 1951. Instr. math. M.I.T., 1946-51, research mathematician Instrumentation Lab., 1951-56, adj. prof. aeros. and astronautics, 1979—; sr. staff mem. Ops. Research Group, Arthur D. Little, Inc., Cambridge, Mass., 1956-58; tech. dir. Apollo Mission devel.; assoc. dir. Instrumentation Lab., 1958-73; assoc. head NASA Program dept. Charles Stark Draper Lab., Inc., 1973—, mem. aerospace safety adv. panel, 1980—. Author: (with J.H. Laning, Jr.) Random Processes in Automatic Control, 1956, Astronautical Guidance, 1964; Mem. editorial com.: Celestial Mechanics, 1968-74. Pres. Project Impact, 1981—; Mem. Lexington (Mass.) Town Meeting, 1956—; mem. Lexington Appropriations Com., 1958-64. Served to lt. (j.g.) Supply Corps USNR, 1945-46. Recipient Louis W. Hill Space Transp. award AIAA, 1972, Mechanics and Control of Flight award, 1978; Superior Achievement award Inst. of Navigation, 1980; Teaching award dept. aeros. and astronautics M.I.T., 1981. Fellow Am. Inst. Aeros. and Astronautics (asso. editor jour., chmn. astrodynamics tech. com. 1978-80, dir. tech. 1979-82); mem. Nat. Acad. Engring., Internat. Acad. Astronautics, Celestial Mechanics Inst., Sigma Xi. Club: Hancock Men's (pres. 1974-76). Subspecialty: Aerospace engineering and technology. Current work: NASA space shuttle. Home: 15 Paul Revere Rd Lexington MA 02173 Office: 555 Technology Sq Cambridge MA 02139

BATTISTA, JERRY JOSEPH, medical physicist; b. Montreal, Que., Can., Jan. 11, 1950; s. Evan and Angelina (Cuccioletta) B.; m. Leigh M. Barton, May 17, 1975; children: Michael Evan, Susan Reed. B.Sc. in Physics, Loyola of Montreal, 1971, M.S., U. Western Ont., 1973; Ph.D. in Med. Biophysics, U. Toronto, 1977. Grad. research Ont. Cancer Inst., Toronto, 1973-77, clin. med. physicist, 1977-79; sr. med. physicist Cross Cancer Inst., Edmonton, Alta., Can., 1979—; asst. prof. med physics U. Alta., 1979—. Contbr. over 25 articles to sci. jours. Mem. Am. Assn. Physicists in Medicine, Can. Assn. Physicists. Subspecialties: Biophysics (physics); Graphics, image processing, and pattern recognition. Current work: Applications of physics and computer technology to medical radiation problems. Digital image processing, radiation therapy planning. Office: Cross Cancer Inst 11560 University Ave Edmonton AB Canada T6G 1Z2

BATTISTE, MERLE ANDREW, chemistry educator, researcher; b. Mobile, Ala., July 22, 1933; s. David Theodore and Flossie (Older) B.; m. Anita Elaine Luise, Mar. 12, 1960; children: Mark Andrew, John Lawrence. B.S., The Citadel, 1954; M.S., La. State U.,-Baton Rouge, 1956; Ph.D., Columbia U., 1959. Research fellow UCLA, 1959-61; asst. prof. U. Fla.-Gainesville, 1961-66, assoc. prof., 1966-70, prof. chemistry, 1970—. Contbr. articles in field to profl. jours. Pres. Gainesville Camellia Soc., 1981-82, v.p., 1982-83. Served to 2d lt. U.S. Army, 1960-61. Predoctoral fellow Am. Cyanamid Co., 1958-59; Alfred P. Sloan Found. research fellow, 1967-69; Fulbright-Hays sr. research scholar, 1974. Mem. Am. Chem. Soc. (alt. councilor 1980-83), Royal Soc. Chemistry, Sigma Xi. Democrat. Subspecialties: Organic chemistry; Synthetic chemistry. Current work: Isolation, identification, and synthesis of insect semiochemicals, alicyclic chemistry, reaction mechanisms, photochemistry, organometallic compounds. Home: 427 SW 41st St Gainesville FL 32607 Office: U Fla Dept chemistry Gainesville FL 32611

BATTISTICH, VICTOR ANTHONY, research psychologist, consultant, educator; b. Sacramento, Sept. 9, 1952; s. Carl Anthony and Marian Rita (Hansen) B.; m. Martha Susan Montgomery, Jan. 6, 1978; 1 dau., Sarah. B.A. in Psychology with high honors, Calif. State U.-Sacramento, 1974; M.A. in Personality, Social Psychology, Mich. State U., 1976, Ph.D., 1979. Sr. research assoc. Ctr. for Evaluation and Assessment dept. psychology Mich. State U., 1978-79; vis. asst. prof. dept. psychology and First Coll., Cleve. State U., 1979-80, participant in planning, devel. and implementation of, 1979-80; sr. research assoc. Child Devel. Project, Devel. Studies Ctr., San Ramon, Calif., 1981—; cons. to numerous state, local agys.; lectr. in field. Contbr. reviews and articles to profl. jours. Mem. AAAS, Am. Psychol. Assn., Internat. Assn. Study Cooperation in Edn., Midwestern Psychol. Assn., Soc. Personality Assessment, Soc. Personality and Social Psychology, Soc. Psychol. Study Social Issues, Phi Kappa Phi. Subspecialties: Social psychology; Personality. Current work: Primary research interests in relationships between personality, social cognition and social behavior; currently involved in major longitudinal study of the development of social attitudes, motives and behavior in children between ages 3 and 10. Office: Developmental Studies Center Child Development Project 130 Ryan Ct Suite 210 San Ramon CA 94583

BATY, DAVID LEE, metallurgist, research specialist; b. Easton, Pa., Apr. 21, 1946; s. Philip Leroy and Janice (Wyant) B.; m. Linda Mae Rheinhart, June 21, 1969; children: Jamison, Daniel, Lauren. B.S. in Metall. Engring, Drexel U., Phila., 1969; S.M. in Metallurgy, MIT, 1971. Sr. research engr. Babcock & Wilcox, Lynchburg, Va., 1975-79, group supr., 1979-82, research specialist, 1982—. Subspecialty: Metallurgy. Current work: Reactive metals, advanced materials, failure analysis. Home: 318 Lake Forest Dr Lynchburg VA 24502 Office: Babcock & Wilcox PO Box 239 Lynchburg VA 24505

BATZEL, ROGER ELWOOD, chemist; b. Weiser, Idaho, Dec. 1, 1921; s. Walter George and Inez Ruth (Klinefelter) B.; m. Edwina Lorraine Grindstaff, Aug. 18, 1946; children—Stella Lynne, Roger Edward, Stacy Lorraine. B.S., U. Idaho, 1947; Ph.D., U. Calif. at Berkeley, 1951. Mem. staff Lawrence Livermore (Calif.) Lab., 1953—, head chemistry dept., 1959-67, asso. dir. for chemistry, 1961-71, asso. dir. for testing, 1961-64, asso. dir. for space reactors, 1966-68, asso. dir. chem. and bio-med. research, 1969-71, dir. lab., 1971—. Served with USAAF, 1943-45. Named to Alumni Hall of Fame U. Idaho, 1972. Fellow Am. Phys. Soc.; mem. Sigma Xi. Club: Commonwealth of Calif. (San Francisco). Subspecialty: Nuclear chemistry. Home: 315 Bonanza Way Danville CA 94526 Office: PO Box 808 Livermore CA 94550

BAUCH, TAMIL DANIEL, civil engineer, researcher, artist; b. Bklyn., Aug. 26, 1943; s. Marks Joseph and Mary (Heller) B. B.C.E., Rensselaer Poly. Inst., 1965. Field supt. trainee, asst. supt. Diesel Constrn. Co., N.Y.C., 1969-70; founder, dir. A.I.R. Design Group (name formerly Egge Research), 1971—; staff cons. Dome East Corp., Hicksville, N.Y., 1973-76, Child Environ. Design Inst., Poughkeepsie, N.Y., 1976-78; major cons. Bio-Energy Systems, Inc., Ellenville, N.Y., 1979—; pres. Aerius Design Group, Inc., 1982—; cons. on solar energy and energy conservation field to major corps. Contbr. to profl. publs. Served to lt. (j.g.) USN, 1965-69. Mem. Mid-Hudson Renewable Energy Assn., Tau Beta Pi, Chi Epsilon. Subspecialties: Solar energy; Energy conservation. Current work: Development of: solar collector technology; energy conservation products; solar heated structures (homes, commercial buildings and greenhouses); experimental architecture—molded building components. Patentee composite constrn. panel. Office: RFD 1 Box 394B Kingston NY 12401

BAUE, ARTHUR EDWARD, surgeon, educator; b. St. Louis, Oct. 7, 1929; s. Arthur Christian and Viola (Wegener) B.; m. Rosemary Dysart, Nov. 24, 1956; children—Patricia Sage, Arthur Christian II, William Dysart. A.B. summa cum laude, Westminster Coll., 1950; M.D. cum laude, Harvard, 1954. Diplomate: Am. Bd. Surgery (dir.), Am. Bd. Thoracic Surgery (dir.). Successively intern, resident, chief resident surgery Mass. Gen. Hosp., Boston, 1954-61; asst. prof. surgery U. Mo. Sch. Medicine, 1962-64; asst. prof., then assoc. prof. surgery U. Pa. Sch. Medicine, Phila., 1964-67; Harry Edison prof. surgery Washington U. Sch. Medicine, St. Louis, 1967-75; surgeon-in-chief, dir. dept. surgery Jewish Hosp., St. Louis, 1967-75; chief of surgery Yale-New Haven Hosp., 1975—; prof., chmn. dept. surgery Yale, 1975—, Donald Guthrie prof. surgery, 1977—; cons. surgery Nat. Bd. Med. Examiners; chmn. NIH surgery B study sect., 1978-82. Chief editor: Archives of Surgery, 1977—; mem. editorial bd.: Am. Jour. Physiology. Mem. alumni council Westminster Coll. Served to capt. USAF, 1959. John and Mary R. Markle scholar acad. medicine, 1963; recipient Research Career Devel. award USPHS, 1964. Mem. Am. Assn. Thoracic Surgery, Am. Coll. Cardiology, Am. Coll. Chest Physicians, A.C.S., Assn. Acad. Surgery, New Eng. Surg. Soc., Internat. Cardiovascular Soc., Soc. Thoracic Surgeons, Soc. U. Surgeons, Soc. Vascular Surgery, Internat. Soc. Surgery, Am. Assn. Surgery Trauma, Am. Assn. Artificial Internal Organs, Am. Physiol. Soc., AMA (editorial bd. jour.), Am., Central, Western surg. assns., Soc. Surgery Alimentary Tract, Alpha Omega Alpha. Subspecialties: Surgery; Cardiac surgery. Current work: Shock, circulatory failure, multiple organ failure, subcellular alterations with aschemia, abnormallities of cellular energetics and the effects of ATP-magnesium chloride. Home: 184 Todd St Hamden CT 06514 Office: 333 Cedar St New Haven CT 06510

BAUER, MICHAEL ANTHONY, computer scientist; b. Dayton, Ohio, Feb. 18, 1948; s. Vincent DeJohn and Stephanie (Talmant) B.; m. Angeline Blonski, May 22, 1976; children: Andrea, Michelle. B.Sc., U. Dayton, 1970; M.Sc., U. Toronto, 1971, Ph.D., 1978. Programmer Wright Patterson AFB, Dayton, Ohio, 1966-70; instr. U. Toronto, 1974; research fellow U. Edinburgh, Scotland, 1974-75; lectr. U. Western Ont., London, 1975-78; asst. prof., 1978-81, assoc. prof., 1981—. Judge London (Ont.) Dist. Sci. Fair, 1980—. Nat. Research Council Can. scholar, 1972-74; recipient Excellence in Teaching award U. Western Ont., 1982; Teaching award Ont. Conf. Univ. Faculty Assns., 1982. Mem. IEEE, Assn. Computing Machinery, Soc. Indsl. and Applied Math, Can. Info. Processing Soc. Subspecialties: Software engineering; Algorithms. Current work: Automated tools, techniques and methods to aid in software specification, design and implementation; interfaces to computer systems and software. Office: Dept Computer Sci U Western Ont London On Canada N6A 5B9

BAUER, RICHARD CARLTON, nuclear engineer; b. Batavia, N.Y., July 15, 1944; s. Willard Ronald and Ethel Ann. (Roth) B.; m. Madeline Joy Amreich, June 28, 1969; children: Jason Todd, Cheryl Robyn. B.S., Clarkson Coll. Tech., 1966; M.Eng., Cornell U., 1968; Ph.D., Carnegie-Mellon U., 1974. Registered profl. engr., Pa.; cert. fallout shelter analyst Fed. Emergency Mgmt. Agy. Engr., sr. engr. Bettis Atomic Power Lab., West Mifflin, Pa., 1968-78, staff engr. to gen. mgr., 1978-79, mgr. reactor system performance analysis, 1979—, tng. lectr., 1975—, sec., mem. reactor safety com., 1975—. Author: Reactor Safety, Systems, Analysis, 1978. Chmn. secondary schs. com. Cornell U., Pitts. sect., 1976-78; regional v.p. Soc. Engrs., 1974-80. N.Y. State regents fellow, 1962-66; AEC spl. fellow, 1967, 68; Bettis Lab. doctoral fellow, 1968-74. Mem. Am. Nuclear Soc., Am. Inst. Chem. Engrs., Pa. Soc. Engrs. (dir. 1981-84, chmn. sustaining assocs. 1981—), Nat. Soc. Prof. Engrs., N.Y. Acad. Scis., Sigma Xi, Omega Chi Epsilon, Tau Beta Pi. Club: Triangle. Subspecialties: Nuclear engineering; Nuclear fission. Current work: My current activities involve nuclear reactor safety, protection analysis and plant operations and lecturing; my interests also include commercial nuclear reactor operational economics, reactor testing, plant testing.

BAUER, RICHARD HENRY, educator, researcher; b. Garrison, N.D., Oct. 5, 1939; s. Richard and Martha (Sayler) B. B.A., U. Mont., Missoula, 1964, M.A., 1965; Ph.D., U. Wash., Seattle, 1970. Instr. psychology U. Houston, 1970-72; asst. prof. Kans. State U., Manhattan, 1977-81, Middle Tenn. State U., 1981—. Contbr. articles in field to profl. jours. Served with U.S. Army, 1957-59. USPHS trainee, 1964-65; NSF trainee, 1967-69; NIMH fellow, 1972-74; NIMH grantee, 1973; Upjohn Co. grantee, 1975. Mem. Am. Psychol. Assn., Midwestern Psychol. Assn., Western Psychol. Assn., N.Y. Acad. Sci., Soc. Neurosci., Psychonomic Soc., AAAS, Fedn. Am. Scientists, Internat. Acad. Research in Learning Disabilities, Internat. Soc. Devel. Psychology, Sigma Xi. Subspecialties: Neuropsychology; Physiological psychology. Office: Department of Psychology Middle Tennessee State University Murfreesboro TN 37132

BAUER, ROBERT WILLIAM, research psychologist, educator; b. Indpls., June 23, 1923; s. Nicholas A. and Gertrude Erma (Lawrence) B.; m. Norma Brackett, Dec. 1950 (div. 1964); children: Robin, Nicholas, Christopher, Daniel; m. Jean Anderson, Dec. 12, 1970. Ph.B., U. Chgo., 1948, Ph.D., 1953. Lic. psychologist, Del.; lic. indsl-orgnl. psychologist, Ky. Staff psychologist Sheridan (Wyo.) VA Hosp., 1954-55; chief psychologist Evansville (Ind.) State Hosp., 1955-58; prof. psychology, chmn. dept. Radford (Va.) Coll., 1958-62; research psychologist U.S. Army Human Engring. Lab., Aberdeen Proving Ground, Md., 1962-74, U.S. Army Research Inst., Ft. Knox, Ky., 1974—. Pres. Greenspace, Inc., Elizabethtown, Ky., 1977-78. Mem. Am. Psychol. Assn., Human Factors Soc., AAAS. Subspecialties:

Human factors engineering; Industrial psychology. Current work: Human factors, training, human subsystem development, research management, teaching. Home: Route 1 Box 359B Cecilia KY 42724

BAUER, SIMON HARVEY, chemist, emeritus educator; b. Kaunas, Lithuania, Oct. 12, 1911; came to U.S., 1921, naturalized, 1927; s. Benzion and Golda (Betten) B.; m. Miriam Rosoff, June 25, 1938; children: Frederick, Deborah, Ross. B.S., U. Chgo., 1931, Ph.D., 1935. Postdoctoral fellow Calif. Inst. Tech., Pasadena, 1935-37; instr. Pa. State U., University Park, 1937-39, Cornell U., Ithaca, N.Y., 1939-41, asst. prof., 1941-46; assoc. prof., 1946-50, prof. phys. chemistry, 1950-77, prof. emeritus, 1977—; cons. Los Alamos Nat. Lab., Argonne Nat. Lab., CALSPAN, Atlantic-Richfield Co., 1944—. Contbr. articles to sci. jours. Recipient Alexander von Humboldt award Humboldt Found., 1979; Guggenheim fellow, 1949; NSF sr. postdoctoral fellow, 1962. Fellow Am. Phys. Soc., AAAS; mem. Am. Chem. Soc., Combustion Inst., Sigma Xi, Phi Beta Kappa. Subspecialties: Physical chemistry; Laser-induced chemistry. Current work: Molecular structure of amorphous, heterogeneous catalysts, dynamics of intramolecular transformations of very fast reactions, gas phase chemiluminescent reactions. Home: 412 Klinewoods Rd Ithaca NY 14850 Office: Dept Chemistr Cornell U Ithaca NY 14853

BAUGHCUM, STEVEN LEE, phys. chemist; b. Atlanta, Dec. 18, 1950; s. George Lee and Henrietta (Stevens) Baughcom. B.S., Emory U., 1972; A.M., Harvard U., 1973, Ph.D., 1978. NRC research assoc. Joint Inst. Lab. Astrophysics Nat. Bur. Standards and U. Colo., Boulder, 1978-80; staff mem. Los Alamos Nat. Lab., 1980—; teaching fellow chemistry Harvard U., 1973-74, 75-76. Contbr. articles in field to profl. jours. Served as 1st lt. USAF, 1976. NSF grad. fellow, 1972-75; recipient Am. Inst. Chemists award, 1972. Mem. Am. Chem. Soc., Am. Phys. Soc., Phi Beta Kappa, Sigma Xi. Subspecialties: Laser photochemistry; Kinetics. Current work: Spectroscopy and reaction kinetics of free radicals at both room temperature and high temperatures studied using laser techniques; laser photochemistry. Home: 359 Cheryl Ave Los Alamos NM 87544 Office: Los Alamos National Lab Los Alamos NM 87545

BAUGHMAN, RAY HENRY, materials scientist; b. York, Pa., Jan. 14, 1943; s. Ray Henry and Ruth Marion (Beers) B.; m. Nancy Ann O'Connor, Feb. 15, 1969; children: Lara Crusan, Heather Leigh, Dana Marie, Rebecca Lynn. B.S., Carnegie-Mellon U., 1964; M.S., Harvard U., 1966, Ph.D., 1971. Staff scientist Allied Chem. Co. (name now Allied Corp.), Morristown, N.J., 1970-73, group leader, 1974-77, mgr. organic materials sci., 1977-79, mgr. splty. polymers, corp. research and devel., 1979—; mem. organizing com. Internat. Conf. on Low Dimensional Condrs., 1980-82; mem. adv. group NATO, Materials Research Council, Dept. Energy. Mem. editorial bd.: Jour. Synthetic Metals, 1978-83; regional editor, 1983—; contbr. numerous articles to profl. publs.; patentee in field (14). Gen. Motors Corp. scholar, 1961-64; NSF fellow, 1964-65. Fellow Am. Phys. Soc. (editorial bd. div. high polymers 1975-77); mem. Am. Chem. Soc., Sigma Xi, Phi Kappa Phi. Democrat. Lutheran. Subspecialties: Polymer physics; Solid state chemistry. Current work: New material synthesis, properties and applications; electronic properties of low dimensional conductors; solid-state reactions; polymer physics and chemistry. Home: 41 Glacier Dr Morris Plains NJ 07950 Office: Allied Corp PO Box 1021R Morristown NJ 07960

BAUM, GARY ALLEN, physics educator, materials scientist; b. New Richmond, Wis., Oct. 2, 1939; s. Herbert Louis and Margaret L. (Peters) B.; m. Paula Zoe VanHoeser, Feb. 22, 1964; children: Geoffrey, Kristin. B.S., U. Wis., 1961; M.S., Okla. State U., 1964, Ph.D., 1969. Research engr. Douglas Aircraft, Santa Monica, Calif., 1963-64; sr. physicist Dow Chem. Co., Golden, Colo., 1964-66; teaching asst. Okla. State U., Stillwater, 1966-69; group leader Inst. Paper Chemistry, Appleton, Wis., 1969-80, sect. head, 1980-83, div. dir., 1983—, prof. physics, 1980—. Contbr. articles to profl. jours. NASA fellow, 1967-69. Mem. Am. Assn. Physics Tchrs., TAPPI (v.p. paper physics com. 1980-83, pres. 1983—). Lutheran. Subspecialties: Composite materials; Condensed matter physics. Current work: Materials science, fiber and paper physics, reprography, pulp and paper mill instrumentation, physics education. Home: 426 Fidelis St Appleton WI 54915 Office: Inst Paper Chemistry PO Box 1039 Appleton WI 54912

BAUM, PETER JOSEPH, research physicist, cons., author; b. Lennox, Calif., June 4, 1943; s. Custer Charles and Persis Eugenia (Fell) B.; m.; 1 dau., Maryann Joy. B.A. in Physics, U. Calif., Santa Barbara, 1965, M.S., U. Nev., Reno, 1967; Ph.D. in Physics (NDEA fellow 1968, NSF fellow 1969), U. Calif., Riverside, 1971; Calif. state teaching cert. astronomy, 1973. Research engr. microelectronics Autonetics Div. N.Am. Rockwell Corp., 1967-68; physicist computer analysis Corona Lab., Naval Weapons Center, 1968-70; research asso. solar-plasma physics Office Sci. Research, 1970-74; Air Force grantee, 1972-74; research physicist solar-terrestrial plasma physics U. Calif., Riverside, 1974—, wind energy, energy sci. program, 1980—; cons. Sandia Labs., Albuquerque, 1973-75; cons. wind energy City of Riverside Public Utilities Dept., 1980—; instr. in astronomy Riverside City Coll., 1975-80. Author: (with A. Bratenahl) Magnetic Reconnection, 1982; contbr. articles to tech. publs. Calif. Inst. Tech. Presdl. Fund grantee, 1970; NSF grantee, 1974—; Calif. Space Inst. grantee, 1981-82. Mem. Am. Phys Soc., Am. Astron. Soc., Am. Geophys. Union (life), Phi Kappa Phi. Subspecialties: Solar physics; High energy astrophysics. Current work: Dir. operation UCR 11 meter terrella experiment; lab. stucies on magnetic reconnection; condr. exptl. and theoretical research in geomagnetic substorms, solar flares, magnetic energy conversion; studies wind energy potential San Gorgonio Pass region, Calif. Home: 11410 Lombardy Ln Sunnymead CA 92388 Office: IGPP Univ Calif Riverside CA 92521

BAUM, STANLEY, radiologist, educator; b. N.Y.C., Dec. 26, 1929; s. Herman and Fannie (Harris) B.; m. Jeanne Masch, June 29, 1958; children: Richard Arthur, Laura Dianne, Carol Lisa. B.A., N.Y. U., 1951; M.D., U. Utrecht, Holland, 1957. Intern Kings County Hosp., N.Y.C., 1957-58; resident in radiology Grad. Hosp., U. Pa., Phila., 1958-61; trainee Nat. Cancer Inst., Bethesda, Md., 1958-61; fellow cardiovascular radiology Stanford (Calif.) U., 1961-62; instr. radiology U. Pa., Phila., 1962-63, asst. prof., 1963-66, assoc. prof., 1966-70, prof., 1970—, Eugen P. Pendergrass prof. radiology, 1977—, chmn. dept. radiology, 1975—; chmn. med. bd. Hosp. of U. Pa., 1983—; chief cardiovascular radiology Mass. Gen. Hosp., Boston, 1971-75; prof. radiology Harvard Med. Sch., Boston, 1971-75; cons. Radiation Effects Research Found., Hiroshima, Japan, 1975—; cardiovascular rev. bd. Am. Heart Assn., 1970—. Editorial bd.: Investigative Radiology, 1970-80, New Eng. Jour. Medicine, 1975-76, Radiology, 1975—, Gastrointestinal Radiology, 1975-79, Jour. Continuing Edn, 1978-80, Postgrad. Radiology, 1980—. Fellow Am. Coll. Radiology, Am. Coll. Cardiology; mem. Soc. Cardiovascular Radiology (pres. 1974-76), Soc. Chmn. Acad. Radiology Depts. (sec.-treas. 1983—). Subspecialty: Diagnostic radiology. Home: 401 W Moreland Ave Chestnut Hill PA 19118 Office: 3400 Spruce St Philadelphia PA 19104

BAUM, STEPHEN GRAHAM, internist, researcher, educational administrator; b. N.Y.C., Apr. 28, 1937; s. Samuel Meyerson and Rosalind (Birns) B.; m.; 1 dau., Laurie R. A.B., Cornell U., 1958; M.D., NYU, 1962. Diplomate: Am. Bd. Internal Medicine (Subspecialty in infectious diseases). Intern Harvard U.-Boston City Hosp., 1962-63, resident, 1963-64; research assoc. NIH, 1964-67; asst. prof. medicine and cell biology Albert Einstein Coll. Medicine, 1968-73, assoc. prof., 1973-78, prof. medicine, cell biology, microbiology and immunology, 1978—, co-dir. infectious diseases div., 1975—, dir. M.D.-Ph.D. program, 1973—. Contbr. articles, chpts. to profl. publs. Served to lt. comdr. USPHS, 1964-67. Recipient Career Scientist award N.Y.C. Health Research Council, 1968-72, Faculty Research award Am. Cancer Soc., 1973-78. Mem. Am. Soc. Clin. Investigation, Harvey Soc., Infectious Diseases Soc. Am. Subspecialties: Cell biology (medicine); Microbiology (medicine). Office: 1300 Morris Park Ave F456 Bronx NY 10461

BAUMAN, JOHN E., JR., chemistry educator; b. Kalamazoo, Mich., Jan. 18, 1933; s. John E. and Teresa A. (Wauchek) B.; m. Barbara Curry, June 6, 1964; children: John, Catherine, Amy. B.S., U. Mich., 1955, M.S., 1960, Ph.D., 1962. Chemist Midwest Research Inst., Kansas City, Mo., 1955-58; research assoc. U. Mich., Ann Arbor, 1958-61; prof. chemistry U. Mo., Columbia, 1961—. Active Mo. Symphony Soc., Columbia Audubon Soc. Recipient Faculty Alumni award U. Mich., 1969, Amoco teaching award, 1975, Purple Chalk award U. Mo., 1980. Mem. Am. Chem. Soc. (nat. lectr), Calorimetry Conf., Mo. Acad. Sci., Sigma Xi, Alpha Chi Sigma. Roman Catholic. Lodges: Kiwanis (pres. Little Dixie 1982-83); KC. Subspecialties: Inorganic chemistry; Thermodynamics. Current work: Thermodynamics of inorganic reactions in solution. Home: 1805 Cliff Dr Columbia MO 65201 Office: 237 Chemistry Bldg Columbia MO 65211

BAUMAN, THOMAS CHARLES, metallurgist, cons.; b. Los Angeles, Nov. 26, 1945; s. Thomas and Geraldine Maxine (Ballou) B.; m. Judith Ann Smith, Apr. 28, 1943 (div.); children: Kimberley Ann, Ann Lorraine, Laura Marie.; m. Joy Lynne Mott, Mar. 8, 1951. B.Sc. in Metall. Engrng. with honors, Calif. Poly. State U., 1974; M.Sc. in Mgmt. Sci. and Energy Systems Engrng, West Coast U., 1978; postgrad. in bus. adminstrn, Pepperdine U., 1982. Registered profl. engr., Calif., 1977. Research asst. Calif. Poly State U., 1972-74; grad. research asst. U. Calif., Berkeley, 1974; metallurgist Aluminum Co. Am., Corona, Calif., 1974; engr. So. Calif. Gas Co., Los Angeles, 1974-76; material application specialist R. M. Parsons Co., Pasadena, Calif., 1977-78; sr. metallurigst Borg-Warner Corp., Los Angeles, 1978, supervising metall. engr., 1979, mgr. dept. metall. engring., 1979-81, mgr. tech. services Energy Systems Devel. Center, 1981-82; gen.mgr. Thermal Electron Corp., Metall. Service Group, Los Angeles, 1982-83; engr., cons. Northrop Corp., 1983—; materials cons., partner METCON Inc., Los Angeles, 1977—; cons. materials application, failure analysis of metallic components. Contbr. articles to profl. jours. Served to staff sgt. USAF, 1966-70; Vietnam. Decorated Air Force Commendation medal. Mem. Am. Soc. Metals (corp. rep.), ASME, AM. Welding Soc., Nat. Assn. Corrosion Engrs. (accredited corrosion specialist 1980), Am.Mgmt. Assn. Am., Internat. Platform Assn., MENSA, Alpha Gamma Sigma. Republican. Roman Catholic. Clubs: Byron Jackson Mgmt. (Los Angeles); Am. Legion, VFW, Can. Legion Pipe Band. Subspecialties: Metallurgical engineering; Materials (engineering). Current work: Producibility of advanced weapon systems, thermal treatment of metals and their mech. behavior; heat treatment; metals; mech. properties; mgmt.; energy; adminstrn. Office: 9600 E Washington Blvd Pico Rivera CA 90660

BAUMANN, GERHARD, endocrinologist, medical educator; b. Basel, Switzerland, Sept. 15, 1941; came to U.S., 1968; s. Oscar and Johanna (Muller) B.; m. Mary Paula Casey, June 25, 1972; 1 dau., Catherine. M.D., U. Basel, 1967. Diplomate: Am. Bd. Internal Medicine. Resident VA Hosp., Bklyn., 1968-69; fellow in endocrinology Peter Bent Brigham Hosp., Boston, 1969-71; med. resident, cons. U. Basel (Switzerland) Med. Ctr., 1971-74; vis. scientist NIH, Bethesda, Md., 1974-77; asst. prof. medicine Northwestern U., Chgo., 1977-82, assoc. prof., 1982—; cons. NIH, 1978-79. Contbr. numerous articles to med. jours. Am. Diabetes Assn. grantee, 1979; NIH grantee, 1980—. Mem. AAAS, Endocrine Soc., Central Soc. for Clin. Research, Am. Fedn. for Clin. Research. Presbyterian. Subspecialties: Endocrinology; Receptors. Current work: Pituitary growth hormone, growth disorders, diabetes, insulin receptors, insulin action. Home: 1004 Sheridan Rd Evanston IL 60202 Office: Northwestern U Med Sch 303 E Chicago Ave Chicago IL 60611

BAUMANN, HEINZ, cancer research scientist; b. Winterthur, Switzerland, July 22, 1947; came to U.S., 1976; s. Armin and Kaethe (Raumer) B.; m. Ursula Schmied, Mar. 4, 1972. M.S., U. Zurich, 1971, Ph.D., 1974. Research asst. U. Zurich, Switzerland, 1970-76, lectr., 1973-76; cancer research scientist Roswell Park Meml. Inst., Buffalo, 1978-79, sr. cancer research scientist, 1979—. Grantee Swiss Nat. Sci. Found., 1976-78; Nat. Cancer Inst., 1979—. Mem. Swiss Soc. Cell Biology, Am. Soc. Biol. Chemists. Subspecialties: Cell and tissue culture; Membrane biology. Current work: Hormonal regulation of normal and malignant liver cells. Synthesis and degradation of glycoconjugates. Function and reconstitiuion of plasma membranes. Home: 305 Bryant St Buffalo NY 14222 Office: Dept Cell and Tumor Biology Roswell Park Meml Inst 666 Elm St Buffalo NY 14263

BAUMANN, ROBERT COILE, aerospace systems executive and engineer; b. Alexandria, Va., Oct. 8, 1926; s. Charles Nelson and Carolyn Jeanette (Litaker) B.; m. Evelyn Jean Ford, Sept. 9, 1949; children: Christine, Robert, William, Elizabeth, Lucinda. B.M.E., George Washington U., 1953. Head satellite structures U.S. Naval Research Lab., Washington, 1953-58; head mech. systems br. Goddard Space Flight Center, Greenbelt, Md., 1958-62, chief spacecraft integration and sounding rocket div., 1962-73, dep. dir. engring., 1973-74, assoc. dir. flight project, 1974—. Contbr. articles to profl. jours. Served with USMC, 1944-51; PTO. Recipient Exceptional Service award NASA, 1969, Outstanding Leadership award, 1976; Engring. Alumni award George Washington U., 1976. Fellow AIAA (assoc.). Subspecialties: Aerospace engineering and technology; Systems engineering. Current work: Management space systems and space vehicle systems and allied ground support equipment. Inventor, patented satellite; inventor satellite controls system. Home: 9308 Woodberry St Seabrook MD 20706 Office: NASA Goddard Space Flight Center Code 400-3 Greenbelt MD 20771

BAUM-BAICKER, CYNTHIA, psychologist; b. Amityville, N.Y., Aug. 7, 1953; d. Milton William and Selma (Goldman) Baum; m. Mark Baum-Baicker, July 4, 1976. A.B., Washington U., 1975; Ph.D., Temple U., 1979. Lic. clin. psychologist, Pa. Staff clin. psychologist Phila. Geriatric Ctr., 1975-79; instr. Temple U., Phila., 1976-77; adj. grad. faculty Villanova U., 1979—; workshop leader Temple U.-Center City, Inst. Awareness, 1979—; dir. research/evaluation Access Ctrs., Inc., Phila., 1980—; cons. research psychologist Busch Ctr., Wharton Sch., U. Pa., Phila., 1981—; workshop leader Pa. Psychol. Assn., 1982. Contbr. articles to profl. jours. Mem. Am. Psychol. Assn., Pa. Psychol. Assn., Phila. Soc. Clin. Psychologists, Orgn. Devel. Network, Am. Soc. Tng. and Devel. NOW, Sigma Xi, Phi Beta Kappa. Current work: Socio-technical systems design; workplace stress reduction through systems redesign; alcohol as purposeful behavior, the use of metacommunication patterns in psychotherapy and consulting. Home: 1715 Lombard St Philadelphia PA 19146 Office: Busch Ctr Wharton Sch U Pa 3733 Spruce St Philadelphia PA 19104

BAUMERT, JOHN HENRY, astronomer; b. Washington, Ind., Dec. 13, 1941; s. John Edward and Rachel Lorene (Wininger) B.; m. Jean Ellen Heck, May 20, 1966; children: John Richard, Jeffrey Michael. B.S., U. Mich., 1966; Ph.D., Ohio State U., 1972. Research assoc. Ohio State U., 1970-72; Morehead Obs. fellow, instr. U. N.C., Chapel Hill, 1972-75; asst. prof. Conn. Coll., New London, 1975-80; sect. mgr. Computer Scis. Corp., Silver Spring, Md., 1980—. Author: (with Joseph Jackson) Pictorial Guide to the Planets, 3d edit, 1981; Contbr. numerous articles to profl. jours. Asst. scoutmaster Boy Scouts Am., 1982—. Mem. Am. Astron. Soc., Astron. Soc. Pacific, AAAS. Democrat. Roman Catholic. Club: American Philatelic Soc. Subspecialties: High energy astrophysics; Optical astronomy. Current work: Processing and analysis of cosmic ray data from the Pioneer-10, -11, Voyager-1 and 2 spacecraft and x-ray data from BHEAO-1 spacecraft; searching for x-ray binary pulsars. Office: Computer Scis Corp Dept 792 8728 Colesville Rd Silver Spring MD 20910

BAUMGARTEN, REUBEN LAWRENCE, chemist, educator, cons.; b. N.Y.C., Nov. 19, 1934; s. Leon and Sonia (Jacobson) B.; m. Iris Marsha, Dec. 22, 1963; children: Lainie Nicole, Steven Craig. B.S. cum laude, CCNY, 1956; M.S., U. Mich., 1958, Ph.D., 1962. Instr. Hunter Coll. (named changed to Lehman Coll. 1968), 1962-66, asst. prof. chemistry, 1966-71, assoc. prof., 1971-77, prof., 1977—, chmn. dept. chemistry, 1978—; cons. Marks Chem. Works. Author: Organic Chemistry—A Brief Survey, 1978. Mem. Am. Chem. Soc., Sigma Xi, Phi Lambda Upsilon. Jewish. Lodge: Free Sons of Israel. Subspecialty: Organic chemistry. Current work: Organic chemistry of nitrogen; Rimini-Simon test; hydroxylamines; Frediel-Crafts reaction catalysts; mechanisms and kinetics. Home: 22 Eagle Rd Edison NJ 08820 Office: Herbert H Lehman Coll Bronx NY 10468

BAUMGOLD, JESSE, biochemist; b. Antwerp, Belgium, June 3, 1949; s. Theodore and Tamar (Riwlin) B. B.A., Oberlin Coll., 1971; Ph.D., U. Rochester, 1976. Staff fellow NIMH, Bethesda, Md., 1976-79, sr. staff fellow, 1979—. Contbr. articles to profl. jours. Alberta Heritage Found. fellow, 1982. Mem. Soc. Neurosci., Am. Soc. Neurochemistry. Subspecialties: Biochemistry (biology); Neurochemistry. Current work: Use of monoclonal antibodies and neurotoxins to study sodium channels and muscarinic receptors in brain and muscle cells. Home: 4570 MacArthur Blvd Apt 108 Washington DC 20007 Office: NIMH Bldg 36 Room ID-02 Bethesda MD 20205

BAUMILLER, ROBERT CAHILL, geneticist, university administrator; b. Balt., Apr. 15, 1931; s. Bernard Joseph and Margaret Christine (Sullivan) B. B.S., Loyola Coll., Balt., 1953; Ph.D., Ph.L., St. Louis U., 1961; B.T., Woodstock Coll., Balt., 1964. Joined Soc. of Jesus, Roman Catholic Ch., 1953, ordained priest, 1965; asst. in medicine Johns Hopkins U. Sch. Medicine, Balt., 1963-67; research assoc. Inst. for Natural Sci., Woodstock (Md.) Coll., 1962-67; asst. prof. biology Georgetown U., Washington, 1967-71, asst. prof. ob/gyn, 1967-71, assoc. prof. ob/gyn, 1971-81, prof. ob/gyn and pediatrics, 1981—, dir. div. genetics dept. ob/gyn, 1974—, dir., 1982—; dir. cytogenetics lab. Georgetown U. Hosp., 1970—; mem. med. adv.bd. Hemophilia Found. Capital Area, 1975—; cons. for genetic counseling Pope John XXIII Med.-Moral Research an Edn. Inst., 1975-80; mem. postdoctoral fellowship evaluation panel NRC, 1976-79; mem med. adv. bd. Childbirth Edn. Assn.; cons. Nat. Clearinghouse for Genetic Diseases, Bur. Community Health Services, HEW Pub. Health Services, HEW Pub. Health Service-Health Services Adminstrn. Assoc. editor: Linacre Quar, 1971-78; assoc. editor: Down's Syndrome, 1978—. Bd. govs. Inst. for Advanced Studies in Criminalistics Forensic Scis. Lab., Washington, 1975-77; mem. adv. com. curriculum devel. project Ctr. for Edn. in Human and Med. Genetics, 1982—; mem. adv. council Archiocesan Parent Assn. for Residential Services for Persons Who Are Mentally Retarded, 1981. NSF fellow, 1961-62. Subspecialties: Genetics and genetic engineering (medicine); Reproductive biology (medicine). Current work: Mutation in amniocytes; genetic effects of drugs on the fetus. Office: Dept Ob/Gyn Georgetown U Washington DC 20007

BAUMSTARK, ALFONS LEOPOLD, chemistry educator, researcher; b. Bleiburg, Austria, May 4, 1948; s. Leopold and Maria (Wanko) B.; m. Barbara Ruth Bullock, Mar. 23, 1969; children: Anna, Maria. B.A., U. Calif.-Riverside, 1970; A.M., Harvard U., 1972, Ph.D., 1974. Postdoctoral fellow Harvard U., Cambridge, Mass., 1974-76; asst. prof. Ga. State U., Atlanta, 1976-82, assoc. prof. chemistry, 1982—. Contbr. articles in field to profl. jours. Recipient Camille and Henry Dreyfus Tchr.-Scholar award Dreyfus Found., 1981-86; Jr. Faculty Teaching award Ga. State U., 1982. Mem. Am. Chem. Soc., Phi Beta Kappa. Subspecialty: Organic chemistry. Current work: Reaction mechanisms, oxidation processes, peroxide chemistry, chemiluminescence. Office: Dept Chemistry Ga State U Atlanta GA 30303

BAUSERMAN, DEBORAH NYLEN, psychologist; b. Grand Rapids, Mich., Dec. 9, 1948; d. Harry Barr and Norah Lillian (Drew) Nylen; m. John Stephen Bauserman, Sept. 6, 1975; children: Trenton Drew, Michael Alexander. B.A., Mich. State U., 1971, M.A., 1973; Ed.D., U. No. Colo., 1977. Lic. psychologist, Va. Behavior modification cons. Denver Bd. Mentally Retarded and Seriously Handicapped, 1973-76; sch. psychologist Clarke County Bd. Edn., Berryville, Va., 1977-78, Grafton Sch., Berryville, 1977-79, dir. evaluation tng. and research, 1979-81; psychologist Clin. Psychology Services, Winchester, Va., 1981—; cons. Grafton Sch., 1982—. Contbr. articles to profl. jours.; author: Parents Preschool Guide, 1983; editorial advisor: Sch. Psychology Rev, 1980—. Bd. dirs. Shelter for Abused Spouses, Winchester, 1981, Winchester Mental Health Assn.; dir. Seminars for Presch. Tchrs., Winchester, 1982; lectr. in field. Essay winner Am. Vets., 1972; Sigma Xi grantee-in-aide, 1977. Mem. Am. Psychol. Assn., Nat. Assn. Sch. Psychologists, Kappa Delta Pi. Lutheran. Subspecialties: Clinical psychology; Tests and measurement. Current work: Developing screening tests for handicapped children, measuring and testing emotional and social devel., psychotherapy, cons. Office: Clin Psychology Services 104 W Cork St Winchester VA 22601

BAUTZ, GORDON THOMAS, biochemical pharmacologist; b. Bridgeport, Conn., June 6, 1942; s. Milton and Janet Bedford (Ballou) B.; m. Delmaue Hannemann, June 12, 1965; children: Tracy, Jennifer. B.S., Salem (W.Va.) Coll., 1964. With U.S. Dept. Fisheries, Milford, Conn., 1964-65; assoc. scientist Mt. Sinai Hosp., N.Y.C., 1965-68; scientist Hoffmann-La Roche Inc., Nutley, N.J., 1968—. Contbr. articles and abstracts to profl. jours. Mem. Am. Soc. for Pharmacology and Exptl. Therapeutics, N.Y. Acad. Scis. Subspecialties: Neuropharmacology; Neurochemistry. Office: 314 Kingsland St Nutley NJ 07110

BAWA, KAMALJIT SINGH, biology educator; b. Kapurthala, Panjab, India, Apr. 7, 1939; came to U.S., 1967; s. Rajinder Singh and Dwarki Devi (Ohri) B.; m. Tshering Wangdi, Sept. 22, 1969; children: Sonia, Ranjit. B.S. with honors, Panjab U., Chandigarh, India, 1960, M.S., 1962, Ph.D., 1967. Research asst. Panjab U., Chandigarh, India, 1962-67, U. Wash., Seattle, 1967-72; research fellow Harvard U., Cambridge, Mass., 1972-74; asst. prof. U. Mass., Boston, 1974-77, assoc. prof., 1977-80, prof. biology, 1980—; mem. La Selva Adv. Com., Orgn. Tropical Studies, 1978—. Editorial bd.: Jour. Arnold Arboretum, 1979—; Contbr. articles to profl. jours. Orgn. Tropical

Studies grantee, 1969, 70; NSF grantee, 1970, 75-79, 83; Charles Bullard and Maria Moors Cabot research fellow Harvard U., 1972; U. Mass. grantee, 1975, 76, 78-82; Smithsonian Instn. grantee, 1977, 83; recipient Chancellor's award U. Mass., 1981. Mem. Internat. Soc. Tropical Ecology, Soc. Am. Naturalists, Assn. Tropical Biology, Ecol. Soc. Am., Soc. Study Evolution, New Eng. Bot. Club. Subspecialties: Evolutionary biology; Population biology. Current work: Evolutionary and population biology of plants, particularly tropical forest trees. Office: U Mass Boston MA 02125

BAXTER, JOANN CRYSTAL, dental educator, researcher; b. Pitts., Jan. 9, 1947; d. John and Helen (Halaja) March; m. Albert Donald Baxter, June 14, 1965 (div. June, 1976); m. John David Hasenmiller, June 5, 1981. D.M.D., U. Pitts., 1977, cert. Splty., 1978, M.D.S., 1979. Cert. specialist prosthodontic dentistry. Pvt. practice dentistry, Pitts., part-time, 1977-79; asst. prof. dentistry U. Pitts., 1979-80; asst. prof. prosthodontics, geriatrics U. Ill.-Chgo, 1980—; cons. VA Continuing Edn., Washington, 1981—, U. Conn., 1982—, Henrotin Hosp., Chgo., 1981, VA Hosp., Wood, Wis., 1982. Mem. Am. Coll. Prosthodontics (chairperson edn. com. 1982—), ADA, Chgo. Dental Assn. (program com. 1982—), Am. Equilibration Soc. (chmn. credentials com. 1981-82), Am. Prosthodontics Soc. Republican. Eastern Orthodox. Subspecialties: Prosthodontics; Gerontology. Current work: Prosthodontic specialist; geriatric nutrition; research geriatric dentristry; geriatric nutrition. Home: 155 N Harbor Dr Apt 1303 Chicago IL 60601 Office: U Ill 801 S Paulina St Chicago IL 60680

BAXTER, JOHN DARLING, physician, educator; b. Lexington, Ky., June 11, 1940; s. William Elbert and Genevive Lockhart (Wilson) B.; m. Ethelee Davidson Baxter, Aug. 10, 1963; children: Leslie Lockhart, Gillian Booth. A.B. in Chemistry, U. Ky., 1962; M.D., Yale U., 1966. Intern, then resident in internal medicine Yale-New Haven Hosp., 1966-68; USPHS research assoc. Nat. Inst. Arthritis and Metabolic Diseases, NIH, 1968-70; Dernham sr. fellow oncology U. Calif. Med. Sch., San Francisco, 1970-72, mem. faculty, 1972—, prof. medicine and biochemistry and biophysics, 1979—; dir. endocrine research Howard Hughes Med. Inst., 1973-82; chief div. endocrinology Moffitt Hosp., 1980—, dir. Metabolic Research Unit, 1981—; attending physician U. Calif. Med. Center, 1972—. Editor textbook of endocrinology and metabolism; Author research papers in field; mem. editorial bd. profl. jours. Recipient George W. Thorn award Howard Hughes Med. Inst., 1978, Disting. Alumni award U. Ky., 1980; grantee NIH, Am. Cancer Soc., others. Mem. Am. Chem. Soc., Am. Soc. Clin. Investigation, Am. Thyroid Assn., Assn. Am. Physicians, Am. Fedn. Clin. Research, Endocrine Soc., Western Assn. Physicians, Western Soc. Clin. Research. Subspecialties: Genetics and genetic engineering (biology); Endocrinology. Current work: Gene structure and regulation; hormone action. Office: 671 HSE U Calif Med Sch San Francisco CA 94143

BAXTER, JUDITH LEE, mathematics educator; b. Marlinton, W.Va., Nov. 8, 1948; d. Ernest Lee and Anne Loraine (Adkinson) B.; m. Stephen Douglas Smith; children: Dallas Kevin Williams, Dawn Krystal Williams. B.S., Ill. Benedictine Coll., 1976; M.S., U. Ill.-Chgo., 1978, postgrad. Faculty U. Ill.-Chgo., 1976—, mem., 1982—. Author: Ultrasonics applied to Biological Media, 1979. Mem. Soc. Indsl. and Applied Math., Am. Soc. Math., Assn. Women in Math., Assn. Women in Sci. (com.). Subspecialty: Applied mathematics. Current work: Elastic scattering applied to biological media concentration fluxinn magma chambers. Office: Univ Ill PO Box 4348 Chicago IL 60680

BAY, ZOLTAN LAJOS, research physicist; b. Gyulavari, Hungary, July 24, 1900; s. Joseph and Julianna (Boszormenyi) B.; m. Julia Herzcegh, July 7, 1947; m. Ilona Lazar, Jan. 6, 1932; children: Martha, Zoltan, Julia. M.S., U. Budapest, 1923; Ph.D., 1926; hon. Doctorate, U. Edinburgh, 1978. Prof. theoretical physics U. Szeged, Hungary, 1930-36; prof. atomic physics Tech. U. Budapest, 1938-48; research prof. electronics George Washington U., Washington, 1948-55; physicist Nat. Bur. Standards, 1955-72; sr. research scientist Am. U., Washington, 1972—. Contbr. articles to profl. jours. Recipient Boyden award Franklin Inst., Phila., 1980. Fellow IEEE (sr.), Am. Phys. Soc.; Mem. Hungarian Acad. Scis. (hon.), Hungarian Phys. Soc. (hon.). Subspecialties: Atomic and molecular physics; Relativity and gravitation. Current work: Constancy of the speed of light from relativity experiments and measurements of radar astronomy; proof that unified time-length measurement system via a defined value of the speed of light is practicable. Pioneer in active nitrogen containing free nitrogen atoms, secondary electron multiplication for particle counting; radar echoes from moon in Budapest, 1946; patentee electroluminiscence, time of flight mass spectrometer. Office: American U Washington DC 20016

BAYBARS, ILKER, business educator; b. Gordes, Manisa, Turkey, Nov. 26, 1947; came to U.S., 1970; s. Tevfik and Emine (Basboga) B. B.S., Middle East Tech. U., Ankara, Turkey, 1969; M.S., Carnegie-Mellon U., 1972, Ph.D., 1979. Instr. Middle-East Tech. U., Ankara, Turkey, 1969-70; instr. Carnegie-Mellon U., 1978-79; acad. dir. Quantitative Skills Summer Inst., 1980; vis. asst. prof. Sch Urban Pub. Affairs, 1979-81; asst. prof. Arad. Sch. Indsl. Adminstrn., 1981—; cons. Par Ajans, Ankara, 1968-70, UN, N.Y.C., 1981—, AT&T, Bedminister, N.J., 1981—, Mellon Bank, Pitts., 1982—. Author: Graph Theory; A Self-Pace Text, 1976. Served with Turkish Army, 1976. UN Traveling grantee, 1981; NSF grantee, 1980-82; recipient Carnegie-Mellon U. Limbach Teaching award, 1981. Mem. Pitts. Turkish Am. Assn., Ops. Research Soc. Am., Inst. Mgmt. Sci., Soc. Indsl. Applications of Math., Assn. Pub. Policy Analysis and Mgmt., Sigma Xi. Subspecialties: Operations research (mathematics); Algorithms. Current work: Optimal design of networks; facility-capicity planning and expansion; optimal design of assembly systems; algorithms; graph theory; mathematical programming. Home: 5133 Forbes Ave Pittsburgh PA 15213 Office: GSIA Carnegie Mellon U 5000 Forbes Ave Pittsburgh PA 15213

BAYER, BARBARA MOORE, pharmacologist, consultant, researcher; b. Henderson, Tex., Dec. 10, 1948; d. C.W. and Elizabeth (Richardt) Moore (Jr); m. Michael J. Bayer, June 24, 1972; 1 son, Alexander Moore. B.S. in chemistry, U. Tex.-El Paso, 1972; Ph.D. in Pharmacology, Ohio State U., 1977. Lab. instr. chemistry U. Tex., El Paso, 1970-72; research assoc. Toxicology Lab. Ohio State U., 1972-76, mem. faculty, 1977; research fellow NIH, Bethesda, 1977-79, staff fellow, 1979-81; spl. asst. for sci. to commr. FDA, Rockville, Md., 1981-83, sci. cons., 1983—; asst. prof. pharmacology Georgetown U., 1982—. Contbr. articles to profl. jours. Mem. AAAS, Found. Advanced Edn. in the Scis., Am. Soc. Pharmacology and Exptl. Therapeutics, Sigma Xi, Phi Kappa Phi. Republican. Roman Catholic. Subspecialties: Cellular pharmacology; Immunopharmacology. Current work: Cell cycle analysis and cellular effects on anti-inflammatory drugs; laboratory research, teaching of medical students, directing research of graduate students, review of manuscripts, consulting food and drug administration.

BAYER, JESSE ABRAHAM, systems engineer; b. N.Y.C., Oct. 26, 1945; s. Irving and Sarah (Dratler) B.; m. Leslie Helen Bayer, Jan. 28, 1968; children: Hannah Michelle, Ariel Ian. B.E.M.E., Cooper Union, 1967; S.M.M.E., MIT, 1969; D.Engring. Sci., Columbia U., 1974. Registered engr. in tng. Mem. tech. staff Bell Labs., Whippany, N.J., 1967—. Active Friends of Orange Public Library, Friends of Turtle Back Zoo. Recipient award Cooper Union Alumni Assn., 1967. Mem. ASME. Subspecialties: Systems engineering; Software engineering. Home: 444 Prospect St South Orange NJ 07079 Office: Bell Laboratories Whippany Rd 1C216D Whippany NJ 07981

BAYER, SHIRLEY ANN, neurobiology researcher, educator; b. Evansville, Ind., Aug. 20, 1940; d. Richard Anthony and Mary Margaret (Memmer) B.; m. Joseph Altman, Dec. 8, 1973. B.A., St. Mary-of-the-Woods (Ind.) Coll., 1963; M.A., Calif. State U.-Fullerton, 1969; Ph.D., Purdue U., 1974. Asst. research scientist Purdue U., West Lafayette, Ind., 1974-81, assoc. research scientist, 1981-82; assoc. prof. anatomy Ind.-Purdue U., Indpls., 1983—; mem. neurology A study sect. NIH. Contbr. articles to profl. jours. NSF grantee; NIH grantee. Mem. Soc. for Neurosci., Internat. Soc. Developmental Neurosci., Am. Assn. Anatomists. Subspecialties: Neurobiology; Morphology. Current work: The development of the nervous system from a morphological viewpoint. Home: 231 Spring Valley West Lafayette IN 47906 Office: 1125 E 38th St Biology Dept Indianapolis IN 46223

BAYFIELD, JAMES EDWARD, physicist, educator; b. Wooster, Ohio, July 4, 1938; s. Edward G. and Esther R. (Zurcher) B.; m. Janet Ann Lafrentz; 1 son, Glenn Edward. B.A., Fla. State U., 1959, M.S. in Math, 1960, Yale U., 1961, Ph.D., 1967. Successively instr., asst. prof., assoc. prof. physics Yale U., New Haven, Conn., 1967-76; asso. prof., then prof. physics U. Pitts., 1976—; prin. sr. investigator research NSF, Dept. Energy; cons Gen. Atomic Co., Stanford Research Ins., Army Research Office, Oak Ridge Nat. Lab. Contbr. articles, revs. to profl. jours. Cottrell grantee, 1975-76; Nat. Bur. Standards grantee, 1975-78. Fellow Am. Phys. Soc. Republican. Episcopalian. Subspecialties: Spectroscopy; Atomic and molecular physics. Current work: Laser spectroscopy of atoms that are distorted by strong electromagnetic fields. Atomic and molecular excitation and ionization induced by strong microwaves and by infrared laser beams. Office: U Pitts 100 Allen Hall Pittsburgh PA 15260

BAYLIS, WILLIAM ERIC, physicist, educator; b. Providence, Nov. 28, 1939; s. Charles A. and Ruth W. (Weage) B.; m. Bobbye Kaye Whitenton, June 10, 1961; children: Evelyn M., Katherine R. B.Sc., Duke U., 1961; M.Sc., U. Ill., 1963; D.Sc., Tech. U. Munich, W.Ger., 1967. Postdoctoral research assoc. Joint Inst. Lab. Astrophysics, Boulder, Colo., 1967-69; asst. prof., then assoc. prof. U. Windsor, Ont., Can., 1969-78, prof. physics, 1978—; vis. scientist Max-Planck-Inst., Gottingen, W. Ger., 1976-77. Contbr. articles to profl. jours. Mem. Can. Assn. Physicists, Am. Phys Soc. Subspecialties: Atomic and molecular physics; Theoretical physics. Current work: Theory of atomic interactions; pseudopotentials, pressure broadening of spectral lines, correlation effects in relativistic hartree fock calculations. Home: RR 2 Harrow ON Canada N0R 1G0 Office: U Windsor Physics Dept Windsor ON Canada N9B 3P4

BAYLY, WARWICK MICHAEL, veterinary medicine eductor; b. Melbourne, Australia, Feb. 1, 1952; s. George Raymond and Jean B.; m. Della Marie Lewis; children: Matthew, Daniel. B.V. Sc. with honors, Melbourne U., 1974; postgrad., Tex. A&M U., 1976-77; M.S., Ohi State U., 1979. Diplomate: Am. Coll. Vet. Internal Medicine. Practice vet. medicine, Claresholm, Atla., Can., 1975-76; resident in large animal surgery Tex. A&M U., 1976-77; resident in equine medicine and surgery Ohio State U., 1977-79; asst. prof. vet. medicine Wash. State U., Pullman, 1979—. Mem. AVMA, Wash. State. Vet. Med. Assn. Subspecialties: Internal medicine (veterinary medicine); Animal physiology. Current work: Equine exercise physiology; diseases of equine cardiopulmonary system; investigation of physiological adaptations to training in the horse and design of different training programs; equine renal disease. Office: Dept Vet Clin Medicine and Surgery Wash State U Pullman WA 99164

BAYM, GORDON ALAN, physicist, educator; b. N.Y.C., July 1, 1935; s. Louis and Lillian B.; m. Lillian Hartmann; children—Nancy, Geoffrey, Michael. A.B., Cornell U., 1956; A.M., Harvard U., 1957, Ph.D., 1960. Fellow Universitetets Institut for Teoretisk Fysik, Copenhagen, Denmark, 1960-62; lectr. U. Calif.-, Berkeley, 1962-63; prof. physics U. Ill., Urbana, 1963—; vis. prof. U. Tokyo and U. Kyoto, 1968, Nordita, Copenhagen, 1970, 76, Niels Bohr Inst., 1976, U. Nagoya, 1979; vis. scientist Academia Sinica, China, 1979; mem. adv. bd. Inst. Theoretical Physics, Santa Barbara, Calif., 1978-83; mem. subcom. theoretical physics, physics adv. com. NSF, 1980-81, mem. phys. adv. com., 1982—; Mem. nuclear sci. adv. com. Dept. of Energy/NSF, 1982—. Author: Lectures on Quantum Mechanics, 1969, Neutron Stars, 1970, Neutron Stars and the Properties of Matter at High Density, 1977, (with L.P. Kadanoff) Quantum Statistical Mechanics, 1962; assoc. editor: Nuclear Physics. Recipient Alexander von Humboldt Found. sr. U.S. Scientist award, 1983; Fellow Am. Acad. Arts and Scis.; Alfred P. Sloan Found. Research fellow, 1965-68; NSF postdoctoral fellow, 1960-62. Fellow Am. Phys. Soc.; mem. Am. Astron. Soc., Internat. Astron. Union., Nat. Acad. Scis. Subspecialties: Theoretical physics; Quantum mechanics. Office: Loomis Lab Physics U Ill 1110 W Green St Urbana IL 61801

BAZERMAN, MAX H(AL), organizational behavior educator, researcher; b. Aug. 14, 1955; m. Marla Felcher. B.S. in Econs, U. Pa., 1976; M.S.O.B., Grad. Sch. Indsl. Adminstrn., Carnegie-Mellon U., 1978, Ph.D., 1979. Instr. Grad. Sch. Indsl. Adminstrn., Carnegie-Mellon U., 1977-79; asst. prof. mgmt. U. Tex.-Austin, 1979-80; asst. prof. organizational behavior Sch. Mgmt., Boston U., 1981—; cons., presentations in field; contbg. reviewer NSF, S.W. Acad. Mgmt. Contbr. articles, chpts. to profl. publs.; contbg. reviewer: Mgmt. Sci, 1980—; spl. assoc. editor, 1982—. Recipient prize for conf. paper presentation Columbia U., 1982; William Larimer Mellon fellow, 1976-79; Acad. Mgmt. Consortium fellow, 1978; Ford Motor Co. grantee, 1978-79; Zoecon Industries Corp. grantee, 1980; U. Tex. Grad. Sch. Bus. Adminstrn. course devel. grantee, 1980; research grantee, 1980; Office Naval Research grantee, 1980-81; Boston U. Sch. Mgmt. research grantee, 1981, 82; NSF grantee, 1981-83; Boston U. Grad. Sch. seed research grantee, 1982. Mem. Acad. Mgmt. (contbg. reviewer 1982—, chmn. symposium 1980, organizer symposium 1981, mem. task force on outstanding contbns. in orgn. and mgmt. theory 1980—), Eastern Acad. Mgmt. (contbg. reviewer), Am. Psychol. Assn. (hon. mention Cattell award div. 14 1982), Organizational Behavior Teaching Soc. Subspecialties: Social psychology; Oranizational psychology. Current work: Research in negotiation and arbitration processes, human judgement in managerial decision making, interlocking directorates, personal control, relationships between organization design and motivation/decision making, competitive bidding. Home: 996 Centre St Jamaica Plain MA 02130 Office: Sch Mgmt Boston U 621 Commonwealth Ave Boston MA 02215

BAZINET, LESTER, biologist, educator; b. Escanaba, Mich., May 5, 1929; s. Archie E. and Verlie (Trombly) B. B.S., No. Mich. U., 1959; M.S., U. Mich., 1961; Ed. D., Nova U., 1977; postgrad., Stanford U., 1969, 72, 79. Biology tchr., Swartz Creek, Mich., 1960-65; instr. biology Villanova (Pa.) U., 1965-66, 67-69; prof. biology Community Coll. Phila., 1970—, head dept., 1973, 74. Author: Laboratory Manual Morphology and Evolution in the Plant Kingdom, 1973, Development and Evaluation of a Biology Laboratory Unit on Cells and Tissues, 1975, The Development and Evaluation of an A-T Program of Instruction for Introductory Biology Laboratory, 1977, Audio-Tutorial Laboratory Guide for Biology 101, 1977. Mich. Bd. Regents scholar, 1947; Backus-Jewett fellow, 1960; NSF fellow, 1969; recipient Lindback award for Disting. Teaching, 1976. Mem. N.Y. Acad. Sci., Am. Bryol. and Lichenol. Soc., Am. Bot. Soc., Phycol. Soc. Am., Mich. Bot. Club, Phila. Bot. Club, Kappa Delta Pi. Subspecialties: Evolutionary biology; Taxonomy. Current work: Taxonomy and systematics; marine algae; teaching biology and botany. Home: 1146 S 8th St Philadelphia PA 19147 Office: 34 S 11th St Philadelphia PA 19107

BEACH, LEE ROY, psychologist, educator; b. Gallup, N.Mex., Feb. 29, 1936; s. Dearl and Lucile Ruth (Krumtum) B.; m. Barbara Ann Heinrich, Nov. 13, 1971. B.A., Ind. U., 1957; M.A., U. Colo., 1959, Ph.D., 1961. Aviation psychologist U.S. Sch. Aviation Medicine, Pensacola, Fla., 1961-63; human factors officer Office of Naval Research, Washington, 1963-64; postdoctoral research U. Mich., Ann Arbor, 1964-66; faculty dept. psychology U. Wash., Seattle, 1966—, prof. psychology, 1974—; cons. VA Med. Center, Seattle, 1979—. Contbr. articles to profl. jours.; author: Psychology: Core Concepts and Special Topics, 1973. Served with USN, 1961-64. Recipient Feldman Research award, 1981; NIMH fellow, 1964-66; NIH grantee, 1979—. Fellow Am. Psychol. Assn.; mem. AAAS. Subspecialties: Cognition; Social psychology. Current work: Research on judgement and decision processes and on metacognition. Home: 2129 2d Ave W Seattle WA 98119 Office: Dept Psychology NI25 Univ Wash Seattle WA 98195

BEACHLEY, NORMAN HENRY, mechanical engineer; b. Washington, Jan. 13, 1933; s. Albert Henry and Anna Garnet (Eiring) B.; m. Marion Ruth Iglehart, July 18, 1959; children: Brenda Ruth, Rebecca Sue, Barbara Joan. B.M.E., Cornell U., 1956, Ph.D., 1966. Mem. tech. staff Hughes Aircraft Co., Culver City, Calif., 1956-57; mem. tech. staff Space Tech. Labs., Redondo Beach, Calif., 1959-63; mech. engring. professorial staff U. Wis., Madison, 1966—, prof. mech. engring., 1978—; cons. Lawrence Livermore Labs., 1978—. Co-author: Introduction to Dynamic System Analysis, 1978. Served with USAF, 1957-59. Sci. and Engring. Research Council Gt. Britain fellow, 1981-82. Mem. ASME, Soc. Automotive Engrs., Sigma Xi. Subspecialty: Mechanical engineering. Research in field of energy storage powerplants for motor vehicles, 1970. Home: 2332 Fitchburg Rd Verona WI 53593 Office: U Wis 1513 University Ave Madison WI 53706

BEACHY, ROGER NEIL, biologist, educator; b. Plain City, Ohio, Oct. 4, 1944; s. Neil and Emma (Kramer) B.; m. Teresa Sophia Brown, Dec. 5, 1946; children: Kathryn Corrine, Kyle Andrew. B.A., Goshen (Ind.) Coll., 1966; Ph.D., Mich. State U., 1973. Postdoctoral assoc. dept. agrl. biochemistry U. Ariz., 1973; NIH postdoctoral fellow dept. plant pathology Cornell U., Ithaca, N.Y., 1973-76; research assoc. U.S. Dept. Agr., Cornell U., Ithaca, 1976-78; asst. prof. biology Washington U., St. Louis, 1978—; cons. in field. Contbr. chpts. to books, articles to profl. jours. NIH, Dept. Energy, NSF, Dept. Agr. grantee. Mem. Am. Soc. Virology, Am. Soc. Plant Physiology, Am. Phytopath. Soc., AAAS. Mennonite. Subspecialties: Molecular biology; Genetics and genetic engineering (agriculture). Current work: Gene structure and expression in plants; plant virology and viral gene expression; developmental biology of seeds. Office: Dept Biology Box 1137 Washington U Saint Louis MO 63130

BEADLE, CHARLES WILSON, mech. engr., educator; b. Beverly, Mass., Jan. 24, 1930; s. Thomas and Jean (Wilson) B.; m. Dorothy E. Beadle, May 5, 1956; children: Steven, Sara, Gordon. B.S., Tufts U., 1951; M.S.E., U. Mich., 1954; Ph.D., Cornell U., 1961. Registered prof. engr., Calif., Mich. With Gen. Motors Research Labs., 1951-54, RCA Research Labs., 1954-57; mem. faculty U. Calif.-Davis, 1961—, prof. mech. engring., 1975—, chmn. dept., 1978—. Contbr. articles to profl. jours. Ford Found. resident in engring. practice, 1965-66. Mem. ASME, AM. Soc. Engring. Edn. Subspecialties: Mechanical engineering; Solid mechanics. Current work: Computer-aided mechanical design.

BEADLES, JOHN KENNETH, biology educator; b. Alva, Okla., Sept. 22, 1931; s. Joseph H. and Ellen Amanda (Applebee) B.; m. Sharon Kay Ruch, Dec. 18, 1955; children: John David, Kristi Diane. B.S., Northwestern State Coll., 1957; M.S., Okla. State U., 1963, Ph.D., 1965. Tchr. sci. Alva (Okla.) Pub. Schs., 1957-62; grad. research asst. Okla. State U., Stillwater, 1963-65; asst. prof. biology Ark. State U., State University, 1965-66, prof. biology, chmn. dept. biol. sci., 1968—; teacher, researcher, adv. Haile Sellassie I U., Ethiopia, Africa, 1966-68; project dir. Ark. State U., 1971-77, cons. fish diseases, 1968—; dir. Flintrol, Inc. Contbr. articles to profl. jours.; assoc. editor: Ark. Acad. Scis. Aquatic Environment, 1973—. Trustee Jonesboro (Ark.) United Way, 1975-80. Served with USN, 1950-54. Named Outstanding Biology Tchr. Bell Telephone Co., 1960; Danforth assoc., 1970—. Mem. Am. Fisheries Soc., Southwestern Assn. Naturalists, Am. Soc. Ichthyologists and Herpetologists, Ark. Acad. Sci. (editorial bd. 1974-75, assoc. editor 1975—, pres. 1981-82), Sigma Xi, Phi Sigma. Lodge: Rotary. Subspecialties: Systematics; Taxonomy. Current work: Fishery biology using liquid scintellation counter and electron microscopy. Office: Ark State U Caraway State University AR 72467 Home: 1111 Thrush Rd Jonesboro AR 72401

BEAKES, JOHN HERBERT, consulting firm executive; b. Balt., Feb. 24, 1943; s. John Herbert and Martha (Ailes) B.; m. Rosemary Brown, June 11, 1966; children: Susan Dawn, Sarah Elizabeth, John Herbert. B.S. in Nuclear Sci, U.S. Naval Acad., 1966; M.S. in Environ. Engring, Johns Hopkins U., 1977. Registered profl. engr., Md., Pa., Tenn. Nuclear submarine officer U.S. Navy, 1966-74; sr. engr. Gen. Physics Corp., Columbia, Md., 1974-77, dir. engring. scis., 1977-79, v.p., 1979-82, sr. v.p., 1982—. Served as lt. comdr. USN, 1966-74. Mem. Am. Nuclear Soc., ASME (exec. com. plant engring. and maintenance div. 1981—). Republican. Presbyterian. Subspecialties: Nuclear engineering; Information systems (information science). Current work: Operability, maintainability and reliability enhancements to power generation facilities. Office: Gen Physics Corp 10650 Hickory Ridge Rd Columbia MD 21044

BEAL, JOHN M., physician; b. Starkville, Miss., 1915. M.D., U. Chgo., 1941. Diplomate: Am. Bd. Surgery (chmn. 1970-71). Intern N.Y. Hosp., N.Y.C., 1941-42, asst. resident surgery, 1942-44, 46-47, surgeon, 1947-48, attending surgeon, 1953-63; chmn. tumor bd. and staff surgeon Wadsworth Gen. Hosp., West Los Angeles, 1949-50, chief surg. service, 1950-53; cons. staff St. John's Hosp., Santa Monica, Cal., 1950-53; instr. surgery Cornell U., Ithaca, N.Y., 1948-49, assoc. prof. clin. surgery, 1953-63; instr. surgery UCLA, 1949-50, asst. prof., 1950-53; now J. Roscoe Miller disting. Northwestern U., chmn. dep. surgery, 1963-82; chmn. dept. surgery Chgo. Wesley Meml. Hosp., 1963-69, Northwestern Meml. Hosp., 1973-82; chief surgery Passavant Meml. Hosp., Chgo., 1963-73. Served to capt. M.C. AUS, 1944-46. Fellow ACS (bd. regents 1973-83, pres. 1982-83); mem. Council of Med. Splty. Socs. (sec. 1978-80), Soc. Univ. Surgeons, Soc. Clin. Surgery, AMA, Am. Surg. Assn. Subspecialty: Surgery. Address: Northwestern U Medical School Chicago IL 60611

BEALE, WILLIAM TAYLOR, research and development company executive, mechanical engr.; b. Chattanooga, Apr 17, 1928; s. David and Katherine (Burris) B.; m. Carol Phyllis, June 27, 1959; children:

Faith, Daniel, John. B.S.M.E., Wash. State U., 1950; M.S.M.E., Calif. Inst. Tech., 1953. Research engr. Wash. State U., 1950-52; engr. NACA, Cleve., 1953-57; asst. prof. mech. engring. Boston U., 1958-60; asst. prof. Ohio U., 1960-62, assoc. prof., 1962-75, adj. prof., 1975—; pres. Sunpower Inc., Athens, Ohio, 1975—; cons. in field. Contbr. numerous articles on Stirling engines to profl. jours. Served with USN, 1945-46. Democrat. Unitarian. Subspecialties: Mechanical engineering; Stirling engines. Current work: Stirling engines. Inventor Stirling engines. Home: Route 6 Box 73 Athens OH 45701 Office: 6 Byard St Athens OH 45701

BEALER, STEVEN LEE, physiologist, researcher; b. Evansville, Ind., July 10, 1949; s. Joseph Fredrick and Mildred Bernice (Biggerstaff) B.; m. Jan Marie Ebert, Aug. 28, 1971; 1 dau., Tracy. Ph.D., U. Wyo., 1976. Postdoctoral fellow U. Iowa, 1976-79; asst. prof. U. Tenn. Health Sci. Ctr., Memphis, 1979—. Mem. Am. Physiol. Soc., Soc. Neurosci. Subspecialty: Physiology (biology). Current work: Body-fluid electrolyte regulation.

BEALMEAR, PATRICIA MARIA, immunologist; b. Dodge City, Kans., Oct. 23, 1929; s. John Morgan and Anna Margaret (Wilson) B. B.S., Mount St. Scholastica Coll., 1949; Ph.D., U. Notre Dame, 1965; postgrad., Purdue U., 1957, U. Kans.-Lawrence, 1959, St. Louis U., 1960. Asst.-assoc. research scientist U. Notre Dame, Ind., 1966-71; assoc. prof. Baylor Coll. Medicine, Houston, 1971-76; assoc. cancer research scientist Roswell Park Meml. Inst., Buffalo, 1976—. Contbr. articles in field to profl. jours. Arthur Schmidt fellow U. Notre Dame, 1963-64; NSF fellow Purdue U., 1957, U. Kans., 1959, U. Notre Dame, 1962-63; NIH fellow U. Notre Dame, 1964-65. Mem. Internat. Assn. Gnotobiology (exec. sec. 1981—), Internat. Soc. Exptl. Hematology (trustee 1975-77), Assn. Gnotobiotics (exec. sec. 1975—), AAAS, Am. Assn. Immunologists, AAUP, Am. Assn. Pathologists, Radiation Research Soc., Soc. Exptl. Biology and Medicine, Transplantation Soc., N.Y. Acad. Scis., Sigma Xi, Kappa Gamma Pi. Roman Catholic. Subspecialties: Immunology (medicine); Marrow transplant. Current work: Elimination of graft-vs host disease following bone marrow transplantation using the germfree mouse as the model system. Office: Roswell Park Meml Inst 666 Elm St Buffalo NY 14263

BEAM, THOMAS ROGER, infectious diseases researcher; b. Elizabeth, N.J., July 12, 1946; s. Thomas Roger and Lillian (Norloff) B.; m. Janice Victoria Niesz, Aug. 15, 1970; children: Nancy Victoria, Thomas Roger. A.B. in Biology, U. Pa., 1968, M.D., 1972. Diplomate: Am. Bd. Internal Medicine. Intern, resident, chief resident SUNY-Buffalo, 1972-75, fellow in infectious diseases, 1975-77, asst. prof. medicine, 1977—; chief infectious diseases Buffalo VA Med. Ctr., 1977—; cons. infectious diseases Erie County Med. Ctr., Buffalo, 1981—, J.N. Adam Devel. Ctr., Perrysburg, N.Y., 1981—, West Seneca (N.Y.) Devel. Ctr., 1982—. Author: Infectious Diseases for the House Officer, 1982; contbr. articles to profl. jours. Recipient Joel S. Miller award U. Pa. Sch. Medicine, 1972; Bristol fellow, 1975. Fellow ACP; mem. Infectious Diseases Soc. Am., Am. Fedn. Clin. Research, Am. Soc. Microbiology, N.Y. Acad. Sci. Subspecialties: Infectious diseases; Internal medicine. Current work: Research into antimicrobial therapy of meningitis; immune defenses in the central nervous system; adult immunizations; infection control; geriatric infections. Office: Buffalo VA Med Center 3495 Bailey Ave Buffalo NY 14215

BEAMER, ROBERT LEWIS, pharmacy educator, researcher; b. Pulaski, Va., June 9, 1933; s. Harold Lee and Mary Overton (Smith) B.; m. Joan Fanning, June 6, 1959; 1 dau., Janelle Elizabeth. Student, Davidson Coll., 1950-52; B.S. in Pharmacy, Med. Coll. Va., 1955, M.S., 1957, Ph.D., 1959. Asst. prof. med. chemistry U. S.C., Columbia, 1959-63, assoc. prof., 1963-68, prof., 1968-72, prof. and assoc. dean, 1972-75, prof., 1975-77, 1978—; vis. prof. biochemistry Cornell U. Med. Coll., N.Y.C., 1977-78; cons. author Pharmat Inc., Lawrence, Kans., 1980-81; panel mem. NSF, Boston, 1968; mem. thesis coms. Sri Venteswara U., India, 1982-83; mem. U.S. Pharm. Conv., Washington, 1980—. Contbr. articles on pharmacology to profl. jours. Active drug abuse edn. U. S.C., 1967-75. William S. Merrell Co. grantee, 1963; NIH grantee, 1964; Smith Kline and French grantee, 1964, 66, 68; NSF grantee, 1969; Eli Lilly and Co. grantee, 1973. Mem. Am. Chem. Soc., Am. Pharm. Assn., N.Y. Acad. Scis., AAAS, Am. Assn. Colls. Pharmacy, Am. Diabetes Assn. (lectr.), 5th Dist. Pharm. Assn. Democrat. Methodist. Club: Torch Internat. Subspecialties: Medicinal chemistry; Biochemistry (medicine). Current work: Heterogeneous and homogeneous catalysis, asymmetric synthesis, enzymes, amino acids, proteins, peptides, enzyme inhibitors, drugs, neurochemistry. Home: 1330 Raintree Dr Columbia SC 29210 Office: U SC Coll Pharmacy Columbia SC 29208

BEAMES, CALVIN G., JR., physiologist; b. Kingston, Okla., Oct. 29, 1930; s. Calvin G. and Grace Loree (Chesnut) B.; m. J. Charlene Smith, June 1, 1952; children: Deborah Lee, Calvin G. III, Rebecca Ann. A.B., N.Mex. Highland U., 1955, M.S., 1956; Ph.D., U. Okla., 1961. Tchr. high sch. biology, Santa Fe, N.Mex., 1956-57; teaching asst. U. Okla., Norman, 1957-60; NIH postdoctoral fellow Rice U., Houston, 1960-62; asst. prof. Okla. State U., Stillwater, 1962-63, assoc. prof., 1963-70, prof. physiology, 1970—. Author papers, monographs and books. Served with U.S. Army, 1950-52. NIH grantee, 1964-74, 75-79; Okla. State U. grantee, 1981-82. Mem. Am. Physiol. Soc., AAAS, Am. Soc. Zoology, Am. Chem. Soc., Am. Soc. Parasitologists, Okla. Acad. Scis. Democrat. Methodist. Subspecialties: Physiology (biology); Physiology (biology). Current work: Transport Mechanisms of epithelium - lipid metabolism. Office: Okla State U Room 402 LSW Stillwater OK 74078

BEAN, CHARLES PALMER, biophysicist; b. Buffalo, Nov. 27, 1923; s. Barton Adrian and Theresa (Palmer) B.; m. Elizabeth Harriman, Sept. 13, 1947; children: Katherine G., Bruce P., Margaret E., Sarah H., Gordon T. B.A., U. Buffalo, 1947; M.A., U. Ill., 1948, Ph.D., 1952. Research scientist Gen. Electric Co., Schenectady, 1951—; adj. asso. prof. Renssalear Poly. Inst., Troy, N.Y., 1957-67, adj. prof., 1978—, Disting. prof., 1983—; adj. prof. SUNY, 1978—, Union Coll., Schenectady, 1981—; guest investigator Rockefeller U., N.Y.C., 1973-76. Contbr. articles to profl. jours.; asso. editor: Biophys. Jour., 1974—. Bd. dirs. Dudley Obs., Albany, N.Y., 1975—, pres., 1983—; Bd. dirs. Bellevue Research Found., Schenectady, 1973—. Served with USAAF, 1943-46. Fellow Am. Phys. Soc.; mem. Biophys. Soc., N.Y. State Acad. Scis., Am. Acad. Arts and Scis., Nat. Acad. Sci. Clubs: Fortnightly (Schenectady); Chemists (N.Y.C.). Subspecialties: Biophysics (biology); Molecular biology. Current work: Study of structure and properties of bio-molecules and biosystems. Patentee in field. Home: 2221 Stoneridge Rd Schenectady NY 12309 Office: Box 8 Schenectady NY 12301:

BEAN, GERRITT POST, chemist, educator; b. Amsterdam, N.Y., Apr. 29, 1929; s. Charles Leslie and Doris Elizabeth (Post) B.; m. Virginia Lewis Brewer, June 9, 1956. B.S., Northeastern U., 1952; Ph.D., Pa. State U., 1956. Research chemist Merck and Co., Inc., Danville, Pa., 1956-57, Althouse Chem. Co., Reading, Pa., 1957-60; asst. prof. chemistry Douglass Coll., Rutgers U., New Brunswick, N.J., 1960-66; vis. prof. U. East Anglia, Norwich, Eng., 1966-67; prof. Western Ill. U., 1967—, chmn. dept. chemistry, 1981—. Author: The Chemistry of Pyrroles, 1977. Mem. Am. Chem. Soc., Chem. Soc. (London). Subspecialty: Organic chemistry. Current work: Heterocyclic organic chemistry. Home: 16 Grandview Dr Macomb IL 61455 Office: Dept Chemistry Western Ill U Macomb IL 61455

BEAN, MICHAEL ARTHUR, cancer researcher, research center administrator; b. Alliance, Nebr., Sept. 18, 1940; s. Kenneth C. and Catherine (Chisolm) B.; m. Susan J. McKay, Jan. 3, 1978. B.A., U. Colo., 1962, M.D. cum laude, 1967. Diplomate: Am. Bd. Pathology. Intern in pathology Yale U., 1967-68; resident in pathology USPHS Hosp., S.I., N.Y., 1968-69; staff pathologist Armed Forces Inst. Pathology, Washington, 1969, Atomic Bomb Casualty Commn., Hiroshima, Japan, 1969-70; clin. research trainee Meml. Sloan Kettering Cancer Ctr., N.Y.C., 1970-72, fellow, 1970-72, assoc., 1970-75; sr. investigator Virginia Mason Research Ctr., Seattle, 1975-77, mem., 1977—, assoc. dir., 1980—; affiliate investigator Fred Hutchinson Cancer Research Ctr., 1975—; affiliate assoc., prof., microbiology and immunology U. Wash., 1976—; ad hoc cons, for adv. coms Nat. Cancer Inst., Am. Cancer Soc. Contbr. numerous chpts., articles to profl. publs. Served to lt. comdr. USPHS, 1968-70. Recipient Meller award Meml. Hosp., 1973, Faculty Research award Am. Cancer Soc., 1978-82. Mem. AAAS, Am. Assn. Cancer Research, Am. Assn. Immunologists, Phi Rho Sigma. Subspecialties: Cancer research (medicine); Immunology (medicine). Current work: Cancer, immunology, immunoregulation, immunotherapy, transplantation immunology, suppressor cells. Office: 1000 Seneca St Seattle WA 98101

BEAN, WILLIAM JOSEPH, JR., virologist; b. Albany, N.Y., Mar. 16, 1945; s. William Joseph and Ruth Elizabeth (Lafferty) B.; m. Dianne Lee Pendleton, Sept. 28, 1968; 1 dau., Shawn Katherine. B.S., U. Maine, 1967, M.S. in Bacteriology, 1969; Ph.D. in virology, Rutgers U., 1974. Research asst. Waksman Inst. Microbiology, Piscataway, N.J., 1974-75; research assoc. St. Jude Children's Research Hosp., Memphis, 1975-77, asst. mem., 1977-82, assoc. mem., 1982—; asst. prof. microbiology U. Tenn., Memphis, 1981—. Served to lt. U.S. Army, 1968-71. Mem. Am. Soc. Microbiology, Am. Soc. Virology. Subspecialties: Virology (biology); Molecular biology. Current work: Genetics of influenza virus; molecular and biological basis of genetic changes in human and animal influenza viruses. Home: 3981 Cheryl Dr Memphis TN 38116 Office: 332 N Lauderdale Memphis TN 38101

BEARD, JACOB THOMAS BARRON, JR., mathematics educator; b. Jacksonville, Fla., Nov. 29, 1940; s. Jacob Thomas Barron and Mamie (Green) B.; m. Marjorie Elizabeth Pybas, Dec. 29, 1961; children: Elizabeth Michelle, Katherine Ann. B.S., Tenn. Technol. U., 1962, M.S., 1967; Ph.D., U. Tenn., 1971. Mathematician TVA, Chattanooga, 1962-65; asst. prof. U. Tex.-Arlington, 1971-76, assoc. prof., 1976-79; prof. math. Tenn. Technol. U., 1979—, chmn. dept. math. and computer sci., 1979-81; vis. assoc. prof. Emory U., 1978-79. Research publs. in field. U. Tenn. non-service fellow, 1969-71; U. Tex.-Arlington grantee, 1972-78. Mem. Am. Math. Soc., Soc. Indsl. and Applied Math., Math. Assn. Am., AAUP, Sigma Xi, Pi Mu Epsilon (charter faculty sponsor Tex. Iota chpt. 1975). Episcopalian. Subspecialties: Algebra and number theory; Linear algebra. Current work: Fields, matrices, polynomials, quadratic residues. Office: Tenn Technol U Dept Math and Computer Sci Cookesville TN 38501

BEARD, JAMES TAYLOR, mechanical engineering educator, consultant; b. Birmingham, Ala., Oct. 1, 1939; s. James Robison and Mary Evelyn (McArthur) B.; m. Kathryn LuClaire Lee, June 5, 1965; children: Rosemary Ann, James David. B.M.E., Auburn (Ala.) U., 1961; M.S., Okla. State U., Stillwater, 1963, Ph.D.M.E. (NDEA fellow), 1965. Registered profl. engr., Va. Research asst. Okla. State U., 1963-65; asst. prof. mech. and aerospace engring. U. Va., Charlottesville, 1965-69, assoc. prof., 1969—, asst. provost, 1972-77, mem. faculty senate, 1983—; ptnr. Assoc. Environ. Consultants, Charlottesville, 1971—; short course dir. EPA, 1972-82; tech. adv. com. Va. Air Pollution Control Bd. Contbr. articles to profl. jours. Elder, clk. of session Westminster Presbyn. Ch.; mem. Albemarle County Democratic Com., 1972—; treas. bd. dirs. Charlottesville Housing Found., 1969—. Recipient Algernon Sydney Sullivan award Auburn U., 1961. Mem. ASME, ASHRAE, Internat. Solar Energy Soc., Sigma Xi, Tau Beta Pi, Phi Kappa Phi, Pi Tau Sigma. Club: Colonnade (Charlottesville). Subspecialties: Mechanical engineering; Solar energy. Current work: Heat transfer, solar energy system analysis, air pollution control, univ. engring. educator, combustion as related to air pollution control. Home: 412 Westmoreland Ct Charlottesville VA 22901 Office: Dept Mech and Aero Engineering U Va Charlottesville VA 22901

BEARD, LEO ROY, civil engineer; b. West Baden, Ind., Apr. 6, 1917; s. Leonard Roy and Barbara Katherine (Frederick) B.; m. Marian Janet Wagar, Oct. 21, 1939 (dec.); children: Patricia Beard Huntzicker, Thomas Edward, James Robert; m. Marjorie Elizabeth Pierce Wood, Aug. 30, 1974. A.A., Pasadena City Coll., 1937; B.S., Calif. Inst. Tech., 1939. Engr. U.S. Army C.E., Los Angeles, 1939-49; engr. Office Chief of Engrs., Washington, 1949-52; chief of Reservoir Regulation, Sacramento, 1952-64; dir. Hydrologic Engring. Center, Davis, Calif., 1964-72; prof. civil engring. U. Tex., Austin, 1972—; dir. (Center for Research in Water Resources), 1972-80; cons. Espey, Huston & Assos., Austin, 1980—; v.p. Internat. Commn. of Water Resource Systems; mem. NRC Water Sci. and Tech. Bd. Editor-in-chief: Water International; Editor: Jour. of Hydrology. Served with USNR, 1945-46. Recipient Meritorious Civilian Service award U.S. Army C.E., 1972. Fellow ASCE (water resources exec. com., Julian Hinds award 1981); mem. Internat. Water Resources Assn. (exec. bd.), Am. Water Resources Assn. (hon.), Am. Geophys. Union (pres. hydrology sect.), Nat. Soc. Profl. Engrs., Internat. Assn. Hydrol. Scis., World Meteorol. Orgn. (chmn. com. on hydrol. design data), U.S. Com. on Irrigation, Drainage and Flood Control, Univs. Council on Water Resources (exec. bd.), Nat. Acad. Engring. Subspecialties: Civil engineering; Hydrology. Current work: All aspects of water resources management. Home: 606 Laurel Valley Austin TX 78746 Office: PO Box 519 Austin TX 78767

BEARDEN, ALAN JOYCE, biophysicist; b. Balt., Nov. 23, 1931; s. Joyce Alvin and Lillian Lavonia (Singleton) B. A.B., Johns Hopkins U., 1950, Ph.D., 1958. Asst. prof. physics Cornell U., Ithaca, N.Y., 1960-64; asst. prof. chemistry U. Calif., San Diego, 1966-68, from lectr. to asso. prof. div. med. physics, Berkeley, 1969-76, prof. biophysics, 1976—, chmn. div. med. physics, 1978-79, chmn. dept. biophysics and med. physics, 1979—. Served with USN, 1950-54. NIH career devel. awardee, 1970-74; NIH fellow U. Calif., San Diego, 1964-66; Pollard lectr. Yale U., 1976. Fellow Am. Phys. Soc. (chmn. div. biol. physics 1978-79); mem. Biophys. Soc., Am. Soc. Photobiology, AAAS. Subspecialty: Biophysics (physics). Research in photosynthesis, laser radiation, bioenergetics, energy transfer in biology. Office: Dept Biophysics and Med Physics U Calif Berkeley CA 94720

BEARDEN, HENRY JOE, physiologist, educator; b. Starkville, Miss., May 12, 1926; s. Henry and Tera (Dunaway) B.; m. Marie Josephine Publick, Oct. 26, 1946; children: John Henry, James Rodney, Jeffrey Lee. B.S., Miss. State U., 1950; M.S., U. Tenn., 1951; Ph.D., Cornell U., 1954. Grad. asst. Cornell U., 1951-54, asst. prof., 1954-60, assoc. prof., 1960; prof., head dairy sci. dept. Miss. State U., 1960—; mem. Miss. Bd. Animal Health; adv. bd. Mid-South Animal Breeders Assn., Am. Dairy Assn. Miss., Food and Nutrition Council Deep South. Author: Applied Animal Reproduction, 1980; contbr. articles to profl. jours. Leader, Pushmataha council Boy Scouts Am., 1958-60, 62-69. Served with USN, 1944-46. Mem. Am. Dairy Sci. Assn., Soc. Study of Reprodn., U.S. Animal Health Assn., So. Assn. Agrl. Scientists, Sigma Xi, Alpha Zeta, Phi Kappa Phi, Gamma Sigma Delta. Republican. Presbyterian. Subspecialty: Animal physiology. Current work: Reproductive physiology in dairy cattle and white-tailed deer. Office: Dairy Sci Dept Drawer DD Mississippi State MS 39762

BEARISON, DAVID J., psychologist, educator; b. N.J., Feb. 22, 1944; s. Clarence and Fay (Soskin) B. B.A., Pa. State U., 1965; fellow, Merrill-Palmer Inst., Detroit, 1965-66; M.A., Clark U., 1968, Ph.D., 1973, in psychiatry, Harvard U. Med. Sch., Boston, 1981-82. Asst. prof., mem. doctoral faculty Grad. Center CUNY, 1973-78, assoc. prof., 1978—, sr. research assoc. Center Advanced Study in Edn., 1978—; fellow in psychiatry Harvard U. Med. Sch., Boston. Contbr. articles to sci. jours.; editorial bd.: Human Devel., 1978—, Sex Roles: A Jour. of Research, 1975—, Genetic Epistemologist, 1980—, Jour. Applied Developmental Psychology, 1982—. Mem. Jean Piaget Soc. (dir. 1978-84), Am. Psychol. Assn. Subspecialty: Developmental psychology. Current work: Research and theories of cognitive development across life-span; childhood psychopathology/applied development psychology/sex role development and socialization. Office: 33 W 42d St New York NY 10036

BEARN, ALEXANDER GORDON, physician, scientist; b. Surrey, Eng., Mar. 29, 1923; came to U.S., 1951; s. Edward Gordon and Rose (Kay) B.; m. Margaret Slocum, Dec. 20, 1952; children—Helen Elliot, Gordon Clarence Frederic. Ed., Epsom Coll.; M.B., B.S., Guy's Hosp., U. London, Eng., 1946; M.D., 1951; M.D. Dr. honoris causa, U. René Descartes, Paris, 1974. House physician Guy's Hosp., 1946-47; house physician, registrar Postgrad. Med. Sch., London, 1948-51; mem. staff Rockefeller Inst., N.Y.C., 1951-64, asso. prof., 1957-64, prof., sr. physician, 1964-66; prof. medicine Cornell U., 1966—, Stanton Griffis Distinguished med. prof., 1976-80, chmn. dept., 1966-77; physician-in-chief N.Y. Hosp., 1966-77; med. dir., bd. dirs., sec.-treas. Russell Sage Inst. of Pathology, 1967-79; sr. v.p. for med. and sci. affairs Merck, Sharp & Dohme Internat., Rahway, N.J., 1979—; Lilly lectr., 1973, Lettsomian lectr., 1976; bd. sci. cons. Sloan Kettering Inst., 1967-74; mem. Commn. Human Resources, Nat. Acad. Scis., 1974-77; chmn. div. med. scis. Assembly Life Scis., 1978—; bd. sci. counselors Nat. Inst. Arthritis, Metabolism and Digestive Diseases, 1976-80; mem. Space Sci. Bd., 1978-79; cons. genetics tng. com., div. gen. med. scis. USPHS, 1961-65, cons. genetics study sect., 1966-70; pres. Royal Soc. Medicine Found., Inc., 1976-78; now dir.; mem. bd. sci. overseers Jackson Lab., Bar Harbor; mem. Inst. Medicine, Nat. Acad. Scis. Editor: Am. Jour. Medicine; co-editor: Progress in Medical Genetics, 1962—; asso. editor: Cecil-Loeb Textbook of Medicine; Contbr. articles to profl. jours. Trustee Rockefeller U., Helen Hay Whitney Found. Served as med. officer RAF, 1947-49. Fellow AAAS, Royal Coll. Physicians (Edinburgh, Scotland), Royal Coll. Physicians (London, Eng.); mem. Nat. Acad. Scis., Am. Philos. Soc., Assn. Am. Physicians, Am. Soc. Clin. Investigation, Am. Soc. Human Genetics (pres. 1971), Genetics Soc. Am., Am. Soc. Biol. Chemists, Soc. Exptl. Biology and Medicine, Harvey Soc. (pres. 1972-73, Harvey lectr. 1975), Harveian Soc. London (Council 1959), Assn. Physicians Great Britain and Ireland, Med. Research Soc. Great Britain, Med. Soc. London, Am. Assn. History Medicine, Sigma Xi (pres. Rockefeller chpt. 1962-63); fgn. asso. Norwegian Acad. Sci. and Letters; hon. mem. Sociedad Medica de Santiago, Sociedad de Biologia de Santiago. Presbyterian. Clubs: The Bath (London, Eng.); Century Assn., Grolier (N.Y.C.); Crail Golf (Scotland). Subspecialty: Genetics and genetic engineering (medicine). Home: 1225 Park Ave New York NY 10028 Office: Merck and Co Inc Rahway NJ 07065

BEASLEY, LEROY B., mathematics educator; b. Shelley, Idaho, July 31, 1942; s. Lawrence Byington and Grace Vivian (Davis) B.; m. Debra Anne Schaefer, July 18, 1975; children: Mia Louise, Lisa Nichelle. B.S., Idaho State U., Pocatello, 1964, M.S., 1966; Ph.D., U. B.C., Vancouver, 1969. Teaching asst. Idaho State U., 1964-66, U. B.C., 1966-69; tchr. Middleton (Idaho) High Sch., 1972-81; asst. prof. Utah State U., Logan, 1981—; research assoc. U. Waterloo, Ont., Can., summers 1972, 73, 77-82. Contbr. articles to profl. jours. Served as capt. U.S. Army, 1970-72; Viet Nam. Mem. Math. Assn. Am., Am. Math. Soc., Soc. Indsl. and Applied Math. Lodge: Masons. Current work: Matrix theory, boolean matrices, linear transformations on matrices, spaces of matrices, combinational group theory. Office: Utah State Univ Dept Math UMC 41 Logan UT 84322

BEASLEY, ROBERT LAWRENCE, farm supply cooperative executive; b. Poplar Bluff, Mo., Mar. 6, 1929; s. C. F. and Zula Mae (McAllister) B.; m. Betty Lou Lepp, July 21, 1951; children: Rob, Ann. B.A. in Journalism, U. Mo., 1952. Reporter, photographer, editor Columbia (Mo.) Tribune, 1951-54, Dubuque (Iowa) Herald, 1954-56, Madison (Wis.) State Jour., 1956-57; with Farmland Industries Inc., Kansas City, Mo., 1957—, v.p., 1971—. Chmn. bd. govs. Kansas City YMCA; bd. dirs. Cablevision of Kansas City. Mem. Coop. League U.S.A. (chmn. 1974-76, 83-84, vice-chmn. 1980), Internat. Coop. Alliance Econ. Bur. (vice chmn. 1980). Office: 3315 N Oak Trafficway Kansas City MO 64116

BEASLEY, WAYNE MACHON, materials science educator, consultant; b. Everett, Mass., May 23, 1922; s. William Francis and Elsie May (Machon) B.; m. Evelyn Harriet Eddy, Feb. 28, 1945; 1 dau., Dawn Linda. S.B., Harvard U., 1945; S.M., M.I.T., 1965. Tchr. math. Dover (N.H.) High Sch., 1946-48; cons., Rochester, N.H., 1948-51; physicist Clarostat mfg. Co., Dover, 1951-55; sr. physicist Metals & Controls Corp., Attleboro, Mass., 1955-57; research prof. U. N.H., Durham, 1957-72, prof. materials sci., 1972—; cons. materials tech. to electronics industry in U.S. and Japan. Contbr. articles to various jours. Mem. budget com., Town of Barrington (N.H.), 1978-81; mem. roads com., 1980-81, planning bd., 1982—, Waste Disposal Study Commn., Barrington, 1975-78. Served to ensign USN, 1943-45; PTO. Mem. Am. Phys. Soc., European Phys. Soc., Internat. Soc. for Stereology, Am. Soc. for Metals (pres. N.H. chpt. 1959-60), N.H. Acad. Sci., Union Concerned Scientists, Sigma Xi. Clubs: Harvard Faculty (Cambridge, Mass.); Harvard (N.H.). Subspecialties: Materials; X-ray crystallography. Current work: Application of quantitative stereological techniques to particle packing; research involving spinodal decomposition transformations. Co-patentee method of elec. noise reduction in potentiometers. Home: 22 Weeks Ln Rochester NH 03867 Office: University of New Hampshire Kingsbury Hall Durham NH 03824

BEASON, ROBERT CURTIS, biology educator; b. Ft. Scott, Kans., May 12, 1946; s. Eugene and Lida Jane (Lawson) B.; m. Julia Ann Farthing, Aug. 10, 1967 (div. 1973); m. Delena Lorraine Sloane, Feb. 14, 1981; 1 son, Zachary Adam Sloane. B.A., Bethany Nazarene Coll., 1968; M.S., Western Ill. U., 1970; Ph.D., Clemson U., 1976. Research scientist U.S. Air Force, Kirtland AFB, N.Mex., 1970-74; grad. research asst. Clemson (S.C.) U., 1974-76; research biologist U.S. Forest Service, Columbia, S.C., 1976; vis. lectr. U. Calif.-Irvine, 1977; vis. asst. prof. Western Ill. U., Macomb, 1977-78; asst. prof. biology SUNY-Geneseo, 1978—; cons. FAA, 1977, NASA. Reviewer: Jour. Field Ornithology, 1975—. U.S. Dept. Interior grantee, 1974; NSF grantee, 1981; Geneseo Found. grantee, 1982. Mem. AAAS, Am. Ornithologists Union (life), Animal Behavior Soc., Ecol. Soc. Am.

Subspecialties: Behavioral ecology; Evolutionary biology. Current work: Evolution of migration; avian migration; orientation andnavigation; avian behavioral ecology; role of vocalizations and behavior in avian speciation. Home: 7673 Dutch St Rd Mount Morris NY 14510 Office: Dept Biology SUNY Geneseo NY 14454

BEATON, ROY HOWARD, nuclear industry exec.; b. Boston, Sept. 1, 1916; s. John Howard and Mary Beaton (LaVoie) B.; m.; children—Constance Beaton Reinholz, Roy Howard. B.S., Northeastern U., 1939, D.Sc. (hon.), 1967; D. Eng., Yale U., 1942. With E.I. duPont, 1942-45, plant tech. supr., 1944-45; asso. prof. chem. engring. U. Kans., Lawrence, 1946; with Gen. Electric Co., 1946—, v.p., gen. mgr. electronic systems div., Syracuse, N.Y., 1968-74, v.p., gen. mgr. energy systems and tech. div., Fairfield, Conn., 1974-75, v.p., gen. mgr. nuclear energy systems div., San Jose, Calif., 1975-77, v.p., group exec. nuclear energy group, 1977-79, sr. v.p., group exec. nuclear energy group, 1979—. Chmn. industry div. United Way Campaign, Santa Clara County, Calif., 1978-79. Fellow Am. Inst. Chemists, AAAS, Nat. Acad. Engring., Am. Ordnance Assn., Am. Nuclear Soc., Am. Inst. Chem. Engrs., IEEE, AIAA, Navy League U.S., Air Force Assn., Soc. Mil. Engrs., Santa Clara County Mfg. Group, Sigma Xi, Tau Beta Pi. Subspecialty: Nuclear engineering. Home: PO Box 1018 Saratoga CA 95070 Office: 175 Curtner Ave San Jose CA 95125

BEATTIE, CRAIG WARREN, pharmacologist, educator; b. Elizabeth, N.J., Oct. 26, 1943; s. Warren Edgar and Margaret Ann (Sasz) B.; m. Jean Blankenship, Nov. 24, 1973. B.S., Fairleigh Dickinson U., 1965, M.S., 1968; Ph.D., U. Del., 1971. Postdoctoral fellow dept. pharmacology U. Ill., 1971-72; sr. research scientist Pharm. div. Penwalt Corp., 1972-73; sr. research biologist endocrine sect. dept. pharmacology Wyeth Labs., 1973-76; asst. prof. pharmacology Bowman-Gray Sch. Medicine, Wake Forest U., 1976-77; assoc. prof. pharmacology in surgery div. surg. oncology U. Ill. Sch. Medicine, 1977-83-, prof. pharmacology in surgery div. surg. oncology, 1983—, assoc. prof. pharmacology, 1977-83; vis. prof. Sch. Biochemistry U. New South Wales, Australia, 1983-84. Contbr. numerous articles and revs. to sci. jours., chpts. to books. Recipient award Pharm. Mfrs. Assn. Found., 1976-77; U. Del. research fellow, 1968-70; NIH postdoctoral fellow, 1970-72. Mem. Am. Soc. Pharmacology and Exptl. Therapeutics, Am. Physiol. Soc., Endocrine Soc., Am. Assn. Cancer Research, Tissue Culture Soc., Soc. for Study Reprodn., Am. Soc. Zoologists, Sigma Xi. Subspecialties: Cancer research (medicine); Receptors. Current work: Endocrine correlates to neoplastic transformation and metastasis. Office: Div Surg Oncology U Ill 370A 840 S Wood St Chicago IL 60612

BEATTIE, MICHAEL STEPHEN, neuroscientist, medical educator; b. Boston, June 20, 1949; s. George Chapin and Nancy Ulrica (Fant) B.; m. Denise Elizabeth Craik, Sept. 14, 1970; 1 dau., Jennifer Bresnahan; m. Jacqueline Connor Bresnahan, June 14, 1975. B.S. in Psychology, U. Calif.-Davis, 1972; M.A. in Neuropsychology, Ohio State U., 1974, Ph.D., 1977; postdoctoral, Mich. State U., 1979. Postdoctoral fellow dept. anatomy Ohio State U., Columbus, 1977-78, asst. prof. depts. neurosurgery and anatomy, 1979—; adv. cons. NIH, 1980. Contbr. articles to sci. jours. Sponsor Union Concerned Scientists, 1982—. NIH fellow, 1977-78; NIH and Ohio State U. Grad. Sch. grantee, 1980—. Mem. AAAS, Soc. Neurosci., Internat. Assn. Study of Pain. Subspecialties: Neurobiology; Neurophysiology. Current work: Development and plasticity of the spinal cord; neuroscience research and teaching; ultrastructural techniques; neuroanatomy; neurophysiology, structure and function relationships. Home: 158 Glencoe Rd Columbus OH 43214 Office: 1645 Neil Ave Rm 410 Columbus OH 43210

BEATY, EARL CLAUDE, physicist; b. Zeigler, Ill., Nov. 6, 1930; s. James Ollious and Cecile Mae (Bayless) B.; m. Velma Earle Cornette, Dec. 21, 1952; children: David, Paul, Neil. A.B., Murray (Ky.) State U., 1952; Ph.D., Washington U., St. Louis, 1956. Physicist Nat. Bur. Standards, Washington, 1956-62; fellow Joint Inst. for Lab. Astrophysics, Boulder, Colo., 1962-81; physicist Nat. Bur. Standards, Boulder, 1981—, Internat. Atomic Energy Agy., 1977-78. Contbr. articles to profl. jours. Fellow Am. Phys. Soc.; mem. Am. Geophys. Union. Subspecialty: Atomic and molecular physics. Current work: Atomic and molecular physics research with emphasis on electron-atom collisions; application of atomic physics to atomic frequency standards; laser frequency measurements. Office: National Bureau Standards Boulder CO 80303

BEATY, TERRI HAGAN, geneticist, educator; b. Austin, Tex., Mar. 28, 1951; d. Leander Jackson and Mary Marie (DeGlandon) Hagan; m. Narlin Bennet Beaty, May 27, 1972; 1 son, Narlin, Jr. B.A. in Biology, U. Tex-Austin, 1972, M.A. in Zoology, 1974; Ph.D. in Human Genetics, U. Mich., 1978. Postdoctoral scholar div. endocrinology and metabolism U. Mich. Med. Sch., Ann Arbor, 1978-79; asst. prof. dept. epidemiology Sch. Hygiene and Public Health, Johns Hopkins U., Balt., 1980—. Contbr. numerous articles to profl. jours. Mem. Genetics Soc. Am., Am. Soc. Human Genetics, Phi Beta Kappa. Subspecialties: Genetics and genetic engineering (biology); Statistics. Current work: Statistical genetics; quantitative human genetics; genetic epidemiology. Office: 615 N Wolfe Room 6029 Baltimore MD 21205

BEAULNES, AURELE, physician, research inst. adminstr.; b. Montreal, Que., Can., Aug. 7, 1928; s. Lucien and Berthe (Courteau) B.; m. Rita Archambault, June 16, 1953; children—Pierre, Marie-Helene, Genevieve. B.A., U. Montreal, 1946, M.D., 1953. Intern Notre Dame, St. Justine, St. Jean-de-Dieu, Montreal, 1952-53; resident Notre Dame Hosp., 1953; asst. prof. pharmacology U. Montreal Faculty Medicine, 1957-59, asso. prof., 1959-62, prof., 1962-66, head dept. pharmacology, 1959-66, vice dean, 1962-64; prof., head dept. pharmacology, dir. basic scis. div. U. Sherbrooke Faculty Medicine, 1966-67; vis. prof. U. Ottawa, Ont., 1967-68; prof. pharmacology and therapeutics, Med. Research Council asso. McGill U., Montreal, 1968-71; sec. Can. Ministry of State for Sci. and Tech., Ottawa, 1971-74, 1971-74; dir. Armand-Frappier Inst., Laval-des-Rapides, Que., 1974—; cons. WHO, 1969-71; MacLaughlin vis. prof. McMaster U. Med. Sch., 1970-71. Author: Le Centre Medical Universitaire, 1966; contbr. articles on pharmacology, physiology, med. health and sci. policy to profl. publs. Decorated Queen Elizabeth II Commemorative medal; knight Order St. Lazarus of Jerusalem; John and Mary R. Markle scholar, 1957-62. Mem. La Corporation professionnelle des medecins du Quebec (license), Pharmacol. Soc. Can., Am. Soc. Pharmacology and Exptl. Therapeutics, Can. Soc. Clin. Investigation, Clin. Research Club Que., Can. Found. Advancement Clin. Pharmacology. Subspecialty: Medical research administration. Office: 531 Des Prairies Blvd Laval-des-Rapides PQ H7N 4Z3 Canada

BEAVEN, MICHAEL ANTHONY, research pharmacologist; b. London, Dec. 4, 1936; U.S., 1962, naturalized, 1968; s. Edward and Phyllis Georgina Dorcas (Barker) B.; m. Vida Helms, Feb. 1, 1964. B.Pharm., Chelsea Coll., U. London, 1959, Ph.D., 1962. USPHS vis. fellow Lab. Chem. Pharmacology, Nat. Heart, Lung and Blood Inst., NIH, Bethesda, Md., 1962-65, research pharmacologist, 1965-68, research pharmacologist exptl. therapeutics br., 1968-77, sect. chief sect. cellular pharmacolog Lab. Cellular Metabolism, 1977-82; dep. chief Lab. Chem. Pharmacology, 1982—; assoc. mem. Darwin Coll., Cambridge (Eng.) U., 1982-83. Contbr. over 100 articles and revs. to profl. jours., also chpts. to books. Mem. Pharm. Soc. Gt. Britain, Am. Soc. Pharmacology and Exptl. Therapeutics. Subspecialties: Pharmacology; Immunopharmacology. Current work: Role of histamine in pathological and physiological reactions: mechanism of inflammation. Office: Bldg 10 Room 8N108 NIH Bethesda MD 20205

BEAVER, PAUL CHESTER, parasitologist, educator; b. Glenwood, Ind., Mar. 10, 1905; s. John Chester and Blanche Emma (Murphy) B.; m. Lela E. West, Oct. 16, 1931; 1 dau., Paula Jean Beaver Chipman. A.B., Wabash Coll., 1928, D.Sc. (hon.), 1963; M.S., U. Ill., 1929, Ph.D., 1935. Diplomate: Am. Bd. Microbiology. Asst. zoology U. Ill., 1928-29, 31-34; instr. zoology U. Wyo., 1929-31; instr. biology Oak Park Jr. Coll., 1934-37; asst. prof. biology Lawrence Coll., 1937-42; biologist Wis. Dept. Health, summer 1940, Ga. Dept. Pub. Health, 1942-45; asst. prof. parasitology Tulane U. Med. Sch., 1945-47, asso. prof., 1947-52, prof., 1952—, head dept. parasitology, 1956-71, William Vincent prof. tropical diseases and hygiene, 1958-76, prof. emeritus, 1976—; dir. Internat. Center Med. Research and Tng. in Colombia, 1967-76; vis. prof. Eastern Mont. Normal Sch., summers 1935-37, Colo. State Coll., 1941, U. Mich., 1954-56, 58, U. Natal Med. Sch., Durban, South Africa, 1957; hon. vis. prof. Universidad del Valle, Cali, Colombia, 1970-76; cons. Ga. Dept. Pub. Health, 1946-53, USPHS Hosp., New Orleans, 1949-72, WHO, 1960-77; mem. com. standards and exams. Am. Bd. Microbiology, 1960-67; mem. commn. parasitic diseases Armed Forces Epidemiol. Bd., 1953-73, dir. commn. parasitic diseases, 1967-73; mem. Am. Found. Tropical Medicine, 1960-66; microbiology fellowships rev. panel NIH, 1960-63; mem. WHO expert com. on intestinal helminths, 1963, temp. adv., 1960, 61, 65, 66, 80, 81, WHO expert panel on parasitic diseases, 1963-77; bd. sci. counselors Nat. Inst. Allergy and Infectious Diseases, NIH, 1966-68; mem. NIH parasitic diseases panel U.S.-Japan Coop. Med. Sci. Program, 1965-69; mem. adv. sci. bd. Gorgas Meml. Inst. Tropical and Preventive Medicine, 1970—. Co-author: Animal Agents and Vectors of Human Disease, rev. edit, Craig & Faust's Clinical Parasitology, rev. edit.; contbg. author: Mitchell-Nelson's Pediatrics; editorial bd.: Am. Jour. Tropical Medicine and Hygiene, 1958-60, 67-70; editor-in-chief, 1960-66, 72—; asso. editor: Am. Jour. Hygiene, 1961-64, Jour. Parasitology, 1965-76, Am. Jour. Epidemiology, 1966—; editorial bd.: Transactions of Am. Micros. Soc, 1966-73, Ceskoslovenska Parasitologie, 1966-72; contbr. articles to profl. jours. Fellow Am. Acad. Microbiology (bd. govs. 1966-75), AAAS; mem. Internat. Filariasis Assn., Am. Soc. Tropical Medicine and Hygiene (councilor 1956-57, v.p. 1958, pres. 1969), Royal Soc. Tropical Medicine and Hygiene, Am. Soc. Parasitologists (councilor 1952-54, 56-59, pres. 1968), Am. Micros. Soc. (v.p. 1953, exec. com. 1955-59, 61-62), Am. Pub. Health Assn., Société Belge de Medicine Tropicale de Parasitologie et de Mycologie, Société de Pathologie Exotique (France; hon.), Sociedad Mexicana de Parasitologia (hon.), New Orleans Acad. Sci., Brazilian Soc. Tropical Medicine (hon.), Sigma Xi, Delta Omega, Alpha Omega Alpha (hon.). Club: Round Table (New Orleans). Subspecialty: Parasitology. Current work: Research and consultation on parasites causing tropical diseases, especially parasites of animals causing disease in man. Home: 1416 Cadiz St New Orleans LA 70115 Office: 1430 Tulane Ave New Orleans LA 70112

BEAVER, WILLIAM THOMAS, pharmacologist, educator; b. Albany, N.Y., Jan. 27, 1933; s. Ralph Alexander and Iona Marian (Hunter) B.; m. Nancy Powell, Dec. 16, 1961; children: Diane Elizabeth, Hilary Alexandra, Roderick William. B.A., Princeton U., 1954; M.D., Cornell U., 1958. Diplomate: Nat. Bd. Med. Examiners. Intern Roosevelt Hosp., N.Y.C., 1958-59; postdoctoral fellow in pharmacology Cornell U. Med. Coll., N.Y.C., 1959-61; clin. asst. in medicine Meml. Hosp., N.Y.C., 1961-68; clin. asst. in wards and outpatient dept., 1963-68; clin. asst. vis. physician James Ewing Hosp., N.Y.C., 1962-68; attending physician Calvary Hosp., N.Y.C., 1964-68; mem. staff dept. anesthesia Georgetown U. Hosp., Washington, 1968—; cons. Washington Home/Georgetown U. Hospice Unit, 1978—; instr. pharmacology Cornell U. Med. Sch., 1961-66, asst. prof., 1966-68; assoc. prof. pharmacology Georgetown U. Schs. Medicine and Dentistry, Washington, 1968-79, prof., 1979—; assoc. prof. anesthesia Georgetown U. Sch. Medicine, 1968-79, prof., 1979—; assoc. mem. Lombardi Cancer Research Center, 1978—; lectr. surgery Johns Hopkins U. Sch. Medicine, Balt., 1980—. Contbr. articles to profl. jours. USPHS tng. grantee, 1959-61; recipient Golden Apple award Med. Class of 1975, 1981; Spl. citation Commr. FDA, 1976. Mem. AAAS, Am. Soc. for Pharmacology and Exptl. Therapeutics, Am. Soc. for Clin. Pharmacology and Therapeutics (dir. 1978-81, vice chmn. sci. awards com. 1979-82, chmn. com. on FDA guidelines for clin. evaluation of analgesic drugs 1980—), AAUP, Internat. Assn. for Study of Pain (founding mem.), Am. Pain Soc. (pub. info. com.), Com. on Problems of Drug Dependence (exec. com. 1980—), Alpha Omega Alpha. Subspecialty: Pharmacology. Current work: Clinical pharmacology of analgesic drugs; design considerations in clinical trials, particularly in relation to agents affecting subjective responses; drug dependence. Office: Dept Pharmacology Georgetown U Schs Medicine and Dentistry 3900 Reservoir Rd NW Washington DC 20007

BECCHETTI, FREDERICK DANIEL, JR., physicist; b. Mpls., Mar. 3, 1943; s. Frederick Daniel and Olga Maxine (Sestini) B. B.S., U. Minn., 1965, M.S., 1968, Ph.D., 1969. Research assoc. Niels Bohr Inst., Copenhagen, 1969-71; research assoc. Lawrence Berkeley Lab., 1971-1973; asst. prof. U. Mich., 1973-76, assoc. prof. physics, 1976-82, prof. physics, 1982—. Contbr. articles to profl. jours. NSF fellow, 1970-71. Mem. Am. Phys. Soc., IEEE, Am. Assn. Physics Tchrs. Democrat. Roman Catholic. Subspecialty: Nuclear physics. Current work: Nuclear models; nuclear reactions; nuclear radiation detectors; heavy ion research. Office: 1049 Randall Lab Dept Physics U Mich Ann Arbor MI 48109

BECHTEL, DONALD BRUCE, research chemist; b. Paterson, N.J., Aug. 1, 1949; s. Joseph F. and Helene J. (Fiedeldey) B.; m. Kathleen Ann Bechtel, Aug. 7, 1971; children: Michael Shannon, Ryan John. B.S., Iowa State U., 1971, M.S., 1974; Ph.D. in Biology, Kans. State U., 1982. Chemist U.S. Grain Mktg. Research Lab., Manhattan, Kans., 1974-77, research chemist, 1977—. Contbr. articles to sci. jours. U.S. Dept. Agr. grantee, 1982. Mem. Am. Soc. Microbiology, Am. Assn. Cereal Chemistry, Bot. Soc. Am. Republican. Lutheran. Club: Optimists (Manhattan). Subspecialties: Food science and technology; Plant physiology (agriculture). Current work: Formation and secretion of storage proteins in cereals. Home: 2928 Gary Ave Manhattan KS 66502 Office: 1515 College Ave Manhattan KS 66502

BECHTEL, STEPHEN DAVISON, JR., engineering company executive; b. Oakland, Calif., May 10, 1925; s. Stephen Davison and Laura (Peart) B.; m. Elizabeth Mead Hogan, June 5, 1946; 5 children. Student, U. Colo., 1943-44; B.S., Purdue U., 1946; Dr. Engring. (hon.), 1972; M.B.A., Stanford U., 1948. Registered profl. engr. N.Y., Mich., Alaska, Calif., Md., Hawaii, Ohio, D.C., Va. With Bechtel Corp., San Francisco, 1941—, pres., 1960-73; chmn. cos. in Bechtel group, 1973-80, chmn., 1980—; dir. IBM Co., So. Pacific Co. Mem., vice chmn. Bus. Council; life counillor, past chmn. Conf. Bd.; mem. Presdl. Commn. Urban Housing, 1967-69, Nat. Indsl. Pollution Control Council, 1970-73, Nat. Productivity Commn., 1971-74, Cost of Living Council, 1973-74, Nat. Commn. Indsl. Peace, 1973-74, Labor-Mgmt. Group, 1974—; mem. policy com. Bus. Roundtable; trustee, chmn. bldg. and grounds com. Calif. Inst. Tech.; mem. pres.'s council Purdue U. Served with USMC, 1943-46. Decorated officer French Legion of Honor; recipient disting. alumnus award Purdue U., 1964; Ernest C. Arbuckle disting. alumnus award Stanford U., 1974; Man of Yr. award Engring. News Record, 1974; Outstanding Achievement in Constrn. award Moles, 1977; Disting. Engring. Alumnus award U. Colo., 1979. Fellow ASCE; mem. Nat. Acad. Engring. (chmn.), Am. Inst. Metall. Engrs., Calif. Acad. Scis. (hon. trustee), Chi Epsilon, Tau Beta Pi. Clubs: Pacific Union (San Francisco); Claremont Country (Oakland, Calif.); Cypress Point (Monterey Peninsula, Calif.); Thunderbird Country (Palm Springs, Calif.); Vancouver (B.C.); Ramada (Houston); Bohemian, San Francisco Golf; Links, Blind Brook (N.Y.C.); Augusta (Ga.) Nat. Golf; York (Toronto); Mount Royal (Montreal). Subspecialty: Civil engineering. Office: 50 Beale St San Francisco CA 94105 Office: Bechtel Power Corp PO Box 3965 San Francisco CA 94119

BECK, AARON TEMKIN, psychiatrist; b. Providence, July 18, 1921; s. Harry S. and Elizabeth (Temkin) B.; m. Phyllis Whitman, June 4, 1950; children—Judith, Daniel, Alice, Roy. B.A., Brown U., 1942; M.D., Yale U., 1946. Mem. faculty U. Pa. Med. Sch., 1954—, prof. psychiatry, 1971—; dir. Center Cognitive Therapy, 1965—; mem. rev. panel NIMH, 1965—, chmn. task force suicide prevention in 70s, 1969-70; bd. dirs. West Philadelphia Community Mental Health Consortium, 1975-77. Author: Depression: Causes and Treatment, 1972, Diagnosis and Management of Depression, 1973, Cognitive Therapy and the Emotional Disorders, 1976, Cognitive Therapy of Depression, 1979. Served as officer M.C. U.S. Army, 1952-54. Recipient award research R.I. Med. Soc., 1948. Mem. Am. Acad. Psychoanalysis (trustee 1970-75), Soc. Psychotherapy Research (pres. 1975-76), Am. Psychiat. Assn. (prize research psychiatry 1979), Royal Coll. Psychiatry, Psychiat. Research Soc., Am. Coll. Psychiatrists, Am. Acad. Psychoanalysis, Assn. Advancement Behavior Therapy, Phila. Soc. Clin. Psychologists (ann. award 1978). Subspecialties: Psychiatry; Cognition. Office: 133 S 36th St Room 602 Philadelphia PA 19104

BECK, BARBARA DORIS, toxicologist; b. Providence, Jan. 16, 1948; d. Irving Addison and Edith Elizabeth (Woodhead) B.; m. Robert E. Bahn, June 29, 1974; 1 son, James Earl. A.B., Bryn Mawr Coll., 1968; Ph.D., Tufts U., 1975. Postdoctoral fellow U. Mass. Med. Sch., Worcester, 1975-76, Harvard U., Cambridge, Mass., 1976-78; instr. Tufts U. Sch. Medicine, Boston, 1978-79; fellow Harvard Sch. Pub. Health, Boston, 1979-81, research assoc., 1981—. Cystic Fibrosis Found. postdoctoral fellow, 1975-76; NIH postdoctoral fellow, 1976-78; Interdisciplinary Programs in Health/Harvard Sch. Pub. Health fellow, 79-81. Mem. AAAS, Am. Thoracic Soc. Subspecialties: Environmental toxicology; Toxicology (medicine). Current work: Prediction of pulmonary toxicity of gas and airborne particles, assessing individual variations in response to toxic agent. Office: Dept Physiology 665 Huntington Ave Boston MA 02115

BECK, DORIS JEAN, microbiology educator, researcher; b. Blissfield, Mich.; d. Willard L. and Mary Ann (Fojtik) DeGroff; m.; children: Bill Lee, Dawn Renee. B.S., Bowling Green State U., 1965; M.S., Mich. State U., 1971, Ph.D., 1974. Prof. microbiology Bowling Green (Ohio) State U., 1974—; research assoc. Yale U., New Haven, 1981-82. Mem. Am. Soc. Microbiology, AAAS, NW Ohio Soc. Electron Microscopy, Sigma Xi. Subspecialty: Microbiology. Current work: Research mutagenesis and DNA repair. Office: Dept Biology Sci Bowling Green OH 43403

BECK, JOHN ROBERT, clinical pathologist, educator; b. Cleve., Sept. 8, 1953; s. J. Edward and Maralyn Janet (Smith) B.; m. Sharon Dombkowski, Aug. 30, 1975; children: J. Benjamin, Stefan A. A.B., Dartmouth Coll., Hanover, 1974; M.D., Johns Hopkins U., 1978. Cert. Am. Bd. Pathology. Intern Mary Hitchcock Meml. Hosp., Hanover, N.H., 1978; resident in pathology Dartmouth-Hitchcock Med. Ctr., Hanover, 1978-80, asst. prof. pathology, 1982—; fellow medicine New England Med. Ctr., Boston, 1981; instr. pathology Tufts U., 1981-82; staff pathologist VA Med. Ctr., White River Junction, Vt., 1982—, Hitchcock Clinic, Hanover, 1982—. Contbr. articles to profl. jours. Recipient Young Investigator award Acad. Clin. Lab. Physcians & Scientists, 1979, 80, 81; Koennecke award Johns Hopkins U., 1978; New Investigator Research award Nat. Library Medicine, 1983—. Mem. Am. Assn. Med. System Informatics, Soc. Med. Decision Making, Am. Fedn. Clin. Research, AAAS, Acad. Clin. Lab. Physicians and Scientists. Republican. Subspecialties: Pathology (medicine); Health services research. Current work: Application of clinical decision analysis to laboratory diagnosis and management; multivariate analysis of new laboratory tests; markov modeling of natural history of disease; also therapeutic apheresis and immunohematology. Home: Honeycomb House RFD route 10 West Lebanon NH 03784 Office: Dartmouth-Hitchcock Med Ctr 2 Maynard St Hanover NH 03756

BECK, KENNETH HAROLD, health and safety educator, psychologist, researcher; b. Phila., Dec. 18, 1950; s. Kenneth Harold and Elizabeth Louise (Voigt) B.; m. Candace Sue Brown, July 16, 1976; 1 son, Kenneth Harold. B.S., Pa. State U., 1972; M.A., Syracuse U., 1975, Ph.D., 1977. Teaching asst. Syracuse U., 1972-75, research asst., 1975-76, instr., 1975-77; postdoctoral fellow U. Conn.-Farmington, 1977-79; asst. prof. health edn. U. Md., 1979—, dir. research, 1980—; test supr. Ednl. Testing Service, Syracuse, N.Y., 1975-77; cons. in field. Author: Dept. Agr. safety/health tng. man, 1981; contbr. articles to profl. jours. NIH postdoctoral fellow, 1977; U. Md. grantee, 1980; NIMH grantee, 1981; dept. Agr. grantee, 1981. Mem. Am. Psyhchol. Assn., EAstern Psychol. Assn. Subspecialty: Social psychology. Current work: Health behavior, fear-arousing communications, risk perception. Office: U Md Safety Edn Center College Park MD 20742

BECK, LEE RANDOLPH, reproduction and immunology educator; b. Chgo., Mar. 7, 1942; s. and Beverly (Deitz) Kolner; m. Marjorie Collisson, Aug. 13, 1966; children: John, Jessica. B.A., U. Dubuque, 1965; M.S., N.Mex. Highlands U., 1966; Ph.D., Wash. State U., 1970; postgrad., Ohio State U., 1970-72. Asst. prof. dept. ob-gyn U. Ala.-Birmingham, 1972-78, assoc. prof., 1978-83, prof., 1983—; cons. Stolle Research and Devel. Corp., Cin., 1972—. Contbr. articles on immunology and contraception to profl. jours. USPHS fellow, 1968, 70. Mem. Am. Fertility Soc., Soc. for Study of Reprodn., Soc. for Advancement of Contraception, Controlled Release Soc. Republican. Subspecialties: Immunology (medicine); Immunobiology and immunology. Current work: Research in reproduction and immunology and controlled drug-delivery systems. Patentee in field. Office: U Ala Dept Ob-Gyn University Station Birmingham AL 35294

BECK, MARY MCLEAN, physiologist, educator; b. Oak Ridge, Tenn., Sept. 14, 1946; d. Clifford Keith and Mary Elizabeth (Lassetter) B.; m. Ron Jackson, 1983. B.A., Westhampton Coll. U., Richmond, 1968; M.S., U. Md., 1976, Ph.D., 1980. Bilingual sec. Max-Planck-Institut fur Plasmaphysik, Garching, W. Ger, 1972-74; grad. research asst. dept. poultry sci. U. Md., 1975-80; asst. prof. physiology U. Nebr., 1980—. Contbr. to profl. jours. Hubbard Farms scholar, 1978. Mem. Soc. Neurosci., N.Y. Acad. Sci., Nebr. Acad. Sci., Poultry Sci. Assn., Sigma Xi. Democrat. Subspecialties: Neurobiology; Neurophysiology. Current work: Neural degeneration, cerebral energy metabolism avian model for seizure disorders, auditory function. Office: University of Nebraska 215 Baker Hall Lincoln NE 68583

BECK, NAMA, medical educator; b. Seoul, Korea, Aug. 26, 1933; came to U.S., 1964; s. Rakjoong and Yebun (Park) B.; m. Irene Whayoung, Apr. 23, 1966; children: Edmund C., Alma S. M.S., Seoul U., 1959, M.D., 1963. Diplomate: Am. Bd. Internal Medicine. Asst. prof. medicine U. Pitts., 1970-75; assoc. prof. U. Tex. Health Sci. Ctr., San Antonio, 1975-1981; prof. medicine U. Calif.-Irvine, 1981—. Contbr. articles to profl. jours. Nat. Inst. on Aging grantee, 1980. Fellow ACP; mem. Am. Soc. Nephrology, Am. Physiol. Soc., Am. Soc. Endocrinology. Subspecialty: Nephrology. Current work: Renal metabolism with emphasis on (1) calcium, phosophorus and parathyroid hormone, (2) water metabolism, (3) aging, and (4) renal vasoactive hormones. Home: 8 Fortuna W Irvine CA 92714 Office: University of California Irvine CA 92717

BECK, NATHAN, biochemical pharmacology educator, research scientist; b. Phila., Nov. 30, 1925; s. Joseph and Freda (Goldhursh) B.; m. Toby D. Ticktin, July 24, 1927; children: Ephraim Eli, Aaron Issar, Adina, Rachel Tzvia, Sara Deborah. B.Sc. in Biochemistry, Pa. State U., 1948; M.S. in Pharmacology, Phila. Coll. Pharmacy and Sci., 1953, Ph.D., 1955. Research fellow Wyeth Inst. Med. Research, Phila., 1950-51; research asst. LaWall Meml. Lab. Pharmacology, Phila. Coll. Pharmacy and Sci., 1952-55; cancer research scientist dept. pharmacology Roswell Park Meml. Inst., Buffalo, 1955-53, sr. cancer research scientist, 1958-61, assoc. in pharmacology, 1961—; research fellow in pharmacology Med. Sch., U. Buffalo, 1956-58, asst. prof. pharmacology, 1956-60, assoc. prof., 1961-63; prof. biochem. pharmacology Sch. Pharmacy and Grad. Sch., SUNY-Buffalo, 1963—, chmn. dept. biochem. pharmacology, 1967-71; prof., acting dir. Sch. Pharmacy, Hebrew U., Jerusalem, 1969-71; dir. Pharmacology Inst., Nat. Council Research, Prime Minister's Office, Jerusalem, 1970-71; UN sci. expert UN Devel. Programme, Pharmacology Inst. Project, Jerusalem, 1976-77. Author books; contbr. numerous articles to profl. publs. Served with USN, 1944-46. Recipient E. K. Frey award German Govt. Fellow Royal Acad. Sci., AAAS, N.Y. Acad. Scis., Internat. Hematology Soc., Soc. Exptl. Biology and Medicine; mem. Internat. Soc. Biochem. Pharmacology (v.p.), Am. Assn. Cancer Research, Am. Chem. Soc., Am. Coll. Clin. Pharmacology, Assn. Gnoto Biology, Am. Heart Assn., Internat. Inflammation Soc., Royal Soc. Medicine, Kinin Klub, Sigma Xi, Rho Chi; hon. mem. Italian Soc. Pharmacology, Israel Pharmacol. Soc. Subspecialties: Pharmacology; Cellular pharmacology. Current work: Biochemical; proteases-protease inhibitors; vasopeptide rinins; shock; angiogenesis; tumor growth and vascularization; fibrinolysis. Office: SUNY Hochstetter 355 Buffalo NY 14260

BECK, NIELS CHRISTIAN, JR., psychology educator, researcher; b. St. Louis, July 29, 1948; s. Niels Christian and Lola Maria (Da Mota) B.; m. Mary Margret Wehrheim, Dec. 26, 1970; children: Mary Margaret, Niels Christian, Laura Joanna. A.B., St. Louis U., 1970, Ph.D. in Clin. Psychology, 1977; M.S., So. Ill. U., 1972. Intern, Tulsa Psychiat. Found., 1973-74; trainee Cardinal Glennon Meml. Hosp. for Children, 1974-75; staff psychologist child devel. unit St. Louis Univ. Hosps., 1974; staff psychologist, cons. inpatient psychiat. unit Mid-Mo. Mental Health Center, Columbia, 1975, dir. outpatient alcoholism and drug abuse services, 1976-78, dir. office program evaluation and applied research, 1978—; clin. instr. med. psychology U. Mo. Sch. Medicine, Columbia, 1975, instr., 1976-78, asst. prof., 1978-83, assoc. prof., 1983—, adj. asst. prof. psychology, 1978—, clin. asst. prof. nursing Sch. Nursing, 1978—. Contbr. articles to profl. jours. Mem. Am. Psychol. Assn., Mo. Program Evaluators Soc., Sigma Xi. Subspecialty: Clinical psychology. Current work: Keeping patients in treatment; the reduction of pain during labor and delivery. Office: 3 Hospital Dr Columbia MO 65201

BECK, PAUL ADAMS, metallurgist, educator; b. Budapest, Hungary, Feb. 5, 1908; came to U.S., 1928, naturalized, 1944; s. Philip O. and Laura (Bardos) B.; m.; children—Paul John, Philip Odon. M.S., Mich. Coll. Mining and Tech., 1929; M.E., Royal Hungarian U. Tech. Scis., 1931; Dr.Min. (hon.), Leoben Inst. Tech., 1979. Metallurgist Am. Smelting & Refining Co., Perth Amboy, N.J., 1937-41; chief metallurgist Beryllium Corp., Reading, Pa., 1941-42; supt. metall. lab. Cleve. Graphite Bronze Co., 1942-45; faculty U. Notre Dame, 1945—, prof. metallurgy, 1949—, head dept. metallurgy, 1950-51; research prof. phys. metallurgy U. Ill., 1951-76. Co-author: The Physics of Powder Metallurgy, 1951, Metal Interfaces, 1952, Recrystallization, Grain Growth and Textures, 1966; Editor: Theory of Alloy Phases, 1956, Electronic Structure and Alloy Chemistry of Transition Elements, 1963; co-editor: Magnetic and Inelastic Scattering of Neutrons by Metals, 1968, Magnetism In Alloys, 1972. Recipient U.S. Scientist award Humboldt Found., 1978, Heyn Meml. award German Metall. Soc., 1980. Fellow Metall. Soc. of AIME (Mathewson Gold Medal award 1952, ann. lectr. 1971, Hume-Rothery award 1974), Am. Soc. Metals (Sauveur Achievement award 1976), Am. Phys. Soc., Hungarian Phys. Soc. (hon.); mem. Nat. Acad. Enging. Subspecialty: Alloys. Current work: Magnetism in alloys, effects of short-range atomic order on magnetic properties of alloys. Office: Metallurgy Bldg U Ill Urbana IL 61801

BECK, ROBERT DONALD, software engineer; b. L.I., N.Y., Aug. 25, 1953; s. Donald Edward and Roberta Evelyn (Winter) B. B.S. in Math., Rensselaer Poly. Inst., 1975, M.S., 1975; M.S.C.S., U. Wis., 1977. Staff software engr. Intel Corp., Hillsboro, Oreg., 1977-83; systems design engr. Sequent Computer Systems, Portland, Oreg., 1983—; adj. instr. Oreg. State U., Corvallis, 1981. Mem. Assn. Computing Machinery, IEEE. Subspecialties: Operating systems; Distributed systems and networks. Current work: Distributed operating systems, office of the future, multiprocessing, human interfaces, computer languages, computer graphics. Office: Sequel Computer Systems 14360 NW Science Park Dr Portland OR 97229

BECK, SIDNEY LOUIS, biologist, educator, researcher; b. N.Y.C., Mar. 28, 1935; s. Sol and Helen (Diner) B.; m. Phyllis Sara Silverstein, Aug. 21, 1955; children: Stephen Howard, Laura Gene. B.S., CCNY, 1955; M.A., U. Kans., 1957; Ph.D., Brown U., 1960. Asst. instr. zoology U. Kans., Lawrence, 1955-57; instr. U. Mich., Ann Arbor, 1960-61, USPHS postdoctoral fellow, 1961-64; NIH spl. fellow and research assoc. in genetics Univ. Coll. London, 1964-65; asst. prof. biology U. Toledo, 1965-68, assoc. prof., 1968-69; prof. Wheaton Coll., Norton, Mass., 1969—, chmn. dept. biology, 1969-83; biologist EPA, Research Triangle Park, N.C., 1975-76; sr. fellow in pathology U. Wash., Seattle, 1983-84; cons., grants reviewer. Contbr. articles to sci. jours. NIH grantee, 1966-69; Toledo Hosp. Inst. grantee, 1969; NSF grantee, 1969-70; EPA grantee, 1977-80. Mem. AAAS, Am. Soc. Zoologists, Am. Genetic Assn., Genetics Soc. Am., Soc. Developmental Biology, Teratology Soc. Subspecialties: Teratology; Animal genetics. Current work: Genetic contributions to the teratogen responses; postnatal skeletal assessment to detect prenatal exposure to potentially teratogenic influences; cytogenetic effects of chronic low-level exposure to methylmercury. Office: Wheaton College Norton MA 02766

BECK, THEODORE RICHARD, chem. engr., cons. and researcher; b. Seattle, Apr. 11, 1926; s. Theodore and Gudmunda Elin (Thorarinsdottir) B.; m. Ruth Elizabeth Schaumberger, Dec. 1, 1951; children: Randi Marie, Maren Louisa. B.S. in Chem. Enging, U. Wash., 1949, M.S., 1950, Ph.D., 1952. Registered profl. engr., Wash. Research engr. duPont Corp., Deepwater, N.J., 1952-54; group leader Kaiser Aluminum & Chem. Corp., Permanente, Calif., 1954-59; sect. head Am. Potash & Chem. Corp., Henderson, Nev., 1959-61; mem. tech. staff Boeing Co., Seattle, 1961-65, sr. basic research scientist, 1965-72; div. mgr. Flow Research, Inc., Kent, Wash., 1972-75; pres. Electrochem. Tech. Corp., Seattle, 1975—; guest lectr. U. Wash., 1963-64, research prof. and affiliate prof., 1972—; cons. Argonne Nat. Lab., 1973-80. Editor: Corrosion div. Jour. Electrochem. Soc, 1971-75, Tutorial Lectures in Electrochem. Engrng. and Tech, 1981. Bd. dirs. N. Cascades Conservation Council, Seattle, 1973-80, Seattle Youth Symphony Orch., 1975-78. Fellow Am. Inst. Chem. Engrs.; mem. AAAA, Am. Chem. Soc., Am. Electroplaters Soc., Electrochem. Soc. (pres. 1975-76, Outstanding Achievement award corrosion div. 1981, hon. mem.), Internat. Soc. Electrochemistry, Nat. Assn. Corrosion Engrs., Sigma Xi, Tau Beta Pi, Phi Lambda Upsilon. Club: Seattle Mountaineers. Subspecialties: Chemical engineering; Corrosion. Current work: Developer electrochem. processes and devices and solution of problems in applied electrochemistry. Research in electrochem. kinetics, elec. double layer, transport properties, current distbn. to support devel. activities. Patentee, publs. in electrochem. engrng. and corrosion fields. Home: 10035 31st Ave N E Seattle WA 98125 Office: Electrochem Tech Corp 3935 Leary Way N W Seattle WA 98107

BECK, WILLIAM NELSON, physicist; b. Chgo., Dec. 16, 1923; s. Frank Spurgeon and Bessie (Dunn) B.; m. Beverly Ann Bell, Feb. 6, 1948; children: Bill, John, Belinda. B.A., Dakota Wesleyan U., Mitchell, S.D., 1946; M.S., U. S.D., Vermillion, 1951. Instr. Dakota Wesleyan U., 1946-50; chief field service engr. Globe Aircraft, Lockport, Ill., 1951-54; physicist Argonne Nat. Lab., Ill., 1954—. Contbr. articles to profl. jours. Served with USN, 1942-45. Mem. Am. Nuclear Soc. (Outstanding Achievement award 1979). Subspecialty: Nuclear fission. Current work: Evaluation of reactor fuel materials; applications of ultrasonics in the field on nondestructive testing; neutron radiography techinques for irradiated materials. Inventor, patentee automatic water timer, acoustic lock. Office: Argonne Nat Lab Bldg 310 9700 S Cass St Argonne IL 60439

BECK, WILLIAM SAMSON, physician, educator, biochemist; b. Reading, Pa., Nov. 7, 1923; s. Myron Paul and Gertrude (Harris) B.; m. Helen Samuels, Oct. 24, 1947; children: Thomas Russell, Peter Dean; m. Hanne Troedsson, July 20, 1964; children: John Christopher, Paul Brooks. B.S., U. Mich., 1943, M.D., 1946; A.M. (hon.), Harvard U., 1971. Diplomate: Am. Bd., Internal Medicine, 1954. Instr., asst. prof. medicine UCLA, 1950-55; fellow in biochemistry NYU Coll. Medicine, 1955-57; mem. faculty dept. medicine Harvard U., 1957—, prof., 1979—, tutor in biochemocal scis., 1957, chmn. M.I.T. Div. Health Sci. and Tech., 1978; dir. clin. labs. Mass. Gen. Hosp., 1957-75, chief hematology unit, 1957-72, dir. hematology research lab., 1957—; mem. adv. council Nat. Inst. Arthritis, Metabolism and Digestive Diseases, NIH, 1971-74; mem. hematology study sect. NIH, 1967-71. Author: Modern Science and the Nature of Life, 1957, Human Design, 1971, Hematology, 3d edit., 1981, (with G.G. Simpson), Life: An Introduction to Biology, 3d edit., 1984; contbr. articles to profl. jours. Served with AUS, 1943-46. Mem. Am. Soc. Biol. Chemists, Am. Soc. Hematology (exec. com. 1980—), Assn. Am. Physicians, Am. Soc. Clin. Investigation, Am. Assn. Cancer Research. Subspecialties: Hematology; Biochemistry (biology). Current work: Biochemistry of DNA synthesis, vitamin B12 and folic acid. Home: 85 Arlington St Winchester MA 01890 Office: Mass Gen Hosp Boston MA 02114

BECKER, ALBERT WALTER, engineering educator; b. Belleville, Ill., Sept. 23, 1922; s. Walter F. and Minnie E. (Helmholtz) B.; m. Alberta A. Schroeder, Jan. 17, 1948; children: Tessa, Herschel, Stephen, Albert Walter, Barbara. Student, Chgo. Tech. Coll., 1946, Ill. Tech. Inst., 1950, Colo. State U., 1966; M.S., So. Ill. U., 1969. Cert. tchr. gen. and vocat. edn., Ill. Plant engr. Nat. Lead Co., St. Louis, 1954-66; mem. faculty So. Ill. U., Carbondale, 1966-69, Belleville (Ill.) Area Coll., 1969—, coordinator electricity/electronics, 1969—, coordinator air conditioning/heating/refrigeration engrng., 1973—; cons. elec. systems for air conditioning, heating, refrigeration, 1973—. Served to lt. (j.g.) USCGR, 1942-45. Mem. ASHRAE, Refrigeration Service Engrng. Soc. Republican. Roman Catholic. Club: Exchange. Lodge: K.C. Subspecialties: Electrical engineering; Solar energy. Current work: Designing new electrical circuits for acir conditioning/heating/refrigeration/solar. Patentee 2-stage horizontal compressors. Office: Dept Indsl Engrng Belleville Area Coll 2500 Carlyle Rd Belleville IL 62221

BECKER, DONALD AUGUST, ecologist, educator, planner, consultant; b. Valley City, N.D., July 27, 1938; s. August F. and Ella M. (Amundson) B.; m. Elaine R. Sandberg, Aug., 20, 1960; children: Michelle, Pamela, Donald. B.S., Valley City State Coll., 1960; M.S. in Biology, U. N.D., 1966; Ph.D. in Ecology, U. N.D., 1968. Assoc. prof. biology Midland Luth. Coll., Fremont, Nebr., 1968-75; environ. specialist Mo. River Basin Commn., Omaha, 1975-81; flood plan studies mgr. Mo. Basin State Assn., Omaha, 1982-83; environ. resource specialist U.S.C.E., Omaha, 1983—; cons. govtl. agys., plant ecology. Author: Mo. River Flood Studies, 1981, 82. U.S. Water Resources Council grantee, 1980-1982; Nat. Park Service grantee, 1982-1984. Mem. Nebr. Acad. Sci., Signa Xi. Subspecialties: Ecology; Resource conservation. Current work: Flood plain management, river ecology, land use and cover data bases, grassland ecology, plant autecology, water and related land resources planning and mangement. Office: US Army Corp Engrs Omaha NE 68102

BECKER, DONALD EUGENE, animal science educator, animal nutritionist; b. Delavan, Ill., Feb. 2, 1923; s. George Edwin and Esther C. (Peters) B.; m. Elsie Hendrickson, Dec. 28, 1949; children: Esther Becker Gasser, Phyllis Becker Boerman, Donald E., Jr., William E., Beth A. B.S., U. Ill.-Urbana, 1945, M.S., 1947; Ph.D., Cornell U., 1949. Asst. U. Ill., 1945-47, asst. prof. dept. animal sci., 1950-54, assoc. prof., 1954-58, prof., 1958—, head dept., 1967—; asst. Cornell U., 1947-49; asst. prof. U. Tenn., 1949-50; mem. agrl. research adv. bd. Eli Lilly, 1969-80. Recipient Funk Recognition award U. Ill., 1972; Edn. award Ill. Pork Producers Assn., 1972. Fellow AAAS; mem. Am. Soc. Animal Sci. (editor 1964-70, pres. 1970-71, recipient, Am. Feed Mfrs. award 1957, Morrison award 1977, fellow award 1982), Council Agrl. Sci. and Tech., Am. Registry of Cert. Animal Scientists. Republican. Methodist. Subspecialty: Animal nutrition. Current work: Animal nutrition related to amino acid metabolism, growth, reproduction, vitamin B-Cobalt nutrition, antibiotics in nutrition and niacin-trytophan nutrition. Office: University of Illinois Department of Animal Science 328 Mumford Hall Urbana IL 61801

BECKER, DONALD PAUL, neurosurgeon, educator; b. Cleve., July 5, 1935; s. Benjamin Harry and Sarah (Justin) B.; m. Maria Appelberg, Sept. 11, 1978; children: Mary Jane, Jennifer, Benjamin, Elizabeth. B.A., Williams Coll., 1957; M.D., Case Western U., 1961. Diplomate: Am. Bd. Neurol. Surgery. Instr., asst. prof. neurosurgery Case Western Res. U., 1967-68; asst. prof. surgery-neurosurgery UCLA Sch. Medicine, 1968-69, assoc. prof., 1969-71; prof., chmn. dept. neurosurgery Va. Commonwealth U., Richmond, 1971—, dir., 1976—. Contbr. articles to profl. jours. Mem. Soc. Neurol. Surgeons, Am. Assn. Neurol. Surgeons, Neurol. surg. Soc. Am. Subspecialty: Neurosurgery. Current work: Pathobiology of acute brain damage. Home: Mount Bernard HLR A31 Maidens VA 23102 Office: MCV Station Box 631 Richmond VA 23298

BECKER, ERNEST I., chemistry educator, cons., researcher; b. Cleve., Aug. 18. 1918; s. Harry and Esther (Cohen) B.; m. Marion Ferris, Dec. 20, 1947; children: Jonathan David, Kenneth Alan, Mark Edward, Robert Neal, Paula Sarah. B.S. in Pharmacy, Western Res. U., 1941, M.S. in Chemistry, 1943, Ph.D., 1946. Instr. Poly. Inst. of Bklyn., 1946, asst. prof., 1947-51, assoc. prof., 1951-56, prof., 1956-70, chmn. dept. chemistry, div. natural scis., 1965-70; prof., chmn. dept. chemistry U. Mass., Boston, 1970—, prof., 1972—; cons. in field. Author 6 books on organometallic syntheses and reactions, 1 book first aid manual for chem. accidents; contbr. numerous articles on chemistry to profl. jours; book reviewer. U.S. Army grantee, 1949-53; NIH grantee, 1950-53; U.S. Army Research Office grantee, 1959-67; Fisher Sci. Co. grantee, 1981-83. Fellow N.Y. Acad. Scis., Chem. Soc. of London; mem. Am. Chem. Soc., AAAS, Am. Inst. Chemists (cert.), Nat. Sci. Tchrs. Assn., New Eng. Assn. Chemistry Tchrs. Subspecialties: Organic chemistry; Synthetic chemistry. Current work: Alkylation of heterocycles polymer systheses, flurorescent monomers and polymers. Home: 32 Oxford Rd Newton MA 02159 Office: Dept Chemistry U Mass Boston MA 02125

BECKER, ERNEST LOVELL, physician, educator; b. Cin., Jan. 13, 1923; s. Ernest Louis and Sarah (Lovell) B.; m. Margaret Webb Thompson, Oct. 22, 1949 (div. 1970); children—James T., Margaret W., Frank L.; m. Eleanor Holden, July 14, 1972. A.B., Washington and Lee U., 1944; M.D., U. Cin., 1948. Diplomate: Am. Bd. Internal Medicine. Asst. pharmacology U. Cin. Coll. Medicine, 1946-47; intern Christ Hosp., Cin., 1948-49; jr. asst. resident Med. Coll. Va. Hosp., Richmond, 1949-50, sr. asst. resident, 1950-51, asst. medicine, 1950-51; instr. physiology N.Y. U. Coll. Medicine, 1951-53; investigator Mt. Desert Island Biol. Lab., Salisbury Cove, Maine, summers 1951-52, 55; asst. prof. medicine Med. Coll. Va., 1955-57, Cornell U. Med. Coll., 1957-62, asso. prof., 1962-69, prof., 1969-78, adj. prof., 1978—; dir. dept. grad. med. evaluation AMA, 1978—; dir. Eugene F. DuBois Pavilion, N.Y. Hosp.-Cornell Med. Center, 1960-73, chief nephrology and hypertension service, dept. medicine, 1967-73; asst. attending physician 2d Cornell med. div. Bellevue Hosp., 1957-68; clin. asst. medicine Meml. Hosp., 1964-81; dir. medicine Beth Israel Med. Center, N.Y.C., 1981—; cons. U.S. Naval Hosp., St. Albans, N.Y., 1960-73; asst. attending physician N.Y. Hosp., 1957-65, asso. attending physician, 1965-69, attending physician, 1969-78; asst. vis. physician James Ewing Hosp.; Sec.-gen. Internat. Co. Nomenclature and Nosology Renal Disease; mem. research adv. com. Health Research Council N.Y.C., also mem. metabolic study sect., 1965-78; chmn. ad-hoc com. establishing criteria chronic renal disease centers Regional Med. Programs, 1972; sci. adv. com. artificial kidney-chronic uremia program Nat. Inst. Arthritis and Metabolic Diseases, 1971-78; mem. steering com. lisiting program specialized clin. services, chmn. kidney research com. Joint Commn. Accreditation Hosps., 1972-73; chmn. adv. com. renal provisions HR-1, Social Security Adminstrn., 1972-73. Editorial cons.: Am. Jour. Medicine, 1971—, Clin. Nephrology, 1971; editorial adv. bd.: Current Contents, 1972—; Contbr. articles to profl. jours. Pres. Nat. Kidney Found., 1970-73. Served to capt. USAF, 1953-55. Recipient Lederle Med. Faculty award, 1960-63; Markle scholar med. scis., 1955-60; WHO fellow, 1974, 78. Fellow AAAS, Royal Soc. Medicine, N.Y. Acad. Scis., Royal Soc. Tropical Medicine and Hygiene, ACP; mem. Am. Soc. Nephrology (finance chmn.), Am. Heart Assn. (exec. com. renal sect. council circulation 1965-67), N.Y. Heart Assn. (bd. dirs. 1965-70), AMA, Soc. Exptl. Biology and Medicine, N.Y. County Med. Soc., N.Y. Med. and Surg. Soc., Chgo. Med. Soc., Am. Physiol. Soc., Harvey Soc., Am. Clin. and Climatol. Assn., Chgo. Soc. Internal Medicine, So. Soc. Clin. Research, Explorers Club N.Y. (pres. 1975-76), Sigma Xi. Clubs: Century Assn., Union (N.Y.C.); Cosmos (Washington). Subspecialties: Internal medicine; Nephrology. Current work: Current career activities include the administrative, teaching and research activities involved in the operations of the dept. medicine at Beth Israel Medical Center. Home: 14 Sutton Pl S New York NY 10022 Office: Dir Medicine Beth Israel Med Center 10 Nathan D Perlman Pl New York NY 10003

BECKER, JAMES THOMPSON, neuropsychologist; b. N.Y.C., July 7, 1953; s. Ernest Lovell and Margaret Webb (Thompson) B.; m. Mary-Amanda Dew, July 12, 1980. B.A., Washington and Lee U., 1975; M.A., Northeastern U., 1977, Johns Hopkins U., 1979, Ph.D., 1980. Lic. psychologist, Mass. Nat. Inst. for Alcohol Abuse and Alcoholism postdoctoral research fellow Boston U. Sch. Medicine, 1980-82; instr. dept. psychiatry U. Conn. Med. Sch., 1982—; research cons. Dew-Becker Assocs. Contbr. articles to profl. jours. Mem. Soc. for Neurosci., Internat. Neuropsychol. Soc., AAAS, Research Soc. on Alcoholism. Episcopalian. Subspecialties: Neuropsychology; Physiological psychology. Current work: Memory and memory disorders, brain damage and relationship to behavior, alcohol abuse and effects on cognitive function, visual information processing. Office: Suite M-90 Boston U Sch Medicine 85 E Newton St Boston MA 02118

BECKER, JOAN ALAINE, university administrator, educator; b. Bklyn., Oct. 17, 1954; d. Lewis Bernard and Lillian (Wolkoff) B.; m. Michael N. Sills, Sept. 5, 1982. B.S., Ohio State U., 1976; M.B.A., George Washington U., 1981. Cert. Am. Registry Radiol. Technologists. Staff technologist, clin. educator St. Vincent's Hosp., Toledo, 1976-77; instr. M.T. Owens Tech. Coll., Toledo, 1977-78; asst. prof., acad. coordinator George Washington U., Washington, 1978—; cons. George Washington U. Med. Ctr., 1978—. Dir. new mem. services United Jewish Appeal Young Leadership div., Washington, 1978—; mem. adv. com. radiol. tech. program North Va. Community Coll., Annandale, 1979-81. Kellogg Found. and Am. Soc. Allied Health Professions grantee, 1982-83. Mem. Am. Soc. Allied Health Profls. (dir. 1981-82), Am. Hosp. Radiology Adminstrs., Am. Soc. Tng. and Devel., Am. Soc. Radiol. Technologists. Subspecialties: Imaging technology; Health services research. Current work: Leadership development of the allied health professional, organizational development in the health care system, impact of group dynamics on quality of health care. Office: George Washington U Med Ctr 2300 Eye St NW Washington DC 20037

BECKER, JOEL LEONARD, psychologist; b. N.Y.C., May 2, 1949; s. Irving and Renee (Katz) B. B.A., Hobart Coll., 1971; M.S., U. Ga., 1973, Ph.D., 1977. Dir. Behavior Assocs., Boston, 1977—; instr. Harvard Med. Sch., Boston, 1978—; intern in clin. psychology Mass. Mental Health Ctr., 1974-75; dir. Boston Inst. for Behavior Therapies, 1981—; ednl. dir. Assn. for Advanced Tng. in Behavioral Scis., Los Angeles, 1978—. Author. various articles in profl. jours. Bd. dirs. Feinald Fund, Waltham, Mass., 1981-82. Mem. Am. Psychol. Assn., Mass. Psychol. Assn. (fellow), Assn. for Advancement Behavior Therapy, Sigma Xi, Phi Kappa Phi. Subspecialties: Behavioral psychology; Cognition. Current work: Cognitive behavioral psychotherapy. Office: 45 Newbury St Boston MA 02116

BECKER, ROBERT EARL, clinical research psychologist, educator; b. Buffalo, Apr. 16, 1945; s. Earl and Connie (Miosi) B.; m. Donna D. Hereid, June 21, 1966; children: Kelly R., Karen I., Robert R. B.A., SUNY-Buffalo, 1966; Ph.D., SUNY-Albany, 1977. Cert. sch. psychologist, lic. psychologist, N.Y. Trainee practicums in clin.

psychology, psychology service VA Hosp., Albany, 1971, Union Coll. Counseling Center, 1971-72; practicum in psychol. assessment, phys. medicine and rehab. Albany Med. Center Hosp., 1973; practicum in phychol. assessment St. Catherin's Day Care Ctr., Albany, 1973; research fellow SUNY-Albany, 1971-72, research asst., 1972-73, predoctoral instr., 1972-74, adj. clin. prof. psychology, 1979-80; asst. project dir. SUNY (State Edn. Dept., Grad. Student Financing Study), Albany, 1974-75; intern in psychology Capital Dist. Psychiat. Ctr. and dept. psychiatry Albany Med. Coll., 1974-75, research psychology and asst. dir. research unit, 1979—; staff psychologist outpatient units (Capital Dist. Psychiat. Ctr.), 1975-76, assoc. psychologist, 1976-79; instr. in psychiatry Albany Med. Coll., 1977-79, asst. prof. psychiatry, 1979-82, assoc. prof., 1982—; cons. in field. Contbr. chpts. to books, articles to profl. publs.; reviewer: Jour. Cons. and Clin. Psychology, 1981—; editorial bd.: Behavior Modification, 1982-83. Served to capt. USAF, 1966-70. NIMH grantee, 1979-82, 82—; NIH grantee, 1981-82, 82-83. Mem. Am. Psychol. Assn., Assn. Advancement Behavior Therapy, Psychol. Assn. Northeastern N.Y. (program chmn. 1981-82). Democrat. Lutheran. Subspecialty: Behavioral psychology. Current work: Clinical research into psychological causes and treatment of depression and anxiety disorders. Home: RD 1 Kinderhook NY 12106 Office: Research Unit Dept Psychiatry Albany Med Coll 47 New Scotland Ave Albany NY 12208

BECKER, ROBERT HOWARD, astrophysicist; b. Newark, Aug. 18, 1946; s. Samuel H. and Mildred (Federbush) B.; m. Joanne Rossi, Aug. 17, 1968. B.A., Lehigh U., 1968; M.S., U. Md., 1970, Ph.D., 1975. Research assoc. NASA, Goddard Space Flight Ctr., Greenbelt, Md., 1975-80; research assoc. Columbia U., 1980-81; vis. asst. prof. Va. Poly. Inst. and State U., 1981—. Contbr. numerous articles to profl. jours. Mem. Am. Astron. Soc., Internat. Astron. Union, ACLU, Nat. Audobon Soc. Subspecialties: 1-ray high energy astrophysics; Radio and microwave astronomy. Current work: Observational astronomy, supernova remnants, binary stars. Home: 803 Crestwood Dr Blacksburg VA 24060 Office: Physics Dept Va Tech Blacksburg VA 24060

BECKER, ROBERT OTTO, orthopaedic surgeon, educator; b. River Edge, N.Y., May 31, 1923; s. Otto and Elizabeth (Blank) B.; m. Lillian Moller, Sept. 6, 1946; children: Lisa, Michael, Adam. B.A., Gettysburg Coll., 1946; M.D., NYU, 1948. Diplomate: Am. Bd. Orthopaedic Surgery. Intern Bellevue Hosp., N.Y.C., 1948-49; resident in orthopaedic surgery Downstate Med. Ctr., SUNY, Bklyn., 1953-56; mem. faculty SUNY Upstate Med. Ctr., Syracuse, 1956—; now prof. orthopaedic surgery; clin. prof. orthopaedic surgery La. State U. Coll. Medicine, Shreveport, 1980—. Author: Mechanisms of Growth Control, 1981, Electromagnetism and Life, 1982; contrb. over 150 sci. articles to profl. publs. Served to 1st lt. U.S. Army, 1951-53. Recipient Middleton award VA, 1960, Disting. Alumnus award NYU Coll. Medicine, 1966, Nicolas Andry award Am. Assn. Bone Joint Surgery, 1979; SUNY Faculty Exchange scholar, 1979. Mem. N.Y. Acad. Scis., AAAS. Republican. Club: Angler's (N.Y.). Subspecialties: Biomedical engineering; Cell and tissue culture. Current work: Electrical control systems in living organisms; bio effects electrical currents; electrical fields; magnetic fields. Home and Office: Star Route Lowville NY 13367

BECKER, STEPHEN ALLAN, astronomer, astrophysicist; b. Evanston, Ill., Sept. 11, 1950; s. John Nicholas and Irene Ann (Wlodarski) B.; m. Wendee M. Brunish, May 30, 1980. B.A., Northwestern U., 1972; M.S., Case Western Res. U., 1974; Ph.D., U. Ill., 1979. Research and teaching assoc. astronomy dept. U. Ill., Urbana, 1979-80; research fellow in physics W.K. Kellogg Radiation Lab., Calif. Inst. Tech., Pasadena, 1980-82; staff mem. Los Alamos (N.Mex.) Nat. Lab., 1983—. Contbr. papers to profl. publs. and confs. Mem. Am. Astron. Soc., Planetary Soc. Roman Catholic. Subspecialties: Theoretical astrophysics; Nuclear physics. Current work: Stellar evolution, star clusters, supernovae, abundance anomalies in meteorites, variable stars, nuclear reactions. Office: Los Alamos Nat Lab P O Box 1663 MSB220 Los Alamos NM 87545

BECKER, WILLIAM NICHOLAS, crop consultant; b. Berwyn, Ill., Mar. 15, 1946; s. Alexander B. and Virginia (Kurako) B.; m. Caroline M. Weiss, June 22, 1968; 1 son, Joshua. A.A., Morton Jr. Coll., 1966; B.S., Western Ill. U., 1969, M.S., 1971; Ph.D., U. Ill., 1976. Dist. mgr. Lincoln Land Crop Pro-Tech, Inc., Springfield, Ill., 1977-79; instr. agr. Lincoln Land Community Coll., Springfield, 1979-81; plant pathologist, field rep. Taralan Corp., Springfield, 1981—. Mem. Am. Phytopathol. Soc., Soc. Nematologists. Lodge: K.C. Subspecialties: Plant pathology; Integrated pest management. Current work: Maximum economic yields, soil microbiology and conservation, mycorrhizal fungi. Home: 1229 W Edwards Springfield IL 62704

BECKERS, JACQUES MAURICE, astronomer; b. Arnhem, Netherlands, Feb. 14, 1934; s. Wilhelmus Bartholomeus and Maria Hubertina (Hermans) B.; m. Gerda Maria van Vuurden, Spr. 7, 1959; children: Christian M., Michael P. Doctorandus Astronomy, U. Utrecht, Netherlands, 1959, Dr. Astronomy, 1964. Commonwealth Sci. and Indsl. Research Orgn. fellow, Sydney, Australia, 1959-62; astrophysicist Sacramento Peak Obs., 1962-79; dir. multiple mirror telescope obs. U. Ariz., Tucson, 1979—. Contbr. articles to profl. jours. Recipient Henryk Arctowski award Nat. Acad. Scis., 1975. Mem. Am. Astron. Soc., Internat. Astron. Union, Optical Soc. Am. Subspecialties: Optical astronomy; Solar physics. Current work: Solar physics, stellar atmospheres, speckle interferometry, astronomical instrumentation. Home: 4871 E Avenida del Cazador Tucson AZ 85718 Office: Steward Observatory U Ariz Tucson AZ 85721

BECKING, RUDOLF WILLEM, forestry educator; b. Blora, Java, Indonesia, Oct. 19, 1922; s. J.H. and Martha P. (Hennig) B.; m. Hilda Louise Scheltema, Oct. 4, 1952; children: Irene Martha, John Rudolph, Tasha Karen. M.S., Agr. U., Wageningen, Netherlands, 1952; Ph.D. in Forest Mgmt, U. Wash., 1954. Registered forester, Calif. Forest officer Dutch Forest Service, Utrecht, 1954-56; asst. prof. forestry U. N.H., Durham, 1956-57, Pa. State U., State College, 1957-58; assoc. prof. forestry Auburn (Ala.) U., 1958-60; prof. forestry Coll. Natural Resources, Humboldt State U., Arcata, Calif., 1960—; expert forest inventory FAO, Indonesia, 1980-81. Author: Pocket Flora Redwood Region, 1982. Mem. Arcata City Council, 1972-75. Mem. Soc. Am. Foresters, Ecol. Soc. Am., Am. Soc. Photogrammetry, Assn. Tropical Ecology, Assn. Tropical Biology, Assn. Tropical Foresters. Subspecialties: Integrated systems modelling and engineering; Ecology. Current work: Environmental assessment, all-aged forestry, tropical rainforest research; bird inventories, ZM plant community inventories. Home: 1415 Virginia Way Arcata CA 95521 Office: Coll Natural Resources Humboldt State U Arcata CA 95521

BECKJORD, ERIC STEPHEN, laboratory executive, energy researcher; b. Evanston, Ill., Feb. 17, 1929; s. Walter Clarence and Mary Amelia (Hitchcox) B.; m. Caroline Wendell Gardner, Feb. 28, 1953; children: Eric. H., Amy W., Charles A., Sarah H. A.B. cum laude, Harvard U., 1947; M.S in E.E, M.I.T., 1956. Devel. engr. Gen. Electric Co., San Jose, Calif., 1956-60, project engr., Pleasanton, Calif., 1960-63; engring. mgr. Westinghouse Electric Corp., Pitts., 1963-70; v.p. Westinghouse Nuclear Europe, Brussels, 1970-73; project dir., mgr. strategic planning-nuclear Westinghouse Electric Corp., Pitts., 1973-75; dep. dir. Office Nuclear Affairs, Fed. Energy Adminstrn., 1975; dir. div. reactor devel. and demonstration ERDA, 1976-77; dir. div. nuclear power devel. U.S. Dept. Energy, 1977-78, coordinator, 1978-80; dep. dir. Argonne Nat. Lab., Ill., 1980—; mem. reactor safety criteria Am. Standards Inst., 1965-68; mem. AEC Water Reactor Core Cooling Task Force, 1966-67. Author: Advances in Nuclear Science, 1962; contbr. articles to profl. jours. Bd. dirs. St. Edmund's Acad., Pitts., 1967-69; bd. dirs. Shady Lane Sch., Pitts., 1968-70; vestryman Calvary Episcopal Ch., Pitts., 1970, 75. Served to lt. (j.g.) USNR, 1951-54. Fellow Am. Nuclear Soc. (v.p., pres.-elect Belgian sect. 1972); mem. IEEE, Sigma Xi. Republican. Subspecialties: Nuclear fission; Electrical engineering. Current work: Energy research and development; nuclear fission, nuclear fusion and fossil research and development; direction of applied programs, oversight of laboratory science and technology. Home: 627 E 6th St Hinsdale IL 60521 Office: Argonne National Laboratory 9700 S Cass Ave Argonne IL 60439

BEDELL, RALPH CLAIRON, psychologist, educator; b. Hale, Mo., June 4, 1904; s. Charles Edward and Jennie (Eaton) B.; m. Stella Virginia Bales, Aug. 19, 1929 (dec. 1968); m. Ann Sorency, Dec. 21, 1968 (dec. 1975); m. Mtra Jervey Hoyle, Feb. 14, 1976; 1 son, Brian Hoyle. B.S. in Edn, Central Mo. State U., 1926; M.A., U. Mo.-Columbia, 1929, Ph.D., 1932. Diplomate: Am. Bd. Profl. Psychology; cert. tchr. Mo. Prof. psychology, dir. guidance N.E. Mo. State U., Kirksville, 1933-37; dean faculty and student personnel Central Mo. State U., Warrensburg, 1937-38; prof. ednl. psychology, counselor edn. U. Nebr., Lincoln, 1938-50; prof., chmn. dept. psychology and edn. American U., Washington, 1950-52; sec.-gen. S. Pacific Commn., Noumea, New Caledonia, 1955-58; dir. counseling and guidance insts. U.S. Office Edn., Washington, 1958-67; prof. edn. U. Mo.-Columbia, 1967-74, prof. emeritus, 1974—; cons. and advisor in edn. Co-author: General Science for Today, 1932, rev. edit., 1936, (with Ralph K. Watkins, with Frank E. Sorenson and Harold E. Wise) Element of Pre-Flight Aeronautics, 1942; dir., editor textbooks for naval aviation cadets in WWII, 1942-45. Served to comdr. USNR, 1941-45. Recipient Alumnus award Central Mo. State U., 1971; award Mo. Guidance Assn., 1971, Mo. Coll. Personnel Assn., 1982. Fellow Am. Psychol. Assn., Royal Soc. Health, Explorers Club (Flay award 1980); mem. Am. Personnel and Guidance Assn. (life), Mil. Order of the World Wars, U. Mo. Alumni Assn. (outstanding service to edn. award 1979), Assn. Counselor Edn. and Supervision (outstanding contribution to counselor edn. award 1967), Phi Delta Kappa, Sigma Tau Gamma (Soc. of The Seventeen). Democrat. Unitarian. Clubs: Columbia (Mo.) Country; Army and Navy (Washington). Subspecialties: Behavioral psychology; Counseling. Current work: Cognition in counseling for career change; cross cultural problems in teacher education between the United States and Thailand. Home: 106 S Ann St Columbia MO 65201 Office: Coll Edn U Mo-Columbia 301 Hill Hall Columbia MO 65211

BEDERSON, BENJAMIN, physicist, educator; b. N.Y.C., Nov. 15, 1921; s. Abraham Michael and Lena (Waxlowsky) B.; m. Betty E. Weintraub, Jan. 20, 1956; children: Joshua Benjamin, Geoffrey Adam, Aron Gregory, Benjamin Boris. B.S., CCNY, 1946; M.S., Columbia U., 1948; Ph.D., NYU, 1950. Research scientist MIT, 1950-52; asst. prof. physics to prof. physics NYU, 1952—, dir.; cons. various govt. and acad. agys. Editor: (with D.R. Bates) Advances in Atomic and Molecular Physics, 1974; Assoc. editor: Atomic Data and Nuclear Data Jour; editor: Phys. Rev. A; contbr. articles to profl. jours. Served with AUS, 1942-46. NSF grantee; Dept. of Energy grantee. Fellow Am. Phys. Soc., N.Y. Acad. Scis., AAAS. Subspecialties: Atomic and molecular physics; Laser research. Current work: Atomic structure and collisions using beams, polarized and state-selected beams, laser interactions with atoms and molecules. Office: NYU 4 Washington Pl New York NY 10003

BEDINGER, JOSEPH ARNOLD, nuclear plant design consultant; b. Portales, N.Mex., July 1, 1916; s. Henry Clay and Betty (Miller) B.; m. Ruth V. Jones, June 6, 1969; 1 dau., Sandra Lynn Bedinger Chase. Student, Eastern N.Mex. Jr. Coll., 1935-38, Valley Coll., 1950-51, UCLA, 1951-55. Nuclear design cons. N.Am. Aviation, Canoga Park, Calif., 1950-55, Westinghouse Corp., Richland, Wash., 1977-79, Wash. Pub. Power Supply System, Richland, 1979-80, Gulf States Utility, St. Francisville, La., 1980-82. Mem. ASME, Am. Soc. Metals, Am. Welding Soc., Am. Nuclear Soc., Soc. Automotive Engrs. Mem. Christian Ch. Subspecialties: Theoretical computer science; Computer-aided design. Current work: Use of a computer system, with graphics, to check the nuclear plant design for interference. Home: 909 S Juniper Escondido CA 92025

BEDNOWITZ, ALLAN LLOYD, chemical physicist; b. Bklyn., Oct. 7, 1939; s. Herman and Letty (Lambor) B.; m. Priscilla Terri Nagler, June 26, 1966 (div. 1979); 1 son, Howard Ian. B.Chem. Engring., Cooper Union, 1960; M.S. in Physics, Poly. Inst. N.Y., 1963; Ph.D. in Chem. Physics, Poly. Inst. N.Y., 1966. Teaching fellow Poly. Inst. N.Y., 1960-63, research fellow, Bklyn., 1963-65, Brookhaven Nat. Lab., Upton, N.Y., 1965-67; mem. research staff IBM Watson Research Center, Yorktown Heights, N.Y., 1967—. Editor: IUCr World Directory, 1981. Vice pres. Hebrew Inst., White Plains, N.Y., 1982-83. Mem. AAAS, Am. Phys. Soc., Am. Crystallog. Assn., Assn. Computing Machinery, Sigma Xi. Subspecialties: Graphics, image processing, and pattern recognition; X-ray crystallography. Office: IMB Watson Research Center PO Box 218 Yorktown Heights NY 10598 Home: PO Box 537 Millwood NY 10546

BEDROSIAN, SAMUEL DER, electrical and systems engineer, educator; b. Marash, Turkey, Mar. 24, 1921; came to U.S., 1922, naturalized, 1952; s. Sahag Der and Zabel B.; m. Agnes Morjigian, Nov. 24, 1951; children—Camille, Gregory. A.B., SUNY, Albany, 1942; M.E.E., Poly. Inst. Bklyn., 1951; Ph.D., U. Pa., 1961. Project engr. Signal Corps Engring. Labs., Ft. Monmouth, N.J., 1946-49, sect. chief, 1949-54, asst. br. chief, 1954-55; systems engr. Burroughs Research Center, Paoli, Pa., 1955-60; mem. research staff U. Pa., 1960-64, asst. prof. elec. engring., 1964-68; asso. prof., 1968-73, prof. elec. and systems engring., 1973—, chmn. dept. systems engring., 1975-80, dir. dual degree MBA/MSE program for, 1977-82; cons. in field; organizer, gen. chmn. 22d M.W. Symposium on Circuits and Systems at Moore Sch., Phila., 1979; NAVELE Research chair prof. Naval Postgrad. Sch., Monterey, Calif., 1980-81. Contbr. numerous articles to profl. jours. Served to 1st lt., Signal Corps U.S. Army, 1943-46; PTO. Recipient Kabakjian award Armenian Students Assn. U.S.A., 1977. Fellow IEEE (guest editor Transactions on Circuits and Systems 1979); mem. Franklin Inst. (asso. editor jour. 1966—, guest editor spl. issue jour. 1973, 76), Sigma Xi, Eta Kappa Nu. Subspecialties: Graphics, image processing, and pattern recognition; Distributed systems and networks. Current work: Application of fuzzy, set theory to image clarification quality criteria and information content; design of computer communication network; application of graph theory to VLSI chip layout. Patentee in field. Home: 35 Bryan Ave Malvern PA 19355 Office: Moore Sch U Pa 200 S 33d St Philadelphia PA 19104

BEEDLE, LYNN SIMPSON, civil engineering educator; b. Orland, Calif., Dec. 7, 1917; s. Granville L. and Carol (Simpson) B.; m. Ella Marie Grimes, Oct. 20, 1946; children: Lynn, Helen, Jonathan, David, Edward. B.S., U. Calif., 1941; M.S., Lehigh U., 1949, Ph.D., 1952. With Todd-Calif. Shipbldg. Corp., Richmond, Calif., 1941; instr. Postgrad. Sch., U.S. Naval Acad.; officer-in-charge Underwater Explosions Research div. Norfolk (Va.) Naval Shipyard, 1941-47; dir. Lehigh U. Fritz Engring. Lab., Bethlehem, Pa., 1960—; prof. civil engring. Lehigh U., 1958-77, Univ. Disting. prof., 1978—, dir. High-Rise Inst., 1983—. Author: Plastic Design of Steel Frames, 1958, (with others) Structural Steel Design, 2d edit, 1974; editor-in-chief: Planning and Design of Tall Buildings, 5 vols, 1978-81; contbr. articles to profl. jours. Served with USNR, 1941-47. Recipient Robinson award Lehigh U., 1952, Hillman award, 1973; E.E. Howard award ASCE, 1963; Research prize, 1956; Silver medal Am. Welding Soc., 1957; Constrn. award Engring. News Record, 1965, 73; Regional Tech. Meeting award Am. Iron and Steel Inst., 1958; T.R. Higgins award Am. Inst. Steel Constrn., 1973; Engr. of Year award Lehigh Valley sect. Nat. Soc. Profl. Engrs., 1977. Fellow ASCE (hon. mem.; dir. 1974-77, dir. Lehigh Valley sect. 1977—, past chmn. structural div. exec. com., past mem. research com.); mem. Structural Stability Research Council (life mem., chmn. 1966-70, dir. 1970—), Welding Research Council, Am. Inst. Steel Constrn., Nat. Acad. Engring., Council on Tall Bldgs. and Urban Habitat (chmn. 1970-76, dir. 1976—), Internat. Assn. Bridge and Structural Engring. (hon.). Presbyn. (elder 1957—). Subspecialty: Civil engineering. Current work: Urban design, plastic design, residual stresses, high-strength bolts, welded plate girders,tall buildings. Home: 102 Cedar Rd Hellertown PA 18055 Office: Fritz Engring Lab Lehigh Univ Bethlehem PA 18015

BEER, STEVEN VINCENT, plant pathologist, educator; b. Boston, July 19, 1941; s. Carl and Vera (Radna) B.; m. Beverly Richardson, June 22, 1963; children: David V., Rachel E., Jennifer S. B.S., Cornell U., 1965; Ph.D., U. Calif.-Davis, 1969. Asst. prof. plant pathology Cornell U., Ithaca, N.Y., 1969-77, assoc. prof., 1977—. Contbr. chpts. to books, articles to profl. jours.; patentee in field. NSF fellow, 1962-64; NDEA fellow, 1965-68; Competitive Research grantee USDA, 1978, 80, 83. Mem. Am. Phytopath. Soc., Am. Soc. Microbiology, AAAS, Sigma Xi. Subspecialties: Plant pathology; Microbiology. Current work: Plant diseases caused by bacteria, particularly fire blight of apple and pear, biological control. Physiology and genetics of phytopathogenic bacteria including molecular biology. Home: 312 Eastwood Ave Ithaca NY 14850 Office: 410 Plant Science Bldg Cornell U Ithaca NY 14853

BEERY, KENNETH EUGENE, food scientist; b. Lancaster, Ohio, Apr. 30, 1943; s. Robert David and Lucille Ester (Scholl) B.; m. Marci Annear, Aug. 22, 1965; children: Kevin, Kendra, Kelli, Kyle. B.S., Ohio State U., 1965; Ph.D., Pa. State U., 1970. With U.S. Dept. Agr., Berkeley, Calif., 1972-75, Union Carbide Corp., Chgo., 1975-76; dir. research Archer Daniels Midland Co., Decatur, Ill., 1976. Editorial adv. bd.: Food Tech. mag, 1975-78; contbr. articles to profl. jours. Served with U.S. Army, 1970-72; to maj. Res. Recipient Honored Grad. Student award Am. Oil Chemists Soc., 1969. Mem. Inst. Food Technologists (chmn. Iowa sect. 1978-79), AAAS, Am. Meat Sci. Assn., Am. Soc. Animal Sci., Am. Mamt. Assn., Am. Assn. Cereal Chemists, Research Soc. Am., Sigma Xi, Alpha Gamma Sigma, Gamma Sigma Delta, Alpha Zeta, Phi Tau Sigma. Subspecialties: Food science and technology; Biomass (agriculture). Current work: Process development and anlaysis, corn, soy, wheat and grain. Office: 4666 Faries Pkwy Decatur IL 62526

BEETON, ALFRED MERLE, limnologist, educator; b. Denver, Aug. 15, 1927; s. Charles Frederick and Edna F. (Smith) B.; m. Mary Eileen Wilcox, July 20, 1945; children—Maureen Ann, Heather Ann, Celeste Nadine; m. Ruth Elizabeth Holland, June 4, 1966; children—Jonathan Eugene, Daniel Paul. B.S., U. Mich., 1952, M.S., 1954, Ph.D., 1958. Fishery biologist U.S. Bur. Comml. Fisheries, Ann Arbor, Mich., 1957-65, chief environ. research, 1960-65; prof. zoology U. Wis.-Milw., 1965-76, asst. dir., 1965-69, assoc. dir., 1969-73, assoc. dean, 1973-76; dir. Gt. Lakes and Marine Waters Center; prof. U. Mich., Ann Arbor, 1976—; lectr. biology Wayne State U., 1957-61; lectr. civil engring. U. Mich., 1961-65; mem. research adv. council Wis. Dept. Natural Resources; mem. water quality criteria com. Nat. Acad. Scis.; cons. U.S. Army C.E., 1967-73, Met. San. Dist. Chgo., 1968-76, EPA, 1973-83; adviser to Smithsonian Instn. on projects in, Ghana, Laos, Yugoslavia, 1972-82; to WHO/Pan Am. Health Orgn. in, Venezuela, 1978; mem. environ. studies bd. NRC, 1976-82, internat. environ. program com., 1977-82. Contbr. chpts. to books; articles Ency. Brit. Mem. Internat. Assn. Theoretical and Applied Limnology, Am. Soc. Limnology and Oceanography (treas. 1962-81), Am. Soc. Zoologists, Internat. Assn. Gt. Lakes Research, Sigma Xi. Subspecialties: Ecology; Limnology. Current work: Eutrophication of the Great Lakes. Limnology of tropical man-made lakes. Home: 2761 Oakcleft St Ann Arbor MI 48103 Office: U Mich Ann Arbor MI 48109

BEGA, ROBERT V., plant pathologist; b. Milford, Mass., Aug. 17, 1928; s. Valino B.; m. Joan Fisher, Feb. 20, 1972; 1 dau.: Cyrise Burger. B.S., U. Calif.-Berkeley, 1954, Ph.D., 1957. With Forest Service, U.S. Dept. Agr., Berkeley, Calif., 1956—, supervisory research plant pathologist, 1962—; lectr. U. Calif., Berkeley, 1957—; assoc. Calif. Agrl. Expt. Sta., 1957—. Author books and numerous Sci articles. Served with AUS, 1945-47. EPA grantee, 1967-72. Mem. Am. Phytopathol. Soc., Am. Forestry Assn., Sigma Xi. Subspecialties: Plant pathology; Integrated pest management. Current work: Biological control, chemical control, diagnosis, biometeorology, fungus, genetics. Home: 635 Spruce St Berkeley CA 94707 Office: PO Box 245 Berkeley CA 94701

BEGGS, JAMES MONTGOMERY, government official; b. Pitts., Jan. 9, 1926; s. James Andrew and Elizabeth (Mikulan) B.; m. Mary Elizabeth Harrison, Oct. 3, 1953; children—Maureen Elizabeth, Kathleen Louise, Teresa Lynn, James Harrison, Charles Montgomery. Student, So. Meth. U., 1942-44; B.S. in Engring, U.S. Naval Acad., 1948; M.B.A., Harvard U., 1955; LL.D. (hon.), Washington and Jefferson Coll., 1972; Dr. Engr. Mgmt., Embry-Ripple U., 1972; LL.D. (hon.), Salem Coll.; D.Engring., U. Ala.; LL.D., Maryville Coll. Commd. ensign U.S. Navy, 1947, advanced through grades to lt. comdr., 1954; resigned, 1954; various mgmt. positions Westinghouse Electric Corp., 1955-68, v.p. Def. and Space Center, 1968; asso. adminstr. advanced research and tech. NASA, 1968-69; undersec. Dept. Transp., 1969-73; mng. dir. Summa Corp., Los Angeles, 1973-74; exec. v.p., dir. Gen. Dynamics Corp., St. Louis, 1974-81; adminstr. NASA, Washington, 1981—; dir. ConRail, Phila., EMC, Inc., Cockeysville, Md. Vice chmn. bd. Howard County (Md.) Charter Bd., 1966-67; mem. Md. Bd. Natural Resources, 1966-67. Mem. Nat. Acad. Pub. Adminstrn., Am. Soc. Pub. Adminstrn., AIAA, Am. Soc. Naval Engrs. Clubs: Burning Treel (Bethesda Md.); Met. (Washington); St. Louis, Bellerive Country (St. Louis). Subspecialties: Mechanical engineering; Aerospace engineering and technology. Office: NASA 400 Maryland Ave SW Washington DC 20546

BEHBEHANI, ABBAS M., clin. virologist, educator; b. Iran, July 27, 1925; came to U.S., 1946, naturalized, 1964s; s. Ahmad M. and Roguia B (Tasougi) B.; m.; children: Ray, Allen, Bita. B.A., Ind. U., 1949; M.S., U. Chgo., 1951; Ph.D., Southwestern Med. Sch., U. Tex., 1955. Asst. prof. Baylor U. Coll. Medicine, Houston, 1960-64; assoc. prof. pathology U. Kans. Sch. Medicine, Kansas City, 1967-72, prof., 1972—. Author two books, numerous articles. Fellow Am. Acad. Microbiology; mem. AAAS, Am. Soc. Microbiology, Soc. Exptl. Biology and Medicine. Moslem. Subspecialties: Virology (medicine); Tissue culture. Current work: Use of tissue culture system in the study of human disease. Home: 5415 Hazen St Kansas City KS 66106 Office:

Dept Pathology and Oncology U Kans Med Center Kansas City KS 66103

BEHNKE, WALLACE BLANCHARD, JR., utility company executive; b. Evanston, Ill., Feb. 5, 1926; s. Wallace Blanchard and Dorothea (Bull) B.; m. Joan Fortune Murphy, Sept. 24, 1949; children: Susan B. Behnke Jones, Ann B., Thomas W. B.S., Northwestern U., 1945, B.S.E.E., 1947. Registered profl. engr., Ill. With Commonwealth Edison Co., Chgo., 1947—, exec. v.p., 1973-80, vice chmn., 1980—, also dir.; bd. dirs., chmn. Atomic Indsl. Forum, Washington, 1982—; dir. Lake View Trust & Savs. Bank, Chgo., 1970—, Paxall, Inc., 1981—, Tuthill Corp., Oak Brook, Ill., 1982—. Bd. dirs. United Way of Chgo, 1972—, Robert Crown Ctr. for Health Edn., Hinsdale, Ill., 1972—, Protestant Found. Greater Chgo., 1974—; trustee Northwestern Meml. Hosp., Chgo., 1970—, Ill. Inst. Tech., 1976—. Fellow IEEE; mem. Western Soc. Engrs. (pres. 1976-7), hon. mem. (1980—), Am. Nuclear Soc., Nat. Acad. Engring. Episcopalian. Clubs: Hinsdale Golf (pres. 1981-82); Chgo., Econ. (sec. 1972), Comml. (Chgo.)). Subspecialties: Electrical engineering; Nuclear engineering. Current work: Electric power production, transmission and distribution; nuclear power, including light water reactor operation and liquid metal fast breeder reactor development. Office: Commonwealth Edison Co PO Box 767 Chicago IL 60690

BEHRENDT, JOHN CHARLES, geophysical researcher; b. Stevens Point, Wis., May 18, 1932; s. Allen Charles and Vivian Elaine (Frogner) B.; m. Donna Miriam Ebben, Oct. 6, 1961; children: Kurt Allen, Marc Russell. Student, Wis. State Coll., 1950-52; B.S., U. Wis., 1954, M.S., 1956, Ph.D., 1961. Asst. seismologist Arctic Inst. N.Am., Ellsworth Station, Antarctica, 1956-58; research assoc. U. Wis., Madison, 1958-64; geophysicist U.S. Geol. Survey, Denver, 1964-72, Monrovia, Liberia, 1964-72, br. chief, geologist in charge, Woods Hole, Mass., 1972-77, geophysicist, Denver, 1977—; advisor Dept. State, Washington, 1977—; cons. NSF, 1981—; mem. panel Joint Oceanographic Inst. Inc., Washington, 1981-82, Nat. Acad. Sci., 1973—. Fellow Geol. Soc. Am., Royal Astron. Soc.; mem. Am. Geophys. Union, Soc. Exploration Geophysicists, Glaciol. Soc., AAAS, Seismol. Soc. Am., Explorers Club. Subspecialties: Geophysics; Sea floor spreading. Current work: Tectonics of passive continental margins (U.S. Atlantic, W. Africa, Antarctica); Mid-plate earthquakes; Antarctic continental tectonics. Office: US Geol Survey MS 966 Federal Center Denver CO 80225

BEHRENS, WILLIAM WOHLSEN, JR., oceanographer, engr., research company executive, educator, retired naval officer; b. Newport, R.I., Sept. 14, 1922; s. William Wohlsen and Nell (Vasey) B.; m. Betty Ann Taylor, June 22, 1946; children—Elizabeth Behrens Garland, William Wohlsen III, Charles Conrad, Susan Raker. B.S., U.S. Naval Acad., 1943; student, U.S. Submarine Sch., 1943, U.S. Naval Nuclear Power Sch., 1955-56, Nat. War Coll., 1963; M.A., George Washington U., 1964; Sc.D., Gettysburg Coll., 1974. Commd. ensign U.S. Navy, 1943, advanced through grades to vice adm., 1972, ret., 1974; combat duty 6 submarine patrols, Pacific, World War II, comdr. four submarines, 1953-63; founding dir. Navy's Nuclear Power Sch., 1955; AEC qualified nuclear chief reactor operator, 1956; special asst. Atomic Energy Comm., 1956-57; dir. NATO nuclear planning strategic plans Office Chief Naval Ops., 1964-66; dep. asst. sec. state, mem. policy planning council, 1966-67; comdr. Amphibious Force, Vietnam, 1967-69; dir. politico-mil. policy Office Chief Naval Operations, 1969-70; oceanographer of Navy, 1970-72; asso. adminstr. and naval dep. NOAA, 1972-74; sr. v.p. J. Watson Noah Inc., Arlington, Va.; pres. Earth Resources Applications Inst., Inc., Washington; sci. adviser Wheeler Industries, Inc., Washington; adj. prof. dept. marine sci. U. South Fla.; Vice pres. Am. Oceanic Orgn., 1976. Trustee SEA Edn. Assn., Boston, 1974-78; dir. Fla. Inst. for Oceanography; mem. Com. of 100, Pinellas County, Fla.; bd. govs. Sci. Center, Inc., Pinellas County, 1975-81; mem. Port Commn., St. Petersburg; exec. bd. St. Petersburg Progress, Inc. Decorated D.S.M., Silver Star, Legion of Merit with 4 gold stars and combat V, Navy Bronze Star with combat V, Army Bronze Star with combat V, Joint Services Commendation medal, Navy Commendation medal, other U.S. and fgn. decorations. Mem. N.Y. Acad. Scis., U.S. Naval Inst. (life), U.S. Naval Acad. Alumni Assn. (life), U.S. Naval Acad. Found., Navy-Marine Corps-Coast Guard Residence Found., Naval Acad. Athletic Assn., Explorers Club, U.S. Navy League, Soc. Am. Mil. Engrs., Inst. Navigation, Marine Tech. Soc., AAAS, IEEE, Am. Geophys. Union, Internat. Oceanographic Found., Internat. Game Fish Assn., Arctic Inst., Am. Soc. Naval Engrs., Smithsonian Instn. Nat. Assos., Ret. Officers Assn. (life). Clubs: N.Y. Yacht (N.Y.C.); U.S. Naval Sailing Assn. (Annapolis, Md.) (commodore 1975); Nat. Propeller, Army Navy (Washington). Subspecialties: Offshore technology; Environmental engineering. Home: 1125 Friendly Way S Saint Petersburg FL 33705 Office: 830 1st St S Saint Petersburg FL 33701 My life has taught me the everlasting value of the oceans to man, not only in the nurturing of mankind through the ages, but now, more than ever, in providing to our burgeoning world populations the essential transport for communications and much-needed goods, as well as the all-important potential of energy food and other key resources from the sea and the environmental rejuvenation so vitally required.

BEHRISCH, HANS WERNER, biology educator, researcher; b. Vienna, Austria, Nov. 26, 1941; came to U.S., 1969; s. Hans Hermann and Erika Hilda (Petzold) B.; m. Kathryn Elizabeth Beauchamp, June 2, 1967; children: Tanya, Erika, Kevin. B.S., U. B.C., 1964, Ph.D., 1969; M.A., Oreg. State U., 1966. Asst. prof. Inst. Arctic Biology, U. Alaska, Fairbanks, 1969-73, assoc. prof., 1973-82, prof. zoochemistry, 1982—. Mem. Am. Physiol. Soc., Am. Soc. Biol. Chemists. Subspecialties: Biochemistry (biology); Physiology (biology). Current work: Metabolic regulation, environmental physiology. Office: Inst Arctic Biology Univ Alaska Fairbanks AK 99701

BEHRMAN, RICHARD ELLIOT, pediatrician, neonatologist, university dean; b. Phila., Dec. 13, 1931; s. Robert and Vivian (Keegan) B.; m. Ann Nelson, Aug. 14, 1954; children: Amy Jane, Michael Jameson, Carolyn Ann, Hillary. A.B., Amherst Coll., 1953; J.D., Harvard U., 1956; M.D. (Univ. scholar), U. Rochester, 1960. Intern Johns Hopkins Hosp., Balt., 1960-61, resident in pediatrics, 1963-65; asst. prof. pediatrics U. Oreg. Sch. Medicine, Portland, 1965-67, asso. prof., 1967-68; prof. U. Ill. Coll. Medicine, Chgo., 1968-71; prof., chmn. dept. Columbia U. Coll. Physicians and Surgeons, N.Y.C., 1971-76, Case Western Res. U. Sch. Medicine, Cleve., 1976-81, dean Sch. Medicine, 1980—; dir. dept. pediatrics Rainbow Babies and Children's Hosp., Cleve., 1976-81; chmn. bd. maternal, child and family research Nat. Acad. Sci., NRC, 1977-80; examiner Am. Bd. Pediatrics. Author: Neonatology: Diseases of the Fetus and Infant, 1973, Neonatal-Perinatal Medicine, 1977; editor: Nelson's Textbook of Pediatrics, 1978, 83; mem.: editorial bd., sect. editor fetal and neonatal medicine Jour. Pediatrics, 1970—; asso. editor: Pediatric Research, 1971-80. Served with USPHS, 1961-63. Whipple scholar, 1960-61; Wyeth pediatric fellow, 1963-65. Fellow Am. Acad. Pediatrics; mem. Soc. Pediatric Research (v.p. 1976-77), Inst. Medicine of Nat. Acad. Scis., Am. Pediatric Soc., Inst. Medicine, Assn. Med. Sch. Pediatric Dept., Perinatal Research Soc. (council 1970-73), Pediatric Travel Club, Soc. Gynecol. Investigation, Sigma Xi. Presbyterian. Clubs: Cleve. Racquet, Century Assn. Subspecialties: Pediatrics; Neonatology. Current work: Neonatal intensive care; fetal physiology; health and education administration. Home: 2871 Courtland Blvd Shaker Heights OH 44122 Office: 2101 Adelbert Rd Cleveland OH 44106

BEICKEL, SHARON LYNNE, psychologist, consultant; b. Hanford, Calif., Mar. 1, 1943; d. William Wayne and Kathleen (Haun) B.; m. W. Oran Hutton, Aug. 8, 1964 (div. 1974). B.S., Eastern Oreg. State Coll., 1965; M.S., U. Oreg., 1970, Ph.D., 1977. Cert. psychologist. Head resident, instr. housing dept. U. Oreg., Eugene, 1966-68, research assoc. dept. human devel., 1968-70, research assoc., 1970-75, psychol. intern, 1975-76, psychologist intern, 1976-77; psychologist Counseling Ctr., Ariz. State U., Tempe, 1978—, pvt. practice psychol. cons., 1978—; cons. McHugh & Assocs., Sequim, Wash., 1981—, sch. dists., Oreg., Calif., Colo., Wash., Hawaii, Kans., 1971-76. Mem. Am. Psychol. Assn., Western Psychol. Assn., Ariz. Psychol. Assn., Ariz. State U. Faculty Women's Assn. Democrat. Subspecialties: Behavioral psychology; Developmental psychology. Current work: Mid-life crisis, Jungian psychology. Home: 1678 Orchard St Eugene OR 97403 Office: Counseling Service Ariz State Univ 112 Agriculture Tempe AZ 85281

BEIRNE, OWEN ROSS, oral and maxillofacial surgeon, educator, researcher; b. Santa Maria, Calif., Jan. 18, 1947; s. Owen and Thelma (Ross) B.; m. Sheryl Martha Schochet, Aug. 23, 1970. B.A., U. Calif.-Berkeley, 1968; D.M.D., Harvard U., 1972; Ph.D., U. Calif.-San Francisco, 1976. Cert. oral and maxillofacial surgery. Resident oral and maxillofacial surgery Harbor-UCLA Med. Ctr., Torrance, Calif., 1976-79; asst. prof. Sch. Dentistry, U. Calif.-San Francisco, 1979—; attending staff U. Calif.-Moffitt Hosp., San Francisco, 1979—, Mt. Zion Med. Ctr., 1981—; cons. San Francisco Gen. Hosp., 1980—, Calif. Children's Service, 1980—. Contbr. articles to profl. publs. U. Calif. Cancer Coordinating Com. grantee; recipient Silver award Harvard Sch. Dental Medicine, 1972. Fellow Am. Soc. Dental Anesthesiology; Mem. AAAS, Am. and Internat. Assn. Dental Research, ADA, Am. Assn. Dental Schs., Calif. Honor Soc., Phi Beta Kappa, Omicron Kappa Upsilon. Subspecialties: Oral and maxillofacial surgery; Cancer research (medicine). Current work: Chemical carcinogenesis and oral cancer, preprosthetic surgery, bone grafting, teaching management of medically compromised dental patient, orthognathic surgery. Office: Dept Stomatology Sch Dentistry Univ Calif San Francisco S-653 San Francisco CA 94143

BEISER, LEO, consulting physicist, researcher, author; b. N.Y.C., Sept. 18, 1924; s. Sigmund N. and Sarah (Weiner) B.; m. Edith Vegotsky, Aug. 31, 1946; children: Helene Ronnie, Steven Scott. B.S. in Physics, Hofstra U., 1964, M.A., 1966; elec. engring. honor grad., R.C.A. Insts., 1948; postgrad. in bus. adminstrn, Alexander Hamilton Ins., 1958. Asst. chief engr. CBS, N.Y.C., 1951-56; project mgr. Polarad Electronics Corp., N.Y.C., 1956-60; staff cons. Gen. Instrument Corp., L.I., N.Y., 1960-61; research mgr. Telechrome Corp., 1961-62; staff research specialist Autometric-Raytheon Corp., 1962-63; sr. staff physicist, dir. Dennis Gabor Labs., CBS Labs., Stamford, Conn., 1963-76; pres., research dir. Leo Beiser Inc., Flushing, N.Y., 1976—; seminar leader; editorial advisor. Author: Advanced Electronic and Electro-Optical Publishing and Printing, 1977; contbr.: Laser Scanning Systems chpt. to Laser Applications, Vol. 2, 1974. Served with USAAF, 1943-46; CBI. Recipient IR-100 award for holofacet optical scanner, 1973. Fellow Soc. Info. Display (recognition award 1978), Soc. Photo-Optical Instrumantation Engrs.; mem. Optical Soc. Am., Soc. Motion Picture and TV Engrs. Subspecialties: Information systems, storage, and retrieval (computer science); Laser data storage and reproduction. Current work: Advanced laser scanning; holographic scanning; data storage and retrieval; optical information handling; three dimensional imaging; holography; imaging technologies. Numerous patents and inventions in laser scanning, notably light scanning system utilizing diffraction optics. Home and Office: 151-77 28th Ave Flushing NY 11354

BEITZ, ALVIN JAMES, neurobiologist, researcher, educator; b. Meadville, Pa., Feb. 16, 1949; s. Albert O. and Margaret (Balint) B.; m. Diane W. Beitz, Sept. 4, 1971; children: Jennifer, Scott, Stacey, Mark, Julie. B.S., Gannon Coll, 1971; Ph.D., U. Minn., 1976. Postdoctoral fellow Harvard U. Med. Sch., Boston, 1976-78; asst. prof. anatomy U. S.C., Columbia, 1978-82, asso. prof., 1982; asst. prof. Sch. Vet. Medicine, U. Minn., St. Paul, 1982—. Contbr. articles to sci. jours. Recipient Bacaner research award Minn. Med. Found., 1978. Mem. Am. Assn. Anatomists, Soc. for Neurosci. Roman Catholic. Subspecialties: Neurobiology; Anatomy and embryology. Current work: Research into the anatomy, biochemistry and pharmacology of pain pathways and the central nervous system pain suppression system. Office: Dept Vet Biology U Minn 1988 Fitch Ave Saint Paul MN 55108 Home: 3517 Vivian Ave Roseville MN 55112

BELADY, LASZLO ANTAL, computer scientist; b. Budapest, Hungary, Apr. 29, 1928; s. Laszlo Armin and Ilona (Janauschek) B.; m. Gisele Fordos, Dec. 18, 1959; children: Christian L., Petra H. Diploma in Aeronautics with honors, Tech. U., Budapest, 1950. Design engr., France, Germany, Hungary, 1950-61; mem. research staff IBM Research, Yorktown Heights, N.Y., 1961-71, sr. mgr. software engring., 1972-81, program mgr. software tech., 1981-83; mgr. software engring. IBM Sci. Inst., Tokyo, Japan, 1983—; vis. prof. U. Calif.-Berkeley, 1971-72; adj. prof. computer scis. N.Y. Poly. Inst., 1980-83. Contbr. articles to profl. jours. Recipient Outstanding Contbn. award IBM, 1969, 73, Invention award, 1971, 73, 77, Meritorious Service award, 1982. Mem. Assn. Computing Machinery, IEEE (sr.). Subspecialties: Software engineering; Operating systems. Current work: Tools and methods to increase programmer productivity and program quality. Patentee in field. Office: IBM Japan Sci Inst 36 Kowa Bldg 5-19 Sanban-Cho Chiyoda-Ku Tokyo Japan 102

BEL BRUNO, JOSEPH J(AMES), chem. physicist; b. Passaic, N.J., June 30, 1952; s. Joseph and Carmella (Nicastro) Bel B.; m. Kathleen Cassidy, Aug. 10, 1980. B.S. in Chemistry, Seton Hall U., 1974, Ph.D., Rutgers U., 1980. Research chemist Am. Cyanamid, Bound Brook, N.J., 1974-76; research assoc. Princeton U., 1980-82; asst. prof. chemistry Dartmouth Coll., 1982—. Contbr. articles to profl. jours. Mem. Am. Phys. Soc., Am. Chem. Soc., Sigma Xi. Roman Catholic. Subspecialties: Laser-induced chemistry; Atomic and molecular physics. Current work: Laser-assisted chem. processes; spectroscopic and dynamical studies of van der Waals complexes; molecular energy transfer; laser spectroscopy. Home: 2 Prospect St Hanover NH 03755 Office: Dept Chemistry Dartmouth Coll Hanover NH 03755

BELDING, HIRAM HURLBURT, IV, psychologist; b. Chgo., Nov. 26, 1942; s. Hiram Hurlburt and Nancee Curry (Reitheimer) B.; m. Margaret Irving, June 25, 1966; children: Wendy Kathleen, Lindsay Cameron. A.A., Riverside City Coll., 1966; B.A. in Psychology, San Jose State U., 1968; M.A. in Human Behavior, U.S. Internat. U., 1971, Ph.D., 1975. Lic. psychologist, S.C., Calif. Electronics technician U.S. Air Force, Denver and Las Vegas, 1960-64; dep. probation officer Riverside County Probation Dept., 1968-70; psychologist USN Med. Ctrs. (Bethesda), Md., 1974-79, S.C., 1974-79, 1974-79; clin. psychologist Psychol. Services of San Diego, 1978—; cons. FBI, 1979—. Served to lt. USNR, 1974-79. Mem. Am. Psychol. Assn., Calif. Psychol. Assn., Acad. Psychologists, San Diego Nat. Register Health Care Providers in Psychology, Assn. Advancement Psychology. Republican. Subspecialty: Behavioral psychology. Current work: Individual and family psychotherapy, hypnosis for early life information and painccontrol. Office: Psychological Services of San Diego 3560 4th Ave San Diego CA 92103 Home: 3664 Curtis St San Diego CA 92106

BELIAN, RICHARD DUANE, physicist; b. Santa Fe, N.Mex., Feb. 23, 1938; s. Charles Paul and Era May (Johnson) B.; m. Mary Reyes, Mar. 3, 1962; children: Richard, Raymon, Anthony. M.S., U. N.Mex., 1967. Staff mem. Los Alamos (N.Mex.) Nat. Lab., 1967—. Contbr. articles to profl. publs. Pres. De Vargas Civitan Club, 1972-63, dist. lt. gov., 1980-81. Served with USAR, 1959-65. Mem. Am. Astron. Soc., Am. Geophys. Union. Democrat. Lodge: Elks. Current work: Magnetospheric dynamics and substorms, magnetospheric ion composition particle energization; X-ray bursters and transients. Home: 2005 Zozobra Ln Santa Fe NM 87501 Office: Los Alamos Nat Lab Los Alamos NM 87545

BELIKOFF, STEVEN WILLIAM, microbiologist; b. Glendale, Calif., Sept. 15, 1952; s. Bill and Irene (Goreyko) B.; m. Debra Carol Anderson, Aug. 25, 1973; children: Erin Lynn, Amy Christine. Student, Fullerton Jr. Coll., 1970-72; B.S., Calif. State U.-Fullerton, 1975. With Hancock/Extracorporeal, Inc., Anaheim, Calif., 1975—, now supr. microbiology lab. Mem. Am. Soc. Microbiology, Am. Soc. Quality Control. Quaker. Subspecialties: Microbiology; Artificial organs. Current work: Sterilization of bioprosthetic heart valves by cold chemicals. Office: 4633 E La Palma Ave Anaheim CA 92807

BELJAN, JOHN RICHARD, university dean, medical educator; b. Detroit, May 26, 1930; s. Joseph and Margaret Anne (Brozovich) B.; m. Bernadette Marie Marenda, Feb. 2, 1952; children: Ann Marie, John Richard, Paul Eric. B.S., U. Mich., 1951, M.D., 1954. Diplomate: Am. Bd. Surgery. Intern U. Mich., 1954-55, resident in gen. surgery, 1955-59; dir. med. services Stuart div. Atlas Chem. Industries, Pasadena, Calif., 1965-66; from asst. prof. to assoc. prof. surgery U. Calif. Med. Sch., Davis, 1966-74, from asst. prof. to assoc. prof. engring., 1968-74, from asst. dean to assoc. dean, 1971-74; prof. surgery, prof. biol. engring. Wright State U. Med. Sch., Dayton, Ohio, 1974-83, dean Sch. Medicine, 1974-81, vice provost Sch. Medicine, 1974-78, v.p. health affairs Sch. Medicine, 1978-81, provost, sr. v.p. Sch. Medicine, 1981-83; prof. surgery, biomed. engring., dean Sch. Medicine Hahnemann U., Phila., 1983—; prof. arts and scis., assoc. v.p. med. affairs Central State U., Wilberforce, Ohio, 1976—; trustee Cox Heart Inst., 1975-77, Drew Health Center, 1977-78, Wright State U. Found., 1975-83; trustee, regional v.p. Engring. and Sci. Inst. Hall of Fame, 1983—; bd. dirs. Miami Valley Health Systems Agy., 1975-82; cons. in field. Author articles, revs., chpts. in books. Served with M.C. USAF, 1955-65. Decorated Commendation medal; Braun fellow, 1949; grantee USPHS, 1967—; NASA, 1968—. Fellow A.C.S., Royal Soc. Medicine; mem. Aerospace Med. Assn., AAUP, Inst. Aeros. and Astronautics, AMA (council on sci. affairs 1978—), Assn. Acad. Surgery, Biomed. Engring. Soc., F.A. Coller, Dayton surg. socs., Flying Physicians Assn., Ohio Med. Soc., Greene County Med. Soc., Montgomery County Med. Soc. (councilor 1974—), IEEE, Instrument Soc. Am., Royal Soc. Medicine, Soc. Air Force Clin. Surgeons, Soc. Internat. de Chirurgie, Phi Beta Kappa, Phi Eta Sigma, Phi Kappa Phi, Alpha Kappa Kappa. Clubs: Mich. Alumni (Outstanding Alumnus award 1976), Racquet (Dayton); University of Washington, Oakwood Fur, Fin and Feather. Subspecialties: Space medicine; Chronobiology. Current work: Engineering and physical principles in biological systems and man; characterization and modelling of biological rhythms and bone repair; human performance and measurement. Home: 29 Oriole Way Moorestown NJ 08057 Office: Hahnemann Univ Broad and Vine Sts Philadelphia PA 19102

BELL, A. EARL, genetics educator, consultant; b. Providence, Ky., May 9, 1918; s. George Edgar and Ida Margaret (Cullen) B.; m. Floris Selina Thorning, June 12, 1941; children: Robert Earl, Diane Louise, Donald Wayne. B.S., U. Ky., 1939; M.S., La. State U., 1941; Ph.D., Iowa State U., 1948. Grad. teaching asst. La. State U., Baton Rouge, 1939-41, Iowa State U., Ames, 1941-42; asst. prof. genetics Purdue U., West Lafayette, Ind., 1946-48, assoc. prof., 1948-51, prof., 1951—; Fulbright scholar Am. Edn. Found., Australia, 1964; genetics cons. Ind. Farm Bur. Coop., Indpls., 1955—; DNA cons. Nat. Agrl. Expt. Stas. Policy and Orgn. Com., Washington, 1977-79; genetics cons. Nat. Acad. Scis.-NRC, Washington, 1968-70; biostats. cons. Office Naval Research, Chgo., 1963. Contbr. articles to profl. jours. Served to comdr. USN, 1942-46; PTO. NSF Sr. fellow, 1963; Guggenheim fellow, 1965. Fellow AAAS; mem. Genetics Soc. Am., Am. Genetics Assn., Internat. Genetics Congress. Subspecialties: Genetics and genetic engineering (agriculture); Gene actions. Current work: Biomass efficiency, genetic mechanism of competition, adaptation to stress, genotype x environment interactions, polygenic mutation rate, selection limits. Home: 708 Sugar Hill Dr West Lafayette IN 47906 Office: Population Genetics Inst Purdue U West Lafayette IN 47907

BELL, BARBARA JEAN, human factors specialist, human reliability analyst; b. Houston, Aug. 28, 1954; d. John Leon and Louise (Sturdivant) B. B.A. in Modern Langs, Tex. A&M U., 1976, M.S. in Indsl. Engring, 1979. Mem. tech staff Sandia Nat. Labs., Albuquerque, 1979-82; prin. research scientist Battelle Columbus (Ohio) Labs., 1983—; course instr. U. Wis.-Madison, 1981-82, JBF Assocs., Knoxville, Tenn., 1982-83, Det Norske Veritas, Oslo, 1982. Author: (with A.D. Swain) A Procedure for Conducting Human Reliability Analysis for Nuclear Power Plants, 1981, 2d edit., 1983. Tchr., leader Del Norte Baptist Ch., Albuquerque, 1979-82. Mem. Human Factors Soc., Am. Nuclear Soc. (exec. council TGHF 1982-85). Republican. Subspecialties: Human factors engineering; Nuclear fission. Current work: Human reliability analysis as part of probabilistic risk assessment, usually for nuclear power generating stations. Office: Risk Assessment Group Battelle Columbus Labs 505 King Ave Columbus OH 43201

BELL, CHAROLETTE RENEE, school psychologist; b. St. Louis, Jan. 23, 1949; d. Jesse Leon and Victoria Larue (Hancock) B.; m.; children: J. Leon III, David Anthony. B.A., Dillard U., 1970; M.A., U. No. Colo., 1973, Ed.D., 1976. Cert. sch. psychologist. G.E.D. instr. Collbran (Colo.) Civilian Job Corps Ctr., 1970-72; sch. psychologist Aurora (Colo.) Schs., 1973-76, Cherry Creek Schs., Englewood, Colo., 1976-78; researcher S.C. State Coll., Orangeburg, 1978—; cons. psychologist S.C. Dept. Social Services, Columbia, 1980—, S.C. Dept. Mental Retardation, 1981—; ednl. psychologist Nat. Assessment of Ednl. Progress, Denver, 1978. Contbg. author: Discipline and Classroom Management, 1980. Chmn. Community Adv. Bd. for Retarded, Orangeburg, 1980—; v.p. Orangeburg Mental Health Assn., Orangeburg, 1982—; founder, chmn. Citizens Against Sexual Assault, Orangeburg, 1979-80. Fellow Am. Psychol. Assn.; mem. Evaluation Network, Am. Assn. Sex. Educators, Counselors and Therapists, Assn. Black Psychologists (co-chmn. nat. testing com. 1981), Adlerian Soc. (v.p. 1982—), Alpha Kappa Alpha. Democrat. Episcopalian. Club:

Altrusa. Subspecialty: Testing and evaluation. Current work: Achievement on standardized tests by limited resource students. Home: PO Box 99 Orangeburg SC 29116 Office: SC State Coll Box 1841 Orangeburg SC 29117

BELL, CHESTER GORDON, computer engineering company executive; b. Kirksville, Mo., Aug. 19, 1934; s. Roy Chester and Lola Dolph (Gordon) B.; m. Gwendolyn Kay Druyor, Jan. 3, 1959; children: Brigham Roy, Laura Louise. B.S. in E.E, M.I.T., 1956, M.S., 1957. Engr. Speech Communication Lab., M.I.T., 1959-60; mgr. computer design Digital Equipment Corp., Maynard, Mass., 1960-66, v.p. engring., 1972—; prof. computer sci. Carnegie-Mellon U., 1966-72; dir. Inst. Research and Coordination Acoustic Music, 1976-81, Computer Mus., 1982—. Author: (with Newell) Computer Structures, 1971, (with Grason, Newell) Designing Computers and Digital Systems, 1972, (with Mudge, McNamara) Computer Engineering, 1978, (with Siewiorek, Newell) Computer Structures, 1982. Recipient 6th Mellon Inst. award, 1972. Fellow IEEE (McDowell award 1975, Eckert-Mauchly award 1982), AAAS; mem. Nat. Acad. Engring., AAAS, Assn. Computing Machinery (editor Computer Structures sect. 1972-78), Eta Kappa Nu. Subspecialties: Computer architecture; Computer engineering. Current work: computer science, computer art, electrical engineering, computer engineering. Home: Page Farm Rd Lincoln MA 01773 Office: 146 Main St Maynard MA 01754

BELL, CLYDE RITCHIE, botany educator; b. Cin., Apr. 10, 1921; s. William Harold and Mary Edith (Spielman) B.; m. Sarah Foushee Fore, Jan. 14, 1943. A.B., U. N.C., 1947, M.A., 1949; Ph.D., U. Calif.-Berkeley, 1953. Instr. dept. botany U. Ill., Urbana, 1953-55; asst. prof. dept. botany U. N.C., Chapel Hill, 1955-59, assoc. prof., 1959-66, prof., 1966—; dir. N.C. Bot. Garden, 1960—. Author: (with W.S. Justice) Wild Flowers of North Carolina, 1968, (with B.J. Taylor) Florida Wild Flowers, 1982, (with others) Manual of the Vascular Flora of the Carolinas, 1968, Vascular Plant Systematics, 1974. Served to lt. USAAF, 1942-45. Decorated Air medal; recipient Silver Seal award Nat. Council State Garden Clubs, 1979. Fellow AAAS; mem. Am. Inst. Biol. Scis. (exec. com. governing bd. 1970-73), Commn. Undergrad. Edn., Biol. Scis. (exec. com. 1970), Bot. Soc. Am. (v.p. 1978), Soc. Study Evolution, Am. Soc. Plant Taxonomists (pres. 1976), Internat. Assn. Plant Taxonomists, Assn. Southeastern Biologists (exec. com. 1963-65). Subspecialties: Evolutionary biology; Reproductive biology. Current work: Reproductive biology of plants, evolution and systematics; pollination biology. Office: U NC 406 Coker Hall Chapel Hill NC 27514

BELL, CORNELIUS, contract aerospace engineer; b. Mpls., July 14, 1946; s. Connie and Lorena (Welch) B.; m. Sherry Lesley, June 1976 (div. 1978); m. Joyce Theresa Biles, Jan. 2, 1981. A.A.S., Acad. Aeronautics, Flushing, N.Y., 1968; student, Poly. Inst. Tech., Bklyn., 1975-78; B.S., N.Y. Inst. Tech., 1971-74. Design engr. Fairchild Republic Co., Farmingdale, N.Y., 1974-78; stress analyst Martin Marietta Co., Denver, 1978-80; mass properties engr. Boeing Co., Seattle, 1980-82; pipe stress analyst Sargent & Lundy, Chgo., 1980-82; mech. engr. Raytheon Co., Wayne, N.J., 1982—; contract engr. Salem Corp., Oak Brook., Ill., 1980-82, Gen. Devices, Denver, 1980. Served with U.S. Army, 1968-71; Viet Nam. Mem. ASME, AIAA, Am. Astron. Soc. Democrat. Roman Catholic. Subspecialties: Mechanical engineering; Aerospace engineering and technology. Office: Electromagnetic System Div Raytheon Co 1578 Route 23 Wayne NJ 07470

BELL, GEORGE IRVING, research administrator, researcher; b. Evanston, Ill., Aug. 4, 1926; s. George Irvine and Hazel Virginia (Seerley) B.; m. Laura Virginia Lotz, Jan. 13, 1956; children: Carolyn, George Irving. B.S., Harvard U., 1947; Ph.D., Cornell U., 1951. Mem. staff Los Alamos Nat. Lab., 1951, group leader, 1974—, div. leader, 1976—; Gordon McKay lectr. Harvard U., 1962-63; mem. Basel (Switzerland) Inst. Immunology, 1979-80; adj. prof. Colo. Med. Ctr., Denver, 1969—; scholar in human biology Eleanor Roosevelt Inst. Cancer Research Denver, 1977—. Author: Nuclear Reactor Theory, 1970; editor: Theoretical Immunology, 1978. Fellow Am. Phys. Soc., Am. Nuclear Soc. (cert. merit 1966), AAAS, Am. Alpine Club (David A. Sowles medal 1981, dir. 1966-71), Groupe de Haute Montagne (Paris), Himalayan Club (India). Subspecialties: Cell and tissue culture; Immunobiology and immunology. Current work: Devise and test mathematical models of cell kinetics and activity in immune response cell-cell adhesion and other problems in cell biology. Home: 794-43d St Los Alamos NM 87544 Office: Los Alamos Nat Lab PO Box 1663 T-DO MS B210 Los Alamos NM 87545

BELL, GREGORY LEE, health physicist, emergency planner; b. Bartlesville, Okla., Nov. 28, 1951; s. Roger Lee and Aileen Beverly (Smith) B.; m. Karen Ann Studer, Jan. 4, 1975 (div. 1982). B.S., Okla. State U., 1974. Mgr. quality assurance Conesco Midcontinent Inc., Hinsdale, Ill., 1975; journeyman technician Fla. Power & Light Co., Miami, 1975-77, sr. plant technician, 1977-80; health physicist Tex. Utilities Generating Co., Glen Rose, 1980—; cons. County Emergency Orgn., 1980—. Mem. Am. Nuclear Soc., Health Physics Soc., Nat. Inst. for Certification of Engring. Technicians. Republican. Methodist. Subspecialty: Nuclear fission. Current work: Emergency planning for a nuclear power facility. Office: Texas Utilities Generating Co PO Box 2300 Glen Rose TX 76031

BELL, JIMMY TODD, research chemist; b. Hazlehurst, Ga., Dec. 17, 1938; s. Willie Leon and Mary (Todd) B.; m. Lucille Hill, Jan. 29, 1961; children: Virgie Carol, Stephen Todd. B.A. in Chemistry, Berry Coll., 1960; Ph.D. in Phys. Chemistry, U. Miss., 1963. Research chemist Oak Ridge Nat. Lab., 1963—. Contbr. articles to profl. jours. Deacon First Baptist Ch., Kingston, Tenn., 1980-82. Mem. Am. Chem. Soc. (counselor 1976-77, sec. 1974-75), Am. Nuclear Soc., AAAS. Republican. Subspecialties: Nuclear fission; Fuels. Current work: Research and development in the areas of nuclear fuel reprocessing, tritium permeation of structural materials, and actinide photochemistry. Patentee nuclear reprocessing tech. Office: Oak Ridge Nat Lab PO Box X Oak Ridge TN 37830

BELL, MARVIN CARL, animal nutritionist, researcher, educator; b. Centertown, Ky., Dec. 24, 1921; s. Marvin Cyril and Ida (Coffman) B.; m. Betty Triplett, Aug. 22, 1948; children: Celia Bell Ferguson, Rachel Bell Burley. B.S. in Agr., U. Ky., 1947, M.S., 1948; Ph.D. in Animal Nutrition, Okla. State U., 1951. Assoc. prof. research and teaching animal husbandry-vet. sci. U. Tenn., Knoxville, 1951-57; assoc. prof. U. Tenn.-AEC-Agrl. Research Lab., Oak Ridge, 1957-65, prof., 1965-74; prof. animal sci. U. Tenn., Knoxville, 1974—; cons. Oak Ridge Nat. Lab., 1965-67. Contbr. chpts., numerous articles to profl. publs. Served to maj. USAFR, 1942-46; ETO. Recipient Calcium Carbonate Co.-Nat. Feed Ingredient Assn. Travel Fellowship award, 1969; travel grantee Internat. Congress Nutrition, Hamburg, Germany, 1966, Internat. Congress Nutrition, Mexico City, 1973, Internat. Congress radiation Research, Evion, France, 1970. Fellow AAAS; mem. Am. Soc. Animal Sci., Am. Inst. Nutrition, Animal Nutrition Research Council, Council Agr. Sci. and Tech., Radiation Research Soc., Sigma Xi, Gamma Sigma Delta (Research award of Merit 1970). Methodist. Subspecialties: Animal nutrition; Toxicology (agriculture). Current work: Mineral nutrition and toxicology in livestock; fallout radiation effects on livestock and protection practices.

BELL, RICHARD, physician, educator, researcher; b. Melbourne, Victoria, Australia, Dec. 24, 1946; came to U.S., 1980; s. Richard Chisholm and Yvonne (Thorpe) B.; m. Catherine Joan Callanan; children: Andrew James, David Hamish, Nicholas Matthew. M.B., B.S., U. Melbourne, 1970. Intern Royal Melbourne Hosp., 1971-72, house officer, 1972-73, registrar in medicine, 1973-76, chief resident, 1976-77, physician, 1977-78; fellow, registrar St. Bartholomews Hosp. and Imperial Cancer Research Fund, London, 1978-80; asst. prof. medicine Univ. Hosp., Boston, 1981—. Fellow Royal Australian Coll. Physicians, Royal Australian Coll. Pathologist; mem. Hematology Soc. Australia, Clin. Oncology Soc. Australia, Brit. Assn. Cancer Research, Am. Soc. Clin. Oncology, Am. Fedn. Clin. Research, Am. Soc. Hematology. Subspecialties: Oncology; Biochemistry (biology). Current work: Glucocorticoid receptors in leukemia; terminal deoxynucleoidyl transferase expression and function; oncogene products. Home: 167 Walnut St Wellesley MA 02181 Office: Boston U Med Ctr 75 E Newton St Boston MA 02118

BELL, ROBIN GRAHAM, biologist, research immunologist; b. Canberra, Australia, July 12, 1942; s. Ronald W and Doris A. B.; children: Julian William, Samuel Dylan. B.Sc. in Zoology with honors, Australian Nat. U., 1967; Ph.D. in Exptl. Pathology, John Curtin Sch. Med. Research, Canberra, 1971. Research fellow dept. microbiology U. Western Australia, Perth, 1971-73; sr. research officer Children's Med. Research Found., Princess Margaret Hosp., Perth, 1973-76; research officer J. A. Baker Inst., Coll. Vet. Medicine, Cornell U., Ithaca, N.Y., 1976-79, asst. prof., 1979—. Australian Commonwealth fellow, 1968; Athelston Saw postdoctoral fellow, 1971; USPHS fellow, 1977. Mem. Am. Assn. Immunologists, Am. Soc. Tropical Medicine and Hygiene, N.Y. Acad. Sci. Subspecialties: Immunobiology and immunology; Gene actions. Current work: Immunology of intestinal infections, particularly with intestinal helminths; genetics of expression of immunity. Office: NY State coll Vet Medicine Ithaca NY 14853

BELL, ROGER ALISTAIR, astronomer, educator; b. Walton-on-Thames, Eng., Sept. 16, 1935; came to U.S., 1963; s. William Ernest and Irene May (Elsley) B.; m. Sylvia Anne Gandine-Stanton, July 16, 1960; children: Alistair M., Andrew C. B.Sc., U. Melbourne, Australia, 1957; Ph.D., Australian Nat. U., 1961, Uppsala U., 1982. Lectr. U. Adelaide, Australia, 1962-63; asst. prof. astronomy U. Md., College Park, 1963-69, assoc. prof., 1969-76, prof., 1976—; program dir. NSF, Washington, 1981—. Contbr. articles to profl. jours. Mem. Royal Astron. Soc., Am. Astron. Soc., Internat. Astron. Union. Anglican. Subspecialties: Theoretical astrophysics; Infrared optical astronomy. Current work: Analysis of stellar spectra and colors. Office: Astron Scis NSF 1800 G St Washington DC 20550

BELLER, DAVID I., immunologist, educator; b. N.Y.C., Apr. 24, 1947; s. George J. and Mary R. (Beller) B.; m. Judy J. Bongiorno; 1 son, Aaron J. B.A., Conn. Wesleyan U., 1969; Ph.D., Princeton U., 1975. Research fellow in pathology Harvard U. Med. Sch., Boston, 1975-77, instr., 1977-80, asst. prof., 1980-83, assoc. prof., 1983—. Assoc. editor: Jour. Immunology, 1981—; contbr. articles to sci. jours. Leukemia Soc. Am. spl. fellow, 1978-79. Mem. Am. Assn. Immunologists. Subspecialties: Immunobiology and immunology; Immunology (medicine). Current work: Research in cell biology of immune function, specifically the regulation of activity of macrophages and T-lymphocytes. Office: Harvard U Med Sch Boston MA 02115

BELLER, WILLIAM STERN, aerospace engineer, government official; b. Cleve., Aug. 28, 1919; s. Harold Ansley and Beatrice Irene (Stern) Beller H.; m. Joan Helen Hirshman, Nov. 19, 1955; children: Nancy Kay, Diane Leslie, Stacy Ann. B.S. in M.E, Ga. Inst. Tech., 1941; M.Aero.E., NYU, 1942. Registered profl. engr., N.Y. Sr. aerodynamicist Hughes Aircraft Co., Culver City, Calif., 1942-45; lectr. math. U. So. Calif., Los Angeles, 1946-48; instr. mech. engring. Poly. Inst. Bklyn., 1948-49; sr. editor Missiles and Rockets, Am. Aviation Publs., Washington, 1957-67; chief ocean programs br. U.S. EPA, Washington, 1967—; coordinator marine affairs, dir. Office Energy, Dept. Natural Resources, P.R., 1973-76; chmn. Caribbean Islands Directorate, UNESCO U.S. Man and the Biosphere Program, U.S. Dept. State, Washington, 1976—; spl. cons. U.S. Ho. of Reps., Washington, 1958, chmn. Interagy. Task Force for Georges Bank, 1983—. Author: (with Eric Bergaust) Satellite, 1956, (with Kurt Stehling) Skyhooks, 1962, The American Arsenal, 1968. Recipient Resolution of Thanks U.S. Ho. of Reps., 1959; Editorial Achievement award Assoc. Bus. Pubs., 1965; Meritorious Service award U.S. Dept. Interior, 1970; Honor Plaque Dept. Natural Resources P.R., 1975. Fellow Marine Tech. Soc. (chmn. coastal zone mgmt. com. 1972-76, Meritorious Service award 1978). Club: Nat. Press (Washington). Subspecialty: Resource management. Current work: Automating environmental decisions, establishing environmental criteria for industrial use of U.S. marine waters, and applying the criteria to an automated ecological and toxicological data base to derive conditions for allowable industrial operations, e.g., oil/gas extraction from outer continental shelf. Home: 2701 Largo Pl Bowie MD 20715 Office: US EPA Washington DC 20460

BELLINA, JOSEPH HENRY, obstetrician, gynecologist; b. New Orleans, Jan. 30, 1942; s. Philip Vincent and Sue Bellina; m.; children: Shawn, Todd. M.D., La. State U., 1965. Diplomate: Am. Coll. Obstetrics and Gynecology. Intern Charity Hosp., New Orleans, 1965-66, resident, 1966-69; research assoc. Electrosci. and Biophysics Research Lab., Tulane U. Sch. Engring., New Orleans, 1978—; mem. adv. bd. Baromed. Research Inst., JoEllen Smith Hosp., 1982; mem. Internat. Confedn. Advisors Third World Countries for Laser Instrumentation, WHO, 1982; gynecologic and reproductive infertility med. adv. bd. Vet. Laser Inst., 1982; dir. Omega Inst. and Laser Research Found., New Orleans, 1974—; clin. prof. La. State U. Sch. Medicine, 1980—. Contbr. articles to sci. jours. Recipient 1st Prize award Dist VII Jr., Fellow Papers, 1970; Prize award AMA Meeting, 1975, Am. Coll. Obstetrics and Gynecology, 1979. Mem. Gynecol. Laser Soc. (founding mem., pres. 1979—), Am. Soc. Laser Medicine and Surgery (pres. 1982), AMA, Am. Coll. Gynecologists, Am. Fertility Soc., Am. Assn. Gynecologic Laparoscopists (prize award 1979), Gynecologic Laser Soc., Am. Soc. Colposcopists and Cervical Pathologists, Jefferson Parish Med. Soc., Orleans Parish Med. Soc. Methodist. Subspecialties: Laser medicine; Obstetrics and gynecology. Current work: Lassers in medicine and surgery as applied to tissue interaction, studying wavelength characteristics and tissue effects, laser research in tissue interaction, developing delivery system for laser microsurgery. Office: 3439 Kabel Dr Suite 7 New Orleans LA 70114

BELLIS, EDWARD DAVID, ecologist, researcher, educator; b. Ridley Park, Pa., June 28, 1927; s. Lloyd Monroe and Margaret (Bachman) B.; m. Janice Carol Swaveley, Oct. 27, 1932; 1 dau., Gayle Ann. B.S., Pa. State Coll., 1951; M.S., U. Okla., 1953; Ph.D., U. Minn., 1957. Instr. U. Ga., Athens, 1957-58; asst. prof. Pa. State U., University Park, 1958-63, assoc. prof., 1963-71, prof. biology dept., 1971—, chmn. interdisciplinary program in ecology, 1980—. Served with U.S. Army, 1945-47. Fellow AAAS; mem. Ecol. Soc. Am., Am. Soc. Naturalists, Am. Soc. Ichthyologists and Herpetologists, Wilderness Soc. Subspecialty: Ecology. Current work: Terrestrial ecosystems, amphibians, mammals. Home: RD 1 Spring Mills PA 16875 Office: Pa State Univ Biology Dept 311 Mueller Lab University Park PA 16802

BELLMAN, RICHARD ERNEST, mathematician; b. N.Y.C., Aug. 26, 1920; m., 1963; 2 children. B.A., Bklyn. Coll., 1941; M.A., U. Wis., 1943; Ph.D., Princeton U., 1946. Instr. electronics, Truax Field, Wis., 1942-43; mathematician Princeton U., 1943-44, fine instr. math., 1946-47, research assoc., asst. prof., 1947-48; math. physicist U.S. Navy Radio and Sound Lab., San Diego, 1944-45, U.S. Army, Los Alamos, 1945-46; assoc. prof. Stanford U., 1948-52; mathematician Rand Corp., 1952-65; prof. math., elec. engring., medicine U. So. Calif., Los Angeles, 1965—, prof. medicine, 1974—; assoc. Center for Study Democratic Instns., 1969—. Mem. Nat. Acad. Engring., Am. Math. Soc., Nat. Acad. Scis. Subspecialty: Applied mathematics. Office: Dept Math Univ So Calif Los Angeles CA 90007

BELLO, JAKE, biomed. researcher, chemistry educator; b. Detroit, Feb. 22, 1928; s. Max and Rose (Galanter) B.; m. Helene Riese, Nov. 25, 1957. B.S., Wayne State U., 1948, Ph.D., 1952. Research fellow Purdue U., Lafayette, Ind., 1952-53, Calif. Inst. Tech., Pasadena, 1953-56; scientist Eastman Kodak, Rochester, N.Y., 1956-58; research assoc. Poly. Inst., Bklyn., 1958-59; prin. scientist Roswell Park Meml. Inst., Buffalo, 1960—; cons. Gen. Foods, Tarrytown, N.Y., 1970-73, NIH, Bethesda, Md., 1970-73. Mem. Am. Chem. Soc., Am. Soc. Biol. Chemists, AAAS, Sigma Xi. Subspecialties: Biophysical chemistry; Molecular biology. Current work: Chemistry and structure of proteins and nucleic acids, interferon inducers. Office: Roswell Park Meml Inst 666 Elm St Buffalo NY 14263

BELLONI, FRANCIS LOUIS, physiology educator; b. N.Y.C., Jan. 23, 1949; s. Francis Peter and Frances Marie (Asaro) B.; m. Susan Marie Kelly, May 1, 1971; 1 son, Benjamin. B.S., Providence Coll., 1970; Ph.D., U. Mich., 1975. Teaching asst. U. Mich., Ann Arbor, 1971-75, postdoctoral scholar, 1975-78; research asst. prof. U. Va., Charlottesville, 1979-81; asst. prof. physiology N.Y. Med. Coll., Valhalla, 1981—. Editorial cons. various scientific jours.; contbr. articles to profl. jours. NIH Nat. Research Service award, 1975-78; NIH New Investigator Research award, 1982—; Am. Heart Assn. grantee, 1980-81, 82—. Mem. Am. Physiol. Soc., Fedn. Am. Socs. Exptl. Biology. Roman Catholic. Subspecialty: Physiology (medicine). Current work: Intrinsic cardiovascular regulatory mechanisms. Office: Dept Physiology Basic Sci Bldg NY Med Coll Valhalla NY 10595

BELLO-REUSS, ELSA N., physiology and medical educator, researcher; b. La Plata, Buenos Aires, Argentina, May 1, 1939; came to U.S., 1972; d. Jose Fernando and Julia M. (Hiriart) Bello; m. Luis Reuss, Apr. 15, 1965; children: Luis F., Alejandro E. B.A., U.Chile, 1957, M.D., 1964. SNS fellow U. Chile, Santiago, 1964-66, instr. exptl. medicine, 1966-72; Fogarty fellow U. N.C., Chapel Hill, 1972-74, Louis Welt fellow, 1975-76; career investigator Am. Heart Assn., 1974-75; asst. prof. physiology and medicine Washington U., St. Louis, 1976—. Contbr. articles to med. jours.; referee: Jour. Lab. and Clin. Medicine, Am. Jour. Physiology. Mem. Internat. Soc. Nephrology, Am. Soc. Nephrology, Am. Fedn. Clin. Research. Subspecialties: Physiology (medicine); Nephrology. Current work: Study of transport properties of isolated perfused renal tubules, electrophysiology of renal tubules. Office: Sch Medicine Dept Physiology Washington U 660 S Euclid St Louis MO 63110

BELLPORT, BERNARD PHILIP, cons. engr.; b. LaCrosse, Kans., May 25, 1907; s. Bernard P. and Louise H. (Groves) B.; m. Elsy V. Johnson, June 11, 1931 (dec. Mar. 1954); children—Louise Bellport Garcia, Bernard Philip; m. Mabelle W. Kandolin, Sept. 26, 1955. B.S. in Mining Engring, Poly. Coll. Engring., Oakland, Calif., 1927. Registered profl. engr., Colo. Mining engr. Western U.S., 1927-28; engr.-geologist St. Joseph Lead Co., 1928-31; with Phoenix Utility Co., 1931-32, Mont. Hwy. Commn., 1932-35, Bur. Reclamation, 1936-72; regional dir. region 2, Calif., 1957-59, asso. chief engr., Denver, 1959-63, chief engr., 1963-70, dir. design and constrn., 1970-72; practice as engring. cons., 1972—; arbitrator Constrn. Arbitration Panel, State of Calif. Recipient Distinguished Service award Dept. Interior; Golden Beaver for engring.; named Man of Year Am. Pub. Works Assn., 1970. Mem. Nat. Acad. Engring., U.S. Commn. Large Dams (chmn. 1971-72), Internat. Commn. Irrigation, Drainage and Flood Control, ASCE (pres. Colo. 1966), Am. Arbitration Assn., Rossmoor Engrs. Club, Internat. Water Resources Assn., Hon. Order Ky. Cols., Chi Epsilon (hon.). Episcopalian. Clubs: Masons (32 deg.), Shriners, Round Hill Country. Subspecialty: Civil engineering. Current work: Construction of major heavy structures; improvement of methods and arbitration of contract disputes. Address: 855 Terra California Dr Apt 4 Walnut Creek CA 94595 In retrospect, I believe my highest contribution to humanity and our nation was and is without question my thirty odd years tenure with the United States Bureau of Reclamation—namely my assistance and guidance in the development of water resources in the arid West and emerging foreign nations to produce water for human consumption and recreation, production of food and fibre, industrial production and, not least, cleanly produced electrical energy.

BELLVILLE, JOHN WELDON, anesthesiology educator, reseracher; b. Wauseon, Ohio, Aug. 7, 1926; s. John Francis and Sarah Frederica (Brose) B.; m. Elaine Ramage, 1956; children: Jon, David, Steven, Paul, Charles; m. Lynne Brandstater, Nov. 6, 1979. A.B., Cornell U., 1948, M.D., 1952. Cert. ship's surgeon USCG, diplomate: Am. Bd. Anesthesiology. Intern Mpls. Gen. Hosp., 1952-53; resident N.Y. Hosp.; Bowen-Harlow Brooks scholar N.Y. Acad. Medicine, 1955; asst. prof. surgery Sloan-Kettering Inst., N.Y.C., 1955-59, assoc., 1960; asst. prof. anesthesia Cornell U. Med. Coll., 1959-60; prof. anesthesiology Stanford (Calif.) U. Sch. Medicine, 1960-71, UCLA Med. Sch., 1971—; mem. subcom. on nat. halothane study NRC, 1963-65. Contbr. numerous articles to profl. jours. Served with USN, 1944-46. Mem. AMA (Cons., evaluator council on drugs), Am. Soc. Anesthesiologists (com. on reserach), Societe Belge d'Anesthesie et de Reanimation (hon.), Sigma Xi. Subspecialties: Anesthesiology. Current work: Control of respiration. Patentee continuous servo-medicator. Home: 926 Malcolm Ave Los Angeles CA 90024

BELMAN, SIDNEY, biochemist; b. Bklyn., May 22, 1926; s. Israel and Sadie (Jacobson) B.; m. Hilda Ann Belman, Apr. 6, 1951; children: Sherry, Vickie. B.S., CCNY, 1948; M.S., Bklyn. Poly. Inst., 1952; Ph.D., N.Y.U., 1958. Research asst. dept. environ. medicine N.Y.U. Med. Ctr., 1949-58, instr., 1958-61, asst. prof., 1961-62, assoc. prof., 1965-76, prof. dept. environ. medicine, 1976—. Mem. AAAS, Am. Soc. Biol. Chemists, Am. Chem. Soc., Environ. Mutagen Soc., N.Y. Acad. Scis., Am. Assn. Cancer Research. Subspecialties: Biochemistry (biology); Cancer research (medicine). Current work:

Mechanisms of carcinogenesis, biochemical effects of tumor promoters, nutritional factors that act to prevent carcinogenesis. Home: 376 Edgewood Ave Teaneck NJ 07666 Office: 550 1st Ave New York NY 10016

BELMONT, JOHN MARK, psychologist, psychology educator; b. Oakland, Calif., Sept. 9, 1940; s. Daniel M. and Josephine (Jacobs) B.; m. Jean Ford, Dec. 20, 1963; children: Cynthia, Jessica. B.A., Reed Coll., 1962; M.A., George Peabody Coll., 1963; Ph.D., U. Ala., 1966. Postdoctoral fellow Yale U., 1966-67, research staff psychologist, 1967-70, research assoc., 1970-72; assoc. prof. U. Kans. Med. Ctr., 1972-82, prof., 1982—; cons. Pres.'s Commn. on Mental Retardation, 1968-69; mem. NIH-HUD Study Sect., 1978-82; cons. Gallaudet Coll., 1970—. Contbr. chpts. to books. Mem. Am Psychol. Assn., AAAS, Am. Assn. Mental Deficiency, Soc. Research in Child Devel. Subspecialties: Developmental psychology; Cognition. Current work: Mental development in retarded, gifted, deaf and normal children with main emphasis on children's self-awareness in problem solving. Home: 2112 W 61st Terrace Mission Hills KS 66208 Office: U Kans Med Center 39th and Rainbow Blvd Kansas City KS 66103

BELZER, FOLKERT OENE, surgeon; b. Soerabaja, Indonesia, Oct. 5, 1930; came to U.S., 1951, naturalized, 1956; s. Peter and Jacoba H. (Gorter) B.; m. Aug. 4, 1956; children—Ingrid J., John B., G. Eric, Paul O. A.B., Colby Coll., Waterville, Maine, 1953; M.A., Boston U., 1954, M.D., 1958. Diplomate: Am. Bd. Surgery. Intern Grace-New Haven Hosp., 1958-59; asst. resident 1960-62; chief resident U. Oreg. Med. Sch., 1962-63, instr. surgery, 1963-64; asst. research surgeon U. Calif. Med. Center, San Francisco, 1964, asst. prof. surgery, 1966-69, asst. prof. ambulatory and community medicine, 1966-69; asst. chief Transplant Service, 1967-69, co-chief, 1969-72, chief, 1972-74, asso. prof. surgery, 1969-72, asso. prof. ambulatory and community medicine, 1969-72, prof. surgery, 1972-74; dir. Exptl. Surgery Labs., 1973-74; sr. lectr. Guys Hosp., London, Eng., 1964-66; prof., chmn. dept. surgery U. Wis., Madison, 1974—. Contbr. articles to med. jours. Recipient Samuel Harvey award as outstanding resident, 1960. Mem. A.C.S., Am., Calif. med. assns., Am. Soc. Transplant Surgeons (pres. 1975), Calif. Soc. Transplant Surgeons (pres. 1970-72), Am., Central surg. assns., Calif. Acad. Medicine, Halsted Soc., Howard Surg. Soc., C. Naffziger Surg. Soc., Madison Surg. Soc., Pacific Coast Surg. Soc., San Francisco Surg. Soc. (chmn. program com. 1973-74), Wis. surg. socs), Nat. Kidney Found. (vice chmn. com. on dialysis and transplantation 1974-76), Société Internationale de Chirurgie, Soc. Vascular Surgery, Soc. Surg. Chairmen, Soc. U. Surgeons, Surg. Biology Club III, Transplantation Soc., Whipple Soc. Republican. Subspecialties: Internal medicine; Transplant surgery. Current work: Organ preservation and transplantation; developed first kidney perfusion preservation unit. Developed method and machine for human kidney preservation. Home: 6105 S Highlands Dr Madison WI 53705 Office: U Wis Center for Health Scis 600 N Highland Ave Madison WI 53792

BENACERRAF, BARUJ, physician, educator; b. Caracas, Venezuela, Oct. 29, 1920; came to U.S., 1939, naturalized, 1943; s. Abraham and Henriette (Lasry) B.; m. Annette Dreyfus, Mar. 24, 1943; 1 dau., Beryl. B. es L., Lycee Janson, 1940; B.S., Columbia U., 1942; M.D., Med. Sch. Va., 1945; M.A., Harvard U., 1970; M.D. (hon.), U. Geneva, 1980; D.Sc., NYU, 1981, Va. Commonwealth U., 1981, Yeshiva U., 1982, U. Aix-Marseille, 1982. Intern Queens Gen. Hosp., N.Y.C., 1945-46; research fellow dept. microbiology Columbia U. Med. Sch., 1948-50; charge de recherches Centre National de Recherche Scientique Hospital Broussais, Paris, 1950-56; asst. prof. pathology NYU Sch. Medicine, 1956-58, asso. prof., 1958-60, prof., 1960-68; chief immunology Nat. Inst. Allergy and Infectious Diseases, NIH, Bethesda, Md., 1968-70; Fabyan prof. comparative pathology, chmn. dept. Harvard Med. Sch., 1970—; pres., chief exec. officer Dana-Farber Cancer Inst., 1980; J.S. Blumenthal lectr. in allergy and immunology, 1980; Sci. adviser immunology WHO; mem. immunology study sect. NIH; mem. Am. med. adv. bd. Am. Hosp., Paris; pres. Fedn. Am. Socs. Exptl. Biology, 1974-75; chmn. sci. adv. com. Centre d'Immunologie de Marseille. Editorial bd.: Jour. Immunology. Trustee, mem. sci. adv. bd. Trudeau Found.; mem. sci. adv. com. Children's Hosp. Boston; bd. govs. Weizmann Inst. Medicine; mem. award com. Gen. Motors Cancer Research Found., also chmn. selection com. Sloan prize, 1980. Served to capt. M.C. AUS, 1946-48. Recipient T. Duckett Jones Meml. award Helen Hay Whitney Found., 1976; Rabbi Shai Shacknai lectr. and prize Hebrew U. Jerusalem, 1974; Waterford award for biomed. scis., 1980; Nobel prize for medicine or physiology, 1980. Fellow Am. Acad. Arts and Scis.; mem. Nat. Acad. Scis., Nat. Inst. Medicine, Am. Assn. Immunologists (pres. 1973-74), Am. Assn. Pathologists and Bacteriologists, Am. Soc. Exptl. Pathology, Soc. Exptl. Biology and Medicine, Brit. Assn. Immunology, French Soc. Biol. Chemistry, Harvey Soc., N.Y. Acad. Scis., Scandinavian Immunol. Soc., Internat. Union Immunology Socs. (pres. 1980—), Alpha Omega Alpha. Subspecialties: Immunobiology and immunology; Immunogenetics. Home: 111 Perkins St Boston MA 02130

BENADE, LEONARD EDWARD, virologist, molecular biologist; b. Evansville, Ind., Nov. 13, 1944; s. Leo Edward and Marietta (Taylor) B.; m. Mary Pat Larsen, Nov. 29, 1975; 1 dau.: Tina Marie. B.A. in Biology with high distinction, U. Va., 1966; M.Ph., George Washington U., 1971, Ph.D. in Biochemistry, 1971. Chemist CIA, Washington, 1971-73; sr. biochemist Envirocontrol, Inc., Rockville, Md., 1974-75; sr. analyst JRB Assocs., McLean, Va., 1975-76; postdoctoral fellow Frederick Cancer Reserch Ctr., 1977-79; sr. staff scientist Meloy (Md.) Labs., Springfield, Va., 1979-81; research scientist Lab. Tumor Virus Genetics, Nat Cancer Inst., Bethesda, Md., 1981; head dept. molecular biology and biotech. Microbiol. Assocs., Bethesda, Md., 1982; head dept. virology Am. Type Culture Collection, Rockville, 1982—; lectr. biology No. Va. Community Coll, Alexandria, 1974-77. Contbr. articles to profl. books and jours. NSF trainee, 1963; Miller Scholar, 1965-66; NASA fellow, 1966-68. Mem. Am. Soc. Microbilogy; mem. Am. Soc. Virology; Mem. N.Y. Acad. Sci., AAAS, Found. Advanced Edn. in Scis., Phi Beta Kappa, Phi Sigma, Alpha Epsilon Delta. Subspecialties: Molecular biology; Cancer research (medicine). Current work: Molecular virology, cancer research, regulation of gene expression. Home: 971 Park Ave Herndon VA 22070 Office: 12301 Parklawn Dr Rockville MD 20852

BENAROYA, HAYM, civil and mechanical engineer; b. Tel Aviv, May 12, 1954; U.S., 1961, naturalized, 1966; s. Alfred and Esther (Alfassi) B. B.E.CE., Cooper Union U.-N.Y.C., 1976; M.S. in Engring, U. Pa., (1977), Ph.D., 1981. Sr. research engr. Weidlinger Assocs., N.Y.C., 1981—; adj. prof. engring. Cooper Union U., N.Y.C., 1981—; dir., founder Eagle Dancer Public Policy Corp., N.Y.C., 1982—. Assoc. editor: jour. Probability and Stats. in Engring. Research and Devel, 1983—; reviewer: Applied Mechanics Rev. Jour, 1982. Mem. AIAA, ASCE, ASME, Soc. Petroleum Engrs., Soc. Engring. Sci., Am. Soc. Engring. Edn. Subspecialties: Theoretical and applied mechanics; Mathematical software. Current work: Applied mechanics, probability and statistics; educational policy; foreign policy; futurist studies. Office: Weidlinger Assocs 333 7th Ave New York NY 10001

BENBOW, ROBERT MICHAEL, biology educator; b. San Pedro, Calif., Nov. 10, 1943; s. Henry Robertson and Betty Lou (Pederson) B.; m. Lena Camilla Persson, Jan. 5, 1975; children: Wystan, Bronwen, Trefor, Evan. B.S., Yale U., 1967; Ph.D., Calif. Inst. Tech., 1972. Helen Hay Whitney postdoctoral research fellow Lab. Molecular Biology, Med. Research Council, Cambridge, Eng., 1972-75; asst. prof. Johns Hopkins U., Balt., 1975-81, assoc. prof. molecular biology, 1981—; mem. cell biology panel NSF, 1980—. Co-author: Biochemistry, 1981. NIH Research Career Devel. awardee, 1976. Mem. Am. Soc. Biol. Chemists, Am. Soc. Cell Biology, Sigma Xi. Republican. Roman Catholic. Clubs: Johns Hopkins, Yale (Balt.). Subspecialties: Molecular biology; Developmental biology. Current work: Control of DNA replication during embryogenesis in the frog, Xenopus laevis; molecular mechanisms of genetic recombination. Home: 1907 Fairbank Rd Baltimore MD 21209 Office: Dept Biology Johns Hopkins 34th and Charles St Baltimore MD 21218

BENCOMO, JOSE ANTONIO, physicist, educator; b. Caracas, Venezuela, Nov. 17, 1946; came to U.S., 1975, naturalized, 1982; s. Jose Antonio and Lucia Maria (Ragundez) B.; m. Blanca de Hoyos, Mar. 27, 1976; 1 son, Jose Antonia III. B.S. in Physics, Central U. Venezuela, 1974; M.S. in Radiation Physics, U. Tex., 1978; Ph.D. in Biophysics, U. Tex., 1982. Asst. techr. physics Central U. Venezuela, Caracas, 1968-74, U. Simon Boliver, 1973-74; research asst. U. Tex.-M.D. Anderson Hosp. and Tumor Inst., 1976-1981; asst. prof. radiology U. Tex., Houston, 1982—. Contbr. articles to profl. jours. Mem. Am. Assn. Physicists in Medicine, Soc. Photo-optical Instrumentation Engrs., Soc. Photog. Scientists and Engrs. Subspecialties: Imaging technology; Biophysics (physics). Current work: Radiology, imaging technology.

BENDER, A(LLAN) DOUGLAS, consulting technologist; b. Iowa City, Iowa, May 26, 1936; s. W. Ralph G. and Lynette Ann (Epperson) B.; m. Cynthia Kane, Dec. 28, 1957; children: Kane Douglas, Kimberly North, Jeffrey Griggs, Jonathan Gibbs. B.A., Williams Coll., 1957; M.S., U. Iowa, 1959; Ph.D., Jefferson Med. Coll., Phila., 1962. With Smith Kline Corp., Phila., 1962-81, v.p. planning and ops., 1971-78, v.p. research and devel. ops., 1978-80, v.p. pre-clin. devel., 1980-81; v.p., dir. corp. research and devel. Ralston Purina Co., St. Louis, 1981-82; pres. Thayer Technologies, Inc., Swarthmore, Pa., 1982—; vis. lectr. Jefferson Med. Coll., 1962-68; assoc. prof. Widener U., Chester, Pa., 1983. Contbr. articles to profl. jours. Mem. Wallingford-Swarthmore Sch. Dist. Bd., 1967-72. Fellow Am. Coll. Cardiology, Gerontol. Soc., Am. Inst. Chemists; mem. Am. Physiol. Soc. Republican. Presbyterian. Subspecialties: Pharmacology; Gerontology. Current work: Technical planning, technology transfer. Home: 401 Thayer Rd Swarthmore PA 19081 Office: Thayer Technologies Inc PO Box 305 Swarthmore PA 19081

BENDER, HARVEY ALAN, Geneticist, Biology educator; b. Cleve., June 5, 1933; s. Oscar and Effie (Goldstein) B.; m. Eileen Teper, June 16, 1956; children: Leslie Carol, Samuel David, Philip Michael. B.A., Case-Western Res. U., 1954; M.S., Northwestern U., 1957, Ph.D., 1959. Diplomate: Am. Bd. Med. Genetics. USPHS fellow U. Calif.-Berkeley, 1959-60; asst. prof. U. Notre Dame, Ind., 1960-64, assoc. prof., 1964-69, prof., 1969—; vis. prof. Yale U., New Haven, 1973-74; adj. prof. Ind. U. Sch. Medicine, Indpls., 1979—; Gosney fellow Calif. Inst. Tech., Pasadena, 1965-66; cons. in field; dir. North Central Ind. Genetics Ctr., South Bend, 1980—; co-chmn. Nat. Genetics Edn. Com., 1978-81. Contbr. articles to profl. jours. Mem. Sickle Cell adv. com. St. Joseph County, Ind., 1981—; chmn. human rights com. No. Ind. State Hosp., 1979—. NIH grantee, 1960-70; Dept. Energy grantee, 1965—; United March of Dimes/Nat. Found. Grantee, 1980—; HHS/Bur. Maternal and Child Health Grantee, 1981—; Health Services grantee, 1966—; NSF grantee, 1978-81; others.; Hew fellow, 1965-67; Soc. for Values in Higher Edn. fellow, 1973-74; others. Fellow AAAS; mem. AAUP, Am. Inst. Biol. Sci., Am. Soc. Human Genetics, Genetics Soc. Am., Radiation Reserach Soc., Soc. for Values in Higher Edn., Soc. for Devel. Biology, Am. Assn. on Mental Deficiency, Sigma Xi (nat. dir.-at-large 1980—, chmn. sci. and safety com. 1981—). Jewish. Subspecialties: Genetics and genetic engineering (biology); Genetics and genetic engineering (medicine). Current work: Developmental genetics; human genetics; biomedical legal ethics. Office: Dept Biology Univ Notre Dame Notre Dame IN 46556

BENDER, MYRON LEE, chemist, educator; b. St. Louis, May 20, 1924; s. Averam Burton and Fannie (Leventhal) B.; m. Muriel Blossom Schulman, June 8, 1952; children: Alec Robert, Bruce Michael, Steven Pat. B.S. with highest distinction, Purdue U., 1944, Ph.D., 1948, D.Sc. honoris causae, 1969; postdoctoral student, Harvard, 1948-49; AEC fellow, U. Chgo., 1949-50. Chemist, Eastman Kodak Co., 1944-45; instr. U. Conn., 1950-51; from instr. to asso. prof. Ill. Inst. Tech., 1951-60; mem. faculty Northwestern U., 1960—, prof. chemistry, 1962—, prof. biochemistry, 1975—; cons. to govt. and industry, 1959—; fellow Merton Coll., Oxford U., 1968; J.S.P.S. vis. lectr., Japan, 1974; vis. prof. U. Queensland, Australia, 1979, Nankai U., China, 1982, univs. Tokyo and Kyoto, Japan, 1982. Recipient Midwest award Am. Chem. Soc., 1972; Sloan fellow, 1959-65; Fulbright Hays disting. prof., Zagreb, Yugoslavia, 1977. Fellow Am. Inst. Chemists; mem. Am. Chem. Soc., AAUP, Chem. Soc. (London), Am. Soc. Biol. Chemists, Assn. Harvard Chemists, AAAS (councilor chemistry sect.), Nat. Acad. Scis., Phi Beta Kappa, Sigma Xi, Phi Lambda Upsilon. Subspecialties: Organic chemistry; Biochemistry (biology). Current work: Mechanisms of enzyme action; mechanisms of action of enzyme models. Home: 2514 Sheridan Rd Evanston IL 60201

BENDER, WELCOME W., molecular biologist; b. Balt., June 28, 1949; s. Welcome W. and Mary V. (Priebe) B. A.B., Harvard Coll., 1971; Ph.D., Calif. Inst. Tech., 1977. Postdoctoral fellow dept. biochemistry Stanford U. Med. Sch., 1977-80; asst. prof. biol. chemistry Harvard Med. Sch., Boston, 1980—. Subspecialties: Genetics and genetic engineering (biology); Developmental biology. Current work: Developmental control of genes in Drosophila.

BENDET, IRWIN JACOB, biologist, educator; b. N.Y.C., May 9, 1927; s. Julius and Anna (Feldman) B.; m. Roslyn M. Miller, June 7, 1937; children: David, Elizabeth. B.S., CCNY, 1949; M.A., U. Mich., 1950; Ph.D., U. Calif.-Berkeley, 1954. Faculty U. Pitts., 1954—, prof. biophysics, 1966—. Contbr. articles to profl. publs. Served with USNR, 1945-46. Mem. Biophys. Soc., Electron Microscope Soc. Am., N.Y. Acad. Scis., AAAS, Sigma Xi. Subspecialty: Biophysics (biology). Current work: Virus, protein and nucleic acid structure; electron microscopy. Home: 1321 Cordova Rd Pittsburgh PA 15206 Office: U Pitts Pittsburgh PA 15260

BENDITT, EARL PHILIP, educator, med. scientist; b. Phila., Apr. 15, 1916; s. Milton and Sarah (Schoenfeld) B.; m. Marcella Wexler, Feb. 18, 1945; children—John, Alan, Joshua, Charles. B.A., Swarthmore Coll., 1937; M.D., Harvard, 1941. Intern Phila. Gen. Hosp., 1941-43; resident pathology U. Chgo. Clinics, 1944; mem. faculty U. Chgo. Med. Sch., 1945-57, asso. prof. pathology, 1952-57; asst. dir. research LaRabida Children's Sanitarium, Chgo., 1950-56; prof. pathology, 1957—; chmn. dept. U. Wash. Sch. Medicine, 1957-81; mem. sci. adv. bd. St. Jude Children's Research Hosp.; cons. USPHS-NIH, 1957-80; Commonwealth Fund fellow, vis. prof. Sir William Dunn Sch. Pathology, U. Oxford, Eng., 1965, Macy faculty scholar, 1979-80, Litchfield lectr., 1980; chmn. bd. sci. counselors adv. com. Nat. Inst. Environ. Health Scis., 1976-79, council mem., 1971-74. Mem. editorial bds. scis. publs. Recipient Med. Alumni award univ. Chgo., 1968; Rous-Whipple award Am. Assn. Pathologists, 1980. Fellow AAAS; mem. Am. Soc. Exptl. Pathology (council 1971-77, sec. treas. 1972-73, pres. 1975-76), Nat. Acad. Scis., Am. Assn. Pathologists and Bacteriologists (council 1972-77), Soc. Exptl. Biology and Medicine, Am. Soc. Cell Biology, Am. Soc. Biol. Chemists, Histochem. Soc. (pres. 1963-64), Phi Beta Kappa, Sigma Xi. Subspecialties: Pathology (medicine); Cell biology (medicine). Current work: Atherosclerosis, environmental chemicals and viruses as causes of heart disease; acute responses to injury and toxic substances. Home: 3717 E Prospect St Seattle WA 98112

BENDIXEN, HENRIK HOLT, physician; b. Frederiksberg, Denmark, Dec. 2, 1923; came to U.S., 1954, naturalized, 1960; s. Carl Julius and Borghild Nicoline (Holt) B.; m. Karen Skakke, Dec. 20, 1947; children—Nils, Birgitte. C.phil., c.m., c.chir. (laudabills), U. Copenhagen, 1951. Diplomate: Am. Bd. Anesthesiologists. Postgrad. tng. in surgery and anesthesia in, Denmark and Sweden, 1951-54, also Danish hosp. ship in, Korea; resident in anesthesia Mass. Gen. Hosp., Boston, 1954-57; mem. anesthesia dept. faculty Mass. Gen. Hosp. and Harvard U. Med. Sch., 1957-69; prof. anesthesia, head dept. U. Calif. Med. Sch., San Diego, 1969-73; prof. anesthesiology, chmn. dept. Columbia U. Coll. Phys. and Surg., 1973—; pres. Mass. Soc. Anesthesiologists, 1966; mem. gen. med. research program-project com. NIH, 1967; dir. center research and tng. anesthesiology Harvard U. Med. Sch., 1968; chmn. com. anesthesia NRC, 1970. Author: Respiratory Care, 1965, also articles, revs., abstracts. Mem. Soc. Critical Care Medicine (pres. 1974), Assn. U. Anesthetists, Am. Soc. Pharmacology and Therapeutics., Am. Physiol. Soc., Assn. Am. Med. Colls., N.Y. Acad. Medicine, N.Y. State, N.Y. County med. socs., AMA, Am. Soc. Anesthesiologists, Am. Heart Assn.; hon. mem. Minn. Surg. Soc., Belgian Soc. Anesthesiologists; corr. mem. Danish Soc. Anesthesiologists. Clubs: Harvard (Boston); University (N.Y.C.). Subspecialty: Anesthesiology. Address: Dept Anesthesiology Columbia Univ Coll Phys and Surg 630 W 168th St New York NY 10032

BENDOW, BERNARD, physicist; b. Portland, Maine, Apr. 30, 1942; s. Lipa and Ides (Golub) B.; m. Mary Joann Shea, May 1, 1977. Ph.D. in Physics, NYU, 1969. Research asst. U. Calif.-San Diego, 1969-71; physicist Air Force Cambridge Research Labs., Hanscom AFB, Mass., 1971-75, 1971-75; physicist and supervisory physicist Rome Air Devel. Ctr., 1975-81; sr. prin. scientist The BDM Corp., Albuquerque, 1981—. Editor: Optical Props Highly Transparent Solids, 1975, Theory of Light Scattering in Condensed Matter, 1976, Physics of Fiber Optics, 1980; contbr. articles in field to profl. jours. NASA trainee, 1964-67; recipient Founder's Day award NYU, 1969. Mem. Am. Phys. Soc., Optical Soc. Am., Am. Ceramic Soc. Subspecialties: Fiber optics; Materials. Current work: Optical materials, lasers, laser-matter interactions, infrared glasses, fiber optics. Office: 1801 Randolph Rd SE Albuquerque NM 87106

BENEDEK, GEORGE BERNARD, physicist; b. N.Y.C., Dec. 1, 1928; m., 1955; 2 children. B.S., Rensselaer Polytech Inst., 1949; M.A., Harvard U., 1951, Ph.D. in Physics, 1954. Mem. staff Joint Harvard-Lincoln Lab. MIT project, 1953-55; research fellow Harvard U., 1955-57, lectr. solid state physics, 1957-58, asst. prof. applied physics, 1958-61, assoc. prof., 1961-65; prof. physics MIT, Cambridge, Mass., 1965—. Fellow Guggenheim Found., 1960, Atomic Energy Research Establishment, Harwell, Eng., 1967. Mem. Am. Inst. Physics (bd. govs. 1971-74), Am. Phys. Soc., Inst. Medicine, Nat. Acad. Sci. Subspecialty: Biophysics (physics). Office: Dept Physics MIT Cambridge MA 02139

BENEDETTO, ANTHONY RICHARD, nuclear and med. sci. educator, cons.; b. El Paso, Dec. 19, 1946; s. Leo and Mabel (Kelly) B.; m. Kathryn McAdams, May 23, 1970; children: Michele Eileen, Noel Kathryn. B.S. in Nuclear Engring, Tex. A&M U., 1968, M. Engring. Health Physics, 1970; M.B.A., Sul Ross State U., 1976. Cert. Am. Bd. Sci. in Nuclear Medicine, Am. Bd. Radiology. Commd. 2d lt. U.S. Army, 1968, advanced through grades to capt., 1974; with (Engr. Reactors Group), 1970-72, 1972-73, 1973-79, ret., 1979; asst. prof. radiology (nuclear medicine) U. Tex. Health Sci. Ctr., San Antonio, 1979—. Contbr. articles on nuclear and med. sci. to books and profl. jours. Trustee Am. Bd. Nuclear Medicine. Mem. Soc. Nuclear Medicine, Am. Assn. Physicists in Medicine, Radiol. Soc. N.Am., Am. Coll. Nuclear Physicians, Health Physics Soc., Am. Coll. Radiology. Methodist. Lodges: Masons; Shriners. Subspecialties: Nuclear medicine; Nuclear engineering. Current work: Nuclear cardiology, image perception, air embolism, chemical dosimetry. Office: Dept Radiology 7703 Floyd Curl Dr San Antonio TX 78284

BENEDICT, ANTHONY GORMAN, electric connector mfg. cor. exec.; b. Sydney, Australia, Dec. 24, 1920; came to U.S., 1956, naturalized, 1961; s. Ralph Payne and Elsie Lincoln (Vandegrift) B.; m. Shirley Marshall, Mar. 6, 1942; children: Merril Genevieve Benedict Rogers, Leeanna Ruth Benedict Mickelson, Ralph Scott. Student, Calif. Inst. Tech., 1936-37; B.Engring., U. Sydney, 1948. Registered profl. engr., Calif. Local govt. engr., New South Wales, 1952; Instr. radio theory U. Sydney, 1942-43; elec. engr. Brisbane Water County Council, Australia, 1948-52; supervising engr. Rural Electricity Authority New South Wales, 1952-55; elec. engr. Utah Constrn. Co., Australia, 1956; systems engr. So. Calif. Edison Co., 1956-59; chief devel. div. Ordnance Assos. Inc., South Pasadena, Calif., 1959-62; mem. tech. staff Jet Propulsion Lab., Pasadena, 1962-68; pres. Arc Assocs. Inc., Mpls., 1969—. Served to flight lt. Royal Australian Air Force, 1939-44. Mem. IEEE, ASME, Am. Def. Preparedness Assn., Sigma Xi. Subspecialties: Mechanical engineering; Electrical engineering. Current work: Product development with emphasis on human factors. Home: 8415 Hitsman Ln Maple Plain MN 55359 Office: Arc Assocs Inc 6409 Goodrich Ave Minneapolis MN 55426

BENEDICT, BRUCE JOHN, nuclear engineer; b. Madison, Wis., Nov. 22, 1940; s. R. Ralph and Dorothy O. (Strauss) B.; m. Bonnie Davenport, July 4, 1974. B.S.E.E., Stanford U., 1964; M.S. in Nuclear Engring, U. Wis., 1970. Prin. engr. Babcock & Wilcox, Lynchburg, Va., 1970—. Treas. First Unitarian Ch., Lynchburg, 1973-75, 80—. Served to capt. USAF, 1964-68. Mem. Am. Nuclear Soc., IEEE. Subspecialties: Nuclear fission; Computer engineering. Current work: Development of computer-based CRT displays and associated software to be used as an aid in plant control and monitoring by nuclear power operators. Office: Babcock and Wilcox Co PO Box 1260 Lynchburg VA 24505

BENEDICT, GEORGE FREDERICK, astronomer; b. Los Angeles, Mar. 17, 1945; s. Frederick and Sarah Alice (Guptill) B.; m. Ann Durr, June 24, 1967; children: Michael Robert, Sarah Ann. B.S. in Physics, U. Mich., 1967; M.S., Northwestern U., 1970, Ph.D., 1972. Research scientist assoc. U. Tex., Austin, 1972-78, research scientist, 1979—, part-time asst. prof., 1977—; cons. in field; adv. bd. Space Telescope Guide Star Selection System. Contbr. articles to profl. jours. NSF grantee, 1980. Mem. Am. Astron. Soc., Internat. Astron. Union. Subspecialties: Optical astronomy; Graphics, image processing, and pattern recognition. Current work: NASA Space telescope astrometry team member. Home: 4105 Hycrest Dr Austin TX 78759 Office: McDonald Obs Univ Tex 16.222 RLM Hall Austin TX 78712

BENEDICT, MANSON, chemical engineer, educator; b. Lake Linden, Mich., Oct. 9, 1907; s. C. Harry and Lena I. (Manson) B.; m. Marjorie Oliver Allen, July 6, 1935; children: Mary Hannah (Mrs. Myran C. Sauer, Jr.), Marjorie Alice (Mrs. Martin Cohn). B. Chemistry, Cornell, 1928; M.S., Mass. Inst. Tech., 1932, Ph.D., 1935. NRC fellow chemistry, 1935-36; research asso. geophysics Harvard, 1936-37; research chemist M.W. Kellogg Co., 1938-43; in charge process design gaseous diffusion plant for uranium-235 Kellex Corp., 1943-46; dir. process development Hydrocarbon Research, Inc., 1946-51; tech. asst. to gen. mgr. AEC, 1951-52; prof. nuclear engring. Mass. Inst. Tech., 1951-69, Institute prof., 1969-73, prof. emeritus, 1973—, head dept. nuclear engring., 1958-71; dir. Burns & Roe, Inc., 1979—; sci. adv. Nat. Research Corp., 1951-58, dir., 1962-67; mem. gen. adv. com. AEC, 1958-68, chmn., 1962-64; mem. Mass. Adv. Council on Radiation Protection; dir. Atomic Indsl. Forum, 1966-72; mem. energy research and devel. adv. council Fed. Energy Adminstrn., 1973-75. Co-editor: Engineering Developments in the Gaseous Diffusion Process, 1949; Co-author: Nuclear Chemical Engineering, 1981. Recipient William H. Walker award Am. Inst. Chem. Engrs., 1947, Founders award, 1965; Indsl. and Engring. Chemistry award Am. Chem. Soc., 1962; Perkin medal Soc. Chem. Industry; Robert E. Wilson award in nuclear chem. engring.; Arthur H. Compton award Am. Nuclear Soc.; Fermi award AEC, 1972; John Fritz medal Engring. Founder Socs., 1974; Nat. Medal Sci., 1975; Henry D. Smyth Nuclear Statesman award Atomic Indsl. Forum, 1979; Washington award Western Soc. Engrs., 1982. Fellow Am. Nuclear Soc. (pres. 1962-63), Am. Acad. Arts and Sci., Am. Philos. Soc., Am. Inst. Chem. Engrs.; mem. Nat. Acad. Scis., Nat. Acad. Engring. (Founders award 1976), Sigma Xi. Clubs: Cosmos (Washington); Weston (Mass.) Golf; Country of Naples (Fla.). Subspecialties: Nuclear fission; Chemical engineering. Current work: Nuclear fuel cycle, nuclear power safety. Home: 2151 Gulf Shore Blvd N Naples FL 33940 Office: Dept Nuclear Engring Mass Inst Tech Cambridge MA 02139

BENHAM, CRAIG JOHN, mathematics educator, physical chemist; b. Chgo., Sept. 1, 1946; s. G. Harvey and Sylvia (Gilling) B.; m. Marcia Bielinski, July11, 1975; 1 dau., Allana. B.A., Swarthmore Coll., 1968; M.A., Princeton U., 1971, Ph.D., 1972. Asst. prof. math. Notre Dame U., 1972-76; research fellow in biology Calif. Inst. Tech., Pasadena, 1976-77; asst. prof. math. Lawrence U., Appleton, Wis., 1977-78, U. Ky., Lexington, 1978-81, assoc. prof., 1981—; cons. Gen Molecular Applications, Inc., Columbus, Ohio, 1982—. NDEA fellow, 1969-72; NSF grantee, 1980—; Alfred P. Sloan fellow, 1982—. Mem. AAAS. Subspecialties: Theoretical chemistry; Molecular biology. Current work: Development of theoretical analysis of conformational properties of superhelically stressed DNA and of how these properties may regulate physiological events (initiation of transcription or replication). Home: 2465 Eastway Dr Lexington KY 40503 Office: Math Dept U Ky Lexington KY 40506

BENJAMIN, ROLAND JOHN, optical engineer; b. Williamsfield, Ill, May 18, 1928; s. Harley Wilson and Beulah Isabelle (Doubet) B.; m. Helen Maxine Cadwell, June 25, 1950; 1 dau.: Nancy Anne Benjamin Kwoh. B.S. in Mech. Engring, U. Ill., 1950; M.S. in Theoretical and Applied Mechanics, 1951; Ph.D., 1955. Stress analyst N.Am. Aviation Missile div., Downey, Calif., 1952-56; mgr. aerospace group Cook Research Labs., Morton Grove, Ill., 1956-67; mgr. optical engring. Bell & Howell Co., Lincolnwood, Ill., 1967-71, dir. engring. consumer products group, 1971-74, chief scientist optical div., 1974—. Contbr. articles to profl. jours. Served with AUS, 1945-47. U. Ill. fellow, 1950-52. Mem. Optical Soc. Am., Soc. Photographic and Instrumentation Engrs. Subspecialties: Optical aspheric surfaces; Theoretical and applied mechanics. Current work: Developing techniques and application for precision machining nonspherical optical surfaces on non-ferrous metals and infrared transmitting materials using natural diamond tools. Office: 7100 McCormick Rd Lincolnwood IL 60645

BENJAMIN, STEPHEN ALFRED, veterinary pathologist; b. N.Y.C., Mar. 27, 1939; s. Frank Joseph and Dorothy (Zweighaft) B.; m. Barbara J. Larson, July 25, 1982; children: Jeffrey, Karen, Susan Douglas, Kristine, Eric. A.B., Brandeis U., 1960; D.V.M., Cornell U., 1964, Ph.D., 1968. Asst. prof. Pa. State U. Sch. Medicine, Hershey, 1967-70; pathologist Inhalation Toxicology Research Inst., Lovelace Found., Albuquerque, 1970-77; assoc. prof. Coll. Vet. Medicine and Biomed. Scis., Colo. State U., 1977-80, prof. pathology and radiation biology, 1981—, dir. Collaborative Radiol. Health Lab., 1977—. Contbr. numerous publs. in field; author profl. reports; contbr. chpts. to books. Mem. AVMA Am. Coll. Vet. Pathologists, Am. Assn. Lab. Animal Sci., Radiation Research Soc., Am. Assn. Pathologists, Internat. Acad. Pathologists, Phi Zeta, Phi Kappa Phi. Subspecialties: Pathology (veterinary medicine); Cancer research (veterinary medicine). Current work: Radiation effects, cancer resarch, immunotoxicology. Office: Collaborative Radiol Health Lab Coll Vet Medicine and Biomed Sci Foothills Campus Fort Collins CO 80523

BENJAMINI, ELIEZER, immunologist, educator; b. Tel-Aviv, Israel, Feb. 8, 1929; s. Kalman and Anna (Feinman) B.; m. Joyce Barbara Kushner, Sept. 12, 1953; children: Etan Marc, Leora Ann. B.S., U. Calif., Berkeley, 1952, M.S., 1954, Ph.D., 1958. Toxicologist Nat. Canners Assn. Reserach Lab., Berkeley, Calif., 1954-56; jr. specialist Citrus Experiment Sta., U. Calif., Riverside, 1957-58; with Kaiser Found. Research Inst., San Francisco, 1959-70, assoc. research scientist, 1960-64, research scientist, asst. dir., 1964-70; prof. immunology dept. med. microbiology and immunology U. Calif. Sch. Medicine, Davis, 1970—; mem. adv. panel Calif. Cancer Research Coordinating Com., 1976-80, NSF, 1977-80. Contbr. chpts. to books. articles to profl. jours. Mem. Am. Assn. Immunologists, Entomol. Soc. Am., Am. Assn. Cancer Research. Subspecialties: Immunobiology and immunology; Cell biology. Current work: Research on the immune response and its manipulation; antigenic determinants of protein with emphasis on the relationship between structure and activity; activation of lymphocytes and macrophages. Office: Dept Med Microbiology and Immunology Sch Medicine U Calif Davis CA 95616

BENNER, DRAYTON CHRIS, planetary scientist; b. Rockville Centre, N.Y., Dec. 19, 1947; s. Percival D. and Dorothy M. (Waaser) B.; m. Beverly Ann Bohaty, Oct. 13, 1972; children: Tamsen, Rodelle, Drayton. B.S in Astronomy with distinction, U. Ariz., 1970, Ph.D. in Planetary Scis., 1979; postgrad. in astronomy, U. Tex., 1970-71. Research assoc. Lunar and Planetary Lab., U. Ariz., Tucson, 1979-80; NRC assoc. NASA-Ames Research Ctr. Space Scis. Div., Moffett Field, Calif.; now research asst. prof. physics Coll. William and Mary, Williamsburg, Va. Contbr. articles to profl. jours. Grantee NASA, 1979, 81-83, NRC, 1980. Mem. Am. Astron. Soc., Astron. Soc. Pacific. Subspecialties: Planetary science; Aeronomy. Current work: Spectrum of methane. Application to solar system (including first detection of Pluto's atmosphere.) Spectroscopy of Earth's stratosphere. Halogen occultation experiment. Office: Dept Physics Coll William and Mary Williamsburg VA 23185

BENNETT, BRUCE ALAN, mech. engr., writer, photographer; b. Paterson, N.J., June 25, 1953; s. Alan Kenneth and Elsie Mae (Conklin) B.; m. Antoinette Ann Bennett, Oct. 26, 1975; 1 son, Patrick Alan. B.S.M.E. (Hon. Estrin scholar), N.J. Inst. Tech., 1975, postgrad., 1982—. Mech. engr. Naval Air Engring. Center, Lakehurst, N.J., 1975-77; project engr. U.S. Testing Co., Hoboken, N.J., 1977-79; contract mgr PHB Weserhutte Inc., Allendale, N.J., 1979-82, dir. participation in trade shows, exhbns., 1980-82; materials mgr. IFE Systems Inc., Mahwah, N.J., 1982—; also free-lance writer and photographer. Mem. Jim Russell Internat. Racing Driver's Club, Mont Tremblant, Can. Recipient First Place award N.J. Student Poetry Competition, 1982; also numerous photography awards. Mem. ASME, Soc. Automotive Engrs. Democrat. Roman Catholic. Subspecialties: Mechanical engineering; Systems engineering. Current work: Project management of large bulk mataerial handling systems/equipment from receipt of order through to commissioning; review of specifications, design coordination, production supervision, purchasing, quality control, particularly of large structural items during fabrication; systems include stackers, reclaimers, material blending systems, shiploaders. Home: 123 Post Ave Hawthorne NJ 07506 Office: 370 Franklin Turnpike Mahwah NJ 07430

BENNETT, CECIL JACKSON, biologist, educator; b. Eau Claire, Wis., Oct. 4, 1927; s. Cecil Hilts and Leah M (Lanam) B.; m. Katherine Wilson, Jan. 22, 1951; children: Scott Jackson, Carroll Anne Bennett Carlson; m. Donna Irene Campbell, June 18, 1974. B. S. in Edn, U. Wis., 1949; Ph.D., 1959; M.A. in Zoology, Washington U., 1953. Asst. prof., then assoc. prof., then prof. biology No. Ill. U., DeKalb, Ill., 1975—; vis. prof. opthalmology Washington U., St. Louis, summers 1964-65. Republican precinct committeeman, 1962-80; adv. bd. DeKalb Airport, 1972-82; comdr. DeKalb Area Composite Squadron, CAP, 1978-81. Served with AUS, 1946. Mem. Soc. Study Evolution, AAAS, Am. Inst. Biol. Scis., Genetics Soc. Am., Am. Genetics Assn., Ill, State Acad. Sci. (past pres., now newsletter editor), Assn. Midwestern Coll. Biology Tchrs (past pres.), Ill, Sci, Tchrs. Assn., AAUP, Am. Fedn. Tchrs. (v.p. local 4100), Exptl. Aircraft Assn. (DeKalb County Corn Chpt.), Sigma Xi, Phi Sigma. Lodge: Masons. Subspecialties: Animal genetics; Gene actions. Current work: Behavior, biochemical, population effects of alleles of white locus in Drosophila melanogaster. Behavior, developmental, population, single gene substitution. Home: P:O Box 364 DeKalb IL 60115 Office: Biol Scis Dept No Ill U KeKalb IL 60115

BENNETT, DONALD RAYMOND, physician, pharmacologist; b. Mishawaka, Ind., Feb. 16, 1926; s. Donald Henry and Arline Ida-Edna (Bailey) B.; m. Rosemary Fritz, Mar. 23, 1928; children: Jeffrey Anthony, Ralph Gregory. B.S. in Zoology, U. Mich., 1949, M.S. in Pharmacology, 1951, M.D., 1955, Ph.D. in Pharmacology, 1958. Diplomate: Am. Bd. Family Practice. Mem. faculty dept. pharmacology U. Mich. Med. Sch., Ann Arbor, 1958-65; mgr. biosci. research Dow Corning Corp., Midland, Mich., 1965-74; resident in family practice Midland (Mich.) Hosp., 1974-75; assoc. dir. div. drugs AMA, Chgo., 1976—. Contbr. articles to profl. jours. Bd. dirs. Midland Hosp., 1970-76; mem. instl. rev. bd. Hinsdale (Ill.) Hosp., 1981—. Served with U.S. Army, 1943-46. Decorated Purple Heart. Mem. Am. Soc. Pharmacology and Exptl. Therapeutics, Am. Soc. Clin. Pharmacology and Exptl. Therapeutics, Soc. Toxicology, Am. Acad. Family Practice, AMA. Subspecialties: Family practice; Pharmacology. Office: AMA 535 N Dearborn St Chicago IL 60610

BENNETT, GEORGE ALAN, nuclear engineer, chemical engineer; b. N.Y.C., Sept. 12, 1918; s. George Edward and Florence (Landsberg) B.; m. Frances Acher, Sept., 1949 (div. Oct. 1964); m. Lillian Vlahos, June 27, 1972 (dec. Oct. 1978); children: George, Leslie. Jr. engr. The Solvay Process Co., Syracuse, N.Y., 1943-44; assoc. scientist Los Alamos Nat. Lab., 1944-46; research engr. Esso Standard Oil, Elizabeth, N.J., 1946-47; assoc. scientist Brookhaven Nat. Lab., Upton, N.Y., 1947-51; engr. N.Am. Aviation, Downey, Calif., 1951-52, Argonne (Ill.) Nat. Lab., 1954—. Mem. Am. Nuclear Soc. Clubs: Am. Spaniel (N.Y.C.); Skyline Cocker (Chgo.). Subspecialties: Nuclear engineering; Chemical engineering. Current work: Breeder reactor development and nuclear fuel cycles including spent fuel reprocessing and waste disposal, program management and planning. Patentee flotation methods for table model reactor, lubricating oil treating process. Office: Argonne National Laboratory 9700 S Cass Ave Argonne IL 60439

BENNETT, H(ENRY) STANLEY, anatomy and cell biology educator; b. Tottori City, Japan, Dec. 22, 1910; s. Henry James and Anna Woodruff (Jones) B.; m. Alice Roosa, July 28, 1935; children: Edith Roosa Page, Anna Bennett McNaught, Henry James, Patience Bennett Berkman. A.B., Oberlin Coll., 1932; M.D., Harvard U., 1936; D.Sc. (hon.), Monmouth Coll., 1962. Intern Johns Hopkins Hosp., 1936-67; fellow NRC, research fellow in anatomy Harvard U., 1937-39, instr. anatomy and pharmacology, 1939-41, assoc. in anatomy, 1941-48; asst. prof. cytology dept. biology MIT, 1945-48; prof. anatomy, head dept. U. Wash., 1948-60; acting chmn. dept. anatomy U. Chgo., 1961-63, prof. biophysics, 1961-65, dean div. biol. scis., 1961-65, prof. anatomy, 1961-69, Robert R. Bensley prof. biol. and med. scis., 1966-69, dir. labs. for cell biology, 1966-69; chmn. dept. anatomy U. N.C., 1969-77, Sarah Graham Kenan prof. biol. and med. scis., 1969-81, prof. anatomy, 1969-81, dir. labs. reproductive biology, 1969-81, prof. emeritus, 1981—; mem. U.S.-Japan Com. on Sci. Cooperation, 1964-76, U.S. chmn., 1969-76; mem. U.S.-Japan Com. on Cooperation in Med. Scis., 1965-71; mem. panel environ. pollution Exec. Office of the Pres. Office of Sci. and Tech., 1964-65, chmn. subpanel on soil contamination, 1965; mem. Pres.'s Adv. Commn. on Narcotics and Drug Abuse, 1962-63; mem. nat. adv. health council USPHS, HEW, 1958-62; mem. adv. com. Nat. Cancer Inst., 1971-75; adv. com. High Voltage EM Shared Resources, 1973-80; mem. space sci. and tech. adv. com. Office of Manned Space Flight, NASA, 1964-70; mem. Internat. Commn. Protection Against Environ. Mutagens and Carcinogens, 1977—; guest lectr. U. Ark., 1973, Tulane U., 1974, Japanese Physiol. Soc., 1976, Japanese Assn. Anatomists, 1981, Hugarian Soc. Anatomists, Histologists and Embryologists, 1982. Bd. editors: Jour. Biophys. and Biochem. Cytology, 1954-60, Microbios and Cytobios, 1968—; assoc. editor: Jour. Histochemistry and Cytochemistry, 1956-59; adv. editor: Internat. Rev. Cytology, 1957-61; mem. editorial bd.: Am. Jour. Anatomy, 1976—. Bd. dirs., mem. adminstrv. council Health Info. Found., 1962-65; bd. dirs. Ill. Soc. Med. Research, 1961-69, Infant Welfare Soc. Chgo., 1961-69, Nat. Fund Med. Edn., 1963-66, Nat. Opinion Research Ctr., 1962-65, Nat. Soc. Med. Research, 1962-74, Brain Research Found., 1964-69; trustee Salk Inst. Biol. Studies, 1963-76. Served to capt. USNR, 1942-46; PTO. Decorated Legion of Merit U.S., Order Sacred Treasure 2d Class Japan; recipient numerous grants. Fellow Japanese Soc. Electron Microscipy, Clin. Electron Microscopy Soc. Japan, Am. Acad. Arts and Scis.; mem. Am. Assn. Anatomists (pres. 1959-60), AAAS, Pan Am. Anat. Soc., Am. Chem. Soc., Am. Physiol. Soc., Am. Soc. Cell Biology, Electron Microscopy Soc. Am., Histochem. Soc., So. Soc. Anatomists, Southeastern Electron Microscopy Soc., Hungarian Soc. Anatomists, Histologists and Embryologists, Coligio Anatomica Brasileiro, Sociedad Mexicana de Anatomía, Chiba Med. Soc., Sigma Xi, Alpha Omega Alpha. Clubs: Cosmos (Washington); Quadrangle (Chgo.); Aesulapian (Boston). Subspecialties: Anatomy and embryology; Cell biology (medicine). Current work: Molecular anatomy of cell membranes: general role of actin and myosin; furthering international cooperation in science. Home: Box 203 Route 5 Jones Ferry Rd Chapel Hill NC 27514 Office: Dept of Anatomy U NC 111 Swing Bldg 217H Chapel Hill NC 27514

BENNETT, JAMES LEROY, pharmacology educator; b. Sioux City, Iowa, June 30, 1939; s. Oscar Leroy and Marjorie (Waterhouse) B.; m. Elaine Larsen, Dec. 7, 1940; children: Angela, Courtney, Aaron, Isaac, Sarah. B.S., Brigham Young U., 1964; Ph.D., Johns Hopkins U., 1972. Mem. faculty Mich. State U., East Lansing, 1974—, prof. pharmacology, 1982—; cons. Edna McConnell Clark Found., N.Y.C., 1976—; mem. study sec. parasitology and tropical medicine NIH, 1982—. Contbr. numerous sci. articles to profl. publs. NIH grantee, 1976—; WHO grantee, 1977—. Mem. Am. Soc. Pharmacology, Am. Soc. Parasitology, Am. Soc. Tropical Medicine. Subspecialties: Cellular pharmacology; Parasitology. Current work: Study of the action of drugs on parasitic helminths. Home: 5788 Montebello St Haslett MI 48840 Office: Pharmacology Dept Mich State U B420 Life Scis East Lansing MI 48824

BENNETT, JESSE HARLAND, research scientist, cons.; b. Lehi, Utah, June 21, 1936; s. Clifford Crosby and Dorothy (Carr) B.; m. Marily Gubler, Oct. 12, 1938; children: Scott Jesse, Cheryl, Teresa. B.S., Utah State U., Logan, 1961, M.S., 1965; Ph.D. (USPHS grantee), U. Utah, Salt Lake City, 1969; grad. cert. environ. toxicology, Center for Environ. Biology, 1969. Research assoc. Washington U., St. Louis, 1969-70; NIH postdoctoral fellow in biophysics Mo. Bot. Garden, St. Louis, 1969-70; research, teaching assoc. U. Utah, Salt Lake City, 1970-74; research scientist Dept. Agr., Beltsville, Md., 1974—, plant physiologist, toxicologist; cons. in field. Contbr. numerous articles on biol. scis. to profl. jours. Mem. numerous panels and coms. Served with USNG, 1954-59. Named Outstanding Grad. Student in Biol. Scis. U. Utah, 1967; USPHS grantee, 1970-73; ERDA grantee, 1977-80; Dept. Agr.-EPA grantee, 1977. Mem. AAAS., Am. Soc. Plant Physiologists (chmn. stress physiology session annual meeting 1978), Soc. for Environ. Toxicology and Chemistry (panel mem., reviewer). Mormon. Subspecialties: Plant physiology (biology); Toxicology (agriculture). Current work: Physiological/metabolic effects, modes of action and control of toxic stresses due to environmental pollutants. Home: 12713 Maple St Silver Spring MD 20904

BENNETT, JOAN WENNSTROM, biologist, educator; b. Bklyn., Sept. 15, 1942; d. John A. and Gertrude E. (Johnson) Wennstrom; m. Frank W. Bennett, Jr., June 5, 1966; children: John Frank, Daniel Edgerton, Mark Bradford. B.S., Upsala Coll., 1963; M.S., U. Chgo., 1964, Ph.D., 1967. Research assoc. dept. biology U. Chgo., 1967-68; So. Regional Research Labs., New Orleans, 1968-70; research assoc. dept. biology Tulane U., New Orleans, 1970-71, asst. prof., 1971-76, assoc. prof., 1976-81, prof., 1981—; cons. on bioethics Lutheran Ch. in Am. Editor: (with K.I. Abroms) Genetics and Exceptional Children, 1981, (with A. Ciegler) Secondary Metabolism and differentiation in Fungi, 1982; Contbr. articles to profl. publs. Mem. AAAS, Genetics Soc. Am., Am. Soc. Microbiology, Soc. Indsl. Microbiology, Mycol. Soc. Am., Soc. Gen. Microbiology, ACLU, Ammesty Internat. Lutheran. Subspecialties: Genetics and genetic engineering (biology); Microbiology. Current work: Genetics of secondary metabolism, especially mycotoxins; human aneuploidy, especially Downs Syndrome. Office: Dept Biology Tulane U New Orleans LA 70118

BENNETT, JOE CLAUDE, med. educator; b. Birmingham, Ala., Dec. 12, 1933; s. Claude and Clara Lucille (Clark) B.; m. Nancy Miller Bennett, June 17, 1958; children—Katherine Diane, Miller, Clark Barton. A.B., Samford U., 1954; M.D., Harvard, 1958. Intern Univ. Hosp., Birmingham, 1958-59, resident, 1959-60; practice medicine specializing in rheumatology; with NIH, 1962-64; sr. research fellow div. biology Calif. Inst. Tech., Pasadena, Calif., 1964-65; asst. prof. dept. medicine, asso. prof. dept. microbiology, asst. dir. div. clin. immunology and rheumatology U. Ala. Med. Sch., Birmingham, 1966-70, prof. dept. medicine, dir. div. clin. immunology and rheumatology, prof., chmn. dept. microbiology, 1970—, dir. multipurpose arthritis center, Disting. Faculty lectr., 1979; mem. Nat. Arthritis Adv. Bd., 1977-80; mem. subsplty. bd. rheumatology Am. Bd. Internal Medicine, 1979—. Author: Vistas in Connective Tissue Diseases, 1968; Editor: Arthritis and Rheumatism, 1975-80. John and Mary R. Markle Found. scholar in acad. medicine, 1965-70; recipient Research Career Devel. award NIH; fellow Arthritis Found., arthritis unit Mass. Gen. Hosp., 1960-62. Mem. A.C.P., AAAS, Am. Assn. Immunologists, Am. Fedn. Clin. Research, Am. Rheumatism Assn. (pres. 1981—, Am. Soc. Biol. Chemists), Am. Soc. Clin. Investigation, Am. Soc. Microbiology, Assn. Am. Physicians, Genetics Soc. Am., N.Y. Acad. Sci., Soc. Exptl. Biology and Medicine, So. Soc. Clin. Investigation, Sigma Xi. Subspecialty: Internal medicine. Current work: Research in rheumatology and immunology. Home: 4236 Antietam Dr Birmingham AL 35213 Office: U Ala in Birmingham Dept Medicine Univ Station Birmingham AL 35294

BENNETT, JOHN ROSCOE, computer co. exec.; b. Sparta, N.C., Sept. 14, 1922; s. Walter and Maggie J. (Brooks) B.; m.; 1 son, John Patrick; m. Barbara Wunderle, Sept. 22, 1973. B.S. in Commerce, U. Va., 1949. Nat. accounts mgr. Burroughs Corp., Washington, 1949-58; sales mgr. data systems Collins Radio, Dallas, 1958-65; v.p. mktg. Applied Data Research, Inc., Princeton, N.J., 1965-70, pres., chief exec. officer, 1970—, chmn. bd., 1981—; chmn. bd., dir. ADR Products, Inc., ADR Services, Inc., Mass. Computer Assos., Inc. Served as 1st lt. USAAF, 1943-46. Mem. Assn. Computer Mgmt. Data Processing Mgmt. Assn., Armed Forces Communications and Electronics Assn., Serpentine Club. Clubs: Bedens Brook (Princeton); Pike Brook Country (Belle Mead, N.J.); Elks. Subspecialty: Computer company management. Office: Applied Data Research Inc Route 206 and Orchard Rd Princeton NJ 08540

BENNETT, KELLY RANDOLPH, family counselor, educator, psychological test researcher; b. Los Angeles, Nov. 30, 1931; s. Charles Randolph and Bernice (Wheil) B.; m. Lois Elon Ferman, Jan. 16, 1953 (div. Nov. 1964); m. Kathleen Marie Lawrence, July 7, 1967; 1 dau., Keri Kristen. Student, Los Angeles City Coll., 1953-56; B.S. in Gen. Engring, UCLA, 1958; postgrad. in Biblical studies, Princeton Theol. Sem., 1962-63, in Hebrew and Archaeology, Hebrew Union Sem., 1963-64; student in psychology, U. Calif.-Riverside, 1964-65; M.S. in Clin. Psycology, Calif. State U., Los Angeles, 1967; Ph.D. in Spl. Edn, UCLA in cooperation with Calif. State U., Los Angeles, 1971. Astronautical engr. McDonnell) Douglas Aircraft Co., 1958-64; trainee Family Counseling and Research Ctrs., 1963-68, counselor, 1968-70; acting dir. Ctr. for Study Spl. Edn., Calif. State U., Los Angeles, 1970-71; instr. Calif. State U., Los Angeles, 1970-71; dir.-owner Palos Verdes Counseling Ctr., Palos Verdes Estates, Calif., 1971-75; pres. Life Themes Inc., Palos Verdes Estates, 1982—; pvt. practice marriage, family and child counseling, Palos Verdes Estates, 1968—; instr. Calif. State U.-Dominguez Hills, 1981—. Author: (with N.D. Berke) Early Readers Measure, 1971, (with W.H. Seaver and G. Ceaglio) Life Themes Inventory, 1981. Recipient Hon. Service award Calif. Congress Parents and Tchrs. Inc., 1975; Office Edn. fellow, 1969-71. Mem. Am. Psychol. Assn., Assn. Holistic Health, Assn. Transpersonal Psychology, Western Psychol. Assn. Republican. Subspecialty: Developmental psychology. Current work: Development of revolutionary holistic measure of self-understanding (Life Themes Inventory). Home: 2 Pear Tree Ln Rolling Hills Estates CA 90274 Office: 2550 Via Tejon Suite 3K Palos Verdes Estates CA 90274

BENNETT, LAWRENCE ALLEN, criminal justice research administrator; b. Selma, Calif., Jan. 4, 1923; s. Allen Walter and Geneva Elinor (Hall) B.; m. Beth J. Thompson, Aug. 12, 1948; children: Yvonne Irene, Glenn Livingston. B.A., Fresno (Calif.) State Coll., 1949; M.A., Claremont (Calif.) Grad. Sch., 1954, Ph.D., 1968.

Lic. psychologist, Calif.; registered psychologist, Ill. Clin. psychologist Deuel Vocat. Inst., Tracy, Calif., 1955-57; supervising psychologist Calif. Med. Facility, Vacaville, 1957-60; dept. supr. clin. psychology Calif. Dept. Corrections, Sacramento, 1960-67, chief research, 1967-76; dir. Center for Study of Crime, Delinquency and Corrections, So. Ill. U., Carbondale, 1976-79, Office of Program Evaluation, Nat. Inst. Justice, Washington, 1979—; cons. Nat. Council on Crime and Delinquency, Davis, Calif., 1968-75, Inst. for Study of Crime and Delinquency, Sacramento, 1968-70, Nat. Inst. Law Enforcement, Washington, 1973-78. Author: (with others) Counseling in Correctional Environments, 1978; Contbr. chpts. to books, articles to profl. jours. Bd. dirs. Calif. Crime Technol. Research Found., 1970-75, Am. Justice Ins., Sacramento, 1970-79; chmn. Yolo County Fair Play Com., Davis., 1960-61; mem. Sacramento Area Mental Health Exec. Group., 1970-73, chmn., 1971-72; mem. Yolo County Mental Health Adv. Bd., Woodland, Calif., 1970-74, chmn., 1971; mem. Calif. Interdepartmental Research Coordinating Com., 1967-76, chmn., 1970. Served with U.S. Army, 1943-45, 50-52; ETO. Mem. Calif. State Psychol. Assn. (dir. 1970-75), Sacramento Valley Psychol. Assn. (pres. 1969-70), Sacramento Statis. Assn. (pres. 1970-71), Assn. Criminal Justice Researchers (dir. 1973-76), Safer Found. (adv. bd. 1978-79). Democrat. Unitarian. Lodge: Elks. Subspecialties: Social psychology; Behavioral psychology. Current work: Criminal justice research including improved systems, advanced planning, offender personality functioning, and experimantal testing of new criminal justice concepts dealing with courts, police and correctional systems. Home: 8380 Greensboro Dr 311 The Rotunda McLean VA 22102 Office: Nat Inst Justice 633 Indiana Ave NW Washington DC 20531

BENNETT, MICHAEL VANDER LAAN, neuroscience educator; b. Madison, Wis., Jan. 7, 1931; s. Martin Toscan and Cornelia (Vander Laan) B.; m. Ruth B. Burman, July 19, 1963; children: Nicholas Toscan, Elena Paula. B.S., Yale U., 1952; D.Phil., Oxford U., 1957. Asst. prof. neurology Columbia U., N.Y.C., 1959-61, assoc prof., 1961-66; prof. anatomy Albert Einstein Coll. Medicine, 1967-74, prof. neurosci., 1974—, chmn. dept., 1982—. Contbr. numerous articles profl. jours. Pepsi Cola scholar, 1948; Rhodes scholar, 1952. Fellow AAAS; mem. Am. Physiol. Soc., Am. Soc. Cell Biology, Biophys. Soc., Corp. Marine Biol. Lab., N.Y. Acad. Scis., Nat. Acad. Scis., Soc. Neurosci., Phi Beta Kappa. Subspecialties: Neurophysiology; Neurobiology. Current work: Synaptic transmission, intercellular communication. Home: 10 Alderbrook Rd Bronx NY 10471 Office: 1300 Morris Park Ave Bronx NY 10461

BENNETT, PETER BRIAN, physiology educator; b. Portsmouth, Hants, Eng., June 12, 1931; s. Charles Risby and Doris Isobel (Peckham) B.; m. Margaret Camellia Rose Waren, July 7, 1956; children: Caroline Susan, Christopher Charles. B.Sc., U. London, 1951; Ph.D., U. Southampton, 1964. Scientist to dept. dir. R.N. Physiol. Lab., Alverstake, Hants, U.K., 1953-72; head pressure physiology Def. and Civil Inst. for Environ. Research, Toronto, Ont., Can., 1966-68; prof. anesthesiology dept. anesthesiology Duke Med. Ctr., Durham, N.C., 1972—, co-dir., 1972-77, dir., 1977—, 1980—, chmn., 1982—; advisor U.S./Japan Diving Physiology Panel, NOAA, Washington, 1977—; dir. N.C. Marine Edn. and Resources Found., Raleigh, 1981—. Author: Aetiology of Compressed Air Intoxication, 1966; editor: Physiology and Medicine of Diving, 1969, 75, 82. Served with RAF, 1951-53. Mem. Undersea Med. Soc. (pres. 1975-76, award 1975), European Undersea Med. Soc., Am. Physiol. Soc., AAAS. Subspecialty: Physiology (medicine). Current work: Laboratory director and research interests in diving physiology and medicine, hypobasic oxygen therapy and mechanisms of general anesthesia, dir. simulated deep "trimix" research dives to record 2250 feet to study and prevent the high pressure nervous syndrome. Home: 3010 Harriman Ave Durham NC 27705 Office: F G Hall Lab Duke Med Center Durham NC 27710

BENNETT, WILLIAM F., SR., agronomist, college dean; b. Plainview, Ark., Jan. 23, 1927; s. John and Minnie (Winner) B.; children: Linda Kay, William F., Jacqueline S. B.S., Okla. State U., 1950; M.S., Iowa State U., 1952, Ph.D., 1958. Cert. profl. agronomist. Extension soil chemist Tex. A&M U., 1957-63; v.p. and agronomist Elcor Chem. Co., Dimmitt, Tex., 1963-68; agronomist, assoc. dean for resident instrn. Coll. Agrl. Scis., Tex. Tech U., Lubbock, 1968—. Co-author: Food and Fiber for a Changing World, 2d edit., 1982, Crop Science and Food Production, 1983; sr. author: Modern Grain Sorghum Production, in press. Served as cpl. USAF, 1945-47. Mem. Soil Sci. Soc. Am., Am. Soc. Agronomy. Subspecialties: Soil chemistry; Plant growth. Current work: Soil fertility, fertilizer use and crop management. Home: 3703 68th St Lubbock TX 79413 Office: Tex Tech U Box 4169 Lubbock TX 79409

BENNETT, WILLIAM RALPH, JR., physicist, educator, university official; b. Jersey City, Jan. 30, 1930; s. William Ralph and Viola Mildred (Schreiber) B.; m. Frances Commins, Dec. 11, 1952; children: Jean, William Robert, Nancy. B.A., Princeton U., 1951; Ph.D., Columbia U., 1957; M.A. (hon), Yale U., 1965; D.Sc. (hon.), U. New Haven. Instr. Yale U., New Haven, 1957-59, assoc. prof. physics and applied sci., 1962-64, prof., 1964-72, C.B. Sawyer prof. engring. and applied sci., 1972—, prof. physics, 1972—, master Silliman Coll., 1981—; mem. tech. staff Bell Telephone Labs., Murray Hill, N.J., 1959-62; cons. Tech. Research Group, Melville, N.Y., 1962-67, Inst. Def. Analyses, 1963-70, Laser Scis. Corp., 1968-71, CBS Labs., 1967-68, AVCO Corp., 1978—; cons. in laser field for numerous indsl. and govt. labs., 1962—. Author: Gas Lasers, 1964, (transl. into Russian) Introduction to Computer Applications, 1976, Scientific and Engineering Problem Solving with the Computer, 1976, The Physics of Gas Lasers, 1977, Atomic Gas Laser Data; A Critical Evaluation, 1979; mem. editorial adv. bd.: Jour. Quantum Electronics, 1965-69; guest editor: Applied Optics, 1965. Bd. dirs. Friends of WFCR in So. Conn., 1974-78. Recipient Western Electric Fund award for outstanding teaching Am. Assn. Engring. Educator, 1977, Outstanding Patent award Research and Devel. Council N.J., 1977; Sloan Found. Fellow, 1963-65; Guggenheim Found. fellow, 1967. Fellow Am. Phys. Soc., Optical Soc. Am., IEEE (Morris Lieberman prize 1965); mem. Sigma Xi (chmn. New Haven chpt. 1976). Subspecialties: Laser medicine; Atomic and molecular physics. Current work: Gas lasers; helium-neon laser; noble gas lasers; ion lasers; metal vapor lasers. Patentee laser field, 1st several gas lasers, helium-neon laser. Home: 71 Wall St New Haven CT 06511 Office: Master's Office Silliman College Yale U New Haven CT 06520

BENNINGHOFF, ANNE STEVENSON, botanist, ecologist, palynologist; b. Shelby, Ohio, Aug. 3, 1942; d. Clayton Cooper and Thelma Ruth (Kunkel) Stevenson; m. William Shiffer Benninghoff, June 14, 1969. B.S., Wittenberg U., 1964; M.S., U. Mich., 1967, postgrad., 1969. Researcher Parke-Davis & Co., Rochester, Mich., 1964-65; research investigator U. Mich., Ann Arbor, 1969—; botany cons. Natural Features Inventory for City of Ann Arbor, 1980-82. Contbr. articles to profl. jours. Sci. vol. Ann Arbor Hands-On Mus. Mem. Bot. Soc. Am., Ecol. Soc. Am., Ohio Acad. Scis., Internat. Assn. Aerobiology, Internat. Assn. Gt. Lakes Research, Sigma Xi. Lutheran. Clubs: Women's Research, Women of Univ. Faculty, Sci. Research, Faculty Women's (U. Mich.), Mich. Bot. (sec. Huron Valley chpt.). Subspecialties: Ecology; Aerobiology. Current work: Airborne particulates in realtion to atmospheric potential gradients, vegetation history of western Lake Erie region, plant community studies, electrostatic studies conducted in Antarctica. Home: 3315 Alton Ct Ann Arbor MI 48105 Office: Dept of Botany U Mich Ann Arbor MI 48109

BENNO, ROBERT HOWARD, neurobiology educator, researcher; b. Dallas, Apr. 4, 1948; s. Martin and Fannie Mae B. B.S., Tulane U., 1970, M.S., 1973; Ph.D., U. Iowa, 1978. Research asst., Tulane U., 1970-71, teaching asst., 1971-73; research asst., teaching asst. U. Iowa, 1973-78; postdoctoral fellow Cornell U. Med. Coll., 1978-81, instr., 1981-82; asst. prof. biology William Paterson Coll., 1982—. Mem. Soc. Neurosci., Sigma Xi, Omicron Delta Kappa. Subspecialties: Neurobiology; Immunocytochemistry. Current work: Development and neurotransmitted regulation in the central nervous system; quantitative immunocytochemical analysis of neurotransmitters in the developing and mature rodent brain teaching anatomy, physiology, neuroscience. Home: 26 E 2d St New York NY 10003 Office: Dept Biology William Paterson Coll Wayne NJ 07470

BENSELER, ROLF WILHELM, biologist, educator; b. San Jose, Calif., Sept. 24, 1932; s. William A. and Ella K (Vaeth) B.; m. Donna Alyce Kirk, Dec. 16, 1961; children: William Paul, Mark Christian. Student, San Jose State U., 1951-53; B.S., U. Calif.-Berkeley, 1957; Ph.D., 1968; M.F., Yale U., 1958. Specialist U. Calif. Agrl. Exptl. Sta., Berkeley, 1958-61; instr. Modesto (Calif.) Jr. Coll, 1961-63; asst. prof., then prof. biol. sci. Calif. State U., Hayward, 1968—; mem. edni. adv. com. East Bay Regional Park Dist., East Bay Mcpl. Utility Dist. Served as petty officer USNR, 1951-58. Mem. Am. Inst. Biol. Scis., Bot. Soc. Am., Am. Soc. Plant Taxonomy, Calif. Bot. Soc., Calif. Acad. Sci., Nat. Wildlife Fedn., Wirehaired Pointing Griffon Club Am., Deutschkurzhaar Verband, Nature Conservancy. Lutheran. Subspecialties: Reproductive biology; Evolutionary biology.

BENSEN, DAVID WARREN, nuclear researcher; b. Paterson, N.J., FEb. 6, 1928; s. David and Jane Elizabeth (Kayheart) B.; m. H. Evelyn Streicher, June 12, 1954; children: Douglas, Debra, Clark, Michele. B.S., Rutgers U., 1954, M.S., 1955, Ph.D., 1958. Chemist, Gen. Electric Co., Richland, Wash., 1958-63; sr. scientist Isotopes, Inc., Westwood, N.J., 1963-66; phys. sci. adminstr. U.S. Naval Radiol. Def. Lab., San Francisco, 1966-67; project mgr. Office CD, Washington, 1967-72, Def. Civil Preparation Agy., Dept. Def., 1973-78; sr. scientist Fed. Emergency Mgmt. Agy., Washington, 1978—. Served with U.S. Army, 1946-49. Mem. Health Physics Soc., Am. Nuclear Soc., Sigma Xi. Subspecialties: Integrated systems modelling and engineering; Ecosystems analysis. Current work: Describe the environment created by the explosion of nuclear weapons on strategic targets and develop civil defense measures to protect people and resources from the effects of nuclear weapons. Home: 4624 Sunflower Dr Rockville MD 20853 Office: Fed Emergency Mgmt Agy 500 C St SW Washington DC 20472

BENSON, ANDREW ALM, biochemistry educator; b. Modesto, Calif., Sept. 24, 1917; s. Carl Bennett and Emma Carolina (Alm) B.; m. Dorothy Dorgan, July 31, 1971; children: Claudia Benson Matthews, Linnea, Bonnie Benson Kumar (dec.). B.S., U. Calif., Berkeley, 1939; Ph.D., Calif. Inst. Tech., 1942; Phil.D. (hon.), U. Oslo., 1965. Instr. chemistry U. Calif., Berkeley, 1942-43, asst. dir. bio-organic group Radiation Lab., 1946-54, assoc. prof. agrl. biol. chemistry Radiation Lab., 1955-60; prof. Pa. State U., 1960-61; prof.-in-residence biophys./physiol. chemistry UCLA, 1961-62; prof. Scripps Instn. Oceanography, U. Calif., San Diego, 1962—; research assoc. OSRD dept. chemistry Stanford U., 1944-45. Contbr. articles on biochem. research to profl. jours. Trustee Found. for Ocean Research, San Diego; mem. adv. council The Costeau Soc., 1976—. Recipient Sugar Research Found. award, 1950, Ernest Orlando Lawrence Meml. award, 1962; Sr. Queen's fellow, Australia, 1979. Mem. Am. Chem. Soc., Am. Soc. Plant Physiologists (Stephen Hales award 1972), Japan Soc. Plant Physiologists), Am. Soc. Biol. Chemists, Nat. Acad. Sci., Am. Acad. Arts and Scis. Subspecialties: Biochemistry (biology); Marine metabolism. Current work: Symbiotic relationships in marine organisms; wax ester synthesis and utilization; spawning Pacific metabolism; arsenic detoxification in marine algae and higher organisms; arsenic biochemistry, nutrition and pesticides; photosynthesis. Home: 6044 Folsom Dr La Jolla CA 92037 Office: Scripps Instn Oceanography A-002 La Jolla CA 92093

BENSON, DAVID MICHAEL, plant pathologist, eductor; b. Dayton, Ohio, Aug. 28, 1945; s. Phillip Wayne and Edna Mae (Yowler) B.; m. Patricia Diane Benson, Jan. 28, 1967; 1 dau., Julie Ann. B.S., Earlham Coll., 1967; M.S., Colo. State U., 1967, Ph.D., 1973. Postdoctoral assoc. dept. plant pathology U. Calif.-Berkeley, 1973-74; asst. prof. plant pathology N.C. State U., Raleigh, 1974-79; assoc. prof., 1979—. Subspecialty: Plant pathology. Current work: Ornamental diseases; ecology of soil-borne plant pathogenic fungi; phytophthora diseases. Office: Dept Plant Pathology NC State U Raleigh NC 27650

BENSON, DONALD WARREN, anesthesiologist; b. Jamestown, N.Y., Aug. 17, 1921; s. George Elver and Elen Johanna (Peterson) B.; m. Marjorie Ann Maulsby, June 8, 1946; children—Brian Wesley, Jane Ellen, Ruth Ann. Student, North Park Coll., Chgo., 1940-42, U. Ill., 1942-43, Notre Dame U., 1943-44; B.S., U. Chgo., 1949, M.D., 1950, Ph.D. in Pharmacology, 1957. Diplomate Am. Bd. Anesthesiology (dir. 1969—). Instr., then asst. prof. anesthesiology U. Chgo., 1953-56; asso. prof. anesthesiology Johns Hopkins Med. Sch., 1956-59, prof., after 1959; anesthesiologist-in-charge Johns Hopkins Hosp., 1956-74; prof., chmn. anesthesiology U. Chgo. Hosps. and Clinics, 1975—. Served to ensign USNR, 1944-46. Mem. Am. Soc. Anesthesiologists, Assn. Univ. Anesthesiologists, Sigma Xi. Subspecialty: Anesthesiology. Home: 5719 S Kenwood Ave Chicago IL 60637 Office: U Chgo Dept Anesthesiology 950 E 59th St Chicago IL 60637

BENSON, PETER HOWARD, gas research institute administrator; b. Martinez, Calif., Aug. 25, 1935; s. Howard Lucius and Alberta Clara (Bothe) B.; m. Lois Johanna Lee, May 23, 1958 (dec. 1962); m. Judith Ellen Patterson, July 6, 1964; children: Scott Conrad, Cydney. B.A., U. Calif.-Berkeley, 1958; M.A., San Francisco State Coll., 1960; Ph.D., U. So. Calif., 1973. Scientist, sr. scientist dept. marine biology Lockheed Ocean Lab., San Diego, 1968-73; mgr. Lockheed Marine Biology Lab., San Diego, 1973-76; gen. mgr. Lockheed Center for Marine Research, Carlsbad, Calif., 1976-77, chief scientist, 1977-78; asst. program mgr. Argonne (Ill.) Nat. Lab., 1978-81; asst. dir. Gas Research Inst., Chgo., 1981—; instr. oceanography UCLA, 1967-68; cons. Argonne Nat. Lab., 1978, 81; plankton specialist Antarctic Oceanology Program, Coop. Study Kuroshio, 1965-68. Contbr. articles to tech. publs. Mem. U.S. Adv. Panel to Interciencia Bioresources Program. Fellow Allan Hancock Found.; mem. Marine Tech. Soc., Phi Kappa Sigma. Subspecialties: Fuels and sources; Biomass (energy science and technology). Current work: Biomass and wastes to methane research and development; biofouling and ocean thermal energy conversion. Home: 12115 87th Ave Palos Park IL 60464 Office: Gas Research Inst 8600 W Bryn Mawr Ave Chicago IL 60631

BENSON, ROYAL HENRY, chemist, researcher; b. Galveston, Tex., Oct. 25, 1925; s. Roy Henry and Kathleen (Bradford) B.; m. Aleta Jo Wooten, Mar. 28, 1970; children: Royal H. III, Tamara K., Melissa A. Sc.B., U. Houston, 1948, M.Sc., 1956. Instr. U. Houston, 1947-48; research chemist M.D. Anderson Inst., Houston, 1948-50; VA, Houston, 1950-56; sr. research chemist Monsanto Co., Texas City, Tex., 1956-65, sr. research specialist, 1965-77, sci. fellow, 1977—. Mem. Am. Chem. Soc. (dir. 1965-70), Am. Nuclear Soc. Democrat. Methodist. Subspecialties: Analytical chemistry; Organic chemistry. Current work: Applications of radioisotopes to technical problems in chemical research and chemical manufacturing processes; development of highly sensitive methods of radioassay of wide applicability. Patentee in field. Home: 1522 19th Ave N Texas City TX 77590 Office: Monsanto Co 211 Bay St Texas City TX 77590

BENSON, SIDNEY WILLIAM, chemistry educator; b. N.Y.C., Sept. 26, 1918; m. Ann M. Benson (dec.); 2 children, 4 stepchildren. A.B. with honors in Chemistry, Physics and Math, Columbia U., 1938; A.M., Ph.D. in Phys. Chemistry and Chem. Physics, Harvard U., 1941. Postdoctoral research fellow Harvard U., Cambridge, Mass., 1941-42; instr. chemistry CCNY, 1942-43; group leader Manhattan Project, Kellex Corp., 1943; asst. prof. chemistry U. So. Calif., Los Angeles, 1943-48, assoc. prof., 1948-51, prof., 1951-64, 1976—; research assoc. Nat. Def. Research Council, Chem. Warfare Service, 1944-46; chmn. dept. kinetics and thermochemistry Stanford Research Inst., Calif., 1963-76; research assoc. dept. chemistry and chem. engring. Calif. Inst. Tech., Pasadena, 1957-58; vis. prof. UCLA, 1959; vis. prof. chemistry U. Ill.-Champaign, 1959; hon. Glidden lectr. Purdue U., West Lafayette, Ind., 1961; lectr. Phillips Petroleum Co., 1964; vis. prof. dept. chemistry and chem. engring. Stanford U., Calif., 1966-70, 71, 73; hon. vis. prof. chemistry and chem. engring. U. Utah, 1971; vis. prof. chemistry NSF sr. postdoctoral fellow U. Paris, 1971-72; hon. vis. prof. chemistry U. St. Andrews, Scotland, 1973; sci. dir. Hydrocarbon Research Inst., U. So. Calif., Los Angeles, 1977; hon. vis. prof. chemistry Tex. A&M U., 1978, U. Paris, 1979, U. Lausanne, Switzerland, 1979; dir. Pyrotech N.V. div. KTI, 1980—; cons. and researcher in chemistry and physics; chmn. com. kinetics of chem. reactions NRC, 1969, 77-80. Author: Syllabus for General College Chemistry, 1948, Chemical Calculations, 1952, The Foundations of Chemical Kinetics, 1960, 1982, Atoms, Molecules and Chemical Reactions: Chemistry from a Molecular Point of View, 1970, Thermo-Chemical Kinetics, 1972, 2d edit., 1976, The Current Contents, 1981; co-author: (with H.E. O'Neal) Kinetic Data on Gas Phase Unimolecular Reactions, 1970; contbr. chpts. to books and articles to profl. jours., to proc. profl. confs., encys.; editorial bd., Elsevier Publishing Co., Amsterdam, 1965—, Combustion and Flame, 1967-71, Combustion and Science Technology, 1973—, Oxidation Communications, 1978—, Reviews of Chemical Intermediates, 1979—, Hydrocarbon Letters, 1980—, Jour. Phys. Chemistry, 1981—; editor-in-chief: Internat. Jour. Chem. Kinetics, 1967-84. Recipient cert. of Merit for Contribution to War Effort Nat. Def. Research Council, 1947; Guggenheim fellow, 1950-51; Fulbright fellow, France, 1950-51; NSF sr. postdoctoral fellow, 1957-58; Japanese Soc. Promotion of Sci. fellow, 1980. Fellow AAAS (chemistry sect. com. 1978-82), Am. Phys. Soc.; mem. Nat. Acad. Scis., Am. Chem. Soc. (exec. com. 1966-69, 73-77, chmn. 1974-75, award in petroleum chemistry 1977, Tolman award 1978), Internat. Council Sci. Unions (chmn. task group chem. kinetics 1969-75), Faraday Soc., Phi Beta Kappa, Pi Mu Epsilon, Phi Lambda Upsilon, Sigma Xi, Phi Kappa Phi (hon.). Subspecialties: Kinetics; Physical chemistry. Current work: Atmospheric chemistry; free radicals; combustion; theory of solutions; chemical kinetics. Patentee measurements of concentrations of gaseous phase elements, 1979, conversion of methane, 1980. Home: 533 Palos Verdes Dr West Palos Verdes Estates CA 90274 Office: U So Calif Los Angeles CA 90089

BENT, SAMUEL WATKINS, computer scientist, educator; b. Gardner, Mass., May 11, 1955; s. Gardner L. and Marie B. B. A.B., Cornell U., 1975; Ph.D., Stanford U., 1982. Asst. prof. computer sci. U. Wis., Madison, 1981-. Mem. Assn. for Computing Machinery, Soc. for Indsl. and Applied Math., Math. Assn. Am. Subspecialties: Algorithms; Theoretical computer science. Current work: Analysis of algorithms; design of efficent data structures. Office: Univ Wis 1210 W Dayton Madison WI 53706

BENTLEY, BARBARA LEE, ecologist, educator; b. Los Angeles, Dec. 14, 1942; d. William John and Mary Lou (Price) Morse; m. C. Ronald Carroll, Apr. 23, 1966 (div. 1978); 1 son, Steven Clay; m. Glenn Downes Prestwich, May 25, 1980; 1 dau., Jocelyn Alyssa. B.A., Willamette U., Salem, Oreg., 1964; M.S., UCLA, 1966; Ph.D., U. Kans.-Lawrence, 1974. Asst. prof. SUNY-Stony Brook, 1973-79, assoc. prof., 1979—, assoc. vice provost grad. studies, 1983—; cons. Brookhaven Nat. Lab., 1977-81, NSF, 1982—. Editor: The Biology of Nectaries, 1982. Bd. dirs. Community Health Plan, Hauppauge, N.Y., 1982—. Mem. Orgn. Tropical Studies (v.p. 1980—, dir.), Assn. Tropical Biology (exec. com. 1982—), Am. Soc. Naturalists, Brit. Ecol. Soc., N.Y. Acad. Scis., Sigma Xi. Subspecialties: Ecology; Species interaction. Current work: Nitrogen fixation and cycling in tropical environments; plant/herbivore interactions. Office: Dept Ecology and Evolution SUNY Stony Brook NY 11794

BENTLEY, CHARLES RAYMOND, geophysicist; b. Rochester, N.Y., Dec. 23, 1929; s. Raymond and Janet Cornelia (Everest) B.; m. Marybelle Goode, July 3, 1964; children: Molly Clare, Raymond Alexander. B.S., Yale U., 1950; Ph.D., Columbia U., 1959. Research geophysicist Columbia U., 1952-56; Antarctic traverse leader and seismologist Arctic Inst. N.Am., 1956-59; project asso. U. Wis., 1959-61, asst. prof., 1961-63, asso. prof., 1963-68, prof. geophysics, 1968—; mem. council Internat. Antarctic Glaciol. Project; chmn. polar research bd. Nat. Acad. Sci.; U.S. mem. working group on solid earth geophysics Sci. Com. on Antarctic Research., U.S. alt. del. Chmn. bd. asso. editors: Am. Geophys. Union Antarctic Research series. Recipient Bellingshausen-Lazarev medal for Antarctic research Acad. Scis. USSR, 1971; NSF sr. postdoctoral fellow, 1968-69; U.S.-U.S.S.R. Acad. Scis. exchange fellow, 1977. Mem. Am. Geophys. Union, Soc. Exploration Geophysicists, AAAS, Internat. Glaciol. Soc., Seismol. Soc. Am., Geol. Soc. Am., Am. Quaternary Assn., Am. Geol. Inst., Am. Polar Soc., AAUP, Phi Beta Kappa, Sigma Xi. Subspecialties: Geophysics. Current work: Study of Antarctic ice sheet and subglacial crystal structure; study of past and future extent of the ice. Research on Antarctic glaciology and geophysics, seismic refraction measurements at sea, magnetotelluric exploration of earth structure. Home: 5618 Lake Mendota Dr Madison WI 53705

BENTLEY, MICHAEL MARTIN, biology educator; b. Detroit, Nov. 2, 1947; s. Carvel Martin and Doris Marian (Krueger) B.; m. Laura Suzanne Jordan, Feb. 1, 1969; children: Mark James, Christopher Martin, Joseph Michael. B.A. in Biology and Geology, Coll. Wooster, 1969, Ph.D., McMaster U., Hamilton, Ont., Can., 1977. Postdoctoral fellow biology dept. U. Calgary, Alta., Can., 1976-79, research assoc., 1980, Alta. Heritage Found. Med. Research scholar, 1981, asst. prof. biology, 1982—. Contbr. numerous sci. articles to profl. publs. McMaster Grad. fellow McMaster U., 1971-76; Alta. Heritage Found. Med. Research grantee, 1981-83; Natural Scis. and Engring. Resarch Council Can. grantee, 1983—. Mem. Genetics Soc. Am., Genetics Soc. Can. Subspecialties: Developmental biology; Gene actions. Current work: Developmental and biochemical genetics of Drosophila; developmental genetics; biochemical genetics; molybdoenzymes; NADP enzymes; gene regulation; gene-enzyme systems; Drosophila. Office: Biology Dept U Calgary Calgary AB Canada T2N 1N4

BENTLEY, ORVILLE GEORGE, government official; b. Midland, S.D., Mar. 6, 1918; s. Thomas O. and Ida Marie (Sandal) B.; m. Enolia J. Anderson, Sept. 19, 1942; children: Peter T., Craig E. B.S., S.D.

State Coll., 1942; M.S. in Biochemistry, U. Wis., 1947, Ph.D., 1950; hon. degree, S.D. State U., 1974. Asst. prof. animal sci. Ohio Agrl. Expt. Sta.; also mem. dept. animal sci. and dept. agrl. biochemistry Ohio State U., 1950-58; dean Coll. of Agr. and Biol. Scis., S.D. State U., 1958-65, Coll. Agr., U. Ill. at Urbana, 1965-82; asst. sec. agr. for sci. and edn. USDA, Washington, 1982—; Mem. com. animal nutrition NRC-Nat. Acad. Scis., 1958-67; mem. Council U.S. Univs. for Rural Devel. in India, 1967-74; mem. ad hoc adv. com. Ill. Inst. for Environmental Quality, 1971; mem. tech. adv. com. on food and agr. U.S. Dept Agr., Viet Nam, 1966; mem. panel Nat. Acad. Scis. to meet mems. Indonesian Acad. Scis., 1968; co-chmn. Agrl. Research Policy Adv. Com., 1973-77; mem. Bd. for Internat. Food and Agrl. Devel., 1976-80. Editorial bd.: jour. Animal Sci, 1956-59; Contbr. articles to profl. jours. Bd. dirs. Am. U. Beirut, Midwest Univs., Consortium for Internat. Activities, 1966-76; chmn. bd. dirs. Farm Found., 1971-78. Served to maj., chem. warfare service AUS, 1942-45. Named Young Man of Year Wooster Jr. C. of C., 1953; recipient Distinguished Alumnus award S.D. State U., 1967. Fellow Am. Soc. Animal Sci. (v.p. midwestern sect. 1963, Am. feed mfrs. award 1958); mem. Am. Chem. Soc., Am. Inst. Nutrition, Am. Soc. Animal Sci., Am. Dairy Sci. Assn., Internat. Union of Nutritional Scis., Farm House (hon.), AAAS (committeeman-at-large 1971-82), Sigma Xi, Phi Kappa Phi. Club: Rotarian. Subspecialties: Agricultural research administration.; Animal nutrition. Current work: Planning coordination and policy guidance for the Agricultural Research Service, Cooperative State Research Service, Extension Service and National Agricultural Library. Home: Concord Ln Rural Route 2 Urbana IL 61801 Office: Dept of Agriculture 14th and Independence Ave SW Washington DC 20250

BENY, NETA, mathematician; b. Tripoli, Libya, Nov. 2, 1945; s. Shalom and Ahuva (Vittori) N.; m. Tamar Cerklevitz, Aug. 15, 1971; children: Itay, Leeor, Maital. B.Sc., Tel Aviv U., 1967, M.Sc., 1971; Ph.D., Carnegie Mellon U., 1977. Asst. prof. No. Ill. U., DeKalb, 1977-80; asst. prof. math Tex. Tech. U., Lubbock, 1980—. Mem. Am. Math. Soc., Soc. Indsl. and Applied Math., N.Y. Acad. Sci., AAAS, Sigma Xi. Subspecialty: Numerical analysis. Current work: Numerical solution of differential equations. Home: 7923 Vicksburg Ave Lubbock TX 79424 Office: Dept Math Tex Tech U Lubbock TX 79409

BENZ, EDWARD JOHN, JR., physician, medical educator, researcher; b. Pitts., May 22, 1946; s. Edward John and Verna Marie (Cuddyre) B.; m. Roberta Jean Fiske, Apr. 23, 1947; children: Timothy Edward, Jennifer Kirsten. B.A. cum laude, Princeton U., 1968; M.D. magna cum laude, Harvard U., 1973. Diplomate: Am. Bd. Internal Medicine, Nat. Bd. Med. Examiners. Intern internal medicine Peter Bent Brigham Hosp., Boston, 1973-74, resident in internal medicine, 1974-75; research assoc. NIH, Bethesda, Md., 1975-78; fellow in hematology Yale U., 1978-79, asst. prof. medicine, 1979-82, assoc. prof. medicine, 1982—, assoc. prof. human genetics, 1983—; mem. sci. adv. bd. Conn. chpt. ARC, Farmington, Conn., 1982—; cons. dept. reprodn. genetics McGee Women's Hosp., Pitts., 1980—; mem. research rev. panels ad-hoc NIH, Bethesda, 1977—. Contbr. articles to profl. jours., chpts. to books; reviewer, editorialist numerous sci. jours., agys., 1976—. Served with USPHS, 1975-78. NIH Research Career Devel. awardee, 1982; NIH, March of Dimes Cooley Anemia Found. N.J. grantee, 1979—. Mem. Am. Soc. Clin. Investigation, Am. Soc. Hematology (chmn. subcom. 1982-83), Am. Fedn. Clin. Research, AAAS. Roman Catholic. Club: Princeton Elm. Subspecialties: Genetics and genetic engineering (medicine); Hematology. Current work: Laboratory research focused on the molecular genetics of hemoglobin diseases and leukemia, utilizing recombinant DNA techniques; clinical care of patients, education. Home: 57 Cindy Lane Guilford CT 06437 Office: Yale U Sch Medicine 812 LCI Bldg 333 Cedar St New Haven CT 06471

BENZER, SEYMOUR, scientist, educator; b. N.Y.C., Oct. 15, 1921; s. Mayer and Eva (Naidorf) B.; m. Dorothy Vlosky, Jan. 10, 1942 (dec. 1978); children: Barbara Ann Benzer Freidin, Martha Jane Benzer Goldberg; m. Carol A. Miller, May 11, 1980. B.A., Bklyn. Coll., 1942; M.S., Purdue U., 1943, Ph.D., 1947, D.Sc. (hon.), 1968, Columbia U., 1974, Yale U., 1977, Brandeis U., 1978, CUNY, 1978, U. Paris, 1983. Mem. faculty Purdue U., 1945-67, prof. biophysics, 1958-61, Stuart distinguished prof. biology, 1961-67; prof. biology Calif. Inst. Tech., 1967-75, Boswell prof. neurosci., 1975—; biophysicist Oak Ridge Nat. Lab., 1948-49; vis. asso. Calif. Inst. Tech., Pasadena, 1965-67. Contbr. articles to profl. jours. Research fellow Calif. Inst. Tech., 1949-51; Fulbright Research fellow Pasteur Inst., Paris, France, 1951-52; sr. NSF postdoctoral fellow, Cambridge, Eng., 1957-58; recipient Award of Honor, Bklyn. Coll., 1956; Sigma Xi research award Purdue U., 1957; Ricketts award U. Chgo., 1961; Gold medal N.Y. City Coll. Chemistry Alumni Assn., 1962; Gairdner award of merit, 1964; McCoy award Purdue U., 1965; Lasker award, 1971; T. Duckett Jones award, 1975; Prix Leopold Mayer French Acad. Scis., 1975; Louisa Gross Horwitz award, 1976; Harvey award Israel, 1977; Warren Triennial prize Mass. Gen. Hosp., 1977; Dickson award, 1978. Mem. Nat. Acad. Scis., Am. Acad. Arts and Scis., Am. Philos. Soc., Harvey Soc., AAAS; fgn. mem. Royal Soc. London. Subspecialties: Genetics and genetic engineering (biology); Neurobiology. Current work: Neurogenetics of drosophila. Home: 2075 Robin Rd San Marino CA 91108

BENZING, DAVID HILL, biology educator, cons., researcher; b. Evanston, Ill., Oct. 13, 1937; s. Kenneth K. and Francis (Hill) B.; m. Lizette M. Bonnifield, Oct. 20, 1979; children: Ellen, Carrie, Warren. B.A., Miami U., Oxford, Ohio, 1959; M.S., U. Mich., 1962, Ph.D., 1965. Mem. faculty Oberlin (Ohio) Coll., 1965—, prof. biology, 1976—; NSF grantee, 1969-1983. Contbr. articles on biology to profl. jours. City councilmen-at-large City of Oberlin, 1978-80. Served with USAR, 1960-66. Mem. Bot. Soc. Am., Assn. Tropical Biology, Sigma Xi. Subspecialties: Plant physiology (biology); Evolutionary biology. Current work: Biology of vascular epiphytic plants, functional morphology and stress physiology of epiphytic members of plant families bromeliaceae araceae and orchidaceae.

BERAN, JO ALLAN, chemistry educator; b. Odell, Nebr., Aug. 24, 1942; s. Ernest E. and Marie J. (Bednar) B.; m. Judith Lynn McGuire, Aug. 15, 1964; children: Kyle, Greg. B.A., Hastings Coll, 1964; Ph.D., U. Kans., 1968. Asst. prof. Tex. A&I U., 1968-71, assoc. prof., 1971-76, prof. chemistry, 1976—, chmn. dept., 1975-81; also cons. Author: Laboratory Manual for General Chemistry, Principles and Structure, 1978, 2d edit., 1982, Student's Guide to Fundamentals of Chemistry, 4th edit, 1980, Laboratory Manual for Fundamentals of Chemistry, 1981, 2d edit., 1984. Grantee Robert A. Welch Found., NSF. Mem. Am. Chem. Soc., Tex. Assn. Coll. Tchrs. Subspecialties: Environmental chemistry; Inorganic chemistry. Current work: Chemical education.

BERANEK, LEO LEROY, business and engineering consultant; b. Solon, Iowa, Sept. 15, 1914; s. Edward Fred and Beatrice (Stahle) B.; m. Phyllis Knight, Sept. 6, 1941 (dec. 11/7/82); children: James Knight, Thomas Haynes. A.B., Cornell Coll., 1936, D.Sc. (hon.), 1946; M.S., Harvard U., 1937, D.Sc., 1940; D.Eng. (hon.), Worcester Poly. Inst., 1971, D.Comml. Sci., Suffolk U., 1979, LL.D., Emerson College, 1982. Instr. physics Harvard, 1940-41, asst. prof., 1941-43, dir. research on sound, 1941-45; dir. Electro-Acoustics and Systems Research Labs., 1945-46; assoc. prof. communications engring. MIT, 1947-58, lectr., 1958-81; tech. dir. Acoustics Lab., 1947-53; pres., dir. Bolt Beranek & Newman, Cambridge, Mass., 1953-69, chief scientist, 1969-71, dir., 1953—; pres., chief exec. officer, dir. Boston Broadcasters, Inc., 1963-79, chmn. bd., 1980-83; part-owner WCVB-TV, Boston, 1972-82; chmn. bd. Mueller-BBM GmbH, Munich, Germany, 1962—. Author: (with others) Principles of Sound Control in Airplanes, 1944, Acoustic Measurements, 1949, Acoustics, 1954, Music, Acoustics and Architecture, 1962; Editor, contbr.: Noise Reduction, 1960, Noise and Vibration Control, 1971; Editor: Noise Control mag, 1954-55; assoc. editor: Sound mag, 1961-63; editorial bd.: Noise Control Engring, 1973-77; Contbr. articles on acoustics, audio and TV communications systems to tech. publs. Mem. Mass. Gov.'s Task Force on Coastal Resources, 1974-77; Charter mem. bd. overseers Boston Symphony Orch., 1968-80, chmn., 1977-80, trustee, 1977—, v.p., 1980-83, chmn. bd. trustees, 1983—; mem. vis. com. Center Behavioral Scis., Harvard U., 1964-70, vis. com. biology and related research facilities, 1971-77, mem. vis. com. physics dept., 1983—; mem. advisory com. mgmt. devel. Harvard Bus. Sch., 1965-71; mem. council for arts Mass. Inst. Tech., 1972—; pres. World Affairs Council Boston, 1975-78, vice chmn. bd., 1979—; trustee Cornell Coll., 1955-71, Emerson Coll., 1973-79; bd. dirs. Boston Opera Co., pres., 1961-63; bd. dirs. Boston 200, 1975-77, United Way Mass. Bay, 1975-80, Flaschner Jud. Inst., 1977-81. Guggenheim fellow, 1946-47; recipient Presdl. certificate of merit, 1948; Cornell Coll. Alumni Citation, 1953; 1st Silver medal le Groupement des Acousticiens de Langue Francaise, Paris, 1966; Abe Lincoln TV award So. Bapt. Conv., 1975; Media award NAACP, 1975. Fellow Acoustical Soc. Am. (Biennial award 1944, exec. council 1944-47, v.p. 1949-50, pres. 1954-55, assoc. editor 1946-60, Wallace Clement Sabine Archtl. Acoustics award 1961, Gold medal award 1975), Nat. Acad. Engring. (dir. marine bd., com. pub. engring. policy, aeros. and space engring. bd.), Am. Acad. Arts and Scis., Am. Phys. Soc., AAAS, Audio Engring. Soc. (pres. 1967-68, Gold medal 1971, gov. 1966-71), IEEE (chmn. profl. group audio 1950-51); mem. Inst. Noise Control Engring. (charter pres. 1971-73, dir. 1973-75), Am. Standards Assn. (chmn. acoustical standards bd. 1956-68, dir. 1963-68), Mass. Broadcasters Assn. (dir. 1973-80, pres. 1978-79, Disting. Service award 1980), Boston Community Media Council (treas. 1973-76, v.p. 1976-77), Cambridge Soc. Early Music (pres. 1963-71, dir. 1961-79), Acad. Disting. Bostonians, Greater Boston C. of C. (dir. 1973-79, v.p. 1976-79, Disting. Community Service award 1980), Phi Beta Kappa, Sigma Xi, Eta Kappa Nu. Episcopalian. Clubs: Mass. Inst. Tech. Faculty, Winchester Country, St. Botolph, Harvard. Subspecialties: Acoustics; Acoustical engineering. Current work: Concert hall acoustics; engineering; building acoustics and noise control. Home and Office: 7 Ledgewood Rd Winchester MA 01890

BERARDI, MATTEO P., mech. engr., educator; b. Milford, Mass., July 6, 1935; s. Matteo and Maria (Fantini) B.; m. Marilyn J. Bevere, June 24, 1935; children: Lori A, Susan M. B.S., Northeastern U., 1960, M.S. in Mech. Engring, 1962. Registered profl. engr., Mass.; real estate broker, Mass.; Notary public. Engr. Am. Sci, & Engring., Inc., Cambridge, Mass., 1962-64; supr. Avco Corp., Wilmington, Mass., 1964-67; engring, mgr. Indsl. Magnetics, Inc., Canton, Mass., 1967-71; asst. chief engr. Stone & Webster Engring., Inc., Boston, 1971—; sr. lectr. Lincoln Coll., Northeastern U., Boston, 1960—. Contbr. articles to profl. jours. Mem. ASME. Roman Catholic. Subspecialties: Metallurgical engineering; Metallurgy. Current work: Design, construction and operation of power and processing plants. Home: 15 Putney Ln Lynnfield MA 01940 Office: 245 Summer St Boston MA 02107

BERCZI, ISTVAN, immunologist, educator; b. Hungary, Nov. 12, 1938; emigrated to Can., 1967, naturalized, 1974; s. Istvan and Ilona (Olah) B.; m. Anna Kovacs, Sept. 3, 1967; children: Stephen, Antony, Anna. D.V.M., Vet. Sch., Budapest, Hungary, 1962; Ph.D. in Immunology, Faculty Medicine, U. Man. (Can), Winnipeg, 1972. Research scientist Hungarian Acad. Scis., Budapest, 1962-67; lectr. dept. immunology Faculty Medicine, U. Man., 1972-74, asst. prof. immunology, 1974-82, assoc. prof., 1982—. Contbr. numerous articles to profl. jours. Grantee Med. Research Council, 1972-82, Nat. Cancer Inst., 1973-80, Heart Found., 1975-77, Arthritis Soc., 1982-84. Mem. Can. Soc. Immunology, Am. Assn. Immunologists, Am. Assn. Cancer Research, Transplantation Soc., N.Y. Acad. Scis. Subspecialty: Immunobiology and immunology. Current work: Immunology, immunopathology, immunoregulation. Home: 62 St Michael Rd Winnipeg MB Canada R2M 2K6 Office: 795 McDermot Ave Winnipeg MB Canada R3E OW3

BERELOWITZ, MICHAEL, endocrinologist, researcher, educator; b. Cape Town, South Africa, June 27, 1944; came to U.S., 1977, naturalized, 1982; s. Alec and Doris (Hirsch) B.; m. Merle Finkenstein, Dec. 22, 1968; children: Niki Michelle, Lael. M.B. Ch.B., U. Cape Town, 1968. Diplomate: Am. Bd. Internal Medicine. Intern Groote Schuur Hosp., Cape Town, 1969, resident in pathology, 1970, resident in nuclear medicine, 1971, resident in internal medicine, 1972-75, fellow in endocrinology, 1975-77; jr. lectr. U. Cape Town, 1977; cons. Groote Schuur Hosp., Cape Town, 1977; attending physician Michael Reese Hosp., Chgo., 1978-81; instr. U. Chgo., 1978-79, asst. prof. internal medicine, 1979-81; assoc. prof. U. Cin., 1981—. Contbr. articles to med. and sci. jours. Recipient Robert M. Kark award Chgo. Soc. Internal Medicine, 1980; Career Develop. award Juvenile Diabetes Found., 1981; Charelick Saloman Trust scholar, 1961. Fellow A.C.P., Coll. Physicians (South Africa); mem. Endocrine Soc., Am. Diabetes Assn., Am. Fedn. Clin. Research, Central Soc. Clin. Research, Internat. Soc. Neuroendocrinology. Jewish. Subspecialties: Internal medicine; Neuroendocrinology. Current work: Physiology of somatostatin, hypothalamic regulation of the pituitary, neuroendocrinology of diabetes and obesity; clinical neuroendocrinology. Home: 9599 Heather Ct Cincinnati OH 45242 Office: U Cin Coll Medicine 231 Bethesda Ave Cincinnati OH 45267

BERENDS, LAWRENCE KEITH, marine scientist, educator; b. Seattle, May 19, 1932; s. David P. and Mable Christine (Todd) B.; m. Claudia Widen, July 19, 1956; children: Kimberly, Sharon, Jeffrey, Stuart. B.A., Wash. State U., 1953; M.S., U. So. Calif., 1955, Ph.D., 1958. Assoc. prof. marine biology U. So. Calif., Los Angeles, 1967-70, prof., 1971—; vis. prof. U. Alaska, Fairbanks, 1970-71; cons. NSF. Contbr. articles to profl. jours. Trustee Los Angeles County Mus. Art. Grantee Nat. Acad. Scis., 1981-83. Mem. AAAS. Subspecialty: marine biology. Office: Werik Lab 1244 S Grand Ave Los Angeles CA 90015

BERG, BENGT HENRIK, steel company executive; b. Stockholm, Mar. 13, 1931; s. Henrik and Elsa (Johansson) B.; m. Maud Margareta Lindbert, Nov. 9, 1957; children: Lena, Anders, Henrik. B.S. in Metall. Engring, Royal Inst. Tech., Stockholm, 1968, Ph.D. in Mettalurgy. Research engr. Sandvik AB, Sandviken, Sweden, 1961-67; mgr. process engring. Sandvik Spl. Metals Corp., Kennewick, Wash., 1967-71; tech. mgr. Sandvik Pacific, Santa Ana, Calif., 1971-74; quality assurance mgr. Sandvik, Inc., Scranton, Pa., 1974-82, mgr. metallurgy and product quality, 1982—; instr. metallurgy Pa. State U. Mem. Metall. Soc. of AIME, Am. Soc. Metals, ASME, Am. Soc. Quality Control. Lodge: Lions. 1982-83. Subspecialties: Metallurgical engineering; Alloys. Home: 511 Old Colony Rd Clarks Summit PA 18411 Office: Po Box 1220 Scranton PA 18501

BERG, CLAIRE M., genetics educator; b. Mt. Vernon, N.Y., Apr. 24, 1937. B.S., Cornell U., 1959; M.S., U. Chgo., 1962; Ph.D., Columbia U., 1966. Predoctoral fellow Oak Ridge Nat. Lab., 1964-66; postdoctoral fellow Hammersmith Hosp., London, 1966-67, U. Geneva, 1967-69; asst. prof. biology U. Conn., 1969-73, assoc. prof., 1973-82, prof., 1982—. Mem. Am. Soc. Microbiology (chmn. genetics and molecular biology sect. 1980-81), AAAS (council del. biol. scis. 1983-86, com. on council affairs 1983-85), Genetics Soc. Am., Am. Women in Sci., Sigma Xi. Subspecialty: Genetics and genetic engineering (biology). Office: U Conn Box U-131 Storrs CT 06368

BERG, DOUGLAS E., microbial geneticist, educator; b. Mt. Vernon, N.Y., Sept. 21, 1943. B.S., Cornell U., 1964; Ph.D. in Genetics, U. Wash., Seattle, 1969. Teachine asst. in genetics Cornell U., Ithaca, N.Y., 1964; grad. fellow. teaching asst. in genetics U. Wash., Seattle, 1964-69; USPHS postdoctoral fellow dept. biochemistry Stanford (Calif.) U. Med. Sch., 1969-71; research assoc. dept. molecular biology U. Geneva, Switzerland, 1971-75, lectr., genetics, 1973; research assoc. dept. biochemistry U. Wis., Madison, 1975-77; asst. prof. dept. microbiology and immunology, also dept. genetics Washington U. Sch. Medicine, St. Louis, 1977-81, assoc. prof., 1981—; Charge de recherche, 1971-75. Editorial bds.: Jour. Bacteriology, Plasmid; contbr. writings to profl. publs. Mem. Genetics Soc. Am., Am. Soc. Microbiology, AAAS. Subspecialties: Genetics and genetic engineering (biology); Molecular biology. Office: Dept Microbiology and Immunology Washington Univ Sch Medicine Saint Louis MO 63110

BERG, EDWARD, geophysicist, educator; b. Trier, Mosel, Germany, Nov. 9, 1928; s. Hatthias and Maria (Gerner) B.; m. Francoise Berg, May 30, 1955; children: Christophe, Frederic. Mathematique/ Generales, U. Rennes, France, 1949; Staats Examen, U. Saarlandes, Saarbrucken, W. Ger., 1953, Diplom Physiker, 1953, Dr. rer. nat., 1955. Instr. Physikalisches Institut, U. Saarlandes, 1954-55; chercheur Institut pour la Recherche Scientifique en Afrique Centrale, Lwiro, D.S. Bukavu, Congo, 1955-59; head seismic and volcanic dept. time service IRSAC, 1959-63; instr. U. Bonn, W. Ger., 1957, U. Calif.-Berkeley, 1961; tchr. physics UN in the Congo, Intitut Nat. des Mines, Bukavu, 1962-63; assoc. prof. geophysics Geophys. Inst., U. Alaska, 1963-67, prof., 1967-72; vis. prof. geophysics Hawaii Inst. Geophysics, U. Hawaii, Honolulu, 1971-72, geophysicist, prof. geophysics, 1972—, acting dir., 1980-81, acting chmn. dept. geology and geophysics, 1981, chmn. dept., 1981—; cons. Dames & Moore, San Francisco, 1969-70; mem. com. on Alaska earthquake, seismological panel Nat. Acad. Scis. Recipient Harry Oscar Wood award Carnegie Instn., Washington, 1963. Mem. Am. Geophys. Union, Seismological Soc. Am., AAAS. Club: Rotary. Subspecialty: Seismology. Current work: Earthquake seismology, coastal deformation. Home: 661 N Kainalu Dr Kaulua HI 96734 Office: Dept Geology and Geophysics U Hawaii 2525 Correa Rd Honolulu HI 96822

BERG, HENRY CLAY, research geologist; b. N.Y.C., Apr. 23, 1929; s. Isidore and Freda (Gottschalk) B.; m. Imogen Jocelyn Siff, Oct. 20, 1951 (div. 1981); 1 dau., Jadine Jocelyn; m. Judith Harriet Graham, June 20, 1982. B.S., Bklyn. Coll., 1951; A.M., Harvard U., 1956. Research geologist Alaskan Geol. Br. U.S. Geol. Survey, Menlo Park, Calif., 1956-82, chief tech. reports unit geol. div., 1965-67, mgr. Alaska Mineral resource assessment program, 1974-80, geologist-in-charge, 1980-82, research geologist, Anchorage, 1982—. Served with U.S. Army, 1952-54. C. T. Broderick scholar Harvard U., 1956. Fellow Geol. Soc. Am., Geol. Assn. Can.; mem. AAAS. Democrat. Jewish. Subspecialties: Geology; Tectonics. Current work: Developing a new model for origin and distribution of mineral deposits in southeastern Alaska based on concept of accretionary tectonics, and to apply this new metallogenic model to a predictive mineral resource appraisal of this remote, geologically complex, and mineral-rich region. Home: 1308 Atkinson Dr Anchorage AK 99504 Office: Alaskan Geology Br US Geol Survey 4200 University Dr Anchorage AK 99508

BERG, HOWARD CURTIS, biology educator; b. Iowa City, Iowa, Mar. 16, 1934; s. Clarence P. and Esther M. (Carlson) B.; m. Mary E. Guyer, Dec. 19, 1964; children: Henry, Alexander, Elena. B.S., Calif. Inst. Tech., 1956; A.M., Harvard U., 1960, Ph.D., 1964. Fulbright fellow Carlsberg Lab., Copenhagen, 1956-57; nat. scholar Harvard Med. Sch., Boston, 1957-59; jr. fellow Harvard Soc. Fellows, Cambridge, Mass., 1963-66; asst. prof., assoc. prof., chmn. bd. tutors in biochem. scis. dept. biochemistry and molecular biology Harvard U., 1966-70; assoc. prof., prof. dept. molecular, cellular and developmental biology U. Colo., Boulder, 1970-79; prof. div. biology Calif. Inst. Tech., Pasadena, 1979—. Recipient Grants NIH, NSF, Am. Heart Assn., Research Corp., 1966—. Mem. Am. Phys. Soc., Biophys. Soc., Am. Soc. Biol. Chemists, Am. Soc. Microbiologists, AAAS, N.Y. Acad. Scis. Subspecialties: Biophysics (biology); Microbiology. Current work: Motility and motile behavior of bacteria, chemotaxis in bacteria, behavior of microorganisms, bacterial flagellar rotation. Home: 1401 Crest Dr Altadena CA 91001 Office: Div Biology 216-76 Calif Inst Tech Pasadena CA 91125

BERG, PATRICIA E., molecular biologist; b. Dubuque, Iowa, Sept. 17, 1943; d. Clifford J. and Dorothy R. (McKibben) Emerson; m. Paul S. Lovett, Jan. 5, 1982; 1 dau., Bridget Berg. A.B. (Univ. scholor, Ill. State scholar), U. Chgo., 1965, 1965; Ph.D. (NIH predoctoral fellow), Ill. Inst. Tech., 1973. Postdoctoral fellow U. Chgo., 1973-78; dir. genetic engring. Bethesda (Md.) Research Labs., 1978-80; expert Lab. Molecular Hematology, Nat. Hear, Lung and Blood Inst., NIH, Bethesda, 1980—. Contbr. articles to profl. publs. Mem. AAAS, Am. Soc. Microbiology, Sigma Xi. Club: Md. Masters Swim Team. Subspecialties: Genetics and genetic engineering (biology); Molecular biology. Current work: Molecular cloning with goal of gene therapy; studying expression of regulated eucaryotic gene using cloning and transfer into cells in culture and into mice. Office: Bldg 10 Room 7D-18 NIH Bethesda MD 20205

BERG, PAUL, educator, biochemist; b. N.Y.C., June 30, 1926; s. Harry and Sarah (Brodsky) B.; m. Mildred Levy, Sept. 13, 1947; 1 son, John. B.S., Pa. State U., 1948; Ph.D. (NIH fellow 1950-52), Western Res. U., 1952; D.Sc. (hon.), U. Rochester, 1978, Yale U., 1978. Postdoctoral fellow Copenhagen (Denmark) U., 1952-53; postdoctoral fellow Sch. Medicine, Washington U., St. Louis, 1953-54, Am. Cancer Soc. scholar cancer research dept. microbiology, 1954-57, from asst. to asso. prof. microbiology, 1955-59; prof. biochemistry Stanford Sch. Medicine, 1959—, Sam, Lula and Jack Willson prof. biochemistry, 1970, chmn. dept., 1969-74; non-resident fellow Salk Inst., 1973; lectr. Weizmann Inst., 1977; disting. lectr. U. Pitts., 1978; Priestly lectr. Pa. State U., 1978; Shell lectr. U. Calif., Davis, 1978; adv. bd. NIH, NSF, M.I.T.; vis. com. dept. biochemistry and molecular biology Harvard U. Contbr. profl. jours.; Editor: Biochem. and Biophys. Research Communications, 1959-68; editorial bd.: Molecular Biology, 1966-69. Served to lt. (j.g.) USNR, 1943-46. Recipient Eli Lilly prize biochemistry, 1959; V.D. Mattia award Roche Inst. Molecular Biology, 1972; Henry J. Kaiser award for excellence in teaching, 1972; Disting. Alumnus award, Pa. State U., 1972; Sarasota Med. awards for achievement and excellence, 1979; Gairdner Found. annual award, 1980; Lasker Found. award, 1980; Nobel award in chemistry, 1980; Sci. Freedom and Responsibility award AAAS, 1982; named Calif. Scientist of Yr. Calif. Museum Sci. and Industry, 1963; Harvey lectr., 1972; Lynen lectr., 1977. Mem. Inst. Medicine, Nat. Acad. Scis. (council 1979), Am. Acad. Arts and Scis., Am. Soc. Biol. Chemists (pres. 1974-75), Am. Soc. Microbiology. Subspecialties: Biophysical chemistry; Biochemistry (biology). Home: 838 Santa Fe Ave Stanford CA 94305 Office: Stanford Sch Medicine Stanford CA 94305

BERG, PAUL CONRAD, educator, educational consultant; b. Hallstead, Pa., July 21, 1921; m. Rosalie Elizabeth Helm, Jan. 26, 1945; Mary Lucetta, Doris Roxanna, Ruth Elizabeth. B.A., Syracuse U., 1949; M.S., Cornell U., 1950, Ph.D., 1953. Assoc. clin. dir. Reading Lab., U. Fla., Gainesville, 1952-56; assoc. prof. edn. SUNY, Fredonia, 1956-57; prof. edn. U. S.C., Columbia, 1957—, John E. Swearingen prof., 1980—; cons. Ford Found. Author: (with others) Art of Efficient Reading, 4th edit, 1983, Skimming and Scanning, 1966; contbr. articles to profl. jours. Tchr. Sunday sch. 1st Baptist Ch., Columbia, 1972—; pres. Wardlaw Club, Columbia, 1963. Served with USAAF, 1942-46; PTO. Recipient Oscar Causey award Nat. Reading Conf., 1978. Mem. Internat. Reading Assn. (dir. 1967-70), Nat. Reading Conf. (pres. 1968), Am. Psychol. Assn. Republican. Subspecialty: Learning. Current work: Work with learning disabled children with reading problems. Home: 6033 Poplar Ridge Rd Columbia SC 29206 Office: Univ of South Carolina Columbia SC 29208

BERG, RAISSA LVOVNA, genetics researcher, cons.; b. St. Petersburg (now Leningrad), Russia, Mar. 27, 1913; d. Lev and Paulina (Katlovker) B.; m. Valentin S. Kirpitshnikov, Dec. 20, 1945 (div.); children: Elisabeth, Maria. Diploma, Leningrad U., 1935, Ph.D., 1939; D.Sc., Inst. Cytology and Genetics, Acad. Sci., USSR, 1964. Sr. scientist Inst. Animal Morphology Acad. Sci. USSR, Moscow, 1939-47; assoc. prof. Leningrad U., 1954-63; head Lab. Population Genetics, Acad Sci. USSR, Novosibirsk, 1963-68, Inst. Agrophysics, Acad. Agrl. Sci., Leningrad, 1968-70; prof. Hertsen Pedagogical Inst., Leningrad, 1968-74; assoc. scientist U. Wis., Madison, 1976-81; vis. prof. Washington U., St. Louis, 1981—; cons. in field. Contbr. articles to profl. jours. Mem. Genetics Soc. Am.; mem. Genetics Soc. Japan; Mem. AAAS, Nat. Geog. Soc. Subspecialties: Genetics and genetic engineering (biology); Plant genetics. Current work: Geography of the spontaneous mutation process in Drosophila populations; mechanisms of mutator gene action; history of genetics USSR. Home: 5696 Kingsbury 403 St Louis MO 63112 Office: Dept Genetics Box 8031 Washington U Sch Medicine Saint Louis MO 63110

BERG, RICHARD ALAN, clinical neurophychologist; b. Albany, N.Y., Mar. 30, 1953; s. Gunter Alfred and Sylvia (Falkow) B. B.A., SUNY-Buffalo, 1974; M.A., U. Houston, 1976, Ph.D., 1980. Adj. faculty social scis. Houston Community Coll., 1971-77; postdoctoral fellow neuropsychology Nebr. Psychiat. Inst., Omaha, 1979-80; instr. psychiatry U. Tenn. Center for Health Scis., Memphis, 1980—; adj. faculty psychology Memphis State U., 1980—; clin. neuropsychologist St. Jude Children's Research Hosp., Memphis, 1980—; cons. Crisis Center, Memphis, 1981—, Fed. Corrections Inst., 1981—. Co-author: Interpretation of Halstead Reitan Neuropsychological Battery, 1981, (with Charles Golden) Interpretation of Luria-Nebraska Neuropsychological Battery, 1982. Mem. Am. Psychol. Assn., Nat. Acad. Neuropsychologists, Internat. Neuropsychol. Soc., Biofeedback Soc. Tenn. Subspecialties: Neuropsychology; Pediatric neuropsychology. Current work: Neuropsychological effects of leukemia and its treatment. Development of rehabilitation strategies for cerebral dysfunction. Office: St Jude Children's Research Hosp 332 N Lauderdale St Memphis TN 38101

BERG, RICHARD ALAN, biochemist, educator; b. Spokane, Wash., Apr. 17, 1945; s. Norris Herbert and Leta Kathern (Peterson) B.; m. Samantha Francesca Curran, Jan. 5, 1981. B.S., U. Chgo., 1967; Ph.D., U. Pa., 1972. Asst. prof. U. Medicine and Dentistry N.J., Piscataway, 1973-81, assoc. prof. biochemistry, 1981—; staff investigator NIH, Bethesda, Md., 1978-80. Recipient Sinsheimer Fund award, 1976-79. Mem. Am. Soc. Biol. Chemists, Am. Fedn. Clin. Research, Am. Chem. Soc., AAAS, Soc. Exptl. Biology and Medicine (editorial bd. procs.). Subspecialties: Biochemistry (biology); Cell and tissue culture. Current work: extracellular matrix; expression of genes for connective tissue proteins; lung injury, inflammation and fibrosis; analytical biochemistry; high pressure liquid chromatography. Home: Route 1 Box 148 Lambertville NJ 08530 Office: U Medicine and Dentistry of NJ Hoes Ln Piscataway NJ 08530

BERGER, ALAN ERIC, mathematician; b. Plainfield, N.J., Sept. 4, 1946; s. Walter H. and Shirley (Schwartz) B. A.B., Rutgers U., 1968; M.S., MIT, 1969, Ph.D., 1972. Research mathematician Naval Surface Weapons Center, Silver Spring, Md., 1972—. Mem. Am. Math. Soc., Soc. Indsl. and Applied Math, AAAS, Am. Meteorl. Soc., Math. Assn. Am. Subspecialties: Numerical analysis; Applied mathematics. Current work: Numerical solution of differential equations. Office: Applied Math Br Naval Surface Weapons Center Silver Spring MD 20910

BERGER, AUDREY MARILYN, psychologist; b. Bklyn., Nov. 2, 1955; d. Alexander and Elaine (Kosloff) B. B.A., SUNY, Binghamton, 1976; M.A., U. Iowa, 1978, Ph.D., 1981. Lic. psychologist, N.Y. Psychology intern U. Rochester Med. Ctr., 1980-81; psychologist Rochester Psychiat. Ctr., 1981—. Contbr. chpts to books, articles to profl. jours. Mem. Am Psychol. Assn., Amnesty Internat., Phi Beta Kappa. Subspecialty: Behavioral psychology. Current work: Human aggression, child abuse. Office: Rochester Psychiatric Center 1600 South Ave Rochester NY 14620

BERGER, EDWARD MICHAEL, geneticist; b. N.Y.C., May 2, 1944; s. Benjamin and Sarah (Handelman) B.; m. Barbara Fritsche, Dec. 6, 1981; m. Deirdre Wallace, Dec. 16, 1968; children: Alexander, Nicholas, Tanya, Matthew. B.A., Hunter Coll., 1965; M.S., Syracuse U., 1967, Ph.D., 1969. USPHS postdoctoral fellow Biol. Labs., Harvard U., 1969-70; dept. biology U. Chgo., 1970-71; asst. prof. biology SUNY, Albany, 1971-75; asst. prof. Dartmouth Coll., Hanover, N.H., 1975-77, assoc. prof., 1977—. Contbr. articles to profl. jours. Embo fellow, 1975, 81; NSF grantee, 1975-77; NIH grantee, 1971—. Mem. Genetics Soc. Am. Subspecialties: Gene actions; Genetics and genetic engineering (biology). Current work: Regulation of gene expression in eucryotic cells. Office: Dept Biology Dartmouth Coll Hanover NH 03755

BERGER, HARVEY JAMES, physician, medical scientist, cardiac radiologist; b. N.Y.C., June 6, 1950; s. Howard H. and Edith (Muskat) B.; m. Wendy S. Wolk, May 16, 1976; 1 son: Eric Michael. A.B., Colgate U., 1972; M.D., Yale U., 1977. Diplomate: Am. Bd. Nuclear Medicine, Nat. Bd. Med. Examiners. Resident radiology, nuclear medicine Yale-New Haven Hosp., 1977-81; asst. prof. diagnostic radiology and medicine, dir. cardiovascular imaging Yale U. Sch. Medicine, 1981—. Contbr. numerous articles to profl. jours., chpts. to books. Recipient Marc Tetalman Research award Soc. Nuclear Medicine, 1981; Meml. award Assn. U. Radiologists, 1979. Fellow Am. Coll. Cardiology, Am. Coll. Chest Physicians, Council Cardiovascular Radiology, Am. Heart Assn.; mem. N. Am. Soc. Cardiac Radiology, Am. Physiol. Soc., Am. Fedn. Clin. Research, Am. Coll. Radiology, Am. Roentgen Ray Soc., N.Y. Acad. Sci., Soc. Thoracic Radiology, Sigma Xi. Clubs: Yale of N.Y., Mary's Assn. (New Haven). Subspecialties: Radiology; Cardiology. Current work: Cardiovascular imaging: nuclear, NMR, digital angiography, ultrasound, image processing. Myocardial ischemia, contractivity. Office: Yale U Sch Medicine 333 Cedar St New Haven CT 06510

BERGER, HENRY, psychiatrist, management consultant; b. Marburg, Germany, Jan. 17, 1947; came to U.S., 1949; s. Oscar and Rose (Engelstein) B. B.A., Columbia U., 1968, M.S., 1970, Ph.D., 1973; M.D., Dartmouth Coll., 1978. Instr. Columbia U., N.Y.C., 1972-73; psychiatrist Dartmouth Coll. and Med. Ctr., Hanover, N.H., 1978-80; clin. affiliate N.Y. Hosp., N.Y.C., 1980—; dir. gerontology unit Gracie Sq. Hosp., N.Y.C., 1982—; instr. Cornell U. Med. Coll., N.Y.C., 1982—; pres. FAMBUS Mgmt. Cons., N.Y.C., 1982—; chmn. adv. bd. Inst. for Health-Weight Scis., N.Y.C. Contbr. articles to profl. lit. Recipient George C. Curtis Award Columbia U., 1968; fellow NIMH, 1968-72, Nat. Council on Alcoholism, 1977. Mem. Am. Psychol. Assn., Am. Psychiat. Assn., Am. Mgmt. Assn., Sigma Xi. Subspecialties: Psychiatry; Neuropsychology. Current work: Interplays between psychiatric and medical illness and between emotional and managerial issues; nonverbal behavior; thanatology; gerontology. Home: 8 Hastings House Hastings-on-Hudson NY 10706 Office: 65 Court St White Plains NY 10601

BERGER, PHILIP JEFFREY, educator; b. Newark, June 28, 1943; s. Philip Graham and Jean Bar (Weller) B.; m. Frances Ann Berger, Jan. 9, 1942; children: Sarah Katherine, Philip Calvin. B.S., Delaware Valley Coll., 1965; M.S., Ohio State U., 1967, Ph.D., 1970. Asst. prof. animal sci. Iowa State U., 1972-78, assoc. prof., 1978-82, prof., 1982—; computer cons. animal prodn. div. FAO, UN, Rome; vis. coop. scientist BARD project, Bet Dagan, Israel. Contbr. to profl. jours. Mem. Am. Dairy Sci. Assn., Am. Soc. Animal Sci., Biometrics Soc., Sigma Xi, Delta Tau Alpha, Gamma Sigma Delta. Republican. Methodist. Subspecialties: Animal breeding and embryo transplants; Statistics. Current work: Research directed toward developing mixed model techniques for sire and cow evaluation; teaching computer techniques for biol. research; population dynamics in animal breeding; research sire and cow evaluation for production traits; breeding program devel.

BERGER, SEYMOUR MAURICE, psychology educator; b. Bklyn., Jan. 7, 1928; s. Leo and Bessie Ida (Okun) B.; m. Sara Marilyn (Nappen); children: Evelyn Joyce, Nancy Faith. B.A., Okla. A&M Coll., 1949; M.A., Columbia Coll., 1950; Ph.D., Cornell U., 1959. Instr. Trinity Coll., 1958-59; instr. Ind. U., 1959-62, asst. prof. psychology, 1962-66, assoc. prof. psychology, 1966-69; prof. psychology U. Mass.-Amherst, 1969—. Contbr. articles to profl. jours., chpts. to books. Served with USNR, 1945-46; Served with USAF, 52-55. NIH Research fellow, 1965-66; Fulbright Sr. research fellow, 1975-76; Research Travel fellow, 1983. Fellow Am. Psychol. Assn.; mem. Eastern Psychol. Assn., Midwestern Psychol. Assn. (program chmn. 1969), Soc. Exptl. Social Psychologists. Subspecialty: Social psychology. Current work: Social learning: motoric and symbolic processes in learning, memory, communication. Home: 459 Flat Hills Rd Amherst MA 01002 Office: Dept Psychology U Mass Amherst MA 01003

BERGER, THEODORE WILLIAM, neuroscientist; b. Lafayette, Ind., June 11, 1950; s. Arvid William and Marian Hildegard (Beyer) B.; m. Terry Lynn Berger, May 20, 1972. B.S. summa cum laude, Union Coll., Schenectady, 1972; Ph.D., Harvard U., 1976. Postdoctoral research assoc. dept. psychobiology U. Calif., Irvine, 1976-77; asst. research psychobiologist, 1977-78; Alfred P. Sloan Found. fellow Salk Inst. Biol. Studies, LaJolla, 1978-79; asst. prof. psychology and psychiatry U. Pitts., 1979-81, assoc. prof., 1982—. Contbr. articles to profl. jours. Recipient James McKeen Cattell award, 1978, Alfred P. Sloan Found. fellow, 1978, McKnight Found. scholar award, 1980; NIMH research career devel. awardee, 1981. Mem. Soc. Neuroscience, N.Y. Acad. Scis., Psychonomic Soc., AAAS, Phi Beta Kappa, Sigma Xi. Subspecialties: Neurobiology; Neurophysiology. Current work: Neurobiological analysis of associative learning research, neuroscience, electrophysiology, neuroanatomy. Office: Dept Psycholog Univ Pitts Pittsburgh PA 15260

BERGERON, CLIFTON GEORGE, ceramic engr., educator; b. Los Angeles, Jan. 5, 1925; s. Lewis G. and Rose C. (Dengel) B.; m. Laura H. Kaario, June 9, 1950; children—Ann Leija, Louis Kaario. B.S., U. Ill., 1950, M.S., 1959, Ph.D., 1961. Sr. ceramic engr. A. O. Smith Corp., Milw., 1950-55; staff engr. Whirlpool Corp., St. Joseph, Mich., 1955-57; research asso. U. Ill., Champaign-Urbana, 1957-61, asst. prof., 1961-63, asso. prof., 1963-67, prof., 1967-78, head dept. ceramic engring., 1978—; cons. A. O. Smith Corp., Whirlpool Corp., Ingraham Richardson, U.S. Steel Corp., Pfaudler Corp., Ferro Corp. Editor, Ann. Conf. on Glass Problems. Served in U.S. Army, 1943-46; ETO. NSF grantee, 1961—. Fellow Am. Ceramic Soc.; mem. AAAS, Nat. Inst. Ceramic Engrs., AAUP, KERAMOS, Am. Soc. Engring. Edn., Sigma Xi. Subspecialties: Ceramic engineering; Ceramics. Research in crystallization kinetics in glass; high temperature reactions. Invented high temperature catalytic coatings for oxidation. Home: 208 W Michigan St Urbana IL 61801 Office: 105 S Goodwin St Urbana IL 61801

BERGESON, HAVEN ELDRED, high energy physicist; b. Logan, Utah, Dec. 22, 1933; m., 1957; children: 6. B.S., U. Utah, 1958, Ph.D., 1962. Physicist Space Sci. Lab., Gen. Electric Co., 1961-63; asst. prof. physics U. Utah, Salt Lake City, 1963-64, asst. research prof., 1964-68, assoc. prof., 1968-75, prof. physics, 1975—, chmn. dept., 1976-80. Mem. Am. Phys. Soc. Subspecialty: High energy physics. Office: Dept Physics U Utah Salt Lake City UT 84112

BERGEY, GREGORY KENT, neurologist, neuroscientist, internist; b. Bryn Mawr, Pa., Nov. 9, 1949; s. Robert Harr and Kathryn Agnes (Schmidt) B.; m. Stefanie Friday Antonakos, Aug. 27, 1972; children: Alyssa Noelle, Alexander Christian. A.B. in Biology, Princeton U., 1971; M.D., U. Pa., 1975. Diplomate: Am. Bd. Internal Medicine. Intern Yale-New Haven Hosp., 1975-76, resident in internal medicine, 1975-77; research assoc. Lab. Devel. Neurobiology, Nat. Inst. Child Health and Human Devel., NIH, Bethesda, Md., 1977-79, 81-82; resident in neurology, fellow neurology dept. Johns Hopkins U. Sch. Medicine and Hosp., Balt., 1979-83; asst. prof. neurology and physiology U. Md. Sch. Medicine, Balt., 1983—. Served with USPHS, 1977-79, 81-82. Mem. Soc. Neurosci., Am. Soc. Neurol. Investigation, ACP, Am. Acad. Neurology. Subspecialties: Neurophysiology; Neurology. Current work: Developmental neurobiology; neuronal cell culture; cellular mechanisms of convulsant and anticonvulsant action; synaptogenesis. Home: 1219 John St Baltimore MD 21217 Office: Neurology Dept U Md Sch Medicine 22 S Greene St Baltimore MD 21201

BERGMAN, MICHAEL, endocrinologist; b. Berlin, Germany, Oct. 2, 1947; came to U.S., 1950, naturalized, 1957; s. Paul and Yetta (Applbaum) B.; m. Anne Geliebter, Aug. 10, 1980; 1 dau., Elana Amy. B.A., Queens Coll., 1969; M.D., U. Louvain, Belgium, 1976. Diplomate: Am. Bd. Internal Medicine. Resident in internal medicine U. Md., Balt., 1976-79; fellow in endocrinology and metabolism Yale U., 1979-80, Albert Einstein Sch. Medicine, Bronx, N.Y., 1980-81; asst. prof. medicine, dir. diabetes sect. N.Y. Med. Coll., Valhalla, 1981—; dir. diabetes clinic Westchester County Med. Center, Valhalla, 1981—. Author: Diabetes Mellitus: Theory and Practice, 1982. Mem. A.C.P., Am. Diabetes Assn., N.Y. Diabetes Assn, Am. Fedn. Clin. Research. Democrat. Jewish. Subspecialty: Endocrinology. Current work: Effect of intensive control of diabetes. Home: 3801 Hudson Manor Terr Riverdale NY 10463 Office: NY Med Coll Dept Medicine Valhalla NY 10595

BERGMANN, STEVEN ROBERT, cardiovascular physiologist, educator; b. N.Y.C., Feb. 4, 1951; s. Paul and Therese (Greenfield) B.; m. Joanne L. Rubin, Aug. 17, 1975. B.A., George Washington U., 1972; Ph.D., Hahnemann Med. Coll., 1978. Grad. teaching fellow Hahnemann Med. Coll., Phila., 1973-77; fellow Washington U. Sch. Medicine, St. Louis, 1977-80, research instr., 1979-80, asst. prof., 1980—. Arbitrator Better Bus. Bur., St. Louis, 1978—. Mem. Am. Fedn. Clin. Research, Am. Heart Assn., Am. Physiol. Soc., Internat. Soc. for Heart Transplantation, Physiol. Soc. Phila., Omicron Delta Kappa. Jewish. Subspecialties: Cardiology; Nuclear medicine. Current work: Positron emission tomography, thrombolytic therapy, heart transplantation. Office: Washington U Sch Medicine Box 8086 660 S Euclid Ave Saint Louis MO 63110

BERGQUIST, JAMES WILLIAM, computer scientist, mathematician; b. Ottumwa, Iowa, Apr. 23, 1928; s. Albin and Lucille (Morrison) B.; m. Madonna T. Dunham, Sept. 15, 1951; children: Catherine, James, Mary, Brian, Thomas, John, Timothy, Joseph, Ann, Robert, Paul. B.S., Iowa State U., 1950; M.S., U. So. Calif., 1955, Ph.D., 1963. Analog computers Lockheed Aircraft, Burbank, Calif., 1951-54; theoretical analyst Gilfillan Bros. (now ITT), Los Angeles, 1955-57; computer scientist IBM, Los Angeles, 1958—; vis. assoc. Calif. Inst. Tech., Pasadena, 1963-67. Chmn, sch. bd. La Canada (Calif.) Unified Sch. Dist., 1971-81; v.p. Catholic Social Service, Los Angeles, 1980-81. Recipient Excellence award IBM, 1970; Pub. Relations award, 1981. Mem. Am. Math. Soc., Math. Assn. Am., Assn. Computing Machinery, Soc. Indsl. and Applied Math. (pres. 1980-81, exec. com. 1979-82). Roman Catholic. Club: Serra (Pasadena) (v.p. 1981-82). Subspecialties: Programming languages; Cognition. Current work: Application of mathematics and computer science to the understanding of cognition and human consciousness. Home: 4705 Daleridge Rd PO Box 1036 La Canada CA 91011 Office: IBM 3424 Wilshire Blvd Los Angeles CA 90010

BERGSTRESSER, PAUL RICHARD, physician, educator; b. Ottawa, Kans., Aug., 24, 1941; s. Karl S. and May H. (Holmes) B.; m. Rebecca L. Baird, Jan. 4, 1969; children: Daniel, Laura. A.B., Coll. Wooster, Ohio, 1963; M.D., Stanford U., 1968. Diplomate: Am. Bd. Dermatology, 1976. Resident in dermatology Stanford (Calif.) U., 1969-70, U. Miami, Fla., 1972-74; asst. prof. dermatology U. Miami (Fla.) Sch. Medicine, 1975-76; asst. prof. internal medicine dept. medicine U. Tex. Health Sci. Center, Dallas, 1976-80, assoc. prof. dermatology and internal medicine, 1980—, vice-chmn. dept. dermatology, 1982—; attending staff in dermatology Children's Med. Center, Dallas, 1977—, Parkland Meml. Hosp., 1976—. Contbr. articles to profl. jours. Served to maj. M.C. U.S. Army, 1970-72. Fellow Am. Acad. Dermatology, ACP; mem. Am. Assn. Immunologists, AAAS, Am. Burn Assn., Am Fedn. Clin. Research, Am. Soc. Photobiology, Internat. Soc. Tropical Dermatology, Soc. Investigative Dermatology, Dallas Coounty Med. Assn., Dallas Dermatological Soc., Tex. Med Assn., S.W.Immunology Club. Subspecialties: Dermatology; Immunology (medicine). Current work: Dermatology, immunology, Langerhans cells, contact hypersensitivity, transplantation immunology. Office: 5323 Harry Hines Blvd Dallas TX 75235

BERGSTROM, GARY CARLTON, plant pathologist, educator; b. Chgo., May 12, 1953; s. Robert Carlton and Virginia Mae (Jensen) B. B.S. in Microbiology, Purdue U., 1975, M.S. in Plant Pathology, 1978, Ph.D., U. Ky., 1981. Teaching asst. Purdue U., West Lafayette, Ind., 1975-76, research asst., 1976-77, U. Ky., Lexington, 1978-81; asst. prof. plant pathology Cornell U., Ithaca, N.Y., 1981—; extension pathologist diseases of field and forage crops State of N.Y. Contbr. articles to profl. jours. Mem. Am. Phytopath. Soc., Gamma Sigma Delta. Subspecialties: Plant pathology; Integrated pest management. Current work: Integrated pest management of field crops; interactions of plant pathogenic fungi with insects and pathogens; physiology of parasitism. Office: 316 Plant Sci Bldg Cornell U Ithaca NY 14853

BERI, AVINASH CHANDRA, chem. physicist, educator; b. Jullunder, India, Oct. 28, 1949; s. Amrit Lal and Lila (Myer) B. B.Sc. with honors in physics, Delhi U., 1969; M.S. in physics, SUNY, Albany, 1974; Ph.D. in Physics, SUNY, Albany, 1979. Postdoctoral fellow dept. chemistry U. Rochester, 1979-80, research assoc., 1980—. Contbr. articles profl. jours. Mem. Am. Phys. Soc., Am. Assn. Physics Tchrs., Sigma Xi. Subspecialties: Laser-induced chemistry; Condensed matter physics. Current work: Theory of optical, magnetic and hyperfine properties of ionic solids, theory of laser-stimulated processes at a solid surface. Home: 504 Suburban Ct 1 Rochester NY 14620 Office: Dept of Chemistry University of Rochester Rochester NY 14627

BERING, EDGAR ANDREW, III, physics educator; b. N.Y.C., Jan. 9, 1946; div. B.A. cum laude, Harvard U., 1967; Ph.D., U. Calif.-Berkeley, 1974. Research scientist, physicist U. Houston, 1974-75, asst. prof., 1975-81, asso. prof., 1981—; astronomy instr. U. St. Thomas, 1982—. Contbr. numerous articles to profl. jours. Vice pres. for Nordic Recreation U.S. Ski Assn., 1965-79; mem. exec. com. Houston regional group Sierra Club, 1980-82; bd. dirs. Gulf Coast Council Fgn. Affairs, 1982—; chmn. vol. activities Houston Shakespeare Festival, 1982; treas. Festival Angels, Inc., 1983. Recipient Antarctica Service medal NSF, 1981. Mem. AAAS, Am. Geophys. Union., Am. Astron. Soc., N.Y. Acad. Sci., Sigma Xi. Club: Harvard (Houston). Subspecialties: Satellite studies; Planetary science. Current work: Mangetospheric physics, electrodynamics of auroral arc. Office: Dept Physics U Houston Houston TX 77004

BERING, EDGAR ANDREW, JR., neurosurgeon; b. Salt Lake City, Feb. 18, 1917; s. Edgar A. and Ilsa Louise (Billing) B.; m. Harriet Aldrich, Nov. 5, 1944; children: Edgar A., Charles C., Harriet Bering Hoder. A.B., U. Utah, 1927; M.D., Harvard U., 1941. Diplomate: Am. Bd. Neurol. Surgery, Am. Bd. Electroencephalpgraphy. Surg. house officer Boston City Hosp., 1941-42; spl. research assoc. dept. phys. chemistry Harvard U. Med. Sch., 1942; asst. in neurosurgery N.Y. Med. Coll., Flower Fifth Ave Hosp., N.Y.C., 1946-48; Mosely Traveling fellow Harvard U. Med. Sch., Nat. Hosp., London, 1948-49; also clin. clk.; resident in neurosurgery Children's Hosp. and Peter Bent Brigham Hosp., Boston, 1949-50; Harvey Cushing fellow Peter Bent Brigham Hosp., 1950-51; practice medicine specializing in neurosurgery, Boston, 1951-64, Easton, Md., 1973—; dir Neurosurg. Research Lab., Children's Hosp. Med. Center, 1952-63, assoc. neurosurgeon, 1955-64; sr. fellow in poliomyelitis NRC, 1951-52; cons. in surgery of nervous system Lemuel Shattuck Hosp., Jamaica Plain, Mass., 1953-55, sr. cons., 1955-63; clin. assoc. in surgery Harvard U. Med. Sch., 1956-59, asst. clin. prof., 1959-65; attending neurosurgeon West Roxbury (Mass.) VA Hosp., 1954-63; vis. lectr. UCLA Med. Sch., 1958; mem. Conf. on Computer Techniques for Biol. Scientists, M.I.T., 1962; vis. scientist Nat. Inst. Neurol. Disease and Blindness, NIH, 1963-65; spl. asst. to dir. for program analysis Nat. Inst. Neurol. Disease and Stroke, 1965-71, chief spl. programs, 1971-74, spl. cons., 1974—; cons. to adv. com. om coagulation components Commn. on Plasma Fractionation and Related Products, 1954; assoc. clin. prof. neurol. surgery Georgetown U. Med. Sch., Washington, 1968—; mem. staff Meml. Hosp., Easton, Md., 1974—, vice chief of staff, 1980-83, chief staff, 1983—; cons. Eastern Shore Hosp. Ctr., Cambridge, Md., 1975—, Johns Hopkins Hosp., Balt., 1975—. Contbr. numerous articles on neurosurgery to profl. jours. Served to comdr. USN, 1942-46. Mem. Am. Acad. of Neurology, AAAS, Am. Assn. for Neurol. Surgeons, Soc. for Neurosci. (founding mem.), Internat. Soc. for Pediatric Neurosurgery (founding mem.), Soc. of Neurology, Psychiatry and Neurosurgery (Argentina), Chilean Soc. for Neurosurgery and Neurology, D.C. Med. Soc., Neurosurg. Soc. of D.C., Neurosurg. Soc. Am., New Eng. Neurosurg. Soc., N.Y. Acad. Scis., Royal Soc. Medicine (London), Scandinavian Neurosurg. Soc., Research Soc. of Neurosurg. Surgeons (founding mem.), Am. Assn. for Neurol. Surgeons (founding mem. Pediatric sect.), Talbot County Med. Soc., Soc. for Neurosci., Md. Neurosurg. Soc., Sigma Xi. Subspecialty: Neurosurgery. Patentee fibrim foam. Home: Creek House Oxford MD 21654 Office: 4 Talbottown Ln Easton MD 21601

BERJIAN, RICHARD A., surgical oncologist; b. N.Y.C., Nov. 11, 1929; s. Parker S. and Elizabeth (Tashjian) B.; m. Sally S. Aljian, Aug. 21, 1960; children: Janice, Leslie, Stephanie. B.S., Ursinus Coll., 1951; D.O., Kirksville Coll. Osteopathy and Surgery, 1955. Diplomate: Am. Bd. Osteo. Surgery. Sr. cancer research surgeon Roswell Park Meml. Inst., Buffalo, 1976—; assoc. prof., chmn. dept. surgery Sch. Osteo. Medicine U. Medicine and Dentistry of N.J., Camden, N.J. Contbr. articles to profl. jours. Mem. Am. Soc. Clin. Oncology. Subspecialties: Oncology; Osteopathy. Current work: Metabolic pathway-melanin synthesis. Carcinoma of gastrointestinal tract. Liver-metastatic disease. Office: 300 Broadway Camden NJ 08103

BERKELHAMMER, JANE, immunobiologist; b. Newburyport, Mass., Apr., 13, 1946; d. David and Sarah Pearl (Barth) B. B.S., Simmons Coll., 1968; M.S., Purdue U., 1970; Ph.D., 1972. Postdoctoral fellow, research assoc. Inst. Cancer Research, Fox Chase, Phila., 1972-75; staff scientist II Frederick (Md.) Cancer Research Ctr., 1975-77; asst. scientist Cancer Research Ctr. and asst. prof. biol. sci. U. Mo., Columbia, 1977-82; asst. prof. depts. medicine and microbiology U. Mo. Sch. Medicine, Columbia, 1982-83; scientist AMC Cancer Research Ctr., Lakewood, Colo., 1983—. Contbr. articles to profl. jours. Nat. Cancer Inst. grantee. Mem. Am. Assn. Cancer Research, Am. Assn. Immunologists, Tissue Culture Assn., AAAS, Sigma Xi. Jewish. Subspecialties: Cancer research (medicine); Immunology (medicine). Current work: Tumor Immunobiology, host response to melanoma; investigation and development of metastasis and therapy of malignant melanoma in various experimental systems. Home: 3465 Mirage Dr Colorado CO 80918 Office: AMC Cancer Research Ctr 6401 W Colfare Ave Lakewood CO 80214

BERKEY, DENNIS DALE, mathematician, educator; b. Wooster, Ohio, May 27, 1947; s. William Bruce and Mary Louise (Schrock) B. B.A., Muskingum Coll., New Concord, Ohio, 1969; M.A., Miami U., Oxford, Ohio, 1971; Ph.D., U. Cin., 1974. Instr. math. Miami U., 1973-74; asst. prof. math. Boston U., 1974-78, assoc. prof. math., 1978—, chmn. dept. math., 1979-83, assoc. v.p. acad. affairs, 1983—. Danforth found. faculty assoc., 1979; recipient Disting. Alumnus award Kappa Mu Epsilon, 1981. Mem. Am. Math. Soc., Math. Assn. Am., Soc. for Indsl. and Applied Math., Assn. for Computing Machinery. Subspecialty: Applied mathematics. Current work: Research in differential equations and control. Managed the development of computer science programs at Boston University. Curriculum development. Home: 71 Arlo Rd Newton MA 02161 Office: Dept Math Boston U 264 Bay State Rd Boston MA 02215

BERKLAND, JAMES OMER, geologist; b. Glendale, Calif., July 31, 1930; s. Joseph Omer and Gertrude Madelyn (Thompson) B.; m. Janice Lark Keirstead, Dec. 19, 1966; children: Krista Lynn, Jay Olin. A.A., Santa Rosa Jr. Coll., 1950; A.B., U. Calif.-Berkeley, 1958; M.S., San Jose State U., 1969; postgrad., U. Calif.-Davis, 1969-72; grad., Bur. Reclamation Soils Sch., Denver, 1967. Registered geologist, cert. engring. geologist, Calif. Phys. sci. technician U.S. Geol. Survey, Menlo Park, Calif., 1958-64; engring. geologist U.S. Bur. Reclamation, Sacramento, 1964-69; cons. geologist, Davis, 1969-72; asst. prof. geology Appalachian State U., Boone, N.C., 1972-73; county geologist Santa Clara County, San Jose, Calif., 1973—; faculty San Jose State U., 1974-75, adj. prof., 1975-76; mem. Earthquake Engring. Research Inst., Berkeley, 1979—, Western Council Engrs., 1980. Advisor Geotech. Adv. Com., San Jose, 1974-76; mem. West Valley Legis. Com., San Jose, 1980—, Safety Coordinating Com. Santa Clara, 1981—. Doyle scholar, 1949-51. Fellow Geol. Soc. Am.; mem. AAAS, Saber Soc. (co-founder, membership chmn. 1973-75, pres. 1976-77), Assn. Engring. Geologists (vice chmn. sect. 1977-78), Santa Clara County Engring. and Arch. Assn. (v.p. 1981-82), New Weather Observer Soc., Nat. Geog. Soc., Calif. Scholarship Fedn. (life), Peninsula Geol. Soc. (treas. 1978-79), Internat. Platform Assn., West Coast Aquatics Club (pub. relations officer 1981-82), Creekside Homeowners Assn. (treas. 1977-78), San Jose Hist. Mus. Assn., Golden Hills Aquatics Club, Sigma Xi. Democrat. Subspecialties: Geology; Species interaction. Current work: Quaternary geology and geological hazards; lowering risks associated with land development; earthquake history, mitigation and prediction. Originator, Earthquake Prediction "Seismic Window Theory'", 1974. Home: 14927 E Hills Dr San Jose CA 95127 Office: 70 W Hedding St San Jose CA 95110

BERKLEY, MARK A., psychology educator; b. N.Y.C., July 11, 1936; s. David Philip and Bette Rachael (Sherman) B.; m. Karen J. Greene, Sept. 15, 1965; children: Lara, Tamara. B.S., Trinity Coll., 1958; M.A., Johns Hopkins U., 1961, Ph.D., 1962. NIMH postdoctoral fellow Brown U., 1962-64; research fellow neurophysiology U. Wash., Seattle, 1964-67; asst. prof. dept. psychology Fla. State U., 1967-71, assoc. prof., 1971-76, prof., 1976—. Assoc. editor: Behavioral Brain Research, 1980—; contbr. articles to profl. jours. Mem. Soc. Neurosci., AAAS, N.Y. Acad. Scis., Assn. for Research in Vision and Ophthalmology. Subspecialties: Neurophysiology; Psychophysics. Current work: Neural substrates of vision, animal and human visual psychophysics. Office: Fla State U 212 Kellogg Research Bldg Tallahassee FL 32306

BERKOF, RICHARD STANLEY, mechanical engineer; b. Bklyn., Mar. 2, 1941; s. Alfred H. and Lillian B.; m. Madeline Barton, June 30, 1962; children: David, Michael, Howard. B.M.E., CCNY, 1962; M.S.M.E., Columbia U., 1963; Ph.D., CUNY, 1969. Registered profl. engr., Pa., N.Y. Air conditioning engr. Syska & Hennessy, Inc., N.Y.C., 1962; design engr. Gibbs & Cox, Inc., N.Y.C., 1963-64; lectr. CCNY, 1965-67; research assoc. Am. Can Co., Princeton, N.J., 1968-73; mgr. advanced tech. Gulf & Western Advanced Devel. and Engring. Ctr., Swarthmore, Pa., 1973-80, mgr. proposals and planning, 1980—; lectr. in field. Contbr. articles to profl. jours.; editor: dynamics of machine systems Mechanism and Machine Theory, 1975—; editor pro tem: Jour. Mech. Design, 1979-80. Mem. planning com. Main Line Forum, Radnor, Pa., 1982—; bd. dirs. Main Line Sch. Night, 1983—. Recipient Machine Design award CCNY and Machinery Mag., 1962; N.Y. State Regents fellow, 1962-63, 64-65, 66-67. Mem. ASME (exec. com. 1976-81, editor spl. features Transactions 1977-79, chmn. Design Engring. Div. 1980-81), AAAS, Am. Nuclear Soc., Soc. Mfg. Engrs. (sr.), Am. Soc. Engring. Edn., Robotics Internat. (sr.). Subspecialty: Mechanical engineering. Current work: Management and staff capabilities in mechanical engineering and solid mechanics; including machine dynamics and kinematics, stress analysis, and vibrations; systems and design; applied mathematics and computer applications. Home: 322 Strathmore Dr Rosemont PA 19010 Office: Gulf and Western Advanced Development and Engring Ctr 101 Chester Rd Swarthmore PA 19081

BERKOVITS, SHIMSHON, computer scientist, educator; b. Berlin, Oct. 5, 1936; U.S., 1950; s. Eliezer and Sali (Bickel) B.; m. Lillian Lea Fortgang, June 6, 1963; children: Rahel, Aliza, Avraham Chaim Zvi, Yonatan Shmuel, Gavriel Yosef, Binyamin David. S.B., MIT, 1957; M.S., U. Chgo., 1960; Ph.D., Northeastern U., Boston, 1973. Programmer/analyst United Aircraft, West Hartford, Conn., 1957-58; mem. tech. staff The Mitre Corp., Bedford, Mass., 1963-70; asst. prof. math. Boston U., 1968-74, U. Lowell, Mass., 1974-80, assoc. prof. computer sci., 1980—; cons. The Mitre Corp., Bedford, Mass, 1977. Bd. dirs. Solomon Schechter Day Sch. Mem. Am. Math. Soc., Assn. Computer Machinery. Subspecialties: Cryptography and data security; Distributed systems and networks. Current work: The study of public key algorithms, their design, implementation, cryptanalysis and applications. Office: Dept Computer Sci U Lowell Lowell MA 01854

BERKOVITZ, LEONARD DAVID, mathematician; b. Chgo., Jan. 24, 1924; s. Judea and Esther (Trop) B.; m. Anna Whitehouse, June 18, 1953; children: Dan, Michael, Kenneth Eugene. B.S., U. Chgo., 1946, M.S., 1948, Ph.D., 1951. AEC postdoctoral fellow Stanford (Calif.) U., 1951-52; research fellow Calif. Inst. Tech., Pasadena, 1952-54; mathematician Rand Corp., Santa Monica, Calif., 1954-62; prof. math Purdue U., West Lafayette, Ind., 1962—; cons. Rand corp., 1962-68. Author: Optimal Control Theory, 1974; assoc. editor: Jour. Optimization Theory and Applications, 1969—; mng. editor: Soc. Indsl. and Applied Math. Jour. on Control, 1982—. Served to 1st lt. USAAF, 1943-46. Mem. Am. Math. Soc., Math Assn. Am., Soc. Indsl. and Applied Math, Phi Beta Kappa. Subspecialties: Applied mathematics. Current work: Optimal control theory; differential games; variational problems related to filter problems. Office: Dept Math Purdue U West Lafayette IN 47907

BERKOWITZ, BARRY A., pharmacologist; b. Brookline, Mass., Dec. 29, 1942; s. Frank and Frances (Richman) B.; m. Barbara Berkowitz, Aug. 4, 1963; children: Lauren, Brian. B.S., Northeastern U., 1964; Ph.D. in Pharmacology, U. Calif.-San Francisco, 1968. Postdoctoral fellow Roche Inst. of Molecular Biology, Nutley, N.J., 1968-70; asst. mem., assoc. mem., 1971-79; research assoc. Roche Inst. Molecular Biology, N.Y.C., 1970-71; assoc. dir. biol. research, dir. pharmacology I, dir. pharmacology Smith Kline & French Labs., Phila., 1979-83; v.p. biol. research and devel. Smith Kline, French Labs., Phila., 1983—; adj. assoc. prof. Cornell U. Med. Coll. Contbr. articles on pharmacology to profl. jours. Recipient Northeastern U. Pres.' award, 1963; Lehn and Fink Gold medal, 1963. Mem. Am. Soc. for Pharmacology and Exptl. Therapeutics. Subspecialty: Pharmacology. Current work: Pharmacology, drug discovery and development; cardiovascular/neuroeffector pharmacology.

BERKOWITZ, JEROME, mathematician; b. Bklyn., Oct. 2, 1928; m., 1954; 2 children. Ph.D., N.Y.U., 1953. Mathematician Reeves Instrument Corp., 1948; research asst. Courant Inst. Math. Sci., NYU, N.Y.C., 1950-56, mem. faculty, 1956—, prof. math., 1965—. Mem. Am. Math. Soc., Math. Assn. Am. Subspecialty: Applied mathematics. Office: Courant Inst Math Sci N Y U 251 Mercer St New York NY 10012

BERKOWITZ, JOSEPH, physical chemist, physicist; b. Kosice, Czechoslovakia, Apr. 22, 1930; s. Michael and Margaret (Stieglitz) B.; m. Nina, Aug. 2, 1958; children: David Barak, Robert Ari. B.Chem. Engring., N.Y.U., 1951; M.A., Ph.D., Harvard U. Jr. chem. engr. Brookhaven Nat. Lab., Upton, N.Y., 1951-52; research assoc. dept. physics U. Chgo., 1956-57; asst. physicist Argonne Nat. Lab., Ill., 1958-60, assoc. physicist, 1960-73, sr. physicist, 1973—; vis. asst. prof. dept. chemistry U. Ill., 1959-60; vis. prof. dept. chemistry Northwestern U., 1973. Author: Phtoabsorption, Photoionization and Photoelectron Spectroscopy, 1979; Contbr. articles to profl. jours. NSF fellow, 1952-53; John Simon Guggenheim/Found fellow, 1965-66. Mem. Am. Chem. Soc., Am. Phys. Soc., Am. Soc. Mass Spectrometry. Subspecialties: Atomic and molecular physics; Physical chemistry. Current work: Photoionization and photoelectron spectroscopy of common and high temperature vapors, structure of molecular ions. Office: Div of Physics Argonne Nat Lab 9700 S Cass Ave Argonne IL 60439

BERKOWITZ, RICHARD L., perinatologist; b. N.Y.C., July 28, 1940; s. Sidney S. and Miriam Rosenfeld; m. Trudy B. M. Svala, Sept. 3, 1972; children: Michael N., Laura S. B.A., Cornell U., 1961; M.P.H., Johns Hopkins U., 1972; M.D., N.Y.U., 1965. Intern Kings County Med. Center, Bklyn., 1965-66; resident in obstetrics and gynecology Cornell Med. Center, N.Y.C., 1968-71; obstetrician/gynecologist, gen. surgeon, Kenya, East Africa, 1972-74; asst. prof. obstetrics and gynecology and pub. health Yale U. Sch. Medicine, New Haven, 1974-79, assoc. prof., 1979-82; chief High Risk Obstetrical Service, Yale New Haven Hosp., 1979-82; prof. obstetrics and gynecology Mount Sinai Sch. Medicine, N.Y.C., 1982—; dir. div. maternal-fetal medicine Mount Sinai Med. Center, N.Y.C., 1982—. Editor: A Handbook for the Use of Medications During Pregnancy, 1980, Critical Care of the Obstetrical Patient, 1983; author: Ultrasonographyiin Obstetrics and Gynecology, 2d edit, 1983; contbr. articles in field to profl. jours. Served with USPHS, 1966-68. Mem. Am. Coll. Obstetricians and Gynecologists, Soc. Perinatal Obstetricians, Am. Inst. Ultrasound in Medicine, Am. Soc. Human Genetics, New Haven Obstetrical Soc., Sigma Xi. Subspecialty: Maternal and fetal medicine. Office: Dept Obstetrics and Gynecology Mount Sinai Sch Medicine 1 Gustave L Levy Pl New York NY 10029

BERLAD, ABRAHAM LEON, engineering educator; b. N.Y.C., Sept. 20, 1921; s. Harry and Celia (Eichen) B. B.A., Bklyn. Coll., 1943; Ph.D., Ohio State U., 1950. Chief combustion fundamentals sect. NASA-Lewis Research Ctr., Cleve., 1951-56; sr. staff scientist Gen. Dynamics Corp., San Diego, 1956-64; mem. tech. staff Gen. Research Corp., Santa Barbara, Calif., 1964-66; faculty mem. SUNY-Stony Brook, 1966—, chmn. dept. mechanics, 1966-69, prof. engring., 1966—, dir., 1974—; vis. prof. U. Calif.-Berkeley, 1963-64, U. Calif.-La Jolla, 1973, Hebrew U., Jerusalem, 1973; cons. NASA, Washington, 1968—, chmn. sci. working group on combustion, 1978—; cons. U.S. Dept. Energy, 1973-82, Nuclear Regulatory Commn., 1980-83, others. Contbr. numerous tech. papers, monographs to profl. lit. Mem. L.I. (N.Y.) Energy Task Force, 1974—. Served with AUS, 1943-46; PTO. E.I. DuPont de Nemours & Co. fellow, 1949; research grantee U.S. Dept. Def., 1960-66, NASA, 1966-83, U.S. Forest Service, 1967-70, U.S. Dept. Energy, 1974-82, others. Mem. Combustion Inst. (bd. dirs. 1974—, editor/program chmn. 11th Internat. Symposium on Combustion 1966, editorial adv. bd. Combustion and Flame 1969—), Am. Phys. Soc., Am. Nuclear Soc. Jewish. Subspecialties: Combustion processes; Gravitational effects on Combustion. Current work: Gravitational effects on combustion (space shuttle experimentation); combustion theory. Patentee energy conversion devices. Office: Coll Engring SUNY-Stony Brook Stony Brook NY 11794

BERLEKAMP, ELWYN RALPH, computer engineer, educator; b. Dover, Ohio, Sept. 6, 1940; s. Waldo and Loretta (Kimmel) B.; m. Jennifer Wilson, 1966; children: Persis, Bronwen, David. B.S. in Elec. Engring; M.I.T., 1962; M.S., 1962; Ph.D. in Elec. Engring, 1964. Mem. Math. Research Center, Bell Telephone Labs., Murray Hill, N.J., 1967-71; asst. prof. elec. engring. U. Calif., Berkeley, 1964-67, prof. math., elec. engring. and computer sci., 1971—; now pres. Cyclotomics, Inc. Author: Algebraic Coding Theory, 1968, Winning Ways, 1982. Recipient award Eta Kappa Nu, 1972. Fellow IEEE (pres. group on info. theory 1973); mem. Nat. Acad. Engring. Subspecialty: Computer engineering. Home: 1836 Thousand Oaks Blvd Berkeley CA 94707

BERLIN, CHESTON MILTON, JR., pediatrics educator; b. Pitts., Mar. 28, 1936; s. Cheston Milton and Gladys (Vance) B.; m. Anne Risher, July 9, 1960; children: Jean, Douglas, Alexander, Gordon. B.A., Haverford Coll., 1958; M.D., Harvard U., 1962. Diplomate: Am. Bd. Pediatrics, 1968. Intern Children's Hosp. Med. Center, Boston, 1962-63, resident in pediatrics, 1965-67; asst. prof. pediatrics U. Ala., 1967-68, Sch. Medicine, George Washington U., Washington, 1968-71, spl. lectr. child health and devel., 1971—; assoc. prof. pediatrics and pharmacology M.S. Hershey Med. Center, Pa. State U., Hershey, 1971-75, prof., 1975—. Contbr. numerous articles on biochemistry, pharmacology and pediatrics to profl jours. Served as sr. asst. surgeon USPHS, 1963-65. Markle scholar in acad. medicine, 1969-74. Mem. Am. Acad. Pediatrics, Am. Soc. Exptl. Pharmacology and Therapeutics, Am. Soc. Clin. Pharmacology and Therapeutics, Am. Pediatric Soc. Subspecialties: Pediatrics; Pharmacology. Current work: Pediatric pharmacology, general pediatrics. Home: 1415 Bradley Ave Hummelstown PA 17036 Office: Dept Pediatrics Hershey Med Center Pa State U PO Box 850 Hershey PA 17033

BERLIN, NATHANIEL ISAAC, physician; b. N.Y.C., July 4, 1920; s. Louis and Gertrude (Sugarman) B.; m. Barbara Ruben, June 14, 1953; children: Deborah Joy, Marc David. B.S., Western Res. U., 1942; M.D., L.I. Coll. Medicine, 1945; Ph.D., U. Calif.-Berkeley, 1949. Intern Kings County Hosp., Bklyn., 1945-46, resident pathologist, 1946-47; Nat. Cancer Inst. postdoctorate research fellow U. Calif., 1948-50, research fellow, 1949-50, research asso., 1950-51, instr., 1951, lectr. and research assoc., 1952-53, lectr., assoc. research med. physicist, 1952-53; Nat. Heart Inst. spl. research fellow Nat. Inst. Med. Research, London, 1953-54; med. officer, analysis br. Effects div. Hdqrs. Armed Forces Spl. Weapons Project, 1954-56; head metabolism service, gen. medicine br. Nat. Cancer Inst., 1956-72, chief gen. medicine br., 1959-61, clin. dir., 1961-71, sci. dir. gen. lab. and clinics, 1969-72, dir. div. cancer biology and diagnosis, 1972-75; dir. Cancer Center, Northwestern U., 1975—, Genevieve B. Teuton prof. medicine, 1975—; vis. scientist Walter Hall Inst. Med. Research, Melbourne, Australia, 1980-81; cons. U.S. Naval Hosp., Bethesda, Md., 1955-65, Armed Forces Spl. Weapons Project, Dept. Def., 1957-59; alumni lectr. Downstate Med. Center, 1966; mem. panel diagnostic applications of radioisotopes in hematology Internat. Com. on Standardization in Hematology, 1964—, chmn., 1974-76; chmn. instnl. rev. bd. Fermi Nat. Lab., 1975—; mem. adv. com. div. blood diseases and resources Nat. Heart, Lung and Blood Diseases Inst., 1975-79; mem. adv. bd. cancer control, State of Ill., 1976—. Editorial adv. bd.: Cancer Letters; mem. editorial bd.: Blood; contbr. articles to med. jours. Mem. med. adv. com. Nat. ARC, 1969-75; trustee Ill. Cancer Council, pres., 1979; bd. dirs. Ill. div. Am. Cancer Soc.; mem. med. adv. bd. Leukemia Research Found., 1976-80; chmn. Leukemia Research Council, 1978-80. Served with AUS, 1943-45; lt. comdr. M.C. USNR, 1954-56; comdr. Res. Recipient Superior Service award HEW; Alumni medal for distinguished service to medicine State U. N.Y. Fellow AAAS, N.Y. Acad. Sci., Internat. Soc. Hematology; mem. Am. Fedn. Clin. Research, Soc. Exptl. Biology and Medicine, Am. Physiol. Soc., Biochem. Soc. (Eng.), Radiation Research Soc., Am. Soc. Hematology (publ. com., pub. issues com.), Assn. Am. Physicians, Am. Soc. Clin. Investigation, Am. Clin. and Climatol. Assn., Western Soc. Clin. Research, Mid-Eastern Soc. Nuclear Medicine (sec.-treas. 1957-60), Am. Soc. Clin. Oncology (legis. liaison com., pub. affairs com.), Am. Soc. Preventive Oncology (pres.), Am. Assn. Cancer Research, Sigma Xi, Alpha Omega Alpha, Zeta Beta Tau, Phi Delta Epsilon. Subspecialties: Cancer research (medicine); Hematology. Current work: Director cancer center. Home: 1448 N Lake Shore Dr Chicago IL 60611 Office: Cancer Center Northwestern U 303 Chicago Ave Chicago IL 60611

BERLIND, ALLAN, biology educator; b. N.Y.C., Dec. 24, 1942; s. Morris and Ruth (Fischer) B.; m. Wendy Prindle, Aug. 18, 1968; children: Lisa, Andrew. B.S., Swarthmore Coll., 1964; Ph.D., Harvard U., 1969. NIH fellow in zoology U. Calif., Berkeley, 1969-71; asst. prof. dept. biology Wesleyan U., Middletown, Conn., 1971-77, assoc. prof., 1977—; vis. scholar Cambridge (Eng.) U., 1976-77; vis. assoc. prof. U. Hawaii, 1980-81. Mem. Soc. Neurosci., Am. Soc. Zoologists. Subspecialties: Neurophysiology; Neuroendocrinology. Current work: Action of neurohormones on simple neuronal systems. Office: Dept Biology Wesleyan U Middletown CT 06457

BERLINER, HANS JACK, computer scientist; b. Berlin, Germany, Jan. 27, 1929; came to U.S., 1937, naturalized, 1943; s. Paul and Theodora (Lehfeld) B.; m. Araxie Yacoubian, Aug. 15, 1969. B.A., George Washington U., 1954; Ph.D., Carnegie Mellon U., 1975. Systems analyst U.S. Naval Research Lab., 1954-58; group head systems analysis Martin Co., Denver, 1959-60; adv. systems analyst IBM, Gaithersburg, Md., 1960-69; sr. research scientist Carnegie-Mellon U., Pitts., 1974—. Editorial bd.: Artificial Intelligence, 1976—. Served with AUS, 1951-53. Awarded title Internat. Grandmaster Corr. Chess, 1968. Mem. Assn. Computing Machinery, Internat. Joint Conf. Artificial Intelligence, U.S. Chess Fedn., Internat. Computer Chess Assn. Subspecialties: Artificial intelligence. Current work: Work in artificial intelligence with emphasis on heuristic search, knowledge representation for making judgments, learning. Among leading chess players U.S., 1950—, N.Y. State champion, 1953, Southwest Open champion, 1960, So. Open champion, 1949, U.S. Open Corr. Chess champion, 1955, 56, 59, World Corr. Chess champion, 1968-72. Developed 1st computer program to defeat a world champion at his own game (backgammon), 1979. Home: 657 Ridgefield Ave Pittsburgh PA 15216

BERLINER, LAWRENCE JULES, chemistry educator; b. Los Angeles, Sept. 18, 1941; s. Abraham Murray and Victoria (Levy) B.; m. Barbara Elaine Anderson, June 30, 1979; children: Allegra Elisabeth, Anders Nathaniel. B.S., UCLA, 1963; Ph.D., Stanford U., 1967. Asst. prof. Ohio State U., Columbus, 1969-76, assoc. prof., 1975-82, prof., 1982—; cons. NIH, Bethesda, Md., 1974—. Editor: Spin Labeling, 1976; 1979, Biological Magnetic Resonance, 1978-83; mem. editorial bd.: Thrombosis Research, 1983-87. NIH grantee, 1979, 82, 83; NSF grantee, 1978. Mem. Am. Chem. Soc., Am. Soc. Biol. Chemistry, AAAS, N.Y. Acad. Scis., Soc. Magnetic Research Medicine. Subspecialties: Biophysical chemistry; Nuclear magnetic resonance (biotechnology). Current work: Magnetic resonance in biology, blood coagulation enzymes, protein structure and function. Office: Ohio State U Chemistry Dept 140 W 18th Ave Columbus OH 43210

BERLINER, MARTHA D., research biologist, educator; b. Antwerp, Belgium, Nov. 18, 1928; came to U.S., 1941; d. A. A. and Frieda (Mandelbaum) Dresner; m. S. Newton Berliner, Sept. 14, 1952; children: Leni Susan, Michael Paul. B.A., Hunter Coll., 1949; M.A., U. Mich., 1950; Ph.D., Columbia U., 1953. Med. microbiologist Lynn Hosp., 1953-60; sr. scientist AVCO Corp., 1958-65; prof. biology

Simmons Coll., 1965-82; sr. research assoc. Harvard Sch. Pub. Health, Boston, 1965-74; program mgr., policy analyst NSF, 1980-82; prof., chmn. dept. biology Va. Commonwealth U., Richmond, 1982—; prof. microbiology and immunology Med. Coll. Va., Richmond, 1982—. Contbr. articles to profl. jours.; editorial bd.: Jour. Applied and Environ. Microbiology, 1974-80. Trustee Mary A. Alley Hosp., 1976-80. Fellow Am. Acad. Microbiology. Subspecialties: Cell and tissue culture; Microbiology. Current work: Plant tissue culture, plant biotechnology, protoplasts. Office: 816 Park Ave Richmond VA 23284

BERLINER, ROBERT WILLIAM, physician, univ. dean; b. N.Y.C., Mar. 10, 1915; s. William M. and Anna (Weiner) B.; m. Leah Silver, Dec. 21, 1941; children—Robert William, Alice (Mrs. James L. Hadler), Henry J., Nancy. B.S., Yale, 1936; M.D., Columbia, 1939. Intern Presbyn. Hosp., N.Y.C., 1939-41; resident physician Goldwater Meml. Hosp., N.Y.C., 1942-43, research fellow 3d div. research service, 1943-44, research asst., 1944-47; asst. medicine N.Y.U. Coll. Medicine, N.Y.C., 1943-44, instr., 1944-47; asst. prof. medicine Columbia, research asso. dept. hosps., N.Y.C., 1947-50; chief lab. kidney and electrolyte metabolism Nat. Heart Inst., NIH, Bethesda, Md., 1950-62, dir. intramural research, 1954-68; dir. lab. and clinics NIH, 1968-69, dep. dir. sci., 1969-73; dean Yale U. Sch. Medicine, New Haven, Conn., 1973—; lectr. George Washington U. Sch. Medicine, 1951-73; professorial lectr. Schs. Medicine and Dentistry, Georgetown U., 1964-73. Editorial bd.: Jour. Clin. Investigation, 1954-59, 61-66, Am. Jour. Physiology, 1956-61, Circulation Research, 1958-63, 65-70. Mem. Am. Physiol. Soc. (pres. 1967-68), Soc. Gen. Physiol., Am. Soc. Clin. Investigation (pres. 1959-60), Soc. for Exptl. Biology and Medicine, Am. Acad. Arts and Scis., Washington Acad. Medicine, Philos. Soc. Washington, Asso. Am. Physicians, Nat., Washington acads. scis., Am. Soc. Nephrology (pres. 1968-69), Harvey Soc., Sigma Xi, Alpha Omega Alpha. Subspecialties: Physiology (medicine); Circulatory system. Home: 36 Edgehill Terr New Haven CT 06511 Office: Office of Dean Yale U Sch Medicine New Haven CT 06510

BERLYNE, GEOFFREY MERTON, nephrologist, researcher; b. Manchester, Eng., May 11, 1931; came to U.S., 1976, naturalized, 1981; s. Charles Solomon and Miriam Hannah (Rosenthal) B.; m. Ruth Selbourne, June 7, 1959; children: Jonathan, Benjamin, Suzannah. M.B.Ch.B. with honors, Manchester U., 1954, M.D., 1966. Lectr. U. Manchester, 1961-62, sr. lectr., 1964-68, reader, 1969-70; prof. medicine and life scis. Negev U., Israel, 1970-79; prof. medicine SUNY-Bklyn., 1976—; chief nephrology sect. Brooklyn VA Med. Center, 1976—. Author: Course in Renal Diseases, 1966, Course on Electrolytes and Body Fluids, 1981. Fellow Am. Coll. Physicians, Am. Coll. Nutrition; mem. Japanese Nephrology Soc. (named distinguished nephrologist 1979), Assn. Physicians of G.B., Am. Fedn. Clin. Research. Republican. Jewish (chmn. 1970-74, pres. synagogue 1982—). Subspecialties: Nephrology; Physiology (medicine). Current work: Biology of trace elements; renal disease and physiology; biology of trace elements silicon and aluminum; renal disease and physiology of terminal and preterminal renal failure. Home: 27 Merrall Dr Lawrence NY 11559 Office: Renal Sect III Bklyn VA Hosp 800 Poly Pl Brooklyn NY 11209

BERMAN, ALAN, physicist; b. Bklyn., Nov. 2, 1925; s. Hyman and Sarah (Levy) B.; m. Charlotte Bernstein, Apr. 28, 1962; children—Julia, Jessica, S. Jonathan, Margaret, James. A.B., Columbia, 1947, Ph.D., 1952. Research scientist Hudson Labs., Columbia, N.Y.C., 1952-57, assoc. dir., 1957-63, dir., 1963-67; dir. research Naval Research Lab., Washington, 1967-82; dean Sch. Marine and Atmospheric Scis. U. Miami, 1982—; mem. Naval Research Adv. Com., 1982. Served with AUS, 1944-46. Recipient Superior Civilian Service award Dept. Navy, 1969, Disting. Civilian Service award Dept. Def., 1973, Robert Dextar Conrad award, 1982; named Disting. Sr. Exec. Pres. of U.S., 1980. Fellow Am. Phys. Soc., Acoutical Soc. Am.; mem. Washington Philos. Soc., Sigma Xi. Subspecialty: Research management. Home: 6645 SW 118th St Miami FL 33156 Office: 4600 Rickenbacker Causeway Miami FL 33149

BERMAN, DAVID ALBERT, pharmacologist, educator; b. Rochester, N.Y., Nov. 4, 1917; s. Samuel Moses and Anna (Newman) B.; m. Miriam Goodman, July 13, 1945; children: Shelly Ann Berman Stone, Judith Berman McCleese. B.S., U. So. Calif., 1940, M.S., 1948, Ph.D., 1951. Instr. U. So. Calif., 1952-54, asst. prof., 1954-58, prof. dept. pharmacology and nutrition, 1963—. Served to sgt. MC. AUS, 1941-45. Recipient Elaine Stevely Hoffman Achievement award, 1971; Kaiser Permanente Teaching award, 1974, 77, 79, 81; Assocs. award for teaching excellence, 1980; Alpha Omega Alpha award, 1982. Mem. Western Pharmacology Soc., Biochemistry Soc., Am. Soc. Pharmacology and Exptl. Therapeutics, Sigma Xi. Subspecialty: Pharmacology. Current work: Biochemical mechanisms of cardiac contractility. Home: 3304 Scadlock Ln Sherman Oaks CA 91403 Office: 2025 Zonal Ave Los Angeles CA 90033

BERMAN, HERBERT L(AWRENCE), fired heater engineering specialist, consultant; b. Bklyn., Jan. 8., 1931; s. Moses and Bertha (Silverman) B.; m. Pearl M., Mar. 30, 1957; children: Stacey Perri, Marcy Eydie. B.Ch.E., Poly. Inst. Bklyn., 1952; postgrad. in engring, N.Y.U., 1956-57. Lic. profl. engr., N.Y., Tex. Refinery engr. Shell Oil Co., Houston, 1952-53; fired heater proposal engr. Petro-Chem Devel. Co., N.Y.C., 1955-60; sr. fired heater proposal engr. Foster Wheeler Corp., N.Y.C., 1960-62; mgr. fired heater proposal engring. Alcorn Combustion Co., N.Y.C., 1962-72; engring. supr. of heat transfer, energy conservation and cost engring. Caltex Petroleum Corp., Dallas, 1972—. Contbr.: article series on fired heaters to Chem. Engring. mag, 1978. Served with Chem. Corps U.S. Army, 1953-55. Fellow Am. Inst. Chem. Engrs. (organizer and lectr. Fired Heater Engring. continuing edn. course); mem. ASME, Am. Assn. Cost Engrs., Am. Petroleum Inst. (subcom. on heat transfer equipment). Subspecialties: Combustion processes; Chemical engineering. Current work: Fired heater engring. U.S. patentee double fired multi-path heater; Can. patentee air cooled fired heater. Home: 7310 Blythdale Dr Dallas TX 75248 Office: PO Box 619500 Dallas TX 75261

BERMAN, JOEL D., mathematics educator; b. Mpls., Jan. 1, 1943; s. Morris and Hilda B. B.A., U. Minn., 1965; Ph.D., U. Wash., 1970. Asst. prof. math. U. Ill. at Chgo., 1970-75, assoc. prof., 1975-82, prof., 1982—. Subspecialties: Algebra; Theoretical computer science. Current work: Lattice theory, universal algebra, computer mathematics. Office: Dept Math U Ill at Chgo Chicago IL 60680

BERMAN, MARK LAURENCE, clinical psychologist, consultant, researcher; b. Los Angeles, Sept. 13, 1940; s. Joseph Erwin and Bernice (Levin) B.; m. Teresa R. Davich, July 3, 1966; children: Alisa Ruth, Joseph Daniel. B.A. in Anthropology, UCLA, 1962; M.A., Ariz. State U., Tempe, 1964, Ph.D. in Psychology, 1969. Cert. psychologist, Ariz. Asst. Prof. Pa. State U., 1968-70; research coordinator U. Wash., Seattle, 1970-72; pvt. practice psychology, Phoenix, 1973—; therapist, evaluator Juvenile Ct. Ctr., Phoenix, 1975-80, Bur. Indian Affairs, 1976—, Child Protective Services, 1977-80, Ariz. State Dept. Corrections, 1977—. Author: (with Naomi A. Reich) Essentials of Clothing Construction, 1971; editor: Motivation and Learning, 1971; contbr. articles to profl. jours. Mem. com. mandatory mediation of joint custody Maricopa County Bar Assn., 1983, ex-officio mem. legal services for the elderly com., 1982. Bur. Edn. Handicapped HEW grantee, 1971; Law Enforcement Assistance Agy. grantee, 1969; Pa. State U. grantee, 1969. Mem. Am. Psychol. Assn., Ariz. State Psychol. Assn. (exec. com. 1982), Maricopa Psychol. Assn. (pres. 1982), Assn. Advancement Behavior Therapy, Nat. Rehab. Assn. Democrat. Jewish. Subspecialty: Behavioral psychology. Current work: Child custody, visitation evaluations; therapy, evaluation of senior persons; behavior therapy for phobias, stress, anxiety; psychological and neuropsychological evaluations; therapy, counseling of hyperactive children and their parents; marital counseling; behavioral treatment of respiratory and other disorders. Home: 8603 N Cardinal Dr Phoenix AZ 85028 Office: 2021 N Central Ave Suite 205 Phoenix AZ 85004

BERMAN, MARLENE OSCAR, research neuropsychologist, consultant, educator; b. Phila., Nov. 21, 1939; d. Paul and Evelyn (Hess) Oscar; m. Michael Brack Berman, June 22, 1963 (div. Feb. 1980); 1 son, Jesse Michael. B.A., U. Pa., 1961; M.A., Bryn Mawr Coll., 1964; Ph.D., U. Conn., 1968; postgrad., Harvard U., 1968-70. Research assoc., clin. investigator Boston VA Med. Center, 1970-76, research psychologist, 1976—; research assoc. Boston U. Sch. Medicine, 1970-72, asst. prof. dept. neurology, 1972-76, assoc., prof., 1976-81, prof., 1982—, prof. psychiatry, dir., 1981—; mem. grant rev. coms. Dept. Health and Human Services; cons. in field. Contbr. articles to profl. jours., chpts. to books. Coordinator Newton (Mass.) Community Schs., 1978-80. Recipient clin. investigator award, VA, 1973-76, research scientist devel. awards Nat. Inst. Neurol. Communicative Disorders and Stroke, 1976-81, Nat. Inst. Alcohol Abuse and Alcoholism, 1981—; USPHS grantee, 1964—. Fellow Am. Psychol. Assn. (sec.-treas. div. 6), Mass. Psychol. Assn. (awards com. 1979-80, nominations com. 1982—); mem. Acad. Aphasia, Soc. for Neurosci., Internat. Neuropsychol. Soc., Psychonomic Soc., Eastern Psychol. Assn., N.Y. Acad. Scis., Com. To Combat Huntington's Disease, Sigma Xi (nat. lectr. 1980, 81). Democrat. Jewish. Subspecialties: Neuropsychology; Physiological psychology. Current work: Research on mechanisms of human brain function and behavioral abnormalities accompanying brain damage, especially due to alcohol abuse; also research on cerebral laterality and cognitive function in aging. Office: Boston VA Med Center 150 S Huntington Ave 14th Floor Boston MA 02130

BERMAN, NANCY E., researcher, educator; b. Paducah, Ky., Aug. 14, 1946; d. Hollis and Mary Willy (Steffy) Johnson; m. Terry L. Berman, Sept. 11, 1970; 1 son, John Brian. B.A., Lawrence U., 1968; Ph.D., M.I.T., 1972. Faculty U. Pa., Phila., 1972-74, Wash. U., St. Louis, 1974-76; faculty Med. Coll. Pa., Phila., 1976—, asso. prof. anatomy and physiology, 1981—. Mem. AAAS, Am. Assn. Anatomists, Soc. for Neurosci. Club: Cajal. Subspecialty: Neurobiology. Current work: Teaching and research in anatomy and physiology of vision. Office: Medical College of Pennsylvania 3200 Henry Ave Philadelphia PA 19129

BERMAN, ROBERT HIRAM, mathematician; b. N.Y.C., Apr. 1, 1948; s. David and Eleanor (Davidson) B. B.S., MIT, 1970, Ph.D., 1975. Research fellow Reading (Eng.) U., 1975-78; research scientist MIT, Cambridge, Mass., 1978—. Fellow Royal Astron. Soc.; mem. Am. Phys. Soc., Soc. Indsl. and Applied Math, Internat. Astronom. Union, Am. Astronom. Soc. Subspecialties: Applied mathematics; Plasma physics. Current work: Computational physics; N-body simulations, plasma kinetic theory, plasma turbulence and stochasticity; applications to development of advanced scientific computing environments; including symbolic computing.

BERMANT, GORDON, experimental psychologist, administrator, educator; b. Los Angeles, Oct. 10, 1936; s. Ira George and Josephine (Wilson) B.; m.; children: Laura, Daniel, Jennifer. B.A., UCLA, 1967; M.A., Harvard U., 1961, Ph.D., 1961. Assoc. prof. dept. psychology U. Calif.-Davis, 1964-69; fellow Battelle (Seattle) Research Ctr., 1969-76; dir. systems devel. div., sr. research psychologist Fed. Jud. Ctr., Washington, 1976—; adv. panelist NSF; cons. various orgns. Author: 6 books including Psychology and the Law, 1976; bds. editors various sci. jours. Fellow Am. Psychol. Assn., Inst. Society, Ethics, and Life Scis.; mem. Am. Psychology-Law Soc. (pres. 1983). Home: 5408 Midship Ct Burke VA 22015 Office: Fed Judicial Ctr 1520 H St NW Washington DC 20005

BERNACKI, RALPH JAMES, pharmacologist, cancer researcher; b. Buffalo, Oct. 9, 1946; s. Roman S. and Emily (Dommer) B.; m. Celeste Agnes, Aug. 16, 1969; children: Rachelle, Gwen. B.S., Rensselaer Poly. Inst., 1968; Ph.D., U. Rochester, 1973. Cancer research scientist I Roswell Park Meml. Inst., Buffalo, from 1972, cancer research scientist V, 1981—; asst. prof. SUNY at Buffalo, 1975-82, assoc. prof., 1982—, dir. pharmacology grad. program, 1982—. Contbr. numerous articles to profl. jours. Nat. Cancer Inst. grantee, 1977-82. Mem. Am. Soc. Pharmacology and Exptl. Therapeutics, Am. Assn. Cancer Research, AAAS, Sigma Xi. Subspecialties: Pharmacology; Cancer research (medicine). Current work: Design and evaluation of potential antitumor agents. Office: Dept Exptl Therapeutics Roswell Park Meml Inst Buffalo NY 14260

BERNARD, PETER SIMON, mechanical engineering educator; b. N.Y.C., Jan. 14, 1950; s. Foster Arnold and Helen Rose (Siegal) B.; m. Betty Jean Chaumont, May 20, 1979; children: Jennifer Louise, Alexander Foster. B.E., CCNY, 1972; M.S., U. Calif., Berkeley, 1973, Ph.D., 1977. Postdoctoral fellow Lawrence Berkeley (Calif.) Lab., 1977; asst. prof. mech. engring. U. Md., College Park, 1977—. Editor: (with Tudor Ratiu) Turbulence Seminar Berkeley, 1976/77, 1977. NSF grad. fellow, 1972-75. Mem. ASME, Soc. Indsl. and Applied Math. Subspecialties: Fluid mechanics; Applied mathematics. Current work: Develop theories of turbulent flow. Office: U Md Dept Mech Engring College Park MD 20742

BERND, PAULETTE SALLY, anatomy educator; b. N.Y.C., Jan.21, 1953; d. Addie and Elizabeth Ann (Drucker) B; m. Steven Matthew Erde, Aug. 7, 1982. A.B., Colgate U., 1975; M.A., Columbia U., 1977, M.Phil., 1978, Ph.D., 1980. Postdoctoral fellow dept. pharmacology NYU Sch. Medicine, N.Y.C., 1980-82; asst. prof. anatomy Mt. Sinai Sch. Medicine, N.Y.C., 1982—. Contbr. articles to profl. jours. Pharm. Mfrs. Assn. Found. fellow, 1980-82. Mem. Soc. Neurosci., N.Y. Soc. Electron Microscopists, AAAS, N.Y. Acad. Scis., Assn. of Women in Sci. Subspecialties: Developmental biology; Cell and tissue culture. Current work: Mechanism of action of nerve growth factor. Office: Dept anatomy Mt Sinai Sch Medicine 1 Gustave Levy Pl New York NY 10029

BERNDT, THOMAS JOSEPH, developmental psychologist, educator; b. South Bend, Ind., June 26, 1949; s. Wilmer Anthony and Gertrude (Lukas) B.; m. Emily Gray Swanson, June 17, 1972; children: William, Edward, Brian. B.A., Harvard Coll., 1971; Ph.D., U. Minn., 1975. Asst. prof. Yale U., 1975-80, assoc. prof., 1980-82; assoc. prof. psychology U. Okla., Norman, 1982—; cons. in field. Contbr. chpts. to books, articles to profl. publs. Spencer Found. grantee, 1971; NSF grantee, 1978; NIMH grantee, 1982; Grant Found. grantee, 1982. Mem. Soc. Research in Child Devel., Am. Psychol. Assn. Subspecialty: Developmental psychology. Current work: Studies of characteristics and consequences of children's friendships; investigation of social development more broadly, including social influence in peer groups and social understanding. Office: U Okla Dept Psychology Norman OK 73019

BERNDT, WILLIAM OSCAR, pharmacologist, toxicologist, univ. adminstr., cons.; b. St. Joseph, Mo., May 11, 1933; s. Oscar Emil and Gertrude Ann (Muthig) B.; m. Bonnie Lou Lampe, Aug. 28, 1954; children: Barbara, Carol, David, Mary, Paul. B.S., Creighton U., 1954; Ph.D., SUNY, Buffalo, 1959. Diplomate: Am. Bd. Toxicology. Instr., asst. prof. dept. pharmacology-toxicology Dartmouth Med. Sch., Hanover, N.H., 1959-68, assoc. prof., then prof., 1968-74; prof., chmn. dept. pharmacology-toxicology U. Miss. Med. Center, Jackson, 1974-82; prof. pharmacology, dean for grad. studies and research U. Nebr. Med. Center, Omaha, 1982—; mem. pharmacology study sect. NIH; cons. in field. Contbr. chpts. to books, articles to profl. jours. Served to capt. Med. Services Corps U.S. Army, 1962-63. Am. Heart Assn. estblished investigator grantee, 1964-69; recipient fed. grants, 1962—. Mem. Am. Soc. Pharmacology and Exptl. Therapeutics, Soc. Toxicology, Am. Heart Assn., AAAS, Am. Nephrology Soc., Internat. Soc. Study of Xenobiotics. Subspecialties: Pharmacology; Nephrology. Current work: Effects of drugs and chemicals on kidney function, specifically studies directed at understanding nephrotoxic events caused by chemicals such as fungal toxins (Mycotoxins)., metals and solvents. Office: 5009 Conkling Hall Univ Nebr Med Center 42d and Dewey Ave Omaha NE 86105

BERNER, ROBERT ARBUCKLE, geochemistry educator; b. Erie, Pa., Nov. 25, 1935; s. Paul Nau and Priscilla (Arbuckle) B.; m. Elizabeth Marshall Kay, Aug. 29, 1959; children: John Marshall, Susan Elizabeth, James Clark. B.S., U. Mich., 1957, M.S., 1958; Ph.D., Harvard U., 1962. Sverdrup fellow U. Calif., San Diego, 1962-63; asst. prof. U. Chgo., 1963-65, assoc. prof., 1965; assoc. prof. geology and geophysics Yale U., 1965-71, prof., 1971—. Author: Principles of Chemical Sedimentology, 1971, Early Diagenesis, 1980, (with Elizabeth K. Berner) Natural Water Cycle, 1983; assoc. editor: Am. Jour. Sci, 1968-80; editor, 1980—. Sloan fellow, 1968-72; Guggenheim fellow, Switzerland, 1972. Fellow Mineral. Soc. Am. (life, award 1971), Geol. Soc. Am., Geochem. Soc. (councillor 1976-79, pres. 1983), Am. Geophys. Union. Club: High Lane (North Haven, Conn.). Subspecialties: Geochemistry; Sedimentology. Current work: Application of physical chemistry to sedimentology, oceanography and soil science; evolution of composition of sediments, atmosphere and oceans. Home: 15 Hickory Hill Rd North Haven CT 06473 Office: Dept Geology and Geophysics Yale U New Haven CT 06511

BERNEY, STUART ALAN, clin. pharmacologist, educator; b. Albany, N.Y., Aug. 8, 1945; s. Morris and Ester B.; m. Mary Helen, Aug. 28, 1975; children: Elizabeth, Joshua. B.S., Union U., 1969; Ph.D., Vanderbilt U., 1975. Lectr. medicine U. Toronto, Can., 1976-77; research asso. dept. pharmacology Vanderbilt U., 1978, research asst. prof. dept. psychiatry, 1978—. Mem. AAAS, Soc. for Neurosci. Republican. Subspecialties: Neuropharmacology; Pharmacokinetics. Current work: Pharm. cons. Home: 5562 Ridge Rd Joelton TN 37080

BERNFELD, PETER HARRY WILLIAM, biochemist; b. Leipzig, Ger., June 1, 1912; came to U.S., 1949, naturalized, 1955; s. Isidor and Elsa (Gutfreund) B.; m. Helen Cecily Kroch, Nov. 21, 1940; children: Michele Marion, Mark Raymond. M.S., U. Leipzig, 1935; Ph.D., U. Geneva, 1937. Research fellow U. Geneva, 1937-39, chief chemist dept. chemistry, 1939-49, privat docent enzymology, 1947-49; asst. prof. dept. biochemistry and nutrition Tufts U. Sch. Medicine, 1949-51, asso. prof., 1951-57, biochemist cancer research unit, 1949-57; sr. v.p., dir. research Bio-Research Inst. and Bio-Research Cons., Cambridge, Mass., 1957—. Editor, contbg. author: Biogenesis of Natural Compounds, 1963, 2d edit. 1967; contbr. articles to profl. jours. Recipient Werner medal Swiss Chem. Soc., 1948. Mem. Am. Soc. Biol. Chemists, Soc. Exptl. Biology and Medicine, Am. Chem. Soc., Am. Inst. Chemists, Am. Assn. Cancer Research, AAAS, Am. Coll. Toxicology, N.Y. Acad. Sci., Sigma Xi. Subspecialties: Biochemistry (biology); Toxicology (medicine). Home: 247 Farm Ln Westwood MA 02090 Office: 9 Commercial Ave Cambridge MA 02141

BERNHARDT, PETER, research botanist, science writer, consultant; b. Bklyn., Oct. 21, 1952; s. Jules and Arlene (Samuels) B. M.S. in Botany, SUNY-Brockport, 1976; Ph.D. (Grad. Union fellow), U. Melbourne, 1982. Vis. prof. botany U. El Salvador, 1975-77; plant collector Mo. Bot. Garden, 1975-77; herbarium technician N.Y. Bot. Garden, 1978; editor Biology of Mistletoes U. Melbourne, Australia, 1981-82; research fellow Plant Cell Biology Research Centre, Melbourne, 1982—; sci. writer Garden Mag., N.Y.; cons. floras of Central Am., Australia, northeastern U.S.; postgrad. scholar U. Melbourne, 1978-81. Contbr. articles on botany to profl. jours. Mem. Australian Pollination Ecologists Soc., Internat. Union for the Conservation of Nature, Torrey Bot. Club. Subspecialties: Evolutionary biology; Species interaction. Current work: Evolutionary ecology (pollen-pistil interactions, interspecific isolation, plant parasite/host interactions, floral ecology).

BERNHEIM, ROBERT ALLAN, chemistry educator; b. Hackensack, N.J., June 8, 1933; s. Fred and Virginia (White) B.; m. Gloria D. Bernheim, Feb. 14, 1959; 1 dau., Britt; m. Marguerite Yevitz, Aug. 11, 1975; 1 son, Kyle. B.S., Brown U., 1955; M.A., Harvard U., 1957; Ph.D., U. Ill., 1959. Postdoctoral fellow Columbia U., N.Y.C., 1959-61; faculty Pa. State U., University Park, 1961—, now prof. chemistry. NSF fellow, 1967-68; Guggenheim fellow, 1974-75; Joint Inst. for Lab. Astrophysics vis. fellow, 1981-82. Mem. Am. Chem. Soc., Am. Phys. Soc. Subspecialty: Physical chemistry. Current work: Molecular spectroscopy, laser spectroscopy. Home: 622 McKee St State College PA 16803 Office: Univ of Pa 152 Davey Lab University Park PA 16802

BERNINGER, VIRGINIA WISE, research psychologist, consultant; b. Phila., Oct. 4, 1946; d. Oscar Sharpless and Lucille (Fike) Wise; m. Ronald William Berninger, Aug. 3, 1968. B.A., Elizabethtown Coll., 1967; M.Ed., U. Pitts., 1970; Ph.D., Johns Hopkins U., 1981. Lic. psychologist, Mass. Tchr. Phila. Pub. Schs., 1967-68, Baldwin-Whitehall (Pa.) Pub. Schs., 1969-72, Frederick (Md.) Pub. Schs., 1972-75, Balt. Pub. Schs., 1975-76; research assoc. Children's Hosp., Boston, 1980-83; mem. spl. and sci. staff New Eng. Med. Ctr., Boston, 1983—; instr. psychology Harvard Med. Sch., 1981-83; asst. prof. rehab. medicine Tufts Med. Sch., 1983—; pvt. ednl., pediatric cons., Wellesley, Mass., 1982—. Contbr. articles to profl. jours. Md. Psychol. Assn. grantee, 1980. Mem. Am. Psychol. Assn., Soc. Research in Child Devel. Subspecialties: Learning; Developmental psychology. Current work: Adaptation of information processing paradigms to clinical assessment of learning problems; use of microprocessors in psychoeducational intervention; basic research in reading acquisition. Home: 6 Northgate Rd Wellesley MA 02181 Office: Box 75K Tufts-New Eng Med Ctr 171 Harrison Ave Boston MA 02111

BERNLOHR, ROBERT WILLIAM, biochemistry educator; b. Columbus, Ohio, Apr. 20, 1933; s. William Frederick and Ruth Elizabeth (Russell) B.; m. Carol Jean Smiley, June 11, 1955; children: David, Timothy, Mark, James. B.S. in Chemistry, Capital U., 1955, D.Sc. (hon.), 1980; Ph.D. in Biochemistry, Ohio State U., 1958. Research assoc. Oak Ridge Nat. Lab., 1958-60; asst. prof. biochemistry Ohio State U., Columbus, 1960-62; asst. prof. to prof. microbiology and biochemistry U. Minn., Mpls., 1962-74; prof. biochemistry and microbiology Pa. State U., University Park, 1974—, head biochemistry, microbiology, molecular and cell biology dept., 1974—; cons. NSF, Washington, 1968-71. Mem. editorial bd.: Jour. Bacteriology, 1965—; contbr. over 100 articles to sci. publs. USPHS grantee, 1962-72; NSF grantee, 1960—; NIH grantee, 1980-83. Mem. AAAS, Am. Soc. Biol. Chemists, Biochem. Soc., Am. Soc. Microbiology (chmn. physiology div. 1972-74), Sigma Xi. Subspecialties: Biochemistry (biology); Microbiology. Current work: Basic research on the regulation of growth, physiology, and metabolism of bacteria; regulation of differentiation; enzymology. Office: Pa State Univ 108 Althouse Lab University Park PA 16802

BERNS, DONALD SHELDON, biophysical chemist, researcher, consultant; b. Bronx, June 27, 1934; s. Benjamin and May (Cohen) B.; m. Sylvia Schleicher, Feb. 5, 1956; children: Brian, Neil, Amy. B.S. in Chemistry, Wilkes Coll., 1955, Ph.D., U. Pa., 1959. Postdoctoral fellow Yale U., New Haven, 1959-61; resident research assoc. Argonne (Ill.) Nat. Labs., 1961-62; adj. prof. chemistry dept. Rensselaer Poly. Inst., Troy, N.Y., 1970—; assoc. prof. biochemistry Albany Med. Coll., 1962-76; research scientist, sr. assoc. prin. Center Labs. and Research, Albany, N.Y., 1962—, chief molecular biology sect., 1982—, chief biophysics lab., 1983—; vis. prof. Hadassah-Hebrew U. Med. Sch., Jerusalem, 1972-73; chmn. promotion and tenure com. N.Y. State Health Bd., Albany, 1981-83. Contbr. numerous articles to profl. jours. Pres. Hebrew Acad. Capitol Dist.; exec. com. Jewish Welfare Fund. U.S. Dept. Environ. fellow, 1972-73; NSF grantee, 1962-74; NIH grantee, 1974-82; recipient travel awards Research Corp., France, Israel, Japan, Portugal, 1964, 68, 74, 75. Mem. Am. Chem. Soc., Biophys. Soc., Am. Soc. Photobiology, Am. Soc. Biol. Chemists, AAAS. Subspecialties: Biophysical chemistry; Biochemistry (biology). Current work: Model systems for energy transduction, membrane structure and function, protein structure and function. Home: 42 Winnie St Albany NY 12208 Office: Ctr Labs and Research NY State Health Dept Empire State Plaza Albany NY 12201

BERNSTEIN, ANNE CAROLYN, psychologist, educator, writer; b. N.Y.C., Apr. 8, 1944; d. Alfred Jacob and Clara (Handelman) B.; m. Conn M. Hallinan, Jan. 16, 1982; 1 son, David Alexander Hallinan; m.; stepchildren: Sean B., Antonio B., Brian D. B.A. magna cum laude with honors in Sociology, Brandeis U., 1965; Ph.D., U. Calif., Berkeley, 1973. Lic. psychologist, marriage, family and child counselor, Calif. Psychologist Family Therapy Inst. Marin, San Rafael, Calif., 1972-73; psychologist, instr. U. Calif., Santa Cruz, 1973-75; contract instr. JFK U., Calif. Sch. Profl. Psychology, Calif. State U., Hayward, 1974-76; program dir. Lone Mountain Coll., also U. San Francisco, 1977-81; psychologist Rockridge Health Plan, Oakland, Calif., 1979—; prof. psychology Wright Inst., 1981—; treas., bd. dirs. Children's Rights Group, San Francisco, 1976—; mem. council Psychotherapy Inst., Berkeley, 1981-82; cons. in field. Author: The Flight of the Stork, 1978; contbr. chpts. to books, articles to profl. publs. USPHS grantee, 1967-71. Mem. Am. Family Therapy Assn. (charter), Am. Psychol. Assn., Am. Orthopsychiat. Assn., Sex Edn. and Info. Council U.S., Am. Assn. Sex Educators, Counselors and Therapists, No. Calif. Family Therapy Assn., Phi Beta Kappa. Subspecialties: Clinical psychology; Developmental psychology. Current work: Family therapy, child development and parent education; family systems medicine. Office: Wright Inst 2728 Durant Ave Berkeley CA 94704

BERNSTEIN, CAROL, molecular biologist, educator; b. Paterson, N.J., Mar. 20, 1941; d. Benjamin and Mina (Regenbogen) Adelberg; m. Harris Bernstein, June 7, 1962; children: Beryl, Golda, Benjamin. B.S. in Physics, U. Chgo., 1961; M.S. in Biophysics, Yale U., 1963; Ph.D. in Genetics, U. Calif,-Davis, 1967. NIH postdoctoral fellow U. Calif.-Davis, 1967-68; research asso. dept. molecular and med. microbiology Coll. Medicine, U. Ariz., 1968-77, adj. asst. prof., 1977-81, adj. asso. prof., 1981—; proposal reviewer NSF; invited speaker Am. Soc. Microbiology Ann. Meeting, 1982. Contbr. articles to profl. jours. NSF grantee, 1975-77, 77-79; Nat. Found. grantee, 1975-76; NIH grantee, 1979-82. Mem. Genetics Soc. Am., Biophys. Soc., Am. Soc. Microbiology, Fedn. Am. Scientists, AAUP (pres.-elect U. Ariz. chpt. 1982), Am. Women in Sci. Democrat. Jewish. Subspecialties: Molecular biology; Microbiology. Current work: Molecular basis of sex and aging; DNA repair in phage T4; mechanisms of recombinational repair, including enzymes involved and phys. intermediates; recombinationl repair during sex to reverse aging. Office: Dept Molecular and Med Microbiology Coll Medicine U Ariz Tuscon AZ 85724

BERNSTEIN, ELLIOT ROY, chemistry educator; b. N.Y.C., Apr. 14, 1941; s. Leonard H. and Geraldine (Goldberg) B.; m. Barbara Wyman, Dec. 19, 1965; children: Jephta, Rebecca. A.B., Princeton U., 1963; Ph.D., Calif. Inst. Tech., 1967. Asst. prof. chemistry Princeton U., 1969-75; assoc. prof. chemistry Colo. State U., Ft. Collins, 1975-79, prof., 1979—; cons. Los Alamos Nat. Lab., 1975 82. Contbr. articles to profl. jours. Mem. Am. Chem. Soc., Am. Phys. Soc., AAAS. Subspecialties: Physical chemistry; Spectroscopy. Current work: Condensed matter science, molecular crystals, cryogenic liquids, phase transitions and critical phenomena, laser spectroscopy, light scattering. Patentee in field. Office: Dept Chemistry Colo State U Fort Collins CO 80523

BERNSTEIN, EUGENE HAROLD, biochemist-virologist, research and devel. labs. exec., researcher; b. N.Y.C., Sept. 26, 1926; s. Max and Anna (Cohen) B.; m. Bernice Alberta, Dec. 21, 1952; children: Andrew Mark, Robert Stuart, Jill Terri. B.S., U.S. Mcht. Marine Acad., 1947, Rutgers U., New Brunswick, N.J., 1951, M.S., 1952, Ph.D., 1955. Postdoctoral Fellow Sloan-Kettering Inst., N.Y.C., 1955-56; enzymdogist biochemistry div. Colgate-Palmolive Co., Piscataway, N.J., 1956-59; owner, dir. Univ. Lab., Inc., Highland Park, N.J., 1959—; vis. prof. Waksman Inst. Microbiology, Rutgers U., 1959-68, 82—. Contbr. articles to profl. jours. Served to ensign USNR, 1944-47. Mem. Am. Chem. Soc., AAAS, N.Y. Acad. Sci., N.J. Acad. Sci., Am. Soc. Microbiology, Am. Inst. Chemists, Am. Cancer Soc., Internat. Assn. Leukemia Research, Phi Beta Kappa, Sigma Xi, Beta Beta Beta. Jewish. Subspecialties: Biochemistry (biology); Virology (biology). Current work: Tumor virology research; application of enzyme immuno assay techniques for detection of virus diseases of poultry. Patentee in field. Home: 901 S Park Ave Highland Park NJ 08904 Office: 810 N 2d Ave Highland Park NJ 08904

BERNSTEIN, HARRIS, educator; b. Bklyn., Dec. 12, 1934; s. Benjamin and Hannah (Simonwitz) B.; m. Carol Bernstein, June 7, 1962; children: Beryl, Golda, Benjamin. B.S., Purdue U., 1956; Ph.D., Calif. Inst. Tech., 1961. Postdoctoral fellow Yale U., New Haven, 1961-63; asst. prof. U. Calif., Davis, 1963-68; assoc. prof. U. Ariz, Tucson, 1968-74, prof., 1974—. Contbr. articles to profl. jours. NIH grantee, 1979—. Fellow Am. Acad. Microbiology; mem Am. Soc. Biol. Chemists, Genetics Soc. Am., AAAS. Subspecialties: Molecular biology; Genetics and genetic engineering (biology). Current work: Mechanism of DNA replication, DNA repair, recombination and mutation; evolution of sexual reproduction; DNA damage as basis of aging. Home: 2639 E 4th St Tucson AZ 85716 Office: Dept Molecular and Med Microbiology Coll Medicine Univ Ariz Tucson AZ 85724

BERNSTEIN, HERBERT JACOB, computer consultant, scientist; b. Bklyn., Sept. 26, 1944; s. David and Dorothy (Ashery-Skupsky) B.; m. Frances C. Turnheim, Apr. 7, 1968; children: Michael Edward, Daniel Julius. B.A. magna cum laude with honors in Math, NYU, 1964, M.S., 1965, Ph.D., 1968. Asst. research scientist NYU, 1968, assoc. research scientist, 1968-70, adj. assoc. prof. math., 1968-69, adj. assoc. prof. computer sci., 1982—; vis. mem. Courant Inst. Math. Scis., 1982-83, sr. research scientist, 1983—; sci. program analyst II Brookhaven Nat. Lab., Upton, N.Y., 1970-71, sr. computer sci. analyst, 1971-74, assoc. scientist, 1974-78, scientist, 1978-83, research collaborator, 1983—; computer cons., prin. Bernstein & Sons, Bellport, N.Y., 1971—. Author: (with N.V. Findler and J.L. Pfaltz) Four High Level Extensions of FORTRAN IV, 1972; contbr. articles to profl. jours. N.Y. State Regents sci. scholar, 1961-64; NFS Coop. fellow, 1964-65, 65-66; NSF fellow, 1966-67. Mem. Am. Math. Soc., Am. Math. Assn., Soc. Indsl. and Applied Math., Am. Crystallographic Assn., Phi Beta Kappa, Pi Mu Epsilon, Sigma Pi Sigma. Subspecialties: Numerical analysis; Distributed systems and networks. Current work: High speed parallel computation, robotics, symbolic manipulation, data acquisition systems, graphics, data communications. Home and Office: 5 Brewster Ln Bellport NY 11713 Office: Courant Inst Math Scis NYU 251 Mercer St New York NY 10012

BERNSTEIN, I. LEONARD, physician, educator; b. Jersey City, Feb. 17, 1924; s. Sydney and Jean B.; m. Miriam Goldman, Aug. 29, 1948; children—David, Susan, Ellen, Jonathan. Student, St. John's U., Bklyn., 1940-41, George Washington U., 1941-43; M.D., U. Cin., 1949. Diplomate: Am. Bd. Internal Medicine, Am. Bd. Allergy and Clin. Immunology. Intern Cin. Gen. Hosp., 1949-50; jr. resident in internal medicine Jewish Hosp., Cin., 1950-51; resident in chest diseases Bellevue Hosp., N.Y.C., 1953-55; fellow in allergy and immunology Northwestern U. Med. Sch., 1955-56; mem. faculty U. Cin. Med. Center, 1956—, clin. prof. medicine, 1971—, dir. allergy clinic, 1971—, dir. allergy tng. program, 1958—, dir., 1958—; trustee faculty council on Jewish affairs U. Cin., 1967—; attending physician Cin. Gen., VA, Jewish hosps. Contbr. articles to med. publns. Trustee Adath Israel Synagogue, 1980—. Served with AUS, 1943-45; as officer M.C. USAF, 1951-53. Grantee Nat. Inst. Allergy and Infectious Disease, 1958—. Fellow Am. Acad. Allergy (pres. 1982—), A.C.P.; mem. Am. Assn. Immunologists, AAAS, Central Soc. Clin. Research, Am. Thoracic Soc., Soc. Occupational and Environ. Health. Subspecialties: Allergy; Immunology (medicine). Current work: Occupational asthma; food allergy; respiratory allergy due to dust, mites and algae. Home: 3117 Esther Dr Cincinnati OH 45213 Office: 8464 Winton Rd Cincinnati OH 45231 Productive basic and clinical research requires not only an innovative spark but also patience and unswerving determination. Personal commitment, ability to organize working time and association with intellectually honest colleagues are prerequisites to success in these fields.

BERNSTEIN, ISADORE ABRAHAM, biochemistry educator, researcher; b. Clarksburg, W.Va., Dec 23, 1919; s. William and Rosa B.; m. Claire Bernstein, Sept. 8, 1942; children: Lynne, Amy. A.B., Johns Hopkins U., 1941; P.D., Western Res. U., 1952. Research assoc. Case Western Res. U., Cleve., 1951-52, sr. instr., 1952-53; research assoc. Inst. Indsl. Health, U. Mich., Ann Arbor, 1953-56, 59-70, instr. biol. chemistry, 1954-57, asst. prof., 1957-61, assoc. prof., 1961-68, prof. dept. dermatology, 1968-71, assoc. prof. dept. indsl. health, 1961-67, prof., 1967-70, prof. dept. biochemistry, 1971—, prof. dept. environ. and indsl. health, 1970—; assoc. dir. research Inst. Environ. and Indsl. Health, 1978—; vis. prof. Osaka (Japan) U., 1963-64, Rockefeller U., N.Y.C., 1977-78; vis. scientist Hebrew U. Jerusalem, 1978. Contbr. numerous articles to sci. jours., chpts. to books; author 4 sci. books. Served to capt. U.S. Army, 1941-46. Decorated Bronze Star; co-recipient Internat. Meml. award for Psoriasis, 1959; recipient Stephen Rothman Meml. award Soc. Investigative Dermatology, 1981; Disting. Faculty Achievement award U. Mich., 1981. Fellow AAAS; mem. Am. Soc. Biol. Chemists, Am. Chem. Soc., Am. Soc. Microbiology, Soc. Investigative Dermatology, Am. Soc. Cell Biology, Soc. Toxicology, Am. Pub. Health Assn., Radiation Research Soc., Am. Inst. Biol. Sci., N.Y. Acad. Scis., Sigma Xi. Subspecialties: Biochemistry (medicine); Toxicology (medicine). Current work: Environmental toxicology, cutaneous biochemistry. Office: U Mich Ann Arbor MI 48109

BERNSTEIN, JAY, pathologist, researcher, educator; b. N.Y.C., May 14, 1927; s. Michael Kenneth and Frances K. B.; m. Carol Irene Kritchman, Aug. 11, 1957; children: John Abel, Michael Kenneth. B.A., Columbia U., 1948; M.D., SUNY-Downstate Med. Ctr., 1952. Diplomate: Am. Bd. Pathology. Intern Peter Bent Brigham Hosp., Boston, 1952-53; resident in pathology Boston Lying-In Hosp., 1953, Free Hosp. for Women, Brookline, Mass., 1954, Children's Med. Ctr., Boston, 1954-56; pathologist Children's Hosp. Mich., Detroit, 1956-62; attending pathologist, assoc. prof. Albert Einstein Coll. Medicine, 1962-68; dir. dept. anatomic pathology William Beaumont Hosp., Royal Oak, Mich., 1969—; vis. prof. pathology Albert Einstein Coll. Medicine, 1974 ; clin. prof. pathology Wayne State U. Sch. Medicine, 1977—; clin. prof. health scis. Oakland U., 1977—; mem. sci. adv. bd. Nat. Kidney Found. and Nat. Kidney Found. of Mich.; mem. WHO Com. on Renal Nomenclature. Mem. editorial bd., contbg. editor pathology: Jour. Pediatrics, 1969—; mem. editorial bd.: Perspective in Pediatric Pathology, 1970; co-editor, 1979—; mem. editorial bd.: Pediatric Pathology, 1982—, Urol. Radiology, 1982—. Served with USN, 1945-46. Mem. Am. Assn. Pathologists, Internat. Acad. Pathology, Am. Soc. Clin. Pathologists, Am. Soc. Pediatrics, Am. Acad. Pediatrics, Soc. Pediatric Research, Am. Soc. Pediatric Nephrology (council 1969), Am. Soc. Nephrology, Pediatric Pathology Club (council 1966-67, pres. 1968). Subspecialties: Pathology (medicine); Nephrology. Office: William Beaumont Hosp 3601 W 13 Mile Rd Royal Oak MI 48072

BERNSTEIN, JERALD J(ACK), neuroscientist; b. Bklyn., Mar. 30, 1934; m.; children: Steven, David. B.S., Hunter Coll., 1955; M.S., U. Mich., 1957, Ph.D., 1959. USPHS fellow NIH, Bethesda, Md., 1959; research biologist Lab. Neuroanat. Sci., Nat. Inst Neurol and Communicative Diseases and Stroke, NIH, Bethesda, 1960 -65; asst. prof. anatomy U. Fla. Coll. Medicine, 1965-69, assoc. prof. anat. sci., 1969-70, assoc. prof. neurosci., 1970-76, prof. neurosci. and ophthalmology, 1976-80; prof. neurosurgery and physiology George Washington U. Sch. Medicine, 1980—; chief Lab. Central Nevous System Injury and Regeneration, VA Med. Center, Washington, 1980—; vis. prof. anatomy Hadassah Med. Sch., Jerusalem, 1978; career research scientist VA, 1980. Author numerous articles, abstracts, and book chpts. on spinal cord regeneration. NIH grantee, 1965-82; VA grantee, 1980—; Office Naval Research grantee, 1982. Mem. Soc. Neurosci., Am. Physiol. Soc., Am. Assn. Anatomists, AAAS, Cajal Club. Subspecialties: Regeneration; Neurobiology. Current work: Spinal cord injury and regeneration.

BERNSTEIN, JERROLD, physician; b. Montreal, Que., Dec. 26, 1936; s. David and Yetta Jacqueline (Hertz) B.; m. Kathleen Marlyn Barber, May 9, 1971; children: Kevin, Patrick, Jennifer. B.Sc., McGill U., 1958, M.D., C.M., 1963; Ph.D., U. Ill., 1967. Pharmacologist, group leader Merck Frosst Labs., Kirkland, Que., 1969-74; dir. clin. investigation Smith Kline Can., St. Laurent, 1974-77; head clin. investigation Ciba Geigy Can., Dorval, 1977-79; sr. dir. anti inflammatory/endocrine sect., Summit, N.J., 1979-82; dir. clin. investigation Lederle Labs., Pearl River, N.Y., 1982—; Vice pres. long-range planning, dir. Mixed Media art supply house, Toronto, 1980—. Contbr. articles to profl. jours. Mem. Soc. Clin. Trials, Am. Chem. Soc., Am. Epilepsy Soc., Can. Assn. Clin. Pharmacology, Internat. Assn. Study Pain, N.Y. Acad. Scis., Soc. Neurosci. Subspecialties: Pharmacology; Neurophysiology. Current work: Clinical trials in neurophysiology. Home: 62 Stanford Ave West Orange NJ 07052 Office: Med Research div Lederle Labs Pearl River NY 10965

BERNSTEIN, RALPH, scientist, electrical engineer, educator; b. Zweibrucken, Germany, Feb. 20, 1933; s. Eleazar L. and Martha (Uhlfelder) B.; m. Leah K., Aug. 23, 1959; children: Elanna, Stuart N., Alexander P. B.S.E.E., U. Conn., 1956; M.S.E.E., Syracuse U., 1960. With IBM Corp., 1956—, adv. engr., Bethesda, Md., 1956-69, sr. engr., Gaithersburg, Md., 1969-79; mem. sci. staff Palo Alto (Calif.) Sci. Center, 1980—; mem. Space Sci. Bd. NRC/Nat. Acad. Scis., 1977-81; cons. NASA Space Sci. Adv. Com. of NASA Adv. Council, 198—82; exec. com. Earth Environ. and Resources Conf., 1974; ex-officio mem. com. earth scis. Space Sci. Bd., 1978-79; lectr. George Washington U., Cornell U., U. Pa., U. Tenn., Cath. U., Am. U., U. Calif.-Santa Barbara, U. Calif.-Davis, U.S. Coast Guard Acad., Stanford U. Editor: Digital Image Processing for Remote Sensing, 1978; contbr. chpts. in books. Pres. Luxmanor Citizens Assn., 1973-75, exec. com., 1975-79; mem. Montgomery County Adv. Com. Sector Plan Devel. and Metro, 1974-77. Served with USAF, 1951-53. Recipient medal for exceptional sci. achievement NASA, 1974, Outstanding Contbn. award IBM, 1974, Project Team Group Achievement award NASA, 1972. Fellow IEEE; mem. AAAS, Am. Soc. Photogrammetry, Soc. Info. Display, Am. Astron. Soc. Democrat. Jewish. Subspecialties: Computer engineering; Graphics, image processing, and pattern recognition. Current work: Digital image processing of earch observation sensor data, image enhancement, convection, information extraction. Patentee in field. Home: 1201 Woodview Terr Los Altos CA 94022 Office: 1530 Page Mill Rd Palo Alto CA 94304

BERNSTEIN, RICHARD BARRY, phys. chemist, educator; b. N.Y.C., Oct. 31, 1923; s. Simon and Stella (Grossman) B.; m. Norma B. Olivier, Dec. 17, 1948; children—Neil David, Minda Dianne, Beth Anne, Julie Lynn. A.B., Columbia, 1943, M.A., 1944, Ph.D., 1948. Mem. research staff SAM Lab., Columbia, 1942-46; asst. prof. chemistry Ill. Inst. Tech., 1948-53, Kilpatrick lectr., 1974; from asst. prof. to prof. chemistry U. Mich., 1953-63; prof. chemistry U. Wis., 1963-66, W.W. Daniells prof., 1966-73; W.T. Doherty prof. chemistry, prof. physics U. Tex., Austin, 1973-77; Higgins prof. natural sci. and chemistry Columbia U., 1977—, chmn. chemistry dept., 1979—; Reilly Centennial lectr. and award honor U. Notre Dame, 1965; Mack lectr. Ohio State U., 1966; FMC lectr. Princeton, 1966; N.Y. State distinguished lectr. Yeshiva U., 1969; Falk-Plaut lectr. Columbia U., 1971; Dreyfus lectr. U. Kans., 1973; vis. prof. Hebrew U., Jerusalem, 1973, Pa. State U., 1974; del. Pontifical Acad. Conf., 1966; mem. adv. bds. chemistry NRC, 1965-69; adv. bd. Army Research Office, 1967-69; Sloan fellow, 1956-60; sr. postdoctoral fellow NSF, 1960-61; Mem. program com. A.P. Sloan Found., 1971-76; mem. ednl. adv. bd. J.S. Guggenheim Found., 1978—; chmn. Office Chemistry and Chem. Tech., NRC, Nat. Acad. Scis., 1974-79. Author books papers in area of chem. physics; editor Chem. Physics Letters, 1978—. Served with C.E. AUS, 1942-44. Fellow Am. Phys. Soc. (chmn. div. chem. physics 1967-68), Am. Acad. Arts and Scis., AAAS; mem. Am. Chem. Soc. (chmn. div. phys. chemistry 1965-66, chmn. div. phys. chemistry Nebr. lectr. award, 1976, Peter Debye award in phys. chemistry, 1981), Nat. Acad. Scis., Sigma Xi. Subspecialty: Physical chemistry. Home: 460 Riverside Dr New York NY 10027 Office: Havemeyer Hall Columbia U New York NY 10027

BERNSTEIN, SANFORD IRWIN, molecular biologist; b. Bklyn., June 10, 1953; s. Harold and Adele Dorothy (Kutner) B. B.S., SUNY-Stony Brook, 1974; Ph.D., Wesleyan U., 1979. Research fellow dept. biology U. Va., Charlottesville, 1979-82; asst. prof. dept. biology San Diego State U., 1983—. Contbr. articles to profl. jours. Muscular Dystrophy Assn. postdoctoral fellow, 1979-82. Mem. Genetics Soc. Am., Sigma Xi. Subspecialties: Developmental biology; Genome organization. Current work: Structure, organization and expression of Drosophila muscle genes. (Recombinant DNA technology). Office: Dept Biology and Molecular Biology Inst San Diego State U San Diego CA 92182

BERNSTEIN, SHELLY COREY, pediatric hematologist, oncologist; b. Chgo., May 4, 1951; s. Maurice H. and Robin (Dorfman) B.; m. Nancy J. Levy, June 23, 1977. A.B., U. Chgo., 1973, Ph.D. in Genetics, 1978, M.D., 1980. Diplomate: Nat. Bd. Med. Examiners. Resident in pediatrics Children's Hosp. Med. Ctr., Boston, 1980-82, fellow in medicine, 1982—; clin. fellow div. pediatrics Dana-Farber Cancer Inst., Boston, 1982—; research fellow in pediatrics Harvard Med. Sch., 1982—; vis. scientist Whitehead Inst. Biomed. Research, Ctr. Cancer Research and dept. biology MIT, 1983—. Contbr. articles to profl. jours. ITT Internat. fellow, Cameroon, 1976-77; recipient award Nat. Research Service award NIH, 1975 77; Med. Alumni prize U. Chgo., 1980. Mem. Am. Soc. Human Genetics, Genetics Soc. Am., Am. Genetic Assn., Mass. Med.Soc., AMA, Sigma Xi. Subspecialties: Pediatrics; Hematology. Current work: Cancer research—investigations on the genetics of cancer. Home: 129 Oakdale Rd Newton Highlands MA 02161 Office: Dana-Farber Cancer Inst 44 Binney St Boston MA 02115

BERREMAN, DWIGHT WINTON, physicist; b. Salem, Oreg., May 30, 1928; s. Joel V. and Sevilla M. (Ricks) B.; m. Patricia A. McDorman, July 31, 1954; children: Diana L. Berreman Davies, Steven L., Ronald L.; m. Dorothy Ellen Wickman, Aug. 13, 1977. B.S., U. Oreg., 1950; M.S., Calif. Tech. U., 1952, Ph.D., 1955. Mem. staff Stanford Research Inst., Menlo Park, Calif., 1955-56; asst. prof. physics U. Oreg., Eugene, 1956-61; physicist, mem. tech. staff Bell Labs., Murray Hill, N.J., 1961—. Contbr. numerous articles to profl. publs. Mem. Am. Phys. Soc., Optical Soc. Am., Soc. Info. Display. Subspecialties: Condensed matter physics; Optic research. Current work: Liquid crystal optics and fluid dynamics in LCD devices; physical optics of layered anisotropic media; x-ray optics. Holder 13 U.S. patents. Office: 600 Mountain Av Room 1D 435 Murray Hill NJ 07974

BERRIOS-ORTIZ, ANGEL, biology educator; b. Cidra, P.R., May 4, 1935; s. Ramon and Carmen (Ortiz) Berrios-O. B.S., U. P.R., 1957; M.S., U. Ill., 1961, Ph.D., 1973. Instr. U. P.R., Mayaguez, 1957-62, asst. prof. biology, 1962-67, assoc. prof., 1967-77, prof., 1977—, asst. head biology dept., 1965-67, acting head, 1967-68, curator entomology and zoology collections, 1973—. Co-author book in field. NSF fellow, 1964. Mem. Entomol. Soc. Am., History of Sci. Soc., Sociedad Colombiana de Entomologia, Sociedad Puerto riquena de Entomologia, Sigma Xi. Roman Catholic. Lodge: KC. Subspecialty: Morphology. Current work: Morphology muscular system in insects (immature); blister beetles. Office: Dept Biology U Puerto Rico Mayaguez PR 00708 Home: 22 Aguadilla St Hato Rey PR 00919

BERRY, CHARLES RICHARD, plant pathologist; b. Morgantown, W.Va., Mar. 8, 1927; s. Homer Hays and Fanny Belle (Clowser) B.; m. Frances Ruth Craig, Dec. 25, 1948; children: Thomas Michael, Joseph Warren. B.A., Glenville State Coll., 1949; M.S., W.Va. U., 1955, Ph.D., 1958. Research plant pathologist U.S. Forest Service, Athens, Ga., 1957—. Served with USN, 1945-46; Served with U.S. Army, 1950-51. Mem. Am. Phytopath. Soc., Am. Agronomy Soc., Soil Sci. Am., Am. Council Reclamation Research. Methodist. Subspecialties: Plant pathology; Resource conservation. Current work: Mycorrhieal research, surface mine reclamation. Home: 179 Tara Way Athens GA 30606 Office: Carlton St Foresty Scis Lab Athens GA 30602

BERRY, CHRISTINE ALBACHTEN, physiologist, educator, researcher; b. San Diego, Nov. 27, 1946; d. Hubert Thomas and Rosella (Allen) Albachten; m. William Dearborn Berry, Sept. 23, 1968; 1 son, Christopher. B.A., Stanford U., 1968; Ph.D., Yale U., 1974. Fellow U. Calif., San Francisco, 1974-76, asst. prof. physiology, 1977-83, assoc. prof. Mem. editorial bd.: Am. Jour. Physiology, 1982—; contbr. chpts. to books, articles to profl. jours. NIH grantee, 1979—. Mem. AAAS, Am. Soc. Nephrology, Am. Physiol. Soc., Internat. Soc. Nephrology. Republican. Subspecialties: Physiology (biology); Physiology (medicine). Current work: Salt and water transport across the proximal tubule of the kidney. Home: 5 Owlswood Rd Tiburon CA 94920 Office: Nephrology Division 1065 HS University of California San Francisco CA 94143

BERRY, DAVID LESTER, toxicologist, researcher; b. Red Bluff, Calif., Mar. 27, 1944; s. Lester Joseph and Hilda Mary (Jessen) B.; m. Linda Susan Converse, June 15, 1974; 1 dau., Angela Megan. B.S., U. Calif.-Davis, 1967; M.S., Utah State U., 1971, Ph.D. (Dow fellow), 1974. NIH postdoctoral trainee U. Wash., Seattle, 1974-76; staff scientist Oak Ridge Nat. Lab., 1976-78; vis. prof. German Cancer Research Ctr., Heidelberg, 1979-80; staff scientist, project leader Western Research Ctr., USDA, Berkeley, Calif., 1980—. Contbr. articles to profl. jours. USDA grantee, 1968-71; Nat. Cancer Inst. grantee, 1976-79, 82; Nat. Acad. Scis. grantee, 1978. Mem. Am. Assn. Cancer Research, Am. Chem. Soc., Am. Soc. Pharmacology and Explt. Therapeutics, Internat. Union Against Cancer, Tissue Culture Assn., No. Calif. Genetic Toxicology Assn. Subspecialties: Cancer research (medicine); Toxicology (medicine). Current work: Chemical Carcinogenesis and genetic toxicology, nutrition and cancer. Office: Toxicology Unit 800 Buchanan St Berkeley CA 94710

BERRY, EDWIN X, atmospheric phys. scientist, research and devel. co. exec.; b. San Francisco, June 2, 1935; s. Edwin F. and Frances A. B.; m. Valerie Stella, Oct. 27, 1973. B.S.E.E., Calif. Inst. Tech., 1957; M.A. in Physics, Dartmouth Coll., 1960; Ph.D. in Atmospheric Physics, U. Nev., Reno, 1965. Research assoc. U. Nev., 1961-72; program mgr. NSF, Washington, 1972-73, Burlingame, Calif., 1973-75; pres. Atmospheric Research & Tech., Inc., Sacramento, 1975—. Contbr. articles, primarily in atmospheric physics, to profl. jours. Mem. Am. Meterol. Soc., Am. Wind Energy Assn. Subspecialties: Meteorologic instrumentation; Wind power. Current work: Devel. battery-powered computers; wind energy evaluation. Patentee vibration sensitive valve operating apparatus, vibration/temperature sensitive valve operating apparatus; designer ART800, low powered instrumentation microcomputer, 1981-82. Office: 6040 Verner Ave Sacramento CA 95841

BERRY, GUY CURTIS, educator; b. Greene County, Ill., May 11, 1935; s. Charles Curtis and Wilma Francis (Wickes) B.; m. Marilyn Jane Montooth, Jan. 9, 1957; children: Susan Jane, Sandra Jean, Scott Curtis. B.S., U. Mich., 1957, M.S., 1958, Ph.D., 1960. Mellon Inst. fellow, 1960-64, sr. fellow, 1964-73; assoc. prof. Carnegie-Mellon U., Pitts., 1966-73, prof. chemistry and polymer sci., 1973—. Contbr. articles to sci. jours. Mem. Am. Chem. Soc., Rheology Soc., AAAS. Subspecialties: Polymer chemistry; Polymer physics. Current work: Rheology and light scattering on polymers and their solutions, properties of mesogenic polymers, branched chains, composites. Office: 4400 5th Ave Pittsburgh PA 15213

BERRY, JAMES FREDERICK, biology educator, ecology consultant; b. Washington, Dec. 22, 1947; s. James F. and Joyce (Drummond) B.; m. Cynthia Marie Valukas, July 31, 1982. B.S., Fla. State U., 1970, M.S., 1973; Ph.D., U. Utah, 1978. Chmist Fla. Dept. Agr., Tallahassee, 1973-74; teaching fellow U. Utah, Salt Lake City, 1974-78; asst. prof. biology Elmhurst (Ill.) Coll., 1978—; project leader ENCAP, Inc., DeKalb, Ill., 1981-82; research assoc. Carnegie Mus. Natural History, Pitts., 1983—. U.S. Office Endangered Species grantee, 1982. Mem. Am. Soc. Zoologists, Soc. Systematic Biology, Soc. Study of Evolution, Am. Soc. Ichthyologists and Herpetologists, Soc. Study of Amphibians and Reptiles. Democrat. Roman Catholic. Subspecialties: Population biology; Ecology. Current work: Evolutionary biology, ecology and systematics of reptiles; particularly turtles of the family Kinosternidae. Home: 248 R Washington Highwood IL 60040 Office: Dept Biology Elmhurst Coll 190 Prospect St Elmhurst IL 60126

BERRY, MICHAEL JAMES, phys. chemist; b. Chgo., July 17, 1947; s. Bernie Milton and Irene Barbara (Lentz) B.; m. Julianne Catherine Elward, Apr. 28, 1967; children: Michael James II, Jennifer Anne. B.S., U. Mich., Ann Arbor, 1967; Ph.D., U. Calif., Berkeley, 1970. Asst. prof. chemistry U. Wis., Madison, 1970-75, assoc. prof., 1975-76; mgr. photon chem. dept. Allied Chem. Corp., Morristown, N.J., 1976-79; Robert A. Welch prof. chemistry, dir. Rice Quantum Inst., Rice U., Houston, 1979—; pres. Antropix Corp., Houston, 1982—. Contbr. articles in field to profl. jours. NSF grantee, 1975—; Air Force Office Sci. Research, 1972-76; grantee Office Naval Research, 1972-74; recipient Fresenius award Phi Lambda Upsilon, 1982; Alfred P. Sloan Found. research fellow, 1975-76; Camille and Henry Dreyfus Found. fellow, 1974-76; John Simon Guggenheim Meml. Found. fellow, 1981-82. Mem. Am. Chem. Soc. (Pure chemistry award 1983), Am. Phys. Soc., Am. Soc. Photobiology, Optical Soc. Am. Subspecialties: Laser-induced chemistry; Laser photochemistry. Current work: Chem. applications of lasers, research and devel. in laser application, anti-entropic applications of sci. and tech. Home: 351 Tealwood St Houston TX 77024 Office: Dept Chemistry Rice U PO Box 1892 Houston TX 77251

BERRY, RICHARD WARREN, geology educator, researcher; b. Quincy, Mass., June 21, 1933; s. George Thomas and Blanche Claudia (Lambert) B.; m. Frances Sue Potter, June 11, 1958 (div. Sept. 1982); children: William, Jeffrey, Thomas. B.S. in Mining Engring, Lafayette Coll., 1955; M.A. in Geophysics, Washington U., St. Louis, 1957; Ph.D. in Geochemistry and Clay Mineralogy, Washington U., St. Louis, 1963. Instr. Trinity Coll., 1959-61; asst. prof. geology San Diego State U., 1961-65, assoc. prof., 1965-71, prof., 1971—, chmn. dept. geology, 1976-80; Fulbright prof. U. Baghdad, Iraq, 1965-66; postdoctoral profl. U. Oslo, 1968-69; cons. in field; NASA summer faculty fellow, Huntsville, Ala., about 1974. Author: (with David Tuer) Petroleum Industry Dictionary, 1980. Mem. San Diego Mayor's Com. on Offshore Drilling, 1979; bd. dirs. Campus br. YMCA, San Diego, Interscholar-Fgn. Exchange Group, San Diego. Royal Norweigan Council for Sci. and Indsl. Research postdoctoral research fellow, Oslo, 1968-69; Bur. Land Mgmt. grantee, 1975-76; NSF grantee, 1974-76. Fellow Geol. Soc. Am. (chmn. program com. 1979-80); mem. Mineral. Soc. Am., Soc. Econ. Paleontologists and Mineralogists, Clay Minral Soc., AAAS. Democrat. Presbyterian. Subspecialties: Mineralogy; Sedimentology. Current work: Clay mineralogy, major cation chemistry of marine sediments to reconstruct paleoenvironments or evaluate diagenesis and maturity regarding petroleum potential. Home: 4501 Collwood Blvd Apt 12 San Diego CA 92115 Office: San Diego State U Dept Geol Scis San Diego CA 92182

BERRY, STEPHEN DANIEL, psychobiology educator, neuroscience researcher; b. chgo., June 14, 1947; s. John Joseph and Kathlyn M. (Murphy) B.; m. Cathyann Therese Baldwin, July 12, 1969; children: Kimberly Anne, Seanna Lyn, Kevin Daniel. B.A. cum laude, U.Notre Dame, 1969; M.A., U. Conn., 1972, PH.D., 1975. Postdoctoral fellow dept. psychobiology U. Calif.-Irvine, 1975-77, research assoc., 1977-79; asst. prof. psychology Miami U., Oxford, Ohio, 1979-81, assoc. prof., 1981—, dir. Psychobiology Lab., 1979—. NIMH research fellow, 1975-77; Miami U. Faculty Research Com. fellow, 1980, 83; grantee, 1979-83; NIMH grantee, 1981-82; Nat. Inst. Occupational Safety and Health investigator, 1982-83. Mem. AAAS, Soc. Neurosci., Psychonomic Soc., Sigma Xi. Democrat. Subspecialties: Psychobiology; Neurophysiology. Current work: Neurobiology of learning and memory; neural correlates of classical conditioning in rabbits; the septo-hippocampal system and behavior. Office: Miami U 112 Benton Hall Oxford OH 45056

BERRY, WILLIAM BENJAMIN NEWELL, paleontologist, mus. dir.; b. Boston, Sept. 1,1931; s. John King and Margaret Elizabeth (Newell) B.; m. Suzanne Foster Spaulding, June 10, 1961; 1 son, Bradford Brown. A.B., Harvard U., 1953, A.M., 1955; Ph.D., Yale U., 1957. Asst. prof. geology U. Houston, 1957-58; asst. prof. to prof. paleontology U. Calif., Berkeley, 1958—; curator Mus. of Paleontology U. Calif., Berkeley, 1960-75, dir., 1975—, chmn. dept. paleontology, 1975—; cons. U.S. Geol. Survey. Author: Growth of a Prehistoric Time Scale, 1968; contbr. numerous articles on stratigraphic and paleontol. subjects to profl. jours.; editor publs. in geol. scis. Guggenheim Found. fellow, 1966-67. Mem. Paleontol. Soc., Geol. Soc. Norway, Internat. Platform Assn., Explorers Club. Subspecialties: Paleoecology; Ocean thermal energy conversion. Current work: Research currently involves investigation of anoxic ocean waters and sediment. Role of organics in formation of metal sulfides forming under anoxic conditions is one aspect of current research. Another is those conditions under which life may have developed in early stages of development of oceans—a time when oceans were anoxic save for surface layers. Home: 1366 Summit Rd Berkeley CA 94708 Office: Dept Paleontology U Calif Berkeley CA 94720

BERRY, WILLIAM BERNARD, electrical engineering educator; b. Shelby, Ohio, July 23, 1931; s. Darwin and Norma (Laux) B.; m. Lois Georgia Langford, June 25, 1955; children: Elizabeth Mava, William Joseph, Mary Suzanne, Thomas James. B.S.E.E., U. Notre Dame, 1953, M.S.E.E., 1957; Ph.D., Purdue U., 1963. Registered profl. engr., Ind. With Goodyear Atomic Corp., Portsmouth, Ohio, 1953; faculty Marquette U., Milw., 1957-61, Purdue U., West Lafayette, Ind., 1961-63, U. Notre Dame, Ind., 1963—, prof. elec. engring., 1970—, asst. dean for research and grad. study, 1974—, program mgr. cold weather transit tech., 1980—. Contbr. articles to profl. jours. Active Boy Scouts Am., 1967-72; instl. dir. United Way, 1977. Served with U.S. Army, 1953-55. ASEE/NASA summer faculty fellow, 1965, 66; NRS postdoctoral asso., 1972-73. Mem. Am. Soc. Engring. Edn., IEEE, Electrochem. Soc., Sigma Xi, Sigma Pi Sigma, Eta Kappa Nu. Roman Catholic. Subspecialties: Electronic materials; Solar energy. Current work: Electronic properties of materials, electron process in thin films, photovoltaics, thermoelectricity. Home: 402 E Pokagon St South Bend IN 46617 Office: Coll of Engring U Notre Dame Notre Dame IN 46556

BERS, DONALD MARTIN, physiology educator; b. N.Y.C., Dec. 13, 1953; s. Harold Theodore and Penny B.; m. Kathryn Eileen Hammond, July 17, 1976; children: Brian Alexander, Rebecca Ann. B.A., U. Colo., 1974; Ph.D., UCLA, 1978. Research and teaching assoc. UCLA Sch. Medicine and Nursing, 1976-78; research fellow UCLA Sch. Medicine, 1978-79, asst. research physiologist, 1980-82; research fellow Edinburgh (Scotland) U. Med. Sch., 1979-80; adj. asst. prof. UCLA Sch. Nursing, Los Angeles, 1981—; asst. prof. physiology U. Calif., Riverside, 1982—; Am. Heart Assn. research fellow, 1978-80. Recipient Lievre award Am. Heart Assn., 1978-79; Brit. Am. fellow, 1980-81. Mem. Biophys. Soc., Internat. Soc. Heart Research, Am. Physiol. Soc., AAAS. Subspecialties: Physiology (medicine); Cardiology. Current work: Mechanisms controlling the contractile force of cardiac muscle, excitation-contraction coupling and transmembrane ion movements in cardiac function. Office: U Calif Div Biomed Scis Riverside CA 92521

BERS, LIPMAN, mathematician, educator; b. Riga, Latvia, May 22, 1914; came to U.S., 1940, naturalized, 1949; s. Isaac and Bertha (Weinberg) B.; m. Mary Kagan, May 15, 1938; children: Ruth, Victor. Dr. Rerum Naturalium, U. Prague, 1938. Research instr. Brown U., 1942-45; asst. prof., asso. prof. Syracuse U., 1945-49; mem. Inst. Advanced Study, 1948-50; prof. N.Y. U., 1950-64, chmn. grad. dept. math., 1959-64; prof. Columbia U., 1964—, chmn. dept. math., 1972-75, Davies prof. math., 1973-82, Davies prof. math. emeritus, 1982, spl. prof., 1982-84; Vis. prof. Stanford U., summer 1955; vis. Miller Research prof. U. Calif. at Berkeley, 1968; chmn. Com. Support on Research on Math. Scis., Nat. Acad. Scis.-NRC, 1966-68; chmn. div. math. scis. NRC, 1969-71; chmn. U.S. Nat. Com. for Math., 1977-81. Author math. books.; Contbr. articles to math. jours. Fulbright fellow, 1959-60; Guggenheim fellow, 1959-60, 79. Fellow Am. Acad. Arts and Scis., AAAS (chmn. math. sect. 1973, 83), Am. Philos. Soc.; mem. Am. Math. Soc. (v.p. 1963-65, Steele prize 1975, pres. 1975-77), Fedn. Am. Scientists (council 1977-79, sponsor 1980—), Nat. Acad. Scis. (chmn. math. sect. 1967-70, chmn. com. on human rights 1979—). Subspecialty: Parasitology. Current work: Riemann surfaces, quasicon formal maps, kleinan groups. Home: 111 Hunter Ave New Rochelle NY 10801 Office: Dept Math Columbia U New York NY 10027

BERSON, JEROME ABRAHAM, chemist; b. Sanford, Fla., May 10, 1924; s. Joseph and Rebecca (Bernicker) B.; m. Bella Zevitovsky, June 30, 1946; children—Ruth, David, Jonathan. B.S cum laude, Coll. City N.Y., 1944; M.A., Columbia U., 1947; Ph.D., 1949; Ph.D. NRC postdoctoral fellow, Harvard U., 1949-50. Asst. chemist Hoffmann-LaRoche, Inc., Nutley, N.J., 1944; asst. prof. U. So. Calif., 1950-53, asso. prof., 1953-58, prof., 1958-63, U. Wis., 1963-69, Yale U., 1969-79, Irénée du Pont prof., 1979—, chmn. dept. chemistry, 1971-74; vis. prof. U. Calif., U. Cologne, U. Western, Ont.; Fairchild Distinguished scholar Calif. Inst. Tech.; cons. Riker Labs., Goodyear Tire & Rubber Co., Am. Cyanamid Co., IBM; mem. adv. panel for chemistry NSF; mem. medicinal chemistry study sect. NIH, 1969-73. Mem. editorial adv. bd.: Jour. Organic Chemistry, 1961-65, Accounts of Chemical Research, 1971-77, Nouveau Journal de Chimie, 1977—, Chem. Revs., 1980-83; contbr. articles to profl. jours. Served with AUS, 1944-46. CBI. Recipient Alexander von Humboldt award, 1980; John Simon Guggenheim fellow, 1980. Fellow Am. Acad. Arts and Scis.; mem. Nat. Acad. Scis., Am. Chem. Soc. (Calif. sect. award 1963, James Flack Norris award 1978, chmn. div. organic chemistry 1971), Chem. Soc. London, Phi Beta Kappa, Sigma Xi, Phi Lambda Upsilon. Subspecialty: Organic chemistry. Current work: Reaction mechanism; synthesis of theoretically significant molecules; characterization of reactive intemediates; high-spin systems. Home: 45 Bayberry Rd Hamden CT 06511 Office: Dept Chemistry Yale U PO Box 6666 New Haven CT 06511

BERTALANFFY, FELIX DIONYSIUS, anatomy educator; b. Vienna, Austria, Feb. 20, 1926; emigrated to Can., 1949, naturalized, 1955; s. Ludwig von and Maria (Bauer) B.; m. Gisele D. Lavimodiere, Jan. 20, 1954. Ph.D., Med. Sch., Vienna, 1945; M.Sc., McGill U., 1951, Ph.D., 1954. Asst. prof. anatomy U. Man., Winnipeg, 1955-58, assoc. prof., 1959-63, prof., 1964—. Contbr. articles to profl. jours. Fellow Royal Micros. Soc. London, Pan Am. Cancer Soc.; mem. Am. Assn. Cancer Research, Am. Assn. Anatomists, Canadian Assn. Anatomists, Internat. Soc. Stereology, Sigma Xi. Roman Catholic. Subspecialties: Anatomy and embryology; Microscopy. Current work: Cancer research; exfoliative cytology; histophysiology of respiratory system, regeneration, cell kinetics, carcinogenesis, histochemistry. Home: 886 Lindsay St Winnipeg MB Canada R3N 1H8 Office: 750 Bannatyne Ave Winnipeg MB Canada R3E OW3

BERTANI, GIUSEPPE, research geneticist, educator; b. Como, Italy, Oct. 23, 1923; s. Carlo and Armida (Seveso) B.; m. Lillian E. Teegarden, July 2, 1954; children: Christofer, Niklas. Dr. Sci. Nat., U. Milan, 1945; hon. Dr., Uppsala (Sweden) U., 1982. Various research positions, 1946-54; sr. research fellow in biology Calif. Inst. Tech., Pasadena, 1954-57; assoc. prof. med. microbiology U. So. Calif., Los Angeles, 1957-60; prof. microbial genetics Karolinska Inst., Stockholm, 1960-83; sr. research scientist Jet Propulsion Lab., Calif. Inst. Tech., 1981—. Author articles in microbial and molecular genetics.; Editorial bds. sci. jours. Mem. European Molecular Biology Orgn., Am. Soc. Microbiology, Genetics Soc. Am., AAAS, others. Subspecialties: Genetics and genetic engineering (biology); Microbiology. Current work: Lysogenic bacteria; bacterial viruses; genetic recombination; methanogenic bacteria; applications of genetics to biotechnology. Office: 4800 Oak Grove Dr Pasadena CA 91109

BERTÉ, FRANK JOSEPH, statistical engineering analysis group executive; b. N.Y.C., June 19, 1941; s. Dominic Joseph and Rose Louise (DeMarco) B.; m. Phyllis Theresa DeCristoforo, June 1, 1968; children: Frank Joseph, Nicholas Sebastian. A.A.S., Bronx Community Coll., 1961; B.M.E., CCNY, 1964; S.M., MIT, 1966, Nuclear Engr., 1967, Ph.D., 1971. Dresden asst. project engr. Commonwealth Edison, Chgo., 1971-73; prin. engr. Combustion Engring., Windsor, Conn., 1973-80, cons., 1980-81, supr., 1981—; pres. Innovative Technology Inc., West Hartford, Conn., 1981—, also dir. Mem. Am. Nuclear Soc. (Student Seminar award 1967). Democrat. Roman Catholic. Subspecialties: Nuclear fission; Statistics. Current work: Application of statistical methods to solution of nuclear engineering problems; application of engineering principles to naval architecture and sail design. Patentee improved sailboat. Office: Combustion Engineering Inc 100 Prospect Hill Rd Windsor CT 06095

BERTOLAMI, CHARLES NICHOLAS, oral biologist, oral and maxillofacial surgeon; b. Lorain, Ohio, Dec. 31, 1949; s. Salvatore Charles and Michela (Orlando) B.; m. Linda Marie Silva, June 26, 1977; children: Michela, Joseph. A.A., Lorain Community Coll., 1969; D.D.S., Ohio State U., 1974; D.Med.Sc., Harvard U., 1979. Diplomate: Am. Bd. Oral and Maxillofacial Surgery. Intern Mass. Gen. Hosp., Boston, 1975-76, asst. resident, 1978-79, chief resident, 1979-80; asst. prof. U. Conn., Farmington, 1980-83, Harvard U. Sch. Dental Sch., 1983—; surgeon Mass. Gen. Hosp., 1983—. Callahan Meml. awardee Ohio Dental Assn., 1974; research fellow Am. Soc. Oral Surgeons, 1975; New Investigator Research awardee Nat. Inst. Dental Research, 1983. Fellow Am. Assn. Oral and Maxillofacial Surgeons; mem. Am. Dental Assn., Hartford Dental Soc., Sigma Xi. Subspecialties: Oral biology; Oral and maxillofacial surgery. Current work: Research is directed toward establishing an analogy at a biochemical level between mammalian wound repair and embryonic development. This has involved a study of wound glycosaminoglycans, collagen, and fibroblasts. Goals include development of more objective methods for quantitating healing and generating an experimental animal model for hypertrophic scarring. Home: 67 Howard St Saugus MA 01906 Office: U Conn Health Ctr Farmington CT 06032

BERTOLOTTI, RAYMOND LEE, biomedical engineer, dentist; b. San Pedro, Calif., July 10, 1943; s. Peter J. and Ethel A. (Vaczi) B.; m. Mary V. Reasbeck, Sept. 22, 1973. Ph.D., U. Wash., 1970; D.D.S., U. Calif., San Francisco, 1978. Lic. dentist, Calif. Nuclear engr. Gen. Electric, San Jose, Calif., 1966-67; ceramic engr. Sandia Nat. Labs., Livermore, Calif., 1970-74; assoc. prof. dental materials U. Calif., San Francisco, 1978—. Subspecialties: Prosthodontics; Biomedical engineering. Current work: Dental materials, implants, biomedical engineering. Office: RL Bertolotti Inc 425 Estudillo Ave San Leandro CA 94577

BERTRAND, JOHN, research and development engineer; b. Athens, Greece, Feb. 6, 1943; came to U.S., 1962, naturalized, 1977; s. Peter and Vasso (Tsoli) B.; m. Daphne Helen Apostoleri, June 30, 1972; 1 dau.: Cleo Vasso. B.E.E., Columbia U., 1966; M.E.E., U. Calif.-Berkeley, 1967; PH.D., Columbia U., 1970. Sr. mem. tech. staff Wavetek Rockland, Inc., Rockleigh, N.J., 1972-81, mgr. software devel., 1981-83, sr, fellow, 1983—; instr. Columbia U., 1973-74, adj. asst. prof., 1974-77, adj. assoc. prof., 1979; group leader digital signal processing group Research Ctr. for Nat. Def., Athens, Greece, 1980-81. Contbr. numerous articles to profl. jours.; Patentee in field. Served with Greek Army, 1970-71. Eugene and Mona Gee fellow, 1966-67. Mem. IEEEE (Sr.), Assn. Computing Machinery, N.Y. Acad. Sci., Acoustical Soc. Am., Sigma Xi, Tau Beta Pi, Eta Kappa Nu. Subspecialties: Computer engineering; Computer architecture. Current work: Digital signal processing with applications in spectrum analysis and speech processing. Home: 211 Radcliff Dr Upper Nyack NY 10960 Office: 245 Livingston St Northvale NJ 07647

BERTRANDO, BERTRAND ROBERT, aerospace engineer; b. Chgo., May 30, 1927; s. Secondino and Nora (Burlando) B.; m. Virginia Prettyman, Jan. 7, 1952; children: Robert, William, Michael. B.A., U. Ill.-Urbana, 1948; M.S., Ill. Inst. Tech., 1950; M.B.E., Claremont Grad. Sch., 1973. Registered profl. engr., Calif. Research engr. Inst. Gas Tech., Chgo., 1948-50, Worcester Gas Light Co., Mass., 1949, Jet Propulsion Lab., Pasadena, Calif., 1953-57; mgr. tech. staff Gen. Electric Co., Santa Barbara, Calif., 1957-63; sr. engring. specialist Aerospace Corp., El Segundo, Calif., 1963—. Served with USMC, 1945-46. Fellow AIAA (assoc., pres. central Calif. sect. 1963); mem. System Safety Soc. (sr.). Clubs: La Cumbre Golf and Country (Santa Barbara, Calif.); Antheneum (Pasadena, Calif.). Subspecialties: Satellite studies; Aerospace engineering and technology. Current work: Design of satellites, system safety, environmental sciences. Home: 4108 Paloma Dr Santa Barbara CA 91310 Office: Aerospace Corp 2350 E El Segundo Blvd El Segundo CA

BERTUGLIA, LYNN ELLEN, mechanicial engineer; b. Albuquerque, Mar. 31, 1953; d. James Harlin and Frances Imogene (Looper) Gibson; m. David Thomas Bertuglia, Feb. 23, 1974; children: Michelle Lynn, Adam Scott. B.S. in Engring, Wichita State U., 1976; postgrad., U. Kans. Registered profl. engr., Kans., N.J., Wash. Assoc. engr. Black & Veatch Cons. Engrs., Kansas City, Mo., 1976-80; mech. engr. Franklin Assos., Ltd., Prairie Village, Kans., 1980—. Mem. Nat. Soc. Profl. Engrs., Kans. Engring. Soc., Mo. Soc. Profl. Engrs., ASME, Soc. Woman Engrs. Republican. Roman Catholic. Subspecialties: Resource management; Biomass (energy science and technology). Current work: Analysis and implementation of alternative solid waste mgmt. techniques. Office: 8340 Mission Rd Suite 101 Prairie Village KS 66206

BERTY, JOZSEF MIHALY, chemical engineering educator, reaction engineering company executive; b. Kispest, Hungary, Oct. 25, 1922; came to U.S., 1957, naturalized, 1962; s. Jozsef and Klara (Morzsanyi) B.; m. Gizella E. Reseky, Mar. 25, 1945; children: Laszlo, Bela, Peter, Annamaria, Orsolya, Imre. Dipl. Ch.E., Tech. U. Budapest, Hungary,

1944, D.Sc., 1950. Registered profl. engr., W.Va., N.Y., Pa. Head dept. Hungarian Oil & Gas Research Inst., Veszprem, Hungary, 1950-56; assoc. prof. Coll. Chem. Engring., Veszprem, 1953-56; vis. prof. Tech. U., Leuna-Merseburg, E.Ger., 1956-57; research assoc. Union Carbide Corp., South Charleston, W.Va., 1957-76; v.p. Berty Reaction Engrs. Ltd., Erie, Pa., 1976—; prof. chem. engring. U. Akron, Ohio, 1982—; cons. engring. burs., chem. and petroleum cos. Contbr. articles to profl. jours. Fellow Am. Inst. Chem. Engrs.; mem. Am. Chem. Soc., Soc. Engring. Sci. Rep. Roman Catholic (extraordinary minister 1977—). Subspecialties: Chemical engineering; Fuels. Current work: Chemical reaction engineering; improved design of chemical reactors; scale-up, preventions of thermal runaways; explosions, statistical experimental design. Patentee in field. Home: 310 W Grandview Blvd Erie PA 16508 Office: Chem Engring Dept U Akron Akron OH 44325

BERWALD, DAVID H., nuclear engr., researcher; b. N.Y.C., Feb. 24, 1952; s. Kiva H. and Rhoda M. (Goldberg) B.; m. Patricia E. Juris. B.S., Cornell U., 1974; M.S. in Engring, U. Mich., 1975, Ph.D., 1977. Research engr. Exxon Corp. Research, Linden, N.J., 1977-79; head fusion and reactor engring. sect. TRW Energy Devel. Group, Redondo Beach, Calif., 1979—. Contbr. articles on nuclear engring. to profl. jours.; referee nuclear tech. fusion: Am. Nuclear Soc. Jour. U. Mich. fellow, 1976, 77; Dept. Energy fellow, 1975. Mem. Am. Nuclear Soc. Subspecialties: Fusion; Nuclear engineering. Current work: Directing inter-disciplinary research group which is involved in the design of future nuclear reactors based upon nuclear fusion as the principal driving mechanism.

BESCH, HENRY ROLAND, JR., pharmacologist, educator; b. San Antonio, Sept. 12, 1942; s. Henry Rol and Monette Helen (Kasten) B.; 1 son, Kurt Theodore. B.Sc. in Physiology, Ohio State U., 1964, Ph.D. in Pharmacology (USPHS predoctoral trainee 1964-67), 1967; USPHS postdoctoral trainee, Baylor U. Coll. Medicine, Houston, 1968-70. Instr. Ob-Gyn Ohio State U. Med. Sch., 1967-68; mem. faculty Ind. U. Sch. Medicine, 1971—; prof. pharmacology and medicine, sr. research asso. Krannert Inst. Cardiology, chmn. dept. pharmacology and toxicology, 1977—, Showalter prof. pharmacology, 1980—; Can. Med. Research Council vis. prof., 1979, investigator fed. grants, mem. nat. panels and coms., cons. in field. Contbr. numerous articles pharm. and med. jours.; mem. editorial bds. profl. jours. Fellow Brit. Med. Research Council, 1970-71; Grantee Showalter Trust, 1975—. Fellow Am. Coll. Cardiology; Mem. AAAS, Am. Assn. Clin. Chemistry, Am. Fed. Clin. Research, Am. Heart Assn., Am. Physiol. Soc., Am. Soc. Biol. Chemists, Am. Soc. Pharmacology and Exptl. Therapeutics, Assn. Med. Sch. Pharmacologists, Biochem. Soc., Cardiac Muscle Soc., Internat. Soc. Heart Research (exec. com. Am. sect.), Nat. Acad. Clin. Biochemistry, N.Y. Acad. Scis., Sigma Xi. Subspecialties: Pharmacology; Toxicology (medicine). Office: 1100 W Michigan St Indianapolis IN 46223

BESIC, FRANK CHARLES, cardiologist; b. Chgo., July 13, 1908; s. Michael and Martha (Juric) B.; m. Anna M. Krotke, Dec. 24, 1936; children: Kathryn Ann, Hallie Gwyn. D.D.S., U. Ill., Chgo., 1934. Assoc., operative dentistry U. Ill., Chgo., 1935-46; chief central clinic Forsyth Dental Center, Boston, 1946-47; research assoc. Zoller Clinic, U. Chgo., 1948-67, chief dental clinics, 1967-73; ind. dental researcher, Manitowoc, Wis., 1973—. Clin. investigator NRC, 1951-55. Fellow Am. Coll. Dentists; mem. Chgo. Dental Soc., Ill. Dental Soc., Am. Dental Assn., N.Y. Acad. Sci., AAAS, Sigma Xi, Omicron Kappa Upsilon. Lodge: Lions. Subspecialties: Cariology; Preventive dentistry. Current work: Method and materials for prevention of caries; degradation of teeth. Home: 1042 W Crescent Dr Manitowoc WI 54220

BESMANN, THEODORE MARTIN, nuclear engineer, researcher, energy policy analyst; b. N.Y.C., Feb. 18, 1949; s. Siegfried and Greta (Kohn) B.; m. Wendy Diane Lowe, Oct. 24, 1982. B.E. in Chem. Engring, NYU, 1970; M.S. in Nuclear Engring, Iowa State U., 1971, Ph.D., Pa. State U., 1976. Mem. devel. staff Oak Ridge Nat. Lab., 1975—; cons. in field. Author: (with R.S. Livingston and others) A Desirable Energy Future, 1982. AEC trainee, 1970. Mem. Am. Nuclear Soc. (treas. local sect. 1980-82, mem. exec. com.sect. 1982—), Am. Ceramic Soc., Sigma Xi. Subspecialties: Nuclear fission; High-temperature materials. Current work: High temperature chemistry of energy-related materials; national and international energy policy assessment. Home: 1720 Tonalea Rd Knoxville TN 37919 Office: Oak Ridge Nat Lab PO Box X Bldg 4501 Oak Ridge TN 37830

BESSE, ARTHUR L., research physicist; b. Cambridge, Mass., Sept. 26, 1919; s. Arthur and Eleanor (Pass) B.; m. Virginia Weston, Aug. 14, 1943; children: Carol., Stephen, Anne. B.S., Harvard U., 1942. Research physicist shock wave studies, optical instruments, theoretical studies of space craft-environ. interaction U.S. Air Force, Bedford, Mass., 1964—. Contbr. articles to profl. jours. Current work: Space craft environment interactions. Developer heat shrinkable plastic films, 1947-64; patentee in field. Office: AFGL PH Hanscom Air Force Base Bedford MA 01730 Home: 30 Lantern Ln Weston MA 02193

BESSMAN, ALICE NEUMAN, medical educator; b. Washington, Nov. 7, 1922; d. Lester and Janet (Nusbaum) N.; m. Samuel Paul, July 3, 1945; children: Joel David, Ellen. B.A., Smith Coll., 1943; postgrad., Washington U. Sch. Medicine, St. Louis, 1944-46; M.D., George Washington U., 1949. Intern George Washington U. Hosp., Washington, 1949-50, resident, 1950-51, fellow, 1951-52, 53-54, Harvard U./Mass. Gen. Hosp., 1952-53; instr. medicine Johns Hopkins U., Balt., 1956-68; assoc. prof. U. So. Calif. Sch. Medicine, Los Angeles, 1969-79, prof., 1979—; chief diabetes Rancho Los Amigos Hosp., Downey, Calif., 1969—. Contbr. articles to med. jours.; reviewer various med. jours. Fellow ACP; mem. Am. Diabetes Assn. (grantee 1978, 83), Los Angeles County Soc. Internal Medicine (bd. dirs. 1978-80), Am. Diabetes Assn. of So. Calif. (research com. 1976-81), Am. Fedn. Clin. Research. Subspecialties: Endocrinology; Internal medicine. Current work: Treatment of diabetes mellitus, nature of infection in diabetes mellitus. Home: 7404 Woodrow Wilson Dr Los Angeles CA 90046 Office: Ortho-Diabetes Service Rancho Los Amigos Hosp 7601 E Imperial Hwy Downey CA 90242

BESSMAN, JOEL DAVID, research hematologist, physician; b. Norfolk, Va., Dec. 22, 1946; s. Samuel Paul and Alice B.; m. Joan Lewis, Mar. 24, 1974; children: Daniel Lewis, Elizabeth Ballin, Matthew Stephen. B.A., Columbia U., 1967, M.D., 1972. Diplomate: Nat. Bd. Med. Examiners, Am. Bd. Internal Medicine. Intern U. So. Calif., Los Angeles, 1972-73, resident in internal medicine, 1975-77; staff assoc. Nat. Cancer Inst., Bethesda, Md., 1973-75; research fellow Johns Hopkins Hosp., Balt., 1977-79; asst. prof. medicine U. Tex. Med. Br., Galveston, 1979—. Author: Automated Blood Counting and Differentials, 1984; contbr. chpts. to books. Served with USPHS, 1973-75. NIH fellow, 1977; Nat. Cancer Inst. grantee, 1982; recipient Research award U. Tex., 1982. Fellow ACP; mem. Am. Soc. Hematology, Am. Fedn. Clinc. Research (session chmn. 1983), Gulf Coast Hematology Soc. (pres. 1982-83). Subspecialties: Hematology; Cytology and histology. Current work: Physiology of blood cell maturation; technology of blood cell measurement; clinical application of flow cytometry. Office: John Sealy MW404 U Tex Medical Branch Galveston TX 77550

BEST, PHILIP ERNEST, physicist, educator, researcher, cons.; b. Perth, Australia, July 28, 1938; s. Ernest Edgar and Anne Eileen (Solomon) B.; m. Laurie Elizabeth Terrill, Feb. 25, 1960; children: David Ernest, Susan Laurie, Karen Margaret. B.S. with first class honors, U. Western Australia, Perth, 1959, Ph.D., 1963. Instr., research assoc. Cornell U., Ithaca, N.Y., 1962-65; exptl. physicist United Aircraft Research Lab., East Hartford, Conn., 1965-70; IMS fellow U. Conn., Storrs, 1970-71, assoc. prof., 1971-79, prof. physics, 1979—; vis. fellow Murdoch U., Western Australia, 1978; cons. on instrumentation. Contbr.: articles to sci. publs. including Ency. of Physics. Mem. Am. Phys. Soc., Inst. of Physics (London), Am. Vacuum Soc., IEEE. Subspecialties: Condensed matter physics; Atomic and molecular physics. Current work: Electron spectroscopy to study surfaces; X-ray physics. Home: 19 Brookside Ln Mansfield Center CT 06250 Office: Physics Dept Univ Conn Storrs CT 06268

BETHE, HANS ALBRECHT, physicist, educator; b. Strassburg, Alsace-Lorraine, July 2, 1906; U.S., 1935; s. Albrecht Theodore and Anna (Kuhn) B.; m. Rose Ewald, 1939; children—Henry, Monica. Ed. Goethe Gymnasium, Frankfurt on Main, U. Frankfurt; Ph.D., U. Munich, 1928; D.Sc., Bklyn. Poly. Inst., 1950, U. Denver, 1952, U. Chgo., 1953, U. Birmingham, 1956, Harvard U., 1958. Instr. in theoretical physics univs. of Frankfort, Stuttgart, Munich and Tubingen, 1928-33; lectr. univs. of Manchester and Bristol, Eng., 1933-35; asst. prof. Cornell U., 1935, prof., 1937-75, prof. emeritus, 1975—; dir. theoretical physics div. Los Alamos Sci. Lab., 1943-46; Mem. Presdl. Study Disarmament, 1958; mem. President's Sci. Adv. Com., 1956-60. Author: Mesons and Fields, 1953, Elementary Nuclear Theory, 1957, Quantum Mechanics of One-and Two-Electron Atoms, 1957, Intermediate Quantum Mechanics, 1964; Contbr. to: books Handbuch der Physik, 1933, Reviews of Modern Physics, 1936-37, Phys. Rev. Recipient A. Cressy Morrison prize N.Y. Acad. Sci., 1938-40; Presdl. Medal of Merit, 1946; Max Planck medal, 1953; Enrico Fermi award AEC, 1961; Nobel Prize in physics, 1967; Nat. Medal of Sci., 1976. Fgn. mem. Royal Soc. London; mem. Am. Philos. Soc., Nat. Acad. Scis. (Henry Draper medal 1968), Am. Phys. Soc. (pres. 1954), Am. Astron. Soc. Subspecialty: Theoretical physics. Office: Lab Nuclear Studies Cornell U Ithaca NY 14853

BETTELHEIM, FREDERICK ABRAHAM, chemist, educator, researcher, author; b. Gyor, Hungary, June 3, 1923; came to U.S., 1951, naturalized, 1961; s. Anton and Elizabeth (Gyarfas) B.; m. Annabelle Ganz, June 8, 1946; 1 son: Adriel. Student, U. Szeged, Hungary, 1943-44; B.S., Cornell U., 1953; M.S., U. Calif., Davis, 1954, Ph.D., 1956. Chemist Agrl. Expt. Sta., Rehovoth, Israel, 1947-51, Ithaca, N.Y., 1951-53; research instr. U. Mass., 1956-57; asst. prof. chemistry Adelphi U., 1957-60, assoc. prof., 1960-63, prof., 1963—; cons. in field. Author: Experimental Physical Chemistry, 1971, (with J. March) Essentials of Chemistry, Organic Chemistry and Biochemistry, 1983; author: (with J. Lee) Laboratory Manual, 1983; editor: Exptl. Eye Research, 1983—; Author also mongraphs. Served to 2d lt. Israeli Def. Army, 1948-51. Recipient Kiss award U. Szeged, 1944, Lalor award Lalor Found., 1959; recipient Disting. Tchr. award Adelphi U., 1979. Mem. Am. Chem. Soc., Fedn. Am. Scientists for Exptl. Biology, Assn. Research in Vision and Ophthalmology, Internat. Soc. Eye Research. Jewish. Subspecialties: Biophysical chemistry; Ophthalmology. Current work: Cataract formation in human lens; studying supramolecular changes in cataractogenesis by physical chemical means: light scattering, birefrignence, differential scanning calorimetry. Office: Adelphi U South St Garden City NY 11530

BETTENHAUSEN, LEE H., nuclear engineer; b. Hazleton, Pa., Dec. 14, 1934; s. William Herman and Gertrude (Bradney) B.; m. Mary Ann Cavalovitch, Sept. 29, 1962; children: Maia Ann, Tod William, Lia Marie. B.S., Pa. State U., 1956; Ph.D., U. Va., 1974. Registered profl. engr., Ohio, Va. Research scientist Battelle Meml. Inst., Columbus, Ohio, 1959-66; lectr. U. Va., Charlottesville, 1971-74; health physicist U.S. E.P.A., Phila., 1974-78; nuclear engr. U.S. NRC, Phila., 1978—. Served to 1st lt. USAF, 1956-59. AEC fellow, 1968-71. Mem. IEEE, Am. Nuclear Soc., Health Physics Soc., Va. Acad. Sci., Delaware Valley Soc. for Radiation Safety (pres. elect 1982-83). Subspecialties: Nuclear engineering; Nuclear fission. Current work: Engineering test programs. Home: 7 Long Lane Malvern PA 19355 Office: US Nuclear Regulatory Commn 631 Park Ave King of Prussia PA 19406

BETTERLEY, DONALD ALAN, research mycologist; b. Colorado Springs, Colo., June 17, 1952; s. Robert Leslie and Joan Madelle (Snavely) B.; m. Margaret Louisa McCollough, Sept. 6, 1975. B.S. in Botany with highest distinction, Colo. State U., 1974; Ph.D. in Botany-Mycology, U. Calif.-Berkeley, 1981. Postdoctoral research botanist U. Calif.-Berkeley, 1980-81; dir. research, research mycologist Spawn Mate, Inc., San Jose, Calif., 1981—; guest lectr. U. Calif.-Berkeley. Mem. Mycol. Soc. Am., Brit. Mycol. Soc., N.Y. Acad. Scis., Internat. Mushroom Soc. Tropics, Bot. Soc. Am., Sierra Club, Sigma Xi, Phi Beta Kappa. Subspecialties: Genome organization; Microbiology. Current work: Fungal genetics, physiology, nutrition and biological control mechanisms as related to mushroom cultivation; scientific photography. Office: 555 N 1st St San Jose CA 95112

BETTS, AUSTIN WORTHAM, retired research company executive; b. Westwood, N.J., Nov. 22, 1912; s. Irving Wilcox and Bessie Harris (Boardman) B.; m. Edna Jane Paterson, Dec. 8, 1934; children: Jerry W., Lee W., Lynn P. B.S., U.S. Mil. Acad., 1934; M.S., Mass. Inst. Tech., 1938. Commd. 2d lt. U.S. Army, 1934, advanced through grades to lt. gen., 1966; dist. engr. Bermuda Dist., U.S. Engr. Dept., 1942-43; engr. 14th Air Force, 1944-45; asso. dir. Los Alamos Sci. Lab., 1946-48; chief atomic energy br. G-4, Dept. Army, 1949-52; exec. to chief research and devel. Dept. Army, 1952-54; mil. exec. to spl. asst. of dir. guided missiles Office Sec. Def., 1957-59; dir. Advanced Research Projects Agy., Office Sec. Def., 1959-61; dir. mil. application AEC, 1961-64; dep. chief research and devel. Dept. Army, 1964-66, chief research and devel., 1966-70; retired, 1970; sr. v.p. S.W. Research Inst., San Antonio, 1971-83. Mem. Assn. U.S. Army, Nat. Security Indsl. Assn., Inst. Environ. Scis., Air Pollution Control Assn., Soc. Research Adminstrs., Soc. Am. Mil. Engrs., Am. Inst. Aeros. and Astronautics, Am. Indsl. Preparedness Assn., Sigma Xi. Presbyterian. Club: Masons. Subspecialties: Systems engineering; Environmental engineering. Current work: Corporate management of multidisciplinary contract research, development and evaluations;and consultant on Army Science Board. Home: 6414 View Point San Antonio TX 78229 Early in my life I was impressed by an autobiographical sketch of a great leader who commented that his goal in life was simply to leave tracks. I took that guidance for my own and have since tried to orient all my major activities toward service, those services to be of such nature that I can look back with pride at the tracks I have left behind me.

BETTS, HENRY BROGNARD, physician, educator; b. New Rochelle, N.Y., May 25, 1928; s. Henry Brognard and Marguerite Meredith (Denise) B.; m. Monika Christine Paul, Apr. 25, 1970. A.B., Princeton, 1950; M.D., U. Va., 1954. Diplomate: Am. Bd. Phys. Medicine and Rehab. Intern Cin. Gen. Hosp., 1954-55; resident, teaching fellow N.Y.U. Med. Center Inst. Rehab. Medicine, N.Y.C., 1958-63; practice medicine, specializing in phys. medicine and rehab., Chgo., 1963—; staff physiatrist Rehab. Inst. Chgo., 1963-64, asso. med. dir., 1964-65, med. dir., 1965-69, v.p., med. dir., 1969-75, exec. v.p., med. dir., 1975—; chmn. dept. rehab. medicine Northwestern U. Med. Sch., 1967—, prof., 1968—; cons. Northwestern Meml. Hosp., Chgo. Contbr. articles to profl. jours. Mem. steering com. United Cerebral Palsy, 1967—; Mem. med. adv. com. Nat. Paraplegia Found., 1969—; bd. dirs. Nat. Com. Arts for Handicapped, 1981—; mem. Gov.'s High Blood Pressure Adv. Bd., 1977—. Served with USNR, 1956-58. Named Physician of Year Ill. Gov.'s Com., 1964; commended by Ill. Gen. Assembly, 1967; cited for meritorious service Pres.'s Com. on Employment of Handicapped, 1965. Mem. Ill. Med. Soc. (chmn. com. on rehab. services), Assn. Acad. Physiatrist (pres. 1968-69), Am. Congress Rehab. Med. (med. adv. com., pres. 1976-77), Mid-m. Soc. Phys. Med. and Rehab. (pres. 1969). Subspecialty: Physical medicine and rehabilitation. Home: 1727 N Orleans Chicago IL 60614 Office: 345 E Superior St Chicago IL 60611

BEUTE, MARVIN KENNETH, plant pathology educator; b. Jenison, Mich., Mar. 3, 1935; s. Martin and Jeanette B.; m. Sherlene Joy Russell, Nov. 11, 1955; children: Douglas Russell, Dana Lynn. A.B., Calvin Coll., 1963; Ph.D., Mich. State U., 1967. Asst. prof. plant pathology N.C. State U., 1968-73, assoc. prof., 1973-78, prof., 1978—. Mem. Am. Phytopath. Soc., Sigma Xi. Subspecialty: Plant pathology. Current work: Teaching in plant pathology, research in epidemiology of peanut pathogens. Home: 4104 Picardy Dr Raleigh NC 27612 Office: 2618 Gardner Hall Raleigh NC 27650

BEUTLER, ERNEST, physician, research scientist; b. Berlin, Sept. 30, 1928; U.S., 1936, naturalized, 1943; s. Alfred David and Kaethe (Italiener) B.; m. Brondelle Fleisher, June 15, 1950; children: Steven Merrill, Earl Bryan, Bruce Alan, Deborah Ann. Ph.B., U. Chgo., 1946, B.S., 1948, M.D., 1950. Intern U. Chgo. Clinics, 1950-51; resident in medicine 1951-53; asst. prof. U. Chgo., 1956-59; chmn. div. medicine City of Hope Med. Center, Los Angeles, 1959-78; chmn. dept. clin. research Scripps Clinic and Research Found., 1978—; clin. prof. medicine U. So. Calif., 1964-79, U. Calif., San Diego, 1979—; mem. hematology study sect. NIH, 1970-74; mem. med. adv. com. ARC, 1972-78. Author 8 books, numerous articles in med. jours.; mem. editorial bds. profl. jours. Mem. med. and sci. adv. council Cystic Fibrosis Found., 1976-78. Served with U.S. Army, 1953-55. Recipient Gairdner award, 1975. Mem. Nat. Acad. Scis., Am. Acad. Arts and Scis., Am. Assn. Physicians, Am. Soc. Clin. Investigation, Am. Soc. Hematology (mem. exec. com. 1968-72, v.p. 1977, pres. 1979), Am. Soc. Human Genetics (mem. exec. com. 1968-72). Jewish. Subspecialties: Hematology; Genetics and genetic engineering (biology). Current work: The biochemistry of genetic diseases, particulary those affecting red blood cells. Home: 4308 Caminito del Zafiro San Diego CA 92121 Office: 10666 N Torrey Pines Rd La Jolla CA 92037

BEUTLER, FREDERICK JOSEPH, info. scientist; b. Berlin, Oct. 3, 1926; U.S., 1936, naturalized, 1943; s. Alfred David and Kaethe (Italiener) B.; m. Suzanne Armstrong, Jan. 6, 1969; children—Arthur David, Kathryn Ruth, Michael Ernest. S.B., Mass. Inst. Tech., 1949, S.M., 1951; Ph.D., Calif. Inst. Tech., 1957. Mem. faculty U. Mich., Ann Arbor, 1957—, prof. info. and control engring., 1963—, chmn. computer info. and control engring., 1970-71, 77—; vis. prof. Calif. Inst. Tech., 1967-68; vis. scholar U. Calif. at Berkeley, 1964-65. Editorial cons.: Math. Rev, 1965-67, 75—; contbr. articles to profl. jours. and books. Bd. dirs. Ann Arbor Civic Theatre, 1976-78. Served with AUS, 1945-46. NSF research grantee, 1971-75, 76-81; Air Force Office Sci. Research grantee, 1970-74, 75-80; NASA grantee, 1959-69. Fellow IEEE; mem. Soc. Indsl. and Applied Math (council 1969-74, mng. editor Jour. Applied Math. 1970-75, editor Rev. 1967-70), Am. Math. Soc., Inst. Math. Statistics, Am. Soc. Engring. Edn. (exec. com. 1967-70), Am. Arbitration Assn., Econ. Club Detroit. Club: Barton Boat. Subspecialties: Systems engineering; Probability. Current work: Queueing systems; properties, characterizations, estimation,prediction, optimization. Home: 1717 Shadford Rd Ann Arbor MI 48104

BEUTLER, LARRY EDWARD, clinical psychologist, educator; b. Logan, Utah, Feb. 14, 1941; s. Edward and Violet Stella (Rasmussen) B.; m. Maria Elena Oro, Feb. 25, 1977; children: Jana Lynne, Kelly Jo, Ian David, Gail Lei. B.S., Utah State U., 1965, M.S., 1966; Ph.D., U. Nebr., 1970. Diplomate: Am. Bd. Profl. Psychology. Asst. prof. Highland div. Duke U., Asheville, N.C., 1970-71, Stephen F. Austin State U., Nacogdoches, Tex., 1971-73; assoc. prof. Baylor Coll. Medicine, Houston, 1973-79; prof., chief psychology U. Ariz. Coll. Medicine, Tucson, 1979—; cons. Houston Police Dept., 1975-79, Cypress-Fairbanks pub. schs., Houston, 1973-79, Tex. Dept. Corrections, 1973-76. Author: Eclectic Psychotherapy, 1983; editor: (with others) Special Problems in Child and Adolescent Behavior, 1978; contbr. numerous articles to profl. jours. Mem. Am. Psychol. Assn. (pres. Div. 12, Sect. II), Soc. Psychotherapy Research, Ariz. Psychol. Assn., Assn. Psychology Internship Ctrs. Democrat. Subspecialty: Behavioral psychology. Current work: Psychotherapy research, sleep disorders, sexual dysfunction. Office: University of Arizona Dept of Psychiatry Tucson AZ 85724

BEUTNER, ERNST HERMAN, microbiologist, educator; b. Berlin, Germany, Aug. 27, 1923; came to U.S., 1923; S. Reinhard Heinrich and Hermine (Aye) B.; m. Gloria Parks, Sept. 9, 1925; children: Ann, Eric, Karen, Jean. Ph.D., U. Pa., 1951. Diplomate: Am. Bd. Med. Microbiology, Am. Bd. Med. Lab. Immunology. Research supr. Sias Labs at Brooks Hosp., Brookline, 1951-55; research assoc. Harvard Sch. Dental Medicine, Boston, 1955-56; prof. microbiology SUNY-Buffalo, 1956—. Editor: Immunopathology of the Skin, 1979, Autoimmunity in Psoriasis; others. Fellow Am. Acad. Microbiology; men. Am. Assn. Immunology, Am. Acad. Dermatology, N.Y. Acad. Sci., Soc. Exptl. Biology and Medicine. Subspecialty: Immunobiology and immunology. Office: Dept Microbiology SUNY Buffalo NY 14214

BEVAN, THOMAS EDWARD, psychologist; b. Millville, N.J., Mar. 11, 1947; s. Edward George and Fola (Zimmerman) B.; m. Alice Alexander, Sept. 11, 1971; children: Lesley, Cynthia. A.B., Dartmouth Coll., 1969; Ph.D., Princeton U., 1973. Psychologist Sci. Applications Inc., Rosslyn, Va., 1973-76; psychologist Environ. Research Inst. Mich., Rosslyn, Va., 1980—, mgr. systems evaluation group, 1980—. Contbr. articles to profl. jours. Bd. dirs. No. Va. Football Ofcls. Assn., 1981—. Served to capt. U.S. Army, 1973-76. Mem. Soc. Neurosci. Subspecialties: Human factors engineering; Remote sensing (atmospheric science). Home: 1207 N Quantico St Arlington VA 22205 Office: 1501 Wilson Blvd Suite 1105 Arlington VA 22209

BEVELACQUA, JOSEPH JOHN, physicist, researcher; b. Waynesburg, Pa., Mar. 17, 1949; s. Frank and Lucy Ann (Cataneo) B.; m. Terry Sanders, Sept. 4, 1971; children: Anthony, Jeffrey, Megan, Peter, Michael. B.S. in Physics, Calif. State Coll., 1970; postgrad., U. Maine, 1970-72; M.S. in Physics, Fla. State U., 1974, Ph.D., 1976. Cert. radiol. shield survey engr., Westinghouse Bettis Atomic Power Lab. Teaching/research asst. U. Maine, 1970-72, Fla. State U., 1973-76; nuclear engr. Bettis Atomic Power Lab., West Mifflin, Pa., 1973, sr. nuclear engr., 1976-78; ops. research analyst U.S. Dept. Energy, Oak Ridge, 1978-80, chief physicist advanced laser isotope separation program, 1980-83; sr. engr. GPU Nuclear Corp. (Three Mile Island Sta.-Unit 2), Middletown, Pa., 1983—; con. U.S. Dept. Energy's Process Evaluation Bd. of Isotope Separation, Washington, 1981-82. Contbr.: articles to profl. jours. including Physical Rev. Letters. Mem. Republican Presdl. Task Force. NSF research asst., 1975-76. Mem. Am. Nuclear Soc., Am. Phys. Soc., Health Physics Soc. Republican.

Lutheran. Club: Oak Ridge Sportsman's. Subspecialties: Nuclear physics; Nuclear engineering. Current work: Theoretical studies of light nuclei, few nucleon transfer reactions, radiation shielding, laser isotope separation, neutron nuclei, symmetry violations in nuclei, grand unification theories, quark models of nuclear forces, nuclear fuel cycle, laser fusion and gravitational collapse of stars; nuclear reactor safety. Home: 19 Merion Ln PO Box 166 Hummelstown PA 17036 Office: GPU Nuclear Corp 3 Mile Island Nuclear Generating Sta PO Box 480 Middletown PA 17057

BEVER, BERLINER MICHAEL, metallurgic engineer, educator; b. Germany, Aug. 7, 1911; s. Rudolf and Maria (Bever) Berliner; m. Marian Gordon, Aug. 26, 1936; children: James G., Thomas G., Mary-Ivers B. Witherby. Dr. iur., Heidelberg, 1934; M.B.A., Harvard U., 1937; M.S., MIT, 1942, Sc.D., 1944. Registered profl. engr., Mass. Staff Mem. dept. metallurgy MIT, Cambridge, 1940-44, instr. to prof., 1944-76, prof. emeritus materials sci. and engring., sr. lectr., 1977—; sr. research assoc. Inst. Econ. Analysis, NYU, 1978—; hon. research assoc. Harvard U., 1966-67. Co-editor: Basic Open Hearth Steelmaking, 2d edit, 1951, Metals Handbook Vol. 8, 8th edit, 1973; cons.: McGraw-Hill Mathematics, Science and Engring. series Environ. Impact Assessment Review; co-editor: Conservation and Recycling; editor-in-chief: Ency. Materials Sci. and Engring; contbr. articles to profl. jours. Mem. Boston Mus. Sci. corp. Recipient Nat. Assn. Secondary Materials Industries, recycling award, 1972. Fellow Am. Soc. Metals, AAAS; mem. AIME (Mathewson Gold medal 1965), Metals Soc. (London), Am. Acad. Arts and Scis.; corr. mem. Berliner Wissenschaftliche Gesellschaft. Clubs: Harvard, Harvard Musical Assn. Subspecialties: Metallurgy; Materials. Current work: Physical metallurgy; application of thermodynamics to metallurgy; calorimetry; deformation of metals and intermetallic compounds; alloy theory and metastable phases; surface hardening of metals; characterization of structures; materials engineering; conservation and recycling of materials. Home: 23 Highland St Cambridge MA 02138 Office: MIT Room 13-5026 Cambridge MA 02139

BEYENBACH, KLAUS WERNER, physiologist, educator; b. Mainz, Rheinland-Pfalz, W.Ger., Mar. 19, 1943; came to U.S., 1962, naturalized, 1968; s. Otto Georg and Maria (Eschenauer) B.; m. Christa-Maria Kuhn, Apr. 13, 1979. B.A., St. Mary's U., San Antonio, 1968; M.A., S.W. Tex. State U., 1970; Ph.D., Wash. State U., 1974. Fellow U. Ariz. Coll. Medicine, 1974-78; asst. prof. physiology Cornell U., 1978-82, assoc. prof., 1982—; prin. investigator Mt. Desert Island Biol. Lab., Salisbury Cove, Maine, 1978—, radiation safety officer, 1982—. Served with USAF, 1962-66. NIH predoctoral fellow, 1972-74; postdoctoral fellow, 1976-78; Nat. Kidney Found. fellow, 1974-76; NIH grantee, 1979—. Mem. Am. Physiol. Soc., Am. Soc. Zoologists, Biophys. Soc., AAAS, Nat. Kidney Found. Democrat. Lutheran. Subspecialties: Physiology (medicine); Physiology (biology). Current work: Epithelial transport physiology; comparative physiology of epithelia. Office: Sect Physiology Cornell U Ihaca NY 14853

BEYER, KARL HENRY, JR., pharmacologist; b. Henderson, Ky., June 19, 1914; s. Karl H. and Lennie M. (Beadles) B.; m. Camille Slobodzian, Nov. 9, 1979; children by previous marriage—Annette Matilda Beyer Mears, Katherine Louise Beyer Cranson. B.S., Western Ky. State Coll., 1935; Ph.M., U. Wis., 1937, Ph.D., 1940, M.D., 1943, Sc.D. (hon.), 1972. Asst. dir. pharmacol. research Sharp & Dohme, 1943-44, dir. pharmacol. research, 1944-50, asst. dir. research, 1950-56; dir. Merck Inst. Therapeutic Research, West Point, Pa., 1956-58, pres., 1961-66; v.p. life scis. Merck Sharp & Dohme, Research Labs., West Point, 1958-66, sr. v.p. research, 1966-73; lectr. Med. Coll. Pa.; vis. prof., guest lectr. U. Wis., 1958, Swedish U. Med. Schs., 1962, Howard U., 1964, Free U. Berlin, 1966; vis. prof. Milton S. Hershey Med. Center, Pa. State U., 1973—, Vanderbilt U. Sch. Medicine, 1973-79; chmn. Cosmetic Ingredient Rev., 1976; bd. sci. advisers Merck Inst., 1973-77, 78—; chmn. bd. Phila. Assn. Clin. Trials, 1980. Author: Pharmacological Basis of Penicillin Therapy, 1950, Discovery, Development and Delivery of New Drugs, 1978; editorial bd.: Clin. Pharmacology and Therapeutics; contbr. articles to profl. jours. Recipient Gairdner Found. award, 1964; Modern Pioneers in Creative Industry award NAM, 1965; Modern Medicine Distinguished Achievement award, 1967; Am. Pharm. Assn. Found. Achievement award, 1967; Distinguished Service award Wis. Alumni Assn., 1968; Lasker award, 1975; Torald Sollmann award, 1978; Catell award Am. Coll. Clin. Pharmacology, 1980. Fellow A.C.P., AAAS, N.Y. Acad. Scis., Royal Acad. of Medicine; mem. Nat. Acad. Scis., Am. Chem. Soc., Am., Phila. physiol. socs., Soc. for Exptl. Biology and Medicine, Phila. Med. Soc., Am. Soc. for Pharmacology and Exptl. Therapeutics (pres. 1964-65), Fedn. Am. Soc. Exptl. Biology (pres. 1965-66), Internat. Soc. Biochem. Pharmacology, Phila. Coll. Physicians, Am. Therapeutic Soc., Soc. Toxicology, Am. Soc. Nephrology, Am. Heart Assn. (hypertension research award 1979, council circulation and renal sect.), Heart Assn. Southeastern Pa., Biol. Abstracts (trustee; treas. 1965-69), Nat. Acad. Scis. (drug research bd. 1964-70), Canadian Pharm. Soc. Subspecialty: Pharmacology. Home: Box 387 Penllyn PA 19422

BEYER-MEARS, ANNETTE, physiologist, researcher; b. Madison, Wis., May 26, 1941; d. Karl H. and Annette (Weiss) Beyer; m. William H. Mears, Jr.; 1 son, Karl. B.S., Vassar Coll., 1963; M.S., Fairleigh Dickinson U., 1973; Ph.D., U. Medicine and Dentistry of N.J., 1977. NIH fellow Cornell U. Med. Sch., N.Y.C., 1963-65; instr. physiology Springside Coll., Phila., 1967-71; teaching asst. dept Physiology U. Medicine and Dentistry N.J., N.J. Med. Sch., 1974-77, NIH fellow dept. ophthalmology, 1978-80, asst. prof. dept. physiology, 1980—, asst. prof. dept. ophthalmology, 1979—. Contbr. articles in field of diabetic lens and kidney therapy to profl. jours. Chmn. admissions No. N.J. area Vassar Coll., 1974-79; mem. minister search com. St. Bartholomew Episcopal Ch., Ridgewood, N.J., 1978, chmn. fund raising, 1978, 79; del. Epis. Diocesian Conv., 1977, 78. Recipient Nat. Research Service award, 1978-80; Research award NIH, 1980—, Found. U. Medicine and Dentistry N.J., 1980, Sigma Xi, 1980. Mem. Am. Physiol. Soc., N.Y. Acad. Scis., Soc. for Neurosci., Assn. for Research in Vision and Ophthalmology, Internat. Soc. for Eye Research, AAAS, Aircraft Owners and Pilots Assn., Sigma Xi. Subspecialties: Neuroendocrinology; Ophthalmology. Current work: Research in diabetic lens and kidney therapy. Office: Dept Physiology U Medicine and Dentistry NJ 100 Bergen St Newark NJ 07103

BEZELLA, WINFRED AUGUST, nuclear engineer; b. Milw., Mar. 1, 1935; s. Martin F. and Dorothy (Kellner) B.; m. Evelyn Virginia Bergmark, June 22, 1968; children: Karen, Bruce. B.S. in Chem. Engring, U. Wis., 1957; M.S. in Nuclear Engring, U. Mich., 1959; Ph.D., Purdue U., 1972. Registered profl. engr., Calif. Engr. Allis Chalmers Mfg. Co., Milw., 1959-64; sr. engr. Westinghouse Electric Corp., Pitts., 1964-68; engr. Argonne Nat. Lab., Ill., 1972—. Author tech. articles and reports. AEC fellow, 1968-70. Mem. Am. Nuclear Soc. Subspecialties: Nuclear engineering; Nuclear fission. Current work: Thermal and hydraulic design of nuclear reactors; fast breeder reactor safety analysis, design of in-reactor safety experiments; reactor engineering, reactor kinetics and probabilistic risk assessment. Office: 9700 S Case Ave Argonne IL 60439 Home: 4123 West End Rd Downers Grove IL 60515

BHAGAT, PHIROZ MANECK, mechanical engineer, educator; b. Poona, India, Oct. 28, 1948; came to U.S., 1970; s. Maneck Phirozshaw and Khorshed Eduljee (Batliwala) B.; m. Patricia Jane Steckler, Oct. 13, 1979; 1 child, Kay. B.Tech., Indian Inst. Tech.-Bombay, 1970; M.S.E., U. Mich., 1971, Ph.D., 1975. Research fellow in applied mechanics Harvard U., Cambridge, Mass., 1975-77; asst. prof. engring. Columbia U., N.Y.C., 1977-81, adj. asst. prof., 1981—; staff engr. Exxon Research & Engring. Co., Florham Park, N.J., 1981-83, sr. staff engr., 1983—. Contbr. articles to profl. jours. Named K.C. Mahindra scholar, 1970; J.N. Tata scholar, 1970; Horace Rackham predoctoral fellow, 1973-74, 74-75. Mem. N.Y. Acad. Scis., Am. Inst. Chem. Engrs., ASME, Combustion Inst., AAAS, Tau Beta Pi, Sigma Xi. Subspecialties: Mechanical engineering; Chemical engineering. Current work: Application of the thermal sciences to model petrochemical processes; combustion; heat transfer; fluid mechanics; thermodynamics; coal conversion. Home: 252 Sinclair Pl Westfield NJ 07090 Office: Exxon Research & Engring Co Florham Park NJ 07932

BHALLA, VINOD K., endocrinologist, biochemist, educator; b. Iahore, India, Aug. 4, 1940; came to U.S., 1968, naturalized, 1981; s. Ial C. and Shanti (Punga) B.; m. Madhu B. Sarin, May 29, 1966; children: Niti, Jyot, Varun. B.S., St. John's Coll., Agra, India, 1962, M.S., 1964; Ph.D., Nat. Chem. Lab., Poona, India, 1968. Research assoc. U. Ga., Athens, 1969-72, Emory U., Atlanta, 1972-74; mem. faculty Med. Coll. Ga., Augusta, 1974—, prof. endocrinology, 1982—; speaker in field. Mem. editorial bd.: Biology of Reproduction, 1978—; reviewer: Endocrinology Jour, 1980, Alcoholism-Clinical and Exptl. Research, 1982—, Andrology Jour, 1982—. NSF grantee, 1976-79; NIH grantee, 1976—. Mem. Am. Soc. Biol. Chemists, Endrocine Soc., Soc. for Study Reprodn., N.Y. Acad. Sci., Am. Fertility Soc., Am. Chem. Soc. Subspecialties: Receptors; Biochemistry (medicine). Current work: Polypeptide hormone action at the testicular level: polypeptide hormone receptors, cAMP, mediation and testosterone production. Home: 3541 Westlake Dr Augusta GA 30907 Office: Med Coll Ga Endocrinology Dept Augusta GA 30912

BHARGAVA, BHARAT KUMAR, computer scientist, educator; b. Agra, India, Oct. 15, 1948; came to U.S., 1970; s. Radha Krishan and Indra (Kanti) B.; m. Shail Bhargava, Sept. 5, 1976; children: Anjali, Anu. B.Sc., Punjab U., 1966; B.E., Indian Inst. Sci., 1969; M.S., Purdue U., 1970, Ph.D., 1974. Project mgr. Ind. U. Med. Sch., Indpls., 1973-76; asst. prof. Cleve. State U., 1976-78; asst. prof. research, Stanford U., summer 1978; asst. prof. computer sci. U. Pitts., 1978—; chmn. IEEE Computer Sci. Conf. on Reliability in Distributed Database and Software Systems, 1981-82. Author: Reliability in Distributed Database Systems, 1982; editor: Concurrency and Reliability in Distributed Systems, 1983. Indian Inst. Tech. highest scholarship, 1969. Mem. Assn. Computing Machinery, IEEE (pub. bd. 1981-82, edn. activity bd. 1981—). Subspecialties: Database systems; Software engineering. Current work: Concurrency control and reliability in distributed database systems, software reliability, intelligent database systems. Home: 1220 Ridgewood Rd Pittsburgh PA 15241 Office: Univ Pitts Computer Sci Dept Pittsburgh PA 15260

BHATHENA, SAM JEHANGIRJI, research chemist; b. Bombay, India, Sept. 18, 1936; s. Jemangirji and Pirojbai (Mistry) B.; m. Pauruchisty S. Bhathena, July 13, 1975. B.Sc. with honors, U. Bombay, 1961, M.Sc., 1964, Ph.D., 1970. Vis. fellow HIH, Bethesda, Md., 1971-73, vis. assoc., 1974; research biochemist VA Med. Ctr., Washington, 1975-82; asst. prof. Georgetown U., Washington, 1979-82; research chemist U.S. Dept. Agr., Beltsville, Md., 1983—. Contbr. chtps. to books. Treas. Zorastrian Assn. Met. Washington, 1979-82; v.p. Assn. Indians in Am., Washington, 1979-82. Mem. Endocrine Soc., Am. Diabetes Assn., N.Y. Acad. Sci., Soc. for Exptl. Biology and Medicine, Am. Fedn. for Clin. Research. Democrat. Subspecialties: Receptors. Current work: Effects of carbohydrate diets on receptors for insulin, glucagon, somatostatin, endorphins and LDL. Home: 11912 Judson Ct Wheaton MD 20909 Office: Carbohydrate Nutrition Lab Beltsville Human Nutrition Center Barc East Beltsville MD 20705

BHATNAGAR, GOPAL MOHAN, biochemist, researcher; b. Lucknow, India, July 15, 1937; came to U.S., 1969; s. Jag Mohan Lal and Gopi B. Ph.D., Lucknow U., 1961. Research scientist Commonwealth Sci. and Indsl. Research Orgn., Melbourne, Australia, 1965-69; research staff scientist Boston Biomed. Research Inst., 1969-73; prin. assoc. Harvard Med. Sch., Boston, 1974-77; asst. prof. Johns Hopkins Med. Sch., Balt., 1977—; research scientist NIH, Balt., 1980—; cons. FDA, 1980—. NIH research grantee, 1975-80. Mem. AAAS, Am. Soc. Biol. Chemists, Soc. Investigative Dermatology. Subspecialties: Biochemistry (medicine); Cell biology (medicine). Current work: Biochemistry and physiology of cardiac and skeletal muscle, epithelial differentiation and keratinization; molecular changes involved in cellular aging. Home: 301 Garden Rd Baltimore MD 21204 Office: Gerontology Research Center Nat Inst Health Baltimore MD 21224

BHATNAGAR, JAGDISH PRASAD, radiol. physicist; b. Sikandrabad, India, Oct. 4, 1938; came to U.S., 1966, naturalized, 1980; s. Kamta Prasad and Kalvati (Devi) B.; m. Usha Bhatnagar, Nov. 6, 1965; children: Atul, Sonika. Sc.D., Johns Hopkins U., 1973. Diplomate: Am. Bd. Radiology, Radiol. Physics, 1977. Research asst., research officer Bhabha Atomic Research Center, 1959-66; lab. asst. Sch. Hygiene and Public Health, Johns Hopkins U., Balt., 1966-70; radiol. phsyicst Boston City Hosp. U. Med. and Boston U. Med. Center, 1970-75; dir. div. radiol. physics dept. radiology Mercy Hosp., Pitts., 1975—; cons. in field. Contbr. articles to profl. jours. Sec. Hindu Temple Soc. N.Am., Monroeville, 1976-78; sec. Chinmaya Mission (W), Pitts., 1978—. Mem. Am. Assn. Physicists in Medicine, Health Physics Soc., N.Y. Acad. Sci., Am. Coll. Radiology, Pa. Radiol. Soc. Subspecialty: Radiological physics. Current work: research in radiation dosimetry. Office: 1400 Locust St Pittsburgh PA 15219 Home: 8890 Willoughby Rd Pittsburgh PA 15237

BHATNAGAR, KUNWAR P., anatomy educator; b. Gwalior, India, Mar. 21, 1934; came to U.S., 1967, naturalized, 1976; s. Narayan Swaroop and Bhagwati (Devi) B.; m. Indu Bhatnagar, 1961; children: Divya, Jyoti. B.Sc., Agra (India) U., 1956; M.Sc., Vikram (India) U., 1958; Ph.D., SUNY, Buffalo, 1972. Asst. prof. anatomy Sch. Medicine, U. Louisville, 1972-78, assoc. prof., 1978—. Editor: Bat Research News, 1982—. Chmn. bd. trustees India Community Found. Louisville, 1982—. Recipient Disting. Community Service award India Assn. Louisville, 1981; Max-Planck-Inst. for Brain Research fellow, 1978. Mem. Am. Soc. Mammalogists (life), Am. Assn. Anatomists, Electron Microscopy Soc. Am., Sigma Xi (pres. 1981-82). Democrat. Hindu. Subspecialties: Anatomy and embryology; Morphology. Current work: Mammalian olfaction; biology of bats; human development; organs of special sense. Office: Dept Anatomy U Louisville Louisville KY 40292

BHATNAGAR, RAVI, chemist, researcher; b. Allahabad, India, Nov. 23, 1950; came to U.S., 1979; s. Ram Ratan and Girija B.; m. Arati Varma, Sept. 12, 1980. Sc.B., U. Saugor, 1968, M.Sc. in Physics, 1970; M. Tech., Indian Inst. Sci., 1972; Ph.D., U. Hull, Eng., 1979. Tech. officer Electronics Corp. India, Hyderabad, 1972-75; faculty intern U. Utah, Salt Lake City, 1979-82, research assoc., 1982—. Contbr. articles to profl. jours. Subspecialty: Laser-induced chemistry. Current work: Current interests include study of intra and intermolecular energy transfer processes in polyatomic molecules, effect of vibrational excitation of the reactants on the reaction rate and other laser induced processes in molecules. Office: U Utah Box 53 Dept Chemistry Salt Lake City UT 84112

BHATT, JAGDISH JEYSHANKER, geologist/oceanographer; b. Umreth, Gujarat, India, Feb. 17, 1939; came to U.S., 1961, naturalized, 1976; s. Jeyshanker J. and Kamala J. B.; m. Meena J. Pandaya, Jan. 22, 1970; children: Amar J., Anita J. B.Sc. with honors, U. Baroda, India, 1961; M.S., U. Wis., 1963; Ph.D., U. Wales, 1972; postgrad., U. Calif., Santa Barbara, 1968-69. Instr. geology Panhandle State U., Goodwill, Okla., 1965-66; postdoctoral research scientist Stanford U., Palo Alto, Calif., 1971-72; asst. prof. geology/oceanography U. Buffalo, 1972-74; prof. geology/oceanography Community Coll. R.I., Warwick, 1974—; instr. Jackson (Mich.) Community Coll., 1964-65; environ. cons. Warwick City, 1977; pvt. cons. marine educator, Warwick, 1976-80. Author: Oceanography-Exploring the Planet Ocean, 1978, Environmentology, 1975, Geochemistry and Geology of South Wales, 1976, Applied Oceanography, 1979, others; contbr. articles to profl. jours. Mem. ocean task force R.I. Dept. Edn., Providence, 1976; soccer coach East Greenwich (R.I.) Soccer Assn., 1980. U. Wales Wolfson doctoral scholar, Cardiff, 1970-71. Mem. Marine Tech. Soc. Subspecialties: Geology; Oceanography. Current work: Geological and geochemical studies of marine environment, marine science and technical education, ocean mining, marine sciences planning and policy. Home: 11 Midlands Dr East Greenwich RI 02818

BHATTACHARJEE, JNANENDRA K., microbiologist, educator; b. Digli, Sylhet, Bangladesh, Feb. 1, 1936; s. Gobinda K. and Kumudkamini B.; m. Tripti Chakrabarty, Aug. 20, 1969; children: Gourab, Mala. B.Sc., M.C. Coll., Bangladesh, 1957, M.Sc. in Botany, Dacca U., Bangladesh, 1959; Ph.D. in Microbiology, So. Ill. U., 1966. Research assoc., asst. mem. Albert Einstein Med. Ctr., Phila., 1965-68; assoc. prof. microbiology Miami U., Oxford, Ohio, 1968-73, prof., 1973—. Contbr. over 40 articles to profl. jours. Recipient Pres.'s award, Pakistan, 1960; Fulbright-Hays award, 1961; NSF grantee, 1969, 71, 75, 78, 79, 81; Eli Lilly and Co. grantee, 1970, 72; S&H Found. grantee, 1973; Ohio Bicentennial award, 1976; Miami U. Faculty Research Com. Research awards. Fellow Am. Acad. Microbiology; mem. Am. Soc. Microbiology, Genetic Soc. Am., AAAS, Sigma Xi (v.p., pres. local chpt.). Subspecialties: Microbiology (medicine); Genetics and genetic engineering (biology). Current work: Lysine biosynthesis, gene expression, regulation in yeast, Saccharomyces cerevisiae. Home: 454 Emerald Woods Dr Oxford OH 45056 Office: Dept Microbiology Miami U Oxford OH 45056

BHATTACHARYA, PRADEEP K., biology educator; b. Dacca, Jan. 12, 1940; s. Debendra Kumar and Manorama B.; m. Anuradha Tiruvury, June 29, 1978; 1 child, Jyotsna. B.Sc., Banaras H. U., Varanasi, India, 1957, M.Sc., 1959; Ph.D., U. Sask., 1966. Lectr. in botany M.L.K. Coll., Balrampur, India, 1959-61; postdoctoral fellow U. Sask., Saskatoon, 1966-67, U. Western Ont., 1967-69, U. Wis.-Madison, 1969-70; lectr. biology Rockford (Ill.) Coll., 1970-73; assoc. prof. biology, chmn. dept. Ind. U.-N.W., Gary, 1973—. Contbr. articles to profl. jours. Mem. Am. Soc. Cell Biology, Am. Inst. Biol. Scis., AAAS, Ind. Acad. Scis., Sigma Xi. Subspecialties: Plant physiology (agriculture); Plant pathology. Current work: Physiology of host parasite relations with particular reference to nucleic acid and protein metabolism. Home: 2626 N Lakeview Apt 3608 Chicago IL 60614 Office: Dept Biology Ind U NW 3400 Broadway St Gary IN 46408

BHATTACHARYA, SYAMAL KANTI, biomedical scientist; b. Calcutta, West Bengal, India, Feb. 13, 1949; came to U.S., 1974, naturalized, 1983; s. Sudhir Chandra and Prabhabati B.; m. Keka Ghoshal, Dec. 11, 1969; children: Sumoulindra T., Julie, Syamal Dave. B.Sc. with honors, U. Calcutta, 1968; B.A. in English Lit, 1969; M.S., Murray State U., 1976; A.M., Washington U., St. Louis, 1978; Ph.D., Memphis State U., 1979. Diplomate: Am. Bd. Bioanalysis. Sr. instr. chemistry Bhabanath Instn., Calcutta, India, 1969-70; research and devel. chemist Swastik Household and Indsl. Products Pvt. Ltd., Bombay, India, 1970-74; sr. research technician Washington U. Med. Sch., St. Louis, 1976-77; research assoc. U. Tenn. Med. Ctr., Memphis, 1979-80, instr. medicine, 1980-82, mem. surgery faculty, 1983—, dir., 1982—, 1982—; teaching asst. Murray (Ky.) State U., 1974-76; research/teaching fellow Washington U., 1976-78; presdl. research fellow Memphis State U., 1978-79. Indian Nat. scholar Govt. India, New Delhi, 1965-69; Govt. India scholar Bank of India, 1974-75; Muscular Dystrophy Assn. Am. grantee, 1983-84; Am. Heart Assn. research grantee, 1983-84; recipient Nat. Research Service award in medicine NIH, 1979-81. Fellow Am. Instn. Chemists (cert. profl. chemist 1980), Indian Chem. Soc.; mem. Royal Soc. Chemistry (chartered chemist 1981), Am. Fedn. Clin. Research, N.Y. Acad. Sci. Club: U. Tenn. Faculty (Memphis). Subspecialties: Clinical chemistry; Neurophysiology. Current work: Mineral metabolism in muscular dystrophy and other neuromuscular diseases, hypertrophic cardiomyopathy, and acute pancreatitis, application of calcium-channel blocking drugs and surgical procedures to ameliorate these conditions by inhibiting intracellular calcium shift through the leaky membranes, nutritional assessment in critical care patients and those on hyperalimentation and total parenteral nutrition. Home: 3750 Marion Ave Memphis TN 38111 Office: Dept Surgery U Tenn Center Health Scis 956 Court Ave Suite 2GO4 Memphis TN 38163

BHATTACHARYYA, ASHIM K., physiologist, educator, researcher; b. Kanpur, Uttar Pradesh, India, July 9, 1936; came to U.S., 1966, naturalized, 1981; s. Viswanath and Asha (Bhattacharya) B.; m. Bani Chattapadhyay, July 10, 1966; children: Rupa, Gopa. B.Sc. with honors, Presidency Coll., Calcutta, India, 1957, M.Sc., 1959; Ph.D., U. Calcutta, 1965. Demonstrator Christian Med. Coll., Ludhiana, Punjab, India, 1964-65; lectr. Krishnath Coll., Berhampore, West Bengal, India, 1965-66; postdoctoral fellow Lab Physiol. Hygiene, U. Minn.-Mpls., 1966-68; assoc. research scientist Clin. Research Ctr., U. Iowa, 1968-74, research scientist, 1974-75; assoc. mem. Sch. Grad. studies La. State U. Med. Ctr., 1975—, asst. prof. pathology and physiology, 1975-80, assoc. prof., 1980—. Contbr. chpt., articles to profl. publs. Trustee Indian Assn. New Orleans, 1982—. Fellow Council on Arteriosclerosis of Am. Heart Assn.; mem. Am. Physiol. Soc., Am. Soc. Clin. Nutrition, Am. Inst. Nutrition, Sigma Xi. Hindu. Subspecialties: Physiology (medicine); Nutrition (biology). Current work: Mechanism and regulation of sterol absorption; sterol and bile acid metabolism; diet and atherosclerosis. Office: Dept Pathology La State U Med Center 1901 Perdido St New Orleans LA 70112

BHATTACHARYYA, GOURI KANTA, statistician; b. Hooghly, West Bengal, India, Jan. 12, 1940; m., 1962; 1 child. B.Sc., U. Calcutta, 1958, M.Sc., 1960; Ph.D., U. Calif.-Berkeley, 1966. Lectr. stats. R.K. Mission, Narendrapur, India, 1961; research officer River Research Inst., Calcutta, India, 1961-63; asst. prof. U. Calif.-Berkeley, 1966; faculty mem. U. Wis.-Madison, 1966—, prof. stats., 1975—, cons., 1967—; prof. Indian Inst. Mgmt., Ahmedabad, 1969-70; vis. prof. Indian Statis. Inst., Calcutta, 1975-76, Sch. Computer and Info. Sci., Syracuse (N.Y.) U., 1979-80. Fellow Am. Statis. Assn., Royal Statis. Soc.; mem. Inst. Math. Stats. Subspecialty: Statistics. Office: Dept Stats U Wis 1210 W Dayton Madison WI 53706

BHATTACHARYYA, PRASHAD *See also* **BHATTACHARYYA, SHANKAR**

BHATTACHARYYA, SHANKAR (PRASHAD BHATTACHARYYA), electrical engineering educator, researcher; b. Rangoon, Burma, June 23, 1946; came to U.S., 1967; s. Nil Kantha and Hem Nalini (Mukherji) B.; m. Carole Jeanne Colgate, Feb. 10, 1971; children: Krishna Lee, Mohadev, Sona Lee. B.Tech. with honors, Indian Inst. Tech., Bombay, 1967; M.S.E.E., Rice U., 1969, Ph.D. E.E., 1971. Asst. prof. Fed. U., Rio de Janeiro, Brazil, 1971-72, assoc. prof., 1972-76, prof., 1976-80; research assoc. NASA, Huntsville, Ala., 1974-75; assoc. prof. elec. engring. Tex. A&M U., College Station, 1980—; chmn. elec. engring. dept. Fed. U., Rio de Janerio, 1978-80. NRC fellow NASA Marshall Space Flight Ctr., 1974-75. Mem. IEEE, Soc. Indsl. and Applied Math. Subspecialties: Computer-aided design; Systems engineering. Current work: Research in area of control theory and control systems design. Home: 2803 Normand College Station TX 77840 Office: Texas A&M Univ Elec Engring Dept College Station TX 77843

BHELLA, HARBANS SINGH, research horticulturist; b. Harsi Pind, Punjab, India, Feb. 2, 1943; came to U.S., 1968, naturalized, 1974; s. Dalip Singh and Charan (Kaur) B.; m. Surjit Kaur, May 17, 1964; children: Kenneth Singh, Paul Singh. B.Sc., Punjab Agrl. U., 1963, M.Sc., 1966; M.S., Oreg. State U., 1970, Ph.D., 1973. Agrl. officer Punjab Agrl. U., Ludhiana, India, 1963-68; grad. research asst. Oreg. State U., Corvallis, 1968-73; dir. research Oreg. Bulb Farms, Gresham, 1974, Chgo. Hort. Soc., Glencoe, Ill., 1974-77; horticulturist Agrl. Research Service, U.S. Dept. Agr., Ames, Iowa, 1977-81, research horticulturist, Vincennes, Ind., 1981—; cons. Punjab Agrl. U., 1976, 80, 1974. Contbr. articles to profl. jours. Recipient Richard P. White award Hort. Research Inst., 1976; Baker Excellence award Iowa State U., 1979. Mem. Am. Soc. Hort. Sci., Internat. Soc. Hort. Sci., AAAS, Smithsonian Inst., Internat. Plant Propagators Soc. Sikh religion. Subspecialties: Plant physiology (agriculture); Soil chemistry. Current work: Developing a comprehensive production system research program for increasing efficiency of crop production, thus improving yield and quality and lowering production costs so they are reasonable to the consumer yet profitable to the farmer. Home: 10 Hazelwood Dr Vincennes IN 47591 Office: Agrl Research Service US Dept Agr 1118 Chestnut St PO Box 944 Vincennes IN 47591

BHUSHAN, BHARAT, engineer, researcher; b. Jhinjhana, India, Sept. 30, 1949; s. Narain and Devi (Vati) Dass; m. Sudha Bhushan, June 14, 1975; children: Ankur, Noopur. B.E. in Mech. Engring. with honors, Birla Inst. Tech. and Sci., India, 1970, M.S., MIT, 1971; M.S. in Mechanics, U. Colo., Boulder, 1973, Ph.D. in Mech. Engring., 1976; M.B.A., Rensselaer Poly. Inst., Troy, N.Y., 1980. Registered profl. engr., Pa. Research staff dept. mech. engring. MIT, 1971-72; cons., expert investigator Automotive Specialists, Denver, 1973-76; program mgr. Mech. Tech., Inc., Latham, N.Y., 1976-80; research scientist SKF Industries, Inc., King of Prussia, Pa., 1980-81; adv. engr. IBM Corp, Tucson, 1981—; propr., mgr. Time Sharing Computer Services India. Contbr. tech. papers to profl. jours. Pres. India Students Assn., 1974-76; bd. dirs. Hindu Temple Soc., Albany, N.Y., 1978-79, India Assn., 1979-80. Govt. of India Merit scholar, 1965-70; Ford Found. fellow, 1970-71; recipient Henry Hess award ASME, 1980, Bert L. Newkirk award ASME, 1983; Alfred Noble prize ASCE, 1981; George Norlin award U. Colo., 1983; Achievement award IBM, 1983. Mem. ASME; mem. IEEE, AIME; Mem. Am. Soc. Lubrication Engrs., Am. Acad. Mechanics, N.Y. Acad. Scis., AAAS, Sigma Xi, Tau Beta Pi. Subspecialties: Solid mechanics; Polymers. Current work: Technical involvement in and management of research and development activities in the field of tribology and materials related to sliding and rolling applications related to computer industry. Patentee in field. Office: IBM Corp 75B/061-1 GPD Lab Tucson AZ 85744

BHUSHAN, VIDYA, psychology educator; b. Meerut, India, Jan. 5, 1928; emigrated to Can., 1969; s. Harnand and Tulsa (Devi) Lal; m. Hem Lata Agrawal, Aug. 7, 1966; children: Vipul, Vikas. B.Sc., Agra (India) U., 1949; M.Sc., 1951; M.Ed., Lucknow (India) U., 1958; Ed.D., Ind. U., 1964. Lectr. R.R. Coll., Pilkhawa, India, 1951-53; prin. A.N. High Sch., Atamanda, India, 1954-57, Hindu Coll., Kandhla, India, 1958-61; research assoc. U. Chgo., 1964-66; asst. prof. psychology U. Hawaii, Honolulu, 1966-69; prof. psychology Laval U., Quebec City, Que., Can., 1969—; Cons. Hawaii Curriculum Ctr., Honolulu, 1966-69, Head Start Evaluation Ctr., 1966-69. Author: Matrix Algebra, 1975, Statistical Methods, 1978. Treas. Boy Scouts Can., Quebec City, 1977-82. Can. Council grantee, 1972, 74; U.S. Office of Edn. grantee, 1968; Que. Minister of Edn. grantee, 1973, 74, 77-81. Mem. Am. Psychol. Assn., Am. Ednl. Research Assn., Can. Psychol. Assn., Can. Soc. for Study of Edn., Can. Ednl. Researchers Assn. Hindu. Subspecialties: Statistics; Learning. Current work: Educational measurement, research design, and statistics. Office: Laval U Faculty of Edn Quebec PQ Canada G1K 7P4 Home: 3358 Montpetit Ste-Foy PQ Canada G1W 2T2

BIANCHERIA, AMILCARE, power systems company executive; b. Worcester, Mass., Apr. 28, 1929; s. Annibale and Aggrepina (Falcioni) BIANCHERIA. A.B. in Chemistry, Clark U., 1952, M.A. in Phys. Chemistry, 1954, Ph.D., 1957. Sr. engr. Westinghouse Electric Co., Pitts., 1957-60, fellow engr., 1962-66, mgr. material and analysis, 1967-68, mgr. fuel irradiation, Madison, Pa., 1968-80, mgr. fuel analysis, 1980—; sr. engr. Nuclear Materials and Equipment Corp., Apollo, Pa., 1960-62. Contbr. articles to profl. jours. Bd. dirs. Blackridge Civic Assn., Pitts., 1974. Mem. Am. Chem. Soc., Am. Nuclear Soc., Am. Ceramic Soc., N.Y. Acad. Scis., Am. Natural History Soc. Club: Toastmasters (Monroeville, Pa.) (area gov. 1970). Subspecialties: Nuclear fission; Materials (engineering). Current work: Nuclear fuel element development; develop computer codes describing nuclear fuel element behavior in a reactor; ceramic fuel and cladding alloy irradiation data analysis. Patentee in nuclear fuel elements field. Home: 1014 Old Gate Rd Pittsburgh PA 15235 Office: Westinghouse Electric Co PO Box 158 Madison PA 15663

BIANCO, CELSO, immunologist; b. Sao Paulo, Brazil, May 23, 1941; came to U.S., 1969, naturalized, 1981; s. Jose Antonio and Paulina (Schor) B.; m. Barbara Mei, Sept. 11, 1977; children: Marco, Christina. M.D., Med. Sch. Sao Paulo, 1966. Diplomate: lic. physician, N.Y. State. Resident in internal medicine Hosp. Sao Paulo, 1968; instr. N.Y. U. Sch. Medicine, 1969-71, asst. prof. pathology, 1971-73; asst. prof. cellular physiology and immunology Rockefeller U., 1974-77; asso. prof. pathology SUNY, Bklyn, 1977-80, prof., 1980-82; investigator Lindsley Friske Kimbal Research Inst., The N.Y. Blood Center, 1982—; dir. research and devel. Greater N.Y. Blood Program, The N.Y. Blood Center, N.Y.C., 1982—; mem. allergy and immunology study sect. NIH; adviser in immunology WHO; mem. sci. adv. com. Trudeau Inst. Med. Research, Saranac Lake, N.Y. Contbr. numerous articles on markers of lymphocyte subpopulations, activation of macrophages, interaction of monocytes and macrophages with plasma proteins to sci. jours. Recipient Research Career Devel. award Nat. Cancer Inst., NIH, 1976-81; named to 1,000 Contemporary Scientist Most Cited, 1965-78, Current Contents, 1981; Leukemia Soc. Am., Inc. scholar, 1975-76. Mem. Am. Soc. Clin. Investigation, Am. Soc. Cell Biology, Am. Assn. Immunologists, Harvey Soc., Sigma Xi. Club: The Douglaston (N.Y.). Subspecialties: Immunology (medicine); Cell biology (medicine). Current work: Research on regulation of monocyte function (phagocytosis, adhesion, microbial killing and tumor killing) by proteins of the complement system and of clotting system,including factor B and fibronectin; devel. new areas in blood transfusion. Office: 310 E 67th St New York NY 10021

BIBER, MICHAEL PETER, neurologist; b. Newark, Oct. 9, 1941; s. Irving and Hilda (Zuckerman) B.; m. Sharlene Janice Hesse, May 19, 1978; children: Sarah Alexandra, Julia Ariel. A .B. with high honors, Oberlin Coll., 1963; M.D., U. Chgo., 1967. Diplomate: Am. Bd. Internal medicine, Am. Bd. Psychiatry and Neurology. Intern Mt. Sinai Hosp., N.Y.C., 1967-68; resident in medicine Boston City Hosp., 1970-72, resident in Neurology, 1972-74; research asst. Children's Hosp., Boston, 1975-78, research assoc., 1978; assoc. neurologist Beth Israel Hosp., Boston, 1978—, dir. Sleep unit, 1978-83. Served with USPHS, 1968-70. Mem. Am. Acad. Neurology, AAAS. Subspecialties: Neurology; Neurobiology. Current work: Clinical research on sleep-related problems such as sleep apnea, narcolepsy, seizures; development of ambulatory monitoring systems. Office: 1269 Beacon St Brookline MA 02146

BIC, LUBOMIR, computer scientist; b. Iglau, Czechoslovakia, Nov. 28, 1951; s. Lubomir and Olga (Treska) B. M.S. in Computer Sci, Tech. U. Darmstadt, W.Ger., 1976, Ph.D., U. Calif.-Irvine, 1979. Cons. DKD, Wiesbaden, W.Ger., 1974-76; research asst. dept. info. and computer sci. U. Calif.-Irvine, 1976-78, lectr., 1978-79; research assoc. computer sci. Siemens AG, Munich, W. Ger., 1979-80; asst. prof. info. and computer sci. U. Calif.-Irvine, 1980—. Author: Micos: A Microprogrammable Computer Simulator, 1982. NSF research grantee, 1982-84. Mem. Assn. Computing Machinery, IEEE. Subspecialties: Computer architecture; Database systems. Current work: Highly distributed asynchronous computer architectures, dataflow systems, intelligent data-retrieval systems, database machines. Office: Dept ICS U Calif Irvine CA 92717

BICE, DAVID E., immunologist; b. Cornville, Ariz., Apr. 8, 1938; s. Virgil J. and Sarah D. (Winona) B.; m. Janet Schmutz, July 1, 1960; children: Cheryl, Patricia, Brent, Robyn, Jon, Brian. B.S., Utah State U., 1962; M.S., U. Ariz., 1964; Ph.D., La. State U. Med. Sch., 1968. Research assoc. La State U. Med. Sch., New Orleans, 1964-69; post doctoral fellow Harvard Med. Sch., Boston, 1969-71; asst. prof. La. State U. Med. Sch., New Orleans, 1971-75; immunologist Inhalation Toxicology Research Inst., Albuquerque, 1975—; mem. lung immunology and immunotoxicology study sects. NIH. Contbr. articles to profl. jours. Am. Cancer Soc. grantee, 1972-1975. Mem. Am. Assn. Immunologists, Am. Thoracic Soc., AAAS. Republican. Mormon. Subspecialties: Immunobiology and immunology; Immunotoxicology. Current work: Basic research in the development of immune defenses in the lung and how they are affected by inhaled toxic materials. Home: 3101 Alcazar NE Albuquerque NM 87110 Office: PO Box 5890 Albuquerque NM 87185

BICHSEL, HANS, physicist; b. Basel, Switzerland, Sept. 2, 1924; came to U.S., 1951, naturalized, 1967; s. Paul and Anna (Blaettler) B.; m. Sue Greenwalt, Sept. 15, 1969; children: Elisabeth Christine, Joseph Oliver. M.A., U. Basel, 1951, Ph.D., 1951. Research asso. Princeton U., 1951-54, Rice Inst., Houston, 1954-57; asst. prof. physics U. Wash., Seattle, 1957-59, from assoc. prof.to prof., 1969-78; from asst. prof. to assoc. prof. U. So. Calif., Los Angeles, 1959-72; assoc. prof. physics U. Calif.-Berkeley, 1969; lectr Aarhus (Denmark) U., 1982; cons. Los Alamos Nat. Lab., 1978-83. Contb. articles to profl. jours. Served with Swiss Army, 1943-45. Mem. Am. Phys. Soc., AAUP, Swiss Phys. Soc. Subspecialties: Cancer research (medicine); Atomic and molecular physics. Current work: Interaction of charged particles with matter; dosimetry for charged particle cancer therapy; detection of relativistic charged particles. Address:: 1211 22d Ave E Seattle WA 98112

BICKART, THEODORE ALBERT, electrical and computer engineering educator, consultant; b. N.Y.C., Aug. 25, 1935; s. Theodore Roosevelt and Edna Catherine (Pink) B.; m. Carol Florence Nichols, June 14, 1958 (div. 1972); children: Karl Jeffrey, Lauren Spencer; m. Frani Rudolph, Aug. 14, 1982; 1 stepdau., Jennifer Anne Cumming. B.E.S., Johns Hopkins U., 1957, M.S.E., 1958, D.Eng., 1960. Instr. Johns Hopkins U., 1958-61; asst. prof. elec. and computer engring. Syracuse U., 1963-65, assoc. prof., 1965-70, prof., 1970—; vis. scholar U. Calif., Berkeley, 1977; Fulbright lectr. Kiev (USSR) Poly. Inst., spring 1981; vis. lectr. Nanjing (People's Republic of China) Inst. Tech., summer 1981; cons. Deft Labs., East Syracuse, N.Y., 1982—. Author: (with Norman Balabanian) Electrical Network Theory, 1969, Linear Network Theory, 1981; contbr.: articles to profl. publs. Linear Network Theory. Served to 1st lt. U.S. Army, 1961-63. Decorated Army Commendation medal; NSF grantee, 1970, 79. Fellow IEEE (Syracuse Sect. Best Paper award 1969, 70, 73, 74, 77); mem. N.Y. Acad. Scis., Am. Math. Soc., Soc. Indsl. and Applied Maths., Assn. Computing Machinery. Democrat. Subspecialties: Computer engineering; Mathematical software. Current work: Analysis and design of electrical networks; control systems analysis; design of algorithms and software for analysis of dynamical systems, including retarded systems; logic design of computer systems. Home: 211 Standish Dr Syracuse NY 13224 Office: Syracuse U Elec and Compuer Engring Link Hall Syracuse NY 13210

BICKEL, JOHN HENRY, utility company executive; b. Chgo., June 23, 1950; s. Francis Anthony and Elaine (Broderick) B.; m. Anne Livingston Stuart, June 9, 1973; children: Blake Francis, Catherine Stuart. B.S., U. Vt., 1972, M.S. in Physics, 1974, Rensselaer Poly. Inst., 1975, Ph.D., 1980. Nuclear engr. Combustion Engring., Windsor, Conn., 1975-79; fellow, adv. com. reactor safeguards Nuclear Regulatory Commn., Washington, 1979-80; sr. nuclear engr. Northeast Utilities Co., Berlin, Conn., 1980—. Chmn. 10th Dist. Va. Conservative Caucus, Fairfax County, 1980-81. Recipient High Merit award Nuclear Regulatory Commn., 1980. Mem. Am. Nuclear Soc. (Engring. Achievement award 1983). Republican. Roman Catholic. Subspecialties: Nuclear engineering; Nuclear fission. Current work: Nuclear reactor safety analysis, simulation of reactor core and containment response, hydrogen generation and control, radiolysis and fission product water chemistry; probabilistic risk assessment. Office: Northeast Utilities Co PO Box 270 Selden St Berlin CT 05401

BICKEL, PETER JOHN, university dean, statistician; b. Bucharest, Roumania, Sept. 21, 1940; came to U.S., 1957, naturalized, 1964; s. Eliezer and P. Madeleine (Moscovici) B.; m. Nancy Kramer, Mar. 2, 1964; children: Amanda, Stephen. A.B., U. Calif., Berkeley, 1960, M.A., 1961, Ph.D., 1963. Asst. prof. stats. U. Calif., Berkeley, 1964-67, asso. prof., 1967-70, prof., 1970—, chmn. dept., 1976-79, dean phys. scis., 1980—; vis. lectr. math. Imperial Coll., London, 1965-66; fellow J.S. Guggenheim Meml. Found., 1970-71; NATO sr. sci. fellow, 1974. Author: (with K. Doksum) Mathematical Statistics, 1976; Assoc. editor: Annals of Math. Statistics, 1968-76; contbr. articles to profl. jours. Fellow Inst. Math. Stats. (pres. 1980), Am. Statis. Assn., AAAS; mem. Royal Statis. Soc., Internat. Statis. Inst. Subspecialty: Statistics. Current work: Asymptotic theory, robust and nonparametric statistics. Office: Dept Statistics Evans Hall Univ of Calif Berkeley CA 94720

BICKFORD, ARTHUR ALTON, vet. pathologist; b. Coventry, Vt., Sept. 23, 1936; s. Robert Lee and Evangeline (Tyler) B.; m. Margaret Nash, July 18, 1960; children: Carolyn F., Patricia S., Andrea K. Student, U. Vt., 1954-56; V.M.D., U. Pa., 1960; postgrad., Colo. State U., 1961-62; M.S., Purdue U., 1964, Ph.D., 1966. Cert. Am. Coll. Vet. Pathologists. Asst. animal pathologist U. Vt., 1960-61; instr. Purdue U., 1962-65, asst. prof., 1965-66, assoc. prof., 1968-1973; head pathology sect. William S. Merrell Co., Cin., 1966-67; head pathology dept. ASPCA Hosp. and Clinic, N.Y.C., 1967-68; pathologist Ayest Research Lab., Chazy, N.Y., 1973-74; extension veterinarian U. Calif., Davis, 1974-79; prof., chmn. dept. vet. pathology U. Mo., Columbia, 1979—. Author: Avian Disease Manual, 1980; Contbr. articles and abstracts to profl. and trade jours. Mem. Am. Vet. Med. Assn., Am. Coll. Vet. Pathologists, Am. Assn. Avian Pathologists, Internat. Acad. Pathologists. Club: Kiwanis. Subspecialty: Pathology (veterinary medicine). Current work: Avian pathology, toxicol pathology, diagnostic pathology. Home: 25 Bingham Rd Columbia MO 62501 Office: Dept of Veterinary Pathology University of Missouri Columbia MO 65211

BICKNELL, EDWARD J., veterinarian, diagnostic pathologist; b. Kansas City, Mo., Jan. 23, 1928; s. Arthur F. and Anne V. B.; m. Phyllis L. Saborowski, Aug. 2, 1952. B.A., U. Mo., 1948; M.S., Kans. State U., 1951, D.V.M., 1960; Ph.D., Mich. State U., 1965. Instr. in diagnostic pathology and clin. medicine Mich. State U., 1960-65, Kans. State U., 1966-67; researcher in diagnostic pathology S.D. State U., 1967-73; extension veterinarian, research scientist U. Ariz., Tucson, 1973—. Contbr. articles to profl. jours. Served to sgt. USAF, 1952-56. NIH grantee, 1964-65. Mem. AVMA, Ariz. Veterinary Med. Assn., Sigma Xi, Phi Kappa Phi, Gamma Sigma Delta, Alpha Zeta, Phi Zeta. Episcopalian. Subspecialty: Pathology (veterinary medicine). Current work: Diagnostic pathology infectious and toxic diseases of domestic animals; animal pathology, infectious diseases, toxicology. Office: PO Box 304 Mesa AZ 85201

BIDLACK, WAYNE ROSS, pharmacology and nutrition educator; b. Waverly, N.Y., Aug. 12, 1944; s. Andrew L. and Vivian P. (Cowles) B.; m. Sandra Marie Weber, June 29, 1968. B.S. in Dairy Sci. and Tech., Pa. State U., 1966; M.S. in Food Sci., Iowa State U., 1968; Ph.D. in Biochemistry, U. Calif.-Davis, 1972. Postdoctoral fellow U. So. Calif., Los Angeles, 1972-74; asst. prof. pharmacology and nutrition Sch. Medicine (U. So. Calif.), 1974-80, assoc. prof., 1980—; pres. Greater Los Angeles Nutrition Council, 1982. Contbr. articles to sci. jours.; editorial bd.: Biochem. Medicine, 1982—, Ann. Revs. Nutrition, 1979-82. Bd. dirs. Calif. Council Against Health Fraud, 1982—. Mem. Am. Dietetic Assn., Am. Inst. Nutrition, Inst. Food Technologists, Am. Soc. Pharmacology and Exptl. Therapeutics, Am. Coll. Toxicology, Soc. Toxicologists. Club: Almansor Men's Golf (sec. 1978-79, pres. 1981). Subspecialties: Nutrition (medicine); Toxicology (medicine). Current work: Hepatic drug and toxicant metabolism, vitamin-mineral metabolism and interaction, drug-nutrient interaction. Home: 900 N Almansor St Alhambra CA 91801 Office: 2025 Zonal Ave Los Angeles CA 90033

BIEGER, GEORGE R., psychologist, educator; b. Phila., July 12, 1946; s. John William and Isabelle (Tayoun) B.; m. N. Karen Trotter, Apr. 11, 1970; 1 son, Geoffrey Trotter. B.S., U.S. Naval Acad., 1968; M.A., U. West Fla., Pensacola, 1980; M.S., Cornell U., 1982, Ph.D., 1982. Tchr. Escambia County Pub. Schs., Pensacola, Fla., 1973-78; research asst. Cornell U., 1978-81; asst. prof. ednl. psychology Bucknell U., Lewisburg, Pa., 1981—; cons. IBM, Poughkeepsie, N.Y., 1981-82, Lewisburg Area Schs., 1982. Contbr. articles to profl. jours. Served to lt. USN, 1968-73. USDA grantee, 1978; Office Naval Research grantee, 1979. Mem. Am. Psychol. Assn., Am. Ednl. Research Assn., AAAS, Cognitive Scis. Soc. (assoc.). Subspecialties: Cognition; Learning. Current work: Human learning, cognitive processing, visual perception, language acquisition and development, psycho-educational research. Office: Bucknell Univ B-110 Coleman Hall Lewisburg PA 17837 Home: RD 3 Box 126 Mifflinburg PA 17844

BIELAWSKI, W. BART, telecommunications executive; b. Kolomyja, Poland, May 22, 1936; s. Waclaw M. and Eleonora (Sklarczyk) B.; m. Sue A. Patterson, Dec. 17, 1960; children: W. Mark, H. Michael. B.S.E.E., Wyzsza Szkola Marynarki Wojennej, Poland, 1957; M.B.A., Ind. U., 1965. Research and devel. engr., asst. chief engr. Sarkes Tarzian, Bloomington, Ind., 1960-65; market devel. specialist, sales and mktg. mgr. Corning Glass Works, N.Y., 1966-77; v.p. sales and mktg. Siecor Optical Cable, Horsehead, N.Y., 1978-81; v.p. Siecor Corp., 1982—; gen mgr. Siecor FiberLAN, 1982—; dir. Gardoc Corp. Contbr. articles to profl. jours. Mem. IEEE, Optical Soc. Am. Subspecialties: Distributed systems and networks; Fiber optics. Current work: Product and applications development in fiber optics. Home: 10200 Roadstead Way West Raleigh NC 27612 Office: PO Box 12726 Research Triangle Park NC 27709

BIELLO, DANIEL ROBERT, physician, medical educator; b. Cleve., Feb. 26, 1947; s. Dante Nicholas and Jeanne Lois (Scott) B.; m. Elizabeth Anne Dumbleton, June 21, 1969; children: David Daniel, Timothy Robert. A.B., Ohio Wesleyan U., 1969; M.D., Case Western Res. U., 1973. Diplomate: Am. Bd. Radiology, Am. Bd. Nuclear Medicine. Intern Mallinckrodt Inst. Radiology, St. Louis, resident in radiology, 1974-78, asst. prof. radiology, 1978-81, assoc. prof., 1982—. Mem. Am. Coll. Radiology, Radiol. Soc. N.Am., Am. Roentgen Ray Soc., Soc. Nuclear Medicine, AMA. Subspecialties: Nuclear medicine; Diagnostic radiology. Current work: Cardiac and pulmonary medicine. Office: Mallinckrodt Inst Radiology 510 S Kinghighway Saint Louis MO 63110 Home: 900 Revere Dr Town and Country MO 63141

BIENENSTOCK, ARTHUR IRWIN, physicist; b. N.Y.C., Mar. 20, 1935; s. Leo and Lena (Senator) B.; m. Roslyn Doris Goldberg, Apr. 14, 1957; children—Eric Lawrence, Amy Elizabeth, Adam Paul. B.S., Poly. Inst. Bklyn., 1955, M.S., 1957; Ph.D., Harvard U., 1962. Asst. prof. Harvard U., 1963-67; mem. faculty Stanford U., 1967—, prof. applied physics, 1972—, vice provost faculty affairs, 1972-77, dir. synchrotron radiation lab., 1978—; mem. U.S. Nat. Com. for Crystallography, 1983—; lectr., cons. in field. Author papers in field. Bd. dirs. No. Calif. chpt. Cystic Fibrosis Research Found., 1970-73, mem. pres.'s adv. council, 1980—; trustee Cystic Fibrosis Found., 1982—. Recipient Sidhu award Pitts. Diffraction Soc., 1968, Disting. Alumnus award Poly. Inst. N.Y., 1977; NSF fellow, 1962-63. Fellow Am. Phys. Soc.; mem. Am. Crystallographic Assn., AAAS, N.Y. Acad. Scis. Jewish. Subspecialties: Condensed matter physics. Current work: The development of synchrotron radiation capabilities and utilizarion of it for the determination of atomic arrangements in amorphous materials. Home: 967 Mears Ct Stanford CA 94305 Office: Synchrotron Radiation Lab Bin 69 Box 4349 Stanford CA 94305

BIENIEK, RONALD JAMES, physics educator, researcher; b. South Gate, Calif., Aug. 30, 1948; s. John Peter and Catherine Margaret (Bieniek) B.; m. Georgia Ann Harman, Sept. 5, 1970; children: Michelle Diane, Geoffrey Thomas. B.S. in Physics (Nat. Merit scholar), U. Calif.-Riverside, 1970; postgrad, M.I.T., 1971; A.M. in History of Sci. (NSF predoctoral fellow), Harvard U., 1973; Ph.D. in Physics, Harvard U., 1975. Engring. trainee N.Am. Rockwell, Anaheim, Calif., summers 1966-69; physicist Naval Undersea Research and Devel. Center, Pasadena, Calif., summer 1970; NSF postdoctoral fellow Joint Inst. Lab. Astrophysics, Boulder, Colo., summer 1975; vis. scholar U. Calif.-Berkeley, summer 1976; asst. prof. astronomy and phys. sci. U. Ill.-Urbana, 1975-81, asst. prof. humanities, 1979; asst. prof. physics U. Mo.-Rolla, 1981—; pub. cons.; researcher. NSF postdoctoral fellow, summers 1975-77; grantee NSF,

1976-78; 78-81, Research Corp., 1982—. Mem. Am. Phys. Soc., History of Sci. Soc., Am. Assn. Physics Tchrs., Phi Beta Kappa, Sigma Xi. Subspecialties: Atomic and molecular physics; Theoretical astrophysics. Current work: Theoretical atomic and molecular collisional physics and astrophysics; atomic and molecular collision theory involving energy redistribution, ionization, spectral line broadening, interstellar medium; physics-astronomy edn. Home: Rural Route 1 Box 432 Rolla MO 65401 Office: Dept Physics Univ Mo Rolla MO 65401

BIER, CHARLES JAMES, chemistry educator, Solar researcher; b. Louisville, June 21, 1945; s. Charles Aloysius and Emilie Julie (Miksik) B.; m. Jerryanne Taber, Aug. 3, 1968; children: Rebecca, Jessica, Jonathan, Sara. B.S. in Chemistry, Providence Coll., 1967; Ph.D., MIT, 1971. Research assoc. Duke U., Durham, N.C., 1971-72; asst. prof. U. North Fla., Jacksonville, 1972-75, Eisenhower Coll., Seneca Falls, N.Y., 1975-76; stockroom curator, instr. chemistry Hollins Coll., Va., 1976-77; assoc., prof. chemistry and environ. studies Ferrum Coll., Va., 1977—. Contbr. articles to profl. jours.; solar inventor. Bd. dirs. Whetstone br. Living Sch., Ferrum, 1978—. Franklin County Peace fellow. Mem. Internat. Solar Energy Soc. Catholic. Club: Franklin County Organic Gardening. Subspecialties: Solar energy; Resource management. Current work: Development technology and systems which use minimum of non-renewable resources to support comfortable, productive and full livelihood. Patentee in field. Home: Route 2 Box 35 Ferrum VA 24088 Office: PO Box 163 Ferrum Coll Ferrum VA 24088

BIER, MILAN, engring. and microbiology educator, consultant; b. Vukovar, Yugoslavia, Dec. 7, 1920; came to U.S., 1946, naturalized, 1950; s. Edmond E. and Ada (Diamant) B.; m. Rhoda M. Solvay, Mar. 29, 1952; children: Vicki M., James J., Robert E. Licence Scis. Chimiques, U. Geneva, 1946; Ph.D., Fordham U., 1950. Asst. prof. Fordham U., N.Y.C., 1950-60; vis. asst. prof. Poly. Inst., Bklyn., 1959-61; head protein chemistry Inst. Applied Biology, N.Y.C., 1951-62; research biophysicist VA Hosp., Tucson, 1962-78; prof. U. Ariz., Tucson, 1962—, head biophys. lab., 1978—; cons. Schering-Plough Corp., Bloomfield, N.J., 1981—, U.S. Space Research Assn., Washington, 1981—, GTE, San Diego, 1983; dir. Am. Biol. Corp., Montclair, N.J. NASA grantee, 1972—. Mem. Am. Chem. Soc., Am. Soc. Biol. Chemists, AAAS, Am. Soc. Artificial Organs. Subspecialties: Biomedical engineering; Biophysical chemistry. Current work: Protein separation technology, membrane processes, electrophoresis, artificial internal organs, plasmapheresis. Patentee in field. Home: 5341 E 7th St Tucson AZ 85711 Office: U Ariz Bldg 20 Tucson AZ 85721

BIERLY, EUGENE WENDELL, meteorologist, science foundation administrator; b. Pittston, Pa., Sept. 11, 1931; m., 1953; 3 children. A.B., U. Pa., 1953; cert., U.S. Naval Postgrad. Sch., 1954; M.S., U. Mich., 1957, Ph.D., 1968. Asst. dept. civil engring. meteorol. labs. U. Mich., 1956-60, asst. research meteorologist dept. engring. mechanics, 1960-63, lectr., 1961-63; meteorologist U.S. AEC, 1963-66; program dir. meteorology div. atmospheric sci. NSF, Washington, 1966-71, coordinator global atmospheric research program, 1971-74, head office climate dynamics, 1974-75, head climate dynamics research sect., 1975-79, dir. div., 1979—; cons. Reactor Devel. Co., 1961-62, Pacific Missile Range, 1962-63. Congl. fellow, 1970-71. Fellow AAAS, Am. Meteorol. Soc.; mem. Air Pollution Control Assn., Royal Meteorol. Soc., Am. Polit. Sci. Assn. Subspecialty: Atmospheric science research management. Office: Div Atmospheric Sc Nat Sci Found 1800 G St N W Washington DC 20550

BIERZYCHUDEK, PAULETTE FRANCINE, biologist; b. Chgo., Aug. 25, 1951; d. Casimir A. and Gladys (Chmielewski) B. Student, U. Chgo., 1969-71; B.A., B.S., U. Wash., 1974; Ph.D., Cornell U., 1981. Asst. prof. biology Pomona Coll., Claremont, Calif., 1980—. Mem. Soc. Study Evolution, Am. Soc. Naturalists, Rocky Mountain Biol. Lab., Sierra Club. Subspecialties: Ecology; Evolutionary biology. Current work: Evolution of plant reproductive strategies, especially evolution of sexual reproduction, plant demography. Office: Dept Biology Pomona Coll Claremont CA 91711

BIGELEISEN, JACOB, educator, chemist; b. Paterson, N.J., May 2, 1919; s. Harry and Ida (Slomowitz) B.; m. Grace Alice Simon, Oct. 21, 1945; children: David M., Ira S., Paul E. A.B., NYU, 1939; M.S., Wash. State U., 1941; Ph.D., U. Calif., Berkeley, 1943. Research scientist Manhattan Dist., Columbia, 1943-45; research asso. Ohio State U., Columbus, 1945-46; fellow Enrico Fermi Inst., U. Chgo., 1946-48; sr. chemist Brookhaven Nat. Lab., Upton, N.Y., 1948-68; prof. chemistry U. Rochester, N.Y., 1968-78, chmn. dept., 1970-75, Tracy H. Harris prof., 1973-78; v.p. research, dean grad. studies SUNY, Stony Brook, 1978-80, Leading prof. chemistry, 1978—; vis. prof. Cornell U., 1953; NSF sr. fellow, vis. prof. Eidgen Techn. Hochschule, Switzerland, 1962-63; chmn. Assembly Math. and Phys. Scis., NRC-Nat. Acad. Scis., 1976-80. Mem. editorial bd.: Jour. Phys. Chemistry. Trustee Sayville Jewish Center, 1954-68. Recipient Nuclear award Am. Chem. Soc., 1958, Gilbert N. Lewis lectr., 1963, E.O. Lawrence award, 1964, Disting. Alumnus award Wash. State U., 1983; John Simon Guggenheim fellow, 1974-75. Fellow Am. Phys. Soc., Am. Chem. Soc., AAAS, Am. Acad. Arts and Sci.; mem. Nat. Acad. Scis., Phi Beta Kappa, Sigma Xi, Phi Lambda Upsilon. Subspecialties: Physical chemistry; Nuclear engineering. Current work: Correlation of isotope chemistry with molecular structure; study of molecular motion in condensed media by isotope fractionation. Research in photochemistry in rigid media, semiquinones, cryogenics, chemistry of isotopes, quantum statistics of gases, liquids and solids. Home: PO Box 217 Saint James NY 11780 As a youth I became interested in a career in science because it offered the opportunity to test ideas and hypotheses objectively by experiment. This unique aspect of science, which differentiates it from all other branches of learning and knowledge, has been a guiding principle both in my professional and my personal life. My career has included research, teaching, administration and public service.

BIGELOW, BRIAN JOHN, psychology educator, researcher; b. Windsor, Ont., Can., Dec. 1, 1947; s. Ernest Peter and Dorothy Joan (Pariset) B.; m. Valerie Anne Baker, Mar. 2, 1948; children: Caragh, Ian. B.A., Windsor U., 1970, M.A., 1971, Ph.D., 1974. Cert. psychologist, Ont. Lectr. Lakehead U., Thunder Bay, Ont., 1973-74, asst. prof., 1974-75; behavioral cons. Lakehead Bd. Edn., Thunder Bay, 1975-77; asst. prof. Laurentian U., Sudbury, Ont., 1978-82, assoc. prof., 1982—. Social Sci. and Humanities Research Council Can. grantee, 1982—. Mem. Am. Psychol. Assn., Ont. Psychol. Assn. (dir. No. region), Can. Psychol. Assn., Soc. Research Child Develop. Anglican. Subspecialties: Developmental psychology; Cognition. Current work: Friendship concepts, social rules charting social understanding in children through documenting expressed social rules. Home: 372 3d Ave Sudbury ON Canada P3B 4A1 Office: Laurentian U Ramsey Lake Rd Sudbury ON Canada P3E 2C6

BIGELOW, CHARLES CROSS, biochemist, university administrator; b. Edmonton, Alta., Can., Apr. 25, 1928; s. Sherburne Tupper and Helen Beatrice (Cross) B.; m. Elizabeth Rosemary Sellick, Aug. 22, 1977; children: Ann K. Bigelow McLean, David C. B.A.Sc., U. Toronto, 1953, M.Sc., 1955; Ph.D., McMaster U., 1957. Postdoctoral fellow Carlsberg Lab., Copenhagen, 1957-59; assoc. Sloan-Kettering Inst. Cancer Research, N.Y.C., 1959-62; asst. prof. chemistry U. Alta., Can., 1962-64, assoc. prof., 1964-65; vis. prof. Fla. State U., Tallahassee, 1965; assoc. prof. biochemistry U. Western Ont. (Can.), London, 1965-69, prof., 1969-74; prof., head biochemistry Meml. U. Nfld. (Can.), St. John's, 1974-76; dean of sci., prof. chemistry St. Mary's U., Halifax, N.S., Can., 1977-79, U. Man. (Can.), Winnipeg, 1979—; vis. prof. U. Toronto, 1973-74; Chmn. Ont. Confedn. Univ. Faculty Assns., 1970-71; pres. Can. Assn. Univ. Tchrs., 1972-73. Contbr. articles on protein structure and denaturation to sci. jours. Pres. N.S. New Democratic party, 1978-79, Man. New Dem. party, 1982—. Grantee NRC Can., Med. Research Bd. govs. U. Western Ont., 1972-73. Council Natural Scis. and Engring. Research Council Can.; Fellow Chem. Inst. Can.; mem. Can. Biochem. Soc., Am. Chem. Soc., Am. Soc. Biol. Chemists, AAAS, Sigma Xi. Club: Southwood Golf and Country (Winnipeg). Subspecialty: Biophysical chemistry. Current work: Protein structure and denaturation. Home: 701 South Dr Winnipeg MB Canada R2G OC2 Office: U Man Winnipeg MB Canada R3T 2N2

BIGELOW, JOHN EALY, radiochemical processing engineer, nuclear engineer; b. Hammond, Ind., Jan. 28, 1929; s. Charles Glenford and Cornelia (Ealy) B.; m. Lauretta Marguerite Meyer, Jan. 28, 1954; children: Holly Rea, Timothy Stuart, Andrew Scott. B.S. in Chem. Engring, Purdue U., 1950, S.M., MIT, 1952, Sc.D., 1956. Jr. devel. engr. Oak Ridge Nat. Lab., 1956-58, devel. engr., 1961—; ops. analyst AEC, Washington, 1958-61. Co-author reference book. Vol. leader Boy Scouts Am., Oak Ridge and Knoxville, 1965-77; commr., chmn. Pellissippi dist., Gt. Smoky Mountain council, 1972-73; bd. dirs. Anderson County (Tenn.) Health Council, 1974—, pres., 1977; bd. dirs. East Tenn. Health Improvement Council, 1976-79, 82—. Mem. Am. Chem. Soc., Am. Nuclear Soc., Sigma Xi (sec. chpt. 1975-82), Tau Beta Pi. Republican. Subspecialties: Chemical engineering; Nuclear engineering. Current work: Have been associated with National Transplutonium Element Production Program for 22 years. Coordinate production at Oak Ridge facilities with requests for such materials to be used in research. Home: 111 Concord Rd Oak Ridge TN 37830 Office: Oak Ridge Nat Lab PO Box X Oak Ridge TN 37820

BIGFORD, THOMAS EDWARD, environmental policy analyst, marine ecologist; b. Lansing, Mich., Aug. 9, 1952; s. John Watkins and Helenmarie (Van Hartesveldt) B.; m. Jacqueline Marie Buckley, June 29, 1974. B.S., Mich. State U., 1974; M.S., U. R.I., 1976, M.M.A., 1977. Aquatic biologist U.S. EPA, Narragansett, R.I., 1974-77; marine ecologist Ctr. for Natural Areas, Washington, 1978-80; environ. policy analyst NOAA, U.S. Dept. Commerce, Washington, 1980—. Mem. Coastal Soc. (sec. 1981-83), Mass. Audubon Soc. Subspecialties: Resource management; Ecology. Current work: Fisheries resource management, marine parks and reserves, environmental regulation. Office: Ecology and Conservation Div NOAA 14th and Constitution Ave NW Washington DC 20230

BIGGS, MAURICE EARL, geophysicist, assistant state geologist; b. Sebree, Ky., Oct. 27, 1921; s. Paul Isidore and Bessie (Hunt) B.; m. Jean Bock, June 14, 1946; children: Nancy June, Donald Charles, David Paul, Judith Ann. A.B., Ind. U.-Bloomington, 1948, M.A., 1950, Ph.D., 1973. With Ind. Geol. Survey, Ind. U., Bloomington, 1952—, head geophysics sect., 1952-59, asst. state geologist, 1959—, now head. Served to capt. U.S. Army, 1941-46. Mem. European Assn. Exploration Geophysicists, Sigma Xi. Subspecialties: Geophysics; Seismology. Current work: Refraction and reflection seismology and correlation of seismic velocities and stratigraphic units. Home: 1213 S Brooks Dr Bloomington IN 47401 Office: 611 N Walnut Dr Bloomington IN 47405

BIGLER, ERIN DAVID, psychologist, educator; b. Los Angeles, July 9, 1949; s. Erin Boley and Natalie (Webb) B.; m. Janet Beckstrom, June 22, 1971; children: Alicia Suzanne, Erin Daniel. B.S., Brigham Young U., 1971, Ph.D., 1974; postgrad., St. Joseph's Hosp., Phoenix, 1977. Lic. psychologist, Ariz., Tex. Postdoctoral fellow St. Josephs Hosp., Phoenix, 1975-77; psychologist Austin (Tex.) State Hosp, 1977-78; asst. prof. U. Tex., Austin, 1977-82, assoc. prof. psychology, 1982—; pvt. practice psychology, Austin, 1977—. Editor: Clin. Neuropsychology, 1979—; author: Diagnostic Clinical Neuropsychology, 1983; contbr. articles to profl. jours. NIH grantee, 1975-77; Hogg Found. grantee, 1981-82. Mem. Am. Psychol. Assn., Nat. Acad. Neuropsychology (exec. com. 1982-83), N.Y. Acad. Scis., Soc. for Neurosci., Sigma Xi. Democrat. Mormon. Subspecialties: Neuropsychology; Psychobiology. Current work: Brain-behavior relations; neurological diagnostics. Office: Austin Neurological Clinic 711-F W 38th St Austin TX 78705

BIGLER, RODNEY E., biophysicist; b. Pocatello, Idaho, Mar. 15, 1941; s. Vance S. and Mildred (Higham) B.; m. Louise Anne Calabrese, Sept. 17, 1977; children—Ronald, Julie. Student, Multnomah Coll., 1962-63; B.S., Portland State Coll., 1966, postgrad., 1966-67; Ph.D., U. Tex., Austin, 1971. Teaching asst. Portland State Coll., 1966-67; teaching asst. U. Tex., 1967-68, research asst., 1968-71; research assoc. Sloan-Kettering Inst., N.Y.C., 1971-73, assoc., 1973-82, asst. mem., 1982—; research collaborator Brookhaven Nat. Lab., 1973-79, 83—; asst. prof. Cornell U. Grad. Sch. Med. Scis., 1974—. Contbr. numerous articles to sci. jours. Served with AUS, 1959-62. NIH-Nat. Heart Lung and Blood Inst. grantee, 1978—; NIH-Nat. Cancer Inst. grantee, 1983—. Subspecialties: Biophysics (physics); Nuclear physics. Current work: Medical diagnostic and therapeutic uses of radiation and radioactive materials research; nuclear medicine. Home: 303 E 71st St Apt 2H New York NY 10021 Office: 1275 York Ave New York NY 10021

BIGLER, WILLIAM C., engineering executive; b. Franklin, Pa., Feb. 27, 1940; s. Joseph G. and Mary (Eggbeer) B.; m. Carole L. Lindemann, June 8, 1963; children: Jeffrey C., Pamela J. B.M.E., Cornell U., 1963, M.M.E., 1964. With Corning (N.Y.) Glass Works, 1964; engr., 1964-70, sr. engr., 1970-72, supr. equipment engring., 1972-74, mgr. equipment engring. and maintenance, 1974-75, plant mfg. engr., 1975-77, project mgr., 1977-82, engring. mgr., 1982—; research asst. Cornell U., 1963-64. Served to 1st lt. C.E. U.S. Army, 1964-66. Decorated Army Commendation medal. Mem. ASME, Nat. League Am. Wheelmen, Corning Roadrunners, Sigma Xi. Republican. Congregationalist. Subspecialties: Mechanical engineering; Systems engineering. Current work: Engineering management in industrial manufacturing; engineering management/project management in industrial manufacturing. Office: Corning Glass Works HP-ME-2 Corning NY 14831

BIGLIERI, EDWARD GEORGE, physician; b. San Francisco, Jan. 17, 1925; s. Ned and (Mignacco) B.; m. Beverly A. Bergesen, May 16, 1953; children—Mark, Michael, Gregg. Student, U. San Francisco, 1942-43, Gonzaga U., 1943-44; B.S. in Chemistry summa cum laude, U. San Francisco, 1948; M.D., U. Calif., 1952. Diplomate: Am. Bd. Internal Medicine (endocrine test com. 1971-76). Intern U. Calif., San Francisco Med. Center, also; VA Hosp., San Francisco, 1952-54, resident, 1954-56; clin. asso. and research physician NIH, 1956-58; also metabolic unit U. Calif., 1958-61, asst. prof. medicine, 1962-65, asso. prof., 1965-71, prof., 1971—; program dir. gen. clin. research, also chief endocrinology service San Francisco Gen. Hosp., 1962—; vis. prof. Monash U., Melbourne, Australia, 1967; cons. Oak Knoll Naval Hosp., Travis AFB; mem. study sect. NIH, 1971-74. Contbr. articles on endocrinology ahd hormones in hypertension to profl. jours. Served to lt. (j.g.) USN, 1944-46. NIH grantee, 1972-73. Mem. Endocrine Soc., A.C.P., Am. Soc. Clin. Investigation, Am. Heart Assn. (council high blood pressure research), Assn. Am. Physicians, Western Assn. Physicians, Am. Fedn. Clin. Research. Subspecialties: Internal medicine; Endocrinology. Current work: Rule of hormones in high blood pressure. Home: 129 Convent Ct San Rafael CA 94901 Office: San Francisco Gen Hosp San Francisco CA 94110 The opportunity to study an ever-inquiring profession, and an incredibly supportive family, are the essential ingredients for sustained growth in the academic arena.

BIKERMAN, MICHAEL, geology educator, researcher; b. Berlin, July 30, 1934; U.S., 1945, naturalized, 1950; s. Jacob Joseph and Valentine (Leiv) B.; m. Viola Adler, June 2, 1956; children: Jennifer, David, Tania. B.S. in Chemistry, Queens Coll., 1954, N.Mex. Inst. Mining and Tech., 1956; M.S., U. Ariz., 1962, Ph.D., 1965. Surveyor/assayer Cinnabar Mine, Yellowpine, Idaho, 1957; instr. Boise (Idaho) Jr. Coll., 1958-60; research asst. U. Ariz.-Tucson, 1960-65; asst. prof. Wichita (Kans.) State U., 1965-67; asst. prof. geology U. Pitts., 1967-70, assoc. prof., 1970—; research assoc. Carnegie Mus. Natural History, Pitts., 1979—. Active Am. Field Service, Pitts., 1975—. Fellow Geol. Soc. Am.; mem. Geochem. Soc., AAAS, Pitts. Geol. Soc. (pres. 1977-78). Subspecialties: Geology; Geochemistry. Current work: K-Ar dating of volcanic rocks in Southwest New Mexico, in conjunction with mapping and field petrologic work; various K-Ar studies elsewhere. Home: 215 Woodhaven Dr Mount Lebanon PA 15228 Office: Dept Geology and Planetary Sci U Pittsburgh Pittsburgh PA 15260

BILITCH, MICHAEL, medical educator, physician; b. Belgrade, Yugoslavia, Feb. 8, 1932; came to U.S., 1949; s. Sasha Alexander and Oona Mary (Ball) B.; m. Mary Jo Ann Minges, June 19, 1956 (dec. 1966); children: Bonnie, Kimberly, David; m. Nancy Ann Neher, Sept. 3, 1967 (div. 1982); children: Kendal, Dawn, Robert, Susan, Douglas. A.B., San Jose State Coll., 1954; M.A., Miami U., 1956; M.D., U. So. Calif., 1960. Teaching asst. San Jose State Coll., 1952-54; teaching asst. Miami U., Oxford, Ohio, 1954-56; research asst. U. So. Calif., Los Angeles, 1958-60, instr. medicine, 1964-67, asst. prof. medicine, 1967-71, assoc. prof., 1971—, dir., 1970—; vis. prof. U. Groningen, Netherlands, 1978; cons. Cardiac Pacemakers, St. Paul, 1978—. Author: A Manual of Cardiac Arrhythmias, 1971; co-author: Heart Block, 1972; contbr. articles to profl. jours. Los Angeles County Heart Assn. fellow, 1965-67. Fellow ACP, Am. Coll. Cardiology; mem. Assn. Advancement of Med. Instrumentation (chmn. pacemaker com. 1971-77), N.Am. Soc. Pacing and Electrophysiology (founding mem., exec. com.), Am. Fedn. Clin. Research, AAAS, N.Y. Acad. Scis., Royal Soc. Health. Subspecialties: Cardiology; Internal medicine. Current work: Implantable cardiac pacemakers, development and clinical testing of new devices and systems, evaluating performance standards; cardiac arrhythmias, evaluation by mapping techniques, management by logical schema. Patentee leadless pacemaker. Home: 1420 San Pablo St Apt C-201 Los Angeles CA 90033 Office: U So Calif Sch Medicine 2025 Zonal Ave Los Angeles CA 90033

BILLEN, DANIEL, radiation biologist, educator; b. N.Y.C., Nov. 27, 1924; s. Morris and Gertrude B.; m. Gertrude Eleanor Berlin; children: Jerome, Rhonda, Robin. B.S., Cornell U., 1948; M.S., U. Tenn., Knoxville, 1949, Ph.D., 1951. Biologist Oak Ridge (Tenn.) Nat. Lab., 1951-57; assoc. prof., staff scientist M.D. Anderson Hosp., Houston, 1957-66; program dir. NSF, Washington., 1961-62; prof. radiation biology U. Fla., Gainesville, 1966-73; dir., prof. biomed. scis. Grad. Sch. Biomed. Scis. U. Tenn., Oak Ridge, 1973-77. Editor-in-chief: Radiation Research, 1979—; contbr. research papers to profl. publs. Served with USAAF, 1943-44. Mem. Radiation Research Soc., Am. Soc. Microbiologists, Am. Soc. Biol. Chemists. Jewish. Club: Elks. Subspecialties: Cell biology; Gene actions. Current work: Research in DNA damage and repair, radiation and chemical hazards in cells. Office: Biology Div Oak Ridge Nat Lab Oak Ridge TN 37830

BILLHIMER, WARD LOREN, JR., microbiologist, educator; b. Wadsworth, Ohio, Jan. 9, 1952; s. Ward Loren and Dorothy Jane (Rager) B.; m. Rosetta Paulette Armbrust, June 15, 1974; children: Shelly, Vincent, Todd, Bryan. B.S., U. Akron, 1973; M.S., 1976. Mgr. microbiology dept. Hill Top Research, Inc., Miamiville, Ohio, 1976-79, dir. ops., microbiology div., 1979-81, mgr. data processing, proprietary clin. div., 1981-82, asst. tech. dir. for clin. safety, clin. investigator, proprietary clin. div., 1981—. Chmn. adminstrv. bd. Trinity United Meth. Ch., Milford, Ohio, 1981—. Recipient Wyandott award Alpha Chi Sigma, 1973. Mem. Soc. Indsl. Microbiology, Am. Soc. Microbiology, AAAS, Environ. Mutagen Soc., Alpha Chi Sigma. Subspecialties: Microbiology; Human safety and efficacy evalutions. Current work: Evaluating consumer products for human safety and efficacy. Home: 1009 Marcie Ln Milford OH 45150 Office: SR 126 Miamiville OH 45147

BILLINGHAM, JOHN, aerospace physician, government research center executive; b. Worcester, Eng., Mar. 18, 1930; m., 1956; 2 children. B.A., Oxford U., 1951, M.A., 1954, B.Medicine, 1954, B.Ch., 1954; D.H.L. (hon.), Hawaii Loa Coll., 1981. Med. research officer aviation physiology RAF Inst. Aviation Medicine, Farnborough, Eng., 1956-63; br. chief environ., physiol., space medicine Johnson Spacecraft Ctr., NASA, Houston, 1963-65; div. chief biotechnol., aerospace medicine Ames Research Ctr., Moffett Field, Calif., 1966-75, chief extraterrestrial research div., chief program for search for extraterrestrial intelligence, 1976—; staff mem. marine sci. council Exec. Office of Pres., 1969. Fellow Aerospace Med. Assn. Subspecialties: Space medicine; Exobiology. Address: 33 Campbell Ln Menlo Park CA 94025

BILLINGSLEY, PATRICK PAUL, mathematical educator; b. Sioux Falls, S.D., May 3, 1925; m., 1953; 5 children. B.S., U.S. Naval Acad., 1948; M.A., Princeton U., 1952, Ph.D. in Math, 1955. NSF fellow Princeton U., 1957-58; asst. prof. stats. U. Chgo., 1958-62, assoc. prof. math. and stats., 1962-67, prof., 1967—, chmn. dept. stats., 1980-83; vis. prof. U. Copenhagen, 1964-65. Editor: Annals of Probability, 1976-78. Fellow Inst. Math. Stats. (pres. 1983), Am. Math. Soc. Subspecialty: Statistics. Office: Dept Stats U Chgo Chicago IL 60637

BILOTTO, GERARDO, neurophysiologist; b. Benevento, Italy, Sept. 21, 1948; s. Giuseppe and Violetta (DePasquale) B.; m. Sandra Swanson, Sept. 4, 1977. B.S., N.Y. Inst. Tech., 1970; M.S., Columbia U., 1972, M.Phil., 1976, Ph.D., 1978. Dir. independent living research Human Resources Center, Albertson, N.Y., 1978-79; postdoctoral research fellow Rockefeller U., N.Y.C., 1979-81; sr. research assoc. Liberty Mut. Ins. Co. Research Center, 1981—; research assoc. dept. orthopedic surgery Children's Hosp., Boston, 1981—, Harvard Med. Sch., 1981—. Contbr. articles to profl. jours. NIH fellow, 1972-77, 79-81. Mem. Soc. Neurosci., IEEE, N.Y. Acad. Scis., AAAS, Sigma Xi. Subspecialties: Neurophysiology; Biomedical engineering. Current work: Vestibular and oculomotor physiology; objective evaluation of human muscle fatigue, human vestibular research and oculomotor responses in patients and normal subjects; muscle fatigue and pain physiology. Office: Children's Hosp Med Center 300 Longwood Ave Neuromuscular Research Lab Boston MA 02115

BING, R.H., educator, mathematician; b. Oakwood, Tex., Oct. 20, 1914; s. Rupert Henry and Lula May (Thompson) B.; m. Mary Blanche Hobbs, Aug. 26, 1938; children—Robert H., Susan Elizabeth,

Virginia Gay, Mary Patricia. B.S., S.W. Tex. State Tchrs. Coll., 1935; M.Ed., U. Tex., 1938, Ph.D., 1945. Tchr. high sch., Tex., 1935-42; instr., then asst. prof. math. U. Tex., 1942-47; mem. faculty U. Wis.-Madison, 1947-73, prof. math., 1952-64, research prof., 1964-68, Rudolph E. Langer prof. math., 1968-73, chmn. dept., 1958-60; acting prof. U. Va., 1949-50; vis. prof. U. Tex., 1971-72, prof., 1973—, Ashbel Smith prof. math., 1979—, chmn. dept., 1975-77; dir. Summer Inst. on Set Theoretic Topology, Madison, 1955; mem. Inst. Advanced Study, Princeton, 1957-58, 62-63, 67. Mem. Nat. Sci. Bd. (1968-74, chmn. div. math.), Nat. Acad. Scis., NRC (1967-69), Conf. Bd. Math. Sci. (chmn. 1965-66), Nat. Acad. Scis. (chmn. math. sect. 1970-73, councilor 1977-80), Math. Assn. Am. (pres. 1963-64, vis. lectr. 1954-55, 61-62, chmn. Wis. sect. 1952), Am. Math. Soc. (councilor 1952-54, 58-60, v.p. 1967-68, pres. 1977-78), AAAS (v.p., chmn. sect. A 1959), Pi Mu Epsilon (vice dir. gen. 1960-63). Presbyn. (elder). Office: Dept Math U Tex Austin TX 78712

BINGHAM, BILLY ELIAS, nuclear power engineer; b. Shelby, N.C., July 31, 1931; s. Gettys David and Mary Lillie (Sain) B.; m. Shirley Ann Bridges, May 26, 1957; children: Gregory Elias, Melissa Dawn. A.A., Gardner-Webb Coll., 1952; B.S. in Nuclear Engring., N.C. State U., 1955; Ph.D., U. Va., 1970. Sr. engr. Babcock & Wilcox, Lynchburg, Va., 1958-67, supervisory engr., 1967-69, mgr. safety and systems analysis, 1969-78, mgr. breeder, 1978-82, adv. engr., 1982—. Lay leader Chestnut Hill United Methodist Ch., Lynchburg, 1973-78; scoutmaster Blue Ridge Mountains council Boy Scouts Am., 1974-79. Served with U.S. Army, 1956-58. Mem. Am. Nuclear Soc. Subspecialties: Nuclear fission; Numerical analysis. Current work: Advanced energy systems. Home: 110 Old Wiggington Rd Lynchburg VA 24502 Office: Babcock & Wilcox PO Box 1260 Lynchburg VA 24505

BINSTOCK, MARTIN H., manufacturing company executive and metallurgist; b. N.Y.C., Jan. 16, 1922; m. Adelaide Binstock, Aug. 31, 1947; children: Cathy, Peter, James. B.S. in Chemistry, Rensselaer Poly. Inst., Troy, N.Y., 1943, M.Metall. Engring., 1948. Mgr. Sylvania Electric Products, Inc., Bayside, N.Y., 1953-56; mgr. Atomics Internat., Canoga Park, Calif., 1956-69; dir. contract mgmt. Kerr-McGee Corp., Oklahoma City, 1969-82; cons., 1982—. Served to lt. (j.g.) USN, 1943-46. Mem. Am. Nuclear Soc., Am. Soc. for Metals. Subspecialties: Fuels; Metallurgical engineering. Current work: Consultant for nuclear fuels development and fabrication. Inventor composite thermoelec. assembly methods, uranium base alloy, lead-uranium oxide nuclear fuel element (patented). Home: 1191 Club House Dr Aptos CA 95003 Office: Pure-Flor Solder Inc 1779 Grant St Santa Clara CA 95050

BIRCHLER, JAMES ARTHUR, geneticist; b. Red Bud, Ill., Feb. 7, 1950; s. James Arthur and Edith Ann (Murphy) B. B.S., Eastern Ill. U., 1972; Ph.D., Ind. U., 1977. Postdoctoral investigator Oak Ridge Nat. Lab., 1977-81; postdoctoral affiliate Roswell Park Meml. Inst., Buffalo, 1981-82; specialist dept. genetics U. Calif., Berkeley, 1982—. Mem. Genetics Soc. Am. Subspecialties: Gene actions; Plant genetics. Current work: Dosage analyses of gene expression. Office: U Calif Berkeley 345 Mulford Hall Berkeley CA 94720

BIRCKBICHLER, PAUL JOSEPH, biochemist; b. Greenville, Pa., Nov. 13, 1942; s. Paul Joseph and Stella Theresa (Brandt) B.; m. Donna Jean Bowser, July 17, 1965; children: Stacey, Marc. B.S., Duquesne U., 1964, Ph.D., 1969. Instr. chemistry Pa. State U., McKeesport, 1968-69; postdoctoral research fellow Brown U., Providence, R.I., 1969-71; vis. scientist Bergen (Norway) U., 1971-72; research assoc. biomed. div. Noble Found., Ardmore, Okla., 1972-77, asst. scientist, 1977-81, assoc. scientist, 1981—. Mem. Am. Soc. Biol. Chemists, Am. Assn. Cancer Research, Tissue Culture Assn., Okla. Acad. Scis. Subspecialties: Membrane biology; Cell and tissue culture. Current work: Membranes of normal and transformed human cells, transglutaminase, isopeptide crosslinks. Office: Noble Found Biomedical Div Route 1 PO Box 2180 Ardmore OK 73402

BIRD, JOHN MALCOLM, geologist; b. Newark, Dec. 27, 1931; s. John Robert and Beryl Elizabeth (Wright) B.; m. Marjorie Ann Kelleher, Apr. 18, 1957; children—Anne Elizabeth, Marsha Jean. B.S., Union Coll., Schenectady, 1955; M.S., Rensselaer Poly. Inst., Troy, N.Y., 1959, Ph.D., 1961. Grad. asst. Union Coll., 1957-58; grad. asst. Rensselaer Poly. Inst., 1958-61; from instr. to assoc. prof. SUNY-Albany, 1961-70; prof., 1970-72, chmn. dept. geol. scis., 1969-72, vis. research prof., 1972-76; research assoc. Dudley Obs., Albany, 1964-72; sr. research assoc. Lamont-Doherty Geol. Obs., Columbia U., 1970-73; prof. geology Cornell U., Ithaca, N.Y., 1972—; Nat. Acad. Scis. Vis. scientist, 1967, disting. vis. scientist Am. Geol. Inst., 1971; chmn. Appalachian working group U.S. Geodynamics Com., 1971-73; disting. vis. lectr. Am. Assn. Petroleum Geologists, 1977-78; cons geotech. engring., mineral exploration. Editor: Plate Tectonics, 1981; assoc. editor: Jour. Geophys. Research, 1971-74; contbr. profl. jours. Served with AUS, 1955-57. Research grantee NSF, 1964, 68, 72-80; Nat. Acad. Scis., 1969; Petroleum Research Inst., 1975-77; Office Naval Research, 1978-80; Nat. Geog. Soc., 1977-78, 80. Fellow Geol. Soc. Am. (chmn. N.E. sect. 1975-76), Canadian Geol. Soc., Explorers Club; mem. Am. Geophys. Union, Sigma Xi, Chi Psi. Subspecialties: Tectonics; Petrology. Current work: Plate tetonics, rock deformation, origin of terrestrial metals, TEM and STEM analysis of minerals, nuclear waste disposal, engring. geology, generation and emplacement of ophiolites, carbon in the earth's mantle, landsat. Home: 1187 Ellis Hollow Rd Ithaca NY 14850 Office: Geol Scis Dept Cornell Univ Ithaca NY 14853

BIRD, ROBERT BYRON, chemical engineering educator, author; b. Bryan, Tex., Feb. 5, 1924; s. Byron and Ethel (Antrim) B. Student, U. Md., 1941-43; B.S. in Chem. Engring, U. Ill., 1947; Ph.D. in Chemistry, U. Wis., 1950, U. Amsterdam, 1950-51; D.Eng. (hon.), Lehigh U., 1972, Washington U., 1973, Tech. U. Delft, Holland, 1977, Sc.D., Clarkson U., 1980. Asst. prof. chemistry Cornell U., 1952-53, Debye lectr., 1973; mem. faculty U. Wis., 1951-52, 53—, prof. chem. engring., 1957—, C.F. Burgess distinguished prof. chem. engring., 1968-72, John D. MacArthur prof., 1982—, Vilas research prof., 1972—, chmn. dept., 1964-68; vis. prof. U. Calif., Berkeley, 1977; D.L. Katz lectr. U. Mich., 1971; W.N. Lacey lectr. Calif. Inst. Tech., 1974; K. Wohl Meml. lectr. U. Del., 1977; W.K. Lewis lectr. MIT, 1982; lectr. Lectures in Sci. Humble Oil Co., 1959, 61, 64, 66; lecture tour Am. Chem. Soc., 1958, 75, Canadian Inst. Chemistry, 1961, 65; cons. to industry, 1965—; mem. adv. panel engring. sci. div. NSF, 1961-64. Author: (with others) Molecular Theory of Gases and Liquids, 2d printing, 1964, Transport Phenomena, 30th printing, 1982, Spanish edit., 1965, Czech edit., 1966, Italian edit., 1970, Russian edit., 1974, Een Goed Begin: A Contemporary Dutch Reader, 1963, 2d edit., 1971, Comprehending Technical Japanese, 1975, Dynamics of Polymeric Liquids, Vol. 1, Fluid Mechanics, Vol. 2, Kinetic Theory, 1977; also numerous research publs.; Am. editor: Applied Sci. Research, 1969—; adv. bd.: Indsl. and Engring. Chemistry, 1970-72; editorial bd.: Jour. Non-Newtonian Fluid Mechanics, 1975—. Served to 1st lt. AUS, 1943-46. Decorated Bronze Star; Fulbright fellow, Holland, 1950; Fulbright lectr., 1958; Guggenheim fellow, 1958; Fulbright lectr., Japan, 1962-63, Sarajevo, Yugoslavia, 1972; recipient Curtis McGraw award Am. Assn. Engring. Edn., 1959, Westinghouse award, 1960. Fellow Am. Phys. Soc. (Otto Laporte lectr. 1980), Am. Inst. Chem. Engrs. (William H. Walker award 1962, Profl. Progress award 1965, Warren K. Lewis award 1974), Am. Acad. Arts and Scis.; mem. Am. Chem. Soc. (chmn. Wis. sect. 1966, unrestricted research grant Petroleum Research Fund 1963), Soc. Rheology (Bingham award 1974), Am. Acad. Mechanics, Brit. Soc. Rheology, Dutch Phys. Soc., Royal Inst. Engrs. (Holland), Nat. Acad. Engring., N.Y. Acad. Scis., Am. Acad. Arts and Scis., Wis. Acad. Scis., Arts and Letters, Phi Beta Kappa, Sigma Xi (v.p. Wis. sect. 1959-60), Tau Beta Pi, Alpha Chi Sigma, Phi Kappa Phi, Omicron Delta Kappa, Sigma Tau. Subspecialty: Fluid mechanics. Current work: Kinetic theory, rheology, and fluid dynamics of polymeric liquids. Transport phenomena and engineering applications. Office: Chem Engring Dept 3004 Engring Bldg 1415 Johnson Dr U Wis Madison WI 53706

BIRD, STEPHANIE JEAN, neuroscientist; b. Los Angeles, Oct. 5, 1948; d. Joseph Lester and Jean Belle (Lay) B.; m. Lawrence W. Shannon, Apr. 17, 1976. A.B., UCLA, 1971; Ph.D., Yale U., 1975. Neuroscientist Johns Hopkins U., Balt., 1976-77, Case Western Res. U., Cleve., 1977-79; neuroscientist Neurosci. Research Program, MIT, Boston, 1979-82, Mellon fellow, 1982—. Editor: Molecular Genetic Neuroscience; contbr. articles to profl. jours. Mem. Kennedy Adv. Com., 1982. Mem. Soc. for Neurosci., Assn. for Women in Sci. Subspecialties: Neuropharmacology; Neurophysiology. Current work: Applications of molecular genetics techniques to neuroscience problems, ethical issues in neuroscience and biotechnology applications. Home: 41 Sleepy Hollow Wrentham MA 02093 Office: Sci Tech Soc MIT Cambridge MA 02139

BIRELY, JOHN HORTON, chemist, research executive; b. Glen Ridge, N.J., Oct. 17, 1939; s. Charles W. and Catharine (Horton) B.; m. Laura Mears, Dec. 21, 1969; m. Molly Cooper, June 16, 1982. B.A. in Chemistry, Yale U., 1961; M.S. in Phys. Chemistry, U. Calif.-Berkeley, 1963, Ph.D., Harvard U., 1966. Postdoctoral research fellow dept. phys. chemistry Cambridge (Eng.) U., 1966-67; asst. prof. chemistry UCLA, 1967-69; mem. tech. staff Aerospace Corp., 1969-74; mem. staff Los Alamos Nat. Lab., 1974—, assoc. dir. chemistry, earth and life scis., 1981—. Mem. Am. Chem. Soc., Am. Phys. Soc., AAAS, CAP. Subspecialties: Physical chemistry; Kinetics. Current work: Management of scientific research in chemistry, earth and life sciences. Office: U Calif Los Alamos Nat Lab PO Box 1663 Los Alamos NM 87545

BIRGIT, HERTEL-WULFF, research scientist. Mem. Scandinavian Assn. Immunology. Subspecialties: Immunology (medicine); Immunogenetics. Current work: Total lymfoid irradiation, effector functions, animal models. Office: Div Immunology Dept Medicin Stanford U Sch Medicine Stanford CA 94305

BIRK, JAMES PETER, chemist, educator, cons, author; b. Cold Spring, Minn., Aug. 21, 1941; s. Albert Mathew and Christina Marie (Theisen) B.; m. S. Kay Gunter, June 21, 1974; 1 dau.: Kara Michelle. B.A., St. John's U., Collegeville, Minn., 1963; Ph.D. (NSF fellow), Iowa State U., 1967. Research assoc. U. Chgo., 1967-68; asst. prof. chemistry U. Pa., Phila., 1968-73, Rhodes-Thompson chair of chemistry, 1972-73; assoc. prof. Ariz. State U., 1973-79, prof., 1979—; cons. to textbook pubs. Author: General Chemistry Laboratory Manual, 1975, 77, 80, Mathematical Background for General Chemistry, 1975, Chemistry Laboratory Manual, 1977, 80; Contbr. numerous articles to profl. jours. Recipient award for distinction in undergrad. teaching Ariz. State U., 1980. Mem. Am. Chem. Soc., Sigma Xi, Phi Kappa Phi. Roman Catholic. Subspecialties: Inorganic chemistry; Educational software. Current work: Inorganic reaction kinetics; computers in chem. edn.; research on kinetics; writing ednl. software; creating instructional aids; textbook writing and reviewing. Home: 1201 E Verlea Dr Tempe AZ 85282 Office: Dept Chemistry Ariz State U Tempe AZ 85287

BIRKHOFF, GARRETT, mathematics educator; b. Princeton, N.J., Jan. 10, 1911; s. George David and Margaret (Grafius) B.; m. Ruth Collins, June 21, 1938; children: Ruth W., John D., Nancy C. A.B., Harvard U., 1932, Soc. of Fellows, 1933-36; postgrad., Cambridge (Eng.), U., 1932-33, hon. degree, U. Nacional Mexico, 1951, hon. degree, U. Lille, 1960, hon. degree, Case Inst. Tech., 1964. Instr. Harvard U., 1936-38, asst. prof., 1938-41, assoc. prof., 1941-46, prof., 1946—; cons. to govt. and pvt. industry; Walker-Ames lectr. U. Wash.; Taft lectr. U. Cin., 1947; Chmn. organizing com. Internat. Congress Mathematicians, 1950. Author: (with S. MacLane) Survey of Modern Algebra, 1941, 4th edit., 1977, Lattice Theory, 1940, 3d edit., 1967, Hydrodynamics, 1950, rev. 1960, Jets, Wakes and Cavities, (with E. Zarantonello), 1957, (with G.C. Rota) Ordinary Differential Equations, 1962, 3d edit., 1978, (with S. MacLane) Algebra, 1967, rev., 1979, Source Book in Classical Analysis, 1973. Guggenheim fellow, 1948. Mem. Am. Math. Soc. (v.p. 1958), Math. Assn. Am. (v.p. 1971-72), Am. Acad. Arts and Scis. (v.p. 1966-68), Am. Philos. Soc., Nat. Acad. Sci., Conf. Bd. Math. Sci. (chmn. 1969-70), Assn. Computing Machinery, Soc. Indsl. Applied Math. (pres. 1967-68), AAAS (chmn. sect. 1979); hon. mem. Sociedad Math. Mex., Acad. Ciencias Lima.; Mem. Soc. of Friends. Subspecialty: Applied mathematics. Office: Dept Math Science Center 1 Oxford St Cambridge MA

BIRKITT, JOHN CLAIR, defense and aerospace corporation executive; b. Inglewood, Calif., Aug. 20, 1941; s. Clair W. and Helene (Gille) B.; m. Constance E. May, June 4, 1966 (div. July 1980); children: Andra Diane, Robert John; m. Linda A. Aylmer, Sept. 13, 1980; 1 dau., Danielle Laurissa. B.S., Calif. State Poly. U. - Pomona, 1969. Engr. Aerojet Mfg. Co., Placentia, Calif., 1969-74; test condr. TRW-Def. and Space Systems Group, San Clemente, Calif., 1974-79, 1979—; plant mgr. Advanced Ground Systems Engring. Co., Long Beach, Calif., 1979. Treas., engr. Riverside County Fire Dept., Lake Elsinore, Calif., 1978—. Served with USMCR, 1959-65. Recipient commendation Minuteman Program Office, Norton AFB, Calif., 1975, Dept. Navy High Energy Laser Project, Washington, 1978; Appreciation for Patriotic Civilian Service cert. Dept. Army, Washington, 1979. Mem. AIAA, Nat. Assn. Watch and Clock Collectors, Music Box Soc. Republican. Subspecialties: High energy chemical lasers; Aerospace engineering and technology. Current work: Planning, execution of complex high energy laser tests, reactant storage, delivery systems, mechanical and optical installations, electronic control and data reduction systems and site control and service equipment. Home: 32536 Ortega Hwy Lake Elsinore CA 92330 Office: TRW-Def and Space Systems Group 33000 Avenida Pico San Clemente CA 92672

BIRKLE, A(DOLPH) JOHN, metall. engr., cons.; b. Chgo., Aug. 26, 1930; s. Adolph and Elizabeth (Lemoch) B.; m. Catherine, Sept. 12, 1957; children: Gregory, Kevin, Eric, Julie. B.S. in Metall. Engring, U. Ill., 1956, M.S., U. Wis., 1958. Sr. research engr. Applied Research Lab., U.S. Steel Corp., Monroeville, Pa., 1958-65; research supr. dept. research Youngstown Sheet & Tube Co., Ohio, 1965-68; mgr. product research and devel. CF&I Steel Corp., Pueblo, Colo., 1968-71; sr. staff engr./sect. head projects, engring. and constrn., project engring. services materials sect. Consumers Power Co., Jackson, Mich., 1971—; mem. NRC and Electric Power Research Inst. Corrosion Adv. Com., Material Property Council Com. on Fracture. Contbr. numerous articles to profl. jours. Served with USAF, 1952-54. Mem. Am. Soc. Metals, Am. Welding Soc., ASTM, ASME (chmn. Working Group on Steam Generator Inservice Inspection, chmn. subgroup on water-cooled systems, instr. boiler and pressure vessel code), Edison Electric Inst. (chmn. metallurgy and piping task force). Subspecialty: Metallurgical engineering. Current work: Nuclear power, energy, availability, product research and devel., inspection nuclear power plants, welding, corrosion, cathodic protection. Patentee in field; numerous inventions. Home: 2340 Vaudermere Dr Jackson MI 42901 Office: 1945 Parnall Rd Jackson MI 49201

BIRMINGHAM, BASCOM WAYNE, retired government official; b. Grand Island, Nebr., June 20, 1925; s. James C. and Stella M. (Sorrels) B.; m. Lois Marie Booth, Sept. 3, 1949; children: Steven W., Janet L. Birmingham Chamberlin. S.B., M.I.T., 1948, S.M., 1951. With Sorrels Supply Co., Poteau, Okla., 1947, Wester Geophys. Soc., Worland, Wyo., 1948, W.R. Holway & Assos., Tulsa, 1948-50; chief processes sect., cryogenics div. Nat. Bur. Standards, Boulder, Colo., 1951-63, chief cryogenics div., 1963-68; dep. dir. Inst. Basic Standards, 1968-77; dir. Boulder Labs., 1977-82; cons. cryogenic engring. Birmingham Assocs., 1982—; Mem. Boulder Zoning Bd., 1964-66; bd. dirs. Community Hosp., 1968-77; v.p. Rocky Mountain Eye Found., 1978—. Author: monograph Technology of Liquid Helium, 1968, Am. editor Cryogenics Jour, 1968-83. Served with USNR, 1944-46. Recipient gold medal service award Commerce Dept., 1953, 71, meritorious service award, 1961, Sci. fellow, 1966-67; Best Paper award Cryogenic Engring. Conf., 1961. Mem. Boulder C. of C. (bd. dirs. 1979-82). Subspecialties: Cryogenics; Mechanical engineering. Current work: Strategic long range planning in cryogenics field. Patentee in field. Home: 5440 White Pl Boulder CO 80303

BIRMINGHAM, THOMAS JOSEPH, physicist; b. Milford, Mass., Sept. 28, 1938; s. Thomas Joseph and Lillian May (Martin) B.; m. Jeannette Marie Grellner, Nov. 27, 1965; children: John, James, Mark. B.S., Boston Coll., 1960; M.A., Princeton U., 1962, Ph.D., 1965. Nat. Acad. Scis.-NRC postdoctoral fellow Goddard Space Flight Ctr., Greenbelt, Md., 1965-66, physicist, 1966—. Contbr. articles to profl. jours. Past pres. Hillandale Sch. PTA; past pres. Hillandale Citizens Assn., Hillandale Swim and Tennis Assn.; past scoutmaster troop Nat. Capitol Area council Boy Scouts Am. NSF fellow, 1960-62. Fellow Am. Phys. Soc.; mem. Am. Geophys. Union, Sigma Xi. Roman Catholic. Subspecialties: Plasma physics; Cosmic rays. Current work: Theoretical plasma physicist working on problems related to planetary magnetospheres, solar wind, cosmic rays, and other space problems. Office: Code 695 NASA Goddard Space Flight Ctr Greenbelt MD 20771

BIRNBAUM, EDWARD R., chemistry educator; b. Bklyn., Oct. 28, 1943; s. David and Sylvia (Deutchman) B.; m. Amy Boule Birnbaum, Apr. 4, 1968; children: David, Eva, Jessica. B.S., Bklyn. Coll., 1964; M.S., U. Ill., 1966, Ph.D., 1968. Asst. prof. chemistry N. Mex. State U., 1968-73, assoc. prof., 1973-78, prof., 1978—. DuPont fellow, 1967. Mem. Am. Chem. Soc., AAAS, Biophysical Soc., Am. Assn. Biol. Chemists, Alpha Chi Sigma. Subspecialties: Spectroscopy; Biophysical chemistry. Current work: Lanthanide spectroscopy in calcium binding proteins, research in the area of lanthanide substitution for calcium in calcium binding biological systems; application of spectroscopic techniques. Office: Department Chemistry New Mexico State University PO Box 3C Las Cruces NM 88003

BIRNBAUM, LINDA SILBER, toxicologist; b. Passaic, N.J., Dec. 21, 1946; d. Gene Robert and Gloria Ruth (Marcus) Silber; m. David Alan Birnbaum, June 25, 1967; children: Bernard, Holly, Lisa. A.B., U. Rochester, 1967; M.S., U. Ill., 1969, Ph.D., 1972. Postdoctoral fellow in biochemistry U. Mass., Amherst, 1973-74; asst. prof. Kirkland Coll., Clinton, N.Y., 1974-75; research scientist Masonic Med. Research Lab., Utica, N.Y., 1975-79; sr. staff fellow Nat. Inst. Environ. Health Scis., Research Triangle Park, N.C., 1979-80; research microbiologist chem. disposition lab. systemic toxicology br. Nat. Toxicology Program, 1981—; adj. assoc. prof. Sch. Pub. Health U. N.C., Chapel Hill. Contbr. articles on toxicology to profl. jours. Mem. Soc. Toxicology, Am. Soc. Pharmacology and Exptl. Therapeutics, Gerontology Soc., Am. Aging Assn., AAAS. Jewish. Subspecialties: Pharmacokinetics; Biochemistry (biology). Current work: Biochemical toxicology, chemical disposition of xenobiotics, biochemistry of aging. Office: PO Box 12233 Research Triangle Park NC 27709

BIRNBAUM, MICHAEL HENRY, psychology educator; b. Los Angeles, Mar. 10, 1946; s. Eugene David and Bessie (Holtzman) B.; m. Bonnie Gail Bruck, July 7, 1968; children: Melissa Anne, Kevin Michael. B.A., UCLA, 1968, M.A., 1969, Ph.D., 1972. Research asst. UCLA, 1968-72; postdoctoral scholar U. Calif., San Diego, 1972-73; asst. prof. psychology Kans. State U., Manhattan, 1973-74, U. Ill., Champaign, 1974-76, assoc. prof., 1976-82, prof., 1982—; cons. Rand Corp., 1979 , Scientific Applications, Inc., 1981 . Contbr. articles to profl. jours. Woodrow Wilson fellow, 1968; NIMH fellow, 1968-69; UCLA Alumni Assn. Dissertation award, 1972. Mem. Am. Psychol. Assn., Psychonomic Soc., Psychometric Soc., Soc. for Math. Psychology, AAAS, Phi Beta Kappa. Club: Internat. Brotherhood of Magicians. Subspecialties: Psychophysics; Social psychology. Current work: Processes by which people compare and combine information; social and psychophysical judgement; ratios and differences of subjective value. Home: 1001 Devonshire Dr Champaign IL 61820 Office: Univ Ill 603 E Daniel St Champaign IL 61820

BIRNBAUM, ZYGMUNT WILLIAM, mathematics educator; b. Lwow, Poland, Oct. 18, 1903; came to U.S., 1937, naturalized, 1943; s. Ignacy and Lina (Nebenzahl) B.; m. Hilde Merzbach, Dec. 20, 1940; children: Ann Miriam, Richard Franklin. LL.M., U. Lwow, 1925, Ph.D. in Math, 1929; postdoctoral research, U. Goettingen, Germany, 1929-31. Math. instr. Gymnasium, Lwow, 1926-29; chief actuary Life Ins. Co. Phoenix in Poland, 1931-36; research biometrician, N.Y.U., 1937-39; mem. faculty U. Wash., Seattle, 1939—, prof. math., 1950—; dir. lab. statis. research, 1948—; vis. prof. Stanford, 1951-52, U. Paris, France, 1960-61, U. Rome, Italy, 1964, Hebrew U., Jerusalem, 1980; cons. Boeing Co., 1956—, HEW, 1963—. Editor: Annals of Math. Statistics, 1967-70. Guggenheim fellow, 1960-61. Fellow Inst. Math. Statistics (pres. 1963-64), Am. Statis. Assn.; mem. Am. Math. Soc., Math. Assn. Am., Soc. Indsl. and Applied Math., Internat. Statis. Inst., AAUP. Subspecialties: Applied mathematics; Statistics. Current work: Industrial and biological applications, reliability, life distributions. Office: Math Dept Univ Wash Seattle WA 98195

BIRSHTEIN, BARBARA KATHRYL, immunologist, educator; b. Clarksburg, W.Va., Dec. 13, 1944. B.S. with high honors in Cellular Biology, U. Mich., 1965; Ph.D. with distinction in Biology, Johns Hopkins U., 1971. Postdoctoral fellow Albert Einstein Coll. Medicine, Bronx, N.Y., 1971-73, asst. prof. dept. cell biology, 1973-79, assoc. prof., 1979—. Mem. Am. Assn. Immunologists. Subspecialty: Immunogenetics. Current work: Regulation of immunoglobulin gene arrangement and expression in cultured cells; characterization of primary structure of mutant immunoglobulins; description of rearranged genes by DNA isolation and sequence studies. Office: Dept Cell Biology Albert Einstein Coll Medicine Bronx NY 10461

BISCHOFF, KENNETH BRUCE, chemical engineer, educator; b. Chgo., Feb. 29, 1936; s. Arthur William and Evelyn Mary (Hansen) B.; m. Joyce Arlene Winterberg, June 6, 1959; children: Kathryn Ann, James Eric. B.S., Ill. Inst. Tech., 1957, Ph.D., 1961. Asst. to asso. prof. U. Tex., Austin, 1961-67; asso. prof., then prof. U. Md., 1967-70; Walter R. Read prof. engring. Cornell U., 1970-76, dir. Sch. Chem.

Engring., 1970-75; Unidel prof. biomed. and chem. engring. U. Del., 1976—, chmn. dept. chem. engring., 1978-82; cons. Exxon Research and Engring., NIH, Gen. Foods Corp., W. R. Grace Co., Kappers Co. Author: (with D.M. Himmelblau) Process Analysis and Simulation, 1968, (with G.F. Froment) Chemical Reactor Analysis and Design, 1979; editor: (with R.L. Dedrick and E.F. Leonard) The Artificial Kidney; Contbr. articles to research publs.; Editorial bd.: Advances in Chemistry Series, 1973-76, 78—, Jour. Bioengring, 1976-80, Jour. Pharmacokin. Biopharmacy, 1975—; assoc. editor: Advanced Chem. Engring., 1982—. Mem. council thrombosis Am. Heart Assn., 1971-81. Recipient Ebert prize Acad. Pharm. Scis., 1972, Profl. Progress award Am. Inst. Chem. Engrs., 1976; Shell Found. fellow, 1959; NSF fellow, 1960; U. Ghent, 1960-61. Fellow AAAS, Am. Inst. Chemists; mem. Am. Inst. Chem. Engrs. (dir. 1972-74, chmn. nat. program com. 1978), Am. Chem. Soc., Am. Soc. Engring. Edn., Am. Soc. Artificial Internal Organs, Engrs. Council for Profl. Devel. (dir. 1972-78), Council Chem. Research (governing bd. 1981-84), Catalysis Soc., AAUP, N.Y. Acad. Scis., Sigma Xi, Tau Beta Pi, Phi Lambda Upsilon, Omega Chi Epsilon, Alpha Chi Sigma. Subspecialties: Chemical engineering; Biomedical engineering. Current work: Pharmacokinetic basis for quantitative risk assessment in toxicology and cancer chemotherapy; kinetic modelilng of complex chemical biochemical processes. Home: Box 81A Benge Rd RD 1 Hockessin DE 19707

BISHOP, ALBERT BENTLEY, III, industrial engineering educator; b. Phila., Apr. 7, 1929; s. Albert Bentley and Sara LeCompte (DesPortes) B.; m. Louise Boyd Squire, Nov. 17, 1951; children: John Albert, Suzanne Squire, James DesPortes. B.E.E., Cornell U., 1951; M.S., Ohio State U., 1953, Ph.D., 1957. Mem. tech. staff Bell Telephone Labs., Allentown, Pa., 1954; instr. indsl. and systems engring. Ohio State U., Columbus, 1954-57, asst. prof., 1957-60, asso. prof., 1960-65, prof., 1965—, chmn. dept., 1974-82; pres. Albert B. Bishop and Assos.; cons. U.S. Army Sci. Adv. Panel, 1975-77. Author: Introduction to Discrete Linear Controls—Theory and Application, 1975; contbr. articles to profl. jours. Mem. Ohio Citizens Council on Health and Welfare, 1973—; active Cub and Boy Scouts Am., 1965-74; mem. council Cornell U., 1974-80, 81—; trustee Buckeye Boys Ranch, 1976—. Served with USAF, 1951-53. Recipient Tech. Person of Yr. award Columbus Tech. Council, 1983. Fellow Am. Soc. Quality Control, Am. Inst. Indsl. Engrs. (nat. engring. economy research com. 1957-60, chmn. Columbus sect. research com. 1957-59, dir. 1976-78, mem. editorial bd. 1960-67, chmn. council acad. dept. heads 1980-81, dir. acad. affairs 1981-82, mem. task force on new technologies 1982—); mem. IEEE, Ops. Research Soc. Am., Inst. Mgmt. Sci., Am. Soc. Engring. Edn., Mil. Ops. Research Soc. (dir. 1969-72, co-chmn. edn. com. 1970-71, chmn. prize com. 1971-72), Tau Beta Pi (nat. exec. council 1966-70), Eta Kappa Nu, Alpha Pi Mu, Sigma Xi, Phi Kappa Phi, Pi Mu Epsilon, Sigma Pi. Episcopalian (vestryman 1958-60, 72-75, 77-80, 83—, jr. warden 1961-63, sr. warden 1963-65, mem. diocesan council 1968-72, nat. conv. del. 1973, 76, 79, 82). Subspecialties: Systems engineering; Industrial engineering. Current work: Industrial and systems engineering engineering educator and consultant. Patentee in field. Home: 1946 W Lane Ave Columbus OH 43221

BISHOP, BEVERLY P(ETTERSON), neurophysiologist; b. Corning, N.Y., Oct. 19, 1922; d. Elof Bernard and Bonnie (Hungerford) Petterson; m. Charles William Bishop, May 2, 1944; 1 son: Geoffrey Craig. B.A. in Math, Syracuse U., 1944; M.A. in Exptl. Psychology, U. Rochester, 1948; Ph.D. in Physiology, U. Buffalo, 1958. Asst. in physiology U. Glasgow, Scotland, 1956-57; inst. in physiology U. Buffalo, 1958-62; asst. prof. physiology SUNY, Buffalo, 1962-67, assoc. prof., 1967-75, prof., 1975—; cons., lectr. in field; mem. NIH study sect., 1980-84. Author: Basic Neurophysiology, 1982; contbr. numerous articles to profl. jours.; lectr. for: audio-visual series Illustrated Lectures in Neurophysiology, 1980. Recipient Dean's award SUNY, Buffalo, 1969, Chancellor's award, 1975; Golden Pen award Am. Phys. Therapy Assn., 1976. Mem. Am. Physiol. Soc., Soc. Neurosci., Am. Thoracic Soc., Am. Congress Rehab. Medicine, Internat. Soc. Electromyography and Kinesiology, Am. Assn. Electromyography and Electro-diagnosis, Phi Beta Kappa. Subspecialties: Neurobiology; Physiology (biology). Current work: Motor control of respiratory muscles; research to determine organization of neural circuitry controlling respiration, mastication and other rhymical motor activities. Home: 508 Getzville Rd Buffalo NY 14226 Office: Dept Physiology Sherman Hall SUNY Buffalo NY 14214

BISHOP, JOHN MICHAEL, microbiology educator; b. York, Pa., Feb. 22, 1936; m., 1959. A.B., Gettysburg Coll., 1957; M.D., Harvard U., 1962. Intern in internal medicine Mass. Gen. Hosp., Boston, 1962-63, resident, 1963-64; research assoc. virology NIH, 1964-66, sr. investigator, 1966-68; asst. prof., assoc. prof. Med. Center, U. Calif., San Francisco, 1968-72, prof. microbiology, 1972—, mem. cancer research coordinating com., 1968—; mem. Calif. div. Am. Cancer Soc., 1969—. NIH research grantee, 1968—; recipient Biomed. Research award Am. Assn. Med. Colls., 1981; Lasker award Basic Med. Research, 1982. Mem. Nat. Acad. Sci., AAAS, Am. Soc. Biol. Chemists, Am. Soc. Microbiology, Am. Soc. Cell Biology. Subspecialties: Virology (medicine); Biochemistry (medicine). Office: Dept Microbiology U Calif Med Center San Francisco CA 94143

BISHOP, PAUL EDWARD, microbiologist; b. Portland, Oreg., Feb. 12, 1940; s. Paul Emory and Elizabeth Ann (Burnett) B.; m. Lola Germania Montesdeoca, June 16, 1968; 1 son: Paul Emory. B.S., Wash. State U., 1964; M.S., Oreg. State U., 1970, Ph.D., 1973. Vol. Peace Corps, Panama, 1964-66; research technician U. Wash., Seattle, 1966-67; research asst. Oreg. State U., Corvallis, 1967-73, research assoc., 1973-75, U. Wis., Madison, 1975-77; research microbiologist Agrl. Research Sta., U.S. Dept. Agr., N.C. State U., Raleigh, 1977—; mem. faculty dept. microbiology, 1977—. USDA Competitive grantee, 1978-80, 81—. Mem. Am. Soc. Microbiology. Democrat. Methodist. Subspecialties: Microbiology; Nitrogen fixation. Current work: Genetics and molecular biology of nitrogen fixation in Azotobacter vinelandii. Home: 127 Brooks Ave Raleigh NC 27607 Office: Dept Microbiology NC State U Raleigh NC 27650

BISHOP, RICHARD STEARNS, exploration geologist, researcher; b. Dowagiac, Mich., Apr. 14, 1945; s. Barton Phelps and Margaret (Stearns) B.; m. Edythe Marie White, Jan. 15, 1971; children: Ryan Barclay, Timothy Clinton. B.S., Tex. Christian U., 1967; M.A., U. Mo.-Columbia, 1969; Ph.D., Stanford U., 1977. Devel. geologist Union Oil Co. of California, New Orleans, 1969-71; research geologist Exxon Prodn. Research Co., Houston, 1975-81; exploration geologist Exxon Co. U.S.A., Houston, 1981-82, project leader, 1982—. Fellow Geol. Soc. Am.; mem. Am. Assn. Petroleum Geologists (Cam Sproule award 1980, chmn. edn. com. 1980-82), Houston Geol. Soc. (editor Bull. 1981-83), Soc. Petroleum Engrs. Subspecialties: Geology; Organic geochemistry. Current work: Assessment of hydrocarbon resources; understanding the accumulation and dispersion of hydrocarbons; diapirism, abnormal pore pressure. Home: 10607 Dunlap St Houston TX 77096 Office: Exxon Co USA PO Box 2180 4550 Dacoma Houston TX 77092

BISPLINGHOFF, RAYMOND LEWIS, business exec.; b. Hamilton, Ohio, Feb. 7, 1917; s. Roscoe Earl and Isabelle (Lewis) B.; m. Ruth Doherty, June 20, 1944 (div.); children—Ross Lee, Ron Sprague. A.E., U. Cin., 1940, M.Sc., 1942; Sc.D., 1963; Sc.D., Swiss Fed. Inst. Tech., 1957; D.Eng. (hon.), Case Inst. Tech., 1965. Registered profl. engr., Mass., Mo. Engr. Aeronca Aircraft Corp., 1937-40, Wright Field, 1940-41; instr. U. Cin., 1941-43; engr. Bur. Aero., Navy Dept., Washington, 1943-46; asst. prof. Mass. Inst. Tech., 1946-48, asso. prof., 1948-53, prof., 1953-62; dir. Office Advanced Research and Tech., NASA, Washington, 1962-63, asso. adminstr., 1963-65, spl. asst. to adminstr., 1965-66; prof., head dept. aeros. and astronautics Mass. Inst. Tech., Cambridge, 1966-68; dean Coll. Engring., 1968-70; dep. dir. NSF, 1970-74; chancellor U. Mo-Rolla, 1974-77; sci. adviser to Mo. Gov., 1975-77; v.p. for research and devel. Tyco Labs., 1977—; cons. Dept. Def.; adminstr. NASA, FAA; Wright Brothers lectr. Inst. Aero. Scis., 1955; Samuel P. Langley lectr. U. Pitts., 1962; 3d Ann. von Karman lectr. AIAA, 1965; vis. prof. U. Fla.; sr. lectr. M.I.T.; chmn. bd. No. Energy Corp.; dir. Allied Research Assos., Allied Systems Ltd., Gen. Aircraft Corp., Mobil-Tyco Corp., Grinnell Corp., Simplex Wire & Cable Corp.; mem. corp. adv. council Eastern Air Lines; chmn. sci. adv. bd. USAF; mem. Nat. Sci. Bd., Def. Sci. Bd.; Pres. Internat. Council Aero. Scis.; chmn. investigative bds. for C-5A and B-1 aircraft USAF; mem. vis. com. Princeton Carnegie-Mellon U. Author: (with others) Aeroelasticity, 1955, Principles of Aeroelasticity, (with H. Ashley), 1962, Solid Mechanics, (with J.W. Mar and T.H.H. Pian), 1966, also numerous profl. papers.; Assos. editor: Jour. of Franklin Inst; cons. editor: McGraw Hill Ency. Technology. Recipient certificate of merit USAF; Sylvanus Reed award Inst. Aero. Scis., 1958; Distinguished Service medal NASA, 1967; Extraordinary Service medal FAA, 1968; Carl F. Kayan medal, 1971; Godfrey L. Cabot award, 1972; Distinguished Service award NSF, 1973; hon. fellow Truman Library Inst. Fellow Am. Acad. Arts and Scis., AAAS, Inst. Aero. Scis., Am. Astronautical Soc., Am. Inst. Aeros. and Astronautics (pres. 1966), Royal Aero. Soc.; mem. Nat. Acad. Engring. (chmn. aeros. space engring. bd., chmn. com. on transp.), Nat. Acad. Sci., Internat. Acad. Astronautics, Engrs. Council for Profl. Devel. (dir.), Engrs. Joint Council (dir.), Sigma Xi, Phi Kappa Phi, Tau Beta Pi. Clubs: Mason., Cosmos (Washington); Engineers (St. Louis); Explorers (N.Y.C.). Subspecialties: Aerospace engineering and technology; Mechanical engineering. Office: Tyco Labs Tyco Park Exeter NH 03833

BISSELL, MICHAEL GILBERT, med. scientist, clin. pathologist; b. Ridgecrest, Calif., Mar. 5, 1947; s. Henry Robert and Margaret Alberta (Encell) Benefiel; m. Sherry Helene Sachs, June 1, 1969 (div.); 1 dau., Cassandra Lyons; m. Sherrie Lynne Lyons, Mar. 27, 1977. B.S. in Math, U. Ariz., 1969; M.D., Stanford U., 1975; Ph.D. in Neurobiology, Stanford U., 1977; M.P.H., U. Calif., Berkeley, 1978. Diplomate: Am. Bd. Pathology. Research fellow NSF undergrad. research project U. Ariz., 1968-69, NIH Med. Scientist Tng. Program, Stanford U., 1971-75; intern in pathology Stanford U. Med. Ctr., 1975-76; resident in clin. pathology Martinez VA Med. Ctr. at U. Calif., Davis, 1978-81, lectr. in med. tech., 1978-81; sr. staff assoc. Lab. Neurochemistry, NIMH, Bethesda, Md., 1981—; sci. adv. Power Plant Siting Program, Minn. Environ. Quality Bd. Mem. ACLU, Soc. Sufi Studies (London), Folklore Soc. Greater Washington, Sierra Club., Coll. Am. Pathologists, Am. Pub. Health Assn., Soc. Neurosci., Tissue Culture Assn., Am. Inst. Med. Climatology. Subspecialties: Biochemistry (medicine); Tissue culture. Current work: Biochemical mechanisms of skeletal muscular hypertrophy; chemistry and dosimetry of negative air ions in vitro. Office: NIH Bldg 36 Room 3D30 Bethesda MD 20205

BISSON, MARY ALDWIN, biology educator; b. Fairfield, Calif., Oct. 24, 1948; d. Francis Joseph and Anne Frances (Tutak) Aldwin; m. Terrence Paul Bisson, Aug. 24, 1968. B.A., U. Chgo., 1970; Ph.D., Duke U., 1976. Research asst. U. Chgo., 1968-70; teaching asst. Duke U., Durham, N.C., 1970-74, research asst. marine lab., Beaufort, N.C., 1974-76; postdoctoral trainee U. N.C., Chapel Hill, 1976-77; Queens fellow U. Sydney, Australia, 1978-80; asst. prof. biology SUNY, Buffalo, 1980—. Australian Research Com. fellow, 1979, 80; NSF grantee, 1982. Mem. AAAS, Am. Soc. Plant Physiologists, Am. Women in Sci., N.Y. Acad. Scis., Sigma Xi. Democrat. Subspecialties: Plant physiology (biology); Biophysics (biology). Current work: Control of active and passive ion transport in marine and freshwater algae with respect to control of osmotic pressure and cytoplasmic pH. Home: 372 Voorhees Ave Buffalo NY 14216 Office: SUNY-Buffalo Dept Biol Scis 1109 Cooke Hall Buffalo NY 14260

BISSON, ROBERT ANTHONY, mineral exploration company executive; b. Laconia, N.H., Feb. 20, 1946; s. Reginald Anthony and Frances Adrienne (Shastany) B.; m. Linda Gale Gaillardetz, Mar. 10, 1964; 1 dau., Teresa Adrienne. Cert. master deep-sea diver, 1967. Technician, research diver Office Francais de Recherche Sous-Marine, Marseilles and Musee d'Oceanographique, Monaco, 1966-67; master diver, conshelf exploration Divcon Internat., London, 1967-68; project dir. conshelf exploration Ocean Sci. & Engring., Inc., Riviera Beach, Fla., 1968-69; project dir. ecol. studies Normandeau Assocs., Inc., Manchester, N.H., 1969-72; pres. Biospheric Cons. Internat., Inc. (now BCI-Geonetics, Inc.), Meredith, N.H., 1972—; v.p., dir. Biospherics Group Ltd., Halifax, N.S., Can., 1973—; pres., dir. BCI Map Corp., Laconia, N.H., 1976-77. Contbr. articles to profl. jours. Program leader N.H. Hydrospace Explorers, Boy Scouts Am., 1975—; trustee Ctr. for Environ. Studies, 1976—; bd. dirs. Lake Winnipesaukee Assn., 1977—, pres., 1978-79. Mem. AAAS, Marine Tech. Soc., Am. Water Works Assn., Newcomen Soc. Club: Bald Peak Colony. Subspecialties: Resource management; Offshore technology. Current work: Developing new water supply business ventures using the unique MESA Program to find and deliver groundwater in both gravels and even more in high-yield bedrock wells under long-term no-risk contracts, with turnkey cities, towns and industry; investigating applications in third world countries for new water supplies and the concept of bulk water delivery by VLCCs and other transport media from coastal wells to dry areas of the world; Co-inventor BCI Geonetics of a high-tech mineral exploration procedures called the Mineral Exploration System (MESA) Program.

BISTRIAN, BRUCE RYAN, internist, educator; b. Southampton, N.Y., Oct. 22, 1939; s. Peter and Mary Laura (Ryan) B.; m. Eleanor Alice Dix, Sept. 3, 1964; children: Tennille Ryan, Jordan Brooke, Britton Perry. B.A., NYU, 1961; M.D., Cornell U., 1965; M.P.H., Johns Hopkins U., 1971; Ph.D., MIT, 1975. Diplomate: Am. Bd. Internal Medicine. Intern Cornell U., N.Y.C., 1965-66; metabolism fellow U.Vt., Burlington, 1968-69, resident in medicine, 1969-70; mem. faculty Harvard U. Med. Sch., Boston, 1975—; clin. assoc. physician research resources div. NIH, 1975-78; lectr. MIT, 1981—. Contbr. over 120 sci. articles to profl. publs. Served to capt. U.S. Army, 1966-68. Nat. Inst. Gen. Med. Scis. grantee, 1977-80; Nat. Inst. Arthritis, Metabolism and Digestive Disease grantee, 1979-83. Fellow Am. Coll. Nutrition; mem. Fedn. Am. Soc. Exptl. Biologists, Am. Soc. Clin. Nutrition, Am. Fedn. Clin. Research, Am. Soc. Parenteral and Enteral Nutrition, Mass. Med. Soc. Presbyterian. Subspecialties: Nutrition (medicine); Biochemistry (medicine). Current work: Protein calorie malnutrition; total parenteral nutrition; nutrition and infection; treatment of obesity. Home: Argilla Rd Ipswich MA 01938 Office: New Eng Deaconess Hosp 194 Pilgrim Rd Boston MA 02215

BISWAL, NILAMBAR, molecular virologist; b. Khamar, Orissa, India, Feb. 20, 1934; s. Palau and Jamuna B.; m. Annapurna Biswal, Mar. 7, 1967; children: Sandip, Subrat, Subrina. Ph.D. in Microbiology, Mich. State U., 1965. Asst. research virologist U. Calif., 1965-67; asst. prof. Baylor Coll., 1968-72, assoc. prof., 1972-82; sr. investigator Balt. Cancer Research Program Nat. Cancer Inst. NIH, Balt., 1978-82; head div. molecular biology U. Md. Cancer Center, Balt., 1982—. Contbr. articles in field to profl. jours. Mem. AAAS, Am. Soc. Microbiology, N.Y. Acad. Sci., Am. Assn. Cancer Research, Sigma Xi. Subspecialties: Virology (biology); Cancer research (medicine). Current work: Isolation and characterization of DNA binding proteins of herpes simplex virus on Viral DNA replication, viral latency and cellular transformation; cloning of transforming genes of HSV. Home: 9704 Kerrigan Ct Randallstown MD 21133 Office: Breseler Research Bldg 655 W Baltimore St Baltimore MD 21201

BITO, LASZLO Z., ocular physiologist, educator; b. Budapest, Hungary, Sept. 7, 1934; s. Jozsef and Marianna (Bonin) B.; m.; children: John, Bucky. B.A. (scholar), Bard Coll., 1959; Ph.D., Columbia U., 1963. Hon. research asst. dept. physiology Univ. Coll. London, 1964-65; instr. ophthalmology Columbia U., N.Y.C., 1965-66, asst. prof., 1967-74, sr. research assoc., 1975-77, assoc. prof. ocular physiology, 1977-80, prof., 1980—. Editorial bd.: Exptl. Eye Research. NIH fellow, 1959-63, 63-65; recipient award Semmelweis Sci. Soc., 1974; NIH grantee, 1964—. Mem. Assn. Research in Vision and Ophthalmology, Internat. Soc. Eye Research, Am. Soc. Pharmacology and Exptl. Therapeutics, Am. Physiol. Soc., N.Y. Acad. Sci. Subspecialties: Comparative physiology; Pharmacology. Current work: Ocular fluid composition; prostaglandins, presbyopia. Office: 630 W 168th St New York NY 10032

BITRAN, JACOB DAVID, physician, oncologist, researcher; b. Thessaloniki, Greece, Sept. 23, 1947; came to U.S., 1952, naturalized, 1957; s. David Jacob and Martha (Faratzi) B.; m. Linda Sue Androw, Dec. 26, 1970; children: Lauren, Dina. B.S., U. Ill.-Chgo., 1967, M.D., 1971. Diplomate: Am. Bd. Internal Medicine. Intern Michael Reese Med. Ctr., Chgo., 1971-72; resident Rush-Presbyn. St. Luke's Hosp. Chgo., 1972-73, Michael Reese Med. Ctr., 1973-75; fellow hermatology-oncology U. Chgo., 1975-77; attending physician Michael Reese Med. Ctr., Chgo., 1977—; clin. asst. prof. medicine U. Chgo., 1977-82, clin. assoc. prof., 1982—. Fellow ACP.; mem. Am. Soc. Clin. Oncology, Am. Soc. Cancer Research, Am. Fedn. Clin. Research, Am. Soc. Hematology. Subspecialties: Oncology; Chemotherapy. Current work: Lung cancer, chemotherapy. Office: Monroe Med SC 104 S Michigan Ave Chicago Il 60603

BITTERMAN, MORTON EDWARD, psychologist, educator; b. N.Y.C., Jan. 19, 1921; s. Harry Michael and Stella (Weiss) B.; m. Mary Gayle Foley, June 26, 1967; 1 dau., Sarah Fleming. B.A., NYU, 1941; M.A., Columbia U., 1942; Ph.D., Cornell U., 1945. Asst. prof. Cornell U., Ithaca, N.Y., 1945-50; assoc. prof. U. Tex., Austin, 1950-55; mem. Inst. for Advanced Study, Princeton, N.J., 1955-57; prof. Bryn Mawr (Pa.) Coll., 1957-70, U. Hawaii, Honolulu, 1970—. Author: (with others) Animal Learning, 1979; editor: Evolution of Brain and Behavior in Vertebrates, 1976; co-editor: Am. Jour. Psychology, 1955-73, Animal Learning and Behavior, 1973-76. Recipient Humboldt prize Alexander von Humboldt Found., Bonn, Ger., 1981; Fulbright grantee; grantee NSF, Office Naval Research, NIMH, Air Force Office Sci. Research, Deutsche Forschungsgemeinschaft. Fellow Soc. Exptl. Psychologists, Am. Psychol. Assn., AAAS; mem. Psychonomic Soc. Subspecialties: Learning; Psychobiology. Current work: Comparative analysis of learning; evolution of intelligence. Home: 229 Kaalawai Pl Honolulu HI 96816 Office: Bekesy Lab of Neurobiology Univ Hawaii 1993 East-West Rd Honolulu HI 96822

BITZER, DONALD LESTER, electrical engineering educator; b. East St. Louis, Ill., Jan. 1, 1934; s. Jess L. and Marjorie B. (Look) B.; m. Maryann Drost, July 2, 1955; 1 son, David. B.S. in Elec. Engring, U. Ill., 1955, M.S., 1956, Ph.D., 1960. Asst. prof. U. Ill. Coordinated Sci. Lab., Urbana, 1960-63, assoc. prof., 1963-67, dir. computer-based edn. research lab., prof. elec. engring., 1967—; cons. Control Data Corp., Gandalf Ltd.; mem. various coms. Nat. Research Council. Contbr. articles on computer-based edn. to profl. jours. Mem. Regional Health Resource Center Bd., Urbana, Mercy Hosp. Health Care Found., Urbana. Recipient Vladimir K. Zworykin award Nat. Acad. Engring., 1973; Bobby C. Connelly award Miami Valley Computer Assn., 1973; co-recipient Data Processing Mgmt. Assn.'s Computer Scis. Man of Year award, 1975. Fellow IEEE; mem. Nat. Acad. Engring., Am. Soc. for Engring. Edn. (Chester F. Carlson award 1981), AAAS, AAUP. Subspecialties: Computer engineering; Graphics, image processing, and pattern recognition. Current work: Computer-based education, communication devices, displays. Patentee in field.

BIZZI, EMILIO, neurophysiologist; b. Rome, Feb. 22, 1933; U.S., 1963, naturalized, 1982; s. Vittorio and Anna (Galeazzi) B.; m. Mary Dale Reeves, May 1, 1943. M.D. summa cum laude with highest honors, U. Rome, 1958. Postdoctoral tng. Inst. Med. Pathology, U. Siena, Italy, 1958-60, Inst. Physiology, U. Pisa, 1960-63; research assoc. Neurophysiol. Lab. dept. zoology Washington U., St. Louis, 1963-64; vis. assoc. sect. physiology Lab. Clin. Sci., NIMH, Bethesda, Md., 1964-66; research assoc. dept. psychology MIT, 1966-67, lectr., 1967-68, assoc. prof. psychology, 1969-72, prof. neurophysiology, 1972-80, Eugene McDermott prof. in brain scis. and human behavior, 1980—, dir. Whitaker Coll., 1983—; sr. investigator Istituto di Richerche Cardiovascolari, U. Milan, Italy, 1968-69; mem. NIH Study Sect. Vision B., 1973-77; mem. adv. bd. Biomed. Engring. Center for Clin. Instrumentation; mem. faculty adv. council Whitaker Coll. Contbr. numerous chpts., articles, abstracts to profl. jours.; corr. asso. commentor: Behavorial and Brain Scis, 1980—; mem. editorial staff: Studies of Brain Function, 1980—; editorial bd.: Brain Theory Newsletter, 1980—, Jour. Motor Behavior, 1981—, Jour. Neurobiology, 1981—. Recipient Alden Spencer award Columbia U. Coll. Physicians and Surgeons, 1978, Whitaker Health Scis. award Mass. Inst. Tech., 1978; Found. Fund for Research in Psychiatry fellow, 1978—. Mem. Am. Acad. Arts and Scis., Italian Physiol. Soc., Internat. Brain Research Orgn., AAAS, Am. Physiol. Soc., Soc. Neurosci., Barany Soc. Subspecialties: Neurophysiology; Neurobiology. Current work: Brain mechanisms in motor control. Office: 79 Amherst St E10-238 Cambridge MA 02142

BJORKHOLM, JOHN ERNST, physicist; b. Milw., Mar. 22, 1939; s. Jack W. and Marion B. (Anderson) B.; m. Mary J. Durbin, June 20, 1964; children: Kristen E., Laura J. B.S.E. in Engring. Physics with highest honors, Princeton U., 1961; M.S., Stanford U., 1962, Ph.D. in Applied Physics, 1966. Mem. tech. staff Electronics Research Lab., Bell Labs., Holmdel, N.J., 1966—. Contbr. numerous articles to profl. jours. NSF fellow, 1961-62; Howard Hughes fellow, 1962-65. Fellow Optical Soc. Am., Am. Phys. Soc.; mem. IEEE (sr.). Subspecialties: Applied physics; Nonlinear optics. Current work: Research using lasers as tools; more specifically, nonlinear optics. Patentee in field. Office: 4C-318 Bell Labs Holmdel NJ 07733

BJORKMAN, OLLE ERIK, plant biologist; b. Jonkoping, Sweden, July 29, 1933; came to U.S., 1964, naturalized, 1978; m. Erik Gustaf and Dagmar Kristina (Svensson) B.; m. Monika Birgit Waldinger, Sept. 24, 1955; children: Thomas N.E., Per G.O. M.S., U. Stockholm, 1957; Ph.D., U. Uppsala, 1960, D.Sc., 1968. Asst. scientist dept. genetics and plant breeding U. Uppsala, 1956-61; research fellow Swedish Natural Sci. Research Council, 1961-63; postdoctoral fellow Carnegie Inst. Wash., Stanford, Calif., 1964-65, mem. faculty, 1966—,

Stanford U., 1967—, prof. biology by courtesy, 1967—; vis. fellow Australian Nat. U., Canberra, 1971-72, 78; sci. adv. Kettering Found., 1976-77; mem. panel world food and nutrition study NRC, 1976; com. carbon dioxide effects Dept. Energy, 1977-82; competitive grants panel Dept. Agr., 1978. Co-author: Experimental Studies of the Nature of Species V, 1971, Physiological Processes in Plant Ecology, 1980; contbr. articles to profl. publns.; editorial bd.: Plant, Cell and Environ, 1978. Recipient Linneus prize Royal Swedish Physiographic Soc., 1977. Mem. AAAS, Nat. Acad. Scis., Am. Soc. Plant Physiologists, Am. Acad. Arts and Scis. Subspecialties: Photosynthesis; Ecology. Current work: Response and adaptation of plants to ecologically diverse habitats. Environmental and biological control of photosynthesis; resource utilization by plants. Home: 3040 Greer St Palo Alto CA 94303 Office: 290 Panama St Stanford CA 94305

BLACK, DAVID CHARLES, research scientist; b. Waterloo, Iowa, May 14, 1943; s. Donald M. and Ruth Arlene (Woltz) B.; m. Karen Marie Winter, Aug. 26, 1967. B.S. in Physics, U. Minn., 1965, M.S., 1967, Ph.D., 1970. NRC postdoctoral fellow space sci. div. NASA Ames Research Ctr., Moffett Field, Calif., 1970-72, research scientist, 1972—, acting chief theoretical studies br., 1976-79, acting dep. chief space sci. div., 1980-81. Subspecialties: Theoretical astrophysics; Planetary science. Current work: Formation of stars and planetary systems; theoretical astrophysics; planetary science; leading NASA efforts to search for other planetary systems. Office: Space Sci Div MS 245-3 Ames Research Center Moffett Field CA 94035

BLACK, FRANKLIN OWEN, physician, med. researcher; b. St. Louis, Aug. 8, 1937; s. Frank and Kathleen Ruth (Scowden) B.; m. Jorita Jenkins, Mar. 30, 1961; children: Owen Brent, Christopher Brian, Jeremy Benjamin. B.A. in Chemistry, S.E. Mo. State U., 1955; M.D., U. Mo., 1963. Diplomate: Am. Bd. Otolaryngology. Intern Mobile (Ala.) Gen. Hosp., 1963-64; resident in surgery Bataan Meml. Hosp. and Loveland Found., Albuquerque, 1964-65; resident in otolaryngology U. Colo. Med. Center, Denver, 1965-68, instr., 1968-69; attending physician in otolaryngology Denver VA Hosp., Denver Gen. Hosp., 1968-69; instr., spl. research fellow dept. surgery U. Colo. Med. Center, 1968-69, asst. prof., 1969, asst. clin. prof., 1970-71; assoc. prof. otolaryngology U. Fla. Coll. Medicine, Gainesville, 1971-74; asso. prof. dept. otolaryngology Eye and Ear Hosp., U. Pitts., 1974-82, dir. div. vestibular disorders, 1974-82, vice chmn. dept. otolaryngology, 1976-81, mem. grad. faculty, 1977-82; mem. med. staff Children's Hosp. Pitts., 1980-82; chief div. neurootology Good Samaritan Hosp. and Med. Center, Portland, Oreg., 1982—; sr. scientist Neurol. Scis. Inst., Portland, 1982—. Contbr. articles to profl. jours. Served to lt. comdr. USNR, 1969-71. Decorated Bronze Star.; NIH fellow, 1968-69; NIH career devel. awardee, 1972-77. Fellow ACS; mem. Assn. Research in Otolaryngology (sec.-treas. 1976-78, pres. 1978-79), Pitts. Neuroscience Soc. (pres. 1977), Am. Acad. Otolaryngology (chmn. com. for research in otolaryngology 1978-82), Pa. Acad. Ophthalmology (program chmn. 1976), Am. Audiology Soc., Am. Council Otolaryngology, AMA, Am. Neurotology Soc., Am. Tinnitus Assn., Am. Nat. Standards Inst., Barany Soc., Internat. Brain Research Orgn., N.Y. Acad. Scis., Otosclerosis Study Group, Pan Am. Assn. Oto-Rhino-Laryngology and Broncho-Esophagology, Pa. Med. Soc., Royal Soc. Medicine, Triological Soc., Soc. Ear, Nose and Throat Advances in Children, Soc. Neurosci., Sigma Xi. Subspecialties: Otorhinolaryngology; Neurophysiology. Current work: Vestibular physiology, vestibular neurophysiology, sensorimotor integration. Office: Good Samaritan Hospital and Medical Center 1015 NW 22d Ave Portland OR 97210

BLACK, JAMES EMMETT, JR., systems engineer; b. Jasper, Ala., May 27, 1951; s. James Emmett and Audra Virginia (Green) B. A.S., Walker Coll., 1971; B.S.E.E., U. Ala., 1974, postgrad., 1974-75; postgrad., U. Houston, 1975-76. Assoc. computer programmer Lockheed Electronics Co., Johnson Space Ctr., Houston, 1975-77; sr. systems analyst nat. accounts Sperry Univac, Blue Bell, Pa., 1977-80; sr. systems engr. Gen. Electric Co., Valley Forge, Pa., 1980—; cons. and researcher in computer communications. Contbr. articles to profl. jours. Mem. IEEE, Assn. Computing Machinery, IEEE Computer Soc., Phi Theta Kappa, Alpha Phi Omega. Baptist. Subspecialties: Distributed systems and networks; Computer architecture. Current work: Inter-computer communications, protocols and interfaces; engaged in research into the nature of computer communications; protocols for development of efficient and intelligent mechanisms for protocol handling. Developer of computer communications systems. Home: PO Box 306 Warrington PA 18976 Office: General Electric Co Space Systems Div PO Box 8555 Philadelphia PA 19101

BLACK, JAMES FRANCIS, radiation technology firm executive; b. Butte, Mont., Jan. 15, 1919; s. William Joseph and Agnes Leona (Tillman) B.; m. Edna Frances Zekevich, Oct. 15, 1955; children: Claudia Edna, Gregory James. B.Sc., U. Calif.-Berkeley, 1940; M.A., Princeton U., 1943, Ph.D., 1943. Group leader Exxon Research and Engring. Co., Linden, N.J., 1945-56, research assoc., 1956-59, sr. research assoc., 1959-67, sci. advisor, 1967-83; pres. Applied Radiation Tech., Inc., Chatham, N.J., 1983—; vice chmn. AEC Adv. Com. Isotopes and Radiation, Washington, 1959-64, N.J. Commn. on Radiation Protection, Trenton, 1958-61; mem. select com. on Nat. Weather Service, NRC, Washington, 1979-80. Contbr. articles to profl. jours. Fellow AAAS, Explorers Club; mem. Am. Nuclear Soc. (chmn. isotopes 1968-69), Am. Chem. Soc., Am. Meteorol. Soc. Republican. Roman Catholic. Clubs: N.Y. Sailing (N.Y.C.); Appalachian Mountain (Boston). Subspecialties: Physical chemistry; Resource management. Current work: Application of chemiluminescence for research on liquid phase kinetics; demonstration of large asphalt coatings for producing clouds and rainfall in desert regions; application of nuclear radiation for producing biodegradeable sulfonate detergents; a thermodynamic approach to long range weather forecasting. Patentee in field. Home: 45 Hall Rd Chatham NJ 07928 Office: Applied Radiation Tech Inc 45 Hall Rd Chatham NJ 07928

BLACK, KATHRYN NORCROSS, psychologist, educator; b. Atlantic, Iowa, Jan. 28, 1933; d. Basil John and Opal Lucille (Weaver) Norcross; m. William C. Black, Feb. 2, 1962 (div. 1970); children: Deirdre Rehnberg, Amanda Katherine. B.A., U. Iowa, 1954, M.A., 1956, Ph.D., 1957. Research fellow NIMH, Bethesda, Md.; assoc. prof. dept. psychol. scis. Purdue U., West Lafayette, Ind., 1965—; cons. Gifted Edn. Research Inst., 1982. Contbr. articles to profl. jours. Fellow Am. Psychol. Assn.; mem. Soc. for Research in Child Devel., Phi Betta Kappa (pres. Purdue chpt. 1978-79), Sigma Xi. Democrat. Unitarian. Subspecialty: Developmental psychology. Office: Psychological Scis Purdue Univ West Lafayette IN 47907 Home: 2004 Edgewood West Lafayette IN 47906

BLACK, PERCY, psychology educator; b. Montreal, Jan. 6, 1922; U.S., 1954; s. Ovido and Rose (Vasilevsky) B.; m. Virginia Arne, June 21, 1951; children: Deborah, David, Elizabeth, Jonathan. B.Sc., Sir George Williams Coll., Montreal, 1944; M.Sc., McGill U., 1946; Ph.D., Harvard U., 1953. Asst. prof. psychology U. N.B., Fredericton, 1951-53; dir. research Social Attitude Survey, Yonkers, N.Y., 1955-67; prof. psychology Pace U., Pleasantville, N.Y., 1967—. Contbg. author: Societies Around the World, 2 vols, 1953; author: The Mystique of Modern Monarchy, 1953; contbr. articles to profl. jours. Fellow AAAS; mem. Am. Psychol. Assn., So. Soc. for Philosophy and Psychology, Internat. Assn. Applied Psychology, Internat. Council Psychologists. Lodge: B'nai B'rith. Subspecialties: Social psychology. Current work: Cognitive and affective configurations in moral and immoral decisions. Office: Pace U Pleasantville NY 10703

BLACK, PERRY, neurosurgeon, educator; b. Montreal, Que., Can., Oct. 2, 1930; s. Ovido and Rose (Vasilevsky) B.; m. Phyllis Naomi Black, June 2, 1963; children—Daniel Ovid, Julie Miriam, Amy Rose. B.Sc., McGill U., 1951, M.D., C.M., 1956. Resident in neurosurgery Johns Hopkins U., 1959-63, from instr. to prof. neurosurgery, 1963-79, asst. prof. then asso. prof. psychiatry and behavioral sci., 1967-79; dir. Lab. Neurol. Scis., Friends Med. Sci. Research Center, 1972-79; chmn. Central Research Authority, 1972-79; prof., chmn. dept. neurosurgery Hahnemann Med. Coll. and Hosp., Phila., 1979—; mem. Neurol. A study sect. NIH, 1973-77. Editor: Drugs and the Brain, 1969, Physiological Correlates of Emotion, 1970, Brain Dysfunction in Children: Etiology, Diagnosis and Management, 1981; asso. editor, internat. neurosurgery editor: Neurosurgery. NIH fellow, 1961-62. Mem. Congress Neurol. Surgeons (chmn. internat. com. 1975—, editor newsletter 1972-74, Disting. Service award 1977), Am. Assn. Neurol. Surgeons, Soc. Neuroscience, Am. Epilepsy Soc., Am. Neurol. Assn., Internat. Assn. Study of Pain, AAAS, AAUP, Am. Soc. Stereotactic and Functional Neurosurgery, Internat. Neurosurgical Forum, Pa. Neurosurgical Soc. Jewish. Subspecialties: Neurology; Physiological psychology. Current work: Neurosurgery; experimental studies on spinal cord injury; recovery of function after brain lesions. Office: Hahnemann Med Coll and Hosp 230 N Broad St Philadelphia PA 19102

BLACK, PETER MCLAREN, neurosurgeon, educator; b. Calgary, Alta., Can., Apr. 3, 1944; s. Thomas Herbert and Elizabeth (Peterson) B.; m. Katharine Cohen, June 15, 1967; children: Winifred, Elizabeth, Katharine, Peter Thomas, Christopher. A.B., Harvard U., 1966; M.D., C.M., McGill U., Montreal, Can., 1970; Ph.D., Georgetown U., 1978. Intern, resident Mass. Gen. Hosp., Boston, 1970-72, resident in neurosurgery, 1975-80; asst. prof. surgery Harvard Med. Sch., Boston, 1980—. Author numerous sci. articles. NIH tchr. investigator awardee, 1982. Mem. Congress Neurol. Surgeons, Research Soc. Neurol. Surgeons, Soc. Neurosci., Am. Assn. Neurol. Surgeons, Am. Fedn. Clin. Research. Subspecialty: Neurosurgery. Current work: cerebrospinal fluid physiology, especially neuroendocrine mechanisms; higher cortical functions. Office: Mass General Hosp Ambulatory Care Center Suite 312 15 Parkman St Boston MA 02114

BLACK, WILLIAM CARTER, JR., high tech. exec., physicist; b. San Diego, Calif., Feb. 25, 1939; s. William Carter and Marguerite E. (MacLaggan) B.; m. Helene Therese Laurent, July 27, 1978. B.A., Pomona Coll., Claremont, Calif., 1961; M.S., U. Ill., 1963, Ph.D., 1967. Asst. prof. physics U. Calif., San Diego, 1967-70; gen. mgr. S.H.E. Corp., San Diego, 1970-72, pres., chief exec. officer, 1972—. Contbr. articles to profl. jours. Mem. Am. Phys. Soc., AAAS, Am. Mgmt. Assn., Sigma Xi. Subspecialties: Superconductors; Bioinstrumentation. Current work: Applications of magnetic sensing technology, establishing subsidiaries for development research and clinical biomagnetic instruments. Home: 13081 Via Esperia Del Mar CA 92014 Office: 4174 Sorrento Valley Blvd San Diego CA 92121

BLACKADAR, ALFRED KIMBALL, meteorologist, educator, researcher, consultant; b. Newburyport, Mass., July 6, 1920; s. Walter Lloyd and Harriett Dodge (White) B.; m. Beatrice Fenner, Mar. 23, 1946; children: Bruce E., Russell L., Thomas A. A.B., Princeton U., 1942; Ph.D., NYU, 1950. Cert. cons. meteorologist Am. Meteorol. Soc., 1973. From instr. to assoc. prof. NYU, N.Y.C., 1946-56; from assoc. prof. to prof. Pa. State U., University Park, 1956—, head dept. meteorology, 1967-81; lectr. Columbia U., N.Y.C., 1951-53. Contbr. articles to profl. jours. Served to maj. AC U.S. Army, 1942-46. Alexander von Numboldt Sr. Scientist, 1973. Fellow Am. Meteorol. Soc. (pres. 1971-72, Charles F. Brooks award 1969), Am. Geophys. Union, AAAS; fgn. mem. Deutsche Meteorolgische Gesellschaft. Baptist. Subspecialties: Micrometeorology; Synoptic meteorology. Current work: Turbulent flow and physical processes applied to the atmospheric and oceanic boundary layers. Home: 805 W Foster Ave State College PA 16801 Office: 503 Walker Bldg University Park PA 16802

BLACKBURN, JACOB FLOYD, computer scientist; b. Newton., N.C., Nov. 27, 1918; s. Julius Walter and Lottie Mae (Lael) B.; m. Beverley England, Mar. 29, 1944; 1 son, Gregg Scott. B.A., Lenoir-Rhyne Coll., 1940; certificate, N.Y.U., 1942; M.A., Duke U., 1947; Ph.D., U. N.C., 1953. Asst. prof. The Citadel, Charleston, S.C., 1947-50; mem. Inst. for Advanced Study, Princeton, N.J., 1953-54; assoc. prof. USAF Acad., 1955-56; various mgmt positions IBM, Cranford, N.J., 1956-77, Geneva, Switzerland, 1956-77, Brussels, Belgium, 1956-77; dir. Office Tech. Policy and Space Affairs, Dept. State, Washington, 1977-79; exec. dir. computer sci. bd. Nat. Acad. Scis., 1980-82; liaison scientist Office Naval Research, London, Eng., 1982—. Contbr. articles to profl. jours. Pres. Brussels Am. Club, 1974; pres. Brussels Toastmasters, 1972-73, Paris Toastmasters, 1975-76, Dresden Condominium Assn., Washington, 1978-82. Served to capt. USAAF, 1941-46; to lt. col. USAF, 1951-56. Decorated Air medal.; Recipient Computer Pioneer award Nat. Computer Conf., 1975; Alumnus of Yr. award Lenoir-Rhyne Coll., 1981. Mem. Assn. Computing Machinery, Am. Math. Soc. Democrat. Club: Grosvenor Sq. Toastmasters (London) (pres. 1983). Subspecialties: Numerical analysis; Applied mathematics. Current work: Reporting on European research in computer science and robotics. Office: 223 Old Marylebone Rd London England NW1 Home: 12A Redcliffe Sq London SW England 10

BLACKBURN, WILL R., academic physician, educator, pediatric-perinatal pathologist; b. Durant, Okla., Nov. 4, 1936; s. Henry and Vicie (McIntosh) B.; m. Hope Hutchins, Aug. 18, 1979. B.S., U. Okla., 1957; M.D., Tulane U., 1961. Med. intern Coll. Physicians and Surgeons, Columbia U., N.Y.C., 1961-62, resident in pathology, 1962-64; instr.-resident U. Colo. Med. Center, Denver, 1964-66; assoc. prof. Pa. State U., Hershey, 1968-74; prof. dept. pathology South Ala. Coll. Medicine, Mobile, 1974—, dir., 1974—; research pathologist Women's and Children's Hosp., Bangkok, Thailand, 1966-68. Contbr. chpts. to books in field. Bd. dirs. Sudden Infant Death Syndrome Program Mobile County Bd. Health, 1975-77. Served to capt. USAF, 1966-68. NIH research grantee, 1975-78. Mem. Soc. for Pediatric Research, Teratology Soc., So. Soc. Pediatric Research, Pediatric Pathology Club, Soc. Exptl. Biology and Medicine. Subspecialties: Teratology; Pathology (medicine). Current work: Experimental teratology; birth defects, human and experimental; research human teratology and mammalian experimental teratology; studies in human and animal allomorphology; experimental fetal surgery. Home: 67 Magnolia Ave Fairhope AL 36532 Office: U South Ala Coll Medicine 2451 Fillingham St Mobile AL 36617

BLACKLOW, NEIL RICHARD, medical educator; b. Cambridge, Mass., Feb. 26, 1938; s. Leo Alfred and Clara Edna (Cumenes) B.; m. Margery Lois Brown, June 2, 1963; children: John Andrew, Peter Douglas. B.A., Harvard U., 1959; M.D., Columbia U., 1963. Intern Beth Israel Hosp., Boston, 1963-64, resident, 1964-65; virologist Nat. Inst. Allergy and Infectious Diseases, 1965-68, 69-71; fellow in infectious diseases Mass. Gen. Hosp.-Harvard U., 1968-69; asst. prof. medicine Boston U., 1971-74, assoc. prof., 1974-76; prof. medicine, molecular genetics and microbiology U. Mass., 1976—, dir. div. infectious diseases, 1976—. Mem. editorial bd., Infection and Immunity, 1981—; assoc. editor: Revs. of Infectious Diseases, 1982—; contbr. articles to profl. jours. Served with USPHS, 1965-68. Med. Found. Boston fellow, 1972-74; grantee in field. Mem. Am. Soc. Clin. Investigation, Infectious Diseases Soc. Am., Am. Assn. Immunologists. Subspecialties: Infectious diseases; Virology (medicine). Current work: Etiology, diagnosis and treatment of viral diseases, virologist, infectious diseases. Office: U Mass Medical School Worcester MA 01605

BLACKMAN, MARC ROY, biomed. researcher, cons. in endocrinology; b. Boston, Mar. 2, 1946; s. Max Abraham and Sue (Weiner) B.; m. Linda Ellen Richman, Aug. 8, 1971; children: David Max, Adam Daniel. A.B., Northeastern U., 1968; M.D., NYU, 1972. Diplomate: Am. Bd. Internal Medicine. Intern and resident in internal medicine Bronx Mcpl. Hosp. Ctr.-Albert Einstein Coll. Medicine, 1972-75; postdoctoral clin. research fellow NIH, Bethesda, Md., 1955-77; asst. prof. Johns Hopkins U. Sch. Medicine, Balt., 1977—; guest scientist Nat. Inst. on Aging NIH, Balt., 1980—. Author: Principles of Ambulatory Medicine, 1982, Adv. Internal Medicine, 1978, also articles. Mem. Endocrine Soc., Am. Fedn. Clin. Research, N.Y. Acad. Sci., AAAS, Alpha Omega Alpha. Democrat. Jewish. Subspecialties: Endocrinology; Neuroendocrinology. Current work: Synthesis and secretion of pituitary glycoprotein hormones and their subunits. Home: 7908 Kentbury Dr Bethesda MD 20814 Office: Dept Medicine Balt City Hosps Baltimore MD 21224

BLACKMORE, PETER FREDERICK, physiology educator, researcher; b. Brisbane, Queensland, Australia, May 27, 1947; came to U.S., 1975; s. Frederick Bertrum Arthur and Ruth Evelyn (Deacon) B.; m. Jill Margaret Quirk, Dec. 11, 1971; children: David James, Heather Anne. B.Sc., U. New South Wales, 1971, Ph.D., 1976. Lab. asst. U. New South Wales, Sydney, 1969-71; research asst. Australia Nat. U., Canberra, 1975; research assoc. Vanderbilt U. Sch. Medicine, 1975-77, asst. prof. physiology, 1978—; research assoc. Howard Hughes Med. INst., 1977-80, assoc. investigator, 1980—. Recipient Commonwealth Postgrad. award Govt. of Australia, 1972-75; Australian-Am. Edn. Found. award, 1975-78. Mem. Am. Soc. Biol. Chemists, Am. Diabetes Assn. Subspecialties: Biochemistry (medicine); Endocrinology. Current work: Elucidating mechanism by which catecholamines, insulin, glucagon, vasopressin and angiotensin II regulate hepatic metabolism. Office: Vanderbilt U Howard Hughes Med Inst 731 Light Hall Nashville TN 37232

BLACKWELDER, RON (FOREST BLACKWELDER), aerospace engineering educator, consultant; b. Pratt, Kans., July 16, 1941; s. Forest A. and Evelyn M. (Meyer) B.; m. Judy P. Tiefeld, Sept. 14, 1965; children: Laura, Sherri. B.S., U. Colo., 1964; postgrad., Technische Hochschule, Munich, W.Ger., 1964-65; Ph.D., Johns Hopkins U., 1970. Teaching asst. Johns Hopkins U., 1966-68, research asst., 1968-70; asst. prof. aerospace engring. U. So. Calif., 1970-75, assoc. prof., 1975-79, prof., 1979—; vis. scientist Max-Planck Inst., Gottingen, W.Ger., 1976-77; cons. in field. Author: Hot-Wire Anemometry, 1981; contbr. numerous articles to profl. jours. Recipient Faculty Service award U. So. Calif., 1975; Guggenheim fellow, 1976-77. Mem. AIAA, Am. Phys. Soc., AAAS. Democrat. Lutheran. Subspecialties: Aeronautical engineering; Mechanical engineering. Current work: Experimentalist in fluid mechanics specializing in turbulence; research interests also include developing instrumentation, signal processing, pattern recognition and advanced technologies in fluid mechanics. Inventor vorticity probe using strain, 1979. Office: Dept Aerospace Engring U So Calif Los Angeles CA 90089

BLACKWELL, HAROLD RICHARD, optical science educator, psychophysiologist; , 1943; 2 children. B.S., Haverford Coll., 1941; M.A., Brown U., 1942; Ph.D. in Psychology, U. Mich., 1947. Research psychologist Polaroid Corp., 1943; research assoc. L.C. Tiffany Found., N.Y.C., 1943-45; dir. vision research labs. U. Mich., Ann Arbor, 1946-58; prof. biophysics Ohio State U., Columbus, 1965-78, dir., 1958—, prof., chair div. sensory biophysics, 1978—; Exec. sec. vision com. NRC U.S. Armed Forces, 1945-55. Mem. Optical Soc. Am. (Lomb medal 1950), Illuminating Engring. Soc. (Gold medal 1972), Assn. Research in Vision and Ophthalmology, Am. Acad. Optics, Psychonomic Soc. Subspecialties: Psychophysics; Ophthalmology. Office: Inst Research Vision Ohio State U 1314 Kinnear Rd Columbus OH 43212

BLACKWELL, RICHARD EDGAR, med. educator, physician, researcher; b. Atlanta, Aug. 4, 1943; s. Hubert L. and Mildred (Perry) B.; m. Elizabeth Croom, Mar. 23, 1968. B.A., Jacksonville U., 1966; M.A., U. South Fla., 1968; Ph.D., Baylor U., 1971, M.D., 1975. Diplomate: Am. Bd. Ob-Gyn. Research fellow Salk Inst., La Jolla, Calif., 1971-72; instr. dept. physiology Baylor Coll. Medicine, Houston, 1975; instr. U. Ala.-Birmingham, 1975-79, asst. prof. ob-gyn, 1979-81, asst. prof. physiology, 1979—, assoc. prof. ob-gyn, 1981—; cons. Sandoz Pharm. Co., East Hanover, N.J., 1980—, Lilly Research Labs., Indpls., 1980—, Searle Pharms., Chgo., 1981—, Ansell & Akwell, Inc., Dothan, Ala., 1979—. Contbr. numerous articles to med. jours. NIH fellow, 1968-71, 1971-72; AID fellow, 1972. Mem. Endocrine Soc., Am. Physiol. Soc., Soc. for Neurosci., Soc. for Study of Reprodn., Am. Fertility Soc. Republican. Clubs: Vestavia County, The Club, Relay House (Birmingham). Subspecialties: Reproductive endocrinology; Microsurgery. Current work: Reproductive neuroendocrinology with emphasis on pituitary cell culture, mechanism of hormone action, diagnosis and treatment of pituitary tumors. Home: 3416 Old Wood Ln Birmingham AL 35243 Office: U Ala-Birmingham Dept Ob-Gyn UAB Sta Birmingham AL 35294

BLADE, RICHARD ALLEN, physics and energy science educator; b. Bartlesville, Okla., July 28, 1938; s. Oscar Carl and Helen (Lutz) B. B.S. in Engring. Physics, U. Colo., Boulder, 1960, Ph.D. in Theoretical Physics, 1964. Mem. faculty U. Colo., Colorado Springs, 1967—; now prof. physics and energy sci.; vis. assoc. Calif. Inst. Tech., Pasadena, 1969-70; research assoc. U. Calif.-Berkeley, 1978; pres. Blade and Assoc., Futuristic Computer Products; prin., owner Nat. Inst. Healing Arts.; Mem. pub. policy com. Am. Inst. Physics. Contbr. articles to profl. publs. NSF grantee, 1979-82; Tecton Corp. grantee, 1980. Mem. Am. Assn. Physics Tchrs. (chmn. profl. concerns com.), Internat. Solar Energy Assn., Am. Wind Energy Assn. Subspecialties: Theoretical physics; Electronics. Current work: Solar energy; computer control applications. Office: U Colo Physics and Energy Sci Dept Colorado Springs CO 80907

BLAESE, (ROBERT) MICHAEL, clinical immunologist, pediatrician; b. Mpls., Feb. 16, 1939; s. Robert Marion and Eva Ruth B.; m. Julianne Eleanor Johnson, June 23, 1962; children: Elise, Kristianne. B.S., Gustavus Adolphus Coll., 1961; M.D., U. Minn., 1964. Intern, Parkland Hosp., Dallas, 1964-65; resident in pediatrics U. Minn., Mpls., 1965-66; clin. assoc. metabolism br. Nat. Cancer Inst., Bethesda, Md., 1966-68, sr. investigator, 1968-74, chief cellular immunology sect. metabolism br., 1974—. Mem. editorial bd.: Jour. Immunology; contbr. chpts. to books, articles to profl. jours. Recipient Mead-Johnson award Am. Acad. Pediatrics, 1980; Wellcome vis. prof.

Royal Soc. Medicine, London, 1980. Mem. Am. Soc. Clin. Investigation, Am. Assn. Immunologists, Soc. Pediatric Research. Lutheran. Subspecialties: Immunology (medicine); Pediatrics. Current work: Immunodeficiency diseases, cellular immunology, development of host defence mechanisms, immunoregulation, tumor immunology. Office: NIH Bldg 10 Rm 6B05 Bethesda MD 20205

BLAINE, EDWARD HOMER, pharmacologist, research institute administrator, educator; b. Farmington, Mo., Jan. 30, 1940; s. Theodore Warren and Tessa Ella (McClanahan) B.; m. Susan Irene Cring, June 19, 1963; children: Jennifer, Marquis Edward. A.B., U. Mo.-Columbia, 1962, A.M., 1967, Ph.D., 1970. Mem. team Green Bay Packers Profl. Football Club, 1962, Phila. Eagles Profl. Football Club, 1963-66; asst. prof. U. Pitts. Sch. Medicine, 1972-77; sr. research fellow Merck Inst. Therapeutic Research, West Point, Pa., 1977-78, sr. investigator, 1978-80, dir. renal pharmacology, 1980; adj. assoc. prof. physiology Temple U. Sch. Medicine; postdoctoral fellow Howard Florey Inst., U. Melbourne, Australia, 1971-73. Contbr. numerous articles to profl. publs. Recipient Grand award Mead Johnson Competition, 1970, Research Career Devel. award NIH, 1975-77; NIH postdoctoral fellow, 1971-73. Me. Am. Physiol. Soc., Soc. Exptl. Pharmacology and Therapeutics, Endocine Soc. Subspecialties: Pharmacology; Physiology (medicine). Current work: Hypertension, salt and water homeostasis, renal pharmacology. Office: Merck Inst Therapeutic Research Bldg 26 Room 208 West Poin PA 19468

BLAIR, CAROL DEAN, microbiologist, virologist; b. Salt Lake City, Jan. 31, 1942; d. Doyle and Beulah (Bond) B.; m. Patrick Joseph Brennan, Sept. 5, 1968; children: Deirdre, Niall, Justin. B.A. magna cum laude, U. Utah, 1964; Ph.D., U. Calif.-Berkeley, 1968; M.A. Juri Officii, U. Dublin, 1975. Am. Cancer Soc. fellow U. Dublin, 1968-70, jr. lectr. then lectr. bacteriology, 1970-75; Am. Cancer Soc. Eleanor Roosevelt vis. fellow Baylor U. Coll. Medicine, Houston, 1972-73; asst. prof. microbiology Colo. State U., 1975-77, assoc. prof., 1977—, asst. dean biomed. curricula Coll. Vet. Medicine Biomed. Sci., 1983—. Contbr. articles to profl. jours. Woodrow Wilson fellow, 1964; NSF fellow, 1964-68; Damon Runyon fellow, 1968. Mem. Am. Soc. Microbiology, Am. Soc. Virology, Soc. Gen. Microbiology, Am. Soc. Tropical Medicine and Hygiene, AAAS, Sigma Xi, Phi Beta Kappa. Democrat. Subspecialties: Molecular biology; Virology (biology). Current work: Structure, replication and expression of animal virus genomes. Office: Dept Microbiolog Colo State U Fort Collins CO 80523

BLAIR, DONALD GEORGE RALPH, biochemist, educator; b. Lloydminster, Sask., Can., Nov. 5, 1932; s. George Alfred and Ester Pearl (Stromstad) B.; m. Joan Louise Sprinsteen, June 27, 1959; children: Eric Donald, Karen Jo. B.Sc. with honors, U. Alta., 1955, M.Sc., 1956; Ph.D., U. Wis., 1961. Lect. biochemistry U. Sask., 1963-76, lectr. cancer research, 1961-63, asst. prof., 1963-67, assoc. prof., 1967-76, assoc. prof. biochemistry, 1976—; mem. staff Sask. Research Unit Nat. Cancer Inst. Can., 1961-70, research assoc., 1970-75. Contbr. articles to profl. jours. Robert Tegler Spl. scholar, 1952, 53, 54; recipient Fred H. Irwin prize in organic chemistry, 1955. Mem. Can. Biochem. Soc., Canadian Oncolgy Soc., Am. Assn. Cancer Research, Am. Soc. Biol. Chemists, AAAS, Sigma Xi. Nazarene. Subspecialties: Biochemistry (medicine); Cancer research (medicine). Current work: Structure and function of RNA polymerases of rodent tissues; teaching biochemistry and cancer research; research on RNA polymerases. Office: Department Biochemistry University Saskatchewan Saskatoon SK Canada S7N OWO

BLAIR, JAMES BRYAN, biochemist, educator, researcher; b. Waynesburg, Pa., May 26, 1944; s. Bryan Luther and Beulah (Richey) B.; m. Daphna Lynn Killen, July 9, 1966; children: Dwight, Tristan. B.S. in Chemistry, W.Va. U., 1966; Ph.D. in Biochemistry, U. Va., 1969. Research assoc. U. Wis., Madison, 1969-72; asst. prof. W.Va. U., Morgantown, 1972-76; assoc. prof., 1976-82, prof., 1982—; cons. biochemistry study sect. NIH, Bethesda, Md., 1982—. NIH grantee, 1973—; recipient career devel. award, 1978-83. Mem. Am. Soc. Biol. Chemists, Am. Chem. Soc., AAAS, Biophys. Soc. Subspecialties: Biochemistry (medicine); Endocrinology. Current work: Hormonal control of carbohydrate metabolism. Office: Dept Biochemistry W Va U Med Center Morgantown WV 26506

BLAIR, WILLIAM FRANKLIN, zoologist; b. Dayton, Tex., June 25, 1912; s. Percy Franklin and Mona (Patrick) B.; m. Fern Antell, Oct. 25, 1933. B.S., U. Tulsa, 1934; M.S., U. Fla., 1935; Ph.D., U. Mich., 1938. Research asso. Lab. Vertebrate Biology U. Mich., 1937-46; mem. faculty U. Tex., 1946—, now prof. zoology.; Mem. adv. panel environmental biology NSF, 1958-62, mem. adv. com. div. biology and medicine, 1967-69; chmn. U.S. nat. com. Internat. Biol. Program, 1968-72, mem. spl. internat. com., 1968-74, v.p., 1969-74; mem. internat. environmental programs Com. NAS/NRC, 1970-76, mem. exec. com., 1971-76; mem. monitoring commn. spl. com. on problems environment Internat. Council Scientific Unions, 1970-72; chmn. adv. panel U.S. Bur. Reclamation, 1971—. Author: The Rusty Lizard, 1960; Sr. author: Vertebrates of the United States, 1957, rev. edit., 1968, Big Biology: The US/IBP, 1977; Editor: Vertebrate Speciation, 1961, Evolution in the Genus Bufo, 1972. Fellow AAAS; mem. Am. Soc. Ichthyologists and Herpetologists (bd. govs. 1951-55, 56-61, 62-63, v.p. 1955), Am. Soc. Mammalogists, Am. Soc. Naturalists (editorial bd. 1957-58), Am. Inst. Biol. Scis. (mem. exec. com. 1969-72, pres. 1972), Am. Soc. Internat. Law (mem. panel internat. law and global environment 1969—), Ecol. Soc. Am. (editorial bd. 1960-62, pres. 1963, chmn. public affairs com. 1967-68), Am. Soc. Zoologists (chmn. ecology sect. 1965), Genetics Soc. Am., Soc. Study Evolution (mem. council 1959-61, pres. 1962, asso. editor 1961-62), Soc. Systematic Zoology (council 1968-71). Subspecialties: Evolutionary biology; Ecology. Current work: Professor emeritus. Home: 5401 E 19th St Austin TX 78721

BLAKE, BRIAN FRANCIS, psychology educator, consultant; b. Jersey City, Aug. 26, 1942; s. Andrew A. and Mary A. (White) B.; m. Ann M. Sicola, Jan. 24, 1965; children: Dristin, Eric, Sean. B.A., St. Peter's Coll., 1964; M.S., Purdue U., 1966, Ph.D., 1969. Asst. prof. St. John's U., N.Y.C., 1969-72, assoc. prof., 1972-73; asst. prof. Purdue U., West Lafayette, Ind., 1973-75, assoc. prof., 1975-79, prof., 1979-81; dir. consumer lindsl. research program and prof. psychology Cleve. State U., 1981—; sr. research cons. Human Affairs Research Center, N.Y.C., 1970-72; market research cons. Pfizer Genetics Land O'Lakes, West Lafayette, Ind., 1977-81, others, 1977-81; program evaluation cons. Fgn. Agr. Services, U.S. Dept. Agr., Washington, 1976-77, AID, Bolivia, 1980; pres. Decision Dynamics, Inc., Cleve., 1982—. Contbr. articles to profl. jours., chpts. to books. Mem. Am. Psychol. Assn., Am. Mktg. Assn., Eastern Psychol. Assn., Midwestern Psychol. Assn., Sigma Xi. Subspecialties: Social psychology. Current work: Consumer psychology, public opinion measurement, market research techniques. Office: Cleveland State U Cleveland OH 44115

BLAKE, RICHARD DOUGLAS, biochemistry educator; b. Greenfield, Mass., Sept. 24, 1932; s. Charles Samuel and Vena Beatrice (Dewey) B.; m. Jeanne Marie Mango, Aug. 16, 1958; children: Hannah Elizabeth, Jonathan Dresser. B.S., Tufts Coll., 1958; M.S., Rutgers U., 1963; Ph.D., Princeton U., 1967. Research asst. Harvard U., Cambridge, Mass., 1958-61; research assoc. Princeton U., 1967-68, research staff, 1968-73; prof. biochemistry U. Maine, Orono, 1973—, also chmn. dept. Contbr. articles to profl. jours. NIH grantee, 1975—. Mem. Biophys. Soc., Am. Soc. Biol. Chemists, N.Y. Acad. Scis., Sigma Xi. Subspecialties: Biophysical chemistry; Molecular biology. Current work: Structure and interactions of nucleic acids; evolution of nucleic acid sequences; dynamics and stability of nucleic acid helixes; effects of water and ions on the structure and stability of nucleic acids. Home: 540 College Ave Orono ME 04473 Office: Dept Biochemistry U Maine Orone ME 04473

BLAKE, RICHARD LEE, astrophysicist; b. Berkeley Springs, W.Va., Mar. 8, 1937; s. Robert A. and Ica Virginia (Yost) B. B.S. in Physics, Rensselaer Poly. Inst., 1959; Ph.D. in Astrophysics, U. Colo., 1968. Astronomer Naval Research Lab., Washington, 1959-63; asst. prof. astrophysics U. Chgo., 1968-74; mem. staff astrophysics Los Alamos (N.Mex.) Nat. Lab., 1974—; cons. x-ray tech. Contbr. numerous articles to profl. jours. Served with USNR, 1959-63. Mem. Am. Astron. Soc., Am. Phys. Soc. Republican. Subspecialties: Solar physics; 1-ray high energy astrophysics. Current work: Studies of solar activity via high resolution x-ray spectroscopy; studies in high energy astrophysics via spectroscopic imaging systems. Home: 448 Paige LPW Los Alamos NM 87544 Office: Los Alamos Nat Lab MSD 410 Los Alamos NM 87545

BLAKE, THOMAS MATHEWS, med. educator; b. Sheffield, Ala., Aug. 4, 1920; s. Jeptha Hill and Edna Austin (Mathews) B. B.A., U. Ala., 1941; M.D., Vanderbilt U., 1944. Diplomate: Am. Bd. Internal Medicine. Intern, resident Vanderbilt U. Hosp., 1944-45, 47-49; resident in medicine Strong Meml. Hosp., Rochester, N.Y., 1949-50; prof. medicine U. Miss. Sch. Medicine, Jackson, 1955—. Served to capt. AUS, 1942-47. Fellow ACP, Am. Coll. Chest Physicians, Am. Coll. Cardiology, Am. Heart Assn. Episcopalian. Subspecialty: Cardiology. Current work: Electrocardiography. Home: 4210 Hanover Pl Jackson MS 39211 Office: Univ Hosp 2500 N State St Jackson MS 39216

BLAKELY, JOHN MCDONALD, educator; b. Scotland, Apr. 8, 1936; came to U.S., 1961, naturalized, 1967; s. James A. and Elizabeth M. (McDonald) B.; m. Nanette Stewart, July 1, 1960; children: Robin Mary, Karen Elizabeth. B.Sc., Glasgow U., 1958, Ph.D., 1961. Research fellow Harvard U., Cambridge, Mass., 1961-63; faculty Cornell U., Ithaca, N.Y., 1963—, prof. materials sci. and engring., 1970—; Guggenheim fellow Cambridge (Eng.) U., 1970-71; vis. scientist Argonne (Ill.) Nat. Lab., 1976; NSF fellow, Berkeley, 1977, cons. in field. Author: Introduction to Properties of Crystal Surfaces, 1973, Surface Physics of Materials, vols. I, II, 1975. Recipient Kelvin prize in physics Glasgow U., 1960; NSF grantee, 1970—; Dept. of Energy grantee, 1979—. Fellow Inst. Physics U.K., Am. Phys. Soc.; mem. Am. Vacuum Soc., AIME. Club: Finger Lakes Runners. Subspecialties: Materials; Condensed matter physics. Current work: Surface science, oxidation of metals, catalysis, photographic materials. Home: 332 Forest Home Dr Ithaca NY 14850 Office: Cornell Univ 312 Bard Hall Ithaca NY 14853

BLAKELY, JOHN PAUL, manager technical publications, chemical engineer, researcher; b. Erie, Pa., Oct. 20, 1922; s. Eugene James, Jr. and Isabelle Marie (Carney) B.; m. Tinque June Spann, Aug. 10, 1944; 1 dau., Carol Tinque Blakely Kaplan. B.S., Washington and Lee U., 1943; M.S., U. Tenn.-Knoxville, 1958. Research chemist Tenn. Eastman Corp., Oak Ridge, 1943-48; research chemist Union Carbide Corp., Oak Ridge, 1948-65, mng. editor, 1965-73, supr. tech. publs., 1973-80, chem. engr., 1980—, editor, author publs. manual, 1982. Editor, author: Nuclear Safety Jour, 1965-73. Pres., bd. Y-12 Fed. Credit Union, Oak Ridge, 1950-73. Mem. Soc. Tech. Communication (sr., outstanding communicator of year E. Tenn. chpt. 1982, dir. 1979-84, chmn. strategic planning com. 1981-84), Am. Nuclear Soc. (chmn. publs. com. 1979-81). Republican. Club: Toastmasters (pres. Knoxville 1951-54). Subspecialties: Information science; publications management. Current work: Effective technical communication by means of standardized and accepted format, terminology, and usage. Home: 9635 Tunbridge Ln Knoxville TN 37922 Office: Union Carbide Corp Nuclear Div Y-12 9201-5 PO Box Y Oak Ridge TN 37830

BLAKLEY, GEORGE ROBERT, JR., mathematician, computer scientist, educator; b. Chgo., May 6, 1932; s. George Robert and Gladys Margaret (Baechle) B.; m. Virginia Clarke, Sept. 7, 1957; children: George Robert III, Cynthia Ellen, Lydia Anne. A.B., Georgetown U., 1954; M.A., U. Md., 1959, Ph.D., 1960. Asst. prof. math. U. Ill. at Urbana, 1962-66; asso. prof. State U. N.Y. at Buffalo, 1966-70; prof. Tex. A&M U., College Station, 1970—, head dept., 1970-78; Office Naval Research postdoctoral research asso. Cornell U., 1960. Nat. Acad. Scis.-NRC postdoctoral fellow Harvard U., 1961. Mem. IEEE (sr. mem.), AAAS, Am. Math. Soc., Assn. Computing Machinery, Math. Assn. Am., Soc. for Indsl. and Applied Math. (vis. lectr. biomath. 1968-70, vis. lectr. applied math. 1979-83), Inst. Math. Statistics, Sigma Xi, Phi Kappa Phi, Pi Mu Epsilon. Subspecialties: Cryptography and data security; Information systems, storage, and retrieval (computer science). Current work: Information theory, computer security, cryptography, combinatorics. Home: 1405 Broadmoor St Bryan TX 77802 Office: Dept Math Tex A and M U College Station TX 77843

BLANCHARD, RAY MILTON, psychiatry institute research psychologist; b. Hammonton, N.J., Oct. 9, 1945; s. Ray Milton and Angelina (Celi) Ruggero. A.B., U. Pa., 1967; M.A., h4U. Ill.-Urbana, 1970; Ph.D., 1973. Cert. psychologist Ont. Bd. Examiners. Psychologist Ont. Correctional Inst., Brampton, Can., 1976-80; research psychologist Gender Identity Clinic, Clarke Inst. Psychiatry, Toronto, Ont., 1980—. Killam fellow Dalhousie U., Halifax, N.S., Can., 1973. Mem. Internat. Acad. Sex Research, Am. Psychol. Assn., Can. Psychol. Assn. Subspecialty: Gender indentity disorders. Current work: Taxonomy of gender identity disorders; psychosocial adjustment of transsexuals; phallometric assessment of sexual anomalies. Home: 32 Shaftesbury Ave Toronto ON Canada M4T 1A1 Office: Gender Identity Clinic Clarke Inst Psychiatry 250 College St Toronto ON Canada M5T 1R8

BLANCHARD, ROBERT OSBORN, plant pathologist, college dean; b. Cumberland Center, Maine, July 5, 1939; s. Nelson and Luella B.; m. Ellen L. Murphy, Aug. 14, 1965; children: Lori, Gregory. A.A.S., So. Maine Vocat. Tech. Inst., 1959; B.S., U. Maine, 1964; M.Ed., U. Ga., 1969, Ph.D., 1971, postdoctoral fellow, 1971-72. Jr. high sch. biology tchr., Windsor, Conn., 1964-67; mem. faculty U. N.H., Durham, 1972—, chmn. botany and plant pathology, 1973-76, chmn. intercoll. biol. scis. orgn., 1976-78, assoc. dean resident instrn., 1982—. Author textbook on tree diseases; contbr. numerous articles to sci. jours. Served in Air N.G., 1956-62. NSF fellow, 1967-68; NDEA fellow, 1968-71; recipient disting. service cert. Order of Patrons of Husbandry, 1978, Faculty Merit award Bd. Trustees, U. N.H., 1979; U.S. Forest Service grantee, 1975—. Mem. Mycol. Soc. Am., Bot. Soc. Am., Am. Phytopath. Soc., N.E. Forest Pest Council, Sigma Xi, Gamma Sigma Delta, Phi Sigma, Grange. Republican. Roman Catholic. Subspecialties: Plant pathology; Biophysics (biology). Current work: Diseases of forest and shade trees; electrophysiology of tree tissues. Home: 2 Meserve Rd Durham NH 03824 Office: U NH Durham NH 03824

BLANCK, A.R., scientist, engineer; b. N.Y.C., Feb. 7, 1925; s. Andrew G. and Anne V. B.; m. Edna H. Ruppert, Nov. 25, 1950; children: Elaine Lois, Evelyn Joyce. B.A., N.Y.U., 1950; M.A., Poly. Inst. Bklyn., 1953; Sc.D., Sussex (Eng.) Inst. Tech., 1959. Project engr. Centro Research Labs., Ossining, N.Y., 1948-53; tech. service engr. Celanese Corp., Summit, N.J., 1953-56; head diaelectics lab W.R. Grace, Clifton, N.J., 1956-58; coordinator materials design engring. U.S. Govt., Dover, N.J., 1958-65; pres., tech. dir. Rutherford Research Corp., N.J., 1956—; mem. faculty Sch. Continuing Edn., N.Y.U., 1965—. Contbr. articles to profl. jours. Mem. Am. Council Ind. Labs., Assn. Cons. Chemists and Chem. Engrs., ASTM, AIEE, IEEE. Baptist. Subspecialties: Materials; Electrical engineering. Current work: Electrical insulation; material engineering; plastics and rubbers. Office: Rutherford Research Drawer 249 Rutherford NJ 07070

BLANCO, VICTOR MANUEL, astronomer; b. Guayama, P.R., Mar. 10, 1918; s. Felipe and Adelfa (Pagan) B.; m. Cicely Woods, June 5, 1943; m. Betty Elaine Fried, Jan. 26, 1971; children: Merida, Victor Carlos, Daniel, Elizabeth, David. B.S., U. Chgo., 1943; M.A., U. Calif., Berkeley, 1947, Ph.D., 1949; Dr. Sci. (hon.), U. P.R., 1976. Assoc. prof. physics U. P.R., 1949-51; prof. astronomy Case Inst. Tech., Cleve., 1951-65; dir. astrometry and astrophysics div. U.S. Naval Obs., Washington, 1965-67; dir. Cerro Tololo Interum. Obs., La Serena, Chile, 1967-80, astronomer, 1980—. Author: (with S. W. McCuskey) Basic Physics of the Solar System, 1961. Served with USAAF, 1941-46. Recipient U.S. NSF Disting. Service medal, 1981. Mem. Internat. Astron. Union, Am. Astron. Soc. Subspecialties: Optical astronomy; Observational optical astronomy. Current work: Galactic structure, magellanic clouds, statistics. Home: 603 Casilla La Serena Chile

BLANDFORD, ROGER DAVID, astrophysics educator; b. Grantham, Eng., Aug. 28, 1949; (married), 1972; 2 children. B.A., Cambridge U., 1970, M.A. and Ph.D., 1974. Bye fellow Magdalen Coll. Cambridge U., 1972-73, research fellow St. John's Coll., 1973-76; asst. prof. theoretical astrophysics Calif. Inst. Tech., Pasadena, 1976-79, assoc. prof., 1979—; mem. Inst. Advanced Study, Princeton, N.J., 1974-75; Parisot fellow U. Calif.-Berkeley, 1975. Fellow Royal Astron. Soc.; mem. Am. Astron. Soc. (Warner prize 1982). Subspecialty: Theoretical astrophysics. Office: Calif Inst Tech Pasadena CA 91125

BLANFORD, GEORGE EMMANUEL, JR., physicist, educator; b. Lebanon, Ky., Sept. 16, 1940; s. George Emmanuel and Catherine Josephine (Hardesty) B.; m. Julianne Blanford, Aug. 7, 1971 (div.); children: Elizabeth Braznell, Peter Emmanuel. B.A., Cath. U. Am., Washington, 1964; M.S. (NSF summer fellow), U. Louisville, 1967; Ph.D., Washington U., St. Louis, 1971. Maitre de conference associe Universite de Clermont, Clermont-Ferrand, France, 1971-73; research assoc. NASA Johnson Space Ctr., Houston, 1973-75; asst. prof. physics U. Houston, Clear Lake City, 1975-78, assoc. prof., 1978—. Contbr. articles to profl. jours. NRC resident research assoc., 1973-1975. Mem. AAAS, Am. Assn. Physics Tchrs., Am. Geophys. Union, Am. Phys. Soc., Phi Beta Kappa, Sigma Xi, Sigma Pi Sigma. Democrat. Roman Catholic. Club: Johnson Space Ctr. Bicycle (pres.). Subspecialties: Planetary science; Planetology. Current work: Cosmic ray and solar particle interactions. Office: 2700 Bay Area Blvd B19 Houston TX 77058

BLANK, LAWRENCE WILLIAM, dentist; b. Los Angeles, Oct. 4, 1944; s. S. Steven and Pearl S. (Zuckerman) B.; m. Deborah Edith Glickman, Aug. 28, 1966; 1 son, Forrest Evan. Student, UCLA, 1962-64; B.S.D., D.D.S., U. Calif., San Francisco, 1968; M.S., George Washington U.; postgrad., Naval Grad. Dental Sch., 1973-74; M.S., U. Mich., 1978. Diplomate: Fed. Services Bd. Gen. Dentistry. Rotating intern U.S. Naval Hosp., Oakland, Calif., 1968-69; commd. lt. U.S. Navy, 1968, advanced through grades to capt., 1981; asst. dental officer U.S. Naval Sta., Subic Bay, Philippines, 1969-71; staff instr. operative and preventive dentistry U.S. Naval Hosp., Phila., 1971-73; asst. dental officer U.S.S. Nimitz, 1974-76; chief 2d yr. div. Nat. Naval Dental Ctr., Bethesda, Md., 1978-82, dir. occlusion courses, 1980—, presdl. dentist, Camp David dentist, 1981—, chmn. gen. dentistry dept., 1982—; gen. dentistry cons. U.S. Navy Bur. Medicine, 1982—. Exec. com. mem. East Bethesda Community Assn., 1978-80; mem. PTA, 1978—. Mem. Am. Dental Assn., Assn. Mil. Surgeons of U.S., U.S. Naval Inst., Acad. Gen. Dentistry (chpt. pres.), Internat. Assn. Dental Research, Am. Assn. Dental Schs., Acad. Operative Dentistry, Acad. Gold Foil Operators, Xi Psi Phi. Subspecialties: Periodontics; General dentistry. Current work: Research in electronic measuring devices in endodontics, Gingival response to composite resin restorations, use of mouth splints to improve athletic performance. Home: 5420 Beech Ave Bethesda MD 20814 Office: 8901 Wisconsin Ave Bethesda MD 20814

BLASGEN, MICHAEL WILLIAM, computer scientist; b. Glendale, Calif., May 18, 1941; s. Harry and Dorothy (McKenna) B.; m. Sharon Walther, Sept. 10, 1965; children: Alexandra, Nicholas. B.S., Harvey Mudd Coll., 1963; M.S., Calif. Inst. Tech., 1964; Ph.D., U. Calif.-Berkeley, 1968. Asst. prof. U. Calif.-Berkeley, 1968-69; research staff IBM, Yorktown Heights, NY, 1969-75, mgr. database systems, San Jose, Calif., 1975-80, mgr. advanced systems tech., Bethesda, Md., 1980-83, mgr. advanced minicomputer systems, Yorktown Heights, N.Y., 1983—. Contbr. articles to profl. jours. Mem. Assn. Computing machinery (recipient, best paper award 1982), IEEE. Subspecialties: Database systems; Distributed systems and networks. Current work: Management of large computer based development; development of new office technology. Home: 33 Whippoorwill Rd Chappaqua NY 10514 Office: IBM Corp 6600 Rockledge St Bethesda MD 20817

BLASINGAME, BENJAMIN PAUL, electronics co. exec.; b. State College, Pa., Aug. 1, 1918; s. Ralph Upshaw and Sue Mae (Combs) B.; m. Ella Mae Perry, Aug. 29, 1942; children—Nancy J. Blasingame Wambach, James P., Margaret A. Blasingame Kramer, John R. B.S. in Mech. Engring, Pa. State U., 1940; Sc.D. in Aero. Engring, M.I.T., 1950. Head astronautics dept. U.S. Air Force Acad., 1958-59; resigned, 1959; gen. mgr. electronics div. Gen. Motors Corp., 1959-70; mgr. Milw. operation Delco Electronics div., 1970-72. Author: Astronautics, 1964. Mgr. Santa Barbara operation, 1972-79; Bd.dirs. Santa Barbara Cottage Hosp., 1977—; chmn. Santa Barbara Metro, Nat. Alliance Bus., 1972-75. Commd. 2d lt. U.S. Air Force, 1941; advanced through grades to col., 1959. Decorated Legion of Merit; recipient Public Service award, NASA, 1969, Public Service medal, 1973. Mem. AIAA, Nat. Acad. Engring., N.Y. Acad. Scis., Internat. Acad. Astronautics, Santa Barbara C. of C. (bd. dirs. 1977—). Unitarian. Club: La Cumbre Country. Subspecialties: Aeronautical instrumentation; Aerospace engineering and technology. Current work: Inertial navigation, guidance and control systems. Patentee in field. Home: 517 Carriage Hill Ct Santa Barbara CA 93110

BLAUSTEIN, MORDECAI P., physiology educator; b. N.Y.C., Oct. 19, 1935; s. Norman and Gertrude (Hellman) B.; m. Ellen Baron, June 21, 1959; children: Laura, Marc. B.A., Cornell U., 1957; M.D. (NIH fellow), Washington U., St. Louis, 1962. Intern Boston City Hosp., 1962-63; med. resident U.S. Navy Med Research Inst., Bethesda, Md., 1963-66; assoc. prof. Washington U. Med. Sch., St. Louis, 1968-75, prof., 1975-80; prof., chmn. dept physiology U. Md. Med. Sch., Balt., 1979—; guest scientist Pharmacology Inst. U. Bern, Switzerland, 1971, Lab. of Marine Biol. Assn. U.K., Plymouth, Eng., 1973. Contbr. articles to profl. jours. USPHS fellow Cambridge (Eng.) U., 1966-68;

NIH fellow, 1966-68; NATO sr. fellow, 1971; NIH grantee, 1969—; NSF grantee, 1973—; Muscular Dystrophy Assn. grantee, 1975—. Mem. Am. Physiol. Soc., Biophys. Soc., Physiol. Soc. U.K. (assoc.), Soc. Gen. Physiologists (councillor 1979-81), Soc. Neurosci. Subspecialties: Physiology (medicine); Neurophysiology. Current work: Regulation of cell calcium in nerve and muscle; physiology of presynaptic nerve terminals; role of salt in etiology of essential hypertension. Office: Dept Physiology MD U Med Sch Baltimore MD 21201

BLAZEY, RICHARD NELSON, research electro optical engineer; b. Rochester, N.Y., Aug. 23, 1941; s. Willard and Frances B.; m. Jacqueline Norris; 3 children. B.E.E., Cornell U., 1964; M.S.E.E., Purdue U., 1966. Project engr. Sperry Gyroscope, Great Neck, N.Y., 1966-69; research assoc. Eastman Kodak, Rochester, 1969—. Mem. IEEE, Optical Soc. Am., Rochester Engring. Soc. (v.p. 1976-77), Eta Kappa Nu. Subspecialties: Laser data storage and reproduction; Acoustics. Current work: Hybrid microelectronics; printed circuits; packaging of electronic components. Patentee in field (7). Office: Bldg 81 Kodak Park Rochester NY 14650

BLEACH, RICHARD DAVID, physicist; b. Ft. Riley, Kans., June 7, 1944; divorced. B.S., Rensselaer Poly. Inst., 1966; Ph.D., U. Md., 1972. Reseach assoc. U. Md., College Park, 1972-76; cons. Naval Research Lab., Washington, 1976-77, research physicist, 1977—. Contbr. articles to profl. journs. Recipient Superior Performance award Naval Research Lab., 1979, 81. Mem. AAAS, Am. Astron. Soc., Am. Phys. Soc., Am. Vacuum Soc., Sigma Xi. Subspecialties: Plasma physics; 1-ray high energy astrophysics. Current work: X-ray spectroscopy of high temperature plasmas and x-ray astronomy. Home: 15909 Pennant Ln Bowie MD 20716 Office: Code 6680B Naval Research Lab Washington DC 20375

BLECHNER, MARK J., psychologist; b. N.Y.C., Nov. 6, 1950; s. Norbert and Hannah (Darmstadter) B. B.A., U. Chgo., 1972; M.S., Yale U., 1975, Ph.D., 1977. Lic. psychologist, N.Y. Researcher Herkias Labs., New Haven, 1974-76; researcher Bell Labs., Murray Hill, N.J., 1977-78; asst. clin. prof. Coll. Physicians and Surgeons, Columbia U., N.Y.C., 1981—. Contbr. articles to profl. jours. Mem. AAAS, Am. Psychol. Assn., N.Y. Acad. Sci., Sigma Xi. Subspecialties: Cognition. Current work: Dreams in borderline personality organization, cognitive development of paranoid perception. Address: 145 Central Park W New York NY 10023

BLECKMANN, CHARLES ALLEN, botanist; b. Evansville, Ind., Nov. 24, 1944; s. Emil Adolph and Mary Louise (Blesch) B.; m. Cecilia Andrews, July 3, 1968; 1 son, Mark. B.A., U. Evansville, 1967; M.S., Incarnate Word Coll., 1971; Ph.D., U. Ariz., 1977. Research assoc. U.S. Dept. Agr. Range Sci. Lab., Tuscon, 1977-78; sr. research botanist Conoco, Inc., Ponca City, Okla., 1978—. Served to capt. USAF, 1967-71. Mem. AAAS, Am. Inst. Biol. Sci., Sigma Xi. Subspecialties: Biological aspects of petroleum production; Water supply and wastewater treatment. Current work: Waste treatment and disposal. Office: Box 1267 Ponca City OK 74601

BLEIBERG, MARVIN JAY, pharmacologist, toxicologist; b. Bklyn., Feb. 19, 1928; s. Harold and Rena (Holzer) B.; m. Beulah Matt, Jan. 17, 1960; children: Lawrence, Robert. B.S., Coll. William and Mary, 1949; postgrad., Ohio State U., 1949-50; Ph.D., Med. Coll. Va., 1957. Diplomate: Am. Bd. Toxicology. Postdoctoral trainee in steroid biochemistry Worcester Found. Exptl. Biology, Shrewsbury, Mass., 1957-58; instr., then asst. prof. pharmacology Jefferson Med. Coll., Phila., 1958-61; rev. scientist FDA, Washington, 1961-62; sr. scientist Melpar, Inc., Falls Church, Va., 1962-64; pharmacologist, sect. head Woodard Research Corp., Herndon, Va., 1963-75; interdisciplinary rev. scientist, div. toxicology Bur. Foods, FDA, Washington, 1975—. Contbr. articles to sci. jours. Mem. Am. Soc. Pharmacology and Exptl. Therapeutics, Am. Chem. Soc., Soc. Toxicology, Biometrics Soc., AAAS, N.Y. Acad. Sci., Soc. Exptl. Biology and Medicine (D.C. chpt.), Phi Beta Kappa, Sigma Xi. Subspecialties: Toxicology (medicine); Pharmacology. Current work: Food additivies; safety evaluation. Home: 3613 Old Post Rd Fairfax VA 22030 Office: 200 C St SW Washington DC 20204

BLEVINS, DALE GLENN, plant physiologist; b. Ozark, Mo., Aug. 29, 1943; s. Vernon Henry and Edna Gertrude (Payne) B.; m. Brenda Jo Graves, Aug. 27, 1967; 1 son, Jeremy Justin. B.S., S.W. Mo. State Coll., 1965; M.S., U. Mo., 1967; Ph.D., U. Ky., 1972. Postdoctoral fellow dept. botany Oreg. State U., Corvallis, 1972-74; asst. prof. botany U. Md., College Park, 1974-77; asst. prof. agronomy U. Mo.-Columbia, 1978-81, assoc. prof., 1981—. Mem. Am. Soc. Plant Physiologists, Am. Soc. Agronomy. Methodist. Subspecialties: Plant physiology (agriculture); Nitrogen fixation. Current work: Ureide (allantoic acid) synthesis in soybeans nodules; allantoic acid metabolism in soybean leaves; potassium and proten synthesis in crop plants; hormones in root nodules. Home: 19 West Parkway Columbia MO 65201 Office: Dept Agronomy U Mo 204 Curtis Hall Columbia MO 65211

BLEVIS, BERTRAM CHARLES, research administrator, physicist; b. Toronto, Ont., Can., Feb. 26, 1932; s. Abraham Harry and Kate (Soskin) B.; m. Rhoda Ain, Apr. 12, 1959; children: Lisa, Barbara, Mark. B.A. with honors, U. Toronto, 1953, M.A., 1954, Ph.D., 1956. Def. sci. officer Def. Research Bd., Ottawa, Ont., 1956-66; sci. liaison officer Can. Def. Research Staff, Washington, 1966-69; dir. space systems Communications Research Centre, Ottawa, 1969-75; dir. internat. arrangements Dept. Communications, Ottawa, 1975-76; dir. gen. space tech. and applications Communications Research Centre, Ottawa, 1977—. Mem. editorial bd.: Space Communication and Broadcasting, North Holland Pub. Co., 1982—; mem. adv. bd.: Telematics and Informatics, Pergamon Press, Ltd., 1983—. Mem. Ottawa Talmud Torah Bd. J.S. McLean scholar Univ. Coll., U. Toronto, 1949-51; Edward Blake scholar, 1950-51. Fellow Can. Aeronautics and Space Inst.; mem. IEEE, Canadian Assn. Physicists. Jewish. Subspecialties: Communications; Satellite studies. Current work: Management of the research and development activities of the Canadian government's Department of Communications in space technology and its applications to the fields of communications, broadcasting, remote sensing and search and rescue. Home: 1997 Neepawa Ave Ottawa ON Canada K2A 3L4 Office: Communications Research Centre PO Box 11490 Sta H Ottawa ON Canada K2H 8S2

BLEYMAN, LEA KANNER, biologist, educator; b. Halle, Ger., Nov. 9, 1936; d. Salomon David and Amalia (Azderbal) Kanner; m. Michael Alan Bleyman, June 15, 1958; 1 dau., Anne. B.A. magna cum laude, Brandeis U., 1958; M.A., Columbia U., 1961; Ph.D., Ind. U., 1966. Research assoc. U. Ill., Urbana, 1964-69, U. N.C., Chapel Hill, 1970-72; assoc. prof. biology Baruch Coll., CUNY, 1973-78, prof., 1979—, chairperson dept. natural scis., 1981-83; vis. research scientst Cornell U., Ithaca, N.Y., summers 1975-80. Asst. editor: Protozoological Actualities, 1979; contbr. articles to sci. jours. Max Planck Inst. fellow, Berlin, summer 1974; NIH fellow, 1961-64. Mem. NOW, AAAS, Am. Assn. Women in Sci., Am. Genetics Assn., Am. Soc. Cell Biology, Genetics Soc. Am., N.Y. Acad. Scis., Soc. Protozoologists (exec. com.), Phi Beta Kappa, Sigma Xi. Subspecialties: Gene actions; Cell biology. Current work: Ciliate genetics; aging and life cycle; nuclear regulation. Office: 17 Lexington Ave Box 502 New York NY 10010

BLIGH, JOHN, physiologist, educator; b. London, July 18, 1922; s. Harry Alfred and Florence Emily (Scott) B.; m. Antonia Doris Roever, Aug. 16, 1952; children: Peter, Bettina. B.Sc. with honors, Univ. Coll. London, 1950, Ph.D., 1952, D.Sc., 1978. Sci. officer Agrl. Research Council, Hannah Research Inst., Scotland, 1952-56; sci. officer Inst. Animal Physiology, Babraham, Cambridge, Eng., 1956-77; prof. physiology, dir. div. life scis., dir. Inst. Arctic Biology, U. Alaska, Fairbanks, 1977—. Author: Temperature Regulation in Mammals and Other Vertebrates, 1973; contbr. articles to profl. jours.; mem. editorial bds. profl. jours. Chmn. bd. govs. Sawston Village Coll., Cambridge, Eng., 1970-77. Served with Brit. Army, 1940-42. Mem. Am. Physiol. Soc., AAAS, Arctic Inst. N.Am., Am. Circumpolar Health Soc., Internat. Biometeorol. Soc., Royal Soc. London (vis. prof. Lima, Peru 1972-73), Physiol. Soc. (U.K.), Can. Physiol. Soc., Inst. Biology (U.K.). Subspecialties: Neurophysiology; Physiology (medicine). Current work: Thermoregulatory and environmental physiology; biological regulation; neurophysiology. Home: SR 20864D Fairbanks AK 99701 Office: Inst Arctic Biology U Alaska Fairbanks AK 99701

BLINCOE, CLIFTON ROBERT, biochemist, educator; b. Odessa, Mo., Nov. 21, 1926; s. Clifton R. and Frankie Florence (Downs) B.; m. Bertha Alice Fisher, Jan. 15, 1948; children: Clyde Joseph, Allen Benjamin, Carl Richard. B.S., U. Mo., 1947, M.A., 1948, Ph.D., 1955. Asst. prof. dairy husbandry U. Mo., 1948-56; prof. biochemistry U. Nev., Reno, 1956—; vis. assoc. prof. Theoretical Chemistry Inst., U. Wis., summer 1969; vis. prof. Trinity Coll., Dublin, 1969-70, Cornell U., 1976-77. Contbr. articles to profl. jours. Fulbright lectr. Trinity Coll., Dublin, 1969-70; Calcium Carbonate Co. travel fellow, 1974. Mem. Am. Chem. Soc., AAAS, Biophys. Soc., Am. Soc. Animal Sci. Subspecialties: Biochemistry (biology); Animal nutrition. Current work: Trace mineral nutrition of domestic ruminants. Office: Div Biochemistry U Nev Reno NV 89557

BLISS, LAWRENCE CARROLL, botany educator; b. Cleve., Nov. 29, 1929; s. Laurence and Ada May (Peterson) B.; m. Gweneth Ruth Jones, Mar. 15, 1952; children: Dwight I., Karen L. B.S., Kent State U., 1951, M.S., 1953; Ph.D., Duke U., 1956. Instr. biology Bowling Green State U., 1956-57; instr. botany U. Ill., 1957-58, asst. prof., 1958-61, asso. prof., 1961-66, prof., 1966-68; prof. dept. botany U. Alta., 1968-78, dir. controlled environ. facility, 1968-78; prof. botany U. Wash., 1978—, chmn. dept. botany, 1978—; cons. in field. Author: (with M. Balbach) Laboratory Manual for General Botany, 6th edit., 1977; editor: Truelove Lowland, Devon Island, Canada: A High Arctic Ecosystem, 1977, (with others) Tundra Ecosystems: A Comparative Analysis, 1981; contbr. articles to profl. jours. Fulbright scholar, 1963-64. Mem. Ecol. Soc. Am. (v.p. 1976-77, treas. 1977-81, pres.-elect 1981-82, pres. 1982-83), Am. Inst. Biol. Sci.; fellow AAAS, Arctic Inst. N. Am.; mem. Can. Bot. Assn., Sigma Xi. Republican. Presbyterian. Subspecialties: Ecology; Ecosystems analysis. Current work: Plant communities, ecophysiology and ecosystem studies of Artic and Alpine encironments especially Canadian High Artic, Mount St. Helens. Home: 1226 NW 175th St Seattle WA 98177 Office: Dept Botany U Wash Seattle WA 98195

BLITZ, LEO, astronomer, educator; b. Krakow, Poland, Oct. 21, 1945; s. Abraham and Maria (Salz) B.; m. Judith Ida Klimpl, Aug. 8, 1971; 1 son: Brian Adam. B.S., Cornell U., 1967; M.A., Columbia U., 1975, M.Phil., 1976, Ph.D., 1979. Postgrad. research fellow U. Calif., Berkeley, 1978-81; asst. prof. astronomy program U. Md., College Park, 1981—. Author undergrad. lab. manual; Contbr. articles to profl. jours. Gen. Research Bd. grantee U. Md., 1982; NATO Sci. Affairs grantee, 1983. Mem. Am. Astron. Soc., Internat. Astron. Union. Subspecialty: Radio and microwave astronomy. Current work: Star formation, galactic structure, molecular clouds. Home: 1530 Red Oak Dr Silver Spring MD 20910 Office: Astronomy Program U Md College Park MD 20742

BLITZER, ANDREW, otolaryngologist; b. Pitts., Apr. 25, 1946; s. Martin Hollander and Lyrene Iris (Lave) B.; m. Patricia Volk, Dec. 21, 1969; children: Peter Morgen, Polly Volk. B.A., Adelphi U., 1967; D.D.S., Columbia U., 1970; M.D., Mt. Sinai Sch. Medicine, 1973. Diplomate: Am. Bd. Otolaryngology. Resident in gen. surgery Beth Israel Hosp., N.Y.C., 1973-74; resident in otolaryngology Mt. Sinai Hosp., N.Y.C., 1974-77; asst. prof. otolaryngology Coll. Phys. & Surg., Columbia U., N.Y.C., 1977-82, assoc. prof. otolaryngology and oral surgery, 1982—; dir. div. head and neck surgery Columbia-Presbyn. Med. Ctr., N.Y.C., 1980—, dir. residency edn., 1978—; lectr. dept. otolaryngology Mt. Sinai Sch. Medicine, N.Y.C., 1977—. Co-Author several books; assoc. editor: Oncology Times; contbr. chpts. to books, articles to profl. jours. Recipient award for excellence Am. Assn. Orthodontists, 1970; Nat. Inst. Neurol. Communicative Disorders and Stroke, grantee, 1978—. Fellow A.C.S., N.Y. Acad. Medicine, Am. Soc. for Head and Neck Surgery, Am. Acad. Facial Plastic and Reconstructive Surgery, Am. Laryngol., Rhinol., and Otol. Soc. Subspecialties: Otorhinolaryngology; Cancer research (medicine). Current work: Histopathological and biochemical responses of connective tissue to cancer of the head and neck. Clinical research concerning malignant fibrous histiocytoma and mucormycosis of the paranasal sinuses. Home: 1136 Fifth Ave New York NY 10028 Office: Dept Otolaryngology Coll Phys & Surg Columbia U 630 W 168th St New York NY 10032

BLIZNAKOV, EMILE GEORGE, scientist; b. Kamen, Bulgaria, July 28, 1926; came to U.S., 1961, naturalized, 1966; s. George P. and Paraskeva B. M.D. Faculty of Medicine, Sofia, Bulgaria, 1953; Dir. Regional Sta. for Hygiene and Epidemiology, chief dist. dept. health, Pirdop, Bulgaria, 1953-55; staff scientist, microbiologist Research Inst. Epidemiology and Microbiology, Ministry of Public Health, Sofia, 1955-59; vis. scientist Gamaleya Research Inst. Epidemiology and Microbiology, Acad. Med. Scis., Moscow, 1958-59; sr. staff scientist, prof. life scis. New Eng. Inst., Ridgefield, Conn., 1961-81, dir. personnel, 1968-74, v.p., 1974-76, pres., 1976-81; exec. dir. research and devel. Libra Research, 1981—; pres., sci. dir. Lupus Research Inst., Rockville, Md., 1981—; dir. Child Safety Corp.; cons. to indsl., pharm. and public relations firms. Contbr. articles to profl. jours. Fannie E. Rippel Found. grantee, 1972-80; G.M. McDonald Found. grantee, 1972-81; Whitehall Found. grantee, 1971-75; Wallace Genetic Found. grantee, 1972-81. Fellow Royal Soc. Tropical Medicine and Hygiene (London); mem. AMA, Am. Fedn. Clin. Research, Am. Soc. Microbiology, N.Y. Acad. Scis., Am. Coll. Toxicology, Am. Soc. Neurochemistry, Reticuloendothelial Soc., Bioelectromagnetic Soc., AAAS. Subspecialty: Medical research administration. Home: 189 Ledges Rd Ridgefield CT 06877 Office: Lupus Research Inst 1300 Piccard Dr Rockville MD 20850

BLOCH, ERICH, electrical engineer; b. Sulzburg, Ger., Jan. 9, 1925; came to U.S., 1948, naturalized, 1952; s. Joseph and Tony B.; m. Renee Stern, Mar. 4, 1948; 1 dau., Rebecca Bloch Rosen. Student, Fed. Poly. Inst., Zurich, Switzerland, 1945-48; B.S. in Elec. Engring, U. Buffalo, 1952. With IBM Corp., 1952—, v.p. gen. mgr., East Fishkill, N.Y., 1975-80, v.p. tech. personnel devel., Armonk, N.Y., 1980—; mem. com. computers in automated mfg. NRC, 1980—; Trustee Marist Coll., Poughkeepsie, N.Y., 1978—. Author. Fellow IEEE; mem. Nat. Acad. Engring., AAAS, Am. Soc. Mfg. Engrs. (hon.). Subspecialty: Computer engineering. Patentee in field. Office: 1000 Westchester Ave Harrison NY 10604

BLOCH, HERMAN SAMUEL, chemist; b. Chgo., June 15, 1912; s. Aaron and Esther (Broder) B.; m. Elaine J. Kahn, July 4, 1940; children—Aaron N., Janet L. (Mrs. Daniel Martin), Merry D. (Mrs. Dobroslav Valik). B.S., U. Chgo., 1933, Ph.D., 1936. With UOP, Inc., Des Plaines, Ill., 1936-77, asso. dir. research, 1964-73, dir. catalysis research, 1973-77, cons., 1977—; Chmn. com. phys. scis. Ill. Bd. Higher Edn., 1969-71. Author. Commr. Housing Authority Cook County, 1966—, chmn., 1971—; chmn. Skokie (Ill.) Human Relations Commn., 1965-71; pres. bd. edn. Skokie Sch. Dist. 68, 1962-63; trustee Skokie Library Bd., 1981—. Recipient E.V. Murphree award indsl. and engring. chemistry Am. Chem. Soc., 1974; Eugene J. Houdry award Catalysis Soc., 1971; I-R 100 award Indsl. Research mag., 1973. Hon. mem. Am. Inst. Chemists (honor scroll 1957); mem. Am. Chem. Soc. (chmn. bd. 1973-77), AAAS (v.p. for chemistry 1970), Nat. Acad. Scis., Soc. Chem. Industry, Nat. Acad. Sci., Ill. Acad. Sci., N.Y. Acad. Sci., Chemists Club Chgo., Phi Beta Kappa (pres. Chgo. 1968-69), Sigma Xi. Subspecialties: Catalysis chemistry; Fuels. Current work: Catalyst design; interaction of catalyst surfaces with reactants;mechanisms of catalytic reactions and of catalyst poisoning; processes for petroleum refining and petrochemicals. Patentee petroleum refining, catalysis, petrochems. Home and Office: 9700 Kedvale Ave Skokie IL 60076

BLOCH, KONRAD, biochemist; b. Neisse, Germany, Jan. 12, 1912; came to U.S., 1936, naturalized, 1944; s. Frederick D. and Hedwig (Steimer) B.; m. Lore Teutsch, Feb. 15, 1941; children—Peter, Susan. Chem.Eng., Technische Hochschule, Munich, Germany, 1934; Ph.D., Columbia, 1938. Asst. prof. biochemistry U. Chgo., 1946-50, prof., 1950-54; Higgins prof. biochemistry Harvard, 1954—. Recipient Nobel prize in physiology and medicine, 1964; Ernest Guenther award in chemistry of essential oils and related products, 1965. Fellow Am. Acad. Scis.; mem. Nat. Acad. Scis., Am. Philos. Soc. Subspecialty: Biochemistry (biology). Office: Dept Biochemistry Harvard U 12 Oxford St Cambridge MA 02138

BLOCH, PETER H., radiological physicist; b. N.Y.C., Nov. 11, 1933; s. Jacob and May (Garfinkle) B.; m. Julie Batten, Dec. 23, 1961 (div.); children: Leonard Mark, Martin Adam, Gabby Rosado. B.S., Mich. State U., 1956; M.S., U. Pa., 1965, Ph.D., 1968. Diplomate: Am. Bd. Radiology. Faculty U. Pa., Phila., 1968—, assoc. prof. med. physics, 1972—; dir. radiotherapy physics U. Hosp., 1978—. Contbr. articles to profl. jours. Served with U.S. Army, 1960-61. Mem. Am. Assn. Physicists in Medicine, Soc. Nuclear Medicine, Sigma Xi. Subspecialties: Cancer research (veterinary medicine); Radiology. Current work: Teaching and research in neutron radiation therapy, x-ray fluorescence analysis of heavy metals in patients in situ. Office: University of Pennsylvania Hospital Dept Radiotherapy 3400 Spruce St Philadelphia PA 19104

BLOCK, A. JAY, physician, medical educator; b. Balt., Apr. 11, 1938; s. Michael and Sylvia (Rosenberg) Lissauer; m. Linda Crone, May 25, 1961; children: Margo Dee, Allison Lee. B.A., Johns Hopkins U., 1958, M.D., 1962. Diplomate: Am. Bd. Internal Medicine.; Lic. physician, Md., Fla. Successively intern, resident, fellow Johns Hopkins Hosp., Balt., 1962-67; instr., then asst. prof. Johns Hopkins Sch. Medicine, 1967-70; asst. prof. medicine U. Fla., Gainesville, 1970-73, assoc. prof., 1973-75, prof., 1975—, chief pulmonary div., 1973—; chief pulmonary sect. VA Hosp., Gainesville, 1970-77. Editor: Tice's Practice of Medicine, 1967-72; dept. editor: Chest Jour, 1977—; mem. editorial bd., 1977-82, Am. Rev. Respiratory Disease, 1981-84. Recipient award for excellence in teaching U. Fla., 1974; research grantee NIH, 1979-81. Mem. Am. Thoracic Soc. (chmn. 1975-76), Fla. Thoracic Soc. (pres. 1977-78), Fla. Med. Assn., Alachua County Med. Soc. (treas. 1979-80), So. Sleep Soc. Democrat. Jewish. Subspecialty: Pulmonary medicine. Current work: Sleep and breathing oxygen therapy, preoperative pulmonary evalua. Home: 1520 NW 68th Terr Gainesville FL 32605 Office: U Florida J Hillis Miller Health Ctr Box J-225 Gainesville FL 32610

BLOCK, LAWRENCE HOWARD, pharmacist, educator; b. Balt., Nov. 14, 1941; s. Harry C. and Dina (Cooper) B.; m. Sharon Kelson, Aug. 4, 1974; children: Hal Zachary, Dana Elayne. B.S. in Pharmacy, U. Md., 1962, M.S., 1966, Ph.D., 1969. Asst. prof. pharmaceutics U. Pitts., 1968-70; asst. prof. pharm. chemistry, pharmaceutics Duquesne U., Pitts., 1970-71, assoc. prof. pharm. chemistry, pharmaceutics, 1971-75, prof. pharmaceutics, 1975—; cons. Thrift Drug Co. div. J.C. Penney Co., Biodecison Lab. Inc., Travenol Labs., Inc. Author numerous chpts. for sci. books and articles for profl. jours. Fellow Am. Found. Pharm. Edn.; mem. Acad. Pharm. Scis., Am. Pharm. Assn., N.Y. Acad. Sci., Soc. Cosmetic Chemists, Sigma Xi. Subspecialties: Pharmacokinetics. Current work: Population pharmacokinetics; individualization of drug therapy; pharmacokinetic-pharmacodynamic correlations; transdermal drug delivery; protein-drug interaction. Office: Duquesne U Sch Pharmacy 441 Mellon Hall Sci Pittsburgh PA 15282

BLOCK, ROBERT CHARLES, nuclear engineering educator; b. Newark, Feb. 11, 1929; s. George and Sue (Ehrenkranz) B.; m. Rita Adler, June 28, 1952; children: Keith, Robin. B.S. in Elec. Engring, Newark Coll. Engring., 1950; M.A. in Physics, Columbia U., 1953; Ph.D. in Nuclear Physics, Duke U., 1956. Elec. engr. Nat. Union Radio Corp., W. Orange, N.J., 1950-51, Bendix Aviation Co., Teterboro, N.J., 1951; physicist Oak Ridge Nat. Lab., 1955-66; vis. scientist Atomic Energy Research Establishment, Harwell, Eng., 1962-63; prof. nuclear engring. and sci. Rensselaer Poly. Inst., 1966—; Vis. scientist Am. Inst. Physics, 1961-67; vis. prof. Kyoto (Japan) U., 1973-74; cons. Gen. Electric Co., 1968-79; cons., mem. nuclear cross sect. adv. com. AEC, 1969-72; vis. physicist Brookhaven Nat. Lab., 1975, mem. vis. com. nuclear energy dept., 1982-85; founder, v.p., treas. Becker, Block, & Harris, Inc., 1981—; mem. U.S. Nuclear Data Com., 1972-74, NRC panel on low and medium energy neutrons, 1977; dir. Gaerttner Linac Lab. Co-author chpt. in book. Japanese Ministry Edn. research grantee, 1973-74. Fellow Am. Nuclear Soc.; Mem. AAAS, AAUP, Am. Phys. Soc., Sigma Xi, Sigma Pi Sigma, Phi Beta Tau, Tau Beta Pi. Subspecialties: Nuclear fission; Nuclear fusion. Current work: Experimental neutron physics; radiation technology for industrial application; nondestructive testing. Research on neutron physics. Home: 114 3d St Troy NY 12180 Office: Rensselaer Poly Inst Troy NY 12181

BLOCK, ROBERT MICHAEL, endodontist, educator, researcher; b. Ann Arbor, Mich., Oct. 15, 1947; s. Walter David and Thelma Violet (Levine) B.; m. Anne Powell Marshall, Sept. 4, 1977. B.A., DePauw U., 1969; D.D.S., U. Mich.-Ann Arbor, 1974; cert. in endodontics, Va. Commonwealth U., 1977; M.S. in Pathology, Va. Commonwealth U., 1978. Diplomate: Am. Bd. Endodontics. Clin. instr. Va. Commonwealth U., 1975-77, instr. pathology, 1977-78; research assoc. endocontics U. Conn.-Farmington, 1975—; vis. sr. scientist Nat. Med. Research Inst., Bethesda, Md., 1976-78; research assoc. McGuire Vets. Hosp., Richmond, Va., 1975-78; vis. research scientist U. Conn.-Farmington, 1978—; lectr. endodontics Flint Community Schs. Contbr. articles profl. jours., chpt. in book. Exec. mem. campaign com. candidate for U. Mich. Bd. Regents, 1980; candidate for Mich. State Bd. Edn., 1982. HEW and NIH summer research fellow, 1970-71; research grantee McGuire Vets. Hosp., 1976-78. Mem. Internat. Assn.

Dental Research (Edward P. Hatton award 1977), Am. Assn. Dental Research, Am. Assn. Endodontists (Meml. Research award 1977), Lapeer Dental Study Club (treas. 1978-82), ADA. Club: Bourben Barrell Hunt (Imlay City, Mich.). Subspecialty: Endodontics. Current work: Biologic responses to endodontic materials. Home: 1322 Wood Krest Dr Flint MI 48504 Office: G 3163 Flushing Rd Suite 212 Flint MI 48504

BLOEMBERGEN, NICOLAAS, physicist, educator; b. Dordrecht, Netherlands, Mar. 11, 1920; came to U.S., 1952, naturalized, 1958; s. Auke and Sophia M. (Quint) B.; m. Huberta D. Brink, June 26, 1950; children—Antonia, Brink, Juliana. B.A., Utrecht U., 1941, M.A., 1943; Ph.D., Leiden U., 1948; M.A. (hon.), Harvard, 1951. Teaching asst. Utrecht U., 1942-45; research fellow Leiden U., 1948; mem. Soc. Fellows Harvard, 1949-51, asso. prof., 1951-57, Gordon McKay prof. applied physics, 1957—, Rumford prof. physics, 1974, Gerhard Gade univ. prof., 1980; vis. prof. U. Paris, 1957, U. Calif., 1965; Lorentz guest prof. U. Leiden, 1973; Raman vis prof. Bangalore, India, 1979. Author: Nuclear Magnetic Relaxation, 1948, Nonlinear Optics, 1965, also articles in profl. jours. Recipient Buckley prize for solid state physics Am. Phys. Soc., 1958, Dirac medal U. New South Wales (Australia), 1983; Stuart Ballantine medal Franklin Inst., 1961; Half Moon trophy Netherlands Club N.Y., 1972; Nat. medal of Sci., 1975; Lorentz medal Royal Dutch Acad., 1978; Frederic Ives medal Optical Soc. Am., 1979; von Humboldt sr. scientist award, Munich, 1980; Nobel prize in Physics, 1981; Guggenheim fellow, 1957. Fellow Am. Phys. Soc., Am. Acad. Arts and Scis., IEEE (Morris Liebmann award 1959, Medal of Honor 1983), Optical Soc. Am.; hon. fellow Indian Acad. Scis.; mem. Nat., Royal Dutch acads. scis. Subspecialties: Spectroscopy; Atomic and molecular physics. Current work: Nonlinear optics and spectroscopy. Office: Pierce Hall Harvard Univ Cambridge MA 02138

BLOMEKE, JOHN OTIS, nuclear engr.; b. Austin, Tex., Jan. 16, 1923; s. John and Maud (Douglas) B.; m. Margaret Boruff, Aug. 25, 1946; children: Hugh, Linda. B.S. in Chem. Engring, U. Tex. - Austin, 1943, M.S., 1947; Ph.D. in Chem. Engring, Ga. Inst. Tech., Atlanta, 1950. Devel. engr. Clinton Labs., Oak Ridge, 1944-46; sr. research staff mem. Oak Ridge Nat. Lab., 1950—. Served with U.S. Army, 1945-46. Fellow Am. Nuclear Soc. Subspecialties: Nuclear fission; Nuclear engineering. Current work: Radioactive waste management research, development and applications. Home: Route 3 Sandy Shore Dr Lenoir City TN 37771 Office: Oak Ridge National Laboratory PO Box X Oak Ridge TN 37830

BLOOM, ARNOLD LAPIN, physicist; b. Chgo., Mar. 7, 1923; s. Louis Simon and Edith (Lapin) B.; m. Barbara Ann Bloom, Jan. 26, 1951; 1 dau: Debra Bloom Fisher. Ph.D., U. Calif. - Berkeley, 1951. Physicist Varian Assocs., Palo Alto, Calif., 1951-61; physicist/chief scientist Spectra-Physics, Inc., Mountain View, Calif., 1961-71; physicist Coherent Inc., Palo Alto, 1971—; lectr. Stanford U., 1969-70. Contbr. articles to profl. jours. Served with U.S. Army, 1943-46. Mem. Am. Phys. Soc., Optical Soc. Am., AAAS. Subspecialties: Thin films; Spectroscopy. Current work: Research in optics - thin film multilayer design; lasers resonator theory and spectroscopy. Patentee in field. Home: 1955 Oak Ave Menlo Park CA 94025 Office: 3210 Porter Dr Palo Alto CA 94304

BLOOM, EDA TERRI, immunologist; b. Los Angeles, May 11, 1945; d. Louis and Lucy (Malin) B. A.B., UCLA, 1966, Ph.D., 1970. Asst. research immunologist UCLA, 1971-1978; asst. prof. U. So. Calif., 1978-80; asst. research immunologist UCLA and VA Wadsworth Med. Center, 1980-81, assoc., 1981—. Mem. AAAS, Am. Assn. Cancer Research, Am. Assn. Immunologists, N.Y. Acad. Scis., Gerontol. Soc. Am., Transplantation Soc., Sigma Xi. Subspecialties: Immunobiology and immunology; Cancer research (medicine). Current work: Tumor and gerontological immunology. Office: VA Wadsworth Med Center Los Angeles CA 90073

BLOOM, FLOYD ELLIOTT, physician, research scientist; b. Mpls., Oct. 8, 1936; s. Jack Aaron and Frieda (Shochman) B.; m. D'Nell Bingham, Aug. 30, 1956 (dec. May 1973); children—Fl'Nell, Evan Russell; m. Jody Patricia Corey, Aug. 9, 1980. A.B. cum laude, So. Meth. U., 1956, M.D., Washington U., St. Louis, 1960; D.Sc. h.c., So. Meth. U., 1983. Intern Barnes Hosp., St. Louis, 1960-61, resident internal medicine, 1961-62; research asso. NIMH, Washington, 1962-64; fellow depts. pharmacology, psychiatry and anatomy Yale Sch. Medicine, 1964-66, asst. prof., 1966-67, asso. prof., 1968; chief lab. neuropharmacology NIMH, Washington, 1968-75, acting dir. div. spl. mental health, 1973-75; commd. officer USPHS, 1974-75; dir. Arthur Vining Davis Center for Behavorial Neurobiology; prof. Salk Inst., La Jolla, Calif., 1975—; mem. Commn. on Alcoholism, 1980-81, Nat. Adv. Mental Health Council, 1976-80. Author: (with J.R. Cooper and R.H. Roth) Biochemical Basis of Neuropharmacology, 1971; editor: Peptides: Integrators of Cell and Tissue Function, 1980; co-editor: Regulatory Peptides. Recipient A. Cressy Morrison award N.Y. Acad. Scis., 1971; A.E. Bennett award for basic research Soc. Biol. Psychiatry, 1971; Arthur A. Fleming award Science mag., 1973; Mathilde Solowey award, 1973; Biol. Sci. award Washington Acad. Scis., 1975; Alumni Achievement citation Washington U., 1980; McAlpin Research Achievement award Mental Health Assn., 1980; Lectr.'s medal College de France, 1979. Fellow Am. Coll. Neuropsychopharmacology (mem. council 1976-78); mem. Nat. Acad. Sci. (chmn. sect. neurobiology 1979-83), Inst. Medicine, Am. Acad. Arts and Scis., Soc. Neurosci. (sec. 1973-74, pres. 1976), Am. Soc. Pharmacology and Exptl. Therapeutics, Am. Soc. Cell Biology, Am. Physiol. Soc., Am. Assn. Anatomists, Research Soc. Alcoholism. Subspecialties: Neurobiology; Neuropharmacology. Current work: Neurotransmitters; identification, characterization, and possible medical and behavioral implications. Home: 1145 Pacific Beach Dr Apt B405 San Diego CA 92109 Office: Salk Inst La Jolla CA 92037

BLOOM, JAMES RICHARD, research plant pathologist, educator; b. Clearfield, Pa., Feb., 20, 1924; s. Raymond V. and Rozella G. (Dunlap) B.; m. June Farwell, Sept. 11, 1947; children: James Richard, Heidi L., Gretchen E., Coralie A. B.S. in Botany, Pa. State U., 1950; Ph.D. in Plant Pathology, U. Wis., 1953. Asst. prof. plant pathology Pa. State U., University Park, 1953-59, assoc. prof., 1960-66, prof., 1966—. Contbr. articles to profl. jours. Served to sgt. U.S. Army, 1943-46. Decorated Purple Heart. Mem. Soc. Nematologists, Am. Phytopathol. Soc. Subspecialties: Plant pathology; Nematology. Current work: Interaction of nematodes witn other plant pathogens. Home: Box 317 Lemont PA 19851 Office: 211 Buckhout Lab University Park PA 16802

BLOOM, PAUL RONALD, soil chemist researcher, educator; b. Mpls., Sept. 19, 1943; s. Ronald Elmer and Marjorie (Carlander) B.; m. Milegua Fernandez Layese, Oct. 11, 1975; children: Benjamin, Francisco. B.A., U. Mont., 1970, M.A.T., 1972; Ph.D., Cornell U., 1978. Asst. prof. soil sci. U. Minn.-St. Paul, 1978—. Precinct chmn. Democratic Farm Labor Party, St. Paul, 1982—. Mem. Soil Sci. Soc. Am., Clay Minerals Soc., AAAS, Internat. Humic Substances Soc., Fedn. Am. Scientists. Subspecialty: Soil chemistry. Current work: Metal ion binding in humic substances; kinetic controls of aluminum in acid soils; soil solution chemistry in calcareous soils; plant nutrient chemistry in flooded peats. Office: Dept Soil Sci U Minn Saint Paul MN 155108

BLOOM, RICHARD FREDRIC, behavior scientist, psychologist; b. Bklyn., Oct. 23, 1931; s. Morris and Mary (Schur) B.; m. Myra R. Thal, Dec. 16, 1956 (div.); children: Laura A.; David T.; m. Susan Davies, Nov. 30, 1981; stepchildren: David Y. Redford II, Hugh T. Redford. B.S., Bklyn Coll., 1953; A.E., Newark Coll. Engring., 1959, M.S., 1962; Ph.D., NYU, 1969. Lic. psychologist, Conn., Fla., Pa. Physicist U.S. Nat. Bur. Standards, Washington, 1953; electronics engr. ITT Fed. Labs., Nutley, N.J., 1955-65; behavioral scientist Dunlap & Assocs. East, Inc., Norwalk, Conn., 1965—; pvt. practice psychologist, New Canaan, Conn., 1970—; co-dir. Psychotherapy & Counseling Assocs., New Canaan, 1981—; research cons. Silver Hill Found. Hosp., New Canaan, 1972-75. Editor: Dunlap Monographs, 1967-70; author tech. reports and research papers. Bd. dirs. Am. Cancer Soc., Darien, Conn., 1981—. Served with U.S. Army, 1953-1955; Korea. Fellow Soc. Clin. and Exptl. Hypnosis; mem. Am. Acad. Forensic Scis., Am. Psychol. Assn., IEEE, Soc. Psychophysiol. Research. Jewish. Subspecialties: Behavioral psychology; Psychotherapy. Current work: Human-machine interactions; psychophysiological research; human behavior; market research; training and evaluation; psychotherapy services to individuals, couples and families; forensic hypnosis. Home: 45 Silvermine Rd New Canaan CT 06840 Office: Dunlap & Assocs East Inc 17 Washington St Norwalk CT 06854

BLOOM, SHERMAN, pathologist, educator; b. Bklyn., Jan. 26, 1934; s. Philip and Sadie (Kaplan) B.; m. Miriam Fishman, Feb. 12, 1960; children: Naomi, Stephanie. B.A., NYU, 1955, M.D., 1960. Cert. Am. Bd. Anat. Pathology. Intern in medicine Kings County Hosp., Bklyn., 1960-61; asst. prof. U. Utah Coll. Medicine, Salt Lake City, 1966-70, assoc. prof., 1970-72, U. South Fla. Coll. Medicine, Tampa, 1973-76, prof. pathology, 1976-77, George Washington U. Coll. Medicine, Washington, 1977—; cons. Sci. Rev., NIH; cardiovascular study sect. NSF, FDA; bd. dirs. Scientists Ctr. Animal Welfare. Contbr. numerous articles to profl. publs. Del. Utah State Democratic party, 1968. NIH fellow, 1962; Dilthey Found. fellow, 1982. Fellow Am. Coll. Nutrition; mem. Internat. Acad. Pathologists, Am. Physiol. Soc., Am. Assn. Pathologists, Internat. Soc. Heart Research. Jewish. Subspecialties: Pathology (medicine); Physiology (medicine). Current work: role of electrolytes in myocardial injury; structural basis for myocardial diastolic stiffness. Office: Dept Pathology Med Ctr George Washington U Washington DC 10037 Home: 7604 Hemlock St Bethesda MD 10817

BLOOMER, RICHARD RODIER, petroleum geologist; b. Norfolk, Va., June 26, 1918; s. Alfred T. and Marie A. (Hefferman) B.; m. Anne Louise Egdorf, June 7, 1948; children: Carol Leigh, Charles Richard. B.S., U. Va., 1940, M.A., 1941; Ph.D., U. Tex., 1949. Instr. geology U. Tex.-Austin, 1941-42, 46-48; research geologist Bur. Econ. Geology, U. Tex., 1948-49; exploration geologist Carter Oil Co., Tulsa, 1949-52; ind. geologist, Abilene, Tex., 1952—; prof. geology Hardin-Simmons U., Abilene, 1967-68; owner R.R. Bloomer & Assocs., Abilene, 1978—; Disting. lectr. Am. Assn. Petroleum Geologists, 1981-82; mem. adv. council U. Tex. Geology Found. Contbr. articles to profl. jours. Served to 1st lt. USAF, 1942-46; CBI. Fellow Geol. Soc. Am.; mem. Am. Assn. Petroleum Geologists (pres. S.W. sect. 1982-83, mem. found.), Soc. Econ. Paleontologists and Mineralogists (v.p. S.W. sect. 1981), Am. Inst. Profl. Geologists (cert. profl. geologist), Sigma Xi. Subspecialties: Sedimentology; Tectonics. Current work: Subsurface channel sandstones. Home: 1141 Elmwood Dr Abilene TX 79605 Office: 132 Devonian Bldg 310 N Willis St Abilene TX 79603

BLOOMFIELD, DANIEL KERMIT, college dean, physician; b. Cleve., Dec. 14, 1926; s. Joseph Bernard and Henrietta (Namen) B.; m. Frances Aub, June 10, 1955; children: Louis, Ruth, Anne. B.S., U.S. Naval Acad., 1947; M.S., Western Res. U., 1954, M.D., 1954. Intern Beth Israel Hosp., Boston, 1954-55, resident, 1955-56, Mass. Gen. Hosp., Boston, 1956-67; research fellow chemistry Harvard U., 1957-59; hon. asst. registrar cardiology Nat. Heart Hosp., London, 1959-60; sr. instr. medicine Western Res. U., Cleve., 1960-64, sr. clin. instr. medicine, 1964-70; dir. cardiovascular research Community Health Found., Cleve., 1964-66; asso. medicine Mt. Sinai Hosp., Cleve., 1966-69; prof. medicine U. Ill. Sch. Basic Med. Sci., Urbana, 1970—; dean Coll. Medicine, 1970-83; Investigator Am. Heart Assn., 1960-64; Bd. dirs. E. Central Ill. Health Systems Agy. Served with USN, 1947-50. Recipient citation for contbns. to med. edn. Ohio Heart Assn., 1964. Mem. Alpha Omega Alpha. Subspecialties: Cardiology; Preventive medicine. Current work: Electrocardiographic screening, ambulatory electrocardiographic monitoring. Home: 103 E Michigan St Urbana IL 61801

BLOOMFIELD, SAUL S., physician, clin. pharmacologist, educator; b. Montreal, June 30, 1925; s. Oscar H. and Tillie S. (Schoilovitch) B.; m. Ellen Steinberg, Jan. 9, 1949; children: Laurence, Patricia, Matthew. B.Sc., McGill U., Montreal, 1946, M.Sc., 1948; M.D., U. Geneva, Switzerland, 1953; M.Sc., U. Montreal, 1965. Practice medicine, Montreal, 1955-63; instr. pharmacology U. Montreal Faculty Medicine, 1963-65; asst. prof. medicine and pharmacology U. Cin. Coll. Medicine, 1965-71, asso. prof., 1971-77, prof., 1977—; vis. scientist Merrell Internat. Research Center, Strasbourg, France, 1972-73; vis. prof. Oxford U., Churchill Hosp., 1980. Contbr. articles to profl. jours. Can. Found. Advancement Therapeutics fellow, 1964-65; NIH fellow, 1966-68; grantee, 1968-72. Mem. Am. Soc. Clin. Pharmacology and Therapeutics, Am. Soc. Pharmacology and Exptl. Therapeutics, Am. Fedn. Clin. Research, AAUP, Can. Med. Assn., N.Y. Acad. Scis., Am. Pain Soc. Subspecialties: Internal medicine; Clinical pharmacology. Current work: Clinical pharmacology of analgesics; controlled clinical trials of investigational new analgesics in various acute pain models. Home: 57 Carpenter's Ridge Cincinnati OH 45241 Office: University Cincinnati Medical Center 5502 Medical Sciences Bldg Cincinnati OH 45267

BLOOMFIELD, VICTOR ALFRED, biochemistry educator; b. Newark, June 10, 1938; s. Samuel G. and Miriam B. (Finkelstein) B.; m. Clara Gail Derber, June 11, 1962. B.S. in Chemistry, U. Calif., Berkeley, 1959; Ph.D., U. Wis., 1962. NSF postdoctoral fellow in chemistry U. Calif., San Diego, 1962-64; asst. prof. chemistry U. Ill., Urbana, 1964-69; asso. prof. U. Minn., St. Paul, 1969-70, asso. prof. biochemistry and chemistry, 1970-74, prof. biochemistry and chemistry, 1974—, head dept. biochemistry, 1979—; mem. biophysics and biophys. study sect. NIH, 1975-79; mem. NSF Biophysics Panel, 1980—. Contbr. articles to profl. jours.; author: (with Crothers and Tinoco) Physical Chemistry of Nucleic Acids, 1974, (with R. Harrington) Biophysical Chemistry, 1975; mem. editorial bd.: Jour. Phys. Chemistry, 1974-78, Ann. Rev. of Phys. Chemistry, 1976-80, Biopolymers, 1977—, Biophys. Jour, 1978-81, Biochemistry, 1980—. Recipient Eastman Kodak award, 1962; A.P. Sloan fellow, 1969-71; NIH Research Career Devel. awardee, 1971-76. Fellow AAAS; mem. Am. Chem. Soc. (sec. biol. chemistry div. 1981-83), Am. Soc. Biol. Chemists, Biophys. Soc. (mem. council 1979—, publs. com. 1981-83). Subspecialty: Biophysical chemistry. Office: 1479 Gortner Ave Saint Paul MN 55108

BLOOR, W(ILLIAM) SPENCER, elec. engr., cons.; b. Trenton, N.J., Oct. 16, 1918; s. W. Harry and Evva (Averre) B.; m. Barbara P. Walters, Jan. 19, 1952; children—William G., Robert S. B.S. in Elec. Engring, Lafayette Coll., 1940, D.Eng. (hon.), 1981. Registered profl. engr., Calif., Pa. With Leeds & Northrup Co., 1940-81, product market devel. mgr., Phila., 1966-68, engring. coordination mgr., North Wales, Pa., 1968-69, mgr. steam and nuclear power systems, 1969-81; cons. in pvt. practice, 1981—; chmn. Engrs. Week Com., Delaware Valley, 1981. Served to lt. USN, 1943-46. Named Engr. of Yr., Delaware Valley, 1980. Fellow AAAS, IEEE (chmn. awards com. Phila. sect. 1980—, chmn. Power Industry Computer Application Conf. 1981), Instrument Soc. Am. (v.p. publs. 1968-70, pres. 1974, chmn. admissions com. 1980—); mem. ASME, Nat. Soc. Profl. Engrs., Nat. Acad. Engring., Phi Beta Kappa, Tau Beta Pi, Eta Kappa Nu. Presbyterian. Subspecialties: Electrical engineering; Systems engineering. Current work: Instrumentation and control systems especially for electric power generating stations. Design and application of control and monitoring systems for electric power generating stas. throughout the world. Home and Office: 1904 Jody Rd Meadowbrook PA 19046

BLOSSER, HENRY GABRIEL, physicist; b. Harrisonburg, Va., Mar. 16, 1928; s. Emanuel and Leona (Branum) B.; m. Priscilla May Beard, June 30, 1951 (div. Oct. 1972); children—William Henry, Stephan Emanuel, Gabe Fawley, Mary Margaret; m. Mary Margaret Gray, Mar. 16, 1973. B.S., U. Va., 1951, M.S., 1952; Ph.D., 1954. Physicist Oak Ridge Nat. Lab., 1954-56, group leader, 1956-68; asso. prof. physics Mich. State U., East Lansing, 1958-61, prof., 1961—; dir. Cyclotron Lab., 1961—; cons. U. Mich., 1960-61, Washington U., St. Louis, 1961-62, Lawrence Radiation Lab., 1962, U. Md., 1962-65, Princeton U., 1965—, others. Bd. dirs. Midwestern Univs. Research Assos., 1960-63. Served with USNR, 1946-48. Recipient Distinguished Faculty award Mich. State U., 1972; NSF postdoctoral fellow, 1966-67; Guggenheim fellow, 1973-74. Mem. Sigma Xi, Phi Beta Kappa, Kappa Alpha Order. Subspecialty: Nuclear physics. Current work: Design of advanced cyclotrons; adaptation of superconductivity to cyclotrons; design of detection devices for nuclear reaction studies. Home: 609 Beech St East Lansing MI 48823 Office: Cycletron Lab Michigan State U East Lansing MI 48824

BLOSSEY, ERICH CARL, chemistry educator; b. Toledo, June 10, 1935; s. Erich Frederick August and Marguerite Florence B.; m. Shirley Ann Stanford, Sept. 6, 1958 (div. Nov. 1978); m. Elizabeth Diane Frye, Aug. 11, 1979; children: Catherine, Christina, Elizabeth. B.S., Ohio State U., 1957; M.S., Iowa State U., 1959; Ph.D., Carnegie-Mellon U., 1963. Postdoctoral fellow Stanford U., 1962-63, Syntex S.A., Mexico City, 1963-64; teaching intern Wabash Coll., Crawfordsville, Ind., 1964-65; sr. postdoctoral fellow U. N.Mex., Albuquerque, 1972-73; prof., head dept. chemistry Rollins Coll., Winter Park, Fla., 1965—; cons. White Labs., Inc., Orlando, Fla., 1977—. Author: PSI Study Guide, 1977; co-author: Solutions Manual, 1977; editor: Solid Phase Synthesis, 1975. Arthur Vining Davis fellow, 1978; Research Corp/NIH grantee, 1972. Fellow Royal Inst. Chemistry; mem. Am. Chem. Soc. (chmn. 1970-71, 81-82), N.Y. Acad. Sci., AAAS (chmn. 1971-72), AAUP. Subspecialties: Organic chemistry; Polymer chemistry. Current work: Polymer-bound organic and biochemical reagents; study of molecules attached to polymers. Patentee in field. Home: 261 W Kings Way Winter Park FL 32789 Office: Rollins Coll Dept Chemistry Winter Park FL 32789

BLOTCKY, ALAN JAY, physicist; b. Omaha, July 5, 1930; s. Paul and Evalyn Sylvia (Meyer) B.; m. Wanda June Richmond, Jan. 20, 1953; children: Steven, Beth. B.S., Carnegie Inst. Tech., 1952; M.S., Creighton U., 1971. Sect. head mass spectrometer maintenance sect. Goodyear Atomic Corp., Portsmouth, Ohio, 1953-55; cons. physicist, mfrs. rep. A.J. Blotcky & Assocs., Omaha, Nebr., 1955-57; reactor supr., research physicist, radiation safety officer VA Med. Ctr., Omaha, 1957—. Contbr. articles to profl. jours. Chief, radiol. def. service Omaha Douglas County Civil Def., 1962—; pres. Operation Bridge, 1972, 77-79. Served with Signal Corps USAR, 1952-77. Recipient VA Chief Med. Dirs. Pub. Service award, 1970. Mem. Am. Nuclear Soc., Soc. Nuclear Medicine, Health Physics Soc., Am. Assn. Physicists in Medicine, Instrument Soc. Am., Sigma Xi. Lodge: Rotary. Subspecialties: Analytical chemistry; Nuclear physics. Current work: Use of nuclear reactor for development of elemental and molecular activation analysis techniques, correlation of values in biological tissue. Office: 4101 Woolworth Ave Omaha NE 68105

BLOUT, ELKAN ROGERS, biological chemistry educator, university dean; b. N.Y.C., July 2, 1919; s. Eugene and Lillian B.; m. Joan E. Dreyfus, Aug. 27, 1939; children: James E., Susan L., William L. A.B., Princeton U., 1939; Ph.D., Columbia U., 1942; A.M. (hon.), Harvard U., 1962, D.Sc., Loyola U., 1976. With Polaroid Corp., Cambridge, Mass., 1943-62, successively research chemist, assoc. dir. research, 1948-58, v.p., gen. mgr. research, 1958-62; research assoc. Harvard U., 1950-52, 56-60; lectr. on biophysics 1960-62, prof. biol. chemistry, 1962—, Edward S. Harkness prof. biol. chemistry, 1964—, head dept. biol. chemistry, 1965-69; dean for acad. affairs Sch. Public Health, 1978—; research assoc. Children's Hosp. Med. Center, Boston, 1950-52, cons. chemistry, 1952—; mem. conseil de surveillance Compagnie Financiére du Scribe, 1975—81; trustee Bay Biochem. Research, Inc., 1973—; mem. exec. com. div. chemistry and chem. tech. NRC, 1972-74, mem. assembly of math. and phys. scis., 1979—; mem. sci. adv. com. Center for Blood Research, Inc., 1972—; also mem. bd. dirs.; mem. research adv. com. Children's Hosp. Med. Center, 1976-80; mem. sci. adv. com. Mass. Gen. Hosp., 1968-71, Research Inst., Hosp. for Sick Children, Toronto, Ont., Can., 1976-79; mem. adv. council dept. biochem. scis. Princeton U., 1974—, mem. adv. council program in molecular biology, 1983—; mem. vis. com. dept. chemistry Carnegie-Mellon U., 1968-72; bd. visitors Faculty Health Scis., SUNY, Buffalo, 1968-70; mem. corp. Mus. Sci.; trustee Boston Biomed. Research Inst.; bd. govs. Weizmann Inst. Sci., Rehovot, Israel, 1978—. Mem. adv. bd.: Jour. Polymer Sci, 1956-62; editorial bd.: Biopolymers, 1963—, Am. Chem. Soc. Monograph Series, 1965-72, Internat. Jour. Peptide and Protein Research, 1978—; editorial adv. bd.: Macromolecules, 1967-70, Jour. Am. Chem. Soc, 1978—; Contbr. articles to profl. jours. Recipient Princeton Class of 1939 Achievement award, 1970; NRC fellow Harvard, 1942-43. Fellow AAAS, Am. Acad. Arts and Scis. (fin. com. 1977—), N.Y. Acad. Arts and Scis., Optical Soc. Am. (past pres. New Eng. sect.); mem. Nat. Acad. Scis., mem. Inst. Medicine (treas. 1980—, fin. com. 1976—, adv. com. USSR and Eastern Europe 1979—), USSR Acad. Scis. (fgn. mem.), Am. Chem. Soc. (nat. councillor 1958-61), The Chem. Soc., Am. Soc. Biol. Chemists (fin. com. 1973—), Biophys. Soc., Assembly of Math. and Phys. Scis. of NRC, Internat. Orgn. Chem. Scis. in Devel. (council 1981—), Am. Chem. Soc. (chmn. fin. com. 1982—), Fedn. Am. Socs. Exptl. Biology (investments adv. com.). Subspecialties: Biophysical chemistry; Nuclear magnetic resonance (chemistry). Current work: Investigation of peptides and proteins using NMR, Raman, CD, IR methods. Patentee in field. Home: 1010 Memorial Dr Cambridge MA 02138 Office: Harvard Med Sch Boston MA 02115 also Harvard Sch Public Health Boston MA 02115

BLUE, TODD IRWIN, technical educator; b. Carbondale, Colo., July 27, 1939; s. Lloyd G. and June (Kirk) B.; m. Judith Ann Olson, June 17, 1967. B.A., U. N.Mex., 1980, M.A., 1982. Numerous positions in constrn. industry, 1964—; instr. civil and survey tech. Albuquerque Tech. Vocat. Inst., 1968—, coordinator constrn. trades programs,

1969—; ednl. cons. U.S. Air Force, Zia Co. Served with USN, 1960-64. Mem. Am. Tech. Edn. Assn., Am. Vocat. Assn. Roman Catholic. Subspecialties: Civil engineering; Solar energy. Current work: Satelite imagery with photogrametric applications to surveying systems and data update, cartographic representations of hypsographic and hydrographic features; solar applications. Home: 5700 Aspen NE Albuquerque NM 87110 Office: 525 Buena Vista SE Albuquerque NM 87106

BLUESTEIN, THEODORE, aeronautical engineer; b. N.Y.C., July 7, 1934; s. Harry and Cecile (Cohen) B.; m. Helen Emma Rothweiler, May 25, 1957; children: Julie Anne, Keith Alan, Nancy Lynn, Mark Andrew. B.S., Pa. State U., 1956; M.S., U. So. Calif., 1958. Aerodynamicist Hughes Aircraft Co., Los Angeles, 1956-61; prin. engr. Raytheon Co., Bedford, Mass., 1961-69, 71-76; program mgr. Motorola Co., Scottsdale, Ariz., 1976-79; head Roland Missile systems engrng. Hughes Aircraft Co., Tucson, 1979-81, sr. project engr. system engrng. lab., 1981—. Pres. El Segundo (Calif.) Democratic Club, 1960; mem. Acton (Mass.) Dem. Town Com., 1963-68; chmn. Raytheon Co. United Way, 1975-76, Walden (Mass.) Br. Am. Cancer Soc., 1975-76. Assoc. fellow AIAA; mem. Pa. State Alumni Assn., Sigma Tau. Clubs: Met. Amateur Radio, Catalina Radio (Tucson); Tucson Repeater Assn. Subspecialties: Aeronautical engineering; Aerospace engineering and technology. Home: 4385 N Paseo Rancho Tucson AZ 85745 Office: Hughes Aircraft Co PO Box 11337 Tucson AZ 85734

BLUM, HOWARD STANLEY, computer and management science educator, consultant; b. Bklyn., June 13, 1942; s. George Joseph and Miriam (Milch) B.; m. Carolyn Stevenson, July 12, 1964 (div. 1975); children: Shayna. B.E.E., Pratt Inst., 1964; M.S., SUNY, 1966; Ph.D., Poly. Inst. N.Y., 1973. Electronic engr. U.S. Naval Applied Sci. Lab., Bklyn., 1964-65; analyst Computer Applications, Inc., N.Y.C., 1968-70; cons. Systems Cons., Inc., N.Y.C., 1970-71; v.p. Digital Simulation Systems, Inc., N.Y.C., 1971-75; mgr. software Olivetti Corp., N.Y.C., 1975; asst. prof. Rutgers Grad. Sch. Bus., Newark, 1975-81; assoc. prof. computer and mgmt. sci. Pace U. Grad. Sch. Bus., N.Y.C., 1981—; cons. U.S. Army, Huntsville, Ala., 1979-81; scientist Bell Labs., Holmdel, N.J., 1978; cons. Xybion Corp., Morristown, N.J., 1977, Olivetti Corp., N.Y.C., 1975-76. NSF fellow, 1966-69; NASA fellow, 1966; USN scholar, 1959-64. Mem. Assn. Computing machinery, Ops. Research Soc. Am., IEEE, Am. Math. Soc. Subspecialties: Operations research (mathematics); Distributed systems and networks. Current work: Teaching development and use of mathematical modeling and computer techniques applied to management and engineering systems. Home: 305 E 86th St Apt 7FW New York NY 10028 Office: Pace U Grad Sch Bus 41 Park Row New York NY 10038

BLUM, KENNETH, pharmacology educator; b. Bklyn., Aug. 8, 1939; m. Arlene Schlessel, June 8, 1963; children: Jeffrey H., Seth H. B.S. in Pharmacy, Columbia U., 1961; M.S. in Med. Sci, N.J. Coll. Medicine, 1965; Ph.D. in Pharmacology, N.Y. Med. Coll., 1968. Research pharmacologist U.S. Vitamin & Pharm. Corp., Yonkers, N.Y., 1964-65; postdoctoral research assoc. S.W. Found. Research and Edn., San Antonio, 1968-70, asst. found. scientist, 1970-71; asst. prof. pharmacology U. Tex. Health Scis. Ctr., San Antonio, 1971-75, assoc. prof., 1975-83, prof., 1983—, chief dept., 1975—; fellow U. Colo. Coll. Pharmacy, Boulder, 1977; mem. council Gordon Research Conf., 1979—; mem. sci. adv. com. Internat. Neurotoxicology Congress, 1979; organizer, participant numerous profl. confs.; ad hoc peer reviewer Nat. Council Alcoholism, 1979-80, VA, 1980-81; cons. Tenn. Bd. Edn., 1970-72, Mo. Bd. Edn., 1971-72, U.S. Office Edn., 1972-74, Nat. Drug Abuse Ctr., 1977-79, others; sci. adviser Nat. Alcohol Research Ctr., U. Calif. Sch. Medicine, Irvine, 1981—; mem. Tex. Commn. on Alcoholism, 1979—; v.p., dir. OPOC Corp., 1982—; owner Western Ventures Brokerage Corp., 1982—; lectr. drug abuse. Host: TV spl. Perspectives of Psychology, 1980-81; Editor-in-chief: Substance and Alcohol Actions/Misuses, 1979—; editorial bd.: Jour. Psychoactive Drugs, 1976—, Clin. Toxicology, 1976—, European Jour. Clin. Toxicology, 1982; mem. publ. com.: Alcoholism: Clin. and Exptl. Research, 1979—; author: (with others) Pharmacological and Toxicological Perspectives on Commonly Abused Drugs, 1978, (with A.H. Briggs) Introduction to Social Pharmacology, 1983, (with others) Folk Healing and Herbal Medicine, 1981, (with L. Manzo) Neurotoxicology, 1983, Handbook of Abusable Drugs, 1983; editor: Alcohol and Opiates: Neurochemical and Behavioral Mechanisms, 1977, (with A.H. Briggs) Social Meanings of Drugs: Principles of Social Pharmacology, 1982, (with L. Manzo) Clinical Toxicology, 1981, 82; contbr. chpts. to books, articles to sci. jours.; appearances on local TV shows. Bd. dirs., founder Bexar County Pharm. Speakers Bur., 1969-71; bd. dirs. Agudas Achim Synagogue, 1973-74, San Antonio Free Clinic, 1971-80; pres. San Antonio Free Clinic, 1973-75; v.p. Drug Inst. San Antonio, 1972-73; chmn. bd., pres. Nat. Found. Addictive Diseases, 1982-83; chmn. Hanging Bottle Alcohol Research Fund, San Antonio, 1982-83; bd. dirs. San Antonio Com. Dangerous Drugs, 1972-73; pres. Job Power Now, Inc., 1978-79. Bigelow research fellow Columbia U., 1961; Am. Found. Pharm. Edn. fellow, 1966-68; NSF fellow, summer 1967; recipient citation of honor Nat. Assn. Retail Druggists, 1970; Career Tchrs. award Nat. Inst. Drug Abuse, 1974-77; citation of honor Alamo Area Council Govts., 1979; cert. of Merit St. Mary's U., San Antonio, 1979; Speaking Service award Lions Club, 1979; Gordon Research award, 1979, 82. Fellow Am. Coll. Clin. Pharmacology; mem. Internat. Soc. Substance and Alcohol Abuse (acting pres., co-founder 1982), Tex. Med. Assn. (ad hoc mem. spl. com. drug and alcohol abuse 1978—), Bexar County Med. Soc. (ad hoc mem. disabled physicians com. 1976—), Italian Drug Abuse Soc., World Fedn. Clin. Toxicology, Internat. Biomed. Research Soc. on Alcoholism (charter), Brain Research Soc. Gt. Britain (hon.), AAAS, AAUP, Am. Soc. Pharmacology and Exptl. Therapeutics, Acad. Med. Educators and Substance Abuse (charter), San Antonio Holistic Health Assn., Neurosci. Soc., Research Soc. Am., Soc. Medicinal Chemistry, Com. of One Thousand, N.Y. Acad. Sci., Research Soc. on Alcoholism (nominating and publ. com. 1978—), Tex. Research Soc. on Alcoholism (charter), Bexar County Pharm. Soc. (hon.), Automobile Club of Italy (Rome). Subspecialty: Neuropharmacology. Home: 3707 Castle Crest San Antonio TX 78230 Office: U Tex Health Sci Center San Antonio TX 78284

BLUM, MANUEL, computer science educator; b. Caracas, Venezuela, Apr. 26; s. Bernardo and Ernestine (Horowitz) B.; m. Lenore, June 30, 1961; 1 son, Avrim. B.S. in Elec. Engring, MIT, 1961, M.S., 1964, Ph.D. in Math, 1964. Research asst. to research assoc. Research Lab. Electronics, MIT, 1960-65, asst. prof. dept. math., 1966-68; vis. asst. prof. to prof dept. elec. engring. and computer scis. U. Calif.-Berkeley, after 1968; chmn. for computer sci., 1977-80. Contbr. articles to profl. jours. Sloan Found. fellow, 1972-73; recipient Disting. Teaching award U. Calif.-Berkeley, 1977. Fellow AAAS, IEEE; mem. Assn. for Computing Machinery, Sigma Xi. Subspecialties: Theoretical computer science; Cryptography and data security. Current work: Theoretical computerscience; cryptography and security; number theory. Office: U Calif 583 Evans Hall Berkeley CA 94720 Home: 700 Euclid Ave Berkeley CA 94708

BLUM, PAUL SOLOMON, physiologist, educator; b. N.Y.C., Jan. 13, 1947; s. Leonard F. and Ethel (Siegel) B.; m. Ann S. Havers, Apr. 6, 1968; 1 dau., Lori. B.A., U. Vt., 1969; Ph.D., 1973. Research assoc. Columbia U., 1974-77; asst. prof. physiology Jefferson Med. Coll, 1977-82, assoc. prof., 1982—. NIH predoctoral fellow, 1971-73; NIMH postdoctoral fellow, 1973-74. Mem. Soc. Neurosci., Am. Physiol. Soc. Subspecialties: Physiology (medicine); Neurophysiology. Current work: Control of the sensory and autonomic function of spinal cord; processing of sensory information. Office: Dept Physiology Jefferson Med Coll Philadelphia PA 19107

BLUMBERG, HAROLD, pharmacologist, educator; b. Fairmont, W.Va., June 19, 1909; s. Michael Meyer and Anna (Goodman) B.; m. Dorothy Drasnin, Sept. 1, 1934; 1 son, Mark Allan. Student, Johns Hopkins U., 1926-27, 28-30, W. Va. U., 1927-28; Sc.D., Johns Hopkins U., 1933. Asst., instr. in pediatrics Johns Hopkins Med. Sch., 1933-36; assoc. biochemist USPHS, 1936-38; research biochemist Johns Hopkins U. Sch. Hygiene and Public Health, 1938-42; assoc. toxicologist U.S. Army Indsl. Hygiene Lab., Balt., 1942-44; sr. biologist Sterling-Winthrop Research Inst., Rensselaer, N.Y., 1944-47; assoc. dir. research Endo Labs., Inc., Garden City, N.Y., 1947-74; research prof. pharmacology N.Y. Med. Coll., Valhalla, 1974-81, emeritus, 1981—; cons. in field. Contbr. articles to profl. jours. Mem. AAAS, Am. Inst. Nutrition, Am. Soc. Pharmacology and Exptl. Therapeutics, Eastern Pain Assn., Internat. Narcotic Research Club, N.Y. Acad. Scis., Soc. Exptl. Biology and Medicine, Soc. Toxicology, Parenteral Drug Assn. (hon.). Subspecialties: Pharmacology; Neuropharmacology. Current work: Narcotic antogonists, addiction, analgesics, toxicology. Patentee in field. Home: 1731 Beacon St Apt 213 Brookline MA 02146

BLUMBERG, JEFFREY BERNARD, toxicologist, research center administrator; b. San Francisco, Dec. 31, 1945; s. Waldo Bernard and Helen (Sherman) B.; m. Dorothy Eleanor Frost, Sept. 4, 1971; children: Nathaniel Frank, Zachary Joseph. B.S. in Psychology, Wash. State U., 1969, 1969, Ph.D., Vanderbilt U., 1974. NIH fellow U. Calgary, Alta., Can., 1975; asst. prof. pharmacology Northeastern U., Boston, 1976-79, dir. toxicology program, 1978-80, assoc. prof. pharmacology and toxicology, 1979-81, head pharmacology dept., 1980-81; scientist I and asst. dir. U.S. Dept. Agr. Human Nutrition Research Ctr. on Aging., Boston, 1981—. Contbr. book reviews and sci. articles to profl. publs. Recipient Coordinated Industry Program award Pharm. Mfrs. Assn., 1979; Cert. of Merit award U.S. Dept. Agr., 1982. Mem. AAAS, Soc. Neuroscis., N.Y. Acad. Scis., Toxicology Forum, Am. Coll. Toxicology. Subspecialties: Toxicology (medicine); Nutrition (medicine). Current work: Animal and clinical drug-nutrient interaction research; research administration. Home: 117 Nonantum St Newton MA 02158 Office: 711 Washington St Boston MA 02111

BLUME, JOHN AUGUST, consulting engineer; b. Gonzales, Calif., Apr. 8, 1909; s. Charles August and Vashti (Rankin) B.; m. Ruth Clarissa Reed, Sept. 14, 1942. A.B., Stanford, 1932, C.E., 1934, Ph.D., 1966. Constrn. engr. San Francisco-Oakland Bay Bridge, 1935-36; individual practice civil and structural engring., San Francisco, 1945-57; pres. John A. Blume & Assos. (Engrs.), San Francisco, 1957-81, chmn., sr. cons., 1980—; sr. engr./scientist URS Corp., San Mateo, Calif., 1981—; Mem., past chmn. adv. council Sch. Engring., Stanford U.; chmn. adv. com. Earthquake Engring. Research Center, U. Calif. at Berkeley.; cons. prof. civil engring. Stanford U.; Past chmn. adv. council Sch. Engring. Stanford U.; Adv. com. Earthquake Engring. Research Center. Author: A Machine for Setting Structures and Ground into Forced Vibration, 1935, Structural Dynamics in Earthquake Resistant Design, 1958, A Reserve Energy Technique for the Design and Rating of Structures in the Inelastic Range, 1960, Dynamic Characteristics of Multistory Buildings, 1969; co-author: Design of Multistory Reinforced Concrete Buildings for Earthquake Motions, 1961, An Engineering Intensity Scale for Earthquakes and Other Ground Motion, 1970, The SAM Procedure for Site-Acceleration-Magnitude Relationships, 1977; Contbr. articles to profl. jours. John A. Blume Earthquake Engring. Center at Stanford U. named in his honor. Mem. Nat. Acad. Engring., Structural Engrs. Assn. Calif. (pres. 1949), Cons. Engrs. Assn. Calif. (pres. 1959), ASCE (hon.; pres. San Francisco sect. 1960, Moisseiff award 1953, 61, 69, Ernest E. Howard award 1962), Seismol. Soc. Am., N.Y. Acad. Scis. (hon. life), Soc. Am. Mil. Engrs., Internat. Assn. Earthquake Engring. (hon.), Earthquake Engring. Research Inst. (hon.; pres. 1977-81), Sigma Xi, Tau Beta Pi. Subspecialties: Civil engineering; Probability. Current work: Structural dynamics; earthquake engineering; earthquake damage: causes, risk of; prevention; evaluation; repair. Home: 85 El Cerrito Ave Hillsborough CA 94010 Office: 130 Jessie St San Francisco CA 94105

BLUMENREICH, MARTIN SIGVART, physician; b. Oslo, Norway, Dec. 1, 1949; came to U.S., 1975; s. Sane and Bluma (Nomberg) B.; m. Patricia Dulman, Dec. 23, 1978. M.D., U. Uruguay, 1975. Diplomate: Am. Bd. Internal Medicine (Med. oncology and hematology). Med. intern Jewish Hosp. and Med. Ctr. of Bklyn., 1975-76, resident in medicine, 1976-78; fellow in med. oncology Meml. Sloan-Kettering Cancer Ctr., N.Y.C., 1978-81; asst. prof. medicine U. Louisville, 1981—. Subspecialty: Chemotherapy. Current work: Clinical research. Office: J Graham Brown Cancer Center 529 S Jackson St Louisville KY 40292

BLUMENTHAL, GEORGE RAY, astrophysicist, educator; b. Milw., Oct. 20, 1945; s. Marcel and Lillian (Banks) B.; m. Deborah Kelly Weisberg, Aug. 14, 1977. B.S., U. Wis.-Milw., 1966; Ph.D., U. Calif.-San Diego, 1971. Sr. Scientist Am. Sci. & Engring.Inc., Cambridge, Mass., 1971-72; asst. prof. astronomy and astrophysics U.Calif.-Santa Cruz, 1972-77, assoc. prof., 1977-83, 1983—. Contbr. articles to profl. jours. NSF fellow, 1966-71. Mem. Am. Astron. Soc. Subspecialties: Theoretical astrophysics; Cosmology. Current work: Research on emission line gas in active galactic nuclei, galaxy formation, nonthermal emission processes, and high energy astrophysics. office: Lick Obs U Calif Santa Cruz CA 95064

BLUMENTHAL, JAMES ALAN, psychology educator; b. Flushing, N.Y., Jan. 23, 1947; s. Arthur B. and Ruth A. (Stanton) B. B.S., U. Pitts., 1969; Ph.D., U. Wash., 1975; postdoctoral fellow, Duke U., 1979. Lic. psychologist, N.C. Asst. prof., intern Duke U., Durham, N.C., 1972-73; psychologist Meml. Hosp., Burlington, N.C., 1975-77; fellow Duke U., 1977-79, assoc. in medicine, 1982—, asst. prof. psychology, 1979—; cons. Seattle Head Start, 1969-70, Seattle Crisis Clinic, 1969-71, Randolph County Community Mental Health Ctr., 1974-75, Duke U. Preventive Approach to Cardiology, 1977—. Editor: Assessment Strategies in Behavioral Medicine, 1982; contbr. articles to profl. jours. NIMH traineeship, 1969-70; recipient Dodson award Duke U., 1978; NHLBI investigator, 1980—. Mem. Am. Psychol. Assn., Am. Psychosomatic Soc., Am. Coll. Sports Medicine, Soc. Behavioral Medicine, AAAS. Subspecialties: Clinical psychology; Biofeedback. Current work: Psychological aspects of health and illness; psychophysiology; cardiac rehabilitation; exercise physiology; aging. Office: Duke U Med Ctr PO Box 3926 Durham NC 27710

BLUMENTHAL, KENNETH MICHAEL, biochemist, educator; b. Chgo., Aug. 24, 1945; s. Maurice and Helen (Silbinger) B.; m. Susan Ina Bernstein, June 18, 1967; children: Elizabeth Jill, Deborah Lynn. B.Sc., U. Wis.-Madison, 1967; Ph.D., U. Chgo., 1971. Postdoctoral fellow UCLA, 1971-74; asst. prof. U. Fla.-Gainesville, 1974-76, U. Cin., 1976-81, assoc. prof. biochemistry, 1981—; grant reviewer S.W. Ohio Heart Ass, 1980—. Contbr. articles in field to profl. jours. Recipient Research Career Devel. award NIH, 1977-82; Research Grants NIH, 1977-85, NSF, 1975-76, 80-82. Mem. Am. Chem. Soc., Am. Soc. Biol. Chemists, Sigma Xi. Subspecialties: Biochemistry (medicine); Neurochemistry. Current work: Interaction of polypeptide neurotoxins with their receptors, structure and function of polypeptide neurotoxins and protein cytolysins. Office: Dept Biol Chemistry U Cin Coll Medicine Cincinnati OH 45267

BLUTH, EDWARD IRA, radiologist, educator; b. N.Y.C., May 3, 1946; s. Abraham and Irene B.; m. Elissa J. Weinberg, Nov. 22, 1970; children: Rachel, Jonathan, Marjorie. A.B. cum laude, U.Pa., 1967; M.D., SUNY Downstate Med. Ctr., 1971. Cert. diagnostic radiology Hosp. U. Pa., 1975., diplomate: Am. Bd. Radiology. Research fellow in history medicine Oxford (Eng.) U. and Wellcome Inst. London, 1968; fellow, asst. to 1st dept. commr. health N.Y.C. Dept. Health, 1969; rotating intern Med. Coll. Va., 1971-72; resident in diagnostic radiology Hosp. U.Pa., 1972-75; chief radiol. services USAF Hosp., Tyndall AFB, Fla., 1975-77; staff radiologist Ochsner Clinic and Ochsner Found. Hosp., New Orleans, 1977—; asst. instr. U. Pa. Sch. Medicine, 1972-75; instr. Tulane U. Sch. Medicine, 1977-78, clin. asst. prof. radiology, 1978-82, clin. assoc. prof., 1982—. Contbr. chpt., numerous articles on radiology and ultrasonography to profl. publs.; author profl. papers. Served to maj. USAF, 1975-77. NIH research fellow. Mem. Am. Roentgen Ray Soc., Radiol. Soc. N.Am., Am. Inst.Ultrasound in Medicine, AMA, So. Med. Assn. (sec. Radiology Sect. 1982), New Orleans Ultrasound Assn. (chmn. membership 1979, pres. 1979-80, v.p. 1983). Subspecialties: Diagnostic radiology; Imaging technology. Current work: Ultrasound of gastrointestinal tract, lower extremities. Office: Ochsner Clinic 1514 Jefferson Hwy New Orleans LA 70121

BLYSTONE, ROBERT VERNON, cell biologist, educator; b. El Paso, Tex., July 4, 1943; s. Edward Vernon and Cecilia (Mueller) B.; m. Donna Moore, Mar. 26, 1964; 1 son, Daniel Vernon. B.S., U. Tex., El Paso, 1965, M.A., 1968, Ph.D., 1971. Instr. U. Tex., El Paso, 1965, teaching asst., Austin, 1965-68, predoctoral fellow, 1968-71; asst. prof. biology Trinity U., San Antonio, 1971-76, assoc. prof., 1976—, dir. electron microscopy labs., 1976—; cons. in field. Contbr. articles to profl. jours. Active Alamo Sci. Fair. NSF fellow, 1976; NSF instrument grantee, 1978. Mem. Am. Inst. Biol. Scis., AAAS, Tex. Soc. Electron Microscopy, Electron Microscopy Soc. Am., Tex. Acad. Sci., Assn. Research in Vision and Ophthalmology, Am. Soc. Zoologists, Am. Microscopy Soc., AAUP. Subspecialties: Cell biology; Developmental biology. Current work: The mechanisms leading to the completion of late lung development; the influence of the adrenal pituitary axis on that development. Home: 2635 Worldland St San Antonio TX 78217 Office: Dept Biology Trinity U San Antonio TX 78284

BLYTHE, WILLIAM BREVARD, physician educator, nephrologist; b. Huntersville, N.C., Sept. 23, 1928; s. William LeGette and Esther Emily (Farmer) B.; m. Gloria Eleanor Nassif, Feb. 4, 1956; children: William LeGette, Anne Dewar, David Samuel Brevard, John Alexander. A.B., U. N.C., 1948, Cert. in Medicine, 1951; M.D., Washington U., St. Louis, 1953. Intern N.C. Meml. Hosp., Chapel Hill, 1953-54, resident, 1954-55; asst. prof. medicine U. N.C. Sch. Medicine, Chapel Hill, 1962-65, assoc. prof. medicine, 1965-70; prof. medicine, 1970—, head div. nephrology, 1973—, assoc. dir., 1965-66, dir., 1966-76, editorial bd., 1982—; bd. govs. U. N.C. Press, 1979—; sci. adv. bd. Nat. Kidney Found., 1975-81; cons. NIH, 1968-72. Editor: jour. The Kidney, 1977. Served to capt. M.C. U.S. Army, 1955-57. Fellow ACP; mem. Am. Physiol. Soc., Am. Clin. and Climatalogic Soc., So. Soc. Clin. Investigation, Alpha Omega Alpha. Democrat. Presbyterian. Club: Chapel Hill Tennis. Subspecialties: Internal medicine; Nephrology. Current work: Renal disease, renal physiology. Home: Hillcrest Circle Chapel Hill NC 27514 Office: Dept Medicine Univ NC Sch Medicine Chapel Hill NC 28514

BOAL, JAN LIST, mathematics educator; b. Canton, Ohio, Oct. 20, 1930; s. Clarence Jan and May Ann (List) B.; m. Bobby Snow, Mar. 21, 1953; children: Robert Kelly, Emily Ann Boal Wert, Virginia Baal Jamieson. B.Mech. Engring., Ga. Inst. Tech., 1954, M.S., 1954; Ph.D., MIT, 1959. Part time instr. math. MIT, Cambridge, 1954-59, instr. in math., 1959-60; asst. prof. math. U. S.C., Columbia, 1960-62, assoc. prof. math., 1962-69; prof. math. Ga. State U., Atlanta, 1969—, dept. chmn., 1969-77; mem., chmn. various dept. coms. and univ. coms., speakers bur., 1978—; NSF-AID cons. summer instr. program, India, 1967-68; vis. com. So. Assn. Colls. and Univs., 1972-74, 81; interviewer, reader Danforth Found., St. Louis, 1973-79. Contbr. articles to profl. publs., papers to profl. confs. Mem. Math. Assn. Am., Nat. Council Tchrs. of Math., Ga. Council Tchrs. of Math., Soc. Indsl. and Applied Math., Am. Sci. Affiliation, Blue Key, Sigma Xi, Phi Kappa Phi (Brierian Cup 1954), Tau Beta Pi, Omicron Delta Kappa, Phi Eta Sigma, Pi Tau Sigma. Baptist. Subspecialties: Applied mathematics; Numerical analysis. Current work: Elementary number theory, problems in teaching and exposition of mathematics. Office: Georgia State Univ University Plaza Atlanta GA 30303

BOATRIGHT, TONY JAMES, space shuttle orbiter/cargo systems engineer, space science/physics researcher; b. Chgo., Sept. 13, 1954; s. James Benjamin and Ethel (Warren) B. B.S. in Space Sci, Fla. Inst. Tech., 1979, M.S. in Space Tech, 1983. Sr. engr. Planning Research Corp., Kennedy Space Center, Fla., 1979-82; elec. systems engr. shuttle launch ops. div. Rockwell Internat., Kennedy Space Center, 1982—; mem. adj. faculty dept. physics and space sci. Fla. Inst. Tech., 1980—; dir. space research Centric Found., Inc., Merritt Island, Fla., 1977; test subject for space stress research NASA/Kennedy Space Center Biomed. Office, 1981—; designer/prin. investigator Space Shuttle Getaway Spl. Expt.: Holography of Organic and Inorganic Processes in Zero-Gravity. Mem. Republican presdl. Task Force, Washington, 1982. Recipient Ga. Soc. Profl. Engrs. award, 1972, award for aerospace research NASA, 1972, First Shuttle Flight Achievement award, 1981, Spacelab I Turnover medal NASA/European Space Agy., 1982, Superior Performance award Rockwell Internat., 1983. Mem. AIAA (editor and mem. council Canaveral sect. 1981—), Astron. Soc. of Pacific, Fla. Acad. Sci., Planetary Soc. Subspecialties: Astronautics; Optical astronomy. Current work: Space shuttle scientific payload integration; research and design in permanent manned space facilities; eventual objective: space shuttle mission or payload specialist. Designer shuttle-supported active satellite/mass capture system. Office: Rockwell Internat Shuttle Launch Ops Div ZL-89 Kennedy Space Center FL 32815

BOAZ, NOEL THOMAS, paleoanthropologist, anthropology educator; b. Martinsville, Va., Feb. 8, 1952; s. Thalma Noel and Elena More Anson (Taylor) B.; m. Dorothy Dechant, June 17, 1978. B.A., U. Va., 1973; M.A., U. Calif.-Berkeley, 1974, Ph.D. in Phys. Anthropology, 1977. Sr. mus. scientist (paleoanthropology) U. Calif.-Berkeley, 1975-77; lectr. anthropology UCLA, 1977-78; adj. asst. prof. anatomy N.Y. U. Sch. Medicine, 1978-81, asst. prof. anthropology, 1978—; dir. Internat. Sahabi Research Project, 1976—, Western Rift Research Expdn., 1982—. Contbr. articles to profl. jours. NSF grantee, 1976, 1980-82; Nat. Geog. Soc. grantee, 1978; Presdl. fellow, 1981; Wenner-Gren Found. grantee, 1982. Mem. Am. Assn Phys. Anthropologists, Am. Anthrop. Assn., AAAS, Royal Anthrop. Inst., Soc. Vertebrate Paleontology, Explorers Club. Democrat.

Episcopalian. Subspecialties: Evolutionary biology; Paleontology. Current work: Paleoanthropology, hominid evolution, taphonomy, primate paleoecology, biostratigraphy, vertebrate paleontology. Office: 25 Waverly Pl Paleoanthropology Lab 901 New York NY 10003

BOCKHOP, CLARENCE WILLIAM, agrl. engr.; b. Paullina, Iowa, Mar. 28, 1921; s. Fred Henry and Sophie Dorothea (Laue) B.; m. Virginia Buhman, July 9, 1949; children—Barbara Lucille, Nancy Jeanne, Bryan William, Karl David. B.S. in Agrl. Engring, Iowa State U., 1943, M.S., 1955, Ph.D. in Agr. Engring. and Theoretical and Applied Mechanics, 1957. Service and edn. mgr. Stewart Co., Dallas, 1948-53; mem. faculty Iowa State U., Ames, 1953-57, 60-80, prof. agrl. engring., 1960-80, head dept., 1962-80; head dept. agrl. engring. Internat. Rice Research Inst., Los Banos, Philippines, 1980—; prof., head dept. agrl. engineering U. Tenn., 1957-60; vis. prof. U. Ghana, 1969-70. Gen. reporter, VIth Internat. Congress Agrl. Engring., Lausanne, Switzerland, 1964; Author articles in field. Served to capt. AUS, 1943-48. Fellow Am. Soc. Agrl. Engrs. (chmn. Tenn. sect. 1958-59, chmn. mid-central sect. 1960-61, chmn. Iowa sect. 1963-64, chmn. edn. and research div. 1966-67, dir. 1973-75); mem. Am. Soc. Engring. Edn. (chmn. agrl. engring. div. 1966-67), Sigma Xi, Gamma Sigma Delta, Phi Kappa Phi, Phi Mu Alpha, Tau Beta Pi. Lutheran. Subspecialty: Agricultural engineering. Office: Internat Rice Research Inst Box 933 Manila Philippines

BOCKNEK, GENE, psychologist, educator; b. N.Y.C., Jan. 21, 1930; s. Philip Samuel and Gladys (Grossman) B.; m. Judith Rice Lehtinen, May 16, 1976; children by previous marriage: Kari, Marilyn, Robert. B.S.S., CUNY, 1950; Ph.D., Boston U., 1959. Lic. psychologist, Mass. Pvt. practice psychology, Boston, 1957—; faculty Boston U., 1962—, now prof. psychology, dir. tng. counseling psychology program, 1978—; cons. Hollingsworth & Vose Co., East Walpole, Mass., 1958—. Author: The Young Adult, 1980; contbr. articles to profl. jours. Co-founder Project Discover, Inc., Framingham, Mass., 1970-72, chmn. bd., 1970-71; co-founder, dir. Gender Identity Service, Inc., Boston, 1972-82. Recipient Cert. of Recognition Project Discovery, 1974; Cert. of Appreciation Cambridgeport Problem Ctr., 1973-76. Fellow Mass. Psychol. Assn.; mem. Am. Psychol. Assn., AAUP, Internat. Soc. Behavioral Devel. Subspecialties: Developmental psychology; Psychotherapy and intervention theory. Current work: Adult Developmental life span psychology; intervention theory; an ego/ developmental approach. Office: Boston U 605 Commonwealth Ave Boston MA 02215

BOCTOR, AMAL MORGAN, biochemical pharmacologist; b. Egypt, Mar. 27, 1935; came to U.S., 1962, naturalized, 1975; s. Morgan and Matheilda (Makar) B.; m. Violette H. Roumie, Feb. 3, 1947; children: Nancy and Carolyn (twins). B.Sc., U. Cairo, Egypt, 1955, M.S., 1961; Ph.D., M.I.T., 1967. Asst. researcher Nat. Research Council, Cairo, 1958-62; assoc. research scientist N.Y.U. Med. Ctr., 1967-71, research asst. prof. dept. pharmacology, 1971-82; sr. scientist Warner-Lambert Parke-Davis Pharm. Research div., Ann Arbor, Mich., 1982—. Contbr. articles to profl. publs. Mem. Am. Soc. Pharmacology and Exptl. Therapeutics, N.Y. Acad. Scis., Internat. Union Pharmacology, Sigma Xi. Coptic Orthodox. Subspecialties: Biochemistry; Pharmacology. Current work: Arachidonic acid metabolism; hormonal regulation of metabolism, enzyme induction and multiple forms, pharmacokinetics, drug receptors, modulation of arachidonic acid metabolism. Home: 725 Peninsula Ct Ann Arbor MI 48105

BODENHEIMER, PETER HERMAN, astrophysics educator; b. Seattle, June 29, 1937; s. Edgar and Brigitte Marianne (Levy) B.; m. Dori Bodenheimer, Sept. 11, 1965; children: Daniel, Debora. A.B., Harvard Coll., 1959; Ph.D., U. Calif.-Berkeley, 1965. Research assoc. Princeton U., 1965-67; asst prof., asst. astronomer U. Calif., Santa Cruz, 1967-70, assoc. prof., assoc. astronomer, 1970-76, prof., astronomer, 1976—; vis. scientist Max Planck Inst. for Astrophysics, Munich, W. Ger., 1976, 78, 81, 82. Contbr. articles to profl. jours. NSF fellow, 1965; NSF grantee, 1970—; NASA grantee, 1970—. Mem. Am. Astron. Soc., Internat. Astron. Union, Royal Astron. Soc. Subspecialty: Theoretical astrophysics. Current work: Numerical hydrodynamical calculations; stellar evolution; star formation; formation and evolution of the solar system and the giant planets; supernovae and supernovae remnants. Office: Lick Observatory University of California Santa Cruz CA 95064

BODENSTEIN, DIETRICH HANS FRANZ ALEXANDER, biology educator; b. Corwingen, East Prussia, Ger., Feb. 1, 1908; came to U.S., 1934, naturalized, 1940; s. Hans and Charlotte (Lilienthal) B.; m. Jean Coon, July 22, 1947; 1 dau. by previous marriage, Evelina (Mrs. William C. Suhler). Student, U. Konlgsberg, Ger., 1926-28, U. Berlin, 1928-33; Ph.D., U. Freiburg, Ger., 1953. Research asst. Kaiser Wilhelm Inst. Biology, Berlin, 1928-33; research assoc. German-Italian Inst. Marine Biology, Rovigno d' Istria, Italy, 1933-34, Stanford (Calif.) U. Sch. Biology, 1934-41; Guggenheim fellow dept. zoology Columbia U., N.Y.C., 1941-43; asst. entomologist Conn. Agrl. Exptl. Sta., New Haven, 1944; insect physiologist, med. div. Army. Chem. Med. Ctr., Md., 1945-57; embryologist gerontology br. Nat. Heart Inst., Balt. City Hosps., 1958-60; Lewis and Clark prof. biology U. Va., Charlottesville, 1960-78, prof. emeritus, 1978—, chmn. dept., 1960-73. Contbr. articles to profl. jours. Recipient sr. U.S. Scientist award von Humboldt Found., 1977. Mem. Am. Acad. Arts and Scis., Am. Soc. Zoologists, Am. Soc. Naturalists, Nat. Acad. Sci., Soc. Biology Brazil (hon.), Sigma Xi. Subspecialties: Developmental biology; Cell biology. Home: 536 Valley Rd Charlottesville VA 22903

BODIAN, DAVID, educator; b. St. Louis, May 15, 1910; s. Harry and Tillie (Franzel) B.; m. Elinor Widmont, June 26, 1944; children—Helen, Marion, Brenda, Alexander, Marc. B.S., U. Chgo., 1931, Ph.D., 1934, M.D., 1937. Asst. in anatomy U. Chgo., 1935-38; NRC fellow medicine U. Mich., 1938; anatomy Johns Hopkins, 1939- 40; asst. prof. anatomy Western Res. U., 1940-41; research on problems poliomyelitis, faculty dept. epidemiology Johns Hopkins, 1942-57, asso. prof. epidemiology, 1946-57, prof. anatomy, dir. dept., 1957-75, prof. neurobiology dept. otolaryngology, 1975—; tech. com. poliomyelitis vaccine USPHS, 1957-64; vaccine adv. com. Nat. Found., 1956-60; cons. NIH, mem. bd. sci. counselors, div. biol. standards, 1957-59; mem. bd. sci. advisers Nat. Inst. of Neurol. Diseases, 1968—. Author: Neural Mechanisms in Poliomyelitis, 1942; Mng. editor: Am. Jour. Hygiene, 1948-57; mem. editorial bds.: Jour. Comparative Neurology; Contbr. science articles profl. jours. Served as lt. USNR, World War II. Recipient E. Mead Johnson award in pediatrics Am. Acad. Pediatrics, 1941. Mem. Am. Assn. Anatomists (pres. 1971-72), Am. Acad. Arts and Scis., Nat. Acad. Scis., Am. Philos. Soc., A.A.A.S., Am. Physiol. Soc., Neurosci. Soc., Assn. Research Nervous and Mental Diseases, Phi Beta Kappa, Sigma Xi. Subspecialties: Neurobiology; Cytology and histology. Current work: At present making an electron misroscopic analysis of the auditory organ of Cort. Researcher on structure and diseases of nervous tissue. Home: 3917 Cloverhill Rd Baltimore MD 21218 Office: 1721 E Madison St Baltimore MD 21205

BODILY, DAVID MARTIN, chemist, educator; b. Logan, Utah, Dec. 16, 1933; s. Levi Delbert and Norma (Christenson) B.; m. Beth Alene Judy, Aug. 28, 1958; children—Robert David, Rebecca Marie, Timothy Andrew, Christopher Mark. Student, Utah State U., 1952-54; B.A., Brigham Young U., 1949, M.A., 1960; Ph.D., Cornell U., 1964. Postdoctoral fellow Northwestern U., Evanston, Ill., 1964-65; asst. prof. chemistry U. Ariz., Tucson, 1965-67; asst. prof. fuels engring. U. Utah, Salt Lake City, 1967-70, asso. prof., 1970-77, prof., 1977—, chmn. dept. mining and fuels engring., 1976—. Contbr. articles to profl. jours. Mem. Am. Chem. Soc. (chmn. Salt Lake sect. 1975), Catalysis Soc. N. Am., Am. Inst. Mining and Metall. Engrs., Sigma Xi. Mormon. Subspecialties: Coal; Fuels. Current work: Structure and chemistry of coal synthetic fuels from coal, oil shale and tar sands. Home: 2651 Cecil St Salt Lake City UT 84117 Office: 320 WBB U Utah Salt Lake City UT 84112

BODIN, ARTHUR MICHAEL, clinical psychologist; b. N.Y.C., July 11, 1932; s. Harry S. and Rose B.; m. Miriam Irene Lifton, June 25, 1961; children: Douglas Adam, Laura June. B.A., Swarthmore Coll., 1954; M.A., NYU, 1957; Ph.D., SUNY-Buffalo, 1966. Lic. psychologist, Calif. Tng. and edn. dir. Mental Research Inst., 1967-69; co-dir. communications and human systems program, 1969-70, research assoc., 1965—; sr. clin. psychologist Emergency Treatment Ctr., 1978—; assoc. clin. prof. dept. psychiatry U. Calif.-San Francisco, 1979—; pvt. practice clin. psychology. Contbr. articles to profl. jours. Recipient Outstanding Contbn. to Psychology award Santa Clara County Psychol. Assn., 1977. Mem. Am. Psychol. Assn. (council reps.), Western Psychol. Assn., Calif. Psychol. Assn. (pres. 1976-77, Disting. Service award 1973), Assn. Family Therapists No. Calif., Assn. Advancement Psychology, Am. Family Therapy Assn. (pres. 1970-72, Achievement award 1981), Am. Psychology-Law Soc. (pres. Santa Clara County). Subspecialty: Clinical psychology. Current work: Director Center for Mediation, Mental Research Institute; workshops on divorce mediation to mental health and law professionals. Office: 555 Middlefield Rd Palo Alto CA 94301

BODINGTON, SVEN HENRY MARRIOT, electrical engineer; b. Vancouver, C., Can., May 22, 1912; s. Spencer Marriot and Martha Lisa (Christoffersen) D.; m. Kathleen Dworak, Nov. 16, 1940; children: Thomas, Susan, Peter. A.B., Stanford U., 1934. Chief elec. engr. Scophony Ltd., London, 1935-41; dept. head ITT Corp., 1941-58, v.p., Nutley, N.J., 1958-59, avionics cons. world hdqrs., N.Y.C., 1969—. Co-author: Avionics Navigation Systems, 1969, Electrical Engineers, 1975. Mem. Mountain Lakes (N.J.) Citizens Adv. Com., 1978-79. Recipient Volare award, 1967; N.J. Outstanding Patent award, 1967. Fellow IEEE (pres. aerospace group 1962-63, 66-67, chmn. fellow com. 1978-79, Pioneer award 1980); mem. Inst. Navigation. Republican. Episcopalian. Club: Mountain Lakes (gov. 1966-70). Subspecialty: Aerospace engineering and technology. Home: 1 Briarcliff Rd Mountain Lake NJ 07046 Office: 320 Oark Ave New York NY 10022

BODLEY, JAMES WILLIAM, biochemistry educator; b. Portland, Oreg., Oct. 7, 1937; s. George William and LaVerna Marie (Linnemann) B.; m. Margaret Joylee Zickuhr, June 6, 1960; children: Camille, Andrea, Cade. B.S., Walla Walla Coll., 1960; Ph.D., U. Hawaii, 1964; postgrad., U. Wash.-Seattle, 1964-66. Acting asst. prof. U. Wash.-Seattle, 1966; asst. prof. U. Minn., Mpls., 1967-71, assoc. prof., 1971-77, prof. biochemistry, 1977—; mem. merit review bd. VA, Washington, 1980-83. Am. Cancer Soc. postdoctoral fellow, 1965; sr. fellow, 1973; NIH research grantee, 1974-87. Mem. Am. Soc. Biol. Chemists, Am. Chem. Soc. Subspecialty: Biochemistry (medicine). Current work: The mechanism of translation of genetic information. Home: 1293 Keston St St Paul MN 55108 Office: U Minn 435 Delaware St SE Minneapolis MN 55455

BODNAR, RICHARD JULIUS, physiological psychologist, educator, researcher; b. N.Y.C., Feb. 21, 1946; s. Julius and Irene (Monette) B.; m. Carol B. Greenman, July 4, 1981; 1 son, Benjamin P.G. B.A., Manhattan Coll., 1967; M.A., CCNY, 1973; Ph.D. in Psychology, CUNY, 1976. Postdoctoral fellow Columbia U. Coll. Physicians and Surgeons, N.Y.C., 1976-78; research scientist N.Y. State Psychiat. Inst., N.Y.C., 1978-79; asst. prof. Queens Coll., CUNY, 1979-82, asso. prof., 1983—. Contbr. articles to sci. jours. Served to capt. USAF, 1967-71. Decorated Bronze Star.; Sigma Xi nat. lectr., 1983-85; NIH grantee, 1978-82. Mem. AAAS, Soc. Neurosci., Eastern Psychol. Assn., Internat. Assn. Study of Pain. Subspecialties: Neuropsychology; Physiological psychology. Current work: Mechanisms in central nervous system and endocrine system that inhibit pain. Office: Queens Coll Flushing NY 11367

BODONYI, RICHARD JAMES, mathematics educator; b. Cleve., Nov. 26, 1943; s. Peter and Rose (Baczay) B.; m. Josette Hovance, Jan. 15, 1966; children: Jami, Lisa, Rebecca. B.S. in Math, Ohio State U., Columbus, 1966, Ph.D., 1973; M.S. in Engring, Case Western Res. U., Cleve., 1970. Aerospace engr. NASA Lewis Research Center, Cleve., 1967-68; asst. prof. aerospace engring. Va. Poly Inst. and State U., Blacksburg, Va., 1974-76; prof. math. Ind. U.-Purdue U. at Indpls., Indpls., 1976—. Mem. Am. Phys. Soc., Soc. Indsl. and Applied Math., Am. Math. Soc., Phi Beta Kappa, Sigma Xi. Democrat. Roman Catholic. Subspecialties: Applied mathematics; Aeronautical engineering. Current work: Viscous flow theory, hydrodynamic stability, aerodynamics, rotating flows, applied mathematics. Office: Ind-Purdue U at Indpls 1125 E 38th St Box 647 Indianapolis IN 64223 Home: 308 Thornberry Dr Carmel IN 46032

BODOR, NICHOLAS STEPHEN, educator; b. Satu Mare, Transylvania, Romania, Feb. 1, 1939; came to U.S., 1968, naturalized, 1976; s. Miklos Sandor and Berta (Horvath) B.; m. Sheryl Lee Reinmann, Feb. 26, 1971; children: Nicole, Erik; 1 child by previous marriage: Miklos. B.S.-M.S. in Organic Chemistry, Bolyai U., Romania, 1959; Dr. Chemistry, Babes-Bolyai U. and Cluj, Supreme Council of Romanian Acad. Sci., 1965. Prin. investigator, group leader Chem. Pharm. Research Inst., Cluj, Romania, 1961-68, 69-70; R.A. Welch postdoctoral fellow U. Tex., Austin, 1968-69, 70-72; sr. research scientist Alza Co., Lawrence, Kan., 1972-73; dir. medicinal chemistry Interx Research Co., Lawrence, 1973-78; adj. prof. U. Kan., Lawrence, 1974-78; prof., chmn. dept. medicinal chemistry U. Fla., Gainesville, 1979—; cons. Key Pharms., Inc., Miami, 1980—, Schering-Plough, Bloomfield, N.J., 1981—, Warner-Lambert, Ann Arbor, Mich., 1983—; v.p., dir. research Pharmatec, Inc., Arlington Heights, Ill., 1983—. Contbr. articles to profl. jours. Mem. Am. Chem. Soc., Am. Pharm. Assn., Acad. Pharm. Sci., AAAS, N.Y. Acad. Sci. Subspecialties: Medicinal chemistry; Organic chemistry. Current work: Design of soft (low toxicity) drugs; brain specific drug delivery; computer assisted drug design; neuropharmacology; prodrugs; MO calculations; organic reactions. Inventor in field. Home: 7211 SW 97th Ln Gainesville FL 32608 Office: J H M Health Center Univ Fla Dept Medicinal Chemistry Box J4 Gainesville FL 32610

BOEHM, FELIX HANS, educator, physicist; b. Basel, Switzerland, June 9, 1924; came to U.S., 1952, naturalized, 1964; s. Hans G. and Marquerite (Philippi) B.; m. Ruth Sommerhalder, Nov. 26, 1956; children: Marcus F., Claude N. M.S., Inst. Tech., Zurich, 1948, Ph.D., 1951. Research assoc. Inst. Tech., Zurich, Switzerland, 1949-52; Boese fellow Columbia U., 1952-53; faculty Calif. Inst. Tech., Pasadena, 1953—, prof. physics, 1961—; Sloan fellow, 1962-64, Niels Bohr Inst., Copenhagen, 1965-66, Cern, Geneva, 1971-72, Laue-Langevin Inst., 1980. Recipient Humboldt award, 1980. Fellow Am. Phys. Soc.; mem. Nat. Acad. Scis. Subspecialties: Particle physics; Plasma physics. Current work: Properties of elementary particles and fields. Research on nuclear physics, nuclear beta decay, neutrino physics, atomic physics, muonic and pionic atoms, parity and time-reversal. Home: 2510 N Altadena Dr Altadena CA 91001 Office: Calif Inst Tech Pasadena CA 91125

BOEHM, ROBERT FOTY, mechanical engineer, researcher; b. Portland, Oreg., Jan. 16, 1940; s. Charles Frederick and Lufteria (Christie) B.; m. Marcia Kay Pettibone, June 10, 1961; children: Deborah, Robert, Christopher. B.S. in Mech. Engring, Wash. State U., Pullman, 1962, M.S., 1964; Ph.D., U. Calif.-Berkeley, 1968. Registered profl. engr., Calif. With Westinghouse Corp., Lima, Ohio, 1961, Lawrence Livermore Lab., Livermore, Calif., 1962, Boeing Aerospace Co., Seattle, 1963, Gen. Electric Co., San Jose, Calif., 1964-66, Jet Propulsion Lab., Pasadena, Calif., 1967; mem. faculty U. Utah, Salt Lake City, 1968—, prof. mech. engring., 1976—, chmn. dept., 1981—; Mem. Utah Solar Adv. Com., Utah Energy Conservation and Devel. Council, 1980—. Contbr. articles to profl. jours. Fellow Am. Soc. Engring. Edn., ASME; Mem. Internat. Solar Energy Soc., Utah Solar Energy Soc., Bonneville Corvair Club. Congregationalist. Subspecialties: Mechanical engineering; Solar energy. Home: 2217 E Bryan Circle Salt Lake City UT 84108 Office: U Utah Mech Engring Dept Salt Lake City UT 84112

BOEKELHEIDE, VIRGIL CARL, chemistry educator, consultant; b. Chelsea, S.D., July 28, 1919; s. Charles Frederick and Eleonor Charlotte (Toennies) B.; m. Caroline Ambler Barrett, Sept. 1, 1945; children: Karl, Anne, Erich. A.B., U. Minn., 1939, Ph.D., 1943. Instr. U. Ill., 1943-45; asst. prof. to prof. chemistry U. Rochester (N.Y.), 1946-60; prof. U. Oreg., Eugene, 1960—; cons. Ciba-Geigy Co. Editorial bd.: Organic Reactions and Organic syntheses; contbr. over 220 articles to sci. jours. Guggenheim fellow, 1953-54; Sloan fellow, 1960-62; Roche fellow, 1962-64; recipient Outstanding Achievement award U. Minn., 1967; Fulbright Disting. prof., 1972; Alexander von Humboldt fellow, 1974-75, 82; Centenary lectr. Royal Soc. Gt. Britain, 1983. Mem. Nat. Acad. Scis., Am. Chem. Soc., Swiss Chem. Soc., German Chem. Soc. Subspecialty: Organic chemistry. Current work: Syntheses, heterocycles, novel aromatic and organometallics; organic metals and organic electrical conductors. Home: 2017 Elk Dr Eugene OR 97403 Office: U Oreg Eugene OR 97403

BOER, KARL WOLFGANG, physicist, engr.; b. Berlin, Mar. 23, 1926; U.S., 1961, naturalized, 1972; s. Karl and Charlotte (Gruhlke) B.; m. Renate Schroder, Apr. 18, 1935; children: Ralf-Reinhard, Katarina Karlotta. Dipl. in Physics, Humboldt U., Berlin, 1949, Dr.rer.nat., 1952; Dr.rer.nat. habil., 1955. Asst., docent, prof., chmn. Humboldt U., 1949-61; dir. lab. for dielectric breakdown German Akademie der Wissenschafleng, Berlin, 1955-61; prof. physics U. Del., 1962-72, dir., 1972-75, chief scientist, 1975—, prof. physics and engrings., 1972—, advisor to pres., 1976—; chmn. SES, Inc., 1972-78, chief scientist, 1978—. Editor 3 jours.; editor in chief: Advances in Solar Energy, 1982; editor 8 conf. procs.; contbr. numerous articles to profl. jours. Recipient Humboldt medal, 1958; Charles J. Abbott award, 1981. Fellow Am. Phys. Soc.; mem. IEEE (sr.), Am. Solar Energy Soc. (chmn. 1976–77, sec. 1978). Subspecialties: Solar energy; Condensed matter physics. Current work: First observer Franz Keldysh effect, light induced modulation of absorption, proposed Bose-Einstein condensation of excitons, conduction mechanism of semiconducting glasses, curve shape factor of jV characteristics DC electroluminescence, hybrid solar house system, negative differential conductivity effects. Patentee in field. Home: Buck Toe Hills Kennett Square PA 19348 Office: Coll of Engineering U of Del Newark DE 19711

BOERBOOM, LAWRENCE EDWARD, physiologist, cardiovascular researcher, educator; b. Marshall, Minn., Apr. 30, 1945; s. Lambert J. and Frances B.; m. Mary T. Bushard, June 15, 1968; children: Peter, Christopher. B.S., Coll. St. Thomas, St. Paul, 1968; M.S., St. Cloud (Minn.) State U., 1972; Ph.D., U. N.D.-Grand Forks, 1975. Research fellow U. Calif., San Francisco, 1975-77; asst. prof. cardiothoracic surgery and physiology Med. Coll. Wis., Milw., 1977—, dir. cardiovascular surg. research, 1977—. Contbr. chpts. to books and articles in field to profl. jours. NIH grantee, 1978-79; USPHS grantee, 1978-79; Med. Coll. Wis. grantee, 1980—; Am. Heart Assn. Wis. grantee, 1981, 83. Mem. Am. Heart Assn., Am. Physiol. Soc., AAAS, Sigma Xi. Subspecialties: Physiology (medicine); Cardiac surgery. Current work: Cardiovascular surgical research. Office: Medical College Wisconsin 8700 W Wisconsin Ave Milwaukee WI 53226

BOERINGA, JAMES ALEXANDER, clinical psychologist; b. Oak Park, Ill., Feb. 26, 1940; s. Joseph S. and Alice (Van Stedum) B.; m. Karin M. Moreland, June 17, 1978; 1 son, Michael Alexander. B.A., Hope Coll., 1965; Ph.D., U. Tex.-Austin, 1979. Researcher Tracor, Inc., Austin, Tex., 1966-67; v.p. Tex Mex, Inc., Austin, 1968-73; psychology intern Houston VA Med. Center, 1977; asst. prof. U. Tex. Med. Sch., Galveston, 1978-82; clin. psychologist VA, Peoria, Ill., 1982—. Contbr. articles in field to profl. jours. Served with U.S. Army, 1959-61. Woodrow Wilson fellow, 1965. Mem. Am. Psychol. Soc., AAAS, Soc. Behavioral Medicine, S.W. Soc. Clin. Hypnosis (sec. 1981-82), Phi Kappa Phi. Subspecialties: Behavioral psychology; Behavioral medicine. Current work: Reliability and validity diagnosis in psychiatric disorders, biochemical basis for psychosis, psychological intervention in health problems. Home: 2504 W Melrose Pl Peoria IL 61604 Office: VA Outpatient Clinic 411 w 7th St Peoria IL 61605

BOERNER, WOLFGANG-MARTIN, radar and electronic engineering educator; b. Finschhafen, Morobe, Papua-New Guinea, July 26, 1937; came to U.S., 1963, naturalized, 1981; s. Martin Ernst and Ilse Louise (Stoss) B.; m. Eileen A. Hassebrock, Dec. 23, 1967; children: Vaughan W., J. Allan, Joanna E. Dipl. Ing., Tech. U. Munich, Germany, 1963; Ph.D. in E.E, U. Pa., 1967. Research engr. Radiation Lab., U. Mich., Ann Arbor, 1967-68; postdoctoral fellow elec. engring. U. Man. (Can.), Winnipeg, 1968-69, asst. prof. elec. engring., 1969-71, assoc. prof., 1971-75, prof., 1975-78, research prof., 1978—; prof., dir. Elec. Engring. and Computer Sci Lab., U. Ill.-Chgo., 1978—; cons. radar U.S. Dept. Def., Chgo., 1980—; pres. Polarimetrics Inc., Northbrook, Ill., 1982; Dir. Advanced Study Inst., NATO, Bad Windsheim, W.Ger., 1983. Contbr. articles to profl. jours. Fulbright fellow, 1963. Mem. IEEE, AAAS, Humboldt Soc., Soc. Exploration Geophysics, Soc. Engring. Sci., Can. Assn. Physicists, Verein Deutscher Ingenieure, Soc. Photog. Instrumentation Engrs., Optical Soc. Am., Sigma Xi. Subspecialties: Electrical engineering; Applied magnetics. Current work: Applied electromagnetics; radar polarimetry; inverse scattering; radar remote sensing; geo-electromagnetics; laser optics, atmospheric optics. Home: 1021 Cedar Ln Northbrook IL 60062 Office: Communications Lab Dept Elec Engring and Computer Sci U Ill 851 S Morgan St PO Box 4348 M St SEO-1141 Chicago IL 60680

BOES, ARDEL J., mathematics educator; b. Wall Lake, Iowa, Sept. 24, 1937; m., 1963; 2 children. B.A., St. Ambrose Coll., 1959; M.S., Purdue U., 1961, Ph.D. in Math, 1966. Instr. math. Marycrest Coll., 1963-64; asst. prof. math. Colo. Sch. Mines, Golden, 1966-69, assoc. prof., 1969—, head dept., 1978—. Subspecialty: Probability. Office: Dept Math Colo Sch Mines Golden CO 80401

BOESHAAR, PATRICIA CHIKOTAS, astronomer; b. Butler Twp., Pa., Sept. 25, 1947; d. Joseph S. and Anna (Geritis) Chikotas; m. John Anthony Tyson, Jan. 23, 1981; 1 son, Kristopher Chikotas Tyson.

Student, Duquesne U., Pitts, 1965-67; B.S., Northwestern State U., Natchitoches, La., 1969; Ph.D., Ohio State U., 1976. Instr., research assoc. U. Wash., 1975-77; asst. prof. U. Oreg., 1977-80; research assoc U. Ariz., 1980-81; asst. prof. physics Rider Coll., Lawrenceville, N.J., 1981—. Contbr. articles to profl. jours. NSF grantee, 1978, 79. Mem. Am. Astron. Soc., Astron. Soc. Pacific, Sigma Xi, Sigma Pi Sigma. Subspecialty: Optical astronomy. Current work: Low luminosity star, spectral classification, stellar distributions. Office: Rider Coll PO Box 6400 Lawrenceville NJ 08648

BOFF, KENNETH RICHARD, research psychologist, consultant; b. Bklyn., Aug. 17, 1947; s. Victor Boff and Ann Yunko; m. Judith Marion Schoer, Aug. 2, 1969; 1 son, Cory Asher. B.A., Hunter Coll., CUNY, 1969, M.A., 1972; M. Phil., Columbia U., 1975, Ph.D., 1978. Research asst. Columbia, U., 1976; research psychologist Air Force Human Resources Lab., Wright-Patterson AFB, Ohio, 1977-80, Air Force Aero Med. Research Lab., 1980—; project dir. Intergrated perceptual info. for designers project USAF, U.S. Army, USN, NASA, 1980—; program mgr. tactical aircraft cockpit devel. and evaluation program, 1980—, research team leader, 1980—. Editor: Handbook of Perception and Human Performance; contbr. articles to profl. jours. Recipient Outstanding Performance award USAF, Wright-Patterson AFB, 1981. Mem. Assn. for Research in Vision and Opthalmology, Am. Psychol. Assn., Human Factors Soc. Subspecialties: Human factors engineering; Sensory processes. Current work: Knowledge-based, information-management systems (articial intelligence); biocybernetic control; flight simulation; visually coupled systems; information transfer effectiveness of three-dimensional displayed information. Home: 3114 Village Ct Dayton OH 45432 Office: Air Force Aerospace Med Research Lab Wright-Patterson AFB OH 45433

BOGARD, DONALD DALE, geochemist; b. Fayetteville, Ark., Feb. 6, 1940. B.S., U. Ark., 1962, M.S., 1964, Ph.D. in Isotope Geochemistry, 1966. NSF research fellow geol. sci. Calif. Inst. Tech., 1966-68; staff scientist planetary and earth sci. div. Johnson Space Ctr., NASA, Houston, 1968—, curator Antarctica meteorite collection, 1978—; Assoc. editor Jour. Geophys. Research Am. Geophys. Union, 1975-77. Fellow Meteoritical Soc.; mem. AAAS, Am. Geophys. Union, Geochem. Soc. Subspecialties: Planetology; Meteorites. Office: NASA Johnson Space Ctr Code SN 7 Houston TX 77058

BOGDONOFF, SEYMOUR MOSES, aeronautical engineer; b. N.Y.C., Jan. 10, 1921; s. Glenn and Kate (Cohen) B.; m. Harriet Eisenberg, Oct. 1, 1944; children: Sondra Sue, Zelda Lynn, Alan Charles. B.S., Rensselaer Poly. Inst., 1942; M.S., Princeton U., 1948. Asst. sect. head fluid and gas dynamics sect. Langley Meml. Aero. Lab., NASA, 1942-46; research assoc. aero. engring. dept. Princeton U., 1946-53, asso. prof., 1953-57, prof., 1957-63, Henry Porter Patterson prof. aero. engring., 1963—, chmn. dept. mech. and aerospace engring., 1974-83, head gas dynamics lab.; cons. aero. engr.; mem. adv. council NASA; mem. sci. adv. bd. Dept. Air Force, 1958-76, 80—. Recipient Exceptional Civilian Service award Dept. Air Force, 1968. Fellow AIAA (dir., Fluid and Plasma Dynamics award 1983); mem. Internat. Acad. Astronautics of Internat. Astronautical Fedn. (corr.), Nat. Acad. Engring., ASME, Am. Phys. Soc., Sigma Xi, Tau Beta Pi. Subspecialty: Aeronautical engineering. Home: 39 Random Rd Princeton NJ 08540

BOGERT, GARY MICHAEL, chemical corporation engineer; b. Santa Monica, Calif., Sept. 12, 1944; s. George Dudley and Dorothy Ruth (Williamson) B.; m. Cynthia Jane Drake, Feb. 14, 1970; children: Juillet Noel, Ian Christopher. B.S., Stanford U., 1966. Systems engr. Boeing Aerospace Co., Seattle, 1971-75; engr.ops. mgr. Exxon Nuclear Co., Richland, Wash., 1975-81; project engr. E.I. DuPont de Nemours & Co., Inc., Wilmington, Del., 1981—. Referee U.S. Soccer Fedn., Richland and Newark, Del., 1980—. Served to lt. Submarine Service USNR, 1966-71; lt. comdr. Res. Mem. Am. Nuclear Soc., Res. Officers Assn. (life). Episcopalian. Subspecialties: Nuclear engineering; Laser-induced chemistry. Current work: Process design and engineering of facilities for laser isotope separation of nuclear materials. Co-inventor dual magnification optical head for electronic access to photographically stored micro-images. Office: Engring Dept E I DuPont de Nemours & Co Inc Louviers Bldg Wilmington DE 19898 Home: RD 1 Box 234A West Grove PA 19390

BOGGESS, ROBERT KEITH, chemistry educator and researcher; b. Morgantown, W.Va., Apr. 26, 1945; s. Thomas Cooper, Sr. and Lavinia (Hill) B.; m. Susan Scott, July 5, 1969; children: Teresa Elaine, Brian Michael. B.S., Emory and Henry Coll., 1967; Ph.D., U. Ala.-Tuscaloosa, 1972. High sch. tchr., Tuscaloosa County, Ala., 1968-69; postdoctoral research U. Va., Charlottesville, 1972-74; asst. prof. Thomas More Coll., Covington, Ky., 1974-77; asst. prof., assoc. prof. chemistry Radford (Va.) U., 1977—. Contbr. writings to research publs. Research grantee Research Corp., 1975, 76-77, Am. Chem. Soc., 1979-80, 81-82, Jeffress Found., 1982-83. Mem. Am. Chem. Soc (chmn. Va. Blue Ridge sect. 1982, exec. com. 1983—). Methodist. Subspecialties: Inorganic chemistry; Analytical chemistry. Current work: Inorganic synthesis, model systems for biologically important molecules, electrochemistry. Home: 2108 Charlton Ln Radford Va 24141 Office: Dept Chemistry Radford Univ Radford VA 24142

BOGGS, GEORGE JOHNSON, psychophysicist, human factors engineer; b. Beckley, W.Va., Mar. 2, 1949; s. William Arnie and Margaret Pearl (Johnson) B.; m. Hilda Mae Beaver, Aug. 22, 1972 (div. 1977). B.S., Marshall U., Huntington, W.Va., 1972, M.S., 1974; Ph.D., Purdue U., 1981. Psychologist Nicholas County Mental Health Center, Summersville, W.Va., 1974-75; dir. psychology Cabell County Bd. Edn., Huntington, 1975-77; research asst. Purdue U., West Lafayette, Ind., 1977-81; mem. tech. staff GTE Labs., Inc., Waltham, Mass., 1981—; cons. U. Md., College Park, 1982. Mem. Am. Inst. Physics, Acoustical Soc. Am., N.Y. Acad. Scis., Am. Psychol. Assn., Sigma Xi. Subspecialties: Psychophysics; Human factors engineering. Current work: Basic and applied research in auditory psychophysics and human factors engineering for telecommunications application. Office: GTE Labs Inc 40 Sylvan Rd Waltham MA 02254 Home: 1909 Stearns Hill Rd Waltham MA 02154

BOGGS, JAMES ERNEST, chemistry educator; b. Cleve., June 9, 1921; s. Ernest Beckett and Emily (Reid) B.; m. Ruth Ann Rogers, June 22, 1948; children: Carol, Ann, Lynne. A.B., Oberlin Coll., 1943; M.S. in Chemistry, U. Mich., 1944, Ph.D., 1953. Asst. prof. dept. chemistry Eastern Mich. U., Ypsilanti, 1949-52; instr. U. Mich. at Ann Arbor, 1952-53; mem. faculty dept. chemistry U. Tex. at Austin, 1953—, assoc. prof., 1958-66, prof., 1966—, asst. dean Grad. Sch., 1958-67, dir. Center for Structural Studies, 1969—, acting dir. Inst. Theoretical Chemistry, 1979—. Contbr. 140 articles to profl. jours. Mem. Am. Chem. Soc., Am. Phys. Soc., Phi Beta Kappa, Sigma Xi, Phi Lambda Upsilon, Gamma Alpha. Subspecialties: Physical chemistry; Theoretical chemistry. Current work: Physical and theoretical chemistry. Research in structural chemistry, microwave spectroscopy, quantum chemistry. Home: 4603 Balcones Dr Austin TX 78731

BOGNER, FRED KARL, research engineer; b. Mansfield, Ohio, July 7, 1939; s. Fred W. and Esther V. (Swarz) B.; m. Mary L. Reynolds, July 11, 1939; children: Fred C., Sharon L. B.S. in Civil Engring, Case Inst. Tech., 1961, M.S. in Engring. Mechanics, 1964, Ph.D., 1967. Mem. tech. staff Bell Telephone Labs., Whippany, N.J., 1967-69; research engr. U. Dayton (Ohio) Research Inst., 1969—, group leader, 1976—. Contbr. articles to profl. jours. in field. Mem. AIAA, Soc. Engring. Sci., Am. Acad. Mechanics. Club: Dayton Squash Racquets Assn. Subspecialties: Solid mechanics; Theoretical and applied mechanics. Current work: Solid mechanics; theoretical and applied mechanics; finite element analysis; stress analysis; nonlinear structural analysis. Home: 9516 Bridlewood Trail Spring Valley OH 45370 Office: Univ Dayton 300 College Park Ave Dayton OH 45469

BOGORAD, LAWRENCE, biology educator; b. Tashkent, U.S.S.R., Aug. 29, 1921; came to U.S., 1922; s. Boris and Florence (Bernard) B.; m. Rosalyn G. Sagen, June 29, 1943; children—Leonard Paul, Kiki M. Lee. B.S., U. Chgo., 1942, Ph.D., 1949. Instr. botany U. Chgo., 1948-51, asst. prof. dept. botany, 1953-57, assoc. prof., 1957-61, prof., 1961-67; prof. biology Harvard U., Cambridge, Mass., 1967—, chmn. dept. biology, 1974-76, dir. Maria Moors Cabot Found., Cambridge, Mass., 1976—, Maria Moors Cabot prof. biology, 1980—; vis. investigator Rockefeller Inst., N.Y.C., 1951-53; mem. com. on sci. and public policy Nat. Acad. Scis., 1977-81; mem. Assembly of Life Scis., NRC; mem. joint council on food and agrl. scis. Dept. Agr., 1978-82. Asso. editor: Bot. Gazette, 1958; editoral com.: Annual Rev. Plant Physiology, 1963-67; editorial bd.: Plant Physiology, 1965-66, Biochimica Biophysica Acta, 1967-69, Jour. Cell Biology, 1967-70, Jour. Applied and Molecular Genetics, 1981—, Plant Molecular Biology, 1981—, Plant Cell Reports, 1981—. Served with AUS, 1943-46. Merck fellow, 1951-53; Fulbright fellow, 1960; recipient Career Research award NIH, 1963. Fellow Am. Acad. Arts and Scis.; mem. Am. Soc. Biol. Chemistry, Am. Soc. Cell Biology, Nat. Acad. Scis. (chmn. botany sect. 1974-77), Am. Soc. Plant Physiologists (pres. 1968-69, Stephen Hales award 1982), AAAS (bd. dirs. 1982), Royal Danish Acad. Scis. and Letters (fgn.), Soc. Developmental Biology (pres. 1984). Subspecialty: Plant physiology (agriculture). Office: Dept of Biology Harvard Univ 16 Divinity Ave Cambridge MA 02138

BOHACHEVSKY, IHOR OREST, mathematician, engineer; b. Sokal, Ukraine, Sept. 7, 1928; U.S., 1948; s. Daniel and Rostyslava S. (Nychay) B.; m. Mary Ulana, May 10, 1969. B.Aero. Engring., N.Y. U., 1956, Ph.D., 1961. Research scientist N.Y. U., 1960-63; aero. research engr. Cornell Aero. Lab., Buffalo, 1963-66; prin. research scientist Avco-Everett Research Lab., Mass., 1966-68; staff mem. Bellcomm, Inc., Washington, 1968-72, Bell Telephone Labs., Murray Hill, N.J., 1972-75, Los Alamos Nat. Lab., 1975—. Contbr. articles to profl. jours. Served to p.f.c. AUS, 1951-53; Korea. Recipient W.O. Bryans medal N.Y. U., 1956, NASA Apollo award Bellcomm, 1969, AT & T cert., 1969; NSF fellow, 1956-69. Mem. N.Y. Acad. Scis., Soc. Indsl. and Applied Math., AIAA, Am. Nuclear Soc., Sigma Xi. Roman Catholic. Subspecialties: Applied mathematics; Fusion. Current work: Identification and solution of technical problems associated with utilization of intertial confinement fusion process including identification and analyses of different applications. Inventor beam heated theta pinch. Home: 829 Pine St Los Alamos NM 87544 Office: Los Alamos National Laboratory PO Box 1663 Los Alamos NM 87545

BOHLIN, JOHN DAVID, space scientist; b. Valparaiso, Ind., Mar. 19, 1939; s. George and Margaret (Etruth) B.; m. Jeanne Ann Caudell, June 26, 1966; children—Matthew David, Timothy John. A.B., Wabash (Ind.) Coll., 1961; Ph.D., U. Colo., 1968. Research asst. High Altitude Obs., Boulder, Colo., 1962-67; research fellow Calif. Inst. Tech., Pasadena, 1967-70; research space physicist Naval Research Lab., Washington, 1970-76; chief solar and heliospheric physics br. NASA, Washington, 1977—. Contbr. articles to profl. jours. Mem. Am. Astron. Soc., Am. Geophys. Union, Internat. Astron. Union, Phi Beta Kappa. Lutheran. Subspecialty: Solar physics. Current work: Science program management. Office: Code SC-7 NASA Hdqrs Washington DC 22546

BOHME, DIETHARD KURT, chemistry educator; b. Boston, June 20, 1941; s. Kurt Friedrich Wilhelm and Maria Kunigunda (Kiesel) B.; m. Shirley Faith Broadway, Dec. 23, 1966; children: Kurt, Kenneth Diethard, Heidi Claire. B.Sc., McGill U., 1962, Ph.D., 1965. Asst. prof. chemistry York U., Downsview, Ont., Can., 1970-74, assoc. prof., 1974-77, prof., 1977—, dir. grad. program in chemistry, 1979—; sr. scientist, vis. fellow Sci. Research Council, U. Warwick, Eng., 1978. Contbr. chpts. to books, also articles to profl. jours. Recipient Rutherford Meml. medal Royal soc. Can., 1981; Sloan Found. fellow York U., 1974. Fellow Chem. Inst. Can. exec. (phys. chemistry div. 1980-83, Noranda lectr. award 1983); mem. Am. Soc. Mass Spectrometry, Combustion Inst. Subspecialties: Physical chemistry; Space chemistry. Current work: Gas-phase ion chemistry, ion energetics, experimental reaction kinetics, physical organic chemistry, astrochemistry, flame-ion chemistry, ionospheric chemistry. Home: 28 Colonsay Rd Thornhill Ont Can L3T 3E8 Office: York U Downsview Ont Can M3J 1P3

BOHN, MARTHA D., neurobiologist, educator; b. Niagara Falls, N.Y., Mar. 30, 1943; d. John W. and Elsie L. (Gregory) Churchill. A.B., Cornell U., 1964; M.S., U. Conn., 1977, Ph.D., 1979. Asst. prof. Cornell U. Med. Coll., N.Y.C., 1980-83, SUNY-Stony Brook, 1983—. Contbr. articles to profl. jours. NIH grantee, 1976-78, 79-81, 82—; NINCDS Research Career Devel. award, 1982. Mem. Soc. Neurosci., AAAS, N.Y. Acad. Scis., Phi Beta Kappa, Phi Kappa Phi. Subspecialties: Neurobiology; Developmental biology. Current work: Endocrine effects on neuronal and glial cell differentiation; specification and plasticity of neurotransmitter phenotype.

BOHR, BRUCE A., industrial engineer; b. Wasau, Wis., May 11, 1950; s. Joseph Peter and Mary Veronica (Sikorski) B.; m. Mary Helen, July 15, 1975; 1 dau., Katherine. B.S.M.E., U. Wis.-Madison, 1973. Engring. summer trainee Deere & Co., Dubuque, Iowa, 1970-72, project engr., 1973; chief engr. Wis. Equipment Corp., Medford, 1973-76; v.p. mfg. Foster Needle Co., Manitowoc, Wis., 1976-78; mgr. product devel. Hamilton Industries, Inc., Two Rivers, Wis., 1978—. Patentee in field. Mem. Soc. Automotive Engrs., ASME (vice chmn. N.E. Wis. 1982-83). Roman Catholic. Subspecialty: Mechanical engineering. Current work: Development of professional equipment and furniture for lab applications. Office: 1316 18th St Two Rivers WI 54241

BOILEAU, OLIVER CLARK, aerospace co. exec.; b. Camden, N.J., Mar. 31, 1927; s. Oliver Clark and Florence Mary (Smith) B.; m. Nan Eleze Hallen, Sept. 15, 1951; children—Clark Edward, Adrienne Lee, Nanette Erika, Jay Marshall. B.S. in Elec. Engring, U. Pa., 1951, M.S., 1953; M.S. in Indsl. Mgmt, M.I.T., 1964. With Boeing Aerospace Co., 1953-80; mgr. Minueteman, then v.p., 1968, pres., 1973-80, Gen. Dynamics Corp., St. Louis, 1980—, also dir. Served with USNR, 1944-46. Sloan fellow, 1963-64. Mem. AIAA, Navy League, Air Force Assn., Am. Def. Preparedness Assn., Assn. U.S. Army, Armed Forces Communications and Electronics Assn., Nat. Aeros. Assn., Nat. Space Club, Naval War Coll. Found., Nat. Acad. Engring. Subspecialty: Aerospace engineering and technology. Office: 7733 Forsyth Blvd Clayton MO 63105

BOIS, PIERRE, medical research organization executive; b. Oka, Que., Can., Mar. 22, 1924; s. Henri and Ethier (Germaine) B.; m. Joyce Casey, Sept. 8, 1953; children: Monique, Marie, Louise. M.D., U. Montreal, Que., 1953, Ph.D., 1957; hon. doctorate, U. Ottawa, Ont., 1982. Research fellow pathology U. Montreal, 1957-58, asst. assoc. prof. dept. pharmacology, 1960-64, prof., head dept. anatomy, 1964-70, dean faculty medicine, 1970-81; pres. Med. Research Council of Can., 1981—; asst. prof. histology, Ottawa, Ont., Can., 1958-60. Contbr. over 130 publs. to profl. jours. Fellow Royal Soc. Can., Royal Coll. Physicians and Surgeons Can.; mem. Am., French Canadian assns. anatomists, N.Y. Acad. Scis., AAAS, Can. Fedn. Biol. Socs., Can. Soc. Clin. Investigation. Subspecialty: Medical research administration. Research and numerous publs. on morphological effects of hormones, histamine and mast cells in magnesium deficiency, muscular dystrophy, exptl. thymic tumors. Office: Med Research Council Can 20th Floor Jeanne Mance Bldg Tunney's Pasture Ottawa ON K1A 0W9 Canada

BOISSE, NORMAN ROBERT, research pharmacologist, consultant; b. Biddeford, Maine, Aug. 18, 1947; s. Paul Emile and Rita (Letendre) B. B.S., U. Conn.-Storrs, 1970; Ph.D., Cornell U. Grad. Sch. Med. Scis., N.Y.C., 1976. Registered pharmacist, Maine. Research asst., postdoctoral trainee dept. pharmacology Cornell U. Med. Coll., N.Y.C., 1976-77; assoc. prof. pharmacology and physiology Northeastern U., 1977—, head sect. pharmacology, 1982—, dir. M.S. in pharmacology program, 1982—; cons. pharmacology; participant Pharm. Mfrs. Assn. visitation program Roche Labs., Nutley, N.J., 1981. Nat. Inst. Drug Abuse grantee, 1979-82. Mem. AAAS, N.Y. Acad. Sci., Soc. Neurosci., Am. Assn. Colls. Pharmacy, Sigma Xi, Rho Chi. Subspecialties: Pharmacology; Neuropharmacology. Current work: Neuropharmacology of benzodiazepine tolerance and dependence. Office: Coll Pharmacy and Allied Health Professions Northeastern U Boston MA 02115

BOISVERT, RONALD FERNAND, computer scientist; b. Manchester, N.H., July 26, 1951; s. Fernand Lucien and Irene Dorothy (Demers) B.; m. Rita Linda Lefebvre, Dec. 30, 1972. B.S., Keene State Coll., 1973; M.S., Coll. William and Mary, 1975, Purdue U., 1977, Ph.D., 1979. Computer programmer Coll. William and Mary, Williamsburg, Va., 1973-74, NASA Langley Research Center, Hampton, Va., 1974-75; teaching asst. Purdue U., West Lafayette, Ind., 1975-76, research asst., 1976-79; computer scientist Nat. Bur. Standards, Washington, 1979—. Mem. Assn. for Computing Machinery, Soc. Indsl. and Applied Math., Internat. Assn. Math. and Computers in Simulation, IEEE Computer Soc. Subspecialties: Numerical analysis; Mathematical software. Current work: Research in numerical methods for solution of partial differential equations. Development of mathematical software. Software development tools. Office: National Bureau Standards Washington DC 20234

BOJADZIEV, GEORGE NIKOLOV, mathematics educator; b. Shumen, Bulgaria, July 11, 1927; emigrated to Canada, 1971; s. Nicolas D. and Luba G. (Kodjabanova) B.; m. Maria S. Cekova, Oct. 14, 1956; children: Luba, Nick. Diploma, U. Sofia, Bulgaria, 1950, Ph.D. in Math., 1957. Prof. Math. U. Mech. and Elec. Engring., Sofia, Bulgaria, 1969-70; vis. prof. Simon Fraser U., Burnaby, B.C., Can., 1971-75, assoc. prof. dept. math., 1975-79, prof., 1979—. Mem. Can. Applied Math. Soc., Soc. Indsl. and Applied Math. Subspecialty: Applied mathematics. Current work: Ordinary and partial differential equations and applications; perturbation methods; nonlinear oscillations; population dynamics. Home: 835 Farmleigh Rd West Vancouver BC Canada V7S 1Z8 Office: Simon Fraser U Burnaby BC Canada V5A 1S6

BOLAND, C. RICHARD, physician; b. Johnson City, N.Y., Oct. 19, 1947; s. Clement Richard and Catherine Jane (Armstrong) B.; m. Patricia Ellen Sweeney, Sept. 4, 1970; children: Tara Sweeney, Maureen Sweeney, Brigid Sweeney. B.A., U. Notre Dame, 1969; M.D., Yale U., 1973. Asst. prof. medicine U. Calif., San Francisco, 1981—. Contbr. articles to profl. jours. Served to lt. comdr. USPHS, 1974-78. Mem. Am. Fedn. for Clin. Research, No. Calif. Soc. Clin. Gastroenterology. Subspecialties: Gastroenterology; Internal medicine. Current work: Tumor markers in human colon. Office: GI Research Lab (151-M2) 4150 Clement St San Francisco CA 94121

BOLAND, J. ROBERT, radiological physicist; b. Denver, June 17, 1928; s. Hugh William and Hilda Josephine (Larson) B.; m. Pita Martinez, Feb. 14, 1963; children: Adriana, Nina Marie. B.S., U. Denver, 1952, M.S., 1958. Health physicist Dow Chem. Co., Rocky Flats, Colo., 1952-53; chemist-toxicologist U. Colo. Med. Ctr., Denver, 1953-58; health physicist Dept. Def., White Sands Missile Range, 1959-63; radiation specialist EG & G Inc., Santa Barbara, Calif., 1963-67; health physicist AEC (now U.S. Dept. Energy), Nev. Test Site, 1967—. Contbr. articles to profl. jours. Served with USN, 1946-48. Mem. Am. Chem. Soc., Am. Nuclear Soc., Health Physics Soc., VFW. Club: Rough Riders (trail boss 1981—). Subspecialty: Radioactive waste disposal. Current work: Developing technology for disposal of high specific activity low-level radioactive waste. Home: 5593 Alfred Dr Las Vegas NV 89108 Office: US Dept Energy PO Box 14100 Las Vegas NV 89114

BOLANOWSKI, STANLEY JOHN, JR., neuroscientist, educator; b. Utica, N.Y., Feb. 22, 1950; s. Stanley John and Stella Agnes (Kielar) B. A.A.S., SUNY-Morrisville, 1970; B.A., Syracuse U., 1973, B.S., 1974, Ph.D., 1981. Asst. prof. Ctr. Brain Research, U. Rochester Sch. Medicine and Dentistry, 1982—, postdoctoral fellow, 1980-82. Contbr. articles to sci. jours. Syracuse U. fellow, 1978-80; Uniroyal grantee, 1972-74; NIH-Nat. Eye Inst. grantee. Mem. AAAS, Soc. Neurosci., Sigma Xi, Eta Kappa Nu. Subspecialties: Neurophysiology; Neuropsychology. Current work: Relating physiology and behavior in the somatosensory and visual systems; psychophysics; neurophysiology; vison; somatosensory; mechanotransduction. Office: U Rochester Sch Medicine Box 605 Rochester NY 14646

BOLCH, WILLIAM EMMETT, JR., engineering educator; b. Lenoir, N.C., Oct. 27, 1935; s. William Emmett and Gladys (Hendrix) B.; m. Sandra Lee Talley, Nov. 7, 1959; children: Wesley Emmett, Elizabeth Talley. B.S. in Civil Engring, U. Tex.-Austin, 1959, M.S., 1963; Ph.D., U. Calif.-Berkeley, 1967. Registered profl. engr., Fla. Asst. prof. environ. engring. U. Fla., Gainesville, 1966-70, assoc. prof., 1970-77, prof., 1977—, acting chmn., 1981—; cons., owner Environ. Radiation Group, Gainesville, Fla., 1981—, Occidental Research Corp., Irvine, Calif., 1979-81, Nuclear Safety Assocs., Bethesda, Md., 1977—, Atomic Indsl. Forum, Washington, 1979. Served to 1st lt. USAF, 1959-62. Mem. Am. Nuclear Soc., Health Physics Soc. Democrat. Methodist. Subspecialties: Environmental engineering; Nuclear engineering. Current work: Health physics, radiation protection, environmental monitoring for radioactivity, radiotracers. Office: Dept Environmental Engring 110 AP Black Hall Gainesville FL 32611

BOLDT, ELIHU A., astrophysicist; b. New Brunswick, N.J., July 15, 1931; s. Joel and Yetta (Miller) B.; m. Yvette, Nov. 25, 1971; children: Adam, Abigail, Jessica. Ph.D., MIT, 1958. Asst. prof. physics dept. Rutgers U., New Brunswick, N.J., 1958-64; astrophysicist Goddard Space Flight Ctr., NASA, Greenbelt, Md., 1964—, also head X-ray astronomy group; adj. prof. physics U. Md., College Park, 1982—. Recipient John C. Lindsay Meml. award Goddard Space Flight Center, 1977; Outstanding Sci Achievement award NASA, 1978. Fellow Am. Phys. Soc.; mem. Am. Astron. Soc., Internat. Astron. Union. Subspecialty: 1-ray high energy astrophysics. Current work: X-ray astronomy. Office: Goddard Space Flight Ct NASA Code 661 Greenbelt MD 20771

BOLE, GILES G., physician, rheumatologist; b. Battle Creek, Mich., July 28, 1928; s. Giles Gerald and Kittie Belle (French) B.; m. Beverley June Wilcox, June 16, 1951; children—David Giles, Elizabeth Ann. Grad., U. Mich., 1949, M.D., 1953. Diplomate: Am. Bd. Internal Medicine; cert. rheumatology. Instr. internal medicine U. Mich., 1959-61, asst. prof., 1961-64, asso. prof., 1964-70, prof., 1970—; asso. physician Rackham Arthritis Research Unit, 1961-69, acting physician-in-charge, 1969-71, physician-in-charge, 1971—, 1975—, program dir., co-investigator Arthritis Found., 1968-69, [illegible] grant in rheumatology USPHS, 1969-76, dir., 1976—; prin. investigator NIH, 1976—; mem. policy adv. com. (Center for Disease Control), 1977-80. Contbr. numerous articles to profl. jours., chpts. in textbooks. Pres. Newport Elementary Parent-Tchr. Orgn., 1967-68; chmn. spl. citizens com. safety Ann Arbor Pub. Schs., 1968-69, citizen rep. intergovtl. com. student safety, 1969-73; trustee Huron River Heights Property Owners Assn., 1970-73. Served to capt. USAF, 1956-58. Postdoctoral research fellow Arthritis and Rheumatism Found., 1961-63. Fellow A.C.P.; mem. Am. Rheumatism Assn. (exec. com., v.p. 1979-80, pres. 1980-81), AAAS, Am. Fedn. Clin. Research, Central Clin. Research Club, Central Soc. Clin. Research (sec.-treas. 1970-75, v.p. 1975-76, pres. 1976-77), Alpha Omega Alpha, Phi Kappa Phi. Subspecialties: Biochemistry (medicine); Rheumatology. Current work: Biochemistry of osteoarthritis and cartilage; clinical study of rheumatic diseases. Home: 2524 Blueberry Ln Ann Arbor MI 48104 Office: R4633 Kresge Univ Michigan Ann Arbor MI 48109

BOLENDER, CHARLES L., prosthodontics educator; b. Iowa City, June 2, 1932; m., 1955; 3 children. D.D.S., U. Iowa, 1956, M.S., 1957. From instr. to assoc. prof. Sch. Dentistry, U. Wash., Seattle, 1959-68, prof. prosthodontics, 1968—, chmn. dept., 1963—. Fellow Am. Coll. Dentists, Acad. Denture Prosthetics; mem. Am. Prosthodontic Soc., ADA. Subspecialty: Prosthodontics. Office: Sch Dentistry U Wash Seattle WA 98195

BOLEY, BRUNO ADRIAN, engineering educator; b. Gorizia, Italy, May 13, 1924; came to U.S., 1939, naturalized, 1945; s. Orville F. and Rita (Luzzatto) B.; m. Sara R. Boley, May 12, 1949; children: Jacqueline, Daniel L. B.C.E., CCNY, 1943, D.Sc. hon.; M. Aero. Engring., Poly. Inst. Bklyn., 1945, D.Sc. in Aero. Engring, 1946. Asst. dir. structural research, aero. engring. dept. Poly. Inst. Bklyn., 1943-48; engring. specialist Goodyear Aircraft Corp., 1948-50; asso. prof. aero. engring. Ohio State U., 1950-52; asso. prof. civil engring. Columbia, 1952-58, prof., 1958-68; Joseph P. Ripley prof. engring., chmn. theoretical and applied mechanics Cornell U., Ithaca, N.Y., 1968-72; dean Technol. Inst., Walter P. Murphy prof. Northwestern U., Evanston, Ill., 1973—; Mem. adv. com. George Washington U., Princeton U., Yale U., FAMU/FSU Inst. Engring.; chmn. Midwest Program for Minorities in Engring., 1975—; bd. govs. Argonne Nat. Lab., 1983—. Author: Theory of Thermal Stresses, 1960, High Temperature Structures and Materials, 1964, Thermoinelasticity, 1970, Crossfire in Professional Education, 1976; also articles, numerous tech. papers.; Editor-in-chief: Mechanics Research Communications; bd. editors: Jour. Thermal Stresses, Bull. Mech. Engring. Edn., Internat. Jour. Computers and Structures, Internat. Jour. Engring. Sci., Internat. Jour. Fracture Mechanics, Internat. Jour. Mech. Engring. Scis., Internat. Jour. Solids and Structures, Jour. Applied Mechanics, Jour. Structural Mechanics Software, Letters in Applied and Engring. Sci., Nuclear Engring. and Design. Recipient Disting. Alumnus award Poly. Inst. N.Y., 1974; Townsend Harris medal, 1981; NATO sr. sci. fellow, 1964-65. Fellow AIAA; hon. mem. ASME (exec. com., pres. applied mechanics div. 1975); Hon. mem. Am. Acad. Mechanics (pres. 1974); mem. Nat. Acad. Engring. (chmn. task force engring. edn. 1979-80), Soc. Engring. Scis. (pres. 1975), Assn. Chairmen Depts. Mechanics (pres. 1970-72), Internat. Assn. Structural Mechanics in Reactor Tech. (adv-gen. 1979-80), Internat. Union Theoretical and Applied Mechanics (sec. Congress com. 1976—), N.Y. Acad. Scis. (named Outstanding Educator of Am. 1971), U.S. Nat. Com. Theoretical and Applied Math. (chmn. 1975—), Ill. Council for Energy Research and Devel. (chmn. 1979—). Subspecialties: Theoretical and applied mechanics; Applied mathematics. Home: 1941 Orrington Ave Evanston IL 60201

BOLEY, DANIEL LUCIUS, computer science educator, researcher, consultant; b. N.Y.C., Mar. 23, 1953; s. Bruno Adrian and Sara Roslyn B.; m. Maria Luigia Gini, Oct. 16, 1982. B.A. summa cum laude, Cornell U., 1974; M.S., Stanford U., 1976, Ph.D., 1981. Research asst. Los Alamos Nat. Lab., summers 1978-79; grad. research asst. IBM Research Lab., Zurich, Switzerland, summer 1980; asst. prof. computer sci. U. Minn.-Mpls., 1981—; invited speaker Lund (Sweden) Workshop on Automatic Control, 1980. NSF fellow, 1974; grantee, 1982. Mem. Soc. Indsl. and Applied Math., Assn. Computing Machinery., Phi Beta Kappa. Subspecialties: Numerical analysis; Mathematical software. Current work: Numerical problems in control theory, linear algebra; Eigenvalue problems; large sparse matrix problems; mathematical analysis of scenes (computer vision). Office: U Minn Dept Computer Sci 207 Church St SE Minneapolis MN 55455

BOLLER, BRUCE RAYMOND, nuclear science and engineering educator, consultant; b. N.Y.C., Jan. 22, 1940; s. Raymond Edward and Alice (Stevens) B.; m. Susan Schoelch Saunders, May 21, 1941; children: John David, Janet Lynn. B.S., Iona Coll., 1961; M.S., U. Pitts., 1964; Ph.D., CUNY, 1970. Research Assoc. City Coll.-CUNY, N.Y.C., 1970-73; asst. prof. Baruch Coll., 1973-75; assoc. prof. nuclear sci., engring. Maritime Coll.-SUNY, Bronx, 1975—; cons. Queensborough Community Coll., Bayside, N.Y., 1978—. Recipient Univ. award SUNY, 1978. Mem. Am. Assn. Physics Tchrs., Am. Geophys. Union, Am. Nuclear Soc. Republican. Subspecialties: Magnetospheric physics and geomagnetism; Nuclear fission. Current work: Magnetospheric substorm onsets, neutron transport and thermal hydraulics in nuclear reactor cores. Home: R D 1 Sunset Ln Oak Ridge NJ 07438 Office: Dept Sci Maritime Coll-SUNY Bronx NY 10465

BOLLINGER, JOHN GUSTAVE, mechanical engineer; b. Grand Forks, N.D., May 28, 1935; s. Elroy William and Charlotte (Kirchner) B.; m. Heidelore Ladwig, Aug. 16, 1958; children: William, Kristin, Pamela. B.S., U. Wis., Madison, 1957; M.S., Cornell U., 1958; Ph.D., U. Wis., 1961. Asst. prof. U. Wis., Madison, 1961-65, asso. prof. dept. mech. engring., 1965-68, prof., 1968—, Bascom prof., 1973—, chmn. dept., 1975-79, dir. data acquisition and simulation lab., 1972-75, dean, 1981—; chmn., dir. Unico Inc.; dir. Rexnard Inc., Nicolet Instrn. Corp. Contbr. articles to profl. jours. Bd. dirs. Madison Gen. Hosp. Fulbright postdoctoral fellow, Aachen, Ger., 1962; vis. Fulbright prof. Cranfield (Eng.) Inst. Tech., 1980-81. Fellow ASME (Gustus L. Larson Meml. award 1976); mem. Am. Soc. Engring. Edn., Soc. Mfg. Engrs. (Research Medal award 1978), Sigma Xi. Club: Mendota Yacht. Subspecialties: Robotics; Systems engineering. Current work: Robot intelligence, computer control of machines and processes, manufacturing systems. Patentee in field. Home: 6117 S Highlands Madison WI 53705 Office: 1513 University Ave Madison WI 53706

BOLLINGER, RICHARD COLEMAN, mathematician; b. Johnstown, Pa., Feb. 1, 1932; s. Robert A. and Mary K. (Coleman) B.; m. Gertrude A Toth, July 3, 1954 (div. Dec. 1977); children: David, Lisa; m. Linda M. May, Aug. 10,1979; children: James, Jennifer, Adam. B.S., U. Pitts., 1954, M.S., 1957, Ph.D., 1961. Sr. mathematician Westinghouse Research, Pitts., 1954-62; asst. prof. math. Pa. State U., University Park, 1962-69; assoc. prof. math. Behrend Coll., Pa. State U., Erie, 1969—. Mem. Math. Assn. Am., Soc. Indsl. and Applied Math., AAAS. Subspecialties: Applied mathematics; Numerical analysis. Current work: Problems involving runs of events, with applications to combinatorics, statistics, reliability theory. Office: Behrend Coll Pa State U Station Rd Erie PA 16563 Home: 4212 Sassafras St Erie Pa 16508

BOLLINI, RAGHUPATHY, electrical engineering educator, consultant; b. Madras, India, Oct. 1, 1941; s. Manavala Choudhry and Padmavathi (Gali) B.; m. Sailarani Nalajala; children: Ramalakshmi, Subhashini. B.E., Madras U., 1962; M.S.E.E., Purdue U., 1967, Ph.D., 1971. Registered profl. engr., Ill. Lectr. Osmania U., 1962; asst. engr. Heavey Engring. Corp., India, 1963-65; research asst. in elec. engring. Purdue U., 1965-71, vis. asst. prof., 1971-73; asst. prof. engring. and tech. So. Ill. U.-Edwardsville, 1973-77, prof., 1977—; cons. U.S. Army C.E., St. Louis Dist. Contbr. articles to profl. publs. Mem. IEEE, Sigma Xi. Hindu. Subspecialty: Computer-aided design. Current work: Electrical energy generation, co-generation, resource conservation, electrical energy conditioning, efficiency in energy conversion. Office: Dept Engring and Tech Edwardsville IL 62026

BOLLON, ARTHUR PETER, geneticist, educator; b. N.Y.C., Dec. 15, 1942; s. Arthur and Emily B; m.; children: Marc, Erica. B.A., C.W. Post Coll., 1965; Ph.D., Rutgers U., 1970. Postdoctoral fellow Yale U., New Haven, Conn., 1970-71; asst. prof. U. Tex. Health Sci. Ctr., Dallas, 1972-78; chmn. dept. molecular genetics, dir. genetic engring. Wadley Insts. Molecular Medicine, Dallas, 1979—; adj. prof. N. Tex. State U.; cons. Diamond Shamrock Corp., Dallas, 1981. Editor: Recombinant DNA Products, 1984; contbr. articles to sci. publs. Mem. adv. bd. Dallas Chamber Orch., Dallas Ballet Barre Assn. Research grantee NSF, 1974-79, NIH, 1980-83, Am. Cancer Soc., 1974-76. Mem. Am. Soc. Biol. Chemists, Genetics Soc. Am., Am. Soc. Microbiologists, N.Y. Acad. Scis., Sigma Xi. Subspecialties: Genetics and genetic engineering (biology); Enzyme technology. Current work: genetic engineering and analysis of animal and yeast genes; gen and chromosome organization and expression. Office: Wadley Insts Molecular Medicine 9000 Harry Hines Blvd Dallas TX 75235

BOLT, BRUCE ALAN, educator; b. Largs, Australia, Feb. 15, 1930; s. Donald Frederick and Arlene (Stitt) B.; m. Beverley Bentley, Feb. 11, 1956; children—Gillian, Robert, Helen, Margaret. B.S. with honors, New Eng. U. Coll., 1952; M.S., U. Sydney, Australia, 1954, Ph.D., 1959, D.Sc. (hon.), 1972. Math. master Sydney (Australia) Boys' High Sch., 1953; lectr. U. Sydney, 1954-61, sr. lectr., 1961-62; research seismologist Columbia, 1960; dir. seismographic stas. U. Calif., Berkeley, 1963—, prof. seismology, 1963—, chmn., 1980—; Mem. com. on seismology Nat. Acad. Scis., 1966-72, mem., 1974-76, 79-81; also chmn. nat. earthquake obs. com.; mem. earthquake and wind forces com. VA, 1971-75; mem. Calif. Seismic Safety Commn., 1978—; earthquake studies adv. panel U.S. Geol. Survey, 1979—, U.S. Geodynamics Com., 1979—. Author, editor textbooks on applied math., earthquakes, geol. hazards and detection of underground nuclear explosions. Recipient H.O. Wood award in seismology, 1967, 72; Fulbright scholar, 1960; Churchill Coll. Cambridge overseas fellow, 1980. Fellow Am. Geophys. Union (mem. geophys. monograph bd. 1971-78, chmn. 1976-78), Geol. Soc. Am., Calif. Acad. Scis. (trustee 1981—), Royal Astron. Soc.; mem. Nat. Acad. Engring., Seismol. Soc. Am. (editor bull. 1965-70, dir. 1965-71, 73-76, pres. 1974-75), Internat. Assn. Seismology and Physics Earth's Interior (exec. com. 1964-67, v.p. 1975-79, pres. 1980—), Earthquake Engring. Research Inst., Australian Math. Soc., Sigma Xi. Club: Univ. Subspecialty: Geophysics. Research on dynamics, elastic waves, earthquakes, reduction geophys. observations; inferences on structure of earth's interior; cons. on seismic hazards. Home: 1491 Greenwood Terr Berkeley CA 94708

BOLTAX, ALVIN, nuclear materials executive; b. N.Y.C., Oct. 20, 1930; s. Jacob and Anne (Kasofsky) B.; m. Barbara Rogen, Sept. 16, 1956 (div. 1978); children: Jay David, Leslie Joan, Nancy Ruth. S.B., MIT, 1951, Sc.D., 1955. Engr. MIT Metall. Lab., Cambridge, Mass., 1950-55; project mgr. Nuclear Metals, Inc., Cambridge, 1955-60; mgr. fuel devel. Westinghouse Astronuclear Lab., Large, Pa., 1960-67; mgr. fuel materials tech. Advanced Reactors div. Westinghouse Electric Corp., Madison, Pa., 1967—. Author: Nuclear Engineering Fundamentals, 1964; editor: Nuclear Applications of Non-Fissionable Ceramics, 1966. Mem. Am. Soc. Metals, Am. Inst. Mining and Metall. Engrs., Am. Nuclear Soc. Subspecialties: Nuclear fission; Materials. Current work: Fast reactor oxide and carbide fuel development including: design, fabrication, irradiation testing, data analysis and fuel performance code development. Home: 129 Academy Ave Pittsburgh PA 15228 Office: Westinghouse Advanced Reactors Div PO Box 10864 Pittsburgh PA 15236

BOLTON, CHARLES THOMAS, astronomer; b. Camp Forrest, Tenn., Apr. 15, 1943; s. Clifford Theordore and Pauline Grace (Voris) B. B.S., U. Ill., 1966; M.S., U. Mich., 1968, Ph.D., 1970. Postdoctoral fellow David Dunlap Obs. U. Toronto, 1970-73, asst. prof. astronomy, 1970-71; instr. astronomy Scarborough Coll., 1971-72; asst. prof. astronomy Erindale Coll., 1972-73, U. Toronto, 1973-76, assoc. prof., 1976-80; prof., assoc. dir. David Dunlap Obs., 1980—. Contbr. articles to sci. jours. Mem. Internat. Astron. Union, Can. Astron. Soc., Am. Astron. Soc., Astron. Soc. Pacific, Royal Astron. Soc. Can. Subspecialties: Optical astronomy; Ultraviolet high energy astrophysics. Current work: Optical observations of stellar x-ray and radio sources, mass transfer in binary star systems, stellar mass loss, stellar atmospheres and envelopes, variable stars. Home: 326 Palmer Ave Richmond Hill ON Canada L4C 1P3 Office: David Dunlap Observatory PO Box 360 Richmond Hill ON Canada L4C 4Y6

BOMBIERI, ENRICO, mathematician; b. Milan, Italy, Nov. 26, 1940; came to U.S., 1977; s. Carlo and Luisa (Cambi) B.; m. Susan Russell, Jan. 21, 1967; 1 dau., Donata. Ph.D., U. Milan, 1963. Prof. U. Cagliari, Italy, 1965, U. Pisa, 1966-74, Scuola Normale Superiore, Pisa, 1974; prof. math. Inst. Advanced Study, Princeton, N.J., 1977—. Recipient Fields medal Internat. Math. Union, Vancouver, B.C., 1974; Balzan prize, Rome, 1981. Mem. Am. Acad. Arts and Scis., Accademia Nazionale Delle Scienze (Italy), Accademia Nazionale dei Lincei (corr. mem., recipient Feltrinelli prize 1976, Balzan prize 1981). Office: Inst Advanced Study Princeton NJ 08540

BONA, CONSTANTIN A(TANASIE), immunologist, microbiologist, educator; b. Turnu, Severin, Rumania, July 5, 1934; came to U.S., 1977; s. Traian and Maria B.; m. Alexandra, Dec., 1968; 1 dau., Monique. M.D., Faculty Medicine Bucharest, Rumania, 1958; Ph.D. in Med. Scis, Postgrad. Sch. Medicine Bucharest, 1965; Docteur es Science Naturelles, Faculty Sci., Paris, 1972. Chief lab. Inst. Cantacuzene, Bucharest, 1961-62; research worker Inserm, Paris, 1967-71; maitre de Recherche Pasteur Inst., Paris, 1971-77; vis. scientist NIH, Bethesda, Md., 1977-79; prof. microbiology Mt. Sinai Med. Center, N.Y.C., 1979—. Author: books, including Lymphocytes and Idiotypes, 1980, Lymphocytic Regulations by Antibodies, 1981; contbr. numerous articles to profl. jours. Greek Orthodox. Subspecialty: Immunobiology and immunology. Current work: Immunology. Patentee in field. Home: 406 E 73d St 5R New York NY 10021 Office: One Gustave L Levy Pl Microbiology New York NY 10029

BONAVIDA, BENJAMIN, educator, research cons.; b. Egypt, Jan. 29, 1940; s. Victor M. and Marie (Lamberti) B.; m. Ofra M. Bourla, Oct. 20, 1972; children: Alain, Raymond. B.A., UCLA, 1964, Ph.D. with distinction, 1968. Postdoctoral fellow Weizmann Inst., Israel, 1968-71; asst. prof. UCLA, 1971-78, assoc. prof., 1978-83, prof., 1983—. Contbr. numerous articles to profl. jours. Dept. Health and Human Services grantee, 1971—. Mem. Am. Assn. Immunologists, Transplantation Soc., Am. Assn. Cancer Research. Subspecialties: Immunobiology and immunology; Membrane biology. Current work: Cell-mediated cytotoxicity, cancer research. Office: UCLA Dept Microbiology and Immunology Los Angeles CA 90024

BOND, CLIFFORD WALTER, virologist; b. Buffalo, Apr. 7, 1937; s. Walter F. and Eva M. B.; m. Pamela Jo Bond, Aug. 12, 1978; 1 dau., Tana E. Student, Cornell U., 1954-55; B.A., SUNY, Buffalo, 1966; postgrad., Case Western Res. U., 1967-69; Ph.D., U. Ky., 1973. Technician Case Western Res. U., 1966-67, research fellow, 1967-69, U. Ky., 1969-73, postdoctoral fellow, 1974; trainee U. Calif., San Diego, 1976-77, asst. research pathologist, 1977-78; asst. prof. microbiology Mont. State U., 1978—. Contbr. in field. Leukemia Soc. Am. fellow, 1976-78. Mem. Am. Soc. Microbiology; mem. Am. Soc. Virology; Mem. AAAS. Subspecialties: Virology (biology); Molecular biology. Current work: Molecular biology of pathogenic viruses. Home: 9552 Cougar Dr Bozeman MT 59715 Office: Department Microbiology Montana State University Bozeman MT 59717

BOND, REBEKAH SUZANNE, corporate psychologist, private practitioner; b. Whitmire, S.C., Dec. 2, 1949; d. Russell Dean and O'Neill (Courtney) Bowling; m. Peter Hamlin Bond, June 19, 1971; children: Kristin, Casey, Jeffrey. B.A., Furman U., 1971; M.A., North Tex. State U., 1973; Ph.D., Tex. Woman's U., Denton, 1978. Lic. psychologist, Tex., Oreg., diplomate: Am. Acad. Behavioral Medicine. Staff psychologist Southwestern Med. Sch., Dallas, 1972-76; cons. psychologist Portland Oreg. Pub. Sch., 1977-80, Portland Dept. Human Resources, 1977-80; pvt. practitioner Pacific Psychol. Assocs., Portland, 1979-80; sr. cons. LWFW, Inc., Dallas, 1981-82; mgr. employee assistance programs Control Data Corp., Dallas, 1982—. Contbr. articles profl. jours. Mem. Am. Psychol. Assn., Tex. Psychol. Assn., Dallas Psychol. Assn., Am. Acad. Behavioral Medicine. Current work: Behavioral medicine, stress-related illness, employee assistance programming. Home: 5013 Clover Valley Dr The Colony TX 75056 Office: Control Data Corp 14801 Quorum Dr Suite 400 Dallas TX 75240

BOND, WALTER DAYTON, chemist; b. Lebanon, Tenn., Mar. 21, 1932; s. Charles Powell and Susie Anna (Shipp) B.; m. Martha Ruth Ricks, Aug. 25, 1955; children: Stephanie, Jeffrey, Rachael. B.S., Middle Tenn. State U., 1953; Ph.D., Vanderbilt U., 1957. Chemist Oak Ridge Nat. Lab., 1957—, group leader, 1960—; guest scientist Swiss Fed. Inst. Reactor Research, Wurenlingen, 1971-72. Contbr. articles to profl. jours., chpts. to books. Mem. Am. Chem. Soc., Am. Nuclear Soc. Episcopalian. Subspecialties: Inorganic chemistry; Physical chemistry. Current work: Research and development studies of the nuclear fuel cycle in regard to fuel reprocessing, waste management and fuel fabrication. Home: 6704 Stonemill Rd Knoxville TN 37919 Office: Oak Ridge National Laboratory PO Box X Oak Ridge TN 37830

BOND, WILLIAM HOLMES, scientist; b. Toronto, Ont., Can., Sept. 20, 1916; s. Ernest Albert and Edna Lucille (Haines) B.; m. Virginia King, Jan. 3, 1949 (dec.); children: Roy Alan, James King, Robert Simpson, Linda Jane. B.S., U. Chgo., 1940, M.D., 1942. Diplomate: Am. Bd. Internal Medicine. Intern Vanderbilt U. Hosp., Nashville, 1942-43; resident in internal medicine Ind. U. Med. Ctr., 1948-51; instr. medicine Ind. U., 1952-55, asst. prof., 1955-60, assoc. prof., 1960-67, prof., 1967—. Contbr. articles on chemotherapy to med. jours. Served to capt. AUS, 1944-46. Fellow ACP; mem. AMA, Am. Soc. Hematology, Central Soc. Clin. Research, Am. Assn. Cancer Research, Phi Betta Kappa, Sigma Xi, Alpha Omega Alpha. Republican. Roman Catholic. Subspecialties: Chemotherapy; Hematology. Current work: Cancer chemotherapy. Home: 4525 W 59th St Indianapolis IN 46254 Office: 1100 W Michigan St Indianapolis IN 46223

BONDY, JONATHAN, computer systems engineer; b. Atlanta, Apr. 29, 1951; s. Phillip Kramer and Sarah Berheim (Ernst) B. B.S. in Math., Haverford Coll., 1973. Cons. to dir. advanced devel. orgn. Burroughs Corp., 1973-74; programmer Ketron, Inc., 1974-77; computer systems engr. Gen. Electric Space Div., King of Prussia, Pa., 1977-79, 82—; project mgr., computer systems engr. Energy Data Systems, Wayne, Pa., 1979-81. Contbr. articles in field to profl. jours. Mem. IEEE, Assn. Computing Machinery, UCSDp-System Users Soc., Phila. Area Computer Soc. Subspecialties: Software engineering; Programming languages. Current work: Software tools, fault-tolerant computing. Home: Box 148 Ardmore PA 19003

BONE, DONALD ROBERT, microbiologist/hybridoma technologist, virologist; b. Minot, N.D., July 13, 1944; s. Lawrence Henry and Ora Marie (Sather) B.; m. Jeri Faye Selleseth, Aug. 27, 1966; children: Aaron, Adam. B.A., Pasadena Coll., 1966; Ph.D. in Microbiology, U. Calif.-Davis, 1971. Postdoctoral fellow Baylor Coll. Medicine, Houston, 1971-74; research virologist Lederle Labs., Pearl River, N.Y., 1974-77; postdoctoral fellow Salk Inst., La Jolla, Calif., 1977-78; asst. prof. U. Okla.-Norman, 1978-81, adj. asst. prof. zoology, 1981—; dir. research and devel. Celtek, Inc., Norman, 1981—; adj. asst. prof. microbiology and immunology U. Okla. Coll. Medicine-Oklahoma City, 1980—. Contbr. articles to profl. jours. Mem. AAAS, Am. Soc. Microbiology. Subspecialties: Hybridoma Technology; Cellular engineering. Current work: The development and characterization of monoclonal antibodies specific for microbial pathogens, hormones, and other molecules of clinical significance. Office: 102 W Eufaula St Norman OK 73069

BONGAR, BRUCE MICHAEL, clinical and consulting psychologist, medical psychologist; b. Madison, Wis., Sept. 20, 1950; s. Larry and Elaine B. B.A. with distinction, U. Wis.-Madison, 1972; M.S., U. So. Calif., 1975, Ph.D., 1977. Lic. psychologist, Calif., Mass., Ariz., Conn., diplomate: Am. Bd. Psychotherapy. Staff clin. psychologist Children's Hosp., Los Angeles, 1977-79, sr. psychologist, 1979-80; pvt. practice clin. and health psychology, Santa Monica, Calif., 1978—, cons. behavioral medicine, 1978—; bd. dirs. Internat. Bd. Medicine and Psychology, Santa Ana, Calif., 1982—; Oral commr. Calif. B. Behavioral Sci. Examiners, 1981, Calif. Bd. Med. Quality Assurance Psychology Examining Com., 1982. Author: (with Lyn Paul Taylor) Clinical Applications in Biofeedback Therapy, 1976. State Wis. fgn. lang. scholar, Mexico City, 1969; State Calif. fellow, 1975-77. Fellow Explorers Club; mem. Soc. Behavioral Medicine (charter), A. Psychol. Assn., Western Psychol. Assn., Calif. State Psychol. Assn., Los Angeles County Psychol. Assn., Brit. Psycol. Soc., Am. Soc. Clin. Hypnosis, Am. Coll. Sports Medicine, Biofeedback Soc. Am., Acad. Psychologists in Marital, Sex, and Family Therapy. Current work: Research in behavioral medicine, health psychology and psychotherapy, clinical interventions in treatment of psychological reaction to phsyical disease, psychosomatic and psychophysiological disorders; stress control and occupational burnout research and consultation; psychotherapeutic interventions in stress control and performance enhancement for executives, managers, and world-class athletes. Office: 2700 Neilson Way Apt 631 Santa Monica CA 90405

BONHAM, HAROLD FLORIAN, JR., economic geologist, consultant, educator; b. Los Angeles, Sept. 1, 1928; s. Harold Florien and Viola Violet (Clopine) B.; m. Sally Mae Reimer, Sept. 7, 1952; children: Cynthia Jean, Douglas Craig, Gary Stephen. A.A., U. Calif.-Berkeley, 1951; B.A., UCLA, 1954; M.S., U. Nev.-Reno, 1963. Geologist So. Pacific Co., 1955-61; grad. research asst. Nev. Bur. Mines, Reno, 1961-62, asst. mining geologist, 1963-67, asso. mining geologist, 1967-74, mining geologist, prof. geology, 1974—; cons., Reno, 1970-82. Contbr. articles to sci. jours. Trustee Palamino Valley Gen. Improvement Dist., Reno. Served with USNR, 1946-49. Fellow Geol. Soc. Am.; mem. Soc. Econ. Geologists. Subspecialty: Geology. Current work: Volconology, precious metal deposits, rare metal deposits in volcanic environments. Home: 2100 Right Hand Canyon Rd Reno NV 89510 Office: Nev Bur Mines and Geology Univ Nev Reno NV 89557

BONNER, JAMES, educator; b. Ansley, Nebr., Sept. 1, 1910; s. Walter Daniel and Grace (Gaylord) B.; m. Ingelore Silberbach, Nov. 10, 1967; children by previous marriage—Joey, James Jose. A.B., U. Utah, 1931; Ph.D., Calif. Inst. Tech., 1934. NRC fellow Univs. Utrecht, Leiden, Zürich, 1934-35; faculty Calif. Inst. Tech., Pasadena, 1935-81, prof. biology, 1946-81; chmn. bd., chief exec. officer Phytogen, Inc., 1981—. Author: Plant Biochemistry, 1950, 3d edit., 1977, Principles of Plant Physiology, 1952, The Next 100 Years, 1957, The Nucleohistones, 1964, The Molecular Biology of Development, 1965, The Next 90 Years, 1967, The Next 80 Years, 1977. Mem. Nat. Acad. Sci. Subspecialties: Genetics and genetic engineering (agriculture); Molecular biology. Current work: Research in gentic engineering for plants. Home: 1914 Edgewood Dr South Pasadena CA 91030

BONNER, JAMES JOSE, educator; b. Los Angeles, May 1, 1950; s. James Frederick and Harriet (Rees) B.; m. Monica Atkinson, Mar. 16, 1974. Ph.D., M.I.T., 1976. Am. Cancer Soc. postdoctoral fellow U. Calif.-San Francisco, 1977-78; asst. prof. biology Ind. U., Bloomington, 1979—. Contbr. articles to profl. jours. Am. Cancer Soc. fellow, 1977-78; NIH research grantee, 1979—. Mem. Genetics Soc. Am., Am. Soc. Cell Biology, Am. Soc. Developmental Biology. Subspecialties: Molecular biology; Gene actions. Current work: Molecular and genetic analysis of hyperthermic stress; mechanisms of induction of heat shock-induced proteins and mechanisms whereby these proteins protect cells from lethality at high temperatures. Office: Dept Biology Ind Univ Bloomington IN 47406

BONNER, JOHN TYLER, biology educator; b. N.Y.C., May 12, 1920; s. Paul Hyde and Lilly Marguerite (Stehli) B.; m. Ruth Anna Graham, July 11, 1942; children: Rebecca, Jonathan Graham, Jeremy Tyndall, Andrew Duncan. Grad., Phillips Exeter Acad., 1937; B.Sc., Harvard U., 1941, M.A., 1942, Ph.D. (Jr. fellow 1942, 46-47), 1947; D.Sc. (hon.), Middlebury Coll., 1970. Asst. to asso. prof. Princeton U., 1947-58, prof., 1958—, chmn. dept. biology, 1965-77, 83-84; lectr. embryology Marine Biol. Lab., Woods Hole, Mass., 1951-52; spl. lectr. U. London, 1957; Bklyn. Coll., 1966; trustee Biol. Abstracts, 1958-63; Mem. bd. editors Princeton U. Press, 1965-68, 71, trustee, 1976-82. Author: Morphogenesis, 1952, Cells and Societies, 1955, The Evolution of Development, 1958, The Cellular Slime Molds, 1959, rev. edit. 1967, The Ideas of Biology, 1962, Size and Cycle, 1965, The Scale of Nature, 1969, On Development, 1974, The Evolution of Culture in Animals, 1980, Lifesite; also scientific papers.; Editor: Growth and Form, 1961, Evolution and Development, 1981; Asso. editor: Am. Scientist, 1961-69; editorial bd.: Am. Naturalist, 1958-60, 66-68, Jour. Gen. Physiology, 1962-69, Growth, 1955—, Differentiation, 1976—. Served from pvt. to 1st lt. USAC, 1942-46; staff aero. med. lab.; Wright Field, Dayton, Ohio. Sheldon traveling fellow, Panama, 1941; Rockefeller traveling fellow, France, 1953; Guggenheim fellow, Scotland, 1958, 71-72; recipient Selman A. Waksman award for contbns. to microbiology Theobold Smith Soc.; NSF sr. postdoctoral fellow, 1963. Fellow Am. Acad. Arts and Scis.; mem. Am. Soc. Naturalists, Soc. Growth and Devel., Mycol. Soc. Am., Am. Philos. Soc., Nat. Acad. Scis., Phi Beta Kappa, Sigma Xi. Subspecialties: Developmental biology; Evolutionary biology. Current work: Delevelopment of cellular slime molds, evolution and development. Home: 148 Mercer St Princeton NJ 08540

BONNER, THOMAS PATRICK, university dean; b. Sharon, Pa., Sept. 29, 1942; s. William and Naomi (Kennedy) B.; m. Mary C. Fullerton, June 3, 1964; children: Anne, Matthew. B.S., Ind. U. Pa., 1964; Ph.D., U. Cin., 1969. NIH fellow U. Notre Dame, 1971; asst. prof. biology SUNY-Brockport, 1971-74, assoc. prof., 1974-79; lectr. Leeds (Eng.) U., 1979-80; dean Alt. Coll., SUNY-Brockport, 1981—. Contbr. articles to profl. jours. Recipient Chester A. Herrick award Midwestern Conf. Parsitology, 1969; NIH postdoctoral fellow, 1971. Mem. Am. Soc. Parasitology, Am. Soc. Cell Biology, Am. Assn. Higher Edn., Sigma Xi. Subspecialties: Parasitology; Cell biology. Current work: Parsitology, developmental biology of nematode parasites, collagen synthesis, ultrastructure. Address: Alternate Coll SUNY Brockport NY 14420

BONNET, JUAN AMEDEE, nuclear engineer, environmental research administrator; b. Santurce, P.R., Apr. 22, 1939; s. Juan A. and Josefa L. (Diez) B.; m. Wally Vargas, Dec. 27, 1963; children: Juan, Carlos, Antonio, Luis, Gerardo, Gabriel. B.S. in Chem. Engring, U. Mich., 1960, Ph.D. in Nuclear Engring, 1971; M.S. in Nuclear Tech, U. P.R., 1961. Registered profl. engr., P.R. Safety and analysis engr. P.R. Water Resources Authority, 1962-67, head nuclear engring. dept., 1971-73, head environ. protection, quality assurance and nuclear divs., 1972-75, asst. exec. dir. planning and engring., 1975-77; dir. Center for Energy and Environment Research, U. P.R., San Juan, 1977—; mem. adj. faculty P.R. Technol. U., 1973-77; adhonorem prof. Sch. Medicine, San Juan, 1979—; adj. prof. Engring. Sch. Mayaguez, P.R., 1979-80; asst. prof. Bayamon Technol. U. Coll., 1980—; cons. Energy and Environ. Engring., 1971—; pres. Profl. Engrs., Architects and Surveyors Exam. Bd. P.R., 1979—. Contbr. articles on energy and environ. matters to profl. publs. Bd. dirs. Rincon Fund Raising, P.R. Soc. Mentally Retarded Children, 1967; mem. Caribbean Islands Directorate of UNESCO-U.S. Com. Men and the Biosphere, 1979—; bd. dirs. U. P.R., 1978—, So. Solar Energy Center, 1979-82. Named Outstanding Young Scientist of P.R. Jaycees, 1978, Disting. Engr. Tau Beta Pi, 1978; Sci. award P.R. Mobil Oil, 1981. Mem. Am. Nuclear Soc., P.R. Inst. Chem. Engrs. (pres. 1977-79), Interam. Confedn. Chem. Engrs. (gen. sec. 1976-78), Internat. Solar Energy Assn., Nat. Soc. Profl. Engrs., P.R. Chemist Soc. (editorial bd. 1979—), Pan Am. Union of Assns. Engrs. (energy com. 1972—), Ateneo de P.R., P.R. Acad. Arts and Scis., Internat. Assn. Hydrogen Energy, Assn. Engrs. and Surveyors (dir. 1967-76), Sigma Xi, Phi Eta Mu. Roman Catholic. Subspecialty: Energy research administration. Home: Calle 1 No H-7 Los Frailes Norte Guaynabo PR 00657 Office: GPO Box 3682 San Juan PR 00936 In Puerto Rico, public needs are so very great that public service is especially challenging and its spiritual rewards are very gratifying. The opportunity to serve humanity enlarges one's spirit.

BONO, PETER RICHARD, computer scientist, consultant, lecturer; b. Honolulu, July 25, 1945; s. Leonard and Pauline Milton (Currier) B.; m. Elaine Marie Handsaker, Aug. 18, 1968; 1 dau., Diane. A.B., Harvard U., 1967; M.S., U. Mich., 1969, Ph.D., 1972. Computer specialist Naval Underwater Systems Center, New London, Conn., 1975-81; v.p. product devel. Athena Systems Inc., Pawcatuck, Conn., 1981—; chmn. tech. com. on computer graphics Am. Nat. Standards Inst., 1979—; chief U.S. del. working group on computer graphics Internat. Standards Orgn., 1980—. Contbr.: chpt. to book, articles to publs. in field including Computer Graphics World. Served to lt. USN, 1972-75. Mem. Assn. for Computing Machinery, Spl. Interest Group Graphics (treas. 1977-79), Sigma Xi, Phi Kappa Phi. Subspecialties: Graphics, image processing, and pattern recognition; Software engineering. Current work: Man-machine interaction; user-friendly systems; non-procedural languages; computer graphics standards; device-independent software tools. Office: Athena Systems Inc N Stonington Profl Ctr Box 410 Rt 2 and 184 North Stonington CT 06359

BONVENTRE, JOSEPH VINCENT, nephrologist, researcher; b. Bklyn., May 13, 1949; s. Vincent and Philomena (Orsino) B.; m. Kristina Cannon; 1 dau., Joanna. B.S. with distinction, Cornell U., 1970; M.D., Harvard U., 1976, Ph.D., 1979. Diplomate: Am. Bd. Internal Medicine. Instr. medicine Harvard Med. Sch., Boston, 1980-81, asst. prof. medicine, 1981—; asst. medicine Mass. Gen. Hosp., Boston, 1981—. Author: Key References in Nephrology, 1982. Recipient NIH new investigator research award, 1980-83. Mem. Am. Soc. Nephrology, Am. Fedn. Clin. Research, Am. Physiol. Soc., Internat. Soc. Nephrology, N.Y. Acad. Scis., Tau Beta Pi, Phi Kappa Phi. Roman Catholic. Subspecialties: Nephrology; Cell biology (medicine). Current work: Cellular aspects of anoxic and ischemic injury; renal concentration mechanisms; cell volume regulation; cellular calcium homeostasis. Home: 1 Sylvan Way Wayland MA 01778 Office: Mass Gen Hosp Fruit St Boston MA 02114

BOODMAN, DAVID MORRIS, systems scientist; b. Pitts., July 3, 1924; m., 1948; 2 children. B.S., U. Pitts., 1944, Ph.D. in Phys. Chemistry, 1950. Sr. staff mem. ops. evaluation group MIT, Cambridge, 1950-60, sr. staff mem. ops. research sect., 1960-72; v.p. Arthur D. Little, Inc., Cambridge, 1972—; ops. analyst, mem. staff Comdr.-in-Chief Pacific Fleet, USN, 1951, ops. devel. force, 1953-54, mem. staff, 1958. Fellow AAAS; mem. Math. Assn. Am., Ops. Research Soc. Am. Subspecialty: Operations research (engineering). Office: Arthur D Little Inc 25 Acorn Park Cambridge MA 02140

BOOKER, HAROLD EDWARD, neurologist; b. Indpls., Sept. 26, 1932; s. Clyde Robert and Evangeline (Stewart) B.; m. Sandra Jeanette Eberly, June 8, 1957; children—Patricia Lynn, Stewart Clyde. M.D., Ind. U., 1957. Intern Marion County Gen. Hosp., Indpls., 1957-58; resident Ind. U. Med. Center, Indpls., 1958-61; asst. prof. neurology U. Wis., Madison, 1964-69, asso. prof., 1968-72, prof., 1972-79; dir. neurology service VA, Washington, 1979—; mem. Nat. Adv. Neurol. and Communicative Disorders and Stroke Council, 1979—; cons. epilepsy br. Nat. Inst. Neurol. and Communicative Disorders and Stroke, 1968-79. Cons. editor: Analytical Toxicology, 1976-79. Bd. dirs. Epilepsy Found. Am., 1977—, mem. profl. adv. bd., 1976—; pres. Wis. Epilepsy Assn., 1974-78; chmn. Wis. Advocacy Coalition, 1977. Served with USAF, 1961-64. Mem. Am. Neurol. Assn., Am. Acad. Neurology, Am. Epilepsy Soc., Am. Electroencephalographic Soc., Soc. Neurosci. Subspecialty: Neurology. Office: 810 Vermont Ave NW Washington DC 20420

BOOKER, HENRY GEORGE, educator, scientist; b. Barking, Essex, Eng., Dec. 14, 1910; came to U.S., 1948, naturalized, 1952; s. Charles Henry and Gertrude Mary (Ratcliffe) B.; m. Adelaide Mary McNish, July 9, 1938; children—John Ratcliffe, Robert William, Mary Adelaide, Alice. Student, Palmer's Sch., Grays, Essex, 1921-30; B.A., Christ's Coll., Cambridge, 1933, Ph.D., 1936; Guggenheim fellow, Cambridge U., 1954-55. Fellow Christ's Coll., 1935-48; sci. officer Ministry Aircraft Prodn., London, 1940-45; lectr. Cambridge U., 1945-48; prof. elec. engring. Cornell U., 1948-65, dir. sch. elec. engring., 1959-63; asso. dir. Center Radio Physics and Space Research, IBM; prof. engring. and applied math., 1962-65; prof. applied physics U. Calif. at San Diego, 1965—. Author: An Approach to Electrical Science, 1959, A Vector Approach to Oscillations, 1965, also sci. papers on radio wave propagation. Jr. intermediate and sr. county scholarships, Essex, Eng., 1920-30; Entrance scholarship Christ's Coll., 1930; Allen scholarship, 1934-35; Smith's prize, 1935; Duddell, Kelvin and instn. premiums Instn. Elec. Engrs., London, 1948-50. Fellow IEEE; mem. Nat. Acad. Scis., Internat. Union Radio Sci. (hon. pres. 1979—), Sigma Xi. Subspecialties: Remote sensing (atmospheric science); Plasma physics. Current work: Electromagnetic theory and radio wave propagation. Home: 8696 Dunaway Dr La Jolla CA 92037 Office: Dept Electrical Engineering and Computer Sci U Calif at San Diego La Jolla CA 92093

BOOKMAN, CHARLES ARTHUR, marine affairs program manager; b. Washington, Mar. 19, 1948; s. George B. and Janet (Schrank) B.; m. Betsy Cheney, Oct. 5, 1977; children: Tyras, Zachary. B.A. in Geography, Columbia U., 1970; M.A. in Marine Affairs, U. R.I., 1974. Research asst. Lamont-Doherty Geol. Obs., Palisades, N.Y., 1970-73; natural resources planner Md. Dept. Natural Resources, Annapolis, 1974-77; cons. Charles Bookman Assoc., Washington, 1977-79; dir. Washington ops. Rogers, Golden & Halpern, Inc., Reston, Va., 1979-80; staff officer Nat. Research Council, Washington, 1980—. Project mgr., author for com.: Productivity Improvements in U.S. Naval Shipbuilding, 1982, Ocean Engineering for Ocean Thermal Energy Conversion, 1982, Marine Salvage in the United States, 1982, Safety & Offshore Oil, 1981. Mem. Soc. Naval Architects and Marine Engrs., Marine Tech. Soc. Subspecialties: Ocean engineering; Offshore technology. Current work: Ocean resource development and operations; maritime commerce; relation of engineering technology to human resources and activities, and science and public policies. Home: 7501 Arden Rd Cabin John MD 20818 Office: Marine Bd Nat Research Council 2101 Constitution Ave Washington DC 20418

BOOKMAN, SUSAN STONE, educator, botanical and ecological researcher; b. Fullerton, Calif., Mar. 17, 1954; d. Donald Sherwood and Charity (Murray) Stone; m. Peter Alan Bookman, Aug. 16, 1980; 1 dau., Marie Elizabeth. B.S. in Biology, U. Colo.-Boulder, 1976; M.S., Wash. State U., 1979, Ph.D. in Botany, 1982. Curatorial asst. Marion Ownbey Herbarium, Pullman, Wash., 1977; teaching asst. Wash. State U., 1977-81, research technician, 1978, grad. research asst., 1981-82; sci. tchr. Tatnall Upper Sch., Wilmington, Del., 1982—. Named Outstanding Teaching Asst. in Biology Wash. State U., 1979; Wash. State U. travel grantee, 1979, 81; NSF grantee, 1980-81. Mem. Am. Soc. Plant Taxonomists, Bot. Soc. Am., Ecol. Soc. Am., Soc. Study Evolution, Phi Beta Kappa, Sigma Xi, Phi Kappa Phi. Subspecialties: Evolutionary biology; Population biology. Current work: Floral morphology in relation to pollination and fertilization; fruit production and abortion; reproductive biology of Ascelpias; control of fruit and seed size and number; pollination biology. Home: 204 Brecks Ln Wilmington DE 19807 Office: Tatnall Sch 1501 Barley Mill Rd Wilmington DE 19807

BOOKSTEIN, ABRAHAM, information science educator; b. N.Y.C., Mar. 22, 1940; s. Alex and Doris (Cohen) B.; m. Marguerite Lindley Vickers, June 20, 1968. B.S., C.C.N.Y., 1961; M.S., U. Calif.-Berkeley, 1966; Ph.D., Yeshiva U.-N.Y.C., 1969; M.A., U. Chgo., 1970. Prof. info. sci. and behavioral scis. U. Chgo., 1970—. Editor: Prospects for Change in Bibliographic Control, 1977; editorial bd.: Library Quar., Info. Processing and Mgmt; dir.: ATLA Religion Indexes; contbr. articles in field to profl. jours. NSF grantee, 1981; DFI (Sweden) grantee, 1982. Mem. Am. Soc. Info. Sci., Assn. Computing Machinery. Subspecialties: Information systems, storage, and retrieval (computer science); Operations research (mathematics). Current work: Application of probability theory to information retrieval systems; bibliometric distributions. Home: 5445 S East View Park Chicago IL 60615 Office: U Chgo Grad Library Sch 1100 E 57th St Chicago IL 60637

BOONE, DONALD H(ERBERT), metall. research scientist, cons., educator; b. Moline, Ill., Aug. 27, 1936; s. W. Donald and E. Eugenia (Duncan) B.; m. Karolin M., Aug. 18, 1956; children: Dana L.; Christina A., Cynthia J. B.S. with high honors, U. Ill., 1957, M.S., 1959, Ph.D., 1962. Supr. Pratt & Whitney Aircraft, East Hartford, Conn., 1961-74; tech. mgr. Airco Temescal, Berkeley, Calif., 1974-78; staff scientist Lawrence Berkeley Lab., U. Calif., Berkeley, 1979—; adj. prof. mech. engring. Naval Postgrad. Sch., 1978—; cons. high temperature materials and coatings. Contbr. numerous articles to profl. jours., confs. Served to 1st lt. C.E. U.S. Army, 1962-63. Mem. Am. Soc. Metals, AIME, ASME, AIEE, Am. Vacuum Soc. Subspecialties: Clad metals and coating technology; High-temperature materials. Current work: High temperature coatings and materials. Patentee in field. Home: 2412 Cascade Dr Walnut Creek CA 94598 Office: Lawrence Berkeley Lab U Calif Bldg 62-351 Berkeley CA 94720

BOOR, MYRON VERNON, psychologist, researcher; b. Wadena, Minn., Dec. 21, 1942; s. Vernon LeRoy and Rosella Katharine (Eckhoff) B. B.S., U. Iowa, 1945; M.S., So. Ill. U.-Carbondale, 1947, Ph.D., 1970; M.S. in Hygiene, U.Pitts., 1981. Cert. psychologist, Kans. R.I. Clin. psychology intern Galesburg (Ill.) State Research Hosp., 1968-69; research psychologist Milw. County Mental Health Ctr., Milw., 1970-72; asst. prof. psychology Ft. Hays State U., 1972-76, assoc. prof., 1976-79; NIMH postdoctoral fellow U. Pitts., 1979-81; psychologist, asst. prof. R.I. Hosp.-Brown U., 1981-. Contbr. articles to profl. jours. USPHS fellow, 1965-67; NIMH postdoctoral fellow, 1979-81. Mem. Am. Psychol. Assn., Am. Assn. Suicidology, Soc. Psychol. Study Social Issues, AAAS. Subspecialties: Behavioral psychology; Social psychology. Current work: Social networks; psychosocial adjustment; psychiatric epidemiology; suicidology; psychopathology; clinical psychology. Home: 825 Pontiac Ave Apt 17302 Craston RI 02910 Office: Dept Psychiatry RI Hosp 593 Eddy St Providence RI 02902

BOOSTROM, EUGENE RICHARD, international development physician, consultant; b. Moline, Ill., June 14, 1941; s. Merlyn E. and Irene M. (DeGreve) B.; m. Rebecca Weber, Apr. 15, 1963 (div. 1967); 1 son, Eric Weber (dec.); m. Irene Anne Jillson, May 1, 1983. B.S., St. Louis U., 1963; M.D., U. Ill, 1968; M.P.H., UCLA, 1970, Dr.P.H., 1974. Lic. physician, N.Y., Calif. Resident, chief resident in preventive medicine UCLA Sch. Medicine, 1969-72; internt. pub. health evaluation adv. UCLA and Am. Pub. Health Assn., Los Angeles and Washington, 1974-77; Academia de Ciencias Médicas, Fisicas y Naturales, Guatemala, 1977-79; systems devel. specialist U. Hawaii Med. Sch., Honolulu, 1979-83; pub. health med. adv. Near East Bur. AID, Washington, 1981—; mem. rural health tech. adv. com. Egyptian Ministry of Health, Cairo, 1977—; cons. WHO, 1981—, Am. Pub. Health Assn., 1973—, AID, 1973—. Contbr. articles to profl. jours. Mem. adv. com. UCLA Child Health Care Prepayment Plan, 1970-73; mem. dean's com. Latin Am. studies UCLA, 1970-72. Served to maj. M.C. U.S. Army, 1972-74. Woodrow Wilson fellow, 1963; NSF fellow, 1963; NASA fellow, 1963; NDEA fellow, 1963; USPHS fellow, 1969-72; recipient cert. of appreciation Peruvian Ministry of Health, 1970. Mem. Am. Psychol. Assn., Am. Pub. Health Assn., Psi Chi, Delta Omega. Subspecialties: Preventive medicine; Health services research. Current work: Design, development, analysis and evaluation of health services systems in developing countries; application of modern technology (e.g., small computers, oral rehydration) and of effective management and decisionmaking techniques to the developmental implementation of cost-effective high-coverage health services; competency-based training for health services providers and managers. Address: NE/TECH/HPN/NS6663/AID Washington DC 20523

BOOTH, JAMES ALBERT, research co. exec.; b. Salem, Ohio, Dec. 14, 1946; s. Kenneth Bishop and Helen Elizabeth (Kelly) B.; m. Anita Jean, Feb. 10, 1974; 1 dau., Jennifer Lynn. B.S., Bowling Green State U., 1968, M.S., 1973; M.S. in Nuclear Engring, Ohio State U., 1974. Registered profl. engr., Ohio. Sr. research engr. Monsanto Research Corp., Dayton, Ohio, 1974-81, mgr. engring. research and devel., 1981—. Contbr.: 18 articles on 19th-20th century physicist to Aca. Am. Ency, 1980. Served to sgt. U.S. Army, 1969-71; Vietnam. Decorated Bronze Star, Army Commendation medal. Mem. Am. Nuclear Soc., Am. Soc. Quality Control (cert. quality engr. 1979), ASME, History of Sci. Soc. Republican. Methodist. Subspecialties: Nuclear engineering; Nuclear fission. Current work: Design of neutron and gamma sources; irradiation effects on stainless steels; quality assurance in nuclear industry. Home: 3141 Westview Dr: Xenia OH 45385 Office: PO Box 8 Sta B Dayton OH 45418

BOOTZIN, RICHARD RONALD, psychologist, educator; b. Milw., Feb. 25, 1940; s. Arnold and Evelyn (Myslis) B.; m. Maris Kay Pittelman, Dec. 27, 1959; children—Deborah Jeanne, Helaine Beth. B.S., U. Wis., 1963; M.S., Purdue U., 1966, Ph.D., 1968. Ward psychologist Palo Alto (Calif.) VA Hosp., 1967-68; mem. faculty Northwestern U., 1968—, prof. psychology, 1978—, chmn. dept., 1980—; vis. asso. prof. Stanford U., 1977-78; prin. investigator various grants, 1971—. Author books, papers, revs. in field; editorial cons. profl. jours. Fellow Am. Psychol. Assn.; mem. Assn. Advancement Behavior Therapy (dir. 1973-77, chmn. publs. bd. 1975-77), Sleep Research Soc. Jewish. Subspecialty: Behavioral psychology. Office: Dept Psychology Northwestern Univ Evanston IL 60201

BOPP, CHARLES DAN, development engineer; b. Decatur, Ill., Feb. 4, 1923; s. Charles D. and Edna (Rybolt) B.; m. Mary E. Mcleod, Mar. 13, 1971. B.S. in Chem. Engring, Purdue U., 1944. Chemist Tenn. Eastman Co., Oak Ridge, 1944-47; devel. engr. Union Carbide Corp., Oak Ridge, 1947—. Fellow Am. Inst. Chemists; mem. Am. Chem. Soc., Am. Ceramic Soc., Am. Inst. Chem. Engrs. (assoc.), Am. Nuclear Soc. Subspecialties: High-temperature materials; Kinetics. Current work: Graphite reaction with gaseous oxidants, diffusion rate of gases in graphite proes. Home: 72 Outer Dr Oak Ridge TN 37830

BOPP, LAWRENCE HOWARD, microbiologist; b. Schenectady, Mar. 10, 1949; s. Charles John, Jr. and Sybil Alfreda (Birch) B.; m. Betty Ann Stauning, June 19, 1971; children: Stacy Sara, Joshua Lawrence. B.S., SUNY, Albany, 1971; M.S., Rensselaer Poly. Inst., 1976, Ph.D., 1980. Asso. staff scientist Gen. Electric Corp. Research and Devel. Center, Schenectady, 1977-80, staff scientist, 1980—. Contbr. article to publ. Mem. Am. Soc. Microbiology, AAAS, Sigma Xi. Methodist. Club: Hudson-Mohawk Road Runners. Subspecialties: Microbiology; Genetics and genetic engineering (medicine). Current work: Bacterial/metal interactions and associated genetics. Patentee.

Office: Gen Electric Research and Devel Center Bldg K-1 Room 3B14 Schenectady NY 12301

BORDEN, ERNEST CARLETON, physician; b. Norwalk, Conn., July 12, 1939; s. Joseph Carleton and Violet Ernette (Lanneau) B.; m. Louise Dise, June 24, 1967; children: Kristin Louise, Sandra Lanneau. A.B., Harvard U., 1961; M.D., Duke U., 1966. Diplomate: Am. Bd. Internal Medicine. Intern Duke U. Med. Ctr., 1966-67; asst. resident in internal medicine Hosp. of U. Pa., 1967-68; med. officer Viropathology Lab., Nat. Communicable Disease Ctr., USPHS, Atlanta, 1968-70; clin. instr. dept. medicine Emory U. Sch. Medicine also Grady Meml. Hosp., Atlanta, 1968-70; mem. med. attending staff Peachtree Hosp., Atlanta; postdoctural fellow oncology div. dept. medicine Johns Hopkins U. Sch. Medicine, Balt., 1970-73; asst. prof. div. clin. oncology and depts. human oncology and medicine Wis. Clin. Cancer Ctr., Univ. Hosps. and Sch. Medicine, U. Wis.-Madison, 1973-79, assoc. prof., 1979-83; prof. Wis. Clin. Cancer Ctr., Univ. Hosps. and Sch. Medicine, U. Wis-Madison, 1983—; chief div. clin. oncology William S. Middleton VA Hosp., 1977-81; cons. staff Madison Gen. Hosp., 1974—. Contbr. articles to profl. jours.; editorial bd.: Jour. Interferon Research, 1980—; editorial bd: Cancer Immunology and Immunotherapy, 1981—, Jour. Biologic Response Modifiers, 1982—, Investigational New Drugs, 1982—. Mem. AAAS, ACP, Am. Soc. Microbiology, Am. Assn. Cancer Research, Am. Fedn. Clin. Research, Eastern Coop. Oncology Group, Am. Soc. Clin. Oncology, Am. Assn. Immunologists. Unitarian. Subspecialties: Cancer research (medicine); Immunopharmacology. Current work: Mechanism of action and clinical application of immunomodulators, particularly interferons. Office: 600 Highland Ave K4/414 CSC Madison WI 53792

BORENSTEIN, JEFFREY MARK, manufacturing executive; b. Newton, Mass., June 5, 1941; s. Milton Conrad and D. Anne (Shapiro) R.; m. Margaret Ruth Beller, Dec. 4, 1977; 1 dau., Danielle Louise. A.B., Harvard U., 1968, A.M., 1971, Ph.D., 1975. Adj. asst. prof. physics U. N.H., 1975-76; exec. v.p. Sweetheart Paper Products Co., Inc., Chelsea, Mass., 1976—; sec. Md. Baking Co. of Atlanta, 1976—. Mem. Am. Inst. Physics. Jewish (trustee 1980). Subspecialty: Particle physics. Current work: Gauge field theory. Home: 200 Highland St West Newton MA 02165 Office: Sweetheart Paper Products Co Inc 191 Williams St Chelsea MA 02150

BORG, IRIS YVONNE, geologist; b. San Francisco, Oct. 6, 1928; d. Thomas Dean and Anna Louisa (Siegl) Parnell; m. Richard John Borg, Nov. 3, 1951; 1 son, Lars E. B.S., U. Calif.-Berkeley, 1951, Ph.D., 1954. Vis. fellow Princeton (N.J.) U., 1954-59; research assoc. geology U. Calif.-Berkeley, 1960-61; geologist Lawrence Livermore Nat. Lab., Calif., 1961—; cons. Shell Devel., Houston, 1955-67; com. mem. Vis. Com. for Earth Scis. M.I.T., Cambridge, Mass., 1978-84; vis. lectr. Am. U. Cairo, Egypt, 1971. Author: Calculated X-Ray Powder Patterns, 1969; editor: Flow and Fracture of Rocks, 1972; contbr. articles in field to profl. jours. Guggenheim fellow, Zurich, 1968-69. Fellow Mineral. Soc. Am. (councillor 1974-77), Geol. Soc. Am.; mem. Am. Geophys. Union (sec. tectnophysics 1961-64), Am. Assn. Petroleum Geologists, Am. Geol. Inst. (mem. governing bd. 1975-77). Democrat. Subspecialties: Mineralogy; Fuels. Current work: Energy assessments with focus on liquid and gaseous hydrocarbons and associated recovery tech. Home: 4681 Almond Circle Livermore CA 94550 Office: Lawrence Livermore Nat Lab East Ave Livermore CA 94550

BORG, RICHARD JOHN, research chemist; b. Los Angeles, Oct. 18, 1925; s. Melvin John and Lillian Winefred (Smith) B.; m. Iris Yvonne Parnell, Nov. 3, 1950; 1 son, Lars Eric. B.Sc., U. Calif.-Berkeley, 1950, M.Sc., 1952; Ph.D., Princeton U., 1957. Chemist U. Calif. Radiation Lab., Livermore, 1952-53; instr. Princeton (N.J.) U., 1957-59; chemist Lawrence Livermore (Calif.) Nat. Lab., 1959—; prof. dept. applied sci. U. Calif.-Davis-Livermore, 1963-75; pres. Calif. Metall. Industr., Livermore, 1973-83; cons. Physics Internat., Oakland, Calif., 1966-67, Energy Fund, N.Y.C., 1967-68, Fansteel Corp., Waukegan, Ill., 1973-75. Editor-in-chief: Mineral Processing and Tech. Review, 1983. Served with USAAF, 1944-46. Fellow Am. Phys. Soc.; mem. Am. Soc., AIME. Subspecialties: Solid state chemistry; Space chemistry. Current work: Thermodynamics and diffusion in solids, Mossbauer spectroscopy-magnetic alloys-magnetic order, cosmo-chemistry-meteoritics. Patentee in field. Home: 4681 Almond Circle Livermore CA 94550 Office: Lawrence Livermore Nat Lab Livemore CA 94550

BORGATTI, ALFRED LAWRENCE, biology educator; b. Medford, Mass., Apr. 28, 1928; s. Joseph James and Angeline (Fortini) B. B.S., Tufts Coll., 1951, Ed.M., 1952; Ph.D., Mich. State U., 1961. Tchr. sci. Parlin Jr. High Sch., Everett, Mass., 1954-58; instr. entomology Mich. State U., East Lansing, 1958-61; mem. faculty dept. biology Salem (Mass.) State Coll., 1961—, now, prof.; cons. Briggs Engrng Co., Norwell, Mass., 1980—. Author: Laboratory Manual for Microbiology and its Application, 1982, Laboratory Manual for Microbiology, 1983, Laboratory Manual for Entomology, 1983. Bd. dirs. Everett Eagles Athletic Assn., 1966—. Served with U.S. Army, 1952-54. Mem. Am. Soc. Microbiology, Entomol. Soc. Am. Subspecialties: Microbiology; Insect pathology. Current work: Insect pathology, biological control of insects by microbial pathogens. Office: Dept Biology Salem State Coll Salem MA 01970

BORGSTEDT, HAROLD HEINRICH, physician, pharmacologist, toxicologist; b. Hamburg, Germany, Apr. 21, 1929; s. Gustav Johannes and Anna Dorothea (Wulf) B.; m. Agneta D. von Rehren, May 11, 1957; children: Eric, Astrid. M.D., U. Hamburg, 1956. Intern Rochester (N.Y.) Gen. Hosp., 1956-57; fellow in anatomy and pharmacology U. Rochester Med. Center, 1957-60, instr., 1960-63, sr. instr., 1963-66, research sr. instr. in anesthesiology, 1963-66, asst. prof. pharmacology and toxicology. anesthesiology, 1966-83; research asst. prof. anesthesiology; v.p. Health Designs, Inc., 1983—; cons. in field. Contbr. articles to profl. jours., books. Mem. Am. Soc. Pharmacology and Exptl. Therapeutics, Soc. Toxicology, Drug Info. Assn., N.Y. Acad. Scis., Sigma Xi. Republican. Unitarian. Subspecialties: Pharmacology; Toxicology (medicine). Current work: Halocarbon toxicity, toxicity of polymers and polymer components, mathmatical modelling and computer prediction of toxicity. Office: 183 Main St E Rochester NY 14604

BORING, ARTHUR MICHAEL, physicist; b. Jacksonville, Fla., Feb. 9, 1936; s. John Rutledge and Elizabeth Anne (Andrews) B.; m. Jennie Lee Negin, Jan. 13, 1968; children: Rachel, Neil; m. Patricia Van Etten Teslin, Apr. 12, 1975. B.S., U. Fla., 1963, M.S., 1965; Ph.D., 1968. Staff mem. chem. and material scis. Los Alamos Nat. Lab. div., 1968-73, 1973-75, 1975-78, 1978—; instr. solid state physics U. N.Mex., 1975; vis. scientist Swedish NRC, 1973. Contbr. articles in field to profl. jours. Served with USAF, 1955-59. Mem. Am. Phys. Soc., Sigma Xi, Sigma Phi Sigma. Democrat. Subspecialties: Condensed matter physics; Electronic materials. Current work: Determination of electronic structure actinides in condensed matter phase.

BORISON, RICHARD LEWIS, neuropsychiatrist, psychiatry educator; b. Boston, Mar. 4, 1950; s. Melville and Faye (Golub) B. B.A., Boston U., 1972; Ph.D. in Pharmacology, Chgo. Med. Sch., 1975; M.D., U. Ill.-Chgo., 1977. Diplomate: Am. Bd. Nat. Med. Examiners. Asst. prof. pharmacology Chgo. Med. Sch., 1975-81; intern Mt. Sinai Hosp., Chgo., 1977-78; resident in psychiatry Ill. State Psychiat. Inst., Chgo., 1977-81; assoc. prof. psychiatry Med. Coll. Ga., Augusta, 1981—, dir. neuropsychopharmacology program, 1981—. Contbr. numerous sci. and psychiat. articles to profl. publs. Mem. Ill. Psychiatry Soc. (first prize 1975), Am. Psychiat. Assn., Soc. Biol. Psychiatry, Soc. Neuroscis. Subspecialties: Psychopharmacology; Neuropharmacology. Current work: Clinical neuropsychopharmacology and extrapyramidal movement disorders. Research on the therapeutics and side effects of psychoactive medication, and the teaching of the rational use of psychoactive agents in man. Office: Downtown VA Med Center Psychiatry Service 116A-D Augusta GA 30910

BORKO, HAROLD, information scientist, psychologist, educator; b. N.Y.C., Feb. 4, 1922; s. George and Hilda (Karpel) B.; m. Hannah Levin, June 22, 1947; children: Hilda, Martin. Student, Coll. City N.Y., 1939-41; B.A., U. Calif. at Los Angeles, 1948; M.A., U. So. Calif., 1949, Ph.D. in Psychology, 1952. System tng. specialist Rand Corp., 1956-57; with System Devel. Corp., Santa Monica, Calif., 1957-68, asso. staff head lang. processing and retrieval staff, 1965-68; instr. psychology U. So. Calif., 1957-65; instr. Sch. Library Service U. Calif. at Los Angeles, 1965-68, prof. Grad. Sch. Library and Info. Sci., 1968—. Author: Computer Applications in the Behavioral Sciences, 1962, Automated Language Processing, 1967, Targets for Research in Library Education, 1973, (with H. Sackman) Computers and the Problems of Society, 1972, (with C. Bernier) Abstracting Concepts and Methods, 1975, Indexing Concepts and Methods, 1978; Asso. U.S. editor: Information Processing and Management, 1963—; editor: Academic Press Library and Information Science series, 1970—; book rev. editor: Jour. Ednl. Data Processing, 1963-75. Served with AUS, 1942-46; to capt., Med. Service Corps AUS, 1950-56. Mem. Am. Soc. for Info. Sci. (pres. 1966), Assn. Computing Machinery, Am. Psychol. Assn., Assn. Am. Library Schs., Am. Soc. Indexers, Phi Beta Kappa, Sigma Xi, Phi Gamma Mu. Subspecialties: Automated language processing; Information systems (information science). Current work: Professor, Graduate school of Library science and information. Home: 11507 National Blvd Los Angeles CA 90064 It is unrealistic to expect a person to decide, at age twenty or thereabout, on a career to be followed for the rest of one's life. One should try to attain as good and as general an education as is possible and not be afraid to change professions. The world is changing, and we must be prepared to change with it; only then can we seize the opportunities presented.

BORLAUG, NORMAN ERNEST, agricultural scientist; b. Cresco, Iowa, Mar. 25, 1914; s. Henry O. and Clara (Vaala) B.; m. Margaret G. Gibson, Sept. 24, 1937; children: Norma Jean (Mrs. Richard H. Rhoda), William Gibson. B.S. in Forestry, U. Minn., 1937, M.S. in Plant Pathology, 1940, Ph.D., 1941; Sc.D. (honoris causa), Punjab (India) Agrl. U., 1969, Royal Norwegian Agrl. Coll., 1970, Luther Coll., 1970, Uttar Pradesh Agrl. U., India, 1971, Kanpur U., India, 1970, Mich. State U., 1971, Universidad de la Plata, Argentina, 1971, U. Ariz., 1972, U. Fla., 1973; L.H.D., Gustavus Adolphus Coll., 1971; LL.D. (hon.), N.Mex. State U., 1973, D.Agr., Tufts U., 1982; others. With U.S. Forest Service, 1935-36, 37, 38; instr. U. Minn., 1941; microbiologist E.I. DuPont de Nemours, 1942-44; research scientist in charge wheat improvement Coop. Mexican Agrl. Program, Mexican Ministry Agr.-Rockefeller Found., Mexico, 1944-60; assoc. dir. assigned to Inter-Am. Food Crop Program, Rockefeller Found., 1960-63; dir. wheat research and prodn. program Internat. Maize and Wheat Improvement Center, Mexico City, 1964-79; assoc. dir. Rockefeller Found., 1964—, cons., 1983—; cons., collaborator Instituto Nacional de Investigationes Agricolas, Mexican Ministry Agr., 1960-64; cons. FAO, North Africa and Asia, 1960; ex-officio cons. wheat research and prodn. problems to govts. in, Latin Am., Africa, Asia.; Mem. Citizen's Commn. on Sci., Law and Food Supply, 1973—, Commn. Critical Choices for Am., 1973—, Council Agr. Sci. and Tech., 1973—; dir. Population Crisis Com., 1971; asesor especial Fundacion para Estudios de la Poblacion A.C., Mexico, 1971—; mem. adv. council Renewable Natural Resources Found., 1973—. Recipient Distinguished Service awards Wheat Producers Assns., and state govts. Mexican States of Guanajuato, Queretaro, Sonora, Tlaxcala and Zacatecas, 1954-60; Recognition award Agrl. Inst. Can., 1966, Instituto Nacional de Tecnologia Agropecuaria de Marcos Juarez, Argentina, 1968; Sci. Service award El Colegio de Ingenieros Agronomos de Mexico, 1970; Outstanding Achievement award U. Minn., 1959; E.C. Stakman award, 1961; named Uncle of Paul Bunyan, 1969; recipient Distinguished Citizen award Cresco Centennial Com., 1966; Nat. Distinguished Service award Am. Agrl. Editors Assn., 1967; Genetics and Plant Breeding award Nat. Council Comml. Plant Breeders, 1968; Star of Distinction Govt. of Pakistan, 1968; citation and street named in honor Citizens of Sonora and Rotary Club, 1968; Internat. Agronomy award Am. Soc. Agronomy, 1968; Distinguished Service award Wheat Farmers of Punjab, Haryana and Himachal Pradesh, 1969; Nobel Peace prize, 1970; Diploma de Merito El Instituto Tecnologico y de Estudios Superiores de Monterrey, Mexico, 1971; medalla y Diploma de Merito Antonio Narro Escuela Superior de Agricultura de la U. de Coahuila, Mexico, 1971; Diploma de Merito Escuela Superior de Agricultura Hermanos Escobar, Mexico, 1973; award for service to agr. Am. Farm Bur. Fedn., 1971; Outstanding Agrl. Achievement award World Farm Found., 1971; Medal of Merit Italian Wheat Scientists, 1971; Service award for outstanding contbn. to alleviation of world hunger 8th Latin Am. Food Prodn. Conf., 1972; named to Oreg. State U. Agrl. Hall of Fame, 1981; numerous other honors and awards from govts., ednl. instns., citizens groups. Hon. fellow Indian Soc. Genetics and Plant Breeding; mem. Nat. Acad. Sci., Am. Soc. Agronomy (1st Internat. Service award 1960, 1st hon. life mem.), Am. Assn. Cereal Chemists (hon. life mem., Meritorious Service award 1969), Crop Sci. Soc. Am. (hon. life mem.), Soil Sci. Soc. Am. (hon. life mem.), Sociedad de Agronomia do Rio Grande do Sul Brazil (hon.), India Nat. Sci. Acad. (fgn.), Royal Agrl. Soc. Eng. (hon.), Royal Soc. Edinburgh (hon.), Hungarian Acad. Sci. (hon.), Royal Swedish Acad. Agr. and Forestry (fgn.), Academia Nacional de Agronomia y Veterinaria (Argentina); hon. academician N.I. Vavilov Acad. Agrl. Scis. Lenin Order (USSR.). Address: Centro Internacional de Mejoramiento del Maiz y del Trigo Apartado Postal 6-641 Londres 40 Mexico City 6 Mexico

BORNMANN, JOHN ARTHUR, chemistry educator; b. Pitts., May 1, 1930; s. John Arthur and Iona Ann (Flanegin) B.; m. Sandra Lee Reel, June 12, 1954; children: Patricia Lee, Carol Ann. B.S., Carnegie Inst. Tech., 1952; Ph.D., Ind. U., Bloomington, 1958; postgrad., Technische Hochschule, Stuttgart, Germany, 1956-57. Research chemist E.I. duPont de Nemours & Co., Parlin, N.J., 1958-60; research assoc. Princeton (N.J.) U., 1960-61; asst. prof. No. Ill. U., Dekalb, 1961-65; prof., chmn. dept. chemistry Lindenwood Coll., St. Charles, Mo., 1965—; cons., owner DAS Co., 1974—. Bd. dirs., treas. Bridgeway Counciling Service. Recipient Fulbright award, 1956. Mem. Am. Chem. Soc., Am. Inst. Chemists, AAAS. Lutheran. Club: Disk Drivers Computer (St. Charles, Mo.). Subspecialties: Physical chemistry; Analytical chemistry. Home: Three Briarwood Ln St Charles MO 63301 Office: Lindenwood Coll St Charles MO 63301

BORNSTEIN, ROBERT A., neuropsychologist; b. Brookline, Mass., Nov. 1, 1948; m. Sandra S., Dec. 26, 1970; 1 dau., Bryanne. B.A., Wayne State U., 1972; M.A., Western Mich. U., 1974; Ph.D., U. Ottawa, Ont., Can., 1981. Neuropsychologist U. Alta. (Can.) Hosp., Edmonton, 1979—; asst. clin. prof. U. Alta., 1979—; cons. Misericordia Hosp., Edmonton, 1981—. Mem. Internat. Neuropsychol. Soc., Can. Neurol. Soc. (assoc.), Am. Psychol. Assn. Subspecialty: Neuropsychology. Office: Neuropsychology Service Dept Psychiatr 1-115 Clin Scis Bldg U Alta Edmonton AB Canada T6G 2G3

BOROCHOFF, ROBERT M., research computer scientist; b. Tulsa, Jan. 24, 1956; s. Stan J. and Jean (Halff) B. B.S. Computer Sci, U.S.C., 1978; M.S.C.S., U. Md., 1980. Computer specialist Nat. Library Medicine, Bethesda, Md., 1982; research computer scientist Fed. Jud. Center, Washington, 1982—. Adv. council Internat. Policy Inst., Washington, 1981—; co-chmn. Human Rights, Internat., Washington, 1982—. Recipient Burroughs Corp. scholarship, 1976. Mem. ACM, IEEE, AAAS, Software Psychology Soc. Democrat. Jewish. Subspecialties: Information systems, storage, and retrieval (computer science); Distributed systems and networks. Current work: Current research and work interests: design and implementation of computerized court management systems for the fed. judiciary. Areas of work include distributed systems, networks, databases and human factors. Home: 4615 North Park Ave Apt 1114 Chevy Chase MD 20815 Office: Fed Jud Ctr 1520 H St NW Washington DC 20005

BOROWITZ, GRACE BURCHMAN, chemistry educator, researcher; b. N.Y.C., Dec. 7, 1934; d. Hyman and Edith (Cohen) Burchman; m. Irving J. Borowitz, Nov. 26, 1959; children: Susan, Debra, Lisa Naomi. B.S., CCNY, 1956; M.S., Yale U., 1958, Ph.D. in Organic Chemistry, 1960. Research chemist Am. Cyanamid, Stamford, Conn., 1960-62; lectr. Yeshiva Coll., N.Y.C., 1967; asst. prof. chemistry Upsala Coll., East Orange, N.J., 1967-73; asst. prof. Ramapo Coll. of N.J., Mahwah, 1973-75, assoc. prof., 1975-80, prof., 1980—. Contbr. articles on chemistry to profl. jours. NSF grantee, 1971; Ramapo Faculty Devel. research grantee, 1975-82; recipient Florence Thomas Faculty Research award Ramapo Coll., 1981. Fellow AAAS; mem. Am. Chem. Soc., N.Y. Acad. Scis. (award 1976), N.J. Acad. Scis., Sigma Xi. Democrat. Jewish. Subspecialty: Organic chemistry. Current work: Synthesis and complexing of ionophores with cations, organometallic and platinum complexes.

BORRESEN, C. ROBERT, psychology educator; b. Chgo., July 12, 1926; s. Kristen and Dagny (Mathiesen) B.; m. Thelma Jasper, Dec. 28, 1966; 1 stepson, Mike Meacham. B.A., Northwestern U., 1953; M.A., U.Mo., 1958, Ph.D., 1968. Lic. psychologist, Kans. Instr. psychology U. Mo., Columbia, 1962-63; asst. prof. psychology Memphis State U., 1964-65; asst. prof. Wichita (Kans.) State U., 1965-73; assoc. prof., 1974—, chmn. dept. psychology, 1978-83; expert witness perception and human factors various law firms. Contbr. articles to profl. jours. Treas. Re-election Com. Republican Meachan, Wichita, 1976, 78, 80, 82; advisor Behavioral Sci. Bd., Topeka, Kans., 1980-82. Served with USN, 1944-46. Wichita State U. grantee, 1976, 78, 80. Mem. Am. Psychol. Assn., Southwestern Psychol. Assn., Midwestern Psychol. Assn., Wichita Area Psychol. Assn., Internat. Sci. Soc. for Polit. Psychology. Subspecialties: Psychophysics; Behavioral psychology. Current work: Empirical research on locus of control; research in political behavior. Home: 2215 Hathway Cir Wichita KS 67226 Office: Dept Psychology Wichita State U Wichita KS 67208

BORSOS, TIBOR, immunologist; b. Budapest, Hungary, Mar. 12, 1927; s. Edmund and Anna B.; m. Ruth Moser, July 17, 1950; children: Michael Bela, David Julian. B.A. cum laude, Cath. U. Am., 1954; Sc.D., Johns Hopkins U. Sch. Hydgiene and Public Health, 1958. Asst. prof. microbiology Johns Hopkins U. Sch. Medicine, 1960-62; sr. scientist Nat. Cancer Inst., NIH, Bethesda, Md., 1962-66, head immunochemistry sect., 1966-71, assoc. chief lab. of immunobiology, 1971-81, acting chief lab. of immunobiology, 1981—. Contbr. articles to profl. jours. Recipient Dir.'s award NIH, 1976, Sr. Exec. Service award, 1981. Mem. Am. Assn. Immunologists. Subspecialties: Immunobiology and immunology; Cancer research (medicine). Current work: Interaction of antibodies with cell surface antigens and complement components and the resulting activation of complement and cytotoxic activity. Office: Nat Cancer Inst Bldg 560 Frederick MD 21701

BORTEN, WILLIAM H., research company executive; b. N.Y.C., Mar. 1, 1935; s. David and Susan B.; m. Judith Sue Becker, Feb. 13, 1957; children: Jeffry, Daniel, Matthew. B.B.A., Adelphi U., Garden City, N.Y., 1957. Controller Avien, Inc., Woodside, N.Y., 1959-63; asst. gen. mgr. Fairchild Industries, Germantown, Md., 1963-71; exec. v.p., treas. Atlantic Research Corp., Alexandria, Va., 1971-80, pres., chief operating officer, 1980—, also dir. Bd. dirs. Jr. Achievement Met. Washington; advisor for bus. and industry State Bd. for Community Colls., Richmond, Va., 1983—; trustee Montgomery County Soc. Crippled Children and Adults., Adelphi U., Garden City, N.Y., 1982—; founder, bd. dirs. Montgomery Village Day Care Ctr., Gaithersburg, Md., 1972—. Mem. Nat. Assn. Accountants. Subspecialties: Aerospace engineering and technology; Electronics. Office: 5390 Cherokee Ave Alexandria VA 22314

BORTENFREUND, ELLEN, biologist, researcher, educator; b. Leipzig, Germany; d. David and Paula (Korber) Borenfreund. B.S., Hunter Coll., 1946; M.S., NYU, 1948, Ph.D., 1957. Research assoc. Sloan Kettering Inst., N.Y.C., 1957-61, assoc., 1961-65, asst. prof., 1961-68, assoc. prof., 1968—, assoc. mem., 1968—; adj. assoc. prof. Rockefeller U., 1981—. Recipient Research Career Devel. award NIH, 1963, named to Hunter Coll. Hall of Fame, 1981. Mem. Am. Soc. Cell Biology, Am. Soc. Biol. Chemists, Am. Assn. Cancer Research, Tissue Culture Assn., Harvey Soc., N.Y. Acad. Sci. Subspecialties: Cell biology; Cell and tissue culture. Current work: Tissue culture systems and toxicology. Office: Rockefeller University 1230 York Ave New York NY 10021

BORZELLECA, JOSEPH FRANCIS, pharmacologist, toxicologist, educator, cons.; b. Norristown, Pa., Oct. 3, 1930; s. Peter and Madeline (Fiorillo) B.; m. Mary Elizabeth Ford, Aug. 9, 1931; children: Joseph F., Paul, David, Michael, Therese Marie, Mark. B.S., St. Joseph U., Phila., 1952; M.S., Thomas Jefferson U., 1954, Ph.D., 1956. Instr., assoc. dept. pharmacology Med. Coll. Pa., Phila., 1956-59; asst. prof. pharmacology Med. Coll. Va., Richmond, 1959-62, assoc. prof., 1962-67, prof., 1967—, head div. toxicology, 1972—; pres. Toxicology and Pharmacology, Inc.; cons. FDA, NIMH, EPA, U.S. Army; chmn. carcinogens standards com. OSHA, Dept. Labor. Mem. editorial bd.: Toxicology and Applied Pharmacology, 1975-78, Jour. Environ. Pathology and Toxicology, 1977—, Pharmacology, 1978—, Jour. Environ. Sci. and Health, 1979—, Pharmacology and Drug Devel, 1980—, Jour. Cutaneous and Ocular Toxicology, 1981—; contbr. articles to profl. jours. Mem. AAAS, Am. Chem. Soc., Am. Coll. Toxicology, Am. Soc. Pharmacology and Exptl. Therapeutics, Soc. Exptl. Biology and Medicine (councilor, program chmn. Southeastern sect.), Soc. Toxicology (chmn. edn. com., sec. 1967-71, councilor 1971-72, pres. 1973-74, chmn. numerous coms.), Va. Acad. Sci. (chmn. med. scis. div.), Sigma Xi. Club: Cosmos (Washington). Subspecialties: Toxicology (medicine); Pharmacology. Current work: Toxicology, safety evaluation, water contaminants, food safety. Home: 8718 September Dr Richmond VA 23229 Office: Med Coll Virginia Box 613 Richmond VA 23298

BOSHELL, BURIS RAYE, physician, educator; b. nr. Phil Campbell, Ala., Oct. 9, 1926; s. Harvey M. and Lela (Alexander) B.; m. Martha

Sue Johnson, June 4, 1951; children—Patty, Thomas Eppinger. B.S., Ala. Polytech. Inst., 1947, postgrad. vet. medicine, 1947-49; postgrad., Med. Coll. Ala., 1949-51; M.D., Harvard U., 1953. Diplomate: Am. Bd. Internal Medicine. Intern Peter Bent Brigham Hosp., Boston, 1953-54, resident, 1954-56, chief med. resident, 1958-59; research fellow in medicine Harvard Med Sch., 1956-58; asst. prof. medicine U. Ala. in Birmingham, 1959-62, asso. prof., 1962-64, prof, 1964—, Ruth Lawson Hanson prof. medicine, 1967—, dir. div. endocrinology and metabolism, 1970—, med. dir. Diabetes Research and Edn. Hosp., 1973—; clin. investigator VA Hosp., 1959-62, chief medicine, 1962-64; med. dir. Central Bancshares of South, Rust Engring.; dir. Central Bank Birmingham.; Vis. prof. U. Mexico, 1975, U. Witwatersrand, Johannesburg, S.Africa, 1975; mem. Nat. Commn. on Diabetes, 1975—, chmn. subcom. on scope, 1975—. Author: The Diabetic at Work and Play, 1973, Diabetes Mellitus-40 Case Studies, 1976; Contbr. articles to profl. jours. Pres. bd. dirs. Diabetes Trust Fund of Ala.; bd. dirs. Diabetes Research Lab. Recipient Sr. U.S. Scientist award Alexander von Humboldt Found., Fed. Republic Germany, 1975. Fellow A.C.P., Am. Coll. Clin. Pharmacology and Chemotherapy; mem. AMA, Ala., Jefferson County med. assns., AAUP, Birmingham Acad. Medicine, Ala. Acad. Sci., Am. Diabetes Assn. (bd. dirs. 1973—), New Eng. Diabetes Assn., N.Y. Diabetes Assn., Ala. Diabetes Assn., Endocrine Soc., Am. Fedn. for Clin. Research, So. Soc. for Clin. Investigation, Am. Soc. for Clin. Pharmacology and Therapeutics (com. mem.), Sigma Xi, Omicron Delta Kappa, Phi Kappa Phi, Gamma Sigma Delta, Tau Kappa Alpha, Alpha Omega Alpha. Subspecialty: Endocrinology. Home: 3017 Old Ivy Rd Birmingham AL 35210 Office: 1808 7th Ave Birmingham AL 35294 My goal in life, at this juncture, is to help educate the diabetic, help relieve his suffering and, if it is possible, participate in the development of a cure.

BOSHKOV, STEFAN HRISTOV, educator, mining engr.; b. Sofia, Bulgaria, Sept. 29, 1918; came to U.S., 1938, naturalized, 1944; s. Hristo and Karla (Lubich) B.; m. Bianca G. Amaducci, Aug 28, 1943; children—Lynn Karla, Stefan Robert. Diploma, Am. Coll., Sofia, Bulgaria, 1938; B.S., Columbia U., 1941, E.M., 1942. Mem. faculty Columbia, 1946—; prof. Henry Krumb Sch. Mines, 1951—, chmn., 1967—, Henry Krumb prof., 1980—; disting. prof., sr. scientist (Fulbright program), Yugoslavia, 1969, guest lectr., Taiwan, China, 1972, 76, OAS, Chile, 1972, USSR, 1974, Poland, 1976, Bulgaria, 1976, Bolivia, 1977, People's Republic of China, 1980, 81; cons. engr., 1950—; mem. internat. organizing com. World Mining Congress, 1962—; chmn. 4th Internat. Conf. Strata Control and Rock Mechanics, N.Y.C.; 1964; mem. adv. com. metal and nonmetallic health and safety standards Dept. Labor, 1978. Mem. editorial bd.: Internat. Jour. Rock Mechanics and Mining Scis, 1964-76. Pres. Benedict Found., 1973—, Harrison (N.Y.) No. 7 Sch. Bd., 1954-63. Served to 1st lt. AUS, 1943-46; CBI. Recipient Boleslaw Krupinski medal State Mining Council Poland, 1980. Mem. AIME (Mineral Industry Edn. award 1980), Am. Arbitration Assn., Sigma Xi. Presbyterian. (trustee 1965-69). Club: Masons. Current work: Strata control and rock mechanics; mine economics; consulting; school administration. Home: 119 White Plains Ave White Plains NY 10604 Office: Mudd Bldg Columbia Univ New York NY 10027

BOSIN, TALMAGE RAYMOND, pharmacologist, educator, researcher; b. Fond du Lac, Wis., Mar. 6, 1941; s. Raymond W. and Edna S. B.; m. Elizabeth W. Witler, July 18, 1971; children: Sara, Catherine. B.S. in Chemistry, Wheaton (Ill.) Coll., 1963; Ph.D. in Organic Chemistry (NIH fellow), Ind. U., 1967. Mem. faculty Ind. U. Sch. Medicine, Bloomington, 1969—, assoc. prof. pharmacology, 1973-78, prof., 1978—. Contbr. articles to profl. jours. NIH postdoctoral fellow, 1967-69; sr. internat. fellow, 1977-78. Mem. Am. Soc. Pharmacology and Exptl. Therapeutics, Soc. Toxicology. Evangelical. Subspecialties: Pharmacology; Cellular pharmacology. Current work: Effects of toxicants on vasoactive agent handling by the pulmonary microvasculature. Home: 1300 Valley Forge Bloomington IN 47401 Office: Ind U Myers Hall Bloomington IN 47405

BOSKEY, ADELE LUDIN, researcher, chemistry educator; b. N.Y.C., Aug. 30, 1943; d. Benjamin and Anne (Monoson) Ludin; m. James B. Boskey, June 30, 1970; 1 dau., Elizabeth R. B.A., Barnard Coll., 1964; Ph.D., Boston U., 1970. Instr. Coll. Liberal Arts, Boston U., 1969-70; post-doctoral trainee Imperial Coll., London, 1970-71, Hosp. Spl. Surgery, N.Y.C., 1971-73, asst. scientist, 1973-79, assoc. scientist, 1980—; asst. prof. biochemistry Cornell U. Med. Coll., N.Y.C., 1975-79, assoc., prof., 1980—. Recipient Career Development award NIH-Nat. Inst. Dental Research, 1975. Mem. Am. Chem. Soc., Am. Crystallographic Assn., Orthopedic Research Soc. (Kappa Delta award 1979, exec. com. 1981-82), Am. Soc Bone and Mineral Research, Internat. Assn. Dental Research, Harvey Soc., Soc. Exptl. Biology and Medicine. Subspecialties: Biophysical chemistry; Orthopedics. Current work: Investigation of mechanism of normal and pathologic calcificatinn with emphasis on the roles of lipids and proteoglycans. Home: 4 Winding Way North Caldwell NJ 07006 Office: Hosp Spl Surgery 535 E 70th St NYC NY 10021

BOSMANN, HAROLD BRUCE, geriatrics educator, university dean; b. Chgo., June 17, 1942; s. Harold William and Gladys (Hoagl) B.; m. Maureen Kathleen O'Pray, Oct. 29, 1966; children: Geoffrey, Katharine. B.A., Knox Coll., 1964; Ph.D., U. Rochester, 1966. NSF postdoctoral fellow Cambridge (Eng.) U., 1967; NIH postdoctoral fellow Salk Inst., La Jolla, Calif., 1968; fellow Sidney Sussex Coll., Cambridge, Eng., 1967; Asst. prof. to prof. pharmacology, toxicology and oncology U. Rochester (N.Y.) Sch. Medicine, 1968-78, chmn., prof., 1978-82; assoc. dean, Martha Betty Semmons prof. geriatrics Coll. Medicine, U. Cin., 1982—, dir. gerontology, 1982—; co-dir. Center on Aging, U. Rochester, 1979-82. Contbr. articles in field to profl. jours. Pres. Civic Assn., Riverton, N.Y., 1978-81, Sycamore Trace, Ohio, 1983; bd. edn. West Henrietta (N.Y.) Sch. Dist., 1981. Recipient Research Career Devel. award Nat. Inst. Gen. Med. Scis., 1969-74; Scholar award Leukemia Soc. Am., 1974-79; Geriatric Medicine Acad. award Nat. Inst. Aging, 1979-82. Mem. Am. Chem. Soc., Biophys. Soc., Am. Soc. Pharmacology and Exptl. Therapeutics, Gerontology Soc. Am., Sigma Xi. Subspecialties: Gerontology; Chemotherapy. Current work: Biology, pharmacology and molecular genetics of aging and oncology, geriatric medicine, biochemical pharmacology. Home: 9643 Ash Ct Cincinnati OH 45242 Office: U Cin Coll Medicine Mail Location 555 231 Bethesda Ave Cincinnati OH 45267

BOSS, ALAN PAUL, astrophysicist; b. Lakewood, Ohio, July 20, 1951; s. Paul and Marguerite May (Gehringer) B.; m. Barbara Carol Woltz, Sept. 8, 1973; m. Catherine Ann Starkie, Aug. 4, 1979. B.S., U. South Fla., 1973; M.A., U. Calif.-Santa Barbara, 1975, Ph.D., 1979. Research asst. U. Calif.-Santa Barbara, 1974-79; resident research assoc. Nat. Acad. Scis./NRC, Ames Research Center, Calif., 1979-81; staff assoc. Carnegie Instn. Washington, 1981-83, staff mem., 1983—. Contbr. articles to profl. jours. Mem. Am. Astron. Soc., Am. Geophys. Union, AAAS. Democrat. Subspecialties: Theoretical astrophysics; Planetary science. Current work: Theory of stellar and planetary system formation. Office: 5241 Broad Branch Rd NW Washington DC 20015

BOSS, KENNETH JAY, biology educator; b. Grand Rapids, Mich., Dec. 5, 1935; s. Orrie and Margaret (Oosting) B. B.A., Central Mich. U., 1957; M.Sc., Mich. State U., 1959; Ph.D., Harvard U., 1963. Research malacologist Dept of Interior, Washington, 1963-66; asst. curator Museum of Comparative Zoology, Harvard U., 1966-70, prof. biology, curator, 1970—. Contbr. to: Synopsis and Classification of Organism, 1983. Subspecialty: Systematics. Current work: Systematics. Office: Museum of Comparative Zoology Harvard College Cambridge MA 02138

BOSTON, JOHN ROBERT, med. computing researcher; b. Evanston, Ill., Oct. 16, 1942; s. John Robert and Elizabeth Louise (Olmstead) B.; m. Carol Lee Dillon, Oct. 23, 1971; children: Christopher, Patrick. B.S.E.E., Stanford U., 1964, M.S.E.E., 1966; Ph.D., Northwestern U., Chgo., 1971. Research assoc. Northwestern U., 1971-72; asst. prof. U. Md., College Park, 1972-75, Carnegie-Mellon U., Pitts., 1975-80; research assoc. prof. dept. anesthesiology and critical care medicine U. Pitts., 1980—. Contbr. articles to profl. jours., chpts. in books. Mem. AAAS, IEEE, Soc. Neuroscis., Sigma Xi. Subspecialties: Biomedical engineering; Neurophysiology. Current work: Acquisition and analysis of sensory evoked potentials. Home: 2004 Carriage Hill Rd Allison Park PA 15101 Office: U of Pitts 1060 J Scaife Hall Pittsburgh PA 15261

BOSTROM, CARL OTTO, physicist, educator; , 1954; m.; 3 children. B.S., Franklin and Marshall Coll., 1956; M.S., Yale U., 1958, Ph.D. in Physics, 1962. Sr. physicist Applied Physics Lab., Johns Hopkins U., Laurel, Md., 1960-64, supervisory space physicist, 1964-74, chief scientist, 1974-78, assoc. head, 1978, head space dept., 1979-80, asst. dir. space systems, 1979, dep. dir. lab., 1979-80, dir. lab., 1980—, instr., 1975—. Mem. AAAS, Am. Phys. Soc., Am. Geophys. Union. Subspecialty: Space physics. Current work: Space particles and fields; satellite instrumentation; magnetospheric physics; solar physics; satellite systems design; research administration. Office: Applied Physics Lab Johns Hopkins U Laurel MD 20707

BOSY, BRIAN JOSEPH, mechanical engineer, consultant; b. Glen Cove, N.Y., Sept. 14, 1952; s. Frank Edward and Marie (Lehner) B.; m. Polly Ann Suda, June 28, 1975. S.B.M.E., MIT, 1975, S.M.M.E., 1975, M.B.A., 1980. Registered profl. engr., Mass. Engr. Westinghouse, Westboro, Mass., 1975-76; engr., sales rep. Stowe-Woodward, Newton, Mass., 1976-79; dir. mktg. and sales Allis Chalmers, Cambridge, Mass., 1980-81; prin. engr. Digital Equipment, Hudson, Mass., 1981—; owner, operator Commonwealth Cons., Inc., 1981—. Contbg. author: Materials and Society, Vol. 5, 1981. Fellow Wunsch Found.-Silent Hoist & Crane Co., 1972, Engring. Products Lab., MIT, 1973, Gulf Oil Corp., 1974; recipient Luis de Florez award MIT, 1973; Clapp and Pollak scholar, 1973. Mem. ASME, Soc. Exptl. Stress Analysis. Republican. Roman Catholic. Subspecialties: Mechanical engineering; Robotics. Current work: Robotics, computer-aided material selection, composites. Home: 37 Swanson Rd Framingham MA 01701 Office: 77 Reed Rd HL02 1D12 Hudson MA 01749

BOTT, RAOUL, educator; b. Budapest, Hungary, Sept. 24, 1923; came to U.S., 1947, naturalized, 1959; s. Rudolph and Margit (Kovacs) B.; m. Phyllis Hazell Aikman, Aug. 30, 1947; children—Anthony, Jocelyn, Renee, Candace. B.Engring., McGill U., 1945, M.Engring., 1946; D.Sc., Carnegie Inst. Tech., 1949, U. Notre Dame, 1980; A.M. (hon.), Harvard, 1959. Mem. Inst. Advanced Study, Princeton, 1949-51, 55-57; instr. math. U. Mich., 1951-52, asst. prof., 1952-55, prof., 1957-59; prof. math. Harvard, 1959-67, Higgins prof. math., 1967-79, Graustein prof. math., 1978—, master Dunster House, 1979—; spl. research network theory, topology and geometry. Editor: Topology. Asso; editor: Annals of Math, 1958—. Recipient Veblen prize in geometry Am. Math. Soc., 1964. Mem. Nat. Acad. Sci., Am. Math. Soc. (council), Am. Acad. Arts and Sci., London Math. Soc. (hon.). Subspecialty: Topology. Home: 935 Memorial Dr Cambridge MA 21613 Office: 1 Oxford St Cambridge MA 02138

BOTTEMA, MURK, optical physicist; b. Velsen, Netherlands, May 23, 1923; came to U.S., 1958; s. Fokke and Eatske (Wijbenga) B.; m. Willy Tielrooy, Mar. 3, 1950; children: Frank, Murk Jan, Nicolaas. Ph.D. in Physics, U. Groningen, Netherlands, 1957. Research assoc. physics dept. U. Groningen, 1957-58, 60-62, Lab. for Astrophysics, Johns Hopkins U., Balt., 1958-60, 62-68; staff scientist, later staff cons. space systems div. Ball Aerospace Systems Div., Boulder, Colo., 1968—. Contbr. articles to profl. jours. Fellow Optical Soc. Am.; mem. Am. Astron. Soc., Am. Geophys. Union. Subspecialties: Aerospace engineering and technology; Optical design. Current work: Conceptual design and performance analysis of optical instruments for space research/space astronomy. Home: 2525 Table Mesa Dr Boulder CO 80303 Office. Ball Aerospace Systems Div PO Box 1062 Boulder CO 80306

BOTTOMLEY, SYLVIA STAKLE, physician; b. Riga, Latvia, Mar. 9, 1934; U.S., 1950, naturalized, 1955; d. John Waldemar and Leontine (Miluns) Stakle; m. Richard Harold Bottomley, June 1958; children: Astrid Elizabeth, Ian Philip. B.S., Okla. State U., 1954; M.D., U. Okla., 1958. Diplomate: Am. Bd. Internal Medicine. Intern Salt Lake County Gen. Hosp., Salt Lake City, 1958-59; resident in medicine U. Okla. Health Sci. Center, 1959-61; fellow in hematology, clin. asst. in medicine U. Okla. Coll. Medicine, Oklahoma City, 1961-64, instr. medicine, 1964-67, asst. prof., 1967-71, assoc. prof., 1971-75, prof., 1975—, assoc. prof. pathology, 1973—; asst. chief hematology VA Med. Center, Oklahoma City, 1969—. Contbr. chpts. to books, articles to profl. jours. NIH grantee, 1961-70. Fellow A.C.P.; mem. Am. Soc. Hematology, Central Soc. for Clin. Research, So. Soc. for Clin. Investigation, Sigma Xi (pres. chpt. 1975-76), Alpha Omega Alpha. Lutheran. Subspecialties: Hematology; Internal medicine. Current work: Research in heme-biosynthesis, erythropoiesis and porphyrias. Office: VA Med Center 921 E 13th St Oklahoma City OK 73104

BOUBEL, RICHARD WILLIAM, engring. educator, cons., inventor; b. Portland, Oreg., Aug. 1, 1927; s. William Francis and Cleo Adden (Link) B.; m. Ruth Abbey, Mar. 28, 1952; children: Jane A., Thomas R., William W., Elizabeth Boubel Nutter. B.S. in Mech. Engring, Oreg. State U., 1953, M.S. M.E., 1954; Ph.D. in Environ. Engring, U. N.C., 1963. Registered profl. engr., Oreg., diplomate: Am. Acad. Environ. Engrs., 1970. Instr. Oreg. State U., Corvallis, 1954-57, asst. prof., 1957-63, asso. prof., 1963-67, prof., 1967—, engr., 1957-58, dir., 1979-82; engr. CH2M Hill, Corvallis, 1955-57; statistician U.S. Bur. Mines, Albany, Oreg., 1963-73; cons. in field. Author: Fundamentals of Air Pollution, 1973, also articles. Mem. Corvallis Airport Commn., 1970-73; edler First United Presbyterian Ch., Corvallis, 1966—. Served in USNR, 1944-46. USPHS fellow, 1960; named Eminent Engr. Tau Beta Pi, 1979. Mem. Air Pollution Control Assn. (Pres. 1978-79), ASME (sect. dir. 1977-78, membership chmn. 1981—), Am. Acad. Enviro. Engrs., Oreg. Pilots Assn. (pres. 1958-59), Pi Tau Sigma. Republican. Subspecialties: Environmental engineering; Biomass (energy science and technology). Current work: Gas turbine and jet engine air pollution emission control; biomass utilization with atmospheric emissions; hydrogen/oxgen direct steam production-combustion and cycle analysis. Patentee energy systems and air pollution samplers (5). Home: 2773 NW Skyline Dr Corvallis OR 97330 Office: Oreg State U Corvallis OR 97331

BOUCK, NOEL PATRICK, microbiology educator; b. San Francisco, Oct. 23, 1936; d. J. Howard Partick; m. G. Benjamin Bouck, Aug. 19, 1961; children: Julie, John, Laurie. B.A., Pomona Coll., 1958; M.A., Columbia U., 1960; Ph.D., Yale U., 1969. Lectr. Albertus Magnum Coll., New Haven, 1970-71; vis. prof. Salk Inst., San Diego, 1977-78; asst. prof. U. Ill., Chgo., 1978-79; asst. prof. dept. microbiology Northwestern U., Chgo., 1979—. Pegram fellow, 1961; Merril fellow, 1967-68. Mem. Am. Soc. Microbiology, Tissue Culture Assn., Genetics Soc. Am., Am. Assn. for Cancer Research. Subspecialties: Cancer research (medicine); Cell and tissue culture. Current work: Genetic analysis of invitro neoplastic transformation induced by chemical carcinogens and DNA viruses. Office: Northwestern 303 E Chicago Ave Chicago IL 60611

BOUDART, MICHEL, chemical engineering educator; b. Belgium, June 18, 1924; came to U.S., 1947, naturalized, 1957; s. Francois and Marguerite (Swolfs) B.; m. Marina D'Haese, Dec. 27, 1948; children: Mark, Baudouin, Iris, Philip. B.S., U. Louvain, Belgium, 1944, M.S., 1947; Ph.D., Princeton U., 1950. Research asso. James Forrestal Research Center, Princeton, 1950-54; mem. faculty Princeton U., 1954-61; prof. chem. engring. U. Calif. - Berkeley, 1961-64; prof. chem. engring. and chemistry Stanford U., 1964-80, William J. Keck prof. chem. engring., 1980—; cons. to industry, 1955—; dir. Catalytica Assos., Inc.; Humble Oil Co. lectr., 1958, Am. Inst. Chem. Engrs. lectr., 1961, Sigma Xi nat. lectr., 1965; chmn. Gordon Research Conf. Catalysis, 1962. Author: Kinetics of Chemical Processes, 1968, (with A. Djéga-Mariadassov) Kinetics of Heterogeneous Catalytic Reactions, 1983; editor: (with J.R. Anderson) Catalysis: Science and Technology, 1981; adv. editorial bd.: Jour. Catalysis, 1964—, Internat. Chem. Engring., 1964—, Advances in Catalysis, 1968—, Catalysis Rev., 1968—, Accounts Chem. Research, 1978—. Belgium-Am. Ednl. Found. fellow, 1948; Procter fellow, 1949; Recipient Curtis-McGraw research award Am. Soc. Engring. Edn., 1962, R.H. Wilhelm award in chem. reaction engring., 1974. Fellow AAAS; mem. Am. Chem. Soc. (Kendall award 1977), Catalysis Soc., Am. Inst. Chem. Engrs., Chem. Soc., Nat. Acad. Sci., Nat. Acad. Engring.; fgn. assoc. Académie Royale de Belgique. Subspecialty: Surface chemistry. Current work: Synthesis, characterization and testing of new catalytic matrerials. Home: 512 Gerona Rd Stanford CA 94305 Office: Dept Chem Engring Stanford Univ Stanford CA 94305

BOUDEWYNS, PATRICK ALAN, psychologist, educator, researcher; b. Des Moines, Jan. 27, 1940; s. Robert Charles and Corinne Alice (Rider) B.; m. Mikell Eileen Sheil Wright, Sept. 2, 1961 (div. 1981); children: Brian, Erin. B.A., Drake U., 1962; M.S., U. Wis.-Milw., 1966, Ph.D., 1968. Lic. psychologist, Ga., N.C.; cert. in psychology Nat. Register Health Service Providers in Psychology. Chief psychology service VA Med. Center, Iowa City, 1970-75; asst. prof. psychology U. Iowa, 1970-75; chief psychology service VA Med. Center, Durham, N.C., 1975-79; assoc. prof. psychiatry Duke U. Med. Center, Durham, 1976-81; chief psychology service VA Med. Center, Augusta, Ga., 1981—; prof. psychiatry (psychology) Med. Coll. Ga., Augusta, 1981—. Research editor: VA Chief Psychologists, 1982; Author: (with B. Shipley) Flooding and Implosive Therapy: Direct Therapeutic Exposure in Clinical Practice, 1983; co-editor: (with F. Keefe) Behavioral Medicine in General Medical Practice, 1982; assoc. editor: Iowa Psychologist jour, 1973-78; editorial Bd.: Jour. Behavioral Medicine, 1979—. Served with USAR, 1958-66. Nat. Def. fellow U.S. Govt., 1966-68; research grantee NIMH, 1980, VA Div. Medicine and Surgery, 1972-79. Mem. Am. Psychol. Assn. (vis. psychologist 1979—, program com. 1975-77), Assn. VA Chief Psychologists (council of reps. 1977-79), Assn. Advancement Behavior Therapy (program com. 1974-75), N.C. Psychol. Assn., Ga. Psychol. Assn. Subspecialties: Behavioral psychology; Biofeedback. Current work: Behavior therapy/behavioral medicine. Office: VA Med Center 2460 Wrightsboro Rd Augusta GA 20910

BOUDOULAS, HARISIOS, cardiologist, educator, researcher; b. Velvendo-Kozani, Greece, Nov. 3, 1935; came to U.S., 1975; s. Konstantinos and Sophia (Manolas) B.; m. Olga Paspati, Feb. 27, 1976; step children: Shophia, Konstantinos. M.D. diploma, Aristoteliam U., 1959, U. Thessaloniri, 1967. Diplomate: Am. Bd. Internal Medicine. Intern Red Cross Hosp.-Athens, Greece, 1960; resident in cardiology U. Thessaloniki, 1962-64; sr. lectr. U. Thessaloniri, Greece, 1973-75; resident Ohio State U., 1970-73; asst. prof. cardiology, 1975-78, assoc. prof. cardiology, 1978-80, prof., 1983—, dir. cardiovascular research, 1983; prof. cardiology Wayne State U., Detroit, 1980-82; dir. cardiovascular research, 1980-82, adj. chief cardiology, 1982; chief diagnostic and tng. cardiovascular ctr. VA Med. Ctr., Allen Park, Mich., 1980-82. Contbr. numerous articles to profl. jours. Recipient disting. research investigator award Am. Heart Assn., 1982. Fellow Am. Coll. Cardiology, Am. Coll. Clin. Pharmacology, Council Clin. Cardiology of Am. Heart Assn., ACP; mem. Central Soc. Clin. Research. Greek Orthodox. Club: President's (Columbus). Subspecialties: Cardiology; Internal medicine. Current work: My research interests covers a side spectrum of cardiovascular medicine. Home: 2108 Mackenzie Dr Columbus OH 43220 Office: Ohio State ept Cardiology 466 W 10th Ave Columbus OH 43210

BOUDREAU, ROBERT DONALD, meteorology educator; b. North Adams, Mass., Mar. 9, 1931; s. Lucien Albert and Rose Elizabeth (Franceschini) B. B.S. in Meteorology, Tex. A&M U., 1962, M.S., 1964, Ph.D., 1968. Cert. cons. meteorologist, cert. instrument flight instr., Colo. Weather forecaster, sgt. USAF, U.S., Eng., 1948-57; research meteorologist Atmospheric Scis. Lab., Ft. Huachuca, Ariz., 1965-68, Deseret Test Ctr., Salt Lake City, 1968-70, Meteorological Satellite Lab., Washington, 1970-71; sr. atmospheric scientist NASA Earth Resources Lab., Bay St. Louis, Mo., 1971-75; prof., head dept. meteorology Metropolitan State Coll., Denver, 1976—; pres. Internat. Meteorology, Aviation and Electronics Inst., Boulder, Colo., 1982—; cons. in field. Contbr. articles to profl. jours. Com. mem. Colo. Republican Party, Edgewater, 1982. NSF fellow, 1962; NDEA fellow, 1962; recipient NASA Skylab award, 1974. Mem. Am. Meteorol. Soc., NEA, Airplane Owners & Pilots Assn., Colo. Pilots Assn. (adv. com. 1979—), P.C. Flyers, Phi Kappa Phi, Sigma Xi, Alpha Eta Rho. Unitarian. Subspecialties: Meteorology; Satellite studies. Current work: Aeronautical and aerospace meteorology, airborne cloud seeding. Home: 2601 Ingalls Apt 102 Edgewater CO 80214 Office: Metropolitan State Coll Dept Meteorology 1006 11th St Denver CO 80204

BOUHANA, JAMES P., computer scientist, educator; b. Detroit, July 28, 1943; s. George J. and Anna B. B.A., Oakland U., Rochester, Mich., 1964; M.S., U. Mass., 1966; Ph.D., U. Wis., 1978. Systems specialist Burroughs Corp., San Diego, 1971-76; asst. prof. Purdue U., West Lafayette, Ind., 1978-80; cons. BGS Systems, Inc., Waltham, Mass., 1979-80; assoc. prof. info. tech. Wang Inst. Grad. Studies, Tyngsboro, Mass., 1980—. Author: (with P.J. Denning and W.F. Tichy) Computer Languages and Architecture, 1984. Named Best Tchr. in Computer Sci. Purdue Student Assn., 1979. Mem. Assn. for Computing Machinery, IEEE Computer Soc., Math. Assn. Am., Computer Measurement Group. Subspecialties: Computer Performance Evaluation; Computer architecture. Current work: Performance evaluation, analytic queueing network models, performance - oriented design. Office: Wang Inst Tyng Rd Tyngsboro MA 01879

BOULGER, FRANCIS WILLIAM, metallurgical engineer; b. Mpls., June 19, 1913; s. Francis J. and Mary (Armstrong) B. Metall. Engr., U. Minn., 1934; M.S. (Battelle fellow), Ohio State U., 1937. With A.P.,

1929-34; engr. Minn. Dept. Hwys., 1935-36; metallurgist Republic Steel Corp., Cleve., 1937; research metallurgist Battelle Meml. Inst., Columbus, Ohio, 1938-45, div. chief, 1945-67, sr. tech. adviser, 1967—; cons. USAF; Materials Adv. Bd. OECD. Author: (with others) Forging Materials and Practices, 1968, Tri-Lingual Dictionary of Production Engineering, 1969, Forging Equipment, Materials and Practices, 1973; also numerous articles. Named Man of Yr. Columbus Tech. Council, 1966; Gold medalist Soc. Mfg. Engrs., 1967; recipient Am. Machinist award, 1975. Fellow Am. Soc. Metals, ASME; mem. AIME (Hunt medal 1955), Soc. Mfg. Engrs., Nat. Acad. Engring., Internat. Inst. for Prodn. Research (pres.), Sigma Xi. Roman Catholic. Subspecialties: Materials processing; Mechanical engineering. Current work: Computerized planing and control of manufacturing processes, particularly metalworking operations. Home: 1816 Harwich Rd Columbus OH 43221 Office: 505 King Ave Columbus OH 43201

BOULPAEP, EMILE LOUIS J.B., medical educator, university administrator; b. Aalst, Belgium, Sept. 15, 1938; came to U.S., 1968; s. Henri Jules and Eulalie J. (de Croes) B.; m. Elisabeth J. Goris, July 25, 1964. B.S., U. Louvain, Belgium, 1958, M.D., 1962, Lic. Sc. in Medicine, 1963; M.A., Yale U., 1979. Asst. in medicine U. Louvain, 1962-66, instr., chief asst., 1966-68; asst. prof. physiology Cornell U. Med. Coll., N.Y.C., 1968-69, Yale U. Sch. Medicine, New Haven, 1969-72, assoc. prof., 1972-79, prof., 1979—, chmn., 1979—; mem. study sect. in gen. medicine NIH, 1976-80; overseas fellow Churchill Coll., U. Cambridge, Eng., 1978. Contbr. articles to med. jours.; mem. editorial bd.: Am. Jour. Physiology, 1976-78; assoc. editor: (1976) Yale Jour. of Biology and Medicine. Bd. dirs. Belgian Am. Ednl. Found., Brussels, N.Y.C., 1971—, pres., New Haven, 1977—; bd. dirs. Fondation Universitaire, Brussels, 1977—, Fondation Francqui, Brussels, 1977—, Belgian Soc. Benevolence, N.Y.C., 1980—, Universitas Ltd., N.Y.C., 1982—. Decorated knight Order of the Crown, comdr. Order of Leopold II, Belgium; recipient Specia prize U. Louvain, 1962; Prix des Alumni Foundation Universitaire, Brussels, 1973; Fellow Branford Coll., Yale U., 1973—. Mem. Am. Heart Assn. (council on kidney in cardiovascular disease), Am. Physiol. Soc., Am. Soc. Nephrology, Yale Club of N.Y.C., Royal Acad. Medicine (Belgium), Club of the Fondation Universitaire (Brussels). Subspecialties: Physiology (medicine); Physiology (biology). Home: 11 Burnt Swamp Rd Woodbridge CT 06510 Office: Yale U Sch Medicine 333 Cedar St New Haven CT 06510

BOUMA, HESSEL, III, educator; b. Chgo., Sept. 16, 1950; s. Hessel and Cornelia (Hoving) B.; m. Ruth Ellen Cooper, June 10, 1972; children: Christopher Michael, Amy Rebecca, Brian Cooper. A.B., Calvin Coll., 1972; Ph.D., U. Tex. Med. Br., Galveston, 1975. Postdoctoral fellow dept. chemistry U. Calif.-San Diego, 1975-78; asst. prof. biology Calvin Coll., Grand Rapids, Mich., 1978-82, assoc. prof., 1982—. NIH fellow, 1975-78; Research Corp. grantee, 1980-81. Mem. Am. Soc. Cell Biology, Sigma Xi. Subspecialties: Cell biology; Gene actions. Current work: Biosynthesis of blood clotting protein-fibrinogen; nature and characteristics of acute-phase inflammatory response. Office: Dept Biolog Calvin Coll 3201 Burton St SE Grand Rapids MI 4950

BOUNDY, RAY HAROLD, chem. engr.; b. Brave, Pa., Jan. 10, 1903; s. George W. and Anetta (Cather) B.; m. Geraldine McCurdy, Nov. 27, 1926; children—Richard Ray, Lois Cather. B.S. in Chemistry, Grove City Coll., Pa., 1924, Case Western Res. U., 1926; M.S. in Chem. Engring, Case Western Res. U., 1930; D.Sc. in Chemistry (hon.), Grove City (Pa.) Coll., 1961. With Dow Chem. Co., Midland, Mich., 1926-68, v.p., dir. research, corp. dir., 1951-68, cons. mgmt. of research, 1968—; vol. Internat. Exec. Service Corps., Taiwan and Iran, 1968—. Co-editor: Its Polymers and Copolymers, 1951; Contbr. articles to profl. jours. Bd. dirs. Grove City Coll., Saginaw Valley State Coll. Recipient Gold medal Indsl. Research Inst., 1967; Alumni Achievement award Case Western Res. U., 1967, Grove City Coll., 1968. Mem. NAM, Modern Pioneers in Creative Industry, Nat. Acad. Engring., Am. Chem. Soc., Am. Inst. Chem. Engrs. Clubs: Kiwanis, Torch, Midland Country. Subspecialties: Chemical engineering; International research management. Patentee in field. Address: 600 S Ocean Blvd Apt 1503 Boca Raton FL 33432

BOURDEAU, JAMES EDWARD, researcher in renal physiology, physician; b. Seattle, Feb. 19, 1948; s. Robert Vincent and Beatrice L. (Oman) B.; m. Susan Gwen Perlman, June 17, 1973; 1 dau., Nicole Rochele. B.S., Northwestern U., 1970, Ph.D., 1973, M.D., 1974. Diplomate: Am. Bd. Internal Medicine. Intern, resident Peter Bent Brigham Hosp-Harvard U., Boston, 1974-76; research assoc. NIH, Bethesda, Md., 1976-79; asst. prof. medicine Northwestern U., Chgo., 1979—. Contbr. chpts. to books, articles to profl. jours. Served as surgeon USPHS, 1976-79. NIH fellow, 1970-72; grantee, 1982—; Hartford Found. fellow, 1980-83. Mem. Internat. Soc. Nephrology, Am. Physiol. Soc., Am. Soc. Nephrology, Am. Fedn. for Clin. Research. Jewish. Club: Midwest Salt and Water. Subspecialties: Physiology (biology); Nephrology. Current work: Epithelial calcium transport and its regulation in the kidney. Home: 900 N Lake Shore Dr Unit 607 Chicago IL 60611 Office: VA Lakeside Med Ctr Room 833 333 E Huron St Chicago IL 60611

BOURGELAIS, DONNA BELLE CHAMBERLAIN, electro-optics company research and development executive, medical laser reseacher; b. Battle Creek, Mich., May 20, 1948; d. Donald Blain and Lela May (Kellogg) Chamberlain; m. Frederick Nelson Bourgelais, Feb. 20, 1935; 1 dau., Laura Chamberlain. B.S. in Elec. Engring, M.I.T., 1970. Prin. research scientist AVCO Everett Research Lab., Mass., 1970-82; dir. research and devel. Laakman Electro-Optics, San Juan Capistrano, Calif., 1982—. Contbr. articles to profl. jours. Mem. Beth Israel Hosp. Laser Clinic, Boston. Mem. Am. Soc. for Lasers in Medicine and Surgery, Optical Soc. Am., AAAS. Subspecialties: Laser medicine; Atomic and molecular physics. Current work: Laser system development for medical use; medical lasers; laser tissue interaction. Patentee vortex stabilized repetitively pulsed flashlamps, pumped laser. Office: Leakmann Electro-Optics 33052 Calle Aviador San Juan Capistrano CA 92677

BOURNE, LYLE EUGENE, JR., psychology educator; b. Boston, Apr. 12, 1932. B.A., Brown U., 1953; M.S., U. Wis.-Madison, 1955, Ph.D. in Psychology, 1956. Asst. prof. psychology U. Utah, 1956-61, assoc. prof., 1961-63; assoc. prof. psychology U. Colo., Boulder, 1963-65, prof., 1965—, dir. Inst. Cognitive Sci., 1980-83, chmn. dept. psychology, 1983—; vis. assoc. prof. U. Alta., summer 1961, U. Calif.-Berkeley, 1961-62; vis. prof. U. Wis., summer 1966, U. Mont., summer 1967, U. Calif.-Berkeley, 1968-69, U. Hawaii, summer 1969, U. Kans. Med. Center, 1973-74, clin. prof. psychology, 1973—; mem. psychobiology rev. panel NSF, 1972-76; cons. VA, 1956—; chmn. editor selectioncom. Phsyiological Psychology, 1980. Assoc. editor: Jour. Exptl. Psycology, 1972-75; editor: Jour. Exptl. Psychology: Human Learning and Memory, 1975-80; author: Human Conceptual Behavior, 1966, (with Ekstrand and Dominowski) The Psychology of Thinking, 1971, (with Ekstrand) Psychology: Its Principles and Meanings, 1973, 4th edit., 1982, Readings in the Principles and Meanings of Psychology, 1973, (with Dominowski and Loftus) Cognitive Processes, 1979; contbr. numerous articles to profl. jours., chpts. to books; cons. editor: Psychol. Reports, 1960-69, (with Dominowski and Loftus) Jour. Exptl. Psychology, 1962-72, Psychol. Monographs, 1965-67, Psychonomic Sci, 1968-70, Jour. Verbal Learning and Verbal Behavior, 1968-73, Jour. Exptl. Child Psychology, 1969-73, Jour. Clin. Psychology, 1973—; acad. editor, Scott, Foresman Pub. Co., 1970—, (with Dominowski and Loftus), Holt, Rinehart and Winston, 1976-80, Charles Merrill Pub. Co., 1980—. Francis Whelan scholar, 1951-53; R.I. State Bd. Edn. fellow, 1950-53; grantee Office Naval Research, AT&T, others. Fellow Am. Psychol. Assn. (council of editors 1975-80, chmn. early career awards cm. 1978-79, exec. com. div. 3 1968-71, div. program com. 1964-67, bd. sci. affairs 1973-76, council of reps. 1972-74); mem. Psychonomic Soc. (governing bd. 1976-81, publs. com. 1979—), Cognitive Sci. Soc., Sigma Xi. Subspecialty: Cognition. Office: U Colo Dept Psychology Box 345 Boulder CO 80309

BOURNIA, ANTHONY, nuclear engineer; b. Warren, Ohio, Jan. 30, 1925; s. Alexander and Helen (Crecre) B.; m. Arlene Mae Lautz, June 7, 1963; children: Michael Anthony, Gregory Anthony. B.S., U. Pitts., 1950, Ph.D., 1961; M.S., U. Idaho-Richland, Wash., 1954. Registered profl. engr., Calif. Devel. chemist Sherwin William Paint Co., Cleve., 1950-51; design engr. Gen. Electric Co., Richland, Wash., 1951-55; lead design engr. Westinghouse Elec. Corp., Pitts., 1955-59, sr. design engr., 1960-61, supr. thermal/hydraulics, 1962-71, project leader environment, 1971-72; sr. design engr. Los Alamos (N.Mex.) Sci. Lab., 1961-62; sr. project mgr. U.S. Nuclear Regulatory Commn., Bethesda, Md., 1972—. Served with U.S. Army, 1943-46; ETO. Mem. Am. Nuclear Soc., Am. Inst. Chem. Engrs., Sigma Tau, Phi Lambda Upsilon. Greek Orthodox. Subspecialties: Nuclear engineering; Nuclear rocketry. Current work: Serve as a project manager for the safety review of nuclear power plant construction permit and operating license application. Office: US Nuclear Regulatory Commn 7920 Norfolk Ave Bethesda MD 20014

BOUTTON, THOMAS WILLIAM, JR., biologist; b. Lakewood, Ohio, Sept. 9, 1951; s. Thomas William and Rosemarie (Prell) B.; m. Janet Eileen Cook, Aug. 23, 1979. B.A. in Biology, St. Louis U., 1973, M.S., U. Houston, 1976; Ph.D in Botany, Brigham Young U., 1979. Research assoc. dept. biology Augustana Coll., Sioux Falls, S.D., 1979-81, asst. prof. biology, 1981-82; research assoc. dept. botany U. Nairobi, Kenya, 1979-81; postdoctoral fellow Stable Isotope Lab., Children's Nutrition Research Ctr. dept. pediatrics Baylor Coll. Medicine, Houston, 1982—, Baylor Coll. Medicine, 1982-83, instr., 1983—. Contbr. articles to profl. publs. Rob and Bessie Welder Wildlife Found. grantee, 1975; Sigma Xi grantee, 1979; Provo City Rotary Club grantee, 1979; Nature Conservancy grantee, 1981. Mem. AAAS, Bot. Soc. Am., Brit. Ecol. Soc., Ecol. Soc. Am., Sigma Xi. Subspecialties: Nutrition (medicine); Mass spectrometry. Current work: Utilization of natural variation in relative abundances of stable isotopes of carbon, oxygen, and hydrogen to investigate environmental relationships; application of stable isotope techniques to problems in human nutrition, such as determination of specific nutrient requirements and nutrient utilization within the body. Home: 5500 N Braeswood Apt 172 Houston TX 77096 Office: 6608 Fannin Med Towers 519 Houston TX 77030

BOUWER, HERMAN, lab. exec.; b. Haarlem, Netherlands, July 11, 1927; came to U.S., 1952, naturalized, 1959; s. Eduard and Trinette (Dusschoten) B.; m. Agnes N. Temminck, Mar. 29, 1952; children—Edward John, Herman (Archie) Gerard, Annette Nancy. B.S., Nat. Agr. U., Wageningen, Netherlands, 1949, M.S., 1952; Ph.D., Cornell U., 1955. Asso. agr. engr. Auburn U., 1955-59; research hydraulic engr. U.S. Water Conservation Lab., Phoenix, 1959-72, dir., 1972—; lectr. groundwater hydrology Ariz. State U.; cons. in field. Author: Groundwater Hydrology, 1978; Contbr. articles in field to profl. jours. OECD fellow, 1964; recipient Superior Service awards U.S. Dept., Agr., 1963, 73. Mem. ASCE (Walter Huber Research prize 1966), Am. Soc. Agr. Engrs., Am. Soc. Agronomy, Nat. Water Well Assn., Dutch Inst. Agr. Engrs. Club: Tempe Racquet and Swim. Subspecialties: Water supply and wastewater treatment; Ground water hydrology. Current work: Water conservation, sewage treatment by filtration through soils and aquifers via groundwater recharge, groundwater pollution, irrigation, drainage. Home: 338 La Diosa St Tempe AZ 85282 Office: 4331 E Broadway St Phoenix AZ 85040

BOVAY, HARRY ELMO, JR., engring. co. exec.; b. Big Rapids, Mich., Sept. 4, 1914; s. Harry E. and Addibelle (Bentley) B.; m. Sue Goldston, Feb. 1, 1977; children—Mark Benson, Susan Stone. C.E., Cornell U., 1936. Jr. engring. aide U.S. C.E., 1936-37; jr. metal insp., project engr. Humble Oil & Refining Co., Baytown, Tex., 1937-45; cons. engr., Houston, 1946-62; pres. Bovay Engrs., Inc., Houston, 1962-73, chmn., bd. chief exec. officer, 1974—; pres. Mid-South Telephone Co., Rienzi, Miss., 1959—, Lamar Telephone Co., Millport, Ala., 1975—; owner Bovista Farms, Somerville, Tenn., 1963—; Mem. Houston Adv. Council Naval Affairs, 1959; mem. Tex. Water Resources Research Adv. Com., 1968-71; mem. adv. com. Coastal Engring. Lab., Tex. A. and M. U., 1969; mem. engring. adv. com. Miss. State U., 1974, 77. Editor: Mechanical and Electrical Systems for Buildings. Pres. Sam Houston Area council Boy Scouts Am., 1963-64; exec. com. South Central region, 1973-76, bd. dirs., 1975-79, v.p., 1980-81, pres., 1981-82; chmn. Houston Commn. Zoning, 1959-60, 1982-83, National Audit Com., 1982-83; bd. dirs. Vis. Nurse Assn. Houston, 1970-75; active United Fund Houston and Harris County. Named Disting. Engr. Tex. Engring. Found. Fellow ASCE, ASHRAE (ASHRAE-ALCO award); mem. Nat. Soc. Profl. Engrs. (pres. 1976), Tex. Soc. Profl. Engrs. (pres. 1967-68), Am. Inst. Cons. Engrs. (past pres. Tex. chpt.), Houston Engring. and Sci. Soc. (past 2d v.p.), Am. Rd. Builders Assn. (mem. exec. com.), Am. Concrete Inst., Am. Wood Preservers Assn., ASTM (councilor 1960-64), Forest Products Research Soc., Tex. Forest Products Mfrs. Assn., ASME (Toulmin medal), Pres.' Assn., Newcomen Soc. N.Am., Nat. Acad. Engring. Episcopalian. Clubs: Houston, Kiwanis, Cosmos, Houston Country, Baton Rouge Country, Cornell of N.Y., Warwick, Petroleum. Subspecialties: Civil engineering; Industrial engineering. Current work: Chairman of board and chief executive officer of professional engineering firm which is staffed with all disciplines of engineering furnishing services to industry, government and financial groups. Home: 2200 Willowick Dr Apt 12-H Houston TX 77027 Office: 5619 Fannin Houston TX 77004

BOVEY, FRANK ALDEN, research chemist; b. Mpls., June 4, 1918; s. John Alden and Margaret Eugenia (Jackson) B.; m. Shirley June Elfman, June 19, 1941 (div. 1980); children: Margaret Bovey Glassman, Peter, Victoria A. B.S., Harvard U., 1940; Ph.D., U. Minn., 1948. With 3M Co., 1942, 48-62, Nat. Synthetic Rubber Corp., Louisville, 1942-45; with Bell Telephone Labs., Murray Hill, N.J., 1962—, head polymer chemistry research dept., 1967—; v.p., dir. Bodel Corp., Mpls.; adj. prof. Stevens Inst. Tech., 1965-67, Rutgers U., 1971—. Author: Effects of Ionizing Radiation on Polymers, 1958, Nuclear Magnetic Resonance, 1969, Polymer Conformation and Configuration, 1969, High Resolution NMR of Macromolecules, 1972, Chain Structure and Conformation of Macromolecules, 1982; also articles.; Asso. editor: Macromolecules, 1968—; editorial bd.: Accounts of Chem. Research, 1968-74, Biopolymers, 1972—. Recipient Outstanding Achievement award U. Minn. Fellow N.Y. Acad. Scis.; mem. Nat. Acad. Scis., Am. Chem. Soc. (Union Carbide award 1958, Minn. award 1962, Witco award polymer chemistry 1969, Nichols medal 1978, Phillips award 1983), Am. Phys. Soc. (High Polymer Physics prize 1974), Am. Soc. Biol. Chemists, Sigma Xi, Phi Lambda Upsilon. Subspecialties: Polymer chemistry; Nuclear magnetic resonance (chemistry). Current work: Structure and dynamics of polymer molecules; application of nuclear magnetic resonance spectroscopy to these areas. Home: 9C Dorado Dr Morristown NJ 07960 Office: Bell Labs Murray Hill NJ 07974

BOWDEN, BRYANT BAIRD, mech. engr.; b. Knoxville, Tenn., Sept. 20, 1945; s. Andrew Jackson and Calberta Ethel (Baird) B.; m. Patricia Tuck, June 11, 1966; children: Elisabeth Ann. David Aaron. B.S. in Mech. Engring, U. Tenn., Knoxville, 1968. Registered profl. engr., Tenn. With Nuclear Div. Union Carbide Corp., Oak Ridge, 1968—, head dept. gas centrifuge mech. engring., 1977-82, mgr. mech. engring., 1982—. Chmn. deacons, minister of music Bible Bap. Ch., Lenoir City, Tenn. Mem. ASME, Tenn. Soc. Profl. Engrs., Nat. Soc. Profl. Engrs. Republican. Club: Gideons Internat. Subspecialties: Mechanical engineering; Fuels and sources. Current work: Mechanical design engineering of equipment and facilities for uranium fuel enrichment; gas centrifuge; gaseous diffusion; laser isotope separation. Patentee gas centrifuge products. Home: Route 1 Box 290 Mountain View Dr Lenoir City TN 37771 Office: Mail Stop 363 Oak Ridge TN 37830

BOWEN, BOBBY J., electrical manufacturing company executive; b. Fort Worth, May 14, 1943; s. J. Raymond and Jenny (Harrison) B.; m. Virginia Leigh, June 4, 1966; children: Lauren, Francine, Chris. B.A., U. Tex., 1965, Ph.D., 1969. Advanced devel. specialist process automation Gen. Electric Co., Waukesha, Wis., successively mgr. products engring., gen. mgr. engring. dept., now gen. mgr. X-ray programs dept., 19——; mem. adv. council Marquette U. Coll. Engring.; chmn. liaison council U. Wis. Coll. Engring. Mem. Ch. of Christ. Current work: Advanced diagnostic imaging systems. Home: 6222 Oakland Hills Rd Nashotah WI 53058 Office: 3000 N Grandview Blvd Waukesha WI 53186

BOWEN, HARVEY KENT, engineering educator, consultant researcher; b. Salt Lake City, Nov. 21, 1941; s. Elmer Joseph and Sarah Bailey (Darley) B.; m. Kathleen Jones, Sept. 10, 1965; children: Natalie, Jennifer, Melissa, Kirsten, Jonathan. B.S., U. Utah, 1967; Ph.D., M.I.T., 1971. Asst. prof. dept. materials sci. and engring. M.I.T., 1970-74, assoc. prof., 1974-75, assoc. prof. dept. elec. engring. and computer sci., 1975-76, prof. depts. materials sci. and elec. engring. and computer sci., 1976-81, Ford prof. engring., dir. materials processing center, 1982—; tech. cons. to corps. Author: (with others) Introduction to Ceramics, 1976; contbr. articles to profl. jours. Mem. Am. Ceramic Soc. (Schwartzwalder award 1971, F.H. Norton award New Eng. sect. 1980, R.M. Fulrath award 1981), Am. Chem. Soc., Am. Phys. Soc., Electrochem. Soc. Republican. Mormon. Subspecialties: Ceramics; Materials. Current work: Materials processing, materials in energy and electronic systems, ceramic materials, electronic materials, capacitors, semiconductor packaging, high temperature materials.

BOWEN, JOHN METCALF, pharmacologist, college dean; b. Quincy, Mass., Mar. 23, 1933; s. Loy John and Marjorie Alice (Metcalf) B.; m. Jean Schmidt, Dec. 27, 1956; children: Mark John, Richard Kelley. D.V.M., U. Ga., 1957; Ph.D., Cornell U., 1960. Asst., then assoc. prof. Kans. State U., Manhattan, 1960-63; assoc. prof. U. Ga., 1963-69, prof., 1969—, assoc. dean research and grad. affairs, dir., 1976—. Mem. AMVA (council biologic and therapeutic agts., drug availability com.), Soc. Neurosci., Am. Soc. Pharmacology and Exptl. Therapeutics, AAAS, Sigma Xi. Subspecialties: Cellular pharmacology; Neurophysiology. Current work: Pharmacologic and toxicologic studies of excitable cells in culture. Office: Coll Vet Medicine Athens GA 30602

BOWEN, LAWRENCE HOFFMAN, chemistry educator; b. Lynchburg, Va., Dec. 20, 1934; s. Charles Wesley, Jr. and Eleanor (Hoffman) B. B.S., Va. Mil. Inst., 1956; Ph.D., MIT, 1961. Asst. prof. chemistry dept. N.C. State U., Raleigh, 1961-65, assoc. prof., 1965-70, prof. chemistry, 1970—. Contbr. articles in field to sci. jour. Served to 1st lt. Chem. Corps, U.S. Army, 1961. Recipient Jackson-Hope medal Va. Mil. Inst., 1956. Mem. Am. Chem. Soc., Am. Phys. Soc., Sigma Xi (young scientist research award 1970). Subspecialties: Physical chemistry; Solid state chemistry. Current work: Mossbauer spectroscopy of iron oxides in synthetic systems and environmental samples, magnetic properties of small particles, solid state chemistry. Office: Dept Chemistry NC State U Raleigh NC 27650

BOWEN, PAUL TYNER, environmental engineering educator; b. Macon, Ga., Sept. 2, 1953; s. Irwin Washington and Nell Elizabeth (Tyner) B.; m. Barbara Jean Amstutz, Sept. 14, 1980; 1 dau., Ashlea Harbison. B.S., Mercer U., 1975; M.S., Clemson U., 1976, Ph.D., 1982. Asst. prof. dept. civil engring. and environ. sci. U. Okla., Norman, 1982—. Chmn. social com. First Baptist Ch., Clemson, S.C., 1982, tchr. coll. Bible study, 1981. Mem. Am. Chem. Soc.(assoc.), ASCE (water quality mgmt. task force sludge conditioning and dewatering), Water Pollution Control Fedn., Am. Water Works Assn., Assn. Environ. Engring. Profs., Internat. Assn. Water Pollution Research and Control, Sigma Xi. Subspecialty: Water supply and wastewater treatment. Current work: Wastewater sludge processing, handling and disposal; sludge conditioning by organic polyelectrolytes, physical-chemical wastewater treatment. Office: Sch Civil Engring and Environ Sci U Okla 202 W Boyd St Rm 334 Norman OK 73069

BOWEN, RAFAEL LEE, dental health health found. adminstr.; b. Takoma Park, Md., Dec. 27, 1925; s. William Tyler and Naomi Ruth (Carroll) B.; m. Rosalie Jean, Aug. 3, 1958; Cheryl Lynn, Heather Jean. D.D.S., U. So. Calif., 1953. Gen. practice dentistry, San Diego, 1953-55; research assoc. Am. Dental Assn. Health Found., Washington, 1956-69, assoc. dir., 1970—, mem. spl. com. on future of dentistry research working group, 1981-82; mem. Wilmer Souder award com., 1981-82, Am. Nat. Standards Com./MD 156 dental materials, instruments and equipment, subcom. on biol. evaluation of dental materials, sec., subcom. on direct filling resins. Contbr. numerous articles on dentistry to profl. jours.; mem. rev. coms.: Jour. Biomed. Materials Research. Served with AUS, 1944-46; ETO. Recipient Hollenback Meml. prize Acad. Operative Dentistry, 1981; Clemson U. award, 1982; Callahan Meml. award Ohio Dental Assn., 1976; Dept. Commerce Cert. of Appreciation, 1971; Am. Acad. for Plastics Research in Dentistry award, 1969. Mem. Am. Dental Assn., Internat. Assn. for Dental Research (dental materials group). Subspecialties: Polymer chemistry; Biomaterials. Current work: Materials and techniques for preventive and restorative dentistry. Patentee in field. Home: 16631 Shea Ln Gaithersburg MD 20877 Office: Am Dental Assn Health Found Research Unit Nat Bur Standards Washington DC 20234

BOWEN, WILLIAM HENRY, dental researcher; b. Enniscorthy, Wexford, Ireland, Dec. 1933; came to U.S., 1973; s. William Henry and Pauline Anna (McGrath) B.; m. Carole Ann Barnes, Aug. 9, 1958; children: William, Kevin, Katherine, Deirdre, David. B.D.S., Nat. U. Ireland, 1955; M.S., U. Rochester, 1959; Ph.D., U. London, 1965; D.Sc., U. Ireland, 1974; D.D.Sc., U. Goteborg, Sweden, 1980. Gen. practice dentistry, 1955-56; research fellow Eastman Dental Center, Rochester, N.Y., 1956-59, chmn. dental research, 1982—; acting chief Nat. Caries Program, Bethesda, Md., 1973-79, chief caries prevention and research br., 1979-82; cons. FDA, 1978, Pan Am. Health Orgn., 1974-78. Editor and author: (with Art Melcher) Biology of Periodontium. Quinten Hogg fellow Royal Coll. Surgeons, London, 1959-62; Nuffield Found. fellow, 1962-65; sr. research fellow, 1965-69;

Sir Wilfred Fish fellow, 1969-73. Club: U.S. Pony (Potomac, Md.) (dist. commr. 1979-81). Subspecialties: Cariology; Preventive dentistry. Current work: Oral microbiology, immunology of dental caries, preventive methods, and oral disease in primates. Office: U Rochester Dept Dental Research 601 Elmwood Ave PO Box 611 Rochester NY 14642

BOWER, GORDON HOWARD, psychologist, educator; b. Scio, Ohio, Dec. 30, 1932; s. Clyde Ward and Mabel (Bosart) B.; m. Sharon Anthony, Jan. 30, 1957; children—Lori, Tony, Julia. B.A., Western Res. U., 1954; Woodrow Wilson fellow, U. Minn., 1954-55; M.S., Yale U., 1957, Ph.D., 1959. Prof. psychology Sch. Humanities and Scis., Stanford U., 1959—, Albert Ray Lang prof., 1975—, chmn. psychology dept., 1978—; fellow (Center for Advanced Study in Behavioral Scis.), 1973; mem. exptl. psychology rev. bd. NIMH, 1968-71; mem. psychology and ednl. process com. (Social Sci. Research Council), 1970-72; grant reviewer NIMH, NSF, Nat. Insts. Edn., Can. Research Council. Editor: The Psychology of Learning and Motivation: Advances in Research and Theory, 1968—; editorial bd.: Jour. Comparative and Physiol. Psychology, 1963-70, Jour. Math. Psychology, 1964-70, Jour. Exptl. Psychology, 1965-72, Jour Exptl. Analysis of Behavior, 1965-69, Jour. Verbal Learning and Verbal Behavior, 1968—, Cognitive Psychology, 1969-75, Cognitive Therapy and Research, 1976-79; contbr. articles to profl. jours. NIMH postdoctoral fellow, 1965. Fellow Am. Psychol. Assn. (exec. com. div. exptl. psychology 1974-76, pres. Div. 3 1975, Disting. Sci. Contbn. award 1979); mem. Western Psychol. Assn., Psychonomic Soc. (bd. govs. 1972-76, chmn. governing bd. 1975-76, chmn publs. com. 1978), Nat. Acad. Scis., Am. Acad. Arts and Scis., Soc. Exptl. Psychologists. Subspecialty: Cognition. Home: 750 Mayfield Ave Stanford CA 94305 Office: Dept Psychology Stanford U Stanford CA 94305

BOWER, NATHAN WAYNE, chemistry educator, researcher; b. Clarinda, Iowa, Nov. 10, 1951; s. Kenneth G. and Julia (Emerick) B. B.A., Wooster Coll., 1973; Ph.D., Oreg. State U.-Corvallis, 1977. Asst. prof. Colo. Coll., Colorado Springs, 1977—. Author: Techniques in Instrument and Experimental Design, 1980. Mem. Am. Chem. Soc., Soc. Applied Spectroscopy, Sigma Xi. Roman Catholic/Soc. of Friends. Subspecialty: Analytical chemistry. Current work: Geochemistry and archaeochemistry/chemometrics and sampling methodologies, nutrition and cancer. Office: Chemistry Dept Colo Coll Colorado Springs CO 80903

BOWERS, CHARLES HATHAWAY, telephonic engineer, nuclear safety engineer; b. Chgo, Oct. 13, 1940; s. Abner Martin and Phyllis (Dodge) B.; m. Marjorie Foster, June 12, 1962; children: Jane Elizabeth, Susan Hathaway. B.S. in Engring, U.S. Naval Acad., 1962; M.S. in Nuclear Engring., (fellow), Ohio State U., Ph.D., 1973. Profl. nuclear engr., Calif. Engr. Knolls Atomic Power Lab., Schenectady, 1970-74; group leader Argonne (Ill.) Nat. Lab., 1974-82; mem. tech. staff Bell Telephone La., Naperville, Ill., 1982—; cons. Adv. Com. Reactor Safety, Washington, 1981—, Electric Power Research Inst., Palo Alto, Calif., 1979-82. Contbr. articles in field to profl. jours. Precinct committeeman Republican party, DuPage County, Ill., 1977-79; pres. bd. dirs. Project Home, Downers Grove, Ill., 1977—; bd. dirs. Kobe Coll. Corp., Evanston, Ill., 1981—. Served to capt. USNR, 1962—. Mem. Am. Nuclear Soc. Presbyterian. Current work: Computer electronic switching system architectural engineer working on the switching systems for Bell Telephone Lab. Home: 9 South 610 Main St Downers Grove IL 60516 Office: Bell Telephone Lab Wheaton-Warrenville Rd Naperville IL 60556

BOWERS, MAYNARD CLAIRE, botany educator; b. Battle Creek, Mich., Nov. 5, 1930; s. Frederick Claire and Elnora Alice (Hard) B.; m. Mary Joan Sweet, Aug. 17, 1951 (div. 1969); children: Maynard, Janet; m. Leenamari Kangas, Aug. 16, 1970; children: Piiamari, Eerik. A.B., Albion Coll., 1956; M.Ed., U. Va., 1960; Ph.D., U. Colo., 1966. Tchr. Flint (Mich.) Pub. Schs., 1956-57, St. Petersburg (Fla.) Pub. Schs., 1957-59; asst. prof. botany Towson (Md.) State Coll., 1960-62; mem. faculty dept. biology No. Mich. U., Marquette, 1966—, prof. botany, 1975—. Editor: Through the Years in Glacier National Park, 1960; contbr. articles to profl. jours. Served with USAF, 1951-52. Mem. Internat. Assn. Bryologists, Internat. Assn. Plant Taxonomists, Internat. Orgn. Plant Biosystematists, Mich. Bot. Club, Sigma Xi (chpts. pres.), Phi Sigma. Republican. Lodge: Masons. Subspecialties: Systematics; Plant growth. Current work: Effects of herbicides and their reversal on non-target species; plant biosystematics. Home: 2 Northwoods Ln Marquette MI 49855 Office: Dept Biology No Mich U Marquette MI 49855

BOWERS, PHILLIP FREDERICK, radio astronomer; b. Huntington, Ind., Dec. 14, 1947; s. Frederick Wallace and Frances Maxine (Everett) B.; m. Katherine Lorimer Woolley, June 27, 1970. B.S. in Astrophysics, Ind. U., 1969; Ph.D. in Astronomy, U. Md., 1977. Research assoc. U. Md., 1977; research assoc. Nat. Radio Astronomy Obs., Charlottesville, Va., 1977-80; research astronomer U. Md. and Naval Research Lab., Washington, 1980—. Contbr. articles to profl. jours. E.O. Hurlburt fellow, 1981—. Mem. Internat. Astron. Union, Am. Astron. Soc., Sigma Pi Sigma. Subspecialty: Radio and microwave astronomy. Current work: Interferometric studies of radio emission from stars. Office: EO Hurlburt Ctr Space Research Naval Research Lab Code 4134-5 Washington DC 20375

BOWERSOX, TODD WILSON, silviculturist; b. Lewistown, Pa., Oct. 4, 1941; s. John I. and Dorothy M. B.; m. Judy C. Loht, Sept. 9, 1961; children: Natalie, Elizabeth, Cortney. B.S., Pa. State U., 1966, M.S., 1968, Ph.D., 1975. Mem. faculty Pa. State U., 1968—, assoc. prof. silviculture, 1981—; cons. to industry U.S. Forest Service, Nat. Park Service, AID. Contbr. articles to profl. jours. Served with USN, 1959-62. Mem. Soc. Am. Foresters, Forest Product Research Soc. Subspecialties: Biomass (agriculture); Resource management. Current work: Cultural practices for hardwood forests; biomass plantations (short rotation intensive culture). Office: 204 Ferguson Bldg University Park PA 16802

BOWES, GEORGE ERNEST, botanist, educator, consultant; b. London, May 22, 1942; U.S., 1968; s. George and Ivy Primrose (Haynes) B.; m. Helen Clifford, Sept. 12, 1970; children: George Edward Aden, Joel Asher. B.Sc., Queen Mary Coll., U. London, 1963, Ph.D., 1967; M.I.Biol. (hon.), Inst. Biology, London, 1968. Lectr. in biology Regents St. Poly., London, 1966-68; high sch. tchr. Inner London Edn. Authority, 1968; vis. research assoc. agronomy dept. agr. U. Ill., 1968-71; Carnegie Instn. research fellow dept. plant biology Stanford (Calif.) U., 1971-72; asst. prof. botany U. Fla., Gainesville, 1973-78, assoc. prof., 1978—. Editor: Aquatic Botany, 1982—, What's New in Plant Physiology, 1978-82; contbr. articles on photosynthesis research to profl. jours. Bd. elders Community Evang. Free Ch. Fla. Dept. Natural Resources grantee, 1976-83; Dept. Agr. Sci. and Edn. Adminstrn. grantee, 1978-84; Gas Research Inst. grantee, 1981-83; NSF grantee, 1982-84. Mem. Am. Soc. Plant Physiologists, Soc. Exptl. Biology, Inst. Biology London, Aquatic Plant Mgmt. Soc., Sigma Xi. Subspecialties: Photosynthesis; Biomass (agriculture). Current work: Pioneer in ribulose bisphosphate carboxylase-oxygenase enzyme and its role in photorespiration; aquatic plant photosynthetic pathways, research use of aquatic plants as alternative energy (biomass) source. Office: 3157 McCarty Hall Dept of Botany U Fla Gainesville FL 32611

BOWHILL, SIDNEY ALLAN, electrical engineering educator; b. Dover, Kent, Eng., Aug. 6, 1927; came to U.S., 1955, naturalized, 1962; s. Sidney Allan and Violet (Clarke) B.; m. Margaret M. McLaughlin, Aug. 22, 1959; children: Allan J.C., Amanda M. B.A., Cambridge (Eng.) U., 1948, M.A., 1950, Ph.D., 1954. Research engr. Marconi's Wireless Telegraph Co. Ltd., Chelmsford, Essex, Eng., 1953-55; assoc. prof. elec. engring. Pa. State U., University Park, 1955-62; prof. elec. engring. U. Ill., Urbana, 1962—; Pres. Aeronomy Corp., Champaign, Ill., 1969—; Chmn. U.S.A. Commn. 3, Internat. Sci. Radio Union; assoc. editor Radio Sci., 1964-67, editor, 1968-72; vice chmn. Internat. Commn. 3, 1969-72, chmn., 1972-75; mem. working group 4 Inter-Union Com. Space Research, 1966-79, co-chmn. panel interactions neutral and ionized atmospheres, 1966-79, chmn. sci. commn. C, 1979—, mem. panel sci. ballooning, 1979-82; convenor, program chmn. Symposium (9th meeting), Vienna, Austria, 1966, Seattle, 1971, Solar-Terrestrial Physics Symposium, São Paulo, Brazil, 1974, Innsbruck, Austria, 1978; mem. com. data interchange and data centers Nat. Acad. Scis., 1967-82, potential contamination and interference from space expts., 1963—, com. polar research, 1967-70, chmn. panel upper atmospheric phys., 1967-70, com. on solar terrestrial research, 1969-79; mem. panel on Jicamarca Radio Obs., 1969—, chmn. panel, 1976-78; mem. Inter-Union Sci. Com. on Solar-Terrestrial Physics, 1967 -, chmn. working group II, 1968-73, chmn. atmospheric phys. programs com., 1974—; chmn. steering com. for middle atmosphere program, 1977—; editorial adv. bd. Jour. Atmospheric and Terrestrial Physics, 1965-81. Contbr. numerous articles to profl. jours. Fellow IEEE (procs. bd. cons., procs. editorial bd. 1965-68), AAAS, Am. Geophys. Union, Am. Astron. Soc., Am. Meteorol. Soc.; mem. Am. Soc. Engring. Edn., Nat. Acad. Engring., Sigma Xi, Sigma Tau, Eta Kappa Nu. Club: Cosmos (Washington). Subspecialties: Aeronomy; Remote sensing (atmospheric science). Current work: Middle-atmosphere physics and chemistry; rocket studies of the D region; meteor radar and MST radar studies of atmospheric dynamics; small-signal processing and detection; computer language development. Home: 2203 Anderson St Urbana IL 61801

BOWMAN, CLEMENT WILLIS, petroleum research executive; b. Toronto, Ont., Can., Jan. 7, 1930; s. Clement Willis and Emily (Stockley) B.; m. Marjorie Elizabeth Greer, Aug. 21, 1954; children: Elizabeth Ann, John Clement. B.A.Sc., U. Toronto, 1952, M.A.Sc., 1958, Ph.D., 1961. Registered profl. engr., Alta. Research engr. Imperial Oil Ltd., Sarnia, Ont., Can., 1960-63; research mgr. Syncrude Can. Ltd., Edmonton, Alta., Can., 1964-69; chem. research mgr. Imperial Oil. Ltd., Sarnia, 1969-72, petroleum research mgr., 1972-75; chmn. Alta. Oil Sands Tech. and Research Authority, Edmonton, 1975—; dir. Alta. Research Council, 1979—. Recipient Meritorious Service medal U. Toronto Alumni Assn., 1977; Queen's 25-Yr. Jubilee medal, 1977. Mem. Can. Soc. Chem. Engring.; fellow Chem. Inst. Can. Subspecialties: Fuels; Fuels and sources. Current work: Heavy Oil, tar sand recovery and upgrading. Office: 10010 106th St 500 Highfield Pl Edmonton AB Canada T5J 3L8

BOWMAN, H(ARRY) FREDERICK, bioengineering educator; b. Terre Hill, Pa., Sept. 16, 1941; s. Willis Stauffer and Hanna Mae (Myer) B. B.S., Pa. State U., 1963; M.S. in Nuclear Engring, MIT, 1966, 1966; Ph.D. in Nuclear Engring, 1968. Project engr. Humble Oil & Refining Co., Linden, N.J., 1963; engr. Stone & Webster Engring. Corp., Boston, 1964; sr. acad. adminstr. Harvard-MIT Div. Health Scis. and Tech., 1969—; asst. prof. mech. engring. Northeastern U., Boston, 1973-79, assoc. prof. biomed. engring., 1979—; lectr. mech. engring. MIT, 1976—; cons. clin. cancer program project rev. com. NIH/Nat. Cancer Inst., 1981—, ad hoc mem. diagnostic radiology study sect., 1983; cons., reviewer NSF, NIH, ASME, others. Author: Non-invasive Transducer for Monitoring Perfusion, 1982; contbr. articles to profl. jours. Bd. dirs. Tuborous Sclerosis Assn. Am., 1981—, South Shore Ctr. for Brain-injured Children, 1974—; mem. Mass. Devel. Disabilities Council; mem. selection com. Am. Field Service, 1973-76. AEC fellow, 1964-67. Mem. ASME, Am. Nuclear Soc., Assn. Advancement Med. Instrumentation. Mem. Ch. of Jesus Christ of Latter-day Saints. Subspecialties: Biomedical engineering; Mechanical engineering. Current work: Bioheat and mass transfer; thermal transducer development; thermal methods to quantify blood flow; hyperthermia; modelling; thermometry; instrumentation; boiling heat transfer. Home: 101 Edwardel Rd Needham MA 02192 Office: Harvard MIT Div of Health and Scis E25-518 MIT 77 Massachusetts Ave Cambridge MA 02139

BOWMAN, JAMES EDWARD, physician, educator; b. Washington, Feb. 5, 1923; s. James Edward and Dorothy (Peterson) B.; m. Barbara Taylor, June 17, 1950; 1 dau., Valerie June. B.S., Howard U., 1943, M.D., 1946. Intern Freedmen's Hosp., Washington, 1946-47; resident pathology St. Lukes Hosp., Chgo., 1947-50; chmn. dept. pathology Provident Hosp., 1950-53, Shiraz (Iran) Med. Center, Nemazee Hosp., 1955-61; vis. prof., chmn. dept. pathology Faculty of Medicine, U. Shiraz, 1959-61; dir. labs. U. Chgo., 1971—, prof. dept. pathology, medicine, com. on genetics and biol. scis., collegiate div., 1972—, dir., 1973—; cons. pathology, div. hosp. and med. facilities HEW, USPHS, mem., 1968, (Health and Hosps. Governing Commn. Cook County), 1969-72; mem. exec. com. hemalytic anemia study group NHLI, NIH, Bethesda, Md., 1973-75; Sabbatical fellow Center for Advanced Study in Behavioral Scis., Stanford U., 1981-82. Contbr. articles profl. jours. Served to capt. M.C. AUS, 1953-55. Spl. research fellow NIH Galton Lab., Univ. Coll., London, 1961-62. Mem. Coll. Am. Pathologists, Am. Soc. Clin. Pathologists, Am. Soc. Human Genetics, Central Soc. Clin. Research, Am. Soc. Hematology, Am. Assn. Phys. Anthropologists, Acad. Clin. Lab. Physicians and Scientists. Subspecialties: Pathology (medicine); Genetics and genetic engineering (medicine). Current work: Population distribution of abnormal hemoglobins; ethical and legal issues in genetics programs. Home: 4929 S Greenwood St Chicago IL 60615 Office: 950 E 59th St Chicago IL 60637

BOWMAN, JOEL MARK, educator; b. Boston, Jan. 16, 1948; s. Henry Frank and Irene (Goodman) B.; m. Barbara Ann Brown, July 15, 1978. A.B., U. Calif., Berkeley, 1969; Ph.D., Calif. Inst. Tech., 1974. Asst. prof. Ill. Inst. Tech., 1974-77, asso. prof., 1977-82, prof. chemistry, 1982—; cons. chemistry div. Argonne (Ill.) Nat. Lab., 1977—. Recipient Alfred P. Sloan Fellowship award, 1977-81; grants from NSF, 1976—, DOE, 1981—; Am. Chem. Soc., 1975-80. Mem. Am. Inst. Physics, Am. Chem. Soc. Subspecialties: Theoretical chemistry; Kinetics. Current work: Gas phase, gas surface chemical dynamics, laser chemistry, scattering theory, combustion and catalysis applications; influence of high powered lasers on chemical interactions. Home: 1437 E 54th St Chicago IL 60615 Office: Dept Chemistry Ill Inst Tech Chicago IL 60616

BOWMAN, MARY BRONWYN, clinical neuropsychologist, consultant; b. Atlanta, July 29, 1957; s. Joseph Merrill and Mary Isabella (Nichols) B. B.S., Georgetown U., 1978; Ph.D., U. Tex.-Austin, 1981. Postdoctoral research fellow NIMH, Princeton, N.J., 1981; clin. neuropsychologist John F. Kennedy Med. Center, Edison, N.J., 1981-82; dir. cognitive rehab. and research The Head Injury Center Lewis Bay, Hyannis, Mass., 1982—; dir. Neurologic Ctr. at Forest Manor, Middleboro, Mass., 1982--; cons. Center for Battered Women, Austin, Tex., 1979-81, Assn. Advancement of Mentally Handicapped, Princeton, N.J., 1981-82. Editor: Jour. Community Psychology, 1980-81. U. Tex.-Austin research grantee, 1980. Mem. Am. Psychol. Soc., Internat. Neuropsychol. Soc. Democrat. Methodist. Subspecialties: Neuropsychology; Neurophysiology. Current work: Research in cognitive rehabilitation, restitution of function, research in etiology of the psychoses, epidemiology of mental illness. Office: 410 Virginia Ave Alexandria VA 22302

BOWMAN, THOMAS EUGENE, engring. educator, cons., researcher; b. Darby, Pa., Aug. 3, 1938; s. Robert Alexander and Nina Adelaide (Hardin) B.; m. Ritva Vappu Helina Torsti, Dec. 18, 1961; children: Paul, Katriina, Markus, Anthony. B.S., Calif. Inst. Tech., 1960, M.S. (NSF coop. fellow), 1961; Ph.D. (Walter P. Murphy research fellow), Northwestern U., 1964. Registered profl. engr., Fla. Research scientist Martin Marietta Corp., Denver, 1963-69; assoc. prof. space tech. Fla. Inst. Tech., Melbourne, 1969-72, assoc. prof. mech. engring., 1972-75, prof., 1975—, head dept., 1978—, dir., 1975—, acting dean, 1982-83; chmn. policy adv. bd. Fla. Solar Energy Ctr.; cons. Sci. Application, Inc., CARE, Gen. Dynamics Corp; vis. lectr. applied math U. Reading, Eng., 1967-68. Contbr. articles to profl. jours. Grantee NASA, 1973-78, AID, 1976-77, State of Fla., 1979-83. Mem. ASME, Am. Soc. Engring. Edn., Am. Solar Energy Soc., Tau Beta Pi. Democrat. Roman Catholic. Subspecialties: Solar energy; Mechanical engineering. Current work: Fluid mechanics, energy conversion, solar energy, aerospace propulsin research, research management, educational administration. Home: 8637 Sylvan Dr Melbourne FL 32901 Office: 150 W University Blvd Melbourne FL 32901

BOWNDS, JOHN MARVIN, mathematician, software engineer, consultant; b. Delta, Colo., Apr. 22, 1941; s. John Ennis and Ruth Pauline (Martin) B.; m. Lynne Marie Newman, Sept. 1, 1962; children: Jennie, Layne, John. A.A., Oakland City Coll., 1961; B.A., Calif. State U.-Chico, 1964; M.A., U. Calif-Riverside, 1967, Ph.D., 1968. Mathematician U.S. Naval Weapons Center, China Lake, Calif., 1964-65, Corona, Calif., 1966-68; prof. math. Rensselaer Poly. Inst., Troy, N.Y., 1971-72, U. Ariz., Tucson, 1968—; Einstein vis. prof. Rensselaer Poly. Inst., 1971-72. Contbr. articles in field to profl. jours. Cons., vol. Pima County (Ariz.) Sheriff Dept., Tucson, 1977—, U.S. Park service, 1979—; Vol. Sheriff's Dept. Search and Rescue, State Ariz., 1972—. Mem. Am. Math. Soc., Soc. Indsl. and Applied Math. Subspecialties: Numerical analysis; Mathematical software. Current work: Mathematical software, algorithms development. Office: Dept Math U Ariz Tucson AZ 85721

BOWSHER, ARTHUR LEROY, geologist; b. Wapakoneta, Ohio, Apr. 29, 1917; s. Dallas and Sallie Lenora (Fox) B.; m. Lanorah Jane Higgins, Oct. 19, 1943 (div. 1965); children: Sally Jane, Arthur LeRoy, Anne Lorraine, Dale Clark; m. Ruth E. Webber, Aug. 29, 1967; 1 dau., Donna Jean DeMars. B.S., U. Tulsa., 1941; postgrad, U. Kans., 1941-42, 46-47. Registered geologist. Assoc. curator U.S. Nat. Mus., Washington, 1947-52; geologist U.S. Geol. Survey, Washington and Fairbanks, Alaska, 1952-57; explorationist Sinclair Oil & Gas Corp., Tulsa and Dallas, 1957-70; staff geologist ARAMCO, Dhahram, Saudi Arabia, 1970-78; chief exploration strategy ONPRA, U.S. Geol. Survey, Menlo Park, Calif., 1978-81; sr. geologist Yates Petroleum Corp., Artesia, N.Mex., 1981—. Contbr. articles to profl. jours. Active Boy Scouts Am., 1929—; mem. fin. com. Schmidtt for Senate, N.M., 1982. Served to capt., C.E. AUS, 1942-46; CBI. Recipient Silver Beaver award Boy Scouts Am., 1977. Fellow Geol. Soc. Am.; mem. Am. Assn. Petroleum Geologist, Sigma Xi, Sigma Gamma Epsilon. Republican. Methodist. Subspecialties: Sedimentology; Fuels. Current work: Petroleum exploration, descriptive paleontology of echinoderma and carbonate deposition. Home: 2707 Gaye Dr Roswell NM 88201 Office: Yates Petroleum Corp 207 S 4th St Artesia NM 88210

BOWYER, ALLEN FRANK, cardiologist, administrator; b. Milw., Aug. 9, 1932; s. Charles Maynard and Mildred Berniece (Haagensen) B.; m. Carolyn Isabel Gramlich, June 5, 1954; children: Sylvia Renee, Susan Rayleen. B.A., Pacific Union Coll., 1955; M.D., Loma Linda U., 1959. Diplomate: Am. Bd. Cardiology. Resident in medicine Rush-Presbyn., St. Luke's Hosp., Chgo., 1960-63, fellow in cardiology, 1963-66; extramural fellow Nat. Heart Inst., Bethesda, Md., 1963-66; instr. in medicine U. Ill. Sch. Medicine, Chgo., 1960-65; asst. prof. Loma Linda (Calif.) U., 1966-72, assoc. prof., 1972-73; prof. medicine W. Va. U., Morgantown, 1973-78, East Carolina U., Greenville, 1978—; research physician Rush-Presbyn. Hosp., 1965-66; dir. cardiovascular research lab. Loma Linda U., 1970-73; dir. cardiac catheterization lab. W.Va. U. Sch. Medicine, 1973-78; dir. cardiac labs. East Carolina U.Sch. Medicine, 1975-81. Author and producer: Computer Graphics Film, Heart Motion by Computer Graphics, 1968 (1st prize for research at Internat. Film Festival 1970), Teaching Heart Function by Computer Graphics, (sci. film award Australia, New Zealand Sci. Film Festival 1970). Pres. Monoghalia County Heart Assn., Morgantown, 1974, 75; trustee Columbia Union Coll., Takoma Park, Md., 1976-78; pres. Pitt County Unit Am. Heart Assn., Greenville, N.C., 1983; dir. W. Va. Heart Assn., Charleston, 1974, 75. Recipient Golden Eagle ward Council on Internat. Non-theatrical Events, 1969, 1970; MacNeal award for Med. Research Rush-Presbyn. Hosp., 1963. Mem. IEEE, Instrument Soc. Am., Am. Fedn. for Clin. Research, Assn. for Computing Machinery, Sigma Xi. Democrat. Adventist. Subspecialties: Cardiology; Graphics, image processing, and pattern recognition. Current work: Applications of computer graphics to clinical medicine and cardiology, development of computer graphics programs to describe cardiac valve and chamber motion and development of information theory methods to medical diagnosis, development of new approaches to artificial intelligence. Office: East Carolina U Sch Medicine Sect Cardiology Dept Medicine Greenville NC 27834 Home: 315 King George Rd Greenville NC 27834

BOXER, LAURENCE ALAN, physician, pediatric hematologist; b. Denver, May 17, 1940; s. Sam and Tillie (Belstock) B.; m. Grace Jordison, Aug. 23, 1969; 1 son; David E.K. B.A., U. Colo., 1961; M.D., Stanford U., 1966. Diplomate: Am. Bd. Pediatrics. Intern, resident in pediatrics Yale New Haven Hosp., 1966-68; resident in pediatrics Stanford (Calif.) U., 1966-68; fellow in hematology Childrens Hosp. Med. Center, Boston, 1972-74; instr. pediatrics Harvard Med. Sch., 1974-75; asst. prof. Ind. U. Med. Sch., Indpls., 1975-82, assoc. prof., 1978-82, prof., 1982; prof., dir. pediatric hematology, oncology U. Mich. Med. Sch., Ann Arbor, 1982—. Contbr. numerous articles to med. jours. Served to maj. AUS, 1969-72. Established investigator Am. Heart Assn., 1978—. Fellow A.C.P.; mem. Am. Soc. Clin. Investigators, Am. Soc. Pathologists, Am. Assn. Immunologists, Soc. Pediatric Research, Am. Soc. Pediatric Research, Am. Soc. Hematology, Phi Beta Kappa, Sigma Xi. Republican. Jewish. Subspecialties: Pediatrics; Hematology. Current work: Mechanisms of phagocytic cell function, host defense.

BOXER, STEVEN GEORGE, chemist, educator; b. N.Y.C., Oct. 18, 1947; s. George E. and Lily (Behar) B.; m. Linda Minivum, Sept. 26, 1977. B.S., Tufts U., 1969; Ph.D., U. Chgo., 1976. Asst. prof. chemistry Stanford U., 1976-82, assoc. prof., 1982—. Contbr. articles to profl. jours. Alfred P. Sloan fellow, 1979-83; AEC fellow, 1972-76. Mem. Am. Chem. Soc., Biophysical Soc., Am. Soc. Photobiology. Subspecialties: Biophysical chemistry; Physical chemistry. Current work: Biophysical chemistry, photosynthesis, magnetic resonance, laser spectroscopy. Office: Dept Chemistry Stanford U Stanford CA 94305

BOYAN, BARBARA DALE, biochemistry educator; b. Boise, Idaho, Sept. 20, 1948; d. Jack R. and Berniece Alice (Elling) B.; m. Gregory Allen Salyers, May 17, 1975 (div. Dec. 1981); m. Don Mauldin Ranly, Apr. 24, 1982; children: Elizabeth Melva, Edmund Don. B.A., Rice U., 1970, M.A., 1974, Ph.D., 1975. Postdoctoral fellow Health Sci. Center, U. Tex.-Houston, 1974-75, research assoc., 1975-77, asst. prof. dept. periodontics, 1977-81; assoc. prof. dept. periodontics Health Sci. Center, U. Tex.-San Antonio, 1981—; cons. in field. Contbr. articles to profl. jours. Religious edn. adviser Temple Emanu El, Houston, 1979-81. NIH postdoctoral fellow, 1974-77; Nat. Acad. Sci.-NRC research assoc., 1974; NIH Research Career Devel. awardee, 1977-82; NIH research grantee, 1974—. Mem. Am. Assn. for Dental Research, Internat. Assn. for Dental Research, Am. Soc. for Bone and Mineral Research, Electron Microscopy Soc. Am., Tex. Soc. Electron Microscopy, Alpha Omega. Jewish. Subspecialties: Oral biology; Biochemistry (medicine). Current work: Study of the mechanisms of mineral formation in normal bones and teeth; study of regulation of mineralization by vitamin D; analysis of how pathologic calcification occurs. Office: Department Periodontics University of Texas Health Science Center 7703 Floyd Curl Dr San Antonio TX 78284

BOYCE, JOSEPH MICHEAL, space sci. researcher, govt. exec.; b. Mesa, Ariz., July 30, 1946; s. James William and Florence Evelyn (Gutherie) B.; m. Catherine E., Jan. 25, 1969; children: Amy, Heather. B.S., No. Ariz. U., 1969, M.S., 1972. Geologist Br. Astrogeology, U.S. Geol. Survey, Flagstaff, Ariz., 1969-77; staff scientist NASA Planetary Geology Program, Washington, 1977-79, discipline chief, 1979—. Contbr. writings to sci. publs., Smithsonian Press, 1982—. Mem. AAAS, Am. Geophys. Union, Geol. Soc. Am., Am. Astron. Union. Subspecialty: Planetology. Current work: Management of science programs, research planetary stratigraphy, impact cratering, surface processes. Office: NASA Hdqrs Code El-4 Washington DC 20546

BOYD, ANN LEWIS, viral oncologist, researcher, educator; b. Shreveport, La., Nov. 15, 1944; d. Fletcher Willard and Bess Juanita (Sherman) Lewis; m. James Pierce Boyd, June 4, 1964 (div.); 1 dau., Kathryn Ann. B.S., Northwestern State U., Natchitoches, La., 1965, M.S., 1968; Ph.D., La. State U., 1971. Postdoctoral researcher Baylor U. Sch. Medicine, 1971-73; prin. scientist Nat. Cancer Inst.-Frederick (Md.) Cancer Research Facility, Litton Bionetics, Inc., 1973—; asso. prof. biology Hood Coll., 1981—. Contrb. articles, chpts. to profl. publs. Bd. dirs. Penn Laurel council Girl Scouts Am.; mem. vestry, jr. warden Harriet Chapel Episcopal Ch., Frederick. Grantee Nat. Cancer Inst. Mem. Am. Soc. Microbiology, Am. Tissue Culture Assn., AAAS, N.Y. Acad. Scis., Sigma Xi, Phi Kappa Phi. Republican. Subspecialties: Cancer research (medicine); Virology (medicine). Current work: Manual microinjectin of mammalian cells with DNA, RNA or protein, using glass capillary micropipettes to study gene expression. Home: 8821 Indian Springs Rd Frederick MD 21701 Office: Nat Cancer Inst-FCRF Bldg 560 PO Box B Frederick MD 21701

BOYD, DONALD LOREN, computer scientist; b. Des Moines, July 7, 1941; s. Loren U. and Faith B. (Rippey) B.; m. Mary Angela Coblentez, June 8, 1969; children: Peter Loren, Sarah Elizabeth. B.A., U. Iowa-Iowa City, 1963, M.S., 1965, Ph.D., 1971. System programmer U. Iowa Computer Ctr., Iowa City, 1965-68, instr. computer sci. dept., 1968-69; asst. prof. computer sci. dept. U. Minn., Mls., 1971-76; research scientist Honeywell Computer Sci. Ctr., Bloomington, Minn., 1976-79, mgr. research, 1979-81, dir., 1981—; cons. Control Data Corp., Mpls., 1973-76; adj. prof. computer sci. dept. U. Minn., Mpls., 1976—. Mem. IEEE, Assn. Computing Machinery, Sigma Xi. Democrat. Subspecialties: Software engineering; Operating systems. Current work: Software technology, computer architecture, software and hardware. Home: 2789 Dean Pkwy Minneapolis MN 55416 Office: Honeywell Inc 10701 Lyndale Ave S Bloomington MN 55420

BOYD, ELEANOR H., researcher, educator; b. Phila., Oct. 7, 1935; d. Herbert V. and Eleanor M. (Wister) Hurbrink; m. Eugene S. Boyd, May 7, 1964. B.A., Wellesley Coll., 1956; Ph.D., U. Rochester, 1968. Asst. toxicologist E.I. du Pont, Newark, Del., 1956-61; postdoctoral fellow, assoc. dept. physiology U. Rochester, 1969-72, asst. prof., 1972—. Contbr. articles to profl. jours. NIH grantee, 1978—. Mem. Am. Soc. Pharmacology and Exptl. Therapeutics, Soc. Neursci., Sigma Xi. Subspecialties: Neurobiology; Pharmacology. Current work: Brain mechanisms of motivated behavior; neurophysiologic basis of neuropharmacologic agents. Office: Dept Pharmacology U Rochester Richester NY 14642 Home: Gillis Rd Victor NY 14564

BOYD, JAMES, consulting geologist; b. Kanowna, West Australia, Dec. 20, 1904; s. Julian and Mary (Innes) Cane; ; s. Julian and Mary (Innes) B.; m. Ruth Ragland Brown, Aug. 17, 1932 (dec. 1979); children: James Brown, Harry Bruce, Douglas Cane, Hudson; m. Clemence D. Jandray, 1980. B.S., Calif. Inst. Tech., 1927; M.Sc., Colo. Sch. Mines, 1932, D.Sc., 1934. Instr. geology Colo. Sch. Mines, Golden, 1929-34, asst. prof. mineralogy, 1934-37, asso. prof. econ. geology, 1938-41, dean faculty, 1946-47; asst. to sec. interior chmn. interdeptl. com. (Resources for Marshall Plan), 1947; dir. Bur. Mines, 1947-51, Def. Minerals Adminstrn., 1950-51; exploration mgr. Kennecott Copper Corp., 1951-55, v.p. exploration, 1955-60; pres. Copper Range Co., 1960-70, chmn. bd. dirs., 1970-71; exec. dir. Nat. Commn. on Materials Policy, Washington, 1971-73; pres. Materials Assos., 1974-78; geologist U.S. Geol. Survey, 1933-34; cons. geology, mining and geophysics, 1935-40; pres., gen. mgr. Goldcrest Mining Co., 1939-40; dir. engrs. Joint Council and United Engring.; Trustees, 1969-71; chmn. com. on mineral research NSF, 1952-57; vice chmn. Engrs. Commn. on Air Resources, 1970-71; chmn. sec. interior's adv. com. on non-coal mine safety, 1971-74; exec. dir. Nat. Commn. Materials Policy; chmn. materials com. Office Tech. Assessment, 1974-79; mem. nat. materials adv. bd. NRC, 1975-77, mem. mineral and energy resources bd., 1977-80, mem. mineral resources bd., 1973-75, chmn. com. on surface mining and reclamation, 1978-79; mem. tech. adv. com. Office of Nuclear Waste Isolation, 1979-82; Bd. dirs. Watergate S. Corp., 1972-79, pres., 1976-77. First reader Carmel Christian Sci. Ch., 1983-85. Served from capt. to col. AUS, 1941-46. Decorated Legion of Merit with oak leaf cluster; recipient Distinguished Service medal Colo. Sch. Mines, 1949; Distinguished Alumni award Calif. Inst. Tech., 1967; Hoover medal, 1975. Mem. Mining and Metall. Soc. Am. (pres. 1960-63), Am. Inst. Mining Engrs. (Rand gold medal 1963, pres. 1969), Nat. Acad. Engring., Am. Soc. Econ. Geologists, Geol. Soc. Am., Am. Inst. Profl. Geologists (Parker Meml. medal 1973, v.p. 1965-66), Soc. Exploration Geophysicists, Australasia Inst. Mining and Metallurgy (hon.), Acad. Polit. Sci. Clubs: Cosmos (Washington); Burning Tree. Subspecialties: Nuclear waste; Resource management. Home and Office: 228 Del Mesa Carmel Carmel CA 93921

BOYD, JEFFREY LYNN, clinical psychologist, researcher, educator; b. Mpls., June 16, 1950; s. Harold McGrath and Verna Grace (Peterson) B.; m. Jeanne Cathryn Blomberg, Aug. 17, 1974; children: Cameron Michael, Caitlin Lainey, Andrew Thomas. B.A., U. Minn., 1972; M.S., Fla State U., Tallahassee, 1976, Ph.D., 1979. Lic. cons. psychologist, Minn. Research psychologist Sch. Medicine, U. So. Calif., Los Angeles, 1978-82, asst. prof., 1980-82; sr. clin. psychologist Hennepin County Mental Health Center, Mpls., 1982—; cons. Los Angeles Police Dept., 1980-82, NIMH, Washington, 1982. Author: Family Therapy for Schizophrenia, 1983; Contbr. articles to profl. jours. Mem. Am. Psychol. Assn. Subspecialties: Behavioral psychology; Psychiatry. Current work: Family behavioral therapy for schizophrenia; delivery of psychological services to police officers; assessment of cognitive, emotional, and behavioral sequelae of brain injuries. Home: 5326 Pleasant Ave S Minneapolis MN 55419 Office: Hennepin County Mental Health Center 619 S 5th St Minneapolis MN 55415

BOYD, JOHN PHILIP, geophysl fluid dynamicist; b. Winchester, Mass., Feb. 21, 1951; s. Hugh Robert and Marjorie Frances (Markham) B. A.B., Harvard U., 1973, S.M., 1975, Ph.D., 1976. Postdoctoral fellow Nat. Center Atomospheric Research, 1976; postdoctoral scholar U. Mich., 1977; asst. prof. atmospheric sci., 1977-82, assoc. prof., 1982—. Contbr. articles to profl. jours.; author sci. fiction stories. Mem. Am. Meteorol. Soc. Roman Catholic. Subspecialties: Meteorology; Oceanography. Current work: Fluid dynamics of the atmosphere and ocean; math. and numerical methods. Home: 3385 Burbank Ann Arbor MI 48105 Office: 2455 Hayward Room 2215 Ann Arbor MI 48109

BOYER, CHARLES BENJAMIN, mechanical engineer, research scientist; b. Urbana, Ohio, June 29, 1924; s. Rufus B. and Lena M. (Wyer) B.; m. Martha Muray, June 22, 1947; children: Janet Elizabeth Boyer Kriebel, Eric Martin. B.S.M.E., Tri-State U., Ind., 1950. Asst. to shop supt. Nat. Automatic Tool Co., Richmond, Ind., 1950-56; engr.-assoc. mgr. Battelle Columbus Labs., Ohio, 1956-74, sr. research scientist, process metallurgy sect., 1976—; supr. facilities design and tech., crucible research center Colt Industries, Pitts., 1974-76; expert hot isostatic processing. Contbr. articles to profl. jours. Served with USN, 1943-46. Recipient Engring. Material Achievement award Am. Soc. Metals, 1979; citation ASME, 1980; cert. of appreciation, 1981. Fellow ASME (chmn. high pressure subcom. 1976-79, assoc. editor Jour. Pressure Vessel Tech. 1976-82), Am. Powder Metallurgy Inst., Assn. Internat. Research and Advancement of High Pressure Sci. and Tech., West Jefferson (Ohio) Community Orgn. Presbyterian. Lodges: Mason; Shriners. Subspecialties: Mechanical engineering; Materials processing. Current work: Field of hot and cold isostatic processing. Patentee in field (9). Office: 505 King Ave Columbus OH 43201

BOYER, HERBERT WAYNE, biochemist; b. Pitts., July 10, 1936. B.A., St. Vincent Coll., Latrobe, Pa., 1958; Ph.D., U. Pitts., 1963; D.Sc. (hon.), St. Vincent Coll., 1981. Mem. faculty U. Calif., San Francisco, 1966—, prof. biochemistry, 1976—; investigator Howard Hughes Med. Inst., 1976—; co-founder, dir. Genentech, Inc., South San Francisco, Calif. Mem. editorial bd.: Biochemistry. Recipient V.D. Mattai award Roche Inst., 1977; Albert and Mary Lasker award for basic med. research, 1980; USPHS postdoctoral fellow, 1963-66. Mem. Am. Soc. Microbiology, Am. Acad. Arts and Scis. Subspecialty: Biochemistry (biology). Address: Dept Biochemistry HSE 1504 Univ Calif San Francisco CA 94143

BOYER, JOHN STRICKLAND, plant physiology educator; b. Cranford, N.J., May 1, 1937; m., 1964; 2 children. M.S., U. Wis., 1961; Ph.D., Duke U., 1964. Vis. asst. prof. botany Duke U., 1964-65; asst. physiology Conn. Agr. Expt. Sta., 1965-66; from asst. prof. to assoc. prof. U. Ill.-Urbana, 1966-73, prof. botany and agronomy, 1973—; plant physiology U.S. Dept. Agr., 1978—; mem. vis. com. Carnegie Instn. Washington, Stanford U. Mem. Am. Soc. Agronomy, Am. Soc. Plant Physiology (Shull award 1977, pres. 1981-82), Sigma Xi. Subspecialties: Plant physiology (biology); Photosynthesis. Office: Dept Botany U Ill Urbana IL 61801

BOYER, PAUL D., biochemist, educator; b. Provo, Utah, July 31, 1918; s. Dell Delos and Grace (Guymon) B.; m. Lyda Mae Whicker, Aug. 31, 1939; children: Gail Anne (Mrs. Denis Hayes), Marjorie Lynne (Mrs. Lukman Clark), Douglas. B.S., Brigham Young U., 1939; M.S., U. Wis., 1941, Ph.D., 1943; D.Sc. (hon.), U. Stockholm, 1974. Instr., research asso. Stanford, 1943-45; asst. prof. to assoc. prof. biochemistry U. Minn., 1946-55, Hill research prof., 1955-63; prof. chemistry U. Calif. at Los Angeles, 1963—, dir. Molecular Biology Inst., 1965—; chmn. biochemistry study sect. USPHS, 1962-67; mem. U.S. Nat. Com. for Biochemistry, 1965-71. Editor: Ann. Rev. of Biochemistry, 1965-70, Biochemical and Biophysical Research Communications, 1968-80, The Enzymes, 1970—; Mem. editorial bd.: Biochemistry, 1969-76, Jour. Biol. Chemistry, 1978-83; Contbr. articles to profl. jours. Recipient Am. Chem. Soc. award in enzyme chemistry, 1955, Tolman award, 1982; Guggenheim fellow, 1955. Fellow Am. Acad. Arts and Scis. (council); mem. Nat. Acad. Sci., Am. Soc. Biol. Chemists (past pres., council mem.), Am. Chem. Soc. (past div. chmn.), Biophys. Soc. Subspecialties: Biochemistry (biology). Current work: Enzyme mechanisms, biological energy transductions, oxidative phosphorylation, photophosphorylation, active transport. Home: 1033 Somera Rd Los Angeles CA 90024

BOYER, PAUL SLAYTON, geology educator, researcher; b. Summit, N.J., Feb. 11, 1942; s. Paul Kenneth and Martha (Slayton) B.; m. Marian Dehmel, May 10, 1967; children: Virginia, Charles. A.B. in Geology, Princeton U., 1964, Ph.D., Rice U., 1970. Asst. prof. Franklin and Marshall Coll., Lancaster, Pa., 1969-71; asst. prof. geology Fairleigh Dickinson U., Madison, N.J., 1971-75, assoc. prof., 1975-81, prof., 1981—, chmn. dept. geol. scis., 1979—. Mem. Harding Twp. Bd. Health, New Vernon, N.J., 1978—; chmn. Harding Twp. Republican Com., 1980—. Mem. Am. Assn. Petroleum Geologists, Soc. Econ. Paleontologists and Mineralogists, Internat. Paleontol. Assn., Internat. Sedimentological Assn., N.J. Acad. Scis. (editor Bull. 1980-83). Episcopalian. Subspecialties: Sedimentology; Paleobiology. Current work: Fossil annelids, sedimentology, regional geology of New Jersey and Pennsylvania. Office: Dept Geol Scis Fairleigh Dickinson U 285 Madison Ave Madison NJ 07940

BOYER, RAYMOND FOSTER, physicist; b. Feb. 1910. B.S. in Physics, Case Western Res. U., 1933, M.S., 1935, D.Sc. (hon.), 1955. With Dow Chem. Co., Midland, Mich., 1935—, asst. dir. phys. research lab., 1945-48, dir. phys. research lab., 1948-52, dir. plastics research, 1952-68, asst. dir. corporate research for polymer sci., 1968-72, research fellow, 1972-75; partner Boyer and Boyer, Midland, 1975—; research affiliate Midland Macromolecular Inst., 1975—; Vis. prof. Case Western Res. U., 1974, adj. prof., 1979, Central Mich. U., 1980; guest Russian Acad. Scis., 1972, 78, 80, Polish Acad. Scis., 1973; Past chmn. Gordon Conf. on Polymers. Contbr. numerous articles to profl. jours. Recipient Swinburne award Plastics Inst., London, 1972. Mem. Am. Chem. Soc. (past chmn. high polymer div., Borden award in chemistry of plastics and coatings 1970), Am. Phys. Soc. (past chmn. high polymer div.), Soc. Plastics Engrs. (Internat. award in polymer engring. and sci. 1968), Nat. Acad. Engring. Subspecialties: Polymer physics; Polymers. Current work: Multiple transitions and relaxations in polymers with emphasis on thermal and dynamic mechanical and statistical data analysis. Research on physics and phys. chemistry of high polymers with emphasis on transitions and relaxations in polymers. Patentee in field. Office: 415 W Main St Midland MI 48640

BOYER, ROBERT ERNST, geologist, educator; b. Palmerton, Pa., Aug. 3, 1929; s. Merritt Ernst and Lizzie Venetta (Reinard) B.; m. Elizabeth Estella Bakos, Sept. 1, 1951; children—Robert M., Janice E., Gary K. B.A., Colgate U., 1951; M.A., Ind. U., 1954; Ph.D., U. Mich., 1959. Instr. geology U. Tex., Austin, 1957-59, asst. prof., 1959-62, asso. prof., 1962-67, prof., 1967—, chmn. dept. geol. scis., 1971-80, dean, 1980—; exec. dir. Natural Scis. Found., 1980—; chmn. exec. com. Geology Found., 1971-80. Author: Activities and Demonstrations for Earth Science, 1970, Geology Fact Book, 1972, Oceanography Fact Book, 1974, The Story of Oceanography, 1975, Solo-Learn in the Earth Sciences, 1975, GEO-Logic, 1976, GEO-VUE, 1978; editor: Tex. Jour. of Sci, 1962-65, Jour. of Geol. Edn, 1965-68. Fellow Geol. Soc. Am., AAAS; mem. Tex. Acad. Sci. (hon. life, pres. 1968), Nat. Assn. Geology Tchrs. (pres. 1974-75), Am. Geol. Inst. (pres. 1983), Am. Assn. Petroleum Geologists, Austin Geol. Soc. (pres. 1975), Gulf Coast Asso. Geol. Soc. (pres. 1977). Subspecialty: Geology. Home: 7644 Parkview Circle Austin TX 78731

BOYER, VINCENT SAULL, utility executive; b. Phila., Apr. 5, 1918; s. Philip A. and Gertrude (Stone) B.; m. Ethel Wolf, June 6, 1942; children: Ruth Ann, Suzanne, Sandra Jean. B.S. in Mech. Engring., Swarthmore (Pa.) Coll., 1939; M.S., U. Pa., 1944; D.Engring. Tech. (hon.), Spring Garden Coll., 1979. With Phila. Electric Co., 1939—, mgr. nuclear power, 1963-65, gen. supt. sta. operating, 1965-67, mgr. electric ops., 1967-68, v.p. engring. and research, 1968-80, sr. v.p. nuclear power, 1980—; mem. adv. com. Electric Power Research Inst., 1978-80. Author papers in field. Served with USNR, 1944-46. Named Engr. of Year of Delaware Valley, 1979. Fellow ASME (chmn. Phila. 1970-71, James M. Landis medal 1981), Am. Nuclear Soc. (pres. 1976-77, honors and award com.); mem. Nat. Acad. Engring., Franklin Inst., Engrs. Club Phila. (George Washington medalist 1982), Edison Electric Inst., Assn. Edison Illuminating Cos., Atomic Indsl. Forum (chmn. com. on Three Mile Island 2 recovery, mem. policy com. on nuclear regulation), Am. Nat. Standards Inst. Clubs: Union League, Aronimink Golf (Phila). Subspecialty: Nuclear fission. Address: 2301 Market St Philadelphia PA 19101

BOYLAN, DAVID RAY, educator, chemical engineer; b. Belleville, Kans., July 22, 1922; s. David Ray and Mabel (Jones) B.; m. Juanita R. Sheridan, Mar. 24, 1944; children—Sharon Rae, Gerald Ray, Elizabeth Anne, Lisa Dianne. B.S. in Chem. Engring, U. Kans., 1943; Ph.D., Iowa State U., 1952. Instr. U. Kans., 1942-43; project engr. Gen. Chem. Co., Camden, N.J., 1943-47; sr. engr. Am. Cyanamid Co., Elizabeth, N.J., 1947; plant mgr. Arlin Chem. Co., Elizabeth, 1947-48; faculty Iowa State U., Ames, 1948—, prof. chem. engring., 1956—, asso. dir., 1959—, dir., 1966—, dean, 1970—. Fellow Am. Inst. Chem. Engrs., Am. Chem. Soc., AAAS; mem. Am. Soc. Engring. Edn., Sigma Xi, Phi Lambda Upsilon, Sigma Tau, Phi Kappa Phi, Tau Beta Pi. Subspecialty: Chemical engineering. Research in transient behavior and flow of fluids through porous media, unsteady state and fertilizer tech., devel. fused- phosphate fertilizer processes, theoretical and exptl. correlation of filtration. Patents and papers in field. Home: 1516 Stafford St Ames IA 50010

BOYLAN, ELIZABETH S., biology educator; b. Shanghai, China, Nov. 29, 1946; d. Nathan M. and Elizabeth (Little) Shippee; m. Robert J. Boylan, Oct. 2, 1971; 1 dau. Elizabeth B. A.B., Wellesley Coll., 1968; Ph.D., Cornell U., 1972. Research assoc. U. Rochester (N.Y.) Sch. Medicine, 1972-73; asst. prof. biology Queens Coll., CUNY, Flushing, N.Y., 1973-78, assoc. prof., 1978-83, prof., 1983—; vis. investigator Sloan-Kettering Inst., N.Y.C., 1979-80; mem. Breast Cancer Task Force, Nat. Cancer Inst., 1980-84; mem. adv. com. Am. Cancer Soc., 1981—. Mem. editorial bd.: Jour. Toxicology Environ. Health, 1981—. NIH grantee, 1975-83. Mem. Am. Assn. Cancer Research, Soc. Developmental Biology, Endocrine Soc., Am. Soc. Zoologists. Subspecialties: Developmental biology; Reproductive biology. Current work: Hormones and fetal development; transplacental carcinogenesis; mammary gland development. Office: Dept Biology Queens Coll CUNY Flushing NY 11367

BOYLAN, STEPHEN P., software engineer; b. Meadville, Pa., Dec. 11, 1954; s. Ralph P. and Shirley E. (Stearns) B. B.S., SUNY-Binghamton, 1978. Software engr. Data Gen. Corp., Westboro, Mass., 1978-80, Wang Labs., Lowell, Mass., 1980—; instr. Chamberlayne Jr. Coll., Boston, 1979-80; data gen. rep. Codasyl Cobol Com., Screen Mgmt. Task Group, Irvine, Calif., 1979-81. Mem. Assn. Computing Machinery. Subspecialties: Programming languages; Database systems. Current work: Distributed office systems supporting word processing, data processing, and business graphics. Office: Wang Labs 1 Industrial Ave Lowell MA 02146

BOYLE, BRIAN JOHN, medical information scientist, consultant artificial intelligence; b. Monmouth, Ill., Aug. 5, 1945; s. Brian Edward and Frances Virginia (Hickman) B. B.S., Harvey Mudd Coll., 1967; Ph.D., U. Calif.-San Francisco, 1976. Sr. systems researcher Berkeley Sci. Labs., Calif., 1969-72; mgr. Varian Instrument Data Systems, Walnut Creek and Palo Alto, Calif., 1972-75; pres., dir. Clin. Systems of Calif., Berkeley, 1975—; dir. Integral Systems Internat., San Francisco, Novon, Inc.; cons., vis. prof., lectr. U. Calif.-Berkeley, Stanford U. Alt. mem. Berkeley City Council, 1969-70; mem. Community Coll. Bd., 1971-72; bd. govs. Harvey Mudd Coll., 1974-77. Served with USNR, 1966-69. Decorated Purple Heart medal, Bronze Star medal. Mem. Am. Mgmt. Assn., Assn. Computing Machinery, IEEE, Am. Soc. Math., Assn. Advancement Med. Instrumentation, Software Underground Club. Democrat. Subspecialties: Biomedical engineering; Artificial intelligence. Current work: Artificial intelligence; cognitive science and human factors applied to medical research and genetic engineering. Patentee high speed continuously variable transmission. Office: 222 Downey St San Francisco CA 94117

BOYLE, MICHAEL DERMOT, immunochemist, educator; b. Belfast, No. Ireland, Jan. 4, 1949; came to U.S., 1974; s. Dermot P. and Joan M. (West) B.; m. Carla E. Boyle, Jan. 27, 1973; children: Kieron, Sarah. B.Sc., U. Glasgow, 1971; Ph.D., U. London, 1974. Expert Nat. Cancer Inst., Bethesda, Md., 1976-80, vis. scientist, 1980; assoc. prof. dept. immunology and med. microbiology dept. pediatrics U. Fla., Gainesville, 1981—. Contbr. articles in field to profl. jours. Mem. Am. Assn. Immunology, Am. Assn. Cancer Research, Fedn. Advancement of Edn. in the Scis. Subspecialties: Immunology (medicine); Microbiology (medicine). Current work: Immunochemistry of complement system; cytotoxic action on complement of tumor cells; bacterial Fc-reactive proteins. Home: 1809 SW 44th Ave Gainesville FL 32608 Office: U Fla PO Box J266 THMHC Gainesville FL 32610

BOYLE, WILLARD STERLING, physicist; b. Amherst, N.S., Can., Aug. 19, 1924; naturalized, 1969; s. Ernest Sterling and Bernice Teresa (Dewar) B.; m. Elizabeth Joyce, June 15, 1946; children—Robert, Cynthia, David, Pamela. B.Sc., McGill U., Montreal, Que., Can., 1947, M.Sc., 1948, Ph.D., 1950. Asst. prof. Royal Mil. Coll., Kingston, Ont., 1951-53; mem. staff Bell Labs., 1953-62, 64-79, exec. dir. semiconductor device devel. div., Allentown, Pa., 1968-75, exec. dir. communications scis. div., 1975-79; sr. partner Atlantic Research Assos., 1980—; dir. space sci. Bellcommunications, 1962-64. Author. Served with Canadian Navy, 1942-45. Recipient Ballantine medal Franklin Inst., Nat. Research Council Can. fellow, 1949. Fellow IEEE (Morris Liebman medal 1974), Am. Phys. Soc.; mem. Nat. Acad. Engring. Current work: Applications of microelectronics technology to industry. Patentee in field; co-inventor charge coupled device and 1st continuously pumped ruby laser. Address: Wallace NS B0K 1Y0 Canada

BOYNE, ALAN FREDERICK, chemist, educator; b. Liverpool, Eng., Jan. 22, 1943; s. Frederick and Isobel (Persil) B. B.Sc. with honors, U. Liverpool, 1965; Ph.D. in Biochemistry, U. Calif.-San Francisco, 1970. Asst. prof. Northwestern U. Med. Sch., Chgo., 1975-79; assoc. prof. U. Md. Med. Sch., Balt., 1979—. Contbr. articles to profl. publs. Grantee Nat. Inst. Neurol. Diseases and Stroke, 1977-80, 80-83. Mem. Soc. Neurosci. Subspecialties: Morphology; Neurobiology. Current work: Applying physical fixation by quick freezing to preservation of metabolites, ions and antigenicity of proteins, particularly in brain for purpose of understanding nerve terminal plasticity and memory. Inventor assembly for quick-freezing tissues for electron miscroscopy.; mfr. quick-freezing invention. Home: 112 E Gittings St Baltimore MD 21230 Office: Dept Pharmacology U Md Med Sch 660 W Redwood St Baltimore MD 21201

BOYNTON, JOHN ELLSWOOD, geneticist; b. Duluth, Minn., June 3, 1938; s. Wayne H. and Dorothy K. (Ellswood) B. B.S., U. Ariz., 1960; Ph.D., U. Calif., Davis, 1966. NIH postdoctoral fellow Inst. Genetics, U. Copenhagen, 1966-68; asst., then asso. prof. Duke U., 1968-76, prof. botany, 1977—. Recipient Campbell award Am. Inst. Biol. Scis., 1967, NIH research career devel. award, 1972-77. Mem. Am. Soc. Cell Biology, Genetics Soc. Am., Am. Soc. Naturalists. Current work: Genetics of chloroplasts and mitochondria. Home: 1808 Woodburn Rd Durham NC 27705 Office: Dept Botany Duke U Durham NC 27706

BOYNTON, ROBERT MERRILL, psychology educator; b. Evanston, Ill., Oct. 28, 1924; s. Merrill Holmes and Eleanor (Matthews) B.; m. Alice Neiley, Apr. 9, 1947; children: Sherry, Michael, Neiley, Geoffrey. Student, Antioch Coll., 1942-43, U. Ill., 1943-45; A.B., Amherst Coll., 1948; Ph.D., Brown U., 1952. Asst. prof. psychology and optics U. Rochester, N.Y., 1952-57, asso. prof., 1957-61, prof., 1961-74, dir. Center for Visual Sci., 1963-71, chmn. dept. psychology, 1971-74; prof. psychology U. Calif., San Diego, 1974—; guest researcher Nat. Phys. Lab., Teddington, Eng., 1960-61; vis. prof. physiology U. Calif. Med. Center, San Francisco, 1969-70. Author: Human Color Vision, 1979; Contbr. articles to profl. jours. Served with USNR, 1943-45. Fellow A.A.A.S., Optical Soc. Am. (dir. at large 1966-69), Am. Psychol. Assn., Nat. Acad. Scis. Subspecialties: Psychophysics; Sensory processes. Current work: Visual science, especially human color vision. Home: 376 Bellaire St Del Mar CA 92014

BOYSE, JOHN WESLEY, research engineer, automobile company executive; b. Saginaw, Mich., Dec. 1, 1940; s. John Wesley and Ellen Elizabeth (Dent) B.; m. Georgia Jeffrey, June 21, 1963; 1 dau., Kyla Lynn. A.A., Bay City Jr. Coll., 1960; B.S.E.E., U. Mich., 1963, 1963, M.S.E., 1964, Ph.D., 1971. Research engr. dept. computer sci. Gen. Motors Research Labs., Warren, Mich., 1971-77, research engr., group leader, 1977-82, asst. head dept. computer sci., 1982—. Contbr. articles to profl. jours. Recipient Charles L. McCuen award Gen. Motors, 1982. Mem. IEEE, Assn. Computing Machinery. Methodist. Subspecialties: Computer-aided design; Graphics, image processing, and pattern recognition. Current work: Computer aided design and manufacturing, geometric modeling, solid modeling. Office: Gen Motors Research Labs Warren MI 48090

BOZDECH, MAREK JIRI, physician, scientist; b. Wildflecken, West Germany, Oct. 12, 1946; came to U.S., 1951; s. Jiri Josef and Zofia (Jadwiga) B.; m. Frances Barclay Craig, Dec. 22, 1967; children: Elizabeth, Andrew, Matthew. A.B., U. Mich., 1967; M.D., Wayne State U., 1972. Diplomate: Am. Bd. Internal Medicine. Intern U. Wis.-Madison, 1972-73, resident in internal medicine, 1973-75; fellow in hematology/oncology U. Calif.-San Francisco, 1975-78; instr. medicine, 1977-78; asst. prof. medicine U. Wis.-Madison, 1978—; dir. hematology lab. U. Wis. Hosp., Madison, 1978-82. Am. Cancer Soc. fellow, 1980, 81, 82; recipient excellence in teaching award U. Wis. Hosp., 1981. Mem. ACP, Am. Soc. Hematology, AAAS. Subspecialties: Marrow transplant; Oncology. Current work: New techniques in eliminating toxicity and increasing efficacy of marrow transplantation in hematologic malignancies by use of immunologic radiotherapeutic techniques; ultrastructural cytochemistry of leukocytes. Home: 6233 Countryside Ln Madison WI 53705 Office: U Wis Clin Sci Ctr H4-538 600 Highland WI 53792

BOZYMSKI, EUGENE MICHAEL, medical educator; b. Mansfield, Ohio, Sept. 29, 1935; s. Alexander Casimir and Clementyna Helen (Zablocki) B.; m. Mary Kay Elizabeth Simon, June 21, 1958; children: Michael Joseph, Mark Edward, David Stuart, Eric Alexander. Grad., John Carroll U., 1956; M.D., Marquette U., 1960. Diplomate: Am. Bd. Internal Medicine. Intern Mercy Hosp., San Diego, 1960-61; resident in internal medicine Marquette U., Milw. County Hosp., Milw., 1961-62, 64-65, sr. resident, instr. internal medicine, 1965-66; NIH postdoctoral fellow in gastroenterology U. N.C. Sch. Medicine, Chapel Hill, 1966-68, instr. in medicine, 1968-69, asst. prof. medicine, 1969-73, assoc. prof. medicine, 1973-78, prof. medicine, 1978—; dir. Clin. Gastrointestinal Motility Lab., 1968-75; co-dir. Digestive Diseases and Nutrition Diagnostic and Treatment Ctr., U. N.C.-N.C. Meml. Hosp., 1975-78, dir., 1978—; mem. Med. Triad Com., 1973—; dir. Inpatient Care Utilization Rev. Com., 1976-80; mem. Durham Orange County Patient Care Com. and Public Relations Com., sec.-treas., 1983—, exec. mem.-at-large, 1979-82; mem. Task Force for Role of Attending Physician, other coms., 1981—; med. dir. N.C. State Hwy. Patrol Med. Program; participating mem. Nat. Coop. Crohn's Disease Study. Contbr. sects. to books, articles to profl. publs. Parish council St. Thomas More Parish, Roman Catholic Ch., 1974-76, Little League coach, 1978-80, 82—. Served to capt. U.S. Army, 1962-64. Cokeman-Beckley Found. scholar, 1954-56; Joseph Collins Found. scholar, 1957-60. Fellow ACP; mem. Fedn. Clin. Research, Am. Gastroent. Assn., So. Gut Club, N.C. Med. Soc. (del.), Gullet Club, Phi Chi. Subspecialties: Gastroenterology; Internal medicine. Current work: Esophageal physiology and diseases, gastrointestinal endoscopy and therapeutic applications, inflammatory bowel disease. Office: U NC Sch Medicine Chapel Hill NC 27514 Home: 407 Lyons Rd Chapel Hill NC 27514

BOZZOLA, JOHN JOSEPH, microbiologist, educator; b. Herrin, Ill., Oct. 22, 1946; s. John Joseph and Angeline (Flabbi) B. B.A., So. Ill U.-Carbondale, 1968, M.S., 1970, Ph.D., 1976. Instr. microbiology Med. Coll. Pa., Phila., 1976-78, asst. prof., 1978—. Recipient NIH young investigator award, 1979. Mem. Am. Soc. Microbiology, Electron Microscopy Soc. Am., Internat. Assn. Dental Research, Phi Kappa Phi, Kappa Delta Pi. Roman Catholic. Subspecialties: Microbiology; Cell biology. Current work: Virulence factors in microbes; subcellular vaccines; diagnostic electron microscopy; clinical microbiology. Patentee rapid viral diagnosis. Office: Med Coll Pa Dept Microbiology Philadelphia PA 19129

BRAATZ, JAMES ANTHONY, research biochemist; b. Balt., July 17, 1943; s. James Anthony and Janet Barbara (Klosek) B.; m. Geraldine Lee Waldecker, May 9, 1964; children: James Anthony III, Ronald Chester, Mary Janet. B.S., Johns Hopkins U., 1968, Ph.D., 1973. Sr. research technician W. R. Grace & Co., Clarksville, Md., 1963-68; postdoctoral fellow Nat. Inst. Arthritis, Metabolism and Digestive Diseases, NIH, Bethseda, Md., 1973-75, sr. investigator, 1975-81, head biochemistry sect., Frederick, Md., 1981—; tech. cons. Abbott Labs., Chgo., 1981—, Hoffmann-LaRoche, Nutley, NJ, 1983—. Editorial adv. bd.: Jour. Biol. Response Modifiers, 1982—. Mem. Am. Chem. Soc., AAAS, Internat. Soc. Oncodevel. Biology and Medicine, Am. Soc. Biol. Chemists. Democrat. Roman Catholic. Subspecialty: Biochemistry (biology). Current work: Biochemical studies on human lung tumor-associated proteins, evaluation of these proteins as serum markers for lung cancer. Patentee in field. Home: 4510 Yates Rd Beltsville MD 20705 Office: NIH Nat Cancer Inst Frederick Cancer Research Facility Bldg 560 Room 31-93 Frederick MD 21701

BRACIALE, THOMAS JOSEPH, pathologist, medical scientist, educator; b. Phila., Oct. 22, 1946; s. Thomas and Rose E. (Pascale) B.; m. Vivian Lam, Aug. 5, 1972; children: Kara, Michael. B.S., St. Joseph's Coll., Phila., 1968; Ph.D., U. Pa., 1974, M.D., 1975. Intern, resident in pathology Washington U., St. Louis, 1975-76, asst. prof. pathology, 1978-82, assoc. prof., 1982—; postdoctoral fellow Australian Nat. U., Canberra, Australia, 1976-78; asst. pathologist Barnes Hosp., St. Louis, 1978-82, assoc. pathologist, 1982—, staff pathologist, 1978—. Author: (with Ada and Yap) Topics in Immunology, 1978, (with Braciale and Andrew) Isolation & Characterization of T Cell Clones, 1982. NIH grantee, 1978—. Mem. Am. Assn. Immunologists, Am. Assn. Pathologists, Australian Immunol. Soc. Subspecialties: Immunobiology and immunology; Infectious diseases. Current work: Analysis of the function and properties of cloned populations of thymus-derived lymphocytes from man and experimental animals. Home: 7018 Maryland St Saint Louis MO 63130 Office: Washington Sch of Medicine Dept Pathology 660 S Euclid Ave Saint Louis MO 63110

BRACIALE, VIVIAN LAM, immunologist, educator; b. N.Y.C., June 6, 1948; d. Wing Ching and Wai Ching (Li) Lam; m. Thomas Joseph Braciale, Jr., Aug. 5, 1972; children: Kara, Michael Stephen, Laura. A.B., Cornell U., 1969; Ph.D., U. Pa., 1973. Postdoctoral fellow U. Pa., Phila., 1974-75; postdoctoral fellow Washington U., St. Louis, 1975-76, research instr. pathology, 1978—; vis. fellow Australian Nat. U., Canberra, 1976-78. Contbr. articles on immunology to profl. jours. N.Y. State regent scholar, 1965-69; recipient Nat. Research Service award NIH, 1976-78. Mem. Am. Assn. Immunologists. Lutheran. Subspecialty: Immunobiology and immunology. Current work: Study of the role of T lymphocytes in the immune response and events in their induction and differentiation, using cloned cell populations. Office: Washington U Sch Medicine 660 S Euclid Ave St Louis MO 63110

BRACKETT, BENJAMIN GAYLORD, veterinary medicine educator; b. Athens, Ga., Nov. 18, 1938; s. Ernest Marshall and Julia Claire (Cook) B.; m. Ann Thornton Crawford, Aug. 22, 1959; children: Laura Ellen, Jeffrey Crawford, David Gregory. B.S.A., U. Ga., 1964, M.S., 1964, D.V.M., 1962, Ph.D., 1966; M.A. (hon.), U. Pa., 1971. Lic. veterinarian Ga., diplomate: Am. Coll. Theriogenologists. Postdoctoral fellow in biochemistry U. Ga., Athens, 1962-66; dir. primate colony, assoc. to prof. Pa. Sch. Medicine, Phila., 1966-74; prof. animal reprodn. U. Pa., Kennett Square, Pa., 1974—; cons. in field. Editor: New Technologies in Animal Breeding, 1981; contbr. articles in field to profl. jours. USPHS fellow, 1964-66; recipient NIH career devel. award, 1970-75. Mem. AVMA (pres. student chpt. 1962, fellow 1962-64), Soc. Study Reprodn. (dir. 1979-81), Internat. Embryo Transfer Soc. (program chmn. 1982—, bd. govs. 1983—), Am. Fertility Soc. Subspecialties: Embryo transplants; Reproductive biology (medicine). Current work: Research on fertilization in vitro, especially in rabbits and cows; teaching animal reproduction, especially students of veterinary medicine and postdoctoral fellows in reproductive biology. Home: 432 Bartram Rd Kennett Square PA 19348 Office: U Pa Sch Veterinary Medicine New Bolton Ctr 382 W Street Rd Kennett Square PA 19348

BRADBURY, NORRIS EDWIN, physicist; b. Santa Barbara, Calif., May 30, 1909; s. Edwin Perly and Elvira C. (Norris) B.; m. Lois Platt, Aug. 5, 1933; children—James Norris, John Platt, David Edwin. B.A., Pomona Coll., 1929, D.Sc., 1951; Ph.D., U. Calif., 1932; LL.D., U. N.Mex., 1953; D.Sc., Case Inst. Tech., 1956. NRC fellow in physics M.I.T., 1932-34; asst. prof. physics Stanford U., 1934-37, asso. prof., 1937-42, prof., 1942-50; prof. physics U. Calif., 1951-70; dir. Los Alamos Sci. Lab., 1945-70. Contbr. tech. articles to phys. revs., jours. Served with USNR, 1941-45; capt. Res. Decorated Legion of Merit. Fellow Am. Phys. Soc.; mem. Nat. Acad. Sci. Episcopalian. Subspecialty: Nuclear physics. Home: 1451 47th St Los Alamos NM 87544

BRADFORD, MARK LEE, environ. co. exec.; b. Burbank, Calif., Jan. 10, 1954; s. Harold Earl and Phyllis (Hewson) B.; m. Julie LaPerle, Aug. 29, 1981. B.S., U. Calif.-Davis, 1977. Tech. writer Franklin Inst. GmbH, Munich, W. Ger., 1978-79; project mgr. Ecology & Environment, Inc., San Francisco, 1979—, Calif. Oil Spill Contingency Plan, 1982. Author: Chemical Pollutants. . ...Mediterranean Sea, 1978. Mem. task force tng. subcom. Assn. of Bay Area Govts., Berkeley, Calif., 1981-82; Mem. Nat. Wildlife Fedn., Spill Control Assn. Am. Subspecialties: Evnrionmental pollution contingency planning; Environmental toxicology. Current work: Fate of pollutants in terrestial and aquatic systems, groundwater pollution, spill cleanup management, contingency planning for oil and hazardous materials spills, hazardous waste site investigations. Home: 1815 Vine St Berkeley CA 94703 Office: Ecology and Environment Inc 120 Howard Suite 640 San Francisco CA 94105

BRADLEY, A. FREEMAN, JR., research and development laboratory director, cardiovascular-anesthesia researcher; b. Tuskegee Institute, Ala., Jan. 14, 1932; s. Arthur Freeman and Marion (Davis) B.; m. Dorothy Shamwell, Oct. 31, 1953; children: Lynn, Karen, Freeman. B.A., Lincoln U., 1953; student, Howard U., 1953-54. Biologist, Nat. Heart Inst., Bethesda, Md., 1954-58; staff research assoc. U. Calif.-San Francisco, 1958-68; specialist Cardiovascular Research Inst., 1968-77, dir. research and devel. lab., 1977—; pres., owner Med. Research Specialties, San Francisco, 1961-69. Co-designer: Po_2 and Pco_2 Electrodes, 1958. Chmn., mem. steering com. U. Calif.-San Francisco Black Caucus, 1968—; bd. dirs. Bay Area Men United, 1980—. Mem. Am. Physiol. Soc., AAAS, N.Y. Acad. Sci., Assn. Advancement Med. Instrumentation, Instrument Soc. Am. (sr.). Democrat. Episcopalian. Subspecialties: Bioinstrumentation; Biomedical engineering. Current work: Cardiovascular and anesthesia research with emphasis on respiratory physiology and neuroanatomy, design of biomedical apparatus. Home: 59 Topaz Way San Francisco CA 94131 Office: U Calif San Francisco Research and Devel Lab 4th and Parnassus Ave San Francisco CA 94143

BRADLEY, JAMES HENRY, computer scientist; b. London, Eng., Mar. 26, 1933; came to U.S., 1963, naturalized, 1976; s. John Henry and Marjorie Florence (Jones) B.; m. Estella Bennett, Dec. 27, 1966. B.A., U. Oxford, Eng., 1954; M.A., U. Toronto, 1960; Ph.D., U. Mich., 1967. Asst. prof. Pa. State U., University Park, 1968; research assoc. McGill U., Montreal, Que., 1968-70; asst. prof. Drexel U., Phila., 1970-76; sr. mem. tech staff Electronic Assocs., West Long Branch, N.J., 1976-82; pres. Atlantic Cons., Atlantic Highlands, N.J., 1982—; cons. Govt. of Can., 1968-70, U.S. Govt., 1965-82. Author: Scientific Software Methods, 1983; Contbr. tech. papers to profl. jours. Mem. Instrument Soc. Am. (sr.), Soc. Computer Simulation, Internat. Assn. for Math. and Computers in Simulation, Soc. for Indsl. and Applied Math. Subspecialties: Numerical analysis; Software engineering. Current work: Realism of simulation. Integro, partial and ordinary differential equations. Scientific methodology for general software. Address: 106 Portland Rd Atlantic Highlands NJ 07716

BRADLEY, JOHN MICHAEL, reading educator; b. Los Angeles, Mar. 5, 1940; s. Earl Chase and Angela Grace (McNamee) B.; m. Nancy Joyce Donaldson, Aug. 11, 1962; children: Christopher Michael, Gia Laverne. A.A., Am. River Coll.-Calif., 1960; B.A., San Jose State Coll., 1962; M.A., Calif. State U.-Sacramento, 1967; Ed.D., U. Pa., 1973. Lectr. edn. U. Pa., Phila., 1968-69; asst. prof. tchr edn. U. Sacramento, Calif., 1969-73; asst. prof. dept. reading U. Ariz., Tucson, 1973-79, assoc. prof., 1979—. Pres. Foothills Homeowners Assn., Tucson, 1982. Mem. Am. Psychol. Assn., Internat. Reading Assn., Nat. Assn. Sch. Psychologists, Phi Delta Kappa, Kapp Delta Pi, Psi Chi. Democrat. Roman Catholic. Subspecialty: Reading-related cognition and perception. Current work: Coordinating a reading clinic and researching text readability, comprehension and interest. Home: 3219 E Table Mountain Rd Tucson AZ 85718 Office: Dept Reading U Ariz Tuscon AZ 85721

BRADLEY, JOHN SAMUEL, geologist; b. Oklahoma City, Feb. 23, 1923; s. John Samuel and Charleen (Holloway) B.; m. Janet Rich, Nov. 24, 1951; children: Anne, Elizabeth, James. B.Geol.Engring., Colo. Sch. Mines, 1948; Ph.D., U. Wash., 1952. Geologist Humble Research, Houston, 1950-54, U. Tex., Port Arkansas, 1954-56, Atlantic Research, Dallas, 1956-64, Amoco Prodn. Co., Tulsa, 1964—. Co-author: Permian Reef Complex, 1953; contbr. articles to profl. jours. Served to 2d lt. C.E. U.S. Army, 1943-46. Fellow Geol. Soc. Am.; mem. Am. Assn. Petroleum Geologists, Am. Geophys. Union, Soc. Econ. Paleontologists and Mineralogists. Democrat. Unitarian. Subspecialties: Geology; Remote sensing (geoscience). Current work: Structural geology and rock mechanics. Patentee in field. Home: 3355 S Braden St Tulsa OK 74135 Office: Amoco Prodn Co PO Box 591 Tulsa OK 74102

BRADLEY, KATHARINE TRYON, arboretum director; b. Washington, Oct. 13, 1920; d. Henry Harrington and Margaret (Ramsay) Tryon; m. Joseph Crane Bradley, June 29, 1956; children: James Watrous, Margaret Tryon Timmerman. A.B. (75th Anniversary scholar, Nancy Skinner Clark fellow), Vassar Coll., 1943; M.S., U. Minn., 1944; Ph.D. (Phoenix fellow), U. Mich., 1953. Instr. biology Bowling Green (Ohio) State U., 1946-48; instr. botany Wellesley (Mass.) Coll., 1951-54; instr. U. Wis.-Madison, 1956-57, dir. univ. arboretum, 1974—. Author profl. papers. Vice-pres. 1st Unitarian Soc., 1973-74, pres., 1977-79; pres. Friends of Madison Pub. Library, 1968-71. Am. Cancer Soc. research fellow, 1954-56. Mem. Am. Assn. Bot. Gardens and Arboreta (dir. 1979-82, chmn. awards com. 1979-82), Am. Inst. Biol. Scis., Natural Sci. for Youth Found., Am. Hort. Soc., Nature Conservancy, Wilderness Soc. Subspecialties: Resource management; Resource conservation. Current work: Administration; natural areas restoration and management. Home: 2805 Sylvan Ave Madison WI 53705 Office: 1207 Seminole Hwy Madison WI 53711

BRADLEY, MICHAEL DOUGLAS, hydrology and water resources educator; b. Wichita, Kans., May 20, 1938; s. Lincoln P. and Nora Belle (Stultz) B.; m. Dorotha M. Bradley, Dec. 23, 1961. B.A., U. N.Mex., 1967; M.P.A., U. Mich., 1968, Ph.D., 1971. Postdoctoral fellow Scripps Instn. Oceanography, La Jolla, Calif., 1971-72; asst. prof. hydrology and water resources U. Ariz., Tucson, 1972-76, assoc. prof., 1976—; vis. prof. Grad. Sch. Architecture and Urban Planning, UCLA, 1980-81; cons. U.S. Army C.E., 1981-82, Hydrogeochem, Inc., 1982-83. Author: The Scientist and Engineer in Court, 1983. Bd. dirs. Tucson Art-Sci. Ctr., 1980-83. Served with USMC, 1956-60. Mem. AAAS, Geoscientists for Internat. Devel., Am. Water Resources Assn. Democrat. Subspecialties: Water resources policy; Resource management. Current work: Water resources policy; water law and rights; Indian water rights; use of scientific information in public decision-making, law and science. Home: 3301 N Christmas Ave Tucson AZ 85716 Office: Dept Hydrology and Water Resources U Ariz Tucson AZ 85721

BRADLEY, WALTER G., neurologist; b. Southampton, Eng., Sept. 1, 1937. B.A., U. Oxford, Eng., 1959, B.Sc., 1961, M.A., B.M., B.Ch., 1963, D.M., 1970. Diplomate: Am. Bd. Psychiatry and Neurology. Registrar in neurology Regional Neurol. Center, Newcastle Upon Tyne, Eng., 1966, M.R.C. research fellow in muscular dystrophy, 1967, Welcome sr. research fellow in clin. sci., hon. lectr. in neurology, 1969-71, sr. lectr., 1971-74; Radcliffe traveling fellow in neuropathology Mass. Gen. Hosp.; research fellow in neuropathology Harvard Med. Sch., Boston, 1968-69; sr. research assoc. Muscular Dystrophy Group Research Labs., Newcastle Gen. Hosp., 1971-74; cons. neurologist Newcastle Univ. Hosp. Group, 1971-74, prof. exptl. neurology, 1974-77; hon. cons. Newcastle Area Health Authority, 1974-77; assoc. neurologist in chief, dir. neuromuscular research labs. New Eng. Med. Center Hosp., 1974—; prof. neurology Tufts U. Sch. Medicine, Boston, 1977-82; co-dir. Muscular Dystrophy Assn. Clinic, Tufts-New Eng. Med. Center, 1980-82, dir. ambulatory services dept. neurology, 1980-82; prof. neurology, chmn. dept. neurology U. Vt. Coll. Medicine, Burlington, 1982—; dir. Peripheral Nerve Pathology Lab., Muscular Dystrophy Clinic of Vt., 1982—. Author, editor 2 books.; Contbr. articles to profl. jours. Muscular Dystrophy Assn. grantee; NIH grantee. Fellow Royal Coll. Physicians (London); mem. Am. Acad. Neurology, Am. Assn. Neuropathologists, Am. Neurol. Assn., Assn. Brit. Neurologists, Assn. Physicians Gt. Britain and Ireland, Boston Soc. Psychiatry and Neurology, N.H.-Vt. Neurol. Soc., Brit. Med. Assn., Brit. Neuropath. Soc., Mass. Assn. Neurologists, Mass. Med. Soc., Vt. Med. Soc., Soc. Neurosci., Societe Francaise Neurologie (corr.), Dutch Soc. Neurology (corr.). Subspecialties: Neurobiology; Genome organization. Current work: Neuromuscular apparatus in health and disease; muscular dystrophy; human neuronal degenerating diseases including ALS and Alzheimer's disease; clinical, therapeutics, morphological and biochemical studies. Home: 19 Brewer Pkwy Burlington VT 05401 Office: Dept Neurology U Health Center U Vt Coll Medicine S Prospect St Burlington VT 05405

BRADNER, WILLIAM TURNBULL, research scientist; b. Short Hills, N.J., Aug. 16, 1924; s. Palmer and Emily (Turnbull) B.; m. Ruth Marie Snyder, June 3, 1951; children: Terry Ellen, Tymm, Philip Russell. B.A., Lehigh U., 1948, M.S., 1949, Ph.D., 1952. Research assoc. Brown U., Providence, 1953-55; research assoc. Cornell U. Med. Sch., N.Y.C., 1956-58, asst. prof., 1958; asst. mem. Sloan Kettering Inst., N.Y.C., 1955-58; sr. research scientist Bristol Labs., Inc., Syracuse, N.Y., 1958-65, asst. dir. pharmacology, 1965-73, asst. dir. microbiology, 1973-77, dir. antitumor biology dept., 1977—; Bd. dirs. Am. Cancer Soc. (Onondaga County unit), Syracuse, 1977-81, v.p., 1980. Contbr. over 100 sci. articles to profl. publs. Served with U.S. Army, 1943-45. Decorated Purple Heart. Mem. Am. Assn. Cancer Research, AAAS, Am. Soc. Microbiology, N.Y. Acad. Scis. Republican. Episcopalian. Subspecialties: Cancer research (medicine); Microbiology (medicine). Current work: Discovery and evaluation of new cancer chemotherapeutic agents from fermentation, natural, and synthetic sources. Patentee is field. Home: 4903 Briarwood Circle Manlius NY 13104 Office: Box 4755 Syracuse NY 13221-4755

BRADSHAW, HOWARD HOLT, management consultant; b. Phila., Feb. 28, 1937; s. Howard Holt and Emojeane (Campbell) B.; m. Loretta Warren Sites, Aug. 13, 1982; children by previous marriage:

Elaine Allen, Howard Holt. B.A., Yale U., 1958; postgrad., Duke U., 1958-60. Cert. mgmt. cons. With Western Electric Co., various locations, 1960-68; head behavioral scis. cons. Celanese Fibers Co., Charlotte, N.C., 1968-71; pres. Orgn. Cons., Inc., Charlotte, 1971—; cons. in field. Author: Personal Power, Self Esteem and Performance, 1982, The Management of Self Esteem, 1981; Contbr. articles to profl. jours. Regional chmn. Constl. Party of Pa., Harrisburg, 1964-66; pres. Coordinated Planning League, Inc., Charlotte, 1972-74. Recipient cert. of appreciation Charlotte Police Dept., 1969, Mecklenburg County Com., 1970. Mem. Inst. Mgmt. Cons., Am. Psychol. Assn., Soc. Indsl. and Orgn. Psychology, Am. Soc. Tng. and Devel., Orgnl Devel. Network. Republican. Presbyterian. Subspecialty: Behavioral psychology. Current work: Research and application of general system theory to improve individual and group performance within hi-tech and other organizations. Design of psychosocial systems to facilitate the effective introduction of robotics, CAD/CAM and other technologies. Office: 1913 Charlotte Dr Charlotte NC 28203 Home: 3619 Maple Glen Ln Charlotte NC 28211

BRADSHAW, JERALD SHERWIN, chemistry educator, researcher; b. Cedar City, Utah, Nov. 28, 1932; s. Sherwin Hinton and Maree (Wood) B.; m. Karen Lee, Aug. 6, 1954; children: Donna Maree, Melinda Caroline. B.S., U. Utah-Salt Lake City, 1955; Ph.D., UCLA, 1963. Postdoctoral fellow Calif. Inst. Tech., Pasadena, 1962-63; research chemist Chevron Research Co., Richmond, Calif., 1963-66; asst. prof. Brigham Young U., Provo, Utah, 1966-69, assoc. prof., 1969-73, prof. chemistry, 1973—. Bd. dirs. Internat. Environment Inc., Salt Lake City, 1978—. Served to lt. USN, 1955-59. Named Prof. of Yr. Brigham Young U., 1975; recipient Maeser research award, 1977, Maeser teaching award, 1982. Mem. Internat. Soc. Heterocyclic Chemistry (bd. dirs. 1982-84), Am. Chem. Soc. (chmn. local sect. 1971), Utah Acad. Sci. Republican. Mem. Ch. of Jesus Christ of Latter-day Saints. Lodge: Kiwanis. Subspecialty: Organic chemistry. Current work: Synthesis of macrocyclic multidentate compounds and polysiloxanes. Patentee in field. Office: Brigham Young U Provo UT 84602 Home: 1616 Oak Ln Provo UT 84604

BRADSHAW, JOHN DAVID, research scientist; b. Miami, Fla., Sept. 4, 1952; s. Lee and Helen (Lantz) B.; m. Deborah K. Fouts, 1971. A.S., Palm Beach Jr. Coll, 1973; B.S., U. Fla., 1975, Ph.D., 1980. Research scientist II Ga. Inst. Tech., Atlanta, 1980—. Subspecialties: Atmospheric chemistry; Analytical chemistry. Current work: Development of analytical instrumentation and methodology; laser spectroscopic methods development; global atmospheric monitoring of trace gases. Contbr. numerous articles to sci. jours. Address: Sch Geophys Scis Ga Inst Tech Atlanta GA 30332

BRADT, HALE VAN DORN, physicist, x-ray astronomer, educator; b. Colfax, Wash., Dec. 7, 1930; s. Wilber Elmore and Norma (Sparlin) B.; m. Dorothy Ann Haughey, July 19, 1958; children—Elizabeth, Dorothy Ann. A.B., Princeton U., 1952; Ph.D. in Physics, M.I.T., 1961. Mem. dept. physics M.I.T., 1961—, prof., 1972—; sci. investigator Small Astronomy Satellite, NASA, 1975-79, High Energy Astronomy Obs., 1977-79. Co-editor: X and Gamma Ray Astronomy, 1973; asso. editor: Astrophys. Jour. Letters, 1974-77. Served with USNR, 1952-54. Recipient Exceptional Sci. achievement medal NASA, 1978. Mem. Am. Astron. Soc. (sec.-treas. high energy astrophysics div. 1973-75, chmn. 1981), Am. Phys. Soc., Sigma Xi. Subspecialties: 1-ray high energy astrophysics; Optical astronomy. Current work: Galactic x-ray source; optical counterparts of x-ray sources. Home: Belmont MAOffice: 37-581 MIT Cambridge MA 02139

BRADT, RICHARD CARL, materials science educator, ceramic engineer; b. St. Louis, Nov. 17, 1938; m., 1960. B.Sc., MIT, 1960; M.S., Rensselaer Poly. Inst., 1965, Ph.D. in Materials Engring, 1967. Research metallurgist Fansteel Metall. Corp., 1960-62; ceramist V-R/Wesson div., 1962-63; research fellow materials engring. Rensselaer Poly. Inst., 1963-67; asst. prof. Pa. State U., University Park, 1967-71, assoc. prof., 1971-75, prof. ceramic sci., 1975-78, head materials sci. and engring., 1978-83; chmn. materials sci. and engring. U. Wash., Seattle, 1983—. Recipient R.M. Fulrath award. Fellow Am. Ceramic Soc. (trustee); mem. Am. Soc. Metals, AIME, Brit. Ceramic Soc. Subspecialties: Ceramics; Composite materials. Current work: Thermal and mechanical properties of ceramics and glass. Office: Dept Materials Sci and Engring Wash U Roberts Hall FB-10 Seattle WA 98195

BRADY, J(OHN) MICHAEL, artificial intelligence research scientist, consultant; b. Liverpool, Eng., Apr. 30, 1945; came to U.S., 1980; s. John and Priscilla M. B.; m. Naomi Friedlander, Oct. 2, 1967; children: Sharon, Carol. B.Sc. in Math. with first class honors, U. Manchester, Eng., 1967, M.Sc. in Pure Math, 1968; Ph.D. in Math, Australian Nat. U., Canberra, 1970. Lectr. in computer sci. U. Essex, Eng., 1970-78, sr. lectr., 1978-80; sr. research scientist Artificial Intelligence Lab., M.I.T., 1980—; mem. com. army robotics NRC; invited speaker profl. cons. Author: The theory of Computer Science: a programming approach, 1977; author, editor: (with others) Robot motion: Planning and control, 1983; co-editor: Artificial Intelligence series, 1978—; editor: Computational Models of Discourse, 1983; editorial bd.: Mathematical Sciences series, 1978; contbg. editor: Computer Vision, 1982; editorial bd.: Artificial Intelligence Jour, 1980—; founding editor: Robotics Research, 1982—. Mem. IEEE. Subspecialties: Artificial intelligence; Graphics, image processing, and pattern recognition. Current work: Computer vision research on representation of visible surfaces; development of computational theory of early processing in reading; robotics research. Office: MIT 321 Artificial Intelligence Lab 545 Tech Sq Cambridge MA 02139

BRADY, LUTHER W., JR., physician; b. Rocky Mount, N.C., Oct. 20, 1925; s. Luther W. and Gladys B. A.A., George Washington U., 1944, A.B., 1946, M.D., 1948. Diplomate: Am. Bd. Radiology (treas. 1974—). Intern Jefferson Med. Coll. Hosp., Phila., 1948-50, resident in radiology, 1954-55; resident radiology Hosp. U. Pa., Phila., 1955-56; fellow Nat. Cancer Inst., 1953-57, 1957-59; practice medicine, specializing in radiation therapy and oncology, Phila.; asst. instr. radiology Jefferson Med. Coll. Hosp., 1954-55, U. Pa., Phila., 1955, instr., 1956-57, asso. radiology, 1957-59; asst. prof. radiology Coll. of Physicians and Surgeons, Columbia U., N.Y.C., summer, 1959; asso. prof. radiology Hahnemann Med. Coll. and Hosp., Phila., 1959-62, prof., 1963—, chmn. dept. radiation therapy, 1970—; asst. prof. radiology Harvard Med. Sch., Boston, 1962-63; mem. med. radiation adv. com. Bur. Radiation Health, HEW, 1971-74; cons. radiation therapy various hosps.; mem. U.S. del. to Interam. Congress Radiology, 1975, Internat. Congress on Radiology, 1981; med. adv. radiation therapy, dir. Pa. Blue Shield, Camp Hill. Author: Tumors of the Nervous System, 1975, Cancer of the Lung, Clinical Applications of the Electron Beam; editor: others Cancer Clin. Trials (Am. Jour. Clin. Oncology); editorial bd.: Cancer; asso. editor: Am. Jour. Roentgenology; contbr. articles on radiation therapy to profl. jours. Bd. dirs. Assn. Artists Equity of Phila., Welcome House, 1974—, Settlement Music Sch., 1973—, Phila. Art Alliance, 1977—; trustee Phila. Mus. Art, also mem. oriental art com., 1974—, chmn. exec. com., 1968-72, mem. print com., 1974—. Served to lt. M.C. USN, 1950-54. Recipient Grubbe award Chgo. Radiol. Soc., 1977. Fellow Am. Coll. Radiology (chmn. commn. radiation therapy 1975—, bd. chancellors 1975—); mem. Radiol. Soc. N.Am. (bd. dirs. 1977—, chmn. refresher course com. com. 1971-75, lectr. 1979), Pa. Radiol. Soc. (dir. 1970-77, councilor to Am. Coll. Radiology 1971-77), Am. Radium Soc. (pres. 1976-77, dir., Janeway lectr. 1980), Am. Cancer Soc. (pres. Phila. div. 1976—, dir. 1968—, exec. com. 1976—, mem. breast cancer task force 1974—), Am. Soc. Therapeutic Radiologists (pres. 1971-72), Assn. U. Radiologists, Am. Roentgen Ray Soc., Am. Assn. for Cancer Research, Radiation Research Soc., Am. Fedn. Clin. Oncologic Soc. (exec. com.), Am. Soc. Clin. Oncology, Phila. Roentgen Ray Soc. (pres. 1976-77, mem. exec. com. 1976-78), Am. Fedn. Clin. Research, Coll. Physicians Phila., James Ewing Soc., Assn. Pendergrass Fellows, Philadelphia County Med. Soc., AMA, Med. Soc. State Pa., Internat. Skeletal Soc., Council Acad. Socs., Soc. Chairmen Acad. Radiation Oncology Programs (pres. 1977), Soc. Chairmen Acad. Radiology Depts. (pres. 1974-75), Gynecologic Oncology Group (exec. com. 1971—), Radiation Therapy Oncology Group (chmn. 1980—), Internat. Club Radiotherapists, Nat. Cancer Inst. (bd. sci. counselors, com. for radiation therapy studies 1971—), Smith-Reed-Russell Soc., Alpha Omega Alpha, Phi Lambda Kappa. Clubs: Merion Cricket, Racquet of Phila. Subspecialties: Cancer research (medicine); Radiology. Current work: Monoclonal antibodies, tumor research (radiation synthesizer, combined modality treatment.). Office: 230 N Broad St Philadelphia PA 19102

BRADY, LYNN ROBERT, pharmacognosist, educator; b. Shelton, Nebr., Nov. 15, 1933; s. Connie E. and Laura M. (Vohland) B.; m. Geraldine Ann Walcott, June 23, 1957. B.S., U. Nebr., 1955, M.S., 1957; Ph.D., U. Wash., 1959. Asst. prof. pharmacognosy U. Wash., 1959-63, asso. prof., 1963-67, prof., 1967—, chmn. dept., 1972-80, asst. dean, 1982—; Chmn. conf. Tchrs. Am. Assn. Colls. Pharmacy, 1966-67. Author: (with others) Pharmacognosy, 6th edit., 1970, 7th edit., 1976, 8th edit., 1981. Fellow Acad. Pharm. Scis. (chmn. sect. pharmacognosy and natural products 1969-70); mem. Am. Pharm. Assn., Am. Soc. Pharmacognosy (pres. 1970-71), Sigma Xi, Rho Chi, Kappa Psi. Subspecialty: Pharmacognosy. Current work: Identification of fungal constituents, plant poisonings and chemotaxonomy. Research in fungal constituents, alkaloid biosynthesis, chemataxonomy. Home: 5815 NE 57th St Seattle WA 98105

BRADY, ROSCOE O., physician; b. Phila., Oct. 11, 1923; s. Roscoe O. and Martha (Roberts) B.; m. Bennett Carden Manning, 1972; 2 sons. Student, Pa. State U., 1941-43; M.D., Harvard, 1947; postgrad., U. Pa., 1948-49. NRC fellow U. Pa., 1948-50, USPHS spl. fellow, 1950-52; sect. chief Nat. Inst. Neurol. Diseases and Blindness, NIH, 1954-67, acting lab. chief neurochemistry, Bethesda, Md., 1967; chief developmental and metabolic neurology br. Nat. Inst. Neurol. and Communicative Disorders and Stroke, 1972—; professorial lectr. George Washington Sch. Medicine, 1963—; faculty Georgetown U. Sch. Medicine, 1967—. Author: (with Donald B. Tower) Neurochemistry of Nucleotides and Amino Acids, 1960, Basic Neurosciences, 1975, also numerous articles. Mem. Am. Soc. Biol. Chemists, Am. Chem. Soc., Am. Acad. Neurology, Am. Acad. Mental Retardation, Soc. Exptl. Biology and Medicine, Am. Soc. Clin. Investigation, Nat. Acad. Sci. Subspecialties: Biochemistry (medicine); Genetics and genetic engineering (medicine). Current work: Therapy of inherited metabolic disorders; molecular basis of metabolic diseases in humans. First demonstration of enzyme system for fatty acid synthesis; biosynthesis of myelin sheath lipids, nature of metabolic defects in Gaucher's disease, Niemann-Pick disease, Fabry's diseases and Tay-Sachs disease; diagnostic tests for Gaucher's Niemann-Pick, Fabry's diseases; control and therapy of lipid storage diseases; metabolism of sphingolipids in neoplastic diseases. Home: 9501 Kingsley Ave Bethesda MD 20814 Office: NIH 9000 Rockville Pike Bethesda MD 20205

BRAGG, ROBERT HENRY, physicist; b. Jacksonville, Fla., Aug. 11, 1919; s. Robert Henry and Lilly Camille (McFarland) B.; m. Violette Mattie McDonald, June 14, 1947; children: Robert Henry, Pamela. B.S., Ill. Inst. Tech., 1949, M.S., 1951, Ph.D., 1960. Asso. physicist research lab. Portland Cement Assn., Skokie, Ill., 1951-56; sr. physicist physics div. Armour Research Found., Ill. Inst. Tech., Chgo., 1956-61; sr. mem., mgr. phys. metallurgy dept. Lockheed Palo Alto Research Lab., Calif., 1961-69; prof. materials sci. U. Calif., Berkeley, 1969—, chmn. dept. materials sci. and mineral engring., 1978-81; faculty sr. scientist Lawrence Berkeley Lab., 1969—; dir. Applied Space Products; cons. IBM, NIH., NSF. Contbr. articles to profl. jours. Pres. Palo Alto NAACP, 1967-68. Served with U.S. Army, 1943-46. Decorated Bronze star (2); Recipient Disting. award No. Calif. sect. Am. Inst. Mining and Metall. Engrs., 1970. Mem. Am. Phys. Soc., Am. Ceramics Soc. (chmn. No. Calif. sect. 1980), AIME (chmn. No. Calif. sect. 1970), Am. Carbon Soc., Am. Soc. Metals, AAUP, No. Calif. Council Black Profl. Engrs., Sigma Xi, Tau Beta Pi, Sigma Pi Sigma., Am. Crystallography Assn. Democrat. Subspecialties: High-temperature materials. Current work: Structure and physical properties of carbon materials, graphitization in hard and soft carbons. Home: 2 Admiral Dr 373 Emeryville CA 94608 Office: Dept Materials Sci and Mineral Engring Univ of Calif Berkeley CA 94720

BRAHAM, ROSCOE RILEY, cloud physicist; b. Yates City, Ill., Jan. 3, 1921; s. Roscoe Riley and Edith L. (Bowman) B.; m. Mary Ann Moll, Mar. 12, 1943; children—Ruth Ann, Nancy Kay, Richard Riley, Jean Lou. B.S., Ohio U., 1942; S.M., U. Chgo., 1948, Ph.D., 1951. Meteorologist U.S. Weather Bur., Chgo., 1947-49; meteorologist U. Chgo., 1949-50, 51-54, assoc. prof., 1954-56, prof., 1957—; meteorologist N.Mex. Inst. Tech., Socorro, 1950-51; dir. Inst. Atmospheric Physics, U. Ariz., Tucson, 1954-56. Trustee Univ. Corp. Atmospheric Research, Boulder, Colo., 1965-67, 73-76, 79—. Served to 1st lt. USAAF, 1942-45. Fellow Am. Meteorol. Soc., Royal Meteorol. Soc.; mem. Am. Geophys. Union, Sigma Xi, Phi Beta Kappa. Republican. Presbyterian. Subspecialties: Micrometeorology. Current work: Researcher in meteorology of snow storms, anthropogenic modification of local weather, physics of natural precipitation. Home: 57 Longcommon Rd Riverside IL 60546 Office: University of Chicago 5734 S Ellis Ave Chicago IL 60637

BRAID, THOMAS HAMILTON, experimental physicist; b. Heriot, Scotland, Dec. 21, 1925; m., 1951; 2 children. B.Sc., U. Edinburgh, Scotland, 1947, Ph.D. in Physics, 1951. NRC Can. research fellow in physics Atomic Energy Can., Ltd., 1950-52; research asst. Princeton (N.J.) U., 1952-55, instr., 1955-56, research assoc., 1956; assoc. physicist Argonne Nat. Lab. (Ill.), 1956-80, physicist, 1980—; vis. physicist Atomic Energy Research Establishment, Eng., 1966-67. Assoc. editor: Jour. Applied Physics, 1974-79; editor: Rev. Sci. Instruments, 1979—. Fellow Am. Phys. Soc. Subspecialties: Nuclear physics; Scientific instrumentation. Office: Argonne Nat Lab S Cass Ave Argonne IL 60439

BRAITMAN, DAVID JEFFREY, neurophysiologist, educator; b. Bklyn., Nov. 16, 1947; s. Solomon and Elizabeth (Schulman) B.; m. Helen Joan Bromberg, July 6, 1949; children: Amy, Jesse, Jacqueline. B.A., Hunter Coll., 1970; M.S., U. Conn.-Storrs, 1974, Ph.D., 1976. Research fellow U. Conn., 1975; research fellow NIMH, Bethesda, Md., 1976-78; physiologist Armed Forces Radio-biology Research Inst., Bethesda, 1978—; adj. asst. prof. Uniformed Services U. of Health Sci., 1983—. Contbr. articles to profl. jours. Pres. East Silver Spring Elem. Sch. PTA. Served with U.S. Army, 1978—. NIMH fellow, 1975; USPHS fellow, 1976. Mem. Soc. Neurosci., AAAS, Psi Chi. Subspecialties: Neurophysiology; Neuropharmacology. Current work: Functional and pharmacological characterization of neurotransmitters and receptors in mammalian central nervous system; physiological basis of learning and memory. Office: Dept of Physiology Armed Forces Radiobiology Research Inst Bethesda MD 20814

BRAKEFIELD, JAMES CHARLES, research engineer; b. Janesville, Wis., Nov. 28, 1944; s. John Henry and Edith Gertrude (Thompson) B.; m. Irene Lopez Casiano, June 10, 1978. B.S., U. Wis., 1966, M.S.E.E., M.S. in Computer Sci, 1972. Teaching asst. Engring. Computing Lab., U. Wis.-Madison, 1963-66; programmer analyst Control Data Corp., Arden Hills, Minn., 1968-70; systems. analyst Tex. Instruments, Austin, Tex., 1973-74; research engr. Tech. Inc., San Antonio, 1975—. Contbr. articles to tech jours. Served with AUS, 1966-68. Mem. IEEE, Assn. Computing Machinery. Subspecialties: Computer engineering; Software engineering. Current work: Computer architectures, computer tools and languages; software engineering for biomedical experiments, image processing, computer instruction set design. Home: 5803 Cayuga St San Antonio TX 78228 Office: 300 Breesport St San Antonio TX 78216

BRAKKE, MYRON KENDALL, research chemist; b. Fillmore County, Minn., Oct. 23, 1921; s. John T. and Hulda Christina (Marburger) B.; m. Betty-Jean Einbecker, Aug. 16, 1947; children—Kenneth Allen, Thomas Warren, Joan Patricia, Karen Elizabeth. B.S., U. Minn., 1943, Ph.D., 1947. Research asso. Bklyn. Bot. Garden, 1947-52; research asso. U. Ill., 1952-55; research chemist U.S. Dept. Agr., Lincoln, Nebr., 1955—; prof. plant pathology U. Nebr., Lincoln, 1955—. Editor: Virology, 1960-66; Contbr. articles to profl. jours. Fellow AAAS, Am. Phytopath. Soc.; mem. Am. Chem. Soc., Electron Microscope Soc., Nat. Acad. Scis., Am. Soc. Microbiology, Sigma Xi, Phi Lambda Upsilon, Gamma Sigma Delta, Alpha Zeta. Subspecialties: Plant virology; Biochemistry (biology). Current work: Purification and characterization of plant viruses, nucleic acids and proteins; formation of mosaics virus-induced mutations in maize; cereal viruses. Office: Room 406 Plant Science Bldg 8-K U Nebr Lincoln NE 68583

BRAMBLE, JAMES HENRY, educator; b. Annapolis, Md., Dec. 1, 1930; s. Charles Clinton and Edith (Rinker) B.; m. Margaret H., June 25, 1977; children—Margot, Tamara, Mary, James. A.B., Brown U., 1953; M.A., U. Md., 1955, Ph.D., 1958. Mathematician Gen. Electric Co., Cin., 1957-59, Naval Ordnance Lab., White Oak, Md., 1959-60; asst. prof., asso. prof., prof. U. Md., 1960-68; prof. Cornell U., Ithaca, N.Y., 1968—; dir. Center Applied Math., 1974-80; cons. Brookhaven Nat. Lab., 1976—; vis. prof. Chalmers Inst. Tech., Göteborg, Sweden, 1970, 72, 73, 76, U. Rome, 1966-67, Ecole Poly., Paris, 1978, Lausanne, Switzerland, 1979, U. Paris, 1981; lectr. in field. Chmn. editorial bd.: Mathematics of Computation, 1975—; Contbr. articles profl. jours. Mem. Am. Math. Soc. (council), Soc. Indsl. and Applied Math. Subspecialties: Applied mathematics; Numerical analysis. Address: 220 Berkshire Rd Ithaca NY 14850

BRAME, EDWARD GRANT, JR., analytical chemist; b. Shiloh, N.J., Mar. 20, 1927; m., 1957. B.S., Dickinson Coll., 1948; M.S., Columbia U., 1950; Ph.D. in Analytical Chemistry, U. Wis., 1957. Asst. chemist Columbia U., 1948-50; research analytical chemist Corn Products Refining Co., 1950-53; asst. U. Wis., 1953-56; research chemist plastics dept. E.I. duPont de Nemours & Co., Wilmington, Del., 1957-64, research chemist elastomer chemistry dept., 1964-79, research assoc. polymer prodn. dept., 1980—. Editor: Applied Spectroscopic Rev; mem. adv.: Analytical Chemistry, 1974-76; editor-in-chief: Practical Spectroscopy; editor: Applications of Polymer Spectroscopy, 1978. Mem. sci. adv. bd. Winterthur Mus., Del.; mem. postdoctoral research associateships evaluation panel NRC, 1974-76. Nat. Acad. Sci. exchange scientist to USSR, 1981. Mem. Am. Chem. Soc., Soc. Applied Spectroscopy (pres. 1978), N.Y. Acad. Sci., Sigma Xi. Subspecialties: Nuclear magnetic resonance (chemistry); Spectroscopy. Office: EI DuPont de Nemours & Co Bldg 353 Rm 325 Expt Sta Wilmington DE 19898

BRAMLAGE, LAWRENCE ROBERT, veterinary surgeon, educator; b. Marysville, Kans., July 12, 1951; s. Bernard Wesley and Geraldine Louise (Lierz) B.; m. Marilyn Sue Burns, May 25, 1974; children: Joey Marie, Matthew Bernard. D.V.M., Kans. State U., 1975; M.S., Ohio State U., 1978. Vet. intern Colo. State U., Ft. Collins, 1975-76; resident Ohio State U., Columbus, 1976-78, asst. prof. vet. surgery, 1978-82, 83—; equine surgery cons. Mem. AVMA, Am. Assn. Equine Practitioners. Democrat. Roman Catholic. Subspecialty: Surgery (veterinary medicine). Current work: Orthopedic surgery in horses. Office: 1935 Coffey Rd Columbus OH 43210

BRAMWELL, FITZGERALD BURTON, chemist; b. Bklyn., May 16, 1945; s. Fitzgerald and Lula (Burton) B.; m. Charlott Burns Bramwell, Aug. 12, 1973; children: Fitzgerald T., Elizabeth B. B.A., Columbia U., 1966; M.S., U. Mich., 1967, Ph.D., 1970. Research chemist Esso Research & Engring., Linden, N.J., 1970-71; asst. prof. chemistry Bklyn. Coll., CUNY, 1971-75, assoc. prof., 1975-80, prof., 1980—, 1980—; mem. grad. fellowship evaluation panel NRC, 1981-82, chmn., 1983, cons. in field. Author: (with G. Wieder, C.J. Shahani, C.R. Dillard) General Chemistry 2 Laboratory Manual, 1975, (with C.J. Shahani, C.R. Dillard) General Chemistry 1 Laboratory Manual, 1975, (with C.R. Dillard, C.J. Shahani, G.M. Wieder) Investigations in General Chemistry Quantitative Techniques and Basic Principles, 1977, Instructor's Guide for Investigations in General Chemistry Quantitative Techniques and Basic Principles, 1978; contbr.: book reviews and articles to prof. jours. Instructor's Guide for Investigations in General Chemistry Quantitative Techniques and Basic Principles. NSF grantee, 1977-78; Bell Telephone Labs. Equipment grantee, 1977. Mem. N.Y. Acad. Sci., Am. Phys. Soc., Am. Chem. Soc., Sigma Xi, Phi Lambda Upsilon. Subspecialties: Physical chemistry; Solid state chemistry. Current work: Structure and properties of organic semiconductors, structures and properties of group IV organometallics electron spin resonance spectroscopy. Office: Dept Chemistry Brooklyn Coll Bedford Ave and Ave H Brooklyn NY 11210

BRANCH, LAURENCE GEORGE, health policy researcher, educator, gerontologist; b. Cleve., Oct. 31, 1944; s. John Howard and Mercedes (Brachle) B.; m. Patricia Mary Skalski, June 24, 1967; children: Kathryn Helen, Carolyn Mercedes, Daniel Laurence. B.A., Marquette U., 1967; M.A., Loyola U., Chgo., 1969, Ph.D., 1971. Program dir. Ctr. Survey Research, Boston, 1973-79; Asst. prof. Harvard Med. Sch., 1978—, Harvard Med. Sch. Sch. Pub. Health, 1980—; exec. com. div. aging Harvard Med. Sch., 1979—; assoc. dir. Geriatric Research and Edn. Clin. Center Boston VA Outpatient Clinic, 1982—; trustee, vice chmn. North Hill Life Care Community, Wellesley, Mass., 1980—; mem. profl. staff Brigham & Women's Hosp., Boston, 1981—; cons. Robert Wood Johnson Found., Princeton, N.J. Mem. editorial bd.: Jour. Gerontology, 1981, Jour. Community Health, 1980, Gerontologist, 1982. Served to maj. USAR, 1968—. Fellow Gerontol. Soc. Am.; Mem. Am. Pub. Health Assn. (sect. chmn. 1983—), Am. Psychol. Assn., AAAS. Subspecialties: Health services research; Gerontology. Current work: Epidemiology of disability, normal aging, gerontology, research health policy, population survey research, evaluation research. Home: 20

Hammondswood Rd Chestnut Hill MA 02167 Office: Harvard Medical School 643 Huntington Ave Boston MA 02115

BRANCO, MARIA DOS MILAGRES, nuclear engineer; b. Almada, Portugal, Feb. 5, 1955; came to U.S., 1970; d. Jose Vieira and Herminia (Torrado) B.; m. George Leo Rorke, Oct. 6, 1979; 1 son, Evan Daniel. B.S. in Mech. Engring., Newark Coll., 1974; M.S. in Nuclear Engring., Cornell U., 1975. Cert. profl. engr., N.J. Asst. engr. L.I. Lighting Co., Hicksville, N.Y., 1975-78, engr., 1978-82, prin. engr., 1982—; officer Met. Nuclear Fuel Group, N.Y.C., 1977. Contbr. articles to profl. jours. Mem. Am. Nuclear Soc. Subspecialties: Nuclear engineering; Numerical analysis. Current work: Computerized 3D core follow and predictive methodology. Office: Long Island Lighting Co 175 E Old Country Rd Hicksville NY 11801

BRANDE, SCOTT, geology educator; b. Altoona, Pa., May 23, 1950; s. Harold and Selma (Stein) B. B.S., U. Rochester, 1972; M.S., Calif. Inst. Tech., 1974; Ph.D., SUNY, Stony Brook, 1979. Asst. prof. geology U. Ala., Birmingham, 1979—. Mem. AAAS, Paleontol. Soc., Geol. Soc. Am., Sigma Xi. Subspecialties: Paleobiology; Sedimentology. Current work: Research in recent history of sea level, quantitative measurement of evolutionary changes. Office: University of Alabama Dept Geology Birmingham AL 35294

BRANDENBURG, JAMES H., otolaryngology educator; b. Green Bay, Wis., July 17, 1930; m., 1954; 4 children. B.A., U. Wis., 1952, M.D., 1956. Intern William Beaumont Gen. Hosp., El Paso, 1957, ear, nose and throat preceptorship, 1958; ear, nose and throat resident Brooke Gen. Hosp., San Antonio, 1961; from asst. prof. to assoc. prof. Med. Sch. U. Wis.-Madison, 1964-72, prof. otolaryngology, 1972—, chmn. dept., 1972—; attending VA Hosp., Madison, 1964—; mem. faculty home study course Am. Acad. Otolaryngology and Ophthalmology, 1967—. NIH grantee, 1974—; recipient cert. of achievement Armed Forces Inst. Pathology, 1964. Fellow Am. Acad. Ophthalmology and Otolaryngology, Am. Laryngol., Rhinol. and Otol. Soc.; mem. Am. Council Otolaryngology, Am. Soc. Head and Neck Surgery, AMA. Subspecialty: Otorhinolaryngology. Address: 1300 University Ave Madison WI 53706

BRANDT, CARL DAVID, virologist, researcher; b. Bridgeport, Conn., Jan. 19, 1928; s. Carl August and Hildur (Wedberg) B.; m. Elsa Lund Erickson, Apr. 25, 1964; children: Karen, Erik. B.S., U. Conn., 1949; M.S., U. Mass., 1951; Ph.D., Harvard U., 1958. Research instr. dept. vet. sci. U. Mass., 1949-52, 54; research virologist Charles Pfizer and Co., Ind. and Conn., 1958-62; assoc. dept. epidemiology Public Health Research Inst., N.Y.C., 1962-66; research assoc. virology sect. Children's Hosp., Washington, 1966-79; sr. research assoc. virology sect. Children's Hosp. Nat. Med. Center, 1979—; instr. in pediatrics Georgetown U. Sch. Medicine, 1966-69; asst. prof. child health and devel. George Washington U. Sch. Medicine, 1969-74, asso. prof., 1974—; cons. virology clin. labs. Children's Hosp. Nat. Med. Center. Contbr. numerous articles, chpts. to profl. publs. Served to 1st lt. USAF, 1952-54. Fellow Am. Acad. Microbiology, Infectious Diseases Soc. Am., Am. Coll. Epidemiology; mem. AAAS, Am. Soc. Microbiology, Soc. Epidemiologic Research, Pan Am. Group for Rapid Viral Diagnosis, Sigma Xi. Subspecialties: Virology (medicine); Epidemiology. Current work: Diagnosis and epidemiology of viral gastroenteritis, viral respiratory disease, rapid viral diagnosis, electron microscopy. Home: 819 E Franklin Ave Silver Spring MD 20901 Office: Virology Sect Children's Hosp 111 Michigan Ave NW Washington DC 20010

BRANDT, GERALD BENNETT, optical physicist; b. Pitts., Apr. 20, 1938; m., 1961; 4 children. A.B., Harvard U., 1960; M.S., Carnegie Inst. Tech., 1963, Ph.D. in Physics, 1966. Engr. Westinghouse Research Labs., Pitts., 1960-66, sr. engr., 1966-70, fellow scientist, 1970-74, mgr. electro-optics, applied physics dept., 1974-81, adv. scientist, 1981—; lectr. Carnegie Mellon U., 1969, 70. Assoc. editor: Jour. Optical Soc. Am, 1972-77. Mem. AAAS, Optical Soc. Am., Am. Phys. Soc., Soc. Photo-Optical Instrumentation Engrs. Subspecialties: Holography; Optical signal processing. Address: 208 W Swissvale Ave Pittsburgh PA 15218

BRANHAM, RICHARD LACY, JR., astronomer, computer scientist; b. Balt., Mar. 19, 1943; s. Richard Lacy and Mary Marlene (Mrkonjich) B.; m. Rosa Guadalupe, July 15, 1972; 1 dau., Maria Tersita. B.S., Georgetown U., 1965; M.A., Harvard U., 1968; Ph.D., Case Western Res. U., 1977. Astronomer Nat. Radio Astronomy Obs., Green Bank, W.Va., 1965, Georgetown Coll. Obs., Washington, 1966, NASA Hdqrs, 1967, U.S. Naval Obs, 1968—. Contbr. writings to profl. publs. NATO grantee, 1966. Mem. Am. Astron. Soc., Internat. Astron. Union, N.Y. Acad. Scis. Democrat. Subspecialties: Astrometry; Graphics, image processing, and pattern recognition. Current work: Fundamental astrometry; minor planet orbits; image processing of astronomical data; research into mathematical techniques for astronomical data reduction. Home: 4805 Ertter Dr Rockville MD 20852 Office: US Naval Obs Astrometry Div Washington DC 20390

BRANNIGAN, GARY G(EORGE), clinical psychologist, consultant; b. Bridgeport, Conn., Mar. 22, 1947; s. George Tierney and Anna (Chisarik) B.; m. Linda Ann Baker, Sept. 7, 1969; children: Marc, Michael. B.A., Fairfield U., 1969; M.A., U. Del., 1973, Ph.D., 1973. Lic. clin. psychologist, N.Y.; cert. sch. psychologist, N.Y. Panel psychologist Office Vocat. Rehab.; Asst. prof. psychology SUNY-Plattsburgh, 1973-76, assoc. prof. psychology, 1976-82; prof. psychology, 1982; cons. No. N.Y. Center, Plattsburgh, 1974-77, Early Infant Intervention Program, 1977—, Essex County Head Start, Elizabethtown, N.Y., 1981—; dir. Psychol. Services Clinic, Plattsburgh, 1975-80; cons. editor Exceptional People Quarterly, 1981—. Co-author: Research and Clinical Applications of the Bender Gestalt Test, 1980; test Preschool and Primary Visual Motor Gestalt Screening Instrument, 1982; editor: Psychoeducational Perspectives: Reading in Educational Psychology, 1982; contbr. articles to profl. jours., chpts. to books, articles to newspapers. Bd. dirs. Plattsburgh Little League, 1980-82; coordinator YMCA Youth Basketball Program, 1981—; coach Plattsburgh Grasshopper Baseball Program, 1980—. Grantee State Conn., 1971-73, SUNY, 1974, HEW, 1976-79. Fellow Soc. for Personality Assessment; mem. Am. Psychol. Assn., Clinton County Assn. for Retarded Children, Clinton County Mental Health Assn. Lodge: Elks. Subspecialties: Developmental psychology; Learning. Current work: Psychological assessment of infants and children; development of intervention strategies for handicapped and high-risk infants and children. Home: 53 Leonard Ave Plattsburgh NY 12901 Office: State Unit NY Plattsburgh NY 12901

BRANNON, H(EZZIE) RAYMOND, JR., oil co. scientist; b. Midland, Ala., Jan. 23, 1926; s. Hezzie Raymond and Cora Mae B.; m. Rita Alice Newville, Oct. 19, 1957; 1 dau., Sarah Elaine. B. Engring. Physics, Auburn (Ala.) U., 1950, M.S., 1951. Research asso. Auburn Research Found., 1951-52; engr. Exxon Prodn. Research Co., Houston, 1952—, research scientist, 1973—. Contbr. articles to profl. jours. Served with USNR, 1943-46. Mem. Am. Phys. Soc., Soc. Petroleum Engrs. of AIME (Disting. Lectr. 1976-77), Soc. Exploration Geophysicists, NRC (marine bd.), Nat. Acad. Engring., Sigma Xi, Phi Kappa Phi, Tau Beta Pi, Sigma Pi Sigma. Republican. Subspecialties: Petroleum engineering; Civil engineering. Current work: Offshore petroleum production systems and drilling vessels; design and operation of offshore structures. Patentee in field. Home: 5807 Queensloch St Houston TX 77096 Office: PO Box 2189 Houston TX 77001

BRANSCOMB, LEWIS MCADORY, physicist; b. Asheville, N.C., Aug. 17, 1926; s. Bennett Harvie and Margaret (Vaughan) B.; m. Margaret Anne Wells, Oct. 13, 1951; children—Harvie Hammond, Katharine Capers. A.B. summa cum laude, Duke U., 1945, D.Sc. (hon.); M.S., Harvard U., 1947, Ph.D., 1949; D.Sc. (hon.), Poly. Inst. N.Y., Clarkson Coll., Rochester U., U. Colo., Western Mich. U., Lycoming Coll., L.H.D., Pace U. Instr. physics Harvard U., 1950-51; lectr. physics U. Md., 1952-54; vis. staff mem. Univ. Coll., London, 1957-58; chief atomic physics sect. Nat. Bur. Standards, Washington, 1954-60, chief atomic physics div., 1960-62; chmn. Joint Inst. Lab. Astrophysics, U. Colo., 1962-65, 68-69; chief lab. astrophysics div. Nat. Bur. Standards, Boulder, Colo., 1962-69; prof. physics U. Colo., 1962-69; dir. Nat. Bur. Standards, 1969-72; chief scientist, v.p. IBM Corp., Armonk, N.Y., 1972—; chmn. commn. atomic and molecular physics and spectroscopy Internat. Union Pure and Applied Physics, 1972-75; mem. JASON div. Inst. Def. Analyses, 1962-69; chmn. gen. com. Internat. Conf. Physics of Electron and Atomic Collisions, 1969-71; U.S. rep. to CODATA com. Internat. Council Sci. Unions, 1970-73; mem.-at-large Def. Sci. Bd., 1969-72; mem. high level policy group sci. and tech. info. Orgn. Econ. Coop. and Devel., 1968-70; mem. Pres.'s Sci. Adv. Com., 1965-68, chmn. panel space sci. and tech., 1967-68; mem. Nat. Sci. Bd., 1978—, chmn., 1980—; mem. Pres.'s Nat. Productivity Adv. Com., 1981-82; mem. standing com. controlled thermonuclear research AEC, 1966-68; mem. adv. com. on sci. and fgn. affairs Dept. State, 1973-74; mem. U.S.-USSR Joint Commn. on Sci. and Tech., 1977-80; chmn. Com. on Scholarly Communications with the People's Republic of China, 1977-80; dir. Gen. Foods Corp., Mobil Corp.; Mem. pres.'s bd. visitors U. Okla., 1968-70; mem. astronomy and applied physics vis. coms. Harvard bd. overseers, 1969—; mem. physics vis. com. M.I.T., 1974-79; mem. Pres.'s Com. Nat. Medal Scis., 1970-72; Bd. dirs. Am. Nat. Standards Inst., 1969-72; trustee Carnegie Instn., 1973—, Poly. Inst. N.Y., 1974-78, Vanderbilt U., 1980—. Editor: Rev. Modern Physics, 1968-73. Served to lt. (j.g.) USNR, 1945-46. USPHS fellow, 1948-49; Jr. fellow Harvard Soc. Fellows, 1949-51; recipient Rockefeller Pub. Service award, 1957-58, Gold medal exceptional service Dept. Commerce, 1961, Arthur Flemming award D.C. Jr. C. of C., 1962, Samuel Wesley Stratton award Dept. Commerce, 1966, Career Service award Nat. Civil Service League, 1968, Proctor prize Research Soc. Am., 1972. Fellow Am. Phys. Soc. (chmn. div. electron physics 1961-68, pres. 1979), AAAS (dir. 1969-73), Am. Acad. Arts and Scis.; mem. Nat. Acad. Scis. (council 1972-75), Nat. Acad. Engring., Washington Acad. Scis. (Outstanding Sci. Achievement award 1959), Nat. Acad. Public Adminstrn., Internat. Astron. Union, Am. Geophys. Union, Am. Astron. Soc., Internat. Union Geodesy and Geophysics, Am. Philos. Soc., Phi Beta Kappa, Sigma Xi. Club: American Yacht (Rye, N.Y.). Subspecialties: Information systems, storage, and retrieval (computer science); Computer engineering. Current work: National science and technology policy, information science and technology, experimental atomic physics. Office: Old Orchard Rd Armonk NY 10504

BRANSOME, EDWIN DAGOBERT, JR., educator; b. N.Y.C., Oct. 27, 1933; s. Edwin D. and Margaretta (Homans) B.; m. Janet Grace Williams, June 27, 1959; children: Edwin D., April Grace. A.B., Yale U., 1954; M.D., Columbia U., 1958. Resident internal medicine Peter Bent Brigham Hosp. and Harvard Med. Sch., Boston, 1958-62, research fellow endocrinology, 1959-61; research asso. biochemistry Columbia U., N.Y.C., 1962-64; assoc. in endocrinology Scripps Clinic, La Jolla, Calif., 1964-66; asst. to assoc. prof. MIT, Cambridge, 1966-70; prof. medicine Med. Coll. Ga., Augusta, 1970—; cons. in field; med. adv. Social Security Administrn., Bur. Hearings & Appeals, 1983—; chmn. endocrinology adv. com. U.S. Pharmacopeia and mem. com. of rev., 1976—. Author: Self Assessment of Current Knowledge in Clinical Endocrinology, Metabolism and Diabetes, 1975, The Current Status of Liquid Scintillation Counting, 1970; contbr. articles to profl. jours. Chmn. bd. Am. Diabetes Assn., Ga., 1980-82. Am. Cancer Soc. Faculty Research asso. award, 1965-70; various research grants, 1964—. Mem. Endocrine Soc., Am. Diabetes Assn., Am. Chem. Soc., Am. Physiol. Soc., Am. Clin. and Climatological Soc. Clubs: Palmetto Golf, Augusta Country. Subspecialties: Internal medicine; Endocrinology. Current work: Research on the potential interaction of drugs and hormones with DNA, prediction of activity of drugs or homones using molecular models. Patentee in field. Home: 621 Magnolia Ln Aiken SC 29801 Office: Med Coll Ga Augusta GA 30912

BRANTON, DANIEL, educator; b. Belgium, Jan. 13, 1932; s. Leopold and Josephine (Bandes) B.; m. Lana Lenore Brennan, Sept. 1, 1956; children—Hilary Claire, Benjamin David. A.B., Cornell U., 1954; M.S., U. Calif. at Davis, 1957; Ph.D., U. Calif. at Berkeley, 1961. Postdoctoral fellow Eidgenoessische Technische Hochschule, Zurich, Switzerland, 1961-63; prof. botany U. Calif. at Berkeley, 1963-73, Miller Found. research prof., 1968-69; prof. biology Harvard, Cambridge, Mass., 1973—; Mem. study sect. in molecular biology NIH, 1974—. Mem. editorial bd.: Jour. of Molecular Biology, 1970-73, Jour. of Cell Biology, 1970-73, Jour. of Membrane Biology, 1973—; Mem. editorial adv. bd.: Protoplasma, 1973—; Contbr. articles to profl. jours. Served with AUS, 1954-56. Recipient N.Y. Bot. Garden prize, 1972; J.S. Guggenheim fellow, 1970-71. Mem. Nat. Acad. Scis., Am. Acad. Arts and Scis., Am. Soc. Cell Biology, Biophys. Soc., Am. Soc. Plant Physiologists, A.A.A.S. Subspecialty: Cell biology. Home: 14 Eliot Rd Lexington MA 02173 Office: 16 Divinity Ave Cambridge MA 02138

BRASE, DAVID ARTHUR, pharmacologist, educator; b. Orange, Calif., May 9, 1945; s. Arthur H. and Helen (Rottmann) B. B.S., Chapman Coll., 1967; Ph.D., U. Va., 1972. Postdoctoral fellow U. Calif.-San Francisco, 1972-76; asst. prof. pharmacology Eastern Va. Med. Sch., Norfolk, 1976—. Contbr. articles to profl. jours. Pharm. Mfrs. Assn. Found. research starter grantee, 1977-78. Mem. Am. Soc. Pharmacology and Exptl. Therapeutics, Soc. Neuroscience. Lutheran. Club: Soc. Paper Money Collectors. Subspecialty: Neuropharmacology. Current work: Elucidation of mechanism of tolerance to opioid drugs. Office: 700 W Olney Rd Rm 3025 Norfolk VA 23507

BRATTAIN, MICHAEL G., biochemistry educator; b. Ponca City, Okla., Oct. 31, 1947; s. Harold G. and June M. (Oxford) B.; m. Diane E. Roche, Sept. 11, 1966; children: Kathleen A., Michael A. B.S., Rutgers U., 1970, Ph.D., 1974. Teaching asst. Rutgers U., New Brunswick, N.J., 1970-72, NSF research intern, 1972-74; research assoc. U. Ala., Birmingham, 1974-75, instr., 1976-77, assoc. scientist, 1976-81, asst. prof. dept. biochemistry and pathology, 1977-81; adj. assoc. prof. dept. biochemistry U. Tex. System Cancer Ctr., M.D. Anderson Hosp. and Tumor Inst., Houston, 1982—; adj. assoc. prof. dept. pharmacology, dir. Bristol-Baylor Lab., Baylor coll. Medicine, Houston, 1982—; asst. dir. sci. ops. Nat. Large Bowel Cancer Project. Contbr. numerous articles to sci. publs. Nat. Cancer Inst. Am. Cancer Soc. grantee, 1983. Mem. Am. Assn. Cancer Researchers, Am. Assn. Pathologists. Subspecialties: Cancer research (medicine); Cell biology (medicine). Current work: cancer research; cell biology. Home: 1002 Willowvale Dr Taylor Lake Village TX 77586 Office: Bristol-Baylor Lab 1200 Moursund Ave Houston TX 77030

BRATTER, THOMAS EDWARD, psychotherapist; b. N.Y.C., May 18, 1939; s. Edward Maurice and Marjorie (Lowell) B.; m. Carole Jaffe, Aug. 25, 1963; children: Edward Philip, Barbara Ilyse. B.A., Columbia U., 1961, M.A., 1963, Ed.M., 1964, Ed.D., 1974. Cons. N.Y.C. Probation Dept., 1976-79, Pelham (N.Y.) Guidance Council, 1973-79, North Castle Police Dept., Armonk, N.Y., 1978—. Author: (with Bassin) Reality Therapy Reader, 1976, (with Forrest) Current Treatment of Substance Abuse, 1983, (with Kolodny, Kolodny and Deep) Surviving Your Adolescents' Adolescence, 1983, Toward a Humanistic Treatment of Substance Abuse, 1983; mem. editorial bd.: Jour. for Specialists in Group Work, 1975-78, Jour. Drug Issues, 1970-78, Corrective and Social Psychiatry, 1971-76, Jour. Reality Therapy, 1980—, Addiction Therapist, 1975—. Mem. Am. Group Psychotherapy Assn. (history com. 1974—), Am. Psychol. Assn. (substance abuse com. 1972—), Am. Acad. Family Therapy, Am. Acad. Psychotherapists. Democrat. Jewish. Subspecialties: Behavioral psychology; Learning. Current work: Individual, family and group psychotherapy with adolescent substance abusers and their families. Home: 88 Spier Rd Scarsdale NY 10583

BRATTON, GERALD ROY, veterinarian, educator; b. San Antonio, Sept. 25, 1942; s. Roy and Ruth (Adams) B.; m. Lyle Myrta, May 5, 1965; children: Tony, Cindy, Alan. B.S., Tex. A&M U., 1965, D.V.M., 1966, M.S., 1970, Ph.D., 1977. Instr. dept. vet. anatomy Tex. A&M U., 1966-70, asst. prof. vet. anatomy, 1970-75, prof., 1982—; assoc. prof. animal sci./vet. medicine U. Tenn., 1975-80, prof., 1980-81. Contbr. articles to profl. jours. Named Outstanding Young Man Brazos County (Tex.) Jaycees, 1975; recipient Norden Disting. Teaching award, 1977. Mem. AVMA (student chpt. Teaching award 1977), Am. Assn. Vet. Anatomists, Am. Assn. Vet. Med. Colls., Phi Zeta, Phi Kappa Phi, Gamma Sigma Delta. Lutheran. Lodge: Kiwanis. Subspecialties: Veterinary anatomy; Veterinary toxicology. Current work: Lead toxicology, protozoal CNS diseases of horse, innervation specifics of muscle of limbs. Office: Dept Vet Anatomy Coll Vet Medicine Tex A&M U College Station TX 77843

BRAUMAN, JOHN I., chemist, educator; b. Pitts., Sept. 7, 1937; s. Milton and Freda E. (Schlitt) B.; m. Sharon Lea Kruse, Aug. 22, 1964; 1 dau., Kate Andrea. B.S., Mass. Inst. Tech., 1959; Ph.D. (NSF fellow), U. Calif., Berkeley, 1963. NSF postdoctoral fellow U. Calif., Los Angeles, 1962-63; asst. prof. chemistry Stanford (Calif.) U., 1963-69, asso. prof., 1969-72, prof., 1972-80, J.G. Jackson-C.J. Wood prof. chemistry, 1980—, chmn. dept., 1979—; cons. in phys. organic chemistry; adv. panel chemistry div. NSF, 1974-78; adv. panel NASA, AEC, ERDA, Research Corp., Office Chemistry and Chem. Tech., NRC. Mem. editorial adv. bd.: Jour. Am. Chem. Soc, 1976—, Jour. Organic Chemistry, 1974-78, Nouveau Jour. de Chimie, 1977—, Chem. Revs, 1978-80, Chem. Physics Letters, 1978-80. Alfred P. Sloan fellow, 1968-70; Guggenheim fellow, 1978-79. Fellow AAAS; mem. Nat. Acad. Scis., Am. Acad. Arts Scis., Am. Chem. Soc. (award in pure chemistry 1973, Harrison Howe award 1976, exec. com. phys. chemistry div.), Brit. Chem. Soc., Sigma Xi, Phi Lambda Upsilon. Subspecialties: Organic chemistry; Physical chemistry. Current work: Structure, reactivity, photochemistry and spectroscopy of gas-phase ions. Electron photodetachment spectroscopy. Reaction mechanisms. Home: 849 Tolman Dr Stanford CA 94305 Office: Dept Chemistry Stanford U Stanford CA 94305

BRAUN, ANDREW GEORGE, research scientist, educator; b. Paris, Feb. 5, 1939; U.S., 1942, naturalized, 1957; s. Stanislaw and Sonia (Beniakonska) B.; m. Helen Osborn, Sept. 12, 1964; children: Rebecca, Stephanie. B.S., M.I.T., 1961; Sc.D., Harvard U., 1970. Damon Runyon postdoctoral fellow, then research assoc. Brandeis U., Waltham, Mass., 1970-75; asst. prof. radiation therapy Harvard U. Med. Sch., 1975—. Mem. AAAS, Biophys. Soc., Radiation Research Soc., Am. Assn. Cancer Research, Am. Soc. Photobiology, Sigma Xi. Subspecialties: Cell biology; Biochemistry (biology). Current work: Cell surface structure and function. Home: 464 Heath St Chestnut Hill MA 02167 Office: 50 Binney St Boston MA 02115

BRAUN, CHARLES LOUIS, chemist, educator; b. Webster, S.D., June 4, 1937; s. Louis F. and Myrene C. B.; m. Kathleen L. Brickel; children: Sarah K., David C. B.S. in Chemistry, S.D. Sch. Mines and Tech., 1959; Ph.D. in Phys. Chemistry, U. Minn., 1963. Successively instr., asst. prof., assoc. prof. chemistry Dartmouth Coll., Hanover, N.H., 1965-77, prof. chemistry, 1977—; cons. Eastman Kodak Corp. Contbr. articles to profl. publs. Served to 1st lt. U.S. Army, 1963-65. Co-recipient Eastman Kodak prize, 1963; NSF grad. fellow, 1961-63; NSF grantee, 1966-81; Dartmouth Coll. faculty fellow, 1969-70; Petroleum Research Found. grantee, 1981—; Dept. Energy grantee, 1983—. Mem. Am. Chem. Soc., Am. Phys. Soc. Subspecialties: Physical chemistry; Solid state chemistry. Current work: Photophysics of organic molecules; luminescence, photoionization and photoconductivity of organic liquids and solids. Conformational dynamics of chain molecules. Office: Dept Chemistry Dartmouth Coll Hanover NH 03755

BRAUN, PHYLLIS CELLINI, microbiology educator; b. Bridgeport, Conn., Jan. 19, 1953; d. Rudolph V. and Rose B. (Nappi) Cellini; m. Kevin F. Braun, July 19, 1975; 1 son, Ryan Christopher. B.S., Fairfield U., 1975; Ph.D., Georgetown U., 1978. Postdoctoral fellow U. Conn., Farmington, 1978-79; instr. Fairfield (Conn.) U. (dept. molecular biology), 1979-80, asst. prof., 1980—; radiation safety officer Fairfield U., 1981—, career counselor, 1980—. Contbr. articles to profl. jours.; author: Limited Proteolysis, 1978. Presdl. scholar Fairfield U., 1971; faculty research grantee, 1980, 81. Mem. Am. Soc. Microbiology, N.Y. Acad. Sci., AAAS, Assn. Women in Sci. Republican. Roman Catholic. Subspecialties: Microbiology; Membrane biology. Current work: Investigating activity and regulation of enzyme Mannan Synthetase in both morphological forms yeast and pseudohyphae of Candida albicans. Home: 30 Hanover Rd Newtown Ct 06470 Office: Fairfield U North Benson Rd Fairfield CT 06430

BRAUN, ROBERT DENTON, chemistry educator; b. Santa Ana, Calif., June 28, 1943; s. Robert Louis and Delma L. (Carpenter) B.; m. Barbara J. Fisher. B.S., U. Colo., 1965; M.S., U. Conn., 1969, Ph.D., 1971. Lectr. U. Mich., 1971-72; vis. asst. prof. U. Ill.-Urbana, 1972-73; asst. prof. chemistry Vassar Coll., Poughkeepsie, N.Y., 1973-76, U. Southwestern La., Lafayette, 1976-79, assoc. prof., 1979—. Author: Introduction to Chemical Analysis, 1982; co-author: Applications of Chemical Analysis, 1982; contbr. articles to profl. jours. Recipient Outstanding Teaching award Amoco Found., 1982. Mem. Am. Chem. Soc., N.Y. Acad. Scis., La. Acad. Scis., Sigma Xi (pres. chpt.). Republican. Subspecialty: Analytical chemistry. Current work: Electrochemistry, electroanalytical methods, general analytical methods, nonaqueous studies, spectrophotometric methods. Home: 115 Strasbourg Dr Lafayette LA 70506 Office: U Southwestern La Dept Chemistry Box 44370 Lafayette LA 70504

BRAUN, STEPHEN HUGHES, consulting clinical psychologist, educator; b. St. Louis, Nov. 20, 1942; s. William Lafon and Jayne Louise (Shellabarger) B.; m. Penny Lee Prada, Aug. 28, 1956; 1 son, Damian Hughes. B.A., Washington U., St. Louis, 1964, M.A., 1965; Ph.D., U. Mo.-Columbia, 1970. Asst. prof. psychology Calif. State U.-Chico, 1970-71; dir. social learning div. Ariz. State Hosp., Phoenix, 1971-74; chief planning and evaluation Ariz. Dept. Health Services, Phoenix, 1974-79; pres. Braun and Assocs., Scottsdale, Ariz., 1979—;

asst. prof. psychology Ariz. State U., 1971-79, asst. prof. criminal justice, 1974-79, asst. prof. public affairs, 1979-82. Contbr. articles, chpts. to profl. publs.; author clin. evaluation system: Behavioral Health Treatment Outcome Evaluation System, 1981 and subsequent revisions; editorial cons. to profl. jours., 1971—. Recipient Research award State of Calif., 1971; USPHS fellow, 1967-70; NIMH grantee, 1971-74. Mem. Am. Psychol. Assn., Sigma Xi. Subspecialties: Clinical psychology; Information systems, storage, and retrieval (computer science). Current work: Development of systems for planning and managing human service programs (e.g., mental health, drug abuse, alcoholism, corrections, social services, physical health); development of systems for evaluating the efficiency and effectiveness of human service programs; development of microcomputer software to facilitate the planning, management and evaluation of human service programs. Home: 6122 E Calle Tuberia Scottsdale AZ 85251 Office: Braun and Assocs 7125 E Second St Suite 110 Scottsdale AZ 85251

BRAUN-MUNZINGER, PETER, Physics educator; b. Heidelberg, Ger., Aug. 26, 1946; s. Theodor and Brigitte (Föhrenbach) Braun-M.; m. Gabriele Huys, July 21, 1973; children: Lena, Karen. Diploma, U. Heidelberg, 1970, Ph.D. in Physics, 1972. Research assoc. Max-Planck Inst., Heidelberg, 1973-76; vis. asst. prof. SUNY-Stony Brook, 1976-78, asst. prof., 1978-80, assoc. prof., 1980-82, prof., 1982—. Conbtr. articles to sci. jours. Mem. Am. Phys. Soc. Subspecialty: Nuclear physics. Current work: Study of nuclear reactions and structure of complex nuclei. Home: 3 Locust Ave Box 2095 Setauket NY 11733 Office: SUNY Stony Brook NY 11794

BRAUNWALD, EUGENE, physician; b. Aug. 15, 1929; s. William and Clare (Wallach) B.; m. Nina Starr, May 23, 1952; children—Karen, Allison, Jill. A.B., N.Y. U., 1949, M.D., 1952. Diplomate: Am. Bd. Internal Medicine and Cardiovascular Diseases. Tng. internal medicine Mt. Sinai, Johns Hopkins hosps., also Columbia, 1952-58; commd. USPHS, 1954, med. dir., 1963; research cardiology and physiology Nat. Heart Inst., 1955-68, chief cardiology dept., 1960-67, clin. dir. inst., 1966-68; clin. prof. medicine Georgetown U., 1966-68; prof., chmn. dept. medicine U. Calif. at San Diego Sch. Medicine, La Jolla, 1968-72; Hersey prof., Blumgart prof., chmn. dept. medicine Peter Bent Brigham Hosp., Beth Israel Hosp., Harvard Med. Sch., Boston, 1972—; Mem. Nat. Heart and Lung Adv. Council, 1975—. Author: Heart Disease, A Textbook of Cardiovascular Medicine; Editorial bd.: Yearbook of Medicine; editor: Principles of Internal Medicine; contbr. numerous sci. articles. Recipient John Abel award research pharmacology; Arthur Fleming award for outstanding fed. service; Outstanding Service award USPHS, 1967; Nylin award Swedish Med. Soc., 1970; Einthoven medal, 1970; Research Achievement award Am. Heart Assn., 1972; Lilly medal Royal Coll. Physicians, 1979. Fellow Am. Coll. Chest Physicians (hon.), A.C.P., Am. Coll. Cardiology (v.p., gov., trustee), Am. Acad. Arts and Scis.; mem. Am. Heart Assn. (v.p., dir. chmn. publs. 1965—, Research Achievement award 1972, chmn. council on clin. cardiology), Nat. Acad. Scis., Am. Fedn. for Clin. Research (pres. 1969-70), Soc. Clin. Investigation (pres. 1974-75), Western Soc. for Clin. Research (pres. 1971-72), Am. Physiol. Soc., Am. Pharmacology Soc., Assn. Am. Physicians, Assn. Profs. Medicine (pres. 1974-75). Subspecialties: Internal medicine; Cardiology. Home: 75 Scotch Pine Rd Weston MA 02193

BRAWN, ROBERT IRWIN, geneticist; b. Mifflinburg, Pa., Jan. 13, 1923; s. James Daniel and Eva Irwin B.; m. Carolyn Louise Hauck, June 13, 1922; children: Terry Brawn Wheatley, Kathy Brawn Aldous. B.S., Pa. State U., 1943; Ph.D., U. Wis., Madison, 1956. Asst. prof. agronomy and genetics McGill U., Montreal, 1949-56, asso. prof., 1956-64, prof., 1964-71; mgr. corn research Funk Seeds Internat., Bloomington, Ill., 1971-81; dir. research and devel. Ciba-Geigy Seeds Ltd., Ailsa Craig, Ont., 1981—; adj. prof. biology Ill. State U., 1973-81; dean students Macdonald Coll., 1966-71. Contbr. articles to profl. jours. Mem. AAAS, Am. Genetics Soc., Am. Soc. Agronomy, Am. Genetics Assn., Am. Soc. Hort. Sci., Genetics Soc. Can., Can. Soc. Agronomy. Subspecialty: Plant genetics. Current work: Researcher in corn breeding and genetics. Home: 107 Farmington Way London ON Canada N6K 3N7 Office: Ciba Geigy Seeds Ltd Ailsa Craig ON Canada NOM 1AO

BRAY, ARTHUR PHILIP, corp. exec.; b. San Francisco, Sept. 23, 1933; s. Arthur T. and Anna F. (Nevin) B.; m. Grace McCarthy, June 16, 1956; children—Bernard, Peter, Erin, Eileen, Mary, Florence. A.A., San Francisco City Coll., 1953; B.S.M.E. with highest honors, U. Calif., Berkeley, 1955. With Gen. Electric Co., 1955—, v.p., gen. mgr. nuclear power systems div., San Jose, Calif., 1978—; dir. Gen. Electric Tech. Services Co.; Mem. dean's adv. council Sch. Engring., San Jose State U. Co-author: Nuclear Power and the Public, 1970. Bd. dirs. San Jose Repertory Co.; bd. regents Bellarmine Coll. Prep., San Jose. Recipient Ernest O. Lawrence Meml. award U.S. Dept. Energy, 1977. Mem. Nat. Acad. Engring., Am. Nuclear Soc., Atomic Indsl. Forum, ASME, Phi Beta Kappa, Tau Beta Pi, Pi Tau Sigma. Republican. Roman Catholic. Subspecialty: Nuclear engineering. Patentee in field. Office: 175 Curtner Ave San Jose CA 95125

BRAY, DONALD JAMES, poultry scientist, animal nutritionist; b. Anamosa, Iowa, Nov. 8, 1923; s. Percy Milne and Minnie Leona (Purcell) B.; m. Harlene Mae Smith, Aug. 18, 1948; children: Donna Bray Johnson, Ronald Lee, Sharon Bray Frank, Richard Dean, Jean Marie. B.S. in Poultry sci, Iowa State U., 1950; M.S. in Poultry Nutrition, Kans. State U., 1952, Ph.D., 1954. Prof. animal poultry Sci. U. Ill., 1954-81; prin. poultry scientist Coop. State Research Service, U.S. Dept. Agr., Washington, 1981—. Served with USNR, 1944-46. Recipient Paul A. Funk award U. Ill., 1972. Mem. AAAS, Am. Inst. Nutrition, Poultry Sci. Assn., Worlds Poultry Sci. Assn. Republican. Presbyterian. Club: Exchange Am. (Urbana, Ill.) (pres. 1972). Subspecialties: Animal nutrition; Animal physiology. Current work: Relationships among avian nutrition and environment; poultry research coordination among state agricultural experiment stations. Home: 10713 Tenbrook St Silver Spring MD 20901 Office: 6032 S Agriculture Washington DC 20250

BRAY, ROBERT WOODBURY, meat science educator; b. Dodgeville, Wis., Oct. 17, 1918; m., 1943; 2 children. B.S., U. Wis., 1940; M.S., Kans. State Coll., 1941, Ph.D. in Animal Husbandry and Biochemistry, 1949. Instr. animal husbandry Coll. Agr. U. Wis.-Madison, 1941-43, from instr. to assoc. prof., 1946-53, prof. meat and animal sci., 1954-63, chmn. dept., 1954-65, asst. dir. Agr. Expt. Sta. and asst. dean Coll. Agr., 1966-67, assoc. dir. Agr. Expt. Sta. and assoc. dean Coll. Agr., 1967—, prof. meat and animal sci., 1980—. Recipient award Nat. Assn. Meat Purveyors, 1978; R.C. Pollock award, 1978. Mem. Am. Soc. Animal Sci. (award 1962), Am. Meat Sci. Assn. (Disting. Research award 1978), Inst. Food Technologists. Subspecialty: Food science and technology. Address: Coll Agr 136 Agr Hall U Wis Madison WI 53706

BREAKIRON, LEE ALLEN, astronomer; b. Arlington, Va., July 26, 1948; s. Philip Lewis and Margaret Elizabeth (Jensen) B.; m. Patricia Joy McDonough, June 14, 1975; 1 son, Jason Lance. B.A. in Astronomy, U. Va., Charlottesville, 1970, M.S., U. Pitts., 1973, Ph.D. in Astronomy (Zaccheus Daniel fellow), 1977. Grad. teaching asst., researcher U. Pitts., 1971-76; postdoctoral research fellow Wesleyan U., Middletown, Conn., 1976-80; spl. asst. to dir. div. astron. scis. NSF, Washington, 1980—. Contbr. articles to profl. jours. Mem. Am. Astron. Soc., Sigma Xi. Subspecialty: Optical astronomy. Current work: Astrometry. Home: 5248 Monroe Dr Springfield VA 22151 Office: 1800 G St NW Washington DC 22050

BRECHER, ARTHUR SEYMOUR, biochemist, educator; b. N.Y.C., Mar. 30, 1928; s. Harry and Mollie (Rudich) B.; m. Laura Alma Lyman, June 19, 1966; children: Benjamin, Sharon. B.S., CCNY, 1948; Ph.D., UCLA, 1956. Postdoctoral appointee Purdue U., Lafayette, Ind., 1956-58; biochemist U.S. FDA, Washington, 1958-60; assoc. research scientist Bklyn. State Hosp., 1960-63; asst. prof. biochemistry George Washington U., Washington, 1963-69; assoc. prof. chemistry Bowling Green State U., Ohio, 1969-75, prof., 1975—. Contbr. articles to profl. jours. Bd. dirs. Wood County Cancer Unit, Wood County Heart Br., N.W. Ohio Heart Assn. (chmn. heart health in the young com.). Recipient Bowling Green State U., research and devel. award, 1974, spl. achievement award, 1975. Fellow AAAS; mem. Am. Soc. Biol. Chemists, Internat. Soc. Neurochemistry, Am. Soc. Neurochemistry, Soc. Exptl. Biology and Medicine, Am. Chem. Soc., Sigma Xi, Phi Lambda Upsilon. Jewish. Subspecialties: Biochemistry (medicine); Neurochemistry. Current work: Isolation and characterization of peptide hydrolases which catabolize hormones; isolation of naturally occurring inhibitors of proteases; interaction of anti-tumor alkylating agents with nucleic acids and proteins. Home: 3317 Brantford Rd Toledo OH 43606 Office: Bowling Green State U Bowling Green OH 43403

BRECHER, KENNETH, astronomer, physicist, educator; b. N.Y.C., Dec. 7, 1943; s. Irving and Edythe (Grossman) B.; m. Aviva Schwartz, Aug. 18, 1965; children: Karen, Daniel. B.S. in Physics, MIT, 1969, Ph.D., 1969. Research physicist U. Calif., San Diego, 1969-72; asst., then assoc. prof. physics M.I.T., Cambridge, Mass., 1972-79; assoc. prof. astronomy and physics Boston U., 1979-81, prof., 1981—. Co-author/co-editor: (with G. Setti) High Energy Astrophysics and Its Relation to Elementary Particles Physics, 1974, (with M. Feirtag) Astronomy of the Ancients, 1979; also numerous articles. Recipient 2d prize Gravity Research Found., 1969; grantee NATO, 1972, NSF, 1974—, NASA, 1981-82; Guggenheim meml. fellow, 1979-80. Mem. Am. Astron. Soc. (sec.-treas. hist. astronomy div.), Am. Phys. Soc., Internat. Astron. Union. Subspecialties: High energy astrophysics; General relativity. Current work: Theoretical high energy astrophysics, relativity, cosmology. Theoretical research aimed at applying and testing the know laws of physics in astrophysical objects of extreme conditions of density, gravitational field.

BRECKON, GARY JOHN, educator, environmental consultant; b. Weiser, Idaho, Apr. 5, 1940; s. John O. and Edna L. B.; m. Mary Elizabeth Carlson, Oct. 2, 1960; children: Tracy Ann Breckon-Metcalf, Gary Glen. B.A., San Francisco State Coll., 1967, M.A., 1969; Ph.D., U. Calif. -Davis, 1975. Botanist, Strybing Arboretum, San Francisco, 1967-68; TV tchr. Sta. KQED, San Francisco, 1969-70; asst. prof. botany U. Wis., Madison, 1974-80; asst. prof. biology U. P.R.-Mayaguez, 1980—. Mem. Am. Assn. Plant Taxonomists, Internat. Assn. Plant Taxonomists, Assn. Topical Biology, Cycad Soc., Ecol. Assn. Am., Soc. Study of Evolution, Sigma Xi. Subspecialties: Systematics; Ecology. Current work: Systematics and ecology of tropical plants. Office: Dept Biology U PR Mayaguez PR 00708

BREEN, JOHN EDWARD, civil engineer, educator; b. Buffalo, May 1, 1932; s. Timothy J. and Alice C. (Keenan) B.; m. Marian T. Killian, June 20, 1953; children: Mary L., Michael T., Dennis P., Sheila A., Sean E., Kerry T., Christopher D. B.C.E., Marquette U., Milw., 1953; M.S. in Civil Engring, U. Mo., 1957; Ph.D., U. Tex., Austin, 1962. Registered profl. engr., Tex., Mo. Structural designer Harnischfeger Corp., Milw., 1952-53; asst. prof. U. Mo., Columbia, 1957-59; mem. faculty U. Tex., Austin, 1959—, prof. civil engring., 1969—, J.J. McKetta prof. engring., 1977-81, Carol Cockrell Curran chair in engring., 1981—; dir. P.M. Ferguson Structural Engring. Lab., Balcones Research Center, 1967—; cons. in field. Contbr. articles to profl. jours. Served to lt. USNR, 1953-56. Recipient Teaching Excellence award Gen. Dynamics Corp., 1971, U. Tex. Student Assn., 1963, Standard Oil Found. Ind., 1968. Fellow Am. Concrete Inst. (bd. direction 1974-77, Wason medal 1972, 83, Raymond C. Reese Research medal 1972, 79, Kelly medal 1981, Raymond Davis lectr. 1978); mem. Nat. Acad. Engring., ASCE, Sigma Xi, Chi Epsilon, Tau Beta Pi. Democrat. Roman Catholic. Club: Austin Yacht (commodore 1977). Subspecialty: Civil engineering. Current work: Research and development in design and construction of reinforced and prestressed concrete structures. Home: 8603 Azalea Trail Austin TX 78759 Office: Dept Civil Engring Univ Texas Austin TX 78712

BREESE, GEORGE RICHARD, pharmacologist, researcher; b. Richmond, Ind., Dec. 27, 1936; s. George Ralph and Bertha Lee (Miller) B.; m. Joan C. Skates, Nov. 11, 1939; children: Charles, Alicia. B.S., Butler U., 1959, M.S., 1961; Ph.D., U. Tenn.-Memphis, 1965. Research assoc. pharmacology tng. program NIMH, Bethesda, Md., 1966-68; asst. prof., assoc. prof. Sch. Medicine U. N.C., Chapel Hill, from 1968, now prof. psychiatry and pharmacology; researcher in field. Contbr. writings to profl. publs. USPHS grantee, 1968—. Mem. Am. Coll. Neuropsychopharmacology, Am. Soc. Pharmacology, Soc. Neurosci. Subspecialties: Pharmacology; Neuropharmacology. Current work: Developmental pharmacology, neuropharmacology, neural interactions serotonin, thyrotropin releasing factor, substance P, dopamine, neurotensin; alcohol studies. Home: 1240 Little Creek Rd Durham NC 27713 Office: U NC Sch Medicine 226 BSRC Chapel Hill NC 27514

BREGMAN, ALLYN AARON, educator; b. Bklyn., Apr. 29, 1941; s. Irving and Rose (Balmages) B.; m. Sybil Oakley Brewster, Jan. 2, 1943; children: Susannah, Naomi. B.S. magna cum laude, Bklyn Coll., 1962; M.S., U. Rochester, 1964, Ph.D., 1968. Asst. prof. biology SUNY, New Paltz, 1967-71, asso. prof., 1971—. Author: Laboratory Investigations in Cell Biology, 1983; Contbr. articles in field to profl. jours. Mem. Am. Soc. for Cell Biology, AAAS, Genetics Soc. Am. Subspecialties: Cell biology; Genome organization. Current work: Teacher, writer in area of cell biology. Home: 11 Prospect St New Paltz NY 12561 Office: State University of New York Dept Biology New Paltz NY 12561

BREGMAN, JACOB ISRAEL, corporate executive; b. Hartford, Conn., Sept. 17, 1923; s. Aaron and Jennie (Katzoff) B.; m. Mona Madan, June 27, 1948; children—Janet, Marcia, Barbara. B.S., Providence Coll., 1943; M.S., Poly. Inst. Bklyn., 1948, Ph.D., 1951. Research chemist Fels & Co., 1947-48; head phys. chem. labs. Nalco Chem. Co., Chgo., 1950-59; supr. phys. chemistry research sect. Armour Research Found., Chgo., 1959-63; asst. dir. chemistry research Ill. Inst. Tech. Research Inst., Chgo., 1963-65, dir. chem. scis., 1965-67; dep. asst. sec. U.S. Dept. Interior, 1967-69; pres. Wapora Inc., 1969-82; v.p. Dynamac Corp., 1983—; Chmn. N.E. Ill. Met. Area Air Pollution Control Bd., 1962-63; chmn. Ill. Air Pollution Control Bd., 1963-67; chmn. adv. bd. on saline water conversion NATO Parliamentarians Conf., 1963; chmn. Water Resources Research Council, 1964-67. Author: Corrosion Inhibitors, 1963, Surface Effects in Detection, 1965, The Pollution Paradox, 1966, Handbook of Water Resources and Pollution Control, 1976; contbr. articles to profl. jours. Mem. plan commn., Park Forest, Ill., 1956-58, trustee, 1958-62; Mem. Md. Democratic State Central Com., 1974-78; treas. Montgomery Dem. Central Com., 1974-76; del. Dem. Conv., 1976. Served with AUS, 1943-46; ETO. Fellow Am. Inst. Chemists; mem. Am. Chem. Soc., Sigma Xi, Phi Lambda Upsilon. Subspecialties: Environmental engineering; Water supply and wastewater treatment. Current work: Environmental activities—air, water hazardous and toxic wastes, environmental impact. Home: 5630 Old Chester Rd Bethesda MD 20814 Office: 11140 Rockville Pike Rockville MD 20852

BREGMAN, JOEL NORMAN, astronomer; b. Bklyn., Sept. 11, 1951; s. Michael and Miriam F. B.; m. Elaine Susan Pomeranz, Oct. 23, 1954. B.S. in Physics, SUNY, Stony Brook, 1973; Ph.D. in Astronomy and Astrophysics, U. Calif., Santa Cruz, 1977. Postdoctoral research asso. Columbia U., N.Y.C., 1977-79, N.Y. U., 1979-81, research asst. prof., 1981—. Mem. Am. Inst. Physics, Am. Astron. Soc. Subspecialty: Theoretical astrophysics. Current work: Theoretical studies in the interstellar and intercluster medium; studies of quasars. Address: 4 Washington Pl New York City NY 10003

BREISCH, ERIC ALAN, anatomist, researcher, educator; b. Woodbury, N.J., Nov. 5, 1950; s. Robert Evan and Victoria Katherine (Habib) B.; m. Nancy Marie Spinelli, June 16, 1973; children: Justin Eric, Adriana Marie. B.A., Temple U., 1972, Ph.D., 1977. Research asst. McNeil Labs., Flourtown, Pa., 1972-73; postdoctoral fellow Temple U. (cardiology sect.), Phila., 1977-79; asst. prof. anatomy U. Calif.-San Diego Sch. Medicine, LaJolla, 1979—; guest lectr. Am. Heart Assn., San Diego, 1980-83, com. mem. heart edn. in young, 1982, com. mem. student research, 1983. Judge Greater San Diego Sci. Fair, 1981-83. Recipient Young Investigators award Deborah Heart and Lung Found., 1978; Am. Heart Assn. grantee-in-aid, 1980-82; NIH grantee, 1983. Mem. am. Physiol. Soc., Am. Assn. Anatomists, Am. Heart Assn., AAAS. Subspecialties: Anatomy and embryology; Microscopy. Current work: Cardiac structure function, hypertrophy, exercise, aging, morphometry, electron microscopy, physiology, blood flow. Home: 13063 Calle de las Rosas San Diego CA 92129 Office: U Calif San Diego Gilman Dr La Jolla CA 92093

BREITBART, YURI JACOB, data base group supervisor; b. Moscow, Nov. 7, 1940; U.S., 1975, naturalized, 1976; s. Jacob and Pesia B.; m. Ekaterina Samsorov, Mar. 3, 1963; children: Orit, Alex. B.A., Moscow Pedagogical Inst., Moscow, 1963, M.S., 1968; D.Sc., Technion, Haifa, Israel, 1973. Lectr. Technion, 1972-75; asst. prof. SUNY-Albany, 1975-78; assoc. prof. U. Wis.-Milw., 1978-79; mgr. data base evaluation ITT Programming Tech. Center, Stratford, Conn., 1979-81; data base group supr. Amoco Prodn. Research, Tulsa, 1981—; cons. IBM Sci. Center Israel, Haifa, 1973-74, Astronautic Corp. Am., Milw., 1978-79. Israel Acad. Sci., Bat-Sheba grantee, 1974; NSF grantee, 1976-78. Mem. IEEE, Computer Soc. IEEE, Assn. Computing Machinery, Spl. Interest Group on Mgmt. of Data. Subspecialties: Database systems; Distributed systems and networks. Current work: Logical data base design. Home: 6052 E 56th Pl Tulsa OK 79135

BREITMAN, THEODORE RONALD, biochemist; b. N.Y.C., Feb. 25, 1931; s. Sol and Bertha (Morell) B.; children: Sara, Linda. B.S., CCNY, 1948; M.S., Ohio State U., 1955, Ph.D., 1958. Postdoctoral fellow Brandeis U., 1958-61; research biochemist Lederle Labs., Pearl River, N.Y., 1961-63, Nat. Cancer Inst., Bethesda, Md., 1963—. Mem. Am. Soc. Biol. Chemists, Am. Assn. Cancer Research, AAAS. Subspecialties: Cell and tissue culture; Cell study oncology. Current work: Terminal differentiation of human leukemic cells promoted by physiological substances such as retinoic acid, prostaglandins and lymphokines. Home: 8194 Inverness Ridge Rd Potomac MD 20854 Office: NIH Bldg 37 Room 5B14 Bethesda MD 20205

BREITSCHWERDT, EDWARD BEALMEAR, veterinarian, educator; b. Balt., Oct. 25, 1948; s. Edward Paul and Sarah Ester (Bealmear) B.; m. Anne Shepherd, May 23, 1981; 1 son, Edward Brett. B.S., U. Md., 1970; D.V.M., U. Ga., 1974. Diplomate: Am. Coll. Vet. Internal Medicine, 1979. Intern U. Mo., Columbia, 1974-75, resident, 1975-77; asst. prof. medicine La. State U., Baton Rouge, 1977-82, asso. prof., 1982; sect. chief Small Animal Clinic, 1979-82; assoc. prof. medicine N.C. State U., Raleigh, 1982—. Contbr. articles to profl. jours. Mem. AVMA, Am. Assn. Vet. Med. Colls., Am. Animal Hosp. Assn. Episcopalian. Subspecialty: Internal medicine (veterinary medicine). Current work: Renal tubular dysfunction, canine dirofilariasis, identification of animal diseases and animal models of human disease. Office: Sch Veterinary Medicine NC State U 4700 Hillsborough St Raleigh NC 27606

BRELAND, HUNTER MANSFIELD, psychologist; b. Mobile, Ala., Aug. 11, 1933; s. Robert Milton and Cora (Peirce) B.; m. Nancy Schacht, Aug. 17, 1968; 1 dau., Alison. B.S.M.E., U. Ala., 1955; M.S., U. Tex.-Austin, 1961; Ph.D., SUNY-Buffalo, 1972. Engr. Gen. Dynamics Corp., Ft. Worth, 1955-59; sr. engr. LTC, Inc., Dallas, 1960-61; vol. Peace Corps, Dominican Republic, 1962-65; adminstr. NSF, Washington, 1965-66; mgmt. cons. Harbridge House, Inc., Boston, 1966-67; research scientist Ednl. Testing Service, Princeton, N.J., 1972—. Author: Selective Admissions in Higher Edn, 1977, Population Validity, 1979, Assessing Student Characters, 1981, Personal Qualities in Administration, 1982. Mem. Am. Psychol. Assn., Am. Ednl. Research Assn., Sociedad Interamericana Psychology, AAAS. Subspecialties: Cognition; Artificial intelligence. Current work: Information science, automated language processing; evaluation of written products through both judgemental and automated techniques. Home: 426 Burd St Pennington NJ 08534 Office: Ednl Testing Service Rosedale Rd Princeton NJ 08541

BREMERMANN, HANS J., mathematics educator, biophysicist; b. Breman, Ger., Sept. 14, 1926; m., 1954. M.A., U. Münster (Ger.), Ph.D. in Math, 1951. Instr. math. U. Münster, 1952; research assoc. Stanford U., 1952-53, vis. asst. prof., 1953-54; asst. prof. U. Münster, 1954-55; mem. staff Inst. Advanced Study, 1955-57, 58-59; asst. prof. U. Wash., 1957-58; assoc. prof. math. U. Calif.-Berkeley, 1959-64, assoc. prof. math. and biophysics, 1964-66, prof. math. and med. physics, 1966—; research fellow Harvard U., 1953; indsl. cons.; mem. exec. com. grad. group biophysics and med. physics U. Calif.-Berkeley, 1964—. Mem. Am. Math. Soc., Austrian Math. Soc., German Soc. Applied Math. and Mechanics, German Math. Assn., Biophys. Soc. Subspecialties: Biophysics (biology); Applied mathematics. Current work: Several complex variables; Schwartz distributions; physics; dispersion relations; renormalization; information theory; limitations of genetic control; evolution processes; self-organizing systems; biological algorithm; complexity theory; mathematical ethology; pattern recognition; model verification; optimization; control; medical applications of nonlinear control; mathematical models of epidemics and host-pathogen dynamics; application to integrated pest control. Office: Dept Math U Calif Berkeley CA 94720

BRENCHLEY, JEAN ELNORA, microbiologist, research exec.; b. Towanda, Pa., Mar. 6, 1944; d. John E. and Elizabeth (Jefferson) B. B.S., Mansfield (Pa.) State Coll., 1965; M.S., U. Calif., San Diego, 1967, Ph.D., 1970. Research assoc. M.I.T., Cambridge, 1970; asst. prof., then assoc. prof. Pa. State U., 1971-77; prof. biology Purdue U., Lafayette, Ind., 1977-81; research dir. Genex Corp., Gaithersburg, Md., 1981—; cons., lectr. Editor: Applied and Environ. Microbiology, 1982—; Contbr. chpts. to books, articles to sci. jours. Vol. Crisis Center, Pals Program. NSF and NIH grantee, 1971-81; Becton&Dickinson lectr., 1979; Am. Soc. Microbiology Found. lectr.,

1973. Mem. NOW, Assn. Women in Sci., Am. Soc. Microbiology, Am. Chem. Soc., Am. Soc. Biol. Chemists, Am. Genetics Soc. Subspecialties: Microbiology; Genetics and genetic engineering (biology). Current work: Microbial genetics; biotechnology. Office: 16020 Industrial Dr Gaithersburg MD 20877

BRENDEL, KLAUS, pharmacologist, chemistry researcher; b. Berlin, Germany, June 14, 1933; came to U.S., 1963; s. Erich Rudolf and Eva (Keysselitz) B.; m. Barbara Brigitte Luge, Nov. 2, 1957; children: Sabine, Katrin, Caroline. B.S., Free U. Berlin, 1955, M.S., 1959, Ph.D., 1962. Research assoc. Free U. Berlin, 1962-63; research assoc. U. Pacific, Stockton, Calif., 1963-64, Duke U., 1964-67, asst. prof. pharmacology, 1967-70; assoc. prof. U. Ariz., 1970-76, prof., 1976—. Mem. editorial bd.: Life Sciences, 1974—, Internat. Jour. Biol. Research Pregnancy, 1980—, Internat. Jour. Clin. Pharmacology, 1980—; contbr. numerous articles to profl. jours. Fellow and grantee in field. Fellow Atherosclerosis Council Am. Heart Assn.; mem. Am. Soc. Pharmacology and Exptl. Therapeutics, Gesellschaft Deutscher Chemiker, Am. Chem. Soc., Sigma Xi. Subspecialties: Cellular pharmacology; Molecular pharmacology. Current work: Molecular toxicology, biomaterials research, physical anthropology, paleo toxicology. Home: 3231 N Manor St Tucson AZ 85715 Office: Dept Pharmacology U Ariz Health Scis Ctr Tucson AZ 85723

BRENNAN, DONALD FRANCIS, aerospace co. mktg. exec.; b. Orange, N.J., Dec. 18, 1934; s. Clifford James and Ethel Margaret (Lally) B.; m. L. Barbara Muhn, May 9, 1970; 1 dau., Jean. B.M.E., U. Detroit, 1957; M.B.A., Fairleigh Dickinson U., 1962. Aero-thermodynamicist Curtiss-Wright, Woodridge, N.J., 1955-60, asst. project engr., 1960-62; engring. adminstr. RCA Astro-Electronics, Princeton, N.J., 1962-67, mktg. rep., 1967-81, mgr. sci. programs, 1981—. First v.p. Green Island Community Assn., Toms River, N.J., 1980-83. Mem. Am. Geophys. Union, Am. Inst. Physics. Roman Catholic. Subspecialties: Aerospace engineering and technology; Satellite studies. Current work: Market development of satellite systems and technology for government and commercial applications. Office: PO Box 800 Princeton NJ 08540

BRENNAN, EILEEN G., plant pathologist, educator; b. Jersey City, N.J., Dec. 21, 1922; s. John J. and Mary (Gallagher) B. B.S., Rutgers U., 1944, M.S., 1946, Ph.D., 1975. Research asst. Cook Coll., Rutgers U., 1946-56, asst. prof., 1956-74, prof., 1974—. Contbr. numerous articles to profl. jours. Mem. Am. Phytopathol. Soc., Air Pollution Control Assn. (recipient Reuben W. Wasser award Mid-Atlantic States sect. 1981), Internat. Soc. Arboriculture, Sigma Xi. Subspecialty: Plant pathology. Current work: Effect of air pollutants and heavy metals on vegetation. Home: 120 Cleveland Ave Colonia NJ 07067 Office: Cook Coll Dept Plant Pathology Rutgers Coll Agr Box 231 New Brunswick NJ 08903

BRENNAN, ROBERT WILLIAM, animal scientist, educator; b. Schenectady, May 24, 1953; s. Robert Doyle and Joan Mary (Allen) B.; m. Marlene Hotaling, June 21, 1974; 1 son, Robert James. B.S., Iowa State U., 1975, M.S., 1978, Ph.D., 1983. Technician Iowa Dept. Transp., 1976-78; instr. animal sci. Iowa State U., 1978-82; asst. prof. animal sci. Findlay Coll., 1982—; cons. computer applications in agr. Contbr. articles to profl. jours. Mem. Am. Soc. Animal Sci., AAAS, Gamma Sigma Delta. Democrat. Roman Catholic. Subspecialties: Animal nutrition; Integrated systems modelling and engineering. Office: Findlay Coll Brewer Hall Findlay OH 45840

BRENNER, BARRY MORTON, physician; b. Bklyn., Oct. 4, 1937; s. Louis and Sally (Lamm) B.; m. Jane P. Deutsch, June 12, 1960; children: Robert, Jennifer. B.S., L.I. U.; M.D., U. Pitts.; M.A., Harvard U. Asst. prof. medicine U. Calif., San Francisco, 1969-72, asso. prof. medicine and physiology, 1972-75, prof. medicine, 1975-76; Samuel A. Levine prof. medicine Harvard U. Med. Sch., Boston; with Peter Bent Brigham Hosp., Boston, 1976—; dir. renal div. Brigham and Women's Hosp., Boston, 1979—; cons. NIH. Contbr. numerous articles to various publs. Recipient research award NIH. Mem. Am. Physiol. Soc., Am. Soc. Clin. Investigation, Assn. Am. Physicians, Western Assn. Physicians, Am. Soc. Nephrology, Am. Soc. Clin. Investigation (councillor, v.p.), Salt and Water Club, Interurban Clin. Club, Alpha Omega Alpha, Phi Sigma. Subspecialties: Physiology (biology); Nephrology. Office: 75 Frances St Boston MA 02115

BRENNER, HOWARD, chemical engineer; b. N.Y.C., Mar. 16, 1929; s. Max and Margaret (Wechsler) B.; m.; children: Leslie, Joyce, Suzanne. B.Ch.E., Pratt Inst., 1950; M.Ch.E., NYU, 1954, Eng.Sc.D., 1957. Instr. chem. engring. NYU, 1955-57, asst. prof. chem. engring., 1957-61, assoc. prof., 1961-65, prof., 1965-66, Carnegie-Mellon U., 1966-77; prof., chmn. dept. chem. engring. U. Rochester, 1977-81; Willard Henry Dow prof. chem. engring. MIT, Cambridge, 1981—; sr. vis. fellow Sci. Research Council, Gt. Britain, 1974. Author: (with J. Happel) Low Reynolds Number Hydrodynamics, 1965, 2d edit., 1973, Russian edit., 1976; contbr. numerous articles to profl. jours.; asso. editor: Internat. Jour. Multiphase Flow, 1973—, (with J. Happel) Internat. Jour. PhysicoChem. Hydrodynamics, 1979—. Fairchild Disting. scholar Calif. Inst. Tech., 1975-76; Named to 11th Ann. Honor Scroll Indsl. Engring. Chemistry div. Am. Chem. Soc., 1961. Fellow Am. Inst. Chem. Engrs., AAAS, Am. Inst. Chem. Engrs. (Alpha Chi Sigma award 1976); mem. Nat. Acad. Engring., Soc. Indsl. and Applied Math., Internat. Assn. Colloid and Interface Scientists, Am. Inst. Civil Engrs., Soc. Rheology (Bingham medal 1980), Soc. Nat. Philosophy, Am. Acad. Mechanics, Am. Chem. Soc. Subspecialty: Chemical engineering. Office: Dept Chem Engring MIT Cambridge MA 02139

BRENT, ROBERT LEONARD, physician, educator; b. Rochester, N.Y., Oct. 6, 1927; s. Charles and Rose (Katz) B.; m. Lillian H. Hoffman, Aug. 21, 1949; children: David A., James R., Lawrence H., Deborah A. A.B., U. Rochester, 1948, M.D. with honors, 1953, Ph.D., 1955. Fellow Nat. Found., Strong Meml. Hosp., 1953-54; intern pediatrics Mass. Gen. Hosp., Boston, 1954-55; chief radiation biology Walter Reed Army Inst. Research, 1955-57; faculty Jefferson Med. Coll., 1955—, prof. radiology, 1962—, also prof. pediatrics, chmn. dept. and dir.; Chmn. med. adv. bd. Nat. Found.; Chmn. med. adv. bd., mem. fertility and maternal health com. FDA.; mem. human embryology study sect. NIH, 1970-74. Editor in chief: Teratology, 1976—. Served with U.S. Army, 1955-57. Travelling fellow Royal Soc. Medicine, 1971-72; vis. fellow FitzWilliam Coll., Cambridge, 1971-72; Recipient Richie meml. prize U. Rochester Med. Sch., 1953; Lindback Found. award for distinguished teaching, 1968; Lady Davis scholar Hadassah Med. Ctr., Jerusalem, 1983-84. Mem. Teratology Soc. (pres. 1967-68), AAAS, Radiation Research Soc., Am. Soc. Exptl. Pathology, Soc. Pediatric Research, Am. Pediatrics Soc., Am. Acad. Pediatrics, Soc. Exptl. Biology and Medicine, Phila. Coll. Physicians, Phila. Pediatric Soc., Am. Assn. Immunology, Soc. Developmental Biology, Congenital Malformations Assn. Japan, European Teratology Soc., Sigma Xi, Alpha Omega Alpha. Subspecialties: Pediatrics; Teratology. Current work: Pediatrics; developmental biology. Office: 920 Chancellor St Philadelphia PA 19107

BRESLAU, NEIL ART, medical educator; b. Bklyn., Sept. 3, 1947; s. Nathan and Henny (Rabinowitz) B.; m. Sharon E. Weidman, June 8, 1969; children: Joshua, Jeremy. B.S., Bklyn. Coll., 1968; M.D., Vanderbilt U., 1972. Diplomate: Am. Bd. Internal Medicine. Intern Med. Center Hosp. of Vt., Burlington, 1972-73; resident L.I. Jewish-Hillside Med. Center, New Hyde Park, N.Y., 1973-74; endocrinology fellow and instr. medicine Upstate Med. Center, Syracuse, N.Y., 1974-78; asst. prof. medicine Southwestern Med. Sch., Dallas, 1978—. Contbr. numerous articles to med. jours. NIH grantee, 1978-81, 1981-84. Mem. Am. Fedn. Clin. Research, Endocrine Soc., Am. Soc. Bone and Mineral Metabolism. Subspecialties: Endocrinology; Internal medicine. Current work: Pathophysiology of parathyroid glands, mechanism of kidney stone formation and prevention and metabolic bone disease. Home: 529 Tiffany Trail Richardson TX 75081 Office: Dept Internal Medicine Southwestern Med Sch 5323 Harry Hines Blvd Dallas TX 75235

BRESLOW, LESTER, physician, educator; b. Bismarck, N.D., Mar. 17, 1915; s. Joseph and Mayme (Danziger) B.; m.; children: Norman, Jack, Stephen; m. Devra J.R. Miller, 1967. B.A., U. Minn., 1935, M.D., 1938, M.P.H., 1941. Diplomate: Am. Bd. Preventive Medicine and Public Health. Intern USPHS Hosp., Stapleton, N.Y., 1938-40; dist. health officer Minn. Dept. Health, 1941-43; chief bur. chronic diseases Calif. Dept. Public Health, Berkeley, 1946-60, chief div. preventive medicine, 1960-65, dir. dept., 1965-68; lectr. U. Calif. Sch. Public Health, Berkeley, 1950-68, prof. public health, 1968—, chmn. dept. preventive medicine and social medicine, 1969-72; dean Sch. Pub. Health, UCLA, 1972-80; dir. study Pres.'s Commn. Health Needs of Nation, 1952; cons. Nat. Cancer Inst., 1981—. Author med. publs.; editor: Ann. Rev. Pub. Health, 1979—; editorial cons.: Jour. Preventive Medicine. Served to capt. U.S. Army, 1943-45. Decorated Bronze Star; recipient Lasker award; Sedgwick medal Am. Pub. Health Assn.; Outstanding Achievement award U. Minn. Fellow Am. Coll. Preventive Medicine (Disting. service award 1976), ACP; fellow AAAS; mem. Am. Heart Assn. (fellow epidemiology sect.), Am. Public Health Assn. (past pres.), Public Health Cancer Assn. (past pres.), Am. Epidemiol. Soc., Internat. Epidemiol. Assn. (past pres.), Am. Cancer Soc. (nat. dir., Calif. dir., chmn. adv. com. on research etiology), Assn. Schs. Public Health (pres. 1973-74), Inst. Medicine, Nat. Acad. Scis. (council 1978-80, chmn. bd. health promotion and disease prevention 1981—). Subspecialties: Preventive medicine; Epidemiology. Current work: Measurement of health and factors influencing it; public health. Home: 10926 Verano Rd Los Angeles CA 90024

BRESLOW, RONALD CHARLES, chemist, educator; b. Rahway, N.J., Mar. 14, 1931; s. Alexander E. and Gladys (Fellows) B.; m. Esther Greenberg, Sept. 7, 1955; children: Stephanie, Karen. A.B. summa cum laude, Harvard, 1952, M.A., 1953, Ph.D., 1955. NRC fellow Cambridge (Eng.) U., 1955-56; mem. faculty Columbia, 1956—, prof. chemistry, 1962-66, S.L. Mitchell prof., 1966—; cons. to industry, 1958—; Mem. medicinal chemistry panel NIH, 1964—; mem. adv. panel on chemistry NSF, 1971—. Editor: Benjamin, Inc, 1962—; Author: Organic Reaction Mechanisms, 1965, 2d edit., 1969; also articles.; Mem. editorial bd.: Organic Syntheses, 1964—, Jour. Organic Chemistry, 1969—, Jour. Bio-organic Chemistry, 1972—, Tetrahedron, 1975—, Tetrahedron Letters, 1975—. Recipient Fresenius award Phi Lambda Upsilon, 1966; Mark Van Doren award Columbia, 1969; Roussel prize, 1978; Centenary lectr. London Chem. Soc., 1972. Fellow Am. Acad. Arts and Scis.; mem. Am. Philos. Soc., Nat. Acad. Scis. (chmn. chemistry div. 1974-77), Am. Chem. Soc. (Pure Chemistry award 1966, Baeheland medal 1969, chmn. div. organic chemistry 1970, Harrison Howe award 1974, Remsen award 1977, J. F. Norris award 1980), Phi Beta Kappa (first marshall 1952). Subspecialties: Organic chemistry; Biophysical chemistry. Current work: Artificial enzymes; bio-mimetic chemistry; anti-aromatic compounds; enzyme mechanisms. Home: 275 Broad Ave Englewood NJ 07631 Office: Dept Chemistry Columbia Univ New York NY 10027

BRETT, BETTY LOU HILTON, biologist; b. Hudson, N.Y., Mar. 25, 1952; d. Donald Fredrick and Ethel Mary (Pickering) Hilton; m. Carlton Eliott Brett, May 20, 1974. B.A., SUNY-Buffalo, 1974; M.A. in Biology, U. Mich., 1977; Ph.D. in Natural Resources, U. Mich., 1981. Postdoctoral research assoc. in biology U. Rochester, N.Y., 1981—. Contbr. articles to profl. jours. NSF grantee, 1978-80. Mem. AAAS, Am. Soc. Ichthyologists and Herpetologists, Soc. Study of Evolution, Systematic Zoology Soc., Genetics Soc. Am. Subspecialties: Evolutionary biology; Species interaction. Current work: Gene flow in fish; sex ratio strategies in mealybugs. Office: Dept Biology U Rochester Rochester NY 14627

BREWER, GEORGE J., geneticist, hematologist, educator; b. Lake County, Ind., Feb. 28, 1930; m. Lucia Feitler; children: Bonnie, Holly, Jeannie, Katie. B.S. in Pharmacy, Purdue U., 1952; M.D., U. Chgo., 1956. Intern, U. Chgo. Clinics, 1956-57, resident, 1957-59; research assoc. dept. medicine U. Chgo., 1959-63, research assoc. dept. human genetics U. Mich., Ann Arbor, 1963-67, asst. prof. internal medicine, 1965-67, assoc. prof. dept. human genetics, 1965-67, research assoc. dept. internal medicine, 1967-76, prof. dept. human genetics, 1970—, prof. dept. internal medicine, 1976—. Author: Genetics, 1983; assoc. editor: Am. Jour. Hematology. Served with M.C., U.S. Army, 1961-63. Recipient Borden award Purdue U., 1949; Disting. Alumni award U. Chgo., 1976. Mem. Am. Fedn. Clin. Research (pres. chpt. 1967-68), Am. Soc. Human Genetics, Am. Soc. Hematology, Central Soc. Clin. Research, Am. Soc. Clin. Investigation, Phi Eta Sigma. Subspecialties: Genetics and genetic engineering (medicine); Gene actions. Current work: Orphan drugs, orphan disease research work. Zinc requirement in humans. Treatment of Wilson's Disease with zinc. Treatment of sickle cell anemia. Erythrocyte metabolism and membranes. Home: 3820 Gensley Rd Ann Arbor MI 48109 Office: U Mich 1241 E Catherine Ann Arbor MI 48109

BREWER, LEO, educator; b. St. Louis, June 13, 1919; s. Abraham and Hannah (Resnik) B.; m. Rose Strugo, Aug. 22, 1945; children—Beth A., Roger M., Gail L. B.S., Calif. Inst. Tech., 1940; Ph.D., U. Calif. at Berkeley, 1943. Mem. faculty U. Calif. at Berkeley, 1946—, prof. phys. chemistry, 1955—; research asso. Lawrence Berkeley Lab., 1943-61, head inorganic materials div., 1961-75, asso. dir. lab., 1967-75; Huffman Meml. lectr. Calorimetry Conf., 1966; Coover lectr. Am. Chem. Soc., 1967; Robert W. Williams lectr. Mass. Inst. Tech., 1963; Henry Werner lectr. U. Kans., 1963; O.M. Smith lectr. Okla. State U., 1964; G.N. Lewis lectr. U. Calif., 1964, Faculty lectr., 1966; Corn Products lectr. Pa. State U., 1970; W.D. Harkins lectr. U. Chgo., 1974; mem. rev. com. reactor chem. div. Oak Ridge Nat. Lab.; research asso. Manhattan Dist., U. Calif., Berkeley, 1943-45; sec. gas subcom. high temperature commn. Internat. Union Pure and Applied Chemistry, 1957-60, asso. mem. commn. on thermodynamics and thermochemistry, 1973—; chmn. materials adv. bd. Com. Investigation Application Plasma Phenomena, 1959-60. Author: (with others) Thermodynamics, 1961; Asso. editor: Jour. Chem. Physics, 1959-63; editorial adv. bd.: Jour. Physics and Chemistry Solids, Progress Inoganic Chemistry, Jour. Chem. Thermodynamics, 1968-77, Jour. High Temperature Sci, Jour. Solid State Chemistry, Jour. Chem. Engring. Data, 1977—, Jour. Phys. Chemistry Reference Data, 1978-81; divisional editor high temperature sci. and tech. div.: Jour. Electrochem. Soc, 1977—. Great Western Dow fellow, 1942; Guggenheim fellow, 1950; recipient Ernest Orlando Lawrence Meml. award, 1961; Distinguished Alumni award Calif. Inst. Tech., 1974. Mem. Nat. Acad. Scis. (exec. com. Office Critical Tables 1961-66), AAUP, AAAS, Am. Acad. Arts and Scis., Am. Chem. Soc. (Leo H. Baekeland award 1953), Electrochem. Soc. (lectr. 1970, Palladium Medalist 1971), Am. Plant Life Soc., ACLU, Cobletz Soc., Combustion Inst., Faraday Soc., Fedn. Am. Scientists, Calif. Assn. Chemistry Tchrs., Internat. Plansee Soc. Powder Metallurgy, Am. Optical Soc., Metall. Soc., Am. Phys. Soc., Calif. Acad. Sci., Calif. Native Plant Soc., Calif. Botanic Soc., Lawrence Hall of Sci., Save Redwoods League, Sierra Club, Sigma Xi, Alpha Chi Sigma, Tau Beta Pi. Subspecialty: Physical chemistry. Home: 15 Vista del Orinda Orinda CA 94563 Office: Dept Chemistry Univ California Berkeley CA 94720

BREWER, RICHARD GEORGE, physicist; b. Los Angeles, Dec. 8, 1928; s. Louis Ludwig and Elise B.; m. Lillian Magidow, Sept. 23, 1954; children: Laurence R., Emily S., Catherine. B.S., Calif. Inst. Tech., Pasadena, 1951; Ph.D., U. Calif., Berkeley, 1958. Instr. Harvard U., 1958-60; asst. prof. UCLA, 1960-63; mem. research staff IBM Corp. Research Lab., San Jose, Calif., 1963-73, IBM fellow, 1973—; cons. prof. applied physics Stanford U., 1977—; adj. prof. Nat. Inst. Optics, Florence, Italy, 1977—; vis. prof. M.I.T., 1968-69, U. Tokyo, spring 1975, U. Calif., Santa Cruz, fall 1976; mem. Calif. Scientist of Year Awards Jury, 1980, 81; mem. com. atomic and molecular physics Nat. Acad. Scis.-NRC, 1974-77; mem. rev. panel for Nat. Bur. Standards, 1981—; mem. com. on recommendations U.S. Army Basic Sci. Research, 1982-85; rev. com. San Francisco Laser Center, 1980-83, AEC-Lawrence Berkeley Lab., 1974. Asso. editor: Optics Letters, 1977-80, Jour. Optical Soc. Am, 1980—. Served with AUS, 1955-57. Recipient Albert A. Michelson Gold medal Franklin Inst., 1979. Fellow Am Phys. Soc. (Joint Council Quantum Electronics 1982-83, O.E. Buckley prize com. 1982), Optical Soc. Am. (chmn. optical physics tech. council 1978-80, com. on fellows and hon. mems. 1981, W.F. Meggers award com. 1981); mem. Nat. Acad. Scis. Subspecialties: Atomic and molecular physics; Condensed matter physics. Current work: Laser spectroscopy; quantum optics. Office: IBM Research Lab 5600 Cottle Rd San Jose CA 95193

BREY, WALLACE SIEGFRIED, chemist, educator; b. Schwenksville, Pa., June 6, 1922; s. Wallace S. and Roxie (Lichty) B.; m. Mary Louise Van Natta, Apr. 7, 1955; children: William W., Paul D. B.S., Ursinus Coll., 1942; Ph.D., U. Pa., 1948. With Warner Co., Phila., 1942-44; faculty DePauw U., Greencastle, Ind., 1948-49, St. Joseph Coll., Phila., 1949-52, U. Fla., Gainesville, 1952—, now prof. chemistry. Author books and tech. papers.; Editor: Jour. Magnetic Resonance. Mem. Am. Chem. Soc. (Fla. award 1981), Am. Phys. Soc., Royal Soc. Chemistry, N.Y. Acad. Scis. Subspecialties: Nuclear magnetic resonance (chemistry); Surface chemistry. Current work: Nuclear magnetic resonance spectroscopy in study of molecular structure and molecular interactions. Office: Dept Chemistry U Fla Gainesville FL 32611

BREYER, NORMAN NATHAN, metallurgical engineering educator; b. Detroit, June 21, 1921; m., 1952; 3 children. B.S., Mich. Tech. Inst., 1943; M.S., U. Mich., 1948; Ph.D., Ill. Inst. Tech., 1963. Aero. research scientist Nat. Adv. Com. Aeros., 1948; chief armor sect. Detroit Tank Arsenal, 1948-52; dir. research cast steels and irons Nat. Roll & Foundry, 1952-54; metallurgist-in-charge armor Continental Foundry & Machine div. Blaw-Knox Co., 1955-57; mgr. tech. projects La Salle Steel Co., 1957-64; assoc. prof. metall. engring. Ill. Inst. Tech., Chgo., 1964-69, prof., 1969—, chmn. dept., 1969—. Mem. AIME, Am. Soc. Metals. Subspecialty: Metallurgy. Office: Dept Metall Engring Ill Inst Tech 3300 S Federal St Chicago Il 60616

BREZENOFF, HENRY EVANS, pharmacologist, educator; b. N.Y.C., July 9, 1940. B.S., Columbia U., 1962; Ph.D., N.J. Coll. Medicine and Dentistry, 1968. Postdoctoral fellow Sch. Medicine UCLA, 1968-69; asst. prof. N.J. Coll. Medicine and Dentistry, Newark, 1970-74, assoc. prof., 1974-80, prof. pharmacology, 1980—, asst. Dean, 1978-80, assoc. dean, 1980—; researcher in field. Mem. Am. Soc. Pharmacology and Exptl. Therapeutics, Soc. for Hypertension, Italian Soc. Pharmacology, AAAS. Subspecialties: Pharmacology; Neuropharmacology. Current work: Role of brain acetylcholine in cardiovascular regulation. Office: 100 Bergen St Newark NJ 07103

BRICK, JOHN, biopsychologist, researcher; b. N.Y.C., Mar. 18, 1950; s. H. C. and V.A. (Carmella) B.; m. Laurie Stockton Krulish, May 1, 1976. B.A., Queens Coll., CUNY, 1973; M.A. in Psychology, SUNY, Binghamton, 1979, Ph.D. in Psychology, 1981. Research asst. Rockefeller U., 1973-76; research asso. Center Alcohol Studies Rutgers U., 1980-82, research specialist, asst. prof., 1982—; cons. dept. medicine and surgery VA, Washington, 1982—. Editor: Stress and Alcohol Use; contbr. over 20 articles to profl. jours. Mem. Soc. Neurosci., AAAS, N.Y. Acad. Sci., Brit. Brain Research Assn., European Brain and Behavior, Sigma Xi. Subspecialties: Psychobiology; neuroscience. Current work: Elucidation of neurochemical and neuroendocrine responses involved in the interaction between stress and to alcohol. Office: Center Alcohol Studies Rutgers U New Brunswick NJ 08903

BRIDGE, HERBERT SAGE, space physics educator; b. Berkeley, Calif., May 23, 1919; m., 1941; 3 children. B.S., U. Md., 1941; Ph.D. in Physics, MIT, 1950. Mem. research staff Los Alamos Sci. Lab., 1943-46; research assoc. cosmic ray research MIT, Cambridge, 1946-50, mem. research staff, 1950-55, research physicist, 1955-65, assoc. dir. ctr. space research, 1965-78, dir. ctr., 1978—, prof. physics, 1965—; vis. scientist European Orgn. Nuclear Research, Geneva, 1957-58. Mem. Am. Geophys. Union, Am. Acad. Arts and Sci. Subspecialty: High energy astrophysics. Office: Ctr Space Research MIT Rm 37-241 Cambridge MA 02139

BRIDGES, ALAN LYNN, computer scientist; b. Knoxville, Tenn., Oct. 10, 1950; s. Elijah Paul and Beuna Flynn B. B.S., Ga. Inst. Tech., 1972, M.S., 1974, postgrad., 1975-78. Asst. research scientist Engring Expt. Sta., Ga. Inst. Tech., Atlanta, 1975-78; asst. product mgr. Humphrey Instrs., San Leandro, Calif., 1978-79; pres., founder ETC West Ltd. dba VeXP Research/Systems, Atlanta, 1978-82; new tech. trg. dir. Gen. Motors Assembly Div., Warren, Mich., 1979-80; product mgr., tng. dir. picture processing systems Via Video, Inc., Cupertoni, Calif., 1982; lead engr., prin. investigator electronics research and devel. dept. Lockheed - Ga. Co., Marietta, Ga., 1982—. Contbg. editor: Computer Tech. Rev.; Contbr. articles in field to profl. jours. Mem. Optical Soc. Am., Soc. Photo-optical Instrumentation Engrs., IEEE; mem. AIAA; Mem. Assn. Computing Machinery, Atlanta Computer Soc.; MEM. Nat. Mgmt. Assn.; Mem. Pi. Sigma. Democrat. Clubs: Radio (Atlanta) (editor 1975-76); Kennehoochee Amateur Radio (Marietta, Ga.) (bd. dirs. 1976-79). Subspecialties: Graphics, image processing, and pattern recognition; Distributed systems and networks. Current work: Videographics systems, man-machine interface, speech recognition/Synthesis, intelligent tutoring systems, computer aided instruction and training, BLSI/VASIC, optical memory systems, display technology, automatic test equipment. Home: 2754 Pine Hill Dr NW Kennesaw GA 30144 Office: Lockheed-Ga Co Electronices Research and Devel Dept 72-95 Zone 316 86 S Cobb Dr Marietta GA 30063

BRIDGES, DONALD NORRIS, nuclear engineer; b. Shelby, N.C., Aug. 13, 1936; s. Torrence Festus and Rhea (Hunt) B.; m. Charlene Kiser, Aug. 30, 1959; children: Denise Ann, David Lynn, Daryl Dean, Donna Michelle. B.C.E., N.C. State U.-Raleigh, 1958, M.S., 1960; M.S. in Nuclear Engring, Ga. Inst. Tech., 1968, Ph.D., 1970.

Registered profl. engr., N.C. Nuclear engr. sr. ops. office AEC, Aiken, S.C., 1970-74; nuclear engr. NRC, Washington, 1974-76; chief nuclear safety br., sr. ops. Dept. Energy, Aiken, 1976-80, chief reactors and materials br., 1980—. Cub master Pack 601, Cub Scouts Am., Merriwether Sch., S.C., 1981—; tchr. Sunday Sch., First Bapt. Ch., North Augusta, S.C., 1980—. Served to lt. USNR, 1962-66. Mem. Am. Nuclear Soc., Naval Res. Assn., Soc. Am. Mil. Engrs. Club: Savannah River Federal Employee (Aiken) (pres. 1977-78). Subspecialties: Nuclear engineering; Nuclear fission. Current work: Use of nuclear reactors for materials production and power production. Home: 1002 Longleaf Ct North Augusta SC 29841 Office: Savannah River Ops Office US Dept Energy Aiken SC 29801

BRIDGES, WILLIAM BRUCE, research electrical engineer, educator; b. Inglewood, Calif., Nov. 29, 1934; s. Newman K. and Doris L. (Brown) B.; m. Carol Ann French, Aug. 24, 1957; children: Ann Marjorie, Bruce Kendall, Michael Alan. B.E.E., U. Calif. at Berkeley, 1956, M.E.E. (Gen. Electric Rice fellow), 1957, Ph.D. in Elec. Engring. (NSF fellow), 1962. Asso. elec. engring. U. Calif., Berkeley, 1957-59, grad. research engr., 1959-61; mem. tech. staff Hughes Research Labs. div. Hughes Aircraft Co., Malibu, Calif., 1960-77, sr. scientist, 1968-77, mgr. laser dept., 1969-70; prof. elec. engring. and applied physics Calif. Inst. Tech., Pasadena, 1977—, Carl F. Braun prof. engring., 1983—, exec. officer elec. engring., 1978-81; lectr. elec. engring. U. So. Calif., Los Angeles, 1962-64; Sherman Fairchild Distinguished scholar Calif. Inst. Tech., 1974-75; chmn. Conf. on Laser Engring. and Applications, Washington, 1971. Author: (with C.K. Birdsall) Electron Dynamics of Diode Regions, 1966; contbr. articles on gas lasers, optical systems and microwave tube to profl. jours.; asso. editor: IEEE Jour. Quantum Electronics, 1977-82, Jour. Optical Soc. Am, 1978-83. Active Boy Scouts Am., 1968-82; bd. dirs. Ventura County Campfire Girls, 1973-76. Recipient L.A. Hyland Patent award, 1969. Fellow IEEE (chmn. Los Angeles chpt. Quantum Electronics and Applications Soc. 1979-81), Optical Soc. Am. (chmn. lasers and electro-optics tech. group 1974-75, bd. dirs. 1982—); mem. Nat. Acad. Engring., Nat. Acad. Scis., Am. Radio Relay League, Phi Beta Kappa, Sigma Xi, Tau Beta Pi, Eta Kappa Nu (One of Outstanding Young Elec. Engrs. for 1966). Lutheran. Subspecialties: Physics research; Microwaves. Current work: Laser device physics; microwave device physics. Inventor noble gas ion laser, patentee in field. Home: 413 W Walnut St Pasadena CA 91001 Office: Calif Inst Tech 128-95 Pasadena CA 91125

BRIDGMAN, CHARLES JAMES, engineering educator; b. Toledo, May 6, 1930; s. Charles D. and Wilhhelmina EstherBelle (O'Neill) B.; m. Lucy Hull, May 15, 1954; children: Kathleen Bridgman McFadden, Stephanie Bridgman Danahy, Charles J., Paula J., Kenneth M., Thomas A. B.S., U.S. Naval Acad., 1952; M.S., N.C. State U., 1958, Ph.D., 1963. Commd. 2d lt. U.S. Air Force, 1952, advanced through grades to capt., 1958; resigned, 1963; asst. prof. Air Force Inst. Tech., Dayton, Ohio, 1963-64, assoc. prof. engring., 1964-68, prof., chmn. com. nuclear engring., 1968—. Contbr. articles to sci. jours. Recipient awards Air Force Inst. Tech. Mem. Am. Nuclear Soc., Am. Assn. Engring. Edn., Health Physics Soc., AAUP, Sigma Xi. Republican. Roman Catholic. Subspecialty: Nuclear engineering. Current work: Nuclear weapon effects, neutral particle transport; computational physics. Office: Air Force Inst Tech Wright Patterson AFB OH 45433 Home: 7362 Natoma Pl Huber Heights OH 45424

BRIDGMAN, JOHN FRANCIS, biologist, educator, environmental consultant; b. Kuling Kiangsu, China, Sept 6, 1925; s. Harold Thomas and Eleanor Mae (Galbraith) B.; m. Beverly Alice May, June 3, 1952; children: John Francis, Paul, Alice, Nancy. B.S., Davidson Coll., 1949; M.S., La. State U., 1952; Ph.D., Tulane U., 1968. Instr. biology Delta State Tchrs. Coll., Cleveland, Miss., 1952, U. Tenn., Martin, 1952-54, Shikoku Christian Coll., Zentsuji, Japan, 1955-72; prof. biology Coll. of Ozarks, Clarksville, Ark., 1972—. Contbr. articles to profl. jours. Served with USAAC, 1943-47. Recipient George Henry Penn award in biology Tuland U., 1968; NIH fellow; NSF fellow. Mem. AAAS, Am. Soc. Parasitologists, Southwestern Assn. Parasitologists, Ark. Acad. Sci., Am. Soc. Zoologists. Presbyterian. Club: Lake Dardanelle Sail. Subspecialties: Parasitology; Species interaction. Current work: Parasitology—ecology and host-parasite relations. Office: Coll of Ozarks CPO Box 501 Clarksville AR 72830

BRIERLEY, CORALE LOUISE, mineral technological company executive, microbiologist; b. Shelby, Mont., Mar. 24, 1945; d. Lloyd R. and Louise (Reinlasoder) Beer; m. James A. Brierley, Dec. 21, 1965. B.S., N. Mex. Inst. Mining and Tech.-Socorro, 1968, M.S., 1971; Ph.D., U. Tex.-Dallas, 1981. Microbiologist planetary quarantine Martin Marietta Corp., Denver, 1968-69; electron microprobe analyst N. Mex. bur. Mines, and Mineral Resources, Socorro, 1970-71, chem. microbiologist, 1971-82; pres. Advanced Mineral Technologies, Inc., Socorro, 1982—; lectr. in field. Contbr. over 35 sci. articles to profl. publs. Grantee in field. Mem. Am. Soc. Microbiology, Can. Soc. Microbiologists, AIME, Soc. Indsl. Microbiology, Am. Inst. Biol. Scis., Soc. Environ. Toxicology and Chemistry. Subspecialties: Microbiology; Water supply and wastewater treatment. Current work: Development of microbiological processes for mining industry and wastewater treatment; microbiology; wastewater treatment; materials processing. Home: 1103 Bullock St Socorro NM 87801 Office: Advanced Mineral Technologies, Inc PO Box 1339 Socorro NM 87801

BRIGGS, ARTHUR BRAILSFORD, JR., computer systems engineering official, artificial intelligence official; b. Hamlet, N.C., Jan. 11, 1944; s. Arthur Brailsford and Margaret Louise (Poston) B.; m. Annesley Rembert Stuckey, Mar. 12, 1966; children: Catherine Margaret, Nancy Annesley. B.S. in Math, Wofford Coll., 1965; M.S. in Ops. Research, Air Force Inst. Tech., Dayton, Ohio, 1975. Computer maintenance officer Offutt AFB, Nebr., 1967-70; computer systems analyst Wright-Patterson AFB, Ohio, 1970-77; resigned, 1977; computer systems engr. Tex. Instruments, Inc., Dallas, 1977, mgr. engring. tng. and edn., Dallas and Lewisville, Tex., 1978-82, mgr. artificial intelligence br., 1983—. Commd. 2d lt. U.S. Air Force, 1966; advanced through grades to capt., 1969. Mem. Ops. Research Soc. Am. (assoc.), Assn. Computing Machinery, Air Force Assn. Republican. Lutheran. Subspecialties: Artificial intelligence; Software engineering. Current work: Expert systems logic programming, knowledge based systems, natural language processing, development of software engineering methodology. Home: 1316 North Park Dr Richardson TX 75081 Office: Texas Instruments Inc PO Box 405 MS 3407 Lewisville TX 75067

BRIGGS, ARTHUR HAROLD, pharmacologist; b. East Orange, N.J., Nov. 3, 1930; s. Arthur H. and Marie (Schoepf) B.; m. Elizabeth Jensen, June 6, 1953; children: Kimberlee, Norman Arthur. B.A., Johns Hopkins U., 1952, M.D., 1956. Diplomate: Am. Bd. Internal Medicine. Intern, then resident in internal medicine Vanderbilt U. Hosp., 1956-58, postdoctoral fellow dept. pharmacology, 1959; asst. prof. pharmacology and medicine U. Miss. Med. Center, 1959-68; prof., chmn. dept. pharmacology U. Tex. Health Sci. Center, San Antonio, 1968—. Author books and sci. articles. USPHS spl. fellow, 1967-68. Mem. Am. Soc. Pharmacology and Exptl. Therapeutics, Soc. Exptl. Biology and Medicine, Sigma Xi. Subspecialties: Molecular pharmacology; Internal medicine. Current work: Pharmacology of alcohol and drugs of abuse. Teacher, researcher, chairman department of pharmacology. Home: 707 Serenade St San Antonio TX 78216 Office: 7703 Floyd Curl Dr San Antonio TX 78284

BRIGGS, WILLIAM BENAJAH, aerospace marketing executive; b. Okmulgee, Okla., Dec. 13, 1922; s. Eugene Stephen and Mary Betty (Gentry) B.; m. Lorraine Hood, June 6, 1944; children: Eugene Stephen, Cynthia Anne, Julia Louise, Spencer Gentry. B.A., Phillips U., 1944; M.S. in M.E, Ga. Inst. Tech., 1948; D.Sc., Phillips U., 1977. Research scientist Nat. Adv. Commn. for Aeros., Cleve., 1948-52; engr. div. planning mgr. Chance Vought/LTV, Dallas, 1952-64; mgr. mktg. space/missiles McDonnell Douglas Astronautics Co., St. Louis, 1964-76, mgr. mktg. fusion energy, 1976-80, dir. mktg. fusion energy, 1980—; mem. NASA Planetary Quarantine Adv. Com., 1967-70. Vice-pres., bd. dirs. Christian Bd. Publs., St. Louis, 1975—; v.p., pres. Disciples Council St. Louis, 1971-75. Served to lt. (j.g.) USN, 1943-47. Assoc. fellow AIAA (dir. 1974-77, nat. v.p. 1974-79, Lindbergh award, 981, sect. service award 1980); mem. Am. Nuclear Soc. Disciple of Christ. Lodge: Masons. Subspecialties: Nuclear fusion; Aerospace engineering and technology. Current work: Fusion energy program development, engineering, writing, lecturing, exhibits, films. Patentee in field. Home: 1819 Bradburn Dr St Louis MO 63131 Office: McDonnell Douglas Astronautics Co E080 Box 516 St Louis MO 63166

BRIGGS, WINSLOW RUSSELL, plant physiologist; b. St. Paul, Apr. 29, 1928; s. John DeQuedville and Marjorie (Winslow) B.; m. Ann Morrill, June 30, 1955; children: Caroline, Lucia, Marion. B.A., Harvard U., 1951, M.A., 1952, Ph.D., 1956. Instr. biol. scis. Stanford (Calif.) U., 1955-57, asst. prof., 1957-62, asso. prof., 1962-66, prof., 1966-67; prof. biology Harvard U., 1967-73; dir. dept. plant biology Carnegie Instn. of Washington, Stanford, 1973—. Author: (with others) Life on Earth, 1973; Asso. editor: Annual Review of Plant Physiology, 1961-72; editor, 1972—; Contbr. articles on plant growth and devel. and photbiology to profl. jours. John Simon Guggenheim fellow, 1973-74. Fellow AAAS; mem. Am. Soc. Plant Physiologists (pres. 1975-76), Calif. Bot. Soc. (pres. 1976-77), Nat. Acad. Scis., Am. Acad. Arts and Scis., Am. Inst. Biol. Scis. (pres. 1980-81), Am. Soc. Photbiology, Bot. Soc. Am., Nature Conservancy, Sigma Xi. Subspecialties: Plant physiology (biology); Photobiology. Current work: Role of light in plant development, interaction of light reactions with hormones in plant development, molecular consequences of photoexcitation of pertinent plant photoreceptor molecules in light-regulated development. Home: 480 Hale St Palo Alto CA 94301 Office: Dept of Plant Biology Carnegie Institution of Washington 290 Panama St Stanford CA 94305 With gifted students, remarkable things are possible.

BRIGHT, HAROLD JOHN, biochemistry and biophysics educator; b. Salisbury, Eng., Aug. 25, 1935; s. Leonard George and Kathleen Rosemund (Greenhall) B.; m. Janice Watson; children: Deborah, Leslie, Christopher. B.A. with honors, Cambridge (Eng.) U., 1957; Ph.D., U. Calif.-Davis, 1961. Mem. Faculty Sch. Medicine, U. Pa., Phila., 1962—, prof. biochemistry and biophysics, 1975—; vis. prof. U. Coll. London, 1971, Oxford (Eng.) U., 1971, Bristol (Eng.) U., Eng., 1971, 79; cons. in field. Editorial bd.: Jour. Biol. Chemistry; contbr. over 80 sci. articles to profl. publs. Rector's warden St. Mary's Episcopal Ch., Phila. USPHS Research Career Devel. grantee, 1967-71; NIH Research grantee, 1963—; NSF Research grantee, 1971-74; Guggenheim fellow, 1971; Fogarty Sr. Internat. fellow, 1979. Home: 689 Meadowbrook Ln Media PA 19063 Office: U Pa Sch Medicine Biochemistry and Biophysics Dept Philadelphia PA 19104

BRIGHTON, JOHN AUSTIN, mechanical engineer, educator; b. Gosport, Ind., July 9, 1934; s. John William and Esther Pauline B.; m. Charlotte L. McCarty, Mar. 20, 1953; children: Jill, Kurt, Eric. B.S., Purdue U., 1959, M.S., 1960, Ph.D., 1963. Draftsman Switzer Corp., Indpls., 1952-56; instr. Purdue U., 1960-63; asst. prof. mech. engring. Carnegie-Mellon U., 1963-65; asst. prof. Pa. State U., 1965-67, asso. prof., 1967-70, prof., 1970-77, Mich. State U., 1977-82, chmn. dept. mech. engring., 1977-82; dir. Sch. Mech. Engring. Ga. Inst. Tech., 1982—; Chmn. Community Sponsors Inc., State College, Pa., 1976-77; chmn. Pre-Trial Alts. Program for First Offenders, State College, 1976-77; bd. dirs. Impression 5. Author: (with Hughes) Fluid Dynamics, 1966. NSF grantee, 1975-77; NIH grantee, 1974-78. Mem. ASME (Engr. of Yr. award for Central Pa. 1977, tech. editor Jour. Biomech. Engring. 1976-79), Am. Soc. Engring. Edn., Am. Soc. Artificial Internal Organs. Subspecialties: Mechanical engineering; Biomedical engineering. Current work: Biofluid mechanics, artificial heart, engineering education and administration. Home: 525 Kenbrook Dr Atlanta GA 30327 Office: Ga Inst Tech Atlanta GA 30332

BRILEY, MARGARET ELIZABETH, nutrition educator; b. Abilene, Tex., Aug. 5, 1929; d. Charles Grant and Ivie Mae (Rape) Willis; m. Clyde Briley, Aug. 19, 1950; children: Kathryn Ann Briley Riddles, Kimberly Susanne. B.S., U. Tex.-Austin, 1950; M.S., Tex. Tech. U., 1969, Ph.D., 1973. Techr. Lockney (Tex.) Ind. Sch. Dist., 1954-68; research Tex. Tech. U., Lubbock, 1969-70, research assoc., 1970-73; asst. prof. Tex. Christian U., Ft. Worth, 1973-74; assoc. prof. U. Tex.-Austin, 1974—; nutrition edn. cons. Edn. Service Ctr., Austin, 1978-82; nutrition cons. Tex. State Nutrition Council, 1982-84. Mem. Am. Inst. Nutrition, Am. Dietetic Assn. (registered dietitian), Tex. Dietetic Assn., Austin Dietetic Assn. (legis. chmn.), Inst. Food Tech., Soc. for Nutrition Edn., Sigma Xi, Phi Upsilon Omicron. Methodist. Subspecialties: Nutrition (biology); Food science and technology. Current work: Nutrition, education, vitamin B6, geriatrics. Home: 1420 Yaupon Valley Austin TX 78746 Office: U Tex-Austin GEA 115 Austin TX 78712

BRILL, THOMAS BARTON, chemistry educator, researcher; b. Chattanooga, Feb. 3, 1944; s. Kenneth Gray and Priscilla (Ritchie) B.; m. Patricia Jahn, Aug. 7, 1967; children: Barbara, Russell. B.S., U. Mont.-Missoula, 1966; Ph.D., U. Minn.-Mpls., 1970. Postdoctoral fellow N.C. State U., Raleigh, 1969-70; asst. prof. U. Del., Newark, 1970-74, assoc. prof., 1974-77, prof. chemistry, 1979—; vis. prof. U. Oreg., Eugene, 1977-78. Mem. editorial bd.: Art Materials Tech, 1981—; author: Light: Its interaction with Art and Antiquities, 1980. Air Force Office Sci. Research grantee, 1973—. Mem. Am. Chem. Soc., Sigma Xi. (nat. lectr. 1984-86). Republican. Episcopalian. Subspecialties: Inorganic chemistry; Physical chemistry. Current work: Thermal decomposition processes, synthesis, structure and reactions of organometallic reactions, nuclear quadrupole resonance spectroscopy, vibrational spectroscopy. Home: 101 Tanglewood Ln Newark DE 19711 Office: U Del Newark DE 19711

BRILL, WINSTON J., microbiology educator, biochemical geneticist; b. London, June 16, 1939; m., 1965; 1 child. B.S., Rutgers U., 1961; Ph.D. in Microbiology, U. Ill.-Urbana, 1965. NIH fellow in biology MIT, Cambridge, 1965-67; prof. bacteriology U. Wis.-Madison, 1967-83; dir. research Cetus Madison Corp., Middleton, Wis., 1983—; mem. panel NSF, 1974-77; mem. recombinant DNA adv. panel NIH, 1980—. Grantee USPHS, 1968—, NIH, 1969—, NSF, 1969—; recipient Alexander von Humboldt award, 1979; Eli Lilly award, 1979. Mem. AAAS, Am. Soc. Microbiology. Subspecialties: Plant genetics; Nitrogen fixation. Office: Cetus Madison Corp 2208 Parview Rd Middleton WI 56532

BRILLIANT, HOWARD MICHAEL, air force engineering officer; b. Balt., Aug. 15, 1945; s. Benjamin and Anne Gertrude (Grodnitzky) B.; m. Arleen H. Blatt, Oct. 22, 1978; 1 dau.: Rachelle Idena. B.S. in Mech. Engring, U. Pitts., 1966; M.S.Engring., U. Mich., 1967, Ph.D., 1971. Registered profl. engr., Colo. Commd. 2d lt. U.S. Air Force, 1966, advanced through grades to maj., 1980; project officer (Air Force Aero. Propulsion Lab.), Wright-Patterson AFB, Ohio, 1970-75, faculty mem., Colo., 1975-80, assoc. prof. aeros., 1978-80, chief laser tech. group, Kirtland AFB, N.Mex., 1980-82, chief new laser concepts br., 1982, chief laser systems devel. sect., 1982—; cons. NASA Dryden Flight Research Ctr., Edwards AFB, Calif., 1976-78. Recipient Sci. Achievement award U.S. Air Force Systems Command, 1973, Commendation medal U.S. Air Force, 1974, Meritorious Service medal, 1979. Mem. AIAA, Sigma Xi. Club: Sierra (Colorado Springs, Colo.) (chmn. group 1978). Subspecialties: Aeronautical engineering; Fluid mechanics. Current work: Transonic and supersonic aerodynamics, propulsion, kinetics, thermodynamics, physics of chemical lasers, chemical laser systems. Home: 4104 Glen Canyon Rd N E Albuquerque NM 87111 Office: AFWL/ARAC Kirtland AFB NM 87117

BRILLINGER, DAVID ROSS, statistician; b. Toronto, Oct. 27, 1937; s. Austin Carlyle and Winnifred Elsie (Simpson) B.; m. Lorie Silber, Dec. 17, 1960; children: Jef Austin, Matthew David. B.A., U. Toronto, 1959; M.A., Princeton U., 1960, Ph.D., 1961. Lectr. math Princeton U. and; mem. tech. staff Bell Labs., 1962-64; lectr. stats. London Sch. Econs., 1964-66, reader, 1966-69; prof. stats. U. Calif., Berkeley, 1970—, chmn. dept., 1979-81. Author: Time Series: Data Analysis and Theory, 1975. Woodrow Wilson fellow, 1959; Bell Telephone Labs. fellow, 1960; Social Sci. Research Council postdoctoral fellow, 1961; Miller prof., 1973; Guggenheim fellow, 1975-76, 82-83. Fellow Am. Statis. Assn., Inst. Math. Stats., AAAS; mem. Internat. Statis. Inst., Inst. Math. Stats., Royal Statis. Soc., Can. Statis. Soc., Seismol. Soc. Am., Bernoulli Soc. Subspecialty: Statistics. Current work: Mainstream statistics, but especially applications in seismology and neurophysiology. Office: Dept Stats U Calif Berkeley CA 94720

BRIMHALL, GEORGE H., geology educator, researcher; b. Santa Monica, Calif., Aug. 24, 1947; s. George and Alice (Traver) B.; m. Mary Jane Patroan, June 19, 1971; children: Lara Claire, Hilary Alyse. Project geologist Anaconda Co., Butte, Mont., 1972-76; asst. prof. Johns Hopkins U., Balt., 1976-78; asst. prof. geology U. Calif.-Berkeley, 1978-80, assoc. prof., 1980-82, prof., 1982—; mem. panel on mineral resources NRC, 1981—. Contbr. articles on econ. geology to profl. jours. Mem. Soc. Econ. Geologists (Lindgren award 1980, mem. research com. 1981—, nominating com. 1982—), Geochem. Soc., Mineral Soc. Am., AIME, Geol. Soc. Am. Subspecialties: Petrology; Geochemistry. Current work: Geological, geochemical and petrological study of the origin of economic metals in the crust; ore deposits, and economic geology. Office: Dept Geology and Geophysics U Calif Berkeley CA 94720

BRIMIJOIN, WILLIAM STEPHEN, pharmacologist, educator; b. Passaic, N.J., July 1, 1942; s. William Owen and Georgiana Grier (Macklin) B.; m. Margaret Ross, June 22, 1964; children: Megan, Owen, Alexander. B.A. in Psychology, Harvard U., 1964, Ph.D. in Pharmacology, 1969. Research assoc. NIH, 1969-71; assoc. cons. in pharmacology Mayo Found., Rochester, Minn., 1971-72, cons. in pharmacology, 1972—; asst. prof. pharmacology Mayo Med. Sch., 1972-76, assoc. prof., 1976-78, prof., 1980—; vis. scientist Karolinska Inst., Stockholm, Sweden, 1978-79. Contbr. articles to profl. jours. Served with USPHS, 1969-71. NIH career devel. awardee, 1975-80; NIH grantee, 1971-74, 74—, 82—. Mem. Soc. Neurosci. (pres. So. Minn. chpt. 1976-78, 81—), Am. Soc. Pharmacology and Exptl. Therapeutics, Am. Soc. Neurochemistry. Subspecialties: Neurobiology; Neurochemistry. Current work: Translocation of proteins in nerve cells (axonal transport), mechanisms of peripheral nerve disease, biology of chloinesterases. Home: 1214 6th St SW Rochester MN 55901 Office: Mayo Clinic 200 1st St SW Rochester MN 55905

BRINCH HANSEN, PER, computer scientist, consultant; b. Copenhagen, Nov. 13, 1938; s. Jrgen and Elsebeth (Ring) Brinch H.; m. Milena Marija Hrastar, Mar 27, 1965; children: Mette, Thomas. M.S.E.E., Tech. U. Denmark, 1963, Dr.techn., 1978. Systems programmer Regnecentralen, Copenhagen, 1963-67, head software devel., 1967-70; research assoc. Carnegie-Mellon U., Pitts., 1970-72; assoc. prof. computer sci. Calif. Inst. Tech., Pasadena, 1972-76; prof. U. So. Calif., Los Angeles, 1976—; cons. Henry Salvatori prof. computer sci., 1982—, cons. Author: RC 4000 Computer and Software, 1969, The Programming Languages: Concurrent Pascal, 1975, Edison, 1981, Operating System Principles, 1973, The Architecture of Concurrent Programs, 1977, Programming a Personal Computer, 1983, also articles. Mem. ACM, IEEE, Internat. Working Group Programming Methodology. Subspecialties: Operating systems; Programming languages. Current work: Personal computers, computer networks, distributed processing. Home: 1351 Pleasant Ridge Altadena CA 91001 Office: U So Calif Los Angeles CA 90089

BRINING, DENNIS LEE, administrator, scientist; b. Gary, Ind., Aug. 15, 1946; s. George Lee and Mary May (Popoff) B.; m. Linda L. Shaw, Sept. 7, 1968; children: Eric Lee, Tamara Suzanne. B.S. in Biology, San Diego State U., 1967, M.S., 1969. Assoc. sr. scientist Lockheed Aircraft Service Co., San Diego, 1969-80; sr. scientist, mgr. Lockheed Ocean Sci. Labs., Carlsbad, Calif., 1980—; marine adv. cons. Saddleback Coll., 1980-82, Fullerton Coll., 1980-82. Mem. Marine Tech. Soc. (council 1980-82), Am. Water Works Assn., Internat. Desalination and Environ. Assn. (chmn. environ. effects session 1981). Republican. Roman Catholic. Subspecialties: Resource management; Environmental monitoring and assessment. Current work: Management, new business development, environmental management, assessment. Home: 1634 Juniper Hill Dr Encinitas CA 92024 Office: Lockheed Ocean Sci Labs 6250 Yarrow Dr #A Carlsbad CA 92008

BRINK, GILBERT O., physicist, educator; b. Los Angeles, May 26, 1929; s. Oscar C. and Leoti (Gilbert) B.; m. Lois M. Fredstrom, Mar. 2, 1957; children: Janet L., David R. B.A., Coll. Pacific, Stockton, Calif., 1953; Ph.D., U. Calif., Berkeley, 1957. Asst. research prof. U. Calif., Berkeley, 1957-59; research physicist Lawrence Radiation Lab., Livermore, Calif., 1959-63; vis. asst. research prof. U. Pitts., 1962-63; prin. physicist Cornell Aero. Lab., Buffalo, 1963-68; assoc. prof. physics SUNY-Buffalo, 1968-80, chmn. dept. physics and astronomy, 1972-74, prof. physics, 1980—; vis. McDonnell Disting. prof. Washington U., St. Louis, summer 1981. Contbr. numerous articles to profl. jours. Mem. Am. Phys. Soc., N.Y. Acad. Sci., Sigma Xi. Subspecialties: Atomic and molecular physics; Spectroscopy. Current work: Laser spectroscopy of atoms and molecules, laboratory astrophysics. Home: 139 Segsbury Rd Williamsville NY 14221 Office: Dept Physics and Astronomy SUNY at Buffalo Amherst NY 14260

BRINK, PETER RICHARDS, physiologist, educator; b. Wellsley, Mass., Oct. 24, 1946; s. Raymond and Mary (Richards) B.; m. Nancy Ann Brink, Sept. 17, 1969; 1 dau., Stephanie. B.A., Quinnipiac Coll., 1969; M.S., So. Conn. State Coll., 1971; Ph.D., U. Ill., 1976. Postdoctoral fellow SUNY at Stony Brook, 1975-77, research asst. prof., 1977-80, asst. prof. anat. scis., 1980—. Author: (with Dewey) Diffusions of Substances Inside Cells, 1981; Contbr. articles to profl. jours. NIH research grantee. Mem. Biophys. Soc., Gen. Physiology Soc., Soc. Neurosci. Subspecialties: Biophysics (biology); Neurophysiology. Current work: Mechanism of gap junctional membrane transport. Effects of heavy water, acclimation or junctional membrane resistance. Office: Anatomical Sci HSC SUNY at Stony Brook Stony Brook NY 11794

BRINK, ROYAL ALEXANDER, educator; b. Woodstock, Ont., Can., Sept. 16, 1897; came to U.S., 1920, naturalized, 1933; s. Royal Wilson and Elizabeth Ann (Cuthbert) B.; m. Edith Margaret Whitelaw, Dec. 27, 1922 (dec. May 1962); children—Andrew Whitelaw, Margaret Alexandra; m. Joyce Hickling, Oct. 19, 1963. B.S.A., Ont. Agrl. Coll., 1919; M.S., U. Ill., 1921; D.Sc., Harvard, 1923; postgrad. (NRC fellow), Institut für Vererbungsforschung, Berlin, (NRC fellow), U. Birmingham, 1925-26, (NRC fellow), Calif. Inst. Tech., 1938-39. Chemist Western Can. Flour Mills, Winnipeg, Man., Can., 1919-20; Emerson fellow in biology Harvard, 1921-22; asst. prof. of genetics U. Wis., 1922-27, asso. prof., 1929-31, prof., 1931-68, emeritus prof. genetics, 1968—, chmn. dept., 1939-51. Editor: Heritage from Mendel, 1967; mng. editor: Genetics, 1952-56; Contbr. numerous research papers to biol. jours. Haight Travelling fellow U. Wis., 1960-61; NSF Sr. Postdoctorate fellow, 1966-67. Fellow AAAS; mem. Am. Genetics Assn., Genetics Soc. Am. (pres. 1957), Bot. Soc. Am., Am. Acad. Arts and Scis., Am. Soc. Naturalists (pres. 1963), Nat. Acad. Scis., Wis. Acad. of Scis., Arts and Letters, Sigma Xi (pres. Wis. chpt. 1940-41), Phi Sigma, Phi Eta. Club: University. Subspecialty: Plant genetics. Home: 4237 Manitou Way Madison WI 53711

BRINKHOUS, KENNETH MERLE, pathologist; b. Clayton County, Iowa, May 29, 1908; s. William and Ida (Voss) B.; m. Frances E. Benton, Sept. 5, 1936; children—William Kenneth, John Robert. Student, U.S. Mil. Acad., 1925; A.B., State U. Iowa, 1929, M.D., 1932; D.Sc., U. Chgo., 1967. Asst. in pathology State U. Iowa, 1932-33, instr., 1933-35, asso. in pathology, 1935-37, asst. prof., 1937-45, asso. prof., 1945-46; prof. pathology U. N.C., Chapel Hill, 1946-61, alumni distinguished prof., 1961—; Mem. Nat. Adv. Heart and Lung Council, 1969-74; chmn. med. adv. council Nat. Hemophilia Found., 1954-73; sec. gen. Internat. Com. Hemostasis and Thrombosis, 1966-78. Bd. editors: Perspectives in Biol. Medicine, 1968—; editor: Archives Pathology and Lab. Medicine, 1974—, Yearbook Pathology Clin. Pathology, 1980—. Served from capt. to lt. col. M.C. U.S. Army, 1941-46; col. Med. Res. Corps, 1946—. Co-recipient Ward Burdick award Am. Soc. Clin. Pathologists, 1941; recipient same, 1963, O. Max Gardner award, 1961; N.C. award, 1969; Internat. Heart Research award, 1969; Murray Thelin award Nat. Hemophilia Found., 1972; Distinguished Achievement award Modern Medicine, 1973; H.P. Smith lectr., 1974. Mem. Nat. Acad. Scis. Inst. of Medicine, Am. Acad. Arts and Scis., Assn. Am. Physicians, Internat. Soc. Thrombosis and Haemostasis (pres. 1971), Am. Assn. Pathologists and Bacteriologists (sec., treas. 1968-71, pres. 1973), Am. Soc. Exptl. Pathology (pres. 1965-66), Fedn. Am. Socs. Exptl. Biology (pres. 1966-67), Univs. Asso. Research and Edn. Pathology (pres. 1964-68). Subspecialties: Hematology; Pathology (medicine). Current work: Blood coagulation and platelets; Thrombosis. Home: 524 Dogwood Dr Chapel Hill NC 27514

BRINSON, DONALD EDWARD, data processing executive; b. Ponca City, Okla., Sept. 6, 1953; s. Merwyn Glen and Mildred Colleen (Good) B. B.S., U. Okla., Norman, 1980. Computer programmer Oscar Rose Jr. Coll., Midwest City, Okla., 1974-78, dir. computing services, 1980—; systems programmer Okla. Tax Commn., Oklahoma City, 1978-80. Mem. Assn. for Computing Machinery, Hewlett-Packard Internat. Users Group (program dir. Central Okla. Regional chpt. 1981-82, pres. 1982-83). Democrat. Methodist. Club: Internat. Order Foresters. Subspecialties: Information systems, storage, and retrieval (computer science); Operating systems. Current work: Operating systems, software engineering, application development methodologies. Home: 308 Draper Dr Midwest City OK 73110 Office: Oscar Rose Jr Coll 6420 SE 15th St Midwest City OK 73110

BRINSTER, RALPH LAWRENCE, veterinary medicine educator; b. Montclair, N.J., Mar. 10, 1932. B.S., Rutgers U., 1953; V.M.D., U. Pa., 1960, Ph.D. in Physiology, 1964. Asst. instr. physiology Sch. Medicine, U. Pa., 1960-61, teaching fellow, 1961-64, instr., 1964-65, from asst. prof. to assoc. prof., 1965-70, prof. physiology, 1970—. Mem. AVMA, Am. Soc. Cell Biology, Soc. Study of Reproduction, Brit. Soc. Study of Fertility, Brit. Biochem. Soc. Subspecialty: Reproductive biology. Office: U Pa Sch Vet Medicine Rm 530 Lippincott Bldg Philadelphia PA 19103

BRIOTTA, DANIEL A., JR., astronomer, computer cons.; b. Springfield, Mass., Nov. 24, 1947; s. Daniel A. and Frances T. (Bruno) B. B.S., M.I.T., 1969; M.S., Cornell U., 1973, Ph.D., 1976. Lect., planetarium dir. U. Wyo., 1975-77; research assoc. Cornell U., Ithaca, N.Y., 1977-82; asst. prof. Ithaca (N.Y.) Coll., 1982—. Mem. Am. Astron. Soc., Astron. Soc. Pacific, Sigma Xi. Subspecialties: Infrared optical astronomy; Laboratory microcomputing. Current work: Teaching, laboratory microcomputing, infrared spectroscopy. Office: Dept Physics Ithaca Coll Ithaca NY 14850

BRISKIN, MADELEINE, geology educator; b. Paris, Sept. 4, 1932; U.S., 1951, naturalized, 1956; d. Michael and Mina (Blevinal) B. B.S., CCNY, 1965; M.S., U. Conn., 1967; Ph.D., Brown U., 1973. Asst.prof. geology U. Cin., 1973-79, assoc. prof., 1979—. Mem. Am. Geophys. Union, N.Y. Acad. Scis., Paleontol. Soc., AAAS, Soc. Econ. Paleontologists and Mineralogists, Engring. Soc. Cin., Sigma Xi. Subspecialties: Geology; Oceanography. Current work: Past circulation of ocean and atmosphere, paleoclimatic study (mechanism of climate), new identification of astronomical quasi-periodicity of 413,000 to 430,000 years in deep sea sediments. Office: Dept Geology U Cin Mail Location 13 Cincinnati OH 45221

BRITO, GILBERTO OTTONI, neuroscientist, physician; b. Rio de Janeiro, Brazil, May 24, 1951; s. Ney P. and Herbene I. (Ottoni) B.; children: Alexandre, Bianca. M.D., State U. Rio de Janeiro, 1974; Ph.D., U. Rochester. Postdoctoral fellow Center for Brain Research, U. Rochester Med. Center, 1981; asst. prof. Center for Brain Research U. Rochester Med. Center, 1982—. Contbr. articles to profl. pubs. Co-investigator NIH. Mem. AAAS, Soc. Neurosci., N.Y. Acad. Scis., Assn. Child Psychology and Psychiatry, Sigma Xi. Subspecialties: Neuropsychology; Neuroendocrinology. Current work: Neuropsychology of memory; hormones and behavior. Supervising students' theses and performing own research. Home: 27 Mountain Rd Rochester NY 14625 Office: Center Brain Research U Rochester Med Center Rochester NY 14642

BRITTEN, ROY JOHN, biophysicist; b. Washington, Oct. 1, 1919; s. Rollo H. and Marion (Hale) B.; m. Barbara H. Hagen (div.); children: Gregory H., Kenneth H. B.S., U. Va., 1940; Ph.D., Princeton U., 1951. Mem. staff dept. terrestrial magnetism Carnegie Inst., Washington, 1951-72; Disting. Carnegie sr. research assoc. in biology Calif. Inst. Tech. and CIW joint appointment, 1972—. Mem. Nat. Acad. Scis., AAAS. Subspecialties: Genome organization; Developmental biology. Current work: Study of evolution of the genome and the mechanism of embryonic development. Home: 498 Abbie Way Costa Mesa CA 92627 Office: 101 Dahlia St Corona Del Mar CA 92625

BROADHURST, JOHN HENRY, physics educator; b. Stoke-on-Trent, Eng., Apr. 27, 1935; came to U.S., 1968; s. Leonard B. and Clarice L. (Robinson) B.; m. I. Jaeger, July 25, 1959; children: Nina Louise, Denise Joan. B.Sc., U. Birmingham (Eng.), 1956, Ph.D., 1959. Grad. researcher minerals engring. U. Birmingham, 1956-59, postgrad. in physics, 1959-61, staff fellow physics, 1962-68; assoc. prof. physics U. Minn., Mpls., 1968-78, prof. physics, 1979—; cons. in field. Contbr. articles to profl. jours. Mem. Am. Phys. Soc. Subspecialties: Nuclear physics; Cell and tissue culture. Current work: Research in applied nuclear physics, in physiology of contractile tissue. Office: U Minn 324 Physics Dept 116 Church St SE Minneapolis MN 55455

BROADWELL, RICHARD DOW, neurocytologist, neuropathologist, consultant, researcher, educator; b. Oak Park, Ill., Nov. 4, 1945; s. Robert and Dorothy Jane (Dow) B. B.A., Knox Coll., 1967; M.S., U. Wis., Madison, 1971, D. Phil., 1974. Staff fellow in neurocytology/neuropathology Nat. Inst. Neurol. and Communicable Diseases and Stroke, NIH, Bethesda, Md., 1974-80; asst. prof. pathology, head Lab. Exptl. Neuropathology, U. Md. Sch. Medicine, 1980—; cons. in field. Contbr. numerous articles, chpts. on brain and neurocytology to profl. publs. Recipient Undergrad. Research award NSF, 1966-67; Japanese Soc. for Promotion of Sci. fellow, 1980-81; NIH Nat. Inst. Neurol. and Communicable Diseases and Stroke grantee, 1982-84. Mem. Neurosci. Soc., Am. Soc. Cell Biology, Histochem. Soc., Washington Electron Microscopy Soc. Republican. Presbyterian. Subspecialties: Neurobiology; Cell biology (medicine). Home: 10401 Grosvenor Pl Unit 1010 Rockville MD 20852 Office: U Md Dept Pathology 10 S Pine St Baltimore MD 21201

BROBECK, JOHN RAYMOND, physiology educator; b. Steamboat Springs, Colo., Apr. 12, 1914; s. James Alexander and Ella (Johnson) B.; m. Dorothy Winifred Kellogg, Aug. 24, 1940; children: Stephen James, Priscilla Kimball, Elizabeth Martha, John Thomas. B.S., Wheaton Coll., 1936, LL.D., 1960; M.S., Northwestern U., 1937, Ph.D., 1939; M.D., Yale U., 1943. Instr. physiology Yale, 1943-45, asst. prof., 1945-48, asso. prof. physiology, 1948-52; prof. physiology, chmn. dept. U. Pa., Phila., 1952-70, Herbert C. Rorer prof. med. scis., 1970-82, prof. emeritus, 1982—. Editor: Yale Jour. Biology and Medicine, 1949-52; chmn. editorial bd.: Physiol. Revs, 1963-72. Fellow Am. Acad. Arts and Scis.; mem. Am. Physiol. Soc. (pres. 1971-72), Am. Inst. Nutrition, Nat. Acad. Scis., Am. Soc. Clin. Investigation, Halsted Soc., Phila. Coll. Physicians, Sigma Xi, Alpha Omega Alpha. Subspecialties: Physiology (medicine); Neurophysiology. Current work: Control of food intake, energy balance, temperature regulation. Home: 224 Vassar Ave Swarthmore PA 19081 Office: U Pa G/3 Philadelphia PA 19104

BROBST, DUANE FRANKLIN, pathologist, educator; b. Medicine Lake, Mont., Oct. 8, 1923; s. Herbert Franklin and Inez (Bell) B.; m. Janice Dean, Nov. 26, 1970; children: Clay, Todd, Amy; m.; 1 stepson, Bill Wasson. A.B., U. So. Calif., Los Angeles, 1949; D.V.M., Wash. State U., Pullman, 1954; Ph.D., U. Wis., Madison, 1962. Asst. prof. to prof. Purdue U., West Lafayette, Ind., 1960-70; prof. vet. pathology Wash. State U., 1970—. Author: (with D.J. Blackmore) Biochemical Values in Equine Medicine, 1981; contbr. chpt. in book. Served with USN, 1943-46. Mem. Am. Coll. Vet Pathologists, Am. Soc. Vet. Clin. Pathologists. Republican. Methodist. Subspecialty: Pathology (veterinary medicine). Current work: Acid-base balance in renal disease. Office: McCoy Hall Washington State University Pullman WA 99164

BROCK, THOMAS DALE, educator; b. Cleve., Sept. 10, 1926; s. Thomas Carter and Helen (Ringwald) B.; m. Katherine Middleton, Feb. 20, 1971; children: Emily Katherine, Brian Thomas. B.Sc. cum laude, Ohio State U., 1949, M.Sc., 1950, Ph.D., 1952. Research microbiologist The Upjohn Co., Kalamazoo, 1952-57; asst. prof. Case Western Res. U., Cleve., 1957-60; asst. to prof. Ind. U., Bloomington, 1960-71; E.B. Fred prof. natural scis. U. Wis.-Madison, 1971—; pres. Sci. Tech. Pubs., Madison, 1974—. Author: Milestones in Microbiology, 1961, Thermophilic Microorganism, 1978, Biology of Microorganisms, 1979, Membrane Filtration, 1983. Fellow AAAS, Am. Soc. Limnology and Oceanography (editorial bd. 1982—), Am. Soc. Microbiology (chmn. 1970-71). Club: Town and Gown. Subspecialties: Microbiology; Ecology. Current work: Microbiol ecology, thermophilic microorganisms, aquatic microbiology, computer-assisted publication; computer modeling of natural processes. Home: 1227 Dartmouth Rd Madison WI 53705 Office: U Wis 1550 Linden Dr Madison WI 53706

BROCKETT, ROGER WARE, applied mathematics educator; b. Wadsworth, Ohio, Oct. 22, 1938; s. Roger Lawrence and Grace Ester (Patch) B.; m. Carolann Christina Riske, Aug. 20, 1960; children: Mark, Douglas, Erik. B.S., Case Inst. Tech., 1960, M.S., 1962, Ph.D., 1964; M.S. (hon.), Harvard U., 1969. Ford postdoctoral fellow MIT, 1963-65, asst. prof., 1965-67, assoc. prof., 1967-69; prof. applied math. Harvard U., 1969—; cons. in field. Author: Finite Dimensional Linear Systems, 1970; contbr. articles to profl. jours.; editor: Systems and Control Letters, 1980—. Recipient Donald P. Eckman award Am. Automatic Control Council, 1967; Guggenheim fellow, 1975-76. Fellow IEEE; mem. Am. Math. Soc., Soc. Indsl. and Applied Math., Tau Beta Pi. Subspecialties: Applied mathematics; Robotics. Current work: Control theory, robotics. Home: 29 Oakland St Lexington MA 02173 Office: Div Applied Scis Harvard U G12b Pierce Hall Cambridge MA 02138

BROCKMAN, RONALD PAUL, physiology educator; b. Sask., Can., Jan. 10, 1945; came to U.S., 1980; s. Herman and Lavina (Helmink) B.; m. Lenora M. Kloppenburg, Aug. 19, 1972. B.A., U. Sask., 1966, D.V.M., 1970, M.Sc., 1972; Ph.D., Cornell U., 1975. Profl. Assoc. U. Sask., Saskatoon, 1975-78; assoc. prof., 1978-80, U. Minn., St. Paul, 1980-83; with Animal Health Div., Alta. Dept. Agr., Edmonton, 1983—. Mem. Am. Physiol. Soc., Can. Physiol. Soc., Sask. Vet. Med. Assn. Roman Catholic. Club: Toastmasters. Subspecialties: Animal physiology; Nutrition (biology). Current work: Regulation of carbohydrate metabolism in ruminant animals, particularly the roles of pancreatic hormones.

BROCKMAN, WILLIAM WARNER, virologist; b. Phila., July 8, 1942; s. Frank Gottlieb and Margaret (Elliott) B.; m. Anne Ferris Brockman, June 3, 1967; children: William Ferris, Theodore Elliott. B.S., Cornell U., 1964, M.D., 1968. Intern Balt. City Hosp., 1968-69, resident, 1969-70; fellow dept. medicine Johns Hopkins U., 1968-70, depts. microbiology and medicine, 1970-74; research assoc. Lab. Molecular Biology, NIH, 1974-76; asst. prof. microbiology U. Mich., 1976-80, assoc. prof., 1980—. Contbr. articles in field to profl. jours. Served with USPHS, 1974-76. Mem. AAAS, Am. Soc. Microbiology. Subspecialties: Virology (biology); Molecular biology. Current work: Molecular genetics of animal viruses; virus induced cellular transformation. Office: Mich Med Sch 6606 Med Sci Bldg II Ann Arbor MI 48109

BRODER, SAMUEL, oncologist, research scientist; b. Feb. 24, 1945; m. Gail Lois Steinmetz, Dec. 26, 1966; children: Karen R., Joanna S. M.D. cum laude, U. Mich., 1970. Diplomate: Am. Bd. Internal Medicine (subsplty. Med. Oncology). Intern and resident in internal medicine Stanford U., Palo Alto, Calif., 1970-72; clin. assoc. metabolism br. Nat. Cancer Inst., NIH, Bethesda, Md., 1972-75, investigator medicine br., 1976, sr. investigator metabolism br., 1976-81, dep. clin. dir. Inst., 1981—, assoc. dir. clin. oncology program, 1981—; cons. Am. Cancer Soc. Contbr. numerous articles to sci. and med. jours. Served in USPHS. Fellow ACP; mem. Am. Soc. Clin. Investigation, Am. Assn. Immunologists, Am. Fedn. Clin. Research. Subspecialties: Cancer research (medicine); Immunology (medicine). Current work: Immunology, cancer, infectious diseases. Office: Nat Cancer Inst 6B-15 Bldg 10 Bethesda MD 20205

BRODERICK, JOHN JOSEPH, radio astronomer, educator; b. Locustdale, Pa., Oct. 14, 1940; s. John Joseph and Margaret Mary (Dougherty) B.; m. Sandy Lynne Bussey, Dec. 10, 1979; children: Rosemarie, Jack. B.S., Pa. State U., 1962; M.A., Brandeis U., 1964, Ph.D., 1970. Research assoc. Nat. Radio Astronomy Obs., Charlottesville, Va., 1969-71, Nat. Astronomy and Ionosphere Ctr., Arecibo, P.R., 1971-74; asst. prof. physics Va. Poly. Inst. and State U., Blacksburg, 1974-80, assoc. prof. physics, 1980—. Contbr. articles to profl. jours. NSF grantee, 1977—. Mem. Internat. Astron. Union, Am. Astron. Soc., Va. Acad. Scis., Internat. Sci. Radio Union, Am. Phys. Soc., Sigma Xi. Roman Catholic. Subspecialty: Radio and microwave astronomy. Current work: Research in compact extragalactic radio sources, astronomy education. Home: 2810 Wellesley Ct Blacksburg VA 24060 Office: Physics Dept Va Poly Inst and State U Blacksburg VA 24061

BRODERICK, PATRICIA ANN, research scientist, educator, consultant; b. N.Y.C., Oct. 2, 1939; d. Patrick and Margaret Theresa (Daly) B. B.S. magna cum laude, St. Thomas Aquinas Coll., Sparkill, N.Y., 1963; M.S. in Biology summa cum laude, Fordham U., 1970; Ph.D. in Pharmacology, St. John's U., 1979. Cert. tchr., N.Y. Joined Order Dominican Sisters, 1958; faculty Cathedral High Sch., Manhattan, N.Y., 1963-73; tchr. Monsignor Scanlan High Sch., Bronx, 1973-75; researcher Rockland Psychiat. Research Inst., Orangeburg, N.Y., 1975-76; teaching fellow pharmacology St. John's U., 1976-78, research in neuropsychopharmacology, 1976-79; sr. pharmacologist Revlon Health Care Group, Tuckahoe, N.Y., 1979-81; research fellow Interdeptl. depts. psychiatry and neurosci. Albert Einstein Coll. Medicine, 1981-84, asst. prof. dept. psychiatry, 1984—; adj. asst. prof. anatomy and physiology CUNY, 1972—; adj. asst. prof. St. Thomas Aquinas Coll., 1974-75; adj. prof. drug use and abuse Coll. New Rochelle, N.Y., 1974-75. Contbr. articles to publs. in field. Recipient Tchr. Leadership award NYU, 1981-84; NIMH fellow, 1981-84; NSF grantee, 1967, 71, 72—; N.Y. State grantee, 1973—. Mem. Soc. Neurosci., Assn. Women in Sci., N.Y. Acad. Sci., Pharm. Soc. State N.Y., Inst. Soc. Ethics and the Life Scis., Union Concerned Scientists, Rho Chi. Subspecialties: Neuropharmacology; Neurochemistry. Current work: Study of brain and behavior; in vivo electrochemical tracings of specific neurotransmitters; dopamine interactions with enkephalin and serotonin in discrete brain regions to elucidate mechanisms of mental disease (anxiety, aggressions, schizophrenia and depression); studies of CNS neurotransmitters in retinal tissue. Office: Albert Einstein College Medicine 1300 Morris Park Ave Bronx NY 10461

BRODHAGEN, THOMAS WARREN, materials specialist; b. Traverse City, Mich., Jan. 27, 1942; s. Donald W. and Florence I. (Kratochvil) B.; m. Kathleen Jo Morris, Aug. 20, 1966; children: Katrin, Thomas Warren. A.A., Northwestern Mich. Coll., 1962; B.S. in Biology, Central Mich. U., 1964, M.S., 1966. Tchr. biology Midland (Mich.) pub. schs., 1964-65, Central Mich. U., Mt. Pleasant, 1966-68; biomed. research biologist Dow Corning Corp., Midland, 1968-69, market and clin. devel. analyst, 1969-72, quality and reliability assurance mgr., 1972-79, sr. devel. specialist, research and devel., 1979—. Trustee Mich. Community Blood Ctr. Theodore Roosevelt Found. grantee for biol. studies, 1966. Mem. Am. Assn. Advancement of Med. Instrumentation, Health Industry Mfrs. Assn., Nat. I.V. Therapy Assn. Subspecialties: Biomaterials; Artificial organs. Current work: Development of silicone and silicone containing materials for biomedical devices and drug delivery systems. Office: Dow Corning Corp 12234 Geddes Rd Hemlock MI 48626

BRODIE, HARLOW KEITH HAMMOND, university chancellor; b. Stamford, Conn., Aug. 24, 1939; s. Lawrence Sheldon and Elizabeth White (Hammond) B.; m. Brenda Ann Barrowclough, Jan. 26, 1967; children: Melissa Verduin, Cameron Keith, Tyler Hammond, Bryson Barrowclough. A.B., Princeton U., 1961; M.D., Columbia U., 1965. Diplomate: Am. Bd. Psychiatry and Neurology. Intern Ochsner Found. Hosp., New Orleans, 1965-66; resident in psychiatry Columbia-Presbyn. Med. Center, N.Y.C., 1966-68; clin. assoc. intramural research program NIMH, 1968-70; asst. prof. psychiatry, dir. gen. clin. research center Stanford U. Med. Sch., 1970-74; prof. psychiatry, chmn. dept. Duke U. Med. Sch., 1974-82, prof. law, adj. prof. psychology, 1981—; psychiatrist-in-chief Duke U. Med. Center, 1974-82; chancellor Duke U., 1982—. Co-author: The Importance of Mental Health Services to General Health Care, 1979, Modern Clinical Psychiatry, 1981; co-editor: American Handbook of Psychiatry, vols. 6 and 7, 1975, 81, Controversy in Psychiatry, 1978; asso. editor: Am. Jour. Psychiatry, 1973-81. Chmn. Durham Area Mental Health, Mental Retardation and Substance Abuse Bd., 1981—; Vice pres., trustee Durham Acad., 1979—. Recipient Psychopharmacology research award Am. Psychol. Assn., 1970, Strecker award Inst. of Pa. Hosp., 1980. Mem. Am. Psychiat. Assn. (sec. 1977-81, pres. 1982-83), Am. Coll. Psychiatrists (chmn. publs. com. 1980—), Inst. Medicine, So. Psychiat. Assn., Royal Coll. Psychiatrists, Soc. Biol. Psychiatry (A.E. Bennet research award 1970), Am. Psychosomatic Soc., Soc. Neurosci. Subspecialties: Pharmacology; Neuropsychology. Home: 63 Beverly Dr Durham NC 27707 Office: 215 Allen Bldg Duke U Durham NC 27706

BRODNER, ROBERT ALBERT, neurosurgeion, researcher; b. New Haven, Aug. 15, 1946; s. Albert Abraham and Louise Margaret (DeStefano) B.; m. Stephanie Margaret, Mar. 20, 1970; children: David C., John J., Christina L. B.S., Fordham U., 1968; M.D., Loyola U. of Chgo. Stritch Sch. Medicine, 1972. Diplomate: Nat. Bd. Med. Examiners. Intern in surgery Mt. Sinai Hosp. N.Y., N.Y.C., 1972-73, resident in neurosurgery, 1973-74, 76-78; research fellow in neurosurgery Yale U., 1975-76; sr. instr. in neurosurgery and staff attending neurosurgeon Hahnemann Med. Coll. and Hosp., 1981—. Contbr. articles to profl. jours. Served to lt. comdr. M.C. USN, 1978-80. Recipient First Place prize N.Y. Soc. Neurosurgery; Resident's Research Award, 1976, 78. Mem. Congress Neurosurgery, Soc. Neurosci., Am. Fedn. Clin. Research, Pa. Neurosurg. Soc., Pa. Med. Soc. Roman Catholic. Club: Waynesborough Country (Paoli, Pa.). Subspecialties: Fetal surgery; Neurosurgery. Current work: Microneurosurgery; fetal intracranial surgery; experimental fetal hydrocephalus; spinal cord injury. Inventor fetal ventriculo-amniotic shunt. Office: Dept Neurosurger Hahnemann Med Coll and Hosp Philadelphia PA 19102

BRODSKY, ROBERT FOX, aerospace engineer; b. Phila., May 16, 1925; s. Samuel H. and Sylvia (Fox) B.; m. Patricia Wess, Jan. 24, 1959; children: Bette W., Robert D., David V., Jeffrey M. B.M.E., Cornell U., 1947; M. in Aero. Engring., N.Y.U., 1948, D.Sc. in Engring, 1950; M.S. in Math, U. N. Mex., 1957. Registered profl. engr., Calif., Iowa. Instr. dept. mech. engring. N.Y.U., 1948-50; sup. theoretical aerodynamics Sandia Corp., Albuquerque, 1950-56; chief aerodynamics Convair/Pomona, 1956-59; with Aerojet-Gen. Corp., 1959-71; chief engr. Space-Gen., El Monte, Calif., 1963-67, mgr., Paris, 1969-70; mgr. systems test Aerojet Electrosystems Co., 1970-71; prof., head dept. aerospace engineering Iowa State U., Ames, 1971-80;

research Aircraft Co., 1978-79; dir. tech. planning TRW Space and Tech. Group, Redondo Beach, Calif., 1980—; cons. in field. Served with USN, 1944-46. Recipient Ednl. Achievement award AIAA/Am. Soc. Engring. Edn. Aerospace Div., 1978; NSF/NATO sr. fellow in sci., 1973. Fellow Inst. Advancement Engring.; mem. Am. Astron. Soc., Nat. Soc. Profl. Engrs., AIAA (ednl. activities com. 1972-80, spacecraft systems tech. com. 1978-82, editorial adv. bd. 1977-80), Am. Soc. Engring. Edn. (chmn. aerospace div., chmn. tech. assessment com.), Am. Soc. Aerospace Edn. (v.p. 1979-80, Univ. Educator of yr. 1979), Sigma Xi. Club: Rotary. Subspecialties: Aerospace engineering and technology; Satellite studies. Current work: Design of planetary orbiters and scientific payloads. Inventor space lifeboat. Home: 401 2d St Hermosa Beach CA 90254 Office: R51 1031 TRW One Space Park Redondo Beach CA 90248

BRODY, BURTON ALAN, physicist, researcher, educator; b. N.Y.C., June 8, 1942; s. Jules and Shirley (Nudriv) B.; m. Susan Simon, Aug. 3, 1980. B.A., Columbia U., 1963; Ph.D., U. Mich., Ann Arbor, 1970. Computer programmer, systems analyst, mgr. OLI Systems and SBM, Inc., N.Y.C., 1978-79; prof. physics Bard Coll., Annandale-on-Hudson, 1970—; vis. scholar in physics Columbia U., 1981—. Author: Electronics from the Ground Up; contbr. articles to profl. jours. Active ACLU, Union Concerned Scientists, Fedn. Atomic Scientists. Mem. Am. Phys. Soc. Subspecialties: Atomic and molecular physics; Low temperature physics. Office: Physics Dept Bard College Annandale on Hudson NY 12504

BRODY, MICHAEL J., pharmacology educator; b. N.Y.C., Aug. 16, 1934; m., 1956; 2 children. B.S., Columbia U., 1956; Ph.D., U. Mich., 1961. From instr. to assoc. prof. U. Iowa, Iowa City, 1961-69, prof. pharmacology, 1969—, assoc. dir. Cardiovascular Ctr., 1974—; cons. Upjohn Co.; mem. research rev. com. Nat. Heart Lung Blood Inst., 1979-83. Assoc. editor: Circulation Research, 1981-86. USPHS spl. fellow, U. Lund (Sweden), 1971. Fellow Council on Circulation of Am. Heart Assn.; mem. Am. Soc. Exptl. Biology and Medicine, Am. Fedn. Clin. Research, Am. Soc. Pharmacology and Exptl. Therapeutics, Am. Physiol. Soc. Subspecialty: Pharmacology. Office: Dept Pharmacology U Iowa Iowa City IA 52242

BRODY, STEVEN, physicist, engineer; b. Phila., Apr. 16, 1952; s. Bernard Irving and Eleanor (Albert) B. B.S., Drexel U., 1974; M.S., MIT, 1977. Intelligence analyst/phys. scientist trainee CIA, Washington, 1970-73; study participant space colonies NASA/Ames Research Center, summer 1975; research asst. MIT Earth & Planetary Sci. Dept., 1975-77; avionics systems engr. Intermetrics, Inc., Cambridge, Mass., 1977-80, program mgr. space shuttle test, Huntington Beach, Calif., 1980—. Recipient award for acad. excellence in sci. Drexel U., 1970, others. Mem. AIAA (chmn. sect. tech. com. for guidance, control, and dynamics), Aircraft Owners and Pilots Assn., The L-S Soc., Planetary Soc. Club: Long Beach Rowing Assn. Subspecialties: Aerospace engineering and technology; Software engineering. Current work: Space Shuttle checkout and operation, space sta. devel., space exploration and industrialization, artificial intelligence, civil air traffic control improvements. Office: 5392 Bolsa Ave Huntington Beach CA 92649

BRODY, THEODORE MEYER, educator, pharmacologist; b. Newark, May 10, 1920; s. Samuel and Lena (Hammer) B.; m. Ethel Vivian Drelich, Sept. 7, 1947; children—Steven Lewis, Debra Jane, Laura Kate, Elizabeth. B.S., Rutgers U., 1943; M.S., U. Ill., 1949, Ph.D., 1952. Mem. faculty U. Mich. Med. Sch., Ann Arbor, 1952-66; prof. pharmacology, chmn. dept. Coll. Medicine, Mich. State U., East Lansing, 1966—; cons. NIH, 1969-73, NIDA, 1975—, Internat. Soc. Heart Research, 1973—; mem. sci. adv. com. Pharm. Mfrs. Assn. Found., 1973—; U.S. rep. Internat. Union Pharmacology, 1973-76; mem. bd. Fedn. Am. Socs. for Exptl. Biology, 1973-76; mem. Com. Sci. Soc. Presidents; NSF Distinguished scholar lectr. U. Hawaii, 1974. Mem. editorial bd.: Jour. Pharmacology and Exptl. Therapeutics, 1965—; specific field editor, 1981; editorial bd.: Research Communications in Chem. Pathology and Pharmacology, Molecular Pharmacology, 1972—, Revs. in Pure and Applied Pharmacol. Sci, 1980—. Served with AUS, 1943-46. Mem. Soc. Pharmacology and Exptl. Therapeutics (John Jacob Abel award 1955, chmn. Abel award com. 1966, mem. council 1969-72, sec.-treas. 1970, pres. elect 1973, pres. 1974), Internat. Soc. Biochem. Pharmacology, Am. Coll. Clin. Pharmacology, Assn. Med. Sch. Pharmacologists, Soc. Toxicology, Am. Soc. Pharmacology and Exptl. Therapeutics (pres. 1974-75, awards com. 1977, chmn. 1978), Soc. Neurosci., Japanese Pharmacology Soc., AAUP, Sigma Xi, Rho Chi, Phi Kappa Phi. Subspecialty: Pharmacology. Home: 842 Longfellow Dr East Lansing MI 48823 Office: Dept of Pharmacology and Toxicology Mich State Univ East Lansing MI 48824

BROHN, FREDERICK HERMAN, biochemistry educator; b. Flint, Mich., Mar. 6, 1940; s. William Henry and Ottilia Caroline (Pleger) B.; m. Margaret Sue Standley, Sept. 13, 1969; children: Karl, Philip, Adam, Keith, Margaret, Carolyne. B.S., U. Mich., 1965; Ph.D., Wayne State U., 1972. Instr. Oakland (Mich.) Community Coll., 1970-72; postdoctoral fellow Rockefeller U., N.Y.C., 1972-75, research assoc., 1975-76; asst. prof. biochemistry N.Y.U. Sch. Medicine, N.Y.C., 1976—. NIH grantee, 1980. Mem. Sigma Xi, Phi Lambda Upsilon. Subspecialties: Biochemistry (medicine); Parasitology. Current work: Biochemistry of parasitic protozoa. Home: 1370 Circle Dr W Baldwin NY 11510 Office: Dept Microbiolog NYU Sch Medicin 550 1st Ave New York NY 10016

BROKAW, CHARLES JACOB, educator, cellular biologist; b. Camden, N.J., Sept. 12, 1934; s. Charles Alfred and Doris Evelyn (Moses) B.; m. Darlene Smith, July 29, 1955; children—Bryce, Tanya. B.S., Calif. Inst. Tech., 1955; Ph.D., King's Coll., Cambridge (Eng.) U., 1958. Research asso. Oak Ridge Nat. Lab., 1958-59; asst. prof. zoology U. Minn., 1959-61; mem. faculty Calif. Inst. Tech., 1961—, prof. biology, 1968—, exec. officer div. biology, 1976-80, asso. chmn. div. biology, 1980—. Contbr. articles to profl. jours. Guggenheim fellow, 1970-71. Mem. Biophys. Soc., Am. Soc. Cell Biology, Soc. Exptl. Biology. Subspecialty: Cell biology. Home: 940 Oriole Dr Laguna Beach CA 92651 Office: Biology Div Calif Inst Tech Pasadena CA 91125

BROKER, THOMAS RICHARD, molecular biologist; b. Hackensack, N.J., Oct. 22, 1944; s. Thomas Oerter and Evelyn Anna (Froetscher) B.; m. Louise Tsi Chow, May 26, 1974. B.A. cum laude in Chemistry and Biology (Nat. Merit scholar), Wesleyan U., Middletown, Conn., 1966, M.A. (hon.), 1982; Ph.D. in Biochemistry (Woodrow Wilson fellow, NSF predoctoral fellow), Stanford U., 1972. Helen Hay Whitney postdoctoral fellow in chemistry Calif. Inst. Tech., 1972-75; sr. staff investigator Cold Spring Harbor (N.Y.), Lab., 1975-79, sr. scientist, 1979-84; assoc. prof. biochemistry U. Rochester (N.Y.) Sch. Medicine, 1984—. Contbr. articles sci. jours. Trustee Cold Spring Harbor Fish Hatchery and Aquarium, 1981-83. Nat. Cancer Inst. grantee, 1977—; NSF grantee, 1977, 82. Mem. Am. Soc. Microbiology, Phi Beta Kappa, Sigma Xi. Subspecialties: Microbiology; Virology (biology). Current work: Chromosome organization and expression, RNA Transciption and splicing, human adenoviruses and papilloma viruses, electron microscopy of nucleic acids, recombinant DNA technology. Office: PO Box 607 601 Elmwood Ave Rochester NY 14642

BROMBERG, ROBERT, aerospace co. exec.; b. Phoenix, Aug. 6, 1921; s. Max and Rae (Lipow) B.; m. Hedwig Ella Remak, Aug. 5, 1943; children—Robin Franklin, Janice Kuntz, Kenneth. B.S., U. Calif., Berkeley, 1943, M.S., 1945, Ph.D., 1951. Asso. engr. U. Calif., Berkeley, 1943-46; faculty UCLA, 1946-54; mem. tech. staff Ramo Wooldridge Corp., Los Angeles, 1954-58; asso. dir. (Astrosci. Lab., Space Tech. Labs.), Los Angeles, 1958-59, dir., 1959-61; dir. mechanics div. TRW-STL, Redondo Beach, Calif., 1961-65, v.p., 1962, TRW Systems, Redondo Beach, 1966-72, gen. mgr. power systems div., 1966-68, gen. mgr. sci. and tech. div., 1968-71, gen. mgr. applied tech. div., 1971-72, v.p. research and engring., 1972—. Contbr. articles to tech. jours. Trustee U. Calif., Los Angeles Found.; mem. engring. adv. council U. Calif. Named Engring. Alumnus of Year U. Calif., Los Angeles, 1969. Fellow AAAS, AIAA; mem. Nat. Acad. Engring., Am. Soc. Engring. Edn., Sci. Research Soc. Am., Phi Beta Kappa, Sigma Xi, Tau Beta Pi. Subspecialty: Aerospace engineering and technology. Patentee in field. Home: 1001 Westholme Ave Los Angeles CA 90024

BROMLEY, STEPHEN C., zoology educator; b. Los Angeles, Aug. 31, 1938; m., 1967; 5 children. B.S., Brigham Young U., 1960; M.A., Princeton U., 1962, Ph.D. in Biology, 1965. Research asst. ornithology Los Angeles County Mus., 1957, 59-60; instr. biology Princeton U., 1964-65; asst. prof. zoology U. Vt., 1965-69; research assoc. Mich. State U., East Lansing, 1969-70, assoc. prof. zoology, assoc. prof. biol. sci., 1970-75, prof., 1975—. Mem. AAAS, Am. Sox. Zoologists. Subspecialty: Developmental biology. Office: Mich State U East Lansing MI 48823

BRONDYKE, KENNETH JAMES, metallurgical engineer; b. Sault Sainte Marie, Mich., June 16, 1922; m. Lois Rose Wiese, Oct. 19, 1946; children: Karen Ann Brondyke McLemore, Donald Scott. B.S. in Metall. Engring., U. Mich., 1948. With Alcoa Co., Alcoa Center, Pa., 1948-84, asst. dir. research, 1970-73, asst. dir. metal processing, 1973-74, asso. dir., prodn., 1974-78; dir. Alcoa Labs., 1978-84; mem. adv. bd. Materials Processing Center, MIT; mem. adv. council dept. materials sci. and engring. U. Pa. Bd. dirs. Citizens Gen. Hosp., New Kensington, Pa. Served with U.S. Army, 1944-46. Am. Soc. Metals fellow, 1971. Mem. AIME, Am. Inst. Mining, Metall. and Petroleum Engrs., Indsl. Research Inst. (rep.), Sci. Research Soc. Am., Dirs. Indsl. Research, Sigma Xi. Clubs: Duquesne, Oakmont Country. Subspecialties: Materials processing; Metallurgy. Home: 727 14th St Oakmont PA 15139 Office: Alcoa Labs Alcoa Tech Center Alcoa Center PA 15069

BRONK, BURT V., physics and microbiology educator, researcher; b. Irwin, Pa., Apr. 28, 1935; s. Charles Zadok and Rose (Gusky) B.; m. Gerri Maura Ash, Nov., 1958; children: Robin Lee, Nina Deborah, Benjamin Ron. B.S. in Engring, Pa. State U., 1955; Ph.D. in Physics, Princeton U., 1965. Research assoc. physics SUNY, 1965-66; asst. prof. physics Queens Coll., 1966-68; research assoc. Brookhaven Nat. Lab., 1968-71, U. Tex.-Dallas, 1981-82; prof. physics and microbiology Clemson U., 1971—. Editor, coordinator: Jour. Comments on Molecular and Cellular Biophysics, 1981—; contbr. chpts. to books and articles to profl. jours. Vice pres. Beth Israel Synagogue, Greenville, S.C., 1978-79. Served to lt. j.g. USN, 1955-58. Mem. Phys. Soc., Biophys. Soc. Subspecialties: Biophysics (physics); Biophysics (biology). Current work: Quantitative aspects of DNA repair, biophysics, hyperthermia, cell kinetics. Office: Dept Physics and Microbiology Clemson U Clemson SC 29631

BRONSON, DAVID LEE, cell biologist/virologist, educator; b. Holland, Mich., Oct. 29, 1936; s. John and Dorothy (Tilden) B.; m. Judith Gunn, Apr. 23, 1970. B.A., Hope Coll., 1963; M.S., Iowa State U., 1966, Ph.D., 1969. Postdoctoral fellow dept. virology and epidemiology Baylor Coll. Medicine, 1969-71; research assoc. dept. microbiology U. Miami-Coral Gables, Fla., 1971-72; with dept. urologic surgery U. Minn., Mpls., 1972-83, asst. prof. urologic surgery, 1974-83; assoc. scientist Southwest Found. for Edn. and Research, San Antonio, 1984—. Contbr. articles to sci. jours. Served with USAF, 1954-57. Mem. AAAS. Subspecialties: Cancer research (medicine); Cell and tissue culture. Current work: Control of cell differentiation in human embryonal carcinoma cells. Office: Southwest Found PO Box 28147 San Antonio TX 78284

BRONSON, FRANKLIN HERBERT, zoology educator; b. Pawnee City, Nebr.; m., 1953; 2 children. B.S., Kans. State U., 1957, M.S., 1958; Ph.D. in Zoology, Pa. State U., 1961. Staff scientist Jackson Lab., 1961-68; assoc. prof. U. Tex.-Austin, 1968-72, prof. zoology, 1972—. Mem. AAAS, Animal Behavior Soc., Soc. Study of Reproduction, Am. Soc. Zoologists, Ecol. Soc. Am. Subspecialty: Reproductive biology. Office: U Tex Austin TX 78712

BRONZAFT, ARLINE LILLIAN, psychology educator; b. N.Y.C., Mar. 26, 1936; d. Morris and Ida (Plant) Cohen; m. Bertram Bronzaft, Oct. 7, 1956; children: Robin, Susan. B.A., Hunter Coll., 1956; M.A., Columbia U., 1958, Ph.D., 1966. Instr. Hunter Coll., N.Y.C., 1958-65, Finch Coll., 1965-67; prof. psychology Lehman Coll., Bronx, 1967—; cons. N.Y.C. Transit Authority, 1977—; mem. Transit Services Characteristics Commn., Transp. Research Bd., Nat. Acad. Scis., 1977—. Contbr. articles to profl. jours. Chmn., mem. Mayor's Subway Service Watchdog Com., N.Y.C., 1970-74; mem. Gov. Carey's Energy Task Force, 1975, Commr.'s Adv. Commn. on Transp., N.Y. State, 1975-79; chmn. Adv. Com. on Edn. in 13 C.D., Bklyn., 1970-73. Recipient Regional Cert. of Appreciation, Region 2 U.S. EPA, 1976; Outstanding Woman of Bklyn. Bklyn chpt. NOW, 1974; City U. Faculty Research awards, 1971, 77; Hunter Coll. research fellow, 1956-58. Mem. Am. Psychol. Assn., Phi Beta Kappa, Sigma Xi. Subspecialties: Environmental psychology; Social psychology. Current work: Noise, human factors in transportation, informational aids in transportation. Home: 505 E 79th St New York NY 10021 Office: Herbert H Lehman Coll Bedford Park Blvd Bronx NY 10468

BROOK, MARX, physics educator, academic administrator, researcher; b. N.Y.C., July 12, 1920; s. Abraham and Esther B.; m. Dorothy; children: Janet, Jimmy, Georgia. B.S., U. N. Mex., 1944; Ph.D., UCLA, 1953. Asst. U. N. Mex., 1943-46; research physicist UCLA, 1947-53, N. Mex. Inst. Mining and Tech., 1954-58, assoc. prof. physics, 1958-60, prof., 1960—, chmn. dept. physics, 1960-68, dir. research and devel. div., 1978—; trustee N. Mex. Tech. Research Found.; mem. N. Mex. Gov.'s Com. on Tech. Excellence; mem. steering com. Sen Dominicis Rio Grande Communications Network. Contbr. numerous articles to profl. publs. Recipient Vis. Scientist award Japan Soc. Promotion Sci., Japan, 1976, Disting. Scientist award N. Mex. Acad. Sci., 1981. Fellow Am. Physics Soc., Am. Meteorol. Soc., AAAS; mem. Am. Assn. Physics Tchrs., Am. Geophys. Union, Royal Meteorol. Soc., Sigma Xi. Subspecialties: Remote sensing (atmospheric science); Gas cleaning systems. Current work: Scrubbers; electrostatic dischargers for helicopters; weather radar; air quality; cloud physics; lightning. Patentee fast-scanning meteorol. radar, means and method for removing airborne particulates from aerosol stream. Home: 1216 North Dr Socorro NM 87801 Office: Research and Devel Div Campus Sta N Mex Inst Mining and Tech Socorro NM 87801

BROOKE, MICHAEL HOWARD, neurologist; b. Leeds, Eng., Mar. 4, 1938; came to U.S., 1959, naturalized, 1966; s. Vincent Howard and Margaret (Craven) B.; (div.)children: Jennifer, Brenda, Mark. B.A., Cambridge U., 1955; M.B., B.Ch., Guy's Hosp., London, 1955. Intern San Francisco Children's Hosp., 1959-60; resident in neurology U. Calif. Med. Center, San Francisco, 1960-64; clin. and research fellow NIH, 1964-68; asso. prof. neurology U. Colo. Med. Center, 1968-75; prof. neurology Washington U. Med. Sch., St. Louis, 1975—, prof. preventive medicine, 1979—; dir. Jerry Lewis Neuromuscular Research Center, 1975—; med. dir. I.W.J. Rehab. Inst., 1979—; rehabilitationist-in-chief Barnes Hosp., St. Louis, 1979—. Author: A Clinician's View of Neuromuscular Diseases, 1977; co-author: Muscle Biopsy: A Modern Approach, 1973; editorial bd.: Muscle and Nerve. Served with USPHS, 1966-68. Fellow Am. Acad. Neurology; mem. Am. Neurol. Assn., Amateur Radio Club. Subspecialty: Neurology. Current work: Basic biochemical mechanisms in normal and diseased muscle; treatment of muscular diseases by exercise; medication and orthoses. Office: Dept Neurology Washington Univ Med Sch St Louis MO 63110 Sometimes, as a physician wrestling with health care, one seems to be more part of the problem than of the solution. Until we develop effective systems to maintain the health of the community with the same enthusiasm that we develop expensive technological machinery, we will never cope with disease in this country. I don't know what the answer is, but I do know that waving "good-bye" to the patient from the doorstep of a multi million dollar hospital is not part of it.

BROOKER, ALAN EDWARD, clinical neuropsychologist; b. Madison, Wis., Jan. 26, 1949; s. Russell Alan and Margaret Theresa (Gorman) B.; m. Mary Elizabeth Naglee, Apr. 15, 1972; children: Jeffrey Alan, Jarrod Russell. A.A., Yuba Coll., 1971; B.A., Chapman Coll., 1971; M.S., Calif. State U., 1975; Ph.D., Kans. State U., 1977. Diplomate: in profl. neuropsychology Am. Bd. Profl. Neuropsychology, in profl. psychotherapy Internat. Acad. Profl. Counseling and Psychotherapy. Commd. officer U.S. Air Force, 1968, advanced thorugh grades to maj.; social actions officer, Malmstrom AFB, Mont., 1972-73; rehab. counselor State Dept. Rehab., Auburn, Calif., 1973-77; clin. psychologist, Wiebaden, W.Ger., 1977-81, clin. neuropsychologist, Travis AFB, Calif., 1982—; fellow in neuropsychology U. Oreg. Med. Sch., Portland, 1981-82; vocat. expert Bur. Hearings and Appeals, Sacramento, 1975-77; neuropsychology cons. State Dept. Rehab., 1981—. Vice pres. Calif. Human Services Orgn., Auburn, 1974-75, Voluntary Action Ctr., South Lake Tahoe, Calif., 1976-77; rep. Air Force Soc. of Clin. Psychologists, Wiesbaden, 1978-81. Recipient Proclamation of Appreciation City of South Lake Tahoe, 1977; Calif. Human Services award, 1977; Mil. Psychology award Am. Psychol. Assn., 1981; Cert. of Appreciation U.S. State Dept., 1981. Mem. Am. Psychol. Assn., Internat. Neuropsychol. Assn., Air Force Soc. Clin. Psychologists (rep. 1978-81). Democrat. Roman Catholic. Subspecialties: Behavioral psychology; Neuropsychology. Current work: Neuropathology; autism; memory; behavioral treatment for psychiatric disorders; teaching in psychology. Home: 115 Lamb St Travis AFB CA 94535 Office: David Grant USAF Med Ctr Bldg 121 Travis AFB CA 4535

BROOKER, GARY, biochemistry educator; b. San Diego, Mar. 24, 1942; m., 1964; 2 children. B.S., U. So. Calif., 1966, Ph.D. in Cardiac Pharmacology, 1968. Asst. prof. medicine Sch. Medicine U. So. Calif., Los Angeles, 1968-72, asst. prof. biochemistry, 1969-72; from assoc. prof. to prof. pharmacology Med. Sch. U. Va., Charlottesville, 1972-82; prof. and chmn. dept. biochemistry Georgetown U. Med. Ctr., Washington, 1982—. Grantee Los Angeles County Heart Assn., 1968-70, Am. Heart Assn., 1970-73, USPHS, 1971-82. Mem. Am. Soc. Biol. Chemists, AAAS, Am. Fedn. Clin. Research, Am. Soc. Pharmacology and Exptl. Therapeutics. Subspecialties: Biochemistry (biology); Pharmacology. Office: Georgetown U Med Ctr Washington DC 20007

BROOKS, CHANDLER MCCUSKEY, educator; b. Waverly, W.Va., Dec. 18, 1905; s. Earle Amos and Mary (McCuskey) B.; m. Nelle Irene Graham, June 25, 1932. A.B., Oberlin Coll., 1928; M.A., Princeton U., 1929, Ph.D., 1931; D.Sc. (hon.), Berea Coll., 1970. NRC fellow, teaching fellow Harvard Med. Sch., 1931-33; instr., then asso. prof. physiology Johns Hopkins Med. Sch., 1933-48; prof. physiology and pharmacology, chmn. dept. L.I. Coll. Medicine, 1948-50; prof. physiology, chmn. dept. State U. N.Y. Downstate Med. Center, Bklyn., 1950-72, dir. grad. edn., 1956-66, dean, 1966-72, , acting pres., 1969-71, distinguished prof., 1971—; vis. prof. Tokyo and Kobe (Japan) med. schs., 1961-62, U. Otago, Dunedin, New Zealand, 1975; vis. scholar U. Aberdeen, Scotland, 1973-74; hon. mem. faculty Catholic U., Santiago, Chile.; Mem. study sects. NIH, 1949-69. Author: (with others) Excitability of the Heart, 1955, Humors, Hormones and Neurosecretions, 1962, (with Kiyomi Koizumi) Japanese Physiology, Past and Present, 1965; Editor: (with P.F. Cranefield) The Historical Development of Physiological Thought, 1959, (with others) Cerebrosphinal Fluid and the Regulation of Ventilation, 1965, The Changing World and Man, 1970, (with H.H. Liu) The Sinoatrial Pacemaker of the Heart, 1972, (with K.K. Koizumi) Integrations of Autonomic Reactions, 1972, Jour. of Autonomic Nervous System, 1978—. Trustee Internat. Found., 1972—, chmn. grants com., 1973—. Decorated Order of Rising Sun 3d class, Japan; cited Internat. Physiol. Congress, 1965; Guggenheim fellow, 1946-48; Rockefeller fellow, 1950; China Med. Bd. N.Y. fellow, 1961-62. Mem. Nat. Acad. Scis., Harvey Soc. (pres. 1965), AAAS (council 1950—), N.Y. Heart Assn. (council 1965—), Am. Soc. Pharmacology and Exptl. Therapeutics, Internat. Brain Research Orgn., Nat. Soc. Med. Research, Royal Soc. Medicine, N.Y. Acad. Scis., Soc. Exptl. Biology and Medicine, Soc. Study Internal Secretions, Soc. Study Nervous and Mental Diseases, Am. Coll. Cardiology, Am. Coll. Pharmacology and Chemotherapy, Am. Inst. Biol. Scis., AMA (spl. affiliate), Am. Physiology Soc., Phi Beta Kappa, Sigma Xi, Alpha Omega Alpha; hon. mem. Nat. Acad. Medicine Buenos Aires, Cardiology Soc. Argentina, biol. socs. Montevideo, Uruguay, Inst. Hist. Medicine and Med. Research New Delhi, Alumni Assn., Coll. Medicine Downstate Med. Center. Subspecialties: Physiology (medicine); Neurophysiology. Current work: Nutritional, medical, educational philanthropy, functions of the autonomic nervous system. Research, publns. central control autonomic system, function of hypothalamaus, motor cortex function in regulation of posture, activity in heart and nerve cells. Home: 623 2d St Brooklyn NY 11215 If there is any one secret to success in learning and teaching, I suggest it is to become interested.

BROOKS, COLIN G., immunologist; b. Bury, Lancashire, Eng., Nov. 15, 1949; s. Horace and Vera B.; m. Nada Kasagic, Dec. 20, 1973; children: Alenka, Anica. B.A., Cambridge (Eng.) U., 1971, M.A., 1976; Ph.D., London U., 1974. Postdoctoral fellow Nottingham (Eng.) U., 1974-80; assoc. Fred Hutchinson Cancer Ctr., Seattle, Wash., 1980—. Mem. Brit. Soc. Immunology, Brit. Transplantation Soc. (founding), Am. Assn. Immunologists. Subspecialty: Immunobiology and immunology. Current work: cellular immunology, tumor immunology and biology; cytotoxic cells; natural immunity; natural killer cells; lymphokines. Office: Program Basic Immunology Fred Hutchinson Cancer Research Center 1124 Columbia St Seattle WA 98104

BROOKS, FREDERICK PHILLIPS, JR., computer science educator; b. Durham, N.C., Apr. 19, 1931; s. Frederick Philips and Octavia Hooker (Broome) B.; m. Nancy Lee Greenwood, June 16, 1956; children: Kenneth Phillips, Roger Greenwood, Barbara Suzanne. B.A., Duke U., 1953; S.M., Harvard U., 1955, Ph.D., 1956. Engr. IBM Corp., Poughkeepsie, N.Y., 1956-59, Yorktown Heights, N.Y., 1959-

60, corp. processor devel. system/360 computer, Poughkeepsie, 1960-64, mgr. devel., 1964-65; prof. U. N.C. at Chapel Hill, 1964-75, Kenan prof., 1975—, chmn. dept. computer sci., 1964—; bd. dirs. Triangle Univ. Computation Center, 1966-83, chmn., 1975-77; bd. dirs. N.C. Ednl. Computing Service, 1965—. Author: The Mythical Man-Month-Essays on Software Engineering, 1975, (with K.E. Iverson) Automatic Data Processing, 1963, Automatic Data Processing, System/360 Edition, 1969; Contbr. articles to profl. jours. Chmn. com. Central Carolina Billy Graham Crusade, 1972-73; trustee Durham Acad., pres., 1977-80; mem. corp. Inter-Varsity Christian Fellowship, 1968-77. Recipient McDowell award IEEE Computer Soc., 1970, Computer Pioneer award IEEE Computer Soc., 1982, Man of Year award Data Processing Mgmt. Assn., 1970; NSF grantee; AEC grantee; NIH grantee; Guggenheim fellow, 1975. Fellow IEEE, Am. Acad. Arts and Scis.; mem. Assn. Computing Machinery (council mem.-at-large 1966-70), Nat. Acad. Engring., NRC (computer sci. and tech. bd. 1977-80). Methodist. Subspecialties: Graphics, image processing, and pattern recognition; Computer architecture. Current work: Man-machine interfaces; computer graphics for interactive moleculear modeling. Inventor (with D.W. Sweeney) Program Interruption System, Alphabetical Read-Out Device. Home: 413 Granville Rd Chapel Hill NC 27514

BROOKS, HAROLD LLOYD, investigative cardiologist; b. Durban, South Africa, July 26, 1932; s. John and Erna (Holt) B.; m. Carolyn M. Brooks, Aug. 26, 1932; children: Mark D., Pamela J. B.A., Oklahoma City U., 1955; M.D., Okla. U., 1963. Med. resident U. Tex. Southwestern Med. Sch., Dallas, 1964-67; research fellow in cardiology Harvard U. Med. Sch., Boston, 1967-69; instr. in medicine, 1969-71; dir. cardiac catheterization and Hecht Research Labs., U. Chgo., 1971-76; Northwestern Mut. prof. cardiology, prof. medicine and pharmacology, dir. cardiology, dir. Heath Exptl. Labs., Med. Coll. Wis., Milw., 1977—; chief of staff Milw. County Med. Complex, 1982—; attending County Gen. Hosp., Wood VA Hosp., Froedtert Meml. Luth. Hosp. Editor: (with J. S. Soin) Nuclear Cardiology for Clinicians, 1980, (with L. I. Bonchek) Management of Patients with Operable Heart Disease, 1981; contbr. numerous articles to profl. jours.; co-dir.: film Nuclear Imaging of the Heart, 1980. Raymund J. Thompson scholar, 1959-60; NIH grantee, 1967, 71-76, 83—; recipient Research award Mead Johnson Labs., 1967-69; Louis Block Research award U. Chgo., 1971-76; numerous others. Fellow Am. Coll. Cardiology, Am. Heart Assn. Circulation Council, Am. Coll. Geriatrics, Am. Coll. Clin. Pharmacology; mem. Am. Fedn. Clin. Research, AAAS, Am. Soc. Pharmacology and Exptl. Therapeutics, Am. Physiol. Soc., Central Soc. Clin. Research, Am. Coll. Sports Medicine, Nat. Assn. VA Physicians, Wis. Heart Assn. (bd. govs. 1980-83), Smithsonian Instn., Milw. C. of C. Club: University (Milw.). Subspecialties: Cardiology; Pharmacology. Current work: Investigative cardiology involving experimental myocardial infarction in experimental animal; right ventricular response to stress; response of normal and ischemic myocardium to experimental pharmacologic intervention, including beta adrenergic blockers, calcium antagonists. Office: 8700 W Wisconsin Ave Milwaukee WI 53226

BROOKS, JAMES ELWOOD, educator; b. Salem, Ind., May 31, 1925; s. Elwood Edwin and Helen Mary (May) B.; m. Eleanore June Nystrom, June 18, 1949; children: Nancy, Kathryn, Carolyn. A.B., DePauw U., 1948; M.S., Northwestern U., 1950; Ph.D., U. Wash., 1954. Research assoc. Ill. Geol. Survey, 1950; geologist Gulf Oil Corp., Salt Lake City, summers 1951-53; instr. geol. scis. So. Meth. U., Dallas, 1952-55, asst. prof., 1955-59, assoc. prof., 1959-62, prof., 1962—, chmn. dept., 1961-70, dean, assoc. provost univ., 1970-72, provost, v.p., 1972-80, pres., 1980-81, Inst. for Study Earth and Man, Dallas, 1981—; cons. geologist firm DeGolyer & MacNaughton, Dallas, 1954-59. Contbr. articles to profl. jours. Trustee Inst. Study Earth and Man, Dallas; bd. dirs. Hockaday Sch., Dallas; exec. bd. Circle Ten Council Boy Scouts Am.; mem. So. Meth. U. Bd. Publs.; bd. vistors DePauw U., 1979-83; bd. dirs. Rangeira Corp. Served with USNR, 1943-46. Fellow Geol. Soc. Am., AAAS, Tex. Acad. Sci.; mem. Am. Assn. Petroleum Geologists, Dallas Geol. Soc., Sigma Xi, Sigma Gamma Epsilon, Sigma Phi. Subspecialties: Geology; Sedimentology. Current work: Geomorphology and statrigraphy of the Qattara Depression, N.W. Egypt leading to and understanding of the origin of the Depression. Home: 7055 Arboreal Dr Dallas TX 75231 Office: Inst Study Earth and Man Box 274 Dallas TX 75275

BROOKS, NORMAN HERRICK, environmental and civil enginner; b. Worcester, Mass., July 2, 1928; s. Charles Franklin and Eleanor Merritt (Stabler) B.; m. Frederika Nelson, Dec. 22, 1948; children: Diana, Alexander, Laura. A.B. magna cum laude, Harvard U., 1949, M.S. in Civil Engring, 1950; Ph.D. summa cum laude in Civil Engring. and Physics, Calif. Inst. Tech., 1954. With Calif. Inst. Tech., Pasadena, 1953—, prof., 1962—, James Irvine prof., 1976—, dir. environ. quality lab., 1974-84; vis. asso. prof. SEATO Grad. Sch. Engring., Bangkok, 1959-60; vis. prof. M.I.T., 1962-63; vis. environ. scientist Scripps Instn. Oceanography, fall 1971; cons. in hydraulics and ocean pollution control; mem. assembly sci. and tech. advisory com. Calif. State Legislature, 1970-73; mem. environ. studies bd. Nat. Acad. Scis., 1973-76. Co-author: Mixing in Inland and Coastal Waters, 1979; contbr. articles to profl. jours. Chmn. Altadena-Pasadena Human Relations Com.; mem. Altadena Planning Adv. Com. Fellow ASCE (chmn. com. on hydrologic transport processes 1975-76, Huber research prize 1959, Collingwood prize 1959, J.C. Stevens award 1959, Rudolph Hering medal 1957, 62, Hilgard hydraulics prize 1970); mem. Nat. Acad. Engring., Nat. Acad. Scis., Am. Geophys. Union, Internat. Assn. Hydraulic Research, Univ. Council on Water Resources, AAAS, Water Pollution Control Fedn., Sigma Xi. Subspecialties: Water supply and wastewater treatment; Fluid mechanics. Current work: Turbulant diffusion of sewage effluents and sewage sludge in the ocean; design of outfall pipes for ocean discharge; transport of sediments by rivers and oceans; sedimentation. Home: 2521 N Santa Anita Ave Altadena CA 91001 Office: Keck Labs Calif Inst Tech Pasadena CA 91125

BROOKS, PHILIP RUSSELL, chemistry educator; b. Chgo., Dec. 31, 1938; s. John Russell and Louise Jane (Seyler) B.; children: Scott, Robin, Christopher, Steven. B.S., Calif. Inst. Tech., 1960; Ph.D., U. Calif., Berkeley, 1964. Research assoc. U. Chgo., 1964; asst. prof. Rice U., 1964-70, assoc. prof., 1970-75, prof., 1975—. Contbr. articles to profl. jours. Active Boys Scouts Am. NSF gellow, 1960-63, 64; Alfred P. Sloan Found. fellow, 1970-74; Guggenheim fellow, 1975-76; Robert A. Welch Found. grantee, 1966—; USAF grantee, 1975-79. Mem. AAAS, Am. Chem. Soc., Am. Phys. Soc. Subspecialties: Physical chemistry; Laser-induced chemistry. Current work: Reactions of state selected molecules. Office: Dept Chemistry Rice U Houston TX 77251

BROOME, DOUGLAS RALPH, JR., space systems program executive; b. Columbia, S.C., Nov. 30, 1936; s. Douglas Ralph and Ollie Bill (Odem) B.; m. Shirley Ruth Blizzard, Aug. 25, 1939; m. Patricia Louise Broome, Nov. 26, 1942; children: Douglas Ralph, Morris L., David G. Schulman, Robert M. Schulman. B.S. in Elec. Engring, The Citadel, Charleston, S.C., 1959; postgrad., Tex. A&M U., College Station, 1964-67, George Mason U., Fairfax, Va., 1981—. With NASA, 1959-70, 74—, dep. div. chief, now mgr., Washington; with Raytheon Co., Sudbury, Mass., 1970-72. Bd. dirs. Alcohol Rehab. Inc., Arlington, Va. Served to 2d lt. U.S. Army, 1959-60. Recipient Apollo Program Mgmt. Team award NASA, 1969, Apollo Achievement award, 1969, Medal for Outstanding Leadership, 1980; Group award of the Presdl. Medal of Freedom Pres. of U.S., 1970. Mem. IEEE. Subspecialties: Aerospace engineering and technology; Information systems, storage, and retrieval (computer science). Current work: Manager of implementation of an advanced, satellite-based, end to end science data acquisition, processing and distribution system. Home: 9312 Arlington Blvd Fairfax VA 22031 Office: 600 Independence Ave SW Washington DC 20546

BROSNIHAN, K. BRIDGET, physiologist, researcher; b. Omaha, June 3, 1941; d. Thomas Timothy and Anna Marie (Evon) B.; m. Tony William Simmons, May 30, 1975; children: Joshua, Jonathan. B.S., Coll. St. Mary, Omaha, 1965; M.S., Creighton U., 1970; Ph.D., Case Western Res. U., 1974. Instr. biology Coll. St. Mary, 1967-68; instr. sci. St. Albert High Sch., Council Blufs, Iowa, 1969-70; postdoctoral fellow Cleve. Clinic, 1974-76, asso. staff, then staff, 1976—; adj. asst. prof. dept. physiology Case Western Res. U., Cleve., 1981—. Contbr. articles to profl. jours., chpts. to books. Heart Assn. N.E. Ohio grantee, 1974-79; NIH grantee, 1978—. Mem. Am. Heart Assn., Am. Physiol. Soc., Soc. Neurosci. Endocrine Soc., Sigma Xi, Am. Women in Sci. Democrat. Roman Catholic. Subspecialties: Neuroendocrinology; Physiology (medicine). Current work: Neurohormones in hypertension. Office: 9500 Euclid Ave Cleveland OH 44106

BROSTOW, WITOLD KONRAD, materials scientist, educator; b. Warsaw, Poland, Mar. 21, 1934; came to U.S., 1979; s. Ludomir Brzostowski and Janina (Dorozinska) Bończa- Brzostowski; m. Anna Zujko, Dec. 15, 1973; children: Gabriel Julian, Diana Paulina; m. Maria Jolanta Garbowska, Aug. 15, 1960 (div. 1973); 1 son, Adam Adrian. M.S., U. Warsaw, 1955, Dr.Sc., 1960; D.Sc., Polish Acad. Scis., 1965. Head div. phys. chemistry Inst. Synthetics, Warsaw, Poland, 1968-69; U.S. Nat. Acad. Sci. vis. scholar Stanford (Calif.) U., 1969-70; NRC vis. scholar U. Ottawa, Ont., Can., 1970-71; attache de recherche U. Montreal, 1971-76; prof. materials sci. and chemistry World Open U., Orange, Calif., 1975—; prof. phys. chemistry Ctr. for Advanced Studies, Mexico City, 1976-79; assoc. prof. materials engring. Drexel U., Phila., 1980—; mem. sci. council Internat. Soc. Bioelectricity, 1981—; cons. Interactive Learning, Inc., Ottawa, Ont., Can., 1981—, Tecnicol, Vienna, Austria, 1981—; vis. prof. Ohio State U., 1979-80, Royal Inst. Tech., Stockholm, 1982-83, Tech. U. Gdansk, 1983. Author: Science of Materials, 1979, Introduccion a la ciencia de los materiales, 1981; editorial bd.: Materials Chemistry and Physics, 1981—, Jour. Bioelectricity, 1982—, Jour. Materials Edn., 1983—; contbr. articles to profl. jours. Brit. Council, U. Reading vis. scholar, 1961-62; German Chem. Soc., U. Halle vis. lectr., 1965; French Minister Fgn. Affairs vis. scholar U Montpellier, 1966; USSR Acad. Sci. vis. lectr., 1967. Fellow Royal Soc. Chemistry; mem. Nat. Acad. Scis. Mexico City, French-Canadian Assn. Advancement Sci., Am. Chem. Soc., Materials Research Soc., Am. Phys. Soc., Am. Soc. Engring. Edn., Soc. Plastics Engrs. Roman Catholic. Club: Keystone Automobile (Phila.). Subspecialties: Polymers; Statistical physics. Current work: Prediction of thermophysical and mechanical properties of polymeric and liquid phases; methods used: statistical physics, computer simulations, theory of information, theory of graphs, general systems and experimental. Home: 217 Valley Rd Media PA 19063 Office: Dept Materials Engring Drexel U Philadelphia PA 19104

BROSTROM, MARGARET ANN, pharmacologist, educator, researcher; b. Chgo., Aug. 13, 1941; d. Ferdinand William and Marion Edmunda (Kane) O'Brien; m. Charles Otto Brostrom, Dec. 18, 1965; 1 son, Arthur Charles. B.A. in Chemistry, Clarke Coll., Dubuque, Iowa, 1963; Ph.D. in Biochemistry, U. Ill., Chgo., 1968. Postdoctoral fellow in biochemistry U. Calif., Davis, 1968-71; teaching asst. in pharmacology U. Medicine and Dentistry N.J.-Rutgers Med. Sch., Piscataway, N.J., 1971-72, instr., 1972-74, asst. prof., 1974-79, assoc. prof., 1979—. Contbr. articles to biol. jours. NIH grantee, 1980—. Mem. Am. Soc. for Pharmacology and Exptl. Therapeutics. Subspecialties: Molecular pharmacology; Cell and tissue culture. Current work: Calcium and cyclic nucleotide metabolism in excitable tissues research; teaching of medical and graduate students; committees associated with medical school functions. Home: 133 Summers Ave Piscataway NJ 08854 Office: Dept Pharmacology PO Box 101 U Medicine and Dentistry NJ-Rutgers Med Sch Piscataway NJ 08854

BROTZMAN, HAROLD GEORGE, biologist, educator; b. Harpursville, N.Y., Sept. 18, 1941; s. Lee Willard and Dorothy Mae (Wentzel) B.; m. Nancy Lee Chase, May 30, 1970; children: Kristine, Keri. B.S., SUNY-Syracuse, 1964; M.S., U. Maine, 1967; Ph.D., U. Iowa, 1972. Postdoctoral fellow U. Mo., Columbia, 1972-76; asst. prof. biology North Adams (Mass.) State Coll., 1976-81, assoc. prof., 1981—. Author: (with Merton F. Brown) Phytopathogenic Fungi: A Scanning Electron Stereoscopic Survey. Recipient Disting. Service award North Adams State Coll., 1980, Faculty Devel. award, 1981. Mem. Am. Inst. Biol. Scis., Bot. Soc. Am., Mycol. Soc. Am. Subspecialties: Mycology; Morphology. Current work: The use of light and scanning electron microscopy in morphologic and life history studies of fungi. Home: 1249 N Hoosac Rd Williamstown MA 01267 Office: N Adams State Coll N Adams MA 01247

BROUGHTON, DONALD BEDDOES, chemical engineer; b. Rugby, Eng., Apr. 20, 1917; came to U.S., 1924, naturalized, 1934; s. Walter and Emily (Beddoes) B.; m. Natalie Waitt, Feb. 20, 1943. Asst. prof. chem. engring. M.I.T., 1943-49; with process div. UOP, Inc., Des Plaines, Ill., 1949—, asso. tech. dir., 1979—. Recipient Alpha Chi Sigma award for chem. engring. research Am. Inst. Chem. Engrs., 1967. Fellow Nat. Acad. Engring., Am. Inst. Chem. Engrs.; mem. Am. Chem. Soc. (Rohm & Haas award in Separation Sci. and Tech. 1983), Am. Petroleum Inst. Subspecialties: Chemical engineering; Physical chemistry. Home: 1639 Hinman Ave Evanston IL 60201 Office: 20 UOP Plaza Des Plaines IL 60016

BROUGHTON, PAUL LEONARD, geologist; b. N.Y.C., Sept. 20, 1946; s. Willie Jackson and Charlotte (Burkhardt) B.; m. Victoria Claire Kinnersly, Sept. 12, 1978; 1 son, Edward James. B.Sc., Am. U., 1968; M.Sc., Va. Poly. Inst., 1971; Ph.D., Cambridge U., 1979. Sr. geologist Sask. Geol. Survey, 1971-82; sr. geologist Husky Oil Co., Calgary, Alta., Can., 1982—. Saks Govt. fellow Cambridge U., 1977-79. Fellow Cambridge Philos. Soc., Explorers Club; mem. Nat. Speleological Soc., Canadian Soc. Petroleum Geologists, Sigma Xi. Subspecialties: Combustion processes; Fuels. Current work: Specialist on enhanced oil recovery, sedimentology of heavy oil/tar sands deposits. Office: Husky Oil Co PO Box 6525 Sta D Calgary AB Canada T2P 3G7 Home: 812 15th Ave SW Apt 202 Calgary AB Canada T2V 066

BROUGHTON, ROGER JAMES, medical researcher, consultant; b. Montreal, Que., Can., Sept. 25, 1936; s. James William and Edith Olwen (Edwards) B.; m. Margaret Joan, Sept. 23, 1959; children: Lynn, Michael, Katherine. M.D., C.M., Queen's U., Kingston, Ont., Can., 1960; Ph.D., McGill U., Montreal, 1967. Fgn. visitor clin. neurophysiology U. Aix-Marseille, France, 1962-64; asst. prof. Montreal Neurol. Inst. and McGill U., 1965-68; assoc. prof. U. Ottawa, Ont., 1968-78, prof., 1978—. Contbr. over 200 articles to profl. jours.; co-author book; editor 2 books in field. Med. Research Council Can. grantee, 1968—. Mem. United Ch. Can. Clubs: Ottawa Athletic, Britannia Yacht (Ottawa). Subspecialties: Neurology; Neuropsychology. Current work: Sleep research, neurological aspects of sleep; psychophysiology; biological rhythms. Patentee automatic sleep analysis system. Home: 175 Bradford St Ottawa ON Canada K2B 5Z2 Office: Ottawa Gen Hosp Ottawa ONCanada K1H 8L6

BROWDER, FELIX EARL, mathematician; b. Moscow, July 31, 1927; s. Earl and Raissa (Berkmann) B.; m. Eva Tislowitz, Oct. 5, 1949; children—Thomas, William. S.B., Mass. Inst. Tech., 1946; Ph.D., Princeton U., 1948. C.L.E. Moore instr. math. Mass. Inst. Tech., 1948-51, vis. asso. prof., 1961-62, vis. prof., 1977-78; instr. math. Boston U., 1951-53; asst. prof. Brandeis U., 1955-56; from asst. prof. to prof. math. Yale U., 1956-63; prof. math. U. Chgo., 1963-72, Louis Block prof. math., 1972—, chmn. dept., 1972-77; vis. mem. Inst. Advanced Study, Princeton U., 1953-54, 63-64; vis. prof. Instituto de Matematica Pura e Aplicada, Rio de Janeiro, 1960, Princeton U., 1968; Fairchild disting. visitor Calif. Inst. Tech., 1967, 78; sr. research fellow U. Sussex, Eng., 1970, 76; vis. prof. U. Paris, 1973, 75, 78. Served with AUS, 1953-55. Guggenheim fellow, 1953-54, 66-67; Sloan Found. fellow, 1959-63; NSF sr. postdoctoral fellow, 1957-58. Fellow Am. Acad. Arts and Scis.; mem. Nat. Acad. Scis., Am. Math. Soc. (editor bull. 1959-68, 78—, council mem. 1959-72, 78—, mng. editor 1964-68, 80, exec. com. council 1979—), Math. Assn. Am., AAAS, Sigma Xi. Home: 5505 S Kimbark Ave Chicago IL 60637

BROWDER, WILLIAM, mathematician, educator; b. N.Y.C., Jan. 6, 1934; s. Earl and Raissa (Berkmann) B.; m. Nancy O'Brien, Jan. 30, 1960; children: Julia, Risa, Daniel. B.S., MIT, 1954; Ph.D., Princeton U., 1958. Instr. U. Rochester, 1957-58; from instr. to assoc. prof. math. Cornell U., 1958-63; prof. math. Princeton U., 1964—, chmn. dept., 1971-73; vis. fellow Math. Inst. and Magdalen Coll., Oxford U. (Eng.), 1978-79; chmn. office Math. Scis. NAS-NRC, 1978—. Guggenheim fellow, 1974-75. Mem. Am. Math. Soc. (v.p. 1977-78), Nat. Acad. Scis. Office: Fine Hall PO Box 37 Princeton NJ 08544

BROWER, JOSEPH GILBERT, computer systems programmer, writer, educator; b. Point Pleasant, N.J., Jan 31, 1949; s. Gilbert Joel and Bertha Sarah (Halmuth) B. B.S., Monmouth Coll., West Long Branch, N.J.; M.S., Rutgers U., 1973. Mem. programming staff AT&T Communications, Piscataway, N.J., 1977—. Author-editor: tech. manual VM/CMS Programmer's Guide, 1983; author: slide-tape VM/CMS Introduction for Users, 1982. Mem. Ocean County Rep. Com., Toms River, N.J., 1980-83. Research intern Rutgers U., 1971; Monmouth Coll. Trustee scholar, 1967-71. Mem. AAAS, Am. Math. Soc., Soc. Indsl. and Applied Math., World Boxing Historians Assn., Rochester Boxing Assn., Monmouth Scholars, Lambda Sigma Tau (life 1971). Republican. Baptist. Subspecialties: Operating systems; Software engineering. Current work: Design, coding, testing, implementation of software subsystems in VM/CMS operating systems: VM/CMS technical training and education, coordination of release and distribution of software components to the VM Standard Operating Environment. Home: 1029 Kerwin St Piscataway NJ 08854 Office: AT&T 30 Knightsbridge Rd Piscataway NJ 08854

BROWN, ARNOLD LANEHART, JR., pathologist, educator, univ. dean; b. Wooster, Ohio, Jan. 26, 1926; s. Arnold Lanehart and Wilda (Woods) B.; m. Betty Jane Simpson, Oct. 2, 1949; children—Arnold III, Anthony, Allen, Fletcher, Lisa. Student, U. Richmond, 1943-45; M.D., Med. Coll. Va., 1949. Diplomate: Am. Bd. Pathology. Intern Presbyn.-St. Luke's Hosp., Chgo., 1949-50, resident, 1950-51, 53-56, asst. attending pathologist, 1957-59; practice medicine specializing in pathology, Rochester, Minn., 1959-78; cons. exptl. pathology, anatomy Mayo Clinic, Rochester, 1959-78, also prof., chmn. dept., 1968-78; prof. pathology U. Wis. Med. Sch., Madison, 1978—, dean, 1978—; mem. nat. cancer adv. council NIH, 1971-72; nat. cancer advisory bd. NIH, HEW, 1972-74; chmn. clearing house on environ. carcinogens Nat. Cancer Inst., 1976-80, chmn. com. to study carcinogenicity of cyclamate, 1975-76; mem. Nat. Com. on Heart Disease, Cancer and Stroke, 1975-79; mem. com. on safe drinking water NRC, 1976-77; mem. award assembly Gen. Motors Cancer Research Found., 1978—; co-chmn. panel on geochemistry of fibrous materials related to health risks Nat. Acad. Scis.-NRC, 1978—. Contbr. articles to profl. jours. Served with USNR, 1943-45, 51-53. Nat. Heart Inst. postdoctoral fellow, 1956-59. Mem. Am. Soc. Exptl. Pathology, Internat. Acad. Pathology, Am. Assn. Pathologists and Bacteriologists, Am. Gastroent. Assn., Electron Microscope Soc. Am., AMA. Subspecialty: Pathology (medicine). Home: 1705 Camelot Dr Madison WI 53705 Office: 610 N Walnut St Madison WI 53706

BROWN, BRUCE LEONARD, microbiologist; b. Newburgh, N.Y., Apr. 24, 1937; s. Leonard Hampton and Ellen Ruby (Chandler) B.; m. Barbara Campbell Brown, June 3, 1962; children: Sara, Kristina. B.S. in Biology, No. Ariz. U., 1960; A.A.S. in Microbiology, Orange County Community Coll., 1958. Research asst. DNA Virus Lab., 1975-79, Research assoc., 1979-81; sr. microbiologist Bethesda Research Labs., Gaithersburgh, Md., 1979—. Contbr. articles in field to profl. jours. Clk. City of Rosemont, Md., 1974-81. Served with U. S. Army, 1960-62. Mem. Tissue Culture Assn., Sigma Xi. Republican. Club: Am. Legion (Brunswick, Md.). Subspecialties: Cell and tissue culture; Virology (biology). Current work: Co-developing monoclonal antibodies to seven viral antigens including herpes simplex viruses and selected biochemical antigens. Home: 3670 Petersville Rd Knoxville MD 21758 Office: PO Box 6009 Gaithersburgh MD 20877

BROWN, BRYAN CONRAD, computer analyst, research; b. Oklahoma City, Apr. 6, 1948; s. Roy Bert and Mattie Ann (Nash) B.; m. Mary Grogan, Aug. 22, 1970; 1 dau., Alicia Rosanna. B.A. in Physics (Gen. Motors scholar 1966-70), Northwestern U., 1970; Ph.D. in Astronomy (NDEA fellow 1970-73), U. Md., College Park, 1976. Teaching asst. astronomy program U. Md., College Park, 1973-76; computer analyst Computer Scis. Corp., Silver Spring, Md., 1976—; ind. researcher in celestial mechanics. Contbr. writings to profl. publs. in field. NSF conf. grantee, 1980; Am. Astron. Soc. grantee, 1979; recipient 3M Co., award, 1970. Mem. Am. Astron. Soc., Royal Astron. Soc., Phi Beta Kappa. Democrat. Club: Navy Records Soc. (London). Current work: Precision orbit determination software systems; orbital motions of Galilean Satellites of Jupiter; microcomputer-based algebraic manipulation systems. Home: 4326 Clagett Rd University Park MD 20782 Office: 8728 Colesville Rd Silver Spring MD 20910

BROWN, BURTON PRIMROSE, cons. engr.; b. Denver, Dec. 5, 1917; s. Burton Primrose and Pauline Marie (Roessler) B.; m. June Aileen Rist, Mar. 11, 1944; children—Nancy, Daniel. B.S. in Elec. Engring, U. Colo., 1939; M.S., U. Vt., 1941. With Gen. Electric Co., 1947-62, 63—, cons. engr. electronic systems div., Syracuse, N.Y., 1963—; asst. dir. def. research and engring. Dept. Def., 1962-63; asst. prof. elec. engring. U. Vt., 1946-47; mem. USAF Sci. Adv. Bd., Army Sci. Adv. Panel. Author. Mem. bd. edn., Baldwinsville, N.Y., 1966-69. Recipient Charles P. Steinmetz award Gen. Electric Co., 1975, Charles A. Coffin award, 1951. Fellow IEEE; mem. Nat. Acad. Engring., Sigma Tau, Sigma Pi Sigma, Eta Kappa Nu. Subspecialties: Systems engineering; Electronics. Current work: Consultant-military defense systems. Patentee in field. Home: 50 Brown St Baldwinsville NY 13027 Office: Gen Electric Co Court St Plant Syracuse NY 13201

BROWN, CONNELL JEAN, animal science educator; b. Everton, Ark., Mar. 6, 1924; s. Clarence Jackson and Winnie Dee (Trammell) B.; m. Erma Dexter, May 19, 1946; children: Craig Jay, Mark Allen.

B.S.A., U. Ark., 1948; M.S., Okla. State U., 1950, Ph.D., 1956. Asst. prof. dept. animal sci. U. Ark., Fayetteville, 1950-57, assoc. prof., 1957-62, prof., 1962—; lectr. Internat. Stockmans Short course, 1980—. Contbr. articles to profl. jours. Served with USAAF, 1943-46; PTO. Recipient Research award Performance Registry Internat., 1977, U. Ark. Coll. Agr. research award, 1981. Fellow AAAS; mem. Am. Soc. Animal Sci. (pres. So. sect. 1975, leadership award So. sect. 1975), Am. Genetics Assn., N.Y. Acad. Sci., So. Assn. Agrl. Scientists (dir.), Sigma Xi, Gamma Sigma Delta. Lodge: Kiwanis. Subspecialties: Animal breeding and embryo transplants; Animal genetics. Current work: Genetic and phenotypic relationships among size, shape and performance traits of beef cattle and other farm animals. Home: 2583 Elizabeth St Fayetteville AR 72701 Office: Dept Animal Sci Univ Ark Fayetteville AR 72701

BROWN, DAVID GORDON, med. physics lab. adminstr., researcher; b. Escondido, Calif., Sept. 6, 1942; s. Gordon Arthur and Ester Isabel (Griswold) B.; m. Yunni Rosa Chang, May 25, 1974; children: Matthew, Timothy. B.S., Stanford U., 1964; M.A., U. Calif., Berkeley, 1965, Ph.D., 1968; M.B.A., George Washington U., 1982. Physicist Bur. Radiol. Health, Rockville, Md., 1968-70; vis. prof. Dalat, Saigon and Hue Univs., 1970-73, Nat. Tsing Hua U., Taiwan, 1973-75; chief med. physics br. Bur. Radiol. Health, FDA, Rockville, Md., 1975—; research Chinese economy. Contbr. articles to sci. jours. Mem. Am. Phys. Soc., Am. Assn. Physicists in Medicine, Health Physics Soc. Republican. Methodist. Club: Road Runners. Subspecialty: Theoretical physics. Current work: Quantification of imaging performance in diagnostic imaging; determination of the limits on dose reduction in diagnostic radiology; description of the Chinese economy. Home: 5921 Holland Rd Rockville MD 20851 Office: 5600 Fishers Ln Rockville MD 20857

BROWN, DONALD DAVID, biology educator; b. Cin., Dec. 30, 1931; s. Albert Louis and Louise (Rauh) B.; m. Linda Jane Weil, July 2, 1957; children: Deborah Lin, Christopher Charles, Sharon Elizabeth. M.S., U. Chgo., 1956, M.D., 1956, D.Sc. (hon.), 1976, U. Md., 1983. Staff mem. dept. embryology Carnegie Instn. of Washington, Balt., 1963-76, dir., 1976—; prof. dept. biology Johns Hopkins U., 1968—. Served with USPHS, 1957-59. Recipient U.S. Steel Found. award for molecular biology, 1973; V.D. Mattia award Roche Inst., 1975; Boris Pregel award for biology N.Y. Acad. Scis., 1976; Ross G. Harrison award Internat. Soc. Developmental Biology, 1981; Bertner Found. award, 1982. Fellow Am. Acad. Arts and Scis., AAAS; mem. Nat. Acad. Scis., Soc. Developmental Biology (pres. 1975), Am. Soc. Biol. Chemists, Am. Soc. Cell Biology, Am. Philos. Soc. Subspecialties: Developmental biology; Cell and tissue culture. Home: 5721 Oakshire Rd Baltimore MD 21209 Office: Carnegie Instn of Washington 115 W University Pkwy Baltimore MD 21210

BROWN, ELLEN RUTH, theoretical physicist, engineer; b. N.Y.C., June 15, 1947; d. Aaron Joseph and Grace (Presser) B. B.S., Mary Washington Coll., 1969; M.S., Pa. State U., 1971; Ph.D. in Theoretical Atomic Physics, U. Va., 1981. Physicist, Naval Weapons Lab., Dahlgren, Va., 1969; instr. physics Lord Fairfax Community Coll., Middletown, Va., 1971-74; fellow NASA, Langley, Va., 1974-75; engr. EG&G Washington Analytical Services Center, Dahlgren, 1979—, head dept. analysis and evaluation, Va., 1982—; v.p. Windy Knoll Enterprises, Inc., Magnolia, Tex., 1981—. First violinist, Community Coll. Orch., Fredericksburg, Va., 1979—; Contbr. articles to profl. jours. NSF fellow, 1973; Gov.'s fellow U. Va., 1975-78; Physics Dept. fellow U. Va., 1978-79. Mem. Am. Phys. Soc., Sierra Club, Sigma Xi. Club: Barry Lee Bressler Sci. (Fredericksburg) (pres. 1979—). Subspecialties: Atomic and molecular physics; Theoretical physics. Current work: Photoionization calculations, optical ray tracing in the atmosphere, meteorological effects on reentry bodies, technical assessment, semiconductor physics. Home: PO Box 1397 Fredericksburg VA 22402 Office: PO Box 552 Dahlgren VA 22448

BROWN, ERIC REEDER, immunologist, educator; b. Cortland, N.Y., Mar. 16, 1925; s. Harold McDaniel and Helen (Seitz) B.; m. Chloe Cassandra Ledbetter, May 11, 1961; children—Carl F., Christopher H.A., Amy Elizabeth French; children by previous marriage—Eric Reeder II, Christine Virginia, Dianne Mary, Daniel K. B.A., Syracuse U., 1949, M.S., 1951; Ph.D. (Nat. Cancer Inst. Fellow), U. Kan., 1957; D.Sc., Quincy Coll., 1966. Instr. U. Ill. Med. Sch., 1957-58; asst. prof. U. Ala., 1958-60, U. Minn. Sch. Medicine, 1960-61; sr. research assoc. Hektoen Inst., Chgo., 1961-67; assoc. prof. Northwestern U. Med. Sch., 1964-68; chmn. dept. microbiology Chgo. Med. Sch., 1967-82; cons. Newport Pharms., Inc., Strategic Med. Research Corp., U. Ill.; med. adviser to Ill. dir. SSS; dir. Lake Bluff Labs., Inc., Chesterton, Ind.; reviewer grants NSF, 1978—, Am. Cancer Soc. fellow, 1960-63, Leukemia Soc. scholar, 1965—; mem. med. adv. bd. Leukemia Research Found. Co-author: Cancer Disemination and Therapy, 1961; author: Textbook of Micromolecular Biology, 1974, Immunobiological Characteristics of Leukemia, 1975, Sailing Made Easy, 1978; contbr. articles to profl. jours. Served with USCGR, 1942-46; served to col. USAF, 1951-55; col. Res. Fellow Am. Inst. Chemists, Am. Acad. Microbiology, Chgo. Inst. Medicine; mem. Royal Soc. Medicine, Histochem. Soc., Internat. Soc. Lymphology, AAUP, Am. Mus. Natural History, Med. Mycol. Soc. of Ams., Soc. Exptl. Biology and Medicine, Res. Officers Assn., Phi Beta Kappa, Sigma Xi, Psi Chi, Phi Sigma. Subspecialties: Oncology; Immunology (medicine). Current work: Epidemiology of cancer; gynecological microbiology; tumor virology. Research on virus etiology of cancer and leukemia. Home: PO Box 335 7704 Camellia Chicago IL 60097 Office: Chicago Med Sch 3333 Greenbay Rd North Chicago IL 60064

BROWN, GEORGE BARREMORE, neuroscientist, educator; b. Ithaca, N.Y., Dec. 5, 1945; s. Barremore B. and Marjorie J. (McBride) B.; m. Paula B. Costello, Dec. 21, 1968; children: Jason M., John Christopher. B.S. cum laude, Tulane U., New Orleans, 1968; Ph.D., Stanford U., 1972. Staff fellow NIH, Bethesda, Md., 1972-75; asst. prof. psychiatry U. Ala., Birmingham, 1975-79, asst. prof. biochemistry, 1975—, dir. cancer/pathology/neuroscis. Gas Chromatography/Mass Spectrometry Lab., 1977—, asso. prof. psychiatry, 1979-83, prof. psychiatry, 1983—, also asso. dir. neuroscis. program. Contbr. articles to profl. publs. NIH grantee, 1978—. Mem. Soc. Neurosci., AAAS. Methodist. Subspecialties: Neurophysiology; Mass spectrometry. Current work: Mechanism of neuronal excitability; plasticity. Applications of mass spectrometry in biomed. sics.; measurement of endogenous compounds, metabolites. Office: Neuroscis Program Univ Stat U Ala Birmingham AL 35294

BROWN, GEORGE HAROLD, radio engineer; b. North Milwaukee, Wis., Oct. 14, 1908; s. James Clifford and Ida Louise (Siegert) B.; m. Julia Elizabeth Ward, Dec. 26, 1932; children: James Ward, George H. (twins). B.S., U. Wis., 1930, M.S., 1931, Ph.D., 1933, E.E., 1942; Dr. Eng. (hon.), U. R.I., 1968. With RCA, 1933-37, 38-73, successively research engr., Camden, Princeton, N.J.; dir. Systems Research Lab., chief engr. Comml. Electronic Products div., Camden, chief engr. indsl. electronic products, 1933-59, v.p. engring., 1959-61, v.p. research and engring., 1961-65, exec. v.p. research and engring., 1965-68, exec. v.p. patents and licensing, 1968-72; dir. Trane Co., 1967-79, cons. engr., 1937—; dir. RCA Global Communications, 1962-71, RCA, 1965-72, RCA Internat., Ltd., 1968-72; Shoenberg Meml. lectr. Royal Instn., 1972; Marconi Centenary lectr. AAAS, 1974. Author: (with R.A. Bierwirth and C.N. Hoyler) Radio Frequency Heating, 1947, And Part of Which I Was, 1982; contbr. articles to sci. jours. Exec. bd. George Washington council Boy Scouts Am.; bd. govs. Hamilton Hosp. Recipient Silver Beaver, Silver Antelope awards Boy Scouts Am.; citation Internat. TV Symposium, Montreux, Switzerland, 1965; DeForest Audion award, 1968; David Sarnoff award for outstanding achievements in radio and TV U. Ariz., 1980. Fellow IEEE (Edison medal 1967), AAAS, Royal Television Soc.; mem. Nat. Acad. Engring., Am. Mgmt. Assn., Sigma Xi, Tau Beta Pi, Eta Kappa Nu (eminent mem.). Clubs: Nassau, Springdale. Subspecialties: Electrical engineering; Electronics. Current work: History of electrical engineering and electronics. Patentee in field. Home: 117 Hunt Dr Princeton NJ 08540

BROWN, GEORGE STEPHEN, physicist; b. Santa Monica, Calif., June 28, 1945; s. Paul Gordon and Frances Ruth (Moore) B.; m. Nohema Fernandez, Aug. 8, 1981; 1 dau.: Sonya. B.S., Calif. Inst. Tech., 1967; Ph.D., Cornell U., 1973. Mem. tech. staff Bell Labs., Murray Hill, N.J., 1973-77; sr. research assoc. Stanford (Calif.) U., 1977-82, adj. prof., 1982; physicist Brookhaven Nat. Labs., Upton, N.Y., 1982—. Contbr. articles to profl. jours. Mem. Am. Phys. Soc. Subspecialties: Condensed matter physics; Atomic and molecular physics. Current work: X-ray scattering, emission and absorption. Office: Brookhaven Nat Lab Upton NY 11973

BROWN, GERALD LAVONNE, medical scientist, psychiatrist, psychoanalyst; b. Athens, Georgia, Mar. 8, 1940; s. Coile Frank and Lillie Rice (Spratlin) B.; m. Margaret Stadler, Mar. 25, 1966; children: Klara, Suzanne, Stefanie, Kristine. A.B., Duke U., 1963, M.D., 1967; postgrad., Washington Psychoanalytic Inst., 1970-83. Intern Queens Med. Center, Honolulu, 1967-68; resident Duke U. Med. Center, Durham, N.C., 1968-71, fellow in research, 1971-72; staff scientist intramural research program NIMH, Bethesda, Md., 1974—; pvt. practice child and adult psychiatry and psychoanalysis, McLean, Va., 1972—. Contbr. over 50 med. articles to profl. publs. Served with USMC, 1961-63; with USN, 1963-74; with USPHS, 1974—. Recipient Residency Achievement award Duke U., 1970; Psychobiol. Research Tng. fellow, 1971-72. Mem. Am. Psychiat. Assn., Am. Acad. Child Psychiatry, Am. Psychoanalytic Assn. Baptist. Subspecialties: Psychopharmacology; Neuropharmacology. Current work: Psychobiology of aggression and suicide; psychobiology of childhood mental illness and sequelae; child and adult psychoanalysis. Office: NIMH Clin Center Room 3N234 Bldg 10 Bethesda MD 20205

BROWN, GLENN HALSTEAD, chemist, educator; b. Logan, Ohio, Sept. 10, 1915; s. James E. and Nancy J. (Mohler) B.; m. Jessie Adcock, May 27, 1943; children—Larry H., Nancy K., Donald S., Barbara J. B.S., Ohio U., 1939; M.S., Ohio State U., 1941; Ph.D., Iowa State U., 1951; D.Sc. (hon.), Bowling Green State U., 1972. Asst. prof. U. Miss., 1941-46, 49-50; instr. Iowa State U., 1946-49; asst. prof. U. Vt., 1950-52; asso. prof. U. Cin., 1952-60; with Kent (Ohio) State U., 1960—, prof. chemistry, head dept., 1960-65; dir. Liquid Crystal Inst., 1965—, dean for research, 1963-69, Regents prof. chemistry, 1968—; Bikerman lectr., 1981. Author: (with F.A. Anderson) Fundamentals of Chemistry, 1944, (with Wollett and Fogelsong) Laboratory Manual for Organic Chemistry, 1944, Record Book for Quantitative Analysis, 1954, (with E. M. Sallee) Quantitative Chemistry, 1963, (with others) Liquid Crystals, 1967, Review of the Structure and Properties of Liquid Crystals, 1970, (with J.J. Wolken) Liquid Crystals and Biological Structures; contbr. articles to sci. jours.; editor: Liquid Crystals 2, Parts I and II, 1969, Photochromism, 1971, Liquid Crystals 3, parts I and II, 1972, Advances in Liquid Crystals, vol. I, 1975, vol. II, 1976, vol. III, 1978, vol. IV, 1979; editor-in-chief: Jour. Molecular Crystals and Liquid Crystals, 1968—. Recipient Morley award in chemistry, 1977; Pres.'s award Kent State U., 1980; 8th Internat. Liquid Crystal Conf. dedicated in his hon., Tokyo, 1980. Fellow Ohio Acad. Sci. (pres. 1960, Distinguished Service award 1966); mem. Am. Chem. Soc. (chmn. Akron sect. 1965, nat. councilor Akron sect., chmn. regional meeting planning com. 1968, Distinguished Service award Akron sect. 1971), Am. Inst. Chemists (chmn. Ohio 1969-71), Am. Crystallographic Assn., AAAS, N.Y. Acad. Scis., Sigma Xi (nat. lectureship 1970), Alpha Chi Sigma, Phi Lambda Upsilon, Omicron Delta Kappa. Methodist. Subspecialty: Liquid crystals. Spl. research X-ray structural studies liquids, concentrated salt solutions, photochromism, liquid crystals. Home: 470 Harvey Ave Kent OH 44240

BROWN, GORDON M(ARSHALL), research optical engineer, consultant; b. Detroit, Feb. 17, 1934; s. Everett J. and Agnes E. (Craig) B.; m. Sharla A. Smith, Aug. 15, 1958; children: Gordon C., Julie Marie. Student in math. (Coll. scholar), Greenville Coll., 1952-54; B.S.M.E., Gen. Motors Inst., 1958; M.S. in Nuclear Engring. (Gen. Motors fellow), U. Mich., 1961. Project engr. Bendix Aerospace Systems, Ann Arbor, Mich., 1961-67; dir. engring. GCO, Inc., Ann Arbor, 1967-71, dir. mfg., 1971-73; prin. research engr., physics dept. Ford Motor Co., Dearborn, Mich., 1973—; cons. optical/digital systems engring. Contbr. articles to profl. publs. Mem. Optical Soc. Am., Am. Soc. Nondestructive Testing (Achievement award 1970). Republican. Methodist. Subspecialties: Optical engineering; Holography. Current work: Holographic interferometry, image processing, optical metrology; development holographic methods and apparatus for vibration analysis of large automotive structures; devel. noncontact distance measurement instruments. Inventor holographic tire tester. Home: 3191 Bluett St Ann Arbor MI 48105 Office: Room S-1024 SRL PO Box 2053 Dearborn MI 48121

BROWN, GREGORY GAYNOR, biochemist, educator; b. Englewood, N.J., Aug. 17, 1948; s. Norbert Hugh and Edith Mary (Krabach) B.; m. Sheila Marie Sullivan, Nov. 19, 1977; 1 dau., Meghan Anne. B.S., U. Notre Dame, 1970; Ph.D., Mt. Sinai Sch. Medicine CUNY, 1977. Postdoctoral fellow SUNY-Stony Brook, 1977-81; asst. prof. biology McGill U., Montreal, Que., Can., 1981—, mem. genetic manipulation research group, 1981—. Contbr. articles to profl. jours. NIH fellow, 1979-81; Natural Scis. and Engring. Research Council Can. grantee, 1981—. Mem. Am. Soc. Cell. Biology, Genetics Soc. Am. Subspecialties: Genetics and genetic engineering (biology); Molecular biology. Current work: Organization and evolution of mitochondrial genetic system; regulation of mitochondrial gene activity and replication. Home: 6666 Fielding Ave #703 Montreal PQ Canada H4V IN6 Office: Dept. Biology McGill U 1205 Ave Dr Penfield Montreal PQ Canada H3A 1B1

BROWN, HANNAH R(EEVA), virologist, electron microscopist; b. Bklyn., Apr. 23, 1934; d. Samuel and Fanny (Cumings) Sussman; m. Morris Brown, July 11, 1954; children: Ilene, Eric, Jess. B.A. in Biology (N.Y. State Regents scholar), Hunter Coll., 1954; M.S. in Microbiology, Rutgers U., 1981. Fellow Hunter Coll., 1955; technician dept. virology N.Y. State Inst. for Basic Research in Devel. Disabilities, S.I., 1968-69, asst. research scientist, 1969-73, research scientist I, 1973-79, research scientist II, 1979—. Contbr. articles to profl. jours., 1973—. NIH grantee, 1981-82. Mem. N.Y. Acad. Scis., N.Y. Soc. Electron Microscopists, Electron Microscope Soc. Am., Am. Soc. Microbiology, AAAS, Council Research Scientists, Phi Beta Kappa, Phi Sigma. Subspecialties: Virology (biology); Immunoelectron microscopy. Current work: Immunoelectron microscopy of virus-infected CNS tissue; ultra-structural localization of measles virus antigens and immunoglobulins in brains of SSPE infected ferrets using immunoperoxidase labeling techniques. Home: 199 Ardmore Ave Staten Island NY 10314 Office: 1050 Forest Hill Rd Staten Island NY 10314

BROWN, HAROLD, scientist, educator, corp. dir., cons., former sec. Def.; b. N.Y.C., Sept., 19, 1927; s. A.H. and Gertrude (Cohen) B.; m. Colene Dunning McDowell, Oct. 29, 1953; children—Deborah Ruth, Ellen Dunning. A.B., Columbia U., 1945, A.M., 1946, Ph.D. in Physics (Lydig fellow 1948-49), 1949; D.Eng., Stevens Inst. Tech., 1964; LL.D., L.I. U., 1966, Gettysburg Coll., 1967, Occidental Coll., 1969, U. Calif., 1969; Sc.D., U. Rochester, 1975, Brown U., 1977, U. of the Pacific, San Francisco, 1978, U. S.C., 1979. Research scientist Columbia U., 1945-50, lectr. physics, 1947-48, Stevens Inst. Tech., 1949-50; research scientist Radiation Lab., U. Calif. at Berkeley, 1950-52, lectr. physics, 1951-52; group leader to dir. Radiation Lab. at Livermore, 1952-61; dir. def. research and engring. Dept. Def., 1961-65; sec. of air force, 1965-69; pres. Calif. Inst. Tech., Pasadena, 1969-77; sec. Def., Washington, 1977-81; Disting. vis. prof. nat. security Sch. Advanced Internat. Studies, Johns Hopkins U., Washington, 1981—; cons., 1981—; dir. AMAX, CBS, IBM, Hoover Universal; mem. Polaris Steering Com., 1956-58; cons., mem. Air Force Sci. Adv. Bd., 1956-61, Pres.'s Sci. Adv. Com., 1958-61; sr. sci. adviser Conf. Discontinuance Nuclear Tests, 1958-59; U.S. del. SALT, Helsinki, Vienna and; Geneva, 1969-77; chmn. Tech. Assessment Adv. Council to U.S. Congress, 1974-77; mem. exec. com. Trilateral Commn., 1973-76. Decorated Medal of Freedom; named One of 10 Outstanding Young Men U.S. Jaycees, 1961; recipient Medal of Excellence Columbia U., 1963; Joseph C. Wilson award in internat. affairs, 1976; award for disting. contbns. to higher edn. Stony Brook Found., 1979. Mem. Nat. Acad. Engring., Am. Phys. Soc., Am. Acad. Arts and Scis., Nat. Acad. Scis., Council on Fgn. Relations, N.Y.C., Phi Beta Kappa, Sigma Xi. Clubs: Bohemian (San Francisco); Calif. (Los Angeles). Subspecialty: National security policy. Office: Johns Hopkins U Sch Advanced Internat Studies 1740 Massachusetts Ave NW Washington DC 20036

BROWN, HARRISON SCOTT, chemist, educator; b. Sheridan, Wyo., Sept. 26, 1917; s. Harrison H. and Agatha (Scott) B.; m. Rudd Owen, Nov. 11, 1949; 1 son, Eric Scott; m. Theresa Tellez, 1975. B.S., U. Calif., 1938, LL.D., 1970; Ph.D., Johns Hopkins U., 1941; LL.D., U. Alta., 1961; Sc.D., Rutgers U., 1964, Amherst Coll., 1966, Cambridge U., 1969. Instr. chemistry Johns Hopkins U., 1941-42; asst. dir. chemistry Clinton Labs., Oak Ridge, 1943-46; research asso. plutonium project U. Chgo., 1942-43; asst. prof. Inst. Nuclear Studies, 1946-48, asso. prof., 1948-51; prof. geochemistry Calif. Inst. Tech., 1951-77, prof. sci. and govt., 1967-77; dir. Resource Systems Inst., East-West Center, Honolulu, 1977—. Author: Must Destruction Be Our Destiny?, 1946, The Challenge of Man's Future, 1954, The Next Hundred Years, 1957, The Cassiopeia Affair, 1968, The Human Future Revisited, 1978, Learning How To Live in a Technological Society (Ishizaka lectures, Japan), 1979. Recipient Lasker Found. award, 1958, N.Y. Acad. Scis. award, 1978. Mem. Nat. Acad. Scis. (fgn. sec. 1962-74, chmn. world food and nutrition study 1975-77), Internat. Council Sci. Unions (pres. 1974-76), Am. Chem. Soc. (award in pure chemistry 1952), Geol. Soc. Am., AAAS (ann. award 1947), Am. Geophys. Union, Phi Beta Kappa, Sigma Xi. Subspecialty: Geochemistry. Current work: Demand for and availability of natural resources. Home: 965 Prospect St Honolulu HI 96822 Office: East-West Center Honolulu HI 96848

BROWN, HERBERT CHARLES, chemistry educator; b. London, May 22, 1912; U.S., 1914; s. Charles and Pearl (Stine) B.; m. Sarah Baylen, Feb. 6, 1937; 1 son, Charles Allan. Asso. Sci., Wright Jr. Coll., Chgo., 1935; B.S., U. Chgo., 1936, Ph.D., 1938, D.Sc., 1968, hon. doctorates, 1968; hon. doctorates, Wayne State U., 1980, Lebanon Valley Coll., 1980, L.I. U., 1980, Hebrew U. Jerusalem, 1980, Pontificia Universidad de Chile, 1980, Purdue U., 1980, U. Wales, 1981. Asst. chemistry U. Chgo., 1936-38; Eli Lilly post-doctorate research fellow, 1938-39, instr., 1939-43; asst. prof. chemistry Wayne U., 1943- 46, asso. prof., 1946-47; prof. inorganic chemistry Purdue U., 1947-59, Richard B. Wetherill prof. chemistry, 1959, Richard B. Wetherill research prof., 1960-78, emeritus, 1978—; vis. prof. U. Calif. at Los Angeles, 1951, Ohio State U., 1952, U. Mexico, 1954, U. Calif. at Berkeley, 1957, U. Colo., 1958, U. Heidelberg, 1963, State U. N.Y. at Stonybrook, 1966, U. Calif. at Santa Barbara, 1967, Hebrew U., Jerusalem, 1969, U. Wales, Swansea, 1973, U. Cape Town, S. Africa, 1974, U. Calif., San Diego, 1979; Harrison Howe lectr., 1953, Friend E. Clark lectr., 1953, Freud-McCormack lectr., 1954, Centenary lectr., Eng., 1955, Thomas W. Talley lectr., 1956, Falk-Plaut lectr., 1957, Julius Stieglitz lectr., 1958, Max Tishler lectr., 1958, Kekule-Couper Centenary lectr., 1958, E. C. Franklin lectr., 1960, Ira Remsen lectr., 1961, Edgar Fahs Smith lectr., 1962, Seydel-Wooley lectr., 1966, Baker lectr., 1969, Benjamin Rush lectr., 1971, Chem. Soc. lectr., Australia, 1972, Armes lectr., 1973, Henry Gilman lectr., 1975, others; chem. cons. to indsl. corps. Author: Hydroboration, 1962, Boranes in Organic Chemistry, 1972, Organic Synthesis via Boranes, 1975, The Nonclassical Ion Problem, 1977; Contbr. articles to chem. jours. Bd. govs. Hebrew U., 1969—. Served as co-dir. war research projects U. Chgo. for U.S. Army, Nat. Def. Research Com., Manhattan Project, 1940-43. Recipient Purdue Sigma Xi research award, 1951; Nichols medal, 1959; award Am. Chem. Soc., 1960; S.O.C.M.A. medal, 1960; H.N. McCoy award, 1965; Linus Pauling medal, 1968; Nat. Medal of Sci., 1969; Roger Adams medal, 1971; Charles Frederick Chandler medal, 1973; Chem. Pioneer award, 1975; C.U.N.Y. medal for sci. achievement, 1976; Elliott Cresson medal, 1978; C.K. Ingold medal, 1978; Nobel prize for chemistry, 1979; Priestley medal, 1981; Perkin medal, 1982; others. Fellow Royal Soc. Chemistry (hon.), AAAS, Indian Nat. Sci. Acad. (fgn.); mem. Am. Acad. Arts and Scis., Nat. Acad. Scis., Chem. Soc. Japan (hon.), Pharm. Soc. Japan (hon.), Am. Chem. Soc. (chmn. Purdue sect. 1955-56), Ind. Acad. Sci., Phi Beta Kappa, Sigma Xi, Alpha Chi Sigma, Phi Lambda Upsilon (hon.). Subspecialties: Organic chemistry; Synthetic chemistry. Current work: Development of new hydrideregeants for selective reductions, hydroboration-kinetics and mechanism,new hydroboration agents, organoboranes, syntheses via organoboranes. Research in phys., organic, inorganic chemistry relating chem. behavior to molecular structure; selective reductions; hydroboration; chemistry of organoboranes. Awarded patents (with others) on preparation of borohydrides, diborane, hydroboration; synthesis of aliphatic derivatives. Home: 1840 Garden St West Lafayette IN 47906

BROWN, JACK HAROLD UPTON, health service administrator, educator, biomedical engineer; b. Nixon, Tex., Nov. 16, 1918; s. Gilmer W. and Thelma (Patton) B.; m. Jessie Carolyn Schulz, Apr. 14, 1943. B.S., S.W. Tex. State U., 1939; postgrad., U. Tex., 1939-41; Ph.D., Rutgers U., 1948. Lectr. physics S.W. Tex. State U., San Marcos, 1943-44; instr. phys. chemistry Rutgers U., New Brunswick, N.J., 1944-45, research assoc., 1944-48; lectr. U. Pitts., 1948-50; head biol. scis. Mellon Inst., Pitts., 1948-50; asst. prof. physiology U.N.C., Chapel Hill, 1950-52; scientist Oak Ridge Inst. Nuclear Studies, 1952; assoc. physiology Emory U. Med. Sch., Atlanta, 1952-58, prof., 1959-60, acting chmn. dept. physiology, 1958-60; lectr. physiology George Washington U. Med. Sch., Georgetown U. Med. Sch., Washington, 1960-65; exec. sec. biomed. engring. and physiology tng. coms. Nat. Inst. Gen. Med. Scis., NIH, Bethesda, Md., 1960-62, chief spl. research br. div. research facilities and resources, 1962-63, acting chief gen. clin. research ctrs. br., 1963-64, asst. dir. ops. div. research facilities and resources, 1964-65; acting program dir. pharmacology/toxicology program Nat. Inst. Gen. Med. Scis., 1966-70, acting dir., 1970; spl. asst. to adminstr. Health Services and Mental Health Adminstrn., USPHS, Rockville, Md., 1971-72, assoc. dep. adminstr. for devel., 1972-73; spl. asst. to adminstr. Health Resources Adminstrn., 1973-78; coordinator S.W. Research Consortium, San Antonio, 1974-78; prof. physiology U. Tex. Med. Sch., San Antonio, 1974-78; prof. environ. scis. U. Tex., San Antonio, 1974-78; adj. prof. health services adminstrn. Trinity U., San Antonio, 1975-78; assoc. provost research and advanced edn., prof. biology U. Houston, 1978—; adj. prof. U. Tex. Sch. Pub. Health, 1978—; adj. prof. pub. adminstrn. Tex. Women's U., 1978—; Fulbright lectr. U. Rangoon, Burma, 1950; cons. in health systems WHO, Oak Ridge Inst. Nuclear Studies, VA, Lockheed Aircraft Co., Drexel Inst. Tech., NASA, Vassar Coll. Author: Physiology of Man in Space, 1963, The Health Care Dilemma, 1977, Integration and Control of Biological Processes, 1978, Politics and Health Care, 1978, Telecommunications in Health Care, 1981; co-author: (with A.B. Barker) Basic Endocrinology, 1966, 2d edit., 1970, (with J.F. Dickson) Future Goals of Engineering in Biology and Medicine, 1968, Advances in Biomedical Engineering, vol. II, 1972, vols. III, IV, 1973, vol. V, 1974, vol. VI, 1976, vol. VII, 1978, (with J.E. Jacobs, L.E. Stark) Biomedical Engineering, 1972, (with D.E. Gann) Engineering Principles in Physiology, vols. I, II, 1973; editor: Blood and Body Functions, 1966, Life into Space, 1968; contbr. articles to sci. jours. Med. adv. bd. Ctr. for Cancer Therapy, San Antonio, 1974—; bd. dirs. S. Tex. Health Edn. Ctr. Served with USNR, 1941. Recipient spl. team award NASA, 1978; Gerard Swope fellow Gen. Electric Co., 1946-48; Fulbright grantee, 1950; NIH grantee, 1950-60; Cancer Soc. grantee, 1958; Damon Runyon Cancer award grantee, 1959; Dept. of Energy grantee, 1980-81. Fellow AAAS, Nat. Acad. Engring., IEEE (mem. joint com. engring. in medicine and biology 1966); mem. Am. Chem. Soc. (sr.), Council Biology Editors, Soc. Research Adminstrn., Cosmos Club, Biomed. Engring. Soc. (pres. 1969-70, dir. 1968-69), Inst. Radio Engrs. (nat. sec. profl. group biomed. engring. 1962-64), N.Y. Acad. Scis., Endocrine Soc., Am. Physiol. Soc. (com. mem. 1959-63), Soc. for Exptl. Biology and Medicine, Sigma Xi (research award 1961, pres. Alamo chpt. 1977-78), Pi Kappa Delta, Phi Lambda Upsilon, Alpha Chi. Subspecialties: Biomedical engineering; Health services research. Current work: Research in biomedical engring. and health services. Developer: capsule manometer, respirator for small animals and basal metabolic apparatus for small animals, dust sampler, apparatus for partioning human lung volumes. Home: 2908 Whisper View San Antonio TX 78230 Office: 4400 Calhoun St Houston TX 77004

BROWN, JAY CLARK, microbiologist; b. Jersey City, June 23, 1942; s. John Robert and Vonna Lee (Lamme) B.; m. Sallie Dietrich, June 26, 1965; children: Jeffrey, Norman, Michael. B.A. with honors, John Hopkins U., 1964; Ph.D. in Biochemistry and Molecular Biology, Harvard U., 1969. Postdoctoral researcher Med. Research Council Lab. of Molecular Biology, Cambridge, Eng., 1969-71; asso. prof. microbiology U. Va. Sch. Medicine, 1971—; mem. adv. com. for personnel in research Am. Cancer Soc. Mem. Am. Soc. Biol. Chemists, Am. Soc. Microbiology. Subspecialties: Microbiology (medicine); Cancer research (medicine). Current work: Research on membrane structure and virus structure. Office: Dept Microbiology U VA Med Center Charlottesville VA 22908

BROWN, JIM MCCASLIN, consultant engineering geologist; b. Mpls., Sept. 29, 1938; s. James McCaslin and Alvi Vieno (Ojennus) B.; m. Dean Naomi Alexander, Sept. 1, 1963; children: Robin, Shelly. B.S., U. Alaska, 1960, M.S., 1963; Ph.D., U. Wis.-Madison, 1968. Cert. profl. geologist, Alaska; registered profl. geologist, cert. engring. geologist, Oreg. Jr. geologist Pan Am. Petroleum Corp., Anchorage, 1960; geologist U.S. Geol. Survey, Fairbanks, Alaska, 1962; sr. asst. geologist Ont. Dept. Mines, Toronto, Can., 1964; asst. prof. geology St. Louis U., 1968-69, Ind. U.-Indpls., 1969-74; sr. engring. geologist R&M Cons., Inc., Anchorage, 1974-83; pres. Arctic Geo-Terrain Cons., Wasilla, Alaska, 1983—. Mem. Valdez (Alaska) City Sch. Bd., 1975-76, treas., 1975-76. Fellow Geol. Soc. Am.; mem. Alaska Geol. Soc., AIME, Soc. Mining Engrs., Am. Inst. Profl. Geologists (pres. Alaska sect. 1981-82), U. Alaska Alumni Assn. (bd. dirs. 1980—). Republican. Subspecialties: Remote sensing (geoscience); Geology. Current work: Geology of Arctic regions including permafrost and the development of resource extraction infrastructures; remote sensing and terrain analysis for resource management and design/construction purposes; geologic hazards including seismic analysis and rock slope stability. Office: Arctic Geo-Terrain Cons PO Box 870366 Wasilla AK 99687

BROWN, JOHN CLIFFORD, scientist, educator, researcher; b. Cullman, Ala., Feb. 23, 1943; s. Robert Christopher and Laura Belle (Roberts) B.; m. Diane Brown, Mar. 19, 1966; children: Aaron Dalton, Rachel Diane. B.S., Auburn U., 1965, M.S., 1967; Ph.D., N.C. State U., Raleigh, 1973. Postdoctoral fellow U. Calif., Berkeley, 1973-76; asst. prof. U. Kans., Lawrence, 1976-82, asso. prof. dept. microbiology, 1982—. Contbr. articles to profl. jours. Mem. Douglas County Amateur Baseball Assn., 1976—. Served to lt. USN, 1965-69. Grantee NIH, Mid-Am. Cancer Center. Mem. Am. Assn. Immunologists, Am. Soc. Microbiology, AAUP, AAAS. Subspecialties: Immunology (medicine); Biochemistry (biology). Current work: Immunochemistry; autoimmune reactions in humoral responses; rheumatoid factor induction and specificity; autoimmune antibody production in response to hyperimmunization with bacterial vaccines; hybridomas. Home: 3201 Ranger Dr Lawrence KS 66044 Office: U Kans Lawrence KS 66045

BROWN, JOHN LAWRENCE, JR., electrical engineering educator; b. Ellenville, N.Y., Mar. 6, 1925; s. John Lawrence and Grace Evelyn (Freer) B. B.S., Ohio U., 1948; Ph.D., Brown U., 1953. Asst. prof. engring. research Ordnance Research Lab., State College, Pa., 1951-53, assoc. prof., 1953-60, prof. and head, 1960-69; prof. elec. engring. Pa. State U., University Park, 1969—; vis. fellow dept. elec. engring. Princeton U., 1973-74; Prince vis. scholar Ariz. State U., Tempe, 1982—. Contbr. articles to profl. jours. Served with U.S. Army, 1943-46. Fellow IEEE; mem. Am. Math. Soc., Acoustical Soc. Am., Soc. for Indsl. and Applied Math., Sigma Xi, Phi Beta Kappa, Eta Kappa Nu, Pi Mu Epsilon. Democrat. Methodist. Club: Univ. Subspecialties: Applied mathematics; Electrical engineering. Current work: Research in statistical communication theory, signal processing, applied mathematics, undergraduate and graduate university instruction. Home: 1431 Curtin St State College PA 16803 Office: Pa State Univ 109 Electrical Engring W University Park PA 16802

BROWN, KEITH IRWIN, educator, researcher; b. Hunter, Kans., Sept. 28, 1925; s. Fabius L. and Antonette P. (Lawson) B.; m. Dorothy J. Johnson, Feb. 4, 1951; children: Phyllis, Roger, Ann, Jane. B.S. in Zoology, Kans. State Coll., 1949, M.S., 1950; Ph.D. in Zoology, U.Wis., 1956. Asst. prof. physiology Okla. State U., 1955-57; asst. prof. poultry sci. Ohio Agrl. Research and Devel.Ctr., Ohio State U., 1957-61, assoc. prof., 1961-64, prof., 1964—, chmn. dept., 1964—. Pres. Wooster Fish, 1970-74; pres. Wooster Interfaith Housing Co., 1973-78; mem. Zoning Bd. Apls. Wooster, 1974-78; pres. Wooster Community Residents Inc., 1978-81. Served with USN. Recipient Nat. Turkey Fedn. Research award, 1965; Laymans award merit Wooster Kiwanis, 1974. Mem. Soc. Study Reproduction, Poultry Sci. Assn., Worlds Poultry Sci. Assn., AAAS, Sigma Xi, Gamma Alpha Sigma. Democrat. Presbyterian. Subspecialties: Animal physiology; Endocrinology. Current work: Reproductive physiology in birds, stress physiology in birds, animal welfare with emphasis on fitness traits in modern production systems. Office: Dept Poultry Sci Ohio Agrl Research and Devel Ctr Wooster OH 44691

BROWN, KENNETH LAWRENCE, chemist, educator; b. Phila., July 6, 1946; s. S. Robert and Lillian (Tentzer) B.; m. Kathleen Thompson, Dec. 21, 1968; children: Elizabeth L., Dana Marin. B.S., U. Chgo., 1968; Ph.D., U. Pa., 1971. NIH postdoctoral trainee U. Pa., 1972-73; NIH postdoctoral fellow U. Calif.-Davis, 1973-75; asst. prof. U. Tex., Arlington, 1975-80, assoc. prof. chemistry, 1980—. Contbr. articles to profl. jours. Woodrow Wilson fellow; NSF fellow; grantee NIH, 1976-79, Petroleum Research Fund, 1976-79, Robert A. Welch Found., 1979-82, 82—. Mem. AAAS, Am. Chem. Soc., Sigma Xi, Phi Beta Kappa. Democrat. Subspecialties: Biochemistry (biology); Organometallic chemistry. Current work: Bioinorganic chemistry of vitamin B-12, organocobalt chemistry.; teaching biochemistry, organic chemistry; research in kinetics, thermodynamics, organocobalt synthesis, nuclear magnetic resonance. Office: Dept Chemistry Box 19065 U Tex at Arlington TX 76019

BROWN, LEONARD FRANKLIN, JR., geological research administrator, educator, consultant; b. Seminole, Okla., June 1, 1928; s. Leonard Franklin and Portia Anna (McLeod) B.; m. Mettie Lee Roots, Dec. 14, 1957 (div. Dec. 1981); 1 dau., Ann Leslie; m. Belle Keith McClelland, Sept. 26, 1982. B.S., Baylor U., 1951; M.S., U. Wis., 1953, Ph.D., 1955. Exploration geologist Chevron, Amarillo, Tex., 1955-57; research scientist Bur. Econ. Geology, U. Tex., Austin, 1957-60, 1966-71, assoc. dir., 1971—, prof. geol. sci., 1971—; assoc. prof. Baylor U., 1960-66; internat. cons., 1971—. Author: (with others) books, including Delta Systems in the Exploration for Oil and Gas, 1969, Stratigraphic Interpretation and Petroleum Potential, 1980; contbr. articles to profl. jours.; editor: Baylor Geol. Studies, 1961-66. Recipient Best Paper award Gulf Coast Assn. Geol. Socs., 1969. Fellow Geol. Soc. Am.; mem. Am. Assn. Petroleum Geologists (disting. lectr. 1971-72, assoc. editor 1974—), Permian Basin Soc. Econ. Paleontologists (1st hon. mem.), Sigma Xi. Subspecialties: Sedimentology; Geophysics. Current work: Basin analysis and seismic stratigraphy of sedimentary basins with emphasis on depositional systems analysis in petroleum and mineral exploration. Office: Bur Econ Geology U Tex Box X UT Sta Austin TX 78712 Home: 9613 Bluecreek Ln Austin TX 78758

BROWN, LOREN DENNIS, physician; b. Des Moines, Feb. 21, 1949; s. Wendell James and Vivian Rose (Young) B.; m. Debra Dee Winders, Feb. 27, 1971; children: Marcus Loren, Melissa Lynn, Katherine Megan. B.A., U. Iowa, 1971; D.O., Coll. Osteopathic Medicine and Surgery, Des Moines, 1974. Diplomate: Nat. Bd. Examiners for Osteopathic Physicians and Surgeons. Intern Des Moines Gen. Hosp., 1974-75; resident internal medicine Chgo. Osteopathic Hosp., 1975-76, Youngstown Osteopathic Hosp., 1976-77; fellow hematology/med. oncology Cg Cleve. Clinic Found., 1977-79; practice medicine specializing in hematology/med. oncology Assoc. Med.Clinic, Des Moines, 1979—; assoc. prof. medicine U. Osteopathic Medicine and Health Scis., Des Moines, 1979—; assoc. investigator North Central Cancer Treatment Group/Iowa Oncology Research Assn., Des Moines, 1980—. Mem. Am. Osteopathic Assn., Am. Coll. Osteo. Internists, Am. Soc. Clin. Oncology, Ia. Soc. Osteopathic Physicians and Surgeons. Roman Catholic. Club: Golf and Country (Des Moines). Subspecialties: Oncology; Hematology. Current work: Private practice in hematology and medical oncology, part-time teaching faculty, participate in National Clinical Cancer Research Trials, investigation of new methods in chemotherapy delivery. Home: 4025 Mary Lynn Dr Des Moines IA 50322 Office: 1440 E Grand Ave Suite 2E Des Moines IA 50316

BROWN, LORIN W., physicist, naval officer; b. Cedar City, Utah, July 19, 1937; s. John M. and Althea (Lund) B.; m. Marilynn Thompson, Apr. 17, 1938; children: Christopher, Alexander, Brita, Benjamin, Anna, Caroline, Catherine. A.S., Coll. So. Utah, 1957; B.S., U. Utah, 1959, Ph.D., 1970. Commd. ensign U.S. Navy, 1961, advanced through grades to capt., 1982; asst. prof. naval sci. U. Utah, Salt Lake City, 1966-67; sta. USS Ranger and USS Enterprise, 1962-66, Cecil Field, Fla., 1970-72, USS Independence, 1970-72; physicist Naval Research Lab., Washington, 1972-76, Naval Weapons Evaluation Facility, Albuquerque, 1976-80; comdg. officer Navy Space Systems Activity, Los Angeles, 1981—. Contbr. articles to profl. jours, reports to naval lit. Decorated D.F.C., Airmedal (14); recipient Rear adm. William S. Parson's award for sci. and tech. achievement Navy League U.S., 1976. Mem. Am. Phys. Soc., Optical Soc. Am., Sigma Pi Mu Epsilon. Mormon. Subspecialties: Laser physics; Fiber optics. Current work: Direction of research and development supporting space staellites for communication, navigation, surveillance, general space science, space-based high energy laser and submarine laser communication satellites. Patentee optical access coupler for fiber bundles.

BROWN, LUCY LESEUR, neuroscientist, researcher; b. Boston, May 14, 1945; d. Ralph Sawyer and Rosemary (Wyman) B. B.A., Washington Sq. Coll., NYU, 1968; PH.D., NYU, 1973. Postdoctoral fellow in neurology Albert Einstein Coll. Medicine, Bronx, N.Y., 1973-76, assoc. in neurology, 1976-79, asst. prof. neurology and neurosci., 1979—. Contbr. articles to profl. jours. Mem. Soc. for Neurosci., AAAS. Subspecialties: Neuropharmacology; Neuroanatomy. Current work: Neurobiology of the motor system, especially the basal ganglia. Effects of dopaminergic drugs on brain metabolism and blood flow. Office: Albert Einstein Coll Medicine Bldg F Room G9 Bronx NY 10461

BROWN, MICHAEL STUART, geneticist; b. N.Y.C., Apr. 13, 1941; s. Harvey and Evelyn (Katz) B.; m. Alice Lapin, June 21, 1964; children: Elizabeth Jane, Sara Ellen. B.A., U. Pa., 1962, M.D., 1966. Intern, then resident in medicine Mass. Gen. Hosp., Boston, 1966-68; served with USPHS, 1968-70; clin. assoc. NIH, 1968-71; asst. prof. U. Tex. Southwestern Med. Sch., Dallas, 1971-74; Paul J. Thomas prof. genetics, dir. Center Genetic Diseases, 1977—. Recipient Pfizer award Am. Chem. Soc., 1976, Passano award Passano Found., 1978, Lounsbery award U.S. Nat. Acad. Scis., 1979; Lita Annenberg Hazen award, 1982. Mem. Nat. Acad. Scis., Am. Soc. Clin. Investigation, Assn. Am. Physicians, Harvey Soc. Subspecialty: Genetics and genetic engineering (medicine). Home: 5719 Redwood Ln Dallas TX 75209 Office: 5323 Harry Hines Blvd Dallas TX 75235

BROWN, NEAL BOYD, geophysicist, educator; b. Moscow, Idaho, Dec. 13, 1938; s. Kenneth Wayne and Ruth Alvina (Boyd) B.; m. Frances Claire Tannian, Aug. 23, 1980; children: Steven Ross Sweet, Kris David, Melody Jo, Michael Scott Sweet, Nathaniel Scott. B.S., Wash. State U.-Pullman, 1961; M.S., U. Alaska, 1966. Research engr. NASA Ames, Moffeit Field, Calif., 1961-62; research scientist Am. Geophys. Soc., Thule, Greenland, 1962-63; asst. engr. Geophys. Inst., U. Alaska, Fairbanks, 1968-69; supr., asst. prof. Poker Flat Research Range, 1971—. Commn. mem. Alaskaland, City of Fairbanks, 1981. Mem. Am. Assn. Physics Tchrs., Am. Geophys. Union, AIAA; Air Force Assn., AAAS, Alaska Assn. Computers in Edn. Subspecialties: Aeronomy; Meteorology. Current work: Implementing coordinated studies of atmospheric phenomena using earthbased, aircraft, balloon, rocket and spacecraft-borne sensors. Home: SR 20802 Fairbanks AK 99701 Office: Geophys Inst U Alaska Fairbanks AK 99701

BROWN, PAUL BURTON, physiologist; b. Panama City, Panama, Nov. 29, 1942; s. Harold E. and Nina E. (Wetzler) B.; m. Sally Ann Brown, Dec. 23, 1942. B.S., M.I.T., 1964; Ph.D., U. Chgo., 1968. Research assoc. Cornell U., 1969-72; neurophysiologist Boston State Hosp., 1972-74; asst. prof. physiology W. Va. U., Morgantown, 1974-78, assoc. prof., 1978-82, prof., 1982—. Editor-in-chief: Jour. Electrophysiol. Techniques; Contbr. articles to profl. lit.; Author: Electronics for the Modern Scientist. Recipient Gellhorn prize for grad. research, 1968; grantee NIH, NSF; Fogarty sr. internat. fellow, 1980-81. Mem. AAAS, N.Y. Acad. Scis., Am. Physiol. Soc., Soc. Neurosci. Subspecialties: Neurobiology; Physiology (biology). Current work: Somatosensory system. Electrophysiology, neuroanatomy of spinal cord neurons responding to low threshold mechano-receptor input. Office: Physiology Dept W Va U Med Center Morgantown WV 26506

BROWN, RICHARD DON, pharmacologist, educator; b. Alexandria, La., Mar. 3, 1940; s. Recie and Jewell Isadoor (Cory) B.; m. Fronie Mae Taylor, June 1, 1963; children: Richard Don. B.S., La. Coll., 1964; M.S., La. State U., 1966, Ph.D., 1968. Instr. pharmacology La. State U., New Orleans, 1968-69, asst. prof. pharmacology, Shreveport, 1969-72, assoc. mem. grad. faculty, 1969-73, assoc. prof., 1972-76, coordinator grad. studies, 1978-81, mem. grad. faculty, 1973—, prof. pharmacology, 1976—; cons. in field. Contbr. articles to profl. jours. So. Med. Assn. grantee, 1970-71; La. State U. Found. grantee, 1973-74; NIH grantee, 1975-78; Hoffman LaRoche grantee, 1976-79; Dooner Pharm. Co. grantee, 1978; Glenbrook Lab. grantee, 1979-81; Parke-Davis grantee, 1981—; recipient Frank R. Blood award Soc. Toxicology, 1976. Mem. Am. Soc. Pharmacology and Exptl. Therapeutics, Soc. Neurosci., Acoustical Soc. Am., Soc. Toxicology, Assn. for Research in Otolaryngology, La. State U. Alumni Assn., N.W. La. Computer Group, Mensa, Sigma Xi, Alpha Chi, Alpha Epsilon Delta. Lodge: Masons. Subspecialties: Pharmacology; Mathematical software. Current work: Auditory pharmacology and toxicology, pharmacokinetics and pharmacodynamics, computer applications software development. Office: PO Box 33932 Shreveport LA 71107 Home: 6630 N Park Cir Shreveport LA 71107

BROWN, ROBERT MICHAEL, psychology educator; b. Seattle, Jan. 17, 1945; s. Robert Bruce and Katherine Elizabeth (Schneider) B.; m. Norma Lynn Andersen, Nov. 12, 1966; children: Stephanie Lynn, Michelle Terese. B.A., Seattle U., 1967; M.Sc., U. Calgary, 1972; Ph.D., U. N.C., 1974. Asst. prof. Seattle U., 1974-77; asst. prof. No. Mich. U., Marquette, 1977-78, U. Wash, Seattle, 1978-82; asst. prof. psychology Pacific Luth. U., Tacoma, 1982—. Cons. reviewer: Prentice Hall, 1982; cons. reviewer: Dorsey Press, 1980-81, Little Brown Press, 1976, Child Devel, Chgo., 1979; Sr. author: Psychology, 1984; contbr. articles to profl. jours.; internat. editorial bd., Carfax Pub. Co., Oxford, Eng., 1982—. Summer camp adminstr. Cath. Youth Orgn., Seattle, 1962-67. Teaching fellow U. N.C., 1972-73; Seattle U. grantee, 1976; U. Wash. grantee, 1979, 80; Pacific Luth. U. grantee, 1982. Mem. AAUP, Am. Psychol. Assn., Soc. Research in Child Devel. Subspecialties: Developmental psychology; Cognition. Current work: Development of cognitive processes in children. Office: Dept Psychology Pacific Luth Univ Tacoma WA 98447

BROWN, ROGER WILLIAM, psychologist, educator; b. Detroit, Apr. 14, 1925; s. Frank Herbert and Muriel Louise (Graham) B. A.B., U. Mich., 1948, Ph.D., 1952; M.A. (hon.), Harvard, 1962, D. Univ., U. York, Eng.; D.Sc., Bucknell U., 1980. Asst. prof. psychology Harvard, Cambridge, Mass., 1952-57, prof. social psychology, 1962—, John Lindsley prof. psychology in memory William James, 1974—, chmn. dept. social relations, 1967-70; asso. prof. psychology Mass. Inst. Tech., Cambridge, 1957-61, prof. social psychology, 1961-62; Chmn. behavioral scis. study sect. NIH, 1961-63. Author: Words and Things, 1958, (with others) New Directions in Psychology, 1962, The Acquisition of Language, 1964, Social Psychology, 1965, Psycholinguistics, 1970, A First Language, 1973, (with R. Herrnstein) Psychology, 1975. Recipient Distinguished Research award Nat. Council Tchrs. English, 1974. Mem. Am. (Distinguished Sci. Contbn. award 1971, G. Stanley Hall award), New Eng. Psychol. Assn. (pres. 1965-66), Eastern Psychol. Assn. (pres. 1971-72), Linguistic Soc. Am., Am. Acad. Arts and Scis., Nat. Acad. Scis. Subspecialties: Cognition; Social psychology. Current work: Psychological causality and language. Home: 100 Memorial Dr Cambridge MA 02142

BROWN, R(OY) LEONARD, computer science educator, scientific computing consultant; b. Montgomery, Ala., Aug. 30, 1949; s. Roy Leonard and Mary Emma (Clark) B. B.S. Cum Laude, Tulane U., 1971; M.S., U. Ill.-Urbana, 1973, Ph.D., 1975. Asst. prof. applied math U. Va., Charlottesville, 1974-79; asst. prof. Drexel U., 1979—; vis. researcher Inst. Computer Applications in Sci. and Engring., Hampton, Va., 1975; cons. in field. Trustee Unitarian Ch., Cherry Hill, N.J., 1982. NASA grantee, 1976-79; Air Force Office Sci. Research grantee, 1979-80. Mem. Assn. Computing Machinery, Soc. Indsl. and Applied Math., Am. Math. Soc., Sigma Xi, Alpha Sigma Phi. Subspecialties: Mathematical software; Algorithms. Current work: Development of interactive software to allow engineers to describe a system of differential equations in mathematical language and let the computer determine and display the appropriate numerical solution. Office: Department of Mathematical Sciences Drexel University Philadelphia PA 19104

BROWN, STEPHEN WOODY, psychologist; b. Cleve., Aug. 3, 1939; s. Joe and Enid (Hirsch) B.; m. Malinda Slugocki, July 27, 1975; children: Kimberly M., David M. B.A., Calif. State U.-Los Angeles, 1962; Ph.D., U. So. Calif., 1966. Asst. prof. Calif. State U.-Dominquez Hills, 1968; asst. prof. U. So. Calif. Med. Sch., Los Angeles, 1972-76; prof. Calif. Sch. Profl. Psychology, Fresno, 1976-80; assoc. prof. Pepperdine U., Los Angeles, 1980—; clin. psychologist Fresno Community Hosp., 1978-80; cons. pub. schs., Los Angeles, Orange, Riverside County, 1968—. Author: Schools and Microcomputers, 1982; Contbr. articles to profl. jours. USPHS postdoctoral fellow, 1969. Fellow Am. Geriatrics Soc.; mem. Am. Psychol. Assn. Subspecialties: Psychometrics; Computer applications in education and psychology. Current work: Educational applications of computers; research in clinical psychology. Home: 23391 Devonshire St El Toro CA 92630 Office: Dept Psychology Pepperdine U Los Angeles CA 90044

BROWN, STEVEN HARRY, corporation health physicist, consultant; b. Phila., Sept. 16, 1948; s. Robert Martin and Vera Ethel (Lipovsky) B.; m. Kathryn Helena Vassi, May 24, 1970; children: Chad, Joshua. A.B.S., Temple U., 1970, B.S., 1971; M.A., West Chester (Pa.) U., 1974. Health physicist Temple U., Phila., 1969-71; tchr. phys. sci. Phila. Sch. Dist., 1971-76; mgr. radiation protection Westinghouse Electric Corp., Lakewood, Colo., 1976-80; mgr. western regional office Radiation Mgmt. Corp., Phila., 1980-82; prin. safety analysis engr. Rockwell Internat., Golden, Col., 1982—; cons. Westinghouse Electric Corp., Lakewood, Colo., 1981, Earth Scis. Inc., Golden, 1982, Radiation Mgmt. Corp., Chgo., 1982. Mem. Nat. Health Physics Soc. (pres. Rocky Mountain chpt. 1982-83), Am. Nuclear Soc., Nat. Mgmt. Assn. Subspecialties: Nuclear engineering; Environmental toxicology. Current work: Failure mode and effects analysis applied to nuclear systems; toxicology of uranium and plutonium; emergency response and probabilistic risk assessment for energy generating systems; radioactive waste reduction via extractive metalurgy. Home: 4673 S Vivian Ct Morrison CO 80465 Office: Rockwell Internat PO Box 464 Golden CO 80401

BROWN, WAYNE SAMUEL, mechanical engineering educator; b. Provo, Utah, Mar. 19, 1928; s. Cleveland W. and Wilmirth H. B.; m. Joyce Fechser, Mar. 4, 1948; children: Karen Brewster, Diane Whittaker, Gary W., Don R., Janet. B.S., U. Utah, 1951; M.S., U. Tenn., 1953; Ph.D., Stanford U., 1960. Registered profl. engr., Utah. Development engineer Oak Ridge Nat. Lab., 1951-53; research engr. Utah Research & Devel. Co., Salt Lake City, 1960-64; asst. prof. U. Utah, Salt Lake City, 1953-57, prof. mech. engring., 1964—, chmn. dept., 1964-70, assoc. dean., 1970-72, dean, 1973-78; pres. Utah Innovation Ctr., Inc., Salt Lake City, 1982—; dir. Terra Tek Inc., Native Plants, Inc., Bunnel Life Systems, Inc.; cons. NSF, 1980—. Editor: Technovation, 1980—. Bd. dirs. United Way Greater Salt Lake, 1978-82, v.p., 1981-82. Served with USMC, 1946-47. Mem. ASME, Soc. Exptl. Stress Analysis. Mem. Ch. of Jesus Christ of Latter-day Saints. Subspecialty: Mechanical engineering. Current work: Research on technical innovation and entrepreneurship. Home: 1630 Arlington Dr Salt Lake City UT 85103 Office: Mech Engring Dept U Utah Salt Lake City UT 84112

BROWN, WILLIAM LACY, genetic supply company executive; b. Arbovale, W.Va., July 16, 1913; s. Tilden L. and Mamie Hudson (Orndoff) B.; m. Alice Hevener Hannah, Aug. 17, 1941; children: Alicia Anne, William Tilden. B.A., Bridgewater Coll., Va., 1936; M.S., Washington U., St. Louis, 1939, Ph.D., 1941. Cytogeneticist Dept. Agr., Washington, 1941-42; dir. maize breeding Rogers Bros. Co., Olivia, Minn., 1942-45; with Pioneer Hi-Bred Internat., Inc., 1945—, v.p., dir. corp. research, Des Moines, 1965-75, pres., 1975-79, chief exec. officer, 1975-81, chmn., dir., 1979—; dir. Am. Farmland Trust, Winrock Internat, Pioneer de Centroamerica, S.A., Pishrow Seed Co., Iowa-Des Moines Nat. Bank; extra-mural prof. botany Washington U., 1957-65; mem. Gov. Iowa Sci. Adv. Council, 1977—. Author papers maize cytogenetics, evolution, germplasm conservation. Bd. regents Nat. Colonial Farm. Trustee Accokeek Found., Washington, Bridgewater (Va.) Coll. Fulbright advanced research scholar Imperial Coll. Tropical Agr., Trinidad, 1952-53; Univ. fellow Drake U., 1981—. Fellow Am. Soc. Agronomy, Iowa Acad. Sci. (dir.); mem. AAAS, Nat. Acad. Scis. (trustee), Am., Can. genetics socs., Am. Genetics Assn., Am. Inst. Biol. Scis., Bot. Soc. Am., Soc. Econ. Botany (Disting. Econ. Botanist award 1980), Phi Beta Kappa, Sigma Xi. Quaker. Clubs: Hyperion Field, Des Moines. Subspecialty: Plant genetics. Home: 6980 NW Beaver Dr Johnston IA 50131 Office: 6800 Pioneer Pkwy Johnston IA 50131

BROWN, WILLIAM SAMUEL, speech and linguistics educator; b. Pottstown, Pa., Apr. 25, 1940; s. William Samuel and Elizabeth (Gallagher) B.; m. Elaine Kay Whitehouse, Aug. 18, 1962; children: William Samuel III, Allen Reed. B.S., Edinboro State Coll., 1962; M.A, SUNY-Buffalo, 1967, Ph.D., 1969. Speech pathologist Crawford County, Meadville, Pa., 1962-65; research asst. SUNY, Buffalo, 1965-68; postdoctoral fellow U. Fla., Gainesville, 1968-70, asst. prof., 1970-76, assoc. prof., 1976-81, prof. speech and linguistics, 1981—; dir. Inst. Advanced Study Communication Processes, Gainesville, 1981—. Contbr. articles to profl jours. Basketball ofcl. Mid-Fla. Ofcls. Assn., Gainesville. Fellow Am. Speech and Hearing Assn., Internat. Soc. Phonetic Scis.; mem. Am. Assn. Phoenetic Scis. (exec. sec. 1977-85), Acoustical Soc. Am., Sigma Xi. Subspecialties: Physiological psychology; Sensory processes. Current work: Direct and conduct research in the communication processes, specifically related to the normal and abnormal physiological processes of human speech. Home: 1517 NW 19th St Gainesville FL 32605 Office: Inst Advanced Study Communication Processes ASB 63 U Fla Gainesville FL 32611

BROWN, WILLIAM TED, medical researcher, physician; b. Missoula, Mont., Feb. 18, 1946; s. C. W. and H. F. B.; m. Barbara Blegen, Mar. 31, 1967 (div. Jan. 1982); 1 dau., Vanessa. B.A., Johns Hopkins U., 1967, M.A., 1969, Ph.D., 1973; M.D. cum laude, Harvard U., 1974. Diplomate: Am. Bd. Internal Medicine, Am. Bd. Clin. Genetics. Intern Roosevelt Hosp., N.Y.C., 1974-75, resident in medicine, 1975-77; Nat. Inst. on Aging postdoctoral fellow in genetics Cornell U. Med. Sch., N.Y.C., 1977-78, asst. prof., 1978-82; chmn. dept. human genetics N.Y. State Inst. Basic Research in Developmental Diseases, N.Y.C., 1982—; asst. attending physician N.Y. Hosp., N.Y.C., 1981—; adj. asst. prof. Rockefeller U., N.Y.C., 1981—; mem. Nat. Inst. Aging Research Resources Adv. Panel. Contbr. chpts., articles on aging, genetics, mental retardation to profl. jours.; editorial bd.: Trisory 21. Recipient Andrew W. Mellon Tchr.-Scientist award, 1979; NSF research fellow Johns Hopkins U., 1967-69. Mem. Am. Soc. Human Genetics, Gerontol. Soc., Am. Fedn. Clin. Research, Harvey Soc., N.Y. Acad. Scis. Subspecialties: Genetics and genetic engineering (medicine); Gerontology. Current work: Genetic diseases affecting aging, mental retardation, fragile x syndrome, and Downs syndrome. Office: NY State Inst Basic Research in Developmental Diseases 1050 Forest Hill Rd Staten Island NY 10314

BROWNE, JAMES CLAYTON, physics and computer science educator, researcher; b. Conway, Ark., Jan. 16, 1935; m., 1959; 3 children. B.A., Hendrix Coll., 1956; Ph.D. in Chemistry, U. Tex., 1960. Asst. prof. physics U. Tex.-Austin, 1960-64; prof. computer sci. Queen's U., Belfast, No. Ireland, 1965-68; prof. physics and computer sci. U. Tex.-Austin, 1968—, research scientist, 1973—; cons. NSF. NSF fellow, 1964-65. Mem. Am. Phys. Soc., ACM, Soc. Indsl. and Applied Math. Subspecialty: Operating systems. Office: Dept Physics U Tex Austin TX 78712

BROWNE, RONALD GREGORY, biomedical scientist, educator; b. San Diego, Mar. 15, 1948; s. Berle Eldon and Shirley Reed (Mathews) B. B.A. in Psychology, U. Calif., San Diego, 1971; M.S. in Biomed. Sci, Health Sci. Center, U. Tex., Houston, 1973; Ph.D. in Biomed, Sci, Health Sci. Center, U. Tex., Houston, 1975. Predoctoral fellow Tex. Research Inst. Mental Sci., Houston, 1971-75; NIH postdoctoral research fellow U. Calif., San Diego, La Jolla, 1975-77; instr. in psychology Miramar Coll., 1977-78; Nat. Inst. Drug Abuse postdoctoral research fellow dept. psychiatry U. Calif.- San Diego Sch. Medicine, La Jolla, 1977-78; research scientist Pfizer, Inc., Groton, Conn., 1978-81, sr. research and scientist, 1982—; adj. assoc. prof. psychology U. R.I., 1982—; vis. assoc. prof. psychology Conn. Coll., 1983-84. Contbr. articles, abstracts, chpts. to profl. publs. Subspecialty: Neuropharmacology. Current work: Drug discovery; neuropeptides and behavior; drug discrimination. Office: Central Research Pfizer Inc Groton CT 06340

BROWNELL, KELLY DAVID, educator; b. Evansville, Ind., Oct. 31, 1951; s. Arnold B. and Margaret E. B.; m. Mary Jo Brownell, Aug. 20, 1977; 1 son, Matthew Joseph. B.A., Purdue U., 1973; M.S., Rutgers U., 1975, Ph.D., 1977. Research assoc. Brown U., Providence, 1976-77; asst. prof. U. Pa., Phila., 1977-82, assoc. prof. dept. psychiatry, 1982—. Assoc. editor: Jour. Applied Behavior Analysis, 1980—; editorial bd.: Jour. Cons. Clin. Psychology, 1979—; author: Partnership Diet Program, 1980, (with others) Annual Review of Behavior Therapy, 1982. NIMH grantee, 1980; N.Y. Acad. Sci. Cattell award, 1977; Young Investigator grantee Nat. Heart, Lung and Blood Inst., 1977. Mem. Am. Psychol. Assn., n. Advancement Behavior Therapy (dir.), Soc. Behavioral Medicine (dir.), AAAS, Acad. Behavioral Med. Research. Subspecialties: Behavioral psychology; Preventive medicine. Current work: Coronary risk reduction; nutrition; exercise physiology. Office: Dept Psychiatry Univ Pa 133 S 36 St Philadelphia PA 19104

BROWNIE, ALEXANDER C., biochemistry educator, researcher; b. Bathgate, Scotland, Mar. 6, 1931. B.Sc., U. Edinburgh, 1952, Ph.D. in Biochemistry, 1955, D.Sc., 1981. Asst. biochemistry U. Edinburgh, 1953-55; Scottish Hosps. Endowment Research Trust fellow Royal Infirmary, Edinburgh, 1955-56; Organon research fellow in pharmacology U. St. Andrews, Scotland, 1956-59; med. research fellow, 1959-62, hon. lectr., 1960-62; USPHS fellow U. Utah, 1962-63; asst. prof. pathology and biochemistry SUNY-Buffalo, 1963-67, assoc. prof. biochemistry, 1967-70, prof. biochemistry, 1970—, chmn. dept., 1976—, research assoc. prof. pathology, 1967—. Mem. Endocrine Soc., Am. Soc. Biol. Chemists. Subspecialty: Biochemistry (biology). Office: Dept Biochemistry 102 Cary Hall SUNY Buffalo NY 14214

BROWNING, CHARLES BENTON, university dean; b. Houston, Sept. 16, 1931; s. Earl William and Emma B.; m. Magda L. Luest, Jan. 14, 1956; children: Charles, Susan, Steve, Karen, Heidi, Gary. B.S., Tex. Tech. U., 1955; M.S., Kans. State U., 1956, Ph.D., 1958. Asst. to full prof. Miss. State U., 1958-66; head dept. dairy sci. U. Fla., Gainesville, 1966-69; dean (Coll. of Agr.), 1969-79, Coll. of Agr., Okla. State U., Stillwater, 1979—; dir. Agrl. Expt. Sta., and Coop. Extension Service. Subspecialty: Animal nutrition. Home: 1002 W Will Rogers St Stillwater OK 74074 Office: 139 Ag Hall Okla State U Stillwater OK 74078

BROWNING, MICHAEL DOUGLAS, neurochemist; b. Dallas, Sept, 10, 1946; s. Walter Acker and Mary Jane (Schuhmacher) B. B.A. in English, U. Tex., 1970; Ph.D. in Biology, U. Calif.-Irvine, 1979. NIMH research trainee U. Calif.-Irvine, 1975-79; Muscular Dystrophy Assn. postdoctoral fellow Yale U. Sch. Medicine, New Haven, 1981-83; asst. prof. dept. molecular and cellular neurosci. Rockefeller U., N.Y.C., 1983—. Contbr. articles to profl. jours. Mem. Soc. for Neurosci., N.Y. Acad. Scis. Subspecialties: Neurobiology; Neurochemistry. Current work: Relationship of protein phosphorylation to synaptic plasticity. Synaptic plasticity, protein phosphorylation, synaptic vesicles, calcium transport, pyruvate dehydrogenase. Office: Molecular and Cellular Neurosci Rockefeller U 1230 York Ave New York NY 10021

BROWNLEE, DONALD EUGENE, II, astronomer, educator; b. Las Vegas, Nev., Dec. 21, 1943; s. Donald Eugene and Geraldine Florence (Stephen) B.; m. Paula Szkody. B.S. in Elec. Engring, U. Calif., Berkeley, 1965; Ph.D. in Astronomy, U. Wash., 1970. Research assoc. U. Wash., 1970-77, asso. prof. astronomy, 1977—; asso. geochemistry Calif. Inst. Tech., Pasadena, 1977—; cons. NASA, 1976—. Author papers in field, chpts. in books. Grantee NASA, 1975. Mem. Internat. Astron. Union, Am. Astron. Assn., AAAS, Meteoritical Soc., Com. Space Research Dust. Subspecialties: Planetary science; Geochemistry. Home: 3118A Portage Bay Pl E Seattle WA 98102 Office: Dept Astronomy Univ Wash Seattle WA 98195

BROXMEYER, HAL EDWARD, biologist, researcher; b. Bklyn., Nov. 27, 1944; s. David and Anna (Gurman) B.; m. C. Beth Broxmeyer, Nov. 22, 1945; 1 son, Eric Jay. B.S., CUNY, 1966; M.S., L.I. U., 1969; Ph.D., N.Y. U., 1973. Lab. specialist Midwood High Sch., Bklyn., 1966-71; research asso. dept. medicine Queen's U., Kingston, Ont., Can., 1973-75; asso. researcher Sloan-Kettering Inst., N.Y.C., 1975-76, research asso., 1976-78, asst. mem., 1978—; asst. prof. Cornell U. Grad. Sch. Medicine, N.Y.C., 1980—; mem. study sect. NIH; reviewer Nat. Cancer Inst., Am. Cancer Soc., NSF, VA, U.S.-Israel Bi-nat. Sci. Found. Contbr. articles to sci. jours.; assoc. editor: Exptl. Hematology. Recipient Spl. Fellow award Leukemia Soc. Am., 1976-78, Scholar award, 1978-83; Mellor award Sloan-Kettering Inst., 1976, 77; Founders Day award N.Y. U., 1973; NIH grantee, 1972-73. Mem. Am. Soc. Hematology (subcom. leukocyte physiology 1981-84, subcom. erythropoietin and cell proliferation 1982—), Internat. Soc. Exptl. Hematology, Am. Assn. Immunologists, Am. Assn. Cancer Research, Reticuloendothelial Soc., Internat. Assn. Comparative Research Leukemia and Related Disorders, Cell Kinetics Soc., N.Y. Acad. Sci., AAAS. Subspecialties: Cancer research (medicine); Hematology. Current work: Regulation of the production of blood cells - normal and leukemia.

BRUCE, BARBARA JEAN, molecular biologist; b. Ithaca, N.Y., Aug. 20, 1942; d. William Fausett and Verna Belle (Updike) B. B.A., Ohio Wesleyan U., 1964; M.S., U. Wis., 1967. Research technologist U. Mich., Ann Arbor, 1967-74; research asst. Upjohn Co., Kalamazoo, 1974—; mem. Upjohn Asst. Symposium Commn., Kalamazoo, 1980—, chmn., 1983—. Contbr. articles to profl. jours. Bd. dirs. Portage (Mich.) Parks, 1980—, chmn., 1983. Mem. Am. Soc. Microbiology., AAAS. Subspecialties: Molecular biology; Microbiology. Current work: Expression of eukaryotic genes in prokaryotes and eukaytotes. Patentee in field. Home: 5122 Old Colony Rd Portage MI 49081 Office: Molecular Biology Upjohn Co 301 Henrietta St Kalamazoo MI 49001

BRUCE, JAMES T., oceanographer; b. Bridgeport, Conn., Oct. 21, 1953; s. Alfred C. and Sheila (Donnelly) B. B.S., Nasson Coll.-Maine, 1976; grad. cert., U. Va., 1982. Oceanographer NOAA/NOS, Rockville, Md., 1976-83. Contbr. reports in field. Recipient unit citation NOAA/NOS, 1981; named outstanding performer NOAA/NOS, 1980, 81. Mem. Internat. Oceanic Found., Marine Tech. Soc., Oceanic Soc. Subspecialty: Oceanography. Current work: Tidal current tracking and recording logistical survey planning data retrieval and analysis observational data reduction and publication; specialized computer program development, trouble-shooting and de-bugging; institutional liason field study investigation; data quality assurance program development. Home: 5913 Halsey Rd Rockville MD 20851 Office: NOAA/NOS C-2112 6001 Executive Blvd Rockville MD 20852

BRUCE, THOMAS ALLEN, educator, physician; b. Mountain Home, Ark., Dec. 22, 1930; s. Rex Floyd and Dora Madeline (Fee) B.; m. Dolores Fay Montgomery, May 28, 1960; children—T.K. Montgomery, Dana Fee. B.S.M., M.D., U. Ark., 1955. Intern Duke Hosp., 1956-57; resident medicine Bellevue Hosp., N.Y.C., 1957, Meml. Center Cancer and Allied Diseases, 1958, Parkland Meml. Hosp., Dallas, 1958-59; cardiopulmonary trainee Southwestern Med. Sch. of U. Tex., 1959-60; cardiac research fellow Hammersmith Hosp. and U. London Postgrad. Med. Sch., London, 1960-61, Harvard Bus. Sch., 1974; instr. to prof. medicine Wayne State U., 1961-68, also asst. dean; prof., head cardiovascular sect. U. Okla. Med. Center, 1968-74; prof. medicine, dean Coll. Medicine, U. Ark. Med. Scis., 1974—; med. dir. Barton Research Inst.; coordinator Sino-Am. Med. Exchange Program. Dir. Comml. Nat. Bank.; Mem. Ark. Commn. on Health Cost Containment; active Friends of the Zoo, Partners of the Ams., Ark. Symphony Assn., Ark. Opera Theater, Ark. Art Center, Ark. Chamber Music Soc., Friends of Library. Recipient Ark. Gov.'s Meritorious Achievement award. Fellow A.C.P., Am. Coll. Cardiology, Internat. Coll. Angiology, Council Clin. Cardiology, Council Arteriosclerosis; mem. Assn. U. Cardiologists, Assn. Am. Med. Colls. Council Deans (chmn. so. council deans 1977-78), Central Soc. Clin. Research, Am. Fedn. Clin. Research, Ark. Heart Assn., Internat. Soc. Heart Research, Soc. for Human Values in Medicine, Am. Rural Health Assn. (nat. cabinet), Ark. Caduceus Club, Old Statehouse Founders Soc., Ark. Med. Soc., Pulaski County Med. Soc. (exec. com.), Ark. Hist. Assn., Little Rock Hist. Soc., Smithsonian Inst. Assos., Sigma Xi, Alpha Omega Alpha. Clubs: Rotary, Cosmos (Washington). Subspecialty: Cardiology. Bruce Soc. Am. Research and publs. on cardiovascular disease including left ventricular function in cardiac denervation, coronary heart disease, myocardial metabolism relating to phospholipids in graded cardiac ischemia, med. edn. with particular reference to rural health care. Home: 4 Hillandale Robinwood Little Rock AR 72207 Office: 4301 W Markham Little Rock AR 72201

BRUCE, WILLIAM ROBERT, physician, educator; b. Hamhung, Korea, May 26, 1929; s. George Findlay and Ellen (Tate) B.; m. Margaret MacFarlane, June 15, 1957; children: Graham Douglas, Lynda Jeanne, Kevin Robert. B.Sc., U. Alta., 1950; Ph.D., U. Sask., 1956; M.D., U. Chgo., 1958. Intern Billings Hosp., Chgo., 1958-59; mem. faculty U. Toronto, Can., 1959, prof. biophysics, 1966—; mem. physics sect. Ont. Cancer Inst., Toronto, 1959-81; dir. Ludwig Inst. for Cancer Research, Toronto, 1981—. Fellow Royal Coll. Physicians (Can.), Royal Soc. Can.; mem. Assn. Cancer Research. Subspecialty: Cancer research (medicine). Research, publns. on X-ray and gamma ray penetration, control red blood cell prodn., action of anti-cancer agts. on normal and tumor cells, sperm prodn., computers in med. records, origins of human cancer. Home: 4 Marshfield Ct Don Mills ON Canada Office: 9 Earl St Toronto ON M4Y 1M4 Canada

BRUCK, DAVID KENNETH, botanist, researcher; b. East Lansing, Mich., Sept. 1, 1953; s. Max and Emily Reva (Zisenwein) B. A.B., U. Calif.-Berkeley, 1977; Ph.D., Cornell U., 1982. Postdoctoral fellow UCLA, 1982—. Contbr. articles to profl. jours. Mem. Bot. Soc. Am., Am. Soc. Plant Physiologists. Subspecialties: Plant growth; Morphology. Current work: Control of plant cell differentiation. Office: UCLA Dept of Biology 405 Hilgard Los Angeles CA 90024

BRUECKNER, KEITH ALLAN, theoretical physicist, educator; b. Mpls., Mar. 19, 1924; s. Leo John and Agnes (Holl) B.; children: Jan Keith, Anthony Leo, Leslie. B.A., U. Minn., 1945, M.A., 1947; Ph.D., U. Calif. at Berkeley, 1950; D.Sc. (hon.), Ind. U., 1976. Prof. physics U. Ind., 1951-55; physicist Brookhaven Nat. Lab., N.Y., 1955-56; prof. physics U. Pa., 1956-59, U. Calif. at San Diego, 1959—, chmn. dept. physics, 1959-61, dean, 1963, dean letters and sci., 1963-65, dean grad. studies, 1965; dir. Inst. Pure and Applied Phys. Scis., 1965-69; v.p., dir. research Inst. Def. Analysis, Washington, 1961-62 (on leave); tech. dir. Helliodyne Corp., San Diego, 1968-69 (on leave), KMS Tech. Center, 1969-70; exec. v.p., tech. dir. KMS Fusion, Inc., Ann Arbor, Mich., 1971-74 (on leave). Co-editor: Pure and Applied Physics series. Served with USAAF, 1943-46. Recipient Dannie Heineman prize for math. physics, 1963. Fellow Am. Phys. Soc., Am. Acad. Arts and Scis.; mem. Nat. Acad. Scis. Club: Am. Alpine. Subspecialties: Theoretical physics; Fusion. Office: Dept Physics U Calif La Jolla CA 92093

BRUELS, MARK C., radiol. physicist; b. Kansas City, Mo., Jan. 2, 1941; s. C.B. and Florence Rose (Collins) B.; m. (div.); children: Christine Clare, Nicholas Raphael. B.A., Mankato State U., 1962; M.S., U. Kans., 1965, Ph.D., 1969. Diplomate: Am. Bd. Radiology. Asst. prof. physics Slippery Rock (Pa.) State Coll., 1969-72; Presdl. intern Naval Aerospace Med. Research Detachment, New Orleans, 1972; postdoctoral fellow M.D. Anderson Hosp., Houston, 1972-73; staff internal medicine Baylor Coll. Medicine, Houston, 1973; radiation physicist VA Hosp., Mpls., 1973-77; radiol. physicist Radiology Consultants, Inc., Omaha, 1977—; clin. assoc. prof. radiology Creighton U., Omaha; cons. in field. Contbr. articles to profl. jours. NSF grantee, 1970-72. Mem. Am. Assn. Physicists in Medicine, Am. Coll. Radiology, Am. Soc. Therapeutic Radiologists, Am. Phys. Soc. Subspecialties: Radiological physics; Imaging technology. Current work: 3D treatment planning; non-linear optimization techniques. Clinical radiationphysics. Home: 5518 Grover St Apt 708 Omaha NE 68106 Office: 8031 W Center Rd Ford Plaza Bldg Suite 209 Omaha NE 68124

BRUGEL, EDWARD WILLIAM, astrophysicist; b. N.Y.C., Nov. 29, 1948; s. Edward William and Muriel Ellen (Flanly) B.; m. Kanistra Wointranant, July 25, 1974 (div.); 1 child, Narinton Wointranont. B.S., SUNY, Stony Brook, 1970; M.S., U. Wash., 1978, Ph.D., 1980. Teaching/research asst. U. Wash., Seattle, 1974-80; research asso., dir. Regional Data Analysis Facility, Internat. Ultraviolet Explorer satellite, Lab. Atmospheric and Space Physics, U. Colo., Boulder, 1981—. Contbr. articles to profl. jours. Served with U.S. Army, 1970-74. NASA grantee, 1980, 81, 83. Mem. Am. Astron. Soc., Astron. Soc. Pacific. Subspecialties: Ultraviolet high energy astrophysics; Optical astronomy. Current work: Research, interstellar medium in regions of recent star formation, ejecta from young objects (Herbig-Haro objects); shock waves in interstellar clouds, symbiotic stars. Home: 708 Mohawk Dr 16 Boulder CO 80303 Office: Lab Atmospheric and Space Physics Campus Box 392 U Colo Boulder CO 80309

BRUHN, ARNOLD RAHN, JR., clin. psychologist; b. Bklyn., Dec. 30, 1941; s. Arnold Rahn and Paula (Muich) B.; m. Arlene C. Palmer, June 23, 1967; children: Alexis, Erika. B.A., U. Portland, 1963, M.A., 1966; M.S., Portland State U., 1972; Ph.D., Duke U., 1976. Psychologist Dorothea Dix Hosp., Raleigh, N.C., 1973-76; psychology intern Duke U Med. Center, Durham, N.C., 1975-76; asst. prof. psychology George Washington U., Washington, 1976-82; staff psychologist Alexandria (Va.) Community Mental Health Center, 1977—; asst. research prof. George Washington Med. Center, Washington, 1980—. Cons. editor: Jour. Personality Assessment, 1976—; author articles. USPSH fellow, 1972-76. Mem. Soc. for Personality Assessment, Am. Psychol. Assn. Christian. Subspecialties: Clinical psychology; Cognition. Current work: Personality theory and personality assessment. Earliest memories. Home: 7820 Glenbrook Rd Bethesda MD 20814 Office: George Washington U Med Center 901 23d St NW Washington DC 20037

BRUMBERGER, HARRY, physical chemist, educator, researcher, consultant; b. Vienna, Austria, Aug. 28, 1926; came to U.S., 1940, naturalized, 1948; s. Leon and Rose (Kraft) B.; m. Vilma, June 21, 1950; children: Jesse, Eva. B.S. in Chemistry, Poly. Inst. Bklyn., 1949, M.S., 1952, Ph.D., 1955. Teaching asst. Poly. Inst. Bklyn., 1951-54; research asso. dept. chemistry Cornell U., 1954-57; asst. prof. phys. chemistry Syracuse U., 1957-62, asso. prof., 1962-69, prof., 1969—, dir. grad. biophysics progrm, 1977-83. Contbr. numerous articles to profl. publs. Served with U.S. Army, 1946-47. Research Corp. grantee, 1959-60; NSF grantee, 1959-61, 61-63, 64-66, 64-65, 72-74, 80-81, 80-85; Army Research Office grantee, 1964-65; Am. Chem. Soc. Petroleum Research Fund grantee, 1965-69; Office Naval Research grantee, 1964-69; Oak Ridge Assoc. Univs. grantee, 1977-78; Syracuse U. grantee, 1979-81. Mem. AAAS, Am. Crystallographic Assn. Subspecialties: Physical chemistry; Catalysis chemistry. Current work: Small-angle x-ray scattering and its applications to study of supported-metal catalysts; biol. macromolecules; others. Office: Dept Chemistry Syracuse U Syracuse NY 13210

BRUMM, DOUGLAS BRUCE, electrical engineer, educator, consultant, researcher; b. Barry County, Mich., Aug. 4, 1940; s. Bruce Dwight and Dorotha Clarice (Green) B.; m. Phyllis Jean Orthner, Aug. 14, 1965; children: Bruce Douglas, Dawn Marie. B.S.E.E., Mich. Technol. U., 1962; M.S.E., U. Mich., 1964, Ph.D., 1970. Registered profl. engr., Mich. Assoc. engr. Raytheon Co., Wayland, Mass., 1962-63; Research assoc. U. Mich., Ann Arbor, 1963-70; assoc. prof. elec. engring. Mich. Technol. U., Houghton, 1970—. Contbr. articles to profl. jours. Mem. IEEE, IEEE Computer Soc., Optical Soc. Am., Sigma Xi. Baptist. Subspecialties: Electronics; Fiber optics. Current work: Fiber optic communication systems, microprocessor-controlled instruments, laser applications, holography. Home: 1140 Rockhouse Rd Calumet MI 49913 Office: Elec Engring Dept Mich Technol U Houghton MI 49931

BRUNDA, MICHAEL JOHN, immunologist; b. Passaic, N.J., Dec. 16, 1950; s. John and Helena (Gawronski) B.; m. Patricia Katherine Ann Mongini, July 28, 1979; 1 dau., Nicole Anna. A.B., U. Rochester, 1971; Ph.D., Stanford U., 1975. Postdoctoral fellow Nat. Jewish Hosp. and Research Center, Denver, 1975-78; immunologist Nat. Cancer Inst., Bethesda, Md., 1978-82; sr. scientist dept. exptl. and applied biology Hoffmann LaRoche, Nutley, N.J. Contbr. articles to profl. jours. Nat. Cancer Inst. postdoctoral fellow, 1977; AAI Travel grantee, 1980. Mem. Am. Assn. Immunologists, Am. Assn. Cancer Research, AAAS. Democrat. Orthodox Christian. Subspecialties: Immunobiology and immunology; Immunopharmacology. Current work: Study of interferons, interleukins and other immunomodulators on natural killer cell activity and macrophage function as antitumor defense mechanism. Office: Dept Immunotherapy Hoffman La Roche Nutley NJ 07110

BRUNER, RALPH CLAYBURN, testing services co. exec., metallurgist; b. Oklahoma City, Apr. 22, 1921; s. Ralph Sylvester and Macil Gladys (Stroup) B.; m. Cicely Louise Fidler, July 3, 1954; children: Martha Ellen, David Ralph. B.A. in Chemistry, Calif. State U. - Fullerton, 1965. Registered profl. engr., Okla. Research engr. N.Am. Aviation, Inc., Los Angeles, 1947-52, supr. metallurgy, Columbus, Ohio, 1952-56; chief chem. lab. Autonetics, Anaheim, Calif., 1956-65; mgr. labs. Rockwell Internat., Tulsa, 1965-76; pres. Metlab Testing Services, Inc., Tulsa, 1976—. Contbr. articles to profl. meetings. Mem. Am. Soc. Metals, ASME, Am. Welding Soc., Am. Foundrymen's Soc. Automotive Engrs. Republican. Roman Catholic. Subspecialties: Metallurgy; Alloys. Current work: Metallurgical forensic analysis. Patentee in field. Office: 6825 E 38th St Tulsa OK

BRUNISH, WENDEE M., astronomer; b. Los Angeles, Dec. 31, 1953; d. Robert and Virginia Florence (Hughes) B.; m. Stephen A. Becker, May 30, 1980. A.B., Vassar Coll., 1975; M.S., U. Ill., 1977, Ph.D., 1981. Teaching and research asst. U. Ill., 1975-81; research asso. astronomy U. So. Calif., Los Angeles, 1981-83; vis. staff mem. Los Alamos Nat. Lab., 1983—. Mem. Am. Astron. Soc. Subspecialties: Theoretical astrophysics; Solar physics. Current work: Evolution of massive stars with mass loss; study of solar oscillations. Office: LANL PO Box 1663 MS B288 Los Alamos NM 87545

BRUNJES, PETER CRAWFORD, developmental psychobiologist; b. Columbus, Ohio, June 19, 1953; s. Thomas H. and Marie E. (Baker) B.; m. Victoria L. Manning, 1976; 1 son, Benjamin. B.S., Mich. State U., 1974; Ph.D., Ind. U., 1979. Postdoctoral fellow U. Ill.-Champaign, 1979-80; asst. prof. psychology U. Va., Charlottesville, 1980—. Mem. Internat. Soc. Developmental Psychobiology, AAAS, Assn. for Chemoreception Scis., Soc. for Neurosci. Subspecialties: Neurobiology; Psychobiology. Current work: Teaching and research on the devel. of brain and behavior, factors influencing the rate and course of early life. Office: University of Virginia Dept Psychology Gilmer Hall Charlottesville VA 22901

BRUNK, WILLIAM EDWARD, astronomer; b. Cleve., Nov. 24, 1928; s. Edgar Rea and Mabel Mowbray (Pearson) B.; m.; 1 dau., Anna Kathryn. B.S., Case Inst. Tech., 1952, M.S., 1954, Ph.D., 1963. Aero. research scientist Lewis Flight Propulsion Lab., NACA, Cleve., 1954-58; aerospace engr. Lewis Research Center, NASA, Cleve., 1958-64; staff scientist for planetary astronomy NASA Hdqrs., Washington, 1964-65, program chief planetary astronomy, 1965-77, discipline scientist planetary astronomy, 1977-82, chief planetary sci. br., 1982—. Fellow AAAS; mem. Am. Astron. Soc., Internat. Astron. Union; Mem. Sigma Xi. Subspecialty: Planetary science. Home: 4515 Willard Ave Chevy Chase MD 20815 Office: Code EL-4 NASA Hdqrs Washington DC 20546

BRUNNER, EDWARD A., anesthesiology educator; b. Erie, Pa., July 18, 1929; m., 1955; 10 children. B.S., Villanova U., 1952; M.D., Hahnemann Med. Coll., 1959, Ph.D. in Pharmacology, 1962. Instr. pharmacology Hahnemann Med. Coll. and Hosp., Phila., 1960-62; instr. anesthesia Sch. Medicine U. Pa., Phila., 1962-65, instr. pharmacology, 1964-65; from asst. prof. to assoc. prof. anesthesia Northwestern U. Med. Sch., Chgo., 1966-71, prof. and chmn. dept., 1971—. Subspecialty: Anesthesiology. Office: Dept Anesthesia Northwestern U Med Sch Chicago IL 60611

BRUNS, DAVID EUGENE, clinical pathologist, researcher; b. St. Louis, Dec. 12, 1941; s. Eugene H. and Ellen E. (Johnson) B.; m. Elizabeth Hirst; children: Elizabeth P., David H. B.S. Chem.E., Washington U., St. Louis, 1963, A.B., 1965; M.D., St. Louis U., 1973. Diplomate: Nat. Bd. Med. Examiners, Va. State Bd. Medicine. Intern lab. medicine Barnes Hosp., St. Louis, 1973-74; fellow exptl. pathology Washington U., St. Louis, 1974-75, instr. pathology, 1973-77; resident lab. medicine Barnes Hosp., St. Louis, 1975-77; asst. prof. pathology U. Va., Charlottesville, 1977-81, assoc. prof., 1981—, assoc. dir. clin. chemistry, 1977—. NIH grantee, 1978—; Am. Cancer Soc. grantee, 1982—. Mem. Assn. Clin. Scientists, Am. Assn. Clin. Chemistry, Am. Assn. Pathologists, Nat. Acad. Clin. Biochemistry, Acad. Clin. Lab. Physicians and Scientists. Roman Catholic. Subspecialties: Clinical chemistry; Pathology (medicine). Current work: Research in clinical enzymology, toxicology of polyethylene glycol, tumor markers, and subcellular calcium metabolism. Office: Dept Pathology U Va Sch Medicine Charlottesville VA 22908 Home: 2516 Woodhurst Rd Carlottesville VA 22901

BRUNS, PETER J., geneticist; b. Syracuse, N.Y., May 2, 1942; s. Hans J. and Ursula M. (Bahr) B.; m. Diane Vargo, Jan. 28, 1967. A.B in Zoology, Syracuse U., 1964; Ph.D. in Cell Biology, U. Ill., 1969. Asst. prof. genetics Cornell U., Ithaca, N.Y., 1969-74, asso. prof. genetics, 1975-82, prof., chmn. sect. genetics and devel., 1982—; vis. scientist Biol. Inst. Carlsburg Found., Copenhagen, Denmark, 1977-78, mem. teaching staff, Copenhagen, 1977. Contbr. writings to sci. publs. in field. John Simon Guggenheim Found. fellow, 1977-78; NSF grantee, 1971-73, 73-76, 77-82; NIH grantee, 1980—. Mem. AAAS, Soc. Protozoology, Genetics Soc. Am. Club: Ithaca Yacht. Subspecialties: Genome organization; Gene actions. Current work: Devel. genetics of Tetrahymena thermophila; genome orgn. and devel. of terminally differentiated somatic nucleus. Home: 7 Nottingham Dr Ithaca NY 14850 Office: 203 Bradfield Hall Cornell Univ Ithaca NY 14853

BRUNSO-BECHTOLD, JUDY KAREN, neurobiologist; b. Cin., Aug. 3, 1950; d. Paul Anchor and Thelma Jean (Crouch) Brunso; m. Robert Edmond Bechtold, June 16, 1973; children: Kressa Brunso, Shane Brunso. B.S. with distinction in Psychology, Duke U., 1972; M.S., Fla. State U., 1974, Ph.D., 1977. Research asst. dept. biology Washington U., St. Louis, 1976-79; fellow dept. anatomy Vanderbilt U., 1979-81, research asst. prof., 1981-83; asst. prof. anatomy Bowman Gray Med. Sch., 1983—. Contbr. articles in field to profl. jours. NIH fellow, 1975-76, 79-81. Mem. Soc. Neurosci., Sigma Xi. Subspecialties: Neurobiology; Developmental biology. Office: Department of Anatomy Bowman Gray Sch Wake Forest U Winston-Salem NC 27103

BRUNSON, BRADFORD IRA, psychologist; b. Sioux City, Iowa, Sept. 4, 1949; s. Ira Wandel and Shirley A. (Branch) B.; m. Cheryl A. Hollenshead, Aug. 25, 1980. B.A., Rutgers U., 1977; M.S., Kans. State U., 1979, Ph.D., 1982. Lic. profl. counselor, Tex.; cert. clin. mental health counselor. Instr. Kans. State U., Manhattan, 1979-81; counseling psychologist Tex. A&M U., College Station, 1981-82; dir. counseling center St. Mary's U., San Antonio, 1982—; dir. Commn. II Am. Coll. Personnel Assn., Falls Church, Va., 1980—. Contbr. articles to profl. jours. Active Boy Scounts Am., Montebello, Calif., 1964. Served with USMC, 1979-81. Mem. Am. Psychol. Assn., Am. Personnel and Guidance Assn., Internat. Acad. Profl. Counseling and Psychotherapy, Am. Coll. Personnel Assn., Am. Mental Health Counselors Assn., Psi Chi, Phi Delta Kappa, Alpha Sigma Lambda. Lutheran. Subspecialties: Counseling psychology; Social psychology. Current work: Type A coronary-prone behavior pattern; depression and learned helplessness; social influence in counseling. Home: 5214 Timberhurst St San Antonio TX 78250 Office: Saint Marys U 1 Camino Santa Maria San Antonio TX 78284

BRUSH, ALAN HOWARD, biology educator; b. Rochester, N.Y., Sept. 29, 1934; s. Martin and Rebecca (Trott) B.; m. Nancy D. Carter, Sept. 1, 1961; children: Lisa, Matthew. B.A., U. So. Calif., 1956; M.A., UCLA, 1957, Ph.D., 1964. Postdoctoral fellow Cornell U., Ithaca, N.Y., 1964-65; asst. prof. dept. biology U. Conn., Storrs, 1965-69, assoc. prof., 1970-76, prof., 1976—; vis. fellow U. Calif.-Berkeley, 1971-72; fellow in protein chemistry, Melbourne, Australia, 1976-77. NIH fellow, 1963-64. Fellow Am. Ornithologists Union (H.B. Tucker award 1963, council 1979-82); mem. Cooper Ornithol. Soc., Brit. Ornithol. Union., Am. Physiol. Soc., Am. Soc. Zoologists. Subspecialties: Evolutionary biology; Morphology. Current work: Comparative molecular design in evolution. Home: 86 Willington Oaks Storrs CT 06268 Office: Biol Scis Dept Univ Conn Storrs CT 06268

BRUSH, F. ROBERT, psychologist, educator; b. Phoenixville, Pa., Nov. 24, 1929; s. Franklin Cotton and Anna (Fox) B.; m. (div.); children: Robert C., Elizabeth W. B.A., Princeton U., 1951; M.A., Harvard U., 1953, Ph.D., 1956. Asst. prof. U. Md., College Park, 1956-59, U. Pa., Phila., 1959-65; assoc. prof. U. Oreg. Med. Sch., Portland, 1965-67, prof., 1967-71; prof. psychology Syracuse (N.Y.) U., 1971-81, Purdue U., West Lafayette, Ind., 1981—; vis. asst. prof. U. Wis., Madison, 1964-65; vis. prof. St. Thomas' Hosp. Med. Sch., London, 1970-71, 72, 76. Editor: Aversive Conditioning and Learning, 1971. NIH grantee, 1967-71. Fellow Am. Psychol. Assn.; mem. Psychonomic Soc., Soc. for Neurosci., Soc. for Endocrinology, Behavior Genetics Assn. Subspecialties: Psychobiology; Neuroendocrinology. Current work: Psychoendocrinology, hormone-behavior relationships, behavioral genetics of aversively motivated behavior. Home: 1118 Becker Dr Zionsville IN 46077 Office: Dept Psychol Scis Purdue U West Lafayette IND 47907

BRUTLAG, DOUGLAS LEE, biochemistry educator, consultant; b. Alexandria, Minn., Dec. 19, 1946; s. Minehart and Cora (Lee) B.; m. Simone C. Manteuil, Oct. 11, 1975; children: Pauline, Benjamin. B.S., Calif. Inst. Tech., 1968; Ph.D., Stanford U., 1972. Research scientist Commonwealth Sci. and Indsl. Research Orgn., Canberra, Australia, 1972-74; asst. prof. biochemistry Stanford U., 1974-80, assoc. prof., 1980—; founder/cons. IntelliGenetics Inc., Palo Alto, Calif., 1980—; cons. NIH. Mellon fellow, 1974-76; NIH Sr. Fogarty Internat. fellow, 1981-82; Henry and Camille Dreyfus Tchr.-Scholar grantee, 1979-84. Mem. Am. Soc. Biol. Chemists, Fedn. Am. Soc. Exptl. Biology. Subspecialties: Molecular biology; Biochemistry (biology). Current work: Genetic engineering, DNA topoisomerases, enzymology. Office: Dept Biochemistry Stanford U Med Center Stanford CA 94305

BRYAN, DAVID A., physicist; b. Austin, Tex., July 29, 1946; s. William C. and Virginia S. (Vedder) B.; m. JoAnn Grossman, Aug. 3, 1974; children: Kenna, Kimberlee, Benjamin, Joshua. B.A., Rice U., 1968; M.S., U. Mo.-Rolla, 1973, Ph.D., 1976. Research asst. U. Mo.-Rolla, 1971-76; lead engr. electro-optics dept. McDonnell Douglas Astronautics Co., St. Louis, 1976—. Contbr. articles to profl. jours. Served to lt. USNR, 1968-71. Mem. Optical Soc. Am., Internat. Soc. Optical Engrs. Subspecialties: Holography; Optical signal processing. Current work: Holographic and other optical studies of non-linear optical materials; photorefractive effect of laser-frequency-doubler materials and electro-optic modulator materials. Home: 3044 Westminister Saint Charles MO 63301 Office: McDonnell Douglas Astronautics Co PO Box 516 Saint Louis MO 63166

BRYAN, GLENN LEVAN, research executive, researcher; b. Canton, Ohio, Jan. 10, 1921; s. Henry Paul and Ella Marie (Kashner) B.; m. Olivia Ruth Judson, Jan. 3, 1943; children: Geoffrey, Bradley, William. B.A., Coll. of Wooster, Ohio, 1942; Ph.D., U. So. Calif., 1951. Research assoc. U. So. Calif., Los Angeles, 1951, dir. electronic personnel research, 1953-59; dir. personnal and tng. research Office Naval Research, Washington, 1959-69, dir. psychol. sci. div., Arlington, Va., 1969-81, assoc. dir. research, life scis., 1981—. Contbr. chpts. to books in field. Pres. Woodhaven Citizens' Assn., Bethesda, 1960's. Served with U.S. Army, 1943-46. Recipient research award Nat. Soc. for Programmed Instrn., Tex., 1964; vis. research scholar Stanford U., 1969. Fellow AAAS; mem. Am. Psychol. Assn., Eastern Psychol. Assn., Human Factors Soc. Subspecialties: Cognition; Learning. Current work: Formulate and direct research programs in biological and psychological sciences. Home: 6205 Poe Rd Bethesda MD 20817 Office: Office Naval Research 800 N Quincy St Arlington VA 22217

BRYAN, JOHN HENRY DONALD, biologist, educator, researcher; b. London, Sept. 18, 1926; s. John and Mary (Barnes) B.; m. Janet Goff, Aug. 23, 1952; children: Mary E., Melissa L. B.Sc., U. Sheffield, Eng., 1947; M.A., Columbia U., 1949, Ph.D., 1952. Lectr. zoology Columbia U., N.Y.C., 1949-50; instr. biology M.I.T., Cambridge, 1951-54; asst. prof. Iowa State U., Ames, 1954-62, assoc. prof., 1962-67; prof. zoology U. Ga., Athens, 1967—; cons. AID, 1968-69. Contbr. articles to profl. jours. McCallum Found. fellow, 1950-52. Fellow AAAS; mem. Am. Soc. Naturalists, AAAS, Am. Soc. Cell Biology, Genetics Soc. Am., Soc. Devel. Biology, Internat. Assn. Torch Clubs (chpt. dir. 1974-77, chpt. pres. 1975-76), Sigma Xi. Subspecialties: Cell biology; Gene actions. Current work: Use of mutants as probes to study normal processes in production of male reproductive cells, and in development of the nervous system; relationship between ultrastructural makeup and functional specializations of cells. Office: U Ga Athens GA 30602

BRYANT, HOWARD CARNES, physicist; b. Fresno, Calif., July 9, 1933; s. Leslie Howard and Enla Beryl (Carnes) B.; m. Mona Patricia Jordan, June 4, 1960; children: Matthew, Clifford, Susannah. B.A., U. Calif.-Berkeley, 1955; Ph.D., U. Mich., 1961. Prof. physics U. N. Mex. Fellow Am. Phys. Soc.; mem. Am. Assn. Physics Tchrs. Subspecialties: Atomic and molecular physics; Particle physics. Current work: Laser photodetachment, solar ponds.

BRYANT, JAMES WINSTON, JR., clinical psychologist; b. Oklahoma City, Feb. 26, 1949; s. James Winston and Elly Elizabeth (Geissler) B.; m. Martha Rosa Moroyoqui, Apr. 10, 1982. M.S., Calif. State U., 1973; M.A., Palo Alto Sch. Profl. Psychology, 1979, Ph.D., 1980. Lic. psychologist, Calif. Chem. dependency therapist Pathways Soc., Inc., Santa Clara, Calif., 1980—; psychiat. asst. Hugh Kohn, Ph.D., San Jose, Calif., 1979-81, O'Connor Hosp., Campbell, Calif., 1981-82; clin. psychologist Palomares Group Homes, Inc., San Jose, 1982—, El Dorado Guidance Ctr., 1982—; pvt. practice clin. psychology, Los Gatos, Calif., 1980—; mem. psychologist panel Dept. Vocational Rehab., San Jose, 1982—; provider W. Bay Health Systems Agy., San Mateo Subarea Adv. Council, 1979-81; examiner San Mateo County Oral Commrs. Bd. Mental Health Services, 1979-81. Contbr. articles to profl. jours. Mem. Am. Psychol. Assn. (assoc.), Calif. Psychol. Assn., Calif. Soc. Clin. Hypnosis (bd. govs. 1981-83), San Jose Soc. Clin. Hypnosis, Santa Clara County Psychol. Assn. Democrat. Current work: L-facility with a chronically-impaired schizophrenic population.

BRYANT-GREENWOOD, GILLIAN DOREEN, anatomy educator; b. Witney, Oxfordshire, Eng., July 20, 1942; came to U.S., 1968, naturalized, 1975; d. Gordon Henry and Joyce (Bartlett) Bryant-G.; m. Frederick Charles Greenwood, July 9, 1976; children: Peter, Kate. B.S., Brunel U. London, 1965; Ph.D., Imperial Cancer Research Fund, Eng., 1968. Mem. staff Imperial Cancer Research Fund, London, 1965-68; postdoctoral fellow dept. biochemistry and biophysics U. Hawaii, Honolulu, 1968-72, asst. prof. dept. anatomy and reproductive biology, 1972-76, assoc. prof., 1976-81, prof., 1981—. Co-editor: Relaxin, 1981. Recipient Career Devel. award NIH, 1972-77, research grantee, 1972—. Mem. Soc. Endocrinology, Endocrine Soc. U.S., Soc. Study Reprodn., Assn. Women in Sci. Subspecialties: Endocrinology; Reproductive biology. Current work: The biochemistry and physiology of the hormone relaxin. Office: U Hawaii Dept Anatomy and Reproductive Biology 1960 East-West Rd Honolulu HI 96822 Home: 949 Koae St Honolulu HI 96816

BRYDEN, MARK PHILIP, psychologist; b. Boston, Nov. 14, 1934; s. Samuel David and Ellen Agnes (Sibley) B.; m. Patricia Mabel Rowe, Nov. 2, 1962; children—Penny Elizabeth, Pamela Joanna. S.B., MIT, 1956; M.Sc., McGill U., 1958, Ph.D., 1961. Research assoc. McGill U., Montreal, 1960-63; asst. prof. psychology U. Waterloo, Ont., 1963-64, assoc. prof., 1964-67, prof., 1967—. Author: Laterality, 1982; editor: Can. Jour. Psychology, 1981—. NRC Can. sr. scientist, 1969-70; Can. Council fellow, 1975-76. Fellow Can. Psychol. Assn. (dir. 1981-83), Am. Psychol. Assn.; mem. Internat. Neuropsychology Soc., Psychonomic Soc. Subspecialties: Neuropsychology; Cognition. Current work: Brain lateralization, handedness, sex hormones and behavior. Office: Dept Psychology U Waterloo Waterloo ON Canada N2L 3Gl

BRYNJOLFFSON, ARI, nuclear physicist; b. Akureyri, Iceland, July 12, 1926; came to U.S., 1965, naturalized, 1970; s. Brynjolfur and Gudrun (Rosinkarsdottir) Sigtryggsson; m. Marguerite Reman, Dec. 22, 1950; children: Ariane, Olaf, Erik, John, Alan. Cand. Phil., U. Copenhagen, 1949, Cand. Mag., 1954, Mag. Scien., 1954; D.Phil., Niels Boht Institut Theoretical and Exptl. Nuclear Physics, 1973; post grad., Advanced Mgmt. Program, Harvard U., 1971. Dir. radiation research Danish Atomic Energy Research Establishment, Roskilde, 1957-65; chief radiation research U.S. Army Natick (Mass.) Lab., 1965-72; dir. food irradiation program, 1972-80, spl. asst. for physics, 1980—; lectr. M.I.T., 1980—. Contbr. articles to profl. jours. Recipient Mollers Found. award for exceptional service to Danish industry, 1965. Mem. Am. Phys Soc., Radiation Research Soc., Am. Nuclear Soc., Am. Soc. Physicists in Medicine, Inst. Food Technologists. Subspecialties: Nuclear physics; Population biology. Current work: Astrophysics, theoretical physics, general theory of relativity. Biological effects of radiation. Home: 7 Bridle Path Wayland MA 01778 Office: 6 Kansas St Natick MA 01760

BRYSK, MIRIAM MASON, dermatology educator, cell biologist, researcher; b. Warsaw, Poland, Mar. 10, 1935; came to U.S., 1947, naturalized, 1952; d. Henry and Betty (Zablocki) M.; m. Henry Brysk, June 5, 1955; children: Judith Brysk Rocher, Helen L. Brysk Mandell. B.A., NYU, 1955; M.S., U. Mich., 1958; postgrad., St. John's U., 1961-62; Ph.D., Columbia U., 1967. Postdoctoral fellow Inst. Muscle Disease, N.Y.C., 1967-69; research biologist U. Calif., San Diego, 1969-71; research asst. prof. U. N.Mex., Albuquerque, 1977-79; asst. prof. dermatology U. Tex. Med. Br., Galveston, 1979-82, assoc. prof. dermatology and microbiology, 1982—. NIH fellow, 1974-77; NIH grantee, 1979—; Robert A. Welch Found. grantee, 1982—. Mem. Am. Soc. Microbiology, Soc. Investigation Dermatology, Am. Acad. Dermatology, Sigma Xi. Democrat. Jewish. Subspecialties: Cell biology (medicine); Cancer research (medicine). Current work: Cell biology and biochemistry of skin cancer, specifically cell surface changes. Home: 32 Colony Park Circle Galveston TX 77551 Office: Department Dermatology lUniversity of Texas Medical Branch Galveston TX 77550

BRYSON, ARTHUR EARL, JR., engring. educator; b. Evanston, Ill., Oct. 7, 1925; s. Arthur Earl and Helen Elizabeth (Decker) B.; m. Helen Marie Layton, Aug. 31, 1946; children—Thomas Layton, Stephen Decker, Janet Elizabeth, Susan Mary. Student, Haverford Coll., 1942-44; B.S., Iowa State U., 1946; M.S., Calif. Inst. Tech., 1949, Ph.D. in Aeros, 1951; M.A. (hon.), Harvard., 1956. With Container Corp. Am., 1947-48, United Aircraft Corp., 1948; research asst. aero. Calif. Inst. Tech., 1949-50; mem. tech. staff Hughes Research & Devel. Labs., 1950-53; mem. faculty Harvard, 1953-68, Gordon McKay prof. mech. engring., 1961-68; mem. faculty Stanford, 1968—, chmn. dept. applied mechanics, 1969-71, chmn. dept. aeros. and astronautics, 1971-79, Paul Pigott prof. engring., 1972—; Hunsaker prof. Mass. Inst. Tech., 1965-66; Mem. nat. com. Fluid Mechanics Films, 1961-68. Author: (with Y.C. Ho) Applied Optimal Control, 1969. Served as ensign USNR, 1944-46. Recipient Rufus Oldenberger medal ASME, 1980. Fellow Am. Inst. Aeros. and Astronautics (asso. editor Jour. 1963-65, bd. dirs. 1965-68, Pendray Award 1968, mechanics and control of flight award 1980); mem. Am. Acad. Arts and Scis., Am. Soc. Engring. Edn. (Westinghouse award 1969), Nat. Acad. Engring. (aero. and space engring. bd. 1970-79), Nat. Acad. Scis., Sigma Xi, Tau Beta Pi. Conglist. Subspecialty: Theoretical and applied mechanics. Home: 761 Mayfield Ave Stanford CA 94305

BRYSON, GEORGE GARDNER, physiologist/psychologist; b. Santa Barbara, Calif., Dec. 16, 1935; s. George Omar and Mary Dewey (Gardner) B. A.B. in Zoology, U. Calif.- Santa Barbara, 1957, Ph.D., 1981; M.A. in Psychology, San Francisco State Coll., 1971. Zoologist Santa Barbara Cottage Hosp. Research Inst., 1957-71, physiologist, 1971-82, physiologist-psychologist, 1982—; teaching asst. dept. psychology U. Calif., Santa Barbara, 1976; rev. cons. Nat. Inst. Child Health and Human Devel., 1977. Contbr. articles to profl. jours. Mem. Am. MENSA Selection Assn. Mem. Am. Assn. Cancer Research. Subspecialties: Neuroendocrinology; Neuropsychology. Current work: Steroid hormones and behavior; neuropharmacology of biogenic amines; basic research on metabolism and on carcinogenesis. Home: 2412 Chapala St Santa Barbara CA 93105 Office: PO Box 689 Santa Barbara CA 93102

BRYSON, REID ALLEN, educator; b. Detroit, June 7, 1920; s. William Riley and Elma (Turner) B.; m. Frances Edith Williamson, June, 13, 1942; children—Anne, William, Robert, Thomas. A.B., Denison U., 1941, D.Sc. (honoris causa), 1971; postgrad., U. Wis., 1941, 46; Ph.D., U. Chgo., 1948. Asst. prof. meteorology and geology U. Wis., 1946-48, asst. prof. meteorology, 1948-50, asso. prof., 1950-56, chmn. dept., 1948-50, 52-54, prof., 1957—; dir. Inst. for Environ. Studies, 1970—; prof. U. Ariz., 1956-57; Mem. various coms. Nat. Acad. Sci.-NRC, 1958—, mem. remote sensing com., 1964-67, mem. com. on mil. geography, 1966-69; mem. (Smithsonian Council), 1976—, sr. cons., 1975—; Trustee Univ. Corp. for Atmospheric Research. Author: Atlas of 500 mb Wind Characteristics for the Northern Hemisphere, 1958, Atlas of Five-Day Normal Sea-Level Pressure Charts for the Northern Hemisphere, 1958, Atlas of 300 mb Wind Characteristics, 1959; Editor: (with F.K. Hare) Climates of North America, 1974, Climates of Hunger, 1977 (Banta medal 1978); Contbr.: articles to profl. jours. Climates of Hunger. Cited by Denison U., 1966. Fellow Am. Meteorol. Soc., Explorers Clubs, Wis. Acad. Scis., Arts and Letters; mem. Wis. Phenological Soc. (past pres.), Soc. Am. Archaeology, Assn. Am. Geographers, Am. Soc. for Limnology and Oceanography, Arts and Letters (pres. 1981), Phi Beta Kappa, Sigma Xi, Phi Kappa Phi (hon.). Subspecialty: Climatology. Current work: Year-or-more-in-advance climate forecasting; paleo climatology. Application of climatology to archaeol. problems; regional and global climatic modification; climatic changes and world food supply; interdisciplinary environmental studies. Home: 11 Rosewood Circle Madison WI 53711

BUBBERS, JOHN ERIC, researcher, immunologist; b. Rockville Centre, N.Y., Mar. 13, 1951; s. John and Sydney Felicia (Browne) B.; m. Sandra Welgreen, May 28, 1977; 1 dau., Emily Julia. B.A., Johns Hopkins U., 1973; Ph.D., Sch. Medicine, 1975. Postdoctoral fellow Albert Einstein Coll. Medicine, N.Y.C., 1975-77, Scripps Clinic and Research Found., La Jolla, Calif., 1977-79; asst. research immunologist UCLA, 1979-82, adj. asst. prof. dept. radiation oncology, 1982—. Mem. Am. Assn. Immunologists, Internat. Assn. Comparative Research Leukemia and Related Disorders. Subspecialties: Cancer research (medicine); Immunogenetics. Office: Dept Radiation Oncology UCLA Sch Medicine Los Angeles CA 90024

BUBE, RICHARD HOWARD, materials scientist; b. Providence, Aug. 10, 1927; s. Edward Neser and Ella Elvira (Baltteim) B.; m. Betty Jane Meeker, Oct. 9, 1948; children: Mark Timothy, Kenneth Paul, Sharon Elizabeth, Meryl Lee. Sc.B., Brown U., 1946; M.A., Princeton U., 1948, Ph.D., 1950. Mem. sr. research staff RCA Labs., Princeton, N.J., 1948-62; prof. materials sci. and elec. engring. Stanford U., 1962—, chmn. dept., 1975—; cons. to industry and govt. Author: A Textbook of Christian Doctrine, 1955, Photoconductivity of Solids, 1960, The Encounter Between Christianity and Science, 1968, The Human Quest: A New Look at Science and Christian Faith, 1971, Electronic Properties of Crystalline Solids, 1974, Electrons in Solids, 1981, Fundamentals of Solar Cells, 1983; also articles; editor: Jour. Am. Sci. Affiliation; editorial bd.: Solid State Electronics; asso. editor: Ann. Rev. Materials Sci., Materials Letters. Fellow Am. Phys. Soc., AAAS, Am. Sci. Affiliation; mem. Am. Soc. Engring. Edn., Internat. Solar Energy Soc., Sigma Xi. Evangelical. Subspecialties: Semiconductors; Electronic materials. Current work: Solid state physics; photo electronic materials and devices with particular attention to semiconductor junction conversion of solar energy into electricity. Home: 753 Mayfield Ave Stanford CA 94305 Office: Dept Materials Sci and Engring Stanford Univ Stanford CA 94305 I find no contradiction or conflict between science and Christian faith, but rather a marvelous compatibility that touches all aspects of life.

BUCCAFUSCO, JERRY JOSEPH, medical educator, researcher; b. Jersey City, Aug. 20, 1949. B.S., St. Peter's Coll., 1971; M.S., Canisius Coll., 1973; Ph.D., U. Medicine and Dentistry N.J., 1980. Postdoctoral fellow Roche Inst. Molecular Biology, Nutley, N.J., 1977-79; asst. prof. pharmacology and psychiatry Med. Coll. Ga., 1979—; research pharmacologist VA Med. Ctr., Augusta, Ga. Contbr. articles, abstracts to profl. jours. Recipient New Investigator award Nat. Inst. Drug Abuse, 1980—; Nat. Heart, Lung and Blood Inst. grantee, 1981-84. Mem. Am. Soc. Pharmacology and Exptl. Therapeutics, Soc. Neurosci., Soc. Exptl. Biology and Medicine, Am. Heart Assn., Southeastern Pharmacol. Soc., AAAS, Sigma Xi. Subspecialties: Pharmacology; Neuropharmacology. Current work: Central regulation of blood pressure; central mechanisms and hypertension; the neurochemistry of acetylcholine; mechanisms of narcotic dependence and withdrawal. Office: Dept Pharmacology Med Coll Ga Augusta GA 30912

BUCHANAN, WILLIAM, dental educator, researcher; b. N.Y.C., May 5, 1950; s. James Francis and Marcella (Nally) B. B.S., Calif. State U.-Long Beach, 1972; D.D.S., UCLA, 1977, M.S., Sch. Dentistry, 1979; specialty cert., Harvard Sch. Dental Medicine, Boston, 1981, M. Md.Sc., Med. Sch., 1981. Research fellow Harvard U. Sch. Dental Medicine, Boston, 1978-81; asst. prof. periodontics U. Miss. Dental Sch., 1981—. Nat. Inst. Dental Research fellow, 1978-81. Mem. ADA, Am. Acad. Periodontology, AAAS, Internat. Assn. Dental Research. Subspecialties: Periodontics; Oral biology. Current work: Periodontal research using immunofluorescent technique for rapid identification of bacterial pathogens at disease sites. Office: U Miss Sch Dentistry 2500 N State St Jackson MS 39216

BUCHER, BRADLEY DEAN, psychologist, educator; b. Astoria, Ill., July 16, 1932; emigrated to Can., 1970; s. Ezra Gibble and Irma Louise (Onion) B. M.A., Ph.D., Princeton U., 1957, U. Pa., 1965. Registered psychologist, Ont. With Analytical Research Group, Princeton, N.J., 1955-56, Statis. Research Group, Princeton, 1956-57; staff, project dir. Weapons Systems Evaluation Group, Pentagon, Washington, 1957-60; asst. prof. psychology UCLA, 1965-70; assoc. prof. psychology U. Western Ont., London, 1970-82, prof., 1982—; cons. in field. Editor: Perspectives in Behavior Modification with Deviant Children, 1974; contbr. articles to profl. jours. Med. Research Council grantee, 1970-73; Ont. Mental Health Found. grantee, 1972-83; Can. Council grantee, 1975-79; NIMH grantee, 1964-65. Mem. Am. Psychol. Assn., Assn. for Advancement Behavior Therapy, Assn. for Behavior Analysis, Behavior Research Therapy Soc. Subspecialties: Behavioral psychology; Developmental psychology. Current work: Behavior modification. Home: 688 Fanshawe Park Rd E London ON Canada N5X 2B9 Office: Dept Psychology Univ Western Ont London ON Canada N6A 5C2

BUCHER, NANCY L.R., biomedical researcher; b. Balt., May 4, 1913; d. John Howard and Lula E. (Langrall) C. A.B., Bryn Mawr Coll., 1935; M.D., Johns Hopkins U., 1943. Intern and fellow in medicine Mass. Meml. Hosp., Boston, 1943-45, clin. fellow, assoc. biologist, biologist, 1945-83; research fellow Harvard U., 1945, assoc. prof. medicine (oncology), 1972-79, assoc. prof. surgery (oncology), 1979-83; research prof. pathology Boston U. Sch. Medicine, 1983—; trustee Worcester Found. Exptl. Biology, Shrewsbury, Mass. Contbr. articles to profl. jours. Grantee in field. Mem. AAAS, Am. Assn. Cancer Research, Am. Assn. Study Liver Diseases, Am. Physiol. Soc., Am. Soc. Biol. Chemists, Am. Soc. Cell Biology, Internat. Assn. Study Liver, N.Y. Acad. Scis., Soc. Developmental Biology, Tissue Culture Assn. Subspecialties: Cell and tissue culture; Biochemistry (medicine). Current work: Mechanisms for pysiological control of tissue growth, specifically liver; liver regeneration; hormones and growth factors in primary hepatocyle cultures. Office: Boston U Sch Medicine 80 E Concord St Boston MA 02118

BUCHHOLZ, DONNA MARIE, microbiologist, immunologist; b. Chgo., May 27, 1950; d. Arthur George and Doris Hedwig (Lewis) B.; m. William E. Hourigan, Oct. 12, 1974. B.S., Quincy (Ill.) Coll., 1972; M.S. in Microbiology-Immunology, U. Ill.-Chgo., 1975, Ph.D., 1978. Postdoctoral research scientist div. biol. and med. research Argonne (Ill.) Nat. Lab., 1978-80; research info. scientist pharm. products div. Abbott Labs., North Chicago, Ill., 1980-82, project mgr. research mgmt. and devel., 1982—. Contbr. articles to profl. jours. Recipient research award Sigma Xi, 1978. Mem. Soc. Indsl. Microbiology, Ill. Soc. Microbiology (councilor exec. com.), AAAS, Assn. Women in Sci., Am. Soc. Microbiology, Am. Assn. Immunologists, Sigma Xi (chpt. pres.). Subspecialties: Microbiology (medicine); Infectious diseases. Current work: Antibiotic discovery and development; microbiology, infectious diseases, immunology, health services, environl Toxicology, hematology. Home: 451 Highland Ave West Chicago IL 60185 Office: Abbott Labs North Chicago IL 60064

BUCHI, GEORGE HERMANN, chemistry educator, consultant; b. Baden, Switzerland, Aug. 1, 1921; came to U.S., 1948, naturalized, 1954; s. George Jakob and Martha (Muller) B.; m. Anne Westfall Parkman, Aug. 20, 1955. D.Sc., Swiss Fed. Inst. Tech., Zurich, 1947; Dr.h.c., U. Heidelberg, 1983. Firestone postdoctoral fellow U. Chgo., 1948-51; mem. faculty MIT, 1951—, asst. prof. chemistry, 1951-56, assoc. prof., 1956-58, prof., 1958-71, Camille Dreyfus prof. chemistry, 1971—. Served with Swiss Army, 1942-45. Mem. Nat. Acad. Scis., Am. Chem. Soc. (Fritzsche award 1958, award for creative work in synthetic organic chemistry 1973), German Chem. Soc., Brit. Chem. Soc., Japanese Chem. Soc., Swiss Chem. Soc. (Ruzicka prize 1957), Pharm. Soc. Japan (hon.). Republican. Subspecialties: Organic chemistry; Synthetic chemistry. Current work: Natural products, carcinogenesis, photochemistry. Office: 77 Massachusetts Ave 18-287 Cambridge MA 02139

BUCHSBAUM, DAVID ALVIN, mathematics educator; b. N.Y.C., Nov. 6, 1929; m., 1949; 3 children. A.B., Columbia U., 1949, Ph.D., 1954. Instr. math. Princeton U., 1953-54, NSF fellow, 1954-55; instr. math. U. Chgo., 1955-56; asst. prof. Brown U., Providence, 1956-59, assoc. prof., 1959-60, Brandeis U., Waltham, Mass., 1960-63, prof. math., 1963—. Guggenheim fellow, 1964-65; NSF fellow, 1960-61. Mem. Am. Math. Soc. Office: Dept Math Brandeis U Waltham MA 02154

BUCHSBAUM, SOLOMON JAN, physicist; b. Stryj, Poland, Dec. 4, 1929; came to U.S., 1953, naturalized, 1957; s. Jacob and Berta (Rutherfoer) B.; m. Phyllis N. Isenman, July 3, 1955; children—Rachel Joy, David Joel, Adam Louis. B.S., McGill U., 1952, M.S., 1953; Ph.D., Mass. Inst. Tech., 1957. Mem. tech. staff Bell Labs., Murray Hill, N.J., 1958-61, dept. head, 1961-65, dir., 1965-68; v.p. Sandia Labs., Albuquerque, 1968-71; exec. dir. Bell Labs., 1971-76, v.p., 1976-79, exec. v.p., 1979—; sr. cons. Def. Sci. Bd., chmn., 1972-77; mem. AEC Controlled Thermonuclear Fusion Com., 1965-72, Pres.'s Sci. Adv. Com., 1970-73, Pres.'s Com. on Sci. and Tech., 1975-76; mem. fusion power coordinating com. ERDA, 1972-76; mem. advisory group sci. and tech. NSF, 1976-77; chmn. Energy Research Adv. Bd., 1978—; mem. Naval Research Adv. Com., 1978—; mem. vis. com. M.I.T., 1977—, mem. corp. devel. com., 1980—; cons. (Office Sci. and Tech.), 1976—. Asso. editor: Revs. Modern Physics, 1968-72, Jour. Applied Physics, 1968-70, Physics of Fluids, 1963-64; Co-author: Waves in Plasmas, 1963; contbr. numerous articles to profl. jours. Trustee Argonne Univs. Assn., 1979—. Moyse traveling fellow, 1953-54; IBM fellow, 1954-56; recipient Anne Molson Gold medal and Sec. Def. medal Outstanding Pub. Service, 1977. Fellow Am. Phys. Soc. (chmn. div. plasma physics 1968, mem. council 1973-76), IEEE, Am. Acad. Arts and Scis., AAAS; mem. Nat. Acad. Engring. (exec. com. 1975-76), Nat. Acad. Scis., Cosmos Club. Subspecialty: Plasma physics. Research in gaseous and solid state plasmas, communications. Patentee in field. Office: Bell Labs Holmdel NJ 07733

BUCHWALD, HENRY, surgeon, educator; b. June 21, 1932; m. Emilie D. Bix, June 6, 1954; children: Jane Nicole, Amy Elizabeth, Claire Gretchen, Dana Alexandra. B.A. summa cum laude, Columbia U., 1954, M.D., 1957; M.S., U. Minn., 1966, Ph.D. in Surgery, 1966. Diplomate: Am. Bd. Surgery. Intern in surgery Columbia Presbyn. Med. Ctr., N.Y.C., 1957-58; resident in surgery U. Minn. Hosp., 1960-66; instr. in surgery U. Minn. Med. Sch., 1966-67, asst. prof. surgery, 1967-70, assoc. prof. surgery, 1970-77, prof. surgery, 1977—, prof. biomed. engring., 1977—; established investigator Am. Heart Assn., 1964-69. Mem. editorial bd.: Jour. Clin. Surgery. Served to capt. SAC USAF, 1958-60. Helen Hay Whitney fellow, 1962-64; Found. for Allergic Diseases fellow, 1956; recipient Essay prize Am. Coll. Chest Physicians, 1957; Shering award, 1957; 1st Clin. Research award Minn. Surg. Soc., 1965; 1st prize research forum Am. Coll. Chest Physicians, 1966; Samuel D. Gross award, 1969; Disting. Service award Am. Assn. Acad. Surgery, 1976. Fellow Am. Surg. Assn., ACS, Soc. Univ. Surgeons, Central Surg. Assn. (program com.), Assn. for Acad. Surgery, Am. Heart Assn., Am. Coll. Cardiology, Soc. for Surgery of Alimentary Tract, Am. Therapeutic Assn.; mem. Minn. Surg. Assn., Mpls. Surg. Assn., Minn. Heart Assn., Am. Assn. History of Medicine, Saint Paul Surg. Soc. (hon.), AAAS, Paleopathology Club, Hennepin County Med. Assn., Minn. State Med. Assn., Am. Coll. Nutrition (chmn. surgery council), Am. Soc. Artificial Internal Organs (program com.), Internat. Study Group on Diabetes Treatment with Implantable Insulin Delivery Devices, Phi Beta Kappa, Alpha Omega Alpha, Phi Lambda Upsilon. Subspecialties: Surgery; Biomedical engineering. Current work: Implantable infusion devices; lipid metabolism; obesity surgery. Office: Minn Sch Medicine Minneapolis MN 55455

BUCK, RICHARD FORDE, electronics engineering educator; b. Enterprise, Kans., Dec. 8, 1921; s. Charles Fay and Ruth Wykoff (Scott) B.; m. Harriet J. Ojers, June 4, 1944; children: David R., Janet H., Paul S., Bryan T., Neal A., Daniel C. B.S. in Physics, U. Kans.-Lawrence, 1942, M.S., Okla. State U.-Stillwater, 1960. Registered profl. engr., Okla. Cereal chemist Flour Mills of Am., Kansas City, Mo., 1939-43; applications engr. Tung Sol Electric, Newark, N.J., 1946-48; mem. faculty Okla. State U., Stillwater, 1960—, projects dir., 1948-60, dir. electronics lab., 1976—; cons. in field. Mem. Sheltered Workshop for Payne County (Okla.), 1967—, pres. bd. dirs., 1972-73, 78-79, life mem., 1980—. Served to 1st lt. U.S. Army, 1942-46. Mem. Okla. Soc. Profl. Engrs. (Pres.'s award 1972), Okla. Assn. Retarded Citizens, Sigma Xi. Democrat. Subspecialties: Electronics; Satellite studies. Current work: Research instrumentation for space and atmospheric exploration. Home: 1301 Westwood Dr Stillwater OK 74074 Office: Electronics Lab Okla State U 1700 W Tyler St Stillwater OK 74078

BUCK, ROSS WORKMAN, communication sciences and psychology educator; b. Sewickley, Pa., Aug. 16, 1941; s. Ross Workman and Ruth (Hadley) B.; m. Marianne Jenney, Dec. 28, 1963; children; William, Marian, Jenney Theodore. B.A., Allegheny Coll., 1963; M.A., U. Wis., Madison, 1965; Ph.D., U. Pitts., 1970. Research assoc. U. Pitts. Med. Sch., 1967-70; asst. prof. Carnegie-Mellon U., Pitts., 1970-74, U. Conn., Storrs, 1974-76, assoc. prof., 1976-81, prof. communication sciences and psychology, 1981—; vis. scholar dept. psychology Harvard U., Cambridge, Mass., 1980-81; vis. scholar aphasia research unit Boston VA Hosp., 1980-81. Author: Human Motivation and Emotion, 1976, Emotion and Nonverbal Behavior, 1983; mem. editorial bd.: Jour. Personal and Social Psychology, 1981—, Jour. Nonverbal Behavior, 1978—. Chmn. social action community of reconciliation University and City Ministries, Pitts., 1970-74. NIMH grantee, 1972. Mem. Am. Psychol. Assn., Internat. Communication Assn., Eastern Psychol. Assn., Midwestern Psychol. Assn. Democrat. Subspecialties: Social psychology; Neuropsychology. Current work: Neural bases of the communication of emotion; social learning and temperament in emotional development and education, emotion, communication, stress and disease. Home: 64 Cedar Swamp Rd Storrs CT 06268 Office: Department Communication Sciences U-85 University of Connecticut Storrs CT 06268

BUCKLAND, MICHAEL KEEBLE, librarian, educator; b. Wantage, Eng., Nov. 23, 1941; came to U.S., 1972; s. Walter Basil and Norah Elaine (Rudd) B.; m. Waltraud Leeb, July 11, 1964; children: Anne Margaret, Anthony Francis. B.A., Oxford (Eng.) U., 1963; postgrad.; diploma in librarianship, Sheffield (Eng.) U., 1965, Ph.D., 1972. Grad. trainee Bodleian Library, Oxford, Eng., 1963-64; asst. librarian U. Lancaster (Eng.) Library, 1965-72; asst. dir. tech. services Purdue U. Libraries, West Lafayette, Ind., 1972-75; dean Sch. Library and Info. Studies, U. Calif.-Berkeley, 1976—; v.p. Ind. Coop. Library Services Authority, 1974-75; vis. scholar Western Mich. U., 1979; vis. prof. U. Klagenfurt, Austria, 1980. Author: Book Availability and the Library User, 1975; co-author: The Uses of Gaming in Education for Library Management, 1976, Reader in Operations Research for Libraries, 1976. Mem. Library Assn. (London), ALA, Am. Soc. Info. Sci., Assn. Records Mgrs. and Adminstrs., Assn. Am. Library Schs., Calif. Library Assn. Office: Sch Library Studies U Calif Berkeley CA 94720

BUCKLER, MICHAEL J., electrical engineer, research executive; b. Washington, Sept. 29, 1949; s. James J. and Marijo (Malloy) B.; m. Carol L. Buckler, June 28, 1969; children: Mickey J., Brendan J. B.S.E.E., M.S.E.E., Ga. Inst. Tech. Mem. tech. staff Bell Telephone Labs., Holmdel, N.J., 1971-75, Atlanta, 1975-80, supr. lightguide applications and field testing group, 1980—. Contbr. articles to tech. jours.; patentee in field. Mem. IEEE, Optical Soc. Am. Roman Catholic. Subspecialty: Fiber optics. Current work: Lightwave communication systems: research, development, implementation and testing. Office: 2000 NE Expressway Norcross GA 30071

BUCKLEY, JOSEPH PAUL, pharmacologist, educator, univ. dean; b. Bridgeport, Conn., Jan. 12, 1924; m. Shirley Elizabeth Shipman, Aug. 16, 1947. B.S., U. Conn., 1949; M.S., Purdue U., 1951, Ph.D., 1952. Asst. prof. pharmacology U. Pitts., 1952-55, assoc. prof., 1955-58, prof., chmn. dept., 1958-73, assoc. dean Sch. Pharmacy, 1969-73; prof. pharmacology U. Houston, 1973—, dean Coll. Pharmacy, 1973—, dir., 1973—; hon. prof. San Carlos U., Guatemala City, Guatemala, 1967. Author: (with Ferrario) Central Actions of Angiotensin and Related Hormones, 1977, Central Nervous System Mechanisms in Hypertension, 1981; contbr. over 200 articles to sci. publs. Served to 2d lt. USAAF, 1943-45. Decorated Air medal with oak leaf clusters; recipient honors achievement award Angiology Research Found., 1965; Am. Found. for Pharm. Edn. fellow, 1950-52; numerous grants NIH, Dept. Def., pharm. industry, 1955—. Mem. Am. Heart Assn. (council for high blood pressure research), Soc. Exptl. Biology and Medicine, Acad. Pharm. Scis., Am. Pharm. Assn. (Eli Lilly award in pharmacodynamics 1966), Am. Soc. Pharmacology and Exptl. Therapeutics, N.Y. Acad. Scis., Sigma Xi. Methodist. Clubs: University (Pitts.); Warwick (Houston). Subspecialty: Pharmacology. Current work: Central nervous system actions of angiotensin with emphasis on relationship to hypertension; mechanisms of action of antihypertensive compounds, vasodilators and antiarrhythmic compounds. Home: 13714 Pebble Brook Houston TX 77079 Office: Coll Pharmacy U Houston 139 SR-2 Houston TX 77004

BUCKLEY, PAGE SCOTT, chemical engineer; b. Hampton, Va., June 23, 1918; s. Walter A. and Byrd E. Buckley; m. Betty Hill, Jan. 29, 1948; children: Ann, Kebba, Judith, Elizabeth. B.A., B.S., Columbia U., 1940; Dr. Eng. (hon.), Lehigh U., 1975. Registered profl. engr., Del., Calif. Asst. to cons. engr., Ottawa, Ont., Can., 1940; tech. asst. Monsanto Co., St. Louis, 1941-49; prin. cons. E.I. DuPont de Nemours & Co., Wilmington, Del., 1949—. Author: Techniques of Process Control, 1964. Mem. Am. Inst. Chem. Engrs., Nat. Acad. Engring., Instrument Soc. Am. Subspecialty: Chemical engineering. Current work: Design advanced control systems for chemical plants. Home: 9 N Kingston Rd Brookside Newark DE 19713 Office: DuPont Co Engring Dept Wilmington DE 19898

BUCKLIN, ANN CONE, marine biologist, researcher in population genetics; b. St. Louis, Oct. 13, 1951; d. Donale H. and Hope L. (Cone) B. A.B., Oberlin Coll., 1975; Ph.D., U. Calif.-Berkeley, 1980. Grad. asst. dept. zoology U. Calif.-Berkeley, 1975-78; postdoctoral scholar Woods Hole (Mass.) Oceanographic Instn., 1980-81, 82—, also guest investigator. Contbr. writings to profl. publs. NATO postdoctoral fellow dept. zoology, U. Reading (Eng.), 1981-82, Marine Biol. Assn. U.K., Plymouth, Eng., 1982; Oberlin alumni fellow, 1980; NSF grantee, 1980. Mem. AAAS, Am. Soc. Zoologists, Genetics Soc. Am., Soc. Study of Evolution, Western Soc. Naturalists. Subspecialties: Population biology; Ecology. Current work: Population genetic correlates of life history variation in marine invertebrates.

BUCKMAN, MAIRE TULTS, medical educator, physician; b. Tartu, Estonia, Sept. 25, 1939; came to U.S., 1949, naturalized, 1954; d. Harald and Kate (Gaag) T.; m.; children: James Harold, Sabrina Ellen. M.D., U. Wash., 1966. Diplomate: Nat. Bd. Med. Examiners; lic. physician, N. Mex. Rotating intern Santa Clara Hosp., San Jose, Calif., 1966-67; resident in endocrinology and metabolism Sch. Medicine, U. N. Mex., Albuquerque, 1971-72, NIH spl. fellow, 1972-74, instr. medicine, 1973-74, asst. prof., 1974-80, assoc. prof., 1980—; chief endocrinology and metabolism, dir. Radioimmunoassay Lab., VA Med. C., Albuquerque, 1980—. Contbr. articles to med. jours. Mem., active speaker Physicians for Social Responsibility, Albuquerque, 1982—. Served as capt. M.C. U.S. Army, 1966-68. Irwin Collison scholar, 1965; VA Career Devel. Program awardee, 1974-80; VA Merit Rev. awardee, 1980-82. Mem. ACP, Am. Fedn. Clin. Research (counselor 1977-80), Endocrine Soc., Western Soc. Clin. Research, Pacific Coast Fertility Soc., West Coast Endocrine Club (chmn. Carmel, Calif. 1978). Subspecialties: Neuroendocrinology; Psychopharmacology. Current work: Prolactin physiology and patholphysiology with special interest in prolactin role in regulation of blood pressure and renal function and its effect on psychological well-being in humans. Home: 2415 Vista Larga Albuquerque NM 87106 Office: University of New Mexico School of Medicine 2211 Lomas St NE Albuquerque NM 87106

BUCY, J. FRED, electronics company executive; b. Tahoka, Tex., July 29, 1928; s. J. Fred and Ethel (Montgomery) B.; m. Odetta Greer, Jan. 25, 1947; children: J. Fred III, Roxanne, Diane. B.A. in physics, Tex. Tech. U., 1951; M.A., U. Tex.-Austin, 1953. With Tex. Instruments Inc., Dallas, 1953—, corporate v.p., 1963-67, corporate group v.p., components, 1967-72, exec. v.p., 1972-75, 1975-76, chief operating officer, 1975—, pres., 1976—; also dir., rep. dir. Tex. Instruments

Japan Ltd.; mem. Tech. Assessment Adv. Council Office of Tech. Assessment Congress of U.S.; Bd. regents Tex. Tech. U. and Tex. Tech. U. Sch. of Medicine, chmn., 1980-82; mem. Harvard U. Vis. Com. Russian Research Center, MIT Physics Vis. Com. Recipient Disting. Engr. award Tex. Tech. U., 1972. Fellow IEEE; mem. Nat. Acad. Engring., Soc. Exploration Geophysicists, Sigma Pi Sigma, Tau Beta Pi. Methodist. Clubs: Cosmos (Washington); Petroleum, Univ., Northwood (Dallas). Subspecialties: Microelectronics; Geophysics. Patentee in field. Office: PO Box 225474 Dallas TX 75265

BUDDINGTON, PATRICIA ARRINGTON, aerospace engineer; b. Takoma Park, Md., Dec. 25, 1950; d. Warren and Elsie (Miller) B. B.S. in Aerospace Engring, Northop Inst. Tech., 1973; postgrad space technology, Flas. Inst. Tech., 1982—. With Air Force Systems Command Rocket Propulsion Lab., Edwards AFB, Calif., 1973-78, facility design engr., 1974-76, primary test engr. magneto-plasma-dynamic electric propulsion thrustor-pulsed plasma, 1976-78; with Boeing Aerospace Co., 1979—; test engr. reaction control system, inertial upper stage Cape Canaveral Air Force Sta., Cocoa Beach, Fla., 1981—. Mem. AIAA, Inst. Environ. Scis. (treas. Pacific N.W. chpt. 1981). Subspecialty: Aerospace engineering and technology. Current work: Advanced propulsion/space applications of technology: space industrialization. Home: 17 B Cape Shore Dr Cape Canaveral FL 32920 Office: Boeing Aerospace Co PO Box 220 M/S FC-51 Cocoa Beach FL 32931

BUDIANSKY, BERNARD, educator; b. N.Y.C., Mar. 8, 1925; s. Louis and Rose (Chaplick) B.; m. Nancy Cromer, Dec. 21, 1952; children: Michael, Stephen. B.C.E., CCNY, 1944; Sc.M., Brown U., 1948, Ph.D., 1950. With NACA, Langley Field, Va., 1944-55, head structural mechanics br., 1952- 55; faculty Harvard, 1955—, Gordon McKay prof. structural mechanics, 1961—; vis. prof. Technion, Haifa, Israel, 1976; Mem. research adv. com. on aircraft structures NASA, 1966-71, also adv. com. on space systems and tech., 1978—; mem. U.S. Nat. Com. on Theoretical and Applied Mechanics, 1970-80; mem. materials research council DARPA, 1968—. Bd. editors: Jour. Math. and Physics, 1961-68; cons. editor: Addison-Wesley Pub. Co, 1962-78, North Holland Pub. Co, 1978—; Author tech. reports. Recipient Townsend Harris medal City Coll. N.Y., 1974; Guggenheim fellow Tech. U. Denmark, 1961. Fellow AIAA (asso. editor Jour. 1963-66), ASME; mem. Nat. Acad. Scis., Nat. Acad. Engring., Am. Acad. Arts and Scis., Royal Netherlands Acad. Arts and Scis. (fgn.), Danish Center for Applied Math. and Mechanics (fgn. mem.), ASCE (von Karman medal 1982), Am. Geophys. Union, Sigma Xi, Tau Beta Pi. Subspecialties: Solid mechanics; Theoretical and applied mechanics. Current work: Fracture mechanics; plasticity; composite materials. Home: 11 DeMar Rd Lexington MA 02173 Office: Pierce Hall Harvard Cambridge MA 02138

BUDINGER, THOMAS FRANCIS, biomedical engineering educator, researcher; , 1965; 3 children. B.S., Regis Coll., Colo., 1954; M.S., U. Wash., 1957; M.D., U. Colo., 1964; Ph.D., U. Calif.-Berkeley, 1971. Analytical chemist Indsl. Labs., 1954; sr. oceanographer U. Wash., 1961-66; physicist Lawrence Livermore Lab. U. Calif.-Berkeley, 1966-67; research physician Donner Lab. and Lawrence Berkeley Lab., 1967-76; H. Miller prof. med. research and group leader research medicine Donner Lab., also prof. elec. engring. and computer sci., 1976—. Mem. AAAS, Am. Geophys. Union, N.Y. Acad. Sci., Soc. Nuclear Medicine. Subspecialties: Imaging technology; Nuclear magnetic resonance (biotechnology). Office: Donner La U Calif Berkeley CA 94720

BUDZYNSKI, ANDREI ZYGMUNT, biochemist, researcher; b. Warsaw, Poland, July 21, 1926; came to U.S., 1968; s. Zygmunt Stefan and Stefania (Kamienska) B.; m. Grazyna Halina Ochniewska, Dec. 31, 1950; children: Lech, Ewa. M.S. in Chemistry, U. Warsaw, 1951; postgrad., U. Vienna, 1959-61; Ph.D. in Biochemistry, Polish Acad. Scis., 1962. Instr. organic chemistry, dept. chemistry U. Warsaw, Poland, 1949-51; dir. dept. chemistry Manufacture of Sera and Vaccines, Warsaw, 1951-56; research chemist dept. radiochemistry Inst. Nuclear Research, Warsaw, 1956-59, asst. prof. dept. radiobiology, 1961-68; Internat. Atomic Energy Agy. fellow dept. radiochemistry U. Vienna, Austria, 1959-61; asst. research biochemist, dept. molecular biology and virus lab. U. Calif.-Berkeley, 1968-70; asst. prof. biochemistry and medicine Temple U. Health Sci. Center, Phila., 1970-78, assoc. prof. biochemistry and medicine, 1978-81, prof. biochemistry, 1981—. Contbr. writings to profl. publs. Mem. Am. Chem. Soc., Am. Heart Assn., Internat. Soc. Thrombosis and Haemostasis, Fedn. Am. Scientists, Am. Soc. Biol. Chemists. Republican. Subspecialties: Biochemistry (medicine); Hematology. Current work: Relation of protein structure to biologic functions. Research focus on enzymes and architectural proteins involved in formation and dissolution of blood clots, especially in man. Office: Temple Univ Sch Medicine 3400 N Broad St Philadelphia PA 19140

BUEHLER, ROBERT JOHN, biochemist; b. Schenectady, Dec. 4, 1947; s. Frederick P. and Frederica R. (Wickert) B.; m.; children: Sandra, Robyn. B.S. in Biology, Boston Coll., 1969; M.A. in Biochemistry, Boston U., 1970; Ph.D., 1975. Research asst. Wellesley (Mass.) Coll., 1973-75; cons. toxicologist Leary Clinic Lab., Boston, 1973-75; dir. research and devel. Damon Diagnostics, Needham Heights, Mass., 1975-82; project leader Corning Med. and Sci., East Walpole, Mass., 1982—. Contbr. articles in field to profl. jours. Served with U.S. Army, 1970-72. USPHS fellow, 1969-70. Mem. Am. Assn. Clin. Chemistry, Clin. Ligand Assay Soc., AAAS. Roman Catholic. Subspecialties: Immunology (medicine); Cancer research (medicine). Current work: Development of in vitro diagnostic test procedures for use in thyroid and oncological medicine; devel. of hybridoma cell lines for production ofmonoclonal antibodies and cell growth factors. Office: 333 Coney St East Walpole MA 02032

BUELOW, FREDERICK HENRY, educator; b. Minot, N.D., Mar. 13, 1929; s. Albert Wilhelm Gustav and Frieda Alvina Adele (Hass) B.; m. Selma Lois Ione Eia, July 21, 1954; children—David Frederick, Diane Louise, Darci Jo, Darin Martin. B.S., N.D. Agrl. Coll., 1951; M.S.E., Purdue U., 1952; Ph.D., Mich. State U., 1956. Faculty agrl. engring. Mich. State U., 1956-66, prof., 1965-66; prof., chmn. dept. agrl. engring. U. Wis.-Madison, 1966—. Served to lt. USAF, 1952-54. NSF grantee, 1963, 69, 70. Mem. Am. Soc. Agrl. Engrs. (Jour. Paper award 1957, dir. 1972-74, 77-79), Am. Soc. for Engring. Edn., Sigma Xi, Gamma Sigma Delta. Lutheran. Club: Kiwanis. Subspecialty: Agricultural engineering. Home: 6401 Landfall Dr Madison WI 53705

BUERGELT, CLAUS DIETMAR, veterinary pathology educator; b. Schmiedeberg, Germany, Sept. 16, 1939; came to U.S., 1972; s. Wilhelm and Elsa (Langer) B.; m. Nancy Kordak, Apr. 12, 1968. D.V.M., Vet. Coll., Hannover, W.Ger., 1965; Ph.D., Cornell U., 1976. Diplomate: Am. Coll. Vet. Pathology. Asst. prof. Yale U., New Haven, 1976-77; assoc. prof. vet. pathology U. Fla., Gainesville, 1978—. Mem. German Vet. Assn., AMVA, Am. Coll. Vet. Pathologists, Internat. Acad. Pathologists, Soc. Environ. Pathologists. Lutheran. Subspecialty: Pathology (veterinary medicine). Current work: Bovine paratuberculosis; macrophage phagocytosis and digestion. 32610

BUETTNER, GARRY RICHARD, chemist, educator; b. Vinton, Iowa, Oct. 11, 1945; s. Albert E. and Marlene L. (Koopman) B.; m. Claudia J. Barber, Oct. 9, 1948; children: Daniel, David, Adam, Elisabeth. B.A., U. No. Iowa, 1967; M.S., U. Iowa, 1969, Ph.D., 1976. Asst. prof. chemistry Wabash Coll., Crawfordsville, Ind., 1978—. Author research publs. Served with USAF, 1969-73. Recipient Nat. Research Service award, 1976-78. Mem. Am. Chem. Soc., Am. Soc. Photobiology; mem. Soc. Free Radical Research; Mem. N.Y. Acad. Scis., Sigma Xi. Subspecialty: Physical chemistry. Current work: Role of superoxide and superoxide dismutase in human health. Magnetic resonance lineshape simulation. Home: 714 S Green St Crawfordsville IN 47933 Office: Dept Chemistry Wabash Coll Crawfordsville IN 47933

BUFF, JAMES STEVE, research scientist; b. Lincolnton, N.C., Nov. 2, 1947; s. Clee Howard and Pauline Susan (Hovis) B.; m. Margaret Anne LeClair, Jan. 2, 1982. B.S. in Physics, U.N.C., Chapel Hill, 1969, Ph.D., U. colo., Boulder, 1974. Research scientist Mass. Inst. Tech., 1974-76; research scientist IBM Watson Research Center, Yorktown Heights, N.Y., 1976-78; vis. asst. prof. Dartmouth Coll., 1978-80; research scientist Mission Research Corp., Albuquerque, 1980—. Contbr. articles to sci. jours. Mem. Am. Astronom. Soc., Am. Phys. Soc., Internat. Astronom. Union, Phi Beta Kappa. Subspecialties: Plasma physics; 1-ray high energy astrophysics. Current work: Inertial fusion. Office: 1720 Randolph Rd SE Albuquerque NM 87106

BUFFORD, RODGER KEITH, psychology educator, clinical psychologist; b. Santa Rosa, Calif., Dec. 23, 1944; s. John Samuel and Evelyn (Rude) B.; m. Kathleen A. Parson, Aug 17, 1968; children: Heather, Brett. B.A., King's Coll., 1966; M.A., U. Ill.-Urbana, 1970, Ph.D., 1971. Lic. psychologist, Va., Ga., Oreg. Asst. prof. Am. U., 1971-76; asst. prof. chmn. dept. psychology Huntington Coll., 19976-77; assoc. prof., dir. clin. tng. Psychol. Studies Inst., 1977-81; psychologist Atlanta Counseling Center, 1980-82; assoc. prof. psychology, chmn. dept. psychology Western Sem., 1982—; bd. dirs. Mental Health Assn., Huntington, Ind., 1976-77. Author: The Human Relfex: Behavioral Psychology in Biblical Perspective, 1981; contbr. articles to profl. jours.; editor: Baker's Ency. of Psychology, 1984; contbg. editor: Christian Assn. Psychol. Studies Bull, 1979—; cons. reviewer: Jour. Psychology and Theology, 1974-81; contbg. editor, 1982—; cons. reviewer jours., pubs., Elder Chapel Woods Presbyterian Ch., Decatur, Ga., 1982—. USPHS trainee, 1967-68, 70-71; Am. U. Faculty research grantee, 1972. Mem. Christian Assn. Psychol. Studies (program chmn. 1982), Am. PSychol. Assn., Am. Sci. Affiliation, Midwestern Psychol. Assn., Southeastern Psychol. Assn. Republican. Subspecialties: Behavioral psychology; Psychology and religion. Current work: Behavioral psychology and biblical principles; psychology of religion; measurement of religious variables; training in psychology and religious perspectives. Home: 19504 Hidden Springs Rd West Linn OR 97068 Office: Dept Psychology Western Sem 5511 SE Hawthorne Blvd Portland OR 97215

BUFTON, JACK LYTLE, elec. engr.; b. Kenmore, N.Y., Jan. 25, 1945; s. Jack Van Hawn and Dorothy Kennedy (Lytle) B.; m. Carol Luttermoser, June 2, 1973; children: Diane, Hobbs, Julie Patterson, Suzanne Patterson. B.S. in Engring. Physics, Lehigh U., 1966; M.S. in Physics, U.Md., 1970; Ph.D. in Elec. Engring, U.Md., 1976. Research and devel. engr. laser remote sensing, and electro-optics NASA/Goddard Space Flight Center, Greenbelt, Mass., 1966—, sec. head electro-optical instrument br., 1980. Contbr. numerous articles on applied optics to jours. Recipient Spl. Achievement award NASA, 1976. Mem. Optical Soc. Am. Subspecialties: Remote sensing (atmospheric science); Planetary atmospheres. Current work: Remote sensing, lidar, sub-millimeter wave tech., CO2 TEA lasers, nonlinear infrared crystals, heterodyne radiometers. Office: Goddard Space Flight Center Code 723 Greenbelt MD 20771

BUJTAS, MARK STEVEN, mechanical engineer; b. Princeton, N.J., Aug. 24, 1952; s. Andrew E. and Anneliese (Kindler) B.; m. Cornelia E., Sept. 7, 1980. B.S.M.E., Newark Coll. Engring., 1974. Asst. engr. U.S. Machine Co., Cedar Grove, N.J., 1974-75; project engr. Fluorocarbon, Pine Brook, N.J., 1975-77, chief engr., 1977—; v.p. Bermag Corp. Internat., 1983—. Mem. ASME (asso.), Soc. Mfg. Engrs. (cert. mfg. engr.), Am. Concrete Inst. (bearing systems com.). Subspecialties: Mechanical engineering; Materials. Current work: Low-friction structural expansion bearings using resin teflon and woven teflon, particularly for use in pot bearings, spherical bearings, and elastomeric backed slide bearings. Office: 18 Commerce Rd Fairfield NJ 07006

BUKHARI, AHMAD IQBAL, microbiology educator; b. Amritsar, Punjab, India, Jan. 5, 1943; came to U.S., 1964, naturalized, 1983; s. Noorul Islam and Sarwar (Begum) B.; m. Christine Karen Morgan, Sept. 6, 1975; children: Yousef Ali, Jaffer Ahmad. B.Sc., U. Karachi, Pakistan, 1961, M.Sc., 1963; M.S., Brown U., 1966; Ph.D., U. Colo., 1970. Postdoctoral fellow Cold Spring Harbor (N.Y.) Lab., 1970-72, sr. staff scientist, 1978—; assoc. prof. microbiology SUNY, Stony Brook, 1978—. Jane Coffin Childs Meml. Fund fellow, 1971-73; NIH grantee, 1975-80. Mem. Am. Soc. for Microbiology, Genetics Soc. Am., AAAS. Subspecialties: Microbiology; Genome organization. Current work: Mechanisms of DNA rearrangements. Office: Cold Spring Harbor Lab Cold Spring Harbor NY 11724

BUKOVSAN, LAURA ANN, biologist; b. Norfolk, Va., Jan. 2, 1940; d. Robert Hunter and Grace Elizabeth (Grabman) Colgin; m. William Bukovsan, Aug. 29, 1964; children: William Hunter, James Richard. B.S., U. Richmond, 1961; M.A., Ind. U., 1963, Ph.D., 1969. Lectr. biology SUNY, Oneonata, 1974-75, asso. prof., 1980—; lectr. Hartwick Coll., Oneonta, 1979. Mem. Genetics Soc. Am., AAAS, Soc. Protozoologists, Sigma Xi. Subspecialties: Microbiology; Gene actions. Current work: Researcher in aging in Tokophrya lemnarum.

BUKOVSAN, WILLIAM, biologist, educator; b. Chgo., June 4, 1929; s. John and Elizabeth (Vitek) B.; m. Laura Ann Colgin, Aug. 29, 1964; children: William Hunter, James Richard. B.S., U. Ill., 1955, M.S., 1958; Ph.D., Ind. U., 1967. Assoc. prof. biology SUNY, Oneonta, 1967-70, prof., 1970—. Author: Laboratory Studies in Animal Biology-Zoology, 1977; contbr. articles to profl. jours. Chmn. Zoning Bd. Appeals Otego, N.Y., 1980—. Served with U.S. Army, 1952-54. Mem. N.Y. Acad. Scis., AAAS, Am. Soc. Zoologists, Sigma Xi. Subspecialties: Endocrinology; Neuroendocrinology. Current work: Endocrinology and cell biology; cell culture, adrenal glands, corticotropin. Office: Dept Biology SUNY Oneonta Oneonta NY 13820

BUKRY, J(OHN) DAVID, geologist; b. Balt., May 17, 1941; s. Howard Leroy and Irene Evelyn (Davis) Snyder. Student, Colo. Sch. Mines, 1959-60; B.A., Johns Hopkins U., 1963; M.A., Princeton U., 1965, Ph.D., 1967; postgrad., U. Ill., 1965-66. Geologist U.S. Army Corp. Engrs., Balt., 1963; research asst. Mobil Oil Co., Dallas, 1965; geologist U.S. Geol. Survey, La Jolla, Calif., 1967—; research assoc. dept. geol. research div. U. Calif.-San Diego, 1970—; cons. Deep Sea Drilling Project, LaJolla, 1967—. Author: Leg I of the Cruises of the Drilling Vessel Glomar Challenger, 1969, Coccoliths from Texas and Europe, 1969, Leg LXIII of the Cruises of the Drilling Vessel Glomar Challenger, 1981; editor: Marine Micropaleontology, 1976—. Mobil Oil, Princeton U. fellow, 1965-67; Am. Chem. Soc., Princeton U. fellow, 1966-67. Fellow AAAS, Geol.Soc. Am., Explorer's Club; mem. Paleontol. Research Inst. (assoc.), Sigma Xi. Club: San Diego Shell. Subspecialties: Oceanography, biochronology; Paleoecology. Current work: Defining new species of phytoplankton for chronologic and paleoecologic analysis of ocean strata. Home: 675 S Sierra Ave Solana Beach CA 92075 Office: US Geol Survey A015 Inst Oceanography La Jolla CA 92093

BULGER, ROGER JAMES, physician, medical school president; b. Bklyn., May 18, 1933; s. William Joseph and Florence Dorothy (Poggi) B.; m. Ruth Ellen Grouse, June 8, 1960; children: Faith Anne, Grace Ellen. A.B., Harvard U., 1955, M.D., 1960; postgrad., Emmanuel Coll., Cambridge (Eng.) U., 1955-56. Intern, resident internal medicine U. Wash. Hosps., 1960-62, 64-65; postgrad. trainee infectious disease and microbiology U. Wash., 1962-63, 65-66; renal and metabolic diseases Boston U., 1963-64; asst. prof., then asso. prof. medicine U. Wash. Med. Sch., Seattle, 1966-70; med. dir. Univ. Hosp., Seattle, 1967-70; prof. community health scis., asso. dean allied health Duke U. Med. Center, 1970-72; exec. officer Inst. Medicine, Nat. Acad. Scis., 1972-76; prof. internal medicine George Washington U. Sch. Medicine, 1972-76; prof. internal medicine, family and community medicine, dean Med. Sch., chancellor Worcester campus U. Mass., 1976-78; pres. U. Tex. Health Sci. Center, Houston, 1978—; mem. report rev. com. Nat. Acad. Scis.; adv. panel nat. health ins. com., ways and means com. U.S. Ho. Reps., 1975-76. Author: Hippocrates Revisited, 1973, also articles, chpts. in books.; Mem. editorial bds. various jours. Bd. dirs. Georgetown U. Lionel de Jersey Harvard fellow, 1955-56. Fellow A.C.P.; mem. Inst. Medicine, Am. Soc. Microbiology, Infectious Disease Soc. Am., Am. Fedn. Clin. Research, Soc. Tchrs. Preventive Medicine, Am. Soc. Nephrology, Soc. Health and Human Values. Subspecialty: Internal medicine. Office: Office of Pres U Tex Health Sci Center at Houston PO Box 20036 Houston TX 77225

BULKLEY, GREGORY BARTLETT, general and thoracic surgeon, medical educator; b. Spokane, Wash., Apr. 28, 1943; s. George J. and Patricia (Bartlett) B.; m. Bernadine Healy, Aug. 13, 1967 (div. 1982); 1 child, Bartlett Anne. A.B., Princeton U., 1965; M.D., Harvard U., 1970. Diplomate: Am. Bd. Surgery. Intern Johns Hopkins U. Hosp., Balt., 1970-72, resident in surgery, 1974-77, staff, 1978—; mem. faculty Johns Hopkins U. Sch. Medicine, Balt., 1977—, assoc. prof. surgery, 1982—, dir. surgical research, 1982—. Author, editor: Measurement of Blood Flow, 1982; contbr. numerous sci. articles to profl. publs. Served in USPHS, 1972-74. NSF fellow, 1963; Am. Heart Assn. grantee, 1979; George & Sadie Hyman Found. grantee, 1979; NIH grantee, 1982. Fellow ACS; mem. Assn. Acad. Surgery (mem. membership com. 1981-83), Am. Physiol. Assn., Am. Gastroenterol. Assn., Soc. Univ. Surgeons. Subspecialties: Surgery; Physiology (medicine). Current work: Control of the splanchnic circulation; pathophysiology; diagonsis and treatment of intestinal ischemic disease; conduct primary original research in these conditions - both basic laboratory work and clinical applications and clinical trials. Office: Johns Hopkins Univ Sch Medicine 600 N Wolfe St Baltimore MD 21205

BULL, BRIAN STANLEY, physician, educator; b. Watford, Hertfordshire, Eng., Sept. 14, 1937; came to U.S., 1954, naturalized, 1960; s. Stanley and Agnes Mary (Murdoch) B.; m. Maureen Hannah Huse, June 3, 1963; children—Beverly Velda, Beryl Heather. B.S. in Zoology, Walla Walla Coll., 1957; M.D., Loma Linda (Calif.) U., 1961. Diplomate: Am. Bd. Pathology. Intern Yale U., 1961-62, resident in anat. pathology, 1962-63; resident in clin. pathology NIH, Bethesda, Md., 1963-65, fellow in hematology and electron microscopy, 1965-66, staff hematologist, 1966-67; research asst. dept. anatomy Loma Linda U., 1958, dept. microbiology, 1959, asst. prof. pathology, 1968-71, assoc. prof., 1971-73, prof., 1973—, chmn. dept. pathology, 1973—, consulting pathologist, 1968—; vis. prof. Institut de Pathologie Cellulaire, Paris, 1972, 74, Royal Postgrad. Med. Sch., London, 1972, U. Wis.-Madison, 1973, U. Ohio, Columbus, 1974, U. Minn., Mpls., 1979. Bd. editors: Jour. Clin. and Lab. Haematology, U.K., 1980—; contbr. chpts. to books and numerous articles to med. jours. Served with USPHS, 1963-67. Nat. Inst. Arthritis and Metabolic Diseases fellow, 1967-68; recipient Daniel D. Comstock Meml. award Loma Linda U., 1961, Merck Manual award, 1961, Mosby Scholarship Book award, 1961; Ernest B. Cotlove Meml. lectr. Acad. Clin. Lab. Physicians and Scientists, 1972. Fellow Am. Soc. Clin. Pathologists, Am. Soc. Hematology, Coll. Am. Pathologists, N.Y. Acad. Scis.; mem. AMA, Assn. Pathology Chmn., Calif. Soc. Pathologists, San Bernardino County Med. Soc., Acad. Clin. Lab. Physicians and Scientists, Am. Soc. Exptl. Pathology, Sigma Xi, Alpha Omega Alpha. Seventh-day Adventist. Subspecialties: Hematology; Pathology (medicine). Current work: Researcher, inventor devices for med. analysis, devel. math. algorithms for quality control, red cell modelling, antocoagulant control and thrombosis prevention. Patentee in field. Home: 24489 Barton Rd Loma Linda CA 92354 Office: Department of Pathology and Laboratory Medicine Loma Linda University School of Medicine Loma Linda CA 92350

BULL, COLIN BRUCE BRADLEY, educator; b. Birmingham, Eng., June 13, 1928; s. George Ernest and Alice Matilda (Collier) B.; m. Diana Gillian Garrett, June 16, 1956; children—Nicholas, Rebecca, Andrew. B.Sc., Birmingham U., 1948, Ph.D., 1951. Geophysicist later chief scientist Brit. N. Greenland Expdn., 1952-56; sr. lectr. physics Victoria U., Wellington, New Zealand, 1956-61; asso. prof. geology Ohio State U., Columbus, 1962-65; prof., dir. Inst. Polar Studies, 1965-69, chmn. dept. geology, 1969-72, dean, 1972—; Vis. fellow geophysics Australian Nat. U., Canberra, 1960; vis. scholar Cambridge (Eng.) U., 1969; vis. prof. Nat. Inst. Polar Research, Tokyo, Japan, 1983; U.S. rep. working group on glaciology Sci. Com. Antarctic Research, 1974—, sec., 1978—. Contbr. articles to profl. jours. Recipient Polar medal Queen Elizabeth, 1954; U.S. Antarctic Service medal, 1974. Fellow Arctic Inst. N. Am. (bd. govs. 1966-72); Fellow Geol. Soc. Am., Royal Soc. Arts; mem. Internat. Glaciology Soc. (council 1974-78), Am. Geophys. Union, Phi Beta Kappa (hon.). Subspecialties: Geophysics; Glaciology. Current work: Stability and dynamics of ice sheets, Antarctica and Greenland. Home: 4187 Olentangy Blvd Columbus OH 43214

BULL, FRANCES ELEANOR, oncologist, medical educator; b. Ann Arbor, Mich., Jan. 10, 1927; d. Hempstead S. and Sarah E. (Carr) B. B.S., U. Mich., 1948, M.D., 1952; M.S., U. Minn., Mpls., 1960. Diplomate: Am. Bd. Internal Medicine. Intern Phila. Gen. Hosp., 1952-53; resident in internal medicine Mayo Clinic, Rochester, Minn., 1954-57; instr. to prof. internal medicine sect. med. oncology U. Mich., 1957—; co-investigator S.W. Oncology Group. Contbr. articles on cancer treatment to profl. jours. Fellow ACP; mem. Am. Soc. Clin. Oncology, Am. Fedn. Clin. Research. Subspecialties: Oncology; Cancer research (medicine). Current work: Clinical research in new drugs for cancer treatment, clinical research, pharmacology, antineoplastic agents. Office: Medical Center Ann Arbor MI 48109

BULL, LEONARD SETH, animal nutrition educator, researcher; b. Westfield, Mass., Jan. 31, 1941; s. Floyd Milton and Erna Magdaline (Preissler) B.; m. Deborah Carol Stevens, May 22, 1971. B.S., Okla. State U., 1963, M.S., 1964; Ph.D., Cornell U., 1969. Cert. animal scientist Am. Soc. Animal Sci., 1978. Fellow med. physiology U. Va., 1968-70; asst. prof. nutrition U. Md., 1970-74, assoc. prof., 1974-75; assoc. prof. nutrition U. Ky., 1975-79; prof. U. Maine, 1979-81; prof. animal sci. U. Vt., 1981-; also chmn. dept.; mem. subcoms. Nat. Acad. Scis.; cons. in field. Contbr. numerous articles to profl. jours. Recipient Moorman Internat. Travel award, 1979. Mem. Am. Dairy Sci. Assn., Am. Soc. Animal Sci., Am. Inst. Nutrition. Republican. Methodist.

Subspecialties: Animal nutrition; Nutrition (medicine). Current work: Research on nutritional energetics and protein metabolism; teaching in animal sci.; adminstrn. of animal scis. academic program. Home: 23 Tanglewood Dr Essex Junction VT 05452 Office: 220 Carrigan Hall Vt Burlington VT 05405

BULL, RICHARD JAMES, toxicologist; b. Stillwater, Okla., Oct. 25, 1940; s. Campbell Carlos and Jullian Georgiana (Reiter) B.; m. Roberta DiAnne Mohaupt, Sept. 1, 1962; children: Greta, Erika, Marta, Jeffrey. B.S. in Pharmacy, U. Wash., 1964; Ph.D. in Pharmacology, U. Calif.-San Francisco, 1971. Registered pharmacist, Wash. Pharmacy intern USPHS, R.I., 1964-65; research chemist, Narragansett, R.I., 1965-67; research pharmacologist EPA, Cin., 1970-77, chief toxicological assessment br., 1977-81, dir. toxicology and microbiology div., 1981—. Contbr. numerous research articles, chpts. to profl. publs. Recipient First Sci. and Technol. Achievement award EPA, 1979. Mem. AAAS, Soc. Toxicology, Am. Soc. Pharmacology and Exptl. Therapeutics, Soc. Neurochemistry, Soc. Neurosci., Sigma Xi. Subspecialties: Pharmacology; Toxicology (medicine). Current work: Toxicological and carcinogenic activities of chemicals that are water pollutants; effects of lead on development of mammalian brain. Office: EPA 26 W Saint Clair Cincinnati OH 45268

BULL, STANLEY RAYMOND, engineer, research administrator; b. Montezuma, Iowa, May 15, 1941; s. Raymond W. and Neola A. B.; m. Diana Lee Maxwell, Sept. 7, 1963; children: Melanie J., Julia D., Jeffrey D. B.S. in Chem. Engring, U. Mo., 1963; M.S., Stanford U., 1964; Ph.D. in Mech. Engring, 1967. Registered profl. engr., Mo. Successively asst. prof., assoc. prof., prof. Coll. Engring., U. Mo., Columbia, 1967-80; Solar Energy Research Inst., Golden, Colo., 1980-81; sr. sci. adviser, 1981-82, mgr. planning and research devel., 1982—; Fulbright-Hays prof. Centre d'Études, Grenoble, France, 1973-74. Contbr. articles to tech. jours. Mem. Am. Soc. Engring. Edn., Internat. Am. Solar Energy Soc., Am. Phys. Soc. Subspecialties: Engineering physics; Solar energy. Current work: Energy systems and resources; research and development resource allocations and plans; evaluation renewable energy technology. Office: 1617 Cole Blvd Golden CO 80401

BULLA, LEE AUSTIN, JR., science educator and administrator, industrial consultant and researcher; b. Oklahoma City, May 1, 1941; s. Lee Austin and Ruthie (Pearce) B.; m. Betty Lee Harrison, Sept. 29, 1962; children: Stephen, Susan, Stacey. B.S. in Biology and Math, Midwestern U., 1965; Ph.D. in Microbiology and Biochemistry, Oreg. State U., 1968. Microbiologist and project leader No. Regional Research Lab., Agr. Research Service, U.S. Dept. Agr., Peoria, Ill., 1968-73; microbiologist Kans. State U., Manhattan, 1973-81, prof. div. biology, 1973-81; prof. dept. bacteriology and biochemistry U. Idaho, Moscow, 1981—; assoc. dean Coll. Agr. and dir. Idaho Agr. Expt. Sta., 1981—; dir. Inst. Molecular and Agr. Genetic Engring., 1981—. Contbr. numerous articles to profl. jours. Bd. dirs. Idaho Research Fdn., 1983. Mem. Am. Soc. Microbiology, AAAS. Lodge: Kiwanis. Subspecialties: Microbiology; Genetics and genetic engineering (agriculture). Current work: Genetic characterization of insecticidal proteins of Bacillus Thuringiensis and regulation of protein synthesis in gram-positive bacilli. Office: Coll Agr U Idaho Moscow ID 83843

BULLAS, LEONARD RAYMOND, microbiologist, educator, researcher; b. Lismore, New South Wales, Australia, Dec. 8, 1929; s. Raymond and Arum Adelaide (Semmens) B.; m. Rosemary Grace Ekdahl, Nov. 30, 1958; children: Roslyn Mary, Graham Leonard. B.S., U. Adelaide, Australia, 1953, M.S., 1958; Ph.D., Mont. State U., 1962. Teaching asst. U. Adelaide, Australia, 1953-58; research asst. Mont. State U., Bozeman, 1958-62; instr. Loma Linda (Calif.) U., 1962-64, asst. prof., 1964-70, asso. prof., 1970-80, prof. microbiology, 1980—; vis. prof. European Molecular Biology Lab., Heidelberg, W. Ger., 1981-82; researcher in field; research contract Office Naval Research, 1965-69. Contbr.: articles to profl. publs. including Jour. of General Microbiology. Basic sci. fellow Loma Linda U. Med. Sch. Alumni Assn., 1981; named basic sci. investigator of year, 1975, 81. Mem. Am. Soc. Microbiology, Genetics Soc. Am., Sigma Xi. Seventh-day Adventist. Subspecialties: Genetics and genetic engineering (biology); Genome organization. Current work: Genetics and molecular biology of genes for Salmonella restriction endonucleases. Office: Dept Microbiology Loma Linda Univ Loma Linda CA 92350

BULLERMAN, LLOYD BERNARD, food scientist, educator; b. Adrian, Minn., June 20, 1939; s. Leonard Theodore and Alma Frances (Voss) B.; m. Kathleen Ann Persing, July 30, 1960; children: Lisa, Michael, Lori, Mark. B.S., S.D. State U., 1961, M.S., 1965; Ph.D., Iowa State U., 1968. Food scientist Green Giant Co., LeSueur, Minn., 1968-70; asst. prof. food sci. and tech. U. Nebr., Lincoln, 1970-75, assoc. prof., 1975-79, prof., 1979—, interim dept. head, 1980-81; cons. in field. NIH predoctoral fellow; NIH health grantee; indsl. grants awardee.; FDA contractee. Mem. Inst. Food Technologists, Am. Soc. Microbiology, Am. Assn. Cereal Chemists, Internat. Assn. Milk Food and Environ. Sanitarians, Sigma Xi, Gamma Sigma Delta. Subspecialties: Food science and technology; Microbiology. Current work: Food microbiology, mycology and mycotoxicology. Food safety, toxicology and safety evaluation. Home: 6701 Amhurst Dr Lincoln NE 68510 Office: Dept Food Sci and Tech U Nebr Lincoln NE 68583

BULLOCH, KAREN, neuroimmunologist; b. Washington, Sept. 18, 1945; d. Douglass Eugene and Grace Edith (Webb) B.; m. Walter William Muryasz, Aug. 22, 1964 (div.); children: Kimberly Lynne, Scott William. Student, MacMurray Coll., Jacksonville, Ill., 1964-66; Ph.D., U Calif.-San Diego, 1981. Research investigator Salk Inst., 1970-78; lectr., postdoctoral fellow SUNY-Stony Brook, 1980-82, research asst. prof. dept. neurology, 1982—. Contbr. articles to profl. jours. NINCDS-NIH grantee, 1982—. Mem. Soc. Neuroscis., N.Y. Acad. Sci., Sierra Club. Democrat. Subspecialties: Immunology (medicine); Neuroimmunology. Current work: Neuroendocrine-immune system integration, autonomic nervous system, humoral and cell mediated immunity, neuroendocrine, lymphocytes, receptor. Office: Dept Neurolog SUNY Stony Brook T12-033 Stony Brook NY 1179

BULLOCK, FRANCIS JEREMIAH, research exec.; b. Brookline, Mass., Jan. 14, 1937; m. Lorraine M. Littig, Aug. 26, 1961; children: Christine, Gregory. B.S., Mass. Coll. Pharmacy, 1958; A.M., Harvard U., 1961, Ph.D., 1963. Research assoc. Harvard U. Med. Sch., Boston, 1963-64, U. Calif. - Berkeley, 1964-65; sr. staff Arthur D. Little, Inc., Cambridge, Mass., 1965-72; mgr. medicinal chemistry Abbott Labs., Chgo., 1972-79; v.p. new drug discovery Schering-Plough, Inc., Bloomfield, N.J., 1979-81, sr. v.p. research ops., 1981—. Mem. Am. Chem. Soc., Chem. Soc. (London), Am. Soc. Pharmacology and Exptl. Therapeutics, Am. Soc. Microbiology. Subspecialties: Medicinal chemistry; Pharmacology. Current work: Management of research directed to discovery of new therapeutics through application of chemistry, pharmacology, microbiology, biochemistry or genetic engineering. Office: 60 Ornge St Bloomfield NJ 07003

BULLOCK, ROBERT MORTON, III, mathematics educator; b. Cin., June 30, 1937; s. Robert Morton and Frieda Delores (Schreiber) B.; m. Evonne Lorton, Feb. 1, 1963; children: Tanja Lynne, Robert Morton, IV. B.S., U. Cin., 1961, M.A., 1962, Ph.D., 1966. Tchr. math. Cin. Country Day Sch., 1962-63; Instr. math. U. Cin., 1963-66; mem. faculty Miami U., Oxford, Ohio, 1966—, assoc. prof. math., 1982—. Contbg. editor: Speaker Builder mag, Peterborough, N.H., 1981—. Served with USAF, 1955-59. Mem. Soc. for Indsl. and Applied Math., Audio Engring. Soc. (assoc.). Subspecialties: Applied mathematics; Audio engineering. Current work: Modelling of loudspeaker systems and crossover networks; model analysis. Home: 1301 Dana Dr Oxford OH 45056 Office: Dept Math and Stats Miami U Oxford OH 45056

BULLOCK, SCOTT VERNE, nuclear power plant administrator; b. Buffalo, Dec. 6, 1947; s. Leland Verne and Joyce (Knapp) B. B.S., U. Rochester, 1976. Constrn. insp. Rochester Gas & Electric, 1967-70, startup tester, 1970-72, quality control insp., 1973-77, quality control inspecion engr., 1977-78, quality control engr., 1978-79, quality assurance engr., adminstr. quality assurance group, 1979—; pres., chief exec. officer SVB Enterprises, Rochester, 1973—. Head dept. United Ch. of Christ, Rochester, 1970-75. Served with U.S. Army, 1970-71. Mem. Am. Nuclear Soc., Am. Soc. Quality Control, Am. Soc. Nondestructive Testing. Republican. Club: Ontario (N.Y.) Country. Lodge: Eagles. Subspecialties: Nuclear fission; Nuclear engineering. Current work: Striving to upgrade standards for safe nuclear power plant operation. Office: Rochester Gas & Electric Corp 89 East Ave Rochester NY 14649

BUNAG, RUBEN DAVID, pharmacology educator; b. Manila, Philippines, June 3, 1931; came to U.S., 1960; s. Maximino Galura and Angelina (David) B.; m. Pros Liongson, Apr. 7, 1956; children: Royce, Karen. M.D., U. Philippines, 1955; M.A., U. Kans., 1962. USPHS internat. postdoctoral research fellow U. Kans., 1960-62; Med. Life Ins. research fellow Case-Western Reserve U., Cleve., 1963-64; research fellow (Cleveland Clinic), 1964-69; asso. prof. pharmacology U. Kans., 1970-75, prof. pharmacology, 1975—; research pharmacologist VA Hosp., Kansas City, Mo., 1970-72. Contbr. numerous articles to sci. jours., abstracts. NIH research grantee, 1971—. Mem. Am. Physiol. Soc., Council for High Blood Pressure Research, Soc. Exptl. Biology and Medicine, Neurosci. Soc. Roman Cath. Subspecialties: Comparative physiology; Molecular pharmacology. Current work: Experimental hypertension, autonomic drugs, contraceptive steroids, experimental diabetes. Home: 9900 Perry Dr Overland Park KS 66212 Office: U Kans Med Ctr Kansas City KS 66103

BUNCH, PHILLIP CARTER, physicist; b. Maryville, Tenn., Sept. 14, 1946; s. Everett J. and Margaret J. (Sutton) B.; m. Margaret Tyler, Dec. 22, 1967; children: Jonathan A., Jessica L. A.B. with honors, U. Chgo., 1969, M.S. in Radiol. Scis, 1971, Ph.D. in Med. Physics, 1975. Research physicist Kodak Research Labs., Rochester, N.Y., 1975-79, sr. research physicist, 1979—; mem. Internat. Commn. on Radiation Units and Measurements Com. on Modulation Transfer Functions of Screen-Film Systems, 1979—. Contbr. articles to profl. jours. NIH grad. fellow, 1969-74. Mem. Am. Assn. Physics in Medicine, Optical Soc. Am., Soc. Photo-Pptical Instrumentation Engrs., IEEE, Soc. Photog. Scientists and Engrs. Unitarian. Subspecialties: Imaging technology; Mathematical software. Current work: Radiologic imaging, image analysis, computer modelling, data analysis, diagnostic radiologic imaging, computers. Home: 129 Landing Rd N Brighton NY 14625 Office: Kodak Research Labs Bldg 81 Rochester NY 14650

BUNCH, WILBUR LYLE, nuclear physicist; b. Pine Bluffs, Wyo., Apr. 24, 1925; s. Bradley K. and Elvira (Carlstrum) B.; m. Margaret M. May, Nov. 24, 1926; children: Larry, Jerry, Jim, Janice, Judy. B.S., U. Wyo., 1949, M.S., 1951. Physicist Gen. Electric Co., Richland, Wash., 1951-64; mgr. Battelle N.W., Richland, 1964-70, Westinghouse Hanford Co., 1970—. Contbg. author: Reactor Handbook, 1964, Engineering Compendium, 1974. Cubmaster Boy Scouts Am., 1961-63; coach Little League, Richland, 1958-62, Westside Ch., Richland, 1964-70. Srved in U.S. Army, 1943-46. Mem. Am. Nuclear Soc. (Best Paper award 1981, div. chmn. 1977-78). Republican. Subspecialties: Nuclear fission; Nuclear engineering. Current work: Interaction of radiation with matter; radiation shielding; radioisotope production; fast breeder reactor technology; fission products; gamma radiation. Home: 2403 Pullen St Richland WA 99352 Office: PO Box 1970 Richland WA 99352

BUNDY, HALLIE FLOWERS, biochemist, educator; b. Santa Monica, Calif., Apr. 2, 1965; d. Douglas and Phyllis (Flowers) B. B.A., Mt. St. Mary's Coll., 1947; M.S., U. So. Calif., 1955, Ph.D. in Biochemistry, 1958. Instr. U. So. Calif. Sch.Medicine, 1959-60; asst. program dir. undergrad. research program NSF, Washington, 1965-66; from asst. prof. to assoc. prof. Mt. St. Mary's Coll., Los Angeles, 1960-65, prof. biochemistry, 1966—. Contbr. articles to profl. jours. USPHS fellow, 1955-58; NSF faculty fellow, 1969; Grad. Women in Sci. Grant-in-Aid awardee, 1974. Mem. Am. Chem. Soc., Pacific Slope Biochem. Conf., AAAS, Grad. Women in Sci., Sigma Xi. Subspecialties: Biochemistry (biology); Enzymology. Current work: Comparative biochemistry of algal, higher plant and animal carbonic anhydrases. Office: 12001 Chalon Rd Los Angeles CA 90049

BUNGAY, HENRY ROBERT, III, biochem. engr., educator, researcher, cons.; b. Cleve., Jan. 22, 1928; s. Henry Robert, Jr. and Marguerite (Callahan) B.; m. Mary Louise Bungay, Feb. 25, 1929; children: Margaret Bungay Rizzner, Henry, James. B.Ch.E. (scholar), Cornell U., Ithaca, N.Y., 1949; Ph.D., Syracuse U., 1954. Registered profl. engr., Va., N.J. Biochemist Eli Lilly, Indpls., 1954-62; prof. san. engring. Va. Poly Inst., 1962-67; bioengring. Clemson U., 1967-72; program mgr. NSF, 1972, ERDA, 1976; v.p., tech. dir. Worthington Biochem. corp., Freehold, N.J., 1973-76; prof. chem. and environ. engring. Rensselaer Poly. Inst., Troy, N.Y., 1976—; cons. to industry. Author: Energy, the Biomass Options, 1981 (Am. Soc. Pubs. award for most outstanding tech. book); adv. bd.: Biotech. and Bioengring, 1972-75; editorial bd.: Applied Biochemistry and Biotech, 1981—; contbr. articles profl. jours. Mem. Am. Chem. Soc. (councillorr 1971-74, chmn, div. microbial and biochem. tech. 1969-70), Am. Soc. Microbiology, Am. Inst. Chem. Engrs. (James Van Lanen Disting. Service award 1981), Sigma Xi. Subspecialties: Biomass (energy science and technology); Chemical engineering. Current work: Microelectrodes for studying mass transfer in microbial aggregates, commercialization of biomass hydrolysis, high-rate continuous culture. Home: 2 Fairlawn Ln Troy NY 12180 Office: Rensselaer Poly Inst 204 Ricketts Bldg Troy NY 12181

BUNGE, CARLOS FEDERICO, physicist, educator; b. Buenos Aires, Mar. 27, 1941; s. Mario Augusto and Julia Delfina (Molina y Vedia) B.; m. Ana Maria Vivier, May 28, 1962; children: Pablo, Veronica, Lucia, Diego. B.Sc. in Chemistry, U. Buenos Aires, 1962, Ph.D., U. Fla., 1966. Asst. prof. Central U. Venezuela, Caracas, 1968-70; prof. U. Sao Paulo, Sao Carlos, Brazil, 1971-76; sr. theoretical physicist, prof. Nat. U. Mex., 1976—. Guggenheim Meml. fellow, 1981-82. Mem. Am. Phys. Soc., Optical Soc. Am. Subspecialties: Atomic and molecular physics; Theoretical chemistry. Current work: Computer programs for atomic and molecular electronic structure calculations, atomic spectroscopy, laser physics, chemical structure. Patentee in field. Home: Edificio 29 104 Villa Olimpica Mexico DFMexico 14020 Office: Instituto de Fisica Apdo 20 364 Mexico DFMexico 01000

BUNN, PAUL AXTELL, JR., oncologist; b. N.Y.C., Mar. 16, 1945; s. Paul Axtell and Elizabeth (Maxwell) B.; m. Camille Ann Bunn, Aug. 17, 1968; children: Rebecca, Kristen, Paul. B.S., Amherst (Mass.) Coll., 1967; M.D., Cornell U., 1971. Diplomate: Am. Bd. Internal Medicine, Nat. Bd. Med. Examiners; Lic. physician, Md. Intern U. Calif., San Francisco, 1971-72, resident in medicine, 1972-73; with USPHS, 1973—; clin. assoc. Nat. Cancer Inst., NIH, Bethesda, Md., 1973-76; sr. investigator med. oncology br. Washington VA Hosp., Nat. Cancer Inst., 1976-81; asst. prof. medicine Georgetown U., 1978-81; chief cellular kinetic sect. navy med. oncology br. Nat. Naval Med. Center, Nat. Cancer Inst., Bethesda, 1981—; assoc. prof. medicine Uniformed Services U. Health Scis., Bethesda, 1981—. Assoc. editor: Cancer Treatment Reports, 1976—; contbr. articles and abstracts to profl. jours. Recipient Howard Hill Mossman award Amherst Coll., 1967, Sondra Lee Shaw Research award Cornell U. Md. Coll., 1971. Assoc. fellow ACP; mem. Am. Soc. Hematology, Am. Assn. Cancer Research, Am. Soc. Clin. Oncology, Cell Kinetic Soc., Am. Fedn. Clin. Research, Internat. Assn. Study of Lung Cancer, Alpha Omega Alpha. Subspecialties: Oncology; Cancer research (medicine). Current work: T cell lymphomas, T cell biology, lung cancer, cell kinetics. Home: 9118 Potomac Ridge Rd Great Falls VA 22066 Office: NCI Navy Medical Oncology Branch NNMC Bethesda MD 20814

BUNNER, ALAN NEWTON, physicist, astronomer, educator; b. St. Catharines, Ont., Can., Jan. 11, 1938; came to U.S., 1960, naturalized, 1979; s. William Kelvin and Freda Helen (Newton) B.; m. Diane Martha Bishop, June 15, 1963 (div. 1971); m. Barbara Lin Ames, July 31, 1973; children: Andrew Ames, Anne Elizabeth. B.A., U. Toronto, 1960; M.Sc., Cornell U., 1966, Ph.D., 1967. Research assoc. Cornell U., Ithaca, N.Y., 1966-67; assoc. scientist, dept. physics U. Wis., Madison, 1967-79, instr. dept. physics, 1971-72; mgr. future astronomy programs Perkin-Elmer Corp., Danbury, Conn., 1979—; mem. proposal evaluation coms. NASA, 1976-82, cons. to facility definition teams, 1979-83. Contbr. articles to profl. jours. Mem. Internat. Astron. Union, Am. Astron. Soc., AIAA, Internat. Soc. Optical Engring. Subspecialties: High energy astrophysics; Aerospace engineering and technology. Current work: X-ray astronomy; space astronomy projects. Project scientist on advanced x-ray astrophysics facility (AXAF) at Perkin-Elmer; project scientist on solar optical telescope. Home: 16 Beech Tree Rd Brookfield Center CT 06805 Office: 100 Wooster Hts Rd Danbury CT 06800

BUNNETT, JOSEPH FREDERICK, chemist, educator; b. Portland, Oreg., Nov. 26, 1921; s. Joseph and Louise Helen (Boulan) B.; m. Sara Anne Telfer, Aug. 22, 1942; children—Alfred Boulan, David Telfer, Peter Sylvester (dec. Sept. 1972). B.A., Reed Coll., 1942; Ph.D., U. Rochester, 1945. Mem. faculty Reed Coll., 1946-52, U. N.C., 1952-58; mem. faculty Brown U., 1958-66, prof. chemistry, 1959-66, chmn. dept., 1961-64; prof. chemistry U. Calif. at Santa Cruz, 1966—; Erskine vis. fellow U. Canterbury, New Zealand, 1967; vis. prof. U. Wash., 1956, U. Würzburg, Germany, 1974; research fellow Japan Soc. for Promotion of Sci., 1979; Lady Davis vis. prof. Hebrew U., Jerusalem, Israel, 1981. Contbr. articles to profl. jours. Trustee Reed Coll., Societá Chimica Italiana (hon.). Fulbright scholar Univ. Coll., London, Eng., 1949-50; Guggenheim fellow, Fulbright scholar U. Munich, Germany, 1960-61. Fellow AAAS; mem. Am. Acad. Arts and Scis., Am. Chem. Soc. (editor jour. Accounts of Chem. Research), Chem. Soc. (London), Internat. Union Pure and Applied Chemistry (chmn. commn. on phys. organic chemistry 1978—, sec. organic chemistry div. 1981—). Subspecialty: Organic chemistry. Home: 608 Arroyo Seco Santa Cruz CA 95060 Office: U of California Santa Cruz CA 95064

BURBIDGE, E. MARGARET, astronomy educator; b. Davenport, Eng., Aug. 12, 1919. B.Sc., U. London, 1939, Ph.D. in Astrophysics, 1943; D.Sc. (hon.), Smith Coll., 1963, U. Sussex, 1970, U. Bristol, 1972, U. Leicester, 1972, City U. London, 1974, U. Mich., 1978, U. Mass., 1978. Acting. dir. U. London Obs., 1943-51; research assoc. astronomy Yerkes Obs., 1951-53; Shirley Farr fellow in astronomy, 1957-59; research fellow in astrophysics Calif. Inst. Tech., Pasadena, 1955-57; assoc. prof. U. Chgo., 1959-62; research assoc. in astrophysics U. San Diego, La Jolla, Calif., 1962-64, prof. astronomy, 1964—; Abby Mauze Rockefeller prof. MIT, Cambridge, 1968; mem. Space Sci. Bd., Nat. Acad. Sci., NRC, 1971-74, mem. astronomy com., 1973-75; mem. NSF astronomy adv. panel, 1972-74; mem. steering group Large Space Telescope, 1973-77; mem. Associated Univs. for Research Astronomy Bd., 1974—; Virginia Gildersleeve prof. Barnard Coll., N.Y.C., 1974. Mem. AAAS (pres. 1982-83), Nat. Acad. Sci., Am. Astron. Soc. (Warner prize 1959, pres. 1976-78), Royal Astron. Soc., Am. Acad. Arts and Scis., Internat. Astron. Union. Subspecialty: Optical astronomy. Office: Dept Physics U San Diego La Jolla CA 92093

BURBIDGE, GEOFFREY, astrophysicist, educator; b. Chipping Norton, Oxon, Eng., Sept. 24, 1925; s. Leslie and Eveline B.; m. Margaret Peachey, 1948; 1 dau. B.Sc. with spl. honors in Physics, Bristol U., 1946; Ph.D., U. Coll., London, 1951. Asst. lectr. U. Coll., London, 1950-51; Agassiz fellow Harvard, 1951-52; research fellow U. Chgo., 1952-53, Cavendish Lab., Cambridge, Eng., 1953-55; Carnegie fellow Mt. Wilson and Palomar Obs., Calif. Inst. Tech., 1955-57; asst. prof. dept. astronomy U. Chgo., 1957-58, assoc. prof., 1958-62, U. Calif. at San Diego, La Jolla, 1962-63, prof. physics, 1963—; dir. Kitt Peak Nat. Obs., Tucson, 1978; Phillips vis. prof. Harvard U., 1968; bd. dirs. Associated Univs. Research in Astronomy, 1971-74; trustee Associated Univs., Inc., 1973-82. Author: (with Margaret Burbidge) Quasi-Stellar Objects, 1967; Contbr. articles to sci. jours. Fellow Royal Soc. London, Am. Acad. Arts and Scis., Royal Astron. Soc.; mem. Am. Phys. Soc., Am. Astron. Soc., Internat. Astron. Union, Astron. Soc. of Pacific (pres. 1974-76). Subspecialties: High energy astrophysics; Cosmology. Office: Kitt Peak Nat Obs PO Box 26732 Tucson AZ 85726

BURCH, JOHN BAYARD, zoology educator; b. Charlottesville, Va., Aug. 12, 1929; m., 1952; 3 children. B.S., Randolph-Macon Coll., 1952; M.S., U. Richmond, 1954; Ph.D., U. Mich., 1959. Research assoc. U. Mich., Ann Arbor, 1958-62, curator mollusks, 1962-75, from asst. prof. to prof. zoology, 1962-77, prof. biology, 1977—. Editor-in-chief: Malacologia; editor: Malacol. Rev. Recipient research awards Va. Acad., 1953, NSF, 1954; USPHS grantee, 1964. Mem. Inst. Malacology (exec. sec.-treas. 1961-62, treas. 1962-67), Soc. Systematic Zoology, Am. Soc. Zoology, Ecol. Soc. Am., Am. Micros. Soc. Subspecialties: Evolutionary biology; Systematics. Office: Dept Biol Sci U Mich Ann Arbor MI 48104

BURCHFIEL, BURRELL CLARK, geology educator; b. Stockton, Calif., Mar. 21, 1934; s. Beryl Edward and Agnes (Clark) B.; m.; children: Brian Edward, Brook Evans. B.S., Stanford U., 1957, M.S., 1958; Ph.D., Yale U., 1961. Prof. geology Rice U., 1961-76, M.I.T., 1977—. Served with U.S. Army, 1958-59. Fellow Geol. Soc. Am.; mem. Geol. Soc. Australia; Mem. Am. Assn. Petroleum Geologists, Am. Geophys. Union. Subspecialties: Tectonics; Geology. Home: 8 Eastern Ave Arlington MA 02174 Office: 54-1010 MIT Cambridge MA 02139

BURCHFIEL, JAMES LEE, research neurophysiologist; b. Los Angeles, Mar. 16, 1941; s. Beryl Edward and Agnes (Clark) B.; m. Rae Demler, Aug. 17, 1963; children: Corin, Kendra. B.S., Stanford U., 1963, Ph.D., 1969. Instr. neuroanatomy U. Md. Sch. Medicine, Balt., 1969-70; research fellow Harvard U. Med. Sch. and Children's Hosp. Med. Center, Boston, 1970-71, research asso. in neurology (neurophysiology), 1971-75, prin. research asso. in neurology, 1975—. Contbr. articles to sci. jours. Served to capt., Med. Service Corps U.S. Army, 1968-70. Nat. Inst. Neurol. and Communicative Disease and Stroke grantee, 1978—. Mem. Soc. for Neurosci., Am. EEF Soc., Eastern Assn. EEG, Am. Epilepsy Soc. Subspecialties: Neurophysiology; Neuropharmacology. Current work: Experimental epilepsy, neurophysiology and pharmacology of developing visual system; computer analysis of electroencephagraphic data. Home: 19 White Pine Rd Newton MA 02164 Office: 300 Longwood Ave Boston MA 02115

BURCHSTED, CLIFFORD ARNOLD, mechanical engineer; b. Braintree, Mass., Aug. 19, 1921; s. Charles Fowles and Emma May (Corey) B.; m. Elizabeth Johns Simmons, Oct. 3, 1943. Student, Ripon Coll., 1943-44; B.S.M.E., Northeastern U., 1948; M.S., U. Tenn., 1959. Registered profl. engr., Tenn. Jr. engr. Tuttle and Bailey Co., New Britain, Conn., 1948; plant engr. Sample-Dureck Co., Chicopee, Mass., 1949-52; with nuclear div. Union Carbide Corp., Oak Ridge, 1952—, sr. engr., 1975—; cons. engr.; guest lectr. Harvard U. Sch. Pub. Health, Oak Ridge Assoc. Univs. Author: Design, Construction and Testing of High Efficiency Air Cleaning Systems, 1970, Nuclear Air Cleaning Handbook, 1976; contbr. articles to profl. jours. Chmn. Citizens Adv. Com., Clinton, Tenn., 1970-80. Served with USAAF, 1942-46. Fellow Royal Soc. Health, ASME (Centennial medal 1980), ASTM; mem. Inst. Environ. Scis. (Seligman award 1974), Sigma Xi, Phi Kappa Phi, Beta Gamma Sigma. Methodist. Lodge: Lions. Subspecialty: Gas cleaning systems. Current work: High efficiency air cleaning for nuclear, manufacturing, biological, pharmaceutical, electronics applications.

BURD, ROBERT M., physician, educator; b. N.Y.C., Aug. 25, 1937; s. David and Ann (Popkin) B.; m. Alice E. Stoller, May 30, 1964; children: Russell J., Stephen J. B.A., Columbia U., 1959, M.D., 1963. Diplomate: Am. Bd. Internal Medicine, (subspecialty in hematology, oncology). Med. intern Albert Einstein Med. Coll., N.Y.C., 1963-64, resident in medicine, 1964-66; fellow in hematology Montefiore Hosp., N.Y.C., 1966-67; physician in pvt. practice, Fairfield, Conn., 1969—; chief hematology/oncology St. Vincents Med. Center, Bridgeport, Conn., 1980—; assoc. prof. clin. medicine Yale U. Med. Center, New Haven, 1981—; program dir. Yale-St. Vincents Oncology Program, New Haven, 1980—. Editor: St. Vincent Med. Bull, 1972—; mem. editorial bd.: Conn. Medicine, 1972-76; contbr. articles to profl. jours. Chmn. profl. edn. Am. Cancer Soc., Bridgeport, 1983, bd. dirs., 1983—; vice pres. Leukemia Soc. Am., Fairfield County, Conn., 1973-75. Served to lt. comdr. U.S. Navy, 1967-69. Ettinger fellow Am. Cancer Soc., 1983. Fellow ACP; mem. Am. Soc. Hematology, Am. Soc. Clin. Oncology, N.Y. Acad. Sci., Am. Soc. Internal Medicine, AMA. Jewish. Subspecialties: Hematology; Oncology. Current work: Cooperative protocols in therapy of lung cancer. Office: 1305 Post Rd Fairfield CT 06430

BURDETTE, WALTER JAMES, surgeon, educator; b. Hillsboro, Tex., Feb. 5, 1915; s. James S. and Ovazene (Weatherred) B.; m. Kathryn Lynch, Apr. 9, 1947; children—Susan, William J. A.B., Baylor U., 1935; A.M., U. Tex., 1936, Ph.D., 1938; M.D., Yale, 1942. Diplomate: Am. Bd. Surgery, Am. Bd. Thoracic Surgery. Intern Johns Hopkins Hosp., 1942-43; Harvey Cushing fellow surgery Yale, 1943-44; resident staff surgery New Haven Hosp., 1944-46; instr., asst., assoc. prof. surgery La. State U., 1946-55; vis. surgeon Charity Hosp. of La., 1946-55; cons. Touro Infirmary and So. Baptist Hosp., 1952-55, Oak Ridge Inst. Nuclear Studies Hosp., 1953-59; vis. investigator Chester Beatty Inst. Cancer Research, Brompton, and Royal Cancer Hosp., London, 1953, Max Planck Institut Fuer Biochemie, Tuebingen, Germany, summer 1955; prof., chmn. dept. surgery U. Mo., 1955-56; prof. clin. surgery St. Louis U. Sch. Medicine, 1956-57; prof., head dept. surgery U. Utah, 1957-65; dir. lab. clin. biology, surgeon-in-chief Salt Lake Gen. Hosp., 1957-65; chief surg. service VA Hosps., Salt Lake City, 1957-65; prof. surgery, asso. dir. U. Tex-M.D. Anderson Hosp. and Tumor Inst., Houston, 1965-72; prof. surgery U. Tex. Sch. Medicine at Houston, 1971-79; adj. prof. pharmacology U. Houston, 1975—; pres. Nat. Biomed. Found., 1972—; cons. Hermann Hosp., Center Pavilion Hosp., 1970—, St. Luke's Hosp., 1975—; Gibson lectr. advanced surgery Oxford U., 1966; vis. prof. U. Oxford, spring 1965; ofcl. U. Congo, summer 1968. Editor, author: Etiology, Treatment of Leukemia, 1958; editor, author: Methodology in Human Genetics, 1962, Methodology in Mammalian Genetics, 1962, Methodology in Basic Genetics, 1963, Primary Hepatoma, 1965, Carcinoma of the Alimentary Tract, 1965, Viruses Inducing Cancer, 1966, Carcinoma of the Colon and Antecedent Epithelium, 1970, Planning and Analysis of Clinical Studies, 1970, Invertebrate Endocrinology and Hormonal Heterophylly, 1974; mem. editorial bd.: Surg. Rounds; contbr. articles to med. and sci. jours. Chmn. genetics study sect., mem. morphology study sect. NIH; cons. Nat. Cancer Inst.; mem. Nat. Adv. Cancer Council, Nat. Adv. Heart Council, Surgeon General's Com. on Smoking and Health; chmn. U.S.A. nat. com. Internat. Union Against Cancer; mem. transplantation com. Nat. Acad. Scis.; chmn. working Cadre on cancer large intestine Nat. Cancer Inst.; elder, deacon Christian Ch. Rockefeller travel fellow, summer 1957. Fellow A.C.S.; mem. Soc. Surgery Alimentary Tract, Am. Assn. Cancer Research (dir.), Am. Cancer Soc. (chmn. research adv. council, mem. council on analysis and projection), Am. Surg. Assn., Soc. Clin. Surgery (treas.), Soc. U. Surgeons, A.M.A., Soc. Exptl. Biology and Medicine, Genetics Soc. Am., AAAS, Western Soc. Clin. Research, Am. Assn. Thoracic Surgery, Transplantation Soc., N.Y. Acad. Sci., Soc. Am. Naturalists, New Orleans, St. Louis, Salt Lake City, Houston surg. socs., Tex. Med. Soc., Harris County Med. Soc., So., Western surg. assns., So. Thoracic Surg. Soc., Peruvian Cancer Soc. (hon.), Am. Assn. for Cancer Research, Soc. for Surgery Alimentary Tract, Am. Soc. Clin. Oncology, Am. Soc. for Cancer Edn., Tex. Surg. Soc., Assn. Yale Alumni in Medicine (exec. com. 1977), Soc. Internat. de Chirurg, Sigma Xi, Alpha Omega Alpha. Subspecialties: Surgery; Cancer research (medicine). Current work: Gene action and modifications for cancer therapy; general and thoracic surgery; genetic research. Home: 239 Chimney Rock Rd Houston TX 77024 Office: Doctors Center 7000 Fannin St Suite 1740 Houston TX 77030

BURDICK, CHARLES KENNETH, army officer, researcher; b. Charleston, W.Va., Jan. 13, 1946; s. Kenneth Welcome and Dorris Donnelle (Adams) Gustad; m. Rosemary Núñez, Aug. 12, 1967; children: Damien Michael, Elizabeth Suzanne, Charles Welcome. Student, Alfred U., 1963-64; B.A., U. South Fla., 1967; M.A., U. Man., Can., 1969; Ph.D., Wash. U., 1975. Psychologist Hastings State Hosp., Ingleside, Nebr., 1969; research technician U.S. Army Med. Research Lab., Ft. Knox, Ky., 1971-72; research asst. Central Inst. Deaf, St. Louis, 1972-75; research psychologist U.S. Army Aeromed. Research Lab., Ft. Rucker, Ala., 1975-79; staff officer U.S. Army Med. Research and Devel. Command, Ft. Detrick, Md., 1979-82; chief biomed. model devel. br. U.S. Army Med. Research Inst. Chem. Def., Aberdeen Proving Ground, Md., 1982—; advanced through ranks to maj., 1982; fellow Inst. Advanced Study of Communication Processes, Gainesville, Fla., 1978-81; sponsor rep. NAS-NRC Com. Hearing Bioacoustics and Biomechanics, Washington, 1980—; grant reviewer, 1979. Contbr. chpts. to books, articles to profl jours. U. Man. grad. fellow, 1968; recipient N.Y. State Incentive award Alfred U., 1963. Mem. AAAS, Acoustical Soc. Am., Am. Psychol. Assn., Assn. Research Otolaryngology, Drug Info. Assn., Sigma Xi, Psi Chi. Democrat. Subspecialties: Sensory processes; Neuropsychology. Current work: Science management; drug development with special interest in in vitro, in vivo and behavioral modeling of drug effects; noise-induced hearing loss. Office: US Army Med Research Inst Chem Def SGRD-UV-DB Aberdeen Proving Ground MD 21010

BURDICK, GLENN ARTHUR, university dean, electrical engineer; b. Pavilion, Wyo., Sept. 9, 1932; s. Stephen Arthur and Mary Elizabeth (McClerg) B.; m. Joyce Mae Huggett, July 14, 1951; children: Stephen Arthur, Randy Glenn. B.S., Ga. Inst. Tech., 1958, M.S., 1959; Ph.D., MIT, 1961. Reg. profl. engr., Fla. Office mgr. Statewide Contractors, Las Vegas, Nev., 1955-56; spl. tool designer Ga. Inst. Tech., Atlanta, 1954-55, instr., 1956-59; sr. mem. research staff Sperry Microwave, Oldsmar, Fla., 1961-65; prof. elec. engring. U. So. Fla., Tampa, 1965—, dean Coll. Engring., 1979—. Invented underground pipeline leak detector, 1956, sail boat mast insulation, 1981. Mem. Tampa Bay Fgn. Affairs Com., 1981—, Pinellas County (Fla.) High Speed Rail Task Force, 1982—, Gov. of State of Fla. Energy Task Force, 1980—; vice chmn. Fla. Task Force for Sci, Energy and Tech. Service to Industry, 1981-82. Tex. Gulf scholar, 1957-58; NSF fellow, 1958-61; Woodrow Wilson fellow, 1958-59. Mem. Fla. Engring. Soc. (Engr. of Yr. award 1981), Internat. Soc. Hybrid Microelectronics (nat. pres. 1974), IEEE (sr. mem., Engr. of Yr. award 1980), Am. Ry. Engring. Assn., N.Y. Acad. Sci., Am. Soc. Engring. Edn. Clubs: Clearwater Tennis (pres. Fla. chpt. 1965, 69), Downtown.). Lodge: Rotary. Subspecialties: Theoretical physics; Electrical engineering. Current work: Engineering education; accident reconstruction. Home: 1000 County Rd 80 Tarpon Springs FL 33589 Office: Univ So Fla Coll Engring Tampa FL 33620

BURDICK, JAMES ALAN, mental health adminstr.; b. Omaha, Oct. 24, 1934; s. Chester Childres and Mary Elma (Remberger) B.; m. Elaine Louise Thomas, Oct. 1, 1960; children: Dakin Robert, Mathew Justin. B.A., James Milikin U., 1956; M.A., CUNY, 1965; Ph.D., U. Man., Can., 1978. Cert. psychologist Ind. Psychophysiologist Schick Pharm., Inc., Seattle, 1967-69; lectr. dept. psychiatry U. Man., Winnipeg, 1969-77; sr. program analyst Dept. Health and Social Devel., Province of Man., Winnipeg, 1974; dir. research and program evaluation Alcoholism Found. of Man., Winnipeg, 1974-77; dir. for program evaluation and research Otis. R. Bowen Ctr., Warsaw, Ind., 1978-81; dir. quality assurance N.Y. State Office of Mental Health-Gowanda Psychiat. Center, Helmuth, 1981—. Served with Signal Corps U.S. Army, 1959-60. USPHS grantee, 1966, 67; Med. Research Council Can. grantee, 1971. Mem. Am. Psychol. Assn., Can. Psychol. Assn., Soc. for Psychophysiol. Research, N.Y. Acad. Scis., N.Y. State Office of Mental Health Assn. Facility Dirs. Subspecialties: Physiological psychology; Psychophysiology. Current work: Temporal and nontemporal variability of physiological systems in humans and its relationship to behavior, intelligence, attitude and personality. Alcoholism research; evaluation of mental health systems. Home: PO Box 236 Helmuth NY 14079 Office: NY State Office Mental Health-Gowanda Psychiat Ctr Helmuth NY 14079

BURFORD, HUGH JONATHAN, pharmacologist, educator; b. Memphis, Aug. 5, 1931; s. Tolbert Hugh and Margaret Elizabeth (Henderson) B.; m. Dorothy Marie Moffett, Dec. 26, 1957; children: Jonathan Mark, Jennifer Lynn. B.S., Millsaps Coll., 1954; M.S., U. Miss., 1956; Ph.D. in Pharmacology, U. Kans., 1962. USPHS fellow in pharmacology Tulane U., New Orleans, 1962-63; asst. Bowman Gray Sch. Medicine, 1963-68; assoc. prof. pharmacology and pharmacy, teaching asst. in basic med. sci. U. N.C., Chapel Hill, 1971—; cons. for pharmacology teaching materials Health Scis. Consortium; cons. Audio Visual (Computer Search) Line; liaison mem. Computer Assisted Teaching System Consortium. Author: HEALER programs and data base for student self-assessment in med. pharmacology, 1978—, also articles. Democratic precinct worker. Served with U.S. Army, 1956-57. Recipient Sloan Faculty Facilitator award U.N.C., 1973-74. Mem. Am. Soc. Pharmacology and Exptl. Therapeutics, Assn. Med. Dental Edn. Health Scis. (founding 1968), Spl. Interest Group in Health Profl. Edn., Am. Edn. Research Assn. (founding 1969), Sigma Xi. Democrat. Baptist. Subspecialties: Pharmacology; Information systems, storage, and retrieval (computer science). Current work: Research in cognitive learning styles and computer instruction in pharmacology for health profession students.

BURGE, CHARLES ARTHUR, emergency planning engineer; b. Honolulu, Feb. 25, 1948; s. Carlos and Mary Chieko (Koyama) B.; m. Rebecca Lynn Johnson, Sept. 7, 1971 (div. Jan. 1978); m. Deborah Simon, Oct. 9, 1982. Student, Canton Community Coll., 1966-67, U. Ill.-Urbana, 1967-68, N.J. Inst. Tech., 1979-82. Engr. ops. dept. Pub. Service Elec. & Gas Co., Salem Generating Sta., Hancocks Bridge, N.J., 1977-78, engr. licensing hdqrs., Newark, 1978-79, engr. emergency plan, Hancocks Bridge, 1979-82, lead engr, emergency planning, 1982—; mem. Prompt Notification Com. KMC, Inc., Washington, 1982—. Mem. nat. com. Republican party, Washington, 1981—, mem. congl. com., 1982; mem. Nat. Trust Historic Preservation, Washington, 1981—. Served with USN, 1969-77. Mem. Am. Nuclear Soc., N.Y. Acad. Scis., Alpha Pi Mu. Republican. Episcopalian. Subspecialties: Nuclear fission; Industrial engineering. Current work: Relationship between technology and the lay person - especially how to present the case for commercial nuclear power. Home: 617 W Landing Rd Mantua NJ 08051 Office: Pub Service Elec and Gas Co PO Box 236 Hancocks Bridge NJ 08038

BURGER, CHRISTIAN PIETER, electrical engineering mechanics educator; b. George, S. Africa, Dec. 30, 1929; came to U.S., 1971; m. Marie Mundy Kenyon, June 30, 1956; children: Marie Elise, Christian David, Robert Johann. B. Sc., U. Stellenbosch, South Africa, 1952; Ph.D., U. Cape Town, South Africa, 1967. Engr. Brush Abboe Co., Eng., 1953-55; tech. engr. Vacuum Oil Co., South Africa, 1956-62; sr. lectr. dept. mech. engring. U. Cape Town, 1962-71; prof. dept. engring. mechanics Iowa State U., Ames, 1971—. Editor: Jour. Exptl. Mechanics, 1978-82. Mem. Soc. Exptl. Stress Analysis (chmn. rev. com. 1978-82, exec. bd. 1982—, R.E. Peterson award 1973, 76), Brit. Soc. Strain Measurement, Am. Soc. Nondestructive Testing. Subspecialties: Solid mechanics; Fracture mechanics. Current work: Experimental stress analysis; bioengineering. Home: 1724 Meadowlane Ave Ames IA 50010 Office: Iowa State U Dept Engring Mechanics 202A Lab Mech Ames IA 50011

BURGER, DIETER, virologist, microbiologist, educator; b. Ulm-Danube, W.Ger., Jan. 1, 1926; came to U.S., 1957; s. Rudolf and Else (Didie) B.; m. Anke David, Sept. 26, 1964; children: Rolf Michael, Jutta Christine. Dr.med.vet., Ludwigs-Maximilian U., Munich, Ger., 1953; M.S. in Vet. Sci, Wash. State U., 1962; Ph.D., U. Wis., 1970. Diplomate: Am. Coll. Vet. Microbiology. Research asst. Inst. Animal Hygiene, U. Munich, 1954-57; asst. prof. dept. vet. microbiology Wash. State U., Pullman, 1964-70, assoc. prof., 1971-75, Prof. dept. vet. microbiology and pathology, 1976—. Contbr. articles to profl. jours. Mem. Am. Coll. Vet. Microbiologists, Am. Assn. Virology, Am. Soc. Microbiology. Subspecialties: Virology (veterinary medicine); Microbiology (veterinary medicine). Current work: Infectious virus-antibody complexes, restriction enzyme analysis of viral DNA. Office: Wash State U Pullman WA 99164

BURGER, RICHARD M., molecular biologist; b. N.Y.C., Mar. 23, 1941; s. Sidney J. and Regina (Biederman) B.; m. Deborah Mellis, Feb. 6, 1977; children: Caludia, Abigail. Student, Brandeis U., 1958-60; B.A., Adelphi Coll., 1962; postgrad., Brandeis U., 1965-68; Ph.D., Princeton U., 1969. Fellow U. Calif.-Berkeley, 1968-71; asst. prof. biology Middle East Tech. U., Ankara, Turkey, 1971-72; assoc. Sloan-Kettering Inst., N.Y.C., 1972-77; asst. prof. biochemistry Cornell U., 1972-77; adj. assoc. prof. H.H. Lehman Coll. CUNY, 1980-81; prin. assoc. Albert Einstein Coll. Medicine, 1977—; research assoc. Columbia U., 1976-77, Haskins Labs., N.Y.C., 1958—. Nat. Cancer Inst. grantee, 1976; Leukemia Soc. Am. spl. fellow, 1981; scholar, 1983. Mem. Am. Soc. Biol. Chemists, Harvey Soc. Subspecialties: Molecular biology; Medicinal chemistry. Current work: Biochemical mechanisms of oxygen activation, drug metabolism and DNA injury, repair and replication. Office: Albert Einstein Coll Medicine 1300 Morris Park Av Bronx NY 10461

BURGESS, ERIC, high technology company executive, writer; b. Stockport, Cheshire, Eng., May 30, 1920; came to U.S., 1956, naturalized, 1962; s. William and Lily B.; m. Lilian Slater, Aug. 9, 1947; children: Janis, Marie, Stephen Roy, Howard John. B.A., Coll. Commerce, Manchester, Eng., 1940; B.Sc., Coll. Tech., Manchester, 1950. Vice-pres. Mellonics Inc., Tucson and Northridge, La., 1959-62; sr. tech. staff mem. Informatics Inc., Sherman Oaks, Calif., 1962-65; dep. dir. Wolf Research & Develop Co., Encino, Calif., 1965-68; sci. correspondent The Christian Sci. Monitor, Boston, 1968-71; pvt. practice cons. space mission, high tech., satellite communications, Santa Rosa and Los Angeles, 1971-81; sr. v.p. corp. devel. Space Microwave Labs. Inc., Santa Rosa, Calif., 1981—; cons. Delta-Vee Inc., San Jose, Calif. Author: Rocket Propulsion, 1952, Frontier to Space, 1954, Rockets & Spaceflight, 1956, Guided Weapons, 1957, Long-Range Ballistic Missiles, 1961, Assault on the Moon, 1966; Author with Bruce Murray: Flight to Mercury, 1977; Author: (with James Dunn) To The Red Planet, 1978, The Voyage of Mariner 10, 1978, (with R.O. Fimmel, James A Van Allen) First to Jupiter, Saturn and Beyond, 1981, By Jupiter, 1982, Celestial Basic, 1982, Pioneer Venus, 1983, (with H.J. Burgess) Timex, Sinclair 1000: Astronomy, 1983; contbr. numerous articles to profl. jours. Fellow Royal Astron. Soc., Brit. Interplanetary Soc. (chmn. 1946-47), AIAA; mem. Nat. Assn. Sci. Writers. Subspecialties: Software engineering; Planetary science. Current work: Planetology, military radar and electronic warfare. Home: 13361 Frati Ln Sebastopol CA 95472 Office: Space Microwave Labs Inc 1255 N Dutton Ave Santa Rosa CA 95401

BURGESS, JOHN HENRY, engineering psychologist, operations research analyst; b. Niagara Falls, N.Y., Aug. 9, 1923; s. John Hersha and Mary Ann (Crandall) B.; m. Sylvia Marie Johnson, June 30, 1965. B.S., Kent State U., 1948; M.A., U. Miami, 1950. Engring. psychologist Bell Aerosystems, Buffalo, 1954-66; ops. researcher Ill. Dept. Mental Health, Decatur, 1966-73; engring. psychologist U.S. Army, Champaign, Ill., 1973-76; psychol. cons. Human Factors Assocs., Syracuse, N.Y., 1976—. Author: System Approaches, 1978, Human Factors Environment, 1980, Forms Design, 1983. Served to staff sgt. AUS, 1943-46. Recipient Loss Prevention award Hartford Nat. Safety Council, 1980. Mem. Am. Psychol. Assn., Human Factors Soc. Subspecialties: Human factors engineering; Operations research (engineering). Current work: Application of aerospace-derived methodologies to domestic and indsl. products and operations.

BURGGREN, WARREN WILLIAM, zoology educator; b. Edmonton, Alta., Can., Aug. 14, 1951; came to U.S., 1978; s. William Arthur Ora and Ella Maureen (Hill) B.; m. Margaret Joan McAdam, Aug. 5, 1972 (div.); 1 dau., Kimberly Alison. B.Sc. with 1st class honors, U. Calgary, 1973; Ph.D., U. East Anglia, Norwich, Eng., 1976. Postdoctoral assoc. U. B.C., Vancouver, 1976-78; asst. prof. zoology U. Mass., Amherst, 1978-82, assoc. prof., 1982—. Author: (with others) Evolution of Air Breathing in Vertebrates, 1981; contbr. articles to profl. jours. NSF grantee, 1980, 83. Mem. Am. Soc. Zoologists, Can. Soc. Zoologists, Soc. Exptl. Biology. Subspecialties: Comparative physiology; Evolutionary biology. Current work: Comparative and environmental animal physiology; physiological adaptations associated with evolution of terrestrial animal life, especially respiration, circulation, osmoregulation. Office: Dept Zoolog U Mass Amherst MA 01003

BURICH, RAYMOND LUCAS, physiology educator; b. Cleve., June 23, 1943; s. Joseph John and Caroline Anne (Swetich) B.; m. Nancy Jane Wannemacher, June 19, 1965; children: Christine Lynn. B.S., Kent State U., 1965, M.S., 1967; Ph.D., Mich. State U., 1970. Assoc. prof. U. Mo., Kansas City, 1971—. Author: Essentials of Applied Oral Physiology, 1983, also articles. Mem. Am. Physiol. Soc., Gerontol. Soc., Internat. Soc. Dental Research. Subspecialties: Physiology (medicine); Oral biology. Home: 10247 Hauser St Lenexa KS 66215 Office: U Mo Dental Sch 650 E 25th St Kansas City MO 64108

BURISH, THOMAS GERARD, clinical psychology educator, researcher; b. Menominee, Mich., May 4, 1950; s. Bennie Charles and Donna Mae (William) B.; m. Pamela J. Zebrasky, June 19, 1976; children: Mark Joseph, Brent Christopher. A.B., U. Notre Dame, 1972; M.A., U. Kans.-Lawrence, 1975, Ph.D., 1976. Asst. prof. psychology Vanderbilt U., 1976-80, assoc. prof., 1980—, dir. clin. tng., 1980—; cons. in field. Editor: Coping with Chronic Disease, 1983. Vice chmn. St. Ann's Sch. Bd., Nashville, 1982. Recipient David Schulman Meml. award in clin. psychology U. Kans., 1975, Madison Sarratt award Vanderbilt U., 1980; Nat. Cancer Inst. grantee, 1979-83; Ctr for Health Services Research grantee, 1980-84. Mem. Am. Psychol. Assn. (com. chmn. 1979-80, co-chmn. div. com. 1981-83), Biofeedback Soc. Am., Biofeedback Soc. Tenn. Roman Catholic. Subspecialties: Behavioral psychology; Biofeedback. Current work: Research in general area of stress and coping, including specific areas of biofeedback, health psychology, and cognitive coping strategies. Home: 625 Brook Hollow Rd Nashville TN 37205 Office: Vanderbilt U Dept Psychology Nashville TN 37240

BURK, CREIGHTON, marine geology educator; b. Laramie, Wyo., Feb. 1, 1929; m., 1949; 3 children. B.Sc., U. Wyo., 1952, M.A., 1953; Ph.D. in Geology, Princeton U., 1964. Field geologist Stanolind Oil & Gas Co., 1952-53; exploration geologist Richfield Oil Corp., 1953-60; chief scientist Am. Miscellaneous Soc., Mohole Project Nat. Acad. Sci., 1962-64; corp. exploration adv. Socony Mobil Oil Co., Inc., 1964-69; chief geologist and mgr. regional geology Mobil Oil Corp., 1969-75; prof. geol. sci., chmn. dept. marine studies, dir. Marine Sci. Inst. U. Tex.-Austin, 1975—; instr. Princeton U., 1960-61; vis. prof. geology and geophys. sci., 1967-75. Mem. U.S. Geodynamics Com., Pres.'s Commn. on Marine Sci. Engring. and Resouces, Nat. Oceanic Adv. Com., 1968—; U.S. del. World Petroleum Congress, 1975; mem. and del. U.S.-USSR Protocol on Oceanography, 1974—. NSF fellow, 1961-62; recipient Frank A. Morgan award, 1956. Mem. Geol. Soc. Am., Am. Geophys. Union, Am. Assn. Petroleum Geologists, Marine Tech. Soc. (v.p. 1974—), Geol. Soc. (London). Subspecialty: Marine geology. Office: Marine Sci Inst U Tex Box 7999 Austin TX 78712

BURK, MARTYN WILLIAM, surgical oncologist, immunologist; b. N.Y.C., Jan. 28, 1942; s. Martin William and Lenore (Koch) B.; m. Kathleen Robinson. B.S., Fordham U.; M.D., SUNY-Bklyn.; Ph.D., U. Minn. Diplomate: Am. Bd. Surgery. Resident in surgery U. Minn. Hosps., 1968-75, chief resident in surgery, 1975-76; asst. prof. surgery UCLA, 1983—, assoc. prof., 1983—. Contbr. articles to med. and sci. jours. Served to lt. comdr. M.C. USN, 1976-78. Mem. Am. Assn. Immunologists, Am. Assn. Cancer Research, Am. Soc. Clin. Oncology,

Soc. Univ. Surgeons, ACS. Subspecialties: Surgery; Cancer research (medicine). Current work: Monoclonal hybridoma antibodies, tumor cell surface antigens, immunotoxins, surgical oncology. Office: 54-140 CHS UCLA Sch Medicine Los Angeles CA 90024

BURK, RAYMOND FRANKLIN, JR., medical educator, researcher; b. Kosciusko, Miss., Dec. 9, 1942; s. Raymond Franklin and Florence Annie (Davis) B.; m. Enikoe Vikor, June 17, 1967; children: Teresa Marie, Stephen Morrison. B.A., U. Miss., 1963; M.D., Vanderbilt U., 1968. Diplomate: Am. Bd. Internal Medicine. Intern. Vanderbilt Hosp., Nashville, 1968-69, resident, 1969-70; asst. prof. internal medicine U. Tex., Health Sci. Ctr., Dallas, 1975-78, assoc. prof. medicine, San Antonio, 1980-82, prof. medicine, 1982—; assoc. prof. medicine La. State U., Shreveport, 1978-80. Contbr. research papers to profl. publs. Served to maj. U.S. Army, 1970-73. NIH research grantee, 1974—. Mem. Am. Soc. Biol. Chemists, Am. Soc. Clin. Investigation. Subspecialties: Gastroenterology; Biochemistry (medicine). Current work: Biochemical function of selenium, role of lipid peroxidation in cell injury. Home: 3706 Hunters Peak San Antonio TX 78230 Office: Dept Medicine U Tex Health Sci Ctr 7703 Floyd Curl Dr San Antonio TX 78284

BURKART, BURKE, geology educator, geologic consultant, researcher; b. Dallas, Feb. 23, 1933; s. Herman Frederick and Velma (Ball) B.; m. Marilyn Caskey, Apr. 2, 1966; children: Patrick Caskey, Michael David. B.S. in Geology, U. Tex.-Austin, 1954, M.A., 1960, Ph.D., Rice U., 1965. Asst. prof. Temple U., Phila., 1965-70; assoc. prof. geology U. Tex.-Arlington, 1970-82, prof., 1982—; Fulbright-Hayes lectr., Ecuador, 1972; cons. Banter Exploration, Ft. Worth, 1981—. Served to 1st lt. USAF, 1955-58. Fellow Geol. Soc. Am.; mem. Am. Assn. Petroleum Geologists, Am. Geophys. Union, Geochem. Soc. Subspecialties: Geochemistry; Tectonics. Current work: Tectonics of Central America and Caribbean; geochemistry of phosphates. Home: 1818 Kenwood Terr Arlington TX 76013 Office: Dept Geology U Tex at Arlington Arlington TX 76019

BURKE, BERNARD FLOOD, physicist, educator; b. Boston, June 7, 1928; s. Vincent Paul and Clare (Brine) B.; m. Jane Chapin Pann, May 30, 1953; children—Geoffrey Damian, Elizabeth Chapin, Mark Vincent, Matthew Brine. S.B., M.I.T., 1950, Ph.D., 1953. Staff mem. terrestrial magnetism Carnegie Instn. of Washington, 1953-65, chmn. radio astronomy sect., 1962-65; prof. physics Mass. Inst. Tech., 1965—; vis. prof. U. Leiden, Netherlands, 1971-72; trustee N.E. Radio Obs. Corp., 1973—, vice chmn., 1975-82, chmn., 1982—; cons. NSF, NASA. Trustee Associated Univs., Inc., 1972—. Recipient Helen Warner prize Am. Astron. Soc., 1963; Rumford prize Am. Acad. Arts and Scis., 1971. Fellow AAAS; mem. Nat. Acad. Scis., Am. Acad. Arts and Scis., Am. Phys. Soc., Am., Royal astron. socs., Internat. Astron. Union, Internat. Sci. Radio Union. Subspecialty: Planetary science. Research on microwave spectroscopy, radio astronomy, galactic structure, antenna design. Home: 10 Bloomfield St Lexington MA 02173 Office: Mass Inst Tech Cambridge MA 02139

BURKE, CARROLL NUTILE, virologist; b. New Haven, Conn., Aug. 16, 1929; d. Albert C. and Matilda C. (Grant) Nutile; m. Ronald C. Burke, Apr. 26, 1952; children: Tara J., Megan P., Alison K. B.S., U. Conn., 1951, M.S., 1959, Ph.D., 1965. Asst. prof. pathobiology U. Conn., 1966-70, assoc. prof., 1970-76, prof., 1976—, head Electron Microscopy Lab., 1966-79, acting dir. Inst. Water Resources, 1976-77, asst. dir. Inst. Water Resources, 1979-80, dir. Inst. Water Resources, 1980—; vis. scientist Eastern Fish Disease Lab. U.S. Dept. Interior, W.Va., 1974-75; co-dir. Water Fowl Field Test Tishomingo Wildlife Refuge, Okla., 1973; lectr. in field. Author: (with G.C. Rovozzo) A Manual of Basic Virological Techniques, 1974; contbr. articles to profl. jours. Mem. Electron Microscope Soc. Am. (examiner for certification 1977, reviewer 1981), Am. Assn. Avian Pathologists, Am. Fisheries Soc., AAAS, Am. Soc. Microbiology, N.Y. Acad. Scis., Sigma Xi (v.p. U. Conn. chpt. 1977, pres. 1978, chmn. nominating com. 1980). Subspecialty: Virology (biology). Home: 507 Bolton Rd Vernon CT 06066 Office: Dept Pathobiology U Conn Storrs CT 06268

BURKE, DEREK CLISSOLD, scientific administrator; b. Birmingham, Eng., Feb. 13, 1930; s. Harold and Ivy Ruby (Clissold) B.; m. Mary Elizabeth, May 21, 1953; children: Elizabeth, Stephen, Rosemary, Virginia. B.Sc. in Chemistry, U. Birmingham, 1950, Ph.D., 1953; LL.D., U. Aberdeen, Scotland, 1982. Research fellow Yale U., New Haven, Conn., 1953-55; mem. sci. staff Nat. Inst. for Med. Research, London, 1955-60; lectr. to sr. lectr. dept. biochemistry U. Aberdeen, 1960-69; prof. biol. scis. U. Warwick, Eng., 1969-82; v.p., sci. dir. Allelix, Toronto, Ont., Can., 1982—. Contbr. articles to profl. jours. Subspecialties: Virology (biology); Cell biology. Current work: Interferon. Office: 6850 Goreway Dr Mississauga ON Canada L4V 1P1

BURKE, DOYLE, resource cons., logging engr., educator; b. Denver, July 2, 1939; s. John R. and Leta M. D.; m. Joanne, Jan. 2, 1960; children: Matthew, Sheryl. B.S., U. Wash., 1964, M.S., 1968. Registered profl. engr., Wash., Oreg.; registered profl. surveyor, Wash.; registered profl. forester, Calif. With U.S. Forest Service, 1964-72, civil engr., Washington, 1970, research engr., Seattle, 1970-72; owner, operator Doyle Burke & Assoc., Seattle, 1970-81; partner Pan Sylvan cons. firm, Seattle, 1981—; prof. logging engring. U. Wash., 1975—. Contbr. articles on logging engring. to profl. jours. Mem. Nat. Soc. Profl. Engrs., Cons. Engrs. Council Wash. Subspecialty: Forestry engineering. Current work: Logging engring. Office: 18021 15th Ave NE Seattle WA 98155

BURKE, EDWARD ALOYSIUS, atomic physicist, educator; b. White Plains, N.Y., Oct. 21, 1929; s. Edward Walter and Nellie Agnes (Cunningham) B.; m. Maria Susan Wildermann, June 11, 1955; children: Joseph, Lawrence, Catherine, Vincent, Gerard, John, Margaret. B.A., NYU, 1954; M.S., Fordham U., 1955, Ph.D., 1959. Assoc. physicist Brookhaven Nat. Lab., Yaphank, N.Y., 1965-66; asst. prof. Montclair State Coll., Upper Montclair, N.J., 1958-59; assoc. prof. physics St. John's U., Jamaica, N.Y., 1959-66; prof. physics Adelphi U., Garden City, N.Y., 1966—; cons. Adelphi Ctr. for Energy Studies, 1982—. Author articles. Served with USAF, 1951. U.S. Navy grantee, 1962-63; U.S. Air Force grantee, 1965-68. Mem. Am. Phys. Soc., Am. Assn. Physics Tchrs., AAUP. Roman Catholic. Subspecialty: Atomic and molecular physics. Current work: Atomic wave function calculations, specifically employing numerial techniques and recently developed numerical methods. Office: Dept Physics Adelphi U Garden City NY 11530

BURKE, JOSEPH ELDRID, materials engineer; b. Berkeley, Calif., Sept. 1, 1914; s. Charles Eldrid and Ruth Enid (Hadcock) B.; m. Kathleen Mary Wilson, Sept. 11, 1939; children: Charles Robert, Margaret (Mrs. Charles Craig Van Decar). B.A., McMaster U.-Hamilton, Ont., Can., 1935; Ph.D., Cornell U., 1940. Chemist Internat. Nickel Co., Bayonne, N.J., 1940-41; chemist Norton Co., Worcester, Mass., 1941-43; group leader Manhattan Dist. Lab., Los Alamos, 1943-46; assoc. prof. metallurgy U. Chgo., 1946-49; mgr. metallurgy Knolls Atomic Power Lab., Gen. Electric Co., Schenectady, N.Y., 1949-54, mgr. ceramics, 1954-72, mgr. planning and resources, 1972-74, mgr. spl. projects, 1974-79, ret., 1979; cons. materials sci. and engring., 1979—; adj. prof. Renssalaer Poly. Inst., Troy, N.Y., 1979-83; mem. eval. panel Nat. Bur. Standards, 1965-70, chmn., 1977-81; vis. com. U. Ill.-Urbana, 1974-79. Co-author: (with A.U. Seybolt) Procedures in Experimental Metallurgy, 1953; editor: (with D.W. White) The Metal Beryllium, 1955; gen. editor: Progress in Ceramic Science, 1964-70, Vols. I-IV; adv. editor: Jour. Nuclear Materials, 1959-80; contbr. articles to profl. jours. Chmn. zoning bd. appeals, Town Ballston, N.Y., 1960-69. Fellow Am. Ceramic Soc. (pres. 1974-75), Am. Soc. Metals, Am. Nuclear Soc.; mem. AAAS, Nat. Acad. Engring., Sigma Xi. Subspecialties: Ceramics; Materials. Current work: Consultant fundamental and applied research metals, ceramics, glass, research administration. Home: 33 Forest Rd Burnt Hills NY 12027

BURKE, LUKE ANTHONY, chemistry educator, researcher; b. N.Y.C., Feb. 18, 1948; s. Luke A. and Virginia (Kelly) B.; m. Bernadette Otte, July 24, 1976; 1 dau., Catheryn. B.S., Fordham U., 1969; M.S., NYU, 1973; D.Sc., Universite de Louvain, Belgium, 1978. NATO postdoctoral fellow Facultes de Namur, Belgium, 1978-79, SUNY-Stony Brook, 1979-80; asst. prof. chemistry Rutgers U., 1980—. NASA faculty fellow, 1982; recipient Jean Stas prize Royal Acad. Arts and Scis., Belgium, 1978. Mem. Am. Chem. Soc., Societe Chimique de Belgique. Club: Chemists (N.Y.C.). Subspecialties: Theoretical chemistry; Organic chemistry. Current work: Theoretical description of reaction mechanisms; electronic properties of polymers and molecules of biological interest. Home: 40 Lincoln Dr PO Laurel Springs NJ 08021 Office: Rutgers University Department of Chemistry 406 Penn St Camden NJ 08102

BURKE, PATRICIA VIRGINIA, biophysicist; b. Washington, June 3, 1942; d. Oscar and Virginia Norgorden. B.A., Pomona Coll., 1964; M.S., Calif. Inst. Tech., 1967, Ph.D., 1971. NIH postdoctoral fellow U. Calif.- Santa Cruz, 1970-72, research assoc., 1972-76, asst. research biophysicist, 1976—. Mem. Am. Assn. Physics Tchrs., Genetics Soc. Am., Am. Soc. Photobiology. Subspecialties: Biophysics (biology); Membrane biology. Current work: Mechanisms of sensory transduction; lipid-protein interactions; genetics of fungus Phycomyces. Office: U Calif 466 Natural Sciences II Santa Cruz CA 95064

BURKE, ROBERT EMMETT, neurophysiologist; b. N.Y.C., July 26, 1934; s. N. Thomas and Margaret K. (Meudt) B.; m. Patricia Donovan, July 16, 1960; children: Mary Kay, David W., Jean E., Christina E. B.S., St. Bonaventure U., 1956; M.D., U. Rochester, 1961. Katherine Whipple postdoctoral scholar U. Rochester Sch. Medicine, 1961; intern Mass. Gen. Hosp., Boston, 1961-62, asst. resident in medicine, 1962-63, asst. resident in neurology, 1963-64; med. officer Lab. Neural Control, Nat. Inst. Neurol, and Communicative Disorders and Stroke, NIH, 1964-76; chief Lab. Neural Control, 1975—. Contbr. numerous articles, revs. essays to profl. jours. Recipient Borden prize U. Rochester Sch. Medicine, 1961; Superior Service award HEW, 1972. Mem. Am. Physiol. Soc., Soc. Neurosci., Internat. Brain Research Orgn., AAAS, Alpha Omega Alpha. Subspecialties: Physiology (biology); Neurophysiology. Current work: Structure and function of spinal cord, spinal cord, physiology, anatomy, neurobiology of motor units, synaptic transmission, dendrites. Office: NIH Bldg 36 Room 5A29 Bethesda MD 20205

BURKE, RONALD JOHN, administrative studies educator; b. Winnipeg, Man., Can., Oct. 22, 1937; s. John Stanley and Anne B.; m. (div.); children: Sharon, Rachel, Jeff. B.A., U. Man., 1960; M.A., U. Mich.-Ann Arbor, 1962, Ph.D., 1966. Teaching fellow U. Mich.-Ann Arbor, 1962-64, research asst., 1964-66; asst. prof. U. Minn., 1966-68, York U., Toronto, 1968-69, assoc. prof. adminstrv. studies, 1969-72, prof., 1972—; cons. Gulf Oil Corp., Toronto, 1980-81, Imperial Oil Co., 1979, Control Data Can., 1971; Bank of Can., Ottawa, 1977. Editor: Canadian Jour. Adminstrv. Scis, 1980—; cons. editor: Jour. Group and Orgn. Studies. Ont. Ministry of Labor grantee, 1981; Imperial Oil Ltd. grantee, 1980; Health and Welfare of Can. grantee, 1977; Donner Found. grantee, 1976. Mem. Canadian Mental Health Assn. (com. chmn.), Canadian Psychol. Assn., Am. Psychol. Assn., Acad. Mgmt., Internat. Assn. Applied Psychology, Adminstrv. Scis. Assn. Can. Subspecialty: Developmental psychology. Current work: Mentoring in organizations, home and work interfaces, coronary heart disease and executive behavior, stress, skill obsolescence and aging. Home: 2 Lafayette Pl Thornhill ON Canada L3T 1GS Office: Faculty of Adminstrv Studies York U 4700 Keele St Downsview ON Canada M2J 1P3

BURKE, WYATT WARNER, psychology educator, consultant; b. DeKalb County, Ala., May 12, 1935; s. Alfred Vernard and Ruby Inez (Gilbert) B.; m. Grace Ann Barbour, Jan. 25, 1958 (div. Sept. 1974); 1 son, Donovan Warner; m. Roberta Joann Luchetti, Oct. 5, 1974; children: Courtney Robyn, Warner Brian. B.A., Furman U., 1957; M.A., U. Tex.-Austin, 1961, Ph.D., 1963. Diplomate: in indsl. organizational psychology Am. Bd. Profl. Psychology. Instr. U. Tex.-Austin, 1962-63; asst. prof. psychology U. Richmond, 1963-66; ctr. dir. NTL Inst., WAshington, 1966-74; cons. orgn. devel., Arlington, Va., 1974-76; prof. mgmt., chmn. dept. mgmt. Clark U., Worcester, Mass., 1976-79; prof. psychology, co-chmn. dept. psychology Tchrs. Coll., Columbia U., N.Y.C., 1979—; cons. in field. Author: Organization Development: Principles and Practices; editor: The Cutting Edge: Current Theory and Practice in Organization Development, Organizational Dynamics, 1979—. Served to 1st lt. U.S. Army, 1958-60. Research contracts NASA, 1980-81, 83. Mem. Am. Psychol. Assn., Acad. Mgmt. (rep.-at-large bd. govs. 1981-83). Democrat. Episcopalian. Club: Pelham (N.Y.) Country. Subspecialties: Organizational psychology; Social psychology. Current work: Management competence-behavioral dimensions; leadership; organizational climate. Home: 235 Pelhamdale Ave Pelham Heights NY 10803 Office: Tchrs Coll Columbia U Box 24 New York NY 10027

BURKET, GEORGE EDWARD, JR., family physician; b. Kingman, Kans., Dec. 10, 1912; s. George Edward and Jessie May (Talbert) B.; m. Mary Elizabeth Wallace, Nov. 12, 1938; children: George Edward III, Carol Sue, Elizabeth Christine. Student, Wichita State U., 1930-33; M.D., U. Kans., 1937. Diplomate: Am. Bd. Family Practice (pres. 1976-78). Intern Santa Barbara (Calif.) Gen. Hosp., 1937-38, resident, 1938-39; grad. asst. in surgery Mass. Gen. Hosp., Boston, 1956-57; practice medicine, Kingman, 1939-73; preceptor in medicine U. Kans. Med. Sch., 1950-73, asso. prof., 1973-78, clin. prof., 1978—. Contbr. articles to profl. jours. Mem. Kingman Bd. Edn., 1946-58; mem. Kans. State Bd. Health, 1960-66. Mem. Kans. Med. Soc. (pres. 1966-67), Am. Acad. Family Physicians (pres. 1967-68, John Walsh Founders award 1979), inst. Medicine, AMA, Assn. Am. Med. Colls., Soc. Tchrs. Family Medicine, Alpha Omega Alpha. Republican. Episcopalian. Clubs: Masons, Shriners, Garden of Gods (Colorado Springs, Colo.); Wichita Country, Wichita. Subspecialty: Family practice. Home: Spring Lake Route 1 Kingman KS 67068 Office: Rainbow Blvd at 39th St Kansas City KS 66103

BURKHARDT, KENNETH JOHN, JR., computer consultant, educator; b. Elizabeth, N.J., Mar. 30, 1945; s. Kenneth J. and Virginia (Morgan) B.; m. Joanne K. Petrock, Sept. 6, 1981; 1 son, Kenneth J. A.B., Cornell U., 1967; M.S., Rutgers U., 1970; Ph.D., U. Wash., 1975. Systems programmer Am. Cyanamid Co., Bound Brook, N.J., 1967-70; cons., pres. K.J. Burkhardt & Co., Quakertown, N.J., 1970—; research faculty U. Wash., 1974-76; asst. prof. elec. engring. dept. Rutgers U., 1976—; cons. maj. Am., European cos. in computer industry. Contbr. articles to tech. jours. Recipient Am. Cyanamid Postgrad. Research award, 1970-72. Mem. Assn. Computing Machinery, IEEE, Sigma Xi. Subspecialties: Computer architecture; System design. Current work: Design and application of computers systems for the office of the future; innovative applications of microprocessors. Home: Quaker Ln Farm Quakertown NJ 08868 Office: Box 420 Quakertown NJ 08868

BURKHART, RICHARD HENRY, mathematician; b. Tacoma, Wash., Dec. 17, 1946; s. Perry Needham and Dorothy Alice (Shoff) B.; m. Saowalak Suntharalak, Apr. 13, 1971; children: Sanyaalak, Sandra, Diana. B.A., Reed Coll., 1969; A.M., Dartmouth Coll., 1974, Ph.D., 1976. Asst. prof. U. N.C., Wilmington, 1976-80; applied mathematician Boeing Computer Services, Tukwila, Wash., 1981—. Mem. Am. Math. Soc., Soc. Indsl. and Applied Math., Math. Assn. Am., Assn. Computing Machinery. Democrat. Unitarian-Universalist. Subspecialties: Algorithms; Graphics, image processing, and pattern recognition. Current work: Development of real-time algorithms for optimization and suboptimal estimation, in an image processing, pattern recognition context. Office: Boeing Computer Services 565 Andover Park W MS 9C-01 Tukwila WA 98188 Home: 11230 SE 325th St Auburn WA 98002

BURKMAN, ALLAN MAURICE, pharmacology educator, researcher, drug evaluation cons.; b. Waterbury, Conn., Apr. 23, 1932; s. Leon Oscar and Ann (Deitcher) B.; m. Katherine Horween, Aug. 8, 1965; children: David Eric, Deborah Rae. B.Sc., U. Conn., 1954; M.Sc., Ohio State U., 1955, Ph.D., 1958. Registered pharmacist, Conn. Asst. prof. pharmacology U. Ill.-Chgo., 1958-63; assoc. prof. pharmacology Butler U., Indpls., 1963-66, Ohio State U., Columbus, 1966-71, prof., 1971—, chmn. div. pharmacology, 1978—; vis. prof. pharmacology U. Utah, Salt Lake City, 1981-82. Mem. editorial rev. bd., referee various profl. and sci. jours.; contbr. numerous articles to profl. and sci. jours., chpts. to books. Recipient R. B. Allen Instructorship award U. Ill., 1963; Mead Johnson Research award, 1965; Am. Found. for Pharm. Edn. fellow, 1957-58; NIH, Office Naval Research, pharm. industry grantee, 1959—. Mem. Am. Soc. Pharmacology and Exptl. Therapeutics, Soc. for Exptl. Biology and Medicine, Am. Pharm. Assn., AAUP. Subspecialties: Pharmacology; Neuropharmacology. Current work: Characterization of drugs acting on the central nervous and cardiovascular systems; emetic and anti-emetic drug evaluation; drug effects on anterior pituitary function. Office: Div Pharmacology Coll Pharmac Ohio State U 500 W 12th Ave Columbus OH 43210

BURKS, THOMAS FRANKLIN, II, pharmacologist; b. Houston, Apr. 3, 1938; s. Jerry Y. and Goldie M. (Riley) B.; m. Dorothy Anne Travis, Sept. 1, 1962; children: Kathleen L., Thomas Franklin III. B.S., U. Tex., Austin, 1962, M.S., 1964; Ph.D., U. Iowa, 1967. Postdoctoral fellow Nat. Inst. Med. Research, London, 1967-68; asst. prof. pharmacology U. N. Mex. Sch. Medicine, 1968-71, asso. prof., 1971, U. Tex. Med Sch., Houston, 1971-74, prof., 1974-77; prof., head dept. pharmacology U. Ariz. Coll. Medicine, 1977—; mem. bd. sci. dirs. Vega Biotechs.; bd. dirs. Gibson-Stephens Inst. Author: books, including Prostaglandins: Organ - and Tissue - Special Actions, 1982; contbr. articles to profl. jours. Recipient John H. Freeman award U. Tex., 1977; USPHS grantee, 1968—. Mem. Am. Soc. Pharmacology and Exptl. Therapeutics, Western Pharmacology Soc., Am. Gastroenterol. Assn. Methodist. Subspecialties: Pharmacology; Neuropharmacology. Current work: Autonomic and neuropharmacology; neuropepetides; ppioids; neuropharmacology of digestive system; sensory function; thermoregulation; intestinal motility. Home: 3763 N Knollwood Circle Tucson AZ 85715 Office: Dept Pharmacology Coll Medicine U Ariz Tucson AZ 85724

BURLEIGH, BRUCE DANIEL, biochemist researcher; b. Augusta, Ga., June 23, 1942; s. Bruce Daniel and Billie Ann (Carter) B.; m. Dorothy Jean Roskos, Sept. 4, 1962; 1 son, Michael Eugene. B.S. in Chemistry, Carnegie-Mellon U., 1964; M.S. in Biochemistry, U. Mich., 1967, Ph.D., 1970. Post-doctoral fellow MRC Lab. Molecular Biology, Cambridge, Eng., 1970-72, sci. staff, 1972-73; asst. prof. biochemistry M.D. Anderson Hosp., Houston, 1973-78, assoc. biochemist, 1978-81; research biochemist research and devel. div. Internat. Minerals and Chem. Corp., Terre Haute, Ind., 1981—; cons. biochemist, endocrinologist, molecular naturalist, 1981—. Contbr. book chpts., numerous articles to profl. jours. Robert A. Welch Found. grantee, Houston, 1979. Mem. Am. Chem. Soc., Am. Soc. Biol. Chemists, Endocrine Soc., N.Y. Acad. Scis., Union Concerned Scientist, AAAS, Tau Beta Pi, Phi Lambda Upsilon, Sigma Xi. Episcopalian. Subspecialties: Biochemistry (biology); Cell and tissue culture. Current work: Peptide and protein chemistry; endocrinology and mechanics of peptide hormone action; molecular and cellular endocrinology; recombinant DNA technology; industrial production of protein and peptide hormones; regulation of animal growth. Home: 381 S 22d St Terre Haute IN 47803 Office: Research and Devel Div International Minerals & Chem Corp PO Box 207 1331 S 1st SE Terre Haute IN 47808

BURLEY, ELLIOTT LENHARD, nuclear engineer, consultant; b. Kenmore, N.Y., Jan. 10, 1925; s. Alan Elliott and Mildred Marie (Lenhard) B.; m. Helen Kathryn Hoskinson, Dec. 30, 1948; children: David, Diana, Molly, Ann. B.S.Ch.E., Cornell U., 1945, B.Ch.E., 1947. Nuclear engr. Gen. Electric Co., Hanford, Wash., 1947-59, San Jose, Calif., 1960-82; cons. nuclear engring., San Jose, Calif., 1983—. Served to lt. (j.g.) USNR, 1943-46. Mem. Am. Inst. Chem. Engrs., Am. Nuclear Soc. Republican. Subspecialties: Nuclear fission; Nuclear engineering. Current work: Nucleare waste treatment; reactor coolant treatment and purification systems. Inventor BWR Steam Dryer System, 1970. Home: 14127 Hydraulic Ridge Close Nevada City CA 95959

BURMEISTER, JOHN LUTHER, chemistry educator, consultant; b. Fountain Springs, Pa., Feb. 20, 1938; s. Luther John and Frieda May (Tielmann) B.; m. Doris Aileen Crawford, June 25, 1960; children: Lisa Anne, Jeffrey Scott. B.S., Franklin and Marshall Coll., 1959; Ph.D., Northwestern U., 1964. Instr. U. Ill., Urbana, 1963-64; asst. prof. U. Del., Newark, 1964-69, assoc. prof., 1969-73, prof. chemistry, 1973—, assoc. chmn, dept. chemistry, 1974—; cons. Control Data Corp., Mpls., 1981—, AMP, Inc., Harrisburg, Pa., 1971-73, Sun Oil Co., Marcus Hook, Pa., 1969-74. Editorial bds.: Inorganica Chimica Acta, Padova, Italy, 1967—, Synthesis and Reactivity in Inorganic and Metal-Organic Chemistry, St. Louis, 1970—. Ruling elder Head of Christiana Presbyn. Ch., Newark, 1969—; pres. Covered Bridge Farms Maintenance Corp., Newark, 1977-79. Recipient Gelewitz Award Northwestern U., 1963, excellence in teaching awards Lindback Found. and U. Del. Alumni Assn., 1968, 79, Chem. Mfrs. Assn., 1981. Mem. Am. Chem. Soc. (inorganic div. sec./treas 1975-77, alt. councilor 1977-79), Royal Soc. Chemistry, Sigma Xi, Phi Lambda Upsilon, Phi Kappa Phi (chpt. pres. 1980-81), Omicron Delta Kappa. Republican. Subspecialties: Inorganic chemistry; Synthetic chemistry. Current work: Coordination chemistry of ambidentate ligands, synthesis of metal complexes in unusual oxidation states, oxidative addition reactions of metal complexes. Home: 1 Carriage Ln Covered Bridge Farms Newark DE 19711 Office: Dept Chemistr Univ Delaware Newark DE 19711

BURNE, RICHARD ALLEN, neurobiologist, educator; b. N.Y.C., Apr. 26, 1951; s. Seymour William and Norma (Stewart) B.; m. Lynn

C. Burne, Aug. 1, 1979. B.S.E.E., Washington U., St. Louis, 1973; Ph.D., U. Rochester Sch. Medicine, 1979. Instr. U. Tex. Health Sci. Center, 1979-81; mem. tech. staff Advanced Tech. Center, Bendix Corp., Columbia, Md., 1981—; asst. instr. physiology U. Rochester, 1975-76; instr. neurosci. U. Tex., 1977-81. Mem. Soc. Neurosci., AAAS. Subspecialties: Neurophysiology; Neurobiology. Current work: Neuroanatomy and neurophysiology of auditory system; electrophysiology of midbrain function; neuronal tract tracing; neurophysiological single cell recording. Office: 9140 Old Annapolis Rd Route 108 Columbia MD 21045

BURNETT, JOHN LAURENCE, geologist, geology educator; b. Wichita, Kans., Aug. 28, 1932; s. Virgil Milton Burnett and Bertha (Van Order) Gambill; m. Annette Saywell, July 2, 1954 (div. 1975); children: John Forrester, Laurence Gregory; m. Joyce Frances Brady, Sept. 6, 1980. B.A., U. Calif.-Berkeley, 1957, M.S., 1960. Registered geologist, cert. engring. geologist, Calif. Staff geologist Calif. Div. Mines and Geology, San Francisco, Redding, and Sacramento, 1958—; instr. geology U. Calif.-Berkeley Extension, 1964-72, Cosumnes River Coll., Sacramento, 1981—; cts. expert Superior Ct. Los Angeles, 1967-72. Co-author: Urban Geology Master Plan for California, 1972. Served with U.S. Army, 1953-55. Fellow Geol. Soc. Am.; mem. Assn. Engring. Geology, Geol. Soc. Sacramento (pres. 1971). Subspecialties: Geology; Exploration geoglogy. Current work: Regional geology Northern California; geology of metallic and industrial mineral deposits; geologic hazards. Home: 8275 Holly Jill Way Sacramento CA 95823 Office: Calif Div Mines and Geology 610 Bercut Dr Sacramento CA 95814

BURNETT, JOHN NICHOLAS, chemistry educator; b. Atlanta, Aug. 19, 1939; s. Joseph Nicholas and Maurine (Morris) B. B.A., Emory U., 1961, M.S., 1963, Ph.D., 1965. Research chemist DuPont Co., Wilmington, Del., 1965-66; research assoc. U. N.C., Chapel Hill, 1966-68; asst. prof. chemistry Davidson (N.C.) Coll., 1968-72, assoc. prof., chmn., 1972-80, prof., chmn., 1980—, assoc. dean faculty, 1980—, Maxwell Chambers prof., 1981—; Mem. com. on sci. instrumentation N.C. Bd. Sci. and Tech., Raleigh, 1983. Contbr. articles to profl. jours. Recipient Thomas Jefferson Teaching award Davidson Coll., 1980; also grantee. Mem. Am. Chem. Soc., AAAS, Royal Chem. Soc. (London), AAUP, Sigma Xi. Republican. Presbyterian. Subspecialty: Analytical chemistry. Current work: Electroanalytical chemistry. Home: PO Box 2182 727 Virginia Ave Davidson NC 28036 Office: Dept Chemistry Davidson Coll Davidson NC 28036

BURNETT, LOUIS ELWOOD, JR., comparative animal physiologist, educator; b. Richmond, Va., Sept. 12, 1951; s. Louis Elwood and Theresa (Morrell) B.; m. Karen Gray, Jan. 20, 1973. B.S., Coll. William and Mary, 1973; Ph.D., U. S.C.-Columbia, 1977. Research biologist EPA, Charleston, S.C., 1977; asst. prof. biology U. San Diego, 1978-82, assoc. prof., 1982—. Contbr. articles to profl. jours. NSF grantee, 1981-82. Mem. Am. Soc. Zoologists, Am. Physiol. Soc., Soc. Exptl. Biology, Southeastern Estuarine Research Soc., Sigma Xi. Subspecialties: Physiology (biology); Comparative physiology. Current work: Respiratory adaptations of animals living in unstable environs; function of gills in respiration and ion regulation using isolated perfused gills. Home: 3814 Martha St San Diego CA 92117 Office: Dept Biology University of San Diego Alcala Park San Diego CA 92110

BURNHAM, DONALD CLEMENS, manufacturing company executive; b. Athol, Mass., Jan. 28, 1915; s. Charles Richardson and Freda (Clemens) B.; m. Virginia Gobble, May 29, 1937; children: David Charles, Joan (Mrs. Robert Graham), John Carl, William Lawrence, Mary Barbara (Mrs. F. David Throop). B.S. in Mech. Engring, Purdue U., 1936, D.Eng. (hon.), 1959, Ind. Inst. Tech., 1963, Drexel Inst. Tech., 1964, Poly. Inst. Bklyn., 1967. With Gen. Motors Corp., 1936-54, asst. chief engr., 1953-54; with Westinghouse Electric Corp., 1954—, group v.p., 1962-63, pres., chief exec. officer, 1963-68, chmn., chief exec. officer, 1969-75, dir.-officer, 1975-80; dir. Mellon Bank (N.A.), Mellon Nat. Corp. Mem. The Bus. Council; life trustee Carnegie-Mellon U.; trustee Carnegie Inst.; bd. dirs. Am. Wind Symphony Orch., Logistics Mgmt. Inst. Served to maj. AUS, World War II. Recipient Outstanding Achievement in Mgmt. award Am. Inst. Indsl. Engrs., 1964. Mem. ASME, Soc. Mfg. Engr. (Hoover Medal award 1978), Soc. Automotive Engrs., IEEE, Nat. Acad. Engring., Am. Assn. Engring. Socs. (Nat. Engring. award 1981). Club: Duquesne (Pitts.). Subspecialties: Industrial engineering; Systems engineering. Current work: Productivity improvement. Home: 615 Osage Rd Pittsburgh PA 15243 Office: Westinghouse Bldg Gateway Center Pittsburgh PA 15222

BURNS, EDWARD R., hematologist, medical educator; b. N.Y.C., Apr. 4, 1951; s. Erwin and Pola B.; m. Helen Silber, July 14, 1977; children: Judah, Joshua, Aaron. B.A., Yeshiva U., 1973; M.D., Albert Einstein Coll. Medicine, 1976. Intern in internal medicine Montefiore Hosp. and Med. Center, Bronx, N.Y., 1976-77, resident in medicine and hematology, 1977-80; instr. NYU Med. Center, N.Y.C., 1980-81; asst. prof. medicine Albert Einstein Coll. Medicine, Bronx, 1982—, asst. prof. lab. medicine, 1982—, dir. hematology lab., 1982—. Contbr. chpts. to books, articles to profl. publs. NIH grantee, 1982. Mem. Am. Soc. Hematology, Am. Fedn. Clin. Research, N.Y. Acad. Scis. Jewish. Subspecialties: Hematology; Microscopy. Current work: Hemostosis; thrombosis; sickle cell anemia; electron microscopy. Office: Albert Einstein Coll Medicine 1300 Morris Park Ave Bronx NY 10461 Home: 70-45 173rd St Flushing NY 11365

BURNS, JACK O'NEAL, JR., astrophysicist; b. Ayer, Mass., Jan. 2, 1953; s. Jack O'Neal and Irene Blanche (Gendron) B.; m. Cathleen Spalding, Nov. 8, 1980. B.S., U. Mass., 1974; M.A., Ind. U., 1976, Ph.D., 1978. Postdoctoral research asso. Nat. Radio Astronomy Obs. (VLA Project), Socorro, N.Mex., 1978-80; asst. prof. astronomy U. N.Mex., Albuquerque, 1980—; cons. computer image processing Sandia Nat. Labs., Albuquerque, 1980—. Contbr. Articles to profl. jours. Sandia Nat. Labs. grantee, 1980-82; NASA grantee, 1980-83; Ind. U. Found. fellow, 1977-78. Mem. Internat. Astron. Union, Internat. Union Radio Sci., Am. Astron. Soc., Royal Astron. Soc. Eng. Democrat. Subspecialties: Radio and microwave astronomy; High energy astrophysics. Current work: Radio, X-ray, optical studies of clusters of galaxies, research on extragalactic radio sources, radio jets, optical spectra of radio galaxies for distance determinations. Home: 12208 La Charles Ave NE Albuquerque NM 87111 Office: Dept Physics and Astronomy U NMex Albuquerque NM 87131

BURNS, JAY, III, physicist and space scientist, educator; b. Lake Wales, Fla., Mar. 22, 1924; s. Jay, Jr. and Harlan (Sheafe) B.; m. Dulcie Evans, Sept. 18, 1948; children: Jay, Wendy, William Scott. B.S., Northwestern U., 1947; M.S., U. Chgo., 1951, Ph.D., 1959. Dir. Chgo. Midway Labs., U. Chgo., 1962-65; assoc. prof. astrophysics Northwe. U., 1965-72; assoc. dir. research Rauland div. Zenith Radio Corp., 1967-68, sr. research cons., 1968-78; dir. Lakeside Labs., Chgo., 1972-75; prof., head dept. physics and space scis. Fla. Inst. Tech., Melbourne, 1976—; cons. U.S. Navy. Contbr. articles sci. jours. Served with USNR, 1944-46. Research grantee U.S. Air Force, U.S. Navy, U. S. Army, NSF. Mem. Am. Phys. Soc. Club: East Coast Cruising Assn. (Indian Harbour Beach, Fla.). Subspecialties: Atomic and molecular physics; Condensed matter physics. Current work: Electron emission from solids, surface physics. Home: 226 Sand Pine Rd North Indialantic FL 32903 Office: Dept Physics and Space Sciences Fla Inst Tech Melbourne FL 32901

BURNS, JOHN JOSEPH, pharmaceutical company executive, pharmacology educator; b. Flushing, N.Y., Oct. 8, 1920; s. Thomas F. and Katherine (Kane) B.; m. Margaret Hitchcock, 1974. B.S., Queens Coll., 1942; M.A., Columbia U., 1948, Ph.D., 1950. With lab. chem. pharmacology Nat. Heart Inst., 1950-60, dep. chief lab., 1957-60; head sec. clin. pharmacology, also adj. asst. prof. biochemistry N.Y. U. research service Goldwater Meml. Hosp., Welfare Island, N.Y., 1950-57; dir. research pharmacodynamics div. Wellcome Research Labs., Burroughs Wellcome & Co. (U.S.A.) Inc., Tuckahoe, N.Y., 1960-66; v.p. for research Hoffmann-LaRoche Inc., Nutley, N.J., 1967—; Vis. prof. pharmacology Albert Einstein Coll. Medicine, 1960-68; adj. prof. pharmacology Cornell U. Med. Coll., 1969—; sr. cons. pharmacology-toxicology programs NIH; chmn. com. problems drug safety Drug Research Bd., 1965-72. Author articles metabolism drugs, vitamins and carbohydrates. Served with AUS, 1944-46. Fellow Am. Inst. Chemists; mem. Inst. Medicine, Nat. Acad. Scis., N.Y. Acad. Scis. (v.p. 1964-65), Am. Soc. Pharmacology and Exptl. Therapeutics (pres. 1972-73), Am. Soc. Biol. Chemists, Am. Inst. Nutrition, Am. Coll. Neuropsychopharmacology, Internat. Union Pharmacology (pres. 1975-78). Subspecialties: Psychopharmacology; Neurology. Home: PO Box 104 Southport CT 06490 Office: Hoffmann-LaRoche Inc Nutley NJ 07110

BURNS, JOSEPH ARTHUR, Astronomer, educator; b. N.Y.C., Mar. 22, 1941; s. John Driscoll and Genevieve Mary (McCarthy) B.; m. Judith Ann Klein, July 1, 1967; 1 son, Patrick Matthew. B.S., Webb Inst., Glen Cove, N.Y., 1962; Ph.D., Cornell U., 1966. Asst. prof. Cornell U., 1966-67, 68-73, asso. prof., 1974-81, prof. space mechanics, 1981—; NRC postdoctoral fellow NASA-Goddard Space Flight Center, Greenbelt, Md., 1967-68; U.S. exchange fellow Schmidt Inst., Moscow, 1973; sr. scientist NASA Ames, Moffett Field, Calif., 1975-76; astronom titular Paris Obs., Meudon, France, 1979; vis. prof. U. Calif., Berkeley, 1982-83. Contbr. numerous articles to profl. jours. Grantee N.Y. Arts Council, 1973, NSF, 1973, NASA, 1975—. Mem. Internat. Astron. Union, AAAS, Am. Astron. Soc., Am. Geophys. Union, Sigma Xi. Subspecialties: Planetary science; Theoretical and applied mechanics. Current work: Solar system dynamics. Office: Cornell U 237 Thurston Hall Ithaca NY 14853

BURNS, MARY ANN THERESA, college administrator; b. Phila., Jan. 24, 1928; d. John Joseph and Anna Marie (McLean) B. A.B., Rosemont Coll., 1949; A.M., U. Pa., 1950, Ph.D. (Fund for Advancement Edn. fellow 1954-55), 1960. Tchr. Springfield (Pa.) High Sch., 1951-60; faculty U. Wis., Milw., 1960-73, asso. prof., 1964-68, prof., 1968-73; dean Wilson Coll., Chambersburg, Pa., 1973-76, Emmanuel Coll., Boston, 1977-79; v.p. acad. affairs, dean Mary Washington Coll., Fredericksburg, Va., 1979—; vis. lectr. Marquette U., 1965; exec. sec. Eta Sigma Phi. Author: Lingua Latina: Liber Primus, 1964, Lingua Latina Liber Alter, 1965; also articles, revs. Mem. Am. Philol. Assn., Archaeol. Inst. Am., Classical Assn. of Middle West and South, Am. Classical League (pres. 1980—). Address: 101 Wilderness Ln Fredericksburg VA 22401

BURNS, ROBERT DONALD, III, engineering company administrator, consultant; b. Ft. Atkinson, Wis., May 31, 1951; s. Robert Donald and Sophia Agnes (Schuelke) B.; m. Bertina Kathleen Jacobson, May.15, 1976; children: Brian, Robert IV, Nicholas. B.S. in Engring, Purdue U., 1973, M.S., 1973, Ph.D., 1976. Student research assoc. Argonne (Ill.) Nat. Lab., 1974; staff Los Alamos (N.Mex.) Sci. Lab., 1976-80, sect. leader, 1980-81; supr. Impell Corp., San Francisco, 1981-82, mgr., 1982—; staff Pres.'s Commn. on Accident at Three Mile Island; cons. U.S. NRC; lectr. U. Ariz.; tech. com. Internat. AEC, Vienna. Mem. Soc. Risk Analysis, Am. Nuclear Soc. Subspecialties: Nuclear engineering; Software engineering. Current work: Analysis of fission product transport during severe degraded core accidents in commercial nuclear power plants. Home: 3749 Morningside Dr El Sobrante CA 94803 Office: Impell Corp 350 Lennon Ln Walnut Creek CA 94598

BURNSIDE, ORVIN CHARLES, agronomy educator, researcher; b. Hawley, Minn., June 9, 1932; s. John Joseph and Sena (Dwyre) B.; m. Delores Sadie Schattschneider, Dec. 22, 1954; children: Bruce Donald, Kristi Lynn. B.S., N.D. State U., 1954; M.S., U. Minn., 1958, Ph.D., 1959. Research asst. U. Minn., St. Paul, 1956-59; asst. prof. U. Nebr., Lincoln, 1959-63, assoc. prof., 1963-66, prof. agronomy, 1966—; cons. U.S. Congress, OTA, EPA, AID, Dept. Agr., Washington, 1970—, U. Berkeley, Oakland, Calif., 1970-72, Agrimark, Inc., Pretoria, S. Africa, 1981—, Farmland Industries, Kansas City, Kans., 1981-82. Served to 1st lt. U.S. Army, 1954-56. Recipient Outstanding Spokesman award N.E. Agrl. Chem. Assn., 1970; Outstanding Agrl. award Ciba-Geigy Corp., 1979. Mem. Weed Sci. Soc. Am. (sec. 1981-83, outstanding research award 1979), Internat. Weed Sci. Soc., Am. Soc. Agronomy, N. Central Weed Control Conf. (pres., hon.), Gamma Sigma Delta (outstanding research award 1981). Republican. Lutheran. Lodge: Kiwanis. Subspecialty: Integrated pest management. Current work: Develop weed control systems for agronomic crops, herbicide physiology and reduced tillage cropping system, weed biology studies. Patentee in field. Office: U Nebr Dept Agronomy Lincoln NE 68583 Home: 6111 Lexington St Lincoln NE 68505

BUROW, DUANE FRUEH, chemist, educator; b. San Antonio, June 12, 1940; s. Martin and Hermine Elizabeth (Frueh) B.; m. Janice M. Bockstahler, Aug. 3, 1969; children: Nathan Allan, Bethanie Kristin. B.A., U. Tex., 1961, Ph.D., 1966. Instr. chemistry St. Edwards U., Austin, Tex., 1964-65; research phys. chemist engring. physics lab. E.I. DuPont de Nemours, Inc., Wilmington, Del., 1966-67; asst. prof. chemistry Mich. State U., 1967-69, U. Toledo, 1969-73, assoc. prof., 1973-78, prof. chemistry, 1978—. Contbr. chpts. to books, articles to profl. jours. Mem. Sylvania Schs. Com. on Gifted Edn., 1979—. NSF grantee, 1974, 77; Dept. Energy grantee, 1978; Ohio Coal Research Labs. Assn. grantee, 1980. Mem. Am. Chem. Soc., Royal Soc. Chemistry (U.K.), Coblentz Soc., Optical Soc. Am., Soc. Applied Spectroscopy, Sigma Xi. Subspecialties: Physical chemistry; Inorganic chemistry. Current work: Chemistry of sulfur dioxide, solution chemistry, application to fossil fuels and minerals processing; molecular spectroscopy, modeling of moleculaoptical properties, application to bio and geochemical systems. Office: Dept Chemistry U Toledo Toledo OH 43606

BURR, ALEXANDER FULLER, physicist, educator; b. Cambridge, Mass., July 18, 1931; s. A.C. and Lily (Fuller) B.; m. Marjorie McKinstry, Aug. 18, 1962; children: Margaret, Catherine, Susan. B.S., Jamestown Coll., 1953; M.A., U. Edinburgh, Scotland, 1958; Ph.D., Johns Hopkins U., 1967. Solid state physicist U.S. Naval Research Lab., Washington, 1965-66; sr. vis. fellow U. Strathclyde, Glasgow, Scotland, 1973; chief research scientist Duntech Industries, Las Cruces, N.Mex., 1979-80; instrumentation expert IAEA, Vienna, 1981, 83; prof. physics N.Mex. State U., Las Cruces, 1966—. Author book, also papers and articles. Served with U.S. Army, 1954-56. Grantee Am. Assn. Physics Tchrs. Mem. Am. Phys. Soc., Am. Assn. Physics Tchrs., AAAS. Subspecialty: Atomic and molecular physics. Home: 2025 O'Donnell Dr Las Cruces NM 88001 Office: Physics Dept Box 3 NMex State U Las Cruces NM 88003

BURR, BALDWIN GWYNNE, energy cons. firm exec.; b. Columbus, Ohio, Dec. 29, 1945; s. Baldwin Gwynne and Edna Mae (Tracey) B.; m. Katherine Moore Harvey, May 27, 1972; 1 dau., Katherine Alexandra. B.A., U. N.Mex., 1969, M.A., 1977. Cert. energy auditor, N. Mex., Ariz., Colo., Fla. Asst. dir. So. Vt. Art Center, Manchester, 1969-70; prin. Godbold Burr and Burr, Albuquerque, 1975-; cons. N.Mex. Energy Inst., Albuquerque, 1978-80; pres. Energy Mgmt. Co., Tome, N.Mex., 1978—; guest lectr. U. N.Mex., Valencia. Author manuals in field of energy conservation. Served with U.S. Army, 1969-71. Mem. Profl. Energy Auditors Assn. N.Mex., Solar Lobby, Greater Albuquerque C. of C., Phi Gamma Delta. Subspecialties: Solar energy; Mathematical software. Current work: Development of microcomputer software for sophisticated energy analysis of buildings. Home: Route 2 Box 816 Los Lunas NM 87031 Office: Box 362 Tome NM 87060

BURR, BROOKS MILO, zoology educator; b. Toledo, Ohio, Aug. 15, 1949; s. Lawrence E. and Beverly Joy (Herald) B.; m. Patti Ann Grubb, Mar. 5, 1977. B.A., Greenville (Ill.) Coll., 1971; M.S., U. Ill., 1974. Ph.D., 1977; Lab. instr. Greenville Coll., 1971-72; research asst. Ill. Natural History Survey, Urbana, 1972-77; asst. prof. zoology So. Ill. U., Carbondale, 1977-81, assoc. prof., 1981—; mem. adv. com. Ill. Dept. Conservation Endangered Species Protection Bd., 1980—. Contbr. articles to profl. jours. NSF grantee, 1977-78; U.S. Dept. Agr. grantee, 1978-79; Ill. Dept. Conservation grantee, 1981-82. Mem. AAAS, Am. Soc. Ichthyologists and Herpetologists, Soc. Systematic Zoology, Am. Fisheries Soc., Biol. Soc. Washington. Methodist. Subspecialties: Systematics; Ecology. Current work: Systematics, life history and zoogeography of native North American fishes; intensive ecological studies of nongame fishes; endangered species; faunal studies. Office: Dept Zoology So Ill U Carbondale IL 62901

BURR, THOMAS JAMES, plant pathologist; b. Oshkosh, Wis., Mar. 19, 1949; s. James Gordon and Marilyn Jane (Faust) B.; m. Judith Ann Bonsall, May 17, 1974; children: Andrew Thomas, Alison Michelle, Holly Nicole. B.S., U. Ariz., 1971, M.S., 1973; Ph.D., U. Calif.-Berkeley, 1977. Asst. prof. plant pathology Cornell U., Geneva, N.Y., 1977—. Contbr. articles to profl. jours. Mem. Am. Phytopath. Soc. Subspecialty: Plant pathology. Current work: Biological and control research on fruit diseases, fruit disease extension. Co-inventor Plant Growth-Promoting bacteria. Home: 91 Highland Ave: Geneva NY 14456 Office: NY State Agrl Expt St Cornell U Dept Plant Pathology Geneva NY 14456

BURRELL, CHARLES FREDERICK, physicist; b. Danville, Pa., Sept. 26, 1942; s. Charles Earl and Helen Gertrude (Conrad) B. B.S., Lehigh U., 1964; M.S., Cornell U., 1967; Ph.D. in Physics, U. Md., 1974. Postdoctoral trainee U. Md., College Park, 1975; physicist Lawrence Berkeley Lab., Berkeley, Calif., 1976—. Contbr. articles to profl. jours. Mem. Am. Phys. Soc., AAAS. Subspecialties: Fusion; Plasma physics. Current work: Research and development of neutral beam injectors for plasma heating in magnetic fusion energy experiments; especially neutral beam diagnostics by optical spectroscopy and laser induced fluorescence. Home: 6526 Kensington Ave Richmond CA 94805 Office: Lawrence Berkeley Lab Berkeley CA 94720

BURRIS, JOSEPH STEPHEN, seed science educator; b. Cleve., Apr. 18, 1942; s. Charles Richard and Catherine (Pravika) B.; m. Judith Kay Burkley, June 8, 1963; children: Jeffrey, John, Jennifer, Jason. B.Sc., Iowa State U., 1964; M.Sc., Va. Poly. Inst., 1965, Ph.D., 1967. Research technician Iowa State U., Ames, 1962-64; NDEA fellow Va. Poly. Inst., Blacksburg, 1964-67, dir., 1967; asst. prof. Iowa State U., Ames, 1968-70, assoc. prof., 1970-76, prof., 1977—. Pres. Ames PTA, 1979-81. Mem. Am. Soc. Agronomy, Crop Sci. Soc. Am. (chmn.), Assn. Ofcl. Seed Analysts (editor). Presbyterian. Subspecialties: Plant physiology (agriculture); Plant growth. Current work: Basic research on the development, maturation, drying, longevity and germination of seeds. Office: Iowa State U Seed Sci Center Ames IA 50011

BURRIS, LESLIE, laboratory administrator, chemical engineer; b. Carmi, Ill., Sept. 1, 1922; s. Leslie O. and Nellie (Varney) B.; m. Mary Elizabeth Bush, Mar. 19, 1944; children: Susan Elizabeth Burris Madeira, James Leslie, Kathryn Lorraine Burris Turner. B.S., U. Colo., 1943; M.S., Ill. Inst. Tech., 1956. Analytical chemist Monscanto Chem. Co., Monsanto, Ill., 1943-45; staff engr. Moncanto Chem. Co., Oak Ridge, 1945-48, Argonne Nat. Lab., Argonne, Ill., 1948-50, group leader, 1950-60, sect. head, 1960-67, project mgr., 1967-70, assoc. dir., 1970-73, div. dir. chem. tech., 1973—. Contbr. articles to profl. jours. Mem. elem. sch. bd., Naperville, Ill., 1961-72, pres., 1964-70; mgr. United Way campaign, Naperville, 1978-79. Mem. Am. Nuclear Soc., Am. Inst. Chem. Engrs., Research Soc. Am. Republican. Methodist. Subspecialties: Nuclear fission; Chemical engineering. Current work: Nuclear fuel cycle-fuel reprocessing and nuclear waste management; management of energy research and development. Patentee in field. Home: 206 Douglas St Naperville IL 60540 Office: Argonne National Laboratory 9700 S Cass Ave Argonne IL 60439

BURRIS, MARTIN JOE, animal breeding science educator; b. Hebron, Nebr., Mar. 30, 1927; s. Sheridan A. and Nellie Hazel (Johnson) B.; m. Helen Ada Storey, Sept. 13, 1953 (dec. Dec. 1982); children: Emilie Burris Dohleman, Lucy Ellen, Ruth Ada, Martin Joe. B.S., U. Nebr., 1949, M.S., 1950; postgrad., Iowa State U., 1949; Ph.D., Oreg. State U., 1953. Asst. prof. U. Ark., Fayetteville, 1953-54; assoc. prof. Va. Poly. Inst., Front Royal, 1954-57; research adminstr. U.S. Dept. Agr., Washington, 1957-66; asso. dir. Mont. Agrl. Expt. Sta., Bozeman, 1966-80; prof. animal sci., animal and range scis. dept. Mont. State U., Bozeman, 1981—. Contbr. articles to profl. jours. Served with AUS, 1945-46. Purdue U. fellow, 1964-65. Mem. Am. Soc. Animal Sci., Genetics Soc. Am. Methodist. Subspecialties: Animal genetics; Animal breeding and embryo transplants. Current work: Genetic environment of livestock; efects of crossbreeding in livestock; death losses. Home: 2503 Spring Creek Dr Bozeman MT 59715 Office: Animal and Range Scis Dept Mont State U Bozeman MT 59717

BURRIS, ROBERT HARZA, biochemist, educator; b. Brookings, S.D., Apr. 13, 1914; s. Edward T. and Mable C. (Harza) B.; m. Katherine Irene Brusse, Sept. 12, 1945; children: Jean Carol, John Edward, Ellen Louise. B.S., S.D. State Coll., 1936, D.Sc., 1966; M.S., U. Wis., 1938, Ph.D., 1940. NRC fellow Columbia U., 1940-41; faculty U. Wis., Madison, 1941—, prof., 1951—; chmn. biochemistry Coll. Agr., 1958-70, W.H. Peterson prof. biochemistry, 1976—. Recipient Charles Thom award Soc. Indsl. Microbiology, 1977; Nat. Medal of Sci., 1980; Guggenheim fellow Cambridge U., 1954. Mem. Am. Chem. Soc., Am. Soc. Biol. Chemistry, Am. Soc. Plant Physiologists (Stephen Hales award 1968, Charles Reid Barnes award 1977, pres. 1960), Japanese Soc. Plant Physiology, Biochem. Soc., AAAS, Am. Soc. Microbiology, Nat. Acad. Scis., Am. Acad. Arts and Scis., Am. Philos. Soc. Subspecialties: Nitrogen fixation; Biochemistry (biology). Current work: Biochemical mechanism of biological nitrogen fixation; role of hydrogenase in nitrogen fixation; associative nitrogen fixation; blue-green algae. Home: 1015 University Bay Dr Madison WI 53705

BURROUGHS, RICHARD H., III, oceanographer, geologist, educator; b. New Haven, July 5, 1946; s. Richard H., Jr. and Mary Drummond (Page) B. A.B. with honors, Princeton U., 1969; Ph.D. (grad. research fellow), M.I.T.- Woods Hole Oceanographic Inst.,

1975. Staff officer Nat. Acad. Scis-NRC, 1974-77; sr. fellow Ecosystems Center, Marine Biol. Lab., Woods Hole, Mass., 1977-79, 81; vis. lectr. Sch. Forestry and Environ. Studies, Yale U., 1979; sci. advisor to dir. Bur. Land Mgmt., Dept. Interior, Washington, 1979-81; researcher John Gray Inst., Lamar U., Beaumont, Tex., 1982-83; asst. prof. grad. program in marine affairs U. R.I., Kingston, 1983—. Contbr. articles to profl. jours. Mem. AAAS, Am. Geophys. Union, Geol. Soc. Am., Sigma Xi. Subspecialties: Resource management; Oceanography. Current work: Management of natural resources. Office: Grad Program Marine Affairs U RI Kingston RI 02881

BURRUS, CHARLES ANDREW, JR., research physicist; b. Shelby, N.C., July 16, 1927; s. Charles Andrew and Velma (Martin) B.; m. Barbara Ione Dunlevy, May 4, 1957; children: Charles Andrew III, Barbara Jean, John Alan. B.S. cum laude in Physics, Davidson Coll., 1950; M.S. in Physics, Emory U., 1951; Ph.D. in Physics (Tex. Co. fellow, Shell Co. fellow), Duke U., 1955. Research assoc. dept. physics Duke U., Durham, N.C., 1954-55; mem. tech. staff Bell Telephone Labs., Holmdel, N.J., 1955—. Contbr. articles on millimeter and submillimeter-wave spectroscopy, techniques and semicondr. devices for lightwave communications, long-wavelength photoemitters and photodetectors for lightwave communications to tech. jours. Served with USNR, 1945-46. Named Disting. Mem. Tech. Staff Bell Telephone Labs., 1982. Fellow AAAS, Am. Phys. Soc., IEEE, Optical Soc. Am. (David Richardson medal 1982). Methodist. Subspecialties: Semiconductors; Fiber optics. Current work: Research on devices and technology for lightwave (optical fiber) communications. Home: 62 Highland Ave Fair Haven NJ 07701 Office: Bell Labs Crawford Hill Lab Holmdel NJ 07733

BURRY, KENNETH ARNOLD, physician; b. Monterey Park, Calif., Oct. 2, 1942; s. Frederick Harvey and Betty Jean (Bray) B.; m. Mary Lou Tweedy, June 5, 1964 (div.); 1 son, Michael Curtis; m. Katherine A. Johnson, Apr. 3, 1982; 1 dau., Lisa Bray. B.A., Whittier Coll., 1964; M.D., U. Calif.-Irvine, 1968. Diplomate: Am. Bd. Ob-Gyn. Intern Orange County Med. Center, Orange, Calif., 1968-69; resident in ob-gyn U. Oreg. Med. Sch., 1971-74; fellow U. Wash., Seattle, 1974-76; asst. prof. Oreg. Health Sci. U., Portland, 1976-81, assoc. prof., 1981—; dir. Oreg. Reproductive Research and Fertility Program, Portland, 1982—. Served to capt. U.S. Army, 1969-71. Fellow Am. Coll. Ob-Gyn; mem. Endocrine Soc., Am. Fedn. Clin. Research, Am. Fertility Soc., Pacific Coast Fertility Soc. Presbyterian. Subspecialties: Obstetrics and gynecology; Reproductive endocrinology. Current work: Reporductive endocrinology, cell differentiation, sex steroid receptors. Office: Oreg Hosp State U 3181 SW Sam Jackson Park Rd Portland OR 97201 Home: 8630 SW Pacer Dr Beaverton OR 97005

BURSEY, MAURICE M., chemistry educator; b. Balt., July 27, 1939; s. Reginald Price and Edna Frances (Moyer) B.; m. Joan Marie Tesarek, Dec. 28, 1970; children: John Thomas Kieran, Sara Helen Moyer. B.A., Johns Hopkins U., 1959, M.A., 1960, Ph.D., 1963. Lectr. Johns Hopkins U., Balt., 1963-64; asst. prof. Purdue U., Lafayette, Ind., 1964-66; asst. prof. chemistry U. N.C., Chapel Hill, 1966-69, assoc. prof., 1969-74, prof., 1974—. Contbr. articles to profl. jours. Recipient various research grants. Fellow Am. Inst. Chemists, Royal Soc. Chemistry (assoc.); mem. Am. Chem. Soc., Am. Soc. for Mass Spectroscopy, Alpha Chi Sigma. Democrat. Roman Catholic. Subspecialty: Analytical chemistry. Current work: Ion formation, decompositions, and reactions with molecules. Analysis of large molecules by mass spectroscopy. Ion structures, expecially of negative ions. Environmental applications of mass spectroscopy. Home: 101 Longwood Ct Chapel Hill NC 27514 Office: U NC Dept Chemistry Chapel Hill NC 27514

BURSTEIN, ELIAS, physicist, educator; b. N.Y.C., Sept. 30, 1917; s. Samuel and Sarah (Plotkin) B.; m. Rena Ruth Benson, Sept. 19, 1943; children—Joanna Bliss, Sandra Joy, Miriam Stephanie. A.B., Bklyn. Coll., 1938; A.M., U. Kans., 1941; postgrad., MIT, 1941-43, Cath. U., 1946-48; D. Tech. (hon.), Chalmers U. Tech., Göteborg, Sweden, 1982. Asst. instr. U. Kans., 1939-41; research asst. M.I.T., 1941-43, research asso., 1943-44; project engr. White Research Assos., Boston, 1944-45; physicist Crystal br. U.S. Naval Research Lab., 1945-58, head semiconductor br., 1958; prof. physics U. Pa., Phila., 1958—, Mary Amanda Wood prof. physics, 1982—; Jubilee vis. prof. physics Chalmers U. Tech., Göteborg, 1981. Editor-in-chief: Solid State Communications, 1969—; co-editor: Comments on Solid State Physics, 1971—. Recipient John Price Wetherill medal Franklin Inst., 1979; Guggenheim fellow, 1980. Fellow Am. Phys. Soc., Optical Soc. Am.; mem. Nat. Acad. Scis., AAAS, Sigma Xi. Club: Cosmos (Washington). Subspecialty: Solid State Physics. Patentee in field. Office: Dept Physics U Pa Philadelphia PA 19104

BURSTEIN, JOSEPH, analyst, researcher; b. Gorky, White Russia, Nov. 24, 1926; came to U.S., 1976; s. Mois and Luybov (Aizenstadt) B. M.S. in Mech. Engring, Moscow Higher Tech. Sch., 1950; Ph.D. in Engring. Sci, Odessa (Ukraine) Technol. Inst., 1966. Diplomate: Engring. diplomate. Various engring. and acad. positions, industry, research, edn., Russia, 1950's-60's; analyst Bendix Corp., Southfield, Mich., 1976-80, SDC/Burroughs Corp., Cambridge, Mass., 1980—. Author: Instrument Complex for Exploration in Wells, 1955, Dynamic Programming in Planning, 1968, Blackout in Mathematical Manuals, 1983. Mem. Soc. Indsl. and Applied Math. Subspecialties: Applied mathematics; Algorithms. Current work: Mathematical optimization. Scientific grounds of lucidity in mathematical manuals. Office: SDC/Burroughs Corp 2500 Colorado Ave Santa Monica CA 90406

BURSTEIN, PAUL HARRIS, physicist; b. Chelsea, Mass., Aug. 28, 1948; s. Sol and Esther (Levine) B.; m. Dorothy P. Peavy, Dec. 7, 1946. B.S., MIT, 1970; Ph.D., U. Wis., 1976. Postdoctoral fellow U. Wis., Madison, 1976; sr. staff scientist Am. Sci. & Enging., Inc., Cambridge, Mass., 1976—. Contbr. articles to profl. jours. Mem. Am. Phys. Soc., Am. Astron. Soc., Optical Soc. Am. Subspecialty: Physics - X-Ray applications. Current work: Non-destructive testing applications of x-rays; high energy x-ray inspection; computed tomography; science management. Patentee in field. Home: 27 Fountain Rd Arlington MA 02174 Office: Fort Washington Cambridge MA

BURT, DAVID REED, med. educator, biomed. researcher; b. East Orange, N.J., Oct. 28, 1943; s. Clifton and Ruth (Wurts) B.; m. Dorothy Ann Schick, June 21, 1968. A.B., Amherst Coll., 1965; Ph.D. in Biophysics, Johns Hopkins U., 1972. Asso. research scientist in biophysics Johns Hopkins U., Balt., 1972-73, postdoctoral fellow in pharmacology, 1973-76; asst. prof. pharmacology U. Md. Sch. Medicine, Balt., 1976-82, asso. prof., 1982—; cons. NIH, NSF. Contbr. articles to profl. jours., chpts. to books. Gen. Motors scholar, 1961-65; Nat. Inst. Neurol. and Communicative Disorders and Stroke fellow, 1974-76; NIMH and NSF grantee. Mem. Am. Soc. Pharmacology and Exptl. Therapeutics, Soc. Neurosci., Am. Soc. Neurochemistry, Endocrine Soc., N.Y. Acad. Scis., Sigma Xi, Phi Beta Kappa. Subspecialties: Neuropharmacology; Neurochemistry. Current work: Neuropeptides, receptor binding, neuroendocrinology, psychopharmacology; thyrotropin releasing hormone, dopamine, acetylcholine, retina, limbic forebrain. Office: U Md Sch Medicine Baltimore MD 21201

BURT, MICHAEL EDWARD, physician, surgeon; b. Newark, Jan. 3, 1948; s. Edward and Shirlee (Muzzio) B.; m. Jacqueline JoAnn Panlone, July 9, 1970; children: Bryan Michael, Lauren Gail. A.B., Rutgers U., 1970; M.D., St. Louis U., 1975; Ph.D., George Washington U., 1981. Diplomate: Nat. Bd. Med. Examiners. Intern dept. surgery U. Wash. Affiliated Hosps., Seattle, 1975-76; jr. asst. resident N.Y. Hosp.-Cornell Med. Center, N.Y.C., 1976-77, resident, 1981-83; clin. assoc. Nat. Cancer Inst. NIH, Bethesda, Md., 1977-79, investigator, 1979-80, cancer expert, 1980-81. Contbr. articles to profl. jours. Served with USPHS, 1977-80. Recipient James Ewing research award Soc. Surg. Oncology, 1980, 81; research award Assn. Acad. Surgery, 1981. Mem. N.Y. Acad. Scis., Am. Fedn. Clin. Research, Am. Assn. Cancer Research, Am. Physiol. Soc., Fedn. Am. Soc. Exptl. Biology. Subspecialties: Surgery; Cancer research (medicine). Current work: Evaluating metabolic host-tumor interactions at the substrate level. Office: NY Hosp-Cornell Med Center 525 E 68th St Dept Surgery New York NY 10021

BURTNER, ROGER LEE, research geologist; b. Hershey, Pa., Mar. 31, 1936; s. Bruce Lemuel and Bernetta Viola (Quigle) B.; m. Carol Ann Spitzer, Aug. 1, 1965; 1 dau., Pamela Sue. B.S., Franklin and Marshall Coll., 1958; M.S., Stanford U., 1959; Ph.D., Harvard U., 1965. Assoc. research geologist Calif. Research Corp. div. Standard Oil Co. California, La Habra, CAlif., 1963-64, research geologist, 1964-68; exploration geologist Standard Oil Co. Texas div. Standard Oil Co. California, Corpus Christi and Houston, 1968-69; research geologist Chevron Oil Field Research Co. div. Standard Oil Co. Calif., La Habra, 1969-74, sr. research geologist, 1974-77, sr. research assoc., 1977—. Founder Orange County Center for Performing Arts, Costa Mesa, Calif., Christ Coll., Irvine, Calif.; pres. Lutheran HIgh Sch. Assn. Orange County, Orange, Calif., 1977-79, v.p., 1979-81; v.p. Prince of Peace Luth. Ch., Anaheim, Calif., 1980—. Fellow Geol. Soc. Am.; mem. Am. Assn. Petroleum Geologists, Soc. Econ. Paleontologists and Mineralogists, Clay Minerals Soc. (councillor 1981-84), Los Angeles Basin Coast Geol. Soc., Microbeam Analysis Soc., Phi Beta Kappa, Sigma Xi. Republican. Subspecialties: Geochemistry; Petrology. Current work: Diagenesis of sandstones and shales; stable isotope geochemistry of water and authigenic minerals in sandstone; K/Ar age dating of authigenic illite in sandstone; geochemistry of $_{129}$I and $_{127}$I; stratigraphic correlation utilizing trace elements. Home: 721 E Harmony Ln Fullerton CA 92631 Office: Chevron Oil Field Research Co PO Box 446 La Habra CA 90631

BURTON, GLENN WILLARD, geneticist; b. Clatonia, Nebr., May 5, 1910; s. Joseph Fearn and Nellie (Rittenburg) B.; m. Helen Maurine Jeffryes, Dec. 16, 1934; children: Elizabeth Ann (Mrs. John Edward Fowler), Robert Glenn, Thomas Jeffryes, Joseph William, Richard Bennett. B.Sc., U. Nebr., 1932, D.Sc. (hon.), 1962; M.Sc., Rutgers U., 1933, Ph.D., 1936, D.Sc. (hon.), 1955. With U.S. Dept. Agr. and U. Ga. at Tifton Expt. Sta., 1936—, prin. geneticist, 1952—, chmn. div. agronomy, 1950-64; Univ. Found. prof. U. Ga., 1957. Mem. Tift County Bd. Edn., 1953-58. Recipient 1st ann. agrl. award So. Seedsmen Assn., 1950; Sears-Roebuck research award, 1953, 60; Superior Service award Dept. Agr., 1955; Disting. Service award, 1980; 1st Ford Almanac Crops and Soils Research award, 1962; named Man of Year in So. Agr. Progressive Farmer, 1954; numerous other awards and citations. Fellow Am. Soc. Agronomy (Stevenson award 1949, John Scott award 1957, v.p. 1961, pres. 1962); mem. Am. Genetic Assn., Am. Soc. Range Mgmt., Nat. Acad. Sci., Sigma Xi, Alpha Zeta, Gamma Sigma Delta. Subspecialties: Plant genetics; Plant physiology (agriculture). Current work: Genetic improvement and germplasm management of Cynodon spp; Paspalum spp. developing superior genetic and physiology research methods. Home: 421 W 10th St Tifton GA 31794

BURTON, HOWARD ALAN, neuropsychologist, computer company executive; b. Bklyn., May 9, 1951; s. Lawrence and Esther P. (Scondutto) B. B.A., SUNY, Stony Brook, 1973, M.A., 1974; Ph.D., St. Louis U., 1978. Lectr. Dowling Coll., Oakdale, N.Y., 1974; data processing cons. Immunology Researching Found., Great Neck, N.Y., 1975; research asst. sleep lab. St. Louise U., 1976; research asso. Md. Psychiat. Research Center, Balt., 1978-81; pres. Synclastic Communications, Inc., Syncom, Inc., Richmond Heights, Mo., 1982—. Contbr. articles to profl. jours. Active Boy Scouts Am., Eagle Scout. Recipient Order of Arrow. Mem. Soc. Neurosci., Brit. Brain Research Assn., European Brain and Behavior Soc. Subspecialties: Neuropsychology; Numerical analysis. Current work: Neural correlates of behavior/information processing and real time data capture and analysis.

BURTON, ROBERT CLYDE, geology educator; b. Borger, Tex., Feb. 27, 1929; s. Earl and Edith (Roman) B.; m. Betty Jean Hill, Oct. 6, 1951; children: Randall L., Roger E., Jana S., Jill E. B.A., Tex. Tech U., 1957, M.S., 1959; Ph.D., U. N.Mex., 1965. Instr. Tex. Tech U., Lubbock, 1958-59; vis. lectr. U. N.Mex., Albuquerque, 1964; prof. geology West Tex. State U., Canyon, 1959—; owner Geomag Surveys, Canyon, 1980—; dir. West Tex. State U. Geology Field Camp, Salida, Colo., 1970-83. Contbr. articles to profl. jours. Served as sgt. USAF, 1948-52. Named Favorite Prof. West Tex. State U., 1963. Mem. Nat. Assn. Geology Tchrs., Panhandle Geol. Soc., Pander Soc., Sigma Xi, Sigma Gamma Epsilon. Presbyterian. Subspecialties: Geology; Paleoecology. Current work: Biostratigraphy, micropaleontology—conodonts, geomagnetics. Office: West Tex State U Canyon TX 79016 Home: 711 Taylor Ln Canyon TX 79015

BURTON, WILLIAM BUTLER, astronomer; b. Richmond, Va., July 13, 1940; s. Joseph Ashby and Denison (Laws) B.; m. Judy Marie Johnson, Mar. 26, 1972; children: Hannah Marie, Benjamin Joseph, Molly Catherine. B.A., Swarthmore Coll., 1962; postgrad. (Fulbright scholar 1962-63, 63-64, Kovalenko scholar 1964), U. Leiden, Netherlands, 1965, Ph.D., 1970. Research asso. Nat. Radio Astromomy Obs., 1971-73, asst. scientist, then scientist, 1973-78; prof. astronomy, chmn. dept. U. Minn., 1978-81; prof. astronomy U. Leiden, 1981—. Mem. Am. Astron. Soc., Internat. Astron. Union, Netherlands Astron. Soc., Sigma Xi. Subspecialty: Radio and microwave astronomy. Current work: Observation radio astronomy with interest focussed on problems relating to galactie structure and the interstellar medium. Address: Sterrewacht U Leiden PO Box 9513 Leiden Netherlands

BURZYNSKI, STANISLAW RAJMUND, internist, cancer researcher; b. Lublin, Poland, Jan. 23, 1943; came to U.S., 1970, naturalized, 1977; s. Grzegorz and Zofia Miroslawa (Radzikowski) B.; m. Barbara Burzynski, Feb. 8, 1979; 1 son, Grzegorz Stanislaw. M.D. with distinction, Med. Acad., Lublin, 1967, Ph.D., 1968. Lic. Tex. State Bd. Med. Examiners, 1973. Teaching asst. Med. Acad., Lublin, 1962-67; intern, 1967-68, resident, 1969-70; research asso. Baylor Coll. Medicine, 1970-72, asst. prof., 1972-77; practice medicine specializing in internal medicine, Houston, 1977—; pres. Burzynski Research Inst., Houston, 1977—. Contbr. articles in field to profl. jours. Nat. Cancer Inst. grantee, 1974-77; West Found. grantee, 1975. Mem. AMA, AAAS, Am. Assn. Cancer Research, Harris County Med. Soc., Polish Nat. Alliance (pres. Houston chpt. 1974-75), Soc. Neuroscience, Tex. Med. Assn., Sigma Xi. Subspecialties: Internal medicine; Cancer research (medicine). Current work: Antineoplastons and their role in the treatment of neoplastic disease.

BUSCH, CHRISTOPHER WILLIAM, scientific company executive; b. Colton, Wash., Apr. 14, 1937; s. Christopher Matthias and Louise Clara (Grieser) B.; m. Dorothy Charvat, June 9, 1960; children: Therese, Joseph, Mark, Maria. B.S.M.E., Gonzaga U., 1960; M.S. M.E., U. Wis., 1962; Ph.D., U. Calif.-Berkeley, 1965. Asst. prof. mech. engring. Wash. State U., Pullman, 1965-66; sect. head TRW Systems, Redondo Beach, Calif., 1966-71; v.p. Sci. Applications, Inc., El Segundo, Calif., 1971-75; pres. Spectron Devel. Labs., Inc., Costa Mesa, Calif., 1975—. Mem. ASME, Am. Inst. Chem. Engrs. Republican. Roman Catholic. Subspecialties: Combustion processes; Aeronautical engineering. Current work: Combustion, aerodynamics, optical measurements, non-destructive testing, mfg. tech.

BUSCH, HARRIS, medical educator; b. Chgo., May 23, 1923; s. Maurice Ralph and Rose Lillian (Feigenholtz) B.; m. Rose Klora, June 16, 1945; children: Daniel Avery, Laura Anne Busch Smolkin, Gerald Irwin, Fredric Neal. B.S., U. Ill., 1944, M.D. with honors, 1946; M.S., U. Wis., 1950, Ph.D., 1952. Intern Cook County Hosp., Chgo., 1946-47; asst. surgeon, sr. asst. surgeon USPHS, 1947-49; postdoctoral fellow Nat. Cancer Inst., 1950-52; asst. prof. biochemistry, internal medicine Yale U., 1952-55; asso. prof., prof. pharmacology U. Ill., 1955-60; prof. biochemistry, chmn. dept. Baylor U. Coll. Medicine, 1960-62, prof. pharmacology, chmn. dept., 1960—, disting. service prof., 1978—, chmn. student promotions com., 1969—, mem. policy planning com., 1972—, dir. Cancer Research Center; vis. prof. U. Chgo., 1968, 71, Northwestern U., 1968, Ga. Med. Coll., 1971, Washington U., St. Louis, 1972, U. Ala., Birmingham, 1972, Ind. U., Indpls., 1972, U. Nev., Reno, 1978, U. Colo., Denver, 1980; cons. lectr. U. Tenn., U. Tex., San Antonio, 1971; disting. lectr. SUNY, Buffalo, 1977; Centennial lectr. U. Ill. Coll. Medicine, 1981; cons. VA, Meth. hosps., both Houston, Bristol-Myers.; mem. adv. com. cell and devel. biology Am. Cancer Soc., 1978—; cancer chemotherapy study sect. USPHS; mem. Nat. Cancer Planning Com., 1971; mem. bd. sci. counselors to div. cancer treatment Nat. Cancer Inst., 1975. Author: Chemistry of Pancreatic Diseases, 1959, An Introduction to the Biochemistry of the Cancer Cell, 1962, Histones and Other Nuclear Proteins, 1965; co-author: Chemotherapy, 1966, The Nucleolus, 1970; editor: Frontiers in Medical Biochemistry, 1962, The Nucleus of the Cancer Cell, 1963, Jour. Phys. Chemistry and Physics, Methods in Cancer Research, vol. I, 1966, vols. II and III, 1967, vol. IV, 1968, Methods in Cancer Research, vol. V, 1970, vol.VI, 1971, vols. VII-IX, 1973, vol. X, 1973, Methods in Cancer Research, vol. XI, 1975, vols. XII and XIII, 1976, Methods in Cancer Research, vols. XIV, XV, 1978, Molecular Biology of Cancer, 1974, Cell Nucleus, Vols. I-III, 1974, IV-VII, 1978, VIII-IX, 1980; editorial bd.: Jour. Cancer Research and Clin. Oncology, Jour. Biol Chemistry, Cancer Investigation, New Drugs, Physiol. Chemistry, Phys. Life Scis. Recipient Outstanding Alumnus award for service to edn. and research U. Ill., 1977, Disting. Faculty award Baylor U. Coll. Medicine, 1982; Baldwin scholar oncology Yale U. Sch. Medicine, 1952-55; scholar cancer research Am. Cancer Soc., 1955. Mem. Am. Soc. Biol. Chemists, Am. Assn. Cancer Research (public issues com. 1977), Am. Chem. Soc., Soc. Pharmacology and Exptl. Therapeutics, Soc. Exptl. Biology and Medicine, Sigma Xi, Alpha Omega Alpha. Subspecialties: Cell study oncology; Pharmacology. Current work: The nucleus of the cancer cell. Home: 4966 Dumfries Dr Houston TX 77096

BUSCH, ROBERT HENRY, research geneticist, educator; b. Jefferson, Iowa, Oct. 22, 1937; s. Henry and Lena Margret (Osterman) B.; m. Mavis Ann Bushman, Nov. 23, 1958; children: Shari Lynne, Todd William. B.Sc., Iowa State U., Ames, 1959, M.Sc., 1963; Ph.D., Purdue U., 1967. Research assoc. Iowa State U., Ames, 1961-63; grad. research asst. Purdue U., West Lafayette, Ind., 1964-66, grad. research instr., 1966-67; from asst. prof. to prof. N.D. State U., Fargo, 1967-78; research geneticist, prof. U.S. Dept. Agr., Agrl. Research Service, U. Minn., St. Paul, 1978—; cons. Nat. Crop Hail Ins. Council, 1976-77. Author: (with W. Kranstad) Wheat in the Peoples' Republic of China, 1977. Research grantee Nat. Crop Hail Ins. Council, 1969-72, U.S. Dept. Agr., 1977—, Minn. Wheat Council, 1978—. Mem. Crop Sci. Soc. Am. (assoc. editor jour. 1976-78), Am. Soc. Agronomy, Sigma Xi, Phi Kappa Phi, Gamma Sigma Delta. Subspecialties: Plant genetics; Gene actions. Current work: Plant breeding, quantitative genetics, breeding methods, disease resistance. Home: 2485 Galtier Circle St Paul MN 55113 Office: Dept Agronomy and Plant Genetics U Minn St Paul MN 55108

BUSCHE, ROBERT MARION, chemical engineer; b. St. Louis, June 14, 1926; s. Ferdinand and Irma (Seim) B.; m. Norma Jean Nickles, Sept. 17, 1950 (div.); children: Robert Eric, David Clay, Kristin Anne, Amy Ellen; m. Emma Elizabeth Ruch, June 21, 1980. B.S. in Chem. Engring, Washington U., St. Louis, 1948, M.S., 1949, D.Sc., 1952; A.S., Oreg. State U., 1945. Process engr. U.S. Bur. Mines, Louisiana, Mo., 1950-53; asst. tech. supt., synthesis gas mfg. E.I. DuPont de Nemours & Co. Inc., Charleston, W. Va., 1953-56; research supr. Sabine River Works, Orange, Tex., 1956-59, Polyolefin Products Exptl. Sta., Wilmington, Del., 1959-63; tech. service supr. Polyolefin Mktg., Chestnut Run Labs., Wilmington, 1963-66; tech. mgr. heat transfer products div. Willow Bank Plant, Wilmington, 1969-74; planning cons. life scis., biotech. & bioenergy resources, central research & devel. dept. DuPont Co. Exptl. Sta., Wilmington, 1974—; adj. prof. chem. engring. U. Pa., Phila., 1983. Contbr. articles to profl. jours.; mem. editorial bd.: Handbook Series in Biosolar Resources, 1978-83. Elder, deacon Presbyterian Ch.; active Boy Scouts Am., 1959—. Recipient Order of Merit Boy Scouts Am., 1970, Silver Beaver, 1974. Mem. Am. Inst. Chem. Engrs., Am. Chem. Soc. (bio-energy council), Soc. Plastics Engrs., Inst. Food Technologists, Phi Eta Sigma, Tau Beta Pi, Sigma Xi, Tau Kappa Epsilon. Republican. Subspecialties: Chemical engineering; Biomass (energy science and technology). Current work: Chemicals from renewable resources via bio conversion; fermentation research planning, engineering evaluations, new separation processes, business development, market research on chemicals and specialty products. Patentee in biotech. field. Home: 2205 Pennington Dr Wilmington DE 19810 Office: Central Research & Devel Dept E I DuPont De Nemours & Co Wilmington DE 19898

BUSECK, PETER R., geochemistry educator; b. Sept. 30, 1935; s. Paul M. and Edith G. (Stern) B.; m. Alice E. Buseck, June 20, 1960; children: Lori, David, Susan, Paul. A.B., Antioch Coll., 1957; M.A., Columbia U., 1959, Ph.D., 1962. Fellow geophys. Lab. Carnegie Inst., Washington, 1961-63; with depts. chemistry and geology Ariz. State U., 1963—; now prof., vis. prof. dept. geology Oxford U., 1970-71, Stanford U., 1979-80. Contbr. articles to profl. jours. NSF fellow, 1970-71; Overseas fellow Churchill Coll., Cambridge U., 1979-80; recipient Corning award, 1975; JEOL award Microbeam Analysis Soc., 1981. Fellow AAAS, Geol. Soc. Am., Mineral. Soc. Am.; mem. Am. Geophys. Union, Geochem. Soc., Microbeam Soc., Soc. Econ. Geologists, Can. Mineral Soc., Air Pollution Control Assn., Am. Assn. Aerosol Research, Electron Microscope Soc. Am. Subspecialties: Geochemistry; Mineralogy. Current work: Solid state geochemistry, mineralogy, electron microscopy, meteoritics, atmospheric chemistry environmental geochemistry/air pollution, geochemical monitoring of active volcanism and geothermal activity. Office: Dept Chemistry Arizona State Tempe AZ 85287

BUSH, C. ALLEN, chemist; b. Rochester, N.Y., Aug. 2, 1938; s. C. Allen and Helen (Roberts) B.; m. Luise A. Graff. Ph.D., U. Calif.-Berkeley, 1965. Prof. chemistry Ill. Inst. Tech., Chgo., 1980—. Subspecialties: Biophysical chemistry; Nuclear magnetic resonance (chemistry). Address: Dept Chemistry Ill Inst Tech 3300 S Federal St Chicago IL 60616

BUSH, DAVID FREDERIC, psychology educator, researcher, consultant; b. Watertown, N.Y., July 12, 1942; s. Frederic R. and Charlotte M. (Ellingworth) B. B.A., U. South Fla., 1965; M.A., U. Wyo., 1968; Ph.D., Purdue U., 1972. Mgmt. trainee Pan Am. World Airways, Cocoa Beach, Fla., 1964-65; instr. Hiram Scott Coll., Scottsbluff, Nebr., 1967-69; assoc. dir. human orgn. sci. Villanova (Pa.) U., 1981—, assoc. prof. psychogy, 1972—; cons. Delaware Valley Transplant, Phila., 1979—; dir. Bush Assocs., Broomall, Pa., 1980—, Life Guidance Services, Broomall, 1977—; cons. MidAtlantic Research Inst., Bethesda, Md., 1975-78. Co-editor: Straight Talk, 1975, Communication in the Consultation, 1982; editor: Human Development, 1975; assoc. editor: Social Sci. Rev, 1977—. Mem. Phila. Orch. Assn., 1976—, People's Light & Theater Co., Malvern, Pa., 1982, Am. Lung Assn., Norristown, Pa., 1982. Recipient Disting. Service award Hiram Scott Coll., 1969; David Ross fellow, 1972; NDEA fellow, 1969-71; E-SU grantee Oxford U., summer 1980. Mem. Internat. Communication Assn. (div. membership chmn. 1980—), Am. Psychol. Assn., Acad. Mgmt., Internat. Assn. Applied Psychology, Midwestern Psychol. Assn. (local rep. 1981—), Phi Kappa Phi (chpt. pres. 1982-83), English-Speaking Union. Club: Phila. Masters Track. Subspecialties: Social psychology; Health psychology. Current work: Gender differences in nonverbal communication; effects of gender differences in speaker on listener's ability to recall information; influences on memory for medical information and health-related decision making. Home: 53 S Greenhill Rd Broomall PA 19008 Office: Villanova U Villanova PA 19085

BUSH, DAVID LYNN, plant breeder, pathologist; b. San Antonio, Oct. 30, 1947; s. Jerry and Mary Elizibeth (Pyron) B.; m. Joy Sue Adams, July 28, 1968. B.S. in Microbiology, Tex. A&M U., 1972, M.S. in Plant Pathology, 1973, Ph.D., 1980. Cert. plant breeder, Tex. Research asst. Tex. A&M U., College Station, 1972-73, research asso., 1973-78; dir. research Custon AG Service, Inc., Loraine, Tex., 1978—. Contbr. articles to profl. jours. Served with U.S. Army, 1970-76. Mem. Am. Soc. Agronomy, Am. Phytopathol. Soc., Crop Sci. Soc. Am., Tex. Seed Trade Assn. (dir. cotton div.). Republican. Baptist. Lodge: Lions. Subspecialties: Plant pathology; Plant genetics. Current work: Development of disease and insect resistant cotton and soybean cultivars with emphasos on resistance to seedling diseases. Home: C E 23 Route 1 Sweetwater TX 79556 Office: PO Box 97 Loraine TX 79532

BUSH, FRANCIS MARION, dental educator; b. Bloomfield, Ky., Sept. 5, 1933; s. Francis Marion and Emily Mae (Arnold) B.; m. Madge Morgan, Dec. 5, 1959. B.S., U. Ky., 1955, M.S., 1957; Ph.D., U. Ga., 1962; D.M.D., U. Ky., 1975. Faculty U. Ky., Lexington, 1955-57, U. Ga., Athens, 1957-61, Samford U., Birmingham, Ala., 1961-64; faculty Med. Coll. Va., Richmond, 1964—, assoc. prof. dentistry, 1975—. Contbr. articles to profl. jours. Served with USNG, 1953-61. Named Outstanding Dental Sch. Tchr. Med. Coll. Va. Dental Sch., 1979, 80. Mem. Am. Equilibration Soc., Internat. Assn. Dental Research, Am. Assn. Dental Schs., Am. Assn. Dental Research, Omicron Kappa Upsilon. Subspecialty: Dental education. Current work: Facial pain, temporomanidibular joint dysfunction, occlusion. Home: 2350 Castlebridge Rd Midlothian VA 23113 Office: Med Coll Va Station PO Box 566 Richmond VA 23298

BUSH, GEORGE EDWARD, physicist, cons. B.S. in Physics, Purdue U., West Lafayette, Ind., 1960; M.S. in Cybernetic Systems, San Jose (Calif.) State Coll., 1975. Physicist Midwestern Univs. Research Assn., Madison, 1960-66, Lawrence Livermore (Calif.) Nat. Lab., 1966—; founder Solar Systems, Inc. Contbr. to: Solar Technology Handbook, 1980. Mem. AAAS, Internat. Solar Energy Soc. Subspecialties: Solar energy; Computer engineering. Current work: Man-machine interface, small computers, data acquisition systems, instrumentation development, work station development, graphics applications, databases. Office: Box 931 Livermore CA 94550

BUSH, RAYMOND SYDNEY, physician; b. Toronto, Ont., Can., Apr. 11, 1931; s. Raymond E. and Alice T. (Hampson) B.; m. Margaret J. Bush, Sept. 5, 1959; children—Catherine, Elizabeth, Jennifer. B.Sc., U. London, 1951; M.D., U. Toronto, 1961, M.A., 1964. Physicist No. Electric, 1952-53, Can. AEC, 1953-55; resident Toronto East Gen. Hosp., 1961-62, Princess Margaret Hosp., Toronto, 1962-64, radiation oncologist, 1966—; dir. Ont. Cancer Inst. (incorporating Princess Margaret Hosp.), 1976—; resident Victoria Gen. Hosp., Halifax, N.S., Ca., 1964-65; Richard's fellow, Ont. Cancer Soc. awardee Christie Hosp., Manchester, Eng., other centers, 1965-66; prof. radiology U. Toronto, 1976—, asso. prof. med. biophysics, 1976—; vis. prof. U. London, 1974, U. Manchester, 1974; Lichfield lectr. Oxford (Eng.) U., 1974. Author: Malignancies of the Ovary, Uterus and Cervix, 1979; mem. editorial bd.: Cancer Nursing, 1978—, Internat. Jour. Radiation Oncology, 1977, Clin. and Investigative Medicine, 1979—. Mem. Acad. Medicine (Toronto), Can. Assn. Radiologists (editorial bd.), Can. Oncology Soc., Nat. Cancer Inst. Can., Can. Cancer Soc., Royal Coll. Physicians and Surgeons Can., Am. Soc. Therapeutic Radiologists, Am. Soc. Clin. Oncology. Subspecialties: Radiology; Oncology. Office: 500 Sherbourne Toronto ON M4X 1K9 Canada

BUSH, SPENCER HARRISON, metallurgist; b. Flint, Mich., Apr. 4, 1920; s. Edward Charles and Rachel Beatrice (Roser) B.; m. Roberta Lee Warren, Aug. 28, 1948; children: David Spencer, Carl Edward. Student, Flint Jr. Coll., 1938-40, Ohio State U., 1943-44, U. Mich., 1946-53. Registered profl. engr., Calif. Asst. chemist Dow Chem. Co., 1940-42, 46; asso. Engrng. Research Inst., U. Mich., 1947-53; research asst. Office Naval Research, 1950-53, instr. dental materials, 1951-53; metallurgist Hanford Atomic Products Operation, Gen. Electric Co., 1953-54, supr. phys. metallurgy, 1954-57, supr. fuels fabrication devel., 1957-60, metall. specialist, 1960-63, cons. metallurgist, 1963-65; cons. to dir. Battelle N.W. Labs., Richland, Wash., 1965-70, sr. staff cons., 1970-83; pres. Rev. & Synthesis Assocs., cons., 1983—; lectr. metall. engrng. Center for Grad. Study, U. Wash., 1953-67, affiliate prof., 1967—; chmn., com. study group on pressure vessel materials Electric Power Research Inst., 1974—; cons. U. Calif. Lawrence Berkeley Labs., 1975—; chmn. com. on reactor safeguards U.S. AEC, 1971; mem. Wash. Bd. Boiler Rules, 1972—; Gillett lectr. ASTM, 1975; Mehl lectr., 1981. Contbr. tech. articles to profl. jours. Served with U.S. Army, 1942-46. Recipient Silver Beaver award Boy Scouts Am.; Am. Foundrymens Soc. fellow, 1948-50; Regents prof. U. Calif., Berkeley, 1973-74. Fellow Am. Nuclear Soc. (adv. editorial bd. nuclear applications 1965-77), Am. Soc. Metals (chmn. program council 1966-67, trustee 1967-69, chmn. fellow com. 1968), ASME (Langer award 1983); mem. AIME (chmn. ann. seminar com. 1967-68), ASTM, Nat. Acad. Engrng., Sigma Xi, Tau Beta Pi, Phi Kappa Phi. Subspecialties: Metallurgical engineering; Nuclear engineering. Current work: Reactor safety, pressure boundary codes and standards, nondestructive examination, fracture mechanics. Home: 630 Cedar Ave Richland WA 99352 Office: PO Box 999 Richland WA 99352:

BUSHAW, THOMAS HENRY, chemist; b. Pullman, Wash., Sept. 9, 1953; s. Donald Wayne and Sylvia Ruth (Lybecker) B. B.A., Carleton Coll., 1975; Ph.D., Purdue U., 1979. With Mobil Exploration and Prodn. Research Lab., Mobil Research & Devel. Corp., Dallas, 1979-82, research chemist, 1982—. Contbr. articles to profl. jours. Mem. Am. Chem. Soc. Democrat. Subspecialties: Analytical chemistry. Current work: Inductively coupled plasma emission spectroscopy; fluorescence spectroscopy; laboratory automation and computer applications to chemistry.

BUSHNELL, ROBERT HEMPSTEAD, solar energy consultant, engineer; b. Wooster, Ohio, May 11, 1924; s. John and Dyllone (Hempstead) B.; m. Martha W. Dicks, Oct. 2, 1965; children: Helen, Orson. B.Sc. in Engrng. Physics, Ohio State U., 1946, M.Sc. in Physics, 1947; Ph.D. in Meteorology (Univ. Corp. Atmospheric Research fellow), U. Wis., 1962. Registered profl. engr., Colo. Physicist Hoover Co., Ohio, 1948-50; engr. Goodyear Aircraft Corp., Ohio, 1950-56, RCA, N.J., 1957-58; research meteorologist Nat. Center Atmospheric Research, Boulder, Colo., 1962-74; cons. solar energy, Boulder, 1974—. Contbr. articles to profl. jours. Served with USNR, 1944-46. Mem. Am. Meteorol. Soc., ASHRAE, Internat. Solar Energy Soc., AAAS, U.S. Metric Assn. (treas. 1975-79). Subspecialties: Climatology; Solar energy. Current work: Climatology of solar irradiation and temperature, response of bldgs. to climate, distbn. of intensity of solar irradiation. Home and Office: 502 Ord Dr Boulder CO 80303

BUSHNELL, WILLIAM RODGERS, plant physiologist, educator; b. Wooster, Ohio, Aug. 19, 1931; s. John and Dyllone (Hempstead) B.; m. Ann Holcomb, Sept. 20, 1952; children: Thomas, John, Mary. B.A., U. Chgo., 1951; B.S., Ohio State U., 1953, M.S., 1955; Ph.D., U. Wis., 1960. Research plant physiologist Agrl. Research Service, U.S. Dept. Agr., St. Paul, 1960—; asst. prof. dept. plant pathology U. Minn., 1966-72, assoc. prof., 1972-73, adj. prof., 1973—. Contbr. articles in field to profl. jours. NSF fellow, 1957-60; NSF grantee, 1964, 66. Mem. Am. Phytopath. Soc., Am. Soc. Plant Physiologists, AAAS. Subspecialties: Plant physiology (agriculture); Plant pathology. Current work: Research on physiology of rust and powdery mildew diseases of cereal plants. Office: University of Minnesota Cereal Rust Lab Saint Paul MN 55108

BUSIS, NEIL AMDUR, neurologist, neurobiologist; b. Pitts., Mar. 4, 1951; s. Sidney Nahum and Sylvia (Amdur) B.; m. Cynthia Dickter, May 24, 1981. B.A. summa cum laude (Nat. Merit scholar), Yale Coll., 1973; M.D., U. Pa., 1977. Diplomate: Nat. Bd. Med. Examiners. Student researcher Lab. Neuropharmacology, NIMH, Washington, summers 1974, 75, Mar. 1976; intern and asst. resident in medicine Johns Hopkins Hosp., Balt., 1977-79; research assoc. Lab. Biochem. Genetics, Nat. Heart, Lung and Blood Inst., NIH, 1979-81; resident dept. neurology Mass. Gen. Hosp., Boston, 1981—. Contbr. articles to profl. jours. Served with USPHS, 1979-81. Recipient Mosby Book award U. Pa. Sch. Medicine. Mem. Am. Acad. Neurology, AAAS, Soc. Neurosci., Amyotrophic Lateral Sclerosis Soc. Am., Phi Beta Kappa, Alpha Omega Alpha. Subspecialties: Neurobiology; Neurology. Current work: Nervous system development and neuronal communication. Office: Mass Gen Hosp Dept Neurology Boston MA 02114

BUSKE, NORMAN L., consulting scientist, energy researcher; b. Milw., Oct. 11, 1943; s. Gilbert and Genevieve (Strutt) B.; m. Patricia Teller, June 10, 1965; children: Heather, Alisyn, Robin; m. Linda S. Josephson, Aug. 25, 1980. B.A. in Physics, U. Conn., 1964, M.A., 1965; M.S. in Oceanography, Johns Hopkins U., 1967. Oceanographer Ocean Sci. & Engrng., Inc., Rockville, Md., 1968-71; prin. Sea-Test Co., Laie, Hawaii, 1972-76; sr. scientist/engr. Van Gulik & Assocs., Lake Oswego, Oreg., 1976-77; dir. research Pacific Engrng. Corp., Portland, Oreg., 1977-78; prin. Search Technical Services, Davenport, Wash., 1978—; cons. Physicians for Social Responsibility, Portland, Oreg., 1980-82. Inventor internal pressure engine. Mem. ASME, ASTM, IEEE, N.Y. Acad. Scis., Soc. Automotive Engrs. Current work: Development of methodologies for fire and accident investigations; energy and nuclear weapons policy; development of low-pressure Brayton expanders for solar energy conversion, and non-lubricated pressure seals. Home and Office: Star Route Box 61 Davenport WA 99122

BUSS, EDWARD GEORGE, geneticist; b. Concordia, Kans., Aug. 28, 1921; s. George Edward and Kathryn (Luginsl) B.; m. Dorothy Ruth; children: Ellen, Norman. B.S., Kans. State Coll., 1943; M.S., Purdue U., 1949, Ph.D., 1956. Research asst. Purdue U., West Lafayette, Ind., 1946-49, instr. poultry sci., 1955-56; asst. prof. Colo. State U., Fort Collins, 1949-55; acting head dept., 1950-55; assoc. prof. poultry sci. Pa State U., State College, 1956-65, prof., 1965—. Served to capt. inf. U.S. Army, 1943-46. NIH research grantee, 1960-79. Mem. AAAS, Am. Inst. Biol. Scis., Am. Genetic Assn., Am. Soc. Zoologists, Genetics Soc. Am., Poultry Sci. Assn., Soc. for Study Reproduction, Phi Eta Sigma. Democrat. Unitarian. Subspecialties: Animal genetics; Reproductive biology. Current work: Parthenogenesis in turkeys, biochem. nature of difference between normal and mutant individuals and between desired and undesired individuals for quantitative traits. Home: 1420 S Garner St State College PA 16801 Office: Pennsylvania State University 2 Agrl Engring Bldg University Park PA 16802

BUSSARD, ROBERT WILLIAM, physicist; b. Washington, Aug. 11, 1928; s. Marcel Julian and Elsa Mathilda (Griesser) B.; m. Dolly H. Gray, 1981; children: Elise Marie Bussard Bright, William Julian, Robert Lee, Virginia Lesley. B.S. in Engring., UCLA, 1950, M.S. in Engring., 1952; A.M. in Physics, Princeton U., 1959, Ph.D. in Physics, 1961. Design engr. Falcon program Hughes Aircraft Co., 1949-51; mech. engr. aircraft nuclear propulsion project Oak Ridge Nat. Lab., 1952-55; alt. group leader nuclear rocket program Los Alamos Sci. Lab., 1955-62, alt. leader laser div., 1971-73; dir. nuclear systems staff, asst. dir. mechanics div. Space Tech. Labs., Thompson-Ramo-Wooldridge, Inc., Redondo Beach, Calif., 1962-64; assoc. mgr. research and engring., corp. chief scientist Electro-Optical Systems div. Xerox Corp., Pasadena, Calif., 1964-69; with CSI Corp., Los Angeles, 1969-70; mgr. Cherokee Assos., Pasadena, Md., 1970-74; asst. dir. div. controlled thermonuclear research U.S. AEC, Washington, 1973-74; founder, pres., chmn. Energy Resources Group, (ERG), Inc., La Jolla, Calif., Alexandria, Va., 1974; Internat. Nuclear Energy Systems Co. (INESCO), Inc., La Jolla and McLean, 1976—; cons. NATO, 1960-64, U.S. Dept. Energy, 1974-78; lectr. UCLA, 1960-69, U. Fla., 1962-64. Author: (with R.D. DeLauer) Nuclear Rocket Propulsion, 1958; author: Fundamentals of Nuclear Flight, 1965; editor: Nuclear Thermal and Electric Rocket Propulsion, 1967; contbr. articles to profl. jours. Fellow AIAA; mem. Am. Phys. Soc., Internat. Acad. Astronautics. Clubs: Princeton (N.Y.C.); Cosmos, Capitol Hill (Washington). Subspecialties: Fusion; Aerospace engineering and technology. Current work: Fusion power development; space, propulsion; space weapons. Patentee space nuclear propulsion, power generation, fusion and fission power, solar power systems. Office: 11077 N Torrey Pines Rd La Jolla CA 92037 The future is constructed in a fashion and to a scale envisioned by those who perceive what it might be, and who work to make these visions happen. At any one time, probably no more than a few thousand dedicated people are actively working to shape the world of tomorrow from the tools, techniques, and ideas of today. I like to think that I have tried to spend my life in this manner, always with the goals of improving the lot of man, and of assisting in ensuring the survival and growth of my people and my country, in order that the freedom of men, as it flourishes under democracy, might be preserved and extended for future generations.

BUSSE, EWALD WILLIAM, physician, educator, researcher; b. St. Louis, Aug. 18, 1917; s. Frederich Eugene and Emily Louise (Stroh) B.; m. Ortrude Helen Schnaedelbach, July 18, 1941; children: Ortrude, Barbara, Richard, Emily. A.B., Westminster Coll., 1938, Sc.D. (hon.), 1960; M.D., Washington U., St. Louis, 1942. Diplomate: Am. Bd. Psychiatry and Neurology. Intern in neurology St. Louis City Hosp., 1942-43; resident U. Colo. Hosps, Denver, 1946-48; instr. to prof. U. Colo., 1946-53; head psychosomatic medicine and EEG Colo. Gen. Hosp., 1946-53; prof. Duke U. Med. Ctr., Durham, N.C., 1953—, chmn. dept. psychiatry, 1953-74, dir. Ctr. for Study of Aging, 1957-70, dean Sch. Medicine, 1974-82, assoc. provost, 1974-82, J.P. Gibbons prof. psychiatry, 1953—, dean emeritus, 1982—; chmn. com. on geriatrics and gerontology VA, 1981-82; chmn. program com. Nat. Council on Aging, 1981—; mem. Pres.'s Biomed. Research Panel. Author: (with D.G. Blazer) Handbook of Geriatric Psychiatry, 1980, Cerebral Manifestations of Episodic Cardiac Dysrhythmics, 1979. Served to maj. M.C. AUS, 1943-46. Decorated Meritorious Service award; recipient award Brookdale Found., 1982; Salmon award N.Y. Acad. Medicine, 1980; Freeman award Gerontol. Soc. Am., 1978; William C. Menninger award ACP, 1971; others. Mem. Am. Psychiat. Assn. (pres. 1971-72, chmn. ethics com.), World Psychiat. Assn. (mem. ethics com.), Internat. Congress of Gerontology (pres. 1985), Internat. Assn. Gerontology (pres. 1983), Am. Geriatrics Soc. (pres. 1975-76), Geront. Soc. Am. (pres. 1967-68). Presbyterian. Clubs: Hope Valley (Durham); Beech Mountain (Banner Elk, N.C.). Lodge: Masons. Subspecialties: Gerontology; Neuropsychology. Current work: Brain changes with aging-impact on behavior, learning, and memory. EEG changes with age—normal and pathologic. Neuropeptides and neurotransmitters in late life. Home: 1132 Woodburn Rd Durham NC 27705 Office: Duke Univ Medical Center PO Box 2948 Durham NC 27710

BUSSE, WILLIAM WALTER, immunology educator, researcher; b. Sheboygan, Wis., Jan. 20, 1941; s. Walter and Bernice B.; m.; 2 children. B.S., U. Wis.-Madison, 1962, M.D., 1966. Diplomate: Am. Bd. Internal Medicine, Am. Bd. Allergy and Clin. Immunology. Intern Cin. Gen. Hosp., 1966-67; resident in internal medicine U. Wis. Hosp., Madison, 1967-68, 70-71; research fellow in medicine, allergy and clin. immunology U. Wis.-Madison, 1971-73; pvt. practice medicine, Mpls., 1973-74 and; clin. instr. U. Minn. Med. Sch., Mpls., 1973-74; asst. prof. medicine U. Wis. Med. Sch., Madison, 1974-78, assoc. prof. and head sect. allergy and clin. immunology, dept. medicine, 1978—. Contbr. chpts. to books, articles to profl. jours. Served to maj. M.C. U.S. Army, 1968-70. Fellow Am. Acad. Allergy, ACP; mem. Am. Assn. Immunologists, Am. Fedn. Clin. Research, Am. Thoracic Soc., Central Soc. Clin. Research. Subspecialties: Allergy; Immunology (medicine). Home: 5510 S Hill Dr Madison WI 53705 Office: 600 Highland Ave Madison WI 53792

BUSSGANG, JULIAN JAKOB, electronic engr.; b. Lwow, Poland, Mar. 26, 1925; came to U.S., 1949, naturalized, 1954; s. Joseph and Stephanie (Philipp) B.; m. Fay Rita Vogel, Aug. 9, 1960; children—Jessica Edith, Julia Claire, Jeffrey Joseph. B.Sc., U. London, 1949; S.M. in Elec. Engring, M.I.T., 1951; Ph.D. in Applied Physics, Harvard U., 1955. Registered profl. engr., Mass. Mem. tech. staff Lincoln Lab., M.I.T., Lexington, 1951-55; mgr. applied research RCA, Burlington, Mass., 1955-62; pres. Signatron, Inc., Lexington, 1962—; vis. lectr. Harvard U., 1964; lectr. Northeastern U., Boston, 1962-65; mem. Mass. del. White House Conf. on Small Bus., 1980. Assoc. editor: Radio Sci, 1976-78; contbr. chpts. to books, also articles. Mem. Town Meeting, Lexington, 1975—; mem. alumni council M.I.T., 1965-72. Served with Free Polish Forces, 1942-46. Fellow IEEE; mem. Research Mgmt. Assn., Smaller Bus. Assn. New Eng., Am. Assn. Small Research Cos. Subspecialties: Electronics; Systems engineering. Current work: Communication theory; information theory. Patentee in field. Office: 12 Hartwell Ave Lexington MA 02173 I was a child-refugee, an adolescent-soldier, a student-immigrant, a young engineer and an adult entrepreneur. In every phase of my life I was blessed with the friendship and support of many wonderful people from various walks of life. Even in the darkest moments I had faith that each of us could improve the world a little.

BUTCHER, FRED RAY, biochemist, educator; b. Rochester, Pa., Aug. 11, 1943; s. Goble S. and Monnie (Gibson) B.; m. Letty J. Lytton, July 16, 1965; children: Allen R., Amy J. B.S., Ohio State U., 1965, Ph.D., 1969. Postdoctoral fellow U. Wis., Madison, 1969-71; asst. prof. Brown U., Providence, R.I., 1971-75, assoc. prof., 1975-78; prof. W. Va. U., Morgantown, 1978—, chmn. dept. biochemistry, 1981—. Contbr. articles to sci. publs. NIH research grantee, 1974—. Mem. Am. Soc. Biol. Chemists. Subspecialties: Biochemistry (medicine); Endocrinology. Current work: Mechanism of intracellular action of cyclic nucleotides and calcium in secretion, hepatic carbohydrate metabolism and differentiation. Home: Rural Route 1 Box 242 Independence WV 26374 Office: W Va U Med Center Morgantown WV 26506

BUTCHER, HARVEY RAYMOND, III, astronomer; b. Salem, Mass., Aug. 3, 1947; s. Harvey Raymond, Jr. and Marilyn (Corning) B.; m. Phillipa Ruth Newton, Apr. 21, 1947; children: Jeremy Robert, Christopher Thomas. B.S., Calif. Inst. Tech., 1969; Ph.D., Australian Nat. U., Canberra, 1974. Research asst. infrared astronomy group Calif. Inst. Tech., Pasadena, 1967-69; research scholar Mt. Stromlo Obs., Australian Nat. U., Canberra, 1970-74; Bart Bok fellow Steward Obs., U. Ariz., Tucson, 1975-76; asst. astronomer Kitt Peak Nat. Obs., Tucson, 1976-79, asso. astronomer, 1979-81, astronomer, 1981-83; prof., dir. Kapteyn Obs. U. Groningen (Netherlands), 1983—. Mem. Am. Astron. Soc., Royal Astron. Soc. Subspecialty: Optical astronomy. Current work: Observational research on evolution of galaxies and on physics of non-thermal radio sources; devel. of advanced instrumental techs. Office: Postbus 800 9700 Ave Groninger Netherlands

BUTCHER, JAMES WALTER, biologist; b. Pa., Feb. 14, 1917; s. Louis and Mary B.; m. Mary Katharine Culley, June 18, 1944; children: Craig, Mary Helen. B.S., U. Pitts., 1943; M.S., U. Minn., 1949, Ph.D., 1951. With Gulf Research Devel. Corp., 1948, Dept. Agr., 1950-52, Minn. Dept. Agr., 1952-57; mem. faculty Mich. State U., East Lansing, 1957—, prof. biology, 1965—, asst., asso. dean research, 1969-74, chmn. dept. zoology, 1974-81, prof., chmn. emeritus, 1981—, acting dean, 1973; cons. in field. Author papers in field, rev. articles. Served with USAAF, 1943-47. Fulbright sr. research scholar U. Vienna, 1966-67; grantee fed. and state govts., also industry. Mem. Ecol. Soc. Am., Am. Zool. Soc., Entomol. Soc. Am., Phi Beta Kappa, Sigma Xi. Subspecialty: Entomology. Home: 1002 Aragon Saint Augustine FL 32086 Office: 203 Natural Sci Bldg Mich State Univ East Lansing MI 48824

BUTCHER, JOHN EDWARD, animal scientist, educator, researcher; b. Belle Fourche, S.D., Aug. 4, 1923; s. James E. and Eva L. (Kirk) B.; m. Virginia O. Butcher, Apr. 28, 1951; children: Joan, Jean, James. B.S., Mont. State U., 1950, M.S., 1952; Ph.D., Utah State U., 1956. Ranch mgr. Mont. State U., 1949-50, mem. coop. extension staff, 1952, mem. range mgmt. faculty, 1952-53; mem. faculty Utah State U., 1955—; asst. prof. animal, dairy and vet. sci., 1955-59, assoc. prof., 1959-67, prof., 1967—; cons. Office Internat. Cooperation and Devel. Contbr. articles on beef cattle and sheep nutrition to profl. jours., 1953—. Mem. Nat. Land Adv. Council, 1982. Served with U.S. Army,

1946, 51. Named Utah Citizen of Yr. Utah Cattlemen's Assn., 1972; NSF fellow, 1954-55, 63; Rockefeller Found. fellow, Mexico, 1965-72. Fellow Am. Soc. Animal Sci., AAAS; mem. Am. Soc. Farm Mgrs. and Rural Appraisers, Am. Registry Cert. Animal Scientists, Soc. Range Mgmt., Am. Inst. Biol. Scis., Council Agrl. Sci. and Tech. Presbyterian. Lodge: Masons. Subspecialties: Animal nutrition; Integrated systems modelling and engineering. Current work: Beef cattle and sheep, from production to consumption; dietary phosphorus requirements of cows; integrated reproductive management on beef and sheep. Home: 1703 E 1030 N Logan UT 84321 Office: Utah State UUMC 48 Logan UT 84322

BUTLER, ANN BENEDICT, neurobiologist; b. Wilmington, Del., Dec. 2, 1945; d. Thomas H. and Arlene C. (Johnson) Benedict; m. Thomas P. Butler, June 8, 1968; 1 dau., Whitney Elizabeth. B.A., Oberlin Coll., 1967; Ph.D., Case Western Res. U., 1971. NIH postdoctoral fellow Brown U., 1971-72, U. Va., 1972-73; asst. prof. anatomy George Washington U., 1973-75; asst. prof. Georgetown U., 1975-79, assoc. prof., 1979—. Contbr. articles, abstracts, chpts. to profl. publs. Mem. Soc. Neurosci. (sec.-treas. Potomac chpt. 1980-81), Am. Assn. Anatomists, Am. Soc. Zoologists. Subspecialties: Comparative neurobiology; Physiological psychology. Current work: Comparative anatomy of central nervous system, particularly relating to evolution of sensory systems and forebrain; anat. and behavioral studies of regenerated olfactory system preparation in rodents. Office: Dept Anatomy Georgetown U Washington DC 20007

BUTLER, JACKIE DEAN, horticulture educator, consultant; b. Raleigh, Ill., Mar. 1, 1931; s. Gilbert Lowry and Winnie Ellen (Braden) B.; m. Dianna Mathis, June 9, 1957; children: Lisa Mathis, John Eric. Student, So. Ill. U., 1949, 54-56; B.S., U. Ill., 1957, M.S., 1959, Ph.D., 1966. Asst. farm adv. U. Ill.-Jerseyville, 1957-58; research asst., assoc. U. Ill., Urbana, 1958-63, instr., 1963-66, asst. prof., 1966-70, assoc. prof., 1970-71, Colo. State U., Fort Collins, 1971-76, prof. horticulture, 1976—; turfgrass cons. Greenscape Ltd., Menlo Park, Calif., 1976-83; turfgrass ccons. Pacific Planners Internat., Palo Alto, Calif., 1976-83, Robert Trent Jones II, Palo Alto, 1976-83. Contbr. articles to profl. jours. Served with USN, 1950-54. Recipient Disting. Service awards Midwest Golf Course Supts., 1971, Chicagoland Golf Course Supts., 1971, Rocky Mountain Turfgrass Assn., 1981; Outstanding Service award Ill. Turfgrass Found., 1969. Mem. Am. Soc. Hort. Sci., Am. Soc. Agronomy, Internat. Turfgrass Soc., Sigma Xi, Phi Alpha Xi, Gamma Sigma Delta, Phi Sigma, Epsilon Sigma Phi. Presbyterian. Current work: Primary efforts have been devoted to water utilization in landscape systems and to turfgrass production in hostile environments. Office: Colo State U Dept Hort Fort Collins CO 80523 Home: 220 S County Rd 5 Fort Collins CO 80524

BUTLER, JAMES EHRICH, research chem. physicist; b. Tenafly, N.J., Nov. 29, 1944; m. Shahla Amoukhteh-Agah. S.B., MIT, 1966; Ph.D., U. Chgo., 1972. Research chem. physicist Naval Research Lab., Washington, 1975—. NASA trainee, 1969-71; NIH fellow, 1972-74. Mem. Am. Phys. Soc., Optical Soc. Am., Sigma Xi. Subspecialties: Physical chemistry; Laser photochemistry. Current work: Reaction dynamics and kinetics, spectroscopy, photochemistry, molecular interaction at surfaces. Office: Code 6110 US Naval Research Lab Washington DC 20375

BUTLER, JAMES PRESTON, physicist, physiologist; b. Boston, Aug. 19, 1945; s. Clay Preston and Delilah Graham (Barber) B.; m. Susan Nowers, Aug. 18, 1973; children: Matthew Preston, Aaron Joseph. B.A., Pomona Coll., 1967; M.A., Harvard U., 1968, Ph.D., 1974. Research assoc. Los Angeles County-U. So. Calif. Med. Ctr., 1969-71; research assoc. Harvard U., Boston, 1971-74, 75-78, asst. prof. dept. physiology, 1978—; vis. scholar Henry Luce Found., Sendai, Japan, 1974-75; cons. Adage, Inc., Billerica, Mass., 1982, Mass. Gen. Hosp., Boston, 1982—. Contbr. articles to profl. jours. Mem. Back Bay Water Table Com., Boston, 1981—. Recipient Young Investigator Pulmonary Research award NIH, 1976, Light Scattering Stereology award, 1982. Mem. Math. Assn. Am. Episcopalian. Subspecialties: Biophysics (physics); Physiology (biology). Current work: Mathematics and physics of lung gas exchange and mechanics, optical dynamic stereology with lasers, mathematics of ill-posed inverse problems, oscillatory continuum mechanics. Home: 176 Coolidge St Brookline MA 02146 Office: Dept Physiolog Harvard Sch Pub Health 665 Huntington Ave Boston MA 02115

BUTLER, JOHN EDWARD, immunologist; b. Rice Lake, Wis., Jan. 10, 1938; s. Edward Walter and Ida (Fredrick) B.; m. (div.); 1 dau., Kirsten D. B.S., U. Wis.-River Falls, 1961; Ph.D., Kans. U., 1965. Ranger, naturalist U.S. Nat. Park Service, Crater Lake, Oreg., 1961-63; teaching asst. U. Kans., 1961-65, acting asst. prof., 1965-67; research biologist U.S. Dept. Agr., Washington, 1967-71; asst. prof. U. Iowa, 1971-74, assoc. prof., 1974-80, prof., 1980—. Contbr. articles to profl. jours., chpts. to books. USPHS trainee, 1965-66; NSF fellow, 1964; Max-Planck Inst. fellow, W.Ger., 1973-74; fellow Fogarty Internat., W.Ger., 1982-83. Mem. Am. Assn. Immunologists, Sigma Xi. Subspecialties: Immunobiology and immunology; Biochemistry (biology). Current work: Secretory immunity; respiratory disease, ELISA immunochemistry, maternal-neonatal immune regulation. Home: 712 11th Ave Coralville IA 52241 Office: U Iowa 3-450 Bowen Sci Bldg Iowa City IA 52242

BUTLER, JOHN J., physician, educator; b. Rochester, N.Y., Oct. 18, 1920; s. John Joseph and Josephine (Fitzgerald) B.; m. Meta Lois Regan, Apr. 3, 1948; children: L Kathleen M., Timothy J., Paul, Anne E., Susan C. B.A., U. Toronto, 1941; M.D., U. Rochester, 1944. Diplomate: Am. Bd. Internal Medicine (subspecialty in hematology and oncology). Intern Grady Meml. Hosp., Atlanta, 1946, resident in internal medicine, 1947-48; assoc. prof. medicine N.J. Coll. Medicine, Newark, 1959-66; assoc. prof. clin. medicine NYU Med. Sch., 1972—; chief hematology, med. oncology Cath. Med. Ctr., Bklyn., 1972—. Mem. ACP, Am. Soc. Hematology, Am. Soc. Clin. Oncology. Club: N.Y. Athletic. Subspecialties: Chemotherapy; Hematology. Home: Gt Meadow Rd Locust Valley NY 11560 Office: Cath Med Ctr 152-11 89th Ave Jamaica NY 11432

BUTLER, LARRY GENE, biochemist; b. Elkhart, Kans., Dec. 14, 1933; s. Benjamin Herman and Leota Marie B.; m. Mary Frances McIntosh, Nov. 6, 1953; children: Scott Daniel, Sarah Beth, Frank Andrew, Anna Ruth. B.S. (NSF fellow) in Chemistry, Okla. State U., 1960; postgrad., U. Minn., 1960-63; Ph.D. (NSF fellow), UCLA, 1964. Chmn. dept. natural scis. Los Angeles Baptist Coll., Newhall, Calif., 1964-65; postdoctoral researcher U. Ariz., Tucson, 1965-66; asst. prof. chemistry Purdue U., W. Lafayette, Ind., 1966-68, asso. prof., 1968-73, prof., 1973—; cons., adv. bd. Jour. Agrl. and Food Chemistry, 1981--. Contbr. numerous articles on biochemistry to profl. jours. Served with U.S. Army, 1954-56. Recipient NIH Research Career Devel. award, 19; recipient Staley Disting. Scientist Lectureship Grand Rapids Bapt. Coll., 1978. Mem. Am. Chem. Soc., Am. Soc. Biol. Chemists. Subspecialties: Biochemistry (biology); Enzyme technology. Current work: Tannins and other plant polyphenols, their chemical characterization and elucidation of the basis for their biological effects as plant protectants, dietary antinutritional factors, and possible carcinogens, structure/function relationships of enzymes. Patentee in field. Home: 2126 Robinhood Ln W Lafayette IN 47906 Office: Dept Biochemistry Purdue U W Lafayette IN 47907

BUTLER, ROBERT NEIL, gerontologist, psychiatrist, writer, educator; b. N.Y.C., Jan. 21, 1927; s. Fred and Easter (Dikeman) B.; m. Diane McLaughlin, Sept. 2, 1950; children: Ann Christine, Carole Melissa, Cynthia Lee; m. Myrna I. Lewis, May 19, 1975; 1 dau., Alexandra Nicole. B.A., Columbia U., 1949, M.D., 1953. Intern St. Lukes Hosp., N.Y.C., 1953-54; resident U. Calif. Langley Porter Clinic, 1954-55, NIMH, 1955-56, research psychiatrist, 1955-62; founder geriatric unit Chestnut Lodge, 1958, adminstr., 1958-59; research psychiatrist Washington Sch. Psychiatry, 1962-76; dir. Nat. Inst. on Aging, NIH, 1976-82; Brookdale prof. geriatics and adult devel. Mt. Sinai Sch. Medicine, N.Y.C., 1982; mem. faculty George Washington U. Med. Sch., Washington, 1962—, Howard U. Sch. Medicine; cons. NIMH, 1967-76, U.S. Senate Spl. Com. on Aging. (Recipient Pulitzer prize for gen. nonfiction 1976); Author: (with others) Human Aging, 1963, (with Myrna I. Lewis) Aging and Mental Health, 1973, Why Survive? Being Old in America, 1975, Sex After Sixty, 1976; Mem. editorial bd.: Jour. Geriatric Psychiatry, Aging and Human Development; Contbr. articles to publs. Sec. Nat. Ballet of Washington, 1962-75; chmn. D.C. Advisory Commn. on Aging, 1969-72; bd. dirs. Nat. Council on Aging. Served with U.S. Maritime Service, 1945-47. Leo Laks award, 1976; McIntyre award, 1977; others. Fellow Am. Psychiat. Assn., Am. Geriatrics Soc. (founding mem.); mem. Group for Advancement Psychiatry (trustee 1974-76), Gerontol. Soc., Forum for Profls. and Execs. (founding). Club: Cosmos (Washington). Subspecialties: Gerontology; Psychiatry. Current work: Senile demontia; aging processes: geriatrics. Home: 3815 Huntington St NW Washington DC 20015 Office: Nat Inst on Aging NIH Bldg 31 Room 5L02 9000 Rockville Pike Bethesda MD 20014 To always stretch the limits of the possible through personal relationships, scholarship, science, writing, action and political activism. To work toward making life a work of art. To do no harm.

BUTLER, SHAHLA, computer scientist; b. Tehran, Iran, May 19, 1946; came to U.S., 1964; d. Ali Asgar and Saeedeh (Zanjani) Amoukhteh; m. James Ehrich, Sept. 14, 1969. B.S., U. Mich., 1968, M.S., 1972; Ph.D., U. Chgo., 1974. Asst. prof. U. Md., College Park, 1974-76; cons. Am. Mgmt. Systems, Arlington, Va., 1976-80; program mgr. Wilson Hill Assocs., Washington, 1980-81; dir. devel. Student Loan Mktg. Assn., Washington, 1981-82; exec. v.p. Planning Analysis Corp., Arlington, 1982—. Mem. IEEE, Assn. Computing Machinery. Subspecialties: Software engineering; Database systems. Home: 2724 Fort Scott Dr Arlington VA 22202 Office: 1000 Wilson Blvd Arlington VA 22202

BUTLER, THOMAS PARKE, physician, researcher, educator; b. Chgo., Apr. 12, 1945; s. Robert Elliott and Barbara Jane (Parke) B.; m. Ann Benedict, June 8, 1968; 1 dau., Whitney Elizabeth. B.A., Oberlin Coll., 1967; M.D., Case Western Res. U., 1971. Diplomate: Am. Bd. Internal Medicine., Am. Bd. Med. Oncology. Intern R.I. Hosp., Providence, 1971-72; research assoc. Nat. Cancer Inst., Bethesda, Md., 1972-74; resident in internal medicine Georgetown U., Washington, 1974-76, fellow in oncology, 1976-78, clin. assoc. prof. medicine, 1982—; practice medicine specializing in oncology/hematology, Arlington, Va., 1978—; clin. staff Vincent T. Lombardi Cancer Ctr., Washington, 1978—; dir. oncology unit Arlington Hosp., 1982—. Bd. dirs. Arlington (Va.) dist. Am. Cancer S oc. Served with USPHS, 1972-74. Recipient Golden Apple for Teaching award Georgetown U., 1981. Fellow ACP; mem. Am. Assn. Cancer Research, Am. Soc. Clin. Oncology, AMA, Am. Soc. Internal Medicine, Sigma Xi. Presbyterian. Subspecialties: Oncology; Hematology. Current work: Private practice of medical oncology and hematology, clinical research per Lombardi Cancer Center, and teaching per Georgetown University Medical School. Office: 1715 N George Mason Dr Arlington VA 22207

BUTLER, THOMAS WARWICK, JR., engring. co. exec.; b. Niagara Falls, N.Y., Oct. 9, 1922; s. Thomas Warwick and Genevieve Margaret (Casey) B.; m. Jeanne E. Lindsey, Aug. 24, 1950; 1 son, Thomas W. B.S., M.S., U. Mich., Ph.D., 1961. Research asso. U. Mich., Ann Arbor, 1951-54, research engr., 1960-62, asso. research engr., 1954-60; dir. engring. and research Mech. Products, Inc., Jackson, Mich., 1962-65; asso. prof., dir. Cooley Electronics Lab., U. Mich., 1965-70, dir., prof., 1970-74; v.p., corp. officer engring. and research AMF, Inc., White Plains, N.Y., 1974—; dir. Fed. Screw Works, Detroit, 1978—; chmn. adv. bd. Applied Research Lab., Pa. State U., 1976—; chmn. Nat. Engring. Consortium seminar on product planning, 1978-79; spl. cons. to dir. NSF, 1975; vis. prof. Mich. State U., 1973; advisor fgn. tech. div. USAF, 1962-65. Author book. Served with USAAF, 1942-45. Mem. N.Y. Zool. Soc., Indsl. Research Inst., IEEE, Am. Soc. Engring. Edn., Soc. Profl. Engrs., Sci. Research Club. Clubs: Greenwich Country, Indian Harbor Yacht; Seabrook Island (Charleston, S.C.). Subspecialties: Electronics; Computer engineering. Patentee in field. Home: 9 Joshua Ln Greenwich CT 06830 Office: 777 Westchester Ave White Plains NY 10604

BUTLER, WILLIAM MANION, physician, medical educator; b. Lexington, Ky., May 14, 1947; s. Benjamin Joseph and Ruth (Manion) B.; m. Lynn Mackie, Sept. 26, 1970; children: Benjamin L., Charles M. A.A., U. Louisville, 1968; M.D., Tulane U., 1972. Intern Charity Hosp., New Orleans, 1972-73, resident, 1973-75; asst. prof. medicine U. S.C., Columbia, 1981—. Contbr. articles to profl. jours. Served with U.S. Army, 1975-81. Fellow ACP; mem. Am. Soc. Hematology, Am. Soc. Clin. Oncology, Am. Fedn. Clin. Research, AAAS. Subspecialties: Hematology; Chemotherapy. Current work: Coagulation, red cell diseases. Home: 4729 Heath Hill Columbia SC 29206 Office: University of South Carolina School of Medicine 3321 Medical Park Rd Columbia SC 29203

BUTTON, KENNETH J(OHN), physicist; b. Rochester, N.Y., Oct. 11, 1922; s. Kenneth P. and Ruth C. (Wagner) B.; m. Margaret Jane Wells, Dec. 22, 1952. B.S., U. Rochester, 1950, M.S. in Physics, 1952. Research physicist MIT, 1952-62, research group leader, 1962-72, sr. scientist, 1972—; organizer, program chmn. Ann. Internat. Conf. on Infrared and Millimeter Waves, 1974-84. Author: Microwave Ferrites and Ferrimagnetics, 1962; Editor: Infrared and Millimeter Waves, vols. 1-12, 1979-84; editor: Internat. Jour. Infrared and Millimeter Waves, 1980-84. Served to sgt. U.S. Army, 1942-46. Decorated Bronze Star with oak leaf cluster.; Recipient Disting. Service award IEEE Microwave Theory and Techniques Soc., 1980, cert. merit, 1981. Fellow IEEE, Am. Phys. Soc. Subspecialties: Condensed matter physics; Electronics. Current work: Experimental semiconductor physics, millimeter and submillimeter wave propagation; semmicondrs; millimeter and submillimeter wave propagation in semiconductors, magnetic materials, ceramics, glasses and liquids. Office: MIT Nat Magnet Lab Cambridge MA 02139

BUYNISKI, JOSEPH PAUL, pharmacologist; b. Worcester, Mass., July 18, 1941; s. Julius M. and Wanda (Waziak) B.; m. Barbara Claire Buyniski, Sept 22, 1973; children: Larissa, Alexei. B.S., U. Cin., 1959-63, Ph.D., 1967; postgrad. cardiovascular program, Bowman Gray Med. Sch., 1967-69. Instr. Bowman Gray Med. Sch., 1968-69; with Bristol Labs., Syracuse, 1969-82, head cardiovascular research, 1969-75, asst. dir. dept. pharmacology, 1975-77, project dir. cardiovascular drugs, 1977-82; dir. dept. pharmacology Pharm. Reserch and Devel. div. Bristol Myers Co., Syracuse, 1982—. Contbr. articles to profl. jours. NIH trainee, 1963-67; fellow, 1967-69; N.C. Heart Assn. grantee, 1968-69. Mem. Am. Soc. Pharmacology and Exptl. Therapeutics, Am. Physiol. Soc., Am. Heart Assn. Republican. Roman Catholic. Clubs: Syracuse Athletic, Syracuse Stamp. Subspecialties: Pharmacology; Cardiology. Current work: Preclinical and clinical cardiovascular research; preclincial gastrointestinal,analgetic and antiarthritic research. Office: Bristol Myers Dept Pharmacology PO Box 657 Syracuse NY 13201

BYARS, T. DOUGLAS, veterinarian; b. San Francisco, Sept. 20, 1943; s. Tandy Douglas and Ruby Jane B.; m. Susan Elizabeth Arnett, May 17, 1968; 1 dau., Carrie Rebecca. A.A., Modesto Jr. Coll., 1965; B.S., Calif. State Poly. U., 1968; D.V.M., U. Calif.-Davis, 1974. Cert. Am. Coll. Vet. Internal Medicine, 1981. Intern U. Ga., 1974-75; pvt. practice vet. medicine, Ocala, Fla., 1975-76, resident veterinarian to pvt. practitioner, Logandale, Nev., 1976-77; assoc. prof. large animal medicine U. Ga., Athens, 1977—; cons. equine veterinarian. Active Big Brother program. Calif. Thoroughbred Breeders grantee, 1973; recipient 1st ann. alumni award U. Calif., 1974. Mem. Am. Assn. Equine Practitioners, Am. Assn. Coagulationists, AVMA, Am. Coll. Vet. Internal Medicine, Miniature Veterinarians Assn. Subspecialty: Internal medicine (veterinary medicine). Current work: Equine blood coagulation and anticoagulation, blood coagulation relating to equine colic syndrome.

BYCK, ROBERT, psychiatrist; b. Newark, Apr. 26, 1933; s. Louis and Lucy Ruth (Landau) B.; m. Susan Elizabeth Wheeler, Aug. 21, 1976; children: Carl, Gillian, Lucas. A.B., U. Pa., 1954, M.D., 1959; M.A. (hon.), Yale U., 1978. Intern U. Calif.-San Francisco, 1959-60; research assoc. NIH, 1960-62; resident in psychiatry Yale U., New Haven, 1969-72; asst. prof. pharmacology and rehab. medicine Albert Einstein Coll. Medicine, Bronx, 1964-69; lectr. pharmacology, resident in psychiatry Yale U. Med. Sch., New Haven, 1969-72; assoc. prof. psychiatry and pharmacology Yale Sch. Medicine, 1972-77, prof., 1977—; cons. N.Y. Zool. Soc., VA Hosp., West Haven, Med Letter on Drugs and Therapeutics. Editor: Cocaine Papers (Sigmund Freud), 1975. Served with USPHS, 1960-63. NIMH career devel. awardee, 1967; Burroughs Wellcome scholar, 1972. Mem. AAAS, Am. Soc. Pharmacology and Explt. Therapeutics, Am. Soc. Clin. Pharmacology and Therapeutics, Am. Coll. Neuropsychopharmacology, Sherlock Holmes Soc. Subspecialties: Psychopharmacology; Neuropharmacology. Current work: Nuclear magnetic resonance imaging. Home: 197 McKinley Ave New Haven CT 06515 Office: Dept Pharmacology Yale U Med Sch New Haven CT 06510

BYERLEE, JAMES DOUGLAS, geophysicist; b. Cairns, Australia; s. Christian Joseph and Marie (Clarkson) B.; m. Merle, Dec. 7, 1954; children: David, Ian, Madonna. B.Sc. with honors in Geology, U. Queensland, Australia, 1963; Ph.D., MIT, 1966. Geophysicist U.S. Geol. Survey, Menlo Park, Calif. Contbr. numerous articles, abstracts to profl. jours. Zinc Corp. scholar, 1961-62; U. Queensland travelling scholar, 1963-66. Mem. Am. Geophys. Union. Democrat. Presbyterian. Subspecialty: Geophysics. Current work: Physical properties of rocks and high temperature and pressure. Office: 345 Middlefield Rd Menlo Park CA 94025

BYERRUM, RICHARD UGLOW, coll. dean; b. Aurora, Ill., Sept. 22, 1920; s. Earl Edward and Florence (Uglow) B.; m. Claire Somers, Apr. 3, 1945; children—Elizabeth, Robert, Mary, Carey. A.B., Wabash Coll., 1942, D.Sc. (hon.), 1967; Ph.D., U. Ill., 1947. Teaching asst. U. Ill., 1942-44; research asso. U.S. Chem. Corps, toxicity dept. U. Chgo., 1944-47; faculty Mich. State U., East Lansing, 1947—, prof. biochemistry, 1957—, acting dir., 1961-62, dean, 1962—. Author: (with others) Experimental Biochemistry, 1956; Editorial bd.: Phytochemistry, 1961—; Contbr. numerous articles to profl. jours. Mem. Project Hope, 1961—; Trustee Mich. Health Council, 1961—, pres., 1966. Travel grantee Internat. Congress Biochemistry, Vienna, 1958, Internat. Congress Biochemistry, Montreal, 1959. Mem. Am. Chem. Soc. (lectr. vis. scientist program, awards com., visitor for com. profl. tng.), N. Central Assn. Colls. and Secondary Schs., A.A.A.S., Am. Soc. Plant Physiologists (trustee, exec. com.), Am. Soc. Biol. Chemists, Soc. Exptl. Biology and Medicine, Mich. Acad. Arts, Sci. and Letters, Phi Beta Kappa (pres. local chpt. 1962), Sigma Xi (awards com., Jr. Research award Mich. State U. chpt. 1958), Phi Kappa Phi (pres. 1968-69), Phi Lambda Upsilon, Alpha Chi Sigma, Beta Theta Pi. Subspecialties: Biochemistry (biology); Plant physiology (biology). Current work: Administration of science college; alkoloid biosynthesis. Patentee cancer tumor inhibiting material. Home: 602 Wildwood Dr East Lansing MI 48823

BYERS, BRECK EDWARD, cell biologist, educator; b. St. Louis, July 4, 1939; s. F. Donald and Melba Constance (Boothman) B.; m. Margaret Read, Nov. 26, 1964; children: Mark Andrew, Carl Bradford. Research fellow dept. biology Harvard U., 1967-68; charge des researches Inst. Molecular Biology, U. Geneva, Switzerland, 1968-70; asst. prof. dept. genetics U. Wash., Seattle, 1970-76, assoc. prof., 1976-80, prof., 1980—. Mem. editorial bd.: Molecular and Cellular Biology; contbr. articles to sci. jours. Recipient NIH career devel. award, 1971-76, research grantee, 1971—. Mem. Genetics Soc. Am., Am. Soc. for Cell Biology. Subspecialties: Cell biology; Genetics and genetic engineering (biology). Current work: Yeast cell division and meiotic recombination. Office: Dept Genetics SK-50 U Wash Seattle WA 98195

BYERS, GEORGE WILLIAM, educator; b. Washington, May 16, 1923; s. George and Helen (Kessler) B. B.S., Purdue U., 1947; M.S., U. Wash., 1949, Ph.D., 1952. Asst. prof. dept entomology U. Kans.-Lawrence, 1956-60, assoc. prof., 1960-65, prof., 1965—, chmn. dept. entomology, 1969-72, assoc. chmn., 1972—; collaborator U.S. Dept. Agr., Washington, 1970—. Contbr. articles to profl. jours.; editor: Systematic Zoology, 1964-67, Jour. Kans. Entomol. Soc, 1958-59, 83—. Served to lt. col. U.S. Army, 1942-72. Rackham fellow, 1952-53; NSF grantee, 1958—. Mem. Entomol. Soc. Am. (editorial bd. 1968-73), Central States Entomol. Soc. (pres. 1958-59). Subspecialties: Taxonomy; Morphology. Current work: Taxonomy and morphology of crane files and Mecoptera. Home: 2215 Princeton Blvd Lawrence KS 66044 Office: Dept Entomology U Kans Lawrence KS 66045

BYERS, JAMES MARTIN, pathologist, chemist, educator; b. Chillicothe, Ohio, Oct. 16, 1944; s. James Martin and Arlene Alberta (Courser) B.; m. Donna Mary Aher, June 15, 1968. A.B., Dartmouth Coll., 1966; M.D., Ohio State U., Columbus, 1975. Diplomate: Am. Bd. Pathology. Intern, fellow pathology Johns Hopkins U. and Univ. Hosp, Balt., 1970-71; asst. resident, fellow pathology Johns Hopkins U. and U. Md. Hosp, 1971-72, asst. resident lab. medicine, fellow pathology, 1973-75; asst. prof. pathology U. Ariz., Tuscon, 1975-81, assoc. prof., 1981—. Fellow Coll. Am. Pathologists (mem. toxicology resource com. 1978—, vice-chmn. 1980—); mem. Am. Soc. Clin. Pathologists (mem. commn. on grad. edn. in pathology 1982—). Subspecialties: Pathology (medicine); Clinical chemistry. Current work: Drug assay, laboratory studies in applied clinical toxicology, characterization of immuno-reactivity of antibody conjugates, medical diagnostic chemical methods development. Office: Dept Pathology U Ariz 1501 N Campbell Ave Tucson AZ 85724

BYLER, JAMES W., plant pathologist; b. Bellefontaine, Ohio, June 19, 1940; s. A. Milford and M. Lorene (Yoder) B.; m. JoAnn Brown. Ph.D. in Plant Pathology, U. Calif.-Berkeley, 1970. Plant pathologist U.S. Forest Service, Region 5, San Francisco, 1971-79, supervisory plant pathologist, Missoula, Mont., 1979—. Mem. Am. Phytopathol.

Soc. Subspecialty: Plant pathology. Current work: Extension forest pathology. Office: Forest Service CFPM Po Box 7669 Missoula MT 59807

BYLUND, DAVID B., pharmacologist; b. Spanish Fork, Utah, Apr. 16, 1946; s. H. Bruce and Rhea (Bowen) B.; m. Elaine C. Thurman, May 27, 1970; children: Carma, Eric, Michelle, Kevin, Jennifer. Student, Calif. Inst. Tech., 1966-68; B.S., Brigham Young U., 1970; Ph.D., U. Calif., Davis, 1974. Postdoctoral fellow Johns Hopkins U. Med. Sch., Balt., 1975-77; asst. prof. pharmacology U. Mo., Columbia, 1977-82, assoc. prof., 1982—. Contbr. articles to profl. jours. Scoutmaster Boy Scouts Am., 1980-81. Mem. Am. Soc. Pharmacology and Exptl. Therapeutics, Soc. Neurosci., AAAS. Mormon. Subspecialties: Neuropharmacology; Molecular pharmacology. Current work: Adrenergic receptors; adenylate cyclase and radioligand binding studies; smooth muscle; hypertension; receptor purification. Office: U Mo Columbia MO 65212

BYRD, JAMES WILLIAM, SR., physicist, educator, cons.; b. Mt. Olive, N.C., Dec. 4, 1936; s. William Kilby and Essie Pearl (Jernigan) B.; m. Marvis Ann Edwards, May 31, 1959; children: James William, John Edward, Jennifer Ann. B.S., N.C. State U., 1959, M.S., 1960; Ph.D., Pa. State U., 1963. Asso. prof. East Carolina U., Greenville, N.C., 1962-64, prof. physics, 1964—, chmn. physics dept., 1965—. Contbr. articles to profl. publs. Grantee NSF, HEW, Dept. Energy. Mem. Am. Phys. Soc., Am. Assn. Physics Tchrs., N.C. Acad. Sci., Sigma Xi. Democrat. Presbyterian. Subspecialties: Energy transport; Solar energy. Current work: Energy transport in radiant beams and subterranean media, solar ponds, mathematical methods. Home: 225 York Rd Greenville NC 27834 Office: Dept Physics East Carolina Univ Greenville NC 27834

BYRD, KENNETH ELBURN, anatomy educator, researcher; b. Phoenix, Nov. 26, 1951; s. James Elburn, Jr. and Alberta Maureen (Demarest) B.; m. Elizabeth Lee Hamilton, May 22, 1976. B.S. in Anthropology, Ariz. State U., 1973; Ph.D. in Phys. Anthropology, U. Wash., 1979. Sr. fellow dept. biol. structure U. Wash., Seattle, 1979-81, postdoctoral fellow dept. physiology and biophysics, 1979-81; adj. asst. prof. dept. physiology U. Tex. Health Sci. Ctr. Dental Br., Houston, 1982—; asst. prof. anatomy Sch. Phys. Therapy, Tex. Woman's U., Houston, 1981—. NIH NRSA/NIDR fellow, 1979; recipient organized research award Tex. Woman's U., 1981, 83; research assoc. award, 1982. Mem. Am. Assn. Anatomists, Am. Assn. Phys. Anthropologists, AAAS, Internat. Assn. Dental Research, Neurosci. Group of Internat. Assn. Dental Research. Democrat. Subspecialties: Morphology; Neurophysiology. Current work: Interaction of craniofacial growth and development with neurophysiological processes during the act of mastication. Office: Sch Phys Therapy Tex Woman's U 1130 MD Anderson Blvd Houston TX 77030

BYRD, LARRY DONALD, research scientist, pharmacologist; b. Salisbury, N.C., July 14, 1936; s. Donald Thomas and Mildren Alexina (Gardner) B.; m. Vivian Corinne Williams, Dec. 23, 1961; children: Kay, Lynn, Renee, Andrew. A.B., East Carolina U., 1962, M.A., 1964; Ph.D., U. N.C., Chapel Hill, 1968, Harvard U. Med. Sch., 1967-70. Assoc. scientist New Eng. Regional Primate Research Center, Southborough, Mass., 1969-74; instr. psychobiology Harvard U., Boston, 1970-73, prin. assoc. in psychiatry, 1973-74; psychobiologist, chmn. div. primate behavior Yerkes Regional Primate Research Center, Emory U., Atlanta, 1974-79, assoc. researchprof., chmn., 1979-80, assoc. research prof., 1980-82, chief div. behavioral biology, 1980—, research prof., 1982—, adj. prof psychology, 1982—, assoc. prof.pharmacology, 1981—; lectr. psychology Ga. Inst. Tech., 1974—; cons. Nat. Center Toxicol. Research, FDA, Jefferson, Ark., 1976-77, Naval Aerospace Med. Research Lab., Pensacola, Fla., 1977, Addiction Research Center, Lexington, Ky., 1979, S.W. Found. Research and Edn., San Antonio, 1977, MIT Press, Cambridge, Mass., 1975, Nat. Inst. Drug Abuse, Rockville, Md., 1979—. Editorial bd.: Jour. Exptl. Analysis of Behavior, 1969-79; assoc. editor, 1970-76; editor: Psychopharmacology Newsletter, 1976-82; cons. editor: Am. Jour. Primatology, 1980—; editorial advisor: Jour. Pharmacology and Exptl. Therapeutics, 1973—, Psychopharmacology, 1976—, Sci, 1973—, Physiology and Behavior, 1980—, Behavioral and Neural Biology, 1981—. Served with AUS, 1954-57. Recipient Outstanding Alumnus award East Carolina U., 1977. Fellow Am. Psychol. Assn. (pres. div. psychopharmacology 1982-83); mem. Am. Soc. Pharmacology and Exptl. Theapeutics, Behavioral Pharmacology Soc., Soc. Exptl. Analysis of Behavior, AAAS, Soc. Neurosci., Ea. Psychol. Assn., Southea. Psychol. Assn., Phi Sigma Pi. Subspecialties: Psychopharmacology; Psychobiology. Current work: Behavioral pharmacology and behavioral physiology. Home: 1026 Viking Dr Stone Mountain GA 30083 Office: Yerkes Regional Primate Research Center Emory University Atlanta GA 30322

BYRNE, BARBARA JEAN, biology educator; b. Baraboo, Wis., Aug. 9, 1941; d. Joseph Patrick and Charlotte Augusta (Graves) McManamy; m. Bruce Campbell Byrne, Feb. 2, 1968; children: Joshua, Jenny. B.A., Blackburn Coll., 1962; M.A., Ind. U., 1963, Ph.D., 1969. Vis. lectr. Ind. U., Bloomington, 1972, vis. scholar, 1981-82; research assoc. Cornell U., Ithaca, N.Y., 1972-74; asst. prof., assoc. prof. biology Wells Coll., Aurora, N.Y., 1974-80, 1974—, dir. summer confs., 1982—. Contbr. papers in field to profl. lit. Research Corp. grantee, 1976-80; NIH grantee, 1979-82. Mem. Genetics Soc. Am., AAAS, Electron Microscopy Soc. Am. Democrat. Subspecialties: Genetics and genetic engineering (biology); Gene actions. Current work: Control of ciliary antigen genes in Paramecium; cloning genes to determine mechanisms of control of family of ciliary antigen genes; interstock crosses to do same. Home: Main St Aurora NY 13026 Office: Dept Biology Wells Coll Aurora NY 13026

BYRNE, JEFFREY EDWARD, research pharmacologist, educator; b. Mpls., July 15, 1939; s. Maurice Charles and Edna Francis (Kinney) B.; m. Janice V. Grove, Feb. 1, 1960 (dec. 1976); children: Christopher, Maura; m. Margaret Ann Kaiser, June 17, 1978; 1 son, Jason. B.A., U. N.D., 1962; M.A., U. S.D., 1964, Ph.D., 1966. Postdoctoral fellow, lectr. U. Man., Winnipeg, Can., 1966-69; sr. scientist Mead Johnson Pharms., Evansville, Ind., 1969-71, sr. investigator, 1971-78; sr. research assoc. Mead Johnson/Bristol-Myers, Evansville, 1978-82; prin. research scientist Bristol-Myers, Evansville, 1982—; assoc. faculty Ind. U. Sch. Medicine, Evansville, 1972-80, U. Evansville Sch. Nursing, 1972-80. Contbr. chpt. to book, articles to jours. in field. Mem. N.Y. Acad. Scis., Am. Heart Assn., AAAS, Am. Soc. Pharmacology and Exptl. Therapeutics, Mensa, Sigma Xi. Republican. Lutheran. Subspecialties: Pharmacology; Physiology (medicine). Current work: Cardiovascular pharmacology - animal experimentation, cardiac arrhythmias, myocardial ischemia, ischemic heart disease, coronary physiology, drug discovery and preclinical development. Home: 5120 New Harmony Rd Evansville IN 47712 Office: Bristol-Myers Pharmaceutical Research 2404 Pennsylvania Ave Evansville IN 47721

BYRNE, JOHN MAXWELL, biological sciences educator; b. Gassaway, W.Va., May 7, 1933; s. George Coble and Margret Cathering (Heater) B.; m. Garnet Ruthe Bobblet, July 8, 1960; 1 dau. Kimberly Ann. B.A., Glenville State U., 1960; M.A., Miami U., Oxofrd, Ohio, 1964, Ph.D., 1969. Asst. prof. Va. Poly. Inst. and State U., Blacksburg, 1969-75; asso. prof. biol. Scis. Kent (Ohio) State U., 1975—. Contbr. articles on biol. scis. to profl. jours. Served with USMC, 1954-57. Mem. Bot. Soc. Am., Am. Inst. Biol. Sci., AAAS, Sigma Xi. Current work: Developmental plant anatomy, root development. Office: Dept Biol Scis Kent State U Kent OH 44242

BYSTRYN, JEAN-CLAUDE, dermatologist; b. Paris, May 8, 1938; m. Marcia Bystryn, Dec. 17, 1947; 1 dau. Anne. B.S., U. Chgo., 1958; M.D., N.Y.U., 1962. Diplomate: Am. Bd. Dermatology. Intern Montefiore Hosp., N.Y.C., 1962-63, resident in medicine, 1973-64; resident in dermatology N.Y.U., 1966-69, USPHS fellow dermatology, 1968-72, mem. faculty, 1970—, assoc. prof. dermatology, 1976—, dir. immunofluroescence lab., 1972—; co dir. Bullous Disease Clinic, Skin and Cancer Clinic, 1974—; attending physician N.Y.U. Hosp. Contbr. articles to profl. jours. Served with USPHS, 1964-66. Ford Found. fellow, 1954; recipient Irma T. Hirschl Career Scientist award, 1979. Mem. Am. Assn. Immunologists, Am. Assn. Cancer Research, Soc. Investigative Dermtology, Am. Acad. Dermatology, Am. Soc. Cell Biology, Task Force on Immunofluorescence Internat. Soc. Tropical Dermatology, Am. Fedn. Clin. Research, N.Y. Dermatol. Soc., Dermatology Found., Skin Cancer Found. (chmn. grant rev. com. 1980—), Dystrophic Epidermolysic Bullosa Found. (adv. bd. 1980-81). Subspecialties: Dermatology; Immunology (medicine). Current work: Melanoma immunology, immunology of blistering diseases of the skin and vitiligo. Office: 530 1st Ave Suite 7F New York NY 10016

CABRERA, BLAS, physicist, educator, researcher; b. Paris, Sept. 21, 1946; s. Nicolas and Carmen (Navarro) C.; m. JoAnn Nelson, Apr. 1 1972; children: Nicolas, Joseph, Blas Jacob. B.S. in Physics, U. Va., 1968, Ph.D., Stanford U., 1974. Research assoc. Stanford U., 1975-78, sr. research assoc., 1979, acting asst. prof., 1980, asst. prof., 1980—. Contbr. articles to sci. publs. Woodrow Wilson fellow, 1968; Churchill fellow Cambridge (Eng.) U., 1968; NSF fellow, 1968-72; Nat. Bur. Standards grantee, 1978-81. Mem. Am. Phys. Soc., Sigma Xi, Sigma Pi Sigma. Subspecialty: Low temperature physics. Current work: Application of cryogenic techniques and devices to study of fundamental physics: e.g., search for magnetic monopoles, determination of Planck's constant divided by the electron mass, and test of general relativity with orbiting gyroscope. Office: Dept Physics Stanford U Stanford CA 94305

CABRERA, EDELBERTO JOSE, immunologist; b. Pinar del Rio, Cuba, Nov. 5, 1944; s. Baltazar Edelberto and Maria Paulina (Chirino) C.; m. Lourdes Elena Rodriguez, Aug. 13, 1944; children: Edward, Michelle. Ph.D., U. Ill., 1972. Research assoc. U. N.Mex., 1972-77; research scientist Norwich Eaton Pharms Inc., N.Y., 1977—; faculty SUNY, Binghamton, 1978-79. Contbr. articles to profl. jours. Mem. Am. Assn. Immunologists, N.Y. Acad. Sci. Roman Catholic. Club: Peaks and Trail Ski. Subspecialties: Immunopharmacology; Infectious diseases. Current work: Immune regulation, immunomodulators. Home: RD 4 Gibbon Rd Norwich NY 13815 Office: Norwich Eaton Pharms Inc subs Procter & Gamble Norwich NY 13815

CACAK, ROBERT KENT, med. physicist, educator, tech. cons. in radiology; b. Fairbury, Nebr., Dec. 13, 1942; s. Frank Louis and Edna (Junker) C.; m. Roxene Frances, June 28, 1963; children: Jody, Kent. B.Sc., U. Nebr., 1965, M.Sc. (NASA fellow), 1967, Ph.D., 1970. NRC Can. postdoctoral fellow U. Western Ont., 1969-72; research asso. and vis. asst. prof. physics U. Md., College Park, 1972-75; med. physicist, asst. prof. radiology U. Colo. Health Scis. Center, Denver, 1975—; v.p. Plenergy Devel., Ltd.; cons. radiology. Contbr. numerous articles to profl. jours. Mem. Am. Assn. Physicists in Medicine, Am. Phys. Soc. Subspecialties: Medical physics; Diagnostic radiology. Current work: Radiation therapy for cancer treatment; diagnostic imaging in radiology; instrumentation in medicine. Inventor in field. Home: 1718 S Nile Ct Aurora CO 80012 Office: 4200 E 9th Ave C-278 Denver CO 80262

CACHAT, JOHN F., electrical engineer, manufacturing company executive; b. Cleve., Sept. 1, 1916; s. Joseph Anthony and Mary B. (Dunn) C.; m. Mary Ann Cachat, Aug. 1, 1940; children: Anne M. Kohler, Anthony J., Beth. B.S.E.E., Case Western Res. U., 1939, M.B.A., 1952. Indsl. sales engr. Illuminating Co., 1945-46; comml. engr. TOCCO Div. Park-Ohio Industries, Cuyahoga Heights, Ohio, 1946-49, dist. engr., 1949-52, dist. mgr., 1952-57, works mgr., 1957-67, v.p., gen. mgr., 1967—; pres. TOCCO Ala., Inc. Patentee in field. Trustee Jr. Achievement Greater Cleve., 1970-79. Served to lt. USNR, 1942-45. Fellow IEEE (past pres. Industry Applications Soc.); mem. Forge Industry Assn. Roman Catholic. Club: Cleve. Yachting (Rocky River, Ohio). Subspecialty: Electrical engineering. Office: 4620 E 71st St Cuyahoga Heights OH 44125

CACIOPPO, JOHN TERRANCE, psychologist, educator, researcher; b. Marshall, Tex., June 12, 1951; s. Cyrus Joseph and Mary Katherine (Kazimour) C.; m. Barbara Lee Andersen, May 17, 1981. Student, Iowa State Univ., Ames, 1969-70; B.S., U. Mo.-Columbia, 1973; M.A., Ohio State U., 1975; Ph.D., 1977. Research asst. Ohio State U., 1974-75, grad. teaching assoc., 1975-76, grad. fellow, 1976-77; asst. prof. psychology U. Notre Dame, 1977-79, U. Iowa, 1979-81, assoc. prof. psychology, 1981—. Author: (with R. Petty) Attitudes and Persuasion, 1981; Editor: Perspectives in Cardiovascular Psychophysiology, 1982, Social Psychophysiology: A Sourcebook, 1983; Contbr.: articles to profl. jours. Social Psychophysiology: A Sourcebook. Old Gold fellow U. Iowa, 1980; NSF grantee, 1979-85; NIH grantee, 1980-81; U. Iowa faculty scholar, 1980-83; Young Psychologists grantee, 1981. Mem. Am. Psychol. Assn., 500. Soc. for Psychophysiol. Research, Midwestern Psychol. Assn., Soc. for Advancement of Social Psychology, Soc. for Exptl. Social Psychology, Sigma Xi, Phi Kappa Phi. Subspecialties: Social psychology; Psychophysiology. Current work: Investigating the elementary operations underlying social influence and attitudinal processes using verbal, behavioral, chronometric, and psychophysiological procedures. Home: Box 144N Rural Route 6 Woodland Heights Iowa City IA 52240 Office: Dept Psychology U Iowa Iowa City IA 52242

CACUCI, DAN GABRIEL, nuclear engineer, researcher; b. Cluj, Romania, May 16, 1948. M.S., Columbia U., N.Y.C., 1973, M. Philosophy, 1977, Ph.D., 1978. Nuclear engring. asso. Brookhaven Nat. Lab., Upton, N.Y., 1975-76; lead engr. Ebasco Services, Inc., N.Y.C., 1976-77; group leader Oak Ridge (Tenn.) Nat. Lab., 1977—; assoc. prof. U. Tenn., Knoxville, 1983—. Contbr. sci. articles to profl. publs. Recipient Merriman Meml. award Columbia Univ., 1977. Mem. AAAS, Am. Nuclear Soc. (nat. planning com. 1983—), N.Y. Acad. Scis., Sigma Xi. Subspecialties: Applied mathematics; Nuclear fission. Current work: Sensitivity and uncertainty analysis; climate impact of CO_2 increase; regular and chaotic motion of dynamical systems; intelligent control systems; A-bomb dose reassessment. Office: Oak Ridge Nat Lab PO Box X Oak Ridge TN 37830

CADDELL, JOAN LOUISE, research physician; b. N.Y.C., Jan. 16, 1927; d. Alfred Mathew and Anna Emily (Mielke) C. B.A., Coll. for Women, U. Pa., 1948; M.S., U. Pa., 1953. Diplomate: Am. Bd. Pediatrics. Fellow in cardiology and endocrinology Children's Hosp., Phila., 1957-59; fellow in cardiology Yale U., 1959-61, instr. in pediatrics, 1961-62; research prof. St. Louis U., 1980-83; guest research worker NIH, Bethesda, Md., 1982—; physician to Tarascan Indians, Michoacan, Mexico, 1958; to Soc. to Protect Children from Cruelty, Phila., 1957-58; lectr. in pediatrics, Rockefeller Found. fellow Makerere U., Kapala, Uganda, 1962-63, U. Ibadan, Nigeria, 1963-65. Contbg. editor: Nutrition Revs, 1979—. NIH fellow, Chiangmai, Thailand, 1969-71; grantee, 1972-75. Fellow Am. Acad. Pediatrics, Am. Coll. Cardiology, Am. Coll. Nutrition, Am. Inst. Nutrition, Am. Soc. Clin. Nutrition; mem. Am. Pediatric Soc., Soc. Exptl. Biology and Medicine. Episcopalian. Subspecialties: Nutrition (medicine); Cardiology. Current work: Metabolism of essential minerals in man and animal models; research in magnesium metabolism, using rat as animal model. Office: Nat Inst Child Health NIH Bldg 10 8 D-48 Bethesda MD 20205

CADDELL, ROBERT MACORMAC, mechical engineer, consultant, researcher, educator; b. Paterson, N.J., Nov. 13, 1925; s. David and Louise (Coutts) C.; m. Doris Louise Nash, June 25, 1954; children: Steven, David, Gary. B.S., Newark Coll. Engring., 1948; M.S., U. Mich., 1951, Ph.D. (NSF sci. faculty fellow 1962, DuPont grantee-in-aid, 1963, Ford Found. faculty devel. grantee 1963), 1963. Registered profl. engr., Mich. Instr. U. Mich., 1952-55, asst. prof. prodn. engring., 1955-56, assoc. prof. mech. engring. and applied mechanics, 1956-63, asso. prof. mech. engring. and applied mechanics, 1963-71, prof., 1971—; cons. materials and mfg. processes, mech. behavior solids, fracture. Author: Deformationand Fracture of Solids, 1980, (with W.F. Hosford) Metal Forming: Mechanics and Metallurgy, 1982; contbr. numerous articles to profl. jours. Served with inf. U.S. Army, 1944-46. Decorated Bronze Star; named hon. editorial adv. bd. mem. Internat. Jour. Mech. Scis., 1981—; recipient Outstanding Tchr. award Coll. Engring., U. Mich., 1982. Mem. ASME, Am. Soc. Metals, ASTM, Research Club U. Mich., Sigma Xi, Tau Beta Pi, Pi Tau Sigma (Outstanding Mech. Engring. Prof. 1981). Club: Ann Arbor (Mich.) Golf and Outing. Subspecialties: Solid mechanics; Fracture mechanics. Current work: Mech. behavior of solids (deformation and fracture) and sheet metal forming. Home: 1840 Mershon Ann Arbor MI 48103 Office: Dept Mech Engring U Mich 2020 GGBL North Campus Ann Arbor MI 48109

CADDEN, JAMES MONROE, research petroleum engineer; b. Williston, S.C., Dec. 19, 1923; s. George Otis and Martha Adilade C.; m. Joanna Craig, Aug. 25, 1947; children: Christine Cadden Macmurphy, Nancy Cadden Wells. B.S. in Chemistry and Math, Tex. A&I U., 1949, 1950; postgrad., U. Houston, 1957. Registered petroleum engr., Tex. Gas engr., Palestine, Tex., 1950, petroleum engr., Venice, La., 1950-52, Mid Continent div. Getty Oil Co., Houston, 1952-57, asst. to sr. v.p. corp. office, 1957-58; lead petroleum engr. (Coastal dist.), Ventura, Calif., 1958-62, dist. engr., 1962-69, mgr. systems and data processing for supply and transp. div., mktg. and mfg. div. and internat. exploration and prodn. and minerals div., 1969-70, mgr. engring. Calif. exploration and prodn. div., 1970-72, mgr. engr., 1972-73, chief engr., Los Angeles, 1973-77, mgr., Houston, 1977—. Served with USN, 1942-45. Mem. AIME, Am. Petroleum Inst. Republican. Episcopalian. Club: Petroleum (Houston). Subspecialty: Petroleum engineering. Current work: Optimization of recovery from energy sources, development of innovative methods of finding, evaluating and using energy sources worldwide. Home: 10219 Briar Dr Houston TX 77042 Office: 10210 Westpark Dr Houston TX 77042

CADDY, GLENN ROSS, clinical psychologist, researcher, author, consultant; b. Sydney, Australia, Apr. 8, 1947; came to U.S., 1972; s. Thomas Ross and Hazel (Hendy) C.; m. Janice Lynn Nafziger; 1 son, Gavin David. B.A. with honors, U. New South Wales, Australia), Sydney, 1968, Ph.D., 1972. Assoc. prof. and dir. Addiction Research and Treatment Center, Old Dominion U., 1976-80; prof. psychology, dir. clin. tng. Nova U., 1980—; cons. in field; clin dir Family Inst. of Broward, Ft. Lauderdale, Fla., 1981—; mem. nat. adv. bd. Post-Doctoral Inst. for Mental Health, Ft. Lauderdale, 1982—; chmn. psychology rev. task force Broward Circuit Ct., Ft. Lauderdale, 1982. Contbr. numerous articles to profl. jours.; Assoc. editor: Internat. Jour. Eclectic Psychotherapy, 1982—; co-editor: Ablex Series Developments in Clinical Psychology, 1982—; editor: The Education of Professional Psychologists, 1982; editorial bd.: Addictive Behaviors, 1978—. Mem. exec. com. Va. Mental Health Assn., Richmond, 1979-80. Served to lt. Australian Army, 1970-71. Recipient Tchrs. Coll. award Alexander Mackey Tchrs. Coll., Sydney, 1966; named Outstanding Youg man U.S. Jaycees, 1981; Australian Govt. Commonwealth scholar, 1966-69; U. New South Wales Commonwealth fellow, 1969-71. Mem. Am. Psychol. Assn., Assn. Advancement Behavior Therapy, Southeastern Psychol. Assn.. Episcopalian. Clubs: U. New South Wales Sports Assn. (life), U. New South Wales Regt. Assn. (Sydney) (life). Subspecialties: Behavioral psychology. Current work: Developing psychological tests for computer-based evaluations; studying work stress and remediation programming; evaluating psychological prevention programs for cardiovascular disease. Office: Nova U Dept Psychology 3301 College Ave Fort Lauderdale FL 33314 Home: 9890 SW 1st Ct Plantation FL 33324

CADOGAN, DONALD ANDREW, psychotherapist, consultant; b. N.Y.C., Apr. 10, 1937; s. Thomas Joseph and Helen (Callahan) C.; m. Maureen Patricia Donnelly, Jan. 22, 1972; 1 dau., Laura Maureen. B.A. in psychology, Fairleigh Dickinson U., 1969, M.A. in Psychology, 1970; Ph.D. in Human Relations, Ind. No. U., 1973. Lic. marriage, family and child counselor, Calif. Clin. psychologist Bergen Pines Hosp., Paramus, N.J., 1970-73; psychotherapist, Rosemead, Calif., 1973-76, pvt. practice psychotherapy, Pasadena, Calif., 1976—; counseling cons. Bell & Howell Co., 1982—. Author: (with others) Family Therapy of Drug and Alcohol Abuse, 1979. Served with AUS, 1956-58. Mem. Am. Psychol. Assn., Calif. Assn. Marriage and Family Therapists, So. Calif. Psychotherapy Affiliation. Subspecialty: Psychotherapy. Home: 1019 S Magnolia St West Covina CA 91791 Office: 550 N Rosemead Blvd Pasadena CA 91107

CADY, PHILIP DALE, civil engineering educator, researcher; b. Elmira, N.Y., June 26, 1933; s. Hugh William and Martha Elizabeth (Budnick) C.; m. Shirley Mary Broadbent, Oct. 1, 1955; children: Stephen, Alan, Joyce, Michael, Thomas. Student, Mansfield (Pa.) State Coll., 1951-53; B.S., Pa. State U., University Park, 1956, M.S., 1964, Ph.D., 1967. Registered profl. engr., Pa.; registered profl. land surveyor, Pa. Engr. Esso Standard Oil. Co., Balt., 1956-57; engr. Lago Oil and Transport Co., Ltd., Aruba, Netherlands Antilles, 1957-62; research asst. Pa. State U., 1962-67, asst. prof. civil engring., 1967-69, assoc. prof., 1969-75, prof., 1975—; cons. in field. Contbr. articles to tech. jours. Vice chmn. College-Harris Joint Sewer Authority, 1972—. Recipient Researcher of Yr. award Pa. State U. Engring. Soc., 1976. Fellow ASCE, Am. Concrete Inst; mem. ASTM (Sanford E. Thompson award 1968), Transp. Research Bd. Subspecialties: Civil engineering; Materials (engineering). Current work: Research on deteriorative mechanisms in concrete constructions; weathering of concrete; freezing and thawing deterioration of concrete; reinforcement corrosion; deleterious aggregates; polymer impregnated concrete. Patentee in polymer impregnation of concrete. Home: Box 158 Lemont PA 16851 Office: Pa State U 212 Sackett Bldg University Park PA 16802

CADY, WALLACE MARTIN, research geologist; b. Middlebury, Vt., Jan. 29, 1912; s. Frank William and Alice Marian (Kingsbury) C.; m. Helen Johanna Raitanen, Jan. 1, 1942; children: John Wallace, Nancy Helen, Norma Louise. B.S., Middlebury Coll., 1934; M.S.,

Northwestern U., 1936; Ph.D., Columbia U., 1944. Asst. in geology Northwestern U., Evanston, Ill., 1934-36, Columbia U., N.Y.C., 1938-40; geol. field asst. U.S. Geol. Survey, Olympia Peninsula, Wash., 1938-40; tutor geology Bklyn Coll., 1939-40; research geologist U.S. Geol. Survey, 1941—; Fulbright lectr. Varonezh State U., USSR, 1975. Fellow Geol. Soc. Am., Am. Geophys. Union, AAAS; mem. Colo. Sci. Soc. (pres. 1974-75), Soc. Econ. Geologists. Subspecialties: Geology; Tectonics. Current work: Geology of the Olympic Peninsula, Northwestern Washington. Home: 348 S Moore St Lakewood CO 80226 Office: US Geol Survey PO Box 25046 Fed Ctr Denver CO 80225

CAFLISCH, RUSSEL EDWARD, mathematics educator; b. Charleston, W.Va., Apr. 29, 1954; s. Edward George and Dorothy Gail (Barrett) C. B.S., Mich. State U., 1975; M.S., NYU, 1977, Ph.D., 1978. Fannie and John Hertz fellow Courant Inst. NYU, N.Y.C., 1978-79, asst. prof., 1983—; asst. prof. math. Stanford (Calif.) U., 1979-82; cons. Los Alamos Nat. Lab., 1979—, Livermore Nat. Lab., 1977—, Bell Labs., Murray Hill, N.J., 1981. Author: Studies in Statistical Mechanics, 1983. Mem. Am. Math. Soc., Soc. Indsl. and Applied Math. Subspecialty: Applied mathematics. Current work: Fluid dynamics and its relation to the Boltzmann equation of kinetic theory, sedimentation theory, inverse scattering, asymptotic analysis. Office: Courant Inst Math Scis New York Univ New York NY 10012 Home: 88 Bleecker St New York NY 10012

CAGIN, NORMAN ARTHUR, physician, medical educator and researcher; b. Dover, N.J., Mar. 7, 1941; s. Martin M. and Selma (Huckman) C.; m. Joyce Ann Hammerstedt, Dec. 1, 1979. B.S., Rutgers U., 1963; M.D., N.Y. Med. Coll., 1967. Diplomate: Am. Bd. Internal Medicine. Asst. prof. N.Y. Med. Coll., N.Y.C., 1974-77; clin. asst. prof. Albert Einstein Coll. Medicine, Bronx, N.Y., 1977-78; clin. asst. prof. medicine Coll. Physicians and Surgeons, Columbia U., N.Y.C., 1978-81, assoc. prof. clin. medicine, 1981—; research assoc. VA, East Orange, N.J., 1978—. Served to lt. comdr. USN, 1972-74. Fellow Am. Coll. Cardiology; mem. N.Y. Cardiology Soc., Am. Fedn. for Clin. Research, N.Y. Acad. Sci., N.Y. Heart Assn. Jewish. Subspecialty: Cardiology. Current work: Investigate the role of the central nervous system in causing digitalis toxicity. Office: 315 W 70th St New York NY 10023

CAHILL, GEORGE FRANCIS, JR., physician, educator; b. N.Y.C., July 7, 1927; s. George Francis and Eva Marion (Wagner) C.; m. Sarah Townsend duPont, Dec. 20, 1949; children: Colleen (Mrs. Thomas P. Remley), Peter duPont, George F. III, Sarah Rhett, Eva Wagner (Mrs. William M. Doll), Elizabeth Anglin. B.S., Yale, 1949; M.D., Columbia U., 1953; M.A., Harvard U., 1966. Intern Peter Bent Brigham Hosp., Boston, 1953-54, resident, 1954-55, 57-58, asso. in medicine, 1962-65, sr. physician, 1983—; research fellow biol. chemistry Harvard U., 1955-57, prof. medicine, 1970—; practice medicine specializing in metabolism, Boston, 1965—; Prin. cons. endocrinology, metabolism VA, 1972-75; investigator Howard Hughes Med. Inst., 1962-68, dir. research, 1978—; mem. research tng. coms. NIH. Contbr. articles to profl. jours. Served with USNR, 1945-47. Recipient Banting medal, U.S., 1971, Eng., 1974, J.P. Hoet award, Belgium, 1973. Mem. Am. Diabetes Assn. (pres. 1975, Lilly award 1965), Endocrine Soc. (Oppenheimer award 1963, Gairdner Internat. award 1979), Nat. Commn. on Diabetes, Am. Soc. Clin. Investigation, Assn. Am. Physicians, Am. Clin. Climatol. Assn., Am. Physiol. Soc., Am. Acad. Arts and Scis. Club: Wellesley Country. Subspecialties: Endocrinology; Physiology (medicine). Home: Upton Pond Stoddard NH 03464 Office: 398 Brookline Ave Boston MA 02215

CAHILL, LAURENCE JAMES, JR., physicist, educator; b. Frankfort, Maine, Sept. 21, 1924; s. Laurence J. and Wilma (Lord) C.; m. Alice Adeline Krieger, Sept. 10, 1949; children: Laurence James III, Thomas G., Daniel A. Student, U. Maine, 1942-43; B.S., U.S. Mil. Acad., 1946, U. Chgo., 1950; M.S., U. Iowa, 1956, Ph.D., 1959. Staff U. Iowa, 1954-59, research asso., 1959; mem. faculty U. N.H., 1959-68, prof. physics, 1965-69; dir. Space Scis. Center, 1966-68; prof. physics U. Minn., Mpls., 1968—, asso. head physics, 1974-77; dir. Space Sci. Center, 1968-74; chief physics NASA Hdqrs., Washington, 1962-63, cons., 1962—; vis. prof. U. Calif. at San Diego, 1965-66; cons. NSF, 1965—. Recipient NASA award for sustained superior performance, 1963; NATO sr. fellow, 1974; vis. scientist Max Planck Inst. Extraterrestrial Physics, W. Ger., 1977-78. Fellow Am. Geophys. Union, Am. Phys. Soc.; mem. AAAS, Sigma Xi. Subspecialties: Space physics; Satellite studies. Current work: Rocket studies of auroral phenomena; hydromagnetic waves in magnetosphere. Research and publs. on measurement by rocket-borne magnetometer of elec. currents in ionosphere, measurement boundary between earth's magnetic field and interplanetary medium, ring current of charged particles encircling earth and causing magnetic storms, hydromagnetic waves. Home: Afton MN 55001 Office: U Minn Dept Physics 116 Church St SE Minneapolis MN 55455

CAHILL, THOMAS ANDREW, physicist; b. Paterson, N.J., Mar. 4, 1937; s. Thomas Vincent and Margery (Groesbeck) C.; m. Virginia Ann Arnoldy, June 26, 1965; children: Catherine Frances, Thomas Michael. B.A., Holy Cross Coll., Worcester, Mass., 1959; Ph.D. in Physics; NDEA fellow, U. Calif., Los Angeles, 1965. Asst. prof. in residence U. Calif., Los Angeles, 1965-66; NATO fellow, research physicist Centre d'Etudes Nucleaires de Saclay, France, 1966-67; prof. physics U. Calif., Davis, 1967—; acting dir. Crocker Nuclear Lab., 1972, dir., 1980—; dir. Inst. of Ecology, 1972-75. Author: (with J. McCray) Electronic Circuit Analysis for Scientists, 1973; Contbr. to profl. jours., articles on physics, applied physics, and air pollution. OAS fellow, 1968. Mem. Am. Phys. Soc., Air Pollution Control Assn., Sigma Xi, Sigma Pi Sigma. Democrat. Roman Catholic. Club: Sierra. Subspecialties: Nuclear physics; Atmospheric chemistry. Current work: Application of nuclear and atomic techniques to analysis of atmospheric particles. Home: 1813 Amador Ave Davis CA 95616 Office: Dept Physics U Calif Davis CA 95616

CAHILL, VERN RICHARD, meat scientist, educator, researcher; b. Tiro, Ohio, May 5, 1918; s. Verrill W. and Marie A. (Galehr) C.; m. Ruth Alice Huber, June 23, 1946; children: Nancy Cahill Haar, Donna Cahill Solovay, Kenneth. B.Sc., Ohio State U., 1941, M.S., 1942, Ph.D., 1955. Mem. faculty Ohio State U., 1946—, asst. prof. meat sci., 1955-56, assoc. prof., 1956-61, prof., 1961—, coordinator meat sci., 1972—, acting chmn. animal sci., 1983. Author: Meat Processing, rev. edit, 1980; contbr. numerous articles to profl. jours. Served to lt. col. U.S. Army. Recipient Educators award Nat. Assn. Meat Purveyors, 1972. Mem. Am. Meat Sci. Assn. (Disting. Teaching award 1967, Signal Service award 1979), Am. Assn. Animal Sci., Inst. Food Technologists, Ohio Meat Processors Assn. (dir.), Ohio Meat Industries Assn. (trustee). Lutheran. Subspecialty: Food science and technology. Current work: Meat quality, comminuted meat processing. Home: 133 Aldrich Rd Columbus OH 43214 Office: 2029 Fyffe Rd Columbus OH 43210

CAHILL, WILLIAM JOSEPH, JR., public utility executive, engineer; b. Suffern, N.Y., June 13, 1923; s. William Joseph and Sophie A. (Scozzafava) C.; m. Edna Kierman, Oct. 3, 1953; children: William E., Kathleen, Madeleine. B.S. in Mech. Engring, Poly. Inst., 1949. Engr. Consol. Edison, N.Y.C., 1949-54, Knolls Atomic Power Lab., Schenectady, N.Y., 1954-56, Consol. Edison, N.Y.C., 1957-60, nuclear plant engr., 1961-68, v.p., 1969-80; sr. v.p. Gulf States Utilities Co., St. Francisville, La., 1980—; chmn. safety and analysis task force Electric Power Research Inst., Palo Alto, Calif., 1978-80. Pres. Queens County Young Republicans; bd. dirs. Rockland County Assn. Retarded. Served with U.S. Army, 1942-46. Mem. ASME, Am. Soc. Profl. Engrs., Am. Nuclear Soc. (dir. 1980-83), La. Nuclear Soc. (chmn. 1982-83). Roman Catholic. Club: City (Baton Rouge). Subspecialties: Mechanical engineering; Nuclear engineering. Current work: Nuclear safety, electrical power technology. Patentee nuclear reactor vessel, self-activated valve, triggerable fuse. Home: PO Box 835 St Francisville LA 70775 Office: Gulf States Utilities Co PO Box 220 St Francisville LA 70775

CAHN, JOHN WERNER, metallurgist, educator; b. Germany, Jan. 9, 1928; came to U.S., 1939, naturalized, 1945; s. Felix H. and Lucie (Schwarz) C.; m. Anne Hessing, Aug. 20, 1950; children—Martin Charles, Andrew David, Lorie Selma. B.S., U. Mich., 1949; Ph.D., U. Calif. at Berkeley, 1953. Instr. U. Chgo., 1952-54; with research lab. Gen. Electric Co., 1954-64; prof. metallurgy Mass. Inst. Tech., 1964-78, adj. prof. materials sci., 1979—; center scientist Nat. Bur. Standards, 1978—; vis. prof. Israeli Inst. Tech., Haifa, 1971-72, 80; cons. in field, 1963—; chmn. Gordon conf. Phys. Metallurgy, 1964; vis. scientist Nat. Bur. Standards, Gaithersburg, Md., 1977; hon. prof. Jiao Tung U., Shanghai, China, 1980—. Guggenheim fellow, 1960; research fellow Japan Soc. for Promotion of Sci., 1981-82; recipient Dickson prize Carnegie Mellon U., 1981. Fellow Am. Acad. Arts and Scis.; mem. Am. Inst. Metall. Engrs., AAAS, Nat. Acad. Scis. Subspecialties: Materials; Condensed matter physics. Current work: Surfaces and interfaces, thermodynamics, phase changes. Home: 6610 Pyle Rd Bethesda MD 20817 Office: Nat Bur Standards Washington DC 20234

CAIN, CHARLES ALAN, bioengineering educator; b. Tampa, Fla., Mar. 3, 1943. B.E.E., U. Fla., 1965; M.S.E.E., MIT, 1967; Ph.D. in Elec. Engring, U. Mich., 1972. Mem. tech. staff Bell Labs., 1965-68; asst. prof. elec. engring. U. Ill.-Urbana, 1972-78, assoc. prof., 1978-83, prof., 1983—, also chmn. Engring. Faculty, also dir. microwave bioengring. lab. Mem. IEEE (sr.). Subspecialty: bioengineering. Office: Dept Elec Engring U Ill Urbana IL 61801

CAIN, J(AMES) ALLAN, geology educator; b. Douglas, Isle of Man, Gt. Brit, July 23, 1935; came to U.S., 1958; s. James Herbert and Margaret Ivy (Moore) C.; m. Leila Nation Scelonge, June 20, 1959 (div. 1978); children: Geoffrey Carleton, Trevor Bradley. B.Sc., U. Durham, Eng., 1958; M.S., Northwestern U., 1960, Ph.D., 1962. Instr. Case Western Res. U., Cleve., 1961-62, asst. prof. geology, 1962-66; assoc. prof. U. R.I., Kingston, 1966-71, prof. geology, 1971—, chmn. dept., 1967—, acting. assoc. dean Coll. Arts and Scis., 1983-84; hon. research assoc. Harvard U., 1973; vis. scientist MIT, Cambridge, 1973, Union Oil of Calif., Ventura and Brea, 1980, Hanna Mining Co., Salt Lake City, 1980. Author: Geology - A Synopsis, Part 1, Physical Geology, 1980, (with E. J. Tynan) Part 2, Historical Geology, 1981, Environmental Geology, (with J. C. Boothroyd), 1983. Vice chmn. bd. trustees Kingston Library, 1970-78; v.p. Providence Opera Theatre, 1978-79; mem. Richmond Conservation Commn., 1983—. Served with RAF, 1953-55. Research grantee NSF, 1964-68, Research Corp., 1962-65, U.S. Navy, 1968-69, U.S. Geol. Survey, 1975-77. Fellow Geol. Soc. Am.; mem. Nat. Assn. Geology Tchrs., AAUP, Sigma Xi. Subspecialties: Petrology; Mineral resources. Current work: Environmental geology; distribution of strategic minerals in New England. Home: Clarke Trail West Kingston RI 02892 Office: Dept Geology U RI Kingston RI 02881

CAIN, STEPHEN MALCOLM, physiology and biophysics educator; b. Lynn, Mass., Oct. 4, 1928; s. Herbert and Eva (Rowe) C.; m. Helen Gladys Allen, Sept. 14, 1951; children: Nancy Ellen Cain Hosmer, Carol Ann. B.S., Tufts U., 1949; Ph.D., U. Fla., 1959. Physiologist U.S. Air Force Sch. Aerospace Medicine, San Antonio, 1959-71; prof. physiology and biophysics U. Ala. in Birmingham, 1971—; cons. Nat. Heart, Lung and Blood Inst., Bethesda, Md. Contbr. chpts. to books, articles to profl. jours. Served with U.S. Army, 1951-53. Nat. Heart, Lung and Blood Inst. grantee, 1971—. Mem. Soc. Exptl. Biology and Medicine, Am. Physiol. Soc., Can. Physiol. Soc., Am. Thoracic Soc., Sigma Xi. Subspecialty: Physiology (medicine). Current work: Oxygen transport to tissues, hypoxia. Home: 3752 Forest Run Rd Birmingham AL 35223 Office: U Ala Dept Physiology and Biophysics Birmingham Al 35294

CAIN, WILLIAM S., psychology educator, researcher; b. N.Y.C., Sept. 7, 1941; s. William H. and June (Stanley) C.; m. Eileen M. Nugent, Jan. 25, 1964; children: Justin, Alison. B.S., Fordham U., 1963; S.M., Brown U., 1966, Ph.D., 1968. Assoc. fellow Pierce Found., New Haven, 1967—; assoc. prof. epidemiology and psychology Yale U., 1969—; v.p. Fragrance Found. Philanthropic Trust, N.Y.C., 1981—. Contbr. numerous articles to profl. publs.; co-editor: Stimulus and Sensation, 1971, Evaluation, Utilization, and Control, 1974. Fellow N.Y. Acad. Scis. (v.p. N.Y.C. 1983—, nat. v.p.), Am. Psychol. Assn.; mem. Assn. for Chemoreception Scis. (exec. chmn.-elect 1982-83). Subspecialties: Sensory processes; Environmental health. Current work: Chemoreception, i.e. smell, taste, irritation; sensory reactions to indoor and outdoor air contaminants; disorders of sense of smell. Home: 135 Royden Rd New Haven CT 06511 Office: John B Pierce Found 290 Congress Ave New Haven CT 06519

CAIRD, JOHN ALLYN, physicist; b. Bklyn., Dec. 24, 1947; s. William John and Ethel (Fountain) C. B.S., Rutgers U., 1969; M.S., UCLA, 1971; Ph.D., U. So. Calif., 1975. Reliability analyst Consol. Edison, N.Y.C., 1968; staff mem., masters fellow, doctoral fellow, staff physicist Hughes Aircraft Co., Culver City, Calif., 1969-76; NSF postdoctoral fellow, Argonne postdoctoral fellow Argonne Nat. Lab., Ill., 1976-78; engring. specialist, sr. scientist Bechtel Group, Inc., San Francisco, 1978-81; staff mem. Los Alamos Nat. Lab., 1981—. Contbr. articles to profl. jours. Mem. IEEE, Am. Phys. Soc., Optical Soc. Am. Club: Ski (Santa Fe). Subspecialties: Laser-induced chemistry; Atomic and molecular physics. Current work: Laser-induced chemistry and spectroscopy, energy-related tech., primarily nuclear fission, currently performing experiments to demonstrate feasibility of molecular laser isotope separation. Office: Los Alamos Nat Lab Mail Stop J-565 Los Alamos NM 87545

CAIRNS, THEODORE LESUEUR, chemist; b. Edmonton, Alta., Can., July 20, 1914; came to U.S., 1936, naturalized, 1945; s. Albert William and Theodora (MacNaughton) C.; m. Margaret Jean McDonald, Aug. 17, 1940; children—John Albert, Margaret Eleanor (Mrs. William L. Etter), Elizabeth Theodora (Mrs. Ernest I. Reveal III), James Richard. B.S., U. Alta., 1936, LL.D., 1970; Ph.D., U. Ill., 1939. Instr. organic chemistry U. Rochester, 1939-41; research chemist central research dept. E.I. duPont de Nemours & Co., Wilmington, Del., 1941-45, research supr., 1945-51, lab. dir., 1951-63, dir. basic scis., 1963-66, dir. research, 1966-67, asst. dir. central research and devel. dept., 1967-71, dir., 1971-79; Regents prof. U. Calif., Los Angeles, 1965-66; mem. adv. bd. Organic Syntheses, 1958—; mem. Pres.'s Sci. Adv. Com., 1970-73, Pres.'s Com. Nat. Medal Sci., 1974-75; chmn. Office of Chemistry and Chem. Tech., NRC, 1979-81. Editorial bd.: Organic Reactions, 1959—, Jour. Organic Chemistry, 1965-69. Recipient award for creative work in synthetic organic chemistry Am. Chem. Soc., 1968; Perkin medal, 1973; Cresson medal Franklin Inst., 1974. Mem. Nat. Acad. Scis., Am. Chem. Soc. (chmn. organic div. 1964-65), AAAS, Sigma Xi, Phi Lambda Upsilon, Alpha Chi Sigma, Phi Lambda Upsilon (hon.). Subspecialty: Organic chemistry. Home: Box 3941 Greenville DE 19807

CALABRESE, VINCENT PAUL, physician; b. Jamaica, N.Y., Sept. 23, 1939; s. Giuseppe O. and Florence (Verderese) C.; m. Linda Metzger, June 24, 1966; children: Gregory Paul, Dana Lynn. A.B., Columbia U., 1961; M.D., Downstate SUNY, 1965. Diplomate: Am. Bd. Psychiatry and Neurology. Intern U. Pitts., 1965-66, resident in internal medicine, 1966-67; resident in neurology Albert Einstein Coll. Med., N.Y.C., 1967-70; asst. prof. neurology Med. Coll. Va., Richmond, 1972-78, assoc. prof., 1978—; staff physician McGuire VA Hosp., Richmond, 1974—; dir. neurochemistry lab., 1972—. Contbr. articles to profl. jours. Served to maj USAF, 1970-72. Fellow NIH, 1979-80. Mem. Am. Acad. Neurology, Soc. Neurosci., Am. Soc. Neurochemistry, Internat. Soc. Neurochemistry, AAAS, N.Y. Acad. Sci. Subspecialties: Neurochemistry; Neuroimmunology. Current work: Myelin and axolemma interactions, isolation of specific proteins from axolemma and interspecies differences; neuroimmunology—demyelinating disease, determination of antigens for abnormal IgA, IgM, IgG produced. Office: Dept Neurology Med Coll Va Box 599 Richmond VA 23298

CALDER, CLARENCE ANDREW, mech. engr., educator, cons., researcher; b. Baker, Oreg., Oct. 30, 1937; s. Clarence Leroy and Viola Mary (Lucas) C.; m. Judy Lee Wood, Dec. 15, 1961; children: Brian, Gregory, Kaylene, Chad, Jared. B.S.M.E., Oreg. State U., 1960; M.S., Brigham Young U., 1962; Ph.D. in Engring. Sci, U. Calif., Berkeley, 1969. Registered profl. engr., Calif., Oreg. Design engr. Boise Cascade Corp., Emmett, Idaho, 1960-61; project engr. Sandia Labs., Albuquerque, 1962-64; research assoc. U. Calif., Berkeley, 1966-69; asst. prof. Wash. State U., 1969-74; research engr. Lawrence Livermore (Calif.) Nat. Labs., 1974-78; assoc. prof. mech. engring. Oreg. State U., 1978—; cons. laser applications in engring., athletic shoe performance. Contbr. articles on exptl. mechanics, laser applications in engring., instrumentation, stress waves in materials, athletic shoe performance to profl. jours. Recipient Best Paper award of recognition Am. Nuclear Soc., 1978. Mem. Soc. Exptl. Stress Analysis (F. G. Tatnal award 1982), ASME, Soc. Engring. Edn. Republican. Mormon. Subspecialties: Mechanical engineering; Solid mechanics. Current work: Research and applications using lasers as sensors and as ultrasonic stress wave generators in non-destructive testing; noncontact material testing using lasers; athletic shoe performance testing and analysis. Home: Route 1 Box 484N Philomath OR 97370 Office: Dept Mech Engring Oreg State U Corvallis OR 97331

CALDWELL, DAVID ORVILLE, physicist, educator; b. Los Angeles, Jan. 5, 1925; s. Orville Robert and Audrey Norton (Anderson) C.; m. (div.); children: Bruce David, Diana Miriam. B.S., Calif. Inst. Tech., 1947; postgrad., Stanford U., 1947-48; M.A., UCLA, 1949, Ph.D., 1953. Instr. dept. physics MIT, 1954-56, asst. prof., 1956-58, assoc. prof., 1958-63; vis. assoc. prof. Princeton U., 1963-64; lectr. U. Calif.-Berkeley, 1964-65; prof. physics U. Calif.-Santa Barbara, 1965—; cons. Lawrence Radiation Lab., U. Calif., 1950-51, 65-70, Am. Sci. and Engring., Cambridge, Mass., 1959-60, Inst. Def. Analyses, 1960-67, Dept. Def., 1966-70; assoc. dir Intercampus Inst. Research at Particle Accelerators. Contbr. chpts. to books, articles to profl jours. Served to 2d lt. USAAF, 1943-46. AEC predoctoral fellow, 1950-52; NSF postdoctoral fellow, 1953-54; NSF sr. postdoctoral fellow, 1960-61; Ford Found. fellow, 1961-62; Guggenheim Found. fellow, 1971-72; research grantee AEC, Energy Research and Devel. Agy., Dept. of Energy, 1966—. Fellow Am. Phys. Soc. Subspecialties: Particle physics; Nuclear physics. Current work: Teaching and research; research generally seeking the fundamental laws of nature; current experiments in two-photon annihilation and double-beta decay. Office: Dept Physics U Calif Santa Barbara CA 93106

CALDWELL, ELWOOD FLEMING, food science educator, researcher; b. Gladstone, Man., Can., Apr. 3, 1923; s. Charles Fleming and Frances Marion (Ridd) C.; m. Irene Margaret Sebille, June 13, 1949; children: John Fleming, Keith Allan; m. Florence Annette Zaz, June 23, 1979. B.Sc., U. Man., 1943; M.A. in Food Chemistry, U. Toronto, 1949; Ph.D. in Nutrition, U. Toronto, 1953; M.B.A., U. Chgo., 1956. Chemist Lake of the Woods Milling Co., Can., 1943-47; research chemist Can. Breweries Ltd., Toronto, Ont., 1948-49; chief chemist Christie, Brown & Co. (Nabisco), Toronto, 1949-51; research assoc. in nutrition U. Toronto, 1951-53; with Quaker Oats Co., Barrington, Ill., 1953-72, dir. research and devel., until 1972; prof., head dept. food sci. and nutrition U. Minn., St. Paul, 1972—; chmn. bd. Dairy Quality Control Inst., Inc., St. Paul, 1972—, R.& D. Assocs. for Mil. Food & Packaging, Inc., San Antonio, 1970-71; chmn. evening programing food sci. Ill. Inst. Tech., Chgo., 1965-69. Contbr. articles to sci. jours. Chmn. North Barrington (Ill.) Bd. Appeals, 1966-69, mayor, 1969-72; vice-chmn. Barrington Area Council Govts., 1972; bd. dirs. Family Guidance Barrington, 1971-72. Recipient cert. of appreciation for civilian service U.S. Army Materiel Command, 1970. Fellow Inst. Food Technologists (Chmn.'s Service award Chgo. sect. 1975); mem. Am. Assn. Cereal Chemists, Am. Home Econs. Assn., Phi Tau Sigma, Sigma Xi, Gamma Sigma Delta. Republican. Lutheran. Club: Minnesota Alumni (Mpls.). Subspecialties: Food science and technology; Nutrition (biology). Current work: Administration of teaching, research and extension in food science and nutrition; direct undergraduate and graduate instruction and graduate research in food science, food technology, nutrition, dietetics, consumer studies. Office: 1334 Eckles Ave U Minn Saint Paul MN 55108

CALDWELL, GLYN GORDON, physician, epidemiologist, virologist, govt. adminstr.; b. St. Louis, Jan. 14, 1934; s. Cecil Gordon and Zelma Mae (Peeler) C.; m. Mary Jean Pandolfo, Aug. 13, 1960; children: Michael, Elizabeth, Thomas. B.S. (Univ. scholar), St. Louis U., 1960; M.S., U. Mo., 1962, M.D., 1966. Intern USPHS Hosp., Brighton, Mass., 1966-67; resident in internal medicine Cleve. Met. Gen. Hosp., 1969-71; commd. sr. asst. surgeon USPHS, 1966, advanced through grades to dir., 1979; with Center for Disease Control, Atlanta, 1967—, Chief field investigations sect. cancer research, 1974-77, chief cancer br., chronic diseases div., 1977-82, dep. dir. epidemiology and environ. service div., 1982—. Contbr. articles to profl. publs. Served to sgt. Signal Corps U.S. Army, 1954-62. Recipient Commendation medal USPHS, 1980. Mem. Am. Soc. Microbiology, Soc. Epidemiologic Research, Am. Soc. Preventive Oncology, USPHS Commd. Officers Assn. Roman Catholic. Lodge: KC. Subspecialties: Epidemiology; Internal medicine. Current work: Cancer epidemiology, including radiation, viral, chem. and genetic oncology.

CALDWELL, WILLARD E., psychologist, psychology educator; b. Flushing, N.Y., July 10, 1920; s. Howard E. and Lillian (Warner) C. A.B. in Psychology, U. Fla.-Gainesville, 1940, M.A., 1941, Ph.D., Cornell U., 1946. Lic. psychologist, D.C., Fla. Prof. psychology, head psychology dept. Mary Baldwin Coll., Staunton, Va., 1946-47; mem. faculty George Washington U., Washington, 1947—, prof. psychology, 1983—; pvt. practice psychol. counseling, Washington. Co-author; co-editor: Principles of Comparative Psychology, 1960; contbr. over 50 articles to profl publs. Mem. Am. Psychol. Assn., D.C. Psychol. Assn., Eastern Psychol. Assn., Internat. Soc. Biometeorology, Sigma Xi. Episcopalian. Club: Cornell (Washington). Subspecialties: Psychobiology; Ethology. Current work: Corona discharge photography applied to living organisms in both animal and humans; general theory of apnea as it applied to Sudden Infant Death Syndrome. Home: 1500 Massachusetts Ave NW Apt 250 Washington DC 20005 Office: George Washington U 316 Bldg GG Washington DC 20052

CALEDONIA, GEORGE ERNEST, research scientist; b. Boston, Nov. 9, 1941; s. George F. and Gilda (Cimmino) C.; m. Diane M., Nov. 6, 1965; children: Karen, Julie, Elizabeth. A.B., Northeastern U., 1965, M.S., 1967. Prin. scientist Avco Everett Research Lab., Mass., 1967-73; prin. scientist Phys. Scis. Inc., Andover, Mass., 1973-80, v.p. research, 1980—, also dir. Contbr. articles to profl. jours. Recipient Marcus O'Day Meml. award USAF Geophysics Lab. Mem. Am. Phys. Soc., Am. Geophys. Union, Am. Chem. Soc. Subspecialties: Aeronomy; Atomic and molecular physics. Current work: Upper atmospheric radiative and kinetic phenomena, high temperature gas radiation properties, electron excitation of gases, kinetic analysis. Office: Research Park Andover MA 01810

CALESNICK, BENJAMIN, physician, clinical pharmacologist; b. Phila., Dec. 27, 1915; s. Samuel and Ida (Lichtenstein) C.; m. Sophie Adele Brenner, Dec. 27, 1921; 1 son, Jay Lee. B.S., St. Joseph's Coll., 1938; M.A., Temple U., 1941; M.D., Hahnemann Med. Coll., 1944. Intern Phila. Gen. Hosp., 1944-45; faculty Hahnemann U., Phila., 1946—; prof. pharmacology and medicine Hahnemann Med. Coll., 1971—. Served to lt. comdr. USN, 1945-54. Subspecialties: Internal medicine; Pharmacology. Current work: Teaching and research in clin. pharmacology. Home: 646 W Springfield Rd Springfield PA 19064 Office: Hahnemann U Dept Pharmacology and Medicine 230 N Broad St Philadelphia PA 19102

CALIFANO, JOSEPH MICHAEL, quality engr.; b. Bklyn., Jan. 1, 1951; s. Joseph Frank and Ann (Ruscitto) C.; m. Theresa Ann Marie Swartout, June 16, 1978; 1 son. Eric Michael. B.A. in Physics, Hunter Coll., CUNY, 1972. Asst. mgr. research and devel. lab. Pall Corp., Glen Cove, N.Y., 1974-75, mgr. biomed. quality control, 1975-77; mgr. quality control Amicon Corp., Lexington, Mass., 1978-79, Gelman Scis. Co., Ann Arbor, Mich., 1979-81; quality engr., 1981—. Mem. Am. Soc. for Quality Control (cert. quality engr.), High I.Q. Soc. San Francisco. Democrat. Roman Catholic. Subspecialties: Statistics; Filtration membranes. Current work: Design, review and validation of products and processes for filtration membranes; statistical analysis of product and process data; design of experiments; establish test methods.

CALIGIURI, JOSEPH FRANK, multinational company executive; b. Columbus, Ohio, Feb. 13, 1928; s. Frank and Angeline J. (Gentila) C.; m. Barbara Jane DeLaney, June 15, 1948; children: Mark, Timothy, Jeffrey, Andrew. B.S.E.E., Ohio State U., 1949, M.S.E.E., 1951. Chief engr. Sperry Gyroscope Co., Great Neck, N.Y., 1966-69; v.p. engring. Guidance and Control Systems, Wedland Hills, Calif., 1969-71; pres., 1971-77; v.p. Litton Industries, Inc., Beverly Hills, Calif., 1971-77, sr. v.p., group head, 1977-81, exec. v.p. and group head, 1981—. Mem. Assn. U.S. Army (chpt. dir., pres.), Nat. Security Indsl. Assn. (trustee), Am. Def. Preparedness Assn. (dir.), Inst. Nav., IEEE, AIAA. Republican. Office: 360 N Crescent Dr Beverly Hills CA 90210

CALIO, ANTHONY JOHN, scientist, govt. ofcl.; b. Phila., Oct. 27, 1929; s. Antonio and Mary Emma (Cappuccio) C.; m. Cheryll Kay Madison, Feb. 28, 1971. B.A., U. Pa., 1953, postgrad., 1953; postgrad., Carnegie Inst. Tech., 1959; Sc.D. (hon.), Washington U., St. Louis, 1974; postgrad. (Sloan fellow), Stanford U., 1974-75. With Westinghouse Electric Corp., Pitts., 1956-59; chief nuclear physics sect. Am. Machine & Foundry Co., Alexandria, Va., 1959-61; v.p., mgr. ops. Mt. Vernon Research Co., Alexandria, 1961-63; mem. electronic research task group Hdqrs. NASA, Washington, 1963-64, chief research engring., Boston, 1964-65, chief instrumentation and systems integration br., Washington, 1965-67, asst. dir. planetary exploration, 1967-68, dir. sci. and applications, Houston, 1969-75, dep. asso. adminstr., Washington, 1975-77, asso. adminstr., 1977—, acting dep. adminstr., 1981—. Served with U.S. Army, 1954-56. Recipient Group Achievement award (2) NASA, 1969, Exceptional Service medal, 1969, Apollo Achievement award, 1970, Exceptional Sci. Achievement medal, 1971, Lunar Sci. Team award, 1973, Distinguished Service medal, 1973, Exec. Performance award, 1976; presdl. rank of Disting. Exec., 1980. Fellow Am. Astron. Soc., AIAA (asso); mem. AAAS, Am. Geophys. Union, N.Y. Acad. Scis. Subspecialties: Satellite studies; Remote sensing (atmospheric science). Current work: Space physics; chemistry engineering. Home: Nat'l Oceanic and Atmospheric Admn. 19th Between E and Constitution Ave NW Washington DC 20230 Office: Office of Asso Adminstr Space and Terrestrial Applications NASA Washington DC 20546

CALKIN, PARKER EMERSON, geology educator; b. Syracuse, N.Y., Apr. 27, 1933; s. Frank G. and Georgia (Spencer) C.; m. Joan A. Chace, Sept. 15, 1955 (div. 1978); children: Mark, Lisa; m. Harriet R. Simons, Feb. 19, 1979. B.S., Tufts U., 1955; M.Sc., U. B.C., 1959; Ph.D., Ohio State U.-Columbus, 1963. Asst. prof. geology Coll. at SUNY-Buffalo, 1963-65; asst. prof. SUNY-Buffalo, 1965-68, assoc. prof., 1968-75, prof. geology, 1975—; dir. Tech. Systems Research, Inc., N.Y., 1981—. Editor, author: series Great Lakes Coastal Geology, 1981, 82. Served to lt. (j.g.) USN, 1955-57. NSF grantee, 1960-64, 66-72, 76-83; SUNY faculty research fellow, 1973; NOAA sea grantee, 1974-80. Fellow Geol. Soc. Am.; mem. Glaciological Soc., N.Y. State Geol. Assn. (pres. 1982), Am. Quaternary Assn., Sigma Xi. Subspecialty: Geology. Current work: Polar geology, particularly geomorphology and glacial geology in Antarctica, Greenland and Alaska. Alaskan work concentrates on Holocene glacial chronology of Brooks Range. Home: 49 Blossom Heath Williamsville NY 14221 Office: Dept Geol Scis SUNY 4240 Ridge Lea Rd Buffalo NY 14226

CALKINS, EVAN, physician, educator; b. Newton, Mass., July 15, 1920; s. Grosvenor and Patty (Phillips) C.; m. Virginia McC. Brady, Sept. 9, 1946; children: Sarah Calkins Oxnard, Stephen, Lucy McCormick, Joan Calkins Bender, Benjamin, Hugh, Ellen Rountree, Geoffrey, Timothy. Grad., Milton Acad., 1939; A.B., Harvard U., 1942, M.D., 1945. Intern, asst. resident medicine Johns Hopkins, 1946-47, 48-50; chief resident physician Mass. Gen. Hosp., 1951-52, mem. arthritis unit, 1952-61; NRC fellow med. scis. Harvard, 1950-51, instr., asst. prof. medicine, 1952-61; practice medicine, specializing in rheumatology, Boston, 1951-61, Buffalo, 1961—; prof. medicine SUNY, Buffalo, 1961—, chmn. dept., 1965-77, head div. geriatrics and gerontology, 1978—; head dept. medicine Buffalo Gen. Hosp., 1961-68; dir. medicine E.J. Meyer Meml. Hosp., 1968-78; head geriatrics service Buffalo VA Med. Center, 1978—; cons. Nat. Inst. Arthritis and Metabolic Diseases Tng. Grants Com., 1958-62, Program Project Com., 1964-68, Nat. Insts. Spl. Study Sect. for Health Manpower, 1969-77, for Behavioral Medicine, 1978—; mem. acad. awards com. Nat. Inst. on Aging, 1979-80. Editor: Handbook of Medical Emergencies, 1945; editor: Practice of Geriatric Medicine, 1983; Contbr. articles to profl. jours. Served to capt., M.C. AUS, 1943-45, 46-48. Fellow A.C.P.; mem. Am. Assn. Pathologists, Gerontol. Assn., Am. Geriatrics Soc., Am. Rheumatism Assn. (pres.), Am. Clin. and Climatological Assn., Am. Soc. Clin. Investigation, Assn. Am. Physicians, Central Soc. for Clin. Research, Soc. Medicine Argentina (hon.), Alpha Omega Alpha. Subspecialties: Internal medicine; Gerontology. Current work: Research on clinical aspects of geriatrics and rheumatology, amyloidosis and curriculum development in geriatric medicine. Home: 3799 Windover Hamburg NY 14075 Office: VA Med Center 3495 Bailey Ave Buffalo NY 14215

CALLAGHAN, OWEN HUGH, biochemistry educator; b. Johannesburg, South Africa, Oct. 2, 1927; came to U.S., 1958, naturalized, 1975; s. Vernon Owen and Flora (Fraser) C.; m. Carol Turner, July 10, 1959; children: Colleen, Maureen, James. M.Sc., U. Witwatersrand, South Africa, 1956; Ph.D., U. Sheffield, Eng., 1958. Asst. prof. biochemistry Chgo. Med. Sch., 1962-66, assoc. prof., 1966-69, prof., 1969—. NIH grantee, 1962-66. Mem. Biochem. Soc., Am. Assn. Immunologists, Chgo. Assn. Immunologists, Sigma Xi. Republician. Mem. Ch. of England. Subspecialties: Biochemistry (medicine); Neurochemistry. Current work: Metabolism of neurotransmitters. Enzyme kinetics of neurotransmitters. Office: U Health Scis-Chgo Med Sch 3333 Green Bay Rd North Chicago Il 60064

CALLAHAN, HUGH JAMES, biochemist, educator; b. Phila., July 9, 1940; s. John and Bridget (Shields) C.; m. AnnaMarie Faragasso, Aug. 12, 1961; children: Michael, Anthony, Joseph, Maria, Bernadette. B.S., St. Joseph's Coll., Phila., 1962; M.S., Ill. Tech., Chgo., 1967; Ph.D., Thomas Jefferson U., Phila., 1970. Research assoc. Evanston (Ill.) Hosp., 1962-66; instr. biochemistry Thomas Jefferson U., 1967-70, asst. prof. biochemistry, 1972-77, assoc. prof., 1977—. Contbr. articles to profl. jours. Mem. Am. Assn. Immunologists. Democrat. Roman Catholic. Subspecialties: Biochemistry (medicine); Immunology (medicine). Current work: Biochemical and immunochemical characterization of cell surface receptors involved in triggering the immune response; biochemical characterization of surface structures involved in urinary bladder anti-bacterial defenses. Home: 129 Worthman Ave Bellmawr NJ 08031 Office: 1020 Locust St Philadelphia PA 19107

CALLAHAN, JOHN EDWARD, geology educator, economic geology researcher; b. Buffalo, Feb. 26, 1941; s. Edward John and Elizabeth (Grimmer) C.; m. Karen Marie McGreery, June 26, 1965; children: Cathleen, Michael. B.A., SUNY-Buffalo, 1963, M.Ed., 1965; M.S., U. N.C.-Chapel Hill, 1968; Ph.D., Queen's U., Kingston, Ont., Can., 1973. Asst. prof. geology Appalachian State U., 1970-75, assoc. prof., 1975-80; prof., 1980—; geologist Br. Exploration Research, U.S. Geol. Survey, Golden, Colo., 1978—. Contbr. articles to profl. jours. Mem. AIME, Can. Inst. Mining and Metall. Engrs., Soc. Exploration Geochemists. Subspecialty: Geology. Current work: Use of heavy minerals in exploration geochemistry. Office: Dept Geology Appalachian State U Boone NC 28608 Home: Rural Route 4 Box 130 A Boone NC 28607

CALLAWAY, JOSEPH, educator, physicist; b. Hackensack, N.J., July 15, 1931; s. Joseph and Sybil Leigh (Mock) C.; m. Mary Morrison, July 30, 1949; children: Joseph A., Paul E., Jessie S. B.S., Coll. William and Mary, 1951; M.A., Princeton U., 1953, Ph.D., 1956. Asst. prof. physics U. Miami, 1954-60; asso. prof. physics U. Calif.-Riverside, 1960-64, prof., 1964-67; prof. physics and astronomy La. State U., 1967-76, Boyd prof. physics and astronomy, 1976—. Author: Quantum Theory of the Solid State, 1974; contbr. articles to sci. jours. Fellow Am. Phys. Soc.; fellow Inst. Physics U.K.; mem. AAAS, European Phys. Soc., Phi Beta Kappa. Democrat. Subspecialties: Condensed matter physics; Atomic and molecular physics. Current work: Theoretical solid state and atomic physics; band theory of solids; theory of ferromagnetism; theory of electron scattering by atoms. Office: Dept Physics La State U Baton Rouge LA 70803

CALLEN, JEFFREY PHILLIP, dermatologist, educator; b. Chgo., May 30, 1947; s. Irwin R. and Rose P. (Cohen) C.; m. Susan M. Manis, Dec. 21, 1968; children: Amy, David. B.S., U. Wis., 1969; M.D., U. Mich., 1972. Diplomate: Am. Bd. Internal Medicine, 1975, Am. Bd. Dermatology, 1977. Intern/ resident in internal medicine U. Mich., 1972-75, in dermatology, 1975-77; asst. clin. prof. U. Louisville Sch. Medicine, 1977-81, assoc. clin. prof., 1982—. Author: Manual of Dermatology, 1980, Cutaneous Aspects of Internal Disease, 1981; editor: Clinics in Rheumatic Disease, 1982; editor-in-chief Dermavision video program; editorial cons.: Dialogues in Dermatology audio tape program. Asst. chief Shawnee Indian Guides, Louisville; bd. dirs. Lousiville Jewish Community Ctr., United Jewish Campaign Louisville. Fellow ACP, Am. Acad. Dermatology (chmn. audio/visual edn. com., task force therapeutic agents, internal medicine symposium); mem. Am. Fedn. Clin. Research, AMA, Am. Rheumatism Assn., Dermatology Found. (chmn. corp fund raising com.). Subspecialties: Dermatology; Internal medicine. Current work: condition in which systemic disease has cutaneous manifestations. Office: Dept Dermatology U Louisville 554 Medical Towers S Louisville KY 40202

CALLENS, E(ARL) EUGENE, JR., government defense aerospace engineering executive; b. Memphis, Mar. 8, 1940; s. Earl Eugene and Sadie Esther (Chipman) C.; m. Barbara Ann Brossette, Sept. 4, 1960; children: Angela Kay, Earl Eugene III, Ryan Lawrence, Eric Lee, Nathaniel Jobe, Amy Elizabeth, Shannon Marie, Adam Spencer. B.S., Ga. Inst. Tech., 1962, M.S., 1964; diploma with honors, Von Karman Inst. for Fluid Dynamics, Rhode-St.-Genese, Belgium, 1967; Ph.D., U. Tenn. Space Inst., Tullahoma, 1976. Registered profl. engr., Tenn. Research assoc. Von Karman Inst. for Fluid Dynamics, 1966-67; instr. Ga. Inst. Tech., Atlanta, 1967-68; research engr. ARO, Inc, Arnold Engring. and Devel. Ctr. Div., Arnold Air Force Station, Tenn., 1968-77, supr. range projects, 1977-78, engring. mgr., 1978-81, Calspan Field Services, Inc., Arnold Air Force Station, 1981—. Contbr. articles in field to profl jours. Pres. Bel Aire PTA, Tullahoma, 1976-77; v.p. Tullahoma Am. Little League, 1977-78; mem. Tullahoma Bd. Edn., 1977-83; mem. exec. bd. Middle Tenn. council Boy Scouts Am., Nashville, 1979-81. Sloan Found. scholar Ga. Inst. Tech., 1958-62; Ford Found. fellow, 1963-66; recipient Von Karman award, 1967. Assoc. fellow AIAA (Gen. H.H. Arnold award 1974, chmn. Tenn. sect. 1982-83); mem. Nat. Mgmt. Assn., ASTM, Sigma Xi. Republican. Mormon. Subspecialties: Aerospace engineering and technology; Fluid mechanics. Current work: Development of advanced ground test facilities and test techniques for reentry vehicles at hypervelocities; technical specialties include aerothermodynamics, hypersonic flow, reentry physics, high speed ablation/erosion and hypervelocity impact phenomena. Home: 110 Redbud Ln Tullahoma TN 37388 Office: VKF/AB MS-440 Calspan Field Services Inc Arnold Air Force Station TN 37389

CALLEWAERT, DENIS MARC, chemist, educator; b. Detroit, Feb. 20, 1947; s. Marcel August and Mary Theresa (Lams) C.; m. Karen Margaret Koehn, Sept. 25, 1971; children: Amy Marlene, Megan Elizabeth. B.S. in Chemistry, U. Detroit, 1969; Ph.D. in Biochemistry, Wayne State U., 1973. Research assoc. Wayne State U., Detroit, 1969-73, postdoctoral assoc., 1973-74; asst. prof. chemistry Oakland U., Rochester, Mich., 1974-80, assoc. prof., 1980—, chmn. interdepartmental biochemistry program, 1980—; cons. in field. Author: (with J. Genyea) Basic Chemistry: General Organic, Biological, 1980; contbr. articles to profl. jours. Leukemia Soc. Am. fellow, 1975-77; Nat. Cancer Inst. awardee, 1980-85. Mem. Am. Chem. Soc., Am. Assn. Immunologists, AAAS, Sigma Xi. Subspecialties: Biochemistry (biology); Immunobiology and immunology. Current work: Function and biochemistry of human natural killer cells; modulation of human immune function; enzyme evolution. Home: 1600 Hosner Rd Oxford MI 48051 Office: Dept Chemistry Oakland U Rochester MI 48063

CALLIHAN, DIXON, federal agency administrator, editor; b. Searbro, W.Va., July 20, 1908; s. Alfred D. and Janie (Dixon) C.; m. Mildred Thompson Martin, Dec. 18, 1982. A.B., Marshall U., 1928, D.Sc., 1964; M.A., Duke U., 1931; Ph.D., NYU, 1933. Instr. CCNY, 1934-46, asst. prof., 1946-48; war research staff Columbia U., N.Y.C., 1942-45, Union Carbide Corp., N.Y.C., Oak Ridge, 1945-73; adminstrv. judge U.S. NRC, Washington, 1963—. Editor: Nuclear Sci. Engring, 1963—. Fellow Am. Phys. Soc., Am. Nuclear Soc.; mem. Sigma Xi. Republican. Subspecialties: Nuclear physics; Nuclear engineering. Home: 102 Oak Ln Oak Ridge TN 37830

CALLIS, PATRIK ROBERT, chemist, educator, researcher; b. Ontario, Oreg., Mar. 17, 1938. B.S. in Chemistry, Oreg. State U., 1960; Ph.D. in Phys. Chemistry, U. Wash., 1965. Postdoctoral research Calif. Inst. Tech., Pasadena, 1966-68; asst. prof. dept. chemistry Mont. State U., Bozeman, 1968-71, assoc. prof., 1971-74, prof., 1974—. Contbr. articles to profl. sci. jours. NSF grantee, 1980-82; NIH grantee, 1983—. Mem. Am. Chem. Soc. Club: Am. Alpine. Subspecialties: Physical chemistry; Spectroscopy. Current work: Electronic structure and laser spectroscopy of aromatic molecules, including biological heterocyclics. Office: Dept Chemistry Mont State U Bozeman MT 59717

CALLOWAY, DORIS HOWES, nutrition educator; b. Canton, Ohio, Feb. 14, 1923; m., 1981; 2 children. B.S., Ohio State U., 1943; Ph.D. in Nutrition, U. Chgo., 1947. Diplomate: Am. Bd. Nutrition, 1951. Intern dietetics Johns Hopkins Univ. Hosp., Balt., 1944; research dietitian dept. medicine U. Ill., 1945; cons. Med. Assocs., Chgo., 1948-51; nutritionist OM Food and Container Inst., 1951-58, head metabolism lab., 1958-59, chief nutrition br., 195-61; chmn. dept. food sci. and nutrition Stanford Research Inst., Calif., 1961-64; prof. nutrition U. Calif.-Berkeley, 1963—, provost, 1981—; Bd. dirs. Am. Bd. Nutrition, 1968-71; mem. panel White House Conf. on Food, Nutrition and Health, 1969; trustee Nat. Council Hunger and Malnutrition in U.S., 1969-71; mem. food and nutrition bd. Nat. Acad. Sci.-NRC, 1972-75; mem. vis. com. MIT, 1972-74; mem. expert adv. panel on nutrition WHO, 1972—; mem. adv. council Nat. Inst. Arthritis and Metabolism and Digestive Diseases, NIH, 1974-77; cons. nutrition div. FAO, 1974-75; mem. adv. council, Nat. Inst. Aging, NIH, 1978-81. Assoc. editor: Nutrition Rev, 1962-68; editorial bd.: Jour. Nutrition, 1967-72, Environ. Biology and Medicine, 1969—, Jour. Am. Dietetic Assn, 1974-77, Interdisciplinary Sci. Rev, 1975—. Mem. Am. Inst. Nutrition (pres. 1982-83), Am. Dietetic Assn. Subspecialty: Nutrition (biology). Office: U Calif Berkeley CA 94720

CALL-SMITH, KATHY MEREDITH, dentist, educator, research consultant; b. Fairborn, Ohio, Mar. 8, 1949; d. John Perry and Geraldine (Barry) Call; m. Fredrick Bond Smith, Apr. 13, 1974; 1 dau., Lauren Anna. B.A., Sarah Lawrence Coll., 1971; D.M.D., Sch. Dentistry Med. Coll. Ga., 1978. Lic. dentist, Ga., N.C. Hosp. resident Med. Coll. Ga., 1978-79, instr. in community dentistry, 1979-80, asst. prof. community dentistry, 1981—, supervising dentist, 1980-82, co-dir., 1982—. Mem. Internat. Assn. Dental Research, Acad. Gen. Dentistry, ADA, NOW, Sierra Club. Republican. Episcopalian. Subspecialties: Cariology; Preventive dentistry. Current work: Clinical field study of filled sealant over class I caries without caries removal or other tooth preparation, sealed vs unsealed amalgams. Home: 510 Coventry Rd Apt 13A Decatur GA 30030 Office: Sch Dentistry Restorative Dept Med Coll Ga Augusta GA 30912

CALVIN, MELVIN, chemist, educator; b. St. Paul, Apr. 8, 1911; s. Elias and Rose I. (Hervitz) C.; m. Marie G. Jemtegaard, 1942; children: Elin, Karole, Noel. B.S., Mich. Coll. Mining and Tech., 1931, D.Sc., 1955; Ph.D., U. Minn., 1935, D.Sc., 1969; hon research fellow, U. Manchester, Eng., 1935-37; Guggenheim fellow, 1967; D.Sc., Nottingham U., 1958, Oxford (Eng.) U., 1959, Northwestern U., 1961, Wayne State U., 1962, Gustavus Adolphus Coll., 1963, Poly. Inst. Bklyn., 1962, U. Notre Dame, 1965, U. Gent, Belgium, 1970, Whittier Coll., 1971, Clarkson Coll., 1976, U. Paris Val-de-Marne, 1977, Columbia U., 1979. With U. Calif., Berkeley, 1937—, successively instr. chemistry, asst. prof., prof., Univ. prof., dir. Lab. Chem. Biodynamics, 1963-80, asso. dir. Lawrence Radiation Lab., 1967-80; Peter Reilly lectr. U. Notre Dame, 1949; Harvey lectr. N.Y. Acad. Medicine, 1951; Harrison Howe lectr. Rochester sect. Am. Chem. Soc., 1954; Falk-Plaut lectr. Columbia U., 1954; Edgar Fahs Smith Meml. lectr. U. Pa. and Phila. sect. Am. Chem. Soc., 1955; Donegani Found. lectr. Italian Nat. Acad. Sci., 1955; Max Tishler lectr. Harvard U., 1956; Karl Folkers lectr. U. Wis., 1956; Baker lectr. Cornell U., 1958; London lectr., 1961, Willard lectr., 1982; Vanuxem lectr. Princeton U., 1969; Disting. lectr. Mich. State U., 1977; Prather lectr. Harvard U., 1980; Dreyfus lectr. Dartmouth Coll., 1981, Berea Coll., 1982; Barnes lectr. Colo. Coll., 1982; Nobel lectr. U. Md., 1982; Abbott lectr. U. N.D., 1983; Gunning lectr. U. Alta., 1983; Eastman prof. Oxford (Eng.) U., 1967-68. Author: (with G. E. K. Branch) The Theory of Organic Chemistry, 1940, Isotopic Carbon, (with others), 1949, Chemistry of Metal Chelate Compounds, (with Martell), 1952, Path of Carbon in Photosynthesis, (with Bassham), 1957, Photosynthesis of Carbon Compounds, 1962, Chemical Evolution, 1969; contbr. articles to chem. and sci. jours. Recipient prize Sugar Research Found., 1950, Flintoff medal prize Brit. Chem. Soc., 1953, Stephen Hales award Am. Soc. Plant Physiologists, 1956, Nobel prize in chemistry, 1961; Davy medal Royal Soc., 1964; Virtanen medal, 1975; Priestley medal, 1978; Am. Inst. Chemists medal, 1979; Feodor Lynen medal, 1983; Oesper award Cin. sect. Am. Philos. Soc., 1981. Mem. Britain's Royal Soc. London (fgn. mem.), Am. Chem. Soc. (Richards medal N.E. sect. 1956, Chem. Soc. Nichols medal N.Y. sect. 1958, award for nuclear applications in chemistry, pres. 1971, Gibbs medal Chgo. sect. 1977, Priestley medal 1978), Am. Acad. Arts and Scis., Nat. Acad. Scis., Royal Dutch Acad. Scis., Japan Acad., Am. Philos. Soc., Sigma Xi, Tau Beta Pi, Phi Lambda Upsilon. Subspecialties: Photochemistry. Current work: Hydrocarbon-biomass; photochemical decomposition of water 1) with H2 product or c02 red 2) with 02 or organic oxide. Home: 2683 Buena Vista Berkeley CA 94708

CALVIN, WILLIAM HOWARD, neurophysiologist, educator, author; b. Kansas City, Mo., Apr. 30, 1939; s. Fred Howard and Agnes (Leebrick) C.; m. Katherine Graubard, Sept. 1, 1966. B.A., Northwestern U., 1961; postgrad., MIT, 1961-62; Ph.D., U. Wash.-Seattle, 1966. Instr. U. Wash.-Seattle, 1967-69, asst. prof., 1969-74, assoc. prof. neurol. surgery, 1974—; vis. prof. Hebrew U., Jerusalem, 1978-79. Author: (with G.A. Ojemann) Inside the Brain, 1980, The Throwing Madonna, 1983. State pres. ACLU, Seattle, 1973-74. Mem. Soc. Neurosci. (program com. 1975-78), Am. Physiol. Soc., Internat. Assn. Study Pain, Biophys. Soc., Internat. Brain Research Orgn. Subspecialties: Neurobiology; Neurophysiology. Current work: Neurophysiology of tic douloureux and epilepsy; evolution of language and bigger brains. Home: 1543 17th Ave E Seattle WA 98112 Office: U Wash R1-20 Seattle WA 98195

CALVO, JOSEPH MARLE, biochemist; b. Seattle, Oct. 11, 1934; s. Marco J. and Rose (Azore) C.; m. Rita A. Rottenberg, May 14, 1942; children: Rachel, Naomi, Sarah. B.A., Whitman Coll., 1956; Ph.D.,

Wash. State U., 1962. Fulbright fellow, Germany, 1956-57, staff, NIH postdoctoral fellow, Cold Spring Harbor, N.Y., 1962-64; asst. prof. Cornell U., Ithaca, N.Y., 1964-70, asso. prof., 1970-79, prof. biochemistry, 1979—. Recipient Chancellor's award for excellence in teaching; Fogarty Sr. Internat. Fellow, 1981. Mem. Am. Soc. Microbiology, Genetics Soc. Am. Subspecialties: Biochemistry (biology); Genetics and genetic engineering (biology). Current work: Regulation of gene expression in bacterial and euccryotes. Office: Cornell Univ Dept Biochemistry Ithaca NY 14853

CAME, PAUL E., biotechnology exec., microbiologist, cell biologist; b. Dover, N.H., Feb. 23, 1937; s. Melvin L. and [illegible] M. ([illegible]) C.; m. Elizabeth J. Came, June 11, 1960; children: Paula, Heather. A.B., St. Anselm's Coll., 1958; M.S., U. N.H., 1960; Ph.D., Hahnemann Med. Coll., 1964. Asst. virologist St. Jude Hosp., Memphis, 1963-65; research assoc. Rockefeller U., 1965-66; dir. virology Schering Corp., Bloomfield, N.J., 1966-75; dir. microbiology Winthrop Research inst., Rensselaer, N.Y., 1975-81; pres., chief exec. officer, dir. sci. affairs HEM Research, Inc., Rockville, Md., 1981—; lectr. in field. Contbr. articles to profl. jours. Served with USNG, 1954-58. Mem. Am. Assn. Immunologists, Am. Soc. Microbiology, N.Y. Acad. Sci., Am. Assn. Cancer Research. Subspecialties: Cell and tissue culture; Genetics and genetic engineering (biology). Current work: Production of glycoproteins and proteins from cultured human cells; manufacture and marketing of tissue culture products. Home: 316 Osborne Rd Loudonville NY 12211 Office: 12220 Wilkins Ave Rockville MD 12852

CAMERMAN, ARTHUR, chemist, pharmacologist, educator, researcher; b. Vancouver, C., Can., Apr. 12, 1939; came to U.S., 1967; s. Philip and Dora (Charkow) C. B.Sc. with honors, U. B.C., 1961, Ph.D. in Chemistry, 1964. NRC of Can. overseas postdoctoral fellow Royal Instn. Gt. Brit., London, 1964-66; research asso. depts. biol. structure and pharmacy U. Wash., Seattle, 1967-71, research asst. prof. depts. medicine and pharmacology, 1971-75, research asso. prof., 1975-81, research prof., 1981—; vis. investigator Howard Hughes Med. Inst., 1971-72. Contbr. numerous articles and abstracts to sci jours. NIH research career devel. awardee, 1974-79; Klingenstein sr. fellow in the neurosci., 1983-86; Internat. League Against Epilepsy research award, 1983; numerous research grants NIH, NSF. Mem. AAAS, Am. Chem. Soc., Am. Crystallographic Assn., Am. Soc. for Neurochemistry, Am. Soc. Pharmacology and Exptl. Therapeutics, Fedn. Am. Scientists. Subspecialties: Crystallography; Molecular pharmacology. Current work: Structure-activity relationships in drugs and biol. molecules; antiepileptic drugs; neurochemistry; anti-cancer drugs. Office: U Wash Neurology Dept RG-20 Seattle WA 98195

CAMERON, ALASTAIR GRAHAM WALTER, astrophysicist; b. Winnipeg, Can., June 21, 1925; came to U.S., 1959, naturalized, 1963; s. Alexander Thomas and Airdrie Edna (Bell) C.; m. Elizabeth Aston MacMillan, June 11, 1955. B.Sc., U. Man., 1947; Ph.D., U. Sask., 1952, D.Sc. (hon.), 1977, A.M., Harvard U., 1973. Asst. prof. physics Iowa State Coll., Ames, 1952-54; asst., asso. and sr. research officer Atomic Energy Can., Ltd., Chalk River, Ont., 1954-61; sr. research fellow Calif. Inst. Tech., Pasadena, 1959-60; sr. scientist Goddard Inst. Space Studies, N.Y., 1961-66; prof. space physics Yeshiva U., 1966-73; prof. astronomy Harvard U., Cambridge, Mass., 1973—; chmn. Space Sci. Bd., 1976-82, Nat. Acad. Scis. Contbr. articles to profl. jours. Mem. Nat. Acad. Scis., Am. Acad. Arts and Scis., World Acad. Art and Sci., Royal Soc. Can., AAAS, Am. Phys. Soc., Am. Geophys. Union, Am. Astron. Soc., Royal Astron. Soc., Internat. Astron. Union, Internat. Assn. Geochemistry and Cosmochemistry, Meteoritical Soc. Club: Cosmos. Subspecialties: Theoretical astrophysics; Planetary science. Office: 60 Garden St Cambrioge MA 02138

CAMERON, DEBORAH JANE, immunologist, microbiologist, educator; b. Hackensack, N.J., Aug. 14, 1949; d. Alan Duncan and Jane Katherine (Bocket) C. B.S., Jackson Coll., 1971; Ph.D., Coll. Physicians and Surgeons, Columbia U., 1977; postgrad., Harvard Med. Sch., 1977-79. Research technician Sloan Kettering Cancer Ctr., N.Y.C., 1971-72; staff assoc. Columbia U. Coll. Physicians and Surgeons, N.Y.C., 1977; research assoc. Harvard Med. Sch., Boston, 1977-79; instr. Med. U. S.C., Charleston, 1979-80, asst. prof., 1980—. NIH - Nat. Cancer Inst. fellow, 1978-79; grantee, 1981-84. Mem. Reticuloendothelial Soc., Am. Assn. Immunologists, Sigma Xi (sec. 1980-81). Republican. Presbyterian. Subspecialties: Cancer research (medicine); Immunopharmacology. Current work: studying macrophage mediated cytotoxicity; effects of sturgeon factors on tumor cells (anti-neoplastic effects); immunochemistry xenogeneic transplantation. Home: 611 E Hobcaw Dr Mt Pleasant SC 29464 Office: Med U SC 171 Ashley Ave Charleston SC 29425

CAMERON, H. RONALD, educator; b. Oakland, Calif., June 30, 1929; s. Sidney H. and Violet N. (Mecklenberg) C.; m. Barbara L. Snook, June 23, 1956; children: William S., Kathryn L. B.S., U. Calif., Davis, 1951; Ph.D., U. Wis., 1955. Faculty plant pathology Oreg. State U., Corvallis, 1955—. Contbr. articles to profl. jours. NSF fellow, 1962; NATO fellow, 1972. Mem. Am. Phytopath. Soc. (treas. 1982—), Sigma Xi. Subspecialties: Plant virology; Plant pathology. Current work: Researcher in isolation and identification of previously unidentified virus-like diseases. Office: Oregon State University Dept Botany and Plant Pathology Corvllis OR 97331

CAMERON, JOHN STANLEY, cardiovascular physiology educator, researcher; b. Chgo., Aug. 23, 1952; s. John H. and Jean (Wallace) C. B.S., Coll. William and Mary, 1974; M.S., U. Mass.-Amherst, 1977, Ph.D., 1979. Fellow Albany (N.Y.) Med. Coll., 1979-81; research asst. prof. cardiovascular physiology U. Miami, 1981—. Sponsor Dade County Div. Exceptional Student Edn., Miami, 1981—. Am. Heart Assn. fellow, 1979-81; grantee, 1982-84; NIH trainee, 1981-82; NIH grantee, 1982—. Mem. Am. Physiol. Soc., Am. Heart Assn., Am. Soc. Zoologists. Subspecialties: Pharmacology; Comparative physiology. Current work: Cardiovascular physiology, pharmacology; role of the autonomic nervous system in arrhythmogenesis. Office: Dept Pharmacology U Miami Sch Medicine PO Box 016189 Miami FL 33101

CAMIOLO, SARAH MAY, cancer researcher; b. Enna, Italy, May 18, 1924; came to U.S., 1924; d. Stephen and Mary (Incardona) C. B.A., Seton Hill Coll., 1946; M.S., Canisius Coll., 1949; Ph.D., American U., 1971. Research chemist surg. research lab. E. J. Meyer Meml. Hosp., Buffalo, 1950-52; silicones div. Union Carbide Corp., Tonawanda, N.Y., 1952-63; chemistry educator Peace Corps, Cameroon, West Africa, 1963-66; research assoc. J. F. Mitchell Med. Inst., Washington, 1967-72; asst. research prof. SUNY-Buffalo dept pharmacology, 1974-76; sr. cancer research scientist Roswell Park Meml. Inst., Buffalo, 1976—; cons. AID, Kerala, India, 1967, Madras, India, 1968. Mem. pastoral council Diocese of Buffalo, 1982—. Michael Reese Blood Center fellow, 1972; United Way of Western N.Y. fellow, 1973. Mem. Am. Chem. Soc., Am. Soc. Biol. Chemists, Internat. Soc. on Thrombosis and Haemostasis, Am. Women in Sci., N.Y. Acad. Scis., Iota Sigma Pi. Subspecialties: Cancer research (medicine); Molecular biology. Current work: Plasminogen activators and the role of these fibrinolytic enzymes in cancers. Home: 120 Lakewood Pkwy Buffalo NY 14226 Office: Roswell Park Meml Inst 666 Elm St Buffalo NY 14263

CAMPBELL, ALLAN MCCULLOCH, educator; b. Berkeley, Calif., Apr. 27, 1929; s. Lindsay and Virginia Margaret (Henning) C.; m. Alice Del Campillo, Sept. 5, 1958; children—Wendy, Joseph. B.S. in Chemistry, U. Calif. at Berkeley, 1950; M.S. in Bacteriology, U. Ill., 1951; Ph.D., 1953, U. Chgo., 1978, U. Rochester, 1981. Instr. bacteriology U. Mich., 1953-57; research asso. Carnegie Inst., Cold Spring Harbor, N.Y., 1957-58; asst. prof. biology U. Rochester, N.Y., 1958-61, asso. prof., 1961-63, prof., 1963-68; prof. biol. sci. Stanford, 1968—; mem. genetics study sect. NIH, 1964-69, mem. DNA recombinant adv. com., 1977-81; mem. genetics panel NSF, 1973-76. Author: Episomes, 1969; co-author: General Virology, 1978; Editor: [illegible] 1969—; editorial bd.: Jour. Bacteriology, 1966-72, Jour. Virology, 1967-75. Served with AUS, 1953-55. Recipient Research Career award USPHS, 1962-68. Mem. Nat. Acad. Scis., Am. Acad. Arts and Scis., Am. Soc. Microbiology, Soc. Am. Naturalists, AAAS. Democrat. Subspecialties: Genome organization; Virology (biology). Current work: Lyosogenny in bacteriophage; biotin biosynthesis in E. coli. Home: 947 Mears Ct Stanford CA 94305 Office: Dept Biol Scis Stanford U Stanford CA 94305 I've always thought that each individual has some contribution to human knowledge that he is uniquely suited to make. So I try to be organized and to avoid doing things that I expect will get done, anyway, by others. And, of course, everything worthwhile requires hard work.

CAMPBELL, CARLOS BOYD GODFREY, physician, army officer; b. Chgo., July 27, 1934; s. Joseph G. Bumzahem and Ruby V. Brown-Campbell; m. Deborah E. Stephens, June 28, 1958 (div.); children: Ellen, Gowan, Kenneth, Christopher; m. Nydia Haydee Gonzalez, Feb. 3, 1979; 1 stepdau., Zinnia. B.S., M.S., M.D., Ph.D., U. Ill. Neuroanatomist Walter Reed Army Inst. Research, Washington, 1964-67, research neurologist, 1979—; asst. to assoc. prof. neural scis., anatomy, and physiology Ind. U., Bloomington, 1967-74; asso. clin. prof. anatomy U. Calif., Irvine, 1975-77; vis. asso. in biology Calif. Inst. Tech., Pasadena, 1976-77; prof., head dept. anatomy U. P.R., Rio Piedras, 1977-79; adj. prof. anatomy Georgetown U. Schs. Medicine and Dentistry, Washington, 1980—; lectr. psychology U. Md., College Park, 1982—; commd. lt. col. M.C. U.S. Army, 1979—. Editor: (with others) Evolution of Brain and Behavior in Vertebrates, 1976; Contbr. articles, abstracts, papers, book revs., forewords to profl. lit. Served to capt. M.C. U.S. Army, 1964-67. NIH, NSF grantee. Mem. AAAS, Cajal Club, Am. Soc. Primatologists, Internat. Primatological Soc., Am. Assn. Anatomists, Soc. Neurosci., Am. Soc. Zoologists, Sigma Xi, Phi Rho Sigma. Episcopalian. Subspecialties: Evolutionary biology; Neurobiology. Current work: Research currently involves central nervous system regulation of respiration. Other areas of interest are comparative neurology and evolutionary biology. Home: 6003 McKinley St Bethesda MD 20817 Office: Walter Reed Army Inst Research Washington DC 20307

CAMPBELL, CATHERINE CHASE, geologist; b. N.Y.C., July 1, 1905; s. John Hildreth and Eliza Robbins C.; m. Ian Campbell, Sept. 16, 1930 (dec. Feb. 1978); 1 son, Dugald Robbins. B.A., Oberlin Coll., 1927, M.A., 1927; M.A., Radcliffe Coll., 1932, Ph.D., 1933. Instr. geology Mount Holyoke Coll., South Hadley, Mass., 1927-29; tech. editor Army Air Force, Pasadena, Calif., 1943-46; head publs. br. U.S. Naval Ordnance Test Sta., Pasadena, Calif., 1946-61; geologist U.S. Geol. Survey, Menlo Park, Calif., 1961—; sec. Naval Interlab. Com. for Editing and Pub., Pasadena, Calif., 1959-60; expert examiner U.S. Civil Service Comm., Pasadena, 1956-61. Editor: Environmental Planning and Geology, 1971 (cert. of appreciation Assn. Engring. Geologists 1972). Recipient Outstanding Performance award USN, 1952, 53, 54, 56 59, Superior Accomplishment award, 1952, 53, 56, 59; Outstanding Performance award U.S. Geol. Survey, 1965, 66; Superior Performance award, 1967; Meritorious Civilian award, 1969; Sp. Achievement award, 1975. Fellow Geol. Soc. Am., Calif. Acad. Scis.; mem. Soc. Tech. Communication, Geosci. Info. Soc., Assn. Earth Sci. Editors, No. Calif. Geol. Soc. Republican. Unitarian. Clubs: Commonwealth, Metropolitan (San Francisco); Athenaeum (Pasadena, Calif.). Subspecialty: Geology. Current work: Editor of reports on environmental geology and engineering geology. Home: 1333 Jones St Apt 906 San Francisco CA 94109 Office: US Geol Survey 345 Middlefield Rd Menlo Park CA 94025

CAMPBELL, CHARLES JOHN, ophthalmologist; b. Steubenville, Ohio, June 24, 1926; m. Mary Catherine McGuigan, July 2, 1955; children—Catherine Mary, Barbara Irene, Charles Arbuthnot III. B.S., Muskingum Coll., 1949; M.D., George Washington U., 1948; M.S. in Optics, U. Rochester, 1951; Med.Sc.D., Columbia U., 1957. Intern George Washington U. Hosp., 1948-49; resident Edward S. Harkness Eye Inst., 1954-57; practice medicine specializing in ophthalmology, N.Y.C., 1957—; dir. Edward S. Harkness Eye Inst., Ophthalmology Service of Columbia-Presbyn. Med. Center, N.Y.C., 1974—, Knapp Meml. Lab. of Physiol. Optics, 1957—; prof., chmn. dept. ophthalmology Columbia U., 1974—. Served with USAF, 1952-54. Mem. Am. Ophthal. Soc., Am. Acad. Ophthalmology and Otolaryngology, A.C.S., Optical Soc. Am., Assn. U. Profs. Ophthalmology, AMA, N.Y. Ophthal. Soc., Assn. for Research in Vision and Ophthalmology, N.Y. State Med. Soc., N.Y. Acad. Scis. Subspecialty: Ophthalmology. Current work: Clinical studies on vitreous and retina. Office: 635 W 165th St New York NY 10032

CAMPBELL, DAVID OWEN, research and development chemist, consultant; b. Merriam, Kans., Nov. 11, 1927; s. Sullivan Gee and Lenore Louise (Engel) C.; m. Barbara Jean Powers, Feb. 26, 1954; children: Evelyn, James, Robert. B.A., U. Kansas City, 1947; Ph.D., Ill. Inst. Tech., 1953. Chemist Oak Ridge (Tenn.) Nat. Lab., 1953-55, 57-72; group leader Oak Ridge Nat. Lab., 1972—; tech. assistance and adv. group Three Mile Island, Harrisburg, Pa., 1981—. Co-author: LWR Nuclear Fuel Cycle, 1981; contbr. numerous articles to profl. jours. Mem. Am. Chem. Soc., Am. Nuclear Soc. (spl. award 1981). Subspecialties: Nuclear fission; Physical chemistry. Current work: Nuclear reactor fuel reprocessing, fission product chemistry, nuclear reactor safety, actinmide element chemistry. Patentee (7). Home: 102 Windham Rd Oak Ridge TN 37830 Office: Oak Ridge Nat Lab PO Box X Oak Ridge TN 37830

CAMPBELL, DONALD LEROY, chemistry educator; b. Waverly, Iowa, July 16, 1940; s. Harold Leroy and Margaret Loretta (McDonough) C.; m. Darlene Janice Mehl, July 18, 1949; 1 dau., Meghan Ann. B.S., Iowa State U., 1967; Ph.D., U. Ill., 1969. Chemist Liquid Carbonic div. Gen. Dynamics Corp., Chgo., 1962-65; asst. prof. chemistry U. Wis.-Eau Claire, 1969-77, assoc. prof., 1977—. Mem. Am. Chem. Soc., Sigma Xi. Subspecialty: Inorganic chemistry. Office: Dept Chemistry U Wis-Eau Claire Eau Claire WI 54701

CAMPBELL, GILBERT SADLER, medical educator; b. Toronto, Ont., Can., Jan. 4, 1924; came to U.S., 1926; s. Gilbert S. and Ellen (Thorson) C.; m. Dorothy Jean Nugent, Sept. 18, 1947 (div. 1960); children: Kathryn, Rebecca, Thomas, William; m. Joan Louise Hancock, Sept. 28, 1961; children: Susan, John. Student, Hampden-Sydney Coll., 1939-40; B.A., U. Va., 1943, M.D., 1946; M.S., U. Minn., 1949, Ph.D., 1954. Diplomate: Am. Bd. Surgery. Asst. prof. surgery U. Minn., Mpls., 1954-58; prof. surgery U. Okla., Oklahoma City, 1958-65; prof., head dept.surgery U. Ark. Med. Sch., Little Rock, 1958—. Contbr. over 150 articles to med. jours. Served to capt. U.S. Army, 1949-51; Korea. Decorated Silver Star (2), Bronze Star (2), Purple Heart; Markle scholar, 1954-59. Mem. Am. Surg. Assn., So. Surg. Assn. (v.p. 1980-81), Internat. Cardiovascular Soc. (v.p. 1973), Southwestern Surg. Congress (pres. 1980), So. Thoracic Surg. Assn. (v.p. 1982). Subspecialty: Surgery. Current work: Cardiovascular and thoracic surgery. Home: 66 River Ridge Rd Little Rock AR 72207 Office: 4301 W Markham Little Rock AR 72205

CAMPBELL, JAMES NORMAN, neurosurgeon, neurophysiologist, educator; b. Royal Oak, Mich., Aug. 5, 1948; s. James Stewart and Lilla Cleo (Upton) C.; m. Regina Helen Anderson, July 31, 1982. B.A., U. Mich., 1969; M.D., Yale U., 1973. Diplomate: Am. Bd. Neurol. Surgery. Resident dept. neurosurgery Johns Hopkins U., Balt., 1973-79, postdoctoral fellow in neurophysiology, 1975-77, asst. prof. neurosurgery, 1979—; neurosurgeon Johns Hopkins Hosp., 1979—. Contbr. articles and abstracts to profl. jours. Recipient research fellowship award NIH, 1975, tchr. investigator award, 1980. Mem. Soc. for Neurosci., Internat. Assn. for Study of Pain, Am. Assn. Neurol. Surgeons, AAAS, Research Soc. for Neurol. Surgery, Phi Beta Kappa. Subspecialties: Neurophysiology; Neurosurgery. Current work: Pain physiology, microneurosurgery, psychophysics; investigation of mechanisms of pain sensation, pain secondary to nerve injury, and also performance of vascular and peripheral nerve neurosurg. procedures. Home: 15 St Georges Rd Baltimore MD 21210 Office: Dept Neurosurgery Johns Hopkins Hosp Baltimore MD 21205

CAMPBELL, JAMES WAYNE, biologist; b. Highlandville, Mo., Mar. 2, 1932; s. Frank Pauline and Mable (Kentling) C.; m. Bonnalie Josephine Oetting, Sept. 4, 1960; children—Heather Anne, James Kentling. B.S., S.W. Mo. State Coll., 1953; M.S., U. Ill., 1955; Ph.D. (USPHS fellow), U. Okla., 1958. Nat. Acad. Sci.-NRC fellow Johns Hopkins U., 1958-59; mem. faculty Rice U., Houston, 1959—, prof. biology, 1970—, chmn. dept., 1974-78; vis. asso. prof., USPHS fellow U. Wis. Med. Sch., Madison, 1964-65; program dir. regulatory biology Div. Biol. and Med. Sci. NSF, 1973-74, dir. div. physiology, cellular and molecular biology, 1979—; cons. to govt., 1969—. Editor: Comparative Biochemistry of Nitrogen Metabolism, 2 vols, 1970; co-editor: Nitrogen Metabolism and the Environment, 1972; contbr. profl. jours. Recipient USPHS Career Devel. award, 1966-70; NSF USPHS grantee, 1960—. Fellow AAAS; mem. Am. Physiol. Soc., Am. Soc. Biol. Chemists, Biochem. Soc. Eng., Am. Soc. Zoologists, Sigma Xi, Phi Sigma, Phi Lambda Upsilon. Subspecialty: Physiology (biology). Address: 2628 Fenwood Rd Houston TX 77005

CAMPBELL, JUDITH LYNN, molecular biology educator; b. New Haven, Mar. 24, 1943; d. John and Marjorie (Hutt) C. B.A., Wellesley Coll., 1965; Ph.D., Harvard U., 1974. Postgrad. Harvard U. Med. Sch., 1969-74, research assoc., 1974-77; asst. prof. Calif. Inst. Tech., Pasadena, 1977—; biochemistry study sect. NIH, Bethesda, Md., 1982—. Research Career Devel. awardee NIH, 1979-84. Mem. Fedn. Am. Socs. Exptl. Biology. Subspecialties: Molecular biology; Genetics and genetic engineering (biology). Current work: Genetic and biochemical analysis of DNA replication, recombination and repair in microorganisms. Office: Calif Inst Tech 1201 E California Blvd Pasadena CA 91125

CAMPBELL, RUSSELL BRUCE, educator; b. Hartford, Conn., Sept. 18, 1952; s. Andrew Burr and Marian Priscilla (Champlin) C. Sc.B., Brown U., 1974, Sc.M., 1974; Ph.D., Stanford U., 1979. Asst. prof. math. Purdue U., West Lafayette, Ind., 1979—. Contbr. articles to profl. jours. Mem. Am. Math. Soc., Genetics Soc. Am., Am. Soc. Naturalists, Sigma Xi. Subspecialties: Evolutionary biology; Applied mathematics. Current work: Selection-migration interaction; implications of natural selection versus neutral drift, sources of genetic variation, phenotypic selection and genetic change. Office: Dept Math Purdue Univ West Lafayette IN 47907 Home: 7 Waterside Ln West Hartford CT 06107

CAMPBELL, STEPHEN LAVERN, mathematics educator; b. Belle Plaine, Iowa, Dec. 8, 1945; s. LaVern L. C. B.A., Dartmouth Coll., 1967; M.S., Northwestern U., Evanston, Ill., 1968, Ph.D., 1972. Instr. Marquette U., Milw., 1970-72; asst. prof. N.C. State U., Raleigh, 1972-76, assoc. prof., 1976-81, prof. math., 1981—. Author: Singular Systems of Differential Equations, 1978, Singular Systems of Differential Equations II, 1982, (with Carl D. Meyer, Jr) Generalized Inverses of Linear Transformations, 1979; editor: Recent Applications of Generalized Inverses, 1982; mem. editorial staff: Linear Algebra and Its Applications, 1981—. Mem. Soc. Indsl. and Applied Math. (editorial staff jour.). Subspecialty: Applied mathematics. Current work: Analytic and numerical solution of implicit differential equations arising in circuit and control theory. Office: Dept Math NC State U Raleigh NC 27650

CAMPBELL, THOMAS COLIN, nutritional science educator, consultant; b. Annandale, N.J., Mar. 14, 1934; s. Thomas McIlwain and Bessie Hoagland (DeMott) C.; m. Karen Lee Margaret, Sept. 1, 1962; children: Nelson, LeAnne, Keith, Daniel, Thomas. B.S., Pa. State U., 1956; M.S., Cornell U., 1957, Ph.D., 1962. From asst. prof. to prof. biochemistry and nutrition Va. Inst. Tech., Blacksburg, 1965-75; prof. nutritional sci. Cornell U., Ithaca, N.Y., 1975—; cons. Contbr. 200 articles to sci. jours. Mem. panels Nat. Acad. Scis. NIH grantee; Nat. Cancer Inst. exchange scholar People's Republic of China, 1981. Mem. Am. Inst. Nutrition, Am. Soc. Pharmacology and Exptl. Therapeutics, Soc. Toxicology. Subspecialties: Nutrition (biology); Cancer research (medicine). Current work: Research on diet and cancer. Office: Cornell U Ithaca NY 14850

CAMPBELL-SMITH, ROSEMARY GILLES, dental sci. educator; b. Rapid City, S.D., Mar. 16, 1939; d. Albert Peter and Anna (Schmitz) Gilles; m. Richard Lee Smith, Aug. 6, 1978; 1 dau. by previous marriage, Christina Lynn Campbell. Cert., Eastman Sch. Dental Hygiene, 1960; B.S.H.E. with high honors, U. Fla., 1964; M.S., U. Miami, 1968, Dr.Arts, 1976. Registered dental hygienist, Fla. Dental hygienist in pvt. practice, Palm Beach County, Fla., 1960-63; chief lab. technician dept. physiology U. Fla., Gainesville, 1965-66; instr. dept. biology U. Miami, Coral Gables, 1968-70; asst. prof. dept. dental hygiene Miami-Dade Community Coll., 1974-77; lead instr. Dental Aux. Programs, Santa Fe Community Coll., 1977-81; cons. Campbell-Smith Cons., Gainesville, 1982—. Author: Head and Neck: What's It All About, 1976; editor: Prophyways/Prophygram, 1982. Counselor Birthright, Inc., Gainesville, 1981-82. Recipient Albert E. Sevenson award for art. sci. and service N.Y. Dental Soc., 1960; Sci. award Eastman Sch. Dental Hygiene, 1960; J. Hillis Miller award U. Fla., 1964; Cancer Assn. award, 1964; Merit Citation Am. Dental Assn. Commn. on Accreditation Report, 1979. Mem. Am. Dental Hygienists Assn. (student liaison Dist. IV 1981—), Fla. Dental Hygienists Assn. (student advisor 1978-81), Internat. Assn. Dental Research, AAAS, Am. Mgmt. Assn., Sigma Xi, Phi Kappa Phi, Phi Lambda Pi. Democrat. Roman Catholic. Clubs: Dental Guild, Dental Wives. Subspecialties: Dental practice consulting; Neuropsychology. Current work: Endagenous pain control/pain control/ pain/endorphins/ dental anxiety reduction. Office: 3609 NW 30 Blvd Gainesville FL 32605

CAMPION, DENNIS ROBERT, research physiologist; b. Janesville, Wis., Oct. 28, 1945; s. Robert E. and Marie (Wilbur) C.; m. Rita Ann Fish, July 5, 1969; children: Anna, Sara, Carrie, Andrea, Amy. B.S., U. Wis.-Madison, 1969, M.S., 1971, Ph.D., 1973. Postdoctoral assoc. Iowa

State U., Ames, 1973-74; research biochemist U.S. Meat Animal Research Ctr., Clay Center, Nebr., 1974-76; research physiologist R. B. Russell Research Ctr.,/U.S. Dept. Agr., Athens, Ga., 1976—; Editorial bd. Jour. Animal Sci., 1979-81. Pres. Friendship Recreation Club, Inc., Watkinsville, Ga., 1982. Mem. Ga. Nutrition Council (publicity chmn. 1982-83, sec. 1983-84), Am. Meat Sci. Assn. (research priorities com. 1979—), Am. Soc. Animal Sci., Am. Inst. Nutrition, Soc. Exptl Biology and Medicine. Subspecialty: Animal physiology. Current work: Growth and development of skeletal muscle of meat animals; factors that regulate myofiber nuclear proliferation and muscle metabolism are identified and characterized. Office: USDA RB Russell Research Ctr College Station Rd Athens GA 30613

CAMPOY, LEONEL PEREZ, nuclear/computer engineer; b. Los Angeles, Dec. 23, 1954; s. Lauro and Consuelo (Pérez) Villegas C.; m. María Paz Fernandez, Aug. 20, 1981. B.S., U. Ariz., 1979, M.S., 1981. Asst. researcher nuclear engring. dept. U. Ariz., Tucson, 1979, Solar Energy Research Facility, 1979-81; nuclear/computer engr. So. Calif. Edison, San Clemente, 1981—. Author: Industrial Solar Process Heat Systems Simulation, 1981; co-author: A Guidebook for Solar Process Heat Applications, 1981. Mem. Am. Nuclear Soc., Soc. Computer Simulation. Subspecialties: Nuclear engineering; Software engineering. Current work: Simulation of nuclear power plants, simulation languages, and industrial solar energy applications. Office: So Calif Edison Co PO Box 128 San Clemente CA 92672

CAMRAS, MARVIN, engineer, inventor; b. Chgo., Jan. 1, 1916; s. Samuel and Ida (Horwich) C.; m. Isabelle Pollack, 1951; children: Robert, Carl, Ruth, Michael, Louis. B.S., Armour Inst. Tech., (now Ill. Inst. Tech.), 1940; M.S., Ill. Inst. Tech., 1942, LL.D., 1968. Registered profl. engr., Ill. Mem. staff Armour Research Found., Chgo., 1940—, asst. physicist, 1940-45, asso. physicist, 1945-46, physicist, 1946-49, sr. physicist, 1949-58, sr. engr., 1958-65, sci. adviser, 1965-69, sr. sci. adviser, 1969—; Chmn. S-4 com. Am. Nat. Standards Inst., 1966, mem., 1966—. Editor: Inst. of Radio Engrs. Transactions on Audio, 1958-63. Recipient Distinguished Service award Ill. Inst. Tech. Alumni Assn., 1948; Achievement award for outstanding contbn. motion picture photography U.S. Camera mag., 1949; John Scott medal, 1955; Ind. Tech. Coll. citation, 1958; Achievement award I.R.E., 1958; Indsl. Research Mag. Product award, 1966; John S. Potts Meml. Gold Medal award Audio Engring. Soc., 1969; merit award Chgo. Tech. Socs., 1973; Alumni medal Ill. Inst. Tech., 1978; named to Hall of Fame, 1981; Inventor of Yr. award Patent Law Assn., Chgo., 1979. Fellow IEEE (sec.-treas. 1951-53, Consumer Electronics award 1964, nat. chmn. profl. group on audio 1953-54 of I.R.E), Acoustical Soc. Am. (patent rev. bd.), AAAS, Soc. Motion Picture and Television Engrs.; mem. Nat. Acad. Engring., Western Soc. Engrs. (Washington award 1979), Physics Club Chgo. (dir. 1969—, pres. 1973-74), Radio Engrs. Club Chgo., Chgo. Acoustic and Audio Group (dir. 1967-68), Audio Engring. Soc. (hon.; gov. 1970—), Midwest Acoustics Conf. (dir. 1969—), Sigma Xi (chpt. pres. 1959-60), Tau Beta Pi, Eta Kappa Nu. Subspecialties: Applied magnetics; Electronics. Current work: Magnetic recording, audio, video, tape. Patentee in field, devels. in wire and tape recorders and stereo sound reproduction, motion picture sound, video recorders. Home: 560 Lincoln Ave Glencoe IL 60022 Office: Technology Center Chicago IL 60616

CANCRO, MICHAEL PAUL, immunologist, educator; b. Washington, Oct. 28, 1949; s. Ciro Anthony and Florence Lilian Meekma; m. Jamie Arlene Robinson, June 26, 1975; 1 dau., Robin Elizabeth. B.S., U. Md., 1973, Ph.D., 1976. Research specialist NIH, Bethesda, Md., 1975-76; postdoctoral fellow U. Pa. Med. Sch., 1976-79, asst. prof., 1979—. Contbr. numerous articles to profl. jours. Asst. scoutmaster Boy Scouts Am., Springfield, Pa. Am. Assn. Immunologists award, 1980; Nat. Cancer Inst. grantee, 1979—. Mem. AAAS, Am. Assn. Immunologists, Am. Soc. Zoologists, Internat. Soc. Developmental and Comparative Immunology. Subspecialties: Immunobiology and immunology; Developmental biology. Current work: Developmental biology of B-lymphocyte repertoire; study of Murine models. Office: Pathology Dept U Pa Med Sch 36th and Hamilton Walk Room 284 G-3 Philadelphia PA 19104

CANDIA, OSCAR A., medical educator; b. Buenos Aires, Argentina, Jan. 4; came to U.S.; s. Jose F. and Luisa (Mitri) C.; m. Blanca F.; children: Roberto, Leticia, Silvina. M.D., U. Buenos Aires, 1959. Asst. prof. U. Louisville, 1965-68; assoc. prof. biophysics, physiology Mt. Sinai Sch. Medicine, N.Y.C., 1977—, prof. ophthalmology, 1978—, dir. small group, 1976—; mem. vision research program com. Nat. Eye Inst., Bethesda, Md., 1979—, cons. cataracts, 1980—. U. Buenos Aires postdoctoral fellow, 1961-63, USPHS career devel. award, 1966-68. Subspecialties: Membrane biology; Ophthalmology. Current work: Mechanisms of ultralaser infrascopy in ophthalmology. Home: 17 Sunny Ridge Rd New Rochelle NY 10804 Office: Mount Sinai Sch Medicine One Gustave Levy Pl New York NY 10021

CANIZARES, CLAUDE ROGER, physicist, educator; b. Tucson, June 14, 1945; s. Orlando and Stephanie (Bolan) C.; m. Jennifer Wilder, Aug. 31, 1968; children: Kristen Elizabeth, Alexander Orlando. B.A., Harvard U., 1967, M.A., 1968, Ph.D., 1972. Staff scientist MIT, 1971-74, asst. prof. physics, 1974-78, assoc. prof. physics, 1978—. Contbr. numerous articles to profl. jours. Alfred P. Sloan Research fellow, 1980. Mem. Am. Astron. Soc., Am. Phys. Soc., Internat. Astron. Union, Phi Beta Kappa, Sigma Xi. Subspecialties: 1-ray high energy astrophysics; Optical astronomy. Current work: X-ray astronomy, satellite experiments, x-ray spectroscopy, plasma diagnostics, extragalactic astrophysics, optical telescopes and instrumentation. Office: MIT 37-501 Cambridge MA 02139

CANNON, J. TIMOTHY, psychologist, educator; b. Scranton, Pa., Aug. 26, 1949; s. Thomas F. and Cecilia A. (Culkin) C.; m.; children: Christina, Sean, Michael. B.S., U. Scranton, 1971; Ph.D., U. Maine, 1977; postdoctoral student, UCLA, 1977-79. Asst. research psychologist UCLA, 1980-81, vis. asst. prof., summer 1981; asst. prof. psychology U. Scranton, 1981—. Contbr. articles to profl. jours. NSF grantee, 1971, 77-79; NDEA fellow, 1972-73. Mem. AAAS, Internat. Assn. Study Pain, Soc. Neurosci. Subspecialties: Neuropsychology; Neuropharmacology. Current work: Functional neuroanatomy, pharmacology and electrophysiology of pain, endogenous pain inhibitory mechanisms and the endorphins. Home: 840 Jefferson Ave Scranton PA 18510 Office: O'Hara Hall U Scranton Scranton PA 18510

CANO, ELMER RAUL, radiotherapist, educator; b. Trujillo, Peru, Feb. 23, 1946; came to U.S., 1974, naturalized, 1979; s. Justo Roman Cano and Zoila Rosa Iglesias; m.; children: Malina Dianah, David Raul. B.S., U. Trujillo (Peru), 1971, M.D., 1972. Intern St. John Hosp., Detroit, 1974-75, resident in therapeutic radiology, 1976-79; fellow in radiotherapy M.D. Anderson Hosp., U. Tex., Houston, 1979-80; assoc. radiotherapist St. Joseph Hosp., Elgin, Ill., 1980-81; asst. prof. radiotherapy Duke U., Durham, N.C., 1981-83; radio therapist St. Francis Hosp., Memphis, 1983—. Mem. AMA., Am. Soc. Therapeutic Radiology, Radiol. Soc. N.Am., Am. Coll. Radiology, Fletcher's Soc. Roman Catholic. Subspecialty: Radiology. Current work: Gynecologic malignancies research. Home: 8520 Mysen Cove Cordova TN 38018 Office: Duke Univ Med Center Box 3275 Durham NC 27706

CANON, RONALD MARTIN, chemical engineer; b. Paducah, Ky., Apr. 3, 1947; s. Charles Curd and Mary Edith (Lax) C.; m. Mary Alice Downey, June 24, 1966; children: Jeffrey, Jason, Rachel. Ph.D., U. Mo.-Rolla, 1974. Devel. engr. Oak Ridge Nat. Lab., 1974-79, group leader resource recovery, 1979-82, mgr. chem. tech. div., 1982-83, mgr. chem. process design, 1983—; cons. in field. Contbr. articles to profl. jours. Mem. Karns Recreation Commn., Knoxville, Tenn., 1981—. Served with U.S. Army, 1970-72. Decorated Bronze Star, Air medal; recipient Indsl. Research IR-100 award Indsl. Research mag., 1978. Mem. Am. Inst. Chem. Engrs., Sigma Xi. Club: Optimist (Knox County, Tenn.). Subspecialties: Chemical engineering; Materials processing. Current work: Separation and recovery of critical metals from alternate ores and wastes; utilizing ion exchange, continuous chromatography, solvent extraction and precipitation techniques; recovery of aluminum from fly ash. Patentee continuous chromatography. Office: Oak Ridge Nat Lab PO Box X Building 1000 Oak Ridge TN 37830

CANONICO, DOMENIC ANDREW, engring. co. exec., lectr., cons., researcher; b. Chgo., Jan. 18, 1930; s. Angelo Anthony and Anna (Contratto) C.; m. Colleen Margaret Jennings, Aug. 27, 1955; children: Judith Canonico Asreen, Mary Carol, Angelo Edward, Domenic Michael, Catherine Ann. B.S. in Metall Engring, Mich. Tech. U., 1951; M.S. in Metall. Engring, Lehigh U., 1961, Ph.D., 1963. Group leader Pressure Vessel Tech. Lab. div. metals and ceramics Union Carbide Corp. at Oak Ridge Nat. Lab., 1965-81; dir. Metall. and Materials Lab., Combustion Engring., Inc., Chattanooga, 1981—. Served with USAF, 1952-53. Fellow Am. Soc. Metals; mem. ASME, Am. Welding Soc. (Rene D. Wasserman award 1977, N.E. Sect. Disting. Service award 1978, nat. Lincoln Gold medal 1980), Sigma Xi. Roman Catholic. Club: Walden (Chattanooga). Subspecialty: Metallurgical engineering. Current work: Research and devel. for current and advanced energy systems. Home: 3 Big Rock Rd Signal Mountain TN 37377 Office: 911 W Main St Chattanooga TN 37402

CANRIGHT, JAMES EDWARD, botany educator; b. Delaware, Ohio, Mar. 1, 1920; s. Ralph Olin and Ruth Evelyn (Edwards) C.; m. Margaret Jean Barnthouse, May 21, 1943; children: J. Douglass, Lawrence, Susan, Margaret Eloise. A.B., Miami U., Oxford, Ohio, 1942; M.A. in Biology, Harvard U., 1947, Ph.D., 1949. From instr. to prof. botany Ind. U., Bloomington, 1949-64; prof. botany, chmn. dept. botany and microbiology Ariz. State U., Tempe., 1964-72, prof. botany, 1972—; vis. prof. Nat. Taiwan U., 1971; hon. research assoc. in palynology Geol. Survey, Victoria, Australia, 1981. Contbr. articles to sci. jours. Mem. Republican Nat. Com. Served to lt. USCGR, 1943-46. Guggenheim fellow, Indonesia and Malaya, 1960-61. Fellow Explorers Club, AAAS; mem. Bot. Soc. Am., Internat. Orgn. Paleobotanists, Am. Inst. Biol. Scis. (exec. com. 1975-80), Am. Assn. Stratigraphic Palynologists (pres. 1979-80). Republican. Methodist. Current work: Palynology of cretaceous and tertiary rocks. Office: Ariz State U Tempe AZ 85287

CANTANZARO, ANTONINO, physician; b. Bklyn., Dec. 20, 1940; s. Ignazio and Anna (Alongi) C.; m. Margaret Ann Mandato, June 30, 1963; children: Andrew Thomas, Brian Edward, Donald Grant, Melissa Lynn. Student, U. Buffalo, 1958-61; M.D., SUNY-Buffalo, 1965. Intern Buffalo Gen. Hosp., 1965-66; resident in medicine Georgetown U. Hosp., 1968-70; fellow pulmonary div. U. Calif.-San Diego Sch. Medicine, 1970-72; fellow exptl. pathology Scripps Clinic and Research Found., 1970-73; asst. prof. dept. medicine U. Calif.-San Diego Sch. Medicine, 1974-82, assoc. prof., 1982—, dir., ; cons. control programs County San Diego, Dept. Pub. Health, San Diego, 1976—; rep. Pub. Health Statutes Recodification Project, Dept. Health Services, 1981—; dir. Ctr. Indochinese Health Edn., 1980—. Mem. com. Am. Lung Assn. Calif., 1979—; mem. TB com. Am. Lung Assn. San Diego and Imperial Counties, 1972—, chmn., 1978—, bd. dirs., 1975—, pres., 1982-83; mem. Resource Devel. Com. of San Diego Organizing Project. Served with USPHS, 1966-68. Mem. Am. Thoracic Soc. (mem. ad hoc com. 1979-80), Am. Coll. Chest Physicians (mem. fungus disease com. 1977—, pres. 1981-83), Med. Mycological Soc. Am., Calif. Thoracic Soc. (mem. membership com. 1979—), San Diego County Soc. Internal Medicine. Subspecialty: Pulmonary medicine. Office: U Calif Spl Center San Diego CA 92103

CANTOR, ELINOR H, pharmacologist, educator; b. Phila., Aug. 1, 1948; d. Joseph and Marcella (Morris) C. B.A. in Biochemistry, U. Pa, 1970; M.Sc. in Pharmacology, Phila. Coll. Pharmacy and Sci., 1973; Ph.D., Med. Coll. Pa., 1979. Postdoctoral fellow Roche Inst. Molecular Biology, Nutley, N.J., 1979-81, asst. mem., 1981—; adj. assoc. prof. dept. biol. scis. Fairleigh Dickinson U., Rutherford, N.J., 1981—. Contbr. articles to profl. jours. Mem. secondary sch. com. U. Pa. Mem. Assn. Women in Sci., Soc. Neurosci., Sigma Xi. Subspecialties: Pharmacology; Molecular pharmacology. Current work: Isolation and purification of membrane receptors with the aim of understanding the nature and regulation of receptor macromolecules. Office: Roche Inst Nutley NJ 07110

CANTRELL, CYRUS DUNCAN, III, physicist; b. Bartlesville, Okla., Oct. 4, 1940; s. Cyrus Duncan and Janet Ewing (Robinson) C.; m. Mary Lynn Marple, Nov. 18, 1972. B.A., Harvard U., 1962; Ph.D., Princeton U., 1968. Asst., then asso. prof. physics Swarthmore Coll., 1967-73; staff mem. Los Alamos Sci. Lab., 1973-79; professeur associe Universite Paris-Nord, 1980; prof. physics, dir. Center Quantum Electronics and Applications, U. Tex. at Dallas, 1980—; cons. in field. Editor: Laser Induced Fusion and X-Ray Laser Studies, 1976; contbr. articles to profl. jours. Fellow Optical Soc. Am., Am. Physic Soc.; mem. Am. Phys. Soc., Optical Soc. Am., IEEE, Am. Chem. Soc. Subspecialties: Laser-induced chemistry; Spectroscopy. Current work: Laser separation of isotopes; infrared multiphoton molecular excitation; coherent laser propagation effects. Patentee in field. Office: U Tex at Dallas PO Box 688 Richardson TX 75080 Home: 2409 Lawnmeadow Dr Richardson TX 75080

CANTRELL, JOSEPH SIRES, JR., chemistry educator; b. Parker, Kans., July 31, 1932; s. Joseph Sires and Alta Fern (Collins) C.; m. Margaret Joyce Herr, Aug. 15, 1958; children: Mark Alan, Kenneth Aaron, Keith Floyd. Ph.D., Kans. State U., 1961. Research chemist Procter & Gamble Co., Cin., 1961-65; asst. prof. Miami U., Oxford, Ohio, 1965-69, assoc. prof., 1969—; cons. in field. Contbr. articles to profl. jours. Asst. scoutmaster Boy Scouts Am., 1979—. Served with U.S. Army, 1954-56. Fellow Ohio Acad. Sci.; mem. Am. Chem. Soc., Am. Crystallographic Assn., Electrochem. Soc., Sigma Xi. Club: Isaac Walton League. Lodge: Masons. Subspecialties: Physical chemistry; X-ray crystallography. Current work: Relationships between structure (x-ray crystallographics) and catalytic activity for transition metal oxides and hydrides. Office: Dept Chemistry Miami U Oxford OH 45056

CANTRELL, WELDON KERMIT, utility company nuclear engineer; b. Columbus, N.C., Feb. 7, 1940; s. Otis Lee and Thelma Bertha (Greene) C. B.S., N.C. State U., 1975. Registered profl. engr., N.C. Teaching asst. N.C. State U., Raleigh, 1975-77; nuclear engr. Carolina Power & Light Co., Raleigh, 1977—. Mem. Am. Nuclear Soc., Nat. Soc. Profl. Engrs. Democrat. Subspecialty: Nuclear fission. Current work: BWR reload design methods. Home: 1217 Hardimont Rd Apt 4 Raleigh NC 27609

CAPANZANO, CHARLES THOMAS, clinical psychologist, mental health administrator; b. N.Y.C., Aug. 2, 1949; s. Salvatore and Josephine Marguerite (Lupo) C. B.A. with honors in Psychology, Dartmouth Coll., 1971; M.A. in Clin. Psychology, U. Windsor, Ont., 1973, Ph.D., 1976. Registered psychologist, N.Y. Cons. Children's Aid Soc., Windsor, 1975-76; project research dir., parental counselor U. Windsor, 1975-76; dir. profl. services Heartline, Inc., Detroit, 1975-76; dir. continuing edn., Twin Tiers br. N.Y.Sch. Psychiatry, Elmira, 1976-78; dir. community services Schuyler and Yates Counties, N.Y., 1978-79, Cortland County, 1979—; sec. exec. com. Conf. Mental Hygiene Dirs., Albany, 1978—; chmn. Central N.Y. Mental Health Dirs., 1980-81, Cortland Area council Central N.Y. Health Systems Agy., Syracuse, 1981—. Chmn. govt. div. United Way of Cortland County, 1981—; treas. Hitchcock Hose Co.-Cortland Fire Dept., 1981—; adv. council Central N.Y. Div. of Youth, 1982—. Regents scholar, 1967; Laidlaw Found. grantee, 1975-76. Mem. Am. Psychol. Assn. Lodge: Rotary. Subspecialty: Psychology clinic. Current work: Parent counseling; cognitive behavior modification; forensic mental health, mental health administration. Home: 1 1/2 Broadway Cortland NY 13045 Office: Cortland County Community Services 60 Central Ave Cortland NY 13045

CAPEN, CHARLES FRANKLIN, astronomer, photographer; b. Gilman, Ill., Jan. 1, 1926; s. Charles Franklin and Lettie (Knoepfel) C.; m. Virginia Watkins, June 18, 1956; children: Mars, Rigel W., Regulus W. Diploma in flight aeros, Spartan Coll. Aeros., 1946, diploma in meteorology, 1951; postgrad. in Math. and Physics, N.Mex. State U., 1952-56; M.S. in Astronomy, U. Ill., 1955; Ph.D. in Astrophysics, Ind. U., 1956. Dir. Smithsonian Astrophys. Obs. Sta., Shiraz, Iran, 1957-60; sr. scientist, resident astronomer Table Mountain Obs., Jet Propulsion Lab., Calif. Inst. Tech., 1961-70, also chief planetary patrol programs in support of Mariner spacecraft missions, 1965-70; planetary astronomer, lectr. Planetary Research Ctr., Lowell Obs., Flagstaff, Ariz., 1970—; research astronomer cons. Braeside Obs., Flagstaff, 1979—; Inst. Planetary Research Obs., Miami, Fla., 1979—; lectr. in field. Contbr. articles to profl. jours., chpts. to books. Served with USAF, 1950-52. Recipient Space Sci. div. award for service as mem. Mariner IV Mars Team Jet Propulsion Lab., Calif. Inst. Tech., 1965; award Inst. Environ. Sci., 1969; Apollo Achievement award NASA; Mariner VI and VII Space Sci. div. award Jet Propulsion Lab., Calif. Inst. Tech., 1969; G. Bruce Blair gold medal Western Amateur Assn., 1970. Mem. Internat. Astron. Union, Am. Astron. Soc., Assn. Lunar and Planetary Observers (dir. Mars sect. 1967—), Am. Geophys. Union, Flagstaff Astron. Soc. Episcopalian. Lodge: Elks. Subspecialties: Planetology; Planetary atmospheres. Current work: Telescopic planetary photography for raw data. Computer program studies of Martian and Jovian meteorology for weather prediction and climatic changes. Study of historic planetary observations for weather and climatic changes on planets and earth. Archivist of classical planetary observations and charts. Pioneer in planetary color photography for sci. and comml. uses; C.F. Capen Planetary Library and Classical Charts Archive, Inst. Planetary Research Obs., named in his honor. Home: 223 W Silver Spruce St Flagstaff AZ 86001 Office: Lowel Obs Mars Hill Flagstaff AZ 86001

CAPEN, RONALD LEROY, biology educator; b. Niles, Mich., Feb. 22, 1942; s. Thomas Wells and Jacqueline Lorraine (Yagman) C.; m. Sigrid Stull, Dec. 17, 1967; children: Quinn, Ingrid, Jody. B.A., U. Calif.-Riverside, 1965; M.A., Harvard U., 1966; Ph.D., U. Calif.-Berkeley, 1971. Biologist Navy Electronics Lab., San Diego, summers 1965, 66; assoc. prof. biology Colo. Coll., Colorado Springs, 1971—; vis. asst. prof. cardiology U. Colo. Health Scis. Ctr., Denver, 1977-78, instr., summer 1980. Nat. research service fellow USPHS, NIH, 1977-78; Woodrow Wilson hon. grad. fellow, 1965; grad. fellow NSF, 1965-71. Mem. Am. Physiol. Soc. Democrat. Subspecialty: Physiology (biology). Current work: Research on control of pulmonary blood circulation. Office: Biology Dept Colo Coll Colorado Springs CO 80903

CAPLAN, JOHN DAVID, industrial research company executive; b. Weiser, Idaho, Mar. 5, 1926; s. Manley Maurice and Kathleen Malaby (Coldwell) C.; m. Loris Elizabeth Green, June 21, 1952; children: Barbara E., Carole E., Nancy B. Student, Stanford U., 1943-44; B.S. in Chem. Engring, Oreg. State U., 1949; M.S. in Mech. engring, Wayne State U., 1955; grad. advanced mgmt. program, Harvard U., 1976. With Gen Motors Research Labs., Warren, Mich., 1949—, dept head, 1963-67, tech. dir. basic and applied scis., 1967-69, exec. dir., 1969—; bd. dirs. Coordinating Research Council, 1970—, pres., 1975-77. Research publs. in field. Served to staff sgt. U.S. Army, 1943-46. Recipient Crompton-Lanchester medal INstn. Mech. Engrs., 1964. Fellow AAAS, Am. Inst. Chem. engrs.; mem. Am. Chem. Soc., Am. Mgmt. Assn., Engring. Soc. Detroit, Soc. Automotive Engrs., Nat. Acad. Engring, Dirs. Indsl. Research (sec. 1980, chmn. 1981). Subspecialties: Chemical engineering; Mechanical engineering. Current work: Engine fuels and combustion; automotive air pollution; research administration. Office: GEN Motors Research Labs Warren MI 48090

CAPLE, RICHARD BASIL, psychology educator, psychotherapist; b. Maxwell, Iowa, Dec. 25, 1930; s. Basil Hunt and Charlotte Martha (Green) C.; m. Anna May Sharp, July 5, 1963; children: Holly Ann, Michael Huntlee. B.A., Cornell Coll., 1953; M.A., Columbia U. Tchrs. Coll., 1956, Ed.D., 1961. Lic. psychologist, Mo. Counselor men N.Mex. State U., 1957-61; dean students Northwestern State Coll., 1961-67; assoc. prof. psychology U. Mo.-Columbia, 1967-72, prof., assoc. dir., 1972-80, prof. edn., counseling psychologist, 1980—; cons. Fulton (Mo.) State Hosp., 1972-73. Contbr. articles to profl. jours., chpts. to books. Pres. Rockbridge PTA, Columbia, Mo., 1976-77; project leader Rockbridge 4-H Club, 1981-82. Served with AUS, 1953-55. Mem. Am. Coll. Personnel Assn., Mo. Coll. Personnel Assn. (pres. 1982-83), Am. Psychol. Assn., Am. Assn. Higher Edn., Mo. Psychol. Assn. Democrat. Unitarian. Subspecialties: Counseling psychology; Social psychology. Current work: Practice of group and individual psychotherapy; research in group dynamics and group psychotherapy; theory building; development of wholistic counseling model and methods in health psychology. Home: RR 4 Box 210AA Columbia MO 65201 Office: U Mo-Columbia 214 Parker St Columbia MO 65211

CAPORAEL, LINNDA ROSE, psychologist; b. N.Y.C., July 28, 1947; d. Francis T. and Rosita (Diaz) C. B.A., U. Calif.-Santa Barbara, 1975, Ph.D., 1979; prostgrad., U. London, 1977-78. Asst. prof. Rensselaer Poly. Inst., Troy, N.Y., 1980—. Fulbright Hayes scholar, U.K., 1977; Ford Found. fellow, 1975-79. Mem. Am. Psychol. Assn., Assn. Computing Machinery, Sigma Xi. Subspecialties: Social psychology; Cognition. Current work: Social dimensions of computing; sociality in decision-making and information processing. Office: Rensselaer Poly Inst Troy NY 12181 Home: 163 Benson St Albany NY 12206

CAPPUCCINO, CARLETON C., endodontist, educator, researcher; b. Boston, Apr. 24, 1947; s. Anthony and Gertrude (Rogers) C.; m. Noreen Coughlan, Aug. 1, 1970. B.A. magna cum laude, Tufts U., Medford, Mass., 1969; D.M.D. cum laude, Harvard U., 1973; cert. in endodontics, Harvard U. and Forsyth Dental Center, Boston, 1975. Diplomate: Am. Bd. Endodontics. Endodontist Thomas, Mellion & Cappuccino, Inc., Warwick, R.I., 1975—; staff assoc. dept. endodontics Forsyth Dental Center, Boston, 1975-78, asst. clin. prof. endodontics, 1978-82, staff assoc., dept. endodontics, 1982—; clin.

instr. in endodontics Harvard Sch. Dental Medicine, 1978-82, asst. clin. prof., 1982—. Author, editor: Textbook of Oral Biology, 1978; Contbr. articles to profl. jours. Mem. U.S. Olympic Com., Boston, 1982—. Recipient cert. of merit Harvard U. Odontological Soc., 1973. Fellow Am. Coll. Stomatologic Surgeons; mem. Am. Assn. Endodontists (editorial bd. jour. 1981—), Am. Dental Assn., Internat. Assn. for Dental Research, R.I. Dental Assn., Boston Cancer Research Assn. Subspecialties: Endodontics; Oral biology. Current work: Patient care; clinical and basic science teaching; histochemistry of the human dental pulp; evaluation of new materials. Office: Thomas Mellion & Cappuccino Inc 265 Jefferson Blvd Warwick RI 02888

CARASSO, ALFRED SAM, mathematician; b. Alexandria, Egypt, Apr. 9, 1939; came to U.S., 1962; s. Samuel and Renee (Ades) C.; m. Beatrice Kozak, June 12, 1964; children: Adam, Rachel. B.S. in Physics, U. Adelaide, Australia, 1960; M.S. in Meteorology, U. Wis.-Madison, 1964; Ph.D. in Math, U. Wis.-Madison, 1968. Asst. prof. math. Mich. State U., East Lansing, 1968-69, U. N.Mex., Albuquerque, 1969-72, assoc. prof. math., 1972-76, prof. math., 1976-81; mathematician Nat. Bur. Standards, math. analysis div. Center for Applied Math., Washington, 1982—; cons. Los Alamos Nat. Lab., 1972-81. Mem. Soc. for Indsl. and Applied Math. Subspecialties: Applied mathematics; Numerical analysis. Current work: Mathematical and numerical analysis of inverse problems and their application in heat conduction, seismology, acoustics, image processing, and electromagnetics. Office: Nat Bur Standards Adminstrn A-302 Washington DC 20234

CARBERRY, JAMES JOSEPH, operations research analyst, consultant; b. Newark, Aug. 8, 1953; s. James Joseph and Alice Reeves (Camp) C. B.S., Stevens Inst. Tech., 1975; M.S., Am. U., 1979. Water analyst N.J. Med. Labs., Bricktown, 1973-74; ops. research analyst Naval Facilities Engring. Command, Alexandria, Va., 1975—. Author: Economic Analysis Handbook, 1980. Mem. Ops. Research Soc. Am., Mil. Ops. Research Soc., Nat. Computer Graphics Assn., Washington Ops. Research/Mgmt. Sci. Council, Pi Lambda Phi. Subspecialties: Operations research (engineering); Graphics, image processing, and pattern recognition. Current work: Developing computer-aided graphics planning system for master planning and natural resource management applications. Home: 2421 Byrd Ln Alexandria VA 22303 Office: Naval Facilities Engring Command 200 Stovall St Alexandria VA 22332

CARBON, MAX WILLIAM, educator; b. Monon, Ind., Jan. 19, 1922; s. Joseph William and Mary Olive (Goble) C.; m. Phyllis Camille Myers, Apr. 13, 1944; children—Ronald Allen, Jean Ann, Susan Jane, David William, Janet Elaine. B.S. in Mech. Engring, Purdue U., 1943, M.S., 1947, Ph.D., 1949. With Hanford works Gen. Electric Co., 1949-55, head heat transfer unit, 1951-55; with research and advanced devel. div. Avco Mfg. Corp., 1955-58, chief thermodynamics sect., 1956-58; prof., chmn. nuclear engring. dept. U. Wis. Coll. Engring., 1958—; group leader Ford Found. program, Singapore, 1967-68; mem. Adv. Com. on Reactor Safeguards. Served to capt. ordnance dept. AUS, 1943- 46. Mem. Am. Nuclear Soc., AAAS, Am. Soc. Engring. Edn., AAUP, Sigma Xi, Tau Beta Pi. Subspecialties: Nuclear fission; Nuclear engineering. Current work: LMFBR safety; nuclear engineering education. Office: Engring Research Bldg U Wis Madison WI 53706

CARBONE, PAUL PETER, cancer researcher; b. White Plains, N.Y., May 2, 1931; s. Antonio and Grace (Cappelieri) C.; m. Mary Iamurri, Aug. 20, 1954; children—David, Kathryn, Karen, Kim, Paul J., Mary Beth, Matthew. Student, Union Coll., Schenectady, 1949-52; M.D., Albany (N.Y.). Med. Coll., 1956. Diplomate: Am. Bd. Internal Medicine. Joined USPHS, 1956; intern USPHS Hosp., Balt., 1956-57, resident in internal medicine, San Francisco, 1958-60; mem. staff Nat. Cancer Inst., NIH, Bethesda, Md., 1960-76, chief medicine br., 1968-72, asso. dir. for med. oncology, div. cancer treatment, 1972-76, dep. clin. dir., 1972-76; clin. prof. Georgetown U. Med. Sch., 1971-76; lectr. hematology Walter Reed Army Inst. Research, 1962-76; prof. medicine and human oncology U. Wis., Madison, 1976—, dir. div. clin. oncology, 1976—, chmn. dept. human oncology, 1977—; dir. Wis. Clin. Cancer Center, 1978—. Contbr. profl. jours. Decorated USPHS Commendation medal; recipient Trimble Lecture award Md. Chirurgical Faculty, 1968; Lasker award clin. cancer chemotherapy, 1972; Rosenthal award for improvement in clin. cancer care, 1977. Mem. A.C.P., Exptl. Hematology Soc., Am. Soc. Clin. Investigation, Am. Soc. Clin. Oncology (pres. 1972-73), Am. Soc. Clin. Investigation, Am. Soc. Hematology, Am. Assn. Cancer Research (pres. 1978-79), Am. Fedn. Clin. Research, AMA, Alpha Omega Alpha. Subspecialties: Internal medicine; Oncology. Current work: Breast cancer, lynmphomas, clinical trials. Home: 6115 N Highlands Ave Madison WI 53705 Office: Univ Hosps 600 Highland Ave Madison WI 53792

CARCATERRA, THOMAS, civil engineer, engineering company executive; b. N.Y.C., Nov. 11, 1922; s. Arcangelo Francesco and Carmela (Giuffre) C.; m. Edna Ruth Law, Feb. 12, 1955; children: Thomas A., Steven M. B.C.E., CCNY, 1943, M.C.E., 1955. Registered profl. engr., N.Y., Fla., Va., Md., Pa., Tex., W.Va., N.J., D.C. Structural designer various firms in, N.Y. and Fla., 1946-55; chief structural engr. Chatelain, Gauger & Nolan, Washington, 1955-58; structural engr. Mills, Petticord & Mills, Washington, 1958-61; ptnr. Smislova & Carcaterra, Rockville, Md., 1961-66; prin. Caracterra & Assocs., Silver Spring, Md., 1966-76; ptnr. Chatelain, Samperton & Carcaterra, Washington, 1976-78; pres. E/A Design Group, Chartered, Washington, 1978—. Served with U.S. Army, 1944-46. Named Man of Yr. Cons. Engrs. Council Met. Washington, 1968, 75. Mem. ASCE, Am. Concrete Inst., Prestressed Concrete Inst., Washington Soc. Engrs. Subspecialty: Structural engineering. Current work: Management of multi-disciplined architectural engineering firm including architectural, structural, mechanical andelectrical engineering. Home: 4210 Isbell St Silver Spring MD 20906 Office: 1625 K St NW Washington DC 20006

CARD, DARRELL HOLDER, consultant radioactive waste management; b. Salem, Utah, Nov. 8, 1926; s. Earl F. and Louise (Holder) C.; m. Wynnette Kartchner, Sept. 1, 1950; children: Darelyn, Dyann, William, Miriam, David, Michael. B.S., Brigham Young U., 1950; M.Engring. Adminstrn., U. Utah, 1967. Chemist U.S. Bur. Mines, Salt Lake City, 1950-52; shift supr. E.I. DuPont de Nemours & Co., Inc., Martinsburg, W.Va., 1952-54; shift supt. Phillips Petroleum Co., Idaho Falls, Idaho, 1954-61; sr. process engr. Thiokel Chem. Corp., Brigham City, Utah, 1961-67; supr. rocket mfg. Northrup Aircraft Co., North Edwards AFB, Calif., 1967-68; mfg. project engr. Lockheed Propulsion Co., Redlands, Calif., 1968-74; project mgr. E.G.&G Idaho Inc., Idaho Falls, 1974-77; program mgr. Ford, Bacon & Davis, Salt Lake City, 1977—. Served with USNR, 1945-46. Mem. Am. Nuclear Soc. Mormon. Lodge: Lions. Subspecialty: Nuclear engineering. Current work: Development of new concepts in radioactive waste management and disposal techniques. Home: 454 N Main St Alpine UT 84003 Office: Ford Bacon & Davis 375 Chipeta Way Salt Lake City UT 84108

CARDAMONE, MICHAEL J., physics educator; b. York, Pa., Dec. 29, 1944; s. Joseph M. and Pauline (Resetar) C.; m. Barbara A. Spitale, Sept. 2, 1967; children: Deborah, Lori. B.S., U. Scranton, 1966; M.S., Pa. State U., 1969, Ph.D., 1970. Postdoctoral fellow U. Guelph, Ont., Can., 1970-71; mem. factulty Pa. State U. (Schuylkill Campus), Schuylkill Haven, 1971—, now assoc. prof. physics. Contbr. articles to profl. jours. NSF grantee, 1981. Mem. Optical Soc. Am., Sigma Xi. Subspecialty: Atomic and molecular physics. Current work: Research on light scattering from organic liquids.

CARDIFF, ROBERT DARRELL, pathologist, educator; b. San Francisco, Dec. 5, 1935; s. George Darrell and Helen (Kohfield) C.; m. Sally Bounds, June 23, 1962; children: Darrell, Todd Trevor, Shelley Lynn. B.S., U. Calif.-Berkeley, 1958; Ph.D., 1968; M.D., U. Calif., San Francisco, 1962. Diplomate: Am. Bd. Anatomical Pathology, 1968. Intern Kings County Hosp., Bklyn., 1962-63; resident in pathology U. Oreg., Portland, 1963-66; NIH spl. fellow U. Calif.-Berkeley, 1966-68; mem. faculty U. Calif. Med. Sch.-Davis, 1971—, prof. pathology, 1977—. Contbr. articles to profl. jours. Served to lt. col. M.C. U.S. Army, 1968-71. NIH grantee. Mem. AAAS, Internat. Acad. Pathology, Am. Soc. Pathology Bacteriologists, Internat. Assn. Breast Cancer Research, Internat. Assn. Comparative Research Leukemia, N.Y. Acad. Scis., Am. Assn. Cancer Research. Subspecialties: Cancer research (medicine); Pathology (medicine). Current work: Research in tumor biology, breast cancer, molecular biology-genetics. Office: Dept Pathology U Calif Med Sch Davis CA 95616

CARDONA, OCTAVIO, astronomer; b. Zacatecas, Zac., Mexico, Nov. 25, 1943; s. Moises and Maria de Jesus (Nunez) C.; m. Laura Judith Saver, July 1, 1954. B.A., Universidad Nacional de Mexico, 1970; Ph.D., U. Colo, 1978. Research asst. Instituto de Astronomia, Universidad Autonoma de Mexico, 1967-71; research asst. Joint Inst. Lab. Astrophysics, U. Colo., 1972-78; research astronomer Instituto Nacional de Astrofisica, Optical y Electronica, Tonanitzintla, Puebla, Mex., 1978—. Mem. Internat. Astron. Union, Am. Astron. Soc., Sociedad Mexicana de Fisica. Subspecialties: Theoretical astrophysics; Optical astronomy. Current work: Stellar atmospheres, numerical models, observational classification of stars, peculiar stars. Home: INAOE Tonantzintla Apartado Postal 51 PueblaMexico

CARDUCCI, BERNARDO JOSEPH, psychology educator; b. Detroit, May 20, 1952; s. Edward and Mary (Bosco) C.; m. Andra Lee Evans, Nov. 22, 1973; 1 dau., Rozana. A.A., Mt. San Antonio Coll., 1972; B.A., Calif. State U., Fullerton, 1974, M.A., 1976; Ph.D., Kans. State U., 1981. Asst. prof. psychology Ind. U., New Albany, 1979—; Mem. rev. bd. Replications in Social Psychology, 1979—; textbook reviewer Holt, Rinehart and Winston, 1981—. Author: Instructor's Manual to Accompany Mehr's Abnormal Psychology, 1983; contbr. articles to profl. jours. Ind. U. S.E. instructional devel. grantee, 1980; Ind. U. S.E. summer faculty fellow, 1981. Mem. Am. Psychol. Assn., Soc. for Personality and Social Psychology, Midwestern Psychol. Assn., Council of Undergrad. Psychology Depts., Southeastern Psychol. Assn., Soc. for Psychol. Study of Social Issues, Psi Chi. Subspecialties: Social psychology; Personality. Current work: The fole of cognitive factors in the attribution of responsibility for domestic and sexual violence and psychophysiological responses to interpersonal stressors; cross-generational perceptions of sexuality and interpersonal attraction. Home: 4002 Summer Pl New Albany IN 47150 Office: Department of Psycholog Indiana University Southeast New Alban IN 47150

CARELLI, MARIO DOMENICO, nuclear engineer; b. Taggia, Imperia, Italy, Feb. 6, 1942; came to U.S., 1969, naturalized, 1982; s. Giuseppe and Rosa (Cichero) C.; m. Maria Sabina Viti, Mar. 29, 1969; children: Eric Viti, Eliana Jennifer. B.Sc., U. Florence, Italy, 1962; Ph.D. in Nuclear Engring. summa cum laude, U. Pisa, Italy, 1966. Registered profl. engr. Italy, Pa. With Westinghouse Advanced Energy Systems Div., Madison, Pa., 1969—, prin. engr., 1974-77, engring. fellow, 1977—; adj. faculty prof. U. Pitts., 1975—, mem. faculty adv. com. energy resources, 1982—. Contbr. over 60 nuclear engring. articles to profl. publs. Recipient Engring. Achievement award Westinghouse, 1983; Research fellow U. Pisa, 1966-67. Mem. Am. Nuclear Soc., Internat. Assn. Hydraulic Research (mem. exec. com. 1979—), ASME. Roman Catholic. Subspecialties: Nuclear engineering; Fluid mechanics. Current work: Nuclear breeder reactors core design; heat transfer and fluid flow research and design; academic teaching on nuclear technology. Patentee flow orificing of breeder core assemblies, 1975, blanket management method for breeder reactors, 1982. Office: Westinghouse Advanced Energy Systems Div PO Box 158 Madison PA 15663 Home: 652 Buckingham Circle Greensburg PA 15601

CAREN, LINDA ANN DAVIS, biology educator; b. Corsicana, Tex., Apr. 17, 1941; d. Doyle Truitt and Elizabeth Nell (Roberts) Davis; m. Robert Poston Caren, Mar. 27, 1963; children: Christopher, Michael. B.S. summa cum laude, Ohio State U., 1962; M.A., Stanford U., 1965, Ph.D., 1967. Nat. Acad. Scis.-NRC postdoctoral fellow NASA Ames Research Ctr., Moffett Field, Calif., 1967; instr. biol. scis. San Jose (Calif.) City Coll., 1976-77, De Anza Coll., Cupertino, Calif., 1977-78; lectr., vis. asst. prof. dept. biology U. Santa Clara, Calif., 1974-80, asst. prof., 1980—; participant NATO Advanced Studies Inst. on Immunotoxicology, Wolfville, N.S., 1982; mem. adv. bd., blood banking specialist Santa Clara chpt. ARC, 1982—. Contbr. articles to profl. jours. Mem. Am. Soc. Microbiology, N.Y. Acad. Scis., Reticuloendothelial Soc., Soc. Exptl. Biology and Medicine, Internat. Soc. Study of Origin of Life, Am. Assn. Immunologists, Phi Beta Kappa, Sigma Xi. (v.p. U. Santa Clara chpt. 1981-83, pres. 1983-85). Subspecialties: Immunobiology and immunology; Immunotoxicology. Current work: Immunotoxicology; effects of environmental substances on the immune system; human granulocyte antigens. Office: Dept Biology U Santa Clara Santa Clara CA 95053

CAREW, JAMES L., geology educator; b. Lydney, Eng., May 2, 1945; came to U.S., 1946; s. Ernest W. and Joyce M. (Andrews) C. A.B., Brown U., 1966; M.A., U. Tex.-Austin, 1969, Ph.D., 1978. Asst. prof. dept. geology Williams Coll., Williamstown, Mass., 1972-75, Rensselaer Poly. Inst., Troy, N.Y., 1975-77; marine scientist Mystic (Conn.) Seaport Mus., 1977-80; assoc. prof. dept. geology Coll. of Charleston, S.C., 1981—. Contbr. articles to profl. jours. Pres. Forest Lakes Civic Club, Charleston, 1982-83. Union Oil Co. fellow, 1969; Cuyler Meml. scholar, 1970. Mem. Geol. Soc. Am., Soc. Econ. Paleontologists and Mineralogists, Paleontol. Soc., Bahamas Nat. Trust, Sigma Xi, Phi Kappa Phi, Sigma Gamma Epsilon. Subspecialties: Paleoecology; Geology. Current work: Pleistocene-Holocene geology and climatology of the Bahamas; Acropora cervicornis growth and calcification rates; permo-carboniferous paleoncology. Office: Dept Geology Coll of Charleston Charleston SC 29244

CAREW, LYNDON BELMONT, JR., nutritionist, educator; b. Lynn, Mass., Nov. 27, 1932; s. Lyndon Belmont and Myrtle Louella (Woodworth) C.; m. Lynn Harrington, July 9, 1960; children: Leslie, Audre. B.S., U. Mass., 1955; Ph.D., Cornell U., 1961. Research asst. and assoc. in animal nutrition Cornell U., Ithaca, N.Y., 1955-61, 65; dir. Colombian Nat. Poultry Program and Animal Nutrition Lab., Rockefeller Found., Bogota, 1961-65; dir. poultry research div. Hess and Clark, Ashland, Ohio, 1966-69; sci. program mgr. Internat. Nutrition Project, U. Vt., Burlington, 1980-82, prof. animal scis., prof. human nutrition and foods, 1969—. Contbr. articles to profl. jours. Pres. Vt. Nutrition Council, 1976-81; mem. Vt. Gov.'s Council Phys. Fitness; vice chmn. Shelburne (Vt.) Conservation Com., 1970-73. Recipient Carrigan Outstanding Teaching award U. Vt., 1981, George V. Kidder Outstanding Faculty award U. Vt., 1983; grantee NSF, U.S. Dept. Agr., Bur. Internat. Food and Agr. Devel., U. Vt., Agway Inc., Muscular Dystrophy Assn., others. Mem. Am. Inst. Nutrition, Soc. Exptl. Biology and Medicine, Endocrine Soc., Poultry Sci. Assn., World Poultry Sci. Assn., Animal Nutrition Research Council, Nutrition Edn. Soc., N.Y. Acad. Scis., Am. Inst. Biol. Scis., AAAS, Nat. Assn. Coll. Tchrs. of Agr., Am. Council Sci. and Health, Council Agrl. Sci. and Tech., Sigma Xi, Gamma Alpha, Phi Kappa Phi. Subspecialties: Animal nutrition; Endocrinology. Current work: Radioimmunoassay, nutrition, thyroid, energy metabolism, growth hormone thyroid receptors, unusualffeedstuffs, poultry nutrition. Nutrition in Latin America, poultry science in Latin America, computer assisted instruction, computers in diet analysis. Home: 57 Collamer Circle Shelburne VT 05482 Office: Bioresearch Lab 655 Spear St South Burlington VT 05401

CARICO, JAMES EDWIN, biology educator; b. Galax, Va., Mar. 20, 1937; s. James Khale and Nita Sue (Cox) C.; m. Nelle Arp, Aug. 21, 1959; children: James Kipp, Edwin Lenn. B.S., East Tenn. State U., 1959; M.S., Va. Poly. Inst. and State U., 1964, Ph.D., 1970. Lab. asst. Va. Poly. Inst. and State U., 1959-64; asst. prof. to prof. biology Lynchburg Coll., 1964—. Mem. Am. Arachnological Soc., Brit. Arachnological Soc., Centre International Documentation Arachnologique, Va. Acad. Sci., Sigma Xi. Subspecialties: Systematics; Ethology. Current work: Systematics and behavior of spiders. Office: Biology Dept Lynchburg College Lynchburg VA 24501

CARINO, FELIPE, JR., computer scientist, tech. cons.; b. Bronx, N.Y., Feb. 5, 1956; s. Felipe and Manuela (Rios) C. B.A. in Computer Sci. and Math, N.Y.U., 1977, M.S., 1979. Mem. tech. staff Bell Labs., Piscataway, N.J., 1977-80; sr. systems programmer Fairchild Test Systems, San Jose, Calif., 1980-81; staff engr. Ampex Corp., Redwood City, Calif., 1981; sr. software engr. Britton-Lee, Berkeley, Calif., 1981, Ford Aerospace and Communications Corp., Palo Alto, Calif., 1982—. Mem. Assn. for Computing Machinery, IEEE, Robot Inst. Am. Democrat. Subspecialties: Database systems; Graphics, image processing, and pattern recognition. Current work: Research into data base theory and machines, especially as applied to pictorial data base management system. Home: 655 S Fairoaks Ave N103 Sunnyvale CA 94086 Office: 3939 Fabian Way Palo Alto CA 94303

CARLEY, JOHN WESLEY, III, industrial psychology consultant; b. Dallas, Dec. 15, 1942; s. John Wesley and Velma Ruth (Miller) C.; m. Pamela A. Reynolds, Jan. 26, 1978. B.S., N. Tex. State U., Denton, 1965, Ph.D., 1970; postgrad., Harvard U., 1979. Dir. Tex. Dept Mental Health and Mental Retardation, Austin, 1964-76, asst. commr., 1976-78, dep. commr., 1978-82; self-employed indsl. cons., Austin, 1982—; pres. Tex. Showdown Games, Austin, 1982—, Greater Southwest Devel. Corp., 1981—. Author: Transportation and Mental Retardation, 1980; Contbr. articles to profl. jours. Fellow Am. Assn. Mental Deficiency; mem. Am. Psychol. Assn., Tex. Assn. Mental Deficiency (pres. 1977-78). Republican. Methodist. Subspecialties: Human factors engineering. Current work: Management, personnel turnover, employee safety. Home: 2106 Raleigh Ave Austin TX 78703 Office: PO Box 5877 Austin TX 78763

CARLO, JAIME RAFAEL, immunomicroscopist, immunologist; b. Utuado, P.R., Sept. 15, 1943; s. Rafael and Carmen Aida (Casellas) C.; m. Judith B. Bragg, Dec. 19, 1964; children: Judson Dupree, Tyler Bragg. B.S., Med. Coll. Ga., 1967; M.S., U. Ga., 1968; Ph.D., U. Md., 1976. Research immunologist Becton-Dickinson, Cockeysville, Md., 1972-73; pres. Immunodiagnostics, Inc., Houston, 1973-74; asst. prof. pathology U. Md., Balt., 1973-76; asst. prof. pathology/medicine Med. Coll. Va., Richmond, 1976-82; asst. prof. med. lab. sci. Northeastern U., Boston, 1982—. Contbr. articles in field to profl. jours. Advisor AIDS (Aquired Immune Deficiency Syndrome) Action Com., Boston, 1982-83. Served to lt. comdr. USN, 1968-72. Med. Coll. Va. grantee-in-aid, 1977; Dept. Health and Human Services grantee, 1977-79; NIH grantee, 1982; recipient Research and Scholarship Devel. award Northeastern U., 1983. Fellow Arthritis Found.; mem. Am. Assn. Immunologists, Am. Soc. Microbiology (pres. Va. br. 1979-81), Am. Soc. Clin. Pathologists, Am. Fedn. Clin. Research, N.Y. Acad. Scis. Democrat. Roman Catholic. Subspecialties: Immunocytochemistry; Immunobiology and immunology. Current work: The characterization of the catabolism of the third component of complement in immunologically mediated tissue injury by immunocytochemical methods. Home: 39 Middlesex Ave Swampscott MA 01907 Office: Northeastern U 360 Huntington Ave Boston MA 02026

CARLQUIST, SHERWIN, botanist, educator; b. Los Angeles, July 7, 1930; s. Robert William and Blanche Helen (Bauer) C. B.A., U. Calif., Berkeley, 1952, Ph.D., 1956. Asst. prof. botany Claremeont (Calif.) Grad. Sch., 1956-61, assoc. prof., 1962-67, prof., 1968—; prof. botany Pomona (Calif.) Coll., 1976—. Author: Comparative Plant Anatomy, 1961, Island Life, 1965, Japanese Festivals, 1965, Hawaii A Natural History, 1970, Island Biology, 1974, Ecological Strategies of Xylem Evolution, 1975. Recipient numerous awards and grants. Subspecialties: Plant anatomy; Insular biology. Current work: Wood anatomy. Wood evolution and interrelationship between function, ecology and structure in wood. Office: Rancho Santa Ana Botanic Garden Claremont CA 91711

CARLSON, CARL EDWARD, petroleum geologist; b. LaCrosse, Kans., Jan. 30, 1922; s. Carl Edward and Laura (Pine) C.; m. Iris May Spielman, Aug. 18, 1947; 1 son, Dwight Jesse. B.A., U. Wyo., 1948, M.A., 1949. Geologist Mobil Oil Corp., Wyo., 1949-52, dist. geologist, Mont., Wyo., 1952-63, dist. exploration supt., Tex., 1963-66, area exploration mgr., 1966-76, geol. cons., 1976-82; pvt. practice petroleum geologist Rocky Mountain region, San Diego, 1982—. Served as lt. (j.g.) USNR, 1942-46; PTO. U. Wyo. scholar, 1940. Fellow Geol. Soc. Am.; mem. Am. Assn. Petroleum Geologists, Rocky Mountain Assn. Petroleum Geologists, Rocky Mountain Oil and Gas Assn., Sigma Xi, Phi Kappa Phi. Republican. Clubs: Am. Cause, Aircraft Owners and Pilots Assn. Subspecialties: Geology; Sedimentology. Current work: Particular interest realtive to stratigraphic entrapment of oil and gas deposits. Home: 4734 Oporto Ct San Diego CA 92124

CARLSON, HAROLD ERNEST, physician, medical educator; b. S.I., May 17, 1943; s. Clarence Herbert and Edith Amelia (Anderson) C.; m. Gabrielle Arakelian, July 2, 1966. B.S. in Chemistry, Rensselaer Poly. Inst., 1964; M.D., Cornell U. Med. Coll., 1968. Diplomate: Am. Bd. Internal Medicine. Intern Barnes Hosp., St. Louis, 1968-69, resident, 1969-70; fellow in metabolism Washington U. Sch. Medicine, St. Louis, 1972-74; asst. chief endocrinology Wadsworth VA Hosp., Los Angeles, 1974-82; asst. prof. medicine UCLA, 1974-79, assoc. prof., 1979-82; chief endocrinology sect. Harry Truman VA Hosp., Columbia, Mo., 1982—; assoc. dir. endocrinology div. , assoc. prof. medicine U. Mo.-Columbia, 1982—. Contbr. numerous articles, abstracts to profl. publs.; contbg. author, editor: Endocrinology, 1983; editorial bd.: Jour. Clin. Endocrinology and Metabolism, 1979-82. Served to lt. comdr. USPHS, 1970-72. VA Merit Rev. research grantee, 1974-83. Mem. Am. Fedn. Clin. Research (sec.-treas. Western sect. 1982-83), Endocrine Soc., Western Soc. Clin. Investigation, Central Soc. Clin. Research. Subspecialty: Endocrinology. Current work: Regulation of pituitary and parathyroid gland function. Office:

Endocrinology Sect Harry Truman VA Hosp 800 Stadium Rd Columbia MO 65201

CARLSON, JOHN B., astronomer, anthropologist; b. Joliet, Ill., June 22, 1945; s. Bernard C. and Mary W. (West) C.; m. Linda C. Landis, May 19, 1979. B.A. in Physics, Oberlin Coll, 1967; M.S. in Astronomy, U. Md., College Park, 1971, Ph.D., 1977. Tinker Found. postdoctoral fellow Yale U., 1977-79; dir., founder Center for Archaeoastronomy, U. Md., 1978—; vis. asst. prof., research assoc. Inst. Phys. Sci. and Tech., U. Md., 1977-81; lectr. Smithsonian Instn. Resident Assoc. Program, 1977—; affiliate asst. prof. astronomy program, adj. asst. prof. anthropology U. Md. Editor-in-chief, founder: Archaeoastronomy, 1977—; assoc. editor: Current Anthropology, 1979—; adv. editor: Jour. History of Astronomy, 1979; contbr. articles profl. jours. Dumbarton Oaks summer research fellow, 1982. Fellow Royal Astron. Soc.; mem. Am. Astron. Soc. (council hist. astronomy div.), Internat. Astron. Union, Soc. Am. Archaeology, AAAS. Subspecialties: Archaeoastronomy; Extragalactic astronomy. Current work: Archaeoastronomy, native Am. Astronomy, pre-Columbian studies, Maya hieroglyphic writing, history of astronomy, extragalactic astronomy. Office: Center for Archaeoastronomy U M College Park MD 20742

CARLSON, JOHN EDWARD, molecular biologist educator; b. DuBois, Pa., July 16, 1952; s. Gust Elwood and Mable Irene (Bundy) C. B.S. cum laude, U. Pitts., 1974; M.S., U. Ill., 1978, Ph.D., 1982. Research asst. U. Pitts., 1974-75; grad. asst. U. Ill., 1975-82; research assoc. plant pathology Kans. State U., 1982-83; research fellow Allelix, Inc., Mississauga, Ont., Can., 1983—. Contbr. articles to profl. jours. Bd. dirs. Ch. of the Brethren, Champaign, 1977-80. Mem. AAAS, Plant Molecular Biology Assn., Am. Assn. Plant Physiologists, Genetics Soc. Am. Subspecialties: Genetics and genetic engineering (agriculture); Molecular biology. Current work: Development of vectors of transferring foreign genes into corn; analysis of mitochondrial DNA of cytoplasmically male-sterile corn for nucleo-cytoplasmic interactions. Office: 6850 Goreway Dr Mississauga ON Canada L4V 1P1

CARLSON, JOSEPH RALPH, neurophysiology educator; b. Kansas City, Kans., May 30, 1945; s. Joseph and Ida Natalie (Dobbie) C.; m. Karen Evans Cody, June 17, 1967; children: Brian, Peter. A.B., U. Calif.-Berkeley, 1967, Ph.D., 1975. Postdoctoral fellow dept. biology U. Oreg., Eugene, 1975-77; mem. faculty dept. zoology Iowa State U., Ames, 1977—. Nat. Inst. Neural and Communicative Disorders and Stroke grantee, 1980-83. Mem. AAAS, Am. Soc. Zoologists, Midwest Neurobiologists. Subspecialties: Neurophysiology; Neuroendocrinology. Current work: Research into mechanisms of the steroid hormone, ecdysone's actions on identified neuros generating the hormonally controlled ecdysis (molting behavior) of insects. Home: 3433 Woodland St Ames IA 50010 Office: Dept Zoology Iowa State U Ames IA 50011

CARLSON, KENNETH T., chemist, educator; b. Douglas, N.D., June 18, 1921; s. Torkel and Clara A. (Jorgens) C.; m. Sylvia G. Aafedt, June 12, 1949; children: Sandra Whitney, David Carlson, Cartor C. B.S., Minot State Coll., 1947; M.A., U. No. Colo., 1953; postgrad., Oreg. State U., 1963-64. Tchr. math. Harvey (N.D.) High Sch., 1947-48; tchr. sci. Carrington (N.D.) High Sch., 1948-50; prin. high sch., Cando, N.D., 1950-54; prof. sci. Mayville (N.D.) State Coll., 1954—. Treas. Mayville Vol. Dept., 1956-63. Served with USAAF, 1942-46. Recipient Service award Mayville State Coll., 1978. Mem. Nat. Sci. Tchrs., Am. Chem. Soc., World Future Soc., N.D. Acad. Sci., Red River Valley Chem. Soc., N.D. Edn. Soc., NEA, Lambda Sigma Tau. Republican. Lutheran. Lodge: Elks. Subspecialties: Organic chemistry. Current work: Science education, environment, energy and the future. Office: Mayville State Coll Mayville ND 58257

CARLSON, MARIAN BILLE, molecular genetics educator, researcher; b. Princeton, N.J., Oct. 19, 1952; d. B.C. and Louise G. (Winston) C.; m. Stephen P. Goff, Oct. 15, 1977. A.B., Radcliffe Coll., Cambridge, Mass., 1973; Ph.D., Stanford U., 1978. Fellow dept. biology M.I.T., Cambridge, 1978-81; asst. prof. human genetics and devel. Coll. Physicians and Surgeons Columbia U., N.Y.C., 1981—. Jane Coffin Childs fellow, 1978-81; recipient Irma T. Hirschl Career Scientist award, 1982. Mem. Genetics Soc. Am. Subspecialties: Genetics and genetic engineering (biology); Molecular biology. Current work: Molecular genetics of yeast. Office: 701 W 168th St New York NY 10032

CARLSON, RICHARD FREDERICK, physics educator; b. St. Paul., June 19, 1936; s. Richard E. and Margaret (Kaercher) C.; m. Sandra Jean Johnson, Sept. 7, 1957; children: Karen Jean, Kristin Ann, Keith Richard. B.S., U. Redlands, Calif., 1957; M.S., U. Minn., 1962, Ph.D., 1964. Acting asst. prof. UCLA, 1963-67; prof. physics U. Redlands, Calif., 1967—. Contbr. articles in field to profl jours. Mem. Am. Phys. Soc., Am. Sci. Assn. Mem. Reformed Ch. Am. Subspecialty: Nuclear physics. Current work: Experimental low to medium energy nuclear physics; nuclear reactions. Office: Dept Physics U Redlands Redlands CA 92373 Home: 318 Marcia St Redlands CA 92373

CARLSON, RICHARD MERRILL, aeronautical engineer; b. Preston, Idaho, Feb. 4, 1925; s. Carl and Oretta C.; m. Venis Johnson, July 26, 1946; children: Judith, Jennifer, Richard. B.S. in Aero. Engring; M.S., U. Wash.; Ph.D. in Engring. Mechanics, Stanford U. Registered profl. engr., Calif. Chief aero.-structures engr. Hiller Aircraft, Menlo Park, Calif., 1949-64; rotary wing div. engr. Lockheed Calif. Co., Burbank, 1964-72; chief adv. systems research U.S. Army Air Mobility R&D Lab., Moffett Field, Calif., 1972-76; dir. Research and Tech. Labs., 1976—; lectr. Stanford U.; designated engring. rep. FAA. Contbr. articles to profl. jurs. Served to lt. (j.g.) AC USN, 1943-46. Consol.-Vultee fellow, 1947; recipient Meritorious Civilian Service award U.S. Army, 1975, 77. Fellow AHS (hon.), AIAA; mem. Swedish Soc. Aeros. and Astronautics, Sigma Xi. Republican. Mem. Ch. Jesus Christ of Latter-day Saints. Lodge: Elks (Preston, Idaho). Subspecialties: Aeronautical engineering; Theoretical and applied mechanics. Current work: Rotary wing design, structural dynamics, performance analysis, research management. Office: US Army Research and Tech Lab Ames Research Ctr Moffett Field CA 94035

CARLSON, ROY WASHINGTON, cons. civil engr.; b. Big Stone, Minn., Sept. 23, 1900; s. John Carlson and Christine (Olson) C.; m. Eleanor Cutler, Sept. 14, 1927; children—Suzan Carlson Dieden, Sally Carlson Johnson. A.B., U. Redlands, Calif., 1922, Sc.D. (hon.), 1951; M.S., U. Calif., Berkeley, 1933; Sc.D., Mass. Inst. Tech., 1939. Asst. prof. physics U. Redlands, 1924-25; testing engr. So. Calif. Edison Co., Los Angeles, 1925-27; test engr. County of Los Angeles, 1927-31, U. Calif., Berkeley, 1931-34; asst. prof. civil engring. Mass. Inst. Tech., 1934-35, asso. prof., 1936-43, U. Calif., Berkeley, 1935-36, research asso., 1945—; with atomic bomb project U. Calif., Los Alamos, 1943-45; cons. civil engring., Berkeley, 1945—. Author tech. papers. Recipient Outstanding Civilian Service award U.S. Army, 1972; Berkeley citation U. Calif., Berkeley, 1980. Fellow ASCE; hon. mem. Am. Concrete Inst. (Wason medal 1935, Turner medal 1967), Concrete Inst. Brazil; mem. ASTM (Dudley medal 1939), Nat. Acad. Engring., Sigma Xi. Subspecialties: Materials; Materials (engineering). Current work: Striving to improve concrete technology. Inventor elec. resistance meters for measuring stress, strain, pressure and temperature. Address: 55 Maryland Ave Berkeley CA 94707

CARLTON, DONALD MORRILL, research and development company executive; b. Houston, July 20, 1937; s. Spencer William and Ruth (Morrill) C.; m. Elaine Yvonne Smith, Jan. 28, 1961; children: Donna Kay, Spencer Frank, Monica Elaine. B.A., U. St. Thomas, Houston, 1958; Ph.D., U. Tex., Austin, 1962. Mem. staff, then group leader Sandia Corp., Albuquerque, 1962-65; with Tracor, Inc., Austin, 1965-69, asst. dir. research, 1968-69; pres., chmn. bd. Radian Corp., Austin, 1969—; dir. Hartford Steam Boiler Insp. and Ins. Co., Interfirst Bank, Austin. Mem. Air Pollution Control Assn., Am. Chem. Soc., Am. Mgmt. Assn. (pres.'s club), Austin C. of C. (past dir.). Subspecialties: Coal; Gas cleaning systems. Current work: Environmental aspects of coal gasification; sulfur dioxide control. Home: 4601 Cat Mountain Dr Austin TX 78731 Office: PO Box 9948 Austin TX 78766

CARLTON-FOSS, JOHN ANDREW, psychology and tech. co. exec.; b. Appleton, Minn., May 4, 1945; s. Harvey J. and Camilla A. (Person) Foss; m. Elizabeth J. Dennison, June 21, 1972; m. Rhona Newcomb Carlton, Aug. 15, 1980. S.B., M.I.T., 1967; S.M., 1969; Ed.M., Harvard U., 1973; Ph.D. in Psychology, Saybrook Inst., 1981. Ind. research and devel. and orgnl. cons., Lincoln, Mass., 1971-75; lead project researcher, adminstr. M.I.T., 1975-77; lead engr., action researcher Energy Investment, Inc., Boston, 1978; pres. Human-Tech. Energy Systems, Inc., Lincoln Center and Weston, Mass., 1978—; instr. physics and energy U. Mass., 1980-81; mgr. Energy Edn. Project, Dept. Energy/Mass. Energy Office, 1980-81; Mem. local congl. energy task force Program Task Force for Suffolk County Project for Reliable and Affordable energy; energy adv. com. Weston Pub. Schs.; com. for ednl. policy M.I.T. Contbr. articles to profl. jours. Recipient William L. Steward, Jr., award M.I.T.; NDEA fellow. Mem. Am. Psychol. Assn., AAAS, ASHRAE (corr. mem. Task Group), Am. Phys. Soc., Orgnl. Devel. Network. Episcopalian. Club: M.I.T. of Boston. Current work: Organizational consulting on human-organizational-technical factors in analysis, design, and implementation of energy and computer-based systems; research on human and organizational factors associated with technology; personal and situational stresses in relation to burnout and preventive medicine; technical inventions in these areas. Office: Box 151 Lincoln Center MA 01773 99 School St Weston MA 02193

CARLYLE, JACK WEBSTER, computer scientist, educator; b. Cordova, Alaska, Feb. 23, 1933; s. Jack Bartley and Helen Beatrice (Havil) C.; m. Sheila Adele Greibach, Mar. 22, 1970; 1 son, Jay Samuel. B.A., U. Wash., 1954, M.S. in Elec. Engring, 1957; M.A., U. Calif., Berkeley, 1961, Ph.D., 1961. Asst. prof. elec. engring. Princeton U., 1961-63; mem. faculty dept. engring. and applied sci. UCLA, 1963—, prof. computer sci., 1980—, chmn. system sci. dept., 1975-80. Contbr. articles to profl. jours. Mem. IEEE Computer Soc. (chmn. tech. com. founds. of computing 1975-78), Assn. Computing Machinery, Inst. Math. Stats., Am. Math. Soc., Soc. Indsl. and Applied Math, Am. Radio Relay League, Phi Beta Kappa, Sigma Xi, Pi Mu Epsilon. Subspecialty: Theoretical computer science. Office: Dept Computer Sci UCLA Los Angeles CA 90024

CARMAN, JOHN G(RIFFITH), plant geneticist; b. Salt Lake City, Nov. 1, 1951; s. Frank Cyrus and Anna Katherine (Griffith) C.; m. Cynthia Ann, Apr. 26, 1976; 4 children. B.S., Brigham Young U., 1979; Ph.D., Tex. A&M U., 1982. Asst. prof. plant sci. Utah State U., 1982—. Contbr. articles to profl. jours. Mem. Crop Sci. Soc. Am., Agronomy Soc. Am., Internat. Assn. Plant Tissue Culture, Plant Molecular Biology Assn. Subspecialties: Plant cell and tissue culture; Plant genetics. Current work: Wide hybridization/germplasm enhancement of cereals and forages; tissue culture applications to crop improvement. Office: Utah State U Plant Sci Dept UMC 48 Logan UT 84322

CARMICHAEL, RICHARD DUDLEY, mathematics educator; b. High Point, N.C., Mar. 13, 1942; s. Charles Hubert and Flossie Mae (Sullivan) C.; m. Doris Jane Aaron, Mar. 26, 1967; 1 dau., Mary Jane Carmichael. B.S., Wake Forest U., 1964; A.M., Duke U., 1966, Ph.D., 1968. Asst. prof. Va. Poly. Inst., Blacksburg, Va., 1968-71; asst. prof. Wake Forest U., Winston-Salem, N.C., 1971-74, assoc. prof., 1974-80, prof. math., 1980—; vis. lectr. U. Calif.-Davis, 1973; vis. assoc. prof. Iowa State U.-Ames, 1978-79. Contbr. articles in field to profl. jours. Mem. Am. Math. Soc., Math. Assn. Am., Soc. Indsl. and Applied Math., N.C. Acad. Sci. Republican. Baptist. Subspecialty: Analysis. Current work: Representation of distbns. as boundary values of holomorphic functions, harmonic analysis and H0P0 spaces, Abelian theorems. Home: 3131 Burkeshore Rd Winston-Salem NC 27106 Office: Dept Math Wake Forest U Winston-Salem NC 27109

CARMICHAEL, STEPHEN WEBB, anatomist, educator; b. Detroit, July 17, 1945; s. Lucien Webb and Sue (Peil) C.; m. Adrienne St. Pierre, Aug. 8, 1970; 1 son, Allen St. Pierre. A.B. with honors in Biology, Kenyon Coll., 1967; Ph.D. in Anatomy, Tulane U., 1971. Instr. anatomy W.Va. U., Morgantown, 1971-72, asst. prof., 1972-75, assoc. prof., 1975-82; assoc. prof. anatomy Mayo Med. Sch., Rochester, Minn., 1982—; UNESCO cons. Calcutta (India) U., 1981—; exchange scientist Hungarian Acad. Sci., 1980. Author: The Adrenal Medulla, Vols. 1, 2 and 3, 1979-83; contbr. over 50 articles to sci. jours. Recipient MacLaughlin award for excellence in teaching basic med. sci. W.Va. U., 1974; NIH grantee, 1974-83; Am. Heart Assn. grantee, 1980-82. Mem. Am. Assn. Anatomists, Soc. Neurosci., Am. Soc. Cell Biology, Electron Micros. Soc. Am. Republican. Episcopalian. Subspecialties: Anatomy and embryology; Cell biology (medicine). Current work: Basic mechanisms of (neuro)secretion, using the adrenal medulla as a model. Office: Anatomy Dept Mayo Clinic Med Scis Bldg 3 Rochester MN 55905

CARNEY, BRUCE WILLIAM, astronomer; b. Guam, Mariana Islands, Nov. 30, 1946; s. William Robert and Anne Elizabeth (Skow) C.; m. Lynn Christopher, Dec. 18, 1971. B.A., U. Calif.-Berkeley, 1969; A.M., Harvard U., 1971, Ph.D., 1978. Carnegie fellow dept. terrestrial magnetism, Washington, 1978-80; asst. prof. astronomy U. N.C., Chapel Hill, 1980—; Kitt Peak Nat. Obs. Users Com., 1980-83. Contbr. articles in field to profl. jours. Served with U.S. Army, 1971-74. NSF trainee, 1969-70; Harvard fellow, 1970-71. Mem. Internat. Astron. Union, Am. Astron. Soc. (Shapley lectr.), Astron. Soc. Pacific., Audubon Soc., Sierra Club. Democrat. Subspecialties: Optical astronomy; Infrared optical astronomy. Current work: Optical photometry, imagery and spectroscopy, studies aimed at stars nd clusters of our galaxy's oldest stars. Home: 201 Carl Dr Chapel Hill NC 27514 Office: U NC Dept Physics and Astronomy Phillips Hall 039 Chapel Hill NC 27514

CARNEY, RICHARD EDWARD, research psychologist, educator, consultant; b. Miami, Fla., Feb. 5, 1929; s. Clifford R. and Johnie Ora (Des Roches) C.; m. Jane Rima Wallace, June 20, 1953; children: Cathleen Jane, Daniel Richard, Bonnie Ann. Student, Fla. State U., 1947-48; I.B.S., U. Wash., 1954, M.S., 1956; Ph.D. in Psychology, U. Mich., 1961. Research asst. U. Wash., Seattle, 1954-55, teaching asst., 1955-58; asst. prof. psychology Drake U., Des Moines, 1958-62, Ind. U.-Gary, 1962-64; assoc. prof. psychology Calif. Western U. (later U.S. Internat. U.), San Diego, 1964-70; founder, pres. Carney Enterprises, Inc., San Diego, 1969—; prof. psychology Eastern Ky. U., Richmond, 1970-72; contract and core faculty Calif. Sch. Profl. Psychology, San Diego, 1972-79; co-founder, v.p. Timao Found. For Research and Devel., San Diego, 1972-80, pres., 1980—; research scientist System Devel. Corp., Santa Monica, Calif., 1974-75; assoc. dir. South San Diego Area Health Edn. Ctr., 1980-82; summer lectr. U. Mich. Extension Service, Flint and Saginaw, 1961; lectr. Roosevelt U., Chgo., 1962-63; summer assessment officer two Peace Corps tng. programs, 1966; cons. Non-Linear Systems, Inc., Solano Beach, Calif., 1967, Educator's Assistance Inst., 1972-74; evaluation cons. Title IX U.S. Office of Edn., 1977-78; cons. various sch. systems and orgns.; speaker profl. confs. Editor/author: Risk Taking Behavior, 1961; co-author: How to Reach Your Goals, 1981; author, co-author psychol. tests, 1976-81; contbr. articles, book revs., chpts. to publs. in field. Active ACLU, Planned Parenthood, Zero Population Growth; bd. dirs. San Diego Hypertension Program, Neighborhood House, Head Start; speaker polit., civic, religious groups. Served with U.S. Army, 1949-52. Grantee Ind. U., 1963, Methodist Ch., 1965, Calif. Western U., 1967, Dept. Def., 1974-75, Title IX U.S. Dept. Edn., 1977-78, USPHS, 1980-82. Mem. Am. Psychol. Assn., Western Psychol. Assn., San Diego Psychol. Assn. (pres. 1967-68), AAAS, AAUP (chpt pres. 1966-67), Soc. Psychologists in Substance Abuse (founding), Sigma Xi, Psi Chi (faculty advisor 1964-70). Democrat. Unitarian. Subspecialties: Physiological psychology; Learning. Current work: Psychophysiological bases of substance abuse and health application of learning-motivation. Inventor Digi-Tutor, 1968. Home: 3955 Alpine Blvd Alpine CA 92001 Office: Timao Found for Research and Devel 2223 El Cajon Blvd #307 San Diego CA 92104

CAROFF, LAWRENCE JOHN, research scientist; b. Windber, Pa., Aug. 26, 1941; s. Lawrence B. and Mary Sylvia (Mattie) C.; m. Velita Worden, Sept. 10, 1959 (div.); children: Michael Andrew, Christine Leigh, Jacqueline. B.S., Swarthmore Coll., 1962; postgrad., UCLA, 1962-63; Ph.D., Cornell U., 1967. Research scientist NASA Ames Research Ctr., Moffett Field, Calif., 1967—, dep. chief, space sci. div., 1982—. Contbr. articles to profl. jours. Mem. Am. Astron. Soc., ACLU, Sigma Xi. Democrat. Subspecialties: Theoretical astrophysics; Infrared optical astronomy. Current work: Basic research in theoretical astrophysics supported by observations in infrared astronomy. Office: NASA Ames Research Center MS 245-1 Moffett Field CA 94035

CAROTHERS, ZANE BLAND, botany educator; b. Phila., Nov. 7, 1924; s. Zane Bl and Louise Ann (Kirn) C.; m. Diane Marie Foxhill, June 28, 1952; children: Bruce Douglas, Robert Dale. B.S. in Biology, Temple U., 1950, M.Ed., 1952; Ph.D., U. Mich., 1958. Instr. botany U. Ky., 1957-59; asst. prof. botany U. Ill., 1959-64, assoc. prof., 1964-76, prof., 1976—. Author: (with others) The Plant World, 4th edit, 1963, 5th edit, 1972; contbr. numerous articles to sci. jours. Served with USAAF, 1943-46. NSF grantee, 1964-66, 68, 76-78, 80—. Mem. AAAS, Am. Bryological and Lichenological Soc., Bot. Soc. Am., Brit. Bryological Soc., Electron Microscope Soc. Am., Internat. Assn. Bryologists, Internat. Soc. Plant Morphologists, Phi Beta Kappa, Phi Kappa Phi, Sigma Xi. Subspecialties: Morphology; Reproductive biology. Current work: Ultrastructural cytology of spermatogenesis in bryophytes and lower vascular plants, sporophyte anatomy of embryophytes. Office: Botany Dept U Ill 505 S Goodwin St Urbana IL 61801

CAROZZI, ALBERT VICTOR, geology educator; b. Geneva, Switzerland, Apr. 26, 1925; came to U.S., 1955, naturalized, 1963; s. Luigi and Anna-Maria (Ferrario) C.; m. Marguerite Peier, July 23, 1949; children: Viviane Marrocco, Nadine B. M.S., U. Geneva, 1947, Dr.Sc. summa cum laude, 1948. Lectr. geology U. Geneva, 1953-57; asst. vis. prof. U. Ill.-Urbana, 1955-56, assoc. prof geology, 1957-59, prof. geology, 1959—; cons. Petroleo Brasileiro, 1969—, Yacimientos Petroliferos Fiscales, 1978—, Philippine Oil Devel. Co., 1970—. Author: Microscopic Sedimentary Petrography, 1972, Carbonate Depositional Models, 1983; editor: Sedimentary Rocks, 1975. Recipient Davy award U. Geneva, 1949, 54; Disting. Lectr. Am. Assn. Petroleum Geologists, 1959. Fellow Geol. Soc. Am.; mem. Soc. Econ. Paleontoogists and Mineralogists (councillor), Internat. Com. History Geol. Scis., U.S. Com. History of Geology, History of Earth Scis. Soc. (pres.-elect). Subspecialties: Sedimentology. Current work: Devel. of depositional models for carbonates with computer and microfacies techniques; experimental simulation of generation of porosity and permeability at depth in major basins; analysis of historical development of geological concepts. Office: Dept Geology NHB 245 U Ill 1301 W Green St Urbana IL 61801 Home: 709 W Delaware St Urbana IL 61801

CARPENTER, BARRY KEITH, chemist, educator, researcher; b. Hastings, Eng., Feb. 13, 1949; came to U.S., 1973; s. George Henry and Gladys May (Reekie) C.; m. Constance Joyce, Jan. 23, 1974. B.Sc., U. Warwick, Eng., 1970; Ph.D., Univ. Coll., London, 1973. NATO postdoctoral fellow Yale U., 1973-75; asst. prof. chemistry Cornell U., 1975-81, assoc. prof., 1981—. Contbr. articles to profl. jours. A.P Sloan fellow, 1980-82. Mem. Am. Chem. Soc., Chem. Soc. (London), AAAS. Subspecialties: Organic chemistry; Theoretical chemistry. Current work: Mechanisms of organic and organometallic reactions; development application of qualitative molecular orbital theories. Office: Dept Chemistry Cornell U Ithaca NY 14853

CARPENTER, CHARLES BERNARD, medical educator; b. Melrose, Mass., Sept. 11, 1933; s. Seymour Charles and Pauline Annette (Freeman) C.; m. Sandra Davis, Aug. 4, 1956; children: Bradford, Scott. A.B., Dartmouth Coll., 1955; M.D., Harvard U., 1958. Diplomate: Am. Bd. Internal Medicine, 1966. Research fellow, assoc., instr. Harvard U., 1962-70, asst. prof., 1970-73; assoc. prof. medicine Peter Bent Brigham Hosp., 1975-80, prof. medicine, 1980—; dir. Immunogenetics Lab., Brigham and Women's Hosp., Harvard U., 1980—; investigator Howard Hughes Med. Inst., 1973-80; mem. exec. com. New Eng. Organ Bank. Contbr. articles to profl. jours.; mem. editorial bd.: Am. Jour. Kidney Diseases, Jour. Human Immunology. Served to lt. USN, 1960-62. NIH research career devel. awardee, 1968-73. Mem. Assn. Am. Physicians, Am. Soc. Clin. Investigation, Transplantation Soc., Am. Soc. Transplant physicians, Am. Assn. Immunologists, Am. Soc. Nephrology, Am. Assn. Clin. Histocompatibility Testing. Baptist. Subspecialties: Immunogenetics; Transplantation. Current work: Immunogenetics and transplanation. Office: 75 Francis St Boston MA 02115

CARPENTER, CHARLES COLCOCK JONES, physician, educator; b. Savannah, Ga., Jan. 5, 1931; s. Charles Colcock Jones and Alexandra (Morrison) C.; m. Sally R. Fisher, Nov. 29, 1958; children—Charles Morrison, Murray Douglas, Andrew Fisher. A.B., Princeton, 1952; M.D., Johns Hopkins, 1956. Diplomate: Am. Bd. Internal Medicine (mem. bd. 1976—, exec. com. 1980—, chmn. 1983-84). Intern Johns Hopkins Hosp., 1956-57, resident, 1957-59, 61-62; practice medicine, specializing in infectious disease, Balt., 1962-73; asst. prof. medicine Johns Hopkins, 1962-67, asso. prof., 1967-69, prof., 1969-73; physician-in-chief Balt. City Hosps., 1969-73; prof., chmn. dept. medicine Case Western Res. Sch. Medicine, 1973—; physician-in-chief Case Western Res. Univ. Hosp., 1973—; dir. Cholera Research Program, Johns Hopkins Center Med. Research and Tng., Calcutta, India, 1962-64; chmn. cholera panel U.S.-Japan Coop. Med. Sci. Program, 1965-72, mem., 1973—; mem. exec. com. Bd. Sci. and Tech. Nat. Acad. Scis., 1981—; mem. adv. com. Sch. Medicine Johns Hopkins U., 1982—; mem. Am. Bd. Med. Spltys., 1982—. Trustee Internat. Center for Infectious Disease Research, Bangladesh,

1979. Served as sr. asst. surgeon USPHS, 1959-61. Fellow ACP; mem. Am. Soc. Clin. Investigation, Assn. Am. Physicians (sec. 1975-81, councillor 1981-86), Infectious Diseases Soc. Am. Subspecialties: Infectious diseases; Internal medicine. Current work: Chairman of department of medicine, with active involvement in clinical investigation, patient care and teaching; major research interest is immune mechanisms in enteric infections. Home: 2720 Dryden Rd Shaker Heights OH 44121

CARPENTER, DAVID ORLO, neurobiologist; b. Fairmont, Minn., Jan. 27, 1937; s. Orlo Ernest and Mae Elizabeth (Poulson) C.; m. Deborah Elizabeth [illegible], Aug. 19, 1961 (div.); children: Amanda, Adam. B.A., Harvard U., 1959, M.D. cum laude, 1964. Research fellow dept. physiology U. Goteborg, Sweden, 1961-62; research asso. dept. physiology Harvard Med. Sch., 1964-65; neurophysiologist Lab. Neurophysiology, NIMH, Bethesda, Md., 1965-73; chmn. neurobiology dept. Armed Forces Radiobiology Research Inst., Def. Nuclear Agy., Bethesda, 1973-80; dir. Center for Labs. and Research, N.Y. State Dept. Health, Albany, 1980—; adj. prof. biol. scis. SUNY, Albany; adj. prof. dept. anatomy Albany Med. Coll.; adj. prof. biology Rensselaer Poly. Inst. Editor books in field; contbr. chpts. to books, articles to profl. jours. Served with USPHS, 1965-70. Recipient Leon Risnick prize for research Harvard U. Med. Sch., 1964. Mem. Internat. Brain Research Orgn., Am. Physiol. Soc., Soc. Gen. Physiologists, Biophys. Soc., Soc. Neuroscience. Subspecialties: Neurophysiology; Biophysics (biology). Current work: Neurobiology of synaptic transmission, neurobiology, public health, physiology, biophysics. Home: 2749 Old State Rd Schenectady NY 12309 Office: Center for Labs and Researc New York State Department Health Alban NY 12201

CARPENTER, GENE BLAKELY, chemistry educator; b. Evansville, Ind., Dec. 15, 1922; s. Leland A. and Juanita L. (Blakely) C.; m. Elizabeth E. Corkum, Apr. 15, 1949; children: Jonathan R., Anne E. B.A., U. Louisville, 1944; Ph.D., Harvard U., 1947. NRC postdoctoral fellow Calif. Inst. Tech., 1947-48; research fellow, 1948-49; instr. chemistry Brown U., Providence, 1949-52, asst. prof., 1952-56, assoc. prof., 1956-63, prof., 1963—; vis. prof. U. Groningen, Netherlands, 1963-64; Fulbright-Hayes lectr. U. Zagreb, Yugoslavia, 1971-72; vis. scientist Oak Ridge Nat. Lab., 1980. Author: Principles of Crystal Structure Determination, 1969. Guggenheim fellow U. Leeds, Eng., 1956-57. Mem. Am. Crystallographic Assn., Am. Chem. Soc. Subspecialties: X-ray crystallography; Physical chemistry. Current work: Small molecule structure determination. Home: 8 Angell Ct Providence RI 02906 Office: Dept Chemistry Brown U Providence RI 02912

CARPENTER, PATRICIA, clinical psychologist, consultant; b. Detroit, May 16, 1920; d. William Henry and Kathryn Virginia (Dix) Humphrey; m. Warren Henry Carpenter, Mar. 29, 1982. A.B., Oberlin Coll., 1941; B.L.S., Western Res. U., 1943; M.A., Wayne State U., 1958, Ph.D., 1961. Lic. psychologist, Mich., diplomate: Am. Bd. Psychology and Psychiatry. Librarian Detroit Pub. Library, 1943-50, UAW-CIO, Detroit, 1950-56; research assoc. Wayne State U., 1957-58; psychology intern Lafayette Clinic, Detroit, 1958-59; psychologist Clinic for Child Study, Wayne County Juvenile Ct., Detroit, 1959-60, psychologist II, 1960-61, psychologist III, 1961-63, dir., 1963-81, Psychol. Services for Youth, Brighton, Mich., 1981—; cons. Wayne County Juvenile Ct. Clinic, 1981—, Livingston County Mental Health, 1967-68; U. Windsor, 1970-81, North Suburban Counseling Ctr., Mt. Clements, Mich., 1974-81, Genesee County Child and Adolescent Clinic, Flint, Mich., 1980-82. Mem. Am. Psychol. Assn., Mich. Psychol. Assn., Sigma Xi, Psi Chi. Subspecialties: Behavioral psychology; Clinical psychology. Current work: Psychotherapy with children, adolescents, especially delinquents; research in delinquency, groups, consulting; psychological testing. Home: 2875 W Coor Lake Rd Howell MI 48843 Office: Psychological Services for Youth 121 W North St Brighton MI 48116

CARPENTER, RICHARD AMON, chemist; b. Kansas City, Mo., Aug. 22, 1926; s. Harry Russell and Ina Marie (Garver) C.; m. Joanne Fisher, Aug. 14, 1948; children: Stephen Russell, Lynne, Wendy. B.S., U. Mo., 1948, M.A., 1949. Chemist Shell Oil Co., Wood River, Ill., 1949-51; asst. mgr. Midwest Research Inst., Kansas City, 1951-58, trustee, 1964-69, 75-79; mgr. Washington office Callery Chem. Co., 1958-64; sr. specialist in sci. and tech. Congl. Research Service, Library of Congress, Washington, 1964-69, chief environmental policy div., 1969-72; exec. dir. Commn. on Natural Resources, NRC, Nat. Acad. Scis., Washington, 1972-77; research asso. East-West Center, Honolulu, 1977—; vis. prof. environ. studies, Dartmouth, 1976. Editor: Assessing Tropical Forest Lands: Their Suitability for Sustainable Uses, 1981; editor: Natural Systems for Development: What Planners Need to Know, 1983; Contbr. articles to profl. jours. Trustee Inst. Ecology, 1979—; Mem. corp. vis. com. dept. civil engring. M.I.T., 1974—; mem. internat. environ. programs com. NRC-Nat. Acad. Scis., 1980—. Served with USAAF, 1945. Fellow Am. Inst. Chemists, AAAS; mem. Am. Chem. Soc., Ecol. Soc. Am., Sigma Xi, Sigma Chi. Presbyn. Club: Cosmos (Washington). Subspecialties: Resource management; Ecosystems analysis. Current work: Methods for acquiring and analyzing information about natural systems to be used in economic development planning and management. Patentee in field. Home: 2419 Halekoa Dr Honolulu HI 96821 Office: 1777 East West Rd Honolulu HI 96848

CARR, DANIEL BARRY, endocrinologist, medical researcher; b. N.Y.C., Apr. 6, 1948; s. Andrew Joseph and Florence (Glassman) C.; m. Justine M. Meehan, Nov. 11, 1978; children: Nora, Rebecca. B.A., Columbia U., 1968, M.A., 1970, M.D., 1976. Diplomate: Am. Bd. Internal Medicine (subsplty. bd. Endocrinology and Metabolism). Intern Columbia-Presbyn. Med. Ctr., N.Y.C., 1976-78; resident med. service Mass. Gen. Hosp., Boston, 1978-79, endocrine fellow, 1979-82, staff physician endocrine unit, 1982—, clin. assoc. physician, clin. research ctr., 1982—, clin. asst. in medicine, 1983—; cons. internal medicine Mass. Eye and Ear Infirmary, 1980-82; instr. medicine Harvard U. Med. Sch., 1982—. Contbr. articles, research reports, essays, revs. to profl. lit. Daland fellow Am. Philos. Soc., 1980—. Mem. AAAS, Am. Fedn. for Clin. Research, Alpha Omega Alpha. Clubs: Columbia Univ. (dir. Boston 1982—), Corinthians.). Subspecialties: Neuroendocrinology; Neurobiology. Current work: Clinical neuroendocrinology of opioid peptides. Office: Mass Gen Hosp Fruit St Boston MA 02114

CARR, EDWARD GARY, psychology educator and consultant; b. Toronto, Ont., Aug. 20, 1947; s. Saul and Anne (Goldsmith) C. B.A., U. Toronto, 1969; M.A., U. Calif.-San Diego, 1970, Ph.D., 1973. Adj. asst. prof. UCLA, 1973-76; asst. prof. SUNY-Stony Brook, 1976-81, assoc. prof., 1981—; cons. psychologist Suffolke Child Devel. Ctr., Smithtown, N.Y., 1976—; adv. bd. May Inst., Chatham, Mass., 1980—; bd. cons. Children's Treatment Program, Binghamton, 1980—; faculty sponsor Fulbright Program, Washington, 1980-82. Author: In Response to Aggression, 1981, How to Teach Sign Language, 1982; contbr. articles to profl. jours. W.W. Found. fellow, 1969; Regents fellow U. Calif.-San Diego, 1969-73; UCLA fellow, 1973-76; recipient Cert. of Commendation Nat. Soc. Autistic Children, 1981. Mem. Am. Psychol. Assn., Assn. for Advancement Behavior Therapy, Soc. Research Child Devel., Nat. Soc. Autistic Children. Subspecialties: Behavioral psychology; Developmental psychology. Current work: Research in experimental child psychopathology including studies of language disorders and analysis and remediation of severe behavior problems. Office: SUNY Dept Psychology Stony Brook NY 11794 Home: 28 Rolling Rd Miller Place NY 11764

CARR, GERALD DWAYNE, botany educator, researcher; b. Pasco, Wash., Apr. 1, 1945; s. Emery Winfield and Lydia Katherine Marie (Schuman) C.; m. Barbara Jean Myers, Aug. 19, 1977; 1 dau., Melissa Ann. B.A., Eastern Wash. U., 1968; M.S., U. Wis.-Milw., 1970; Ph.D., U. Calif.-Davis, 1975. Asst. prof. botany U. Hawaii, Honolulu, 1975-81, assoc. prof., 1981—; cons. on rare and endangered plant species. Contbr. numerous articles on botany to profl. jours. Served with USAR, 1963-67. NDEA fellow, 1968-70; NSF trainee, 1970-74; NSF grantee, 1979—; recipient U. Hawaii research award, 1975-77. Mem. Soc. for Study of Evolution, Am. Soc. Plant Taxonomy, Internat. Assn. Plant Taxonomists, Calif. Bot. Soc., Bot. Soc. Am., Northwest Sci. Assn., Hawaiian Bot. Soc. (Pres. 1976), Sigma Xi. Subspecialties: Evolutionary biology; Systematics. Current work: Biosystematics, chromosome evolution, adaptive radiation in insular ecosystems, speciation. Home: 47024 C Hui Iwa Pl Kaneohe HI 96744 Office: Dept Botan U Hawaii 3190 Maile Way Honolulu HI 96822

CARR, LAURENCE A., pharmacologist; b. Ann Arbor, Mich., Mar. 21, 1942; s. Hollis and Virginia (Finkbeiner) C.; m. Jeanne M. Carr, Aug. 15, 1964; children: L. Alan, Rachel M. B.S., U. Mich., 1965; M.S., Mich. State U., 1967, Ph.D., 1969. Asst. prof. pharmacology U. Louisville, 1969-75, assoc. prof., 1975-81, prof., 1981—. Assoc. editor: Clin-Alert, Sci. Editors, Louisville, 1972—. Fulbright scholar, Paris, 1980-81. Mem. Am. Soc. Pharmacology and Exptl. Therapeutics, Soc. Neurosci., Internat. Soc. Neuroendocrinology, Sigma Xi, AAAS. Democrat. Methodist. Subspecialties: Neuropharmacology; Neuroendocrinology. Current work: Effects of drugs on development of central monoamine systems in the neonate. Office: U Louisville Louisville KY 40292

CARR, RALPH W., mathematics educator; b. St. Paul, Apr. 19, 1946; s. Charles W. and Betty J. (Westman) C.; m. Linda M. Berscheid, June 3, 1979. B.A., Carleton Coll., 1968; Ph.D., U. Wis.-Madison. Asst. prof. math. St. Cloud (Minn.) State U., 1977—. Served with U.S. Army, 1969-72; Vietnam. NSF grad fellow U. Wis., 1968. Mem. Am. Math. Soc., Math. Assn. Am., Soc. Indsl. and Applied Math., AAAS, N.Y. Acad. Scis. Unitarian/Universalist. Subspecialties: Applied mathematics; Differential and integral equations. Current work: Volterra integro-differential equations. Home: 1821 11th Ave S Saint Cloud MN 56301 Office: Dept Math and Computer Sc St Cloud State Saint Cloud MN 5630

CARR, WILLIAM HOGE, JR., nuclear fuel reprocessing engineer, safety and licensing consultant; b. Princeton, W.Va., Jan. 31, 1921; s. William H. and Bertalee (Sackett) C.; m. Joy Hammitt, 1940 (div.); children: Melissa Carr Sullivan, Anne Carr Wackeen; m. Florence Louise Smith, Mar. 28, 1948; children: Robert A., William W. B.S. in Chem Engring, Va. Tech., 1943. Registered profl. engr., Tenn., S.C. Jr. engr. SAM Lab, Columbia U., N.Y.C., 1943-44; prodn. supr. Oak Ridge Gaseous Diffusion Plant, 1944-46; pilot plant engr. Oak Ridge Nat. Lab., 1946-49; devel. engr., 1954-70; research engr. and lab. comdg. office U.S. Army Chem. Corps, Md., Utah, Korea and Japan, 1949-53; devel. engr. Allied Gen. Nuclear Services, Barnwell, S.C., 1970—. Contbr. articles in field to profl. jours. Pres., chmn. bd. trustees Oak Ridge Hosp., 1964-70; vice chmn., bd. assocs. Hiwassee Coll., Tenn., 1968-70; bd. dirs. ORNL Credit Union, 1963-70. Served to capt. U.S. Army, 1949-53. Mem. Am. Chem. Soc., Am. Inst. Chem Engrs., Am. Nuclear Soc. Republican. Methodist. Lodges: Masons; Lions. Subspecialties: Chemical engineering; Nuclear engineering. Current work: Nuclear fuel reprocessing; primary emphasis on radioactive waste processing, storage, and disposal. Patentee diaphragm pumping system. Office: Allied-Gen. Nuclear Services PO Box 847 Barnwell SC 29812

CARRANO, ANTHONY VITO, cytogeneticist, researcher; b. N.Y.C., Mar. 22, 1942; s. Anthony and Geraldine Agnes (Salerno) C.; m. Elizabeth Patricia Hnatow, June 20, 1964; children: Christopher, Scott. B.S. in Chemistry, Rensselaer Poly. Inst., 1964; M.Bioradiology, U. Calif., Berkeley, 1970, Ph.D. in Biophysics, 1972. Postdoctoral fellow Argonne Nat. Lab., Ill., 1972-73; biomed. scientist Lawrence Livermore Nat. Lab., Calif., 1973—, sect. leader cell biology and mutagenesis, biomed. scis. div., 1980—; adj. prof. San Jose (Calif.) State U., 1980—. Served to capt. USMC, 1964-68. Mem. Environ. Mutagen Soc. (treas. 1983—), Am. Soc. Human Genetics, Am. Soc. Cell Biology, Radiation Research Soc., AAAS, Genetics Soc. Am. Subspecialties: Genome organization; Cancer research (medicine). Current work: Mechanisms and significance of cytogenetic damage; organization and structure of the mammalian chromosome; relation between mutation/cytogenetic damage and cancer. Office: PO Box 5507 Livermore CA 94550

CARRANO, RICHARD A, research pharmacologist; b. Bridge port, Conn.; s. Alfred J. and Irene E. (Genci) C.; m. Linda Lane; children: Beth Ayn, Richard, Karin, Roxanne. B.S. in Pharmacy, U. Conn., 1963, M.S. in Pharmacology and Pathology, 1965, Ph.D. in Pharmacology and Biochemistry, 1967. Research pharmacologist ICI Am., Inc., Wilmington, Del., 1966-67; supr. gen. pharmacology Stuart Pharm. div. ICI U.S., Wilmington, 1968-74; mgr. pharmacology/toxicology Adria Las., Inc., Wilmington, 1974-77, dir. preclin. research, Columbus, Ohio, 1977—; adj. assoc. prof. dept. microbiology Ohio State U., 1978—. Contbr. articles to profl. jours. Am. Pub. Health Assn. fellow, 1964-66. Mem. Western Pharmacology Soc., Acad. Pharm. Scis., Am. Pharm. Assn., Am. Chem. Soc., N.Y. Acad. Scis., Internat. Inflammation Club, Environ. Mutagen Soc., Am. Soc. Pharmacology and Exptl. Therapeutics. Subspecialties: Pharmacology; Toxicology (medicine). Current work: Development of new therapeutic agents in preclinical support of IND/NDA applications. Office: PO Box 16529 Columbus OH 43216

CARRAWAY, KERMIT LEE, biology educator; b. Utica, Miss., Mar. 1, 1940; s. Kermit L. and Louise (Greer) C.; m. Coralie Ann Carothers, May 26, 1962; children: Kermit Lyell, Kirsten Leigh. B.S., Miss. State U., 1962; Ph.D., U. Ill.-Urbana, 1966. Research fellow U. Calif.-Berkeley, 1966-68; asst. prof. Okla. State U., Stillwater, 1968-71, assoc. prof., 1971-75, prof., 1975-78, regents prof., 1978-81; prof. biology, chmn. U. Miami Sch. Medicine, 1981—; mem. molecular cytology study sect. NIH, Bethesda, Md., 1975-78. Mem. Am. Soc. Cell Biology, Am. Soc. Biol. Chemists, Sigma Xi (chpt. lectr. 1980). Democrat. Subspecialties: Biochemistry (biology); Cell biology. Current work: Cell surface structure/function relationships, membrane-cytoskeleton interactions, membrane proteins, membrane enzymes, glycoproteins. Home: 6465 SW 112th St Miami FL 33156 Office: U Miami Sch Medicine 1600 NW 10th Ave Miami FL 33101

CARRIER, E. BERNARD, microbiologist, university dean; b. Ferriday, La., Dec. 26, 1929; s. Dewey H. and Ella L. (Bunch) C.; m. Pauline Erma Peak, Jan. 31, 1951; children: Ernest Bernard, Gail Christine Carrier Mason. B.S., Southeastern La. U, 1953; M.S., La. State U., 1961, Ph.D., 1963. Cert. specialist in pub. health bacteriology, Am. Soc. Microbiology. Med. lab. technologist Baton Rouge Gen. Hosp., 1953-60; mem. faculty Southeastern La. U., Hammond, 1961-74; dir. La. Office Fed. Affairs, Baton Rouge, 1974-77; asst. sec. La. Dept. Culture, Baton Rouge, 1977-80; dean Grad. Sch., Southeastern La. U., 1980—. Served in USAF, 1950-54. Mem. Am. Soc. Microbiology, La. Acad. Scis., Sigma Xi. Democrat. Baptist. Subspecialties: Microbiology; Immunobiology and immunology. Current work: Medical microbiology, immunology and general microbiology. Home: Rt 6 Box 527 Denham Springs LA 70726 Office: 100 W Dakota St Hammond LA 70402

CARRIER, GEORGE FRANCIS, applied mathematics educator; b. Millinocket, Maine, May 4, 1918; s. Charles Mosher and Mary (Marcaux) C.; m. Mary Casey, June 30, 1946; children: Kenneth, Robert Mark. B.S. in Mech. Engring, Cornell U., 1939, Ph.D., 1944. From asst. prof. to prof. Brown U., 1946-52; Gordon McKay prof. mech. engring. Harvard U., 1952-72, T. Jefferson Coolidge prof. applied math., 1972—; mem. council Engring. Coll., Cornell U. Co-author: Functions of a Complex Variable, 1966, Ordinary Differential Equations, 1968, Partial Differential Equations, 1976; assoc. editor: Quar. Applied Math. Former trustee Rensselaer Poly. Inst., Troy, N.Y. Recipient Von Karman prize ASCE, 1977. Fellow Am. Acad. Arts and Scis.; hon. fellow Brit. Inst. Math. and Its Applications; hon. mem. ASME (Timoshenko medal 1978, Centennial medal 1980); mem. Nat. Acad. Scis. (award applied math. and mumerical analysis 1980), Soc. Indsl. and Applied Math. (Von Karman prize 1979), Nat. Acad. Engring., Am. Philos. Soc. Subspecialties: Fluid mechanics; Combustion processes. Current work: Dynamics of Tsunaml and atmospheric vortices; growth and propogation of large fires; phenomena in internal combustion engines; centrifuge phenomena. Office: Pierce Hall 311 Harvard Univ Cambridge MA 02138

CARRIGAN, CHARLES ROGER, research geophysicist; b. Pasadena, Calif., Sept. 7, 1949; s. Charles Francis and Alyce (Krosley) C.; m. Suzann Lundin, Feb. 21, 1976; 1 dau., Alisa Lynn. B.A., UCLA, 1971, M.S., 1973, Ph.D., 1977. Research fellow dept. geodesy and geophysics Cambridge U., 1977-78, 79; NATO fellow, 1978-79; research geophysicist Inst. Geophysics and Planetary Physics, UCLA, 1979-80; mem. tech. staff (Geophysics Research Div. 1541, Sandia Nat. Labs.), Albuquerque, 1980—; Chancellor's intern fellow UCLA, 1971, Chancellor's dissertation fellow, 1975; NATO postdoctoral fellow, 1977. Mem. Am. Geophys. Union, Sigma Xi. Club: Bible Study Fellowship (Albuquerque). Subspecialties: Geophysics; Fluid mechanics. Current work: Natural convection in geophysical systems, laboratory studies of convective stability, fluid mechanics of volcanism, geophysical fluid dynamics, crustal volcanism, thermal convection. Office: Sandia Nat Labs PO Box 5800 Geophysics Research Div (1541) Albuquerque NM 87185

CARROLL, BARBARA ANNE, radiology educator; b. Beaumont, Tex., Oct. 20, 1945; s. Theron Demp and Annette Ione (Anderson) C. B.A., U. Tex.-Austin, 1967; M.D., Stanford U., 1972. Intern Stanford U., 1972-73, resident in radiology, 1973-76, fellow in ultrasound, 1974-76, instr. radiology, 1976-78; asst. prof., 1978—; cons. NIH, Diasonica, Inc., Fremont, Calif. Nat. Cancer Inst. awardee, 1981-83. Mem. Am. Inst. Ultrasound in Medicine, Am. Coll. Radiology, Soc. Radiologists in Ultrasound. Subspecialties: Diagnostic radiology; Imaging technology. Current work: Development of ultrasound contrast agent; development of new, high frequency ultrasound equipment. Office: Dept Radiology Stanford Med Sch Pastuer Dr Palo Alto CA 94305

CARROLL, DANA, biochemist, educator; b. Palm Springs, Calif., Sept. 2, 1943; s. William Robert and Harriet (Dana) C.; m. Susan Slade, June 25, 1966; children: Adam Slade, Jessica Ann. B.A. in Chemistry, Swarthmore Coll., 1965; Ph.D., U. Calif.-Berkeley, 1970. Postdoctoral fellow Beatson Inst. Cancer Research, Glasgow, Scotland, 1970-72, Carnegie Instn. of Washington, Balt., 1972-75; asst. prof. microbiology, adj. asst. prof. biology U. Utah, Salt Lake City, 1975-81, assoc. prof. cellular, viral, and molecular biology, adj. assoc. prof. biology, 1981—. Contbr. articles to profl. jours. Recipient Ivy award Swarthmore Coll., 1965; Jane Coffin Childs postdoctoral fellow, 1970-72; USPHS fellow, 1966-70, 73-75; Am. Cancer Soc. scholar, 1983. Mem. Am. Soc. Microbiology, AAAS, Sigma Xi, Phi Beta Kappa. Subspecialties: Genetics and genetic engineering (biology); Molecular biology. Current work: Eukaryotic gene structure and function; mechanisms of genetic recombination; recombinant DNA technology. Office: Dept CVM U Utah Med Sch Salt Lake City UT 84132

CARROLL, DAVID STEWART, physician, oncologist; b. N.Y.C., Aug. 28, 1947; s. David Stewart and Alice Elizabeth (Knierim) C. B.A., Colgate U., 1969; M.D., Columbia U., 1973. Med. intern Roosevelt Hosp., N.Y.C., 1973-74, med. resident, 1974-76, hematology fellow, 1976-77, dir. solid tumor oncology, 1979—; oncology fellow Meml. Sloan Kettering, N.Y.C., 1977-79; assoc. clin. medicine Columbia U., N.Y.C., 1979—. Mem. Am. Coll. Physicians, Am. Soc. Clin. Oncology, N.Y. County Med. Soc., N.Y. Cancer Soc., Soc. for Study of Blood, Cancer and Acute Leukemia Group B. Subspecialties: Chemotherapy; Hematology. Office: 16 E 90th St New York NY 10028

CARROLL, DYER EDMUND, mechanical engineer; b. Boston, June 4, 1921; s. Jeremiah Charles and Grace Mildred (Rice) C.; m. Betty Wilder, Nov. 23, 1921; children: Dyer Edmund, Nancy Wilder Carroll O'Brien. Cert. M.E., Lowell Inst., 1942; M.E., Northeastern U., 1953, B.A. in Engring. and Mgmt, 1963. Registered profl. engr., Mass. Supr. research Mut. Boiler and Machine Ins. Co., Boston, 1948-51, 54-65; chief metall. engr. Factory Mut. Engring. Corp., Norwood, Mass., 1965-68; pres. Carroll Engrs., Inc., Andover, Mass., 1968—. Chmn. Bd. of Appeal, Stoneham, Mass., 1962-64; mem. Stoneham Park Commn., 1953-56; active Little League, Boy Scouts Am. Served with USN, 1942-46. Mem. ASME, Am. Soc. for Non-destructive Testing, ASTM, Am. Soc. for Metals, Am. Legion. Republican. Roman Catholic. Subspecialties: Metallurgical engineering; Mechanical engineering. Current work: Accident investigation and reconstruction—industrial, aircraft, automotive, marine. Materials and design consultant. Training of destructive, non-destructive and industrial quality control and assurance personnel. Home: 89 Spring St Stoneham MA 02180 Office: 200 Andover St Ballardvale MA 01810

CARROLL, FELIX ALVIN, JR., chemistry educator; b. High Point, N.C., Aug. 17, 1947; s. Felix Alvin and Addie R. (Doss) C.; m. L. Carol Crutchfield, July 15, 1972; 1 dau., Heather Elaine. B.S. with highest honors in Chemistry, U. N.C., 1969; Ph.D. in Organic Chemistry, Calif. Inst. Tech., 1973. Polymer synthesis chemist Burlington Industries Research Center, Greensboro, N.C., summers, 1968,69; asst. prof. chemistry Davidson (N.C.) Coll., 1972—80, assoc. prof., 1980—. Contbr. articles on chemistry to profl. jours. Mem. Am. Chem. Soc., Inter-Am. Photochem. Soc., AAAS, N.C. Acad. Sci. Club: Wildcat Investment (Davidson). Subspecialties: Photochemistry; Organic chemistry. Current work: Kinetics and mechanisms of photochemical reactions, applications of heavy atom effects, chemical conversion and storage of solar energy. Office: Chemistry Dept Davidson Coll Davidson NC 28036

CARROLL, JOHN STEPHEN, psychology educator, researcher; b. Bklyn., Nov. 5, 1948; s. Hyman Benjamin and Estelle (Silverman) C.; m. Helaine Dankner, June 13, 1970; children: Michael David, Deborah Ann. S.B. in Physics, M.I.T., 1970; M.A. in Social Psychology, Harvard U., 1972, Ph.D., 1973. Asst. prof. psychology

Carnegie Mellon U., 1973-78; assoc. prof. Loyola U., Chgo., 1978-83; assoc. prof. Sloan Sch. Mgmt. MIT, Cambridge, 1983—; vis. assoc. prof. Grad. Sch. Bus. U. Chgo., 1981-82; cons. Decision Assocs., Chgo., 1982—. Editor: (with J.W. Payne) Cognition & Social Behavior, 1976, (with Frieze and Bar-Tal) New Approaches to Social Problems, 1979. NSF fellow, 1970-72; NSF grantee, 1975-77; NIMH grantee, 1978—. Fellow Am. Psychol. Assn.; mem. Soc. Exptl. Social Psychology, Am. Psychology Law Soc. (dir. 1982—). Subspecialties: Social psychology; Cognition. Current work: Research on decision processes of criminal justice personnel (judges, parole boards), attribution theory, decision sciences. Office: Sloan Sch Mgmt MIT 50 Memorial Dr Cambridge MA 02139

CARROLL, LEE FRANCIS, elec. engr., cons.; b. Berlin, N.H., Oct. 14, 1937; s. Alton Francis and Mary Elizabeth (Cushing) C.; m. Judith A. Magoun, Apr. 9, 1960; children: Shawn, Pamela, Bruce. B.S.E.E., Northeastern U., 1960. Registered profl. engr., Maine, N.H., Vt., Mass., N.Y., Pa., Va., La., Tex. Maintenance engr. Am. Optical Co., Southbridge, Mass., 1964-65; elec. engr. Ga. Pacific Corp., Lyons Falls, N.Y., 1965-66; chief elec. engr. Brown Co., Berlin, N.H., 1966-70, Wright Pierce Barnes & Wyman, Topsham, Maine, 1970-73; propr. L.F. Carroll, P.E. Elec. Cons., Gorham, N.H., 1973—. Commr. Water and Sewer Dept., Gorham; treas. Gorham Congl. Ch.; pres. Gorham Devel. Corp. Mem. Nat. Soc. Profl. Engrs., IEEE, Illumination Engring. Soc., Nat. Fire Protection Assn. Subspecialty: Electrical engineering. Current work: Co-generation electrical design-utility interface coordination; alternate fuel (wood chip-waste) power plants. Office: 1 Exchange St Gorham NH 03581

CARROLL, PAUL T., pharmacologist, pharmacology educator; b. San Francisco, Oct. 22, 1943; s. Paul T. and Emmadell (Schroeder) C.; m. Carolyn Sue Carroll, June 15, 1968; children; Craig, Christopher. A.B., U. Calif.-Berkeley, 1966; M.A., San Jose State Coll., 1969; Ph.D., U. Md., 1973. Postdoctoral student Johns Hopkins U., 1974-76; Scientist Alcon Labs., Ft. Worth, Tex., 1969-70; asst. prof. pharmacology U. R.I., 1976-81; assoc. prof. pharmacology Tex. Tech Med. Sch., 1981—. Contbr. articles to sci. jours. Recipient NSF grants, 1981, 82—. Mem. Am. Soc. Pharmacology and Exptl. Therapeutics, Soc. Neurosci., Western Pharmacology Soc., Rho Chi. Subspecialties: Pharmacology; Neurochemistry. Current work: Subcellular origin of acetylcholine release in brain; characterization of membrane-bound choline o-acetyltransferase in brain; mechanisms of acetylcholine release. Home: 7906 Lynnhaven St Lubbock TX 79423 Office: Dept Pharmacology Tex Tech Univ Health Sci Center 4th St Lubbock TX 79430

CARROLL, ROBERT BUCK, educator; b. Wellsburg, W. Va., June 27, 1940; s. Ralph Edwin and Arta Maxine (Buck) C.; m. Ruthalee Markle, June 27, 1963; children: Tiffany Diane, Stanford Brent, Bradley Buck. B.S., W.Va.U., 1962, M.S., 1964; Ph.D., Pa. State U., 1971. Instr. Pa. State U., State College, 1970-71; extension plant pathologist U. Del., Newark, 1971-77, asso. prof. plant pathology, 1977—. Contbr. articles in field to profl. jours. Mem. Nat. Assn. Colls. and Tchrs. of Agr., Am. Phytopath. Soc., N.Y. Acad. Scis. Republican. Mem. Ch. of Christ. Subspecialties: Plant pathology; Integrated pest management. Current work: Researcher, tchr. utilization of ridge regression analysis in plant pathology. Home: 98 Cambridge Rd Surrey Ridge Elkton MD 21921 Office: University of Delaware 214 Agricultural Hall Newark DE 19711

CARROLL, ROBERT BYERS, virology educator; b. Washington, Oct. 10, 1940; s. Frank Samuel and Maryhelen (Byers) C.; m. Suzanne Elaine Edmonds, Aug. 28, 1965; children: Douglas Edmonds, Margot Buder, Monica Byers. B.A. in Zoology, U. Wash., 1962; Ph.D. in Biochemistry, U. Cin., 1970. Asst.prof. pathology NYU Med. Sch., N.Y.C., 1976-81, assoc. prof., 1981—. Leukemia Soc. Am. fellow Imperial Research Fund Labs., 1973-75; Leukemia Soc. Am. scholar, 1981-85. Mem. Am. Soc. Microbiology, Am. Soc. Biol. Chemists, Harvey Soc., N.Y. Acad. Scis., Sigma Xi. Subspecialties: Biochemistry (biology); Virology (biology). Current work: Mechanisms of viral carcinogenesis; molecular activities of SV4O T antigen and the host cell cycle control; p53 tumor antigen; protein-protein and protein-nucleic acid interactions. Office: 550 1st Ave New York NY 10016

CARSON, DENNIS ANTHONY, physician; b. N.Y.C., May 31, 1946; s. Edward M. and Rita (Brown) C.; m. Sandra Coler, May 31, 1970; children: Dora, Rebecca, Joseph. B.A., Haverford Coll., 1966; M.D., Columbia U., 1970. Diplomate: Am. Bd. Internal Medicine. Intern, resident in medicine U. Calif., San Diego, 1970-72; clin. assoc. Nat. Inst. Arthritis and Metabolic Diseases, 1972-74; postdoctoral fellow U. Calif., San Diego, 1974-76; asst. mem. Scripps Clinic and Research Found., LaJolla, Calif., 1976-80, assoc. mem., 1980—. Served to lt. cmdr. USPHS, 1972-74. Mem. Am. Soc. Clin. Investigation, Am. Assn. Immunologists, Am. Rheumatologic Assn. Subspecialties: Immunobiology and immunology; Biochemistry (biology). Current work: Clinical immunology, cancer research, arthritis. Office: 10666 N Torrey Pines Rd LaJolla CA 92037

CARSON, GEORGE STEPHEN, computer scientist; b. Lakewood, Ohio, Dec. 7, 1948; s. Sylvester and Madelyn Frances (Melson) C.; m. Brenda Geraldine Whaley, Feb. 7, 1969; children: Stephen, Elizabeth. B.S., U. Tenn., 1970; Ph.D., U. Calif.-Riverside, 1975. Mem. tech. staff B-1 Div. Rockwell Co., 1976-77, GTE Labs., Northlake, Ill., 1977-78; sr. assoc. prin. engr. Harris Govt. Electronics Systems Div., 1978-81; pres. GSC Assocs., Hawthorne, Calif., 1981—; tchr. UCLA Extension, 1981—. Served to capt. U.S. Army, 1974-75. Mem. IEEE (assoc. editor), IEEE Computer Soc., Am. Math. Soc., Assn. Computing Machinery, Math. Assn. Am., AAAS, Am. Nat. Standards Inst. (mem. computer graphics standards com.), Sigma Xi. Subspecialties: Distributed systems and networks; Graphics, image processing, and pattern recognition. Current work: Computer systems engring., graphics, databases, distributed systems, real-time systems. Office: 13254 Jefferson Ave Hawthorne CA 90250

CARSON, HAMPTON LAWRENCE, genetics educator; b. Phila., Nov. 5, 1914; s. Joseph and Edith (Bruen) C.; m. Meredith Shelton, Aug. 14, 1937; children: Joseph II, Edward Bruen. A.B., U. Pa., 1936, Ph.D., 1943. Instr. dept. zoology U. Pa., Phila., 1938-42; mem. faculty dept. biology Washington U., St. Louis, 1943-70, prof. biology, 1956-70; prof. genetics U. Hawaii, Honolulu, 1970—; vis. prof. biology U. Sao Paulo, Brazil, 1951, 77. Author: Heredity and Human Life, 1963; contbr. articles to profl. jours. Trustee B.P. Bishop Mus., Honolulu. Fulbright research scholar zoology U. Melbourne, Australia, 1961. Mem. Nat. Acad. Scis., Am. Acad. Arts and Scis., Genetics Soc. (pres. 1982), Soc. Study of Evolution (pres. 1971), Am. Soc. Naturalists (pres. 1973), AAAS, Phi Beta Kappa, Sigma Xi. Subspecialties: Evolutionary biology; Genetics and genetic engineering (biology). Current work: Genetics and evolution. Office: U Hawaii Honolulu HI 96822 Home: 2001 Ualakaa St Honolulu HI 96822

CARSON, JAMES MATTHEW, mech. engr.; b. Camden, N.J., Feb. 14, 1944. B.S., U.S. Air Force Acad., 1966; M.S., Drexel U., 1972, Ph.D., 1978. Registered profl. engr., Colo. Commd. 2d lt. USAF, 1966, advanced through grades to capt., 1969; engr. (Air Force Materials Lab.), Wright-Patterson AFB, Ohio, 1966-70, asst. prof., 1972-76, ret. active duty, 1976; now maj. USAFR; research engr. DuPont, Seaford, Del., 1978-80; sr. research engr. N. Mex. Engring. Research Inst., U. N. Mex., 1980—. Contbr. articles to profl. jours., confs. Mem. ASME, Tau Beta Pi, Pi Tau Sigma. Subspecialties: Theoretical and applied mechanics; Materials. Current work: Applications of applied mechanics techniques to solve civil engring. research problems; current techniques include signal analysis, stats. and exptl. methods. Home: 902 Ganado Ct SE Albuquerque NM 87123 Office: U N Mex N Mex Engring Research Inst Box 25 Albuquerque NM 87131

CARSON, PAUL ELBERT, pharmacology educator; b. Champaign, Ill., Feb. 18, 1925; s. Paul E. and Flo J. (Crowder) C.; m. Mary C. Silvans, Feb. 21, 1947; children: Jan Bates, Jeffrey, Amy. Student, U. Chgo., 1941-44; M.D., Harvard U., 1947. Intern Presbyn. Hosp., Chgo., 1947-48, resident, 1948-50, research fellow, 1948-50; chief dept. biochemistry Atomic Bomb Casualty Commn., Hiroshima, Japan, 1950-52; physician, scientist dept. biochemistry Brookhaven Nat. Lab., Upton, N.Y., 1952-55; research assoc. dept. medicine U. Chgo., 1955-62, asst. to assoc. prof., 1962-71; prof., chmn. dept. pharmacology Rush Med. Coll., 1975—, dir. sect. pharmacogenetics, 1971-74; sr. attending physician Rush-Presbyn.-St. Luke's Med. Ctr., Chgo., 1971—; research assoc. Army Malaria Research Project, 1955-57, sci. dir., 1964-69, dir., 1969-75; vis. scientist Galton Lab. of Human Genetics and Biometry, Univ. Coll., London, Eng., 1968-69. Discoverer G-6-PD deficiency; contbr. numerous articles to profl. jours. Bd. dirs. Schweppe Found., 1972—; mem. med. adv. com. Met. Chgo. chpt. March of Dimes, 1972-80; mem. Midwest Com. of Drug Investigation, 1972-82, chmn., 1977-80; mem. peer review research com. Chgo. Heart Assn., 1982—. Recipient Research Career Devel. award NIH, 1961-71; Disting. Service award Med. Alumni Assn. U. Chgo., 1981. Mem. Am. Soc. Pharmacology and Exptl. Therapeutics, Am. Soc. Clin. Pharmacology and Therapeutics, Central Soc. for Clin. Research, Am. Soc. Clin. Investigation, Am. Soc. for Human Genetics, Am. Soc. Tropical Medicine and Hygiene, Royal Soc. Tropical Medicine and Hygiene, Red Cell Club. Unitarian. Subspecialties: Pharmacology; Genetics and genetic engineering (medicine). Current work: Pharmacogenetics, malaria, drug and cell metabolism, role of human investigation in scientific and social advancement. Office: 1725 W Harrison Chicago IL 60612

CARSON, PAUL LANGFORD, radiologist, educator. B.S., Colo. Coll., 1965; Ph.D., U. Ariz., 1972. Diplomate: Am. Bd. Radiology. Instr. dept. radiology U. Colo. Sch. Medicine, Denver, 1972—, asst. prof., 1973-78, assoc. prof., 1978-81; assoc. prof., dir. radiol. physics and engring. dept. radiology U. Mich. Hosps., Ann Arbor, 1981—. Subspecialties: Imaging technology. Current work: Physics of Diagnostic ultrasound, new technology, nuclear magnetic resonance imaging, general radiological physics.

CARSON, STEVEN, educator, toxicologist, pharmacologist, cons., researcher; b. Bklyn., Oct. 17, 1925; s. David and Rebecca (Kraiewicz) C.; m. (married), Aug. 22, 1948; children: Ellen J., Susan L. Carson Friedman. B.S., Washington U., St. Louis, 1948; M.S. in Sci. Edn, N.Y.U., 1950, Ph.D., 1958. With Pub. Health Research Inst., N.Y.C., 1948; with Endo Labs., Inc., N.Y.C., 1950-58, Food & Drug Research Labs., Inc., 1958-72, Biometrics Testing, Inc., Englewood Cliffs, N.J., 1972-75, Toxi Con Assocs., N.Y.C., 1975-79; assoc. prof. St. John's U. Coll. Pharmacy and Allied Professions, N.Y.C., 1979—. Served with U.S. Army, 1942-46. Fellow Am. Inst. Chemists, Royal Soc. Health, Soc. Cosmetic Chemists (Lit. award 1964); mem. Am. Soc. Pharmacology and Exptl. Therapeutics, Clin. Soc. Exptl. Therapeutics, Harvey Soc., Internat. Soc. Immunopharmacology, Soc. Toxicology, European Soc. Toxicology. Club: Chemists (N.Y.C.). Subspecialties: Toxicology (medicine); Pharmacology. Current work: Effects of drugs, foods, cosmetics, and chemicals on behavioural responses of tissues, organs, and organisms; relevance to ultimate safe use in/by man. Home: PO Box 373 Ryder Sta Brooklyn NY 11234 Office: St John's U Coll Pharmacy Jamaica NY 11439

CARSON, STEVEN DOUGLAS, biochemist, educator; b. Bartlesville, Okla., Apr. 9, 1951; s. Harvey Arthur and Evelyn (Rule) C.; m. Sharon McLaren; 1 son, Shawn Kevin. B.A., Rice U., 1973; Ph.D., U. Tex.-Galveston, 1978. Asst. in chemistry U.S. EPA, Houston, 1972-73; research asst. U. Tex. Med. Br., Galveston, 1973-78; postdoctoral fellow Yale U., New Haven, 1978-82, research assoc., 1982, lectr., 1982-83; asst. prof. U. Colo., Denver, 1982—; chemist Pathology Lab., VA, Denver, 1982—. Robert Welch Found. fellow, 1970-71; NSF fellow, 1978-79; NIH fellow, 1979-82. Mem. Am. Soc. Biol. Chemists, AAAS, Am. Soc. Human Genetics, N.Y. Acad. Sci. Subspecialties: Biochemistry (biology); Membrane biology. Current work: Structure-function studies of membrane proteins and coagulation factors; production of monoclonal antibodies for immunochemical assays and characterization of protein antigens. Home: 18361 E Hawaii Pl Aurora CO 80017 Office: U Colo Health Sci Ctr Dept Pathology 4200 E 9th Ave Denver CO 80262

CARSON, VIRGINIA ROSALIE GOTTSCHALL, biologist; b. Pitts., Jan. 22, 1936; d. Walter Carl and Rosalie Madelaide (Paulin) G.; m. John Richard Carson, June 12, 1960; children: Margaret Rosalie, Kenneth Robert. Student, Swarthmore Coll., 1953-57; B.A., Calif. State U., Los Angeles, 1960, M.A., 1965; Ph.D., UCLA, 1970. Research aide Calif. Inst. Tech., Pasadena, 1958-60; asst. clin. research chemist Magaw Labs., Glendale, Calif., 1960-64; mental health trainee Brain Research Inst. UCLA, 1965-69; NIH postdoctoral trainee, 1972-74; asst. prof. biology Chapman Coll., Orange, Calif., 1971-77, assoc. prof., 1977-83, prof., 1983—, chairperson div. natural scis., 1983—; assoc. prof. So. Calif. Coll. Optometry, Fullerton, 1979—; assoc. research pharmacologist U. Calif., Irvine, 1981—, asst. research pharmacologist, 1972-81. Contbr. articles to profl. jours. Recipient Outstanding Faculty Mem. award Chapman Coll., 1979-80; Chapman Coll. research fellow, 1983—. Mem. AAAS, Am. Pharmacol. Assn., Am. Soc. Pharmacology and Exptl. Therapeutics, IEEE, Soc. Neuroscience, Iota Sigma Pi. Republican. Presbyterian. Subspecialties: Physiology (biology); Neuropharmacology. Current work: Physiological behavioral and neurochemical effects of alcohol and aging.

CARSONS, STEVEN ERIC, research physician, educator; b. N.Y.C., Nov. 24, 1950; s. Theodroe and Adele (Heilbraun) C.; m. Lesley Diane Freedberg, Oct. 31, 1975; 1 dau., Cara Frances. A.B., NYU, 1972; M.D., N.Y. Med. Coll., 1975. Cert. Am. Bd. Internal Medicine. Intern Maimonides Hosp., Bklyn., 1975-76, resident in medicine, 1976-78; fellow in rheumatology SUNY, Bklyn., 1978-80, asst. prof. medicine, 1980—; physician in charge clin. immunology L.I. Jewish Hosp., New Hyde Park., 1982—; cons. rheumatology St. John's Episc. Hosp., Bklyn., 1980-82, VA Hosp., 1980—. Renee Carhart Amory fellow N.Y. Arthritis Found., 1979; Lupus Found. Am. grantee, 1982; recipient clin. assoc. award Pub. Health Service NIH, 1980. Mem. Am. Fedn. Clin. Research, Am. Rheumatism Assn., N.Y. Rheumatism Assn. Subspecialties: Immunology (medicine); Biochemistry (medicine). Current work: The role of the extracellular matrix in modulating immune effector cell function. Home: 86 Muttontown Rd Syosset NY 11791 Office: LI Jewish Hosp Room 1153 270-05 76th Ave New Hyde Park NY 11042

CARSRUD, ALAN LEE, consulting research psychologist; b. Denver, July 23, 1946; s. George Edward and Clara Lee (Jones) C.; m. Karen Banks, Sept. 9, 1976; 1 son, Nichel David Victor. Asst. prof. psychology SUNY-Brockport, 1973-75; asst. prof. Tex. A&M U., 1975-77; coordinator psychol. services and research Travis State Sch., Tex. Dept. Mental Health/Mental Retardation, Austin, 1977-79; research assoc., cons. psychologist Robert Helmreich, Inc. (indsl. psychologists), Austin, 1980—; research assoc. dept. psychology U. Tex.-Austin, 1981—, lectr. dept. mgmt., 1982—; cons. in field, 1978—; mem. adv. com. on supervision standards Tex. Bd. Examiners Psychologists, 1979-80, mem. adv. com. on profl. and ethical exams., 1981-82. Author: (with others) Study Guide and Instructor's Manual for Lindzey,Hall, and Thompson: Psychology, 2d edit, 1978; contbr. articles to profl. publs. NDFA fellow, 1968-72; Eastern Psychol. Assn. grantee, 1971; U. N.H. Teaching and Learning Council grantee, 1972; grantee Am. Psychol. Assn. and Nat. Council Psi Chi, 1973-77; Research Found. SUNY grantee-in-aid, 1974; faculty research fellow, 1974; Research Found. Tex. A&M U. mini-grantee, 1977; Tex. Dept. Mental Health and Mental Retardation grantee, 1979. Mem. Am. Psychol. Assn. (grantee 1979), Eastern Psychol. Assn., Southeastern Psychol. Assn., Acad. Mgmt., Sigma Xi, Alpha Kappa Delta. Subspecialties: Social psychology; Applied social/personality psychology. Current work: Effects of competition and personality traits on complex task performance; relationship of group think to personality characteristics of decision makers, or achievement motivation, scholastic aptitude and personality to academic performance; social, psychological and environmental variables affecting outcome of psychotherapeutic treatment; analysis of complex organizational behavior and its relationship to personality factors. Home: 8911 Briardale Dr Austin TX 78758 Office: Dept Mgmt U Tex Austin TX 78712

CARTER, BENJAMIN DUDLEY, nuclear engineer; b. Lake Wales, Fla., Apr. 16, 1952; s. Bobby and Shirley (Cresse) C.; m. Sandra D. Earl, June 15, 1974; 1 son, Brian Dudley. B.S., U. Fla., 1974, M.E. Nuclear, 1977. Research asst. U. Fla., Gainesville, 1974-79, 8081; sr. research engr. RTS Labs., Inc., Gainesville, 1979-80; sr. resident startup engr. Combustion Engring., Windsor, Conn., 1981—. Inventor nuclear pumped helium-neon laser, ENRAD device. Van Dorn scholar, 1973. Mem. Am. Nuclear Soc., Tau Beta Pi. Democrat. Subspecialties: Nuclear engineering. Current work: Startup, low power physics and power acention testing-nuclear power plants. Home: 2683 Cupid St New Orleans LA 70114 Office: 1000 Prospect Hill Rd Windsor CT 06095

CARTER, FORREST LEE, chemist; b. Indpls., 1930. A.B., Harvard U., 1951; Ph.D. in Chemistry, Calif. Tech. U., 1956. Sr. chemist Westinghouse Research Lab., 1957-64; mem. staff Naval Research Lab., Washington, 1964—. Noyes fellow Calif. Tech. U., 1955-56. Subspecialty: Solid state chemistry. Office: Naval Research Lab Code 6175 Washington DC 20375

CARTER, HERBERT EDMUND, univ. ofcl.; b. Mooresville, Ind., Sept. 25, 1910; s. George Benjamin and Edna (Pidgeon) C.; m. Elizabeth Winifred DeWees, Aug. 30, 1933; children—Anne Winsett, Jean Elizabeth. A.B., DePauw U., 1930, Sc.D., 1952; A.M., U. Ill., 1931, Ph.D., 1934, Sc.D., 1974; Sc.D., U. Ind., 1974; L.H.D., Thomas Jefferson U., 1975. Instr. chemistry U. Ill., 1933-35, asso., 1935-37, asst. prof., 1937-43, asso. prof., 1943-45, prof., 1945-71, acting dean grad. coll., 1963-64, head dept. chemistry and chem. engring., 1954-67, vice chancellor for acad. affairs, 1967-71; coordinator interdisciplinary programs U. Ariz., Tucson, 1971-77, head dept. biochemistry, 1977-81; research fellow Office Arid Lands Studies, 1981—; Mem. Pres.'s Com. on the Nat. Medal of Sci., 1963-66; mem. nat. sci. bd. NSF, 1963-76, chmn., 1970-74; mem. Citizens Commn. Sci., Law and Food Supply.; Mem. exec. com. div. chemistry and chem. tech. NRC, 1949-55, 57-68. Mem. editorial bd.: Bio Chem. Preparations; editor-in-chief, Vol. I.; Contbr. to tech. publs. Trustee Assn. Univs. for Argonne, 1980—, Nutrition Found., 1972—. Awarded Rector Scholarship, Rector Fellowship DePauw U., Eli Lilly & Co.; Annual award ($1,000 and bronze medal to biochemist under 35 years of age showing promise in research), 1943; Am. Oil Chemists Soc. award in lipid chemistry, 1966. Mem. Am. Chem. Soc. (dir., asso. editor Bio-Chemistry 1961—, recipient William H. Nichols medal N.Y. sect., also Spencer award Kansas City sect. 1969), Am. Inst. Nutrition (sec. 1945-47), Am. Soc. Biol. Chemists (editorial bd. 1951-60, editorial com. 1963-66, pres. 1956-57), Nat. Acad. Scis. (chmn. section biochemistry 1963-66, mem. council 1966-69), Blue Key, Phi Beta Kappa, Sigma Xi, Phi Eta Sigma, Lambda Chi Alpha, Gamma Alpha, Alpha Chi Sigma. Democrat. Presbyn. Subspecialties: Biochemistry (biology); Food science and technology. Current work: Consultant to programs in nutritional sciences and in biochemistry.

CARTER, JEFF CROSSETT, computer scientist; b. Washington, Apr. 12, 1952; s. William Allen and Jean (Crossett) C.; m. Harriet Haynes, Jan. 2, 1955; 1 son: David. Student, Clarkson Coll. Tech., 1974-77; B.S. in Elec. and Computer Engring, Rochester Inst. Tech., postgrad., 1978—. Tech. specialist Xerox Corp., Rochester, N.Y., 1977—. Mem. IEEE, Computer Soc., Assn. for Computing Machinery. Subspecialties: Distributed systems and networks; Operating systems. Current work: Development of O.S. for distributed computer controls. Engaged in development of high level language, architecture and communications of the O.S. environment for distributed computer based control systems. Office: Xerox Corp 800 Phillips Rd Webster NY 14580

CARTER, LELAND LAVELLE, nuclear engineer, physicist; b. Oberlin, Kans., Nov. 27, 1937; s. LaVelle Wilford and Della Belle (Kathka) C.; m. Gerry L. Lindley, June 7, 1958; children: Carol Lynn, Michael Lee, Linda Marie, Wayne Mark. B.A., N.W. Nazarene Coll., Nampa, Idaho, 1961; M.S., U. Wash., Seattle, 1964, Ph.D., 1969. Scientist Battelle Nat. Lab., Richland, Wash., 1962-65; nuclear engr. Los Alamos Nat. Lab., 1969-77, alt. group leader, 1973-77, cons., 1978—; fellow engr. Hanford Engring. Devel. Lab., Richland, Wash., 1977—. Author: Particle-Transport Simulation with the Monte Carlo Method, 1975. AEC trainee, 1965-69. Mem. Am. Nuclear Soc. (mem. exec. com. 1980—). Republican. Nazarene. Subspecialties: Nuclear engineering; Numerical analysis. Current work: Neutronics and shielding for fusion facilities and fast breeder reactors and the development and application of the Monte Carlo method; previously was alternate group leader of the Monte Carlo Group at Los Alamos National Laboratory. Home: 2417 Michael Ave Richland WA 99352 Office: Hanford Engineering Development Laboratory PO Box 1970 Richland WA 99352

CARTER, MARY KATHLEEN, pharmacologist; b. Franklinton, La., July 11, 1922; d. Elijah Augustus and Ora Victoria (Kemp) C. B.A., Newcomb Coll. Tulane U., 1949, M.S., 1953; Ph.D., Vanderbilt U., 1955. Postdoctoral fellow U. Kans., 1955-57; instr. Tulane Med. Sch., 1957-59, asst., then asso. prof., 1959-73, prof., 1973—. Contbr. articles to profl. jours. USPHS postdoctoral awardee, 1955-57; sr. research career awardee, 1957-61. Mem. Am. Soc. Pharmacology and Exptl. Therapeutics, Soc. Exptl. Biology and Medicine, Am. Soc. Nephrology. Democrat. Baptist. Subspecialties: Pharmacology; Nephrology. Current work: Renal function studies related to effects of autonomic nervous system agents on electrolyte excretion. Home: 3021 Jena St New Orleans LA 70125 Office: 1430 Tulane Ave New Orleans LA 70112

CARTER, ROBERT LEROY, engineer, educator; b. Leavenworth, Kans., Aug. 22, 1918; s. Joseph LeRoy and Viola Elizabeth (Hayner)

C.; m. Jewell M. Long, June 3, 1941; children: Roberta, Benjamin, Judy Meadows, Frederick, Camille Ronchetto. B.S. in Engring. Physics, U. Okla., Norman, 1941; Ph.D., Duke U., 1949. Registered profl. engr., Mo. Tesing technician Eastman Kodak Co., Rochester, N.Y., 1940-42; physicist Tenn. Eastman Co., Oak Ridge, 1945-46; engring. group leader, research specialist Atomics Internat., Canoga Park, Calif., 1949-63; vis. scientist Los Alamos Sci. Lab., 1968-69; prof. elec. engring. and nuclear engring. U. Mo., Columbia, 1962—; commr. Gov.'s Low Level Radioactive Waste Task Force, Jefferson City, Mo., 1981—. Contbr. articles to profl. jours. Trustee Mo. Soc. Profl. Engrs. Edn. Found., Jefferson City, 1982—. Served with AUS, 1942-45; PTO. Mem. Am. Phys. Soc., Nat. Soc. Profl. Engrs., Mo. Soc. Profl. Engrs. (pres. chpt. 1977-78, state dir. 1978-80), Am. Nuclear Soc. (mem. nat. program com. 1973-77), Am. Soc. Engring. Edn. Republican. Methodist. Subspecialties: Nuclear fission; Nuclear fusion. Current work: Examining variety of applications of cryogenic techniques to engineering practice. Home: 1311 Parkridge Dr Columbia MO 65201 Office: Dept Elec Engring U Mo Columbia MO 65211

CARTER, TIMOTHY HOWARD, microbiologist, educator; b. Los Angeles, Nov. 6, 1944; s. Everett and Cecile (Doudna) C.; m. Jocklyn Armstrong, Dec. 31, 1976; 1 son, Benjamin. A.B., Harvard Coll., 1966; Ph.D., Princeton U., 1972. Postdoctoral fellow Pa. Plan to Develop Scientists in Med. Research, U. Pa., Phila., 1972-73; postdoctoral fellow Mattheson Found., 1973-74, Nat. Cancer Inst., 1974-75, Columbia U. Coll. Physicians and Surgeons, N.Y.C., 1973-75; asst. prof. Pa. State U. Med. Sch., Hershey, 1975-78; asst. prof. dept. biol. scis. St. John's U., N.Y.C., 1978-81, assoc. prof., 1981—; cons. in field; sci. adv. bd. Nuclear and Genetic Tech., Inc., 1982—. Contbr. chpts. to books, articles to jours.; also musician. Research grantee Nat. Cancer Inst., NIH, 1978-82. Mem. Am. Soc. Microbiology; mem. Am. Soc. Biol. Chemists; Mem. Am. Soc. Virology, N.Y. Acad. Scis., Sigma Xi. Subspecialties: Molecular biology; Virology (biology). Current work: Regulation of gene expression and action; molecular biology of adenovirus replication; mechanism of viral oncogenesis; interaction of viruses with carcinogens, growth regulators, tumor promoters. Office: Dept Biol Scis St John's Univ New York NY 11439

CARTER, WILLIAM DOUGLAS, geologist, consultant; b. Keene, N.H., Apr. 24, 1926; s. William Ambrose and Laura (Tuckerman) C.; m. Mary Jane Shannon, Sept. 10, 1950; children: Cindy Jean, Judy Lynn, Katherine Ann, William Douglas. A.B., Dartmouth Coll., 1949; postgrad., Johns Hopkins U., 1951, U. Colo., 1956. Geol. field asst. U.S. Geol. Survey, Fairbanks, Alaska, 1948-50, geologist, Grand Junction, Colo., 1951-57, mining geologist, Santiago, Chile, 1957-62, commodity geologist, Washington, 1962-65, remote sensing geologist, 1965-71, asst. program mgr. research, Reston, Va., 1972-82; geol. adv. AID, Santiago, 1957-62, remote sensing advisor, Washington and Costa Rica, 1975, InterAm. Devel. Bank, Washington, UN, 1982, China, 1982; co-leader internat. geol. program on remote sensing and mineral exploration, 1976-82. Contbr. articles to profl. jours. Active Boy Scouts Am., Reston, 1978. Served to cpl. USAAF, 1944-45. Fellow Geol. Soc. Am.; mem. Geol. Soc. Washington (sec. 1968), Am. Assn. Petroleum Geologists (lectr. continuing edn. program 1975—), Soc. Econ. Geologists, Soc. Exploration Geophysicists. Club: Fairfax (Va.) Jubilaires. Subspecialties: Geology; Remote sensing (geoscience). Current work: Mapping geologic structures and spectral discrimination of rock types, ore deposits, and energy resources from space platforms.

CARTER, WILLIAM EUGENE, research geodesist; b. Steubenville, Ohio, Oct. 16, 1939; s. Donald W. and Helen (Martin) C.; m. Marilyn Johnson, Jan. 16, 1961; children: Terri Lynn, Merri Sue, Pamela. B.S., U. Pitts., 1961; M.S., Ohio State U., 1965; Ph.D., U. Ariz., 1973. Commd. 2d lt. USAF, 1961, advanced through grades to capt., 1965; geodetic officer, 1961-69, research geodesist, 1969-72; research assoc. U. Hawaii, 1972-76; chief gravity astronomy and satellite div. (Nat. Geodetic Survey), 1977-81, chief advanced tech. br., 1981—. Contbr. 40 articles to profl. jours. Mem. Am. Astron. Soc., Internat. Astron. Union, Am. Geophys. Union, Internat. Union Geodesy and Geophysics. Subspecialties: Geodesy; Radio and microwave astronomy. Current work: Research and development of advanced techniques for geodesy. Home: 19004 Oxcart Pl Gaithersburg MD 20879 Office: Nat Geodetic Survey N/CG114 Rockville MD 20852

CARTER, WILLIAM HAROLD, research physicist; b. Houston, Nov. 17, 1938; s. William Henry and Fannie Augusta (Simpson) C.; m.; children: William Harold, Elizabeth Lee. B.S.E.E., U. Tex., Austin, 1962, M.S.E.E., 1963, Ph.D., 1966. Instr. elec. engring. U. Tex.-Austin, 1966; research assoc. in physics U. Rochester, 1966-70; research physicist Naval Research Lab., Washington, 1971—; vis. research fellow U. Reading, Eng., 1976-77; prof. elec. engring. U. Nebr., Lincoln, 1981-82. Assoc. editor: Jour. Optical Soc. Am, 1980-85; contbr. articles to profl. jours. Served to capt. U.S. Army, 1967-69. Fellow Optical Soc. Am.; fellow Internat. Soc. Optical Engring. (chmn. tech. council); mem. Am. Phys. Soc., Sigma Xi, Tau Beta Pi, Eta Kappa Nu. Club: Cosmos (Washington). Lodge: Masons. Subspecialties: Laser research; optics research. Current work: Optical communications systems design; developing theoretical models for describing the effects of interference phenomena on radiometry and radiative transfer by use of coherence theory. Office: Naval Research Lab Code 7740 Washington DC 20375

CARTLIDGE, EDWARD SUTTERLEY, mech. engr.; b. Trenton, N.J., Feb. 5, 1945; s. Leon James and Agnes Jean (Cinkay) C.; m. Marilyn Spinuzza, July 21, 1979. B.S. in Marine Engring, U.S. Mcht. Marine Acad., 1968; M.S. in M.E, N.J. Inst. Tech., 1971; M.B.A., Temple U., 1982. Registered profl. engr., Pa., Ill., Wis., Minn., Calif. Marine engr. Seatrain Lines, 1968-69; performance engr. Foster Wheeler Corp., Livingston, N.J., 1969-71; cons. engr. Fluor, Sargent & Lundy, and Kuljian Corp., 1971-75; chief engr. Gimpel Corp., Langhorne, Pa., 1976-79; sr. research and devel. engr. Yarway Corp., Bluebell, Pa., 1976-79; sr. project engr. Merck, Sharp & Dohme, West Point, Pa., 1982—. Served to lt. comdr. USNR, 1968—. Mem. Nat. Soc. Profl. Engrs. (chpt. pres.), Pa. Soc. Profl. Engrs. (Young Engr. of Yr. 1980), ASME, Instruments Soc. Am., Soc. Mfg. Engrs., Am. Soc. Metals, Soc. Naval Architects and Marine Engrs. Christian. Subspecialties: Fluid mechanics. Current work: State-of-the-art pharmaceutical tablet manufacturing plant. Patentee in field.

CARTNER, JOHN AUBREY, maritime consulting company executive; b. Jacksonville, N.C., Nov. 6, 1947; s. John Alexander and Anna Gertrude (Hardison) C.; m. Tanya Lynn Morris, Feb. 18, 1978; children: Christian W. J., Natalie V. O. B.S., U.S. Merchant Marine Acad., 1969; M.Sc., U. Ga., 1974, Ph.D., 1975; M.B.A., Ga. State U., 1979. Research scientist U.S. Army Research Inst., Alexandria, Va., 1976-78; asst. prof. bus. adminstrn. U. Ga. System-Columbus, 1978-79; dir. marine transp. cons. Grumman Data Systems, Bethpage, N.Y., 1979-81; v.p. IMA Resources, Inc., Washington, 1981; prin. Phillips Cartner & Co., Inc., Washington, 1981—, mng. prin., 1981—; mem. Supts.'s Council, U.S. Merchant Marine Acad., 1982—. Contbr. articles to profl. jours., 1975—; author various monographs, reports, 1974—. Adv. U.S. Merchant Marine Acad. Found., 1980—, Christ Ch. Found., Alexandria, Va., 1982—. Served to lt. USNR, 1965-73. Recipient Supt.'s trophy U.S. Merchant Marine Acad., 1982. Mem. Soc. Naval Architects and Marine Engrs., Am. Psychol. Assn., Inst. Navigation, N.Y. Acad. Scis., Sigma Xi. Republican. Episcopalian. Clubs: Downtown Athletic, Whitehall (N.Y.C.). Subspecialties: Naval architecture and marine engineering; Offshore technology. Current work: Applications of state-of-the-art technology in computers and information processing to command, control and communications in maritime settings; applications of creative paradigms to engineering solutions in the maritime environment; transfer of commercial maritime technology to military applications offshore. Office: Phillips Cartner & Co Inc 1629 K St NW Washington DC 20006

CARTWRIGHT, DAVID CHAPMAN, technical administrator, researcher; b. Mpls., Dec. 2, 1937; s. Arvid Chapman and Elizabeth (Swain) C.; m. Carole Roth, July 10, 1965; children: Scott, Alison. B.S. in Physics, Hamline U., 1962; M.S. in Applied Mechanics, Calif. Inst. Tech., Pasadena, 1964; Ph.D. in Chem. Physics and Physics, Calif. Inst. Tech., Pasadena, 1968. Staff mem. Space Scis. Lab. The Aerospace Corp., El Segundo, Calif., 1969-74; staff mem., group leader Los Alamos Nat. Lab., 1975-80, dep. div. leader, 1980—. Contbr. articles in field to profl. jours. Served with U.S. Navy, 1955-58. NSF fellow, 1961-62; NATO fellow, 1967-68. Mem. Am. Phys. Soc., Am .Geophys. Union. Republican. Subspecialties: Atomic and molecular physics; Aeronomy. Current work: Electron impact processes; spectral properties of atoms and molecules; atmospheric processes. Office: PO Box 1663 MS J563 Los Alamos NM 87545

CARTWRIGHT, KEROS, geologist; b. Los Angeles, July 25, 1934; s. Eugene E. and Charlotte (Searle) C.; m. Jennifer E. Moberley, Mar. 9, 1962; children: Sylvia, Jennifer, David, Bridget. B.A., U. Calif.-Berkeley, 1958; M.S., U. Nev.-Reno, 1961; Ph.D., U. Ill.-Urbana, 1973. Cert. prof. geologist, Ind. Geologist Ill. State Geol. Survey, Champaign, 1961—; cons. in field. Assoc. editor: Jour. of Hydrology, 1982—; contbr. numerous geol. articles to profl. publs. Fellow Geol Soc. Am. (chmn. hydrogeology div. 1978-79, assoc. editor 1980-83); mem. Am. Geophys. Union (assoc. editor 1975-81), Nat. Water Well Assn., Am. Water Resources Assn. Club: Explorers (fellow). Subspecialties: Hydrogeology; Geophysics. Current work: Developed the use of soil temperature in groundwater exploration. Home: 310 W John St Champaign IL 61820 Office: Ill State Geol Survey Nat Resources Bldg 615 E Peabody Dr Champaign IL 61820

CARUCCIO, FRANK THOMAS, geology educator; b. N.Y.C., Sept. 7, 1935; s. Nicholas and Genoveffa (DiNucci) C.; m. Gwendelyn Geidel, Oct. 23, 1976; 1 son, Nicholas Charles. B.S., CCNY, 1957; M.S., Pa. State U., 1961, Ph.D., 1967. Research asst. Gulf Oil Corp., N.Y.C., 1956-58; post-doctoral research fellow dept. geology Pa. State U., University Park, 1967-68; asst. prof. geology SUNY-New Paltz, 1968-70, assoc. prof., 1970-71; assoc. prof. geology U. S.C., Columbia, 1971-82, prof., 1982—; ptnr. CARGEID, Columbia, 1978—. Served with U.S. Army, 1959-61. EPA grantee, 1974, 80; NSF grantee, 1979; Office Surface Mining grantee, 1981. Fellow Geol Soc. Am.; mem. Am. Geophys. Union, AAAS, Nat. Waterwell Assn., Sigma Xi. Episcopalian. Subspecialties: Geology; Hydrogeology. Current work: Acid mine drainage, prediction and mitigation; environmental hydrogeology; ground water quality and resource evaluation. Home: 3823 Edinburgh Rd Columbia SC 29204 Office: Dept Geology U SC Columbia SC 29208

CARUSO, FRANK LAWRENCE, educator; b. Hackensack, N.J., Nov. 18, 1949; s. Victor and Ruby (Akre) C.; m. Barbara Ann Steller, Dec. 27, 1975; 1 dau., Emily. B.A., Gettysburg Coll., 1971; M.S., U. Mass., 1974; Ph.D., U. Ky., 1978. Asst. prof. U. Maine, 1979—. Contbr. articles to profl. jours. Grantee in field. Mem. Am. Phytopathol. Soc., N.Y. Acad. Scis., AAAS, Sigma Xi, Gamma Sigma Delta. Subspecialties: Plant pathology; Plant physiology (agriculture). Current work: Physiology of parasitism, mechanisms of disease resistance.

CARUTHERS, MARVIN HARRY, chemistry educator; b. Des Moines, Feb. 11, 1940; s. Harry A. and Eva D. (Schultz) C.; m. Jennie Mary Smoley, Oct. 9, 1971; children: Jonathan, Andrew. B.S., Iowa State U.-Ames, 71962; Ph.D., Northwestern U., 1968. Postdoctoral assoc. U. Wis., Madison, 1968-70; sr. research scientist MIT, Cambridge, 1970-72; asst. prof. U. Colo., Boulder, 1973-77, assoc. prof., 1977-79, prof. chemistry, 1979—; cons. Applied Biosystems, Foster Ci.v, Calif., 1980—, Applied Molecular Genetics, Thousand Oaks, Calif., 1980—, Monsanto Co., St. Louis, 1980-81. Contbr. articles in field to profl jours. Recipient Research Career Devel. award NIH, 1975-80; Guggenheim fellow, 1981; Grants NIH/NSF/Am. Chem. Soc., 1973-83. Mem. Am. Chem. Soc., AAAS, Am. Soc. Biol. Chemists, Phi Lambda Upsilon. Subspecialties: Biochemistry (biology); Genetics and genetic engineering (agriculture). Current work: Nucleic acid chemistry and the study of gene control regions. Office: U Colo Dept Chemistry Boulder CO 80309

CASALI, PAOLO, immunologist; b. Pizzighettone, Italy, Nov. 20, 1947; came to U.S., 1979; s. Angelo and Luigia (Savoldi) C.; m. Segolene Ede, Oct. 17, 1981. M.D., U. Milan, Italy, 1973. Research asst. WHO Immunology Research and Tng. Center, U. Geneva, 1976-78; asst. prof. dept. immunology U. Genoa, Italy, 1979, Scripps Clinic and Research Found., La Jolla, Calif., 1980. Recipient Priz Bizot U. Geneva, 1977. Mem. Am. Assn. Immunologists, Am. Fedn. for Clin. Research, N.Y. Acad. Sci., Am. Soc. for Microbiology. Subspecialty: Infectious diseases. Current work: Immunology of viruses. Home: 2222 Calle Frescota La Jolla CA 92037 Office: Dept Immunology Scripps Clinic and Research Found 10666 N Torrey Pints Rd La Jolla CA 92037

CASASENT, DAVID PAUL, elec. engring. educator, cons.; b. Washington, Dec. 8, 1942; s. Harold Kane and Delta (Fletchal) C.; m. Paula T. Timko, Feb. 14, 1977; children: Candace, Erin, Tod, Jon. B.S. in Elec. Engring, U. Ill., 1964, M.S., 1965, Ph.D., 1969. Research asst. Digital Computer Lab., U. Ill., 1965-69; prof. elec. engring. Carnegie-Mellon U., Pitts., 1969—, George Westinghouse prof. elec. engring., 1980—; tchr. numerous short courses on signal and image processing; cons. to numerous indsl. and govt. agys., including task force on automatic target recognition Def. Sci. Bd., 1982. Author 3 books on electronics and data processing; contbr. chpts. to 12 books, over 200 articles to profl. jours. Active Mt. Lebanon Civic League. Recipient best paper award AIAA, 1979. Fellow IEEE (pres. Pitts. electron devices chpt. 1971-72, best paper award 1976), Optical Soc. Am. (pres. Pitts. chpt. 1975-77), Soc. Photo-Optical Instrumentation Engrs. (numerous coms.). Democrat. Roman Catholic. Subspecialties: Optical image processing; Optical signal processing. Current work: Real-time hybrid optical/digital data processing. Patentee in field. Office: Dept Elec Engring Carnegie-Mellon U Pittsburgh PA 15213

CASCIANO, DANIEL ANTHONY, research cell biologist; b. Buffalo, Mar. 1, 1941; m., 1964; 2 children. B.S., Canisius Coll., 1962; Ph.D. in Cell Biology, Purdue U., 1971. Research asst. tissue culture Roswell Park Meml. Inst., 1963-64; research asst. cell biology Purdue U., 1965-66, asst. microbiology, 1969; investigator biochemistry U. Tenn., 1971-73; research biologist Nat. Ctr. Toxicol. Research, Jefferson, Ark., 1973-79, dir. div. mutagenesis research, 1979—. Subspecialty: Cell biology. Office: Nat Ctr Toxicol Research Jefferson AR 72079

CASE, DELVYN CAEDREN, JR., hematologist, clin. researcher; b. Brownwood, Tex., Jan. 15, 1945; s. Delvyn Caedren and Dorothy Nellie (Dul) C.; m. Carole Ann Case, Aug. 1, 1970; children: Delvyn Caedren III, Wendy Nadia, Keith William. A.B. cum laude, Brown U., 1967; M.D., Jefferson Med. Coll., 1971. Diplomate: Am. Bd. Internal Medicine, Am. Bd. Hematology, Am. Bd. Oncology. Intern, resident in internal medicine Cornell U., N.Y.C., 1971-74; fellow in hematology and oncology Meml. Sloan-Kettering Cancer Center, N.Y.C., 1974-76; attending physician Maine Med. Center, Portland, 1976—; clin. researcher Found. for Blood Research, 1979—; assoc. Sidney Farber Cancer Inst., 1978—; instr. Tufts U., 1976-77; sr. clin. instr. U. Vt., 1978-80, clin. asst. prof., 1980—; cons. hematologist Ventrex Corp., Portland, 1980—; bd. dirs. Found. Blood Research, Scarborough, Maine, 1979—. Contbr. articles on hematology and immunology to profl. jours. Recipient 2d prize for cancer research Meml. Sloan-Kettering Cancer Center, 1976. Fellow ACP; mem. Am. Soc. Hematology, Am. Soc. Clin. Oncology, Am. Assn. Cancer Research, Sigma Xi. Subspecialties: Hematology; Oncology. Current work: Myeloma, Waldenstrom's macroglobucinemia, leukemia, lymphoma, M-components, chemotherapy. Office: 180 Park Ave Portland ME 04102

CASE, MARY ELIZABETH, educator; b. Crawfordsville, Ind., Dec. 10, 1925; d. Ralph Thomas and Leila Luckenbill (Sharar) C. B.A., Maryville Coll., 1947; M.S., U. Tenn., 1950; Ph.D., Yale U., 1957. Research asso. Yale U., New Haven, Conn., 1957-72, lectr., 1965-72; assoc. prof. genetics U. Ga., Athens, 1972—. Contbr. articles in field to profl. jours. Subspecialty: Genetics and genetic engineering (biology). Current work: Neurospace molecular genetics. Office: University of Georgia Molecular and Population Genetics Dept Athens GA 30602

CASE, RONALD MARK, biology educator; b. Wausau, Wis., Oct. 7, 1940; s. Frederick William and Edith (Guenther) C.; m. Sheri Rae Hulliberger, Aug. 17, 1963; children: Douglas Mark, Michael William. A.B., Ripon Coll., 1962; M.S., U. Ill., 1964; Ph.D., Kans. Stae U, 1971. Instr. U. Mo.-Columbia, 1971-72; mem. faculty dept. natural resources U. Nebr., Lincoln, 1972—, prof. wildlife biology, 1980—. Served to capt. U.S. Army, 1964-66. Mem. Wildlife Soc., Ecol. Soc. Am., Am. Inst. Biol. Scis., Sigma Xi. Subspecialties: Population biology; Ecology. Current work: Teaching and research in wildlife related matters; research in canids and range rodents. Office: Dept Forestry Fish and Wildlife U Nebr Lincoln NE 68583

CASE, TED JOSEPH, biology educator; b. Sioux City, Iowa, July 19, 1947; m. Benita L. Epstein, Nov. 28, 1981; children from previous marriage: Ted Nathan, John Dustin. B.S., U. Redlands, 1969; Ph.D., U. Calif.-Irvine, 1973. Asst. prof. biol. scis. Purdue U., West Lafayette, Ind., 1976-78; asst. prof. to assoc. prof. biology U. Calif.-San Diego, La Jolla, 1978—; chmn. U. Calif. Natural Land and Water Res., 1980—. Editor: Island Biogeography in the Sea of Cortez, 1983; assoc. editor: U. Calif. Publs. in Entomology, 1981—. Subspecialties: Evolutionary biology; Population biology. Office: Dept Biology U Calif San Diego Box 109 La Jolla CA 92093

CASEY, FRANCIS B., immunopharmacologist; b. Cambridge, Mass., Jan. 21, 1940; s. Francis and Marion (Osborn) C.; m. Garbrielle M. Bernier, June 16, 1966; children: Monica, Michele. B.S., Norwich U., 1961; M.S., U. Vt., 1966; Ph.D., U. N.Mex., 1969. Research scientist U.S. Army Inst. Environ. Medicine, Natick, Mass., 1969-72; research investigator Squibb Inst. Med. Research, Princeton, N.J., 1972-79; group leader Sterling-Winthrop Research Inst., Rensselaer, N.Y., 1979—. Author articles. Served to capt. U.S. Army, 1969-72. Decorated Army Commendation medal; NIH predoctoral fellow, 1967-69. Mem. Am. Assn. Immunologists, N.Y. Acad. Scis., Sigma Xi. Subspecialty: Immunopharmacology. Current work: Drug discovery and pathophysiologic studies related to allergic and inflammatory diseases. Emphasis on archadonic acid metabolism, pulmonary function and cell biology. Office: Dept Pharmacology Sterling-Winthrop Research Inst Rensselaer NY 12144

CASEY, MICHAEL ALLEN, anatomist, med. researcher; b. Tacoma, Wash., Feb. 27, 1952; s. William H. and Doris I. (Conner) C.; m. Cynthia Ann Myers, Aug. 17, 1971; 1 son, Michael Allen. B.S., Eastern Ky. U., 1975, M.S., 1977; Ph.D., Tulane U., 1981. Research assoc. dept. anatomy Boston U. Sch. Medicine, 1981—. Contbr. articles to profl. publs. Grantee Cancer Assn. Greater New Orleans, 1980. Mem. Soc. Neurosci. Subspecialties: Neurobiology; Cytology and histology. Current work: Cytological changes in aged and diseased nervous system; neurobiology, neuropathology, neurocytology, age-related neuron loss, electron microscopy, light microscopy, auditory system.

CASHDOLLAR, KENNETH LEROY, physicist; b. Pitts., May 6, 1947; s. Chester Leroy and Grace E. (Magee) C. B.S. in Physics, Dickinson Coll., Carlisle, Pa., 1969; M.S. in Astronomy, U. Wis.-Madison, 1973; postgrad., U. Pitts., 1976-77. Asst. in research Yale U., New Haven, 1969-71; research assoc. Lowell (Mass.) Tech. Inst. Research Found., 1971-72; project asst. U. Wis.-Madison, 1972-73; physicist U.S. Bur. Mines Pitts. Research Ctr., 1973—. Contbr. articles to profl. jours. Active Western Pa. Conservancy. Mem. Optical Soc. Am., Am. Phys. Soc., Soc. Photo-Optical Instrumentation Engrs., Combustion Inst., Am. Astron. Soc., Astron. Soc. of Pacific, ASTM, Phi Beta Kappa. Subspecialties: Combustion processes; Infrared spectroscopy. Current work: Flammability of dusts and gases, applied electro-optics, optical probes for monitoring dust clouds, multi-channel infrared pyrometers. Patentee in field. Office: PO Box 18070 Cochrans Mill Rd Pittsburgh PA 15236

CASHIN, KENNETH DELBERT, chemical engineering educator; b. Lowell, Mass., May 10, 1921; s. Arthur Henry and Adaline (Breck) C.; m. Wilma Faye Cozad, May 14, 1944; children: Arthur H., Wilma Linda, David Graham. B.S., Worcester Poly. Inst., 1947, M.S., 1948; Ph.D., Rensselaer Poly. Inst., 1955. Registered profl. engr., Mass. Asst. prof. chem. engring. U. Mass.-Amherst, 1948-55, assoc. prof., 1955-63, prof., 1963-68, 70—, assoc. dean, 1978-81, exec. officer chem. engring. dept., 1982—; head dept. chem. engring. U. Petroleum and Minerals, Dhahran, Saudi Arabia, 1968-70. Served with USN, 1944-46. NSF grantee, 1956-57, 78-80. Mem. Am. Inst. Chem. Engrs., Am. Chem. Soc., Am. Soc. Engring. Edn., Sigma Xi, Tau Beta Pi. Republican. Subspecialties: Chemical engineering; Combustion processes. Current work: Spectroscopic studies of radicals produced in combustion processes, including flame velocity measurements. Home: 55 Goodell St Belchertown MA 01007 Office: Dept Chem Engring U Mass Amherst MA 01003

CASLAVSKA, VERA BARBARA, chemistry researcher; b. Charudim, Czechoslovakia; came to U.S., 1966; d. Vilem and Vera (Kudrnkova) Novak; m. Jaroslav L. Caslavsky, Dec. 25, 1952; 1 dau., Veronika. M.S., Charles U., Prague, 1957, C.Sc., 1965. Research asst. Czech Acad. Sci., Prague, 1957-61; research assoc. Mining Inst., Prague, 1962-65, Pa. State U., 1966-69; asst. staff mem. Forsyth Dental Ctr., Boston, 1969—. Recipient award Czech Acad. Sci., 1961. Mem. Internat. Assn. Dental Research, Sigma Delta Epsilon. Subspecialties: Preventive dentistry; Inorganic chemistry. Current work: Chemistry of fluoride interactions with human enamel, topical fluoridation to effect caries prevention, crystal growth, structure and properties of glasses, reduction processes of iron ores. Patentee in field. Office: Forsyth Dental Center 140 Fenway St Boston MA 02115

CASLER, LAWRENCE, psychology educator, researcher; b. Portland, Oreg., Jan. 26, 1932; s. David H. and Fyrne (Levinson) C. B.A., Harvard U., 1953, M.A., 1954; Ph.D., Columbia U., 1962. Staff editor for psychology Internat. Ency. Social Scis., N.Y.C., 1962-63; asst. prof. CCNY, 1963-66; lectr. L.I.U., Bklyn., 1965; research assoc. S.I. Mental Health Soc., N.Y., 1965-66; assoc. prof. SUNY-Geneseo, 1966-68, prof. psychology, 1968—; cons. Theleme exptl. community, Akron, N.Y., 1971-72; scholar-in-residence U. Del., Newark, 1976; mem. postdoctoral fellowship evaluation panel NRC, 1976. Editorial bd.: Developmental Psychlogy, 1970-75; author: Is Marriage Necessary?, 1974; contbr. articles to profl. jours. Mem. ACLU. Served in U.S. Army, 1955-56. Grantee in field; recipient Chancellor's award SUNY-Albany, 1975; sr. exchange scholar SUNY-Moscow State U., 1980. Mem. Am. Psychol. Assn., Parapsychol. Assn., AAUP, Com. for Elimination of Death (adv. council 1968-78), Com. for Extended Lifespan, Soc. Advancement of Good English (founder, pres. 1978—). Subspecialties: Behavioral psychology; Social psychology. Current work: Alternatives to marriage; relationships between hypnosis and parapsychology; psychological approaches to healthy logevity. Office: SUNY Geneseo NY 14454

CASPARI, ERNEST W(OLFGANG), biologist, educator; b. Berlin, Germany, Oct. 24, 1909; s. Wilhelm and Gertrud (Gerschel) C.; m. Hermine Bertha Abraham, Aug. 16, 1938. Ph.D., U. Gottingen, Ger., 1933; M.A., Wesleyan U., 1951; Dr. rer. nat. (hon.), U. Giessen, Ger., 1982. Asst. U. Gottingen, 1933-35; asst. in microbiology U. Istanbul Med. Sch., 1935-38; fellow, later asst. prof. Lafayette Coll., 1938-44; asst. prof. U. Rochester, 1944-46, prof., 1960-75, prof. emeritus, 1975—; chmn. dept., 1960-65; assoc. prof. Wesleyan U., 1946-47, prof., 1949-60; mem. staff dept. genetics Carnegie Instn., 1947-49. Author: (with W.A. Ravin) Genetic Organization, 1969; editor: Advances in Genetics, Vols. 10-21, 1961-82, Genetics, 1968-72. Fellow Center Advanced Study in Behavioral Sci., 1956-57, 65-66; recipient Sr. U.S. Scientist award A. von Humboldt Found., 1981. Fellow AAAS, Am. Acad.; mem. Genetics Soc. Am. (pres. 1966), Am. Soc. Naturalists (v.p. 1961, hon. mem.), Am. Soc. Zoologist, Behavior Genetics Soc. Subspecialties: Genetics and genetic engineering (biology); Gene actions. Current work: Genetic basis of learning ability in fish. Office: Dept Biology U Rochester Rochester NY 14627

CASPERSON, LEE WENDEL, elec. engr., educator; b. Portland, Oreg., Oct. 18, 1944; s. Rudolph Oliver and Effie Marie (Dahlman) C.; m. Susan Diane Lunnam, Oct. 18, 1974; children: Julie Diane, Janet Marie. B.S., M.I.T., 1966; M.S. in Elec. Engring, Calif. Inst. Tech., 1967; Ph.D. in Elec. Engring. and Physics, Calif. Inst. Tech., 1971. Asst. prof. engring. and applied sci. UCLA, 1971-76, assoc. prof. engring. and applied sci., 1976-80, prof., 1980—; vis. prof. physics U. Auckland, New Zealand, 1981; cons. aerospace industries. Contbr. articles to profl. jours. Mem. Optical Soc. Am., IEEE, AAAS, Sigma Xi. Republican. Lutheran. Subspecialties: Electrical engineering; Laser research. Home: 2571 Westwood Blvd Los Angeles CA 90064 Office: UCLA 7731D Boelter Hall Los Angeles CA 90024

CASPERSON, RICHARD L., engineering company executive, cons.; b. McKeesport, Pa., Apr. 9, 1940; s. Robert E. and Beulah B. (Overturf) C.; m., June 10, 1967; children: Mylee M., Lea G. B.S. in Civil Engring. with honors, Colo. State U., 1963; M.S. in Mechanics, U. Colo., 1966. Sr. engr., lead engr. Martin-Marietta Aerospace, Denver, 1965-67, 71-72; scientist amd project mgr. Aerojet Nuclear Co., Idaho Nat. Engring. Lab., Idaho Falls, 1972-76; pres. Energy Engring. Group, Inc., Golden, Colo., 1977—; vis. faculty U. Colo.; cons. nuclear and alternative energy areas and computer applications in engring. scis. Contbr. articles in field of aerospace, nuclear safety and solar energy articles to profl. jours. Colo. Energy Research Inst. fellow, 1977-78; recipient Outstanding Faculty award U. Colo., 1980; U.S. Dept. Energy grantee, 1977-82. Mem. AAAS, ASCE, Internat. Solar Energy Soc., ASHRAE, Wilderness Soc., Audubon Soc., Sigma Tau, Chi Epsilon. Democrat. Club: Sierra. Subspecialties: Solar energy; Theoretical and applied mechanics. Current work: Mathematical optimization in building energy design; solar energy applications; numerical and computer methods in engineering analysis and design. Home: PO Box 10 2051 Miner St Idaho Springs CO 80452 Office: Energy Engring Group Inc 602 Park Point Dr Suite 213 Golden CO 80401

CASSEDAY, JOHN HERBERT, neurobiology educator; b. Pasadena, Calif., Aug. 11, 1934; s. John Herbert and Marjorie Marie (Hoban) C.; married; children: Patrick, Tara. B.A., U. Calif.-Riverside, 1960; M.A., Ind. U., 1963, Ph.D., 1970. Postdoctoral fellow dept. psychology Duke U., 1970-72, asst. prof., 1972-80, assoc. med. research prof. otolaryngology, 1979—, assoc. prof. psychology, 1980—. Contbr. articles to profl. jours. Served with U.S. Army, 1954-56. NSF grantee, 1972, 77—; NIH grantee, 1975—. Mem. AAAS, Acoustical Soc. Am., Soc. Neurosci., Assn. for Research in Otolaryngology. Subspecialties: Comparative neurobiology; Behaviorism. Current work: Neuroanatomy and neurophysiology of central auditory system, comparative neuroanatomy. Office: Duke U Med Ctr PO Box 3843 Durham NC 27710

CASSENS, ROBERT GENE, animal science educator; b. Morrison, Ill., June 10, 1937; s. Ludwig P. and Ethel A. C.; m. Dessa Trautwein, June 11, 1960; children: Martha, Martin. B.S., U. Ill., 1959; M.S., U. Wis., 1960, Ph.D., 1962. Asst. prof. U. Wis-Madison, 1964-67, assoc. prof., 1967-71, prof. meat and animal sci., 1971—, chmn. dept., 1980—. Contbr. articles to sci. jours. Romnes fellow, 1976; Inst. Food Tech. research grantee, 1967; Am. Soc. Animal Sci. research grantee, 1969; Am. Meat Sci. Assn. grantee, 1971. Mem. Inst. Food Tech., Am. Soc. Animal Sci., Am. Meat Sci. Assn., AAAS, Am. Chem. Soc. Club: Gyro. Subspecialty: Food science and technology. Current work: Muscle Metabolism, nitrites. Office: Dept Meat and Animal Sci U Wis Madison WI 53706

CASSIDY, MARTIN MACDERMOTT, oil company executive, geologist; b. N.Y.C., Mar. 29, 1933; s. George Livingston and Mary-Light (Schaeffer) C.; m. Jo Reesor, June 10, 1955; children: Cathy-Jo, Brandt, Caroline. B.A. cum laude, Harvard U., 1955, postgrad., 1960-62; M.S., U. Okla., 1962. Cert. petroleum geologist. Gordon McKay teaching fellow Harvard U., Cambridge, Mass., 1960-62; geologist Amoco Prodn. Co., Corpus Christi-Houston, Tex., 1962-69; chief geologist Amoco Libya Oil Co., Tripoli, 1969-73; staff geology div. ops supr. Amoco Prodn. Internat. Co., Chgo. and Houston, 1973-82; div. ops. supr. Amoco UK Exploration Co., London, 1982—. Served to 1st lt. USAF, 1956-58; Korea. Mem. Am. Assn. Petroleum Geologists, Houston Geol. Soc. (treas. 1968, 2d v.p. 1969). Republican. Episcopalian. Subspecialties: Fuels; Geology. Current work: Exploration for oil and gas. Office: Amoco UK Exploration 1 Stephen St Tottenhmm Court Rd London United Kingdom WIP IPJ Home: Expatriate London UK PO Box 4381 Houston TX 77210

CASSIDY, WILLIAM ARTHUR, geochemistry educator; b. N.Y.C., Jan. 3, 1928; m., 1961; 3 children. B.S. in Geolgoy, U. N.Mex., 1952; Ph.D. in Geochemistry, Pa. State U., 1961. Mem. staff seismic computation Superior Oil Co., Calif, 1952-53; Fulbright fellow U. South Australia, Calif., 1953-55; research scientist meteoritics Lamont Geol. Obs., Columbia U., 1961-67; assoc. prof. U. Pitts., 1968-81, prof. geology and planetary sci., 1981—. NSF grantee, 1976—. Mem. Am. Geophys. Union, Meteoritical Soc. Subspecialty: Planetology. Office: Dept Geology and Planetary Sci U Pitts Pittsburgh PA 15260

CASSILETH, PETER ANTHONY, hematologist, oncologist, educator; b. N.Y.C., Aug. 16, 1937; s. Lee H. and Geraldine A. (Weisman) C.; m. Barrie Rabinowitz, Dec. 28, 1958; children: Jodi, Wendy, David, Gregory. B.A., Union Coll., Schenectady, N.Y., 1958; M.D., Columbia U., 1962. Diplomate: Am. Bd. Internal Medicine. Intern Columbia-Presbyn. Med. Ctr., N.Y.C., 1962-63, resident in medicine, 1963-66, fellow in hematology, 1965-66, 68-69; asst. prof. medicine U. Pa., 1969-73, assoc. prof., 1973-79, prof., 1979—, assoc. chief hematology oncology sect. Hosp., 1976—. Author: Practical Approaches to Hematology & Oncology, 1982; editor: Clinical Care of the Terminal Cancer Patient, 1982. Served to capt. U.S. Army, 1966-68. Fellow ACP; mem. Am. Soc. Hematology, Am. Soc. Clin. Oncology, Phila. Coll. Physicians, Am. Fedn. Clin. Research, Phi Beta Kappa, Alpha Omega Alpha. Subspecialties: Hematology; Oncology. Current work: Clinical studies in leukemia, autologous transplantation. Office: Hospital University of Pennsylvania 3400 Spruce St Philadelphia PA 19104

CASSIN, SIDNEY, physiologist, educator; b. Chelsea, Mass., June 8, 1928; s. Morris and Minnie (Cogan) C.; m. Barbara Ellnore Covin, Sept. 3, 1950; children: Robin, Nancy, Lisa, Kim. B.A., N.Y.U., 1950, M.A., 1954; Ph.D., U. Tex., 1957. Asst. prof. physiology U. Fla. Coll. Medicine, Gainesville, 1960-66, assoc. prof. physiology, 1966-69, prof. physiology, 1969—. NIH grantee, 1961—. Mem. Am. Physiol. Soc., Sigma Xi. Subspecialty: Physiology (medicine). Current work: Perinatal pulmonary circulation, prostaglandins, fetal lung liquid formation. Home: 1405 NW 35th Way Gainesville FL 32605 Office: Coll Medicine Dept Physiology U Fla Gainesville FL 32610

CASSINELLI, JOSEPH PATRICK, astronomer, educator; b. Cin., Aug. 23, 1940; s. Herbert John and Louise (Schlottmann) C.; m. Mary LeFever, May 4, 1962; children: Joseph Michael, Carolyn Marie, Mary Kathleen. B.S. in Physics, Xavier U., Cin., 1962, M.S., U. Ariz., 1965; Ph.D. in Astronomy, U. Wash., 1970. Research engr. Boeing Co., Seattle, 1965-66; postdoctoral research assoc. Joint Inst. for Astrophysics, U. Colo., 1970-72; postdoctoral fellow U. Wis., Madison, 1972-73, asst. prof. astronomy, 1973-77, assoc. prof., 1977-81, prof., 1981—; vis. research scientist Astron. Inst., Utrecht, Netherlands, 1975; Langley Abbott research fellow Center for Astrophysics, Cambridge, Mass., 1981. Mem. Am. Astron. Soc., Internat. Astron. Union. Subspecialties: Theoretical astrophysics; Ultraviolet high energy astrophysics. Current work: Stellar atmospheres, coronae, stellar winds, radiation transfer, structure of very massive stars, analyses of x-Ray and ultraviolet satellite observations. Office: Washburn Obs 475 N Charter St Madison WI 53706

CASTELFRANCO, PAUL ALEXANDER, botany educator; b. Florence, Italy, Oct. 16, 1921; came to U.S., 1939; s. Giorgio and Matilde (Forti) C.; m. Marie Sander, Jan. 23, 1954; children: Ann, John. A.B., U. Calif.-Berkeley, 1943, M.S., 1950, Ph.D., 1954; S.T.B., Harvard U., 1957. Postdoctoral fellow Sch. Medicine Tufts U., Boston, 1957-58; faculty U. Calif.-Davis, 1958—; now prof. botany, also botanist Expt. Sta. Exec. editor: Archives of Biochemistry and Biophysics, 1981-82; mem. editorial bd.: Plant Physiology, 1976-81. NIH spl. fellow U. Milan, Italy, 1962-63; Guggenheim fellow, Bristol, Eng., 1973-74. Mem. Am. Soc. Plant Physiologists, Am. Soc. Photobiology, Am. Soc. Biol. Chmists. Subspecialties: Plant physiology (agriculture); Developmental biology. Current work: Tetrapyrrole metabolism in plants; chlorophyll biosynthesis; cell organelle biogenesis and physiology. Home: 511 7th St Davis CA 95616 Office: Dept Botany U Calif Davis Davis CA 95616

CASTELLINO, FRANCIS JOSEPH, univ. dean; b. Pittston, Pa., Mar. 7, 1943; s. Joseph Samuel and Evelyn Bonita C.; m. Mary Margaret Fabiny, June 5, 1965; children—Kimberly Ann, Michael Joseph, Anthony Francis. B.S., U. Scranton, 1964; M.S., U. Iowa, 1966, Ph.D. in Biochemistry, 1968. Postdoctoral fellow Duke U., Durham, N.C., 1968-70; mem. faculty dept. chemistry U. Notre Dame, Ind., 1970—, prof., 1977—, dean, 1979—. Contbr. articles to profl. jours. NIH fellow, 395201968-70. Fellow N.Y. Acad. Scis.; mem. AAAS, Am. Heart Assn., Am. Chem. Soc., Am. Soc. Biol. Chemistry. Roman Catholic. Subspecialties: Biochemistry (biology); Biophysical chemistry. Current work: Structure and function of components of the blood coagulation and fibrinolytic systems. Office: College of Science University of Notre Dame Notre Dame IN 46556

CASTELLO, JOHN DONALD, biology educator; b. Paterson, N.J., May 1, 1952; s. Dominick and Catherine (Ackaway) C.; m. Jody Lynn, May 22, 1982. B.A., Montclair State Coll., 1973; M.S., Wash. State U., 1975; Ph.D., U. Wisc., 1978. Asst. prof. dept. environ. and forest biology SUNY, Syracuse, 1978—. Mem. Am. Phytopath. Soc., AAAS, Sigma Xi, Phi Kappa Phi. Subspecialties: Plant pathology; Immunology (agriculture). Current work: Researcher in enzyme linked immunosorbent assay techniques for use in detecting and characterizing viruses and mycoplasmas infecting woody perennials. Office: State University of New York Dept Environ Forest Biology Syracuse NY 13210

CASTER, WILLIAM OVIATT, nutritionist, educator; b. Topeka, Dec. 7, 1919; s. Charles A. and Rena May (Oviatt) C.; m. Lora Elizabeth Joos, Apr. 3, 1943; children: Charles A., A. Bruce, John D., Donald M. B.A., U. Wis., 1942, M.S., 1944; Ph.D., U. Minn., 1948. Biochemist nutrition br. USPHS, Brattleboro, Vt., 1948-51, Chamblee, Ga., 1948-51; asst. prof. physiol. chemistry U. Minn. Med. Sch., Mpls., 1951-63; prof., nutrition U. Ga., Athens, 1963—; mem. Gov's. Adv. Bd. Atomic Devel., Minn., 1957-59, Gov's. Adv. Bd. Hunger and Malnutrition, Ga., 1972, NASA/NRC Adv.Bd. Nutrition and Radiation, Chgo., 1960-62; pres. Ga. Nutrition Council, 1982-83. NIH/Nat. Heart Inst. spl. research fellow, 1956-61. Mem. Am. Soc. Biol. Chemists, Am. Inst. Nutrition, Nutrition Soc., Soc. Environ. Geochemistry and Health (editorial bd. 1978-82, councilor 1980-82). Democrat. Methodist. Subspecialties: Nutrition (medicine); Applied mathematics. Current work: study of eating habit patterns relating health to diet. Home: 155 Devonshire Dr Athens GA 30606 Office: U Ga 201 Dawson Hall Athens GA 30602

CASTLE, CHARLES HILMON, physician, educator; b. Eupora, Miss., Feb. 15, 1928; s. Hays D. and Ruby Ward (Bowen) C.; m. Carol Mae Losee, July 16, 1954 (div. Aug. 1981); children: Cy, Chris, Candace; m. Elaine Marie Litton, Sept. 19, 1981. B.S., U. Miss., 1948; M.D., Duke U., 1951. Instr. medicine Sch. Medicine, U. Utah, Salt Lake City, 1959-61, asst. prof., 1961-65, assoc. prof., 1965-70, prof., 1970—. Served to maj. USAF, 1954-56. Fellow ACP, Am. Coll. Cardiology, Am. Heart Assn. (council clin. cardiology). Baptist. Subspecialties: Internal medicine; Cardiology. Current work: Cardiology; echocardiography.

CASTRO, ALBERTO, medical educator; b. San Salvador, Nov. 15, 1933; U.S., 1952; s. Alberto Lemus and Maria Emma (de la Cotera) C.; m. Jeris Adelle Goldsmith, Oct. 19, 1956; children: Stewart, Sandra, Albert, Richard, Juan. B.S., U. Houston, 1958; M.T., Jeff Davis Hosp.-Baylor U., 1958; Ph.D., U. El Salvador, 1962; M.D., Cetec U., Santo Domingo, Dominican Republic, 1982. Prof., head basic scis. dept. U. El Salvador, San Salvador, 1958-66; NIH sr. research fellow U. Oreg. Med. Sch., Portland, 1966-70, asst. prof. pediatrics, 1969-73; sr. scientist Papanicolaou Cancer Research Inst., Miami, Fla., 1973-75; assoc. prof. dept. pathology, dept. medicine U. Miami Sch. Medicine, 1975-77, prof., 1977—, dir. hormone research lab., 1974—, dir. researchLatin Am. program, 1981—, coordinator exec. program com. internat. tech. transfer and tng. program, 1979—. Contbr. chpts. to books, more than 200 articles to profl. jours. Fellow Acad. Clin. Lab. Physicians and Scientist, Am. Inst. Chemists, Acad. Biochemistry (charter), Royal Soc. Health; mem. N.Y. Acad. Scis. Democrat. Roman Catholic. Subspecialties: Immunobiology and immunology; Endocrinology. Current work: Research in biotechnology, immunology (hybrodomas, immunochemistry), hormone mechanisms and receptors. Home: 6275 SW 123d Terr Miami FL 33156 Office: Dept Pathology U Miami Sch Medicine Miami FL 33101

CASTRO, GILBERT ANTHONY, physiology and cell biology educator; b. Pt. Arthur, Tex., Apr. 24, 1939; s. Richard N. and Pauline (Torres) C.; m. Georgia Fay Hazleton, Dec. 24, 1961; children: Theresa Kay, Mitzi Rene. B.S., Lamar U., 1961; M.S., U. Ark.-Fayetteville, 1963; Ph.D., U. Tex.-Galveston, 1966. Fellow U. Mass., Amherst, 1966-68; asst. prof. parasitology U. Okla., Oklahoma City, 1968-71, assoc. prof., 1971-72, U. Tex., Houston, 1972-77, prof., 1977—; mem. study sect. NIH, Bethesda, md., 1981-82, Fogarty Internat. Ctr., Bethesda, 1981-82. Mem. editorial bd.: Jour. Parasitology, 1980—, Exptl. Parasitology, 1980—; contbr. chpts. to books and articles in field to profl. jours. Recipient D.B. Williams award Tex. chpt. Am. Soc. Microbiology, 1965; Career Devel. award NIH, 1975; Disting. Alumnus award U. Tex., 1979. Mem. Am. Soc. Parasitologists (mem. council 1979-82), Am. Physiol. Soc., Am. Soc. Tropical Medicine and Hygiene. Subspecialties: Parasitology; Physiology (medicine). Current work: Physiological aspects of the host immune response to enteric parasites. Office: U Tex Med Sch PO Box 20708 Houston TX 77080

CASTRO, JOSEPH RONALD, physician, oncology researcher; b. Chgo., Apr. 9, 1934; m. Barbara Ann Kauth, Oct. 12, 1957. B.S. in Natural Sci, Loyola U.-Chgo., 1956, M.D., 1958. Diplomate: Am. Bd. Radiology, 1964. Intern Rockford (Ill.) Meml. Hosp.; resident U.S. Naval Hosp., San Diego; assoc. radiotherapist and assoc. prof. U. Tex.-M.D. Anderson Hosp. and Tumor Inst., 1967-71; prof. radiology/radiation oncology U. Calif. Sch. Medicine, San Francisco, 1971—, vice-chmn. dept. radiation oncology, 1980—; dir. particle radiotherapy Lawrence Berkeley Lab., Calif., 1975—; mem. program project rev. com. NIH/Nat. Cancer Inst. Cancer Program, 1982—. Author sci. articles. Past pres., chmn. bd. trustees No. Calif. Cancer Program, 1980-83. Served to lt. comdr., M.C. USN, 1956-66. Recipient Teaching award Mt. Zion Hosp. and Med. Center, San Francisco, 1972. Mem. Rocky Mountain Radiol. Soc. (hon.), Am. Coll. Radiology, Am. Soc. Therapeutic Radiology. Subspecialties: Oncology; Cancer research (medicine). Current work: Clinical research in radiation oncology. Office: Bldg 55 Lawrence Berkeley Lab Berkeley CA 94720

CASTRONOVO, FRANK PAUL, JR., radiopharmacologist, pharmacist, health physicist, radiation safety officer; b. Newark, Jan. 2; s. Frank Paul and Edna Viola (Weingartner) C.; m. Judith Anne Belli, Apr. 3, 1977; children: Jessica Belli, Elizabeth Frances. Pharmacy B.S., Rutgers U., Newark, 1962, M.S. in Health Physics, 1963-64; Ph.D., Johns Hopkins U., 1970. Registered pharmacist, 1965; cert. Med. Nuclear Physics Am. Bd. Radiology, 1978; Nuclear Med. Scientist, 1979, diplomate: (Radiopharms./Radiochem. Sci.) Am. Bd. Sci. in Nuclear Medicine, 1979. Staff pharmacist Terry's Drug, Verona, N.J., 1962-63; health physics trainee Brookhaven Nat. Lab., Upton, N.Y., 1964; research scientist in radiopharms. Squibb Inst. for Med. Research, New Brunswick, 1964-65; part time staff pharmacist Johns Hopkins Hosp., Balt., 1965-70, part time radiopharmacy technologist, 1965-70; asst. physicist in radiology Mass. Gen. Hosp., Boston, 1970-80, radiopharmacy supr. in radiology, 1970-75, asso. radiopharmacologist in radiology, 1980—, health physicist, 1980, radiation safety officer, 1980—; vis. scientist M.I.T., 1975—, adj. asso. prof. radiopharmacology, 1975-77; adj. clin. asso. prof. radiopharmacology Mass. Coll. Pharmacy, 1977—; asst. prof. radiology Harvard U. Med. Sch.; radiopharmacy radiation safety cons. various regional hosps.; dir., lectr. hosp. courses in field. Contbr. chpts. to books, articles to profl. publs. Recipient Anthony De Rosa Meml. Award Rutgers Coll. Pharmacy, 1959, parenteral research award Parenteral Drug Assn., 1969. Mem. Am. Pharm. Assn., Health Physics Soc., Soc. Nuclear Medicine, Am. Assn. Physicists in Medicine, AAAS, Am. Soc. Hosp. Pharmacists, Internat. Radiation Protection Assn., Ethnopharmacology Soc., Radiopharm. Sci. Council (bd. dirs. 1977-80, chmn. quality control com. 1978), Am. Coll. Radiology, N.Y. Acad. Scis., Soc. for Risk Analysis. Subspecialties: Nuclear medicine; Pharmacology. Current work: Radiopharmacology, devel. new radiopharms., med. health physics (patient and personnel radiation doses), ethnopharmacy. Inventor 125I Phenylphosphonic Acid. Home: 36 Gedlick Rd Burlington MA 01803 Office: Mass Gen Hosp Radiology Boston MA 02114

CASWELL, HAL, marine biologist; b. Los Angeles, Apr. 27, 1949; s. Herbert Hall and Ethel (Preble) C. B.S. with high honors, Mich. State U., 1971, Ph.D., 1974. Research assoc. Mich State U., East Lansing, 1974-75; asst. prof. U. Conn., Storrs, 1975-79, assoc. prof., 1979-82; assoc. scientist Woods Hole (Mass.) Oceanographic Inst., 1981—. Contbr. articles to profl. jours. NSF grantee, 1974, 76, 78, 81, 1982. Mem. Ecol. Soc. Am., Brit. Ecol. Soc., N.Y. Acad. Sci., AAAS, Sigma Xi. Subspecialties: Theoretical ecology; Population biology. Current work: Life history theory; evolutionary demography, especially of plants and marine invertebrates; community structure and succession; toxic substance effects on populations. Home: 302 Thomas Landers Rd East Falmouth MA 02536 Office: Woods Hole Oceanographic Inst Water St Woods Hole MA 02543

CASWELL, LYMAN RAY, chemistry educator; b. Omaha, Sept. 29, 1928; s. Omar and Emma Effie (Richardson) C.; m. Loretta Lynda Kinnard, July 17, 1964; children: Lyman Randolph, Timothy Omar, Amy Elizabeth. B.S., Ind. U., 1949, M.A., 1950; Ph.D., Mich. State U., 1956. Asst. prof. chemistry Ohio No. U., Ada, 1955-56; head dept. chemistry Upper Iowa U., Fayette, Iowa, 1956-61; asst. prof. chemistry Tex. Woman's U., Denton, 1961-64, assoc. prof., 1964-68, prof., 1968—, chmn. dept. chemistry, 1967-70, 78-82. Contbr. articles to profl. jours. Robert A. Welch Found. grantee, 1963-78. Fellow Am. Inst. Chemists, AAAS; mem. Am. Chem. Soc., AAUP, Soc. Coll. Sci. Tchrs. (charter), Am. Philatelic Soc. Subspecialty: Organic chemistry. Current work: Nucleophilic displacement reactions of aromatic polyhalides; measurement of dipole moments; solvent effects in absorption spectrophotometry. Home: 2217 Burning Tree Ln Denton TX 76201 Office: Dept Chemistry Tex Woman's U Box 23973 TWU Station Denton TX 76204

CATACOSINOS, PAUL ANTHONY, geology educator; b. N.Y.C., Sept. 29, 1933; s. Anthony and Elpeniki (Cocotos) C.; m. Joan Diane Cook, Jan. 16, 1958; children: Alice E., Andrew C., Diana N. B.A., U. N.Mex., 1957; M.S., 1962; Ph.D., Mich. State U., 1972. Petroleum exploration geologist Mt. Fuel Supply Co., Salt Lake City, 1962-66, Consumers Power Co., Jackson, Mich., 1967-69; prof. geology Delta

Coll., Univ. Center, Mich., 1969—; v.p. exploration Rom Energy Corp., Birmingham, Mich., 1982-83. Contbr. articles to profl. jours. Served to lt. (j.g.) USN, 1956-60. Recipient Bergstein award Delta Coll., 1972. Mem. Am. Assn. Petroleum Geologists, Geol. Soc. Am., Am. Astron. Soc., Sigma Xi. Subspecialty: Stratigraphy. Current work: Oil and gas exploration in the Mich. Basin; origin of the Mich. Basin. Office: Dept Geology Delta Coll University Center MI 48710

CATALANO, MICHAEL ALFRED, rheumatologist, researcher, educator; b. Manhattan, N.Y., Jan. 25, 1947; s. Michael V. and Lillian E. (Troise) C. B.S. magna cum laude, Boston Coll., 1968; M.D., Yale U., 1972. Diplomate: Am. Bd. Internal Medicine. Intern in medicine U. Ky. Med. Ctr., 1972-73, resident in internal medicine, 1973-74, fellow in clin. rheumatology, 1975; fellow Scripps Clinic, La Jolla, Calif., 1975-78; clin. assoc. Gen. Clin. Research Ctr., 1978-80; assoc. prof. internal medicine U. Hawaii, 1980—; staff rheumatologist Arthritis Ctr. Hawaii, 1980—, dir., 1981—; bd. dirs. Arthritis Found., Honolulu, 1980—. Contbr. articles to profl. publs. Mem. quality assurance com. Beach Area Community Clinic, sAn Diego, 1978, co-dir., 1978, vol. physician, 1975-81. Recipient Clin. Assoc. Physician award Gen. Clin. Research Center Program, NIH, 1978-80; Teamsters Union scholar, 1964-68. Mem. Am. Rheumatism Assn., Am. Fedn. Clin. Research, Nat. Jesuit Honor Soc. Subspecialties: Rheumatology; Immunology (medicine). Current work: urrent work: Rheumatic disease immunology and epidermiology; comprehensive arthritis care. Home: 666 Prospect St Honolulu HI 96813 Office: U Hawaii John A Burns Sch Medicine Arthritis Ctr Hawaii 347 N Kuakini St Honolulu HI 96817

CATE, MARTIN EDWARD, industrial engineering educator; b. Garden City, Kans., Dec. 12, 1934; s. Matthew Sheldon and Dolores Jane (Perkins) C.; m. Janice L. Fischer, April 20, 1964; children: Kenneth, Edwina. B.S.I.E., Lehigh U., 1954, M.S.O.R., 1955; Ph.D., Ga. Inst. Tech., 1958. Elec. engr. Emerson Electric Co., St. Louis, 1958-63, mgr. microwave lab., 1963-66; asst. prof. engring. U. Mo., Rolla, 1966-70; assoc. prof. U. Wash., Seattle, 1970-75, prof., 1975—; dir. Almac/Stroum Electronics Corp. Contbr. articles to profl. jours. Fellow IEEE; mem. Am. Soc. Quality Control, Inst. Indsl. Engrs., Ops. Research Soc. Am., Tau Beta Pi, Alpha Pi Mu. Subspecialties: Industrial engineering; Operations research (engineering). Office: Werik Bldg 7030-15 NW Seattle WA 98117

CATES, LINDLEY ADDISON, JR., medicinal chemist, educator, pharm. cons. and researcher; b. Chgo., Nov. 20, 1932; s. Lindley Addison and Alice Jewett (Gilbert) C.; m. Ruth Elizabeth Cates, Oct. 27, 1932; children: Catherine, Douglas. B.S. in Pharmacy, U. Minn., 1954; M.S., U. Colo., 1958, Ph.D., 1961. Lic. pharmacist, Tex. Commd. 2d lt. U.S. Air Force, 1954, advanced through grades to capt., 1956; ret., 1965; instr. in pharmacy U. Colo, 1958-61; asst. to prof. pharmacy U. Houston, 1961—, chmn. dept. medicinal chemistry and pharmacognosy, 1973-83; vis. prof. Baylor U., Houston, 1970-72, U. London, 1972; pres. Syncates Assocs., Inc.; v.p. Gt. So. Labs., Inc. Contbr. articles profl. jours. Research grantee NIH, USPHS, Robert A. Welch Found. Mem. Am. Pharm. Assn., Am. Assn. Cancer Research, Acad. Pharm. Scis., Am. Chem. Soc., Am. Assn. Colls.of Pharmacy. Methodist. Lodge: Masons. Subspecialty: Medicinal chemistry. Current work: Design and synthesis of potential anticancer, anticonvulsant and anti-fertility agts. Patentee in field.

CATES, REX GORDON, plant ecologist; b. Vernon, Tex., Aug. 30, 1943; s. George A. and Evelyn (Gordon) C.; m. Kay Roberts, Aug. 5, 1965; children: Todd, Michele, Brian, Tanya, Shawn. B.S., Utah State U., 1965, M.S., 1968; Ph.D., U. Wash., 1971. Research assoc. U. Wash., Seattle, 1972-75; asst. prof. biology U. N.M., Albuquerque, 1975-78, assoc. prof., 1978—; mem. ecology adv. panel NSF, Washington, 1978-81, mem. ecosystem adv. panel, 1978-81. Exec. officer Boy Scouts Am., Albuquerque, 1978-80. NSF grantee, 1976—. Mem. Am. Soc. Naturalists, Ecol. Soc. Am. Democrat. Mormon. Subspecialties: Ecology; Species interaction. Current work: Research in plant-herbivore interactions, stress physiology of trees, outbreaks of forest insects, and defensive natural product chemistry of plants. Office: Chemical Ecology Lab Dept Biology U NMex Albuquerque NM 87131 Home: 1709 Ridgecrest Dr SE Albuquerque NM 87108

CATHOU, RENATA EGONE, scientist, consultant; b. Milan, Italy, June 21, 1935; d. Egon and Stella Mary Egone; m. Pierre-Yves Cathou, June 21, 1959. B.S., MIT, 1957, Ph.D., 1963. Postdoctoral fellow, research assoc. in chemistry MIT, Cambridge, Mass., 1962-65; research assoc. Harvard Med. Sch., 1965-69, instr., 1969-70; research assoc. Mass. Gen. Hosp., 1965-69, instr., 1969-70; asst. prof. dept. biochemistry Tufts U. Sch. Medicine, 1970-73, assoc. prof., 1973-78, prof., 1978-81; vis. prof. dept. chemistry UCLA, 1976-1977; ind. cons., writer, 1981—; mem. adv. panel NSF, 1974-75; mem. bd. sci. counselors Nat. Cancer Inst., 1979-83. Mem. editorial bd.: Immunochemistry, 1972-75; contbr. chpts. to books, articles to profl. jours. NIH predoctoral fellow, 1958-62; sr. investigator Arthritis Found., 1970-75; grantee Am. Heart Assn., 1969-81, USPHS, 1970-81. Mem. Am. Soc. Biol. Chemists, Am. Assn. Immunologists, AAAS, Biophys. Soc., Am. Soc. for Phtobiology, N.Y. Acad. Sci. Subspecialties: Biochemistry (biology); Immunobiology and immunology. Current work: Writing in the biological sciences for scientists and laymen; consulting in biochemistry and immunology. Home: 23 Partridge St Lexington MA 02173 Office: 430 Marrett Rd Lexington MA 02173

CATLIN, B. WESLEY, microbiologist, educator, researcher; b. Mt. Vernon, N.Y., June 26, 1917; s. Harold Burd and Abby Faber (Dunning) C.; m. Lew S. Cunningham, Feb. 15, 1954. M.A., UCLA, 1943, Ph.D., 1947. Postdoctoral fellow Carnegie Instn. Washington, Cold Spring Harbor, N.Y., 1948-50; mem. faculty Med. Coll. Wis. (name formerly Marquette U. Sch. Medicine), Milw., 1950—, assoc. prof. microbiology, 1955-65, prof., 1965—; mem. bd. sci. counselors Nat. Inst. Allergy and Infectious Diseases, Bethesda, Md., 1977-80; mem. Neisseriaceae subcom. Internat. Com. on Systematic Bacteriology, 1964—, sec., 1966-78. Mem. editorial bd.: Jour. Bacteriology, 1972-75; contbr. articles to sci. jours., 8 chpts. to books. Recipient ann. citation Nat. Bd. Med. Coll. Pa., 1978, Disting. Service award Med. Coll. Wis., 1980, Pasteur award Ill. Soc. Microbiology, 1983; NIH grantee, 1954—. Mem. Am. Soc. for Microbiology, Genetics Soc. Am., Soc. for Study Evolution, AAAS, Soc. for Gen. Microbiology (London). Subspecialties: Microbiology; Genome organization. Current work: Genetics and physiology of Neisseria; microbiology teaching and research.

CATTELL, RODERIC GEOFFREY GALTON, computer research scientist; b. Urbana, Ill., May 4, 1953; s. Raymond Bernard and Alberta Karen (Schuettler) C.; m. Nancy Ellen Worner, Aug. 11, 1973 (div. Oct. 1982); 1 son, Eric. B.S., U. Ill.-Urbana, 1974; Ph.D., Carnegie-Mellon U., 1978. Research assoc. Carnegie-Mellon U., Pitts., 1978; research staff mem. Xerox Palo Alto (Calif.) Research, 1978—. Author: Formalization and Automatic Generalization of Code Generators, 1982; contbr. articles in field to profl. jours. Mem. IEEE, Assn. Computing Machinery (outstanding Ph.D. thesis award 1978). Democrat. Subspecialties: Database systems; Programming languages. Current work: Development of database user interfaces and high level data models. Office: Xerox Research Centers 3333 Coyote Hill Rd Palo Alto CA 94304

CAUGHEY, DAVID ALAN, engineering educator; b. Grand Rapids, Mich., Mar. 5, 1944; s. Carl William and Margaret (Arman) C.; m. Linda Criss Jones, June 28, 1969; children: Elizabeth Roma, Amanda MeeAe. B.S.E., U. Mich., 1965; M.A., Princeton U., 1967, Ph.D., 1969. Research scientist McDonnell Douglas Research Labs., St. Louis, 1971-75; asst. prof. Cornell U., Ithaca, N.Y., 1975-80, assoc. prof., 1980—; cons. McDonnell Douglas Corp., St. Louis, 1975—. Recipient Excellence in Teaching award Cornell Soc. Engrs., 1976-77. Mem. AIAA (Lawrence Sperry award 1979). Subspecialties: Aeronautical engineering; Fluid mechanics. Current work: Development of computational techniques to predict aerodynamic forces on flight vehicles. Home: 125 Pearl St Ithaca NY 14850 Office: Dept Aero Engring Cornell U Ithaca NY 14853

CAULK, DAVID ALLEN, mechanical engineer; b. Mpls., Sept. 24, 1950; m. Sharon F., June 10, 1972; children: Rebecca, Sarah, Deborah. B.S., Rensselaer Poly. Inst., 1972; M.S. (NSF fellow), U. Calif., Berkeley, 1974, Ph.D., 1976. Staff research engr. Gen. Motors Research Labs., Warren, Mich., 1976—. Contbr. numerous articles to profl. jours. Mem. ASME, Sigma Xi. Subspecialties: Theoretical and applied mechanics; Fluid mechanics. Current work: Fluid mechanics, polymer processing, metal plasticity, industrial research in composites processing and manufacturing. Home: 4303 Auburn Dr Royal Oak MI 48072

CAUSER, GARY LEE, statistician, applied mathematician; b. Johnstown, Pa., Sept. 27, 1953; s. James Wilbur and Helen Lucille (Elliott) C. B.S., Indiana U. of Pa., 1975, M.S. in Math., 1976, Ohio State U., 1979; M.S. in Computer Sci., Johns Hopkins U., 1983. Research systems analyst and statistician Systems Research Labs., Dayton, Ohio, 1978-79; statis. engr. Corning Glass Works (N.Y.), 1979-81; statistician research div. W.R. Grace Co., Columbia, Md., 1981—. Author statis. computing programs. Mem. Am. Statis. Assn., Am. Soc. Quality Control. Subspecialties: Statistics; Applied mathematics. Current work: Statistical and mathematical consulting activities for research chemists and chemical engineers; consulting for manufacturing facilities. Home: 3352-L N Chatham Rd Ellicott City MD 21043 Office: 7379 Route 32 Columbia MD 21044

CAVALLERO, JOSEPH JOHN, microbiologist, educator, cons.; b. Lawrence, Mass., Mar. 18, 1932; s. John and Salvatrice (Zappala) C.; m. Kathleen Frances Kraus, Dec. 2, 1972; children: Theresa Margaret, Sandra Marie, Elizabeth Camille, Danielle Kay. Student, Merrimack Coll., 1948-50; B.S., Tufts Coll., 1952; M.S., U. Mass., 1954; Ph.D., U. Mich., 1966. Pub. health sanitarian Hartford (Conn.) Health Dept., 1954-55, 57-61; teaching assoc. in microbiology U. Mass., 1961-62; research virologist Med. Research Labs. Charles Pfizer & Co., Groton, Conn., 1966-67; research assoc. in epidemiology U. Mich., 1967-70; microbiologist diagnostic immunology tng. br. Center Diseases Control, Atlanta, 1971—; cons. Pan Am. Health Orgn., Colombia, 1976, 77, Brazil, 1977; asst. prof. dept microbiology and immunology Emory U. Med. Sch., Atlanta, 1982—; lectr. Morehouse Sch. Medicine. Author: (with others) Serodiagnosis of Mycotic Diseases, 1977; author, co-author lab. manuals, 1971—; contbr. articles, chpts. to profl. publs. Served with Med. Service Corps U.S. Army, 1955-57. Mem. Am. Soc. Microbiology, Am. Assn. Immunologists, N.Y. Acad. Sci., Sigma Xi. Democrat. Roman Catholic. Subspecialties: Immunology (medicine); Infectious diseases. Current work: Diagnostic immunology; teaching, training of laboratory personnel in diagnostic immunology. Home: 1325 Balsam Dr Decatur GA 30033 Office: 1600 Clifton Rd 1-B-206 Atlanta GA 30333

CAVALLI-SFORZA, LUIGI LUCA, geneticist; b. Genova, Italy, Jan. 25, 1922; s. Pio and Attilia Manacorda; m. Alba Maria Ramazzotti, Jan. 12, 1946; children: Matteo, Francisco, Tomaso, Violetta. Degree in medicine, U. Pavia, Italy, 1944; M.A., U. Cambridge, Eng., 1950. Asst. in research U. Cambridge, 1948-50; dir. research in microbiology Istituto Sieroterapico Milanese, 1950-57; prof. genetics U. Parma, Italy, 1958-62; prof., chmn. genetics dept. U. Pavia, 1962-71; dir. internat. lab. of genetics and biophysics, Naples, Italy, 1964, vice dir., dir. Pavia sect., 1965-70; prof. genetics Stanford U., Calif., 1970—. Author: Statistical Methods in Biology, 1958, (with W. Bodmer) The Genetics of Human Populations, 1971, Genetics Evolution and Man, 1976, also numerous articles. Fgn. assoc. Nat. Acad. Scis. Subspecialty: Genetics and genetic engineering (biology). Office: Dept Genetics Stanford Univ Med Ctr Stanford CA 94035

CAVALLO, TITO, pathology educator; b. Sao Paulo, Brazil, Feb. 9, 1936; came to U.S., 1968, naturalized, 1977; s. Fiore and Carmen (Martins) C.; m. Anita Hahn, Jan. 28, 1966; children: Alexander L., Charles A. Student, Oswaldo Cruz Coll., Sao Paulo, 1953-55; M.D., U. Sao Paulo, 1963; postgrad., Harvard U., 1968-71. Intern Hospital das Clinicas, U. Sano Paulo Sch. Medicine, 1963; resident Peter Bent Brigham Hosp., Boston, 1968-69, Mallory Inst. Pathology, 1969-72; instr. pathology U. Sao Paulo (Brazil) Sch. Medicine, 1964-68, Harvard U. Med. Sch., Boston, 1971-72; assoc. prof. pathology U. Pitts. Sch. Medicine, 1972-77; prof. pathology U. Tex. Med. Br., Galveston, 1977—, dir. renal immunopathology div.; cons. U. Pitts. Health Ctr. Hosps., Pitts., 1972-77. Recipient Dr. Menotti Sainati award U. Sano Paulo Sch. Medicine, 1963; Golden Apple award Student AMA, 1975. Mem. Internat. Acad. Pathology, Am. Soc. Nephrology, Am. Assn. Pathologists, AAAS, Internat. Soc. Nephrology. Subspecialties: Pathology (medicine); Nephrology. Current work: Immunologic renal disease, glomerular permeability, inflammation. Office: U Tex Med Br 201 Keiller Bldg Galveston TX 77550

CAVE, WILLIAM THOMPSON, JR., medical educator and researcher; b. Washington, Oct. 17, 1942; s. William Thompson and Dorothy Mae (Cleary) C.; m. Jacqueline Clarice Cave, May 4, 1968; children: Catherine, John, Christopher. B.A. cum laude, Kenyon Coll., Gambier, Ohio, 1963; M.D., Yale U., 1967. Diplomate: Am. Bd. Internal Medicine. Intern U. Va. Hosp., 1967-68; resident in internal medicine U. Va., 1968-69, fellow in endocrinology, 1972-75; instr. U. VA., 1975-77, asst. prof., 1977; asst. prof. dept. internal medicine and cancer center U. Rochester, N.Y., 1977-83, assoc. prof., 1983—; assoc. chmn. dept. internal medicine St. Mary's Hosp., Rochester, 1979-83. Contbr. in field. Bd. dirs. Rochester Regional Am. Diabetes Assn. Served to maj. USAR, 1969-72. NIH fellow, 1974-75; NIH grantee, 1978-79, 81—. Mem. Endocrine Soc., Am. Thyroid Assn., Am. Assn. Cancer Research, Soc. Exptl. Biology and Medicine, A.C.P., Am. Fedn. Clin. Research, Am. Diabetes Assn., N.Y. Acad. Scis., Sigma Xi. Subspecialties: Endocrinology; Cancer research (medicine). Current work: Nutritional and hormonal factors that influence mammary tumor growth, medical researcher, consulting endocrinologist. Office: Endocrine Unit Saint Mary's Hosp 89 Genesee St Rochester NY 14611

CAVINESS, VERNE STRUDWICK, JR., neurologist; b. Raleigh, N.C., July 25, 1934; s. Verne Strudwick and Alice Hill (Webb) C.; m. Madeline Viva Harrison, June 2, 1962; children: Gwendoline Angela, Alison Chantal. B.A., Duke U., 1956; D.Phil., Oxford U., 1960; M.D., Harvard U., 1962. Diplomate: Am. Bd. Neurology and Psychiatry. Intern Mass. Gen. Hosp., Boston, 1962-63, resident in medicine, 1963-64, resident in neurology, 1964-67, staff, Boston, 1969—, neurologist, 1982—; mem. faculty Harvard Med. Sch., 1969—, assoc. prof., 1976-82, prof., 1982—; dir. Southard Lab., Eunice Kennedy Shriver Center, Waltham, Mass. Served with USAF, 1967-69. NIH fellow, 1969-71; Joseph P. Kennedy Meml. Found. fellow, 1971-74. Mem. Am. Neurol. Assn., Am. Acad. Neurology, AAAS, Soc. Neurosci. Subspecialties: Neurology; Neurobiology. Current work: Anatomic studies of developing cerebral cortex.

CAVINS, JOHN ALEXANDER, physician, research scientist; b. Terre Haute, Ind., Feb. 18, 1929; s. Alexander W. and Grace Lillian (Erickson) C.; m. Myrtle I, McCleod, June 24, 1961; children: Scott A., Bonnie H., Susie G. A.B., Amherst Coll., 1950; M.D., Johns Hopkins U., 1954. Intern Univ. Hosp., Columbus, Ohio, 1954-55, resident in medicine and hematology, 1955-59; research assoc. Harvard U. Med. Sch., Cambridge, Mass., 1963-70; assoc. prof. Ind. U. Sch. Medicine, Indpls., 1970-72, clin. assoc. prof. medicine, 1975—; chief oncology and hematology dept. St. Vincent Hosp., Indpls., 1974—, dir. cancer research, 1974—; pvt. practice medicine specializing in oncology and hematology, Indpls., 1972—; cons. Beta Med. Pharm. Co., Indpls., 1982—. Contbr. sci. articles to profl. publs. Served to lt. comdr. USN, 1955-62. Mem. Internat. Soc. Hematology, Am. Soc. Hematology, Am. Assn. for Cancer Research, AAAS, Am. Soc. Clin. Pharmacology and Therapeutics. Republican. Presbyterian. Subspecialties: Cancer research (medicine); Hematology. Current work: Cancer immunology and immunotherapy, cancer research, cancer medicine, and clinical trials. Home: 6202 N Sherman Dr Indianapolis IN 46220 Office: 8220 Naab Rd Indianapolis IN 46260

CAWLEY, CHARLES NASH, environmental scientist, educator, researcher; b. Shreveport, La., Aug. 21, 1937; s. Charles P. and Carnall (Nash) C. B.A., U. Okla., 1970, M.A., 1970; M.S., U. Tex.-Dallas, 1975, Ph.D., 1978. Project leader TWU Research Inst., Denton, Tex., 1964-73; gen. ptnr. Southwest Textile Lab., Denton, 1973-77; asst. prof. design and environ. analysis Cornell U., Ithaca, N.Y., 1978—. Mem. Soc. for Computer Simulation, Am. Nuclear Soc., Health Physics, Am. Public Health Assn., Sigma Xi. Subspecialties: Environmental effects of energy technologies; Environmental toxicology. Current work: Modelling environmental tritium, tritium dose following chronic exposure, modelling tritium half-life in man, consequences of energy development. Home: 211 Hudson Ithaca NY 14850 Office: Cornell U Design and Environmental Analysis Ithaca NY 14853

CAZES, ALBERT N., computer specialist, educator; b. Cairo, July 30, 1942; U.S., 1961; s. Nessim B. and Julia J. (Chaky) C.; m. Rose V. Chimchirian, Aug. 20, 1972; children: Caroline,Annabelle. B.S., CUNY, 1969; M.A., Hunter Coll., 1975; Ph.D., Poly. Inst., 1982. Computer specialist N.Y.C.-D.S.S.-O.D.P., 1970—. Served with U.S. Army, 1965-67. Mem. Am. Math. Assn., Am. Math. Soc., Soc. Indsl. and Applied Math., Sigma Xi. Democrat. Jewish. Subspecialties: Information systems, storage, and retrieval (computer science); Applied mathematics. Current work: Office automation, ordinary differential equations, inverse boundary problems, Hill's equations. Home: 269 Ocean Pkwy Brooklyn NY 11218

CECH, THOMAS ROBERT, chemistry educator; b. Chgo., Dec. 8, 1947; m., 1970. B.A., Grinnell Coll., 1970; Ph.D., U. Calif.-Berkeley, 1975. Nat. Cancer Inst. fellow molecular biology MIT, 1975-77; asst. prof. chemistry U. Colo., Boulder, 1978-82, assoc. prof., 1982-83, prof., 1983—. NIH grantee, 1978—; NIH research career devel. grantee, 1980—. Mem. Am. Soc. Biol. Chemists. Subspecialties: Photochemistry; Molecular biology. Office: Dept Chemistry U Colo Boulder CO 80309

CEDERBAUM, ARTHUR I., biochemist; b. Bklyn., Oct. 5, 1943; s. Nathan and Lillian (Knobel) C.; m. Arlene Dunayer, July 9, 1967; children: Neil, Robert. B.S., Bklyn. Coll. Pharmacy, 1966; Ph.D., Rutgers U., 1971. Research assoc. Mt. Sinai Sch. Medicine, N.Y.C., 1971-73, research asst. prof., 1973-76, asst. prof., 1976-79, assoc. prof. biochemistry, 1979—. Assoc. editor: Alcoholism: Clin. Exptl. Research, 1981—; mem. editorial bd.: Substance and Alcohol Abuse/Misuse, 1981—. Research Career Devel. awardee Nat. Inst. Alcohol Abuse and Alcoholism, 1975—; USPHS research grantee, 1978-81, 77—, 78—. Mem. Am. Soc. Biol. Chemists, Research Soc. Alcoholism, Am. Chem. Soc., Soc. Free Radical Research, N.Y. Acad. Scis. Club: Soccer (North Brunswick, N.J.) (coach 1980—). Subspecialties: Biochemistry (biology); Biochemistry (medicine). Current work: Regulation and metabolic effect of alcohol and acetaldehyde on the lives, generation of, and role of oxygen radicals in biological reactions. Home: 1314 Aaron Rd North Brunswick NJ 08902 Office: Mount Sinai Sch Medicine 1 Gustave Levy Pl New York NY 10029

CEDERQVIST, LARS LENNART, obstetrician, gynecologist, educator, researcher; b. Helsingborg, Sweden, Aug. 13, 1935; came to U.S., 1968; s. Lennart Frans Vilhelm and Astrid Teresa (Nilsson) C.; m. Marianne Stahlmalm, Aug. 22, 1943; children: Fredrik, Marcus, Christian. B.Med., U. Lund, Sweden, 1957, B.Phil., 1959, M.D., 1964. Diplomate: Am. Bd. Ob-Gyn. Resident in ob-gyn U. Lund, 1965-68; intern in pathology N.Y. Hosp.-Cornell Med. Ctr., N.Y.C., 1974-75, resident in ob-gyn., 1968-69, 72-73, asst. prof. ob-gyn., 1973-78, assoc. prof. clin. ob-gyn., 1978—; NIH postdoctoral fellow Cornell U. Med. Sch., N.Y.C., 1969-72. Assoc. editor: Am. Jour. Reproductive Immunology, 1980—; contbr. numerous articles, chpts. to profl. publs. Recipient Tchr.-Scientist award Mellon Found., 1976. Mem. AMA, N.Y. Acad. Scis., Am. Assn. Immunology, Am. Soc. Human Genetics, Am. Coll. Obstetricians and Gynecologists, Soc. Gynecologic Investigation. Subspecialties: Maternal and fetal medicine; Immunogenetics. Current work: Reproductive immunology, genetics, prenatal diagnosis, fetal infection, maternal-fetal medicine. Office: 570 E 70 MO31 New York NY 10021

CELLA, ALEXANDER, testing laboratory exec.; b. Paterson, N.J., Feb. 4, 1935; s. Alexander and Frances C.; m. Elizabeth Ann Cella, Sept. 27, 1959; children: Ellen, Kathy, Heidi. B.S.M.E., Calif. Poly. State U., 1958. Engr. Swepco Tube, 1959-63; v.p. engring. Pipeco Steel Corp., 1963-70; pres. Universal Testing Labs., Cedar Grove, N.J., 1971—. Served with USAF. Mem. Am.Nuclear Soc., ASME (cert.), ASTM, Am. Soc. Metals. Subspecialty: Nuclear fission. Current work: Nuclear steam generators. Patentee in field.

CENSOR, YAIR, educator, researcher; b. Rehovot, Israel, Nov. 29, 1943; s. Emanuel and Else (Simon) C.; m. Erga Goldfinger, Aug. 21, 1966; children: Aviv, Nitzan, Keren. B.S., Technion, Haifa, Israel, 1967, M.S., 1969, D.Sc., 1975. Research fellow Technion, Haifa, 1975-76; lectr. U. Haifa, 1976-77, sr. lectr., 1979—; research asst. prof. SUNY-Buffalo, 1977-79; research assoc. Med. Image Processing Group, Buffalo, 1977-79; vis. assoc. prof. U. Pa., Phila., 1982-83. Mem. Israeli Union of Math., Soc. Indsl. and Applied Math. Subspecialties: Applied mathematics; Imaging technology. Current work: Mathematical optimization theory; numerical analysis; optimization theory techniques in image reconstruction from projections. Office: Dept Radiology U Pa Hosp 3400 Spruce St G1 Philadelphia PA 19104 Home: 25 Gilboa St Haifa Israel 32716 Office: Univ Haifa Dept Math Mount Carmel Haifa Israel 31999

CENZER, DOUGLAS ALFRED, mathematics educator; b. Detroit, Nov. 15, 1947; s. Alfred V. and Hattie (Czaczkowski) C.; m. Pamela Scharstein, May 21, 1970; 1 son, Michael. B.S. in Math., Mich. State U., 1968, Ph.D., U. Mich., 1972. Research asst. Nat. Security

Adminstrn., Ft. Meade, Md., 1967; asst. prof. U. Fla., Gainesville, 1972-77, assoc. prof., 1977—; vis. assoc. prof. North Tex. State U., Denton, 1981-82; referee NSF, Washington, 1978—, Trans. Am. Math. Soc., Providence, 1982—. Reviewer: Math. Revs, 1976—, Jour. Symbolic Logic, 1981—; contbr. articles on math. to profl. jours. Mem. exec. com. United Faculty of Fla., Gainesville, 1977-79. Ford scholar, 1966-68; NSF fellow, 1968-72; grantee, 1974-76; U.Fla. grantee, 1978. Mem. Am. Math. Soc., Assn. for Symbolic Logic, Soc. for Indsl. and Applied Math., Phi Beta Kappa, Pi Mu Epsilon. Democrat. Current work: Analytic sets, computational complexity, game theory, inductive definability. Home: 2439 NW 12th Pl Gainesville FL 32605 Office: U Fla Dept Math 201 Walker Hall Gainesville FL 32611

CEPERLEY, PETER HUTSON, physics educator, researcher; b. Charleston, W.Va., Sept. 5, 1945; s. Florian Fairchild and Ellen Axon (Rodes) C.; m. Mary Christine Eakin, June 22, 1975; 1 son, Daniel. B.S., U. Mich., 1967; Ph.D., Stanford U., 1973. Research assoc. Stanford (Calif.) U., 1972-76; vis. asst. prof. George Mason U., Fairfax, Va., 1977-79, asst. prof. physics, 1979—. Contbr. articles in field to profl. jours. Mem. Am. Phys. Soc., Internat. Solar Energy Soc., Acoustical Soc. Am. Democrat. Subspecialties: Energy conversion devices; Accelerator physics. Current work: Development an acoustical driven stirling engine with no moving parts. Office: Physics Dep George Mason U Fairfa VA 22030

CERA, LEE MARIE, veterinarian; b. Chgo., June 24, 1950; d. Ernest J. and Gloria E. (Bonet) C. B.A., St. Xavier U., 1971; B.S., U. Ill, 1973, D.V.M., 1975; postgrad., U. Chgo., 1975—. Vet. asst. BevLab Vet. Hosp., Blue Island, Ill., 1974-75, emergency clinician, 1975-77; equine cons. Pine Bluff Animal Hosp., Joliet, Ill., 1975-77; instr. St. Xavier Coll., 1975-77; resident, trainee in comparative pathology U. Chgo.-Lincoln Park Zoo, 1975-77; resident in comparative medicine and pathology A.J. Carlson Animal Research Facility, Pritzker Sch. Medicine, U. Chgo., 1975-78, chief clin. and lab. services biol. scis. div., 1978-80, acting dir., 1980-81, dir. animal care, 1981—; course asst. comparative pathophysiology and med. histology U. Chgo., 1977-80; asst. program dir. Charles Louis Davis D.V.M. Found. for Vet. Pathology, 1977-82, program dir., 1982—. Contbr. articles to profl. jours. Mem. med. adv. bd. Lincoln Park Zoo., Wyler Children's Hosp. Recipient award Lake County Humane Soc., 1975, Pfizer Co., 1974. Mem. Avma, Midwest Vet. Pathology Assn., Wildlife Disease Assn., Am. Assn. for Lab. Animal Sci., Am. Animal Hosp. Assn., Am. Soc. for Lab. Animal Practitioners, Ill. Soc. Med. Research (V.p. 1981—). Roman Catholic. Subspecialty: Pathology (veterinary medicine). Current work: Laboratory animal science, burn rehabilitation. Office: Box 144 Chicago IL 60637

CERAMI, ANTHONY, biochemist; b. Newark, Oct. 3, 1940; m. Helen Vlassara, May 1, 1981; children: Carla, Ethan. B.S., Rutgers U., 1962; Ph.D., Rockefeller U., 1967. Asst. prof. Rockefeller U., 1969-72, asso. prof., head lab. med. biochemistry, 1972—, co-dir. M.D./Ph.D. program, 1978-80, prof., 1978—. Contbr. articles on biochemistry to profl. jours. Mem. Am. Soc. Biol. Chemists, Am. Soc. Hematology, Am. Soc. Pharmacology and Exptl. Therapeutics. Subspecialties: Biochemistry (biology); Parasitology. Current work: Application of chemistry to problems of metabolism, e.g. diabetes, aging and parasitic diseases. Home: 430 E 63rd St 12A New York NY 10021 Office: 1230 York Ave New York NY 10021

CERCEO, JOHN MICHAEL, physicist, sci. corp. exec.; b. Washington, July 1, 1933; s. John G. and Marie L. C.; m. Aline V. Cook; children: Michael, Andrea, Alyssa, Michelle, Anthony. B.S. cum laude, Cath. U., 1957, M.S. in Physics, 1961, Ph.D., Pacific Western U., 1981. Scientist, advanced research div. AMF, 1961-64, Gen. Techs. Corp., 1964-65; tech. group leader TRACOR Labs., 1965-70; sr. scientist VERTEX Corp., 1970-72; sr. scientist, pres. ISC, Inc., 1972-75, v.p., partner, 1977—, also dir.; ind. cons., 1975-77; pres. AFL, Inc., 1977—, also dir. Contbr. tech. papers and articles to profl. lit. Served with U.S. Army, 1954-55. Mem. AAAS, IEEE, Optical Soc. Am., Philos. Soc. of Washington, Acoustical Soc. Am., Am. Inst. Physics, AIAA, Potomac Chase Citizens Assn. Subspecialties: Acoustics; Applied mathematics. Current work: Computer systems analyses/applications; executive technical director, programs mgr. Home: 13305 Moran Dr Travilah MD 20878 Office: 8478 Tyco Rd Vienna VA 22180

CERINI, COSTANTINO PETER, virologist, researcher; b. Phila., Nov. 19, 1931; s. Joseph and Rita Lillian (Cruciani) C.; m. Lydia G. De Angelis, June 18, 1960; children: Angela, Christina. B.A., LaSalle Coll., Phila., 1953; M.S., Lehigh U., 1960, Ph.D., 1964. Research virologist Am. Cyanamid Corp., Pearl River, N.Y., 1964-70; group leader Lederle Labs., Pearl River, 1970-75, sr. research virologist, 1975—. Contbr. articles to profl. publs. Served with U.S. Army, 1955-57. Mem. N.Y. Acad. Scis., Am. Soc. Microbiology, AAAS, Sigma Xi. Roman Catholic. Subspecialties: Virology (biology); Immunobiology and immunology. Current work: Development of vaccines for infectious diseases. Patentee in field. Office: Lederle Labs Pearl River NY 10965

CERMAK, IVAN ANTHONY, research and development executive; b. Povazska Bystrica, Czechoslovakia, Dec. 23, 1940; s. Anton Josef and Maria (Lysa) C.; m. Joan Elizabeth Hinch, Aug. 15, 1964. B.Eng. (E.E.), McGill U., 1963, M.Eng., 1967, Ph.D. (E.E.), 1969. Mem. tech. staff Bell Labs., Holmdel, N.J., 1969-70, tech. supr., 1970-78, tech. dept. head, Denver, 1978-81; corporate v.p., dir. Advanced Tech. Center, ITT, Shelton, Conn., 1981—; cons. Contbr. articles to sci.-tech. jours. Served to capt. RCAF, 1959-66. Mem. IEEE (editorial bd. Spectrum), U.S. C. of C. Subspecialties: Electrical engineering; Computer engineering. Current work: Systems, particularly telecommunications and office systems. Home: 27 Adams Dr Huntington CT 06484 Office: 1 Research Dr Shelton CT 06484

CERNI, TODD ANDREW, atmospheric scientist, educator; b. Milw., Apr. 28, 1947; s. Andrew J. and Margaret L. C. B.S. in Physics, Marquette U., Milw., 1969, M.S., Ind. U., 1971; Ph.D. in Atmospheric Sci, U. Ariz., Tucson, 1976. Research assoc. U. Wyo., Laramie, 1976-79, asst. prof. dept. atmospheric sci., 1979—. Contbr. articles to profl. jours. Mem. Am. Phys. Soc., Am. Meteorol. Soc. Subspecialties: Meteorology; Meteorologic instrumentation. Current work: Research in cloud physics, instrumentation, radiative transfer. Patentee in field. Office: Atmospheric Sci Dept U Wyo Laramie WY 82071

CERNY, JOSEPH, III, nuclear chemist, educator; b. Montgomery, Ala., Apr. 24, 1936; m., 1959; 2 children. B.S., U. Miss., 1957; Ph.D. in Nuclear Chemistry, U. Calif.-Berkeley, 1961. From asst. prof. to assoc. prof. U. Calif.-Berkeley, 1975-79, prof. chemistry, 1971—; research chemist Lawrence Berkeley Lab., 1971—, head nuclear sci. div. and assoc. dir., 1979—; cons. U.S. Army Research Office, 1963-65. Guggenheim fellow Oxford (Eng.) U., 1969-70; vis. fellow Australian Nat. U., Canberra, 1975; recipient E.O. Lawrence award AEC, 1974. Fellow Am. Phys. Soc.; mem. Am. Chem. Soc., Fedn. Am. Scientists, AAAS. Subspecialty: Nuclear chemistry. Office: Lawrence Berkeley Lab Bldg 88 Berkeley CA 94720

CERRETO, MARY CHRISTINE, pediatrics psychology educator; b. Putnam, Conn., Apr. 29, 1951; d. Sebastian S. and Viola (Bonneville) C.; m. David Louis Coulter, Aug. 7, 1982. A.B., Conn. Coll., 1973; Ph.D., U. Wash., 1976. Lic. psychologist, Tex. Asst. prof. pediatrics U. Tex. Med. Br., Galveston, 1977-80, assoc. prof., 1981—; asst. prof. pediatrics Vanderbilt U. Med. Sch., Nashville, 1980-82; cons. Joseph P. Kennedy Jr. Found., Washington, 1979—. Editor: Jour. Children's Health Care, 1980—. Grantee Adminstrn. for Children, Youth and Families, 1978-81, Tex. Dept. Community Affairs, 1979-81, W. T. Grant Found., 1980, March of Dimes, 1980. Mem. Am. Assn. on Mental Deficiency (v.p. psychology 1980-83), Am. Psychol. Assn., Soc. for Research in Child Devel., So. Soc. for Pediatrics Research. Democrat. Roman Catholic. Subspecialties: Behavioral psychology; Developmental psychology. Current work: Research on brothers and sisters of handicapped children and children's understanding of health and illness, training behavioral pediatrics. Office: Dept Pediatric U Tex Med Br Galveston TX 77550

CETAS, THOMAS CHARLES, physicist, educator; b. Petoskey, Mich., Aug. 25, 1941; s. Carl N. and Rae L. (Jolls) C.; m. Betty Lou Schultz, June 15, 1963; children: Amanda, Justin, Melissa. Student, Cedarville (Ohio) Coll., 1959-61; A.B. cum laude, Hope Coll., 1963; Ph.D., Iowa State U., Ames, 1970. Fellow Nat. Measurement Lab., Sydney, New South Wales, Australia, 1970-73; physicist Nat. Bur. Standards, Gaithersburg, Md., 1973-75, Bur. Radiol. Health, Rockville, Md., 1975; research assoc. radiology U. Ariz., Tucson, 1975-78, asst. prof. radiology and elec. engring., 1978-82, assoc. prof., 1982—. Recipient Research Grants Nat. Cancer Inst., 1976—, Technicon Corp., 1979—, Bur. Radiol. Health, 1976, 78. Mem. Radiation Research Soc., Am. Phys. Soc., Am. Assn. Physicists in Medicine, Bioelectromagnetics Soc., Sigma Xi. Baptist. Subspecialties: Medical physics; Biomedical engineering. Current work: Phys. apsects of hyperthermia for cancer therapy, hyperthermia, thermometry, thermal dosimetry, microwave heating, radiofrequency, ultrasound. Patentee in field. Home: 7520 N San Lorenzo St Tucson AZ 85704 Office: Radiation Oncology Div Health Sci Center Tucson AZ 85724

CEZAIRLIYAN, ARED, physicist; b. Istanbul, Turkey, May 9, 1934; came to U.S., 1957, naturalized, 1971; s. Onnik and Valantin (Mangassarian) C.; m. Sylvia Papazian, May 24, 1970; 1 son, Brent. B.S.M.E., Robert Coll., Istanbul, 1957; M.S.M.E., Purdue U., 1960, Ph.D., 1963. Research engr. Purdue U., West Lafayette, Ind., 1960-63; research physicist Nat. Bur. Standards, Washington, 1963—; cons. CINDAS/Purdue U., 1968—. Founding editor: Internat. Jour. Thermophysics, 1980—; contbr. articles to profl. jours. Recipient Silver medal Dept. Commerce, 1975, Gold medal, 1980, Engr. of Yr. award Nat. Soc. Profl. Engrs., 1982; Gold medal French High Temperature Soc., 1982; named Disting. Engring. Alumnus Purdue U., 1978. Mem. Am. Phys. Soc., ASME (Heat Transfer Meml. award 1981), ASTM, Internat. Thermophysics Congress (chmn.), Washington Philos. Soc. Subspecialties: High-temperature materials; Condensed matter physics. Current work: Research in developing advanced techniques for measurement of properties of substances at high temperatures by rapid dynamic pulse-heating techniques of millisecond and microsecond resolution. Home: 12 Tapiola St Rockville MD 20850 Office: Nat Bur Standards Washington DC 20234

CHA, SE DO, physician; b. Seoul, Korea, Dec. 17, 1942; came to U.S., 1966, naturalized, 1977; s. Young Sun and Hee Joo (Chang) C.; m. Elsa Jane Greene, Dec. 21, 1974. M.D., Yon Sei U., 1966. Diplomate: Am. Bd. Internal Medicine. Cardiologist Roger Williams Gen. Hosp., Providence, 1973-75; cardiologist Deborah Heart and Lung Center, Browns Mills, N.J., 1975—, asst. dir. adult cardiac catheterization lab., 1975—; instr. Brown U., Providence, 1973-75. Contbr. articles to profl. jours. Fellow Am. Coll. Angiology, Soc. for Cardiac Angiography, ACP; mem. AMA, Fedn. Clin. Research. Subspecialties: Cardiology; Internal medicine. Current work: Valvular heart disease, especially tricuspid valve disease. Intracardiac phonocardiography, primary pulmonary hypertension. Home: RD 4 Box 268 Hartford Rd Medford NJ 08055 Office: Deborah Heart and Lung Center Trenton Rd Bronws Mills NJ 08015

CHABOT, JEAN FINCHER, biological researcher; b. Chgo., June 19, 1944; d. Chalmers Hudson and Bridget (O'Donnell) Fincher; m. Brian Frank Chabot, July 29, 1967; children: Carolyn, Michael. B.A., Lake Forest Coll., 1965; Ph.D., Duke U., 1970. Lect. U. N.H., 1973; research assoc. food sci. dept. Cornell U., Ithaca, N.Y., 1973-77, research assoc. ecology and systematics, 1977-78, sr. research assoc. div. biol. scis., 1979—; research assoc. Boyce Thompson Inst., 1980-81. Mem. Bot. Soc. Am., Am. Soc. Plant Physiology, Ecol. Soc. Am., Electron Microsocopy Soc. Am., Grad. Women in Sci., Phi Beta Kappa, Sigma Xi. Subspecialty: Cell biology. Current work: Cell biology. Office: Cornell U Ecology and Systematics Corson Bldg Ithaca NY 14853

CHACE, MILTON, mechanical engineering educator; b. Takoma Park, Md., Mar. 19, 1934; m., 1958; 3 children. B.E.P., Cornell U., 1957; M.S.M.E., U. Mich., 1962, Ph.D., 1964. Engr. AEC, 1957-59; mech. engr. Atomic Power Devel. Assocs., 1959-61; staff engr. systems devel. div. IBM, 1964-67; assoc. prof. mech. engring. U. Mich., Ann Arbor, 1967-75, prof., 1975—, dir., 1971—, Mech. Dynamics, Inc. Mem. ASME. Subspecialties: Computer-aided design; Theoretical and applied mechanics. Address: 3265 N Maple Rd Ann Arbor MI 48103

CHACKO, GEORGE KUTTY, biochemist, educator; b. Kottarakkara, Kerala, India, Feb. 15, 1933; came to U.S., 1962, naturalized, 1967; s. Kurian and Sosamma (Abraham) C.; m. Thankamma Joseph, Jan. 11, 1962; children: Jacob, Joseph, George. B.Sc., U. Kerala, 1956, M.Sc., 1958; Ph.D., U. Ill., 1966. Postdoctoral research assoc. U. Wash., Seattle, 1966-67, U. Ariz., Tucson, 1967-68; asst. prof. biochemistry Med. Coll. Pa., Phila., 1968-74, assoc. prof., 1974—. Wright fellow, 1963. Mem. Am. Chem. Soc., Am. Physiol. Sco., AAAS, Soc. Exptl. Biology and Medicine., Am. Soc. Biol. Chemists. Subspecialties: Biochemistry (medicine); Membrane biology. Current work: Metabolism of lipids and lipoproteins; lipoprotein-membrane interactions; membrane structure and function. Home: 1131 Cleveland Rd Blue Bell PA 19422 Office: Med Coll Pa Philadelphia PA 19129

CHADHA, KAILASH CHANDRA, cancer scientist, researcher, microbiologist; b. Churu, Rajasthan, India, July 1, 1943; came to U.S., 1969, naturalized, 1975; s. Ralla Ram and Budhwanti Devi C.; m. Anju, May 22, 1971; children: Sonia, Priya. B.Sc., U. Rajasthan, Jaipur, 1962; M.Sc., Indian Agrl. Research Inst., New Delhi, 1964; Ph.D., U. Guelph, Ont., Can., 1968. NRC Can. postdoctoral fellow Can. Dept. Agr., Vineland, Ont., 1969-70; cancer research scientist I, Roswell Park Meml. Inst., Buffalo, 1971-74, cancer research scientist III, 1974-78, cancer research scientist IV, 1978—; assoc. prof. microbiology SUNY-Buffalo, 1978—. Contbr. articles to profl. jours. NIH grantee, 1980-82; Am. Cancer Soc. grantee, 1979-82. Mem. N.Y. Acad. Scis., Am. Soc. Microbiology. Subspecialties: Microbiology; Animal virology. Current work: Microbiology, virology, interferons, herpes viruses, others. Home: 72 Princess Ln West Seneca NY 14224 Office: 666 Elm St Buffalo NY 14263

CHAFFIN, DON B., industrial engineering educator; b. Sandusky, Ohio, Apr. 17, 1939; m., 1966; 1 child. B.I.E., Gen. Motors Inst., 1962; M.S.I.E., U. Toledo, 1964; Ph.D., U. Mich., 1967. Jr. draftsman Mack Iron Steel Co., Ohio, 1955-57; quality control engr. New Departure div. Gen. Motors Corp., Ohio, 1960-62; project engr. Micrometrical div. Bendix. Corp., Southfield, Mich., 1963-64; asst. prof. phys. medicine U. Kans., 1967-68; asst. prof. indsl. engring. U. Mich., Ann Arbor, 1968-70, assoc. prof. indsl. and bioengring., 1970-77, prof. indsl. and ops. engring., 1977—; cons. Bendix Corp., 1964. Grantee Western Electric Co., 1967-71, NASA, 1970-71, Aerospace Med. Research Labs., 1970-71, Nat. Inst. Occupational Safety and Health, 1971-72. Mem. Human Factors Soc., Nat. Soc. Profl. Engrs., Am. Inst. Indsl. Engrs., Biomed. Engring. Soc., Brit. Ergonomics Research Soc. Subspecialties: Biomedical engineering; Human factors engineering. Office: Coll Engring U Mich Ann Arbor MI 48105

CHAGANTI, RAJU SREERAMA KAMALASANA, geneticist; b. Samalkot, India, Mar. 12, 1933; s. Sanyasi Raju and Seeta Siromani (Vallury) C.; m. Seeta R. Kurada, Nov. 20, 1939; children: Seeta, Sara. B.Sc. with honors, Andhra U., Waltair, India, 1954, M.S., 1955; Ph.D., Harvard U., 1964. Lectr. botany Andhra U., 1961-67; mem. sci. staff Med. Reserch Council Brit., 1967-71; research assoc. to assoc. investigator lab. human genetics N.Y. Blood Cente, N.Y.C., 1971-76; assoc. mem. and head lab. cancer genetics and cytogenetics Meml. Sloan-Kettering Cancer Center, N.Y.C., 1976—; asst. prof. genetics Cornell U. Grad. Sch. Med. Scis., N.Y.C., 1974—; cons. N.Y. Blood Center. Contbr. over 150 articles to sci. and tech. jours. Mem. Am. Soc. Human Genetics, Harvey Soc., Genetics Soc. Am., Indian Soc. Human Genetics. Subspecialties: Cancer research (medicine); Genetics and genetic engineering (medicine). Current work: Human genetics; role of heredity in origin of birth defects and cancer. Office: 1275 York Ave New York NY 10021

CHAISSON, ERIC JOSEPH, astrophysicist; b. Lowell, Mass., Oct. 26, 1946; s. Louis Joseph and Marion Loretta (Brennan) C.; m. Lola Judith Eachus, May 1, 1976. B.S., U. Lowell, 1968; M.A., Harvard U., 1969, Ph.D., 1972. Fellow Nat. Acad. Scis., Smithsonian Astrophys. Obs., 1972-74; asst. prof. Harvard U., 1974-79, assoc. prof., 1979-82; prof. astrophysics Haverford (Pa.) Coll., 1982—; mem. NASA Working Group on Extraterrestrial Intelligence; Shapley vis. prof. Am. Astron. Soc.; mem. sci. adv. com. Boston Mus. Scis., Hayden Planetarium; sci. adv. com. Public Broadcasting System series Search for Solutions; chmn. com. on public edn. Harvard-Smithsonian Obs. Author: Cosmic Dawn: The Origins of Matter and Life, 1981; Contbr. numerous articles to profl. jours. Served to capt. USAF, 1969-70. Sloan fellow, 1977; recipient Bok prize Harvard U., 1977, Smith prize, 1978; Phi Beta Kappa Writing prize, 1981. Mem. AAAS, Am. Astron. Soc., Am. Assn. Physics Tchrs., Fedn. Am. Scientists, Internat. Union Radio Sci., Internat. Astron. Union, Authors Guild, Am. Inst. Physics. (sci. writing award 1981). Subspecialties: Radio and microwave astronomy; Cosmology. Home: 803 Stoke Rd Villanova PA 19085 Office: Haverford Coll Haverford PA 19041

CHAKRABARTI, CHUNI LAL, chemistry educator; b. Patuakhali, India, Mar. 1, 1920; m., 1962; 1 child. B.Sc., U. Calcutta, 1941; M.Sc., U. Birmingham, 1960; Ph.D. in Chemistry, Queen's U., Belfast, 1962, D.Sc., 1980. Supervisory chemist Metal & Steel Factory (Govt. India), 1941-45; chemist in-charge Mines and Indsl. Dept. (Govt. Burma), 1945-52; chief chemist and mgr. Mineral Resources Devel. Corp., 1952-59; vis. asst. prof. and fellow La. State U., 1963-65; group leader research ctr. Noranada Mines Ltd., 1965; asst. prof. Carleton U., Ottawa, Ont., Can., 1965-67, assoc. prof. analytical and inorganic chemistry, 1967-76, prof. chemistry, 1976—. Recipient Gerhard Herzberg award Spectroscopy Soc. Can., 1977; Fisher award Chem. Inst. Can., 1981. Fellow Chem. Inst. Can.; mem. Am. Chm. Soc., Brit. Soc. Analytical Chemistry. Subspecialty: Analytical chemistry. Office: Dept Chemistry Carleton U Ottawa ON Canada K1S 5B6

CHAKRABARTY, ANANDA MOHAN, microbiologist; b. Sainthia, India, Apr. 4, 1938; s. Satya Dos and Sasthi Bala (Mukherjee) C.; m. Krishna Chakraverty, May 26, 1965; children—Kaberi, Asit. B.Sc., St. Xavier's Coll., 1958; M.Sc., U. Calcutta, 1960, Ph.D., 1965. Sr. research officer U. Calcutta, 1964-65; research asso. in biochemistry U. Ill., Urbana, 1965-71; mem. staff Gen. Electric Research and Devel. Center, Schenectady, 1971-79; prof. dept. microbiology U. Ill. Med. Center, 1979—. Editor: Genetic Engineering, 1977. Named Scientist of Year Indsl. Research Mag., 1975. Mem. Am. Soc. Microbiology, Soc. Indsl. Microbiology, Am. Soc. Biol. Chemists. Subspecialties: Genetics and genetic engineering (biology); Microbiology. Current work: Toxic chemical pollution, genetically engineered microorganism for secondary oil recovery. Home: 206 Julia Dr Villa Park IL 60181 Office: Dept Microbiology U Ill Med Center 835 S Wolcott St Chicago IL 60612

CHAKRABORTY, DEV PRASAD, radiol. physicist, educator, researcher; b. Calcutta, India, Mar. 26, 1948; s. Nani Gopal and Maya Usha (Bhattacharyya) C.; m. Nupur Mukherjee, Nov. 19, 1973; 1 child, Rimi. M.S., U. Delhi, India, 1970; Ph.D., U. Rochester, 1977. Grad. asst. U. Rochester, N.Y., 1970-77; research investigator U. Pa., Phila., 1977-79; postdoctoral fellow U. Ala., Birmingham, 1979-81, instr. diagnostic radiology, 1981-82, asst. prof., 1982—. Contbr. writings to profl. publs. in field. Mem. Am. Assn. Physicists in Medicine. Subspecialties: Diagnostic radiology; Imaging technology. Current work: Hyperthermia, imaging tech., mammography, computers. Office: Diagnostic Radiology 619 S 19th St Birmingham AL 35233

CHAKRAVARTI, ARAVINDA, genetics educator; b. Calcutta, India, Feb. 6, 1954; s. Satya Sharan and Bani C.; m. Shukti, Nov. 22, 1978. B.Stat. with honors, Indian Statis. Inst., 1974; Ph.D., U. Tex.-Houston, 1979. Postdoctoral fellow U. Wash., Seattle, 1979-80; asst. prof. human genetics and biostats. U. Pitts., 1980—, mem., 1982—. Contbr. articles to profl. jours. Recipient Nat. Sci. Talent Search award, 1970; Jagadis Bose Nat. Sci. Talent Search award, 1970; UNESCO prize, 1971; Merit scholar U. Tex., 1977; Elsa Pardee Found. grantee, 1982-83; United Way grantee, 1982; NIH grantee, 1982. Mem. Am. Statis. Assn. (chpt. v.p. 1981-83), Am. Soc. Human Genetics, Indian Soc. Human Genetics, Genetics Soc. Am. Subspecialties: Genetics and genetic engineering (biology); Cancer research (medicine). Current work: Using statistical data analysis and molecular data to elucidate the genetics of human disease. Home: 343 Hastings St #2 Pittsburgh PA 15206 Office: U Pitts Pittsburgh PA 15261

CHAKRAVARTY, INDRANIL, research scientist; b. New Delhi, India, Jan. 7, 1954; came to U.S., 1971; s. Sunil Kumar and Monica (Bagchi) C. B.S.E.E., NYU, 1974; M.Eng., Rensselaer Poly. Inst., 1976, Ph.D., 1982. Research asst. Image Processing Lab., Rensselaer Poly. Inst., 1975-80, instr. dept. elec. and systems engring., 1980-81; mem. profl. staff Schlumerger-Doll Research Ctr., Ridgefield, Conn., 1981—. Mem. IEEE, Assn. for Computing Machinery, Spl. Interest Group Computer Graphics, Sigma Xi. Subspecialties: Graphics, image processing, and pattern recognition; Computer engineering. Current work: 3D image synthesis, image realism, computer vision and robotics, computer aided design. Home: 5 Cottonwood Lane Ridgefield CT 06877 Office: Schlumberger-Doll Research Ctr Old Quarry Rd Ridgefield CT 06877

CHALEFF, ROY SCOTT, research scientist; b. N.Y.C., Oct. 30, 1947; m., 1976. B.A., Amherst Coll., 1968; M.Phil., Yale U., 1970, Ph.D. in Biology, 1972. Fellow in biology Brookhaven Nat. Lab., Upton, N.Y., 1972-74; sr. sci. officer applied genetics John Innes Inst., Norwich,

Eng., 1974-76; asst. prof. plant breeding Cornell U., Ithaca, N.Y., 1976-80; research scientist exptl. sta. E.I. DuPont de Nemours & Co., Inc., Wilmington, Del., 1980—. Author: Genetics of Higher Plants: Applications of Cell Culture, 1981. AID grantee, 1978-80. Mem. Genetics Soc. Am. Subspecialties: Plant genetics; Plant cell and tissue culture. Office: EI DuPont de Nemours & Co Exptl Sta Wilmington DE 19899

CHALLONER, DAVID REYNOLDS, university health affairs administrator; b. Appleton, Wis., Jan. 31, 1935; s. Reynolds R. and Marion (Below) C.; m. Jacklyn Anderson, Aug. 30, 1958; children: David H., Laura R., Britt D. B.S., Lawrence U., 1956; M.D., Harvard U., 1961. Diplomate: Am. Bd. Internal Medicine. Intern Presbyn. Hosp., N.Y.C., 1961-62; resident in internal medicine U. Wash., Seattle, 1965-67; prof. medicine Ind. U., Indpls., 1967-74; dean sch. medicine St. Louis U., 1974-82; v.p. health affairs U. Fla., Gainesville, 1982—; cons. Eli Lilly and Co., Indpls., 1968-80. Recipient alumni award Harvard Med. Sch., 1961; Dr. William Beaumont award AMA, 1982. Mem. Am. Fedn. Clin. Research (pres. 1975-76), Assn. Am. Physicians, Inst. Medicine (membership com. 1982-84). Clubs: Cosmos, Washington. Subspecialties: Endocrinology; Internal medicine. Office: U Fla PO Box J-14 JHMHC Gainesville FL 32610

CHALMERS, JOHN H., JR., educator; b. St. Paul, Minn., Mar. 5, 1940; s. John H. and Jane Amelia (Bedson) C. A.B., Stanford U., 1962; Ph.D., U. Calif.-San Diego, 1968. Postdoctoral fellow U. Wash.-Seattle, 1968-71; postdoctoral fellow U. Calif.-Berkeley, 1971-73; research fellow Merck Sharpe & Dohme Research Labs., Rahway, N.J., 1973-75; asst. prof. Baylor Coll. Medicine, Houston, 1976—. NIH Predoctoral and Postdoctoral fellowships, 1966-67, 69-71; NSF fellowship, 1963-67. Mem. AAAS, Genetics Soc. Am., Computer Music Assn., Loglan Inst., Sigma Xi. Subspecialties: Molecular biology; Biochemistry (biology). Current work: Molecular biology of tumor rejection and animal tumor viruses, biochemical genetics of yeast. Patentee in field. Office: Biosyne Corp 2210 Maroneal Suite 302 Houston TX 77030

CHALMERS, THOMAS CLARK, physician, ednl. and research adminstr.; b. Forest Hills, N.Y., Dec. 8, 1917; s. Thomas Clark and Elizabeth (Ducat) C.; m. Frances Crawford Talcott, Aug. 31, 1942; children—Elizabeth Ducat (Mrs. Daniel G. Wright), Frances Talcott, Thomas Clark, Richard Matthew. Student, Yale, 1936-39; M.D., Columbia, 1943. Diplomate: Am. Bd. Internal Medicine. Intern Presbyn. Hosp., N.Y.C., 1943-44; research fellow N.Y. U. Malaria Research Unit, Goldwater Meml. Hosp., N.Y.C., 1944-45; resident Harvard Med. Services of Boston City Hosp., 1945-47; asst. physician Thorndike Meml. Lab., 1947-53; chief med. services Lemuel Shattuck Hosp., Boston, 1955-68; asst. chief med. dir. for research and edn. VA, Washington, 1968-70; asso. dir. clin. care NIH, also dir. clin. center NIH, Bethesda, Md., 1970-73; pres. (Mt. Sinai Med. Center), 1973—; prof. medicine, dean Mt. Sinai Sch. Medicine, N.Y.C., 1973—; lectr. medicine Harvard; prof. medicine Tufts U., 1961-68, George Washington U., 1970-73; mem. ethics adv. bd., spl. cons. NIH, HHS, 1980. Contbr. numerous articles profl. jours. Bd. dirs. New Eng. Home for Little Wanderers, 1960-65; bd. regents Nat. Library Medicine, 1978-79. Served as capt., M.C. AUS, 1953-55. Mem. Am. Assn. Study Liver Diseases (pres. 1959), Am. Clin. and Climatol. Assn., A.C.P., Am. Fedn. Clin. Research, Am. Gastroent. Assn. (pres. 1969), Am. Soc. Clin. Investigation, Assn. Am. Physicians, N.Y. Acad. Medicine, Inst. Medicine of Nat. Acad. Scis., Eastern Gut Club. Subspecialty: Gastroenterology. Current work: Methodology of clinical trials. Office: Mt Sinai Med Center One Gustave L Levy Pl New York NY 10029

CHAMBERLAIN, CHARLES FRANKLIN, marine geologist, consultant; b. Columbus, Ohio, Sept. 7, 1946; s. George Victor and Margaret Lytle C.; m. Janet A. Mathers, Aug. 7, 1970; 1 dau., Circe Ann. B.S., St. Joseph's Coll., 1970; M.S., Boston Coll., 1976. Research assoc. Skidaway Inst. Oceanography, Savannah, Ga., 1973-78; instr. Calif. State U., Northridge, Calif., 1978-79; sr. oceanographer Interstate Electronic, Anaheim, Calif., 1979-80; marine geologist Ertec Western, Inc., Long Beach, Calif., 1980-81; mgr. offshore ops. Mesa, Inc., Northridge, Calif., 1981—; lectr. Marymount; Los Angeles. Contbr. articles in field to marine geologic jours. Mem. Marine Tech. Soc. (bd. dirs. 1982—), Soc. Exploration Geophysicists, Soc. Econ. Paleontologists and Mineralogists, Am. Geophys. Union, Sigma Xi. Subspecialties: Sedimentology; Offshore technology. Current work: Direct offshore geophysical data acquisition. Office: Mesa 2 Inc 4250 Pennsylvania Ave La Crescenta CA 91214

CHAMBERLAIN, JOSEPH WYAN, astronomer, educator; b. Boonville, Mo., Aug. 24, 1928; s. Gilbert Lee and Jessie (Wyan) C.; m. Marilyn Jean Roesler, Sept. 10, 1949; children: Joy Anne, David Wyan, Jeffrey Scott. A.B., U. Mo., 1948, A.M., 1949; M.S., U. Mich., 1951, Ph.D., 1952. Project sci. aurora and airglow USAF Cambridge Research Center, 1951-53; research asso. Yerkes Obs., Chgo., 1953-55, asst. prof., 1955-59, asso. prof., 1959-60, prof., 1961-62, asso. dir., 1960-62; asso. dir. planetary scis. div. of Kitt Peak Nat. Obs., 1962-70, astronomer, planetary scis. div., 1970-71; dir. Lunar Sci. Inst., Houston, 1971-73; prof. dept. space physics and astronomy Rice U., Houston, 1971—; Cons. Stanford Research Inst., NASA. Author: Physics of the Aurora and Airglow, 1961, Theory of Planetary Atmospheres, 1978; Editor: Revs. of Geophysics and Space Physics, 1974-80; editorial bd.: Planetary Space Sci. Recipient Warner prize Am. Astron. Soc., 1961; Alfred P. Sloan research fellow, 1961-63. Fellow Royal Astron. Soc. (fgn.), A.A.A.S., Am. Geophys. Union (councilor 1968-70); mem. Am. Astron. Soc. (councilor 1961-64, chmn. div. planetary scis. 1969-71), Am. Phys. Soc., Internat. Astron. Union, Internat. Union Geodesy Geophysics, Internat. Sci. Radio Union, NRC, Assembly Math. and Phys. Scis. (exec. com. 1973-78), Nat. Acad. Sci. (chmn. geophysics sect. 1972-75). Subspecialty: Planetary atmospheres. Office: Dept Space Physics and Astronomy Rice U Houston TX 77251

CHAMBERLAIN, OWEN, nuclear physicist; b. San Francisco, July 10, 1920; m., 1943 (div. 1978); 4 children; m. June Steingart, 1980. A.B. (Cramer fellow), Dartmouth Coll., 1941; Ph.D., U. Chgo., 1949. Instr. physics U. Calif., Berkeley, 1948-50, asst. prof., 1950-54, asso. prof., 1954-58, prof., 1958—; civilian physicist, Manhattan Dist., Berkeley, Los Alamos, 1942-46. Guggenheim fellow, 1957-58; Loeb lectr. at Harvard U., 1959; Recipient Nobel prize (with Emilio Segrè) for physics, for discovery anti-proton, 1959. Fellow Am. Phys. Soc., Am. Acad. Arts and Scis.; mem. Nat. Acad. Scis. Subspecialty: Nuclear physics. Address: Physics Dept U Calif Berkeley CA 94720

CHAMBERS, JAMES VERNON, food science educator, consultant; b. Pekin, Ill., Mar. 12, 1935; s. Hershel O. and Clela Belle (Stoops) C.; m. Shirley Ann Clarke, Sept. 1, 1957; 1 dau., Deborah Evelyn. B.Sc., Ohio State U., 1961, M.Sc. in Dairy Tech, 1966, Ph.D. in Food Sci, 1972. Corp. microbiologist Ross Labs., Columbus, Ohio, 1961-69; lab. dir. div. foods, dairy and drugs Ohio Dept. Agr., 1969-71; asst. prof. food sci. U. Wis.-River Falls, 1972-74; assoc. prof., extension specialist Purdue U., West Lafayette, Ind., 1974—; cons. indsl. wastewater mgmt. Contbr. articles to sci. jours.; co-author extension manual series on water and wastewater mgmt. Served in USN, 1953-58. Recipient award Nat. Ice Cream Retailers Assn., 1979, Linder Dairy Ctrs., 1979. Mem. Internat. Assn. Milk, Food and Environ. Sanitarians, Am. Dairy Sci. Assn., N.Y. Acad. Scis., Water Pollution Control Fedn., Am. Soc. Microbiology, Inst. Food Technologists. Subspecialties: Food science and technology; Water supply and wastewater treatment. Current work: Medical microbiology in mastitis; food by-product utilization; environmental microbiology. Office: Purdue U West Lafayette IN 47907

CHAMBERS, KENTON LEE, botanist, educator, researcher; b. Los Angeles, Sept. 27, 1929; s. Maynard Macy and Edna Georgia (Miller) C.; m. Henrietta Laing, June 21, 1958; children: Elaine Patricia, David Macy. A.B. with highest honors, Whittier Coll., 1950; Ph.D., Stanford U., 1955. NSF postdoctoral fellow UCLA, 1955-56; instr. botany Yale U., New Haven, 1956-58; asst. prof., 1958-60; assoc. prof. botany Oreg. State U., Corvallis, 1960-65, prof., 1965—, curator herbarium, 1960—; program dir. systematic biology NSF, Wahington, 1967-68. Contbr. articles to sci. jours. Mem. Bot. Soc. Am., Am. Soc. Plant Taxonomists, Soc. Study of Evolution, AAAS, Am. Inst. Biol. Scis., Calif. Bot. Soc., Soc. Systematic Zoology, Assn. Tropical Biology. Subspecialties: Evolutionary biology; Systematics. Current work: Evolutionary genetics and chromosome cytology of higher plants; quantitative DNA variation in natural populations of annual species of plants; polyploidy and hybridization in natural populations of annual plants. Home: 3220 NW Lynwood Circle Corvallis OR 97330 Office: Oreg State U Corvallis OR 97331

CHAMBERS, LAWRENCE PAUL, aerospace engineer; b. Fairmont, W.Va., Oct. 22, 1937; s. Merritt Edward and Elsie Naomi (Crow) C.; m. Carol Ann Erdman, Oct. 28, 1967; children: Kristin Lynn, Robert Jason. B.S. in Physics, W.Va. U., 1960; postgrad. in fin. mgmt, George Washington U., 1961-62. Group engr. Martin-Marietta Corp., Walt., 1960-65; aerospace engr. NASA Goddard Space Flight Ctr., Greenbelt, Md., 1965-68; aerospace mgr. NASA Hdqrs., Washington, 1968-74, project dir., 1975-77, program mgr. flight programs, 1978—. Bd. dirs. Community Assn., 1979-81, treas., 1971. Recipient Skylab Group Achievement cert. Office Manned Space Flight, 1975; Exceptional Performance award NASA, 1976; Cosmos Achievement award, 1975, 77, 81. Republican. Methodist. Subspecialties: Biomedical engineering; Space medicine. Current work: Manage the design and development of all human life sciences experiments to be flown on Shuttle/Spacelab missions. Office: Life Scis Div NASA Fed Office Bldg 10B 600 Independence Ave SW Washington DC 20546

CHAMBERS, WILLIAM ALBERT, physicist, electronics engineer; b. Claytonville, Ill., May 24, 1933; s. William O. and Doris M. (Kramer) C.; m. Caroline Elizabeth Fuller, May 6, 1960; children: Lawrence Alan, Robert Edward, Katherine Ann. B.S. in Elec. Engring, Valparaiso Tech. Inst., 1961, Purdue U., 1959, M.A., U. So. Calif., Los Angeles, 1966. Mem. tech. staff Hughes Aircraft Co., Culver City, Calif., 1959-68; supr. Electro Optics Bell Aerospace Textron, Buffalo, 1968-73, chief engr. optics tech., 1973—. Contbr. articles on high energy laser optics to profl. jours. Served with U.S. Army, 1953-55. Recipient Hughes Masters Fellowship award Hughes Aircraft Co., 1961. Mem. Optical Soc. Am., Sigma Pi Sigma, Pi Mu Epsilon. Club: Buffalo Turnverein. Current work: Design and analysis of resonators for high energy laser, development of optical diagnostic techniques for high energy chemical lasers. Patentee in field. Home: 593 Deerfield North Tonawanda NY 14120

CHAMBLESS, DIANNE L., clin. psychologist, educator; b. Montgomery, Ala., Feb. 26, 1948; d. Alexander Donnelly and Donna Lenora (Marrison) C. B.A., Newcomb Coll., 1969; M.A., Temple U., 1972, Ph.D., 1979. Lic. psychologist, Pa., Ga., D.C.; cert. Nat. Register Health Service Providers in Psychology. Psychotherapist Feminist Therapy Collective, Phila., 1972-78; research psychologist Temple U. Med. Sch., Phila., 1976-78, asst. prof., 1981-82, U. Ga., Athens, 1978-82; asst. prof. psychology Am. U., Washington, 1982—; Cons. research agoraphobia program Temple U. Med. Sch., 1980-81, 82—. Editor: Agoraphobia, 1982; editor: newsletter Feminist Behaviorist, 1982. Mem. Am. Psychol. Assn., Assn. for Advancement of Behavior Therapy, Assn. for Women in Psychology, Soc. for Psychotherapy Research. Subspecialty: Clinical psychology. Current work: Research and treatment of agoraphobia and anxiety disorders, human sexuality, and psychological problems of women. Home: Apt UT-02 4100 Massachusetts Ave NW Washington DC 20016 Office: Dept Psychology American U Massachusetts and Nebraska Aves NW Washington DC 20016

CHAMBLISS, GLENN HILTON, microbiologist; b. Jasper, Tex., Feb. 14, 1942; s. Steven Elbert and Rosalee (Allen) C.; m. Lois Diane Derouen, Aug. 14, 1965; 1 dau., Christine Noel. B.A. in Zoology, U. Tex.-Austin, 1965; M.A. in Microbiology, Miami U., Oxford, Ohio, 1967, Ph.D., U. Chgo., 1972. Jane Coffin Childs Meml. Fund for Med. Research postdoctoral fellow U. Paris, Orsay, France, 1972-73, Phillipe Found. postdoctoral fellow, 1973-74; asst. prof. dept. bacteriology U. Wis.-Madison, 1974-79, assoc. prof. bacteriology, 1979—. Co-editor: Spores VII, 1978, Ribosomes: Structure, Function and Genetics, 1980; contbr. chpts. to textbooks, articles to profl. jours. Bd. dirs. Madison Audubon Soc., 1979—. Recipient Orton K. Stark award microbiology dept. Miami U., 1967, Phi Sigma award Phi Sigma Soc., 1967. Mem. Am. Soc. Microbiology, N.Y. Acad. Scis., Sigma Xi. Unitarian. Subspecialty: Microbiology (medicine). Current work: Research in regulation of protein synthesis in bacterium bacillus subtilis, also enzymology of carbohydrate utilization by bacteria of genus bacillus. Home: 2205 Van Hise Ave Madison WI 53705 Office: 1550 Linden Dr Madison WI 53706

CHAMBLISS, JOE PRESTON, aerospace engineer; b. Kerrville, Tex., Nov. 5, 1947; s. Preston Ray and Marian Joyce (Ganz) C.; m. Marcia Jeanelle Jeanes, Aug 19, 1978. B.S., U. Tex.-Austin, 1970; M.S in Aero. Engring, Rice U., 1972, Rice U., 1975. Engr. McDonnell Douglas, St. Louis, 1970; engr. Lockheed Electronics, Houston, 1974-76, Rockwell Internat., 1976-79; sr. staff engr. McDonnell Douglas Tech. Services Co., Houston, 1979—. Contbr. articles to tech. jours. Mem. AIAA, L-5 Soc., Nat. Space Inst., Fedn. Am. Scientists, Am. Astron. Soc., Am. Phys. Soc., ACLU. Club: Rockwell Flying. Subspecialties: Theoretical and applied mechanics; Aerospace engineering and technology. Current work: Advanced life support system design and integration engineering. Home: 15030 St Cloud St Houston TX 77062 Office: McDonnell Douglas Tech Services Co 16811 Space Center Houston TX 77062

CHAMPE, SEWELL PRESTON, microbiologist, educator; b. Montgomery, W.Va., Nov. 24, 1932; s. Sewell J. and Janice M.; m. Gertrud G. Graubart, Aug. 3, 1959 (div. 1968); children: Mark, Peter; m. Pamela Chambers, June 28, 1969. B.S., M.I.T., 1954; Ph.D., Purdue U., 1959. Asst. prof. biology Purdue U., West Lafayette, Ind., 1962-66, assoc. prof. biology, 1966-69; prof. microbiology Rutgers U., New Brunswick, N.J., 1969—. Contbr. articles to sci. jours. NIH grantee, 1963—. Mem. Am. Soc. Microbiology, Am. Soc. Biol. Chemists. Subspecialties: Microbiology; Gene actions. Current work: Genetics of fungal development; gene expression during development. Home: 17 Beech Ln Edison NJ 08820 Office: Waksman Inst Microbiology Rutgers U New Brunswick NJ 08903

CHAMPLIN, KEITH S(CHAFFNER), electrical engineering educator; b. Mpls., Aug. 20, 1930; m., 1954; 2 children. B.S., U. Minn., 1954, M.S., 1955, Ph.D., 1958. From asst. prof. to assoc. prof. U. Minn., Mpls., 1958-66, prof. elec. engring., 1966—; Exchange prof. Sorbonne, Paris, 1963. Mem. IEEE, Am. Phys. Soc. Subspecialties: 3emiconductors; Microwave electronics. Address: 5437 Elliot Ave Minneapolis MN 55417

CHAN, ARTHUR WING KAY, research scientist, educator; b. Hong Kong, June 24, 1941; came to U.S., 1969, naturalized, 1976; s. Yut-Fat and Shin-Yee (Yung) C.; m. Shirley Pou, Feb. 27, 1967; 1 son, Alvin Mark. B.Sc. with 1st class honors, Australian Nat. U., 1966, Ph.D., 1969. Postdoctoral fellow chemistry dept. Washington U., St. Louis, 1969-71, postdoctoral research asso. pharmacology dept., 1971-73; research scientist III Research Inst. on Alcoholism, Buffalo, 1974-76, research scientist IV, 1976-79, research scientist V, 1979—; research asst. prof. pharmacology SUNY, Buffalo, 1974-81, research asso. prof., 1982—. Contbr. articles to profl. jours. Nat. Inst. Alcohol Abuse and Alcoholism research grantee, 1975-77; N.Y. State Health Research Council grantee, 1976-77, 78-80; Nat. Inst. Drug Abuse grantee, 1980-81. Mem. Am. Soc. for Pharmacology and Exptl. Therapeutics, Research Soc. on Alcoholism, Sigma Xi. Subspecialties: Pharmacology; Psychopharmacology. Current work: Biochemical and pharmacological effects of alcohol and other drugs, as well as their interactions. Office: 1021 Main St Buffalo NY 14203

CHAN, CHIU YEUNG, mathematics educator, researcher; b. Hong Kong, Feb. 28, 1941; came to U.S., 1969; s. Hak Tan and Ching Yuen (Lam) C.; m. Mui Lai Tania Lee, May 6, 1970; children: Andy Sung-Kin, Gary Sung-Hong. B.S.Gen., U. Hong Kong, 1964, B.S. Spl., 1965; M.S., U. Ottowa, 1967; Ph.D., U. Toronto, Ont., Can., 1969. Asst. prof. Fla. State U., Tallahassee, 1969-74, assoc. prof., 1974-81, prof., 1981-83; prof. math. U. Southwestern La., Lafayette, 1982—. Contbr. math. research papers to profl. publs. Mem. Am. Math. Soc., Soc. Indsl. and Applied Math., Am. Acad. Mechanics, Southeastern Atlantic Sect. So. Indsl. and Applied Math. Subspecialty: Partial differential equations. Current work: Free boundary problems; demography; Sturmian theory; nuclear reactor kinetics; heat conduction and radiation; integral equations; mathematical modeling; extremum principles. Home: 110 Charles Reed Ave Lafayette LA 70503 Office: Dept Math and Stats U Southwestern La Lafayette LA 70504

CHAN, DONALD PIN KWAN, orthopaedic surgery educator; b. Rangoon, Burma, Jan. 21, 1937; came to U.S., 1968; s. Charles Y.C. and Josephine (Golamco) C.; m. Dorothy S. Lau, July 31, 1966; children: Joanne, Elaine. M.B., B.S., U. Rangoon, 1960. Diplomate: Am. Bd. Orthopaedic Surgery. Intern U. Hong Kong, 1960-61, surg. trainee, 1961-68; resident in orthopedic surgery U. Vt. Hosps., 1968-71; Scoliosis Research Soc. traveling fellow, 1971-72; asst. prof. orthopedics U. Rochester, 1972-80, assoc. prof., 1980—, chief sect. spine surgery, 1983. Contbr. chpts. in books; designer spine instruments for surgery. Recipient Moire Topography award Easter Seals Soc., 1979, Phase Locked Interferometry award Scoliosis Research Soc., 1979, award for spinal cord injury model system HEW, 1981, 82, 83. Fellow ACS, Scoliosis Research Soc., Am. Acad. Orthopedic Surgeons; mem. Eastern Orthopaedic Assn., AMA. Roman Catholic. Subspecialty: Orthopedics. Current work: Spine surgery and research related to disorders of spine. Office: School of Medicine and Dentistry University of Rochester Rochester NY 14642

CHAN, IU-YAM, chemistry educator; b. Hong Kong, Apr. 1, 1940; U.S., 1962; s. Shu Kwan and Choi Gee (Chow) C.; m. Sophia R. Su, Aug. 18, 1973; 1 son: Raymond. B.Sc., Cheng Kung U., Tainan, Taiwan, 1961; Ph.D., U. Chgo., 1969. Research asst. U. Leiden, The Netherlands, 1968-71; asst. prof. Brandeis U., 1971-77, assoc. prof., 1977—; vis. scientist IBM Research Lab., Yorktown Heights, N.Y., 1977-78. Contbr. articles to sci. jours. Union Carbide Fellowship awardee, 1964; Gustavus F. Swift Fellowship awardee, 1965; William Rainey Harper Fellowship awardee, 1966; recipient numerous research grants. Mem. Am. Phys. Soc., Am. Chem. Soc., Sigma Xi. Subspecialty: Condensed matter physics. Current work: Utilization of the fluorescence line-narrowing technique in studying phonon-assisted energy transfer; interaction of an electronically excited molecule with phonons in molecular crystals; supersonic jet spectroscopy of large molecules. Home: 100 Hickory Rd Weston MA 02193 Office: Dept Chemistry Brandeis U 415 South St Waltham MA 02254

CHAN, KWAN MING, geology educator, consultant; b. Hong Kong, July 30, 1935; s. Cheuk Fai and Sau Ying C.; m. Karen Kung-Mei, July 9, 1965; children: Karl, Ken, Kim. B.Sc. with spl. honors, U. Hong Kong, 1960; Ph.D., U. Liverpool, Eng., 1966. Research officer Hong Kong Dept. Agr. and Fisheries, 1961-68; asst. scientist Woods Hole Oceanographic Instn., Mass., 1968-69; prof. Calif. State U.-Long Beach, 1969—; vis. scientist Commonwealth Sci. Indsl. Research Orgn., Sydney, Australia, 1961, Maritime Safety Agy., Tokyo, 1967; vis. prof. Nat. Taiwan U., Taipei, 1975. Pres. Huntington Beach (Calif.) Chinese Sch., 1977; treas. Chinese Am. Faculty Assn., 1979. Rotary Club scholar, 1959; UNESCO fellow, Paris, 1967. Mem. AM. Soc. Limnology and Oceanography, Am. Chem. Soc., Royal Inst. Chemistry (assoc.). Democrat. Subspecialties: Oceanography; Analytical chemistry. Current work: Oceanography, chemistry, education, Chinese culture. Home: Box 605 Westminster CA 92684 Office: 1250 Bellflower Blvd Long Beach CA 90840

CHAN, PAK HOO, biochemist, researcher; b. Canton, Kwangtong, China, Apr. 11, 1942; came to U.S., 1968, naturalized, 1979; s. Sik-Kee and Hong-King (Leung) C.; m. Helen Shang Chu, Mar. 17, 1974; children: Tammy Yuen-Wah, Olivia Yee-Wah, Goldie Yan-Wah. B.S., Chinese U., Hong Kong, 1964; M.A., UCLA, 1970, Ph.D., 1972. Postdoctoral fellow U. Calif.-Berkeley, 1972-74, Stanford U., 1974-75; asst. research biochemist U. Calif.-San Francisco, 1975-80, assoc. research biochemist, 1980—; cons. clin. fellow, 1975—. Author: Symposium Brain Edema, 1982, Symposium Neural Membranes, 1982. Recipient Yale-New Haven Scholarship award, 1964; NIH grantee, 1978—. Mem. Am. Soc. Neurochemistry, Soc. Neurosci., Internat. Soc. Neurochemistry, N.Y. Acad. Scis., Am. Fedn. Clin. Research. Subspecialties: Neurochemistry; Biochemistry (medicine). Current work: Biochemistry and molecular aspects of neurological disorders, e.g., brain edema, stroke; biochemical mechanisms. Home: 635 Jackson St Albany CA 94706 Office: U Calif Sch Medicine San Francisco CA 94143

CHAN, PHILLIP C., biochemistry educator; b. Amoy, Fujian, China, June 14, 1928; came to U.S., 1949, naturalized, 1965; s. Hwa-Liong and Kim-Luan (Tan) C.; m. Joan Mildred Pon, June 2, 1965; 1 dau., Melinda. B.S., Monmouth Coll., 1952; M.A., Columbia U., 1953, Ph.D., 1957. Postdoctoral fellow Johns Hopkins U., Balt., 1957-59; med. research fellow Max-Planck Inst. for Cell Chemistry, Munich, W.Ger., 1959-60; from asst. prof. to prof. biochemistry SUNY Downstate Med. Ctr., Bklyn., 1960—. Contbr. articles to profl. jours. Mem. Am. Soc. Biol. Chemists, Am. Chem. Soc., N.Y. Acad. Scis. Subspecialty: Biochemistry (biology). Current work: Biochemical reactions involving reactive oxygen species including oxygen free radicals, excited states of molecular oxygen, ozone, peroxides and metal oxygen complexes. Office: SUNY Downstate Med Ctr 450 Clarkson Ave Brooklyn NY 11203

CHAN, PO CHUEN, cell biologist, cancer research scientist; b. Ichong, Hupeh, China, May 13, 1935; came to U.S., 1960, naturalized, 1975; s. Tin Wa and Pui Wa (Ho) C.; m. Lillian Mak, Oct. 16, 1961; children: Yola, Vella. B.A., Internat. Christian U., Tokyo, 1960; M.A., Columbia U., 1963; Ph.D., N.Y.U., 1967. Research assoc. Sloan-Kettering Inst. Cancer Research, N.Y.C., 1967-70; assoc. mem. Am. Health Found., Valhalla, N.Y., 1971-78; cancer research scientist IV Roswell Park Meml. Inst., Buffalo, 1978—. Contbr. articles to profl. jours. Nat. Cancer Inst. grantee, 1980-81. Mem. Am. Assn. Cancer Research, N.Y. Acad. Sci., Am. Physiol. Soc., Tissue Culture Assn., Am. Soc. Zoologists. Subspecialty: Cancer research (medicine). Current work: Mammary carcinogenesis, relationship between dietary fat and mammary carcinogenesis. Home: 323 Patrice Terr Williamsville NY 14221

CHAN, SEK KWAN, research physicist; b. Canton, Kwangtung, China; s. Yuk Lun and Shun (Ying-Ho) C.; m. Ching Ping Mark, May 18, 1971; 1 dau., Nancy. B.Engring., U. New South Wales, Australia, 1966; M.Engring. Sci., U. Sydney, 1968; Ph.D., U. Toronto, 1973. Research physicist C-L-L Inc., McMasterville, Que., Can., 1973-78, 81—; sr. research scientist I.C.I., Stevenston, Scotland, 1979-81. Contbr. articles to books and jours. in field. Utias fellow U. Toronto, 1968; M.H. Beatty fellow, 1969; NRC Can. scholar, 1970; grantee, 1976. Mem. AIAA, Can. Aeros. and Space Inst. Subspecialties: Explosives; Condensed matter physics. Current work: Shock and detonation wave phenomena, shock tube design and application, explosive sensitivity: shock, heat, impact, friction. Explosives ignition, reaction kinetics, transition to detonation. Explosives hazard prediction and evaluation. Office: C-I-L Explosives Research Lab McMasterville PQ Canada J3G 1T9

CHAN, SHIH HUNG, mech. engr., cons. and researcher; b. Changhwa, Taiwan, Nov. 8, 1943; came to U.S., 1964, naturalized, 1976; s. Chan Ping and Liao (Fu) Zon; m. Shirley Shih, Apr. 15, 1944; children: Bryan, Erick. M.S. in Mech. Engring, U. N.H., 1966; Ph.D. (Ehrman fellow), U. Calif., Berkeley, 1969. Registered profl. engr., Wis. Asst. to assoc. prof. N.Y.U., 1969-73; assoc. prof. Poly. Inst. N.Y., 1973-74; mem. research staff Argonne Nat. Lab., 1974-75; prof., chmn. dept. mech. engring. U. Wis.-Milw., 1975—; cons. Contbr. heat transfer and nuclear reactor safety articles to tech. jours. NSF grantee, 1970-82. Mem. Am. Nuclear Soc., ASME. Subspecialties: Fluid mechanics; Nuclear fission. Current work: Radiative heat transfer, thermal hydraulic of nuclear reactor safety, fouling and scale in heat exchange. Home: 3416 W Meadowview Ct Mequon WI 53092 Office: Dept Mech Engring U Wis PO Box 784 Wilwaukee WI 53201

CHAN, TAK HANG, educator, chemist; b. Hong Kong, June 28, 1941; s. Ka King and Ling Yee (Yick) C.; m. Christina W.Y. Hui, Sept. 6, 1969; children: Juanita Y., Cynthia S. B.A., U. Toronto, 1962; M.A., Princeton U., 1963, Ph.D., 1965. Research assoc. Harvard U., 1965-66; asst. prof. McGill U., Montreal, Que., Can., 1966-71, assoc. prof., 1971-77, prof. chemistry, 1978—. Contbr. articles to profl. jours. Recipient Merck, Sharp and Dohme award Chem. Inst. Can., 1982. Mem. Chem. Inst. Can., Am. Chem. Soc., Royal Soc. Chemistry. Subspecialties: Organic chemistry; Synthetic chemistry. Current work: synthesis of organic compounds, new organic reactions, new reagents, silicon chemistry. Office: 801 Sherbrooke St West Montreal PQ Canada H3A 2K6

CHAN, W. Y., pharmacologist, educator; b. Shanghai, China, Dec. 1, 1932; came to U.S., 1952, naturalized, 1968; m. Beatrice Ho Chan, June 11, 1961; children: Mina, Jennifer. B.A., U. Wis., Madison, 1956; Ph.D. in Pharmacology, Columbia, U., 1961. Research asso. to asst. prof. biochemistry Cornell U. Med. Coll., N.Y.C., 1960-67, asst. prof. to asso. prof. pharmacology, 1966-76, prof., 1976—, acting chmn., 1982—; mem. basic pharmacology adv. com. Pharm. Mfrs. Assn. Found., 1973-80; mem. study sect. NIH, 1977, cons., 1981. Contbr. articles to profl. jours. Recipient NIH research career devel. award, 1968-73; Irma T. Hirschl Career Scientist, 1973-77; NIH grantee, 1965—. Mem. Am. Soc. Pharmacology and Exptl. Therapeutics, Soc. Study of Reprodn., Soc. Exptl. Biology and Medicine, Harvey Soc., N.Y. Acad. Sci., AAAS. Subspecialties: Pharmacology; Endocrinology. Current work: Pharmacology of neurohypophys hormones and polypeptides; uterine and renal actions of oxytocin and prostaglandins; pathophysiology and pharmacology of dysmenorrhea. Office: Cornell U Med Coll New York NY 10021

CHAN, WAI-YEE, biochemical geneticist, laboratory administratory, educator; b. Kwangzhou, China, Apr. 28, 1950; came to U.S., 1974; s. Kui and Fung-Hing (Wong) C.; m. May-Fong Sheung, Sept. 3, 1976; children: Connie Hai-Yee. B.Sc. (1st hons.), Chinese Univ. Hong Kong, 1974; Ph.D., U. Fla., 1977. Teaching asst U. Fla., 1974-77; fellow U. Okla., Oklahoma City, 1977-78, research assoc., 1978-79, asst. prof., 1979-82; assoc. prof., 1982—; staff affiliate Okla. Children's Meml. Hosp., Oklahoma City, 1979—; vis. scientist U. Wash., Seattle, 1982; sci. dir. trace metals lab. State Okla. Teaching Hosp., Oklahoma City, 1982—; cons. VA Med. Ctr., Oklahoma City. Editor: Metabolism of Trace Metals in Man: Developmental Biology and Genetic Implications, Volume I and II, 1983; consulting editor: Jour of Am. Coll. Nutrtion, 1982—; contbr. articles to profl. jours. NATO fellow, 1979; recipient Tak Shing prize for biochemistry Chinese U. Hong Kong, 1973; NIH grantee, 1983. Mem. Am. Soc. Biol. Chemists, Soc. Pediatric Research, Am. Inst. Nutrition, N.Y. Acad. Sci., Biochemical Soc. Eng., Am. Soc. Human Genetics, Nutrition Soc. Eng. Lodge: Rotary. Subspecialties: Genetics and genetic engineering (medicine); Nutrition (medicine). Current work: Molecular biology of genetic diseases; cellular polyamine metabolism and biochemical nutrition and metabolism of trace metals. Home: 8725 Raven Ave Oklahoma City OK 73132 Office: University of Oklahoma Health Sciences Center Oklahoma City OK 73190

CHANANA, ARJUN DEV, biomedical researcher; b. LyallPur, Punjab, India, Nov. 6, 1930; came to U.S., 1963, naturalized, 1975; s. Bhagat R. and Vidya V. (Sachdev) C.; m. Judith Taylor, Sept. 9, 1963; children: Arun Paul, Nina Lee. M.D., Swai Man Singh Med. Coll., Jaipur, Rajasthan, 1955. Intern Irwin Hosp., New Delhi, India, 1955-57; resident in surgery Bolton (Eng.) Dist. Gen. Hosp., 1960-63; with Brookhaven Nat. Lab., Upton, N.Y., 1963—, sr. scientist, 1979—; cons. researcher Nassau County Med. Ctr., East Meadow, N.Y., 1974—; guest prof. immunopathology Med. Faculty Senate, U. Berne, Switzerland, 1971-72; Mem. NIH study sect. U.S. Dept. HHS, Bethesda, Md., 1974-79. Contbr. over 100 sci. articles to profl. publs. Fellow Royal Coll. Surgeons London, Royal Coll. Surgeons Edinburgh; mem. Transplantation Soc., Am. Soc. Hematology, Am. Assn. Pathologists, Soc. Exptl. Biology and Medicine, AAAS, Radiation Research Soc. Subspecialties: Environmental toxicology; Hematology. Current work: Pulmonary immunobiology; hemopoletic cell proliferation; transplantation immunology. Office: Brookhaven Nat Lab Bldg 490 Upton NY 11973

CHANCE, BRITTON, educator; b. Wilkes Barre, Pa., July 24, 1913; s. Edwin M. and Eleanor (Kent) C.; m. Jane Earle, Mar. 4, 1938 (div.); children: Eleanor, Britton, Jan, Peter; m. Lilian Streeter Lucas, Nov. 1956; children: Margaret, Lilian, Benjamin, Samuel; stepchildren—Ann Lucas, Gerald B. Lucas, A. Brooke Lucas, William C. Lucas. B.S. and M.S., U. Pa., 1936, Ph.D. (E.R. Johnson Found. fellow), 1940, U. Cambridge, 1942, D.Sc., 1952; M.D. (hon.), Karolinska Inst., Stockholm, 1962, Semmel Weis U., Budapest, 1976; D.Sc., Med. Coll. Ohio, 1974, Hahnemann Coll. and Hosp., 1977. Asst. prof. biophysics U. Pa., 1940-48, prof., chmn., 1949—, acting dir., 1940-41, dir., 1949—, Eldridge Reeves Johnson prof. biophysics and phys. biochemistry, 1949, 77—; dir. Inst. Structural and Functional Studies, 1982—; staff Mass. Inst. Tech., 1941-46; Cons. NSF, 1952-55; mem. Pres.'s Sci. Adv. Com., 1959-60; mem. advisory council Nat. Inst. Alcohol Abuse and Alcoholism, 1971-75; mem. molecular control working group Nat. Cancer Inst., 1973—. Author: (with F.C. Williams, V. Hughes, E.F. McNichol, David Sayre) Waveforms, 1949, Electronic Time Measurements, (with R.I. Hulsizer, E.F. McNichol, F.C. Williams), 1949, Energy-linked Functions of Mitochondria, 1964, (with Q.H. Gibson, R. Eisenhardt, K.K. Lonberg-Holm) Rapid Mixing and Sampling Techniques in Biochemistry, 1964, (with R.W. Estabrook, J.R. Williamson) Control of Energy Metabolism, 1965, (with R.W. Estabrook, T. Yonetani) Hemes and Hemoproteins, 1966, (with others) Probes of Structure and Function of Macromolecules and Enzymes, 1971, Alcohol and Aldehyde, Vol. I, 1974, II, III, 1977, Tunneling in Biological System, 1979; rev. articles Advances in Enzymology, Vo. 12, 1951, Vol. 17, 1956, Ann. Rev. of Biochemistry, 1952, 70, 76, The Enzymes, Vol. II. Part 1, 1952, Vol. XIII, 1976, Ann. Rev. Plant Physiology, 1958, 68; Bd. editors: Physiol. Revs, 1951-54, FEBS Letters, 1973-75, BBA Reviews, 1972—, Photobiochemistry and Photobiophysics, 1979—; Contbr.: articles to Am., Brit., Swedish, German and Japanese Jours. Presdl. lectr. U. Pa., 1975; Julius L. Jackson Meml. lectr. Wayne State U., 1976; Da Costa oration Phila. County Med. Coll., 1976; Recipient Paul Lewis award for enzyme chemistry, 1950; Pres.'s Certificate of Merit for services, 1941-45, as staff mem. Radiation Lab. of M.I.T., 1950; Guggenheim fellow, Stockholm, 1946-48; Harvey lectr., 1954; Phillips lectr., 1955, 65; Pepper lectr., 1957; Exchange scholar to, USSR, 1963; Genootschapps medal Dutch Acad. Scis., 1965; Heineken medal, 1970; Keilin medal Brit. Biochem. Soc., 1966; Harrison Howe award, 1966; Franklin medal, 1966; Overseas fellow Churchill Coll., 1966; Herter lectr. N.Y. U., 1968; Pa. award for excellence in life scis., 1968; Nichols award N.Y. sect. Am. Chem. Soc., 1970; Phila. sect. award, 1969; Redfearn lectr., 1970; Gairdner award, 1972; Post-Congress Festschrift, Stockholm, 1974; Semmelweis medal, 1974; Nat. medal Sci., 1974. Fellow Am. Phys. Soc., IEEE (Morlock award 1961), AAAS, Am. Inst. Chemists; mem. Internat. Union Pure and Applied Biophysics (pres. 1972-75), Chem. Soc., Royal Soc. Arts, Biochem. Soc. Eng., Am. Soc. Biol. Chemists, Am. Philos. Soc., Am. Acad. Arts and Sci., Nat. Acad. Sci., Am. Physiol. Soc., Soc. Gen. Physiologists (council 1957-60), Am. Inst. Physics, Soc. for Neurosci., Biophys. Soc. (council 1959-62), Swedish Biochem. Soc., Royal Swedish Acad. Scis., Royal Acad. Arts and Scis., Sweden, Bavarian Acad. Scis., Acad. Leopoldina DDR, Max-Planck Gesellschaft fü Forerung der Wissenschaften (fgn.), Argentine Nat. Acad. Sci., Royal Soc. London (fgn.), Harvey Soc., Sigma Xi, Tau Beta Pi. Clubs: Corinthian Yacht (Phila.); St. Anthony. Subspecialties: Biophysics (biology); Nuclear magnetic resonance (biotechnology). Current work: Structure of metalloenzymes and proteins by x-ray techniques; reaction kinetics of unstable intermediates at low temperatures; tunneling phenomena; 31-P NMR of animal models and humans. Holder numerous patents on automatic steering devices, also spectrophotometric devices, radar circuitry. Gold medal winner (yachting) 1952 Olympics. Home: 4014 Pine St Philadelphia PA 19104

CHANCE, RONALD RICHARD, chemicals and electronics research executive, physical chemist; b. Memphis, July 24, 1947; s. Roy Lauerne and Betty Jean (Jackson) C.; m. Judy Smith, Aug. 21, 1967; children: Kristin Carol, Ronald Richard. B.S. in Chemistry, Delta State U., Cleveland, Miss., 1970; Ph.D., Dartmouth Coll., 1974. Research mgr. Allied Corp., Morristown, N.J., 1940—. Contbr. numerous articles to profl. jours. Mem. Am. Phys. Soc., Am. Chem. Soc. Subspecialties: Polymers; Physical chemistry. Patentee in field. Office: Allied Corp Columbia Rd and Park Ave Morristown NJ 07960

CHANDLER, JERRY LEROY, biochemical geneticist, toxicologist, sci. policy advisor; b. Little Falls, Minn., Sept. 14, 1940; s. Bert Emery and Blanche Anastasia (Drellack) C.; m. Donna Kay Fitzgerald, June 7, 1969; children: Bert, Jerry-David. A.A., Brainerd Jr. Coll., 1960; B.S., Okla. State U., 1963, Ph.D. in Biochemistry, 1968. Instr. biochemistry dept. Okla. State U., Stillwater, 1969, research assoc., 1973-74; chemistry group leader Central Lab. for Mutagenicity, Freiburg, W.Ger., 1969-72; scientist Nat. Inst. for Occupational Safety and Health, Rockville, Md., 1975-76, sr. scientist, 1976-82, sr. sci. advisor office of dir., 1982—, criteria documentation sect. chief, 1977-79, sci. advisor div. criteria documentation, 1980; faculty Found. for Advanced Edn. in Scis., NIH; lectr. in field. Contbr. articles in field to profl. publs. Chmn. Cub Scouts troop Nat. Capital Area Council Boy Scouts Am. Served as commd. officer USPHS, 1975—. NSF research grantee, 1973. Mem. Am. Indsl. Hygiene Assn., Am. Statis. Assn., Inst. for Theol. Encounter with Sci. and Tech., Soc. for Risk Analysis, USPHS Commd. Officers Assn. Roman Catholic. Club: Langley (Va.). Subspecialties: Environmental toxicology; Genetics and genetic engineering (medicine). Current work: Quantitative molecular origins of health and disease; government health policy science advisor, quantitative risk assessment, molecular dynamics of health and disease processes, occupational health standards. Home: 7412 Churchill Rd McLean VA 22101 Office: 5600 Fishers Ln Rockville MD 20857

CHANDLER, JOHN CHRISTOPHER, nuclear engineer; b. Gulfport, Miss., Oct. 31, 1946; s. Thomas Alfred and Jo Beth (Majure) C., Jr.; m. Lynne Blair Bratcher, Aug. 1, 1967 (div. 1976); 1 dau., Gwendolyn Amy; m. Jean Marie Vaz, June 29, 1980. B.S., Miss. State U., 1968; M.S. in Nuclear Engring, U. Wash., 1980. Nuclear engr. Ingalls Nuclear Ships Co., Pascagoula, Miss., 1968; advanced engr. Westinghouse Hanford Co., Richland, Wash., 1973-79; nuclear engr. Middle South Services, New Orleans, 1979-80; sr. engr. Exxon Nuclear Co., Inc., Richland, 1980—. Contbr. articles to profl. jours. Advisor CAP, Richland, 1973-76. Served to capt. USAF, 1968-73. Mem. Am. Nuclear Soc. Subspecialties: Nuclear engineering; Nuclear fission. Current work: Determination of safety requirements for nuclear reactor operations; establishment of operating limits to assure reactor safety; analytical determination of accident results. Office: Exxon Nuclear Co Inc 2101 Horn Rapids Rd Richland WA 99352

CHANDRA, ASHOK KUMAR, computer scientist; b. Allahabad, India, July 30, 1948; s. Harish and Sushila C.; m. Mala, Sept. 17, 1974; children: Ankur, Anuj. B.Tech., Indian Inst. Tech., 1969; M.S., U. Calif.-Berkeley, 1970; Ph.D., Stanford U., 1973. Mem. research staff IBM Thomas J. Watson Research Center, Yorktown Heights, N.Y., 1973-83, mgr. theoretical computer sci., 1981-83; tech. adv. office of v.p. and chief scientist IBM, Armonk, N.Y., 1983—. Editor: Jour. Computing Soc. Indsl. and Applied Math, 1982; contbr. articles to profl. jours. Recipient Pres.'s Gold medal Indian Inst. Tech., 1969; IBM Outstanding Innovation award, 1980; Invention Achievement award, 1977, 81. Mem. ACM, Soc. Indsl. and Applied Math. Subspecialties: Theoretical computer science; Algorithms. Current work: Theoretical computer science, VLSI testing, magnetic bubbles. Patentee magnetic bubble tech. (2). Office: PO Box 218 Yorktown Heights NY 10598

CHANDRA, RAMESH, med. physicist, educator; b. Nakur, U.P., India, June 9, 1942; came to U.S., 1963, naturalized, 1982; s. Brij Bhushan and Prakash Vati (Goel) Gupta; m. Mithilesh Aron, Aug. 19, 1966; children: Anurag, Ritu. B.Sc., Agra U. India, 1959; M.Sc., Allahabad U. India, 1961; Ph.D., Boston U., 1968. Asst. prof. dept. radiology N.Y.U. Med. Center, N.Y.C., 1970-74, assoc. prof., 1974-77, prof., 1977—. Author: Introductory Physics of Nuclear Medicine, 1976; contbr. articles to profl. jours. Mem. Soc. Nuclear Medicine, Am. Assn. Physicists in Medicine, Health Physics Soc. Hindu. Subspecialties: Nuclear medicine; Nuclear medicine. Current work: Application of radiation in medicine. Home: 41 Pine St Ardsley NY 10502 Office: 560 1st Ave Dept Radiology New York NY 10016

CHANDRA, SATISH, cancer biophysicist; b. Ghazipur, India, May 6, 1927; came to U.S., 1958, naturalized, 1973; s. Shyamlal and Lilawati (Srivastava) Vidyarthi; m. Shimmi P. Chandra, Mar. 9, 1953; children: Arun, Piyush, Arti. B.S., U. Allahabad, India, 1947, M.S., 1949; Ph.D., U. Toronto, 1958. Lectr. physics U. Allahabad, India, 1950-55; research fellow Ont. Cancer Inst., Toronto, 1957-58, Sloan Kettering Inst., N.Y.C., 1958-61; sr. research assoc. U. Buffalo, 1961-63; asst. prof. biophysics Post Grad. Inst. of Med. Edn. Research, Chandigarh, India, 1963-65; research biologist John L. Smith Meml. Center for Cancer, Maywood, N.J., 1966-71; viral oncologist Mercy Hosp., Chgo., 1971-75; assoc. prof. microbiology Abraham Lincoln Sch. Medicine, Chgo., 1973-78; cancer biophysicist in therapeutic radiology VA Hosp., Hines, 1976—; prof. anatomy Loyola U. Stritch Sch. Medicine, 1978—; cons. in field of hyperthermia; cons. UN Devel. program, 1982. Bd. dirs. Am. Cancer Soc., Bergen County, N.J., 1968-71. McGee-Gilchrist Fund fellow U. Toronto, 1956-57; Damon Runyon Fund for Cancer fellow, 1958-61; recipient quality award VA, 1978. Mem. Assn. of Indians in Am., Am. Assn. for Cancer Research, AAAS, Radiation Research Soc. Subspecialties: Cancer research (medicine); Cell biology (medicine). Current work: Cancer research, radiation biology, hyperthermia. Office: VA Edward Hines Jr Hosp Hines IL 60141

CHANDRASEKHAR, SUBRAHMANYAN, theoretical astrophysicist; b. Lahore, India, Oct. 19, 1910; came to U.S., 1936, naturalized, 1953; m. Lalitha, Madras, India, Sept. 1936. M.A., Presidency Coll., Madras, 1930; Ph.D., Trinity Coll., Cambridge, 1933, Sc.D., 1942; Sc.D., U. Mysore, India, 1961, Northwestern U., 1962, U. Newcastle Upon Tyne, Eng., 1965, Ind. Inst. Tech., 1966, U. Mich., 1967, U. Liege, Belgium, 1967, Oxford (Eng.) U., 1972, U. Delhi, 1973, Carleton U., Can., 1978, Harvard U., 1979. Govt. India scholar in theoretical physics Cambridge, 1930-34; fellow Trinity Coll., Cambridge, 1933-37; research asso. Yerkes Obs., Williams Bay and U. Chgo., 1937, asst. prof., 1938-41, assoc. prof., 1942-43, prof., 1944-47, Disting. Service prof., 1947-52, Morton D. Hull Disting. Service prof., 1952—; Nehru Meml. lectr., Padma Vibhushan, India, 1968. Author: An Introduction to the Study of Stellar Structure, 1939, Principles of Stellar Dynamics, 1942, Radiative Transfer, 1950, Hydrodynamic and Hydromagnetic Stability, 1961, Ellipsoidal Figures of Equilibrium, 1969, The Mathematical Theory of Black Holes, 1983; Mng. editor: The Astrophysical Jour, 1952-71; Contbr. various sci. periodicals. Recipient Bruce medal Astron. Soc. Pacific, 1952, gold medal Royal Astron. Soc., London, 1953; Rumford medal Am. Acad. Arts and Scis., 1957; Nat. Medal of Sci., 1966, Nobel prize, 1983. Fellow Royal Soc. (London) (Royal medal 1962); mem. Nat. Acad. Scis. (Henry Draper medal 1971), Am. Phys. Soc. (Dannie Heineman prize 1974), Am. Philos. Soc., Cambridge Philos. Soc., Am. Astron. Soc., Royal Astron. Soc. Club: Quadrangle (U. Chgo.). Subspecialties: General relativity; Theoretical astrophysics. Address: Lab for Astrophysics and Space Research 933 E 56th St Chicago IL 60637

CHANEY, WILLIAM REYNOLDS, forestry educator; b. McAllen, Tex., Dec. 2, 1941; s. Harold Glen and Mary (Reynolds) C.; m. Joann Judith Simon, Aug. 24, 1966; children: Brandon Chaney, Carey Chaney. B.S., Tex. A&M U., 1964; Ph.D., U. Wis., 1969. Research assoc. U. Wis.-Madison, 1969-70; prof. dept. forestry Purdue U., West Lafayette, Ind., 1970—. Mem. Am. Soc. Plant Physiologists. Subspecialties: Biomass (agriculture); Plant physiology (agriculture). Current work: The Physiology of mycorrhizae and applications in forestry. Office: Dept Forestry and Natural Resources Purdue U West Lafayette IN 47906

CHANG, ALBERT FUWU, computing analyst, educator; b. Pingtung, Taiwan, Dec. 30, 1934; came to U.S., 1960; s. Ming and Fei-chi (Lin) C.; m. Joan R. Yang, Dec. 30, 1964. B.S., Nat. Taiwan U., 1957; M.S., U. Houston, 1963, Calif. Inst. Tech., 1964; Ph.D., U. So. Calif., 1979. Mem. profl. staff Calif. Inst. Tech., Pasadena, 1965—; cons. Dr. Maxwell C. Cheung & Assocs., Irvine, Calif., 1977—. Mem. Soc. Indsl. and Applied Math., Assn. Computing Machinery. Subspecialties: Mathematical software; Numerical analysis. Current work: Mathematical software development, microcomputer application, numerical analysis. Office: California Institute Technology 1201 E California Blvd Pasadena CA 91125

CHANG, B(YUNG) JIN, optical systems engring. co. exec., elec. engr.; b. Danyang, Korea, Sept. 26, 1941; came to U.S., 1968, naturalized, 1974; s. Yung S. and Ahzie (Chon) C.; m. Sharon O. Hong, Dec. 27, 1969; children: Jane Y., Michael. Ph.D. in Elec. Engring, U. Mich., 1974. Research engr. ERIM, Ann Arbor, Mich., 1971-79; v.p., chief scientist Kaiser Optical Systems, Inc. subs. Kaiser Aerospace & Electronics Corp., Ann Arbor, 1979—. Contbr. articles to profl. jours. Mem. IEEE, Optical Soc. Am., Soc. Info. Display, Soc. Photo-optical Instrumentation Engrs. Methodist. Lodge: Ann Arbor Rotary. Subspecialties: Holography; Optical signal processing. Current work: Holographic optical elements, dichromated gelatin, interferometry, optical systems design, optical signal processing. Patentee in field. Home: 1495 Folkstone Ct Ann Arbor MI 48105 Office: 6087 Jackson Rd Ann Arbor MI 48106

CHANG, CHI KWONG, chemistry educator; b. Nanking, China, Dec. 25, 1947; m. Bonnie H. Chang, Dec., 1971; children: Timothy, Daniel. B.Sc., Fu Jen Cath. U., Taipei, Taiwan, 1969; Ph.D., U. Calif.-San Diego, 1973. Research assoc. U. Calif.-San Diego, La Jolla, 1974; postdoctoral fellow U. B.C., Vancouver, 1975-76; asst. prof. chemistry Mich. State U., East Lansing, 1976-79, assoc. prof., 1979-82, prof., 1982—; collaborating scientist Brookhaven Nat. Lab., Upton, L.I., N.Y., 1977—. Contbr. articles to profl. jours. Sloan fellow, 1980-84; recipient Camille and Henry Dreyfus Tchr.-Scholar award, 1981—. Mem. Am. Chem. Soc. Subspecialties: Organic chemistry. Current work: Synthesis of porphyrins and macrocyclic ligans, biological oxygen binding and activation, models of metalloenzymes, photosynthesis, electrocatalysis. Home: 660 Tarleton Ave East Lansing MI 48823 Office: Dept Chemistry Michigan State U East Lansing MI 48824

CHANG, CHING-JER, educator, educator, researcher; b. Hsin-Chu, Taiwan, Oct. 17, 1942; came to U.S., 1968, naturalized, 1980; s. Ting-Ian and A-wei (Lai) C.; m. Shu-fang Kuo Chang, Dec. 25, 1977; children: son Philip S-s, Sylvia W-y. Ph.D., Ind. U., 1972. Asst. prof. medicinal chemistry Purdue U., 1973-78, assoc. prof., 1978—. Contbr. articles to profl. jours. NIH grantee, 1973-82; Ind. Elks grantee, 1973-80; Am. Cancer Soc. grantee, 1979-80. Mem. Am. Chem. Soc., Chem. Soc. London, Am. Assn. Cancer Research, Phytochem. Soc. N.Am., Am. Soc. Pharmacognosy, Am. Pharm. Assn. Buddhist. Subspecialties: Medicinal chemistry; Phytochemistry. Current work: Chemical modifications of nucleic acids; modes of action of bioactive compounds; isolation, structural determination and biosynthesis of natural products; biomedical spectroscopy. Home: 217 Cedar Hollow Ct West Lafayette IN 47906 Office: Dept Med Chemistry Purdue U West Lafayette IN 47907

CHANG, ERNEST SUN-MEI, endocrinologist, educator; b. Berkeley, Calif., Dec. 7, 1950; s. Shu Chi and Helen (Fong) C. A.B., U. Calif., Berkeley, 1973; Ph.D., UCLA, 1978. Postdoctoral fellow dept. biochemistry U. Chgo., 1978; asst. prof. animal sci. U. Calif., Davis, 1978—; Asst. prof. Bodega Marine Lab., Bodega Bay, Calif., 1978—. Am. Cancer Soc. grantee, 1979-82; Calif. Sea Grant Coll. Program grantee, 1981-84. Mem. AAAS, Am. Soc. Zoologists, Tissue Culture Assn., World Mariculture Soc., Crustacean Soc. Subspecialties: Endocrinology; Developmental biology. Current work: Effects of hormones on crustacean and insect development. Office: Bodega Marine Lab PO Box 247 Bodega Bay CA 94923

CHANG, JAE CHAN, hematologist, oncologist; b. Chong An, Korea, Aug. 29, 1941; came to U.S., 1965, naturalized, 1976; s. Tae Whan and Kap Hee (Lee) C.; m. Sue Young Chung, Dec. 4, 1965; children: Sung-Jin, Sung-Ju, Sung-Hoon. M.D., Seoul Nat. U., 1965. Diplomate: Am. Bd. Internal Medicine, Am. Bd. Pathology. Intern Ellis Hosp., Schenectady, 1965-66; resident in medicine Harrisburg (Pa.) Hosp., 1966-69; fellow hematology and oncology U. Rochester (N.Y.) Med. Ctr., 1970-72; chief hematology sect. VA Hosp., Dayton, Ohio, 1972-75; dir. oncology unit, chief hematology and oncology Good Samaritan Hosp. and Health Ctr., Dayton, 1975—; clin. prof. medicine Wright State U. Sch. Medicine, Dayton, 1980—. Trustee Montgomery County Soc. Cancer Control, 1976—, Community Blood Ctr.; mem. med. adv. com. Leukemia Soc. Am., Dayton chpt., 1977—. NIH grantee, 1970-72. Mem. Am. Fedn. Clin. Research, ACP, Am. Soc. Hematology; mem. Am. Soc. Clin. Oncology; Mem. Am. Assn. Cancer Research. Roman Catholic. Subspecialties: Hematology; Oncology. Current work: Biochemical behavior of cancer and malignant cells and chemotherapy of hematologic malignancy and solid tumors. Home: 1122 Wycliffe Pl Dayton OH 45459 Office: 2222 Philadelphia Dr Dayton OH 45406

CHANG, JAMES C., chemistry educator; b. Shanghai, China, Aug. 8, 1930. B.S., Mt. Union Coll., 1957; Ph.D., UCLA, 1964. Mem. faculty dept. chemistry U. No. Iowa, Cedar Falls, 1964—, prof., 1974—. Fellow Iowa Acad. Sci.; mem. Am. Chem. Soc., Chinese Am. Chem. Soc., Sigma Xi. Subspecialty: Inorganic chemistry. Current work: Nucleophilic substitution reactions of transition-metal complexes. Office: Dept Chemistry U No Iowa Cedar Falls IA 50614

CHANG, JI YOUNG, metallurgist; b. Sudongmyon Hamyang Kun, Kyungnam, South Korea, Apr. 18, 1932; came to U.S., 1958, naturalized, 1972; s. Jae Bouk Chang and Kyung Douk (Ieshil) Lee; m. Ki Chung, Apr. 3, 1965; children: Grace Saechung, Gloria Younchung, Gail Hyochung. B.S., Seoul (Korea) Nat. U., 1955; M.S., U. Tenn.-Knoxville, 1964, postgrad., 1965-69; Ph.D., Inha U., Inchun, Korea, 19. Registered profl. engr., Pa. Research engr. Sci. Research Inst., Seoul, 1955-58; sr. engr. Atomic Energy Research Inst., Seoul, 1959-63; cons. Oak Ridge Nat. Lab., 1963-67, devel. engr., 1967-70; sr. and prin. engr. Westinghouse Advanced Reactors Div., Madison, Pa., 1970-81, fellow engr., 1981—. Contbr. articles to profl. publs. Mem. Am. Soc. Metals, Am. Nuclear Soc., Korean Nuclear Soc., Internat. Metall. Soc., Korean Engrs. and Scientists Assn. (pres. chpt. 1976-77). Republican. Presbyterian. Subspecialties: Metallurgical engineering; High-temperature materials. Current work: Thermal fatigue and self-welding behavior of high-temperature materials in liquid metals. U.S., Korean patentee in field. Home: 3308 Hermar Dr Murrysville PA 15668 Office: Westinghouse Advanced Energy Systems Div PO Box 158 Madison PA 15663

CHANG, JOSEPH YUNG, research scientist; b. Nanking, China, Jan. 30, 1932; came to U.S., 1954, naturalized, 1968; s. Chun-Yen and Pi-Kuan C.; m. Edna H. Li, Dec. 21, 1963. B.S., Tainan (Taiwan) Coll. Engring., 1953; M.S., U. Notre Dame, 1957, Ph.D., 1958. Postdoctoral research assoc. U. Notre Dame, South Bend, Ind., 1958-61; project scientist Research div. Philco Corp., Blue Bell, Pa., 1961, N.Y.U., N.Y.C., 1961-63; sr. research scientist Grumman Aerospace Corp. Research and Devel. Ctr., Bethpage, N.Y., 1963—. Mem. Am. Chem. Soc., Am. Nuclear Soc., AAAS, Sigma Xi. Subspecialties: Fusion. Current work: Fusion plasma effects on materials, radiation effects on optical materials, thermoluminescence and radiation induced prompt luminescence, basic radiation effects mechanisms. Home: 16-D Cold Spring Hills Rd Huntington NY 11743 Office: Grumman Aerospace Corp Research and Devel Center Mail Stop A01-26 Bethpage NY 11714

CHANG, KENNETH SHUEH-SHEN, microbiologist, med. researcher; b. Taipei, Taiwan, Jan. 3, 1929; came to U.S., 1967, naturalized, 1973; s. Chi-Lin and Pi-Lien (Chen) C.; m. Janet Tasi-Fong Lee, Jan. 9, 1952; children: Susan, Judy, Kathy, Robert, William. M.D., Nat. Taiwan U., 1951; Dr. Med. Sci., Tokyo U., Japan, 1960. Teaching asst. Nat. Taiwan U., 1951-55, lectr., 1955-59, assoc. prof., 1959-64, prof., 1964-67; sr. virologist Flow Labs., Rockville, Md., 1967-69; chief viral oncogenesis sect. Lab. Cell Biology, Nat. Cancer Inst., Bethesda, Md., 1969—; vis. scientist UCLA Med. Sch., 1962-64. Contbr. articles to profl. jours. WHO fellow, 1956; Nat. Acad. Sci. fellow, 1962-64. Mem. Am. Assn. Cancer Research, Am. Soc. Microbiology, Am. Assn., Immunologists, Internat. Assn. Comparative Research on Leukemia and Related Diseases. Subspecialties: Cancer research (medicine); Virology (medicine). Current work: Cancer immunology and virology. Office: Nat Inst Health Bethesda MD 20205

CHANG, KWEN-JEN, biochemist, pharmacologist, educator; b. Taiwan, Oct. 21, 1943; s. Hwa-chi and Hsing-Mei (Chou) C.; m. Lan Yui Chiang, July 25, 1969; children: Linie, Emily. B.S., Nat. Taiwan U., 1966; PH.D., SUNY-Buffalo, 1972. Postdoctoral fellow Johns Hopkins U., 1972-75; research scientist Burroughs Wellcome Co., Research Triangle Park, N.C., 1975—; adj. prof. Duke U. Med. Ctr., 1980—. Contbr. articles to profl. jours. NIH postdoctoral fellow, 1972-74; research fellow Am. Heart Assn., 1974-75. Mem. Am. Soc. Pharmacology and Exptl. Therapeutics, Am. Soc. Biol. Chemists, Soc. Neurosci., N.Y. Acad. Scis. Subspecialties: Neurobiology; Biochemistry (biology). Current work: Opiates, endorphins, neuropeptides, hormones, biochemistry and pharmacology of receptors.

CHANG, LOUIS WAI-WAH, pathology educator, educational administrator, researcher, consultant; b. Hong Kong, July 1, 1944; came to U.S., 1962, naturalized, 1977; s. Ernest Yung-Pao and Jeanny Ming-Wei (Ma) C.; m. Dora Tao-Loo Lin, July 14, 1968; children: Michael, Jennifer-Michelle. B.A. in Chemistry, U. Mass.-Amherst, 1966; M.S. in Anatomy, Tufts U., 1969; Ph.D. in Pathology, U. Wis.-Madison, 1972. Instr. pathology U. Wis.-Madison, 1972-73, asst. prof., 1973-76, dir. histopathology lab., 1973-76; assoc. prof. U. Ark. Med. Sch., Little Rock, 1977-80, prof., 1980—, dir. grad. studies, 1977-80, dir. exptl. pathology, 1980—; cons. to govt. agys. Editor: Jour. Histotechnology, 1976—, Neurotoxicology, 1979—, Neurobehavioral Toxicology and Teratology, 1981—; contbr. chapts. to books, articles to profl. jours. Tres. Chinese Community Assn., Little Rock, 1982-83. Recipient golden Forcep award Nat. Soc. Histotechnology, 1976; NIH grantee, 1973-78; NSF grantee, 1977-80; EPA grantee, 1981—. Mem. Am. Assn. Neuropathologists, Am. Assn. Pathologists, Soc. Neuroscience, Soc. Toxicology, Soc. Toxicopathology. Roman Catholic. Subspecialties: Pathology (medicine); Environmental toxicology. Current work: Environmental toxicology/pathology; heavy metal and drug toxicology; neuropathology/toxicology; developmental toxicology/teratology; histochemistry; electron microscopy. Office: University of Arkansas Medical School 4301 W Markham St Little Rock AR 72201

CHANG, PETER HON-YOU, computer scientist, researcher; b. Cheron County, Kiangsu, China, June 29, 1946; came to U.S., 1969; s. Chin-Yee and Tsing-Fong C.; m. Johanna Chung-Ling, June 26, 1971; children: Mary Kon-Hua, Margaret Kon-Tsing, Lawrence Kon-Luen. B.S., Nat. Tsing-Hua U., 1968; M.S., U. Iowa, 1971; Ph.D., U. Minn., 1975. Research assoc. U. Alta. (Can.), Edmonton, 1975-77, U. Western Ont. (Can.), London, 1977-79; vis. asst. prof. Iowa State U., Ames, 1979-80; assoc. prof. dept. math. computer sci. U. Nebr.-Omaha, 1980-83; computer software cons. AGS Computers Inc., Denver, 1983—; cons. Anderson Cons. Service, Omaha, 1982-83. Bd. dirs. Omaha Am. Chinese Club, 1980-82. Mem. Soc. Indsl. and Applied Math., Am. Math. Soc., Internat. Assn. Math. Modelling, Nebr. Chpt. Am. Statis. Assn. Roman Catholic. Subspecialties: Software engineering; Mathematical software. Current work: UNIX operating system, C language programming, CAD-CAM telecommunication, analytical and numerical solutions of nonlinear partial differential equations applied to models of gas dynamics, vibrating string, optics, combustion, biophysics, traffic flows. Office: AGS Computers Inc 1900 Wazee Denver CO 80202 Home: 301 E Malley Dr #105 Northglenn CO 80233

CHANG, RAYMOND SHEN-LONG, pharmacologist; b. Taipei, Taiwan, Oct. 22, 1943; came to U.S., 1971; s. Sen Der and Pao (Yang) C.; m. Sing Tai Chuang, Aug. 25, 1973; children: Henry, Amy. Ph.D., U. Miami, 1977. Postdoctoral fellow Johns Hopkins U., 1977-79; sr. research pharmacologist Merck, Sharp & Dohme, Rahway, N.J., 1979—. Contbr. articles to profl. jours. Mem. Soc. for Neurosci. Subspecialties: Neuropharmacology; Neurochemistry. Current work: Drug and neurotransmitter receptors. Office: PO Box 2000 Rahway NJ 07065

CHANG, RICHARD LI-CHAI, research chemist; b. Hupei, China, Jan. 23, 1932; s. Kaitsen and K.F. C.; m. Tsailing Chang, Feb. 8, 1961; children: Wan B., Ben Y. B.S. in Agrl. Chemistry, Nat. Taiwan Chung Hsing U., 1957; M.S. in Biochemistry and Nutrition, Utah State U., 1964. Research asst. dept. pharmacology Sch. Medicine, N.Y. U., N.Y.C., 1964-67; research biochemist Burroughs Wellcome and Co. (U.S.A.), Inc., Tuckahoe, N.Y., 1967-70; scientist dept. biochemistry Schering Corp., Bloomfield, N.J., 1970-74; scientist dept. biochemistry and drug metabolism Hoffmann-La Roche, Inc., Nutley, N.J., 1974-79, sr. scientist, 1979—. Contbr. numerous articles to sci. jours. Mem. Am. Soc. Pharmacology and Exptl. Therapeutics. Subspecialties: Cancer research (veterinary medicine); Pharmacology. Current work: Chemical carcinogenesis by polycyclic aromatic hydrocarbons. Home: 107 Konner Ave Pine Brook NJ 07058 Office: Kingsland St Nutley NJ 07110

CHANG, TE-WEN, microbiologist, researcher, educator; b. Nanchang, Kiangsi, China, Oct. 12, 1920; came to U.S., 1949, naturalized, 1960; m. Diana Tan, June 22, 1952; children. M.D., Nat. Central U. Med. Coll., Chengdu, China, 1945. Intern Gen. Hosp. Chengdu, 1944-45; resident in internal medicine Univ. Hosp., Nanking, 1946-49; researcher in virology U. Kans. Med. Ctr., 1951-52; asst. in medicine Boston U. Sch. Medicine, 1952-54, instr., 1954-57; sr. instr. Tufts U. Sch. Medicine, 1957-59, asst. prof. medicine and microbiology, 1959-67, assoc. prof., 1968—; cons. in field. Contbr. chpts., numerous articles, abstracts to profl. publs. Served to capt. Chinese Army, 1944-46. Mem. AAAS, Am. Fedn. Clin. Research, N.Y. Acad. Scis., Am. Soc. Microbiology, Am. Acad. Microbiology, Infectious Disease Soc. Am. Subspecialties: Microbiology; Infectious diseases. Current work: Bacterial toxins, herpes virus and chemotherapy. Home: 55 Deer Path Ln Weston MA 02193 Office: 171 Harrison Ave Boston MA 02111

CHANG, THOMAS MING SWI, physician, medical scientist; b. Swatow, Kwantang, China, Apr. 8, 1933; s. Henry Sue-Yue and Frances Hue-Soo (Lim) C.; m. Lancy Yuk Lan, June 21, 1958; children: Harvey, Victor, Christine, Sandra. B.Sc., McGill U., 1957, M.D., C.M., 1961, Ph.D., 1965, F.R.C.P.(C), 1972. Intern Montreal (Que.) Gen. Hosp., 1961-62; research fellow McGill U., Montreal, 1962-65, lectr. dept. physiology, 1965, asst. prof., 1966-69, asso. prof., 1969-72, prof. physiology, 1972—, prof. medicine, 1975—, dir. artificial organs research unit, 1975-79, dir. artificial cells and organs research center, 1979—; practice medicine specializing in med. scis., Montreal, 1962—; staff Royal Victoria Hosp.; hon. cons. Montreal Chinese Hosp., 1970—; cons. Montreal Children's Hosp., 1979—; Med. Research Council fellow, 1962-65, scholar, 1965-68, career investigator, 1968—. Inventor artificial cells; Author: Artificial Cells, 1972, Biomedical Application of Immobilized Enzymes and Protiens, Vols. I and II, 1977, Artificial Kidney, Artificial Liver and Artificial Cells, 1978, Hemoperfusion-Kidney and Liver Supports and Detoxification, 1980, Hemoperfusion, 1981, Past Present and Future of Artificial Organs, 1983; sect. editor: Internat. Jour. Artificial Organs, 1977—; asso. editor: Jour. Artificial Organs, 1977—; editorial bd.: Jour. Biomaterial Med. Devel. and Orgn, 1972—, Jour. Membrane Soc, 1975—, Jour. Bioengring, 1975-79, Jour. Enzyme and Microbial Tech, 1978—. Fellow Royal Coll. Physicians Can.; mem. Biophysic Soc., Am., Canadian physiology socs., Internat. Soc. Artificial Organs (trustee), Can. Soc. Artificial Organs (pres. 1980-82), Canadian Med. Assn. Subspecialties: Artificial organs; Enzyme technology. Current work: Research (laboratory and clinical) and teaching and administration; artificaial cells, artificaial organs, micro-encapsulation, immobilised enzymes, immobilised cells, absorbents, detoxification, biotechnology. Office: Artificial Cells and Organs Research Centre McGill U 3655 Drummond St Montreal PQ H3G 1Y5 Canada

CHANG, TU-NAN, physics educator; b. Hangzhou, China, Apr. 20, 1945; came to U.S., 1967, naturalized, 1979; s. Hao and Pe-ven (Yu) C.; m. Hsiao-lin Chang, Nov. 1, 1945; children: Amy Chia-Mae, Angela Chia-Loh. B.S., Tunghai U., Taiwan, 1966; M.S., U. Calif.-Riverside, 1969, Ph.D., 1972. Research assoc. U. Chgo., 1973-75; asst. prof. physics U. So. Calif., Los Angeles, 1975-81, assoc. prof., 1981—. Contbr. articles to profl. jours. NSF research grantee, 1978—. Mem. Am. Phys. Soc. Subspecialties: Atomic and molecular physics; Theoretical physics. Current work: Atomic and molecular physics, many-body interactions in atomic transitions, photon-atom interactions. Office: Physics Dept U So Calif Los Angeles CA 90089-1341

CHANG, WEI, med. nuclear physicist, researcher; b. Szechun, China, Oct. 20, 1945; came to U.S., 1969, naturalized, 1982; s. Cheng-Jen and Hui-Yin (Chueh) C.; m. Ching Lin, Jan. 5, 1974; 1 child, Brian Lin. B.S., Nat. Taiwan U., Taipei, 1968; Ph.D., SUNY-Buffalo, 1976. Diplomate: Am. Bd. Sci. in Nuclear Medicine, 1980. Physicist Miriam Hosp., Providence, 1975-76; med. nuclear physicist Loyola U. Med. Center, Maywood, Ill., 1977-82, assoc. prof. radiology, 1981-82; med. physicist, assoc. prof. radiology U. Md. Med. Sch., Balt., 1983—. Mem. Soc. Nuclear Medicine, Am. Assn. Physicist in Medicine. Subspecialties: Imaging technology; Bioinstrumentation. Current work: Computer processing of nuclear medicine images, emissions computed tomography; collimator tomographic techniques. Home: 309 Ridgemoor Dr Willowbrook IL 60521 Office: Dept Radiology U Md 222 Greene St Baltimore MD 21201

CHANG, WILLIAM SHEN CHIE, electrical engineering educator; b. Nantung, Kiangsu, China, Apr. 4, 1931; s. Tung Wu and Phoebe Y.S. (Chow) C.; m. Margaret Huachen Kwei, Nov. 26, 1955; children: Helen Nai-yee, Hugh Nai-han, Hedy Nai-lin. B.S.E., U. Mich., 1952, M.S.E., 1953; Ph.D., Brown U., 1957. Lectr., research asso. elec. engring. Stanford, 1957-59; asst. prof. elec. engring. Ohio State U., 1959-62, asso. prof., 1962-65; prof. dept. elec. engring. Washington U., St. Louis, 1965-76, chmn. dept., 1965-71; dir. Applied Electronic Scis. Lab., 1971-79, Samuel Sachs prof. elec. engring., 1976-79; prof. dept. elec. engring. and computer scis. U. Calif., San Diego, 1979—. Author: Principles of Quantum Electronics, 1969; Contbr. articles to profl. jours. Mem. Am. Optical Soc., Am. Phys. Soc., AAUP, IEEE. Subspecialty: Electrical engineering. Research on quantum electronics and optics. Home: 763 Santa Olivia Solana Beach CA 92075

CHANG, WILLIAM WEI-LIEN, pathologist, histologist, researcher, educator; b. Taipei, Taiwan, China, Feb. 7, 1933; came to U.S., 1970; s. Symonds Tung-Lan and Grace Yun-Luei (Chen) C.; m. Delphine Li-Fen, Oct. 23, 1965; children: Phyllis, Bernice, Albert. M.D., Nat. Taiwan U., 1958; M.Sc., Ohio State U., 1966; Ph.D., McGill U., 1970. Cert. Am. Bd. Pathology. Intern Buffalo Gen Hosp., 1961-62; resident in pathology Ohio State U. Hosp., Columbus, 1962-66; fellow in pathology Queen's U., Kingston, Ont., Can., 1966-67; research fellow in anatomy McGill U., 1967-70; asst. prof. anatomy Mt. Sinai Sch. Medicine, N.Y.C., 1970-73, assoc. prof., 1973-79; assoc. prof. pathology W.Va. U., 1979—; staff pathologist W.Va. U. Hosp., 1979—. Served to 2d lt. Chinese Air Force, 1958-60. Recipient Excellence in Teaching award Mt. Sinai Sch. Medicine, 1976. Mem. Am. Assn. Pathologists, Internat. Acad. Pathology, Am. Soc. Cell Biology, Am. Assn. Anatomists, AAAS. Subspecialties: Pathology (medicine); Cancer research (medicine). Current work: Experimental colonic carcinogenesis with emphasis on histogenesis and modifications of carcinogenesis; histogenesis of various types of cancers in relation to organogenesis. Home: 117 Bakers Dr Morgantown WV 26505 Office: Dept Pathology W Va U Med Ctr Medical Center Dr Morgantown WV 26506

CHANIN, LOREN MAXWELL, electrical engineering educator; b. Roland, Man., Can., Aug. 14, 1927; m., 1950; 3 children. B.S., U. Man., 1949; M.S., U. N.Mex., 1951; Ph.D. in Physics, U. Pitts., 1959. Research engr. Westinghouse Corp., 1951-59; research scientist, sect. head Honeywell, Mpls., 1959-65; assoc. prof. elec. engring. U. Minn., Mpls., 1965-68, prof., 1968—; cons. U.S. Bur. Mines. Mem. Am. Phys. Soc., IEEE, European Phys. Soc. Subspecialty: Plasma physics. Office: Dept Elec Engring U Minn Minneapolis MN 55455

CHANMUGAM, GANESAR, physicist, astronomer, educator, researcher; b. Colombo, Sri Lanka, Oct. 24, 1939; s. Paul K. and Savundramanie (Canapathipillai) C.; m. Prithiva S. Kanagasundram, Aug. 22, 1966; children: Ravi, Suresh. B.Sc. in Math. with honors, U. Ceylon, Colombo, 1961, B.A., Cambridge U. (Eng.), 1963; Ph.D. in Physics, Brandeis U., 1970. Instr. physics U. Mass., Amherst, 1963-64; research fellow Institut d'Astrophysique, Université de Liège, Belgium, 1969-71; research assoc. dept. physics and astronomy La. State U., Baton Rouge, 1971-72, asst. prof., 1972-78, assoc. prof., 1978-83, prof., 1983—; vis. fellow Joint Inst. Lab. Astrophysics, U. Colo., Boulder, 1979-80. Contbr. writings to publs. in field. Grantee NSF, 1976, 77-85, NASA, 1981-82, 82-83; recipient award So. Regional Edn. Bd., 1982. Fellow Royal Astron. Soc. (U.K.); mem. Am. Phys. Soc., Am. Astron. Soc., Internat. Astron. Union. Subspecialty: Theoretical astrophysics. Current work: Theoretical astrophysics, magnetic degenerate stars (white dwarfs and neutron stars). Office: Dept Physics and Astronomy La State U Baton Rouge LA 70803

CHANNIN, DONALD JONES, research scientist; b. Evanston, Ill., Aug. 29, 1942; s. Nathaniel S. and Eleanor L. (Jones) C.; m. Elizabeth F. Bezerra, June 23, 1979; 1 dau., Jennifer L. B.S., Case Western Res. U., 1964; Ph.D., Cornell U., 1970. Mem. tech. staff RCA Labs., Princeton, N.J., 1970—. Contbr. articles to profl. jours. Mem. IEEE (sr.), Optical Soc. Am., Am. Phys. Soc. Subspecialties: Fiber optics; Semiconductors. Current work: Design and analysis of fiber optic systems, research on optoelectronic components, support for optical communication applications. Office: RCA Labs Princeton NJ 08540

CHANNON, STEPHEN R., physicist; b. Washington, Aug. 13, 1947; m., 1969; 1 child. B.S., Ohio State U., 1970; Ph.D. in Physics, Rutgers U., 1981. Physicist Fusion Energy Corp., Princeton, N.J., 1974-77, cons. physicist-programmer, 1981-82, dir. theory programs, 1982—; vis. assoc. prof. U. Buenos Aires, 1982—. Mem. Am. Phys. Soc. Subspecialty: Fusion. Office: Fusion Energy Corp PO Box 2005 Princeton NJ 08540

CHAO, BEI TSE, mechanical engineering educator; b. Soochow, China, Dec. 18, 1918; came to U.S., 1948, naturalized, 1962; s. Tse Yu and Yin T. (Yao) C.; m. May Kiang, Feb. 7, 1948; children: Clara, Fred Roberto. B.S. in Elec. Engring. with highest honor, Nat. Chiao-Tung U., China, 1939; Ph.D. (Boxer Indemnity scholar), Victoria U., Manchester, Eng., 1947. Asst. engr. tool and gage div. Central Machine Works, Kunming, China, 1939-41, asso. engr., 1941-43, mgr. tool and gage div., 1943-45; research asst. U. Ill., Urbana, 1948-50, asst. prof. dept. mech. engring., 1951-53, asso. prof., 1953-55, prof., 1955—, head thermal sci. div., 1971-75, head dept. mech. and indsl. engring., 1975—, asso. mem., 1963-64; cons. to industry and govtl. agys., 1950—; Russell S. Springer prof. mech. engring. U. Calif.-Berkeley, 1973; mem. reviewing staff Zentralblatt für Mathematik, Berlin, 1970-82; mem. U.S. Engring. Edn. Del. to Visit People's Republic of China, 1978; mem. adv. screening com. in engring. Fulbright-Hays Awards Program, 1979-81, chmn., 1980, 81; mem. com. U.S. Army basic sci. research NRC, 1980-83. Author: Advanced Heat Transfer, 1969; contbr. numerous articles on mech. engring. to profl. jours.; tech. editor: Jour. Heat Transfer, 1975-81; mem. adv. editorial bd.: Numerical Heat Transfer, 1977—. Recipient Outstanding Tchr. award Ill. Mech. Engring. Alumni, 1978. Fellow ASME (Blackall award 1957, Heat Transfer award 1971); mem. Am. Soc. Engring. Edn. (Outstanding Tchr. award. 1975, Western Electric Fund award 1973), Nat. Acad. Engring., Soc. Engring. Sci., Chiao-Tung U. Alumni Assn. (pres. Mid-West sect. 1975-76), Sigma Xi, Tau Beta Pi, Pi Tau Sigma. Subspecialties: Mechanical engineering; Fluid mechanics. Current work: Thermal hydraulics in nuclear reactors; multiphase flows; particlate removal from engine exhaust and fluidized bed dynamics. Home: 704 Brighton Dr Urbana IL 61801 Office: 148 Mechanical Engineering Bldg Univ Ill Urbana IL 61801

CHAO, JIATSONG JASON, nuclear engineer, research project manager; b. Venden, Shanton, China, Dec. 15, 1948; came to U.S., 1972; s. Gwei-Hai and Chui-Yun Tsoong C.; m. Lily Ni, May 26, 1974; 1 son, Neal. B.S., FuJen Catholic U., Taipei, Taiwan, 1971; M.A., U. Tex.-Austin, 1974; Ph.D., M.I.T., 1979. Research assoc. M.I.T., 1976-79; sr. scientist Sci. Applications Inc., Oak Brook, Ill., 1979-81; project mgr. Electric Power Research Inst., Palo Alto, Calif.,

1981—, Argonne (Ill.) Nat. Lab., 1981—, Gen. Electric, Sunnyvale, Calif., 1981—, Energy, Inc., Idaho Falls, 1981—. Contbr. articles to profl. jours. Mem. Am. Nuclear Soc., ASME. Subspecialties: Nuclear fission; Nuclear fusion. Current work: Thermal hydraulic and neutronic design of fusion reactor blanket magnetohydrodynamics, two-phase heat transfer nuclear plant analyses for pressurized thermal shock and steam generator tube break events. Patentee fusion blanket design. Home: 1841 Newcastle Dr Los Altos CA 94022 Office: Electric Power Research Institute 3412 Hillview Ave Palo Alto CA 94303

CHAO, SHUI LIN, laser physicist; b. Chekiang, China, Jan. 24, 1935; s. Tze Ding and Lian Hwa (Chiang) C.; m. Cecilia Ching-Haw Ho, Aug. 17, 1967; 1 dau., Cindy. B.S., Chinese Naval Inst. Tech., 1956; M.S., U. Md., 1967; Ph.D., U. Rochester, 1973. Research engr. GTE Lab., Bayside, N.Y., 1967-68; research physicist Gen. Laser Corp., Natick, Mass., 1969-70; research scientist Atlantic Research Corp., Alexandria, Va., 1972-75; optical scientist TRW Systems, Redondo Beach, Calif., 1975-80, optics sect. head, 1980—. Contbr. articles to profl. jours. Mem. adv. com. Pasadena City Coll. Mem. Optical Soc. Am., Soc. Photo-Optics Instrumentation Engrs. Subspecialty: High energy laser systems. Home: 30804 Rue de la Pierre Rancho Palos Verdes CA 90274 Office: 1 Space Park Redondo Beach CA 90278

CHAPIN, CHARLES EDWARD, geologist; b. Porterville, Calif., Oct. 25, 1932; s. William Frank and Gladys Lillian (Mitchell) C.; m. Carol Ruth Giles, June 11, 1958; children: Giles Mathew, John Edward, Laura Ann. B.Geol.Engring., Colo. Sch. Mines, 1954, D. Sc., 1965. Exploration geologist J.M. Huber Corp., Borger, Tex., 1955-58; devel. geologist Lucky Mc Uranium Corp., Riverton, Wyo., 1958-59; asst. prof. U. Tulsa, 1964-65; asst. and assoc. prof. N.Mex. Inst. Mining & Tech., Socorro, 1965-70, head geosci. dept., 1968-70; sr. geologist N. Mex. Bur. Mines, Socorro, 1970—; cons. Western Nuclear Inc., 1967-69, Rampart Exploration, 1978—; chmn. exec. com. Rio Grande Rift Consortium, 1982—; chmn. program com. Internat. Symposium on Rio Grande Rift, 1977-78. Author, co-editor: Field Guide. . . Datil-Mogollon Volcanic Field, N.Mex, 1978, Ash-Flow Tuffs, 1979; contbr. articles to profl. jours. Cubmaster Boy Scouts Am., Socorro, 1968-70. Served with U.S. Army, 1955-57. Recipient Van Diest Gold medal Colo. Sch. Mines, 1980. Fellow Geol. Soc. Am.; mem. N.Mex. Geol. Soc., Am. Geophys. Union, Soc. Exploration Paleontologists and Mineralogists, Rocky Mountain Assn. Geologists. Subspecialties: Tectonics; Volcanology. Current work: Tectonics of Southern Rocky Mountain-Rio Grande rift area; volcanic geology of the N.E. Datil-Mogollon volcanic field, N.Mex.; Tallahassee Creek uranium dist., Colo. Home: 204 Grant Ave Socorro NM 87801 Office: New Mex Bur Mines Campus Station Socorro NM 87801

CHAPMAN, LLOYD WILLIAM, physiology educator; b. Pasadena, Calif., Jan. 5, 1938; s. and Gilbert Wilford and Helen (Moldt) C.; m. Eleanor Edwards, Aug. 10, 1968; children: Judith Anne, David Nathaniel. B.A., UCLA, 1961; Ph.D., U. So. Calif., 1971. Instr. U. So. Calif., Los Angeles, prof. physiology, 1973-78; mem. faculty Coll. Osteopathic Medicine of the Pacific, Pomona, Calif., 1978—, assoc. prof. physiology, 1982—; research assoc. Cedars-Sinai Med. Center, Los Angeles, 1974—, U. Calif., Riverside, 1981-83. Mem. bd. dirs. Jasper Sch. PTA, Alta Loma, Calif., 1980—. Mem. Am. Physiol. Soc. Democrat. Subspecialty: Physiology (medicine). Current work: Cardiovascular and fluid and electrolyte research. Office: Coll Osteopathic Medicine of the Pacific 309 Pomona Mall East Pomona CA 91766 Home: 6167 Topaz St Alta Loma CA 91701

CHAPMAN, ORVILLE LAMAR, chemist, educator; b. New London, Conn., June 6, 1932; s. Orville Carmen and Mabel Elnora (Tyree) C.; m. Faye Newton Morrow, Aug. 20, 1955 (div. 1980); children: Kenneth, Kevin; m. Susan Elizabeth Parker, June 15, 1981. B.S., Va. Poly. Inst., 1954; Ph.D., Cornell U., 1957. Prof. chemistry Iowa State U., 1957-74; prof. chemistry UCLA, 1974—; Cons. Mobil Chem. Co. Recipient John Wilkinson Teaching award Iowa State U., 1968, award Nat. Acad. Scis., 1974; Founders prize Tex. Instruments; George and Freda Halpern award in photochemistry N.Y. Acad. Scis., 1978. Mem. Am. Chem. Soc. (award in pure chemistry 1968, Arthur C. Cope award 1978, Midwest award 1978, Havinga medal 1982). Subspecialty: Organic chemistry. Home: 1213 Roscomare Rd Los Angeles CA 90024 Office: Dept Chemistry U Calif 405 Hilgard Ave Los Angeles CA 90024

CHAPMAN, RUSSELL LEONARD, phycologist, cytologist, educational administrator; b. Bklyn., May 30, 1946; s. Russell Hood and Helen Theresa (Egnotas) C.; m. Melanie Ripperton, June 28, 1969; children: Christopher John, Timothy Sean. A.B., Dartmouth Coll., 1968; M.S., U. Calif., Davis, 1970, Ph.D. (NSF fellow), 1973. Asst. prof. dept. botany La. State U., Baton Rouge, 1973-77, assoc. prof., 1977-83, prof., 1983—, assoc. dean Coll. Arts and Scis., 1979—. Contbr. numerous articles on phycology and cytology to profl. jours. Active Found. for Hist. La. Recipient Amoco Found. Outstanding Undergrad. Teaching award, 1978; La. State U. Alumni Fedn. Disting. Faculty award, 1981. Mem. Phycological Soc. Am., Brit. Phycological Soc., Internat. Phycological Soc., Bot. Soc. Am., Am. Soc. Cell Biology, Electron Microscopy Soc. Am., Sigma Xi, Phi Kappa Phi. Republican. Episcopalian. Subspecialties: Cell biology; Evolutionary biology. Current work: Ultrastructure and phylogeny green algae, cellular research, ultrastructure, phycology. Home: 9120 Sage Circle Baton Rouge LA 70809 Office: Dept Botany La State U Baton Rouge LA 70803

CHAPPEL, SCOTT CARLTON, research scientist, educator; b. Syracuse, N.Y., Apr. 22, 1950; s. Robert W. and Betty A. (Bachman) C.; m. Claudia B. Lippman, Aug. 15, 1976; children: Benjamin, Jessica. B.S., Pa. State U., 1972; Ph.D., U. Md., 1976. Asst. scientist Oreg. Primate Ctr., Beaverton, 1977-79; asst. prof. U. Pa., Phila., 1979-83; assoc. scientist U. Mich., Ann Arbor, 1984—. NIH grantee, 1979. Mem. Soc. Study Reprodn., Endocrine Soc., Am. Physiol. Soc., N.Y. Acad. Sci. Subspecialties: Endocrinology; Neuroendocrinology. Current work: Neuroendocrine regulation of gonadotropin synthesis and release. Home: Geddes Lake Ann Arbor MI 48105 Office: Univ Pa Dept Ob-Gyn 576 Dulles Hosp Philadelphia PA 19104

CHAPPELL, CHARLES RICHARD, scientist, physicist; b. Greenville, S.C., June 2, 1943; s. Gordon Thomas and Mabel Winn (Ownbey) C.; m. Barbara Lynne Harris, May 15, 1968; 1 son, Christopher Richard. B.A. in Physics, Vanderbilt U., Nashville, 1965; Ph.D. in Space Sci, Rice U., Houston, 1968. Research scientist Lockheed Research Lab., Palo Alto, Calif., 1968-72, staff scientist, 1972-74; chief magnetospheric physics br. NASAMarshall Space Flight Ctr., Huntsville, Ala., 1974-80, chief solar terrestrial physics div., 1980—. Contbr. articles on magnetospheric physics to profl. jours. Recipient John Underwood award in physics Vanderbilt U., 1965, Alfred P. Sloan fellow, 1961-65; NASA traineeship Rice U., 1965-68; Exceptional Sci. Achievement award NASA, 1981. Mem. Am. Geophys. Union, Internat. Assn. Geomagnetism and Aeronomy, Phi Beta Kappa, Phi Eta Sigma. Methodist. Subspecialties: Plasma physics; Satellite studies. Current work: Satellite measurement of the composition and energy of low energy plasma in the Earth's magnetospheric space environment including the regions of the plasmasphere, ionosphere, plasma trough, and polar cap. Office: Solar Terrestrial Physics Div Space Science Lab Marshall Space Flight Center Huntsville AL 35812

CHAPPELL, GARY ALAN, software designer; b. Independence, Mo., Jan. 24, 1954; s. Jesse Earl and Eunice Mildred (Ralston) C. B.S. in Computer Sci, U. Mo.-Rolla, 1976, 1976. Grad. teaching asst. Stanford U., 1976-77; graphics programmer Lawrence Livermore Lab., Calif., 1977-79; prin. investigator research and devel. Def. and Space Systems Group TRW, Redondo Beach, Calif., 1979; product mgr. Comtal Image Processing, Inc., Pasadena, Calif., 1979-80; mgr. user interface devel. Tymshare, Inc., Cupertino, Calif., 1980-82; software mgr.-graphics Qubix Graphic Systems, Saratoga, Calif., 1982—. Mem. Assn. Computing Machinery, Soc. Info. Display, Am. Chem. Soc., Mensa. Subspecialties: Graphics, image processing, and pattern recognition. Current work: Design and implementation of interactive, graphics-oriented human-computer interfaces for non-technical computer system users; research into software and hardware components of highly interactive graphics systems. Home: 2220 Homestead Ct #204 Los Altos CA 94022 Office: 18835 Cox Ave Saratoga CA 95070

CHAPPELL, ROBERT PAUL, dental educator; b. Des Moines, Iowa, Nov. 28, 1918; s. William and Nellie Margaret (Florer) C.; m. Penelope Alyce Nelson, Apr. 20, 1979. B.S., U. Ill.-Chgo., 1948; D.D.S., 1950. Lic. dentist, Mo. Commd. 2d lt. U.S. Army, 1941, advanced through grades to lt. col., 1964; served, U.S. and E.T.O., 1941-46, dental officer, U.S.A., France, Eng., 1952-67; asst. prof. U. Mo., Kansas City, 1967-70, assoc. prof., 1970-80, prof., 1980—, chmn. dept. dental materials, 1974—. Contbr. articles to profl. jours. Mem.(life) Retired Officers Assn., Alexandria, Va., Mil. Order of World Wars., Washington. Univ. Mo. faculty grantee, 1972, 74. Fellow Internat. Coll. Oral Implantology; mem. Internat. Assn. Dental Research, Am. Assn. Dental Schs. Republican. Subspecialties: Implantology; Biomaterials. Current work: oral implantology, endosteal, biomaterials. Office: School of Dentistry Univ Mo 650 E 25th St Kansas City MO 64108

CHARACKLIS, WILLIAM GREGORY, microbial chemical process and environmental engineering researcher, consultant; b. Annapolis, Md., Aug. 21, 1941; s. Gregory Arthur and Artemis (Batayannis) C.; m. Nancy Crowley, Aug. 30, 1964; children: Gregory William, Erin Elizabeth. B.E.S., Johns Hopkins U., 1964, Ph.D., 1970; M.S.Ch.E., U. Toledo, 1967. Lic profl. engr., Tex. Research engr. Olin-Mathieson Chem. Corp., New Haven, 1964-65; grad. research asst. U. Toledo, 1965-67, Johns Hopkins U., 1967-70; asst. prof. Rice U., 1970-74, assoc. prof., 1974-78, prof., 1978-79; prof. environ. engrin. Mont. State U., 1979—; prin. W. G. Characklis Cons. Engrs., Inc., Bozeman, Mont., 1981—. Merck fellow, 1972; NSF fellow, 1977-78. Mem. Am. Inst. Chem. Engrs., Am. Soc. Microbiology, Internat. Assn. Water Pollution Research, Water Pollution Control Fedn., Cooling Tower Inst. Club: Karl Marx Athletic and Social (Bozeman). Subspecialties: Chemical engineering; Water supply and wastewater treatment. Current work: Microbial and chemical process engineering. Inventor in field. Home: 516 W Cleveland Bozeman MT 59715 Office: Mont State U Coll Engring Bozeman MT 59717

CHARAN, NIRMAL BISWAS, physician, researcher; b. Ranikhet, India, Dec. 7, 1946; came to U.S., 1977; s. Isaac Albert and Salomi (Singh) C.; m. Lalita Clive, June 20, 1973; children: Ankur, Neev. B.Sc., Hindu Coll., Moradabad, India, 1964; M.B.B.S., Christian Med. Coll., Ludhiana, India, 1969. Intern Christian Med. Coll. Hosp., Ludhiana, India, house physician, 1970, resident in internal medicine, 1970-74; registrar in internal medicine Auckland (New Zealand) Hosp., 1974-77; fellow in pulmonary U. Wash., Seattle, 1977-80, asst. prof., 1981—; staff physician VA Med. Ctr., Boise, 1980—, dir., 1980—. Author: Drug Therapy in Elderly, 1983; cons editor: Jour. Chest, 1982; contbr. articles to profl. jours. NIH fellow, 1979-80; VA grantee, 1981—. Fellow A.C.P., Am. Coll. Chest Physicians; mem. Am. Thoracic Soc., Am. Physiol. Soc., Am. Fedn. Clin. Research. Methodist. Subspecialties: Internal medicine; Pulmonary medicine. Current work: Lung fluid balance, lung mechanics and role of bronchial circulation in health and disease. Home: Apt 24A VA Medical Center Boise ID 83702

CHARGAFF, ERWIN, biochemistry educator; b. Austria, Aug. 11, 1905; came to U.S., 1928, naturalized, 1940; s. Hermann and Rosa C.; m. Vera Broido; 1 son, Thomas. Dr. Phil., U. Vienna, 1928; Dr. phil. h.c, U. Basel, 1976; Sc.D. (hon.), Columbia, 1976. Research fellow Yale, 1928-30; asst. U. Berlin, Germany, 1930-33; research asso. Inst. Pasteur, Paris, France, 1933-34; faculty Columbia, 1935—, prof. biochemistry, 1952-74, prof. emeritus, 1974—, chmn. dept. biochemistry, 1970-74; vis. prof., Sweden, 1949, Japan, 1958, Brazil, 1959, Coll. de France, 1965, Naples, Palermo, Cornell, 1966, Stazione biologica, 1969. Author: Essays on Nucleic Acids, 1963, Voices in the Labyrinth, 1977, Heraclitean Fire, 1978, Das Feuer des Heraklit, 1979, Unbegreifliches Geheimnis, 1980, Bemerkungen, 1981, Wernungstafeln, 1982, Kritikder Zukunft, 1983; numerous articles in field, other lit. work in English and German.; Editor: The Nucleic Acids, 3 vols, 1955, 60. Guggenheim fellow, 1949, 58; recipient Pasteur medal Soc. Biol. Chemistry, Paris, 1949; Neuberg medal Am. Soc. European Chemists, 1958; Bertner Found. award, Houston, 1965; C.L. Mayer prize French Acad. Scis., 1963; Dr. H.P. Heineken prize Netherlands Acad. Scis., 1964; Gregor Mendel medal German Acad. Scis. Leopoldina, 1973; Nat. medal Sci., 1975; medal N.Y. Acad. Medicine, 1980; Disting. Service award Columbia U., 1982. Fellow Am. Acad. Arts and Scis.; mem. Nat. Acad. Scis., Am. Philos. Soc.; fgn. mem. Royal Swedish Physiographic Soc., German Acad. Scis. Leopoldina. Subspecialty: Biochemistry (medicine). Home: 350 Central Park W New York NY 10025

CHARLESWORTH, EDWARD ALLISON, clinical psychologist, author, publishing corporation executive; b. New Orleans, AMar. 23, 1949; s. Albert Ernest and Wilma Nadine (Wright) C.; m. Robin Elaine Rupley, Dec. 7, 1974. B.S., U. Houston, 1974, M.S., 1978, Ph.D., 1980. Diplomate: Am. Acad. Behavioral Medicine; lic. psychologist, Tex. Pres. Stress Mgmt. Research Assos., Inc., Houston, 1977—; research assoc. Baylor Coll. Medicine, Houston, 1970-81, psychologist, instr., 1981-82; dir. Willowbrook Psychol. Assos., Houston, 1982—; chmn., chief editor Biobehavioral Press, Houston, 1980—. Author: recordings Stress Management Trng. Program, 1977, Stress Mgmt. and Relaxation Program, 1981; co-author: Stress Management: A Conceptual and Procedural Guide, 1980, Stress Management: A Comprehensive Guide to Wellness, 1982. Founder, sponsor Forest Lake Teens in Action, Houston, 1978-80; instr. Cypress Fairbank Sch. Dist. Wellness Program, Houston, 1982. Trainee NIMH, 1976-78, VA. Hosp, 1978. Mem. Am. Psychol. Assn., Tex. Psychol. Assn., Harris County Biofeedback Soc., Am. Soc. Biofeedback Clinicians. Democrat. Presbyterian. Club: Mensa. Subspecialties: Behavioral psychology; Preventive medicine. Current work: Research: behavioral treatments for hypertension, addictive disorders and stress related problems; clinical work, psychotherapy, hypnotherapy, biofeedback. Home: 11803 Moorcreek Houston TX 77070 Office: Willowbrook Psychological Assos 10603 Grant #204 Houston TX 77070

CHARNES, ABRAHAM, mathematics educator; b. Hopewell, Va., Sept. 4, 1917; m., 1950; 3 children. A.B., U. Ill., 1938, M.S., 1939, Ph.D. in Math, 1947. Office Naval Research fellow U. Ill., 1947-48; from asst. prof. to assoc. prof. math. Carnegie Inst. Tech., Pitts., 1948-52, assoc. prof. indsl. adminstrn., 1952-55; prof. math., dir. research dept. transp. and indsl. mgmt. Purdue U., West Lafayette, Ind., 1955-57; research prof. applied math. and econs. Northwestern U., Evanston, Ill., 1965-68, Walter P. Murphy prof., 1968-78; Jesse H. Jones prof. biomath., gen. bus. and mgmt. sci., prof. math., gen. bus. and computer sci., dir. Ctr. Cybernetic Studies, U. Tex.-Austin, 1973—. Editor: Jour. Inst. Mgmt. Sci, 1954—. Fellow AAAS, Econometric Soc.; mem. Ops. Research Soc. Am., ACM, Inst. Mgmt. Sci. (v.p. 1958, pres. 1960). Subspecialty: Applied mathematics. Office: U Tex BEB 203E Austin TX 78712

CHARPIE, ROBERT ALAN, physicist; b. Cleve., Sept. 9, 1925; s. Leonard Asbury and Dorothy (McLean) C.; m. Elizabeth Downs, July 12, 1947; children—Richard Alan, Carol Elizabeth, David Wayne, John Robert. B.S. with honors, Carnegie Inst. Tech., 1948, M.S., 1949, D.Sc. in Theoretical Physics, 1950; D.H.L., Denison U., 1965; D.Sc., Alderson-Broaddus Coll., 1967; LL.D., Marietta Coll., 1975; D.Sc., Boston Coll., 1982. With Westinghouse Electric Corp., 1947-50; with Oak Ridge Nat. Lab., 1950-51, tech. asst. to research dir., 1952-54, asst. research dir., 1954-58, dir. reactor div., 1958-61; mgr. adv. devel. Union Carbide Corp., 1961-63, gen. mgr. devel. dept., 1963-64, dir. tech., 1964-66, pres. electronics div., 1966-68; pres. Bell & Howell Co., Chgo., 1968-69; pres., dir. Cabot Corp., Boston, 1969—; trustee Mitre Corp., Boston, 1966-82, chmn., 1972-82; dir. 1st Nat. Bank, Boston, Champion Internat. Corp., Schlumberger Ltd., Northwest Airlines, Inc.; sec. gen. adv. com. AEC, 1959-63; mem. Nat. Sci. Bd., 1969-76; sci. sec., editor-in-chief proc., also asst. U.S. mem. 7 nation adv. com. 1st Internat. Conf. Peaceful Uses Atomic Energy, 1955; coordinator U.S. fusion research exhibit, 2d Conf., 1958; chmn. invention and innovation panel U.S. Dept. Commerce, 1965-67. Gen. editor: Internat. Monograph Series on Nuclear Energy, 1955-60; editor: Progress Series in Nuclear Energy, 1955-60, Jour. Nuclear Energy, 1955-60. Mem. Oak Ridge Bd. Edn., 1957-61; pres. Byram Hills Central Sch. Dist., 1966-68; Trustee Carnegie Inst. Tech., 1962—. Named one of Ten Outstanding Young Men U.S. Jr. C. of C., 1955; recipient Alumni Merit award Carnegie Inst. Tech., 1957. Fellow Am. Phys. Soc., Am. Nuclear Soc. (dir.); mem. N.Y. Acad. Sci., Nat. Acad. Engring., Sci. Research Soc. Am., Sigma Xi, Tau Beta Pi, Phi Mu Epsilon. Subspecialty: Theoretical physics. Home: 45 Ridgeway Rd Weston MA 02193 Office: 125 High St Boston MA 02110

CHARUDATTAN, RAGHAVAN, plant pathologist, educator; b. Tanjore, Tamilnadu, India, Apr. 7, 1942; s. Venkatarama and Sarada (Srinivasan) Raghavan; m. Dharini Mani, Mar. 15, 1952; 1 child, Savitar Kartika. Ph.D., U. Madras, India, 1968. Research asst. U. Madras, 1963-68; research plant pathologist U. Calif., Davis, 1968-70; research assoc. U. Fla., Gainesville, 1970-73, asst. prof., 1973-78, assoc. prof. plant pathology, 1978—; mem. tech. com. S-136 Coop. Regional Research Project. Editor: (with H.L. Walker) Biological Control of Weeds with Plant Pathogens, 1982; contbr. articles to profl. jours.; assoc. editor: Plant Disease, 1980-82. NIH grantee, 1968-70. Mem. Am. Phytopathological Soc., Mycological Soc. Am., Aquatic Plant Mgmt. Soc., Weed Sci. Soc. Am. Hindu. Subspecialties: Plant pathology; Weed science. Current work: Biological control of weeds with plant pathogens and the development of microbial herbicides. Office: Dept Plant Pathology U Fla Gainesville FL 32611

CHARYK, JOSEPH VINCENT, satellite telecommunications company executive; b. Canmore, Alta., Can., Sept. 9, 1920; came to U.S., 1942, naturalized, 1948; s. John and Anna (Dorosh) C.; m. Edwina Elizabeth Rhodes, Aug. 18, 1945; children: William R., J. John, Christopher E., Diane E. B.Sc. in Engring. Physics, U. Alta., 1942, LL.D., 1964; M.S., Calif. Inst. Tech., 1943, Ph.D., 1946; D.Engring. (hon.), U. Bologna, 1974. Sect. chief Jet Propulsion Lab., Calif. Inst. Tech., 1945-46, instr. aeros., 1945-46; asst. prof. aeros. Princeton, 1946-49, asso. prof., 1949-55; dir. aerophysics and chemistry lab., missile systems div. Lockheed Aircraft Corp., 1955-56; dir. aero. lab. Aeronutronic Systems, Inc. subs. Ford Motor Co., 1956-58, gen. mgr. space tech. div., 1958-59; asst. sec. Air Force (for research and devel.), 1959, under sec., 1960-63; pres., dir. Communications Satellite Corp., 1963—, chief exec. officer, 1979—, chmn., 1983—; chmn. bd. Comsat Gen. Corp., Environ. Research and Tech., Inc., Satellite TV Corp.; mem. partners' com. Satellite Bus. Systems; dir. Am. Security Corp., Abbott Labs.; mem. corp. C. S. Draper Lab., Inc. Fellow AIAA, IEEE; mem. Nat. Acad. Engring., Internat. Acad. Astronautics, Nat. Inst. Social Scis., Nat. Space Club, Sigma Xi. Clubs: 1925 F Street, Chevy Chase, Burning Tree, Met. Subspecialty: Telecommunications. Home: 5126 Tilden St NW Washington DC 20016

CHASE, CLEMENT GRASHAM, geophysics and geology educator; b. Phoenix, Mar. 27, 1944; s. Clement Kelsey and Bertha Louise (Grasham) C.; m. June Louise McClay, June 11, 1966; children: Anne, Katherine. Asst. prof. geophysics and geology U. Minn.-mpls., 1970-75, assoc. prof., 1975—. Author: The Evolving Earth, 1975; assoc. editor: Jour. Geophys. Research, 1983—. NSF fellow Cambridge, 1970. Mem. Am. Geophys. Union, Geol. Soc. Am. Subspecialties: Geophysics; Tectonics. Current work: Global gravity and geoid interpretation, mantle isotopic systems as kinematic tracers, plate tectonic motions, mechanics of continental faulting. Office: Dept Geology and Geophysics U Minn 310 Pillsbury Dr SE Minneapolis MN 55455

CHASE, GRAFTON D., chemistry educator, consultant; b. Camden, N.J., May 2, 1921; s. Walter Ezra and Margaret May (Wieland) C.; m. Alice Louise Delamater, Sept. 6, 1952; children: Deborah, Leslie, Grafton D. B.S., Phila. Coll. Pharmacy and Sci., 1943; M.A., Temple U., 1951, Ph.D., 1955. With Crown Can Co., 1945-47; faculty Phila. Coll. Pharmacy and Sci., 1946—, prof. chemistry, 1965—, chmn. dept. chemistry, 1981—; cons. FDA, 1975—. Editor: Remington's Pharmaceutical Sciences, 1965—, Principles of Redioisotope Methodology, 3d edit, 1969. Served with USN, 1942-45. Recipient honor scroll Phila. Inst. Chemists, Am. Inst. Chemists. Mem. Am. Chem. Soc.; mem. Am. Assn. Clin. Chemists; Mem. Soc. Nuclear Medicine, Clin. Ligand Assay Soc., Sigma Xi, Rho Chi. Republican. Subspecialties: Physical chemistry; Nuclear physics. Current work: Antibody binding mechanisms, receptor site binding mechanisms. Home: 316 E Lancaster Pk Philadelphia PA 19151 Office: Phila Coll Pharmacy and Sci 43d and Kingsessing Mall Philadelphia PA 19104

CHASE, JOHN WILLIAM, molecular biologist, educator; b. Balt., May 30, 1944; s. John Webster and Ethel Adele (Mewshaw) C.; m. Anne Christine Wright, July 1, 1967; children: Kristen Lynnette, Kimberly Anne. A.B., Drew U., Madison, N.J., 1966; Ph.D., Johns Hopkins U., 1971. NIH fellow Harvard U. Med. Sch., Boston, 1971-75, Am. Cancer Soc. fellow, 1972-73; asst. prof. Albert Einstein Coll. Medicine, Bronx, N.Y., 1975-81, assoc. prof. molecular biology, 1981—; cons. NIH, Bethesda, Md., 1978—; established investigator Am. Heart Assn., Dallas, 1978-83. Contbr. numerous sci. articles to profl. publs. NIH grantee, 1976—; Am. Cancer Soc. grantee, 1977-78. Mem. Am. Soc. Biol. Chemists, Am. Soc. Microbiologists. Subspecialties: Biochemistry (biology); Enzyme technology. Current work: Biochemical and genetic studies of DNA replication; recombination; repair mechanisms. Home: 90 Perth Ave New Rochelle NY 10804 Office: Albert Einstein Coll Medicine 1300 Morris Park Ave Bronx NY 10461

CHASE, MERRILL WALLACE, educator, immunologist; b. Providence, Sept. 17, 1905; s. John Whitman and Bertha H. (Wallace)

C.; m. Edith Steele Bowen, Sept. 5, 1931 (dec. 1961); children: Nancy Steele (Mrs. William W. Cowles), John Wallace, Susan Elizabeth; m. Cynthia Hambury Pierce, July 8, 1961. A.B., Brown U., 1927, Sc.M., 1929, Ph.D., 1931, Sc.D. honoris causa, 1977, M.D., U. Münster, West Germany, 1974. Instr. biology Brown U., 1931-32; staff mem. Rockefeller Inst. Med. Research, 1932-65; prof. immunology and microbiology, head lab. immunology and hypersensitivity Rockefeller U., 1956-79; med. adv. council Profl. Ednl. and Research Task Force, Asthma and Allergy Found. Am., 1955-82. Editor: (with C.A. Williams) Methods in Immunology and Immunochemistry, Vol. 1, 1967, Vol. 11, 1968, Vol. III, 1970, vols. IV, 1977, and V, 1976. Hon. fellow Am. Acad. Allergy (distinguished service award 1969), Am. Coll. Allergists; fellow Am. Acad. Arts and Scis.; mem. Am. Assn. Immunologists (pres. 1956-57), Am. Soc. Microbiology (program chmn. 1959-61), AAAS, Harvey Soc., N.Y. Acad. Scis., N.Y. Allergy Soc. (hon.), Nat. Acad. Sci. Republican. Universalist-Unitarian. Subspecialties: Allergy; Dermatology. Spl. research hypersensitivity to simple chem. allergens, studies Kveim antigen in sarcoidosis, studies tuberculins and mycobacterial antigens. Office: Rockefeller U 1230 York Ave New York NY 10021

CHASE, ROBERT ARTHUR, surgeon, educator; b. Keene, N.H., Jan. 6, 1923; s. Albert Henry and Georgia Beulah (Bump) C.; m. Ann Crosby Parker, Feb. 3, 1946; children—Deborah Lee, Nancy Jo, Robert N. B.S. cum laude, U. N.H., 1945; M.D., Yale, 1947. Diplomate: Am. Bd. Surgery, Am. Bd. Plastic Surgery. Intern New Haven Hosp., 1947-48, asst. resident, 1949-50, sr. resident surgery, 1952-53, chief resident surgeon, 1953-54; mem. faculty Yale Sch. Medicine, 1948-54, 59-62, asst. prof. surgery, 1959-62; mem. faculty U. Pitts., 1957-59, resident plastic surgeon, also teaching fellow, 1957-59; attending surgeon VA Hosp., W. Haven, Conn., 1959-62, Grace New Haven Community Hosp., 1959-63; prof., chmn. dept. surgery Stanford Sch. Medicine, 1963-74, Emile Holman prof. surgery, 1972—; prof. surgery U. Pa., 1974-77; attending surgeon Pa. Hosp., Hosp. U. Pa., Grad. Hosp., Phila., 1974-77; pres., dir. Nat. Bd. Med. Examiners, Phila., 1974-77; prof. anatomy Stanford (Calif.) U., 1977—; Cons. plastic surgery Christian Med. Coll. and Hosp., Vellore, S. India, 1962; cons. to surgeon gen. USAF, 1970—; Benjamin K. Rank prof. Australasian Coll. Surgeons, 1974. Author: Atlas of Hand Surgery; Editor: Videosurgery, 1974—; editorial bd.: Med. Alert Communication; Contbr. articles to profl. jours. Served to maj. M.C. AUS, 1949-57. Recipient Francis Gilman Blake award Yale Sch. Medicine, 1962, Henry J. Kaiser award Stanford U. Sch. Medicine, 1978-79. Fellow A.C.S., Australasian Coll. Surgeons (hon.); mem. Am. Soc. Plastic Surgery, Calif. Acad. Medicine (pres.), San Francisco Surg. Soc., Am. Surg. Assn., Santa Clara County, Conn. med. socs., Am. Soc. Surgery Hand, Am. Soc. Cleft Palate Rehab., Am. Assn. Surgery Trauma, Plastic Surgery Research Council, AMA, Soc. Clin. Surgery, Western Surg. Assn., Pacific Coast Surg. Soc., Am. Assn. Plastic Surgery, Am. Cancer Soc. (clin. fellowship com.), Found. Am. Soc. Plastic and Reconstructive Surgery (dir.), Soc. Univ. Surgeons, Inst. Med. (exec. com. 1976), Nat. Acad. Scis., Am. Soc. Most Venerable Order Hosp., St. John of Jerusalem, Halsted Soc., South African Soc. Surgery Hand (hon.), South African Soc. Plastic and Reconstructive Surgery (hon.), Sigma Xi. Subspecialties: Computers in Medicine; Surgery. Current work: Medical director man-made interface; computer-imaging interest. Home: 797 N Tolman Ln Stanford CA 94305 Office: Dept Surgery Stanford U Stanford CA 94305

CHASE, SHERRET SPAULDING, research botanist, educator; b. Toledo, June 30, 1918; s. Clement E. and Helen (Kelsey) C.; m. Catherine Compton, Nov. 27, 1943; children: Catherine, Sherret E., W. Compton, Helen, Alice. Student, U. Ariz., 1935-36; B.S., Yale U., 1939; Ph.D., Cornell U., 1947. Assoc. prof. botany and plant pathology Iowa State U., Ames, 1947-53; research geneticist, dir. fgn. seed ops. Dekalb AgResearch, Ill., 1954-66; Bullard fellow, Cabot fellow Harvard U., Cambridge, Mass., 1966-68; prof. biology SUNY-Oswego, 1968-80; dir. plant breeding and cell biology Internat. Plant Research Inst., San Carlos, Calif., 1980—; dir., founding pres. Catskill Ctr. Conservation and Devel., 1969—. Contbr. articles to profl. jours.; pioneer new methods of corn breeding. Bd. dirs. Mid-Hudson Pattern for Progress, Inc., 1972—, Assn. Protection Adirondacks, Inc., 1972—, Hanford Mills Mus., Inc., 1973-80, Ulster County Arts Council, 1974-80, Ecol. Catskill Cultural Ctr., 1974-80. Served to 1st lt. AC U.S. Army, 1942-45; ETO. Decorated D.F.C. Fellow AAAS; mem. N.Y. Acad. Sci., Am. Soc. Agronomy, Am. Forestry Assn., Genetics Soc. Am., Am. Inst. Biol. Scis., Am. Genetic Assn., Bot. Soc. Am., Sigma Xi, Phi Kappa Phi, Gamma Sigma Delta. Democrat. Subspecialty: Genetics and genetic engineering (agriculture). Current work: Economic botany, genetics and plant breeding. Office: IPRI 853 Industrial Rd San Carlos CA 94070

CHASEN, SYLVAN HERBERT, computer applications consultant, educator; b. Richmond, Va., May 19, 1926; s. Nathan and Hanna (Pass) C.; m. Catherine Hudlow, Mar. 25, 1946; children: Deborah Wyatt, Dianne Lipsey, Jane Morrison, Susan. Student, Va. Poly. Inst., 1943-44; B.S. in Engring, Ga. Inst. Tech., 1946, B. Chem. Engring., 1946; M.S., Emory U., 1951, 1951. Math. instr. Ga. Inst. Tech., Atlanta, 1946-50; head computer facility Naval Air Test Ctr., Patuxent, Md., 1951-58; dir. advanced computing CAD and interactive graphics Lockheed-Ga. Co., Marietta, 1958—; pres. Center CAD/CAM Tech., Inc.; cons. Author: Geometric Principles and Procedures for Computer Graphics Applications, 1978, The Guide for the Evaluation and Implementation of CAD/CAM Systems, 1980, 2d edit., 1983. Served as ensign USN, 1944-46. Recipient Outstanding Contbns. award Gov. Md., 1957; Disting. Contbns. award Soc. Mfg. Engrs., 1982. Mem. ASME, Soc. Mfg. Engrs., SIGGRAPH, NCGA. Subspecialties: Graphics, image processing, and pattern recognition; Computer-aided design. Current work: Interactive computer graphics and computer-aided design; evaluation, consulting in high technology with emphasis on interactive computer applications. Home: 760 Starlight Ct NE Atlanta GA 30342 Office: PO Box 76042 Atlanta GA 30328

CHASIN, MARK, pharm. co. exec.; b. N.Y.C., Feb. 20, 1942; s. Philip J. and Florence (Friedman) C.; m. Rena Bleiweiss, June 19, 1963; children: Jeffrey, Larry, Marni. A.B. in Chemistry, Cornell U., 1963; Ph.D. in Biochemistry (Am. Cancer Soc. fellow), Mich. State U., 1967. NIH trainee, 1964-67; sr. research investigator Squibb Inst., Princeton, N.J., 1967-74; dir. biochem. research Ortho Pharm. Corp., Raritan, N.J., 1974-78, dir. intercorp. new product devel., 1978-80, dir. clin. devel., 1980-81, dir. research info. services, 1981—. Editor: Methods in Cyclic Nucleotide Research, 1972; contbr. articles to profl. jours. Mem. Am. Soc. Biol. Chemists, Am. Soc. for Pharmacology and Exptl. Therapeutics, Am. Soc. Clin. Pharmacology and Therapeutics. Subspecialties: Biochemistry (medicine); Pharmacology. Current work: Cyclic nucleotides, control of hormonal action.

CHASSON, ROBERT LEE, physics educator; b. Cin., May 30, 1919; s. Mayer Leon and Fanny (Kondritzer) C.; m. Frances Jean Bray, Nov. 20, 1971; children by previous marriage: Barbara, William. A.B., U. Calif.-Berkeley, 1940, M.A., 1950, Ph.D., 1951. Asst. prof. physics U. Nebr., Lincoln, 1951-56, assoc. prof., 1956-59, prof., 1959-62, chmn. dept., 1956-62; prof. physics U. Denver, 1962—, chmn. dept. physics, div. head research inst., 1962-76; mem. adv. panel atmospheric sci. NSF, 1968-70; trustee Univ. Corp. Atmospheric Research, Boulder, Colo., 1970-79; mem. geophysics research bd. Nat. Acad. Scis., 1973-76. Author, editor works in field. Served with AUS, 1943-45. Guggenheim Found. fellow, 1962-63; Rockefeller Found. fellow, 1978-79. Fellow Am. Phys. Soc., AAAS; mem. Am. Geophys. Union, AAUP (chpt. pres. 1960-61), Sigma Xi. Democrat. Subspecialties: Cosmic ray high energy astrophysics; Solar physics. Current work: Cosmic rays and geophysics, solar-terrestrial relationships, interplanetary fields and particles, high energy astrophysics. Home: 3255 S Albion St Denver CO 80222 Office: Dept Physics U Denver Denver CO 80208

CHASSY, BRUCE MATTHEW, molecular biologist; b. Ft. Jackson, S.C., Oct. 22, 1942; s. Gilbert and Rose (Yellin) C.; m. Joan Ruth Trettan, Jan. 6, 1977; children by previous marriage: Lee Matthew, Lisa Marie. A.B., San Diego State U., 1962; Ph.D., Cornell U., 1965. Staff fellow NIH, Bethesda, Md., 1968-70, research chemist, 1970—; lectr. Am. U., Washington, 1968-72. Contbr. over 75 sci. articles to profl. publs. NSF fellow, 1960-62; NIH fellow, 1963-65. Mem. Am. Soc. Biol. Chemists, Am. Soc. Microbiology, Soc. Indsl. Microbiology, AAAS, N.Y. Acad. Scis., Sigma Xi. Subspecialties: Genetics and genetic engineering (medicine); Biochemistry (medicine). Current work: Gene cloning; molecular biology, lactose matabolism plasmids; sugar transport systems. Office: NIH Room 532 Bldg 30 Bethesda MD 20205

CHASTAIN, GARVIN, educator; b. Ft. Worth, Tex., Feb. 23, 1945; s. Garvin Dunn and Bertha Pearl (Parrish) C.; m. Gloria Jean Pollard, Nov. 21, 1975; 1 child by previous marriage, Ross Calvert. Head computer instruction Durhams Coll, Austin, Tex., 1976-77; research scientist Human Resources Research Orgn., Ft. Hood, Tex., 1977-78; asst. prof. Boise (Idaho) State U., 1978-82, assoc. prof. psychology, 1982—; Bd. dirs. Univ. Scholastic Assn., Boise, 1982—; mem. Ida. Com. of Creation/Evolution, Boise, 1982—. Contbr. articles to profl. jours. Vice pres. Rockies, Freedom From Religion Found., Madison, Wis., 1980-82. Faculty research grantee Boise State U., 1982. Mem. Psychonomic Soc., Am. Psychol. Assn., Rocky Mt. Psychol. Assn., Phi Kappa Phi, Psi Chi. Libertarian. Subspecialties: Cognition; Psychophysics. Current work: Human exptl. psychology; visual info. processing; pattern recognition; spatial attention; reading; visual signal detection. Home: 1819 Tendoy Dr Boise ID 83705 Office: Boise State Univ 1910 University Dr Boise ID 83725

CHATTERJEE, SATYA NARAYAN, medical educator, surgeon; b. Calcutta, India, Dec. 31, 1934; came to U.S., 1973; s. R. N. and J. C.; m. Patricia Sheppard, Sept. 26, 1964; children: Sharmila, Shalini, Arun. Inter.Sci., Scottish Ch. Coll., Calcutta, 1951; M.B.B.S., R.G. Kar Med. Coll., Calcutta, 1957. Diplomate: Brit. Bd. Surgery. Registrar dept. surgery North Staff Royal Infirmary, Stoke, Eng., 1963-64; med. supt. A.R.T. Co., Margherita, India, 1965-69; registrar transplant unit U. Edinburgh, Scotland, 1970-73; clin. and research fellow U. So. Clif. Med. Sch., Los Angeles, 1973-75, asst. prof. surgery, 1975-77; assoc. prof. surgery U. Calif.-Davis, Sacramento, 1977—; cons. surgeon Martin Luther King Gen. Hosp., Los Angeles, 1975-77; dir. renal transplant unit No. Calif. Tissue Bank, San Jose, 1977-78. Editor: Surgical Clinics of North America, 1978, Manual of Renal Transplantation, 1979, Renal Transplantation: A Multidisciplinary Approach, 1980, Organ Transplantation, 1982. R.G. Kar Med. Coll. scholar, 1954. Fellow ACS; mem. Am. Soc. Transplant Surgeons, Assn. Acad. Surgery, Transplantation Soc., Royal Soc. Medicine, Western Assn. Transplant Surgeons (pres. 1979-80). Hindu. Club: Tennis (Davis). Subspecialties: Transplant surgery; Transplantation. Current work: Transplantation immunology, clinical renal transplantation, infectious complications of transplantation, viral infections in transplantation, immunologic monitoring, and tissue typing. Home: PO Box 3102 El Macero CA 95618 Office: 4301 X St Sacramento CA 95817

CHATTERJEE, SUNIL KUMAR, cancer research scientist; b. Calcutta, India, Aug. 7, 1940; came to U.S., 1966, naturalized, 1981; s. Bhupendra Nath and Parimal Bala (Banerjee) C.; m. Malaya Chatterjee, Oct. 25, 1972; children: Indranil, Sumana. B.S., U. Calcutta, 1959, M.S., 1961, Ph.D., 1966. With U. Pa., 1966-68, Inst. Cancer Research, Phila., 1968-70; research assoc. U. Calcutta, 1970-71; asst. research officer Max-Planck Institut, Gottingen, W. Ger., 1972-73; sr. cancer research scientist Roswell Park Meml. Inst., Buffalo, 1973—. Contbr. articles and abstracts to profl. jours. Mem. AAAS, Am. Assn. Cancer Research, N.Y. Acad. Scis. Democrat. Hindu. Subspecialties: Biochemistry (biology); Oncology. Current work: Biochemistry of cancer cell, glycoprotein biosynthesis, membrane changes in malignacy, markers for cancers, cancer metastasis. Home: 152 Telfair Dr Williamsville NY 14221 Office: 666 Elm St Buffalo NY 14263

CHATTOPADHYAY, SOMNATH, mech. engr., researcher, educator; b. Howrah, West Bengal, India, Feb. 25, 1946; came to U.S., 1969, naturalized, 1981; s. Bireswar and Shakti C.; m. Mandira, July 2, 1971; children: Somak, Parama. Ph.D. in Mechanics, Princeton U., 1974. Registered profl. engr., Pa. Research asst. Princeton U., 1969-71, instr., 1971-73; postdoctoral fellow Purdue U., 1973-74; sr. engr. Westinghouse Electric Corp., Pitts., 1974-78, lead engr., steam generator, 1979—; cons. Princeton U., 1974; mem. adj. faculty Pa. State U. Contbr. articles to profl. jours. Govt. India nat. scholar, 1962-67. Mem. ASME, Soc. Engring. Sci., Am. Acad. Mechanics, ASTM. Subspecialties: Mechanical engineering; Theoretical and applied mechanics. Current work: Dynamics, structural stability, composites, electromagnetic interaction, structural mechanics in reactor technology, fatigue design of structures. Home: 1005 Crestwood Dr North Huntingdon PA 15642

CHATURVEDI, ARVIND KUMAR, toxicologist, pharmacologist; b. Azamgarh, India, July 29, 1947; s. Vijai Narain and Sursari Devi (Pandey) C.; m. Mira Shukla, June 14, 1970; children: Priyanka, Vivek. B.S., Gorakhpur U., 1966; M.S., Banaras Hindu U., 1968; Ph.D., Lucknow U., 1972. Research fellow Lucknow U., India, 1968-74; research assoc. Vanderbilt U., Nashville, 1974-77, research instr., 1977-78; acting state toxicologist N.D. State U., Fargo, 1980, asst. prof. toxicology, 1978-82, assoc. prof., 1982—; asst. toxicologist Office N.D. State Toxicologist, 1978—. Am. Cancer Soc. grantee, 1979—; NIH grantee, 1980-82. Mem. Indian Pharmacol. Soc. (life); mem. Indian Acad. Neurosci. (life); Mem. Internat. Soc. Biochem. Pharmacology, Am. Soc. Pharmacology and Exptl. Therapeutics, Soc. Environ. Toxicology and Chemistry, Internat. Soc. Study Xenobiotics, Soc. Exptl. Biology and Medicine, Rho Chi. Subspecialties: Toxicology (medicine); Pharmacology. Current work: Drug toxicity on male reproductive system, influence of toxic agents on calcium transport, devel. of choline acetyltransferase inhibitors, analytical toxicology. Office: Dept Toxicology ND State U Coll Pharmacy PO Box 5195 Farg ND 58105

CHAUDHARI, ANSHUMALI, biochemist, researcher, educator; b. Allahabad, Uttar Pradesh, India, June 22, 1947; came to U.S., 1975; s. Krishna Sahai and Shanti Devi (Verma) C.; m. Suman Lata, Feb. 7, 1973; children: Swetanshu, Shruti. B.S., U. Lucknow, India, 1966, M.S., 1968, Ph.D., 1976. Jr. research fellow U. Lucknow, Uttar Pradesh, India, 1968-72, sr. research fellow, 1972-75; vis. fellow Nat. Inst. Environ. Health Scis., Research Triangle Park, N.C., 1975-77; research assoc. Wayne State U., Detroit, 1977-80; adj. asst. prof. medicine UCLA, 1980—; vis. fellow Fogarty Internat. Center, NIH, 1975; sr. research fellow Indian Nat. Sci. Acad., 1972, Indian Council Med. Research, 1974; advanced research fellow Am. Heart Assn., 1980. Mem. Am. Soc. Pharmacology and Exptl. Therapeutics, Internat. Soc. Biochem. Pharmacology. Subspecialties: Biochemistry (medicine); Molecular pharmacology. Current work: Evaluation of role of prostaglandins in pulmonary and renal functions, prostaglandin endoperoxide synthetase-dependent metabolism of xenobiotics and elucidation of physiologic and pathphysiologic significance of renal prostaglandin receptors. Home: 14607 Erwin St #105 Van Nuys CA 91411 Office: Dept Medicine UCLA 700 Tiverton Los Angeles CA 90024

CHAUDHURI, TAPAN KUMAR, physician; b. Calcutta, West Bengal, India, Nov. 25, 1944; came to U.S., 1967, naturalized, 1978; s. Taposh K. and Bulu R. Chowdbury; m. Chhanda Sen, Mar. 4, 1980; children: Lakshmi, Madhu, Krishna. I.Sc., U. Calcutta, 1961, M.D., 1966. Intern NRS Med. Coll. Hosp., Calcutta, 1966-67, South Side Hosp., Pitts., 1968-69; NRS Med. Coll. Hosp., 1967, W. Va. U. Hosp., Morgantown, 1970-71; chief nuclear medicine VA Med. Center, Hampton, Va., 1974—; prof. nuclear medicine Eastern Va. Med. Sch., Norfolk, 1979—. Recipient Gold and Silver medals U. Calcutta, 1961-66. Fellow ACP, Am. Coll. Gastroenterology; mem. Soc. Nuclear Medicine. Subspecialty: Nuclear medicine. Current work: Medical research, administrator, educator. Home: 304 Rudisill Rd Hampton VA 23669 Office: VA Med Center Hampton VA 23667

CHAVE, KEITH ERNEST, marine geochemist, edcuator; b. Chgo., Jan. 18, 1928; s. Ernest J. and Winnifred M. (Carruthers) C.; m. Edith H. Hunter, May 19, 1969; children: Alan D., Warren T. Ph.B., U. Chgo., 1948, M.S., 1950, Ph.D., 1951. Research geochemist Calif. Research Corp., LaHabra, 1952-59; prof. geology Lehigh U., 1959-67; prof. U. Hawaii, 1967—, assoc. dean research, 1976-77; pres. Palau Marine Research Inst., Koror, Palau, Hawaii, 1979—. Contbr. numerous articles to profl. jours. Sr. U.S. Scientist Alexander von Humboldt Stiftung, Kiel, W.Ger., 1974. Fellow AAAS; mem. Sigma Xi. Subspecialties: Oceanography; Geochemistry. Current work: Seawater-mineral interactions, teaching graduate students. Office: University of Hawaii Honolulu HI 96822

CHAVIN, WALTER, biologist, educator; b. N.Y.C., Dec. 6, 1925; s. Isidor and Fanny (Kesch) C. B.S., CCNY, 1946; M.S., NYC, 1949; Ph.D., NYU, 1954. Instr. biology CCNY, 1946-47, U. Ariz., 1949-51; research specialist Am. Mus. Natural History, N.Y.C., 1951-53; asst. prof. Wayne State U., Detroit, 1953-60, assoc. prof., 1960-65, prof. dept. biol. scis., 1965—, prof. radiobiology Sch. Medicine, 1975-80; research assoc. Argonne Nat. Lab., 1955-58; NSF sr. postdoctoral fellow Sorbonne, Paris, 1960-61; cons. AEC, 1955-58, Dept. Interior, 1963-68. Author: Responses of Fish to Environmental Changes, 1973. Fellow AAAS, N.Y. Acad. Sci.; mem. Sigma Xi (research award 1968). Subspecialties: Endocrinology; Cancer research (medicine). Current work: Control mechanisms in cells and changes producing abnormal growth or function, the impact of hormones upon cellular control mechanisms and the effects of environmentally induced hormone alterations on cellular control mechanisms. Home: 1368 Joliet Pl Detroit MI 48207 Office: Wayne State U 5104 Gullen Mall Detroit MI 48202

CHAVKIN, LEONARD THEODORE, mfg. co. exec.; b. N.Y.C., Dec. 2, 1925; s. Charles I. and Sadie (Greenspan) C.; m. Betty Schein, May 28, 1944; children—Dana, Charles, Andrea, Matthew. B.S., Columbia U., 1944; M.S., Phila. Coll. Pharmacy and Sci., 1947; Ph.D., N.Y. U., 1960. Mem. faculty Columbia U. Coll. Pharmacy, 1947-59, trustee, 1965-70; research dir. Bristol-Myers Co., Hillside, N.J., 1959—. Author articles indsl. pharmacy. Mem. bd. health, Mountainside, N.J., 1973-77. Served with AUS, 1944-46. Mem. Soc. Cosmetic Chemists, Indsl. Research Inst., Assn. Research Dirs. Subspecialty: Pharmaceutical Research Management. Home: 340 W Dudley Ave Westfield NJ 07090 Office: 1350 Liberty Ave Hillside NJ 07207

CHEAL, MARYLOU, psychobiologist; b. St. Clair County, Mich.; d. Marion Louis and Leda Eleanor Shaw (Martin) Fast; m. James Cheal, Mar. 13, 1946; children: Thomas James, Catheryn Leda, Robert David. B.A. with honors, Oakland U., Rochester, Mich., 1969; Ph.D. in Psychology, U. Mich., Ann Arbor, 1973. Research investigator dept. zoology U. Mich., 1973-75, dept. oral biology, 1975-76; postdoctoral fellow McLean Hosp., Belmont, Mass., 1976-77, asst. psychologist, 1977-81; assoc. psychologist Neuropsychology Lab., 1981-83; mem. research faculty dept. psychology Ariz. State U., Temple, 1983—; lectr. psychology Harvard U. Med. Sch., also mem. summer sch. faculty. Contbr. articles and revs. to profl. jours. Charles A. King fellow, 1976-77; Howard U. and Rockefeller Found. award, 1980; grantee McLean Hosp., 1976-78, NIMH, 1977-78, Scottish Rite Schizophrenia Research Program, 1977-79, McLean Hosp., 1979-81. Mem. AAAS, Am. Psychol. Assn., Assn. Women in Sci., Women in Neurosci. (com.), Eastern Psychol. Assn., Soc. Neurosci., Assn. Chemoreception Sci., Sigma Xi. Subspecialties: Psychobiology; Physiological psychology. Current work: Neural mechanisms of attention and habituation, aging,psychopharmacology, gerbils, brain, olfactory, sex differences, reprodn., stereotypy, dopamine, acetylcholine scopolamine, apomorphine, catecholamines, odor preferences, norepinephrine, clonidine, animal restraint. Home: 127 E Loma Vista Temple AZ 85282 Office: Lab Animal Care Program Ariz State U Temple AZ 85287

CHECHIK, BORIS E., immunologist, leukemia researcher; b. Kislovodsk, USSR, Mar. 31, 1931; emigrated to Can., 1973, naturalized, 1977; s. Eber L. and Eva (Grinberg) C.; m. Batia Jurbinski, Nov. 21, 1956; 1 dau., Miriam Sternberg. M.D., 2d Moscow Med. Sch., 1956; Ph.D., Acad. Med. Sci., P.A. Gertzen Oncol. Inst., Moscow, 1963. Gen. practice medicine, Moscow, 1956-59; scientist, then sr. scientist P.A. Gertzen Oncological Inst., 1959-71; scientist Hebrew U., Jerusalem, Israel, 1971-73; research assoc. Hosp. for Sick Children, Toronto, Ont., Can., 1974-76; research investigator Mt. Sinai Hosp., Toronto, 1977-80, sr. scientist, 1981—; assoc. prof. immunology U. Toronto, 1981—. Contbr. articles to profl. jours. Nat. Cancer Inst. Can. grantee, 1977—; Leukemia Research Fund grantee, 1977—. Mem. Internat. Assn. Comparative Research on Leukemia and Related Diseases, Am. Assn. Cancer Research, Can. Soc. Immunology. Subspecialties: Cancer research (medicine); Immunology (medicine). Current work: Cancer research, hematology, immunochemistry, immunomorphology, leukemia-associated antigens, differentiation of lymphoid cells. Discovered 1st human thymus-leukemia-assoc. antigen, 1967-68, established enzymatic nature, 1980, developed detection methods, 1979-81. Office: Dept Research Mt Sinai Hosp 600 University Ave Toronto ON Canada M5G 1X5

CHEDID, ANTONIO, pathologist, educator, researcher; b. Barranquilla, Colombia, May 5, 1939; came to U.S., 1966; s. Aziz Antonio and Maria (Turbay) C.; m. Hoda Abi-Rached, Sept. 14, 1974; children: Anthony John, Marie-Claude, Erica Houda. B.S., Coll. of Barranquilla, 1954; M.D., U. Madrid, 1962. Diplomate: Am. Bd. Pathology. Intern Columbus Hosp., Chgo., 1967-68; resident in pathology Michael Reese Hosp., Chgo., 1968-72; instr. pathology Pritzker Sch. Medicine, U. Chgo., 1972-73; asst. prof. pathology U. Cin. Coll. Medicine, 1973-76; assoc. prof. pathology Chgo. Med. Sch., North Chicago, Ill., 1976—; pathol. cons. coop. study of alcoholic hepatitis VA, 1979—. Am. Cancer Soc. grantee, 1974. Mem. Am. Assn. Pathologists, Fedn. Am. Socs. Exptl. Biology, Internat. Acad.

Pathology, Am. Assn. for Study of Liver Diseases. Subspecialties: Pathology (medicine); Hepatology. Current work: Cell enzyme markers during differentiation and experimental carcinogenesis as well as chronobiology.

CHEH, HUK YUK, electrochemist, engineering educator; b. Shanghai, China, Oct. 27, 1939; s. Tze Sang and Sue Lang (Che) C.; m. An-li, July 26, 1969; children: Emily, Evelyn. B.A.Sc. in Chem. Engring, U. Ottawa, Can., 1962; Ph.D., U. Calif.-Berkeley, 1967. Mem. tech. staff Bell Telephone Labs., N.J., 1967-70; asst. prof. chem. engring. Columbia U., N.Y.C., 1970-73, assoc. prof., 1973-79, prof., 1979—, Ruben-Viele prof., 1982—, chmn. dept., 1980—; program dir. NSF, 1978-79; vis. research prof. Nat. Tsinghua U., Taiwan, 1977. Contbr. articles to sci. jours. Recipient Harold C. Urey award, 1980. Mem. Am. Inst. Chem. Engrs., N.Y. Acad. Scis., Electrochem. Soc., Am. Electroplaters Soc., Sigma Xi. Subspecialties: Chemical engineering; Catalysis chemistry. Current work: Electrochemistry and surface chemistry, high temperature processes, heterogeneous catalysis, transport processes. Office: Columbia U New York NY 10027

CHEN, CHEN-HO, biologist, educator; b. Kiangsu-China, Sept. 9, 1929; s. Tung-Hou and Jane-Ou (Ku) C.; m. Jane-Ru Po, July 22, 1960; children: Winnie, Frankie, Judie. B.S., Nat. Taiwan U., 1954; M.S., La. State U., 1960; Ph.D., S.D. State U., 1964. Postdoctoral fellow S.D. State U., Brookings, 1963-64; research assoc. Argonne Nat. Lab., Ill., 1964-65; lectr., acting head Hong Kong Bapt. Coll., 1965-68; from asst. prof. to prof. biology and plant sci. S.D. State U., 1968—; Mem. Bot. Soc. Am., S.D. Acad. Sci. Contbr. articles to profl. jours. Subspecialties: Plant cell and tissue culture; Genetics and genetic engineering (agriculture). Current work: Plant cell and tissue culture; cytogenetics; cell biology. Home: 114 Gilley Ave Brookings SD 57006 Office: SD State U Brookings SD 57007

CHEN, CHIH PING, engineer; b. Kwangtung, China, June 12, 1938; s. Thai and Kieu (Mach) Tran; m. Virginia Wu-jen Wu, June 15, 1968; 1 dau., Christine. B.S., Cheng-Kung U., Taiwan, 1961; M.S., Pa. State U., 1965; Ph.D., Carnegie-Mellon U., Pitts., 1970. Registered profl. engr., Pa., N.J. Project engr. Carnegie Inst. Tech., Pitts., 1965-70; lead engr. Gibbs & Hill, Inc., N.Y.C., 1970-73; sr. staff engr. Pub. Service Electric & Gas Co., Newark, N.J., 1973-77; staff nuclear engr. Power Authority of State N.Y., N.Y.C., 1977-79; project mgr. Battelle Meml. Inst., Columbus, Ohio, 1979—; cons. Inst. Nuclear Energy Research, Taiwan, 1978. Contbr. articles to profl. jours. Treas. Columbus Chinese Sch., 1980-81. NSF grantee, 1965; NDEA fellow, 1970. Mem. Am. Nuclear Soc. (mem. standards com.), ASME, N.Y. Acad. Sci., Nat. Soc. Profl. Engrs., Sigma Xi. Subspecialties: Mechanical engineering; Nuclear engineering. Current work: Managed projects to develop and license a deep geologic repository for disposal of high-level nuclear wastes. Experience in nuclear power which includes engineering and design, operations supports, licensing, safety and engineering analyses. Research in magnetohydrodynamics. Home: 254 Chinkapin Way Westerville OH 43081 Office: Battelle Meml Inst 505 King Ave Columbus OH 43201

CHEN, CHIN, earth science educator, oceanography researcher; b. Foochow, Fukien, China, Apr. 15, 1927; came to U.S., 1955, naturalized, 1969; s. Y.C. and N.M. (Lin) C.; m. Concordia Chao, July 2, 1960; children: Marie H.M., Albert C. B.S., Nat. Taiwan U., China, 1952; M.S., Wayne State U., 1957; Ph.D., Boston U., 1962. Jr. geologist Taiwan (China) Geol. Survey, 1952-55; adj. staff Boston Coll., 1961-62; research assoc. Lamont-Doherty Geol. Observatory, Columbia U., 1962-69; prof. earch sci. Western Conn. State U., 1969—; cons. Dobex Internat. Ltd., W. Nyack, N.Y., summer, 1981; Lamont-Doherty Geol. Observatory, summer 1980. Author: Economic Geology and Mineral Resource of China, 1984. Recipient Sohio Scholarship award, 1958. Fellow Geol. Soc. Am.; mem. Paleontol. Research Inst. Roman Cath. Subspecialties: Paleoecology; Deep-sea biology. Current work: Pteropod-globigerina ooze of deep sea sediments and stratigraphy; sea level change in South China Sea; petroleum geology and energy resources in People's Republic of China. Home: Mountain Pass Box 34 RD 6 Hopewell Junction NY 12533 Office: Western Conn State U 181 White St Danbury CT 06810

CHEN, CHING JEN, engineering educator, administrator, research scientist, consultant; b. Taipei, Taiwan, July 6, 1936; s. I. Sung and T. (Yen) C.; m. Ruei Man, Aug. 14, 1965; children: Sandra, Anthony. Diploma, Taipei Inst. Tech., 1957; M.S. in Mech. Engring., Kans. State U., 1962; Ph.D., Case Western Res. U., 1967. Design engr. Ta-Tung Grinding Co., Taipei, Taiwan, 1959-60; asst. prof. U. Iowa, Iowa City, 1967-70, assoc. prof., 1970-77, prof., 1977-82, chmn., prof. energy div., 1982—, chmn. dept. mech. engring., 1982—, sr. research scientist Iowa Inst. Hydraulic Research, 1970—; sci. advisor U.S. Army Weapons Command, Rock Island, Ill., 1973; cons. Westinghouse Research Lab., Pitts., 1973-79, Jet Propulsion Lab., Pasadena, Calif., 1973, Argonne (Ill.) Nat. Lab., 1979—. Author: Vertical Turbulent Buoyant Jets, 1979; contbr. monograph to profl. publ. Old Gold fellow U. Iowa, Iowa Found., 1968; recipient sr. U.S. scientist award Alexander von Humboldt Fund., W.Ger., 1974. Mem. ASME, AIAA, Am. Phys. Soc., ASCE, Sigma Xi. Subspecialties: Fluid mechanics; Combustion processes. Current work: Heat transfer, computational fluid mechanics, turbulent flows, numerical analysis. Home: 7 Heather Dr Iowa City IA 52240 Office: Univ Iow EB 2216 Iowa City IA 52242

CHEN, CHING-NIEN, government health institution chemist; b. Checkiang, China, Jan. 22, 1945; came to U.S., 1973, naturalized, 1980; s. Ko-Fei and Ju-Chuen (Ma) C. B.S., Nat. Taiwan U., 1967; M.S., Nat. Tsing-Hua U., 1969; Ph.D., SUNY-Stony Brook, 1980. Instr. Fu-jen U., Taipei, Taiwan, 1968-70, Nat. Tsing-Hua U., Hsinchu, Taiwan, 1969-73; vis. fellow NIH, Bethesda, Md., 1980-81, expert chemist, 1981—. Mem. Soc. Magnetic Resonance in Medicine, Sigma Xi (assoc.). Subspecialties: Nuclear magnetic resonance (chemistry); Biomedical engineering. Current work: Nuclear magnetic resonance imaging (zeugmatography) to obtain 3D NMR signal distribution in objects - one application is to image the internal structure of human body as in CT, but with added dimension of physiologically related chemical information. Office: NIH Bldg 13 3W-13 Bethesda Md 20205

CHEN, CHI-PO, pharmacologist, educator; b. Taiwan, May 14, 1940; m. Pai-Fo Wang, Apr. 20, 1969; children: Joyce, Eric. B.S., Kachsiung Med. Coll., Taiwan, 1963; M.S., Queen's U., Can., 1969; Ph.D., U. Ky., 1973. Research asso. Vanderbilt U., Nashville, 1973-74; asst. prof. U. Conn., Storrs, 1974-80; assoc. prof. Ohio Coll. Podiatric Medicine, Cleve., 1980—; researcher in field. Mem. Am. Soc. for Pharmacology and Exptl. Therapeutics, Am. Pharm. Assn., Acad. of Pharm. Sci., N.Y. Acad. Scis. Subspecialties: Pharmacology; Cellular pharmacology. Current work: Hepatic uptake and clearance of therapeutic agents, drug interactions, biological membrane transport. Office: 10515 Carnegie Ave Cleveland OH 44106

CHEN, CHONG-MAW, life science educator; b. Taoyuan, Taiwan; s. Jen-ho and Chao (Wu) C.; m. Gong-rong Chuang, Jan. 29, 1967; children: Sharon, Alice, Howard. B.S., Nat. Taiwan Normal U.-Taipei, 1959; Ph.D., U. Kans.-Lawrence, 1967. Mem. faculty U. Wis.-Parkside, Kenosha, 1971—, prof. life sci., 1977—, mem., 1981—, dir. lab. plant biochemistry, 1972—. Contbr. numerous sci. articles to profl. publs. Recipeint Teaching Excellence award U. Wis.-Parkside, 1972, 78; NSF grantee, 1972—; NIH grantee, 1976—. Mem. Am. Soc. Biol. Chemists, Am. Chem. Soc., Am. Soc. Plant Physiologists, AAAS, Sigma Xi. Subspecialties: Molecular biology; Plant growth. Current work: Nucleic acids and protein synthesis; plant hormones; crown-gall tumors. Home: 2701 Village Green E Racine WI 53406 Office: Univ Wis-Parkside Dept Life Sci and Biomed Research Inst Kenosha WI 53141

CHEN, ER-PING, laboratory executive; b. Ping-Liang, Kansu, China, May 19, 1944; came to U.S., 1967, naturalized, 1977; s. Shen-Huang and Tze-Yu (Chou) C.; m. Regina Chi Chen, Mar. 31, 1973; children: Candice, Benjamin. B.S., Nat. Chung-Hsing U., Taichung, Taiwan, 1966; M.S., Lehigh U., 1969, Ph.D., 1972. Asst. prof. Lehigh U., Bethlehem, Pa., 1972-77, assoc. prof., 1977-78; mem. tech. staff Sandia Nat. Labs., Albuquerque, 1978—. Author: Cracks in Composites, 1981; contbr. articles in field to profl. jours. Mem. ASME, Soc. Exptl. Stress Analysis, Soc. Engring. Sci., Chinese Inst. Engrs., Sigma Xi. Subspecialties: Fracture mechanics; Solid mechanics. Current work: Fracture of composites analysis, nuclear power plant safety analysis, fracture and fragmentation in rock. Home: 1128 Bernalillo St SE Albuquerque NM 87123 Office: Div 1522 Sandia Nat Labs Albuquerque NM 87185

CHEN, HO SOU, physicist; b. Nov. 24, 1932; m.; 2 children. B.S., Nat. Taiwan U., 1956; M.S., Brown U., 1963; Ph.D. in Applied Physics, Harvard U., 1967. Mem. tech. staff Bell Telephone Labs., 1968-70; asst. prof. chemistry Yeshiva U., 1970-71; physicist Allied Chem. Corp., Morristown, N.J., 1971-72; mem. tech. staff Bell Labs., Murray Hill, N.J., 1972—. Recipient George W. Morey award Am. Ceramic Soc., 1978. Subspecialties: Amorphous metals; Ceramics. Address: 600 Mountain Ave Murray Hill NJ 07974

CHEN, HOFFMAN HOR-FU, chemistry educator; b. Hualien, China, Aug. 5, 1941; came to U.S., 1969; s. Sing-Wong and Youg-Mei (Ray) C.; m. Ay-Ming Yuam, May 5, 1973; children—Jeffrey, Jennifer, Justin. B.S., Tamkang U., 1963; Ph.D., U. Tex., 1976. Research scientist U. Tex., Austin, 1976; research asso. Okla. State U., Stillwater, 1977-78; assoc. prof. chemistry Grambling State U., La., 1979—. Robert Welch Found. fellow, 1972-76; NIH fellow, 1980. Mem. Am. Chem. Soc., Beta Kappa Chi. Presbyterian. Subspecialties: Organic chemistry; Catalysis chemistry. Current work: Researcher in homogeneous and heterogeneous catalysis especially as applied to new friedelcrafts reactions and fischer tropsch chemistry. Home: 1600 Haskell Dr Ruston LA 71270 Office: Grambling State University Dept Chemistry Grambling LA 71245

CHEN, HOLLIS CHING, electrical engineer, educator, researcher; b. Chekiang, China, Nov. 17, 1935; came to U.S., 1960, naturalized, 1971; s. Yu-Chao and Shui-tan C.; m. Donna Liu, Sept. 3, 1961; children: Desiree, Hollis Tao. B.S., Nat. Taiwan U., 1957; M.S., Ohio U., 1961; Ph.D., Syracuse U., 1965. Asst. prof. Syracuse (N.Y.) U., 1962-65; asst. prof. elec. engring. Ohio U., Athens, 1965-67, assoc. prof., 1967-75, prof., 1975—; mem. Commn. B., U.S. Nat. Com., Internat. Union Radio Sci. Contbr. numerous articles to profl. jours. Mem. AAAS, IEEE (sr., best paper and best corr. awards 1967), Soc. Indsl. and Applied Math., Math. Assn. Am., Optical Soc. Am., Am. Soc. for Engring. Edn., Sigma Xi, Eta Kappa Nu. Subspecialties: Computer engineering; Plasma. Current work: Electromagnetic wave propagation and excitation, mocrowave, antennas, applied mathematics. Home: 1 Ball Dr Athens OH 45701 Office: Elec Engring Dept Ohio U Athens OH 45701

CHEN, HSIANG TSUN, physicist; b. Taiwan, Oct. 10, 1947; came to U.S., 1973; s. Chin Fa and Pan C.; m. Ching-Min Chen, Mar. 18, 1973; children: Hong Hsiang, Kaivin Hong. Ph.D., U. Ga., 1980. Sr. scientist Machlett Labs., Inc., Stamford, Conn., 1979-83, Ortho Diagnostic Systems, Inc., Raritan, N.J., 1983—. Served to 2d lt. China Air Force, 1971-72. Mem. Optical Soc. Am. Subspecialties: Optical signal processing; Computer engineering. Current work: Application of experience in physics, optics, electronics, computer engineering to design and build microprocessor based diagnostic equipment of medical imaging and clinical chemistry. Home: 5 Carriage Way Belle Mead NJ 08502

CHEN, HSIEN-JEN JAMES, research neuroendocrinologist, reproductive biologist; b. Chin Hsien, Kwangtung, China, June 19, 1931; came to U.S., 1971; s. Yi-Chung and Yueh-mei (Kuo) C.; m. Teresa Hui-li Ho, Aug. 1, 1971; 1 son, Edward Joseph. B.S., Nat. Taiwan Normal U., Taipei, 1965; M.s., Nat. Taiwan U. Med. Coll., Taipei, 1969; Ph.D., Mich. State U., 1976. Postdoctoral fellow Mt. Sinai Hosp., Toronto, Ont., Can., 1976-78; NIH postdoctoral fellow U. Tex. Health Sci. Ctr.-San Antonio, 1978-79, chief Brain Peptide Research Lab., Barrow Neurol. Inst., Phoenix, 1979—. Endocrine Soc. grantee, 1980; Am. Physiol. Soc. grantee, 1983. Mem. Am. Physiol. Soc., Endocrine Soc., Soc. Neurosci. Subspecialties: Neuroendocrinology; Reproductive biology (medicine). Current work: Research in hypothalamic control of pituitary hormones, brain peptide control of pituitary hormones and reproduction. Home: 2844 E Cortez Phoenix AZ 85028 Office: Barrow Neurol Inst Saint Joseph's Hosp and Med Center 350 W Thomas Rd Phoenix AZ 85028

CHEN, KIRK CHING SHYONG, biochemistry educator; b. Taichung, Taiwan, Nov. 30, 1941; came to U.S., 1968; s. Chi-Nan and Yen-Yang (Pang) C.; m. Cindy C.Y. Wang, Mar. 18, 1972; children: Lucy, Megan. B.S. in Agrl. Chemistry, Nat. Taiwan U., 1965, M.S. in Biochemistry, 1968, Ph.D., U. Okla., 1972. Postdoctoral fellow Rockefeller U., 1972-75, Lab. of Molecular Biology, Med. Research Council, Cambridge, Eng., 1975-76; asst. prof. pathobiology U. Wash., 1976-80, assoc. prof., 1980—. Served to 2d lt. Chinese Army, 1965-66. Helen Hay Whitney Found. postdoctoral fellow, 1973-76. Mem. Harvey Soc., Am. Soc. Biol. Chemists. Subspecialties: Biochemistry (medicine); Microbiology (medicine). Current work: Structural and functional studies on surface antigens from pathogenic microorganisms, degradation of beta-lactam antibiotics by microorganisms. Home: 543 NE 79th St Seattle WA 98115 Office: Dept of Pathobiolog University of Washington Seattl WA 98195

CHEN, LINDA HUANG, biochemistry and nutrition educator; b. Tokyo, Mar. 22, 1937; U.S., 1960; d. Chun mu and Chiung tien (Lin) Huang; m. Boris Yuen-jien Chen, Dec. 23, 1960; children: Audrey Huey-wen, Lisa Min-yi. B.S. in Pharmacy, Nat. Taiwan U., 1959; Ph.D. in Biochemistry, U. Louisville, 1964. Research assoc. U. Louisville, 1964-66, asst. prof., 1967-72, assoc. prof., 1972-79; prof. U. Ky., Lexington, 1979—; mem. cancer research manpower rev. com. Nat. Cancer Inst., Bethesda, Md., 1980-84. Author: Nutritional Biochemistry, 1973; contbr. articles on nutritional biochemistry to profl. jours. Mem. Am. Inst. Nutrition, Am. Soc. Clin. Nutrition, Gerontol. Soc. Am., Inst. Food Technologists. Subspecialties: Nutrition (medicine); Gerontology. Current work: Nutrition and aging, nutrition and cancer, nutrient drug interaction, nutrient interactions. Home: 531 South Bend Dr Lexington KY 40503 Office: U Ky Lexington Dept Nutritiona and Food Sci Lexington KY 40506

CHEN, MICHAEL MING, engr., educator; b. Hankow, China, Mar. 10, 1933; came to U.S., 1953, naturalized, 1965; s. Kwang Tzu and Hwei Chuing (Deng) C.; m. Ruth Hsu, Oct. 15, 1961; children—Brigitte (dec.), Derek, Melinda. B.S., U. Ill., 1955; S.M., M.I.T., 1957, Ph.D., 1961. Sr. staff scientist research and devel. Avco Corp., Wilmington, Mass., 1960-63; asst. prof. engring. and applied sci. Yale U., 1963-69; asso. prof. mech. engring. N.Y. U., 1969-73; prof. mech. engring. and bioengring. U. Ill., Urbana-Champaign, 1973—; cons. A.D. Little Co., NIH, Argonne Nat. Lab. Asso. editor: Jour. Biomech. Engring; contbr. to profl. publs. Mem. ASME, Am. Phys. Soc., Sigma Xi, Phi Kappa Phi, Tau Beta Pi, Pi Tau Sigma. Subspecialties: Biomedical engineering; Mechanical engineering. Home: 311 Eliot Dr Urbana IL 61801 Office: 144 MEB Univ Ill 1206 W Green St Urbana IL 61801

CHEN, RICHARD YUAN ZIN, anesthesiologist, educator; b. Chia-Yi, Taiwan, Feb. 7, 1947; came to U.S., 1972, naturalized, 1977; s. L.C. and Elizabeth (Pi-Aru) C.; m. Chang Chiu, July 14, 1972; children: Jason T.S., Gregory T.M. M.D., Nat. Taiwan U.-Taipei, 1971. Diplomate: Am. Bd. Anesthesiology. Intern in surgery Maimonides Med. Center, Bklyn., 1972-73; resident in anesthesiology Albert Einstein Sch. Medicine, Bronx, N.Y., 1973-74, Columbia-Presbyterian Med. Center, N.Y.C., 1974-75; NIH research fellow Coll. of Physicians and Suregons, Columbia U., N.Y.C., 1975-76, asst. prof. anesthesiology, 1976—. Fellow Am. Coll. Anesthesiologists; mem. Am. Physiol. Soc., Am. Soc. Anesthesiologists, N.Y. Acad. Scis., Internat. Anesthesia Research Soc. Subspecialties: Anesthesiology; Psychophysiology. Current work: Circulatory physiology; blood volume, blood rheology; cerebral circulation; hypothermia; circulatory effects of anesthetics; anesthetic actions; neurohum control of circulation. Office: Coll of Physicians and Surgeons Columbia Univ 630 W 168th St New York NY 10032

CHEN, SHIH-YEW, nuclear engineer, consultant; b. Taichung, Taiwan, June 12, 1948; came to U.S., 1971; s. Teh-Sheng and Shiu-Mei (Liu) C.; m. Mei-Ying Wang, July 4, 1973; 1 son: Frederick Wey-Min. B.S., Nat. Tsing Hua U., 1970; M.S., U. Ill.-Urbana, 1973, Ph.D., 1978. Nuclear engr. Sci. Applications, Inc., Schaumberg, Ill., 1977-79, ETA Engring., Inc., Westmont, Ill., 1979-81; sr. staff engr. Franklin Research Ctr., Phila., 1981—; nuclear cons. Argonne (Ill.) Nat. Lab., 1979-81. Mem. Am. Nuclear Soc., Health Physics Soc., Sigma Xi. Subspecialties: Nuclear engineering; Nuclear fission. Current work: Research involves radiation protection and shielding, nuclear reactor safety, radiological engineering and health physics. Home: 109 Collins Dr Cherry Hill NJ 08003 Office: Franklin Research Ctr 20th St at Race St Philadelphia PA 19103

CHEN, TAO-SENG, physicist, educator; b. Hunan, China, July 25, 1938; s. Yeh-Chieu and Zu-chuan (Yang) C.; m. Yun-Jane Dai, Apr. 20, 1967; 1 son, David. B.S., Taiwan Cheng Kung U., 1962; M.S., Nat. Tsing Hua U., 1965; Ph.D., U. Tex., 1971; postdoctoral, Harvard U. Med. Sch., 1978-80. Research assoc. physicist U. Tex., Austin, 1972-76, instr., 1976-78; med. physicist Harvard U. Med. Sch., Boston, 1978-80; asst. prof. radiology W.Va. U., 1980-81; chief med. physicist, asst. prof. dept. radiation oncology Vanderbilt U., Nashville, 1981—. Contbr. articles to profl. jours. Recipient Robert Welch Found. award, 1971-73; NIH research grantee, 1978-80. Mem. Am. Phys. Soc., Am. Assn. Physicists in Medicine, Sigma Xi. Subspecialties: Radiology; Condensed matter physics. Current work: High energy photon and electron radiation oncology. Home: 144 Cottonwood Dr Franklin TN 37064 Office: Dept Radiation Oncology Vanderbilt U Nashville TN 37232

CHEN, THERESA, SHANG-TSING, toxicologist, researcher; b. Taipei, Taiwan, Oct. 26, 1944; came to U.S., 1967, naturalized, 1976; d. Chiung Lin and Chin Shiu (Su) C.; m. Walter Michael Williams, Dec. 16, 1972; children: Robert, Richard. B.S in Pharmacy, Nat. Taiwan U., 1967; Ph.D. in Pharmacology, U. Louisville, 1971. Postdoctoral fellow U. Chgo., 1971-72; research assoc. U. Pitts., 1974-77; sr. pharmacologist Travenol Labs., Inc., Morton Grove, Ill., 1977-79; asst. prof. pharmacology and toxicology U. Louisville, 1979—, also researcher. Contbr. articles to profl. publs. Ky. Heart Assn. fellow, 1973-74. Mem. Am. Soc. Pharmacology and Exptl. Therapeutics, N.Y. Acad. Scis., Soc. Toxicology, Soc. Exptl. Biology and Medicine, Sigma Xi (U. Chgo. chpt.). Subspecialties: Cellular pharmacology; Toxicology (medicine). Current work: Biochem. toxicology and drug metabolism.

CHEN, TUNG-SHAN, educator; b. Chungking, China, Apr. 17, 1939; came to U.S., 1962, naturalized, 1976; s. Sze-Chen and Mary M. (Chen) Lin; m. Yolanda Chu Chen, Dec. 26, 1964; children: Andy, Lynn. B.S., Nat. Taiwan U., 1960; M.S., U. Calif., Berkeley, 1964, Ph.D., 1969. Research asst. U. Calif., Berkeley, 1962-69; asst. prof. food sci. and tech. Calif. State U., Northridge, 1969-73, asso. prof., 1973-78, prof., 1978—; vis. asso. prof. UCLA, 1974. Contbr. articles to profl. jours. Pres. San Fernando Valley Chinese Cultural Assn., 1974-75; bd; dirs. Greater Los Angeles Nutrition Council, 1978-81. Mem. AAAS, Am. Chem. Soc., Am. Dietetic Assn. (Plan IV rep. 1980—), Am. Inst. Chemists (fellow), Inst. Food Technologists. Subspecialties: Food science and technology; Nutrition (biology). Current work: Improving and developing methodology for vitamin analysis and assessing stability of nutrients during food processing. Office: Calif State Univ 18111 Nordhoff St Northridge CA 91330

CHEN, WAI-FAH, civil engrineer, researcher, consultant, educator; b. Chekiang, China, Dec. 23,1936; came to U.S., 1961, naturalized, 1972; s. Yu-Chao and Shui-Da (Hsia) C.; m. Lily Lin-Lin Hsuan, June 11, 1966; children: Eric, Arnold, Brian. B.S.C.E., Cheng-Kung U., Tainan, Taiwan, 1959; M.S.C.E., Lehigh U., 1963; Ph.D., Brown U., 1966. Profl. engr., Taiwan, 1959. Asst. prof. civil engring. Lehigh U., 1966-71, assoc. prof., 1971-75, prof., 1975-76; prof. structural engring. Sch. Civil Engring., Purdue U., West Lafayette, Ind., 1976-80, head structural engring., prof., 1980—; cons. to oil co. Author: books, the most recent being Plasticity in Reinforced Concrete, 1982, Constitutive Equations for Engineering Materials, vol. 1, 1982, vol. 2, 1984, Tubular Members in Offshore Structures, 1984, Nonlinear Analysis in Geotechnical Engineering, 1984. Co-recipient James F. Lincoln Arc Welding Found. award, 1972, 74, 81; Am. Iron and Steel Inst. grantee, 1966-76; NSF grantee, 1968—; Nat. Coop. Hwy. Research Program grantee, 1972-75; Can. Steel Industries Constrn. Council grantee, 1973-74; Am. Petroleum Inst. grantee, 1973-75; Dept. Energy grantee, 1974-76; Naval Constrn. Bn. Center grantee, 1977-79; Bechtel Group grantee, 1983-85. Mem. ASCE, Internat. Assn. Bridge and Structural Engrs., Am. Acad. Mechanics, Structural Stability Research Council, Earthquake Engring. Research Inst., Am. Concrete Inst., ASME. Subspecialties: Civil engineering; Solid mechanics. Current work: Structural engineering and structural mechanics-inelastic behavior of structures and mathematical modeling of materials; plasticity and elasticity theories applying to metal, soil, concrete and rock mechanics; computer analysis of structures. Home: 1021 Vine St West Lafayette IN 47906 Office: Sch Civil Engring Purdue U West Lafayette IN 47907

CHEN, WAI-KAI, electrical engineering and computer science educator, consultant; b. Nanking, China, Dec. 23, 1936; came to U.S., 1959; s. You-Chao and Shui-Tan (Shen) C.; m. Shirley Shiao-Ling, Jan. 13, 1939; children: Jerome, Melissa. B.S., Ohio U., 1960, M.S., 1961; Ph.D, U. Ill.-Urbana, 1964. Asst. prof. Ohio U., 1964-67, assoc. prof., 1967-71, prof., 1971-78, disting. prof., 1978-81; prof., head dept. elec. engring. and computer sci. U. Ill.-Chgo., 1981—; vis. assoc. prof. Purdue U., 1970-71. Author: Applied Graph Theory, 1970, Theory and Design of Broadband Matching Networks, 1976, Applied Graph

Theory: Graphs and Electrical Networks, 1976, Active Network and Feedback Amplifier Theory, 1980, Linear Networks and Systems, 1983; editor: Brooks/Cole Series in Electrical Engineering, 19; assoc. editor: Jour. Circuits, Systems and Signal Processing, 1981-86. Recipient Lester R. Ford award Math. Assn. Am., 1967; Research Inst. fellow Ohio U., 1972. Fellow IEEE, AAAS; mem. Soc. Indsl. and Applied Math., Assn. Computing Machinery, Tensor Soc. Gt. Britain, Sigma Xi, Phi Kappa Phi, Pi Mu Epsilon, Eta Kappa Nu. Subspecialties: Electrical engineering; Integrated circuits. Current work: Electrical circuits, filters, signal processing; applied graph theory, broadband matching, feedback amplifier design. Office: Dept of Electrical Engineering and Computer Science University of Illinois PO Box 4348 Chicago IL 60680

CHEN, WANG-PING, geophysics researcher and educator; b. Taipei, Taiwan, Nov. 28, 1951. B.Sc., Nat. Taiwan U., 1974; Ph.D., MIT, 1979. Research asst. MIT, Cambridge, 1974-79, postdoctoral assoc., 1979-80, research assoc., 1980-81; asst. prof. geophysics U. Ill.-Urbana, 1981—. Contbr. articles to profl. jours. NSF grantee, 1982—. Mem. Am. Geophys. Union, Seismol. Soc. Am. Subspecialties: Geophysics; Tectonics. Current work: Earthquake source processes and the physical properties of the lithosphere, regional earth structures, faulting, tectonics. Office: Dept Geology U Ill 1301 W Green St Urbana IL 61801

CHEN, WAYNE H., electrical engineer, university dean; b. Soochow, China, Dec. 13, 1922; came to U.S., 1947, naturalized, 1957; s. Ting Li and Yung-Chin (Hu) C.; m. Dorothy Teh Hou, June 7, 1957; children: Avis Shirley and Benjamin Timothy (twins). B.S. in E.E, Nat. Chiao Tung U., China, 1944; M.S., U. Wash., 1949; Ph.D., 1952. Registered profl. engr., Fla. Electronic engr. cyclotron project Applied Physics Lab. U. Wash., 1949-50, asso. in math, 1950-52; mem. faculty U. Fla., Gainesville, 1952—, prof. elec. engring., 1957—, chmn. dept., 1965-73, dean Coll. Engring., dir. Engring. and Indsl. Expt. Sta., 1973—; vis. prof. Nat. Chiao Tung U., Nat. Taiwan U., spring 1964; vis. scientist Nat. Acad. Scis. to USSR, 1967; mem. tech. staff Bell Tel. Labs., summers 1953, 54, cons., 1955-60; mem. tech. staff Hughes Aircraft Co., summer 1962; vis. prof. U. Caraboho, Venezuela, summer 1972. Author: The Analayis of Linear Systems, 1963, Linear Network Design and Synthesis, 1964, The Robotosyncrasies (pseudonym Wayne Hawaii), 1976, The Year of the Robot, 1981. Recipient Fla. Blue Key Outstanding Faculty award, 1960, Outstanding Publs. award Chia Hsin Cement Co. Cultural Fund, Taiwan, 1964, Tchr.-Scholar award U. Fla., 1971. Fellow IEEE; mem. AAUP, Am. Soc. Engring. Edn., Fla. Engring Soc., Nat. Soc. Profl. Engrs., Blue Key, Sigma Xi, Sigma Tau, Eta Kappa Nu, Tau Beta Pi, Epsilon Lambda Chi, Omicron Delta Kappa, Phi Tau Phi., Phi Kappa Phi. Club: Rotary. Subspecialties: Applied mathematics; Electrical engineering. Patentee in field. Home: 2065 19th Ln NW Gainesville FL 32605 Office: Coll Engring U Fla Gainesville FL 32611

CHEN, YI-HSIANG (ALAN), medical educator, researcher; b. Taipei, Taiwan, Feb. 20, 1939; s. Chi-Tsan and Yueh-Kwei (Chang) C.; m. Yu-Hsi (Judy) Chang, Oct. 21, 1967; children: Jennie, Josephine, Jacqueline. M.D., Nat. Taiwan U., 1964; Ph.D., U. Colo., 1970. Diplomate: Am. Bd. Internal Medicine, 1975, Hematology, 1976, Med. Oncology, 1981. Intern U. Miss. Med. Center, Jackson, 1970-71; resident U. Ill. Hosp., Chgo., 1971-72, fellow, 1972-74; asst. prof. U. Ill. Coll. Medicine, Chgo., 1974-80, assoc. prof., 1980—; vis. prof. Taipei Med. Coll., 1982. Contbr. articles to sci. and med. jours. Served to 2d lt. Chinese Army Med. Corps, 1964-65. Fellow ACP.; Mem. Am Soc. Hematology, Am. Assn. Immunologists, Central Soc. Clin. Research, Am. Soc. Clin. Oncology, Nat. Taiwan U. Med. Coll. Alumni Assn. N.Am. (treas. 1982), N.Am. Taiwanese Profs. Assn. Subspecialties: Immunology (medicine); Cell study oncology. Current work: Immunological dysfunctions and tumor immunity in lymphoid neoplasia, and the proliferation and differentiation of cancer cells. Home: 528 Hunter Ct Wilmette IL 60091 Office: 840 S Wood Chicago IL 60612

CHEN, YING-CHIH, physicist; b. Chung-King, China, July 7, 1948; came to U.S., 1971; s. Po-Shia and Tsun-Ming (Nie) C.; m. Ching Ling Chen, Nov. 1, 1975; 1 dau., Jennifer. B.S., Nat. Taiwan U., 1970; Ph.D., Columbia U., 1978. Scientist Optical Info. Systems, Elmsford, N.Y., 1979—; research assoc. Columbia U., N.Y.C., 1978-79. Contbr. articles to profl. jours. Mem. Am. Phys. Soc., IEEE, Optical Soc. Am. Subspecialties: Semiconductor lasers; Fiber optics. Current work: research and development of semiconductor lasers for light wave communications. Home: 67 MacArthur Ave Closter NJ 07624 Office: 350 Executive Blvd Elmsford NY 10523

CHEN, YU MIN, research biochemist; b. I-shing, Kiangsu, China, Dec. 23, 1922; came to U.S., 1954, naturalized, 1976; s. Mu Fan and Yueh Wah (Pan) C.; m. Jia-chun Nei, July 11, 1964; children: Peter P.T., Plato P.T. B.S. in Agrl. Chemistry, Nat. Chekiang U. China, 1946; M.S. in Biochemistry, Nat. Taiwan U., 1950; Ph.D. in Biochemistry and Nutrition, U. So. Calif., 1960. Assoc. prof. biochemistry Nat. Taiwan U., 1960-63; fellow Inst. Chemistry Academia Sinica, Taiwan, 1962-63; research scientist Wayne State U., 1963-68, Mich. Cancer Found., Detroit, 1968-73; prin. investigator Wayne State U., 1973-80; sr. research biochemist Providence Hosp., Southfield, Mich., 1981—. Contbr. articles in field to profl. jours. NIH grantee, 1971-74, 75-78. Mem. Am. Assn. Cancer Research, China Chem. Soc., Chinese Assn. Agrl. Chemistry, Sigma Xi. Subspecialties: Biochemistry (medicine); Immunology (medicine). Current work: Tyrosinase isozymes and inhibitors in vertebrate melanogenesis; serum tyrosinase in malignant disease, its immunological aspects and their applications for early detection in certain cancers. Home: 30105 High Valley Rd Farmington Hills MI 48018

CHENEY, DARWIN LEROY, pharmacologist; b. Burley, Idaho, Sept. 1, 1940; s. Harold and Lila (Banner) C.; m. Bonnie Gray, Sept. 14, 1964; childrn—Darin Lynn, David Lance, Allison, Anamarie, Dale Lawrence. B.S., Brigham Young U., 1965, M.S., 1968; Ph.D. in Pharmacology, Stanford U., 1971. NIMH postdoctoral fellow St. Elizabeth's Hosp., Washington, 1971-72, Pharmacol. Research Tng. asso. fellow, 1972-74; pharmacologist, lab. preclin. pharmacology NIMH, Wasington, 1974-75; acting sect. chief sect. molecular pharmacodynamics at St. Elizabeth's Hosp., Washington, 1975-78, sect. chief molecular pharmacodynamics, lab. preclin. pharmacology, 1978—. Editorial adv. bd.: Neuropharmacology; Contbr. numerous articles to sci. jours. Mem. Am. Soc. Pharmacology and Exptl. Therapeutics, Soc. Neurosci. Republican. Mem. Ch. Jesus Christ of Latterday Saints. Subspecialty: Molecular pharmacology. Current work: Modulation of acetylcholine, gamma-aminobutyric acid, glutamate turnover rates in brain nuclei using gas chromatography-mass fragmentography; dynamic relationships between various putative neurotransmitters.

CHENG, ANDREW FRANCIS, physicist; b. Princeton, N.J., Oct. 15, 1951; s. Sin I. and Jean (Sing) C.; m. Linda Sun Hu, Nov. 24, 1979. C.B., Princeton U., 1971; Ph.D., Columbia U., 1977. Postdoctoral fellow Bell Labs., Murray Hill, N.J., 1976-78; asst. prof. physics Rutgers U., Piscataway, N.J., 1978-83; sr. staff Applied Physics Lab., John Hopkins U., Laurel, Md., 1983—. Mem. Am. Phys. Soc., Am. Astron. Soc., Am. Geophys. Union, N.Y. Acad. Sci. Subspecialty: Theoretical astrophysics. Current work: Theoretical studies on physics of magnetospheres. Office: Applied Physics Lab Johns Hopkins U. Laurel MD 20707

CHENG, CHIA-CHUNG, chemist, researcher, educator; b. Tientsin, China, May 5, 1925; s. Kuo-Liang and Chui-yuen (Chien) C.; m. Katherine Cheng, May 30, 1953; children: Amy Yuwei, Anna Yumin, Alice Yuray, Audrey Yuhui. B.S., Nat. U. Chekiang, China, 1948; M.A., U. Tex, Austin, 1951, Ph.D. in Organic Chemistry, 1954. Chemist, munitions plant, Chungking, China, 1944-45; research assoc. N.Mex. Highlands U., 1954-57; research assoc. dept. chemistry Princeton U., 1957-59; head medicinal chemistry sect., prin.adv. in chemistry Midwest Research Inst., Kansas City, Mo., 1959-78; dir. Mid-Am. Cancer Center, Kansas City, 1978-81; prof. pharmacology, toxicology and therapeutics U. Kans. Med. Center, Kansas City, 1981—, dir. drug devel. lab., 1981—. Contbr. articles to profl. jours. NIH grantee. Mem. Am. Chem. Soc., Am. Assn. Cancer Research, Internat. Soc. Heterocyclic Chemistry, Sigma Xi, Phi Lambda Upsilon. Club: Chinese Club of Greater Kansas City (1st pres.). Subspecialties: Synthetic chemistry; Medicinal chemistry. Current work: Design and synthesis of anticancer and antimalarial agents. Patentee in field. Office: U Kans Med Center Kansas City KS 66103

CHENG, DAVID, physicist; b. Chungking, China, July 21, 1941; s. Elbert C.Y. and Emma (Shah) C.; m. Jennifer H.F. Jong, Dec. 24, 1982; 2 children. B.S., U. Calif.-Berkeley, 1962, M.A., 1963, Ph.D., 1965. Staff physicist Lawrence Berkeley Lab., 1965-67; physicist Brookhaven Nat. Lab., Upton, N.Y., 1967-69; mem. tech. staff Bell Labs., Murray Hill, N.J., 1969-74; project leader Xerox Research Center, Palo Alto, Calif., 1974—; pres. Computers Machine Control, Inc., San Jose, Calif., 1981—; cons. in field. Contbr. articles to profl. jours.; patentee in field. Mem. Optical Soc. Am., IEEE, Am. Phys. Soc., Sigma Xi. Subspecialties: Laser data storage and reproduction; Systems engineering. Current work: Laser-data storage and reproduction; computer control of machine tool. Office: 3333 Coyote Hill Rd Palo Alto CA 94304 Home: 974 Sherman Oaks Dr San Jose CA 95128

CHENG, EDWARD TEH-CHANG, nuclear engineer, researcher; b. Ping Tung, Taiwan, Nov. 23, 1946; came to U.S., 1972; s. Sui Ping and Tsai Yin (Li) C.; m. Shu-Ching Lai, June 4, 1973; children: Eric, Wendy. B.S., Nat. Tsing Hua U., 1969, M.S., 1971; M.S., U. Wis., 1974, Ph.D., 1976. Research assoc. U. Wis., Madison, 1976-78; staff engr. GA Techs. Inc., San Diego, 1978—. Mem. Am. Nuclear Soc. (sec./treas. fusion energy div. 1980-82). Subspecialties: Nuclear engineering; Nuclear fusion. Current work: Fusion neutronics, fusion blanket engineering and technology, fusion-fission hybrid, nuclear data. Home: 14090 Recuerdo Dr Del Mar CA 92014 Office: GA Techs Inc PO Box 85608 San Diego CA 92138

CHENG, H(WEI)-H (SIEN), soil science educator; b. Shanghai, China, Aug. 13, 1932; came to U.S., 1951, naturalized, 1961; s. Chi-Pao and Anna (Lan) C.; m. Jo Yuan, Dec. 15, 1962; children—Edwin, Antony. B.A., Berea Coll., 1956; M.S., U. Ill., 1958, Ph.D., 1961. Cert. profl. soil scientist. Research assoc. Iowa State U., Ames, 1962-64, asst. prof., 1964-65; asst. prof. dept. agronomy and soils Wash. State U., Pullman, 1965-71, assoc. prof., 1971-77, prof., 1977—, assoc. dean, 1982—; vis. scientist Julich Nuclear Research Ctr., W. Ger., 1972-73, 79-80, Academia Sinica, Taipei, Republic of China, 1978, Fed. Agrl. Research Center, Braunschweig, W.Ger., 1980. Fulbright scholar, 1962-63. Fellow Am. Soc. Agronomy, Soil Sci. Soc. Am.; Mem. AAAS, Am. Chem. Soc., Soc. Environ. Toxicology and Chemistry. Methodist. Subspecialties: Soil chemistry; Environmental toxicology. Current work: Transformation of nitrogen, pesticides, and organic matter in the soil-water environment; impact of organics on enviromental quality; development of 15N and 14C tracer methodology for soils research; nitrogen availability and fertilization. Home: NW 305 Joe St Pullman WA 99164 Office: Dept Agronomy and Soils Wash State U Pullman WA 99164

CHENG, KWOK-TSANG, physicist; b. Hong Kong, May 5, 1950; U.S., 1972; s. Sau-Mou and Kam-Mui (Chan) C.; m. Guang-Meei Doris, Aug. 30, 1975. B.Sc. (Hong Kong Govt. scholar), Chinese U. of Hong Kong, 1971; Ph.D., Notre Dame U., 1977. Postdoctoral employee Argonne (Ill.) Nat. Lab., 1977-80, asst. physicist, 1980—. Contbr. articles to physics jours. Mem. Am. Phys. Soc. Subspecialties: Atomic and molecular physics; Theoretical physics. Current work: Theoretical atomic physics, atomic structure calculations, relativistic many-body theory. Home: 400 Ascot Dr Park Ridge IL 60068 Office: Argonne Nat Lab Physics Bldg 203 Argonne IL 60439

CHENG, MEI-FANG HSIEH, psychobiology educator; b. Keelung, Taiwan, Nov. 24, 1938; d. Chao-Chin and Ai Chu Hsieh; m. Wen-Kwei Cheng, 1963; children: Suzanne, Po-Yuan, Julie. B.S., Nat. Taiwan U., 1958; M.A., U. Oreg., 1961; Ph.D., Bryn Mawr Coll., 1965. Postdoctoral and research assoc. U. Pa., 1965-68; asst. research prof. Inst. Animal Behavior, Rutgers U., Newark, 1969-72, assoc. prof., 1973-78, prof., 1979—; mem. edn. and research rev. com. NIMH; mem. ad hoc rev. com. NSF. Contbr. chpts. to books, articles to profl. jours. Fulbright scholar, 1959-63; NIMH career devel. awardee, 1974-79, 79-84. Mem. Soc. Animal Behavior, Soc. Neuroscience, N.Y. Acad. Scis., Internat. Soc. Psychoneuroendocrinology, AAAS. Subspecialties: Neuroendocrinology; Neurobiology. Current work: Hormonal and neural mechanisms of reproductive behavior. Office: Inst Animal Behavior Rutgers U 101 Warren St Newark NJ 07102

CHENG, YUNG-CHI, pharmacology and medicine educator; b. London, Dec. 29, 1944; U.S., 1968; s. J.Y.W. and Lucy C.; m. Elaine H.C., June 14, 1969; children: Pei Kwen, Pei Drin. B.S. in Chemistry, Tunghai (Taiwan) U., 1966; Ph.D. in Biochem. Pharmacology, Brown U., 1972. Research assoc. Brown U., Providence, 1972; mem. postdoctoral research staff pharmacology dept. Yale U. Sch. Med., New Haven, 1972-73, research assoc., 1973-74; sr. cancer research scientist dept. exptl. therapeutics Roswell Park Meml Inst., Buffalo, 1974-76, assoc. cancer research scientist, 1976-77, cancer research scientist V, 1977-79; asst. prof. dept. pharmacology Roswell Park div. Grad. Sch. SUNY-Buffalo, 1974-77, assoc. prof., 1977-79; prof. dept. pharmacology and medicine U. N.C.-Chapel Hill, 1979—; mem. study sect. exptl. therapeutics Nat. Cancer Inst., NIH, 1980-84; mem. sci. adv. bd. G. D. Searle Pharm. and Co., 1981—; cons. antivirus program Syntex U.S.A., Inc., 1981—. Contbr. numerous articles on pharmacology to profl. jours.; assoc. editor: Methotrexate Update, 1983—; co-editor: Herpesvirus, 1982. Am. Leukemia Soc. scholar, 1976-81; recipient Rhoads Meml. award Am. Assn. of Cancer Research, 1981. Mem. Am. Microbiology Soc., Am. Soc. Pharmacology and Exptl. Therapeutics, Sigma Xi. Subspecialties: Cancer research (medicine); Molecular pharmacology. Current work: Anti-viral, anti-cancer drugs, biology and biochemistry of cancer cells and viral infected cells. Home: 2462 Sedgefield Dr Chapel Hill NC 27514 Office: 917 FLOB 231 H U NC 2462 Sedgefield Dr Chapel Hill NC 27514

CHENNAULT, MADELYN JOANNE, psychology educator, reseacher; b. Atlanta, July 15, 1934; d. Benjamin Quillion and and Othello (Jones) C.; m.; children: Eugene Chaires Majaied, Bennel Mosby, Lyssa Sampson. B.S., Morris Brown Coll., Atlanta, 1957; M.A., U. Mich., 1961; Ed. S, Ind. U.-Bloomington, 1965, Ed.D., 1966; postgrad., U. Ga., 1970-72. Tchr. 3d grade R.L. Craddock Sch., Atlanta, 1957-59, Caldwell Sch., Fontana, Calif., 1959-60; asst. prof. psychology Albany (Ga.) State Coll., 1962-64, Atlanta U., 1966-67; prof. edn. and psychology Ft. Valley State Coll., Ga., 1967—; B.E. May prof. psychology Morehouse Coll., Atlanta, 1979—, chairperson dept. psychology, 1979—; cons., mgr. O & M Properties, Atlanta, 1974-82. Mem. Am. Psychol. Assn., Southeastern Psychol. Assn., Am. Pub. Health Assn., Council Exceptional Children, Am. Acad. Behavioral Medicine, Alpha Kappa Alpha. Club: Link's Inc. Subspecialties: Psychobiology; Social psychology. Current work: Principal investigator Middle Georgia Community Hypertension Intervention Program. Office: Fort Valley State College State College Dr Fort Valley GA 31030 Home: 1930 Honeysuckle Ln SW Apt 32 Atlanta GA 30311

CHENOWETH, PHILIP ANDREW, consulting geologist; b. Chgo., Aug. 21, 1919; s. Joseph Gayne and Helen (Burton) C.; m. Marilyn Myers, Apr. 11, 1952; children: Kathryn, Amelia. B.A., Columbia Coll., 1946; M.A., Columbia U., 1947, Ph.D., 1949. Cert. Am. Inst. Profl. Geologists. Instr. Amherst (Mass.) Coll., 1949-51; asst. dist. geologist Sinclair Oil and Gas Co., Ardmore, Okla., 1951-54; assoc. prof. U. Okla., Norman, 1954-60; research assoc. Sinclair Research Labs., Tulsa, 1960-68; cons., Tulsa, 1968—. Contbr. articles to profl. jours. Chmn. Higher Edn. Subcom., Goals for Tulsa. Served with U.S. Army, 1941-45. Recipient Best Tchr. award U. Okla., 1956. Fellow Geol. Soc. Am.; mem. Am. Assn. Petroleum Geologists, Oklahoma City Geol. Soc., Rocky Mountain Assn. Geologists; hon. mem. Tulsa Geol. Soc. Subspecialty: Geology. Inventor oil spill removal method. Home: 5828 62d Pl Tulsa OK 74136 Office: 1000 Petroleum Club Bldg Tulsa OK 74119

CHEO, PEN CHING, plant virologist; b. Ho-fei, Anhwei, China, Mar. 28, 1919; came to U.S., 1947, naturalized, 1965; s. Hsiu Cheng and Yee (Dah) C.; m. Helen Chen, Aug. 16, 1949. Ph.D., U. Wis., 1951. Research assoc. U. R.I., Kingston, 1952-53; research fellow Calif. Inst. Tech., Pasadena, 1957-61; asst. prof. U. Wash., Wenatchee, 1961-66; plant pathologist Los Angeles State and County Arboretum, Arcadia, 1966-67, chief research div., 1967—. Contbr. articles to profl. jours. Mem. Am. Phytopathology Soc., Sigma Xi. Subspecialties: Plant virology; Plant pathology. Current work: Physiology of viral resistance in plant, virus degradation in soil. Home: 144 S Catalina Apt 9 Pasadena CA 91106 Office: 301 N Baldwin St Arcadia CA 91006

CHERBAS, PETER THOMAS, biology educator; b. Bryn Mawr, Pa., Mar. 26, 1946; s. Thomas and Virginia (Stamboolian) C.; m. Lucy Fuchsman, June 14, 1968; 1 dau., Katherine. B.A., Harvard U., 1967, Ph.D., 1973. Research fellow Cambridge (Eng.) U., 1973-74; research fellow Harvard U., Cambridge, Mass., 1974-77, asst. prof. biology, 1977-81, assoc. prof., 1981—. Mem. Soc. Devel. Biology, Am. Genetics Soc. Subspecialties: Developmental biology; Molecular biology. Current work: Steroid hormone action in Drosophila; endocrinology/steroids; Drosophila; insect biochemistry; molecular biology; recombinant DNA. Office: 16 Divinity Ave Cambridge MA 02138

CHERMOL, BRIAN HAMILTON, psychologist; b. Bryn Mawr, Pa., June 24, 1944; s. John Thomas and Esther Louise (Hamilton) C.; m. Judy S. Williams, July 7, 1965 (div. 1972); 1 dau., Sherry L.; m. Jackie L. Burroughs, Feb. 7, 1975 (dec. 1975); m. Annie L. Mahone, Jan. 8, 1983; 1 dau., Laurie A. B.A., Park Coll., 1970; M.A., U. Mo., 1972; Ph.D., U. S.C., 1978. Lic. psychologist, Tex., Ala. Psychologist Ga. VA Hosp., Augusta, 1974; commd. 2d lt. U.S. Army, 1962, advanced through grades to lt. col., 1981—; staff psychologist Psychology Service, Walter Reed Army Med. Ctr., Washington, 1976-78, Community Mental Health, Ft. Rucker, Ala., 1978-80, chief psychiatry dept., 1980-81; chief behavorial sci. specialty br. Acad. Health Sci., Ft. Sam Houston, Tex., 1981—; cons. White House Staff, 1977-78, Washington Police Dept., 1976-78, U.S. Helicopter Team, 1981, Army Safety Ctr., 1981. Contbr. articles to profl. jours. Decorated Silver Star, Bronze Star medal; named Man of the Yr. Columbus C. of C., 1963. Mem. Am. Psychol. Assn., Assn. for Advancement Psychology, Am. Soc. Clin. Hypnosis. Republican. Subspecialties: Behavioral psychology; Social psychology. Current work: Determine best methods for identifying and treating casualties generated by nuclear war; assess new teaching formats and procedures, aviation psychological research. Home: 8120 Scottshill San Antonio TX 78209 Office: Acad Health Scis Rm 287 Fort Sam Houston TX 78234

CHERNICK, MICHAEL ROSS, mathematical statistician; b. Havre de Grace, Md., Mar. 11, 1947; s. Jack and Norma Leonia (Weiner) C. B.S., SUNY-Stony Brook, 1969; M.A., U. Md., 1973; M.S., Stanford U., 1976, Ph.D., 1978. Mathematician U.S. Army, Aberdeen Proving Ground, Md., 1969-74; math. statistician Oak Ridge Nat. Lab., 1978-80, Aerospace Corp., Los Angeles, 1980—; adj. prof. Calif. State U.-Fullerton, 1982—; vis. lectr. U. Calif.-Santa Barbara, 1981. Referee: Annals of Probability, 1981, Zeitschrift Wahrscheinlichkeit Theorie, 1982; reviewer: Math. Reviews, 1982. Stanford U. fellow, 1974. Mem. Am. Statis. Assn., Bernoulli Soc., Inst. Math. Statistics, Soc. Indsl. and Applied Math. Democrat. Jewish. Subspecialties: Statistics; Probability. Current work: Develop statistical procedures for detecting outliers in multivariate and time series data. Statistical methods and probability models for satellite and other space systems. Home: 20920 Anza Ave Apt 208 Torrance CA 90503 Office: Aerospace Corp PO Box 92957 M4-955 Los Angeles CA 90009

CHERNICOFF, DAVID PAUL, hematologist, oncologist, educator; b. N.Y.C., Aug. 3, 1947; s. Harry and Lillian (Dobkin) C.; m. Jean Bogart, June 20, 1970; children; William, Jacki. A.B., U. Rochester, 1969; D.O., Phila. Coll. Osteopathic Medicine, 1973. Cert. internal medicine, hematology/oncology Nat. Bd. Osteopathic Examiners; cert. Am. Osteopathic Bd. Internal Medicine. Intern Rocky Mountain Hosp., Denver, 1973-74; resident Community Gen. Osteopathic Hosp., Harrisburg, PA., 1974-76; fellow Cleve. Clinic, 1976-78; asst. prof. medicine Chgo. Coll. Osteopathic Medicine, 1978-82, assoc. prof., 1982—; staff physician hematology-oncology Chgo. Osteopathic Med. Ctr., 1978-82, dir. sect. hematology oncology, 1982—; Olympia Fields (Ill.) Osteopathic Med. Ctr., 1982—; sr. investigator Eastern Coop. Oncology Group, Chgo., 1980-83; mem. adv. bd. Hospice Suburban South, Olympia Fields. Contbr., articles to profl. jours. Trustee Ill. Cancer Council, Chgo.; bd. dirs. Chgo. unit Am. Cancer Soc. Mem. Am. Soc. Clin. Oncology, Am. Osteopathic Assn., AMA. Subspecialties: Oncology; Cancer research (medicine). Current work: Cancer treatment (chemotherapy) and clinical cancer research protocols. Office: Chicago Osteopathic Medical Center 5200 S Ellis Ave Chicago IL 60615

CHERNOFF, AMOZ IMMANUEL, hematologist, govt. adminstr.; b. Malden, Mass., Mar. 17, 1923; s. Isaiah and Celia (Margolin) C.; m. Renate R. Fisher, Jan. 25, 1953; children—David F., Susan N., Judith A. B.S. in Chemistry with honors, Yale U., 1944, M.D. cum laude, 1947. Diplomate: Am. Bd. Internal Medicine. Med. intern Mass. Gen. Hosp., Boston, 1947-48; asst. resident in medicine Barnes Hosp., St. Louis, 1948-49; fellow in hematology Michael Reese Hosp., Chgo., 1949-51, asst. dir. hematology research lab., 1950-51; A.C.P. fellow Washington U. Sch. Medicine, St. Louis, 1951-52; USPHS spl. research fellow, 1952-53, instr. in medicine, 1952-54, asst. prof., 1954-56; asso. prof. medicine Duke U., 1956-58; chief sect. hematology VA Hosp., Durham, N.C., 1956-58; research prof. U. Tenn. Meml. Research Center, Knoxville, 1958-79, dir., 1964-77; asso. vice chancellor for acad. affairs Center Health Scis., 1977-79; prof. medicine Coll. Medicine, Memphis, 1966-79; med. dir. Cystic Fibrosis Found., Atlanta, 1975-77; dir. div. blood diseases and resources Nat. Heart Lung and Blood Inst., NIH, Bethesda, Md., 1979—; cons. med. program devel. Contbr. articles to profl. jours. Served with U.S. Army, 1943-45. Recipient Campbell award Yale U. Sch. Medicine, 1947, Research Career award USPHS, 1962-77. Fellow A.C.P.; mem. Am. Soc. Clin. Investigation, Am. Soc. Hematology, Internat. Soc. Hematology, Central Soc. Clin. Research, So. Soc. Clin. Investigation, Soc. Exptl. Biology and Medicine, Am. Fedn. Clin. Research, Sigma Xi, Alpha Omega Alpha. Subspecialties: Internal medicine; Hematology. Current work: Director of NIH's extramural programs of support of hematologic research, particulary in red cell, heoglobin area, in thrombosis and hemostasis, and in blood banking. Home: 9417 Copenhaver Dr Potomac MD 20854 Office: NIH Fed Bldg Room 518 Bethesda MD 20205

CHERNOFF, HERMAN, educator; b. N.Y.C., July 1, 1923; s. Max and Pauline (Markowitz) C.; m. Judith Ullman, Sept. 7, 1947; children—Ellen Sue, Miriam Cheryl. B.S., CCNY, 1943; Sc.M., Brown U., 1945, Ph.D., 1948. Research asso. Cowles commn. for research in econs. U. Chgo., 1948-49; asst. prof. U. Ill., Urbana, 1949-51, asso. prof., 1951-52, Stanford (Calif.) U., 1952-56, prof. stats., 1956-74; prof. applied math. M.I.T., Cambridge, 1974—. Author: (with L.E. Moses) Elementary Decision Theory, 1959, Sequential Analysis and Optimal Design, 1972. Recipient Townsend Harris medal CCNY Alumni Soc., 1981. Mem. Nat. Acad. Scis., Inst. Math. Stats. (pres. 1967-68), Am. Acad. Arts and Scis. Subspecialty: Statistics. Research in large sample theory, optimal design of expts., sequential analysis, pattern recognition. Home: 75 Crowninshield Rd Brookline MA 02146

CHERRY, JOHN PAUL, physical science administrator; b. Rhinbeck, N.Y., Jan. 31, 1941; s. John and Susan (Borovsky) C.; m. Janet Carroll, June 20, 1941; children: Jamie Paulette, Janine Collette. B.S., Furman U., 1963; M.S., W.Va. U., 1966; Ph.D., U. Ariz., 1970. Research chemist U.S. Dept. Agr., New Orleans, 1970-72; supervisory research chemist So. Regional Research Ctr., New Orleans, 1976-82; assoc. dir. Eastern Regional Research Ctr., Phila., 1982—; research assoc. dept. biochemistry and biophysics Tex. A&M U., College Station, 1972-73; asst. prof. dept. food sci. Exptl. Sta., U. Ga. Coll. Agr., Experiment, 1973-76; cons. and lectr. in field. Contbr. chpts. to books, articles to profl. jours. Research grantee in field. Mem. Inst. Food Technologists, Am. Chem. Soc., Am. Assn. Cereal Chemists, Assn. Ofcl. Analytical Chemists, Am. Peanut Research and Edn. Assn., Am. Oil Chemists Soc., Smithsonian Instn., Planetary Soc., Sigma Xi. Methodist. Subspecialties: Food science and technology; Biochemistry (biology). Current work: Planning, directing and evaluating research programs in agriculture. Patentee in phys. scis. Home: 360 Dundee Dr Blue Bell PA 19422 Office: Eastern Regional Research Ctr Agrl Research Servic US Dept Agr 600 E Mermaid L Philadelphia PA 19118

CHERRY, ROBERT NEWTON, JR., health physicist, army officer, physicist; b. Bowling Green, Ky., Oct. 6, 1946; s. Robert Newton and Lolita Violet (Tomes) C.; m. Janet Marie Reichenbach, May 31, 1969; children: Christopher Patrick, Gregory Allan. B.S., U. Mich., 1968, M.S., 1972, Ph.D., 1975. Cert. Am. Bd. Health Physics, 1981. Commd. 2d lt. U.S. Army, 1969, advanced through grades to maj., 1983; served, Netherlands, Vietnam, 1969-71; asst. prof. physics Hamilton Coll., Clinton, N.Y., 1975-77; health physicist Army Environ. Hygiene Agy., 1977-78, 79-80, Enewetak Atoll, 1978-79; health physics officer Brooke Army Med. Center, Ft. Sam Houston, 1980—; vis. research assoc. U. Rochester, N.Y., 1976-77. Contbr. articles to profl. publs. Decorated Bronze Star Air Medal. Mem. Am. Phys. Soc., Health Physics Soc., Am. Assn. Physicists in Medicine, Assn. U.S. Army. Subspecialty: Medical health physics. Current work: Radiation protection officer; support tng. of residents and fellows; participant research activities involving ionizing radiation. Home: 12115 Los Cerdos San Antonio TX 78233 Office: Health Physics Office Brooke Army Med Center Fort Sam Houston TX 78234

CHERYAN, MUNIR, food engineer, biotechnologist, consultant; b. Cochin, Kerala, India, May 7, 1946; came to U.S., 1968; s. Parathuparambil C. and Aley (Mathulla) C.; m. Leela S. Sundararajan, Jan. 26, 1972; children: Sapna, Anura. B.Tech. with honors, Indian Inst. Tech, Kharagpur, West Bengal, 1968; M.S., U. Wis., 1970, Ph.D., 1974. Research asst. chem. engr. U. Wis., Madison, 1969-70, research asst. food sci., 1970-74, postdoctoral fellow food sci., 1974-75; research assoc. INTSOY, U.Ill., Urbana, 1975-76, asst. prof. engring., 1976-80, assoc. prof., 1980—. Ill. Soybean Program Operating Bd. grantee, 1977-80; U.S. Dept. Agr. grantee, 1976-84. Mem. Inst. Food Technologists, Am. Inst. Chem. Engrs., Am. Chem. Soc., Am. Assn. Cereal Chemists, Sigma Xi. Subspecialties: Food science and technology; Enzyme technology. Current work: Membrane processing, enzyme technology, heat transfer and thermal process design, biochemical reactor engineering, functional properties and interactions of food constituents. Inventor protein hydrolyzats, protein blends. Office: U Ill Dept Food Scis 1302 W Pennsylvania Ave Urbana IL 61801

CHESNEY, MARGARET ANN, psychologist, researcher; b. Balt., May 28, 1949; d. Robert William and Margaret Montgomery (Renton) C.; m. George William Black, Dec. 15, 1979. B.A., Whitman Coll., Walla Walla, Wash., 1971; M.S., Colo. State U., Ft. Collins, 1973, Ph.D., 1975. Lic. psychologist, Calif. Psychol. asst. Colo. State U., 1971-74, intern, 1974-75; vis. asst. prof. U. No. Colo., Greeley, 1975; postdoctoral fellow dept. psychiatry Temple U., Phila., 1975-76; health psychologist SRI Internat., Menlo Park, Calif., 1976-79, dir., sr. health psychologist, 1979—; mem. grant review com. NIMH, Rockville, Md., 1979-82; mem. U.S. del. to USSR, Nat. Heart, Lung and Blood Inst., Bethesda, Md., 1982; sci. cons. Stanford U., 1980—. Contbr. chpts. to books, articles to profl. jours. Fellow Behavior Therapy and Research Soc.; mem. Acad. Behavioral Med. Research, Am. Psychol. Assn. (membership chmn. div. health psychology 1980-82, program chmn. 1982), Soc. Behavioral Medicine (chmn. local arrangements 1979). Subspecialties: Behavioral psychology; Preventive medicine. Current work: Engaged in research on biobehavioral aspects of chronic disease with focus of role of behavioral or psychosocial factors in etiology, treatment and prevention of illness. Office: Dept Behavioral Medicine SRI International 333 Ravenswood Ave Menlo Park CA 94025 Home: 440 W Oakwood Blvd Redwood City CA 94061

CHESNEY, ROBERT HAROLD, microbiologist, educator; b. Paterson, N.J.; s. David Martin and Goldie (Heiman) C.; m. Ellen Lucas, Aug. 19, 1972; 1 dau., Dana Llewellyn. B.A., U. Va., Charlottesville, 1968, Ph.D., 1974. Postdoctoral fellow Emory U., Atlanta, 1974-77; research assoc. U. Ga., Athens, 1977-78; asst. prof. microbiology Rutgers U., Newark, 1978—. Contbr. articles to jours. NIH postdoctoral fellow, 1975-77. Mem. AAAS, Am. Soc. Microbiology, Sigma Xi, Phi Sigma. Jewish. Subspecialties: Genetics and genetic engineering (biology); Microbiology. Current work: Asparaginase genetics, gene mapping, gene cloning, Bacteriophages P1 and P7, Agrobacterium tumefaciens, tumorigenesis. Office: Dept Zoology and Physiology Rutgers U Newark NJ 07102

CHESSON, EUGENE, JR., civil engineering educator, consultant; b. Sao Paulo, Brazil, Dec. 1, 1928; s. Eugene and Mary Josie (Foy) C.; m. Marilyn Ryder Hershey, Aug. 21, 1954; children: Christopher Eugene, David Anson. B.S.C.E., Duke U., 1950; M.S., U. Ill.-Urbana, 1956, Ph.D., 1959. Registered profl. engr., Ill., Del. Refinery engr. Standard Oil Ind., Whiting, 1953; research asst., research assoc. civil engring. dept. U. Ill.-Urbana, 1953-59, asst. prof., 1959-62, assoc. prof., 1962-66; prof. civil engring. U.Del., Newark, 1966—; pres. Chesson Engring., Inc., Newark, 1981—. Contbr. articles in field to profl. jours. Mem. Nat. Defense Exec. Reserve, U.S. Dept. Transp., 1973—. Served to lt. (j.g.), C.E. USN, 1950-53. Named Outstanding Young Faculty Mem. dept. Civil Engring. U. Ill., 1962, Del. Outstanding Engr. Del. soc. Profl. Engrs., 1981. Fellow ASCE (pres. local sect. 1982-83); mem. Am. Soc. Engring. Edn. (W. E. Wickenden award 1981), Am. Inst. Steel Constrn., Nat. Soc. Profl. Engrs. Republican. Presbyterian. Subspecialty: Civil engineering. Current work: Structural steel design, application of minicomputers to engineering design, forensic engineering and failure analysis, offshore industrial port-islands. Office: U Del 130 DuPont Hall Newark DE 19716 Home: 30 Bridle Brook Ln Newark DE 19711

CHESTNUT, HAROLD, systems engineering, consultant; b. Albany, N.Y., Nov. 25, 1917; s. Harry and Dorothy (Schulman) C.; m. Erma Ruth Callaway, Aug. 24, 1944; children—Peter Callaway, H. Thomas, Andrew T. B.S. in Elec. Engring, Mass. Inst. Tech., 1939, M.S., 1940; D.E. (hon.), Case Western Res. U., 1966, Villanova U., 1972. With Gen. Electric Co., 1940—; cons. systems engr., aeros. and ordnance dept. Advanced Tech. Lab., Schenectady, 1956-66; mgr. Research and Devel. Center, 1966-71. Editor: Systems Engring. and Analysis, John Wiley and Sons, 1965; author: Servomechanisms and Regulating Systems Design, Vol. I, 1951, Vol. II, 1955, Systems Engineering Tools, 1965, Systems Engineering Methods, 1967; editor: Jour. Automatica, 1961-67. Mem. commn. sociotech. systems NRC, 1975-78. Case Western Res. U. Centennial scholar, 1980. Fellow IEEE (v.p. tech. activities 1970-71, v.p. regional activities 1972, pres. 1973, exec. com. 1967-75), AAAS, Instrument Soc. Am.; mem. Nat. Acad. Engring., Internat. Fedn. Automatic Control (pres. 1957-58), World Federalists Assn. (bd. dirs. 1980—), Am. Automatic Control Council (pres. 1962-63). Subspecialty: Systems engineering. Current work: Improving international stability; presently president SWIIS Foundation, Inc. involves systems engineering. Home: 1226 Waverly Pl Schenectady NY 12308 Office: PO Box 8 Schenectady NY 12301

CHESWORTH, ROBERT HADDEN, nuclear research and development company executive, consultant; b. Artesia, Calif., Sept 27, 1929; s. Willard John and Kathryn Elizabeth (Hadden) C.; m. Gwenavere Spratling, Sept. 20, 1952; children: Robert A., David L., James H., Philip D., Kathryn E., Richard M., Karen L. Student, U. Redlands, 1947-49; B.S. in Chem. Engring, U. Calif.-Berkeley, 1951; postgrad. in nuclear engring, U. Calf.-Berkeley, 1957-59. Process engr. Gen. Electric Co., Richland, Wash., 1951-54; research engr. Atomics Internat., Canoga Park, Calif., 1954-56; tech. dir. Aerojet-Gen. Corp., San Ramon, Calif., 1956-71, cons., Sacramento, 1971-74; mgr. mktg. Gen. Atomic Co., San Diego, 1971-77; v.p. mktg. Norman Engring. Co., Los Angeles, 1977-79; dir. Triga reactors Ga. Techs. Inc., San Diego, 1979—; mem. com. NASA, Washington, 1969-71. Mem. Am. Nuclear Soc. (sec. 1967-68), Am. Inst. Chem. Engrs., AAAS, Alpha Xi Sigma. Republican. Mormon. Subspecialties: Nuclear engineering; Chemical engineering. Current work: Nuclear research reactor design and development; medical radioisotope development and applications. Home: 252 Calle Florecita Escondido CA 92025 Office: GA Techs Inc PO Box 85608 San Diego CA 92138

CHEUNG, HERBERT CHIU-CHING, biochemistry educator, researcher; b. Canton, China, Dec. 19, 1933; came to U.S., 1950, naturalized, 1965; s. Kun Hui and Tak Woo (Yuen) C.; m. Daisy S. Lee, Feb 12, 1966; children: Sharon, Melissa. A.B., Rutgers U., 1954, Ph.D., 1961; M.S., Cornell U., 1956. Research chemist Allied Chem. Corp., Morristown, N.J., 1963-66; sr. fellow U. Calif. Med. Ctr.-San Francisco, 1966-69; assoc. prof. biophysics U. Ala.-Birmingham, 1969-75, prof., 1975-82, prof. biochemistry, 1982—; mem. phys. biochemistry rev. group NIH, 1980-81, mem. phys. biochemistry study sect., 1981—; mem. rev. com. Ala. Heart Assn., 1981-83. Recipient Research Career Devel. award NIH, 1972-77; NIH grantee, 1970—; NSF grantee, 1974, 78-80. Mem. Am. Soc. Biol. Chemists, Biophys. Soc., Sigma Xi. Subspecialties: Biophysics (biology); Biophysical chemistry. Current work: Molecular mechanism of energy transduction in muscle, structure-function relationship of proteins, particularly calcium binding proteins; fluorescence; physical studies of biomembranes; transporting ATPases. Office: U Ala in Birmingham 520 CHSB Birmingham AL 35294

CHEUNG, JOHN YAN-POON, educator; b. Hong Kong, Oct. 13, 1950; came to U.S., 1965, naturalized, 1983; s. Paul To-Kwong and Mei-Shin (Lau) C.; m. Rose Kwan-Fun, Aug. 6, 1977; children—Christina Phoebe, Jonathan Paul. B.S. in Math, Oreg. State U., 1969; Ph.D. in E.E, U. Wash., 1975. Sr. engr. Boeing Computer Service, Seattle, 1977-78; asst. prof. U. Okla., Norman, 1978—; reviewer IEEE, Choice, Brooks/Cole Pub. Co. Contbr. articles to profl. jours. NSF grantee, 1981-83; United Engring. Trustees, Inc. grantee, 1981-82. Mem. Acoustical Soc. Am., IEEE, Assn. Computing Machinery, Eta Kappa Nu. Mem. Christian Ch. Subspecialties: Computer engineering; Computer architecture. Current work: Computer applications in signal processing, biomedical instrumentation, parallel processing as applied to algorithms, computer arch. and very large scale integration (VLSI) designs. Office: 202 W Boyd St Rm 219 Norman OK 73019 Home: 1000 W Imhoff St Norman OK 73069

CHEUNG, LIM HUNG, exploration geophysicist; b. Hong Kong, Feb. 18, 1953; came to U.S., 1971, naturalized, 1982; s. Shu Yu and Wai (Kwan) C.; m. Carol T., Nov. 22, 1954. B.S., Calif. Inst. Tech., 1975; Ph.D., U. Md., 1980. Undergrad. research asst. Calif. Inst. Tech., 1972-75; research asst. U. Md., 1975-80, NASA Goddard Space Flight Center, Greenbelt, Md., 1975-80; astrophysicist Harvard-Smithsonian Center for Astrophysics, Cambridge, Mass., 1980-81; research geophysicist Gulf Research & Devel. Co., Pitts., 1981—. Contbr. numerous articles to profl. jours. Active ARC, Pitts. Mem. Am. Phys. Soc., Soc. Exploration Geophysicists, Am. Astron. Soc. Subspecialties: Geophysics; Acoustics. Current work: Geophysics; seismology; signal analysis; computer applications; oil and gas exploration. Office: PO Box 2038 Pittsburgh PA 15230

CHEVALIER, ROBERT LOUIS, pediatrics educator and researcher; b. Chgo., Oct. 25, 1946; s. Frank Charles and Marion Helen (Jahnke) C.; m. Janis Julia Slezak, Dec. 23, 1970; 1 dau., Juline Ariane. B.S. in Zoology, U. Chgo., 1968, M.D., 1972. Diplomate: Am. Bd. Pediatrics (subcert. in pediatric nephrology). Intern, residence in pediatrics N.C. Meml. Hops., Chapel Hill, 1972-75; asst. prof. pediatrics U. Va., Charlottesville, 1978-83, assoc. prof., 1983—, chief div. pediatric nephrology, 1978—. Louis Welt scholar U. N.C., Chapel Hill, 1975; research grantee NIH, 1979; established investigator Am. Heart Assn., 1983. Mem. Am.Soc. Nephrology, Internat. Soc. Nephrology, Am. Fedn. Clin. Research, Nat. Kidney Found., Am. Soc. Pediatric Nephrology. Subspecialties: Pediatrics; Nephrology. Current work: Investigation of developmental renal physiology using micropuncture techniques. Home: 113 Wildflower Dr Charlottesville VA 22901 Office: U Va Dept Pediatrics Box 386 Charlottesville VA 22908

CHEVILLE, NORMAN FREDERICK, pathologist, researcher; b. Iowa, Sept. 30, 1934; s. Fred M. and Lucille H. C.; m. Beth M. Clark, June 22, 1958; children. D.V.M., Iowa State U., 1959; Ph.D., U. Wis., 1964. Research assoc. U. Wis., 1961-63; with Nat. Animal Disease Center, Ames, Iowa, 1963—, vet. med. officer, 1963-65; chief Pathology Research lab., 1965—; prof. pathology Iowa State U., 1966—. Research numerous publs. in field; author: Cytopathology of Viral Diseases, 1975, Cell Pathology, 1976. Pres. Assn. Handicapped Children, Ames; bd. dirs. First United Methodist Ch. Served to capt. U.S. Army, 1959-61. Recipient Outstanding Achievement award; Alumni Merit award Iowa State U., 1980. Mem. Am. Coll. Vet. Pathologists, Internat. Acad. Pathology, AAAS, Sigma Xi, Phi Zeta, Phi Kappa Phi. Subspecialty: Pathology (veterinary medicine). Current work: Infectious diseases of domestic animals. Office: Pathology Research Lab Nat Animal Disease Center PO Box 70 Ames IA 60010

CHEW, CATHERINE STRONG, physiology educator; b. Savannah, Ga., July 6, 1942; d. William Harmon and Pegge Catherine (Bradley) Strong; m. Frank Ellis Chew, Sept. 25, 1969; 1 son, Frank Ellis. B.S., Armstrong State Coll., Savannah, 1970; Ph.D., Emory U., Atlanta, 1970-77. Horse trainer, instr. Sa-Hi Stables, Savannah, 1960-65; research technician Emory U., 1970-74, research assoc., 1977-78, NIH postdoctoral fellow, 1979-80; asst. prof. physiology Morehouse Sch. Medicine, Atlanta, 1980—; lectr. Internat. Congress of Physiol. Sci., Budapest, Hungary, 1980, Montpellier, France, 1979, Porter Found., Atlanta, 1978, 80, 82, 83; instr. Ga. Heart Assn., 1975. Author: Hormone Receptors in Digestion and Nutrition, 1969, Advances in Physiological Sciences, 1980; Contbr. articles to profl. jours. Mem. ednl. policy com. Paideia Sch., Atlanta, 1981—; mem. Candler Park Neighborhood Orgn., Atlanta, 1977—, CAUTION, Atlanta, 1982—. Mem. AAUP, Am. Physiol. Soc., N.Y. Acad. Scis., Am. Women in Sci., AAAS. Democrat. Episcopalian. Subspecialties: Physiology (medicine); Cell biology (medicine). Current work: Gastrointestinal physiologist studying cellular mechanism of action of neural, hormonal and paracrine factors controlling gastric (parietal cell) acid secretion and chief cell pepsinogen secretion. Home: 607 Page Ave NE Atlanta GA 30307 Office: Morehouse Sch Medicine 720 Westview Dr SW Atlanta GA 30310

CHEWNING, JUNE SPANGLER, energy labor force specialist; b. Atlanta, July 27, 1925; d. George McClannahan and Esther (Ward) Spangler; m. Bernard P. Chewning, June 25, 1955; children: B. Peter, Pamela A. B.A., Am. U., 1950; A.A., McNeese State U., 1946; degree, Ga. Inst. Tech., 1982. Editor/writer Dept. Navy, Washington, 1949-52; econ. analyst Library of Congress, Washington, 1952-63, Dept. Def., Arlington, Va., 1963-66; edn. and tng. analyst U.S. AEC, Germantown, Md., 1966-72; sr. manpower analyst Dept. Energy, Washington, 1972—; cons. numerous career guide publs., 1971—; cons. IAEA, Vienna, 1979-80. Editor: Industrial Noise Control, 1963; author govt. publs.; contbr. articles to profl. jours. Co-founder Federally Employed Women, Inc., 1968, CHANGE, 1980; dept. chmn., deacon National City Christian Ch., Washington, 1979-82; bd. dirs. various civic orgns. Recipient Superior Performance award Library of Congress, 1950; Spl. Achievement award ERDA, 1976. Mem. Am. Nuclear Soc., Atomic Indsl. Forum, AAAS, Nuclear Energy Women, Women in Energy, Am. U. Alumni Assn. (v.p. 1974-75), Women's Equity Action League, FEW, Phi Theta Kappa. Democrat. Mem. Disciples of Christ Ch. Subspecialties: Employment aspects of energy production; Nuclear fission. Current work: Projections of needs for and supply of nuclear-trained personnel, adequacy of training, evaluation of education and training delivery modes. Transfer of nuclear technology to developing nations. Home: 3637 Appleton St NW Washington DC 20008 Office: US Dept Energy 1000 Independence Ave SW Washington DC 20585

CHHABRA, RAJENDRA SINGH, pharmacologist, toxicologist; b. India, Mar. 4, 1939; came to U.S., 1970; s. Jagdish S. and Vidya (Rajpal) C.; m. Swarn K., Mar. 25, 1966; children: Videsha, Punit. B.V.Sc. & A.H., Vikram U., India, 1962; Ph.D., U. London, 1970. Diplomate: Am. Bd. Toxicology, 1981. Asst. research officer Vet. Coll., India, 1962-66; pharmacologist Biorex Labs., London, 1966-67; sr. staff fellow Nat. Inst. Environ. Health Sci., Research Triangle Park, N.C., 1970-77, supervisory [pharmacologist, 1978—. Contbr. articles to profl. jours. Mem. Soc. Toxicology, Am. Soc. Pharmacology and Exptl. Therapeutics. Subspecialty: Toxicology (medicine). Current work: Metabolism of drugs; toxicologic and carcinogenic eval. of chems. Office: PO Box 12233 Research Triangle Park NC 27709

CHI, DAVID S., immunologist, educator, researcher; b. Taiwan, July 7, 1943; s Ching Hsing and Ah Chu (Chen) C. B.S., Nat. Chung-Hsing U., Taiwan, 1965; M.A., U. Tex., Galveston, 1974, Ph.D., 1977. Cert. clin. lab. dir. Am. Bd. Bioanalysis, 1981. Farm dir. Taiwan Sugar Corp., 1968-69; asst. mgr. Chai-Tai Enterprise Co., Ltd., Taiwan, 1969-70; postdoctoral fellow and research scientist N.Y.U. Med. Ctr., N.Y.C., 1977-80; assoc. prof. and clin. lab. dir. E. Tenn. State U., Johnson City, 1980—, chief div. biomed. research, 1981—. Contbr. articles to sci. jours. Recipient New Investigator Research award Nat. Cancer Inst., 1980. Mem. Am. Soc. Microbiology, Reticuloendothelial Soc., Am. Assn. Immunologists, Soc. Exptl. Biology and Medicine, N.Y. Acad. Scis., Am. Soc. Zoologists, Harvey Soc., Internat. Soc. Developmental and Comparative Immunology, Sigma Xi. Subspecialties: Immunobiology and immunology; Immunology (medicine). Current work: Research in cellular immunology, tumor immunology and clinical immunology. Home: 419 W Locust St Johnson City TN 37601 Office: E Tenn StateU Johnson City TN 37614

CHI, MYUNG SUN, biochemical nutritionist, health researcher, educator; b. Ik San, Korea, Sept. 28, 1940; came to U.S., 1969, naturalized, 1982; s. Won Gill and Soon Rae (Lee) C.; m. Eun Ja Kim, June 13, 1970; children: Linda, Carolyn. B.S., Kon-Kuk U., Seoul, Korea, 1966; M.S., U. Minn-Mpls., 1972, Ph.D., 1975. Agrl. cons. Korean Farm & Fishery Developint Corp., Seoul, 1967-68; research asst. U. Minn.-Mpls., 1969-74, research assoc., 1974-78; asst. prof. human nutrition Alcorn State U., Lorman, Miss., 1978-81; assoc. prof. Lincoln U., Jefferson City, Mo., 1981—, project leader hypertension research, 1981—. Lay leader Korean United Methodist Ch. Minn., St. Paul, 1976-78; mem. administrv. bd. First United Methodist Ch., Jefferson City, Mo., 1983. U.S. Dept. Agr. grantee, 1979—; recipient Best Research award 1890 Land-Grant Colls., 1980, 82. Mem. Am. Inst. Nutrition, Poultry Sci. Assn. Methodist. Subspecialties: Nutrition (medicine); Biochemistry (medicine). Current work: Nutritional effects in lipid metabolism; arterial wall metabolism; hormones related to hypertension. Home: 917 Airview Dr Jefferson City MO 65101 Office: Lincoln University Jefferson City MO 65101

CHIANG, CHAO-WANG, mechanical engineering educator, consultant, researcher; b. Anu-hui, China, May 21, 1925; s. Tsao-Chun and Ying (Chen) C.; m. Carole Tien; children: Wilfred, Manfred, Anne, Cathy. Ph.D., U. Wis.-Madison, 1960. Profl. engr., Colo.; Prof. mech. engring. U. Denver, 1964-74; prof. mech. engring. dept. S.D. Sch. Mines and Tech., Rapid City, 1974—, head dept., 1974-82; cons. in field. Contbg. author publs. in field. Mem. ASME, Am. Soc. Engring. Edn., Sigma Xi, Pi Tau Sigma. Subspecialties: Solar energy; Mechanical engineering. Current work: Solar energy, heat transfer, fluid mechanics. Home: 913 St Patrick St Rapid City SD 57701 Office: SD Sch Mines and Tech 500 E St Joe St Rapid City SD 57701

CHIANG, CHIN LONG, biostatistician, educator; b. Chekiang, China, Nov. 12, 1916; came to U.S., 1946, naturalized, 1963; s. Tse Shang and (Chen) C.; m. Fu Chen Shiao, Jan. 21, 1945; children: William S., Robert S., Harriet W. B.A. in Econs, Tsing Hwa U., 1940; M.A., U. Calif.-Berkeley, 1948, Ph.D. in Stats, 1953. Teaching asst. U. Calif., Berkeley, 1948, research asst., 1950-51, asso., 1951-53, instr., 1953-55, asst. prof. biostatistics, 1955-60, assoc. prof., 1960-65, prof., 1965—, chmn. div. measurement scis., 1970-75, chmn. faculty Sch. Public Health, 1975-76, chmn., 1970—, co-chmn. group in biostats., 1971—; vis. prof. U. Mich., 1959, U. Minn., 1960, 61, Yale U., 1965-66, Emory U., 1967, U. Pitts., 1968, U. Wash., 1969, U. N.C., 1969, 70, U. Tex., 1973, Vanderbilt U., 1975, Harvard U., 1977; cons. WHO, HEW, NIH, others. Author: Introduction to Stochastic Processes in Biostatistics, 1968, Life Table and Mortality Analysis, 1978, An Introduction to Stochastic Processes and Their Applications, 1979, The Life Table and its, Applications, 1984; Asso. editor: Biometrics, 1972-75, Math. Biosciences, 1976—; editorial bd.: WHO World Health Statis. Quar, 1979—. Nat. Heart Inst. fellow, 1959-60; Fulbright sr. lectr., 1964. Fellow Am. Statis. Assn., Inst. Math. Stats., Am. Public Health Assn., Royal Statis. Soc. London; mem. Internat. Statis. Inst., Biometric Soc. Democrat. Subspecialty: Statistics. Home: 844 Spruce St Berkeley CA 94707 Office: School Public Health U Calif Berkeley CA 94720

CHIANG, CHWAN-KANG, physicist; b. China, Jan. 18, 1943. Ph.D., Mich. State U., 1974. Research investigator U. Pa., Phila., 1974-78; physicist Nat. Bur. Standards, Washington, 1978—. Mem. Am. Phys. Soc., Am. Chem. Soc. Subspecialty: Condensed matter physics. Current work: Electrical properties; conducting polymers. Office: Nat Bur Standards B 320 Polymer Washington DC 20234

CHIANG, FU-PEN, mechanical engineering educator; b. Checkiang, China, Oct. 10, 1936; s. Chien-lo and Lien-yin (Mao) C.; m. Charlotte Chen-yi Chen, June1, 1963; children: Brian (dec.), Ted, Michelle. B.S. in Civil Engring, Nat. Taiwan U., 1953-57; M.S., U. Fla., 1963; Ph.D. in Engring. Sci. and Mechanics, 1966. Civil engr., 1958-62; asst. prof. mech. engring. SUNY-Stony Brook, 1967-70, assoc. prof., 1970-73, prof., 1974—; vis. prof. Swiss Fed. Inst. Tech., Lausanne, 1973-74; sr. vis. fellow dept. physics Cavendish Lab., U. Cambridge, Eng., 1980-81; cons. Army Material and Mechanics Research Ctr., Army Missile Command, Grumann Aerospace Corp., and; others. Contbr. articles to profl. jours. Postdoctoral fellow Cath. U. Am.; NSF grantee, 1968-73, 76-83; Office Naval Research grantee, 1982-84. Fellow Soc. Exptl. Stress Analysis; mem. Optical Soc. Am., Soc. Photo-Optical Instrumentation Engrs., Am. Acad. Mechanics, AAAS, ASME. Subspecialties: stress analysis; Solid mechanics. Current work: Development of optical stress analysis techniques such as laser speckles techniques, holographic interferometry, white light speckle techniques, moire methods, electron speckle and acoustic speckle techniques. Office: Dept Mech SUNY Stony Brook NY 11794

CHIANG, JOHN YOUNG LING, biochemistry educator, researcher; b. Hangchew, CheKiang, China, July 29, 1947; came to U.S., 1970, naturalized, 1980; s. Ming-ming and Ya-Jung (Huang) C.; m. Lisa H. Kang, Aug. 3, 1973; children: Eric, David. B.S., Chung-Hsing U., Taichung, Taiwan, 1969; M.S., SUNY-Albany, 1973, Ph.D., 1976. Postdoctoral scholar U. Mich. Med. Sch., Ann Arbor, 1976-78; asst. prof. biochemistry Northeastern Ohio U., Rootstown, 1978-83, assoc. prof., 1983—. Contbr. articles to profl. jours. NIH fellow, 1977-78; Pharm. Mfrs. Assn. Found. grantee, 1982-83; Am. Heart Assn. grantee, 1980-82; NIH grantee, 1983—. Mem. AAAS, Am. Soc. Biol. Chemists. Subspecialty: Biochemistry (medicine). Current work: Basic biochemical research in studying the induction and regulation of liver detoxication enzymes. Home: 1173 Erin Dr Kent OH 44240 Office: Northeastern Ohio University College of Medicine 4209 State Route 44 Rootstown OH 44272

CHIANG, JOSEPH FEI, chemist, educator, researcher; b. Hunan, China, Feb. 22, 1938; s. K. K. and W. S. (Ma) C.; m. Nancy S. Chang; children: Calvin, Amy. B.S., Tunghai U., 1960; M.S., Cornell U., 1965, Ph.D., 1967. Research fellow Cornell U., Ithaca, N.Y., 1967-68; prof. chemistry SUNY at Oneonta, 1968—; vis. prof. U. Chgo., 1983. Contbr. articls on chemistry to profl. jours. Nat. Acad. Scis. exchange scholar to Hungarian Acad. Scis., 1980; NIH fellow, 1975-76; research fellow Harvard U., 1978. Mem. Am. Chem. Soc., Am. Phys. Soc., Sigma Xi. Republican. Roman Catholic. Subspecialties: Physical chemistry; Biophysical chemistry. Current work: Molecular structure by spectroscopic techniques, X-ray cyrstallography, van der waals molecules, molecular quantum mechanics. Home: 16 Suncrest Terr Oneonta NY 13820 Office: Dept Chemistry SUNY Oneonta NY 13820

CHIANG, PETER K., research chemist; b. Hong Kong, Oct. 20, 1941; s. Wing K. and Kwei Y. (Lee) C.; m. Sabrina C. Hung, Sept. 30, 1941; children: Michelle, Stephanie. B.S., U. San Francisco, 1965; M.Sc., U. Alta., 1967, Ph.D., 1971. Postdoctoral fellow Johns Hopkins U., 1971-72; vis. fellow Nat. Inst. Child Health and Human Devel., NIH, 1972-74; sr. staff fellow NIMH, 1974-81; research chemist Walter Reed Army Inst. Research, Washington, 1982—. Mem. Am. Soc. Biol. Chemists, Am. Soc. Pharmacology and Exptl. Therapeutics, N.Y. Acad. Scis. Roman Catholic. Subspecialties: Biochemistry (medicine); Cellular pharmacology. Current work: Methylation, inhibitors of methylation, cellular chemistry. Patentee in field. Office: Walter Reed Army Inst Research Washington DC 20012

CHIANG, TUNG CHING, computer scientist; b. Canton, Kwangtung, China, Jan. 9, 1941; came to U.S., 1964, naturalized, 1973; s. Wu Chi and Wei Shi (Lee) C.; m. Angela R. H. Feng, Apr. 11, 1976; children: Emily, Evelyn. B.S.E.E., Cheng Kung U., Taiwan, 1963; M.S.E.E., U. N.D., 1966; Ph.D., U. Calif.-Berkeley, 1972. Engr. Sperry Univac, Mpls., 1966-68; teaching asst. U. Calif.-Berkeley, 1968-72; mem. tech. staff Bell Labs., Piscataway, N.J., 1972—. Recipient Disting. mem. award Bell Labs., 1982. Mem. Assn. Computing Machinery, Sigma Xi, Eta Kappa Nu. Subspecialties: Database systems; Software engineering. Current work: Database management systems, database machines, distributed databases, application to videotex systems. Home: 110 North Berkeley Heights NJ 07922 Office: Bell Labs 6 Corporate Pl Piscataway NJ 08854

CHIASSON, ROBERT BRETON, veterinary scientist, educator; b. Griggsville, Ill., Oct. 9, 1925; s. Placid Nelson and Marie Anna C.; m. Frances Marguirite Kientzle, Oct. 24, 1944; children: Phyllis, Robert, Sarah, John, William, Mary, Annette, Laura. A.B., Ill. Coll., 1949; M.S., U. Ill., 1950; Ph.D., Stanford U., 1955. Spl. supr. Ill. State Mus., Springfield, 1950-51; instr. U. Ariz., Tucson, 1951-55, asst. prof., 1956-60, assoc. prof., 1960-65, prof., 1965—; vis. prof. U. Calif.-Berkeley, summer 1963; Fulbright prof. biology U. Sci. and Tech., Kumasi, Ghana, 1969-70; vis. prof. research Poultry Research Centre, U. Edinburgh, Scotland, 1976-77. Contbr. articles to profl. jours. Bd. dirs. Ariz. Consumers Council, 1973—; vice chmn. Air Quality Control Bd., Pima County, Ariz. Served with U.S. Army, 1943-46; ETO. Am. Cancer Soc. grantee; NIH grantee; NSF grantee; Am. Nature Assn., Am. Physiol. Soc., Dept. Agr. grantee. Mem. Am. Physiol. Soc., Am. Soc. Zoologists, Ariz. Acad. Sci., N.Y. Acad. Scis., World Assn. Vet. Anatomists, Fed. Socs. for Exptl. Biology, Sigma Xi. Subspecialties: Animal physiology; Morphology. Current work: Comparative vertebrate anatomy; endocrinology. Home: 6941 Calle Jupiter Tucson AZ 85710 Office: U Ariz Tucson AZ 85721

CHIBUCOS, THOMAS ROBERT, family and child studies educator, researcher, writer; b. Chgo., Apr. 14, 1946; s. Gus and Jennie (Lalla) C.; m. Pamela Elizabeth Perry, May 15, 1982; m. Mary Clare Gilbert, Jan. 21, 1967 (div. Feb. 1976); children: Thomas, Marcus, Elise. B.A. in Psychology, No. Ill. U., 1969, M.A., 1970; Ph.D. in Developmental Psychology, Mich. State U., 1974. Research assoc. Nat. Inst. Edn., Washington, 1973-75; asst. prof. psychology No. Ill. U., 1975-76, asst. prof. child devel., 1976-80, assoc. prof., coordinator div. family and child studies, 1981—; day care cons. Growing Place Day Care, DeKalb, 1979-81; expert witness Kane County States Atty., Wheaton, Ill., 1982. Guest editor: spl. issue Infant Mental Health Jour, 1980; contbr. chpt. to book, articles to profl. jours. Coach Am. Youth Socer Orgn., DeKalb, Little Leauge, DeKalb; mem. DeKalb Human Relations Commn., 1982. Mich. Office Edn. grantee, 1972-83; Ill. Office Edn. grantee, 1977. Mem. Am. Psychol. Assn., Soc. Research in Child Devel., Ill. Assn. Infant Mental Health (co-founder, dir. 1981—), Internat. Assn. Infant Mental Health. Subspecialties: Developmental psychology; Child development and family studies. Current work: Social policy, children and families, child mistreatment, fathers and infants, life span development, family violence, research methodology. Office: No Ill U 209 Wirtz Hall DeKalb Il 60115

CHIEN, ROBERT TIENWEN, elec. engr.; b. Wusih, Kiangsu, China, Nov. 20, 1931. B.S. in Elec. Engring, U. Ill., 1954, A.M. in Math, 1957, Ph.D. in Elec. Engring, 1958. Mem. research staff, then group mgr. Thomas J. Watson Research Center, IBM Corp., 1958-64; mem. faculty U. Ill., 1965—, prof. elec. engring., 1966—, dir. coordinated sci. lab., 1975—. Author book, over 100 articles. Fellow IEEE; mem. Assn. Computing Machinery. Subspecialty: Electrical engineering. Patentee in field. Office: 1101 W Springfield Ave Urbana IL 61801

CHIGNELL, COLIN FRANCIS, pharmacologist; b. London, Apr. 7, 1938; s. Francis George and Elsie Mary (Lee) C.; m. Anke K. Kreienbring, Nov. 19, 1966; children: Kimberly, Kevin. B.Pharm. with honours, U. London, 1959, Ph.D., 1962. Vis. fellow Lab. Chemistry, Nat. Inst. Arthritis and Metabolic Disease, NIH, Bethesda, Md., 1962-64; vis. assoc. Lab. Chem. Pharmacology, Nat. Heart, Lung and Blood Inst., 1964-69, research pharmacologist, 1969-74, research pharmacologist pulmonary br., 1974-77; chief Lab. Environ. Biophysics Nat. Inst. Environ. Health Scis., Research Triangle Park, N.C., 1977—; adj. prof. pharmacology U. N.C., Chapel Hill, 1978—. Editor: Methods in Pharmacology Vol. 2, 1972; mng. editor: Jour. Biochem. and Biophys. Methods; contbr. numerous articles to sci. and tech. jours. Mem. Am. Soc. Pharmacology and Exptl. Therapeutics (J.J. Abel award 1973), Am. Soc. Biol. Chemists, Biophys. Soc., Am. Soc. Photobiology, AAAS, Pharm. Soc. Gt. Britain. Lutheran. Subspecialties: Molecular pharmacology; Medicinal chemistry. Current work: Molecular mechanisms of chemical toxicity, toxicology, spectroscopy, biochemical pharmacology, molecular pharmacology. Home: 128 Bruce Dr Cary NC 27511 Office: Lab Environ Biophysics Nat Inst Environ Health Scis PO Box 12233 Research Triangle Park NC 27709

CHILDERS, JOHN STEPHEN, psychologist, psychological educator; b. Elizabeth City, N.C., Aug. 10, 1946; s. Earl Stephen and Norma (Houchin) C.; m. Beth Austin, Feb. 13, 1966; children: John Stephen, Amy Suzanne. B.A., E. Carolina U., 1968, M.A., 1972; Ed. D., N. C. State U., 1983. Lic. psychol. assoc., N.C. Psychologist adult services Coastal Plain Mental Health Center, Greenville, N.C., 1969-71; acting chmn. Inst. Mental Health, Pitt Tech. Inst., Greenville, 1971-72; dir. testing, asst. prof. psychology E. Carolina U., 1972—; cons. numerous pub. sch. systems, N.C. area, 1970—; vice-chmn. N.C. State Dept. Pub. Instrn., 1978-81; area coordinator various test developing cos., N.C., 1972—. Contbr. articles to profl. jours. Mem. Greenville Mayor's Adv. Council on Leisure Activities, 1980-81; mem. Domiciliary Home Community Adv. Com., Pitt County, N.C., 1982—; adv. Real Crisis Intervention Center, Greenville, 1971—. Am. Guidance Services Inc. grantee, 1979—; N.C. Mental Health Authority grantee, 1973-74. Mem. Am. Psychol. Assn., Southeastern Psychol. Assn. Republican. Methodist. Clubs: Civitan, Greenville Sports. Subspecialties: Developmental psychology; Social psychology. Current work: Research projects include reliability and validity research on the Kaufman-Assessment Battery for Children; the Vineland Adaptive Behavior Scales and Child Behavior Checklist; researching social networks inpact on elderly. Home: 1101 Johnston St Greenville NC 27834 Office: Psychology Dept East Carolina U Greenville NC 27834

CHILDRESS, DUDLEY STEPHEN, biomedical engineering educator; b. Cass County, Mo., Sept. 25, 1934; m., 1959; 2 children. B.S., U. Mo.-Columbia, 1957, M.S., 1958; Ph.D. in Elec. Engring, Northwestern U., 1967. From instr. to asst. prof. elec. engring. U. Mo.-Columbia, 1959-63; research asst. physiol. control systems lab. Northwestern U., Evanston, Ill., 1964-66, from asst. prof. to assoc. prof. elec. engring. and orthopedic surgery, Chgo., 1972-77; prof. elec. engring. and orthopedic surgery research Med. Sch., Chgo., 1977—, dir. prosthetics research lab., 1971—, dir. rehab. engring. program, 1972—; Mem. com. prosthetics research and devel. Nat. Acad. Sci.-NRC, 1969-72, mem. subcom. design, 1970-73, chmn. upper-extremity prosthetics panel, 1971-73; mem. applied physiol. and bioengring. study sect. NIH, 1974-78. Nat. Inst. Gen. Med. Sci. grantee, 1970-75; recipient Goldenson award United Cerebral Palsy Found. Mem. AAAS, IEEE, Biomed. Engring. Soc., Rehab. Engring. Soc. N.Am., Internat. Soc. Prosthetics and Orthotics. Subspecialty: Biomedical engineering. Office: Prosthetics Research Lab Room 1441 345 E Superior St Chicago IL 60611

CHILDS, BARTON, educator, physician; b. Chgo., Feb. 29, 1916; s. Robert William and Katherine Sayles (Barton) C.; m. Eloise L.B. MacKie, Mar. 29, 1950; children—Anne Lloyd, Lucy Barton. A.B., Williams Coll., 1938; M.D., Johns Hopkins, 1942. Successively intern, asst. resident, resident pediatrics Johns Hopkins Hosp., 1942-43, 46-48; research fellow Children's Hosp., Boston, 1948-49; Commonwealth Fund fellow Univ. Coll., London, Eng., 1952-53; mem. faculty Johns Hopkins Sch. Medicine, 1949—, prof. pediatrics, 1962—; Mem. NIH Cons. Coms., 1959-63, 63-67, 67-69, 70-74, 78—. Served to capt., M.C. AUS, 1943-46. John and Mary Markle scholar, 1953-58; Grover F. Powers Distinguished scholar, 1960-62; recipient Research Career award NIH, 1962, Meade Johnson award pediatrics, 1959. Mem. Am. Pediatric Soc., Soc. Pediatric Research, Am. Acad. Pediatrics, Am. Soc. Human Genetics, Genetics Soc. Am., Inst. Medicine, Am. Acad. Arts and Scis. Subspecialty: Gene actions. Home: 1019 Winding Way Baltimore MD 21210

CHILDS, JEFFREY JOHN, computer engr.; b. Des Moines, Oct. 11, 1956; s. John Edwin and Dorothy Louise (Graham) C.; m. Debbie Banwart, June 24, 1978. B.S.E.E., Iowa State U., 1978; postgrad., U. Minn., 1981—. Machinist Hull Industries, Iowa, 1974-76; computer technician Cyclone Computer Lab., Iowa State U., Ames, 1976-78; software design engr., computer systems div. Hewlett Packard, Cupertino, Calif., 1978-80; hardware design engr. Def. Systems div. Sperry Univac, St. Paul, 1980-83, Star Techs., Inc., Brooklyn Park, Minn., 1983—. Mem. IEEE, Assn. for Computing Machinery, Eta Kappa Nu, Tau Beta Pi. Presbyterian. Club: Apple Computer (St. Paul). Subspecialties: Computer engineering; Computer architecture. Current work: Design and implementation of array processor architectures for real-time processing. Home: 1272 Easter Ln Eagan MN 55123 Office: 7101 Northland Circle Suite 102 Brooklyn Park MN 55428

CHILDS, JOSEPH EDWIN, systems analyst; b. Manchester, Conn., May 17, 1954; s. Frederick Lewis and Jean (Picaut) C.; m. Karen Furlong, July 3, 1975. B.S., Colo. State U., 1976, M.S., 1981. Temporary faculty dept. computer Sci. Colo. State U., Ft. Collins, 1980-81; v.p. APCOMP, Inc., Ft. Collins, 1981—; instr. Larimer County Vocat.-Tech., Ft. Collins, 1980—; Cons. Larimer County Vocat.-Tech. Center. Contbr. articles to profl. jours. Mem. Assn. for Computing Machinery. Subspecialties: Algorithms; Resource management. Current work: Numerical modelling; air resource management; simulation modelling; applied data management; computer graphics; software systems, systems design. Office: 1311 S College Ave Suite 4 Main Level Fort Collins CO 80524

CHILDS, WILLIAM JEFFRIES, physicist; b. Boston, Nov. 9, 1926; s. Paul Dudley and Clemence d'Espaigne (Jeffries) C.; m. Jean Mallory, June 17, 1951; children: Linton Jeffries, Lee Tracy. A.B., Harvard U., Cambridge, Mass., 1948; M.S., U. Mich., Ann Arbor, 1949, Ph.D., 1956. Physicist Argonne (Ill.) Nat. Labs., 1956—, sr. physicist, 1980—; vis. prof. U. Bonn., W. Ger., 1972-73. Contbr. articles to profl. jours. Served with AUS, 1944-46. Mem. Am. Phys. Soc., Am. Optical Soc. Subspecialty: Atomic and molecular physics. Current work: Laser and radiofrequency spectroscopy of atomic and molecular beams, fine and hyperfine structure. Home: 539 Fairview Ave Glen Ellyn IL 60137 Office: 9700 S Cass Ave Argonne IL 60439

CHILINGARIAN, GEORGE VAROS, educator; b. Tbilisi, Ga., July 22, 1929; s. Varos and Klavdia (Gorchakova) C.; m. Yelba Maria Salmeron, June 12, 1953; children: Modesto George, Eleanore Elizabeth, Mark Steven. B.E., U. So. Calif., 1949, M.S., 1950, Ph.D., 1954; Dr. Honoris Causa, Kensington U., 1977, Pacific Western U., 1978, Pepperdine U., 1979, Academiya Studiorum, Italy, 1960. With Wright-Patterson AFB, Dayton, 1954-56; faculty U. So. Calif., Los Angeles, 1956—, prof. petroleum engring., 1970—. Contbr. numerous articles to profl. jours.; author 32 books on petroleum engring. and geology. Served to col. USAFR, 1954—. Recipient numerous profl. awards. Republican. Subspecialties: Petroleum engineering; Geology. Home: 101 S Windsor Blvd Los Angeles CA 90004 Office: Dept Petroleum Engring Univ So Calif Los Angeles CA 90007

CHILTON, NEAL W(ARWICK), dental biostatistician, consultant; b. N.Y.C., June 24, 1921; s. Benjamin Bernard and Bertha (Warich) C.; m. Naomi Lilian Alexander, Dec. 28, 1947; children: Peninah, Jonathan, Abigail, Seth, Miriam. B.S., CCNY, 1939; D.D.S., N.Y. U., 1943; M.P.H., Columbia U., 1946. Diplomate: Am. Bd. Periodontology, Am. Bd. Endodontics. Asst. prof. pharmacology N.Y.U., N.Y.C., 1944-46; adj. prof. Columbia U., N.Y.C., 1948-56, sr. research assoc., 1960—; clin. prof. periodontics Temple U., Phila., 1957-75; research prof. oral medicine U. Pa., Phila., 1975-79, research cons., 1979—; Lady Davis prof. preventive dentistry Hebrew U., Jerusalem, Israel, 1977. Author: Design and Analysis in Dental and Oral Research, 2nd edit, 1982; contbr. articles to profl. publs. Served to maj. U.S. Army, 1953-55. NIH grantee, 1963—; contract, 1966, 81; Office Naval Research grantee, 1947. Fellow Am. Coll. Dentists; mem. Am. Acad. Periodontology, Am. Acad. Endodontics, Am. Acad. Oral Pathology, European Orgn. for Caries Research, Internat. Assn. Dental Research, Phi Beta Kappa, Sigma Xi, Omicron Kappa Upsilon. Democrat. Subspecialties: Periodontics; Statistics. Current work: Clinical trails of oral preventive and therapeutic agents; dental biostatistics; research consultant; epidemiology. Home: 2975 Princeton Pike Lawrenceville NJ 08648 Office: Univ Pennsylvania 4001 Spruce St Philadelphia PA 19104

CHIN, FRANCIS YUK-LUN, computing science educator, researcher; b. Hong Kong, Apr. 8, 1948; emigrated to Can., 1968, naturalized, 1980; s. Chuen and Suk-Ching (Cheng) C.; m. Priscilla Mui-Kuen Poon, June 21, 1977; children: Jerome Bokmen, Vivian Waimen. B.A. in Sci, U. Toronto, 1972; M.Sc., Princeton U., 1974, M.A., 1974, Ph.D., 1976. Asst. prof. U. Md., 1975-76; asst. prof. dept. computing sci. U. Alta., Edmonton, 1976-80, assoc. prof., 1980—; asst. prof. U. Calif.-San Diego, La Jolla, 1980-81; tech. cons. Alta. Research Council, Edmonton, 1979. Mem. IEEE, Assn. Computing Machinery, Soc. Indsl. and Applied Math., Canadian Info. Processing Soc., Sigma Xi. Subspecialties: Database systems; Algorithms. Current work: Distrubuted systems, data security, database, protocols, parallel algorithms, geometrical complexity. Office: University of Alberta Department of Computing Science Edmonton AB Canada T6G 2H1 Home: 10939-38 Ave Edmonton AB Canada T6J OK6

CHIN, GILBERT YUKYU, metallurgist; b. Kwangtung, China, Sept. 21, 1934; s. George Shee Ng and Liawah (Gee) C.; ; s. George Shee Ng (father Am. citizen) and C.; m. Ginie Wong, June 26, 1960; children—Patrick Ken, Michael Philip, Grace Fay, Karen Jean. S.B., MIT, 1959, Sc.D., 1963. Mem. tech. staff Bell Telephone Labs, Murray Hill, N.J., 1962—, head phys. metallurgy and crystal growth research dept., 1973-75, head phys. metallurgy and ceramics research and devel. dept., 1975—. Author. Recipient Achievement award Chinese Inst. Engrs. of U.S.A., 1980, 83. Fellow Inst. Metall. Engrs. (Mathewson Gold medal 1974); mem. Am. Soc. Metals, Nat. Acad. Engring., Metall. Soc. Am., Am. Ceramics Soc., Magnetics Soc. of IEEE, N.Y. Acad. Scis., AAAS, Sigma Xi, Tau Beta Pi, Phi Lambda Upsilon. Episcopalian. Subspecialties: Metallurgy; Materials (engineering). Current work: Methallurgy of magnetic alloys and development of magnetically soft, semi-hard and hard alloys; mechanical properties of materials; management of research and development program on electronic materials including metals, ceramics and semiconductors. Patentee in field. Office: Bell Telephone Labs Mountain Ave Murray Hill NJ 07974

CHIN, HONG W., radiation oncologist; b. Seoul, Korea, May 14, 1935; came to U.S., 1974; s. Jik H. and Woon K. (Park) C.; m. Soo J. Cheung, Dec. 27, 1965; children: Richard, Helen, Ki. M.D., Seoul Nat U., 1962, Ph.D., 1974. Diplomate: Am. Bd. Radiology. Radiation oncologist, asst. prof. dept. radiation medicine U. Ky. Med. Ctr., Lexington, 1979—. Contbr. med. articles to profl. jours. Mem. AMA, Am. Coll. Radiology, Am. Soc. Therapeutic Radiologists, AAAS, Can. Assn. Radiologists, N.Y. Acad. Scis., Radiation Research Soc., Radiol. Soc. N.Am., Sigma Xi. Subspecialties: Oncology; Cancer research (medicine). Current work: Researcher in cancer treatment, especially brain tumors. Office: U Ky 800 Rose St Lexington KY 40536

CHIN, JIN H., research engineer; b. Toi Shun, Kwang Tung, China, Oct. 15, 1928; came to U.S., 1948; s. Bing Suey and Ru Ju (Liau) C.; m. Jane E. Heng, Sept. 11, 1960; children: Goodwin R., Kingsley N. B.S. in Chemistry, Stanford U., 1950; M.S.E. in Chem. Engring, U. Mich., 1951, Ph.D., 1955. Research assoc. U. Mich., Ann Arbor, 1955-57; heat transfer specialist Gen. Electric Co., Evandale, Ohio, 1957-60; research specialist Lockheed Missiles and Space Co., Inc., Sunnyvale, Calif., 1960-62, staff engr., 1962-80, sr. staff engr., 1980—. Eastman Kodak fellow, 1951; Rackham fellow, 1952, 53; recipient Cost Improvement award Lockheed Missiles and Space Co., Inc., 1973. Mem. AIAA, AAAS. Subspecialties: Chemical engineering; Fluid mechanics. Current work: Advance computer analysis of heat transfer, fluid flow, and thermodynamics. Home: 727 Christine Dr Palo Alto CA 94303 Office: Lockheed Missiles and Space Co Inc 1111 Lockheed Way Sunnyvale CA 94086

CHIN, WILLIAM WAIMAN, molecular biologist, physician; b. N.Y.C., Nov. 20, 1947; s. James Gampoy and Yoke Ting (Chu) C.; m. Denise Jean-Claude. A.B., Columbia U., 1968; M.D., Harvard U., 1972. Diplomate: Am. Bd. Internal Medicine. Research assoc. NIH, Bethesda, Md., 1974-76; instr. medicine Harvard Med. Sch., Boston, 1979-81, asst. prof., 1981—; assoc. investigator Howard Hughes Med. Inst., Boston, 1979—. Served to lt. comdr. USPHS, 1974-76. Mem. AAAS, Thyroid Assn., Endocrine Soc., Am. Fedn. Clin. Research, Am. Soc. Biol. Chemistry., Am. Physiol. Soc. Subspecialties: Molecular biology; Neuroendocrinology. Current work: Regulation of expression of genes encoding the subunits of the glycoprotein hormones (TSH, LH, FSH, CG). Office: Lab Molecular Endocrinology Bulfinch 3 Mass Gen Hosp Boston MA 02114

CHINCARINI, GUIDO LUDOVICO, educator, astronomer; b. Venice, Italy, Jan. 24, 1938; came to U.S., 1965; s. Ludovico and Maria-Luisa (Bonaldi) C.; m. Ioanna Manousoyannaki, Mar. 27, 1980; children: Ludwig Boris, Wolfgang Maximillian. M.S., Lyceum of Sci., Venice, 1956; Ph.D., U. Padua, Italy, 1960. Astronomer U. Padua, 1961-68, Lick Obs. U. Calif., 1964-67, U. Bonn., Ger., 1968-69; research assoc. NASA, Houston, 1969-71, Wesleyan U., Conn., 1969-71; research scientist McDonald Obs., Austin, Tex., 1971-74; vis. asst. prof. astronomy U. Okla., 1975, vis. assoc. prof., 1976-77, assoc. prof., 1977-79, prof., 1979—; chmn. astrophysics U. Naples, Italy, 1976; chmn. astronomy U. Bologna, 1976; assoc. European So. Obs., Garching bei Munchen, W.Ger., 1978—. Contrb. articles in field to profl. jours. Served with Nat. Def., 1967-68. NASA grantee, 1980-83; NSF grantee, 1983-84. Mem. Internat. Astron. Union, Am. Astron. Soc., Italian Astron. Soc., Sigma Xi. Subspecialties: Cosmology; Optical astronomy. Current work: Study of the large scale distribution of matter in the universe; interaction between the interstellar medium of galaxies and intracluster Galaxies. Home: 1223 Cruce St Norman OK 73069 Office: Physics and Astronomy University of Oklahoma Norman OK 73019

CHINN, KENNETH SAI-KEUNG, research chemist; b. Hong Kong, Dec. 20, 1935; came to U.S., 1951; s. Edward K. and Sinn-tai (Zan) C.; m. Marie L. Leung, Aug. 22, 1964; 1 son, Stephen A. B.S., Washington U., St. Louis, 1957, postgrad., 1957-59. Research chemist U.S. Army Med. Research Ctr., Denver, 1959-68; research biol. Chemist U.S. Naval Med. Research Ctr., Taipei, Taiwan, 1968-74; research chemist U.S. Army, Dugway, Utah., 1974—; cons. in field. Served with U.S. Army, 1959-61. Mem. Am. Nutrition Soc., Colo.-Wyo. Acad. Sci., S.E. Asian Ministers of Edn. Orgn. Subspecialties: Biochemistry (medicine); Nutrition (medicine). Current work: Effects of nutrition regarding nutrient utilization, growth, neurometer development, resistance to infectious disease as well as various aspects of biochemical, clinical and nutritional interaction, various aspects of chemical and biological warfare agents (worldwide). Home: 499 Country Club Dr Stansbury Park UT 84074 Office: US Army Dugway Proving Ground Dugway UT 84022

CHINTAPALLI, HEMANTHA KUMAR, systems programmer; b. Mudinepalli, India, May 25, 1951; came to U.S., 1974; s. Ramachandra Rao and Anasuya Devi (Nutakki) C.; m. Nirmala Kumari Yernent, June 16, 1976. B.Tech., Andhra U., India, 1971, M.Tech., 1974; M.C.S., Tex. A&M U., 1975. Programmer City of Houston, 1976-78; systems programmer Tex. Instruments Inc., Houstin and Austin, 1978-81, IBM Corp., White Plains, N.Y., 1981—. Mem. IEEE, Assn. for Computing Machinery. Subspecialties: Operating systems; Distributed systems and networks. Current work: Operating systems and communications software, with special interest in computer architecture. Home: 100 Hospital Plaza 411 Paterson NJ 07503 Office: 1 Corporate Park Dr White Plains NY 10604

CHIORAZZI, NICHOLAS, physician; b. Weehawken, N.J., Oct. 2, 1945; s. Joseph P. and Mary L. (Ippolito) C.; m. Mary Lorraine Dziadowicz, June 19, 1971; children: Anne, Michael. B.A., Coll. Holy Cross, 1966; M.D., Georgetown U., 1970. Diplomate: Am. Bd. Internal Medicine. Intern, resident Cornell Cooperating Hosps., N.Y.C., 1970-74; postdoctoral fellow in immunology dept. pathology Harvard Med. Sch., Boston, 1974-76, Rockefeller U., 1976-77, asst. prof., assoc. physician, N.Y.C., 1972-82, assoc. prof., physician, 1982—. Mem. ACP, Am. Assn. Immunologists, Am. Fedn. Clin. Research. Subspecialties: Immunology (medicine); Internal medicine. Current work: Clinical investigator of medical problems relating to immunology. Home: 18 Bliss Ave Tenafly NJ 07670 Office: Rockefeller U 1230 York Ave New York NY 10021

CHIOU, JIUNN PERNG, mechanical engineer, consultant; b. Nanking, China, July 29, 1933; came to U.S., 1958, naturalized, 1969; s. Chiou Cho Y. C. and Wang Mong L.; m. (married), Aug. 15, 1965; children: Derek, Jeff. B.S., Nat. Taiwan U., China, 1954; M.S., Oreg. State U., 1960; Ph.D., U. Wis., Madison, 1964. Mech. engr. Keelung Harbor Bur., Taiwan, 1955-58; engring. specialist; AiResearch Mfg. div. Garrett Corp., Los Angeles, 1964-69; assoc. prof. mech. engring. U. Detroit, 1969-78, prof., 1978—; cons. Contbr. articles to profl. jours. Mem. ASME, Soc. Automotive Engineers, Internat. Solar Energy Soc., Sigma Xi., Tau Beta Pi, Pi Tau Sigma. Subspecialties: Mechanical engineering; Solar energy. Current work: Research and development in heat transfer, heat exchanger, solar energy, applied methematics, computer simulation and modelling.

CHIPMAN, ERIC GEORGE, scientist; b. Kingston, Ont., Canada, July 29, 1944; s. Robert Avery and Lois Margaret (Retallack) C.; m. Susan Elizabeth Fitzgerald, Aug. 26, 1966. A.B. in Astronomy, Harvard U., 1965, M.A., 1969, Ph.D., 1972. Research assoc. Lab. Atmospheric and Space Physics U. Colo., 1971-78; program scientist Solar Physics Office NASA, Washington, 1978-83; staff scientist Space Telescope Sci. Inst. Computer Sci. Corp., 1983—. Contbr. articles to profl. jours. NASA grantee, 1977. Mem. Am. Astron. Soc., Am. Geophys. Union. Home: 2606 S Joyce St Arlington VA 22202

CHIPMAN, GORDON LEIGH, JR., government executive; b. Newport News, Va., May 12, 1942; s. Gordon L. and Arta (Leon) C.; m. Geraldine J. Frankhauser, Dec. 3, 1965. B.S. in Elec. Engring, U. Nebr., 1965. Engr. Westinghouse Elec. Corp., Pitts., 1970-72; sect. leader U.S. NRC, Washington, 1972-79; tech. cons. U.S. Ho. of Reps., Washington, 1979-81; dep. asst. sec. breeder reactor programs U.S. Dept. Energy, Washington, 1981. Served as lt. USN, 1965-70. Recipient nat. def. service medal, spl. achievement award U.S. NRC, 1979. Mem. Am. Nuclear Soc., Scis. and Engrs. for Secure Energy. Republican. Subspecialties: Nuclear fission; Nuclear engineering. Current work: Develop breeder reactor technology to readiness for industrialization; develop and provide nuclear power systems for space applications. Home: 6953B Linganore Rd Frederick MD 21701 Office: Asst Sec Nuclear Energy Dep Asst Sec Breeder Reactor Programs Washington DC 20545

CHISHOLM, MALCOLM HAROLD, chemistry educator, researcher; b. Bombay, India, Oct. 15, 1945; came to Eng., 1946; s. Angus MacPhail and Gweneth (Robey) C.; m. Cynthia Ann Brown, May 1, 1982; m. Susan Patricia Sage, Oct. 1968 (div. Apr. 1978); 1 son, Calum. B.Sc. with spl. honors, London U., 1966, Ph.D. in Inorganic Chemistry, 1969, D.Sc. (hon.), 1980. Session lectr. U. Western Ont., Can., London, 1970-72; asst. prof. Princeton (N.J.) U., 1972-78; assoc. prof. Ind. U., Bloomington, 1978-80, prof. chemistry, 1980—; cons. to industry; sci. councilor to Sen. Birch Bayh of Ind., 1979-80; editorial cons. Inorganic Reactions, 1980—. Am. Editor: Polyhedron Reports, 1982—; editor/author: Reactivity of Metal-Metal Bonds, 1981, Inorganic Chemistry: Toward the 21st Century, 1983; contbr. over 150 articles to sci. jours. Sci. Research Council Eng. fellow, 1966-69; Alfred P. Sloan Found. fellow, 1976-78; Dreyfus Found. Tchr.-scholar, 1979-84; recipient Corday-Morgan medal Royal Soc. Chemistry, Eng., 1979. Mem. Am. Chem. Soc. (Akron sect. award 1981, sec.-treas. Princeton sect. div. inorganic chemistry 1976, alt. councilor So. Ind. sect. 1979—). Club: Ind. Univ. Subspecialties: Inorganic chemistry; Organometallic chemistry. Current work: Transition metal chemistry; metal-metal bonds; catalysis. Home: 900 S High St Bloomington IN 47401 Office: Ind U Dept Chemistry Bloomington IN 47405

CHITALEY, SHYAMALA DINKAR, paleobotanist, educator, researcher; b. Nasik, Maharashtra, India, Feb. 15, 1918; came to U.S., 1978; d. Ganesh S. and Srarswati G. (Joglekar) Dixit; m. Dinkar V. Chitaley, Jan. 20, 1934; children: Avinash, Aniruddha. B.S., Coll. of Sci., Nagpur, India, 1942, M.S., 1945; Ph.D. (Ida Smith Internat. fellow), U. Reading, Eng., 1955. Curator paleobotany Cleve. Mus. Natural History, 1980—; lectr. botany Inst. of Sci., Nagpur, India, 1948-61, asso., 1961-74, prof., 1974-76; researcher fossils. Contbr. articles on paleobotany to profl. jours. Univ. Grants Commn. (India) grantee, 1955,57,59,69,72,74; recipient Inventions Promotion Bd. (India) Prize, 1968. Mem. Internat. Orgn. Paleobotany, Internat. Assn. Angiosperm Paleobotany, Am. Assn. Botany, Am. Assn. Paleobotany, Internat. Assn. Palynology, Thoreau Soc. (Cleve.), Fossil Soc. (Cleve.). Subspecialties: Sedimentology; Evolutionary biology. Current work: Plant fossils from Devonian and Cretaceous-Tertiary, curator plant fossils, researcher plants from Cleveland shale and spores from Cleveland shale correlated with carbon content. Inventor preservation of plant specimans in color and slide, 1968. Office: Wade Oval University Circle Cleveland OH 44106

CHITTENDEN, MARK EUSTACE, JR., aquatic biologist, educator; b. Jersey City, N.J., July 30, 1939; s. Mark Eustace and Margaret Beaumont (Neil) C.; m. Susan Rae Morrison, Jan. 20, 1968; children: Laura Lynne, Julie Anne. B.A., Hobart Coll., 1960; M.S., Rutgers U., 1965, Ph.D. in Aquatic Biology, 1969. Fisheries biologist N.J. Div. Fish and Game, 1960-64; research fellow dept. environ. sci. Rutgers U., 1964-67, research asst., 1967-68, research assoc. fisheries, 1968-69; asst. prof. marine fisheries Coll. William and Mary and U. Va., 1969-72; assoc. marine scientist Va. Inst. Marine Sci., Gloucester Point, 1969-72; asst. prof. dept. wildlife and fisheries sci. Tex. A&M U., College Station, 1973-77, assoc. prof., 1977—. Contbr. articles to profl. jours. Mem. Am. Fisheries Soc., Am. Soc. Ichthyologists and Herpetologists, AAAS, Gulf and Caribbean Fisheries Inst., Am. Inst. Fishery Research Biologists. Subspecialties: Ecology; Oceanography. Current work: Research and teaching in marine fish and fisheries ecology and population dynamics. Office: Dept Wildlife and Fisheries Tex A&M U College Station TX 77843

CHITWOOD, DAVID JOSEPH, zoologist; b. Balt., Apr. 29, 1950; s. John Allen and Dolores Rita (Popoli) C. Ph.D., U. Md., College Park, 1980. Postdoctoral nematologist Insect Physiology Lab., Agrl. Research Ctr., U.S. Dept. Agr., Beltsville, Md., 1981—. Contbr. articles to profl. jours. Mem. AAAS, Soc. Nematologists, Am. Phytopath. Soc. Subspecialties: Plant nematology; Biochemistry (biology). Current work: Biochemistry, physiology and endocrinology of plant parasitic nematodes. Office: US Dept Agr B-467 BARC-E Beltsville MD 20705

CHIU, JEN-FU, biochemist, educator; b. Tungshi, Taiwan, China, Sept. 30, 1940; came to U.S., 1972, naturalized, 1979; s. Kuo-Fun and Chien-Leon (Yu) C.; m. Lucia Chi-Kai, May 30, 1970; children: Rosaleen I-Hsuen, Cynthia I-Tyng. B.Pharm., Taipei Med. Coll., 1964; M.Sc. in Biochemistry, Nat. Taiwan U., 1967; Ph.D., U. B.C., Can., 1972. Project investigator M.D. Anderson Hosp. and Tumor Inst., Houston, 1972-74; asst. biochemist, 1974-75; asst. prof. biochemistry Vanderbilt U., Nashville, 1975-78; assoc. prof. biochemistry U. Vt., Burlington, 1978—, dir. biochemistry grad. studies, 1982—. Contbr. articles to profl. jours. Can. Med. Research Council student, 1968-72; Rosalie B. Hite fellow, 1972-73; NIH grantee, 1975—. Mem. Am. Soc. for Cell Biology, Am. Assn. Cancer Research, Can. Biochem. Soc., Biophys. Soc., Am. Soc. Biol. Chemists, Sigma Xi. Subspecialties: Gene actions; Cancer research (medicine). Current work: Biochemistry of chromatin and gene regulation; biochemistry of cancer. To study the alteration of genome organization and gene expression in cells during neoplastic transformation by using hybridoma and recombinant DNA techniques. Home: 65 East Terr South Burlington VT 05401 Office: Dept Biochemistry U Vt Coll Medicin Burlington VT 05405

CHIU, PETER JIUNN-SHYONG, pharmacologist; b. Miao-Li, Taiwan, June 9, 1942; came to U.S., 1967; s. Uh-Tsuen and May-May (Hsu) C.; m. Peggy Tsui-Fang, Aug. 30, 1967; children: Vivian, Faye, Peter. B.S., Taipei Med. Coll. Sch. Pharmacy, 1964; M.S., Nat. Taiwan U., 1966; Ph.D., Columbia U., 1972. Postdoctoral research fellow Sch. Medicine U. Pa., 1972-74; sr. scientist Schering Corp., Bloomfield, N.J., 1974-77, prin. scientist, 1977-82, sr. prin. scientist, 1982—. Contbr. articles to profl. jours. Served with Nat. Chinese Army, 1966-67. Mem. Am. Soc. Pharmacology and Exptl. Therapeutics, Am. Soc. Nephrology, Internat. Soc. Nephrology, Am. Soc. Microbiology, Am. Fedn. Clin. Research, AAAS. Subspecialties: Pharmacology; Nephrology. Current work: Research and development of antihypertensive and antiulcer drugs; renal safety evaluation of antihypertensive and antiinfective drugs. Home: 9 Edwin Rd Morris Plains NJ 07950 Office: 60 Orange St Bloomfield NJ 07003

CHIUTEN, DELIA FUNG SHE, internist, med. educator; b. Philippines, Apr. 24, 1943; came to U.S., 1972; d. Juan and Constancia (Fungshe) C. B.S., U. San Carlos, Philippines, 1966; M.D., U. Philippines, 1971. Resident in surgery Victoria (B.C., Can.) Vets. Hosp., 1972; rotating intern Mt. Sinai Hosp. Services, City Hosp. at Elmhurst, N.Y., 1972-73, resident in medicine, 1973-75, chief resident in medicine, 1975; fellow in hematology Montefiore Hosp. and Med. Center, Bronx, N.Y., 1975-76; fellow in oncology Balt. Cancer Research Center, Nat. Cancer Inst., U. Md. Hosp., 1976-77; vis. asso. Nat. Cancer Inst., NIH, Bethesda, Md., 1977-79; faculty asso. and instr. in devel. therapeutics U. Tex. System Cancer Center M.D. Anderson Hosp. and Tumor Inst., Houston, 1979-80, asst. internist, asst. prof. medicine, 1980—; cons. Office Health Resources Opportunity Program, USPHS, HEW, 1981-82. Mem. Pan Asian Women's Orgn., Am. Soc. Clin. Oncology, Am. Assn. Cancer Research, Internat. Assn. Study Lung Cancer, Am. Fedn. Cancer Research, AAAS, Am. Soc. Internal Medicine, Am. Soc. Microbiology, Tex. Soc. Internal Medicine. Subspecialties: Cancer research (medicine); Chemotherapy. Current work: Evaluation of new antitumor drugs in lung cancer and other tumors in man. Office: 6723 Bertner Ave LP3 003 Houston TX 77030

CHO, ALFRED Y., electrical engineer, educator; b. Peking, China, July 10, 1937; s. Edward I-Lai and Mildred (Chen) C.; m. Mona Lee Willoughby, June 11, 1943; children: Derek, Diedre, Brynna, Wendy. B.S. in Elec. Engring, U. Ill.-Urbana, 1960, M.S., 1961, Ph.D., 1968. Research physicist Ion Physics Corp., Burlington, Mass., 1961-62; mem. tech. staff TRW-Space Tech. Labs., Redondo Beach, Calif., 1962-65; research asst. U. Ill.-Urbana, 1965-68, vis. prof. elec. engring., research prof., 1977-78, adj. prof. elec. engring., adj. prof., 1978—; mem. tech. staff Bell Labs., Murray Hill, N.J., 1968—. Contbr. numerous articles to profl. jours. Recipient Disting. Tech. Staff award Bell Labs., 1982. Fellow IEEE (Morris N. Liebmann award 1982); mem. Am. Phys. Soc. (Internat. prize for new materials 1982), Am. Vaccum Soc., Electrochem. Soc. (Electronic Div. award 1977), N.Y. Acad. Scis., AAAS, Sigma Xi, Tau Beta Pi, Eta Kappa Nu, Sigma Nu. Subspecialties: Electronic materials; Condensed matter physics. Current work: Thin film growth of semiconductor, metal and insulator for microwave and opto-electronics device applications. Patentee in field; developed crystal growth tech called molecular beam epitaxy. Office: Bell Labs Murray Hill NJ 07974

CHO, ARTHUR KENJI, pharmacologist, educator; b. Oakland, Calif, Nov. 7, 1928; s. Iwao and Mary Yoshiko (Takata) C.; m. Sachiko Yoshida, Aug. 16, 1953; children: David, Nancy. B.S., U. Calif., Berkeley, 1952; M.S., Oreg. State U., 1953; Ph.D., UCLA, 1958. Research chemist Don Baxter Inc., Glendale, Calif., 1961-65; research pharmacologist Nat. Heart Inst., Bethesda, Md., 1965-70; from asso. to prof. dept. pharmacology UCLA, 1970—. Author abstracts, articles and book chpts. on drug metabolism, neurochemistry and medicinal chemistry. Mem. Am. Chem. Soc., Am. Soc. Pharmacology and Exptl. Therapeutics, Soc. Toxicology. Subspecialties: Molecular pharmacology; Medicinal chemistry. Current work: Drug metabolism, mechanisms of nitrogen oxidation, adrenergic mechanisms. Home: 3393 Colbert Ave Los Angeles CA 90066 Office: Dept Pharmacology U Calif Los Angeles CA 90024

CHO, BYUNG-RYUL, veterinary microbiology educator; b. Seoul, South Korea, Feb. 3, 1926; s. Tai E. and Ah R. (Kim) C.; m. Jung-Sook Kim, Nov. 26, 1948; children: Sang, Young, Yungmi. D.V.M., Seoul Nat. U., 1950; M.S., U. Minn., 1959, Ph.D., 1961. Research assoc. dept. vet. microbiology Wash. State U., Pullman, 1964-67, asst. prof., 1967-72, assoc. prof., 1972-76, prof., 1976—. Contbr. articles to profl. jours. Mem. AVMA, Am. Assn. Avian Pathologists (P.P. Levine award 1977), Am. Soc. Microbiology, AAAS, World Vet. Poultry Assn. Methodist. Subspecialties: Virology (veterinary medicine); Cancer research (veterinary medicine). Current work: Virus diseases of birds with major emphasis on virus-induced neoplastic diseases. Home: NW 1405 Deane St Pullman WA 99163 Office: Dept Vet Microbiology-Pathology Wash State U Pullman WA 99164

CHO, CHAIDONG, pathologist, educator; b. Suwon, South Korea, Dec. 26, 1935; came to U.S., 1965; s. Baik Hyun and Chung Sook (Shin) C.; m. Ihmyung Moon, Mar. 21, 1965; children: Theresa M., Irene S. M.D., Seoul Nat. U., 1960, M.Med. Sci., 1962. Diplomate: Am. Bd. Pathology. Rotating intern Charleston (W.Va.) Gen. Hosp., 1965-66; resident VA Hosp., Hines, Ill., 1966-70; asst. prof. dept. pathology Upstate Med. Center, SUNY-Syracuse, 1974-81, assoc. prof. dept. pathology, 1982—; chief chemistry sect. Lab Service VA Med. Center, Syracuse, 1974—. Contbr. writings to profl. pubis. in field. Pres. Korea Assn. Syracuse, 1982—. Served to capt. Korean Army, 1962-65. Co-investigator, grants NIH, VA. Fellow Coll. Am. Pathologists, Am. Soc. Clin. Pathologists; mem. Am. Assn. Pathologists, Internat. Acad. Pathology U.S.-Can. Div. Subspecialty: Pathology (medicine). Current work: Metabolic modification of liver injury, hepatic and gastrointestinal pathology. Home: 4768 Edgeworth Dr Manlius NY 13104 Office: Lab Service MA Med Center 800 Irving Ave Syracuse NY 13210

CHO, JOON HO, nuclear engr.; b. Seoul, Korea, Oct. 13, 1949; came to U.S., 1975, naturalized, 1979; s. Nam Chul and Hee (Sook) C.; m. Hanna Kim, July 4, 1980. B.S., Seoul Nat. U., 1972; M.E., Tex. A&M U., 1976; M.S. in Engring, M.I.T., 1980. Grad. Asst. Tex. A&M U., College Station, 1976-77; research asst. M.I.T., Cambridge, 1979-80; nuclear engr. Stone & Webster Engring. Corp., Boston, 1980—. Mem. Am. Nuclear Soc., Korean Scientists and Engrs. in Am. Presbyterian. Subspecialties: Nuclear fission; Mechanical engineering. Current work: Nuclear safety analysis following a loss of coolant accident of nuclear power plant. Home: 199 Massachusetts Ave 415 Boston MA 02115 Office: Stone and Webster Engring Corp 245 Summer St Boston MA 02107

CHO, SOUNG MOO, energy company executive, engineering educator; b. Seoul, Korea, Oct. 1, 1937; came to U.S., 1962, naturalized, 1975; s. Song-Nyung and Mi-Ok (Lee) C.; m. Ki Sun Kim, Mar. 6, 1965; children: Rose, Richard, Karen, Grace, Christopher, Andrew. B.S., Seoul Nat. U., 1960; M.S., U. Calif.-Berkeley, 1964, Ph.D., 1967. Registered profl. engr., N.J. Researcher Korean Atomic Energy Research inst., Seoul, 1960-61; research asst./specialist U. Calif.-Berkeley, 1962-67; engring. specialist Garrett Corp., Los Angeles, 1967-69; staff cons. Energy Tech. Engring. Ctr., Canoga Park, Calif., 1969-73; engring. mgr. Foster Wheeler Energy Corp., Livingston, N.J., 1973-80, dep. dir. engring., 1980—; adj. prof. Stevens Inst. Tech., Hoboken, N.J., 1978—, Fairleigh Dickinson U., Teaneck, N.J., 1975—; cons. lectr. to various univs., utilities, profl. socs., 1975—. Contbr. articles to profl. jours.; reviewer various jours. Recipient Republic Korea Presdl. award, 1960; Fulbright travel scholar, 1962. Mem. ASME (exec. com. nuclear engring. div. 1982—, chmn. nuclear heat exchanger com. 1980-82), Am. Nuclear Soc. Subspecialties: Mechanical engineering; Nuclear engineering. Current work: Liquid metal fast breeder reactor engineering, advanced heat transfer device design, thermal/fluid science education. Home: 1051 Vail Rd Parsippany NJ 07054 Office: Foster Wheeler Energy Applications Inc 110 S Orange Ave Livingston NJ 07039

CHOATE, JERRY RONALD, zoologist, museum director, educator; b. Bartlesville, Okla., Mar. 21, 1943; s. C.W. and Alice Joyce (Cox) Marks; m. Rosemary Fidelis Walker, Apr. 13, 1963; 1 son, Judd Randolph. B.A. in Biology, Pittsburg (Kans.) State U., 1965; Ph.D. in Zoology, U. Kans.-Lawrence, 1969. Cert. wildlife biologist. Asst. prof. U. Conn.-Storrs, 1969-71; lectr. Yale U., New Haven, 1970; asst. prof. Ft. Hays State U., Hays, Kans., 1971-76, assoc. prof., 1976-80, prof. dept. zoology, 1980—, dir. museums, 1980—; cons. Kans. Fish and Game, 1975—. Assn. Systematics Collections, Lawrence, 1976—, Sunflower Electric Coop., Hays, 1980. Contbr. articles to sci. jours. Coach Little League baseball, Hays, 1980, Jr. League football, Hays, 1979, 80. NSF grantee, 1967-82; recipient grants, research contracts various orgns. and agencies. Mem. Am. Soc. Mammalogists (rec. sec. 1974—), Assn. Systematics Collections (sec. 1976-78), Southwestern Assn. Naturalists (pres. 1979-80), other profl. assns. Republican. Subspecialties: Systematics; Evolutionary biology. Current work: Systematics, evolution, biogeography of mammals. Home: Route 1 Victoria KS 67671 Office: Fort Hays State Univ 600 Park St Hays KS 67601

CHOBANIAN, ARAM VAN, medical scientist, cardiovascular physician; b. Pawtucket, R.I., Aug. 10, 1929; s. Vahan and Marina (Arsenian) C.; m. Jasmine Chobanian, June 5, 1955; children: Karin, Lisa, Aram. A.B., Brown U., 1951; M.D., Harvard U., 1955.

Diplomate: Am. Bd. Internal Medicine. Intern, resident Univ. Hosp., Boston, 1955-59; cardiovascular research fellow Boston U., 1959-62, asst. prof. medicine, 1964-68, assoc. prof., 1968-71, prof., 1971—, prof. pharmacology, 1975—, dir., 1974—. Trustee Armenian Library and Mus. of Am. Served to capt. USAF, 1955-57. Fellow ACP, Am. Coll. Cardiology, Council on Arteriosclerosis and Council on Hypertension; mem. Am. Heart Assn. (chmn. elect Council on High Blood Pressure Research), Assn. Am. Physicians, Am. Soc. Clin. Investigation, Am. Physiol. Soc., Phi Beta Kappa, Sigma Xi, Alpha Omega Alpha. Subspecialties: Cardiology; Cell biology. Current work: Research activities concerned with studies on vascular metabolism and the effects of hypertension hyperlipoproteinemia and diabetes on inducing vascular injury; also research involves role of sympathetic nervous system in hypertension. Office: 80 E Concord St Boston MA 02118

CHODOROW, MARVIN, physicist, educator; b. Buffalo, July 16, 1913; s. Isidor and Lena (Cohen) C.; m. Leah Ruth Turitz, Sept. 19, 1937; children: Nancy Julia, Joan Elizabeth. B.A., U. Buffalo, 1934; Ph.D., Mass. Inst. Tech., 1939; LL.D., U. Glasgow, 1972. Research asso. Pa. State Coll., 1940-41; instr. physics Coll. City N.Y., 1941-43; sr. project engr. Sperry Gyroscope Co., 1943-47; faculty Stanford, 1947—, prof. physics, 1947-54, prof. applied physics and elec. engring., 1954-78, Barbara Kimball Browning prof. applied physics, 1975-78, prof. emeritus applied physics and elec. engring., 1978—; dir. Edward L. Ginzton Lab., 1959-78, chmn. dept. applied physics, 1962-69; Cons. Def. Dept., Rand Corp.; vis. lectr. Ecole Normale Superieure, Paris, France, 1955-56; vis. research asso. U. Coll. London, 1969-70. Co-author: Fundamentals of Microwave Electronics, 1964; Contbr. articles to profl. jours. Fulbright fellow Cambridge (Eng.) U., 1962-63. Fellow IEEE (W.R.G. Baker award 1962, Lamme medal 1981), Am. Acad. Arts and Scis., Am. Phys. Soc.; mem. Nat. Acad. Scis., Nat. Acad. Engring., Am. Assn. Physics Tchrs., AAAS, AAUP, Phi Beta Kappa, Sigma Xi. Designed 1st klystron for microwave relay systems, 1946, 1st megawatt klystron, 1949, 1st megawatt traveling-wave tube, 1952-57. Home: 809 San Francisco Terr Stanford CA 94305 Office: Edward L Ginzton Lab Stanford U Stanford CA 94305

CHOI, SEUNG HOON, nuclear engr., cons.; b. Korea, Aug. 12, 1943; came to U.S., 1973, naturalized, 1981; s. Maeng Soon and Hwa Jin (Lee) C.; m. Carol Changsoon Kim, Nov. 28, 1970; children: Tina Eunhye, Joanne Mindulla, Brian Myongwon. B.S., Han Yang U., Seoul, Korea, 1969; cert., Westinghouse Nuclear Tng. Sch., Zion, Ill., 1974; postgrad., Northwestern U., 1974-78. Registered profl. engr., Mich., N.H. Nuclear engr. Korea Electric Power Corp., Seoul, 1969-73; nuclear process engr. Sargent & Lundy Engrs., Chgo., 1974-77; nuclear systems engr. Gilbert/Commonwealth Co., Jackson, Mich., 1977-79; sr. engr. Yankee Atomic Electric Co., Framingham, Mass., 1979—. Served with Korean Army, 1964-65. Mem. ASME, Am. Nuclear Soc. Subspecialties: Nuclear engineering; Nuclear fission. Current work: Nuclear power plant engring., design and operation. Home: 6 Hawthorne Circle Northboro MA 01532 Office: 1671 Worcester Rd Framingham MA 01701

CHOI, YONG SUNG, immunologist, educator; b. Kwangju, Korea, Sept. 11, 1936; came to U.S., 1962, naturalized, 1972; m. Ilzit Zemjanis, Aug. 27, 1966. B.M.S., Seoul Nat. U., 1957, M.D., 1961; Ph.D., M.S., U. Minn., 1965. Med. fellow pediatics U. Minn., Mpls., 1967-70; research assoc. Salk Inst., La Jolla, Calif., 1967-69; asst. prof. pediatics and biochemistry U. Minn., Mpls., 1969-73; mem. staff and prof. Sloan-Kettering Inst., Rye, N.Y., 1973—. Contbr. numerous articles to med. jours. NIH fellow, 1969-70; Am. Cancer Soc. grantee, 1970; USPHS grantee, 1971. Mem. Am. Assn. Exptl. Pathology, Harvey Soc., Am. Assn. Immunologists, N.Y. Acad. Sci., Sigma Xi. Subspecialties: Biochemistry (biology); Immunobiology and immunology. Current work: Biochemical and immunological studies of lymphocyte membrane proteins, cellular and molecular mechanism of human B-lymphocyte differentiation. Home: 9 Megan Ln Armonk NY 10504 Office: Sloan -Kettering Inst for Cancer Research 145 Boston Post Rd Rye NY 10580

CHOLVIN, NEAL ROBERT, educator; b. Chippewa Falls, Wis., Sept. 8, 1928; s. Elmer Frank and Olympia (Elkow) C.; m. Valerie Perkins, June 25, 1957; children—Brooke Diane, Craig Steven, Mark Douglas. B.S., Wayne State U., 1949; D.V.M., Mich. State U., 1954, M.S., 1958; Ph.D., Iowa State U., 1961. From instr. to asso. prof. dept. vet. surgery and medicine Mich. State U., 1955-63; mem. faculty Iowa State U., 1959—, USPHS postdoctoral research fellow, 1960-61, prof. vet. physiology and pharmacology, 1963—, prof. biomed. engring., 1963—, chmn. biomed. engring. program, 1963-74, 80—, chmn. dept. vet. anatomy, pharmacology and physiology, 1974-79; USPHS spl. research fellow U. Wash., 1971-72. Asso. editor: Lab. Animal Sci, 1967-75. Mem. Am. Physiol. Soc., Soc., AAAS, Am. Heart Assn. (pres. Iowa affiliate 1978-80), Am. Vet. Med. Assn., Sigma Xi. Subspecialties: Biomedical engineering; Surgery (veterinary medicine). Current work: Surgical efficay studies on experimental animals to support applications to FDA for markering approval of surgical devices; biomechanics research on wound healing and fastening. Research in cardiopulmonary control mechanisms, hemodynamics, vascular dynamics. Home: 215 Hickory Dr Ames IA 50010

CHOPPIN, GREGORY ROBERT, chemist; b. Eagle Lake, Tex., Nov. 9, 1927; s. Gilbert P. and Nellie M. C.; m. Ann Mary Warner, June 9, 1951; children: Denise, Suzanne, Paul, Nadene. B.S. maxima cum laude, Loyola U., New Orleans, 1949, D.Sc. (hon.), 1969; Ph.D., U. Tex., Austin, 1953. Research scientist Radiation Lab., U. Calif., Berkeley, 1956-59; prof. chemistry Fla. State U., Tallahassee, 1956—, chmn. dept., 1968-77; research scientist Centre d'Etude Nucleaire, Mol, Belgium, 1962-63; vis. scientist European Transuranium Inst., Karlsruhe, W.Ger., 1979-80. Author: Experimental Nuclear Chemistry, 1961, Nuclei and Radioactivity, 1964, (with B. Jaffe) Chemistry, 1965, 3d edit., 1973, (with L. Summerlin) Chemistry, 1978, 2d edit., 1982, (with R. Johnson) Introductory Chemistry, 1972, (with R. Rydberg) Nuclear Chemistry, 1980; contbr. articles in field to physics and chemistry jours. Served with U.S. Army, 1946-48. Recipient Mfg. Chemist Nat. award, 1979; Alexander von Humboldt Stiftung U.S. sr. sci. award, 1979. Mem. Am. Chem. Soc. (Fla. sect. award 1973, So. Chemist award Memphis sect. 1971), AAAS, Sigma Xi. Roman Catholic. Subspecialties: Inorganic chemistry; Nuclear fission. Current work: The behavior of the lanthanide and actinide elements in aquatic systems, the complexation and separation of these elements and their behavior in the environment. Home: 3290 Longleaf Dr Tallahassee FL 32304 Office: Dept Chemistry Fla State U Tallahassee FL 32306

CHOPPIN, PURNELL WHITTINGTON, virologist, physician; b. Baton Rouge, July 4, 1929; s. Arthur Richard and Eunice (Bolin) C.; m. Joan Harriet Macdonald, Oct. 17, 1959; 1 dau., Kathleen Marie. Student, La. State U., 1946-53, M.D., 1953. Cert. Am. Bd. Internal Medicine. Intern Barnes Hosp., St. Louis, 1953-54, resident in medicine, 1957-58; fellow Rockefeller U., 1957-59, research assoc., 1959-60, asst. prof., 1960-62, 1962-64, assoc. prof., 1964-70, prof., 1970—, Leon Hess prof., 1980—, assoc. physician, 1960-62; assoc. physician, 1962-64, physician, 1964-70, sr. physician, 1970—; chmn. virology study sect. NIH; mem. nat. adv. council Nat. Inst. Allergy and Infectious Diseases, NIH; mem. bd. sci. cons. Meml. Sloan Kettering Cancer Center. Contbr. articles in field to profl. jours. Served with USAF, 1954-56. Nat. Found. fellow, 1954-56; grantee in field; recipient Howard Taylor Ricketts award U. Chgo., 1978. Mem. Nat. Acad. Sci., Assn. Am. Physicians, Am. Clin. Research, Am. Soc. Microbiology, Am. Soc. Cell Biology, Am. Assn. Immunologists, Soc. Exptl. Biology Medicine, Harvey Soc., Infectious Diseases Soc. Am., Royal Soc. Medicine Found. (dir. 1979—). Club: Century (N.Y.C.). Subspecialties: Virology (biology); Infectious diseases. Current work: Virus structure, replication, and mechanisms of pathogenesis virology, infectious diseases, membrane biology. Home: 530 E 72d St Apt 7G New York NY 10021 Office: The Rockefeller University 1230 York Ave New York NY 10021

CHOU, ALBERT CHUNG-HO, biochemistry educator; b. Hunan, China, Mar. 14, 1944; came to U.S., 1967; s. Cheng-chih and Liang (Chou) C.; m. Iris Ching-ho-Lee, Dec. 19, 1970; 1 dau.: Lynne. B.S., Nat. Taiwan U., 1966; M.S., W.Va. U., 1969; Ph.D., Mich. State U., 1973. Research assoc. St. Louis U., 1975-78, asst. research prof., 1978-80; asst. prof. nutrition Tulane U., New Orleans, 1980—; asst. dir. research Touro Infirmary, New Orleans, 1980—. Mem. Am. Soc. Biol. Chemists, Am. Fedn. Clin. Research, Sigma Xi. Subspecialties: Biochemistry (medicine); Hematology. Current work: Role of ferriheme in hemolytic anemias; vitamin E and iron metabolism; hormone and drug receptors; purification and characterization of enzyme. Home: 7254 Cornell Ave Saint Louis MO 63130 Office: Tulane Univ Dept Nutrition 1430 Tulane Ave New Orleans LA 70112

CHOU, GEORGE *See also* **CHOU, IIH-NAN**

CHOU, IIH-NAN (GEORGE CHOU), cell biologist, biochemist; b. Taiwan, China, Apr. 12, 1943; came to U.S., 1967, naturalized, 1979; s. Chou Ming-Ho and Yeh Hwang-Chiao; m. Denise Kuo-Howere, Sept. 8, 1968; children: Jerome, Wendy. B.S., Nat. Taiwan U., 1966; Ph.D., U. Ill., 1971. Instr. Harvard Med. Sch., Boston, 1975-77, asst. prof., 1977-80; asst. biochemist Mass. Gen. Hosp., Boston, 1977-80; asst. prof. Boston U. Sch. Medicine, 1980-82, assoc. prof., 1982—; also mem. Hubert H. Humphrey Cancer Research Ctr. Contbr. articles to sci. jours. Recipient Nat. Research Service award NIH, 1974-75; NIH postdoctoral fellow, 1972-74; Am. Cancer Soc. scholar, 1977-80. Mem. Am. Soc. Cell Biology, Am. Soc. Microbiology, N.Y. Acad. Scis., AAAS. Subspecialties: Cell and tissue culture; Cancer research (medicine). Current work: Laboratory research in cell biology and cancer; role of the cytoskeleton in growth control; plasminogen activators; mechanisms of cell activation. Teaching graduate course in cell biology and medical microbiology laboratory. Home: 151 Albemarle Rd Newton MA 02160 Office: Dept Microbiology Boston U Sch Medicine Boston MA 02118

CHOU, LIU-GEI, microbiologist; b. Chungking, China, Apr. 26, 1944; s. P.G. Harold and Teh-Hsien (Liu); m. Chyau-Bin Grace Chou, Nov. 28, 1947. B.Sc., Nat. Taiwan U., 1967; M.Sc., Ohio State U., 1970, Ph.D., 1973. Area mgr. Pacific region Buckman Labs., Inc., Memphis, 1974—. Mem. Am. Phytopathol. Soc., TAPPI. Subspecialties: Plant pathology; Microbiology. Current work: Agricultural and industrial microorganisms control. Office: 1256 N McLean Blvd Memphis TN 38108

CHOU, SHELLEY NIEN-CHUN, neurosurgeon, medical educator; b. Chekiang, China, Feb. 6, 1924; s. Shelley P. and Tse-tsun (Chao) C.; m. Jolene Johnson, Nov. 24, 1956 (div. 1977); children: Shelley T., Dana, Kerry; m. remarried, 1979. B.S., St. John's U., Shanghai, China, 1946; M.D., U. Utah, 1949; M.S., U. Minn., 1954, Ph.D., 1964. Diplomate: Am. Bd. Neurol. Surgery (mem. bd.). Resident U. Minn. Hosps., 1950-55; practice medicine, specializing in neurosurgery, Salt Lake City, 1955-58, Bethesda, Md., 1959, Mpls., 1960—; clin. asst. Coll. Medicine U. Utah, 1956-58; vis. scientist Nat. Insts. Neurol. Diseases and Blindness NIH, 1959; mem. faculty U. Minn., 1960—, asso. prof. neurosurgery, 1965-68, prof. neurosurgery, 1968—, head dept. neurosurgery, 1974—; mem. Am. Bd. Neurol Surg., 1974-79. Contbr. numerous articles to profl. jours.; Publs. on studies of intracranial lesions using radioactive angiography techniques; malformations of cerebral vasculature; neurol. dysfunctions of urinary bladder. Mem. AMA, A.C.S., Congress Neurol. Surgery, Soc. Neurol. Surgeons (pres. 1978-79), Am. Acad. Neurol. Surgery, Soc. Nuclear Medicine, Am. Assn. Neurol. Surgeons (bd. dirs. 1980-83), Neurosurg. Soc. N.Am. (pres. 1977-78), N.Y. Acad. Medicine, Forum Univ. Neurosurgeons (pres. 1968-69), AAAS, Phi Rho Sigma. Subspecialties: Neurosurgery; Neurophysiology. Current work: Cerebrovasospasn, cerebral andeurysms, cerebral A-V malformation,spinal deformity, bladder physiology. Home: 12 S Long Lake Trail North Oaks MN 55110 Office: B-590 Mayo Meml 420 SE Delaware St Minneapolis MN 55455

CHOU, SHYAN-YIH, nephrologist, internist, researcher; b. Taipei, Taiwan, Aug. 7, 1941; came to U.S., 1968, naturalized, 1980; s. En-Truen and Lin-Oh (Lin) C.; m. Wanda Louie, Dec. 7, 1974; children: Janet, Denise. M.D., Nat. Taiwan U., Taipei, 1966. Diplomate: Am. Bd. Internal Medicine. Intern Brookdale Hosp. Med. Center, Bklyn., 1968-69, resident in Medicine, 1969-70; Nat. Kidney Found. fellow, 1970-73, asst. attending physician, 1973-74, assoc. attending physician, 1974-77, attending physician, 1977—, physician-in-charge, 1977—; asst. prof. medicine SUNY-Downstate Med. Center, Bklyn., 1978—. Contbr. sci. articles to profl. publs. Fellow ACP; mem. Am. Soc. Nephrology, Internat. Soc. Nephrology, Am. Fedn. Clin. Research, Am. Heart Assn., AAAS, Am. Physiol. Soc. Subspecialties: Nephrology; Comparative physiology. Current work: Hormonal factors regulating medullary blood flow; role of medullary hemodynamics in regulating renal sodium excretion. Office: Brookdale Hosp Med Center Nephrology Div Linden Blvd and Brookdale Plaza Brooklyn NY 11212

CHOU, TING-CHAO, biochemical pharmacologist, theoretical biologist; b. Hsin-Chu, Hu-Ko, Taiwan, China, Sept. 9, 1938; came to U.S., 1965, naturalized, 1976; s. Chao-Yun and Sheng-Mei (Chen) C.; m. Dorothy Tsui-Shin Tseng, June 26, 1965; children: Joseph Tsin-I; Julia Hsin-Ya. B.S., Kaoshiung Med. Coll, Taiwan, 1961; M.S., Nat. Taiwan U., Taipei, 1965; Ph.D., Yale U., 1970. Postdoctoral fellow Johns Hopkins U. Sch. Medicine, Balt., 1969-72; asst. prof. biology Cornell U. Grad. Sch. Med. Scis., N.Y.C., 1972-77, asst. prof. pharmacology, 1977-79, assoc. prof. pharmacology and therapeutics, 1979—; assoc. mem. Sloan-Kettering Inst. Cancer Research, Meml. Sloan-Kettering Cancer Center, N.Y.C., 1979—, assoc., 1972-78; research asst. pharmacology Yale U. Sch. Medicine, New Haven, 1969; teaching asst. pharmacology Nat. Taiwan U. Coll. Medicine, Taipei, 1964-65. Nat. Cancer Inst. research grantee NIH, USPHS, 1975-78, 80-83, 83—; cancer research grantee Am. Cancer Soc., 1976-80. Mem. AAAS, Am. Assn. Cancer Research, Am. Soc. Pharmacology and Exptl. Therapeutics, Am. Soc. Preventive Oncology, Am. Soc. Biol. Chemists, Sigma Xi. Club: Yale (Princeton, N.J.). Subspecialties: Molecular pharmacology; Cancer research (medicine). Current work: Pharmacology of cancer chemotherapeutic agents, metabolism of drugs, mechanism of action of drugs, evaluation of selective effects and toxicity of drugs, theoretical biology of dose-effect relationships, receptor theory, evaluation of synergism, antagonism and additivism of multiple drugs, and low-dose risk assessment of toxic substances and carcinogens. Office: Sloan-Kettering Inst Cancer Research 1275 York Ave New York NY 10021

CHOVER, JOSHUA, mathematics educator; b. Detroit, Mar. 26, 1928; m., 1952. Ph.D. in Math, U. Mich., 1952. Research mathematician Bell Telephone Labs., 1952-56; from instr. to assoc. prof. U. Wis.-Madison, 1956-65, prof. math., 1965—, chmn. dept., 1977-79; mem. Inst. Advanced Study, Princeton, N.J., 1955-56. Mem. Am. Math. Soc. Subspecialty: Probability. Office: Dept Math U Wis Madison WI 53706

CHOVNICK, ARTHUR, geneticist; b. N.Y.C., Aug. 2, 1927; s. Herman and Fannie (Hutkin) C.; m. Elinor Joy Mosher, June 7, 1949; children: Lisa, Benjamin. A.B., Ind. U., 1949, M.A., 1950; Ph.D., Ohio State U., 1953. Instr. zoology U. Conn., 1953-57, asst. prof., 1957-59; asst. dir. Biol. Lab., Cold Spring Harbor, N.Y., 1959-60, lab. dir., 1960-62; prof. genetics and cell biology U.Conn., 1962—; mem. genetics study sect. NIH, 1972-76. Assoc. editor: Genetics, 1972—, Genetical Research, 1981—. Mem. Genetics Soc. Am. (treas., bd. dirs. 1981—). Subspecialties: Gene actions; Genome organization. Current work: Researcher in organization and control of gene expression during development in higher organisms. Office: Dept Genetics U Conn Storrs CT 06268

CHOW, CHUEN-YEN, aerospace engineering educator; b. Nanchang, Kiangsi, China, Dec. 5, 1932; came to U.S., 1956, naturalized, 1972; s. Pan-Tao and Huey-Ching (Yang) C.; m. Julianna H.S. Chen, June 26, 1960; children: Chi Hui, Chi Tu, Chi An. B.S.M.E., Nat. Taiwan U., Taipei, 1954; M.S.A.E., Purdue U., 1958; S.M. in Aeros. and Astros, M.I.T., 1961; Ph.D., U. Mich., 1964. Asst. prof. U. Notre Dame, 1965-67, assoc. prof., 1967-68; assoc. prof. aerospace engring. scis. U. Colo., Boulder, 1968-76, prof., 1976—; Disting. Vis. prof. U.S. Air Force Acad., 1979-80. Author: (with A.M. Kuethe) Foundations of Aerodynamics, 3d edit, 1976, An Introduction to Computational Fluid Mechanics, 1979; contbr. articles to profl. jours. USAF grantee, 1980—; NASA grantee, 1982—; NSF grantee, 1967-68, 71-73. Mem. AIAA, Sigma Xi, Sigma Gamma Tau. Subspecialty: Aeronautical engineering. Current work: Fluid dynamics, computational fluid mechanics, unsteady aerodynamics. Home: 345 Seminole Dr Boulder CO 80303 Office: Dept Aerospace Engring Sci U Colo Campus Box 429 Boulder CO 80309

CHOW, LOUISE TSI, molecular biologist, biochemistry educator; b. Hunan, China, Sept. 30, 1943; came to U.S., 1965; d. David and Jane (Lee) C.; m. Thomas Richard Broker, May 26, 1974. B.S., Nat. Taiwan U., Taipei, 1965; Ph.D., Calif. Inst. Tech., 1973. Postdoctoral fellow U. Calif. Med. Ctr., San Francisco, 1973-74, Calif. Inst. Tech., 1974-75; postdoctoral fellow Cold Spring Harbor (N.Y.) Lab., 1975-76, staff investigator, 1976-77, sr. staff investigator, 1977-79, sr. scientist, 1979-83; assoc. prof. biochemistry U. Rochester Sch. Medicine and Dentistry, 1984—, mem. Cancer Ctr., 1984—. Author: (with T.R. Broker) books, including The Electron Microscopy of Nucleic Acids, 3d edit, 1979; contbr. articles, chpts. to profl. publs. NIH grantee, 1977—; Am. Cancer Soc. grantee, 1982. Mem. Am. Soc. Microbiology, Am. Soc. Virology. Subspecialties: Molecular biology; Virology (biology). Current work: Chromosome organization and expression, RNA splicing; electron microscopical heteroduplex techniques for visualization of nucleic acids; recombinant DNA technology. Office: Biochemistry Dept U Rochester Sch Medicine and Dentistry 601 Elmwood Ave Rochester NY 14642

CHOW, MICHAEL HUNG-CHUN, pedodontist, educator; b. Hong Kong, Aug. 9, 1952; U.S., 1971; s. Shun-Ching and Tsun-Ming (Chiang) C.; m. Anita Ling-Han Liu, June 11, 1976; children; Bridget, Denise. A.B., Ind. U., 1973; D.D.S., Northwestern U., 1977; M.M.Sc., Harvard U., 1982. Pvt. practice dentistry, Nashua, N.H., 1982—; instr. oral pediatrics Tufts U., Boston, 1979—; mem. assoc. staff Tufts-New Eng. Med. Center, Boston, 1979—; mem. staff Lawrence (Mass.) Gen. Hosp., 1982—, Meml. Hosp., Nashua, 1982—, St. Joseph Hosp., 1982—. Contbr. articles in field to profl. jours. Recipient Founder's Day award Ind. U., 1972, 73. Mem. Internat. Assn. Dental Research, ADA (postdoctoral student cons. 1979-81), Am. Assn. Orthodontists, Am. Acad. Pedodontics. Subspecialties: Dental growth and development; Orthodontics. Current work: Osteoclast activating factor, bone tissue culture, phorbol esters, lymphocyte-macrophage interactions. Home: 7 Whittier Rd Lexington MA 02173

CHOW, PAO-LIU, mathematics educator, researcher; b. Fujian, China, Nov. 28, 1936; came to U.S., 1962; s. Fa-cheng and Tse-jung (Lin) C.; m. Chien-Jen Huo, June 19, 1965; children: Lawrence, Lily. B.S., Nat. Cheng-Kung U., Taiwan, 1959; M.S., Rensselaer Poly. Inst., 1964, Ph.D., 1967. Asst. prof. Rensselaer Poly. Inst., 1967-68; asst. prof. N.Y. U., 1968-72; assoc. prof. math. Wayne State U., 1972-77, prof., 1977—; vis. scholar U. Calif.-Berkeley, 1978; vis. prof. Nat. Inst. Research Computer and Automation, Le Chesnay, France, 1979. Editor: Multiple Scattering and Wave Propagation in Random Media, 1981. NSF grantee, 1970; NASA grantee, 1974—; U.S. Army Research Office grantee, 1976—. Mem. Am. Math. Soc., Soc. Indsl. and Applied Math. Subspecialties: Applied mathematics; Theoretical and applied mechanics. Current work: Stochastic partial differential equations, differential equations, probability, wave propagation, turbulence, stability theory, filtering and control theory, fluid dynamics, inverse problems, methods in applied mathematics. Home: 19621 Hickory Leaf Ln Southfield MI 48076 Office: Wayne State University Detroit MI 48202

CHOW, PAUL CHUAN-JUIN, physics educator; b. Beijing, China, Aug. 1, 1926; came to U.S., 1955; s. C.K. and Lily P. (Shen) C.; m. Vera Chow, June 25, 1965; children: Maria, Theresa, Teh-Han. B.S., U. Calif.-Berkeley, 1960; Ph.D., Northwestern U., 1965. Vis. asst. prof. U. So. Calif., Los Angeles, 1965-67; research scientist U. Tex.-Austin, 1967-68; asst. prof. physics Calif. State U.-Northridge, 1968-72, assoc. prof., 1972-80, prof., 1980—; cons. Control Data Corp., Mpls., 1981. Bd. dirs. San Fernando Valley Chinese Cultural Assn., Northridge, 1979-81, China Inst. Calif. State U.-Northridge, 1983—. Mem. Am. Phys. Soc., Sci. and Tech. Center Chinese-Ams. in So. Calif. (bd. dirs. 1981—, pres. 1983—). Subspecialty: Theoretical physics. Current work: Computer applications in physics, computer aided instructions. Office: Dept Physics Calif State U Northridge 18111 Nordhoff St Northridge CA 91330

CHOW, TSU SEN, scientist; b. China, Nov. 8, 1939; came to U.S., 1964, naturalized, 1974; s. Kong and Helen (Chen) C.; m. Shang Mei Tang, June 10, 1967; 1 son, Albert. B.Sc., Nat. Cheng Kung U., Taiwan, 1962; M.E., Rensselaer Poly. Inst., 1966; Ph.D., Carnegie-Mellon U., 1968. Teaching asst. Nat. Cheng Kung U., 1963-64; research assoc. chemistry U. N.C.-Chapel Hill, 1968-72; sr. scientist, mem. research staff Xerox Corp., Webster, N.Y., 1972—. Contbr. articles sci. jours., chpts. in books. Served to 2d lt. Chinese Army, 1962-63. Recipient Spl. Merit award Xerox Corp., 1981. Mem. Am. Phys. Soc., Am. Chem. Soc., Soc. Rheology, Am. Acad. Mechanics. Democrat. Subspecialties: Polymer physics; Composite materials. Current work: Basic and applied research in physical properties of polymers and composites, thin film surface and adhesion, photoreceptors. Home: 1608 Brattleboro Dr Webster NY 14580 Office: Xerox Webster Research Center 800 Phillips Rd W 114 Webster NY 14580

CHOWDHURY, PARIMAL, physiologist, educator; b. Chittagong, Bangladesh, Dec. 31, 1940; s. Paresh Nath and Kiranbala C.; m.

Pranati, Nov. 14, 1971; children: Parag, Pritam. B.S., Chittagong Govt. Coll., Bangladesh, 1960; M.S., Dacca U., Bangladesh, 1962; Ph.D., McGill U., Montreal, Can., 1970. Lectr. in biochemistry Calcutta (India) Med. Coll., 1965-67; instr. medicine U. Medicine and Dentistry, Newark, 1970-76, asst. prof. medicine, 1976-80; asst. prof. physiology U. Ark., Little Rock, 1980—, asst. prof. toxicology, 1981—; chemist Standard Chem Corp., Chittagong, Bangladesh, 1964; lect. Bipra Das Pal Chowdhury Inst. Tech., West Bengal, India, 1964-65. Author: Structure and Function of Biopolymers, 1968; contbr. articles to profl. publs. Mem. Kallol N.J. Inc., Friends of India, Ark. (exec.). NIH grantee, 1975-80, 78-81, 1980-82, 82-87. Mem. AAAS, Am. Chem. Soc., AAUP, Sigma Xi. Subspecialties: Environmental toxicology; Physiology (biology). Current work: Pulmonary toxicology (mechanism of lung injury); effect of smoking and environmental pollutants on gastrointestinal distribution, secretion and release of gastrointestinal hormones. Home: 5 New Haven Ct Little Rock AR 72207 Office: Dept Physiology U Ark Med Scis 4301 W Markham Little Rock AR 72205

CHRIST, DUANE MARLAND, systems engineer; b. Lakota, Iowa, Jan. 5, 1932; s. George Andrew and Esther Gertrude (Franke) C.; m. Lily Esther Shih, Sept. 14, 1963; 1 son, Wesley Anzo. B.S., Iowa State U.-Ames, 1953; M.A., U. Minn., 1960. Sci. programmer United Aircraft Corp., Hartford, Conn., 1960-63; sr. assoc. programmer IBM Corp., N.Y.C., 1964-68, staff programmer, 1968-72, staff instr., 1973-76, adv. systems engr., 1976-82, sr. systems engr., 1982—. Served to 1st lt. USAF, 1953-56. IBM resident study fellow, 1966-68. Mem. Assn. Computing Machinery, Soc. Indsl. and Applied Math., Math. Assn. Am. Lutheran. Subspecialties: Systems engineering; Distributed systems and networks. Current work: Network architecture. Office: IBM Corp 77 Water St New York NY 10005

CHRISTEN, ARDEN GALE, dental educator and researcher; b. Lemmon, S.D., Jan. 25, 1932; s. Harold John and Dorothy Elizabeth (Taylor) Deering; m. Joan Ardell Akre, Sept. 10, 1955; children: Barbara, Penny, Rebecca, Sarah. B.S., U. Minn.-Mpls., 1954, D.D.S., 1956; M.S.D., Ind. U.-Indpls., 1965; M.A., Ball State U., 1973. Lic. dentist, S.D., Minn., Ind. Chief oral diagnosis, tng. officer Lackland AFB, Tex., 1965-70; base dental surgeon Zaragoza Air Base, Spain, 1970-73, Bentwaters Air Base, Eng., 1973-75; Air Force preventive dentistry officer Sch. Aerospace Medicine, Brooks AFB, Tex., 1975-78, chief dental research, Brooks AFB, 1978-80; chmn. dept. preventive dentistry Ind. U.,-Indpls., 1981—; mil. cons USAF, 1973-75; sr. med. service cons. Surgeon Gen. USAF, 1974-80; spl. cons. preventive dentistry asst. Surgeon Gen. for Dental Services, Washington, 1975-80; dental cons. Dow Chem. Co., Indpls., 1982—. Author: (with others) Primary Preventive Dentistry, 1982; contbr. numerous articles to profl. jours. Bd. dirs. Bexar County chpt. Am. Cancer Soc., 1976-80; mem. Ind. div. Pub. Edn. Standing Com., Indpls., 1980; bd. dirs. Marion County chpt. Am. Cancer Soc., 1980—. Served to col. USAF, 1956-80. Decorated Meritorious Service Medal with 2 oak leaf clusters, Legion of Merit. Fellow Am. Coll. Dentists; mem. ADA, Am. Acad. Oral Pathology, Internat. Assn. Dental Research, Am. Acad. History of Dentistry. Lutheran. Subspecialty: Preventive dentistry. Current work: Preventive dentistry, motivation, psychology, clinical oral pathology, dental history, quit-smoking programs, tobacco effects on the mouth, smokeless tobacco, nicotine. Home: 7112 Sylvan Ridge Rd Indianapolis IN 46240 Office: Oral Health Research Institute Indiana University School of Dentistry 415 Lansing St Indianapolis IN 46202

CHRISTENFELD, ROGER MICHAEL, epidemiologist, mental health services planner; b. N.Y.C., Jan. 3, 1936; s. Bernard and Sylvia (Weiss) C.; m. Elizabeth Vincent-Daviss, Sept. 11, 1960; children: Timothy, Nicholas, Thomas. B.A., Harvard U., 1957; M.A., Oxford U., Eng., 1965; Ph.D., U. Mich.-Ann Arbor, 1968. Sr. psychologist USPHS, Columbia, Mo., 1963-70; research assoc. Columbia U., 1970—; adj. assoc. prof. SUNY-New Paltz, 1973—; planning dir. N.Y. State Office Mental Health, Wingdale, 1976—; lectr. in field. Contbr. chapts. to books and articles to profl. jours. Pres. Retired Sr. Vol. Program, Poughkeepsie, N.Y., 1980-82; v.p. Assn. Sr. Citizens, Poughkeepsie, 1982—. NIMH grantee, 1970-75; NSF grantee, 1973-75. Mem. World Assn. Social Psychiatry (dir. 1968—), Am.Psychol. Assn., Am. Sociol. Assn., World Fedn. Mental Health. Club: Harvard (pres. Poughkeepsie 1979—). Subspecialties: Social psychology; Epidemiology. Current work: Mental health administration, services planning and consultation. Home: 103 S Hamilton St Poughkeepsie NY 12601 Office: New York State Office of Mental Health Station A Wingdale NY 12594

CHRISTENSEN, CLARK GARDNER, astronomer; b. Spanish Fork, Utah, June 17, 1943; s. Howard J. and Edna (Gardner) C.; m. Vicki Marie Enders, Sept. 5, 1967; children: Maren, Dedra Ann, Clark Glen, Melissa, Julie Marie, Nathan James. B.S., Brigham Young U., 1966; Ph.D., Calif. Inst. Tech., 1972. Asso. Calif. Inst. Tech., Pasadena, 1973; asst. prof. physics, astronomy Brigham Young U., Provo, Utah, 1972-78, assoc. prof., 1978—. Contbr. articles to profl. jours. NSF fellow, 1966-70; ARCS Found. fellow, 1971. Mem. Am. Astron. Soc., Astron. Soc. Pacific. Republican. Mem. Ch. of Jesus Christ of Latter Day Saints. Subspecialty: Optical astronomy. Current work: Stellar photometry, metal poor stars, chem. evolution of the galaxy, galaxian luminosity function. Home: 3268 Navajo Ln Provo UT 84604 Office: Brigham Young Univ 410 ESC Provo UT 84602

CHRISTENSEN, JAMES, physician, researcher, educator; b. Ames, Iowa, Jan. 4, 1932; s. Leo Martin and Eva I. (Patterson) C.; m. Carol S. Asbury, July 26, 1958; children: Laura E., Martha A., J. Martin. B.A., U. Nebr.-Lincoln, 1953; M.S. in Medicine, U. Nebr.-Omaha, 1957, M.D., 1957. Intern in medicine Alameda County Hosps., Oakland, Calif., 1957-58; field med. office USPHS, Albuquerque, 1958-60; resident in medicine U. Iowa, Iowa City, 1960-63, fellow in gastroenterology, 1963-65, asst. prof. medicine, 1966-69, assoc. prof. medicine, 1969-72, prof. medicine, 1972—, dir. div. gastroenterology-hepatology, 1972—; vis. instr. dept. pharmacology U. Alta., Can., 1965-66. Editorial bd.: Jour. Gastroenterology, 1969-74, Am. Jour. Physiology, 1972-76, Gastrointestinal and Liver Physiology, Am. Jour. Physiology, 1979—; editorial council: Rendiconti di Gastroenterologia, 1971—; reviewer: Jour. Clinical Investigation, 1970—, New Eng. Jour. Medicine, 1970—, Jour. Applied Physiology, 1970—, Am. Jour. Digestive Diseases, 1970—, Clinical Research, 1970—, Gastroenterology, 1970—; editor: Gastrointestinal Motility, 1979; assoc. editor: textbook Physiology of the Digestive Tract, vols. 1 and 2, 1981. Markle scholar in acad. medicine, 1965-70; USPHS research career devel. awardee, 1969-74; research grantee, 1980—. Fellow ACP; mem. AAUP, Am. Fedn. Clin. Research, Am. Gastroent. Assn., Am. Physiol. Soc., Central Soc. Clin. Research, Iowa Clin. Med. Soc., Am. Soc. Clin. Investigation, Central Clin. Research Club, Iowa Found. Med. Care, Am. Inst. Biol. Scis., Assn. Am. Physicians, Am. Assn. for Study Liver Diseases, Brit. Soc. Gastroenterology, Am. Motility Soc. (pres. 1982—), Midwest Gut Club, Iowa State Med. Soc., AMA, Soc. for Exptl. Biology and Medicine, Phi Beta Kappa, Sigma Xi, Alpha Omega Alpha. Subspecialty: Gastroenterology. Current work: Research in gastrointestinal motility, smooth muscle physiology in gut and autonomic neurophysiology in the gut. Home: 1532 Rochester Ave Iowa City IA 52240 Office: Dept Internal Medicine U Iowa Hosps Iowa City IA 52242

CHRISTENSEN, J(AMES) ROGER, microbiologist; b. Des Moines, Oct. 28, 1925; s. Aksel Nels and Florence (Williams) C.; m. (married), Sept. 8, 1951; children: David, Sarah. B.S., Iowa State Coll., 1949; Ph.D., Cornell U., Ithaca, N.Y., 1953. Fellow in biophysics U. Colo. Med. Sch., Denver, 1953-55; mem. faculty U. Rochester Sch. Medicine and Dentistry, N.Y., 1955—, prof. microbiology, 1970—. Served with inf. U.S. Army, 1943-46. Nat. Cancer Inst. fellow, 1980-81. Mem. Am. Soc. Microbiology, Am. Soc. Virology, Genetics Soc. Am. Subspecialties: Virology (biology); Gene actions. Current work: Molecular genetics of bacteriophage T1. Home: 105 Greenaway Rd Rochester NY 14610 Office: U Rochester Med Center Microbiology Dept Box 672 Rochester NY 14642

CHRISTENSEN, LARRY WAYNE, chemistry educator, chemical researcher; b. Elkhart, Ind., Sept. 1, 1943; s. Gerald K. and Joyce L. (Chisle) C.; m. Bonita L. Stark, Aug. 20, 1964; children: Kimberly, Kara, Kathie. B.A., Goshen Coll., 1965; Ph.D., Purdue U., 1969. Research assoc. U. Fla., Gainesville, 1969-70; asst. prof. chemistry Houghton Coll., N.Y., 1970-72, assoc. prof., 1972-75, prof., chmn., 1976—; vis. research fellow U. Ariz., Tucson, 1975-76; sci. advisor to U.S. congressman Western Dist. N.Y., 1977-79. Contbr. articles to profl. jours. Chmn. Voluntary Service Program, Houghton, N.Y., 1981-83. David Ross fellow, 1968; NSF fellow, 1970; Research Corp. grantee, 1974-75. Mem. Am. Chem. Soc., AAAS, Union Concerned Scientists, Sigma Xi, Phi Lambda Upsilon. Mennonite. Subspecialties: Organic chemistry; Synthetic chemistry. Current work: Synthesis of novel organic compounds, development of new synthetic methods, research on biomedical implants (heartpacer). Office: Houghton College Houghton NY 14744

CHRISTENSEN, MARY LUCAS, virologist, lab. adminstr., researcher; b. St. Louis, Oct. 18, 1937; d. Kermit and Margaret Isabelle (Lucas) C. B.A. in Bacteriology, U. Iowa, 1959, M.S. in Microbiology, 1961, Ph.D., Northwestern U., 1974. Research virologist Wyeth Labs., Phila., 1961-65, Abbott Labs., North Chicago, Ill., 1965-68; chief clin. virology lab. Northwestern U., 1969-71; research fellow Nat. Cancer Inst., Northwestern U., 1974-78, asst. prof. pathology and pediatrics, 1978—; dir. virology lab. Children's Meml. Hosp., 1978—; guest lectr. in field. Author: Basic Laboratory Procedures in Diagnostic Virology, 1977, Microbiology for Nursing and Allied Health Students, 1982; contbr. articles in field to profl. jours. Mem. Am. Soc. Microbiology, Ill. Soc. Microbiology, AAAS, U. Iowa Alumni Assn., Northwestern U. Alumni Assn., Iota Sigma Pi, Gamma Phi Beta. Episcopalian. Subspecialties: Virology (medicine); Biochemistry (medicine). Current work: Biochemistry of tumor viruses; rapid lab. diagnosis of virus infections; tumor virus research; author books in microbiology and related health fields. Home: 900 N Lake Shore Dr Apt 1905 Chicago IL 60611 Office: 2300 Children's Plaza Chicago IL 60614

CHRISTENSEN, NIKOLAS IVAN, geophysicist, educator; b. Madison, Wis., Apr. 11, 1937; s. Ivan Rudolph and Alice Evelyn (Ethen) C.; m. Karen Mary Luberg, June 18, 1960; children—Kirk Nathan, Signe Kay. B.S., U. Wis.-Madison, 1959, M.S., 1961, Ph.D., 1963. Research fellow in geophysics, Harvard, 1963-64; asst. prof. geol. scis. U. So. Calif., 1964-66; prof. U. Wash., 1966—; Mem. Pacific adv. panel Joint Oceanographic Instns. for Deep Earth Sampling, Seattle, 1973-75, mem. igneous and metamorphic petrology panel, 1973-75, mem. ocean crust panel, 1974-77; mem. adv. panel on oceanography NSF, 1976-78; mem. adv. panel on continental lithosphere NRC, 1979—; mem. adv. panel Internat. Assn. Geodesy, 1980—. Contbg. author: Geodynamics of Iceland and the North Atlantic Area, 1974; Contbr. numerous articles to profl. jours. NSF grantee, 1968-80. Fellow Geol. Soc. Am.; mem. Am. Geophys. Union, Seismol. Soc. Am. Subspecialties: Geophysics; Tectonics. Current work: High pressure physics, perology of the Earth's interior, structural petrology, marine geophysics. Research on nature of Earth's interior. Home: 30 Bridlewood Circle Kirkland WA 98033 Office: Dept of Geol Scis U of Wash Seattle WA 98195

CHRISTENSEN, VERN LEE, poultry science and physiology educator; b. Moroni, Utah, June 25, 1945; s. Charles W. and Mary R. (Christensen) C. B.S., Utah State U., 1971; M.S., Brigham Young U., 1974; Ph.D., U. Mo., 1978. Research asst. Brigham Young U., 1973-75, U. Mo., 1975-78; assoc. prof. poultry sci. and physiology N.C. State U., 1978—. Served to capt. U.S. Army, 1971-73. Mem. Poultry Sci. Assn., AAAS, Am. Soc. Zoologists, Sigma Xi. Mormon. Subspecialties: Animal physiology; Animal breeding and embryo transplants. Current work: Embryonic respiration of avian species and reproductive physiology of avian species; incubation, fertility, hatchability, reproduction. Home: 7415 Post Oak Rd Raleigh NC 27609 Office: NC State U Dept Poultry Science PO Box 5307 Raleigh NC 27650

CHRISTIAN, BARRY THEODORE, clinical child psychologist, behavioral scientist; b. Erie, Pa., Sept. 11, 1951; s. Archie Theodore and Alma Jean (Bundy) C.; m. Lillian Schwartz, Aug. 10, 1974. B.S. in Psychology, Edinboro State Coll., 1973, M.A., Austin Peay State U., 1975; Ed.S. in Counseling, U. Mo., 1977; Ph.D. in Psychology, U. Mo., 1981. Cert. psychologist, N.Mex.; cert. examiner in sch. psychology; cert. psychol. counselor. Psychol. technician Harriet Cohn Mental Health Ctr., Clarksville, Tenn., 1973-75; psychology intern U. Mo. Med. Ctr., Columbia, 1978-79; trng. cons.-lectr. emergency med. services, 1978-79; instr. ednl. psychology U. Mo., 1979-80; pvt. practice psychol. counseling Columbia, 1980-81; child clin. psychologist Coop. Edn. Services of N.Mex., Albuquerque, 1981—; research cons. Big Bros. Am., Columbia, 1975-76; behavior modification cons. Woodhaven Learning Ctr., Columbia, 1977-78. Contbr. articles to profl. jours. Mem. Am. Psychol. Assn., N.Mex. Personnel and Guidance Assn., Nat. Assn. Sch. Psychologists, Nat. Rifle Assn., N.Mex. Shooting Sports Assn. Republican. Mem. Ch. of Nazarene. Club: Zia Rifle and Pistol (Albuquerque). Subspecialties: Behavioral psychology; Learning. Current work: Applied behavior analysis in family and educational settings; learning theory applications in psychotherapy; cognitive behavior therapy research; behavior modification in the classroom. Office: Cooperative Ednl Services of NMex 208 Carlisle St NE Albuquerque NM 87108

CHRISTIAN, CAROL ANN, astronomer; b. Cin., Dec. 28, 1950; d. Robert H. and Marjorie (Ruff) C.; m. Patrick G. Waddell, Oct. 17, 1981. B.S., U. Dayton, 1972; M.A., Boston U., 1974, Ph.D., 1978. Postdoctoral lectr. Yale U., 1978-79; postdoctoral fellow Kitt Peak Nat. Obs., Tucson, 1979-81; resident astronomer U. Hawaii at Can. France Hawaii Telescope Corp., Kamuela, 1981—. Contbr. articles to profl. jours. Mem. Exptl. Aircraft Assn., Am. Astron. Soc.; mem. Soc. Photo-optical Instrumentation Engrs.; Mem. Sigma Xi. Club: Kailua Kona Sailing. Subspecialties: Optical astronomy; Graphics, image processing, and pattern recognition. Current work: Galactic astronomy, spectrophotometry, abundance analysis, star clusters, image processing, digital detectors. Office: CFHT Box 1597 Kamuela HI 96743

CHRISTIAN, GARY DALE, chemistry educator; b. Eugene, Oreg., Nov. 25, 1937; s. Roy and Edna (Trout) C.; m. Suanne Byrd Colbourne, June 17, 1961; children: Dale Brian, Carol Jean. B.S., U. Oreg., 1959; M.S., U. Md., 1962, Ph.D., 1964. Research analytical chemist Walter Reed Army Inst. Research, Washington, 1961-67; asst. prof. chemistry U. Md., College Park, 1965-66, U. Ky., Lexington, 1967-70, assoc. prof., 1970-72; prof. U. Wash., Seattle, 1972—; indsl. cons. Author: Instrumental Analysis, 1978, Analytical Chemistry, 3d edit, 1980, Atomic Absorption Spectroscopy, 1970; Mem. editorial bd. various jours. Fulbright scholar, Belgium, 1978-79; recipient medal Free U. Brussels, 1978. Mem. Am. Chem. Soc. (chmn. Puget Sound sect. 1982), Soc. for Applied Spectroscopy (chmn. Pacific N.W. sect. 1979-80), Spectroscopy Soc. Can., Am. Inst. Chemists. Republican. Subspecialties: Analytical chemistry; Clinical chemistry. Current work: Clinical chemistry, enzyme analysis, competitive protein binding, electroanalytical chemistry, atomic spectroscopy, multicomponent fluorescence analysis, flow injection analysis, chromatography detectors. Home: 7827 NE 12th St Medina WA 98039 Office: Dept Chemistry U WAsh Seattle WA 98195

CHRISTIAN, JOHN EDWARD, educator; b. Indpls., July 12, 1917; s. George Edward and Okel Kandus (Waltz) C.; m. Catherine Ellen Spooner, July 23, 1948; 1 dau., Linda Kay. B.S., Purdue U., 1939, Ph.D., 1944. Control chemist Upjohn Co., 1939-40; faculty Purdue U., Lafayette, Ind., 1940—, prof. pharm. chemistry, 1950-59, head dept. radiol. control, 1956-59, prof. bionucleonics, head dept., 1959—; chmn. adminstrv. com. Trace Level Research Inst., 1960—; dir. Inst. for Environmental Health, 1965—; head Sch. Health Scis., 1979—, Hovde Disting. prof., 1979—; vis. prof. radiation therapy Ind. U. Sch. Medicine, 1970—; Harvey Washington Meml. lectr. Purdue U., 1955; Edward-Kremers Meml. lectr. U. Wis., 1956; vis. lectr. U. Tex., 1959, Taylor U. Ann. Sci. Lecture Series, Upton, Ind., 1960; Julius A. Koch Meml. lectr. U. Pitts., 1961. Asso. editor: Radiochem. Letters. Mem. revision com. U.S. Pharmacopeia, 1950-60, mem. adv. panel on radioactive drugs, 1960-70; adv. com. isotope distbn. AEC, 1952-58, mem. med. adv. com., 1967-75; mem. radiation and chem. def. sect. Ind. Dept. Civil Def., 1954—; vice chmn. Radiation Control Adv. Commn., Ind., 1958—; mem. exec. com. Ind. Comprehensive Health Planning Council, 1972-76; mem. adv. com. radiopharms. FDA, 1970-75; mem. Ind. Gov.'s Pesticide Council, 1970-73; Alumni research councilor Purdue Research Found., 1964—; mem. Ind. Environmental Mgmt. Bd., 1972—, Nat. Energy Policy Task Force, Dept. Energy, 1981—. Recipient award Chilean Iodine Ednl. Bur., 1956; Julius Sturmer award Phila. Coll. Pharmacy and Sci., 1958; Leather medal Purdue U., 1971. Fellow AAAS (past sec. and chmn. pharm. sci. sect., mem. council), Ind. Acad. Sci.; mem. Am. Assn. Colls. Pharmacy (past mem. exec. com., chmn. conf. tchrs., chmn. conf. grad. study and grad. tchrs., chmn. com. study grad. edn. in pharmacy), Am. Chem. Soc. (past chmn. Purdue sect.), Am. Pharm. Assn. (Ebert medal 1957, Justin L. Powers Research Achievement award 1963, past chmn. sci. sect.), Acad. Pharm. Sci. (past v.p.), Ind. Pharm. Assn., Am. Pub. Health Assn., A.M.A. (spl. affiliate), Am. Nuclear Soc., Am. Soc. Bacteriology, Health Phys. Soc., AAUP, Sigma Xi (past pres. Purdue chpt., research award Purdue chpt. 1950), Rho Chi, Phi Lambda Upsilon, Sigma Pi Sigma., Eta Sigma Gamma, Gamma Sigma Delta. Subspecialties: Environmental toxicology; Nuclear physics. Current work: Environmental toxicology and health; radiological health; analytical methods; environmental fate of toxicants. Home: 1301 Woodland Ave West Lafayette IN 47906 Office: Sch Health Scis Bionucleonics Dept Purdue U West Lafayette IN 47907

CHRISTIAN, JOHN JERMYN, biologist, educator; b. Scranton, Pa., Apr. 12, 1917; s. John Oren and Margaret Adams (Jermyn) C.; m. Constance Koons, June 26, 1944 (div. 1958); children: John Jermyn, Patricia E.; m. Patricia Hart, Nov. 6, 1958. A.B., Princeton U., 1939; Sc.D., Johns Hopkins U., 1954. Exptl. med. physiologist Naval Med. Research Inst., Bethesda, Md., 1951-59; assoc. dir. Lab. Comparative Pathology, Phila. Zool. Soc., 1959-62; mem. Albert Einstein Med. Center Research Labs., Phila., 1962-70; prof. biology SUNY, Binghamton, 1970—. Author articles. Served with U.S. Navy, 1944-46. Recipient Merrer award Ecol. Soc., 1957. Mem. Endocrine Soc., Am. Soc. Mammalogists, Am. Ornithol. Union., Am. Soc. Exptl. Biology and Medicine, Am. Assn. Pathologists, N.Y. Acad. Scis., Wildlife Soc. Subspecialties: Behavioral ecology; Population biology. Current work: Population endocrinology, population regulation. Home: Box 24 Starlight PA 18461 Office: SUNY Binghamton NY 13901

CHRISTIAN, JOHN THOMAS, consulting civil engineer; b. N.Y.C., Nov. 2, 1936; s. Thomas Douglas and Evelyn Catherine (Maestri) C.; m. Lynda Ballou Gregorian, June 8, 1960; children: Douglas Arthur, Shirin Lynda. B.S., MIT, 1958, M.S., 1959, Ph.D., 1966. Registered profl. engr., Mass., Maine. Pvt. geotech. engring. cons., Cambridge, Mass., 1966-73; asst. prof. civil engring. MIT, Cambridge, 1966-70, assoc. prof., 1970-73; geotech. cons. Stone & Webster Engring. Corp., Boston, 1973-76, cons. engr., 1976-79, sr. cons. engr., 1979—; mem. com. on mechanics of layered media Univs. Council for Earthquake Engring. Research, Transp. Research Bd., 1978-80. Co-author, editor: (with C. S. Desai) Numerical Methods in Geotechnical Engineering, 1977. Served to 1st lt. USAF, 1959-63. Fellow ASCE (exec. com. geotech. engring. div. 1981—, also numerous other coms., tech. council on computer practices 1974—, named outstanding news corr. 1979); mem. Boston Soc. Civil Engrs. (recipient Desmond Fitzgerald medal 1974), Internat. Soc. Soil Mechanics (Found. Engring.), Earthquake Engring. Research Inst. (mem. research com. 1975-79), Seismol. Soc. Am., Brit. Geotech. Soc., Assn. for Computing Machinery. Club: Boston Racquet. Subspecialties: Civil engineering; Software engineering. Current work: Applications of computer and numerical methods to geotechnical and earthquake engineering; offshore structures; design and construction of energy facilities including dams, nuclear power plants, petrochemical facilities; development of production computer software. Home: 23 Fredana Rd Waban MA 02168 Office: Stone & Webster Engring Corp PO Box 2325 Boston MA 02107

CHRISTIANS, CHARLES JOHN, animal science educator; b. Parkersburg, Iowa, Apr. 15, 1934; s. Dick and Johanna (Franken) C.; m. Betty Lou Anderson, Sept. 14, 1957; children: John Charles, Linda Lou, Neil Allen, Cheryl Kay. B.S., Iowa State U., 1955; M.S., State U., 1958; Ph.D., Okla. State U., 1961. Teaching asst. N.D. State U., Fargo, 1955-58, Okla State U., Stillwater, 1958-61; assoc. prof. Miss. State U., Starkville, 1961-64; prof. dept. animal sci. U. Minn., Mpls., 1964—; instr. Worldwide Coll. Auctioneering, 1980—, Zenton, Japanese Pork Products, Tokyo, 1981-82. Author: Angus Bloodlines, 1958; editor procs.: Nat. Swine Improvement Fedn, 1982—. Served with USAR, 1955—. Recipient State 4-H Alumni award Iowa State U., 1966; Minn. Pork Producers Assn. award, 1974; named Outstanding Tchr. U. Minn., 1972; Outstanding Alumnus Internat. Farm House Fraternity, 1981. Mem. Soc. Animal Sci. (pres. 1971-72), Am. Inst. Biol. Sci., Iowa State U. Alumni Assn. (pres.), Minn. Farm House (dir.), Nat. Swine Improvement Fedn. (sec.-treas.), Sigma Xi, Epsilom Sigma Phi. Republican. Lutheran. Subspecialties: Animal breeding and embryo transplants; Animal genetics. Current work: Large animal genetic engineering. Home: 536 Inca Ln New Brighton MN 55112

CHRISTIE, STEPHEN ROLLAND, plant pathologist; b. Dunedin, Fla., Nov. 5, 1929; s. William Joseph and Sybil (Rouseau) C.; m. Mayumi Yoshida, Jan 11, 1969 (div.); 1 son, Stephen Yoshio. A.A., U. Fla., 1955. Lab. technician I dept. food sci. U. Fla., Gainesville, 1958-60, lab. technician II plant pathology dept., 1960-64, sr. lab. technician, 1964-68, electron microscope technician I, 1968-73, plant pathologist III, 1973—. Contbr. articles to profl. jours. Served with USN, 1948-50. Democrat. Subspecialty: Plant virology. Current work:

Characterization of plant viruses by electron microscopy, serology, and physical and chemical techniques; description and classification of new viruses; investigation of virus byproducts such as viral induced inclusions. Office: Plant Virus Lab Plant Pathology Dept U Fla Gainesville FL 32611

CHRISTLIEB, ALBERT RICHARD, physician, medical educator, researcher; b. Boston, July 24, 1935; s. Albert Rudolph and Marion L. Reimer C.; m. Shirley J. Nichols, July 23, 1960; children: Pamela, Gregory, Scott, Jeffrey. A.B., Williams Coll, 1957; M.D., Tufts U., 1961. Diplomate: Am. Bd. Internal Medicine. Intern Balt. City Hosps., 1961-62, resident, 62-63, New Eng. Med. Center Hosps., Boston, 1965-66; assoc. dir. hypertension unit Peter Bent Brigham Hosp., Boston, 1968-69; mem. staff New Eng. Deaconess Hosp., Boston, 1969—, 1969—, med. dir., 1977-82; assoc. in medicine Brigham and Women's Hosp., Boston, 1971—; assoc. prof. medicine Harvard U. Med. Sch., Boston, 1978—; chmn. med. adminstrv. bd. New Eng. Deaconess Hosp., 1982—. Contbr. numerous articles to med. jours. Trustee Joslin Diabetes Center, 1977, Greater Boston Diabetes Soc., 1979, New Eng. Deaconess Hosp., 1982. Served to capt. USAF, 1963-65. NIH grantee; Nat. Heart and Lung Inst. grantee, 1970-76. Fellow ACP, Am. Coll. Cardiology, Council for High Blood Pressure Research of Am. Heart Assn., Am. Fedn. Clin. Research. Republican. Subspecialty: Internal medicine. Current work: Research in hypertension as it relates to diabetes mellitus. Office: Joslin Clinic 1 Joslin Pl Boston MA 02215 Home: 20 River Glen Rd Wellesley MA 02181

CHRISTMAN, ARTHUR CASTNER, JR., scientific advisor; b. North Wales, Pa., May 11, 1922; s. Arthur Castner and Hazel Ivy (Schirmer) C.; m. Marina Ilia Diterichs, Apr. 17, 1945; children: Candace Lee Cupps, Tatiana Marina Harvey, Deborah Ann Clark, Arthur C. III, Keith Ilia, Cynthia Ellen. B.S. in Physics, Pa. State U., 1944, M.S., 1950. Teaching asst. dept. physics Pa. State U., State College, 1943-44, grad. asst., 1946-48; instr. dept. physics George Washington U., Washington, 1948-51; physicist ops. research Johns Hopkins U., Chevy Chase, Md., 1951-58; sr. physicist Stanford Research Inst., Menlo Park, Calif., 1958-62, head ops. research group, 1962-64, dept. mgr., 1965-67, dir. dept., 1968-71, dir. tactical weapons systems, 1971-75; sci. adv. to dep. chief staff Tng. & Doctrine Command, Ft. Monroe, Va., 1975-81, sci. adv. to comdg. gen., 1982—; cons. in field. Contbr. articles in field to profl. jours. Fellow AAAS; mem. Am. Phys. Soc., Ops. Research Soc. Am., Sigma Xi, Sigma Pi. Republican. Am. Baptist. Subspecialties: Operations research (mathematics). Current work: Advisor on technical quality and analytical soundness of training and doctrine command. Home: 102 Sherwood Dr Williamsburg VA 23185 Office: Sci Advisor ATCGS HQTRADOC Fort Monroe VA 23651

CHRISTMAN, JUDITH KERSHAW, biochemistry educator; b. Teaneck, N.J., Apr. 8, 1941; d. James and Ruth (Niederer) Kershaw; m. Donald A. Christman, June 6, 1959. A.B., NYU, 1962; Ph.D. in Biochemistry, Columbia U., 1967. Postdoctoral fellow N.Y. Blood Ctr., 1967-71; asst. mem. Inst. Muscle Disease, N.Y.C., 1971-74; asst. prof. pediatrics Mt. Sinai Sch. Medicine, N.Y.C., 1974-75, assoc. prof. pediatrics, 1975-80, research prof. pediatrics, 1980—, assoc. prof. biochemistry, 1977-82, prof. biochemistry, 1982—. Contbr. articles to profl. jours.; assoc. editor: Cancer Research, 1981—. N.Y. Heart Assn. grantee, 1977-81; NIH grantee. Mem. Harvey Soc., Am. Soc. Cell Biology, Am. Soc. Biol. Chemists, Am. Assn. Cancer Research, Sigma Xi. Subspecialties: Biochemistry (biology); Genetics and genetic engineering (biology). Current work: DNA methylation and regulation of gene expression, regulation of cloned hepatitis genes in mammalian cells, effects of tumor promoters on differentiation of cultured leukemic cells. Office: Dept Biochemistry 1 Gustave Levy Plaza New York NY 10029

CHRONIC, JOHN, geologist; b. Tulsa, June 3, 1921; s. Byron John and Pansy Lee (Whitehead) C.; m. Carol A. Williams, June 18, 1982; children by previous marriage: Emily Ann, Felice Jane, Lucy Marylka, Susan Elizabeth. B.S. with honors, U. Tulsa, 1942; profl. cert, U. Chgo., 1943; M.S., U. Kans., 1947; Ph.D., Columbia U., 1949. Instr. U. Mich., Ann Arbor, 1949-50; asst. prof. to prof. emeritus U. Colo., Boulder, 1950-80; chief geologist Scarth Oil & Gas Co., Amarillo, Tex., 1980-81; exploration mgr. Evans Exploration Co., Houston, 1981—; cons. Houston Oil Internat., 1981—; sr. geologist Keplinger & Assocs., Houston, 1981-83; chief geologist Nuclear Geophysics Inc., 1983—; instr. U. Edinburgh, Haile Selassie U., U. P.R.; cons. Oceanic Exploration Co., Australia, 1969-70; mus. assoc. Houston Mus. Sci., 1982—. Co-author: Prairie, Peak and Plateau, 1972. Served to 1st lt. USAF, 1942-46. Fellow Geol. Soc. Am., Geol. Soc. London, AAAS; mem. Am. Assn. Petroleum Geologists (assoc. editor jour. 1982-85), Soc. Petroleum Engr., Soc. Exptl. Mineralogy and Paleontology, Geol. Soc. Malaysia and Greece. Democrat. Subspecialties: Geology; Paleontology. Current work: Research and devel. of gamma ray spectrometer for oil and gas exploration. Office: 6620 Harwin St Houston TX 77076

CHU, ANN MARIA, physician, researcher; b. Hong Kong; came to U.S., 1965, naturalized, 1978; d. Paul C. and Lai-Chun (Pansie) Ho) C. I.Sc., U. Calcutta, 1960; M.D., U. Sask., Can., 1965. Intern Med. Coll. Va., 1965-66, resident, 1966-69; fellow M.D. Anderson Hosp. and Tumor Inst., 1969-71; asst. radiotherapist Mass. Gen. Hosp.-Harvard U. Med. Sch., 1971-78; radiotherapist Tufts New England Med. Center Hosp., Boston, 1979-81, dir. residency tng. program, 1979-81; dir. clin. radiotherapy, vice chmn. dept. radiotherapy U. Louisville, 1981—; vis. radiobiologist Gray Lab. Mt. Vernon Hosp., Northwood, Eng., 1977; vis. scientist TNO Radiobiological Inst., Rijswijk, Netherlands, 1977. Contbr. articles in profl. jours. Mem. acad. adv. com. Am. Assocs. Ben-Gurion U. Negev. Recipient Physicians Recognition award. Mem. AMA, Am. Coll. Radiology, Am. Soc. Therapeutic Radiologists, Radiol. Soc. North Am., Am. Radium Soc., New Engl. Soc. Radiotherapeutic Oecologists, Mass. Med. Soc., Am. Assn. Cancer Edn., Radiation research Soc., Am. Assn. Cancer Research, AAAS, Am. Soc. Clin. Oncology, Am. Med. Womens Assn., British Inst. Radiology, N.Y. Acad. Sci., Am. Med. Chinese Assn., Ky. Radiol. Soc., Jefferson County Med. Assn., Ky. Med. Assn., Greater Louisville Radiol. Soc. Subspecialties: Oncology; Cancer research (medicine). Current work: The diagnosis and treatment of cancer patients with radiation oncol.; edn. and tng. of physicians specializing in this field; cancer related research activities. Office: 529 S Jackson St Louisville KY 42008

CHU, BARBARA C.F., biochemist; b. Shanghai, China, Nov. 13, 1942; d. Wen Hsiung and Chia Jui (Chang) C. B.Sc., U. Calif., Berkeley, 1964; Ph.D., Cambridge (Eng.) U., 1971. Asst. research biochemist U. Calif., San Diego, 1978—; research fellow Scripps Clinic and Research Found., La Jolla, Calif., 1974-78. Contbr. articles to profl. jours. European Molecular Biology Orgn. fellow, 1969; Shell Internat. fellow, 1973-74; Nat. Cancer Inst. grantee, 1979-82. Mem. Am. Assn. Cancer Research, AAAS. Democrat. Subspecialties: Chemotherapy; Cellular pharmacology. Current work: Attachment of chemotherapeutic agents to high molecular weight carriers to increase selectivity of these agents for tumor cells and to increase overall therapeutic effectiveness. Home: 13716 Ruetta Le Parc Del Mar CA 92014 Office: Dept Medicine U Calif La Jolla CA 92093

CHU, BENJAMIN, chemist, educator; b. Shanghai, China, Mar. 3, 1932; came to U.S., 1953, naturalized, 1967; s. Charles and Gladys (Chen) C.; m. Louisa King, Mar. 30, 1959; children: Peter, Joanne, Laurence. B.S. magna cum laude, St. Norbert Coll., 1955; Ph.D., Cornell U., 1959. Vis. scientist Brookhaven Nat. Lab., summer 1957; research assoc. Cornell U., 1958-62; asst. prof. chemistry U. Kans., 1962-65, asso. prof., 1965-68; prof. chemistry SUNY, Stony Brook, 1968—, chmn. dept., 1978—. Contbr. articles to profl. jours. Recipient Humboldt award for sr. U.S. scientists, 1976-77; Disting. Achievement award in natural sci. St. Norbert Coll., 1981; Alfred P. Sloan research fellow, 1966-68; John Simon Guggenheim fellow, 1968-69. Fellow Am. Inst. Chemists; mem. Am. Chem. Soc., Am. Phys. Soc., AAAS, N.Y. Acad. Scis., Sigma Xi, Phi Lambda Upsilon. Roman Catholic. Subspecialties: Physical chemistry; Polymer physics. Current work: Rayleigh, Brillouin and Raman scattering; small angle x-ray scattering and small angle neutron scattering; static and dynamical properties of macromolecular solutions and colloidal suspensions; critical opalescence and spinodal decomposition. Home: 27 View Rd Setauket NY 11733 Office: Dept Chemistry SUNY Stony Brook NY 11794

CHU, ERNEST HSIAO-YING, genetics educator; b. China, June 3, 1927; came to U.S., 1949, naturalized, 1962; s. Homing and Sitseng (Tang) C.; m. Nien-Si Liu, Aug. 14, 1954; children: Clara, David, Wellington. B.Sc., St. John's U., Shanghai, China, 1947, M.S., 1951; Ph.D., U.Calif.-Berkeley, 1954. Research asst. dept. botany Yale U., 1954-56, research assoc., 1954-59; lectr. in anatomy Yale U. Sch. Medicine, 1958-59; biologist Oak Ridge Nat. Lab., 1959-72; prof. zoology U. Tenn., 1967-72; prof. human genetics U. Mich. Med. Sch., 1972—. Mem. Genetics Soc. Am., Am. Soc. Cell Biology, Am. Soc. Human Genetics, Radiation Research Soc., Environ. Mutagen Soc., Tissue Culture Assn. Subspecialties: Gene actions; Cell and tissue culture. Current work: Somatics cell genetics, mutation research. Offfice: 1137 E Catherine St Ann Arbor MI 48109

CHU, JAMES CHIEN HUA, physicist; b. Nanking, China, July 6, 1948; came to U.S., 1974; s. Tau-tsun and Yu (Auyang) C.; m. Sherry H.J. Yuan, May 20, 1974; children: Michael C.Y. Ph.D., U. Tex., Houston, 1978. Research asst. U. Tex., Southwestern Med. Sch., Dallas, 1973-74; fellow U. Tex., Houston, 1974-78; asst. prof. U. Pa. Sch. Medicine, Phila., 1978—; cons. in field. Rosalie B. Hite fellow U. Tex. Cancer Center, 1975-78. Contbr. articles to profl. jours. Mem. Am. Assn. Physicists in Medicine, Am. Soc. Therapeutic Radiologists. Subspecialties: Radiology; Biomedical engineering. Current work: Radiaton therapy, dosimetry, imaging. Office: 3400 Spruce St Philadelphia PA 19104

CHU, NAI-SHIN, medical educator, neurologist, neuroscience researcher; b. Taiwan, Apr. 28, 1937; came to U.S., 1964; s. Han-Yau and Ping-Yi C.; m. Shiu-Yuan, Dec. 23, 1967; children: Eric, Curran. M.D., Nat. Taiwan U., 1963; Ph.D., U. Mich., 1969. Diplomate: Am. Bd. Psychiatry and Neurology. Intern U. Colo., Denver, 1972-73, resident in neurology, 1973-76; vis. scientist NIMH, Bethesda, Md., 1970-73; asst. prof. neurology U. Calif.-Irvine, 1976-80, assoc. prof., 1980—, dir. seizure clinic, 1976—; mem. profl. adv. bd. Orange County (Calif.) Epilepsy Soc., 1976—. Rackham fellow, 1966-69; NIH Fogarty fellow, 1970-73; also recipient research career devel award, 1980-85. Mem Am. Acad. Neurology, Am. Epilepsy Soc., Internat. Soc. Biochem. Research on Alcoholism. Subspecialties: Neurology; Neuropharmacology. Current work: Effects of alcohol on the brain, epilepsy and clinical neurology. Current efforts are concentrated on the actions of alcohol on the single brain cells and the damaging effects of chronic alcoholism on the brain stem. Home: 32 Sycamore Creek Irvine CA 92715 Office: Univ of California Irvine Medical Center 101 City Drive South Orange CA 92668

CHU, TING L., elec. engr., educator; b. Beijing, China, Dec. 26, 1924; m. Shirley S. Yu, Sept. 6, 1954; children: Dennis, Dora, Daniel. B.S., Cath. U. Peking, 1945, M.S., 1948; Ph.D., Washington U., 1952. Asst. prof. Duquesne U., Pitts., 1952-55, assoc. prof., 1955-56; research scientist, fellow scientist, mgr. electronic materials Westinghouse Research Labs., Pitts., 1956-67; prof. elec. engring. So. Meth. U., Dallas, 1967—; cons. Poly Solar, Inc., Garland, Tex. Contbr. articles to profl. jours. NSF, NASA, Dept. Energy grantee. Mem. Electrochem. Soc., IEEE, Am. Soc. Engring. Edn., AAUP. Subspecialties: Electronic materials; 3emiconductors. Current work: Electronic materials and devices, including photovoltaic solar energy conversion, growth and characterization of crystals and films and fabrication and characterization of junction devices, dielectric-semiconductor devices. Patentee in field. Home: 12 Duncannon Ct Dallas TX 75225 Office: So Meth U Dallas TX 75275

CHU, WESLEY WEI-CHIN, computer science educator; b. Shanghai, China, May 5, 1936; s. Loon Fay and Yen Yen (Chung) Yau; m. Julia M. Nee, Dec. 27, 1960; children; Milton, Christina. B.S. in Elec. Engring, U. Mich., 1960, M.S., 1961, Ph.D., Stanford U., 1966. Electronic circuit designer computer dept. Gen. Electric Co., Phoenix, 1961-62; computer designer IBM Corp., Menlo Park and San Jose, Calif., 1964-66; research in computer communications and networking Bell Telephone Labs., Holmdel, N.J., 1966-69; prof. computer sci. UCLA, 1969—; cons. in computer communications, distributed processing and distributed data bases. Contbr. articles to profl. jours. Fellow , IEEE (Meritorious Service award 1983); mem. Assn. Computing Machinery. Subspecialties: Distributed systems and networks; Computer architecture. Current work: Research and development of distributed processing and distributed data base systems. Office: UCLA Dept Computer Sci 3731 Boelter Hall Los Angeles CA 90024

CHU, WILLIAM TONGIL, physicist; b. Seoul, Korea, Apr. 16, 1934; came to U.S., 1953, naturalized, 1968; s. Yohan and Sunbok (Choi) C.; m. Insoo La, June 16, 1962; children: Joan Inyul, Jean Suyul. B.S., Carnegie Inst. Tech., 1957, M.S., 1959, Ph.D., 1963. Research asso. Brookhaven Nat. Lab., 1963-64; asst. prof. physics Ohio State U., Columbus, 1964-70; asso. prof. radiation scis. Loma Linda (Calif.) U. Sch. Medicine, 1975-78, prof., 1978-79; scientist III div. accelerators and fusion research Lawrence Berkeley Lab., U. Calif., 1979—. Contbr. articles in field to profl. jours. Mem. Am. Phys. Soc., Radiation Research Soc., Am. Assn. Physicists in Medicine, Korean Scientists and Engrs. Am., Sigma Xi, Tau Beta Pi. Republican. Subspecialties: Particle physics; Biophysics (physics). Current work: Radiation physics, heavy-ion applications in biomed. research. Home: 3282 Ameno Dr Lafayette CA 94549 Office: 64-230 Lawrence Berkeley Lab Berkeley CA 94720

CHUAN, RAYMOND LU-PO, science-technology consultant, atmospheric aerosol researcher; b. Shanghai, China, Mar. 4, 1924; came to U.S., 1941; s. Peter Shao-wu and Katherine (Tao) C.; m. Norma Nicoloff, Dec. 22, 1951 (dec. 1973); children: Jason, Alexander; m. Eugenia Nishimine Sevilla, Apr. 23, 1982. B.A., Pomona Coll., 1944; M.S., Calif. Inst. Tech., 1945, Ph.D., 1953. Dir. engring. center U. So. Calif., Los Angeles, 1957-64; pres. Celesco, South Pasadena, Calif., 1964-68; mgr. advanced tech. Atlantic Research Corp., Costa Mesa, Calif., 1968-72; staff scientist Celesco Industries, Costa Mesa, 1972-76; Brunswick Corp., Costa Mesa, 1976—; trustee Sequoyah Sch., Pasadena, 1958-71; cons. NASA, Hampton, Va., 1978—. Assoc. fellow AIAA (Minta Martin Award 1953); mem. Am. Phys. Soc. Subspecialties: Aerospace engineering and technology. Current work: Gas detection, fire detection, acoustic arrays (underwater), characterization and transport of stratospheric aerosols. Inventor high speed ground transp. fire detection system. Home: 19471 Sandcastle Ln Huntington Beach CA 92648 Office: Brunswick Corp Costa Mesa CA 92626

CHUANG, HANSON YII-KUAN, pathology educator; b. Nanking, China, Sept. 24, 1935; came to U.S., 1963, naturalized, 1972; s. Wai-Ching and Yah-Fang (Chang) C.; m. Lucy Wen-Hwa Tai, Apr. 2, 1966; children: Philip Duen-Ho, Helen Duen-Fang. B.S., Nat. Taiwan U., Taipei, 1958; Ph.D., U. N.C., Chapel Hill, 1968. Postdoctoral fellow Johns Hopkins U., Balt., 1968-71; instr. U. N.C., Chapel Hill, 1972-73, asst. prof., 1973-75, Brown U., Providence, 1975-77, U. South Fla., Tampa, 1977-79; research assoc. prof. U. Utah, Salt Lake City, 1979—. Author: Replacement of Renal Function by Dialysis, 1978, 83, Textbook of Hemostasis and Thrombosis, 1982. NIH grantee, 1978—. Mem. Am. Chem. Soc., Am. Assn. Pathology, Internat. Soc. Artificial Organs, N.Y. Acad. Sci. Subspecialties: Biochemistry (medicine); Biomaterials. Current work: Thrombosis and hemostasis, blood-biomaterial interaction, blood coagulation, blood banking, platele biochemistry and function, artificial organs. Home: 3427 E Brockbank Dr Salt Lake City UT 84124 Office: U Utah 50 N Medical Dr Salt Lake City UT 84132

CHUANG, HENRY NING, energy mgmt. cons., researcher, aero. engr.; b. Nanking, China, July 5, 1937; came to U.S., 1960, naturalized, 1972; m. Molley M. Chuang, 1965; children: Susan, Leah, Philip. B.S.M.E., Nat. Taiwan U., 1958; M.S. in Aero. Engring, U. Md., 1962; Ph.D., Carnegie Inst. Tech., 1966. Registered profl. engr., Ohio. Instr. mech. engring. dept. U. Dayton, Ohio, 1965-66, asst. prof., 1966-69, assoc. prof., 1969-78, prof., 1978—; dir. Energy Analysis and Diagnostic Center, 1980-82, energy coordinator, 1977-80; public speaker. Author: How Much Insulation is Enough, 1977. Recipient award for outstanding profl. achievement Affiliate Socs. Council of Engring. and Sci. Found. of Dayton, 1978. Mem. ASME, Am. Soc. Engring. Edn., ASHRAE. Subspecialties: Energy management; Mechanical engineering. Current work: Energy conversion, energy mgmt., power generation, direct energy conversion and battery tech., energy auditing and mgmt. Home: 5361 Red Coach Rd Centerville OH 45429 Office: KL 121K University of Dayton Dayton OH 45469

CHUANG, HENRY YING HUANG, computer science educator; b. Hsi-lo, Taiwan, Jan. 12, 1934; s. C.T. and Y.C. (Liao) C.; m. Shirley H. Ou, June 5, 1960; children: Peggy, Susie, William, Jenny, Mary. B.S., Taiwan U., 1956; M.S., Chiao Tung U., 1960; Ph.D., N.C. State U., 1966. Instr. Chiao Tung U., Hsinchu, Taiwan, 1961-63; research engr. Washington U., St. Louis, 1966-73, asst. prof., 1966-72, assoc. prof., 1972-73; assoc. prof. computer sci. U. Pitts., 1973—. Contbr. articles to profl. jours. NSF grantee, 1972, 74. Mem. IEEE (sr.), Assn. Computing Machinery. Subspecialties: Computer architecture; Software engineering. Current work: Research in reliable computer design, including hardware and software; VLSI and high performance computer architecture. Office: Dept Computer Sci U Pittsburgh Alumni Hall Pittsburgh PA 15260

CHUANG, RONALD Y(AN-LI), pharmacologist, biochemist; b. Feb. 12, 1940; s. Ching Wha C.; m. Linda F., July 8, 1967; children: Ann, Katherine, Teddy. M.S., U. Calif., Davis, 1966, Ph.D., 1970. Asst. prof. pharmacology Duke U., 1972-76; asst. biochemist Calif. Primate Research Center, U. Calif., Davis, 1976-78, asst. prof. pharmacology, 1981—; asst. prof. biochemistry Oral Roberts U., 1978-81. Contbr. articles to profl. jours.; editorial bd., mem. adv. bd.: Jour. Molecular Pharmacology, 1979—. NIH postdoctoral fellow, 1971-72; NIH grantee, 1974-78, 82-85. Mem. Am. Soc. Biol. Chemists, Am. Assn. Cancer Research. Subspecialties: Biochemistry (biology); Enzyme technology. Current work: Control mechanism of gene expression in eukaryotic cells; mechanism of mode of action of cancer chemotherapeutic agts; recombinant DNA researh in animal virus. Home: 1521 Brown Dr Davis CA 95616 Office: Dept Pharmacology U Calif Davis CA 95616

CHUANG, TZE-JER, physicist; b. Chiayi, Taiwan, July 19, 1943; came to U.S., 1968, naturalized, 1976; s. Len-shen and Su (Jean) C.; m. Jenny H., June 24, 1974; childrn: Jessica Z., Jonathan Y. Sc.M., Duke U., 1970; Ph.D., Brown U., 1975. Registered profl. engr., Pa. Sr. engr. Westinghouse Electric Corp., Pitts., 1974-80; physicist Nat. Bur. of Standards, Washington, 1980—. Mem. ASME, Am. Ceramic Soc., Am. Soc. Metals, Sigma Xi. Democrat. Subspecialties: Fracture mechanics; High-temperature materials. Current work: Basic research of fracture properties of crystalline solids, particularly in high temperature range. Office: Fracture and Deformation Div Center Materials Sci Nat Bur Standards Washington DC 20234

CHUBB, WALSTON, nuclear engineer, consultant; b. Washington, July 23, 1923; s. Robert Walston and Irene (Sylvester) C.; m. Carolyn Elizabeth Carpenter, June 16, 1951; children: Walston, Catherine Louise. B.A., Harvard U., 1944; B.S., U. Mo.-Rolla, 1948, M.S., 1949. Asst. engr. Brush Beryllium Corp., Luckey, Ohio, 1949-51; research fellow Batelle Meml. Inst., Columbus, 1951-72; prin. engr. Westinghouse Electric Corp., Monroeville, Pa., 1972—. Contbr. articles to profl. jours. Served with USNR, 1944-46. Ludlow-Saylor Wire Co. fellow, 1948-49. Mem. Am. Soc. Metals, Am. Nuclear Soc., Nat. Soc. Profl. Engrs., Sigma Xi. Subspecialties: Materials (engineering); High-temperature materials. Current work: Application of high temperature material science to the design of fuels for nuclear reactors; design of mathematical codes to describe the behavior of fuels in reactors, interpretation of radiochemical content of primary reactor water as an indicator of the condition of the reactor core. Patentee in field. Home: 3450 MacArthur Dr Murrysville PA 15668 Office: Westinghouse Electric Corp PO Box 3912 Monroeville PA 15146

CHUDNOVSKY, DAVID VOLF, mathematician; b. Kiev, USSR, Jan. 22, 1947; came to U.S., 1978, naturalized, 1983; s. Volf Gersh and Malka (Vienberg) C. Dipl. Math., Kiev State U., 1969; Ph.D. in Math, Inst. Math., Kiev, 1972. Research fellow Inst. Mechanics, Ukrainian Acad. Scis., Kiev, 1969-76; research fellow Centre de Mathematique, Ecole Polytechnique, France, 1977-78; vis. prof. Center of Nuclear Energy, Saclay, France, 1979, 80; charge de recherche CNRS, Ecole Normale Superieure, Paris, 1981—; research assoc. dept. math. Columbia U., 1978—. Contbr. numerous articles to sci. jours. Ed. math., math. physics books. John Simon Guggenheim fellow, 1981. Mem. Am. Math. Soc., Am. Phys. Soc., Math. Assn. Am. Jewish. Subspecialties: Applied mathematics; Theoretical physics. Current work: Theoretical mathematics: General topology, partial differential equations, Hamiltonian systems; mathematical physics: field theories, quantum systems. Office: Dept Math Columbia Univ New York NY 10027

CHUEY, CARL FRANCIS, botany educator; b. Youngstown, Ohio, Mar. 19, 1944; s. Joseph F. and Marcella L. (Scheetz) C.; m.; 1 son, Matthew C. B.S., Youngstown U., 1966; M.S., Ohio U., 1969; Ed.D., Ohio Christian Coll., 1971. Instr. biol. scis. Youngstown State U., 1967-74, asst. prof., 1974-81, assoc. prof., 1981—, curator herbarium, 1967—. Author: The First 150 Years - The Scheetz Family America, 1971; editor: Laitsch, 1976—. Mem. Jackson-Milton Bd. Edn., North Jackson, Ohio, 1970-74, v.p., 1972-74; precinct committeeman

Democratic Party, 1972-74; mem. Environ. Rev. Com., Youngstown, 1975—. Mem. Am. Fern Soc. (life), Am. Soc. Plant Taxonomists, Ohio Acad. Sci. (life), Brit. Pteridological Soc., Ohio Forestry Assn. Roman Catholic. Subspecialties: Taxonomy; Systematics. Current work: Pteridophyte distribution. Home: 214 Wildwood Dr Youngstown OH 44512 Office: Dept Biol Scis Youngstown State U 410 Wick Ave Youngstown OH 44555

CHUGHTAI, GUL MUHAMAD, med. physicist; b. Chak Khanewal, Pakistan, May 10, 1943; came to U.S., 1971, naturalized, 1977; s. Bahadur Khan and Noorjehan (Bahadur) C.; m. Sarwat Gul, Mar. 1, 1970; children: Tabinda, Farhan. M.S. in Physics, Punjab U., Pakistan, 1967; Ph.D., Pacific W.U., 1982. Radiation physicist Georgetown U., Washington, 1975-77; sr. med. physicist Pakistan Atomic Commn., Islamabad, 1978-80; commd. capt. U.S. Air Force, 1980; chief med. physics services Malcolm Grow USAF Med. Ctr., Andrews AFB, Washington, 1980—; research instr. Uniformed Services U. of Health Sci., Bethesda, Md., 1982—. Mem. Am. Assn. Physicists in Medicine. Current work: Study of thermoluminescent dosimetry; ionizing radiation for treatment cancer; treatment plg. with computer for external interstitial and intra cavitory radiation therapy treatment. Home: 7518 Abbington Dr Oxon Hill MD 20745 Office: Dept Radiology MGMC Andrews AFB Washington DC 20331

CHULICK, EUGENE THOMAS, research engineer; b. Jackson, Calif., Jan. 8, 1944; s. Michael Frances and Anna (Valko) C.; m. Gail Patricia Henderson, Aug. 9, 1966; children: Nicole Ann, Jeffrey Eugene. B.S., U. of Pacific, Stockton, Calif., 1961-65; Ph.D., Washington U., St. Louis, 1965-69. Instr. Tex. A & M U., College Station, 1973, research assoc., 1969-74; sr. research engr. Babcock & Wilcox, Lynchburg, Va., 1974-77, group supr., 1977-79, sect. mgr., 1979-81; venture mgr. Alliance, Ohio, 1981—; tech. expert IAEA, Athens, Greece, 1972, Sao Paulo, Brazil, 1973. Contbr. numerous articles to profl. publs. Mem. Nat. Acad. Sci. (mem. subcom. 1982-84), Am. Chem. Soc., Am. Phys. Soc., Am. Nuclear Soc., AAAS, Sigma Xi. Subspecialties: Water supply and wastewater treatment; Nuclear fission. Current work: Water supply purification; pollution measurement; electrical energy production. Office: Babcock & Wilcox 1562 Beeson St Alliance OH 44601 Home: 2147 Vixen St NW North Canton OH 44720

CHUMBLEY, ALVIN BRENT, orthodontist; b. Jamestown, Ky., Mar. 3, 1949; s. Morris Marvin and Avis Jean (Bradshaw) C. D.M.D., U. Ky., 1974, M.S.D., 1980. Cert. orthodontist. Dental dir. dentist Mountain Comprehensive Health Corp., Hazard, Ky., 1974-78; practice orthodontics, Corbin, Ky., 1980—; guest lectr. U. Ky., Lexington, 1982; lectr. profl. conf., 1981. Vice pres. Cancer Soc., Corbin, 1981-82; active Big Brothers-Big Sisters, Corbin, 1981-82; organist Central Baptist Ch., Corbin, 1981-82. Recipient various dental awards U. Ky. Mem. Internat. Assn. Dental Research, ADA, Ky. Dental Assn., Southeastern Ky. Dental Assn., Ky. Orthodontic Soc., So. Am. Orthodontic Assn., Am. Assn. Orthodontists. Republican. Baptist. Lodge: Lions. Subspecialty: Orthodontics. Current work: Biochemical and mechanical aspects of bone resorption in orthodontic tooth movement. Office: 206 S Main St Corbin KY 40701 Home: Barton Mill Village Corbin KY 40701

CHUN, SUN WOONG, government energy executive; b. Seoul, Korea, Mar. 30, 1934; s. Eun L. and Bok S. C.; m. Inmook Lee, Aug. 19, 1961; children: Felicia, Edward. B.Ch.E., Ohio State U., 1959, M.S. in Chem. Engring, 1959, Ph.D., 1964. Research scientist Union Camp Corp., Princeton, N.J., 1964-67; chem. engr. Gulf Research & Devel. Co., Pitts., 1967-75; dir. Pitts. Energy Tech. Center, Dept. Energy, 1975—; mem. Pitts. Fed. Exec. Bd.; pres. Internat. Energy. Agy.'s 1983 Internat. Coal Sci. Conf. Contbr. numerous articles on refining and synthetic fuels to profl. publs. Recipient Benjamin G. Lamme award Coll. Engring., Ohio State U., 1981. Mem. Am. Inst. Chem. Engr., Am. Chem. Soc., Am. Mgmt. Assn., Sigma Xi. Subspecialties: Chemical engineering; Coal. Current work: Coal technology, petroleum refining, combustion. Patentee in field. Office: PO Box 10940 Pittsburgh PA 15236

CHUNG, CHIEN, nuclear science educator, consultant; b. Tokyo, Sept. 7, 1950; s. Han-Po and Wing-Jing (Fan) C.; m. Ching Ling Lee, June 22, 1973; children: Martin, Brian. B.Sc., Nat. Tsing Hua U., Hsinchu, Taiwan, 1972; Ph.D., McGill U., Can., 1980. Teaching asst. McGill U., Montreal, Que., 1975-79, postdoctoral fellow, 1980; guest research assoc. Brookhaven Nat. Lab., Upton, N.Y., 1980-83; research assoc. U. Md., College Park, 1980-83; assoc. prof. nuclear sci. Nat. Tsing Hua U., Hsinchu, 1983—. Contbr. articles to profl. publs. Recipient Winkler award McGill U., 1982; McGill U. scholar, 1975-80. Mem. Am. Nuclear Soc., Am. Phys. Soc., Am. Chem. Soc. (affiliated mem.). Subspecialties: Nuclear fission; Nuclear engineering. Current work: Front-line research on basic and applied nuclear science: analytical, isotopes production, nuclear reaction, fission and structure. Office: Inst Nuclear Sci Nat Tsing Hua U Tsinchu Taiwan 300

CHUNG, CHUNG-TAIK, medical educator, physician; b. Tae-Gue, Korea, Jan. 11, 1942; came to U.S., 1970, naturalized, 1977; s. Woon Yong and Boon-Jo (Kim) C.; m. Nam Sook Chung, May 10, 1969; children: Michael, Robert, Terrine. M.D., Yonsei U., Seoul, Korea, 1967. Diplomate: Am. Bd. Radiology; lic. physician, N.Y., Pa., Calif. Intern H.S. Martland Hosp., Newark, 1970-71; resident in radiation therapy SUNY Upstate Med. Center-Syracuse, 1971-74, instr. radiology, 1974-75, asst. prof., 1976-81, assoc. prof., 1981—, assoc. prof. otolaryngology, 1982—; staff radiotherapist Wilkes-Barre (Pa.) Gen. Hosp., 1975-76; cons. radiotherapist Crouse Irving Meml. Hosp., VA Hosp., Community Gen. Hosp., St. Joseph's Hosp. Contbr. articles to med. jours. Served with Republic of Korea Air Force, 1967-70. Mem. Am. Coll. Radiology, Am. Soc. Therapeutic Radiologists, Radiol. Soc. N.Am., Central N.Y. Radiol. Soc., N.Y. State Med. Soc., Korean Soc. Central N.Y. (pres. 1974-75), Yonsei U. Coll. Medicine Alumni Assn. Central N.Y. (sec. 1973-80). Club: Pompey (Pompey, N.Y.). Subspecialties: Oncology. Current work: Radiation oncology. Home: 4853 Candy Ln Manlius NY 13104 Office: SUNY Upstate Medical Center 750 E Adams St Syracuse NY 13210

CHUNG, HEE MOK, nuclear engr., material scientist; b. Okchun, Chungbuk, Korea, Jan. 26, 1941; came to U.S., 1966; s. Chin Wook and In Hah (Choi) C.; m. Haijung W. Lee, Sept. 9, 1972; children: Gina, Joanne. B.S., Seoul Nat. U., 1963; Ph.D., U. Pa., 1972. Research assoc. Yale U., New Haven, 1972-74; asst. metallurgist Argonne (Ill.) Nat. Lab., 1974-78, project mgr., 1980-82, prin. investigator metallurgist, 1978—; cons. Dept. Energy, 1979-82, U.S. Nuclear Regulatory Commn., 1977—. Recipient cert. of award U.S. Dept. Energy, 1979. Mem. AIME, Am. Nuclear Soc. ASTM. Methodist. Subspecialties: Clad metals and coating technology; Nuclear fission. Current work: Nuclear reactor fuel behavior and performance, related safety aspects, advanced alloy design and performances in energy systems, materials corrosion, surface phenomena, irradiation effects. Developer Chung-Kassner Criterion in reactor safety, 1979, Chung-Thomas effect for reactor accident fuel behavior, 1982. Office: Argonne Nat Lab 9700 S Cass Ave Argonne IL 60439 Home: 1709 Warbler Dr Naperville IL 60540

CHUNG, MELVIN CHUNG-HING, anatomist, educator; b. Honolulu, Feb. 11, 1935; s. Harry Su-Lung and Rosaline Tam (Tom) C.; m. Jane Ching-An Hsia, Aug. 21, 1965; children: Mark K. S., Mona M.-L. B.A., U. Nebr., Lincoln, 1957, M.S., 1960; Ph.D. in Anatomy, U. Calif., Berkeley, 1971. Instr. life scis. San Francisco City Coll., 1964-66; instr. U. Rochester, 1971-73, asst. prof. anatomy, 1973-78; assoc. prof. anatomy Med. Coll. Va., Va. Commonwealth U., 1978—. Contbr. articles to profl. jours. Human Growth Found. grantee, 1976-78; NIH grantee, 1976-77, 77-80. Mem. Am. Soc. Anatomy, Endocrine Soc., Neurosci. Soc., N.Y. Acad. Sci., Va. Acad. Sci., So. Soc. Anatomy, AAAS, Sigma Xi. Subspecialties: Neuroendocrinology; Reproductive biology. Current work: Hypothalamic release of neuropeptides influencing pituitary gonadotrophic and thyrotrophic function and growth; target hormone feedback on pituitary and hypothalamic activity. Home: 12301 Roaring Brook Ct Richmond VA 23233

CHUNG, YOUNG SUP, geneticist, educator; b. Inchon, Korea, Jan. 27, 1937; s. Tai Hyun and Ip Boon (Kim) C.; m. Inhi Angelica Kim, May 29, 1965; children: Rex Kyuchang, Lucille Yoonhi. B.Sc., Purdue U., 1960; Ph.D., Giessen (W.Ger.) U., 1964. Research assoc. Frankfurt (W.Ger.) U., 1964-65; research assoc., postdoctoral fellow Palo Alto (Calif.) Med. Research Found., 1965-69; asst. prof. genetics U. Montreal, Que., Can., 1970-73, assoc. prof., 1974-78, prof., 1979—; vis. prof. Institut Pasteur, Paris, 1980-81, Bochum U., W. Ger., 1980-81. Contbr. articles to sci. internat. and nat. jours. Research grantee Natural Scis. and Engring. Research Council Can., 1971—, German Acad. Exchange Service, 1980, Mission de Ordre, 1980, Fonds Barre, 1981, U. Montreal, 1982. Mem. Am. Soc. Microbiology, Genetics Soc. Am., AAAS, N.Y. Acad. Scis., Societe de genetique et cytologie du Can., Assn. canadienne-francaise pour l'avancement des Sciences. Subspecialties: Gene actions; Molecular biology. Current work: DNA repair, ozone, metagenesis, mapping of genes. Home: 3241 Appleton Ave Montreal PQ Canada H3S 1L6 Office: Dept Biol Scis U Montreal Montreal PQ Canada H3C 3J7

CHURCH, ALLEN CHARLES, neuroscientist; b. Binghamton, N.Y., Aug. 1, 1950; s. William Allen and Margaret Eletheare C.; m. Jennifer Lindsay, Oct. 30, 1976; 1 son, Joel. B.A., Harpur Coll., 1972; M.A., SUNY, Binghamton, 1974, Ph.D., 1976. Fellow Jackson Lab., Bar Harbor, Maine, 1976-78; fellow dept. psychiatry and pharmacology U. Pa., Phila., 1978-80, asst. prof. anatomy, 1980; sr. staff fellow NIMH, St. Elizabeth's Hosp., Washington, 1980—. Mem. AAAS, Soc. for Neurosci., N.Y. Acad. Sci. Subspecialties: Neurobiology; Neuropharmacology. Current work: Researcher in interactions of neurotransmitters and their receptors. Office: NIMH St Elizabeth's Hospital Adult Psychiatry Branch Washington DC 20032

CHURCH, ALONZO, educator; b. Washington, June 14, 1903; s. Samuel Robbins and Mildred Hannah Letterman (Parker) C.; m. Mary Julia Kuczinski, Aug. 25, 1925 (dec. Feb. 1976); children—Alonzo, Mary Ann, Mildred Warner. A.B., Princeton, 1924, Ph.D., 1927; D.Sc., Case Western Res. U., 1969. Faculty Princeton, 1929-67, prof. math., 1947-61, prof. math. and philosophy, 1961-67; prof. philosophy and math. UCLA, 1967—. Author: Introduction to Mathematical Logic, vol. I, 1956; Editor: Jour. Symbolic Logic, 1936-79; contbr. articles to math. and philos. jours. Mem. Am. Acad. Arts and Scis., Assn. Symbolic Logic, Am. Math. Soc., AAAS, Nat. Acad. Scis., Brit. Acad. (corr.), Am. Philos. Assn. (pres. Pacific div. 1973-74). Subspecialty: Mathematical Logic. Current work: Mathematical logic, philosophy of mathematics. Address: Dept Philosophy U Calif Los Angeles CA 90024

CHURCH, DAVID ARTHUR, physicist; b. Berlin, N.H., Apr. 3, 1939; s. Andrew Ivan and Barbara Brown (Holmes) C.; m. Diane Claire Burnham, Sept. 15, 1963; children: Kirin Alene, Aran Holmes. A.B., Dartmouth Coll., 1961; M.S., U. Wash., 1963, Ph.D., 1969. Research assoc. U. Bonn, W. Ger., 1969, U. Mainz, W.Ger., 1969-71, U. Ariz., Tucson, 1971-72; physicist Lawrence Berkely (Calif.) Lab., 1972-75; asst. prof. to assoc. prof. physics Tex. A&M U., College Station, 1975—. Contbr. articles profl. jours., chpts. in books. Recipient Precision Measurement Grant award Nat. Bur. Standards, 1981; research grantee Dept. Energy, Research Corp., NSF. Mem. Am. Phys. Soc. Presbyterian. Subspecialties: Atomic and molecular physics. Current work: Coherence spectroscopy and anistropic excitation of fast ions, slow ion storage and spectroscopy, multi-charged ion charge transfer. Home: 1810 Langford St College Station TX 77840 Office: Physics Dept Texas A&M University College Station TX 77843

CHURCH, EUGENE LENT, research physicist; b. Yonkers, N.Y., July 30, 1925, s. Wallace L. and Wilhelmina L. (Dinger) C.; m. Anne Richardson Meirs, May 15, 1948; children: Rebecca Meirs, David Lent. A.B., Princeton U., 1948; Ph.D., Harvard U., 1953. With Dept. Def., 1952—; sr. physicist Frankford Arsenal, Phila., 1971-77; phys. scientist U.S. Armament Research and Devel. Command, Dover, N.J., 1977—; guest physicist Argonne Nat. Lab., 1952-55, Brookhaven Nat. Lab., 1955-59; vis. scientist Niels Bohr Inst., Copenhagen, 1959-61; guestphysicist Brookhaven Nat. Lab., 1961-71. Contbr. articles to profl. jours. Served with USN, 1944-46. Fellow Am. Phys. Soc., AAAS, Soc. Photo-Optical, Instrumentation Engrs.; mem. Am. Optical Soc., IEEE. Republican. Presbyterian. Club: Princeton (N.Y.C.). Subspecialty: Optical engineering. Current work: Properties and metrology of high performance optical surfaces; precision machining; digital signal processing. Office: US Armament Research and Devel Command Bldg 455 Dover NJ 07801

CHURCHILL, STUART WINSTON, chemical engineering educator; b. Imlay City, Mich., June 13, 1920; s. Howard Heenan and Faye Erma (Shurte) C.; m. Donna Belle Lewis, Feb. 22, 1946 (div.); children: Stuart Lewis, Diana Gail, Cathy Marie, Emily Elizabeth; m. Renate Ursula Treibmann, Aug. 3, 1974. B.S. in Math, U. Mich., 1942, 1942, M.S., 1948, Ph.D., 1952; M.A. honoris causa, U. Pa., 1972. Technologist Shell Oil Co., 1942-46; tech. supr. Frontier Chem. Co., 1946-47; mem. faculty U. Mich., 1949-67, prof. chem. engring., 1957-67, chmn. dept. chem. and metall. engring., 1962-67; Carl V.S. Patterson prof. chem. engring. U. Pa., 1967—; chmn. region 2 edn. and accreditation com. Engrs. Council Profl. Devel., 1961-65, mem. nat. council, 1965-71, exec. com., 1968-71; cons. heat transfer and combustion. Recipient S. Reid Warren, Jr. award for distinguished teaching U. Pa., 1976; Max Jakob Meml. award for heat transfer ASME/Am. Inst. Chem. Engrs., 1979; Japan Soc. for Promotion of Sci. grantee, 1977. Fellow Am. Inst. Chem. Engrs. (nat. council 1962-64, pres. 1966, Profl. Progress award 1964, William H. Walker award 1969, Warren K. Lewis award 1978, Founders award 1980); mem. Nat. Acad. Engring., Combustion Inst., Am. Chem. Soc., Verein Deutscher Ingenieure (Corr. mem.), Sigma Xi, Phi Kappa Phi, Phi Lambda Upsilon (award U. Mich. chpt. 1961), Tau Beta Pi. Unitarian. Subspecialties: Chemical engineering; Combustion processes. Current work: Interpretation and use of rate data, correlation, natural convection in enclosures, combustion in a fefractory tube, reaction kinetics,pollutioncontrol. Home: 137 Pole Cat Rd Glen Mills PA 19342

CHURGIN, LAWRENCE S., prosthodontist; b. Newark, Apr. 26, 1923; s. Leopold and Anna (Gray) C.; m. Judith F. Schonfeld, June 21, 1950; children: Gary, Carol, Erica. B.A., N.Y.U., 1948, D.D.S., 1947, M.S.D., 1968. Asst. research scientist N.Y.U. Coll. Dentistry, 1969-71, asst. prof. material sci., 1971-72; asst. prof. occlusion N.J. Dental Sch., Newark, 1972-74, assoc. prof. occlusion, 1974-77, assoc. prof. fixed prosthodontics, 1978—; pvt. practice prosthodontics, Bloomfield, N.J.; cons. Warner-Lambert, Morris Plains, N.J., 1975-80; dir. Group Health Ins. N.J., East Brunswick, 1975—. Served to capt. Dental corps U.S. Army, 1951-52; Korea. Fellow Greater N.Y. Acad. Prothodontics, Am. Coll. Dentists; mem. Am. Coll. Prosthodontics, Research Soc. Am., Am. Assn. for Dental Research, Am. Dental Assn. Subspecialty: Prosthodontics. Current work: Teach graduate prosthodontics; research activity in the area of material science and its clinical applications in prosthodontics. Office: 301 Belleville Ave Bloomfield NJ 07003

CHURNET, HABTE GIORGIS, geology educator, researcher; b. Shewa, Ethiopia, May 9, 1946; came to U.S., 1975; s. Churnet Argaw and Yeshewa (Mebrat) Tilahun; m. Enat Negussie, Jan. 17, 1983. B.Sc., Haile Selassie I U., Addis Ababa, Ethiopia, 1969; M.Sc., Leeds (Eng.) U., 1972; Ph.D., U. Tenn., 1979. Tchr. Ras Abate Boyalew High Sch., Hossana, Ethiopia, 1967-68; lectr. Haile Selassie I U., 1971-75; asst. prof. geology U. Tenn.-Chattanooga, 1980-83, assoc. prof. geoscis., 1983—; research-investigator dept. geoscis. U.S. Geol. Survey, Chattanooga, 1980—. Author: (in Ethiopian): Techet, 1974. Brit. Council scholar, 1970. Mem. AAAS, Geol. Soc. Am. Subspecialties: Geology; Petrology. Current work: Ore deposition, sedimentary processes. Home: 731-M Mansion Circle Chattanooga TN 37405 Office: Dept Geoscis 615 McCallie Ave Chattanooga TN 37402

CHUSED, THOMAS M., immunologist, physician; b. St. Louis, Mar. 29, 1940; s. Joseph J. and Marie Irene (Steinberg) C.; m. Judith A. Chused, June 28, 1965; children: Amy Elizabeth, Nicholas Fingert. B.A., Harvard U., 1962, M.D., 1967. Diplomate: Am. Bd. Internal Medicine. Intern Cleve. Met. Gen. Hosp., 1967-68, resident, 1968-69; clin. assoc. Nat. Inst. Arthritis and Metabolic Diseases, NIH, Bethesda, Md., 1969-72; sr. investigator Nat. Inst. Dental Research, 1972-79, Nat. Inst. Allergy and Infectious Diseases, NIH, Bethesda, 1979—. Contbr. articles to profl. jours. Served to lt. comdr. USPHS, 1969-72. Mem. Am. Assn. Immunologists, Am. Rheumatism Assn. Subspecialties: Immunobiology and immunology; Immunology (medicine). Current work: Pathogenesis of autoimmune disease, mechanism of lymphocyte activation, immunoregulation. Home: 1805 Randolph St NW Washington DC 20011 Office: NIH Bldg 5 Bethesda MD 20205

CHUTE, DOUGLAS LAWRENCE, educator; b. Toronto, Ont., Aug. 22, 1947; s. Andrew L. and Helen (Reid) C.; m.; children: Jesse Robert, Deborah Evans, Andrew Lawrence. B.A., U. Western Ont., 1969; M.A., U. Mo., 1971, Ph.D., 1973. Lic. psychologist, Ont. Research fellow NASA, Columbia, Md., 1972-73; asst. prof. U. Houston, 1973-77; lectr. U., Otago, Dunedin, N.Z., 1977-80; assoc. clin. prof. McMaster U., Hamilton, Ont., 1980—; supr. neuroscis. U. Toronto, Scarborough, Ont., 1980—. Editorial bd.: Clin. Neuropsychology, 1978—; author, editor: Drug Discrimination and State Dependent Learning, 1978; author: Introduction to Surgery in Neuroscience, 1974, General Experimental Psychology, 2nd edit, 1977. Group com. chmn. Boy Scouts, Can., 1981—. Mem. Am. Psychol. Assn., Canadian Psychol. Assn., Soc. for Neursci., Canadian Assn. Neursci., Sigma Xi. Subspecialties: Neuropharmacology; Neuropsychology. Current work: Role of cyclic nucleotides and protein phosphorylation in memory rehab. in severe neurotrauma. Office: Univ Toronto 1265 Military Trail Scarborough ON Canada M1C 1A4

CHYTIL, FRANK, biochemist; b. Prague, Czechoslovakia, Aug. 28, 1924; came to U.S., 1965, naturalized, 1971; s. Frantisek and Ruzena (Vitouskova) C.; m. Lucie Scheinost, Nov. 26, 1949; children—Frank, Anna, Helena. M.S., Sch. Chem. Tech., Prague, 1949, Ph.D., 1952; C.Sc., Czechoslovak Acad. Sci., Prague, 1956. Research biochemist Charles U., Prague, 1949-51; research fellow Inst. Human Research, Prague, 1952-63; sr. scientist Czechoslovak Acad. Sci., Prague, 1956-64; sr. research fellow Brandeis U., Waltham, Mass., 1964—, sr. research assoc., 1965-66; head sect. enzymology S.W. Found. Research and Edn., San Antonio, 1966-69; mem. faculty Vanderbilt U., 1969—, prof. biochemistry, 1975—; adj. assoc. prof. U. Tex., San Antonio, 1968-69. Editor: Vitamins and Hormones, 1983; mem. editorial bd.: Analytical Biochemistry, 1980—, Jour. Biol. Chemistry, 1982—; contbr. profl. jours. USPHS grantee, 1967—. Mem. Am. Chem. Soc., Am. Soc. Biol. Chemists, Am. Inst. Nutrition (chmn. (nomenclature) 1982, 83, Osborne Mendel award 1983), Endocrine Soc., Sigma Xi. Subspecialties: Biochemistry (biology); Animal nutrition. Current work: Mechanism of vitamin A action in nonvisual tissues. Address: 914 Lynnwood Blvd Nashville TN 37205 Office: Vanderbilt U Sch Medicine Nashville TN 37232

CIACCIO, LEONARD LOUIS, chemistry and environmental science educator, researcher, consultant, inventor; b. N.Y.C., June 21, 1924; s. Benjamin Joseph and Accursia (Bacchi) C.; m. Eva Agostini, July 1, 1946 (div.); children: Gloria Sebastian, Luke, Dominic, Imelda Vetter, Rita; m. mad, Mae Margaret Searles, June 18, 1983. B.S magna cum laude, St. Peters Coll., 1951; M.S. in Chemistry, Poly. Inst. N.Y., 1956, Ph.D., 1962. Lab. technician Charles Pfizer, N.Y.C., 1942-43, 46-48, analytical research co-mgr., N.Y.C. and Groton, Conn., 1953-60; research chemist Allied Chems. & Dye Corp., N.Y.C., 1951-53; mgr. tech. services T. J. Lipton, Inc., Englewood Cliffs, N.J., 1960-64; mgr. water and waste research Wallace & Tierman, Belleville, N.J., 1964-69; mgr. environ. sci. GTE Labs., Inc., N.Y.C., 1969-72; prof. environ. sci. and chemistry Ramapo Coll., 1972—; cons. to tech. industries; researcher U. Que., Can., 1982. Contbr. articles to profl. jours.; editor: Water and Water Pollution Handbook, 4 vols, 1971-73. Served to sgt. Chem. Warfare Service U.S. Army, 1943-46; PTO. N.J. Dept. Environ. Protection grantee, 1982-83. Mem. Am. Chem. Soc., AAAS, N.Y. Acad. Scis., Internat. Assn. Water Pollution Research, Water Pollution Control Fedn., Air Pollution Control Assn., Sigma Xi, Phi Lambda Upsilon. Subspecialties: Analytical chemistry; Environmental analysis of air and water. Current work: Analysis of air, water and soil pollutants (identification of organic substances in airborne particulate matter); electroanalytical methods of trace analysis; effects of trace metals and complexing agents in water ecology; development of automated instrumentation. Patentee method and apparatus for automated measurement of energy oxygen, method and composition for cleaning dairy equipment. Home: 124 Kiel Ave Butler NJ 07405 Office: 505 Ramapo Valley Rd Mahwah NJ 07430

CIANCIO, SEBASTIAN GENE, pharmacology and periodontology educator; b. Jamestown, N.Y., June 21, 1937; m., 1963; 1 child. D.D.S., U. Buffalo, 1961. Diplomate: Am. Bd. Periodontology. Fellow in pharmacology and periodontology SUNY Sch. Dentistry, Buffalo, 1963-65, from asst. prof. to assoc. prof. periodontology, chmn. dept., 1965-74, assoc. prof. pharmacology, 1965-73, prof. periodontology and chmn. dept., 1973—, clin. prof. pharmacology, 1973—; Cons. VA Hosp., Buffalo, 1970—, Pharm. Mfrs. Assn., 1973—, U.S. Pharmacopeae, 1975—, Am. Cyanamid Corp., 1978—, Dupont Co., 1981—; dir. Chautauqua Dental Congress, 1979. Grantee United Health Found., Western N.Y., 1965-66, 70-71, Nat. Inst. Dental Research, 1967-69, Merrill Nat. Labs., 1973—. Fellow Internat. Coll. Dentists; mem. ADA (chmn. council dental therapeutics 1976-78), Internat. Assn. Dental Research, Am. Assn. Dental Research (pres.

pharmacology, toxicology and therapeutics group 1979), Am. Acad. Periodontology, Nat. Soc. Med. Research (dir. 1979—). Subspecialties: Periodontics; Dental materials. Office: SUNY Dental Sch Buffalo NY 14214

CIANCUTTI, MARK ALAN, statistician, mathematics, educator; b. Pitts., May 29, 1948; s. Louis Amadio and Laura Marie (Kuhlman) C. B.S. in Math, Carnegie-Mellon U., 1970, M.S.I.A. in Bus, 1973, Ph.D. in Stats, 1981; M.A. in Math, U. Mich., 1971. Auditor Arthur Young & Co., N.Y.C., 1973-75; asst. prof. Pa. State U.- Monaca, 1981—. Mem. Soc. Indsl. and Applied Math. Club: Shotokan Karate of Am. (Pitts) (instr., 1st deg. black belt). Subspecialties: Operations research (mathematics); Statistics. Current work: Optimal transformations of data, differential geometry, combinatorial geometry, statistics. Home: 1483 Elizabeth Blvd Pittsburgh PA 15221 Office: Pa State U Brodhead Rd Monaca PA 15061

CIAPPENELLI, DONALD JOHN, chemistry educator; b. Worcester, Mass., Dec. 4, 1943. B.Sc., U. Mass., 1966; M.Sc., Brandeis U., 1967, Ph.D. in Chemistry, 1971. Dir. chemistry dept. Brandeis U., Waltham, Mass., 1972-77, Harvard U., Cambridge, Mass., 1977—; chmn. bd. Cambridge Lab. Cons., Inc.; dir. Acton Corp., Phillips Tech. Mem. Am. Chem. Soc., AAAS. Office: Dept Chemistry Harvard U Cambridge MA 02138

CIARALDI, THEODORE PAUL, biochemist, educator; b. Ft. Lee, Va., Sept. 7, 1950; s. Americo Sam and Jeanette (Bonaldi) C. B.S., U. Md., 1972; Ph.D., U. Rochester, 1977. Postdoctoral fellow Stanford U. Med. Sch., 1977-78; postdoctoral fellow U. Colo. Med. Center, Denver, 1978-79, research assoc., 1979-82, instr. biochemistry, 1982—. Juvenile Diabetes Found. grantee, 1982. Mem. Am. Diabetes Assn., Endocrine Soc., Am. Fedn. for Clin. Research, Sigma Xi. Democrat. Subspecialties: Receptors; Biochemistry (medicine). Current work: Relationship between hormone-receptors and regulation of cellular transport systems. Regulation of hormone receptors and transport systems in pathiophysiological states. Office: U Colo Med Center 4200 E 9th Ave Denver CO 80220

CIARANELLO, ROLAND DAVID, psychiatry educator; b. Schenectady, Feb. 27, 1943; s. Roland Victor and Carmella (Vertucci) C.; m. Nancy Rogers, June 29, 1968; 1 dau.: Andrea. B.S., Union Coll., 1965; M.D., Stanford U., 1970. Postdoctoral fellow Stanford U., 1970-71; research assoc. NIMH, 1971-74; resident in psychiatry and child psychiatry Stanford U., 1974-78, asst. prof. psychiatry, 1978-81, assoc. prof., 1981—; dir. lab of developmental neurochemistry dept. psychiatry, 1976—. Contbr. numerous articles to sci. jours. A. E. Bennett Neuropsychiatry awardee, 1968; NIMH Research Career Devel. awardee, 1978—; John Merck faculty fellow, 1982—. Mem. AAAS, Am. Psychiat. Assn., Psychiat. Research Soc., Am. Acad. Child Psychiatry, Am. Coll. Neuropsycho pharmacology, Am. Soc. Pharmacology and Exptl. Therapeutics, Internat. Soc. Psychoneuroendocrinology, Sigma Xi. Subspecialties: Neurochemistry; Psychopharmacology. Current work: Biochemical and molecular genetics of neurotransmitter enzymes and reeptors. Home: 996 Wing Pl Stanford CA 94305 Office: Dept Psychiatry Stanford Med Center Stanford CA 94305

CIARLONE, ALFRED EDWARD, pharmacologist, consultant; b. Reading, Pa., May 2, 1932; s. Jack and Minnie (D'Agostino) C.; m. Jo Ann Nina Zuccaro, July 18, 1959; children: Lisa Anne, Mark David. Student, Kutztown State Coll., 1953-55, Albright Coll., Reading, Pa., 1954; D.D.S., U. Pitts., 1959, Ph.D., 1973. Lic. dentist, Pa. Gen. practice dental resident VA Hosp., Pitts., 1959-60; pvt. practice gen. dentistry, Reading, Pa., 1960-69; Nat. Inst. Dental Research trainee U. Pitts., 1969-73; assoc. prof. oral biology-pharmacology Sch. Dentistry, Med. Coll. Ga., 1973-77; asst. prof. pharmacology Sch. Medicine, Med. Coll. Ga., 1977-82, prof., 1982—, assoc. prof. pharmacology, 1979—; prof. oral biology and pharmacology Sch. Grad. Studies, Med. Coll. Ga., 1982—; also cons.; pharmacology test preparer Nat. Bd. Dental Examiners. Author: (with L.P. Gangarosa Sr. and A.H. Jeske) Pharmacotherapeutics in Dentistry; textbook, 1983, also numerous articles, abstracts and book revs.; Cons., adv. editorial bd.: Jour. of ADA. Served with USAF, 1951-52; to maj. USAFR, 1963-69. Nat. Inst. Dental Research grantee, 1978-81. Mem. Am. Soc. Pharmacology and Exptl. Therapeutics, Internat. Assn. Dental Research, Am. Assn. Dental Schs., Ga. Acad. Sci. Subspecialties: Neuropharmacology; Oral biology. Current work: Neuropharmacology of CNS; mineralization of oral hard tissues. Fluorometry, dopamine, norepinephrine, serotonin; calcium, phosphorus, fluoride, enamel, dentin. Home: 25 Plantation Hills Dr Evans GA 30809 Office: Med Coll Ga Sch Dentistry Dept Oral Biology-Pharmacology 1120 15th St Augusta GA 30912

CIBOSKY, WILLIAM, ocean engineer, aerospace engineer; b. Haverhill, Mass., June 25, 1933; s. Edward and Mary (DiPucchio) C.; m. Beverlee Johnson, June 7, 1958 (div. 1978); children: Stephen, Stacia, Dawn, John Courtney, Thomas; m. Marlene Lois Johnson, Feb. 14, 1980. B.S., U.S. Mil. Acad., 1958; M.S., Stevens Inst. Tech., 1964; postgrad. student, Northwestern U., 1967-69. Commd. 2d lt. U.S. Army, 1951, advanced through grades to capt.; project engr., Kwajalein, Mich., 1964-66; ret., 1966, sr. project engr., Wilmington, Mass., 1966-72, project mgr. ocean surveillance, Portsmouth, R.I., 1972-75; mgr. bus. devel. Rockwell Internat., Anaheim, Calif., 1975-77; progrm mgr. Fiber Optics, ITT, Roanoke, Va., 1977-78; project mgr. TRW, Ocean and Energy Systems, Redondo Beach, Calif., 1978-80, project engr. reentry tech. project, San Bernadino, Calif., 1980-81, project engr. undersea surveillance program, McLean, Va., 1982—. Mem. Air Force Assn., Marine Tech. Soc., Soc. Naval Architects and Marine Engrs., U. S. Mil. Acad. Alumni Assn. Club: Harbor View Recreation Assn. (bd dirs. 1982). Subspecialties: Ocean engineering; Aerospace engineering and technology. Current work: Ocean engineering; hydrodynamics; marine design; ocean testing; aerospace engineering; advanced reentry concepts and missile systems. Patentee deep ocean, multi-leg non-rotating moors. Home: 10725 Greene Dr Lorton Va 22079 Office: TRW 7600 Colshire Dr McLean VA 22102

CIMENT, MELVYN, mathematician, lawyer; b. Bronx, N.Y., Sept. 23, 1941; s. Jack and Regina (Moskowitz) C.; m. Barbara Ann Kagan, July 3, 1966; children: Ethan Joseph, Daniel Isaac. B.S., U. Miami, 1962; M.S., NYU, 1964, Ph.D., 1968; J.D., Am. U., 1978. Bar: Bar: D.C 1979, Fla. 1979, Md. 1979. Mathmatician Nat. Bur. Standards, Washington, 1977—; asst. prof. math. U. Mich., Ann Arbor, 1969-72; mathematician Naval Surface Weapons Ctr., Silver Spring, Md., 1972-77, Nat. Bur. Standards, Washington, 1977—, program analyst, 1981-82; program dir. applied math. NSF, 1983—; Congl. fellow, 1980-81; mem. organizing com. workshop on large scale computing NSF, 1982-83. Mem. Soc. Indsl. and Applied Math. (govt. affairs com. 1982-84, Washington editor SIAM News 1983-85), AAAS, ABA, Computer Law Assn. Jewish. Subspecialties: Applied mathematics; Numerical analysis. Current work: Numerical solution of partial differential equations; large scale scientific computing—policy and requirements. Home: 11712 Kemp Mill Rd Silver Spring MD 20902 Office: Nat Bur Standards Washington DC 20234

CINADER, BERNHARD, scientist, immunologist, educator; b. Vienna, Austria, Mar. 30, 1919; s. Leon and Adele (Schwarz) C.; m.; 1 dau., Agatha. B.Sc., U. London, 1945, Ph.D., 1948, D.Sc., 1958. Research asst. Jenner Meml. student Lister Inst. Preventive Medicine, London, 1945-46, Beit Meml. fellow, 1949-53; fellow immunochemistry Inst. Pathology, Western Res. U., Cleve., 1948-49; prin. sci. officer, dept. exptl. pathology Inst. Animal Physiology, Babraham Hall, Cambridge; also hon. lectr. biochemistry dept. U. Coll., London, 1955-58; head subdiv. immunochemistry, div. biol. research Ont. (Can.) Cancer Inst., Toronto, 1958-69; assoc. prof. depts. med. biophysics and pathol. chemistry U. Toronto, 1958-67, prof. dept. med. biophysics, 1967—, prof. dept. med. cell biology, 1969—; dir. Inst. Immunology, 1971-81; mem. governing body U. Toronto, 1980—; vis. prof. U. Man., 1967, U. Alta., 1968, U. Sask, 1970, U. Western Ont., 1972, U. Bombay, 1981; chmn. immunology con. Biol. Council Can., 1967—; mem. WHO Expert Adv. Panel on Immunology, 1970—; chmn. adv. bd. Internat. Immunology Tng. and Research Center, Amsterdam, 1975-80; mem. adv. bd. dept. basic and clin. immunology Med. U. S.C., 1974; mem. adv. bd. Research in Immunology and Immunobiology, 1972-74; chmn. nomenclature com. WHO/Internat. Union Immunol. Socs., 1980—; lectr. numerous instns., profl. meetings, confs. and seminars. Editor: Antibody to Enzymes - A Three Component System, 1964, Antibodies to Biologically Active Molecules, 1967, Regulation of the Antibody Response, 1968, Immunological Response of the Female Reproductive Tract, 1976, Immunology of Receptors, 1976-77; Series editor: Receptors and Ligands in Intercellular Communication, 1983—; editorial bd.: Immunochemistry, 1965-70; editorial bd. Immunology, Serology, Transplantation sect., Excerpta Medica Found., 1966—; editorial bd.: Can. Jour. Biochemistry, 1967-71, Immunol. Methods, 1970-74, Bolletino dell-istituto sieroterapico Milanese, 1972—, Immunol. Communication, 1973—, Jour. Immunogenetics, 1973—, Immunology Letters, 1978—, Jour. Receptor Research, 1979—, Asian Pacific Jour. Allergy and Immunology, 1983—; contbr. articles to numerous profl. publs.; also catalogues and articles on Canadian Indian art. Recipient Old Student prize, London, 1944; medal Société de Chimie Biologique, Paris, 1954; Pfizer fellow Institut de Recherches Cliniques de Montreal, 1972; Jubilee medal, Ottawa, 1977; Ignác Semmelweis medal, Budapest, 1978. Fellow Royal Inst. Chemistry (U.K.), Royal Soc. Can. (Thomas W. Eadie medal 1982), N.Y. Acad. Scis.; mem. Internat. Union Immunol. Socs. (chmn. 1970—, pres. 1969-74), Can. Soc. Immunology (pres. 1967-69, 79-81), Nat. Com. Immunology (chmn. 1981—), Can. Fedn. Biol. Socs. (chmn. 1976-77), Internat. Council of Sci. Unions (mem. council and assembly 1980—). Subspecialty: Immunology (agriculture). Current work: Cellular immunology, immunigenetics and gerontology. Home: 73 Langley Ave Toronto ON M4K 1B4 Canada Office: Inst Immunology Rm 4366 Med Scis Bldg U Toronto Toronto ON M5S 1A8 Canada

CINTRON, GUILLERMO BO., cardiologist, educator; b. San Juan, P.R., Mar. 28, 1942; s. Guillermo Cintron-Ayuso and Rosa Luz (Silva) C.; m. Susan H. Beans, Aug. 8, 1968; children: Guillermo Carlos, Francisco Jaime, Michael Antonio. B.S. cum laude, U. P.R.-Rio Piedras, 1963; M.D., Loyola U., Chgo., 1967. Diplomate: Am. Bd. Internal Medicine. Intern George Washington U. Hosp., Washington, 1967-68; resident in internal medicine VA Hosp and Georgetown U. Hosp., Washington, 1970-72; fellow in cardiology VA Hosp., Washington, 1972-74; dir. CCU, VA Hosp., San Juan, 1975-83; chief cardiology sect. VA Hosp., Tampa, Fla., 1983—; asst. prof. medicine U. P.R. Sch. Medicine, San Juan, 1975-82, assoc. prof., 1982-83; assoc. prof. medicine U. So. Fla., Tampa, 1983—; bd. dirs. P.R. Cardiovascular Ctr., San Juan, 1980-82; cons. cardiology Roosevelt Rds. Naval Hosp., Ceiba, P.R., 1978-83. Mem. editorial bd.: Bull. P.R. Med. Assn, 1968-70; contbr. articles to profl. publs. Team physician Pee Wee Football League, San Juan, 1982. VA coop studies program research grantee, Washington, 1976-80. Fellow Am. Coll. Cardiology, ACP (treas. P.R. chpt. 1980-82, bd. dirs. 1980-82), Am. Heart Assn. (mem. council clin. cardiology 1978—, dir. P.R. chpt. 1979-81); mem. Dominican Republic Cardiology Soc. (hon.), Am. Fedn. Clin. Research. Roman Catholic. Subspecialties: Cardiology; Internal medicine. Current work: Myocardial infarction; coronary artery disease. Home: 14908 Lake Forest Dr Lutz FL 33549 Office: VA Hosp Cardiology Dept 111-A 13000 N 30th St Tampa FL 33612

CIPOLLA, SAM JOSEPH, physics educator; b. Chgo., July 24, 1940; s. Joseph and Florence M. (Mistretta) C.; m. F. Virginia Stover, Jan. 12, 1939; children: Mark, Karen. B.S., Loyola U., Chgo., 1962; M.S., Purude U., 1965, Ph.D., 1969. Adminstrv. asst. Purdue U., West Lafayette, Ind., 1968-69, research assoc., 1969; asst. prof. Creighton U., Omaha, 1969-73, assoc. prof., 1973-83, prof., 1983—; vis. prof. U. Nebr., 1982-83; cons. in field. Contbr. articles to profl. jours. NSF grantee, 1970, 82; Research Corp. grantee, 1972, 74, 79, 80, 81. Mem. Am. Phys. Soc., Am. Assn. Physics Tchr., Nebr. Acad. Sci. Democrat. Roman Catholic. Subspecialty: Radiation physics and radiation dosimetry. Current work: Heavy ion-atom collision studies at low energies in solids; efficiency response determinations of semiconductor radiation detectors. Home: 2917 S 116 Ave Omaha NE 68144

CISNE, JOHN LUTHER, biostratigrapher, paleontologist, researcher, educator; b. Summit, N.J., Apr. 27, 1947; s. Luther Elmore and Georgia Lee (Johnson) C.; m. Robin Hope Fisher, July 22, 1978; 1 son, Joel Edwin. B.S., Yale U., 1969; Ph.D., U. Chgo., 1973. Asst. prof. dept. geol. scis. and div. biol. scis. Cornell U.-Ithaca, 1973-79, assoc. prof., 1979—; trustee Paleontol. Research Instn., Ithaca, 1976—, asst. sec., treas., 1980—. Contbr. articles to profl. jours. NSF grantee, 1975—. Fellow AAAS; mem. Geol. Soc. Am., Internat. Paleontol. Assn., Paleontol. Soc., Soc. Econ. Paleontologists and Mineralogists. Subspecialties: Stratigraphy; Paleontology. Current work: Gradient analysis of fossil assemblages as applied to paleobathymetry, stratigraphic correlation by sea level curve and paleobathymetric mapping; Morphometric study of evolution. Office: Dept Geol Scis Cornell U Ithaca NY 14853 Home: 115 Oak Hill Rd Ithaca NY 14850

CIVIN, CURT INGRAHAM, physician, cancer researcher; b. Syracuse, N.Y., May 29, 1949; s. Chester J. and Vivian L. (Rutes) C.; m. Nancy L. Banks, June 7, 1970; children: Joshua I., Marcus V. B.A. magna cum laude, Amherst Coll, 1970; M.D. cum laude, Harvard U., 1974. Diplomate: Am. Bd. Pediatrics (pediatric hematology). Intern, then resident in pediatrics Children's Hosp. Med. Center, Boston, 1974-76; clin. asso. immunology br. Nat. Cancer Inst., NIH, 1976-78, investigator pediatric oncology br., 1978-79; asst. prof. Johns Hopkins U. Med. Sch., 1979—. Author numerous papers, reports in field. Served with USPHS, 1976-79. Recipient Oscar E. Shotte award, 1970, Soma Weiss prize, 1974; grantee NIH, Tobacco Research Council, Am. Cancer Soc., Blood Products, Inc.; fellow Am. Cancer Soc., 1980-83. Mem. AAAS, Am. Soc. Clin. Oncology, Am. Soc. Hematology, Am. Assn. Cancer Research, Internat. Soc. Exptl. Hematology, Johns Hopkins U. Immunology Council. Democrat. Jewish. Club: Variety. Subspecialties: Cell biology (medicine); Cell study oncology. Current work: Research in hematopoiesis, granulocyte function, clin. and research activities on leukemia, monoclonal antibodies, hematopoiesis, hybridomas, cell surface antigens, hematology, leukemia, immunology. Office: Oncology 3-121 Johns Hopkins Hosp 600 N Wolfe St Baltimore MD 21205

CIVJAN, SIMON, dental educator, researcher, consultant; b. Linkuva, Lithuania, May 25, 1920; came to U.S., 1940, naturalized, 1945; s. Haim and Sonia Rebecca (Blumberg) C.; m. Velta Lilia Jaunarajs, Nov. 6, 1946; children: Ralph Haime, Neal Gabriel. B.Chem. Engring. with honors, U Fla., 1944; D.D.S., U. Md., 1954; M.S. in Dental Materials, Georgetown U., 1963. Enlisted in U.S. Army, 1944, advanced through grades to col., 1971; dental officer, chief clinician various locations, 1954-63; chief div. dental materials U.S. Army Inst. Dental Research, Washington, 1963-73, dir., 1973-74; dep. comdr. for dental activities U.S. Army Med. Dept. Activity, Landstuhl, W.Ger., 1974-77, Ft. Leonard Wood., Mo., 1977-80; ret., 1980; prof. U. Tex. Dental Br., Houston, 1980—; U.S. Army cons. standards com. on dental materials and devices ADA, Chgo., 1969-74. Contbr. to: Improving Dental Practice through Preventive Measures, 1975; contbr. articles to profl. jours. Decorated Legion of Merit with oak leaf cluster. Fellow Am. Coll. Dentists, Internat. Coll. Dentists; mem. ADA, Am. Assn. Dental Schs., Federation Dentaire Internationale, Assn. Mil. Surgeons U.S., Internat. Assn. Dental Research. Jewish. Subspecialties: Dental materials; Biomaterials. Current work: Teaching dental students, undergraduate and graduates; dental materials science and research methodology; research in physical, chemical and manipulative properties of dental materials; and act as consultant to individual dentists and dental organizations or groups. Home: 5734 Indigo St Houston TX 77096 Office: U Tex Health Sci Center at Houston Dental Br 6516 John Freeman Ave Houston TX 77030

CLAFLIN, ROBERT MALDEN, educator; b. Flint, Mich., Nov. 11, 1921; s. Robert Hugh and Kathryn Elizabeth (Ruhl) C.; m. Barbara Ellen Garrison, June 21, 1957; children—Deborah Ann, Blair Lawrence, Kathryn Elizabeth. D.V.M., Mich. State U., 1952; M.S., Purdue U., 1956, Ph.D., 1958. Faculty Purdue U., Lafayette, Ind., 1952—, prof. vet. pathology, 1959—, head dept. vet. microbiology, pathology and pub. health, 1959—. Mem. AVMA, Internat. Acad. Pathology, Conf. Research Workers Animal Diseases N.A., Sigma Xi, Phi Zeta, Phi Kappa Phi. Subspecialties: Microbiology (veterinary medicine); Solar physics. Current work: Pathology and academic administration. Home: 706 Carrolton Blvd West Lafayette IN 47906 Office: Purdue U Lafayette IN 47907

CLAGETT, JAMES ALBERT, immunologist, immunopathologist, educator; b. Frederick, Md., Nov. 6, 1942; s. Albert Washington and Catherine Gertrude (Howes) C.; m. Carolyn Appleget, June 4, 1964; 1 dau., Allison Beth. B.A., DePauw U., 1964; M.S., U. Nebr., 1966, Ph.D., 1970. Postdoctoral fellow in cellular immunology Scripps Clinic and Research Insts., LaJolla, Calif., 1970-73; assoc. prof. depts. microbiology-immunology and periodontics Schs. Medicine and Dentistry, U. Wash., Seattle, 1978—; cons. Immune Responses, Inc., Seattle. Mem. editorial bd.: Jour. Dental Research; contbr. articles to profl. jours. USPHS fellow, 1970-73; USPHS grantee, 1973—. Mem. Am. Assn. Immunologists, Am. Soc. Microbiology, Sigma Xi. Subspecialties: Immunology (medicine); Pathology (medicine). Current work: Research into cellular basis for recruitment, activational and proliferation of lymphocytes and macrophages in the lung. Office: Dept Periodontics Sch Dentistry and Medicine U Wash Seattle 98195

CLANTON, JEFFREY ALAN, nuclear pharmacist, educator, researcher; b. Evansville, Ind., Nov. 3, 1953; s. Esly Arthur and Gladys Marie (Seaton) C.; m. Pamela Bean, Aug. 28, 1976; 1 dau., Jennifer Leigh. B.S. in Pharmacy, Samford U., 1976; M.S. in Radio-pharmacy, U. So. Calif., 1977. Lic. pharmacist, Nev., Tenn. Teaching asst. U. So. Calif., 1976-77; intern in radiopharmacy U. Utah, 1977-78; research instr. Vanderbilt U., 1978-79, assoc. in radiology, 1979—; chief radiopharmacy services Vanderbilt U. Med. Ctr., 1981—. Mem. CAP, Birmingham, Ala., 1976 Mem. CAP, Los Angeles, 1977. Mem. Am. Pharm. Assn. (chmn. communications com. nuclear pharmacy sect.), Soc. Nuclear Medicine, Soc. Magnetic Resonance Imaging, Internat. Aerobatic Club (pres. Springfield, Tenn. chpt. 1980-82), Omicron Delta Kappa. Subspecialties: Nuclear medicine; Imaging technology. Current work: Developing and testing radiopharmaceuticals and nuclear magnetic resonance imaging contrast agents. Co-inventor paramagnetic metal compounds for physiol. use as Nuclear Magnetic Resonance contrast agts. Office: Div Nuclear Medicine Vanderbilt U Med Center Nashville TN 37232

CLAPP, JAMES LESLIE, univ. adminstr., civil engr.; b. Madison, Wis., Mar. 14, 1933; s. Don Jay and Stelia (Anderson) C.; m. Susan DeForest Randolph, June 3, 1961; children—Lee William, Leonard Jay, Don Randolph. B.S.C.E., U. Wis., Madison, 1956, B.S.N.S., 1956, M.S., 1961, Ph.D., 1964. Registered profl. engr., Wis.; registered land surveyor, Wis. Constrn. engr. Drave Corp., Pitts., 1959-60; engr. Wis. Hwy. Commn., 1960-61; instr., research asst. U. Wis., Madison, 1961-64, asst. prof. civil engring., 1964-68, asso. prof., 1968-70, prof., dir. environ. monitoring group, 1970-78; dean engring. and sci., U. Maine, Orono, 1978—; cons. in field. Served to 1st lt. USMC, 1956-59. Recipient Congl. medal for Antarctic research, 1967; Steiger award U. Wis., 1967; Polygon award, 1972, 73, 75. Mem. Am. Soc. Engring. Edn., Wis. Soc. Profl. Engrs., Maine Soc. Profl. Engrs., Nat. Soc. Profl. Engrs., ASCE, Am. Congress Surveying and Mapping (N.E. sect. Outstanding Educator 1977), Am. Soc. Photogrammetry. Subspecialty: Remote sensing (geoscience). Research on remote sensing water resources, 1968-78, on land info. systems, 1970—, on environ. impact coal fire power plants, 1975-78. Home: 14 Park St Orono ME 04473 Office: 101 Barrows Hall U Maine Orono ME 04469

CLARE, GEORGE HADLEY, systems integration manager, safety engineer; b. Hornell, N.Y., June 11, 1950; s. Don Waldo and Mary Louise (Gigee) C.; m. Carol Margaret Clarke, Sept. 6, 1969 (div. Jan. 1981); m. Annette Bindgen, July 3, 1983. B.S., Cornell U., 1972, M.Engr., 1974. Engr. licensing, systems engring. Advanced Reactors div. Westinghouse Co., Madison, Pa., 1974-75, Oak Ridge, 1975-79, mgr. licensing, 1980-83; Westinghouse rep. Fast Reactor Safety Tech. Mgmt. Center, Argonne, Ill., 1979-80. Mem. Am. Nuclear Soc. Unitarian-Universalist. Subspecialties: Nuclear fission; Nuclear engineering. Current work: Advanced reactor systems and safety system concepts, especially for liquid metal fast breeder reactor plants. Office: Westinghouse Co Advanced Reactors Div 120 Jefferson Circle PO Box W Oak Ridge TN 3783

CLARK, ALAN FRED, materials scientist; b. Milw., June 29, 1936; m., 1957; 4 children. B.S., U. Wis., 1958, M.S., 1959; Ph.D. in Nuclear Sci, U. Mich., 1964. Nat. Acad. Sci.-NRC research assoc. low temperature physics Nat. Bur. Standards, Boulder, Colo., 1964-66, staff physicist low temperature physics and materials sci., 1966-78, chief thermophys. properties of solids, 1978-80, chief superconductors magnetic materials, 1981—; asst. prof. materials sci. colo. State U., Ft. Collins, 1965-68; mem. NASA-USAF-FAA panel on titanium combustion, 1975—; mem. magnetic energy storage panel U.S. Dept. Energy, 1977—; chmn. Internat. Cryogenic Materials Conf. Bd., 1979-83. Tech. editor: Rev. Sci. Instruments, 1974-76; adv. editor: Cryogenics, 1977—. Recipient Superior Accomplishment award Nat. Bur. Standards, 1967. Mem. Am. Nuclear Soc., Am. Phys. Soc., Combustion Inst., Internat. Cryogenic Materials Conf. Subspecialties: Superconductors; Magnetic physics. Office: Nat Bur Standards Boulder CO 80303

CLARK, ALAN LEE, research engineer; b. Vicksburg, Mich., Aug. 15, 1951; s. Loren Herbert and Anne Eleanor (Bishop) C. B.S., U. Mich.-Ann Arbor, 1979. Programmer U. Mich.-Ann Arbor, 1972-74; sr. systems analyst MIS Internat., Inc., Southfield, Mich., 1974-81; research engr. Ford Motor Co., Dearborn, Mich., 1981—; vis. sr. engr. P.A.P., U. Rochester, N.Y., 1982. Orch. mgr. Ann Arbor Comic Opera

Guild, 1982. Mem. Assn. Computing Machinery. Subspecialties: Graphics, image processing, and pattern recognition; Geometric modelling of solids. Current work: Evaluation/enhancement of state-of-the-art solid modelling systems and applications, evaluation/enhancement of user interfaces to solid modelling systems. Home: 3155 Dolph St Ann Arbor MI 48103 Office: Ford Motor Co SCI Research Lab MS E-1134 PO Box 2053 Dearborn MI 48121

CLARK, ALLAN HERSH, mathematician, univ. adminstr.; b. Cin., July 16, 1935; s. Elmer Edward and Dorothy Mildred (Hersh) C.; m.; children—Edward Dunn, Geoffrey Allan, Nathaniel Hersh. B.S., Mass. Inst. Tech., 1957; M.A., Princeton U., 1959, Ph.D., 1961. Instr. Brown U., 1961-63, asst. prof. math, 1963-66, asso. prof., 1966-70, prof., 1970-75, chmn. dept. math., 1971-73; prof., dean Sch. Sci., Purdue U., West Lafayette, Ind., 1975—; vis. mem. Inst. for Advanced Study, 1965-66; guest prof. Aarhus U., Denmark, 1970-71. Author: Elements of Abstract Algebra, 1971. Trustee Univ. Research Assn. Mem. Am. Math. Soc., AAUP, Am. Assn. Higher Edn. Subspecialty: Operations research (mathematics). Current work: Algebraic topology, chomology theories. Office: Sci Adminstrn Mathematics Bldg Purdue U West Lafayette IN 47907

CLARK, BRIAN OLIVER, geophysicist; b. New Castle, Pa., July 19, 1948; s. Stanley Kenneth and Melba Sunshine (Brickner) C.; m. Wendy Carol Ronas., March 26, 1970; children: Justin, Aaron. B.S., Ohio State U., 1970; M.S., Harvard U., 1971, Ph.D., 1976. Instr. physics Brandeis U., Waltham, Mass., 1976-78, asst. prof., 1979; mem. profl. staff, program leader Schlumberger-Doll Research, Ridgefield, Conn., 1979—. Mem. Am. Phys. Soc., AAAS, Phi Beta Kappa. Subspecialties: Petroleum engineering; Geophysics. Current work: Electromagnetic sensing of petroleum. Office: Old Quarry Rd Ridgefield CT 06877

CLARK, DAVID DELANO, physicist; b. Austin, Tex., Feb. 10, 1924; s. David Lee and Grace (Delano) C.; m. Gladys Braunstein, Dec. 27, 1949; children: Marcia Susan, Gordon Richard, Janet Mirella. Student, U. Tex., 1941-42, 46; A.B. in Physics, U. Calif. at Berkeley, 1948; Ph.D. (AEC predoctoral fellow 1950-52), U. Calif. at Berkeley, 1953. Research asso. Brookhaven Nat. Lab., 1953-55, vis. scientist, summers, 1957, 58, 64, 73, fall 79; mem. faculty engring. physics Cornell U., 1955—, asso. prof., 1958-65, prof., 1965—, dir. Ward Lab. Nuclear Engring., 1960—; cons. Gen. Atomic Co., summer 1959; vis. prof. Tech. U., Munich, W. Ger., 1976. Contbr. articles to profl. jours. Served with USAAF, 1942-46. Euratom fellow Euratom Research Center, Ispra, Italy, 1962; Guggenheim Found. fellow Niels Bohr Inst., 1968-69. Mem. Am. Phys. Soc., Phi Beta Kappa, Sigma Xi. Subspecialties: Nuclear physics; Nuclear fission. Current work: Neutron reactions with nucles; especially neutron emission and absorption involving highly unstable or isomenic nuclides. Home: 105 Needham Pl Ithaca NY 14850

CLARK, DONALD ELDON, materials scientist; b. Watertown, S.D., Dec. 24, 1936; s. Eldon Orris and Bertha (Brandon (Andersen)) C.; m. Beverly Jean McInroy, May 31, 1959; children: David O., Deborah A. A.B., U. S.D., 1961; Ph.D., Iowa State U., 1965. Research asst. AEC Lab., Ames, Iowa., 1961-65; research chemist Miami Valley Labs., Procter and Gamble Co., Cin., 1965-66; asst. to assoc. prof. chemistry Western Ill. U., Macomb, 1966-74; sr. scientist Westinghouse Hanford Co., Richland, Wash., 1974-80; sr. chemist Brookhaven Nat. Lab., Upton, N.Y., 1980-82; lead project mgr. Battelle, Columbus, Ohio, 1982—; AEC faculty appointee Battelle Pacific N.W. Labs., 1969-70, Douglas United Nuclear, 1972; Vis. staff U. Calif. Los Alamos Sci. Labs., 1971. Served with U.S. Army, 1954-57. Recipient Twin Cities medal Am. Inst. Chemists, 1961. Mem. Am. Chem. Soc., Am. Nuclear Soc., Health Physics Soc., ASTM, Sigma Xi. Club: VFW. Lodge: Elks. Subspecialties: Physical chemistry; Materials. Current work: Materials development and research projects related to geologic disposal of nuclear wastes. Home: 1974 Keswick Dr Columbus OH 43220 Office: Project Mgmt Div Battelle 505 King Ave Columbu OH 43201

CLARK, EDWARD ALAN, immunogeneticist, educator; b. Long Beach, Calif., Sept. 3, 1947; s. Elliott Goss and Iris Evelyn (Price) C.; m. Yukika Tanaka, July 14, 1975; children: Tomas Dylan, Sashya Sabina Tanaka. B.S., UCLA, 1969, Ph.D., 1977. Staff research asso., renal transplant coordinator UCLA, 1970-74, postgrad. research asst., 1974-77; hon. staff research asst. Univ. Coll., London, 1977-79; asst. prof. genetics Primate Center, U. Wash., Seattle, 1979—; sr. scientist Genetic Systems Corp., Seattle, 1981—; affiliate Fred Hutchinson Cancer Center, Seattle, 1980—. Contbr. articles to profl. jours. Univ. Regents scholar, 1965-69; Inter-Sci. Research Found. prize, 1977; Edna A. Old Meml. fellow Cancer Research Inst. N.Y., 1977-79. Mem. AAAS, Physicians for Social Responsibility, Zero Population Growth., Am. Assn. Immunologists. Subspecialties: Immunogenetics; Infectious diseases. Current work: Immunogenetics, natural killer cells, monoclonal antibodies, Epstein-Barr virus, human B cells. Office: Genetic Systems Corp 3005 1st Ave Seattle WA 98121

CLARK, FRANK OLIVER, astronomer, educator; b. Augusta, Ga., Feb. 27, 1943; s. Anson L. and Gwendolyn C. (Peebles) C.; m. Deane Flanders, 1970; children: Rachel Suzanne, Juliana Elizabeth. A.S. with honors, So. Tech. Inst., 1963, B.S., Augusta Coll., 1969; M.S., U. Va., 1971, Ph.D., 1973. Field engr. IBM, Atlanta, 1963-67; NRC postdoctoral research assoc. Nat. Bur. Standards, Washington, 1973-75; asst. prof. astronomy U. Ky., Lexington, 1975-79, assoc. prof., 1979—. Contbr. numerous articles to profl. jours. NDEA fellow, 1969-72; NRC fellow, 1973-75; research grantee NSF, 1977-81, Research Corp., 1976-82, U. Ky. Research Found., 1975-82. Mem. Internat. Astron. Union, Am. Astron. Soc., Ky. Assn. Physics Tchrs., Blue Grass Astron. Soc. Democrat. Presbyterian. Subspecialties: Radio and microwave astronomy; Remote sensing (atmospheric science). Current work: Star formation, interstellar medium, interstellare molecules, circumstellar masers. Home: 337 Melbourne Way Lexington KY 40502 Office: Dept Physics and Astronom U K Lexington KY 40506

CLARK, GEORGE ALFRED, JR., biology educator; b. Camden, N.J., May 6, 1936; s. George Alfred and Emily Elizabeth (Fox) C.; m. Nancy Barnes, June 17, 1961; 1 son, Kevin Douglas. B.A., Amherst Coll., 1957; Ph.D., Yale U., 1964. Acting instr. U. Wash., Seattle, 1964-65; asst. prof. U. Conn., Storrs, 1965-70, assoc. prof., 1970-82, prof. biology, 1982—; state ornithologist State of Conn., 1981—. Mem. Northeastern Bird Banding Assn. (pres. 1981—). Subspecialties: Morphology; Ethology. Current work: Research on integumental structure in relation to behavior in birds. Office: Biol Scis Group U Conn Storrs CT 06268

CLARK, GEORGE EUGENE, dentist, naval officer, researcher; b. Akron, Ohio, July 2, 1938; s, George Gerald and Juanita Gwendolyn (McKellar) C.; m. (div.); children: Michael Brian, Kevin Douglas. D.D.S., Ohio State U., 1962; M.S. in Biochemistry, George Washington U., 1968, Ph.D., 1976. Commd. lt. U.S. Navy, 1962, advanced through grades to capt., 1979; chief periodontal research div. Naval Med. Research Inst., Bethesda, Md., 1972-76; head dental caries br. Naval Dental Research Inst., Gt. Lakes, Ill., 1976-78, chief oral diseases div., 1978-80, dir. clin. investigation dept., 1980-81, dir. sci. investigation dept., 1981-82, comdg. officer, 1982—. Cubmaster pack 271, Boy Scouts Am., Libertyville, Ill., 1977-80; v.p. Liberty Road and Track Club, 1982—. Fellow Internat. Coll. Dentists; mem. ADA, Am. Assn. for Dental Research (sec.-treas. Chgo. sect. 1981-82), Internat. Assn. for Dental Research. Subspecialties: Cariology. Current work: Commanding officer and administration of research dental institute; epidemiology of and effects of caries on the dental pulp; production of toxins in dental caries and their effects on the dental pulp. Home: 937 S 4th Ave Libertyville IL 60048 Office: Naval Dental Research Inst Naval Base Bldg 1-H Great Lakes IL 60088

CLARK, GEORGE WHIPPLE, educator, physicist; b. Evanston, Ill., Aug. 31, 1928; s. Robert Keep and Margaret (Whipple) C.; m. Elizabeth Kister, Dec. 18, 1954; children—Katherine, Jacqueline. B.A., Harvard, 1949; Ph.D., Mass. Inst. Tech., 1952. Instr. physics Mass. Inst. Tech., 1952-54, asst. prof., 1954-60, asso. prof., 1960-65, prof., 1965—; bd. dirs., past mem. vis. com., past mem. space telescope inst. council Asso. Univs. for Research in Astronomy. Mem. Nat. Acad. Scis. (mem. space astronomy subcom. of space sci. bd., mem. astron. survey com., chmn. panel high energy astrophysics), Am. Acad. Arts and Scis., Internat. Astron. Union (nat. com.), Am. Astron. Soc., Am. Phys. Soc. (past chmn. div. cosmic physics). Subspecialty: 1-ray high energy astrophysics. Research on high energy astronomy. Home: 177 Gardner Rd Brookline MA 02146 Office: Mass Inst Tech Cambridge MA 02139

CLARK, HUGH KIDDER, physicist, consultant; b. St. Louis, Jan. 22, 1918; s. Arthur Henry and Persis Thorndike (Kidder) C.; m. Marie Theresa Folsom, May 26, 1942; children: Lawrence Arthur, Barbara Alice. A.B., Oberlin Coll., 1939; Ph.D., Cornell U., 1943. Research assoc. Harvard U., Cambridge, Mass., 1943-45; research chemist E. I. duPont de Nemours & Co., Waynesboro, Va., 1945-62, research assoc., Aiken, S.C., 1962—. Contbr. articles to profl. jours. Pres. Aiken (S.C.) Community Concert Assn., 1958-59. Fellow Am. Nuclear Soc. (standards com. 1962—, chmn. nuclear criticality safety div. 1969-70, achievement award 1976); mem. Am. Chem. Soc., AAAS. Presbyterian. Subspecialties: Nuclear physics; Physical chemistry. Current work: Nuclear criticality safety assessments. Home: 225 Lakeside Dr Aiken SC 29801 Office: E I duPont de Nemours & Co Savannah River Lab Aiken SC 29808

CLARK, JAMES DERRELL, veterinarian; b. Atlanta, Mar. 8, 1937; s. Dallas Virgil and Ruby Irene (Deaton) C.; m. Martha Elizabeth Doster, Aug. 6, 1960; children: Michele Elizabeth, Melissa Elaine, Matthew Derrell. D.V.M., U. Ga., 1961, M.S., 1963; D.Sc., Tulane U., 1967. Diplomate: Am. Coll. Lab. Animal Medicine, 1967. Research asst. Coll. Vet.Medicine, U. Ga., Athens, 1961-62; asst. prof. Tulane U. Sch. Medicine, New Orleans, 1967-72; dir. animal resources Lab. Animal Medicine U. Ga., Athens, 1972—, assoc. prof. med. microbiology, 1972—; cons. AAALAC, Chgo., 1970—, Am. McGaw, Milledgeville, Ga., 1976—. Contbr. chpts. to books, articles to profl. jours. Served to capt. USAF, 1963-65. NIH fellow, 1962-63, 65-67. Mem. AVMA, Am. Coll. Lab. Animal Medicine, Am. Assn. Lab. Animal Sci., Am. Assn. Zoo Veterinarians, Am. Assn. Vet. Med. Colls., Assn. New Orleans Veterinarians, North Ga. Vet. Med. Assn. Baptist. Subspecialties: Laboratory Animal Medicine; Toxicology. Current work: Aflatoxicosis, laboratory animal care and use. Home: Box 186 Bogart GA 30622 Office: Animal Resources Coll Vet Med U Ga Athens GA 30602

CLARK, JAMES HENRY, biology educator; b. Earlington, Ky., June 17, 1932; s. Henry H. and Emma Louise (Peyton) C.; m. Janis L. Hendrix, Sept. 18, 1957; children: James Gregory, Tricia Lynn. B.S., Western Ky. U., 1959; M.S., Purdue U., 1966, Ph.D., 1968. Asst. prof. Purdue U., Lafayette, Ind., 1970-73; assoc. prof. Baylor U. Coll. Medicine, Houston, 1973, prof. cell biology, 1973—; cons. in field. Author: (with E.J. Peck Jr.) Female Sex Steriods: Receptor and Function, 1979; contbr. over 150 sci. articles to profl. publs. NIH grantee, 1970—; Am. Cancer Soc. grantee, 1972-78. Mem. AAAS, Endocrine Soc., Soc. Study of Reprodn., Sigma Xi. Subspecialties: Receptors; Cell biology (medicine).

CLARK, JOHN DESMOND, anthropology educator; b. London, Eng., Apr. 10, 1916; U.S., 1961; s. Thomas John Chown and Catherine (Wynne) C.; m. Betty Cable Baume, Apr. 30, 1938; children: Elizabeth Ann (Mrs. David Miall Winterbottom), John Wynne Desmond. B.A. Hons, Cambridge U., 1937, M.A., 1942, Ph.D., 1950, Sc.D., 1974. Dir. Rhodes-Livingstone Mus., No. Rhodesia, 1938-61; prof. anthropology U. Calif., Berkeley, 1961—, faculty research lectr., 1979; Raymond Dart lectr. Inst. for Study Man, Africa, 1979; Sir Mortimer Wheeler lectr. Brit. Acad., 1981. Author: The Stone Age Cultures of Northern Rhodesia, 1950, The Prehistoric Cultures of the Horn of Africa, 1954, The Prehistory of Southern Africa, 1959, Prehistoric Cultures of Northeast Angola, 1963, Distribution of Prehistoric Culture in Angola, 1966, The Atlas of African Prehistory, 1967, Kalambo Falls Prehistoric Site, Vol. I, 1968, Vol. II, 1973, The Prehistory of Africa, 1970. Served with Brit. Army, 1941-46. Decorated comdr. Order Brit. Empire; comdr. Nat. Order Senegal; Huxley medallist Royal Anthrop. Inst., London, 1974; Ad personam internat. Gold Mercury award, Addis Ababa, 1982. Fellow Am. Acad. Arts and Scis., Brit. Acad., Royal Soc. S. Africa, Soc. Antiquaries London, AAAS; mem. Pan-African Congress Prehistory, Geog. Soc. Lisbon, Istituto Italiano di Preistoria e Protostoria, Body Corporate Livingstone Mus. Zambia. Subspecialty: Paleoanthropology. Office: Dept Anthropology U Calif Berkeley CA 94720

CLARK, JOHN HAMILTON, chemist, educator; b. San Gabriel, Calif., Nov. 22, 1949; s. Charles Warren and Nellie May (Hamilton) C.; m. Piyanud Ruth Hussey, June 12, 1971; 1 dau., Cynthia Alison. A.B. in Chemistry and Physics with highest honors, U. Calif., Santa Barbara, 1971, Ph.D., 1976. J. Robert Oppenheimer research fellow Los Alamos Sci. Lab., 1976-79, asst. group leader laser photochemistry, 1979; asst. prof. chemistry U. Calif., Berkeley, 1979—. Contbr. numerous articles to sci. jours. Nat. Merit scholar, 1967-71; Regents' scholar, 1967-71; Bank of Am. scholar, 1967; President's Undergrad. fellow, 1970-71; NSF Summer fellow, 1970; Chancellor's Sci. fellow, 1971-72; Charles Kofoid Eugenics fellow, 1972-73; Camille and Henry Dreyfus tchr.-scholar, 1981—; Alfred P. Sloan Research fellow, 1982—. Mem. Am. Chem. Soc., Am. Phys. Soc., AAAS, Inter-Am. Photochem. Soc., Soc. Applied Spectroscopy, Optical Soc. Am., Phi Beta Kappa, Sigma Xi. Subspecialty: Laser-induced chemistry. Current work: Picosecond laser spectroscopy, ultrafast reaction kinetics, chemical dynamics of reactions in solution, laser photochemistry. Patentee in field.

CLARK, JOHN MAGRUDER, JR., biochemist, educator; b. Ithaca, N.Y., June 10, 1932; s. John M. and Emily (Blood) C.; m. Lucie T. Welles, Dec. 19, 1957; children: Theodore W., Jennifer T. B.S., Cornell U., 1954; Ph.D., Calif. Inst. Tech., 1958. Instr. U. Ill.-Urbana, 1958-60, asst. prof., 1960-66, assoc. prof., 1966-79, prof. biochemistry dept., 1979—. Author, editor: textbook Experimental Biochemistry, 1964; author 2d edit., 1976. Mem. Am. Soc. Biol. Chemists, Am. Chem. Soc. Subspecialties: Biochemistry (biology); Molecular biology. Current work: Mechanism of protein biosynthesis, nucleic acid structure and function. Office: Biochemistry Dep U Ill 1209 W California Urban IL 61801

CLARK, JON D(ENNIS), computer science educator, researcher; b. Detroit, Feb. 8, 1946; s. James H. and Ellen F. (Much) C. B.A., Mich. State U., 1968; M.B.A., Eastern Mich. U., 1972; Ph.D., Case Western Res. U., 1977. Process engr. Hoover Ball & Bearing Co., Ann Arbor, Mich., 1968-70, programmer/analyst, 1970-71; faculty Coll. Bus., North Tex. State U., Denton, 1979—. Author: Data Base Selection, Design and Administration, 1980, Computer System Selection, 1981; monograph Physical Data Base Record Design, 1979; asst. editor: Performance Evaluation Rev., Record. Mem. Assn. for Computing Machinery. Subspecialties: Database systems; Information systems, storage, and retrieval (computer science). Current work: Cognitive science applied to information systems problems. Office: North Texas State Univ Coll Business Denton TX 76203

CLARK, JULIA BERG, pharmacologist, educator; b. Moline, Ill., June 7, 1940; d. Leslie Willard and Margaret Vivian Freeman; m.; children: Margaret Louise, Brian Alexander. B.A., Radcliffe Coll., 1962; Ph.D., Ind. U., 1966. Research assoc. Harvard U., 1966-67; staff fellow NIH, 1967-69; research assoc. dept. pharmacology Ind. U., Indpls., 1970-71, asst. prof., 1971-76, assoc. prof., 1976—, dir. grad. studies, 1977—, dir., 1978—. Author: (with S. F. Queener, V.B. Karb) The Pharmacological Basis of Nursing Practice, 1982. Mem. AAAS, Am. Diabetes Assn., Am. Fedn. Clin. Research, Am. Soc. Pharmacology and Exptl. Therapeutics, Endocrine Soc., Soc. Exptl. Biology and Medicine, Sigma Xi. Subspecialties: Pharmacology; Cellular pharmacology. Current work: Hormonal regulation of metabolism research. Office: Dept Pharmacology 635 Barnhill Dr Indianapolis IN 46223

CLARK, MELVILLE, JR., physicist, electrical and nuclear engineering company executive; b. Syracuse, N.Y., Dec. 19, 1921; s. Melville and Dorothy Drew (Speich) C. S.B., MIT, 1943, postgrad., 1943-44; postgrad., U. N.Mex., 1945-46, Princeton U., 1946; M.A., Harvard U., 1947, Ph.D., 1949. Mem. staff Radiation Lab. MIT, 1942-45; mem. staff Manhattan Dist. U. Calif., Los Alamos, 1945-46; physicist Brookhaven Nat. Lab., Upton, N.Y., 1949-53, Radiation Lab. U. Calif., Livermore, 1953-55; sr. engring. specialist Sylvania Electric Products, Waltham, Mass., 1962-64; sr. cons. scientist AVCO, Wilmington, Mass., 1964-67; sr. scientist NASA, Cambridge, Mass., 1967-70; sr. devel. engr. Thermo Electron, Waltham, Mass., 1970-73; sr. cons. engr. Combustion Engring., Windsor, Conn., 1973-81, sr. tech. strategist, 1973-81; pres. Melville Clark Assocs., Wayland, Mass., 1955—; assoc. prof. MIT, 1955-62. Author: (with Rose) Plasmas and Controlled Fusion, 1961, (with Hansen) Numerical Methods of Reactor Analysis, 1964; transl., editor: (with B. Daniel) Introduction to the Theory of Ionized Gases, 1960; contbr. articles to profl. jours. NRC fellow, 1946-49; Hercules Powder Co. fellow, 1946. Mem. Am. Phys. Soc, Am. Inst. Physics, AAAS, Acoustical Soc. Am., Fusion Power Assocs., IEEE, Boston Computer Soc., Sigma Xi. Subspecialties: Electrical engineering; Electronics. Current work: Fusion, musical acoustics, mathematical physics, numerical analysis, electronics. Patentee in field. Office: 8 Richard Rd Wayland MA 01778

CLARK, MERVIN LESLIE, educator, physician, clinical pharmacologist; b. Balt., May 18, 1921; s. Harry and Kate (Simons) C.; m. Lenore Meyers, Aug. 20, 1949; children: Lawrence, Ellen, Andrew, Kathryn. B.S. in Chemistry, Va. Poly. Inst., 1942; M.D., Northwestern U., 1948. Intern Wesley Meml. Hosp., Chgo., 1948-49; resident in medicine City Receiving Hosp., Detroit, 1949-51, VA Hosp., Oklahoma City, 1953-55; gen. practice medicine, mem. Exptl. Therapeutics Unit, U. Okla. Sch. Medicine, Norman, 1955-56; asst. prof. dept. medicine, 1956-62, asso. prof., 1962-69, prof., 1969—, adj. prof. dept. psychiatry and behavioral scis., 1979—, sect. head clin. pharmacology dept. medicine, 1970-75; prof. medicine Health Scis. Center; chief medicine, dep. supr. med. services Central State Griffin Meml. Hosp., Hayden H. Donahue Mental Health Inst., Norman, 1980—; cons., mem. staff VA Hosp., Oklahoma City, part-time, 1971—. Contbr. articles on pharmacology to profl. jours; mem. editorial bd.: Psychopharmacology Communications, 1977-81. Served to capt. U.S. Army, 1944-45, 1951-53. NIMH grantee, 1957-77; Okla. Dept. Mental Health, 1981. Fellow Am. Coll. Neuropsychopharmacology; mem. Central Soc. for Clin. Research, So. Soc. for Clin. Investigation, Am. Soc. for Pharmacology and Exptl. Therapeutics, Am. Soc. for Clin. Pharmacology and Therapeutics, N.Y. Acad. Scis., AAAS, Okla. County Med. Assn., Okla. Med. Assn., AMA, Cleveland County Med. Assn., Sigma Xi, Alpha Omega Alpha. Subspecialties: Internal medicine; Psychopharmacology. Current work: Schizophrenia, drug and other treatments, clinical evaluation and experimental design, pharmacokinetics mechanisms. Home: 1019 Mockingbird Ln Norman OK 73071 Office: PO Box 151 Norman OK 73070

CLARK, PAMELA ELIZABETH, planetary geoscientist, educator; b. Troy, N.Y., Apr. 26, 1951; s. Frederick Earl and Elizabeth Smyth C. B.A., St. Joseph Coll. for Women, 1973; Ph.D., U. Md., 1979. Lab. asst./instr. biochem./geochem. chromatography, organizer/editor newspaper, coordinator/biology subgroup dir./field studies investigator NSF project St. Joseph Coll. for Women, West Hartford, Conn., 1970-73; teaching asst. U. Md., College Park, 1973-74; research asst. lunar geochemistry geology Goddard Space Flight Center, Greenbelt, Md., 1974-79; geologist remote sensing sci. U.S. Geol. Survey Astrogeol. Br., Flagstaff, Ariz., 1977-78; resident research assoc. planetary geochemistry/geophysics Nat. Acad. Scis., Jet Propulsion Lab., NASA, Pasadena, Calif., 1980-82; asst. prof geoscis. Murray State U., Ky., 1982—; prin. investigator Planetary Geoscis. Program NASA, 1983—; cons. planetary radar NASA-Jet Propulsion Lab., 1983—. Conbtr. articles in field to profl. jours. Mem. Am. Geophys. Union, AAAS, Geol Soc. Am., Los Angeles Fedn. Scientists., Sigma Xi. Club: Sierra. Subspecialties: Remote sensing (geoscience); Planetology. Current work: Geochem. classification of planetary terrains, planetary radar data analysis, devel. correlation techniques for related remote sensing of different resolutions and wavelengths, devel. text on planetary applications remote sensing, synthesis geol. data for comparison major terrains of terrestrial planets.

CLARK, RAYMOND LOYD, plant pathologist; b. Tacoma, Jan. 23, 1935; s. James Earl and Anna Mary (Knauf) C.; m. Nancy Lee Claussen, Sept. 11, 1955; children: James Paul, Michael Loyd, Daniel Lee, Rebecca Jean. B.S., Wash. State U., 1957, Ph.D., 1961. Research asst. plant pathology dept. Wash. State U., 1957-61; research plant pathologist U.S. Dept. Agr., Agrl. Research Service, Prosser, Wash., 1961-65, Ames, Iowa, 1965—. Bd. dirs. Gilbert Community Sch. Dist., 1976—, pres. bd. dirs., 1981-82. Mem. Am. Phytopath. Soc., Iowa Acad. Sci., Sigma Xi. Lodge: Kiwanis. Subspecialties: Plant pathology; Plant virology. Current work: Host plant resistance to disease; plant germplasm, maintenance and utilization. Home: 3721 Dawes Dr Ames IA 50010 Office: Plant Introduction Sta Ames IA 50011

CLARK, RICHARD LEFORS, systems engineer, consultant; b. Aberdeen, S.D., Oct. 29, 1936; s. Robert Mongomery and Marion Shook C.; m. Barbara Louise Battersby, Mar. 28, 1980. B.S., Jackson State Coll., 1968, M.A., 1972; B.A., Pacific Western U., 1974, M.S., 1975, Ph.D., 1978. Field service technician Honeywell Co., Mpls., 1957-58; engr. quality assurance Martin Co., Cape Canaveral, Fla., 1958-59; reconnaissance systems engr. Gen. Dynamics, Rochester, N.Y., 1959-66; pvt. practice reconnaissance systems engr., San Diego, 1966-71; supr. Graco, Inc., Mpls., 1971-74; cons. systems engr., San Diego, 1975—; cons. Future Concepts, Winter Haven, Fla., 1981—;

lectr. World Trade Center, Los Angeles, 1983—; cons. Mut. U.F.O. Network, Sequin, Tex., 1981—; field reporter Psychical Research Found., Chapel Hill, N.C., 1980—. Author: Two Dimensions of the Time Domain, 1978; contbr. articles to profl. jours. Served with U.S. Army, 1954-57. Psychical Research Found. fellow, 1981; U.S. Psychotronic Assn. fellow, 1982. Mem. Am. Nuclear Soc., Am. Math. Soc., Math. Assn. Am., Soc. Indsl. and Applied Math., Am. Chem. Soc., AAAS, Mensa. Republican. Episcopalian. Subspecialties: Systems engineering; Theoretical physics. Current work: Theoretical and systems research in Maxwellian gravitational physics, as applied to deactivating and controlling nuclear devices and alternate energy systems devel. Inventor Fusion containment system. Home: 4813 Crown Point Dr P-3 San Diego CA 92109

CLARK, RONALD DUANE, chemistry educator; b. Hollywood, Calif., Nov. 21, 1938; s. Marvin Ansel and Elsie Susana (Appel) C.; m. Rosalind Estelle Proell, Sept. 9, 1967; children: Jennifer, Roger, Kenneth, Stephanie. B.S., UCLA, 1960; Ph.D., U. Calif.-Riverside, 1964. Research chemist Standard Oil Co. Ohio, Warrensville, 1965-69; prof. chemistry N.Mex. Highlands U., Las Vegas, 1969—, chmn. div. sci. and math., 1977—; dir. SWC Co., Inc., Palmdale, Calif. Author: Chemistry - The Science and the Scene, 1975; contbr. articles to profl. jours. Grantee NASA, 1974-75, NSF, 1979-81, Dept. Energy, 1980-82. Mem. Am. Chem. Soc. Republican. Subspecialties: Organic chemistry; Coal. Current work: Coal fly ash formation, stereochemistry, microcomputers in science education. Patentee in field. Home: 509 Dora Ceveste Las Vegas NM 87701 Office: N Mex Highlands U Las Vegas NM 87701

CLARK, ROY WHITE, chemistry educator; b. Oak Park, Ill., Nov. 11, 1930; s. Roy E. and Clara W. (Moore) C.; m. Suma Jane Maupin, Mar. 19, 1955; children: Stephen (dec.), Kathy, Samuel. B.S., Middle Tenn. State Coll., 1957; M.S., La. State U., 1959, Ph.D., 1966. Instr. Middle Tenn. State U., Murfreesboro, 1955-56, prof., 1963—; instr. La. Tech. U., Ruston, 1959-61, La. State U., Baton Rouge, 1962-63. Co-author: Concepts of General Chemistry, 1967. Mem. exec. com. Rutherford County Democratic Party, Murfreesboro, 1976—. Served with USAF, 1948-52. Mem. Am. Chem. Soc., Am. Assn. Physics Tchrs. (chmn. Tenn. sect. 1981-82), AAAS, AAUP, Tenn. Acad. Sci. Subspecialty: Physical chemistry. Current work: Applications of continuous wave n.m.r.; physics and chemistry of modern weapons. Home: 1315 Lakeshore Dr Murfreesboro TN 37130 Office: Dept Chemistry and Physics Middle Tenn State U Murfreesboro TN 37132

CLARK, RUDOLPH ERNEST, applied mathematics educator, biostatistics consultant; b. Springfield, Mass., Feb. 2, 1931; s. Rudolph Franklin and Nellie Francis (Peterson) C.; m. Renata Marie Einstein, Dec. 18, 1954; children: Diana Christine, Michael Scott. B.A., U. N.H., 1955; M.A., U. Iowa, 1958; Ph.D., 1958. Lic. psychologist, Mass. Psycho-physiologist U.S. Army Inst. Environ. Medicine, Natick, Mass., 1961-64, dir. human factors, 1964-66; assoc. prof. Northeastern U., 1966-68, Tufts U. Sch. Dental Medicine, 1968-75, prof., chmn. div. biostats., 1975—, dir. grad. edn. in oral biology, 1983; sci. dir. Biostats. Assocs., Inc., Northboro, Mass., 1970—; cons. Ctr. Tropical Disease, Lowell, Mass., 1981—, New Eng. Nuclear, Billerica, Mass., 1980—. Contbr. articles to sci. jours., chpts. to books. Mem. council on religious edn. 1st Parish Unitarian Ch., Framingham, Mass., 1970-74. Served to 1st lt. MSC AUS, 1958-61. U. N.H. scholar, 1950-55; named Disting. Mil. Grad. ROTC, U. N.H., 1955. Mem. Biometric Soc., Am. Statis. Assn., AAAS, Psychonomic Soc. Club: Appalachian Mountain (Boston). Subspecialties: Applied mathematics; Oral biology. Current work: Mathematics, psychology, oral biology. Home: 36 Alexander Rd Hopkinton MA 01748 Office: Tufts U 1 Kneeland St Boston MA 02111

CLARK, SANDRA HELEN BECKER, geologist, researcher; b. Kansas City, Mo., July 27, 1938; d. LuVern John and Mildred (File) Becker; m. Allen LeRoy Clark, Sept. 29, 1955 (div. 1977); children: Brett Harlan, Holly Lin. Student, Iowa State U., 1956-60; B.S., U. Idaho, 1963, M.S., 1964, Ph.D., 1968. Field and teaching asst. Coll. Mines, U. Idaho, 1963-66; geologist Cominco Am., Inc., Spokane, Wash., 1966-67, Alaska br. U.S. Geol. Survey, Menlo Park, Calif., 1967-72, staff geologist, office mineral resources, Washington, 1972-75, equal opportunity officer, Reston, Va., 1976-80, geologist, eastern mineral resources, 1980—; coordination staff mem. Dept. Interior Arctic Gas Systems Project, Washington, 1974-75; chmn. women geoscientists com. Am. Geol. Inst., 1979. Contbr. articles to profl. publs.; author maps. Fellow Geol. Soc. Am.; mem. Geol. Soc. Washington, Sigma Xi, Phi Kappa Phi. Subspecialty: Geology. Current work: Geology and resources of barite; Mississippi-Valley-type lead and zinc deposits; geology of Chugach Mountains, Alaska. Office: US Geol Survey Nat Center MS 954 Reston VA 22092

CLARK, THOMAS ALAN, astronomer, educator; b. Coalville, Leicestershire, Eng., Mar. 14, 1938; s. Walter Joseph and May (Neale) C.; m. Jean Dennis, Aug. 11, 1960; children; Gillian Anne, David Andrew. B.Sc. in Physics, U. Leeds, Eng., 1959, Ph.D. in Cosmic Ray Physics, 1963. NRC postdoctoral fellow U. Calgary, Alta., Can., 1962-64, sessional lectr., 1965, asst. prof., 1970-71, assoc. prof., 1971-81, prof., 1981—, co-dir. Rothney Astrophys. Obs., 1981—; Killam resident fellow, 1979; lectr. in physics, tutor Univ. Coll., London, 1965-69. Contbr. sci. articles to profl. jours. Fellow Inst. Physics, Explorers Club; mem. Can. Astron. Soc., Am. Astron. Soc., Can. Assn. Physicists. Anglican. Subspecialties: Infrared optical astronomy; Solar physics. Current work: Infra-red solar studies from balloon altitudes; stratospheric constituent measurement; infra-red astronomy; fourier transform spectroscopy; balloon astronomy. Office: Dept Physics U Calgary 2500 University Dr NW Calgary AB Canada T2N 1K4

CLARK, WILLIAM R., immunologist, educator; b. Detroit, Aug. 18, 1938; s. Russell William and Elma Iona (Hamilton) Wootton; m. Edith Lee Anne Clark, Oct. 31, 1966; m. Erica Robin Clark, July 29, 1981. B.S. in Zoology, UCLA, 1963; M.S. in Chemistry, U. Ill., 1965; Ph.D. in Biochemistry, U. Wash., 1968. Mem. faculty UCLA, 1970—, prof. immunology, 1975—, head, 1975—, asst. dir., 1976—. Author: Experimental Foundations of Modern Immunology, 1980. Recipient USPHS Career Devel. award, 1975-80. Mem. Am. Assn. Immunologists, Biophys. Soc., AAAS. Subspecialties: Transplantation; Immunobiology and immunology. Current work: Transplantation immunology. Office: Molecular Biology Inst UCLA Los Angeles CA 90024

CLARKE, GEORGE ALTON, chemist, educator and univ. adminstr.; b. N.Y.C., Apr. 4, 1933; s. Cecil M. and Linda M. (Dicks) C.; m. Jan Avery, July 16, 1966; children: Jill, Kristin. B.S., CCNY, 1955; Ph.D., Pa. State U., 1960. Research assoc. Columbia U., 1960-62; asst. prof. chemistry SUNY, Buffalo, 1962-68; assoc. prof. Drexel U., 1968-71; assoc. prof. chemistry U. Miami, Coral Gables, Fla., 1971—, assoc. dean, 1978—. Chmn. sub-com. and bd. dirs. Am. Heart Assn. Greater Miami, 1972—, bd. dirs., v.p. Fla. affiliate, 1976—. Mem. Am. Chem. Soc., Am. Phys. Soc. Subspecialties: Physical chemistry; Theoretical chemistry. Current work: Fluorescence of bio-molecules in complex enviornments (Micelles): a study of physical effects. Office: U Miami Office of Dean 234 Ashe Coral Gables FL 33124

CLARKE, JOHN, physicist, educator; b. Cambridge, Eng., Feb. 10, 1942; s. Victor Patrick and Ethel May (Blowers) C.; m. Grethe Pedersen, Sept. 15, 1979; 1 dau., Elizabeth Jane. B.A., U. Cambridge, 1964, M.A., 1968, Ph.D., 1968. Postdoctoral scholar U. Calif.-Berkeley, 1968-69, asst. prof. physics, 1969-71, assoc. prof., 1971-73, prof., 1973—. Contbr. articles to profl. jours. Research fellow Christ's Coll., Cambridge, 1968; Alfred P. Sloan Found. fellow, 1970; Adolph C. and Mary Sprague Miller Found. fellow, 1975; John Simon Guggenheim Meml. Found. fellow, 1977. Fellow AAAS; mem. Am. Inst. Physics. Subspecialties: Condensed matter physics; Low temperature physics. Current work: Superconducting devices based on Josephson tunneling; nonequilibrium effects in superconductors; noise in solids, especially at low temperatures; magnetotellurics. Office: Dept Physics U Calif Berkeley CA 94720

CLARKE, JOHN TERREL, astrophysicist, researcher; b. Chgo., Mar. 4, 1952; s. Terrel Edward and Catherine Evelyn (Carr) C. B.S., Denison U., 1974; M.A., Johns Hopkins U., 1978, Ph.D., 1980. Tchr. The Congl. Prep. Sch., Falls Church, Va., 1974-75; grad. teaching asst. Johns Hopkins U., 1975-77, grad. research asst., 1977-80; research physicist U. Calif., Berkeley, 1980—. Contbr. articles to profl. jours. Forbush fellow Physics Dept., Johns Hopkins U., 1980. Mem. Am. Astron. Soc., Am. Geophys. Union, Am. Phys. Soc. Subspecialties: Planetary science; Optical astronomy. Current work: Observational astrophys, engaged in UV planetary spectroscopy and studies of optical counterparts of celestial x-ray sources. Office: Space Scis Lab Univ Calif Berkeley CA 94720

CLARKE, LARRY DENMAN, aerospace company executive; b. Eng., 1925. Degree, Osgoode Hall Law Sch., 1949. Mem. staff DeHavilland Aircraft; founder, chmn., chief exec officer Spar Aerospace, Toronto, 1967—. Served with Royal Navy. Subspecialty: Aerospace engineering and technology. Office: Spar Aerospace Ltd Royal Bank Plaza Toronto ON Canada M5J 2J2

CLARKE, ROBERT FRANCIS, nuclear physics educator, consultant; b. Mpls., Mar. 20, 1915; s. Charles Patrick and Maurine Elizabeth (Clark) C.; m. Charlotte Adele Radwill, July 24, 1966; children: Robert, Carol, David. B.S., U. Fla., 1948; M.S., U. Ariz., 1971; grad., USAF Air Tactical Sch., Air Command and Staff Coll., Air War Coll., U.S. Army Command and Gen. Staff Coll., Indsl. Coll. Armed Forces. Meteorologist U.S. Weather Bur., Washington, 1940-42, 52-58; supervisory electronics engr. Dept. Army, Fort Huachuca, Ariz., 1956-58, nuclear physicist, 1958-62; aerospace engr. NASA, Lewis Research Center, Cleve., 1962-66; physicist optics Hughes Aircraft Co., Tuscon, Ariz., 1966-68; cons., dir. North Star Internat. Metals, Tucson, 1973-75, 79—; radiol. defense officer Fed. Emergency Mgmt. Agy., Tucson, 1980—. Staff officer sr. programs CAP, Tucson, Ariz., 1978—, Patriotic and Civic Coordinating Council Tucson. Served to col. USAF, 1942-75; PTO. Recipient Grad. Scholarship NSF, 1969; Disting. and Outstanding Service awards Mil. Order World Wars, 1982; Scholarships U. Chgo., 1932, U. Minn., 1934. Mem. IEEE, AIAA, AAUP, Am. Meteorol. Soc. (pres. So. Ariz. chpt. 1982-83), Am. Nuclear Soc., N.Y. Acad. Scis., Ariz.-Nev. Acad. Scis., Assn. Former Intelligence Officers, Ret. Officers Assn. (n.p. Tucson chpt.), Assn. U.S. Army (pres. 1980-81), Mil. Order World Wars (dir. 1981-83), VFW (community services chmn. 1982). Lodges: Kiwanis; Elks; Odd Fellows. Subspecialties: Nuclear physics. Current work: Consultant, physical processes involved in metallic ore extraction and processing using new technology, conducting public education programs relative to nuclear fission and fusion power generation. Home: 5846 E S Wilshire Dr Tucson AZ 85711 Office: North Star Internat Metals Inc 35 N Camino Espanol Tucson AZ 85716

CLARKE, STEVEN GERARD, biochemist, educator; b. Los Angeles, Nov. 19, 1949; s. Gerard Theodore and Ann (Rose) C.; m. Catherine Freitag, Dec. 19, 1982. B.A., Pomona Coll., 1970; Ph.D., Harvard U., 1976. Miller fellow dept. biochemistry U. Calif.-Berkeley, 1976-78; asst. prof. chemistry and biochemistry UCLA, 1978-83, assoc. prof., 1983—. A.E. Sloan Found. fellow, 1982-84. Mem. Am. Soc. Biol. Chemists. Subspecialty: Biochemistry (biology). Current work: Protein chemistry, role of protein methylation reactions in cell function, protein-lipid interactions. Office: Dept Chemistry and Biochemistry UCLA Los Angeles CA 90024

CLAUS, RICHARD O., elec. engr., educator, cons.; b. Balt., May 29, 1951; s. Otto R. and Anita L. (Richter) C. B.E.S., Johns Hopkins U., 1973, Ph.D., 1977. Asst. prof. elec. engring. Va. Poly. Inst. and State U., Blacksburg, 1977-81, assoc. prof., 1981—; cons. to pvt. industry. Contbr. articles to profl. jours. Recipient C. Holmes MacDonald Outstanding Teaching award Eta Kappa Nu, 1982. Mem. IEEE (Outstanding Br. Counselor award 1979, chmn. Va. Mountain sect. 1982-83), Optical Soc. Am., Soc. Photo-Optical Instrumentation Engrs., Am. Soc. Engring. Edn. Club: Blind Buckers Saddle (Blacksburg). Subspecialties: Electrical engineering; Acoustics. Current work: Applied optics, ultrasonics; nondestructive evaluation; acoustooptics research, education and consulting. Patentee in field. Office: Elec Engring Dept Va Poly Inst and State U Blacksburg Va 24061

CLAUSING, ARTHUR MARVIN, engring. educator, cons., researcher; b. Palatine, Ill., Aug. 17, 1936; s. Arthur Henry Fred and Emma Marie Sophia (Opfer) C.; m. Willa Louise Spence, Dec. 19, 1964; children: Erin, Kimberly. B.S., Valparaiso (Ind.) U., 1958; M.S., U. Ill., Urbana, 1960, Ph.D., 1963. Assoc. prof. mech. engring., dir. solar energy program U. Ill., Urbana, 1979—; cons. Electric Power Research Inst., Office Energy Related Inventions, Machinenfabrik Angsburg-Nurnberg Neue Technologies, Ill. Power Co. Contbr. articles to profl. jours. Recipient Standard Oil award for heat transfer lab. devel., 1968. Mem. ASME, Am. Solar Energy Soc., Internat. Solar Energy Soc., ASHRAE. Lutheran. Club: Ill. Track. Subspecialties: Mechanical engineering; Solar energy. Current work: Heat transfer, cryogenic modeling, solar energy, numerical methods, performance monitoring. Home: 613 Hessel Blvd Champaign IL 61820 Office: 1206 W Green St Urbana IL 61801

CLAXTON, LARRY DAVIS, genetic toxicologist; b. Chattanooga, June 17, 1946; s. Carl Woods and Margaret Jane (Davis) C.; m. Betty Reed, May 29, 1971; children: Meredith, Matthew. B.S., Middle Tenn. State U., 1967; M.S., Memphis State U., 1971; Ph.D., N.C. State U., 1980. Asst. Oak Ridge (Tenn.) Nat. Lab., 1971-72; biologist Nat. Inst. Environ. Health Sci., Mutagenesis Br., Research Triangle Park, N.C., 1972-77; research biologist Genetic Toxicology Div., EPA, Research Triangle Park, 1977—. Editorial bd.: Environ. Mutagenesis; Contbr. articles to profl. jours. Recipient Bronze medal EPA, 1980. Mem. Environ. Mutagen Soc., Genotoxicity and Environ. Mutagen Soc. (pres. 1983-84), AAAS, Genetic Soc. Am., Am. Soc. Risk Analysis, Genetic Toxicology Assn., Beta Beta Beta, Gamma Beta Phi. Mem. Ch. of Christ. Club: Toastmasters. Subspecialties: Environmental toxicology; Genetics and genetic engineering (biology). Current work: Development genetic systems for detecting mutagenic/carcinogenic effects of environmental substances, their mechanism of action and statistical methods. Home: 5121 Huntingdon Dr Raleigh NC 27606 Office: EPA MD 68 Research Triangle Park NC 27711

CLAY, CHARLES GEORGE, JR., computer scientist; b. Kingston, N.Y., Oct. 8, 1954; s. Charles George and Evelyn Bernice (Miller) C. B.Engring., Stevens Inst. Tech., 1976; M.S. in Computer Scis, 1976. Student assoc. IBM, Kingston, 1974; computer systems engr. Space div. Gen. Electric Co., King of Prussia, Pa., 1976-81, Software project engr., Lanham, Md., 1981—. Recipient Batchelor award Stevens Inst. Tech., 1976; Gen. Electric Profl. Recognition award, 1979, 83. Mem. Assn. for Computing Machinery, SIGACT, SIGARCH, SIGPLAN, SIGNUM. Subspecialties: Software engineering; Mathematical software. Current work: Image processing, numeric processing methods. Bulk image processing software engineering management. Home: 8201 Mandan Ct Greenbelt MD 20770 Office: 4701 Forbes Blvd Lanham MD 20706

CLAY, GEORGE A, pharmacologist, pharmaceutical company executive; b. Cambridge, Mass., June 14, 1930; s. George W. and Mary A. (Reynolds) C.; m. Hollie F. Hicks, May 15, 1965; children: Karen B., G. Thomas, Lauren R., Sarah E. A.B., Dartmouth Coll., 1961; M.A., Boston U., 1964, Ph.D., 1968. Staff fellow NIH, Bethesda, Md., 1967-70; asst. prof. pharmacology Bowman Gray Sch. Medicine, Wake Forest (N.C.) U., 1970-72; research investigator G.D. Searle & Co., Skokie, Ill, 1972-74, head sect. central nervous system pharmacology, 1974-82, asst. dir. gastroenterology clin. research, 1982—. Mem. Am. Soc. Pharmacology and Exptl. Therapeutics, AAAS, Sigma Xi. Republican. Roman Catholic. Subspecialties: Gastroenterology; Neuropharmacology. Current work: Clinical research in gastroenterology, neuropharmacology; conduct clinical research trials in gastroenterology. Office: 4901 Searle Pkwy Skokie IL 60077

CLAYMAN, LEWIS, oral and maxillofacial surgeon, educator; b. Bklyn., Sept. 3, 1947; s. Irwin and Esther (Small) C.; m. Minou Rouhani, July 3, 1976; children: Eric Harald, Lara Marie. B.S., Bklyn. Coll., 1968; D.M.D., Harvard Sch. Dental Medicine, 1972; M.S., Wayne State U., 1978; M.D., Hahnemann Med. Coll., 1978. Diplomate: Am. Bd. Oral and Maxillofacial Surgery. Resident oral and maxillofacial surgery Sinai Hosp., Detroit, 1972-76; resident anesthesiology U. Pa., Phila., 1978-79; asst. prof. surgery, chief div. oral and maxillofacial surgery Marshall U. Sch. Medicine, Huntington, W. Va., 1979—; examiner sect. on Oral and Maxillofacial Surgery, Oral Surgery Licensing Sect. W. Va. Bd. Dental Examiners, 1981-82. Contbr. articles to profl. publs. Huntington Clin. Found. grantee, 1981-83. Fellow Am. Assn. Oral and Maxillofacial Surgeons, Am. Dental Soc. of Anesthesiology, Internat. Assn. Oral Surgeons; mem. Chalmers Lyons Acad. Oral Surgeons, N.Y. Acad. Scis. Democrat. Jewish. Club: Harvard. Lodge: Rotary. Subspecialties: Oral and maxillofacial surgery; Laser medicine. Current work: Application of lasers in bone surgery, microsurgical transfers of bone. Home: 2217 S Inwood Dr Huntington WV 25701 Office: 2828 1st Ave Huntington WV 25701

CLAYTON, DAVID LAWRENCE, laboratory director, ocean engineer; b. Santa Monica, Calif., Sept. 23, 1952; s. Robert Ashton and Marion (Mackie) C.; m. Theresa Garvey, Aug. 21, 1982. B.S., U. So. Calif., 1975; M.S. in Ocean Engring., Fla. Inst. Tech., 1977. Technician mech. Harbor Br. Found., Inc., Fort Pierce, Fla., 1976-77, mech. engr., 1977-79, program mgr., 1979-80, research and devel. mgr., 1980-81; dir. Link Engring. Lab., 1981—. Dist. commr. Indian River dist. Boy Scouts Am., 1977-79; dist. chmn. Gulf Stream council, 1979—, mem. exec. com., 1979—. Served with USN, 1970-73. Recipient Dist. award of merit Gulf Stream council Boy Scouts Am., 1981. Mem. AIAA, Soc. Naval Architects and Marine Engrs. (com. mem. 1980—), ASME (com. mem. 1982-87), IEEE, AAAS, N.Y. Acad. Scis., Sigma Xi. Clubs: Pelican Yacht (Fort Pierce, Fla.); Westside Tennis (Vero Beach, Fla.) (1977—). Subspecialties: Ocean engineering; Computer-aided design. Current work: Submersible technology, oceanographic instrumentation, remotely-operated vehicles (design and operation), computer-aided design, engineering management. Home: 4351 Second Sq SW Vero Beach FL 32960 Office: Harbor Branch Found Inc Rural Route 1 Box 196 Fort Pierce FL 33450

CLAYTON, PAULA JEAN, psychiatry educator; b. St. Louis, Dec. 1, 1934; m., 1958; 3 children. B.S., U. Mich., 1956; M.D., Washington U., St. Louis, 1960. Intern St. Luke's Hosp., St. Louis, 1960-61; asst. resident and chief resident psychiatry Barnes and Renard Hosp., St. Louis, 1961-65; from instr. to assoc. prof. psychiatry Sch. Medicine Washington U., 1965-74, prof., 1974—; cons. Malcolm Bliss Mental Health Ctr., St. Louis, 1972—, dir. tng. and research, 1975; dir. psychiat. inpatient service Barnes and Renard Hosp., 1975-81; prof., head dept. psychiatry U. Minn. Med. Sch., Mpls., 1981—. Fellow Am. Psychiat. Assn.; mem. Psychiat. Research Soc., Assn. Research in Nervous and Mental Diseases, Am. Psychopath. Assn., Soc. Biol. Psychiatry. Subspecialty: Psychiatry. Office: Dept Psychiatry U Minn Minneapolis MN 55455

CLAYTON, RODERICK KEENER, biophysicist, educator; b. Tallinn, Estonia, Mar. 29, 1922; s. John Heber and Helena (Mullerstein) C.; m. Betty Jean Compton, June 28, 1944; children—Roderick Dale, Ann Keener. B.S., Calif. Inst. Tech., 1947, Ph.D., 1951. Merck postdoctoral fellow Stanford, 1951-52; asso. prof. physics U.S. Naval Postgrad. Sch., Monterey, Calif., 1952-57; NSF sr. postdoctoral fellow Oxford, Eng., Trondheim, Norway, 1957-58; sr. biophysicist Oak Ridge Nat. Lab., 1958-62; vis. prof. microbiology Dartmouth, 1962-63; sr. investigator C.F. Kettering Research Lab., Yellow Springs, Ohio, 1963-66; prof. biology, biophysics Cornell U., Ithaca, N.Y., 1966—; Instr. Marine Biol. Lab., Woods Hole, Mass. Author: Molecular Physics in Photosynthesis, 1965, Light and Living Matter, 1970, Photosynthesis: Physical Mechanisms and Chemical Patterns, 1981; Contbr. articles to profl. jours. Pres. Carmel P.T.A., 1955-56, Willowbrook P.T.A., Oak Ridge, 1960-61. Served to 1st lt. USAAF, 1942-46. Decorated Air medal with oak leaf cluster; John Simon Guggenheim fellow, 1973-74, 80-81. Fellow Am. Acad. Arts and scis., A.A.A.S.; mem. Am. Soc. Biol. Chemists, Biophys. Soc., Nat. Acad. Scis., Soc. Gen. Physiologists, Am. Soc. Plant Physiologists, Sigma Xi. Subspecialties: Plant physiology (agriculture); Biophysics (physics). Home: 111 Brandon Pl Ithaca NY 14850

CLEELAND, CHARLES SAMUEL, clinical psychologist, researcher, educator; b. Jacksonville, Ill., Sept. 23, 1938; s. Joseph C. and Charlotte S. (Swanson) C.; m. Lynne Mary Schultheiss, Dec. 31, 1981; 1 dau., Sarah. B.A., Wesleyan U., 1960; Ph.D., Washington U., St. Louis, 1966. Instr. U. Wis. Med. Sch.-Madison, 1966-67, asst. prof. neurology, 1967-72, assoc. prof. neurology, 1972—. Contbr. articles to profl. jours. NIH grantee, 1980—; Nat. Cancer Inst. grantee, 1979—; Robert Wood Johnson grantee, 1982—. Fellow Am. Psychol. Asns.; mem. Am. Acad. Neurology. Subspecialties: Behavioral psychology; Neuropsychology. Current work: Pain research and treatment, behavioral treatment in chronic disease, neuropsychology.

CLELAND, CHARLES CARR, psychologist, educator; b. Murphysboro, Ill., May 15, 1924; s. Homer W. and Stella (Carr) C.; m. Betty Lou Woodburn, July 18, 1948. B.S., So. Ill. U., 1950, M.S., 1951; Ph.D., U. Tex.-Austin, 1957. Lic. psychologist, Tex. Chief psychologist Lincoln (Ill.) State Sch., 1956-57, Austin State Sch., 1957-59; supt. Abilene (Tex.) State Sch., 1959-63; prof. spl. edn. and ednl. psychology U. Tex.-Austin, 1963—. Author: Mental Retardation, 1969, 2d edit., 1978, Profound Retardation, 1979, Exceptionalities, 1982; contbr. articles to profl. jours. Bd. dirs. Child Guidance Center, Austin, 1966-67. Served with USAAF, 1943-46; PTO. Recipient Disting. Psychologist award Tex. Psychol. Assn., 1980; Edn. award Am. Assn. Mental Deficiency, 1978. Fellow AAAS, Am. Psychol. Assn., Am. Assn. for Mental Deficiency (v.p. psychology div. 1973); mem. Tex.

Psychol. Assn. (pres. 1962-63). Republican. Presbyterian. Club: Headliners (Austin). Subspecialty: Developmental psychology. Current work: Creativity, origins of; non-verbal communication; mental retardation (profound). Patentee in field. Home: 3427 Monte Vista Austin TX 78731 Office: Univ Texas EDB408A Austin TX 78712

CLELAND, ROBERT ERSKINE, plant physiologist, educator; b. Balt., Apr. 30, 1932; s. Ralph Erskine and Elisabeth (Shoyer) C.; m. Mary Love, Sept. 2, 1957; children: Thomas Andrew, Alison Anne. B.A. in Chemistry, Oberlin Coll, 1953; Ph.D. in Biochemistry, Calif. Inst. Tech., 1957. NIH postdoctoral fellow U. Lund, Sweden, 1957-58, King's Coll., London, 1958-59; asst. prof. U. Calif.-Berkeley, 1959-64; assoc. prof. botany U. Wash, 1964-68, prof., 1968—. Research numerous publs. in field. Guggenheim fellow, 1967-68; NSF sr. postdoctoral fellow, 1975-76. Fellow AAAS; mem. Am. Soc. Plant Physiologists (sec. 1971-73, pres. 1974-75), Bot. Soc. Am., Japanese Soc. Plant Physiologists. Subspecialties: Plant physiology (biology); Plant growth. Current work: Study of mechanism of control of plant growth, especially by hormones; mechanical properties of plant cell walls. Home: 7834 56th Pl NE Seattle WA 98115 Office: U Wash Dept Boty KB-15 Seattle WA 98195

CLELAND, W(ILLIAM) WALLACE, biochemistry educator; b. Balt., Jan. 6, 1930; s. Ralph E. and Elizabeth P. (Shoyer) C.; m. Joan K. Hookanson, June 18, 1967; children: Elsa E., Erica E. A.B. summa cum laude, Oberlin Coll., 1950; M.S., U.Wis.-Madison, 1953, Ph.D., 1955. Postdoctoral fellow U. Chgo., 1957-59; asst. prof. U. Wis.-Madison, 1959-62, assoc. prof., 1962-66, prof., 1966, M. J. Johnson prof. biochemistry, 1978, Steenbock prof. chem. sci., 1982. Contbr. numerous articles on biochemistry to profl. jours. Served with M.C. U.S. Army, 1955-57. NIH, NSF grantee, 1960—. Mem. Am. Acad. Arts and Scis., Am. Soc. Biol. Chemists, Am. Chem. Soc., Sigma Xi. Subspecialty: Biochemistry. Current work: The use of enzyme kinetic studies to determine enzyme Mechanisms. Office: U Wis Dept Biochemistry Madison WI 53706

CLEM, JOHN R., physics educator, consultant; b. Waukegan, Ill., Apr. 24, 1938; s. Gilbert D. and Bernelda M. (Moyer) C.; m. Judith A. Paulsen, Aug. 27, 1960; children: Paul G., Jean A. B.S., U. Ill.-Urbana, 1960, M.S., 1962, Ph.D., 1965. Research assoc. U. Md., College Park, 1965-66; vis. research fellow Tech. U., Munich, Germany, 1966-67; asst. prof. physics Iowa State U., Ames, 1967-70, assoc. prof., 1970-75, prof., 1975—, chmn. dept., 1982—; cons. Argonne (Ill.) Nat. Lab., 1971-76, Brookhaven Nat. Lab., Upton, N.Y., 1980-81, Oak Ridge (Tenn.) Nat. Lab., 1981, IBM Watson Research Ctr., Yorktown Heights, N.Y., 1982; vis. staff mem. Los Alamos (N.Mex.) Sci. Lab., 1971—; guest prof. U. Tuebingen, West Germany, 1978. Contbr. articles in field to profl. jours. Fulbright sr. research fellow, 1974-75. Fellow Am. Phys. Soc., AAUP, AAAS, Sigma Xi, Tau Beta Pi, Phi Kappa Phi. Demcrat. Presbyterian. Subspecialties: Condensed matter physics; Low temperature physics. Current work: Theoretical research on the electrical and magnetic properties of superconductors, especially the electrodynamic behavior of current-carrying superconductors subjected to magnetic fields. Inventor superconducting magnetic shielding apparatus and method, 1982. Home: 2217 Ferndale Ave Ames IA 50010 Office: Dept Physics Iowa State U Ames IA 50011

CLEMENS, DONALD FAULL, chemistry educator; b. Dorer, Ohio, Aug. 14, 1929; s. John William and Ruth A (Faull) C.; m. Martha Kay Lemmon, July 2, 1950; children: Richard D., Nancy K., Barbara J., Rebecca S., Margaret A. B.S., Fla. So. Coll., Lakeland, 1961; M.S., U. Fla.-Gainesville, 1963, Ph.D., 1965. Assoc. prof. East Carolina U., Greenville, 1965-69, prof., 1969—; vis. Whitehurst Assocs. Inc., New Bern, N.C., 1981—. Bd. dirs. Wesley Found., Greenville. N.C. Energy Inst. grantee, 1980. Fellow Am. Inst. Chemists; mem. N.C. Inst. Chemists (sec. 1978—), Am. Chem. Soc., N.C. Acad. Sci., Sigma Xi. Methodist. Lodges: Kiwanis; Masons. Subspecialties: Inorganic chemistry; Fuels and sources. Current work: Development of improved bright dip baths for aluminum; sugar acid sequestering agents for calcium and iron. Patentee peat fuel slurry, products and processes from sweet potatoes, phesphoric acid brightening reagent for aluminum. Office: East Carolina U Dept Chemistry Greenville NC 27834 Home: 1701 Sulgrave Rd Greenville NC 27834

CLEMENS, JAMES ALLEN, neuroendocrinologist; b. Windsor, Pa., Feb. 4, 1941; s. Allen Victor and Helen M. (Smeltzer) C.; m. Clare Hess, Sept. 25, 1964; children: James C., Amy J. B.S., Pa. State U., 1963, M.S., 1965; Ph.D. in Physiology, Mich. State U., 1968, NIH postdoctoral fellow dept. anatomy UCLA, 1968-69; research adv. Eli Lilly & Co., Indpls., 1969—. Contbr. numerous articles to profl. publs. Mem. Endocrine Soc., Soc. Neurosci., Internat. Soc. Neuroendocrinology, Am. Soc. Pharmacology and Exptl. Therapeutics, Sigma Xi. Subspecialties: Neuropharmacology; Neurobiology. Current work: Neurophysiology; effects of drugs on brain in vivo electrochemistry; endocrinology. Patentee in field. Office: Lilly Research Labs Eli Lilly & Co 307 E McCarty St Indianapolis IN 46285

CLEMENS, WILLIAM ALVIN, paleontology educator, curator; b. Berkeley, Calif., May 15, 1932; m., 1955; 4 children. B.A., U. Calif.-Berkeley, 1954, Ph.D. in Paleontology, 1960. From asst. prof. to assoc. prof. zoology U. Kans., 1961-67; also assoc. curator higher vertebrate fossils; assoc. prof. U. Calif.-Berkeley, 1967-71, prof. paleontology, 1971—, also cur. mammals; vis. prof. Miller Inst., 1982—. NSF fellow, 1960-61, 68-69; Guggenheim fellow, 1974; Alexander von Humboldt fellow, 1978-79. Mem. Soc. Systematic Zoology, Soc. Vertebrate Paleontology, Geol. Soc. Am., Paleontol. Assn., Zool. Soc. London. Subspecialty: Paleontology. Office: U Calif Berkeley CA 94720

CLEMENS, WILLIAM JENKINS, psychology educator, researcher; b. Leesburg, Va., Feb. 11, 1947; s. John William and Mary Morton (Riddle) C.; m.; children: Carolyn, Jennifer. A.B., U. N.C., 1968; Ph.D., U. Tenn., 1972. Asst. prof. psychology Coll. of Cape Breton, Sydney, N.S., Can., 1971-77, assoc. prof. psychology, 1978—, chmn. dept. psychology, 1979—; asst. research psychologist, dept. psychiatry UCLA, 1977; vis. research fellow U. Hull, Eng., 1977-78; mem. exec. N.S. Confedn. Univ. Faculty Assns., 1975-77, 81-83. Contbr. articles, abstracts in field to publs.; editorial cons. publs. in field. Grantee St. Francis Xavier U., 1971-74, Nat. Research Council Can., 1975-79, Natural Sci. and Engring. Research Council, 1979—. Mem. Am. Psychol. Assn., Can. Psychol. Assn., AAAS, Soc. for Psychophysiol. Research, N.Y. Acad. Scis. Episcopalian. Subspecialties: Learning; Physiological psychology. Current work: Autonomic self-control, biofeedback, visceral perception. Office: Univ Coll of Cape Breton PO Box 5300 Sydney NS Canada B1P 6L2

CLEMENTE, CARMINE DOMENIC, anatomist; b. Penns Grove, N.J., Apr. 29, 1928; s. Ermanno and Caroline (Friozzi) C.; m. Juliette Vance, Sept. 19, 1968. A.B., U. Pa., 1948, M.S., 1950, Ph.D., 1952; postdoctoral fellow, U. London, 1953-54. Asst. instr. anatomy U. Pa., 1950-52; faculty U. Calif. at Los Angeles, 1952—, prof., 1963—, chmn. dept. anatomy, 1963-73, dir. brain research inst., 1976—; hon. research asso. Univ. Coll., U. London, 1953-54; cons. Sepulveda VA Hosp., NIH; mem. med. adv. panel Bank Am.-Giannini Found.; chmn. sci. adv. com., mem. bd. dirs. Nat. Paraplegia Found. Author: Aggression and Defense: Neurol Mechanisms and Social Patterns, 1967, Physiological Correlates of Dreaming, 1967, Sleep and the Maturing Nervous System, 1972, Anatomy, An Atlas of the Human Body, 1975, 2d edit., 1981; editor: Gray's Anatomy, 1973—, also Exptl. Neurology; asso. editor: Neurol. Research; contbr. articles to sci. jours. Recipient award for merit in sci. Nat. Paraplegia Found., 1973. Mem. Pavlovian Soc. N.Am. (Ann. award 1968, pres. 1972), Brain Research Inst. (dir. 1976—), Am. Physiol. Soc., Am. Assn. Anatomists (v.p. 1970-72, pres. 1976-77), Am. Acad. Neurology, Am. Acad. Cerebral Palsy, Am. Neurol. Assn., AMA-Assn. Am. Med. Colls. (exec. com. 1978—, disting. service mem. 1982), Council Acad. Socs. (adminstrv. bd. 1973-81, chmn. 1979-80), Assn. Anatomy Chairmen (pres. 1972), Biol. Stain Commn., Inst. Medicine of Nat. Acad. Scis., Internat. Brain Research Orgn., Med. Research Assn. Calif. (dir. 1976—), N.Y. Acad. Sci., Nat. Bd. Med. Examiners, Nat. Acad. Sci. (mem. com. neuropathology, BEAR coms.), Japan Soc. Promotion of Sci. (Research award 1978), Sigma Xi. Democrat. Subspecialties: Anatomy and embryology; Regeneration. Current work: Regeneration of nerve fibers in brain and spinal cord; brain mechanisms related to wakefulness and sleep; neuroanatomical substrates related to behavior. Home: 11737 Bellagio Rd Los Angeles CA 90049

CLEMENTI, ENRICO, theoretical chemist; b. Trento, Italy, Nov. 19, 1931; s. Ambrogio and Laura (Marzari) C.; m. Hildegard Cornelius, 1956 (div. 1972). Ph.D., U. Pavia, Italy, 1954; Postdoctal student, Politecnic Inst. Milano, Fla. State U., Tallahassee, U. Calif., Berkeley, U. Chgo. Mem. staff IBM Research, San Jose, Calif., 1961-65, mgr. research, 1965-69, IBM fellow, 1961-74; mgr. Calcolo Chimico-Montedison, Italy, 1974-79; IBM fellow, mgr. Dept. D55, tech. advisor sci., engring. computation, Poughkeepsie, N.Y., 1979—. Contbr. numerous articles to sci. jours. Mem. Am. Phys. Soc., Am. Chem. Soc., N.Y. Acad. Sci. Subspecialties: Theoretical chemistry; Biophysical chemistry. Current work: Computational methods to solve biophysical and biochemical problems; use of quantum theory, statistical mechanics, fluid dynamics. Home: 104 Beechwood Ave Poughkeepsie NY 12601 Office: IBM PO Box 390 Bldg 701 Dept D5 Poughkeepsie NY 12602

CLEMENTS, GREGORY LELAND, physics educator, microcomputer programmer, astronomer; b. Lincoln, Nebr., Apr. 5, 1949; s. Dwight L. and Marjory R. (Horstman) C.; m. Pamela A. Clements, Jan. 5, 1974; children: Christina, James, John, Brian. B.A. in Physics, U. Iowa, 1971, M.S. in Astronomy, 1976, Ph.D. in Physics, 1978. Asst. prof. Dickinson Coll., 1978-82; systems mgr. Softec Inc., Iowa City, 1982-83; asst. prof. Midland Lutheran Coll., Fremont, Nebr., 1983—. Mem. Am. Assn. Physics Tchrs. Subspecialties: Optical astronomy; Software engineering. Current work: College teaching. Home: 749 N Clarkson Fremont NE 68025 Office: Midland Coll Fremont NE 68025

CLEMENTS, JOHN ALLEN, physiologist; b. Auburn, N.Y., Mar. 16, 1923; s. Harry Vernon and May (Porter) C.; m. Margot Sloan Power, Nov. 19, 1949; children—Christine, Carolyn. M.D., Cornell U., 1947. Research asst. dept. physiology Cornell U. Med. Coll. N.Y., 1947-49; commd. 1st lt. U.S. Army, 1941, advanced through grades to capt., 1951; asst. chief clin. investigation br. (Army Chem. Center), 1951-61; asso. research physiologist U. Calif. at San Francisco, 1961-64, prof. pediatrics, 1964—, mem. staff, 1961—; career investigator Am. Heart Assn., 1964—; mem. group in biophysics and med. physics U. Calif. at Berkeley, 1969—; cons. Surgeon Gen. USPHS, 1964-68, Surgeon Gen. U.S. Army, 1972-79; sci. counselor Nat. Heart and Lung Inst., 1972-75; Bowditch lectr. Am. Physiol. Soc., 1961; 2d ann. lectr. Neonatal Soc., London, 1965; Distinguished lectr. Can. Soc. Clin. Investigation, 1973. Mem. editorial bd.: Jour. Applied Physiology, 1961-65, Am. Jour. Physiology, 1965-72, Physiol. Reviews, 1973—, Jour. Developmental Physiology, 1979—; assoc. editor: Am. Rev. Respiratory Diseases, 1973-79; chmn. publs. policy com.: Am. Thoracic Soc., 1982—. Recipient Dept. Army Research and Devel. Achievement award, 1961; Modern Medicine Distinguished Achievement award, 1973; Howard Taylor Ricketts medal and award U. Chgo., 1975; Mellon award U. Pitts., 1976; Calif. medal Am. Lung Assn. of Calif., 1981; Trudeau medal Am. Lung Assn., 1982; Internat. award Gairdner Found., 1983. Hon. fellow Am. Coll. Chest Physicians; mem. N.Y. Acad. Scis., Western Assn. Physicians, Western Soc. Clin. Research, Perinatal Research Soc. (councillor 1973-75), Nat. Acad. Scis., Am. Lung Assn. (hon., life). Subspecialties: Physiology (medicine); Neonatology. Current work: Pulmonary surfactant respiratory distress syndrome. Office: U Calif Sch Medicine Cardiovascular Research Inst 3d and Parnassus Ave San Francisco CA 94143

CLEMMER, EDWARD JOSEPH, psychologist, psycholinguist; b. Ft. Wayne, Ind., June 28, 1948; s. Benjamin Othmar and Rita Cecilia (Weaver) C.; m. Kathleen Ann Herber, Dec. 29, 1971; children: Kenneth Benjamin, James Anthony, Andrew John, Stephen Joseph. Student, Loras Coll., Dubuque, Iowa, 1966-67; B.A., St. Francis Coll., Ft. Wayne, 1970; postgrad., New Sch. Social Research, N.Y.C., 1970-71; M.S., St. Louis U., 1973, Ph.D., 1975. Teaching asst. Parks Coll., Cahokia, Ill., 1975; asst. prof. orthodontics St. Louis U., 1976-80, asst. prof. psychology, 1976-80, Met. Coll., St. Louis, 1979-80, SUNY-Oswego, 1980-81, Ind. U.-Purdue U.-Ft. Wayne, 1981—. Contbr. in field. Beaumont Fund grantee, 1979; IPFW grantee, 1982. Mem. Am. Psychol. Assn., Midwestern Psychol. Assn., Eastern Psychol. Assn. Roman Catholic. Subspecialties: Cognition; Psycholinguistics. Current work: The temporal analysis of language performance; cognitive processes in schizophrenic speech, oral reading skills and rhetoric in young children, dramatists, radio and television broadcasters. Home: 3933 Hedwig Dr Fort Wayne IN 46815 Office: Indiana University Purdue University 2101 Coliseum Blvd E Fort Wayne IN 46805

CLENDENIN, JAMES EDWIN, physicist, engr.; b. Paris, Tenn., June 10, 1939; s. Lane Edwin and Margurite Helen (Odom) C.; m. Judith P., Dec. 24, 1962; 1 dau., Anne Elizabeth. B.E.E., U. Va., 1962; Ph.D. in Physics, Columbia U., 1975. Research asso. in physics Yale U., New Haven, 1975-78; physicist-engr. Stanford (Calif.) Linear Accelerator Center, 1979—. Served to lt. USN, 1962-66. Mem. Am. Phys. Soc., IEEE. Subspecialties: Particle physics; Nuclear physics. Current work: High energy physics, accelerator physics, atomic physics. Office: SLAC Bin 12 PO Box 4349 Stanford CA 94305

CLEVELAND, SIDNEY EARL, psychologist, clinical psychology educator; b. Boston, Jan. 22, 1919; s. Herbert C. and Edith (Willey) C.; m. Marjorie Spacht, Nov. 27, 1942; children: John A., Carol T., Mark E., Sarah D. A.B., Brown U., 1941; M.A., U. Nebr., 1942; Ph.D., U. Mich., 1950. Lic. psychologist, Tex. Staff psychologist VA Med. Center, Houston, 1950-57, asst. chief psychologist, 1957-62, chief psychology service, 1962—; clin. prof. psychology Baylor Coll. Medicine, 1957—; pvt. practice psychology, Houston, 1961—; cons. Houston Police Dept., 1967-68, U.S. Surgeon Gen., Washington, 1965, NIMH, 1962, NYU Postgrad. Sch. Medicine, 1966. Co-author: Body Image and Personality, 1958, 68; contbr. numerous articles to profl. jours. Served to comdr. USNR, 1942-46. Fellow Am. Psychol. Assn.; mem. Assn. VA Chief Psychologists, Houston Psychol. Assn. (pres. 1955-56), Southwest Psychol. Assn. Subspecialty: Clinical psychology. Current work: Medical psychology; psychosomatic medicine; body image; cardiac rehabilitation. Home: 12021 Tall Oaks Houston TX 77024 Office: VA Med Center Houston TX 77211

CLEWELL, DON BERT, microbiology educator, research scientist; b. Dallas, Sept. 5, 1941; s. Dayton H. and Vesta Jean (Rapp) C.; m. Lynda Lee Robinson, Oct. 23, 1945; children: Amy, Anna. A.B., Johns Hopkins U., 1963; Ph.D., Ind. U., 1967. Asst. prof. microbiology Sch. Dentistry and Medicine, U. Mich., Ann Arbor, 1970-73, assoc. prof., 1973-77, prof., 1977—, assoc research scientist, 1977-80, research scientist, 1980—; Burroughs Wellcome vis. prof. U. Rochester, N.Y., 1982. Contbr. over 100 sci. articles and abstracts to profl. publs. Recipient Research Career Devel. award Nat. Inst. Allergy and Infectious Diseases, 1975-80; NIH-Nat. Cancer Inst. fellow U. Calif.-San Diego, 1967-70. Mem. Am. Soc. Microbiology, Am. Assn. Biol. Chemists, AAAS, Am. Chem. Soc., Internat. Assn. Dental Research, Sigma Xi. Subspecialties: Microbiology (medicine); Genetics and genetic engineering (medicine). Current work: Research in bacterial plasmids, conjugation, microbiology, streptococcal plasmids, sex pheromones, transposons, drug resistance, molecular genetics. Home: 1841 Alhambra Dr Ann Arbor MI 48103 Office: U Mich 300 N Ingalls Bldg Ann Arbor MI 48109

CLEWS, HENRY MADSION, aircraft engr.; b. Phila., Nov. 19, 1944; s. M. Madison and Margaret (Strawbridge) C.; m. Henrietta Thompson, Aug. 20, 1966; children: Alex, Margaret, Leta, Charlotte. B.S.M.E., U. Pa., 1967; postgrad. in areo. engring, N.C. State U., 1968-69. Project engr., chief engr. Bensen Aircraft Corp., Raleigh, N.C., 1967-70; project engr., test pilot Thurston Aircraft Corp., Sanford, Maine, 1970-72; founder, pres. Solar Wind, Inc., East Holden, Maine, 1972-76; chief design engr. Enertech Corp., Norwich, Vt., 1976—. Author: Electric Power from the Wind, 1972. Subspecialty: Wind power. Current work: Design of state of the art wind energy conversion devices. Designed the Enertech 1500, 1800, 4000, also the Enertech 44, wind generators. Office: Enertech Corp PO Box 420 Norwich VT 05055

CLIFFORD, ALAN FRANK, chemist, educator, consultant, researcher; b. Natick, Mass., June 8, 1919; s. Arthur Woodbury and Elva Elisabeth (Buck) C.; m. Shirley Catherine Mittleman, Aug. 20, 1949; children: Abbie Louise Clifford Wysor, Philip Alan. A.B. in Chemistry, Harvard U., 1941; M.S., U. Del., Ph.D., 1949. Chemist Kankakee Ordnance Works, 1941-43; Manhattan Project, U. Chgo. Metall. Lab., 1943-45; devel. chemist Du Pont Expti. Sta., 1945-46; instr. U. Del., 1947-49; asst. prof. Ill. Inst. Tech., 1949-52; asst. prof., then assoc. prof. Purdue U., 1953-66; prof. chemistry Va. Poly. Inst., 1966—, head dept. chemistry, 1966-81; cons., lectr. in field. Author: Inorganic Chemistry of Qualitative Analysis, 1961; contbr. chpt., numerous articles to profl. jours.; contbg. editor: Van Nostrand's Internat. Ency. of Chem. Sci, 1964. Served with Ill. State Militia, 1942-43. Guggenheim fellow, 1951-53; Grantee NSF, Am. Chem. Soc. Petroleum Research Fund, Army Research Office, Sherwin Williams Co., Research Corp., 3M Co., Battelle Devel. Corp., Va. Poly. Inst. Mem. Am. Chem. Soc., AAAS, Royal Chem. Soc. (Eng.), Soc. Chemistry and Industry (Eng.), Triangle (chpt. adv.), Sigma Xi, Phi Lambda Upsilon, Alpha Chi Sigma. Presbyterian. Subspecialties: Inorganic chemistry. Current work: Organic chemistry without carbon (chemistry derived from NESF3); inorganic fluorine chemistry; hydrogen fluoride solvent system; inorganic polymers; homogenous catalysis; rare earths; Mossbauer spectroscopy; interhalogen chemistry. Patentee in field. Office: Va Poly Inst Dept Chemistry Blacksburg VA 24061

CLIFFORD, MARGARET LOUISE, health center administrator, counselor; b. Lakeland, Fla., Dec. 13, 1920; d. Thomas Saxon and Beatrice (Tillie) C.; m. Charles Robert Davis, Apr. 4, 1950 (div. June 1974); children: Daniel Thomas Davis, Kelly Owen Davis. B.A., Chapman Coll., 1950; M.S., San Diego State U., 1972; Ph.D., Union Grad. Sch., Cin., 1976. Cert. sch. psychologist, Calif., Fla.; lic. mental health counselor, Fla. Tchr. elem. schs., Hanford, Cuyama and Blythe, Calif., 1950-68; columnist Daily Midway Driller, Taft, Calif., 1955; owner, operator Marge Davis Sch. Dance, Blythe, 1961-64; psychologist, dance inst. Peace Corps, Kingston, Jamaica, W.I., 1973-76; psychologist Apalachee Community Mental Health Center, Quincy, Fla., 1977-80; coordinator elderly services Beth Johnson Community Mental Health Center, Orlando, Fla., 1980—; pvt. practice counseling; guest speaker Fla. So. Coll., 1981-82, Rollins Coll., 1982. Contbr.: articles to La Femme Newspaper. Organizer, pres. Widowed Person Service of Orange County, Fla., 1981—. Served with USN WAVES, 1943-45. Mem. Fla. Council for Community Mental Health (pres. chpt. 1979-80), Orange County Citizens Adv. Council on Aging (sec. 1982-83), Am. Psychol. Assn., Fla. Council on Aging, Am. Personnel and Guidance Assn., Am. Assn. Ret. Persons, Art Therapy Assn. Democrat. Subspecialties: Developmental psychology, Behavioral psychology. Current work: Counseling services to elderly. Home: 223 N Central St Winter Garden FL 32808 Office: Beth Johnson Community Mental Health Center 2804 Belcoe Dr Orlando FL 32808

CLIFTON, DAVID GEYER, research chemist; b. Pomeroy, Ohio, Mar. 20, 1924; s. A.R. and Helen (Geyer) C.; m. Anna Marie. B.A., Miami U., Oxford, Ohio, 1948, M.S., 1950; Ph.D., Ohio State U., 1955. Research chemist E.I. duPont de Nemours & Co., Del., 1955-56; postdoctoral fellow Ohio State U., Columbus, 1956-57; research chemist Los Alamos Nat. Lab., 1957-63, Gen. Motors Def. Research Lab., Santa Barbara, Calif., 1963-68, Los Alamos Nat. Lab., 1968—. Contbr. articles to sci. jours. Served to 1st lt. USAF, 1943-46; PTO. Mem. Am. Chem. Soc., Am. Phys. Soc., Am. Nuclear Soc., AAAS. Subspecialties: Physical chemistry; Thermodynamics. Current work: Plutonium chemistry. Home: 352 Cheryl Ave Los Alamos NM 87544 Office: PO Box 1663 Los Alamos NM 87545

CLIFTON, RODNEY JAMES, engineering educator, consultant; b. Orchard, Nebr., July 10, 1937; s. James Edward and Minnie Gertrude (Williamson) C.; m. MercaDee Bonde, Dec. 28, 1958; children: Mark Bradford, Jeffrey John, Gregg Andrew, Anne Michelle. B.S. in Civil Engring, U. Nebr.-Lincoln, 1959; M.S., Carnegie-Mellon U., 1961, Ph.D., 1964. Registered profl. engr., R.I. Engr. trainee Paxton & Vierling Steel Co., Omaha, 1959-60; fellow Brown U., 1964-65, asst. prof. engring., 1965-68, assoc. prof., 1968-71, prof., 1971—, chmn. div. engring., 1974-79; cons. Sandia Labs., Albuquerque, Terra Tek Inc., Salt Lake City. Assoc. editor: Jour. Applied Mechanics, 1981; mem. editorial bd.: Jour. Mechanics and Physics of Solids, 1982—; contbr. numerous articles to profl. jours. NDEA fellow, 1960-63; NSF fellow, 1971-72. Fellow Am. Acad. Mechanics; mem. ASCE, Soc. Engring. Sci. (pres. 1982-83), AAAS., ASME. Presbyterian. Subspecialties: Solid mechanics; Theoretical and applied mechanics. Current work: Laser interferometry, dymanic plasticity, hydraulic fracturing, dislocation dynamics, plate impact, shear bands. Home: 18 Starbrook Dr Barrington RI 02806 Office: Brown U Div Engineering Providence RI 02912

CLINARD, FRANK WELCH, JR., materials scientist, researcher; b. Winston Salem, N.C., Aug. 4, 1933; s. Frank Welch and Hazel Helen (Hauser) C.; m. Elva Adams Hyatt, Apr. 2, 1968. B.M.E., N.C. State U., Raleigh, 1955; M.S. in Metall. Engring, 1957; Ph.D. in Materials Sci, Stanford U., Palo Alto, Calif., 1965. Mem. staff Sandia Corp., Albuquerque, 1957-61; research asst. Stanford U., 1961-64; mem. staff Los Alamos Nat. Lab., 1964-77, sect. leader, 1977—; adj. prof. materials sci. U. N.Mex., Albuquerque, 1967-70. Contbr. research papers, articles to profl. jours. Bd. dirs. county pub. TV Orgn., Los

Alamos, 1981, 82; mem. ACLU, Amnesty Internat. Mem. Am. Soc. for Metals (chmn. local chpt. 1969-70), AAAS, Am. Ceramic Soc., Am. Nuclear Soc., Sigma Xi, Pi Tau Sigma, Tau Beta Pi, Phi Kappa Phi. Clubs: Sports Car del Valle Rio Grande (Los Alamos) (pres. 1967); Sandia Gun (Albuquerque)). Subspecialties: Materials; Nuclear engineering. Current work: Leader radiation effects section; research on radiation damage in ceramics for fusion reactors; research on radiation effects in nuclear waste materials. Home: 2940 Arizona Ave Los Alamos NM 87544 Office: Los Alamos Nat Lab Los Alamos NM 87545

CLINE, THOMAS LYTTON, astrophysicist; b. Peking, China, May 14, 1932; s. Warren Williams and Helen (Thomas) C.; m. Marjorie Hart, Aug. 7, 1954; children: Judith L., Karen B., Marcia V. B.A. in Math, Hiram Coll., 1954; Ph.D. in Physics, M.I.T., 1961. Astrophysicist, sr. scientist NASA Goddard Space Flight Center, Greenbelt, Md., 1961—. Contbr. articles to profl. jours. Recipient John Lindsay Meml. award Goddard Space Flight Center, 1980, medal for exceptional sci. achievement NASA, 1981. Fellow Am. Phys. Soc.; mem. Am. Astron. Soc., Washington Acad. Scis., Washington Philos. Soc., Exptl. Aviation Assn. Subspecialty: Gamma ray high energy astrophysics. Current work: Cosmic gamma ray bursts, research in astrophysics of gamma ray transients. Home: 13708 Sherwood Forest Dr Silver Spring MD 20904 Office: Laboratory for High Energy Astrophysics Code 661 Goddard Space Flight Center Greenbelt MD 20771

CLINTON, JAMES MICHAEL, quality assurance scientist; b. Framingham, Mass., May 30, 1950; s. J. Earl and Dorothea Ann (Keller) C.; m. Diane Reazin, Oct. 2, 1976; children: Patricia, Stephanie. Student, Bryan & Stratton Sch. Bus., Boston, 1969-70, Quinsigamond Community Coll., 1971-72; B.A., U. Mass., 1974; M.S., Ind. U., 1977. Cert. quality engr. Research technician, asst. Ind. U., 1976-78; supr. microbiology, immunology quality assurance, radiation safety officer Bio-Dynamics div. Boehringer Mannheim Diagnostics, Indpls., 1978-81; quality assurance devel. scientist Ames Div. Miles Labs., Elkhart, Ind., 1981—. Contbr. article to profl. jours. Active St. Vincent de Paul Soc. Mem. Am. Soc. Microbiology, Am. Soc. Quality Control, Clin. Ligand Assay Soc. Democrat. Roman Catholic. Club: YMCA. Subspecialties: Immunology (medicine); Microbiology (medicine). Current work: Devel. quality assurance systems for new immunoassay products. Home: 10 Manchester Ln Elkhart IN 46514 Office: 1127 Myrtle St Elkhart IN 46515

CLOPINE, GORDON ALAN, consulting geologist, educator; b. Los Angeles, Nov. 28, 1936; s. Walter Gordon and Sara Elizabeth (Donahue) C.; m. Sara Rose Lapinski, Mar. 2, 1979; children: William, Susan, Russell, Cynthia. B.S., U. Redlands, Calif., 1958; M.S., U. Houston, 1960. Registered geologist, Calif.; cert. profl. geol. scientist, Calif. Pres. Clopine Geol. Services (cons. geologists), Redlands, 1961—; prof. San Bernardino Valley Coll., San Bernardino, Calif., 1961—, dean instrn., 1978-81; lectr. U. Redlands, 1961—; mem. extension faculty U. Calif.-Riverside, 1965—, field leader geol. field studies and natural environment series. Author numerous reports and studies on geol. hazards in So. Calif. and San Andreas Fault Zone. Pres. San Bernardino County Mus. Assn., 1972. Fellow Geol. Soc. Am.; mem. Am. Inst. Profl. Geologists (cert. profl. geol. scientist). Republican. Subspecialties: Geology; Sedimentology. Current work: Geologic field studies; lecturer and researcher on San Andreas fault zone in Southern California; geologic hazards investigation. Home: 13093 Burns Ln Redlands Ca 92373 Office: San Bernardino Valley Coll 701 S Mt Vernon San Bernardino CA 92410

CLOSS, GERHARD LUDWIG, chemist, educator; b. Wuppertal, Germany, May 1, 1928; came to U.S., 1955, naturalized, 1966; s. Ludwig and Maria (Pfeiffer) C.; m. Liselotte Else Pohmer, Aug. 17, 1956. Student, U. Würzburg, 1949-52; Ph.D., U. Tubingen, 1955. Research fellow Harvard, 1955-57; asst. prof. chemistry U. Chgo., 1957-61, asso. prof., 1961-63, prof., 1963—, A.A. Michelson Distinguished Service prof., 1974—; Editorial adviser for chemistry Ency. Brit., 1964-77. Contbr. articles on phys. organic chemistry to profl. jours. A.P. Sloan Found. fellow, 1962-66. Fellow AAAS; mem. Nat. Acad. Scis., Am. Chem. Soc., Am. Acad. Arts and Scis., Chem. Soc. (London). Subspecialties: Organic chemistry; Physical chemistry. Current work: Reactive intermediates; photosynthesis; charge transfer; magnetic resonance. Home: 12655 Timberlane Dr Palos Park IL 60464

CLOUD, JAMES DOUGLAS, electrical engineer; b. Dover, Ohio, Mar. 29, 1928; s. Joseph Douglas and Vieleta (Stiffler) C.; m. Virginia Jane, Aug. 9, 1946; children: Joy, Jay Lee, Paul Franklin, Dayne Bryan, Blake Darron. B.E.E., Purdue U., 1951, M.E.E., 1952; cert., UCLA, 1973. Mgr. systems engring., analysis lab Hughes Aircraft Co., El Segundo, Calif., 1965-67, 70-73, asst. program. mgr. engring., mfg., 1967-68, assoc. mgr. systems labs., 1968-69, advanced systems labs, 1969-70, defense systems div., 1973-76, mgr. tech. div., 1976-79, group v.p., div. mgr. electro-optical, data systems group, 1979—; chmn. bd. Santa Barbara Research Ctr., 1979—. Served with USN, 1946-48. Recipient Surveyor V spl. award Hughes Aircraft Co., 1966. Assoc. fellow AIAA; sr. mem. IEEE; mem. Eta Kappa Nu, Tau Beta Pi, Sigma Pi Sigma. Subspecialty: Aerospace engineering and technology. Current work: engineering management. Office: Hughes Aircraft Co 2000 E El Segundo St El Segundo CA 90245

CLOUD, PRESTON, biogeology educator emeritus; b. West Upton, Mass., Sept. 26, 1912; s. Preston E. and Pauline L. (Wiedemann) C.; m. Janice Gibson, 1972; children by previous marriage: Karen, Lisa, Kevin. B.S., George Washington U., 1938; Ph.D., Yale U., 1940. Instr. Mo. Sch. Mines and Metallurgy, 1940-41; research fellow Yale U., 1941-42; geologist U.S. Geol. Survey, 1942-46, 48-61, 74-79, chief paleontology and stratigraphy br., 1949-59; research geologist, 1959-61, 74-79; asst. prof., curator invertebrate paleontology Harvard U., 1946-48; prof. dept. geology and geophysics U. Minn., 1961-65, chmn., 1961-63; faculty geology dept. UCLA, 1965-68; prof. bio-geology and environ. studies dept. geol. scis. U. Calif., Santa Barbara, 1968-74, prof. emeritus, 1974—; vis. prof. U. Tex., 1962, 78; H.R. Luce prof. cosmology Mt. Holyoke Coll., 1979-80; Sr. Queens fellow Baas-Becking Geobiology Lab., Canberra, Australia, 1981; Nat. Sigma Xi lectr., 1967; Emmons lectr. Colo. Sci. Soc.; Bownocker lectr. Ohio State U.; French lectr. Pomona Coll.; Dumaresq-Smith lectr. Acadia Coll., N.B., Can.; A.L. DuToit Meml. lectr. Royal Soc. and Geol. Soc. of South Africa; internat exchange scholar Research Council Can., 1982; hon. vis. prof. U. Ottawa (Ont. Can.), 1982; mem. governing bd. NRC, 1972-75; mem. Pacific Sci. Bd., 1952-56, 62-65; del. internat. sci. congresses; cons. to govt., industry, founds. and agys. Author: Terebratuloid Brachiopoda of the Silurian and Devonian, 1942, (with Virgil E. Barnes) The Ellenburger Group of Central Texas, 1948, (with others) Geology of Saipan, Mariana Islands, 1957, Environment of Calcium Carbonate Deposition West of Andros Island, Bahamas, 1962, Resources and Man, 1969, Cosmos, Earth and Man, 1978; also articles; editor and co-author: (with others) Adventures in Earth History, 1970. Recipient A. Cressey Morrison prize natural history, 1941; Rockefeller Pub. Service award, 1956; U.S. Dept. Interior Distinguished Service award and gold medal, 1959; medal Paleontol Soc. Am., 1971; Lucius W. Cross medal Yale U., 1973; Penrose medal Geol. Soc. Am., 1976, J.S. Güggenheim fellow, 1982-83. Fellow Am. Acad. Arts and Scis. (com. on membership 1978-80, council 1980-83); mem. Am. Philos. Soc., Nat. Acad. Scis. (com. on sci. and pub. policy 1965-69, mem. council 1972-75, exec. com. 1973-75, chmn. com. on resources and man 1965-69, chmn. ad hoc com. nat. materials policy 1972, chmn. study group on uses of underground space 1972, chmn. com. mineral resources and environment 1972-73, chmn. com. geology and climate 1977, chmn. sect. geology 1976-79, mem. assembly math. and phys. scis. 1976-79, C.D. Walcott medal 1977, Polish Acad. Scis., fgn. assoc.), Geol. Soc. Am. (council 1972-75), Paleontol. Soc. Am., Paleontol. Soc. India (hon.), AAAS, Am. Soc. Naturalists, Geol. Soc. Belgium (hon. fgn. corr.), Phi Beta Kappa, Sigma Xi, Sigma Gamma Epsilon. Current work: Biological, sedimentological, geochemical and atmospheric processes in earth history; crustal evolution. Field work 6 continents and 2 oceans. Home: 400 Mountain Dr Santa Barbara CA 93103 Office: Dept Geol Scis U Calif Santa Barbara CA 93106

CLOUET, DORIS HELEN, research scientist, administrator; b. New Haven, Conn., Mar. 5, 1919; s. Napoleon Arthur and Helen (Roarke) C. B.A., Albertus Magnus Coll., New Haven, 1940; Ph.D., U. Rochester, 1949. Asst. prof. biochemistry Vanderbilt U., 1950-54; research scientist Psychiat. Inst. State of N.Y., N.Y.C., 1956-70; asst. dir. Drug Abuse Lab., Bklyn., 1970—; prof. psychiatry Downstate Med. Sch., 1977—; chmn. biochemistry N.Y. Acad. Sci., 1980-82. Editor: Narcotic Drugs: Biochemistry Pharmacology, 1971, (with D.H. Ford) Tissue Responses to Addictive Drugs, 1974; contbr. chpts. to books and articles in field to profl. jours. USPHS fellow, 1954-56; NIMH grantee, 1967-70, NIDA, 1970—. Mem. Am. Soc. Pharmacology and Exptl. Therapeutics, Am. Soc. Biol. Chemistry. Subspecialties: Neurochemistry; Neuropharmacology. Current work: Molecular mechanisms of action of drugs of abuse; molecular mechanisms of behavior. Home: 71-12 Courtland Ave Stamford CT 06902 Office: New York State Div Substance Abuse Services 80 Hanson Pl Brooklyn NY 11217

CLOUGH, DAVID WILLIAM, molecular geneticist; b. Schenectady, Dec. 27, 1952; s. David W. and Constance Marie (Patnaude) C. B.S., U. Ariz., 1974; Ph.D., Med. Coll. Wis., 1979. Research fellow Eye Inst. of Med. Coll. Wis., 1979; fellow in microbiology and molecular genetics and dept. medicine Harvard U., Harvard Med Sch., Boston, 1979-81; instr. genetics Ctr. for Genetics, U. Ill., Chgo., 1981—. Contbr. articles to profl. jours.; referee profl. jours. Mem. AAAS. Subspecialties: Genetics and genetic engineering (biology); Gene actions. Current work: Investigations concerning regulation of expression of viral and cellular genes in mammalian cells. Office: 808 S Wood St Chicago IL 60612

CLOUGH, RAY WILLIAM, JR., educator; b. Seattle, July 23, 1920; s. Ray William and Mildred (Nelson) C.; m. Shirley Claire Potter, Oct. 30, 1942; children—Douglas Potter, Allison Justine, Meredith Anne. B.S. in Civil Engring, U. Wash., 1942; M.S., Calif. Inst. Tech., 1943; S.M., Mass. Inst. Tech., 1947; Sc.D., MIT, 1949; D.Tech. (hon.), Chalmaers U., Goteborg, Sweden, 1979, U. Tron d heim, (Norway). Registered profl. engr., Wash. Faculty U. Calif.-Berkeley, 1949—, prof. civil engring., 1959—, chmn. div. structural engring. and structural mechanics, 1967-70, dir., 1973-76, Nish Kran prof. structural engr., 1983—; Cons. in field, 1953—; Mem. Nat. Acad. Scis.-Nat. Acad. Engring. adv. com. Environmental Sci. Services Adminstrn., 1967-70; mem. dynamics panel Nat. Acad. Scis. adv. bd. on hardened electric power system, 1964-70; mem. U.S. C.E. Structural Design Adv. Bd., 1967—. Served to capt. USAAF, 1942-46. Fulbright fellow Ship Research Inst., Trondheim, Norway, 1956-57; Overseas fellow Churchill Coll., Cambridge (Eng.) U., 1963-64; hon. researcher Laboratorio Nacional de Engenharia Civil, Lisbon, Portugal, 1972; Fulbright fellow Tech. U. Norway, Trondheim, 1972-73. Fellow ASCE (chmn. engring. mechanics div. 1964-65, Research award 1960, Howard award 1970, Newmark medal 1979, Moissieff medal 1980); mem. Structural Engrs. Assn. No. Calif. (dir. 1967-70), Earthquake Engring. Research Inst. (dir. 1957-60, 70-73), Seismol. Soc. Am. (dir. 1970-73), Nat. Acad. Scis., Nat. Acad. Engring., Det Kongelige Norske Videnskabers Selskab. Subspecialty: Civil engineering. Current work: Research in methods of analysis of dynamic structural response to earthquakes; shaking table & study of nonlinear structural dynamic behavior. Home: 576 Vistamont Ave Berkeley CA 94708

CLOWES, ALEXANDER WHITEHILL, vascular surgeon, researcher; b. Boston, Oct. 9, 1946; s. George H.A. (Sr.) and Margaret (Jackson) C.; m. Monika Meyer, May 4, 1980. A.B., Harvard U., 1968, M.D., 1972. Diplomate: Am. Bd. Surgery. Resident in surgery Case Western Res. U., Cleve., 1972-74, 76-79; research fellow in pathology Harvard Med. Sch., Boston, 1974-76; fellow in vascular surgery Peter Bent Brigham Hosp., Boston, 1979-80; asst. prof. surgery U. Wash., Seattle, 1980—; attending surgeon Seattle hosps., Seattle Pub. Health Hosp., Harborview Med. Center, Univ. Hosp., Seattle VA Hosp., 1980—. NIH Research Career Devel. awardee, 1983. Mem. Assn. for Acad. Surgery, Henry Harkins Soc. Subspecialties: Surgery; Cell biology (medicine). Current work: Atherosclerosis, arterial injury, and repair, arterial endothelial and smooth muscle cell proliferation, mechanisms of arterial graft failure. Home: 702 Fullerton Ave Seattle WA 98122 Office: Univ Wash Sch Medicine RF-25 Seattle WA 98195

CLUFF, LEIGHTON EGGERTSEN, physician, found. exec.; b. Salt Lake City, June 10, 1923; s. Lehi Eggertsen and Lottie (Brain) C.; m. Beth Allen, Aug. 19, 1944; children—Claudia Beth, Patricia Leigh. Student, U. Utah, 1941-44; M.D. with distinction, George Washington U., 1949. Intern Johns Hopkins Hosp., 1949-50, asst. resident, 1951-52; asst. resident physician Duke Hosp., 1950- 51; vis. investigator, asst. physician Rockefeller Inst. Med. Research, 1952-54; fellow Nat. Found. Infantile Paralysis, 1952-54; mem. faculty Johns Hopkins Sch. Medicine; staff Johns Hopkins Hosp., 1954-66, prof. medicine, 1964-66, physician, head div. clin. immunology, allergy and infectious diseases, 1958-66; prof., chmn. dept. medicine U. Fla., 1966-76; exec. v.p. Robert Wood Johnson Found., 1976—; U.S. del. U.S.-Japan Coop. Med. Sci. Program, 1972-81; mem. council drugs A.M.A., 1965-67; mem. NRC-Nat. Acad. Sci. Drug Research Bd., 1965-71; mem. expert adv. panel bacterial diseases (coccal infection) WHO; mem. council Nat. Inst. Allergy and Infectious Diseases, 1968-72; cons. FDA; tng. grant com. NIH, 1964-68. Author, editor books on internal medicine, infectious diseases, clin. pharmacology.; Contbr. articles to profl. jours. Markle scholar med. scis., 1955-62; recipient Career Research award NIH, 1962. Mem. Inst. Medicine-Nat. Acad. Scis., Am. Soc. Clin. Investigation, Assn. Life Scis.-Nat. Acad. Scis., Assn. Am. Physicians, Soc. Exptl. Biology and Medicine, Am. Assn. Immunologists, Am. Fedn. Clin. Research, Harvey Soc., N.Y. Acad. Sci., Infectious Disease Soc. Am. (pres. 1973), So. Soc. Clin. Investigation, A.C.P. (Fla. gov. 1975-76, Mead-Johnson postgrad. scholar 1954-55, Ordronaux award med. scholarship 1949), Am. Clin. and Climatological Assn., Alpha Omega Alpha. Subspecialties: Internal medicine; Pharmacology. Home: 7 Beechtree Ln Princeton NJ 08540

CLUM, GEORGE ARTHUR, psychologist, educator; b. Johnson City, N.Y., May 12, 1942; s. George Albert and Theresa (Jankiewicz) C.; m.; children: Gretchen Ann, Christina Lee. B.S., Scranton U., 1963; M.S., St. John's U., 1965, Ph.D., 1968. Diplomate: in clin. psychology Am. Bd. Profl. Psychology; lic. psychologist, Va. Psychology intern Bethesda (Md.) Naval Hosp., 1965-66; research psychologist Neuropsychiat. Research, San Diego, 1966-69; asst. prof. Va. U., Charlottesville, 1969-75; assoc. prof. to prof. psychology Va. Tech., Blacksburg, 1975—; cons. St. Albans Hosp., Redford, Va., 1976-80, Cooper House, Blacksburg, 1977-82. Contbr. articles to profl. jours. Served to lt. USN, 1965-69. Mem. Am. Psychol. Assn., Southeastern Psychol. Assn. Roman Catholic. Current work: Health psychology - research in assessment and treatment of pain; prediction and treatment of parasuicide. Home: 406 Murphy St Blacksburg VA 24060 Office: Va Poly Inst and State U Blacksburg VA 24061

COAKLEY, STELLA MELUGIN, plant pathologist, educator; b. Modesto, Calif., Sept. 1, 1947; d. John Bannister and Alice Dora (Caulkins) Melugin; m. James Alexander Coakley, Sept. 18, 1971; children: Sarah Christine, Miriam Alice, Martha Vey. A.A., Modesto Jr. Coll., 1967; B.S., U. Calif., Davis, 1969, M.S., 1970, Ph.D., 1973. Instr. dept. biol. scis. U. Denver, 1973; vis. asst. prof. Instr. dept. biol. scis., 1973-75; adj. asst. prof. U. Denver, 1977-81, assoc. research prof., 1981—; postdoctoral fellow advanced study program Nat. Ctr. Atmospheric Research, Boulder, Colo., 1975-76. Contbr. articles to profl. jours. NSF trainee, 1969-72; NSF grantee. Mem. Am. Phytopath. Soc., AAAS, Am. Meteorol Soc. Subspecialties: Plant pathology; Climatology. Current work: Investigation of the effect of climatic variation on the occurrence of plant diseases; emphasis on fungal diseases on wheat. Office: Nat Ctr Atmospheric Research PO Box 3000 Boulder CO 80307

COAN, RICHARD WELTON, psychologist, educator; b. Martinez, Calif., Jan. 24, 1928; s. Otis Welton and Esta Dorothy (Wilson) C.; m. Signa Carolyn Roswall, Dec. 8, 1950 (div. 1972); children: Lisa Cooper, Cynthia; m. Susan Anne Chamberlain, June 9, 1979; 1 dau., Angela. A.A., Los Angeles City Coll., 1946; A.B., U. Calif.-Berkeley, 1948, M.A., 1950; Ph.D., U. So. Calif., 1955. Instr. Los Angeles City Coll., 1950-55; research assoc. U. Ill.-Urbana, 1955-57; asst. prof. U. Ariz., Tucson, 1957-60, assoc. prof., 1960-64, prof., 1964—. Author: The Optimal Personality, 1974, Hero, Artist, Sage, or Saint?, 1977, Psychologists, 1979, Psychology of Adjustment, 1983. Fellow Am. Psychol. Assn.; mem. Soc. Multivariate Exptl. Psychology, Assn. Humanistic Psychology. Subspecialties: Personality; Psychology of science. Current work: Psychology symbolism and myths, evolution of consciousness, evolution of human personality. Home: 2136 N Marion Blvd Tucson AZ 85712 Office: Psychology Dept U Ariz Tucson AZ 85721

COATE, BARRIE DOUGLAS, horticulturist, consultant; b. Juneau, Alaska; s. Carl Douglas and Ruth (Carmichael) C.; m. Bernice Frances Gryba, 1962; children: Richard D., Wesley D.; m. Carol Ann Riehl. Student, San Jose State Coll., 1952-53, San Jose City Coll., 1954-56. Supt. Saratoga Hort. Found., 1956-63; br. mgr. Pacific Nurseries, Inc., Colma, Calif., 1963-69; mgr. Western Tree Nurseries, Inc., Gilroy, Calif., 1969-71; v.p., gen. mgr. Barrier's Trees & Shrubs, Aptos, Calif., 1971-78; dir. hort. Saratoga (Calif.) Hort. Found., 1978—. Author: Selected Native Plants in Color, 1980, A Success List of Water Conserving Plants, 1982, Selected California Native Plants with Commercial Sources, 1979. Mem. Calif. Water Conservation Com. Served with inf. U.S. Army, 1952-54. Mem. Internat. Plant Propagators Soc., Internat. Soc. Arboriculture, Soc. Cons. Arborists. Club: Midori Bonsai (San Jose, Calif.). Subspecialties: Resource conservation; Plant growth. Current work: Search, evaluation for introduction of new ornamental, drought-tolerant plants for mid-California. Office: PO Box 308 Saratoga CA 95070

COATES, CLARENCE LEROY, JR., educator, research engineer; b. Hastings, Nebr., Nov. 5, 1923; s. Clarence Leroy and Mildred (Creighton) C.; m. Henrietta Hoff, Jan. 1, 1943; children: Catherine Anne, Christopher John; m. Lila M. Mustola, Mar. 5, 1969; 1 son, Randall Lee; m. Henrietta Coates, July 17, 1972. B.S. in Elec. Engring, U. Kans., 1944, M.S., 1948; Ph.D., U. Ill., 1954. Instr. elec. engring. U. Kans., 1946-48; instr., then asso. prof. elec. engring. U. Ill., 1948-56; research scientist Gen. Electric Research Labs., 1956-63; prof. elec. engring. U. Tex., 1963-71, chmn. dept., 1964-66, dir., 1967-71, (Coordinated Sci. Lab.); prof. U. Ill., 1971-72; head Sch. Elec. Engring., Purdue U., 1973—; Cons. NSF, 1969-70, mem. sci. info. council, 1972-75; mem. research adv. com. NASA, 1971-76. Author: Threshold Logic; Cons. editor, Blaisdell Pub. Co., 1968-70; Contbr. articles in field to profl. jours. Served as officer USNR, 1944- 46. Fellow IEEE (v.p. publ. activities, dir. 1971-72), AAAS; mem. Sigma Xi, Phi Kappa Phi, Tau Beta Pi, Eta Kappa Nu, Sigma Tau. Subspecialty: Computer engineering. Home: 116 Glen Court West Lafayette IN 47906

COATES, GARY JOSEPH, educator, designer, consultant; b. Annapolis, Md., July 20, 1947; s. Joseph Leonard and Claire (Robertson) C.; m. Julie Taylor, Sept. 6, 1969; 1 son, Jason Christopher. B. magna cum laude in Environ. Design, N.C. State U., 1969, M.Arch., 1971. Asst. prof. dept. design and environ. analysis N.Y. State Coll. Human Ecology Cornell U., Ithaca, N.Y., 1971-77; assoc. prof. dept. architecture Kans. State U., Manhattan, 1977—; dir. Univ. for Man appropriate tech. program; cons. in field. Editor: Alternative Learning Environments, 1974, Resettling America-Energy, Ecology and Community, 1981. N.C. Design Found. fellow, 1970-71; Danforth Found. assoc. mem., 1980-86. Mem. AIA (assoc.), Am. Solar Energy Soc., Kans. Soc. AIA, Phi Eta Sigma, Phi Kappa Phi. Democrat. Subspecialty: Solar energy. Current work: Passive solar energy systems for heating and cooling buildings, landscape planning for energy conservation and food production, community energy planning. Designer UFM passive solar addition. Home: 315 North 15th St Manhattan KS 66502 Office: Kans State U Dept Architectur Seaton Hall Manhattan KS 66506

COATS, DOUGLAS A., research physicist. B.S., Calif. Poly. U., 1975; M.S., Scripps Inst. Oceanography, 1979, Ph.D., 1982. Research assoc. Scripps Inst. Oceanography, La Jolla, Calif., 1975-82; sr. research physicist Exxon Prodn. Research Co., Houston, 1982—. Mem. Am. Geophys. Union, Seismological Soc. Am., Earthquake Engring. Research Inst. Subspecialties: Petroleum engineering; Oceanography. Current work: Criteria development for the earthquake-resistant design of offshore structures. Office: Exxon Production Research Co PO Box 2189 Houston TX 77001

COBB, FIELDS WHITE, JR., forest pathologist; b. Key West, Fla., Feb. 16, 1932; s. Fields W. and Alice Mae (Presson) C.; m. Octavia H. Smith, May 24, 1958; children: Cynthia Leigh, David Fields, Stephen Lewis. B.S., N.C. State U., 1955; M.Forestry, Yale U., 1956; Ph.D., Pa. State U., 1963. Research forester U.S. Dept. Agr. Forest Service, Lake City, Fla., 1955-57, plant pathologist, Gulfport, Miss., 1957; instr. plant pathology Pa. State U., State College, 1958-63; asst. prof. U. Calif., Berkeley, 1963-70, asso. prof., 1970-82, prof. plant pathology, 1982—. Mem. Am. Phytopath. Soc., Soc. Am. Foresters. Subspecialties: Plant pathology; Ecology. Current work: Biology and control of forest tree diseases; interactions among forest diseases, humans, insects and other components of the forest ecosystem. Home: 821 Reliez Station Rd Lafayette CA 94549 Office: Dept Plant Pathology U Calif Berkeley CA 94720

COBB, WILLIAM THOMPSON, research scientist; b. Spokane, Wash., Nov. 10, 1942; s. Elmer Jean and Martha Ella (Napier) C.; m. Sandra L. Hodgson, Aug. 30, 1964; children: Michael R., Melanie S. Cobb Kaye, Megan A., William Thompson. B.A., Eastern Wash. U., 1964; Ph.D., Oreg. State U., 1973. Cert. profl. agronomist. Mgr./

agronomist Sun Royal Co., Royal City, Wash., 1970-74; sr. scientist Lilly Research Labs., Kennewick, Wash., 1974-78, research scientist, 1978—; instr. plant pathology Columbia Basin Coll., 1971, 73, 75, 77. Mem. Royal City (Wash.) Sch. Bd., 1973-74; scoutmaster, 1972-73. Served to 1st lt. AC U.S. Army, 1964-66. Mem. Am. Soc. Agronomy, Weed Sci. Soc. Am., Am. Phytopath. Soc., Sigma Xi. Subspecialties: Plant pathology; Weed science. Current work: Pesticide research and development. Home: 815 S Kellogg Kennewick WA 99336

COBLITZ, DAVID BARRY, research engr.; b. Ashtabula, Ohio, Oct. 1, 1949; s. Sandford E. and Leah Pearl (Shapiro) C.; m. Sandra Gay Tischler, Dec. 26, 1976; children: Brian Andrew, Evan Daniel. B.S. in Physics, Case Inst. Tech., 1971; M.S. in Engring, U. Rochester, 1973. Computer programmer Wheeler Mfg. Corp., Ashtabula, 1966-71; engr. optics research dept. Xerox Corp., Webster, N.Y., 1972; staff engr. direct research and devel. visually coupled systems McDonnell Douglas Electronics Co., St. Charles, Mo., 1973-83; sr. program engr. DCS Corp., Alexandria, Va., 1983—. Contbr. articles to profl. jours. Mem. Assn. Computing Machinery. Subspecialties: Aerospace engineering and technology; Graphics, image processing, and pattern recognition. Current work: Computer generated image system design, sensor system simulation, display system design, vision, human engineering, training techniques, artificial intelligence, pattern recognition. Patentee in field. Office: DCS Corp 1055 N Fairfax Alexandria VA 22314

COBURN, HERBERT DIGHTMAN, JR., mech. design engr.; b. N.Y.C., Nov. 5, 1919; s. Herbert Dightman and Miriam (Ware) C.; m. Julia Mae Ledbetter, July 29, 1944; children: Herbert Bryant, Randall Nye. Student, Friends Sem., N.Y.C., 1936-39, Newark Coll. Engring., 1939-42; B.S.M.E., So. Meth. U., 1947. Chem. lab. asst. Philip Stroughton & Co., N.Y.C., 1936-39; plate grainer Photoplate Co., Newark, 1939-42; results engr. Southwestern Electric Service Co., Jacksonville, Tex., 1947-50, Tex. Electric Service Co., Monahans, 1950-52; mech. design engr., geophys.-exploration equipment Tex. Instruments Inc., Dallas, 1952—. Active Circle Ten council Boy Scouts Am., 1958-75; active Southwood Meth. Ch., Dallas, 1958-82, including chmn. bd., 1959-60. Served with U.S. Army, 1942-45. Recipient Order of Arrow award Boy Scouts Am., 1962. Mem. ASME. Subspecialties: Mechanical engineering; Petroleum engineering. Current work: Design, development, prototype construction and evaluation of geophysical exploration equipment. Patentee in field. Home: 3427 S Ravinia Dallas TX 75233 Office: PO Box 225621 M/S 3904 Dallas TX 75265

COCHE, ERICH HENRY ERNST, psychologist; b. Nijmegen, Netherlands, June 24, 1941; came to U.S., 1968; s. Erich Johannes Maximilian and Frieda Sophie (Moellmann) C.; m. Judith Abbe Milner, Oct. 16, 1966; children: Raymond Erich, Juliette Laura. Ph.D., U. Bonn, 1968. Diplomate: Am. Bd. Profl. Psychology. Teaching asst. Tchrs. Coll., Bonn, Ger., 1966-68; clin. psychologist Friends Hosp., Phila., 1968-74, dir. psychol. services and research, 1974-81; pvt. practice psychology, Phila. Contbr. articles to profl. jours. Vice pres. Mental Health Assn. Southeastern Pa., 1982-84. Fellow Pa. Psychol. Assn., Soc. Personality Assessment; mem. Am. Psychol. Assn., Phila. Soc. Clin. Psychologists (pres. 1977-78). Subspecialties: Clinical psychology; Behavioral psychology. Current work: Psychotherapy outcome and process research. Address: 2037 Delancey Pl Philadelphia PA 19103

COCHRAN, KENNETH WILLIAM, pharmacologist, virologist, educator; b. Chgo., Nov. 2, 1923; s. Kenneth William and Mabel Alice (Hoffmann) C.; m. Martha Louise Wells, May 10, 1945; children: Kenneth W., Kimberly Wells Cochran Nelson. Diploma, Wright Jr. Coll., Chgo., 1940-42; S.B., U. Chgo., 1943, Ph.D., 1950. Diplomate: Am. Bd. Med. Microbiology, Am. Bd. Indsl. Hygiene. Research asst. toxicity lab. U. Chgo., 1947-52; instr., research assoc. dept. pharmacology U.S. Air Force Radiation Lab., U. Chgo., 1946-52; mem. faculty U. Mich., Ann Arbor, 1952—, prof. epidemiology, 1968—, assoc. prof. pharmacology, 1975. Contbr. articles to profl. jours. Served to 1st lt. AUS, 1942-46. Mem. Am. Soc. Pharmacology and Exptl. Therapeutics, Am. Soc. Microbiology, Soc. Exptl. Biology and Medicine, Mycological Soc. Am., Sigma Xi, Delta Omega. Subspecialties: Virology (medicine); Toxicology (medicine). Current work: Virus Chemotherapy, toxic and beneficial effects of higher fungi. Home: 3556 Oakwood Ann Arbor MI 48104 Office: Dept Epidemiology Sch Public Health U Michigan 109 Observatory Ann Arbor MI 48109

COCKERHAM, COLUMBUS CLARK, geneticist, educator; b. Mountain Park, N.C., Dec. 12, 1921; s. Corbett C. and Nellie Bruce (McCann) C.; m. Joyce Evelyn Allen, Feb. 26, 1944; children: Columbus Clark Jr., Jean Allen, Bruce Allen. B.S., N.C. State Coll., 1943, M.S., 1949; Ph.D., Iowa State Coll., 1952. Asst. prof. biostats. U. N.C., Chapel Hill, 1952-53; mem. faculty N.C. State U., Raleigh, 1953—, prof. stats., 1959-72, William Neal Reynolds prof. stats. and genetics, 1972—; mem. genetics study sect. NIH, 1965-69; cons. adv. com. protocols for safety evaluation FDA, 1967-69. Author papers population and quantitative genetics, plant and animal breeding.; Editor, assoc. editor: Theoretical Population Biology, 1975—; editorial bd.: Genetics, 1969-72; asso. editor: Am. Jour. Human Genetics, 1978-80. Served with USMCR, 1943-46. Recipient N.C. award in sci., 1976, Oliver Max Gardner award, 1980; grantee Nat. Inst. Gen. Med. Scis., 1960—. Fellow Am. Soc. Agronomy; mem. Nat. Acad. Scis., AAAS, Am. Soc. Animal Sci., Am. Soc. Naturalists, Biometric Soc., Genetics Soc. Am., Am. Soc. Human Genetics, Sigma Xi, Gamma Sigma Delta (award merit 1964), Phi Kappa Phi. Subspecialties: Animal genetics; Plant genetics. Home: 2110 Coley Forest Pl Raleigh NC 27607 Office: Dept Statistics NC State Univ Raleigh NC 27650

CODY, WILLIAM JAMES, JR., numerical analyst, consultant; b. Melrose Park, Ill., Nov. 28, 1929; s. William James and Ruth Muriel (Helton) C.; m. Joanne Bond Search, Aug. 8, 1953; children: Patricia N., Steven J., William D., Richard M., Elizabeth L. B.S., Elmhurst Coll., 1951; Sc.D. (hon.), 1977; M.S., U. Okla., 1956. Asst. mathematician Argonne (Ill.) Nat. Lab., 1959-66, assoc. mathematician, 1966-79, sr. mathematician, 1979—; cons. Internat. Math. and Statis. Libraries, Inc., Houston, 1973-81. Author: Software Manual for Elementary Functions, 1980. Served with USAF, 1951-53. Mem. Am. Math. Soc., Math. Assn. Am., Soc. Indsl. and Applied Math., Assn. Computing Machinery, IEEE Computer Soc., Signum (dir. 1969-72, 76-78). Subspecialties: Numerical analysis; Mathematical software. Current work: Approximation and computation of special functions, preparation and testing of mathematical software, computer arithmetic. Office: Argonne Nat Lab 9700 S Cass Ave Argonne IL 60439

COE, JOHN EMMONS, research immunologist; b. Evanston, Ill., Sept. 1, 1931; s. Emmons Sylvester and Lillian Elizabeth (Beckman) C.; m. Nancy Rowland, June 18, 1954; children: Kristine Wing Coe Sutton, Anne Lindstrom, Paul Rowland. B.A., Oberlin Coll., 1953; M.D., Hahnemann Med. Coll., 1957. Intern U. Ill. Research and Ednl. Hosp., Chgo., 1957-58; resident in medicine U. Colo. Med. Ctr., Denver, 1958-60; surgeon USPHS, NIH, Rocky Mountain Lab., Hamilton, Mont., 1960-63; fellow dept. pathology Scripps Clinic and Research Found., LaJolla, Calif., 1963-65; med. officer Nat. Inst. Allergy and Infectious Diseases, NIH, Rocky Mountain Lab., Hamilton, 1965—; affiliate prof. dept. microbiology U. Mont., Missoula, 1966—. Bd. dirs. Mill Lake Irrigation Dist. Mem. Am. Assn. Immunologists, Alpha Omega Alpha. Clubs: Hamilton (Mont.) Lacrosse., Handball. Subspecialties: Immunology (medicine); Infectious diseases. Current work: Selective induction of antibody production in immunoglobulin classes of rodents; immunochemistry and pathophysiology of pentraxins. Home: NW 986 Orchard Dr Hamilton MT 59840 Office: Rocky Mountain Lab Hamilton MT 59840

COELHO, JAIME BERNARDINO, physician, pharmaceutical executive; b. Buenos Aires, Argentina, Nov. 13, 1931; s. Jaime M. and Maria M. (Bilbao) C.; m. Mary M. Conrow, Feb. 17, 1962; children: Daniel J., Christian A., Sarah E. M.D., U. Buenos Aires, 1956. Postgrad. physician U. Buenos Aires, 1957-60, physician, 1963-67; fellow in medicine Columbia U., N.Y.C., 1960-63, research physician, 1967-77; dir. clin. research Ayerst Labs., N.Y.C., 1977—; NRC career investigator, Buenos Aires, 1963-67. Contbr. articles to sci. publs. Mem. Am. Physiol. Soc., Am. Soc. Nephrology, AMA (Physician's recognition award 1980), Internat. Soc. Nephrology, Am. Heart Assn. Subspecialties: Nephrology; Cardiology. Current work: renal physiology; cardiovascular drug development; clinical pharmacology. Office: Ayerst Labs 685 3d Ave New York NY 10017

COFFEY, BRIAN JOSEPH, physicist, cons.; b. Washington, Mar. 19, 1951; s. Francis A. and Marion (Wolberg) C.; m. Lillian D. Valchar, June 24, 1978; 1 son, Alexander. B.A., Columbia U., 1972, M.A., 1974, M.Phil., 1975, Ph.D., 1980. Postdoctoral fellow City Coll. of CUNY, N.Y.C., 1980—; cons. Philips Labs., Briarcliff Manor, 1981—. Contbr. articles to profl. publs. Mem. Am. Phys. Soc. Subspecialties: Theoretical physics; Non-linear optics. Current work: Free electron lasers; unstable resonators; coop. optical phenomena; electronic properties of heterstructures; patern recognition, silicon device simulation.

COFFEY, TIMOTHY, physicist; b. Washington, June 27, 1941; s. Timothy and Helen (Stevens) C.; m. Paula Marie Smith, Aug. 24, 1963; children: Timothy, Donna, Marie. B.S. in Elec. Engring. (Cambridge scholar 1958), MIT, 1962; M.S. in Physics, U. Mich., 1963, Evening News Assn. fellow, 1964, Ph.D., 1967. Research physicist Air Force Cambridge Research Lab., 1964; theoretical physicist EGG, Inc., Boston, 1966-71; head plasma dynamics br., then supt. plasma physics div. Naval Research Lab., Washington, 1971-80, asso. dir. research for gen. sci. and tech., 1980-83, dir. research, 1983—. Recipient award Naval Research Lab., 1974, 75. Fellow Am. Phys. Soc.; mem. Am. Inst. Physics, AAAS, N.Y. Acad. Scis., Internat. Union Radio Sci. Office: 4555 Overlook Ave SW Washington DC 20375

COFFIN, JOHN MILLER, microbiologist, educator, researcher; b. Boston, Apr. 20, 1944; s. Louis Fussel and Mary Elizabeth (McCarthy) C.; m. Marion Claire Szurek, June 22, 1968; children: Erica Mary, Heather Rachel. B.A., Wesleyan U., 1967; Ph.D., U. Wis., Madison, 1972. NIH predoctoral research fellow U. Wis., 1967-72; postdoctoral fellow Institut fur Molekularbiologie, Universitat Zurich, Switzerland, 1972-75; asst. prof. Tufts U. Sch. Medicine, 1975-78, assoc. prof. molecular biology and microbiology, 1978-82, prof., 1982—; mem. virology study sect. NIH, 1981-84. Contbr. numerous publs. in field; editorial bd.: Jour. Virology, 1979—, Virology, 1979—. Recipient Faculty Research award Am. Cancer Soc., 1978-83; Jane Coffin Childs postdoctoral fellow, 1972-72. Mem. Am. Soc. Microbiology, Am. Soc. Virology. Subspecialties: Virology (biology); Genome organization. Current work: Molecular biology of RNA tumor viruses; retroviruses; genetic engineering viral oncology; viral genetics; genome structure. Home: 116 Old Oaken Bucket Rd Scituate MA 02066 Office: 136 Harrison Ave Boston MA 02111

COFFIN, LOUIS FUSSELL, JR., mech. engr.; b. Schenectady, Aug. 30, 1917; s. Louis Fussell and Laura C. (Glen) C.; m. Mary Elizabeth McCarthy, Apr. 24, 1943; children—John, Sarah (Mrs. Joseph Fitzgerald), Laura (Mrs. Thomas Koch), Robert, Patricia (Mrs. Jeffrey Mullen), Deborah (Mrs. Patrick Higgins), Louis Fussell III, Margaret. B.S., Swarthmore (Pa.) Coll., 1939; Sc.D., Mass. Inst. Tech., 1949. From asst. to asst. prof. mech. engring. Mass. Inst. Tech., 1939-49; research asso., then supr. mech. metallurgy Knolls Atomic Power Lab., Gen. Electric Co., 1949-54, mech. engr. corporate research and devel., Schenectady, 1954—; adj. prof. mech. engring. Rensselaer Poly. Inst., Troy, N.Y., 1955-60, Union Coll., Schenectady, 1965—; vis. fellow Clare Hall Cambridge U., 1976. Author. Recipient Alfred E. Hunt award Am. Soc. Lubrication Engrs., 1958; award excellence Carborundum Co., 1974; Clayton lectr. Inst. Mech. Engrs., London, 1974; Coolidge fellow, 1974. Fellow ASME (Nadai award 1979), Am. Soc. Metals (Albert Sauveur Achievement award 1980), ASTM (chmn. E9 com. on fatigue 1974—, Dudley award 1975, award of merit 1978); mem. Nat. Acad. Engring., Am. Inst. Metall. Engrs. (Disting. Career award 1978), Sigma Xi, Pi Tau Sigma, Sigma Tau. Subspecialties: Fracture mechanics; High-temperature materials. Current work: Mechanism, materials and environmental aspects of fracture of structural materials under cyclic and static loadings, especially fatigue and stress corrosion. Patentee in field. Home: 1178 Lowell Rd Schenectady NY 12308 Office: Corporate Research and Devel Gen Electric Co PO Box 8 Schenectady NY 12301

COFFINO, PHILIP, molecular biology and medical educator; b. N.Y.C., Sept. 7, 1942. B.A., U. Calif.-Berkeley, 1966; Ph.D., Einstein U., 1971, M.D., 1972. Intern U. Calif.-San Francisco, 1972-73; fellow in biochemistry, 1973-74, asst. prof. medicine and microbiology, 1974-77, assoc. prof., 1979—. Mem. Am. Assn. Biol. Chemistry, Am. Soc. Clin. Investigation. Subspecialties: Molecular biology; Genetics and genetic engineering (biology). Current work: Somatic cell genetics; hormone action; regulation of cell growth. Office: Medicine and Microbiology Dept Univ Calif 412 Sci Bldg San Francisco CA 94143 Home: 1523 Cole St San Francisco CA

COGBILL, CHARLES LIPSCOMB, III, nuclear engineer; b. Rochester, N.Y., Nov. 25, 1946; s. Charles Lipscomb and Elton Francis (Mills) C.; m. Hortensia Martin Sanchez, Mar. 28, 1981. B.A.S., Troy State U., 1980. Registered profl. engr., N.H. Engr./sr. engr. B Westinghouse Electric Co., Zion, Ill., 1974-78, sr. engr. B., Dothan, Ala., 1978-80, sr. engr., Almaraz, Spain, 1980-82, Seabrook, N.H., 1982—. Contbr. articles in field to profl. jours. Served with USN, 1965-74. Mem. Am. Nuclear Soc. Republican. Subspecialties: Nuclear engineering; Human factors engineering. Current work: Mechanical engineering (startup) work at nuclear power plants. Office: Seabrook Sta Startup Dept PO Box 700 Seabrook NH 03874

COHEN, ALLEN BARRY, physician, educator; b. Ft. Wayne, Ind., July 14, 1939; s. Samuel A. and Dorothy (Weisse) C.; m. Geraldine Ellen Stein, Oct. 24, 1939; children: Rachel, Deborah. B.A., George Washington U., 1960, M.D., 1963; Ph.D., U. Calif.-San Francisco, 1972. Diplomate: Am. Bd. Internal Medicine. Intern Strong Meml. Hosp., Rochester, N.Y., 1963-64; resident Washington U. Sch. Medicine, St. Louis, 1964-66; Asst. prof. medicine U. Calif., San Francisco, 1971-75; prof. medicine and physiology Temple U. Med. Sch., Phila., 1975—. Contbr. numerous articles to sci. jours. Served as surgeon UPHS, 1966-68. Recipient Am. Coll. Chest Physicians Cecile Mayer research award, 1972. Mem. Am. Soc. Clin. Investigation, Am. Soc. Biol. Chemists, Am. Physiol. Soc., Am. Lung Assn. Jewish. Subspecialties: Pulmonary medicine; Cell biology (medicine). Current work: experimental emphysema; mechanisms of lung diseases.

COHEN, ARTHUR DAVID, geologist, consultant; b. Wilmington, Del., Feb. 26, 1942; s. Herman and Anna Mary (Stein) C.; m. Mary Jo Purcell, June 7, 1970; children: Benjamin, Jonathan. B.S., U. Del., 1964; Ph.D., Pa. State U.-State College, 1968. Assoc. prof. So. Ill. U., Carbondale, 1969-74; geologist U.S. Geol. Survey, Reston, Va., 1974-75; prof. U. S.C., Columbia, 1975—; vis. staff mem. Los Alamos Nat. Lab., 1982—; dir. Organic Sediments Research Center, U. S.C., Columbia, 1975—; cons. U.S. Dept. Justice, Dept. Energy, 1979-82. Contbr. over 80 articles to profl. jours. Research grantee NSF, 1969-82, Dept. Energy, 1979-82, Nat. Geographic Soc., 1980-81, Sea Grant Consortium, 1980. Fellow Geol. Soc. Am. (chmn. coal div. 1976); mem. ASTM (chmn. peat classification com. 1979—), Sigma Xi, Phi Beta Kappa, Phi Kappa Phi. Subspecialties: Coal; Geology. Current work: Origin of coal, development of models to aid in coal mining and coal quality evaluation, peat resource evaluation, geology and paleoecology of modern and ancient swamps. Home: 4 Inca Ln Los Alamos NM 87544 Office: Los Alamos Nat Lab MS D 462 Los Alamo NM 87545

COHEN, DAVID HARRIS, neuroscientist; b. Springfield, Mass., Aug. 26, 1938; s. Nathan Edward and Sylvia (Golden) C.; m. Anne Helena Remmes, Jan. 17, 1981; children from previous marriage: Bonnie, Daniel, Ian. B.A., Harvard U., 1960; Ph.D., U. Calif., Berkeley, 1963. Postdoctoral fellow in physiology UCLA, 1963-64; asst. prof. physiology Western Res. U., Cleve., 1964-68; asso. prof. physiology U. Va., 1968-71, prof. physiology, dir. neurosci., 1971-79; leading prof., chmn. neurobiology and behavior SUNY, Stony Brook, 1979—; mem. study sects. and spl. rev. group NIH; mem. adv. groups NSF; vis. prof. numerous univs. Author books, articles in field of neurosci. Mem. Soc. Neurosci. (pres. 1981-82), Am. Physiol. Soc., Nat. Soc. Med. Research (dir.), Pavlovian Soc. (pres. 1978-79), Am. Assn. Anatomists. Jewish. Subspecialties: Neurobiology; Neurophysiology. Current work: Cellular basis of memory; neural control of the cardiovascular system. Office: Dept Neurobiology SUNY Stony Brook NY 11794

COHEN, EDGAR ALLAN, JR., mathematician; b. Charleston, S.C., Aug. 29, 1938; s. Edgar Allan and Elizabeth (Sternberger) C.; m. Joy Anne Bashlow, Nov. 23, 1975. A.B., Duke U., 1960; M.S., U. Cin., 1964, Ph.D., 1968. Applied mathematician Naval Surface Weapons Center, Silver Spring, Md., 1968—. Asst. scoutmaster Nat. Capital Area council Boy Scouts Am., Silver Spring, 1969-74. Served U.S. Army, 1961-62. Mem. Soc. for Indsl. and Applied Math., Assn. for Computing Machinery, Md. Entomol. Soc., Bot. Soc. Washington, Md. Ornithol. Soc. Democrat. Jewish. Subspecialties: Probability; Graphics, image processing, and pattern recognition. Current work: Signal processing as applied to target detection, stochastic modeling for minefield theory, pattern recognition and image processing as applied to target detection. Home: 5454 Marsh Hawk Way Columbia MD 21045 Office: Naval Surface Weapons Center New Hampshire Ave White Oak Silver Spring MD 20910

COHEN, EDWARD, cons. engr.; b. Glastonbury, Conn., Jan. 6, 1921; s. Samuel and Ida (Tanewitz) C.; m. Elizabeth Belle Cohen, Dec. 19, 1948 (dec. June 1979); children—Samuel, Libby, James; m. Carol Suzanne Kaleb, Jan. 11, 1981. B.S. in Engring, Columbia U., 1945, M.S. in Civil Engring, 1954. Registered profl. engr., N.Y., Conn., Fla., Ga., Md., N.J., Pa., D.C., Okla., Va., Wis., Del., Nat. Council Engring. Examiners. Engring. aide Conn. Hwy. Dept., 1940-41; asst. engr. East Hartford Dept. Pub. Works, 1942-44; structural engr. Hardesty & Hanover, N.Y.C., 1945-47, Sanderson & Porter, 1947-49; lectr. architecture Columbia, 1948-51; with Ammann & Whitney (cons. engr.), N.Y.C., 1949—, partner, 1963-74, sr. partner, 1974-76, mng. partner, 1977—; exec. v.p. in charge bldg., transp., communications, mil. projects Ammann & Whitney, Inc., 1963-78, chmn., chief exec. officer, 1978—; v.p. Ammann & Whitney Internat. Ltd., 1963-73; pres. Safeguard Constrn. Mgmt. Corp., 1973—; cons. to govt. and industry.; Stanton Walker lectr. U. Md., 1973; adv. com. Urban and Civil Engring. U. Pa., 1974—; mem. engring. council Columbia U., 1975—; dir. Concrete Industry Bd., 1976—, pres., 1978-79; concrete specialist European Concrete Com. Contbr. manuals to profl. assns., articles to profl. jours.; Co-editor Structural Concrete Handbook, 1981. Bd. dirs. Cejwin Youth Camps, 1972—; trustee Hall of Sci., N.Y.C., 1976—; mem. Bklyn. Bridge Centennial Commn., 1981. Recipient Illig medal Columbia, 1946, Egleston medal Columbia U., 1981; Patriotic Civilian Service award Dept. of Army, 1973. Fellow ASCE (Ridgway award 1946, Civil Engring. State of the Art award 1974, Raymond Reese award 1976, Earnest Howard award 1983, v.p. Met. sect. 1978-79, pres. 1980, chmn. reinforced concrete research council 1980—), N.Y. Acad. Scis. (hon. life, Laskowitz Aerospace research award 1970, vice chmn. engring. sect. 1975-77, chmn. 1977-79), Am. Concrete Inst. (hon. mem., dir. 1966-76, v.p. 1970-72, pres. 1972-73, chmn. bldg. code requirements for reinforced concrete 1963-71, Wason medal 1956, Delmar Bloem award 1973), Am. Cons. Engring. Council; mem. N.Y. Assn. Cons. Engrs. (dir. 1981—), Nat. Acad. Engring., Am. Welding Soc., Am. Nat. Standards Inst. (chmn. minimum design loads for bldgs. and other structures 1968—), ASCE Performance of Structures Research Council (chmn. com. long term observations 1972-76), N.Y. Concrete Constrn. Inst. (pres. tall bldg. council 1975—), Am. Ordnance Assn., Internat. Assn. Bridge and Structural Engrs., Internat. Bridge and Turnpike Assn., Moles, Sigma Xi, Chi Epsilon (hon.), Tau Beta Pi. Jewish religion. Clubs: Engineers (N.Y.C.) (dir. 1974-75); Wings.). Subspecialties: Civil engineering; Aerospace engineering and technology. Current work: Development of seismic building codes,dynamic response of structures advancing knowledge of reinforced concrete, wind force evaluation antenna structures. Home: 56 Chestnut Hill Roslyn NY 11576 Office: 2 World Trade Center New York NY 10048 Do not give up personal integrity for any apparent "practical" advantage. Strive successful projects but do not seek personal credit. Make no adverse judgments of people unless it is an active consideration in a necessary decision. Judge people by their actions, not their words.

COHEN, EDWIN, psychologist; b. N.Y.C., Aug. 26, 1924; s. Paul and Clara (Lobel) C.; m. Myrna Skalovsky, Aug. 22, 1949 (div. 1951); 1 dau., Kaye Suzanne Kramer; m. Judith Barnett, Aug. 21, 1958; children: Rebecca, Deborah. A.B., Cornell U., 1944; M.S., U. Okla., 1949, Ph.D., 1955. Lic. psychologist, N.Y. State; cert. Soc. Automotive Engrs. Project dir. Psychol. Research Assocs., Washington, 1953-55; tng. specialst The RAND Corp., Santa Monica, Calif., 1955-56; project dir. Ednl. Research Corp., Cambridge, Mass., 1956-58; staff scientist link flight simulation div. The Singer Co., Binghamton, N.Y., 1958—; adj. faculty Simmons Coll., Boston, 1957-58, SUNY-Binghamton, 1968-74; mem. N.Y. State Bd. Psychology, 1977—. Contbr. articles to tech. jours. Officer, trustee Temple Concord, Binghamton, 1970-73. Served with USN, 1944-46. Mem. Am. Psychol. Assn., Human Factors Soc. Republican. Jewish. Subspecialties: Human factors engineering; Systems engineering. Current work: Training, simulation and training equipment, visual perception as related to visual simulation, control/display optimization, human performance measurement, safety. Inventor. Home: 5 Crestmont Rd Binghamton NY 13905 Office: Singer Co Link Flight Simulation Div Binghamton NY 13902

COHEN, GLENN MILTON, ear researcher, consultant; b. Elizabeth, N.J., Sept. 8, 1943; s. Hyman and Alice (Pollack) C.; m. Rona Gail Zwillman, June 4, 1968. B.A., Rutgers U., 1965; Ph.D., Fla. State U., Tallahassee, 1970. Asst. research scientist NYU Med. Sch., 1970-72; asst. prof. biol. scis. Fla. Inst. Tech., Melbourne, 1972-78, assoc. prof., 1978—. Contbr. articles to profl. jours. NASA Research fellow, 1976, 77; NIH grantee, 1977. Mem. Assn. Research in Otolaryngology, Soc. Neursci., Electron Microscopy Soc. Am., Am. Soc. Zoologists. Subspecialties: Prosthetics; Otorhinolaryngology. Current work: Development of ear. Effects of drugs on senses of hearing and balance; motion sickness. Home: PO Box 3451 Indiatlantic FL 32903 Office: Dept Biol Scis Fla Inst Tech 150 W University Blvd Melbourne FL 32901

COHEN, HARLEY, civil engineering educator, researcher; b. Winnipeg, Man., Can., May 12, 1933; s. Joseph and Ettie (Gilman) C.; m. Estelle Brodsky, Dec. 25, 1956; children: Brent, Murray, Carla. B.Sc. with honors, U. Man., Winnipeg, 1956; Sc.M., Brown U., 1958; Ph.D., U. Minn.-Mpls., 1964. Research engr. Boeing Co., Seattle, 1958-60; prin. scientist Honeywell Inc., Mpls., 1960-64; asst. prof. aero. and engring. mechanics U. Minn., 1964-65; prof. civil engring. U. Man., 1965-83, Univ. disting. prof., 1983—; James L. Record prof. U. Minn.-Mpls., 1979; Killam vis. scholar U. Calgary, Alta., Can., 1982. Mem. editorial bd.: Utilitas Mathematics, 1975—, Iran Jour. Sci. Tech, 1975—; contbr. numerous articles to profl. jours. Mem. Acad. Mechanics, Soc. Engring. Sci., Soc. Natural Philosophy. Subspecialties: Theoretical and applied mechanics; Applied mathematics. Current work: Nonlinear wave propagation in rods, shells and membranes. Home: 55 Tanoak Park Dr Winnipeg MB Canada R2V 2W6 Office: Department Civil Engineering University Manitoba Winnipeg MB Canada R3T 2N2

COHEN, IRA LARRY, psychologist, therapist, researcher; b. Phila., Nov. 15, 1947; s. Meyer and Pauline (Smogar) C.; m. Chaya Frohman, June 7, 1970; children: Micole, Mirit. B.A., Temple U., 1969; M.S., Rutgers U., 1972, Ph.D., 1974; postgrad., NYU Med. Ctr., 1975-77. Adj. asst. prof. Lehman Coll., CUNY, 1974; instr. Seton Hall U., 1974, Kean Coll., 1974-77, NYU Med. Ctr., N.Y.C., 1975-80, research asst. prof. psychiatry, 1981—; research scientist Inst. Basic Research, S.I., N.Y., 1981—; behavior therapist Ctr. Counseling & Behavior Therapy, N.Y.C., 1977-78, Group for Psychol. Counseling, 1978-79; consulting psychologist Assn. Advancement Blind and Retarded, St. Albans, N.Y., 1980—. Contbr. chpts. to books and articles to profl. jours. NIMH fellow, 1975. Mem. Am. Psychol. Assn., AAAS, Am. Assn. Mental Deficiency, N.Y. State Psychol. Assn. Subspecialties: Behavioral psychology; Developmental psychology. Current work: Behavioral and learning characteristics of autistic and other developmentally disabled persons; behavior therapy with these same populations; psychopharmacological treatment with same populations. Office: Inst Basic Research 1050 Forest Hill R Staten Island NY 10314

COHEN, IRWIN, chemistry educator; b. Cleve., Feb. 28, 1924; s. Louis M. and Sadie (Dorsky) C.; m. Beatrice Lewin, Dec. 30, 1945; children: Martin, Richard, David. A.B., Western Res. U., 1944, M.S., 1948, Ph.D., 1950. Asst. prof. chemistry Youngstown (Ohio) State U., 1949-53, asso. prof., 1953-58, prof., 1958—, dir. individualized curriculum program, 1971-73; coordinator Youngstown B.S./M.D. program (Ohio) State U., 1974-77; coordinator non-traditional programs non-traditional programs, 1978-81. Contbr. articles on chemistry to profl. jours. Served with USN, 1944-46. Recipient Disting. Prof. award Youngstown State U, 1965, 79; NSF grantee, 1967, 1967,68. Mem. Am. Chem. Soc., AAUP, Ohio Edn. Assn., Sigma Xi, Alpha Epsilon Delta, Phi Beta Kappa, Phi Kappa Phi. Jewish. Subspecialties: Organic chemistry; Theoretical chemistry. Current work: Relation of molecular orbital calculations to chemical covalent bond order and charge distribution, using density differences, localized orbitals and population analysis. Home: 45 Melrose Ave Youngstown OH 44512 Office: Youngstown State U Youngstown OH 44555

COHEN, JAMES SAMUEL, physicist; b. Houston, July 29, 1946; s. Herman and Jimmie Ruth (Harrington) C.; m. Marion Fay Daniel, Dec. 28, 1968; children: Stephen James, Christy Lynn. B.A., Rice U., 1968, M.A., 1970, Ph.D., 1973. Staff mem. Los Alamos (N.Mex.) Nat. Lab., 1972—; vis. asso. prof. Rice U., Houston, 1979-80. Contbr. articles in field to profl. jours. Recipient H. A. Wilson Prize Rice U., 1973. Mem. Am. Phys. Soc., Phi Beta Kappa, Sigma Xi. Subspecialties: Atomic and molecular physics; Theoretical physics. Current work: Research in basic atomic and molecular physics, laser kinetics and muon physics. Home: 330 Valle del Sol Los Alamos NM 87544 Office: Los Alamos Nat Lab T-12 MS-J569 Los Alamos NM 87545

COHEN, JEROME BERNARD, materials science educator; b. Bklyn., July 16, 1932; s. David I. and Shirley Anne C.; m. Lois Nesson, Sept. 15, 1957; children: Elissa Diane, Andrew Neil. B.S., Mass. Inst. Tech., 1954, Sc.D., 1957. Sr. scientist materials AVCO Corp., Wilmington, Mass., 1958-59; mem. faculty Northwestern U., 1959—, prof. materials sci. and engring., 1965—, chmn. dept. materials sci. and engring., 1973-78, Frank C. Engelhart prof., 1974—, fellow Center Teaching Professions, 1971—, prof. Technol. Inst., 1983—; sci. liaison officer Office Naval Research, London, 1966-67; cons. to govt. and industry. Mem. bd. Author: Diffraction Methods in Materials Science, 1966; co-author: Diffraction from Materials, 1978; Co-editor: Local Atomic Arrangements Studied by X-Ray Diffraction, 1967, Jour. Applied Crystallography, Modulated Structures, 1979. All-Star coach Glencoe (Ill.) Hockey Assn., 1974-77. Served as 1st lt. AUS, 1959. Fulbright fellow U. Paris, 1957-58; recipient Tech. Inst. Teaching award Northwestern U., 1976. Fellow Am. Inst. Metall., Mining and Petroleum Engring. (Hardy Gold medal 1960), Am. Soc. Metals (Henry Marion Howe medal 1981); mem. Am. Soc. Engring. Edn. (George C. Westinghouse award 1976), Am. Ceramic Soc., Am. Crystallographic Assn., Royal Instn. Gt. Britain, AAUP, Sigma Xi, Tau Beta Pi, Alpha Sigma Mu, Phi Lambda Upsilon. Jewish. Subspecialty: Metallurgy. Patentee in field. Home: 362 Jackson Ave Glencoe IL 60022 Office: 2145 Sheridan Rd Evanston IL 60201

COHEN, JOEL RALPH, microbiologist, biology educator; b. Chelsea, Mass., Oct. 20, 1926; s. Julius Meyer and Pearl (Mankin) C.; m. Marilyn Roberta Lezar, Sept. 7, 1947; children: Robert Neil, Deborah Ellen, Peter Alan. B.S., U. Mass., 1949, M.S., 1950, Ph.D., 1975. Specialist, registered microbiologist Nat. Registry of Microbiology; cert. clin. lab. specialist Nat. Cert. Agy. for Med. Lab. Personnel. Microbiologist Baystate Med. Ctr., Springfield, Mass., 1950-68, chief clin. labs., 1960-68; assoc. prof. biol. sci. Springfield Coll., 1968-75, chmn. biology dept., 1969-81; prof. biol. sci., 1975—; cons. VA Med. Ctr., Springfield Mcpl. Hosp., Springfield Dept. Pub. Health, Baystate Med. Ctr. Contbr. articles to profl. jours. Served to col. USAR, 1943-80. Fellow AAAS, Am. Pub. Health Assn. (sec. lab. sect. 1979-82), Royal Soc. Health, Am. Acad. Microbiology; mem. N.Y. Acad. Scis, Am. Soc. Microbiology, Am. Assn. Blood Banks, Sigma Xi. Subspecialty: Microbiology. Current work: History of development of public health laboratory methods. Office: Springfield College 263 Alden St Springfield MA 01109 Home: 14 Inglewood Ave Springfield MA 01119

COHEN, JUDITH GAMORA, astronomer; b. N.Y.C., May 5, 1946; d. Isidore and Nettie (Marshak) C.; m. Gaston Araya, 1973. B.A., Radcliffe Coll., 1967; M.S., Calif. Inst. Tech., 1969, Ph.D., 1971. Asst. astronomer Kitt Peak Nat. Obs., Tucson, 1974-77, asso. astronomer, 1977-79; asso. prof. astronomy Calif. Inst. Tech., 1979—; mem. staff Palomar Obs. Miller fellow, 1972-74. Mem. Internat. Astron. Union, Am. Astron. Soc. Subspecialties: Optical astronomy; Infrared optical astronomy. Office: 105-24 California Institute Technology Pasadena CA 91125

COHEN, KARL PALEY, nuclear energy consultant; b. N.Y.C., Feb. 5, 1913; s. Joseph M. and Ray (Paley) C.; m. Marthe H. Malartre, Sept. 20, 1938; children: Martine-Claude Lebouc, Elisabeth M. Brown, Beatrix Josephine Cashmore. A.B., Columbia U., 1933, M.A., 1934, Ph.D., 1937; postgrad., U. Paris, 1936-37. Research asst. to Prof. H. C. Urey Columbia U., 1937-40; dir. theoretical div., SAM Manhattan project, 1940-44; physicist Standard Oil Devel. Co., 1944-48; tech. dir. H.K. Ferguson Co., 1948-52; v.p. Walter Kidde Nuclear Lab., 1952-55; cons. AEC, sr. sci. Columbia U., 1955-56; mgr. advance engring. atomic power equipment dept. Gen. Electric Co., 1956-65, gen. mgr. breeder reactor devel. dept., 1965-71, mgr. strategic planning, nuclear energy div., 1971-73, chief scientist, nuclear energy group, 1973-78; cons. prof. Stanford U., 1978-81. Author: The Theory of Isotope Separation as Applied to Large Scale Production of U-235, 1951; contbr. articles to profl. jours. Recipient Energy Research prize Alfried Krupp Found., 1977; Chem. Pioneer award Am. Inst. Chemists, 1979. Fellow Am. Nuclear Soc. (pres. 1968-69, dir.), AAAS; mem. Nat. Acad. Engring., Am. Phys. Soc., Cactus and Succulent Soc., Phi Beta Kappa, Sigma Xi, Phi Lambda Upsilon. Subspecialties: Nuclear engineering; Nuclear fission. Current work: Reactor safety and nuclear power economics and their interaction. Office: 928 N California Ave Palo Alto CA 94303

COHEN, MARLENE LOIS, research scientist, educator; b. New Haven, May 5, 1945; d. Abraham David and Jeanette (Bader) C.; m. Jerome Herbert Fleisch, August 8, 1976; children: Abby Faye, Sheryl Brynne. B.S. in Pharmacy, U. Conn., 1968; Ph.D. in Pharmacology, U. Calif., San Francisco, 1973. Registered pharmacist, Conn., Calif. Postdoctoral fellow Roche Inst. Molecular Biology, Nutley, N.J., 1973-75; sr. pharmacologist Lilly Research Labs., Eli Lilly and Co., Indpls., 1975-80, research scientist, 1980—; adj. asst. prof. pharmacology Ind. U. Sch. Medicine, Indpls., 1976-82, adj. asso. prof., 1982—; ad hoc reviewer for sci. jours. Mem. editorial bd.: Jour. Clin. and Exptl. Hypertension, 1978—, Procs. Soc. Exptl. Biology and Medicine, 1979—; contbr. articles, revs. and abstracts to profl. jours., also chpts. to books. Mem. Soc. for Exptl. Biology and Medicine, Am. Soc. Pharmacology and Exptl. Therapeutics, Indpls. Children's Mus., Alpha Lambda Delta, Phi Kappa Phi, Rho Chi. Subspecialties: Pharmacology; Neuropharmacology. Current work: Smooth muscle function with emphasis on treatment of hypertension; role of neuronal innervation, enkephalins and angiotensin converting enzyme in smooth muscle function. Office: Cardiovascular Pharmacology Dept Lilly Research Labs Indianapolis IN 46285

COHEN, MARSHALL HARRIS, radio astronomer, educator; b. Manchester, N.H., July 5, 1926; s. Solomon and Mollie (Epstein) C.; m. Shirley Kekst, Sept. 19, 1948; children—Thelma, Linda, Sara. B.E.E., Ohio State U., 1948, M.S., 1949, Ph.D., 1952. Research asso. Ohio State U., 1950-54; faculty Cornell U., 1954-66; prof. applied electrophysics U. Calif., San Diego, 1966; vis. asso. prof. radio astronomy Calif. Inst. Tech., Pasadena, 1965, prof. radio astronomy, 1968—, exec. officer for astronomy, 1981—. Contbr. articles to profl. jours. Co-recipient Rumford medal Am. Acad. Arts and Sci., 1971; John Simon Guggenheim Meml. Found. fellow, 1960-61, 80-81. Fellow A.A.A.S.; mem. Am. Astron. Soc., Astron. Soc. of Pacific (dir. 1970-73), Internat. Astron. Union, Internat. Sci. Radio Union (U.S. chmn. Commn. V 1969-72). Subspecialty: Planetary science. Current work: Very-long baseline interferometry; extragalactic sources. Office: Astronomy Dept Calif Inst of Tech Pasadena CA 91125

COHEN, MARSHALL JAY, research physicist; b. Detroit, Aug. 29, 1949; s. Abraham and Ronnie (Shur) C.; m. Deborah Duchin, Aug. 11, 1974; 1 dau., Jessica Rachel. B.S., U. Mich.-Ann Arbor, 1971; Ph.D., U. Pa.-Phila., 1975. Research assoc. U. Pa.-Phila., 1975-77; mem. tech. staff Rockwell Internat. Sci. Center, Thousand Oaks, Calif., 1977-80, sect. mgr., 1980-82; sr. research physicist Chevron Research Co., Richmond, Calif., 1982—. Contbr. articles in field to profl. jours. Mem. IEEE, Am. Phys. Soc. Jewish. Subspecialties: Condensed matter physics; Solar energy. Current work: Materials science and applications of electrically conducting polymers. Inventor epitaxial polymer films, 1975, poly (sulfur-nitride) solar cells, 1980, plastic solar cells, 1982. Office: Chevron Research Co PO Box 1627 Richmond CA 94801 Home: 236 McNear Dr San Rafael CA 94901

COHEN, MARTIN, research astronomer; b. Prestwich, Manchester, Eng., July 27, 1948; came to U.S., 1970; s. Harold and Mildred (Glynn) C.; m. Barbara Anne Freda, July 4, 1982. B.A. with honors, Clare Coll., U. Cambridge, Eng., 1969, Ph.D., 1972. Research astronomer U. Calif., Berkeley, 1972-75, research astronomer, 1976-79; research fellow Royal Observatory, U. Edinburgh, Scotland, 1975; sr. research asso. NRC, NASA, Ames Research Center, Moffett Field, Calif., 1979-81; vis. asso. research astronomer Radio Astronomy Lab., U. Calif., Berkeley, 1981—; contract astronomer NASA Ames Research Center, Moffett Field, Calif., 1981—. Author: In Quest of Telescopes, 1980; Contbr. numerous articles to tech. jours. Recipient Clare Coll., Cambridge U. Thomas Greene Silver Cup award, 1969. Mem. Internat. Astron. Union, Am. Astron. Soc. Subspecialties: Infrared optical astronomy; Radio and microwave astronomy. Current work: Star formation; late stellar evolution; bipolar nebulae; T Tauri stars; Herbig-Haro objects; VLA observations of stars; interactive software for two-dimensional IR CCD image processing; bipolar nebulae. Home: 3801 Laguna Ave Oakland CA 94602 Office: Radio Astronomy Lab 601 Campbell Hall Univ Calif Berkeley CA 94720

COHEN, MARVIN SIDNEY, experimental psychologist; b. Walterboro, S.C., Dec. 15, 1946; s. Mordecai and Dorothy (Gelson) C.; m. Susan Mary Allan, June 15, 1980; 1 son, Isaac William. B.A., Harvard U., 1968, Ph.D., 1980; M.A., U. Chgo., 1971. Research psychologist Harvard U., Cambridge, Mass., 1971-79; analyst to sr. analyst Decision Sci. Consortium, Inc., Falls Church, Va., 1979-82, 83; program mgr. Maxima Corp., Bethesda, Md., 1982. Mem. Brain and Behavioral Scis., Am. Psychol. Assn., Human Factors Soc., Inst. Mgmt. Sci., Am. Soc. Naval Engrs., Sigma Xi. Subspecialties: Cognition; Information systems, storage, and retrieval (computer science). Current work: Design and testing of computer-based decision support systems; implementation of Bayesian and non-Bayesian prescriptive models of reasoning; tailoring systems to individual cognitive styles and shortcomings. Home: 1257 Beverly Rd McLean VA 22101 Office: Decision Sci Consortium Inc 7700 Leesburg Pike Suite 421 Falls Church VA 22043

COHEN, MELVIN JOSEPH, neuroscientist; b. Los Angeles, Sept. 28, 1928; s. Samuel and Bessie (Firman) C.; m. Catherine Black, Dec. 27, 1963; children: Frank M., Linn C., Sarah R., Samuel D. B.A., UCLA, 1949, M.A., 1952, Ph.D., 1954; student, U. Calif., Berkeley, 1950-51. Instr. biology Harvard, 1955-57; asst. prof. to prof. biology U. Oreg., 1957-69; prof. biology Yale, 1969—. Contbr. articles to profl. jours. NSF postdoctoral fellow, Stockholm, 1954-55; Guggenheim fellow, Oxford, Eng., 1965. Mem. Nat. Acad. Scis., Internat. Brain Research Orgn., Soc. Gen. Physiologists (pres. 1976-77), Am. Soc. Zoologists, Soc. Neurosci. Home: 15 Salt Meadow Ln Madison CT 06443 Office: Dept Biology Yale New Haven CT 06511

COHEN, MICHAEL PAUL, mathematical statistician; b. San Mateo, Calif., July 8, 1947; s. Herman Charles and Evadna Fern (Tull) C. B.A., U. Calif.-San Diego, 1969; M.A., UCLA, 1971, Ph.D., 1978. Statistician, Office Prices and Living Conditions, Bur. Labor, Stats., U.S. Dept. Labor, Washington, 1979—. Mem. Am. Statis. Assn., Inst. Math. Stats., Am. Math. Soc., Soc. Indsl. and Applied Math., Am. Assn. Pub. Opinion Research. Club: Calif. State Soc. (Washington). Subspecialties: Statistics; Applied mathematics. Current work: Variance estimation for the U.S. Consumer Price Index, composite estimation for small areas, applied survey design research, properties of finite population estimators. Office: Office Prices Bur Labor Statistics 600 E St NW Room 5217 Washington DC 20212

COHEN, MORREL HERMAN, physicist, biologist, educator; b. Boston, Sept. 10, 1927; s. David and Rose (Kemler) C.; m. Sylvia Zwein, June 18, 1950; children: Julie, Robert, Daniel, Lisa. B.S. in Physics, Worcester Poly. Inst., 1947, D.Sc. (hon.), 1973; M.A. in Physics, Dartmouth Coll., 1948, Ph.D., U. Calif.-Berkeley, 1952. Mem. faculty U. Chgo., 1952-81, assoc. prof. physics, 1957-60, prof., 1960-81, prof. theoretical biology, 1968—, Louis Block prof. physics and theoretical biology, 1972-81, mem. com. developmental biology, 1973-74, publs. bd., 1969-70; acting dir. James Franck Inst., 1965-66, dir., 1968-71; dir. materials research lab. NSF, 1977-81; sr. sci. advisor corp. research scis. lab. Exxon Research and Engring. Co., 1981—; cons. govt. and industry, 1953—; vis. scientist NRC, Can., 1960, Xerox Corp., 1978; Shrum lectr. Simon Fraser U., 1973; assoc. Clare Hall U. Cambridge, Eng., 1973—; vis. prof. U. Va., 1976, Kyoto U., 1979; mem. adv. panel electrophysics NASA, 1962-66; mem. adv. com. Nat. Magnet Lab., 1963-66; mem. rev. com. solid state sci. and metallurgy div. Argonne Nat. Lab., 1964-67, chmn., 1966, bd. govs., 1982—; chmn. Gordon Conf., 1968, 4th Internat. Conf. Armorphous and Liquid Semicondrs., 1971; mem. adv. com. Inst. Amorphous Studies, 1982—; mem. Army Basic Research Com., 1979—, mem. steering com., 1980—; mem. adv. com. dept. physics U. Tex., Austin, 1982—. Author articles physics of solids, liquids, gases, theoretical and developmental biology.; assoc. editor: Jour. Chem. Physics, 1960-63; mem. editorial bd.: advanced physics monograph series McGraw-Hill Co., 1963-70; editorial bd.: The Physics of Condensed Matter, 1962-74; publs. bd., U. Chgo., 1969-70; bd. editors: Jour. Statis. Physics, 1970-75. AEC fellow, 1951-52; Guggenheim fellow, 1957-58; NSF sr. postdoctoral fellow, Rome, 1964-65; NIH spl. fellow, 1972-73. Fellow Am. Phys. Soc. (council 1978-82, chmn. solid state physics div. 1970, div. councillor 1978-82); mem. AAAS, Am. Inst. Physics, Nat. Acad. Scis., Sigma Xi (nat. lectr. 1966). Subspecialties: Biophysics (physics); Condensed matter physics. Home: 1100 Crim Rd Bridgewater NJ 08807 Office: PO Box 45 Linden NJ 07036

COHEN, MORTON I., physiology educator; b. N.Y.C., July 11, 1923; s. Oscar and Bessie (Beller) C.; m. Priscilla J. Goettler, Dec. 16, 1958; children: Sibyl A., Eve L. B.S., CCNY, 1942; A.M., Columbia U., 1950, Ph.D., 1957. Instr. Albert Einstein Coll. Medicine, Bronx, N.Y., 1957-59, asst. prof., 1959-68, assoc. prof., 1968-73, prof., 1973—. Served with USAAF, 1945-46. Mem. Am. Physiol. Soc., Soc. for Neurosci., European Neurosci. Assn. Subspecialty: Neurophysiology. Current work: Genesis of respiratory rhythm, central pattern generators. Office: Dept Physiology Albert Einstein Coll Medicine 1300 Morris Park Ave Bronx NY 10461

COHEN, NICHOLAS, immunologist, educator, consultant; b. N.Y.C., Nov. 20, 1938; s. Saris and Frances Edith (Pakett) C.; m. Jayne Sevin, July 1, 1962; children: Jaime, Jessica; m. Catharina Johanna, Oct. 23, 1974; children: Misha, Mark. A.B., Princeton U., 1959; Ph.D., U. Rochester, 1966; postdoctoral scholar, UCLA, 1965-67. Asst. prof. microbiology (immunology) U. Rochester (N.Y.) Sch. Medicine and Dentistry, 1967-73, assoc. prof., 1973-80, prof., 1980—, dir. div. immunology, 1980—; mem. Basel (Switzerland) Inst. Immunology, 1975-76; vis. prof. Agrl. U. Wageningen (Netherlands); mem. peer rev. panels and study sects. NIH and NSF; cons. Mem. editorial bds. profl. jours.; also editor books; contbr. articles to profl. jours. NIH grantee, 1967—; recipient Donald R. Charles Meml. award U. Rochester Dept. Biology, 1966; Fulbright scholar, 1982-83. Mem. Am. Soc. Zoologists, Transplantation Soc., Am. Assn. Immunologists, Internat. Soc. Developmental and Comparative Immunology, ACLU, Sigma Xi. Subspecialties: Immunogenetics; Neuroimmunology. Current work: Evolution of immunity, ontogeny of immunity, transplantation immunology, psychoneuroimmunology. Office: U Rochester Med Center Box 672 Rochester NY 14642

COHEN, NORMAN EDWARD, software development company manager; b. Louisville, Dec. 28, 1947; s. Seymour and Shirley C.; m. Jessica L. Field, Aug. 25, 1968; 1 son, Geoffrey. B.A., Cornell U., 1970; M.S., Va. Poly. Inst. and State U., 1980. Programmer Gen. Electric Co., Syracuse, N.Y., 1980-82; mgr. graphics applications Genigraphics Corp., Liverpool, N.Y., 1982—. Recipient Teaching Excellence awards Va. Poly. Inst. and State U., 1979, 80. Mem. Assn. Computing Machinery, IEEE Computer Soc. (assoc.). Subspecialties: Graphics, image processing, and pattern recognition; Software engineering. Current work: Interactive, full-color design stations for producing presentation and business graphics, recording digital images on film, computer animation. Office: Genigraphics Corp PO Box 591 Liverpool NY 13088 Home: 4520 Waltham Dr Manlius NY 13104

COHEN, PAUL JOSEPH, mathematician, educator; b. Long Branch, N.J., Apr. 2, 1934; s. Abraham and Minnie (Kaplan) C.; m. Christina Martha Karls, 1963; children—Eric, Steven, Charles. Student, Bklyn. Coll., 1950-53; M.S., U. Chgo., 1954, Ph.D., 1958. Instr. U. Rochester, 1957-58, Mass. Inst. Tech., 1958-59; fellow Inst. Advanced Study, Princeton, 1959-61; mem. faculty Stanford, 1961—, prof. math., 1964—. Recipient Fields medal Internat. Math. Union, 1966; Bocher prize Am. Math. Soc., 1964; Research Corp. award Research Corp., 1964; Nat. Medal Sci., 1967. Proved impossibility of demonstrating continuum hypothesis from the axims of set theory. Office: Dept Math Stanford U Stanford CA 94305

COHEN, PAUL S., microbiology educator; b. Boston, Jan. 20, 1939; s. Solomon and Frieda (York) C.; m.; children: Matthew, Dan, Allison. A.B., Brandeis U., 1960; M.S., Boston U., 1962, Ph.D., 1964. Nat. Cancer Inst. postdoctoral trainee St. Jude Children's Research Hosp., Memphis, 1964-66; asst. prof. microbiology U. R.I., Kingston, 1966-69, assoc. prof., 1969-75, prof., 1975; vis. assoc. prof. biology MIT, 1973; vis. research scientist Ciba-Geigy Research Div., Basel, Switzerland, 1981. Contbr. articles to sci. jours.; editorial bd.: European Jour. Clin. Microbiology, 1981—, Antimicrobial Agtrs. and Chemotherapy, 1978—. NSF grantee. Mem. Am. Soc. Biol. Chemists, Am. Soc. Microbiology, Sigma Xi. Subspecialties: Microbiology; Molecular biology. Current work: Regulation of vesicular stomatitis virus protein synthesis, the molecular basis of E coli colonization of the mammalian colon. Home: Glen Rock Rd West Kingston RI 02892 Office: U RI Kingston RI 02881

COHEN, PAUL SHEA, chemistry educator; b. N.Y.C., June 18, 1938; s. Irving and Grace (Bromberg) C.; m. Brenda H. Cooperman, June 18, 1960; children: Stephen, Mara. B.A., Bklyn. Coll., 1960; M.S., U. Ill., 1963; Ed.D., Temple U., 1970. Asst. prof. chemistry Frostburg

(Md.) State Coll., 1963-65; asst. prof. chemistry Trenton (N.J.) State Coll., 1965-73, assoc. prof., 1973—. Author: Preparatory Chemistry, 1983; contbr. articles to profl. jours. Bd. dirs. Sci. Camp, Trenton, 1977—, Inst. for Gifted and Talented Services, Trenton, 1980—; mem. Environ. Commn., Ewing, N.J., 1980—. Mem. Sigma Zeta. Subspecialties: Inorganic chemistry. Current work: Calcium analysis, education of gifted children in the sciences. Office: Dept Chemistry Trenton State Coll CN 550 Trenton NJ 08625

COHEN, PHILIP, hydrogeologist; b. N.Y.C., Dec. 13, 1931; s. Isadore and Anna (Katz) C.; m. Barbara Sandler, Dec. 26, 1954; 1 son, Jeffery. B.S. cum laude, CCNY, 1954; M.S., U. Rochester, 1956. With U.S. Geol. Survey, 1956—, asso. chief land info. and analysis office, Reston, Va., 1975-78, asst. chief hydrologist water resources div., Reston, 1978-79, chief hydrologist water resources div., 1979—. Contbr. numerous articles on geology and hydrology to profl. jours. Recipient Ward medal Coll. City, N.Y., 1954; Meritorious Ser. award Dept. Interior, 1975, Disting. Ser. award, 1979. Fellow Geol. Soc. Am.; mem. Am. Water Resources Assn., Am. Geophys. Union, Am. Inst. Profl. Geologists, Sigma Xi. Subspecialties: Hydrology; Ground water hydrology. Current work: Science manager. Office: 709 Geol Survey Reston VA 22092

COHEN, PHILIP LAWRENCE, physician, research scientist, educator; b. N.Y.C., Aug. 22, 1948; s. Morris L. and Jean S. C.; m. Deborah Lawrence, Nov. 8, 1974; children: Andrew, Lisa. B.S., CCNY, 1968; M.D., Yale U., 1972. Diplomate: Am. Bd. Internal Medicine. Intern Presbyn. Hosp., N.Y.C., 1972-73, resident, 1975-76; research assoc. Nat. Inst. Arthritis and Infectious Diseases, NIH, 1973-75; fellow U. Tex. Health Sci. Ctr., Dallas, 1976-77, asst. prof. medicine, 1977-79; asst. prof. medicine and bacteriology and immunology U. N.C., Chapel Hill, 1979—; practice medicine, specializing in rheumatology, Chapel Hill, 1979—. Served with USPHS, 1973-75. Arthritis Found. fellow, 1976-77; sr. investigator, 1982—; NIH clin. investigator, 1977-1979. Fellow ACP; mem. Am. Assn. Immunologists, Am. Rheumatism Assn. Subspecialty: Immunology (medicine). Current work: Cellular mechanism of autoimmune diseases. Office: Univ North Carolina 932 FLOB 231H Chapel Hill NC 27514 Home: 119 Meadowbrook Dr Chapel Hill NC 27514

COHEN, PHILIP PACY, biochemist, educator; b. Derry, N.H., Sept. 26, 1908; s. David Harris and Ada (Cottler) C.; m. Rubye Herzfeld Tepper, June 15, 1935; children: Philip T., David B., Julie A., Milton T. B.S., Tufts Coll., 1930; Ph.D., U. Wis., 1937, M.D., 1938; D.Sc. (hon.), U. Mex., 1979. NRC fellow, Sheffield, Eng., 1938-39, Yale, 1939-40, instr., 1940-41; intern Wis. Gen. Hosp., 1941-42; asst. prof. clin. biochemistry U. Wis., 1942-45, asso. prof. physiol. chemistry, 1945-47, prof., 1947—, chmn. dept. physiol. chemistry, 1948-75, H.C. Bradley prof., 1968—, acting dean, 1961-63; Chmn. com. on growth, mem. exec. com. div. med. scis. NRC, 1954-56; bd. sci. counselors Nat. Cancer Inst., 1957-59, chmn., 1959-61; mem. physiol. chemistry study sect. NIH, 1959-62, nat. adv. cancer council, 1963-67, mem. adv. com. to dir., 1966-70. Mem. Nat. Adv. Arthritis and Metabolic Disease Council, 1970-74; adv. com. biology and medicine AEC, 1963-71; mem. adv. com. on med. research Pan Am. Health Orgn., 1967-75; mem. Nat. Commn. on Research, 1978-80; hon. mem. Med. Sch. Faculty, U. Chile. Commonwealth Fund fellow Oxford U., Eng., 1958. Fellow A.A.A.S.; mem. Nat. Acad. Scis., Am. Soc. Biol. Chemists (treas. 1951-56), Am. Chem. Soc., Biochem. Soc. (Eng.)., Sigma Xi; hon. mem. Harvey Soc., Chiba Med. Soc. (Japan), Argentina Biochem. Soc., Nat. Acad. Med. (Mex.), Japanese Biochem. Soc. Subspecialties: Biochemistry (biology); Molecular biology. Current work: Enzyme biosynthesis and regulation. Office: 587 Medical Sciences Bldg U Wis Madison WI 53706

COHEN, RICHARD S., physicist; b. Pitts., Sept. 20, 1946; s. John J. and Harriet H. (Goldstein) C. B.A., Princeton U., 1968; M.A., Columbia U., 1970, M.Phil., 1971, Ph.D., 1977. Resident research assoc. Nat. Acad. Scis. Inst. Space Studies, N.Y.C., 1977-79; research assoc. Columbia U., 1979—. Contbr. articles to profl. jours. Mem. Am. Physiol. Soc., Am. Astron. Soc., IEEE. Subspecialty: Radio and microwave astronomy. Current work: Millimeter-wave radio astronomy, galactic structure. Office: Institute for Space Studies 2880 Broadway New York NY 10025

COHEN, ROBERT EDWARD, chemical engineering educator, consultant; b. Oil City, Pa., Jan. 21, 1947; s. David M. and Minnie (Magdovitz) C.; m. Jane Woodman, Nov. 18, 1978; 1 dau., Genevieve E. B.S., Cornell U., 1968; M.S., Calif. Inst. Tech., 1970, Ph.D., 1972; postdoctoral student, Oxford (Eng.) U., 1972-73. Asst. prof. chem. engring. M.I.T., 1973-77, assoc. prof., 1977-82, prof., 1982—; cons. Hercules Inc., Wilmington, Del., 1980—, Xerox Corp., Webster, N.Y., 1980—. Named Dreyfus Found. Tchr.-Scholar, 1977. Mem. Am. Inst. Chem. Engring., Am. Chem. Soc., Soc. Rheology, British Soc. Rheology, N.Y. Acad. Scis. Subspecialties: Polymer engineering; Chemical engineering. Current work: Physics and chemistry of polymers. Co-patentee polymer blends; process for fluorinating polymers surfaces. Office: Dept Chem Engring Mass Inst Tech Bldg 66 Room 554 Cambridge MA 02139

COHEN, ROBERT JAY, biochemist; b. Milw., May 31, 1942; s. Harry and Mildred (Muellen) C.; m. Carol N. Neal, Feb. 25, 1968; children: Benjamin, Jonathan, Deborah. B.S., U. Wis., 1964; Ph.D., Yale U., 1969. Postdoctoral fellow biology div. Calif. Inst. Tech., Pasadena, 1969-71; asst. prof. dept biochemistry and molecular biology U. Fla., Gainesville, 1971-76, assoc. prof., 1976—. Contbr. articles to profl. jours. Mem. Hazardous Material Com., City of Gainesville, 1980—. NIH predoctoral fellow, 1966-69; postdoctoral fellow, 1969-71; grantee, 1972-79; NSF grantee, 1980—. Mem. Am. Chem. Soc., Biophys. Soc., Assn. for Research in Vision and Ophthalmology, Am. Soc. Biol. Chemists. Subspecialties: Biochemistry (biology); Biophysics (biology). Current work: Biochemistry, biophysics and genetics of sensory processing in Phycomyces; theoretical biology; phys. biochemistry of nucleic acids. Office: Dept Biochemistry and Molecular Biolog U Fla Coll Medicine Gainesville FL 32610

COHEN, RONALD ALEX, dentist; b. Flushing, N.Y., Sept. 16, 1944; s. Joseph Nathan and Rose (Rutkay) C.; m. Anita Sharon Liedarson, Oct. 26, 1969; children-Jill Stacie, Jodi Nicole. A.A., Queensborough Community Coll., 1964; B.A., Qunnipiac Coll., 1968; M.S., Adelphi U., 1970; D.D.S., N.Y.U., 1974. Tchr., N.Y.C., 1968-71; dentist Am. Dental Assn., Chgo., 1971—; instr Coll. Dentistry N.Y.U., 1978-80; lectr., U.S., 1979—. Sec. Bayside (N.Y.) Democratic Club, 1966. Fellow 11th Dist. Dental Soc., Acad. Gen. Dentistry; mem. Queens Acad. Gen. Dentistry (v.p. 1979-80), Am. Acad. Oral Medicine, Am. Acad. Periodontics, Coral Springs C. of C., Alpha Omega, Alpha Epsilon Pi. Subspecialties: Periodontics; Prosthodontics. Current work: preventive dentistry, periodontics, iatrogenic dentistry lectrs. Home: 10717 NW 19th St Coral Springs FL 33065 Office: 7305 W Sample Rd Coral Springs FL 33065

COHEN, SAMUEL M., pathologist; b. Milw., Sept. 24, 1946; s. David A. and Harriett (Goldman) C.; m. Janet L. Olson, Jan. 27, 1968; children—Sheri Lyn, Benjamin Aaron, Daniel Eric, Erica Ann. B.S., U. Wis.-Madison, 1967, M.D., 1972, Ph.D., 1972. Diplomate: Am. Bd. Pathology, 1976. Resident in pathology St. Vincent Hosp., Worcester, Mass., 1972-75, staff pathologist, Mass., 1975-81; asso. prof. U. Mass. Med. Sch., Worcester, 1977-81; prof., vice chmn. dept. pathology and lab medicine U. Nebr. Med. Ctr., Omaha, 1981—; vis. prof. Nagoya (Japan) City U. Med. Sch., 1976-77; mem. chem. pathology study sect. Nat. Cancer Inst. Assoc. editor: Cancer Research, 1982—; contbr. articles to profl. jours. Mem. Am. Assn. Cancer Research, Am. Assn. Pathologists, Japanese Cancer Assn., AAAS. Subspecialties: Pathology (medicine); Cancer research (medicine). Current work: Chemical carcinogenesis, urinary bladder carcinogenesis, carcinogen metabolism, computer modeling, diagnostic electron microscopy. Home: 2818 S 101 St Omaha NE 68124 Office: U Nebr Med Ctr 42d and Dewey Ave Omaha NE 68105

COHEN, SHELDON AVERY, research psychologist, consultant, educator; b. Detroit, Oct. 11, 1947; s. Harry F. and Ruth (Shapiro) C. Ph.B., Monteith Coll., Wayne State U., 1969; Ph.D., N.Y.U., 1973. Asst. prof. U. Oreg., 1973-78; assoc. prof., 1978-82; prof. psychology Carnegie-Mellon U., Pitts., 1982—. Co-Author: Behavior, Health and Environmental Stress, 1984; contbr. chpts., articles to profl. publs.; editor: (with L. Syrme) Social Support and Health, 1984. NIMH trainee, 1969-73; grantee NSF, 1974-77, 77-80, 80-83, Nat. Inst. Environ. Health Scis., 1978-80, Nat. Heart, Lung and Blood Inst., 1982-84. Fellow Am. Psychol. Assn.; mem. Acad. Behavioral Medicine, Internat. Commn. on Biol. Effects of Noise (exec. com. 1978-83), Soc. Exptl. Social Psychology, Soc. Psychol. Study Social Issues, Environ. design Research Assn., Acad. Behavioral Medicine. Subspecialty: Social psychology. Current work: Effects of environmental and psychological stresses on health and behavior; role of social support systems in protecting people from stress-induced illness. Office: Dept Psychology Carnegie-Mellon U Pittsburgh PA 15213

COHEN, STANLEY NORMAN, educator, geneticist; b. Perth Amboy, N.J., Feb. 17, 1935; s. Bernard and Ida (Stolz) C.; m. Joanna Lucy Wolter, June 27, 1961; children: Anne, Geoffrey. B.A., Rutgers U., 1956; M.D., U. Pa., 1960. Intern, Mt. Sinai Hosp., N.Y.C., 1960-61; resident Univ. Hosp., Ann Arbor, Mich., 1961-62; clin. asso. arthritis and rheumatism br. Nat. Inst. Arthritis and Metabolic Diseases, Bethesda, Md., 1962-64; sr. resident in medicine Duke U. Hosp., Durham, N.C., 1964-65; Am. Cancer Soc. postdoctoral research fellow Albert Einstin Coll. Medicine, Bronx, 1965-67, asst. prof. devel. biology and cancer, 1967-68; mem. faculty Stanford (Calif.) U., 1968—, prof. medicine, 1975—, prof. genetics, 1977, chmn. dept. genetics, 1978—; mem. com. recombinant DNA molecules Nat. Acad. Sci.-NRC, 1974; mem. com. on genetic experimentation Internat. Council Sci. Unions, 1977—. Mem. editorial bd.: Jour. Bacteriology, 1973-79; asso. editor: Plasmid, 1977—. Served with USPHS, 1962-64. Recipient Burroughs Wellcome Scholar award, 1970; V.D. Mattia award Roche Inst. Molecular Biology, 1977; Albert Lasker basic med. research award, 1980; Wolf prize, 1981; Josiah Macy Jr. Found. faculty scholar, 1975-76; Guggenheim fellow, 1975. Mem. Nat. Acad. Sci., Am. Acad. Arts and Sci., Am. Soc. Biol. Chemists, Genetics Soc. Am., Am. Soc. Microbiology, Am. Soc. Pharmacology and Exptl. Therapeutics, Am. Soc. Clin. Investigation, Phi Beta Kappa, Sigma Xi. Subspecialties: Gene actions; Genome organization. Office: Dept Genetics S-337 Stanford U Sch Medicine Stanford CA 94305

COHEN, WAYNE ROY, medical educator; b. N.Y.C., Apr. 27, 1946; s. Eugene Mark and Helen (Paul) C.; m. Sharon Rose Ominski, Aug. 24, 1980; 1 son by previous marriage, Aaron. A.B., U. Rochester, 1967; M.D., Boston U., 1971. Diplomate: Am. Bd. Ob-Gyn. Asst. prof. Harvard Med. Sch., Boston, 1978-82; assoc. prof. ob-gyn Albert Einstein Coll. Medicine, N.Y.C., 1983—. Co-editor: Management of Labor, 1983; contbr. articles to profl. jours. Fellow Am. Coll. Ob-Gyn; mem. Am. Fedn. Clin. research, Soc. for Exptl. Biology and Medicine. Jewish. Subspecialties: Maternal and fetal medicine; Reproductive biology (medicine). Current work: Fetal sympathoadrenal function; management of parturition. Home: Braxmar Dr N Harrison NY 10528 Office: North Central Bronx Hosp 3424 Kossuth Ave Bronx NY 10467

COHEN-ADDAD, JEAN-PIERRE, physicist, educator, consultant; b. Dieppe, Normandie, France, Mar. 27, 1939; s. Robert Charles and Suzanne (Hodencq) Cohen-A.; m. Claudine Daniele Franckel, Dec. 22, 1960; children: Sylvie, Nicolas. Docteur es Science, U. Grenoble, France, 1966. Asst. U. Grenoble, 1960-62, mâitre assistant, 1962-67, prof. physics, 1970—; mem. tech. staff Bell; Telephone Labs., Murray Hill, N.J., 1967-69. Mem. Am. Phys. Soc., Society Française Physique, N.Y. Acad. Sci. Subspecialties: Polymer physics; Polymers. Current work: Slow dynamical processes related to polymer molecule diffusion; bio-information; nuclear magnetic resonance. Home: 25 Bis Cours Berriat 38000 Grenoble France Office: U Grenoble 38041CDX Grenoble France

COHN, DAVID V(ALOR), biological chemistry educator, research consultant; b. N.Y.C., Nov. 11, 1926; s. Ralph and Clara (Schenkman) C.; m., 1947; children: Robert Warren, Emily. B.S., CCNY, 1948; Ph.D., Duke U., 1952; postgrad., Western Res. U., 1953. Mem. faculty U. Kans. Sch. Medicine, Kansas City, 1953—, now prof. biochemistry, assoc. dean research, 1974-82; assoc. chief staff for research devel. VA Med. Ctr., Kansas City, Mo., 1953-82; prof. biochemistry U. Mo.-Kansas City, 1971-82; v.p. research and devel. Immuno Nuclear Corp., Stillwater, Minn., 1982, sci. cons., 1983; research prof. oral biology and biochemistry U. Louisville Sch. Medicine, Sch. Dentistry, 1984—; pres. Internat. Conf. on Calcium Regulating Hormones; mem. bd. sci. counselors Nat. Inst. Dental Research, Bethesda, Md., 1980. Editor: Hormonal Regulation of Calcium Metabolism, 1981; contbr. articles to profl. jours. Served with USN, 1945-46. USPHS grantee, 1957—; VA grantee, 1975—; Am. Cancer Soc. grantee, 1959-60. Mem. Am. Soc. Biol. Chemists, Am. Chem. Soc., AAAS, Gordon Research Conf. Chem. and Biol. of Bones and Teeth (chmn. 1974). Jewish. Subspecialties: Endocrinology; Biochemistry (medicine). Current work: Calcium metabolism, parathyroid gland biosynthesis and secretion, bone cell growth, differentiation and hormone responsivity. Home: 1238 Wyncrest Ct Arden Hills MN 55112

COHN, JAY BINSWANGER, psychiatrist, neurologist, educator; b. Pelham, N.Y., Feb. 2, 1922; s. Louis Marbe and Beatrice (Binswanger) C.; m. Sally Cohn, Oct. 8, 1945; children: Jo Ann, Laurie Marbe. B.A., Amherst Coll., 1942; M.D., Yale U., 1945; Ph.D., U. Calif.-Irvine, 1974; J.D., Am. Coll. Law, Anaheim, Calif., 1982. Intern St. Elizabeth's Hosp., Washington, 1945-46; resident Cleve. Receiving Hosp., 1951-53; lectr. Western Research Grad. Sch., Cleve., 1954-59; clin. instr. psychology UCLA, 1959-65; asst. clin. prof. U. So. Calif., 1965-67; assoc. prof. U. Calif.-Irvine, 1967-71, clin. prof., 1971—, prof. psychology, 1974-; dir. psychopharm. lab., 1979—; prof. UCLA, 1982—, Am. Coll. Law, Anaheim, Calif., 1980—. Contbg. editor: Methods and Findings in Pharmacology, 1980; reviewing editor: Am. Jour. Forensic Psychiatry, 1982. Fellow So. Calif. Psychiat. Soc.; mem. Am. Acad. Psychoanalysis, Soc. Calif. Psychoanalytic Soc., AMA, AAUP. Subspecialty: Comparative physiology. Current work: Research activity: cognition, electroencephalography, psychological testing, psychopharmacology, alcohol effects. Office: Dept Psychiatry Alcohol Research Unit University of California at Los Angeles Los Angeles CA 90024 Home: 1315 S Roxbury Dr Suite 302 Los Angeles CA 90035

COHN, MAJOR LLOYD, neurologist, clinic adminstr., researcher, cons. toxicology; b. N.Y.C., Oct. 29, 1927; s. Isidore and Pauline (Bustein) C.; m. Marthe Hoffnung, Feb. 9, 1958; children—Stephan Jacques, Remi Benjamin. M.D., U. Geneva, Switzerland, 1956; Ph.D. (Am. Cancer Soc. fellow) in Biochemistry and intermediate Metabolism, U. Pitts., 1969. Intern Bklyn. Jewish Hosp., 1956-57; resident Washington U., St. Louis, 1957-58, Cornell U. Med. Sch. - Mem. Hosp. Cancer and Allied Diseases, 1959-60; Am. Cancer Soc. fellow asso. attending in medicine Montefore Hosp., N.Y.C., 1961-62; asst. prof. Sloan-Kettering Inst. and Meml. Hosp. for Cancer, 1960-61; U. Pitts. Sch. Medicine, 1969-79; asso. prof. UCLA Charles R. Drew Med. Sch., 1979—, dir. anesthesiology research. Contbr. articles on neuropharmacology to profl. jours. Served with USN, 1945-46. Recipient Henry L. Moses award, 1978. Fellow Royal Soc. Medicine; mem. Soc. for Neurosci., Internat. Assn. for Study of Pain, Internat. Soc. Psychoneuroendocrinology, Am. Soc. for Pharmacology and Exptl. Therapeutics, Am. Coll. Toxicology. Subspecialties: Internal medicine; Neurochemistry. Current work: Neurochemistry, neuropharmacology, neurotoxicology, biochemistry, neuroendocrinology. Home: 4015 Exultant Dr Rancho Palos Verdes CA 90274 Office: Charles R Drew Med Sch 1621 E 120th St Los Angeles CA 90059

COHN, MILDRED, biochemist; b. N.Y.C., July 12, 1913; d. Isidore M. and Bertha (Klein) C.; m. Henry Primakoff, May 31, 1938; children—Nina, Paul, Laura. B.A., Hunter Coll., 1931; M.A., Columbia, 1932, Ph.D., 1938; Sc.D. (hon.), Med. Coll. Pa., 1966, Radcliffe Coll., 1978, Washington U., St. Louis, 1981. Research asst. biochemistry George Washington U. Sch. Medicine, 1937-38; research asso. Cornell U., 1938-46, Washington U., 1946-50, 51-58, asso. prof. biol. chemistry, 1958-60; asso. prof. biophysics and phys. biochemistry U. Pa. Med. Sch., 1960-61, prof., 1961-78, Benjamin Rush prof. physiol. chemistry, 1978-82, prof. emerita, 1982; sr. mem. Inst. Cancer Research U. Pa., 1982—. Editorial bd.: Jour. Biol. Chemistry, 1958-63, 67-72. Established investigator Am. Heart Assn., 1953-59, career investigator, 1964-78. Recipient Garvan medal, Cresson medal., Nat. Medal of Sci. Mem. Am. Philos. Soc., Nat. Acad. Scis., Am. Chem. Soc., Harvey Soc., Am. Soc. Biol. Chemists (pres. 1978), Am. Biophys. Soc., Am. Acad. Arts and Scis., Phi Beta Kappa, Sigma Xi. Subspecialties: Biophysical chemistry; Nuclear magnetic resonance (chemistry). Current work: Nuclear magnetic resonance of enzyme mechanisms and in vivo phosphate metabolism. Office: Inst Cancer Research 7701 Burholme Ave Philadelphia PA 19111

COHN, NATHAN, cons., former engring. co. exec.; b. Hartford, Conn., Jan. 2, 1907; s. Harris and Dora Leah (Levin) C.; m. Marjorie Kurtzon, June 30, 1940; children—Theodore Elliot, David Leslie, Anne Harris, Amy Elizabeth, Julie Archer. S.B., M.I.T., 1927; D.Eng. (hon.), Rennsalear Poly. Inst., 1976. With Leeds & Northrup Co., Phila., 1927-72, mgr. market devel. div., 1955-58, v.p. tech. affairs, 1958-65, sr. v.p. tech. affair, 1965-67, exec. v.p. research and corp. devel., 1967-72, dir., 1963-75; cons. mgmt. and tech. of measurement and control, Jenkintown, Pa., 1972—; dir. AEL Industries Inc., Alkco Mfg. Co., Weinschel Engring. Co., Milton Roy Co., Modular Comptar Systems, Parlex Corp.; gen. partner Network Systems Devel. Assos.; pres. Nat. Electronics Conf., 1950; mem. NRC; exec. bd. Found. Instrumentation, Edn. and Research, 1962-64; del. congress Internat. Fedn. Automatic Control, 1960, 63, 66, 69, 72, 75, 78, 81, chmn. tech. com. on applications, 1969-72; chmn. U.S. organizgin com. 1975 World Congress, mem. tech. coms. on computers, systems, mem. com. on social effects of automation, 1975—; mem. vis. com. libraries M.I.T., 1964-69, mem. vis. com. philosophy, 1972-74. Contbr. articles to profl. jours., chpts. to books, textbook. Bd. dirs., v.p. Eagleville (Pa.) Hosp. and Rehab. Center. Fellow IEEE (life, Lamme medalist 1968, chmn. fellow com. 1974-76, chmn. awards bd. 1977-78, mem. Centennial com. 1979—); chmn. Intersoc. Hoover Medal (Bd. of Award 1978-81), Instrument Soc. Am. (v.p. industries and scis. 1960-61, sec. 1962, pres. 1963, Sperry medalist 1968, hon. mem. award 1976), AAAS, Franklin Inst. (life, Wetherill medalist 1968, mem. bd. mgrs. 1971—, chmn. bd. mgrs. 1971-75); mem. Nat. Acad. Engring., Engrs. Joint Council (exec. bd. 1975-78, commn. on internat. relations 1978-79), Am. Assn. Engring. Socs. (council for internat. affairs 1980-81), Indsl. Research Inst., Engrs. Council Profl. Devel. (vis. com. curriculum accreditation), Sci. Apparatus Makers Assn. (exec. bd. 1961-62, 66-73, pres. 1969-71, SAMA award 1978), Nat. Soc. Profl. Engrs. (Engr. of Yr. Delaware Valley 1968, State of Pa. 1969), Sigma Xi, Tau Beta Pi, Eta Kappa Nu, Pi Lambda Phi. Jewish. Club: Rydal (Phila.). Subspecialties: Electrical engineering; Systems engineering. Current work: Automatic control and control performance evaluation of interconnected electric power systems. Patentee electric power systems controls.

COHN, STANTON HARRY, medical research investigator, educator; b. Chgo., Aug. 25, 1920; s. Harry S. and Ethel (Goldberg) C.; m. Sylva M. Dushkes, Mar. 20, 1923; children: Avra, Haldan, Holly Cara, Evan. B.S., U. Chgo., 1946, M.S., 1949; Ph.D., U. Calif.-Berkeley, 1952. Biochemist Argonne (Ill.) Nat. Lab., 1946-49; head internal toxicity br. U.S. Naval Radiol. Def. Lab., San Francisco, 1949-58; head. med. physics dept. Brookhaven Nat. Lab., Upton, N.Y., 1958—; prof. medicine and clin. physiology SUNY-Stony Brook, 1978—. Author: Non-invasive measurements of bone mass and their clinical application, 1981; contbr. over 250 articles to profl. publs. Mem. Am. Physiol. Soc., Radiation Research Soc., Sigma Xi. Subspecialties: Medical physics; Physiology (medicine). Current work: Application of nuclear technologies to medical problems.

COHN, VICTOR HUGO, pharmacology educator, consultant; b. Reading, Pa., July 9, 1930; s. Victor Hugo and Kathleen (Morrow) C.; m. Marlene Kaschmann, Dec. 20, 1954; children: David Nathan, Wendy, Jonathan Charles. B.S., Lehigh U., 1952; A.M., Harvard U., 1965; Ph.D., George Washington U., 1961. Neuropharmacologist VA Research Labs. in Neuropsychiatry, Pitts., 1955-57; pharmacologist Nat. Heart Inst., Bethesda, Md., 1957-61; asst. prof. pharmacology George Washington U., 1961-65, assoc. prof., 1965-71, prof., 1971—, acting chmn., 1970-71, 78-79; cons. in field of drug abuse edn.; mem. Bd. on Toxicology and Environ. Health Hazards, Nat. Acad. Sci., 1980-83; advisor on drug policy The White House, 1979-80. NSF fellow, 1952-55. Fellow AAAS; mem. Am. Soc. Pharmacology and Exptl. Therapeutics, N.Y. Acad. Sci., Internat. Soc. Biochem. Pharmacology, Histamine Club, Assn. for Med. Edn. and Research in Substance Abuse, Sigma Xi, Tau Beta Pi. Subspecialties: Pharmacology; Psychopharmacology. Current work: Alcohol and other drug abuse; factors influencing drug metabolism; histamine metabolism; fluorometric methods of analysis. Office: 2300 I St NW Washington DC 20037

COHN, ZANVIL A., physician; b. N.Y.C., Nov. 16, 1926; s. David and Esther (Schwartz) C.; m. Fern R. Dworkin, Dec. 19, 1949; children: David, Ellen. B.S., Bates Coll., 1949; M.D. summa cum laude, Harvard U., 1953. Mem. staff Rockefeller Inst. (now Rockefeller U.), N.Y.C., 1958—, prof., sr. physician, 1966—; Co-dir. Joint Rockefeller U.-Cornell U. Med. Coll. M.D.-Ph.D. program, 1973-78; adj. prof. medicine Cornell U., 1977—. Contbr. articles to profl. jours. Mem. Commn. on Radiation and Infection, Armed Forces Epidemiology Bd.; mem. study sect. on immunology NIH; cons. Nat. Cancer Inst. Recipient Boylston Medal Harvard U., 1961; Basic Sci. award Am. Soc. Cytology, 1970; Fifth Ann. Squibb award Am. Soc.

Infectious Diseases, 1972; 7th Ann. Research award Samuel Noble Found., 1982. Mem. Nat. Acad. Sci. Subspecialties: Cell biology (medicine); Infectious diseases. Current work: Cellular physiology and immunology. Office: 1230 York Ave New York NY 10021

COKELET, GILES ROY, biophysics educator, researcher; b. N.Y.C., Jan. 7, 1932; s. Roy Sylvester and Anna Mary (Trippel) C.; m. Sarah Drew, June 15, 1963; children: Becky, Bradford Roy. A.A., Pasadena City Coll., 1953; B.S., Calif. Inst. Tech., 1957, M.S., 1958; Sc.D., M.I.T., 1963. Research Engr. Dow Chem. Co., Williamsburg, Va., 1958-61; asst. prof. M.I.T., Cambridge, 1963-64, Calif. Inst. Tech., Pasadena, 1964-68; assoc. prof. Mont. State U., Bozeman, 1968-76, prof., 1976-78; prof. radiation biology and biophysics U. Rochester, N.Y., 1978—; ad hoc cons. NIH, 1970—; cons. Cordis-Dow Co., 1977. Assoc. editor: Advances in Chemical Engineering, 1969-81; editor: Erythrocyte Mechanics and Blood Flow, 1980; contbr. articles to profl. jours. Served with U.S. Army, 1954-55. Recipient Sr. U.S. Scientist award A. von Humboldt-Stiftung, Cologne, W.Ger., 1981-82. Fellow AAAS; mem. Am. Inst. Chem. Engrs., Mirocirculatory Soc. (council 1981—), Soc. Rheology, European Microcirculatory Soc. Subspecialties: Biomedical engineering; Biophysics (physics). Current work: Blood rheology and hemodynamics; microcirculatory blood flow; mass transport phenomena in biological systems, including the microcirculation. Office: U Rochester 601 Elmwood Ave Rochester NY 14642 Home: 62 Burrows Hills Dr Rochester NY 14625

COLAHAN, PATRICK TIMOTHY, vet. surgeon; b. Klamath Falls, Oreg., May 31, 1948; s. Robert Martin and Maggie A. (Lovelady) C.; m. Carlye Ann Baker, July 28, 1973. B.S., U. Calif.-Davis, 1970, D.V.M., 1974. Diplomate: Am. Coll. Vet. Surgeons. Intern N.Y. Vet. Coll., Ithaca, 1974-75; resident U. Calif. Sch. Vet. Medicine, Davis, 1975-77, lectr., 1977-78; asst. prof. vet. surgery U. Fla. Coll. Vet. Medicine, Gainesville, 1978—, chief large animal surgery service, 1979—. Mem. League Conservative Voters, Gainesville, 1982—. Recipient award for proficiency in vet. clin. medicine Upjohn Co., 1974. Mem. Am. Assn. Equine Practitioners, Vet. Orthopedic Soc., AVMA, Am. Forestry Assn., Phi Zeta. Democrat. Roman Catholic. Club: Commonwealth of Calif. (San Francisco). Subspecialties: Surgery (veterinary medicine); Biomedical engineering. Current work: General and orthopedic surgery; biomechanical analysis of equine musculoskeletal and respiratory systems. Home: 7716 SW 53d Pl Gainesville FL 32608 Office: Dept Surg Scis Coll Vet Medicine U Fla Box J-116 JHMHC Gainesville FL 32610

COLASANTI, BRENDA KAREN, pharmacologist, educator; b. Charleston, W.Va., Dec 5, 1945; d. Harry Gordon and Mary Louise (Moore) Frame; m. Louis Colasanti, Jr., Sept. 21, 1968. B.A. in Zoology, W.Va. U., 1966, Ph.D. in Pharmacology, 1970. NIMH postdoctoral fellow in neuropsychopharmacology Mt. Sinai Sch. Medicine, N.Y.C., 1970-72; asst. prof. ophthalmology and pharmacology W.Va. U., Morgantown, 1972-76, asso. prof., 1976-80, prof., 1980—. Contbr. articles to sci. jours.; contbr.: (Craig and Stitzel) 8 chpts. to Modern Pharmacology, 1982. NIH grantee, 1974-80, 82—; acad. investigator award, 1977-80. Mem. Am. Soc. Pharmacology and Exptl. Therapeutics, Assn. for Psychophysiol. Study of Sleep, Am. Soc. for Neurochemistry, Assn. for Research in Vision and Ophthalmology, AAAS, Sigma Xi. Subspecialties: Ophthalmology; Neuropharmacology. Current work: Neurochem. mechanisms of psychotropic drug action; role of autonomic input in the intraocular pressure lowering effects of drugs. Office: Dept Pharmacology and Toxicolog WVa U Med Cente Morgantown WV 26506

COLBERG, MAGDA, research psychologist; b. San Juan, P.R., July 25, 1936; d. Carlos and Maria I. (Colberg) Munoz-Santaella. A.B., Colegio Universitario del Sagrado Corazon, San Juan, 1957; Ph.D., Georgetown U., 1965; postgrad., Princeotn U., 1969-71. Asst. liaison officer OAS, Washington, 1965-69; research psychologist U.S. Office Personnel Mgmt., Washington, 1974—. Contbr. articles to profl. jours. Recipient award for Superior Accomplishment U.S. Office Personnel Mgmt., 1981. Mem. Am. Philos. Assn., Assn. Symbolic Logic, Am. Psychol. Assn., Am. Math. Soc., Internat. Assn. Applied Psychology. Club: Princeton. Current work: Special interest in event predictive induction in psychometrics. Home: 4850 Connecticut Ave NW Washington DC 20008 Office: 1900 E St NW Washington DC 20405

COLBERT, EDWIN H., paleontologist, museum curator; b. Clarinda, Iowa, Sept. 28, 1905; s. George Harris and Mary (Adamson) C.; m. Margaret Mary Matthew, July 8, 1933; children: George Matthew, David William, Philip Valentine, Daniel Lee, Charles Diller. Student, N.W. Mo. State Tchrs. Coll., 1923-26; B.A., U. Nebr., 1928, Sc.D., 1973; A.M., Columbia U., 1930, Ph.D., 1935; Sc.D., U. Ariz., 1976. Student asst. Univ. Museum, U. Nebr., 1926-29; univ. fellow Columbia U., 1929-30, lectr. dept. zoology, 1938-39, prof. vertebrate paleontology, 1945-69, prof. emeritus, 1969—; research asst. Am. Museum Natural History, 1930-32, asst. curator, 1933-42, acting curator, 1942, curator, 1943, chmn. dept. amphibians and reptiles, 1943-44, curator of fossil reptiles and amphibians, 1945-70, chmn. dept. geology and paleontology, 1958-60, chmn. dept. vertebrate paleontology, 1960-66, curator emeritus, 1970—; curator vertebrate paleontology Mus. No. Ariz., Flagstaff, 1970—. Author: Evolution of the Vertebrates, 1955, 69, 80, Millions of Years Ago, 1958, Dinosaurs, 1961, (with M. Kay) Stratigraphy and Life History, 1965, The Age of Reptiles, 1965, Men and Dinosaurs, 1968, Wandering Lands and Animals, 1973, The Year of the Dinosaur, 1977, A Fossil Hunter's Notebook, 1980, Dinosaurs: An Illustrated History, 1983; also sci. papers and monographs. Recipient John Strong Newberry prize, Columbia U., 1931; Daniel Giraud Elliot medal Nat. Acad. Sci., 1935; medal Am. Mus. Natural History, 1970. Fellow AAAS, Geol. Soc. Am., Paleontol. Soc. (v.p. 1963), N.Y. Zool. Soc.; mem. Soc. Vertebrate Paleontology (sec.-treas. 1944-46, pres. 1946-47), Soc. Mammalogy, Soc. Ichthyology and Herpetology, Soc. for Study Evolution (editor 1950-52, v.p. 1957, pres. 1958), Nat. Acad. Sci., Sigma Xi. Subspecialties: Paleobiology; Evolutionary biology. Current work: Vertebrate paleontology: evolution and post distribution of vertebrates; evolution: morphological evolution in vertebrates, especially reptiles. Office: Museum of No Arizona Route 4 Box 720 Flagstaff AZ 86001 The paramount factor in the development of my scientific career has been a love of original research. Research is creative, and there is true satisfaction in doing creative things.

COLBORN, JOSEPH NELSON, mech. engr.; b. Indpls., Feb. 7, 1937; s. Joseph Judson and Katherine Elizabeth (Lemons) C.; m. Janet Devure Wells, Aug. 24, 1960; children—Joseph Jefferson, Heather Jene. B.S.M.E., Gen. Motors Inst., 1960; M.S.M.E. (E. B. Newill fellow), Purdue U., 1963. Detail engr. Detroit Diesel Allison div. Gen. Motors Corp., Indpls., 1961-62, project engr, 1962-66, group leader, 1966-70, sr. projet engr., 1970-81; head dept. heat transfer Teledyne CAE, Toledo, 1982—. Mem. ASME (vice chmn. Central Ind. sect.). Methodist. Subspecialties: Aeronautical engineering; Aerospace engineering and technology. Current work: Gas turbine heat transfer research and devel. related to air-cooled components. Office: 1330 Laskey R PO Box 6971 Toledo OH 43612

COLBY, NATHANIEL FRED, nuclear corporation engineer; b. Bennington, Vt., July 5, 1936; s. Nathaniel Henry and Ruth Marjorie (Dudley) C.; m. Mary Louise Grenfell, June 6, 1959; children: Susan, Nathaniel R., Lindsey, Robert J.D. B.S., U.S. Mil. Acad., 1959; M.S., U. Mo.-Rolla, 1972; M.B.A., Fairleigh Dickinson U., 1983. Commd. 2d lt. U.S. Army, 1959; advanced through grades to lt. col.; research coordinator Def. Nuclear Agy., 1973-74; exec. asst. to dep. dir. for sci. and tech. Nuclear Agy., 1975, ret., 1981; sr. engr. GPU Nuclear Corp., Parsippany, N.J., 1981—. Decorated Silver Star medal, Bronze Star medal with cluster, Legion of Merit. Mem. Am. Nuclear Soc. Republican. Episcopalian. Subspecialty: Nuclear engineering. Home: 12 Baker Ave Dover NJ 07801 Office: GPU Nuclear Corp 100 Interpace Pkwy Parsippany NJ 07054

COLE, CHARLES NORMAN, JR., molecular biologist; b. N.Y.C., Oct. 28, 1946; s. Charles N. and Betty Jane (Heldman) C.; m. Elizabeth Ryan, June 15, 1969; children: Noah Jonthan, Ethan Chalres. A.B., Oberlin Col., 1968; Ph.D., M.I.T., 1972. Postdoctoral fellow M.I.T., 1972-73, instr., 1973; postdoctoral fellow Stanford U., 1974-77; asst. prof. human genetics Yale U. Med. Sch., New Haven, 1977—. Mem. AAAS, Am. Soc. Microbiology, Union Concerned Scientists, Sigma Xi. Subspecialties: Gene actions; Molecular biology. Current work: Regulation of gene expression in eucaryotes. Home: 220 Canner St New Haven CT 06511 Office: Dept Human Genetics Yale U Med Sch 333 Cedar St New Haven CT 06510

COLE, JACK WESTLEY, physician; b. Portland, Oreg., Aug. 28, 1920; s. Alva Warren and Louise (Shafer) C.; m. Ruth Adele Kraft, Dec. 22, 1943; children—Deborah, Linda, Douglas, John. A.B., U. Oreg., 1941; M.D., Wash. U., 1944; M.A., Yale, 1966. Mem. faculty Western Res. U. Sch. Medicine, 1952-63; prof., chmn. dept. surgery Hahnemann Med. Coll. and Hosp., 1963-66; Ensign prof. surgery Yale U. Sch. Medicine, 1966—, chmn. dept. surgery, 1966-74, Josiah Macy Jr. faculty scholar, 1974-75, dir. div. oncology and cancer center, 1975—; cons. various hosps. Eleanor Roosevelt Internat. Cancer Research fellow, 1962. Mem. Am. Surg. Assn., Halsted Soc., Soc. Surgery of Alimentary Tract, Am. Soc. Cell Biology, Soc. Cryobiology. Subspecialty: Oncology. Research and publs. on histochemistry, cytochemistry, carcinogenesis; studies dealing with cellular kinetics in normal and abnormal intestinal epithelium. Home: Prospect Ct Woodbridge CT 06525 Office: 333 Cedar St New Haven CT 06510

COLE, KENNETH DEAN, designer, consultant; b. Peoria, Ill., July 3, 1951; s. Don R. and Emma Jayne C.; m. Jeanette Eileen O'Brien, May 2, 1975. Student, Sacramento City Coll., 1968-69, Am. River Coll., 1970. With Bechtel Corp., 1969-73, 76-78, Fluro M & M, 1973-76; ind. cons. to Detroit Edison Co., Pacific Gas & Electric Co. and Chevron Chem. Co., 1978-81; owner, operator Interim Tech. Assocs., Pinole, Calif., 1981—. Mem. John Anderson Presdl. Campaign, 1980. Mem. ASME (affiliate). Subspecialties: Mechanical engineering.

COLE, KENNETH S(TEWART), biophysicist, educator; b. Ithaca, N.Y., July 10, 1900; s. Charles Nelson and Mabel (Stewart) C.; m. Elizabeth Evans Roberts, June 29, 1932 (dec. 1966); children: Roger Braley, Sarah Roberts. A.B., Oberlin Coll., 1922; Sc.D. (hon.), 1954; Ph.D., Cornell U., 1926; Sc.D. (hon.), U. Chgo., 1967, M.D., U. Uppsala, 1967. Fellow NRC, Harvard U., 1926-28; research fellow Gen. Edn. Bd., Leipzig, 1928-29; asst. prof. physiology Columbia, 1929-37, asso. prof., 1937-46; Cons. physicist Presbyn. Hosp., N.Y.C., 1929-46; fellow Guggenheim Found., Inst. Advanced Study, Princeton, 1941-42; prin. biophysicist, metall. lab. U. Chgo., 1942-46, prof. biophysics and physiology, 1946-49; tech. dir. Naval Med. Research Inst., Bethesda, 1949-54; chief lab of biophysics Nat. Inst. Neurol. Diseases and Blindness, NIH, Bethesda, Md., 1954-66, sr. research biophysicist, 1966-77, scientist emeritus, 1978—; regents prof. U. Calif., Berkeley, 1963-64, prof. biophysics, 1965-77, adj. prof. neurosci., San Diego, 1980—; vis. prof. physiology and biophysics U. Tex. Med. Br., Galveston, 1972—; Priestley lectr. Pa. State Coll., 1939; Tennent lectr. Bryn Mawr Coll., 1941. Mem. bd. Biol. Lab., Cold Spring Harbor, 1940-45; Trustee Marine Biol. Lab., Woods Hole, Mass., 1947-55, 56-64, emeritus, 1966—. Decorated Order of So. Cross, Brazil; recipient Nat. Medal of Sci., 1967. Fellow Am. Acad. Arts and Scis., Am. Phys. Soc., AAAS, N.Y. Acad. Sci.; mem. Nat. Acad. Scis., Am. Physiol. Soc. (council 1963- 65), Société philomatique Paris, Soc. Gen. Physiologists, Biophys. Soc. (council 1957-62, pres. 1963), Royal Soc. London (fgn.), Sociedade de Bologia de Chile (hon.), Sociedade Brasileira de Biologia (hon.), Sigma Xi, Alpha Epsilon Delta (hon.), Epsilon Chi (hon.). Club: Cosmos (Washington). Office: 2404 Loring St San Diego CA 92109

COLE, MICHAEL, psychologist, educator; b. Los Angeles, Apr. 13, 1938. B.A. with highest honors in Psychology, UCLA, 1959; Ph.D. in Psychology (Woodrow Wilson fellow 1959-60, Ford Found. fellow 1960-62), Ind. U., 1962. Research asst. UCLA, 1958-59, Ind. U., 1959-62; exchange scholar, State Dept. grantee Moscow U., 1962-63; research asso. Inst. Math. Studies in Social Scis., lectr. Stanford U., 1963-64; asst. prof. Yale U., 1964-66; asso. prof. U. Calif., Irvine, 1966-69, Rockefeller U., 1969-75, prof., 1975-78; prof. psychology U. Calif., San Diego, 1978—, also dir. lab. comparative human cognition and coordinator communications program; mem. for psychology Joint Soviet-Am. Commn. in Social Scis. Author: (with others) The Cultural Context of Learning and Thinking, 1971, (with S. Scribner) Psychology of Literacy, 1981; others; author introductions, forewords; contbr. articles to profl. publs.; editor: (with others) Developing Child Series, Soviet Psychology, 1969—; guest editor spl. issue: Am. Psychologist, 1977; cons. editor to jours. Recipient Research Scientist Devel. award USPHS, 1969-74 (hon. 1970-74); Behavioral Sci. award N.Y. Acad. Sci., 1978; Van Leer-Jerusalem Found. fellow, 1975—. Mem. AAAS, Am. Psychol. Assn., Psychonomic Soc., Am. Anthrop. Assn., Librarian Research Assn., Council on Anthropology and Edn., Social Sci. Research Council, Soc. Research in Child Devel., Am. Acad. Arts and Sci., Phi Beta Kappa, Sigma Xi. Subspecialty: Cognition. Office: U Calif at San Diego Center for Human Info Processing Lab Comparative Human Cognition La Jolla CA 92093

COLE, THOMAS EARLE, nuclear engineer; b. Winter Park, Fla., Dec. 13, 1922; s. Henry Earle and Lizzie Bell (Perrine) C.; m. Jean Holden, Feb. 24, 1944; children: Henry Earle, Edmund Platt. B.S., Rollins Coll., 1946. Various staff assignment Oak Ridge Nat. Lab., 1946-73, spl. assignment, 1973-74, mgr., 1974-77, group leader various programs, 1977-81, mgr., 1980—, group leader reactor systems analysis, 1981; cons. in field of nuclear research reactors to various U.S. and fgn. cos., 1955-65. Contbg. author: Ann. Rev. of Nuclear Science-Technology of Research Reactors, 1963. Served to lt. j.g. USNR, 1943-46; PTO. Mem. Am. Nuclear Soc. (charter), Am. Phys. Soc., Sigma Xi. Subspecialties: Nuclear engineering; Nuclear fission. Current work: Nuclear reactor systems analysis and safety studies; studies of the role of nuclear reactor electric power plants in the U.S. electric utility industry, e.g., energy parks, advanced concepts, etc. Home: 103 Disston Rd Oak Ridge TN 37830

COLEMAN, ANNETTE WILBOIS, biologist; b. Des Moines, Feb. 28, 1934; d. Fred J. and Agnes M. (Dunshee) Wilbois; m. John R. Coleman, July 26, 1958; children: Alan, Benjamin, Suzanne. B.A., Barnard Coll., 1955; Ph.D., Ind. U., 1958. Postdoctoral fellow Johns Hopkins U., Balt., 1958-61; research asst. U. Conn., Storrs, 1961-65; research assoc. Brown U., Providence, 1963-71, asst. prof., 1971-80, assoc. prof., 1980—. Contbr. articles to profl. jours. Mem. N.Y. Acad. Scis., Phycological Soc. Am., Am. Bot. Soc., Soc. Protozoologists. Subspecialties: Cell biology; Microbiology. Current work: Cell biology, reproduction and development. Home: 101 Keene St Providence RI 02906 Office: Bio-Med Div Brown U Providence RI 02912

COLEMAN, C. NORMAN, physician, educator, researcher; b. N.Y.C., Jan. 24, 1945; s. Samuel A. and Minna (Kramer) C.; m. Karolynn Forsburg, May 25, 1970; children: Gabrielle, Keith. B.A. summa cum laude, U. Vt., 1966; M.D., Yale U., 1970. Intern and resident U. Calif.-San Francisco, 1970-72; fellow in med. oncology Nat. Cancer Inst., Bethesda, Md., 1972-74; resident in radiation therapy Stanford (Calif.) U. Med. Sch., 1975-78, asst. prof. radiology and medicine, 1978-83, assoc. prof. radiology and medicine, 1983—; prin. investigator NIH grants Stanford-No. Calif. Oncology Group, 1978—, Stanford-Radiation Therapy Oncology Group, 1980—. Served to lt. comdr. USPHS, 1972-74. Recipient Newell award Stanford U., 1978; Am. Cancer Soc. jr. faculty clin. fellow, 1979-82; Moule fellow Stanford U., 1980-83. Fellow ACP; mem. Am. Coll. Radiology, Am. Assn. Cancer Research, Am. Soc. Therapeutic Radiologists, Am. Soc. Clin. Oncology, Radiation Research Soc., Phi Beta Kappa, Alpha Omega Alpha. Subspecialties: Cancer research (medicine); Oncology. Current work: Study of late effects of cancer treatment, e.g. secondary cancers; study of chemical modifiers of chemotherapy and radiotherapy, sensitizers of cancer to therapy and protectors of normal tissues; cancer treatment. Home: 1336 Marilyn Dr Mountain View CA 94040

COLEMAN, JAMES R., psychologist, neuroscientist; b. Kansas City, Mo., Oct. 12, 1946; s. James D. and Marion (Rice) C.; m. Yolanda G. Salitrero, Dec. 18, 1971. B.A. in Zoology, UCLA, 1969, Ph.D. in Psychology, 1974. Postdoctoral trainee Duke U., Durham, N.C., 1974-76; asst. prof. psychology U. S.C., Columbia, 1977-81, assoc. prof., 1981—, adj. asst. prof. physiology, 1978-81, adj. assoc. prof., 1981—. Contbr. articles to profl. jours. NIH grantee, 1979-82; Deafness Research Found. grantee, 1979-83. Mem. AAAS, Am. Assn. Anatomists, Assn. Research in Otolaryngology, Internat. Soc. Developmental Psychobiology, Soc. Neuroscience. Subspecialties: Neurophysiology; Developmental neuroscience. Current work: Developmental neuroscience; development of central auditory systems; aging of central auditory systems. Office: Dept Psychology U SC Columbia SC 29208

COLEMAN, JAMES STAFFORD, government energy research administrator; b. Cleelum, Wash., May 8, 1928; m. (divorced). B.S., Wash. State U., 1950; Ph.D. in Phys. Chemistry, MIT, 1953. Mem. staff Los Alamos Nat. Lab., 1953-67; chemist div. Research, AEC, 1967-69; tech. advisor Office of Gen. Mgr., 1969-75; asst. div. dir. Div. Phys. Research, Energy Research & Devel. Adminstrn., 1975-77; div. dir. Office of Energy Research, Dept. Energy, Germantown, Md., 1977—, also div. dir. Mem. AAAS, Sigma Xi. Subspecialty: Energy Research Management. Office: Engring Math & Geoscis Div Dept of Energy Route 270 Germantown MD 20545

COLEMAN, JOSEPH EMORY, biophysics and biochemistry educator; b. Iowa City, Iowa, Oct. 11, 1930; s. George Hopkins and Leah Estelle (Rose) C.; m. Phoebe Newman, Apr. 29, 1961; children: Michael Newman, Samuel Hopkins, Julia Heath. B.A., U. Va., 1953, M.D., 1957; Ph.D., MIT, 1964. Intern Peter Bent Brigham Hosp., Boston, 1957-58, resident medicine, 1963-64; asst. prof. molecular biophysics and biochemistry Yale U., New Haven, 1964-66, assoc. prof., 1966-75, prof., 1975—, chmn. dept. molecular biophysics and biochemistry, 1976-82. Mem. Am. Soc. Biol. Chemists, Am. Chem. Soc., Biophys. Soc. Subspecialties: Biochemistry (biology); Biophysics (biology). Current work: Enzymology (metalloenzymes), DNA binding proteins, mechanisms of gene expression. Home: 764 Lambert Rd Orange CT 06477 Office: C-139 SHM 333 Cedar St Box 3333 New Haven CT 06510

COLEMAN, MARILYN RUTH ADAMS, poultry science educator, consultant; b. Lancaster, S.C., Mar. 27, 1946; d. Coyte and Jill J.D. (Lyon) Adams; m. George Edward Coleman, III, Jan. 27, 1968; children: Jill Ann Marie, George Edward. Student, S.C. Med. Coll., 1967; B.S. in Biology, U. S.C., 1968; postgrad., U. Va., 1971, 72, Va. Poly. Inst., 1972; Ph.D. in Physiology, Auburn U., 1976. Teaching asst. U. S.C., 1967-68; research technician in animal sci. Va. Poly. Inst. and State U., Blacksburg, 1968, teaching asst. in biology, 1970-72; biology tchr., basketball coach Brunswick County Schs., 1968-69; research asst. in poultry sci. Auburn U., 1973-76; asst. prof. poultry sci. Ohio Agrl. Research and Devel. Ctr., Ohio State U., Columbus, 1977-82; owner MAC Assocs., Upper Arlington, Ohio, 1974—; Resource person in scis. Upper Arlington Schs., 1977—; cons. Bird House, Cin. Zoo, 1978; mem. research com. Columbus Zoo, 1979—; welcome chmn. Upper Arlington French Exchange Program, 1979—. Grantee NSF, 1967, 71, 72. Mem. Poultry Sci. Assn. (session chmn. Southeastern sect. 1980), World Poultry Sci. Assn. (life), Am. Physiol. Soc., Chem. Biomed. Environ. Research Group (chmn. seminar series), Auburn U. Alumni Assn. (life), Va. Poly. Inst. Alumni Assn., U. S.C. Alumni Assn., Sigma Xi, Phi Sigma, Gamma Sigma Delta. Republican. Baptist. Subspecialties: Animal physiology; Reproductive biology. Current work: Incubation and reproductive problems of poultry; effect of environment and management on reproductive performance and hatchability of poultry. Home and Office: 2532 Zollinger Rd Columbus OH 43221

COLEMAN, MICHAEL MURRAY, polymer science educator; b. Herne Bay, Eng., Jan. 24, 1938. B.Sc., Borough Poly., Eng., 1968; M.S., Case Western Res. U., 1971, Ph.D. in Polymer Sci., 1973. Assayer, Rhokana Corp. Ltd., Zambia, 1955-61; analytical chemist Johnson Mathey Ltd., Eng., 1963-64; research chemist polymers Revertex Ltd., Eng., 1968-69, E.I. du Pont deNemours & Co., 1973-75; asst. prof. Pa. State U., University Park, 1975-78, assoc. prof. polymers, 1978-82, prof., 1982—, program chmn. polymer sci., 1976—, acting head dept. materials sci. and engring., 1983—. Mem. Am. Chem. Soc., Royal Inst. Chemistry, Am. Phys. Soc. Subspecialty: Polymer chemistry. Office: Dept Materials Sci Pa State U University Park PA 16802

COLEMAN, MORTON, oncologist, hematologist; b. Norfolk, Va., Sept. 15, 1939; s. Isadore and Bessie (Levine) C.; m. Joyce Goodman, May 30, 1968; children: Ingrid Alexandra, Benjamin Lee, Abigail Rachael. A.A., Coll. William and Mary, 1958; B.A., Johns Hopkins U., 1959; M.D., Med. Coll. Va., 1963. Diplomate: Am. Bd. Internal Medicine, 1971, subsplty. in hematology, 1972, in oncology, 1973. Asst. prof. Cornell U. Med. Coll., N.Y.C., 1970-74, assoc. prof., 1974—; asst. attending physician N.Y. Hosp., 1970-74, assoc., 1974—, assoc. dir. chemotherapy service, 1970-83; chmn. new agts. com. Cancer and Leukemia Group B, 1975-82; med. chmn. Fund for Blood and Cancer Research, 1976—. Served to lt. comdr. USN, 1965-67. Fellow ACP; mem. Am. Soc. Clin. Oncology, Am. Soc. Hematology, AAAS, Am. Fedn. Clin. Research, Am. Radium Soc., Internat. Soc. Hematology, Internat. Soc. Memostasis, AMA, N.Y. Cancer Soc., Harvey Soc., N.Y. Acad. Sci., Soc. Study Blood, Am. Assn. Cancer Research, Alpha Omega Alpha, Sigma Zeta. Subspecialties: Hematology; Chemotherapy. Current work: Clin. research in oncology/hematology. Office: 525 E 68th St New York NY 10021

COLEMAN, PAUL D., neurobiologist; b. N.Y.C., Dec. 2, 1927; s. Aaron Barnett and Martha (Michaels) C.; m.; children from previous marriage: Laura, Paul. A.B. magna cum laude, Tufts U., 1948; Ph.D., U. Rochester, 1953. Asst. prof. Tufts U., Medford, Mass., 1956-59;

computer ctr. assoc. M.I.T., Cambridge, Mass., 1957-59; spl. fellow Johns Hopkins U. Med. Sch., Balt., 1959-62; assoc. prof. physiology U. Md. Med. Sch., Balt., 1962-67; prof. anatomy U. Rochester (N.Y.) Med. Sch., 1967—. Contbr. articles to profl. jours. Served to 1st lt. U.S. Army Res., 1953-56. NIH spl. fellow, 1959-62; NSF grantee, 1957-67; NIH grantee, 1962—. Mem. Soc. Neuroscience, Am. Assn. Anatomy, Am. Psychol. Assn. Club: Rochester Yacht. Subspecialties: Neurobiology; Anatomy and embryology. Current work: Quantitative neuroanatomical studies of aging brain. Office: Univ Rochester Med Ctr Box 603 Rochester NY 14642

COLEMAN, RONALD L., toxicology educator; b. Wellington, Tex., Aug. 20, 1934; s. J. Leon and Mary Emalyne (Blevins) C.; m.; children: Christy, John Edward, Dennis. B.S., Abilene Christian Coll., 1956; postgrad., Tex. A&M U., 1958; Ph.D. in Biochemistry, U. Okla., 1963. Asst. prof. biochemistry U. Okla., Oklahoma City, 1964-69, assoc. prof. environ. health, 1969-73, prof. environ. health, 1975—, chmn. dept., 1982—; owner, lab. dir. Environ. Cons., Oklahoma City, 1970—. Served with U.S. Army, 1958-60. Mem. Am. Coll. Toxicologists, Am. Indsl. Hygiene Assn., Am. Conf. Govt. Indsl. Hygienists, Soc. Environ. Geochemistry and Health, Am. Chem. Soc. Subspecialties: Environmental toxicology; Toxicology (medicine). Current work: Toxicology of metals/biological monitoring; water and wastewater; occupational health. Office: Dept Environ Health U Okla Health Scis Ctr PO Box 26901 Oklahoma City OK 73190

COLEMAN, SIDNEY RICHARD, physicist, educator; b. Chgo., Mar. 7, 1937; s. Harold Albert and Sadie (Shanas) C. B.S., Ill. Inst. Tech., 1957; Ph.D., Calif. Inst. Tech., 1962. Research fellow dept. physics Harvard U., 1961-63, asst. prof., 1963-66, assoc. prof., 1966-69, prof., 1969—; vis. prof. U. Rome, Italy, 1968, Princeton U., 1973, Stanford U., 1979-80; partner Advent Pubs. Recipient prize for physics lectures Ettore Majorana Centre Sci. Culture; Boris Pregel award N.Y. Acad. Scis. Fellow Am. Phys. Soc., Am. Acad. Arts and Sci., Nat. Acad. Sci.; mem. LILAPA. Subspecialties: Particle physics; Theoretical physics. Current work: Quantum field theory, symmetry principles, magnetic monopoles. Home: Unit 12 1 Richdale Ave Cambridge MA 02140 Office: Lyman Lab Harvard U Cambridge MA 02138

COLES, EMBERT HARVEY, JR., veterinary medicine educator, researcher, consultant; b. Garden City, Kans., Oct. 12, 1923; s. Embert Harvey and Neva Ann (Blanchard) C.; m.; children: Charles David, Kay, Ann. D.V.M., Kans. State U., 1945, Ph.D., 1958; M.S., Iowa State U., 1946. Instr. Iowa State U., 1945-48; pvt. practice vet. medicine, 1949-54; asst. prof. clin. pathology dept. vet.medicine Kans. State U., 1954-59, assoc. prof., 1960-63, prof., 1964—, head dept., 1964-70, head dept. lab. medicine, 1972-82, dean vet. medicine, 1970-72, dir. fed. research programs, 1982—, chief of party, Zaria, Nigeria, 1970-72. Author: Veterinary Clinical Pathology, 1967, 3d edit., 1980; contbr. numerous articles to profl. jours. Named Outstanding Tchr. in Vet. Medicine Upjohn Co., 1982; NIH grantee, 1954-60, 56-64; Mark L. Morris Found. grantee, 1963. Mem. AVMA, Kans. Vet. Med. Assn., Am. Soc. Microbiology, Confrerie of Research Workers in Animal Diseases. Congregationalist. Lodge: Rotary. Subspecialties: Pathology (veterinary medicine); Microbiology (veterinary medicine). Current work: Teaching veterinary medicine, clinical pathology and immunology; research in clinical immunology; service in clinical immunology. Office: Dept Lab Medicine Kans State U Manhattan KS 66506

COLES, GARY JOHN, research psychologist; b. Bottineau, N.D., Nov. 23, 1944; s. John Anthony and Audrey Mae (Jirikowic) C.; m. Patricia Ellen McIntosh, Aug. 2, 1981. B.A., St. Olaf Coll., 1966; M.A., U. N.D., 1968; Ph.D., 1970; M.B.A., U. Mo., 1981. Lic. cons. psychologist, Minn. Assoc. research scientist Am. Insts. Research, Palo Alto, Calif., 1970-73, research scientist, 1973-75, sr. research scientist, 1976-79; program analyst Food and Nutrition Service, U.S. Dept. Agr., Washington, 1979-80; research analyst Bus. Incentives, Inc., Mpls., 1982—. Mem. Am. Psychol. Assn., Am. Mktg. Asssn., Multiple Linear Regression Spl. Interest Group. Subspecialties: Applied research psychology; Statistics. Current work: Program evaluation, marketing research, applied statistical analysis. Home: 1305 Berry Ridge Rd Eagan MN 55123 Office: Bus Incentives 7630 E Bush Lake Rd Edina MN 55435

COLGATE, STIRLING A., physicist; b. N.Y.C., Nov. 14, 1925; m. Rosemary Williamson; children: Henry, Sarah, Arthur. B.A. in Physics, Cornell U., Ithaca, N.Y., 1948, Ph.D., 1952. With Lawrence Radiation Lab., Berkeley, Calif., 1951-52, Livermore, Calif., 1952-64; electron and accelerator physicist; physicist nuclear weapons and tests, 1955; staff Controlled Thermonuclear Fusion project, 1955-64; tech. adviser Conf. Discontinuance Nuclear Weapons Tests, Geneva, 1959; pres. N.Mex. Inst. Mining and Tech., Socorro, 1965-74; sr. fellow, physicist, spl. research on controlled thermonuclear fusion, astrophysics, atmospheric physics Los Alamos Nat. Lab., 1976—; partner Richard M. Colgate (patent devel.), 1958—; Mem. nuclear panel Sci. Adv. Bd., 1959-61; adv. com. fluid mechanics NASA, 1960-62; cons. ballistic missile div. USAF, 1960-62; cons. Def. Atomic Support Agy., 1962-64; mem. adv. com. environ. scis. NSF, 1967; mem. Nat. Acad. Sci. panel on space plasma physics, 1977-79, panel on physics of sun, 1979—; chmn. panel on physics of sun Space Sci. Bd., 1980-81. Trustee-at-large Asso. Univs., 1970-73, Aura-Kitt Peak, 1973-78, Space Sci. Bd., 1976-79. Fellow Am. Phys. Soc.; mem. Am. Astron. Soc., Sigma Xi. Subspecialty: Plasma physics. Home: 4616 Ridgeway Los Alamos NM 87544 Office: MS 275 Los Alamos Nat Lab Los Alamos NM 87545

COLIGAN, JOHN ERNEST, biochemist, chemist, consultant; b. Canonsburg, Pa., July 3, 1944; s. John Baptist and Rose (Koval) C.; m. Nelda Olivia Clodfelter, Apr. 13, 1968. B.A., Wabash Coll., 1966; M.S., Ind. U. Med. Sch., Indpls., 1968, Ph.D., 1971. Jr. research scientist City of Hope Nat. Med. Center, Duarte, Calif., 1971-73, asst. research scientist, 1973-75; asst. prof. Rockefeller U., 1975-77; research chemist Nat. Inst. Allergy and Infectious Disease, Bethesda, Md., 1977—. Editor: Surveys in Immunologic Research, 1981—; assoc. editor: Jour. Immunology. Mem. Am. Assn. Immunologiss, Am. Assn. Cancer Research, Fedn. Am. Socs. for Exptl. Biology, Phi Beta Kappa, Sigma Xi. Subspecialties: Biochemistry (biology); Immunobiology and immunology. Current work: Structure-function relationships of membrane proteins. Home: 10913 Broad Green Terr Potomac MD 20854 Office: Nat Inst Allergy and Infectious Disease 9000 Rockville Pike Bethesda MD 20205

COLINVAUX, PAUL ALFRED, ecologist, author; b. St. Albans, Eng., Sept. 22, 1930; came to U.S., 1959; m. L. Hillis, July 10, 1961. B.A., Cambridge U., 1956, M.A., 1960, cert. agr., 1956; Ph.D., Duke U., 1962. Research biologist Yale U., New Haven, 1963-64; faculty Ohio State U.-Columbus, 1964—, prof. ecology, 1972—; mem. ecology adv. com. NSF, Washington, 1979-82; chmn. sponsors Inst. Ecology, Indpls., 1977-79; mem. U.S. Internat. Quateternary Assn. Nat. Commn., Washington, 1977—. Author: Introduction to Ecology, 1973, Why Big Fierce Animals are Rare, 1978 (Award 1978), The Fates of Nations, 1980; actor/narrator/writer: TV series What Ecology Really Says, 1973. Concilor Charles Darwin Found. for Galapagos Isles, 1972—. Served to lt. Brit. Army, 1951. NSF research grantee, 1963—; Guggenheim fellow, 1972; NATO fellow, 1963; Tansley lectr British Ecol. Soc., 1981. Fellow Explorers Club, Arctic Inst. N.Am.; mem. Ecol. Soc. Am., Am. Soc. Limnology and Oceanography, Am. Soc. Naturalists. Subspecialties: Ecology; Evolutionary biology. Current work: Environment of ice age earth using sediments of ancient lakes in Amazonia, Andes, Galapagos, Alaska, Siberia, China, and Ohio, ecological causes of war and history of human populations. Office: Ohio State U 484 W 12th Ave Columbus OH 43210 Home: 319 S Columbia Ave Columbus OH 43209

COLLARD, SNEED BODY, biol. oceanographer, educator; b. Denver, July 11, 1939; s. Sneed B. and Margaret E. (Payne) C.; m. Suzanne S. Spencer, Feb. 14, 1981; children by previous marriage: Sneed, Gidon, Meghann, Tyler. B.A. magna cum laude (Regents scholar), U. Calif., Santa Barbara, 1965, M.A. (Ford Found. fellow, Woodrow Wilson fellow, NIH fellow, NDEA fellow), 1966, Ph.D. (USPHS fellow), 1968. Asst. prof. zoology U. Calif., Santa Barbara, 1968; NSF postdoctoral fellow Woods Hole Oceanographic Instn. and Harvard U., 1968-69; assoc. prof. biology U. West Fla., Pensacola, 1969—; vis. investigator Steinich Marine Lab., Israel, 1969-70, Australian Inst. Marine Sci., 1981; cons. NASA, Corps of Engrs., Nat. Park Service; advisor Fla. Dept. Natural Resources; participant research cruises and expdns. to, Red Sea, Mediterranean, Caribbean, Coral Seas, North and South Atlantic and Paciic Oceans, Gulf of Mex.; vis. lectr. Marine Environ. Scis. Consortium, Dauphin Island, Ala., summers 1972, 73; speaker sci. meetings. Editorial bd.: N.E. Gulf Sci, 1979—; Contbr. articles to profl. jours. Served with USAF, 1956-60. Grantee Smithsonian Instn., Sea Grant, NSF, Fla. Inst. Oceanography. Mem. Am. Soc. Limnology and Oceanography, Am. Inst. Biol. Scis., Am. Soc. Parasitologists, Am. Soc. Ichthyology and Herpetology, Assn. Southea. Biologists, Fla. Acad. Scis. (chmn. 1974-75), AAAS, Pensacola Area C. of C., Fla. Inst. Oceanography (adv. bd.), N.W. Fla. Zool. Soc. (gov. 1971-72). Democrat. Subspecialties: Deep-sea biology; Parasitology. Current work: Open Sea biology, neuston, zooplankton, marine symbioses, corals, biology of sharks. Home: 10 Main St PO Box 116 Bagdad FL 32539 Office: University of West Florida Biology Dept Pensacola FL 32540

COLLEN, MORRIS FRANK, physician; b. St. Paul, Nov. 12, 1913; s. Frank Morris and Rose (Finkelstein) C.; m. Frances B. Diner, Sept. 24, 1937; children—Arnold Roy, Barry Joel, Roberta Joy, Randal Harry. B.E.E., U. Minn., 1934, M.B. with distinction, 1938, M.D., 1939. Diplomate: Am. Bd. Internal Medicine. Intern Michael Reese Hosp., Chgo., 1939-40; resident Los Angeles County Hosp., 1940-42; chief med. service Kaiser Found. Hosp., Oakland, Calif., 1942-53, chief of staff, San Francisco, 1953-61; med. dir. Permanente Med. Group (West Bay Div.), 1953-79; dir. Med. Methods Research, 1962-79, Tech. Assessment, 1979—; chmn. exec. com. Permanente Med. Group, Oakland, Calif., 1953-73; dir. Permanente Services, Inc., Oakland, 1958-73; lectr. Sch. Public Health, U. Calif., Berkeley, 1966-78; lectr. info. sci. U. Calif., San Francisco, 1970—; lectr. U. London, 1972, Harvard U., 1974, Stanford Med. Center, 1973, 75, Johns Hopkins U., 1976, others; cons. Bur. Health Services, USPHS, 1965-68, chmn. health care systems study sect., 1968-72, adv. com. demonstration grants, 1967; adv. U.S. VA, 1968; cons. European region WHO, 1968-72, USAF Med. Fitness Program, 1968, Pres.'s Biomed. Research Panel, 1975; adv. com. Automated Multiphasic Health Testing, 1970; mem. Pres.'s Ad Hoc Panel on Prevention and Personal Health Care, 1971; discussant Nat. Conf. Preventive Medicine, Bethesda, Md., 1975; mem. com. on tech. in health care Nat. Acad. Sci., 1976; mem. adv. group Nat. Commn. on Digestive Diseases, U.S. Congress, 1978; mem. adv. panel to U.S. Congress Office of Tech. Assessment, 1980-82; mem. peer rev. adv. group Dept. Def. program, 1978-83; program chmn. Internat. Conf. Med. Informatics, Tokyo, 1980. Author: Treatment of Pneumococcic Pneumonia, 1948; author: Hospital Computer Systems, 1974, Multiphasic Health Testing Systems, 1977; editor: Permanente Med. Bull, 1943-53, Lecture Notes in Med. Informatics, 1943-53; mem. editorial bd.: Diagnostic Medicine; contbr. articles to med. jours., chpts. to med. books. Johns Hopkins U. Centennial Scholar, 1976. Fellow ACP, Am. Coll. Cardiology, Am. Coll. Chest Physicians; mem. Pan Am. Med. Assn. (council mem.), Internat. Fedn. Med. Electronics, AMA, Am. Fedn. Clin. Research, Inst. Medicine Nat. Acad. Sci., IEEE, Soc. Biomed. Computing, Salutis Unitas (v.p. 1972), Soc. Advanced Med. Systems (pres. 1973), Ops. Research Soc. Am., Nat. Acad. Practioners in Am., Alpha Omega Alpha, Eta Kappa Nu, Tau Beta Pi. Subspecialty: Medical Research Administration. Home: 4155 Walnut Blvd Walnut Creek CA 94596 Office: 3451 Piedmont Ave Oakland CA 94611

COLLETT, MARC STEPHEN, molecular biologist; b. Detroit, May 23, 1951. B.S., U. Mich., 1973, Ph.D., 1977. Postdoctoral researcher U. Colo., Denver, 1977-80; asst. prof. microbiology U. Minn., Mpls., 1980—; sr. scientist Molecular Genetics, Inc., Minnetonka, Minn., 1980—. Recipient Wilson Stone award M.D. Anderson Hosps., Houston, 1980; Damon Runyon-Walter Winchell fellow, 1977-79; Leukemia Soc. Am. fellow, 1979-80; Nat. Cancer Inst. grantee, 1981-83. Mem. Am. Soc. Microbiologists, Am. Soc. Virology. Subspecialties: Molecular biology; Cell study oncology. Current work: Laboratory research addressing the molecular mechanisms of viral oncogenesis; application of recombinant DNA technology genetic engineering in the generation of useful vaccine and related products in the animal health care industry. Office: Molecular Genetics Inc 10320 Bren Rd E Minnetonka MN 55343

COLLIER, JOHN WALTER, mech. engr., chem. co. inspection and testing adminstr.; b. Ocala, Fla., Jan. 8, 1941; s. Jessie Wilburn and Blanche Adelia (East) C.; m. Shirley Ann Moore, July 27, 1963; children: Leslie Dawn, Michael Shawn. Student in math. (Univ. football scholar), Harding U., 1959-62; B.S. in Tech. (Dow Loan scholar), U. Houston, 1978. Cert. engring. technician Inst. Cert. Engring. Technicians,1974. Test technician Engring. Test Services, Inc., Houston, 1963-66; devel. engr. Magnaflux Corp., Houston, 1966-68; v.p. NDT Services, Inc., Houston, 1968-71; pres. Inspection Engrs. Inc., Houston, 1971-73; supr. Tex. div. Dow Chem. U.S.A., Freeport, 1973—. Contbr. articles to profl. jours. Vice pres. Brazosport sect. S.W. Basketball Ofcls. Assn., Freeport, Tex., 1974—; sec. Brazosport Affiliated Bd. Ofcls., 1978—; mem. City of West Columbia (Tex.) Parks Bd., 1981—. Served with U.S. N.G., 1958-66. Named Outstanding Student student sect. Soc. Mfg. Engrs., 1978. Mem. Am. Soc Nondestructive Testing, ASME. Republican. Mem. Ch. of Christ. Current work: Ultrasonic flaw detection in weldments with computer applications; nondestructive testing—ultrasonics, radiography, penetrants, magnetic particle, eddy current, leak detection, and visual inspections. Home: 601 Kirby West Columbia TX 77486 Office: Dow Chem USA Tex Div B-2615 Freeport TX 77541

COLLIER, LINDA LEE, comparative ophthalmology researcher and educator, veterinary ophthalmic pathologist; b. Galena Park, Tex., Apr. 29, 1950; d. Joseph M. and Lela P. (Anderson) C. B.S. with highest honors, U. Maine-Orono, 1971; D.V.M. with distinction, Cornell U., 1975; Ph.D., Wash. State U., 1979. NIH postdoctoral fellow, dept. vet. microbiology and pathology Wash. State U., Pullman, 1976-79, research assoc., 1979; asst. prof. dept. vet. pathology and dept. ophthalmology U. Mo.-Columbia, 1980—. Contbr. articles to sci. jours. NIH fellow, 1976-79; grantee, 1981—. Mem. Assn. Research in Vision and Ophthalmology, AAAS, AVMA, Am. Genetics Assn., Vizsla Club Am., Nat. Wildlife Fedn., Phi Kappa Phi, Phi Zeta. Subspecialties: Pathology (veterinary medicine); Ophthalmology. Current work: Ophthalmic pathology; the major part of my research involves studying development of and pathologic changes in the visual system of the feline animal model of Chediak-Higashi syndrome, using autoradiographic, histologic, ultrastructural and enzyme cytochemistry techniques. Office: U Mo W213 Vet Med Bldg Columbia MO 65211

COLLIN, ROBERT EMANUEL, electrical engineering educator; b. Donalda, Alta., Can., Oct. 24, 1928; came to U.S., 1958, naturalized, 1964; s. Knute Emanuel and Hannah (Hanson) C.; m. Kathleen Patricia Smith, Sept. 15, 1952; children: Patricia Ann, Linda Marie, David Robert. B.S. in Engring. Physics, U. Sask., Can., 1951; Ph.D., Imperial Coll., U. London, Eng., 1954. Sci. officer Canadian Def. Research Bd., 1954-58; faculty Case Western Res. U., 1958—, prof. elec. engring., 1965—, chmn. elec. engring. and applied physics dept., 1978-82. Author: Field Theory of Guided Waves, 1960, (with R. Plonsey) Principles and Applications of Electromagnetic Fields, 1961, Foundations for Microwave Engineering, 1966; contbr., editor: (with F.J Zucker) Antenna Theory, 2 vols., 1969. Recipient Jr. Achievement award Cleve. Tech. Socs. Council, 1964. Fellow IEEE (sr. mem., chmn. Que. subsect. 1956-57); mem. Sigma Xi (v.p. Case Inst. Tech. chpt. 1966-67), Eta Kappa Nu. Subspecialties: Electrical engineering; Electronics. Current work: Antennas, microwave devices. Home: 1041 West Mill Dr Highland Heights OH 44143 Office: 10900 Euclid Ave Cleveland OH 44106

COLLINS, ANITA MARGUERITE, research geneticist; b. Allentown, Pa., Nov. 8, 1947; d. Edmund and Virginia (Hunsicker) C. B.S., Pa. State U., 1969; M.Sc., Ohio State U., 1972, Ph.D., 1976. Instr. biology Mercyhurst Coll., Erie, Pa., 1975-76; research geneticist Agrl. Research Service, U.S. Dept. Agr., Bee Breeding and Stock Ctr. Lab., Baton Rouge, 1976—. Mem. Assn. Women in Sci. (chpt. pres. 1981), Am. Genetic Assn., Entomol. Soc. Am., Animal Behavior Soc., Sigma Xi. Subspecialties: Animal genetics; Animal breeding and embryo transplants. Current work: Behavior genetics of defensive behavior of the honey bee, especially africanized bee. Office: Route 3 Box 82B Ben Hur Rd Baton Rouge LA 70808

COLLINS, ARLENE RYCOMBEL, microbiologist; b. Buffalo, Jan. 2, 1940; d. Alex and Jean (Krzyzaniak) Rycombel; m. Charles R. J. Collins, Oct. 9, 1965. B.S. cum laude, D'Youville Coll., 1961; M.A., SUNY-Buffalo, 1964, Ph.D., 1967. USPHS postdoctoral fellow Marquette U. Med. Coll. Wis., Milw., 1969-70, SUNY-Buffalo, 1967-68, asst. prof., 1971-78, assoc. prof. microbiology, 1978—; vis. investigator Scripps Clinic and Research Inst., LaJolla, Calif., 1980-81. Mem. Am. Soc. Virology, Am. Women in Sci., Soc. Gen Microbiology. Subspecialties: Microbiology; Cellular engineering. Current work: Human hybridoma technology; preparation of heterophile and viral monoclonal antibodies from patients with infectious mononucleosis. Home: 24 Hendricks Blvd Amherst NY 14226 Office: SUNY-Buffalo 218 Sherman Hall Buffalo NY 14220

COLLINS, FRANK GIBSON, mechanical engineering educator; b. Chgo., Feb. 20, 1938; s. Forrest Gibson and Elizabeth (Freeman) C.; m. Sarah Ruth Knight, May 13, 1960; children: James Forrest, Pamela Ruth. B.S.C.E., Northwestern U., 1961; Ph.D.M.E., U. Calif.-Berkeley, 1968. Registered profl. engr., Tex. Asst. prof. aerospace engring. U. Tex., Austin, 1968-74; assoc. prof. aerospace engring. U. Tenn. Space Inst., Tullahoma, 1974-81, prof., 1981—; cons. to industry. Contbr. articles to scholarly jours. Bd. deacons 1st Christian Ch., 1978-81; treas. N.W. Austin Civic Assn., 1973-74. NSF fellow, 1961-63; NSF grantee, 1974-75; recipient award for paper ASCE, 1961. Mem. AIAA (sec. 1981-92), Am. Phys. Soc., Sigma Xi, Tau Beta Pi, Pi Mu Epsilon, Order of Engr., Sigma Gamma Tau. Lodge: Kiwanis. Subspecialties: Fluid mechanics; Aerospace engineering and technology. Current work: Wind tunnel testing, viscous interactions, flow instability, space processing of materials, computational fluid mechanics. Office: University of Tennessee Space Institute Tullahoma TN 37388 Home: 1703 Country Club Dr Tullahoma TN 37388

COLLINS, FRANK MILES, microbiologist, researcher; b. Adelaide, South Australia, Mar. 30, 1928; s. Frank Vernon and Ethelwyn Rollison (Littler) C.; m. Lorna Fay Hannaford, May 24, 1952; children: William Mark, Michael James. M.Sc., U. Adelaide, 1952, Ph.D., 1960, D.Sc., 1976. Lectr. bacteriology U. Adelaide, 1954-60, asst. prof. microbiology, 1961-65; assoc. mem. Trudeau Inst., Saranac Lake, N.Y., 1965-68, mem., 1969—; mem. study sects. NIH, 1976-80, 81—. Fellow Am. Acad. Microbiologists; mem. Soc. Gen Microbiology (U.K.), Am. Soc. Microbiology, Reticuloendothelial Soc., Am. Assn. Immunologissts, Am. Thoracic Soc. Republican. Presbyterian. Lodge: Rotary. Subspecialties: Infectious diseases; Microbiology (medicine). Current work: Cellular mechanisms of antibacterial immunity to facultative intracellular parasites; antituberculous and antityphoid immunity; role of T-cells in DTH and CMI. Office: Trudeau Inst Algonquin Ave Saranac Lake NY 12983

COLLINS, J.G., neuroscientist; b. Pitts., Jan. 15, 1947; s. John Francis and Katherine (Mulroy) C.; m. Jane Kohler, July 10, 1971; children: Sean, Katherine, Margaret. B.S., U. Pitts., 1969, Ph.D., 1979. Predoctoral fellow in pharmacology U. Pitts., 1973-78; postdoctoral assoc. in anesthesiology Yale U., 1978-80, instr. anesthesiology, lectr. pharmacology, 1980-81, asst. prof. anesthesiology and pharmacology, 1981—. Mem. Am. Soc. Anesthesiologists, Soc. Neurosci, Am. Pain Soc., Internat. Assn. Study of Pain. Subspecialty: Pharmacology. Current work: Drug effects on pain transmission systems in the central nervous system. Office: Dept Anesthesiology Yale U School of Medicine 333 Cedar St New Haven CT 06510

COLLINS, JIMMY HAROLD, biochemistry educator; b. Gaffney, S.C., Dec. 9, 1948; s. James P. and Edith J. (Phillips) C.; m. Carol J. Adkins, June 21, 1972 (div. 1977); m. Helena Swanljung, Mar. 16, 1979; children: Johan, Peter, Henrik, Erik. B.S., Duke U., 1971; Ph.D., U. Tex.-Austin, 1977. Staff fellow NIH, Bethesda, Md., 1977-82; asst. prof. dept. biochemistry Sch. Medicine, U. Pitts., 1982—. Contbr. articles to profl. jours. Research grantee Health Research and Services Found. Pitts., 1983. Mem. Am. Soc. Biol. Chemists, Am. Soc. Cell Biology, Biophys. Soc., AAAS. Democrat. Subspecialties: Biochemistry (biology); Cell biology. Current work: Non-muscle cell motility; biochemistry of contractile proteins in epithelial cells; cytoskeleton; protein phosphorylation. Home: 5654 Darlington Rd Pittsburgh PA 15217 Office: Dept Biochemistry Sch Medicine U Pitts Pittsburgh PA 15261

COLLINS, JOHN H., biochemistry educator; b. Peabody, Mass., Sept. 6, 1942. A.B. in Chemistry, Northeastern U., 1965; Ph.D. in Biochemistry, Boston U., 1970. Research assoc. Boston Biomed. Research Inst., 1969-74, staff scientist, 1974-76; asst. prof. Baylor Coll. Medicine, Houston, 1976-77, U. Cin., 1977-79, assoc. prof., 1979—. Grantee in field. Mem. Am. Soc. Biol. Chemists, Biophys. Soc., Am. Chem. Soc., N.Y. Acad. Scis., AAAS. Subspecialties: Biochemistry (medicine); Pharmacology. Current work: Protein structure, contractile and membrane proteins. Office: Dept Pharmacology and Cell Biophysics Univ Cin Coll Medicine Mail Location 57 231 Bethesda Ave Cincinnati OH 45267

COLLINS, MARGARET STRICKLAND, zoology educator; b. Institute, W.Va., Sept. 4, 1922; d. Rollins Walter and Luella (Bowling) James; m. Bernard E. Strickland, July 5, 1942 (div. 1949); m. Herbert L. Collins, Aug. 1951 (div. 1963); children: Herbert Louis Jr., James

Joseph. B.S., W.Va. State Coll., 1943; Ph.D., U. Chgo., 1949. Instr. zoology Howard U., Washington, 1947-50, asst. prof., 1950-51, prof., from, 1978—, now ret; prof., zoology Fla. A&M U., Tallahassee, 1951-63; prof. biology Fed. City Coll., Washington, 1969-78. Editor: Science and the Question of Human Equality, 1981. Bd. dirs. Nat. Assn. So. Poor, Norfolk, Va., 1979—. Mem. Entomol. Soc. Washington (pres. 1982—), Ecol.Soc. Am., N.Y. Acad. Scis., Earthwatch. Democrat. Subspecialties: Behavioral ecology; Entomology. Current work: Defensive behavior in South American termites; termite ecology; species abundance in virgin and disturbed tropical rain forests. Home: 1642 Promrose Rd NW Washington DC 20012 Office: Dept Entomology Smithsonian Inst NHB Washington DC

COLLINS, MICHAEL THOMAS, microbiology educator; b. St. Paul, Jan. 2, 1949; s. Thomas Wellington and June Claire (Anderson) C.; m. Jeannette McDonald, June 6, 1981; children: Christopher, David, Katrina. B.S., U. Minn., 1970, D.V.M., 1972; Ph.D., U. Ga., 1976. Asst. prof. dept. microbiology Colo. State U., Ft. Collins, 1976-81, assoc. prof., 1981-83; assoc. prof. dept. pathobiol. scis. U. Wis.-Madison, 1983—; mem. faculty aquavet program Woods Oceanographic Instn., 1979—; cons. Colo. Assn. Continuing Med. Lab. Edn., 1979. Contbr. articles to profl. jours. NIH fellow, 1974; Marshall Found. fellow, 1981. Mem. AVMA, Am. Soc. Microbiology, Am. Coll. Vet. Microbiology, Conf. Research Workers in Animal Diseases, Wildlife Disease Assn. Roman Catholic. Subspecialties: Microbiology (veterinary medicine); Microbiology (medicine). Current work: Pathogenesis of bacterial infections of food producing animals on Pasteurella; immunology, serology and antigenic structure of Legionella.

COLLINS, PAUL FRANCIS, nuclear engineer, consultant; b. N.Y.C., Feb. 28, 1927; s. John Patrick and Estelle (Levitre) C.; m. Mary E. Campbell, Apr. 11, 1959. B.S. in Mgmt. Engring, Rensselaer Poly. Inst., 1952. Quality control supr. DuPont, Aiken, S.C., 1952-58, reactor supr., 1958-63, reactor dept. instr., 1963-65; examiner reactor operators AEC, Bethesda, Md., 1965-69; br. chief operator licensing NRC, Bethesda, 1969-82; nuclear cons. KMC, Inc., Washington, 1982—. Author: Operator Requalification Program, 1972. Served to 2d lt. U.S. Army, 1952-53. Mem. Am. Nuclear Soc. (treas. reactor ops. div. 1980-81, sec. 1981-82). Republican. Roman Catholic. Club: R1Kenwood Country (Bethesda). Lodges: KC; Am. Legion. Subspecialties: Nuclear engineering; Human factors engineering. Current work: Nuclear power plant operations. Home: 4920 Sentinel Dr Apt 205 Bethesda MD 20816 Office: KMC Inc 1747 Pennsylvania Ave Washington DC 20006

COLLINS, WILLIAM EDWARD, aviation psychology research administrator, researcher; b. Bklyn., May 16, 1932; s. William Edward and Loretta Agnes (Brasier) C.; m. Corliss Jean Allen, June 20, 1970; 1 dau. Corliss Adora. B.S., St. Peter's Coll., 1954; M.A., Fordham U., 1956, Ph.D., 1959. Lic. psychologist, Okla. Psychol. research asst. Fordham U., 1954-56, teaching fellow, 1958, research asst., 1958-59; research psychologist Aviation Psychology Lab., FAA Civil Aeromed. Inst., Oklahoma City, 1961-63, chief sensory integration sect., 1963-65, supr., 1965—; adj. assoc. prof. psychology U. Okla.-Norman, 1963-70, adj. prof., 1970—; adj. assoc. prof. research psychology dept. psychiatry and behavioral scis. U. Okla. Health Scis. Center, Oklahoma City, 1965-70; adj. prof., 1970—; mem. Nat. Acad. Sci.-NRC Com. on Vision, 1963—, mem. exec. council, 1973-81; mem. Nat. Acad. sci.-NRC Com. on Hearing, Bioacoustics and Biomechanics, 1963—; appearances before House Sub-Com. on Pub. Health and Environ., 1971; mem. rating panel Interagy Bd., U.S. Civil Service Examiners for State Okla., 1967—; judge Okla. State Sci. and Engring. Fair, Ada, 1980, 81, 82; mem. Okla. Bd. Examiners Psychologists, 1981-84, chmn., 1982-84; evaluator proposals NSF, 1968—, HeW, 1971—; lectr. in field. Contbr. chpts., numerous articles to profl. publs.; research numerous prsentations in field. Served to 1st lt. Med. Services Corps U.S. Army, 1959-61. Recipient award for employee invention FAA, 1966, Sustained Superior Performance award, 1966, 67, Spl. Achievement award, 1971, 83, Quality Performance award, 1964, 69, 70, 74, Outstanding Performance rating, 1966-71, 74, 81, 83. Fellow AAAS, N.Y. Acad. Scis., Am. Psychol. Assn. (abstractor Psychol. Abstracts 1962—, citation 1973), Aerospace Med. Assn. (assoc., Raymond F. Longacre award 1971, presdl. exec. com. 1982-84, exec. council 1982—, editorial bd. Aviation, Space and Envir. Medicine 1974—, assoc. editor 1980—); mem. Assn. Aviation Psychologists (pres. 1974-75), Barany Soc., Okla. acad. Sci. (vis. scientist program 1966), Okla. Psychol. Assn. (vice chmn. awards com. 1980, 82, chmn. sci. program com. ann. meeting 1977). Subspecialties: Behavioral psychology; Sensory processes. Current work: Effects of motion on perceptual and motor skills; vestibular-visual interactions; alcohol and drug effects on orientation, vestibular function and performance. Home: 8900 Sheringham Dr Oklahoma City OK 73132 Office: FAA Civil Aeromed Inst AAC-118 PO Box 25082 Oklahoma City OK 73125

COLLIPP, BRUCE GARFIELD, ocean engineer; b. Niagara Falls, N.Y., Nov. 7, 1929; s. Platon Garfield and Audrey Marie (Otto) C.; m. Pricilla Jane Milbury, Dec. 25, 1954; children: Gary, Richard. B.S. in Marine Transp, MIT, 1952, M.S. in Naval Architecture and Naval Engring, 1954. Lic. profl. engr., ship's engr. With Shell Oil Co., Houston, 1954—; sr. staff engr., 1966—; lectr. U. Tex., Austin, 1970—, Internat. Assn. Drilling Contractors Tng. Ctr., Houston, 1970—. Author: Buoyancy Stability and Trim, 1966; invented semi submersible drilling rig., 1956; contbr. over 50 tech. articles to profl. publs. Chmn. bd. dirs. Hidden Coves, Point Blank, Tex., 1980—. Served with USNR, 1951-60. Recipient Silver cup MIT, 1979; Holley medal ASME, 1979; Industry award. Offshore Tech. Conf., 1971, 81. Mem. Am. Petroleum Inst., Marine Tech. Soc., Sigma Psi, Phi Sigma Phi. Republican. Presbyterian. Subspecialties: Ocean engineering; Fuels. Current work: Offshore rigs; platforms; operations oceanography. Patentee in field. Office: Shell Oil Co Box 2099 Houston TX 77001

COLLISON, BETTY CHRISTINE, researcher; b. Balt., July 17, 1952. B.S., Towson State U., Balt., 1974; M.A.S., Johns Hopkins U., 1979; postgrad., U. Md. Dental Sch., 1983—. Research asst. psychology Johns Hopkins U., 1975-77, research asst. immunology, 1977-79; commd. officers summer tng. extern in immunology Nat. Inst. Dental Research, NIH, Bethesda, 1980-81; research assoc. U. Md. Dental Sch., Balt., 1982—. Contbr. articles to profl. publs. Research scholar U. Md., 1982. Mem. Acad. Gen. Dentistry, Am. Assn. Dental Research, ADA (student clinician). Subspecialty: Immunology (medicine). Current work: Dental student conducting research investigating neutrophil dysfunction and abnormalities in juvenile periodontitis and other periodontitis patients. Home: 3321 Chestnut Ave Baltimore MD 21211 Office: University of Maryland Dental School 666 W Baltimore St Baltimore MD 21201

COLLMAN, JAMES PADDOCK, educator; b. Beatrice, Nebr., Oct. 31, 1932; m. (married). B.Sc., U. Nebr., 1954, M.S., 1956; Ph.D. (NSF fellow), U. Ill., 1958. Instr. chemistry U. N.C., Chapel Hill, 1958-59, asst. prof., 1959-62, asso. prof., 1962-67; prof. chemistry, Stanford, 1967—; Frontiers in Chemistry lectr., 1964, Nebr. lectureship, 1968; Venable lectr. U. N.C., 1971; Edward Clark Lee lectr. U. Chgo., 1972; vis. Erskine fellow U. Canterbury, 1972; Plenary lectr. French Chem. Soc., 1974; Dreyfus lectr. U. Kans., 1974; distinguished inorganic lectr. U. Rochester, 1974; Reilley lectr. U. Notre Dame, 1975; William Pyle Philips lectr. Haverford Coll., 1975; Merck lectr. Rutgers U., 1976; FMC lectr. Princeton, 1977; Julius Steiglitz lectr. Chgo. sect. Am. Chem. Soc., 1977; Pres.'s Seminar Series lectr. U. Ariz., 1980; Frank C. Whitmore lectr. Pa. State U., 1980; Plenary lectr. 3d IUPAC Symposium on Organic Synthesis, 1980; Brockman lectr. U. Ga., 1981. Guggenheim fellow, 1977-78. Mem. Am. Chem. Soc. (Calif. Sect. award 1972, soc. award in inorganic chemistry 1975), N.Y. Acad. Sci., Chem. Soc. (London), Nat. Acad. Sci., Am. Acad. Arts and Scis., Phi Beta Kappa, Sigma Xi, Phi Lambda Upsilon, Alpha Chi Sigma. Subspecialties: Inorganic chemistry; Catalysis chemistry. Current work: Biomimetic chemistry, oxyen carriers, catalytic oxygenation, electrocatalysis, metal-metal bonds, molecular metals. Office: Stanford U Stauffer II Stanford CA 94305

COLMAN, ROBERT WOLF, educator, physician; b. N.Y.C., June 7, 1935; s. Jack K. and Miriam (Greenblatt) C.; m. Roberta Fishman, June 16, 1957; children: Sharon V., David S. A.B., Harvard U., 1956, M.D., 1960; M.A., U. Pa., 1973. Diplomate: Am. Bd. Internal Medicine. Intern Boston City Hosp., 1960-61; resident Beth Israel Hosp., Boston, 1961-62, Barnes Hosp., St. Louis, 1964-65, fellow in hematology, 1965-67; asst. prof. Harvard Med. Sch., Boston, 1967-73; assoc. prof., 1973, U. Pa., Phila., 1973-75, prof. medicine, 1975-78, Temple U. Coll. Medicine, Phila., 1978—, dir., 1979—. Editor: Hemostasis and Thrombosis, 1982; mem. editorial bd.: Blood, 1982—, Thrombosis Research, 1983—. Bd. govs. Am. Heart Assn., S.E. Pa., Phila., 1982—. Served with USPHS, 1962-64. Fellow ACP; mem. Am. Soc. Clin. Investigation, Internat. Soc. Hemostatis and Thrombosis, Internat. Soc. Hematology, Assn. Am. Physicians, Am. Soc. Biol. Chemists, Am. Soc. Hematology, Am. Physiol. Soc., Am. Assn. Pathology, Sigma Xi, Phi Beta Kappa, Alpha Omega Alpha. Democrat. Jewish. Subspecialties: Hematology; Biochemistry (medicine). Current work: Biochemistry of coagulation proteins, plasma proteolytic inhibitors, plasma proteolytic enzymes, cell biology of platelets including membrane receptors, role of prostoglandins, cyclic AMP, and ADP. Home: 9 Rose Valley Rd Moylan PA 19065 Office: Thrombosis Research Center Temple U Sch Medicine 3400 N Broad St Philadelphia PA 19140

COLOTLA, VICTOR ADOLFO, psychologist, educator; b. Mexico City, May 8, 1944; s. Adolfo and Josefina (Espinosa) C.; m. Xochitl Gallegos, Sept. 18, 1967; children: Ian Rolando, Eileen Vivian. Grad. in psychology, Nat. U. Mex., 1966; M.A., U. Toronto, 1969; Ph.D., York U., Toronto, 1973. Registered psychologist, Ont. Prof. psychology Nat. U. Mexico, Mexico City, 1973-74; psychologist Toronto Western Hosp., 1974-75; coordinator program on behavioral pharmacology Nat. U. Mex., 1976-80, chief dept. psychophysiology, 1980-81, head grad. studies, 1981—. Co-editor: Modificacion de Conducta: Applicaciones a la Biomedicina, 1980; author articles. Recipient various scholarships and grants. Mem. Sociedad Mexicana de Analisis de la Conducta (pres. 1977-78), Sociedad Mexicana de Psicologia (sec. sci. events 1979—), Am. Psychol. Assn., Eastern Psychol. Assn., N.Y. Acad. Scis. Club: Terranova (Mexico City). Subspecialties: Behavioral toxicology; Learning. Current work: Effects of organic solvents on behavior; biofeedback; history of psychology in Mexico. Home: Zaragoza 3 Chimalcoyotl Mexico City Mexico 14630 Office: Nat U Mex Ciudad Universitaria Mexico City Mexico 04510

COLSON, ELIZABETH FLORENCE, anthropologist; b. Hewitt, Minn., June 15, 1917; d. Louis H. and Metta (Damon) C. B.A., U. Minn., 1938, M.A., 1940; M.A., Radcliffe Coll., 1941; Ph.D. (A.A.U.W. Traveling fellow), 1945, Brown U., 1978. Asst. social sci. analyst War Relocation Authority, 1942-43; research asst. Harvard, 1944-45; research officer Rhodes-Livingstone Inst., 1946-47, dir., 1948-51; sr. lectr. Manchester U., 1951-53; asso. prof. Goucher Coll., 1954-55; research asso. African Research Program, Boston U., 1955-59, part-time, 1959-63; prof. anthropology Brandeis U., 1959-63, U. Calif. at Berkeley, 1964—; Lewis Henry Morgan lectr. U. Rochester, 1973. Author: The Makah, 1953, Marriage and the Family Among The Plateau Tonga, 1958, Social Organization of the Gwembe Tonga, 1960, The Plateau Tonga, 1962, The Social Consequences of Resettlement, 1971, Tradition and Contract, 1974; Jr. Author (Secondary Education and the Formation of an Elite), 1980; Sr. editor: Seven Tribes of British Central Africa, 1951. Fellow Center Advanced Study Behavioral Scis., 1967-68; Fairchild fellow Calif. Inst. Tech., 1975-76. Fellow Am. Anthrop. Assn., Brit. Assn. Social Anthropologists, Royal Anthrop. Inst. (hon.); mem. Nat. Acad. Sci., Am. Acad. Arts and Scis., Phi Beta Kappa. Subspecialty: Behavioral ecology. Current work: Longitudinal study of changing adaptation in Gwembe District, Zambia. Office: Dept Anthropology U Calif Berkeley CA 94720

COLSON, STEVEN DOUGLAS, chemical physicist; b. Idaho Falls, Idaho, Aug. 16, 1941; m., 1962; 6 children. B.S., Utah State U., 1963; Ph.D., Calif. Inst. Tech., 1968. Asst. prof. Yale U., New Haven, 1968-73, assoc. prof., 1973-80, prof. chemistry, 1980—, jr. faculty fellow, 1972-73; mem. Nat. Research Council Adv. Bd. to U.S. Army Research Office, 1972-75. Assoc. editor: Jour. Chem. Physics, 1980—. Subspecialties: Physical chemistry; Spectroscopy. Office: Dept Chemistry Yale U 225 Prospect St New Haven CT 06520

COLTMAN, CHARLES ARTHUR, JR., med. oncologist; b. Pitts., Nov. 7, 1930; s. Charles Arthur and Sarah Margaret (Carman) C.; m. Eleanore Gobel, Jan. 13, 1951; children: Charles Arthur III, Douglas Clair, Rodney Reed, Susan Dale. B.S., U. Pitts., 1952, M.D., 1956; M.M.S., Ohio State U., 1963. Diplomate: Am. Bd. Internal Medicine. Intern Del. Hosp., Wilmington, 1956-57, resident pathology, 1957; jr. asst. resident, asst. medicine Ohio State U. Hosp., 1959-60, asst. resident, asst. medicine, 1960-61, sr. asst. resident hematology, asst. instr. medicine, 1961-62, chief med. resident, demonstrator medicine, 1962-63, attending physician, instr. medicine, 1963; clin. asso. prof. medicine U. Tex. Health Sci. Center, San Antonio, 1970-75, clin. prof. medicine, 1975-77, asso. prof., dir. clin. med. oncology sect., div. oncology, dept. medicine, 1977—, prof. medicine, 1978—; med. dir. Cancer Therapy and Research Center, San Antonio, 1977—; chief med. oncology sect. med. service Audie Murphy VA Hosp., San Antonio, 1977—; Am. Cancer Soc. prof. clin. oncology, 1978—; chmn. sci. adv. bd. Cancer Therapy and Research Found. S. Tex., 1976—. Mem. editorial bd.: Seminars in Oncology, 1979—, Investigational New Drugs: The Jour. New Anti-Cancer Agts, 1982—; Contbr. articles in field to profl. jours. Served to col., M.C. USAF, 1971-77. Decorated Legion of Merit; recipient Stitt award Assn. Mil. Surgeone, 1970, Harold Brown award USAF, 1971, Phillip Kyle Meml. Lecture Soc. Air Force Physicians, 1977. Fellow ACP; mem. Am. Fedn. Clin. Research, Am. Soc. Hematology, Am. Assn. Cancer Research, Am. Soc. Clin. Oncology, Soc. Air Force Physicians (pres. 1974,75), Assn. Mil. Surgeons U.S., Central Soc. Clin. Research, S.W. Oncology Group (chmn. 1981—), Tex. Med. Assn., Bexar County Med. Soc. Club: San Antonio Research (pres. 1966). Subspecialty: Cancer research (medicine). Office: 4450 Medical Dr San Antonio TX 78229

COLTMAN, JOHN WESLEY, physicist; b. Cleve., July 19, 1915; s. Robert White and Louise (Tyroler) C.; m. Charlotte Waters Beard, June 10, 1941; children—Sally Louise, Nancy Jean. B.S. in Physics, Case Inst. Tech., 1937; M.S., U. Ill., 1939, Ph.D. in Physics, 1941. Research scientist Research Labs. Westinghouse Electric Corp., Pitts., 1941-49, mgr. electronics and nuclear physics dept., 1949-60, asso. dir. research labs., 1960-64, research dir. central research labs., 1964-74, dir. research and devel. planning, 1977-80; mem. adv. group on electron devices Dept. Def., 1958-62; mem. Naval Intelligence Sci. Adv. Com., 1971-73, NRC Commn. on Human Resources, 1977-80. Contbr. articles to profl. jours. Recipient Longstreth medal Franklin Inst., 1960; Roentgen medal Remscheid, W. Ger., 1970; Gold medal Radiol. Soc. N.Am., 1982. Fellow Am. Phys. Soc., IEEE; mem. Nat. Acad. Engring., Am. Musical Instrument Soc. Republican. Presbyterian. Subspecialties: Acoustics; Optical image processing. Current work: Acoustical behavior of musical wind instruments. Inventor x-ray image amplifier, scintillation counter. Home: 3319 Scathelocke Rd Pittsburgh PA 15235

COLUMBIA, TIMOTHY FRANCIS, mech. engr., glass mfg. adminstr.; b. Hanover, N.H., Sept. 16, 1948; s. Francis Thomas and Vera Helen (Borry) C.; m. Barbara Anne Lawrence, Dec. 18, 1971; children: Meade Lauren, Jonathan Lawrence. B.S.M.E., Marquette U., 1971. Project engr. optical waveguides Corning Glass Works, Wilmington, N.C., 1976-78, supr. equipment engring., 1978-81, supr. process engring., 1981-82, supr. spl. plant projects, 1982—. Served with Submarine Service USN, 1971-76. Mem. ASME, U.S. Naval Inst. Subspecialty: Fiber optics. Current work: Manufacture of optical waveguides. Home: 205 Brookshire Ln Wilmington NC 28403 Office: 310 College Rd Wilmington NC 28405

COLVIN, BURTON HOUSTON, mathematician, government administrator; b. West Warwick, R.I., July 12, 1916; s. Asa Burton and Sara Elsie (Houston) C.; m. Lois Ann Scholes, Dec. 22, 1947; children: Daniel Burton, David Walter, Thomas Alan. A.B., Brown U., 1938, A.M. in Math. (Grand Army of Republic fellow), 1939; Ph.D. in Math. (Univ. fellow), U. Wis., 1943. Instr. math. and mechanics, dept. math. U. Wis., Madison, 1943, instr. math., and asst. prof. math., 1946-51; tech. aide nat. def. research com. Office Sci. Research and Devel., 1944-45; cons. applied mathematician phys. research staff Boeing Co., Seattle, 1951-55, supr. math. analysis group, 1955-58; with Boeing Sci. Research Labs., Seattle, 1958-72, head math. research lab., 1958-70, acting head info. scis. lab., 1966-70, head math. and info. scis. lab., 1970-72; chief div. applied math. Nat. Bur. Standards, Dept. Commerce, Washington, 1972-78; dir. Center for Applied Math., 1978—; NSF lectr., 1957; mem. council Conf. Bd. Math. Scis., 1964, 70-77, chmn., 1975-77; adv. bd. Sch. Math. Study Group, 1963-71, chmn., 1965-66; chmn. computer sci. adv. com. Stanford U., 1970-71. Recipient Silver Medal award Dept. Commerce, 1978, Gold medal award Dept. Commerce, 1981, Presdl. Meritorious Rank award, 1980. Fellow AAAS (council 1965-67, vice-chmn. commn. on sci. edn. 1968-72, chmn. task force on tech. edn. 1968-69); mem. Soc. Indsl. and Applied Math. (vis. scientist lectr. 1962-63, trustee 1962-65, 67-70, 78-80, pres. 1971-72), AAAS (council 1965-67, vice-chmn. commn. on sci. edn. 1968-72, chmn. task force on tech. edn. 1968-69), Math. Assn. Am. (vis. lectr. 1963-65), Am. Math. Soc., NEA, Inst. Math. Stats., Assn. Women in Math., Nat. Council Tchrs. Math, Phi Beta Kappa, Sigma Xi. Subspecialty: Applied mathematics. Office: Nat Bur Standards Center for Applied Math Washington DC 20234

COLWELL, RITA ROSSI, microbiology educator; b. Beverly, Mass., Nov. 23, 1934; d. Louis and Louise (Di Palma) Rossi; m. Jack H. Colwell, May 31, 1956; children: Alison, Stacie. B.S., Purdue U., 1956, M.S., 1958; Ph.D., U. Wash.-Seattle, 1961. Asst. research prof. marine microbiology U. Wash., Seattle, 1961-64; guest scientist div. applied biology NRC of Can., Ottawa, 1961-63; vis. asst. prof. biology Georgetown U., Washington, 1963-64; asst. prof. biology, 1964-66, assoc. prof., 1966-72; prof. microbiology U. Md., College Park, 1972—, dir. sea grant program, 1977—; acting dir. Ctr. Environ. and Estuarine Studies, U. Md., College Park, 1980-81. Subspecialties: Microbiology; Bacteriology. Current work: Marine and estuarine microbial ecology; survival of pathogens in the aquatic environment; temperature and high pressure effects on marine bacteria; applications of computers in biology and medicine. Home: 5010 River Hill Rd Bethesda MD 20816 Office: U Md Microbiology Dept College Park MD 20742

COMBS, CLAUD STEVE, health industry co. exec.; b. Macon, Ga., June 12, 1952; s. Claud L. and Virginia (Cheatham) C.; m. Cynthia Ham, Nov. 4, 1952; children: Joshua, Matthew. B. Elec. Engring. Tech., So. Tech. Inst., Marietta, Ga., 1975. Engr. Carolina Med. Electronics, Inc., King, N.C., 1976-79; dir. quality assurance and regulatory affairs Healthdyne, Inc., Marietta, 1979—. Mem. Assn. Advancement Med. Instrumentation, Am. Soc. Quality Control (biomed. div.). Subspecialties: Bioinstrumentation; Anesthesiology. Current work: FDA-approved research of high frequency ventilator health care, bio instrumentation, regulatory affairs, product quality assurance, high frequency ventilation, bio technology. Office: 2253 Northwest Pkwy Marietta GA 30067

COMBS, GERALD FUSON, health science administrator; b. Olney, Ill., Feb. 23, 1920; s. Lloyd Roscoe and Ina Roe (Fuson) C.; m. Lily McMaster Jams, Mar. 26, 1943; children: Gerald F., Lawrence L., John W., Gregory L. B.S., U. Ill.-Urbana, 1940; Ph.D., Cornell U., 1948. Diplomate: Am. Bd. Human Nutrition. Grad. research asst. Cornell U., Ithaca, N.Y., 1940-41, 46-48; prof. nutrition U. Md., College Park, 1948-69; dep. dir. nutrition program Health Services Adminstrn., Rockville, Md., 1969-71; nutrition and food sci. coordinator U.S. Dept. Agr., Washington, 1971-73; head dept food and nutrition U. Ga., Athens, 1973-75; dir. nutrition program Nat. Inst. Arthritis, Diabetes and Digestive and Kidney Diseases, NIH, Bethesda, Md., 1975—; adv. com. Food and Agr., Rome, 1960-63; mem. com. on animal nutrition Internat. Union Nutrition Sci., 1960-65. Editorial bd.: Jour. Nutrition, 1960-63, Poultry Sci. Jour, 1958-62. Mem. Poultry Sci. Assn., AAAS, Soc. Exptl. Biology and Medicine, N.Y. Acad. Sci., Am. Inst. Nutrition (council 1965-68), Fedn. Am. Soc. Exptl. Biology (pres., bd. 1978-80), Sigma Xi, Alpha Zeta, Phi Kappa Phi. Republican. Methodist. Subspecialties: Nutrition (medicine); Animal nutrition. Home: 10750 Kinloch Rd Silver Spring MD 20903 Office: NIADDK NIH 5333 Westbard Ave Bethesda MD 20205

COMBS, LEON LAMAR, chemistry educator; b. Meridian, Miss., Sept. 19, 1938; s. Leon Lamar and Roberta (Weems) C.; m. Martha Carol Combs, Feb. 17, 1962; 1 son, Jeffrey Lamar. B.S., Miss. State U., 1961; Ph.D., La. State U., 1968. Polymer chemist Devoe and Reynolds, Louisville, 1961-64; asst. prof. chemistry and physics Miss. State U., 1967-71, asso. prof., 1971-75, prof., 1975 —, head dept. chemistry, 1981—; vis. prof. quantum chemistry U. Uppsala, Sweden, 1977-78. Contbr. articles to profl. jours. Tchr. Sunday sch. United Methodist Ch., 1978-81, Faith Baptist Ch., 1981—. Esso fellow, 1965-66. Mem. Am. Chem. Soc., Am. Phys. Soc., Sigma Xi, Phi Kappa Phi, Phi Lambda Upsilon. Subspecialty: Theoretical chemistry. Current work: Quantum and statistical mechanics of molecular interactions. Office: Miss State U Drawer CH Mississippi State MS 39762

COMINS, NEIL FRANCIS, astronomer; b. N.Y.C., May 11, 1951; s. Francis Malcolm and Pearl Murian (Finkelstein) C.; m. Suzanne Rodrique. B.S., Cornell U., 1972; M.S., U. Md., 1975; Ph.D. in Astronomy, Univ. Coll., Cardiff, Wales, 1978. Registered Emergency Med. Technician. Research technician La. State U., 1975; asst. prof. physics and astronomy U. Maine, Orono, 1978—; cons. Am. Coll. Testing Service. Contbr. articles to profl. assn. papers, proceedings. Recipient Rotary Internat. fellowship, 1976; NASA-ASEE Summer Faculty fellowship, 1980, 81. Fellow Royal Astronom. Soc.; mem. Am.

Astronom. Soc., Mt. Desert Island Astronom. Assn. (acting pres., v.p.), Canadian Astronom. Soc., Nat. Wildlife Fedn., Sigma Xi. Subspecialty: Theoretical astrophysics. Current work: Modeling of galaxy dynamics and evolution galaxies, computer simulations, N-Body programs, stochastic self-propagating star formation, galactic evolution. Holder copyright Stellar 28 constellation games. Home: 95 Silver Rd Bangor ME 04401 Office: U Maine 314 Bennett Hall Orono ME 04469

COMMISSO, FRANKLYN W., biologist, educator, curator; b. White Plains, N.Y., July 14, 1948; s. Frank R. and Elsie M. (Rowe) C.; m. Mary Elizabeth Jerry, Aug. 14, 1976. B.A., Pace U., 1970; M.A., Queens Coll., 1973; Ph.D., Fordham U., 1981. Teaching asst. U. Conn., Storrs, 1970-71; lab. technician Pace U., Pleasantville, N.Y., 1966-70, lab. supr., 1971-75, adj. prof., 1971—; curator Natural History Mus., 1974—; adj. asst. prof. Coll. New Rochelle, N.Y., 1980—, Marymount Coll., 1982—, Manhattanville Coll., 1983—; researcher dept. ornithology Am. Mus. Natural History, 1971—. Contbr. chpts. to books. Mem. Am. Ornithologists Union, Am. Mus. Natural History, Nat. Wildlife Fedn., Moravian Music Found., Sigma Xi, Tri Beta. Democrat. Lutheran. Subspecialties: Evolutionary biology; Morphology. Current work: Analysis of myology and osteology of avian hindlimb with associated functional aspects correlated to niche utilization, species interaction, taxonomy and evolution. Home: 78 Elmore Ave Croton-on-Hudson NY 10520 Office: 861 Bedford Rd Pleasantville NY 10570

COMP, PHILIP CINNAMON, physician, med. researcher; b. Kewanee, Ill., Feb. 28, 1945; s. Franklin Howard and Alberta (Cinnamon) C.; m. Carol Winter, June 21, 1974; children: Vanessa, Justin. B.A., Reed Coll., 1967; M.D., U. Wash., 1971; Ph.D. in Biochemistry, U. Okla., 1977. Resident in medicine Hosp. of U. Pa., Phila., 1971-74; research fellow U. Okla., Oklahoma City, 1974-76, asst. prof. medicine, 1976-82, assoc. prof., 1982—; dir. Adult Hemophilia Program, State of Okla., Oklahoma City, 1981—, Thrombosis and Coagulation Lab., State of Okla. Teaching Hosps., 1981—. Weyerhaeuser Found. scholar, 1963-67. Mem. A.C.P., Am. Soc. Hematology, Am. Soc. for Clin. Research, N.Am. Mycological Soc. Democrat. Subspecialty: Hematology. Current work: Thrombotic disease; fibrinolysis; blood coagulation. Home: 12740 St Andrews Terr Oklahoma City OK 73120 Office: Dept Medicine U Okla PO Box 26901 Oklahoma City OK 73190

COMPAAN, ALVIN DELL, physics educator, researcher; b. Hull, N.D., June 11, 1943; s. William and Dena (DeJong) C.; m. Mary Han, Oct. 6, 1946; children: Timothy, Kristina, Deanne, David. A.B., Calvin Coll., 1965; M.S., U. Chgo., 1966, Ph.D., 1971. Assoc. research scientist N.Y.U., 1971-73; asst. prof. physics Kans. State U., 1973-77, assoc. prof., 1977-81, prof., 1981—. Contbr. articles to profl. jours. NDEA fellow, 1967-68; NSF trainee, 1968-70; Humboldt fellow, 1982-83. Mem. Am. Phys. Soc., AAAS, Materials Research So., Optical Soc. Am., Sigma Xi. Subspecialties: Condensed matter physics; Spectroscopy. Current work: Laser interactions with semiconductors, coherent Raman spectroscopy. Office: Department Physics Kansas State University Manhattan KS 66506

COMPANION, AUDREY LEE, chemistry educator; b. Tarentum, Pa., Aug. 19, 1932; d. August and Mabel (McFall) C. B.S., Carnegie Inst. Tech., 1954, M.S., 1956, Ph.D., 1958. Instr. Ill. Inst. Tech., Chgo., 1958-60, asst. prof., 1960-65, assoc. prof., 1965-75; assoc. prof. chemistry U. Ky., Lexington, 1975-76, prof., 1976—, dir. grad. studies, dept. chemistry, 1977-81. Author: Chemical Bonding, 2d edit, 1979. NSF grantee; Am. Chem. Soc. grantee; others. Mem. Am. Chem. Soc. (alt. councillor 1975-79), Am. Phys. Soc., AAUP, Sigma Xi. Clubs: Blue Grass (pres. 1981-82), Bus. and Profl. Women's (Lexington)). Subspecialties: Theoretical chemistry; Physical chemistry. Current work: Computer simulations, storage of hydrogen in inter-metallics, hydrogen in non-metals. Home: 949 Lily Dr Lexington KY 40504 Office: Dept Chemistry U Ky Rose St Lexington KY 40506

COMPERE, CLINTON LEE, physician; b. Greenville, Tex., Feb. 17, 1911; s. Edward L. and Clara (Davison) C.; m. Katharine Gram, Mar. 31, 1949; children: Clinton Lee, Mary Katherine. B.S., U. Chgo., 1936, M.D., 1937. Diplomate: Am. Bd. Orthopaedic Surgery. Intern Henry Ford Hosp., Detroit, 1938-39; resident Blodgett Meml. Hosp., Grand Rapids, Mich., 1939-40; practice medicine specializing in orthopaedic surgery, Chgo., 1946—; mem. sr. attending staff Chgo. Wesley Meml. Hosp., 1949—, chief staff, 1964-66; acad. dir. Prosthetic Research Center, Chgo., 1955—, Prosthetic-Orthotic Edn., 1958—; dir. Rehab. Engring. Center, 1972—; cons. 5th Army Hdqrs., 1947—; cons. amputee clinics Regional Office VA, 1947—; assoc. prof. orthopaedic surgery Northwestern U. Med. Sch., 1954-65, prof., 1965—, Edwin Ryerson prof., chmn. dept. orthopaedic surgery, 1978-80; vice chmn. bd. Rehab. Inst., Chgo.; mem. med. adv. com. Ill. Div. Vocational Rehab.; sec.-treas. Orthopedic Research and Edn. Found., 1972-78; med. dir. Ill. State Med. Drs. Services, 1980—. Co-author: Fracture Treatment, 1937, also articles. Served to lt. col., M.C. AUS, 1940-46. Recipient citation Pres.'s Com. Employment Physically Handicapped, 1959; Profl. Achievement award U. Chgo., 1979. Mem. Am. Acad. Orthopaedic Surgeons (sec. 1959-62, pres. 1963-64), Ill., Chgo. med. socs., A.M.A., A.C.S., Am., 20th Century orthopaedic assns., Chgo. Orthopaedic Soc., Clin. Orthopaedic Soc., Ill. Soc. Med. Research, Internat. Soc. Orthopaedic Surgery and Traumatology, Alpha Omega Alpha. Subspecialties: Orthopedics; Physical medicine and rehabilitation. Home: 2397 Demaret Dr Dunedin FL 33528 Office: 233 E Erie St Chicago IL 60611

COMPTON, MARK MELVILLE, research endocrinologist, physiology educator; b. Richmond, Va., Nov. 21, 1953; s. Archie M. and Betty I. (Sparkhall) C.; m. Kathy Yoho, June 23, 1979. B.S. in Biology, Va. Poly. Inst. and State U., 1975, M.S. in Poultry Sci, 1978, Ph.D. in Animal Sci, 1980. Lab. technician Va. Poly. Inst. and State U., 1975-76; postdoctoral research assoc., instr. Med. Coll. Va., 1979-83; postdoctoral research assoc. U.N.C., 1983—; participant NSF U.S. Antarctic Research Program, McMurdo, 1974-75. Mem. Poultry Sci. Assn., AAAS, Sigma Xi, Phi Sigma, Gamma Sigma Delta. Republican. Presbyterian. Subspecialties: Endocrinology; Cell biology (medicine). Current work: Molecular endocrinology of glucocorticoid induced lymphocytolysis. Office: Med Coll Va Dept Physiology Richmond VA 23298

COMPTON, W. DALE, physicist; b. Chrisman, Ill., Jan. 7, 1929; s. Roy L. and Marcia (Wood) D.; m. Jeanne C. Parker, Oct. 14, 1951; children: Gayle Corinne, Donald Leonard, Duane Arthur. B.A., Wabash Coll., 1949; M.S., U. Okla., 1951; Ph.D., U. Ill., 1955; D.Eng. (hon.), Mich. Technol. U., 1976. Physicist U.S. Naval Ordnance Test Sta., China Lake, Calif., 1951-52, U.S. Naval Research Lab., Washington, 1955-61; prof. physics U. Ill. at Urbana, 1961-70, dir. coordinated sci. lab., 1965-70; dir. chem. and phys. scis., exec. dir. sci. research staff, v.p. research Ford Motor Co., Dearborn, Mich., 1971—; Mem. Presdl. Commn. for Award of Medal of Sci., 1979—; mem. vis. com. Nat. Bur. Standards, 1975—, chmn. vis. com., 1979. Author: (with J.H. Schulman) Color Centers in Solids, 1962; Editor: Interaction of Science and Technology, 1969. Bd. dirs. Mich. Cancer Found.; bd. govs. Cranbrook Inst. Sci. Fellow Am. Phys. Soc., AAAS, Washington Acad. Scis.; mem. Research Soc. Am., Nat. Acad. Engring. Subspecialties: Condensed matter physics; Materials. Current work: Solid state physics (automotive); engineering and automotive systems. Home: 5565 Forman Dr Birmingham MI 48010

COMROE, JULIUS HIRAM, JR., educator; b. York, Pa., Mar. 13, 1911; s. Julius Hiram and Mollie (Levy) C.; m. Jeanette Wolfson, June 30, 1936; 1 dau., Joan Von Gehr. A.B., U. Pa., 1931, M.D., 1934, D.Sc. (hon.), 1978; Commonwealth Fund fellow, Nat. Inst. Med. Research, London, 1939; M.D. (hon.), Karolinska Inst., Stockholm, 1968; D.Sc., U. Chgo., 1968. Intern Hosp. of U. Pa., 1934-36; instr. in pharmacology U. Pa. Med. Sch., 1936-40, asso., 1940-42, asst. prof., 1942-46; prof. physiology and pharmacology U. Pa. Grad. Sch. Medicine, 1946-57; prof. physiology, dir. Cardiovascular Research Inst., U. Calif. Med. Center, San Francisco, 1957-73, Herzstein prof. biology, 1973-78; chmn. 1st Teaching Inst. Asso. Am. Med. Colls., 1953, also chmn., 1961 Inst; chmn. Physiology Study Section, 1955-58; mem. bd. sci. counselors Nat. Heart Inst., 1957-61; mem. Nat. Adv. Mental Health Council, 1958-62, Nat. Adv. Heart Council, 1963-67, Nat. Adv. Heart and Lung Council, 1970-74, Pres.'s Panel Heart Disease, 1972; mem. adv. com. to dir. NIH, 1976-78; Cons. med. research div. CWS, 1944-46. Author: Physiological Basis for 02 Therapy, 1950, Methods in Medical Research, Vol. 2, 1950, The Lung: Clinical Physiology and Pulmonary Function Tests, 1955, 62, Physiology of Respiration, 1964, 74, Pulmonary and Respiratory Physiology (Dowden), 1976, Retrospectroscope-Insights to Medical Discovery, 1977, Exploring the Heart, 1983; Editor: Physiology for Physicians, 1963-66, Circulation Research, 1966-70, Ann. Rev. Physiology, 1971-75; asso. editor: Am. Rev. Respiratory Disease, 1973-79; mem. editorial bd.: Proc. Nat. Acad. Scis, 1977—. Recipient Am. Physiol. Soc. Travel award, 1938, Research Achievement award Am. Heart Assn., 1968, Coll. medal Am. Coll. Chest Physicians, 1970, Trudeau award, 1974, Wiggers award, 1974, Gold Heart award Am. Heart Assn., 1975; Kovalenko medal Nat. Acad. Scis., 1976; Sci. Contbns. award ACP, 1977; Daggs award Am. Physiol. Soc., 1977; medal U. Calif., San Francisco, 1978; Eugenio Morelli award Accademia dei Lincei, Rome, 1979; Abraham Flexner award in med. edn. Am. Assn. Med. Colls., 1979. Fellow Am. Coll. Cardiology (hon.), Royal Coll. Physcians (London), Royal Soc. Medicine (London); mem. Assn. Am. Physicians, Am. Physiol. Soc. (pres. 1960-61), Am. Soc. for Pharmacology and Exptl. Therapeutics (councilor 1953-56), Nat. Acad. Scis. (mem. bd. medicine 1967-70), Inst. Medicine (exec. com. 1970), Am. Acad. Arts and Scis., Harvey Soc. (hon. mem.), Am. Soc. for Clin. Investigation, Phi Beta Kappa, Sigma Xi, Alpha Omega Alpha. Subspecialties: Physiology (medicine); Pulmonary medicine. Home: 555 Laurent Rd Hillsborough CA 94010 Office: Cardiovascular Research Inst U Cal Med Center San Francisco CA 94143

CONAN, ROBERT JAMES, JR., chemistry educator, consultant; b. Syracuse, N.Y., Oct. 30, 1924; s. Robert J. and Helen M. (O'Brien) C. B.S., Syracuse U., 1945, M.S., 1947; Ph.D., Fordham U., 1950. Instr. Fordham U., N.Y.C., 1947-49; asst. prof. Le Moyne Coll., Syracuse, N.Y., 1949-54, assoc. prof., 1954-58, prof. chemistry, 1958—; cons. Carrier Corp., Syracuse, 1949-63, Owl Wire and Cable Co., Oneida, N.Y., 1952—, Edison Audio Archives, Syracuse, N.Y., 1972—. Com. mem. Onondaga Lake Sci. Council, Syracuse, N.Y., 1964-65. Mem. Am. Chem. Soc. (chmn. Syracuse sect. 1958, 72, mem., nat. councillor 1982—); Mem. Tech. Club Syracuse (pres. 1982, plaque award 1982). Republican. Roman Catholic. Subspecialties: Physical chemistry; Surface chemistry. Current work: Physical properties of liquids and solutions, properties of gases, wetting phenomenon. Home: 263 Robineau Rd Syracuse NY 13207 Office: Le Moyne Coll Syracuse NY 13214

CONCANNON, JAMES THOMAS, psychopharmacologist, educator; b. Shamokin, Pa., Apr. 18, 1951; s. Theodore Patrick and Mary Dolores (Sokoloskie) C. B.S., St. Joseph's Coll., Phila., 1973; M.A., SUNY, Binghamton, 1975, Ph.D., 1977. USPHS fellow, postdoctoral research asso. Kent (Ohio) State U., 1977-79; asst. prof. psychology U. R.I., Providence, 1979-80; research instr. dept. pharmacology Northeastern Ohio Univs. Coll. Medicine, Rootstown, 1980-82, research asst. prof., 1982—. Contbr. articles to profl. jours. Nat. Inst Alcohol Abuse and Alcoholism fellow, 1976-77; NIH grantee, 1980-81; Occupational Health grantee, 1982—. Mem. Soc. for Neurosci., Am. Psychol. Assn. Internat. Soc. for Developmental Psychobiology, Soc. for Stimulus Properties of Drugs, AAAS, Eastern Psychol. Assn., Midwestern Psychol. Assn., Sigma Xi. Subspecialties: Neuropharmacology; Pharmacology. Current work: Animal models of psychiatric and neurological problems; behavorial pharmacology. Office: Northeastern Ohio Univs Coll Medicine State Route 44 Rootstown OH 44272

CONCORDIA, CHARLES, electrical engineer, consultant; b. Schenectady, June 20, 1908; s. Francis G. and Susie Elizabeth (Decker) C.; m. Frances Butler, Dec. 18, 1948. Sc.D. (hon.), Union Coll., 1971. With Gen. Electric Co., Schenectady, 1926-73, in electric utility systems engring., 1936-73, applications engr., 1936-49, in aircraft devel., 1941-45, cons. engr., 1949-73; cons. electric power systems engring., Venice, Fla., 1973—; lectr. various univs. Author: Synchronous Machines, 1951; contbr. 120 articles to profl. jours. Recipient Lamme medal Am. Inst. Elec. Engrs., 1961; Coffin award Gen. Electric Co., 1942; Steinmetz award, 1973; named Engr. of Yr. Profl. Engrs. Soc., 1963. Fellow IEEE, ASME, AAAS; mem. Assn. Computing Machinery (founding mem.), Conf. Internationale des Grands Reseaux Electriques a Haute Tension, Nat. Acad. Engring., Nat. Soc. Profl. Engrs., Sigma Xi, Tau Beta Pi. Republican. Presbyterian. Clubs: Venice Yacht, Mohawk Golf. Subspecialties: Systems engineering; Numerical analysis. Current work: Electric power system dynamic performance and control: system modeling, reliability, operating problems, major breakdowns (blackouts), stability improvement, design and planning. Patentee in field (6). Home and Office: 629 Alhambra Rd Apt 402N Venice FL 33595 I am told that I tend never to take anything on faith, but that I try to find the truth without fear or favor. If there is a talent that I have, it is the ability to abstract the simple essentials from complex problems; and I have never had an uninteresting job.

CONCUS, PAUL, mathematician; b. Los Angeles, June 18, 1933; s. Wulf and Flora (Malin) C.; m. Celia Gordon, Mar. 22, 1959; children: Marian, Adriane. B.S., Calif. Inst. Tech., 1954; A.M., Harvard U., 1955, Ph.D., 1959. Applied mathematician IBM Corp., Oakland, Calif., 1959-60; staff sr. scientist Lawrence Berkeley (Calif.) Lab., 1960—; adj. prof. U. Calif., Berkeley, 1978—; vis. prof. U. Di Trento, Italy, 1977, 78; cons. Gen. Electric Co., San Jose, Calif., 1973-77, Lockheed Research Lab., Palo Alto, Calif., 1961-70. Sr. vis. fellow Sci. Research Council, Gt. Britain, 1971. Mem. Soc. Indsl. and Applied Math. (jour. editor 1979—), Tau Beta Pi. Subspecialties: Applied mathematics; Numerical analysis. Current work: Numerical solution of partial differential equations; capillary fluid mechanics; petroleum reservoir simulation. Office: Lawrence Berkeley Lab U Calif Berkeley CA 94720

CONDAL, ALFONSO RAMON, geophysicist, astronomer, researcher; b. Santiago, Chile, Sept. 22, 1939; came to U.S., naturalized, 1975; s. Alfonso Angel and Isabel (Beretta) C. Diploma math., U. Chile, Santiago, 1965; M.Sc., U. Alaska Coll., 1971; Ph.D. in Astronomy, U. B.C., Vancouver, 1979. Research, tchr. Inst. Physics Cath. U., Santiago, 1971-71; researcher Max-Planck Inst. fur Astronomie, Heidelberg, W. Ger., 1980-81, Atmospheric Environ. Service, Toronto, Ont., Can., 1982—. Contbr. articles to profl. jours. Mem. Am. Astron. Soc., Canadian Astron. Soc. Subspecialties: Remote sensing (atmospheric science); Optical astronomy. Current work: Developing of computer assisted image analysis of visual and infrared data from NOAA satellites. Office: 4905 Dufferin St Toronto ON Canada M3H 5T4

CONDON, JAMES JUSTIN, radio astronomer; b. New Orleans, Apr. 15, 1945; s. Justin Jerome and Jean Louise (Rodger) C.; m. Marlene Ann Cabana, Apr. 27, 1979. B.A., Cornell U., 1966, Ph.D. in Astronomy, 1972. Research assoc. Nat. Radio Astronomy Obs., Charlottesville, Va., 1972-74, vis. asst. scientist, 1977-78, scientist, 1980—; asst. prof. physics Va. Poly. Inst. and State U., 1974-77, assoc. prof., 1978-79. Alfred P. Sloan Found. fellow, 1977-78. Mem. Am. Astron. Soc., Internat. Astron. Union. Subspecialty: Radio and microwave astronomy. Current work: Radio emission from spiral galaxies, QSO's, low-frequency variability. Office: NRAO Edgemont Rd Charlottesville VA 22901 Home: 3785 Skyline Crest Dr Charlottesville VA 22901

CONE, EDWARD J., analytical chemist, educator, cons.; b. Mobile, Ala., Sept. 17, 1942; s. Melvin M. and Geanie R. (McCrory) C.; m. Eleanor E. Bray, Feb. 16, 1963; children: Randall, Jennifer. B.S., Mobile Coll., 1967; Ph.D., U. Ala., Tuscaloosa, 1971. Chemist Shell Oil Co., Norco, La., 1966-67; postdoctoral fellow U. Ky., 1971-72; research chemist Nat. Inst. Drug Abuse, Addiction Research Center, Lexington, Ky., 1972—, sect. chief chemistry, 1980—; adj. prof. Sch. Pharmacy, U. Ky., 1975—. Contbr. articles to profl. jours. Mem. Am. Chem. Soc., Am. Soc. Pharmacology and Exptl. Therapeutics, Internat. Soc. for Study of Xenobiotics. Subspecialties: Analytical chemistry; Pharmacology. Current work: Chemistry and metabolism of drugs of abuse drug metabolism, mechanisms of action, medicinal chemistry. Office: Nat Inst Drug Abuse Addiction Research Center PO Box 12390 Lexington KY 40583

CONE, ROBERT EDWARD, immunologist, educator; b. Bklyn., Aug. 18, 1943; s. Joseph and Ruth C.; m. Michele Joy Nash, Aug. 22, 1967; children: Jennifer, Laura. B.S., Bklyn Coll., 1964; M.S., Fla. State U., 1967; Ph.D., U. Mich., 1970. Postdoctoral fellow Walter and Eliza Hall Inst. for Med. Research, Melbourne, Australia, 1971-73, Basel (Switzerland) Inst. for Immunology, 1973-74; asst. prof. depts. surgery and pathology Yale U., 1974-80, asso. prof., 1980—; mem. histocompatibility com. New Eng. Organ Bank. Author articles. F.G. Novy fellow, 1968; Horace Rackham fellow, 1969-70; Damon Runyon postdoctoral fellow, 1971-73; Upjohn Co. grantee, 1979, 80. Mem. AAAS, Am. Assn. Immunologists, Internat. Cell Research Orgn. Jewish. Subspecialties: Transplantation; Membrane biology. Current work: Structure and function of lymphocyte membranes and receptors.

CONE, RUFUS LESTER, physicist, educator; b. Statesboro, Ga., Apr. 28, 1944; s. Rufus Lester and Louise (Lipford) C.; m. Margaret Nelson Van Horn, Aug. 12, 1967. B.S., Ga. Inst. Tech., 1966, M.S., 1967; M.Phil., Yale U., 1968, Ph.D., 1971. Asst. prof. physics U. Ga., Athens, 1971-74; asst. prof. physics Mont. State U., Bozeman, 1974-77, assoc. prof., 1977—; cons. coherent laser spectroscopy Bell Labs., Murray Hill, N.J., 1980—; vis. fellow Clarendon Lab., U. Oxford, Eng., 1983. Contbr. articles to profl. jours. NSF grantee, 1974—; Research Corp. grantee, 1974-75; Mont. State U. Research/Creativity grantee, 1980-82. Mem. Am. Phys. Soc., Optical Soc. Am., IEEE, Yale Sci. and Engring. Assn., Nat. Rail Hist. Subspecialties: Spectroscopy; Condensed matter physics. Current work: Nonlinear optics and laser spectroscopy. Laser spectroscopy of condensed matter, nonlinear optical processes, interionic interactions, rare earth insulators. jOffice: Dept Physic Mont State U Bozeman MT 59717

CONERY, JOHN SIMPSON, computer scientist; b. Tacoma, Feb. 26, 1952; s. Patrick Daniel and Elaine Margaret (Simpson) C. B.A. with honors in Psychology, U. Calif.-San Diego, 1976; M.S. in Computer Sci, U. Calif-Irvine, 1979; Ph.D. in Info. and Computer Sci., U. Calif.-Irvine, 1983. Programmer, research asst. U. Calif.-San Diego, 1971-74, research asst. phonetics lab., 1975-76; programmer audiology dept. VA Hosp., La Jolla, Calif., 1975-76; sr. systems programmer Pertec Computer Corp., Irvine, 1976-78; teaching asst., instr., research asst. U. Calif.-Irvine, 1978-83; asst. prof. computer and info. sci. U. Oreg., 1983—; cons. in field. Mem. Assn. for Computing Machinery, AAAS. Subspecialties: Programming languages; Computer architecture. Current work: Logic programming, functional languages, dataflow architecture, database technology. Home: 511 W 27th Ave Eugene OR 97405 Office: Dept Info and Computer Sci U Oreg Eugene OR 97403

CONEY, CHARLES CLIFTON, educator; b. Kingsport, Tenn., Aug. 24, 1949; s. Charles Herbert and Emma Sue (Snell) C.; m. Sallie Ruth Dingus, Nov. 24, 1974; 1 dau., Sonia Lorraine. B.S., East Tenn. State U., 1977, M.S., 1980. Research asst. East Tenn. State U., Johnson City, 1977-79, teaching asst., 1979-80; instr. biology U. S.C., Conway, 1980-83; collection mgr. malacology sect. Natural History Mus., Los Angeles, 1983—. Contbr. articles to profl. jours. Mem. Paleontological Soc., Paleontological Research Inst., Am. Malacological Union, S.C. Acad. Sci. Subspecialties: Evolutionary biology; Ecology. Current work: Ontogeny and phylogeny of molluscan bivalve gill, with emphasis on comparative microanatomy of bivalve gill structure, leading to an understanding of mode and tempo of bivalve gill devel. and speciation among convergent bivalve shell homeomorphs. Office: Malacology Sect Natural History Mus 900 Exposition Blvd Los Angeles CA 90007 Office: Univ SC Coastal Carolina Coll Conway SC 29526

CONGER, BOB VERNON, plant scientist, educator; b. Greeley, Colo., July 2, 1938; s. Vernon Fred and Florence Violet (Pierce) C.; m. Donna Dee Russell, June 5, 1960; children: Gregory, Rhonda, Stephen, Michael. B.S. in Agronomy, Colo. State U., 1963; Ph.D. in Genetics, Wash. State U., 1967. Asst. prof. Wash. State U., 1967-68; asst. prof. plant and soil sci. U. Tenn., Oak Ridge and Knoxville, 1968-73, assoc. prof., 1973-78, prof., Knoxville, 1978—. Author: Cloning Agricultural Plants Via in Vitro Techniques, 1981; assoc. editor: Environ. Exptl. Botany; editor: CRC Critical Revs. in Plant Scis; contbr. articles to profl. jours. NASA predoctoral trainee Wash. State U., 1964-67; U.S. Dept. Agr. grantee, 1979-81, 82—. Mem. AAAS, Am. Genetic Assn., Am. Soc. Agronomy, Tissue Culture Assn., Crop Sci. Soc. Am., Internat. Assn. Plant Tissue Culture, Internat. Soc. Plant Molecular Biology, Sigma Xi, Gamma Sigma Delta, Phi Kappa Phi. Democrat. Methodist. Subspecialties: Plant cell and tissue culture; Plant genetics. Current work: Cell and tissue culture, cytogenetics, breeding of cool season forage grasses. Office: Dept Plant and Soil Sci U Tenn Knoxville TN 37996

CONIGLIO, JOHN GIGLIO, biochemistry educator; b. Tampa, Fla., July 21, 1919; s. Guiseppe and Maria (Giglio) C.; m. Carmen Sylvia Moreno, Dec. 27, 1942; children: John William, Robert Freeman, David Martin. B.S. in Chemistry, Furman U., 1940; Ph.D. in Biochemistry, Vanderbilt U., 1949. Tchr. secondary schs., Kershaw, S.C., 1940-41; chemist E.I. duPont, Childersburg, Ala., 1942-43; chemist, supr. Tenn. Eastman Corp., Oak Ridge, 1944-45; asst. prof., assoc. prof. Vanderbilt U., Nashville, 1950-63, prof., 1963—. Contbr. numerous articles on biochemistry to profl. jours. Mem. Am. Soc. Biol.

Chemists, Am. Inst. Nutrition, AAAS, Soc. Exptl. Biology and Medicine. Subspecialties: Biochemistry (biology); Biochemistry (medicine). Current work: Lipid chemistry and metabolism, biochemistry of lipids of reproductive organs, biosynthesis and metabolism of polyunsaturated lipids. Home: 202 Lauderdale Rd Nashville TN 37205 Office: Dept Biochemistry Vanderbilt U Nashville TN 37232

CONLEY, CAROLYNN LEE, aerospace technologist; b. Burbank, Calif., Oct. 10, 1948; d. William Charles and Dorothy Marie (Gibson) Faubion. B.S., UCLA, 1971, M.S., 1977; M.B.A., U. Houston, 1980. Research geochemist [illegible], Houston, 1977; research geochemist NASA-Johnson Space Ctr., Houston, 1977, aerospace technologist, 1978—. Cyclethon vol. Am. Heart Assn., Clear Lake City, Tex., 1980, 82; Peach Creek Camp cons. Jan Jacinto Girl Scout council, Houston, 1982; reporter Tex. Dept. Pub. Safety, Houston, 1982-83. Recipient Superior Performance award NASA-Johnson Space Center, 1981, Suggestion award, 1982, Outstanding Speaker award, 1982, Quality Increase award, 1982. Mem. AIAA (vice-chmn. ops. 1981-82, newsletter assoc. editor 1982-84), Am. Bus. Women's Assn., Soc. Women Mgrs. Republican. Presbyterian. Club: Toastmasters (Houston) (v.p. adminstrn. 1982-83). Subspecialties: Aerospace engineering and technology; Systems engineering. Current work: Flight activity officer develop mission crew activity plan for space shuttle flights integrating vehicle, crew and payload requirements; responsible to flight director for implementation and control of the realtime mission crew activity plan. Office: NASA Johnson Space Center Mail Code NH4 Houston TX 77058

CONLEY, JAMES FRANKLIN, research geologist, educator; b. Logan, N.C., Dec. 28, 1931; s. Thaddeus Lafayette and Mallie Marie (Morehead) C.; m. Ellen Marie Reedy, Dec. 23, 1954; children: James Franklin, John Horace, David Loren. B.A., Berea Coll., 1954; M.S., Ohio State U., 1956; Ph.D., U. S.C., 1982. Cert. profl. geologist Am. Inst. Profl. Geologists. Geologist N.C. Div. Mineral Resources, Raleigh, 1956-65; geologist Va. Div. Mineral Resources, Charlottesville, 1965—, research geologist, 1956—; geologist U. Va., 1982—. Contbr. articles to profl. publs. Fellow Geol. Soc. Am.; mem. Am. Inst. Mining Engrs. (chmn. Va. sect. 1981-82), Am. Inst. Profl. Geologists (pres. Va. sect. 1983), Am. Assn. Petroleum Geologists, Potomac Geophys. Soc. Methodist. Subspecialties: Geology; Tectonics. Current work: Structural geology. Home: 1614 Trailridge Rd Charlottesville VA 22903 Office: Va Div Mineral Resources PO Box 3667 Charlottesville VA 22901

CONLON, BARTHOLOMEW FREDERICK, JR., mechanical engineer; b. Alexandria, Va., Aug. 14, 1931; s. Bartholomew Frederick and Maggie Lee (Hughes) C.; m. Ruth Alice Tompkins, Oct. 3, 1953; children: Philip, David, Mark, Bart III, Patrick, Charles. B.S., U. Va., 1961. Registered profl. engr., Va. Staff supr. Newport News Shipbuilding & Dry Dock Co., Va., 1961-62; mech. design engr. Gen. Electric Co., Waynesboro, 1962-79, mgr. mech. design engring., 1979—; instr. Blue Ridge Community Coll., 1982. Troop leader Explorer Scouts, 1964-69. Served to cpl. U.S. Army, 1953-55; Korea. Mem. ASME (Centennial award 1980). Roman Catholic. Subspecialty: Mechanical engineering. Current work: Development and design of high technology products in area of high speed printing mechanisms. Patentee belt tracking system, tape reel drive mechanism. Home: Rt 1 Box 13 Waynesboro VA 22980 Office: Gen Electric Co Waynesboro VA 22980

CONLY, JOHN FRANKLIN, engineering educator; b. Ridley Park, Pa., Sept. 11, 1933; s. Harlan and Mary Jane (Roberts) C.; m. Jeanine Therese McDonough, Apr. 15, 1967; children: Mary Ann, John Paul. B.S., U. Pa., 1956, M.S., 1958; Ph.D., Columbia U., 1962. Research asst. Columbia U., N.Y.C., 1959-62; asst. prof. San Diego State U., 1962-65, assoc. prof., 1965-69, prof. aerospace engring. and engring. mechanics, 1969—, chmn. dept., 1971-74, 80—; also dir. Wind Tunnel, 1978—. Contbr. articles to profl. jours. Guggenheim Found. fellow, 1958; Am. Soc. Engring. Edn.-NASA fellow, 1970. Assoc. fellow AIAA (San Diego chmn. 1970). Republican. Episcopalian. Subspecialties: Aeronautical engineering; Fluid mechanics. Current work: Experimental aerodynamics. Home: 6478 Bonnie View Dr San Diego CA 92119 Office: Dept Aero Engring San Diego State U San Diego CA 92182

CONN, P. MICHAEL, research scientist; b. Oil City, Pa., May 12, 1949; s. Robert H. and Frances G. C. B.S., U. Mich., 1971; M.S., N.C. State U., 1973; Ph.D., Baylor Coll. Medicine, 1976; postdoctoral, Nat. Inst. Child Health and Human Devel., NIH, 1978. Assoc. prof., sr. fellow in aging, dept. pharmacology Duke U. Med. Ctr., Durham, N.C., 1978—. Editor: Cellular Regulation of Secretion, 1982, Methods in Enzymol. Neuroendocrine Peptides, 1983, The Receptors, 1983; editor: Jour. Clin. Endocrinology and Metabolism, 1983—; bd. dirs.: Am. Jour. Physiology, 1982—, Molecular Pharmacology, 1980—, Endocrinology, 1982—. Recipient research career devel. award NIH, 1979. Mem. Endocrine Soc., Soc. Study Reproduction, Am. Soc. Cell Biology. Subspecialties: Neuroendocrinology; Receptors. Current work: Research in mechanism of action of gonadotropin releasing hormone in the pituitary. Inventor novel forms of contraception. Office: Dept Pharmacology Duke Univ Med Center Durham NC 27710

CONN, REX BOLAND, JR., physician, educator; b. Marengo, Iowa, Aug. 3, 1927; s. Rex Bol and Helena Dorothea (Schoenfelder) C.; m. Victoria Grace Sellens, Dec. 28, 1950; children—Elizabeth Marian, Victoria Anne, Mary Catherine. B.S., Iowa State U., 1949; M.D., Yale U., 1953; B.Sc., U. Oxford, Eng., 1955; M.S., U. Minn., 1960. Prof. pathology, dir. clin. labs. W.Va. Med. Center, Morgantown, 1960-68; prof. lab. medicine, dir. dept. Johns Hopkins Med. Instns., Balt., 1968-77; prof. pathology and lab. medicine, dir. clin labs. Emory U., Atlanta, 1977—; Mem. pathology tng. com. NIH, 1972-73; cons. Walter Reed Army Med. Center, 1972-77. Co-editor: Current Diagnosis, 1980, Yearbook of Pathology and Clinical Pathology, 1980. Served with USNR, 1945-46. Mem. Coll. Am. Pathologists, Am. Soc. Clin. Pathologists (dir. 1975-81), Acad. Clin. Lab. Physicians and Scientists (pres. 1972—). Subspecialties: Pathology (medicine); Biochemistry (medicine). Current work: Diagnostic laboratory medicine, clinical enzymology, medical instrumentation, medical computing. Home: 2505 Greenglade Rd Atlanta GA 30345 Office: 1364 Clifton Rd Atlanta GA 30322

CONNELLY, JOHN PETER, pediatrician; b. Boston, May 12, 1926; s. Thomas Joseph and Bridget T. (Finnigan) C.; m. Martha T. Cronin, June 24, 1950; children: Maureen, Marie, Eileen, Martha, Cathleen, John, Michael. B.S., Boston Coll., 1951; M.D., Georgetown U., 1955. Diplomate: Am. Bd. Pediatrics. Intern Royal Victoria Hosp., Montreal, Que., Can., 1955-56; resident in pediatrics Johns Hopkins Hosp., 1957-58; resident Mass. Gen. Hosp., Boston, 1956-68, 61-62, asst. pediatrician, 1961-64, chief, 1963-64, chief ambulatory div., 1964-69, pediatrician to children;s service, 1967-73; exec. dir. Bunker Hill Health Ctr., 1967-73; vis. physician Boston Hosp. for Women, 1961-73; chief of pediatrics Foster G. McGaw Hosp., Loyola U., Chgo., 1972-76; cons. dir. Regional Center for Study of Sudden Infant Death Syndrome, 1976-81; dir. div. health services research and devel. Am. Acad. Pediatrics, Evanston, Ill., 1976—; mem. faculty Harvard Med. Sch., 1957-73, assoc. prof. pediatrics, 1969-73; vis. lectr. Harvard U. Sch. Pub. Health, Harvard Ctr. Community Health and Med. Care; prof., chmn. dept. pediatrics Loyola U.-Stritch Sch. Medicine, Chgo., 1972-76; sr. lectr. U. Chgo. Sch. Social Service Adminstrn. and Policy, 1979—; prof. community health scis. U. Ill. Sch. Pub. Health, 1980—; vis. scholar Northwestern U. Ctr. Health Services Policy Research, 1981—. Contbr. chpts. to books, articles to profl. jours. Trustee Nat. Sudden Infant Death Syndrome Found., 1975—; bd. dirs. Mass. Soc. Prevention of Cruelty to Children, 1966-73; mem. Ill. Sudden Infant Death Syndrome Study Commn., 1975—. Served to rear adm. M.C. USNR, 1944—. Decorated Naval Commendation medal, others; recipient Cahill medal Georgetown U.; citation for humanitarianism Commonwealth of Mass., 1973, numerous others. Mem. Soc. Med. Cons. to Armed Forces, Naval Res. Assn. (life), Am. Fedn. Clin. Research, Assn. Ambulatory Pediatric Services, Irish and Am. Pediatric Soc. (pres. 1977-78), Royal Coll. Medicine, Am. Irish Found., AMA, Chgo. Med. Soc., Ill. Med. Soc., Mass. Med. Soc. Res. Officers Assn. U.S., Harvard Med. Alumni Assn. (assoc.), Riverside C. of C. Roman Catholic. Club: St. Mary's Holy Name Soc. (Riverside). Lodge: Lions. Subspecialties: Health services research; Pediatrics. Current work: Health services research on maternal and child health related projects; i.e., sudden infant death syndrome. Home: 147 Herrick Rd Riverside IL 60546 Office: Am Acad Pediatrics 1801 Hinman St Evanston IL 60204

CONNELLY, WILL ARTHUR, electronics research company executive, researcher; b. Chgo., June 25, 1930; s. Will Henry and Jane Katherine (Kaye) C.; m. Aileen M. Silvera, Feb. 15, 1953. Tech. rep. George Davis Sales Co., Los Angeles, 1953-57; dir. mktg. Roberts Electronics Corp., Hollywood, Calif., 1957-59; pres. Connelly Sales Co., La Mirada, Calif., 1959-63; v.p. Tracor Marine, Inc., Miami, 1964-70, Southcom Internat., Escondido, Calif., 1970-71; pres. U.S. Technology Corp., Ft. Lauderdale, Fla., 1972—. Author: Musician's Guide to Independent Record Production, 1981; contbr. articles to profl. jours. Mem. exec. com. Broward County (Fla.) Republican Party, Ft. Lauderdale, 1978; pres. Hot Jazz and Alligator Gumbo Soc., Ft. Lauderdale, 1980. Served with USAAF, 1947-50. Mem. Audio Engring. Soc., U.S. Naval Inst., Marine Tech. Soc., Am. Radio Relay League. Subspecialties: Microelectronics; Offshore technology. Current work: Invention, evolution of precision electronic navigation and positioning system. Patentee precision navigation. Home: PO Box 21518 Fort Lauderdale FL 33335 Office: US Technology Corporation PO Box 21518 Fort Lauderdale FL 33335

CONNER, JERRY POWER, physicist; b. Sherman, Tex., Mar. 20, 1927; s. Vester Fulton and Maude Allene (Power) C.; m. Annie Lela Lee, Sept. 7, 1947; children: Susan, Patrick, Kim. Ph.D., Rice U., 1952. Mem. staff Los Alamos Nat. Lab., 1952-74, group leader, 1974-79, asst. div. leader, 1979—. Served with USNR, 1944-46. Recipient Wilson award Rice U., 1952. Mem. Am. Phys. Soc., Am. Geophys. Union, Am. Astron. Soc. Subspecialties: High energy astrophysics; Space project management. Current work: Detection and analysis of radiation in space, space project management. Office: Los Alamos Nat Lab Mail Stop D446 Los Alamos NM 87545

CONNER, ROSS F., social psychologist, evaluation researcher; b. Denver, Aug. 23, 1946; s. Ross D. and Mary Ann C. B.A., Johns Hopkins U., 1968; M.A., Northwestern U., 1973, Ph.D., 1974. Instr. Northwestern U., Evanston, Ill., 1972-74; asst. prof. U. Calif.-Irvine, 1974-80, assoc. prof., 1980—, assoc. dir. social ecology program, 1982—; cons. in field; evaluation specialist Peace Corps, Washington, 1980-81. Co-author: Attorneys as Activists, 1979; editor: Methodological Advances in Evaluation Research, 1981; contbr. articles on evaluation research to profl. jours. Planning commr. Planning commn. City of Laguna Beach, Calif., 1982—. Mem. Evaluation Research Soc. (council 1982—), Am. Psychol. Assn., AAAS, Evaluation Network. Subspecialties: Social psychology; Evaluationresearch. Current work: Utilization of social science and evaluation research data in public policy decisionmaking, use of experimental and control groups in program evaluation. Office: Social Ecology Program U Calif Irvine CA 92717

CONNETT, RICHARD JAMES, physiology educator, researcher; b. York, Nebr., Sept. 11, 1943; s. Robert Paul and Janice M. (Brugh) C.; m.; children: Denise, Elise. B.A., Park Coll., 1963; Ph.D., Duke U., 1969. Assoc. U. Rochester (N.Y.) Med. Sch., 1973-74, asst. prof. physiology, 1974-81, assoc. prof. physiology, 1981—; mem. NIH rev. com., study workshop. Mem. Am. Physiol. Soc., Am. Biophys. Soc., AAAS, Sigma Xi. Episcopalian. Club: Am. Recorder Soc. (pres. Rochester chpt. 1978-81). Subspecialties: Biochemistry (biology); Physiology (biology). Current work: Studies on metabolic regulation of skeletal muscle at rest and during exercise. Home: 2167 Westfall S Rd Rochester NY 14618 Office: U Rochester Med Sch PO Box 642 Dept Physiology 601 Elmwood Ave Rochester NY 14642

CONNEY, ALLAN HOWARD, pharmacologist; b. Chgo., Mar. 23, 1930; s. Leo Younkers and Celia (Gasway) C.; m. Diana Locke, Sept. 5, 1954; children: Michael Raymond, Steven Herbert. B.S. in Pharmacy, U. Wis., 1952, M.S. in Oncology, 1954, Ph.D., 1956. Registered pharmacist, Ill. Research asst. McArdle Lab., Madison, Wis., 1952-56; guest investigator Nat. Heart Inst., Bethesda, Md., 1957-58, pharmacologist, 1958-60; head dept. biochem.pharmacology Burroughs Wellcome & Co., Tuckahoe, N.Y., 1960-70; dir. dept. biochemistry Hoffman-LaRoche, Inc., Nutley, N.J., 1970-71, dir. dept. biochemistry and drug metabolism, 1971—, assoc. dir. exptl. therapeutics, 1979—; adj. prof. Rockefeller U.; lectr. Columbia U.; vis. disting. prof. Rutgers U. Contbr. numerous articles to profl. jours. Mem. Am. Soc. Biol. Chemists, Nat. Acad. Scis., Am. Soc. Pharmacology and Exptl. Therapeutics, Am. Assn. Cancer Research (G.H.A. Clowes award lectr. 1981), Soc. Toxicology, Inc., Acad. Pharm. Scis., Am. Pharm. Assn., AAAS, N.Y. Acad Scis. Subspecialties: Cancer research (medicine); Biochemistry (medicine). Current work: Chemical Carcinogenesis, factors influencing metabolism and action of drugs, steroids and chemical carcinogens. Office: 340 Kingsland St Bldg 86 Nutley NJ 07110

CONNOLLY, LEO PAUL, astronomer, educator; b. Sebastopol, Calif., Jan. 1, 1947; s. Leo F. and Virginia U. (Trigeiro) C.; m. Jacqueline B. Bonanno, July 14, 1973; 1 son: Richard. B.A., U. Calif.-Berkeley, 1969; Ph.D. in Astronomy, U. Ariz., 1975. Astronomy and physics instr., dir. MacLean Obs., Sierra Nevada Coll., Incline Village, Nev., 1973-78; assoc. prof. astronomy Southeast Mo. State U., Cape Girardeau, 1978—; also researcher. Contbr. numerous articles to sci. publs. NSF Sci. Faculty Profl. Devel. awardee, 1979. Mem. Am. Astron. Soc., Astron. Soc. of Pacific. Subspecialty: Optical astronomy. Current work: Photometry of pulsating stars. Home: 913 N Missouri Ave Cape Girardeau MO 63701 Office: Dept Physics Southeast Mo State U Cape Girardeau MO 63701

CONNORS, GERARD JOSEPH, psychologist; b. Bklyn., Mar. 30, 1952; s. Gerard James and Mary (Woods) C.; m. Kathy Lang, May 28, 1977. B.A., U. Kans., 1974; M.S., George Peabody Coll., 1976; Ph.D., Vanderbilt U., 1980. Lic. psychologist, Tex. Psychology intern VA Med. Ctr., Brockton, Mass., 1980-81, research psychologist, 1981-82; asst. prof. U. Tex. Med. Sch., Houston, 1982—; coordinator alchol outpatient services U. Tex. Med. Sch., 1982—. Contbr. articles to profl. jours.; author: monograph 4th Spl. Report to U.S. Congress on Alcohol & Health, 1981. Nat. Inst. on Alcohol Abuse and Alcoholism clin. trainee, 1977-78; research trainee, 1978-80. Mem. Am. Psychol. Assn., Assn. for Advancement Behavior Therapy, Southeastern Psychol. Assn. Subspecialties: Behavioral psychology; Clinical psychology. Current work: Behavior therapy, assessment and treatment of alcohol abuse and nondrug factors influencing drug responses. Home: 7950 N Stadium #231 Houston TX 77030 Office: U Tex Med Sch 6410 Fannin Suite 600 Houston TX 77030

CONOLLY, MATTHEW ELLIS, physician, pharmacology researcher; b. London, June 12, 1940. M.B., B.S., London U., 1963, M.D., 1972. House surgeon, physician Westminster Hosp., London, 1963-64; research registrar Postgrad. Med. Sch., London U., 1966-71; postdoctoral fellow Vanderbilt U., Nashville, 1971-72; cons. physician Royal Postgrad. Med. Sch., London, 1972-77; prof. medicine and pharmacology UCLA Sch. Medicine, 1977—. Recipient Pharm Mfrs. Found. award, 1978-81; research awardee in bronchial asthma NIH, 1982—. Fellow ACP; mem. Royal Coll. Physicians (London). Subspecialties: Internal medicine; Pharmacology. Current work: Adrenergic function in health and disease. Office: UCLA Sch Medicine Los Angeles CA 90024

CONOMY, JOHN PAUL, neurologist, health cons., researcher; b. Cleve., July 31, 1938; s. John Paul and Marie Elizabeth (Bimbea) C.; m. Jeannette Melchior, Oct. 19, 1963; children: John, Lisa, Christopher. B.S. cum laude, John Carroll U., 1960; M.D., St. Louis U., 1964. Diplomate: Am. Bd. Psychiatry and Neurology (examiner). Intern, house officer St. Louis U., 1964-65; resident in neurology Univ. Hosps., Cleve., 1965-68; fellow in neuropathology Cleve. Met. Gen. Hosp., 1968; research fellow neuroanatomy U. Pa., 1970-71; staff neurologist Scott and White Found., 1971-72; asst. prof. neurology Case Western Res. U., 1972-75; chmn. dept. neurology Cleve. Clinic Found., 1975—; cons. health care allocation; attending physician Univ. Hosps., Cleve., 1968; Highland View Hosp., Cleve., 1968; assoc. neurologist Hosp. U. Pa., 1970; sr. staff neurologist Scott and White Clinic and Hosp., Temple, Tex., 1971-72; clin. attending neurologist Parkland Hosp., Dallas, 1971-72; clin. instr. neurology U. Tex. Southwestern Med. Sch., Dallas, 1971-72; attending neurologist Univ. Hosps., Cleve., 1972—; cons. neurologist VA Center, Temple, 1971, VA Hosp., Cleve., 1972; examiner Am. Bd. Neurol. Surgery, 1976—; cons. NSF, 1977; mem. physician evaluation bd. WhittakerCorp., 1980—; sci. adv. bd. Communicative Disorders Found; vis. prof. various univs.; lectr. Reviewer, NIH, 1978, Postgrad. Medicine, 1975—, Neurology, 1977—, Cleve. Clinic Quar, 1977, Neurosurgery, 1979—, Am. Jour Physiology, 1980-81; contbr.: articles and abstracts to profl. jours. Am. Jour Physiology. Served to capt. M.C. USAF, 1968-70. Decorated Air Force Commendation medal; recipient Frances Grogan prize in psychiatry, 1964; Reinberger Found. grantee, 1978—; NIH grantee. Fellow Am. Acad. Neurology, ACP; mem. Am. Neurol. Assn., Soc. Neurosci., AAAS, Am. EEG Soc., Soc. Neurosci. (past pres. Cleve. chpt.), Am. Assn. History of Medicine, Ohio Med. Assn., Cleve. Acad. Medicine, No. Ohio Neurol. Soc., Assn. Research in Nervous and Mental Diseases, Soc. Clin. Neurologists, Assn. Univ. Profs. Neurology, Internat. Assn. Study of Pain, AMA, Am. Assn. Neurol. Surgeons, AAAS, Royal Soc. Medicine, Cleve. Med. Library Assn. (trustee). Roman Catholic. Subspecialty: Neurology. Current work: Behavorial aspects of neurology and neurophysiology related to hypertension, stroke and peripheral nerve illness.

CONRAD, BRUCE PHILLIPS, mathematician; b. Ann Arbor, Mich., July 2, 1943; s. John Phillips and Charlotte Elizabeth (Merchant) C.; m. Rebecca Kay Smith, Dec. 22, 1964; children: Clinton Phillips, Esther Charlotte, Jessica Louise, Rosemary McCutcheon. B.S., Harvey Mudd Coll., 1964; Ph.D., U. Calif.-Berkeley, 1964-69. Asst. prof. math Temple U., Phila., 1969-74, assoc. prof., 1974—. Subspecialty: Applied mathematics. Current work: Differential equations; bifurcation theory; stability theory. Home: 147 Pelham Rd Philadelphia PA 19119 Office: Dept Math Temple U Philadelphia PA 19122

CONRAD, HANS, educator, metall. engr.; b. Konradstahl, Germany, Apr. 19, 1922; came to U.S., 1926, naturalized, 1944; s. Henry K. and Martha Ann (Bader) C.; m. Emma Ann Bort, June 10, 1944; children—Sandra Joy, Roberta Lee, Gary Richard. Student, Washington and Jefferson Coll., 1940-42; B.S. in Metall. Engring, Carnegie Inst. Tech., 1943; M.Eng., Yale, 1951, D.Eng., 1956. Research metallurgist Chase Copper & Brass Co., Waterbury, Conn., 1953-55; supervisory engr. Westinghouse Research Labs., Churchill Boro, Pa., 1955-59; sr. research specialist Atomics Internat., Canoga Park, Calif., 1959-61; head dept. physics Aerospace Corp., El Segundo, Calif., 1961-64; tech. dir. Franklin Inst. Research Labs., Phila., 1964-67; prof., chmn. dept. metall. engring. and materials sci., asso. dir. Inst. Mining and Minerals Research, U. Ky., Lexington, 1967-80; prof., head dept. materials engring., dir. minerals and materials research programs N.C. State U., 1981—. Contbr. articles to profl. jours. and books. Recipient U. Ky. Research award, 1971; U.S. Sr. Scientist award Alexander von Humboldt-Stiftung, 1974; Japan Soc. Promotion Sci. vis. prof., 1976. Fellow Am. Soc. Metallurgy; mem. Am. Inst. M.E., Am. Soc. Metals, Am. Soc. Testing and Materials, Sigma Xi, Tau Beta Pi. Subspecialty: Metallurgy. Home: 205 Glasgow Rd Cary NC 27511

CONRAD, NICHOLAS, research engr.; b. Ankara, Turkey, June 8, 1950; s. Collins and Alla (Shengalia) C.; m. Deborah Jansen, Apr. 11, 1951. B.S. in Aerospace Engring, Tex. A&M U., 1971, M.S. in Interdisciplinary Engring, 1973, Ph.D., 1976; M.S. in Petroleum Engring, U. Houston, 1982. Registered profl. engr., Tex. Sr. engr. Halliburton Services, Duncan, Okla., 1976-78; research scientist Getty Oil Co., Houston, 1978-81, research engr., 1981—. Contbr. articles to profl. jours. Mem. Soc. Petroleum Engrs., ASME. Presbyterian. Subspecialties: Petroleum engineering; Solid mechanics. Current work: Well completions, formation damage.

CONSIGLI, RICHARD ALBERT, virologist; b. Bklyn., Mar. 2, 1931; s. Benjamin M. and Maria (Corchia) C.; m. Barbara J. Seel, Jan. 29, 1938; children: Linda, Joanne, Maria. B.S., Bklyn. Coll., 1954; M.A., U. Kans., 1956, Ph.D., 1960; postdoctoral, U. Pa., 1960-63. Instr. U. Kans., 1959-60; prof. Kans. State U., 1963—. Contbr. numerous articles to profl. jours. NIH research grantee, 1961, 68-73; recipient Grad. Faculty award Kans. State U., 1976. Fellow Am. Acad. Microbiology; mem. Am. Soc. Microbiology, AAAS, Am. Assn. Cancer Research, Soc. Exptl. Biology and Medicine, Am. Soc. Virology. Roman Catholic. Subspecialty: Microbiology (medicine). Current work: Tumor virology, biochemistry of virus proteins. Office: Kans State U Manhattan KS 66506

CONSROE, PAUL FRANCIS, pharmacologist, toxicologist; b. Cortland, N.Y., Oct. 18, 1942; d. Frank J. and Johanna (Musenga) C.; m. Robin L. Maish, June 13, 1981. B.S., Union U., Albany, N.Y., 1966; M.S., U. Tenn., 1969, Ph.D., 1971. Registered pharmacist, N.Y., Calif. Assoc. prof. pharmacology U. Ariz., Tucson, 1971-81, prof. pharmacology and toxicology, 1981—. Contbr. articles to sci. and med. jours. Alcohol Drug Abuse and Mental Health Adminstrn. grantee, 1971-81; Nat. Inst. Neurol. Communicative Diseases and Stroke, 1979—. Mem. Soc. Neuroscience, Western Pharmacology Soc., Am. Soc. Pharmacology and Exptl. Therapeutics, Sigma Xi. Roman Catholic. Subspecialties: Molecular pharmacology; Psychopharmacology. Current work: Pharmacology and toxicology of drugs of abuse; neuropharmacology, psychopharmacology. Home:

9351 E Helen St Tucson AZ 85715 Office: Coll Pharmacy U Ariz Tucson AZ 85721

CONTE, RICHARD JOSEPH, energy engineer; b. Phila., Nov. 18, 1947; s. Alfred Angelo and Susan (Pastore) C.; m. Donna Marie DeSesa, June 17, 1972; children: Richard, Christopher, Daniel, Kevin. B.A. in Physics, LaSalle Coll., 1969; M.S. in Energy Engrng, U. Pa., 1981. Reactor insp. U.S. Nuclear Regulatory Commn., King of Prussia, Pa., 1976-79; sr. resident insp. Unit 1, Three Mile Island, Middletown, Pa., 1979—. Served to lt. U.S. Navy, 1970-76. Mem. Am. Nuclear Soc., Ams. for Energy Independence, Am. Physics Soc. Republican. Roman Catholic. Subspecialties: Nuclear fission; Nuclear engineering. Current work: Manager for regulatory inspection program for startup of the undamaged reactor subsequent to the accident at Three Mile Island. Office: US Nuclear Regulatory Commn PO Box 311 Middletown PA 17057

CONWAY, JOHN G., physicist; b. Pitts., May 16, 1922; s. John George and Irene M. (Clifford) C.; m. Florence M. Bittner, May 10, 1922; children: John George III, Jane M., Michael F., Ann S., Kathleen M., Patrick K., Caroline M. B.S. in Physics and Engrng, U. Pitts., 1944. Mem. staff Los Alamos Sci. Lab., 1944-46; research assoc. U. Pitts., 1946-47, CNRS, Orsay, France, 1973-74, 79-80; with Lawrence Berkeley Lab., 1947—, sr. staff scientist, 1950—. Contbr. articles to profl. jours. Mem. El Cerrito (Calif.) City Council, 1958-63; mayor of, El Cerrito, 1961. Named Outstanding Man of Yr. El Cerrito Jaycees, 1959; recipient Louis A. Strait award No. Calif. Soc. for Spectroscopy, 1978. Fellow Optical Soc. Am. (Wm. F. Meggers award 1980); mem. Am. Phys. Soc., Soc. for Applied Spectroscopy, AAAS. Roman Catholic. Subspecialties: Atomic and molecular physics; Condensed matter physics. Current work: Optical spectroscopy of the actinide elements, both free ion and crystals. Home: 1153 King Dr El Cerrito CA 94530 Office: Lawrence Berkeley Lab 1 Cyclotron Rd Berkeley CA 94720

CONWAY, JOHN RICHARD, biology educator; b. Cin., Feb. 28, 1943; s. John Richard and Beverly (Bishop) C.; m. Sharon Diane Croft, June 14, 1969; 1 son, John Richard. B.S. in Zoology, Ohio State U., 1965; M.A., U. Colo., 1968, Ph.D. in Biology, 1975. Instr. biology U. Colo., Colorado Springs, 1972-75; instr. Regis Coll., Denver, 1972; asst. prof. biology Marycrest Coll., Davenport, Iowa, 1976-78, Elmhurst (Ill.) Coll., 1978—; cons. World Book Ency., 1979, Alan Landsburg Prodns., 1981. Contbr. articles to profl. jours. Mem. Ill. Acad. Sci., Nat. Assn. Biology Tchrs. Current work: Biology of honey ants in the genus Myrmecocystus. Home: 263 Highland St Elmhurst IL 60126 Office: Elmhurst Coll 190 Prospect Elmhurst IL 60126

CONWAY, KENNETH EDWARD, plant pathologist, educator; b. Phila., June 7, 1943; s. William Herbert and Marie Louise (Fretz) C.; m. Cynthia Anne Gary, June 22, 1968; children: Kenna Anne, Deanna Lynne, Heather Alene. B.A., SUNY-Potsdam, 1966; M.S., SUNY Coll. Forestry, 1968; Ph.D. (NDEA fellow), U. Fla., 1973. Asst. research scientist U. Fla., Gainesville, 1973-78; asst. prof. plant pathology Okla. State U., Stillwater, 1978-82, asso. prof., 1982—. Contbr. articles to sci. jours. Mem. Am. Phytopathol. Soc., Mycol. Soc. Am., Okla. Acad. Scis., Sigma Xi, Beta Beta Beta. Lodge: Stillwater Rotary. Subspecialties: Plant pathology; Microbiology. Current work: Biological control of soil borne diseases; mycologist; diseases of horticultural crops. Patentee method and composition for controlling water hyacinth. Office: Dept Plant Pathology Okla State U Stillwater OK 74078

CONWAY, LYNN ANN, computer scientist, electrical engineer; b. Mount Vernon, N.Y., Jan. 2, 1938. B.S., Columbia U., 1962, M.S.E.E., 1963. Mem. research staff computer architecture IBM Corp., 1964-69; sr. staff engr. Memorex Corp., 1969-73; mem. research staff Digital Systems Architecture, 1973-77; mgr. LSI Systems Area, 1977-80; research fellow, mgr. VLSI Systems Design Area, Xerox Palo Alto Research Center, 1980-1983; computer research mgr. Defense Advanced Research Projects Agency, 1983—; cons. Systems Industries, 1973-74; vis. assoc. prof. elec. engring. and computer sci. MIT, 1978-79. Mem. IEEE, Assn. Computing Machinery, AAAS. Subspecialties: Artificial intelligence; Computer architecture. Office: 1400 Wilson Blvd Arlington VA 22209

CONWAY, RICHARD WALTER, computer scientist, educator; b. Milw., Dec. 12, 1931; s. Ralph Walter and Tennie May (Mitchell) C.; m. Edythe Davies, Aug. 29, 1953; children: Kathryn Dimiduk, Ralph, Evan. B.M.E., Cornell U., 1954, Ph.D., 1958. Instr. to assoc. prof. computer sci. Cornell U., 1956-64, prof. computer sci., 1965— Author: numerous books, including Theory of Scheduling, 1967, Introduction to Programming, 3d edit, 1979, Programming for Poets, 1979. Mem. Assn. Computing Machinery, Tau Beta Pi. Subspecialties: Database systems; Software engineering. Current work: Programming environments, office systems. Office: Cornell U 408 Upson Hall Ithaca NY 14853

CONWAY, THOMAS PATRICK, immunologist; b. Chgo., Dec. 31, 1941; s. Thomas John and Nora (McWalter) C.; m. Patricia Ann Walsh, Oct. 26, 1974; children: Felicia Yvonne, Erin Colleen. B.S., Loyola U., Chgo., 1963; M.S., U. Ill., 1966, Ph.D., 1969. USPHS fellow Scripps Clinic and Research Found., La Jolla, Calif., 1969-75; research assoc. Willian Beaumont Hosp., Royal Oak, Mich., 1975-77; asst. prof. dept. biol. scis. No Ill. U., DeKalb, 1978—; adj. asst. prof. dept. microbiology Wayne State U., Detroit, 1975-77. Contbr. articles to sci. jours. Served to capt., M.S.C. U.S. Army, 1969-71. USPHS trainee, 1967-69; Am. Cancer Soc. fellow, 1979-80. Mem. AAAS, Am. Soc. Microbiology, Am. Assn. Immunologists, Sigma Xi. Subspecialties: Immunobiology and immunology; Membrane biology. Current work: Expression, synthesis and degradation of membrane proteins, especially B2 microglobulin in normal, activated and transformed lymphocytes. Office: Dept Biol Scis No Ill DeKalb IL 60115

CONWAY, THOMAS WILLIAM, biochemistry educator; b. Aberdeen, S.D., June 6, 1931; s. James L. and Agnes (Mullen) C.; m. Mary Patricia Leadon, July 7, 1957; children: Catherine Ann, James M. B.S., Coll. of St. Thomas, 1953; M.A., U. Tex.-Austin, 1955, Ph.D., 1962. Guest investigator Rockefeller U., N.Y.C., 1962-64; asst. prof., assoc. prof., prof. biochemistry U. Iowa, Iowa City, 1964—. Co-author: Biochemistry: A Case-Oriented Approach, 1983. Served toi 1st lt. USAF, 1953-58. Mem. Am. Soc. Biol. Chemists., Am. Chem. Soc., Am. Soc. Microbiology, Sigma Xi. Subspecialties: Molecular biology; Biochemistry (medicine). Current work: Control of protein synthesis in virus-infected cells, action of interferon. Home: 245 Hutchinson Ave Iowa City IA 52240 Office: Dept Biochemistry U Iowa Iowa City IA 12242

CONWAY, WILLIAM SCOTT, research plant pathologist; b. Philipsburg, Pa., Aug. 12, 1943; s. John William and Eleanor Eunice (Hickson) C.; m. Robin Buff Tepper, June 10, 1979. B.S., Pa. State U., 1965; M.S., U.N.H., 1974, Ph.D., 1976. Asst. prof. Delhi (N.Y.) Agrl. & Tech. Coll., 1976-79; research plant pathologist Beltsville (Md.) Agrl. Research Center, 1979—. Served to lt. U.S. Army, 1967-69. Decorated Bronze Star, Silver Star. Mem. Am. Phytopath. Soc. Lutheran. Subspecialty: Plant pathology. Current work: Researcher in postharvest pathology of fruit. Office: Beltsville Agricultural Research Center Beltsville MD 20705

CONWAY DE MACARIO, EVERLY, research scientist; b. Buenos Aires, Argentina, Apr. 20, 1939; d. Delfin E. and Maria G. (Benatuil) Conway; m. Alberto J. L. Macario, Mar. 16, 1963; children: Alex, Everly. Ph.D. in Pharmacy, Nat. U. Buenos Aires, 1960, 1962. Research fellow Nat. Acad. Medicine Argentina, Buenos Aires, 1962-63; head lab. oncology and immunology Argentinian Assn. Against Cancer, Buenos Aires, 1966-77; chief immunology Sch. Medicine, U. Buenos Aires, 1967-68; research fellow dept. tumor biology Karolinska Inst., Stockholm, 1969-71; sr. research scientist lab. cell biology NRC Italy, Rome, 1971-73; vis. scientist Internat. Agy. Research on Cancer, WHO, Lyon, France, 1973-74, Brown U., Providence, 1974-76; research scientist Lab. Medicine Inst., N.Y. State Dept. Health, Albany, 1976—, supr. pregnancy test proficiency testing program, 1978-80; grant reviewer nat. and internat. agys. Contbr. chpts. to books, articles to profl. jours. Recipient Prof. J. M. Mezzadra award Nat. U. Buenos Aires, 1969; travel grantee French Soc. Immunology, 1974, Am. Assn. Immunologists, 1977; Gold medal Argentinian Soc. Biochemistry, 1980; Hans Osterman Found. grantee, Sweden, 1969; Sir Samuel Scott of Yews Trust grantee, Sweden, 1970; Winifred Cullis grantee Internat. Fedn. Univ. Women, 1972; NATO grantee, 1975-81; U.S. Dept. Energy grantee, 1981. Mem. Argentinian Soc. Biochemistry, Scandinavian Soc. Immunology, Italian Assn. Immunologists, French Soc. Immunology, Am. Assn. Immunologists, Am. Soc. Microbiology. Subspecialties: Immunobiology and immunology; Immunology (medicine). Current work: Immunology of methanogenic bacteria; methanogenic bacteria, monoclonal antibodies, bacterial surface immunochemistry. Home: 18 Carriage Rd Delmar NY 12054 Office: Empire State Plaza E-225 Albany NY 12201

CONWELL, ESTHER MARLY, physicist; b. N.Y.C., May 23, 1922; d. Charles and Ida (Korn) C.; m. Abraham A. Rothberg, Sept. 30, 1945; 1 son, Lewis J. B.A., Bklyn. Coll., 1942; M.S., U. Rochester, N.Y., 1945; Ph.D., U. Chgo., 1948. Lectr. Bklyn. Coll., 1946-51; mem. tech. staff Bell Telephone Labs., 1951-52; physicist GTE Labs., Bayside, N.Y., 1952-61, mgr. physics dept., 1961-72; vis. prof. U. Paris, 1962-63; Abby Rockefeller Mauze prof. M.I.T., 1972; prin. scientist Xerox Corp., Webster, N.Y., 1972-80, research fellow, 1981—; cons., mem. adv. com. engring. NSF, 1978—. Author: High Field Transport in Semiconductors, 1967, also research papers; editorial bd.: Jour. Applied Physics; proc.: IEEE. Fellow IEEE, Am. Phys. Soc. (sec.-treas. div. condensed matter physics 1977—); mem. Soc. Women Engrs. (Achievement award 1960), Nat. Acad. Engring. Subspecialties: Condensed matter physics; Semiconductors. Current work: Basic research in condensed matter physics, electrical engineering, semiconductors. Patentee in field. Office: 800 Phillips Rd Webster NY 14580

COOK, CHARLES J., industrial planner; b. West Point, Nebr., Oct. 2, 1923; m., 1945; 2 children. B.S., U. Nebr., 1948, M.A., 1950, Ph.D., 1953; participant, Advanced Mgmt. Program, Harvard Grad. Sch. Bus., 1968. Research assoc. U. Nebr., 1953-54; physicist SRI Internat., 1954-56, mgr., 1956-62, dir., 1962-69, exec. dir. phys. sci., 1969-76, v.p. office research ops., 1976-77, sr. v.p. research ops., 1977-81; mgr. internat. planning Bechtel Group, Inc., San Francisco, 1981—; instr. San Jose City Coll., 1957-58, Foothill Coll., 1959-62; sr. research assoc. Queen's U., Belfast, 1962-63. Mem. Am. Phys. Soc., Am. Assn. Physics Tchrs., AIAA, Am. Inst. Physics, Am. Defense Preparedness Assn. Subspecialties: Research Management; Technology transfer. Office: Bechtel Group Inc 50 Beale St PO Box 3965 San Francisco CA 94119

COOK, CLARENCE EDGAR, biochemist; b. Jefferson City, Tenn., Apr. 27, 1936; s. Edgar M. and Lillie G. (Hodge) C.; m. Gail O'Connor McKee, June 1, 1957; children: David Grey, Lisa O'Connor, Kevin McKee. B.S., Carson-Newman Coll., 1957; Ph.D. in Organic Chemistry, U. N.C., 1961; postdoctoral, Cambridge (Eng.) U., 1961. Chemist, Research Triangle Inst., Research Triangle Park, N.C., 1962-64, sr. chemist, 1964-78, group leader, 1968-71, asst. dir., 1971-75, dir. life scis. and bioorganic chemistry, 1975-80, dir. bioorganic chemistry, 1980—, v.p. chemistry and life scis., 1983—. Contbr. articles to profl. jours. Am. Chem. Soc. Petroleum Research Fund fellow, 1961. Mem. Am. Soc. Pharmacology and Exptl. Therapeutics, Am. Chem. Soc., AAAS; fellow N.Y. Acad. Scis. Subspecialties: Organic chemistry; Pharmacology. Current work: Development of immunoassay methodology—enantioselective antibodies; metabolism of xenobiotics, especially drugs of abuse; synthesis of steroids as contraceptives. Office: PO Box 12194 Research Triangle Park NC 27709

COOK, DANIEL WALTER, rehabilitation research scientist, educator; b. Urbana, Ill., Dec. 31, 1945; s. Arthur Elwood and Mary Kay (Meade) C.; m. Becky Ray Childers, Aug. 10, 1968; 1 dau., Carrie Anne. B.A., So. Ill. U., 1967; M. Ed., U. Mo., 1969, Ph.D., 1974. Lic. psychologist, Ark.; cert. rehab. counselor. Rehab. specialist Fulton (Mo.) State Hosp., 1969-71; research fellow Regional Rehab. Research Inst., Columbia, Mo., 1972-74; counseling psychologist U. Mo., Columbia, 1973-74; assoc. prof. rehab. edn. U. Ark., Fayetteville, 1979—; sr. research scientist Ark. Rehab. Research and Tng. Ctr., Fayetteville, 1974—; cons. state, fed. agys. Author: (with P. Cooper) Rehabilitation Evaluation Research, 1979, (with B. Bolton) Rehabilitation Client Assessment, 1980; contbr. articles to profl. jours., chpts. to books. Recipient research award U. Ark., 1979; HEW research fellow, 1972-74. Mem. Am. Psychol. Assn., Am. Rehab. Counseling Assn. (membership chmn. 1977-79, research award 1977-80, superior service award 1980). Subspecialties: Physical medicine and rehabilitation; Behavioral psychology. Current work: Spinal cord injury; computer based diagnosis; research methodology; program evaluation. Home: 2781 Stagecoah Fayetteville AR 72701 Office: U Ark 346 N West Fayetteville Ark 72701

COOK, JOHN W., research astrophysicist; b. Selma, Ala., Oct. 25, 1946; s. John W. and Veda Ray (McInturff) C. B.S., M.I.T., 1967; M.A., CCNY, 1970; Ph.D., Dartmouth Coll., 1976. Skylab Workshop postdoctoral astrophysicist High Altitude Obs., Washington, 1976-78; resident Naval Research Lab., Washington, 1976-78; research astrophysicist solar physics br. Space Sci. Div. Naval Research Lab., 1978—. Contbr. writings in field to profl. publs. Served with U.S. Army, 1969-71. Mem. Am. Phys. Soc., Am. Astron. Soc. (and solar div.), Internat. Astron. Union. Subspecialty: Solar physics. Current work: Physics of solar atmosphere; analysis of high spatial and spectral resolution ultraviolet spectra; solar UV variability; cool stars. Office: Code 4163 Naval Research Lab Washington DC 20375

COOK, PAUL FABYAN, biochemistry educator; b. Ware, Mass., Aug. 2, 1946; s. Fabyan Henry and Almina Carrie (Dragon) C.; m. Sandra Joanna Urba, May 17, 1969; 1 dau., Karen Michelle. B.A., Our Lady of the Lake Coll., San Antonio, 1972; Ph.D., U. Calif.-Riverside, 1976. NIH postdoctoral fellow U. Wis.-Madison, 1976-80; asst. prof. La. State U. Med. Ctr., New Orleans, 1980-82; asst. prof. dept. biochemistry North Tex. State U. (Tex. Coll. Osteo. Medicine), Denton, 1982—. Contbr.: articles to profl. jours. including Analytical Biochemistry. Served with USAF, 1966-70. Ad hoc reviewer NIH grants, 1981—; U.Calif. Regents predoctoral fellow, 1975-76; NIH Research Career Devel. awardee, 1983—; grantee Research Corp., New Orleans, 1981-82, NIH, 1981—. Mem. Am. Chem. Soc., AAAS, Biophys. Soc., Am. Soc, Biol. Chemists, Sigma Xi. Subspecialties: Biochemistry (biology); Biophysical chemistry. Current work: Determination of enzyme mechanism; steady state kinetics; isotope effects. Home: 504 Chisolm Trail Denton TX 76201 Office: Dept Biochemistry North Tex State U Tex Coll Osteo Medicine Denton TX 76203

COOK, ROBERT C., theoretical chemist; b. New Haven, June 5, 1947; s. Russell C. and Tensia (Veazey) C.; m.; children: Andrew, Daniel. B.S., Lafayette Coll., Easton, Pa., 1969; M.Ph., Yale U., 1971; Ph.D., 1973. Asst. prof. Lafayette Coll., 1973-81; research scientist Lawrence Livermore Lab., Livermore, Calif., 1981—. Contbr. articles to profl. jours. Mem. Am. Phys. Soc., Am. Chem. Soc., Sigma Xi. Subspecialties: Polymer physics; Theoretical chemistry. Current work: Physics and chemistry of polymeric materials, computer simulation of polyeric systems, dynamics and failure of polymeric materials. Office: Lawrence Livermore Lab L-338 Livermore CA 94550

COOK, STUART DONALD, physician, educator; b. Boston, Oct. 23, 1936; s. Martius and Nina (Schwartzman) C.; m. Josepha Emdin, June 26, 1960; children—Andrew, Peter, Jonathan. A.B., Brandeis U., 1957; M.S., U. Vt., 1959, M.D., 1962. Diplomate: Am. Bd. Psychiatry and Neurology. Intern Upstate Med. Center, Syracuse, N.Y., 1962-63, resident in neurology, 1965-67, chief resident, 1967-68; instr. dept. neurology Albert Einstein Coll. Medicine, Bronx, N.Y., 1968-69; asst. prof. neurology Coll. Physician and Surgeons, Columbia U., 1969-71; prof. medicine N.J. Med. Sch., Newark, 1971—, chmn. dept. neuroscis., 1972—; chief neurology service VA Med. Center, East Orange, N.J., 1971—; vis. scientist div. virology Nat. Inst. Med. Research, London, 1977-78; cons. HEW. Contbr. articles to profl. jours. Served with USN, 1963-65. Mem. Am. Acad. Neurology (S. Weir Mitchell award 1968), Am. Assn. Neuropathologists, AAUP, Am. Fedn. Clin. Research, Assn. Univ Profs. Neurology, Harvey Soc., N.Y. Acad. Sci., Sigma Xi, Alpha Omega Alpha. Subspecialties: Neurology; Neuroimmunology. Current work: Epidemiology, virology, immunology of multiple sclerosis, inflammatory neuropathies. Home: 26 Dogwood Dr Morristown NJ 07960 Office: VA Med Center East Orange NJ 07019

COOK, WILLIAM JOHN, mechanical engineer; b. Des Moines, Apr. 12, 1929; m., 1953; 2 children. B.S., Iowa State U., 1957, M.S., 1959, Ph.D., 1964. Instr., asst. prof., assoc. prof. Iowa State U., Ames, 1959-76, prof. mech. engring., 1976—. Mem. Am. Soc. Engring., ASME, AIAA. Subspecialty: Mechanical engineering. Office: Dept Mech Engrin Iowa State U Ame IA 50010

COOKE, ALLAN ROY, physician, medical educator; b. Lismore, Australia, Apr. 30, 1936; came to U.S., 1964; s. Lindsay and Elsie Ellen (Vidler) C.; m.; children: Ian Russell, Colin Roy. M.B., B.S., U. Sydney, Australia, 1958, M.D., 1970. Intern Royal Prince Alfred Hosp., Sydney, 1958-59, resident in internal medicine, 1960-63; mem. faculty U. Iowa, Iowa City, 1970-76, prof. medicine, 1974-76, U. Kans., Kansas City, 1976; sr. physician Univ. Hosp., 1976—. Contbr. articles to profl. publs. Fellow A.C.P., Australian Coll. Physicians; mem. Am. Soc. Clin. Investigation, Am. Physiol. Soc., Am. Fedn. Clin. Research (pres. 1975-76). Subspecialties: Gastroenterology; Psychophysiology. Current work: Gastric emptying; drug damage to stomach. Home: 6436 High Dr Mission Hills KS 66208 Office: Dept Medicine U Kansas Med Center Rainbow Blvd Kansas City KS 66103

COOK-IOANNIDOUS, LESLIE PAMELA, mathematics educator and researcher; b. Kingston, Ont., Can., Aug. 23, 1946; d. Leslie Gladstone and Alfreda Mary Cook; m. George Ioannidis, 1972; 1 son, Alexander. B.A., U. Rochester, 1967; M.S., Cornell U., 1969, Ph.D., 1971. NATO postdoctoral fellow U. Utrecht, Netherlands, 1971-72; research asst. and instr. Cornell U., Ithaca, N.Y., 1972-73; adj. asst. prof. UCLA, 1973-75, asst. prof., 1975-80, assoc. prof., 1980—. Contbr. articles to profl. jours. NDEA Title IV fellow, 1967-70; NSF grad. fellow, 1970-71; NSF grantee, 1977—. Mem. Soc. Indsl. and Applied Math., Am. Math. Assn., AIAA, Soc. Women Engrs., Am. Women in Math., Sigma Xi. Subspecialty: Applied mathematics. Current work: Asymptotics and singular perturbations, tronsonic aerodynamics, biomathematics. Office: Dept Math UCLA Los Angeles CA 90024

COOKSON, ALBERT ERNEST, tel. and tel. co. exec.; b. Needham, Mass., Oct. 30, 1921; s. Willard B. and Sarah Jane (Jack) C.; m. Constance J. Buckley, Sept. 10, 1949; children—Constance J., William B. B.E.E., Northeastern U., 1943; M.E.E., Mass. Inst. Tech., 1951; Sc.D., Gordon Coll., 1974. Group leader Research Lab. Electronics, Mass. Inst. Tech., 1947-51; lab. dir. ITT Fed. Labs., Nutley, N.J., 1951-59, v.p., dir. operations, Paramus, N.J., 1959-62; pres. ITT Intelcom, Falls Church, Va., 1962-65; dep. gen. tech. dir. Internat. Tel. and Tel. Corp., N.Y.C., 1965-66, v.p., tech. dir., 1966-68, sr. v.p., gen. tech. dir., 1968—; chmn. bd. ITT Interplan; dir. Internat. Standard Electric, ITT Industries.; Mem. Def. Communications Satellite Panel; adviser research and engring. on def. communications satellite systems Dept. Def.; mem. indsl. panel sci. and tech. NSF.; Mem. Fairfax County Econ. and Indsl. Devel. Com., 1962-65; mem. nat. council Northeastern U.; mem. pride com. U. Hartford, 1973-76; elec. engring./computer adv. bd. Mass. Inst. Tech., 1977—. Served with USNR, 1943-46. Fellow IEEE; mem. Armed Forces Communications and Electronics Assn., Am. Mgmt. Assn., Am. Inst. Aeros. and Astronautics, Electronic Industries Assn., Sigma Xi, Tau Beta Pi. Subspecialties: Electrical engineering. Patentee frequency search and track system. Home: 2 Baywater Dr Darien CT 06820 Office: 320 Park Ave New York NY 10022

COOL, RODNEY LEE, physicist, educator; b. Platte, S.D., Mar. 8, 1920; s. George E. and Muriel (Post) C.; m. Margaret E. MacMillan, June 21, 1949; children: Ellen, John, Mary Lee, Adrienne. B.S., U. S.D., 1942; M.A., Harvard U., 1947, Ph.D., 1949. Research physicist Brookhaven Nat. Lab., Upton, L.I., N.Y., 1949-59, dep. chmn. high energy physics, 1960-64, asst. dir. high energy physics, 1964-66, asso. dir., 1966-70, sec. high energy adv. com., 1960-67, chmn., 1967-70; prof. exptl. high energy physics Rockefeller U., N.Y.C., 1970—; Mem. policy com. Stanford Linear Accelerator Center, 1962-67, 76-80; mem. Asso. Univs. High Energy Panel, Asso. Univs., Inc., 1963-70; mem. Walker panel, com. on sci. and pub. policy Nat. Acad. Sci., 1964; mem. Princeton-Pa. Accelerator Sci. Com., 1966-68; mem. high energy physics adv. panel AEC, 1967-70; chmn. physics adv. com. Nat. Accelerator Lab., 1967-70; mem. adv. panel for physics NSF, 1970-73; sci. asso. European Center Nuclear Research, 1973—; trustee Univs. Research Assn., 1977-83; mem. rev. com. Argonne Univs. Assn., 1978-80. Co-editor: Advances in Particle Physics, vols. I and II, 1968; contbr. articles to profl. jours. Served to maj., Signal Corps AUS, 1942-46. Decorated Bronze Star medal. Fellow Am. Phys. Soc. (program cons. div. particles and fields 1968-70); mem. Nat. Acad. Scis., Phi Beta Kappa, Sigma Xi. Subspecialty: Particle physics. Current work: Research in experimental high energy particle physics. Home: 450 E 63d St New York NY 10021 Office: Rockefeller University New York NY 10021

COOMBE, JOHN RAYMOND, engineering executive; b. Trenton, N.J., Sept. 26, 1926; s. John Raymond and Elizabeth (Tholander) C.; m. Kathleen Marie Jennings, Feb. 22, 1958; children: John, Mary, Elizabeth, James. B.A., Washington and Jefferson U., 1951. Engr. Gen. Electric Co., Schenectady, 1951-59; physicist Tech. Ops., Burlington, Mass., 1959-60; supr. Alco Products, Schenectady, 1960-62; mgr. Westinghouse Co., Pitts., 1962-72; asst. chief licensing Stone &

Webster, Boston, 1972—. Mem. Am. Nuclear Soc. (nat. program com., standards com.). Subspecialties: Nuclear engineering; Cryogenics. Current work: Power plant and advanced concept licensing activities. Patentee in field. Office: Stone & Webster 245 Summer St Boston MA 02107

COOMBS, CLYDE HAMILTON, psychologist, educator; b. Paterson, N.J., July 22, 1912; s. Clyde and Mildred (Horandt) C.; m. Lolagene Convis, Sept. 1, 1939; children: Steven, Douglas. A.B., U. Calif., Berkeley, 1935, M.A., 1937; Ph.D., U. Chgo., 1941; D.S.S. (hon.), U. Leiden, Netherlands, 1975. Instr. psychology U. Chgo., 1939-41, research asso. biophysics, 1939-41; research psychologist Adj. Gen.'s Office, War Dept., 1941-43; mem. faculty dept. psychology U. Mich., Ann Arbor, 1947—, Disting. Univ. prof., 1978—; vis. prof. Harvard U., Boston, 1948-49; vis. research prof. U. Amsterdam, Netherlands, 1955-56; cons. Dept. Army, 1957-59; lecture tour various univs. in, Europe, 1964-65; vis. prof. U. Colo., Boulder, 1965, U. Wash., Seattle, 1967; cons. VA Hosp., Ann Arbor, 1968-69; vis. prof. U. Western Australia, 1969, Central U. Venezuela, 1970, Inst. Psychology, Academia Sinica, Beijing, China, 1981; mem. U.S.-USSR Interacad. seminar math. psychology, Tbilisi, Russia, 1979; vis. research scholar U. London, 1978; mem. com. biometry and epidemiology NIH, 1971. Author: (with R.M. Thrall and R.C. Davis) Decision Processes, 1954, (with R.C. Kao) Nonmetric Factor Analysis, 1955, A Theory of Data, 1964, (with R.M. Dawes and A. Tversky) Mathematical Psychology: An Elementary Introduction, 1970; cons. editor: Psychol. Rev, 1953-58; book rev. editor: Psychometrics, 1951-54. Served from capt. to maj. U.S. Army, 1943-46. Decorated Legion of Merit; recipient Fulbright award, 1955-56, Fulbright-Hayes award, 1975; named Disting. Sr. Faculty lectr. U. Mich., 1980. Fellow Am. Psychol. Assn. (pres. div. 5 1958-59, chmn. bd. sci. affairs 1960-62), Am. Acad. Arts and Scis., Am. Statis. Assn. (hon.); mem. Soc. Math. Psychology (pres. 1977-78), Psychometric Soc. (pres. 1955-56), Nat. Acad. Scis., Psychol. Assn. Spain (hon.). Home: 3419 Daleview Dr Ann Arbor MI 48103 Office: 580 Union Dr Dept Psychology Univ Mich Ann Arbor MI 48109

COON, MINOR JESSER, biological chemistry educator; b. Englewood, Colo., July 29, 1921; s. Minor Dillon and Mary (Jesser) C.; m. Mary Louise Newburn, June 27, 1948; children: Lawrence R., Susan L. B.A., U. Colo., 1943; Ph.D., U. Ill., 1946; D.Sc. (hon.), Northwestern U., 1983. Postdoctoral research asst. in biochemistry U. Ill., 1946-47; instr. dept. physiol. chemistry U. Pa., 1947-49, asst. prof., 1949-53; asso. prof., 1953-55; prof. dept. biol. chemistry U. Mich. Med. Sch., 1955—, chmn. dept., 1970—; research fellow dept. pharmacology N.Y. U., 1952-53; research fellow Fed. Poly. Inst., Zürich, Switzerland, 1961-62; cons. Oak Ridge Inst. Nuclear Studies, 1956-58; mem. adv. council Life Ins. Med. Research Fund, 1960-65; mem. biochem. study sect. NIH, 1963-66, research career award com., 1966-70. Editor-in-chief: Biochemical Preparations, 1962, Microsomes, Drug Oxidations and Chemical Carcinogenesis, 1980; mem. editorial bds.: Biochemistry, 1971-74, Molecular Pharmacology, 1972—, Jour. Biol. Chemistry, 1976—, Proc. Sci. Conf. on Cytochrome P-450: Structural and Functional Aspects, 1980; Contbr. articles to profl. jours. Recipient Distinguished Faculty achievement award U. Mich., 1976; William C. Rose award in biochemistry and nutrition, 1978; Bernard B. Brodie award in drug metabolism, 1980; Disting. Faculty lectureship award in biomed research U. Mich., 1980. Fellow N.Y. Acad. Scis.; mem. Am. Chem. Soc. (award in enzyme chemistry 1959), Am. Soc. Biol. Chemists (sec. 1981—), Am. Soc. Pharmacology and Exptl. Therapeutics, Biophys. Soc., AAAS, Assn. Med. Sch. Depts. Biochemistry (pres. 1974-75), Internat. Union Biochemistry (chmn. oxygenases interest group 1981—), Am. Inst. Biol. Scis., Am. Soc. Microbiology, Am. Oil Chemists Soc., Nat. Acad. Scis., Phi Beta Kappa, Sigma Xi, Phi Kappa Phi, Alpha Chi Sigma, Phi Lambda Upsilon. Subspecialty: Biochemistry, medicine. Home: 1901 Austin Ave Ann Arbor MI 48104 Office: Dept Biol Chemistry U Mich Ann Arbor MI 48109

COONEY, CHARLES LELAND, engineering educator; b. Phila., Nov. 9, 1944; s. Leland E. and Jean B. (Bader) C.; m. Margaret Mary Reiser, Jan. 28, 1978; children: Matthew Leland, Charles Brendan Reiser. B.S. in Chem. Engring, U. Pa., 1966; S.M. in Biochem. Engring, MIT, 1967, Ph.D., 1970. Chem engr. DuPont Co., Wilmington, Del, 1966; cons. in fermentation tech. E.R. Squibb Co., New Brunswick, N.J., 1970; asst. prof. biochem. engring. MIT, 1970-76, assoc. prof. biochem. engring., 1976-82, prof. chem. and biochem. engring., 1982—. Author: (with A.C. Olson) Immobilized Enzymes in Food and Microbial Processing, 1974, (with others) Fermentation and Enzyme Technology, 1979; contrbr. articles and papers to profl. jours. Mem. Am. Chem. Soc., Am. Inst. Chem. Engrs., Am. Soc. for Microbiology, Inst. Food Tech., Soc. for Indsl. Microbiology. Subspecialties: Chemical engineering; Enzyme technology. Current work: Fermentation technoloy, computer control of biological processes, bioconversion of renewable resources to fuels and chemicals, bioreactor design and operation, chemical engineering of genetically engineered organisms. Patentee in field. Office: Room 66-46 MIT 77 MassachusettsAv Cambridge MA 02139

COONEY, DAVID ANTHONY, physician; b. Arlington, Mass., June 10, 1938; s. Harold N. and Mildred (Hale) C. A.B. cum laude, Holy Cross Coll., 1959, M.D., Georgetown U., 1962. Intern Buffalo Gen. Hosp., 1963-64; com. med. officer USPHS, 1964—; postdoctoral fellow dept. pharmacology Yale U. Med. Sch., New Haven, 1971-72; staff scientist, la. toxicology Nat. Cancer Inst., NIH, 1977-79, head biochemistry sect., lab. med. chemistry and biology, div. cancer treatment, 1979—. Contbr. articles to profl. jours. Mem. Soc. Toxicology, Am. Assn. Cancer Research, Am. Soc. Pharmacology and Exptl. Therapeutics, Alpha Omega Alpha. Subspecialty: Cancer research (medicine). Current work: Development of effective new anticancer drugs; antimetabolite drug development. Office: Lab Medicinal Chemistry and Pharmacology Nat Cancer Inst Bethesda MD 20205

COONEY, MARION KATHLEEN, educator; b. Mercedes, Tex., Feb. 2, 1920; d. Albert John and Marie (Jansen) C. B.A., Coll. St. Benedict, 1939; M.S., U. Minn., 1953, Ph.D., 1962. Med. technologist Fairview Hosp., Mpls., 1940-43; bacteriologist Minn. Dept. Health, Mpls., 1943-45, bacteriologist/virologist, 1945-53; chief, lab. sect. of virus and rickettsial diseases, 1953-65; asst. prof. dept. preventive medicine U. Wash., Seattle, 1966-70, asst. prof. dept. pathobiology, 1970-72, asso. prof., 1972-78, prof., 1978—. Contbr. articles in field to profl. jours. Mem. Am. Soc. Microbiology, Am. Assn. Immunologists, Soc. Exptl. Biology and Medicine, Am. Public Health Assn., N.Y. Acad. Sci. Democrat. Roman Catholic. Subspecialties: Virology (medicine); Infectious diseases. Current work: Researcher, tchr. med. virology, infectious disease epidemiology and methodology. Office: University of Washington F262C Health Sciences Bldg Seattle WA 98195

COONTS, HARVEY LEE, petroleum engr.; b. Weslaco, Tex., Mar. 8, 1934; s. Charles Francis and Ruth Jean (Smith) C.; m. Rosa Lee Knowles, Apr. 26, 1955; children: Diana, Joy, Elaine, Janice. B.S. in Petroleum and Natural Gas Engring, Tex. A&I U., 1958. Registered profl. engr., Colo., Tex., Calif. Petroleum engr. Findlay Engring., Alice, Tex., 1958-59; state petroleum engr. State of Utah, Salt Lake City, 1959-64; sr. petroleum engr. Sunset Internat., Beverly Hills, Calif., 1964-66; staff petroleum engr. Union Pacific R.R., Wilmington, Calif., 1966-70; pres. Coonts Petroleum Engring., Inc., Littleton, Colo., 1970—. Contbr. articles to profl. jours. Mem. Soc. Petroleum Engrs., AIME. Subspecialty: Petroleum engineering. Current work: Consulting in petroleum engineering.

COOPER, ALAN DOUGLAS, chemist, educator; b. Lynn, Mass., Apr. 4, 1942; s. Everett James and Helen Eunice (Marble) C.; m. Susan Marie Brown, Sept. 10, 1967; children: Deborah, Stephen, Christina. B.S. in Chemistry, Tufts U., 1964; A.M., Boston U., 1966, Ph.D., 1968. Postdoctoral fellow in biochemistry Johns Hopkins U., Balt., 1968-70; vis. scientist Worcester Found. for Exptl. Biology, 1980; assoc. prof. chemistry Worcester (Mass.) State Coll., 1970-83, prof. chemistry, 1983—. Contbr. articles on chemistry to profl. jours. Mem. Town Meeting, Auburn, Mass., 1974—. NSF grantee, 1979-80. Mem. Am. Chem. Soc. (chmn.-elect central Mass. sect.), AAAS, New Eng. Assn. Chemistry Tchrs., Beta Beta Beta. Democrat. Roman Catholic. Subspecialties: Biochemistry (biology); Biophysical chemistry. Current work: Nucleotide sequences of actin genes, recombinant DNA, sequencing of actin genes, evolution of actin genes, orgins of life, synthesis and characterization of proteinoids. Office: Worcester State Coll Worcester MA 01602

COOPER, ARTHUR WELLS, forester, educator, researcher; b. Washington, Aug. 15, 1931; s. Gustav Arthur and Josephine Phelps (Wells) C.; m. Jean Farnsworth, Aug. 30, 1953; children: Paul Arthur, Roy Alan. B.A., Colgate U., 1953, M.A., 1955; Ph.D., U. Mich., 1958. Asst. prof. N.C. State U., Raleigh, 1958-63, assoc. prof., 1963-68, prof. botony, 1968-71, prof. forestry, 1976-80, head dept. forestry, 1980—; dep. dir. Dept. Conservation and Devel. State of N.C., Raleigh, 1971; asst. sec. Dept. Natural and Econ. Resources, 1971-76; trustee N.C. Nature Conservancy, 1977—; chmn. bd. trustees Inst. Ecology, Washington, 1982—. Mem. N.C. Coastal Resources Commn., Raleigh, 1976—; chmn. com. scientists U.S. Forest Service, Washington, 1977-79, 82. Recipient Conservation award Am. Motors, 1972; Environ. award Sol Feinstone, Syracuse, N.Y., 1982. Fellow AAAS; mem. Ecol. Soc. Am. (pres. 1981), N.C. Acad. Sci. (pres. 1978-79), Soc. Am. Foresters (chmn. N.C. div. 1983-84). Democrat. Subspecialties: Forestry; Ecology. Current work: Forest ecology, application of ecol principles to renewable resource mgmt. Home: 719 Runnymeade Rd Raleigh NC 27607 Office: Dept Forestry NC State U Raleigh NC 27650

COOPER, CHARLES DEWEY, physics educator; b. Whittier, N.C., Jan. 11, 1924; s. Grady T. and Mary L. (Howell) C.; m. Corrie W. Johnson, Dec. 21, 1946; children: Norma, Claire, Edward. B.S., Berry Coll., 1944; M.A., Duke U., 1948, Ph.D., 1950. Asst. prof. physics U. Ga., 1950-56; assoc. prof., 1956-61, prof., 1961—; research fellow Harvard U., 1954-55; cons. Oak Ridge Nat. Lab. Contbr. articles to profl. publs. Served with USN, 1944-46. Recipient Micheal research award U. Ga., 1958. Mem. Am. Phys. Soc., Am. Assn. Physics Tchrs., Sigma Xi. Subspecialties: Atomic and molecular physics; Spectroscopy. Current work: Multiphoton ionization and angular distribution of photoelectrons resulting from laser light interaction with matter. Home: 4235 Barnett Shoals Rd Athens GA 30605 Office: Physics Dept U G Athens GA 30602

COOPER, DAVID STEPHEN, physician; b. N.Y.C., Jan. 26, 1948; s. Harold Coleman and Brenda (Brenner) C.; m. Ellen Grace Shiff, Aug. 23, 1970; 1 son, Jonathan. B.A., Johns Hopkins U., 1969; M.D., Tufts U., 1973. Diplomate: Am. Bd. Internal Medicine. Intern, resident Barnes Hosp., St. Louis, 1973-76; fellow Mass. Gen. Hosp., Boston, 1976-78, clin. asst. in medicine, 1978-80, asst. in medicine, 1980—; asst. prof. medicine Harvard Med. Sch., Boston, 1980—. Author: Your Thyroid, 1982. NIH New Investigator grantee, 1981. Fellow A.C.P.; mem. Am. Thyroid Assn., Endocrine Soc., Am. Fedn. for Clin. Research, Alpha Omega Alpha. Subspecialty: Endocrinology. Current work: Thyroidology, pituitary physiology, clinical endocrinology. Office: Thyroid Unit Mass Gen Hosp Boston MA 02114

COOPER, DAVID YOUNG, biochemist, pharmacologist, science historian; b. Henderson, N.C., Aug. 14, 1924; s. James and Frances (Chatham) C.; m. Cynthia Laughlin, Aug. 6, 1955; children: Lucy, Allison. B.S., U. N.C., 1946; M.D., U. Pa., 1948. Diplomate: Am. Bd. Surgery. Intern Hosp. U. Pa., Phila., 1948-49, resident in surgery, 1952-57; asst. prof. U. Pa. Med. Sch., Phila., 1957-60, assoc. prof., 1960-68, prof., 1968—. Contbr. articles to profl. jours. Served in USN, 1943-46, 49-51. Mem. Am. Physiol. Soc., Am. Soc. Pharmacology and Exptl. Therapeutics, Am. Soc. Biol. Chemistry, Endocrine Soc., Sigma Xi, others. Episcopalian. Club: Racquet (Phila.). Subspecialties: Biochemistry (medicine); Toxicology (medicine). Current work: Biochemical pharmacology; chemical carcinogenesis; experimental microsurgery. Office: U Pa Med Sch 36th and Spruce Sts Philadelphia PA 19104

COOPER, EDWARD SAWYER, internal medicine educator, consultant; b. Columbia, S.C., Dec. 11, 1926; s. Henry Howard and Ada Crosland (Sawyer) C.; m. Jean Marie Wilder, Dec. 2, 1951; children: Lisa Marie Cooper Hudgins, Edward Sawyer, Jan Ada, Charles Wilder. A.B., Lincoln U., Pa., 1946; M.D., Meharry Med. Coll., Nashville, 1949; M.S., U. Pa., 1972. Diplomate: Nat. Bd. Med. Examiners, Am. Bd. Internal Medicine. Intern Phila. Gen. Hosp., 1949-51, resident in medicine, 1951-54, NIH fellow in cardiology, 1956-57, pres. med. staff, 1969-71, co-dir., 1968-74, chief of med. service, 1973-76; prof. medicine U. Pa., 1973—; dir. Blue Cross of Greater Phila., 1975—; adv. bd. Hypertension Detection and Followup Program, Phila., 1974—. Trustee Am. Found. Negro Affairs, 1969—. Served to capt. USAF, 1954-56. Fellow ACP (govs. adv. bd.), Phila. Coll. Physicians (mem. council), Am. Coll. Chest Physicians; mem. Am. Heart Assn. (chmn., dir.), Alpha Omega Alpha. Democrat. Methodist. Subspecialty: Internal medicine. Current work: Stroke and hypertension. Home: 6710 Lincoln Dr Philadelphia PA 19119 Office: University of Pennsylvania Hospital 3400 Spruce St Philadelphia PA 19119

COOPER, FRANKLIN SEANEY, speech scientist; b. Robinson, Ill., Apr. 29, 1908; s. Frank A. and Myrtle Alma (Seaney) C.; m. Frances Edith Clem, Feb. 14, 1935; children: Robert Craig, Alan Kent. B.S. in Engring. Physics, U. Ill., 1931; Ph.D. in Physics, Mass. Inst. Tech., 1936; D.Sc. (hon.), Yale U., 1976. Teaching and research asst. U. Ill., 1931-34, Mass. Inst. Tech., 1934-36; research engr. Gen. Electric Research Labs., 1936-39; asso. research dir. Haskins Labs., New Haven, 1939-55, 75—, pres., research dir., 1955-75; liaison officer, then sr. liaison officer OSRD, 1941-46; vis. com. dept. modern langs. Mass. Inst. Tech., 1949-65; adv. com. research div. Coll. Engring., N.Y. U., 1949-65; bd. dirs. Center Applied Linguistics, 1968-74; adj. prof. phonetics Columbia U., 1955-65; adj. prof. linguistics U. Conn., 1969-80, vis. prof., 1980—; sr. research asso. linguistics Yale U., 1970-76; fellow Calhoun Coll., 1971-80, asso. fellow, 1980—; adv. panel on White House tapes U.S. Dist. Ct. D.C., 1973-74; chmn. communicative scis. interdisciplinary cluster President's Biomed. Research Panel, 1975; mem. adv. council Nat. Inst. for Neurol. and Communicative Disorders and Stroke, NIH, 1978-81. Author papers, book chpts. speech processing, perception and prodn., aids for blind and deaf, biophysics, high-voltage/high vacuum engring. Recipient Presdl. Certificate of Merit, 1948; honors of assn. Am. Speech and Hearing Assn., 1966; Warren medal Soc. Exptl. Psychology, 1975; Fletcher-Stevens award Brigham Young U., 1977. Fellow IEEE (Pioneer award speech communication 1972), Acoustical Soc. Am. (Silver medal speech communication 1975); mem. Nat. Acad. Engring., Sigma Xi. Congregationalist. Club: Cosmos (Washington). Subspecialties: Psychophysics; Automated language processing. Current work: Speech science; production and perception of speech; speech analysis and synthesis by computer; experimental linguistics. Home: 5 Parsell Ln Westport CT 06880 Office: 270 Crown St New Haven CT 06511

COOPER, GEORGE DAVID, psychologist, educator; b. Hagerstown, Md., July 7, 1935; s. George E. and Mary E. (Longanecker) C. A.B., Shepherd Coll., 1957; Ph.D., Duke U., 1962. Lic. psychologist, Md., Va., W.Va. Chief psychologist Petersburg (Va.) Tng. Sch., 1961-63; sch. psychologist Springfield Youth Ctr., Sykesville, Md., 1963-64; clin. psychologist V.A.C., Martinsburg, W.Va., 1964-73; assoc. prof. George Mason U., Fairfax, Va., 1973-79; clin. dir. Glaydin Sch., Leesburg, Va., 1979—; cons. Bur. Crippled Children, Richmond, Va., 1959-61; bd. dirs. A.I.D., Inc., Martinsburg, 1974-75; cons. Am. Research Inst., Washington, 1975-76, D.C. Task Force on Women and Alcoholism, 1979-80. Contbr. articles to profl. jours. Duke U. scholar, 1957-59; U.S. VA assistantship, 1958-62. Fellow Md. Psychol. Assn.; Mem. W.Va. Psychol. Assn., Am. Assn. on Mental Deficiency, Am. Psychol. Assn., Kappa Delta Pi, Pi Delta Epsilon. Republican. Methodist. Lodge: Kiwanis. Subspecialties: Cognition; Sensory processes. Current work: Cognitive and humanistic psychotherapy. Office: Route 3 PO Box 334 Leesburg VA 22075 Home: 3010 Berridge Dr Shepherdstown WV 25443

COOPER, GEORGE, IV, physician, educator; b. Charlottesville, Va., June 25, 1942; s. George Jr. and Juliet (Paine) C.; m. Elizabeth Louise Roemig, Sept. 12, 1981; children: George Franklin, William. B.A., Williams Coll., 1964; M.D., Cornell U., 1968. Diplomate: Am. Bd. Internal Medicine. Intern/resident Case Western Res. U., Cleve., 1968-71; research assoc. physiology Mayo Clinic, Rochester, Minn., 1971-73; cardiology fellow Duke U., Durham, N.C., 1973-75; asst. prof. medicine U. Iowa, Iowa City, 1977-81; assoc. prof. medicine and physiology Temple U., Phila., 1981—. Served as lt. comdr. USN, 1975-77. NIH grantee, 1983. Fellow Soc. Clin. Research; mem. Am. Fedn. Clin. Research, Am. Heart Assn. (circulation council fellow, Basic Sci. Council), Internat. Soc. Heart Research (cardiovascular fellow), Am. Physiol. Soc. (circulation fellow), Cardiac Muscle Soc.. Subspecialties: Cardiology; Physiology (medicine). Current work: energetics and mechanics of cardiac contraction; load regulation of the myocardium. Office: Temple U Dept Cardiology 3401 N Broad St Philadelphia PA 19140

COOPER, HERBERT ASEL, medical educator, pediatric hematologist, experimental pathologist; b. Grand Junction, Colo., Feb. 21, 1938; s. Herbert I. and Inez F. (Samples) C.; m. Karen R. Groe, June 7, 1963; children: Christopher Scott, Kevin Andrew. B.A. in Chemistry, U. Kans., 1960, M.D., 1964. Diplomate: Am. Bd. Pediatrics. Intern Charles T. Miller Hosp., also Children's Hosp., St. Paul, 1965; resident in pediatrics, 1965-67; in pediatric hematology/oncology Mayo Clinic, Rochester, Minn., 1969-71; postdoctoral fellow U. N.C., Chapel Hill, 1971-74, asst. prof. pathology and pediatrics, 1974-78, assoc. prof., 1978—; vis. research prof. Theodor Kocher Inst., Bern, Switzerland, 1977-78; asst. med. dir. clin. coagulation lab. dept. hosp. labs. N.C. Meml. Hosp., Chapel Hill, 1974-78, assoc. dir. share lab. service, 1980—; mem. exec. com. Specialized Center for Research Hemostasis and Thrombosis, Chapel Hill, 1980—; sci. advisor Comprehensive Hemophilia Diagnostic and Treatment Center, Chapel Hill, 1975—; cons. Chapel Hill Bd. Edn. Contbr. to numerous publs. in field. Mem. and sec. exec. bd. Chapel Hill United Fund. NIH Research Career Devel. awardee, 1975-80. Mem. Am. Assn. Pathologists, Soc. for Pediatric Research, Am. Soc. Hematology, Internat. Soc. Hemostasis and Thrombosis, Orange County Med. Soc. Lutheran. Subspecialties: Biochemistry (medicine); Hematology. Current work: Hemostasis and thrombosis; structure function of Factor VIII; Hemophilia A and von Willebrand's disease; biochemistry of blood coagulation; platelets, platelet membrane receptors. Office: Department of Pathology University of North Carolina School of Medicine 722 PCEB 228-H Chapel Hill NC 27514 Home: 111 Springhill Forest Chapel Hill NC 27514

COOPER, (HOWARD) GORDON, computer peripheral co. exec., researcher; b. Joliet, Ill., Feb. 16, 1927; s. Howard Gordon and Jennie (Paarlberg) C.; m. Lacy Ellen Underwood, June 2, 1953; children: Mary, John. Student, Ill. Inst. Tech., 1943-45; B.S., U. Ill., 1949, M.S., 1950, Ph.D., 1954. Mem. tech. staff Bell Telephone Labs., Murray Hill, N.J., 1954-62, Whippany, N.J., 1962-70; mgr. corp. research Recognition Equipment Inc., Dallas, 1970—. Contbr. articles to profl. jours. Mem. U.S. Senatorial Bus. Adv. Com., 1980—; mem. Republican Presdl. Task Force, 1981. Served with USN, 1945-46. Mem. Optical Soc. Am., Pattern Recognition Soc., Am. Phys. Soc., IEEE, World Future Soc., Optical Soc. Tex., Sigma Xi, Tau Beta Pi, Sigma Tau. Methodist. Club: Brookhaven. Subspecialties: Graphics, image processing, and pattern recognition; Computer engineering. Current work: Optical scanning, image processing, pattern recognition, image displays, info. processing, electronic printing, computer systems, robotics. Patentee in field. Home: 1005 Sierra Pl Richardson TX 75080 Office: PO Box 222307 Dallas TX 75222

COOPER, JACK ROSS, pharmacologist; b. Ottawa, Ont., Can., July 26, 1924; s. Harry and Jean (Levine) C.; m. Helen Achbar, Aug. 14, 1951; children: Marilyn, Sheila, Nancy. B.A., Queen's U., 1948; M.A., George Washington U., 1952, Ph.D., 1954. Asst. prof. pharmacology Yale U. Sch. Medicine, 1958-63, asso. prof., 1963-71, prof., 1971—; Mem. adv. bd. U.S.-Israel Binat. Sci. Found. Editorial bd.: Jour. Neurochemistry. Served with RCAF, 1945. Mem. Am. Soc. Pharmacology and Exptl. Therapeutics, Internat. Soc. Neurochemistry., Soc. Neuroscis. Subspecialties: Molecular pharmacology; Neurochemistry. Current work: Cholinergic mechanisms; thiamin in nervous tissue. Home: 11 Jenick Ln Woodbridge CT 06525 Office: 333 Cedar St New Haven CT 06510

COOPER, JOHN ALLEN DICKS, med. educator; b. El Paso, Tex., Dec. 22, 1918; s. John Allen Dicks and Cora (Walker) C.; m. Mary Jane Stratton, June 17, 1944; children—Margaret Ann, John Allen Dicks, Patricia Alison, Randolph Arend Stratton. B.S. in Chemistry, N.Mex.State U., 1939, LL.D. (hon.), 1971; Ph.D. in Biochemistry, Northwestern U., 1943, M.D., 1951, D.Sc. (hon.), 1972; D.Honoris Causa, U. Brasil, 1958; D.Sc. (hon.), Duke U., 1973, Med. Coll. Ohio, Toledo, 1974, Med. Coll. Wis., 1978, N.Y. Med. Coll., 1981, D.Med. Sci, Med. Coll. Pa., 1973. Intern Passavant Meml. Hosp., Chgo., 1951, mem. attending staff, 1955-69; mem. faculty Northwestern U., 1943-69, prof. biochemistry, 1957-69, asso. dean, 1959-63, dean scis., 1963-69, mem. faculty, 1955-69, Georgetown U., 1970—; prof. practice of health policy Duke, 1973-78; vis. prof. U. Brasil, 1956, U. Buenos Aires, 1958; dir. radioisotope service VA Research Hosp., Chgo., 1954-65, cons. in research, 1954-69; adviser to adminstr. AID, U.S. Dept. State, 1966-71; Mem. bd. of pub. health advisers State of Ill., 1962-69; mem. Ill. Legis. Com. Atomic Energy, 1964-69; mem. policy adv. bd. Argonne Nat. Lab., 1957-63, mem. review com. divs. biol. and med. research and radiol. physics, 1958-63, chmn. review com., 1958-62; mem. com. on licensure AEC, 1956-69, cons. div. edn. and tng., 1963; mem. adv. council on health research facilities NIH, 1965-69; organizing com. Pan Am. Fedn. of Assn. Med. Colls., 1962-64, treas., 1963-76; adv. com. personnel for research Am. Cancer Soc., 1962-66;

cons. commr. food and drugs FDA, 1965-70; spl. cons. to dir. NIH, 1968-70; cons. to div. physician and health professions edn. Bur. Health Manpower Edn. NIH, 1970-73; mem. adv. com. instnl. relations NSF, 1967-71; cons. edn. and tng. surgeon gen. U.S. Navy, 1972-73; mem. Inst. Medicine, Nat. Acad. Scis., 1972—; chmn. Fedn. Assns. Schs. Health Professions, 1972; mem. spl. med. adv. group VA, 1981—; Mem. alumni council Northwestern U.; Mem. bd. higher edn., Ill., 1964-69; chmn. Gov.'s Sci. Adv. Council, State Ill., 1967-69; mem. council Asso. Midwest Univs., 1963-68, v.p., bd. dirs., 1964-65, pres., bd. dirs., 1965-66; v.p., bd. trustees Argonne Univs. Assn., 1965-68; bd. dirs. Nat. Fund Med. Edn., 1970-79. Editor: Jour. Med. Edn, 1962-71. Served to 1st lt., San. Corps AUS, 1945-47. Recipient Outstanding Alumni award N.Mex. State U., 1960; Alumni medal Northwestern U., 1976; John and Mary R. Markle scholar in acad. medicine, 1951-56. Mem. Am. Soc. Biol. Chemists, Assn. Am. Med. Colls. (del. numerous confs., mem. various coms., pres. 1969—), AMA, AAAS, Central Soc. Clin. Research, Chgo. Inst. Medicine, Am. Hosp. Assn. (hon.), Asociación Venezolana Para el Avance de la Ciencia (hon.), Sigma Xi, Alpha Omega Alpha. Clubs: Cosmos (Washington); Tavern (Chgo.). Subspecialty: Medical education. Current work: Medical Administration. Home: 4118 N River St Arlington VA 22207 Office: No 1 DuPont Circle NW Washington DC 20036

COOPER, LEON N., physicist, educator; b. N.Y.C., Feb. 28, 1930; s. Irving and Anna (Zola) C.; m. Kay Anne Allard, May 18, 1969; children—Kathleen Ann, Coralie Lauren. A.B., Columbia, 1951, A.M., 1953, Ph.D., 1954, D.Sc., 1973; D.Sc. hon. degrees, U. Sussex, Eng., 1973, U. Ill., 1974, Brown U., 1974, Gustavus Adolphus Coll., 1975, Ohio State U., 1976, U. Pierre et Marie Curie, Paris, 1977. NSF postdoctoral fellow, mem. Inst. for Advanced Study, 1954-55; research asso. U. Ill., 1955-57; asst. prof. Ohio State U., 1957-58; asso. prof. Brown U., Providence, 1958-62, prof., 1962-66, Henry Ledyard Goddard U. prof., 1966-74, Thomas J. Watson Sr. prof. sci., 1974—; co-dir. Center for Neural Sci.; lectr. Summer Sch., Varenna, Italy, 1955; vis. prof. Brandeis Summer Inst., 1959, Bergen Internat. Sch. Physics, Norway, 1961, Scuola Internatzionali Di Fisica, Erice, Italy, 1965, L'Ecole Normal Supèrieure, Centre Universitaire Internationale, Paris, 1966, Cargèse Summer Sch., 1966; cons. indsl., ednl. orgns. Author: Introduction to The Meaning and Structure of Phsyics, 1968; Contbr. articles to profl. jours. Alfred P. Sloan Found. research fellow, 1959-66; John Simon Guggenheim Meml. Found. fellow, 1965-66; Recipient Comstock prize Nat. Acad. Scis., 1968, Nobel prize, 1972. Fellow Am. Phys. Soc., Am. Acad. Arts and Scis.; mem. Am. Philos. Soc., Nat. Acad. Scis., Phi Beta Kappa, Sigma Xi. Subspecialty: Theoretical physics. Home: 49 Intervale Rd Providence RI 02906

COOPER, MERRI-ANN, consulting psychologist, researcher; b. N.Y.C., Dec. 22, 1946; d. Isidore and Florence (Koplick) C.; m. Stephan Kessler, Aug. 11, 1965 (div. 1970). B.A., Bklyn. Coll., 1967; Ph.D., U. Chgo., 1974. Teaching asst. U. Chgo., 1968-70, research asst., 1969-72; asst. prof. Ill. State U., 1972-76; postdoctoral student U. Minn., 1976-78; personnel researcher Personnel Decisions Research Inst., Mpls., 1977-78; personnel research Hennepin County, Mpls., 1978-79; sr. research scientist Advanced Research Resources Orgn., Bethesda, Md., 1979—; project dir. FBI, Washington, 1982—, U.S. Air Force, 1982—; Dept. Labor, 1981, Edison Electric Inst., 1979-81. Contbr. articles to profl. jours.; test developer interviews, written measurement tests, phys. tests, 1978—. U. Chgo. Scholarship awardee, 1968; Ill. State U. grantee, 1974-76. Mem. Am. Psychol. Assn., Acad. Mgmt., Personnel Testing Council, Am. Bus. Women's Assn., NOW. Dem. Jewish. Current work: Test development validation; evaluation of organizational interventions; organizational research; training research; organizational diagnosis. Home: 4515 Willard Ave Chevy Chase MD 20815 Office: Advanced Research Resources Organization 4330 East-West Hwy Bethesda MD 20814

COOPER, NORMAN STREICH, pathologist, medical educator; b. N.Y.C., Dec. 23, 1920; s. Samuel and Edith (Streich) C.; m. Evelyn Fickler, Apr. 13, 1945; 1 son, Jonathan Samuel. B.A., Columbia Coll., 1940; M.D., U. Rochester, 1943. Diplomate: Am. Bd. Pathology. Intern, resident pathology N.Y. Hosp., N.Y.C., 1944-46, intern in medicine, 1948-49; scientist Oak Ridge Nat. Lab., 1947-48; instr. microbiology NYU Sch. Medicine, N.Y.C., 1949-51, instr. pathology, 1951-54, asst. prof., 1954-56, assoc. prof., 1956-67, prof. pathology, 1967—; chief lab. service VA Med. Ctr., N.Y.C., 1967—; founding pres. Assn. VA Lab. Service Chiefs, 1974-76. Editorial bd.: Clin. Immunology and Immunopathology, 1979-82. Vice-pres., treas. Arthritis Found., N.Y. chpt., 1978—. Served to capt. U.S. Army, 1946-48. Mem. Am. Assn. Immunologists, Am. Assn. Pathologists, Internat. Acad. Pathology, Am. Rheumatism Assn., Transplantation Soc. Clubs: Saltaire (N.Y.) Yacht; Racquet (East Hampton, N.Y.). Subspecialties: Immunology (medicine); Cell and tissue culture. Current work: Investigation of metabolic inhibitor, derived from mycoplasma, of the mutiplication of murine plasmacytoma cells in vitro and in vivo. Home: 333 E 68th St New York NY 10021 Office: NYU/VA Med Ctr 408 1st Ave New York NY 10010 Home: 333 E 68th St New York NY 10021

COOPER, RALPH SHERMAN, physicist; b. Newark, June 25, 1931; s. Morris David and Fay Bella (Gottfried) C.; m. Sandra Lenore Kleeman, Jan. 30, 1956; children: Laurie, Brett. B.Ch.E., Cooper Union, N.Y.C., 1953; M.S. in Physics, U. Ill., 1954, Ph.D., 1957. Mem. staff Los Alamos (N.Mex.) Sci. Lab., 1957-65, assoc. leader laser div., 1969-75; chief scientist McDonnel Douglas, Richland, Wash., 1965-69; dep. dir. research and devel. div. Physics Internat., San Leandro, Calif., 1975-82; dir. comml. systems Inesco, Inc., La Jolla, Calif., 1982—. Contbr. articles to profl. jours. Recipient Young Author prize Electrochem. Soc., 1956. Mem. Am. Phys. Soc., Am. Nuclear Soc. (mem. editorial adv. com. 1964-65), AIAA (mem. nuclear propulsion com. 1963-65), Soc. Photo-Optical Instrument Engrs., Am. Mgmt. Assn. Subspecialties: Nuclear fusion; Laser-induced chemistry. Current work: Magnetic fusion reactor design, X-ray microlithography, laser fusion, laser isotope separation, nuclear engineering, space power. Co-inventor, patentee lasers for isotope separation. Home: 10323 Rue Finisterre San Diego CA 92131 Office: Inesco Inc 11077 N Torrey Pines Rd La Jolla CA 92037

COOPER, REGINALD RUDYARD, orthopaedic surgeon, educator; b. Elkins, W.Va., Jan. 6, 1932; s. Eston H. and Kathryn (Wyatt) C.; m. Jacqueline Smith, Aug. 22, 1954; children—Pamela Ann, Douglas Mark, Christopher Scott, Jeffrey Michael. B.A. with honors, W.Va. U., 1952, B.S., 1953; M.D., Med. Coll. Va., 1955; M.S., U. Iowa, 1960. Diplomate: Am. Bd. Orthopedic Surgeons (examiner 1968—). Orthopedic surgeon U.S. Naval Hosp., Pensacola, Fla., 1960-62; asso. in orthopedics U. Iowa Coll. Medicine, Iowa City, 1962-65, asst. prof. orthopaedics, 1965-68, asso. prof. orthopedics, 1968-71, prof. orthopedics, 1971—, chmn. orthopedics, 1973—; research fellow orthopedic surgery Johns Hopkins Hosp., Balt., 1964-65; exchange fellow to Britain for Am. Orthopedic Assn., 1969. Trustee Nat. Easter Seals Research Found., 1977-81, chmn., 1979-81. Served to lt. comdr. USNR, 1960-62. Mem. Iowa, Johnson County med. socs., Orthopedic Research Soc. (sec.-treas. 1970-73, pres. 1974-75), Am. Acad. Orthopedic Surgeons (Kappa Delta award for outstanding research in orthopedics 1971), Canadian, Am. Orthopedic assns., Am. Acad. Orthopedic Surgeons (dir. 1973-74), N.Y. Acad. Sci., Assn. Bone and Joint Surgeons, AMA, Am. Rheumatism Assn., Am. Fedn. Clin. Research, Am. Acad. Cerebral Palsy, Am. Acad. Orthopedic Surgeons (chmn. exams. com. 1978-82, sec. 1982). Subspecialty: Orthopedics. Current work: Ultrastructure of musculoskeletal system. Home: 201 Ridgeview Ave Iowa City IA 52240

COOPER, RICHARD ALAN, hematologist; b. Milw., Sept. 23, 1936; s. Peter and Annabelle (Schlomovitz) C.; m. Jaclyn Koppel, June 22, 1958; children—Stephanie, Jonathan. B.S., U. Wis., 1957; M.D., Washington U., St. Louis, 1961. Intern Harvard U. Med. Services, Boston City Hosp., 1961-63, resident in medicine, 1965-66; fellow in hematology Thorndike Meml. Lab., Boston City Hosp., 1966-69; asst. prof. medicine Harvard Med. Sch., 1969-71; chief hematology div. Thorndike Meml. Lab. and Harvard Med. Services, Boston City Hosp., 1969-71; prof. medicine, dir. Cancer Center, chief hematology-oncology sect. U. Pa., Phila., 1971—. Mem. editorial bd.: Blood, 1979—, Lipid Research, 1983—. Served with USPHS, 1963-65. NIH grantee. Mem. Am. Soc. Hematology, Am. Soc. Clin. Oncology, Am. Fedn. Clin. Research, Am. Soc. Clin. Investigation, Phi Beta Kappa, Alpha Omega Alpha. Subspecialty: Hematology. Current work: Cell membrane structure; membrane lipids; cellular differentiation. Office: 3400 Spruce St Philadelphia PA 19104

COOPER, ROBERT ARTHUR, JR., physician, educator; b. St. Paul, Aug. 27, 1932; s. Robert Arthur and Theodora (Yarborough) C.; m. June Lorraine Spalty, Aug. 29, 1969; children—Robert Arthur, III, Timothy Rychner, Theodore Thomas. A.B., U. Pa., 1954; M.D., Jefferson Med. Coll., Phila., 1958. Intern Moffitt Hosp., U. Calif., 1958-59, resident in pathology, 1959-62; chief resident in pathology Women's Free Hosp. and Boston Lying-in Hosp., Harvard, 1962-63; teaching fellow Harvard Med. Sch., 1962-63; from asst. prof. to prof. Pathology U. Oreg. Med. Sch., 1963-69; mem. faculty U. Rochester (N.Y.) Med. Sch., 1969—, asso. dean curricular affairs, asso. prof. pathology, 1969-72, prof. pathology, dir. surg. pathology, 1972-75, prof. oncology in pathology, dir. cancer center, 1974—; cons. subcom. on comprehensive cancer centers Nat. Cancer Adv. Bd., Nat. Cancer Inst., 1976-78, mem. breast cancer treatment com. (breast cancer task force), div. cancer biology, 1974-77, mem. cancer center support grant rev. com., 1978—, chmn., 1981—; mem. spl. study sect. cancer epidemiology NIH, 1972; bd. dirs. United Cancer Council, Inc., v.p., 1981—; bd. dirs. Monroe County unit Am. Cancer Soc., 1976-80, mem. profl. edn. com. N.Y. State div., 1974-78; cons. Population Council, Rockefeller U., 1972—; mem. Lasker award jury Lasker Found., 1977. Author articles, chpts. in books.; Asso. editor: Internat. Jour. Radiation Oncology, Biology and Physics, 1974—; mem. editorial bd.: Cancer Clin. Trials, 1977—. Bd. dirs. Brighton Little League, 1973-77, vice commnr., 1975-76. Recipient Allan J. Hill Teaching award U. Oreg. Med. Sch., 1966, 67, 69; named 2d Year Tchr. of Year U. Rochester Med. Sch., 1973; Lester P. Slade civic achievement award Real Estate Bd. of Rochester, 1980; research fellow Am. Cancer Soc., 1960-61; grantee Nat. Cancer Inst., 1975—. Mem. Eastern Ski Assn., Alpha Omega Alpha. Clubs: Hunt Hollow Ski (Naples, N.Y.); Genesee Valley (Rochester, N.Y.). Subspecialties: Pathology (medicine); Cancer research (medicine). Current work: Radiation pathology-toxicology/ultrastructure. Home: 555 Clover Hills Dr Rochester NY 14618 Office: U Rochester Cancer Center 601 Elmwood Ave Box 704 Rochester NY 14642

COOPER, ROBERT CHAUNCEY, environmental health educator; b. San Francisco, July 4, 1928; m., 1956. B.S., U. Calif.-Berkeley, 1952; M.S., Mich. State U., 1953, Ph.D., 1958. Asst. prof., assoc. prof. pub. health Sch. Pub. Health, U. Calif.-Berkeley, 1958-74; prof. environ. health service, dir. San. Engring. and Environ. Health Research Lab., 1980—. Mem. AAAS, Am. Soc. Microbiology, Water Pollution Control Fedn., Internat. Assn. Water Pollution Research. Subspeciality: Water supply and wastewater treatment. Office: Sch Pub Health U Calif Berkeley CA 94720

COOPER, SHELDON MARK, immunology educator; b. N.Y.C., Dec. 5, 1942; s. Alex and Sylvia (Silverman) C.; m. Amy Diane Freedman, Nov. 21, 1966; 1 son, Jonas Eric. B.S. cum laude, Hobart Coll., 1963; M.D., NYU, 1967. Diplomate: Am. Bd. Internal Medicine and Rheumatology. Intern, resident Kings County Hosp., Bklyn., 1967-69; resident in internal medicine Los Angeles County Hosp., 1969-70; asst. prof. medicine, clin. immunology and rheumatic disease U. So. Calif., 1974-79, assoc. prof. medicine, research coordinator clin. immunology and rheumatic disease sect., 1980—; assoc. prof. medicine, dir. rheumatology and clin. immunology unit U. Vt., 1982—; attending staff in medicine Med. Ctr. Hosp. Vt., Burlington. Contbr. articles profl. jours. Exec. com. Vt. chpt. Arthritis Found., 1982, med. and sci. com., 1982—. Served to maj. USAF, 1972-73. NIH fellow, 1971-72; NIH grantee, 1978-81, 1975-78; organizing com. 15th Leucocyte Culture Conf., Asilomar, Calif., 1982. Mem. Am. Assn. Immunologists, Am. Fedn. Clin. Research, Reticuloendothelial Soc., Am. Rheumatism Assn., AAAS, Phi Beta Kappa. Subspecialties: Immunology (medicine); Internal medicine. Current work: Research, teaching, patient care. Home: Barstow Rd Shelburne VT 05482 Office: Univ Vt Given Bldg D301 Burlington VT 05405

COOPER, STUART LEONARD, chemical engineering educator; b. N.Y.C., Aug. 28, 1941; s. Jacob and Anne (Bloom) C.; m. Marily Portnoy, Aug. 29, 1965; children: Gary, Stacey. B.S. in Chem. Engring, MIT, 1963; Ph.D., Princeton U., 1967. Asst. prof. chem. engring. U. Wis., Madison, 1967-71, assoc. prof., 1971-74, prof., 1974—; vis. assoc. prof. U. Calif., Berkeley, 1974; vis. prof. Technion, Haifa, Israel, 1977. Editor: Multiphase Polymers, 1979, Biomaterials Interfacial Phenomena and Interactions, 1982. Trustee Argonne (Ill.) Univ. Assn., 1975-81. Fellow Am. Phys. Soc.; mem. Am. Inst. Chem Engrs., Am. Chem. Soc., Am. Soc. Artificial Internal Organs, AAAS, Soc. for Biomaterials. Subspecialties: Chemical engineering; Biomaterials. Current work: Polymer science and engineering, structure-property relations of polyurethanes, mechanical and dielectric spectroscopy, x-ray scattering, rheo-optical characterization, EXAFS analysis of ion containing polymers, biomaterials, blood-material interactions, protein and platelet absorption to artifical surfaces. Office: Department of Chemical Engineering University of Wisconsin Madison WI 53706

COOPER, THEODORE, pharmaceutical company executive, physician; b. Trenton, N.J., Dec. 28, 1928; s. Victor and Dora (Popkin) C.; m. Vivian Cecilia Evans, June 16, 1956; children—Michael Harris, Mary Katherine, Victoria Susan, Frank Victor. B.S., Georgetown U., 1949; M.D., St. Louis U., 1954, Ph.D., 1956. USPHS fellow St. Louis U. Dept. Physiology, 1955-56; clin. asso. surgery br. Nat. Heart Inst., Bethesda, Md., 1956-58; faculty St. Louis U., 1960-66, prof. surgery, 1964-66; prof., chmn. dept. pharmacology U. N.Mex., Albuquerque, 1966-68, on leave, 1967-69; assoc. dir. artificial heart, myocardial infarction programs Nat. Heart Inst., Bethesda, 1967-68; dir. Nat. Heart and Lung Inst., 1968-74; dep. asst. sec. for health HEW, 1974-75, asst. sec. health, 1975-77; dean Med. Coll., Cornell U., N.Y.C., 1977-80; provost for med. affairs Cornell U., 1977-80; exec. v.p. Upjohn Co., Kalamazoo, 1980—; mem. USPHS Pharmacology and Exptl. Therapeutics Study Sect., 1964-67; Bd. overseers Meml. Sloan-Kettering Cancer Center. Author: (with others) Nervous Control of the Heart, 1965, Heart Substitutes, 1966, The Baboon in Medical Research, Vol. II, 1967, Factors Influencing Myocardial Contractility, 1967, Acute Myocardial Infarction, 1968, Advance in Transplantation, Prosthetic Heart Valves, 1969, Depressed Metabolism, 1969; Editorial bd.: Jour. Pharmacology and Exptl. Therapeutics, 1965-68, 77—, Circulation Research, 1966-71; editor: Supplements to Circulation, 1966-71; sect. co-editor for: Jour. Applied Physiology, 1967-70; Contbr. numerous articles med. jours. Recipient Borden award, 1954; Albert Lasker Spl. Public Service award, 1978; Ellen Browning Scripps medal, 1980. Mem. Am. Soc. Pharmacology and Exptl. Therapeutics, Am. Physiol. Soc., Soc. Exptl. Biology and Medicine, Am. Soc. Clin. Investigation, Am. Fedn. Clin. Research, Am. Soc. Artificial Internal Organs, Internat. Cardiovascular Soc., Am. Coll. Chest Physicians, AAUP, Am. Coll. Cardiology, AAAS, Sigma Xi. Subspecialties: Pharmacology; Cardiology. Current work: Pharmaceutical business executive. Discoverer new techniques of denervating heart which have helped delineate role of nerves in heart, on its ability to function under a wide variety of circumstances, and on its ability to respond to drugs. Home: 3656 Woodcliff Dr Kalamazoo MI 49008 Office: Upjohn Co 7000 Portage Rd Kalamazoo MI 49001

COOPER, WILLIAM SECORD, information science educator; b. Winnipeg, Man., Can., Nov. 7, 1935; came to U.S., 1965; m. Helen Clare Dunlap, July 22, 1964. B.Sc., Principia Coll., 1956; M.SC., MIT, 1959; Ph.D., U. Calif.-Berkeley, 1964. Alexander von Humboldt scholar U. Erlangen, Ger., 1964-65; asst. prof. info. sci. U. Chgo., 1966-70; assoc. prof. info. sci. U. Calif.-Berkeley, 1971-76, prof., 1976—; acting dir. Inst. Library Research, 1970-71; Miller prof. Miller Inst., Berkeley, 1975-76. Hon. research fellow Univ. Coll., London (Eng.), 1977-78. Mem. AAAS, Assn. Symbolic Logic, Am. Soc. Info. Sci., Am. Statis. Assn., ACM. Current work: Symbolic logic; foundation of language; information storage and retrieval; decision theory, evolutionary theory. Office: Sch Library and Info Studies U Calif Berkeley CA 94720

COOPS, MELVIN (STERLINE), inorganic chemist, researcher, nuclear chemist, consultant; b. Sonoma, Calif., Apr. 21, 1930; s. Arthru August and Aloha (Millard) C. B.S. in Chemistry, U. Calif.-Berkeley, 1952. Staff chemist Lawrence Livermore Nat. Lab., Livermore, Calif., 1952—; dir. Identra, Inc., Saratoga, Calif. Contbr. numerous articles to profl. jours. Mem. Am. Nuclear Soc., Am. Chem. Soc. Republican. Subspecialties: Inorganic chemistry; Nuclear fission. Current work: Research and develpment on pyrochemistry as a method for processing nuclear fuels and actinide metals. Office: Lawrence Livermore National Laboratory PO Box 808 L-360 Livermore CA 94550

COPE, MICHAEL KEITH, physiology educator; b. Battle Creek, Mich., July 31, 1947; s. Stephen Van Duyne and Vivian Elaine (Keith) C.; m. Sharon Marie Bezzeg, Sept. 9, 1972. B.S. cum laude, U. Mich.-Dearborn, 1969; Ph.D., Ohio State U.-Columbus, 1975. Instr. Ohio No. U., Ada, 1972-73, Ohio Wesleyan U., Delaware, 1973-75; asst. prof. W.Va. Sch. Osteo. Medicine, Lewisburg, 1975-79, assoc. prof., 1979-80, assoc. prof., chmn. physiology, 1980—. Treas. Troop 70 Boy Scouts Am., 1981-82. Mem. Am. Heart Assn., Chmn. Depts. Physiology, Sigma Xi. Presbyterian. Subspecialties: Physiology (medicine); Osteopathy. Current work: Exercise in chronic obstructive pulmonary disease rehabilitation, measurement of the cranial rhythmic impulse. Home: Boggs Rd Frankford WV 24938 Office: W VA Sch Osteopathic Medicine 400 N Lee St Lewisburg WV 24901

COPEL, SIDNEY LEROY, psychologist; b. Phila., Aug. 29, 1930; s. Manuel and Anne (Snyder) C.; m. Joan Danzig, Dec. 28, 1959; children: Valerie, Kenneth. B.A., Temple U., 1951, M.Ed., 1953, Ed.D., 1958. Staff psychologist Sch. Clinic, Camden, N.J., 1954-56; lectr. psychology Bryn Mawr (Pa.) Hosp., 1965-66; cons. psychologist Hahnemann Hosp., Phila., 1966-67; lectr. psychology Jefferson Sch. Nursing, Phila., 1964-69; cons. psychologist Chester (Pa.) pub. schs., 1966-73; adminstr. psychol. clinic Deverevy Found., Devon, Pa., 1956-68; chief psychologist Youth Psychoterapy Ctr., Bryn Mawr, 1968-71; pvt. practice psychology, Bryn Mawr, 1968—. Author: Psychodiagnostic Study of Children and Adolescents, 1967, Advanced Psychopathology and and Research in Interpretation; editor: Behavior Pathology of Children and Adolescents, 1973. Fellow Pa. Psychol. Assn., Soc. Personality Assessment; mem. Am. Psychol. Assn., AAAS, Phila. Soc. Clin. Psychologist. Current work: Research in projective techniques and application to clin. problems. Home: 615 E Manoa Rd Havertown PA 19083 Office: 800 Summit Grove Ave Bryn Mawr PA 19010

COPELAND, JAMES CLINTON, enzyme manufacturing company executive, microbial geneticist; b. Chgo., Nov. 15, 1937; s. Wallace J. and Ann T. (Tuka) C.; m. Ella Grace Greene, Apr. 30, 1960; children: Catherine, Carolmarie, Christina, James C., Jeffrey. B.S., U. Ill.-Urbana, 1959; M.S., U. Tenn.-Knoxville, 1962; Ph.D. (NIH fellow), Rutgers U.-New Brunswick, N.J., 1965; M.B.A., U. Chgo., 1981. Sci. asst. biology div. Oak Ridge Nat. Labs., 1960-62; Am. Cancer Soc. postdoctoral fellow Albert Einstein Coll. Medicine, 1963-67, assoc. geneticist Div. Biology and Medicine, Argonne (Ill.) Nat. Lab., 1967-72; assoc. prof. microbiology Ohio State U., 1972-77; dir. biotech. CPC Internat., Argo, Ill., 1977-81; founder, pres., chief exec. officer Enzyme Tech. Corp., Ashland, Ohio, 1981—, also dir.; dir. Genon Corp. Contbr.: articles to profl. jours. Ency. Brit. Book of Yr., 1968-71; editor: Microbiol. Genetics Bull., 1973-77. Recipient Career Devel. award NIH, 1973-78; Oak Ridge Inst. Nuclear Studies fellow, 1961. Mem. AAAS, Ohio Acad. Sci., Sigma Xi. Clubs: Torch (Chgo.); Country of Ashland. Subspecialties: Genetics and genetic engineering (biology); Enzyme technology. Current work: Microbiology, genetic engineering, germentation, enzymes, biotechnical process development. Office: 783 US 250 E Route 2 Ashland OH 44805

COPELAND, WILLIAM D., metallurgy educator; b. Colorado Springs, Colo., Mar. 16, 1934; m., 1959; 3 children. B.A., Carleton Coll., 1956; Ph.D., U. Minn., 1966. Prof. metallurgy Colo. Sch. Mines, Golden, 1966—, dean, 1972—. Am. Council Edn. adminstrv. fellow, 1970-71. Mem. Am. Soc. Metals, Metall. Soc., AIME. Subspecialty: Metallurgy. Office: Colo Sch Mines Golden CO 80401

COPES, JOHN CARSON, III, consulting mechanical engineer; b. Baton Rouge, Sept. 6, 1923; s. John Carson and Beatrix (Lyons) C.; m. Edith Estelle Givens, Mar. 4, 1944; 1 son, John Carson IV. B.S., La. State U., 1947. Registered profl. engr. La. Mech. design engr. Esso Research Labs., 1947-50; design engr. Ethyl Corp., 1950-56; head mech. sect. Frederic R. Harris, Inc, 1956-61; chief engr. Arthur G. Keller Inc., 1961-65; prin. John C. Copes, Cons. Engrs., Baton Rouge, 1965—. Chmn. Livingston parish dist. Boy Scouts Am., 1951-52. Served with U.S. Army, 1943-45. Decorated Purple Heart., Bronze Star. Mem. La. Engring. Soc. (James M. Todd Tech. Accomplishment medal 1977), Nat. Soc. Profl. Engrs., ASME, SAR. Democrat. Episcopalian. Lodges: Masons; Shriners. Subspecialty: Mechanical engineering. Current work: Mechanical products development. Patentee split mech. seal inserts. Home: 2750 McConnell Dr Baton Rouge LA 70809 Office: 1956 Wooddale Ct Baton Rouge LA 70806

COPPOC, GORDON LLOYD, pharmacology educator; b. Larned, Kans., Nov. 11, 1939; m. Harriet Kagay, June 9, 1962; children: Laura Jean, Elizabeth Ann. B.S. cum laude, Kans. State U., 1963; D.V.M., Ph.D., Harvard U., 1968. Research assoc. U. Chgo., 1969-71; asst. prof. Sch. Vet. Medicine, Purdue U., West lafayette, Ind., 1971-73, assoc. prof., 1973-77, prof., 1977&, prof., head vet. physiology and pharmacology, 1979—; adj. prof. pharmacology Ind. U. Sch. Medicine, 1983—; Contbg. editor Jour. Vet. Med Edn., 1973-75; cons.

Custom Chem. Labs., Inc., Indpls., 1982—, others. Judge sci fairs.; Bd. dirs. Bach Chorale Singers, Inc., 1975-79, pres., 1977-79. Served to capt. USAF, 1967-69. Grantee Showalter Trust, USPHS, Eli Lilly & Co., others. Mem. , Am. Assn. Vet. Med. Colls. (publs. com.; chmn. nominating com. 1975, nat. sec. 1975-77, chmn. long range planning com. 1975-78, chmn. 1980-81), Am. Acad. Vet. Pharmacology and Therapeutics (sect.-treas. 1979-81, 81-83). Baptist. Subspecialties: Database systems; Cancer research (veterinary medicine). Current work: Comparative oncology program; renal disease models; pharmacokinetics application determining dose schedules and analysis of toxicoses; microcomputers in education. Office: Sch Vet Medicine Purdue U Lynn Hall West Lafayette IN 47907

COPPOLA, RICHARD, biomed. researcher, computer technologist; b. N.Y.C., May 27, 1947; s. Henry and Ann (Lenick) C.; m. Elisabeth R. Curtz, Apr. 1, 1978; children: Henry, Rebecca. B.S., M.I.T., 1968; D.Sc., George Washington U., 1978. Engr., scientist IBM, 1968; mem. tech. staff TRW, 1969; asst. USPHS, from 1970, sr. engr., 1981—; researcher Lab. of Psychology, NIMH, Bethesda, Md., 1970—. Contbr. numerous articles on electrophysiology, EEG, computer processing of neurophysiol. data to profl. jours. Mem. Hist. Takoma Soc., Takoma Park, Md. Recipient Service plaque USPHS, 1981. Mem. AAAS, Soc. Neurosci., IEEE, Sigma Xi. Subspecialties: Neurophysiology; Sensory processes. Current work: Brain research, especially as related to attention and sensory processing, EEG, evoked potentials, attention, brain mapping. Office: NIMH Lab Psychology Bethesda MD 20205

CORAN, ARNOLD GERALD, pediatric surgeon, educator; b. Boston, Apr. 16, 1938; s. Charles and Anne (Cohen) C.; m. Susan Williams, Nov. 17, 1960; children—Michael, David, Randi Beth. B.A. cum laude, Harvard U., 1959, M.D., 1963. Diplomate: Am. Bd. Surgery, Am. Bd. Thoracic Surgery. Intern Peter Bent Brigham Hosp., Boston, 1963-64, resident in surgery, 1964-68, chief surg. resident, 1969; resident in surgery Children's Hosp. Med. Center, Boston, 1965-66, sr. surg. resident, 1966, chief surg. resident, 1968; instr. surgery Harvard, Cambridge, Mass., 1967-69; asst. clin. prof. surgery George Washington U., 1970-72; head physician pediatric surgery Los Angeles County-U. So. Calif. Med. Center, 1972-74; asst. prof. surgery U. So. Calif., 1972-73, asso. prof., 1973-74; prof. surgery U. Mich., Ann Arbor, 1974—; head sect. pediatric surgery U. Mich. Hosp., 1974—. Contbr. numerous articles in field to profl. jours. Served to lt. comdr. MC AUS. Fellow ACS; mem. Am. Acad. Pediatrics, Soc. Univ. Surgeons, Am. Pediatric Surg. Assn., Western, Central surg. assns. Subspecialty: Surgery. Current work: Neonatal and pediatric metabolism; shock and nutrition. Home: 3450 Vintage Valley Rd Ann Arbor MI 48105 Office: Mott Children's Hosp Room F7516 Box 66 Ann Arbor MI 48109

CORBASCIO, NICOLA ALDO, physician, pharmaceutical company executive; b. Castellana, Italy, Mar. 21, 1928; came to U.S., 1954, naturalized, 1961; s. Vincenzo and Caterina (Zinza) C.; m. Elise Margareta Holgerson, Nov. 5, 1965; children: Sebastian, Matthias, Catherine. B.A., Horatius Flaccus Lycee, 1947; M.D., U. Bari, 1953; D.Sc., U. Pa., 1958. Intern U. Bari Med. Clinic, 1953-54; Fulbright scholar U. Pa. Med. Sch., 1954-55; fellow therapeutic research U. Hosp. U. Pa.-Phila., 1954-56; instr. pharmacology U. Pa. Med. Sch., 1956-59; asst. research pharmacologist U. Calif. Med. Ctr., 1959-65; assoc. med. dir. Miles Labs., Westhaven, Conn., 1982—; prof. pharmacology Coll. Physicians and Surgeons, U. Pacific, 1963-68, prof., chmn. dept. pharmacology, 1968-76; dir. med. affairs Nordic countries Rhone Poulenc A/S Neuropharmacologist Napa State Hosp., Imola, Calif., 1980—. Contbr. articles to sci. jours.; Co-author: Interactions of Drugs and Anesthetics, 1981. Mem. Am. Soc. Clin. Pharmacology, Western Pharmacol. Soc., AMA, Calif. Med. Assn., Alameda-Contra Costa Med. Soc. Subspecialty: Molecular pharmacology. Current work: Cardiovascular pharmacology; psychopharmacology. Patentee cardiovascular pharmacology. Office: Miles Labs 400 Morgan Ln West Haven CT 06516

CORBATÓ, FERNANDO JOSÉ, educator; b. Oakland, Calif., July 1, 1926; s. Hermenegildo and Charlotte (Jensen) C.; m. Isabel Blandford, Nov. 24, 1962 (dec. July 1973); children: Carolyn Suzanne, Nancy Patricia; m. Emily S. Gish, Dec. 6, 1975; stepchildren: David Lawrence Gish, Jason Charles Gish. Student, UCLA, 1943-44; B.S. in Physics, Calif. Inst. Tech., 1950, Ph.D., Mass. Inst. Tech., 1956. With Computation Center, Mass. Inst. Tech., 1956-66, dep. dir., 1963-66; head computer systems research group of project MAC Mass. Inst. Tech., 1963-72, co-head div., 1972-73, co-head automatic programming div., 1972-73, mem. faculty, 1962—, prof. computer sci. and engring., 1965—, asso. dept. head for computer sci. and engring., 1974-78, 83—, Cecil H. Green prof. computer sci. and engring., 1978-80, dir. computing and telecommunication resources, 1980-83; Mem. computer sci. and engring. bd. Nat. Acad. Sci., 1971-73. Co-author: The Compatible Time Sharing System, 1963, Advanced Computer Programming, 1963. Served with USNR, 1944-46. Fellow IEEE (W.W. McDowell award 1966, Computer Pioneer award 1982), AAAS, Am. Acad. Arts and Sci., Nat. Acad. Engring.; mem. Assn. Computing Machinery (council 1964- 66), Am. Fedn. Info. Processing Socs. (Harry Goode Meml. award 1980), Am. Phys. Soc., Sierra Club, Sigma Xi. Subspecialties: Operating systems; Software engineering. Current work: computer operating systems, architecture, networks and user interfaces. Home: 88 Temple St West Newton MA 02165 Office: 545 Technology Sq Cambridge MA 02139

CORBEIL, LYNETTE BUNDY, microbiologist; b. Vancouver, C., Can., May 19, 1938; d. Geert H. and Olive C. (Bundy) Keur; m. Robert Roland Corbeil, Dec. 14, 1963; children: Jacqueline Anne, Thomas Etienne. D.V.M., Ont. Vet. Coll., U. Toronto, 1962; Ph.D., Cornell U., 1974. Assoc. prof. immunology Kans. State U., 1978-79; assoc. prof. pathology U. Calif.-San Diego, 1979-82; assoc. prof. microbiology and pathology Coll. Vet. Medicine, Wash. State U., Pullman, 1982—, dir. animal diagnostic lab., 1979-82. Contbr. articles to profl. jours. NIH fellow, 1974; NIH grantee, 1978-82. Mem. AVMA, Can. Vet. Medicine Assn., B.C. Vet. Med. Assn., Am. Assn. Immunology, Can. Soc. Immunology, Am. Soc. Microbiology. Subspecialties: Microbiology (veterinary medicine); Infectious diseases. Current work: Immunity to gram negative bacteria, mucosal immunity. Office: Dept Vet Microbiology and Pathology Wash State U Pullman WA 99164

CORDELL, BRUCE MONTEITH, geophysicist, planetary scientist; b. Shelby, Mich., Sept. 10, 1949; s. Carl C. and Ruth M. C.; m. Lee Clark, Aug. 3, 1977. B.S., Mich. State U., East Lansing, 1971; M.S., UCLA, 1973; Ph.D., U. Ariz., 1977. Weizmann research fellow Calif. Inst. Tech., Pasadena, 1977-78; asst. prof. physics and earth scis. Central Conn. State Coll., New Britain, 1978-80; asst. prof. physics and geology Calif. State Coll., Bakersfield, 1980—. Contbr. articles in field to profl. jours. Mem. Am. Geophys Union, Am. Astron. Soc., Soc. Exploration Geophysicists. Republican. Subspecialties: Geophysics; Planetary science. Current work: Origin/evolution of earth and planets, climatic change mechanisms, So. Calif. tectonics, planetary geology and manned planetary exploration. Office: Physics Dept Calif State Coll Bakersfield CA 93309

CORDELL, ROBERT JAMES, geologist, research co. exec.; b. Quincy, Ill., Jan. 7, 1917; s. Vail R. and Gertrude (Robison) C.; m. Frances Regina Sparacio, Sept. 20, 1942; children: Victor V., David M., Margaret L. B.S., U. Ill., 1939, M.S., 1940; Ph.D., U. Mo., 1949. From instr. to asst. prof. U. Mo., Columbia, 1946-47, Colgate U., Hamilton, N.Y., 1947-51; research paleontologist, research geologist, sr. research geologist Sun Oil Co., Abilene, Tex., 1951-55, mgr. geol. research, Richardson, Tex., 1955-63; sr. sect. mgr., sr. research scientist, sr. profl. geologist Sun Co., Richardson, 1963-77; pres. Cordell Reports Inc., Dallas, 1977—; mem. exec. bd. Potential Gas Co., Colorado Springs, Colo., 1976—. Co-editor: Problems of Oil Migration, 1981; contbr. articles to profl. jours. Bd. dirs. Richardson (Tex.) Community Concerts Inc., 1958-70, pres., 1963-64; bd. dirs. Richardson Symphony Inc., 1962-82, pres., 1970-71, chmn. bd., 1971-72; area chmn. James Collins Campaign for U.S. Senate, Richardson, 1982. Fellow Geol. Soc. Am., AAAS; mem. Am. Assn. Petroleum Geologists (gen. chmn. nat. cov. 1975, Spl. Service award 1975), Dallas Geol. Soc. (hon. life, pres. 1977-78, Spl. Service award 1975, Research Publ. award 1980), Soc. Econ. Paleontologists and Mineralogists, Paleontological Soc. Republican. Episcopalian. Subspecialties: Geology; Fuels. Current work: Determining petroleum (oil and gas) potential of various regions based on geological interpretations. Home: 305 West Shore Dr Richardson TX 75080 Office: Cordell Reports Inc Suite 727 13771 N Central Expressway Dallas TX 75243

CORDER, MICHAEL PAUL, physician, educator; b. Zanesville, Ohio, Jan. 20, 1940; s. Thurman Edward and Dorothy (Shipps) C.; m. Sue Tawney, June 20, 1962; m. Dorothy M. Smith, Aug. 6, 1970; children: Anita, Wendy, Jennifer. B.S., Capital U., 1961; M.D., Ohio State U., 1965. Diplomate: Am. Bd. Internal Medicine, subcert. in med. oncology. Resident in internal medicine Letterman Army Med. Center, 1965-69, fellow in oncology, 1971-75; asst. prof. U. Iowa Hosps., 1975-78, assoc. prof., 1978-83; chmn., dir. oncology/hematology Kern Med. Center, Bakersfield, Calif., 1983—; cons. in field. Contbr. articles to profl. jours., chpts. to books. Served with M.C. U.S. Army, 1964-75; to col. Res., 1980—. Fellow in oncology Nat. Cancer Inst., NIH, 1969-71. Fellow ACP, Internat. Soc. Hematology; mem. Am. Soc. Hematology, Am. Assn. Cancer Research, Am. Soc. Clin. Oncology, Am. Soc. Preventive Oncology. Subspecialties: Oncology; Algorithms. Current work: Med. decision making, cancer clin. trials. Home: 12430 Cattle King Bakersfield CA 93306 Office: Dept Medicine U Kern Med Center Bakersfield CA 93305

CORDERO, JULIO, systems engineer; b. San Jose, Costa Rica, Jan. 10, 1923; came to U.S., 1944, naturalized, 1961; s. Julio R. and Maria C. (Fonseca) C.; m. Claire Cox, Oct. 26, 1963; 1 child, Astrid Cox Cordero. B.S. in Meteorology, Inter-Am. Meteorol. Inst., Colombia, 1943; B.S.A.E., Wayne State U., 1948; M.S. in Aerodynamics, U. Minn., 1951. Meteorologist Pan-Am. World Airways, 1944; scientist U. Minn., Mpls., 1951-53; sr. scientist Fluidyne Engring. Corp., Mpls., 1953-61, AVCO Research and Devel., Wilmington, Mass., 1961-68; sr. systems analysis ANSER, Falls Church, Va., 1969-75; chief engr. magnetohy drodynamics MIT, Cambridge, 1975-81; tech. staff AVCO SD, Everett, Mass., 1981—; cons. Mobil Oil Corp., N.Y.C., 1970. U.S. Weather Bur. scholar, 1943; Kales scholar, 1976. Mem. Am. Phys Soc., AIAA, AAAS, Sigma Xi. Club: MIT Faculty. Subspecialties: Systems engineering; Plasma. Current work: Large laser systems integration. Home: 23 Mohawk St Danvers MA 01923 Office: 2385 Revere Beach Pkwy Everett MA 02149

CORDES, JAMES MARTIN, astronomy educator; b. Conneaut, Ohio, Dec. 3, 1949; s. John E. and Gladys (Ross) C. Ph.D., U. Calif.-San Diego, 1975. Postdoctoral research assoc. U. Mass., Amherst, 1975-79; asst. prof. Cornell U., Ithaca, N.Y., 1979—. Recipient Calif. Inst. Tech. Presdl. Fund. grant, 1980-82, Alfred P. Sloan Found. fellow, 1983-85. Mem. Am. Astronom. Soc., N.Y. Astronom. Soc., N.Y. Acad. Sci., Internat. Astronom. Union. Subspecialties: Radio and microwave astronomy; High energy astrophysics. Current work: Studies of pulsars and extragalactic radio sources, wave propagation in random media, signal processing techniques. Home: 939 E State St Ithaca NY 14850 Office: Space Sci Bldg Cornell U Ithaca NY 14853

CORDINGLEY, GARY EDWARD, neurologist, researcher; b. Chgo., Nov. 29, 1949; s. Robert Vincent and Louise Catherine (Nichol) C.; m. Catherine Janette Arthur, May 26, 1974; children: Sarah, Amanda. B.S., Purdue U., 1971; Ph.D., Duke U., 1976, M.D., 1977. Med. intern U. Mich. Hosps., Ann Arbor, 1977-78; resident in neurology Columbia-Presbyterian Med. Center, N.Y.C., 1978-81; pharmacology research asso. NIH, Bethesda, Md., 1981—. Contbr. articles to med. jours. Mem. Am. Acad. Neurology, Soc. Neurosci. Subspecialties: Neurology; Neuropharmacology. Current work: Clin. and basic synaptic pharmacology of the basal ganglia. Office: Lab Preclin Studies NIAAA 12501 Washington Ave Rockville MD 20852

CORDOVA, FRANCE ANNE-DOMINIC, astrophysicist; b. Paris, France, Aug. 5, 1947; d. Frederick Ben and Joan Frances (McGuinness) C. B.A. magna cum laude, Stanford U., 1969; Ph.D., Calif. Inst. Tech., 1979. Postdoctoral research asst. Calif. Inst. Tech., Pasadena, 1979; staff scientist, project leader Los Alamos (N.Mex.) Nat. Lab., 1979—; cons. Lawrence Berkeley Radiation Lab., Calif., 1979-80; mem. adv. council NSF, 1981—. Contbr. articles to various sci. books, mags., sci. jours. and popular sci. jours. U. Calif. Regents scholar, 1965; Ford Found. research grantee, 1968-69; NATO fellow, 1982. Mem. Am. Astron. Soc., New Mexicans for Space Exploration. Subspecialties: Ultraviolet high energy astrophysics; 1-ray high energy astrophysics. Current work: Observational astronomy: X-ray, ultraviolet, optical, infrared, and radio measurements of high energy sources using satellites and ground-based telescopes. Office: Los Alamos Nat Lab M S D436 Los Alamos NM 87545

CORDOVA-SALINAS, MARIA ASUNCION, physiology educator; b. Punta Arenas, Magallanes, Chile, May 14, 1941; came to U.S., 1972; d. Miguel and Maria Asuncion (Requena Aizcorbe) Cordova-Santana; m. Carlos Francisco Salinas, July 27, 1963; children: Carlos Miguel, Claudio Andres, Maria Asuncion. B.S., U. Chile, 1958, D.D.S., 1965; cert., Johns Hopkins U., 1974. Faculty U. Chile, Valparaiso, 1965-74; postdoctoral fellow Johns Hopkins U., Balt., 1972-74; vis. scientist N.Y. Med. Coll., N.Y.C., 1974; faculty dept. pharmacology Med. U. S.C., Charleston, 1975-79, instr. physiology, 1980—; pvt. practice dentistry, Valparaiso, 1965-72. Coordinator Circulo Hispanoamericano, Charleston, 1978—; bd. dir. Iglesia Hispanica, Charleston, 1983. Pan Am. Health Orgn./WHO fellow, 1972-74; named Guest of Honor City of Mayaguez, P.R., 1979. Mem. Am. Physiol. Soc. Subspecialties: Physiology (medicine). Current work: Oral physiology and oral diagnosis. Office: Dept Physiology Med Univ SC 171 Ashley Ave Charleston SC 29425

CORDUNEANU, CONSTANTIN C., mathematics educator, researcher; b. Iasi, Moldavia, Romania, July 26, 1928; came to U.S., 1978; s. Costache and Aglaia (Anitoaie) C.; m. Alice Olga Vultur, July 23, 1949. Diploma in Math, U. Iasi, 1951, D.Math., 1956. Instr. U. Iasi, Romania, 1949-55, asst. prof., 1955-62, assoc. prof., 1962-67, prof., 1968-78; prof. dept. math. U. Tex.-Arlington, 1979—; vis. prof. U. R.I.- Kingston, 1967-68, 73-74, 80, U. Tenn.-Knoxville, 1978-79. Author: Almost Periodic Functions, 1968, Principles of Differential and Integral Equations, 1971, 77, Integral Equations and Stability of Feedback Systems, 1973; assoc. editor: jours. including Math. System Theory, 1967-75, Revue Roumaine Pure Applied Math, 1973-78, Nonlinear Analysis, 1977—, Jour. Integral Equations, 1979—, Libertas Mathematica, 1981—. Served with Romanian Army, 1952. Recipient research prizes Romanian Acad., Bucharest, 1963, Ministry of Edn., Bucharest, 1965; research fellow Inst. Math. Romanian Acad., Iasi, 1954-59, 63-67. Mem. Am. Math. Soc., Math. Assn. Am., Soc. Indsl. and Applied Math., Am. Romanian Acad. Arts and Scis. Christian Orthodox. Subspecialty: Applied mathematics. Current work: Differential and related equations. Home: 812 S Collins Ave Arlington TX 76010 Office: Univ Tex-Arlington S Cooper St Arlington TX 76019

COREY, JOYCE YAGLA, chemistry educator; b. Waverly, Iowa, May 26, 1938; d. Clifford Durwood and Laura Gail (Laird) Yagla; m. Eugene Ray Corey, Sept. 8, 1962. B.S. in Chemistry, U. N.D., 1960, M.S., 1961; Ph.D., U. Wis., 1964. Faculty Villa Madonna Coll., Covington, Ky., 1964-68; faculty U. Mo.-St. Louis, 1968—, prof. chemistry, 1980—. Recipient AMOCO Good Teaching Award, 1977. Mem. Am. Chem. Soc., Chem. Soc. London, Sigma Xi. Subspecialties: Inorganic chemistry; Synthetic chemistry. Current work: Organometallic compounds of Groups III and IV. Synthesis and structural characterization. Silicon analogs of psychotropic drugs. Home: 49 Bellerive Acres Saint Louis MO 63121 Office: Dept Chemistry Univ Mo St Louis 8001 Natural Bridge Rd Saint Louis MO 63121

CORFF, NICHOLAS J., architect, consulting computer systems developer; b. Oklahoma City, June 30, 1942; s. Nicholas C. and Barbara E. (Geirk) C. B.Arch., Stanford U., 1966, M.Arch., 1972. Dir., Workshops on Polit. and Social Issues Stanford U., 1970-72, instr. Micronesia seminar, 1971-72, founder Stanford Oceania Ctr., 1972; cons. Campbell & Assocs., Menlo Park, Calif., 1972-73; dir. planning div. Parsell Yeager, Inc., Seattle, 1973-74; pres. chief exec. officer Corff & Shapiro, Inc., Seattle, 1974-77; pres., chief exec. officer CDS Assocs., Ltd., Seattle, 1977-81, Corgraphics Corp., 1981-82, chmn. bd., 1983; cons., 1983—. Author works in field. Vol. Peace Corps, Palau, Micronesia, 1966-67. Served as officer C.E., U.S. Army, 1968-69. Mem. Inst. Island Research and Assistance (pres. 1977—), Assn. Computing Machinery, Nat. Computer Graphics Assn., Northwest Computer-Aided Mapping Assn. (vice chmn. 1982—), Palau Forum. Subspecialties: Graphics, image processing, and pattern recognition; Artificial intelligence. Current work: Computer aided design, computer based graphics analysis, graphic-based language design, island research; founder, key concept developer of computer graphics systems corporation; consulting, writing, architecture. Home: 511 Malden Ave E Seattle WA 98112

CORI, CARL FERDINAND, educator, biochemist; b. Prague, Czechoslovakia, Dec. 5, 1896; came to U.S., 1922, naturalized, 1928; s. Carl I. and Maria (Lippich) C.; m. Gerty Theresa Radnitz, Aug. 5, 1920 (dec. 1957); m. Anne FitzGerald Jones, Mar. 23, 1960; 1 son, Carl Thomas. Student, Gymnasium, Austria, 1906-14; M.D., German U. Prague, 1920, U. Trieste, Italy, 1974; Sc.D., Yale, Western Res. U., 1947, Boston U., 1948, Cambridge U., Eng., 1949, U. Granada, Spain, 1966, Brandeis U., 1965, Monash U., Melbourne, Australia, 1966, Washington U., St. Louis, 1967, St. Louis U., 1967, Gustavus Adolphus Coll., 1963. Asst. in pharmacology U. Graz, Austria, 1920-21; biochemist State Inst. for Study Malignant Disease, Buffalo, 1922-31; prof. pharmacology and biochemistry Washington U. Sch. Medicine, 1931-66; cons. biochemistry, vis. lectr. Mass. Gen. Hosp., Harvard U. Med. Sch., Boston, 1966—, dir. Enzyme Research Lab., 1966—; mem. faculty Harvard Med. Sch., 1966—. Contbr. articles, chiefly on carbohydrate metabolism and enzymes of animal tissues to Am. sci. jours.; Mem. editorial bd.: Biochimica et Biophysica Acta. Recipient Nobel Prize in medicine and physiology, 1947; Willard Gibbs medal Am. Chem. Soc., 1948; Sugar Research Found. award, 1947, 50; Lasker award, 1946; Squibb award, 1947; St. Louis award. Mem. Nat. Acad. Scis.; hon. mem. Harvey Soc.; mem. Am. Soc. Biol. Chemists, Am. Chem. Soc. (Mid-West award 1946), A.A.A.S., Royal Soc. London, Am. Philos. Soc., Sigma Xi. Subspecialties: Gene actions; Membrane biology. Current work: Gene repression in rat-mouse hepatic hybrid cells.

CORIELL, LEWIS L., pediatrician, educator, medical research administrator; b. Sciotoville, Ohio, June 19, 1911; m., 1936; 3 children. B.A., U. Mont., 1934; M.A., U. Kans., 1936, Ph.D. in Bacteriology, 1940, M.D., 1942. Asst. instr. botany U. Mont., 1934; from asst. instr. to instr. bacteriology U. Kans., 1934-40; instr. pediatrics U. Pa., Phila., 1946-49, assoc. prof. immumological pediatrics, 1949-63, prof. pediatrics, 1963—; NRC fellow in virus disease Inst. Med. Research, Camden, N.J., 1947-49, dir. inst., 1955—; med. dir. Camden Mcpl. Hosp. for Contagious Disease, 1949-61; pediatrician Cooper Hosp., 1949—; sr. physician Children's Hosp., 1954—; cons. Phila. Naval Hosp., 1956-66. Mem. AAAS; assoc. mem. AMA, Am. Soc. Microbiology, Assn. Mil. Surgeons U.S., Soc. for Pediatric Research. Subspecialty: Medical research administration. Office: Inst for Med Research Copewood St Camden NJ 08103

CORK, DOUGLAS JAMES, biology educator; b. St. Paul, May 5, 1950; s. Willis L. and Ruth M. C.; m. Margaret Seavy; 1 dau., Sarah Margaret. B.S., U. Ariz., 1972, M.S., 1974, Ph.D., 1978. Teaching asst. U. Ariz., 1972-74, research asst., 1974-78; asst. prof. U. Miss., 1979-80, Ill. Inst. Tech., 1980—; indsl. microbiologist, biochem. engr. AquaTerra Biochemistry Corp. Am., Dallas, 1979-80. Contbr. articles to profl. jours. Mem. Am. Inst. Chem. Engrs., Am. Chem. Soc., Soc. Indsl. Microbiology, Am. Soc. Microbiology. Subspecialties: Biomass (energy science and technology); Biomass (agriculture). Current work: Developed two microbial desulfurization bioprocesses which are based upon the fundamental microbial ecology and physiology of photoautotrophic sulfur producers, acetate degrading sulfate reducers and acetate and H_2 degrading methanogens. Patentee in field. Office: Dept Biology Illinois Inst Tech Chicago IL 60616

CORK, LINDA K. COLLINS, pathology educator; b. Texarkana, Tex., Dec. 14, 1936; m. (div); 2 children. B.S., Tex. A. and M. U., 1969, D.V.M., 1970; Ph.D., Wash. State U., 1974; diploma, Am. Coll. Vet. Pathologists, 1975. Asst. prof. vet. pathology U. Ga., 1974-76; asst. prof. pathology div. computer medicine Sch. Medicine, Johns Hopkins U., Balt., 1976—. Mem. editorial bd.: Vet. Pathology, 1977—. Co-investigator Amytrophic Lateral Sclerosis Soc., 1977-80; mem. Com. Animal Models and Genetic Stocks, NRC, 1978—; prin. investigator Kroc Found., 1978-80. NIH fellow Wash. State U., 1970-74; prin. investigator, inst. research grantee Johns Hopkins U., 1976-77, Nat. Multiple Sclerosis Soc., 1977. Mem. Am. Coll. Vet. Pathologists, Am. Assn. Neuropathologist, Am. Assn. Pathology. Subspecialty: Pathology (veterinary medicine). Office: Dept Pathology Johns Hopkins U Baltimore MD 21205

CORKIN, SUZANNE, psychology educator; b. Hartford, Conn., May 18, 1937; m. Charles Corkin II, Sept. 8, 1962; children: J. Zachary, Jocelyn H., Damon L. B.A., Smith Coll., 1959; M. Sc., McGill U., Can., 1961, Ph.D., 1964. Lic. psychologist, Mass. Research assoc. dept. psychology MIT, Cambridge, 1964-77, 1964-79, lectr. dept. psychology, 1977-79, princ. research scientist, 1979-81, sr. investigator, 1979—, assoc. prof., 1981—; cons. psychologist, neurosurgery service Mass. Gen. Hosp., Boston, 1975—. Editor: Alzheimer's Disease: A Report of Progress in Research, 1982. Mem. Aging Assn., AAAS, Am. Pain Soc., Am. Psychol. Assn., Boston Soc. Psychiatry and Neurology, Eastern Pain Assn., Eastern Psychol. Assn., Gerontol. Soc., Internat.

Assn. for Study of Pain, Internat. Neuropsychol. Symposium, Soc. for Neurosci., Sigma Xi. Subspecialty: Neuropsychology. Home: 152 Deacon Haynes Rd Concord MA 01742 Office: MIT E10-003A 77 Massachusetts Ave Cambridge MA 02139

CORLEY, KARL COATES, JR., physiologist, researcher; b. Washington, Oct. 23, 1935; s. Karl C. and Isabel M. (Southgate) C.; m. Mary C. Coulson, Dec. 27, 1970; children: Crystal, Hugh. B.S., Trinity Coll., Hartford, Conn., 1958; Ph.D., U. Rochester, 1963. Research assoc. U. Rochester Ctr. Brain Research, 1962-63; vis. scientist Walter Reed Med. Ctr., Washington, 1963; research asso. Merck Inst., West Point, Pa., 1963-65; asst. prof. physiology Med. Coll Va., Richmond, 1965-80, assoc. prof., 1980—. NIH grantee, 1971-81. Mem. Am. Psychology Assn., Am. Physiol. Soc., Soc. Neurosci., N.Y. Acad. Scis., Sigma Xi. Subspecialties: Neurophysiology; Psychophysiology. Current work: Research in autonomic physiology. Office: Physiology and Biophysics Dept Box 125 Med Coll VA Richmond Va 23298

CORLEY, RONALD BRUCE, immunologist, educator, researcher; b. Durham, N.C., Oct. 22, 1948; s. Charles B. and Barbara (Murph) C.; m. Janice O. Grier, Oct. 18, 1980; 1 son, Evan Grier. B.S., Duke U., 1970, Ph.D., 1975. Mem. scientist Basel (Switzerland) Inst. Immunology, 1975-77, vis. scientist, 1977, 79, 80; asst. prof. Duke U. Med. Ctr., Durham, N.C., 1977-82, mem., 1978—, assoc. prof. immunology, 1982—. Recipient Scholar awarr Leukemia Soc. Am., 1979-84; NIH grantee, 1980. Mem. Am. Assn. Immunologists, Am. Assn. Clin. Hisocompatibility Testing. Subspecialties: Cellular engineering; Cell biology (medicine). Current work: Intercellular communication among lymphocytes, especially role of one class of T lymphocytes, called helper T cells, to regulate immune responses by other T cells and by B lymphocytes. Home: 1404 Auburndale Dr Durham NC 27713 Office: Div Immunology Duke Med Ctr Box 3010 Durham NC 27710

CORNELL, ROBERT WITHERSPOON, mechanical engineer, consultant; b. Orange, N.J., Aug. 16, 1925; s. Edward S. and Helen L. (Lawrence) C.; m. Patricia Delight Plummer, June 24, 1950; children: Richard W., Delight W., Elizabeth P., Roberta S. B.E. in M.E. with high honors, Yale U., 1945, M.E., 1947, D.Eng. in M.E, 1950; grad., U.S. Naval acad., 1945. Registered profl. engr., N.Y. State, Conn. With Naval Research Lab., Washington, 1945-46; instr. math. New Haven Jr. Coll., 1947-48; Engr. Pratt & Whitney Aircraft, East Hartford, Conn., 1947; with Hamilton Standard div. United Techologies Corp., Windsor Locks, Conn., 1948—, now chief applied mechanics and aerodynamics; instr. engring. Hillyer Coll., Hartford, Conn., 1955; prin. Cornell Cons., West Hartford, 1973—. Contbr. articles to tech. jours. Republican dist. committeeman; bd. dirs., past pres. West Hartford Taxpayers Assn.; mem. coms. West Hartford Sch. Task Force. Served in USNR, 1943-46. Mem. Yale Sci. and Engring. Assn. (award 1969, dir., treas.), ASME, Sigma Xi, Tau Betz Pi. Congregationalist. Clubs: Hartford Golf, Yale. Subspecialties: Mechanical engineering; Aerospace engineering and technology. Current work: Manage technical group specializing in aerodynamics, acoustics, applied mechanics, vibrations and structure; advanced work on gear dynamics, stressing and strength. Patentee in field (8), including space suit joint and prop-fan multi-bladed propeller. Home: 40 Belknap Rd West Hartford CT 06117 Office: Hamilton Standard Bradley Field Rd Windsor Locks CT 06096

CORNETT, JAMES BRYCE, research microbiologist; b. Orange, Calif., Apr. 30, 1945; s. Royce Wesley and Margaret Lititia (Glenn) C. B.A., U. Calif., Riverside, 1967; M.S., U. Calif.-Davis, 1969; Ph.D., U. Ariz., 1973. NIH postdoctoral fellow Temple U. Sch. Medicine, Phila., 1973-75, research asst. prof., 1977-78; sr. research biologist Sterling-Winthrop Research Inst., Rensselaer, N.Y., 1978-81, group leader microbiology dept., 1981-83, sect. head, 1981—. Contbr. articles to profl. jours. Mem. Am. Soc. Microbiology (nat. councillor). Subspecialty: Microbiology (medicine). Current work: The development of antimicrobial agents for clinical use; drug development. Office: Sterling Winthrop Research Inst Columbia Turnpike Rensselaer NY 12144

CORNETTE, JAMES LAWSON, mathematics educator; b. Bowling Green, Ky., May 8, 1935; s. James Percival and Mary Elizabeth (Lawson) C.; m. Jimmye Carolyn Christian, Apr. 21, 1962; children: James Terrill, Frances Elizabeth. B.S., W.Tex. State Coll., 1955; M.A., U. Tex.-Austin, 1959, Ph.D., 1962. Asst. prof. Iowa State U., Ames, 1962-66, assoc. prof., 1966-70, prof. math., 1970—. Fulbright lectr. U.S. Dept. State, Kuala Lumpur, Malaysia, 1973-74. Mem. Math. Assn. Am. (bd. govs. 1977-80), Am. Math. Soc., Soc. Indsl. and Applied Math. Lodge: Kiwanis. Subspecialties: Applied mathematics; Evolutionary biology. Current work: Mathematical models of population genetics and animal physiology. Office: Iowa State Univ Ames IA 50011 Home: 2814 Torrey Pines Circle Ames IA 50010

CORNO, LYN, educational psychologist, educator; b. Williams, Ariz., May 25, 1950; d. Edward Eugene and Verla (Hornbacher) C. B.A., Ariz. State U., 1972; M.A., Stanford U., 1977, Ph.D., 1978. Research asst. SWRL Edn. Research and Devel., Los Alamitos, Calif., 1972-74, Ctr. for Ednl. Research, Stanford, Calif., 1974-77; research asst. prof. edn. Stanford U., 1977-82; assoc. prof. edn. Tchrs. Coll., Columbia U., 1982—; cons. Calif. Dept. Edn., Sacramento, 1980-82; asst. dir. Center for Ednl. Research, Stanford, 1977-78; cons. computer corps., 1981—. Contbr. articles to profl. jours. Spencer Found. grantee, 1979-81; Calif. Dept. Edn. grantee, 1980-82; Nat. Inst. Edn. grantee, 1981-82. Mem. Am. Psychol. Assn., Am. Ednl. Research Assn., AAAS, Nat. Soc. Study of Edn., Phi Delta Kappa. Democrat. Episcopalian. Subspecialties: Learning; Cognition. Current work: Study of self-regulated learning; cognitive effects of media and technology; effective instructional systems. Office: Tchrs Coll Columbia Univ 120th and Broadway New York NY 10027

CORNYN, JOHN JOSEPH, computer consultant, physicist; b. Pitts., Dec. 20, 1944; s. John Joseph and Leanore Marie (Steiner) C. B.S., Carnegie Inst. Tech., 1967; M.S. in Physics, Carnegie-Mellon U., 1969, U. Md., 1974. Tech. asst. Westinghouse Electric, Pitts., 1965-67; instr. physics Carnegie-Mellon U., Pitts., 1967-69; research physicist Naval Research Lab., Washington, 1969-76, Naval Ocean Research/Devel. Activity, Bay St. Louis, Miss., 1976-81; pres., founder Third Wave Data Systems, Inc., St. Petersburg, Fla., 1981—. Recipient Edison Meml. Scholarship Naval Research Lab., 1969. Mem. IEEE, Assn. Computing Machinery, Ind. Computer Cons. Assn. Subspecialties: Numerical analysis; Acoustics. Current work: Microcomputer software development, primarily for engineering, science and investment applications, interactive computer graphics. Office: Third Wave Data Systems Inc 3210 9th St N PO Box 7206 Saint Petersburg FL 33734

CORRADINO, ROBERT A., biomedical researcher, educator; b. Lancaster, Pa., Aug. 6, 1938; s. Nicholas and Minnie M. C.; m. (div.); 1 son, Tony R. B.S., State Coll., Millersville, Pa., 1960; M.S., Purdue U., 1962; Ph.D., Cornell U., 1966. Research assoc. Wyeth Labs., Radnor, Pa., 1962-64; research assoc. Cornell U., Ithaca, N.Y., 1966-72, sr. research assoc., 1972-80, asso. prof. physiology, 1980—. Editor: Functional Regulation at the Cellular and Molecular Level, 1983; contbr. 70 articles to profl. jours. Served with USMC, 1958-64. Recipient Research Career Devel. award NIH, 1975-80. Mem. Am. Physiol. Soc., Am. Inst. Nutrition, Endocrine Soc., Soc. Exptl. Biology and Medicine, AAAS, Sigma Xi. Subspecialties: Physiology (biology); Endocrinology. Current work: Study of the regulation of induction of a specific intestinal calcium-binding protein and the role of this protein in the cholecalciferol (vitamin D030)-stimulated calcium transport. Home: 17-2E Lansing Apts N Ithaca NY 14850 Office: Cornell U 720 VRT Ithaca NY 14853

CORREIA, JOHN ARTHUR, physicist, educator; b. Brookline, Mass., June 8, 1945; s. Arthur Francis and Mary Elizabeth (Kenneally) C.; m. Dorothy Ellen Reddington, Sept. 8, 1967; 1 son, Gabriel. B.S., Lowell Tech. Inst., 1967, Ph.D., 1973. Research fellow dept. medicine Mass. Gen. Hosp., 1973, research fellow in radiology, 1973-75, asst. physicist, 1975-80, assoc. physicist, 1983—; assoc. dir. Mass. Gen. Hosp. Physics Research Lab., 1982—; research fellow Harvard Med. Sch., 1973, research fellow in radiology, 1973-75, research assoc., 1975-79, asst. prof. radiology, 1979-82, assoc. prof. radiology, 1983—. Contbr. chpts. to books, articles to profl. jours. Mem. Am. Nuclear Soc., Am. Assn. Physicists in Medicine, Am. Heart Assn., Soc. Nuclear Medicine. Subspecialty: Imaging technology. Current work: Medical image processing, positron emission tomography, cerebral blood flow and metabolism measurements, physiological modelling. Office: Mass Gen Hospnl Dept Radiology Boston MA 02114

CORRSIN, STANLEY, educator, fluid dynamicist; b. Phila., Apr. 3, 1920; s. Herman and Anna (Schor) C.; m. Barbara Daggett, Sept. 25, 1945; children—Nancy Eliot, Stephen David. B.S., U. Pa., 1940; M.S., Calif. Inst. Tech., 1942, Ph.D. in Aero, 1947; Docteur honoris causa, Université Claude Bernard, France, 1974. Research and teaching asst. aero. Calif. Inst. Tech., 1940-45, instr., 1945-47; asst. prof. aero. Johns Hopkins, 1947-51; asso. prof., 1951-55, prof. mech. engring., chmn. dept., 1955-60, prof. fluid mechanics, 1960—. Recipient distinguished alumnus citation U. Pa. Sch. Engring., 1955. Fellow Am. Phys. Soc. (chmn. div. fluid dynamics 1964), Am. Acad. Arts and Scis., ASME, Nat. Acad. Engring.; mem. Am. Inst. Aeros. and Astronautics, AAUP (pres. Johns Hopkins chpt. 1964-65), AAAS, Phi Beta Kappa, Sigma Xi, Tau Beta Pi, Pi Tau Sigma. Subspecialties: Fluid mechanics; Biomedical engineering. Home: Riderwood MD 21139 Office: Johns Hopkins U Baltimore MD 21218

CORRY, ANDREW FRANCIS, utility co. exec.; b. Lynn, Mass., Oct. 28, 1922; s. Andrew Francis and Julia Agnes (Gaynor) C.; m. Mildred M. Dunn, Sept. 16, 1950 (dec. 1977); children—Andrea, Janice, James. B.S. in E.E, M.I.T., 1947; postgrad, Advanced Mgmt. program Harvard U., 1966. Registered profl. engr., Mass. With Boston Edison Co., 1947—, asst. to exec. v.p., 1969-72, dir. engring., planning, nuclear and systems ops., 1972-75, v.p. engring. and distbn., 1975-79, sr. v.p., 1979—. Served with Signal Corps U.S. Army, 1942-46. Fellow IEEE, Nat. Acad. Engring. Roman Catholic. Club: Ancient Order of Hibernians. Subspecialty: Electrical engineering. Home: 34 Randlett Park West Newton MA 02165 Office: 800 Boylston St Boston MA 02199

CORSO, JOHN FIERMONTE, psychologist, consultant; b. Oswego, N.Y., Dec. 1, 1919; m. Josephine A. Solazzo, Feb. 8, 1943; children: Gregory Michael, Douglas Jerome, Christine Ann. B.Ed., SUNY-Oswego, 1942; M.A., U. Iowa, Iowa City, 1948, Ph.D., 1950. Lic. psychologist, N.Y. Chief sound and vibration sect. Psychology Br. Army Med. Research lab., Ft. Knox, Ky., 1950-51; chief human factors office Rome Air Devel. Center, Griffiss AFB, Rome, N.Y., 1951-52; prof., dir. human factors research program Pa. State U., 1952-62; prof., dir. dept. psychology St. Louis U., 1962-63; prof., chmn. dept. psychology SUNY-Cortland, 1963-80, Disting. prof., 1973—; engring. psychology cons. HRB-Singer, State College, Pa., 1952-62; staff psychologist U.S. Naval Tng. Device Ctr., Port Washington, N.Y., 1959. Author: The Experimental Psychology of Sensory Behavior, 1967, Aging Sensory Systems and Perception, 1981; contbr. chpts. to books. Served to capt. U.S. Army, 1942-46. NSF grantee, 1955-58; Pa. State U. grantee, 1956-61; Nat. Inst. Neurol. Diseases and Blindness grantee, 1960-63; SUNY grantee, 1965-80; Nat. Inst. Child Health and Human Devel. fellow, 1969-70. Fellow Am. Psychol. Assn., AAAS, Human Factors Soc.; mem. Eastern Psychol. Assn., Midwestern Psychol. Assn., Acoustical Soc. Am., N.Y. State Psychol. Assn., Pa. Psychol. Assn., AAUP, Pa. Acad. Sci., N.Y. Acad. Sci., Internt. Soc. Cybernetic Medicine, Internat. Soc. Audiology, Psychonomic Soc., Sigma Xi, Psi Chi. Roman Catholic. Subspecialties: Sensory processes; Aging. Current work: Research in experimental psychology, especially sensory processes, psychophysics, and perception in older adults; psychoacoustics; audiology; musical acoustics; human factors; research design and administration. Home: Cosmos Hill Rd RD #4 Cortland NY 13045 Office: Department of Psychology State Univeristy of New York Cortland NY 13045

CORWIN, HAROLD GLENN, JR., astronomer; b. Pasadena, Calif., June 5, 1943; s. Harold Glenn and Eleanor Aida (Burbank) C.; m. Kathleen Chaloner Castellini, Dec. 21; 1977. B.A., U. Kans., 1965, M.A., 1967; Ph.D., U. Edinburgh, Scotland, 1981. Research scientist assoc. dept. astronomy U. Tex., Austin, 1971-76; research astronomer dept. astronomy U. Edinburgh, Scotland, 1976-81; research scientist assoc. dept. astronomy U. Tex., Austin, 1981—, cons., 1977-81, lectr., 1982; cons. dept. astronomy UCLA, 1977-81. Author: (with others) Second Reference Catalogue of Bright Galaxies, 1976. Active Austin Civic Chorus. Served to capt. USAF, 1967-71. NSF grantee, 1982. Mem. Internat. Astron. Union, Royal Astron. Soc., Am. Astron. Soc., Astron. Soc. Pacific, AAAS, Sigma Xi. Subspecialty: Optical astronomy. Current work: Galaxy catalogues, galaxy photometry, large-scale structure of the universe. Office: Dept Astronomy U Tex Austin TX 78712

COSCINA, DONALD VICTOR, psychiatric institute administrator, researcher, educator; b. New Britain, Conn., Oct. 11, 1943; emigrated to Can., 1970, naturalized, 1971; s. Victor Joseph and Edna Agnes (Prills) C.; m. Elizabeth Deen Bridgen, June 25, 1966; children: David Victor, Lynn Elizabeth. B.A., U. Vt., 1965; M.A., Bucknell U., 1967; Ph.D., U. Chgo., 1971. Registered psychologist, Ont. Vis. scientist in neurochemistry Clarke Inst. Psychiatry, Toronto, Ont., Can., 1970-71, research scientist in neurochemistry, 1971-77, head biopsychology, 1977—; assoc. prof. psychology U. Toronto, 1977—, assoc. prof. psychiatry, 1980—; mem. initial rev. group in neuropsychology NIMH, Washington, 1982-86. Editor: Anorexia Nervosa, Recent Developments in Research, 1983; editor: Pharmacological Biochem. Behavior, 1973—, Neurosci. and Biobehavior Rev, 1977—, Nutrition and Behavior, 1981—. Recipient Research Fund award Clarke Inst. Psychiatry, 1974; Bucknell Univ. scholar, 1965-67; USPHS trainee U. Chgo., 1967-70. Mem. AAAS, Am. Psychol. Assn., Soc. Neurosci., Eastern Psychol. Assn. Roman Catholic. Subspecialties: Physiological psychology; Neurochemistry. Current work: Neuroanatomical and neurochemical determinants of behavior; abnormal feeding and body weight in humans (e.g., obesity, anorexia nervosa); recovery of function in central nervous system. Home: 3075 Council Ring Rd Mississauga ON Canada L5L 1N7 Office: Clarke Inst Psychiatry 250 College St Toronto ON Canada M5T 1R8

COSMAN, BARD, plastic surgeon, sculptor; b. Bklyn., Nov. 10, 1930; s. Max and Cornelia (Kaps) C.; m. Madeleine Pelner, Sept. 7, 1958; children: Marin, Bard Clifford. A.B., Columbia U., 1952, M.D., 1955. Diplomate: Am. Bd. Plastic Surgery. Intern Roosevelt Hosp., N.Y.C., 1955-59; resident Presbyn. Hosp., N.Y.C., 1959-61; instr. in surgery Columbia U., 1963-67, asst. clin. prof. surgery, 1967-70, assoc. clin. prof. surgery, 1970-73, assoc. prof. surgery, 1973-77, prof. clin. surgery, 1977-82, prof. clin. surgery in anatomy and cell biology, 1982—; pres., exec. bd. N.Y. Physicians Art Assn., 1979—. Abstracts editor, contbr.: articles Cleft Palate Jour; contbr.: Plastic and Reconstructive Surgery Jour; sculptor: Madeleine (first prize AMA Physicians Art Show 1961). Served to lt. comdr. USN, 1961-63. Recipient Curtis Gold Medal Columbia U., 1947, Janeway Prize, 1955. Fellow Am. Assn. Plastic Surgeons; mem. N.Y. Regional Soc. Plastic and Reconstructive Surgery (pres. 1979-80), N.Y. State Med. Soc. (chmn. Sect. on Plastic and Reconstructive Surgery 1973), Am. Cleft Palate Assn., Am. Soc. Plastic and Reconstructive Surgeons, Am. Soc. for Laser Medicine and Surgery (founding). Democrat. Jewish. Subspecialties: Surgery; Laser medicine. Current work: Argon laser treatment of port-tine stain hemangiomas. Home: 32 Knickerbocker Rd Tenafly NJ 07670 Office: Columbia Univ Med School 630 W 168th St New York NY 10032

COSMIDES, GEORGE JAMES, pharmacologist/toxicologist, consultant; b. Pitts., July 23, 1926; s. James and Catherine (Palogaris) C.; m. Nasia Murlas, Sept. 12, 1948; 1 dau., Leda. B.S., U. Pitts., 1952; M.S., Purdue U., 1954, Ph.D. in Pharmacology, 1956. Registered pharmacist, Pa. Sr. scientist Smith Kline & French Labs., Phila., 1956-57; asst. prof. pharmacology U. R.I., Kingston, 1957-59; pharmacologist NIMH, Bethesda, MD., 1959-63; dir. pharmacology-toxicology Nat. Inst. Gen. Med. Scis., Bethesda, 1963-74; dep. dir. div. specialized info. services Nat. Library of Medicine, Bethesda, 1974—; cons. Contbr. articles to profl. jours. Mem. Citizens for Good Govt., Rockville, Md., 1962—; pres. Montgomery County (Md.) PTA, 1963—; mem. parish council St. George's Ch., 1970—. Served with inf. U.S. Army, 1944-46. Decorated Bronze Star; recipient Disting. Alumnus award U. Pitts., 1966; Disting. lectr. AAAS, 1971; others. Mem. Am. Soc. Pharmacology and Exptl. Therapeutics, Soc. Toxicology, Am. Soc. Info. Sci. Subspecialties: Pharmacology; Toxicology (medicine). Current work: Online handling of specialized info. in med. scis.; toxicology info.; database systems; distributed systems and networks; info. systems, storage and retrieval; environ. health/toxicology. Home: 639 Crocus Dr Rockville MD 20850 Office: NIH 8600 Rockville Pike Bethesda MD 20209

COSNER, RAYMOND ROBERT, aeronautical engineer; b. Charleston, W.Va., Dec. 18, 1949; s. Robert Ronald and Winona Harriet (Hinkley) C.; m. Mary Elizabeth Stuesse, May 23, 1979. B.S., Purdue U., 1972, M.S., 1972; Ph.D., Calif. Inst. Tech., 1975. Cons. Analytic Services, Inc., Falls Church, Va., 1972; research asst. Calif. Inst. Tech., Pasadena, 1972-75; sr. engr. propulsion McDonnell Aircraft Co., St. Louis, 1975-80, lead engr. propulsion, 1980—. Area chmn. Caltech Alumni Fund, St. Louis, 1981—. Assoc. fellow AIAA (chmn. coms. 1976-81, outstanding young profl. engr. 1980); mem. Tau Beta Pi, Sigma Xi (assoc.). Club: McDonnell Douglas Railroad (St. Louis). Subspecialties: Aeronautical engineering; Fluid mechanics. Current work: Development of computational fluid dynamics procedures for application to design; development of future fighter aircraft. Home: 12765 Castlebar Dr Creve Coeur MO 63141 Office: McDonnell Aircraft Co Saint Louis MO 63166

COSSINS, EDWIN ALBERT, educational administrator, researcher; b. Havering, Essex, Eng., Feb. 28, 1937; s. Albert Joseph and Elizabeth Henrietta (Brown) C.; m. Lucille Jeannette Salt, Sept. 1, 1962; children: Diane Elizabeth, Carolyn Jane. B.Sc. with 1st class honors, U. London, 1958, Ph.D. in Plant Biochemistry, 1961, D.Sc., 1981. Research assoc. Purdue U., 1961-62; asst. prof. botany and plant biochemistry U. Alta., Edmonton, 1962-65, assoc. prof., 1965-69, prof., 1969—, acting head dept. botany, 1965-66, assoc. dean sci., 1983, McCalla research prof., 1982-83; chmn. grant selection com. for cell biology and genetics NRC Can., 1976-77; invited prof. U. Geneva, Switzerland, 1972-73. Assoc. editor: Can. Jour. Botany, 1969-75; contbr. articles to profl. jours. Recipient Centennial medal Govt. Can., 1967; NRC Can. grantee, 1963—. Fellow Royal Soc. Can. (life); mem. Can. Soc. Plant Physiologists (pres. 1976-77), Am. Soc. Plant Physiologists, Japanese Soc. Plant Physiologists. Club: Royal Glenora (Edmonton). Subspecialties: Biochemistry (biology); Plant physiology (biology). Current work: Folate biochemistry with emphasis on plant systems. Basic research in plant biochemistry; training of graduate students at doctoral level; university teaching; university administration. Office: Dept Botany U. Alberta Edmonton AB T6G 2E9

COSTA, MAX, toxicologist/pharmacologist, researcher; b. Cagliari, Italy, Jan. 10, 1952; came to U.S., 1957, naturalized, 1962; s. Erminio and Anna (Marazzi) C.; m. Elizabeth R. Costa, June 15, 1974. B.S. in Biology, Georgetown U., 1974; Ph.D. in Pharmacology, U. Ariz., 1976. Asst. prof. lab medicine U. Conn., 1977-79; asst. prof. pharmacology U. Tex., Houston, 1980-81, assoc. prof., 1981—; speaker, community cons. toxicol. problems; cons. EPA, AMAX, Inc., Nickel Producers Environ Research Assn. Reviewer profl. jours.; contbr. abstracts, articles and revs. to profl. jours. Recipient Young Envrion. Scientist award NIH, 1977-81; EPA grantee; Nat. Cancer Inst. grantee; Nat. Inst. Environ. Health Scis. grantee. Mem. Am. Soc. Biol. Chemists, Am. Soc. Pharmacology and Exptl. Therapeutics, Am. Soc. Cell Biology, AAAS, N.Y. Acad. Scis., Soc. Toxicology, Am. Assn. Cancer Research. Subspecialties: Biochemistry (biology); Gene actions. Current work: Mechanisms of metal carcinogenesis. Office: Dept Pharmacology U Tex Med Sch Houston TX 77025

COSTANZA, MARY E., physician; b. Quincy, Mass., Feb. 21, 1937; d. Fred P. and Clara (Zottoli) C. A.B. magna cum laude, Radcliffe Coll., 1958; M.A., U. Calif., Berkeley, 1963; M.D., U. Rochester, 1968. Diplomate: Am. Bd. Internal Medicine, 1972. Resident in medicine Tufts-New Eng. Med. Ctr., Boston, 1968-70, fellow in med. oncology, 1970-72, instr. medicine, 1972-74, asst. prof., 1974-79; assoc. prof. medicine U. Mass. Med. Sch., Worcester, 1979—, dir. div. oncology, 1980—. Contbr. articles to profl. jours. Woodrow Wilson fellow, 1961-63; recipient Lange book award, 1968. Fellow ACP; mem. Am. Assn. Clin. Oncologists, Am. Assn. Cancer Research, Am. Fedn. Clin. Research, Am. Assn. Edn. Subspecialties: Internal medicine; Oncology. Current work: Oncology chemotherapy, cancer research, clinical cancer education. Office: 55 Lake Ave N Worcester MA 01605

COSTANZO, RICHARD MICHAEL, physiologist, educator, researcher; b. Bklyn., July 18, 1947; s. William H. and Agatha (Maraventano) C.; m. Linda S. Schupper, July 3, 1971; children: Daniel, Rebecca. B.S. in Biology, SUNY, Stony Brook, 1969; Ph.D. in Physiology, SUNY Upstate Med. Sch., Syracuse, 1975. NIH postdoctoral fellow Rockefeller U., N.Y.C., 1975-77; research asso. dept. physiology N.Y.U. Sch. Medicine, N.Y.C., 1977-78, instr., 1978-79; asst. prof. dept. physiology Med. Coll. Va., Richmond, 1979—. Contbr. articles to profl. jours. Served with USAF Res., 1971-77. Nat. Inst. Neurol. Communicative Disease and Stroke grantee, 1981—. Mem. Physiol. Soc., Soc. for Neurosci., Assn. for Chemoreception Scis., N.Y. Acad. Scis., Sigma Xi. Subspecialty: Neurophysiology. Current work: Research on olfaction.

COSTEA, NICOLAS VINCENT, physician, researcher; b. Bucharest, Romania, Nov. 10, 1927; came to U.S., 1957; s. Nicolas and Florica (Ionescu) C.; m. Ileana Paunescu, Apr. 20, 1973. B.A., Nat. Coll., Bucharest, 1946; M.S., U. Paris, 1949; M.D., 1956. Intern St. Francis

Hosp., N.Y.C., 1956-57; resident L.I. Jewish Hosp., 1957-59; fellow in hematology Tufts U., 1959-62; dir. clinic Pratt Clinic, Boston, 1962-63; clin. investigator VA West Side Med. Ctr., Chgo., 1963-68; chief hematology U. Ill., Chgo., 1968-70, prof. medicine, 1970-72; chief hematology-oncology UCLA-VA Hosp., Sepulveda, 1972—; prof. UCLA, 1972—; vis. prof. Nat. Acad. Scis., 1972. Contbr. numerous chpts., articles to profl. publs. Recipient Lederle award Lederle Industries, 1966. Mem. Am. Soc. Hematology, Am. Soc. Immunology, N.Y. Acad. Scis., Western Soc. Clin. Research. Subspecialties: Immunology (agriculture); Hematology. Current work: Antibody mediated hemolysis; biochemistry of immune competent cells. Home: 3651 Terrace View Encino CA 91436 Office: VA Med Center UCLA Sepulveda CA 91343

COSTELLO, WALTER JAMES, research scientist, educator; b. San Antonio, Tex., Nov. 5, 1945; s. James Edward and Ella Lee (Garvin) C.; m. Carol A. Jones, Nov. 19, 1966 (div.); 1 son, Trevor Durant; m. Mary D. Boone, June 27, 1981; 1 dau., Tirzah Mary. B.A., Trinity U., San Antonio, 1966, M.S., 1969; Ph.D., Boston U., 1978; postgrad., Yale U., 1978-81. Trench supr. Am. Expdn. to Hebron, Jordan, 1966; asst. prof. biomedical scis. Ohio U., Athens, 1981—. Contbr. articles to profl. jours. Cub den leader Boy Scouts Am., 1979-81; judge Athens Sci. Fairs. Served to lt. U.S. Army, 1970-72; Served to maj. USAR, 1972. Muscular Dystrophy Assn. fellow, 1978-79; NIH fellow, 1979-81; Steps-toward-Independence fellow, 1982; Muscular Dystrophy Assn. grantee, 1983—. Mem. AAAS, Am. Soc. Zoologists, Soc. Neuroscience, N.Y. Acad. Sci., Sigma Xi. Republican. Subspecialties: Neurobiology; Developmental biology. Current work: Neurogenetics; development of neuromuscular systems; research in developmental neurobiology, education in biological sciences. Office: Zoological and Biomed Sciences Coll of Osteopathic Medicine Ohio U Athens OH 45701

COTA, HAROLD MAURICE, environmental engineer, educator; b. San Diego, Apr. 16, 1936; s. Florencio Moisa and Doris Alberta (Wright) C.; m. Judith Jane Pritchard, June 30, 1959; children: Rebecca Jo, Cynthia Jane, Michael Maurice. Student, San Diego State Coll., 1954-56; B.S., U. Calif.-Berkeley, 1959; M.S., Northwestern U., 1960; Ph.D., U. Okla., 1966. Diplomate: Am. Acad. Environ. Engrs.; registered profl. engr., Calif. Research engr. Lockheed Missle & Space Co., Sunnyvale, Calif., 1960-62; grad. and research asst. U. Okla., Norman, 1962-66; prof. environ. engring. Calif. Poly State U., San Luis Obispo, Calif., 1966—, dir. air pollution control tng. program, 1969—; dir. Environ. Research Found., 1979—. Mem. Calif. Regional Water Quality Control Bd., 1970—; active Boy Scouts Am. EPA tng. grantee, 1969-82. Mem. Air Pollution Control Assn., Am. Inst. Chem. Engrs., Am. Indsl. Hygiene Assn. Methodist. Subspecialties: Environmental engineering; Chemical engineering. Current work: Modeling environmental problems - air, water, noise. Office: Civil and Environmental Engring Dept Calif Poly State U San Luis Obispo CA 93407

COTLER, SHERWIN BARRY, clin. psychologist; b. Chgo., Mar. 2, 1941; s. Leo and Bessie (Lustig) C.; m.; children: Stacy R., Lisa M. B.A., Calif. State U., 1964, M.A., 1966; Ph.D., Wash. State U., 1970. Lic. psychologist, Calif., Wash. Community mental health psychologist Los Angeles County, 1970-75; clin. psychologist Family Guidance Center, Buena Park, Calif., 1975-77; instr. Calif. State U., Long Beach and Fullerton, 1970-81; psychologist Western State Hosp., Steilacoom, Wash., 1981—; pvt. practice psychology, Calif. and Wash., 1971—; cons. in field. Author: (with Julio Gerra) Assertion Training, 1976; author: comml. cassette Self-Relaxation Training, 1976. USPHS fellow, 1967-68. Mem. Am. Psychol. Assn., Western Psychol. Assn., Long Beach Psychol. Assn. (sec. 1972-73), Sigma Xi, Phi Kappa Phi. Subspecialties: Behavioral psychology; Cognition. Current work: Stress management; psychopathology; professional training; legal offenders; anxiety and depression; human factors. Address: 2838 Madrona Beach Rd NW Olympia WA 98502

COTT, DONALD W(ING), engineering analyst, power systems consultant; b. San Francisco, July 17, 1941; s. Bill and Eula Vesta (Sullivan) C.; m. Sandra Bales, Jan. 20, 1963; children: Craig, Nathan, Jamalea, Kamalea. B.S. in Mech. Engring, Okla. State U., 1964, M.S., 1966; Ph.D. in Aerospace Engring, U. Tenn., 1969. Asst. prof. N.C. State U., Raleigh, 1969-72; assoc. prof. Clemson (S.C.) U., 1972-75; research scientist Reynolds Metals Co., Sheffield, Ala., 1975-77; sr. engring. analyst Mountain States Energy, Inc., Butte, Mont., 1977—; gen. mgr. High County Engring., Inc., Butte, 1981—; also dir.; cons. Reynolds Metals Co., 1978, M. Jones & Assocs., Arlington, Va., 1977, R.S. Noonan, Inc., Greenville, S.C., 1975; Army Missile Command Redstone Arsenal, Ala., 1982—. Recipient award outstanding faculty publ. of year N.C. State U., 1971; NDEA grad. fellow, 1964-66. Mem. ASME, Systems Analysis Com. of ASME, AIAA. Baha'i. Subspecialties: Plasma; Numerical analysis. Current work: Magnetohydrodynamic energy conversion, applications to utility power generation and military pulsed systems, numerical analysis of MHD generator physics, analysis of advanced thermodynamic power cycles. Home: Route 1 150 Moose Creek Rd Butte MT 59701 Office: Mountain States Energy Inc Box 3767 Butte MT 59702

COTTER, DAVID JAMES, biology educator; b. Glens Falls, N.Y., July 24, 1932; s. Harold Francis and Helen Marie (Maher) C.; m. Joann Wood, Aug. 21, 1953; children: David Barry, Mark Wood, John Walter. B.S., U. Ala., Tuscaloosa, 1952, A.B., 1953, M.S., 1955; Ph.D., Emory U., 1958. Research assoc. Emory U., Atlanta, 1957-58; asst. prof. U. Montevallo, Ala., 1958-62, assoc. prof., 1962-66; research assoc. U. Ga., Savanna River Ecology Lab., S.C., 1966, 67, 68; prof., chmn. biology dept. Ga. Coll., Milledgeville, 1966—; staff biologist Office Biol. Edn., Washington, 1969-70. Contbr. articles in field to profl. jours. Panelist Ga. Council Arts and Humanities, Atlanta, 1979-81. Recipient Art Service award Milledgeville - Baldwin County Allied Arts Commn., 1981; grantee NSF, AEC, others, 1958—. Mem. Ecol. Soc. Am., Assn. Southeastern Biologists (v.p. 1973, pres. 1974), Ga. Acad. Sci. (counselor-at-large 1977-80), Ala. Acad. Sci. (v.p. 1966, editor newsletter 1962-66), Guilde Natural Sci. Illustrators, Ga. Climatologists, Ga. Conservancy. Roman Catholic. Subspecialties: Ecology; Evolutionary biology. Current work: Test banks and manuscript reviews for biology textbooks. Home: 1652 Pine Valley Rd Milledgeville GA 31061 Office: Biology Dept Milledgeville GA 31061

COTTER, WILLIAM BRYAN, JR., anatomist, educator, researcher; b. Hartford, Conn., May 8, 1926; s. William Bryan and Johanna (Sumeriva) C.; m. Alice Wadsworth Wendt, Sept. 11, 1948; children: William B., Daniel T., Ellen W., Elizabeth W., Sarah D., Andrew R. Student, New Britain (Conn.) Tchrs. Coll., 1946-47; B.A. with honors, Wesleyan U., 1959, M.A., 1951; Ph.D., Yale U., 1956. Postdoctoral teaching fellow Med. Coll. S.C., 1959-60; Asst. prof. biology Coll. Charleston, S.C., 1955-56, asst. prof., 1957-59; asst. prof. biology Conn. Wesleyan U., 1956-57; asst. prof. anatomy U. Ky., Lexington, 1960-66, assoc. prof., 1967-73, prof., 1974—. Served in USAAF, 1943-45. Mem. Am. Assn. Anatomists, Am. Soc. Zoologists, Soc. Study of Evolution, Genetics Soc. Am., Am. Soc. Human Genetics, Sigma Xi. Subspecialties: Evolutionary biology; Anatomy and embryology. Current work: Physiological genetics; specifically, behavioral modification, courtship behavior, melanism, peiotropy, polymorphic population maintenance. Home: 1005 Lane Allen Rd Lexington KY 40504 Office: Rose St The Medical Center Lexington KY 40536

COTTON, EILEEN GIUFFRE, educator; b. Oakland, Calif., Apr. 23, 1947; d. Leonard and Helen Marie (Weiss) Giuffre; m. Chester Christie Cotton, Apr. 3, 1971. B.A., Calif. State U. - Hayward, 1968; M.A., Calif. State U.-Sacramento, 1976; Ph.D., U. Md., 1979. Tchr. Hayward (Calif.) Unified Sch. Dist., 1969-70, Rescue Union Sch. Dist., Calif., 1970-72, Chico (Calif.) Unified Sch. Dist., 1972-76; program dir. Crownsville Hosp. Ctr., Md., 1976-77; grad. asst. U. Md., College Park, 1977-78; assoc. prof. dept. edn. Calif. State U., Chico, 1978—. Author: A: Guide to Re-Entry, 1977; contbr. articles to profl. jours. Treas. Women's Faculty Assn., 1982. Mem. Am. Psychol. Assn., Internat. Assn. Applied Psychology, Internat. Reading Assn., Am. Ednl. Research Assn., Assn. for Supervision and Curriculum Devel., Calif. Reading Assn., Kappa Delta Pi, Phi Delta Kappa. Subspecialties: Learning; Educational psychology. Current work: Psychology learning. Office: Calif State U Dept Edn Chico CA 95929

COTTON, FRANK ALBERT, chemist, educator; b. Phila., Apr. 9, 1930; s. Albert and Helen (Taylor) C.; m. Diane Dornacher, June 13, 1959; children: Jennifer Helen, Jane Myrna. Student, Drexel Inst. Tech., 1947-49; A.B., Temple U., 1951, D.Sc. (hon.), 1963; Ph.D., Harvard U., 1955; Dr. rer. Nat. (hon.), Bielefeld U., 1979, D.Sc., Columbia U., 1980, Northwestern U., 1981, U. Bordeaux, 1981, St. Joseph's U., 1982, U. Louis Pasteur, 1982, U. Valencia, 1983, Kenyon Coll., 1983, Technion-Israel Inst. Tech., 1983. Instr. chemistry M.I.T., 1955-57, asst. prof., 1957-60, assoc. prof., 1960-61, prof., 1961-71; Robert A. Welch Distinguished prof. chemistry Tex. A&M U., 1971—; Cons. Am. Cyanamid, Stamford, Conn., 1958-67, Union Carbide, N.Y.C., 1964—. Author: (with G. Wilkinson) Advanced Inorganic Chemistry, 4th edit, 1980, Basic Inorganic Chemistry, 1976, Chemical Applications of Group Theory, 2d edit, 1970, (with L. Lynch and C. Darlington) Chemistry, An Investigative Approach; Editor: Progress in Inorganic Chemistry, Vols. 1-10, 1959-68, Inorganic Syntheses, Vol. 13, 1971, (with L.M. Jackman) Dynamic Nuclear Magnetic Resonance Spectroscopy, (with R.A. Walton) Multiple Bonds between Metal Atoms. Recipient Michelson-Morley award Case Western Res. U., 1980, Nat. Medal of Sci., 1982. Mem. Nat Acad. Scis., Am. Soc. Biol. Chemists, Am. Chem. Soc. (awards 1962, 74, Baekeand medal N.J. sect. 1963, Nichols medal N.Y. sect. 1975, Pauling medal Oreg. and Puget Sound sect. 1976, Kirkwood medal N.Y. sect. 1978, Gibbs medal Chgo. sect. 1980), Am. Acad. Arts and Scis., Royal Danish Acad. Scis. and Letters (hon.), N.Y. Acad. Scis. (hon. life), Göttingen Acad. Scis. (corr.), Royal Soc. Chemistry (hon.), Societa Chimica Italiana (hon.). Subspecialties: Inorganic chemistry; X-ray crystallography. Home: Route 2 Box 285 Bryan TX 77801 Office: Tex A and M U College Station TX 77843

COTTON, JOHN L., psychology educator; b. Beloit, Wis., May 11, 1952; s. Alfred L. and Louise A. (Parker) C.; m. Laura Molitor, July 19, 1975. B.A., U. Wis., 1974; M.A., U. Iowa, 1976, Ph.D., 1979. Temporary asst. prof. Iowa State U., Ames, 1978-80; vis. asst. prof. U. Iowa, Iowa City, 1980-81; asst. prof. psychology, organizational behavior Purdue U., West Lafayette, Ind., 1981—. Contbr. articles to profl. jours. U. Iowa teaching/research fellow, 1975-78. Mem. Am. Psychol. Assn., Acad. Mgmt., Soc. Personality and Social Psychology. Current work: Cognitive inferences in organizations, temperature and aggression, organization turnover, realistic job previews, philosophy of science. Home: 941 Southland Dr N Lafayette IN 47905 Office: Purdue U 710 Krannert West Lafayette IN 47907

COTTRELL, STEPHEN F., biology educator; b. Trenton, N.J., Apr. 17, 1943; s. Russell and Frances (Denton) C.; m. Jean E. Doublsky, June 29, 1971. B.A., Wilkes Coll., 1965; M.S., Rutgers U., 1969, Ph.D., 1970. Postdoctoral fellow U. Chgo., 1970-72, Ind. U., Bloomington, 1972-73; research assoc. NASA-Ames Research Ctr., Moffett Field, Calif., 1973-75; asst. prof. biology Bklyn. Coll., CUNY, 1975-80, assoc. prof., 1981-82, prof., 1983—. Contbr. articles to profl. jours. NASA grantee, 1976, 82; NIH grantee, 1977—; Dept. Energy grantee, 1980-82; U.S. Dept. Agr., 1981-83. Mem. AAAs, Gerontol. Soc. Am., AAUP. Subspecialties: Cell biology; Molecular biology. Current work: Research areas include biological mechanisms of agine, ethanol production from biomass, mitochondrial autonomy and biogenesis, organelle formation during the cell cycle. Office: Dept Biology Bklyn Coll CUNY Brooklyn NY 11210

COUCHMAN, PETER ROBERT, polymer scientist, materials scientist; b. Barnet, Hertfordshire, Eng., Jan. 5, 1947; came to U.S., 1969; s. Clifford Robert and Pamela Joyce (Izzett) C.; m. Signe Mary Lund, July 3, 1977; 1 son, David Peter. B.Sc., U. Surrey, Eng., 1969; M.S., U. Va., 1972, Ph.D., 1976. Postdoctoral fellow U. Bristol, Eng., 1973; postdoctoral fellow U. Mass., Amherst, 1976, 77; asst. prof. polymer sci. Case Western Res. U., 1978; asst. prof. mechanics and materials sci. Rutgers U., 1978-81, assoc. prof., 1981—. Contbr. articles to sci. jours. Mem. Am. Phys. Soc., Am. Chem. Soc., N.Am. Thermal Analysis Soc., Sigma Xi. Subspecialties: Polymers; Polymer physics. Current work: Structure-property relations; transition and phase behavior of polymeric materials; multicomponent polymeric materials; surface tension of polymers and their solutions. Office: Dept Mechanics and Materials Sci Rutgers U PO Box 409 Piscataway NJ 08854

COULL, BRUCE CHARLES, marine ecology educator; b. N.Y.C., Sept. 16, 1942; s. Charles and Ida Louise (Lind) C.; m. Judith Mapletoft, June 3, 1967; children: Brent Andrew, Robin Lind. B.S., Moravian Coll., 1964; M.S., Lehigh U., 1966, Ph.D., 1968. Postdoctoral fellow Duke U., Beaufort, N.C., 1968-70; asst. prof. Clark U., Worcester, Mass., 1970-73; assoc. prof. marine ecology U. S.C.-Columbia, 1973-78, prof., 1978—, dir., 1982—; cons. Mediterranean Sorting Ctr., Khereddine, Tunisia, 1970, Tex. Instruments, Inc., 1976-78, Roy F. Weston, Inc., 1979-80, U.S. EPA, 1980-81. Editor: Ecology of Marine Benthos, 1977, (with Tenore) Marine Benthic Dynamics, 1980; Contbr. articles to profl. jours. NSF grantee, 1972—; Fulbright-Hays research scholar, 1981; recipient Russell research excellence award U. S.C., 1982. Mem. Internat. Assn. Meiobenthologists (chmn. 1974-76), Am. Microscopical Soc. (exec. com. 1978-81), Am. Soc. Zoologists, Ecol. Soc. Am., Sigma Xi. Methodist. Subspecialties: Ecology; Oceanography. Current work: Marine benthic ecology research, particularly ecology and role of meiobenthos. Home: 2856 Stepp Dr Columbia SC 29204 Office: U SC Marine Sci Program Columbia SC 29208

COULSON, PATRICIA BUNKER, medical biology researcher, educator; b. Kankakee, Ill., Apr. 27, 1942; d. Francis and Whilamine (Kammann) Bunker; m. James H. Coulson, Jan. 29, 1965; children: Christina, Pamela. Ph.D., U. Ill., 1970. Asst. prof. dept. zoology U. Tenn., Knoxville, 1972-78; assoc. prof. E. Tenn. State U. Coll. Medicine, Johson City, 1978-81; assoc. prof. depts. ob.-gyn. and med. biology Coll. Medicine, U. Tenn., Knoxville, 1981—. Mem. Am.Soc. Cell Biologists, Am. Soc. Endocrinology, Soc. Study of Reprodn., Am. Assn. Clin. Chemistry. Subspecialties: Reproductive biology; Physiology (biology). Current work: Reproductive endocrinology; mechanism of hormone action. Home: 7417 Sheffield Dr Knoxville TN 37919 Office: Room 216-R 1924 Alcoa Hwy Knoxville TN 37920

COULTER, PHILIP WYLIE, physicist; b. Phenix City, Ala., Apr. 19, 1938; s. Leonard Alton and Winslow Lanae (Gullatt) C.; m. Peggy Mullins, Nov. 23, 1960; children: Pippa Elayne, Kathryn Alicia. B.S., U. Ala., 1959, M.S., 1961; Ph.D., Stanford U., 1965. Research assoc. U. Mich., 1965-66; research assoc. U. Calif., Irvine, 1967-68, asst. prof., 1967-71; assoc. prof. U. Ala., University, 1971-76, prof. dept. physics and astronomy, 1976—, dept. chmn., 1982—; cons. Oak Ridge Nat. Lab. Contbr. articles to profl. jours. Mem. Am. Phys. Soc., Sigma Xi. Subspecialties: Atomic and molecular physics; Theoretical physics. Current work: Electron-atom scattering; optical model and close-coupling calculations, electron-atom scattering in laser field. Office: Dept Physics and Astronomy U Ala University AL 35486

COUNCIL, EDWARD LATIMER, software engr.; b. Princeton, N.J., Mar. 18, 1956; s. Harold Edward and Sarah Latimer (Skinner) C. B.S.E.E., Bradley U., 1978, postgrad., 1979-80. Assoc. engr. Midwest Engring., Creve Coeur, Ill., 1981; software engr. Motorola Microsystems, Phoenix, 1981—. Sound technician, tech. dir. sound Scottsdale Community Theatre, 1981—. Mem. IEEE, Assn. Computing Machinery. Club: Motorola Computer (v.p.). Subspecialties: Software engineering; Programming languages. Current work: Software quality assurance, software engineering; programming language development; microprocessor applications.

COUNSELMAN, CHARLES CLAUDE, III, planetary scientist, corporation executive; b. Balt., Apr. 27, 1943; s. Charles Claude, Jr. and Catherine Louise (Roloson) C.; m. Eleanor Frey, June 25, 1966; children: Catherine Marie, Charles Boyd. B.S.E.E., MIT, 1964, M.S.E.E., 1965, Ph.D. in Instrumentation, 1969. Asst. prof. planetary sci. MIT, 1969-74, assoc. prof., 1974-82, prof., 1982—. Contbr. articles sci. and tech. jours. Recipient medal for exceptional sci. achievement NASA, 1980, Carl Pulfrich prize, 1983. Mem. Internat. Assn. Geodesy, Internat. Astron. Union, Am. Geophys. Union, IEEE. Subspecialties: Radio and microwave astronomy; Planetary science. Current work: Geodesy and astrometry via radio interferometry. Inventor macrometer interferometric surveyor. Home: 123 Radcliffe Rd Belmont MA 02178 Office: Massachusetts Institute of Technology Room 54 620 Cambridge MA 02139

COUNSELMAN, CLARENCE JAMES, agricultural research corporation administrator; b. West Palm Beach, Fla., July 4, 1925; s. Clifford and Victoria C.; m. Marion Helseth, July 9, 1949; children: Michael, Jenine, Lynn C. Garriott, Steven. B.S., Auburn U., 1952, M.S., 1953. Prodn. mgr. Big Springs Hatchery, Albertville, Ala., 1953-55; project leader State of Fla., Vero Beach, 1955-57; asst. mgr. Mobay (then Vero Beach Labs.), Vero Beach, 1957-63; mgr. CIBA-GEIGY Corp., Vero Beach, 1963—. Contbr. to profl. confs. Served with USN, 1942-46. Decorated Purple Heart. Mem. Am. Inst. Biol. Sci., Am. Phytopath. Soc., Am. Soc. Photogrammetry, Entomol. Soc. Am. (cert. entomologist), Fla. Hort. Soc., Internat. Platform Assn., N.Y. Acad. Scis., Soc. Nematology, Phi Kappa Phi. Democrat. Subspecialties: Biochemistry (biology); Polymer chemistry. Current work: Administration and research. Patentee chlordimeform with organophosphate compounds. Office: PO Box 1090 Vero Beach FL 32960

COUNTS, WAYNE BOYD, chemist, educator, cons.; b. Prosperity, S.C., Oct. 27, 1936; s. Clarance Boyd and Nell (Long) C.; m. Mary Grace, Sept. 24, 1962; children: Wayne Boyd, Alicia Anne, Cynthia Lynn. B.S., Furman U., 1958; Ph.D., U. N.C., Chapel Hill, 1963. NIH fellow, Bethesda, Md., 1966-69; prof. Lincoln Meml. U., Harrogate, Tenn., 1969; assoc. prof. chemistry Ga. Southwestern Coll., 1969-79, prof., 1979—. Mem. Am. Chem. Soc., Sigma Xi. Baptist. Subspecialties: Organic chemistry; Synthetic chemistry. Current work: Synthesis of hetrocyclic and novel organic compounds. Home: 402 Sharon Dr Americus GA 31709 Office: Ga Southwestern Coll Americus GA 31709

COUPER, JAMES RILEY, chemical engineer, educator; b. St. Louis, Dec. 10, 1925; s. James G. and Annetta (Riley) C.; m. Fanny D. Collins, Sept. 5, 1953 (div.); children: Geoffrey, Kathleen; m. Maribelle Wyton, Aug. 12, 1979. B.S.Ch.E., Washington U., St. Louis, 1949, M.S. Ch.E., 1950, D.Sc., 1957. Registered profl. engr., Mo., Ark. Research engr. Mo. Portland Cement, St. Louis, 1951-52; sr. engr. Organic Chem. Engring. dept. Monsanto, St. Louis, 1952-58; prodn. supr. J.F. Queeny plant, St. Louis, 1958-59; assoc. prof. U. Ark., Fayetteville, 1959-65, prof. dept. chem. engring., 1965—, chmn. dept., 1968-79; also cons. Contbr. articles to profl. pubs. Served to lt. USNR, 1944-62. Mem. Am. Inst. Chem. Engrs., Am. Chem. Soc., Am. Soc. Engring. Edn., Soc. Rheology, Tau Beta Pi, Omega Chi Epsilon, Alpha Chi Sigma, Sigma Chi. Episcopalian. Subspecialty: Chemical engineering. Current work: Process economics and process design, polymer rheology. Office: U Ark E-227 Dept Chem Engring Fayetteville AR 72701

COURANT, ERNEST DAVID, physicist; b. Goettingen, Germany, Mar. 26, 1920; came to U.S., 1934, naturalized, 1940; s. Richard and Nina (Runge) C.; m. Sara Paul, Dec. 9, 1944; children: Paul N., Carl R. B.A., Swarthmore Coll., 1940; M.S., U. Rochester, 1942, Ph.D., 1943; M.A. (hon.), Yale U., 1962. Scientist Atomic Energy Project, Montreal, Que., Can., 1943-46; research asso. physics Cornell U., 1946-48; mem. staff Brookhaven Nat. Lab., 1947—, sr. physicist, 1960—; Brookhaven prof. physics Yale U., 1962-67; vis. prof. Yale, 1961-62; prof. physics and engring. State U. N.Y. at Stony Brook, 1967—; vis. asst. prof. Princeton, 1950-51; cons. Gen. Atomic div. Gen. Dynamics Corp., 1958-59; vis. physicist Nat. Accelerator Lab., 1968-69. Fulbright research fellow Cambridge (Eng.) U., 1956. Fellow Am. Phys. Soc., AAAS; mem. Nat. Acad. Scis., N.Y. Acad. Scis. (Boris Pregel prize 1979). Subspecialties: Particle physics; Theoretical physics. Current work: Dynamics of particle accelerators. Co-originator strong-focusing particle accelerators. Home: 109 Bay Ave Bayport NY 11705

COURNAND, ANDRE F., physiologist; b. Paris, France, Sept. 24, 1895; came to U.S., 1930, naturalized, 1941; s. Jules and Marguerite (Weber) C.; m. Sibylle Blumer (dec. 1959); children: Muriel, Marie-Eve, Marie Claire; m. Ruth Fabian, 1963 (dec. 1973); m. Beatrice Bishop Berle, 1975. B.A., Sorbonne U., Paris, 1913, P.C.B. in Sci, 1914; M.D., U. Paris, 1930; Dr. h.c., U. Strasbourg, 1957, U. Lyon, 1958, U. Brussels, 1959, U. Pisa, 1960, Columbia U., 1965, U. Brazil, 1965, U. Nancy, 1969; D.Sc., U. Birmingham, 1961, Gustavus Adolphus Coll., 1963. Prof. emeritus medicine Coll. Phys. & Surg., Columbia. Served with French Army, 1915-19. Decorated Croix de Guerre (France); recipient Laureate (silver medal), faculty medicine U. Paris; Andrea Retzius silver medal Swedish Soc. Internal Medicine; Lasker award USPHS; winner (with Dr. Dickinson W. Richards and Dr. Werner Forssman) of 1956 Nobel Prize in medicine and physiology; recipient Jiminez Diaz prize, 1970. Fellow Royal Soc. Medicine; mem. Nat. Acad. Scis. U.S.A., de l'Academie Nationale de Medecine (fgn.) (France), Academie Royale de Medecine de Belgique, Am. Physiol. Soc., Assn. Am. Physicians, Brit. Cardiac Soc., Swedish Soc., Internal Medicine, Soc. Medicale Hopitaux de Paris, Academie des Sciences, Institut de France (fgn. mem.). Clubs: Century Assn., Am. Alpine. Subspecialties: Physiology (medicine). Current work: Since retirement, I have published on history of science and on ethics and psychology of scientists. Home: 142 E 19th St New York NY 10003

COURSEY, BERT MARCEL, chemist, editor; b. Birmingham, Ala., Mar. 27, 1942; s. Paul B. and Velma (Anderson) C.; m. Rebecca J. Davis, June 6, 1965; children: Steven, Matthew, Johnathan. B.S. in Chemistry, U. Ga., 1965, Ph.D., 1970. Chemist Southeastern Rubber

Mfg. Co., Athens, Ga., 1964-67; research chemist Nat. Bur. Standards, Washington, 1972—; vis. scientist Central Bur. for Nuclear Measurements of the European Community, Geel, Belgium, 1983-84. Editor: Internat. Jour. Applied Radiation and Isotopes, 1977—; co-editor: Technetium-99m, 1982; contbr. articles to profl. jours. Served to capt. C.E. U.S. Army, 1969-71. Petroleum Research Fund fellow, 1968. Mem. Am. Chem. Soc., Am. Nuclear Soc. (div. exec. com. 1982-84), AAAS. Democrat. Methodist. Club: Frederick (Md.) Steeple Chasers (pres. 1978-79). Subspecialties: Physical chemistry; Nuclear medicine. Current work: New applications for radioisotopes in nuclear medicine, nuclear physics, cosmogenic and terrestrial dating, as well as in nuclear fuel cycle. Home: 8607 Imagination Ct Walkersville MD 21793 Office: Nat Bur Standards Bldg 245 Room C114 Washington DC 20234

COURTHEOUX, RICHARD JAMES, quantitative methods consultant, lecturer; b. Rochester, N.Y., Mar. 25, 1949; s. David and Lillian (Altman) C.; m. Perri Orenstein, June 20, 1976; children; Suzanne, Karen. B.S. in Engring, Yale U., 1972; M.S. in Computer Sci, Weizmann Inst., Rehovot, Israel, 1975; M.B.A., U. Chgo., 1978. Cons. Kestnbaum & Co., Chgo., 1978—; lectr. Northwestern U., Evanston, Ill., 1980—, DePaul U., Chgo., 1978. Developer, author: Database Mgmt. System, 1980. Budget dir. Hillel Found., U. Chgo., 1976—. Mem. Assn. Computing Machinery, Direct Mktg. Assn., Sierra Club. Jewish. Subspecialty: Database systems. Current work: Database systems; design and analysis of marketing databases; mathematical models of stock option pricing. Home: 1700 E 56th Apt 2009 Chicago IL 60637 Office: Kestnbaum & Co 221 N LaSalle Chicago IL 60601

COURTNEY, JOHN CHARLES, nuclear engineering educator, consultant; b. Washington, June 11, 1938; s. Julian Warner and Marie Loretta (Vernon) C.; m. Peggy Joy Roberts, June 29, 1969. B.S. in Civil Engring, Cath. U. Am., 1960, M.S. in Nuclear Engring, 1962, Ph.D. in Engring, 1965. Registered profl. engr., Calif.; cert. health physicist. Asst. project engr. Allis Chalmers Mfg. Co., Washington, 1962-63; cons. Oak Ridge Nat. Lab., 1965; research officer USAF, Sacramento, 1965-68; supr. radiation shielding Aerojet Nuclear Systems Co., Sacramento, 1968-71; prof. La. State U., Baton Rouge, 1971—; cons. in field. Editor: A Handbook of Radiation Shielding Data, 1976; contbr. articles to profl. jours. Served to 1st lt. USAF, 1965-68. Mem. Am. Nuclear Soc. (tech. achievement award 1980), Health Physics Soc., Air Force Assn., Scientists and Engrs. for Secure Energy. Democrat. Roman Catholic. Subspecialties: Nuclear engineering; Nuclear fission. Current work: Nuclear radiation safety; radiation protection, training and nuclear safety analysis. Office: Nuclear Sci Ctr Baton Rouge LA 70803

COURTNEY, JOHN VINCENT, physician, radiologist; b. Belfast, Ireland, Aug. 4, 1948; came to U.S., 1975; s. John Vincent and Catherine Patricia (Rogers) C.; m. Ann Margaret Kieran, Oct. 5, 1973; 1 dau., Rachel elanie. M.B., B.Ch., BAO, Univ. Coll., Dublin, 1974. Intern St. Lawrence's Hosp., Dublin, 1974-75; resident in radiology U. Chgo., 1976-79, asst. prof. radiology, 1979—. Mem. Chgo. Med. Soc., Ill. Med. Soc., AMA, Radiol. Soc. N. Am. Roman Catholic. Subspecialty: Diagnostic radiology. Current work: Cancer of esophagus, lung cancer, mammography. Office: Dept Radiolog Univ Chicago Chicago IL 60637

COURTNEY, KENNETH RANDALL, basic med. researcher; b. Snohomish, Wash., Dec. 24, 1944; s. Herbert Wesley and Olga Lena (Salvadalena) C.; m. Janet Claire Courtney, June 7, 1966; children: Edward, Mark, Jesse. B.S., Wash. State U., 1967; M.S., U. Wash., 1968, Ph.D., 1974; postgrad., U. Colo. Med. Center, 1974-75. Instr. math. Centralia (Wash.) Coll., 1967-70; research asso. Stanford (Calif.) U. Med. Center, 1975-77; pharmacologist SRI Internat., Menlo Park, Calif., 1976-78; sr. investigator Palo Alto (Calif.) Med. Found., 1978—. Contbr. articles and abstracts to sci. publs. Mem. Biophys. Soc., Soc. for Neurosci., Am. Soc. for Pharmacology and Exptl. Therapeutics, Phi Beta Kappa, Phi Kappa Phi. Subspecialties: Physiology (medicine); Cellular pharmacology. Current work: Local anesthetic and antiarrhythmic drug mechanisms in nerve and heart muscle, sodium channels. Office: Palo Alto Med Found 860 Bryant St Palo Alto CA 94301

COUSE, NANCY LEE, geneticist; b. Syracuse, May 3, 1941; d. Charles Richard and Rosa Elizabeth (Saupe) C.; m. George Albert Desborough, Aug. 10, 1966. B.S., Cornell U., 1962; Ph.D., U. Wis., Madison, 1966. Research assoc. U. Wis., 1966; staff fellow NIH, Bethesda, Md., 1966-67; postdoctoral fellow U. Colo. Med. Ctr., Denver, 1967-69; asst. prof. biol. scis. U. Denver, 1969—; cons. Denver Research Inst. Contbr. articles to profl. jours. NSF grantee, 1979-81; EPA grantee, 1977-79; Research Corp. grantee, 1970. Mem. Am. Genetic Assn., Genetics Soc. Am., AAAS, Am. Inst. Biol. Scis., Am. Soc. Microbiology, Soc. Coll. Sci. Tchrs., Sigma Xi. Subspecialties: Gene actions; Environmental toxicology. Current work: Researcher in physiology of higher fungi, enzymological responses of microbes to environmental chemicals. Patentee in improved grid for use in counting colonies of bacteria present in discrete areas of a spiral deposition pattern. Home: 2164 Zang St Golden CO 80401 Office: Dept Biology U Denver Denver CO 80208

COUSER, WILLIAM GRIFFITH, medical educator, academic administrator, nephrologist; b. Lebanon, N.H., July 11, 1939; s. Thomas Clifford and Winifred Priscilla (Ham) C. B.A., Harvard U., 1961, M.D., 1965; B.M.S., Dartmouth Coll. Med. Sch., 1963. Diplomate: Am. Bd. Internal Medicine. Intern Moffitt Hosp. / U. Calif. Med. Ctr., San Francisco, 1965-66, 1966-67; resident Boston City Hosp., 1969-70; asst. prof. medicine U. Chgo., 1972-73; asst. prof. Boston U., 1973-77, assoc. prof., 1977-82; prof., 1982, U. Wash., 1982—, head div. nephrology, 1982—; mem. sci. adv. bd. Kidney Found. Mass., Boston, 1974-82; mem. research grant com. Nat. Kidney Found., N.Y.C., 1981—; mem. merit rev. bd for nephrology VA, Washington, 1981—; mem. exec. com. Council on Kidney in Cardiovascular Disease, Am. Heart Assn., Dallas, 1982. Contbr. numerous articles, chpts., abstracts to profl. publs.; editorial bd.: Kidney Internat, 1982. Served as capt. U.S. Army, 1967-69; Vietnam. Recipient Research Career Devel. award NIH, 1975-80; Nat. Kidney Found. fellow, 1971; NIH fellow, 1973; grantee, 1974—. Fellow ACP; mem. Am. Soc. Clin. Investigation, Am. Soc. Nephrhology, Internat. Soc. Nephrology, Am. Assn. Pathologists, Western Assn. Physicians. Subspecialties: Internal medicine; Nephrology. Current work: Immunologic mechanisms of kidney disease. Office: U Wash Div Nephrology Box RM 11 Seattle WA 98195

COUTTS, ROBERT LAROY, psychologist; b. Royal, Nebr., Nov. 5, 1928; s. Hazen Coutts and Anita (Morrison) Coutts C.; m. Sue Webb, Feb. 17, 1952 (div. Oct. 1973); children: Robert LaRoy, William C., Candila S., R. Christopher; m. Mary E. Reutinger, Jan. 13, 1974. B.S., Fla. State U., 1951, M.S., 1954, Ph.D., 1962. Asst. prof. guidance Western Ill. U., Macomb, 1962-64; prof. psychology SUNY, Oswego, 1964-66; assoc. prof. psychology Parsons Coll., Fairfield, Iowa, 1966-73; psychologist Mesa County Mental Health Center, Grand Junction, Colo., 1973-78; pvt. practice, Colorado Springs, 1979—. Author: Love and Intimacy: A Psychological Approach, 1973. Mem. Am. Pyschol. Assn., Colo. Psychol. Assn., El Paso County Psychol. Assn. (dir. 1981-83), Phi Delta Kappa. Lodge: Lions. Subspecialty: Neuropsychology. Current work: Obtaining baseline data regarding brain dysfunction; assessment of specific response to psychotropic medications. Home: 3752 Quiet Circle Colorado Springs CO 80917 Office: Colorado Springs Psychol Center 308 W Fillmore St #200 Colorado Springs CO 80907

COVENEY, RAYMOND MARTIN, JR., geology educator, researcher; b. Marlboro, Mass., Oct. 15, 1942; s. Raymond Martin and Rita Marie (Brani) C.; m. Anne Marie Keating, Feb. 22, 1965; children: Christina, Maureen, David. B.S., Tufts U., Medford, Mass., 1964; M.S., U. Mich., 1968, Ph.D., 1972. Teaching fellow U. Mich., Ann Arbor, 1966-71, Rackham predoctoral fellow, 1971-72; exploration geologist N.J. Zinc Co., Hanover, N.Mex., 1968; mine geologist Dickey Exploration Co., Alleghany, Calif., 1969-71; asst. prof. geology U. Mo.-Kansas City, 1972-77, assoc. prof., 1977—. Contbr. articles to profl. jours. Served as lt. (j.g.) USNR, 1964-66. Fellow Geol. Soc. Am.; mem. Geochem. Soc., Soc. Econ. Geologists, Mineral. Soc. Am., Assn. Mo. Geologists, AAAS, Sigma Xi. Subspecialty: Geology. Current work: Geology, geochemistry and mineralogy of ore deposits, especially gold ores and metalliferous black shales of Pennsylvanian age. Home: 5640 Charlotte St Kansas City MO 64110 Office: Dept Geoscis U Missouri 5100 Rockhill Rd Kansas City MO 64110

COVERT, EUGENE EDZARDS, educator, engineer; b. Rapid City, S.D., Feb. 6, 1926; s. Perry and Eda (Edzards) C.; m. Mary Solveig Rutford, Feb. 22, 1946; children: David H., Christine J., Pamela M., Steven P. B.S., U. Minn., 1946, M.S., 1948; Sc.D., MIT, 1958. Registered profl. engr., Mass.; chartered engr., U.K. Preliminary design group USNADS, Johnsville, Pa., 1948-52; mem. staff MIT Aerophysics Lab., 1952—, asso. dir. aerophysics lab., 1963—, asso. prof. aeronautics and astronautics, 1963-68, prof., 1968—; cons. Bolt, Beranek & Newman, Inc., Hercules, Inc., MIT Lincoln Lab., U.S. Army Research Office; chief scientist USAF, 1972-73; mem. panel Naval Aeroballistic Adv. Com., 1965-75; mem., chmn. USAF Sci. Adv. Bd.; chmn. Power, Energetics and Propulsion panel Adv. Group for Aerospace Research and Devel. NATO; dir. Megatech Inc., Billerica, Mass., Sverdrup ARO Inc., Tullahoma, Tenn. Served with USNR, 1943-47. Recipient Exceptional Civilian Service award USAF, 1973, Univ. Educator of Yr. award Am. Soc. Aerospace Edn., 1980, Pub. Service award NASA, 1980. Fellow AIAA, Royal Aero. Soc.; mem. AAAS, N.Y. Acad. Sci., Nat. Acad. Engring., Sigma Xi. Subspecialties: Fluid mechanics; Aeronautical engineering. Current work: Unsteady fluid mechanics, in particular as relates to turbulence, separation characteristics for application to stability, turbines, compressors and turbo pumps. Office: Mass Inst Tech 77 Massachusetts Ave Cambridge MA 02139

COVINGTON, EDWARD ROYALS, chemistry educator, consultant; b. Meridian, Miss., Jan. 6, 1925; s. James Howard and Lolah Lillian (Lucius) C.; m. Dorothy Louise Warnack, Oct. 17, 1944 (div. 1969); 1 son, James Edward; m. Janet Elaine Ferguson, Mar. 22, 1975; 1 son, Andrew Royals. A.B. in Chemistry, Emory U., 1948, M.A., 1951, Ph.D. in Organic Chemistry, 1954. Research chemist, staff scientist, film dept. DuPont Co., Old Hickory, Tenn., 1953-63, Richmond, Va., 1953-63, staff chemist fabrics and finishes dept., Old Hickory, 1968-70; research assoc. Ethyl Corp., Baton Rouge, 1963-64; sr. chemist So. Research Inst., Birmingham, Ala., 1964-68; prof. chemistry U. Tenn., Nashville, 1970-79, Tenn. State U., 1979—; cons. polymer and organic chemistry. Served to 2d lt. USAAF, 1943-45; ETO. Recipient Outstanding Tchr. award U. Tenn. Gen. Alumni Assn., 1973. Republican. Presbyterian. Subspecialties: Organic chemistry; Polymer chemistry. Current work: Teaching organic and polymer chemistry; establishment of polymer research program. Inventor/co-inventor coatings and copolymers. Office: Tennessee State University 10th and Charlotte Sts Nashville TN 37203

COVIT, ANDREW B., internist; b. Mineola, N.Y., Mar. 30, 1954; s. Harold and Gertrude (Lindenbaum) C.; m. Michelle Masarsky, July 3, 1978. B.A. in Biology, Hofstra U., 1975; M.D. magna cum laude, SUNY Downstate Med. Ctr., 1979. Diplomate: Am Bd. Internal Medicine. Intern N.Y. Hosp., Cornell U. Med Ctr., N.Y.C., 1979-80, asst. jr. and sr. physician, 1980-82, fellow in nephrology, 1982-1984. Contbr. articles to sci. publs. Cornell U. Cardiovascular Ctr. Research assoc., 1983-84. Mem. Am. Fedn. Clin. Research, ACP (assoc.), Alpha Omega Alpha. Subspecialties: Nephrology; Cardiology. Current work: Neuro-humoral factors in congestive heart failure; hemodynamics and renal function in hypertension. Home: 445 E. 68th St New York NY 10021 Office: Cornell U Med Ctr Box 205 525 E 68th St New York NY

COWAN, ALAN, pharmacology educator; b. Selkirk, Scotland, July 6, 1942; s. Robert Lindsay and Jean (Anderson) C.; m. Fiona Forsyth, Oct. 11, 1968; children: Christopher, Nicky. B.Sc., U. Glasgow, Scotland, 1964; Ph.D., U. Strathclyde, Glasgow, 1968. Sect. leader dept. pharmacology Reckitt & Colman, Hull, Eng., 1968-76; research asst. prof. dept. pharmacology Temple U. Sch. Medicine, Phila., 1977-79, asst. prof., 1979-82, assoc. prof., 1983—. Mem. Am. Soc. Pharmacology and Exptl. Therapeutics, Brit. Pharmacology Soc. Subspecialties: Pharmacology; Neuropharmacology. Current work: Pharmacology of drugs of abuse; behavioral pharmacology of peptides; thermopharmacology. Office: Dept Pharmacology Temple Univ Sch Medicine 3400 N Broad St Philadelphia PA 19140

COWAN, DARREL SIDNEY, geoscience educator; b. Los Angeles, Feb. 26, 1945; s. Cedric Donald and Dorothy Amelia (Kilman) Cowan S. B.S., Stanford U., 1966, Ph.D., 1972. Geologist Shell Oil Co., Houston, 1971-74; asst. prof. U. Wash., Seattle, 1974-79, assoc. prof., 1979—. Mem. editorial bd.: Geology, 1982—. Fellow Geol. Soc. Am.; mem. Am. Geophys. Union. Subspecialty: Tectonics. Current work: Mesozoic and early Cenozoic tectonic evolution of western Cordillera; tectonic processes and structural styles in subduction zones; origin of chaotic melanges. Office: Dept Geol Scis U Wash Seattle WA 98195

COWAN, JOHN JAMES, astronomy educator; b. Washington, Apr. 3, 1948; s. John Robert and Anna Wise (Vick) C.; m. Linda Demetry Cowan, May 29, 1971. B.A. in physics, George Washington U., 1970; M.S., Case Inst. Tech., Cleve., 1972; Ph.D. in astronomy, U. Md., College Park, 1976. Teaching asst. U. Md., 1971-74; research asst. NASA Goddard Space Flight Center, 1972-75; instr. U. Md., 1975, research asst., 1976; fellow Center for Astrophysics Harvard U., 1976-79; asst. prof. astronomy U. Okla., 1979—. Contbr. articles to profl. jours. Am. Astron. Soc.-NASA grantee, 1980. Mem. Am. Astron. Soc., Internat. Astron. Union, Sigma Xi, Phi Beta Kappa, Sigma Pi Sigma, Pi Mu Epsilon. Subspecialties: Theoretical astrophysics; Planetary science. Current work: Theoretical astrophysics; stellar evolution and nucleosynthesis, computer modeling, cometary physics; astron. observations; radio galaxies, optical spectropotometry and spectroscopy. Home: 2813 Meadow Ave Norman OK 73069 Office: Department of Physics and Astronomy Norman OK 73019

COWAN, ROBERT DUANE, research physicist; b. Lincoln, Nebr., Nov. 24, 1919; s. Ralph Ellis and Florence Athey (Eller) C.; m. Dorothy Mabel Martinson, July 6, 1944; children: Nancy Jean Cowan Lemons, Charles Eller, Gerald Stanley, Marjorie Sue, Cowan Larson. B.A., Friends U., Wichita, Kans., 1942; Ph.D., Johns Hopkins U., 1946, U. Lund, Sweden, 1982. NRC fellow U. Chgo., 1946-47, research assoc., 1947-48; prof. physics Friends U., 1948-51; staff mem. Los Alamos Nat. Lab., 1951-82, fellow, 1982—; adj. prof. U. N.Mex., 1958, 64, 72; vis. prof. Purdue U., 1971; Fulbright lectr., Lima, Peru, 1958-59; cons. Appleton Lab., Eng., 1977, Internat. Astron. Union, 1968—. Author: The Theory of Atomic Structure and Spectra, 1981; contbr. articles profl. jours. Fellow Am. Phys. Soc., Optical Soc. Am. Subspecialties: Atomic and molecular physics; Theoretical physics. Current work: Theory of atomic structure and spectra, computer calculation of atomic spectra. Home: 4493 Trinity Dr Los Alamos NM 87544 Office: Los Alamos National Laboratory PO Box 1663 Los Alamos NM 87545

COWAN, WILLIAM MAXWELL, neurobiologist; b. Johannesburg, South Africa, Sept. 27, 1931; s. Adam and Jessie Sloan (Maxwell) C.; m. Margaret Sherlock, Mar. 31, 1956; children: Ruth Cowan Eadon, Stephen Maxwell, David Maxwell. B.Sc., Witwatersrand U., Johannesburg, 1951, B.Sc. (Hons.), 1952; D.Phil., Oxford U., 1956, B.M., B.Ch., 1958, M.A., 1959. From demonstrator to lectr. anatomy Oxford U., 1953-66; fellow Pembroke Coll., 1958-66; vis. prof. anatomy Washington U. Med. Sch., St. Louis, 1965-66, prof., chmn. dept., 1968-80; asso. prof. U. Wis. Med. Sch., Madison, 1966-68; research prof., dir. Weingart Lab. Devel. Neurobiology, Salk Inst. Biol. Studies, La Jolla, Calif., 1980; mem. Inst. Medicine, Nat. Acad. Scis., 1978; fgn. asso. Nat. Acad. Scis., 1981. Editor-in-chief: Jour. Neurosci.; editor: Am. Revs. Neurosci. Fellow Am. Acad. Arts and Scis., Royal Soc. (London); mem. Internat. Brain Research Orgn. (exec. council), AAAS, Anat. Soc. Gt. Britain and Ireland, Royal Micros. Soc., Am. Assn. Anatomists, Soc. Neurosci. (pres. 1977-78), Sigma Xi, Alpha Omega Alpha. Subspecialty: Neurobiology. Current work: The development of the brain and the organization of the visual system. Home: 1230 Avocet Ct Cardiff CA 92007 Office: Salk Inst PO Box 85800 San Diego CA 92138

COWBURN, DAVID ALAN, scientist; b. Sale, Cheshire, Eng., July 13, 1945; came to U.S., 1970; s. Thomas Allen and Grace Muriel (Seagrief) C.; m. Elizabeth Stoner, Sept. 25, 1977; children: Adam, Leah. B.Sc. with honors, U. Manchester, 1965; Ph.D., U. London, 1970, D.Sc., 1981. Research assoc. Coll. Phys. and Surgs., Columbia U., N.Y.C., 1970-73; asst. prof. Rockefeller U., N.Y.C., 1973-79, assoc. prof., 1979—. Contbr. articles to profl. jours. U.K. Ministry Edn. state scholar Manchester U., 1962-65; NIH research grantee, 1977—; Am. Heart Assn. research grantee, 1982-84. Mem. Am. Chem. Soc., Am. Soc. Biol. Chemistry, Soc. Magnetic Resonance in Medicine. Subspecialties: Nuclear magnetic resonance (chemistry); Nuclear magnetic resonance (biotechnology). Current work: The application of spectroscopic techniques, particularly nuclear magnetic resonance, to studies of biomolecular confirmation, intracellular communication and other biomedical problems. Home: 500 E 63d St New York NY 10021 Office: Rockefeller U 1230 York Ave New York NY 10021

COWETT, RICHARD MICHAEL, pediatrician, educator; b. N.Y.C., Sept. 20, 1942; s. Allen Abraham and Sylvia (Kazin) C.; m. Katherine Manz, June 25, 1966; children: Beth Ellen, Allison Ann, Allen Manz. B.A., Lawrence Coll., 1964; M.D., U. Cin., 1968; M.A. ad eundum, Brown U., 1982. Diplomate: Am. Bd. Pediatrics. Intern Cleve. Met. Gen. Hosp., 1968-69, resident, 1969-70; resident in medicine Children's Hosp. Med. Ctr., Boston, 1970-71; teaching fellow in pediatrics Harvard Med. Sch., 1970-71; fellow in neonatology Women and Infants Hosp., Providence, 1973-76; assoc. physician, 1976—, physician in charge, 1981—; asst. prof. pediatrics Brown U., Providence, 1976-81, assoc. prof., 1981—; assoc. physician R.I. Hosp., Providence, 1976—; Mem. adv. council Gifted Student Program, Providence public schs., 1981—. Contbr. chpts. to books, articles to profl. jours. Served to maj. USAF, 1971-73. NIH fellow, 1974-76; Research Career Devel. awardee, 1980—. Fellow Am. Coll. Nutrition, Am. Fedn. for Clin. Research; mem. Am. Acad. Pediatrics, R.I. Acad. Pediatrics, Soc. for Pediatric Research, Am. Inst. Nutrition. Jewish. Subspecialties: Pediatrics; Neonatology. Current work: Evaluation of carbohydrate homeostasis in the perinatal-neonatal period using stable isotopes. Evaluation of glucose kinetics in the pregnant woman and her newborn to determine how the pregnant woman changes her metabolic milieu during pregnancy to provide for the fetus and subsequently how the newborn establishes maturation of glucose homeostasis. Home: 125 Woodbury St Providence RI 02908 Office: Dept Pediatrics Women and Infants Hosp RI 50 Maude St Providence RI 02908

COWIE, LENNOX LAUCHLAN, educator; b. Jedburgh, Scotland, Oct. 18, 1950; came to U.S., 1970, naturalized, 1980; s. James Reid C. B.Sc., Edinburgh U., 1970; postgrad., Glasgow U., 1970-71; Ph.D., Harvard U., 1976. Research assoc. Princeton (N.J.) U., 1976-78, research staff mem., 1978-79, research astronomer, assoc. prof. astrophys. sci., 1979-81; assoc. prof. physics M.I.T., Cambridge, 1980—. Assoc. editor: Astrophys. Jour. Letters; Contbr. articles to profl. jours. Kennedy scholar, 1971-73; Fairchild Disting. scholar, 1980. Mem. Internat. Astron. Union, Royal Astron. Soc., Am. Astron. Soc., Am. Phys. Soc. Subspecialties: Theoretical astrophysics; Optical astronomy. Current work: Gas dynamics of interstellar and intergalactic gas. Office: MIT 6-209 Physic Dept Cambridge MA 02139

COWIN, STEPHEN CORTEEN, biomedical engineering educator, consultant; b. Elmira, N.Y., Oct. 26, 1934; s. William Corteen and Bernice (Reidy) C.; m. Martha Agnes Eisel, Aug. 10, 1956; children: Jennifer Marie, Thomas Burrows. B.S.E. in Civil Engring. (Md. State scholar, Ambrose Howard Carner scholar), Johns Hopkins U., 1956, M.S.C.E. (Univ. fellow), 1958; Ph.D. in Engring. Mechanics, Pa. State U., 1962. Registered profl. engr., La. Prof. mech. engring. Tulane U., 1969-77, prof. mechanics dept. biomed. engring., 1977—, adj. prof. orthopedics, 1978—, prof.-in-charge, 1974-75, chmn. applied math. program, 1975-79; Sci. Research Council Gt. Brit. sr. vis. fellow U. Strathclyde, 1974, 80; vis. research prof. Instituto de Matematica, Estatistica e Ciencia da Computanao, Universidade Estadual de Campinas, Brazil, 1978; participant U.S. Nat. Acad. Scis. interacad. exchange program with Bulgaria, 1983. Editor: (with M. Satake) Continuum Mechanical and Statistical Approaches in the Mechanics of Granular Materials, 1978, Mechanics Applied to the Transport of Granular Materials, 1979, (with M.M. Carroll) The Effects of Voids on Material Deformation, 1976; assoc. editor: Jour. Applied Mechanics, 1974-82, Jour. Biomech. Engring, 1982—; editorial adv. bd.: (with M.M. Carroll) Handbook of Materials, Structures and Mechanics, 1981—, Handbook of Bioengineering, 1981—. Served to capt. AUS, 1957-64. Research grantee NSF, NIH, U.S. Army Research Office, Edward G. Schlieder Found. Fellow AAAS, ASME; mem. Am. Acad. Mechanics, Orthopedic Research Soc., Soc. Rheology, Soc. Natural Philosophy (treas. 1977-79), Soc. Engring. Sci., Math. Assn. Am., N.Y. Acad. Scis., Sigma Xi. Subspecialties: Theoretical and applied mechanics; Biomedical engineering. Current work: Mechanics of materials with microstructure, continuum mechanics, biomechanics, bone remodeling, granular materials. Home: 1545 Exposition Blvd New Orleans LA 70118 Office: Dept Biomedical Engineering Tulane U New Orleans LA 0118

COWLES, JOE RICHARD, biologist, educator; b. Edmonson County, Ky., Oct. 29, 1941; s. Otis and Mannie (Rountree) C.; m. Barbara S. Cowles, June 5, 1965; children: Richard W., Daniel M. B.S., Wstern Ky. U., 1963; M.S., U. Ky., 1965; Ph.D., Oreg. State U., 1968. Postdoctoral assoc. Purdue U., West Lafayette, Ind., 1968-69, U. Ga., Athens, 1969-70; asst. prof. U. Houston, 1970-76, assoc. prof.,

1976-82, prof. biology, 1982—, chmn. dept., 1981—. Contbr. articles to sci. jours. Grantee NSF, Dept. Energy, U.S. Dept. Agr., NASA, 1972—. Mem. Am. Soc. Plant Physiology, AAAS, Sigma Xi. Democrat. Baptist. Subspecialty: Plant physiology (biology).

COWLEY, ANNE PYNE, astronomer, educator; b. Boston, Feb. 25, 1938; d. Charles Crosby and Elizabeth (Brown) Pyne; m. Charles R. Cowley, June 9, 1960; children: David M., James R. B.A., Wellesley Coll., 1959; M.A., U. Mich., 1961, Ph.D., 1963. Research assoc. U. Chgo., 1963-67; research scientist U. Mich., Ann Arbor, 1968-83, prof. astronomy, 1983, Ariz. State U., Tempe, 1983—. Contbr. articles to profl. publs. Mem. Am. Aston. Soc. (past v.p.), Internat. Astron. Union, Astron. Soc. of Pacific. Subspecialty: Spectroscopy. Current work: Research in astronomical spectroscopy, especially of x-ray sources. Office: Dept Physics Arizona State U Tempe AZ 85281

COWPLAND, MICHAEL CHRISTOPHER JOHN, electronic engineer; b. Bexhill, Sussex, Eng., Apr. 23, 1943; s. George Ronald and Marjorie Ann (Plackitt) C.; m. Darlene Ann McDonald, Sept. 14, 1968; children: Paula, Christine. B.Sc.Eng., Imperial Coll., London, 1964; M. Eng., Carleton U., Ottawa, Can., 1968, Ph.D., 1973. Mem. sci. staff Bell No. Research, Ottawa, 1964-68; mgr. integrated circuit design Microsystem Internat., Ottawa, 1968-73; pres., chief exec. officer Mitel Corp., Ottawa, 1973—. Mem. Mensa. Club: Rideau (Ottawa). Subspecialty: Telecommunications management. Holder more than 100 patents. Office: Mitel Corp 350 Legett Dr Ottawa ON Canada K2K 1Y5

COX, ARTHUR NELSON, astrophysicist; b. Van Nuys, Calif., Oct. 12, 1927; s. Arthur Hildreth and Sarah (Nelson) C.; m. Joan Frances Ellis, Oct. 21, 1973; children by previous marriage: Bryan, Kay, Sally, Charles, Edward. B.S., Calif. Inst. Tech., 1948; M.A., Ind. U., 1952, Ph.D., 1953, D.Sc. (hon.), 1973. Staff Los Alamos (N. Mex.) Sci. Lab., summer, 1947, 48, 49, astrophysicist on staff, 1953—; lab. fellow Los Alamos (N.Mex.) Sci. Lab., 1983—; program dir. adviser NSF, 1973-74; vis. prof. UCLA, 1966; staff Arco-Everett Research Lab., 1960-61. Contbr. articles to profl. jours. NSF fellow, 1952-53; Harvard U. fellow, summer 1952; NATO sr. fellow in sci., 1968; Fulbright research scholar, Belgium, 1968-69. Mem. Am. Astron. Soc., AAAS, Internat. Astron. Union, Sigma Xi. Subspecialties: Theoretical astrophysics; Atomic and molecular physics. Current work: Stellar opacities, stellar structure, evolution, stability, pulsation, atmospheres. Home: 1700 Camino Redondo Los Alamos NM 87544 Office: MS B288 PO Box 1663 Los Alamos NM 87545

COX, BRIAN MARTYN, pharmacologist; b. Sutton, Surrey, Eng., Nov. 3, 1939; s. Leslie Henry and Rosalie (Potbury) C.; m. Mavis Thornley, Sept. 5, 1964; children: Helen, Adrian, Graham. Diploma in Tech. of Pharmacology, Chelsea Coll., London, 1962; Ph.D. in Pharmacology, U. London, 1965. Nicholas research fellow St. Mary's Hosp. Med. Sch., London, 1962-65; lectr. pharmacology Chelsea Coll., 1965-74; dir. lab. research group Addiction Research Found., Palo Alto, Calif., 1974-77, assoc. dir., 1977-81; cons. assoc. prof. pharmacology Stanford U., 1978-81; prof. pharmacology Uniformed Services U., Bethesda, Md., 1981—; researcher, cons. Contbr. articles to profl. jours. Recipient Research Scientist award USPHS, 1978-81. Mem. Brit. Pharmacol- Soc., Am. Soc. Pharmacology and Exptl. Theapeutics, Soc. Neurosci., Internat. Narcotics Research Conf. Subspecialties: Neurochemistry; Neuropharmacology. Current work: Properties and functions of neuropeptides, especially endogenous opioids peptides,opioid receptor mechanisms and opiate drug tolerance and dependence. Office: Dept Pharmacology Uniformed Services U 4301 Jones Bridge Rd Bethesda MD 20814

COX, CHARLES SHIPLEY, oceanography educator; b. Paia, Maui, Hawaii, Sept. 11, 1922; s. Joel Bean and Helen Clifford (Horton) C.; m. Maryruth Louise Melander, Dec. 23, 1951; children: Susan, Caroline, Valerie, Ginger, Joel. B.S., Calif. Inst. Tech., 1944; Ph.D., U. Calif.-San Diego, 1954. Research Asst. to prof. dept. oceanography U. Calif.-San Diego, La Jolla, 1954—. Fellow Am. Geophys. Union, Royal Astronomical Soc.; mem. AAAS. Subspecialties: Geophysics; Oceanography. Current work: Research on oceanic internal waves and turbulence; research on electrical interactions between the ocean and underlying rocks. Office: Scripps Inst Oceanography U Calif-SanDiego Mail Stop A030 La Jolla CA 92093

COX, DONALD CODY, microbiology educator; b. Peoria, Ill., Mar. 31, 1936; s. Cody Alfred and Helen Jean (Beitel) C.; m. Nancy Mae Deniston, July 10, 1941; children: Kathleen Ann, Brian Cody. B.A. in Biology, Northwestern U., 1958; Ph.D. in Epidemiology, U. Mich., Ann Arbor, 1965. Asst. prof. U. Okla., 1965-70, assoc. prof., 1970-78, acting chmn. dept. botany and microbiology, 1976-77; prof., chmn. dept. microbiology Miami U., Oxford, Ohio, 1978—; indsl. cons. Contbr. articles profl. jours. Am. Cancer Soc. research grantee, 1969-71; Damon Runyon Fund Cancer Research grantee, 1971-72. Mem. Am. Soc. Microbiology, AAAS, Sigma Xi. Lodge: Rotary. Subspecialties: Virology (biology); Cancer research (medicine). Current work: The therapeutic effects of human viruses on neoplastic diseases. Home: 3631 Pamajera Dr Oxford OH 45056 Office: Miami U 204 Upham Hall Oxford OH 45056

COX, DONALD STEPHEN, periodontist, immunologist, educator; b. Buffalo, Dec. 24, 1948; s. John Cecil and Ann (Cherup) C.; m. Molly Joslyn Gushue, July 17, 1971. B.S., Syracuse U., 1971; D.D.S., SUNY-Buffalo, 1975; D.M.Sc., Harvard U., 1980. Research assoc. Forsyth Dental Center, Boston, 1979-80; asst. prof. SUNY-Stony Brook, 1980—. Recipient Univ. award N.Y. State, 1980-82; Nat. Research award NIH, 1975-80; Am. Fund for Dental Health grantee, 1979-81; NIH grantee, 1983—. Mem. AAAS, N.Y. Acad. Sci., Internat. Assn. for Dental Research, Northeast Soc. Periodontists, Am. Acad. Periodondology, Sigma Xi. Subspecialties: Immunology (medicine); Periodontics. Current work: Secretory immune system—methods of control of antibody amounts and affinity in external secretions. Immunochemistry of IgA. Office: Dept Periodontics Sch Dental Medicine SUNY Stony Brook NY 11794

COX, G. STANLEY, biochemist, educator; b. Roswell, N.Mex., Jan. 26, 1946; s. Jack W. and Lucille Ruby (Smith) C.; m. Cheryl Lynn Whitson, Sept. 2, 1967; children: Joseph Stanley, Kimball Leigh. B.S., N.Mex. State U., 1968; Ph.D., U. Iowa, 1973. Grad. research asst. U. Iowa, Iowa City, 1968-72; postdoctoral fellow Roche Inst. Molecular Biology, Nutley, N.J., 1972-74; staff fellow NIH, Bethesda, Md., 1974-76; vis. scientist dept. biochemistry U. Iowa, Iowa City, 1982; asst. prof. dept. biochemistry and biophysics Iowa State U., Ames, 1976-83; assoc. prof. dept. biochemistry U. Nebr. Med. Ctr., Omaha, 1983—. Contbr. articles to profl. jours. NIH grantee, 1977-80, 82—; Population Council grantee, 1977-78. Mem. Am. Soc. for Microbiology, Sigma Xi, Phi Kappa Phi, Phi Lambda Upsilon. Subspecialties: Molecular biology; Biochemistry (biology). Current work: Regulation of gene expression in mammalian cells; tumor molecular biology. Office: Dept Biochemistr U Nebr Med Ctr Omaha NE 68105

COX, HOLLACE LAWTON, JR., physical chemist; b. Oak Park, Ill., Nov. 17, 1935; s. Hollace Lawton and Frances Marian (Murray) C.; m. Sue Burdon, June 25, 1983. A.B., U. Rochester, 1959; Ph.D., Ind. U., 1967. Mem. tech. staff Tex. Instruments, Inc., Dallas, 1967-70; Robert A. Welch postdoctoral fellow Baylor U. dept. chemistry, Waco, Tex., 1970-73; Robert A. Welch postdoctoral fellow dept. pathology M.D. Anderson Hosp. and Tumor Inst., Houston, 1973-75, postdoctoral fellow med. physics tng. program, 1975-76; instr. Washington U. Sch. Medicine, St. Louis, 1976-77; assoc. prof. U. Kans. Med. Sch., Kansas City, 1977-80, U. Louisville Sch. Medicine, 1980-83; assoc. prof. dept. elec. engring. U. Louisville, 1983—. Robert A. Welch fellow, 1973-75; NSF grantee, 1980-82. Mem. Am. Phys. Soc., Optical Soc. Am., Am. Assn. Physicists in Medicine, N.Y. Acad. Sci., Sigma Xi, Sigma Pi Sigma, Phi Lamba Upsilon. Epsicopalian. Subspecialties: Physical chemistry; Laser-induced chemistry. Current work: Laser-induced photochemistry studies in cells, laser wave guide application for studying cells and development of thin film techniques for biological studies. Office: Dept Elec Engring WS Speed Bldg U Louisville Louisville KY 40292

COX, JEROME ROCKHOLD, JR., elec. engr.; b. Washington, May 24, 1925; s. Jerome R. and Jane (Mills) C.; m. Barbara Jane Lueders, Sept. 2,1951; children—Nancy Jane Cox Battersby, Jerome Mills, Randall Allen. S.B., Mass. Inst. Tech., 1947, S.M., 1949, Sc.D., 1954. Mem. faculty Washington U., St. Louis, 1955—, prof. elec. engring., 1961—, prof. biomed. engring. in physiology and biophysics, 1965—, dir., 1964-75, chmn. computer labs., 1967—, program dir. tng. program tech. in health care, 1970-78, chmn. dept. computer sci., 1975—, sr. research asso., 1975—; co-chmn. computers in cardiology conf. Inst. Medicine, Nat. Acad. Scis., 1974—; cardiology adv. com. Nat. Heart and Lung Inst., 1975-78; mem. epidemiology biostatistics and bioengring. cluster President's Biomed. Research Panel, 1975. Editorial bd.: Computers and Biomed. Research, 1967—; asso. editor, IEEE, trans. biomed. engring., 1969-71. Served with U.S. Army, 1943-44. Fellow Acoustical Soc. Am., IEEE; mem. Biomed. Engring. Soc., Biophys. Soc., Assn. Computing Machinery, Sigma Xi, Eta Kappa Nu, Tau Beta Pi. Subspecialties: Computer architecture; Biomedical engineering. Author, patentee air traffic control, computerized tomography. Office: Dept Computer Sci Box 1045 Washington Univ St Louis MO 63130

COX, MARY E., physicist, educator; b. Detroit, Nov. 11, 1937; d. Willis H. and Dorothy E. (Nicholls) Buckles; m. Kendall B. Cox, July 1, 1961; 1 son, Kendall M. B.A., Albion Coll., 1959; M.A., U. Mich., 1961. Instr. physics U. Mich.-Flint, 1966-71, asst. prof. physics, 1971-76, assoc. prof., 1976-82, chmn. dept. physics and astronomy, 1976-77, acting dir. computing ctr., 1972-73, acting assoc. dean curriculum and program devel. Coll. Arts and Scis., 1980, assoc. dean, 1980-81, acting dean, 1981-82; tutor physics Sommerville Coll., U. Oxford, Eng., 1977-78. Contbr. articles to profl. jours. NSF grantee, 1977-78; NIH grantee. Mem. Am. Assn. Physics Tchrs. (exec. council Mich. sect. 1974-75, pres. Mich. sect. 1973-74), Optical Soc. Am., Soc. Photo-Optical Instrumentation Engrs. Subspecialties: Holography; Biomedical engineering. Current work: Applications of coherent optics to problems in biomedical areas. Office: Dept Physics U Mich-Flint Flint MI 48503

COX, PAUL ALAN, biologist; b. Salt Lake City, Oct. 10, 1953; s. Leo A. and Rae (Gabbitas) C.; m. Barbara Ann Wilson, May 21, 1975; children: Emily Ann, Paul Matthew, Mary Elisabeth. B.S., Brigham Young U., 1976; M.Sc., U. Wales, 1978; A.M., Harvard U., 1978; Ph.D., 1981. Teaching fellow Harvard U., Cambridge, Mass., 1977-81; Miller Research fellow Miller Inst. Basic Research in Sci., Berkeley, Calif., 1981-83; asst. prof. Brigham Young U., Provo, Utah, 1983—; ecologist Utah Environ. Council, Salt Lake City, 1976; staff ecologist Utah MX Coordination Office, Salt Lake City, 1981. Recipient Bowdoin prize; Danforth Found. fellow, 1976-81; Fulbright fellow, 1976-77; NSF fellow, 1977-81. Mem. Brit. Ecol. Soc., N.Y. Acad. Scis., Am. Soc. Naturalists, Assn. for Tropical Biology, AAAS, New Eng. Bot. Club. Mormon. Subspecialties: Evolutionary biology; Population biology. Current work: Evolution of plant breeding systems; tropical plant ecology; Polynesian ethnobotany; pollination ecology. Office: Dept Botany Brigham Young U Provo UT 84602

COX, RACHEL DUNAWAY, psychologist, researcher in personality; b. Murray, Ky., Jan. 20, 1904; d. Enoch T. and Khadra (Fergeson) Dunaway; m. Reavis Cox, Feb. 18, 1928; children: David J., Rosemary Cox Masters. B.A., U. Tex., 1925; M.A., Columbia U., 1930; Ph.D., U. Pa., 1943. Diplomate: in clin. psychology.; Lic. psychologist, Pa. Reporter, feature writer N.Y. Herald Tribune, N.Y.C., 1926-31; from lectr. to assoc. prof. Bryn Mawr (Pa.) Coll., 1944-55, prof., 1955-71, vis. prof., 1976, dir., 1944-70, researcher on adult personality, 1973—; pvt. practice clin. psychology, Wallingford, Pa., 1973—. Author: Counselors and Their Work, 1944, Youth into Maturity, 1970. Bd. dirs. Ctr. for Early Childhood Services, Phila., 1980-82, mem. adv. bd., 1977-80; bd. mgrs. Sleighton Sch., Darling, Pa., 1955-80, Devel. Ctr. for Autistic Children, Phila., 1960-80. Grant Found. grantee, 1963-70, 73—. Fellow Pa. Psychol. Assn. (Disting. Service award 1975); mem. Am. Psychol. Assn., Nat. Assn. Social Workers, Soc. for Projective Techniques (past nat. sec.), LWV. Republican. Presbyterian. Club: Media Book (Delaware County, Pa.). Subspecialties: Developmental psychology; Clinical. Current work: Research in the adult personality with major emphasis on the norm. Home and office: 219 Sykes Ln Wallingford PA 19086

COX, STEPHEN KENT, atmospheric sciences educator; b. Galesburg, Ill., Sept. 2, 1940; s. Joseph E. and Ruth B. (Burroughs) C.; m. Phyllis A. Cox, Dec. 23, 1961; children: Mark Joseph, Elizabeth. B.A., Knox Coll., Galesburg, Ill., 1961; M.S., Ph.D., U. Wis.-Madison, 1967. Research meteorologist ESSA, Madison, Wis., 1965-67; research scientist Space Sci. and Engring. Ctr., U. Wis.-Madison, 1967-69; mem. faculty Colo. State U., Ft. Collins, 1969—, prof. atmospheric scis., 1977—. Contbr. over 100 sci. articles to profl. publs. Mem. Am. Meteorol. Soc. Subspecialties: Meteorology; Remote sensing (atmospheric science). Office: Atmospheric Scis Dept Colo State U Fort Collins CO 80523

COYLE, PATRICIA K., neurology educator; b. N.Y.C.; d. Daniel E. and Eileen J. (Tuohy) C. B.S., Fordham U., 1970; M.D., Johns Hopkins U., 1974. Diplomate: Am. Bd. Psychiatry and Neurology. Intern N.Y. Hosp.-Cornell U., 1974-75; asst. resident, fellow neurology Johns Hopkins U., 1975-77; chief resident, fellow, 1977-78, USPHS postdoctoral fellow in neurovirology and immunology, 1978-80; asst. prof. neurology SUNY-Stony Brook, 1980—. Goldberger fellow AMA, 1971; MS Soc. predoctoral fellow, 1973; Biomed. Research grantee SUNY-Stony Brook, 1980; Nat. MS Soc. grantee, 1981; Kroc Found. grantee, 1981; VA Research grantee, 1981, 82; Tchr.-investigator devel. award NINC DS, 1983. Mem. Am. Acad. Neurology, AAAS, Am. Fedn. Clin. Research, Am. Soc. Microbiology, N.Y. Acad. Scis., Soc. neurosci., Phi Beta Kappa, Alpha Omega Alpha. Subspecialties: Neurology; Neuroimmunology. Current work: Neuroimmune disorders, multiple sclerosis, immune complex diseases. Office: Dept Neurology HSC T12 SUNY Stony Brook NY 11794

COYNE, GEORGE VINCENT, astronomer; b. Balt.; s. Francis Verbrick and Elizabeth Ann (Brune) C. Ph.D. in Astronomy, Georgetown U., 1962. Joined Soc. of Jesus; ordained priest Roman Catholic Ch.; dir. Vatican Obs., Rome; adj. prof. dept. astronomy U. Ariz., Tucson. Mem. Internat. Astron. Union, Am. Astron. Soc. Subspecialty: Optical astronomy.

COZZARELLI, FRANCIS ANTHONY, engineering educator, researcher; b. Jersey City, Apr. 8, 1933; s. Nicholas and Katherine (Meluso) C.; m. Kathleen Isabella Burke, June 7, 1958; children: Catherine, Isabelle, Delia, Julia, John, Claire. B.S. in Mech. Engring, Stevens Inst. Tech., 1955, M.S. in Applied Mech, 1958, Ph.D., Poly. Inst. N.Y., 1964. Stress analyst Gibbs and Cox, Inc., N.Y.C., 1955-57; instr. dept. engring. sci. Pratt Inst., 1957-62, asst. prof., 1961-62; instr.div. interdisciplinary studies and research SUNY-Buffalo, 1962-66, assoc. prof. div. interdisciplinary studies and research, dept. elec. engring. and engring. sci., dept. engring. sci., 1966-71, prof. dept. engring. sci., 1971-80, dir. grad. studies dept. engring sci., aerospace engring. and nuclear engring., 1971-74, 78-80, prof. dept. civil engring., 1980-82, prof. dept. mech. and aerospace engring., 1982—; vis. prof. Lab. Applied Mechanics, Technische Hogeschool, Delft, The Netherlands, 1968-69; vis. scientist Joint Research Ctr. (Euratom materials div.), Ispra, Italy, 1977-; vis. prof. Inst. Mechanics, Politecnico de Milano, Milan, Italy, 1976-77. Reviewer: Jour. Applied Mechanics. NSF fellow, 1968-69; Fulbright-Hays Research grantee, 1976-77; NSF grantee, 1967-72, 80-83; SUNY Research Found. grantee, 1963, 1974-75; Office Naval Research grantee, 1964-82. Mem. ASME, Engring. Sci. Soc., Am. Acad. Mechanics, Sigma Xi, Tau Beta Pi. Democrat. Roman Catholic. Subspecialties: Solid mechanics; Tectonics. Current work: Geophysical solid mechanics, theory of creep rupture, thermal and irradiation creep, inelastic stress analysis, viscoelasticity, inelastic wave propagation, stochastic structural mechanics, continuum mechanics. Home: 393 Starin Ave Buffalo NY 14216 Office: SUNY Dept Mech and Aerospace Engring Furnas 1012 Buffalo NY 14260

CRABB, CHARLES FREDERICK, elec. engr.; b. Plainfield, N.J., July 3, 1948; s. Fred Topping and Dorothy (Higgins) C.; m. Helen Macarof, July 22, 1979. Student, Cornell U., 1966-68; B.S. magna cum laude, U. N.C., 1976; M.S., Boston U., 1980. Researcher U. N.C., Greensboro, 1977; digital design engr. Prime Computer, Natick, Mass., 1978-79; researcher Boston U., 1979-81; sr. design engr. Computervision Corp., Bedford, Mass., 1981-83; design engr. IPL Systems, Inc., Waltham, Mass., 1983—. Served with USMC, 1969-73. Mem. IEEE, Assn. Computing Machinery, Union Concerned Scientists, Sigma Pi. Subspecialties: Computer architecture; Computer engineering. Home: 44 Church St Hopkinton MA 01748 Office: 360 2d Ave Waltham MA 02254

CRABTREE, ROBERT HOWARD, chemistry educator, consultant; b. London, Apr. 17, 1948; U.S., 1977; s. Arthur and Marguerite Marie (Vaniere) C. B.A., Oxford U., 1970; M.A. (hon.), 1977; D.Phil., Sussex U., 1973. Postdoctoral fellow CNRS, Gif-s-Yvette, France, 1973-75, attach de recherche, 1975-77; asst. prof. Yale U., New Haven, 1977-82, assoc. prof. chemistry, 1982—; cons. Am. Cyanamid, Stamford, Conn., 1978—, Proctor & Gamble, Cin., 1981—. Contbr. articles in field to profl. jours. Govett scholar New Coll., Oxford U., 1966-70; DuPont young faculty fellow, 1977; A.P. Sloan Found. fellow, 1981; C and H Dreyfus Found. tchr.-scholar, 1982. Mem. Royal Soc. Chemistry, Am. Chem. Soc., N.Y. Catalysis Soc. Democrat. Clubs: Oxford and Cambridge Rifle Assn., Oxford U. Rifle (capt. 1969-70). Subspecialties: Catalysis chemistry; Inorganic chemistry. Current work: Research in homogeneous catalysis and organometallic chemistry, alkane activation, hydrogenation, hydride complexes, colloidal metals, halocarbon complexes. Home: 131 Maple St New Haven CT 06520 Office: Yale U Chemistry Dept 225 Prospect St New Haven CT 06520

CRADDOCK, CAMPBELL, geologist, educator; b. Chgo., Apr. 3, 1930; s. John and Bernice (Campbell) C.; m. Dorothy Dunkelberg, June 13, 1953; children—Susan Elizabeth, John Paul, Carol Jean. B.A., DePauw U., 1951; M.A., Columbia U., 1953, Ph.D., 1954. Geologist Shell Oil Co., N.Mex., Tex., Colo., Wyo., 1954-56; asst. prof. U. Minn., Mpls., 1956-60, asso. prof., 1960-67; prof. geology U. Wis., Madison, 1967—, chmn. dept., 1977-80; leader Antarctic geologic field research programs, 1959-69, 80, Alaskan geologic field research programs, 1968-81, Svalbard field programs, 1977—; cons. C.E., AUS, 1957-58, N. Star Research Inst., 1965-68, Dept. State, 1976; vis. scientist N.Z. Geol. Survey, 1962-63; lectr. Nanjing and Beijing univs., China, 1981; chmn. panel geology and geophysics NRC, 1967-71, mem. polar research bd., 1978—; U.S. mem. working group on geology Sci. Com. on Antarctic Research, 1967—, chmn. group, 1973-80; co-chief scientist Leg 35, Deep Sea Drilling Project, Antarctica, 1974; chmn. Antarctic panel Circum-Pacific Map Project, 1979—; cons. Phillips Petroleum Co., 1980. Editor: Antarctic Geoscience, 1982; Co-editor: Geologic Maps of Antarctica, Folio 12, Antarctic Map Folio Series, Am. Geog. Soc., 1970, Initial Reports of the Deep Sea Drilling Project, Vol. 35, 1976; Contbr. articles sci. jours. Higgins fellow, 1951-52; NSF fellow, 1952-53; research grantee, 1957—; recipient U.S. Antarctic Service medal, 1968; Bellingshausen-Lazarev medal Soviet Acad. Scis., 1970; Alumni citation DePauw U., 1976. Fellow Geol. Soc. Am. (chmn. sect. 1982-83, books editor 1982—); mem. Internat. Union Geol. Scis. (mem. commn. on structural geology 1968-76, mem. com. on tectonics 1976—, del. Sci. Com. on Antarctic Research 1974—, mem. com. on geologic map of world 1974—), Polar Research Bd. (1978-82), Internat. Union Geol. Scis. (v.p. for Antarctica 1979—), Am. Geophys. Union, Am. Assn. Petroleum Geologists, Phi Beta Kappa, Sigma Xi. Subspecialty: Geology. Mailing Address: 1109 Winston Dr Madison WI 53711 Office: Dept Geology and Geophysics U Wis Madison WI 53706

CRAGLE, RAYMOND GEORGE, animal physiologist; b. Orangeville, Pa., Feb. 28, 1926; m., 1950; 3 children. B.S., N.C. State Coll., 1951, M.S., 1954; Ph.D., U. Ill., 1957. Asst. prof., assoc. prof., prof. physiology and nutrition Agrl. Research Lab., AEC, U. Tenn., 1957-68; vis. prof. Lab. Genetics, U. Wis., Madison, 1968-69; prof. dairy sci., head dept. Va. Poly. Inst. and State U., 1970-78; dir. Agrl. Expt. Sta., U. Ill., 1978—. Mem. Am. Dairy Sci. Assn., Am. Soc. Animal Sci., Am. Inst. Nutrition, AAAS. Subspecialty: Animal physiology. Office: Agrl Expt Sta U Ill 1301 W Gregory Dr Blacksburg VA 24061

CRAIG, BURTON MACKAY, chemist; b. Vermilion, Alta., Can., May 29, 1918; s. Walter Alexander and Mary Jessie (Baillie) C.; m. Inez Gladys Guttormson, July 5, 1945; children—Wayne Keith, Cheryl Lynne. B.Sc.A., U. Sask., 1944, M.Sc., 1946; Ph.D., U. Minn., 1950. Research officer Prairie Regional Lab., Saskatoon, Sask., 1950-69, asso. dir., 1969-70, dir., 1970—. Fellow Chem. Inst. Can.; mem. Am. Oil Chemists Soc., AAAS, Research Mgrs. Assn. Can., Sigma Xi; hon. life mem. Agrl. Inst. Can. Lutheran. Subspecialties: Plant cell and tissue culture; Nitrogen fixation. Current work: Plantcell, plant tissue culture; cryopreservation; nitrogen fixation, photosynthesis; cell biology; mutagenesis. Home: 423 Lake Crescent Saskatoon SK Canada S7H 3A3 Office: Prairie Regional Lab Nat Research Council Can 110 Gymnasium Rd Saskatoon SK Canada S7N 0W9

CRAIG, CHARLES ROBERT, pharmacologist, educator; b. Buckhannon, W.Va., Jan. 24, 1936; s. Thorne C. and Martha (Blakeslee) C.; m. Margaret A. Craig, June 12, 1960; 1 son, Gary Lynn. B.S., Glenville State Coll., 1959; Ph.D., U. Wis., 1964. Sr. investigator GD Searle & Co., Chgo., 1964-65; asst. prof. pharmacology W.Va. U., Morgantown, 1966-71, assoc. prof., 1971-75, prof., 1975—. Editor: (with others) Modern Pharmacology, 1982; Contbr. articles, abstracts to profl. jours. Served with U.S. Army, 1954-57. Mem. Am. Soc.

Pharmacology and Exptl. Therapeutics, Soc. Neurosci., Am. Soc. Neurochemistry, Can. Coll. Neuropsychopharmacology, Sigma Xi. Democrat. Methodist. Subspecialties: Neuropharmacology; Neurochemistry. Current work: Pharmacological and morphological studies of experimental epilepsy. Cobalt-induced epilepsy, transmitters, amino acids, anticonvulsants. Home: 1315 Cherry Ln Morgantown WV 26505 Office: Dept Pharmacology and Toxicology WVa U Med Center Morgantown WV 26506

CRAIG, DONALD M., data processing exec.; b. Ayrshire, Scotland, Aug. 2, 1949; s. John M. and Edith (Thomson) C. Student, U. Western Ont., 1967-69. Systems programmer Ampex Can., Toronto, Ont., 1972-73; digital systems cons. Canadian Broadcasting Corp., Montreal, Que., 1973-78; mgr. software Central Dynamics Ltd., Montreal, Que., 1978-81, mgr. research and devel. lab., Ottawa, Ont., 1981—. Mem. Assn. Computing Machinery, Soc. Motion Picture and Television Engrs. Subspecialties: Operating systems; Graphics, image processing, and pattern recognition. Current work: Digital NTSC and PAL TV signal processing, systems engineering management, high speed signal procssing architecture design, software engring. Office: 1775 Courtwood Crescent Ottawa ON Canada K2C 3J2

CRAIG, HARMON, geochemist, oceanographer; b. N.Y.C., Mar. 15, 1926; s. John Richard and Virginia (Stanley) C.; m. Valerie Kopecky, Sept. 27, 1947; children—Claudia Campbell, Cynthia Camilla, Karen Constance. M.S., U. Chgo., 1950, Ph.D., 1951. Research asso. Enrico Fermi Inst. Nuclear Studies, U. Chgo., 1951-55; with Scripps Instn. Oceanography, U. Calif., San Diego, 1955—; now prof. geochemistry and oceanography; Guggenheim fellow U. Pisa, 1962-63; chief scientist on 12 oceanographic expdns.; expdn. leader UN (FAO) Lake Tanganyika Expdns., 1973, 75; chief scientist on 16 oceanographic expdns.; expdn. leader UNDP Phase-2 Geothermal Survey of Ethiopia, 1976; mem. Haicheng earthquake study del., China, 1976, Yangbajain geothermal field study, Tibet, 1980. Contbr. numerous articles to profl. publs.; editor: Earth and Planetary Sci. Letters, 1971—. Served with USN, 1944-46. Recipient V. M. Goldschmidt medal Geochem.Soc., 1979, A.L. Day medal Geol. Soc. Am., 1983. Mem. Nat. Acad. Scis., Am. Acad. Arts and Scis., Am. Geophys. Union., Explorers Club. Subspecialties: Geochemistry; Oceanography. Current work: Rare gases, especially helium-3, deep ocean circulation, oceanic basalts and xenociths, atmospheric gases, polar ice cores, earthquake prediction. Home: 8553 La Jolla Shores La Jolla CA 92037 Office: Scripps Instn Oceanography A-020 U Calif San Diego La Jolla CA 92093

CRAIG, PAUL LAWRENCE, psychologist; b. Oklahoma City, Nov. 20, 1953; s. John Odra and Edith (Margritz) C.; m. Carol Ann Hult, Dec. 23, 1979; 1 dau., Marie Hult Craig. B.S., Nebr. Wesleyan U., 1975; M.S., U. Wyo., 1978, Ph.D., 1980. Lic. psychologist, Alaska. Psychology intern U. Minn. Health Sci. Ctr., Mpls., 1979-80; postdoctoral fellow U. Wash. Med. Sch., Seattle, 1980-81; program dir., psychologist Community Mental Health Ctr., Homer, Alaska, 1981—; dir. mental health planning Alaska Mental Health Dirs. Assn., 1982—; adv. council Kenai Peninsula Community Coll., 1982—. Editorial adv. bd.: Clin. Neuropsychology, 1980—; Contbr. articles to profl. jours. NIMH fellow, 1976; Clark scholar, 1979. Mem. Alaska Psychol. Assn., Am. Psychol. Assn., Internat. Neuropsychol. Soc., Nat. Acad. Neuropsychologists, C. of C. Methodist. Subspecialties: Behavioral psychology; Neuropsychology. Current work: Behavioral principles applied to rural community mental health services in cross-cultural environment. Home: PO Box 2906 Homer AK 99603 Office: PO Box 2274 Homer AK 99603

CRAIG, RICHARD G., geology educator; b. Wilmington, Del., June 3, 1949; s. Murray Breeze and Elinore Mary (Russell) C.; m., 1982; 1 dau., Lara. B.A., Dickinson Coll., 1971; M.S., Pa. State U., 1976, Ph.D., 1979. Asst. prof. dept. geology Kent (Ohio) State U., 1978-82, assoc. prof., 1982—; cons. in field. Contbr. articles to profl. jours.; editor: Future Trends in Geomathematics, 1982, Applied Geomorphology, 1982. Mem. AAAS, Am. Geophys. Union, Geol. Soc. Am. Subspecialties: Geology; Mathematical software. Current work: Computer simulations of geologic processes in landform evolution. Office: Dept Geology Kent State U 116A McGilvery Hall Kent OH 44242 Home: 955 Meloy Rd Kent OH 44240

CRAIG, ROBERT BRUCE, science manager, consultant; b. Washington, Apr. 22, 1944; s. Glenn Horace and Evelyn May (Fiddler) C.; m. Judy Gail Nelson, Apr. 5, 1965; children: Kimberlie, Cassandra. B.S., U. Calif., Davis, 1970, M.A., 1972, Ph.D., 1974. Research assoc. Oak Ridge (Tenn.) Nat. Lab., 1974-75, research staff, group leader, 1975-78, program mgr., 1979-81; dept. mgr. HDR Scis., Knoxville, Tenn., 1981-82, v.p., Santa Barbara, Calif., 1982—. Author: Coastal Ecology, 1972, Water Resources Regions, 1980. Mem. Ecol. Soc. Am. Nuclear Soc. Democrat. Subspecialties: Ecology; Fuels and sources. Current work: Environmental effects of advanced energy systems, population biology. Home: 1465 Crestline Dr Santa Barbara CA 93105 Office: HDR Scis 804 Anacapa St Santa Barbara CA 93101

CRAIGHEAD, JOHN J., ecology educator; b. Washington, Aug. 14, 1916. B.A., Pa. State U., 1939; M.S., U. Mich., 1940, Ph.D., 1950. Biologist N.Y. Zool. Soc., 1947-49; dir. survival tng. for armed forces U.S. Dept. Def., 1950-52; wildlife biologist, leader Mont. Coop. Wildlife Research Unit Bur. Sport Fisheries and Wildlife, 1952-77; prof. zoology and forestry U. Mont., Missoula, 1952—. Recipient LaGorce gold medal Nat. Geol. Soc., 1979; grantee AEC, NASA, U.S. Forest Service, Mont. Fish and Game Dept. Mem. Wildlife Soc. (Einarsen award 1977, v.p. 1962-63), AAAS, Wilderness Alliance. Subspecialties: Wildlife biology; Population biology. Office: 5125 Orchard Ln Missoula MT 59801

CRAIK, REBECCA LYNN, researcher, educator; b. East Liverpool, Ohio, Sept. 25, 1948; d. Donald Hugh and Betty Jane C. B.A., Case Western Res. U., 1970; M.S., Duke U., 1972; Ph.D., Temple U., 1981. Lic. phys. therapist, Pa., N.C. Clinician Temple U., Phila., 1972-73; researcher U. Pa., 1981-83, Moss Hosp., Phila., 1973-82; tchr. Allied Health Sch., Temple U., 1981—, Phys. Therapy Sch., Beaver Coll., 1983—; adj. instr. physiology Temple U. Sch. Medicine, 1982—. Contbr. articles to profl. jours. Recipient award Easter Seal Soc., 1966; Rehab. Service Adminstrn. trainee, 1970-72; others. Mem. AAAS, Am. Phys. Therapy Assn., Internat. Soc. Prosthetics and Orthotics, Internat. Soc. Biomechanics, Soc. Behavioral Kinesiology, Soc. Neurosci., World Congress Phys. Therapy. Subspecialties: Biomedical engineering; Neurophysiology. Current work: Facilitation of movement patterns with elec. stimulation in humans; neural plasticity of somatosensory cortex in the rat. Office: Rehab Engring Center Moss Rehab Hosp 12th and Tabor Rd Philadelphia PA 19141

CRAIN, CHESTER RAYMOND, statistician; b. St. Louis, Apr. 17, 1944; s. Chester and Mary Louise (Landers) C.; m. Barbara Hope Fagnan, Sept. 2, 1967; 1 dau., Michelle C. Pidot. A.B., Knox Coll., 1965; M.A., U. Calif.-Riverside, 1967; Ph.D., U. N. Mex., 1974. Biostatistician Inhalation Toxicology Research Inst., Albuquerque, 1974-76; research statistician Schering Research Ctr., Bloomfield, N.J., 1976-80; statistician Knoll Pharm., Whippany, N.J., 1980; sr. statistician McNeil Pharm., Spring House, Pa., 1980-81; sr. biostatistician Miles Pharm., West Haven, Conn., 1982-83; dir. statis. services Boots Pharms., Shreveport, La., 1983—. Contbr. articles to profl. jours. Mem. Am. Statis. Assn., East Coast Pharm. Stats. Soc. (pres. 1980-81), Am. Soc. Quality Control, Biometric Soc., Math. Assn. Am., Soc. Clin. Trials, Phi Beta Kappa, Pi Mu Epsilon. Unitarian. Subspecialty: Statistics. Office: 3003 Knight St Suite 222 Shreveport LA 71105

CRAIN, CULLEN MALONE, elec. engr.; b. Goodnight, Tex., Sept. 10, 1920; s. John Malone and Margaret Elizabeth (Gunn) C.; m. Virginia Raftery, Jan. 16, 1943; children—Michael Malone, Karen Elizabeth. B.S. in Elec. Engring, U. Tex., Austin, 1942, M.S., 1947, Ph.D., 1952. From instr. to asso. prof. elec. engring. U. Tex., 1943-57; group leader communications and electronics Rand Corp., Santa Monica, Calif., 1957-69, asso. head engring. and applied scis., 1969—; cons. to govt., 1958—. Author numerous papers in field. Pres. Austin chpt. Nat. Exchange Club, 1954, Santa Monica chpt., 1975. Served with USNR, 1944-46. Fellow IEEE; mem. Nat. Acad. Engring. Subspecialties: Electrical engineering; Electronics. Current work: Current activities directed toward military communications and electronics applications and architecture of future systems. Inventor microwave atmospheric refractometer. Home: 463 17th St Santa Monica CA 90402 Office: 1700 Main St Santa Monica CA 90406

CRAIN, STANLEY M., neuroscience educator, researcher; b. N.Y.C., Feb. 5, 1923; s. Rubin and Bertha (Lipson) C.; m. Bea Heller, Aug. 18, 1946; children: Steven, Michael. Radiol. research Health Physics div. Argonne Nat. Lab., 1945-47; electrophysiology researcher dept. neurology Columbia U., 1950-57; cell physiology researcher, head sect. nerve tissue culture lab Abbott Labs., Chgo., 1957-61; asst. prof. anatomy dept. neurology Columbia U., 1961-65; assoc. prof. dept. physiology Albert Einstein Coll. Medicine, 1965-69, prof., 1969—, prof. dept. neurosci., 1974—; neurosci researcher Rose F. Kennedy Ctr. for Research in Mental Retardation and Human Devel., 1974—. Contbr. articles to profl. jours. Served with U.S. Army, 1942-45. Recipient Grass fellow award Marine Biol. Lab., Woods Hole, 1957; Nat. Inst. Neurol. Diseases and Blindness Research Career Devel. awardee, 1961-65; Joseph P. Kennedy Found. for Mental Retardation scholar awardee, 1965-75. Mem. Soc. Neurosci., Am. Physiol. Soc., Am. Assn. Anatomists, Am. Soc. Cell Biology, Internat. Soc. Devel. Neurosci., Tissue Culture Assn., AAAS. Subspecialties: Neurobiology; Cell and tissue culture. Current work: Electrophysiologic and cytologic research on tissue cultures of fetal mammalian brain and spinal cord networks during development of nerve cell functions in vitro. Home: 10 Linden Terr Leonia NJ 07605 Office: Dept Neurosci Albert Einstein Coll Medicine 1300 Morris Park Ave Bronx NY 10461

CRAINE, ERIC RICHARD, research scientist; b. Harlan, Ky., June 21, 1946; s. Eugene Richard and Carol Dawn (Ward) C.; m. Margaret Helen Zavala, Sept. 18, 1976; children: Patrick Richard, Jennifer Carol. B.S. in Math., Physics, Astronomy, U. Okla., Norman, 1968; Ph.D. in Astronomy, Ohio State U., 1973. Research fellow Ohio State U. Radio Obs., Columbus, 1973-74; project scientist Steward Obs. U. Ariz., Tucson, 1975—; sr. scientist Energy/Environ. Research Group, Inc., Tucson, 1980—; dir. Electro-Optical Test Facility, Bell Aerospace, Tucson. Author: Quasistellar and BL Lacertae Objects, 1977, Near Infrared Photographic Sky Survey, 1980; contbr. numerous articles to profl. jours. Grantee in field. Mem. Am. Astron. Soc., Internat. Astron. Union, Soc. Photo-Optical Instrumentation Engrs. Republican. Presbyterian. Club: Tucson Sailing. Subspecialties: Infrared optical astronomy; Solar energy. Current work: Optical infrared sky survey astronomy; electro-optical instrumentation, infrared sources, molecular clouds and BL Lacertae objects. Office: Steward Observatory University of Arizona Tucson AZ 85721

CRALL, JAMES MONROE, plant pathologist, plant breeder; b. Monongahela, Pa., July 13, 1914; s. James Shelby and Margaret Bureau (Rabe) C.; m. Duronda Stanberry, Dec. 22, 1943; children: Cynthia Ann Crall Smith, James Stanberry. Student, Washington and Jefferson Coll., 1934-35; B.S., Purdue U., 1939; M.A., U. Mo., 1941, Ph.D., 1948. Asst. prof. dept. botany and plant pathology Iowa State U., Ames, 1948-52; prof., dir. Agrl. Research Center, Inst. Food and Agrl. Scis., U. Fla., Leesburg, 1952-77, prof., plant pathologist, 1977—. Bd. dirs. United Appeal, Leesburg, 1960-69, v.p., 1965, pres., 1966. Served with USAAF, 1942-46; CBI. Mem. AAAS, Council Agrl. Sci. and Tech., Am. Phytopath. Soc., Mycological Soc. Am., Am. Soc. Hort. Scis., Sigma Xi, Gamma Sigma Delta, Gamma Alpha. Democrat. Episcopalian. Lodge: Kiwanis. Subspecialties: Plant genetics; Plant pathology. Current work: Development of new watermelon cultivars. Watermelon breeding, with particular emphasis on disease resistance in high quality shipping-type watermelons. Home: PO Box 321 Leesburg FL 32748 Office: PO Box 388 Leesburg FL 32748

CRAM, DONALD JAMES, chemistry educator; b. Chester, Vt., April 22, 1919; s. William Moffet and Joanna (Shelley) C.; m. Jane Maxwell, Nov. 25, 1969. B.S., Rollins Coll., 1941; M.S., U. Nebr., 1942; Ph.D. (Nat. Research fellow), Harvard, 1947, U. Uppsala, 1977. Research chemist Merck and Co., 1942-45; asst. prof. chemistry UCLA, 1947-50, asso. prof., 1950-56, prof., 1956—; chem. cons. Upjohn Co., 1952—, Union Carbide Co., 1960-81, Eastman Kodak Co., 1981—; State Dept. exchange fellow to Inst. de Quimica, Nat. U. Mex., summer 1956; guest prof. U. Heidelberg, Germany, summer 1958; guest lectr., South Africa, 1967; Centenary lectr. Chem. Soc. London, 1976. Author: (with S.H. Pine, J.B. Hendrickson and G.S. Hammond) Organic Chemistry, 1960, 4th edit., 1980, Fundamentals of Carbanion Chemistry, 1965, (with John H. Richards and G.S. Hammond) Elements of Organic Chemistry, 1967, (with J.M. Cram) Essence of Organic Chemistry, 1977; Contbr.: chpts. to Applications of Biochemical Systems in Organic Chemistry; also articles in field of host-guest complexation chemistry, carbonium ions, stereochemistry, mold metabolites, large ring chemistry. Named Young Man of Yr. Calif. Jr. C. of C., 1954, Calif. Scientist of Yr., 1974; recipient award for creative work in synthetic organic chemistry Am. Chem. Soc., 1965, Arthur C. Cope award, 1974; Herbert Newby McCoy award, 1965, 75; award for creative research organic chemistry Synthetic Organic Chem. Mfrs. Assn., 1965; Am. Chem. Soc. fellow, 1947-48; Guggenheim fellow, 1954-55. Mem. Am. Chem. Soc., Nat. Acad. Scis., Am. Acad. Arts and Scis., Chem. Soc. (Eng.), Sigma Xi, Lambda Chi Alpha. Club: San Onofre Surfing. Subspecialty: Organic chemistry. Current work: Host-guest complexation chemistry; biomimetic chemistry; designed ionosphere chemistry; chiral regonition chemistry. Home: 1250 Roscomare Rd Los Angeles CA 90077

CRAMBLETT, HENRY GAYLORD, pediatrician, virologist, educator; b. Scio, Ohio, Feb. 8, 1929; s. Carl Smith and Olive (Fulton) C.; m. Donna Jean Reese, June 16, 1960; children: Deborah Kaye, Betsy Diane. B.S., Mt. Union Coll., 1950; M.D., U. Cin., 1953. Diplomate: Am. Bd. Pediatrics, Am. Bd. Microbiology. Clin. research asso. Nat. Inst. Allergy and Infectious Diseases, Clin. Center, Bethesda, Md., 1955-57; faculty State U. Iowa, 1957-60, asst. prof., 1958-60; faculty Bowman Gray Sch. Medicine, 1960-64, prof. pediatrics, 1963-64, dir. virology lab., 1960-64; prof. pediatrics Ohio State U., Columbus, 1964—, prof. med. microbiology, 1966—, exec. dir. Children's Hosp. Research Found., 1964-73, chmn. dept. med. microbiology, 1966-73, dean Coll. Medicine, 1973-80, acting v.p. for med. affairs, 1974-80, v.p. health scis., 1980-83, Warner M. and Lora Kays Pomerene chair in medicine, 1982—; Mem. Ohio Med. Bd.; chmn. com. on cert., subcert. and recert. Am. Bd. Med. Specialists; mem. coms. on written exam., comprehensive qualifying evaluation program Nat. Bd. Med. Examiners; chmn. Accreditation Council Continuing Med. Edn.; mem. adv. com. on undergrad. med. evaluation; pres. Fedn. State Med. Bds.; bd. dirs. Ohio State U. Hosp., 1979-80. Recipient Hoffheimer prize U. Cin., 1953, Eben J. Carey award in anatomy, 1950. Fellow Am. Acad. Microbiology, AAAS; mem. Infectious Diseases Soc. Am., So. Soc. Pediatric Research (past pres.), Soc. Pediatric Research, Am. Pediatric Soc., Am. Acad. Pediatrics, Midwest Soc. Pediatric Research, Soc. Exptl. Biology and Medicine, Am. Soc. Microbiology, Alpha Omega Alpha. Subspecialties: Pediatrics; Infectious diseases. Research, publs. on etiologic assn. virus infections in illnesses of infants and children, estimation of importance of various viruses in morbidity and mortality in pediatric age group. Home: 2480 Sheringham Rd Columbus OH 43220 Office: 200 Adminstrn Center 370 W 9th Ave Ohio State U Columbus OH 43210

CRAMER, DONALD VERNON, pathology educator; b. Long Beach, Calif., Oct. 22, 1941; s. Neil Allen and Mary Josephine (Lewis) C.; m. Elizabeth Joy Lindley, July 31, 1971; children: Matthew, Kristen. B.S., U. Calif.-Davis, 1964, D.V.M., 1966; Ph.D., Harvard u., 1978. Cert. Am. Coll. Vet. Pathologists. Instr. pathology U. Pitts., 1971-72, asst. prof., 1972-78, assoc. prof., 1978—. Alexander von Humboldt Found. fellow, 1980-81; recipient Research Career Devel. award NIH, 1978-83. Mem. Am. Assn, Pathologists, Am. Assn. Immunologists, Internat. Transplantation Soc. Subspecialties: Immunogenetics; Pathology (veterinary medicine). Current work: Structure and function of the major histocompatibility complex, immunogenetics, genetic monitoring, transplantation biology. Home: 603 Lexington Ave Pittsburgh PA 15215 Office: University of Pittsburgh School of Medicine 3550 Terrace St Pittsburgh PA 15261

CRAMPTON, GEORGE HARRIS, psychologist, educator; b. Spokane, Wash., Nov. 20, 1926. B.S., Wash. State U., 1949, M.S., 1950; Ph.D., U. Rochester, 1954. Prof. psychology Wright State U., Dayton, Ohio, 1971—. Served to col. U.S. Army; ret. Fellow Am. Psychol. Assn.; mem. Soc. Neurosci. Subspecialties: Neurophysiology; Sensory processes. Current work: Vestibular system. Office: Dept Psychology Wright State U Dayton OH 45435

CRANBERG, LAWRENCE, physicist, consultant; b. N.Y.C., July 4, 1917; s. Hyman and Fannie (Rubinstein) C.; m. Charlotte Mount, Oct. 31, 1953; children: Alexis Mount, Nicole. B.S., CCNY, 1937, M.S. in Edn.; A.M. in Physics, Harvard U., 1950; Ph.D., U. Pa., 1949. Sr. physicist Signal Corps Engring. Labs., 1940-50; mem. staff Los Alamos Sci. Lab., 1950-63; prof. physics, dir. physics accelerator lab. U. Va., Charlottesville, 1963-71; v.p. research and devel. Accelerators, Inc., Austin, Tex., 1971-73; pres. TDN, Inc., Austin, 1974—, Tex. Fireframe Co., 1975—; cons. physicist, 1973—; Charter mem. Austin Soc. to Oppose Pseudosci., 1982—. Editor: Nuclear and Chem. Waste Mgmt., 1979—; contbr. articles to sci. jours. Guggenheim fellow, 1958. Fellow Am. Phys. Soc.; mem. AAUP (emeritus). Club: Men's Garden (Austin). Subspecialties: Combustion processes; Nuclear engineering. Current work: Solid-fuel combustion; high-level nuclear waste disposal; neutron therapy. Home and Office: 1205 Constant Springs Dr Austin TX 78746

CRANDALL, ELBERT WILLIAMS, chemistry educator; b. Normal, Ill., Nov. 4, 1920; s. Elbert Williams and Estella (Baker) C.; m. Betty Jo Baber, June 2, 1951; children: Paula, Patricia, Linda, William. B.Ed., Ill. State U., 1942; M.A., U. Mo.-Columbia, 1948, Ph.D., 1950. Analytical chemist Allied Chem., Henderson, Ky., 1940-42; research chemist U.S. Rubber Co., Detroit, 1951-52; prof. chemistry Pittsburg (Kans.) State U., 1952—. Mem., pres. Pittsburg (Kans.) Safety Council, 1973, Pittsburg Pride Com., 1979, 82. Served with USN, 1942-44. Mem. Am. Chem. Soc. Methodist. Lodge: Kiwanis. Subspecialties: Organic chemistry; Polymer chemistry. Current work: Mechanism of epoxy resin crosslinking. Office: Pittsburg State U Pittsburg KS 66762 Home: 704 Twin Lakes Pittsburg KS 66762

CRANDALL, MARJORIE ANN, mycologist; b. Bklyn., Mar. 15, 1940; d. George Henry and Beatrice Louise (Hart) Krautter; m. Jack Kenneth Crandall, Feb. 2, 1960 (div. 1974); 1 dau., Laura Lynn. B.S., Cornell U., 1961; Ph.D., Ind. U., 1968. Lectr. Ind. U., Bloomington, 1971-74; instr. Bklyn. Coll., 1974; asst. prof. U. Ky., Lexington, 1974-80; assoc. research prof. Harbor-UCLA Med. Center, Torrance, 1980—, NSF vis. prof. for women in sci. and engring., 1982-83. Contbr. articles to profl. jours. Mem. Am. Soc. Microbiology, Genetics Soc. Am., Med. Mycol. Soc. Am., Assn. Women in Sci., NOW, Sierra Club. Democrat. Unitarian. Subspecialties: Microbiology (medicine); Gene actions. Current work: Genetics of virulence in the pathogenic yeast Candida albicans. Home: 23930 Los Codona Apt 115 Torrance CA 90505 Office: Harbor-UCLA Med Center E-5 Torrance CA 90509

CRANDALL, WILLIAM BROOKS, university research found. exec.; b. Andover, N.Y., Feb. 243, 1921; s. Ezekiel Rogers and Harriet Louise (Brooks) C.; m. Mary Burdick, Aug. 22, 1942; children: John David, Mary Jean, Donald William. B.S., Coll. Ceramics Alfred U., 1942, M.S., 1944. Dir. Office Naval Research Alfred U., N.Y., 1946-63, asst. to assoc. prof., 1946-63, dir. indsl. liaison, 1974-76; dir. Alfred U. Research Found., 1976—; v.p. research Pfaudler Co., Sybron Corp., Rochester, N.Y., 1963-70; dir. ceramic research Ill. Inst. Tech.Research Inst., Chgo., 1970-74. Contbr. articles to profl. jours. Mem. energy adv. com. Rep. Lundine's Energy Policy Com.; chmn. council Union Univ. Ch. Fellow Am. Ceramic Soc.; mem. Nat. Inst. Ceramic Engrs, Nat. Council Univ. Research Adminstrs., Soc. Univ. Patent Adminstrs. Subspecialties: Materials; Clad metals and coating technology. Current work: Ceramic materials and products, glass fiber products, glass on steel, and waste and by-product utilitzation. Patentee in ceramics, glass, chem. igniter, tech. transfer. Home: RD 1 Alfred Station NY 14803 Office: Alfred U Alfred NY 14802

CRANE, HORACE RICHARD, educator, physicist; b. Turlock, Calif., Nov. 4, 1907; s. Horace Stephen and Mary Alice (Roselle) C.; m. Florence Rohmer LeBaron, Dec. 30, 1934; children—Carol Ann, Janet (dec.), George Richard. B.S., Calif. Inst. Tech., 1930, Ph.D., 1934. Research fellow Calif. Inst. Tech., 1934-35; mem. faculty U. Mich., Ann Arbor, 1935—, prof. physics, 1946—, chmn. dept. physics, 1965-72, George P. Williams Univ. prof., 1972-78, emeritus, 1978—; Research asso. (radar) Mass. Inst. Tech., 1940-41; physicist Carnegie Inst. Washington, 1941; project dir., proximity fuze project U. Mich., 1941-43, atomic energy project, 1943-45; cons. NDRC, 1941-45; mem. standing com. on controlled thermonuclear research AEC, 1969-72; Vice pres. Midwestern Univs. Research Assn., 1956-57, pres., 1957-60; mem. policy bd. Argonne Nat. Lab., 1957-67; Bd. govs. Am. Inst. Physics, 1964-71, chmn., 1971-75; mem. Commn. on Human Resources, 1977-80, Council for Internat. Exchange of Scholars, 1977-80. Contbr. sci. articles to profl. mags. Recipient Davisson-Germer prize, 1967; Henry Russel lectr., 1967; Distinguished Alumni medal Cal. Inst Tech., 1968; Distinguished Service award U. Mich., 1957. Fellow Am. Phys. Soc., AAAS, Am. Acad. Arts and Scis.; mem. Nat. Acad. Scis., Am. Assn. Physics Tchrs. (pres. 1965, Oersted medal 1977), Sigma Xi. Clubs: Research Univ. of Mich. (pres. 1956-57); Science Research (U. Mich.) (v.p. 1946-47, pres. 1947-48. Subspecialty: Nuclear physics. Inventor of Race Track, a modified form of synchrotron for nuclear studies, 1946; made early discoveries in field of artificially produced radioactive atoms, 1934-39; measurements of magnetic moment of free electron, 1950. Home: 830 Avon Rd Ann Arbor MI 48104

CRANE, JOSEPH LELAND, botany and plant pathology educator, mycologist; b. Wilmot, N.H., Jan. 9, 1935; s. William and Hazel (Jones) C. B.S., U. Maine, 1961; M.S., U. Del., 1964; Ph.D., U. Md., 1967. Asst. mycologist Ill. Natural History Survey, 1967-72, assoc. mycologist, 1972-79, mycologist, 1979—; assoc. prof. botany U. Ill.-Champaign-Urbana, 1976—, assoc. prof. plant pathology, 1978—. Mem. Mycol. Soc. Am., Bot. Soc. Am., Brit. Mycol. Soc., Société Mycologique de France, Brit. Lichen Soc. Subspecialties: Systematics; Plant pathology. Current work: Systematics of ascomycetes and fungi imperfecti. Office: Ill Natural History Survey 172 Natural Resources Bldg 607 E Peabody St Champaign IL 61820

CRANEFIELD, PAUL FREDERIC, educator, physician; b. Madison, Wis., Apr. 28, 1925; s. Paul Frederic and Edna (Rothnick) C. Ph.B., U. Wis., 1946, Ph.D., 1951; M.D., Albert Einstein Coll. Medicine, 1964. Fellow biophysics Johns Hopkins U., 1951-53; from instr. to assoc. prof. physiology State U. N.Y. Downstate Med. Center, N.Y.C., 1953-62; research fellow psychiatry Albert Einstein Coll. Medicine, 1960-64; exec. sec. com. publs. and med. information, editor bull. N.Y. Acad. Medicine, 1963-66; adj. assoc. prof. pharmacology Columbia Coll. Physicians and Surgeons, 1964-75, adj. prof., 1975—; assoc. prof. Rockefeller U., 1966-75, prof., 1975—. Author: (with Hoffman) The Electrophysiology of the Heart, 1960, Paired Pulse Stimulation of the Heart, 1968, (with C. McC. Brooks) The Historical Development of Physiological Thought, 1959, The Way In and the Way Out, 1974, The Conduction of the Cardiac Impulse, 1975, Claude Bernard's Revised Edition of his Introduction a L'Etude de la Médicine Expérimentale, 1976; also numerous articles.; Editor: Jour. Gen. Physiology, 1966—; mem. editorial bd.: Circulation Research Spl. Collections; cons. editor: Internat. Microform Jour. Legal Medicine, 1969-77. Chmn. bd. dirs. LaMama Exptl. Theatre Club, 1965-69; chmn. bd. dirs. Circle Repertory Co., 1970-76, The Working Theatre; trustee Milton Helpern Library Legal Medicine. Fellow N.Y. Acad. Medicine, Internat. Acad. History of Medicine; mem. Am. Physiol. Soc., Biophys. Soc., Am. Assn. History Medicine, Episcopal Actors Guild. Clubs: Century, Players, Nat. Arts (N.Y.C.); Coffee House; Cosmos (Washington); Savile (London). Subspecialties: Biophysics (biology); Electrphysiology. Current work: Electrophysiology of heart, electrophysiological basis of cardic arrythmias. Home: 310 E 9th St New York NY 10003 Office: 1230 York Ave New York NY 10021

CRANG, RICHARD FRANCIS EARL, botanist, research center adminstr.; b. Clinton, Ill., Dec. 2, 1936; s. Richard Francis and Clara Esther (Cummins) C.; m. Linda L., Aug. 10, 1958 (div.); children: Steven E., Douglas E. B.S., Eastern Ill. U., 1958; M.A., U. S.C., 1962; Ph.D., U. Iowa, 1965. Asst. prof. biology Wittenberg U., 1965-69; assoc. prof. biol. sci. Bowling Green State U., 1969-74, prof., 1974-80; prof. botany U. Ill., Urbana-Champaign, 1980—, dir., 1980—; adj. prof. anatomy Med. Coll. Ohio, 1974-80; vis. scientist, botany Cambridge (Eng.) U., 1978-79. Research, numerous publs. in field, 1967—. Mem. Statewide Democratic Support Group, Ill. Recipient Outstanding Faculty Research Recognition award Bowling Green State U., 1973, 75; Paint Research Inst. grantee, 1976-83; NSF grantee, 1981-83. Mem. AAAS, Bot. Soc. Am. Mem. Christian Ch. (Disciples of Christ). Subspecialties: Cell and tissue culture; Microbiology. Current work: Inhibitation of biodegredational fungi involving studies with electron microscopy; air pollution studies. Home: 2903 Deske Ct Champaign IL 61820 Office: 905 S Goodwin Av 74A Bevier Hall Urbana IL 61801

CRANIN, ABRAHAM NORMAN, oral surgeon, researcher; b. Bklyn., June 17, 1927; s. Samuel Leonard and Henrietta C.; m. Marilyn Sunners, June 14, 1953; children: Jonathan, Andrew, Elizabeth. A.B., Swarthmore Coll., 1947; D.D.S., NYU, 1951; cert., Mt. Sinai Hosp., 1952, 53. Assoc. attending oral surgeon Mt. Sinai Hosp., N.Y.C., 1961—; practice dentistry specializing in oral surgery; chief oral surgery Greenpoint Hosp., Bklyn., 1961-63; attending oral surgeon Community Hosp., Bklyn., 1962-72; assoc. clin. prof. Mt. Sinai Sch. Medicine, N.Y.C., 1974—; clin. prof. oral and maxillofacial surgery NYU, 1975—; dir. dental and oral surgery Brookdale Hosp. Med. Ctr., Bklyn., 1965—; cons. Nat. Patent Devel. Corp., N.Y.C., 1964-70; dir. Soc. for Biomaterials, San Antonio, 1973—; cons. oral surgeon Bklyn. Devel. Ctr., 1973—. Editor-in-chief: Jour. Oral Implantaology Quar, 1973 (Gold Key 1974), Jour. Biomed. Materials and Research, 1978 (cert. 1981). Pres. Informed Citizens Com. Hewlett Bay Park, N.Y., 1976. Served to ensign USN, 1950-51. Named Man of Yr. Fedn. of Jewish Philanthropies, Bklyn., 1973; honoree United Jewish Appeal, Brookdale Hosp., 1980; recipient award Soc. for Biomaterials, 1978; award of honor Met. Conf. Hosp. Dentists, 1982. Fellow Am. Acad. Implant Dentistry (pres. 1971), Am. Dental Soc. Anesthesiology, Royal Soc. Health, Internat. Coll. Dentistry, Brazilian Soc. Oral and Maxilofacial Surgery (hon., Rene Lefort medal 1980); hon. mem. Japanese Soc. Implant Dentistry (medal and plaque 1980). Clubs: Woodmere Bay (Bay Park, N.Y.); Woodmere (N.Y.). Subspecialties: Implantology; Oral and maxillofacial surgery. Current work: Basic and clinical research, dental, oral and maxillofacial implants, sintered titanium, hydroxyapatite, chrome-alloy anchors, Brookdale bar implant. Home: Copper Top Hewlett Bay Park NY 11557 Office: Brookdale Hosp Med Center Brookdale Plaza Brooklyn NY 11212

CRASEMANN, BERND, physicist, educator; b. Hamburg, Germany, Jan. 23, 1922; came to U.S., 1946, naturalized, 1955; s. Pablo Joaquin and Hildegard Carlota (Vorwerk) C.; m. Jean Millicent McEown, June 7, 1952. A.B., U. Calif. at Los Angeles, 1948; Ph.D., U. Calif. at Berkeley, 1953. With Lavadora de Lanas S.A., Viña del Mar, Chile, 1941-46; asst. prof. physics U. Oreg., Eugene, 1953-58, asso. prof., 1958-63, prof., 1963—, chmn. dept., 1976—; Guest asso. physicist Brookhaven Nat. Lab., Upton, N.Y., 1961-62; vis. prof. U. Calif. at Berkeley, 1968-69, Université Pierre et Marie Curie, Paris, 1977; cons. Lawrence Radiation Lab., 1954-68, physicist, 1968-69; mem. com. on atomic and molecular sci. NRC/Nat. Acad. Scis., 1976—; vis. scientist NASA Ames Research Center, 1975-76. Author: (with J.L. Powell) Quantum Mechanics, 1961; Editor: Atomic Inner-Shell Processes, 1975; mem. editorial bd.: Phys. Rev. C, 1978; Contbr. articles to sci. jours. Mem. region XIV selection com. Woodrow Wilson Nat. Fellowship Found., 1959-61, 62-68. Recipient Ersted award for distinguished teaching U. Oreg., 1959; NSF research grantee, 1954-64; U.S. AEC grantee, 1964-72; NASA grantee, 1972-79; AFOSR grantee, 1979—. Fellow Am. Phys. Soc. (chmn. div. electron and atomic physics 1981-82, councillor 1983-86); mem. Am. Assn. Physics Tchrs. (pres. Oreg. sect. 1956-57), Sierra Club, ACLU, Phi Beta Kappa. Subspecialties: Atomic and molecular physics; Nuclear physics. Current work: Atomic inner-shell processes; applications of synchrotron radiation to atomic physics; interface of atomic and nuclear physics. Office: Dept Physics U Oreg Eugene OR 97403

CRATTY, LELAND EARL, JR., chemist; b. Oregon, Ill., June 3, 1930; s. Leland E. and Kathryn E. (Prugh) C.; m. Carol R. Gardiner, June 9, 1956 (dec.); children: Paul Donald, Sarah Louise, Susan; m. Margaret B. Mason, May 28, 1982. Sc.B., Beloit Coll., 1952; Ph.D., Brown U., 1957. Research chemist Linde Air Products, Buffalo, 1956-58; vis. fellow Mellon Inst., Pitts., 1964-65, Ames (Iowa) Lab., 1969-70; faculty Hamilton Coll., Clinton, N.Y., 1958—, prof. chemistry, 1973—; vis research faculty U. Pa., 1982-83. Pres. Clinton (N.Y.) A Better Chance Program, 1976-77, Hamilton Coll. Sewer Commn., 1978—. Mem. Am. Chem. Soc., AAAS, Phi Beta Kappa, Sigma Xi. Subspecialties: Inorganic chemistry; Catalysis chemistry. Current work: Fast ion conduction in solids, cluster metal catalysis. Office: Hamilton Coll Clinton NY 13323

CRAVALHO, ERNEST GEORGE, educator; b. San Mateo, Calif., Feb. 25, 1939; s. Ernest and Naomi Marie (Dameral) C.; m.; children—Mark Andrew, Lisa Louise. B.S., U. Calif., Berkeley, 1961, M.S., 1962, Ph.D., 1967. Asst. prof. mech. engring. M.I.T., 1967-70, asso. prof., 1970-76, prof., 1976, Matsushita prof. mech. engring. in medicine, 1976—, asso. dir., 1977—, (Whitaker Coll. Health Scis., Tech. and Mgmt.), 1978—. Mem. editorial bd.: Jour. of Cryobiology, 1974. Served with USMC, 1961-67. Mem. ASME, Soc. Cryobiology. Subspecialty: Cryogenics Office: MIT Bldg E25-335 Cambridge MA 02139

CRAVEN, DONALD EDWARD, physician, educator; b. Omaha, Jan. 13, 1944; s. Orvin William and Florence Elizabeth (Waite) C.; m. Margaret Anne McClave, Aug. 27, 1966; m. Dianne Munson, Sept. 10, 1983. B.A., Wesleyan U., 1966; M.D., Albany Med. Coll., 1970. Med. intern Albany (N.Y.) Med. Ctr., 1971; med. resident McGill U., Montreal, Can., 1971-74; infectious disease fellow Boston U., 1974-76; research assoc. Bur. Biologics, /FDA/NIH, Bethesda, Md., 1976-79; asst. prof. medicine Boston U., 1979—, assoc. prof., 1983—; assoc. prof. epidemiology and stats. Boston U. Sch. Pub. Health, 1983—; epidemiologist Boston City Hosp., 1979—; cons. Merck Sharpe & Dohme, West Point, Pa., 1982-83. Author: Internal Medicine, 1983; contbr. articles to sci. jours. Served to sr. surgeon USPHS, 1976-79. Fellow ACP, Royal Coll. Physicians; mem. Infectious Diseases Soc., Am. Soc. Microbiology, Am. Fedn. Clin. Research. Subspecialties: Infectious diseases; Epidemiology. Current work: nosocomial infections; gram negative sepsis, bacterial adherence; meningococcal vaccines; hospital epidemiology. Home: 17 Ellery Sq Cambridge MA 02138 Office: Boston City Hosp 818 Harrison Ave Boston MA 02118

CRAVEN, JOHN PINNA, civil engineering educator; b. Bklyn., Oct. 30, 1924; s. James McDougal and Mabel (Pinna) C.; m. Dorothy Drakesmith, Feb. 4, 1951; children: David John, Sarah Johannah. B.S. in Civil Engring, Cornell U., 1946, M.S., Calif. Inst. Tech., 1947; Ph.D., U. Ia., 1951; J.D., George Washington U., 1959. Hydrodynamicist David Taylor Model Basin, 1951-59; chief scientist U.S. Navy Spl. Projects Office, 1959-71, project mgr. deep submergence systems project, 1965-67, chief scientist project, 1967-70; vis. prof. polit. sci. and naval architecture Mass. Inst. Tech., 1969-70; dean marine programs U. Hawaii, Honolulu, 1970-81; marine affairs coordinator State Hawaii, 1970-76, 77—; dir. Law of Sea Inst., 1977—; adj. prof. Herbert M. Humphrey Inst., 1983—. Served with USNR, 1943-46. Recipient Meritorious Civilian Service award Navy Dept., 1953, Distinguished Civilian Service award, 1960; Fleming award U.S. C. of C., 1960; William S. Parsons award Navy League, 1966; Distinguished Civilian Service award Dept. Def., 1969; Lockheed award Menne Tech. Soc., 1982. Mem. Nat. Acad. Engrs. Presbyterian. Subspecialty: Civil engineering. Home: 4921 Waa St Honolulu HI 96821 Office: Univ of Hawaii Honolulu HI 96822 Adaptation to opportunity and adversity with the goal of increasing mankind's rewards and reducing his catastrophes—with malice toward none and compassion for all.

CRAWFORD, BRYCE LOW, JR., chemist, educator; b. New Orleans, Nov. 27, 1914; s. Bryce Low and Clara Hall (Crawford) C.; m. Ruth Raney, Dec. 21, 1940; children: Bryce, Craig, Sherry Ann. A.B., Stanford U., 1934, M.A., 1935, Ph.D., 1937; Nat. Research fellow, Harvard U., 1937-39. Instr. chemistry Yale U., 1939-40; asst. prof. U. Minn., Mpls., 1940-43, assoc. prof., 1943-46, prof. phys. chemistry, 1946-82, Regents' prof. chemistry, 1982—, chmn. dept., 1955-60, dean grad. sch., 1960-72; Mem. Grad. Record Exam. Bd., 1968-72; chmn. Council Grad Schs. in U.S., 1962-63; pres. Assn. Grad. Schs., 1970; dir. research on rocket propellants under Div. 3 Nat. Def. Research Com., 1942-45. Editor: Jour. Phys. Chemistry, 1970-80. Trustee Midwest Research Inst., 1963—. Guggenheim fellow, 1950-51, 72-73; Fulbright grantee Oxford, 1951, Oxford, Tokyo, 1966; recipient Presdl. Cert. of Merit. Mem. Am. Chem. Soc. (dir. 1969-77, Priestley medal 1982, Pitts. Spectroscopy award, Ellis Lippincott award), Optical Soc. Am., AAAS, AAUP, Am. Phys. Soc., Nat. Acad. Scis. (council 1975-78, home. sec. 1979—), Coblentz Soc., Am. Philos. Soc., Am. Acad. Arts and Scis., Phi Beta Kappa, Sigma Xi, Phi Lambda Upsilon, Alpha Chi Sigma. Episcopalian. Clubs: Campus, Cosmos. Subspecialties: Physical chemistry; Molecular spectroscopy. Specialist in molecular structure and molecular spectra. Home: 1545 Branston St Saint Paul MN 55108 Office: Molecular Spectroscopy Lab U Minn 207 Pleasant St SE Minneapolis MN 55455

CRAWFORD, DAVID LIVINGSTON, astronomer; b. Tarentum, Pa., Mar. 2, 1931; s. William L. and A. Blanche (Livingston) C.; m. Mary Louise Mueller, Apr. 4, 1962; children: Christine Bayze, Deborah, Lisa. Ph.D., U. Chgo., 1958. Asst. prof. Vanderbilt U., 1958-59; astronomer Kitt Peak Nat. Observatory, Tucson, 1960—; project mgr. 4-meter Telescopes, 1962-72, assoc. dir., 1970-73. Contbr. numerous articles to sci., tech. jours. Chmn. Outdoor Lighting Com., City of Tucson, 1980—. Mem. Am. Astron. Soc. (chmn. light pollution com.), Internat. Astron. Union, Astron. Soc. Pacific, Soc. Illuminating Engrs. Subspecialties: Optical astronomy; Optical engineering. Current work: Large telescope design, astronomy instrumentation, photoelectric photometry, galactic structure, star clusters, roadway lighting and light trespass. Home: 3545 N Stewart St Tucson AZ 85716 Office: KPNO Box 26732 Tucson AZ 85726

CRAWFORD, E. DAVID, physician; b. Cin., June 6, 1947; s. Edward G. and Gertrude (Wagner) C.; m. Barbara L. Schobor, June 28, 1969; children: Mike, Marc, Ryan. M.D., U. Cin., 1973. Diplomate: Am. Bd. Urology. Intern Good Samaritan Hosp., Cin., 1973-74, resident in urology, 1974-77; assoc. prof. U. N.Mex., Albuquerque, 1981-83, U. Miss., 1983—; chmn. (S.W. Oncology Group), 1980—. Editor: Genitourinary Oncology, 1982. Fellow ACS; mem. Am. Urology Assn. Subspecialties: Urology; Chemotherapy. Current work: Testis and bladder cancer. Home: 115 SW Allen Brandon MS 39042

CRAWFORD, MICHAEL HOWARD, cardiologist, educator; b. Madison, Wis., July 10, 1943; s. William Henry and A. Kay (Keller) C.; m. Janis Rae Kirschner, June 23, 1968; children: Chelsea, Susan, Dinah Jaye, Stuart Michael. A.B., U. Calif.-Berkeley, 1965; M.D., U. Calif.-San Francisco, 1969. Diplomate: Am. Bd. Internal Medicine (cardiovascular diseases). Intern U.Calif. Hosp., San Francisco, 1969-70; resident, 1970-71, Beth Israel Hosp., Boston, 1971-72, U. Calif.-San Diego, 1972-74, asst. prof. medicine, 1974-76, U. Tex. Health Sci. Ctr., San Antonio, 1976-78, assoc. prof., 1978-82, prof., 1982—, dir. cardiac non-invasive labs., San Antonio, 1976—, co-dir. div. cardiology, 1983—. Contbr. articles to med. jours. Pres. Am. Heart Assn., 1982, bd. dirs., 1978. Recipient Paul Dudley White award Assn. Mil. Surgeons of U.S., 1981; grantee Am. Heart Assn., VA NIH, 1976—. Fellow Am. Coll. Cardiology, ACP; mem. Am. Soc. Echo cardiography (bd. dirs. 1981—), Assn. Univ. Cardiologists, So. Soc. Clin. Investigation. Club: Canyon Creek (San Antonio). Subspecialties: Cardiology; Internal medicine. Current work: Non-invasive evaluation of left ventricular performance during physiologic and pharmacologic stress, mechanisms of tissue infury in myocardial infarction. Home: 113 N Oak San Antonio TX 78432 Office: Univ Tex Health Sci Ctr 7703 Floyd Curl Dr San Tonio TX 78284

CRAWFORD, WILLIAM ARTHUR, geology educator; b. Norman, Okla., Mar. 25, 1935; s. Francis Weldon and Mildred (Crall) C.; m. Maria Luisa Buse, Aug. 28, 1963. B.S., Kans. State U., 1957; M.S., U. Kans., 1960; Ph.D., U. Calif., Berkeley, 1965. Instr. U. Calif., Berkeley, 1965; asst. prof. geology Bryn Mawr (Pa.) Coll., 1965-72, assoc. prof., 1972-81, prof., 1981. Contbr. Articles to profl. jours. Served to capt. U.S. Army. Fellow Geol. Soc. Am.; mem. Mineral. Soc. Am., Nat. Assn. Geology Tchrs., Phila. Geol. Soc. (pres. 1982-84). Republican. Episcopalian. Subspecialties: Geochemistry; Petrology. Current work: Petrological and geochemical studies of igneous and metamorphic rock suites in southeastern Pennsylvania, Egypt and southeastern Alaska. Home: 131 Pennsylvania Ave Bryn Mawr PA 19010 Office: Dept Geology Bryn Mawr Coll Sci Center New Gulph Rd Bryn Mawr PA 19010

CRAWLEY, JACQUELINE N., neurobiologist; b. Phila., June 14, 1950; d. Samuel and Miriam Lerner. B.A., U. Pa., 1971; Ph.D., U. Md., 1976. Postdoctoral research fellow Yale U. Sch. Medicine, 1976-79; pharmacology staff fellow NIMH, Bethesda, Md., 1979-81, chief unit on behavioral neuropharmacology, 1983—; neurobiologist Central Research Life Scis. Program, E.I. DuPont de Nemours & Co., Wilmington, Del., 1981-83. Subspecialty: Neuropharmacology. Current work: Neuropsychopharmacology, peptides; animal models for anxiety and depression. Home: 204 E Delaware Ave Wilmington DE 19809 Office: DuPont Co 500 Ridgeway Ave Glenolden PA 19036

CRAWLEY, PAUL F., nuclear engineer; b. Carthage, N.Y., Nov. 23, 1943; s. Richard F. and Alice M. (Franks) C.; m. Lorraine J. Campbell, Jan. 28, 1967; children: Kevin, Sean. J.T. A.B. cum laude, Kenyon Coll., 1965; M.S., Carnegie-Mellon U., 1967. Scientist, Bettis Atomic Power Lab., West Mifflin, Pa., 1967-72; nuclear engr. Middle South Services, New Orleans, 1972; scientist Bettis Atomic Power Lab., 1972-74; sr. nuclear engr. Boston Edison Co., 1974-78; nuclear supr. Ariz. Pub. Service, Phoenix, 1978-83, mgr. nuclear fuel mgmt., 1983—. Chmn. Litchfield Little League, 1981-82. Mem. Am. Nuclear Soc. Republican. Roman Catholic. Subspecialties: Nuclear fission; Nuclear engineering. Current work: Optimization of nuclear fuel management strategies consistent with commercial nuclear power plant reactor engineering operational needs. Home: 300 W Llano Dr Litchfield Park AZ 85340 Office: Ariz Pub Service Co PO Box 21666 Phoenix AZ 85036

CRAWSHAW, RALPH, psychiatrist; b. N.Y.C., July 3, 1921. A.B., Middlebury (Vt.) Coll., 1943; M.D., N.Y. U., 1947. Diplomate: Nat. Bd. Med. Examiners, Am. Bd. Psychiatry and Neurology. Intern Lenox Hill Hosp., N.Y.C., 1947-48; resident Menninger Sch. Psychiatry, Topeka, 1948-50, Oreg. State Hosp., Salem, 1950-51; practice medicine specializing in psychiatry, Washington, 1954; staff psychiatrist C.F. Menninger Meml. Hosp., Topeka, 1954-57; asst. chief VA Mental Hygiene Clinic, Topeka, 1957-60; staff psychiatrist Community Child Guidance Clinic, Portland, Oreg., 1960-63; founder, clinic dir. Tualatin Valley Guidance Clinic, Beaverton, Oreg., 1961-67; pvt. practice medicine, specializing in psychiatry, Portland, 1960—; mem. staff Holladay Park Hosp., 1961—; lectr. dept. child psychiatry U. Oreg. Med. Sch., 1961-63, asso. clin. prof. dept. psychiatry, 1976; lectr. Sch. Social Work, Portland State U., 1964-67; founder Benjamin Rush Found., 1968, pres., 1968—; founder Friends of Medicine, 1969, Ct. of Man, 1970, Club of Kos, 1974. Contbr. editor: AMA Jour. of Socio-Econs, 1972-75; Columnist: Prism mag, 1972-76, The Pharos, 1972—, Portland Physician, 1975; Contbr. articles to med. jours. Cons. Bur. Hearings and Appeals, HEW, 1964—, Albina Child Devel. Center, Portland, 1965-75, HEW Region 8 Health Planning, 1979; mem. Inst. Medicine, Nat. Acad. Sci., 1978, Oreg. Health Coordinating Council, 1979; Mem. Gov.'s Adv. Com. on Mental Health, 1966-72; ad hoc com. Nat. Leadership Conf. on Am. Health Policy, 1976, Gov.'s Adv. Com. on Med. Care to Indigent, 1976—; trustee Millicent Found., 1964-67, Multnomah Found. for Med. Care, 1977; vis. scholar Center for Study Democratic Instns., 1969, Jack Murdock Charitable Trust, 1977; U.S.-USSR exchange scholar, 1973. Served with AUS, 1943-46; to lt., M.C. USN, 1951-54. Named Oreg. Dr./Citizen of Yr., 1978; U.S.-USSR exchange scholar, 1973, 79. Fellow Am. Psychiat. Assn.; mem. AMA, Nat. Med. Assn., Oreg. Med. Assn. (trustee 1972—), Multnomah County Med. Soc. (pres. 1975), Royal Soc. Medicine, Inst. of Medicine of Nat. Acad. Sci., Am. Psychol. Assn., N.Pacific Soc. Neurology and Psychiatry, Soc. for Psychol. Study Social Issues, Western European Assn. Aviation Psychology, Am. Med. Writers Assn., AAAS, Portland Psychiatrists in Pvt. Practice (pres. 1971), Alpha Omega. Subspecialty: Psychiatry. Current work: Health services research; impaired physician; allocation of health resources. Address: 2525 NW Lovejoy St Suite 404 Portland OR 97210

CRAY, SEYMOUR R., computer designer; b. Chippewa Falls, Wis., 1925. B.S.E.E., U. Minn., 1950, 1950. Computer scientist Univac, until 1957; founder Control Data Corp., 1957-72. Subspecialty: Computer architecture. Founder Cray Research Inc., Mendota Heights, Minn., 1972—, designer Cyber 205, Cray-1, Cray X-MP and other computer systems. Office: Cray Research Inc 1440 N Sand Dr Mendota Heights MN 55120

CREAGAN, ROBERT JOSEPH, energy scientific consultant; b. Rockford, Ill., Aug. 24, 1919; s. Paul Thomas and Mary (Sulivan) C.; m. Irene Marie Tkacs, Aug. 21, 1948; children: Susan, Robert II, Mary, Tim. B.S. in Engring, Ill. Inst. Tech., Chgo., 1942; M.S. in Physics, Yale U., 1943; Ph.D., 1948. Mgr. nuclear engring. Westinghouse Electric Corp., Pitts., 1949-57, asst. tech. dir. nuclear engring., 1961-69, project mgr., 1969-74, dir. tech. assessment, 1974-79; cons. Research and Devel. Ctr., 1978—; dir. nuclear program Bendix Corp., Detroit, 1957-61; dir. Atomic Power Devel. Assocs., Detroit, 1958-61; cons. Power Reactor Devel. Corp., Detroit, 1961-65, Yankee Atomic Electric Corp., Rowe, Mass., 1965-68. Author: Nuclear Engineering, 1948; Contbr. articles to profl. jours. Served as lt. USNR, 1943-46; PTO. Fellow Am. Nuclear Soc. (chmn. power div. 1965-66); mem. Nat. Acad. Engring. Club: Gateway Heights (Monroeville, Pa.) (pres. 1976-81). Subspecialties: Agricultural engineering; Fuels. Current work: Nuclear power development; advanced energy development energy, power and fuel economics. Home: 2305 Haymaker Rd Monroeville PA 15146 Office: Westinghouse Electric Corp 1310 Beulah Rd Pittsburgh PA 15146

CREAGER, JOE SCOTT, geology and oceanography educator; b. Vernon, Tex., Aug. 30, 1929; s. Earl Litton and Irene Eugenia (Keller) C.; m. Barbara Clark, Aug. 30, 1951; children: Kenneth Clark, Vanessa Irene. B.S., Colo. Coll., 1951; postgrad., Columbia U., 1952-53; M.S., Tex. A&M U., 1953, Ph.D., 1958. Asst. prof. Sch. Oceanography, U. Wash., Seattle, 1958-61, assoc. prof., 1962-66, asst. chmn., 1964-65, prof. oceanography, 1966—, prof. geol. sci., 1981—, assoc. dean arts and scis. for earth and planetary scis., assoc. dean research, 1966—; program dir. oceanography NSF, 1965-66; chief scientist numerous oceanographis expdns. to Arctic and sub-arctic including Leg XIX of Deep Sea Drilling project, 1959—; vis. geol. scientist Am. Geol. Inst., 1962, 63, 65; U.S. Nat. coordinator Internat. Indian Ocean Expedition, 1965-66; vis. scientist, program lectr. Am. Geophys. Union, 1965-72; Battelle cons., advanced waste mgmt.,

1974; cons. U.S. Army C.E., 1976, U.S. Depts. Interiors and Commerce, 1975; exec. sec., exec. com., chmn. planning com. Joint Oceanographic Insts. Deep Earth Sampling, 1970-72, 76-78. Editorial bd.: Internat. Jour. Marine Geology, 1964—; assoc. editor: Jour. Sedimentary Petrology, 1963-67; asst. editor: Quaternary Research, 1970-82; contbr. articles in field to profl. jours. Skipper, Sea Scout Ship Boy Scouts Am., Bryan, Tex., 1957; coach Little League Baseball, Seattle, 1964-71, sec., 1971; cons. sci. curriculum Northshore Sch. Dist., 1970; mem. Seattle Citizens Shoreline Com., 1973-74, King County Shoreline Com., 1980. Served with U.S. Army, 1953-55. Colo. Coll. scholar, 1949-51; NSF grantee, 1962-82; ERDA grantee, 1962-64; U.S. Army C.E. Grantee, 1975-82; Office Naval Research grantee; U.S. Dept. Commerce grantee; U.S. Geol. Survey grantee. Fellow Geol. Soc. Am., AAAS; mem. Internat. Assn. Quaternary Research, Am. Geophys. Union, Internat. Assn. Sedimentology, Soc. Econ. Paleontologists and Mineralists, Marine Tech. Soc. (sec.-treas. 1972-75), Sigma Xi, Beta Theta Pi, Delta Epsilon. Club: Explorers. Subspecialties: Oceanography; Sedimentology. Current work: Shallow-water marine sediment transport mechanics; sea-level changes; recent marine sedimentary stratigraphy. Home: 6320 NE 157th St Bothell WA 98011 Office: U Wash Sch Oceanography WB-10 Seattle WA 98195

CREDIDIO, STEVEN GEORGE, psychologist, educator; b. Bronx, Sept. 5, 1949; s. George G. and Josephine (Persiano) C. B.S., Fordham U., 1971; M.A., U. Detroit, 1974, Ph.D., 1975. Diplomate: Am. Bd. Profl. Psychology; Lic. psychologist, Calif., N.Y.; cert. biofeedback therapist Biofeedback Cert. Inst. Am. Lestr. Marygrove Coll., Detroit, 1975-76; psychologist F.D.R. VA Med. Ctr., Montrose, N.Y., 1976-80; clin. assoc. prof. Fuller Grad. Sch., Pasadena, 1980—; adj. prof. Pepperdine U., Los Angeles, 1982—; psychologist VA Outpatient Clinic, Los Angeles, 1980—; dir. Biofeedback Headache Clinic, 1977-80, Stress Mgmt. Clinic, Los Angeles, 1980—. U. Detroit fellow, 1972-73. Mem. Am. Psychol. Assn., Biofeedback Soc. Am., Calif. Psychol. Assn., Western Psychol. Assn., Assn. for Advancement of Psychology, Psi Chi. Subspecialties: Biofeedback; Psychophysiology. Current work: Biofeedback, meditation, stress management, psychophysiological a. Home: 12772 Pacific Ave #5 Los Angeles CA 90066 Office: VA Outpatient Clinic 425 S Hill St Los Angeles CA 90013

CREECH, RICHARD HEARNE, physician; b. Boston, Apr. 6, 1940; s. Hugh J. and E. Marie (Hearne) C.; m. Charlotte E. Goetz, Dec. 28, 1963; children: Susan Marie, Nancy Elizabeth. A.B., Johns Hopkins U., 1961; M.D., U. Pa., 1965. Diplomate: Am. Bd. Internal Medicine (Subspecialty in med. oncology, hematology). Intern, resident in medicine Hosp. of U. Pa., 1965-67; clin. assoc. lab. molecular pharmacology Nat. Cancer Inst., Bethesda, Md., 1967-70; fellow in hematology and immunology Hosp. of U. Pa., 1970-71; chief med. oncology service Phila. Gen. Hosp., U. Pa. Service, 1971-72; assoc. attending physician Am. Oncologic Hosp., Fox Chase Cancer Ctr., Phila., 1972—. Contbr. articles to profl. jours. Served with USPHS, 1967-70. Fellow ACP; mem. Am. Soc. Clin. Oncology, Am. Assn. Cancer Research, AMA, Coll. Physicians Phila. Republican. Episcopalian. Subspecialties: Cancer research (medicine); Oncology. Current work: Development of minimally toxic but maximally active chemotherapy regimens for patients with breast cancer; determination of respective roles of multi-drug chemotherapy and radiation therapy in treatment of localized small cell lung cancer. Office: Am Oncologic Hosp Fox Chase Cancer Ctr Central and Shelmire Aves Philadelphia PA 19111

CREED, DAVID, chemistry educator; b. Colchester, Eng., Sept. 22, 1943; came to U.S., 1968; s. Kenneth Wilfred and Joyce Eleanor (Smith) C.; m. Sherry Kay Bain, July 29, 1976; 1 son, Benjamin. B.Sc., U. Manchester, Eng., 1965, M.Sc., 1966, Ph.D., 1968. Research fellow U. Tex.-Dallas, 1968-71, Welch fellow, 1973-77; SRC fellow Royal Instn., London, 1971-72; lectr. U. Warwick, Eng., 1972-73; asst. prof. chemistry U. So. Miss., Hattiesburg, 1977-80, assoc. prof., 1980—. Mem. Chem. Soc. (London), Am. Chem. Soc., Am. Soc. Photobiology, Inter-Am. Photochem. Soc. Subspecialties: Photochemistry; Solar energy. Current work: Organic photochemistry in relation to solar energy conversion and photobiology. Home: 136 Pinehills Dr Hattiesburg MS 39401 Office: U So Miss Box 5043 Hattiesburg MS 39406

CREESE, IAN NIGEL RICHARD, pharmacologist, educator; b. Bristol, Eng., Apr. 4, 1949; s. Douglas Ernest and Marjorie Florence (Hannell) C.; m. Paula Anne Tallal, July 21, 1972. B.A., Queens' Coll., Cambridge (Eng.) U., 1970, M.A., 1972, Ph.D., 1973. Postdoctoral fellow Johns Hopkins Med. Sch., 1973-79; asst. prof. U. Calif., San Diego, 1978-81, asso. prof., 1981—; mem. study sect. NIMH, 1979-84; cons. in field. A.P. Sloan fellow, 1978-82; NIMH research award, 1978-85. Mem. Soc. Neurosci., AAAS, European Soc. Neurosci., Am. Soc. Pharmacology and Exptl. Therapeutics, Am. Soc. Neurochemistry; mem. Am. Coll. Neuropsychopharmacology; Mem. Internat. Narcotics Research Club. Subspecialties: Neuropharmacology; Regeneration. Current work: Psychopharmacology, drug and neurotransmitter receptors.

CREIGHTON, DONALD JOHN, biochemist, educator; b. Stockton, Calif., Jan. 25, 1946; s. William H. and Helen K. (Norberg) C.; m. Arlene G. Keh, Aug. 16, 1969; children: Diane K., Christine Gail. B.S. magna cum laude, Calif. State U.-Fresno, 1968; Ph.D. in Biol. chemistry, UCLA, 1972. Postdoctoral fellow in biochemistry Inst. Cancer Research, Phila., 1972-75; assoc. prof. chemistry U. Md., Balt., 1975—. Contbr. articles to profl. jours. Grantee Am. Cancer Soc., 1980, 82, Research Corp., 1976, Petroleum Research Fund, 1976, NIH, 1979, 83. Mem. Am. Chem. Soc., Sigma Xi. Democrat. Subspecialty: Biochemistry (medicine). Current work: Protein structure and enzyme mechanisms. Office: U Md Baltimore County 5401 Wilkens Ave Catonsville MD 21228

CRENSHAW, DAVID BROOKS, educator, animal scientist; b. Columbia, Mo., May 15, 1945; s. Joe Perry and A. Karleen (Brooks) C.; m. Sherry Gail Quisenberry, Mar. 26, 1966; children: Denise Renee, David Keith, Joe Perry. B.S. in Agr, U. Mo., Columbia, 1968, M.S. in Animal Husbandry, 1969, Ph.D. in Animal Genetics, 1972. Asst. prof. animal sci. Tex. A&I U., Kingsville, 1972-76, asso. prof., 1976-82, prof., 1982—. Contbr. articles to profl. jours. Named Outstanding Prof. Coll. Agr. Tex. A&I U., 1981; recipient numerous research grants. Mem. Am. Soc. Animal Sci. Democrat. Mem. Disciples of Christ. Lodge: Javelina Kiwanis of Kingsville (pres. 1978). Subspecialties: Animal breeding and embryo transplants; Animal genetics. Current work: manipulation of postpartum estrus period in range beef cow and heifer, cytogenetic effects of drugs and feed additives on chromosomes in domestic animals. Home: 2111 Colorado St Kingsville TX 78363 Office: Box 156 Kingsville TX 78363

CRERAR, DAVID ALEXANDER, geochemistry educator; b. Toronto, Ont., Can., July 23, 1945; came to U.S., 1969; s. Louis Alexander and Dorothy Mary (Biehl) C.; m. Scotia W. MacRae, Aug. 14, 1971. B.Sc., U. Toronto, 1967, M.Sc., 1969; Ph.D., Pa. State U., 1974. Research asst. U. Toronto, 1967-69, Pa. State U., State College, 1969-74; asst. prof. geochemistry Princeton U., 1974-80, assoc. prof., 1980-83, prof., 1983—. Contbr. articles to profl. jours. Shell Found. disting. term prof., 1980. Mem. Geochem. Soc., Soc. Econ. Geologist (Lindgren award 1982). Subspecialties: Geochemistry; High temperature chemistry. Current work: Geochemistry: hydrothermal, environmental, sedimentary. Home: 50 Patton Ave Princeton NJ 08540 Office: Princeton U Dept Geology Guyot Hall Princeton NJ 08544

CRESS, DEAN ERVIN, geneticist; b. Birmingham, Ala., Apr. 29, 1949; s. Glen H., Jr. and Dorothy Dean (Earnst) C.; m. Pamela Brackett, Dec. 1, 1951. B.S. in Zoology, U. Tenn., 1971, Ph.D. in Biochemistry, 1976. Postdoctoral research assoc. dept. biology U. Utah, Salt Lake City, 1976-79; research geneticist U.S. Dept. Agr., Beltsville, Md., 1979—; faculty U.S. Dept. Agr. Grad. Sch., Washington. Contbr. articles to publs. in field. Am. Cancer Soc. postdoctoral fellow, 1976-78; U.S. Dept. Agr. competitive research grantee, 1981—; recipient U.S. Dept. Agr. Sci. and Edn. Dir.'s Award, 1982. Mem. Am. Soc. Microbiology, Am. Soc. Plant Physiologists. Subspecialties: Genetics and genetic engineering (agriculture); Developmental biology. Current work: Regulation of gene expression in higher plants and development control of plant growth. Office: Beltsville Agrl Research Center Beltsville MD 20705

CREVELING, CYRUS ROBBINS, pharmacologist, educator; b. Washington, May 30, 1930; s. Cyrus Robbins and Edith Lois (Hill) C.; m. Cornelia Mills Rector, Aug. 6, 1935; children: Victoria, Diana. A.A., George Washington U., 1952, B.S., 1954, M.S., 1955; Ph.D. (Am. Diabetes fellow), Harvard U., 1964. Instr. George Washington U., 1954-58; Nat. Heart Inst. grantee, 1958-62; research assoc. Harvard Sch. Medicine, Mass. Gen. Hosp., 1962-64; pharmacologist Nat. Inst. Arthritis Metabolics Digestive Kidney Diseases, Lab. Bioorganic chemistry, NIH, Bethesda, Md., 1980—; adj. prof. Howard U. Med. Sch., Washington, 1969—; mem. div. research grants, study sect. A Pharmacology and Exptl. Therapeutics; mem. NSF neurobiology rev. adv.; mem. project adv. group FDA div. nutrition; spl. examiner Bd. Civil Service Examiners; Wash. Acad. Scis. fellow, 1977; chmn. awards com. Found. for Advanced Edn. in the Scis., 1979—. Contbr. articles on pharmacology to profl. jours.; editor: Transmethylation, 1978, Biochemistry of S-Adeno-sul Methionine and Related Compounds, 1982. Active BCC-YMCA, 1964—; mem. Pres.'s Club, Howard U., 1979—. Mem. Soc. Exptl. Biology and Medicine (pres. D. C. Chpt. 1978-79, recipient disting. scientist award 1979), AAAS, Am. Soc. Pharmacology and Exptl. Therapeutics, Am. Chem. Soc., Chem. Soc. Washington, Gordon Conf. on Cylic Nucleotides and Catecholamines, Soc. Neurosci. Republican. Methodist. Clubs: Howard U. Century (Washington); Catecholamine (pres. 1980-81). Subspecialties: Molecular pharmacology; Neuropharmacology. Current work: Pharmacology of biogenic amines; biosynthesis and degradation of biogenic amines; immunolocalization of biogenic amines and related biosynthesis and degredative enzymes. Office: Lab Bioorganic Chemistr NIADDK NIH Bethesda MD 20205

CREWE, ALBERT VICTOR, physicist, research adminstr.; b. Bradford, Yorkshire, Eng., Feb. 18, 1927; came to U.S., 1955, naturalized, 1961; s. Wilfred and Edith Fish (Lawrence) C.; m. Doreen Blunsdon, Apr. 9, 1949; m.; children—Jennifer, Sarah, Elizabeth, David. B.S. in Physics, U. Liverpool, Eng., 1947, Ph.D., 1951; hon. degrees, Lake Forest Coll., 1972, U. Mo., 1972, Elmhurst Coll., 1972. Asst. lectr. U. Liverpool, Eng., 1950-52, lectr., 1952-55; research asso. U. Chgo., 1955-56, asst. prof., 1956-58, asso. prof., 1958-63; prof. dept. physics and Enrico Fermi Inst., 1963-71, dean phys. scis. div., 1971-81; also William Rather Disting. Service prof. physics and biophysics; dir. particle accelerator div. Argonne Nat. Lab., 1958-61, dir., 1961-66. Chmn. Chgo. Area Research and Devel. Council. Recipient Outstanding Local Citizen in Field of Sci. award Chgo. Jr. Assn. Commerce and Industry, 1961; Outstanding New Citizen of Year award Citizenship Council Chgo., 1962; award for outstanding achievement in field of sci. Immigrant's Service League, 1962; Man of Year in Research award Indsl. Research, Inc., 1970; Michelson medal Franklin Inst., 1977; Duddell medal Inst. of Physics, 1980. Fellow Am. Phys. Soc., Am. Nuclear Soc.; mem. Sci. Research Soc. Am., Electron Microscopy Soc. Am. (Disting. Service award 1976), N.Y. Microscope Soc. (Abbe award 1979), Am. Acad. Arts and Scis., Nat. Acad. Scis. Subspecialty: Nuclear physics. Nuclear physics research using particle accelerators. Devel. particle accelerator, external beams from cyclotrons and synchrotron design, devel. electron microscopes. Home: 63 Old Creek Rd Palos Park IL 60464

CREWS, DAVID P., behavorial endocrinologist; b. Jacksonville, Fla., Apr. 18, 1947; s. Sidney Walker and Anne (Pafford) C. B.A., U. Md., 1969; Ph.D., Rutgers U., 1973. Research asst. Bur. Social Sci. Research, Washington, 1968-69; NIMH predoctoral trainee Inst. Animal Behavior, Rutgers U., 1969-73; research zoologist dept. zoology U. Calif., Berkeley, 1973-75; assoc. Mus. Comparative Zoology, Harvard U., Cambridge, Mass., 1975-81, lectr. biology and psychology, 1975-76, asst. prof. biology and psychology, 1976-78, assoc. prof. biology and psychology, 1978-81; sr. assoc. prof. zoology and psychology U. Tex., Austin, 1982—. Contbr. numerous articles to sci. jours. NIMH research scientist devel. awardee, 1977—; Sloan fellow, 1978-80. Fellow AAAS; Mem. Am. Soc. Zoologists, Soc. Study of Reprodn., Soc. Neurosci., Animal Behavoir Soc., Endocrine Soc., Sigma Xi, Psi Chi. Subspecialties: Ethology; Evolutionary biology. Current work: Psychoneuroendocrinology; comparative endocrinology; reproductive biology of lower vertebrates. Office: Dept Zoology U Tex Austin TX 78712

CRICK, FRANCIS HARRY COMPTON, biologist, educator; b. June 8, 1916; s. Harry and Annie Elizabeth (Wilkins) C.; m. Ruth Doreen Dodd, 1940 (div. 1947); 1 son; m. Odile Speed, 1949; 2 daus. B.Sc., Univ. Coll., London; Ph.D., Cambridge U. Scientist Brit. Admiralty, 1940-47, Strangeways Lab., Cambridge, 1947-49; biologist Med. Research Council Lab. of Molecular Biology, Cambridge, 1949-77; Kieckhefer Disting. prof. Salk Inst. for Biol. Studies, San Diego, 1977—, nonresident fellow, 1962-73; vis. lectr. Rockefeller Inst., N.Y.C., 1959; vis. prof. chemistry dept. Harvard U., 1959, vis. prof. biophysics, 1962; fellow Churchill Coll., Cambridge, 1960-61, UCLA, 1962; Warren Triennial prize lectr. (with J.D. Watson), Boston, 1959; Korkes Meml. lectr. Duke U., 1960; Henry Sidgewick Meml. lectr., Cambridge, 1963, Graham Young lectr., Glasgow, 1963; Robert Boyle lectr. Oxford U., 1963; Vanuxem lectr. Princeton U., 1964; William T. Sedgwick Meml. lectr. MIT, 1965; Cherwell-Simon Meml. lectr. Oxford U., 1966; Shell lectr. Stanford U., 1969; Paul Lund lectr. Northwestern U., 1977; Dupont lectr. Harvard U., 1979. Author: Of Molecules and Men, 1966, Life Itself, 1981; contbr. papers and articles on molecular and cell biology to sci. jours. Recipient Prix Charles Leopold Mayer French Academies des Sciences, 1961; Research Corp. award with J.D. Watson, 1961; Gairdner Found. award, 1962; Nobel Prize for medicine (jointly), 1962; Royal Medal Royal Soc., 1972; Copley Medal, 1976; Michelson-Morley award, 1981. Fellow AAAS, Royal Soc.; mem. Am. Acad. Arts and Scis. (fgn. hon.), Am. Soc. Biol. Chemistry (hon.), U.S. Nat. Acad. Scis. (fgn. assoc.), German Acad. Sci., Am. Philos. Soc. (fgn. mem.), French Acad. Scis. (assoc. fgn. mem.). Subspecialties: Cell biology; Neurobiology. Office: Salk Inst Biol Studies PO Box 85800 San Diego CA 92138

CRIDER, ANDREW BLAKE, psychology educator, medical consultant; b. Cleve., June 11, 1936; s. Blake and Doris (Towne) C.; m. Anne Horrocks, Apr. 25, 1964; children: Juliet Gage, Jonathan Andrew. B.A., Colgate U., 1958; M.S., U. Wis., 1960; Ph.D., Harvard U., 1964. Lic. psychologist, Mass. Research assoc. Harvard U., 1964-68; asst. prof. psychology Williams Coll., 1968-71, assoc. prof., 1971-77, prof., 1977—; dir. biofeedback Berkshire Med. Ctr., Pittsfield, Mass., 1981—. Author: Schizophrenia, 1979, Psychology, 1983; contbr. chpts. to books, articles to profl. jours. Bd. dirs. North Berkshire Mental Health Assn., North Adams, Mass., 1982—. Fulbright scholar, 1959; NIMH grantee, 1964-74. Mem. Am. Psychol. Assn., Soc. Psychophysiol. Research, Biofeedback Soc. Am., AAAS. Subspecialties: Biofeedback; Neurochemistry. Current work: Clinical biofeedback for stress-related disorders; research on neurochemistry of schizophrenia. Home: 770 Hancock Rd Williamstown MA 01267 Office: Williams College Williamstown MA 01267

CRIM, FORREST FLEMING, JR., chemistry educator; b. Waco, Tex., May 30, 1947; s. Forrest Fleming and Almanor Adair (Chapman) C.; m. Joyce Ann Wileman, June 13, 1969; 1 dau., Tracy. B.S., Southwestern U., Georgetown, Tex., 1969; Ph.D., Cornell U., 1974. Staff mem. Western Electric, Princeton, N.J., 1974-76; staff mem. Los Alamos Nat. Lab., 1976-77; asst. prof. chemistry U. Wis.-Madison, 1977-82, assoc. prof., 1982—. Alfred P. Sloan Found. fellow, 1981; Camille and Henry Dreyfus Found. scholar, 1982. Mem. Am. Chem. Soc., Am. Phys. Soc., AAAS. Subspecialties: Kinetics; Laser-induced chemistry. Current work: Molecular dynamics; state to state chemistry, chemical physics. Home: 432 N Few St Madison WI 53703 Office: University of Wisconsin 1101 University Ave Madison WI 53706

CRIM, GARY ALLEN, dental educator, researcher; b. Louisville, July 13,1949; s. John William and Ruby Mae (Willis) C. Student, Ind. U., 1967-69; student, U. Ky., 1969-70, D.M.D., 1974; M.S.D., Ind. U. Sch. Dentistry-Indpls., 1981. Practice gen. dentistry, Milton, Ky., 1974-77; asst. prof. restorative dentistry U. Louisville Sch. Dentistry, 1977—. Recipient Nat. Research Service award Nat. Inst. Dental Research, 1980. Mem. Internat. Assn. Dental Research, ADA, Acad. Operative Dentistry, Acad. Internat. Dental Studies, Am. Assn. Dental Schs., Omicron Kappa Upsilon. Democrat. Methodist. Club: Gideons (Clarksville, Ind.) (v.p. club 1982—). Subspecialties: Biomaterials; Operative dentistry. Current work: Microleakage of composite restorative resins; cavity preparations for Class II posterior composite resins. Home: 9810 Vieux Carre Dr 3 Louisville KY 40223 Office: U Lousville Sch Dentistry 501 S Preston Louisville KY 40292

CRIM, JOE WILLIAM, zoology educator; b. Wichita Falls, Tex., Nov. 19. 1945; s. Ed Franklin and Sarah Rose (Butcher) C.; m. Lana Rene Emmons, Nov. 3, 1975. A.B., U. Calif., Berkeley, 1968, M.A., 1972, 1974. Research scientist Am. Med. Ctr., Denver, 1974-75; research assoc. U. Wash., Seattle, 1975-78; asst. prof. U. Ga., Athens, 1978—; ad hoc reviewer NSF, 1980-83. Precinct committeeman Democratic Party, Seattle, 1976-77. NSF grantee, 1980, 82; NIH fellow, 1971-74, 1975-77. Mem. Am. Soc. Zoologists (sec. div. comparative endocrinology 1981-83). Subspecialties: Endocrinology; Immunocytochemistry. Current work: Function and evolution of vertebrate endocrine systems; comparative neurobiology of hypophysiotropic and selected other peptides; neurosecretion in modern and primitive fishes. Office: U Ga Dept Zoology Athens GA 30602

CRIMINALE, WILLIAM OLIVER, JR., mathematics educator; b. Mobile, Ala., Nov. 29, 1933; s. William Oliver and Vivian Gertrude (Sketoe) C.; m. Ulrike Irmgard Wegner, June 7, 1962; children: Martin Oliver, Lucca. B.S., U. Ala., 1955; Ph.D., Johns Hopkins U., 1960. Asst. prof. Princeton (N.J.) U., 1962-68; asso. prof. U. Wash., Seattle, 1968-73, prof. oceanography, geophysics, applied math., 1973—, chmn. dept. applied math., 1976—; cons. Aerospace Corp., 1962-65, Boeing Corp., 1968-72, AGARD, 1967-68, Lennox Hill Hosp., 1967-68; guest prof., Can., 1965, France, 1967-68, Germany, 1973-74, Sweden, 1973-74, Nat. Acad. exchange scientist, USSR, 1969, 72. Author: Stability of Parallel Flows, 1967; Contbr. articles to profl. jours. Served with U.S. Army, 1961-62. Boris A. Bakmeteff Meml. fellow, 1957-58; NATO Postdoctoral fellow, 1960-61; Alexander von Humboldt Sr. fellow, 1973-74. Mem. AAAS, Am. Phys. Soc., Am. Geophys. Union, Fedn. Am. Scientists, Soc. Indsl. and Applied Math. Subspecialty: Applied mathematics. Home: 1635 Peach Court E Seattle WA 98712 Office: Applied Math FS-20 U Wash Seattle WA 98195

CRIPPEN, RAYMOND C., cons. chemist; b. Bronx, N.Y., Mar.1, 1917; s. CharlesH. and Betty B. (Brixner) C.; m. Helen L. Wolf, July 5, 1941; children: Lawrence J., Judith A. B.S., Iowa State U., 1939; M.S., Johns Hopkins U., 1948; postgrad., U. Md., 1949-54, U. Del., 1954-61; Ph.D., St. Thomas Inst., 1970. Group leader DuPont, Ft. Madison, Iowa, 1940-45; fellow Johns Hopkins U., 1945-48; dir. Crippen Labs. Inc., div. Foster D. Snell, Inc., Balt., 1949-61; group leader Atlas Chem. Co. Inc., Wilmington, Del., 1961-66; sect. head Stauffer Chem. Co., Adrian, Mich., 1966-68, Richardson-Merrell, Cin., 1968-70; instr. N. Ky. State U., 1970-75; pres. Crippen Labs., Inc., New Castle, Del, 1975—. Author 2 books in field. Mem. Am. Chem. Soc., Soc. Applied Spectroscopy, ASTM, Fedn. Analytical Chemists, Chromatography Discussion Group. Republican. Methodist. Clubs: Cityside, Inc., Dickinson Theatre Organ Soc. (Wilmington, Del.). Subspecialties: Analytical chemistry; Biochemistry. Current work: Analytical chemistry, biochemistry, product development. Home: 1806 Lovering Ave Wilmington DE 19806 Office: 4027 New Castle Ave New Castle DE 19720

CRIPPS, DEREK J., physician; b. London, Sept. 17, 1928; m., 1963; 4 children. M.B., B.S., U. London, 1953, M.D., 1965; M.S., U. Mich., 1961. Diplomate: Am. Bd. Dermatology. Intern in medicine London Hosps., 1953-54; resident dermatology Med. Center U. Mich., 1959-62; sr. registrar Inst. Dermatology Eng., 1962-65; asst. prof. medicine U. Wis., Madison, 1965-68, assoc. prof. dermatology, 1968-72, prof. dermatology, chmn. dept., 1972—; mem. study com. NIH, 1969—. NIH grantee, 1966—. Mem. British Dermatol. Assn., Am. Acad. Dermatology, Am. Fedn. Clin. Research, Soc. Investigative Dermatology, ACP. Subspecialty: Dermatology. Office: Med Center U Wis 600 Highland Ave Madison WI 53792

CRISP, MICHAEL DENNIS, research scientist, corp. exec.; b. Elmhurst, Ill., Apr. 27, 1942; s. Richard and Eleanor M. (Luebbers) C.; m. Mellisa Brown, Aug. 28, 1966 (dec.); 1 son, Kevin Michael. A.B., Bradley U., 1964; M.S., Washington U., St. Louis, 1968, Ph.D., 1968. Research assoc. Columbia U., 1968-70; sr. scientist Owens-Ill., Inc., Toledo, 1970-76; energy legis. asst. to Senator Howard H. Baker, Washington, 1977; profl. staff U.S. Senate Com. Commerce, Sci. and Transp., Washington, 1978-80; dir. Fed. Tech. Liaison, Woens-Ill., Inc., Washington, 1980—. Contbr. articles to profl. jours. OSA, AAAS Congl. Sci. and Engring. fellow, 1976-77; NDEA fellow, 1964-67. Mem. Optical Soc. Am., Am. Phys. Soc., AAAS. Lutheran. Subspecialties: Atomic and molecular physics; Laser physics. Current work: Interaction of light and matter. Patentee in field. Office: 1700 K St NW #1107 Washington DC 20006

CRISS, CECIL M., chemist; b. Wheeling, W.Va., Apr. 22, 1934; s. Cecil M. and Anna V. (Reece) C.; m. Laura A. Criss, Aug. 18, 1958; children: Cecil M. III, Laura A. M. A.B., Kenyon Coll., 1956; Ph.D., Purdue U., 1961. Research assoc. Purdue U., 1961; asst. prof. U. Vt., Burlington, 1961-65, U. Miami, Coral Gables, 1965-70, assoc. prof.,

1970-76, prof. chemistry, 1976—; vis. scientist U. Lund (Sweden), 1977-78; vis. research prof. San Diego State U., 1978-79; program officer chem-div. NSF, 1982-83. Mem. Am. Chem. Soc. (award Fla. 1977), Chem. Soc. (London), AAAS, Calorimetry Conf., Sigma Xi, Phi Lambda Upsilon. Episcopalian. Subspecialties: Physical chemistry; Thermodynamics. Current work: Thermodynamic properties of aqueous and nonaqueous electrolytic solutions, especially ionic entropies, volumes, heat capacities, and compressibilities. Home: 4910 San Amaro Dr Coral Gables FL 33146 Office: Dept Chemistry U Miami Coral Gables FL 33124

CRISTOL, STANLEY JEROME, chemistry educator; b. Chgo., J. 14, 1916; s. Myer J. and Lillian (Young) C.; m. Barbara Wright Swingle, June 1957; children: Marjorie Jo, Jeffrey Tod. B.S., Northwestern U., 1937; M.A., UCLA, 1939, Ph.D., 1943. Research chemist Standard Oil Co., Calif., 1938-41; research fellow U. Ill., 1943-44; research chemist U.S. Dept. Agr., 1944-46; asst. prof., then asso. prof. U. Colo., 1946-55, prof., 1955—, Joseph Sewall disting. prof., 1979—, chmn. dept. chemistry, 1960-62, grad. dean, 1980-81; vis. prof. Stanford U., summer 1961, U. Geneva, 1975, U. Lausanne, Switzerland; with OSRD, 1944-46; adv. panels NSF, 1957-63, 69-73, NIH, 1969-72. Author: (with L.O. Smith, Jr.) Organic Chemistry, 1966; editorial bd., Chem. Revs., 1957-59, Jour. Organic Chemistry, 1964-68; contbr. research articles to sci. jours. Guggenheim fellow, 1955-56, 81, 82; recipient James Flack Norris award in phys.-organic chemistry, 1972. Fellow AAAS, Chem. Soc. London; mem. Am. Chem. Soc. (chmn. organic chemistry div. 1961-62, adv. bd. petroleum research fund 1963-66, council policy com. 1968-73), AAUP, Colo.-Wyo. Acad. Sci., Nat. Acad. Scis., Phi Beta Kappa, Sigma Xi, Phi Lambda Upsilon. Subspecialties: Organic chemistry; Photochemistry. Current work: Organic reaction mechanisms; reaction of excited states; energy and electron transfer; stereochemistry of excited state reactions. Home: 2918 3d St Boulder CO 80302 Office: U Colo Boulder CO 80309

CRITES, JOHN LEE, zoology educator; b. Wilmington, Ohio, July 10, 1923; s. Wilfred John and Mildred (Baker) C.; m. Phyllis Naomi Steelquist, July 21, 1946; children: Jill Ann, Robert Hilton. B.Sc., U. Idaho, 1949, M.Sc., 1951; Ph.D., Ohio State U., 1956. Instr. dept. zoology Ohio State U., Columbus, 1955-59, asst. prof., 1959-63, assoc. prof., 1963-67, prof., 1967—, chmn. zoology, 1981—, assoc. dir., Put-in Bay, Ohio, 1970-80; cons. biology USAID-NSF Aligarh, U.P., India, 1964. Contbr. articles to profl. jours. Served with U.S. Army, 1942-45; PTO. Fellow Ohio Acad. Sci.; mem. Am. Soc. Parasitology (council), Wildlife Disease Assn., Helminthological Soc. Washington (editorial bd.), Sigma Xi. Subspecialties: Parasitology; Ecology. Current work: Parasites of fishes and wild animals. Home: 1019 Lansmere Ln Columbus OH 43220 Office: Ohio State Dept Zoology 1735 Neil Ave Columbus OH 43210

CRITOPH, EUGENE, laboratory executive, consultant; b. Vancouver, C., Can., Mar. 29, 1929; m. Mary Elizabeth Ivens, Feb. 9, 1952; children: Christopher M., S. Bard, E. Mark, Boyd. B.A.Sc., U. B.C., 1951, M.A.Sc., 1957. Research officer Atomic Energy Can. Ltd. Chalk River Nat. Lab., Chalk River, Ont., Can., 1953-67, br. head, 1967-75, div. head, 1975-79, v.p., gen. mgr., 1979—. Mem. Am. Nuclear Soc., Can. Nuclear Soc., Can. Assn. Physicists. Subspecialty: Nuclear fission. Current work: Research and development in nuclear power; particularly in nuclear engineering and reactor physics. Home: 4 Darwin Crescent Deep River ON Canada K0J 1P0 Office: Atomic Energy Can Ltd Chalk River ON Canada K0J 1J0

CROAT, THOMAS BERNARD, research botanist; b. St. Marys, Iowa, May 23, 1938; s. Oliver Theodore and Irene Mary (Wilgenbusch) C.; m. Patricia Swope, Sept 4, 1966; children: Anne Irene, Thomas Kevin. B.A., Simpson Coll., 1962; M.S., U. Kans., 1966, Ph.D., 1967. Tchr. pub. schs., V.I., 1962-63, Knoxville, Iowa, 1963-64; research assoc. Ctr. Biology of Natural Systems, Washington U., St. Louis, 1967-70, adj. asst. prof., 1970-82, adj. assoc. prof., 1982—; botanist Mo. Bot. Gardens, St. Louis, 1970—, P.A. Schulze curator, 1982—; adj. prof. U. Mo., St. Louis, 1974—. Author: Flora of Barro Colorado Island, 1978. Pres., 15th Ward Republican Club, 1981-83; treas. Holy Family Sch. Home and Sch. Assn., 1980-81. Served in U.S. Army, 1956-58. NDEA scholar, 1958-62; NDEA fellow, 1964-67; NSF grantee, 1972-74, 77-80, 80-83; Nat. Geog. Soc. grantee, 1973, 77; Sigma Xi grantee, 1975. Mem. Am. Soc. Plant Taxonomists, Flora Neotropica Orgn. (commn.), Internat. Aroid Soc. (hon. dir.), Assn. Tropical Biology, Internat. Soc. Plant Taxonomists, Bot. Soc. Am. Roman Catholic. Subspecialty: Systematics. Current work: Systematics of Araceae; floristic revisions and monographic revisions of neotropical Araceae, especially Anthurium. Home: 4043 Parker Saint Louis MO 63116 Office: PO Box 299 Saint Louis MO 63116

CROCE, CARLO MARIA, molecular geneticist; b. Milan, Italy; m.; 1 child. M.D., U. Rome, 1969. Scientist Wistar Inst. Anatomy and Biology, Phila., 1970-71, research assoc., 1971-74, assoc. mem., 1974-76, prof., 1976-80, assoc. dir., inst. prof., 1980—; vis. scientist Carnegie Inst., 1978-79; mem. Mammalian Genetics Study Sect., NIH, Bethesda, Md., 1979-83. Subspecialties: Genetics and genetic engineering (biology); Molecular biology. Office: Wistar Inst Anatomy and Biology 36th St at Spruce Philadelphia PA 19104

CROCKER, JENNIFER, psychology educator; b. Hartford, Conn., Sept. 3, 1952; d. David Curtis and Marion (Snyder) C.; m. Jeffrey L. Walters, Aug. 19, 1972 (div. 1983). B.A., Mich. State U., East Lansing, 1975-; Ph.D., Harvard U., 1979; student, Reed Coll., 1970-71. Asst. prof. psychology Northwestern U., Evanston, Ill., 1979—; summer scholar Center for Advanced Study in Behavioral Scis., Palo Alto, Calif., 1982. Contbr. articles to profl. jours.; editor: Jour. Am. Behavioral Scientist, 1984. Mem. Am. Phsychol. Assn., Soc. for Psychol. Study of Social Issues, Midwestern Psychol. Assn., Soc. Personality and Social Psychology. Club: Women in Research. Subspecialty: Social psychology. Current work: Psychology of stereotyping, social perception, cognitive processes in depression. Office: Dept Psychology Northwestern U Evanston IL 60201

CROFF, ALLEN GERALD, nuclear engineer; b. Flint, Mich., July 19, 1949; s. Gerald Webster and Anna Catherine (Deninger) C.; m. Linda Irene Wildman, Aug. 25, 1973. B.S. in Chem. Engring, Mich. State U., East Lansing, 1971; M.S. in Nuclear Engring, MIT, 1974; M.B.A., U. Tenn., 1981. Research engr. chem. tech. div. Oak Ridge (Tenn.) Nat. Lab., 1974-80, mgr. engring. analysis and planning, 1980—. Author tech. reports and papers. Mem. Am. Nuclear Soc. (mem. program com. fuel cycle div. 1981—), Soc. for Risk Analysis, Beta Sigma Gamma. Subspecialties: Nuclear fission; Chemical engineering. Current work: Nuclear fuel cycles with emphasis on waste management; advanced energy systems analysis; technical planning and analysis; economics. Home: 980 W Outer Dr Oak Ridge TN 37830 Office: Oak Ridge Nat Lab Bldg 4500 N PO Box X Oak Ridge TN 37830

CROFT, BARBARA YODER, physicist; b. Port Chester, N.Y., Aug. 11, 1940; d. Paul Henry and Harriet Franch (Postle) Yoder; m. Joseph Edward Croft, Dec. 15,1977. B.A., Swarthmore Coll., 1962; M.A., Johns Hopkins U., 1964, Ph.D., 1967. Sr. scientist Johnston Labs., Inc., Balt., 1967-69; instr. radiology U. Va., Charlottesville, 1969-72, asst. prof., 1972—; cons. radiopharm. adv. com. FDA. Author: Basics of Radiopharmacy, 1978. Mem. Am. Chem. Soc., Am. Assn. Physicists in Medicine, Soc. Nuclear Medicine, Am. Coll. Nuclear Physicians, Pattern Research Soc., Sigma Xi. Episcopalian. Subspecialty: Nuclear medicine. Current work: Single photon emission computed tomography; computerized radiology. Home: Route 2 Box 565 Scottsville VA 24590 Office: U Va Dept Radiology Box 170 Charlottesville VA 22908

CROFT, THOMAS A(RTHUR), research executive; b. Denver, Feb. 15, 1931; s. Edwin T. and Pearl M. (Jordan) C.; m. Rachel Marie Whitman, Apr. 3, 1965; children: Andrew A., Steven T., Rachel E. B.A., Dartmouth Coll., 1953, M.S., 1954; Ph.D. in Elec. Engring, Stanford U., 1964. Elec. engr. Convair Astronautics, Gen. Dynamics Corp., 1957-59; research assoc. radio propagation Radiosci. Lab. Stanford U., 1959-68, adj. prof., 1968-75, cons., 1965—; dir. research Tech. for Communications Internat., Mountain View, Calif., 1982—; cons. Applied Tech. Inc., 1965—, Barry Research Corp., 1968—, ESSA, 1969—. Served to lt. USN, 1954-57. Mem. Internat. Sci. Radio Union (assoc. mem. commn. 3 1966—, assoc. editor Radio Sci.), Am. Geophys. Union, AAAS (div. planetary scis.), IEEE (sr.), Soc. Motion Picture and TV Engrs. Subspecialties: Electronics; Graphics, image processing, and pattern recognition. Current work: Ionespheric physics; simulation of performance of complex high freequency radar systems by means of digital computer wave propagation analysis and synthesis; use of artificial intelligence methods to control HF skywave communication. Office: Tech for Communications Internat 1625 Stierlin Rd Mountain View CA 94043

CROMBIE, DOUGLASS DARNILL, govt. ofcl.; b. Alexandra, N.Z., Sept. 14, 1924; came to U.S., 1962, naturalized, 1967; s. Colin Lindsay and Ruth (Datnill) C.; m. Pa uline L.A. Morrison, Mar. 3, 1952. B.Sc., Otago U., Dunedin, N.Z., 1947, M.Sc., 1949. N.Z. nat. research fellow Cavendish Lab., Cambridge, Eng., 1958-59; head radio physics div. N.Z. Dept. Sci. and Industry Research, 1961-62; chief spectrum utilization div., chief low frequency group Inst. Telecommunications Scis., Dept. Commerce, Boulder, Colo., 1962-71, dir. inst., 1971-76; dir. Inst. Telecommunication Scis., Nat. Telecommunications and Info. Adminstrn., 1976-80; chief scientist Nat. Telecommunication and Info. Agy., 1980—. Served with N.Z. Air Force, 1943-44. Recipient Gold medal Dept. Commerce, 1970, citation, 1972. Mem. IEEE, Nat. Acad Engring., Union Radio Sci. Internat. Subspecialties: Telecommunications Systems; Electromagnetic Wave Propogation. Current work: Efficiency of radio frequency spectrum use by communication systems. Propagation of medium frequency radio signals to great distances. Selective fading of microwave signals. Home: 1441 Mariposa Ave Boulder CO 80302 Office: 325 S Broadway Boulder CO 80302

CRONHOLM, LOIS S., college dean, biology educator; b. St. Louis, Aug. 15, 1930; s. Fred and Emma (Tobias) Kisslinger; m. Stuart E. Neff, Apr. 11, 1975; children: Judith Frances, Peter Foster; m. James Norman Cronholm, Aug. 11, 1964 (div. 1973). B.A., U. Louisville, 1962, Ph.D., 1966. Asst. prof. U. Louisville, 1973-76, assoc. prof., 1976-80, prof. biology, 1980—, dean, 1979—. Chmn. Jefferson County-Louisville Human Relations Commn., 1968-73; mem. nominating com. Met. United Way, Louisville, 1981—; chmn. Econ. Devel. Council, Jefferson County, Ky., 1982-83; mem. Leadership Louisville, 1982-83. HEW fellow, 1963-66, 66-69; recipient Research award Dept. Interior, 1975-78, NOAA, 1976-77, Dept. Interior, 1977-82. Subspecialties: Microbiology; Water supply and wastewater treatment. Current work: Efficiency of package plants in wastewater treatment, ecology of Histoplasma capsulatum and its relationships to roosting birds. Home: 9811 Gandy Rd Louisville KY 40272 Office: Coll Arts and Scis U Louisville Louisville KY 40292

CRONIN, JAMES WATSON, educator, physicist; b. Chgo., Sept. 29, 1931; s. James Farley and Dorothy (Watson) C.; m. Annette Martin, Sept. 11, 1954; children: Cathryn, Emily, Daniel Watson. A.B., So. Methodist U., (1951); Ph.D., U. Chgo. Asso. Brookhaven Nat. Lab., 1955-58; mem. faculty Princeton, 1958-71, prof. physics, 1965-71, U. Chgo., 1971—; Loeb lectr. physics Harvard U., 1967. Recipient Research Corp. Am. award, 1967; John Price Wetherill medal Franklin Inst., 1976; E.O. Lawrence award ERDA, 1977; Nobel prize for physics, 1980; Sloan fellow, 1964-66; Guggenheim fellow, 1970-71, 82-83. Mem. Am. Acad. Arts and Scis., Nat. Acad. Sci. Subspecialty: Particle physics. Participant early devel. spark chambers; co-discover CP-violation, 1964. Home: 5825 S Dorchester St Chicago IL 60637

CRONIN, MICHAEL JOHN, neuroendocrine physiologist, educator; b. Los Angeles, Nov. 19, 1949; s. Gilbert F. and Dorothy (Fahey) C.; m. Anna Marie Schneider, Feb. 24, 1979. B.S., Loyola U., Los Angeles, 1971; Ph.D. in physiology, U. So. Calif., 1976; postdoctoral fellow, U. Calif.-San Francisco., 1976-79. Asst. prof. physiology U. Va. Sch. Medicine, Charlottesville, 1979—. Contbr. articles to profl. jours. Recipient Los Angeles Mayor's Spl. Service award, 1970, Community Service award, 1970; Alumni award Loyola U., 1976; NIH grantee, 1981—. Mem. Am. Physiol. Soc., Endocrine Soc., Internat. Soc. Neuroendocrinology, Soc. Neurosci. Roman Catholic. Subspecialties: Neuroendocrinology; Physiology (biology). Current work: Mechanism of action of hypothalamic brain hormones on anterior pituitary; pituitary and neural cell physiology; second messenger biochemistry; receptor identification; tumor model systems; hormone secretory dynamics.

CRONKITE, EUGENE PITCHER, physician; b. Los Angeles, Dec. 11, 1914; s. Clarence Edgar and Anita (Pitcher) C.; m. Elizabeth Erna Kaitschuk, Aug. 17, 1940; 1 dau., Christina Elizabeth. A.B., Stanford U., 1936, M.D., 1940; D.Sc. (hon.), L.I. U., 1962. Intern Stanford U. Hosps., San Francisco, 1939-40, resident in medicine, 1941-42; commd. lt. (s.g.) U.S. Navy, 1942, advanced through grades to rear adm., 1969, ret., 1954; head hematology Naval Med. Research Inst., Bethesda, Md., 1945-54; sr. scientist med. dept. Brookhaven Nat. Lab., 1954—, chmn., 1967-79; prof. medicine Health Sci. Center, SUNY, Stony Brook, 1979—. Contbr. articles to med. jours. Recipient Alfred Benzon award, Denmark, 1969; Ludwig Heilmeyer medal, W.Ger., 1974; Semmelwiess award, Hungary, 1975; Alexander von Humboldt sr. scientist award, W.Ger., 1977. Mem. Am. Soc. Hematology (pres. 1970), Internat. Soc. Exptl. Hematology (pres. 1976), U.S. Nat. Acad. Scis., Am. Soc. Clin. Investigation, Assn. Am. Physicians, Am. Soc. Hematology, Am. Assn. Physiologists. Subspecialties: Hematology; Radiation Biology. Current work: Regulation of blood cell production, effects of radiation on man and mammals. Office: Med Dept Brookhaven Nat Lab Upton NY 11973

CRONSHAW, JAMES, cell biologist, educator; b. Lancashire, Eng., Mar. 11, 1933; came to U.S., 1962; s. William and Edith (Wilkinson) C.; m. Patricia Birwhistle, Sept. 1, 1956; 1 dau., Caroline Anne. B.S. in Botany, U. Leeds, (Eng.), 1954, Ph.D., 1957, D.Sc., 1973. Demonstrator dept. botany U. Leeds, 1955-57; research officer, div. forest products Commonwealth Sci. and Indsl. Research Orgn., Melbourne, Australia, 1957-62; demonstrator dept. botany U. Melbourne, 1957-62; asst. prof. dept. biology Yale U., 1962-65; assoc. prof. dept. biol. scis. U. Calif-Santa Barbara, 1965-71, prof., 1971—. Contbr. articles to profl. jours. NSF grantee, 1962-64, 65-66, 67-69, 69-72, 75-78, 80-82; NIH grantee, 1962-67; Santa Barbara Med. Found. Clinic grantee, 1974-76; Dickson Blanchard Pathology Group grantee, 1978; U.S. Dept. Interior grantee, 1976-80. Mem. Am. Soc. Cell Biology, Bot. Soc. Am., Electron Microscopy Soc. Am., So. Calif. Soc. for Electron Microscopy, AAAS, Soc. Exptl. Biology, Sigma Xi. Subspecialty: Cell and tissue culture. Current work: Enzyme cytochemistry of phloem and electron microscopy of the adrenal cortex. Office: Dept Biol Scis U Calif Santa Barbara CA 93106

CROOKE, PHILIP SCHUYLER, III, mathematics educator; b. Summit, N.J., Mar. 10, 1944; s. Philip and Emma (Kaucky) C., Jr.; m. Barbara E Carey, Aug. 31, 1968; children: Philip Alexander, Cornelia Elizabeth. B.S., Stevens Inst. Tech., 1966; Ph.D., Cornell U., 1970. Asst. prof. math. Vanderbilt U., Nashville, 1970-76, assoc. prof., 1976—; vis. fellow Cornell U., Ithaca, N.Y., 1982. Mem. Soc. Indsl. and Applied Math. Subspecialties: Applied mathematics; Chemical engineering. Current work: Mathematical modelling of fermentation systems; partial differential equations. Home: 611 Cantrell Ave Nashville TN 37215 Office: Vanderbilt U Box 6205 Sta B Nashville TN 37235

CROOKE, STANLEY THOMAS, pharmacologist, pharmaceutical company executive, educator; b. Indpls., Mar. 28, 1945; s. Robert Ellison and Catherine Elizabeth C.; m. Nancy Ann Alder, Aug. 29, 1964; 1 son, Evan. B.S., Butler U., 1966; Ph.D., Baylor U. Coll. Medicine, 1971, M.D., 1974. Postdoctoral fellow Baylor Coll. Medicine, Houston, 1971-72, intern, 1974-75, dir. Bristol-Baylor Molecular Pharmacology Lab., 1976-80, asst. prof. dept. pharmacology, 1976-79, assoc. prof., 1979-82, prof., 1982—; asst. dir. med. research Bristol Labs., Syracuse, N.Y., 1975-76, assoc. dir. med. research, 1976-77, assoc. dir. research and devel., 1977-79, v.p., assoc. dir. research and devel., 1979-80; clin. instr. dept. medicine Upstate Med. Center, Syracuse, 1976-77, clin. asst. prof., 1977-79, clin. assoc. prof., 1979-80; v.p. research and devel. Smith Kline & French Labs., Phila., 1980-82, pres. research and devel., 1982—. Contbr. articles to profl. jours. Recipient Research award So. Med. Assn., 1973-74; Julius W. Sturmer Meml. Lecture award Rho Chi, 1981; USPHS fellow, 1968-71, 71-72; Nat. Cancer Inst. awardee, 1973-74; named Outstanding Faculty Mem. Baylor U. Coll. Medicine; other awards. Mem. AAAS, Am. Assn. Cancer Research, Am. Soc. Microbiology, Cancer and Acute Leukemia Group B, Am. Soc. Clin. Pharmacology and Therapeutics, Am. Soc. Pharmacology and Exptl. Therapeutics, Am. Soc. Clin. Oncology, Coll. Physicians Phila. Subspecialty: Molecular biology. Current work: Molecular pharmacology, antitumor biology, clinical research, anticancer and other drug development. Office: PO Box 7929 F-124 Philadelphia PA 19101

CROSBY, MARSHALL ROBERT, systematic botanist, researcher; b. Jacksonville, Fla., June 3, 1943; s. Robert Gilbert and Anne (Respess) C.; m. Carol Anderson, June 10, 1947; children: Matthew Turner, Sara Elizabeth. B.S., Duke U., 1965, Ph.D., 1969. Curator of cryptogams Mo. Bot. Garden, St. Louis, 1968-74, chmn. dept. botany, 1974-79, dir. research, 1977—; research assoc. dept. botany Washington U., St. Louis, 1968-69, adj. asst. prof. botany dept. biology, 1970-73, faculty assoc. dept. biology, 1974-79, adj. prof. biology, 1980—; hon. assoc. curator of Bryophyta Museo Nacional de Costa Rica, 1978—; mem. commons steering com. U. Mo.-St. Louis, adj. prof. biology, 1983—; mem. adv. com. Projecto Flora Amazonica NSF, 1977—. Editor: Annals of the Mo. Bot. Garden, 1969-74; assoc. editor: Bryologist, 1974-78; editor: Monographs in Systematic Botany from the Mo. Bot. Garden, 1978—; mem. editorial. com.: Systematic Botany Monographs, 1978—; co-editor: Herbarium News, 1981—; Contbr. articles to profl. jours. Mem. Am. Bryological and Lichenological Soc. (bus. mgr. 1972-79, sec.-treas. 1973-79), Bot. Soc. Am., Brit. Bryological Soc., Internat. Assn. for Plant Taxonomy (mem. com. for Bryophyta 1975—, sec. 1976—, mem. gen. com. bot. nomenclature 1976—), AAAS, Am. Soc. Plant Taxonomists (councilor 1982—), Internat. Assn. Bryologists (mem. council 1979—), Sigma Xi. Subspecialty: Taxonomy. Current work: Systematics and taxonomy of mosses with emphasis on tropical groups. Office: Missouri Botanical Garden PO Box 299 Saint Louis MO 63166

CROSLEY, DAVID RISDON, physical chemist; b. Webster, City, Iowa, Mar. 4, 1941; m., 1963; 1 child. B.S., Iowa State U., 1962; M.A., Columbia U., 1963, Ph.D., 1966. Research assoc. physics Joint Inst. Lab. Astrophysics, 1966-68; asst. prof. phys. chemistry U. Wis., Madison, 1968-75; project leader Ballistic Research Labs., Aberdeen Proving Ground, Md., 1975-79; program mgr. Molecular Physics Lab., SRI Internat., Menlo Park, Calif., 1979-80, sr. chem. physicist, 1979—. Joint Inst. Lab. Astrophysics fellow, 1966-68. Mem. Am. Phys. Soc., AAAS, Combustion Inst. Subspecialties: Physical chemistry; Laser-induced chemistry. Office: Molecular Physics Lab SRI Internat Menlo Park CA 94025

CROSS, KENNETH JAMES, computer system designer and programmer; b. Muskegon, Mich., Sept. 7, 1951; s. Charles G. and Irene (Cunniff) C.; m. Nancy Kay Irland, Nov. 20, 1971; children: Heather, Adam. B.E.E., Gen Motors Inst., 1974; M.S.E.E., U. Colo., 1974. Engr. Buick Motor Div., Flint, Mich., 1969-74; devel. engr. Oak Ridge (Tenn.) Nat. Lab., 1974-81; dir. software Perceptics Corp., Knoxville, Tenn., 1981—; cons. Tech. for Energy, Knoxville, 1980-81, Digital Decision Systems, 1981; instr. Roane State Community Coll., Harriman, Tenn., 1979-80, Instrument Soc. Am., Knoxville, 1978-79. Mem. Assn. Computing Machinery, IEEE, Digital Equipment Computer Users Soc. (chmn. 1975-81). Subspecialties: Graphics, image processing, and pattern recognition; Software engineering. Current work: Development of advanced image processing systems, including pattern recognition. Home: 1101 Albemarle Ln Knoxville TN 37923 Office: Perceptics Corp 221 B Clark St Knoxville TN 37921

CROSS, RALPH EMERSON, mech. engr.; b. Detroit, June 3, 1910; s. Milton Osgood and Helen (Heim) C.; m. Eloise Florence Fountain, June 18, 1932; children—Ralph Emerson, Carol (Mrs. Peter G. Wodtke), Dennis W. Student, Mass. Inst. Tech., 1933; D.Eng. (hon.), Lawrence Inst. Tech., 1977. Vice pres. Cross Co., Fraser, Mich., 1932-67, pres., gen. mgr., 1967-79, chmn., 1979—; chmn. bd. Cross & Trecker, Bloomfield Hills, Mich., 1979—; chmn. bd., pres. Cross Internat. A.G., Fribourg, Switzerland, 1965-68; pres. Cross Export Corp., 1972-80; dir. Kearney & Trecker Corp., Milw., Roberts Corp. Asst. adminstr. Bus., Def. Services Adminstrn. U.S. Dept. Commerce, 1954; spl. cons. to asst. sec. Air Force for Material, 1955-59; Mem. corp. Econ. Devel. Corp. Greater Detroit, 1968-73, Mich. Blue Shield, 1969-74; mem. corp. devel. com. Mass. Inst. Tech., 1970—; mem. Iranian Joint Bus. Council, 1975-76; trustee Lawrence Inst. Tech., 1979—; pres. SME Edn. Found., 1979—. Recipient Engring. citation Am. Soc. Tool Engrs., 1956; Corp. Leadership award Mass. Inst. Tech., 1976. Mem. Nat. Acad. Engring., Nat. Machine Tool Builders Assn. (pres. 1975), Soc. Automotive Engrs., Soc. Mfg. Engrs. (hon.), Engring. Soc. Detroit. Clubs: Detroit Athletic, Lochmoor. Subspecialty: Mechanical engineering. Current work: Machine tools. Home: 50 N Deeplands Rd Grosse Pointe Shores MI 48236 Office: 17801 Fourteen Mile Rd Fraser MI 48026

CROSS, RICHARD JAMES, JR., chemist, educator; b. N.Y.C., June 28, 1940; s. Richard James and Margaret Whittemore (Lee) C.; m. Ann Marie Dyman, Oct. 11, 1969; children: Donna L., John D. B.S., Yale U., 1962; postgrad., U. Calif., Berkeley, 1962-63; Ph.D., Harvard U., 1966. Postdoctoral fellow U. Fla., 1965-66; asst., then assoc. prof. Yale U., New Haven, 1966-79, prof. chemistry, 1979—. Contbr. numerous articles to profl. jours. Mem. Am. Chem. Soc., Am. Phys. Soc., Am. Soc. Mass Spectrometry. Democrat. Subspecialty: Physical chemistry. Current work: Research on dynamics of chemical reactions, theory of molecular scattering. Office: Yale Chemistry Box 6666 New Haven CT 06511

CROSS, RICHARD LESTER, biochemist, educator; b. Hoboken, N.J., Aug. 26, 1943; s. Richard L. and Ann Jane (Walls) C.; m. Donna Woolfolk, Dec. 17, 1971; 1 dau., Emily Woolfolk. B.A., Hartwick Coll., 1966; Ph.D., Yale U., 1970. Postdoctoral fellow UCLA, 1970-73; asst. prof. SUNY Upstate Med. Ctr., Syracuse, 1973-78, assoc. prof., 1978-82, chmn. biochemistry, 1982—; mem. adv. panel NSF, 1981—. Mem. editorial bd.: Jour. Bioenergetics and Biomembranes, 1983—; Contbr. sci articles to profl. publs. NIH research grantee, 1976—. Mem. Am. Soc. Biol. Chemists, Am. Biophys. Soc. Subspecialties: Biochemistry (biology); Biochemistry (medicine). Current work: Energy transducing membrane systems: the mechanism and regulation of ATP synthesis by oxidative phosphorylation and photophosphorylation. Office: Biochemistry Dept SUNY Upstate Med Ctr 766 Irving Ave Syracuse NY 13210

CROSSLEY, D.A., JR., ecologist, entomology educator; b. Kingsville, Tex., Nov. 6, 1927; s. D.A. and Eugenia (Baird) C.; m. Nettie Keirsey, Sept. 15, 1950 (dec. 1959); 1 dau., Mary Eugenia; m. Dorothy Money. B.A., Tex. Tech U., 1949, M.S., 1951; Ph.D., U. Kans., 1956. Ecologist Oak Ridge Nat. Lab., 1956-67; prof. entomology U. Ga., 1967—. Contbr. articles to profl. jours. Recipient Creative Research medal U. Ga., 1982. Mem. Entomol. Soc. Am., Ecol. Soc. Am., Am. Soc. Naturalists. Subspecialties: Ecology; Ecosystems analysis. Current work: Nutrient cycling in terrestrial ecosystems; roles of consumers in ecosystems; decomposition regulation; insect ecology; agroecosystems. Office: Department Entomology University Georgia Athens GA 30601

CROTHERS, DONALD MORRIS, biochemist, educator; b. Fatehgarh, India, Jan. 28, 1937; came to U.S., 1939, naturalized, 1937; s. Morris K. and Eunice F. C.; m. Leena Kareoja, June 24, 1960; children—Nina H., Kristina A. B.S., Yale U., 1958; B.A., Cambridge U., 1960; Ph.D., U. Calif.-, San Diego, 1963. NSF postdoctoral fellow Max Planck Inst., Gottingen, Germany, 1963-64; asst. prof. Yale U., New Haven, 1964-68, asso. prof., 1968-71, prof. chemistry and molecular biophysics and biochemistry, 1971—, chmn. dept. chemistry, 1975-81; chmn. biophysics, biophysical chemistry B study sect. NIH, 1972-76; co-chmn. nucleic acids Gordon Conf., 1975. Author: Physical Chemistry of Nucleic Acids, 1974, Physical Chemistry with Application to the Life Sciences, 1979; Mem. editorial bd.: Jour. Molecular Biology, 1971-75, Nucleic Acids Research, 1973-82, Biochemistry, 1975-78, Biopolymers, 1977—; Contbr. articles to profl. jours. Recipient Sci. and Engring. award Yale U., 1977; Alexander von Humboldt Sr. Scientist award, 1981; Mellon fellow Clare Coll. Cambridge U., 1958-60; Guggenheim fellow, 1978. Mem. Biophys. Soc. (council 1979-82), Am. Soc. Biol. Chemists. Subspecialty: Biophysical chemistry. Office: Dept Chemistry Yale U New Haven CT 06520

CROUCH, ROSALIE KELSEY, biochemistry educator, researcher; b. Norfolk, Va., Nov. 26, 1941; d. Denham Arthur and Elodia (Yancey) Kelsey; m. William Ellsworth Crouch, June 3, 1967; children: Katherine Anne, Richard William. A.B., Randolph Macon Woman's Coll., 1963; M.S., Lehigh U., 1965; Ph.D., Yeshiva U., 1972. Asst. prof. Med. U. S.C., Charleston, 1975-78, assoc. prof., 1978-82, prof., 1982—. Contbr. articles to profl. jours. Bd. dirs. Charleston Civic Ballet, 1983. Mem. Biophys Soc., Am. Chem. Soc., Sigma Xi. Subspecialties: Biochemistry (biology); Synthetic chemistry. Current work: Vision chemistry, oxygen toxicity. Home: 50 Church St Charleston SC 29401 Office: Med Univ SC 171 Ashley Ave Charleston SC 29425

CROUTHAMEL, CARL EUGENE, nuclear scientist, consultant; b. Lansdale, Pa., Dec. 25, 1920; s. Mervin Dorn and Helen Ruth (Kile) C.; m. Marion Griffith Phillips, Sept. 9, 1944; children: Steven John, Kevin Griffith, David Roger. B.S., Eastern Nazarene Coll., Quincy, Mass., 1941; M.A., Boston U., 1942; Ph.D., Iowa State U., Ames, 1950. Sr. scientist Agronne (Ill.) Nat. Lab., 1950-73; cons. Exxon Nuclear Co., Richland, Wash., 1973—, Electric Power Research Inst., Palo Alto, Calif. Author: Regenerative EMF Cells, 1967, Applied Gamma-Ray Spectrometry, 1970; contbr. articles to profl. jours. Served to lt. USNR, 1943-46. Mem. Am. Chem. Soc., Am. Nuclear Soc. Lutheran. Subspecialties: Nuclear fission; Nuclear engineering. Current work: Light water reactor fuel design and performance verification and consulting. Home: 71 Park St Richland WA 99352 Office: Exxon Nuclear Company Inc 2101 Horn Rapids Rd Richland WA 99352

CROVITZ, HERBERT FLOYD, psychologist; b. Providence, May 21, 1932; s. Jack and Natalie (Turick) C.; m. Elaine Kobrin, 1957 (div. 1971); children: Gordon, Deborah, Sara. A.B., Clark U., 1953, M.A., 1954; Ph.D., Duke U., 1960. Research psychologist VA Hosp., Durham, N.C., 1961—; prof. med. psychology, dept. psychiatry Duke U. Med. Ctr., Durham, 1963—, adj. prof. dept. psychology, 1963—. Author: Galton's Walk, 1970; cons. editor: Jour. Exptl. Psychology: Gen, 1976—. Mem. Am. Psychol. Assn., Psychonomic Soc., Internat Neuropsychology Soc., Optical Soc. Am., N.C. Psychol. Assn. Subspecialties: Cognition; Neuropsychology. Current work: Research on recovery of memory functions after head injury, other loss of consciousness; improving of all classes of cognitive abilities, consciousness. Home: 16 Glenmore Dr Durham NC 27707 Office: VA Hosp Fulton St Durham NC 27705

CROW, EDWIN LOUIS, mathematical statistician, consultant; b. Browntown, Wis., Sept. 15, 1916; s. Frederick Marion and Alice Blanche (Cox) C.; m. Eleanor Gish, June 13, 1942; children: Nancy Rebecca, Dorothy Carol Crow-Willard. B.S. summa cum laude, Beloit Coll., 1937; Ph.M., U. Wis., 1938, Ph.D., 1941; postgrad., Brown U., 1941, 42, U. Calif.-Berkeley, 1947, 48, Univ. Coll. London, 1961-62. Instr. math. Case Sch. Applied Sci., Cleve., 1941-42; mathematician Bur. Ordnance, Navy Dept., Washington, 1942-46, U.S. Naval Ordnance Test Sta., China Lake, Calif., 1946-54; cons. statistics Boulder Labs., U.S. Dept. Commerce, Boulder, Colo., 1954-73; statistician Nat. Ctr. Atmospheric Research, Boulder, 1975-82; cons. stats. Nat. Telecommunications and Info. Adminstrn., Boulder, 1974—; instr. math. extension div. UCLA, China Lake, 1947-54; adj. prof. math. U. Colo., Boulder, 1963-81; lectr. stats. Met. State Coll., Denver, 1974. Author: Statistics Manual, 1960; assoc. editor: Communications in Statistics, 1972—, Jour. Am. Statis. Assn, 1967-75; contbr. articles to profl jours. Survey statistician Boulder Valley Sch. Dist., Boulder, 1971-72; founder, pres. Boulder Tennis Assn., 1967-69, 82. Recipient Outstanding Publ. award Nat. Telecommunications and Info. Adminstrn., 1980; Bronze medal U.S. Dept. Commerce, 1970. Fellow Royal Statis. Soc., Am. Statis. Assn. (council mem. 1959-60, 68-69), AAAS; mem. Am. Math Soc., Math. Assn. Am., Inst. Math. Stats., Soc. Indsl. and Applied Math., Sigma Xi, Phi Beta Kappa. Democrat. Unitarian. Clubs: U.S. Tennis Assn., Colo. Mountain, Harvest House Sporting Assn. (Boulder). Subspecialty: Statistics. Current work: Statistical standards for data communication systems, statistical methods for weather modification. Home: 605 20th St Boulder CO 80302 Office: ITS N3 Nat Telecommunications and Info Adminstrn 325 Broadway Boulder CO 80303

CROW, JAMES FRANKLIN, educator; b. Phoenixville, Pa., Jan. 18, 1916; s. H. Ernest and Lena (Whitaker) C.; m. Ann Crockett, Aug. 9, 1941; children—Franklin, Laura, Catherine. A.B., Friends U., 1937; Ph.D., U. Tex., 1941. Instr., then asst. prof. zoology Dartmouth, 1941-48; faculty U. Wis., 1948—, prof. genetics, 1954—, chmn. dept. med. genetics, 1958-63, 65-71, acting dean sch. medicine, 1963-65; Chmn. genetics study sect. NIH, 1965-68. Author: Genetics Notes, 7th edit, 1976, Introduction to Population Genetics Theory, 1970, also articles. Mem. Nat. Acad. Scis. (chmn. com. genetic effects atomic radiation 1960-63, 70-72, chmn. com. chem. environ. mutagens 1980—), Genetics Soc. Am. (pres. 1960), Am. Soc. Human Genetics (pres. 1963). Subspecialty: Genetics and genetic engineering (biology). Home: 24 Glenway Madison WI 53705

CROW, LOWELL THOMAS, psychology educator; b. Teague, Tex., Dec. 14, 1931; s. Marler and Ruby Lena (Overturf) C.; m. Dorothy Louise McNeil, Aug. 17, 1956; children: Pamela Anne, Lowell Thomas. B.S., U. S.C., 1957, M.A., 1959; Ph.D., U. Ill., 1962. Asst. prof. psychology Western Wash. State Coll., 1962-67, assoc. prof., 1968-72; prof. psychology Western Wash. U., 1972—; asst. prof. psychology U. S.C., 1967-68; Mem. sci. adv. council Lic. Beverage Industries, Inc., 1962, 68. Contbr. articles to profl. jours., chpts. to books. Bd. dirs. Whatcom County Citizens Council on Alcoholism, 1966-70. Served with USAF, 1951-54. Recipient award Nat. Inst. Alcohol and Alcohol Abuse, 1971-73. Mem. Am. Pyschol. Assn., Soc. Neurosci., Psychonomic Soc., Sigma Xi. Subspecialties: Physiological psychology; Neuropharmacology. Current work: Behavioral effects of alcohol, alcoholism. Office: Western Washington Universit Dept of Psychology Bellingham WA 98225

CROWDER, LARRY B., zoology educator; b. Fresno, Calif., June 9, 1950; s. Earl M. and L. Jean (Mayer) C.; m. Judith Ann Morris, Apr. 13, 1973; 1 dau., Emily Joy. B.A., Calif. State U.-Fresno, 1973; M.S., Mich. State U., 1975, Ph.D., 1978. Lectr. Calif. State U.-Fresno, 1973; grad. asst. Mich. State U., East Lansing, 1973-78; postdoctoral fellow, 1978; project assoc. U. Wis.-Madison, 1978-80, asst. research scientist, 1980-82; asst. prof. zoology N.C. State U., Raleigh, 1982—. Contbr. articles to profl. jours. Mem. Am. Fisheries Soc., Am. Soc. Limnology and Oceanography, Ecol. Soc. Am., Sigma Xi, Phi Kappa Phi. Democrat. Presbyterian. Subspecialties: Behavioral ecology; Species interaction. Current work: Resource use and species interactions, community ecology, aquatic ecology, bioenergetics, fish community ecology, competition and predation. Office: Dept Zoology NC State U Raleigh NC 27650

CROWE, DENNIS TIMOTHY, JR., veterinary surgeon, researcher; b. Milw., Nov. 21, 1946; s. Dennis Timothy and Anna Mae (Persen) C.; m. Deborah Gene Coulson, Jan. 2, 1971; children: Michael, Kristin. D.V.M., Iowa State U., 1972. Diplomate: Am. Coll. Veterinary Surgeons. Intern Colo. State U., 1972-73; resident in surgery Ohio State U., 1973-76; chief surgery Westcott Hosp. and Animal Emergency Room, Detroit, 1976-78; asst. prof. surgery Kans. State U., Manhattan, 1978-80, U. Ga., Athens, 1980—, dir. Shock Trauma team, 1981—, co-dir. intensive care unit, 1981—, asst. chief gen. surgery, 1982—; lectr. in field. Contbr. numerous articles to profl. jours. Veterinary Med. Experiments Sta. grantee, 1981-83; recipient G.G. Graham award in clin. veterinary medicine Iowa State U., 1972. Mem. Am. Veterinary Med. Assn., Am. Coll. Veterinary Surgeons, Am. Animal Hosp. Assn., Veterinary Critical Care Soc. Methodist. Subspecialties: Surgery (veterinary medicine); Traumatology-Critical Care. Current work: Research in shock (hemorrhagic, septic) abdominal counterpressure, cardiopulmonary rescuscitation; primarily involved in the teaching and training of students residents and interns in the field of small animal general surgery and critical care and in related research. Home: 630 Sandstone Dr Athens GA 30605 Office: U Ga H316 Veterinary Teaching Hospital Department of Small Animal Medicine and Surgery College of Veterinary Medicine Athens GA 30602

CROWELL, JOHN CHAMBERS, educator, geologist; b. State College, Pa., May 12, 1917; s. James W. and Helen H. (Chambers) C.; m. Betty Marie Bruner, Nov. 22, 1946; 1 dau., Martha Lynn. B.S. in Geology, U. Tex., 1939; M.A. in Meteorology, UCLA, 1946; Ph.D. in Geology, UCLA, 1947; D.Sc. (hon.), U. Louvain, Belgium, 1966. Jr. geologist Shell Oil Co., Inc., 1941-42; mem. faculty UCLA, 1947-67, prof. geology, 1960-67, chmn. dept., 1957-60, 63-66; prof. geology U. Calif., Santa Barbara, 1967—; chmn. Office of Earth Scis., NRC/Nat. Acad. Scis., 1979-82; nat. lectr. Sigma Xi, 1981-82. Served to capt. USAAF, 1942-46. Fellow Geol. Soc. Am., Am. Acad. Arts and Scis.; mem. Am. Assn. Petroleum Geologists, Am. Geophys. Union, AAAS, Am. Inst. Profl. Geologists, Nat. Acad. Scis. Subspecialty: Tectonics. Spl. research structural geology, tectonics, interpretation sedimentary rocks, studies San Andreas fault system, tectonics Calif., ancient glaciation, continental drift. Home: 727 Avenida Pequenia Santa Barbara CA 93111

CROWELL, RICHARD LANE, virologist, immunologist, educator; b. Springfield, Mo., Sept. 27, 1930; s. Thomas Rolla and Addie Malinda (Lane) C.; m. Arlene M. Prell, June 27, 1953; children: Steven R., Kathleen M., Barbara L., Wendy J. B.A., U. Buffalo, 1952; M.S., U. Minn., 1954, Ph.D., 1958. Diplomate: Am. Acad. Microbiology. Instr. microbiology U. Minn. Med. Sch., Mpls., 1958-60; asst. prof. microbiology and immunology Hahnemann U. Sch. Medicine, Phila., 1960-64, assoc. prof., 1964-71, prof., 1971—, chmn. dept., 1979—, dir. Contbr. articles to profl. jours. Recipient Lindback prize for teaching Hahnemann U. Sch. Medicine, 1968; Research Career Devel. awardee NIH, 1962-72. Mem. Am. Soc. for Microbiology (pres. br. 1974-76), Am. Soc. Immunologists, AAAS, Soc. Exptl. Biology and Medicine, Soc. for Gen. Microbiology, Am. Soc. Virology, Sigma Xi. Presbyterian. Subspecialties: Virology (medicine); Immunology (medicine). Current work: Mechanisms of early events in coxsackievirus infections of cells; characterization of virus proteins and cellular receptors; skeletal muscle and heart diseases by coxsackieviruses. Office: 230 N Broad St Philadelphia PA 19102

CROWLE, ALFRED JOHN, immunologist; b. Mexico City, Apr. 15, 1930; s. Alfred C. and Hazel Araminta (Mason) C.; m. Clarice Marjorie Futrelle, Oct. 22, 1954; children: Nelson Frederick, Cynthia Nanette. A.B., San Jose State Coll., 1951; Ph.D., Stanford U., 1954. Instr., U. Colo. Sch. Medicine, Denver, 1956-59, asst. prof., 1959-65, asso. prof., 1965-75, prof., 1974—; research microbiologist Webb-Waring Lung Inst., Denver, 1956-59, head div. immunology, 1965—; cons. U.S. Army, NSF, others. Author: Immunodiffusion, 2d edit., 1973, Delayed Hypersensitivity in Health and Disease, 1962; also articles. Active Boy Scouts Am., 1966-71. Nat. Tb Assn. fellow, 1953-55; Nat. Acad. Scis.-NSF postdoctoral fellow, 1955; James Alexander Miller fellow N.Y. Tb and Health Assn., 1960-61; grantee NIH, NSF, others. Mem. Reticuloendothelial Soc., Am. Assn. Immunologists, AAAS, AAUP, Am. Soc. Microbiology, Soc. Exptl. Biology and Medicine. Clubs: Cherry Creek Gun (pres. 1967-70), Colorado Mountain (head rock climbing sch. Denver group 1977-82, dir.). Subspecialties: Immunology (medicine); Infectious diseases. Current work: Immunodiffusion and crossed immunoelectrophoresis; applications in clinical medicine; bacteria, macrophages, lymphokines, lymphocytes: tests and mechanisms of cellular immunities to infectious diseases; tuberculosis and immunization: mechanisms and control of tuberculoimmunity in man and animals. Office: B-122 E 9th Ave Denver CO 80262

CROWLEY, JOHN JAMES, biostatistics educator; b. San Diego, Feb. 20, 1946; m., 1969; 3 children. B.A., Pomona Coll., 1968; M.S., U. Wash., 1970, Ph.D., 1973. Fellow biostats. Stanford U., 1973-74; asst. prof. stats. and human oncology U. Wis., Madison, 1974-79, assoc. prof., 1979-81; assoc. prof. biostats. U. Wash., 1981—; assoc. mem. Fred Hutchinson Cancer Research Center, 1981—. Mem. Am. Statis. Assn., Biometric Soc., Inst. Math. Statistics, AAAS. Subspecialty: Statistics. Office: Dept Stats U Wis 1210 W Dayton St Madison WI 53706

CROWLEY, JOSEPH MICHAEL, electrical engineer; b. Phila., Sept. 9, 1940; m., 1963; 5 children. B.S., MIT, 1962, M.S., 1963, Ph.D., 1965. Guest scientist Max Planck Inst., Gottingen, 1965-66; asst. prof. U. Ill., Urbana, 1966-71, assoc. prof., 1971-79, prof. elec. engring., 1979—; NATO fellow, 1965-66; cons. Xerox Corp., 1968—; pres. Joseph M. Crowley, Inc., 1981—. Mem. Am. Phys. Soc., IEEE, Electrostatic Soc. Am., Soc. Info. Display. Subspecialties: Electrical engineering; Fluid mechanics. Office: Dept Elec Engring U Ill Urbana IL 61801

CROWLEY, LAWRENCE G., surgeon, med. educator; b. Newark, 1919. M.D., Yale, 1944. Diplomate: Am. Bd. Surgery. Surg. intern New Haven Hosp., 1944-45, resident in surgery, 1947-53; fellow vascular research Emory U. Hosp., 1949; practice medicine, specializing in surgery, 1949—; attending surgeon Kaiser Found. Hosp., Los Angeles, 1953-59, cons., 1959-64; attending tumor surgeon Los Angeles County Hosp., 1954-64; mem. surg. staff Pomona Valley Community Hosp.; attending surgeon Palo Alto-Stanford Hosp.; chief surgeon Palo Alto VA Hosp., 1964-72; from instr. in surgery to asst. clin. prof. U. So. Cal., 1951-64; asso. prof. surgery Stanford, 1964-69, prof., 1969-73; asso. dean for planning Sch. Medicine, 1972-73, dep. dean, 1977-79, chief staff, 1977-78, acting v.p. and dean, 1979-80, v.p. med. affairs, acting dean, 1980—; pres. Stanford U. Hosp., 1981—; prof. surgery, dean med. sch. U. Wis., Madison, 1973-77. Served to capt. M.C.; Served to capt. AUS, 1945-47. Fellow A.C.S.; mem. AMA, Western, Pacific Coast surg. assns. Subspecialties: Surgery; Medical administration. Office: Stanford U Sch Medicine Stanford CA 94305

CROWN, BARRY MICHAEL, psychologist, educator; b. Waukegan, Ill., June 23, 1943; s. Frank Edward and Sara (Babel) C.; m. Sheryl Joyce Lowenthal, Oct. 31, 1982. B.A., U. Miami, 1965; M.Ed., Fla. Atlantic U., 1966; Ph.D., Fla. State U., 1969. Research coordinator Fla. Gov.'s Office, Tallahassee, 1967-69; fellow in psychiatry Mass. Gen. Hosp., Boston, 1969-70; asst. prof. psychology U. Ill. Med. Sch., Chgo., 1970-71; asst. prof. psychiatry U. Miami Med. Sch., 1971-73; assoc. prof. psychology Fla. Internat. U., Miami, 1973-78, courtesy prof., 1978—; assoc. dir. Nat. Council Drug Abuse, Chgo., 1972—; clin. dir. NIMH Nat. Drug Abuse Tng. Ctr., Miami, 1972-74. Recipient award for disting. contrbns. Am. Ontoanalytic Assn., 1976; Sci. Achievement award Nat. Council Drug Abuse, 1975; NIMH fellow, 1969. Fellow Royal Soc. Health; mem. Am. Psychol. Assn., Soc. Behavioral Medicine, Am. Psychology and Law Soc., Southeastern Psychol. Assn. Democrat. Jewish. Club: Tiger Bay. Subspecialties: Neuropsychology; Behavioral psychology. Current work: Developing diagnostic instruments in neuropsychology; applying behavioral concepts in psychosomatic medicine. Office: 7800 Red Rd Suite 310 South Miami FL 33143

CRUESS, RICHARD LEIGH, surgeon, university dean; b. London, Ont., Can., Dec. 17, 1929; s. Leigh S. and Martha A. (Peever) C.; m. Sylvia Crane Robinson, May 30, 1953; children: Leigh S., Andrew C. B.A., Princeton U., 1951; M.D., Columbia U., 1955. Diplomate: Am. Bd. Orthopedic Surgery. Intern Royal Victoria Hosp., Montreal, Que., 1955-56, resident surgery, 1956-57, N.Y. Orthopedic Hosp., 1959-60, asst. resident orthopedic surgery, 1960-61, resident orthopedic surgery, 1961-62, Annie C. Kane fellow orthopedic surgery, 1961-62; research asso. depts. orthopedic surgery and biochemistry Columbia U., N.Y.C., 1962-63; John Armour Travelling fellow, 1962-63, Am.-Brit.-Can. Travelling fellow, 1967, practice medicine specializing in orthopedic surgery, Montreal, 1963—; orthopedic surgeon Royal Victoria Hosp., orthopedic surgeon-in-charge, 1968-81, asst. surgeon-in-chief, 1970-81; chief surgeon Shriner's Hosp. for Crippled Children, Montreal, 1970-82; prof. surgery McGill U., Montreal, 1970—, chmn. div. orthopedic surgery, 1976-81, dean faculty medicine, 1981—; hon. cons. orthopedic surgery Queen Elizabeth Hosp., 1972—; mem. clin. grants com. Med. Research Council, 1972-75. Contbr. articles on surgery to profl. jours.; editorial bd.: Jour. Internat. Orthopedics, 1976—, Jour. Bone and Joint Surgery, 1977—, Current Problems in Orthopedics, 1977—. Served to lt. M.C., USN, 1957-59. Fellow Royal Coll. Physicians and Surgeons Can. (chief examiner orthopedic surgery 1970-72), ACS, Am. Acad. Orthopedic Surgeons; mem. Can. Orthopedic Assn. (sec. 1971-76, pres. 1977-78), Can. Orthopedic Research Soc. (pres. 1971-72), Am. Orthopedic Research Soc. (pres. 1975-76), Am. Orthopedic Assn., Assn. Orthopedic Surgeons Province Que. (treas. 1971-72), Société Française de Chirurgie Orthopedique, McGill Osler Reporting Soc. Subspecialty: Orthopedics. Home: 526 Mount Pleasant Ave Montreal PQ H3Y 3H5 Canada Office: 3655 Drummond St Montreal PQ H3G 1Y6 Canada

CRUICKSHANK, MICHAEL JAMES, marine mining engineer, educator; b. Glasgow, Scotland, July 9, 1929; came to U.S., 1961; s. John Norman and Elizabeth Helen May (Slimmon) C.; m. Beatrix I. Collick, July 11, 1953; children: James I., Francesca; m. Victoria J. White, Dec. 29, 1978; children: Michael D., Alexandra. A.C.S.M., Camborne Sch. Mines, Cornwall, U.K., 1953; M.Sc. in Mining, Colo. Sch. Mines, 1962; Ph.D. in Oceanography, U. Wis.-Madison, 1978. Chartered engr., U.K. Dist. engr. Anglo-Greek Magnesite Co., Euboea, Greece, 1953-55; mine mgr. Muirshiel Barytes Co., Muirshiel, Scotland, 1955-61; research specialist Lockheed Missiles and Space Co., Sunnyvale, Calif., 1965-66; research mining engr. U.S. Bur. Mines and NOAA, Tiburon, Calif., 1966-73, U.S. Geol. Survey, Washington, 1973-82; vis. prof. mining Columbia U., N.Y.C., 1981-82; marine mining engr. U.S. Minerals Mgmt. Service, Reston, Va., 1982—; cons. UN, 1973—, Nat. Acad. Scis., 1976—. Contbr. numerous articles to profl. publs., chpts. to boods. Served to capt. Brit. Royal Marines, 1955-62. Recipient Richard Pearce Gold medal Camborne Sch. Mines, 1953, Spl. Achievement award U.S. Geol. Survey, 1974; Sloan fellow Stanford U., 1975. Mem. Soc. Mining Engrs., Inst. Mining and Metallurgy (Travellilng scholar 1952), Marine Tech. Soc. (founding mem. 1963), World Dredging Assn. (pres. Pacific chpt. 1971-73). Current work: Mineral resources and technology development on the outer continental shelf. Home: PO Box 295 Hamilton VA 22068 Office: US Minerals and Mgmt Service 647 National Center Reston VA 22092

CRUIKSHANK, DALE PAUL, astronomer; b. Des Moines, Aug. 10, 1939; s. Paul Cecil and Bette Helen (Jones) C.; m. Nooria Noor, May 1, 1964; m. Susan Frank, Oct. 4, 1975; children: Paul Shammim, Mark Tammim, Jeffrey Frank. B.S. in Physics, Iowa State U., 1961; M.S. in Geolgoy, U. Ariz., Tucson, 1965; Ph.D. in Planetary Geology, U. Ariz., Tucson, 1968. Research assoc. Lunar and Planetary Lab., U. Ariz., 1961-70; asst. astronomer Inst. for Astronomy, U. Hawaii, 1970-73, assoc. dir., 1974-76, assoc. astronomer, 1973-80, astronomer, 1980—; supervising scientist Mauna Kea Obs, 1974-76; dir. Canada-

France-Hawaii Telescope Corp., 1980—; mem. team Voyager IRIS, 1981—; mem. numerous NASA working groups and coms. Contbr. articles to profl. jours., books. Recipient NASA Group Achievement award for Voyager spacecraft activites, 1981. Mem. Internat. Astron. Union, Am. Astron. Soc., Astron. Soc. Pacific, Am. Geophys. Union, Soc. Sci. Exploration. Subspecialties: Infrared optical astronomy; Planetary science. Current work: Remote sensing of planetary bodies—planets, satellites, comets and asteroids—by astronomical and geological techniques. Cosmochemistry of the solar system by astronomical observations. Studies of terrestrial volcanic gases by spectroscopic field techniques and study of the magmatic gas composition of active volcanoes. Planetary surfaces and atmospheres. Office: Inst for Astronomy Univ Hawaii 2680 Woodlawn Dr Honolulu HI 96822

CRUSBERG, THEODORE CLIFFORD, biology and biotechnology educator; b. Meriden, Conn., Feb. 23, 1941; s. Clifford Herbert and Frances (Blankenburg) C.; m. Marie Morris, Sept. 2, 1966; children: Heidi, Gretchen, Nils. B.A., U. Conn., 1963; M.S., Yale U., 1964; Ph.D., Clark U., 1968. Postdoctoral NIH trainee Tufts U. Med. Sch., Boston, 1968-69; assoc. prof. biology and biotech. Worcester (Mass.) Poly. Inst., 1969—; cons. in field; co-dir. Water Quality Resource Study Group, Worcester, 1972—. Editor.: Solid Waste Management, 1974, Proceedings, Water Quality and the Public Health, 1983. Resource person Dept. Pub. Wks., Solid Waste and Resource Recovery, Worcester, 1983. NSF grantee, 1970; Mass. Bd. Higher Edn. grantee, 1972. Mem. AAAS, New Eng. Soc. Electron Microscopy, Sigma Xi. Subspecialties: Genetics and genetic engineering (medicine). Current work: Drinking water quality and public health; genetics of senescence of podospora anserina; recombiant DNA methods; electron microscopy of cytoskeletal proteins. Home: 9 Hilltop Cir Worcester MA 01609 Office: Worcester Poly Inst Dept Biology and Biotech Worcester MA 01609

CRUTY, MICHAEL ROBERT, physicist; b. Johnson City, N.Y., June 11, 1941; s. Michael and Anna (Halaburka) C. B.S., SUNY-Stonybrook, 1963; M.S.E.E., Clarkson Coll. Tech., 1971, Ph.D., 1972. Diplomate: Am. Bd. Radiology. Research assoc. U. San Francisco, 1972-74; radiol. biophysicist Lawrence Berkeley Lab., 1974-76; research assoc. U. Wis., Madison, 1976-78; assoc. med. physicist Swedish Am. Hosp., Rockford, Ill., 1978-81; dir. med. physics Lakeland (Fla.) Regional Med. Ctr., 1981—. Contbr. articles to profl. jours. Served to lt. (j.g.) USNR, 1964-68. Mem. Am. Assn. Physicists in Medicine, Am. Coll. Radiology, Health Physics Soc., Fla. Radiol. Soc., Sigma Pi Sigma. Subspecialty: Medical physics. Current work: Research in clinical radiological physics, radiation safety. Office: 1400 Lakeland Hills Blvd Lakeland FL 33802

CRUZ, JOSE BEJAR, engring. educator; b. Bacolod City, Philippines, Sept. 17, 1932; came to U.S., 1954, naturalized, 1969; s. Jose P. and Felicidad (Bejar) C.; m. Patria Cunanan, June 23, 1953; children—Fe E., Ricardo A., Rene L., Sylvia C., Loretta C. B.S. in Elec. Engring. summa cum laude, U. Philippines, 1953; M.S., Mass. Inst. Tech., 1956; Ph.D., U. Ill., 1959. Registered profl. engr., Ill. Instr. elec. engring. U. Philippines, Quezon City, 1953-54; research asst. Mass. Inst. Tech., 1954-56, vis. prof., 1973; instr. U. Ill., Urbana-Champaign, 1956-59, asst. prof., 1959-61, asso. prof., 1961-65, prof. elec. engring., 1965—; research prof. Coordinated Sci. Lab., 1965—; asso. mem. Center Advanced Study, 1967-68; vis. asso. prof. U. Calif. at Berkeley, 1964-65; vis. prof. Harvard, 1973; pres. Dynamic Systems; mem. theory com. Am. Automatic Control Council, 1967; gen. chmn. Conf. on Decision and Control, 1975. Author: (with M.E. Van Valkenburg) Introductory Signals and Circuits, 1967, (with W.R. Perkins) Engineering of Dynamic Systems, 1969, Feedback Systems, 1972, System Sensitivity Analysis, 1973, (with M.E. Van Valkenburg) Signals in Linear Circuits, 1974; Asso. editor: Jour. Franklin Inst, 1976—, Jour. Optimization Theory and Applications, 1981; series editor: Advanced in Large Scale Systems Theory and Applications; Contbr. articles fields network theory, automatic control systems, system theory, sensitivity theory of dynamical systems, large scale systems and dynamic games to sci., tech. jours. Recipient Purple Tower award Beta Epsilon U., Philippines, 1969, Curtis W. McGraw Research award Am. Soc. for Engring. Edn., 1972. Fellow IEEE (chmn. linear systems com., group on automatic control 1966-68, asso. editor Trans. on Circuit Theory 1962-64, adminstrv. com. Control Systems Soc. 1966-75, 78-80, pres. 1979, fellow com. 1970-73, chmn. awards com. Control Systems Soc. 1973-75, ednl. activities bd. 1973-75, editor Trans. on Automatic Control 1971-73, mem. tech. activities bd. 1979-81, edn. med. com 1977-79, v.p. fin. and adminstrv. activities Control Systems Soc. 1976-77, dir. 1980—, vice-chmn. publs. bd. 1981, chmn. panel of tech. editors 1981, chmn. TAB periodicals com. 1981, chmn. PUB Soc. publs. com. 1981); mem. Soc. Indsl. and Applied Math., AAUP, AAAS, U.S. Nat. Acad. Engring., Philippine-Am. Acad. Sci. and Engring. (founding), Sigma Xi, Phi Kappa Phi, Eta Kappa Nu. Subspecialties: Systems engineering; Electrical engineering. Current work: Control systems, sensitivity analysis, decentralized and theoretical decision-making. Home: 2014 Silver Ct W Urbana IL 61801 Office: Coordinated Sci Lab U Ill Urbana IL 61801

CRYER, DENNIS ROBERT, medical scientist, educator; b. Dearborn, Mich., Mar. 30, 1944; s. Earl Wilton and Marguerite Gladys (Root) C.; m. Leilani Chen, June 10, 1967 (div.); children: Jonathan Eric. B.A. with honors, Johns Hopkins U., 1968; M.D., Albert Einstein Coll. Medicine, 1977. Resident in pediatrics Children's Hosp., Phila., 1977-79, asst. chief resident physician, 1980-81, fellow in human genetics, 1981-83; clin. asst. prof. pediatrics, 1983—; postdoctoral fellow in pathology U. Pa. Sch. Medicine, Phila., 1979-80. Contbr. articles to profl. jours. Mem. Genetics Soc. Am., Am. Soc. Microbiology, Alpha Epsilon Delta. Subspecialties: Biochemistry (medicine); Genetics and genetic engineering (medicine). Current work: Biochemistry and genetics of hyperlipidemia and atherosclerosis; developing new methods to study normal and abnormal lipid metabolism and elucidate mechanisms of atherosclerosis using stable isotopes and gas chromatography-mass spectrometry. Office: Children's Hosp Phila Di Genetics 34th St and Civic Center Blvd Philadelphia PA 19104

CSINOS, ALEXANDER STEPHEN, plant pathology educator; b. Tillsonburg, Ont., Can., Jan. 28, 1948; s. Alexander Joseph and Elizabeth (Ribar) C.; m. Lucia Veronica Csinos, June 23, 1973; 1 dau., Alexa Nicole. B.S. in Agr, U. Guelph, 1972; Ph.D., U. Ky., 1977. Vis. asst. prof. Coastal Plain Expt. Sta., U. Ga.-Tifton, 1977-78, asst. prof., 1978-81, assoc. prof., 1981—. Contbr. articles to profl. jours. Mem. Ga. Assn. Plant Pathologists, Am. Phytopathol. Soc., Am. Peanut Research and Edn. Assn., Ga. Vegetable Growers Assn., Tobacco Disease Council, Gamma Sigma Delta. Subspecialty: Plant pathology. Current work: Investigative research on control of soil-borne diseases of tobacco and peanuts. Office: Dept Plant Pathology Coastal Plain Experiment Station PO Box 748 Tifton GA 31793

CUADRA, CARLOS ALBERT, information scientist, management consultant; b. San Francisco, Dec. 21, 1925; s. Gregorio and Amanda (Mendoza) C.; m. Gloria Nathalie Adams, May 3, 1947; children: Mary Susan Cuadra Nielsen, Neil Gregory, Dean Arthur. A.B. with highest honors in Psychology, U. Calif., Berkeley, 1949, Ph.D. in Psychology, 1953. Staff psychologist VA, Downey, Ill., 1953-56; with System Devel. Corp., Santa Monica, Calif., 1957-78, mgr. library and documentation systems dept., 1968-70, mgr. edn. and library systems dept., 1971-74; gen. mgr. SDC Search Service, 1974-78; founder Cuadra Assos., Santa Monica, 1978—. Contbr. articles to profl. jours.; Editor: Ann. Rev. of Info. Sci. and Tech, 1964-75. Mem. Nat. Commn. Libraries and Info. Sci., 1971—. Served with USN, 1944-46. Recipient Merit award Am. Soc. Info. Sci., 1968, Best Info. Sci. Book award, 1969; named Disting. Lectr. of Year, 1970; received Miles Conrad award Nat. Fedn. Abstracting and Indexing Services, 1980. Mem. Info. Industry Assn. (bd. dirs., Hall of Fame award 1980). Home: 13213 Warren Ave Los Angeles CA 90066 Office: 2001 Wilshire Blvd Suite 505 Santa Monica CA 90403

CUATRECASAS, PEDRO MARTIN, research pharmacologist; b. Madrid, Sept. 27, 1936; U.S., 1947, naturalized, 1955; s. Jose and Martha C.; m. Carol Zies, Aug. 15, 1959; children: Paul, Lisa, Diane, Julia. A.B., Washington U., St. Louis, 1958, M.D., 1962. Intern, then resident in internal medicine Osler Service, Johns Hopkins Hosp., 1962-64, asst. physician, 1972-75; clin. asso., clin. endocrinology br. Nat. Inst. Arthritis and Metabolic Diseases, NIH, 1964-66; spl. USPHS postdoctoral fellow Lab. Chem. Biology, 1966-67, med. officer, 1967-70; professorial lectr. biochemistry George Washington U. Med. Sch., 1967-70; asso. prof. pharmacology and exptl. therapeutics, asso. prof. medicine, dir. div. clin. pharmacology, Burroughs Wellcome prof. clin. pharmacology Johns Hopkins U. Med. Sch., 1970-72, prof. pharmacology and exptl. therapeutics, asso. prof. medicine, 1972-75; v.p. research, devel. and med. Wellcome Research Labs.; dir. Burroughs Wellcome Co., Research Triangle Park, N.C., 1975—; adj. prof. Duke U. Med. Sch., 1975—; adj. prof., mem. adv. com. cancer research program U. N.C. Med. Sch., 1975—; bd. dirs. Burroughs Wellcome Fund. Editor: Receptors and Recognition Series, 1975, Jour. Solid-Phase Biochemistry, 1975-80; editorial bd.: Jour. Membrane Biology, 1973, Internat. Jour. Biochemistry, 1973, Molecular and Cellular Endocrinology, 1973-77, Biochimica Biophysica Acta, 1973-79, Life Scis., 1978—, Neuropeptides, 1979—, Jour. Applied Biochemistry, 1978—, Cancer Research, 1980-81, Jour. Applied Biochemistry and Biotech., 1980—, Toxin Revs., 1981—, Biochem. Biophys. Research Communications, 1981—; contbr. articles to profl. jours. Recipient John Jacob Abel prize in pharmacology, 1972, Laude prize Pharm. World, 1975. Mem. Am. Soc. Biol. Chemists, Nat. Acad. Scis., Inst. Medicine of Nat. Acad. Scis., Am. Soc. Pharmacology and Exptl. Therapeutics (Goodman and Gilman award 1981), Am. Soc. Clin. Investigation, Am. Soc. Clin. Research, Spanish Biochem. Soc., Md. Acad. Scis. (Outstanding Young Scientist of Year 1970), Am. Cancer Soc., Endocrine Soc., Am. Chem. Soc., Am. Diabetes Assn. (Eli Lilly award 1975), Sigma Xi. Subspecialties: Cell biology (medicine); Molecular pharmacology. Home: 626 Kensington Dr Chapel Hill NC 27514 Office: 3030 Cornwallis Rd Research Triangle Park NC 27709

CUBICCIOTTI, DANIEL, research chemist; b. Phila., June 28, 1921; s. Daniel and Ida (Orecchia) C.; m. Lois de Roos, Dec. 25, 1948; children: Daniel III, Roger, Kelly. B.S., U. Calif.-Berkeley, 1942, Ph.D., 1946. Asst. Prof. Ill. Inst. Tech., Chgo., 1948-51; research chemist N.Am. Aviation Corp., Downey, Calif., 1951-55; sci. fellow Stanford Research Inst., Menlo Park, Calif., 1955-72, sr. scientist, 1974-80; tech. specialist Gen. Electric Co., Pleasanton, Calif., 1972-74; project mgr. Electric Power Research Inst., Palo Alto, Calif., 1980—; chmn. chem. panel Air Force Office Sci. Research, Washington, 1964-74; chmn. high temperature chemistry Gordon Conf., Kingston, R.I., 1966; chmn. com. on high temperature chemistry NRC, Washington, 1971-74. Contbr. articles to profl. jours. Mem. Electrochem Soc. (div. editor Jour. 1975—), Am. Chem. Soc., Am. Inst. Metall. Engrs., Am. Nuclear Soc. (Best Paper award Nat. Sci. Div. 1976). Subspecialties: High temperature chemistry; Corrosion. Current work: Research on chemical behavior of fission products in nuclear accidents and on corrosion of structural materials in nuclear power systems. Home: 1125 Las Flores Los Altos CA 94022 Office: Electric Power Research Inst Box 10412 Palo Alto CA 94303

CUDABACK, DAVID DILL, astronomer, lecturer, consultant; b. Napa, Calif., Jan. 18, 1929; s. Walter Harold and Luella Matilda (Dill) C.; m. Dorothea Jean, Aug. 16, 1953; 1 dau.: Cynthia Nova. B.A. in Physics, U. Calif., Berkeley, 1951, Ph.D. in Astronomy, 1962. Physicist Lawrence Berkeley Lab., 1950-53, 54-58; Los Alamos Sci. Lab., 1953-54; research assoc. Stanford Electronics Lab., 1958-62; research astronomer U. Calif., Berkeley, 1962—; assoc. dir. White Mountain Research Sta., 1972-80. Contbr. articles to profl. publs., (with Richard o'Hanlon) astron. sculptures; installed, San Raphael, Palo Alto, Berkeley, Calif. Mem. Am. Astron. Soc., Astron. Soc. of the Pacific, Internat. Astron. Union, Internat. Sci. Radio Union, AAAS. Subspecialties: Infrared optical astronomy; Radio and microwave astronomy. Current work: Studies of interstellar dust and star formation with radio and infrared techniques, development novel instruments for these studies, development high altitude observatories. Home: 6639 Longwalk Dr Oakland CA 94611 Office: Astronomy Dept Univ Calif Berkeley CA 94720

CUDKOWICZ, LEON, physician, medical educator, physiology educator; b. Lodz, Poland, Jan. 18, 1923; came to U.S., 1969, naturalized, 1978; s. Maurice and Masza (Malinska) C.; m. (div.); children: Alexander, Penelope. M.B.B.S., U. London, 1946, M.D., 1951. Intern King's Coll. Hosp., U. London; resident St. Thomas Hosp., U.London; assoc. prof. medicine and physiology Dalhousie U., Halifax, N.S., 1960-69; prof. medicine Jefferson Med. Coll., 1969-72; prof. medicine, physiology Hahnemann Med. Coll., 1973-75; prof. dept. medicine, chmn. Wright State U., 1975-77; prof. dept. medicine, chmn. dept. King Faisal U., Dammam, Saudi Arabia, 1978-81; prof. medicine U. Cin. Med. Center, 1981—. Author: Human Bronchial Circulation in Health and Disease, 1968; Contbr. numerous articles to profl. jours., chpts. to books. Served to capt. Royal Army, 1948-49. Canadian Soc. Clin. Investigation Schering awardee, 1973; Orgn. Am. States Fellowship awardee, 1982. Fellow Am. Coll. Cardiology, Am. Coll. Physicians, Royal Coll. Physicians, Royal Soc. Medicine. Subspecialties: Cardiology; Pulmonary medicine. Current work: High altitude physiology. Office: U Cin Bethesda Ave Cincinnati OH 45267

CUDWORTH, KYLE MCCABE, astronomer, educator; b. Mpls., June 7, 1947; s. Kyle Gilmore and Jane McCabe (Irvine) C. B. Physics, U. Minn., 1969; Ph.D., U. Calif., Santa Cruz, 1974. Asst. prof. U. Chgo. (Yerkes Obs.), Williams Bay, Wis., 1974-81, asso. prof., 1981—. Contbr. articles to profl. jours. Alfred P. Sloan Found. fellow, 1980. Mem. Am. Astron. Soc., Astron. Soc. Pacific, Internat. Astron. Union, Am. Sci. Affiliation. Mem. Calvary Community Ch. Club: Quadrangle (Chgo.). Subspecialty: Optical astronomy. Current work: Stellar motions, distances, and photometry; star clusters; planetary nebulae. Office: Yerkes Obs Box 258 Williams Bay WI 53191

CUELLAR, ORLANDO, biologist, educator; b. San Antonio, Sept. 6, 1934; s. Antonio and Juana (Lopez) C.; m. Gloria Quiroga, May 25, 1960; children: Leticia, Graciela, Carolina. B.A., U. Tex., Austin, 1964; M.S., Tex. Tech U., 1965; Ph.D., U. Colo., 1969. Asst. prof. dept. biology U. Utah, Salt Lake City, 1972-77, assoc. prof., 1978—. Served with USAF, 1952-56. USPHS fellow, 1967-69; postdoctoral fellow, 1969-71; Ford Found./Nat. Acad. Sci. postdoctoral fellow, 1980-81. Mem. Am. Soc. Naturalists, Am. Assn. Ichthyologists and Herpetologists, AAAS, Am. Soc. Zoologists, Assn. Animal Behavior. Subspecialties: Population biology. Current work: Lizard ecology, behavior, reprodn., parthenogenesis; research on demography, foraging, mating, competition, tissue transplantation, cytogenetics, reproductive effort. Home: 4835 Quail Point Rd Salt Lake City UT 84117 Office: Dept Biology Univ Utah So Biology Rm 203 Salt Lake City UT 84112

CUGINI, EDWARD THOMAS, mech. engr.; b. Pitts., Aug. 28, 1930; s. Addison Gregory and Antonette Pearl (Misivich) C.; m. Martha Jean Peacher, Nov. 5, 1953; children: Gregory, Thomas, Joel, Matthew. B.S., Calif. U.-Long Beach, 1963. Project engr. Shaffer Tool Works, Brea, Calif., 1965-67, chief engr., 1967-69; engr. Grant Oil Tool, Los Angeles, 1969-75; chief engr. Shafco Industries Inc., Anaheim, Calif., 1975—. Served with USN, 1951-55; Korea. Mem. ASME, Am. Soc. Metals. Lutheran. Subspecialty: Mechanical engineering. Current work: Design pressure vessels used during drilling and production of oil, gas and steam wells. Patentee in field. Home: 429 Catalpa Ave Brea CA 92621 Office: 2850 E Coronado St Anaheim CA 92806

CUKOR, GEORGE, medical microbiologist, researcher, educator; b. Szolnok, Hungary, Mar. 16, 1946; came to U.S., 1957, naturalized, 1962; s. Andor and Lili (Vamos) C.; m. Adrienne G., Aug. 2, 1969; children: Michael, Daniel. B.A. (N.Y. State Regents scholar), Bklyn. Coll., 1968; Ph.D. (USPHS tng. grantee), Rutgers U., 1973. Postdoctoral fellow Boston U. Sch. Medicine, 1973-76; instr. U. Mass. Med. Sch., 1976-77, asst. prof., 1977-80, assoc. prof. medicine and molecular genetics and microbiology, 1980-84; research microbiologist E.I. DuPont de Nemours & Co., 1984—; dir. Hosp. Diagnostic Virology Lab., U. Mass. Med. Center, 1978-83. Contbr. numerous articles, abstracts to profl. publs.; editorial bd.: Jour. Clin. Microbiology, 1982—. Nat. Cancer Inst. postdoctoral research fellow, 1974-76. Mem. Am. Soc. Microbiology, Am. Assn. Immunologists, AAAS, N.Y. Acad. Scis., Am. Soc. Virology. Subspecialties: Infectious diseases; Virology (medicine). Current work: Immunology, pathogenesis and rapid diagnosis of infectious disease; research on gastroenteritis viruses, cystic fibrosis, immunoassays, monoclonal antibodies, bacterial toxins. Home: 7 Kensington Heights Worcester MA 01602 Office: New Eng Nuclear/DuPont Med Diagnostics 601 Treble Cove Rd North Billerica MA 01862

CULLARI, SALVATORE SANTINO, psychologist, researcher; b. Caroniti, Catanzaro, Italy, Apr. 1, 1952; came to U.S., 1955, naturalized, 1968; s. Carmelo and Carmela (Cullari) C. B.A., Kean Coll., 1974; M.A. with honors, Western Mich. U., 1976, Ph.D., 1981. Lic. psychologist (ltd.), Mich.; lic. psychologist, Pa. Assoc. dir. Kalamazoo Learning Village, 1976-77; cons. Lansing (Mich.) Dept. Mental Health, 1978-79; staff psychologist Coldwater (Mich.) Regional Center, 1979-80; research psychologist Anova Research Assn., Kalamazoo, 1980-82; psychol. services coordinator White Haven (Pa.) Center, 1982—. Co-author booklet on bulimia, 1982. Mich. Dept. Mental Health grantee, 1981. Mem. Am. Psychol. Assn., Am. Assn. Mental Deficiency, Assn. Advancement Behavior Therapy, Assn. Behavior Analysis, Mich. Mental Health Assn., Kalamazoo Inst. Arts, AAAS. Subspecialty: Behavioral psychology. Current work: Eating disorders (bulimarexia); radical behaviorism; hemispheric laterality and retarded; eastern religions; psychopharmacology; use of video games for training; artificial intelligence; communal living. Office: Danville State Hosp Danville PA 17821 Home: RD2 PO Box 750 Danville PA 17821

CULLEN, DANIEL EDWARD, mathematician, consultant; b. Oak Park, Ill., Feb. 16, 1942; s. Kenneth Arthur and Ruth (Voltz) C.; m. Paula Bramsen, Aug. 24, 1963; children: Sean, Erik. B.S., Stanford U., 1963; M.A. in Math, U. Ill., 1964; Ph.D. in Applied Math, Washington U., St. Louis, 1967. Tech. staff mem. Bell Telephone Labs., Naperville, Ill., 1967-68; mem. staff MATHEMATICA Inc., Princeton, N.J., 1968—, asst. dir. ops. research, 1969-76; v.p. MATHTECH, Inc., Princeton, 1976—, assoc. dir. mgmt. and tech. studies, 1976-79, dir. mgmt. scis., 1979—; cons. govt. research including Depts. Def., Energy, Transp., Interior, Labor, NASA, NSF, FCC, N.J. Reapportionment Commn., N.J. Dept. Labor and Industry, sch. dists.; cons. fin. planning, environ. analysis, credit monitoring, mktg. and risk analysis, prodn. and distbn., personnel, research and devel., econs.; bds. dirs. various firms. Contbr. articles to profl. jours. Pres. bd. trustees Chapin Sch., Princeton, 1982—. Mem. Fin. Mgmt. Assn., Inst. Mgmt. Scis., Air Pollution Assn., Am. Math. Soc., Ops. Research Soc. Am. (Lanchester Prize Com. 1976, long range planning com. 1977), AAAS, Math. Programming Soc., Soc. Indsl. and Applied Math. Subspecialties: Operations research (engineering); Operations research (mathematics). Current work: Use of quantitative methods to solve management, planning, financial and operational problems of business and government. Home: 980 Stuart Rd Princeton NJ 08540 Office: MATHTECH Subs MATHEMATICA PO Box 2392 Princeton NJ 08540

CULLEN, DONALD LEE, rapidly solidified metals co. exec., physicist; b. Shortcreek Twp., Ohio, Dec. 28, 1940; s. James and Grace Virginia (Huffman) C.; m. Hedy Louise, June 24, 1943; children: Michelle Lynn, Christine Kelley. B.S.E.P., Ohio State U., 1970, M.B.A., 1980. Project engr. instrument div. Reliance Electric Co., Columbus, Ohio, 1963-69; engring. mgr. Autech Corp., Columbus, 1969-74, exec. v.p., 1974-76, pres., 1976-81, Transmet Corp., Columbus, 1981—; cons. electro-optics, lasers; dir. Harrison Enterprises, Morrison Electronics, Digitronics; speaker. Contbr. articles on application of low powered lasers to profl. jours. Mem. Brookside Civic Assn. Served with USN, 1959-63. Mem. Instrument Soc. Am., Optical Soc. Am., IEEE, Am. Phys. Soc., Soc. Mfg. Engrs. Republican. Subspecialties: Materials processing; Systems engineering. Current work: Electro-optics and rapidly quenched metals; devel. of product and markets of rapidly solidified metals. Patentee in field. Office: 4290 Perimeter Dr Columbus OH 43228

CULLER, FLOYD LEROY, JR., chem. engr.; b. Washington, Jan. 5, 1923; s. Floyd LeRoy Culler; m. Della Hopper, 1946; 1 son, Floyd LeRoy III. B. Chem. Engring. cum laude, Johns Hopkins, 1943. With Eastman Kodak and Tenn. Eastman at Y-12, Oak Ridge, 1943-47; design engr. Oak Ridge Nat. Lab., 1947-53, dir. chem. tech. div., 1953-64, asst. lab. dir., 1965-70, dep. dir., 1970-77; pres. Electric Power Research Inst., Palo Alto, Calif., 1978—; research design chem. engring. applied to atomic energy program, chem. processing nuclear reactor plants, energy research. Mem. sci. adv. com. Internat. Atomic Energy Agy., 1974—. Chmn. 1st Municipal Planning and Zoning Commn., Oak Ridge. Recipient Ernest Orlando Lawrence award, 1964; Atoms for Peace award, 1969; Robert E. Wilson award in nuclear chem. engring., 1972; Engring. Achievement award E. Tenn. Engrs. Joint Council, 1974. Fellow Am. Nuclear Soc. (dir. 1973-80, spl. award 1977), Am. Inst. Chemists, AAAS, Inst. Chem. Engrs.; mem. Am. Chem. Soc., Nat. Acad. Engring. Subspecialty: Energy science and engineering. Home: 1385 Corinne Ln Menlo Park CA 94025 Office: 3412 Hillview Ave Palo Alto CA 94303

CULLINGFORD, HATICE S(ADAN), chemical engineer, consultant; b. Konya, Turkey, June 10, 1945; d. Ahmet and Emine (Kadayifciglu) Harmanci. B.S. with high honors, N.C. State U., 1969, Ph.D., 1974. Reactor engr. AEC, Washington, 1973-75, spl. asst., 1975; mech. engr. Dept. Energy, Washington, 1975-78; mem. staff Los Alamos Nat. Lab., 1978-82; mem. Fusion Power Assocs., Gaithersburg, Md.,

1981—; organizing mem. 3d, 4th, 5th, 6th Alt. Energy Sources Confs., Miami, Fla., 1979-83. Contbr. articles to profl. jours. Mem. Tex. Round Table on Hazardous Waste, Houston, 1982—; Vol. income tax asst. ARC and IRS, Houston, 1983. Recipient Spl. Achievement award ERDA, 1976, Inventor award Los Alamos Nat. Lab., 1982, Women's badge Tau Beta Pi, 1968; Cities Service fellow, 1969-72. Mem. Am. Nuclear Soc. (sec.-treas. fusion energy div. 1983-84), Am. Inst. Chem. Engrs. (chmn. low pressure processes and tech. 1981—, rep. to Engrs.' Council Houston 1983—), Am. Chem. Soc., Am. Vacuum Soc., Internat. Assn. Hydrogen Energy (hon.), Phi Kappa Phi, Pi Mu Epsilon. Club: No. N. Mex. Chem. Engrs. (chmn./organizer 1980-82). Subspecialties: Nuclear fission; Nuclear fusion. Current work: Fluid flow, heat/mass transfer, hydrogen systems, and low-pressure processes and technology. Co-inventor method and apparatus for storing hydrogen isotopes. Home: 404 Oak Harbor Dr Houston TX 77062

CULLUM, JANE KEHOE, applied mathematician; b. Norfolk, Va., Sept. 17, 1938; d. William Kenneth and Mildred Grace (Moss) Kehoe; m. Clifton David Cullum, Jr., Dec. 20, 1959; children: Christina Grace, David Kenneth. B.S. in Chem. Engring, Va. Poly. Inst., 1960, M.Sc. in Math, 1962; Ph.D. in Applied Math, U. Calif., Berkeley, 1966. Mem. research staff IBM Research, Yorktown Heights, N.Y., 1966—, mgr. math. scis. dept., 1979-82, sr. mgr. math. scis. dept., 1982—. Recipient Lillian Moller Gilbreith award Soc. Women Engrs., 1959. Mem. Soc. Indsl. and Applied Math. (sec. 1972-76, v.p. 1980-83), IEEE (control systems soc. adminstrv. com. 1978-80, 82—), Math. Assn. Am. Subspecialty: Numerical analysis. Current work: The analysis and development of numerical algorithms for the solution of large, difficult engineering and scientific problems: emphasis on eigenvalue computations. Office: IBM Research PO Box 218 Yorktown Heights NY 10598

CULP, LLOYD ANTHONY, cell biologist, educator, researcher; b. Elkhart, Ind., Dec. 23, 1942; s. Robert Eugene and Genevieve (Murdock) C.; m. Margaret Mary, July 17, 1965; children: Robert Joseph, Catherine Anne. B.S. in Chemistry, Case Inst. Tech., 1964; Ph.D. in Biochemistry, MIT, 1969. Postdoctoral trainee in virology Harvard U. Med. Sch., Boston, 1969-71; asst. prof. microbiology Case Western Res. U. Sch. Medicine, 1972-77, assoc. prof., 1977-83, prof. molecular biology, 1983—; cons., lectr. in field. Contbr. numerous articles to profl. jours., texts, 1969—. Recipient Career Devel. award Nat. Cancer Inst., 1974-79; Pinney scholar, 1973-75; NIH grantee, 1972—; Am. Cancer Soc. grantee, 1974-80. Mem. AAAS, Am. Soc. Cell Biology, N.Y. Acad. Scis., Soc. Complex Carbohydrates, Sigma Xi. Subspecialties: Cell and tissue culture; Cell study oncology. Current work: Adhesion of normal or malignant fibroblasts or neuronal cells to extracellular matrices. Office: Dept Microbiology Case Western Res U Sch Medicine Cleveland OH 44106

CUMBERBATCH, ELLIS, mathematics educator; b. Oldham, Eng., Apr. 19, 1934; came to U.S., 1958; s. Richard and Isabella (Mellor) C.; m. Barbara Mary Blears, July 24, 1957; children: Guy, Louis, Evelyn. B.S., Manchester (Eng.) U., 1955, Ph.D., 1958. Research fellow Calif. Inst. Tech., Pasadena, 1958-60; assoc. research scientist Courant Inst. Math., N.Y.C., 1960-61; lectr. in math. Leeds (Eng.) U., 1961-64; assoc. prof. math. Purdue U., West Lafayette, Ind., 1964-68, prof. math., 1968-81, Claremont (Calif.) Grad. Sch., 1981—. Mem. Soc. Indsl. and Applied Math. Subspecialties: Applied mathematics; Fluid mechanics. Current work: Wave propagation, asymptotic analysis. Home: 644 W 10th St Claremont CA 91711 Office: Dept Math Claremont Grad Sch Claremont CA 91711

CUMMINGS, CHARLES WILLIAM, physician, educator; b. Boston, Nov. 16, 1935; s. Harry Blanchard and Madge (Frey) C.; m. Macon Lee Howard, Dec. 20, 1958; children—Charles William, Lee Blanchard, Evelyn Howard. A.B., Dartmouth Coll., 1957; M.D., U. Va., 1961. Intern Mary Hitchcock Meml. Hosp., Hanover, N.H., 1961-62; resident otolaryngology Harvard U. Med. Sch., 1965-68; practice medicine specializing in otolaryngology, Seattle, 1978—; asso. prof. otolaryngology Upstate Med. Sch., SUNY, Syracuse, 1976-78; prof., chmn. dept. otolaryngology U. Wash. Med. Sch., Seattle, 1978—. Contbr. sci. articles to profl. jours. Served to capt., M.C. USAF, 1963-65. Mem. A.C.S., Soc. Head and Neck Surgeons, Am. Soc. for Head and Neck Surgery, Soc. U. Otolaryngologists, Assn. Acad. Depts. Otolaryngology, Triological Soc., Laryngological Soc., Bronchoesophagological Soc. Episcopalian. Subspecialty: Otorhinolaryngology. Current work: Head and neck surgical reconstruction. Office: RL-30 Dept Otolaryngology U Wash Med Sch Seattle WA 98195

CUMMINGS, EDWARD MARK, psychologist; b. Honolulu, May 19, 1950; s. Edward Mark and Miriam (Gilchrist) C.; m. Mary Lorraine Cummings, Mar. 25, 1972. B.A., Johns Hopkins U., 1972; M.A., UCLA, 1973, Ph.D., 1977. Lectr. UCLA, 1977-79; sr. staff assoc. NIMH, Bethesda, Md., 1979—; cons. NBC-TV, N.Y.C., 1978, Alfred Pub., Los Angeles, 1977. Author, editor: Development of Aggression and Altruism. Johns Hopkins U. scholar, 1968-72; NIMH grantee, 1972-76; U. Calif. Regents grantee, 1976-79; Soc. for Research in Child Devel. grantee, 1982; Fellow Internat. Soc. for Research on Aggression. Mem. Am. Psychol. Assn., Soc. for Research in Child Devel., Johns Hopkins U. Alumni Assn., MacArthur Found. Network, AAAS, Internat. Soc. Research in Behavioral Devel. Republican. Roman Catholic. Subspecialty: Developmental psychology. Current work: Aggression in young children, the impact of stress on devel. of aggression, parental psychopathology and aggression, anger in the home and aggression, day care and emotional devel. Home: 6009 67th Ave Apt 4 Riverdal MD 20737 Office: Lab Developmental Psychology NIMH Bldg 15K 9000 Rockville Pike Bethesda MD 20205

CUMMINGS, GARTH ELLIS, nuclear engineer; b. Oakland, Calif., Jan. 31, 1934; s. Ellis N. and Dorothy M. (Boyd) C.; m. Shirley E. Wolfe, Nov. 10, 1956; children: Gregg A., Jill L. B.S. in Mech. Engring., U. Calif.-Berkeley, 1956, M.S. in Engring. Sci., 1959; Ph.D., U. Calif.-Davis, 1978. Registered profl. engr., Calif. Nuclear/mech. engr. Lawrence Livermore Lab., Livermore, Calif., 1956—, chief reactor ops. Livermore PoolType Reactor, 1963-67, leader nuclear systems group, 1974-79, leader engring. mechanics sect., 1979-81, dep. program leader nuclear systems safety program, 1981—; systems analyst AEC reactor safety study Lawrence Livermore Lab./AEC, 1972-74. Chmn. com. Boy Scouts Am., Danville, Calif., 1977-80; adviser De Molay, Walnut Creek Calif., 1979-81. Mem. Am. Nuclear Soc. (sect. chmn. 1971-72), AAAS, Soc. Risk Analysis. Subspecialties: Nuclear engineering; Mechanical engineering. Current work: Nuclear reactor safety research and technical application with specialization in probabilistic risk assessment, seismic safety, project and general management. Home: 1551 Harlan Dr Danville CA 94526 Office: Lawrence Livermore National Laboratory PO Box 808 Livermore CA 94550

CUMMINGS, JACK ALAN, educational psychology educator; b. Phila., Aug. 31, 1953; s. Roy John and Flora (Marks) C.; m. Linda A. Frith, July 29, 1977; children: Scott, Megan, Jennifer. A.B., Ohio U., 1975; M.Ed., Indiana U. Pa., 1977; Ph.D., U. Ga., 1980. Mem. programming staff Pinehill Rehab. Ctr., Phila., 1976-77; asst. prof. ednl. psychology Ind. U., Bloomington, 1980—; cons. on measurement Am. Guidance Service, Circle Pines, Minn., 1979—. Contbr. articles to profl. jours. Spencer Found. grantee, 1981-83; Proffitt Found. grantee, 1982-83; U.S. Dept. Edn. grantee, 1982—. Mem. Am. Psychol. Assn., Nat. Assn. Sch. Psychologists. Current work: Psychometric investigations of instruments for the assessment of academic and cognitive abilities. Home: 4861 N Hite Dr Bloomington IN 47401 Office: Inst for Child Study Ind U Bloomington IN 47405

CUMMINGS, MARTIN MARC, physician, sci. adminstr.; b. Camden, N.J., Sept. 7, 1920; s. Samuel and Cecelia (Silverman) C.; m. Arlene Sally Avrutine, Sept. 27, 1942; children—Marc Steven, Lee Bernard, Stuart Lewis. B.S., Bucknell U., 1941, D.Sc., 1969; M.D., Duke, 1944; D.Sc., U. Nebr., Emory U.; L.H.D., Georgetown U., 1971; M.D. (hon.), Karolinska Inst., 1972. Diplomate: Am. Bd. Microbiology. Intern, resident Boston Marine Hosp., 1944-46; resident Tb Grasslands Hosp., Valhalla, N.Y., 1946-47; dir. Tb evaluation lab. Communicable Disease Center, USPHS, Atlanta, 1947-49; instr. medicine Emory U. Sch. Medicine, 1948-50, asso. medicine, 1950-52, asst. prof., 1953; chief Tb sect., also dir. Tb research lab. VA Hosp., Atlanta, 1949-53; dir. research services VA Central Office, Washington, 1953-59; spl. lectr. microbiology George Washington U. Sch. Medicine, 1953-59; prof. microbiology, chmn. dept. Okla. U. Sch. Medicine, 1959-61; chief Office Internat. Research, NIH, USPHS, 1961-63; dir. Nat. Library of Medicine, 1964—; asso. dir. for research grants NIH, 1963-64; chmn. com. med. research Nat. Tb Assn., 1958-59; chmn. panel Sarcoidosis NRC-Nat. Acad. Scis., 1958-60. Author: (with Dr. H.S. Willis) Diagnostic and Experimental Methods in Tuberculosis, 1952; Contbr. chpt. on: Tubercle Bacilli, Diagnostic Procedures and Reagents, 1950. Served with AUS, 1943-44. Recipient Exceptional Service award VA, 1959; Distinguished Service award HEW, 1968; Rockefeller Pub. Service award, 1973; Disting. Achievement award Modern Medicine, 1976; Disting. Service award Am. Coll. Cardiology, 1978; John C. Leonard award Assn. Hosp. Med. Edn., 1979. Sr. mem. Am. Soc. Clin. Investigation, Am. Fedn. Clin. Research; mem. Am. Clin. and Climatol. Assn., AAAS (dir.). Subspecialties: Information systems (information science); Information systems, storage, and retrieval (computer science). Current work: Develop and distribute medical information and data bases. Home: 11317 Rolling House Rd Rockville MD 20852 Office: Nat Library Medicine Bethesda MD 20209

CUMMINGS, MICHAEL RICHARD, biological science educator; b. Chgo., July 7, 1941; s. Mark J. and Margaret E. (Lamping) C.; m. Lee Ann Mirek, Apr. 16, 1966; children: Brendan, Kerry. B.A., St. Mary's Coll., 1963; M.S., Northwestern U., 1965, Ph.D., 1968. Mem. faculty U. Ill., Chgo., 1969—, assoc. prof. biol. scis., 1972—; research assoc. prof. Inst. Study of Devel. Disabilities, Chgo., 1980—. Author: (with W.S. Klug) Concepts of Genetics, 1983; cons. editor, Charles Merrill Pub. Co., Columbus, Ohio, 1982—; contbr. numerous sci. articles to prof. publs. Mem. Genetics Soc. Am., Soc. for Devel. Biology, Am. Soc. Human Genetics, Am. Soc. Cell Biology, Nat. Assn. Down Syndrome, Down Syndrome Congress. Clubs: Jackson Park Yacht (Chgo.); Gladstone Yacht (Gladstone, Mich.). Subspecialties: Genetics and genetic engineering (biology); Genome organization. Current work: s: Molecular basis of Down Syndrome; genome organization; gene action; molecular biology. Office: U Ill Dept Biol Scis PO Box 4348 Chicago IL 60680

CUMMINGS, SUE CAROL, chemistry educator; b. Dayton, Ohio, Apr. 24, 1941; s. Carl Leroy and Kathryn Irene (Hurlow) C.; m. H. Glynn Marsh, June 15, 1979. B.A., Northwestern U., Evanston, Ill., 1963; M.S., Ohio State U.- Columbus, 1965, Ph.D., 1968. Chemist Dow Chem. Co., Midland, Mich., 1965-66; vis. research assoc. Aerospace Research Labs., Wright Patterson AFB, Ohio, 1968-69; asst. prof. chemistry dept. Wright State U., Dayton, Ohio, 1969-73, assoc. prof., 1973-77, prof., 1977—, acting assoc. dean, 1976-77. Named Outstanding Scientist Engring. and Sci. Found. and Affiliate Socs., 1977. Fellow AAAS; mem. Am. Chem. Soc. (sec., mem. bd. Dayton sect. 1974-76), Phi Beta Kappa, Iota Sigma Pi. Subspecialties: Inorganic chemistry; Synthetic chemistry. Current work: Synthesis characterization of transition metal complexes, activation and catalytic reactivities of compleses in presence of small molecules, bioinorganic, synthetic oxygen carriers. Office: Dept Chemistry Wright State U Dayton OH 45435

CUMMINS, JOSEPH EDWARD, geneticist, educator; b. Whitefish, Mont., Feb. 5, 1933; s. Harold Geddes and Catherine Cynthia (Edwards) C.; m. (widowed); 1 dau.: Lili Frances. B.S., Wash. State U., 1955; Ph.D., U. Wis., 1962. Postdoctoral fellow U. Edinburgh, Scotland, 1962-64, McArdle Lab. for Cancer, 1964-66, Karolinska Inst., Stockholm, 1969; faculty Rutgers U., 1966-67, U. Wash., 1967-71; asso. prof. genetics U. Western Ont., London, 1972—, also researcher. Contbr. articles to profl. publs. Recipient awards Nat. Sci. and Engring. Research Council Can., 1974, 1982. Mem. Canadian Genetics Soc., Am. Genetics Soc., Environ. Mutagen Soc., Brit. Biochem. Soc., Am. Soc. Cell Biology. Subspecialties: Genetics and genetic engineering (medicine); Toxicology (medicine). Current work: Environ. mutagenesis and carcinogenesis, molecular biology of fungi. Home: 738 Wilkins St London ON Canada N6C 4Z9 Office: U Western Ont London ON Canada N6A 5B7

CUNNINGHAM, BRUCE ARTHUR, biochemist; b. Winnebago, Ill., Jan. 18, 1940; s. Wallace Calvin and Margaret Wright (Clinite) C.; m. Katrina Sue Susdorf, Feb. 27, 1965; children—Jennifer Ruth, Douglas James. B.S., U. Dubuque, 1962; Ph.D., Yale U., 1966. NSF postdoctoral fellow Rockefeller U., N.Y.C., 1966-68, asst. prof. biochemistry, 1968-71, assoc. prof., 1971-77, prof. molecular and devel. biology, 1978—. Editorial bd.: Jour. Biol. Chemistry, 1978-83. Camille and Henry Dreyfus Found. grantee, 1970-75; recipient Career Scientist award Irma T. Hirschl Trust, 1975. Mem. Am. Soc. Biol. Chemists, Am. Assn. Immunologists, Am. Chem. Soc., Harvey Soc. Democrat. Lutheran. Subspecialties: Developmental biology; Molecular biology. Research on structure and function of molecules on cell surfaces. Office: 1230 York Ave New York NY 10021

CUNNINGHAM, CHARLES GODVIN, geologist; b. Springfield, Vt., Dec. 5, 1940; s. R. Robert and Mildred (Wilson) Heydt; m. Cheryl Lee Cunningham, Aug. 10, 1968; children: Wendy, Betsy. A.A., Norwalk Community Coll., 1964; A.B., Amherst Coll., 1967; M.S., U. Colo., 1969; Ph.D., Stanford U., 1973. Asst. prof. Syracuse (N.Y.) U., 1973-74; research geologist U.S. Geol. Survey, Denver, 1974—; supervisory geologist Central Mineral Resources Br., 1974—. Contbr. articles to profl. jours. Fellow Geol. Soc. Am.; mem. Soc. Econ. Geologists (research com. 1979-81), Yellowstone-Bighorn Research Assn., Colo. Sci. Soc. Current work: Specialist in fluid inclusion goethermometry and geobarometry of ore deposition. Office: Denver Federal Ctr PO Box 25046 MS 905 Denver CO 80225

CUNNINGHAM, CLARENCE MARION, chemistry educator, computer programming consultant; b. Cooper, Tex., July 24, 1920; s. Willie Lee and Naoma Mae (Stokes) C.; m. Janet Ruth Kohl, Oct. 14, 1922; children: Elizabeth Jane, Daniel Marvin, Steven Charles, Margaret Helen. B.S., Tex. A&M U., 1942; M.S., U. Calif.-Berkeley, 1958; Ph.D., Ohio State U., 1954. Teaching asst. U. Calif., 1946-48; instr. chemistry Calif. State Poly. Inst., 1948-49; research asst. Ohio State U., 1949-51; cryogenic engr. AEC program H. L. Johnston, Columbus, Ohio, 1951-54; asst. prof. chemistry Okla. State U., 1954-59, assoc. prof., from 1959, now prof.; cons. Author: (with Jones) Electrolytic Conductance of Lithium Bromide in Acetone and Acetone-Bromosuccinic Acid Solutions, 1976, (with More) A Student's Guide to Independent Study for Petrucci's General Chemistry, 1977. Sec. Payne County (Okla.) Democratic Com., 1962-65, 75-77; scoutmaster Will Rogers council Boy Scouts Am., 1967—; bd. dirs. Stillwater (Okla.) Neighborhood Nursery, 1967—. Served with U.S. Army, 1942-46; to col. USAR, 1946-72. Recipient Silver Beaver award Boy Scouts Am., 1977; Petroleum Research grantee, 1955-60; NASA grantee, 1964-70. Mem. Am. Chem. Soc., Am. Phys. Soc., AAAS, AAAUP. Quaker. Lodge: Kiwanis. Subspecialties: Physical chemistry; Nuclear magnetic resonance (chemistry). Current work: Association constants from the chemical shift in the nuclear magnetic resonance spectrum of solutions. Home: 924 Lakeridge Ave Stillwater OK 74074 Office: Dept Chemistry Okla State U Stillwater OK 74078

CUNNINGHAM, EARLENE BROWN, biochemistry educator; b. Cleve., Aug. 27, 1930. B.S., U. Ill., 1949; M.S., UCLA, 1951; Ph.D., U.S.C., 1954. Research assoc. Ind. U. Sch. Medicine, Indpls., 1954-59; asst. prof. Howard U. Coll. Medicine, Washington, 1959-63, assoc. prof., 1963-64; research fellow U. Calif.-Berkeley, 1964-68; lectr. U. S.C., Columbia, 1968-71; assoc. prof. Med. U. S.C., Charleston, 1971-78, U. Medicine and Dentistry N.J.-N.J. Med. Sch., Newark, 1978—. Recipient Lederle Med. Faculty award Howard U. Coll. Medicine, 1961-63; NIH fellow, 1964-68. Mem. Sigma Xi. Subspecialty: Biochemistry (medicine). Current work: Biochemical regulation involving cell membrane components and mediated by calcium and/or cyclic AMP. Office: Dept Biochemistr U. Medicine and Dentistry NJ-NJ Med Sch 100 Bergen St Newark NJ 07103

CUNNINGHAM, GLENN NORMAN, chemist, educator; b. Spring City, Tenn., Sept. 13, 1940; s. Teed and Margret L. (McDaniel) C.; m. Mildred L. Clawson, Mar. 2, 1962; children: Darrell G., Randall C. B.S., U. Tenn., 1961; M.S., N.C. State U., 1964, Ph.D., 1966. Postdoctoral fellow U. Tex., Austin, 1966-68; asst. prof. chemistry N.E. La. State U., 1968-69, U. Central Fla., Orlando, 1969-72, assoc. prof. chemistry, 1972-78, prof. chemistry, 1978—; mem. research com. Fla. chpt. Am. Cancer Soc., 1979—. Contbr. articles to profl. jours. Bd. dirs. John Young Sci. Center, 1979—, sec., exec. com., 1982—. Co-recipient Outstanding Researcher of Yr. award U. Central Fla., 1978; NSF grantee, 1977-79; NIH grantee, 1979-82; Dow Chem. Co. grantee, 1981-83; NSF grantee, 1983. Mem. Am. Chem. Soc., Genetics Soc. Am., AAAS, Electrophoresic Soc., Sigma Xi. Democrat. Baptist. Subspecialties: Biochemistry (biology); Molecular biology. Current work: Biochem. genetics, enzymology; biochemistry of differentiation in eukaryotics; mode of action of insecticides; enzyme histochemistry. Home: PO Box 801 Oviedo FL 32765 Office: Chemistry Dept U Central Fla Orlando FL 32816

CUNNINGHAM, HARRY NORMAN, JR., biology educator; b. Imperial, Pa., Mar. 7, 1935; s. Harry N. and Flora F. (Broadhurst) C.; m. Louise Gittins, Dec. 26, 1957; children: Elisa L., John C. B.S., U. Pitts., 1955, M.S., 1960, Ph.D., 1966. Instr. Mt. Union Coll., Alliance, Ohio, 1959-61; asst. prof. Thiel Coll., Greenville, Pa., 1963-67; asst. prof. biology Behrend Coll. of Pa. State U., Erie, 1967-72, assoc. prof., 1972—; cons. Aquatic Ecology Assocs., Pitts., 1974-78. Mem. Am. Soc. Mammalogists, Ecol. Soc. Am., Pa. Acad. Sci. Subspecialties: Ecology; Species interaction. Current work: Effects of heavy metals on small mammal populations. Office: Dept Biology Behrend Coll of Pa State U Station Rd Erie PA 16563

CUNNINGHAM, JOHN EDWARD, laboratory executive; b. Chgo., Mar. 18, 1920; s. Edward Joseph and Marion (Fergus) C.; m. Dolores Mori, Sept. 2, 1944; children: Diane Cunningham Jameson, John E., LaVonne Cunningham Spicer, Joan Cunningham Wofford, Robert Joseph. B.S., U. Ill., 1944; postgrad., Ohio State U., 1944-45; M.S., U. Tenn., 1950, 1968-72. Tech. engr. U.S. Army C.E., Oak Ridge, 1944-46; assoc. metallurgist Clinton Labs., Oak Ridge, 1946-48; supr. metals and ceramics div. Oak Ridge Nat. Lab., 1948-55, asst. dir. metals and ceramic div., 1955-68, assoc. dir., 1968—, coordinator honors and awards, 1979—. Editor: Nuclear Engineering Materials Handbook Series, 1977-84. Pres. Oak Ridge chpt. Tenn. Vols. for Life; bd. dirs., treas. Prisoners Aid Soc. Tenn., Oak Ridge; mem. Health Council Anderson County, Oak Ridge. Fellow Am. Nuclear Soc. (cert. of merit 1965, exceptional service award 1980), Am. Soc. Metals (silver cert. 1968); mem. ASTM, AIME, Sigma Xi. Republican. Roman Catholic. Lodge: KC. Subspecialties: Nuclear fission; Materials (engineering). Current work: Engineering management, high temperature testing, ceramic fabrication, electron microscopy-materials science, alloy development, x-ray diffraction scattering. Home: 106 Mountainview Ln Oak Ridge TN 37830 Office: Oak Ridge Nat Lab PO Box X Oak Ridge TN 37830

CUPCHIK, WILLIAM, psychologist; b. Montreal, June 5, 1940; s. David and Chana (Trifskin) C.; m. Gila Gladys Holtzman, Aug. 20, 1961 (div. 1972); 1 son, Jeffrey Wayne. B.Engring., McGill U., 1961; B.A., Carleton U., 1963; M.Ed., U. Toronto, 1970, Ph.D., 1979. Registered psychologist, Ont. Navigational systems design engr. Computing Devices of Can., Bells Corners, Ont., 1961-62; tchr., guidance counselor Ottawa Bd. Edn., 1963-66, North York Bd. Edn., 1966-69; attendance counselor Etobicoke Bd. Edn., Ont., 1969-72; staff psychologist forensic service Clarke Inst. Psychiatry, Toronto, Ont., 1979—; guest faculty Order of Mt. Mary Immaculate, Lafayette, Calif., 1977-82; lectr. Sch. Continuing Studies, U. Toronto, 1980-83; cons., group leader career counseling Sun Oil Co. Can., 1980-83. Originator, author psychotherapeutic procedure, reintrojection therapy, 1983; author: Clinical Uses of Mental Imagery, 1983; originator, inventor: navigational guidance system Map Display Unit and Vector Adder, 1962. Mem. Am. Group Psychotherapy Assn., Ont. Group Psychotherapy Assn. (treas. 1980-81), Internat. Transactional Analysis Assn., Am. Psychol. Assn., Ont. Psychol. Assn., Internat. Council Psychologists. Jewish. Subspecialty: Psychotherapy. Current work: Clinical uses of mental imagery procedures; reintrojection therapy development; criminal activities of middle and upper class ethical majority. Office: 250 College St Toronto ON Canada M5T 1R8 Home: 276 Waverly Rd Toronto ON Canada M4L 3T6

CUPPAGE, FRANCIS EDWARD, pathologist, researcher, educator; b. Cleve., Aug. 17, 1932; s. Frank Edward and Eunice (Bartels) C.; m. Virginia Lee Bartch, Aug. 18, 1956; children: Lisa, Peter, Sharon. B.S., Case Western Res. U., 1954; M.S. in Pathology, Ohio State U., 1959, P.D., 1959. Diplomate: Am. Bd. Pathology. Intern U. Hosp. of Cleve., 1959-60, resident, 1960-64; instr. Case Western Res. U., 1964-65; asst. prof. Ohio State U., 1965-67; asst. prof. pathology U. Kans.-Kansas City, 1967-69, assoc. prof., 1969-73, prof., 1973—; cons. pathologist VA, Kansas City, Mo., 1967—; mem. study sect. NIH, 1976-79. Ch. leader Lutheran Ch. in America, Kansas City, Kans., 1967—. Recipient Alumni Teaching award U. Kans.-Kansas City, 1972; NIH Fogarty sr. fellow, 1979; NIH grantee, 1968-80; Kidney Found. grantee, 1982. Mem. Kansas City Soc. Pathology (sec.-treas. 1980-81, pres. 1981-82), Internat. Acad. Pathology, Internat. Soc. Nephrology, Am. Soc. Nephrology, Alpha Omega Alpha. Subspecialty: Pathology (medicine). Current work: Toxic injury to kidney. Office: U Kans Med Center 39th and Rainbow Blvd Kansas City KS 66103

CURD, JOHN GARY, immunologist, researcher; b. Grand Junction, Colo., July 2, 1945; s. H. Ronald and Edna (Hegsted) C.; m. Karen Wendel, June 12, 1971; children: Alison, Jonathan, Edward. B.A.

magna cum laude, Princeton U., 1967; M.D. cum laude (Univ. nat. scholar), Harvard U., 1971. Diplomate: Am. Bd. Internal Medicine (rheumatology, allergy and immunology). Intern Med. Service, Mass. Gen. Hosp., Harvard U. Med. Sch., Boston, 1971-72, asst. resident, 1972-73; research assoc. Lab. Chem. Biology, Nat. Inst. Arthritis, Metabolism and Digestive Diseases, NIH, Bethesda, Md., 1973-75; postdoctoral fellow div. rheumatology Sch. Medicine, U. Calif.-San Diego, La Jolla, 1975-77; postdoctoral fellow dept. molecular immunology Research Inst., Scripps Clinic, La Jolla, 1977-78, asst. mem. II dept. molecular immunology and dept. clin. research, 1978—; assoc. dir. Gen. Clin. Research Ctr., Scripps Clinic and Research Found., La Jolla, 1981-83; asst. clin. prof. U. Calif.-San Diego, La Jolla, 198[illegible]; active San Diego Rheumatism Soc., established investigator Am. Heart Assn., 1983—. Research numerous publs. in field. Served with USPHS, 1973-75. Helen Hay Whitney fellow, 1975-78; NIH clin. investigator, 1978-81. Mem. Am. Acad. Allergy, Am. Rheumatism Assn., Am. Assn. Immunologists. Republican. Methodist. Subspecialties: Internal medicine; Immunology (medicine). Current work: Complement research; arthritis research. Office: 10666 N Torrey Pines Rd La Jolla CA 92037

CURIALE, MICHAEL STEVEN, molecular geneticist, researcher; b. Chgo., Jan. 20, 1953; s. Samuel Robert and Amelia Beth (Gilliatt) C.; m. Lynn Ann Reimer, July 17, 1977. Student, Ferris State Coll., 1972; B.S., U. Ill.-Chgo., 1974; Ph.D., Oreg. State U., 1980. Research assoc. dept. molecular biology and microbiology Tufts U. Sch. Medicine, 1980—. Mem. Am. Soc. Microbiology, Genetics Soc. Am., Sigma Xi. Subspecialty: Genetics and genetic engineering (biology). Current work: Genetic organization and mechanism of microbial tetracycline resistance. Home: 23 Sullivan St Boston MA 02129 Office: Tufts U Sch Medicine Dept Microbiology 138 Harrison Ave Boston MA 02111

CURL, ROBERT FLOYD, JR., educator; b. Alice, Tex., Aug. 23, 1933; s. Robert Floyd and Lessie (Merritt) C.; m. Jonel Whipple, Dec. 21, 1955; children—Michael, David. B.A., Rice U., 1954; Ph.D. (NSF fellow), U. Cal. at Berkeley, 1957. Research fellow Harvard, 1957-58; asst. prof. chemistry Rice U., Houston, 1958-63, asso. prof., 1963-67, prof., 1967—; master Lovett Coll., 1968-72; Vis. research officer NRC Can., 1972-73; vis. prof. Inst. for Molecular Sc., Okazaki, Japan, 1977. Contbr. articles profl. jours. Alfred P. Sloan fellow, 1961-63; NATO postdoctoral fellow, 1964; recipient Clayton prize Instn. Mech. Engrs., London, 1958. Mem. Am. Chem. Soc., Phi Beta Kappa, Sigma Xi. Methodist. Subspecialties: Physical chemistry; Spectroscopy. Current work: Laser spectroscopy of transient molecules. Home: 1824 Bolsover St Houston TX 77005

CURNOW, RANDALL THOMAS, pharmacologist, educator; b. Omaha, May 11, 1942; s. Leonard Thomas and Bernice May (Vanderbeek) C.; m. Mabel R. Simpson, Aug. 3, 1963; children: Randall Thomas, Stacey Lynn, Robert Thomas. M.D., U. Omaha, 1967. Diplomate: Am. Bd. Internal Medicine. Intern in internal medicine U. Minn., 1967-68, resident, 1968-69; fellow in endocrinology and metabolism, 1969-71; asst., then assoc. prof. internal medicine and pharmacology U. Va. Sch. Medicine, Charlottesville, 1974-82, prof., 1982—. Served to maj. M.C. U.S. Army, 1971-74. Recipient Schering Found. career devel. award, 1975-78; NIH research career devel. award, 1978—. Fellow ACP; mem. Am. Assn. Biol. Chemists, Am. Soc. Pharmacology and Exptl. Therapeutics, Alpha Omega Alpha (pres. 1967). Subspecialties: Biochemistry (biology); Endocrinology. Current work: Insulin action; phosphoprotein phosphatases. Office: Depts Medicine/Pharmacology U Va Box 419 Charlottesville VA 22908 Home: 326 Carrsbrook Dr Charlottesville VA

CURNUTTE, BASIL, JR., physicist, educator, researcher; b. Portsmouth, Ohio, Mar. 1, 1923; s. Basil and Lula Alafair (Cooper) C.; m. Mary Leete Lukemire, June 10, 1945; children; William Basil, Gregory Mark. B.S., U.S. Naval Acad., 1945; Ph.D. (NSF fellow), Ohio State U., 1953. Grad. asst. Ohio State U., 1950-51, Univ. research scholar, 1951, research assoc., 1953; asst. prof. physics Kans. State U., 1954-55, assoc. prof., 1956-64, prof., 1964—; physicist Lawrence Radiation Lab., 1961; vis. scientist Am. Inst. Physics, 1965-71; vis. prof. Ariz. U., 1968. Research, publs. in atomic and molecular spectroscopy. Served to lt. USN, 1945-49. Fellow Am. Phys. Soc., Optical Soc. Am.; mem. AAAS, Am. Assn. Physics Tchrs., Kans. Acad. Sci. Subspecialties: Atomic and molecular physics; Infrared spectroscopy. Current work: Research in atomic spectroscopy on fast ion beams, Mossbauer Spectroscopy, applied x-ray spectroscopy. Home: 2074 College Heights Manhattan KS 66502 Office: Dept Physics Kans State U Manhattan KS 66506

CURRAH, WALTER EVERETT, mech. engr., cons.; b. Tacoma, Wash., Apr. 10,1936; s. Walter Everett and Gayle Ann C. B.S. in Mech. Engring, Wash. State U., 1964. Owner, prin. Currah Enterprises, Tacoma, Wash., 1969—. Contbr. articles to profl. publs. Mem. ASME. Subspecialties: Analytical chemistry; Atomic and molecular physics. Current work: Writing level 3 stress programs for computer analysis of wind-driven power plant structures. Patentee in field. Home: 4401 S 6th St Tacoma WA 98409 Office: PO Box 2929 Tacoma WA 98409

CURRAN, BRUCE HOWLETT, med. physicist; b. New Haven, Oct. 26, 1951; s. Lawrence T. and Barbara (Howlett) C. A.B., Dartmouth Coll., 1973, M.E., 1982. Sr. programmer Mary Hitchcock Meml. Hosp., Hanover, N.H., 1973-74, sr. engring. analyst, 1974-78; med. physicist Tufts New Eng. Med. Center, Boston, 1978—, instr., 1978—. Contbr. articles to profl. jours. Mem. Am. Assn. Physicists in Medicine, Assn. Computing Machinery, IEEE, Health Physics Soc. Subspecialties: Medical physics; Biomedical engineering. Current work: Hyperthermia, computers in radiology; radiation therapy, ultrasound, digital radiography, cardiology, nuclear medicine. Office: 171 Harrison Ave Boston MA 02111

CURRAN, DENNIS PATRICK, chemistry educator; b. Easton, Pa., June 10, 1953; s. William Vincent and Jane C.; m. Suzanne S. Holcomb, Nov. 24, 1979. B.S., Boston Coll., 1975; Ph.D., U. Rochester, 1979. NIH postdoctoral fellow U. Wis., Madison, 1979-81; asst. prof. chemistry U. Pitts., 1981—. Dreyfus Found. grantee, 1981—. Mem. Am. Chem. Soc. Subspecialties: Synthetic chemistry; Organic chemistry. Current work: Development of new methods for organic synthesis. Total synthesis of complex natural products. Office: Dept Chemistry U Pittsburgh Pittsburgh PA 15260

CURRIE, MALCOLM RODERICK, scientist, aerospace executive; b. Spokane, Wash., Mar. 13, 1927; s. Erwin Casper and Genevieve (Hauenstein) C.; m. Sunya Lofsky, June 24, 1951; children—Deborah, David, Diana; m. Barbara L. Dyer, Mar. 5, 1977. A.B., U. Calif. at Berkeley, 1949, M.S., 1951, Ph.D., 1954. Research engr. Microwave Lab., U. Calif. at Berkeley, 1949-52, elec. engring. faculty, 1953-54; lectr. U. Calif. at Los Angeles, 1955-57; research engr. Hughes Aircraft Co., 1954-57, v.p., 1965-66; head electron dynamics dept. Hughes Research Labs., Culver City, Calif., 1957-60, dir. physics lab., Malibu, Calif., 1960-61, asso. dir., 1961-63, v.p., dir. research labs., 1963-65, v.p., mgr. research and devel. div., 1965-69; v.p. research and devel. Beckman Instruments, Inc., 1969-73; dir. def. research and engring. Office Sec. Def., Washington, 1973-77; v.p. missile systems group Hughes Aircraft Co., Canoga Park, Calif., 1977-83, exec. v.p., 1983—; mem. Def. Sci. Bd. Author articles. Served with USNR, 1944-47. Decorated comdr. Legion of Honor, France; named nation's outstanding young elec. engr. Eta Kappa Nu, 1958, one of 5 outstanding young men of Calif. Calif. Jr. C. of C., 1960. Fellow IEEE, AIAA; mem. Nat. Acad. Engring., Am. Phys. Soc., Phi Beta Kappa, Sigma Xi, Lambda Chi Alpha. Subspecialty: Aerospace engineering and technology. Patentee in field. Home: 28780 Wagon Rd Agoura CA 91301 Office: Hughes Aircraft Co 200 N Sepulveda Blvd Mail Sta C2/A103 El Segundo CA 90245

CURRIE, VIOLANTE EARLSCORT, physician, med. oncologist; b. Chgo.; d. Clarence Clifton and Violante Earlscort (Robertson) C. M.D., U. Ill., 1967. Intern R.I. Hosp., Providence, 1967-68, med. resident, 1968-70; fellow med. oncology Meml. Hosp. Cancer and Allied Diseases, N.Y.C., 1970-71; research fellow Sloan-Kettering Inst., N.Y.C., 1970-71; now research assoc.; fellow in medicine Cornell U., 1970-71; now asst. prof.; fellow in med. oncology Thomas Jefferson U., Phila., 1971-72, asst. attending physician, 1972-73, chief oncology outpatient facility, 1972-73; now asst. attending physician Meml. Hosp. Cancer and Allied Diseases, N.Y.C.; Bd. dirs. Harlem Breast Exam. Center; cons. to vestry St. Philip Episcopal Ch. Contbr. articles to profl. jours. Am. Cancer Soc. fellow, 1970-71; Nat. Cancer Inst. clin. grantee, 1971-72. Mem. Am. Med. Women's Assn., Am. Fedn. Clin. Research, Am. Soc. Clin. Oncology, N.Y. Acad. Scis., Am. Assn. Cancer Research. Subspecialties: Chemotherapy; Cancer research (medicine). Current work: Testing new anticancer agts. Office: 1275 York St New York NY 10021

CURRIER, WILLIAM WESLEY, biochemist, educator, researcher; b. Seattle, Sept. 18, 1947; s. Walter S. and Angeline B. (Clarke) C.; m. Janice K. Cameron, Mar. 22, 1969; 1 son, Reid B. B.S. in Chemistry, U. Wash., 1969; Ph.D. in Biochemistry, Purdue U., West Lafayette, Ind., 1974. Postdoctoral fellow dept. plant pathology Mont. State U., 1974-77; asst. prof. biochemistry, dept. microbiology and biochemistry U. Vt., 1977-82, assoc. prof., 1982—. Contbr. numerous articles to profl. jours. Mem. AAAS, Am. Soc. Plant Physiologists, Sigma Xi (pres. Vt. chpt. 1982-83). Mem. Chs. of Christ. Subspecialties: Nitrogen fixation; Plant pathology. Office: Dept Microbiology and Biochemistry U Vt 115 Hills Bldg Burlington VT 05405

CURRO, FREDERICK ANTHONY, dental educator, dentist; b. Bklyn., Apr. 10, 1943; s. Joseph and Josephine C.; m. Linda Goulet, Sept. 12, 1971; 1 son, Matthew Lawrence. B.S., St. John's U., 1966; D.M.D., Tufts U., 1972; Ph.D., Ohio State U., 1976. Asst. prof. U. Tex., Houston, 1972-73, SUNY, Buffalo, 1973-74; clin. assoc. prof. NYU, N.Y.C., 1977—; assoc. prof. Fairleigh Dickinson U., Hackensack, N.J., 1977-83, prof., 1983—, chmn., 1979—; clin. dir. David B. Kriser Oral Facial Pain Center, N.Y.C., 1979—; cons. Colgate-Palmolive, Piscataway, N.J., 1979—, Block Drug Co., Jersey City, 1980—. Author: The Dental Clinics of North America, 1978, Prostaglandins in Health and Disease, 1982, Nutrition in Oral Health and Disease; editor: CRC Handbook Series of Experimental Dentistry. NIH fellow, 1974. Mem. Internat. Assn. Dental Research, Am. Assn. Dental Schs., Am. Pain Soc., Pharmacology, Toxicology and Therapeutics Group (pres.-elect 1982—). Roman Catholic. Subspecialties: Pharmacology; Anesthesiology. Current work: Evaluating the effects of autonomic drugs on subtypes of Alpha adrenergic receptors. Home: 3 Powell Rd Emerson NJ 07630 Office: Fairleigh Dickinson U 110 Fuller Pl Hackensack NJ 07601

CURRY, MARY GRACE, environmental scientist; b. New Orleans, June 16, 1947; d. Clyde Lalio and Gladys Ruth (Ehret) C. B.S., U. New Orleans, 1969, M.S., 1971; Ph.D., La. State U., 1973. Cert. environ. profl., La. Vis. asst. prof. botany La. State U., Baton Rouge, 1974; environ. scientist VTN La., Inc., Metairie, 1974-79; environ. impact officer Parish of Jefferson, Metairie, 1979—. Contbr. articles to profl. jours. Mem. La. Acad. Scis., La. Environ. Profls. Assn. (pres. 1979, exec. council 1977—), Am. Inst. Biol. Scis., Bot. Soc. Am., Assn. Southeastern Biologists, Ecol. Soc. Am., So. Appalachian Bot. Soc., Coastal Soc., Gretna Hist. Soc., Jefferson Histo. Soc. La., D.A.R., U.D.C., Sigma Xi. Democrat. Roman Catholic. Subspecialties: Ecology; Taxonomy. Current work: Taxonomy, ecology and distbr. of freshwater leeches; vascular plant taxonomy and ecology; coastal zone and wetland mgmt. Home: 3404 Tolmas Dr Metairie LA 70002 Office: 3330 N Causeway Blvd Room 303 Metairie LA 70002

CURTIN, DAVID YARROW, educator, chemist; b. Phila., Aug. 22, 1920; s. Ellsworth Ferris and Margeretta (Cope) C.; m. Constance O'Hara, July 1, 1950; children—Susan McLean, Kathy Gardner, David Ferris, Jane Yarrow. A.B., Swarthmore Coll., 1943; Ph.D., U. Ill., 1945. Pvt. asst. Harvard, 1945-46; instr., then asst. prof. chemistry Columbia, 1946-51; mem. faculty U. Ill., Urbana, 1951—, prof. chemistry, 1954—, head div. organic chemistry, 1963-65; Vis. lectr. Inst. de Quimica, Mexico, summer 1955, U. Tex., 1959; Reilly lectr. U. Notre Dame, 1960. Mem. editorial bd.: Organic Reactions, 1954- 64; adv. bd., 1965—; mem. bd. editors: Jour. Organic Chemistry, 1962-66. Mem. Am., Brit., Swiss chem. socs., Nat. Acad. Sci., Am. Crystallographic Assn. Subspecialties: Organic chemistry; Solid state chemistry. Spl. research organic reaction mechanisms, stereochemistry, exploratory organic chemistry, reactions in solid state. Home: 3 Montclair Rd Urbana IL 61801

CURTIS, BILL, programming executive, software engineering scientist; b. Meridian, Tex., Sept. 3, 1948; s. Willard Holt Curtis and Virginia (White) Stedman; m. Janell Johnston, Jan. 3, 1981. B.A., Eckerd Coll., 1971; M.A., U. Tex.-Austin, 1974; Ph.D., Tex. Christian U., 1975. Grad. research fellow Inst. Behavioral Research, Tex. Christian U., 1972-75; research asst. prof. U. Wash., 1975-76; staff psychologist Weyerhaeuser Co., Tacoma, 1977; mgr. software mgmt. research Gen. Electric Co., Arlington, Va., 1978-80; mgr. programming trends analysis ITT Programming, Stratford, Conn., 1980—. Author: Human Factors in Software Development, 1981, The Human Element in Programming, 1983; editor: Jour. of Systems and Software, 1980, Human Computer Interaction, 1983. Bd. dirs. Canoe Brook Lake Tax Dist., Trumbull, Conn., 1982. Mem. IEEE, Assn. Computing Machinery, Am. Psychol. Assn., Human Factors Soc., Psychometric Soc. Presbyterian. Subspecialties: Software engineering; Cognition. Current work: Advancing knowledge about cognitive and industrial psychological aspects of programming; refining software engineering measurements; developing artificial intelligence systems. Home: 921 Belvin San Marcos TX 78666 Office: ITT Programming Tech Center 1000 Oronoque Ln Stratford CT 06497

CURTIS, CHRISTOPHER MICHAEL, editor; b. N.Y.C., May 7, 1934; m. Jean Getchell, Sept. 30, 1961; children—Christopher Michael, Hilary Ann, Hans Peter Kahn. B.A., Cornell U., 1957, postgrad., 1959-63. Asso. editor Atlantic Monthly, Boston, 1963—. Served with AUS, 1957. Clubs: St. Botolph, Union Boat (Boston). Subspecialties: Sedimentology; Petroleum engineering. Current work: Geological consulting for petroleum industry, particulary stratigrophic analysis for exploration. Home: 668 Bedford St Concord MA 01742 Office: 8 Arlington St Boston MA 02116

CURTIS, GEORGE DARWIN, ednl. research adminstr., cons.; b. Galveston, Tex., Apr. 30, 1928; s. Darwin Agustus and Frances (Gymer) C.; m. Alice Ann Kidd, June 10, 1951 (div. 1963); children: Carol, Paul, Ann, Ted; m. Peggy Ann Plummer, Mar. 20, 1977. B.S. in Physics, N. Tex. U., 1952; postgrad., U. Hawaii, 1972-73. Jr. researcher Mobil Oil Co., Dallas, 1951-52; jr. engr. LTV, Inc., Dallas, 1952-55, sr. engr., 1956-60; sr. scientist LTV Research Center, Honolulu, 1960-68; sr. specialist Control Data Corp., Honolulu, 1968-70; research mgr. U. Hawaii, Honolulu, 1979—; cons. R.A. Darby & Assocs., Honolulu, 1973-78, State of Hawaii, 1970-72; cons., dir. Tiki Gems, Inc., Honolulu, 1969—; lectr. in field. Co-author book chpt. Bd. dirs. ARC, Honolulu, 1975-77; squadron comdr. CAP, Honolulu, 1980—; election officer State of Hawaii, Honolulu, 1974—. NSF grantee, 1980—. Mem. Marine Tech. Soc. (state chmn. 1982-83), IEEE (sr. mem., state chmn. 1978-79), Acoustical Soc. Am., AAAS. Republican. Episcopalian. Subspecialties: Systems engineering, Alternate energy sources. Current work: Photovoltaic and wind energy, instrumentation, technical program management. Inventor core resistivity cell, 1952. Home: 47-363C Hui Iwa Kaneohe HI 96744 Office: U Hawaii 2540 Dole St Honolulu HI 96822

CURTIS, GRAHAM RAY, geologist, mining executive; b. Redfield, Iowa, Jan. 20, 1927; s. Francis Willard and Belva Faith (Graff) C.; m. Barbara L. Wheeler, Aug. 13, 1952; children: Kathleen R., Mark S., Carl M. B.A., U. Iowa, 1951, postgrad., 1951. Registered profl. geologist. Staff geologist Gulf Oil Corp., Mont., Wyo., Kans., La., Nev., Utah, 1952-68, Denver, 1968-77; pres. Gold Cup Exploration Co., Denver, 1977—, Brent Mining Colo. Corp., 1979—, Brenwest Mining Ltd., 1980—, Can. Corp., Calgary, 1980—; chmn. bd. Brent Mining Co., Brenwest Mining Ltd., Gold Cup Exploration Inc. Editor: Oilfields of Kansas, 1959. State del. Republican Com. Jefferson County, Colo., 1978, county del., 1976-81. Served as sgt. AUS, 1945-47. Fellow Geol. Soc. Am.; mem. Am. Assn. Petroleum Geologists, Am. Inst. Profl. Geologists, Rocky Mountain Assn. Geologists, Mont. Geol. Soc., AAAS, Kans. Geol. Soc. Republican. Club: Denver Coal (pres. 1978-79). Subspecialties: Geology; Tectonics. Current work: Energy and mineral development and exploration in the Rocky Mountain States. Home and office: 11880 Swadley Dr Lakewood CO 80215

CURTIS, KENT KRUEGER, science administrator, consultant, researcher; b. Charles City, Iowa, Jan. 24, 1927; s. James Hubert and Lydia Ethel (Krueger) C.; m. Sidnee Smith, 1956; children: Greta, Christian, Sandra; m. Herta Kley, June 7, 1971; children: Celia, Katherine. B.S. in Math, Yale U., 1948; M.A. in Physics, Dartmouth Coll., 1950; postgrad., U. Calif.-Berkeley, 1955. Head div. math. and computing Lawrence Berkeley Lab., Berkeley, Calif., 1955-67; head computer sci. sect. NSF, Washington, 1967—; cons. Livermore Nat. Lab., Los Alamos Nat. Lab., Swedish Tech. Devel. Union; lectr. U. Calif.; research scientist Courant Inst. Math. Scis., N.Y.C. Contbr. articles to profl. jours. Disting. lectr. Computer Soc. of IEEE. Mem. Assn. Computing Machinery, IEEE, AAAS. Subspecialties: Computer science research administration; Computer engineering. Current work: Research and education in computers and their applications. Support research in computer science and computer engineering. Office: Div Math and Computer Scis Nat Science Found Washington DC 20550

CURTIS, LORENZO JAN, physics educator; b. St. Johns, Mich., Nov. 4, 1935; s. Lorenzo F. and Grace C.; m. Maj. R., Nov. 29, 1971. B.S., U. Toledo, 1958; M.S., U. Mich., 1961, Ph.D., 1963. Registered profl. engr., Ohio. Faculty U. Toledo, 1963—, prof. physics and astronomy, 1973—; docent U. Lund, Sweden, 1976-79. Contbr. articles to profl. jours. Mem. Am. Phys. Soc., Am. Assn. Physics Tchrs., Optical Soc. Am., European Phys. Soc., Swedish Phys. Soc. Subspecialty: Atomic and molecular physics. Current work: Teaching and research in the atomic structure of heavy and highly ionized systems. Office: U Toledo Dept Physics and Astronomy Toledo OH 43606

CURTIS, RONALD SANGER, computer scientist, educator; b. Claremont, N.H., Nov. 1, 1950; s. Harding Sanger and Dorothy (Therrien) C. B.A. in Math, Keene State Coll., 1972, M.S., U. N.H., 1974. Tchr. sci. Windsor (Vt.) High Sch., 1974-76; research asst. SUNY, Buffalo, 1976-79; chmn. dept. computer sci. Canisius Coll., Buffalo, 1979-83. Contbr. articles to profl. jours. Mem. Assn. Computing Machinery, IEEE, Kappa Delta Pi. Subspecialties: Distributed systems and networks; Programming languages. Current work: Distributed systems, microcomputer networks, network computers, debugging environments. Home: PO Box 43 Centereach NY 11720 Office: SUNY-Stony Brook Stony Brook NY 11794

CURTISS, ROY, III, biology educator; b. May 27, 1934; m. Josephine Clark, Dec. 28, 1976; children: Brian, Wayne, Roy IV, Lynn, Gregory Clark, Eric Garth. B.S. in Agr., Cornell U., 1956; Ph.D. in Microbiology, U. Chgo., 1962. Instr., research asst. Cornell U., 1955-56; jr. tech. specialist Brookhaven Nat. Lab., 1956-58; fellow microbiology U. Chgo., 1958-60, USPHS fellow, 1960-62; biologist Oak Ridge Nat. Lab., 1963-72; lectr. microbiology U. Tenn., 1965-72, lectr., Oak Ridge, 1967-69, prof., 1969-72, assoc. dir., 1970-71, interim dir., 1971-72; Charles H. McCauley prof. microbiology U. Ala., Birmingham, 1972—83; sr. scientist Inst. Dental Research, 1972—83, Comprehensive Cancer Center, 1972—83; dir. molecular cell biology grad. program, 1973—82; acting chmn. dept. microbiology Comprehensive Cancer Center, 1981—82; dir. Cystic Fibrosis Research Center, 1981—83; prof., chmn. dept. biology Washington U., St. Louis, 1983—; vis. prof. Instituto Venezolana de Investigaciones Científicas, 1969, U. P.R., 1972, U. Católica de Chile, 1973, U. Okla., 1983; Mem. NIH Recombinant DNA Molecule Program Adv. Com., 1974-77, NSF Genetic Biology Com., 1975-78; mem. NIH Genetic Basis of Disease Rev. Com., 1979—83, chmn., 1981—83. Contbr. articles to profl. jours.; Editor: Jour. Bacteriology, 1970-76. Mem. Oak Ridge City Council, 1969-72. Fellow Am. Acad. Microbiology; mem. Genetics Soc. Am., Soc. Gen. Microbiology, Am. Soc. Microbiology (parliamentarian 1970-75, dir. 1977-80), N.Y. Acad. Scis., AAAS, Council Advancement Sci. Writing (dir. 1976—82, v.p. 1978—82), Sigma Xi. Subspecialties: Genetics and genetic engineering (biology); Microbiology. Current work: Genetic and biochemical mechanisms of bacterial pathogenicity. Home: 6065 Lindell Blvd Saint Louis MO 63112 Office: Dept Biology Washington University Saint Louis MO 63130

CUSHING, JIM MICHAEL, mathematics educator; b. North Platte, Nebr., Mar. 20, 1942; s. Harry Joseph and Lorraine Frances (Rohr) C.; m. Dagmar Maria Cushing, July 30, 1971; children: Alina Stephanie, Lara Jennifer. B.A. magna cum laude, U. Colo., 1964; Ph.D., U. Md., 1968. Asst. prof. math. U. Ariz., Tucson, 1968-73, assoc. prof., 1974-81, prof., 1982—. Author: Integrodifferential Equations and Delay Models in Population Dynamics, 1979; contbr. articles to profl. jours. IBM fellow, 1971-72; Humboldt fellow, 1976-77; NSF grantee, 1979-82. Mem. Am. Math. Soc., Soc. Indsl. and Applied Math., Phi Beta Kappa. Subspecialties: Applied mathematics; Theoretical ecology. Current work: Derivation and analysis of mathematical models which describe the dynamics of population growth and interaction with its environment. Office: Dept Math U Ariz Tucson AZ 85721

CUSHMAN, DAVID WAYNE, research biochemist; b. Indpls., Nov. 15, 1939; s. Wayne B. and Mildred M. (Coffin) C.; m. Linda L. Kranch, July 31, 1964; children: Michael, Laura. B.A., Wabash Coll., 1961; Ph.D., U. Ill., 1966. Research investigator Squibb Inst. for Med. Research, Princeton, N.J., 1966-69, sr. research investigator, 1969-73, research fellow, 1973-78, sr. research fellow, 1978-83, asst. dept. dir.,

1983—. NSF fellow, 1961-63. Mem. Am. Soc. Pharmacology and Exptl. Therapeutics, Am. Chem. Soc. (Alfred Burger award in medicinal chemistry 1982), AAAS, N.Y. Acad. Sci., Am. Soc. Biol. Chemists, Phi Beta Kappa, Delta Phi Alpha, Sigma Xi. Subspecialties: Biochemistry (biology); Molecular pharmacology. Current work: Design of drugs acting via inhibition of pathophyiologically important enzyme systems; renin-angiotensin system and blood pressure control; opiate peptides and analgesia. Home: RD 1 20 Lake Shore Dr Trenton NJ 08648 Office: Squibb Inst for Med Research PO Box 4000 Princeton NJ 08540

CUSHMAN, PAUL, JR., physician; b. N.Y.C., Feb. 4, 1930; s. Paul and Cordelia (Hepburn) C.; m. Paulette Bessire, Apr. 4, 1959; children: Paul III, Clare Hepburn. B.A., Yale U., 1951; M.D., Columbia U., 1955. Diplomate: Am. Bd. Internal Medicine. Intern Barnes Hosp., St. Louis, 1955-56; resident Strong Meml. Hosp., Rochester, N.Y., 1956-57, St. Lukes Hosp., N.Y.C., 1960-61; attending physician, 1961-77; instr. Columbia U., 1962-72, asst. prof. medicine, 1972-77; assoc. prof. medicine, pharmacology and psychiatry Med. Coll. Wis., 1977-82, Med. Coll. Va., 1982; dir. substance abuse McGuire VA Hosp., Richmond, 1982—; cons. in field. Contbr. numerous articles to profl. jours.; assoc. editor: Am. Jour. Drug Alcohol Abuse, 1975—, Advances Substance Abuse, 1980—. Served as capt. USAF, 1957-59. Recipient Henry E. Sigerist Found. award, 1967; Caleb Fiske Found. prize, 1973; Career tchr. in substance abuse award Nat. Inst. Drug Abuse, 1979-82. Fellow A.C.P.; mem. AAAS, Am. Physiol. Soc., Am. Soc. Clin. Pharmacology Therapeutics, N.Y. Acad. Medicine, Endocrine Soc. Episcopalian. Clubs: Union, Church (N.Y.C.); Westwood (Richmond). Subspecialties: Internal medicine; Pharmacology. Current work: Substance abuse topics, including endocrine, immunological public health and clinical pharmacology of drugs of abuse; also methadone maintenance programs: evaluation and use. Home: 1826 Park Ave Richmond VA 23220 Office: McGuire Vets Hosp Richmond VA 23249

CUSUMANO, JAMES ANTHONY, research and engring. co. exec.; b. Elizabeth, N.J., Apr. 14, 1942; s. Chalres Anthony and Carmella Madeline (Catalano) C.; m. (married); 1 dau.: Doreen Ann. B.A. (univ. scholar), Rutgers U., 1964, Ph.D., 1968. Dir. catalyst research Exxon Research and Engring. Co., Linden, N.J., 1967-74; pres. Catalytica Assocs., Inc., Santa Clara, Calif., 1974—, also dir. Author: Catalysis in Coal Conversions, 1979, also articles. Recipient award Continental Oil Co., 1964. Mem. Am. Chem. Soc., Am. Phys. Soc., Am. Inst. Chem. Engrs., AAAS, N.Y. Acad. Scis., Sigma Xi. Subspecialties: Catalysis chemistry; Surface chemistry. Current work: Developer new catalytic processes for chemical and energy industries. Patentee in catalysis and surface chemistry fields. Home: 3376 Villa Robleda Dr Mountain View CA 94040 Office: Catalytica Assos Inc 3255 Scott Blvd Bldg 7 E Santa Clara CA 95051

CUTKOMP, LAURENCE KREMER, entomology educator; b. Wapello, Iowa, Jan. 24, 1916; s. Fred Morgan and Glen (Kremer) C.; m. Martha Jaques, June 15, 1939 (div. 1965); children: Kay Cutkomp Bahan, Terry Cutkomp Ostovar, Kent, Lee Cutkomp Ross. B.A., Iowa Wesleyan Coll., Mt. Pleasant, 1936; postgrad., Iowa State U., 1937-38; Ph.D., Cornell U., 1942. Research assoc. U. Minn., St. Paul, 1945-46; entomologist TVA, Wilson Dam, Ala., 1946-47; prof. entomology U. Minn., St. Paul, 1947—; cons. Internat. Atomic Energy Agy., 1965-67, Minn. Mining and Mfg. Mem. Entomol. Soc. Am. (pres. North Central br. 1962), AAAS, Sigma Xi, Gamma Sigma Delta (pres. 1972-73). Subspecialties: Toxicology (agriculture); Chronobiology. Current work: Mode of action of insecticides; insecticides on ATPase activity; insect rhythms. Home: 2418 Doswell Ave Saint Paul MN 55108 Office: University of Minnesota 1980 Folwell Ave Saint Paul MN 55108

CUTLER, IRVING, geography educator; b. Chgo., Apr. 11, 1923; s. Zelig and Frieda (Wopner) C.; m. Marian Horovitz, Aug. 31, 1951; children: Daniel, Susan. A.A., Herzl Jr. Coll., 1942; student, Dartmouth Coll., 1943; M.A., U. Chgo., 1948; Ph.D., Northwestern U., 1964. Investigator U.S. Dept. Labor, 1946-47; tchr. high sch., Bridgman, Mich., 1948-50, Chgo., 1950-59; transp. economist C.E. U.S. Army, 1957; cons., film script writer Ginn & Co., 1964-65; cons. OEO, Kankakee County, Ill., 1965-66, Jour. Films, 1967-71, Gov.'s Task Force on Future of Ill., 1979-80; prof. geography Chgo. State U., 1961—, chmn. dept., 1974—. Author: The Chicago-Milwaukee Corridor, 1964, (with Brian J.L. Berry) Chicago: Transformations of an Urban System, 1976, Urban Geography, 1978, (with others) Urban Communities, 1982, Chicago: Metropolis of the Mid-Continent, 1982; editor: The Chicago Metropolitan Area: Selected Geographic Readings, 1970; mem. editorial bd.: Urban Issues, 1980—; contbr. chpts. to books, articles to profl. jours. Bd. dirs. Chgo. Jewish Hist. Soc., 1977—; mem. edn. com. Anti-Defamation League, 1983—; mem. Save the Dunes Council, 1982—; mem. adv. bd. Explore Your Am. Program, 1976-80. Served to lt. (j.g.) USN, 1943-46; PTO. Haas research grantee, 1964. Mem. Internat. Geographers Union, Ill. Geog. Soc., Geog. Soc. Chgo. (bd. dirs 1976—, v.p. 1982—), Nat. Council for Geog. Edn., Assn. Am. Geographers. Subspecialty: Ecology. Current work: Geographic, environmental, and human aspects of Chicago area. Home: 3217 Hill Ln Wilmette IL 60091 Office: 95th and King Dr Chicago IL 60628

CUTLER, NEAL R., research psychiatrist; b. St. Louis, Sept. 3, 1949; s. Harry and Frances (Zimmerman) C.; m. Beth Leslie Dunitz, Aug. 28, 1977; children: Alexander Jay, Samantha Blair. B.A., St. Louis U., 1971, M.D., 1975. Diplomate: Am. Bd. Psychiatry and Neurology. Intern St. John Mercy Med. Center, St. Louis, 1975-76; resident in psychiatry U. Calif.-Irvine, 1976-78; staff psychiatrist sect. on psychobiology 3-West Clin; Research Unit Biol. Psychiatry Br., NIMH, Bethesda, Md., 1978-80; staff psychiatrist adult psychiatry br. Div. Spl. Mental Health Research NIMH, St. Elizabeth's Hosp., Washington, 1980-82; chief sect. on aging Nat. Inst. Aging, Bethesda, 1981—, dep. clin. dir., 1981—; expert div. neuropharmacol. agts. FDA, 1981—; clin. asst. prof. psychiatry George Washington U., Washington, 1982—; asst. prof. pharmaceutics U. Md., College Park. Tech. editor: Clin. Psychiatry News, 1980—; contbr. articles to profl. jours. Mem. AMA, Am. Psychiatry Assn., Washington Psychiat. Soc., Assn. Psychophysiol. Study of Sleep, ACP, Va. Med. Soc., Fairfax County Med. Soc., N.Y. Acad. Scis., Gerontol. Soc., Am. Geriatrics Soc., Biol. Psychiatry, Am. Soc. Clin. Pharmacology and Exptl. Therapeutics. Subspecialties: Psychopharmacology; Gerontology. Home: 4550 N Park Chevy Chase MD 20815 Office: Nat Inst on Aging Bethesda MD 20205

CUTLER, RHODA, behavioral scientist, educator; b. N.Y.C.; d. Samuel and Sophie (Petrushinsky) C. A.B., Hunter Coll.; M.A., NYU, Ph.D., 1967; postgrad., Yeshiva U. Counseling psychologist N.Y. Assn. for New Americans, N.Y.C., 1947-60; clin. psychologist intern Inst. Phys. Medicine and Rehab., NYU-Bellevue Med. Ctr., N.Y.C., 1960-61; asst. instr. psychiatry N.Y. Med. Coll., N.Y.C., 1961-63; sch. psychologist Bur. Child Guidance, N.Y.C., 1964-65; psychol. cons. Div. Adoption Service, Bur. Child Welfare, N.Y.C., 1965—; psychotherapist Mental Health Cons. Ctr., N.Y.C., 1967-70; research assoc. prof. Inst. for Devel. Studies, Sch. Edn., NYU, 1966-68; vis. assoc. prof. NYU Sch. Edn., 1969; asst. prof. psychiatry N.Y. Med. Coll., N.Y.C., 1968-69; dir. research Child Devel. Ctr., N.Y.C., 1969-70; asst. clin. prof. psychiatry Mt. Sinai Sch. Medicine, Beth Israel Med. Ctr., N.Y.C., 1970—. Contbr. articles to profl. jours. Mem. Am. Psychol. Assn., N.Y. Psychol. Assn., N.Y. Soc. Clin. Psychologists, N.Y. Acad. Sci., Soc. Clin. and Exptl. Hypnosis, Council for Nat. Register of Health Service Providers in Psychology. Home: 230 E 88 St Apt 10A New York NY 10028 Office: Mt Sinai Sch Medicine and Beth Israel Med Center 10 Nathan D Perlman Pl New York NY 10003

CUTTLER, JERRY MILTON, nuclear physicist-engineer; b. Toronto, Ont., Can., Feb. 11, 1942; s. Sam and Annie (Weiner) C.; m. Vera Nikolic, June 5, 1967; children: Sandra, Shai. B.A.Sc., U. Toronto, 1964; M.Sc., Technion-Israel Inst. Tech., 1968, D.Sc., 1971. Research and devel. engr. Israel AEC Soreq Nuclear Research Ctr., Yavne, Israel, 1964-67; research and devel. engr., lab. mgr. dept. nuclear sci. Technion, Haifa, Israel, 1967-71; tech. mgr. Seforad-Applied Radiation Ltd., Jordan Valley, Israel, 1971-74; supr. reactor control Atomic Energy of Can. Ltd., Mississauga, Ont., 1974-78, mgr., 1979-82, asst. to v.p. engring., 1982; engring. mgr. Bruce B, 1983—. Contbr. publs. to profl. lit. Mem. Assn. Profl. Engrs. Province Ont., Am. Nuclear Soc., Can. Nuclear Soc. Subspecialties: Nuclear engineering; Nuclear physics. Current work: Technical management of nuclear engineering, design and development in control and instrumentation for nuclear power plants. Patentee fast neutron spectrometer, self-powered neutron and gamma-ray flux detector. Home: 1188 Vanier Dr Mississauga ON Canada L5H 3X1 Office: Atomic Energy of Can Ltd Sheridan Park Mississauga ON Canada L5K 1B2

CWYCYSHYN, WALTER, mechanical engineer, consultant; b. Detroit, Aug. 10, 1927; s. John and Julia (Bush) C.; m. Rebecca J., July 25, 1933; children: Bradley, David. B.M.E., U. Detroit, 1951. Registered profl. engr., Mich. Supr. Gen. Motors Tech. Ctr., Warren, Mich., then, devel. engr., staff engr., now supr. developing. Served with USAAF, 1945-47. Mem. Soc. Mfg. Engrs. Subspecialties: Robotics; Mechanical engineering. Current work: Robotics—development engineering, systems, manufacturing engineering. Patentee in fields of mechanics and control. Office: APMES General Motors Technical Center Warren MI 48090

CYPESS, RAYMOND HAROLD, medical educator; b. Bklyn., July 27, 1940. B.S., Bklyn. Coll., 1961; B.V.S., U. Ill., 1965, D.V.M., 1967; Ph.D., U. N.C., 1971. Research asst. U. Ill., Urbana, 1965-67; fellow Sch. Pub. Health U. N.C., Chapel Hill, 1967-69, NIH fellow, 1969-70; asst. prof. to assoc. prof. parasitology and epidemiology U. Pitts., 1970-76; prof. microbiology, dir. diagnostic lab. N.Y. State Coll. Vet. Medicine, Cornell U., Ithaca, N.Y., 1977—, chmn. dept. preventive medicine, 1978—, dir. clin. microbiology lab., 1978—; cons. diseases Sch. Medicine, U. Pitts., 1971-77, lectr., 1973—, Fogarty Internat. fellow, 1975; mem. tropical medicine and parasitology study sect. NIH, 1977—; adj.prof. Grad. Sch. Pub. Health, U. Pitts., 1977—; adj.prof. preventive medicine Upstate Med. Sch., Syracuse, 1978—. Recipient career devel. award Nat. Inst. Allergy and Infectious Disease, 1975. Mem. Am. Soc. Parasitologists, Am. Soc. Tropical Medicine and Hygiene, Tchrs. Vet. Medicine Assn., AVMA, Sigma Xi. Subspecialties: Preventive medicine (veterinary medicine); Parasitology. Office: Dept Preventive Medicine Cornell U Ithaca NY 14853

CYR, REGINALD JOHN, electronics engineering executive; b. Caribou, Maine, Dec. 24, 1933; s. Dennis Merchant and Regina Agnes (Trusty) C.; m. Clare Marie Tardif, Dec. 26, 1953; children: Roxanne M., Philip Dennis, Scott M., Krista M. B.S. in E.E., magna cum laude, U. Maine, 1959; postgrad., UCLA, 1962, 66. Design engr. Bendix Electrodynamics Div., Sylmar, Calif., 1959-63; v.p., dir. engring. Aquasonics Engrs., 1963-65; staff engr. to chief engr. Bendix Corp., Sylmar, 1966-68; dir. engring. EMS div. Marine Resources, Inc., Northridge, Calif., 1968-73; pres. Sonatech Inc., Goleta, Calif., 1973—, also dir. Served with USN, 1951-54. Recipient IR 100 award, 1971. Mem. Acoustical Soc. Am., Inst. Navigation. Democrat. Roman Catholic. Subspecialties: Electronics; Acoustical engineering. Current work: Development of acoustic navigation systems, command and control, acoustic telemetry, acoustic sensors—for government, offshore petroleum and deep sea mining. Patentee in field. Home: 1442 Crestline Dr Santa Barbara CA 93117 Office: Sonatech Inc 449 Kellogg Way Goleta CA 93117

CZARNECKI, CAROLINE MARY ANNE, veterinary anatomy educator; b. Detroit, Aug. 3, 1929; s. Daniel and Pauline (Panas) C. B.S., Bemidji State U., 1950; M.A., U. No. Iowa, 1960; Ph.D., U. Minn., 1967. Instr. high schs., Eagle Bend, Minn., 1950-51, Williams, Minn., 1951-53, 54-59, Warroad, Minn., 1953-54, Robbinsdale, Minn., 1960-62; asst. prof. vet. anatomy U. Minn., 1967-71, assoc. prof., 1971-76, prof., 1976—. Contbr. to profl. jours. NIH-USPHS trainee, 1962-67; recipient Norden Disting. Teaching award, 1971. Mem. AAAS, AAUP, Am. Assn. Vet. Anatomy, World Assn. Veterinary Diseases, Am. Assn. Anatomy, Conf. Research Workers in Animal Diseases. Subspecialties: Veterinary anatomy; Histology. Current work: Investigation of drug induced cardiomyopathy using an avian model; teaching the microanatomy of domestic animals to veterinary students. Home: 2159 S Rosewood Ln Roseville MN 55113 Office: Dept Vet Biology 1988 Fitch Ave Saint Paul MN 55108

CZERNIK, DANIEL EDWARD, mechanical engineer; b. Chgo., July 7, 1936; s. Edward Albert and Frances Alvina (Stanek) C.; m. Christina Cecilia Barry, July 30, 1963; children: Michele, Brian, Bradley. B.S. in Mech. Engring, U. Ill., 1959, M.S., 1960; M.B.A., Northwestern U., 1972. Registered profl. engr., Ill. Product engr. Argonne Nat. Lab., Ill., 1960-63; asst. chief engr. Victor Mfg. & Gasket Co., Chgo., 1963-69; dir. product engring. Fel-Pro Inc., Skokie, Ill., 1969—. Contbr. numerous sci. articles to profl. publs. Sec. Timberlake Civic Assn., Hinsdale, Ill., 1981-83. Served with USMC, 1963. Mem. ASME, Soc. Automotive Engrs., Nat. Soc. Profl. Engrs. Republican. Roman Catholic. Subspecialties: Mechanical engineering; Theoretical and applied mechanics. Current work: Gasket engineering; director of product engineering of gaskets for engines and equipment. Patentee in field. Office: 7450 N McCormick Blvd Skokie IL 60076

CZYZEWSKI, HARRY, consulting metallurgical and mechanical engineer, consulting firm executive; b. Chgo., Feb. 13, 1918; s. Leon and Irene (Mierczynska) C.; m. Wilma E. Hood, Nov. 27, 1943; children: Sharon, Bettina, Marie. M.S. in Metall. Engring, U. Ill., 1949. Registered profl. engr. Oreg. Alaska, Calif., Wash. Instr. war tng. program Bradley U., 1942-45; research metallurgist Caterpillar Tractor Co., 1941-46; asst. prof. phys. metallurgy research U. Ill., 1947-51; pres. Metall. Engrs., Inc., Portland, Oreg., 1946-69, MEI-Charlton, Inc., 1946—; chmn. indsl. devel. com. Associated Oreg. Industries, 1965; mem. Oreg. Gov.'s Manpower Coordinating Com., 1968-71. Contbr. numerous articles to profl. jours.; U.S., Can. patentee hard facing alloys. Recipient Pres.'s Citation for extraordinary service Cons. Engrs. Council Oreg., 1979, 82, Alumni Honor award/Gallo medal U. Ill., 1979. Mem. Am. Council Ind. Labs. (nat. pres. 1978-80), Am. Cons. Engrs. Council (pres. Oreg. chpt. 1960, Honor and award Engring. Excellence Awards Competition 1977), Am. Foundrymen's Soc. (pres. Oreg. chpt. 1955), Profl. Engrs. Oreg. (pres. 1959). Subspecialties: Metallurgical engineering; Theoretical and applied mechanics. Current work: Tribology; fracture mechanics; environment—enhanced failure of structures; metal structure geriatrics; fit-for-purpose evaluations. Home: 1966 NW Ramsey Crest Portland OR 97229 Office: 2233 Sw Canyon Rd Portland OR 97201

DABKOWSKI, KRZYSZTOF JERZY, mechanical engineer; b. Warsaw, Poland, Mar. 11, 1944; s. Emil J. and Mieczyslawa M. (Rozmarynowska) D.; m. Halina R. Rarog, Oct. 2, 1976; 1 son, Adrian. M.S.B.S. in mech. Engring, Warsaw Poly., 1970. Project engr. Ursus Tractor Co., Warsaw, 1970-72; mgr. engring. methods sect. Nat. Research Ctr. for Machine Industry Devel., Warsaw, 1972-77; designer/draftsman Superior Welding & Mfg., Niagara Falls, N.Y., 1977-80; designer Bell Aerospace of Can., Grand Bend, Ont., 1980; mfg. engring. mgr. Watson Industries, Jamestown, N.Y., 1981-82; asst. sr. engr. IBM, Endicott, N.Y., 1982—. Recipient Polish Ministry of Machine Industry award, 1974. Mem. ASME. Club: IBM Country. Subspecialties: Mechanical engineering; Computer engineering. Home: 232 Woodgate Ln Vestal NY 13850 Office: 1701 North St T91 Endicott NY 13760

DACEY, GEORGE CLEMENT, laboratory administrator; b. Chgo., Jan. 23, 1921; s. Clement Anthony Dacey and Helyn MacLachan; m. Anne Zeamer, June 20, 1954; children: Donna Lynn, John Clement, Sarah Anne. B.S. in E.E, U. Ill., 1942; Ph.D. in Physics, Calif. Inst. Tech., 1951. Research engr. Westinghouse Research Labs, East Pittsburgh, 1942-45; mem. tech. staff transistor research Bell Telephone Labs, 1952-55, head transistor devel., 1955-58, dir. solid state electronics research, 1958-61, exec. dir. telephones div., 1963-68, v.p. customer equipment devel., 1968-70, v.p. transmission systems, 1970-79, v.p. ops. systems, 1979-81; pres. Sandia Nat. Labs., 1981—; v.p. research Sandia Corp., Albuquerque, 1961-63; dir. Perkin-Elmer Corp., Norwalk, Conn., 1st N.Mex. Bankshare Corp. Contbr. articles on transistor physics, lasers to tech. jours. Mem. exec. bd. Monmouth council Boy Scouts Am., 1970-75; bd. dirs. Monmouth Mus., 1972—. Recipient distinguished alumnus award U. Ill. Elec. Engring. Alumni Assn., 1970. Fellow IEEE, Am. Phys. Soc.; mem. Nat. Acad. Engring., Sigma Xi, Phi Kappa Phi, Tau Beta Pi, Eta Kappa Nu. Subspecialties: 3emiconductors; Solid state physics. Patentee transistors. Home: 1201 Cuatro Cerros TR SE Albuquerque NM 87123 Office: Sandia National Laboratories Albuquerque NM 87185

DACHILLE, FRANK, geochemistry educator; b. N.Y.C., Sept. 15, 1917. B.Ch.E., CCNY, 1939; Ph.D. in Geochemistry, Pa. State U., 1959. Mem. research and devel. staff resin lab. U.S. Indsl. Chems., 1939-41, plant constrn. and chem. supr., 1941-43, asst. plant mgr., 1943-55; asst. prof. dept. geochemistry Pa. State U., University Park, 1955-59, research assoc. geochemist, 1959-62, asst. prof. to assoc. prof., 1963-75, prof., 1975—, cons. div. undergrad. studies, 1974—. Mem. Am. Chem. Soc., Geochem. Soc., Meteoritic Soc., Am. Geophys. Union, Mineral Soc. Am. Subspecialty: Geochemistry. Office: Pa State U Deike Bldg University Park PA 16802

DAEHLER, MARK, research physicist; b. Cedar Rapids, Iowa, Mar. 21, 1934; s. Max and Mary Gertrude (Bingham) D. B.A., Coe Coll., 1955; M.A., U. Wis., Madison, 1957, Ph.D., 1966. Mem. Staff Los Alamos Sci. Lab., 1966-68; guest scientist Max Planck Inst. for Plasma Physics, Garching, W.Ger., 1968-71; research physicist Naval Research Lab., Washington, 1971—. Contbr. articles to profl. jours. Mem. IEEE, Optical Soc. Am., Am. Astron. Soc., Am. Phys. Soc. Subspecialties: Infrared optical astronomy; Optical astronomy. Current work: Prediction of inonospheric properties relevant to high frequency communications; establishment of global HF frequency management systems. Home: 900 Massachusetts Ave NE Washington DC 20002 Office: Naval Research Lab Code 4180 Washington DC 20375 Office: Naval Research Lab Code 4181 Washington DC 20375

DAFNY, NACHUM FRENKEL, neuroscientist, educator; b. Tel Aviv, Mar. 5, 1934; U.S., 1969, naturalized, 1974; s. Nathan and Zelda Frenkel; m. Dita Mirkin Dafny, June 15, 1969; children: Galit, Leemore, Hadar. B.Sc., Hebrew U., Jerusalem, 1964, M.Sc., 1965, Ph.D., 1968. Fellow Calif. Inst. Tech., Pasadena, 1969, Brain Research Inst., UCLA, 1970, Coll. Physicians and Surgeons, Columbia U., N.Y.C., 1971-72; prof. neurobiology and anatomy U. Tex. Med. Sch., Houston, 1972—. Contbr. over 100 articles to sci. publs. Mem. Soc. for Neurosci., Am. Physiol. Soc., Am. Soc. Pharmacology and Exptl. Therapeutics, Am. Assn. Anatomy, Am. Neuroendocrinology Soc. Subspecialties: Neurobiology; Neurochemistry. Current work: Neurophysiology, neuropharmacology and neuroendocrinology of drug addiction, pain and peptides. Office: Dept Neurobiology and Anatomy U Tex Medical School Houston TX 77030

DAHL, ALAN RICHARD, toxicologist; b. Ottawa, Ill., Feb. 24, 1944; s. Henry Victor and Mildred Sylvia (Bicknase) D.; m. Mary Louise Frieberg, July 24, 1982; children: Erica, Douglas. A.B., Princeton U., 1966; Ph.D. in Chemistry, U. Colo., 1971. Cert. Am. Bd. Toxicology. Faculty U. Munich (W. Ger.) and Tech. U., Berlin, 1971-73, Northwestern U., 1973-74; with USPHS, 1974-77; toxicologist Lovelace Biomed. and Environ. Research Inst., Albuquerque, 1977—. Contbr. articles to profl. jours. Mem. Am. AAAS, Chem. Soc., Soc. Toxicology, Am. Coll. Toxicology, Sigma Xi. Subspecialties: Toxicology (agriculture); Inorganic chemistry. Current work: Biochemical toxicology of the respiratory tract. Home: 3337 Wilway NE Albuquerque NM 87106 Office: Lovelace Biomedical and Environmental Research Inst Albuquerque NM 87185

DAHL, BILLIE EUGENE, range mgmt. educator; b. Grady County, Okla., Oct. 24, 1929; s. John O. and Grace (Lusk) D.; m.; children: Desirae, Arlyn Craig, Kelly Christine. B.S., Okla. A&M Coll., 1951; M.S., Utah State U., 1953; Ph.D., U. Idaho, 1966. Range mgr. Bur. Land Mgmt., Rawlins, Wyo., 1953-56; range research Colo. Agrl. Expt. Sta., Akron, 1956-66; grad. fellow U. Idaho, 1962-64; assoc. prof. range mgmt. Colo. State U., 1967; prof. Tex. Tech. U., 1967—; cons. in field. Contbr. to profl. jours. Mem. Soc. Range Mgmt., Sigma Xi. Republican. Lutheran. Club: Toastmasters. Subspecialties: Resource conservation; Resource management. Current work: Grazing management, range improvements, sand stabilization; teach range management and research grazing management, seedling establishment of range forages; sandstabilization and range anaimal diet and nutrition. Home: 43 Highland Dr Route 3 Slaton TX 79364 Office: Range & Wildlife Management Texas Tech U Lubbock TX 79409

DAHLBERG, RICHARD CRAIG, nuclear engineer; b. Astoria, Oreg., July 23, 1929; s. W. A. and Ruth (Slotte) D.; m. Patricia Ravage, June 2, 1957; children: Robin, Julia, Andrea. B.S. in Physics, U. Oreg., 1951; M.S., U. Mich., 1953; Ph.D. in Nuclear Engring, Rensselaer Poly. Inst., 1964. Cert. Profl. engr., Calif. Mgr. nuclear engring. Knolls Atomic Power Co., Schnectady, N.Y., 1957-65; mem. engring. staff General Atomic Co., San Diego, 1965-70, mgr. nuclear engring., 1970-75, dir. fuel engring., 1975-81, dir. gen. engring., 1981—. Pres. LaJolla Shores Assn., 1975-77; pres. LaJollans Inc., 1978-80. Fellow Am. Nuclear Soc.; mem. AAAS, Phi Beta Kappa, Sigma Xi, Phi Eta Sigma. Democrat. Subspecialties: Nuclear fission; Nuclear engineering. Office: GA Technologies Inc PO Box 85608 San Diego CA 92138 Home: 7733 Esterel Dr La Jolla CA 92037

DAHLMAN, GEOFFREY EDWIN, metall. engr.; b. chgo., Jan. 20, 1948; s. Edwin John and Patricia (Cronin) D. B.S. in Metall.Engring, Purdue U., 1970. Student engr. Youngstown Sheet & Tube, East Chicago, Ind., 1970; with Santa Fe Ry., Topeka, 1970—, mgr. reliability engring., 1974-75, mgr. appld research, 1975-78, mgr. research and tests, 1978—. Bd. dirs. Topeka YMCA. Mem. ASME,

Am. Soc. Nondestructive Testing, Am. Ry. Engring. Assn. Lutheran. Lodge: Eagles. Subspecialties: Metallurgy; Metallurgical engineering. Current work: Alloy ry. wheel devel.; rail specifications; failure analysis, finite element analysis; heat treatment procedures; welding procedures. Home: 4105 NE Seward St Topeka KS 66616 Office: 1001 N Atchison Topeka KS 66616

DAHLSTROM, DONALD ALBERT, equipment manufacturing company executive; b. Mpls., Jan. 16, 1920; s. Raymond Estin and Dora Adina (Bloomgren) D.; m. Betty Cordelia Robertson, Dec. 4, 1942; children: Mary Elizabeth, Donald Raymond, Christine Dora, Stephanie Lou, Michael Jeffrey. Student, Macalester Coll., 1937-39; B.S. in Chem. Engring, U. Minn., 1942, Ph.D., Northwestern U., 1949. Petroleum engr. Internat. Petroleum Co., Ltd., Negritos, Peru, 1942-45; from instr. to asso. prof. chem. engring. Northwestern U., 1946-56; with Eimco Corp., Palatine, Ill., 1952—, v.p., dir. research and devel., 1960—, also dir.; v.p. research and devel. Envirotech Corp., Salt Lake City, 1969—; v.p., dir. Erco-Environtech, 1974—; sr. v.p. research and devel. Eimco Process Equipment Co., 1981—; dir. Process Engrs., Inc.; Am. mem. internat. sci. com. 6th Internat. Mineral Processing Congress, 1963; mem. adv. council on engring. NSF. Contbr. to handbooks. Mem. State Air Conservation Com. State Utah, 1971-78, vice chmn., 1977-78; Mem. sch. bd. dist. 110, Deerfield, Ill., 1959-61; pres. Riverwoods Residents Assn., 1962-63; chmn. bd. Northwestern YMCA, 1950-52; trustee Village of Riverwoods, 1966-69. Served with USNR, 1945-46. Recipient Merit award Northwestern U., 1965. Mem. Am. Inst. Chem. Engrs. (dir. 1960-62, v.p. 1963, pres. 1964-65, chmn. environ. div. 1971, Founders award 1972, Environ. award 1977), Am. Inst. Mining, Metall. and Petroleum Engrs. (chmn. minerals benefication div. 1963-64, bd. dirs. soc. mining engrs. 1965—, pres. soc. mining engrs. 1974-75, dir. 1973—, Rossiter W. Raymond award 1952, Richards award 1976, Krumb lectr. 1980, Taggart award 1983), Am. Chem. Soc., Nat. Acad. Engring., Water Pollution Control Fedn., Canadian Inst. Mining and Metallurgy, The Filtration Soc. (London), Air Pollution Control Assn., Mining and Metall. Soc. Am. (dir. Engrs. Council Profl. Devel.), Nat. Acad. Engrs., Sigma Xi (Holgate award Northwestern U. chpt. 1949), Phi Lambda Upsilon, Tau Beta Pi (nat. pres. 1958-62). Presbyterian. Subspecialties: Chemical engineering; Metallurgical engineering. Current work: Development of improved and new types of equipment and processes in liquid-solid separation and mineral processing. Home: 5340 Cottonwood Ln Salt Lake City UT 84117 Office: Eimco Process Equipment Co Box 300 Salt Lake City UT 84110

DAIGNEAULT, ERNEST ALBERT, pharmacology educator; b. Holyoke, Mass., Aug. 16, 1928; s. Lucien and Jeanne Rose (Debien) D.; m.; children: Renee Hennessey, Celesete, David, Brian. Instr., then asst. prof. U. Tenn., 1957-60; asst. prof. La. State U., 1960-63, assoc. prof., 1963-69, prof., 1969-77; prof., chmn. dept. Quillen Dishner Coll. Medicine, East Tenn. State U., 1977—. Author: Pharmacology of Hearing, 1982. NIH grantee, 1965-69; NSF grantee, 1969-70. Mem. Soc. Pharmacology and Exptl. Therapeutics, Soc. Nuclear Medicine, Acoustical Soc. Am., AAAS, Sigma Xi. Roman Catholic. Lodge: Rotary. Subspecialties: Neuropharmacology; Otorhinolaryngology. Current work: Auditory pharmacology, clinical pharmacology. Office: Dept Pharmacology East Tenn State U Johnson City TN 37601 Home: 104 Hillside Rd Johnson City Tn 37614

DAILEY, BENJAMIN PETER, chemist, educator; b. San Marcos, Tex.; s. Benjamin Peter and Anna Clementine (Waldo) D.; m. Beverley Elizabeth Holmes, June 30, 1945; children: Peter, William, Stephen, Stephen. B.S., S.W. Tex. State U., 1938; M.A., Tex., 1940; Ph.D., Tex., 1942. Postdoctoral fellow Harvard U., 1946-47; asst. prof. then assoc. prof. chemistry Columbia U., 1947-57, prof., 1957—. Contbr. chpts., numerous articles to profl. publs. Sloan fellow; Ernest Kempton Adams fellow; NSF sr. postdoctoral fellow; Guggenheim fellow. Mem. Am. Chem. Soc., Am. Phys. Soc., Am. Acad. Arts and Scis. Subspecialties: Physical chemistry; Nuclear magnetic resonance (chemistry). Current work: Oriented molecules in nuclear magnetic resonance. Patentee in field. Home: 440 Riverside D Apt 82 New York NY 10027 Office: Chandler 868 Dept Chemistr Columbia U New Yor NY 10027

DAILEY, JOHN WILLIAM, pharmacologist, educator, researcher; b. Keyser, W.Va., Feb. 10, 1943; s. Michael Joseph and Katherine (Getty) D.; m. Mary Tiller, Jan. 13, 1973; children: John William, David Ryan. A.A., Potomac State Coll., 1963; B.S. (A.M. Lichtenstein scholar in pharmacy), U. Md., 1966; Ph.D. (NIH fellow), U. Va., 1971. Registered pharmacist, Md., Va. Am. Scandinavian Assn. fellow in neuropharmacology Karolinska Inst., Stockholm, 1970-72; research asso. in cardiovascular pharmacology Cleve. Clinic Found., 1972-73; asst. prof. pharmacology George Washington U. Med. Sch., Washington, 1973-76, La. State U. Med. Sch., Shreveport, 1976-78, asso. prof., 1978—. Contbr. articles to profl. jours. Named Best Prof. of Pharmacology La. State U. Med. Center, 1976; 11 grants for sci. studies. Mem. AAAS, Am. Soc. Pharmacology and Exptl. Therapeutics, Am. Heart Assn., Western Pharm. Soc., Soc. for Neurosci., Sigma Xi. Democrat. Roman Catholic. Subspecialties: Pharmacology; Molecular pharmacology. Current work: Neurotransmitter synthesis and metabolism in relation to disease states and drug action. Home: 9908 Village Green Dr Shreveport LA 71115 Office: Dept Pharmacology La State U Med Center PO Box 33932 Shreveport LA 71130

DAILEY, ROBERT ARTHUR, animal scientist, educator, researcher; b. Charles Town, W.Va., May 20, 1945; s. James Hersel and Esther Pauline (Butler) D.; m. Jean Emily Heinz, Aug. 15, 1971; children: Brian Robert, Eric James. B.S., W.Va. U., 1967; M.S., U. Wis.-Madison, 1969, Ph.D., 1973. Research asst. U. Wis., 1967-69, 71-73; instr. Emory U., 1975-77, postdoctoral researcher, 1973-75; asst. prof. animal sci. W.Va. U., 1977-80, assoc. prof., 1980—, assoc. animal scientist, 1980—. Conbr. articles to profl. jours. Served with U.S. Army, 1969-81. Recipient Jr. Faculty award of Merit Gamma Sigma Delta, 1981. Mem. Am. Soc. Animal Sci., Endocrine Soc., Am. Dairy Sci. Assn., Soc. Study Reproductive Physiology. Democrat. Methodist. Subspecialties: Animal physiology; Neuroendocrinology. Current work: Reproductive physiology, measurement hormones, folliculogenesis studies, hypothalamic function. Office: G040 Agrl Sci W Va Morgantown WV 26506

DAILY, FAY KENOYER, research botanist; b. Indpls., Feb. 17, 1911; d. Frederick and Camellia Thea (Neal) Kenoyer; m. William A. Daily, June 24, 1937. A.B., Butler U., 1935, M.S., 1952. Lab. technician Eli Lilly & Co., Indpls., 1935-37, Abbott Labs., North Chicago, Ill., 1939, William S. Merrell Co., Reading, Ohio, 1940-41; lubrication chemist Indpls. propellor plant Curtiss Wright Corp., 1945; lectr. botany Butler U., Indpls., 1947-49, instr. immunology and microbiology, 1957-58, lectr. microbiology, 1962-63, mem. herbarium staff, 1949—. Contbr. articles to profl. jours. Ind. Acad. Sci. grantee. Mem. Am. Inst. Biol. Sci., Phycol. Soc. Am., Bot. Soc. Am., Internat. Phycol. Soc., Ind. Acad. Sci., Sigma Xi, Phi Kappa Phi, Sigma Delta Epsilon. Methodist. Subspecialties: Morphology; Taxonomy. Current work: Charophytes, extant and fossil research. Patentee in field. Home: 5884 Compton St Indianapolis IN 46220

DAILY, JAMES WALLACE, engineering educator, consultant; b. Columbia, Mo., Mar. 19, 1913; s. Wallace Edgar and Marjory Isabel (McGrath) D.; m. Sarah Vanderlip Atwood, Sept. 10, 1938; children: John Wallace, Sarah Anne Vanderlip (Mrs. Charles Rosenberg). A.B., Stanford U., 1935; M.S., Calif. Inst. Tech., 1937, Ph.D., 1945. Registered profl. engr. Test engr. Byron Jackson Co., Berkeley, Calif., 1935; research asst. hydraulics Calif. Inst. Tech., 1936-37, research fellow, mgr. hydraulic machinery lab., 1937-40, instr. mech. engring., 1940-46; hydraulic engr. OSRD, Navy Research Projects, 1941-46; asst. prof. hydraulics M.I.T., 1946-49, asso. prof., 1949-55, prof., 1955-64; prof. engring. mechanics, chmn. dept. U. Mich., 1964-72, prof. fluid mechanics and hydraulic engring., 1972-81, prof. emeritus, 1981—; vis. prof. Tech. U. of Delft, Netherlands, 1971; vis. scientist Electricite de France Centre de Recherches et d'Essais, Paris, 1971; mem. U.S. del. water resources specialists to People's Republic of China, 1974; vis. prof. East China Coll. Hydraulic Engring., Nanking, 1979; domestic and internat. cons. various firms. Author: (with D.R.F. Harleman) Fluid Dynamics, (with R.T. Knapp and F.G. Hammitt) Cavitation; Contbr. tech. articles Am., fgn. jours. Mem. sch. com. Town of Arlington, Mass., 1959-65. Recipient Naval Ordnance Devel. award, 1945. Mem. Nat. Acad. Engring., Internat. Assn. Hydraulic Research (hon. mem., pres. 1967-71, mem. Council 1963-65, 71-77), ASCE, ASME (hon.), Japan Soc. C.E. (hon.), Internat. House of Japan, Sigma Xi, Tau Beta Pi, Chi Epsilon. Conglist. Club: Cosmos (Washington). Subspecialties: Mechanical engineering; Fluid mechanics. Current work: Hydraulic engineering: pumps and turbines. Home: 2968 San Pasqual St Pasadena CA 91107

DAI-SHU-HO, chemical machinery educator; b. Beijing, China, May 16, 1923; s. Ming-Zheng Dai and Wei-Xin Xu; m. Jia-Zhen Chu, Oct., 1947; children: Dai, Qi. B.S., Nat. Central U., Chungking, China, 1946. Mem. faculty Nat. Central U., Nanjing, China, 1947-49, U. Nanjing, 1950-52; lectr. Nanjing Inst. Tech., 1953-58; sr. lectr. Nanjing Inst. Chem. Tech., 1959-77, prof. chem. machinery, 1978—, vice chmn. dept. chem. machinery, 1958-77, chmn. dept., 1978—; cons. Xerox Co., Rochester, N.Y., 1981-82; vis. prof. U. Rochester, 1980-82. Author: (with others) Silicate Industrial Equipment, 1958, Cement Product Euipment, 1959, Process Equipment Design, 1961,65, 80. Mem. Chinese Soc. Chem. Engring. (councilor 1978—), Chinese Soc. Chem. Machinery (councilor 1979—), Jiangsu Province Br. Chinese Soc. Chem. engring. (councilor 1978—), Am. Soc. Engring. sci. Subspecialties: Materials (engineering); Mechanical engineering. Current work: Research on dynamics generation of dislocations from a crack tip and on reliability of process equipment. Office: Nanjing Inst Chem Tech 5 New Model Rd Nanjing Jiangsu Province People's Republic of China 210009

DAJANI, ESAM ZAPHER, pharmacologist, pharmaceutical company executive; b. Jaffa, Palestine, May 30, 1940; came to U.S., 1958, naturalized, 1971; s. Zapher Rageb and Mamdouha (Moustapha) D.; m. Najwa Said Beidas, July 14, 1964; children: Mona, Zapher, Nora. Ph.D. in Pharmacology, Purdue U., 1968. Sr. pharmacologist Rohm & Haas, Phila., 1968-72; sr. research investigator G. D. Searle & Co., Chgo., 1972-74, group leader, 1974-79, sect. head, 1979-80, chmn. digestive disease com., 1974-80, asst. dir. gastroenterology clin. research, 1980-82, assoc. dir. gastroenterology clin. research, 1982—. Contbr. numerous articles to profl. jours. Mem. Am. Soc. Pharmacology and Exptl. Therapeutics, Am. Gastroent. Assn., Am. Coll. Gastroenterology, Am. Fedn. Clin. Research, Soc. Exptl. Biology and Medicine, Arab Am. Univ. Grads. Subspecialties: Pharmacology; Gastroenterology. Current work: Prostaglandins; gastrointestinal secretion and motility; pharm. research and devel. Patentee gastrointestinal pharamcology and therapeutics. Office: PO Box 5110 Chicago IL 60680

DAKSHINAMURTI, KRISHNAMURTI, biochemistry educator; b. Vellore, India, May 20, 1928; s. S.V. and Kamakshi Sundara (Deekshadar) Krishnamurti; m. Ganga Bhavani Venkataraman, Aug. 28, 1961; children: Shyamala Lalitha, Sowmya Sowmitri. Ph.D., U. Rajputana, India, 1957. Research assoc. U. Ill., 1958-61, MIT, Cambridge, 1961-62; assoc. dir. research St. Joseph Hosp., Lancaster, Pa., 1963-65; assoc. prof. biochemistry Faculty of Medicine, U. Man., Winnipeg, Can., 1965-73, prof., 1973—; vis. prof. dept. cell biology Rockefeller U., N.Y.C., 1974-75; mem. Internat. Commn. for Clin. Correlates of Vitamin Deficiencies. Contbr. articles to profl. jours. Recipient Borden award Nutritional Soc. Can., 1973. Fellow Royal Soc. Chemistry, Can. Neuropsychopharmacology; mem. Am. Soc. Biol. Chemists, Internat. Brain Research Orgn., Internat. Soc. for Neurochemistry, Soc. for Neurosci., Am. Inst. Nutrition, Biochem. Soc. (London). Hindu. Subspecialties: Biochemistry (medicine); Neurochemistry. Current work: Metabolic control mechanisms; metabolic function of biotin; regulation of lipogenesis; neurobiology of pyridoxine. Office: Dept Biochemistry Faculty Medicine U Man Winnipeg MB Canada R3E 0W3

DAKSS, MARK LUDMER, applied physicist; b. N.Y.C., Mar. 1, 1940; s. Joseph Walter and Rose (Ludmer) D.; m. Sheryl Judith (Cooper), Nov. 4, 1973; children: Jonathan, Alison. B.E.E., Cooper Union, N.Y.C., 1960; A.M. in Physics, Columbia U., 1962, Ph.D., 1966. Research asst. Columbia U., N.Y.C., 1962-66; research staff mem. IBM Research Ctr., Yorktown Heights, N.Y., 1966-71; mem. tech. staff GTE Labs., Inc., Waltham, Mass., 1971—; lectr. State-of-the-Arts engring. program Northeastern U., Boston, 1974—. Contbr. articles to profl. jours. Eugene Higgins fellow, 1966-67; Columbia U. Pres. fellow, 1968-69, 70-71; Raytheon fellow, 1969-70. Mem. Optical Soc. Am., IEEE Quantum Electronics and Applications Soc. Subspecialty: Fiber optics. Current work: Developing state-of-art techniques for optical fiber measurements and studying light propagation in fibers. Patentee in field. Office: 40 Sylvan Rd Waltham MA 02254

DALAL, NAR SINGH, chemistry educator, researcher; b. Rewarikhera, Panjab, India, May 11, 1941; came to U.S., 1978; m. Jyotsna Deshmukh, July 10, 1965; children: Najma, Geeta. B.S., Panjab U., 1962, M.S., 1963; Ph.D., U. B.C., Vancouver, 1971. Research assoc. NRC, Ottawa, Ont., Can., 1976-78; assoc. prof. chemistry W.Va. U., 1978—. Contbr. articles to profl. jours. Killam fellow, 1971-73. Mem. Am. Chem. Soc. (chmn. No. W. Va. and Ohio Valley sect. 1981-83). Subspecialties: Nuclear magnetic resonance (chemistry); High temperature chemistry. Current work: Nuclear magnetic resonance, electron spin resonance cooperative phase transitions, ferroelectrics, coal science and technology. Office: Chemistry Department West Virginia University Morgantown WV 26506

DALBY, (JOHN) THOMAS, psychologist; b. Oshawa, Ont., Can., Feb. 25, 1953; s. John Thomas and Marion Cecelia (Kinlin) D.; m. Deborah Lynn Dutton, Nov. 19, 1971; children: Krista Faith, Meagan Carmel. B.A. with honors, York U., 1975; M.A., U. Guelph, 1976; Ph.D., U. Calgary, 1979. Cert. psychologist, Alta., diplomate: Am. Acad. Behavioral Medicine, Nat. Register of Health Service Providers in Psychology. Fellow Hosp. for Sick Children, Toronto, Ont., Can., 1976-77; research assoc. U. Calgary (Alta., Can.) Med. Sch., 1979-82; clin. psychologist Calgary Gen. Hosp., 1982—; research coordinator, neurology Alta. Children's Hosp., Calgary, 1979-82. Contbg. editor: Alberta Psychology, 1981-82; cons. editor: Jour. Pediatric Psychology, 1979; author: Introduction to Learning Disabilities: Student Handbook, 1983; contbr. articles to profl. publs. in field. Bd. Assn. for Children with Learning Disabilities, 1978. Can. Council doctoral fellow, Calgary, 1978; Social Scis. and Humanities Research Council Can. doctoral fellow, Calgary, 1979. Mem. Am. Psychol. Assn., Can. Psychol. Assn., Psychologist's Assn. Alta., Nat. Acad. Neuropsychologists, Soc. Pediatric Psychology. Roman Catholic. Subspecialties: Neuropsychology; Behavioral psychology. Current work: Brain dysfunction in children and adults, forensic psychology, behavioral medicine, learning disorders. Home: 4 Varshaven Pl NW Calgary AB Canada T3A 0E1 Office: Calgary Gen Hosp 841 Centre Ave E Calgary AB Canada T2E 0A1

DALGARNO, ALEXANDER, astronomy educator; b. London, Eng., Jan. 5, 1928; s. William and Margaret (Murray) D.; m. Barbara W.F. Kalle, Oct. 31, 1937 (div.); children: Penelope, Rebecca, Piers, Fergus; m. Emily K. Izsak, June 23, 1972. B.Sc., U. London, 1947, Ph.D., 1951; M.A. (hon.), Harvard U., 1967; D.Sc. (hon.), Queen's U. Belfast, 1980. Lectr., Queen's U., Belfast, No. Ireland, 1951-56, reader, 1956-61, prof. math. physics, 1961-67, dir. computation lab., 1961-66; prof. astronomy Harvard U., 1967—, Phillips prof., 1977—, chmn. dept., 1971-76; asso. dir. Center for Astrophysics Harvard U., 1973-80; acting dir. Harvard Coll. Obs., 1971-73; research scientist Smithsonian Astrophys. Obs., Cambridge, Mass., 1967—. Editor: Astrophys. Jour. Letters, 1973—; contbr. articles to profl. jours. Recipient Hodgkins medal Smithsonian Instn., 1977. Fellow Royal Soc., Phys. Soc. (London), Am. Phys. Soc. (Davisson-Germer award 1980); mem. Am. Geophys. Union, Am. Acad. Arts and Scis., Royal Astron. Soc., Internat. Acad. Astronautics (corr. mem.). Subspecialties: Theoretical astrophysics; Metallurgical engineering. Current work: Study of atomic and molecular phenomena in astrophysical, atmospheric and laboratory plasmas. Home: 244 Franklin St Newton MA 02158

DALLA-FAVERA, RICCARDO, medical scientist; b. Legnano, Italy, Dec. 30, 1951; came to U.S., 1978; s. Luciano and Leda (Perissinotto) Dalla-F. M.D., U. Milan, Italy, 1976. Cert. Italian Bd. Hematology. Vis. fellow Nat. Cancer Inst., Bethesda, Md., 1978-81, vis. assoc., 1981-82; asst. prof. pathology, assoc. mem. Kaplan Cancer Ctr., NYU Sch. Medicine, N.Y.C., 1982—. Contbr. articles to profl. jours. Leukemia Soc. Am. spl. fellow, 1982—. Mem. N.Y. Acad. Scis., AAAS, Am. Assn. Microbiology. Subspecialties: Cancer research (medicine); Hematology. Current work: Cancer research, cancer genetics, oncogenes and leukemia. Home: 300 E 33d St Apt 18D New York NY 10016 Office: Dept Pathology NYU Med Ctr 550 1st Ave New York NY 10016

DALLMAN, JOHN CLAY, research engineer; b. Chgo., Oct. 19, 1947; s. Herman H. and Marie Barbara (Westermeir) D.; m. Patricia Roberts, Apr. 10, 1976. B.S. in Math. and Physics, St. Procopius Coll., Lisle, Ill., 1965-69; M.S. in Nuclear Engring, U. Ill., 1972-72, Ph.D., 1978. Nuclear steam supply engr. Combustion Engring., Inc., Windsor, Conn., 1972-73; mem. staff Los Alamos Nat. Lab., N.Mex., 1978—. Author: Investigation of Separated Flow Model, 1979. Pres., bd. dirs. Los Alamos YMCA, 1982-83. Mem. Soc. Physics Students, Am. Nuclear Soc. Roman Catholic. Subspecialties: Fluid mechanics; Nuclear engineering. Current work: Experimental investigation of two-phase flow and heat transfer phenomena related to liquid metal fast breeder reactor safety problems. Inventor application of ultrasonics to measurement of flowing liquid films. Office: Los Alamos Nat Lab Box 1663 Los Alamos NM 87545

DALVI, RAMESH R., toxicologist, educator, cons.; b. Bombay, India, Nov. 8, 1938; s. Rajaram S. and Sumitra R. (Sawant) D.; m. Rekha R. Jadhav, Jan. 22, 1969; children: Rajan, Samir. B.Sc. with honors, U. Bombay, 1962, B.Sc.Tech., 1964, M.Sc., 1967; Ph.D., Utah State U., Logan, 1972. Diplomate Am. Bd. Toxicology. Research fellow Univ. Grants Commn., New Delhi, 1964-67; biochemist Hindustan Lever, Ltd., Bombay, 1967; sci. research officer Bhabha Atomic Research Ctr., Bombay, 1967-69; grad- research fellow Utah State U., 1969-72; postdoctoral fellow Vanderbilt U., Nashville, 1972-74; asst. prof. to prof. toxicology Tuskegee (Ala.) Inst., 1974—; cons. Nat. Acad. Scis. Editorial bd., internat. adv. bd.: Tropical Veterinarian, 1982—; contbr. articles to profl. jours., chpt. in book. Recipient award So. Regional Edn. Bd., 1975, numerous research grants, 1975—. Mem. Soc. Toxicology, Am. Coll. Vet. Toxicologists, Am. Chem. Soc., Am. Soc. Vet. Physiol. Pharmacology, AAAS, Inst. Food Tech., Internat. Soc. Study of Xenobiotics, Am. Assn. Vet. Med. Colls., Pharm. Soc. Japan, Sigma Xi. Subspecialties: Toxicology (medicine); Environmental toxicology. Current work: Analytical and diagnostic toxicologic service to veterinarians, toxicologic research, especially cytochrome P-450 mediated metabolism of toxic substances. Home: 1243 Ferndale Dr Auburn AL 36830 Office: School of Vet Medicine Tuskegee Inst Tuskegee AL 36088

DALY, JOHN ANTHONY, research administrator; b. N.Y.C., Oct. 7, 1937; s. Anthony C. and Ethel E. (Braunton) D.; m. Patricia Ann Cross, Oct. 7, 1966; 1 son, John P. B.S., UCLA, 1959; M.S.E.E., U. Calif.-Berkeley, 1962; Ph.D., U. Calif.-Irvine, 1975. Vol. Peace Corps, Chile, 1965-67; sr. research engr. McDonnell Douglas Corp., Newport Beach, Calif., 1962-65, 68-70; dep. dir. research project WHO, Cali, Colombia, 1970-73; dir. health sector analysis Office Internat. Health, HEW, Rockville, Md., 1973-76; dir. sci. and tech. policy Office of Sec., AID, Washington, 1976-82, Nat. Acad. Scis. program coordinator, office sci. adv., 1982—; tutor U. Md., 1982—; adj. prof. U. Valle, Cali, Colombia, 1971-73; instr. U. Calif.-Irvine, 1968; cons. Ford Found., 1967. Coordinator: book series Syncrisis: The Dynamics of Health, 1974-76. Mem. AAAS, Am. Pub. Health Assn., Ops. Research Soc. Am., Inst. Mgmt. Sci., Tau Beta Pi, Phi Eta Sigma. Democrat. Subspecialties: Operations research (engineering); Artificial intelligence. Current work: Manage international, small-grants, innovative-research program stressing underutilized technologies of potential economic value. Home: 14205 Bauer Dr Rockville MD 20853 Office: AID Washington DC 20523

DAMASHEK, MARC, astronomer; b. N.Y.C., Feb. 19, 1948; s. David and Sylvia (Greenberg) D.; m. Nina Semansky, Oct. 31, 1975; 1 dau., Laurel Jane. B.A., Amherst Coll., 1968; M.S., Stanford U., 1971; Ph.D., U. Mass., 1983. Research engr. Clarke Sch. for the Deaf, Northampton, Mass., 1978-83; sci. programmer, analyst Nat. Radio Astronomy Obs., Green Bank, W.Va., 1978-83; systems analyst Space Telescope Sci Inst., Balt., 1983—. Contbr. articles to profl. jours. Subspecialties: Radio and microwave astronomy; Digital signal processing. Current work: Digital signal processing for astronomical pattern recognition; software engineering for space telescope. Office: Space Telescope Sci Inst Homewood Campus Baltimore MD 21218

D'AMBROSIO, STEVEN MARIO, pharmacologist, educator, researcher; b. Phila., May 7, 1949; s. Mario and Louise (Cerino) D'A.; m. Ruth Elaine Gibson, Mar. 21, 1981. B.S., St. Joseph's U., Phila., 1971; Ph.D., Tex. A&M U., 1975. Research assoc. Brookhaven Nat. Lab., Upton, N.Y., 1975-78; asst. prof. Ohio State U., Columbus, 1978-81, assoc. prof. depts. radiology and pharmacology, 1981—, dir. div. radiobiology, dept. radiology, 1980—; cons. in field. Contbr. articles to profl. jours. EPA and NIH grantee, 1978—. Mem. AAAS, Am. Soc. Photobiology, Am. Soc. Pharmacology and Exptl. Therapeutics, Ohio Fedn. Aging Research, Am. Assn. Cancer Research, N.Y. Acad. Scis. Subspecialties: Toxicology (medicine); Genetics and genetic engineering (medicine). Current work: DNA damage, repair and replication as related to cancer, aging and degenerative disease process. Molecular mechnanisms and ways to alter these processes through molecular biology and genetic enegineering. Office: Div Radiobiology 450 W 10th Ave Columbus OH 43210

DAME, RICHARD FRANKLIN, marine science educator, researcher; b. Charleston, S.C., Nov. 16, 1941; s. Richard Franklin and Lawrence May (Heiser) D.; m. Amanda M. Roberts, Apr. 29, 1967; children: Caroline LaRoche, Elizabeth Stewart. B.S., Coll. Charleston, 1964; M.A., U. N.C., Chapel Hill, 1966; Ph.D., U. S.C., Columbia, 1971. Tchr. St. Andrew's High Sch., Charleston, S.C., 1966-68; prof. marine sci. U. S.C.-Coastal Carolina, Conway, 1971—; research assoc. Baruch Inst., Georgetown, S.C., 1971—. Author: Oceans and Man: Study Notes, 1980; editor: Marsh Estuarine Systems Simulation, 1979. Dir. Litchfield Beaches Property Assn., Pawleys Island, 1973-76; mem. vestry Trinity Episcopal Ch., Myrtle Beach, S.C., 1979-81. Baurch fellow, 1970-71; NSF grantee, 1977—; Sea Grant Program grantee, 1981-82. Mem. Am. Soc. Limnology and Oceanography, Ecol. Soc. Am., Estuarine Research Fedn., Nat. Shellfisheries Assn., Myrtle Beach Tennis. Subspecialties: Ecosystems analysis; Ecology. Current work: The analysis of estuarine and coastal shallow water environments at the ecosystem level with specific interest in oyster reefs and the physical and biological processes controlling them and which they control. Office: University South Carolina Coastal Carolina College PO Box 1954 Conway SC 29526

DAMIANOV, VLADIMIR BLAGOI, mechanical engineer, researcher; b. Sofia, Bulgaria, Sept. 19, 1938; came to U.S., 1971, naturalized, 1976; s. Blagoi Petrov and Bona Krasteva D.; m. Millie Melanoff, Sept. 16, 1973; 1 son: William. B.S., U. Sofia, Bulgaria, 1961, M.S., 1967. Mgr. research and devel. dept. Ctr. Lift Trucks, Sofia, Bulgaria, 1961-70; project engr. Modern Tool & Die Products Inc., Cleve., 1972-74; chief engr., prodn. mgr. Canton Stoker Corp., Canton, Ohio, 1974-77; sr. project engr. McNeil Akron Inc., Akron, Ohio, 1977-81; sr. devel. engr. Goodyear Aerospace Corp., Akron, Ohio, 1981—. Mem. ASME, ASTM. Eastern Orthodox. Clubs: Businessmen (v.p. 1980, 81), Fort Island Swim (bd. dirs. 1980-82). Subspecialties: Mechanical engineering; Fluid mechanics. Current work: Mechanical engineering; gas centrifugeuranium enrichment. Patentee field lift trucks. Home: 3021 Morewood Rd Fairlawn OH 44313

DAMJANOV, IVAN, pathologist, educator; b. Subotica, Yugoslavia, Mar. 31, 1941; s. Milenko and Ana (Pavkovic) D.; m. Andrea Zivanovic, 1964; children: Nevena, Ivana, Milena. M.D., U. Zagreb, Yugoslavia, 1964, Ph.D., 1970. Asst. in pathology U. Zagreb, 1969-73; asst. prof. U. Conn., 1973-77; prof. pathology Hahnemann U. Sch. Medicine, Phila., 1977—. Contbr. articles to profl. jours. Mem. Am. Assn. Pathologists, Am. Assn. Cancer Research. Subspecialties: Pathology (medicine); Teratology. Current work: Experimental pathology; developmental biology; teratology.

DAMON, WILLIAM V.B., psychology educator, writer; b. Brockton, Mass., Nov. 10, 1944; s. Philip Arthur and Helen (Meyers) D.; m.; children: Jesse, Maria. B.A., Harvard U., 1967; Ph.D., U. Calif.-Berkeley, 1973. Lic. psychologist, Calif. Prof. psychology Clark U., Worcester, Mass., 1973—; mem. study sect. NIMH, 1981—; cons. State of Mass., 1976, State of Calif., 1977, 79. Author: Social World of the Child, 1977, Social and Personality Development, 1983; editor in chief: New Directions for Child Devel, 1978—; editorial bd.: Social Cognition, 1981—. Trustee Bancroft Sch., 1981—. Research grantee Spencer Found., 1980—, Carnegie Corp., 1976-80; Mellon Found. travelling fellow, 1978. Mem. Am. Psychol. Assn., AAUP, Soc. Research in Child Devel., Jean Piaget Soc. (bd. dirs.). Democrat. Roman Catholic. Subspecialties: Developmental psychology; Social psychology. Current work: Social and personality development of children and adolescents. Home: 20 Hillside Terr Belmont MA 02178 Office: Clark University 950 Main St Worcester MA 01610

DAMUTH, JOHN ERWIN, marine geologist; b. Dayton, Ohio, Nov. 22, 1942; s. Jason Donald and Maxine Sarah (Simpson) D.; m. Patricia Jane Keenen, Oct. 9, 1971. B.S. with distinction in Geology, Ohio State U., 1965; M.A., Columbia U., 1968, Ph.D., 1973. Grad. research asst. Lamont-Doherty Geol. Observatory, Palisades, N.Y., 1965-73, research scientist, 1973-74, research assoc., 1974-82, sr. research assoc., 1982-83; research geologist Mobil Research and Devel. Corp., 1983—; instr. stream ecology N.J. High Sch. Adult Edn., 1977-83; cons. in field. Contbr. articles in fields to profl. jours. Mem. Trout Unlimited, 1976—, dir., 1979-81. Fellow Geol. Soc. Am.; mem. Am. Assn. Petroleum Geologists, Sigma Xi. Subspecialties: Geology; Oceanography. Current work: Broad spectrum of marine geology problems concerning sedimentation processes and paleosedimentation history of continental margins and the deep sea floor. Home: 725 Robin Ln Coppell TX 75019 Office: Mobil Research and Devel Corp Dallas Research Div PO Box 819047 Dallas TX 75381

DANA, MARTIN P., systems engineer, mathematician; b. Longview, Wash., Sept. 23, 1943; s. Roger E. and Ellen L. (Pershall) D.; m. Jacqueline Lan Hing Fung, Aug. 19, 1973; 1 child, Jean-Daniel. B.A. in Humanities, U. Wis., 1966; Ph.D. in Math, Wash. State U., 1971. Asst. prof. dept. math. Wash. State U., Pullman, 1971-72; sr. scientist Hughes Aircraft Co., Fullerton, Calif., 1972—. NASA fellow, 1967-70. Mem. Am. Math. Soc., Soc. Indsl. and Applied Math. Subspecialties: Aerospace engineering and technology; Applied mathematics. Current work: Optimal estimation (Kalman filtering) and multiple sensor surveillance and identification systems analysis. Home: 899 Quail Circle Brea CA 92621 Office: Hughes Aircraft Co PO Box 3310 Fullerton CA 92634

DANA, RICHARD CHARLES, physiologist, psychobiology educator; b. Bethesda, Md., Jan. 29, 1946; s. Charles Arthur and Grace Marie (Reier) D.; m. Sharon Louise Salisbury, July 8, 1972. B.A., U. Calif., San Diego, 1972; Ph.D., U. Okla., 1977. Research fellow U. Calif., San Francisco, 1977-78, Columbia U., N.Y.C., 1978-79, U. Calif., San Diego, 1979-80, research neurophysiologist, Irvine, 1980-83, instr., 1981-83, Calif. Coll. Medicine, Orange, 1982; research psychologist U. Calif., Berkeley, 1983—. Contbr. articles to profl. jours. Served with USN, 1967-70. U. Okla. research grantee, 1976. Mem. Soc. Neurosci. Republican. Presbyterian. Subspecialties: Neuroendocrinology; Neurophysiology. Current work: Control of learning and memory by hormones, in particular how steroids and enkephalins modulate brain functions. Home: 1489 C Del Rio Circle Concord CA 94518 Office: Dept Psychobiology Univ of Calif Irvine CA 92717

DANA, STEPHEN WINCHESTER, geology educator, consultant; b. Kelley's Island, Ohio, Apr. 10, 1920; s. William Jay and Rhea Florence (Brown) D.; m. Jane Major, Jan. 25, 1952; children: Patricia, Julia, Stephanie, Diana, Janie. A.B. in Geology, Oberlin Coll., 1940, M.S., U. So. Calif., 1942; Ph.D. in Geophysics, Calif. Inst. Tech., 1944. Prodn. geologist Shell Oil Co., Los Angeles, 1944-45; geophysicist Standard Oil Co. Calif., San Francisco, 1945; prof. geology U. Redlands, Calif., 1945—, chmn. dept. geology, 1945—; geologist U.S. Geol. Survey, Washington, 1948-49; cons. geophysicist United Geophys. Co., Pasadena, Calif., 1949-53; cons. So. Calif. Edison Co., Los Angeles, 1957-62, U.S. Geol. Survey, Garden Grove, Calif., 1962-65, San Bernardino Valley Mcpl. Water Dist., Calif., 1967-80. Contbr. articles to profl. jours. Served with U.S. Army, 1942. Fellow Geol. Soc. Am.; mem. Am. Assn. Petroleum Geologists, Phi Beta Kappa, Sigma Xi, Sigma Gamma Epsilon, Phi Kappa Phi. Presbyterian. Subspecialties: Geophysics; Geology. Current work: Petroleum geology, ground water geology, gravimetric location of faults. Home: 1914 Verde Vista Dr Redlands CA 92372 Office: Univ Redlands Colton Ave Redlands CA 92373

DANAHER, BRIAN GRAYSON, health behavior change consultant, media education specialist; b. Atlanta, June 16, 1949; s. Eugene Ignacious and Betty LaVerne (Kefauver) D.; m. Kathleen Ellen Horrall, Sept. 8, 1973. B.A., Stanford U., 1971; M.S., U. Oreg., 1974, Ph.D., 1976. Postdoctoral researcher Stanford U., 1976-78; asst. prof. UCLA, 1978-80; pres. B.G. Danaher & Assocs., Inc., Pasadena, Calif., 1980—; cons. Control Data Corp., 1980—, U.S. Army Med. Corps, Ft. Sam Houston, Tex., 1982—; assoc. research dir. Maxicare Research and Ednl. Found., Hawthorne, Calif., 1980—. Author: Become an Ex-Smoker, 1978; contbr. articles profl. jours. Fellow Am. Heart Assn. Council on Epidemiology; mem. Am. Psychol. Assn., Assn. Advancement Behavior Therapy, Am. Publ. Health Assn., Phi Beta Kappa. Subspecialties: Behavioral psychology; Computer-based instruction. Current work: Search for cost-effective delivery systems for health behavior change technology including mass media (television), self-help methods and computer-based and managed instructional approaches. Home: 159 Glen Summer Rd Pasadena CA 91105 Office: Brian G Danaher & Associates Inc 130 S Euclid Ave Suite 2 Pasadena CA 91101

DANCIK, BRUCE PAUL, forest geneticist; b. Chgo., Dec. 27, 1943; s. Charles John and Sophie Marie (Zaleske) D.; m. Deborah Bloomfield, Mar. 5, 1947. B.S., U. Mich., 1965, M.F., 1967, Ph.D., 1972. Asst. prof. Saginaw (Mich.) Valley Coll., 1972-73; asst. prof. U. Alta., Edmonton, Can., 1973-77, assoc. prof., 1977—; chmn. Forestry panel Environ. Council of Alta., 1977-81. Editor: Canadian Jour. Forest Research, 1981—; assoc. editor: Forestry Chronicle, 1976—; Contbr. articles to profl. jours. Recipient Forestry Achievement award Canadian Inst. Forestry, 1979. Mem. Canadian Inst. Forestry, Soc. Am. Foresters. Club: Flyfishers. Subspecialties: Plant genetics; Evolutionary biology. Current work: Population genetics, differentiation, molecular genetics, evolution of trees. Office: Dept Forest Sci Univ Alta Edmonton AB Canada T6G 2G6

DANDO, WILLIAM ARTHUR, educator; b. Newell, Pa., June 13, 1934; s. Carl Frederick and Myrtle Jane (Foster) D.; m. Caroline Zaporowsky, July 19, 1958; children: Christina Elizabeth, Lara Margaret, William Arthur. B.S., Calif. State U., 1959; M.A., U. Minn., 1960, Ph.D., 1969. Vis. instr. U. Man. (Can.), Winnipeg, summer 1961; instr. to asst. prof. U. Md., College Park, 1965-75; asso. prof. to prof. U. N.D., Grand Forks, 1975—; hon. prof. The Chinese U. of Hong Kong, 1981-82; dir. Am. Indians and the Natural Scis. project NSF, 1981-82; Seminaro Conjuncto CIAF and UNDIRS, 1981; dir. N.D. Drought Project, 1980-81, Cyclic Water Levels and Land Use Problems in the Devils Lake Basin project, 1977-78. Contbr. articles to profl. jours. Chmn. bd. Univ. Luth. Ch., 1979-80, Christus Rex Campus Ministry, 1974-81; v.p. N.D. Luth. Campus Ministry Assn., 1982—. Served with USAF, 1954-56. United Bd. grantee, Hong-Kong-Phillipines-China, 1981-82; Fulbright-Hays research fellow, Romania, 1972-73; Tozer fellow, 1963; named Illustrious Alumni Calif. State U., 1976; Danforth Assos., 1970; Excellance in Teaching award U. Md., 1969. Mem. Assn. Am. Geographers, Nat. Council for Geog. Edn., Great Plains Rocky Mt. Div. Am. Assn. Geographers, N.D. Acad. Sci., Sigma Xi. Club: Jefferson-Lafayette Hunting. Lodge: Moose. Subspecialties: Climatology; Remote sensing (atmospheric science). Current work: Agroclimatology, natural and cultural hazards, land use applications of modern tech. to studies of drought and famine; climate and food prodn.; remote sensing - computer applications to world food problems. Home: 2602 5th Ave N Grand Forks ND 58201 Office: Dept Geography Remote Sensing Univ ND Grand Forks ND 58202

DANES, ZDENKO FRANKENBERGER, physicist, educator, cons.; b. Prague, Czechoslovakia, Aug. 25, 1920; s. Zdenko and Eleonora (Rebensteiger von Blankenfeld) Frankenberger; m. Marie V. Hankova, Jan. 20, 1945; children: Peter, Ellen. Ph.D. in Math./Physics, Charles U., Prague, 1949. Designer Vilnes Electronics, Prague, 1942-45; with Czechoslovak Nat. Geophys. Servey, Prague, 1945-47; asst. prof. Charles U., 1948-50; geophysicist Gulf Research & Devel. Co., Pitts., 1952-59, Boeing Co., Seattle, 1959-62; prof. physics U. Puget Sound, Tacoma, 1962—; pres. Danes Research Assocs., Tacoma, 1978—. Contbr. articles to sci. jours. Served in Czechoslovak Army, 1947-48. Mem. Am. Geophys. Union, Soc. Exploration Geophysicists, Czechoslovak Soc. Arts and Scis. in Am. Subspecialties: Geophysics; Applied mathematics. Current work: Gravimetric exploration. Office: U Puget Sound Tacoma WA 98416

DANET, BURTON NORMAN, clinical psychologist, psychoanalyst, consultant; b. Springfield, Mass., July 23, 1939; s. Benjamin Hyman and Lillian Rose (Sosner) D.; m. Marsha Lynn Wilensky, Oct. 23, 1965. B.A., Yale U., 1960; M.A., U. Minn., 1964, Ph.D., 1967; cert. psychoanalysis, William Alanson White Inst. Psychiatry, Psychoanalysis and Psychology, 1973. Pvt. practice psychoanalysis and psychotherapy, N.Y.C., 1967—, Stamford, Conn., 1967—; chief psychologist Riverdale (N.Y.) Mental Health Center, 1971-82; cons. clin. psychologist Good Samaritan Hosp., Suffern, N.Y., 1980-82, Big Sisters, Inc., Bronx, N.Y., 1980—, Abbot House, Irvington, N.Y., 1979—, Howard Beach (N.Y.) Child Guidance and Family Counseling Center, 1981—. NIMH postdoctoral fellow White Inst., N.Y.C., 1969-71; U. Minn. Med. Sch. doctoral research grantee, 1966-67; Hamm Meml. fellow Hamm Clinic, 1964-65; USPHS fellow U. Minn. Hosps., 1965-66. Mem. Nat. Register Health Service Providers in Psychology, Am. Psychol. Assn., William Alanson White Psychoanalytic Soc., N.Y. Soc. Clin. Psychologists, Westchester County Psychol. Assn., Psychologists Fairfield County. Current work: Personal psychotherapeutic principles; body image as reflected in human figure drawings; countertransference in the psychotherapeutic relationship; unidentified clinical depression in the elderly; pitfalls of three roles played by the clinical psychologist. Home: 145 Don Bob Rd Stamford CT 06903 Office: 115 E 87th St Suite 32F New York NY 10028

DANG, RICHARD KAOYU, scientist; b. Nanking, China, Aug. 6, 1948; came to U.S., 1972, naturalized, 1978; s. Peter H. C. and Ay-Fang (Huang) D. B.S., Chung-Yuan Coll., Taiwan, 1970; M.S., N.Y.U., 1974, Ph.D., 1979. Research asst. dept. physics N.Y.U., 1975-79; research assoc. Joint Inst. for Lab. Astrophysics, Boulder, Colo., 1979-81; project engr. EMR Photoelectric, Princeton, N.J., 1981—. Mem. Am. Phys. Soc. Subspecialty: Atomic and molecular physics. Current work: Applied research and engineering ; computer computation and modeling; electron optics system design; laser and optics system. Home: 2-04 Fox Fun Dr Plainsboro NJ 08536 Office: PO Box 44 Princeton NJ 08540

DANG, VI DUONG, chemical engineering educator, researcher; b. Sept. 12, 1943; s. Hon-Xuong and Huu-Boi (Luu) D.; m. Shu-Chen Wu, June, 1971; children: Angela, Albert. B.S., Nat. Taiwan U., 1966; M.S., Clarkson Coll. Tech., 1968, Ph.D., 1971. Registered profl. engr., N.Y. Chem. engr. P.R. Mallory Inc., Burlington, Mass., 1973-74; chem. engr. Brookhaven Nat. Lab., Upton, N.Y., 1974-80; assoc. prof. chem. engring. Catholic U. Am., 1980—. Contbr. articles to profl. jours. Mem. Am. Nuclear Soc., Am. Inst. Chem. Engrs., Am. Chem. Soc., AAAS, N.Y. Acad. Sci. Subspecialties: Chemical engineering; Nuclear engineering. Current work: Heat and mass transfer, energy conversion. Home: 6304 Phyllis Ln Bethesda MD 20817 Office: The Catholic University of America Washington DC 20064

D'ANGELO, GAETANO (GUY D'ANGELO), chemist; b. Bklyn., June 8, 1942; s. Carmine and Rose (Carlino) D'A; m. Nancy K. Lesniewski, Aug. 9, 1969. B.A., Hunter Coll., 1964. Chemist, D.H. Litter Labs., N.Y.C., 1964-66; tech. specialist dept. chemistry SUNY-Stony Brook, 1966-70, research asst. dept. physiology, biophysics Health Sci. Ctr., 1970—. Contbr. articles to profl. jours. Served to comdr. U.S. Coast Guard Aux., 1973. Fellow Am. Inst. Chemists; mem. Am. Chem. Soc. (safety insp. N.Y. sect. 1976—), Am. Cetacean Soc. (pres. N.Y. chpt. 1981). Roman Catholic. Subspecialties: Biochemistry (biology); Comparative physiology. Home: 10 Wesley St Center Moriches NY 11934 Office: Dept Physiology & Biophysics Health Sci Ctr Stony Brook NY 11794

D'ANGELO, GUY *See also* **D'ANGELO, GAETANO**

DANGLER, EDWARD, industrial program manager, marine resources consultant; b. N.Y.C., Mar. 6, 1929; s. Jonas Joseph and Rose (Leibowitz) D.; m. Yvonne Chateland, June 19, 1954; children: Josiane, Chantal. B.S. in Marine Sci, SUNY Maritime Coll., 1949, San Diego State U., 1960; M.B.A., Calif. Western U., 1962; postgrad., U. Calif., Berkeley, 1963-64. Marine transp. officer Isthmian Steamship Co., N.Y.C., 1949-53; sr. systems engr. Gen. Dynamics Corp., San Diego, 1957-63; program mgmt. specialist NASA, Mountain View, Calif., 1963-66; sr. staff engr. Lockheed Missiles Co., Sunnyvale, Calif., 1966-82; dep. program mgr. Lockheed Austin (Tex.) Div., 1982—; tech. advisor U.S. Del. to UN Conf. on Law of the Sea, Switzerland, 1977-82. Contbr. articles to legal publs. Served to lt. USN, 1953-56. Hon. grad. N.Y. Marine Soc., 1949; recipient certs. of appreciation Soc. Logistics Engrs., 1981, Am. Soc. Metals, 1981. Mem. Marine Tech. Soc., Law of Sea Inst., Oceanic Soc., Navy League (dir. 1968—), Naval Res. Assn. (bd. advisor 1980), Assn. Former Intelligence Officers (bd. San Francisco). Jewish. Club: Onion Creek Country (Austin). Lodge: Masons. Subspecialties: Ocean engineering; Offshore technology. Current work: Study of rational development of ocean resources, including deep sea bed minerals, energy and transportation and concommitant legal regimes. Home: 10915 Crown Colony Dr Austin TX 78747 Office: Lockheed Austin Div 2124 E St Elmo Dr Austin TX 78744

DANIEL, MICHAEL ANDREW, microbiologist, cons., researcher; b. Austin, Tex., Mar. 11, 1952; s. Wayne Alvin and Peggy Jane (Riddle) D. B.S. in Microbiology, Mich. State U., 1974; M.S. in Biology, Ill. Inst. Tech., 1980. Lab. dir. Windsor Med. Assocs. S.C., Riverside, Ill., 1975-77; supr. Baxter Travenol Labs., Round Lake, Ill., 1977-78, supt., 1978-80, systems analyst, Deerfield, Ill., 1980-82, program mgr., 1982; mgr. quality assurance Novalor Med. Corp., Oakland, Calif., 1982—; cons. on fin. systems, microbiology and med. products. Contbr. articles to sci. jours. Mem. Am. Soc. for Quality Control (cert. quality engr.), Am. Soc. Microbiology, Am. Acad. Microbiology (registered microbiologist). Subspecialties: Artificial organs; Teratology. Current work: Behavorial teratology, gen. toxicology, immunochemistry, computer information systems, management information and analysis systems, inplantable medical device engineering, microbiology and good manufacturing practice consulting. Office: Novalor Med Corp 7799 Pardee Ln Oakland CA 94621

DANIEL, MUTHIAH D., microbiologist; b. Sri Lanka, July 4, 1927; s. Santhiapillai Nathaniel and Theresa Thayalama (Vethaparanam) M.; m. Gloria Molly Jeyapackiam, Mar. 25, 1931; children: Amirtham, Thersa, Daniel. Grad. in Vet. Sci, Bengal Vet. Coll., Calcutta, 1951; Ph.D., U. Wis., 1966. Veterinarian Colombo (Sri Lanka) Mcpl. Council, 1951-61; prin. assoc. molecular genetics and microbiology Harvard Med. Sch., New Eng. Regional Primate Research Ctr., Southboro, Mass., 1966—. Recipient award Am. Assn. Lab. Animal Sci., 1969. Mem. Am. Assn. Microbiolgy, Am. Assn. Immunologists, Am. Assn. Cancer Research, Internat. Leukemia Assn. Subspecialties: Animal virology; Cancer research (medicine). Current work: Spontaneous diseases of non-human primates; herpes virus induced malignant lymphoma and leukemia; effects of antiviral agents on herpes viruses. Office: 1 Pine Hill Rd Southboro MA 01772

DANIEL, SAMUEL HENDERSON, III, utility company executive, chemist; b. Dublin, Tex., Jan. 7, 1945; s. Samuel Henderson and Bebe Joyce (Barnett) D.; m. Carol Lyn Wright, Dec. 28, 1982. B.S., Tarleton State U., 1967; Ph.D., Tex. A&M U., 1971. Mem. staff chemistry Va. Tech. Inst., Blacksburg, Va., 1971-73; mgr. tech. services Radiation Mgmt. Corp., Phila., 1973-76, mem. staff Scott & White Clinic, Temple, Tex., 1976-78; radiochemist, engr., chemist Tex. Utilities Generating Co., Glen Rose, Tex., 1978—. Contbr. papers to profl. lit. Stage mgr. Temple Civic Theater, 1978. Fellow Am. Inst. Chemists; mem. Am. Chem. Soc. (chmn. 1977-78), ASME (utility subcom. N.Y.C. 1981—), Soc. Nuclear Medicine, Health Physics Soc. (steering com. 1982—, pres. Northeast Tex. chpt. 1982-83), Nat. Assn. Corrosion Engrs., Am. Nuclear Soc. Subspecialties: Nuclear power plant chemistry; Corrosion. Current work: Control of corrosion and minimization of radioactive contamination. Home: 1303 3d St Granbury TX 76048

DANIEL, THOMAS HENRY, laboratory director, researcher; b. Rochester, N.Y., Oct. 24, 1942; s. Henry B. and Margaret E. (Keller) D.; m. Barbara L. MacLachlan, June 30, 1967 (div. Aug. 1978); 1 dau., Sara L.; m. Alice E. Dean, Dec. 31, 1979; 1 son, Michael H. B.S., MIT, 1964; M.S., U. Hawaii, 1973, Ph.D., 1978. Asst. oceanographer Research Corp. U. Hawaii, Honolulu, 1975-78; research scientist Lockheed Ocean Systems, Sunnyvale, Calif., 1979-82; lab. dir. Natural Energy Lab Hawaii, Kailua-Kona, 1982—; tchr. U. Hawaii Continuing Edn. And Community Service, Honolulu/Kailua/Kona, 1978, 83; lectr. high sch. students Lockheed Mgmt. Assn., Sunnyvale, 1979-82; chmn. jr. ocean systems exec. com. Lockheed, Sunnyvale, 1982. Mem. Tech. and Soc. Com., Palo Alto, Calif., 1981-82. Mem. Am. Geophys. Union, Am. Meterol. Soc., Marine Tech. Soc., Assn. Unmanned Vehicle Systems. Quaker. Clubs: Lockheed Sailing (Sunnyvale) (commodore); Am. Radio Relay League (Newington, Conn.)). Subspecialties: Ocean thermal energy conversion; Ocean engineering. Current work: Investigating ocean and solar alternate energy sources, OTEC biofouling, corrosion and biofouling countermeasures experiments, alternate uses of lab's deep (2000') seawater pumping system, artificial upwelling aquaculture, solar ponds. Home: 73-1308 Ka Imi Nani Dr Kailua-Kona HI 96740 Office: Natural Energy Lab Hawaii PO Box 1749 Kailua-Kona HI 96740

DANIEL, WILLIAM LOUIS, geneticist; b. Wyandotte, Mich., Sept. 20, 1942; s. Lafayette Stephen and Dorothy Rose (Reidy) D.; m. Mary Louise Stace, June 20, 1964; children: Mark W., Jennifer M., Kathryn S. B.S., Mich. State U., 1964, Ph.D., 1967. Asst. prof. genetics Ill. State U., 1967-71, assoc. prof., 1971-72; asst. prof. genetics U. Ill., Urbana, 1972-77, assoc. prof., 1977—; cons. Regional Health Resources Ctr., Urbana. Mem. Am. Genetics Soc., AAAS, Genetics Soc. Am., Am. Soc. Human Genetics, N.Y. Acad. Scis., Phi Kappa Phi. Subspecialties: Gene actions; Genome organization. Current work: Researcher in genetic regulation of mammalian arylsulfatase expression. Home: 2502 Bedford Dr Champaign IL 61820 Office: Dept Genetics and Deve Univ Ill Urbana IL 61801

DANIELLI, JAMES FREDERIC, scientist, educator, editor; b. Wembley, Eng., Nov. 13, 1911; s. James Frederic and Helena (Hollins) D.; m. Mary Guy, Jan. 4, 1937; children—Richard, Corinne. Ph.D., London (Eng.) U., 1933, Cambridge (Eng.) U., 1942; D.Sc., London U., 1938, Gent (Belgium) U., 1956, Med. Coll. Pa., 1970, Worcester Poly. Inst., 1972. Fellow Princeton, 1933-35, St. John's Coll., Cambridge (Eng.) U., 1942-45; physiologist Marine Biol. Assn., 1946; reader cell physiology Royal Cancer Hosp., 1946-49; prof. zoology, chmn. dept. King's Coll., London, Eng., 1949-61; prof. medicinal chemistry and biochem. pharmacology State U. N.Y. at Buffalo, 1962-65, chmn. dept. biochem. pharmacology, 1962-65; prof. theoretical biology, dir. Center for Theoretical Biology, 1965-74, provost faculty natural sci. and math., 1967-69, asst. to pres., 1969-74; prof. Worcester (Mass.) Poly. Inst., 1974-80, emeritus prof., 1980—; vis. research prof. Salk Inst., 1973-75; cons. various indsl. firms, pubs., govt. orgns. Author: Permeability of Natural Membranes, 1942, Cell Physiology and Pharmacology, 1952, Cytochemistry, 1953; Editor: Symposia Soc. for Exptl. Biology, 1946-56, Symposia Internat. Soc. for Cell Biology, 1950-60, Internat. Rev. Cytology, 1951—, Jour. Theoretical Biology, 1960—, Gen. Cytochem. Methods, 1958-65, Progress in Surface and Membrane Sci, 1962-79, Jour. Social and Biol. Structures, 1978—; Contbr. articles to profl. jours. Fellow Royal Soc.; mem. Inst. Biology (past sec.), Biochem. Soc., Physiol. Soc., Soc. for Exptl. Biology (past sec.), Am. Soc. Cell Biology, Internat. Soc. for Cell Biology (past sec.), Am. Inst. Biol. Scis., Internat. Soc. for Study of Origin of Life. Subspecialty: Cell biology. Office: Danielli Assos Inc 185 Highland St Worcester MA 01609

DANIELS, CAROLE ANGELA, engineer, consultant; b. Phila., Nov. 10, 1945; d. Arthur Joseph and Angela (Traveline) D.; m. Lewis Earl Sloter, II, June 2, 1978; 1 son, Lewis Earl III. B.S., Drexel U., 1968, M.S., 1973, Ph.D., 1977. Registered profl. engr., Tex. Sr. engr. Westinghouse Electric Co., Madison, Pa., 1974-77; vis. prof. U. Tex.-Arlington, 1977-79; v.p. John Lummus Co., Dallas, 1977-82; cons. Computer and Materials, Dallas, 1977—; assoc. Gen. Energy Assocs., Inc., Cherry Hill, N.J., 1982. Author: In Situ Examination of the Small Steam Generator Model, 1976. Judge sci. fairs Dallas Ind. Sch. Dist., 1979—; lectr. Nuclear Energy Women, Fort Worth, Tex., 1978-82. Mem. AIME, Am. Nuclear Soc., Am. Soc. for Metals, Assn. Traditional Artists, Alpha Sigma Mu. Subspecialties: Materials; Software engineering. Current work: Engineering applications of polymers and development of computer software for architectural application. Address: 7211 Little Canyon Rd Dallas TX 75249

DANIELS, RAPHAEL SANFORD, environ. engr.; b. N.Y.C., Apr. 23, 1931; s. Alfred Jeremiah and Miriam (Lipson) D.; m. Varda Lea Gross-Shmorak, Dec. 18, 1970; children: Maia, Elan. B.S.C.E., CCNY, 1953; M.S. in Sanitary Engring. and Nuclear Engring, M.I.T., 1959. Radiol. health engr. USPHS, Washington, also Idaho Falls, Idaho, 1959-62; from jr. tech. assoc. to tech. assoc. NUS Corp., Washington, also Rockville, Md., 1962-69, sr. tech. assoc., Rio de Janeiro, 1969-72, tech. dir., asst. gen. mgr., sr. exec. cons., Rockville, 1972-81; project engr. Bechtel Nat. Inc., Gaithersburg, Md., 1981-82, dep. dir. tech. planning, 1982—. Served to 1st lt. USAF, 1954-57. Mem. Am. Nuclear Soc., Health Physics Soc., AAAS, Sigma Xi, Chi Epsilon. Subspecialty: Environmental engineering. Current work: Provide adequate and timely technical plans for recovery of Three Mile Island Unit 2. Office: Bechtel Nat Inc 15740 Shadygrove Rd Gaithersburg MD 20877

DANIELS, RAYMOND BRYANT, soil scientist; b. Adair County, Iowa, Feb. 15, 1925; s. Arthur Rex and Minnie Eva (Davis) D.; m. Irene Mae Byles, Feb. 26, 1945; children: Bryant Rex, Lynda Sue. B.S., Iowa State U., 1950, M.S., 1955, Ph.D., 1957. Registered profl. geol. scientist; cert. profl. soil scientist. With Soil Conservation Service (various locations), 1953-80, research soil scientist, Raleigh, N.C., 1960-77, dir. soil survey investigations, Washington, 1977-80; vis. prof. soil sci. dept. N.C. State U., Raleigh, 1981—. Served to lt. USN, 1951-53. Recipient Superior Service award U.S. Dept. Agr., 1977. Fellow Geol Soc. Am.; mem. Soil Sci. Soc. Am., Am. Quaternary Assn., Carolina Geol. Soc., N.C. Soil Sci. Soc. Presbyterian. Subspecialties: Soil science. Current work: Relationship of geomorphology and soils; current research effect of geomorphic processes on soil productivity. Home: 9112 Leesville Rd Raleigh NC 27612 Office: Dept Soil Science NC State U Raleigh NC

DANIELS-HETRICK, BARBARA ANN, plant pathologist, consultant; b. San Francisco, Sept. 10, 1951; d. Wolfgang Rudolph and Rosine Elizabeth (Meek) Daniels-H. B.A., Ohio Wesleyan U., 1973; M.S., Wash. State U., 1975; Ph.D., Oreg. State U., 1978. Postdoctoral research assoc. U. Calif.-Riverside, 1978-80; asst. prof. Kans. State U., 1980—. Grantee Kans. Corn Commn., U.S. Forest Service. Mem. Am. Phytopath. Soc. Democrat. Methodist. Subspecialties: Plant pathology; Microbiology. Current work: Mycorrhizal fungi. Office: Dept Plant Pathology Kans State U Manhattan KS 66506

DANIELSON, NEIL DAVID, chemistry educator; b. Ames, Iowa, July 25, 1950; s. Gordon Charles and Dorothy Elisabeth (Thompson) D.; m. Elizabeth Moore, Aug. 4, 1979. B.S., Iowa State U., 1972; M.S., U. Nebr., 1974; Ph.D., U. Ga., 1978. Asst. prof. chemistry Miami U., Oxford, Ohio, 1978-83, assoc. prof., 1983—. Contbr. articles to profl. jours. Am. Chem. Soc. grantee, 1981; Cottrell Research Corp. grantee, 1981. Mem. Am. Chem. Soc., Soc. Applied Spectroscopy, Sigma Xi. Subspecialties: Analytical chemistry; Polymer chemistry. Current work: High performance liquid chromatography, gas chromatography fluorescence, modification of fluorecarbon polymers. Home: 507 Edgehill Dr Oxford OH 45056 Office: Dept Chemistry Miami U Oxford OH 45056

DANNA, KATHLEEN JANET, molecular biology educator; b. Beaumont, Tex., Aug. 21, 1945; d. William Eugene and Elsie Pearl (Fisher) D.; m. Richard Lloyd Kautz, Nov. 27, 1976. B.S. in Chemistry, N.Mex. Inst. Mining and Tech., 1967; Ph.D. in Microbiology, Johns Hopkins U., 1972. Postdoctoral research fellow Rijksuniversiteit, Gen, Belgium, 1972-73, MIT, 1973-75; assoc. prof. molecular, cellular and devel. biology U. Colo.-Boulder, 1975—. Contbr. articles to profl. jours., chpts. to books. Sec., Rocky Mountain chpt. Am. Rock Garden Soc. Recipient Wilson H. Stone Meml. award for biomed. research, 1973; research grantee NIH, Nat. Cancer Inst., NSF, Milheim. Mem. Am. Soc. Microbiology, Johns Hopkins Med. and Surg. Assn. Subspecialties: Virology (biology); Molecular biology. Current work: Regulation of gene expression in mammalian cells, molecular biology of animal viruses, mechanism by which tumor viruses transform cells and cause tumors in animals. Home: 4020 Greenbriar Blvd Boulder CO 80303 Office: Dept MCD Biology Campus Box 347 U Colo Boulder CO 80309

DANNENBERG, JOSEPH J., chemistry educator, cons.; b. N.Y.C., Apr. 21, 1941. A.B., Columbia U., 1962; Ph.D., Calif. Inst. Tech., Pasadena, 1967. NIH fellow Centre de Mecanique Ondulatoire Appliquee, Paris, 1966-67; research assoc. Columbia U., 1967-68; asst. prof. chemistry Hunter Coll., CUNY, 1968-72, assoc. prof., 1973-77, prof., 1978—; vis. prof. U. Pierre et Marie, Paris, 1974-75; cons. Mem. Am. Chem. Soc., Royal Soc. Chemistry, European Acad. Scis., Arts and Letters (corr.). Subspecialties: Organic chemistry; Theoretical chemistry. Address: Dept Chemistry Hunter Coll 695 Park Ave New York NY 10021

DANNENBRING, GARY LEE, psychology educator; b. Iowa City, June 21, 1948; s. Forrest Glen and Marjorie (Bone) D.; m. Jane Kivitt, May 24, 1969 (dec. July 1981); children: Aaron, Sarah, Heather. B.S., Iowa State U., 1970; M.Sc., McGill U., 1971, Ph.D., 1974. Research assoc. McGill U., Montreal, Que., Can., 1974-76; asst. prof. St. Francis Xavier U., Antigonish, N.S., Can., 1976-80, assoc. prof., 1980—; vis. prof. Iowa State U., 1982-83. Natural Scis. and Engring. Research Council Can. grantee, 1977-80, 78, 80-83. Mem. Am. Psychol. Assn., Can. Psychol. Assn. Subspecialties: Cognition; Human factors in computing. Current work: Memory processes, especially semantic memory; current human factors research: effect of computer response time on user performance and satisfaction. Office: Dept Psychology Iowa State U Ames IA 50011

DANNER, DAVID LEE, computer software company executive, industrial engineer; b. Milw., July 3, 1948; s. George Wilson and Hazel B. (Damisch) D.; m. Pamela Ann Beck, Sept. 30, 1972; children: Constance Lynne, Laurel Beck. B.S. in Indsl. Engring, Purdue U., 1970, M.S., 1970; B.S. in Oceanography, George Washington U., 1972; Ph.D., Cath. U. Am., 1982. Registered profl. engr., Wis. Asst. foreman Automatic Electric Co. subs. Gen. Telephone & Electronics, 1968; indsl. engr. IBM, Endicott, N.Y., 1970; teaching asst. Purdue U., Lafayette, Ind., 1969-70; mem. tech. staff Informatics, Inc., Rockville, Md., 1972-74; mgr. systems applications Ultrasystems, Inc., Washington, 1974-75; pres., chmn. bd. Ideamatics, Inc., Washington, 1975—. Contbr. articles to profl. jours. Treas. Mt. Pleasant Adv. Neighborhood Com., Washington, 1976-80. Served with USMC, 1970-72. Decorated Am. Spirit Honor medal. Mem. Am. Inst. Indsl. Engrs. (1st v.p. Nat. Capital chpt. 1980-81, pres. 1981-82, dir. 1982-83), Purdue U. Alumni Assn., Tau Beta Pi. Subspecialties: Software engineering; Water supply and wastewater treatment. Current work: Development of microprocessor product software implementation techniques, microprocessor software applications, mathematical modelling of physical systems, urban non-point source pollution control. Office: 1806 T St NW Washington DC 20009

DANNER, DAVID WILLIAM, clergyman, educator, researcher; b. Anderson, Ind., Oct. 12, 1940; s. Donald Golden and Charlotte D.; m. Elizabeth Beck Reichardt, Sept. 27, 1966. B.S., Coll. Wooster, Ohio, 1962; student, Edinburgh (Scotland) U., 1960-61; B.D., Princeton Theol. Sem., 1965; M.Ed., Temple U., 1970, Ed.D., 1974. Permanent cert. English, Pa. Ordained to ministry United Presbyterian Ch. U.S.A., 1965; Asst. minister First Presbyn. Ch., Easton, Pa., 1965-66, St. John's United Presbyn. Ch., Devon, Pa., 1966-68; English tchr. Tredyffrin-Easttown Schs., Wayne, Pa., 1968-74; assoc. prof. Presbyn. Sch. Christian Edn., Richmond, Va., 1974-81; dir. Introduction to Ministry Project United Presbyn. Ch., N.Y.C., 1981—; adj. prof. Temple U. Sch. Edn., Phila., 1970-74, Va. Commonwealth U. Richmond, 1976-80; vis. fellow Princeton (N.J.) Theol. Sem., 1981—. Contbr. articles to profl. jours. Recipient Middler Preaching award Princeton Theol. Sem., 1964; research grantee United Presbyn. Ch. and Presbyn. Ch, U.S., 1977. Mem. Am. Psychol. Assn., Assn. Profs. and Researchers in Religious Edn., Religious Research Assn., Assn. for Childhood Edn. Internat. (Richmond br. pres. 1978-80), Assn. Presbyn. Ch. Educators (research grantee 1977). Republican. Subspecialties: Developmental psychology; Social psychology. Current work: Career development; research on career developmental patterns of clergy and of religious educators. Home: 117 N Lincoln Ave Newton PA 18940 Office: Vocation Agy United Presbyterian Ch Room 430 475 Riverside Dr New York NY 10115

DANTZIG, GEORGE BERNARD, operations research and computer science educator; b. Portland, Oreg., Nov. 8, 1914; s. Tobias and Anja (Ourisson) D.; m. Anne Shmuner, Aug. 23, 1936; children: David Franklin, Jessica Rose, Paul Michael. A.B., U. Md., 1936; M.A., U. Mich., 1937; Ph.D., U. Calif.-Berkeley, 1946; hon. degrees, Technion, Israel, 1973, Linkoping (Sweden) U., 1975, U. Md., 1976, Yale U., 1978, Louvain (Belgium) U., 1983, Columbia U., 1983. Jr. statistican U.S. bur. Labor Stats., Washington, 1937-39; chief combat analysis Statis. Control U.S. Air Force, Pentagon, 1941-46, math. advisor, 1946-52; research mathematician Rand Corp., Santa Monica, Calif., 1952-60; prof., head Ops. Research Ctr., U. Calif.-Berkeley, 1960-66; prof. ops. research and computer sci. Stanford U., 1966—; sect. head Inst. Applied Systems Analysis, Laxenburg, Austria, 1972. Author: (with Thomas L. Saaty) Linear Programming and Extensions, 1963, Compact City, 1973; contbr.: articles to profl. jours. Compact City. Recipient Exceptional Civilian Service medal U.S. War Dept., 1944; U.S. Nat. Medal of Sci., 1975; John von Neumann Theory prize Ops. Research/Mgmt. Sci., 1975. Mem. Econometric Soc., Inst. Math. Stats., Inst. Mgmt. Sci. (pres. 1966), Math. Programming Soc., Am. Math. Soc., Ops. Soc. Am., Inst. Mgmt. Scis., IEEE (hon.), Nat. Acad. Sci. (award in applied math. and numerical analysis 1977), Am. Acad. Arts and Scis. Subspecialties: Numerical analysis; Operations research (mathematics). Current work: Mathematical models for planning and scheduling the activities of large scale enterprises; formulation, optimization and applications to real world problems. Office: Department of Operations Research Stanford University Stanford CA 94305

DANTZLER, WILLIAM HOYT, physiologist, researcher; b. Mt. Holly, N.J., Aug. 25, 1935; s. Marion Adam and Margaret Helen (Hoyt) D.; m., Aug. 22, 1959; children: Amy Louise, Kurt Nicol. A.B., Princeton U., 1957; M.D., Columbia U., 1961; Ph.D., Duke U., 1964. Diplomate: Nat. Bd. Med. Examiners. Intern U. Wash. Hosp., Seattle, 1961-62; asst. prof. pharmacology Columbia U., 1964-68; assoc. prof. physiology U. Ariz., Tucson, 1968-74, prof., 1974—; sr. mem. Ariz. Research Lab., 1981—; vis. prof. Institut fur Physiologie, Innsbruck, Austria, 1976; mem. adv. panel regulatory biology NSF, 1973-76; mem. physiology study sect. NIH, 1980-82; mem. sci. adv. bd. Nat. Kidney Found., 1980-83. Mem. editorial bd.: Am. Jour. Physiology, 1973-76, 78-82; assoc. editor, 1981—; mem. editorial bd.: Renal Physiology, 1978—; contbr. articles to profl. jours. NSF grantee, 1965—; NIH grantee, 1973—. Fellow AAAS; mem. Am. Physiol. Soc. (sec. renal sect. 1978-79, chmn. 1979-81), Am. Soc. Nephrology, Am. Soc. Zoologists, Soc. Gen. Physiologists. Subspecialty: Comparative physiology. Current work: Comparative renal physiology with emphasis on renal tubular transport of organic molecules, the regulation of glomerular filtration rate and regulation of excretion of inorganic ions and water. Office: Dept Physiology Coll Medicine U Ariz Tucson AZ 85723 Home: 1601 Entrada Octava Tucson AZ 85718

DAO, KIM C., polymer engineer, scientist; b. Vietnam, May 14,1941; came to U.S., 1975, naturalized, 1982; s. Uyen V. and Tinh R. (Nguyen) D.; m. Hilary N., May 3, 1968; children: Kim H., Travis K. B.S. (N.Z. Govt. Colombo Plan scholar), U. Auckland, N.Z., 1966, M.S., 1968, Ph.D., 1971. Assoc. prof. U. Saigon, Vietnam, 1971-75; research engr. SRI Internat., Menlo Park, Calif., 1975-79; project mgr. Gen. Electric Co., Louisville, 1979—. Contbr. articles on polymer research and devel. to profl. jours. Pres. Vietnamese Assn., Louisville, 1982; bd. dirs. Asian-Am. Coalition, Louisville, 1982. Mem. Soc. Plastics Engrs., ASME. Subspecialties: Polymer engineering; Mechanical engineering. Current work: Characterization studies of polymer systems in areas of morphology, rheology and mechanical behavior; polymer formulation and alloying research; fracture/failure analysis; microscopy technique; computer-aided materials testing and modeling. Office: Gen Electric Co Appliance Park Louisville KY 40225

DAOUST, DONALD ROGER, pharm. and toiletries co. exec.; b. Worcester, Mass., Aug. 13, 1935; s. G. Arthur and Alice Anne (Lavallee) D.; m. Johanna K. Kalinoski, May 30, 1959; children: Donna Jean, Stephen Michael, Sandra Marie. B.A., U. Conn., 1957; M.S., U. Mass., 1959, Ph.D., 1962. Sr. research microbiologist Merck Sharp & Dohme Research Labs., Rahway, N.J., 1962-70, research fellow, 1970-72; mgr. biol. quality control Merck Sharp & Dohme, West Point, Pa., 1972-75; dir. quality control Armour Pharm. Co., Kankakee, Ill., 1975-76, v.p. quality assurance-regulatory compliance, Phoenix, 1976-78; corp. v.p. quality control Carter-Wallace Inc., Cranbury, N.J., 1978—. Contbr. articles to sci. jours., chpt. to book. Bd. dirs. South Plainfield (N.J.) Jaycees, 1968-69, pres., 1969-70; mem. South Plainfield Boro Council, 1970-72; v.p. programs George Washington council Boy Scouts Am., 1981, v.p., treas., 1982. Recipient Disting. Service award South Plainfield Jaycees, 1969; named Outstanding Young Man of N.J. N.J. Jaycees, 1970. Mem. Am. Soc. Microbiology, AAAS, Parenteral Drug Assn., Proprietary Assn., Am. Soc. Quality Control, Pharm. Mfrs. Assn. Roman Catholic. Lodge: Lions. Subspecialty: Microbiology. Patentee on fermentation process for producing physostigmine. Home: 8 Fairway Dr Cranbury NJ 08512 Office: PO Box 1 Cranbury NJ 08512

D'APPOLONIA, ELIO, civil engineer; b. Coleman, Alta., Can., Apr. 14, 1918; came to U.S., 1946, naturalized, 1959; s. Joseph S. and Constance (Piccinni) D'A.; m. Violet Mary D'Apollonia, May 2, 1942; children: David, Kenneth, Michael, Linda, Mark. B.S., U. Alta., 1942, M.S., 1946; Ph.D., U. Ill., 1948; D. Engring. (hon.), Carnegie-Mellon U., 1983. Cons. U.S. Army C.E., Alaska/No. Can., 1942-45; research asso. U. Ill., 1946-48; asst. prof. civil engring. Carnegie-Mellon U., 1948-56; pres., chmn. bd. D'Appolonia Cons. Engrs., Inc., Pitts., 1956—. Contbr. tech. articles to profl. jours. Recipient Keefer medal Engring. Inst. Can., 1948; William Metcalf award for outstanding engring. achievement Engrs. Soc. Western Pa., 1981; Disting. Alumnus award U. Ill., 1981; Disting. Service award Deep Founds. Inst., 1983. Mem. ASCE (Middlebrooks award 1969, Civil Engr. of Year, Pitts. sect. 1972), Nat. Acad. Engring., ASTM, Nat., Pa. socs. profl. engrs., Internat. Soc. Rock Mechanics, Internat. Assn. Bridge and Structural Engrs., Am. Underground Assn., Internat. Soc. Soil Mechanics and Found. Engrs., U.S. Nat. Com. Tunneling Tech., Deep Foundations Inst., Internat. Commn. Large Dams, Am. Inst. Cons. Engrs., Am. Water Resources Assn., Assn. Engring. Geologists. Republican. Roman Catholic. Club: Edgewood Country. Subspecialty: Civil engineering. Home: 1177 McCully Dr Pittsburgh PA 15235 Office: 10 Duff Rd Pittsburgh PA 15235

DARBY, CLIFTON FLOYD, computer scientist; b. Corpus Christi, Tex., Nov. 22, 1947; s. Clifton Wilbourn and Margie Alice (Floyd) D.; m. Betty Ann Boone, Dec. 30, 1978. Student, U. Tex., 1965-70. Cert. computer programmer. Computer operator/programmer Control Data Corp., Rockville, Md., 1971-72; system programmer IRS, Washington, 1972-80; computer scientist Computer Scis. Corp., Herndon, Va., 1980—. Mem. Assn. for Computing Machinery, IEEE (affiliate). Subspecialties: Operating systems; Software engineering. Current work: Large-scale transaction processing systems. Home: 1307 Winterbourne Ct Herndon VA 22070 Office: 3001 Centreville Rd Herndon VA 22070

DARBY, DENNIS ARNOLD, geology educator, researcher; b. Pitts., Oct. 31, 1944; s. William Arnold and Marjorie (Miller) D.; m. Michele Leonardi, June 10, 1972. B.S., U. Pitts., 1966, M.S., 1968; Ph.D., U. Wis., 1971. Research assoc. U. Wis., Madison, summer 1971; asst. prof. Hunter Coll., CUNY, 1971-74; instr. NYU, 1973; asst. prof. geophys. sci. Old Dominion U., 1974-77, assoc. prof., 1978—, chmn. dept. geophys. sci., 1978-81; coordinator/dir. CAUSE Project, 1976-79; cons. in field. Author: lab. book Introduction to Geology, 1980. NSF fellow, 1967; NSF grantee, 1976-79; NOAA grantee, 1981. Fellow Geol. Soc. Am.; mem. Soc. Econ. Paleontologists and Mineralogists, Nat. Assn. Geology Tchrs., Am. Geophys. Union, Va. Acad. Sci. (chmn. sect. 1978-79), Sigma Xi (pres. club 1976-77). Subspecialties: Sedimentology; Petrology. Current work: Provenance stratigraphy and depositional environments of Late Pleistocene, Coastal Plain of Va. and Colombia and Arctic Ocean basin; clay mineralogy and trace element geochemistry-pollution studies. Office: Old Dominion U Dept Geophys Sci Norfolk VA 23508

DARBY, JOHN LITTRELL, nuclear engineer; b. Florence, Ala., June 5, 1950; s. Albert Wright and Lucile (Guin) D.; m. Melanie Patrice Hook, June 20, 1978. B.S., Birmingham So. U., 1972; M.S., U. Wis.-Madison, 1973; Ph.D., 1976. Registered profl. engr. Scientist Sandia Labs, Albuquerque, 1976-82; sr. scientist SEA Inc., Albuquerque, 1982—. Mem. Am. Nuclear Soc., Inst. Nuclear Materials Mgmt. (chmn. phys. security subcom. 1979—), Am. Soc. Indsl. Security, Phi Beta Kappa. Subspecialties: Nuclear engineering; Nuclear physics. Current work: Reactor safety, technical security, nuclear physics, systems analysis. Home: 2300 Dietz Pl NW Albuquerque NM 87107 Office: SEA Inc PO Box 3722 Albuquerque NM 87190

DARDEN, CHRISTINE MANN, aerospace engineer; b. Monroe, N.C., Sept. 10, 1942; d. Noah Horace, Sr. and Desma (Chaney) Mann; m. Walter Lee Darden, Jr., June 13, 1963; children: Jeanne Oletia, Janet Christine. B.S., Hampton Inst., 1962; M.S., Va. State Coll., 1967; D.Sc., George Washington U., 1983—. Math. tchr. Brunswick County Schs., Lawrenceville, Va., 1962-63, Portsmouth (Va.) City Schs., 1964-65; instr. math. Va. State Coll., Petersburg, 1966-67; data analyst Langley Research Ctr., Hampton, Va., 1967-73, aerospace engr., 1973—. Contbr. articles to pubs. Ordained elder Carver Meml. Presbyterian Ch., Newport News, Va., 1980—; program chmn. Nat. Tech. Assn. Student Symposium, Langley Research Center, 1980, 82. HEW research grantee, 1965-66; named Outstanding Alumna Hampton Inst., 1982. Mem. Nat. Tech. Assn. (v.p. 1978-82), AIAA, Alpha Kappa Alpha (treas. 1980-81). Subspecialties: Fluid mechanics; Applied mathematics. Current work: Basic research in analytic methods used to design and analyze wing-body configurations for supersonic flight. Home: 1028 Barry Ct Hampton VA 23666 Office: NASA Langley Research Center Hampton VA 23665

DARDIRI, AHMED HAMED, veterinarian, govt. adminstr., educator; b. Cairo, Egypt, Mar. 10, 1919; came to U.S., 1946, naturalized, 1958; s. Hamed Ahmed and Ameena (Reedy) D.; m. Lucille B., April 7, 1951. D.V.M., U. Cairo, 1940, M.V.Sc., 1946; M.Sc., Mich. State U., 1947, Ph.D. in Microbiology, 1950. Dir. Poultry Research Experiment Sta., Cairo, 1940-46; mem. Egyptian Edn. Mission to U.S., 1946-50; sr. lectr. Cairo U., 1950-55; research assoc. in animal pathology U. R.I., 1955-56, asst. prof., 1956-59, assoc. prof., 1959-61; prin. research veterinarian Plum Island Animal Disease Center, Dept. Agr., Greenport, N.Y., 1961—, also lab. dir. diagnostic investigations; cons. Am. Tech. Aid to Egypt, Cairo, 1951-55; adj. prof. U. R.I. and U. Pa., Phila. Contbr. numerous articles to U.S., fgn. profl. jours. Merck and Shope Research grantee, 1950. Mem. AVMA, N.Y. Acad. Assn. Avian Diseases, U.S. Animal Health Assn., Internat. Soc. Microplasmology, Wildlife Diseases, Am. Soc. Microbiology. Lodge: Rotary. Subspecialties: Microbiology (veterinary medicine); Animal virology. Current work: Veterinary science, microbiology, immunology, foreign animal disease research and diagnosis.

DARLINGTON, SIDNEY, educator, electrical engineer; b. Pitts., July 18, 1906; s. Philip Jackson and Rebecca Taylor (Mattson) D.; m. Joan

Gilmer Raysor, Apr. 24, 1965; children: Ellen Sewall, Rebecca Mattson. B.S. magna cum laude, Harvard U., 1928, MIT, 1929; Ph.D. in Physics, Columbia U., 1940. Mem. tech. staff Bell Telephone Labs., Murray Hill, N.J., 1929-71, head dept., 1960-71; ret., 1971; adj. prof. elec. engring. U. N.H., Durham, 1971—; cons. in field, 1971—; Mem. U.S. commn. VI Internat. Sci. and Radio Union, 1959-75, del. gen. assemblies, 1960, 63, 66, 69. Author. Recipient Medal of Freedom U.S. Army. Fellow IEEE (Edison medal, Medal of Honor 1981), AIAA; mem. Nat. Acad. Engring., Nat. Acad. Scis., Phi Beta Kappa. Club: Appalachian Mountain. Subspecialties: Electrical engineering; Systems engineering. Current work: Design techniques for digital filters. Patentee in field. Home: 8 Fogg Dr Durham NH 03824 Life is a chancy business; my blessings have far outnumbered my adversities.

DARNALL, DENNIS WAYNE, protein biochemist, chemistry educator; b. Glenwood Springs, Colo., Dec. 14, 1941; s. Harvey Glen and Lois Marie (Coleman) D.; m. Judy Marcell, May 31, 1963; children: Nichol, Beth. B.S. in Chemistry, N.Mex. Inst. Mining and Tech., 1963, Ph.D., Tex. Tech. U., 1966. NIH postdoctoral fellow Northwestern U., 1966-68; asst. prof. chemistry N.Mex. State U., 1968-72, assoc. prof., 1972-74, prof., 1974—; cons. biomaterials, protein chemistry. Author: Methods for Determining Metal Ion Environments in Protein, 1980; contbr. articles to profl. jours., chpts. to books. NIH Career Devel. awardee, 1971-76; recipient Westhafer award for research N.Mex. State U., 1978. Mem. Am. Chem. Soc., Am. Soc. Biol. Chemists, AAAS. Republican. Subspecialties: Biophysical chemistry; Biochemistry (medicine). Current work: Metal ions in biology, biosorption of metal ions from water, metallo-proteins and enzymes, biotechnology, biochemistry. Home: 1501 W University Ave Las Cruces NM 88001 Office: Dept Chemistry New Mex State U Las Cruces NM 88003

DARNELL, JAMES EDWIN, JR., cell biologist; b. Columbus, Miss., Sept. 9, 1930; s. James Edwin and Helen (Hopkins) D.; m. Jane Roller, 1957; children—Christopher, Robert, Jonathan. B.A., U. Miss., 1951; M.D., Washington U., St. Louis, 1955. Intern Barnes Hosp., St. Louis, 1955-56; virologist USPHS, 1956-60; spl. postdoctoral fellow Pasteur Inst., Paris, 1960-61; asst. prof., then asso. prof. biology Mass. Inst. Tech., 1961-64; prof. cell biology and biochemistry Albert Einstein Coll. Medicine, 1964-68; Kempner prof. biology, chmn. dept. Columbia U., 1968-72; Vincent Astor prof. molecular cell biology Rockefeller U., 1973—. Co-author: General Virology, 1967, 2d edit., 1977. Mem. Nat. Acad. Scis., Am. Acad. Arts and Scis., Am. Soc. Microbiology, Soc. Biol. Chemists. Democrat. Subspecialty: Cell biology. Home: 96 Edgewood Ave Larchmont NY 10538 Office: Rockefeller U 1230 York Ave New York NY 10021

DARON, HARLOW H., biochemist; b. Chgo., Oct. 25, 1930; s. Garman H. and Gulah V. (Hoover) D.; m. Carol Fields, Aug. 8, 1969; children: Ruth Ann, Leslie Susan, Barbara Phyllis, Charles Edward. B.S., U. Okla., Norman, 1956; Ph.D., U. Ill., Urbana, 1961. Postdoctoral fellow dept. biology Calif. Inst. Tech., 1961-63; asst. prof. biochemistry Tex. A&M U., 1963-67; asst. prof. animal and dairy scis. Auburn U., 1967-70, assoc. prof., 1970-82, prof., 1982—. Served with USN, 1951-54. Mem. AAAS, Am. Soc. Biol. Chemists, Am. Soc. Microbiology, N.Y. Acad. Scis. Subspecialties: Biochemistry (biology); Enzyme technology. Current work: Enzymology. Office: Dept Animal and Dairy Scis Auburn U Auburn AL 36849

D'ARRIGO, JOSEPH SALVATORE, educator; b. N.Y.C., Apr. 4, 1946; s. Joseph Richard and Elizabeth (Medici) D'A.; m. Sachie Aoki, July 28, 1977; 1 son, Paul Sakichi. B.A., Queens Coll., 1967; Ph.D., UCLA, 1972. Postdoctoral fellow U. Utah, Salt Lake City, 1972-73, research instr., 1973-75; asst. prof. physiology U. Hawaii, Honolulu, 1975-77, assoc. prof., 1977—; vis. fellow Australian Nat. U., Canberra, 1982-83; cons. Cavitation-Control Tech., Kaneohe, 1979—. NIH fellow, 1972; NSF grantee, 1975; NASA grantee, 1980. Mem. Biophys. Soc., Am. Physiol. Soc., AAUP, Am. Chem. Soc., AAAS. Democrat. Roman Catholic. Subspecialties: Surface chemistry; Biophysics (biology). Current work: Research is being conducted on stabilization of gas microbubbles in aqueous media by naturally occurring surfactants and in oil-based media by various analogues of biological surfactants. Industrial applications of this research are also being pursued. Inventor bubble monitor, 1976. Office: Dept Physiology U Hawaii Honolulu HI 96822

DARSEY, JEROME ANTHONY, research scientist; b. Houma, La., Aug. 26, 1946; s. Elmer Joseph and Arline (Houghton) D. B.S., La. State U., 1970, Ph.D., 1982. Tchr. Terrebonne High Sch., Houma, La., 1970-74, grad. teaching asst. La. State U., Baton Rouge, 1974-81, postdoctoral research assoc., 1982—. Mem. Am. Phys. Soc., Am. Chem. Soc., AAAS, Am.Soc. Physics Tchrs., Planetary Soc., Phi Lambda Upsilon. Republican. Roman Catholic. Club: First Day Cover Soc. (Cheyenne, Wyo.). Subspecialties: Polymer chemistry; Polymer physics. Current work: Conformational characterization of macromolecules especially biopolymers and structural, conductive and magnetic properties of conductive polymers, chemistry and physics of liquid crystals. Office: Chemistry Dept La State U Baton Rouge LA 70803

DAS, JAGANNATH PRASAD, psychology educator; b. Puri, Orissa, India, Jan. 20, 1931; emigrated to Can., 1968; s. Biswanath and Nilomoni (Mohanty) D.; m. Gita Dasmohapatra, 1955; children: Satya, Sheela. B.A. with honors, Utkal U., 1951; M.A., Patna U., 1953; Ph.D., U. London, 1957. Cert. psychologist, Alta. Dir. Ctr. Study of Mental Retardation, U. Alta., Edmonton, 1972—, prof. ednl. psychology, 1970—. Author: Verbal Conditioning and Behavior, 1969, Simultaneous and Successive Cognitive Processes, 1979, Intelligence and Learning, 1981, others. Recipient award for Best Article Internat. Reading Assn., 1979; Nuffield Found. fellow, 1972; Kennedy Found. fellow, 1963-64. Fellow Am. Psychol. Assn., Can. Psychol. Assn. Subspecialties: Cognition; Developmental psychology. Current work: Cognitive processing in retarded and learning disabled children, their assessment and remediation of handicap by methods derived from neuropsychology. Home: 11724-38A Ave Edmonton AB Canada T6J L9 Office: Univ Alberta 6-123 Education N Edmonton AB Canada

DAS, PANKAJ K., scientist, educator; b. Calcutta, W. Bengal, India, June 15, 1937; came to U.S., 1964, naturalized, 1975; s. Upendra N. and Susama (Paul) D.; m. Virginia Van Kirk, July 29, 1967; children: Andrea, Joshua. B.Sc., U. Calcutta, 1957, M.Sc., 1960, Ph.D., 1964. Instr. Poly. Inst. Bklyn., 1964-65, asst. prof., 1965-68; assoc. prof. U. Rochester, 1968-74, Rensselaer Poly. Inst., 1974-77, prof., 1977—; vis. prof. elec. engring. OAS, Mexico City, 1972-73. Contbr. numerous articles to profl. jours. NSF grantee, 1975—. Mem. Am. Phys. Soc., IEEE, Optical Soc. Am., Am. Soc. Non-destructive Testing, Acoustical Soc. Am. Subspecialties: Semiconductors; Integrated circuits. Current work: Signal processing devices such as SAW and CCD, acousto-optic devices, non-destructive testing using elastic waves and ultrasonic imaging. Home: 15 Johnson Rd Cohoes NY 12047 Office: Rensselaer Poly Inst Troy NY 12181

DASGUPTA, AARON, mechanical engineer, researcher; b. Calcutta, India, Nov. 20, 1943; came to U.S., 1965, naturalized, 1969; s. Krishna Prosad and Amita (Sen) D.; m. Runu Biswas, Mar. 9, 1972; children: Elora, Debraj. B.Tech. with honors, Indian Inst. Tech., Kharagpur, 1963; M.Eng., Tech. U. N.S. (Can.), Halifax, 1968; Ph.D., Va. Poly. Inst., 1975. Registered profl. engr., Ill., Va., Tenn., N.Mex. Assoc. engr. Heavy Engring. Corp., Dhurwa, Bihar, India, 1963-65; electronics engr. Can. Marconi Corp., Montreal, Que., 1967-69; prodn. engr. Whittaker AMTD Corp., Gardena, Calif., 1969-70; design engr. Kingsport Press Inc., Tenn., 1973-75; stress analyst Sundstrand Aviation, Rockford, Ill., 1975; sr. mech. engr. Ballistic Research Lab., Aberdeen, Md., 1975—; environ. cons. El Paso Electric Co., Tex., 1971-72; design cons. Comml. Fabrication Co., Mt. Airy, N.C., 1972-73; cons. Ministry of Def., Harwell, Eng., 1980-81, U.S. Army C.E., Balt., 1980-82, Harry Diamond Lab., Adelphi, Md., 1979-80; systems cons. U.S. Army, 1976—. Contbr. articles to tech. jours.; editor: Dynamics of Ballistic Impact, 1980. State adv. U.S. Congressional Adv. Bd., Washington, 1982. Mem. Soc. Engring. Sci. (co-chmn. 1981-82), N.Y. Acad. Sci., Am. Acad. Mechanics (reviewer 1978-81, Quality Performance award 1982), Sigma Xi. Hindu. Club: Cosmopolitan (treas. 1967-68). Subspecialties: Mechanical engineering; Theoretical and applied mechanics. Current work: Structural dynamics; penetration mechanics; fracture mechanics; blast damage on structures; nuclear effects; numerical analysis and simulation; containment design. Home: 104 John St Perryville MD 21903 Office: U S Army Ballistic Research Lab Terminal Ballistic Div Aberdeen Proving Ground Aberdeen MD 21005

DASH, SANFORD MARK, consulting aerospace scientist; b. N.Y.C., May 26, 1943; s. Jack and Rachel (Calamar) D.; m. Barbara Gaile Held, Dec. 27, 1964; children: David, Kenneth, Jonathan, Naomi. B.M.E., CCNY, 1964; M.S., NYU, 1966, Ph.D., 1969. Research scientist NYU Aerospace Labs., N.Y.C., 1967-70, Advanced Tech. Labs., Jericho, N.Y., 1970-73; sr. research scientist Gen. Applied Sci. Labs., Westbury, N.Y., 1973-77; cons. Aero. Research Assocs., Princeton, N.J., 1977-80; tech. dir. Sci. Applications, Inc., Princeton, 1980—; adj. prof. aero. dept. Dowling Coll., N.Y., 1969-72; guest lectr. U. Tenn. Space Inst., Tullahoma, 1979-80. Contbr. over 50 articles to various jours. Bd. dirs. Jewish Community Center, Mercer County, Ewing, N.J., 1979-81. Recipient Founder's Day award NYU, 1970; cert. of recognition NASA, 1975. Mem. AIAA, Tau Beta Pi, Pi Tau Sigma, Sigma Gamma Tau. Subspecialties: Fluid mechanics; Aerospace engineering and technology. Current work: Research in fluid dynamics of high speed, multiphase, reacting flows as occur in rocket and aircraft exhausts and nozzle combustors. Developed U.S. standard rocket plume model, 1st 3D exhaust flowfield model, NATO target signature model. Home: 1218 Ward Dr Yardley PA 19067 Office: Science Applications Inc 1101 State Rd Bldg N Princeton NJ 08540

DASHEIFF, RICHARD MITCHELL, neurologist; b. Washington, July 9, 1951; s. Stanley and Alaine (Perry) D.; m. Sandra Jean Kulansky, June 16, 1974. B.S. in Physics, U. Md., 1972, 1972, M.D., 1976. Intern U. Md. Hosp., 1976-77; resident in neurology Duke U., 1977-80, postdoctoral fellow, 1980-82; dir. Francis M. Forster Epilepsy Center, Middleton VA Hosp., Madison, Wis., 1982—. Contbr. articles to profl. jours. Nat. Inst. Neurol. and Communicative Disorders and Stroke grantee, 1980-82. Mem. Am. Acad. Neurology, AAAS, Am. EEG Soc., Am. Epilepsy Soc., AMA, Soc. for Neurosci. Subspecialties: Neurology; Neurobiology. Current work: Epilepsy, kindling. Office: Francis M Forster Epilepsy Center VA Hosp 2500 Overlook Terr Madison WI 53705

DASHEN, ROGER FREDERICK, physics educator; b. Grand Junction, Colo., May 5, 1938. A.B., Harvard U., 1960; Ph.D. in Physics, Calif. Inst. Tech., 1964. Research assoc. theoretical physics Calif. Inst. Tech., Pasadena, 1964-65, asst. prof., 1965-66, prof., 1966-69; mem. staff theoretical physics Inst. Advanced Study, Princeton, N.J., 1966-69, prof., 1969—; vis. prof. Princeton U., 1966; mem. Sci. Adv. Com. Stanford (Calif.) Linear Accelerator Ctr., 1968-72; cons. SRI Internat., 1966—, Los Alamos Sci. Lab., 1974—, Brookhaven Nat. Lab., 1977-78; Fermi Nat. Accelerator Lab., 1972-75. Sloan Found. fellow, 1966-73. Subspecialty: Theoretical physics. Office: Inst Advanced Study Princeton NJ 08540

DASHMAN, THEODORE, biochemist; b. Bklyn., Oct. 7, 1928; s. Louis and Yetta (Cohen) D.; m. Eleanor Greif, May 19, 1957; children: Gina, Suzanne, Adrienne. B.S., Bklyn. Coll., 1950; M.S., Fla. State U., 1964; Ph.D., NYU, 1977. Research assoc. St. Barnabas Hosp., Bronx, N.Y., 1964-67; scientist Sandoz Pharms. Inc., Hanover, N.J., 1967; sr. scientist Hoffman-La Roche Inc., Nutley, N.J., 1967—. Contbr. numerous articles to profl. publs. Mem. adv. bd. on arts Teaneck (N.J.) Mcpl. Council, 1980—. Served with U.S. Army, 1953-55. Mem. Am. Chem. Soc. (councilor North Jersey sect.), AAAS, Am. Soc. Pharmacology and Exptl. Therapeutics. Subspecialties: Biochemistry (biology); Molecular pharmacology. Current work: Metabolism of therapeutic agents, enzyme inhibition. Home: 163 Pinewood Pl Teaneck NJ 07666 Office: Hoffman-La Roche Inc Nutley NJ 07110

DATTA, RATNA, med./health physicist; b. Jamshedpur, India, Oct. 10, 1943; came to U.S., 1973, naturalized, 1978; d. Birendranath and Basanti (Roy) Sarkar; m. Sobhendranath Datta, Jan. 23, 1973. B.S., Ranchi U., India, 1963; M.S., Calcutta U., 1966, Ph.D., 1971. Postdoctoral fellow Bonner Lab., Rice Lab., Houston, 1973-74; NIH sr. postdoctoral fellow M.D. Anderson Hosp., Houston, 1974-76; asst. prof. U. Tex. Health Sci. Center, San Antonio, 1976-78; chief radiol. physicist, asst. prof. dept. radiology La. State U., Shreveport, 1978—. Contbr. articles to profl. jours. Ranchi U. scholar, 1961-63; Atomic Energy scholar Govt. of India, 1963-66; Saha Inst. fellow, 1966-71, 71-73; NIH fellow, 1974-76; others. Mem. Am. Assn. Physicists in Medicine, Am. Assn. Therapeutic Radiologists, Health Physics Soc., Assn. Med. Physicists of India. Subspecialty: Medical/health physicist. Current work: Research on effect of radiation on bone healing; comparative study of different dosimetry devices, electron beam treatment planning using CT scan, treatment planning with split beam technique. Office: 1541 Kings Hwy Dept Radiology Shreveport LA 71130

DATTA, SUBHENDU K(UMAR), mechanical engineering educator, consultant; b. Calcutta, India, Jan. 15, 1936; came to U.S., 1968; s. Srish Chandra and Prabhabati (Ghosh) D.; m. Bishakha Roy, May 10, 1966; 1 child, Kinshuk. B.S., Presidency Coll., Calcutta, 1954; M.S., Calcutta U., 1956; Ph.D., Jadavpur U., Calcutta, 1962. Asst. prof. Indian Inst. Tech., Kanpur, 1965-67; asst. prof. U. Colo., Boulder, 1968-69, assoc. prof., 1969-73, prof. mech. engring., 1973—; cons. NBS, Boulder, 1979—. Editor: Earthquake Ground Motion and Its Effects on Structures, 1982, Procs. U.S. Nat. Congress Applied Mechanics, 1974; contbr. articles to publs. in field. Fulbright fellow, 1962; faculty fellow U. Colo., 1972; NSF grantee, 1970—. Mem. ASME, Soc. Indsl. and Applied Math., Am. Acad. Mechanics, Soc. Engring. Sci. Subspecialties: Theoretical and applied mechanics; Applied mathematics. Current work: Elastic wave scattering and ultrasonic nondestructive evaluation, earthquake engineering, mechanics of composite materials. Home: 6252 Old Stage Rd Boulder CO 80302 Office: Univ Colo ECOT 4-6 Campus Box 427 Boulder CO 80309

DATTA, SURJIT KUMAR, immunologist; b. Adampur Doaba, Punjab, India, Jan. 1, 1935; s. Jagdish Chander and Jaswanti Devi (Sharma) D.; m. Evelyn B. Dochy, Dec. 17, 1971. B. V. Sc. and A.H., Punjab U., 1957; M.V.Sc., Agra U., 1963; Ph.D., U. Louvain, Belgium, 1970. Assoc., then asst. prof. Coll. Vet. Medicine, Hissar, India, 1963-73; research fellow microbiology U. Brisbane, Australia, 1973-74; inst., research assoc. div. exptl. biology Baylor Coll. Medicine, 1974-78, research asst. prof., 1978—; NIH co-investigator, 1978—. Contbr. articles to profl. jours. Mem. Am. Assn. Cancer Research, Am. Assn. Immunology, Am.Soc. Microbiology, Internat. Soc. Exptl. Hematology. Subspecialties: Cancer research (medicine); Transplantation. Current work: Natural and immune resistance to cancer, natural resistance to marrow and lymphona grafts. Office: Inst Med Research Bennington VT 05201

DATZ, I. MORTIMER, systems engineer, consultant; b. N.Y.C., Feb. 11, 1928; s. A. Mark and Lillian (Barkin) D.; m. Gerd Elin Alme-Torkildsen, Apr. 30, 1956. B.S., CCNY, 1950; postgrad., U. Bergen, Norway, 1955. Sect. head NASA, Goddard Space Flight Ctr., Greenbelt, Md., 1959-61; group leader Army Strategy and Tactics Analysis Group, Bethesda, Md., 1961-63; div. head Maritime Adminstrn., Washington, 1963-64; tech. adv. Naval Ship Research and Devel. Ctr., Annapolis, Md., 1964-79; ind. cons., Annapolis, 1979—; cons. Band, Lavis & Assocs., Severna Park, Md., 1980-81, Arinc Research Corp., Annapolis, 1980, Research Facilities Corp., Alexandria, Va., 1981—. Author: Planning Tools for Ocean Transportation, 1971, Power Transmission and Automation for Ships and Submersibles, 1975; contbr. articles to profl. jours. Recipient research stipend The Geophysics Inst., Bergen, Norway, 1954; stipend Woods Hole Oceanographic Instn., 1949. Fellow AAAS; mem. Soc. Naval Architects and Marine Engrs., Ops. Research Soc. Am., Am. Def. Preparedness Assn., N.Y. Acad. Scis. Clubs: Am. Scandinavian Found. (Washington); U.S. Naval Inst. Subspecialties: Operations research (engineering); Systems engineering. Current work: Operations research analyses as applied to naval architecture and marine engineering, development of conceptual design models structured to explore alternative ship concepts and to provide an assessment of their capabilities and economics. Home and Office: 700 Americana Dr Annapolis MD 21403

DATZ, SHELDON, physicist; b. N.Y.C., July 21, 1927; s. Jacob and Clara (Green) D.; m. Roslyn Gordon, Aug. 25, 1948; children: William Lawrence, Joan Ellen; m. Jonna Holm, Jan. 23, 1973. B.S., Columbia U., 1950, M.A., 1951; Ph.D., U. Tenn., 1960. Technician SAM Labs., Columbia U., Manhattan Project, 1943-45; physics dept. Columbia U., 1946-51; research chemist Oak Ridge (Tenn.) Nat. Lab., 1951-60, assoc. dir. chemistry div., 1965-75, group leader chem. dynamics, 1960-81, sect. chief atomic physics, physics div., 1981—; cons. Gen. Atomics, 1958-62, Republic Aviation Plasma Propulsion Lab., 1959-62. Author: Atomic Collisions in Solids, 1975, Electronic and Atomic Collisions, 1982; contbr. articles to profl. jours. Served with USN, 1945-46. Fulbright sr. research fellow, 1962-63; Union Carbide Corporate research fellow, 1980—. Fellow Am. Phys. Soc.; mem. Am. Chem. Soc., AAAS. Subspecialties: Atomic and molecular physics; Kinetics. Current work: High energy atomic collision physics, particle solid interactions. Office: Oak Ridge Nat Lab PO Box X Oak Ridge TN 37830

DAU, PETER CAINE, clinical immunologist, researcher; b. Fresno, Calif., Feb. 25, 1939; s. Julius Jensen and Elva (Carne) D.; m. Barbara Joan Berry, Jan. 20, 1965; children: Birgitt, Kirstin, Mairikke; m. Leslie Kenton, July 30, 1959 (div. 1963); 1 son, Branton. B.A., Stanford U., 1960, M.D., 1964. Diplomate: Am. Bd. Psychiatry and Neurology. Intern Los Angeles County Gen. Hosp., 1964-65; resident neurology U. Wis., Madison, 1965-66, U. Chgo., 1966-68, fellow immunology, 1968-69; research asst. Deutsches Krebsforschungs Centrum, Heidelberg, Germany, 1972-74, U. Calif., San Francisco, 1974-75, asst. prof. medicine, 1975-81; assoc. prof. neurology Northwestern U., Chgo., 1982—; head allergy/immunology Evanston (Ill.) Hosp., 1982—; cons. Cutter Labs., Inc., Berkeley, Calif., 1978-81, Cobe Labs., Inc., Denver, 1979-83, Alpha Therapeutics Corp., Los Angeles, 1981-83. Editor: Plasmopheresis and the Immunobiology of Myesthemia Gravis, 1979. Served to maj. USAF, 1969-71. Recipient Peter Bassoe award Chgo. Neurol. Soc., 1969; awards Muscular Dystrophy Assn., 1976-82; NIH, 1982. Mem. Am. Assn. Immunologists. Republican. Subspecialties: Neuroimmunology; Immunology (medicine). Current work: Plasmapheresis/plasma exchange therapy, human autoimmunity, antibody production, immunosuppressive therapy. Home: 635 Rosewood Ave Winnetka IL 60093 Office: Evanston Hosp 2650 Ridge Ave Evanston IL 60201

DAUBEN, WILLIAM GARFIELD, chemist, educator; b. Columbus, Ohio, Nov. 6, 1919; s. Hyp J. and Leilah (Stump) D.; m. Carol Hyatt, Aug. 8, 1947; children—Barbara, Ann. Edward. A.B., Ohio State U., 1941; A.M., Harvard, 1942; Ph.D., 1944; Ph.D. hon. degree, U. Bordeaux, France, 1980. Austin fellow Harvard, 1941-42, teaching fellow, 1942-43, research asst., 1943-45; instr. U. Calif. at Berkeley, 1945-47, asst. prof. chemistry, 1947-52, asso. prof., 1952-57, prof., 1957—; lectr. Am.-Swiss Found., 1962; pres. Organic Reactions, Inc., 1967; mem. med. chem. study sect. USPHS, 1959-64; mem. chemistry panel NSF, 1964-67; mem. Am.-Sino Sci. Cooperation Com., 1973-76; mem. assembly math. and phys. scis. NRC, 1977-80. Mem. bd. editors: Jour. of Organic Chemistry, 1957-62; bd. editors: Organic Syntheses, 1959-67; bd. dirs., 1971—; editor-in-chief: Organic Reactions, 1967—; Contbr. articles profl. jours. Recipient award Calif. sect. Am. Chem. Soc., 1959; Guggenheim fellow, 1951, 66; sr. fellow NSF, 1957-58; Alexander von Humboldt Found. Fellow, 1980. Fellow London, Swiss chem. socs.; mem. Am. Chem. Soc. (chmn. div. organic chemistry 1962-63, councilor organic div. 1964-70, mem. council publ. com. 1965-70, mem. adv. com. Petroleum Research Fund 1974-77, Ernest Guenther award 1973), Nat. Acad. Scis. (chmn. chemistry sect. 1977-80), Am. Acad. Arts and Scis., Phi Beta Kappa, Sigma Xi, Phi Lambda Upsilon, Phi Eta Sigma, Sigma Chi. Club: Bohemian. Subspecialties: Organic chemistry; Photochemistry. Current work: Photochemistry, lasers; synthetic chemistry. Home: 20 Eagle Hill Berkeley CA 94707

DAUBERT, RAYMOND LEO, systems development engineer, automation and robotics consultant; b. Gt. Bend, Kans., June 12, 1948; s. Elmer James and Florine (Hermann) D.; m. LeElda Mary Kechely, Apr. 13, 1974; children: Shane Evans, Shauna René, Shelley Rae. B.S. in Physics, Ft. Hays State U., 1973. Maintenance supr. Johnson & Johnson, Chgo., 1973-74; nuclear engr. Westinghouse Electric, Idaho Falls, Idaho, 1974-76, nuclear fuel supr., 1976-77; systems integration engr. Westinghouse Hanford, Richland, WAsh., 1977-78, systems devel. engr., 1978—. Served with U.S. Army, 1966-69; Germany. Mem. Am. Nuclear Soc. (sessions chmn. 1983-84). Subspecialties: Robotics; Operations research (engineering). Current work: Design and development of automated equipment for remote testing of fusion simulators, nuclear fuel fabrication, and nuclear waste packaging. Home: 2514 S Anderson Kennewick WA 99336 Office: Westinghouse Hanford PO Box 19780 Richland WA 99352

DAUGHERTY, ROBERT EUGENE, state ofcl., agrl. economist, cons.; b. Strong City, Okla., Feb. 27, 1922; s. John Lee and Ada Ann (Snider) D.; m. Thelma Bernadine Schroeder, Mar. 8, 1943; children: Renee Ann, Mary Denise Daugherty Tye, Bobbie Dee. B.S., Okla. A&M Coll., 1943; M.S., Okla. State U., 1959. Asst. county agrl. agt., Creek County, Okla., 1947-49; instr. in animal husbandry Okla. State Inst. Tech., 1949-51, head div. agr., 1952-53; extension livestock specialist Okla. State U., 1953-58, extension livestock mktg. economist, 1958-77, asso. prof. agrl. econs., 1960-77; dir. internat. market devel.

Okla. Dept. Agr., Stillwater, 1977—; livestock and feed mktg. expert, Tehran, Iran, 1973-75; cons. agrl. devel. and mktg.; dir. Central States Milk Goat Mktg. Coop.; mem. Dist. Export Council-North Tex. and Okla., Okla. Gov.'s Internat. Adv. Team, Okla. Gov.'s Waterway Adv. Team. Served with inf. U.S. Army, 1943-46. Decorated Purple Heart. Mem. So. Agrl. Econs. Assn., Western Agrl. Econs. Assn., Am. Assn. Agrl. Economists. Democrat. Roman Catholic. Subspecialties: Agricultural economics; Animal nutrition. Current work: Devel. sophistication in livestock and internat. market devel.; location and devel. agri-industry potential to process and market internationally; devel. internat. demand contacts for U.S. produced products. Home: 1013 Brown Ave Stillwater OK 74074 Office: Okla Dept Agr Room 505 Agrl Hall Okla State U Stillwater OK 74078

DAVENPORT, ALAN GARNETT, engineering educator; b. Madras, India, Sept. 19, 1932. B.A., Cambridge (Eng.) U., 1954, M.A., 1958; M.A.Sc., U. Toronto, 1957; Ph.D. in Engring, Bristol (Eng.) U., 1961; D.Sc. (hon.), U. Lourain. Lectr. engring. U. Toronto, 1954-57; asst. Nat. Research Council Can., 1957-58, Bristol U., 1958-61; assoc. prof. engring. U. Western Ont., London, Can., 1961-66, prof., 1966—; dir. Boundary Layer Wind Tunnel Lab.; cons. design World Trade Ctr., N.Y.C., 1964-65. Recipient Alfred Nobel prize, 1963. Mem. ASCE o3(com. tall bldgs.), Engring. Inst. Can. (Duggan prize 1962, Gzowski medal 1963), Royal Meterol. Soc., Internat. Assn. Bridge and Structural Engrs., Royal Soc. Can. Subspecialty: Structural engineering. Office: Dept Engring U Western Ontario London ON Canada N6A 5B8

DAVENPORT, WILLIAM DANIEL, JR., oral pathology and anatomy educator, research laboratories administrator; b. Corinth, Miss., Apr. 19, 1947; s. William Daniel and Flora Louise (Phillips) D. B.S., U. Miss., 1969, M.S., 1971; Ph.D., Med. Coll. Ga., 1976. Instr. Ark. State U., 1971-72; instr. U. Miss. Med. Center, Jackson, 1975-77, asst. prof., 1977-82; asst. prof. oral pathology and anatomy La. State U. Sch. Dentistry, 1982—, dir. research Labs., 1982—. Named Basic Sci. Tchrs. of Yr. U. Miss. Sch. Dentistry, 1980; Nat. Endowment for Humanities fellow U. Calif.-San Francisco, 1979. Mem. Am. Assn. Anatomists, Electron Microscopy Soc. Am., History of Medicine Soc., Am. Assn. Dental Schs., Internat. Assn. Dental Research, Eastern Soc. Tchrs. Oral Pathology. Republican. Baptist. Subspecialties: Oral biology; Anatomy and embryology. Current work: Surface ultrastructural changes in endothelial cells in diet/environment induced cardiovascular diseases; cell surface changes in metastatic/non-metastatic cancer cells after chemotherapy in vitro as diagnostic aid in prognosis. Home: 4201 Teuton St Apt 27 Metairie LA 70002 Office: Dept Oral Pathology La State U Med Center Sch Dentistry 1100 Florida Ave New Orleans LA 70119

DAVES, GLENN DOYLE, JR., chemistry educator, research scientist; b. Clayton, N.Mex., Feb. 12, 1936; s. Glenn Doyle and Edna Lee (Parker) D.; m. Pamela Gannarelli, Sept. 5, 1959; children: Laura Lee, Anne Kathryn, Glenn Graham. B.S., Ariz. State U., Tempe, 1959; Ph.D., M.I.T., 1964. Research scientist Midwest Research Inst., Kansas City, Mo., 1959-61, Stanford Research Inst., Menlo Park, Calif., 1964-67; prof. dept. chemistry Oreg. Grad. Center, Beaverton, 1967-81; prof., chmn. dept. chemistry Lehigh U., Bethlehem, Pa., 1981—; affiliate mem. dept. biochemistry Oreg. Health Scis. U., Portland, 1968-81. Contbr. articles to profl. jours. Mem. Am. Chem. Soc., Am. Soc. Mass Spectrometry, Internat. Soc. Heterocyclic Chemistry, AAAS. Democrat. Subspecialties: Organic chemistry; Synthetic chemistry. Current work: Chemistry of molecules with important biological activities. Home: RD 3 Box 96 Fire Ln Bethlehem PA 18015 Office: Dept Chemistry Lehigh U Bethlehem PA 18015

DAVID, EDWARD EMIL, JR., elec. engr., bus. exec.; b. Wilmington, N.C., Jan. 25, 1925; s. Edward Emil and Beatrice (Liebman) D.; m. Ann Hirshberg, Dec. 23, 1950; 1 dau., Nancy. B.S., Ga. Inst. Tech., 1945; M.S., Mass. Inst. Tech., 1947, Sc.D., 1950; D.Engring. (hon.), Stevens Inst. Tech., 1971, Poly. Inst. Bklyn., 1971, U. Mich., 1971, Carnegie-Mellon, 1972, Lehigh U., 1973, U. Ill. at Chgo., 1973, Rose-Hulman Inst. Tech., 1978. Exec. dir. research Bell Telephone Labs., Murray Hill, N.J., 1950-70; sci. adviser to Pres. Nixon; dir. Office Sci. and Tech., Washington, 1970-72; exec. v.p. Gould, Inc., 1973-77; ind. cons., 1977; v.p. Exxon Corp., N.Y.C., 1978-80; pres. Exxon Research and Engring. Co., Florham Park, N.J., 1977—; prof. elec. engring. Stevens Inst. Tech.; dir. Materials Research Corp., Orangeburg, N.Y.; cons. Nat. Security Council, 1974—; mem. def. sci. bd. Dept. of Def., 1974-75; chmn. Nat. Task Force on Tech. and Soc.; U.S. rep. to NATO Sci. Com.; mem. research adv. bd. Dept. Energy. Author: (with Dr. J.R. Pierce) Man's World of Sound, 1958, (with Dr. J.R. Pierce and W.A. van Bergeikj) Waves and the Ear, 1960, (with Dr. J.G. Truxal) The Man-Made World, 1969 (Lanchester prize Operations Research Soc. Am. 1971); Contbr. articles profl. jours. Mem. Bicentennial adv. com. Chgo. Mus. Sci. and Industry, 1974-75; mem. adv. bd. Office of Phys. Scis., NRC, 1976—; mem. Pres.'s Commn. on Nat. Medal of Sci., 1975-78; mem. vis. com. to div. phys. scis. U. Chgo., 1976—; mem. adv. council Humanities Inst., 1976—; trustee Aerospace Corp., 1974-81, chmn. bd. trustees, 1975-81; mem. corp. Mass. Inst. Tech., 1974—; bd. dirs. Summit (N.J.) Speech Sch., 1967-70; mem. Marshall Scholarships Adv. Council. Served with USNR, 1943-46. Recipient George W. McCarty award Ga. Inst. Tech., 1958, award Summit Jr. C. of C., 1959; Am. Soc. M.E. award merit, 1971; Harold Pender award Moore Sch. U. Pa., 1972; N.C. award, 1972. Fellow I.E.E.E., Acoustical Soc. Am., Am. Acad. Arts and Scis., AAAS (dir. 1974-75, 77-80, 80—, pres. 1977-78, chmn. bd. dirs. 1979-80), Audio Engring. Soc.; mem. Nat. Acad. Sci., Assn. Computing Machinery, Engring. Soc. Detroit, Nat. Acad. Engring. Subspecialty: Electrical engineering. Patentee in field. Office: Exxon Research and Engring Co 180 Park Ave PO Box 101 Florham Park NJ 07932

DAVID, GARY SAMUEL, immunologist; b. Aurora, Ill., Oct. 2, 1942; s. Samuel Matthew and Elizabeth Irene (Youngen) D.; m. Denise Nakamura, June 14, 1979. B.S., U. Ill., 1964, Ph.D., 1968. Postdoctoral fellow City of Hope Med. Ctr., Duarte, Calif., 1968-69; research assoc. Salk Inst. Biol. Studies, San Diego, 1970-71; research fellow Scripps Clinic and Research Found., LaJolla, Calif., 1971-77; dir. immunochemistry, prin. scientist Hybritech Inc., San Diego, 1978—. Contbr. articles to profl. jours. NIH grantee, 1974-78. Mem. Am. Assn. Immunologists, Tissue Culture Soc., Radioassy Soc. So. Calif., AAAS. Democrat. Subspecialties: Immunobiology and immunology; Hybridoma technology. Current work: Immunology, protein chemistry, cancer research, cell biology; director immunochemistry research, hybridoma technology and monclonal antibody manipulation. Office: 11085 Torreyana Rd San Diego CA 92121

DAVID, HERBERT ARON, statistics educator; b. Germany, Dec. 19, 1925. B.Sc., U. Sydney, 1947; Ph.D. in Stats, U. London, 1953. Sr. lectr. stats. U. Melbourne, Australia, 1955-57; prof. Va. Polytech. Inst., 1957-64, Sch. Pub.Health, U. N.C., Chapel Hill, 1964-72; dir., head statis. lab., dept. stats. Iowa State U., Ames, 1972—, disting. prof., 1980—. Editor: Jour. Biometrics, 1967-72. Fellow Am. Statis. Assn., Inst. Math. Stats.; mem. Internat. Statis. Inst., Biometric Soc. (pres. 1982-83). Subspecialty: Statistics. Office: Dept Stats Iowa State U Ames IA 50011

DAVIDA, GEORGE I., electrical engineering and computer science educator; b. Baghdad, Iraq, Aug. 2, 1944. B.S., U. Iowa, 1967, M.S., 1969, Ph.D. in Elec. Engring, 1970. Asst. prof. theoretical sci. U. Wis.-Madison, 1970-75; assoc. prof. U. Wis.-Milw., 1975-78, assoc. prof. computer sci., 1979-80; program dir. NSF, 1978-79; prof. Ga. Inst. Tech., Atlanta, 1980-81; prof. elec. engring. and computer sci. U. Wis.-Milw., 1981—; prin. investigator NSF grants, 1972—; mem. Pub. Cryptography Study Group, Am. Council Edn., 1979-81. Mem. Assn. Computing Machinery, IEEE (sr. mem.). Subspecialty: Cryptography and data security. Office: Dept Elec Engring & Computer Sci U Wis Milwaukee WI 53201

DAVIDSEN, ARTHUR FALNES, astrophysicist, educator, researcher; b. Freeport, N.Y., May 26, 1944; s. Andrew and Anna (Falnes) D.; m. Anita Clare Saltz, June 4, 1966; children: Andrew, Austin. A.B., Princeton U., 1966; M.A., U. Calif., Berkeley, 1972, Ph.D., 1975. Asst. prof. physics Johns Hopkins U., 1975-78, assoc. prof., 1978-80, prof., 1980—; dir. Assn. Univs. for Research in Astronomy, Inc., 1979—; mem. Mgmt. and Ops. Working Group in Space Astronomy, NASA, 1980—. Contbr. articles to profl. jours. Served to lt. (j.g.) USNR, 1968-71. Alfred P. Sloan Found. research fellow, 1976-80. Mem. Am. Astron. Soc. (councilor 1981-84, Helen B. Warner Prize 1979), Royal Astron. Soc., Internat. Astron. Union. Club: Explorers (N.Y.C.). Subspecialty: Ultraviolet high energy astrophysics. Home: 4338 N Charles St Baltimore MD 21218 Office: Dept of Physics Johns Hopkins University Baltimore MD 21218

DAVIDSON, ARNOLD B., psychopharmacologist; b. Phila., June 5, 1930; s. Yochanan and Beatrice (Abrams) D.; m. Carole Wolfson, June 15, 1952; children: Steven, Andrew. M.A., Bklyn. Coll., 1953; Ed.D., Temple U., 1964. Sr. investigator Smith Kline & French Labs., Phila., 1955-73; asst. dir. pharmacology dept. Hoffman-La Roche Inc., Nutley, N.J., 1973-81, asst. dir. pharmacology, neuropsychiatry dept., 1982—; instr. Peirce Jr. Coll., Phila., 1966-69. Contbr. sci. articles to profl. publs. Served with U.S. Army, 1953-55. Fellow Am. Psychol. Assn.; mem. Am. Soc. Pharmacology and Exptl. Therapeutics, Soc. Neurosci., Behavioral Pharmacology Soc., AAAS. Subspecialties: Psychopharmacology; Neuropharmacology. Current work: Research on the effects of drugs on the behavior and psychiatric disorders of man and animals. Patentee in field. Office: Hoffman-La Roche Inc Bldg 115 5th Floor Nutley NJ 07110

DAVIDSON, DONALD W., biology educator; b. Prairie Grove, Ark., June 8, 1936; s. Emmett and Eugenia Charlotte D.; m. Ruthanna Marie Davis, Oct. 26, 1969; children: Charlotte, Ruth Ellen. B.A., U. Minn., 1959; Ph.D., Rutgers U., 1963. Research asst. Wilderness Research Found., Ely, Minn., summers 1937, 38; botany asst. U. Minn., Duluth, 1956-59; grad. asst. Rutgers U., 1959-63; asst. prof. U. Ala., Tuscaloosa, 1963-65; asst. prof. biology U. Wis.-Superior, 1965-67, assoc. prof., 1967-79, prof., 1979—. NSF grantee, 1963-64, 69. Mem. Ecol. Soc. Am., Am. Inst. Biol. Scis., Nature Conservancy, Torrey Bot. Club, Sigma Xi. Methodist. Subspecialties: Evolutionary biology; Population biology. Current work: Forest ecology of Western Superior Basin. Office: Dept Biology U Wis Superior WI 54880

DAVIDSON, ERIC HARRIS, devel. and molecular biologist; b. N.Y.C., Apr. 13, 1937; s. Morris and Anne D. B.A., U. Pa., 1958; Ph.D., Rockefeller U., 1963. Research asso. Rockefeller U., 1963-65, asst. prof., 1965-71; asso. prof. devel. molecular biology Calif. Inst. Tech., Pasadena, 1971-74, prof., 1974—. Author: Gene Activity in Early Development, 2d edit, 1976. NIH grantee, 1965—; NSF grantee, 1972—. Subspecialties: Developmental biology; Molecular biology. Research, numerous publs. on DNA sequence orgn., gene expression during embryonic devel., gene regulation. Office: Div Biology Calif Inst Tech Pasadena CA 91125

DAVIDSON, EUGENE ABRAHAM, biochemistry educator, endl. adminstr.; b. N.Y.C., May 27, 1930; s. Jack and Sophie Miriam (Deutsch) D.; m. Alice A. Howell, Jan. 25, 1952; children: Mark, Robin, Steven, Ellen. B.S., UCLA, 1950; Ph.D., Columbia U., 1955. Research assoc. U. Mich., Ann Arbor, 1955-58; asst. prof. Duke U., Durham, N.C., 1958-62, assoc. prof., 1962-65, prof. biol. chemistry, 1965-67; prof. and chmn. biol. chemistry Hershey (Pa.) Med. Ctr., Pa. State U., 1967—; cons. USPHS, 1963-68, VA basic sci. rev. group, 1979—, Warner Lambert, 1981-83, Alcide, 1982—. Contbr. numerous articles to profl. jours. Mem. Am. Assn. Cancer Research, Am. Chem. Soc., Am. Soc. Biol. Chemists, Soc. Complex Carbohydrates (pres.-elect 1982), Assn. Med. Sch. Depts. Biochemistry (pub. affairs rep.). Subspecialties: Cancer research (medicine); Biochemistry (biology). Current work: Biochemistry. Home: 131 E High St Hummelstown PA 17036 Office: Milton S Hershey Med Center Pa State U PO Box 850 Hershey PA 17033

DAVIDSON, GILBERT, research and devel. exec.; b. Omaha, June 10, 1934; s. Mike and Reva (Plotkin) D.; m. Barbara Berger, June 8, 1958; children: Amy, Marc, Sharra. Ph.D., M.I.T., 1959. Research asso. Ecole Polytechnique, Paris, 1959-60; v.p. Am. Sci. and Engring., Cambridge, Mass., 1960-72, Infrared Industries, Waltham, Mass., 1972-77, Photometrics, Woburn, Mass., 1977—. Contbr. articles to profl. jours. Fulbright scholar, 1959-60. Mem. Am. Phys. Soc., Am. Vacuum Soc., Optical Soc. Am., Am. Geophys. Union. Subspecialties: Remote sensing (atmospheric science); Atomic and molecular physics. Current work: Remote sensing of atmosphere under a variety of meterol. conditions. Patentee in field. Home: 23 Exmoor Rd Newton Centre MA 02159 Office: 4 Arrow Dr Woburn MA 01801

DAVIDSON, JAMES BLAINE, ocean engineering educator, acoustical consultant; b. Oklahoma City, Nov. 10, 1923; s. Richard Blaine and Bessie Lowrance (Green) D.; m. Anna Ruth Cox, Dec. 18, 1948; children: Annette Jan Davidson Board, Jeannette Ann Davidson McCallum, James Blaine. Student, U. Okla., 1941-43; B.S., U.S. Naval Acad., 1946, U.S. Naval Postgrad. Sch., 1952; M.S., UCLA, 1953. Commd. ensign U.S. Navy, 1946, advanced through grades to comdr., 1962; dir. undersea programs (Office of Naval Research), Washington, 1962-67, ret., 1967; vis. prof. Norwegian Tech. Inst., Trondheim, 1981-82; prof. ocean engring. Fla. Atlantic U., 1967—; cons. Metallgesellschaft A.G., Frankfurt, Ger., 1975-80, DWG & Assocs., Orkanger, Norway and Hamburg, Ger., 1982—. Royal Norwegian Council Sci. and Tech. Research grantee, 1981. Mem. Am. Soc. Engring. Edn. (chmn. ocean engring. 1975-77, sec.-treas. engring. acoustics 1977—), Acoustical Soc. Am., Marine Tech. Soc. Republican. Presbyn. Subspecialties: Ocean engineering; Acoustical engineering. Current work: Computer applications in underwater acoustics, acoustical test tanks (sound absorbing linings); manganese nodule mining; underwater submarine transport. Home: 1190 SW 11th St Boca Raton FL 33432 Office: Fla Atlantic U Dept Ocean Engring Boca Raton FL 33431

DAVIDSON, RICHARD LAURENCE, geneticist, educator; b. Cleve., Feb. 22, 1941. B.A., Case Western Res. U., 1963, Ph.D., 1967. Asst. prof. Harvard Med. Sch., Boston, 1970-73, assoc. prof. microbiology and molecular genetics, 1973-81; research assoc. human genetics Children's Hosp. Med. Ctr., Boston, 1970-81; dir. ctr. genetics and Benjamin Goldberg prof. genetics U. Ill. Med. Ctr., Chgo., 1981—; U.S. Air Force Office Research-NRC fellow, 1967-68, Ctr. Molecular Genetics, Paris, 1967-70; co-dir. Cell Cult Ctr., MIT, Boston, 1975-81; mem. mammalian genetics study sect. NIH, 1975-81; mem. human cell biologY adv. panel NSF, 1973-75. Editor-in-chief: Somatic Cell Genetics. Mem. AAAS, Tissue Culture Assn., Cell Biology Assn. Subspecialty: Genetics and genetic engineering (biology). Office: Ctr for Genetics U Ill Med Ctr Chicago IL 60612

DAVIDSON, RONALD CROSBY, physicist, educator; b. Norwich, Ont., Can., July 3, 1941; s. William Crosby and Annie Beatrice (Caley) D.; m. Jean Farncombe, May 18, 1963; children—Cynthia Christine, Ronald Crosby. B.Sc., McMaster U., 1963; Ph.D., Princeton U., 1966. Mem. faculty dept. physics U. Md., 1968-78; vis. scientist Los Alamos Sci. Lab., 1974-75; asst. dir. for applied plasma physics Office of Fusion Energy Dept. Energy, Washington, 1976-78; prof. physics, dir. Plasma Fusion Center MIT, Cambridge, Mass., 1978—; cons. Sci. Applications, Inc. Author: Methods in Nonlinear Plasma Theory, 1972, Physics of Nonneutral Plasmas, 1974. Ford Found. fellow, 1963-64; Imperial Oil fellow, 1963-66; Sloan Research Found. fellow, 1970-72. Fellow Am. Phys. Soc. (vice chmn. div. applied plasma physics); mem. Fusion Power Assn. (dir.), Sigma Xi. Subspecialty: Plasma physics. Home: 179 Morse Rd Sudbury MA 01776 Office: 167 Albany St Cambridge MA 02139

DAVIDSON, SCOTT, computer scientist; b. Bayside, N.Y., Nov. 14, 1951; s. Lawrence I. and Selma (Sherman) D.; m. Martiscia Stouffer, May 20, 1978; 1 dau.: Helen Martiscia. B.S., MIT, 1973; M.S., U. Ill.-Urbana, 1977; Ph.D., U. Southwestern La., 1980. Mem. research staff Western Electric Engring. Research Center, Princeton, N.J., since 1980—, sr. mem., 1982—. Mem. IEEE, Assn. Computing Machinery (chmn. tech. com. on microprogramming 1983), IEEE Computer Soc. (com. microprogramming). Subspecialties: Computer architecture; Computer-aided design. Current work: Research in microprogramming, hardware testing. Home: 27 E Delaware Ave Pennington NJ 08534 Office: Western Electric Engring Research Ctr PO Box 900 Princeton NJ 08540

DAVIE, EARL W., biochemistry educator; b. Tocoma, Oct. 25, 1927. B.S., U. Wash., 1950, Ph.D. in Biochemistry, 1954. Asst. prof. to assoc. prof. biochemistry Case Western Res. U., Cleve., 1956-62; assoc. prof. biochemistry U. Wash., Seattle, 1962-66, prof., 1966—, dept. chmn., 1977—; Nat. Found. Infantile Paralysis fellow Mass. Gen. Hosp., Boston, 1954-56; NSF, Commonwealth Fund fellow Inst. Molecular Biology, Geneva, 1966-67; mem. Nat. Bd. Med. Examiners; mem. research rev. com. Nat. ARC Blood Program, 1973-77; mem. Med. and Sci. Adv. Council Nat. Hemophilia Found, 1974-79; mem. hematology study sect. NIH, 1975-79. Mem. Am. Acad. Sci. Am. Chem. Soc., Am. Soc. Biol.Chemistry (sec. 1975-78), Am. Soc. Hematology, AAAS. Subspecialty: Biochemistry (medicine). Office: Dept Biochemistry U Wash Sch Medicine Seattle WA 98195

DAVIES, ALBERT OWEN, endocrinologist, educator; b. Salt Lake City, Jan. 2, 1952; s. Clyde A. and Meda R. D.; m. Debra Ann Watts, Dec. 17, 1971; children: Matthew, Jason, Melissa, Stephanie. B.S., U. Utah, 1971, M.D., 1975. Diplomate: Am. Bd. Internal Medicine (subcert. in endocrinology). Intern and resident in internal medicine Duke U. Med. Ctr., Durham, N.C., 1975-77, fellow in endocrinology, 1977-80, asst. prof. medicine, 1981—, dir. Med. ICU, 1982—. Contbr. articles to med. jours. Nat. Research Service awardee Heart, Lung and Blood Inst., 1978-80; Clin. Investigation awardee, 1980-85. Mem. Am. Fedn. Clin. Research, Soc. Critical Care Medicine, Am. Heart Assn. (grantee-in-aid 1981-84), Shock Soc. Subspecialties: Critical care; Receptors. Current work: Metabolic derangements in critical illness. Home: 3 Barrington Pl Durham NC 27705 Office: Duke U Med Ctr Erwin Rd Durham NC 27710

DAVIES, DAVID KEITH, geologist; b. Barry, Eng., Oct. 10, 1940; came to U.S., 1966, naturalized, 1973; s. Buller T. and Muriel G. (Champ) D.; m. Ruth Margaret Mary Gilbertson, Dec. 12, 1964; children: Mark James, John Philip. B.S., U. Wales, 1962, Ph.D., 1966; M.S., La. State U., 1964. Asst. prof. Tex. A. and M. College Station, 1966-68, asso. prof., 1968-70, asst. dean, 1968-70; prof. U. Mo., Columbia, 1970-77; chmn. dept. geoscis., dir. Reservoir Studies Inst., Tex. Technol. U., Lubbock, 1977-80; pres. David K. Davies & Assos., Inc., Houston, 1980—. Contbr. articles to profl. jours. Mem. Planning and Zoning Commn. Columbia, Mo., 1979-80. Recipient A. I. Levorsen Meml. award Am. Assn. Petroleum Geologists, 1978. Fellow Geol. Soc. Am.; mem. Am. Assn. Petroleum Geologists, Soc. Econ. Paleontologists and Mineralogists, Soc. Petroleum Engrs. of AIME, Phi Kappa Phi. Subspecialties: Sedimentology; Mineralogy. Current work: Devel. of new techinques for well stimulation and electric log interpretation in order to optimize oil and gas prodn. Home: 2210 Long Valley Kingwood TX 77339 Office: 1410 Stonehollow Dr Kingwood TX 77339

DAVIES, GEOFFREY, Chemistry educator, cons., researcher; b. Stoke-on-Trent, Eng., Feb. 6, 1942; s. Frank and Alice Ada (Boulton) D.; m. Elizabeth Florence Gardiner, Jan. 8, 1965; children: Warwick Harvey, Russell Howard, Claire Elizabeth. B.S., U. Birmingham, Eng., 1963, Ph.D., 1966. Postdoctoral fellow Brandeis U., 1966-68; research assoc. Brookhaven Nat. Lab., 1968-69; I.C.I. fellow U. Kent, Eng., 1969-71; asst. prof. chemistry Northeastern U., Boston, 1971-77, assoc. prof., 1977-81, prof., 1981—. Contbr. numerous articles on chemistry to profl. jours. Ringing master Ch. of the advent Episcopal Ch., 1971—; mgr. Bell Restoration Project Old North Ch., Boston, 1981—. Recipient Excellence in Teaching award Northeastern U., 1981; Dreyfus Found. grantee, 1978. Mem. Am. Chem. Soc. Republican. Lodge: Masons. Subspecialties: Catalysis chemistry; Inorganic chemistry. Current work: Mechanisms of metal-catalyzed reactions of small, abundant molecules, especially dioxygen. Office: Chemistry Dept Northeastern U Boston MA 02115

DAVIES, IVOR KEVIN, psychologist, consultant; b. Birmingham, Warwichshire, Eng., Dec. 19, 1930; came to U.S., 1970; s. Howard and Selina (Stockton) D.; m. Shirley Diana Winyard, Feb. 24, 1966; children: Simon Winyard, Michelle Winyard. B.A., U. Birmingham, Eng., 1952, M.A., 1953, Dip.Edn., 1955; M.S., U. Ill., 1954; Ph.D., U. Nottingham, 1967. Sr. lectr., chmn. dept. behavioral sci. RAF U., Cranwell, 1968-72; prof. edn. Ind. U., Bloomington, 1972—. Author: Programmed Learning in Perspective, 1962, What is Programmed Learning, 1965, The Management of Learning, 1971,82, Competency Based Learning, 1972, Objectives in Curriculum Design, 1976, 81, Contributions to an Educational Technology, vol. 2, 1978, Instructional Technique, 1981. Served to lt. col. RAF, 1955-71. Fellow Brit. Psychol. Soc., Coll. Preceptors (Eng.); mem. Am. Ednl. Research Assn., Am. Soc. Tng. and Devel., Nat. Soc. Performance and Instrn. Club: RAF (London). Subspecialties: Cognition; Human factors engineering. Current work: Analysis of human performance, with special emphasis to the reduction of error in work situations. Home: 2447 Rock Creek Dr Bloomington IN 47401 Office: Ind U Bloomington IN 47405

DAVIES, MERTON EDWARD, planetary scientist; b. St. Paul, Sept. 13, 1917; s. Albert Daniel and Lucile Francis (McCabe) D.; m. Margaret Louise Darling, Feb. 10, 1946; children: Deidra Louise Davies Stauff, Albert Karl, Merton Randel. A.B., Stanford U., 1938. Group leader math. lofting Douglas Aircraft Co., El Segundo, Calif., 1940-48; mem. sr. staff The Rand Corp., Santa Monica, Calif., 1948; U.S. observer inspected stas. under terms of Antarctic Treaty. Author: (with Bruce Murray) The View from Space, 1971, (With others) Atlas of Mercury, 1978; contbr. articles to tech. jours. Recipient Antarctic Service medal U.S. Navy, 1967, George W. Goddard award Soc. Photo-Optical Instrumentation Engrs., 1966. Asso. fellow AIAA; mem. Am. Astron. Soc., Am. Soc. Photogrammetry (hon. mention Talbert Abrams award 1973), Am. Geophys. Union, AAAS. Subspecialties: Planetary science; Remote sensing (geoscience). Current work: Geodetic control on planets and satellites, photogrammetry, coordinate systems of planets and satellites. Patentee spinning panoramic camera. Home: 1414 San Remo Dr Pacific Palisades CA 90272 Office: 1700 Main St Santa Monica CA 90406

DAVIES, PETER JOHN, plant physiology educator; b. Sudbury, Middlesex, Eng., Mar. 7, 1940; came to U.S., 1966; s. William Bertram and Ivy Doreen (Parmentier) D.; m. Linda Kay DeNoyer, Aug. 2, 1976; children: Kenneth DeNoyer, Caryn Parmentier. B.Sc. with honors, U. Reading, Eng., 1962; M.S., U. Calif.-Davis, 1964; Ph.D., U.Reading, 1966. Instr. Yale U., New Haven, 1966-69; asst. prof. Cornell U., Ithaca, N.Y., 1969-75, assoc. prof., 1975-83, prof., 1983—. Author: (with others) The Life of the Green Plant, 1980, Control Mechanisms in Plant Development, 1970. Pres. bd. dirs. Ithaca Childcare Ctr., 1982-83. Mem. Am. Soc. Plant Physiology, Internat. Plant Growth Substance Assn. Subspecialties: Plant growth; Plant physiology (biology). Current work: Use of genetic lines in investigation of control of plant growth and development; physiology of plant senescence; analysis of plant hormones and their roles in plant growth and development. Office: Plant Biology Cornell Univ Plant Sci Bldg Ithaca NY 14853

DAVIES, PHILIP, pharmacologist, educator; b. Carmarthen, Wales, Dec. 7, 1943; came to U.S., 1975; s. Tudor and Martha (Ceridwen) D.; m. Meril Jones, Dec. 22, 1967; 2; children. B.Pharmacy, U. Wales, Cardiff, 1964; Ph.D., Welsh Nat. Sch. Medicine, Cardiff, 1967. Research assoc. NYU Sch. Medicine, 1968-70; mem. sci. staff Med. Research Council, London, 1970-75; with dept. immunology Merck, Sharp & Dohme Research Labs., Merck Inst. for Therapeutic Research, Rahway, N.J., 1975—, sr. dir. dept., 1982—; adj.prof. Rutgers U. Contbr. numerous articles to profl. publs. Mem. Am. Fedn. Clin. Research, Am. Assn. Immunologists, Biochem. Soc. Subspecialties: Immunobiology and immunology; Cellular pharmacology. Current work: Hose defense mechanisms and their disorders in inflammatory diseases. Office: Dept Immunology Merck Sharp & Dohme Research Labs PO Box 2000 Rahway NJ 07065

DAVIESS, STEVEN NORMAN, consulting geologist; b. Cedar Rapids, Iowa, Jan. 25, 1918; s. Harry Marston and Mary Alice (Davidson) C.; m. Frances Ober, May 14, 1944; children: Norman Frederick, Frank Arthur. B.A., UCLA, 1940, M.A., 1942. Cert. petroleum geologist, registered geologist, Calif. Geologist U.S. Geol. Survey, U.S., Cuba, Alaska, 1942-46, Gulf Oil Corp (various locations), 1946-60; exploration mgr. Gulf Oil Co., Spain, 1960-67, Gulf Mineral Resources Co., Denver, 1967-78; cons. geologist, Englewood, Colo., 1978—. Contbr. articles to profl. jours. Standard Oil Calif. fellow, 1941. Fellow Geol. Soc. Am., Geol. Soc. London., Geol. Soc. South Africa; mem. Am. Assn. Petroleum Geologists, Sigma Xi. Club: Valley Country. Subspecialties: Geology; Tectonics. Current work: Petroleum Geology. Address: 3961 S Dexter St Englewood CO 80110

DAVILA, ENRIQUE, medical educator, physician; b. Bogota, Colombia, Dec. 28, 1948; came to U.S., 1974; s. Enrique Patricio Davila and Elvira (Davila) D. Bachiller, Gimnasio Campestre, Bogota, 1965; M.D., Nat. U. Colombia, Bogota, 1973. Diplomate: Am. Bd. Internal Medicine, Am. Bd. Hematology and Oncology. Intern Nassau Hosp., Mineola, N.Y., 1974-75; resident in medicine U. Miami, Fla., 1975-78, chief med. resident, 1980-81, asst. prof. medicine, 1981—, asst. prof. oncology, 1982—; fellow in hematology/oncology U. Pa., Phila., 1978-80; cons. Pan Am. Health Orgn., 1982—. Mem. Am. Fedn. for Clin. Research, ACP, Sociedad Colombiana de Hematologia, Am. Soc. Clin. Oncology, Southeastern Cancer Study Group. Subspecialties: Chemotherapy; Hematology. Current work: Phase I trials of chemotherapy; continuous infusion chemotherapy; primary and secondary cancer of liver; biochemical modulators; esophageal cancer. Home: 575 Crandon Blvd Apt 512 Key Biscayne FL 33149 Office: Department of Oncology University of Miami 1475 NW 12th Ave Miami FL 33136

DAVILA, JORGE M., dentist; b. Catavi, Potosi, Bolivia, Mar. 3, 1930; came to U.S., 1967, naturalized, 1979; s. Mamerto and Balvina (Michel) D.; m. Martha Rodrigo, Oct. 4, 1958; children: Jorge Antonio, Juan Carlos. D.D.S., U. Cochabamba, Bolivia, 1953; Pedodontist, Eastman Dental Ctr., 1970, M.S., U. Rochester, 1972. Asst. prof. U. Cochabamba, 1966-67; assoc. researcher Eastman Dental Ctr., Rochester, N.Y., 1972-73; asst. prof. U. Md., Balt., 1973-75; sr. clin. and research assoc. Eastman Dental Ctr., 1975—; dir. dental dept. Bolivian Mining Corp., 1954-56; dir. dental dispensary C.N.S.S., Cochabamba, 1966-67; dir dental services Monroe Devel. Ctr., Rochester, 1975—. Contbr. articles to profl. jours.; co-author: A New Technique Restoration of Fractured Teeth, 1973. Mem. Internat. Assn. Dental Research, Am. Acad. Pedodontics, Acad. Dentistry for Handicapped, Am. Soc. Dentistry for Children. Democrat. Roman Catholic. Club: Midvale. Subspecialties: Preventive dentistry. Current work: Use of plastic resins in dentistry; dentistry for handicapped; mgmt. of difficult patient. Home: 120 Boniface Dr Rochester NY 14620 Office: 625 Elmwood Ave Rochester NY 14620

DAVIS, BERNARD DAVID, med. scientist; b. Franklin, Mass., Jan. 7, 1916; s. Harry and Tillie (Shain) D.; m. Elizabeth Menzel, June 19, 1955; children—Franklin A., Jonathan H., Katherine J. A.B., Harvard, 1936, M.D., 1940. Intern, fellow Johns Hopkins Hosp., 1940-41; commd. officer USPHS, 1942-54; successively assigned NIH, Columbia, Pub. Health Research Inst. of N.Y., Rockefeller Inst., and charge; USPHS Tb Research Lab. at Cornell U. Med. Sch., 1947-54; prof. pharmacology, chmn. dept. N.Y. U. Med. Sch., 1954-57; prof. bacteriology, chmn. dept. Harvard, 1957-68, Adele Lehman prof. bacteriology and immunology, 1963-68, Adele Lehman prof. bacterial physiology, dir. bacterial physiol. unit, 1968—; div. com. for biology, medicine NSF, 1954-57; mem. med. adv. bd. Hebrew U., 1956-70; fellow Center for Advanced Study in Behavioral Scis., 1973-74. Trustee Worcester Found. for Exptl. Biology. Recipient Waksman medal Soc. Am. Bacteriologists, 1952. Mem A.A.A.S., Am. Soc. Biol. Chemists, Nat. Acad. Sci., Am. Acad. Arts and Scis. (v.p. 1977-79), Inst. of Medicine, Am. Soc. Microbiology, Soc. Gen. Physiology (pres. 1964-65), Harvey Soc., Phi Beta Kappa, Sigma Xi, Alpha Omega Alpha. Subspecialty: Microbiology (medicine). Home: 23 Clairemont Rd Belmont MA 02178

DAVIS, BILL DAVID, biologist, educator; b. Junction City, Kans., July 22, 1937; s. Harry Oliver and Frances D.; m. Nelda Strnad; children: Paul David, Timothy Charles. B.S. in Edn, Emporia State Tchrs. Coll., 1959, M.S., 1960; Ph.D., Purdue U., 1965. Assoc. prof. biology Rutgers U., New Brunswick, N.J., 1965—; assoc. chmn. for personnel, chmn. dept. biol. scis. Douglass Coll. Contbr. numerous articles on plant physiology to profl. jours. Mem. AAAS, Bot. Soc. Am., Am. Soc. for Plant Physiologist, Soc. of Devel. Biology, Plant Tissue Culture Assn. Subspecialties: Plant growth; Cell and tissue culture. Current work: Plant developmental physiology, regulation of enzyme activities during seed germination and seedling development (peas, beans), plant tissue culture. Home: 31 Harrison Ave Milltown NJ 08850 Office: Dept Biol Scis Rutgers Univ New Brunswick NJ 08903

DAVIS, BRIAN KEITH, biologist; b. Sydney, New South Wales, Aus., May 15, 1937; came to U.S., 1962; S. and Ruby (Constance) D.; m. Nelida Rodrigo Villanueva, Aug. 3, 1963. B.Sc., U. New South Wales, 1958, Ph.D., 1962, D.Sc., 1982. Teaching fellow Sch. Wool and Pastoral Sci. U. New South Wales, Sydney, Aus., 1958-61; Ford Found. fellow Worcester Found. Exptl. Biology, Shrewsbury, Mass., 1962-63; research fellow Harvard U., Cambridge, Mass., 1963-65; sci. officer MRC Lab. Molecular Biology, Cambridge, U.K., 1965-66; research fellow McGill U., Montreal, Can., 1966-69; scientist Worcester Found. Exptl. Biology, Shrewsbury, Mass., 1970-78; research prof. SUNY-Stony Brook, 1978-82; dir. Research Found. So. Calif., San Diego, 1983—. Contbr. articles in field to profl jours. Served with Australian Armed Forces, 1955-57. Australian Atomic Energy Commn. grantee, 1961; AID grantee, 1974-75; Nat. Inst. Child Health & Devel. grantee, 1973—. Mem. Am. Physiol.Soc., Biophys. Soc., Soc. Exptl. Biology & Medicine, N.Y. Acad. Sci., Am. Soc. Cell Biology, Fedn. Am. Scientists. Subspecialties: Reproductive biology; Biophysics (biology). Current work: The biochemistry of fertilization, especially concerning the molecular mechanism of sperm capacitation and decapacitation; biophysics of template-directed polymerization, especially concerning kinetics complexity and far from equilibrium thermodynamics; polymer gel drug delivery systems. Home: PO Box 2135 Del Mar CA 92014 Office: Research Found So. Calif PO Box 2504 La Jolla CA 92038

DAVIS, BRIAN KENT, geneticist; b. Laramie, Wyo., Dec. 2, 1939; s. Preston John Colver and Myrtle Ann (Barrett) D.; m. Janet Anne Pettingill, June 2, 1962; children: Christopher M., Catherine M., Casandra N. B.A., U. Wis., 1962, M.A. (NSF grad. fellow), 1963; Ph.D., U. Wash., 1970. NIH tng. grantee, 1966-70; research fellow U. Calif., San Diego, La Jolla, 1970-72; asst. prof. Va. Poly. Inst. and State U., Blacksburg, 1973-80, Coll. Medicine, King Faisal U., Saudi Arabia, 1980-81; research assoc. Allied Corp., Morristown, N.J., 1981—. Contbr.: articles to sci. jours., chpt. to Dermatoglyphics-50 Years Later, 1980. Am. Cancer Soc. grantee, 1970-72. Mem. Environ. Mutagen Soc., Genetics Soc. Am., Am. Genetic Assn., Internat. Dermatoglyphics Assn. Subspecialties: Toxicology (agriculture); Genetics and genetic engineering (biology). Current work: Genetic toxicology, recombinant DNA. Office: Allied Corp PO Box 1021R Morristown NJ 07960

DAVIS, CHARLES HARGIS, information scientist, university dean; b. Tell City, Ind., Sept. 23, 1938; s. Charles Alban and Ruth Elizabeth (Hargis) D. B.S. (State Merit scholar), Ind. U., 1960, A.M., 1966, Ph.D., 1969; postgrad. (German Govt. Fellow), U. Munich, W. Ger., 1960-61. Asst. editor Chem. Abstracts Service, Columbus, Ohio, 1962-65; chem. info. specialist Ind. U. Aerospace Research Applications Center, 1965-66; dir. systems Ind. U. ERIC Clearinghouse on Reading, 1967-69; asst. prof. library sci. Drexel U., 1969-71; asso. prof. U. Mich., 1971-76; prof., dean Faculty of Library Sci., U. Alta. (Can.), Edmonton, 1976-79, Grad. Sch. Library and Info. Sci., U. Ill., Urbana-Champaign, 1979—; speaker and condr. workshops and seminars in field; cons. in field; cons. editor Greenwood Press, Westport, Conn., 1974-79; pres. Can. Council Library Schs., 1978-79. Author: Illustrative Computer Programming for Libraries: Selected Examples for Information Specialists, 1974, 2d edit., 1981, (with James E. Rush) Information Retrieval and Documentation in Chemistry, 1974, Guide to Information Science, 1979; contbr. numerous articles, revs., bibliographies and columns to profl. publs. NSF research grantee, 1959-60. Mem. AAAS, Am. Chem. Soc., ALA (chmn. Library Research Round Table 1978-79), Am. Soc. Info. Sci. (chmn. Ind. U. student chpt. 1967-68, chmn. Ind. chpt. 1968-69, treas. Delaware Valley chpt. 1971, chmn. Mich. chpt. 1974-75, chmn. Western Can. chpt. 1978-79), Assn. Computing Machinery, Assn. Am. Library Schs. (chmn. research com. 1976-78), Phi Lambda Upsilon, Beta Phi Mu. Subspecialty: Library and information science education administration. Office: Grad Sch Library and Info Sci 410 DKH U Ill at Urbana-Champaign 1407 Gregory Dr Urbana IL 61801

DAVIS, DAVID, plant pathologist; b. Warsaw, Poland, Dec. 29, 1920; s. Samuel and Sarah D.; m. Alice Lifshitz, Apr. 30, 1946; children: Raime, Sharon. B.S., Cornell U., 1945; Ph.D., U. Ill., 1950. Plant pathologist Conn. Exptl. Agrl. Sta., New Haven, 1950-55, Merck & Co., Rahway, N.J., 1956-60, N.Y. Bot. Garden, Bronx, 1960-70; dir. research Phyta Labs., Ghent, N.Y., 1970—. Bd. dirs. Spencertown Acad. Assocs. Served to 1st lt. USAF, 1943-45. Decorated Air medal, Purple Heart; NIH grantee, 1960-69. Mem. Am. Inst. Biol. Scis., Am. Phytopath. Soc., Soc. Nematologists. Subspecialty: Plant pathology. Current work: Plant disease control. Address: Tipple Rd Ghent NY 12075

DAVIS, DAVID GERHARDT, research physiologist; b. Dickinson, N.D., July 21, 1935; s. Zachary Floyd and Evelyn Catherine (Overbeck) D.; m. Phyllis Ann Miller, Feb. 18, 1972; children: Blair David, Anthony Eric. B.Sc., N.D. State U., 1960, M.Sc., 1962; Ph.D., Wash. State U., 1965. Grad. asst. N.D. State U., Fargo, 1960-62, Wash. State U., Pullman, 1962-65; AEC fellow U. Minn., Mples., 1965-67; plant physiologist U.S. Dept. Agr., Fargo, 1967—; research fellow Weizmann Inst. Sci. Contbr. articles to profl. jours. Leader Boy Scouts Am., Fargo, 1969-71. Served with USAF, 1956-59. AEC fellow, 1965-67; summer fellow, 1963. Mem. Am. Soc. Plant Physiologists, Weed Sci. Soc. Am., Scandinavian Soc. Plant Physiologists, Internat. Assn. Plant Tissue Culture, N.D. Acad. Sci., Sigma Xi. Subspecialties: Plant physiology (biology); Plant cell and tissue culture. Current work: Plant growth and development, control of plant processes, weed control, plant tissue culture of crop and weed species. Office: USDA SE/AR Metabolism and Radiation Research Lab State Univ Station Fargo ND 58105 Home: RR 2 Box 62 Fargo ND 58102

DAVIS, DONALD CRAWFORD, physician, pharmacologist; b. Atlanta, Oct. 28, 1940; s. A. Crawford and Margaret May (Langford) D.; m. Jolene Remick, Oct. 27, 1942; children: Lisa Gail, Keith Crawford. B.S., U. Ga., 1962, M.S., 1964; Ph.D., Vanderbilt U., 1969; M.D., Emory U., 1976. Diplomate: Am. Bd. Internal Medicine. Postdoctoral fellow NIH, 1969-72; asst. prof. medicine Emory U., 1979—, ptnr., 1981—; attending physician Emory U. Hosp., Grady Meml. Hosp. Contbr. articles to profl. jours. Mem. Am. Soc. Pharmacology; mem. ACP; Mem. Exptl. Therapeutics, Med. Assn. Ga., DeKalb Med. Soc., Sigma Xi. Subspecialties: Internal medicine; Pharmacology. Current work: Faculty responsibility, medicine and pharmacology. Office: 1365 Clifton Rd NE Atlanta GA 30322

DAVIS, DONALD DEAN, psychology educator, researcher, consultant; b. Mt. Clemens, Mich., Oct. 31, 1950; s. William J. and Dena (Saurini) D. Student, Internat. Inst. Voor Sociale Geshledenis, Amsterdam, Netherlands, 1971-72; B.S., Central Mich. U., 1973, M.A., 1977; Ph.D., Mich. State U., 1982. Instr. psychology and sociology Mid-Mich. Community Coll., Harrison, 1973-76; econ. and community devel. cons. John Ruggles & Assocs., Mt. Pleasant, Mich., 1976-77; asst. dir. Center for Evaluation and Assessment, Mich. State U., East Lansing, 1979-81, instr. psychology, 1981-82; asst. prof. psychology Old Dominion U., Norfolk, Va., 1982—; research cons. various orgns., 1976—. Contbr. writings to profl. publs. Chmn. bd. dirs. Karate Insts. Mich., Detroit, 1974-76; advisor bd. dirs. Karate Insts. Am., Cleve., 1974-76. NIMH trainee, 1977; research grantee Mich. Office Services to Aging, 1979-82; named instr. of year Karate Insts. Am., 1976. Mem. Am. Psychol. Assn., Am. Sociol. Assn., AAAS, Acad. Mgmt., Gerontol. Soc. Am. Subspecialty: Organizational and community psychology. Current work: Innovation and change in organizations; science and social policy. Office: Psychology Dept Old Dominion Univ Hampton Blvd Norfolk VA 23508

DAVIS, GEORGE HAMILTON, cons. geologist; b. Detroit, July 25, 1921; s. George Edward and Edith Blair (Wright) D.; m. Elizabeth Hessler, Nov. 2, 1961; 1 son, George Robert; m. Mildred Yates, July 29, 1949 (dec. May 1961). B.S., U. Ill., Urbana, 1942; postgrad., UCLA, 1946-47. Registered geologist, Calif.; lic. engrng. geologist, Calif. Geologist U.S. Geol. Survey, Sacramento and Washington, 1948-66; 1st officer IAEA, Vienna, Austria, 1966-68; research hydrologist U.S. Geol. Survey, Reston, Va., 1968-79, asst. dir., 1979-81; cons. geologist, Silver Spring, Md., 1981—; mem. Commn. on Shore Mineral Devel., NRC, 1982—, 1982—, UNESCO Working Group on Water and Energy, Paris, 1976-80. Editor: Jour. Hydrology, 1977—, Water Resources Research, 1970-76. Mem. com. Boy Scouts Am., 1975-79. Served as sgt. AUS, 1942-46; ETO. Recipient Meritorious Service award Dept. Interior, 1974, Disting. Service award, 1979. Fellow Geol. Soc. Am. (vice chmn. hydrogeology div. 1980-82, Meinzer award 1972), Internat. Water Resources Assn. (treas. 1976-82); mem. Am. Geophys. Union (editor 1970-76), Am. Inst. Profl. Geologists (pres. sect. 1983), Assn. Engrng. Geologists. Republican. Presbyterian. Subspecialties: Ground water hydrology; Hydrogeology. Current work: Hydrogeology, isotope hydrology, land subsidence, water for energy development. Address: 10408 Insley St Silver Spring MD 20902

DAVIS, GEORGE KELSO, nutrition educator; b. Pitts., July 2, 1910; s. Ross Irwin and Jennie Lovinia (Kelso) D.; m. Ruthanna Wood, Jan. 25, 1936; children: Dorothy Jeanne, Mary Ellen, Ruthanna Marie, Virginia Kay, Robert Wyatt, George William Ross. B.S., Pa. State U., 1932; Ph.D., Cornell U., 1937. Grad. asst. Cornell U., Ithaca, N.Y., 1933-37; asst. prof. Mich. State U., East Lansing, 1937-42; prof. U. Fla., Gainesville, 1942-79, prof. emeritus, 1979—, dir. nuclear scis., 1960-65, dir. biol. scis., 1965-70, dir. research, 1970-75. Editor: Proc. XII Internat. Nutrition Congress, 1982. Recipient Borden award Am. Inst. Nutrition, 1964; Spencer award Am. Chem. Soc., 1980; Scientist of Yr. award Mus. Sci. and Industry, Tampa, 1981; Disting. Alumnus award Pa. State U., 1982. Fellow Am. Inst. Nutrition (pres. 1975-76); mem. Am. Chem. Soc. (chmn. Fla. sect. 1954), Soc. Environ. Geochem. and Health (pres. 1976), Internat. Congress Nutrition (pres. 1981), Nat. Acad. Scis. Democrat. Presbyterian. Clubs: Athenaeum, Gainesville (sec.-treas. 1982-83). Subspecialties: Animal nutrition; Biochemistry (biology). Current work: Factors influencing the metabolism of copper, zinc, magnesium, calcium and phosphorus; special emphasis on enzyme systems. Office: U Fla Gainesville FL 32611

DAVIS, GRAYUM LLOYD, mfg. design engr.; b. Marshall, Tex., Jan. 13, 1943; s. Grayum Debs and Jewel (Brown) D.; m. Dorothy Jean Carter, Oct. 28, 1966; children: Grayum Lee, Randal Lloyd. B.S.M.E., Tex. A&M U., 1966, M.S.M.E., 1970; M.B.A. in Fin, U. Houston, 1974. Registered profl. engr., Tex. Design engr. Cameron Iron Works, Inc., Houston, 1966-67, sr. engr., 1970-75, chief engr., Livingston, Scotland, 1975-77, Houston, 1977-80, chief devel. engr. of subsea pipleline equipment, 1980—. Active Cub Scouts Am., 1978-81. Served to 1st lt. C.E. U.S. Army, 1967-69. Decorated Bronze Star. Mem. ASME, Tex. Soc. Profl. Engrs. Republican. Methodist. Club: Optimists. Subspecialties: Mechanical engineering; Materials (engineering). Current work: Development of state of the art quick pipeline connection systems for subsea pipelines and drilling and production platform risers. Home: 9218 Mauna Loa Ln Houston TX 77040 Office: PO Box 1112 Houston TX 77251

DAVIS, HASKER PAT, psychologist, educator, researcher; b. West Plains, Mo., June 22, 1946; s. Hasker H. and B. Agnes (Pattello) D.; m.; 1 dau.: Christy Marie. A.B., U. Calif.-San Diego, 1975; M.A., U. Calif.-Berkeley, 1978, Ph.D., 1980. Postdoctoral fellow in cerebral metabolism Cornell U. Med. Coll., N.Y.C., 1980-81; asst. prof. psychology St. John's U., Jamaica, N.Y., 1981—. Contbr. articles and revs. on physiol. bases of memory to sci. jours. Served with USN, 1966-69. Nat. Inst. on Aging postdoctoral fellow, 1981. Mem. AAAS, Am. Psychol. Assn., Western Psychol. Assn., Eastern Psychol. Assn., Soc. for Neurosci., Sigma Xi. Subspecialties: Psychobiology; Neuropharmacology. Current work: Psychopharmacology of learning and memory dysfunction in aging. Home: 470 Broome St New York NY 10013 Office: Dept Psychology St John's U Jamaica NY 11439

DAVIS, HOWARD TED, chemical engineering educator; b. Hendersonville, N.C., Aug. 2, 1937; s. William Howard and Gladys Isabell (Rhodes) D.; m. Eugenia Asimakopoulos, Sept. 18, 1960; children: William Howard III, Maria Katherine. B.S., Furman U., 1959; Ph.D., U. Chgo., 1962. NSF postdoctoral fellow Free U. Brussels, 1962-63; asst. prof. U. Minn., Mpls., 1963-65, asso. prof., 1965-68, prof. chem. engring. and chemistry, 1968—, head dept. chem. engring. and materials sci., 1980—, chmn. exec. com. Microelectronic and Info. Scis. Ctr., 1982-84. Editor books.; Contbr. articles to tech. jours. Indsl. cons. NSF fellow, 1959-62; NSF postdoctoral fellow, 1962-63; Sloan Found. fellow, 1968-70; Guggenheim fellow, 1969-70. Mem. Am. Inst. Chem. Engring., Am. Chem. Soc., Am. Phys. Soc., Soc. Petroleum Engrs., Sigma Xi. Subspecialties: Chemical engineering; Physical chemistry. Current work: Teaching, research and administration. Home: 1822 Mt Curve Ave Minneapolis MN 55403

DAVIS, JAMES IVEY, physicist; b. Repton, Ala., Apr. 9, 1937; s. James Ivey and D.; m. (div.); 1 dau., Melinda. B.S., Calif. Inst. Tech., 1962; M.S. in Physics, UCLA, 1965, Ph.D., 1969. Dept. mgr. Hughes Aircraft Co., Culver City, Calif., 1970-74; program dir. Lawrence Livermore Nat. Lab., Livermore, Calif., 1974—, dir. laser isotope separation program, 1975—; instr. advanced optics U. Ghana, 1966-67. Contbr. articles to profl. jours. Served with U.S. Army, 1953-56. Hughes Aircraft Co. master fellow, 1963-65; NSF fellow, 1969-70. Mem. Am. Phys. Soc., AAAS. Subspecialties: Plasma physics; Laser-induced chemistry. Current work: Director of high technology programs; research and development for high technology applications: laser isotope separation, laser fusion; laser photochemistry, laser radar, high energy laser weapons, low energy laser systems for countermeasures and intelligence gathering. Office: PO Box 5508 L-466 Livermore CA 94550

DAVIS, JAMES NORMAN, neurologist, pharmacology researcher; b. Dallas, Oct. 24, 1939; s. Moses and Ruth (Grossman) D.; m. Frances Isabel Cantor, May 1, 1965; children: Amanda, Adam, Joanna. B.A., Cornell U., 1961, M.D., 1965. Diplomate: Am. Bd. Neurology and Psychiatry. Intern Bellevue Hosp., N.Y.C., 1965-66; research assoc. Lab. Chem. Pharmacology, Nat. Heart Inst., NIH, Bethesda, 1966-68; resident Duke U., 1968-69, asst. prof., 1972-77, assoc. prof. medicine and pharmacology, 1977-80, prof., 1980—; resident in neurology Cornell U.-N.Y. Hosp., 1969-72, North Shore Hosp., 1971; instr. neurology Cornell U., 1969-71. Contbr. articles to profl. jours. Served with USPHS, 1966-68. Mem. Am. Neurol. Assn., Soc. Clin. Investigation, Am. Soc. Pharmacology and Exptl. Therapeutics, Am.

Acad. Neurology, Soc. Neurosci. Democrat. Jewish. Subspecialties: Neurology; Regeneration. Current work: Catecholamine neuronal plasticity; catecholmaine neuronal responses to brain injury and adrenergic cholinergic interactions. Home: 6 Harvey Pl Durham NC 27705 Office: Duke University PO Box 3850 Durham NC 27710

DAVIS, JAMES (OTHELLO), physician, educator; b. Tahlequah, Okla., July 12, 1916; s. Zemry and Villa (Hunter) D.; m. Florrilla Louise Sides, Dec. 27, 1941; children: Janet Ruth, James Lawrence. M.A. in Zoology, U. Mo., 1939, Ph.D., 1942, B.S. in Medicine, 1943; M.D., Washington U., 1945. Intern Barnes Hosp., St. Louis, 1945-46; investigator Lab. Kidney and Electrolyte Metabolism, Nat. Heart Inst., Bethesda, Md., 1949-57, chief sect. on exptl. cardiovascular disease, 1957-66; asso. prof. physiology Temple U. Sch. Medicine, Phila., 1955-56; vis. asso. prof. physiology Johns Hopkins Sch. Medicine, 1961-64; vis. prof. physiology U. Va. Sch. Medicine, 1964; prof., chmn. dept. physiology U. Mo. Sch. Medicine, Columbia, 1966—. Mem. editorial bd.: Am. Jour. Physiology, 1961-63, 66-69, Endocrinology, 1962-65, Circulation Research, 1962-66, 71-76, 78-81, Hypertension, 1979-80. Served with AUS, 1943-45; Served with USPHS, 1946-66. Recipient AMA Golden Apple award for teaching U. Mo., 1968; Sigma Xi Research award U. Mo., 1971; Modern Medicine Distinguished Achievement award, 1973; Alumni gold medal U. Mo., 1973; Volhard award, 1974; CIBA award for hypertension research, 1975; Carl T. Wiggers award, 1976; citation of merit U. Mo. Sch. Medicine, 1981. Mem. Am. Heart Assn. (mem. med. adv. council, vice chmn. council for high blood pressure research 1970-72, chmn. council 1972-74), Am. Physiology Soc. (council 1974-78, steering com. circulation group 1978-81, pres. circulation sect. 1981), Endocrine Soc., Soc. Exptl. Biology and Medicine, Nat. Inst. Health Extramural Program, Assn. Physiology Dept. Chairmen (council 1971-74), Inter-Am. Soc. Hypertension (council 1978-80), Internat. Soc. Hypertension (pres. 1980-82), Sigma Xi, Alpha Omega Alpha. Subspecialties: Physiology (biology); Nephrology. Current work: Roles of kidney and adrenal cortex in pathogenesis of hypertension and heart failure. Home: 612 Maplewood Dr Columbia MO 65201

DAVIS, JOEL STEPHEN, scientific analyst; b. Pitts., June 12, 1948; s. Emanuel and Dorothy (Winans) D. B.S. in Physics, MIT, 1970; M.S. in Astro-Geophysics, U. Colo., 1976. Research asst. Lab. Atmospheric and Space Physics, Boulder, Colo., 1974-76; staff scientist Sci. Applications, Inc. (Crystal City office), Arlington, Va., 1976-79, tech. mgr., Dayton, Ohio, 1979-81, sr. analyst and prin. investigator, 1981—. Contbr. articles to profl. jours. Vol. Children's Med. Ctr., Dayton, 1980—. Recipient 2d prize Tactical Weapons Def. Competition, 1967. Mem. Am. Astron. Soc., Nat. Def. Preparedness Assn., Am. Meteorol. Soc., Air Force Assn., Naval Inst., Strategic Inst. Jewish. Club: Dayton Strategic Game and Gourmet Soc. Subspecialties: Climatology; Operations research (engineering). Current work: Analysis of impact of weather on military systems; computer models of laser sysrems, electro-optical sensors, rapid fire cannons. Office: 1010 Woodman Ave #200 Dayton OH 45432 Home: 624A Dodge Ct Dayton OH 45431

DAVIS, JOHN MIHRAN, surgeon, educator; b. N.Y.C., Aug. 13, 1946; s. Drought Delaney and Ruth Radcliff (Kalaidjian) D.; m. Marlene Morgan, Oct. 13, 1973; 1 son, Nicholas Mihran. B.A., Columbia Coll., N.Y.C., 1968; M.D., Wayne State U., Detroit, 1972. Diplomate: Am. Bd. Surgery. Resident in surgery N.Y. Hosp., 1972-77; fellow NIH, Bethesda, Md., 1977-79; asst. prof. surgery Cornell U. Med. Coll., 1979—; prin. investigator NIH, 1981. Mem. soccer adv. com. Columbia Coll., 1977—; mem. alumni council Collegiate Sch., N.Y.C., 1981—. Mem. Surg. Infection Soc. (charter), Am. Burn. Assn., Am. Fedn. Clin. Research, N.Y. Cancer Soc., N.Y. State Med. Soc. Subspecialties: Surgery; Immunology (medicine). Current work: Neutrophil function in burn patients and patients with abdominal sepsis. Office: Cornell Univ Med Coll 1300 York Ave New York NY 10021

DAVIS, JOHN MOULTON, physicist; b. Nottingham, Eng., Aug. 28, 1938; s. John Henry and Gladys Winifred (Moulton) D.; m. Margery Grady, July 10, 1976. B.Sc. with 1st class honors, U. Leeds, 1960, Ph.D., 1964. Research assoc. M.I.T., 1964-70; sr. scientist Am. Sci. & Engring. Inc., Cambridge, Mass., 1970-72, staff scientist, 1972-74, sr. staff scientist, 1974—. Contbr. articles to profl. jours. Mem. Republican Town Com., Lexington, Mass., 1972-76. Recipient Skylab Achievement award NASA, 1974, Sounding Rocket award, 1977. Mem. Am. Geophys. Union, Am. Astron. Soc. Subspecialties: Solar physics; 1-ray high energy astrophysics. Current work: Study of solar corona through analysis of X-ray observations. Home: 386 Winchester St Newton MA 02161 Office: Am Sci & Engring Inc Fort Washington Cambridge MA 02139

DAVIS, JOHN ROWLAND, research adminstrator; b. Mpls., Dec. 19, 1927; s. Roland Owen and Dorothy (Norman) D.; m. Lois Marie Falk, Sept. 4, 1947; children—Joel C., Jacque L., Michele M., Robin E. B.S., U. Minn., 1949, M.S., 1951; postgrad., Purdue U., 1955-57; Ph.D., Mich. State U., 1959. Registered profl. engr., Calif., Oreg. Hydraulic engr. U.S. Geol. Survey, Lincoln, Nebr., 1950-51; instr. Mich. State U., 1951-55; asst. prof. Purdue U., 1955-57; lectr. U. Calif. at Davis, 1957-62; hydraulic engr. Stanford Research Inst., South Pasadena, Calif., 1962-64; prof. U. Nebr., Lincoln, 1964-65, dean, 1965-71; prof., head dept. agrl. engring. Oreg. State U., Corvallis, 1971-75, dir. Agrl. Expt. Sta., asso. dean Sch. Agr., 1975—, instl. athletic rep., 1972—; mem. governing bd. Water Resources Research Inst., 1975—; dir. Western Rural Devel. Center, 1975—, Agrl. Research Found., Jackman Inst.; cons. Stanford Research Inst., Dept. Agr., Consortium for Internat. Devel.; dir. Engrs. Council Profl. Devel., 1966-72; pres. Pacific-10 Conf., 1978-79. Contbr. articles to profl. jours. Served with USNR, 1945-46. Fellow Am. Soc. Agrl. Engrs. (dir. 1971-73, agrl. engr. of year award Pacific Northwest region 1974); mem. AAAS, Nat. Coll. Athletic Assn. (v.p. 1979-83, sec.-treas. 1983—). Subspecialties: Agricultural engineering; Resource conservation. Current work: Irrigation engineering; soil erosion conrtol. Home: 2940 NW Aspen St Corvallis OR 97330

DAVIS, LAWRENCE CLARK, biochemist, educator, researcher; b. London, Aug. 16, 1945; U.S., 1946; s. George Hawkins and Olive Edwina (Clark) D.; m. Linda Ann Wiles, July 22, 1967; children: Colin, Jennie Lynn, Steven. B.S., Haverford Coll., 1966; Ph.D., Yeshiva U., 1970. Postdoctoral fellow U. Wis., 1970-71, 73-75, energy-related postdoctoral fellow, 1975; research assoc. Norwich (Conn.) State Hosp., 1971-73; asst. prof. biochemistry Kans. State U., Manahattan, 1975-81, assoc. prof., 1981—. Contbr. articles to profl. jours. Active Manhattan (Kans.) Area Energy Alliance, Manhattan Area Resettlement Com.; mem. Kans. Bach Choir, Manhattan. NIH grantee, 1975-84; NSF grantee, 1975-81; Research Corp. grantee, 1975-77. Mem. AAAS, Am. Soc. Microbiology, Biophysics Soc., Am. Soc. Biol. Chemists, Am. Soc. Plant Physiology. Democrat. Lutheran. Subspecialties: Nitrogen fixation; Genetics and genetic engineering (agriculture). Current work: Mechanism of enzyme nitrogenase, associating protein systems, improving nitrogenase function in association with legumes. Home: 3419 Womack Way Manhattan KS 66502 Office: Willard Hall Kans State U Manhattan KS 66502

DAVIS, LEONARD GEORGE, research neurochemist; b. Chgo., Nov. 23, 1946; s. Willard George and Mildred Ann (Poole) D.; m. Penny Suzanne Barber, Aug. 30, 1969; 1 dau., Robin. B.S., U. Ill., 1969; M.S. (scholar), Northwestern U., 1974; Ph.D., U. Ill. Med. Sch., Chgo., 1977. Mem. research staff med. research div. William Beaumont Army Hosp., El Paso, Tex., 1975-76; mem. research staff neurochemistry unit Mo. Inst. Psychiatry, St. Louis, 1976-80; research scientist central research dept. E.I. du Pont de Nemours & Co., Wilmington, Del., 1980—. Contbr. numerous articles to sci. jours. Bd. dirs. New Castle County Mental Health Assn. NIMH scholar, 1975-76; Nat. Inst. Drug and Alcohol grantee, 1980; NIHM grantee, 1980; Nat. Inst. Aging grantee, 1979-81. Mem. Soc. for Neurosci. (pres. chpt. 1983), Am. Soc. Neurochemistry, N.Y. Acad. Scis. Subspecialties: Neurochemistry; Neurobiology. Current work: Understanding the role of neuropeptides in brain function and their molecular mechanisms of action. Office: EI du Pont de Nemours & C Exptl Sta Wilmington DE 19898

DAVIS, LEROY WELLINGTON, metall. and chem. engr., engring. tech. co. exec.; b. Cleve., May 29, 1901; s. George Embury and Jessie Eunice (Wellington) D.; m. Ruth Miller Durbs, Oct. 4, 1950 (dec. 1970); children: Gilbert George (dec.) Linda Elaine. B.Sc. in Chem. Engring, Case Inst. Tech., Cleve., 1922, Metall. Engr., 1931. Metallurgist Aluminum Co. of Am., Cleve., 1926-42, asst. div. chief. metallurgist, Pitts., 1942-52; tech. supt. Kaiser Aluminum, Halethrope, Md., 1952-55, Erie, Pa., 1952-55; div. mgr. Harvey Aluminum, Torrance, Calif., 1955-57, asst. dir. research and devel. div., 1958-70; pres. Nev. Engring. and Tech. Corp., Long Beach, Calif., 1971-80, chmn. bd., 1980—. Author: (with S.W. Bardstreet) Metal and Ceramic Matrix Composites, 1971; contbr. numerous articles to profl. publs. Active Heritage Found.; mem. Republican Congressional Com., 1978-82. Recipient Commendation Am. Ceramic Soc., 1982. Mem. ASTM, Am. Soc. Metals, Am. Soc. Nondestructive Testing, Nat. Assn. Corrosion Engrs., Am. Def. Preparedness Assn., Am. Ceramic Soc., AIME. Baptist. Club: Univ. (Erie). Subspecialties: Composite materials; Alloys. Current work: Metal matrix composites information center; composites technology development and evaluation, and non-destructive testing. Patentee in field. Home: 21462 Pacific Coast Hwy 58 Huntington Beach CA 92646 Office: 2225 E 28th St Bldg 511 Long Beach CA 90806

DAVIS, LLOYD EDWARD, veterinarian, educator, researcher; b. Akron, Ohio, Aug. 23, 1929; s. Roger Q. and Myrtle Elva (Burke) D.; m. Thelma L. Brunty, June 4, 1953; m. Carol A. Neff, Sept. 25, 1972; children: Mark E., Kimberly A. D.V.M., Ohio State U., 1959; Ph.D. in Pharmacology, U. Mo., 1963. Lic. veterinarian, Ill., Mo. Instr. then asst. prof., assoc. prof. U. Mo., Columbia, 1959-69; prof. Ohio State U., Columbus, 1969-72; vis. prof. U. Nairobi, Kenya, 1972-74; prof. Colo. State U., Ft. Collins, 1974-78; prof. clin. pharmacology U. Ill., Urbana, 1978—; mem. U.S. Pharmacopeia Gen. Com. Revision; sci. adv. Mercenene Med. Corp. Contbr. numerous articles in field of clin. pharmacology to profl. jours. Served with USN, 1950-53. Recipient Disting. Teaching award U. Mo., 1968. Mem. Am. Soc. Pharmacology and Exptl. Therapeutics, Am. Acad. Vet. Pharmacology and Therapeutics (pres. 1976-78), Am. Soc. Clin. Pharmacology and Therapeutics, Soc. Exptl. Biol. Medicine, Sigma Xi, Phi Zeta (pres. 1982). Subspecialties: Pharmacology; Internal medicine (veterinary medicine). Current work: Clinical evaluation of drugs, drug therapy, influence of disease on disposition and fate of drugs in animal patients. Home: 621 W Hill St Champaign IL 61820 Office: 1102 W Hazelwood Dr Urbana IL 61801

DAVIS, MARK HEZEKIAH, JR., elec. engr.; b. Knoxville, Tenn., Oct. 5, 1948; s. Mark Hezekiah and Grace Carson (Owens) D.; m. Susan Nakamura, July 14, 1977; 1 dau.: Michelle Grace. B.S. in E.E, U. Tenn., 1972, M.S., 1973. Devel. engr. Westinghouse Electric Corp.-U.S. AEC, Pitts. and Oakridge, 1969-76; sr. research engr. N.L. Petroleum Service, Houston, 1977-79; mgr. research and devel. Advanced Ocean Systems div. Hydril Corp., Houston, 1980-81; engring. mgr. Schlumberger Corp., Sugarland, Tex., 1981-82; dir. electronics devel. Tech. for Energy Corp., Knoxville, Tenn., 1982—. Pres. N.W. Houston United Civic Assn., 1980-81. Robert Miller scholar, 1971; U. Tenn. Nat. Alumni scholar, 1972; U.S. AEC grantee, 1973. Mem. IEEE (sr.), Am. Soc. Engring. Edn., Optical Soc. Am., Electro-Chem. Soc., Marine Tech. Soc., Soc. Photo-Optical Instrumentation Engrs. Subspecialties: Ocean engineering; Fiber optics. Current work: Fiber optic sensors and communications systems and high temperature electronics in geosci. Home: PO Box 118 Knoxville TN 37901 Office: Tech for Energy Corp One Energy Ctr Pellissippi Pkwy Knoxville TN 37922

DAVIS, MICHAEL, psychopharmacologist, educator, researcher; b. Bronxville, N.Y., Nov. 14, 1942; s. Pearce and Lucia Banks (Bates) D.; m. Linda Shaffer, Dec. 30, 1967; children: Nathaniel, Alexander. B.A., Northwestern U., 1965; Ph.D., Yale U., 1969. Research assoc. dept. psychiatry Yale U., New Haven, 1969-70, asst. prof., 1970-75, assoc. prof., 1975—. Contbr. numerous articles to sci. jours. NIMH career devel. award, 1975—. Mem. Am. Psychol. Assn., Soc. for Neurosci., Soc. for Psychophysiol. Research. Subspecialties: Neuropharmacology; Neuropsychology. Current work: Measure how drugs affect the brain and behavior.

DAVIS, MICHAEL JAY, plant pathologist, educator; b. Denver, Mar. 9, 1947; s. Jay Edward and Joan Roberta (Dirmeyer) D.; m. Carol Ann Freeman, June 21, 1980; 1 son, Christopher Michael. B.S., Colo. State U., 1973, M.S., 1975; Ph.D., U. Calif., Berkeley, 1978. Asst. prof. dept. pathology Rutgers U., New Brunswick, N.J., 1979-81; asst. prof. plant pathology Agrl. Research and Edn. Center, U. Fla., Ft. Lauderdale, 1981—. Contbr. articles to sci. jours. U.S. Dept. Agr. grantee, 1980, 82. Mem. Am. Phytopath. Soc., Am. Soc. Microbiology, U.S. Fedn. Culture Collections, Sigma Xi, Gamma Sigma Delta. Subspecialties: Plant pathology; Genetics and genetic engineering (agriculture). Current work: Etiology, epidemiology and control of plant disease caused by bacteria; plant pathology, phytobacteriology, fastidious vascular plant pathogens of plants, plant disease resistance, genetic engineering. Home: 10121 NW 21st Ct Pembroke Pines FL 33023 Office: 3205 SW College Ave Fort Lauderdale FL 33314

DAVIS, OSCAR F., psychiatrist, research pharmacologist; b. Oak Park, Ill., June 19, 1928; s. Maurice H. and Lena (Wippman) D.; m. Doris Suzanne Koller, Apr., 1951; children: Scott, Elizabeth, Susan, Karen. Ph.D., Loyola U. Med. Sch., Chgo., 1954, M.D., 1958. Diplomate: Am. Bd. Psychiatry and Neurology. Intern Michael Reese Hosp., Chgo., 1958-59; resident U. Ill. Hosp., 1959-62, U. Ill. Inst. Juvenile Research, 1961-63; research pharmacologist, assoc. prof. pharmacology and psychiatry U. Health Scis. Chgo. Med. Sch., 1963-70; pvt. practice medicine specializing in child and adult psychiatry, Winnetka, Ill., 1965—; assoc. prof. psychiatry Northwestern U.; attending staff Northwestern Meml. Hosp., Evanston (Ill.) Hosp. Contbr. numerous articles on medicine, psychiatry, exptl. pathology and pharmacology to nat., internat. profl. jours. Subspecialties: Psychiatry; Molecular pharmacology. Current work: Relationships between neurochemistry-psychopharmacology and psychoanalytic psychiatry; psychomatic consideration in infant development. Office: 930 Wesley Doctors Tower Northwestern Meml Hosp 251 E Chicago Ave Chicago IL 60611 Office: 6 Elm St Winnetka IL 60093

DAVIS, PHILIP KEITH, engineering mechanics and materials educator; b. Effingham, Ill., Aug. 29, 1931; s. Charles E. and Mary D.; m. Elizabeth Ann Lasater, June 8, 1930; children: Lisa Dawn, Sheila Marie. B.S. in Mech. Engring, U. Tex.-Austin, 1958, M.S., 1959; M.S.E. in Engring. Mechanics, U. Mich., 1963, Ph.D., 1963. Registered profl. engr., Ill. Teaching fellow dept. engring. mechanics U. Mich.-Ann Arbor, 1960-63, instr., 1963-64; asst. prof. Coll. Engring. and Tech., So. Ill. U.-Carbondale, 1964-67, assoc. prof., 1967-71, prof., 1971—, chmn. dept. engring. mechanics and materials, 1971-78, 79—, acting dean, 1978-79. Contbr. articles to profl. jours. Served with USAF, 1951-55. Ford Found. faculty devel. fellow, 1960-63. Mem. Am. Soc. Engring. Edn., ASME, Soc. Engring. Sci., Am. Acad. Mechanics, Soc. Mining Engrs. Subspecialties: Fluid mechanics; Biomedical engineering. Current work: Fluid mechanics, vibrations, biomechanics, rheology and hydrocyclones. Home: Rural Route 4 Box 199 Carbondale IL 62901 Office: Dept Engring Mechanics and Materials So Ill U Carbondale IL 62901

DAVIS, RAYMOND, JR., chemist; b. Washington, Oct. 14, 1914; s. Raymond and Ida Rogers (Younger) D.; m. Anna Marsh Torrey, Dec. 4, 1948; children: Andrew Morgan, Martha Safford, Nancy Elizabeth, Roger Warren, Alan Paul. B.S., U. Md., 1937, M.S., 1939; Ph.D., Yale U., 1942. Chemist Dow Chem. Co., Midland, Mich., 1938-39, Monsanto Chem. Co., Dayton, Ohio, 1946-48; with Brookhaven Nat. Lab., Upton, N.Y., 1948—, now sr. chemist; adj. prof. dept. astronomy U. Pa. Contbr. articles to profl. jours. Served with USAAF, 1942-46. Recipient Boris Prejel prize N.Y. Acad. Scis., 1955; Comstock prize Nat. Acad. Scis., 1978; award for nuclear applications in chemistry Am. Chem. Soc., 1979. Mem. Am. Phys. Soc., Am. Geophys. Union, Am. Astron. Soc., AAAS, Meteoritical Soc., Nat. Acad. Scis. Subspecialties: Solar physics; Space chemistry. Current work: Studying the neutrino radiation from the sun, cosmic radiation, and particle physics. Office: Brookhaven Nat Lab Upton NY 11973

DAVIS, ROBERT GENE, plant scientist; b. Doddsville, Miss., Mar. 2, 1932; s. Robert Jeff and Osa (McCarty) D.; m. Dorothy Hobart, Jan. 29, 1954; children: Sondra, Elizabeth, Carl, David. B.S. in Agr, Miss. State U., 1953, M.S. in Plant Pathology, 1968, Ph.D., La. State U., 1970. Research plant pathologist Miss. Agr. and Forest Expt. Sta., Stoneville, 1971-82; prof. dept. plant pathology and weed sci. Miss. State U., 1981-82; owner Davis Research Co. (cons. and agrl. research firm), Avon, 1982—. Served with U.S. Army, 1955-57. Mem. Am. Phytopath. Soc., Weed Sci. Soc. Am., Sigma Xi, Phi Kappa Phi. Subspecialty: Plant pathology. Address: PO Box 359 Avon MS 38723

DAVIS, ROBERT HARRY, physiology educator; b. Wilkes-Barre, Pa., July 16, 1927; s. Cyril and Clara (Umlah) D.; m. Irene Casper, May 12, 1953; children: Sally, Susan. B.S., King's Coll., Wilkes-Barre, 1950; M.S., Newark Coll., 1955; Ph.D. in Zoology, Rutgers U., 1958. Assoc. prof. physiology Villanova (Pa.) U., 1963-67, Hahnemann Med. Coll., Phila., 1967-75; prof. physiology Pa. Coll. Podiatric Medicine, Phila., 1975—. Contbr. articles to profl. jours. Served with U.S. Navy, 1946. Recipient Wm. J. Stickles award Am. Coll. Ob-Gyn. Mem. Am. Physiol. Soc., others. Subspecialties: Animal physiology; Nutrition (biology). Current work: Hormone isolation. Home: 307 Abrams Mill Rd King of Prussia PA 19406

DAVIS, ROBERT JAMES, astronomer; b. Omaha, Oct. 26, 1929; s. Harry Cleve and Margaret Louise (Homan) D.; m. Ruth Cinnamon, May 16, 1953; children: Carolyn (Mrs. Norman Hargis), Deborah (Mrs. Thomas Mossberg), Paul, Elizabeth. A.B., Harvard U., 1951, M.A., 1956, Ph.D., 1960. With Smithsonian Astrophys. Obs., 1955—, physicist, 1958—. Served with USNR, 1951-54. Recipient Smithsonian Sustained Outstanding Service award, 1968. Mem. Internat. Soc. Philos. Enquiry (historian 1975—), Am. Astron. Soc., Internat. Astron, Union, Sigma Xi. Congregationalist. (sr. deacon 1972-75, moderator 1980-81). Subspecialties: Optical astronomy; Ultraviolet high energy astrophysics. Current work: Stellar and extragalactic spectroscopy; astrophysical data files stars; galaxies, galaxy clusters; spectroscopy; data management. Home: 307 Pleasant St Belmont MA 02178 Office: 60 Garden St Cambridge MA 02138

DAVIS, RODNEY JAMES, technical consultant; b. Mt. Holly, Vt., Sept. 14, 1925; s. Clyde and Pauline E. (Priest) D.; m. Phyllis E. Clark, Sept. 3, 1948; children: Melanie, Hunter, Melissa, Hugh. B.S., U. N.H., 1948, M.S., 1950; Ph.D., Iowa State U., 1954. Vis. prof. Knoxville (Tenn.) Coll., 1966-71; problem leader Oak Ridge Nat. Lab., 1954-70, group leader, 1970-73; resource cons. environ. services NUS Corp., Gaithersburg, Md., 1973-81, sr. exec. cons. maj. projects, 1981—. Served with AUS, 1944-46. Mem. Am. Mgmt. Assn., Am. Nuclear Soc., Research Soc. Am. Subspecialties: Nuclear fission; Surface chemistry. Current work: Fission product transport and mass balance in TMI-2 accident; nuclear waste management and repository siting; environmental impact assessment. Home: 18745 Walker's Choice Rd Gaithersburg MD 20879 Office: NUS Corp 910 Clopper Rd Gaithersburg MD 20878

DAVIS, ROSS, neurol. surgeon; b. Sydney, Australia, Oct. 30, 1931; came to came toU.S., 1966, naturalized, 1971; m. (married); children. B.Sc., U. Sydney, 1954, M.B., B.S., 1957, M.D., 1964. Resident Royal Prince Alfred Hosp., Sydney, Australia, 1957, 60; research fellow neurophysiology U. Sydney, 1958-59, Australian Nat. U., Canberra, 1960; researcher in neurophysiology, visual and ceyebellar systems Neurophysiol. Lab., New South Wales Dept. Pub. Health, Sydney, 1961-66; resident in neurosurgery U. Mich., 1967-71; practice medicine specializing in neurosurgery; asst. prof. U. Miami, Jackson Meml. Hosp., VA Hosp., 1971-74; chmn. dept. neurol. surgery Mt. Sinai Med. Ctr., Miami Beach, Fla., 1974—; assoc. prof. biomed. engring. U. Miami, 1979—; mem. staff Variety Children's Hosp., Miami, 1974-82. Contbr. numerous articles on neurophysiology, neuropharmacology, neurosurgery to profl. jours. Mem. Soc. Neurosci., ACS. Subspecialties: Neurosurgery; Neurophysiology. Current work: Neuroaugmentation; brain and spinal cord stimulators to reduce spasticity and seizures and improve motor skills. Office: Mt Sinai Med Center 213 Greenspan Bldg 4300 Alton Rd Miami Beach FL 33140

DAVIS, STARKEY DEE, pediatrics educator, researcher; b. Atlanta, Jan. 29, 1931; s. Sidney Rush and Veda Louise (Nichols) D.; m. Patsy Baker, Mar. 7, 1957 (div. Jan. 1979); children: Michael, Kimberly, April; m. Kathryn Maegli, Apr. 3, 1982. B.A., Baylor U., 1953, M.D., 1957. Diplomate: Am. Bd. Pediatrics. Intern Confederate Meml. Med. Ctr., Shreveport, La., 1957-58; resident Baylor U. Med. Sch., Houston, 1960-62; asst. prof. pediatrics U. Wash., Seattle, 1965-69, assoc. prof., 1969-73, prof., 1973-75; prof. pediatrics Med. Coll. Wis., Milw., 1975—. Contbr. numerous articles to med. jours.; co-editor: Infections in Children, 1982. NIH grantee, 1973—. Fellow Infectious Diseases Soc. Am.; mem. Am. Assn. Immunologists. Club: Milw. Yacht. Subspecialty: Microbiology (medicine). Current work: Pathogenesis of experimental pseudomonas keratitis. Office: Milw Children's Hosp PO Box 1997 Milwaukee WI 53201

DAVIS, THOMAS EDWARD, medical oncologist; b. Ft. Monroe, Va., July 21, 1943; s. Thomas O. and Marlynn D.; m. Amy Gibson, June 7, 1969; children: Matthew, Sarah. B.A., Johns Hopkins U., 1965, M.D., 1969. Diplomate: Am. Bd. Internal Medicine, also Sub-Bd. Med. oncology. Intern Osler med. service Johns Hopkins U., Balt., 1969-70, resident, 1970-72, fellow in med. oncology, 1972-74; mem. faculty U. Wis. Med. Sch., Madison, 1974—, assoc. prof. human ocology and medicine, 1980—; assoc. dir. clin. programs Wis. Clin.

Cancer Center, 1982—; exec. officer Eastern Coop. Oncology Group, 1977—. Author 60 papers on cancer diagnosis and treatment. Mem. Am. Soc. Clin. Ocology, Am. Assn. Cancer Research, ACP. Subspecialties: Oncology; Chemotherapy. Current work: Cancer chemotherapy.

DAVIS, THOMAS PAUL, pharmacology educator; b. Los Angeles, Jan. 13, 1951; s. Joseph Jefferson and Margaret (Moran) D.; m. Alecia Anne Kiehn, June 17, 1971; children: Melissa Catherine, Rebecca Marie. B.S., Loyola U.-Los Angeles, 1973; M.S., U. Nev., 1975; Ph.D., U. Mo., 1978. Grad. research asst. physiology U. Mo., Columbia, 1975-78; analytical chemist Abbott Diagnostics Div., Abbott Labs., North Chgo., 1978-80; adj. asst. prof. pharmacology U. Ariz., Tucson, 1980-81, asst. prof., 1981—; dir. analytical chemistry, mass spectrometry, lab., 1980—; cons. Assocs. Lab. Medicine, Tucson, 1982—. Contbr. chpt. to book and articles in field to profl. jours. Recipient D.B. Dill award U. Nev., 1974-75. Mem. Soc. Neurosci., Am. Physiol. Soc., Am. Chem. Soc., Assn. Ofcl. Analytical Chemists, Am. Soc. Pharmacology and Exptl. Therapeutics, Sigma Xi. Democrat. Roman Catholic. Subspecialties: Analytical chemistry; Neuropharmacology. Current work: Teaching and research in peptide biochemistry and neuropharmacology; biochemical pharmacology of drugs and metabolites; neuropharmacology of opioid peptides in gut and brain. Home: 7701 N Lundberg Tucson AZ 85741 Offic: U Ari Coll Medicine Tucson AZ 85724

DAVIS, THOMPSON ELDER, JR., psychologist; b. Calhoun, Ga., Sept. 4, 1940; s. Thompson Elder and Anna May (Carper) D.; m. Margaret Louise Campbell, Sept. 7, 1963 (div. 1983); 1 son, Thompson Elder III. Student, King Coll., Bristol, Tenn., 1958-61; B.S., E. Tenn. State Coll., 1962; M.S., U. Ga., 1973, Ph.D., 1974. Tchr., librarian Habersham County Bd. Edn., Alto, Ga., 1962-64; auditor Fed. Res. Bank, Atlanta, 1964-66, So. Services, Inc., 1966-67; asst. dir. internat. auditing div. U. Ga., Athens, 1967-68; mgmt. cons., Athens, 1968-69; clin. psychology intern Johns Hopkins U. Med. Sch., Balt., 1971-72, pub. psychology pilot intern, 1971-72; dir. psychol. services Glass Mental Health Ctr., Balt., 1974-81; v.p. ops. Behavioral Cons. Service, Inc., Balt., 1974-81; pres. Med. Data Fin. Corp. Inc., Balt., 1979-81; dir. psychol. research Glass Mental Health Found., Inc., Balt., 1977-81; pvt. practice clin. psychology, Lutherville, Md., 1979—; Bd. dirs. Friends Med. Sci. Research Ctr., Inc., 1978-80. King Coll. scholar, 1958-59; USPHS grantee, 1972-73; William Cooper Walker scholar, 1971-72. Mem. Am. Psychol. Assn., Md. Psychol. Assn., AAAS, Balt. Assn. Cons. Psychologists, Nat. Register Health Service Providers in Psychology, Psi Chi. Republican. Presbyterian. Subspecialty: Clinical psychology. Current work: Cognitive behavior therapy. Home: Left Bank 10220 Davis Ave Woodstock MD 21163 Office: York Pl 1204 York Rd Suite 14 Lutherville MD 21163

DAVIS, WILBUR MARVIN, pharmacology and toxicology educator, researcher; b. Calumet City, Ill., Apr. 13, 1931; s. Lester and Gladys (Cyphers) D.; m. Sandra Smith, Nov. 23, 1956; children: Brian Lee, Catherine Lee. B.S. in Pharmacy, Purdue U., 1952, M.S., 1953, Ph.D., 1955. From asst. prof. to prof. Coll. Pharmacy, U. Okla., Norman, 1955-64; prof. pharmacology U. Miss. Sch. Pharmacy, Oxford, 1964—, chmn. dept., 1964-83; cons. NIH. Contbr. articles on pharmacology to profl. jours. Recipient Borden award, 1951; Lalor Found. Research award, 1958. Mem. Am. Soc. for Pharmacology and Exptl. Therapeutics, Soc. Toxicology, Soc. for Neurosci., AAAS, Am. Pharm. Assn., Acad. Pharm. Scis., Am. Assn. Colls. Pharmacy. Baptist. Subspecialties: Neuropharmacology; Toxicology (medicine). Current work: Behavioral mechanisms and toxicologic actions of abuse drugs; teaching and research in toxicology and pharmacology. Home: 308 Lewis Ln Oxford MS 38655 Office: Sch Pharmacy U Miss 303 Faser Hall University MS 38677

DAVISON, KENNETH LEWIS, scientist; b. Hopkins, Mo., Dec. 27, 1935; s. Harlan R. and Hilda E. (Mendenhall) D.; m. Joyce Yvonne Schmitt, Sept. 8, 1957; children: Jeanette, Kenneth, Kathryn. B.S., U. Mo., 1957; M.S., Iowa State U., 1959, Ph.D., 1961. Research specialist in plant and animal nutrition Cornell U., 1961-65; research physiologist U.S. Dept. Agr.-Agrl. Research Service Metabolism and Radiation Research Lab., Fargo, N.D., 1965—. Mem. AAAS, Am. Inst. Nutrition, Am. Soc. Animal Sci., Am. Dairy Sci. Assn. Methodist. Subspecialties: Animal nutrition; Animal physiology. Current work: Metabolism of agricultural chemicals by animals. Home: 2860 N 2d St Fargo ND 58102 Office: 1605 W College St Fargo ND 58105

DAVISON, MARK EDWARD, applied mathematician; b. Charleston, W.Va., Oct. 6, 1952; s. James Morgan and Helen June (Baldwin) D.; m. Kathleen Elise Galloway, June 7, 1980. B.S., M.I.T., 1974; M.A., U. Calif.-Berkeley, 1977, Ph.D., 1979. Asst. prof. math. Iowa State U., 1979-82; assoc. mathematician Ames Lab. Iowa State U., 1979—; assoc. Daniel H. Wagner, Assocs., Paoli, Pa., 1980. Contbr. articles to profl. jours. NSF fellow, 1974; Office Naval Research grantee, 1982. Mem. Soc. Indsl. and Applied Math. Subspecialties: Applied mathematics; Imaging technology. Current work: Computational methods for inverting acoustic and elastic scattering data; medical tomography; applications to nondestructive testing, seismology, ocean acoustics, neutron scattering, nuclear medicine. Office: Ames Lab Iowa State U 136 Metallurgy Ames IA 50011

DAVISON, SYDNEY GEORGE, educator; b. Stockport, Eng., Sept. 6, 1934; s. Wilfrid and Sara H. (Warrington) D.; m. Prudence G. Allonby, Mar. 28, 1959; children: Terry, Symon, Timothy Scott. B.Sc., U. Manchester, Eng., 1958, M.Sc., 1962, Ph.D., 1964, D.Sc., 1982. Prof. physics Clarkson Coll., Potsdam, N.Y., 1970-72, Bartol Research Found., Swarthmore, Pa., 1972-74; prof. applied math. U. Waterloo, Ont., 1974—, prof. physics, 1982—; dir. Guelph-Waterloo Surface Sci. and Tech. Group. Editor: Progress in Surface Science; Contbr. articles to nat. and internat. jours. Nat. Sci. and Engring. Research Council Can. grantee, 1965-73, 75—. Fellow Am. Phys. Soc., Inst. Physics (U.K.). Anglican. Subspecialties: Surface chemistry; Condensed matter physics. Current work: Quantum theory of solid surfaces; chemisorption, surface states, disordered systems; quantum electrochemistry. Home: 313 Hiawatha Dr Waterloo ON Canada N2L 2V9 Office: Applied Math Dept U Waterloo Waterloo ON Canada N2L 3G1

DAWES, COLIN, oral biology educator; b. Nelson, Lancashire, Eng., Sept. 22, 1935; emigrated to Can., 1964; s. Harry and Gladys (Nowell) D.; m. Margaret McLeod, Dec. 28, 1960; children: Roger, Elizabeth, Martin, Richard, Susan. B.S., U. Manchester, 1956, B.D.S., 1958; Ph.D., U. Durham, 1962. Research fellow Harvard Sch. Dental Medicine, Boston, 1962-64; assoc. prof. U. Man., Can., Winnipeg, 1964-72, prof. oral biology, 1972—; mem. Med. Research Council Can., 1977-83. Editor: Jour. Dental Research, 1983. Research grantee NRC Can., 1964-69, Med. Research Council, 1969—. Mem. Can. Dental Assn., Can. Assn. Dental Research (pres. 1978-79), Internat. Assn. Dental Research, European Orgn. Caries Research. Anglican. Subspecialties: Oral biology; Cariology. Current work: Factors influencing flow rate and composition of saliva; the role of saliva in oral health. Office: Faculty Dentistry Univ Manitoba 780 Bannatyne Ave Winnipeg MB Canada R3E 0W3

DAWID, IGOR BERT, biologist; b. Czernowitz, Romania, Feb. 26, 1935; came to U.S., 1960, naturalized, 1977; s. Josef and Pepi (Druckmann) D.; m. Keiko Naito Ozato, Apr. 5, 1976. Ph.D., U. Vienna, 1960. Fellow dept. biology MIT, 1960-62; fellow dept. embryology Carnegie Instn. of Washington, Balt., 1962-66, mem. staff, 1966-78; chief devel. biochemistry sect. Lab. Biochemistry, Nat. Cancer Inst., Bethesda, Md., 1978-82; chief lab. molecular genetics Nat. Inst. Child Health and Human Devel. (NIH), Bethesda, 1982—; vis. scientist Max Planck Inst. for Biology, 1964-67; asst. prof. to prof. dept. biology Johns Hopkins U., 1967-78. Editor: Devel. Biology, 1971-75; editor: Cell, 1977—; editor-in-chief: Devel. Biology, 1975-80; adv. editor, 1980—. Mem. Am. Soc. Biol. Chemists, Am. Soc. Cell Biology, Soc. Devel. Biology, Internat. Soc. Devel. Biologists, AAAS, Nat. Acad. Sci. Subspecialties: Developmental biology; Molecular biology. Office: 9000 Rockville Pike Bethesda MD 20205

DAWSON, DAVID HENRY, research forester, cons.; b. Brillion, Wis., Sept. 21, 1919; s. George Edmund and Mabel Elizabeth (Stewart) D.; m. Minka Anne Hofstra, May 6, 1950; children: Jennifer, Dean, Cary, Cecily. B.S. in Forestry, Mich. State U., 1948, D.Forestry, 1957. Internat. forestry cons., 1970-82; chief forstry scis. lab. U.S. Forest Service, Rhinelander, Wis., 1972-82, program mgr. intensive culture forestry research, 1972-82, prin. geneticist, Bottineau, N.D., 1962-67, research cons., Rhinelander, 1982—. Contbr. articles on tree genetics, intensive culture forestry, shelterbelts to profl. jours. Elder Presbyn. Ch. Served to maj. AUS, 1946-46. Decorated D.S.C., Bronze Star, Purple Heart; recipient Superior Service award Dept. Agr., 1975. Mem. Soc. Am. Foresters, AAAS, Internat. Soc. Tropical Foresters. Lodge: Kiwanis. Subspecialties: Plant genetics; Biomass (energy science and technology). Current work: Resource policy, afforestation, short-rotation, intensive forestry for fiber and energy. Home and Office: Box 1695 Star Route 2 Rhinelander WI 54501

DAWSON, EARL BLISS, biochemist, consultant; b. Perry, Fla., Feb. 1, 1930; s. Bliss and Linnie Estella (Calliham) D.; m. Winnie Ruth Isbell, Apr. 10, 1951; children: Barbara Gail, Patricia Ann, Robert Earl, Diana Lynn. B.A., U. Kans., 1955; M.A., U. Mo.-Columbia, 1960; Ph.D., Tex. A. and M. U., 1964. Instr. U. Tex. Med. Br., Galveston, 1963-65, asst. prof., 1965-68, assoc. prof. biochemistry, 1968—; cons. NIH/State Dept., Guatamala City, Guatamala, 1965, NIH, Tex. Health Dept., 1968-71, NIH, Nat. Inst. Childhood Diseases, 1970-76, NIH/EPA, 1974-76. Author: Nutritional Evaluation of Population of Central America and Panama, 1969, Ten State Nutrition Survey, 1974, Effect of Water Borne Nitrates on the Environment of Man, 1977, Effect of Water Borne Fluoride on the Environment of Man, 1977; contbr. articles to profl. jours. Scoutmaster Boy Scouts Am., LaMarque, Tex., 1968—. Served with USNR, 1947-52. NSF scholar, 1962; NIH grantee, 1963; Moody Found. grantee, 1964. Mem. Am. Inst. Nutrition, Am. Soc. Clin. Nutrition, Am. Soc. Exptl. Biology and Medicine, N.Y. Acad. Sci. Baptist. Lodge: Masons. Subspecialties: Nutrition (medicine); Biochemistry (medicine). Current work: Trace metal and vitamin metabolism. Office: Dept Obstetrics and Gynecology U Tex Med Br Galveston TX 77550 Home: 15 Chimney Corners LaMarque TX 77568

DAWSON, GEORGE EUGENE, JR., mechanical engineer; b. Houston, Dec. 29, 1951; s. George Eugene and Elizabeth Barbara (Feindeisen) D.; m. Monica Martin, July 10, 1976. B.S. in Metallurgy, U. Houston, 1977. Engr. technician WKM Valve, Houston; v.p. engring. AT-Tech. Lab., Houston, to 1977; lab. mgr. Brown & Root, Inc., Houston, 1977-79; br. mgr. Shilstone Engring., Houston, 1979-70; pres., gen. mgr. J&J Machine & Welding, Houston, 1980-82; div. mgr. Bas-Tex Corp., Houston, 1982—; lectr. U. Houston. Mem. ASME, Nat. Soc. Profl. Engrs., Tex. Soc. Profl. Engrs., Am. Soc. Metals, Am. Welding Soc. Republican. Methodist. Subspecialties: Materials; Metallurgy. Current work: Computer aided manufacturing of machined components and heat processing equipment. Office: PO Box 40082 Houston TX 77240

DAWSON, GLYN, mental retardation center administrator, medical educator, researcher; b. New Mills, Eng., Mar. 24, 1943; came to U.S., 1967; s. John Thomas and Winifred Arden (Flemming) D.; m. Sylvia Angela Richards, July 2, 1966; children: Philip, Kenneth. B.S.C., U. Bristol, Eng., 1964, Ph.D., 1967. Asst. prof. pediatrics U. Chgo., 1969-75, assoc. prof., 1976—; acting dir. Joseph P. Kennedy Jr. Mental Retardation Research Ctr., Chgo., 1982—; cons. Nat. Tay Sachs Assn. Guggenheim fellow, 1979; NICMD grantee, 1972—; NIMH grantee, 1979—; recipient Research Career Devel. award, 1975-79. Mem. AAAS, Soc. Complex Carbohydrates, Am. Soc. Neurochemistry, Am. Soc. Biol. Chemistry. Subspecialties: Neurochemistry; Biochemistry (medicine). Current work: Receptor-mediated mechanism of action of opiates and serotonin, lysosomal storage disease, membrane biochemistry (leukemias, myelinating cells). Office: Dept Pediatrics U Chicago 950 E 59th St Chicago IL 60637

DAWSON, JAMES CLIFFORD, geologist, environmentalist, educator; b. Toronto, Apr. 19, 1941; s. Clifford and Winnifred (Tadman) D.; m. Caroline Weiss, June 12, 1971. B.A., UCLA, 1965, M.S., 1967; Ph.D., U. Wis.-Madison, 1970. Asst. prof. geology SUNY, Plattsburgh, 1970-74, assoc. prof. geology, 1974-80; chmn. dept. earth sci., 1975-76, prof. environ. sci., 1980—, dir., 1976-82. Mem. Am. Assn. Petroleum Geologists, AAAS, Am. Geophys. Union, Internat. Assn. Sedimentologists, Nat. Assn. Geology Tchrs., Soc. Econ. Paleontologists and Mineralogists, Adirondack Conservancy, Adirondack Council, Adirondack Mountain, Assn. Protection Adirondacks, Appalachian Trail Conf., Clinton County Hist. Assn., Environ. Planning Lobby, Nat. Audubon Soc., Nat. Wildlife Fedn., Nat. Resources Def. Council, Pine Bush Hist. Preservation Project, Vt. Natural Resources Council, Sierra Club, Sigma Xi. Subspecialties: Sedimentology; Resource management. Home: Birchwood Dr Peru NY 12972 Office: Ctr Earth and Environ Sci SUNY Plattsburgh NY 12901

DAWSON, JEFFREY ROBERT, immunology educator; b. Lakewood, Ohio, Oct. 5, 1941; s. Robert Eugene and Elva Rose (Lincks) D.; m. Linda Elizabeth Issler, June 14, 1964; children: Amy Elizabeth, Mary Catherine, Michael Jeffrey. B.S. in Biology, Rensselaer Poly. Inst., 1964; Ph.D. in Biochemistry, Case Western Res. U., 1969; postgrad. (NIH fellow), Duke U. Med. Center, 1969-71. Instr. Immunology Duke U., 1971-72, asso. in immunology, 1972-74, asst. prof., 1974-78, asso. prof., 1978—, mem. scientist, 1978—. Contbr. to books and articles in field to profl. jours. Eleanor Roosevelt Internat. fellow, 1982—; Am. Cancer Soc. grantee, 1976-78; NIH grantee, 1980—. Mem. Brit. Soc. Immunology, Am. Assn. Immunologists, Am. Assn. Cancer Research, Japanese Cancer Soc. Democrat. Episcopalian. Subspecialties: Immunology (medicine); Cancer research (medicine). Current work: Mechanism of natural killing and modulation of natural killing with pharmacologic reagents. Home: 902 Clarion Dr Durham NC 27705 Office: Division of Immunology Duke University Medical Center Box 3010 Durham NC 27710

DAWSON, JOHN MYRICK, plasma physics educator; b. Champaign, Ill., Sept. 30, 1930; s. Walker Myrick and Wilhelmina Emily (Stephan) D.; m. Nancy Louise Wildes, Dec. 28, 1957; children: Arthur Walker, Margaret Louise. B.S., U. Md., 1952, M.S., 1954, Ph.D., 1957. Research physicist Plasma Physics Lab. Princeton U., 1956-73, head theoretical group, 1965-73; prof. plasma physics UCLA, 1973—; dir. Center for Plasma Physics & Fusion Engring., 1973-76; cons. in field; John Danz lectr. U. Wash., 1974; guest Russian Acad. Scis., 1971; invited lectr. Inst. Plasma Physics, Nagoya, Japan, 1972. Contbr. articles in field to profl. jours. Recipient Exceptional Sci. Achievement award TRW Systems, 1977; James Clerk Maxwell prize in Plasma Physics, 1977; named Calif. Scientist of the Year, 1978. Fellow AAAS, Am. Phys. Soc. (chmn. plasma div. 1970-71); mem. Nat. Acad. Scis., N.Y. Acad. Scis., N.J. Acad. Scis., Sigma Pi Sigma, Phi Kappa Phi, Sigma Xi. Unitarian. Subspecialty: Plasma physics. Patentee in field. Home: 359 Arno Way Pacific Palisades CA 90272 Office: University of California 405 Hilgard Ave Los Angeles CA 90024

DAWSON, WILLIAM RYAN, educator; b. Los Angeles, Aug. 24, 1927; s. William Eldon and Mary (Ryan) D.; m. Virginia Louise Berwick, Sept. 9, 1950; children: Deborah, Denise, William. Student, Stanford, 1945-46; B.A., UCLA, 1949, M.A., 1950, Ph.D., 1953; D.Sc., U. Western Australia, 1971. Faculty zoology U. Mich., Ann Arbor, 1953—, prof., 1962—, D.E.S. Brown prof. biol. scis., 1981—, chmn. div. biol. scis., 1974—, dir. mus. zoology, 1982—; Lectr. Summer Inst. Desert Biology, Ariz. State U., 1960-71; researcher Australian-Am. Edn. Found., U. Western Australia, 1969-70; mem. Speakers Bur., Am. Inst. Biol. Sci., 1960-62; Mem. adv. panel NSF environ. biology program, 1967-69; mem. adv. com. for research NSF, 1973-77, adv. panel regulatory biology program, 1979-82; mem. R/V Alpha Helix New Guinea Expdn., 1969; chief scientist R/V Dolphin Gulf of Calif. Expdn., 1976; mem. R/V Alpha Helix Galapagos Expdn., 1978. Editorial bd.: Condor, 1960-63, Auk, 1964-68, Ecology, 1968-70, Ann. Rev. Physiology, 1973-79, Physiol. Zoology, 1976—; co-editor: Springer-Verlag Zoophysiology and Ecology series; asso. editor: Biology of the Reptilia, 1972. Served with USNR, 1945-46. USPHS Postdoctoral Research fellow, 1953; Guggenheim fellow, 1962-63; Recipient Russell award U. Mich., 1959, Distinguished Faculty Achievement award, 1976. Fellow AAAS, Am. Ornithol. Union (Brewster medal 1979); mem. Am. Soc. Zoologists, Am. Physiol. Soc., Ecol. Soc. Am., Cooper Ornithol. Soc. (hon., Painton award 1963), Internat. Soc. Biometeorologists, Phi Beta Kappa, Sigma Xi, Kappa Sigma. Subspecialty: Comparative physiology. Current work: Physiology of temperature adaptation in reptiles, birds and mammals. Home: 1376 Bird Rd Ann Arbor MI 48103

DAY, CALVIN LEE, JR., dermatologist, cancer researcher; b. Karnes City, Tex., July 4, 1951; s. Calvin Lee and Theresa Louise (Hoffman) D.; m. Regina Carroll, Jan. 8, 1972; 1 dau., Berica. B.S. summa cum laude, Tex. A&M U., 1973; M.D. with honors, U. Tex. Southwestern Med. Sch., 1976. Diplomate: Am. Bd. Dermatology. Research fellow in phys. chemistry Tex. A&M U., 1971; research fellow in biochemistry U. Tex. Southwestern Med. Sch., Dallas, 1973; intern in medicine Mass. Gen. Hosp., Boston, 1976-77, resident in medicine, 1977-78, resident in dermatology, 1978-79, 79-80, clin. fellow in dermatology, 1980-81, clin. and research fellow in dermatology, 1981-82; clin. fellow in medicine Harvard U., 1976-77, 77-78, clin. fellow in dermatology, 1978-79, 79-80, 80-81, research fellow in dermatology, 1981-82; fellow in chemosurgery dept. dermatology N.Y.U., 1982-83; clin. asst. prof., dir. skin cancer surgery div. dermatology U. Tex. Med. Sch., San Antonio, 1983—. Contbr. articles, chpts. to profl. publs.; editor: Melanoma Letter, Skin Cancer Found., 1982—. Pres. Tex. Skin Cancer Found., 1983—. Southwestern Med. Found. scholar, 1972-73; Robert Wood Johnson scholar, 1975-76; Tex. Merit scholar, 1975-76; Julia Ball Lee fellow, 1970-71; NIH research grantee, 1978, 80-82. Fellow N.Y. Acad. Scis.; mem. Am. Fedn. Clin. Research, AMA, Alpha Omega Alpha, Phi Kappa Phi, Phi Eta Sigma. Subspecialties: Cancer research (medicine); Dermatology. Current work: Malignant melanoma; microscopically controlled surgery for skin cancers.

DAY, EUGENE DAVIS, immunologist, educator; b. Cobleskill, N.Y., June 24, 1925; s. Emmons Davis and Alice Dorothy (McCartey) D.; m. Shirley Warner, Sept. 14, 1946; 1 son, Eugene Davis. B.S., Union Coll., 1949; Ph.D., U. Del., 1952. Research assoc. Jackson Meml. Lab., Bar Harbor, Maine, 1952-54; sr., then assoc. cancer research scientist Roswell Park Meml. Inst., Buffalo, 1954-62; assoc. prof. Duke U. Med. Ctr., 1962-64, prof., 1964—. Author books; contbr. articles to profl. jours. Served with AUS, 1943-46. Fellow An. Acad. Microbiology; Mem. Am. Assn. Immunologists, Am. Soc. Neurochemistry, Am. Soc. Microbiology, AAUP, Am. Assn. Cancer Research, Sigma Xi. Subspecialties: Neuroimmunology; Immunobiology and immunology. Current work: Immunochemistry of myelin basic protein. Home: 2727 McDowell St Durham NC 27705 Office: Duke U Med Center Box 3045 Durham NC 27710

DAY, FRANK P., JR., ecologist; b. Bristol, Va., July 12, 1947; s. Frank P. and Sarah Elizabeth (Francisco) D.; m. Jean Ann Waldrop, Aug. 30, 1969; children: Jennifer Ann, Melissa Sue, Frank Christopher. B.S., U. Tenn., 1969; M.S., U. Ga., 1971, Ph.D., 1974. Instr. U. Ga., 1974; asst. prof. biol. scis. Old Dominion U., Norfolk, 1974-80, assoc. prof., 1980—, dir. ecol. scis. Ph.D. program, 1981—. Contbr. articles to profl. jours. NSF grantee, 1977, 79. Mem. Ecol. Soc. Am., Bot. Soc. Am., Assn. Southeastern Biologists, Torrey Bot. Club, Am. Inst. Biol. Scis., others. Subspecialties: Ecology; Ecosystems analysis. Current work: Nutrient cycling and decomposition dynamics in wetland ecosystems. Office: Dept Biol Scis Old Dominion U Norfolk VA 23508

DAY, MICHAEL HARDY, systems engineer, physicist, educator; b. St. Louis, Apr. 4, 1950; s. Arthur A. and Grace (Hardy) D.; m. Rose C. Day, Feb. 14, 1972; 1 son, Matthew B. B.Sc. in Physics, U. Mo., Columbia, 1972, Ph.D., U. Wis., Madison, 1977. Research assoc. U. Sussex, theoretical physics div. Atomic Energy Research Establishment, Harwell, Eng., 1978-80; asst. prof. physics Kans. State U., Manhattan, 1980-82; systems engr. MTS Bell Labs., Holmdel, N.J., 1982—. Mem. Am. Phys. Soc. Subspecialties: Atomic and molecular physics; Theoretical physics. Current work: Government communication, telecommunication networks. Office: 4 M 312 AT&T Bell Labs Crawford Corners Rd Holmdel NJ 07733

DAY, STACEY BISWAS, physician, educator, health tech. and info. sci. cons.; b. London, Dec. 31, 1927; s. Satis B. and Emma L. (Camp) D.; m. Ivana Podvalova, Oct. 18, 1973; 2 children. M.D., Dublin, Ireland, 1955; Ph.D. in Exptl. Surgery, McGill U., 1964; D.Sc., U. Cin., 1970. Med. educator various univs., 1955-66; research dir. phase I and II Hoechst Inc., Cin., 1966-67; regional med. dir. for New Eng. Hoffman La Roche Co., 1967-69; educator, cons., 1969—; v.p. Mario Negri Found. (sci. research), 1975-80, Internat. Found. Biosocial Devel. and Human Health, 1979—; cons. in health tech. and informatics; tech. control, communications, behaviorism and social/community control; prof. N.Y. Med. Coll.; cons. Cross River State, Nigeria. Author, editor 50 books; founder: Bioscis. Communications, 1974, Monograph Publs. in Health Communications and Biospychosocial Health; contbr. numerous articles to profl. jours. Served with Brit. Army, 1946-49. Recipient medals and prizes Royal Coll. Surgeons, Ireland, Royal Coll. Surgeons, Eng.; Ciba fellow, 1964. Fellow Japan Soc. BioPsycho Soc. Health (most disting. fellow), Zool. Soc. London; mem. AMA, Harvey Soc., numerous others. Mem. Ch. of England. Subspecialties: Behaviorism; Information systems (information science). Current work: Technology; information; culture and high technology control; biopsychosocial control; behaviorism and technocracy; technology and transition; parasymphatetic way.

Home: 6 Lomond Ave Spring Valley NY 10977 Office: Calabar Nigeria

DAY, WILLIAM HARTWELL EVELETH, computer scientist, educator, researcher; b. Boston, June 23, 1937; s. Chester Morrill and Elizabeth (Eveleth) D.; m. Cecil Beasley, Aug. 23, 1958; children: Geoffrey, Hartwell, John Bradford. A.B., Harvard U., 1958; M.S., Washington U., St. Louis, 1972, D.Sc., 1975. Staff programmer IBM Corp., Poughkeepsie, N.Y., 1960-70; asst. prof. So. Meth. U., Dallas, 1975-79, Meml. U. Nfld., St. John's, 1979-83, assoc. prof., 1983—; vis. lectr. Soc. for Indsl. and Applied Math., Phila., 1982-84; mem. sci. [illegible] Joint European Meeting Psychometrics and Classification [illegible] Jouy en Josas, France, 1982-83. Contbr. articles to profl. jours. Natural Scis. and Engring. Research Council Can. grantee, 1980, 81, 82, 83. Mem. Assn. Computing Machinery, Classification Soc. (council 1980-82), Math. Assn. Am., Soc. Systematic Zoology, Sigma Xi. Subspecialties: Taxonomy; Algorithms. Current work: Computational problems in numerical taxonomy. Quantitative comparison of classifications. Consensus and distance measures. Design and analysis of algorithms. Home: 5 Willicott Lane St John's NF Canada A1C 1L8 Office: Dept Computer Sci Memorial Univ Newfoundland St John's NF Canada A1C 5S7

DEA, PHOEBE K., chemistry educator; b. Canton, China, June 17, 1946; came to U.S., 1964, naturalized, 1977; K.H. and and H.K. (Lau) Wong; m. Frank J. Dea, Dec. 23, 1967; children: Denise I., Melvin C. B.S., UCLA, 1967; Ph.D., Calif. Inst. Tech., 1972. Postdoctoral research fellow ICN Pharms., Irvine, Calif., 1972-74, head biophys. and analytical chemistry, 1974-76; asst. prof. chemistry Calif. State U.-Los Angeles, 1976-79, assoc. prof., 1979—. Contbr. articles to profl. jours. Named Outstanding Prof. Calif. State U., 1981; NIH grantee, 1977—. Mem. Am. Chem. Soc. Subspecialties: Biophysical chemistry; Analytical chemistry. Current work: Research in nuclear magnetic resonance spectroscopy; membrane structure and function. Office: Calif State U 5151 State University Dr Los Angeles CA 90032

DEACON, JAMES EVERETT, biology educator, environmental consulting company executive; b. White, S.D., May 18, 1934; s. James William and Mona Louise (Everett) D.; m. Maxine Shirley, Aug. 15, 1954; children: Cynthia Doris Deacon Williams, David Everett. B.S., Midwestern State U., 1956; Ph.D., U. Kans., 1960. Prof. biology U. Nev, Las Vegas, 1960—; adj. prof. Internat. Coll. of Cayman Islands, 1983—; pres. Environ. Cons., Inc., 1972—. Contbr. ecol. and conservational articles to profl. publs. Cons. Nev. State Council on Arts. Grantee NSF, EPA, Nat. Park Service. Fellow AAAS, Am. Inst. Fisheries and Biology, Ariz.-Nev. Acad. Sci.; mem. Nev. Wildlife Fedn., Nat. Wildlife Fedn., Dessert Fisheries Council, Am. Soc. Ichthyologists and Herpetologists, Am. Fisheries Soc. Club: Cambridge Raquet (Las Vegas). Subspecialties: Ecology; Resource management. Current work: Ecology of desert fishes; interpretive natural history; habitat requirements of desert fishes; effects of introduced fishes on native fishes; interpretive natural. Home: 2772 Quail Ave Las Vegas NV 89120 Office: U Nev Biol Scis Dept Las Vegas NV 89154

DEAN, DONALD HARRY, molecular biology educator; b. San Diego, Nov. 20, 1942; s. Clarence Ray and Mary Lucille (Lee) D.; m. Janice Elaine Webb, July 18, 1969; 1 son, Samuel Edmond. B.S., Tex. Christian U., 1965, M.S., 1968; Ph.D., U. Mich., 1972. Postdoctoral fellow Washington U., St. Louis, 1972-73, Brandeis U., Waltham, Mass., 1973-75; asst. prof. molecular biology Ohio State U., Columbus, 1975-80, assoc. prof., 1980—; dir. Bacillus Genetic Stock Ctr., Columbus, 1978—. Author Gene Structure and Expression, 1980; editor: Microbial Genetics Bull, 1977-82. NIH grantee; NSF grantee, 1978. Mem. Am. Soc. Microbiology, Soc. Insect Pathology. Democrat. Subspecialties: Genetics and genetic engineering (biology); Molecular biology. Current work: Bacteriophage genetics, temperate bacteriophage of bacillus; bacillus genetics especially genetics of sporulation; genetics engineering of insect pathogenic bacteria.

DEAN, EDWIN BECTON, lead cost analyst; b. Danville, Va., Feb. 7, 1940; s. Edwin Becton and Lois Pearl (Campbell) D.; m. Deirdre Anne Jacovides, Aug. 16, 1964; chilren: Jennifer, Kristin, Brian. B.S. in Physics, Va. Poly. Inst., 1963, M.S. in Math, 1965; postgrad., George Washington U., 1974-77. Assoc. engr., technician Applied Physics Lab., Johns Hopkins U., Laurel, Md., 1959-63; mathematician, physicist, elec. engr., ops. research analyst Naval Surface Weapons Ctr., Silver Spring, Md., 1964-79; rep. First Investors Corp., Arlington, Va., 1971—; bus. counselor Gen. Bus. Services, Virginia Beach, Va., 1979—; computer security specialist NAVSUP, Norfolk, Va., 1982-83; tech. resource mgr. NASA Langley Research Ctr., Hampton, Va., 1983—. NASA fellow, 1963-65. Mem. IEEE, Assn. Computing Machinery, Am. Def. Preparedness Assn., Computer Security Inst., Phi Kappa Phi, Sigma Pi Sigma, Pi Mu Epsilon. Club: Tidewater Apple Worms (Virginia Beach). Subspecialties: Operations research (mathematics); Graphics, image processing, and pattern recognition. Current work: Applying differential geometry to optimization, graphics, image processing, pattern recognition, artificial intelligence, business forecasting, economics, numerical analysis and system measurement. Home: 2412 Whaler Ct Virginia Beach VA 23451 Office: NASA Langley Research Ctr Hampton VA 23665

DEAN, JACK HUGH, immunologist, immunotoxicologist; b. Joplin, Mo., Dec. 6, 1941; s. Hugh H. and Lora L. (Lappie) D.; m. Suellen Zeek, July 12, 1961; children: Carl A., John M., Matthew C. B.S., Calif. State U.-Long Beach, 1964, M.S., 1968; Ph.D., U. Ariz., 1972. Research immunologist Litton Bionetics, Inc., Kensington, Md., 1972-76, head immunology dept., 1976-79; head immunology sect. Nat. Toxicology Program, Nat. Inst. Environ. Health Sci., NIH, Research Triangle Park, N.C., 1979-81; head immunotoxicology and cell biology sect. Chem. Industry Inst. Toxicology, Research Triangle Park, 1981-83, head dept. cell biology, 1983—; adj. assoc. prof. pathology Duke U., Durham, N.C., 1981—; adj prof. toxicology U. N.C., Chapel Hill, 1982—; cons. NIH, 1977-79, EPA, 1982—, Nat. Inst. Environ. Health Scis., 1982—. Assoc. editor: Internat. Jour. Immunopharmacology, 1981—, Jour. Biol. Response Modifiers, 1982—. Mem. Am. Soc. Microbiology, Am. Assn. Cancer Research, Am. Assn. Immunologists, Reticuloendothetial Soc. (assoc. editor jour. 1981—). Presbyterian. Subspecialties: Immunology (medicine); Immunotoxicology. Current work: Effect of chemicals and drugs on immune system; mechanisms of chemical induced immune alterations, mechanisms of immune resistance to infection agents and neoplastic cells. Home: Route 7 Box 251E Chapel Hill NC 27514 Office: Chem Industry Inst Toxicology PO Box 12137 Research Triangle Park NC 27709

DEAN, JACK LEMUEL, plant pathologist; b. Keota, Okla., Mar. 15, 1925; s. Delarvin and Beulah (Prentice) D.; m. Norma Ruth Henderson, Sept. 3, 1949; children: Robert Dale, Paul Randal. Student, Tulane U., 1943-45; B.S., Okla. State U., 1949, M.S., 1951; Ph.D., La. State U., 1966. Plant pathologist USDA, Meridian, Miss., 1951-66, Canal Point, Fla., 1966—; mem. grad. faculty dept. plant pathology U. Fla., Gainesville, 1982—. Contbr. articles to profl. jours. Served with USN, 1943-46. Mem. Am. Phytopath. Soc., Am. Inst. Biol. Scis., Internat. Soc. Sugarcane Technologists. Subspecialty: Plant pathology. Current work: Sugarcane diseases—resistance. Research on disease resistance in sugarcane—cooperation with plant breeders to produce disease resistant clones of sugarcane. Home: 92 Dayton Rd Lake Worth FL 33463 Office: US Sugarcane Field Sta Star Route Box 8 Canal Point FL 33438

DEAN, JUDITH CAROL HICKMAN, medical research; b. Baton Rouge, Oct. 19, 1943; d. Cecil Lamar and Johnnie Louise (Efferson) Hickman. B.S., Northwestern State U., La., 1965; M.S., U. Central Ark., 1973; doctoral candidate, U. Ariz., 1980—. Registered nurse, La., Ariz. Psychiat. nurse Mercy Hosp., Urbana, Ill., 1966-67; chief nurse rehab. and extended care Central La. State Hosp., Pineville, 1967-69; instr. nursing U. Ark., Little Rock, 1970-73; asst. dir. nursing for medicine U. Ariz. Health Sci. Ctr., Tucson, 1974; clin. specialist in [illegible] VA Med. Ctr., Tucson, 1976-81; research assoc. internal medicine U. Ariz. Cancer Ctr., Tucson, 1981—; mem. council collegiate edn. for nursing So. Regionl Edn. Bd., 1972-73; vol. abstractor Am. Nurses Found., 1972-74. Author, lectr., cons. Mem. profl. edn. com. Ariz. div. Am. Cancer Soc., 1978—; mem. tumor registry adv. bd. Ariz. Dept. Health Services, 1980—. Walter Teagle Fond. scholar, 1961-65. Mem. Am. Soc. Clin. Oncology, Am. Psychol. Assn. Am. Personnel and Guidance Assn., Oncology Nursing Soc. (charter), Sigma Theta Tau, Kappa Delta Pi. Subspecialties: Cancer research (medicine); Onoclogy Counseling Psychology. Current work: Psychological impact of cancer and cancer treatment(s), special interest areas sexuality issues for the cancer patient, family dynamics, and prevention of alopecia. Office: 1501 N Campbell Ave Tucson AZ 85724

DEAN, RAYMOND S., neuropsychologist, educator, researcher; b. Troy, N.Y., Sept. 24, 1946; s. Stanley R. and Helen W. D.; m. Nancy Ellen Connors, Dec. 22, 1968; children: Heather M., Whitney E. B.S., SUNY-Albany, 1972, B.A., 1973, M.S., 1974; Ph.D., Ariz. State U., Tempe, 1978. Diplomate: , Nat. Acad. Neuropsychologists. Asst. prof. U. Wis.-Madison, 1978-80, U. N.C., Chapel Hill., 1980-81; asst. prof. Washington U. Sch. Medicine, St. Louis, 1980—, dir. neuropsychology lab., 1980—. Editor: Clin Neuropsychology, 1980—; contbr. articles to profl. jours. Fellow Nat. Acad. Neuropsychologists (sec. 1982—, editor bull. 1981—); mem. Psychol. Assn., Soc. Pediatric Psychologists. Subspecialties: Neuropsychology; Pediatrics. Current work: Research in neuropsychological aspects of children's learning disorders. Home: 7410 Stanford Ave University City MO 63110 Office: Dept Psychiatry Washington Univ Sch Medicine 216 Kingshighway Saint Louis MO 63110

DEAN, REGINALD LANGWORTHY, III, research scientist; b. New Haven, Aug. 20, 1953; s. Reginald Langworthy and Mary Ellen (Keenan) D.; m. Luan Kulof, May 9, 1975; 1 dau., Courtney Meghan. B.A., Denison U., 1976; M.A., Western Mich. U., 1978. Asst. scientist Warner Lambert/Park-Davis, Ann Arbor, Mich., 1978-79; research scientist div. med. research Am. Cyanamid Co. Lederle Labs., Pearl River, N.Y., 1979—. Contbr. articles to profl. jours. Mem. Soc. Neurosci., Am. Aging Assn., AAAS. Roman Catholic. Subspecialties: Neuropharmacology; Gerontology. Current work: Development of animal models to accurately mimic age-related memory impairments. Home: 4 Beatrice Ln New City NY 10956 Office: Lederle Labs/Bldg 56B Room 119 Pearl River NY 10965

DEAN, RICHARD ANTHONY, nuclear technologies corporation executive; b. Bklyn., Dec. 22, 1935; s. Anthony David and Ann (Mylod) D.; m. Sheila Elizabeth Grady, Oct. 5, 1957; children: Carolyn, Julie, Drew. B.Mech. Engring., Ga. Inst. Tech., 1957; M.S., U. Pitts., 1963, Ph.D., 1970. Registered profl engr., Calif. Mgr. thermal and hydraulic design Westinghouse Corp., Pitts., 1959-70; v.p. Gulf United Nuclear Fuels Corp., Elmsford, N.Y., 1971-73; div. dir. GA Techs. Inc., San Diego, 1974-76, v.p., 1976—; cons. U.S. Congress Office Tech. Assessment, Washington, 1982—. Served to 1st lt. U.S. Army, 1957-59. Mem. ASME (chmn. nuclear fuels tech. subcom. 1973-76), Am. Nuclear Soc. (nat. planning commn. 1981), Phi Kappa Phi, Tau Beta Pi, Pi Tau Sigma. Republican. Roman Catholic. Lodge: KC. Subspecialties: Nuclear fission; Mechanical engineering. Current work: Pioneer in development of commercial nuclear power; advanced the fundamental understanding of boiling heat transfer. Patentee advanced nuclear fuel. Home: 6699 Via Estrada La Jolla CA 92037 Office: GA Techns Inc 10955 John Jay Hopkins Dr San Diego CA 92037

DEAN, RICHARD H., surgery educator; b. Radford, Va., June 16, 1942; s. Howard Lee and Minnie (Yates) D.; m. Marcella Blaylock, June 22, 1974; children: Richard Lancaster, Harrison Blaylock, Howard Lee Alexander. B.A., Va. Mil. Inst., 1964; M.D., Med. Coll. Va., Richmond, 1968. Intern Vanderbilt Hosp., 1968-69, resident in surgery, 1969-73, chief surg. resident, 1973-74, dir. surg. intensive care unit, 1975—, head div. vascular surgery, 1978—, prof. surgery, 1981—, co-clin. dir. hypertension ctr., 1981—, chmn. profl. program, 1982—; fellow in vascular surgery Northwestern U., 1974-75. Author, editor: (with others) Vascular Disease of Childhood, 1983. NIH grantee, 1976-86; Fellow ACS Southeastern Surg. Congress. Mem. So. Surg. Assn., Halsted Soc., Soc. Vascular Surgery, Soc. Univ. Surgeons, Internat. Cardiovascular Soc., Assn. Acad. Surgery, So. Assn. Vascular Surgery, Internat. Soc. Surgery, Pan-Pacific Surg. Assn., So. Med. Assn., Nashville Surg. Soc., Tenn. Med. Assn., Nashville Acad. Medicine, AMA, H. William Scott Jr. Soc. Episcopalian. Subspecialty: Vascular Surgery. Current work: Hypertension, atherosclerosis, renal function, cerebrovascular disease, liver failure. Home: 5350 Hillsboro Rd Nashville TN 37215 Office: Vanderbilt University Hospital T-2104 21st Ave at Garland St Nashville TN 38232

DEAN, ROBERT CHARLES, JR., mechanical engineer, business executive, educator; b. Atlanta, Apr. 13, 1928; s. Robert C. and Ruth (Andrew) D.; m. E. Nancy Hayes, Sept. 22, 1951; children: Margaret S., James C., Elizabeth S., Martha A., Charles E. B.S., M.S., MIT, 1949, Sc.D., 1954. Project engr. Ultrasonic Corp., 1949-51; head advanced engring. dept. Ingersoll-Rand Co., 1956-60; dir. research Thermal Dynamics Cor., 1960-61; dir. Ecol. Sci. Corp., 1968-70; co-founder, pres. Creare Inc., 1961-75; Ecol. Research Corp., 1968-70; co-founder, chmn. bd., prin. engr. Creare Innovations Co., 1976-79; founder 1979, since pres. Verax Corp., Hanover, N.H.; asst. prof. mech. engring. MIT, 1951-56; prof. engring. Thayer Sch. Engring., Dartmouth Coll., 1960—; mem. turbine and compressor subcom. NACA, 1954-55. Author numerous articles and patentee in field; editor: Jour. Fluid Engring., 1973-79. Recipient Gold medal Pi Tau Sigma, 1953, Master Designer award Product Engring. mag., 1967. Fellow ASME (chmn. hydraulics div. 1962-63, dir. Turbomachinery Inst. 1968—, Thurston lectr. 1977, Fluids Engring. award 1979); mem. Nat. Acad. Engring., Tau Beta Pi. Subspecialties: Biomedical engineering; Mechanical engineering. Current work: Research and development of advanced bioprocessing for manufacture of biochemicals. Home: Hawk Pine Hill Norwich VT 05055 Office: PO Box B-1170 Hanover NH 03755

DEAN, STEPHEN ODELL, association executive, consultant; b. Niagara Falls, N.Y., May 12, 1936; s. Stephen Odell and Marian (Gammon) D.; m. Elizabeth Alice Wiles, July 21, 1962 (div. 1976). B.S., Boston Coll., 1960; S.M., M.I.T., 1962; Ph.D., U. Md.-College Park, 1971. Physicist AEC, Washington, 1962-68; research physicist U.S. Naval Research Lab., Washington, 1968-72; div. dir. AEC, ERDA, Dept. Energy, Washington, 1972-79; div. mgr. Sci. Applications, Inc., Gaithersburg, Md., 1979—; pres., chief exec. officer Fusion Power Assn., Gaithersburg, 1979—. Editor: Prospects for Fusion Power, 1981. Recipient Research Publ. award U.S. Naval Research Lab., 1972; Spl. Achievement award AEC, 1976, U.S. ERDA, 1977. Mem. Am. Nuclear Soc. (exec. com. 1981-84), Am. Phys. Soc., Am. Soc. Assn. Execs. Subspecialties: Fusion; Laser fusion. Current work: Fusion energy development. Office: Fusion Power Assocs 2 Professional Dr Suite 249 Gaithersburg MD 20879

DEAN, THOMAS SCOTT, architect, engring. educator, cons.; b. Sherman, Tex., July 6, 1924; s. Lura Cecil and Lucille (Scott) D.; m. Jan Marie Irvine, June 1, 1945; 1 son, Thomas Scott. B.S., N. Tex. State U., Denton, 1947, M.S., 1949; Ph.D. (fellow), U. Tex., Austin, 1953. Registered architect. Registered profl. engr., Tex., 1953. With Thomas Scott Dean AIA, Dallas, 1950-60; guest lectr. U. Tex., Austin, 1960-64; from assoc. prof. to prof. Okla. State U., Stillwater, 1964-76; prof. architecture and engring. U. Kans., Lawrence, 1976—; vis. prin. lectr. N.E. London Poly., 1973. Author: Engineering Education, 1975, Thermal Storage, 1979, Accumulation de Chaleur, 1980, How to Solarize Your House, 1981; contbr. articles profl. jours.; works include 2000 houses, chs., sci. bldgs. Served with C.E. AUS, 1942-43. Recipient 3 awards of merit House and Home, 1954, 55; NASA-Stanford fellow Am. Soc. Engring. Edn., 1974. Mem. AIA, ASHRAE, Internat. Solar Energy Soc., Am. Soc. Engring. Edn. Subspecialties: Solar energy; Energy conservation. Current work: Solar heat assistance small buildings, teaching, research, energy conserving buildings consultant. Home: 1304 Raintree Pl Lawrence KS 66044 Office: 335 Art and Design Kans Lawrence KS 66045

DEARING, WILLIAM HILL, physician, educator, researcher; b. Memphis, Dec. 3, 1908; s. William H. and Theresa Irene (Trenham) D.; m. Laura Edith Wintersteen, Aug. 29, 1936; children: Jane Dearing Kearney, John Charles (dec.), Carl Baylor. B.A., U. Pa., 1930, M.D., 1934, M.A., 1934; Ph.D., U. Minn., 1941. Diplomate: Am. Bd. Internal Medicine. Intern Geisinger Meml. Hosp., Danville, Pa., 1934-35, assoc. in medicine, 1935-36; cons. in medicine Mayo Clinic, Rochester, Minn., 1936-76, head med. sect., 1955-67, bd. govs., 1955-60, emeritus cons. in medicine, 1977—; prof. medicine Mayo Grad. Sch. Medicine, U. Minn., 1962-76, Mayo Med. Sch., Rochester, 1973-76, prof. medicine emeritus, 1977—. Contbr. articles to profl. jours. Recipient Cert. of Merit U. Minn., 1977. Fellow ACP; mem. AMA (chmn. sect. gastroenterology 1952-73, ho. of dels. 1963-73), Am. Gastroenterology Assn. (officer, bd. govs. 1962-67), Am. Fedn. Clin. Research, Central Soc. Clin. Research, Central Clin. Research Club, AAAS, N.Y. Acad. Sci. Presbyterian. Subspecialties: Internal medicine; Gastroenterology. Current work: Research in gatroenterology. Home: 4505 S Ocean Blvd Apt 808 Highland Beach FL 33431

DE BAKEY, MICHAEL ELLIS, surgeon; b. Lake Charles, La., Sept. 7, 1908; s. Shaker Morris and Raheeja (Zorba) DeB.; m. Diana Cooper, Oct. 15, 1936; children—Michael Maurice, Ernest Ochsner, Barry Edward, Denis Alton, Olga Katerina; m. Katrin Fehlhaber, July 1975. B.S., Tulane U., 1930, M.D., 1932, M.S., 1935, LL.D., 1965; Docteur Honoris Causa, U. Lyon, France, 1961, U. Brussels, 1962, U. Ghent, Belgium, 1964, U. Athens, 1964; D.H.C., U. Turin, Italy, 1965, U. Belgrade, Yugoslavia, 1967; LL.D., Lafayette Coll., 1965; M.D. (hon.), Aristotelean U. of Thessaloniki, Greece, 1972; D.Sc., Hahnemann Med. Coll., 1973, numerous others. Diplomate: Nat. Bd. Med. Examiners, Am. Bd. Surgery, Am. Bd. Thoracic Surgery. Intern Charity Hosp., New Orleans, 1932-33, asst. surgery, 1933-35, U. Strasbourg, France, 1935-36, U. Heidelberg, Germany, 1936; instr. surgery, Tulane U., 1937-40, asst. prof., 1940-46, assoc. prof., 1946-48; prof. surgery, chmn. dept. Baylor U., 1948—, v.p. med. affairs, 1968—; chief exec. officer Baylor Coll. Medicine, 1968-69, pres., 1969—, chancellor, 1979—; dir. Nat. Heart and Blood Vessel Research and Demonstration Center, Baylor Coll. Medicine, 1975—; surgeon-in-chief Ben Taub Gen. Hosp., 1963—; sr. attending surgeon Meth. Hosp.; cons. surgery VA, St. Elizabeth's, M.D. Anderson, St. Luke's, Tex. Children's Hosps.; clin. prof. surgery U. Tex. Dental Br., Houston; cons. Tex. Inst. Rehab. and Research, Brooke Gen. Hosp., Brooke Army Med. Center, Ft. Sam Houston, Tex.; cons. surgery Walter Reed Army Hosp., Washington.; mem. med. adv. com. sec. def., 1948-50; chmn. com. surgery NRC, 1953, mem. exec. com., 1953; mem. com. med. services Hoover Commn.; chmn. bd. regents Nat. Library Medicine, 1959; past mem. nat. adv. heart council NIH; mem. Nat. Adv. Health Council, 1961-65, Nat. Adv. Council Regional Med. Programs, 1965—, Nat. Adv. Gen. Med. Scis. Council, 1965, Program Planning Com., Com. Tng., Nat. Heart Inst., 1961—; mem. civilian health and med. adv. council Office Asst. Sec. Def.; chmn. Pres.'s Commn. Heart Disease, Cancer and Stroke, 1964. Author: (with Robert A. Kilduffe) Blood Transfusion, 1942; author: (with Gilbert W. Beebe) Battle Casualties, 1952, (with Alton Ochsner) Textbook of Minor Surgery, 1955, (with T. Whayne) Cold Injury, Ground Type, 1958, A Surgeon's Visit to China, 1974, (with A.M. Gotto) The Living Heart, 1977; editor: Yearbook of Surgery, 1958-70; chmn. adv. editorial bd.: Medical History of World War II. Mem. Tex. Constl. Revision Commn., 1973. Served as col. Office Surgeon Gen. AUS, 1942-46; now col. Res.; cons. to Surgeon Gen., 1946—. decorated Legion of Merit, 1946; Rudolph Matas award, 1954; Independence of Jordan medal 1st class; Merit Order of Republic 1st class, Egypt; comdr. Cross of Merit Pro Utiliate Hominum Sovereign Order Knights of Hosp. of St. John of Jerusalem in Denmark; Hektoengold medal AMA; Internat. Soc. Surgery Distinguished Service award, 1957; recipient Modern Medicine award, 1957, Roswell Park medal, 1959; A.M.A. Distinguished Service award, 1959; Leriche award Internat. Soc. Surgery, 1959; Great medallion U. Ghent, 1961; Grand Cross of Order Leopold, Belgium, 1962; Albert Lasker award for clin. research, 1963; Order of Merit, Chile, 1964; St. Vincent prize med. scis. U. Turin, 1965; Orden del Libertador Gen. San Martin, Argentina, 1965; Centennial medal Albert Einstein Med. Center, 1966; Gold Scalpel award Internat. Cardiology Found., 1966; Distinguished Service prof. Baylor U., 1968; Distinguished Faculty award, 1973; Eleanor Roosevelt Humanities award, 1969; Civilian Service medal sec. def., 1970; USSR Acad. Sci. 50th Anniversary Jubilee medal, 1973; Phi Delta Epsilon Disting. Service award, 1974; La Madonnina award, 1974; 30 Yr. Service award Harris County Hosp. Dist., 1978; Knights Humanity award honoris causa Internat. Register Chivalry, Milan, 1978; Diploma de Merito Caja Costarricense de Seguro Social, San Jose, Costa Rica, 1979; Disting. Service plaque Tex. Bd. Edn., 1979; Britannica Achievement in Life award, 1979; Medal of Freedom with Distinction Presdl. award, 1969; Disting. Service award Internat. Soc. Atherosclerosis, 1979; Centennial award ASME, 1980; Marian Health Care award St. Mary's U., 1981; numerous others; named Dr. of Year Med. World News, 1965, Med. Man of Year, 1966, Humanitarian Father of Year award, 1974, Tulane U. Alumnus of Year, 1974, Tex. Scientist of Yr., Tex. Acad. Sci., 1979. Fellow A.C.S. (Ann. award Southwestern Pa. chpt. 1973), Inst. of Medicine Chgo. (hon.); mem. Am. Coll. Cardiology (hon. fellow), Royal Soc. Medicine, Halsted Soc., Am. Heart Assn., So. Soc. Clin. Research, AAAS, Southwestern Surg. Congress (pres. 1952), Soc. Vascular Surgery (pres. 1953), AMA, Am. Surg. Assn. (Disting. Service award 1981), So. Surg. Assn., Western Surg. Assn., Am. Assn. Thoracic Surgery (pres. 1959), Internat. Cardiovascular Soc. (pres. 1958, pres. N.Am chpt 1964), Mexican Acad. Surgery, hon.), Soc. Clin. Surg., Soc. Univ. Surgeons, Internat. Soc. Surgery, Soc. Exptl. Biology and Medicine, Hellenic Surg. Soc. (hon.), Bio-med. Engring. Soc. (dir. 1968), Houston Heart Assn. (mem. adv. council 1968-69), Sociedad Nacional de Cirugia (Cuba), C. of C., Sigma Xi, Alpha Omega Alpha. Democrat.

Episcopalian. Club: University (Washington). Subspecialties: Surgery; Cardiac surgery. Current work: Cardiovascular disease. Office: Baylor Coll Medicine 1200 Moursund Ave Houston TX 77030

DEBARI, VINCENT (ANTHONY), research chemist, research laboratory administrator, educator; b. Jersey City, Feb. 1, 1946; s. Vincent and Josephine (Buzzanco) DeB.; m. Margaret A. Danning, Feb. 28, 1970; children: Michele, Christopher, Jillane. B.S. in Chemistry, Fordham U., 1967, M.S., Newark Coll. Engring., (now N.J. Inst. Tech.), 1970; Ph.D. in Biochemistry, Rutgers U.-Newark, 1981. Research chemist Witco Chem. Corp., Oakland, N.J., 1967-73; research chemist Renal Lab., St. Joseph's Hosp. and Med. Center, Paterson, N.J., 1973-81, dir. renal lab., 1981—, mem. affiliate med. staff in medicine and pathology, 1981—; chmn. blood cells sect. 5th Internat. Conf. Histochemistry and Cytochemistry, Bucharest, Romania, 1976; cons. Rutgers U.; adj. faculty Seton Hall U., St. Peter's Coll. Contbr. articles to profl. publs. Founder, mem. research and grants com. Lupus Erythematosus Found. N.J., 1979-83, first chmn., 1979-81, bd. dirs., 1983—. Lions Charity Found. grantee, 1978—; Lupus Erythematosus Found. grantee, 1978—. Mem. Am. Chem. Soc., Am. Assn. Clin. Chemistry, Am. Fedn. Clin. Research, Biophys. Soc., N.Y. Acad. Scis., AAAS, Am. Soc. Clin. Pathology, Sigma Xi. Roman Catholic. Subspecialties: Biochemistry (biology); Clinical chemistry. Current work: Immunochemistry, specifically characterization of auto-antibodies and biochemistry of phagocytic cells; other interests are cation-polyanion interactions and metabolic regulation. Home: 32 Jacksonville Rd Pompton Plains NJ 07444 Office: Renal La St Joseph's Hosp and Med Center 703 Main St Paterso NJ 07503

DEBLASIO, RICHARD, research engineer; b. Norristown, Pa., Nov. 29, 1941; s. Anthony and Louise (Boccarro) DeB.; m. Olivia Ann Crouch, Apr. 6, 1968; children: Richard Andrew, Olivia Catherine. B.S. in Elec. Engring, U. Santa Clara, 1968-72; cert., MIT, 1977. Researcher, technician Stanford U., 1965-72; project engr. Underwriters Lab., Santa Clara, Calif., 1972-74; nuclear engr. U.S. AEC, Washington, 1974-78; group mgr. advanced systems research Solar Energy Research Inst., Golden, Colo., 1978—; cons. Teledyne Isotopes Corp., Palo Alto, Calif., 1967-70, Nat. Nuclear Corp., Palo Alto, 1968, Custom Nuclear Corp., 1968; project acceptance testing leader U.S./Saudi Solar Projects, Golden, Colo., 1981, Saudi Arabia, 1981. Served with USAF, 1961-65. Decorated Air Force Commendation medal. Mem. IEEE (chmn. photovoltaic systems com. 1980—, mem. standards photovoltaic coordinating com. 1980—), Am. Nuclear Soc. Roman Catholic. Subspecialties: Solar energy; Electrical engineering. Current work: Manager for advanced photovoltaic systems research which includes conceptual systems research on thin film solar cells, modules and systems including economic analysis and field testing. Home: 2581 Scorpio Dr Colorado Springs CO 80906 Office: Solar Energy Research Inst 1617 Cole Blvd Golden CO 80401

DEBONS, ALBERT FRANK, medical scientist and educator; b. Bklyn., Nov. 4, 1929; s. Phillip and Carmela Gufrieda D.; m. Evelyn Ruth Quade, Aug. 27, 1955; children: Jeanne, Elizabeth, Matthew. B.S., Syracuse U., 1953; M.S., George Washington U., 1955, Ph.D., 1958. Med. research assoc. Brookhaven Nat. Lab., Upton, L.I., N.Y., 1958-61; prin. scientist and asst. chief Radioisotope Service, VA Hosp., Birmingham, Ala., 1961-64; asst. prof. depts. physiology and medicine U. Ala. Med. Ctr., Birmingham, 1961-64; sr. scientist VA Med. Ctr., Bklyn., 1964—; cons., reviewer grant proposals NIH, NSF, 1970—. Author 32 sci. publs. Research grantee NIH-NIAMD, 1963—; VA Med. Research Funds, 1964—. Mem. Soc. Nuclear Medicine, Endocrine Soc., Am. Fedn. Clin. Research, N.Y. Lipid Research Club. Democrat. Roman Catholic. Subspecialties: Neuroendocrinology; Neurophysiology. Current work: Research activities have been centered on (a) understanding the mechanisms underlying hormone action (insulin, thyrotropin, thyroxine) on target organs and (b) elucidation of the mechanism by which the central nervous system regulates caloric homeostasis (food intake). Home: 34 Bloomingdale Rd Staten Island NY 10309 Office: Nuclear Medicine VA Med Ctr 800 Poly Pl Brooklyn NY 11209

DEBYE, NORDULF W. G., chemistry educator; b. Bad Ullersdorf, Czechoslovakia, Aug. 2, 1943; s. Gerhard and Mayon (Debye) Saxinger; m. Kristina E. Darwe, June 24, 1968; 1 son, Tyson. B.A., Rice U., 1965; Ph.D., Cornell U., 1970. NRC postdoctoral research assoc. Nat. Bur. Standards, 1970-72; asst. prof. chemistry Colo. Women's Coll., Denver, 1972-76; vis. lectr. Towson State U, Balt., 1975-76, asst. prof. chemistry, 1976-83, assoc. prof., 1983—. Contbr. articles to profl. jours. Mem. Am. Chem. Soc., N.Y. Acad. Scis., Am. Assn. Physics Tchrs., Sigma Xi. Democrat. Subspecialties: Physical chemistry; Thermodynamics. Current work: Calorimetry, structural organometallic chemistry, spectroscopy. Office: Dept Chemistry Towson State U Baltimore MD 21204

DECKER, ARTHUR JOHN, optical physicist; b. Butte, Mont., Oct. 16, 1941; s. Lester Paul and Louise Constance (Kraft) D.; m. Marilyn Anne, Nov. 7, 1970; 1 son, Bruce L. B.S., U. Wash., 1963; M.A., U. Rochester, 1966; Ph.D., Case Western Res. U., 1977. Electronics engr. Lewis Research Ctr., NASA, Cleve., from 1966, now optical physicist. Contbr. articles to profl. jours.; author NASA papers on use of optical measurement methods for aerospace applications. Mem. Optical Soc. Am. Subspecialties: Holography; Aerospace engineering and technology. Current work: Three-dimensional imaging methods for display and measurement of properties of fluids and structures: holographic cinematography, interferometry, spectroscopy. Patentee in field. Office: 21000 Brookpark Rd MS 77-1 Cleveland OH 44135

DECKER, BRUCE MICHAEL, electrical engineer, systems specialist; b. Waterbury, Conn., June 9, 1955; s. Lloyd S. and Erma B. (Miller) D.; m. Holly Jean Hatch, May 28, 1977; children: Stephen Paul, Mark Timothy. B.S. in engring, U. Conn.-Storrs, 1977, M.S., 1979. Engr. in tng., Conn. Mfg. engr. DuPont, Newtown, Conn., 1979-82; research engr. U. N.H., 1982; mem. tech. staff RCA, Burlington, Mass., 1983—. Mem. IEEE, Sigma Xi. Republican. Subspecialties: Artificial intelligence; Computer engineering. Current work: Robotics; automation; productivity improvement; testing. Office: RCA Government Systems Division PO Box 588 Burlington MA 01803

DECKER, DAVID ARNOLD, oncology researcher; b. Wyandotte, Mich., June 14, 1948; s. James Edward and Viola (Cordes) D.; m. Veronica Decker, Mar. 25, 1952; children: Brian David, Pal David, Marc David. Student, Eastern Mich. U., 1966-70; M.D., Wayne State U., 1973. Diplomate: Am. Bd. Internal Medicine. Intern Mayo Clinic, Rochester, Minn., 1974, resident, 1975-77; Instr medicine Mayo Med. Sch., 1978; asst. prof. medicine Mich. State U., 1979-81, dir. fellowship program hematology and oncology, 1979-81; head hematology, oncology sect. Ingham County Hosp., 1979-81; asst. prof. oncology Wayne State U., 1982—. Contbr. articles to profl. jours. Bd. dirs. Ingham County unit Am. Cancer Soc. Mem. Am. Fedn. Clin. Research, Am. Soc. Clin. Oncology, Am. Assn. Cancer Reserach, ACP. Home: 3825 Burkoff St Troy MI 48084 Office: 3990 John R Detroit MI 48201

DECKER, WALTER JOHNS, toxicologist, cons., researcher, educator; b. Tannersville, N.Y., June 13, 1933; s. H. Russell and Leola (Coons) D.; m. Barbara Allan Hart, Aug. 19, 1961; children: Karl Hart, Reid Johns, Sam Travis. A.B., SUNY, Albany, 1954, M.A., 1955; Ph.D., George Washington U., 1966. Commd. 2d lt. U.S. Army, 1955, advanced through grades to lt. col., 1970; research asst. Walter Reed Army Inst. Research, Washington, 1955-56, research biochemist, 1957-60; chief indsl. hygiene U.S. Army Med. Lab., Japan, 1956-57; asst. chief clin. research William Beaumont Army Med. Center, El Paso, 1965-71; chief chemistry dept. U.S. Army Med. Lab., San Antonio, 1971-75; ret., 1975; assoc. prof. pharmacology and toxicology, also dept. pediatrics U. Tex. Med. Br., Galveston, 1976-83; cons. in toxicology, 1983—; mem. toxicology data bank peer rev. com. Nat. Library of Medicine. Assoc. editor: Clin. Toxicology; mem. editorial bd.: Jour. Toxicology and Environ. Health; contbr. articles to profl. jours. Decorated Legion of Merit, others; recipient Aesculapius award Tex. Med. Assn., 1977. Fellow Am. Acad. Clin. Toxicology, Am. Acad. Forensic Scis., AAAS, Am. Inst. Chemists; mem. Am. Chem. Soc., Soc. Toxicology, Sigma Xi. Episcopalian. Subspecialties: Toxicology (medicine); Neuroimmunology. Current work: Methods of detection and treatment of poisoning. Home and Office: 10741 Lemonade El Paso TX 79924

DECKER, WAYNE LEROY, meteorologist, educator; b. Patterson, Iowa, Jan. 24, 1922; s. Albert Henry and Effie (Holmes) D.; m. Martha Jane Livingston, Dec. 29, 1943; 1 dau., Susan Jane. B.S., Central Coll., Pella, Iowa, 1943; postgrad., UCLA, 1943-44; M.S., Iowa State U., 1947, Ph.D., 1955. Meteorologist U.S. Weather Bur., Washington and Des Moines, 1947-49; mem. faculty U. Mo. at Columbia, 1949—, prof. meteorology, 1958-67, prof., chmn. dept. atmospheric sci., 1967—; chmn. com. climatic fluctuations and agrl. prodn. NRC, 1975-76; bd. dirs. Council for Agrl. Sci. and Tech., 1978—, mem. exec. com., 1981—. Mem. Am. Meteorol. Soc., Internat. Soc. Biometeorology, Am. Geophys. Union, Am. Agronomy Soc., Sigma Xi, Gamma Sigma Delta. Subspecialties: Climatology. Current work: Graduate education and research on the impacts of weather and clilmate variabilities on agricultural production. Home: 1007 Hulen Dr Columbia MO 65201 Office: 701 Hitt St Columbia MO 65211

DECORA, ANDREW W., energy consultant, educator. B.S. in Chemistry, U.Wyo., 1950, M.S., 1957, Ph.D., 1962. Formerly with U.S. Bur. Mines, U.S. Energy Research Devel. Adminstrn.; with U.S. Dept. Energy, Laramie (Wyo.) Energy Tech. Center, to 1975, dir., 1975-81; adj. prof. chemistry, chem. engring. U. Wyo., Laramie, 1975—; cons. in field; chmn. UN Oil Shale & Tar Sands Panel, 1982; lectr. in field to profl. confs. Contbr. chpts. to books and articles in field to profl. jours. Served to cl. U.S. Army MC, 1951-53. Mem. Am. Chem. Soc. (midwest rep. 1967-68), Am. ASTM (com. pres. 1974), AAAS, Colo. Acad. Scis., Wyo. Acad. Scis., U. Wyo. Alumni Assn. (disting. alumnus 1978), Aircraft Owners and Pilots Assn., NRA, Nat. Muzzle Loaders Rifle Assn., Sigma Xi (chmn. 1971-72), Gamma Sigma Epsilon. Clubs: U. Wyo. Cowboy Joe, Toastmasters. Subspecialty: Energy consulting. Current work: Research in energy, environment, management and education. Office: Laramie Energy Tech Ctr PO Box 3395 University Station Laramie WY 82071

DEDERICK, JUDITH GARRETTSON, psychologist, educator; b. St. Louis, Apr. 4, 1943; d. John Alexander and Elizabeth (Painter) Garrettson; m. Warren Emery Dederick, Aug. 3, 1975; 1 dau., Elizabeth Jane. A.B., Vassar Coll., 1965; M.A., Tchrs. Coll., Columbia U., 1966; Ph.D., Columbia U., 1969. Lic. psychologist, N.Y., N.J. Research asst. Tchrs. Coll., Columbia U., N.Y.C., 1965-68; instr., then asst. prof., then assoc. prof. Hunter Coll., CUNY, 1968-81; prof. psychology Hunter Coll., 1981—; psychologist in pvt. practice, Westfield, N.J., Bklyn., cons. forensic psychology. Contbr. articles to profl. jours. Heft-Patterson fellow, 1966-68; NDEA Title IV scholar, 1966-69; Internat. Inst. Humanistic Edn. fellow, 1978. Mem. Am. Psychol. Assn., N.J. Psychol. Assn., N.J. Acad. Psychology, Phi Beta Kappa. Presbyterian. Subspecialty: Clinical-developmental psychology. Current work: Psychodiagnostics and psychotherapy of children and adults; designing ways in which families in process of divorce can resolve conflicts involving children in non-adversarial ways; discovering reliable methods for making custody decisions when disputed in divorce. Home: 211 N Chestnut St Westfield NJ 07090

DEDIU, MIHAI VIRGIL, computer scientist; b. Iasi, Romania, Nov. 6, 1943; came to U.S., 1978, naturalized, 1983; s. Virgil and Ana (Condurache) D.; m. Sofia Scarlat, July 22, 1964; children: Ovidiu, Horatiu. M.S., U. Bucharest, 1966; Ph.D., Inst. Math., Bucharest, 1972. Researcher Inst. Math., Bucharest, 1967-75; asst. prof. U. Bucharest, 1973-74; researcher Inst. Physics, Bucharest, 1976-76, U. Turin, Italy, 1977-78, Case Western Res. U., Cleve., 1979—; pres. Dr. Dediu Research Inst., Cleve., 1981—. Author: Mathematical Software, 1984. Mem. Republican Presdl. Task Force, Washington, 1981. Mem. Soc. Indsl. and Applied Math., Am. Math. Soc., Math. Assn. Am., Math. Modelling Assn., N.Y. Acad. Scis., Library Computer and Info. Scis. Subspecialties: Mathematical software; Programming languages. Current work: Mathematical software, programming languages, graphics systems analysis, applied numerical analysis, algorithms, artificial intelligence, database systems, information systems, operating systems. Home: 1242 Cook Ave Lakewood OH 44107 Office: Case Western Res U University Circle Cleveland OH 44106

DE DUVE, CHRISTIAN RENE, educator, scientist; b. Thames-Ditton, Eng., Oct. 2, 1917; s. Alphonse and Madeleine (Pungs) de D.; m. Janine Herman, Sept. 30, 1943; children: Thierry, Anne, Francoise, Alain. M.D., U. Louvain, Belgium, 1941; M.Sc., 1946; M.Sc. Dr. honoris causa, univs. Turin, Leiden, Lille, Sherbrooke, Ghent, Liège, Catholic U. Chile, Université René Decartes, Paris, Gustavus Adolphus Coll., St. Peter, Minn., U. Rosario (Argentina), U. Aix-Marseille II. Prof. physiol. chemistry U. Louvain Med. Sch., 1947; prof. biochem. cytology Rockefeller U., N.Y.C., 1962-74, Andrew W. Mellon prof., 1974—. Recipient Prix des Alumni, 1949, Prix Pfizer, 1957, Prix Francqui, 1960, Prix Quinquennal Belge des Sciences Médicales, 1967; Belgium; Gairdner Found. Internat. award merit, Can., 1967; Dr. H.P. Heineken Prijs, Netherlands, 1973; Nobel prize, 1974. Mem. Royal Acad. Medicine, Royal Acad. Belgium, Am. Chem. Soc., Biochem. Soc., Am. Soc. Biol. Chemists, Pontf Acad. Sci., Am. Soc. Cell Biology, Deutsche Akademie der Naturforscher Leopoldina, Soc. Chim. Biol., Soc. Belge Biochim., Sigma Xi; fgn. mem. Am. Acad. Arts and Scis.; fgn. asso. Nat. Acad. Scis., Académie des Scis. de Paris, Academie des Sciences d'Athene. Subspecialties: Biochemistry (biology); Cell biology. Current work: Basic cellular and molecular biology and its applications in medicine, therapeutics and biotechnology. Home: 80 Central Park W New York NY 10023 Office: Rockefeller U York Ave and 66th St New York NY 10021

DEEDWANIA, PRAKASH CHANDRA, cardiologist, educator; b. Ajmer, India, Aug. 28, 1948; came to U.S., naturalized, 1971; s. Gokul C. and Paras D. (Garg) D.; m. Catherine E. Deedwania, June 26, 1977. M.B.B.S., U. Rajasthan, India, 1969. Diplomate: Am. Bd. Internal Medicine, Am. Bd. Cardiology, Am. Bd. Pulmonary Disease. Intern Coney Island Hosp.-Downstate Med. Ctr., 1971-72; resident in medicine VA Hosp., Mt. Sinai, N.Y., 1972-74; fellow in cardiology U. Ill., 1975-77; chief cardiology VA Med. Ctr. U. Calif.-San Francisco, 1980—, dir. med. intensive care unit/CCU, 1980—, chmn. critical care com., 1982—, assoc. clin. prof. medicine, 1980—. Research grantee Merck Sharpe and Dohme, West Pitts., Pa., 1982—, Marion Labs., Kansas City, Mo., 1982—, Miles Labs., West Haven, Conn., 1982—. Fellow Am. Coll. Cardiology, Am. Heart Assn., ACP, Am. Coll. Chest Physicians; mem. Am. Fedn. Clin. Research, Central Valley Heart Assn. Subspecialties: Cardiology; Internal medicine. Current work: Role of myocardial imaging techniques in diagnosis of cardiac patients; newer agents in heart failure and myocardial ischemia; treatment of hypertension; coronary risk factors and prevention of coronary disease. Office: VA Medical Center 2615 E Clinton Ave Fresno CA 93711

DEEN, JAMES ROBERT, nuclear engineer, researcher; b. Dallas, Mar. 1, 1944; s. James Young and Dorothy Faye (Looney) D.; m. Katy James Pavlidou, Aug. 14, 1971; children: Dorothy, Christina, David, Joshua. B.Engr. Sci., U. Tex.-Austin, 1966, M.S. in Mech. Engring, 1970, Ph.D., 1973. Profl. nuclear engr., Calif. Sr. engr. Gen. Electric Co., San Jose, Calif., 1972-76; nuclear engr. Argonne (Ill.) Nat. Lab., 1976—. Contbr. articles to publs. Mem. Am. Nuclear Soc., Sigma Xi. Subspecialty: Nuclear engineering. Current work: Conversion studies for all research and test reactors from use of high to low enriched fuel. Various neutronic analysis techniques used in studies. Home: 593 Cambridge Way Bolingbrook IL 60439 Office: Argonne Nat Lab 9700 S Cass Ave Argonne IL 60439

DEEPAK, ADARSH, meteorologist; b. Sialkot, India, Nov. 13, 1936. B.S. Delhi (India) U, 1956; M.S., 1959; Ph.D. in Aerospace Engring, U. Fla., 1969. Lectr. physics Delhi U., 1959-63; instr. phys. sci. U. Fla., 1965-68, research assoc. physics, 1970-71; NRC fellow Marshall Space Flight Ctr., NASA, 1972-74; research assoc. prof. physics and geophys. sci. Old Dominion U., 1974-77; pres. Inst. Atmospheric Optics and Remote Sensing, Hampton, Va., 1977—, Sci. & Tech. Corp., Hampton, 1979—; cons. engring. sci. Wayne State U., Detroit, 1970-72; mem. staff NASA Tech. Workshop, 1975; adj. prof. physics Coll. William and Mary, 1979-80; leader U.S. del. Internat. Workshop Application Remote Sensing Rice Prodn., India, 1981. NSF travel grantee, India, 1976. Mem. Optical Soc. Am., Am. Metereol. Soc., Air Pollution Control Assn., Am. Geophys. Union, AAAS. Subspecialties: Meteorology; Remote sensing (geoscience). Office: Inst Atmospheric Optics PO Box P Hampton Va 23666

DEERE, DON UEL, consulting engineering geologist; b. Corning, Iowa, Mar. 17, 1922. B.S., Iowa State U., 1943; M.S., U. Colo., 1949; Ph.D. in Civil Engring., U. Ill., 1955. Jr. mine engr. Phelps Dodge Corp., 1943-44; mine engr. exploration dept. Potash Co. Am., N.Mex., 1944-47; asst. prof. to assoc. prof. civil engring. Coll. A&M., P.R., 1946-50, head dept., 1950-51; ptnr. Found. Engring. Co., P.R., 1951-55; assoc. prof. to prof. civil engring. and geology U. Ill., Urbana-Champaign, 1955-76; prof. civil engring. U. Fla., Gainesville, 1976-83; cons engring. geologist, 1983—. Mem. AAAS, Geol. Soc. Am., Am. Geophys. Union, ASCE. Subspecialty: Civil engineering. Address: 6834 S W 35th Way Gainesville FL 32608

DEERING, REGINALD ATWELL, molecular biologist, educator, researcher; b. Brooks, Maine, Sept. 21, 1932; s. Raymond Atwell and Sibyl Louise (Tibbetts) D.; m. Anne-Lise Dahlsrud, Oct. 20, 1956; children: Eric, Mark, Linda, Norman. B.S., U. Maine, Orono, 1954; Ph.D., Yale U., 1958. Asst prof. So. Ill. U., Carbondale, 1957-58; research assoc. U. Oslo, Norway, 1958-59, Yale U., New Haven, 1959-61; asst. prof., assoc. prof. N.Mex. Highlands U., Las Vegas, 1961-64; assoc. prof. to prof. molecular and cell biology Pa. State U., University Park, 1964—; vis. prof. Stanford U., 1974-75. Contbr. articles to nat. and internat. jours. Recipient various research grants. Mem. Am. Soc. Biol. Chemists, Am. Soc. Microbiology, Am. Soc. Photobiology, Biophys. Soc. Subspecialties: Molecular biology; Genetics and genetic engineering (biology). Current work: Research in enzymatic repair of DNA, mutagenesis, development in lower eycaryotes, cloning of repair genes. Office: 201 Althouse Lab University Park PA 16802

DEESE, JAMES EARLE, psychologist, educator; b. Salt Lake City, Dec. 14, 1921; s. Thomas D. and Serena Jane (Johnson) D.; m. Ellin Ruth Krauss, Dec. 24, 1948; children—Elizabeth Ellin, James Lawrence. A.B., Chapman Coll., 1944; A.M., Ind. U., 1946, Ph.D., 1948. From asst. prof. to prof. psychology Johns Hopkins, Balt., 1948-72; Commonwealth prof. U. Va., 1972-80, Hugh Scott Hamilton prof., 1980—; vis. prof. U. Calif. at Berkeley, 1958-59. Cons. editor: Psychol. Rev; Author: (with E.K. Deese) Psychology of Learning, 5th edit, (with S.H. Hulse and H. Egeth), 1980, Principles of Psychology 2d edit, (with W. Lado and R. Goodale), 1975, The Structure of Associations in Language and Thought, 1965, General Psychology, 1967, Psycholinguistics, 1969, Psychology as Science and Art, 1972, (with L.B. Szalay) Subjective Meaning and Culture, 1978; contbg. author: Introduction to Psychology (C.T. Morgan), 1955; Asso. editor: Jour. Exptl. Psychology, 1963-68; editor: Psychol. Bull, 1968-74. Fellow Soc. Exptl. Psychologist, Am. Psychol. Assn. (dir.); mem. Eastern Psychol. Assn. (pres. 1966-67), Linguistic Soc. Am., AAAS (v.p. 1971-72). Democrat. Clubs: 14 West Hamilton Street (Balt.); Greencroft, Colonnade (Charlottesville); Cosmos (Washington). Subspecialties: Cognition; Neuropsychology. Current work: Language, reading, speech, language disabililties. Home: 1829 Westview Rd Charlottesville VA 22903

DEEVEY, EDWARD SMITH, JR., biologist; b. Albany, N.Y., Dec. 3, 1914; s. Edward Smith and Villa (Augur) D.; m. Georgiana Baxter, Dec. 24, 1938 (dec. Jan. 1982); children: Ruth (Mrs. Lehmann), Edward Brian, David Kevin.; m. Dian R. Hitchcock, Jan. 22, 1983. B.A., Yale, 1934, Ph.D., 1938. Instr. biology Rice Inst., Houston, 1939-43; research asso. biology Woods Hole Oceanographic Instn., 1943-46; asst. prof. biology Yale, 1946-51, asso. prof., 1951-57, prof., 1957-68; Dir. Geochronometric Lab., 1951-62; Killam research prof. biology Dalhousie U., Halifax, N.S., Can., 1968-71; grad. research prof. U. Fla., 1971—; curator paleoecology Fla. State Mus., 1971—; Sect. head environ. and systematic biology NSF, 1967-68; mem. Fisheries Research Bd., Canada, 1969-71, Nat. Acad. Scis. NSF sr. postdoctoral fellow; Fulbright travel grantee U. Canterbury, Christchurch, New Zealand, 1964-65; Recipient Fulbright research award, Denmark, 1953-54; Guggenheim fellow, Denmark, 1953-54. Fellow AAAS, Geol. Soc. Am.; mem. Am. Soc. Limnology and Oceanography (pres. 1974—), Ecol. Soc. Am. (pres. 1970), Am. Anthrop. Assn., Soc. Am. Archaeology, Am. Soc. Naturalists. Subspecialty: Ecology. Current work: Historical ecology, paleolimnology, biogeochemistry. Home: 1702 SW 35th Pl Gainesville FL 32608 also Sheldon Pl Pine Orchard Branford CT 06405

DE FANTI, THOMAS ALBERT, computer science educator; b. N.Y.C., Sept. 18, 1948; s. Charles Leonard and Madeline (Kaiser) De F. B.A., Queens Coll., 1969; M.S., Ohio State U., 1970, Ph.D., 1973. Asst. prof. U. Ill., Chgo., 1973-78, assoc. prof. dept. elec. engring. and computer sci., 1978—, dir., 1976—; pres. Real Time Design, Inc., Chgo., 1981—. Editor: Siggraph Video Rev, 1980—. Mem. Assn. Computing Machinery, Spl. Interest Group on Graphics (sec. 1977-81, chmn. 1981—). Subspecialties: Graphics, image processing, and pattern recognition; Programming languages. Current work: Personal real-time interactive video graphics/computer graphics/computer based learning/computer art. Office: Dept Elec Engring U Ill Box 4348 Chicago IL 60680

DEFIEBRE, CONRAD WILLIAM, microbiologist; b. Bklyn., Jan. 19, 1924; s. Conrad William and Barbara (Benisch) deF.; m. Harriet M. Hamm, Oct. 26, 1946; children—Conrad, Henry, Jeremy, Timothy, David, Christopher. B.S., Rensselaer Poly. Inst., 1949; M.S., U. Wis.,

1950, Ph.D., 1952. Asst. bacteriologist U. Wis., 1949-52; research microbiologist duPont, Wilmington, Del., 1952-61; research dir. Wilson Labs. div. Wilson Pharm. & Chem. Corp., Chgo., 1961-67, v.p. research, 1967-69; with Ross Labs. div. Abbott Labs., Columbus, Ohio, 1969—, v.p. research and devel., 1971—; chmn. bd. infant formula council Abbott Labs., 1976-77, 81—; 2d vice chmn. Central Ohio Rehab. Ctr., Goodwill Industries, 1980. Served with AUS, 1943-46. Recipient award of merit Ohio State U., 1981. Fellow Am. Inst. Chemists; mem. Am. Chem. Soc., Am. Soc. Microbiology, Am. Inst. Biol. Scis., Sigma Xi. Roman Catholic. Clubs: Worthington Hills Country Businessmen's Athletic. Subspecialty: Microbiology. Home: 8000 Fairway Dr Worthington OH 43085 Office: 625 Cleveland Ave Columbus OH 43216

DEFILIPPO, FELIX CARLOS, physicist; b. Buenos Aires, Argentina, July 24, 1943; came to U.S., 1979; s. Miguel and Angela (Portuese) D.; m. Nuria Maria Hernandez, Jan. 13, 1967; children: Ernesto Antonio, Eduardo Pablo. M.Physics, Instituto Balseiro, Bariloche, Rio Negro, Argentina, 1967, Ph.D., 1978. Aux. prof. U. Buenos Aires, 1968-69; researcher AEC Argentina, Buenos Aires, 1968-79; research asst. prof. U. Tenn., Oak Ridge Nat. Lab., 1979—; lectr. Nat. U. Cordoba, Argentina, 1972, Peruvian Inst. Nuclear Energy, Lima, 1978-79. Contbr. articles in field to profl. jours. IAEA fellow, Vienna, 1974. Mem. Am. Nuclear Soc. Subspecialties: Nuclear fission; Nuclear engineering. Current work: Nondestructive assay of spent fuel, noise analysis of nuclear reactors, mathematical modeling of nuclear power plants, in general applied physics. Home: 102 Westwind Dr Oak Ridge TN 37830 Office: Oak Ridge Nat Lab PO Box X Bldg 3500 Oak Ridge TN 37830

DE FRIES, JOHN CLARENCE, behavioral genetics educator, institute administrator; b. Delrey, Ill., Nov. 26, 1934; s. Walter C. and Irene Mary (Lyon) DeF.; m. Marjorie Jacobs, Aug. 18, 1956; children: Craig, Brian, Catherine Ann. B.S., U. Ill.-Urbana, 1956, M.S., 1958, Ph.D., 1961. Asst. prof. U. Ill., 1961-63, 64-66, assoc. prof., 1964-67; research fellow U. Calif., Berkeley, 1963-64; assoc. prof. behavioral genetics and psychology U. Colo., 1967-70, prof., 1970—, dir., 1981—. Author: (with G.E. McClearn) Introduction to Behavioral Genetics, 1973, (with Plomin and McClearn) Behavioral Genetics: A Primer, 1980; co-founder: Behavior Genetics jour, 1970. Served to 1st lt. U.S. Army, 1957-65. Grantee in field. Mem. Behavior Genetics Assn. (sec. 1974-77, pres. 1982-83), AAAS. Current work: Genetics of specific cognitive abilities and reading disability. Home: 1725 Ithaca Dr Boulder CO 80303 Office: Inst Behavioral Genetics U Colo Boulder CO 80309

DEGENKOLB, HENRY JOHN, consultant structural engineer; b. Peoria, Ill., July 13, 1913; s. Gustav J. and Alice (Emmert) D.; m. Anna Alma Nygren, Sept. 9, 1939; children: Virginia A. Degenkolb Craik, Joan A. Degenkolb Boain, Marion S. Degenkolb Hune, Patricia H. Degenkolb Blanton, Paul H. B.S. in Civil Engring., U. Calif., Berkeley, 1936. With various engring firms, 1936-46; chief engr., then partner John J. Gould & H.J. Degenkolb, engrs., San Francisco, 1946-61; pres. H.J. Degenkolb & Assocs., San Francisco, 1961-79, chmn. bd., 1980—; lectr. U. Calif. extension, 1947-58; mem. Calif. Bldg. Standards Commn., 1971—, Calif. Seismic Safety Commn., 1975-77, Presdl. Task Force Earthquake Hazard Reduction, 1970-71; mem. engring. criteria rev. bd. Bay Conservation and Devel. Commn., 1970-76; Trustee Cogswell Coll., San Francisco. Contbr. profl. publns. Hon. mem. ASCE (Moiseiff award 1953, Ernest E. Howard award 1967); mem. Am. Concrete Inst., Am. Cons. Engrs. Council, Cons. Engrs. Assn. Calif., Structural Engrs. Assn. Calif. (pres. 1958), Earthquake Engring. Research Inst. (pres. 1974-78), Forest Hills Assn. (pres. 1957). Club: San Francisco Engineers. Subspecialty: Civil engineering. Current work: Structural engineering; earthquake engineering. Home: 95 Linares Ave San Francisco CA 94116 Office: 350 Sansome St San Francisco CA 94104

DEGHETT, VICTOR JOHN, psychology educator, researcher; b. N.Y.C., May 26, 1942; s. Victor and Mary (Gorman) De G.; m. Stephanie Coyne, Aug. 2, 1980. B.A., U. Dayton, 1964; Ph.D., Bowling Green State U., 1972. Instr. psychology U. Dayton, 1966-67; instr. SUNY-Potsdam, 1971-72, asst. prof., 1972-77, assoc. prof., 1977—. Contbr. articles to profl. jours. Recipient Chancellor's award for excellence in teaching SUNY, 1982. Mem. Am. Soc. naturalists, Am. Soc. Mammalogists, Animal Behavior Soc. (chmn. edn. com. 1974-80, mem. editorial bd. 1982-), Sigma Ci. Democrat. Subspecialties: Ethology; Sociobiology. Current work: Field and laboratory research on the social behavior and the development of behavior in the meadow role. Office: SUNY Dept Psychology Potsdam NY 13676 Home: RD 4 Route 72 Potsdam NY 13676

DEGIOVANNI-DONNELLY, ROSALIE F., research biologist, educator, lectr.; b. Bklyn., Nov. 22, 1926; d. Frank and Rose Emily (Quartuccio) DeGiovanni; m. Edward Frances Donnelly, Sept. 23, 1961; children: Edward Francis, Francis M. B.A., Bklyn. Coll., 1947, M.A., 1953; Ph.D., Columbia U., 1961. Chief allergy lab. Univ. Hosp., N.Y.C., 1950-52; research asso. Sch. Pub. Health, Columbia U., N.Y.C., 1952-54, research asst. dept. biochemistry, 1954-62; chief microbial genetics lab. Bionetics Research Labs., Falls Church, Va., 1963-67; research biologist FDA, Washington, 1968—; assoc. prof., lectr. microbial genetics George Washington U., Washington, 1968—, mem. recombinant DNA research com. Contbr. research papers to sci. publs. Recipient FDA merit award, 1971. Mem. Am. Soc. Microbiology, AAAS, Environ. Muttagen Soc., N.Y. Acad. Scis., Sigma Xi. Democrat. Roman Catholic. Club: Highlands. Subspecialties: Molecular biology; Genome organization. Current work: Mechanisms of mutation, mode of action of mutagens, transposons as agents for mutation and evolution of genomes. Home: 1712 Strine Dr McLean VA 22101 Office: FDA Genetic Toxicology HFF 166 200 C St Washington DC 20204

DEGNAN, JOHN JAMES, III, physicist; b. Phila., Dec. 10, 1945; s. John James and Ruth Deloris (Vece) D.; m. Adele Susan Henry, June 27, 1969; children: Adam John, Andrew Paul. B.S., Drexel U., 1968; M.S., U. Md., 1970, Ph.D., 1979. Student trainee NASA Goddard Space Flight Center, Greenbelt, Md., 1964-68, research physicist, 1968-72, sr. physicist, 1972-79, head advanced electro-optical instrument sect., 1979—; assoc. mem. Adv. Group on Electron Devices, Working Group D, 1980. Trustee Scholarship Drexel U., 1963. Recipient Quality Increase award NASA, 1974, Spl. Achievement award, 1976. Mem. Optical Soc. Am., Am. Inst. Physics, Sigma Pi Sigma, Sigma Pi. Roman Catholic. Current work: Laser ranging and altimetry, infrared heterodyne spectroscopy dye lasers, ultrashort pulse Nd: YAG lasers, waveguide carbon dioxide lasers, laser communications. Office: Code 723 NASA Goddard Space Flight Center Greenbelt MD 20771

DE GROAT, WILLIAM CHESNEY, pharmacologist; b. Trenton, N.J., Oct. 18, 1938; s. William Chesney and Margaret (Welch) deG; m. Dorothy Marion Albertson, June 13, 1959; children: Allyson, Cynthia, Jennifer. B.Sc., Phila. Coll. Pharmacy and Sci., 1960, M.Sc., 1962; Ph.D., U. Pa., 1965, postgrad. (NSF fellow), 1965-66. Riker Internat. Pharmacology fellow, NSF fellow Australian Nat. U., Canberra, 1966-67, vis. research fellow, 1967-68; mem. faculty U. Pitts. Med. Sch., 1968—, prof. pharmacology, 1977—, prof. psychology, 1982—; mem. pharmacology test com. Nat. Bd. Med. Examiners, 1979-82; U.S. rep. Internat. Union Physiol. Scis. Commn. for Autonomic Nervous System, 1982—. Editorial bd.: Jour. Pharmacology and Exptl. Therapeutics, 1975-83, Jour. Autonomic Nervous System, 1979-83, Jour. Neurourology and Urodynamics, 1982, Am. Jour. Physiology, 1983; contbr. articles to profl. jours. Mem. Pitts. Neurosci. Soc., AAAS, Am. Soc. Pharmacology and Exptl. Therapeutics, Soc. Neurosci., N.Y. Acad. Scis., Urodynamics Soc., Internat. Brain Research Orgn., Sigma Xi. Subspecialties: Neuropharmacology; Neurophysiology. Current work: Autonomic physiology and pharmacology; neural control of urinary tract function and large intestine; synaptic transmission, neuropeptides. Office: Dept Pharmacology Med Sch Univ Pittsburgh Pittsburgh PA 15261

DEGROOT, LESLIE JACOB, medical educator; b. Ft. Edward, N.Y., Sept. 20, 1928. B.S., Union Coll., 1948; M.D., Columbia U., 1952. Intern, asst. resident in medicine Presbyn. Hosp., N.Y.C., 1952-54; health physician Nat. Cancer Inst., 1954-55; physician U.S. Mission, Afghanistan, 1955-56; resident Mass. Gen. Hosp., Boston, 1957-58; asst. Harvard Med. Sch., 1958-59, instr., 1959-62, assoc., 1962-66; assoc. prof. exptl. medicine MIT, 1966-68, assoc. dir. dept. nutrition and food sci. Clin. Research Ctr., 1966-68; prof. endocrinology Pritzker Sch. Medicine, U. Chgo., 1968—, chief sect. thyroid study unit, 1968—; clin. and research fellow medicine Mass. Gen. Hosp., Boston, 1956, 58-60, asst., 1960-64, asst. physician, 1964-66. Nat. Cancer Inst. clin. fellow, 1954-55. Mem. Assn. Am. Physicians, Am. Thyroid Assn., Endocrine Soc., Am. Soc. Clin. Investigation, Am. Fedn. Clin. Research. Subspecialty: Endocrinology. Office: Univ Chgo Med Ctr Box 138 950 E 59th St Chicago IL 60637

DEGROOT, RICHARD DOUGLAS, computer scientist and educator; b. Austin, Tex., Mar. 13, 1951; s. Val Richard and Margaret Ann (Huffaker) DeG.; m. Katherine Dianne Karol, Aug. 30, 1980. B.S. cum laude in Math., U. Tex.-Austin, 1976, Ph.D. in Computer Sci., 1981. Computer programmer Computation Ctr., U. Tex.-Austin, 1974-78, dept. computer scis., 1978-82; mem. tech. staff Computer Automation, Austin, 1978-82; research staff mem. IBM Thomas J. Watson Research Ctr., Yorktown Heights, N.Y., 1981—; adj. lectr. Pace U., White Plains, N.Y., 1982—; dir. NYU, Manhattanville, 1983—. Contbr. articles to publs. in field. Mem. Assn. Computing Machinery, IEEE (v.p. tech. com. on computer langs.). Subspecialties: Operating systems; Computer architecture. Current work: Manager parallel symbolic processors; logic programming; highly parallel multiprocessor systems; distributed operating systems; semantic networks; expert systems; natural language processing; concurrent programming languages. Inventor.

DE HAAN, HENRY JOHN, research psychologist; b. St. Clair County, Ill., Nov. 23, 1920; s. Henry John and Fanny (Haislip) de H.; m. Mary J. Farrell, Oct. 22, 1943. A.B., Washington U., St. Louis, 1942, M.A., 1949; Ph.D. in Psychology, U. Pitts., 1960. Postdoctoral trainee in physiol. psychology VA Hosp., Coatesville, Pa., 1960-62; research scientist George Washington U., 1962-64; research psychologist Armed Forces Radiobiology Research Inst., Bethesda, Md., 1965-69, Dept. Army, U.S. Army Research Inst., Alexandria, Va., 1969—; faculty mem. Dept. Agr. Grad. Sch., 1967-77. Contbr. articles on research psychology to profl. jours. Served with USN, 1942-44. Mem. AAAS, Am. Psychol. Assn., Internat. Neuropsychology Soc., Internat. Primatological Soc., Psychonomic Soc., Soc. for Neurosci., Sigma Xi. Subspecialties: Perception; Psychophysics. Current work: Speech perception, analysis, and synthesis; time-compressed speech; machine recognition of speech, visual perception, biological factors in perception and behavior. Home: 5403 Yorkshire St Springfield VA 22151 Office: 5001 Eisenhower Ave Alexandria VA 22333

DE HAAS, HERMAN, biochemistry educator; b. Northbridge, Mass., Jan. 6, 1924; s. Nicholas and Elizabeth (Jongsma) De H.; m. Eugenie T. Lee, Dec. 20, 1951; children: David Alan, N. Elisabeth, Pieter Frank, Timothy John. B.S., Westminster Coll., 1947; M.S., U. Mich., 1951, Ph.D. in Biol. Chemistry, 1955. Asst. prof., then assoc. prof. chemistry Westminster (Pa.) Coll., 1955-59; asst. prof. biochemistry U. Maine, Orono, 1959-62, assoc. prof., 1962-71, prof., 1971—; mem. staff Maine Agrl. Expt. Sta., 1959-80. Mem. Am. Chem. Soc., Sigma Xi. Subspecialties: Biochemistry (biology); Biochemistry (medicine). Current work: Protein nutritional quality as affected by processing; identification of biochemical entities in marine species. Office: Dept Biochemistry U Maine 202 Hitchner Hall Orono ME 04469

DEHGAN, BIJAN, horticulture educator, researcher; b. Shiraz, Iran, Mar. 4, 1939; came to U.S., 1961, naturalized, 1981; s. Kamal and Arfaa (Hamidi) D.; m. Nancy Dumars, Feb. 2, 1964; children: Ramine, Michael, Daria. B.S., Pahlavi U., 1960; cert. Am. lang. Columbia U., 1962; B.S., U. Calif.-Davis, 1965, M.S., 1972, Ph.D., 1976. Staff research assoc. dept. botany U. Calif.-Davis, 1965-71, lectr. environ. horticulture, 1972-78; asst. prof. dept. ornamental horticulture U. Fla., Gainesville, 1978—, researcher, mem., 1980—. Contbr. articles on horticulture and botany to profl. jours. Mem. Bot. Soc. Am., Am. Soc. Plant Taxonomists, Internat. Assn. Plant Taxonomists, Linnean Soc. London, Am. Hort. Soc., Sigma Xi. Subspecialties: Biomass (agriculture); Taxonomy. Current work: Energy farming-production of hydrocarbons from plant biomass, taxonomy of the genus Jatropha L. (Euphoribiaceae) and Cycadales. Home: 5720 NW 57th Way Gainesville FL 32606 Office: U Fla 2519 HS/PP Bldg Gainesville FL 32611

DEHMELT, HANS GEORG, physics educator; b. Germany, Sept. 9, 1922; U.S., 1952, naturalized, 1962; s. Georg Karl and Asta Ella (Klemmt) D.; m.; 1 son, Gerd. Grad., Graues Kloster, Abitur, 1940; Ph.D. summa cum laude, U. Goettingen, 1950. Postdoctoral fellow U. Goettingen, Germany, 1950-52, Duke U., Durham, N.C., 1952-55; vis. asst. prof. U. Wash., Seattle, 1955, asst. prof. physics, 1956, asso. prof., 1957-61, prof., 1961—; cons. Varian Assos., Palo Alto, Calif., 1956-76. Contbr. articles to profl. jours. Recipient Humboldt prize, 1974; award in basic research Internat. Soc. Magnetic Resonance, 1980; NSF grantee, 1958—. Fellow Am. Phys. Soc. (Davisson-Germer prize 1970); mem. Am. Acad. Arts and Scis., Nat. Acad. Scis. Subspecialties: Atomic and molecular physics; Particle physics. Current work: Single elementary/atomic particle at rest in space, massed magnetic moments of electron/positron, ultimate laser frequency standard, particle identity. Home: 1600 43d Ave E Seattle WA 98112 Office: Physics Dept FM 15 U Wash Seattle WA 98195

DE HODGINS, OFELIA CANALES, materials scientist, engineer; b. Mexico City, Oct. 25, 1943; U.S., 1974, naturalized, 1978; s. Fernando and Leana (Del Olmo) Canales Rocha; m. Garry Hodgins, Aug. 24, 1974; 1 child, Alfonso Sidarta. B.S. in physics, U. Mexico, 1972; M.S. in Physics and Materials Sci., 1977; Ph.D., 1985. Jr. researcher Lab. Ultracentrifuges Nuclear Inst. Mexico, 1971-73, sr. researcher, 1973-76; vis. research fellow U. Va., 1973; prof. math. Physics and Lab. Physics High Sch. Instituto Freinet de Mexico, Mexico City, 1971-72; asst. prof. thermodynamics U. Mexico, 1972-73; asst. prof. math. Instituto Politecnico Nacional, Mexico City, 1973-74; grad. research asst. dept. physics U. Va., 1974-75, grad. research asst. dept. materials sci., 1976-79, grad. research asst. dept. nuclear engring. and engring. physics, 1979—; staff negr. Internat. Bus. Machines Corp., Poughkeepsie, N.Y., 1981—. Contbr. articles to profl. jours. Dorothea Buck fellow, 1975-76; recipient Nat. Prize of Sci., Mexico, 1974. Mem. Mexican Soc. Physics, Am. Soc. Metals, Am. Nuclear Soc., Electron Microscopy Soc. Am., Soc. Exptl. Stress Analysis, Sigma Xi. Subspecialty: Microchip technology (materials science). Current work: Solid state physics, metallurgy, polymers, electronics integrated circuits, reactor pressure vessel steels.

DEHORITY, BURK ALLYN, rumen microbiologist, educator; b. Peoria, Ill., Sept. 3, 1930; s. Harry A. and Marie B. D.; m. Barbara June, July 5, 1953; children: Katherine, Christine, Sue Ellen, Burk Joel. B.A., Blackburn Coll., 1952; M.S., U. Maine, 1954; Ph.D., Ohio State U., 1957. Asst. prof. dept. animal scis. U. Conn., 1957-59; with Ohio Agr. Research and Devel. Ctr., Ohio State U., 1959—, prof. dept. animal sci., 1970—, assoc. chmn. dept., 1981—. Contbr. to profl. jours. Mem. Am. Soc. Animal Sci., Am. Soc. Microbiology, Am. Dairy Sci. Assn. Lutheran. Subspecialties: Animal nutrition; Microbiology. Current work: Rumen microbiology, studies on the rumen bacteria responsible for breakdown of forages, types and their isolation and classification, synergism between bacterial species; rumen protozoa. Home: 708 Kieffer St Wooster OH 44691 Office: Dept Animal Science Ohio Agr Research and Devel Center Wooster OH 44691

DEIBEL, MARTIN ROBERT, JR., biochemist; b. Columbus, Ohio, Apr. 21, 1949; s. Martin R. and Edith (Marolt) D.; m. Megan Gower, May 19, 1979. B.S., Ohio State U., 1971, Ph.D., 1977. Postdoctoral fellow U. Ky., Lexington, 1977-79, sr. research assoc., 1979-80, research asst. prof. biochemistry, 1980—; researcher Ohio State U., Columbus, 1971-77. Contbr. articles to profl. jours. NIH grantee, 1982. Mem. Am. Soc. Biol. Chemists, AAAS, Phi Beta Kappa. Subspecialty: Biochemistry (medicine). Current work: Biochemical studies of terminal deoxynucleotidyl transferase; an enzyme found in lymphoid tissues and in the blood of some leukemic patients; active site and enzymatic studies of the enzyme; physical biochemistry. Office: Dept Biochemistry U Ky Sch Medicine 800 Rose St Lexington KY 40536

DEIBEL, ROBERT HOWARD, bacteriology educator; b. Chgo., Dec. 20, 1924. M.S., U. Chgo., 1952, Ph.D., 1962. Bacteriologist Am. Meat Inst. Found., 1952-64; assoc. prof. bacteriology Cornell U., Ithaca, N.Y., 1964-66, U. Wis.-Madison, 1966-69, prof., 1969—. Mem. Am. Soc. Microbiology, Can. Soc. Microbiology, Brit. Soc. Gen. Microbiology. Subspecialty: Bacteriology. Office: Dept Bacteriology U Wis Madison WI 53706

DEIBEL, RUDOLF, medical virologist; b. Berlin, Apr. 27, 1924; U.S., 1962, naturalized, 1968; s. Rudolf and Margarete (Bernhoft) D.; m. Waltraut Rosenbrock, Oct. 4, 1957; children: Rudolf, Stephan, Christiane. Cand. med., Humboldt U., Berlin, 1946-51; Dr. med., Albert Ludwigs U., Freiburg, Ger., 1953; Bd. in Pediatrics, Landes Aerztekammer Baden-Wuerttemberg, Ger., 1961. Intern Stadt, Wenckebach Krankenhaus, Berlin-Tempelhof, Ger., 1953-54; resident Inst. Pathology, U. Freiburg, 1955-56, Children's Hosp., Freiburg, 1956-61; assoc. med. virologist Div. Labs. and Research, N.Y. State Dept. Health, Albany, 1962-67; dir. Virus Labs., Ctr. for Labs. and Research, 1967—; prof. microbiology and pediatrics Albany Med. Coll., 1977—; mem. faculty Union U. Mem. Am. Pub. Health Assn., Am. Soc. Microbiology, Am. Assn. Immunologists, Capital dist. Pediatric Soc., Am. Soc. Tropical Medicine and Hygiene, N.Y. State Assn. Pub. Health Labs., Am. Soc. Virology., N.Y. Acad. Sci. Subspecialties: Virology (medicine); Pediatrics. Current work: Pathology and immunology of virus infections.

DEININGER, PRESCOTT LEONARD, biochemistry educator, molecular genetics researcher; b. Northampton, Mass., Nov. 5, 1951; s. Whitaker Thompson and Harriet Sarah (Prescott) D.; m. Celia Ann Hemphill, Feb. 10, 1979; 1 dau., Emily Sarah. A.B. in Chemistry, U. Calif.-Santa Cruz, 1973, Ph.D., U. Calif.-Davis, 1978. Postdoctoral assoc. U. Calif.-San Diego Med. Sch., La Jolla, 1978-80, Med. Research Council Lab. of Molecular Biology, Cambridge, Eng., 1980-81; asst. prof. biochemistry La. State U. Med. Ctr., New Orleans, 1981—. NATO fellow, 1981; NIH grantee, 1981-84; New Orleans Cancer Crusaders grantee, 1982; C.A.G.N.O. grantee, 1983. Mem. Am. Chem. Soc., AAAS. Democrat. Subspecialties: Genome organization; Virology (biology). Current work: Structure and functions of human repetitive DNA sequences using recombinant DNA and genetic engineering technologies, viral genetics to help study eucaryotic gene regulation. Office: La State U Med Ctr 1901 Perdido St New Orleans LA 70112

DEISSLER, ROBERT GEORGE, fluid dynamacist; b. Greenville, Pa., Aug. 1, 1921; s. Victor Girard and Helen Stella (Fisher) D.; m. June Marie Gallagher, Oct. 7, 1950; children: Robert Joseph, Mary Beth, Ellen Ann, Ann Marie. B.S., Carnegie Inst. Tech., 1943; M.S., Case Inst. Tech., 1948. Research engr. Goodyear Aircraft Corp., Akron, Ohio, 1943-44; aero. research scientist NASA Lewis Research Ctr., Cleve., 1947-52, chief fundamental heat transfer br. center, 1952-70, staff scientist fluid physics, 1970—. Contbr. articles to profl. jours. Served to lt. (j.g.) USNR, 1944-46. Recipient Max Jacob Meml. award ASME/Am. Inst. Chem. Engrs., 1975; NACA/NASA Exceptional Service award, 1957; Outstanding Publ. award, 1978. Fellow AIAA (award 1975), ASME (Heat Transfer Meml. award 1964); mem. Am. Phys. Soc., Soc. Natural Philosophy, Sigma Xi. Roman Catholic. Subspecialties: Fluid mechanics; Heat Transfer. Current work: Theoretical turbulence incl. numerical solutions, turbulent heat transfer, vortex flows, atmospheric and astrophys. flows, thermal radiation, heat transfer in powders. Home: 4540 W 213 St Fairview Park OH 44126 Office: NASA Lewis Research Center 21000 Brookpark Rd Cleveland OH 44135

DEITERS, JOAN ADELE, chemistry educator; b. Cin., Apr. 28, 1934; d. Alfred Harry and Rose Catherine (Rusche) D. B.A., Coll. Mt. St. Joseph, Ohio, 1963; Ph.D., U. Cin., 1967. Joined Sisters of Charity, 1952; mem. chemistry faculty Coll. Mt. St. Joseph, 1968-78; assoc. prof. chemistry Vassar Coll., 1978—. Fellow Sigma Xi; mem. Am. Chem. Soc. Democrat. Roman Catholic. Subspecialty: Inorganic chemistry. Current work: Molecular mechanics calculations of structures and reaction mechanisms of phosphorus compounds. Home: Our Lady of Lourdes Convent 29 N Hamilton St Poughkeepsie NY 12601 Office: Vassar College PO Box 143 Poughkeepsie NY 12601

DE JONG, RUDOLPH H., anesthesiologist, educator, cons.; b. Amsterdam, Holland, Aug. 10, 1928; came to U.S., 1946, naturalized, 1953. B.A., Stanford U., 1950, M.D., 1954. Diplomate: Am. Bd. Anesthesiology. Intern Barnes Hosp., St. Louis, 1954-55; resident San Francisco Gen. Hosp., 1955-56, Virginia Mason Hosp., Seattle, 1956-57; asst. prof. anesthesiology U. Calif., San Francisco, 1961-65; prof. anesthesiology and pharmacology U. Wash., Seattle, 1965-76; sr. editor Jour. AMA, Chgo., 1976-78; Saltonstall prof. for research in anesthesiology Tufts U., 1978-82; prof. anesthesiology U. Cin. Coll. Medicine, 1982—; sr. cons. Pain Control Center, 1982—. Author: Physiology and Pharmacology of Local Anesthesia, 1974, Local Anesthetics, 1978; contbr. articles to profl. jours. Served to capt., M.C. U.S. Army, 1957-61. Recipient research career devel. award NIH, 1966-76. Mem. Am.Soc. Anesthesiology, Am. Soc. for Regional Anesthesiology, Pain Soc. Subspecialties: Anesthesiology; Neuropharmacology. Current work: Effects of local and general anesthetics of spinal cord and cerebral function. Home: 981A Paradrome St Cincinnati OH 45202 Office: 234 Goodman St ML 586 Cincinnati OH 45267

DEKKER, EUGENE EARL, biochemistry educator and researcher; b. Highland, Ind., July 23, 1927; s. Peter and Anne (Hendrikse) D.; m. Harriet Ella Holwerda, July 5, 1958; children: Gwen E., Paul D., Tom R. A.B., Calvin Coll., 1949; M.S., U. Ill., 1951, Ph.D., 1954. Instr. U. Louisville Med. Sch., 1954-56; instr. U. Mich. Med. Sch., Ann Arbor, 1956-58, asst. prof., 1958-65, assoc. prof., 1965-70, prof. dept. biol. chemistry, 1970—, assoc. chmn., 1975—. Served with USN, 1945-46. Mem. Am. Chem. Soc., Am. Soc. Biol. Chemists, Am. Soc. Plant Physiologists, Sigma Xi, Phi Lambda Upsilon. Mem. Christian Reformed Church. Subspecialties: Biochemistry (medicine); Plant physiology (biology). Current work: Biochemistry of novel nitrogen-containing compounds in mammals, plants, and bacteria; enzymology; enzyme structure, active sites and mechanisms. Home: 2612 Manchester Rd Ann Arbor MI 48104 Office: Dept Biol Chemistry Univ Mich Med Sch Ann Arbor MI 48109

DE LANEROLLE, NIHAL CHANDRA, neuroanatomy and neuroethology researcher, educator; b. Colombo, Sri Lanka, Apr. 16, 1945; came to U.S., 1974; s. Leslie Barnes and May Adelaide (Jayawardena) de L. B.Sc., U. Ceylon, 1967; D.Phila., U. Sussex, Eng., 1972; B.A., U. Cambridge, Eng., 1974, M.A., 1981. Asst. lectr. zoology U. Ceylon, 1967-69; postdoctoral research fellow in behavioral physiology and psychopharmacology U. Minn., St. Paul, 1975-78; research fellow in biol. psychiatry Yale U. Sch. Medicine, 1978-79, research asso., 1979-82, asst. prof. neurosurgery and neuroanatomy, 1982—; vis. asst. prof. psychology Wesleyan U., Middletown, Conn., 1981-82. Contbr. articles to profl. jours. Amyotrophic Lateral Sclerosis Soc. Am. grantee, 1982. Mem. Soc. for Neurisci., Assn. for Study Animal Behavior, Animal Behavior Soc., N.Y. Acad. Sci., AAAS. Episcopalian. Subspecialties: Neurobiology; Ethology. Current work: Chemically defined neural circuits in brainstem and neocortex of mammals; neural basis of vocalization and emotion in birds and mammals. Office: Sect Neurosurgery Yale U Sch Medicine 333 Cedar St New Haven CT 06510

DE LA NOUE, JOEL JEAN-LOUIS, biologist, university administrator, researcher; b. Ferryville, Tunisia, Mar. 18, 1938; s. Jean and Genevieve (Adelus) de la N.; m. Christiane Laboissonniere, Dec. 26, 1959; children: Eric, Philippe; m. Helene Raymond, July 16, 1980; 1 dau., Marie-Eve. B.Sc., U. Laval., Can., 1960, D.Sc., 1968; postgrad., U. Sheffield, Eng., 1968-69, U. Wash., 1969-70. Asst. prof. biology dept. U. Laval, Que., 1964-70, adj. prof., 1970-72, assoc. prof., 1972-78, prof., 1978—, dir., 1978—, founding pres. faculty union, 1973-77. Contbr. articles to profl. jours. Recipient medal Vice-Gov. Province Que., 1960; Natural Scis. and Engring. Research Council Can. grantee, 1970—; Ministry Edn. Que. grantee, 1972—. Mem. AAAS, Assn. canadienne-francaise pour l'advancement des sciences, Assn. des biologistes du Que., Canadian Inst. Food Sci. and Tech. Subspecialties: Water supply and wastewater treatment; Nutrition (biology). Current work: Biotechnological recycling with integrated food chains, aquaculture, research, teaching, administration, consulting. Co-inventor automatic particle collector for material in suspension in water. Home: 1082 Dijon St Ste-Foy PQ Canada G1W 4M4 Office: Local 1312 Pavillon Comtois U Laval Ste-Foy PQ Canada G1K 7P4

DELAP, JAMES HARVE, chemist; b. Carbondale, Ill., Feb. 6, 1930; s. Harve Eugene and Adena Rosetta (Harriss) D.; m. Clara Prudence Todd, Mar. 29, 1959; children: Carolyn Adena, Mary Amelia, Margaret Jane, James Todd. B.A., So. Ill. U., Carbondale, 1952; postgrad., U. Calif.-Berkeley, 1954-55; M.A., Duke U., 1959, Ph.D., 1960. Research chemist Chemstrand Corp., Durham, N.C., 1960-62; asst. prof. Stetson U., Deland, Fla., 1962-68, assoc. prof., 1968-71, prof. chemistry, 1971—. Served with U.S. Army, 1952-54. Recipient McEniry award Stetson U., 1982; James B. Duke fellow Duke U., 1957-60; Fulbright lectr., 1970-71, 78-79. Mem. Union Concerned Scientists, Am. Chem. Soc., Gamma Sigma Epsilon. Democrat. Presbyterian. Subspecialties: Physical chemistry; Laser photochemistry. Current work: Photochemistry, photodegradation; luminiscence. Home: 1103 N Boston Ave Deland FL 32720 Office: Stetson U Box 8277 Deland FL 32720

DE LA TORRE, JACK CARLOS, physician, brain researcher; b. Paris, France; s. Rafael and Maria C. (Parodi) de la T.; m. Florinda Bayod, June 30, 1962. B.S., Am. U., Washington, 1961; Ph.D. in Anatomy, U. Geneva, 1968; M.D., U. Mex., Juarez, 1979. Asst. prof. neurosurgery and psychiatry U. Chgo. Sch. Medicine, 1969-75, assoc. prof., 1975-79; assoc. prof. neurosurgery U. Miami Sch. Medicine, 1979-82, dir. research, 1979-82; assoc. prof. neurosurgery, dir. neurosurgery research Northwestern U. Med. Sch., Chgo., 1982-83; assoc. prof. neurosurgery, head exptl. surgery U. Ottawa Health Scis., (Ont. Can.), 1983—. Author: Dynamics of Brain Monoamines, 1972; editor-in-chief: Biological Actions and Medical Applications of Dimethyl Sulfoxide, 1983; translator: Histology of the Nervous System (Ramon y Cajal), vol. I, 1983; contbr. articles to profl. jours. Grantee NIH, 1970-73, U. Chgo., 1970-72, Purer Found., 1979-80, Paralysis Cure Research Found., 1981-82, Am. Paralysis Assn., 1983-84. Mem. Internat. Brain Research Orgn., Am. Acad. Neurology, Soc. Neurosci., AAUP, N.Y. Acad. Scis., AAAS, Cajal Club. Subspecialties: Neurobiology; Neurophysiology. Current work: Regeneration of central nervous system; pathophysiology of head and spinal cord trauma and cerebral stroke; therapy of brain ischemia/trauma; stress and brain neurotransmitters. Patentee in control of bacteria. Office: 451 Smyth Rd Ottawa ON K1H 8M5 Canada

DELAUBENFELS, DAVID JOHN, geography educator; b. Pasadena, Calif., Dec. 5, 1925; s. Max Walker and Beth (Jones) de L.; m. Gudrun Josephine Erickson, Dec. 21, 1954; children: Eric Arthur, Lucia Beth de Laubenfels Sweetland, Evelyn Jo, Marion Jean; m. Linda Elaine Price, Dec. 27, 1973. A.A., Pasadena Jr. Coll., 1947; A.B., Colgate U., 1949; A.M., U. Ill., 1950; Ph.D., 1953. Postdoctoral fellow Johns Hopkins U., Balt., 1955-56; asst. prof. U. Ga., Athens, 1953-58, assoc. prof., 1958-59, Syracuse (N.Y.) U., 1959-71, prof., 1971—. Author: Mapping the World's Vegetation, 1975, Gymnospermes, Fasc. 4-Flore de la Nouvelle Caledonie et Dependancies, 1972, A Geography of Plants and Animals, 1970; contbr. articles on geography of plants and animals to profl. jours. Served with AUS, 1944-46. Travel grantee in field. Mem. Assn. Am. Geographers, Am. Geography Soc., Ecol. Soc. Am., Bot. Soc. Am., Internat. Soc. Plant Morphologists, Sigma Xi. Subspecialties: Biogeography; Taxonomy. Current work: Geographic differentiation of vegetation systems, taxonomy and ecology of tropical conifers. Office: Dept Geography Syracuse U Syracuse NY 13210

DE LAUER, RICHARD D., govt. ofcl., former aerospace co. exec.; b. Oakland, Calif., Sept. 23, 1918; s. Michael and Matilda (Giambruno) DeL.; m. Ann Carmichael, Dec. 6, 1940; 1 son, Richard Daniel. A.B., Stanford, 1940; B.S., U.S. Naval Postgrad. Sch., 1949; Aero. Engr., Calif. Inst. Tech., 1950, Ph.D., 1953. Structrual designer Glenn L. Martin Co., Balt., 1940-42; design engr. Northrop Co., Hawthorne, Calif., 1942; commd. ensign USN, 1942, advanced through grades to comdr., 1958; assignments in, U.S., 1943-58, ret., 1966; lab. dir. Space Tech. Labs., El Segundo, Calif., 1958-60, Titan Program dir., 1960-62, v.p., dir. ballistic missile program mgmt., 1962-66; v.p., gen. mgr. systems engring. and integration div. TRW Systems Group, Redondo Beach, Calif., 1966-68, v.p., gen. mgr., 1968-70; exec. v.p. TRW, Inc., Redondo Beach, 1970-81; also dir.; undersec. for research and engring. Dept. Def., Washington, 1981—; dir. Ducommen, Inc., Los Angeles, Cordura, Inc., Chgo.; Vis. lectr. U. Calif., Los Angeles.; Chmn. Nat. Alliance Businessman, 1968-69; chmn. Region IX, 1970; mem. Def. Sci. Bd., Dept. Def. Author: (with R.W. Bussard) Nuclear Rocket Propulsion, 1958, Fundamentals of Nuclear Flight, 1965. Trustee U. Redlands. Fellow Am. Inst. Aeros. and Astronautics, Am. Astron. Soc.; mem. Nat. Acad. Engring., AAAS, Aerospace Industries Assn. (gov.), Sigma Xi. Subspecialty: Research engineering administration. Home: 1101 S Arlington Ridge Rd Arlington VA 22202 Office: Dept Def Research and Engring Office of Undersec The Pentagon Washington DC 20301

DEL BENE, JANET ELAINE, chemist, educator; b. Youngstown, Ohio, June 3, 1939; d. Anthony Joseph and Elizabeth Josephine (Pastier) Del B. B.S. summa cum laude, Youngstown State U., 1963, A.B., 1965; Ph.D., U. Cin., 1968. Postdoctoral fellow Theoretical Chemistry Inst., U. Wis., 1968-69; NIH postdoctoral fellow Mellon Inst., 1969-70; asst. prof. chemistry Youngstown State U., 1970-73, assoc. prof., 1973-76, prof., 1976—; research prof. molecular pathology and biology Northeastern Ohio Univs. Coll Medicine, 1977—; mem. grad. faculty Kent State U. Mem. Girard (Ohio) Bd. Health, 1982—. Research numerous publs. in field. Am. Chem. Soc.-Petroleum Research Fund starter grantee, 1971-74; Camille and Henry Dreyfus Found. tchr-scholar grantee, 1974-79; NIH grantee, 1974-77, 80-83. Mem. Am. Chem. Soc., AAAS, N.Y. Acad. Scis., Sigma Xi, Phi Kappa Phi, Iota Sigma Pi (Agnes Fay Morgan Research award 1972). Roman Catholic. Subspecialty: Theoretical chemistry. Current work: Molecular orbital studies of hydrogen bonding, protonation, and lithium ion assn. Home: 871 N Ward Ave Girard OH 44420 Office: Dept Chemistry Youngstown State U Youngstown OH 44555

DEL CERRO, MANUEL, physician, educator; b. Buenos Aires, Argentina, Aug. 20, 1931; came to U.S., 1964; s. Manuel and Julia (Caceres) del C.; m. Constancia Clotilde Nunez, May 17, 1958; children: Alicia, Marilu. B.A., Nat. Coll., Buenos Aires, 1951, B.S., 1951; med. diploma, U. Buenos Aires, 1958. Sr. instr. histology U. Rochester, N.Y., 1958-61, assoc. prof., 1961-64, research assoc., 1965-69, sr. research assoc., 1969-71, assoc. prof., 1971-79, assoc. prof. dept. anatomy, 1976—, dept. ophthalmology, 1980—; cons. in field. Contbr. articles to profl. jours. Mem. N.Y. State Health Research Council, 1981-84. Served with Argentine Army, 1952. Nat. Eye Inst. grantee, 1978-81, 81—; N.Y. State Health Research Council grantee, 1980-81. Mem. Internat. Brain Research Orgn., Assn. Research in Vision and Ophthalmology, Am. Assn. Neuropathologists. Roman Catholic. Subspecialties: Neuroimmunology; Ophthalmology. Current work: Research in experimental ophthalmology, dealing with eye neuroimmunology and neuropathology. Interest in developmental neurobiology and retinal transplants. Home: 14 Tall Acres Dr Pittsford NY 14534 Office: University of Rochester Medical School PO Box 605 Rochester NY 14642

DELCOMYN, FRED, neurobiologist, educator; b. Copenhagen, June 4, 1939; U.S., 1947, naturalized, 1960; s. Niels T. and Erna A. (Svendsen) D.; m. Nancy Ann Nigg., Dec. 14, 1969; children: Julia C.M., Michael T.W., Erik A.W. B.S., Wayne State U., 1962; M.S., Northwestern U., 1964; Ph.D., U. Oreg., 1969. Research assoc. dept. zoology U. Glasgow, Scotland, 1969-71, lectr., 1971-72; asst. prof. dept. entomology U. Ill., Urbana, 1972-77, assoc. prof., 1977—; vis. prof. dept. physiology U. Alta. (Can.), Edmonton, 1980-81. Mem. editorial adv. bd.: Jour. Comparative Physiology, Heidelberg, W.Ger., 1979—; editorial assoc.: Brain and Behavioral Sci, London, 1981—; contbr. articles to profl. jours. USPHS trainee, 1964-68; USPHS fellow, 1969; HHS grantee. Mem. AAAS, Soc. Exptl. Biology, Soc. Neurosci. Subspecialties: Comparative neurobiology; Neurophysiology. Current work: The role of sensory information in coordination during walking in insects. Office: Dept Entomology U Ill 505 S Goodwin Urbana IL 61801

DELEEUW, J.H., aerospace engineering educator; b. Amsterdam, Netherlands, Jan. 4, 1929. Dipl. Eng., Delft U. Tech., 1953; M.S., Ga. Inst. Tech., 1952; Ph.D. in Aerophysics, U. Toronto, 1958. Research engr. aerodynamics Nat. Aero. Research Inst., Netherlands, 1952-53; asst. prof. to prof. aerospace engring. Inst. Aerospace Studies, U. Toronto, 1958-76, asst. dir. research, 1970-76, prof., dir., 1976—. Mem. AIAA, Can. Aero. and Space Inst. Subspecialty: Aerospace research administration. Office: Inst Aerospace Studies U Toronto Toronto ON Canada M5S 1A1

DELEVORYAS, THEODORE, botanist, educator; b. Chicopee Falls, Mass., July 22, 1929; s. Basil John and Sophie (John) Dulchinos D.; m. Nancy Lou Foster, June 23, 1956 (div. Dec. 1978); children: Matthew Torrey, Christopher Theodore; m. Cecilia Ann Dean, Aug. 14, 1981. B.S., U. Mass., 1950; M.S., U. Ill., 1951, Ph.D., 1954; M.A. (hon.), Yale, 1968. Postdoctoral fellow NRC, U. Mich., Ann Arbor, 1954-55; asst. prof. botany Mich. State U., East Lansing, 1955-56; instr. botany Yale, New Haven, 1956-58, asst. prof., 1958-60, asso. prof. biology, 1962-68, prof. biology, 1968-72; asso. prof. botany U. Ill., Urbana, 1960-62; prof. botany U. Tex., Austin, 1972—, chmn. dept. botany, 1974—. Author: Morphology and Evolution of Fossil Plants, 1962, Plant Diversification, 1966, (with others) Morphology of Plants and Fungi, 1980; Contbr. numerous articles to profl. jours. Fellow Linnean Soc. London; mem. Bot. Soc. Am. (treas. 1967-72, v.p. 1973, pres. 1974), Paleontol. Soc., Palaeontol. Assn., Am. Inst. Biol. Scis., Internat. Assn. Plant Taxonomy, Internat. Soc. Plant Morphologists, Torrey Bot. Club, Am. Inst. Biol. Scis. (mem. bd. govs. 1975-77), Internat. Orgn. Paleobotany (pres. 1978-81), Phi Beta Kappa, Phi Kappa Phi. Club: Austin Yacht. Subspecialties: Botany; Paleontology. Current work: Morphology and evolution of mesozoic plants, especially cycadophytes and conifers. Home: 4204 Zuni Dr Austin TX 78759

DEL GUERCIO, LOUIS RICHARD MAURICE, surgeon, educator; b. N.Y.C., Jan. 15, 1929; s. Louis and Hortense (Ardengo) Del G.; m. Paula Marie Helene de Vautibault, May 18, 1957; children—Louis, Francsca, Paul, Catherine, Maria, Michelle, Christopher Anthony. B.S., Fordham U., 1949; M.D., Yale U., 1953. Diplomate: Am. Bd. Surgery, Am. Bd. Thoracic Surgery. Intern Columbia-Presbyn. Med. Center, N.Y.C., 1953-54; resident St Vincent's Hosp., N.Y.C., 1954-58, Cleve. City Hosp., 1958-60; practice medicine specializing in thoracic surgery, 1960—; mem. faculty Albert Einstein Coll. Medicine, N.Y.C., 1960-71, assoc. prof., 1966-70, prof. surgery, 1970-71, dir., 1967-71; clin. prof. surgery N.J. Coll. Medicine, Newark, 1971-76; prof. surgery N.Y. Med. Coll., N.Y.C., 1976—, chmn. dept., 1976—; surgeon Lincoln Hosp.; chmn. surgery Met. Hosp. Center, Westchester County Med. Center, 1976—; cons. surgeon other hosps.; mem. surg. study sect. NIH, 1970-74; mem. com. on shock NRC-Nat. Acad. Scis., 1969-71; mem. merit rev. bd. VA, 1971-74; mem. council cardiovascular radiology Am. Heart Assn., 1973—; mem. health care tech. study sect. Dept. Health and Human Services, 1980—; cons. Nat. Center Health Services Research, 1980—. Author: (with B.G. Clarke) Urology, 1956; author: The Multilingual Manual for Medical History Taking, 1972, (with S.G. Hershey, R. McConn) Septic Shock in Man, 1971; editor-in-chief: Critical Care Monitor, 1980—; contbr. (with S.G. Hershey, R. McConn) articles to med. jours. Served with Mcht. Marine, 1946-47; served with AUS, 1949-50. Recipient award in medicine Fordham U. Alumni Assn., 1974, Gold award Am. Acad. Pediatrics, 1973, Alpha Omega Alpha Faculty award N.Y. Med. Coll., 1982; Am. Thoracic Soc fellow, 1959-60; grantee Health Research Council N.Y., 1965-71, NIH, 1962-71. Fellow A.C.S.; mem. Am. Trauma Soc. (founding mem.), Soc. Critical Care Medicine (founding mem., pres. 1976), Am. Surg. Assn., Am. Physiol. Soc., Soc. Univ. Surgeons, Equestrian Order of Holy Sepulchre of Jerusalem. Subspecialties: Bioinstrumentation; Biomedical engineering. Current work: Surgical physiology;physiological assessment and mointoring of high-risk and critically ill or injured patient. Patentee in field. Home: 14 Pryer Ln Larchmont NY 10538 Office: NY Medical College Munger Pavilion Valhalla NY 10595 Adaptability and the determination of what is possible are the keys to personal success and contentment.

DELISI, CHARLES, biophysicist; b. N.Y.C., Dec. 9, 1941; s. Jack and Phyllis (Colameo) DeL.; m. Lyn E. Moskowitz, Aug. 11, 1968; children: Jacqueline, Daniel. B.A., CCNY, 1963; Ph.D. in Physics, N.Y.U., 1969. Postdoctoral fellow in chemistry and biophysics Yale U., 1969-72; sr. lectr. engring. and applied sci., 1971-72; elec. engr. Sperry Gyroscope, 1963-65; physicist theoretical div. Los Alamos Nat. Lab.-U. Calif., 1972-77; chief theoretical immunology NIH, Bethesda, Md., 1977—, chief lab. math. biology, 1982—, spl. asst., 1978-79. Author 5 books in field; contbr. numerous articles to profl. jours. Recipient Gordon Research Conf. award, 1979. Mem. AAAS, Am. Assn. Immunologists, Biophysics Soc. Subspecialties: Biophysics (biology); Cell biology. Current work: Immunology, structural biology, mathematical biology. Home: 7700 Persimmon Tree Ln Bethesda MD 20817 Office: NIH Bethesda MD

DELISLE, CLAUDE A(RMAND), physicist; b. Quebec City, Que., Can., Nov. 15, 1929; s. Isidore and Marie (Girard) D.; m. Charlotte Tardif, June 7, 1958; children: Christine, Marie-France, Simon, Josee, Eric. B.A., U. Montreal, 1951; B.Sc., Laval U., 1958, D.Sc., 1963. Asst. prof. dept physics Laval U., 1962-63; research assoc. Inst. of Optics, Rochester, N.Y., 1963-65; asst. prof. dept. physics Laval U., Quebec, 1965-69, assoc. prof., 1969-73, prof., 1973—, dir. Lab. for Research in Optics and Laser, 1977-80; cons. Ministry of Edn., Province of Que., 1973-74. Contbr. articles to sci. jours. Nat. Research Council Can. scholar, 1958-62. Fellow Optical Soc. Am.; mem. Can. Assn. Physicists, French Can. Assn. for Advancement of Scis. Subspecialty: Optics Research. Current work: Interferometry, bistability, optical instrumentation, coherence. Office: Dept Physics Laval U Quebec PQ Canada G1K 7P4

DELIYANNIS, PLATON CONSTANTINE, mathematics educator; b. Athens, Greece, Aug. 21, 1931. Diploma in Engring, Nat. Tech. U., Athens, Greece, 1954; M.S., U. Chgo., 1955, Ph.D. in Math, 1963. Research asst. U. Chgo., 1955-56; instr. to asst. prof. math. Ill. Inst. Tech., Chgo., 1961-65, asst. prof. math., 1969-74, chmn. dept., 1975-80, assoc. prof., 1974—; dir. Ctr. Advanced Studies, Greek Atomic Energy Commn., 1966-69. Mem. Am. Math. Soc., Math. Assn. Am., N.Y. Acad. Scis., Soc. Indsl. and Applied Math. Subspecialties: Topology; Applied mathematics. Office: Dept Math Ill Inst Tech 3300 S Federal St Chicago IL 60616

DELL, RALPH B., pediatrician, medical educator; b. Mountain Village, Alaska, July 31, 1935; s. Elwyn B. and Elizabeth (Bishop) D.; m. Kathryn M. Bownass, June 17, 1957; children: Laura F., Kenneth R.; m. Karen Kramer Hein, May 21, 1983; step-children: Ellan T. Hein, Molly A. Hein. B.A., Pomona Coll., 1957; M.D., U. Pa., 1961. Diplomate: Am. Bd. Pediatrics. Intern Children's Hosp. Med. Ctr., Boston, 1961-62, resident in pediatrics, 1962-63, Columbia-Presbyterian Med. Ctr., Babies Hosp., N.Y.C., 1965; NIH vis. fellow in pediatrics Columbia U. Coll. Physicians and Surgeons, N.Y.C., 1963-65, 66, assoc. in pediatrics, 1966-67, asst. prof. pediatrics, 1967-72, assoc. prof., 1972-78, prof., 1978—, dep. chmn. pediatrics dept., 1979-82; asst. attending Babies Hosp. and Vanderbilt Clinic, N.Y.C., 1967-72, assoc. attending, 1972-78, attending, 1978—, dep. chemn. pediatric service, 1979-82; Mem. adminstrv. com. Bioengring. Inst., Columbia U., 1974—, mem. faculty, 1972—, mem. bioengring. com., 1972-75; career scientist Health Research Council, N.Y.C., 1972-75. Contbr. over 50 sci. articles to profl. publs. Recipient Career Devel. award Nat. Inst. Gen. Med. Scis., 1966-71; Fogarty Sr. Internat. Research fellow NIH, 1975-76. Mem. Am. Fedn. Clin. Research, Am. Acad. Pediatrics, Am. Soc. Nephrology, Am. Physiol. Soc., Am. Soc. Pediatric Nephrology, Am. Soc. Clin. Investigation, Assn. Computing Machinery, Harvey Soc., N.Y. Acad. Sci., Soc. Pediatric Research. Subspecialties: Pediatrics; Biomedical engineering. Current work: Clinical physiology and nutrition; biostatistics, biomathematics, data processing. Home: 116 Pinehurst Ave New York NY 10033 Office: Babies Hosp S Room 116 Columbia Univ Coll Physicians and Surgeons 630 W 168th St New York NY 10032

DELLA-FERA, MARY ANNE, neuroscientist, veterinarian, cons., educator; b. Wilmington, Del., Mar. 29, 1954; d. Vincent W. and Mary (Rickel) D-F. B.S. in Biology, U. Del., 1975; V.M.D. (NIH trainee), U. Pa., 1979; Ph.D. in Anatomy, U. Pa., 1980. NIH fellow U. Pa., 1980-81, research asso., 1981-82, asst. prof. medicine, 1982; research specialist Monsanto Co., St. Louis, 1982—; cons. Recipient Lindsley prize Grass Foundation and Society for Neuroscience, 1981; Alfred P. Sloan Found. fellow, 1981—; recipient Sigma Xi award, 1980. Mem. Am. Vet. Med. Assn., Soc. Neurosci., Am. Physiol. Soc., AAAS. Subspecialties: Neuroendocrinology; Gastroenterology. Current work: Role of brain peptides in body weight regulation and feeding behavior; research on function and mechanism of action of brain peptides. Office: Monsanto Co 800 N Lindbergh Blvd Saint Louis MO 63167

DELLA-LATTA, PHYLLIS, microbiologist, researcher; b. N.Y.C., Feb. 16, 1946; d. Michael and Sylvia (Thompson) Giglio; m. George Della-Latta, Aug. 17, 1968. B.S (regents Scholar), St. John's U., 1966, M.S. in Microbiology, 1968; Ph.D., N.Y. U., 1979. Cert. lab. dir., N.Y. State. Research microbiologist N.Y. U. Dental Coll., 1966-67, Coney Island Hosp., Bklyn., 1967-71; chief supervising microbiologist L.I. Jewish/Queens Hosp. Center, Jamaica, N.Y., 1977-81, asst. attending microbiologist, 1981—, asst. dir. clin. microbiology lab; instr. in field. Contbr. articles to jours. in field. Research soc. grad. fellow, 1972-78. Mem. Am. Soc. Microbiology, N.Y. Acad. Scis. (councillor, adv. com. microbiology sect.), Am. Pub. Health Assn., AAAS, Assn. for Women in Sci., Sigma Xi. Subspecialties: Microbiology; Microbiology (medicine). Current work: Plasmid biology, clinical research. Office: Queens Hospital Center 82-68 164th St Jamaica NY 11432

DELLARIO, DONALD J(OSEPH), university research administrator, research educator; b. Pittston, Pa., July 20, 1946; s. Cataldo Anthony and Antoinette Marie (Viola) D.; m. Nancy Jane Friedley, May 22, 1982. A.B., Catholic U. Am., 1968; M.S., U. Scranton, 1970; Ph.D., U. Wis.-Madison, 1974. Lic. psychologist, cert. rehab. counselor. Dir. undergrad. studies, dept. human resources U. Scranton, 1974-78; dir. rehab. edn. and tng. U. Hawaii, 1978-80; assoc. dir. research Boston U. Rehab. Research and Tng. Center in Mental Health, 1980—. Mem. Am. Psychol. Assn., Am. Ednl. Research Assn., Am. Personnel and Guidance Assn., Nat. Rehab. Assn., Am. Rehab. Counseling Assn. Subspecialty: Psychiatric rehabilitation. Current work: Psychiatric rehabilitation, rehabilitation change, program evaluation. Office: Boston U Rehab Research and Tng Center in Mental Health 1019 Commonwealth Boston MA 02215

DELL-OSSO, LOUIS FRANK, med. scientist, educator; b. Bklyn., Mar. 16, 1941; s. Frank and Rose (Perrone) Dell-O. B.S. in Elec. Engring, Bklyn. Poly. Inst., 1961; Ph.D., U. Wyo., 1968. Bioengring.

cons. Westinghouse Research and Devel. Labs., Pitts., 1966-67, sr. bioengr., 1967-70; asst. prof. biomed. engring. and surgery U. Miami, Fla., 1970-72, asst. prof. neurology, 1972-75, assoc. prof. neurology, 1975-79, prof. neurology, 1979-80; co-dir. ocular motor neurophysiology lab. VA Med. Ctr., Miami, 1972-80; prof. neurology and biomed. engring. Case Wetern Res. U., Cleve., 1980—; dir. ocular motor neurophysiology lab. VA Med. Ctr., Cleve., 1980—. Contbr. articles on neurology and biomed. engring. to profl. jours. Bd. dirs. Vineland Galloway Civic Assn., 1973-76. Mem. IEEE (sr.), Profl. Group on Engring. in Medicine and Biology (chmn. Miami chpt.), AAAS, Assn. Research in Vision and Ophthalmology. Subspecialties: Biomedical engineering; Neuro-ophthalmology. Current work: Study of normal and abnormal human ocular motor control. Computer modeling of brainstems control systems involved in normal and pathological eye movements. Disease diagnosis. Home: 2356 Tudor Dr Cleveland Heights OH 44106 Office: VA Med Ctr Ocular-Motility 127A Cleveland OH 44106

DELOFFRE, BERNARD THIERRY, engineer; b. Enghien-Les-Bains, Val D'oise, France, Jan. 6, 1935; s. Henri Frederic and Marie Madeleine (Kister) D.; m. Jeanne LePelley Du Manoir, May 19, 1962 (div. 1979); 1 son, Renaud; m. Guillemette Anne Dibart De La Villetanet, Feb. 13, 1981; 1 dau., Samantha. Dipl. Eng., Ecole Polytechnique, Paris, 1958, Ecole Nat Superieure de l'Armement, Paris, 1961. Chief engr. Cabinet Norbert Beyrard, Paris, 1964-67; dep. dir. to dir. Centre Nat d' Etudes Spatiales, Centre Spatial Guyanais, Kourou, France, 1967-73; exec. sec. Symphonie Satellite Program, Paris, 1973-75; dir. Spacelab. program European Space Agy., Paris, 1975-76; program dir. Ceusot-Loire Enterprises, Paris, 1976-78; dir. gen. Satel Conseil, Montrouge, France, 1979—; exec. sec. European Cons. Satellite Orgn., Paris, 1982—. Served with French Army, 1961-64. Recipient silver gilt medal Centre Nat. d'Etudes Spatiales, 1975, symponie commemorative medal, 1978; Spacelab commemorative medal European Space Agy., 1980; decorated officer Order of Merit, France, 1981. Mem. Assn. Aeronautique Astronautique de France, AIAA. Roman Catholic. Clubs: Aero de France, de l'Espace (Paris). Subspecialties: Satellite studies; Aerospace engineering and technology. Home: 2 Bis Villa Mequillet Neuilly France 92200 Office: Satel Conseil 5 Rue Louis Lejeune Montrouge France 92128

DELONG, LANCE ERIC, physics educator; b. Denver, Nov. 12, 1946; s. Robert Earl and Svea Virginia (Selander) DeL.; m. Michele Denise Arranaga, Dec. 30, 1977 (div. 1983). B.A., U. Colo., Boulder, 1968; M.S., U. Calif.-San Diego, 1970, Ph.D., 1977. Asst. prof. physics U. Va., Charlottesville, 1977-79; asst. prof. U. Ky., Lexington, 1979-83, assoc. prof., 1983—. Cottrell research grantee Research Corp., 1982; NSF internat. travel grantee, 1982; recipient research contract U.S. Dept. Energy, 1981—. Mem. Am. Phys. Soc., Am. Assn. Physics Tchrs., Materials Research Soc. Subspecialties: Condensed matter physics; Low temperature physics. Current work: Electronic and magnetic properties of solids at low temperatures, high pressures and high magnetic fields, especially the superconducting and magnetic properties of metals. Home: 241 B Medlock Rd Lexington KY 40502 Office: Dept Physics and Astronomy U Ky Lexington KY 40506

DE LORGE, JOHN OLDHAM, research psychologist, administrator; b. Jacksonville, Fla., Dec. 14, 1935; s. Albert Augustus and Eunice Eugenia (Hopkins) de L.; m. Janet Gail Francis, Sept. 10, 1960; children: Celeste Lynn, Claire Francis, Albert Louis. B.A., Jacksonville U., 1960; M.A., Hollins Coll., 1962; Ph.D., U. N.C., 1964. Instr. U. N.C., 1964-65; research psychologist Evansville State Hosp., 1965-66; asst. prof. U. South Ala., 1966-69; research psychologist Naval Aerospace Med. Research Inst., Pensacola, Fla., 1969—; cons. EPRI, Palo Alto, Calif., 1978—; adj. prof. U. West Fla., 1972—, Pensacola Jr. Coll., 1970-73; vis. prof. U. Rochester, 1979-82. Contbr. numerous articles to profl. jours. Regional dir. Fla. Wildlife Fedn., 1981. Served with U.S. Army, 1953-56. NIMH Research grantee, 1966, 67. Mem. Bioelectromagnetics Soc. (dir. 1982—), N.W. Fla. Psychol. Assn. (pres. 1975-76), Am. Psychol. Assn., Psychonomic Soc., AAAS, Sigma Psi. Democrat. Episcopalian. Club: Pensacola Runners. Subspecialties: Behavioral psychology; Psychobiology. Current work: The effects of biophysical agents on behavior. Home: 4533 St Nazaire Rd Pensacola FL 32505 Office: Naval Aerospace Med Research Lab NAS Pensacola FL 32508

DELOS, JOHN BERNARD, chem. physicist, educator; b. Ann Arbor, Mich., Mar. 24, 1944; s. John S. and Katherine (Petruccione) D.; m. Sue Ellen Steere, May 29, 1965; children: Peter, Gregory, Rebecca. B.S. in Chemistry, U. Mich., 1965; Ph.D. in Phys. Chemistry, M.I.T., 1970. Research assoc. U. Alta., 1970-71, U. B.C., 1970-71; asst. prof. physics Coll. of William and Mary, 1971-77, assoc. prof. physics, 1977—; vis. scientist FOM (Netherlands) Inst. Atomic and Molecular Physics; cons. Naval Surface Weapons Lab., 1981; vis. scientist Oak Ridge Nat. Lab., 1982. Contbr. articles to profl. jours. NSF research grantee, 1976—. Mem. Am. Phys. Soc., Soc. Natural Philosophy, Fedn. Am. Scientists. Subspecialties: Atomic and molecular physics; Theoretical chemistry. Current work: Atoms, molecules and their collisions; theoretical research on electronic transitions in atomic collisions; atoms in fields; dynamics of molecular vibrations. Office: Physics Dept Coll William and Mary Williamsburg VA 23185

DEL REGATO, JUAN ANGEL, radio-therapeutist and oncologist, educator; b. Camaguey, Cuba, Mar. 1, 1909; came to U.S., 1937, naturalized, 1941; s. Juan and Damiana (Manzano) del R.; m. Inez Johnson, May 1, 1939; children: Ann Cynthia del Regato Jaeger, Juanita Inez del Regato Peters. John Carl. Student, U. Havana, Cuba, 1930; M.D., U. Paris, France, 1937, Laureat, 1937; Dr.S. (honoris causa), Colo. Coll., 1969; D.Sc. (honoris causa, ad gradum), Hahnemann Med. Coll., 1977, Med. Coll. Wis., 1981. Diplomate: Am. Bd. Radiology (trustee 1975—, historian 1976—). Asst. Radium Inst., U. Paris, 1934-37, Chgo. Tumor Inst., 1938; radiotherapeutist Warwick Cancer Clinic, Washington, 1939-40; research Nat. Cancer Inst., Balt., 1941-43; chief dept. radiotherapy Ellis Fischel State Cancer Hosp., Columbia, Mo., 1943-48; dir. Penrose Cancer Hosp., Colorado Springs, Colo., 1949-73; prof. clin. radiology U. Colo. Med. Sch., 1950-74; prof. radiology U. South Fla., 1974—; David Gould lectr. Johns Hopkins U., 1983. Author: (with L.V. Ackerman, M.D.) Cancer; Diagnosis Treatment and Prognosis, 1947, 54, 62, 70, (with H.J. Spjut), 1977; Editor: Cancer Seminar, 1960—; Contbr. articles to profl. jours. Decorated Order of Carlos Finlay of Cuba; Order Francisco de Miranda Republic of Venezuela; Béclère medal à titre exceptionnel, 1980; recipient Gold medal Radiol. Soc. North Am., 1967, Inter-Am. Coll. Radiology, 1967, Am. Coll. Radiology, 1968; Gold plaque, 1975; Grubbe gold medal Ill. Radiol. Soc., 1973; Prix Bruninghaus French Acad. Medicine, 1979; Disting. Scientist award U. South Fla. Coll. Medicine, 1980; named Disting. Physician VA, 1974; Disting. physician VA, 1974—. Mem. Nat. Adv. Cancer Council, Bethesda, Md. (1967-71); mem. med. adv. com. Milheim Found., Denver.; Fellow Am. Coll. Radiology (bd. chancellors; chmn. commn. radiation therapy, com. awards and honors); mem. A.M.A., Nat. Acad. Medicine of France (Laureat 1948), Radiol. Soc. N.Am. (v.p. 1959-60, Arthur Erskine lectr. 1978), Am. Roentgen Ray Soc., Am. Radium Soc. (v.p. 1963-64, treas. 1966-68, pres. 1968-69, chmn. exec. com. 1971-72, historian 1969—, Janeway gold medal 1973), Assn. Am. Med. Colls., Internat. Club Radiotherapists (pres. 1962-65), Inter-Am. Coll. Radiology (pres. 1967-71, U.S. counselor 1971-79), Am. Soc. Therapeutic Radiologists (sec. 1958-68, historian 1968—, pres. 1974-75, chmn. bd. dirs. 1975-76, gold medal 1977), Fedn. Clin. Oncologic Socs. (pres. bd. dirs. 1976-77); hon. mem. Rocky Mountain, Pacific N.W., Tex., Oreg., Minn. radiol. socs., radiol. socs. Cuba, Mex., Panama, Ecuador, Peru, Paraguay, Can., Argentina, Buenos Aires (Argentina), Am. Inst. Radiology (historian 1978—). Subspecialties: Radiology; Psychophysiology. Current work: Radiophysiology; clinical radio therapy; cancer. Home: 3101 Cocos Rd Carrollwood Tampa FL 33618 Office: Dept Radiology U South Florida Coll Medicine Tampa FL 33618 also VA Med Center 13000 N 30th St Tampa FL 33612

DE LUCA, CARLO JOHN, biomed. engr., researcher; b. Bagnoli del Trigno, Italy, Oct. 12, 1943; came to U.S., 1973, naturalized, 1982; s. John and Josephine De L.; m. Christine M. Rafferty, June 11,1982. B.A.Sc., U. B.C. (Can.), Vancouver, 1966; M.Sc. in Biomed. Engring, U. N.B., Can., 1968; Ph.D. (Ont. Govt. fellow), Queen's U., Kingston, Ont., Can., 1972. Lab. instr. in elec. engring. U. N.B., 1967-68, lectr. in computing sci., 1968-69; lectr. in biomed. engring. Queen's U., 1969-70, lab. instr. in anatomy, 1970-71, lectr. in anatomy, 1971-72, asst. prof. anatomy, 1972-73; research assoc. in orthopedic surgery Harvard U. Med. Sch., Boston, 1973-79, prin. research assoc. in orthopedic surgery, 1979—; lectr. in mech. engring. M.I.T., 1973—; research assoc. in orthopedic surgery Children's Hosp. Med. Ctr., Boston, 1973—, dir., 1980—; adj. assoc. prof. biomed. engring. Boston U., 1977—; project dir. Liberty Mut. Research Ctr., Hopkinton, Mass., 1973—; affiliated scientist New Eng. Regional Primate Ctr., Southboro, Mass., 1977—; research mem. Harvard-M.I.T. Div. Health Sci. and Tech., 1978—. Contbr. chpts., numerous articles to profl. publs.; editor: Procs. of the Fourth Congress of I.S.E.K, 1979; editorial bd.: Jour. Rehab. Research and Devel., 1978—, Jour. Motor Behavior, 1981—. Mem. IEEE (sr.), Internat. Soc. Electrophysiol. Kinesiology (sec. gen. 1976-70, sec. 1980-84), Can. Med. and Biol. Engring. Soc. (sr.), AAAS, Soc. Neurosci., Orthopaedic Research Soc., Biomed. Engring. Soc., Rehab. Engring. Soc. N.Am., Sigma Xi. Club: Harvard (Boston). Subspecialties: Biomedical engineering; Neurophysiology. Current work: Motor control of normal and abnormal muscles; motor unit properties of muscles; control of electro-mech. prostheses; objective evaluation of muscle fatigue in humans; myoelectric biofeedback; interaction of sensory and motor systems in muscle spasticity. Patentee apparatus for interfacing to anatomic signal sources, method and apparatus for interfacing to anatomic signal sources; co-patentee methods and apparatus for interfacing to nerves, monitoring myoelectric signals.

DE LUCA, LUIGI MARIA, research biochemist; b. Maglie, Lecce, Italy, Feb. 25, 1941; s. Antonio and Elena (Toma) De L.; m. Silvana Matilde Mendola, June 30, 1965; children: Nicholas, Mara. Dr. Organic Chemistry, U. Pavia, Italy, 1964. Cert. Italian Bd. Chemists. Dir. clin. lab. U. Milan Poly. Hosps., 1965; research assoc. M.I.T., 1965-69, instr., 1969-71; research chemist lung cancer br. Nat. Cancer Inst., Bethesda, Md., 1971-73, chief differentiation control sect., 1972-75, chief differentiation control sect. lab. exptl. pathology, 1975-81, chief differentiation control sect. lab. cellular carcinogenesis and tumors promotion, 1981—; chmn. program com. Am. Inst. Nutrition. Contbr. numerous articles to profl. jours. Recipient Mead Johnson award, 1979; award Japan Soc. Promotion Sci.; Lions Club award City of Maglie. Mem. Am. Soc. Biol. Chemists, AAAS, Fedn. Am. Sosienties, Soc. Complex Carbohydrates, N.Y. Acad. Scis. Subspecialty: Biochemistry (biology). Current work: Prevention of carcinogenesis by nutrients.

DELUCA, MARLENE A(NDEREGG), biochemist, educator; b. La Crosse, Wis., Nov. 10, 1936; d. Ruben H. and Yerda T. (Harris) Anderegg; m. William D. McElroy, Aug. 22, 1967; 1 son: Eric Gene McElroy. B.S., Hamline U., 1958; Ph.D., U. Minn., 1962. Asst. prof. biology Johns Hopkins U., 1965-69; asst. prof. biochemistry Georgetown U., 1969-72; assoc. prof. chemistry U. Calif., San Diego, 1974-78, prof., 1978—. Contbr. numerous articles to prof. jours.; contbg. editor: Methods in Enzymology, 1978, Basic Chemistry and Analytical Applications, 1981. NIH grantee, 1967—; NSF grantee, 1966—. Mem. Fedn. Am. Socs. Exptl. Biology, Am. Chem. Soc. Subspecialties: Biochemistry (biology); Biochemistry (medicine). Current work: Bioluminescence, enzyme mechanisms. Office: Dept Chemistr U Calif at San Diego La Jolla C 92093

DELUCIA, ANTHONY JOHN, surgery and physiology educator, research administrator; b. Riverside, Calif., July 9, 1948; s. Vincent Luke and Lucy Katherine (Schettino) DeL.; m. Deborah Jean Gray, Sept. 24, 1977. B.A., U. Calif.-Riverside, 1970; Ph.D., U. Calif.-Davis, 1974. Postgrad. research engr. U. Calif.-Davis, 1975-75; staff research assoc., 1975-76; asst. prof. La. State U. Med. Ctr., 1976-77; asst. prof. surgery and physiology East Tenn. State U., 1977-80, assoc. prof., 1980—, dir. surg. research dept. surgery, 1977—. Contbr. articles, abstracts to profl. publs. NSF fellow, 1970-71. Mem. Am. Coll. Sports Medicine, Am. Thoracic Soc., Am. Soc. Primatologists, Am. Physiol. Soc., Nat. Soc. Med. Research, Internat. Soc. Primatologists, Phi Beta Kappa. Democrat. Roman Catholic. Club: Faculty Investment (sec. 1983—). Subspecialties: Environmental toxicology; Physiology (medicine). Current work: Effects of cigarette smoking on cardiorespiratory function; health effects of air pollutants; effects of drugs on lung. Office: Dept Surgery East Tenn State U Box 19 750A Johnson City TN 37614 Home: 902 Millercrest Dr Johnson City TN 37601

DELUCIA, FRANK CHARLES, physicist; b. St. Paul, June 21, 1943; s. Frank Charles and Muriel Ruth (Rinehart) DeL.; m. Shirley Ann Wood, June 25, 1966; children: Frank Charles, Elizabeth Ann. B.S., Iowa Wesleyan Coll., 1964; Ph.D., Duke U., 1969. Instr., research assoc. Duke U., 1969-71, asst. prof. physics, 1971-76, assoc. prof., 1976-83, prof., 1983—. Mem. Am. Phys. Soc., Optical Soc. Am., IEEE. Subspecialties: Atomic and molecular physics; Electrical engineering. Current work: Millimeter and submillimeter waves, molecular spectroscopy, quantum electronics. Office: Dept Physics Duke U Durham NC 27706

DELUISI, JOHN JAMES, atmospheric physicist; b. Little Falls, N.Y., Oct. 4, 1980; s. John Joseph and Anna Marie (Mercaldo) DeL.; m. Prudy Marlene Williams, Aug. 30, 1959; children: John James, Barbara, Rebecca, Nannette, Michael. B.S., SUNY-Albany, 1957, M.S., 1960; M.A.T., Brown U., 1961; Ph.d., Fla. State U., 1967. Tchr. physics, chemistry and biology, public high sch., East Grenbush, N.Y., 1959-61; atmospheric research scientist Nat. Center Atmospheric Research, Boulder, Colo., 1967-75, NOAA Air Resources Labs., Boulder, 1975—, chief aerosols and radiation monitoring, 1977—. Contbr. articles to profl. jours. Served with USN, 1948-49, 51-53. Recipient High Sch. sci. Tchr. Fellowship award NSF, 1960, 61, 62. Mem. Am. Geophys. Union. Roman Catholic. Lodge: Elks. Subspecialties: Remote sensing (atmospheric science); Climatology. Current work: Radiative effects of stratospheric aerosols from 1982 eruption of El Chichon Volcano in Mexico; atmospheric research concerning aerosol effects on the earth's radiation balance and remote sensing of ozone in stratosphere.

DELUSTRO, FRANK ANTHONY, immunologist, researcher, educator; b. Bklyn., May 8, 1948; s. Frank, Jr. and Yolanda (Lombardi) DeL.; m. Barbara Mary Cervini, May 4; 1974. B.S., Fordham U., 1970; Ph.D., Upstate Med. Sch., SUNY, Syracuse, 1976. Postdoctoral fellow Med. U. S.C., Charleston, 1976-78, instr., 1978-80, asst. prof. immunology, 1980-83; mgr. immunology Collagen Corp., 1983—. Mem. Am. Assn. Immunologists, Reticuloendothelial Soc., Soc. Exptl. Biology and Medicine (editorial bd. 1979—). Roman Catholic. Subspecialties: Immunology (medicine); Immunobiology and immunology. Current work: Exploration of immune mechanisms of connective tissue disease: monocyte stimulation of fibrosis, and autoimmunity to components of basement membrane. Office: Collagen Corp 2500 Faber Pl Palo Alto SC 9430 Home: 928 Wright Ave Unit 1001 Mountain Valley CA 94043

DEL VALLE, FRANCISCO RAFAEL, food science researcher, educator, consultant; b. Laredo, Tex., Oct. 19, 1933; s. Roberto and Margarita (Canesco) Del V.; m. Estela Alicia Urrutia, Aug. 6, 1961; children: Estela Margarita, Francisco Roberto. S.B. in Chemistry, MIT, 1954, M.S. in Biochem. Engring, 1956, S.M. in Chem. Engring, 1957, Ph.D. in Food Sci. and Tech, 1965. Prof. chem. engring. and food sci. Instituto Tecnologico y da Estudios Superiores de Monterrey, Mex.) and; Guaymas, 1961-77; dir. Instituto Chihuahuense de Investigacion y Desarollo de la Nutricion, Chihuahua, Mex., 1970-80; prof. Universidad Autonoma de Chihuahua, 1977—; pres. Fundacion de Estudios Alimentarios y Nutricianoles, 1980—; cons. to industry. Contbr. articles to tech. jours. Recipient Nat. Prize in Technology Pres. Mex., 1977; Nat. prize in Sci. and Tech. Banco Nacional de Mex., 1970, 81. Mem. Academia de la Investigacion Cientifica, Academia Nacional de Ingenieria, Asociacion de Tecnicos en Alimentos de Mex., Inst. Food Technologists, Sociedad Latinoamericana de Nutricion. Subspecialties: Food science and technology; Nutrition (biology). Current work: Research, development, evaluation and industrial-commercial production of high-nutrition, low-cost foods from plant, marine and fresh-water fish sources. Developed processes for prodn. of quick-salted fish cakes, soy-oats infant formula, other soy-oats high-nutrition, low-cost foods. Office: Apartado Postal Chihuahua ChihuahuaMexico

DELVILLANO, BERT C., JR., Immunologist; b. Phila., Apr. 9, 1943; s. Bert C. and Beulah A. DelV.; m. Anne M., Nov. 16, 1963; 1 dau., Diane M. B.A., Lehigh U., 1965; Ph.D., U. Pa., 1971. Postdoctoral fellow Scripps Clinic, La Jolla, Calif., 1971-73, asst., 1973-75; mem. staff Cleve. Clinic, 1975-80; dir. product devel. Centocor, Malvern, Pa., 1980—. Contr. articles in field to profl. jours. Basic O'Connor grantee, 1973-75; Nat. Cancer Inst. grantee, 1975-80. Mem. Am. soc. Microbiology, Am. Assn. Immunology. Subspecialties: Cancer research (medicine); Immunology (medicine). Current work: Developed CA 19-9 RIA for detection of pancreatic and other GI cancers.

DE MARCO, THOMAS JOSEPH, periodontist, dean; b. Farmingdale, N.Y., Feb. 12, 1942; s. Joseph Louis and Mildred Nora (Cifarelli) De M.; children—Todd Gordon, Kristin Alice, Lisa Anne. B.S., U. Pitts., 1962; D.D.S., 1965; Ph.D., certificate in Periodontology, Boston U., 1968; cert. in fin. planning, Coll. Fin. Planning, Denver, 1976. Certificate in clin. hypnosis. Practice dentistry specializing in periodontics, Cleve., 1968—; mem. staff Met. Gen. Hosp., Cleve., Univ. Hosp., VA Hosp.; asst. prof. periodontics and pharmacology Case-Western Res. U., 1968-70, asso. prof., 1970-73, prof., 1973—, asso. dean, 1972-76, dean, 1976—. Author review books in dentistry, also articles on periodontology, pharmacology, fin. planning. Grantee Air Force Office Sci. Research, 1969, Upjohn Co., 1970, Columbus Dental Mfg. Co., 1971. Mem. Am. Acad. Periodontology, Internat. Assn. Dental Research, Am. Soc. for Preventive Dentistry (past pres. Ohio chpt.). Subspecialty: Periodontics. Home: 14435 Hunting Hill Dr Novelty OH 44072 Office: 2123 Abington Rd Cleveland OH 44106

DEMAREST, DAVID STEELE, psychologist; b. Sioux Falls, S.D., Sept. 22, 1954; s. Angus Raymond and Dona Jean (Thomson) D. B.A., U. Va., 1976; M.S., Va. Commonwealth U., 1980. Lic. psychologist, S.D. Clin. psychologist Va. Treatment Ctr. for Children, Richmond, 1979-80, Redfield (S.D.) State Hosp. and Sch., 1980-82; clin. psychologist Iowa Meth. Med. Ctr., Des Moines, 1982—. Coach Southside YMCA Football/Basketball, Des Moines, 1982; referee W. Des Moines Soccer League, 1982, Basketball League, 1982. Mem. Am. Psychol. Assn., Am. Assn. Mental Deficiency, Assn. Retarded Citizens, Midwestern Psychol. Assn., S.D. Psychol. Assn., W. Des Moines Soccer League. Subspecialties: Neuropsychology; Behavioral psychology. Current work: Neuropsychol. assessment of and psychotherapy with head trauma and cardiovascular accident victims; family education training workshops for stroke victims; biofeedback therapist. Home: 1800 Grand Ave #271 West Des Moines IA 50265 Office: Dept Psychology Iowa Meth Med Center 1200 Pleasant St Des Moines IA 50309

DEMAREST, HAROLD HUNT, JR., geology educator; b. N.Y.C., Dec. 20, 1946; s. Harold Hunt and Eleanor Frances (Conklin) D.; m. Merry A. Bouzis, Mar. 16, 1968; 1 dau., Joan Elaine. B.A., Reed Coll., 1969; M.A., Columbia U., 1971; Ph.D., UCLA, 1974. Research asst. Lamont-Doherty Geol. Obs., Palisades, N.Y., 1967-71; from research asst. to postdoctoral scholar UCLA, 1971-75; research assoc. U. Chgo., 1975-79; asst. prof. geology Oreg. State U., Corvallis, 1979—; vis. staff mem. Los Alamos Sci. Lab., 1972-75. Contbr. articles to profl. jours. Bd. dirs. Phoenix Sch., Chgo., 1977-78; precinct committeeman Dem. Party, Corvallis, 1982—. Mem. Am. Geophys. Union, Soc. Exploration Geophysicists, AAAS, Nat. Assn. Geology Tchrs., NOW (pres. Corvallis chpt. 1983), Sigma Xi. Subspecialties: Geophysics; Condensed matter physics. Current work: Mineral physics, geostatistics, portfolio analysis, game theory and new ideas that are so revolutionary that they cannot be described in scientific compendiums. Home: 3621 NW Sylvan Dr Corvallis OR 97330 Office: Dept Geology Oreg State U Corvallis OR 97331

DE MARIA, ANTHONY J., electrical engineer; b. Santa Croce, Italy, Oct. 30, 1931; came to U.S., 1935; s. Joseph and Nicolina (Daddona) De M.; m. Katherine M. Waybright, Aug. 29, 1953; 1 dau., Karla Kay. B.S. in Elec. Engring, U. Conn., 1956, Ph.D., 1965; M.S., Rensselaer Poly. Inst., 1960. Acoustic research engr. Anderson Lab., West Hartford, Conn., 1956-57; magnetic research engr. Hamilton Standard Div. United Techs. Corp., Windsor Locks, Conn., 1957-58; scientist United Techs. Research Center, East Hartford, Conn., 1958—; instr. in electronics U. Hartford, 1957-60; adj. prof. physics Rensselaer Poly. Inst. Grad. Center, Hartford, 1970-77; lectr. in lasers UCLA, 1974—; mem. Dept. Def. Adv. Group on Electronic Devices, 1977-80, chmn., 1980—; mem. evaluation com. on electromagnetic tech. Nat. Bur. Standards, 1977-79; mem. Center Elec. and Electronic Engring., 1979—. Author: Lasers, Vol. III, 1972, Vol. IV, 1976; Contbr. articles to profl. jours. Mem. Air Force Sci. Adv. Bd., 1981—. Recipient Disting. Alumnus award U. Conn., 1978, Disting. Engring. award U. Conn., 1983, Davies medal and award Rensselaer Poly. Inst., 1980. Fellow IEEE (editor Jour. Quantum Electronics, Morris N. Liebman meml. award 1980), Optical Soc. Am. (v.p. 1979, pres. 1981); mem. Am. Phys. Soc., Nat. Acad. Engring., Conn. Acad. Scis. and Engring. Subspecialty: Laser physics. Office: United Techs Research Center 400 Main St East Hartford CT 06108

DEMARQUE, PIERRE, astronomer, educator; b. Fes, Morocco, July 18, 1932; m. Marlene J. Demarque, Sept. 15, 1958; children—Jerome, Annick. B.Sc., McGill U., Montreal, Que., Can., 1955; M.A., U.

Toronto, Ont., Can., 1957, Ph.D., 1960; M.A. (hon.), Yale U., 1968. Munson prof. natural philosophy and astronomy Yale U., New Haven. Office: PO Box 6666 New Haven CT 06511

DEMAS, NICHOLAS GEORGE, engineering educator, researcher; b. Cin., Oct. 24, 1934; s. George Nicholas and Helen Margaret (Kappas) D.; m. Barbara Lamonettin, Nov. 8, 1959; children: Helen, George, Constandina. B.S. West Liberty State Coll., 1957; M.S., U. Miss., 1962; Ph.D., U. Wis.-Madison, 1971. Scientist in nuclear design and analysis Bettis Atomic Power Lab., West Mifflin, Pa., 1957-62, sr. scientist in nucclear design, 1971-74; physicist for nuclear studies and investigations P.R. Water Resources Authority, Santurce, 1966-68; sr. scientist in nuclear design Westinghouse Nuclear Fuels Div., Monroeville, Pa., 1974-77; assoc. prof. engrng. Tenn. Technol. U., 1977—. Fellow Am. Nuclear Soc.; mem. IEEE (sr.). Republican. Lodge: Rotary. Subspecialties: Nuclear fission; Nuclear fusion. Current work: Teaching and research in nuclear fission and fusion engineering. Home: 1040 Wilson Ave Cookeville TN 38501 Office: Tenn Technol U Cookeville TN 38501

DE MAYO, PAUL, chemistry educator; b. London, Eng., Aug. 8, 1924; m. Mary Turnbull, May 28, 1949; children—Ann, Philip. B.Sc., U. London, 1944, Ph.D., 1954; D.es-Sc., U. Paris, 1970. Asst. lectr. Birkbeck Coll., London, 1954-55; lectr. U. Glasgow, Scotland, 1955-57, Imperial Coll., London, 1957-59; prof. chemistry U. Western Ont., Can., London, 1959—, dir. photochemistry unit, 1969-72. Author: Mono and Sesquiterpenoids, 1959, The Higher Terpenoids, 1959, numerous publs. in chem. lit.; editor: Molecular Rearrangements, vol. 1, 1963, vol. II, 1964, Rearrangements in Ground and Excited States, Vol. 1-3, 1980; editorial bd.: Nouveau Journal de Chimie. Recipient Merck, Sharp & Dohme Lecture award Chem. Inst. Can., 1966, Centennial medal Govt. Can., 1967, medal Chem. Inst. Can., 1982. Fellow Royal Soc. Can., Royal Soc. (London). Subspecialties: Photochemistry; Organic chemistry. Current work: Photochemistry of adsorbed molecules; surface photochemistry; organic chemisry of semi-conductors . Office: Chemistry Dept U Western Ont London ON Canada N6A 5B7

DEMEIO, JOSEPH LOUIS, virologist; b. Hurley, Wis., Sept. 9, 1917; s. Emanuel and Jennie (Varda) DeM.; m. Agnes V. Steinbach, Oct. 15, 1941; children: Bonnie Ann. B.S., Marquette U., 1950, M.S., 1954; Ph.D., U. Wis.-Madison, 1958. Virologist Naval Med. Research Unit #4, Great Lakes, Ill, 1950-57; asst. chief diagnostic reagents Ctr. for Disease Control, Atlanta, 1957-59; virologist Nat. Drug Co., Swiftwater, Pa., 1959-77; assoc. mgr. Salk Inst., Swiftwater, 1978—. Served in U.S. Army, 1942-45; PTO. Mem. Am. Assn. Immunologists, Soc. Exptl. Biology and Medicine. Roman Catholic. Subspecialties: Virology (biology); Immunobiology and immunology. Current work: Antigenic relationships among myxoviruses. Home: RD 7 Box 7206 Stroudsburg PA 18360 Office: PO Box 250 Swiftwater PA 18370

DEMEO, V(INCENT) JAMES, JR., geographer, researcher, educator; b. Miami, Fla., Jan. 14, 1949; s. Vincent James and Dorothy DeM. B.S., Fla. Internat. U., Miami, 1975; M.A., U. Kans., Lawrence, 1979, Ph.D., 1983. Pres. Natural Energy Co., Delray Beach, Fla., 1974-77; instr. geography and meteorology dept. U. Kans., Lawrence, 1977-81; asst. prof. geography-geology dept. Ill. State U., Normal, 1981—; dir. Orgone Biophys. Research Lab., 1978; lab. and research asst. Tinicum Med. Research Found., Bucks County, Pa., 1976—. Contbr. articles to profl. jours. Mem. S. Fla. Citizens United Against a Radioactive Environ., 1973-77, Radioactive-Free Kans., 1977-78, Lawrence Assn. Parents and Profls. for Safe Alternatives in Childbirth, 1978-81; cons. solar energy home loan program Appropriate Tech. Resource Ctr., Lawrence, 1980-81. Mem. Assn. Am. Geographers, Assn. Arid Lands Studies, Am. Meteorol. Soc. Subspecialties: Desertification studies; Ecosystems analysis. Current work: Arid lands research, sex-economy, technique of cloudbusting. Office: Geography-Geology Dept Ill State U Normal IL 61761

DEMETER, STEVEN, neurologist; b. Budapest, Hungary, Jan. 12, 1947; U.S., 1957, naturalized, 1965; s. Arpad and Ilona (Wiesner) D. B.S., Bklyn. Coll., 1969; M.D., N.Y. Med. Coll., 1973. Diplomate: Am. Bd. Neurology and Psychiatry, 1979. Intern Beth Israel Med. Center, N.Y.C., 1973-74; resident in neurology Albert Einstein Coll. Medicine, Bronx, 1974-77; instr. neurology N.Y. Med. Coll., N.Y.C., 1977-79; fellow asso. in behavioral neurology U. Iowa, Iowa City, 1979-81; fellow Center for Brain Research, U. Rochester, N.Y., 1981—; instr. neurology U. Rochester, N.Y., 1981—. Mem. Am. Acad. Neurology, Nat. Soc. Med Research, AAAS, N.Y. Acad. Scis., Soc. for Newuosci. Subspecialties: Neurology; Neurobiology. Current work: Teaching and research in anatomical and functional bases of higher brain function in primates including man. Office: University of Rochester Medical Center PO Box 605 Rochester NY 14642

DEMETRAKOUPOULOS, GEORGE EVANGELOS, physician, executive consultant; b. Piraeus, Greece, Oct. 22, 1947; came to U.S., 1972, naturalized, 1980; s. Evangelos and Alexandra (Ziogas) D. M.D., U. Athens, 1971, D.Sc., 1975; M.P.H., Harvard U., 1978. Diplomate: Am. Bd. Pediatrics, Am. Bd. Clin. Nutrition, Greek Bds. Pediatrics. Intern Waltham (Mass.) Hosp., 1972-73; resident medicine Children's Hosp. Med. Ctr., Boston, 1973-75; clin. fellow pediatrics Harvard Med. Sch., Boston, 1974-75, research fellow dept. microbiology and molecular genetics, 1975-77, research assoc. dept. microbiology and molecular genetics, 1977-79; clin. assoc. pediatric oncology br. Nat. Cancer Inst., NIH, Bethesda, Md., 1979-80, expert, sr. staff clin. oncology program, 1989-82; practice medicine specializing in clin. nutrition and oncology, Bethesda, 1982—; pres. Nutrilife Corp. Am., Bethesda, 1982; examiner Am. Bd. Nutrition, 1982. Author: Guide to Good Nutrition and Better Health for the Later Years, 1979; editor, co-author: Handbook of Clinical Nutrition, 1983. H. C. Ernst fellow Harvard Med. Soc., 1975-76; Foggarty Internat. scholar, 1979-80; postdoctoral fellow NIH, 1976-78; Travel award Internat. Congress Nutrition, 1981. Mem. Internat. Soc. Leukemia Research, Am. Soc. Clin. Oncology, Am. Soc. Clin. Nutrition, Am. Acad. Pediatrics, AMA, Am. Coll. Nutrition, Am. Inst. Nutrition, Greek Med. Assn., N.Y. Med. Assn., Mass. Med. Soc., Harvard Med. Sch. Alumni Assn. (assoc.), Children's Hosp. Boston Alumni Assn., Harvard Sch. Pub. Health Alumni Assn. Greek Orthodox. Subspecialties: Nutrition (medicine); Oncology. Current work: Director of medical nutrition center.

DEMILLO, RICHARD A., information and computer science educator, researcher, consultant; b. Hibbing, Minn., Jan. 26, 1947; s. Herman and Lorraine K. DeM.; m. Diane Hanson; children: Alan, Gina, Andrew. B.A. in Math, Coll. St. Thomas, St. Paul, 1969; Ph.D. in Info. and Computer Sci, Ga. Inst. Tech., 1972. Teaching asst., programmer analyst dept. quantitiative methods Coll. St. Thomas, 1967-69; teaching asst. info. and computer sci. Ga. Inst. Tech., 1969-72; assoc. prof. info. and computer sci., 1976-81, prof., 1981—; asst. prof. elec. engrng. and computer sci. U. Wis.-Milw., 1972-76; cons.; vis. mathematician Math. Research Center, U. Wis., Madison, 1977. Author: Foundations of Secure Computation; contbr. articles profl. jours. Grantee NSF, 1973-83, U.S. Army Research Office, 1975-83, Office Naval Research, 1981-84, Office Sec. of Def. Mem. Assn. Computing Machinery, Am. Math. Soc., Math. Assn. Am., Soc. Indsl. and Applied Math., AAAS, Assn. Symbolic Logic. Subspecialties: Cryptography and data security; Software engineering. Current work: Computational complexity and algorithms, cryptograph and cryptographic protocols, software testing and reliability, protection of proprietary software. Home: 2508 Woodwardia Rd Atlanta GA 30345 Office: Information and Computer Sci Ga Inst Tech Atlanta GA 30332

DE MIRANDA, PAULO, pharmacologist; b. Goa, India; came to U.S., 1961, naturalized, 1973; m. Elizabeth Carmel Conlon, July 9, 1966; children: Gwen, David, Cynthia, Kevin. M.S., U. Wis., 1963; Ph.D. in Pharmacology, Marquette U., 1966. Instr. in pharmacology Marquette U. Sch. Medicine, 1966; vis. asst. prof. pharmacology Universidad del Valle, Cali, Colombia, 1966-69; sr. research scientist Wellcome Research Labs., Research Triangle Park, N.C., 1969-78, sr. scientist, group leader, 1978—. Contbr. numerous articles, abstracts to profl. publs. Mem. AAAS, Am. Soc. Pharmacology and Exptl. Therapeutics. Subspecialties: Pharmacology; Biochemistry (medicine). Current work: Nucleic acid antagonists, antiviral chemotherapy, drug metabolism. Home: 4828 Radcliff Rd Raleigh NC 27609 Office: 3030 Cornwallis Rd Research Triangle Park NC 27709

DEMIS, D. JOSEPH, dermatologist, pharmacologist; b. N.Y.C., Aug. 19, 1929. B.S., Union College, 1950; Ph.D., U. Rochester, 1953; M.D., Yale U., 1957. Diplomate: Am. Bd. Dermatology. Intern Madigan Hosp., Tacoma, 1957-58; resident Walter Reed Gen. Hosp., Washington, 1958-61; chief dept. dermatology Walter Reed Inst. Research, Washington, 1961-64; dermatologist-in-chief Barnes Hosp., St. Louis, 1964-67; prof. medicine, dir. dermatology Washington U., St. Louis, 1964-67; prof. dermatology Albany (N.Y.) Med. Coll., 1967—; cons. NIH study sect., 1960-68, 68-72. Author: Clinical Dermatology, 10th edit, 1982. Bd. dirs. Albany Symphony Orch., 1977, Van Rennselaer Mansion Com. Served to maj. U.S. Army, 1957-64. Decorated Army Cross; recipient USPHS Career Devel. award, 1963; Microcirculatory Soc. Travel award, 1964; Angiology Soc. award, 1968. Mem. Am. Dermatology Assn., Am. Soc. Pharmacology, Phi Beta Kappa, Sigma Xi. Club: Fort Orange (Albany). Subspecialties: Dermatology; Pharmacology. Current work: MAST cell disease.

DEMMERLE, ALAN MICHAEL, electronic engineer; b. Port Jefferson, N.Y., Nov. 4, 1933. B.S., Carnegie Inst. Tech., 1955; M.S., Columbia U., 1958. Engr. circuit design Westinghouse Electric, 1955-56; engr. U.S. Naval Research Lab., 1957-60; engr. telemetry procesing Goddard Space Flight Ctr., NASA, 1960-66; chief computer systems lab. NIH, Bethesda, Md., 1966—; dir. Aspin Research Inst., 1981—. Subspecialty: Information systems, storage, and retrieval (computer science). Office: NIH Room 2035 Bldg 12 A Bethesda MD 20205

DEMOS, PETER THEODORE, physics educator; b. Toronto, Ont., Can., 1941. B.Sc., Queen's U., Ont., Can., 1941; Ph.D. in Physics, MIT, 1951; LL.D. (hon.), Trent(Ont.) U., 1981. Instr. math. and physics Queen's U., 1941-42; mem. Ballistics Research staff Nat. Research Lab., Ont., 1942-44; mem. staff Can. Army Research Establishment, Quebec, 1944-46; asst. prof. physics MIT, Cambridge, Mass., 1946-51, mem. staff, 1951-52, lectr., assoc. dir. lab., 1952-61, dir., 1961-75, prof. physics, 1961—, dir., 1975—. Mem. AAAS, Am. Phys. Soc., Am. Acad. Arts and Scis. Subspecialty: Particle physics. Office: Dept Physics MIT Cambridge MA 02139

DE MOSS, RALPH DEAN, microbiologist, educator; b. Danville, Ill., Dec. 29, 1922; s. Guy and Ruby (Walker) DeM; m. Patricia H. Day, June 2, 1946 (dec.); children: Susan L., G. Newton, Guy R., Kurt S.; m. Shirley R. Siedler, Nov. 22, 1975. A.B., Ind. U., 1948, Ph.D., 1951; student, Clemson Coll., 1943, St. Louis U., 1943-44. AEC postdoctoral fellow Brookhaven Nat. Lab., 1951-52; asst. prof. McCollum-Pratt Inst., Johns Hopkins U., 1952-56; asso. prof. microbiology U. Ill., Urbana, 1956-59, prof. microbiology, 1959—, head dept. microbiology, 1971—; mem. microbiology tng. com. NIH, 1967-69, chmn., 1969-71, mem. microbial chemistry study sect., 1976-80, chmn., 1978-80; mem. biomed. scis. panel NRC, 1976-80. Editor: Jour. Bacteriology, 1965-70. Served with AUS, 1942-46; ETO. Mem. Am. Soc. Microbiology, Am. Acad. Microbiology, Am. Soc. Biol. Chemists, Soc. Gen. Microbiology. Subspecialties: Microbiology; Biochemistry (biology). Research and publs. in microbial biochemistry and physiology. Home: 801 Harmon Urbana IL 61801 Office: Dept Microbiology 131 Burrill Hall U Ill 407 S Goodwin Urbana IL 61801

DEMOTT, DIANA L(YNN), nuclear engineer; b. Coffeyville, Kans., June 18, 1952; d. Marion Dean and Sharylon Joan (O'Brien) DeM. B.S. in Nuclear Engring, Tex. A&M U., 1974. Field engr. Gen. Electric Co., Atlanta, 1974-76; quality assurance engr. Westinghouse Hanford, Richland, Wash., 1976-79, systems engr., 1979-81, licensing engr. Clinch River breeder reactor project, Oak Ridge, 1981-83, system engr., 1983—. Contbr. tech. paper to profl. publ. Mem. Soc. Women Engrs. (sr.), Am. Nuclear Soc., ASME, Instrument Soc. Am. (standard com.). Methodist. Subspecialty: Nuclear engineering. Home: 1328 Candlewick Knoxville TN 37922 Office: Westinghouse 120 S Jefferson Circle Oak Ridge TN 37830

DEMPSEY, JOHN PATRICK, civil engineering educator; b. Te Kuiti, King Country, N.Z., Oct. 13, 1953; s. Basil John and Elizabeth Sefton (Adlam) D.; m. Marsha Yvette Shifman, Sept. 1, 1980; 1 dau., Megan Constance. B.E. with first honors, Auckland U., 1975, Ph.D., 1978. Researcher dept. civil engring. Northwestern U., Evanston, Ill., 1978-80; asst. prof. dept. civil and environ. engring. Clarkson Coll. Tech., Potsdam, N.Y., 1980—. Assoc. mem. ASME, ASCE, Soc. Indsl. and Applied Math., Am. Acad. Mechanics, Soc. Engring. Sci. Subspecialties: Fracture mechanics; Theoretical and applied mechanics. Current work: Theoretical and applied mechanics. Office: Clarkson Coll Tech Dept Civil and Environ Potsdam NY 13676

DENBURG, JEFFREY LEWIS, neuroscience educator; b. Bklyn., Oct. 5, 1944; s. Alan and Violet (Moskowitz) D.; m. Jutta Drost, July 3, 1970 (div. 1982). B.A., Amherst Coll., 1965; Ph.D., Johns Hopkins U., 1970. Postdoctoral fellow Cornell U., Ithaca, N.Y., 1970-72; research fellow Australian Nat. U., Canberra, 1972-77; asst. prof. U. Iowa, Iowa City, 1977-83, 1983—. NIH Research Career Devel. award, 1979—. Subspecialties: Neurobiology; Developmental biology. Current work: Role of macromolecules in formation of connections between neurons during the development of the nervous system. Home: 1141 E College St Iowa City IA 52240 Office: Univ Iowa Dept Zoology Iowa City IA 52242

DENDY, JOEL EUGENE, JR., numerical analyst; b. Mar. 24, 1945; s. Joel Eugene and Lucile Flora (Warren) D.; m. Leslie Ann Helfrich, June 15, 1968; children: Miranda, Julian. B.A., Rice U., 1967, Ph.D., 1971. Asst. prof. math. U. Denver, 1971-73; mem. staff Los Alamos Nat. Lab., 1973—. Woodrow Wilson fellow, 1968; NSF trainee, 1969-71. Mem. Soc. Indsl. and Applied Math., Phi Beta Kappa. Subspecialty: Applied mathematics. Current work: Research in numerical solution of partial differential equations, multigrid method. Home: 2877 Woodland Rd Los Alamos NM 87544 Office: Los Alamos Nat Lab MS B284 Los Alamos NM 87545

DENEAU, GERALD ANTOINE, pharmacologist, educator; b. Oxford, Mich., May 9, 1928; s. Clare Richard and Catherine (McCallum) D.; m. Loise Elaine Peters, June 4, 1977; m. Irene Whittle, Sept. 5, 1952; children: Janet, Frances, Leslie. B.A., U. Western Ont., 1950, M.Sc., 1952; Ph.D. in Pharmacology, U. Mich., 1956. Instr., asst. prof. pharmacology U. Mich., 1956-65; research scientist So. Research Inst., Birmingham, Ala., 1965-71; assoc. prof. pharmacology U. Calif.-Davis, 1971-74, U. Ala., Birmingham, 1965-71; research scientist N.Y. State OASA Labs., Bklyn., 1974—; assoc. prof. psychiatry Downstate Med. Center, SUNY- Bklyn., 1980—; mem. study sect. on narcotic addiction and drug abuse Nat. Inst. Drug Abuse, 1967-71. Contbr. articles to profl. jours.; mem.: editorial bd. Jour. Pharmacology and Explt. Therapeutics, 1966-75. Mem. Am. Soc. Pharmacology and Exptl. Therapeutics, AAAS, Royal Soc. Medicine, N.Y. Acad. Scis. Subspecialties: Psychopharmacology; Neuropharmacology. Current work: Experimental drug dependence, physiological dependence, tolerance, psychotoxicity. Office: 80 Hanson Pl Brooklyn NY 11217

DENIO, ALLEN A(LBERT), phys. chemist, educator; b. Lowell, Mass., June 6, 1934; s. Albert A. and Ethel (Lawson) D.; m. Valerie S., June 20, 1959; children: Thomas, Susan, Richard. B.S., Lowell Technol. Inst., 1956, M.S. in Textile Chemistry, 1957, U. N.H., 1960, Ph.D., 1962. Chemist Dow Chem. Co., Midland, Mich., summer 1956; chemist dept. textile fibers E. I. duPont de Nemours & Co., Inc., Wilmington, Del., 1957-58, research chemist, 1962-64; mem. faculty U. Wis., Eau Claire, 1964—, asst. prof. phys. chemistry, 1964-68, assoc. prof., 1968-73, prof., 1973—; vis. prof. U. Wis.-Madison, 1969-70, U. Del., Newark, 1978-79; cons. to industry, pubs. Active Wis. Democratic party. Served with USAFR, 1957-63. Mem. Am. Chem. Soc., AAAS, Sigma Xi. Club: Indianhead Track (Eau Claire). Subspecialties: Physical chemistry; Polymer chemistry. Current work: Properties of polymer films containing metal atoms; revision phys. chemistry textbook. Home: 433 McKinley Ave Eau Claire WI 54701 Office: Dept Chemistry U Wis Eau Claire WI 54701

DENKOWSKI, GEORGE CARL, psychologist, consultant; b. Lodz, Poland, Sept. 23, 1942; s. George Sigmund and Gertrude (Frischholtz) C.; m. Kathryn M. Scharsu, July 11, 1978. B.S., Kent State U., 1964; M.S., U. Akron, 1975, Ph.D., 1977. Lic. psychologist, Tex. Spl. edn. tchr. Summit County Bd. Mental Health and Mental Retardation, Akron, Ohio, 1973-74, counseling psychologist, 1977-78; psychology intern U. Akron, 1974-77; asst. program dir. Harris County Bd. Mental Health and Mental Retardation, Houston, 1978-80; dir. adolescent residential services Tex. Dept. Mental Health and Mental Retardation, Fort Worth, 1980—; cons. Tex. Dept. Corrections, Huntsville, 1982, Tex. Council on Crime and Delinquency, Austin, 1982; designer community correctional group homes for violent mentally retarded adolescents 1982. Contbr. articles to profl. jours. Mem. Am. Psychol. Assn., Assn. Behavior Analysis, Am. Assn. Mental Deficiency, Southwestern Psychol. Assn., Tarrant County Psychol. Assn., Assn. Advancement of Psychology, Phi Sigma Kappa. Subspecialties: Behavioral psychology; Rehabilitation psychology. Current work: Design, direction and evaluation of community-based and institutional rehabilitation programs for adolescent and adult mentally retarded offenders. Office: Tex Dept Mental Health and Retardation 5000 Campus Dr Fort Worth TX 76109

DENNING, DOROTHY ELIZABETH, computer science educator, researcher; b. Grand Rapids, Mich., Aug. 12, 1945; d. Cornelius Lowell and Helen Dorothy (Watson) Robling; m. Peter James Denning, Jan. 24, 1974. B.A., U. Mich., 1967, M.A., 1969; Ph.D., Purdue U., 1975. Asst. research mathematician Radio Astronomy Obs., U. Mich., 1967-69; systems programmer Computer Center, U. Rochester, 1969-72, instr. in elec. engring., 1971-72; asst. prof. computer scis. Purdue U., 1975-81, assoc. prof., 1981-83. Author: Cryptography and Data Security, 1982. IBM fellow, 1974-75; NSF grantee, 1981-83. Mem. Assn. Computing Machinery, IEEE Computer Soc., Sigma Xi, Phi Kappa Phi. Subspecialty: Cryptography and data security. Current work: Study methods of protecting sensitive data stored in computer systems. Office: SRI Internat 333 Ravenswood Ave Menlo Park CA 94025

DENNING, PETER JAMES, computer scientist; b. N.Y.C., Jan. 6, 1942; s. James Edwin and Catherine M. (Manton) D.; m. Dorothy Elizabeth Robling, Jan. 24, 1974; children—Anne, Diana. B.E.E., Manhattan Coll., 1964; M.S. in Elec. Engring. (NSF fellow 1964-67), MIT, 1965, Ph.D., 1968. Asst. prof. elec. engrng. Princeton U., 1968-72; assoc. prof. computer scis. Purdue U., 1972-75, prof., 1975-84, head dept., 1979-83; dir. Research Inst. Advanced Computer Sci. NASA Ames Research Center, Mountain View, Calif., 1983; lectr. Author: Profl. Devel. Seminars, 1968—; textbooks, also numerous research papers. Recipient Outstanding Faculty award Princeton U. Engring. Assn., 1971, Best Paper award Am. Fedn. Info. Processing Socs., 1972. Fellow IEEE; mem. Assn. Computing Machinery (pres. 1980-82, editor-in-chief Computing Surveys 1977-79, Best Paper award 1968, editor communications of ACM 1983—, Recognition of Service award 1974), N.Y. Acad. Scis., Sigma Xi, Eta Kappa Nu, Tau Beta Pi. Subspecialty: Theoretical computer science. Home: 30 Bear Gulch Dr Portola Valley CA 94025 Office: Research Inst Advanced Computer Sci NASA Ames Research Center Moffett Field CA 94035

DENNIS, ANTHONY JOSEPH, laboratory manager, research scientist; b. Springfield, Ohio, May 31, 1948; s. Sebastian Angelo and Kate (Schneider) D.; m. India R. Shinn, Sept. 28, 1974. B.Sc., Ohio State U., 1970, Ph.D., 1973. Research assoc. Ohio State U., 1971-73; research scientist (Battelle's Columbus (Ohio) Labs.), 1973-74, sr. research scientist, 1974-76, assoc. sect. mgr., 1976-79, sect. mgr., 1979—; reviewer NSF, NRC, Nat. Cancer Inst. Am. Heart Assn. young investigator grantee, 1971-73; Am. Cancer Soc. young investigator grantee, 1971-73. Mem. Am. Soc. Microbiology, Reticuloendothelial Soc., Planetary Soc., L-5 Soc., Am. Soc. Natural History, Smithsonian Assocs., Sigma Xi. Subspecialties: Genetics and genetic engineering (biology); Immunobiology and immunology. Co-inventor monoclonal antibody diagnostic applications. Office: 505 King Ave Columbus OH 43201

DENNIS, JACK BONNELL, computer science educator; b. Elizabeth, N.J., Oct. 13, 1931. S.B., S.M., MIT, 1954, Sc.D. in Elec. Engring, 1958. Asst. prof. elec. engrng. MIT, Cambridge, Mass., 1954-58, instr. to assoc. prof., 1958-69, prof., 1969—. Mem. IEEE., Assn. Computing Machinery. Subspecialty: Theoretical computer science. Office: Dept Elec Engring MIT Cambridge MA 02139

DENNIS, MELVIN BEST, JR., veterinarian, researcher; b. Bellingham, Wash., Dec. 4, 1937; s. Melvin Best and Agnes Genevieve (Bruskl) D.; m. Sheila Ann Afflebach, Sept. 25, 1965; children: Andrew Martin, John Raymond, Michael Harrison. B.S., D.V.M., Wash. State U. Lic. veterinarian, Wash., Calif. Small animal vet. practice, Seattle, 1965-71; research assoc. nephrology U. Wash., Seattle, 1971-77, research asst. prof., dept. medicine and dept. lab. medicine, 1977-80, asst. prof. div. animal medicine, adj. asst. prof. dept. medicine, 1980—; cons. Seattle VA Hosp., 1981—; charter trustee Found. for Vet. Med. Research for Wash., pres., 1973. Served to capt. Vet. Corps U.S. Army, 1962-64. Recipient outstanding surgeon award Wash. State U. Coll. Vet. Medicine, 1961. Mem. AVMA, Am. Soc. for Artificial Internal Organs, Am. Assn. Lab. Animal Sci., Am. Acad. Surg. Research (founding). Roman Catholic. Subspecialties: Surgery (veterinary medicine); Biomedical engineering. Current work: Blood access, animal modeling, hemodialysis, laboratory animal medicine; researcher animal models of human disease and artificial organs. Inventor fistula catheter for hemodialysis; developer animal models for uremia, gastric ulcers and liver necrosis; developer extracorporeal enzyme reactor for cancer therapy in dogs. Home: 21819 13th Ave S Seattle WA 98188 Office: Div Animal Medicine SB-42 U Wash Seattle WA 98195

DENNISH, GEORGE WILLIAM, cardiologist, consultant, researcher; b. Trenton, Feb. 14, 1945; s. George William and Mary Ann (Bodnar) D.; m. Kathleen Macchi, June 28, 1969; children: Andrew Stewart, Brian George, Michael John. B.A. magna cum laude, Seton Hall U., 1967; M.D., Jefferson Med. Coll., Thomas Jefferson U., 1971. Diplomate: Am. Bd. Internal Medicine. Intern in medicine Naval Hosp., Phila., 1971-72, sr. intern, 1971, resident, 1973, sr. asst. resident, 1973-74; fellow in cardiology Naval Regional Med. Ctr., San Diego, 1976-77, staff cardiologist, 1976-77, dir. coronary care unit, 1977-78; sr. staff cardiologist Scripps Meml. Hosps., La Jolla and Encinitas, Calif., 1978—, chief of medicine, Encinitas, 1983-85, dir. spl. care units, La Jolla, 1980-82; asst. clin. prof. medicine U. Calif.-San Diego, 1976—; lectr. various topics, with emphasis on calcium channel blockers in cardiovascular disease. Contbr. articles to profl. jours. Served to lt. comdr. USN, 1971-78. Recipient Physician's Recognition award AMA, 1982; decorated knight Order of Holy Sepulcher. Fellow Am. Coll. Cardiology, ACP; mem. Am. Coll. Chest Physicians, Clin. Council of Am. Heart Assn., Am. Coll. Angiology, Am. Coll. Pharmacology, N.Am. Soc. Pacing and Cardiac Electrophysiology, Am. Fedn. Clin. Research, Am. Soc. Internal Medicine, Hobart Amory Hare Soc., Alpha Epsilon Delta, Delta Epsilon Sigma. Republican. Roman Catholic. Club: Cathedral Canyon Country (Palm Springs, Calif.). Subspecialties: Cardiology; Critical care. Current work: Cardiovascular pharmacology. Office: Specialty Med Clinic La Jolla and San Diego 9844 Genesee Ave Suite 400 La Jolla CA 92037 1087 Devonshire Dr Suite 100 Encinitas CA 92024

DENNISON, BRIAN KENNETH, astronomer, educator; b. Louisville, Aug. 14, 1949; s. Kenneth George and Earlene Muriel (Burnett) D.; m. Mira Pahlic, July 17, 1976. B.S. in Physics, U. Louisville, 1970; M.S., Cornell U., 1974, Ph.D. in Astronomy, 1976. Asst. planetarium lectr. U. Louisville, 1969-71; research asst. Cornell U., 1971-76; research assoc. Va. Poly. Inst. and State U., Blacksburg, 1976-77, asst. prof. physics, 1977-82, assoc. prof., 1982—; researcher extragalactic radio astronomy, astronomy teaching. Contbr. articles to profl. jours. Research grantee NASA, 1978-79, Research Corp., 1978. Mem. Am. Astron. Soc., Internat. Astron. Union. Subspecialties: Radio and microwave astronomy; High energy astrophysics. Current work: The physics of compact extragalactic radio sources, including quasars, studies of the intracluster medium of clusters of galaxies. Office: Physics Dept Virginia Polytechnic Institute Blacksburg VA 24061

DENNY, FLOYD WOLFE, JR., pediatrician; b. Hartsville, S.C., Oct. 22, 1923; s. Floyd Wolfe and Marion Elizabeth (Porter) D.; m. Barbara H. Denny, Apr. 27, 1946; children: Rebecca E., Mark W., Timothy P. B.S., Wofford Coll., 1944; M.D., Vanderbilt U., 1946. Diplomate: Am. Bd. Pediatrics. Intern Vanderbilt Hosp., Nashville, 1946-47, resident in pediatrics, 1947-48; instr. pediatrics U. Minn., 1951-52, asst. prof., 1952-53; asst. prof. pediatrics Vanderbilt U. Sch. Medicine, 1953-55; asst. prof. preventive medicine and pediatrics Western Res. U. Sch. Medicine, 1955-60, asso. prof. preventive medicine, 1960; prof. Sch. Medicine, U. N.C., Chapel Hill, 1960—, chmn. dept. pediatrics, 1960-81; vis. scholar dept. epidemiology Sch. Pub. Health and Child Devel. Inst., 1977-78; vis. worker Med. Research Council Clin. Research Centre, London, 1970-71; mem. Commn. on Streptococcal and Staphylococcal Diseases Armed Forces Epidemiol. Bd., 1954-72, dep. dir., 1959-63; mem. Commn. on Acute Respiratory Diseases Armed Forces Epidemiol. Bd., 1960-73, dep. dir., 1963-67, dir., 1967-73; mem. Inst. Medicine, Nat. Acad. Scis., 1981-86. Mem. editorial bd.: Am. Rev. Respiratory Diseases, 1971-74; mem. publs. com.: Jour. Infectious Diseases, 1973-78; Contbr. articles to med. jours. Served to maj. M.C. U.S. Army, 1948-51. Mem. Am. Acad. Pediatrics, Am. Assn. Immunologists, Am. Epidemiology Soc., Am. Fedn. Clin. Research, Am. Pediatric Soc. (pres. 1980-81), Am. Soc. Clin. Investigation, Am. Soc. Microbiology, Am. Thoracic Soc., Assn. Am. Physicians, Infectious Diseases Soc. Am. (pres. 1979-80), Soc. Exptl. Biology and Medicine, Soc. Pediatric Research (pres. 1968-69), So. Soc. Clin. Research, So. Soc. Pediatric Research, Phi Beta Kappa, Alpha Omega Alpha. Subspecialties: Pediatrics; Epidemiology. Current work: Epidemiology of acute respiratory tract infections. Home: Route 10 Box 56 Chapel Hill NC 27514 Office: Dept Pediatrics 535 Burnett-Womack Bldg 229-H U NC Sch of Medicine Chapel Hill NC 27514

DENNY, WILLIAM FRANCIS, II, mathmetical sciences educator, consultant; b. Shreveport, Feb. 14, 1946; s. William Francis and Ouida Marie (Harris) D. B.S., La. Tech. U., 1968; M.S., U. Okla.-Norman, 1970, Ph.D., 1974. Instr. U. Okla., 1973-75; asst. prof. dept. math. sci. McNeese State U., Lake Charles, La., 1975-81, assoc. prof., 1981—, dir. academic computing ctr., 1983—, acting head dept. math. sci., 1983; cons. in field. Mem. Soc. Indsl. and Applied Math. Assn. Computing Machinery, Am. Math. Soc., Math. Assn. Am., Sigma Xi. Democrat. Methodist. Subspecialties: Mathematical software; Applied mathematics. Current work: Using computers to study real life situation thru the use of mathematical models. Home: 2229 Bancroft St Apt 3 Lake Charles LA 70605 Office: McNeese State U Dept Math Sci Lake Charles LA 70609

DENOBLE, VICTOR JOHN, psychologist, researcher; b. Woodside, N.Y., Dec. 3, 1949; s. Victor John and Camela Ann (Paleramo) DeN.; m. Caroline, Aug. 19, 1978; 1 dau., Jennifer Caroline. B.A., Adelphi U., 1971, M.A., 1974, Ph.D., 1976. Sr. research scientist SUNY Down State Med. Ctr., N.Y.C., 1974-77, research assoc. dept. biopsychology, 1976-77; research assoc. U. Minn., 1978-80, postdoctoral research fellow, 1979-80; research scientist Philip Morris USA Research and Devel, Richmond, Va., 1980—; adj. assoc. prof. SUNY; adj. asst. prof. CUNY. Contbr. articles to profl. jours. Nat. Inst. Drug Abuse grantee, 1979-82; Walker Found. Minn. grantee, 1980. Mem. Am. Psychol. Assn., Soc. Neurosci., AAAS. Subspecialties: Behavioral psychology; Psychobiology. Current work: Behavioral pharmacology; physiological psychology; psychopharmacology; operant conditioning. Office: Philip Morris USA Research and Devel PO Box 26583 Richmond VA 23261

DENSEN, PAUL MAXIMILLIAN, health adminstr.; b. N.Y.C., Aug. 1, 1913; s. Charles Edwin and Carrie (Weinberg) D.; m. Elizabeth A. Reed, Dec. 19, 1939; children—Rebecca E. (Mrs. John Rothfuss), Peter. A.B., Bklyn.Coll., 1934; D.Sc., Johns Hopkins, 1939; M.A. (hon.), Harvard, 1968. From instr. to asso. prof. preventive medicine Vanderbilt U. Med. Sch., 1939-46; chief div. med. research statistics VA, Washington, 1946-49; asso. prof., then prof. biometry Grad. Sch. Pub. Health, U. Pitts., 1949-54; dir. div. research and statistics Health Ins. Plan Greater N.Y., 1954-59; dept. commr. N.Y.C. Dept. Health, 1959-66; dept. adminstr. N.Y.C. Health Services Adminstrn., 1966-68; dir. Harvard Center Community Health and Med. Care, 1968—; prof. community health Harvard Sch. Pub. Health, 1968—. Fellow Am. Statis. Assn., Am. Pub. Health Assn., AAAS; mem. Am. Epidemiol. Soc., Inst. Medicine. Subspecialties: Epidemiology; Health services research. Current work: Evolution of long-term care activities; data needed for long-term care policy formation. Home: PO Box 304 Sandown NH 03873 Office: 643 Huntington Ave Boston MA 02115

DENT, JAMES NORMAN, biology educator; b. Marlin, Tenn., May 10, 1916; s. James Rolandus and Alta Anne (Norman) D.; m. Valgerda Nielsen, Dec. 27, 1945 (div. 1972); children: Julie Anne Dent Carlyle, Martha Elizabeth. A.B., U. Tenn., 1938; Ph.D., Johns Hopkins U., 1941. Asst. prof. biology Marquette U., Milw., 1945-46; asst. prof. U. Pitts., 1946-49; assoc. prof. biology U. Va., Charlottesville, 1949-57, prof., 1957—; cons. Oak Ridge Nat. Lab., 1955-70; USPHS spl. research fellow Harvard U., 1969-70; Fulbright lectr. Banaras Hindu U., 1976, U. Calcutta, 1976; vis. fellow U. Calif.-Berkeley, 1977. Guggenheim Found. fellow St. Andrews U., Scotland, 1959-60. Mem. AAAS, Am. Soc. Zoologists, Am. Assn. Anatomists. Subspecialty: Systematics. Current work: Developmental physiology; comparative endocrinology. Home: 1940 Thomson Rd Charlottesville VA 22903 Office: Dept Biology U Va Gilmer Hall Charlottesville VA 22901

DENTINGER, MARK PETER, neurologist; b. Rochester, N.Y., June 22, 1945; s. John Cyril and Mary Louise (Peters) D.; m. Nancy Louise Tufano, Aug. 28, 1965; children: Adam, Aaron, Kelleen. B.A. magna cum laude in Chemistry, St. John Fisher Coll., 1967; M.D. cum laude, Albany Med. Coll., 1971. Diplomate: Am. Bd. Neurology and Psychiatry. Intern, Resident in neurology Albany (N.Y.) Med. Coll., 1971-74, assoc. prof. neurology, 1974—; chmn. edn. com.; staff neurologist VA Med. Ctr., Albany, 1974—. Contbr. articles in field to profl. publs. Mem. Am. Acad. Neurology, Delta Epsilon Sigma, Alpha Omega Alpha. Democrat. Roman Catholic. Subspecialties: Neurology; Regeneration. Current work: Experimental neuropathology (electronmicroscopy) of central nervous system: myelin, trauma, axon reaction. Home: 10 Hartwood St Colonie NY 12205 Office: VA Med Ctr Holland Ave Albany NY 12208

DENYSYK, BOHDAN, government executive, physicist; b. Kornberg, W. Germany, Feb. 13, 1947; came to U.S., 1949; s. John and Maria (Zelenewich) D.; m. Halina B. Bubela, June 28, 1969; children: Maria H., Danna L., Adrienne Y., Alexis M. B.S., Manhattan Coll., 1968; M.S., Cath. U. Am., Washington, 1973; Ph.D., Union Exptl. Colls. and Univs., 1981. Physicist Naval Weapons Lab., Dahlgren, Va., 1968-72; biophysicist Naval Medicine Research Inst., Bethesda, Md., 1972-75; physicist, group mgr. Naval Surface Weapons Ctr., White Oak, Md., 1975-78; physicist, dept. head. E G & G, Inc., Rockville, Md., 1978-81; dep. asst. sec. U.S. Dept. Commerce, Washington, 1981—; pres. DLR Assocs., Arlington, Va., 1972-81; cons. Republican Nat. Com., 1980-81, NSF, 1982. Dir. pub. relations Ukrainian Nat. Info. Service, 1978-81. Navy fellow, 1969-71; N.Y. Regents scholar, 1964-68. Mem. AIAA, Am. Def. Preparedness Assn., Am. Phys. Soc. Subspecialty: Fluid mechanics. Current work: Flow field for vehicles reentering into atmosphere from space flight, with special emphasis on computational techniques. Home: 1301 19th Rd S Arlington VA 22202 Office: US Dept Commerce 14th & Constitution Ave NW Washington DC 20230

DEPALMA, ROBERT ANTHONY, endodontist; b. Orange, N.J., July 2, 1941; s. Germano Frederick and A. Lily (Rende) DeP.; m. Mary Lynn McNair, Feb. 7, 1981; 1 son, Robert Anthony II; m.; 1 stepdau., Melissa Ann Mauriell. A.B. in Sci, Villanova U., 1963; D.D.S., W.Va. U., 1968; cert. in endodontics, N.J. Dental Sch., 1972. Diplomate: Am. Bd. Endodontics. Practice gen. denistry, Livingston, N.J., 1968-72, practice specializing in endodontics, Boca Raton, Fla., 1972—; cons. endodontics St. Mary's S. Barnabas, Orange Meml., Newark Beth Israel hosps., N.J., 1968-72, Boca Raton Community Hosp., 1972—; lectr. seminars. Contbr. articles sci. jours. Pres. Boca Raton Fraternal Order Police, 1976-78. Fellow Am. Coll. Dentists, Am. Acad. Oral Medicine; mem. ADA, N.J. Acad. Medicine, Fla. Dental Soc., Am. Hosp. Assn., Am. Assn. Hosp. Dentists, Am. Assn. Endodontics, South Palm Beach County Dental Soc., Cath. Hosp. Assn., Internat. Hosp. Assn., Am. Assn. Dental Radiology, Am. Assn. Clin. Oral Pathology, Internat. Assn. Dental Research, Brit. Endodontic Soc. Lodge: Kiwanis. Subspecialty: Endodontics. Current work: Oral pathology, immunology and microbiology, bone pathology. Office: 2351 N Federal Hwy Boca Raton FL 33432

DEPAULO, JOSEPH RAYMOND, JR., psychiatrist; b. Charleston, W.Va., May 21, 1946; s. Joseph Raymond and Mary Catherine (Wilson) DeP.; m. Elizabeth Ratterman, June 8, 1970; children: Marianne, margaret. B.S. magna cum laude, Xavier U., Cin., 1968; M.D., John Hopkins U., 1972. Diplomate: Am. Bd. Psychiatry and Neurology. Intern, fellow dept. medicine Johns Hopkins Hosp., 1972-73, fellow dept. psychiatry and behavioral scis., 1973-77; resident in psychiatry Balt. City Hosps., 1973-74, Henry Phipps Psychiat Clinic and Johns Hopkins Hosp., 1974-76; Maudsley exchange resident, London, 1975; chief resident in psychiatry Henry Phipps Psychiat. Clinic and Johns Hopkins Hosp., 1976-77; asst. prof. psychiatry Johns Hopkins U., 1977—; staff psychiatrist Johns Hopkins and Balt. City hosps., 1977—; dir. affective disorders sect. Continuous Treatment Clinic, Phipps Clinic outpatient dept. Johns Hopkins Hosp., 1977—; acting chief dept. psychiatry Balt. City Hosps., 1981—. Contbr. abstracts and articles to profl. jours., chpts. in books. Mem. Am. Psychiat. Assn., Md. Psychiat. Soc., Soc. Neurosci., Johns Hopkins Med. and Surg. Soc., Alpha Sigma Mu. Roman Catholic. Subspecialties: Psychopharmacology; Psychiatry. Current work: Clinical psychopharmacology of affective disorders, clinical effects of lithium carbonate. Home: 504 Overbrook Rd Baltimore MD 21212 Office: Dept of Psychiatry Baltimore City Hosps Baltimore MD 21224

DE PENA, ROSA G., meterology educator, cons.; b. Bairamcea, Romania, Sept. 1, 1921; came to U.S., 1967, naturalized, 1976; d. Marcos and Esther (Persky) Gotzulsky; m. Jorge A. Pena, Nov. 30, 1946; children: Adriana I., Edith A. Ph.D. in Chemistry, U. Buenos Aires, Argentina, 1944. Instr. U. Buenos Aires, 1955-66, assoc. prof., 1966-67; research assoc. Pa. State U., University Park, 1964-73, assoc. prof. meterology, 1973-78, prof., 1978—. Contbr. articles to profl. jours. Mem. AAAS, Am. Geophys. Union, Am. Meterol. Soc. Subspecialty: Atmospheric chemistry. Current work: Dry and wet deposition; scavenging processes, chemistry of precipitation. Office: Pa State U 516 Walker Bldg University Park PA 16802 Home: 827 Wheatfield Dr State College PA 16801

DE PREE, ROBERT WILSON, computer engineer, system architect; b. Milw., Oct. 17, 1955; s. Hugh Wilson and Frances Irene (Kiefer) De P.; m. Patricia Ann Roan, Aug. 30, 1975. Student, Ga. Inst. Tech., 1973-75; B.S.E.E., U. Fla., 1977, postgrad., 1977-80; postgrad., Fla. Inst. Tech., 1980—. Asst. in engring. Ctr. Info. Research, Gainesville, Fla., 1977-80, assoc. in engring., asst. dir., 1980—; lead engr. Harris Corp., Melbourne, Fla., 1980-83; pres. Decisionware, Melborne, 1983—. Reviewer: Internat. Jour. Computer and Info. Sci.; system co-architect info. retrieval system, Telebrowsing, 1975-77, med. diagnostic system, Mediks, 1977-79, decision support system, Automatic Typewriter Identification, 1979-80, distributed computer system, Harris 9000 Series, 1980-82. Mem. Assn. Computing Machinery, IEEE (task groups on data bases and operating systems), IEEE Computer Soc., Tau Beta Pi. Subspecialties: Database systems; Graphics, image processing, and pattern recognition. Current work: Intelligent database systems, decision support systems, pattern recognition. Office: Harris Word Processing Div PO Box 2400 Melbourne FL 32935

D'ERCOLE, AUGUSTINE JOSEPH, pediatric endocrinologist; b. Salt Lake City, Mar. 20, 1944; s. Augustine Dominic and Susan Margaret (Gonnella) D'E.; m. Virginia Louise Weyant, Apr. 8, 1972; children: Ethan Marc, Jed Daniel. A.B., U. Notre Dame, 1965; M.D., Georgetown U., 1969. Diplomate: Nat. Bd. Med. Examiners, Am. Bd. Pediatrics. Mem. pediatric house staff Tufts-New Eng. Med. Center Hosps., Boston., 1969-72; fellow in pediatric endocrinology U. N.C., Chapel Hill, 1974-77, asst. prof. pediatrics, 1977-81, assoc. prof., 1981—; vis. prof. several med. schs. Author 1 book, contbr. chpts. to books, articles to profl. jours. Mem. Task Force for Newborn Screening for Congenital Hypothyroidism, State of N.C. Served with USPHS, 1972-74. Jefferson-Pilot fellow in acad. medicine, 1979-83; NIH grantee, 1981—; March of Dimes grantee, 1978—. Mem. Lawson Wilkins Pediatric Endocrine Soc., Soc. Pediatric Research, Am. Acad. Pediatrics, Endocrine Soc., Soc. for Devel. Biology. Subspecialties: Endocrinology; Pediatrics. Current work: Regulation of fetal growth; peptide growth factors. Office: Dept Pediatrics U NC Sch Medicine 509 Clin Scis Bldg 229H Chapel Hill NC 27514

DERDERIAN, GEORGE, physicist, research consultant; b. Rochester, N.Y., Nov. 19, 1922; s. Sirkes G. and Sogoma (Bogoshian) D.; m. Alice Joan Kenney, May 30, 1953; children: Gregory, Jeanne, Susan, Elizabeth. B.S. in Physics, Queens Coll., 1947, M.S., N.Y. U., 1951. Instr. Pratt Inst., Bklyn, 1947-50; physicist Evans Signal Lab., Belmar, N.J., 1950-55, Republic Aviation, Farmingdale, N.Y., 1955-59; adj. prof. Hofstra U., 1956-67; research engr. Sperry, Great Neck, N.Y., 1960-64; head phys. sci. lab. Naval Tng. Equipment Ctr., Orlando, Fla., 1964-80; cons. electro-optics and visual simulation tech., Maitland, Fla., 1980—; bd. dirs. Laser Inst. Am., 1978-80; mem. Army Laser Adv. Group. Contbr. numerous articles to profl. jours. Served with U.S. Army, 1943-46. Decorated Bronze Star medal; recipient Chemistry Tchr.'s award Am. Chem. Soc., 1941, Rockefeller Service award State of N.Y., 1959, Ten. Yrs. Outstanding Tchr. plaque Hofstra U., 1967, Outstanding Service award Laser Inst. Am., 1978. Fellow AAAS, Brit. Interplanetary Soc.; mem. Am. Optical Soc., Navy Laser Group, Sigma Xi (exec. com. chpt. 1969-78, v.p. chpt. 1968-69, pres. chpt. 1969-72, 77-78), Sigma Pi Sigma. Subspecialties: Optical image processing; Laser data storage and reproduction. Current work: Electro-optical systems; applications of optics and lasaers to training systems. Patentees in electro-optics tech. Home and Office: 921 Gillis Ct Maitland FL 32751

DERENIAK, EUSTACE LEONARD, physicist, educator, consultant; b. Standish, Mich., Dec. 29; s. Peter N. A. and Julia Pauline (Dziuban) D.; m. Barbara Catherine, Aug. 31, 1968; children: Teresa D., Andreana M. Ph.D. in Optical Scis, U. Ariz., 1976. Research assoc. Rockwell Internat., Anahiem, Calif., 1965-72, materials scientist, 1972-74; sr. engr. Ball Aerospace, Boulder, Colo., 1974-78; asst. prof. optics U. Ariz., 1978—; cons. astronomy. Mem. Optical Soc. Am. (pres. Tucson sect.). Roman Catholic. Subspecialties: Infrared physics; 3emiconductors. Current work: Charge transfer devices, infrared physics, radiometry. Office: 528 N Martin Tucson AZ 85719

DE RICHEMOND, ALBERT LEO, engineering mechanics specialist; b. Bryn Mawr, Pa., Oct. 17, 1950; s. John Francis and Joan Marie (Lappin) de R.; m. Jean Ann Rollo, Dec. 30, 1972; children: Annelise Rollo, Jeannine Rollo. B.S., Pa. State U., 1972; M.S., Va. Poly. Inst., 1974; postgrad., Drexel U., 1976-79. Registered profl. engr. Pa. Structural analysis engr. Re-Entry and Environ. Systems div. Gen. Electric, Phila., 1974-75; lab. supr. Pa. Crusher Corp., Broomall, 1976-79; sr. devel. engr. Fuller Co., Bethlehem, Pa., 1980—. Mem. Am. Acad. Mechanics, ASME, Mensa. Subspecialties: Theoretical and applied mechanics; Solid mechanics. Current work: Finite element analysis; failure analysis. Home: 41 Blythewood Rd Doylestown PA 18901 Office: PO Box 2040 Bethlehem PA 18001

DERR, VERNON ELLSWORTH, physicist, educator; b. Balt., Nov. 22, 1921; s. William Edward and Edith May D.; m. Mary Louise Van Atta, Mar. 6, 1943; children: Michael, Kathy, Louise, Carol. A.B., St. Johns Coll., Annapolis, Md.; Ph.D., Johns Hopkins U., 1959. Instr. St. John's Coll., 1947-48; research asst. Johns Hopkins U., 1950, jr. instr., 1948-59, research assoc., 1951-59; adj. prof. Rollins Coll., Winter Park, Fla., 1959-67; prin. scientist Martin-Marietta, Orlando, Fla., 1959-67; sr. scientist ESSA Labs. U.S. Dept. Commerce, Boulder, Colo., 1967-73; supervisory physicist Envrion. Research Labs., NOAA, U.S. Dept. Commerce, Boulder, 1974-81, dep. dir., 1981-83; dir. Environ. Research Labs., NOAA, U.S. Dept. Commerce, 1983—; adj. prof. U. Colo. Contbr. articles to profl. jours. Served with U.S. Army, 1942-46. Mem. Optical Soc. Am., Am. Geophys. Union, IEEE, Internat. Union Radio Sci., Am. Meteorology Soc. Subspecialties: Remote sensing (atmospheric science); Climatology. Office: 325 Broadway St Boulder CO 80301

D'ERRICO, ALBERT PASQUALE, JR., psychologist; b. Dallas, Feb. 27, 1941; s. Albert and Carol (Whitney) D'E. A.B., Southwestern at Memphis, 1965; M.A., Whittier Coll., 1969; Ph.D., U. Ga., 1976. Lic. psychologist, Tenn. Psychologist Northeastern La. U., Monroe, 1974-76; diagnostic adminstr. Shelby County and State of Tenn., Memphis, 1976—; cons. in field. Author: On the State of Inerrancy, 1982; contbr. articles to profl jours. Law Enforcement Assistance Adminstrn. grantee, 1976; State of Tenn. and Shelby County grantee, 1977-80. Mem. Am. Psychol. Assn. Subspecialties: Cognition; Behavioral psychology. Current work: Cognition; formal operations intelligence and achievement behavioral psychology; psychological assessment and prediction of behavior by computer. Address: PO Box 1 Elberton GA 30635

DERTOUZOS, MICHAEL LEONIDAS, computer scientist, electrical engineer; b. Athens, Greece, Nov. 5, 1936; came to U.S., 1954, naturalized, 1965; s. Leonidas Michael and Rosana G. (Maris) D.; m. Hadwig Gofferje, Nov. 21, 1961; children—Alexandra, Leonidas. B.S. in E.E, U. Ark., 1957, M.S., 1959; Ph.D., MIT, 1964. Head research and devel. Baldwin Electronics, Inc., 1958-60; research asst. MIT, Cambridge, 1960-64, asst. prof., 1964-68, assoc. prof., 1968-73, prof., 1973—, dir., 1974—; founder, chmn. bd. Computek, Inc., 1968-74; cons. in computers to industry. Author: Threshold Logic: a Synthesis Approach, 1966, (with Athans, Spann and Mason) Systems, Networks and Computation: Multivariable Methods, 1974, Systems, Networks and Computation: Basic Concepts, 1972, (with Clark, Halle, Pool and Wiesner) The Telephone's First Century—and Beyond, 1977, The Computer Age; A Twenty Year View, 1979; Contbr. articles profl. jours. Trustee Athens Coll., Greece, 1973—; chmn. bd. Boston Camerata, 1976—; dir. Cambridge Soc. Early Music, 1974-75. Recipient Terman Internat. Edn. award Am. Soc. Engring. Edn., 1975; Ford postdoctoral fellow, 1964-66; Fulbright scholar, 1954. Fellow IEEE (Thompson best paper prize 1968); mem. Sigma Xi, Tau Beta Pi, Pi Mu Epsilon. Greek Orthodox. Subspecialties: Graphics, image processing, and pattern recognition; Personal computers. Current work: Personal computers, graphics. Patentee in field. Home: 15 Bernard Ln Waban MA 02168 Office: 545 Technology Sq Cambridge MA 02139

DESAIAH, DURISALA, neuropharmacologist, educator, researcher, cons.; b. Gowravaram, India, Nov. 22, 1944; came to U.S., 1970; d. Durisala and Veeramma D. (Kasineni) Hanumaiah; m. Nirmala D. Desaiah, June 14, 1964; 1 son: Rao V.H. B.Sc., Osmania U., India, 1962, M.Sc., 1964, Ph.D., 1969. Research fellow U. Minn., 1970-73, Miss. State U., 1973-75; research assoc. U. Miss. Med. Center, Jackson, 1975-77, instr. dept. pharmacologist, 1977-78, asst. prof. dept. neurology, 1978-81, assoc, prof., 1981—. Contbr. numerous articles and abstracts to profl. publs. Nizam Trust travel grantee to U.S., 1970; NIH research grantee, 1980—; FASEB Vis. Scientist, 1980—. Mem. Am. Soc. Pharmacoloy and Exptl. Therapeutics, Soc. Toxicology, Soc. for Neurisci., N.Y. Acad. Scis., Sigma Xi. Subspecialties: Neuropharmacology; Toxicology (medicine). Current work: Membrane pharmacology, neuropharmacology, neurotoxicology, neuromuscular and neurological disorders. Office: Dept Neurology U Miss Med Center Jackson MS 39216

DE SALVA, SALVATORE JOSEPH, pharmacologist, toxicologist; b. N.Y.C., Jan. 14, 1924; s. Nicola Carol and Frances Agnes (Caldarella) De S.; m. Elaine Mae Radloff, June 14, 1948; children: Salaine Claire De Salva Bonanne, Christopher Joseph, Stephanie De Salva Farrelly, Steven William, Gregory Vincent, Peter Nicholas, Philip Anthony, Deirdre De Salva Berry. B.S., Marquette U., 1947, M.S., 1949; postgrad., U. Ill., Chgo., 1951-53; Ph.D., Stritch Sch. Medicine, Loyala U., Chgo., 1958. Research and teaching asst. Marquette U., Milw., 1947-49; research biochemist Milwaukee County Gen. Hosp., 1954; instr. U. Ill., 1951-52; asst. prof. Chgo. Coll. Optometry, 1951-53; pharmacologist Armour Pharm. Lab., 1953-59; sect. head Colgate Palmolive Co., Piscataway, N.J., 1959-66, sr. research assoc., 1966-72, mgr., 1972-76, assoc. dir. research for pharmacology and toxicology, 1976-83, dir. research for pharmacology and toxicology, 1983—; lectr. Loyola U., 1957-59. Editor: Symposium for Bio Medical Electronic Instrumentation, 1965; contbr. articles to profl. jours. Mem. Park Forest (Ill.) Mosquito Abatement Program, 1952-55, Franklin Twp. (N.J.) Sch. Bd., 1969-70, Somerset (N.J.) Bd. Health, 1965-67, Cath. Youth Orgn., Somerset, 1960-67; v.p. Cedar Hill Swim Club, Somerset; active Boy Scouts Am., Somerset, 1965-67; trustee Franklin Twp. Day Care Center, 1969; mem. technician tng. com. N.J. Council for Research and Devel., Rutgers U., 1969-72. Served with USN, 1942-46. Mem. AAAS, Soc. Exptl. Biology and Medicine, Am. Soc. Pharmacology and Exptl. Therapeutics, Soc. Toxicology, Internat. Union Pharmacology (toxicology sect.), N.Y. Acad. Scis., Sigma Xi. Roman Catholic. Subspecialties: Toxicology (medicine); Pharmacology. Current work: Pharmaco-toxicology of fluorides, sequestering agents and surfactants. Patentee in field. Office: 909 River Rd Piscataway NJ 08854

DESANTIS, MARK EDWARD, neurobiology educator; b. Vineland, N.J., May 9, 1942; s. O.J. and Ellice (Baier) DeS.; m. Gail M. Chambers, July 5, 1969; 1 son, Michael Kevin. B.S. in Biology, Villanova U., 1963; M.S. in Anatomy, Creighton U., 1966, Ph.D., UCLA, 1970. Research assoc. Naval Air Sta., Pensacola, Fla., 1970-71; instr. anatomy Georgetown U., Washington, 1971-72, asst. prof., 1972-77, assoc. prof., 1977-78; assoc. prof. dept. biol. scis. U. Idaho, Moscow, 1978—; grants reviewer NSF, NIMH. Contbr. articles to profl. jours. Scoutmaster Boy Scouts Am., 1983. Mem. AAAS, Am. Assn. Anatomists, Soc. Neurosci., Am. Soc. Zoologists. Roman Catholic. Subspecialties: Neurobiology; Regeneration. Current work: Structure and function of cells in the nervous system particulary those undergoing degenerative and regenerative changes. Office: Dept Biol Scis U Idaho Moscow ID 83843

DE SANTO, DANIEL FRANK, research engineer; b. New Rochelle, N.Y., June 21, 1930; s. Daniel and Millie (Spilka) De S.; m. Gaynelle Hager, Oct. 16, 1965. B.Aero. Engring., NYU, 1952, M.Aero. Engring., 1953, Dr.Engring. Sci., 1961. Research asst. Coll. Engring., NYU, 1952-56; instr. Manhattan Coll., N.Y.C., 1956; flight test research engr. Grumman Aircraft Engring. Corp., Bethpage, N.Y., 1956-59; assoc. research scientist, adj. asst. prof. Coll. Engring., NYU, 1959-63; prin. aerodynamist Cornell Aero. Lab., Buffalo, 1963-70; sr. engr. Westinghouse Research and Devel. Ctr., Pitts., 1970—. Recipient Alexander Klemin award Daniel Guggenheim Sch. Aeros. NYU, 1952; Juan de la Cierva fellow, 1952-53; NSF fellow, 1954-55. Mem. ASME, AIAA, Sigma Xi, Tau Beta Pi. Subspecialties: Mechanical engineering; Aerospace engineering and technology. Current work: Unsteady flows, flow-induced vibration; design, analysis and testing of dynamic scale models of apparatus; development of scaling laws to predict full-scale dynamics behavior. Patentee in field. Home: 15 Morris St Export PA 15632 Office: Westinghouse Research and Devel Center Beulah Rd Pittsburgh PA 15235

DESANTO, JOHN ANTHONY, mathematical physics researcher, applied mathematics educator; b. Wilkes-Barre, Pa. B.S. magna cum laude in Physics, Villanova U., 1962, M.A. in Math, 1962; M.S. in Physics, U. Mich., 1963, Ph.D., 1967. Research physicist Naval Research Lab., Washington, 1967-81; sr. scientist, program mgr. in theoretical physics Electro Magnetic Applications, Inc., Denver, 1981-82; research prof. math. U. Denver, 1982-83, Colo. Sch. Mines, 1983—. Editor: Ocean Acoustics, 1979, (with others) Mathematical Methods and Applications of Scattering Theory, 1980. Recipient Publ. award Naval Research Lab., 1971; Woodrow Wilson fellow, 1962; NSF fellow, 1962-67. Fellow Acoustical soc. Am.; mem. Am. Phys. Soc., Soc. Indsl. and Applied Math., IEEE, INternat. Union Radio Sci. Subspecialties: Theoretical physics; Acoustics. Current work: scattering from rough surfaces; ocean acoustic propagation and inversion problems. Home: 7692 S Saulsbury Ct Littleton CO 80123 Office: Dept Math Colo Sch Mines Golden CO 80401

DESCHAMPS, NICHOLAS HOWARD, energy co. exec.; b. Richmond, Va., Nov. 12, 1937; s. Lawrence Francis and Dorothy Ellis (Lawrence) DesC.; m. Rebecca Moles, Jan. 23, 1944; children: Nikki Eleaine, Douglas Howard. B.S.M.E., Va. Poly. Inst., 1962, Ph.D., 1966. Registered profl. engr., Va., N.J. Mgr. engr. Sanders Assocs., Inc., Nashua, N.H., 1966-71; v.p. engring. Donbar Devel. Corp., N.Y.C., 1971-74; pres. DesChamps Labs., East Hanover, N.J., 1974—. Ford Found. fellow, 1963-65; NSF fellow, 1965. Mem. ASME, ASHRAE, Tau Beta Pi, Pi Tau Sigma. Republican. Presbyterian. Club: Lions (past pres.). Subspecialties: Fluid mechanics; Mechanical engineering. Current work: Research and development and manufacture of energy recovery and energy recycling systems. Patentee in field. Home: 31 Independence Dr Whippany NJ 07981 Office: Box 440 East Hanover NJ 07936

DESCHNER, ELEANOR ELIZABETH, biologist; b. Jersey City, Oct. 18, 1928; d. Fred and Anna (Sichler) D. B.A., Notre Dame Coll. of S.I., 1949; M.S., Fordham U., 1951, Ph.D., 1954. Head lab. digestive tract carcinogenesis Meml. Sloan-Kettering Cancer Center, N.Y.C., 1980—; assoc. prof. medicine/radiology Cornell U. Med. Coll., N.Y.C., 1976—; assoc. radiobiologist dept medicine Meml. Hosp., N.Y.C., 1978—. Contbr. articles to profl. jours. Nat. Cancer Inst. grantee; Am. Cancer Soc. grantee; NIH grantee. Mem. Am. Assn. Cancer Research, Am. Gastroent. Assn., Royal Soc. Medicine, Am. Soc. Cell Biology, Cell Kinetics Soc., Genetics Soc. Am., Am. Inst. Biol. Sci., AAAS, Sigma Xi, Kappa Gamma Pi. Republican. Roman Catholic. Club: Bus./Profl. Women's. Subspecialties: Cell study oncology; Gastroenterology. Current work: Cell proliferation in human/animal gastrointestinal cancer. Address: 1275 York Ave New York NY 10021

DESELM, HENRY (HAL) RAWIE, botany and ecology educator, researcher; b. Columbus, Ohio, Nov. 1, 1924; s. Ralph Emerson and Helen (Rawie) DeS.; m. Mary Elizabeth Hersee, June 11, 1948; children: Diane DeSelm Overcast, Richard Lowell. B.S. in Agr, Ohio State U., 1948, 1949, M.S. in Botany, 1950, Ph.D. in Botany (Plant Ecology), 1953. Profl. ecologist, Ecol. Soc. Am. Instr. botany dept. Ohio State U., 1953-54; instr. biology dept. Middle Tenn. State Coll., 1954-56; instr., research assoc. botany dept. U. Tenn., Knoxville, 1956-62, asst. prof. to assoc. prof. botany dept., 1962-73, prof. botany and ecology, 1973—; cons. engring. and environ. firms. Contbr. articles to profl. jours. Served as 1st lt USMCR, 1943-46. Recipient grants, research contracts, 1966—. Fellow AAAS, Tenn. Acad. Sci., mem. various sci. orgns. Democrat. Unitarian-Universalist. Subspecialties: Ecology; Species interaction. Current work: Vegetation ecology, plant communities of Southeastern U.S. Home: 424 Hillvate Turn W Knoxville TN 37919 Office: Botany Dept U Tenn Knoxville TN 37996-1100

DESHMUKH, VINOD DHUNDIRAJ, neurologist, neurophysiologist, surgeon, educator; b. Wani, Dist. Yeotmal, India, July 31, 1938; came to U.S., 1974, naturalized, 1982; s. Dhundiraj Govind and Leela Keshav (Shekdar) D.; m. Sunanda Vasant Deodhar, May 8, 1940; children: Abhijit, Asvin, Rahul. I.Sc., Sir Parashurambhau Coll. U. Poona, India, 1957, M.B.B.S., 1962; M. Surgery, U. Bombay, 1966; Ph.D., U. Glasgow, 1972. Intern Civil Hosp., Thana, Bombay, India, 1962; resident J. J. Hosp., Bombay, 1963-66, Nat. Hosp. Neurol. Diseases, London, 1967-68, Inst. Neuroscis., U. Glasgow, Scotland, 1969-72; cons. clin.neurophysiology Central Middlesex Hosp., London, 1973-74; asst. prof. neurology Baylor Coll. Medicine, Houston, 1974-76, dir., 1974-76; asst. prof. U. Tex., Houston, 1976-78; clin. asst. prof. U. Fla., 1978—, dir. edn., 1981—; practice medicine specializing in neurology, Jacksonville, Fla., 1978—; mem., tchr. Vedant Edn. Soc., Jacksonville, 1982. Contbr. articles to profl. jours. Local rep. Krishnamurtl Info. Ctr., Jacksonville, 1981-82. Recipient B.J. Med. Coll. Poona 1st prize in physiology, 1958; St. Georges Hosp. Bombay P.W. Shikhare prize, 1964-65. Mem. Soc. Neuroscis. Am. Acad. Neurology, Am. Med. EEG Assn., Am. Assn. Electromyography and Electrodiagnosis, AMA, Duval County Med. Soc., Jacksonville C. of C. Subspecialties: Neurophysiology; Neurology. Current work: Study of attentional energy and its modulations; two distinct modes of attention: mnemic and free modes, its theoretical implications and practical applications in human life; integrative models of observer, observed and the ultimate universal energy. Home: 3600 Rustic Ln Jacksonville Fl 32217 Office: 3599 University Blvd S #601 Jacksonville FL 32216

DESJARDINS, CLAUDE, physiologist, researcher; b. Fall River, Mass., June 13, 1938; s. A. L. and M. J. (Mercier) D.; m. Jane Elizabeth Campbell, Dec. 30, 1962; children: Douglas, Marc, Anne. B.S., U. R.I., 1960; M.S., Mich. State U., 1962, Ph.D., 1967; fellow, Jackson Lab., Bar Harbor, Maine, 1967-69. Asst. prof. physiology Okla. State U., Stillwater, 1969-70, assoc. prof. physiology, 1970-72, U. Tex.-Austin, 1972-74, prof. physiology, 1974—; cons. NIH, NSF, NASA, FDA, VA, 1974—. Editorial bd.: Biology of Reproduction, 1978-82, Endocrinology, 1982—; assoc. editor: Am. Jour. Physiology: Endocrinology and Metabolism, 1982—. Recipient Nat. Research Sci. award NIH, 1982. Mem. Soc. Study of Reproduction (pres. 1982-83), Am. Physiol. Soc., Endocrine Soc., Soc. Exptl. Medicine and Biology, Am. Assn. Anatomists., Soc. Neurosci. Subspecialties: Physiology (medicine); Neuroendocrinology. Current work: Endocrinology; control systems affecting secretion of reproductive hormones; hypothalamo-hypophyseal-gonadal interactions; testicular function and physiology of the male reproductive system. Home: 3513 Highland View Dr Austin TX 78731 Office: Inst Reproductive Biology Patterson Labs Univ Tex Austin Austin TX 78712

DESLATTES, RICHARD DAY, JR., physicist; b. New Orleans, Sept. 21, 1931; s. Richard and Lillian (Lee) D. B.S., Loyola U. of S., 1952; Ph.D., Johns Hopkins U., 1959. Dir. Div. Physics, NSF, Washington, 1980-81; chief quantum metrology group and sr. research fellow Nat. Bur. Standards, Washington, 1978—. Recipient Gold medal Dept. Commerce, 1979, Silver medal, 1967. Fellow Am. Phys. Soc., AAAS. Subspecialty: Atomic and molecular physics. Address: 610 Aster Blvd Rockville MD 20850

DE SMITH, DONALD ALBERT, computer software designer and researcher; b. Apr. 11, 1953. B.S., U. Mich., 1975, M.S., 1977; M.S. in Info. Sci, Eastern Mich. U., 1980. Programmer U. Mich., Ann Arbor, 1977-79, research assoc., 1979-82; v.p. Computerized Office Services, Inc., Ann Arbor, 1982—. Mem. IEEE, Assn. Computing Machinery. Subspecialties: Database systems; Operating systems. Current work: Research interest in distributed database systems—both software design and database design; also interests in laser printing applications. Office: Computerized Office Services Inc 313 N First St Ann Arbor MI 48103

DESOER, CHARLES AUGUSTE, electrical engineer; b. Ixelles, Belgium, Jan. 11, 1926; came to U.S., 1949, naturalized, 1958; s. Jean Charles and Yvonne Louise (Peltzer) D.; m. Jacqueline K. Johnson, July 21, 1966; children—Marc J., Michele M., Craig M. Ingenieur Radio-Electricien, U. Liege, Belgium, 1949, D.Sc. (hon.), 1976; Sc.D. in Elec. Engring, M.I.T., 1953. Research asst. M.I.T., 1951-53; mem. tech. staff Bell Telephone Labs., Murray Hill, N.J., 1953-58; assoc. prof. elec. engring. and computer scis. U. Calif., Berkeley, 1958-62, prof., 1962—, Miller research prof., 1970-71. Author: (with L. A. Zadeh) Linear System Theory, 1963, (with E. S. Kuh) Basic Circuit Theory, 1969, (with M. Vidyasagar) Feedback Systems: Input Output Properties, 1975, Notes for a Second Course on Linear Systems, 1970, (with F. M. Collier) Multivariable Feedback Systems; contbr. numerous articles on systems and circuits to profl. jours. Served with Belgian Arty., 1944-45. Decorated Vol.'s medal; recipient Best Paper prize 2Joint Automatic Control Conf., 1962, Univ. medal U. Liege, 1970, Disting. Teaching award U. Calif., Berkeley, 1971, Prix Montefiore Inst. Montefiore, 1975; award for outstanding paper Control Systems Soc., 1981, IEEE, 1979; Guggenheim fellow, 1970-71. Fellow IEEE (Edn. medal 1975), AAAS; mem. Nat. Acad. Engring., Am. Math. Soc., Math. Assn. Am., Soc. Indsl. and Applied Math. Current work: Systems: controls and circuits. Office: Dept Elec Engring and Computer Sci U Calif Berkeley CA 94720

DESPOMMIER, DICKSON DONALD, medical educator, parasitologist, researcher; b. New Orleans, June 5, 1940; s. Roland Medd and Beverly (Wood) D.; m. Judith Ann Forman, Aug. 5, 1963; children: Bruce, Bradley. B.S., Fairleigh Dickinson U., 1962; M.S., Columbia U., 1964; Ph.D., U. Notre Dame, 1967. Asst. prof. pub. health Columbia U., N.Y.C., 1971-75, assoc. prof., 1975-77, prof. pub. health and microbiology, 1982—; cons. NIH, 1980—, Gen. Foods Corp., 1976, Cordis Corp., 1973-74. Author: Parasitic Diseases, 1982, A Compendium of Parasite Life Cycles, 1983. Recipient Career Devel. award Nat. Inst. A.I.D., 1971-75; named Tchr. of Yr. Columbia U., 1980, 81, 83; Disting. Tchr. award Med. Coll. Ohio, 1980. Mem. AAAS, Am. Soc. Parasitologists, Am. Soc. Tropical Medicine and Hygiene, Harvey Soc., N.Y. Soc. Tropical Medicine (pres. 1980). Club: Trout Unltd. (Oradel, N.J.) (dir. 1976-78). Subspecialties: Parasitology; Infectious diseases. Current work: Isolation and characterization of protection-inducing antigens from Trichinella spiralis. Office: 630 W 168th St New York NY 10032

DESSER, KENNETH BARRY, cardiologist, researcher; b. N.Y.C., Mar. 24, 1940; s. George and Sarah Ruth (Kaplan) D.; m. Carmen Yvonne Fletcher, Sept. 3, 1981; children: Brett Karen, Lori Helen. B.A., NYU, 1961; M.D., N.Y. Med. Coll., 1965. Diplomate: Am. Bd. Internal Medicine. Intern Beth Israel Med. Ctr., N.Y.C., 1965-66, resident, 1968-70; cardiology fellow Inst. for Cardiovascular Diseases, Phoenix, 1970-72; now asst. dir; dir. cardiovascular edn. Good Samaritan Med. Ctr., Phoenix, 1973—; cons. in field. Contbr. numerous articles to med. jours.; mem. editorial bds.: Am. Jour. Cardiology, 1980-82, Jour. Am. Coll. Cardiology, 1983—. Served to capt. M.C. U.S. Army, 1966-68; Vietnam. Recipient Best Research Project award Beth Israel div. Mt. Sinai Sch. Medicine, 1966. Fellow ACP, Am. Coll. Cardiology, Am. Coll. Chest Physicians, Internat. Coll. Angiology; mem. Am. Fedn. for Clin Research, N.Y. Acad. Scis. Subspecialties: Cardiology; Internal medicine. Current work: Genetics of mitral valve prolapse, Dopplar ultrasonic shifts in cardiovascular disease. Applied Doppler ultrasonic flowmeter for human study. Home: 77 E Missouri St Phoenix AZ 85012 Office: Inst for Cardiovasvular Disease Good Samaritan Med Center 1003 E McDowell Rd Phoenix AZ 85006

DESSY, RAYMOND EDWIN, chemistry educator; b. Reynoldsville, Pa., Sept. 3, 1931; b. Raymond John and Martha Ellen (Orr) D.; m. Annabelle Lee, Sept. 8, 1959. B.A., U. Pitts., 1953, Ph.D., 1956. Postdoctoral fellow, instr. Ohio State U., Columbus, 1956-57; asst. prof. U. Cin., 1957-61, assoc. prof., 1961-66; prof. chemistry Va. Poly. Inst. and State U., Blacksburg, 1966—. Author books and numerous articles. Sloan fellow, 1961-64; recipient award Sigma Xi, 1961. Mem. Am. Chem. Soc. Subspecialties: Analytical chemistry; Distributed systems and networks. Current work: Instrument design and automation including design of new detectors for use in analytical chemistry, laboratory automation using networks. Office: Va Poly Inst and State U 325 Davidson Hall Blacksburg VA 24061

DETORRES, CORY DELGADO, psychologist, organizational consultant; b. N.Y.C., Aug. 3, 1942; d. Frank Joseph and Mildred (Kahn) Cohen; m. Fernando Delgado de Torres, June 25, 1966 (div. 1970); m. Tibor St. John de Cholnoky, Mar. 15, 1975; 1 son, Eric. A.B., Barnard Coll, 1964; Ph.D., Temple U., 1979. Lic. clin. psychologist, Pa. Staff assoc. Eastern Inst. Transactional Analysis and Gestalt, Phila., 1973-75; staff assoc. Laurel Inst., Inc., Phila., 1975-78; founding assoc. Phila. Profl. Assocs., Phila., 1978-79; dir. tng. and consultations Access Centers, Inc., Phila., 1979-82; dir. Cory de Torres Assocs., Phila., 1982—; cons. Influence Tng. Systems, Phila., 1980—. Contbr. articles to profl. jours. Mem. Am. Psychol. Assn., Pa. Psychol Assn., Phila. Soc. Clin. Psychology, Internat. Transactional Analysis Assn., Soc. Neuro-Linguistic Programming (cert. trainer). Subspecialty: Neurolinguistic programming. Current work: Interpersonal power and conflict resolution; interpersonal communications technology. Office: Cory de Torres Assocs 1900 Spruce St Philadelphia PA 19103

DETRICK, CARL ANTHONY, nuclear engineer; b. Monmouth, Ill., June 21, 1943; s. Charles Ariel and Mary Kathryn (Sandstrom) D.; m. Linda Claudia Robeson, Dec. 18, 1964; children: Paige Courtney, Catherine Elaine. B.S., Western Ill. U., 1965; M.S., U. Va., 1967; Ph.D., Carnegie-Mellon U., 1973. Engr./sr. engr. Westinghouse Electric Corp. Bettis Atomic Power Lab., West Mifflin, Pa., 1967-73, supr., 1973-74, mgr., 1974—. Mem. Am. Nuclear Soc., Sigma Xi. Republican. Presbyterian. Subspecialties: Oceanography; Nuclear fission. Current work: Ocean particle transport and modeling, deep ocean biological activity and sedimentation processes, radioactive waste management. Home: 738 Marvle Valley Dr Bethel Park PA 15102 Office: Bettis Atomic Power Lab Westinghouse Electric Corp PO Box 79 West Mifflin PA 15122

DETTBARN, WOLF-DIETRICH, pharmacologist, educator; b. Berlin, Ger., Jan. 30, 1928; s. Erwin B. and Maria M. (Conrady) D.; m. Christine A. Keune, Sept. 15, 1960; children: Donata-Andrea, Henning Christian. M.D., U Gottingen, Ger., 1953. Intern. Univ. Clinic, Gottingen, 1953-54; research assoc. biology dept. Ciba Co., Basel, Switzerland, 1954-55, Physiology Inst., U. Saarland, Hamburg, Germany, 1955-58; research assoc. dept. neurology Coll. Physicians and Surgeons, Columbia U., N.Y.C., asst. prof., 1961-67, assoc. prof., 1967-68; prof. pharmacology Med. Sch., Vanderbilt U., Nashville, 1968—; cons. U.S. Army Med. Research and Devel. Command, Nat. Acad. Sci.; corp. mem. Marine Biol. Lab., Woods Hole, Mass. Contbr. articles to profl. jours. Mem. Am. Physiol. Soc., Am. Soc. Pharmacology and Exptl. Therapeutics, Am. Soc. Neurochemistry, Soc. Gen. Physiologists, Soc. Neurosci. Subspecialties: Neuropharmacology; Neurochemistry. Current work: Neurotrophic regulation of muscle; organophosphates.

DETTERMAN, ROBERT LINWOOD, nuclear engineer; b. Norfolk, Va., May 1, 1931; s. George William and Jenneille (Watson) D.; m. Virginia Armstrong, Apr. 19, 1958; children; Janine, Patricia, William Arthur. B.S., Va. Poly. Inst., 1953; Ph.D. in Nuclear Engring, Oak Ridge Sch. Reactor Tech., 1954. Test dir. Foster Wheeler, N.Y.C., 1955-59; sr. research engr. Atomics Internat., Canoga Park, Calif., 1959-62, chief projects engr., 1962-68, dir. bus. devel., 1968—; chmn. space safety com. Atomic Indsl. Forum, N.Y.C., 1968; nuclear cons. Danish Govt., Riso, 1960. Author: Livermore Pool-Type Reactor, 1961; author: Safe Disposal of Reactors, 1965. Trustee Morris Animal Found., Arabian Horse Trust; adv. Calif. State Poly. U.; pres. Bio-Gin, Inc., Thousand Oaks. Mem. Am. Nuclear Soc. (dir. aerospace div. 1965-70), Atomic Indsl Forum, Tau Beta Phi, Phi Kappa Phi, Eta Kappa Nu. Republican. Clubs: Magic Cast, SPVA (Los Angeles). Subspecialties: Nuclear fission; Microchip technology (engineering). Current work: Nuclear reactor development, liquid metal fast breeder, space systems, computer applications. Patentee central rod activator, 1955, mechanism termination nuclear reactor, 1965. Home: 120 Colt Ln Thousand Oaks CA 91360 Office: Rockwell International Corp 8900 DeSoto Ave Canoga Park CA 91304

DETWEILER, DAVID KENNETH, veterinary physiologist, educator; b. Phila., Oct. 23, 1919; s. David Rieser and Pearl Irene (Overholt) D.; m. Inge E. A. Kludt, Feb. 2, 1965; children: Ellen, Diane, Judith, David, Inge, Kenneth. V.M.D., U. Pa., 1942, M.S., 1949; Sc.D. (hon.), Ohio State U., 1966, M.V.D., U. Vienna (Austria), 1968, D.M.V., U. Turin (Italy), 1969. Asst. instr. physiology and pharmacology Sch. Vet. Medicine, U. Pa., Phila., 1942-43, instr., 1943-45, assoc. in physiology, pharmacology, 1945-47, asst. prof., 1947-51, assoc. prof., 1951-62, , chmn. dept. vet. med. scis., 1956-68, dir. comparative cardiovascular studies unit, 1960—, prof., head lab. physiology and pharmacology, 1962-68, prof., head lab. physiology, 1968—, prof. faculty arts and scis., 1968—, chmn. grad. group comparative med. scis., 1971—; mem. Inst. Medicine, Nat. Acad. Scis., 1974—; cons. cardiovascular toxicology, 1950—. Contbr. numerous articles to various publs. Guggenheim fellow, 1955-56; Recipient Gaines award and medal Am. Vet. Med. Assn., 1960; D.K. Detweiler prize in cardiology established in his honor German Group of World Vet. Med. Assn., 1982. Fellow AAAAS; mem. Am. Physiol. Soc., Am. Assn. Vet. Physiology and Pharmacology (pres.), N.Y. Acad. Scis., Am. Vet. Med. Assn., Council Basic Scis., Am. Heart Assn., Acad. Vet. Cardiology (pres.), Am. Coll. Vet. Internal Medicine (cardiology group), Phi Zeta. Subspecialties: Animal physiology; Internal medicine (veterinary medicine). Office: Sch Vet Medicine 3800 Spruce St Philadelphia PA 19104

DETWEILER, STEVEN LAWRENCE, physicist, educator; b. Yonkers, N.Y., Sept. 4, 1947; s. Joseph H. and Catherine A. (Lawrence) D.; m. Nancy J. Logan, Sept. 4, 1971; children: Catherine S., David L. A.B., Princeton U., 1969; Ph.D., U. Chgo., 1974.

Postdoctoral fellow U. Md., College Park, 1974-76; Richard Chase Tolan research fellow, part-time instr. Calif. Inst. Tech., Pasadena, 1976-77; asst. prof. physics Yale U., New Haven, 1977-82; assoc. prof. physics U. Fla., 1982—. Contbr. articles to profl. publs. Mem. Am. Phys. Soc., Am. Astron. Soc., Soc. Gen. Relativity and Gravitation, Sigma Xi. Subspecialties: General relativity; Theoretical physics. Current work: Research involving black holes, gravitational waves and unification of forces of nature. Office: Williamson Hall U Fla Gainesville FL 32611

DETWYLER, THOMAS ROBERT, geographer, educator, environmentalist; b. Jackson, Mich., Aug. 10, 1938; s. Robert E. and Gladys (Bassett) D. B.S., U. Mich., 1960; Ph.D., Johns Hopkins U., 1966. Asst. prof. geography U. Mich., Ann Arbor, 1966-72, assoc. prof., 1972-78; assoc. prof. geography and environ. sci. Willamette U., Salem, Oreg., 1978-81, dir. environ. sci. program, 1980-81; prof. geography U. Wis.-Stevens Point, 1981—, chmn. dept. geography/geology, 1981—; vis. scholar U. Hawaii, Honolulu, 1972-73. Author: Man's Impact on Environment, 1971; editor: Urbanization and Environment, 1972; contbr. numerous articles to profl. jours. Fulbright fellow, N.Z., 1961-62; NDEA fellow, 1963-65; Beaumont fellow, 1965-66. Mem. Assn. Am. Geographers, AAAS, Nat. Council Geog. Edn. Current work: Political economy of environmental quality and resource use; energy, environment and social change; decentralization. Home: 401 Indiana Ave Stevens Point WI 54481 Office: Dept Geography/Geology U Wis Stevens Point WI 54481

DEUPREE, ROBERT GASTON, hydrodynamicist; b. Washington, Aug. 5, 1946; s. Robert Gaston and Mildred (Avery) D.; m. Janet Hammersley, June 22, 1968; children: Alexander, Michael. B.A., U. Wis., Madison, 1968; M.S., U. Colo., Boulder, 1970; Ph.D., U. Toronto, 1974. Postdoctoral fellow dept. astrophys. scis. Princeton U., 1974-75; postdoctoral asst. Los Alamos Sci. Lab., 1975-77; asst. prof. dept. astronomy Boston U., 1978-80; mem. staff Los Alamos Nat. Lab., 1980—. Contbr. articles to sci. jours. NSF grantee, 1978-79, 79-80. Mem. Am. Astron. Soc., Internat. Astron. Union. Subspecialties: Nuclear shock effects; Theoretical astrophysics. Current work: Hydrodynamics of stellar interiors; shock response of different geological media to nuclear explosions. Home: 1981 41st St Los Alamos NM 87544 Office: ESS-5 MS F-665 Los Alamos Nat Lab PO Box 1663 Los Alamos NM 87545

DEUSCHLE, KURT WALTER, physician, educator; b. Kongen, Germany, Mar. 14, 1923; came to U.S., 1924, naturalized, 1949; s. John and Marie (Schaefer) D.; m. Jeanne Magagna, 1975; children by previous marriage—Kurt J., Sally, James. B.S. cum laude, Kent State U., 1944; M.D., U. Mich., 1948. Intern Colo. Gen. Hosp., Denver, 1948-49; resident medicine, fellow oncology Upstate Med. Center of State U. N.Y. at Syracuse, 1950-52, instr. medicine, 1954-55; asst. prof. pub. health and preventive medicine Cornell Med. Coll., 1955-60; prof., chmn. dept. community medicine U. Ky., 1960-68; Ethel H. Wise prof., chmn. and dir. dept. community medicine Mt. Sinai Sch. of Medicine of City U. N.Y., 1968—; Merrimon lectr. U. N.C., Chapel Hill, 1975; vis. prof. U. Lagos, Nigeria, 1977; mem. tech. bd. Milbank Meml. Fund; mem. Tb control adv. com. Center Disease Control Dept. HEW; cons. manpower intelligence NIH; mem. Inst. Medicine of Nat. Acad. Scis., Washington; mem. rural health systems del. to China, 1978. Author: (with J. Adair) The People's Health: Anthropology and Medicine in a Navajo Community, 1970; Contbr. to: (ed. John Norman) Medicine in the Ghetto, 1969, Community Medicine: Teaching, Research and Health Care, (ed. Lathem and Newberry), 1970. Served with AUS, 1943-46. Commonwealth Fund sr. health fellow, 1966-67. Fellow Am. Coll. Preventive Medicine (past pres., Distinguished Service award 1975); mem. Am. Pub. Health Assn. (award for excellence in domestic health 1975), Am. Thoracic Soc., Assn. Tchrs. Preventive Medicine, Internat. Epidemiol. Assn., Alpha Omega Alpha. Subspecialty: Preventive medicine. Home: 1212 Fifth Ave New York NY 10029 Office: Fifth Ave and 100th St New York NY 10029

DEUTSCH, DALE GEORGE, toxicologist, neurobiochemist, educator; b. Liberty, N.Y., Feb. 22, 1943; s. Bernard and Sylvia (Nathanson) D.; m. Lou Charnon, July 2, 1946. B.A., SUNY-Buffalo, 1965; Ph.D., Purdue U., 1972. Postdoctoral fellow U. Colo., 1972-73; research assoc. U. Chgo., 1973-76; research asst. SUNY-Stony Brook, 1976-80, research asst. prof., 1980-81, asst. prof. pathology and biochemistry, head toxicology research on psychoactive drugs, 1981—; cons. in toxicology. Contbr. articles to profl. publs. Grantee USPHS, Nat. Inst. Drug Abuse. Mem. Am. Soc. Neurochemistry, Soc. Neurosci., Electron Microscopy Soc. Am., AAAS, Suffolk County Mental Health Assn., Am. Assn. Clin. Chemistry. Subspecialties: Toxicology (medicine); Neurochemistry. Current work: Effect of psychoactive drugs upon brain function. Research in atherosclerosis and fibrinolysis. Home: 21 Stony Brook Ave Stony Brook NY 11790 Office: University Hosp SUNY Stony Brook NY 11794

DEVANEY, JOSEPH JAMES, physicist; b. Boston, Apr. 29, 1924; s. Joseph Patrick and Madeline Elinor (Darragh) D.; m. Marjorie Ann Jones, Sept. 9, 1954; 1 dau., Kathleen. B.S., MIT, 1947, Ph.D., 1950. Research asst. MIT, Cambridge, 1942, 46, 47, 48; staff physicist Los Alamos Nat. Lab., 1950—; adj. prof. math. U. N Mex., Los Alamos, 1956-59, adj. prof. physics, 1959-70. Contbr. articles in field to profl. jours. Adv. Gov.'s Policy Bd.-Pollution, Santa Fe, N.Mex., 1969-70; mem. County and State Central Co., Los Alamos, Santa Fe, N.Mex., 1953-71; co-founder Anti-Smog Fedn., N.Mex., 1967; nat. patrolman Nat. Ski Patrol, 1953-79; water safety/first aid instr. ARC, Los Alamos, 1952-79. Served with USCG, 1944-45; Served with AUS, 1942-44. Recipient Scholarships M.I.T., 1941, 42, 45, 46, 47; AEC fellow M.I.T., 1947-50. Mem. Am. Phys. Soc., Am. Nuclear Soc. Club: Ski (Los Alamos) (Patrol leader 1962-63). Subspecialties: Theoretical physics; Laser fusion. Current work: Physics for particle transport, air pollution, solar energy, comparative hazards of energy production, laser fusion, nuclear physics, mathematical physics, defense of the U.S. and Free World. Home: 4792 Sandia Dr Los Alamos NM 87544 Office: Los Alamos Nat Lab MS-B226 X6 Los Alamos NM 87545

DEVAULT, DON CHARLES, biophysicist, educator; b. Battle Creek, Mich., Dec. 10, 1915; s. Ralph Pulliam and Ruth C. (Tenney) DeV.; m. Roberta Jeanette Baird, May 15, 1948; children: James Ralph (dec), Julie Ann. B.S. in Chemistry, Calif. Inst. Tech., 1937; Ph.D., U. Calif.-Berkley, 1940. Jr. research assoc. in chemistry Stanford U., 1940-42; instr. Inst. for Nuclear Studies, U. Chgo., 1946-48; assoc. prof. chemistry and physics U. of Pacific, Stockton, Calif., 1949-58; cons. elec. engring. Bipphys. Electronics Co. and Bionic Instruments Co., New Hope and Bala Cynwyd, Pa., 1959-63; asst. prof. biophysics Johnson Found., U. Pa., Phila., 1967-77; vis. asso. prof. biophysics, dept. physiology and biophysics U. Ill.-Urbana, 1977—. Co-editor: Tunneling in Biological Systems, 1979; Contbr. chpts. to books, articles to profl. jours. NSF grantee, 1967—. Mem. Biophys. Soc., Am. Soc. Biol. Chemistis, Am. soc. for Photobiology, Soc. for Social Responsibility in Sci., Phi Beta Kappa, Sigma Xi. Subspecialties: Biophysics (biology); Photosynthesis. Current work: Electron transfer in biological meterials including low temperature (liquid helium) and high pressure (several thousand atmospheres). Also energy transduction from electron transport. Home: 1206 Northwood Dr N Champaign IL 61820 Office: Dept Physiology and Biophysics U Ill 524 Eurrill Hall 407 S Goodwin Ave Urbana Il 61801

DEVAY, JAMES EDSON, plant pathologist, educator; b. Mpls., Nov. 23, 1921; s. James Henry and Sarah M. (Edson) DeV.; m. Mary Alice Bambach, Dec. 27, 1947; children: Susan, Mary Elizabeth, Sally Jane, Joseph, Paula, Michael. B.S., U. Minn., 1949, Ph.D., 1953. Asst. prof. to assoc. prof. dept. plant pathology U. Minn., 1953-57; asst. prof. U. Calif.-Davis, 1957-59, assoc. prof., 1959-65, prof., 1965—, chmn. dept, 1980—. Served with USN, 1942-46. Fellow AAAS, Am. Phytopathology Soc.; mem. Am. Mycol. Soc., Scandinavian Soc. Plant Physiology. Democrat. Roman Catholic. Subspecialties: Plant pathology; Plant physiology (agriculture).

DEVEREUX, WILLIAM PATRICK, aerospace co. adminstr.; b. Yonkers, N.Y., Mar. 21, 1923; s. William Thomas and Mary Cecelia (McCormack) D.; m. Theodora Anna Desider, May 30, 1955; 1 son, Lawrence M. A.B., Woodstock Coll., 1946, Ph.D., 1947. Instr. Canisius Coll., Buffalo, 1947-50; tech. instr. Bell Aircraft Corp., Buffalo, 1952-56; physicist Farrant Optical Co., Inc., Bronx, N.Y., 1956-59; scientist, physicist Gen. Dynamics-Elec. Boat, Groton, Conn., 1960-62; prin. engr. optics Kollsman Instrument Corp., Syosset, N.Y., 1962-71; mgr. electrooptics design Ball Aerospace Systems Div., Boulder, Colo., 1971—. Contbr. articles to profl. jours. Mem. Optical Soc. Am., Soc. Photo-Instrumentation Engrs. Democrat. Roman Catholic. Lodge: KC. Subspecialties: Aerospace engineering and technology; Optical engineering. Current work: Design of instrumentation for space astronomy and navigation. Patentee in field. Home: 805 Agate St Broomfield CO 80020 Office: Ball Aerospace Systems Div PO Box 1062 Boulder CO 80306

DEVINCENZI, RONALD GEORGE, dental consultant; b. San Francisco, Oct. 11, 1930; s. George Louis and Elvyra (DeLuca) DeV.; m. Donna Rita Vondra, June 27, 1953; children—Mark, Ronald George, Paul, Robert, Maria, Dianna, Andrea. B.S., Creighton U., 1952, D.D.S., 1956. Pvt. practice dentistry, Monterey, Calif., 1960-71; ptnr. Monterey Peninsula Dental Group, 1971-79; research investigator Calif. Found. Dental Health, Los Angeles, 1976-79; Western regional dir. Am. Dental Examiners, Monterey, 1980—; dir. Calif. Dental Service Corp., 1976-77; cons. socio-econ. research Calif. Found. Dental Health, Los Angeles, 1976-79. Served to capt. AUS, 1955-60. Named Dentist of Yr. Monterey Bay Dental Soc., 1974. Fellow Am. Coll. Dentists, Am. Inst. Oral Biology; mem. Calif. Dental Assn. (trustee 1976-79), ADA (del. 1974-79), Internat. Assn. Dental Research, Am. Assn. Pub. Health Dentists. Republican. Roman Catholic. Current work: Dental health insurance administration; dental health care evaluation, health policy research. Home: 1141 Wildcat Canyon Rd Pebble Peach CA 93953

DEVITA, VINCENT THEODORE, JR., oncologist; b. Bronx, N.Y., Mar. 7, 1935; s. Vincent Theodore and Isabel DeV.; m. Mary Kay Bush, Aug. 3, 1957; children: Teddy (dec.), Elizabeth. B.S., Coll. William and Mary, 1957; M.D., George Washington U., 1961. Diplomate: Nat. Bd. Med. Examiners, Am. Bd. Internal Medicine. Intern U. Mich. Med. Center, Ann Arbor, 1961-62; resident in medicine George Washington U. Med. Service D.C. Gen. Hosp., 1962-73; sr. resident in medicine Yale New Haven Med. Center, 1965-66; clin. asso. lab. chem. pharmacology Nat. Cancer Inst. NIH, Bethesda, Md., 1963-65, mem. staff, 1966—, chief med. br., 1971-74, dir. div. cancer treatment, 1974-81, clin. dir. inst., 1975-81, dir., 1981—; mem. faculty George Washington U. Med. Sch., 1971—, prof. medicine, 1975—; mem. expert advisory panel WHO, 1976; mem. Lasker Award Jury, 1976; chmn. Com. French-Am. Agreement on Cancer Treatment Research, 1976—; vis. prof. Stanford U. Med. Sch., 1972; 1st ann. Clowes lectr. Roswell Park Meml. Inst. Buffalo, 1973. Contbr. numerous articles to med. jours. Served with USMCR, 1955-61. Tobacco Research Industry fellow, 1959; recipient Albert and Mary Lasker Med. Research award, 1972; Superior Service award HEW, 1975; Esther Langer Found. award, 1976; Alumni medallion Coll. William and Mary, 1976; Jeffrey Gottliev award, 1976; decorated Oren del Sol en el Grando Oficial, Peru, 1970. Fellow A.C.P.; mem. Am. Soc. Clin. Oncology (chmn. program com. 1972, dir. 1973-76, pres. 1977-78), Am. Cancer Soc., Am. Soc. Hematology, Am. Assn. Cancer Research (dir. 1976-79), AMA, Am. Fedn. Clin. Research, Am. Soc. Clin. Investigation, Soc. Surg. Oncology, Smith-Reed-Russel Med. Soc., Alpha Omega Alpha. Subspecialties: Cancer research (medicine); Hematology. Office: Bldg 31 Room 11A52 9000 Rockville Pike Bethesda MD 20014

DEVLIN, FRANK JOSEPH, physical chemist, researcher; b. Dublin, Ireland, Oct. 1, 1949; s. Michael Joseph and Mary (Cunningham) D.; m. Ann Marie Jones, Feb. 20, 1980; 1 son, Michael. B.Sc. with honors, Univ. Coll. Dublin, 1972, Ph.D., 1977. Spectroscopic researcher dept. chemistry Univ. Coll. Dublin, 1972-77; chem. physics research dept. chemistry U. So. Calif., 1977—, now research assoc. Mem. Optical Soc. Am. Roman Catholic. Subspecialties: Physical chemistry; Spectroscopy. Current work: Research in CD/MCD/VCD spectroscopies; development of spectroscopic instrumentation; design and development of cryogenic and superconducting magnet systems. Office: Dept Chemistry U So Calif Los Angeles CA 90089

DE VOLPI, ALEXANDER, research physicist; b. N.Y.C., Feb. 28, 1931; s. Paul Bonaventura and Bertha (Gaber) De V.; m. Judith Carol Klaye, Jan. 12, 1978. B.A., Washington and Lee U., 1953; M.S., Va. Poly Inst., 1958, Ph.D., 1967; postgrad., Internat. Inst. Nuclear Sci. and Engring., Argonne, Ill., 1958-60. Assoc. physicist applied/reactor physics/engring. divs. Argonne Nat. Lab., 1960-71, physicist reactor analysis and safety div., 1971—, div. mgr. diagnostics sect., since 1973—; cons. De Volpi, Inc., Hinsdale, Ill., 1979—; bd. dirs. Fuel Centers Corp., South Bend, Ind., 1973-76. Author: Proliferation, Plutonium and Policy, 1979, (with others) Born Secret, 1981. Co-chmn. Concerned Argonne Scientists, 1969—; bd. dirs. Alliance to End Repression, Chgo., 1973—, DuPage chpt. ACLU, 1970-76. Served to lt. comdr. USNR, 1953-56. Mem. Am. Phys. Soc., Am. Nuclear Soc., AAAS. Subspecialties: Nuclear fission; Instrumentation. Current work: Cineradiographic instrumentation for nuclear-reactor safety applications; denaturing and safeguards for fissile materials. Patentee hodoscope radiography, electronic timing circuit. Home: 10 Kingery Quarter Hinsdale IL 60521 Office: Argonne Nat Lab 9700 S Cass Ave Argonne IL 60439

DEVORE, DALE PAUL, biochemist, researcher; b. Phillipsburg, N.J., Mar. 31, 1943; s. David Henry and Anna Elizabeth (Paul) DeV.; m. Sandra Bernice Grebowiec, Dec. 27, 1965; children: Mychelle, Braden. B.A., Rutgers U., 1966, M.S., 1972, Ph.D., 1973. Research asst. Rutgers U., New Brunswick, N.J., 1966-72; research scientist Battelle Meml. Inst., Columbus, Ohio, 1972-74, prin. research biochemist, 1974-79; research specialist Riker Research div. 3M Corp., St. Paul, 1979-81, sr. research specialist biochemistry, mgr. collagen products labs., 1981—; mem. faculty U. Minn., 1981—. Contbr. chpts. to books, articles to profl. jours. Commr. Vadnais Heights (Minn.) Planning Commn., 1981—. Mem. Am. Rheumatism Assn., Internat. Assn. for Dental Research, Soc. for Biomaterials, Midwest Connective Tissue Assn., Sigma Xi, Alpha Zeta. Subspecialties: Biochemistry (medicine); Artificial organs. Current work: Biochemistry of connective tissue. Investigation of connective tissue degradation in rheumatoid arthritis. Development of collagen-based biomedical implants. Inventor collagen reconstitution. Home: 4334 Greenhaven Circle Vadnais Heights MN 55110 Office: 3M Center St Paul MN 55144

DEVORE, DUANE THOMAS, oral surgeon, lawyer, educator; b. Park Ridge, Ill., May 14, 1933; s. Jacques Joseph and Nellie (Liddil) DeV.; m. Patricia Kilgarriff; 1 dau., Katherine Margaret. D.D.S., Loyola U., Chgo., 1956; Ph.D., U. London, 1975; J.D., U. Md.-Balt., 1979. Diplomate: Am. Bd. Oral Surgery, Am. Bd. Forensic Odontology, Md 1980, D.C 1980; lic. dentist, Ill., Ga., Md. Individual practice dentistry specializing in oral surgery, Savannah, Ga., 1960-67, Balt., 1971—; hon. Research fellow London Hosp., 1968-71; mem. faculty U. Md. Dental Sch., Balt., 1971—, prof. oral and maxillofacial surgery, 1976—; pvt. practice legal and med. cons., Balt., 1971—. Mem. Am. Heart Assn. (dir. Balt. chpt. 1973-78), Charles Village Civic Assn. (dir. 1982-83), ADA, ABA, Am. Assn. Dental Research (councillor 1975—). Subspecialties: Oral and maxillofacial surgery; Microsurgery. Current work: Microvascular surgery; bone replacement materials; clinical resea. Home: 2701 N Calvert St Baltimore MD 21218 Office: U Md Dental Sch 666 W Baltimore St Baltimore MD 21201

DEVORE, PAUL WARREN, techology educator, consultant; b. Parkersburg, W.Va., July 18, 1926; s. Harry and Eleanor Sarah (Dunn) D.; m. Eleanor Jean Condron, Apr. 7, 1952; children: Michelle Ann, Phillip Charles. B.S., Ohio U., 1950; M.A., Kent State U., 1954; Ed.D. (Fellow 1959-60), Pa. State U., 1961; postdoctoral fellow, U. Md., 1965-66. Asst. prof. engring. Grove City Coll., 1953-56; asst. prof. indsl. arts and tech. SUNY-Oswego, 1956-59, dir. div. indsl. arts and tech., 1961-67; prof. program for study of tech. W.Va. U., 1967-73, prof., chmn., 1973—; cons. Davis Publs. Author: Technology: An Introduction, 1980, Introduction to Transportation, 1983. Served with USNR, 1944-46. Recipient Outstanding Tchr. award U. W.Va., 1971; research award Phi Delta Kappa, 1978. Mem. AAAS, Soc. History of Tech., Transp. Research Forum, Energy Conservation Coalition. Subspecialty: Regenerative technical systems. Current work: Biotechnology—variables related to behavior of technical systems and sustainable futures. Home: 668 Colonial Dr Morgantown WV 26505 Office: West Virginia University 609 Allen Hall Morgantown WV 26506

DEVOUS, MICHAEL DAVID, SR., nuclear medicine/cardiovascular specialist, image processing researcher, educator; b. Chgo., Apr. 9, 1949; s. John Leonard and Mary Ruth (Dickens) D.; m. Priscilla Gish, July 13, 1968; children: M. Adrienne, Michael David. A.B., Washington U., St. Louis, 1970; Ph.D., Tex. A&M U., 1976. Teaching asst. Tex. A&M U., 1970-71, research assoc., 1976-78; asst. prof. U. Ill.-Champaign, 1978-81; asst. prof. nuclear medicine U. Tex. Health Sci. Ctr., Dallas, 1981—. Author: Collected Poetry, 1982. Pres. Bryan (Tex.) Sch. Bd., 1977-78. Calif. Scholastic Fedn. scholar, 1966-70; NSF fellow, 1971-72; Tex. A&M U. fellow, 1972-74; Welch fellow, 1974-76. Mem. Soc. Nuclear Medicine, Am. Heart Assn. (research chmn. Ill. affiliate 1979-80), Am. Physiol. Soc., Am. Fedn. Clin. Research, IEEE, N.Y. Acad. Sci., Soc. Magnetic Resonance Imaging (charter), Sigma Xi. Democrat. Roman Catholic. Subspecialties: Nuclear medicine; Cardiology. Current work: Ischemic heart disease; heart and brain blood flow; medical image processing; nuclear magnetic resonance imaging and metabolic measurements; non-invasive diagnosis; cardiovascular and cerebrovascular physiology. Home: 539 Melody Ln Richardson TX 75081 Office: Nuclear Medicine Center U Tex Health Sci Center 5323 Harry Hines Blvd Dallas TX 75235

DEVRIES, DAVID LEE, psychologist, research center executive; b. Holland, Mich., Aug. 11, 1943; s. Martin and Catherine (Vanderleek) DeV.; m. Martha Carol Christian, Aug. 21, 1965; children: David Todd, Mark Christian, Matthew Bishop. B.A., Calvin Coll., 1965; M.S., U. Ill.-Urbana, 1967, Ph.D., 1970. Assoc. research scientist Ctr. Social Orgn. Schs., Johns Hopkins U., 1970-75, asst. prof. psychology and social relations, 1970-75; research psychologist Ctr. for Creative Leadership, Greensboro, N.C., 1975-76, dir. research, 1976-80, v.p. research, 1980-82, v.p. research and programs, 1982, exec. v.p., 1982—. Author: (with others) Teams-games-tournament: The team approach, 1980, Performance appraisal on the line, 1981; contbr. articles to profl. publs.; cons. editor: Research in Higher Edn, 1974-82; reviewer profl. jorus. Mem. Am. Psychol. Assn., Acad. Mgmt., Am. Ednl. Research Assn. Lodge: Rotary. Subspecialty: Behavioral psychology. Current work: Work group structures, performance appraisal, managerial work simulations, human resource management, employee attitudes. Office: Center Creative Leadership 5000 Laurinda Dr Greensboro NC 27402

DE VRIES, GEORGE HENRY, neurobiologist; b. Paterson, N.J., Dec. 22, 1942; s. Henry and Jeanette (Greydanus) DeV.; m. Helen T. De Vries, Aug. 21, 1965; children: Jori Elizabeth, James Thomson. B.S. in Zoology cum laude, Wheaton Coll, 1964; Ph.D. in Biochemistry, U. Ill. Med. Ctr., Chgo., 1969. Postdoctoral fellow Albert Einstein Coll. Medicine, 1969-71; asst. prof. biochemistry Med. Coll. Va., 1972-75, assoc. prof., 1975—; Fogarty Found. fellow, Nice, France, 1980. Contbr. numerous articles to profl. jours. Bd. dirs. Multiple Sclerosis Assn., Richmond. Roche Found. fellow, 1975-76; NIH grantee, 1973; NSF grantee, 1976-78; Kroc Found. grantee, 1982—. Mem. Am. Soc. Neurochemistry (council), Internat. Soc. Neurochemistry, Soc. Neurosci., Am. Soc. Cell Biology. Presbyterian. Club: Highland Hills Swim. Subspecialties: Neurobiology; Biochemistry (biology). Current work: Research in mechanisms of myelination, neuronal influences in dymelinating disease, isolation of plasma membrane. Home: 2329 Tuscora Rd Richmond VA 23235 Office: Dept Biochemistry Med Coll Va Richmond VA 23298

DE VRIES, KENNETH LAWRENCE, mechanical engineer, educator; b. Ogden, Utah, Oct. 27, 1933; s. Sam and Fern (Slater) DeV.; m. Kay M. DeVries, Mar. 1, 1959; children—Kenneth, Susan. A.B., Weber State Coll., 1953; B.S., U. Utah, 1959, Ph.D., 1962. With Convair, Fort Worth, 1957; mem. faculty U. Utah, Salt Lake City, 1961—, prof. mech. engring., 1969-76, prof. dept. mech. and indsl. engring., 1976—, chmn. dept., 1970-81; head polymer program NSF, 1975-76; mem. Utah Council Sci. and Tech., 1973-77. Author: Analysis and Testing of Adhesive Bonds, 1977; contbr. articles on polymers, dental materials, rock mechanics, adhesive design to profl. jours. Mem. ASME, Am. Phys. Soc., Internat. Soc. Dental Research, Am. Chem. Soc., ASTM, Material Soc. Mormon. Subspecialties: Mechanical engineering; Materials (engineering). Current work: Research in material failure, polymers and composites. Home: 1466 Penrose St Salt Lake City UT 84103 Office: 3008 Mech Engring Bldg U Utah Salt Lake City UT 84112

DEVRIES, WILLIAM CASTLE, surgeon, educator; b. Bklyn., Dec. 19, 1943; s. Henrick and Kathryn Lucille (Castle) DeV.; m. Ane Karen Olsen, June 12, 1965; children—Jon, Adrie, Kathryn, Andrew, Janna, William, Diana. B.S., U. Utah, 1966, M.D., 1970. Intern Duke U. Med. Center, 1970-71, resident in cardiovascular and thoracic surgery, 1971-79; asst. prof. surgery U. Utah; chmn. div. cardiovascular and thoracic surgery; chief thoracic surgery Salt Lake VA Hosp. Recipient Wintrobe award, 1970. Mem. A.C.S., Utah Med. Assn., AMA, Intermountain Thoracic Soc., Salt Lake Surg. Soc., Utah Heart Assn., Assn. VA Surgeons, Utah Lung Assn., Alpha Omega Alpha. Mormon. Subspecialties: Artificial organs; Cardiac surgery. Office: 50 N Medical Dr Salt Lake City UT 84132

DEWAR, MICHAEL JAMES STEUART, chemistry educator; b. Ahmednagar, India, Sept. 24, 1918; came to U.S., 1959, naturalized, 1980; s. Francis and Nan (Keith) D.; m. Mary Williamson, June 3,

1944; children: Robert Berriedale Keith, Charles Edward Steuart. B.A., Oxford (Eng.) U., 1940, D.Phil., 1942, M.A., 1943. Imperial Chem. Industries fellow Oxford U., 1945; phys. chemist Courtaulds Ltd., 1945-51; prof. chemistry, head dept. Queen Mary Coll., U. London, Eng., 1941-59; prof. chemistry U. Chgo., 1959-63; Robert A. Welch prof. chemistry U. Tex., 1963—; Reilly lectr. U. Notre Dame, 1951; Tilden lectr. Chem. Soc. London, 1954; vis. prof. Yale U., 1957; Falk-Plaut lectr. Columbia U., 1963; William Pyle Phillips visitor Haverford Coll., 1964, 70; Arthur D. Little vis. prof. MIT, 1966; Marchon vis. lectr. U. Newcastle (Eng.), 1966; Glidden Co. lectr. Kent State U., 1967; Gnehm lectr. Eldg. Technische Hochschule, Zurich, Switzerland, 1968; Barton lectr. U. Okla., 1969; Disting. vis. lectr. Yeshiva U., 1970; Kahlbaum lectr. U. Basel, Switzerland, 1970; Benjamin Rush lectr. U. Pa., 1971; Kharasch vis. prof. U. Chgo., 1971; Phi Lambda Upsilon lectr. Johns Hopkins U., 1972; Firth vis. prof. U. Sheffield, 1972; Foster lectr. SUNY-Buffalo, 1973; Five Colls. lectr., Mass., 1973; Sprague lectr. U. Wis., 1974; Disting. Bicentennial prof. U. Utah, 1976; Bircher lectr. Vanderbilt U., 1976; Pahlavi lectr., Iran, 1977; Michael Faraday lectr. U. No. Ill., 1977; Priestley lectr. Pa. State U., 1980; cons. to industry. Author: Electronic Theory of Organic Chemistry, 1949, Hyperconjugation, 1962, Introduction to Modern Organic Chemistry, 1965, Computer Compilation of Molecular Weights and Percentage Compositions of Organic Compounds, 1969, The Molecular Orbital Theory of Organic Chemistry, 1969, The PMO Theory of Organic Chemistry, 1975; also articles. Recipient Harrison Howe award Am. Chem. Soc., 1961, S.W. regional award, 1978; Robert Robinson Lecture, Chem. Soc., 1974; G.W. Wheland Meml. medal U. Chgo., 1976; Evans award Ohio State U., 1977; hon. fellow Balliol Coll., 1974. Fellow Royal Soc. (Davy medal 1982), Am. Acad. Arts and Scis., Chem. Soc. London; mem. Am. Chem. Soc., Nat. Acad. Sci., Sigma Xi. Home: 6808 Mesa Dr Austin TX 78731

DEWAR, ROBERT LEITH, research physicist; b. Melbourne, Australia, Mar. 1, 1944. B.Sc., U. Melbourne, 1965, M.Sc., 1967, Ph.D. in Astrophysics Sci., Plasma Physics, 1970. Fellow, Ctr. Theoretical Physics, dept. physics and astronomy U. Md., College Park, 1970-71; research assoc. Plasma Physics Lab., Princeton (N.J.) U., 1971-73, mem. research staff, 1977-79, research physicist, 1979-81, prin. research physicist, 1981—; research fellow dept. theoretical physics Research Sch. Phys. Sci., Australian Nat. U., 1974-76, sr. research fellow, 1976-77. Fellow Am. Phys. Soc.; mem. Australian Inst. Physics. Subspecialty: Plasma physics. Office: Plasma Physics Lab Princeton U PO Box 451 Princeton NJ 08540

DEWART, DOROTHY BOARDMAN, clinical psychologist, consultant, researcher; b. Boston, Aug. 19, 1948; d. Thomas Dennie and Dorothy (Potter) Boardman. B.A. in Psychology and Sociology, Salem Coll., 1972; M.A. in Clin. Psychology, Xavier U., 1976; postgrad., U. Cin., 1975-77; Ph.D. in Clin. Psychology, Temple U., 1981. Research asst. Cin. Ctr. Devel. Disorders, 1974-75, psychology trainee, 1975, U. Cin., 1975-76, Rollman' Psychiat. Inst., Cin., 1975-76; psychologist Cen. Psychiat. Clinic, Cin., 1976-77; asst. dir. Pyschol. Services, Clermont Gen. and Tech. Coll., U. Cin., 1976-77; supr. Psychol. Services Ctr., Temple U., 1978, clin. asst., 1980, psychology intern, 1980-81, staff psychologist, dept. psychiat. and chronic pain clinic, 1981-82; cons. maxillo-facial pain clinic, 1981—, clin. instr. dept. psychiatry, 1981—. Profl. com. Wissahickon Hospice, Chestnut Hill, Pa. Research Incentive Fund grantee Temple U. Hosp., 1981—; Pew Meml Trust grantee, 1981—; recipient NIMH assistantship, 1977-78. Mem. Am. Psychol. Assn., Phila. Soc. Clin. Psychologists, Am. Pain Soc. Subspecialties: Diagnosis and control of chronic pain; Nutrition (biology). Current work: Psychotherapy with patients who present significant physical illness; research in effects of tryptophan on chronic pain; supervise and teach psychiatry residents, psychology interns, consultant to physicians. Home: 748 St George's Rd Philadelphia PA 19119 Office: Dept Psychiatry Temple U Hosp 3401 N Broad St Philadelphia PA 19140

DEWERD, LARRY A., med. physicist, educator; b. Milw., July 18, 1941; s. Anthony L. and Dorothy M. (Heling) DeW.; m. Vada M. Anderson, Sept. 14, 1963; children: Scott, Mark, Eric. B.S., U. Wis.-Milw., 1963, M.S., 1965; Ph.D., 1970. Research assoc. U. Wash., Seattle, 1970-72, research asst. prof. mining, metall. and ceramic engring., 1972-75; vis. asst. prof. U. Wis.-Madison, 1975-76, clin. asst. prof. radiology-med. physics, 1976-79, clin. assoc. prof., 1979—, dir., 1981—. Contbr. articles to profl. jours. Mem. Am. Assn. Physicists in Medicine, Am. Phys. Soc., Health Physics Soc., Soc. Photographic-Instrumentation Engrs. Subspecialties: Medical physics; Radiology. Current work: Diagnostic and therapeutic medical physics. Patentee in field. Home: 13 Pilgrim Circle Madison WI 53711 Office: UNIVAVE Med Physics U Wis 1570 MSC/1300 Madison WI 53706

DEWEY, THOMAS GREGORY, chemistry educator; b. Pitts., June 2, 1952; s. Ralph Cooper and Frances (Kiefer) D.; m. Cynthia Miller, Sept. 4, 1982. B.Sc., Carnegie-Mellon U., 1974; M.S., Rochester (N.Y.), 1977, Ph.D., 1979. Asst. prof. chemistry U. Denver, 1981—. Contbr. articles to profl. jours. NIH fellow, 1979-81; Cottrell Research grantee, 1981-83. Mem. Am. Chem. Soc., Biophys. Soc. Subspecialties: Biophysical chemistry; Biophysics (biology). Current work: Research interest concern the structure and function of membrane bound proteins involved in ion transport. Office: U Denver Dept Chemistry Denver CO 80208

DE WIT, MICHIEL, physicist; b. Amsterdam, Netherlands, June 6, 1933; came to U.S., 1951, naturalized, 1968; s. Louis Willem and Marianne (Carels) DeW.; m. Catharine C. Courtney, Dec. 27, 1957; children: Kirsten, Deirdre, Seth, Damiane. B.S., Ohio U., 1954; Ph.D., Yale U., 1960. Mem. tech. staff Tex. Instruments Inc., Dallas, 1959-79, sr. mem. tech. staff, 1979—; occasional lectr. Contbr. articles to profl. publs. Mem. Am. Phys. Soc., Optical Soc. Am., IEEE, AAAS. Subspecialties: Integrated circuits; Atomic and molecular physics. Current work: Analog MOS integrated circuit design; research and development work. Patentee in field. Office: MS 369 PO Box 225621 Dallas TX 75265

DE WIT, ROLAND, research physicist; b. Amsterdam, Netherlands, Feb. 28, 1930; came to U.S., 1950, naturalized, 1961; s. Louis Willem and Marianne (Carels) deW.; m. Gloria Winkel, Aug 22, 1954; children: Bruce, Keith. B.S. in Math, Ohio U., Athens, 1953; Ph.D. in Physics, U. Ill.-Urbana, 1959. Asst. research physicist U. Calif.-Berkeley, 1959-60; physicist Nat. Bur. Standards, Washington, 1960—. Contbr. articles on physics to profl. jours. Gulf Oil fellow, 1957-58; recipient D.C. Certificate of award, 1965, Spl Achievement award, 1971, 1976. Mem. Am. Phys. Soc., Internat. Soc. for Stereology, Washington Acad. Scis., Metal. Soc. of AIME, Am. Soc. Metals, ASME, Md. Inst. Metals, ASTM, Am. Acad. Mechanics, Soc. Natural Philosophy, Phi Beta Sigma, Phi Beta Kappa. Democrat. Subspecialties: Fracture mechanics; Materials. Current work: Theoretical and experimentql research in fracture mechanics, using computers, applied to failure prediction. Home: 11812 Tifton Dr Potomac MD 20854 Office: Fracture and Deformation Div Nat Bur Standards Washington DC 20234

DEWITT, BRYCE SELIGMAN, physics educator; b. Dinuba, Calif., Jan. 8, 1923. B.S., Harvard U., 1943, M.A., 1947, Ph.D. in Physics, 1950. Mem. staff Inst. Advanced Study, Princeton, N.J., 1949-50; Fulbright fellow Tata Inst. Fundamental Research, India, 1951-52; sr. physicist Radiation Lab., U. Calif., 1952-55; research prof. physics U. N.C., Chapel Hill, 1956-59, prof., 1960-64, Agnew Hunter Bahnson Jr. prof., 1964-72, dir. research, 1956-72; prof. physics U. Tex., Austin, 1973—, dir. ctr. relativity, 1973—; Fulbright lectr., France, 1956, Japan, 1964-65; mem. Intenat. Com. Relativity and Gravitation, 1959-72; mem. fellows panel NRC, 1974—; Guggenheim vis. research fellow All Souls Coll., Oxford, Eng., 1975-76; group leader Inst. Theoretical Physics, U. Calif.-Santa Barbara, 1980-81; cons. in physics. NSF sr. fellow, 1964. Fellow Am. Phys. Soc.; mem. Explorer's Club. Subspecialty: Relativity and gravitation. Office: Dept Physics U Tex Austin TX 78712

DE WOLF, DAVID ALTER, elec. engr., educator; b. Dordrecht, Netherlands, July 23, 1934; came to U.S., 1962, naturalized, 1979; s. Marinus and Esfira (Frigind) de W.; m. Hilfman, Aug. 22, 1958 (div. Apr. 1964); children: Naomi, Jiska; m. Peggy L. Lumpkin, May 9, 1975. B.Sc., U. Amsterdam, 1955, Doctorandus in Theoretical Physics, 1959; D.Tech., U. Eindhoven, Netherlands, 1968. Research scientist Nuclear Def. Lab., Edgewood Arsenal, Md., 1962; mem. tech. staff David Sarnoff Research Center, RCA Labs., Princeton, N.J., 1962-82; prof. elec. engring. Va. Poly. Inst. and State U., Blacksburg, 1982—. Asso. editor: Jour. Optical Soc. Am, 1969-81; author articles. Served to 1st lt. Royal Dutch Army, 1960-61. Recipient Achievement award RCA Labs., 1976. Mem. IEEE, Optical Soc. Am., Netherlands Phys. Soc., Internat. Union Radio Sci. (mem. commns. B, C and F of U.S. nat. com.), AAAS, Sigma Xi. Subspecialties: Electrical engineering; Theoretical physics. Current work: Electromagnetic wave propagation, optics, diffraction, electron optics, scattering of waves. Patentee in field. Home: 200 Craig Dr Blacksburg VA 24060 Office: Dept Elec Engring Va Poly Inst and State U Blacksburg VA 24061

DEWYS, WILLIAM DALE, physician, cancer researcher; b. Zeeland, Mich., Sept. 14, 1939; s. Peter and Jennie (Morslnk) DeW.; m. Alice Grace Schut, June 10, 1961; children: Alisa Kay, William Dale, Pamela Jane. B.S., Calvin Coll., Grand Rapids, Mich., 1960; M.D. cum laude, U. Mich., 1964. Diplomate: Am. Bd. Internal Medicine; Am. Bd. Med. Oncology. Asst. prof. internal medicine U. Rochester, 1971-73; assoc. prof. internal medicine Northwestern U., Chgo., 1973-78, prof., 1978-79, chief sect. med. oncology, 1973-79; head nutrition sect. clin. investigations br. Nat. Cancer Inst., Bethesda, Md., 1979-82, chief clin. investigations br., 1980-82, assoc. dir. prevention program, 1982—. Contbr. articles to med. jours. Served to sr. surgeon USPHS, 1966-68. Fellow ACP; mem. Am. Assn. Cancer Research, Am. Soc. Clin. Oncology, Soc. Clin. Trials, Alpha Omega Alpha, Phi Kappa Phi. Subspecialties: Cancer research (medicine); Nutrition (medicine). Current work: Clinical trials in cancer prevention; physiologic studies of cancer cachexia; clinical trials of supportive care in cancer. Home: 6830 Hillmead Rd Bethesda MD 20817 Office: 8300 Colesville Rd Bethesda MD 20205

DEY, SUDHANSU K., physiologist, educator; b. Calcutta, W. Bengal, India, Nov. 8, 1944; s. Satish C. and Surabala (Aich) D.; m. Anjana, Mar. 2, 1970; 1 child, Maruti K. B.S., Calcutta U., India, 1965, M.S., 1968, Ph.D., 1972. Lectr. Calcutta (India) U., 1970-72; postdoctoral fellow Kans. U. Med. Center, Kansas City, 1973-77, asst. prof., 1977-81, assoc. prof., 1981—. Contbr. articles to sci. jours. NIH research grantee, 1979-83. Mem. Soc. Study Reproduction, Soc. Study Fertility, Am. Physiol. Soc., AAAS; mem. Endocrine Soc.; Mem. Sigma Xi (sec. chpt.). Subspecialties: Developmental biology; Reproductive biology. Current work: the physiology of embryo-uterine interaction during early pregnancy. Home: 8806 W 72nd Terr Shawnee Mission KS 66204 Office: U Kans Med Center 39th and Rainbow Kansas City KS 66103 Home: 8806 W 72nd Terr Shawnee Mission KS 66204

DEYOUNG, DAVID SPENCER, astrophysicist, observatory director; b. Colorado Springs, Colo., Nov. 29, 1940. B.A. magna cum laude, U. Colo., 1962; Ph.D., Cornell U., 1967. Sci. staff mem. Los Alamos Nat. Lab., 1967-69; astronomer Nat. Radio Astronomy Obs., Charlottesville, Va., 1969-80, Kitt Peak Nat. Obs., Tucson, 1980—, assoc. dir., 1982—; mem. adv. bd. Aspen Ctr. Physics; chmn. radio astronomy experiment selection panel NASA Deep Space Network. Contbr. articles in field to profl. jours. Mem. Am. Phys. Soc., Am. Astron. Soc., Internat. Astron. Union., Internat. Union Radio Sci. Subspecialty: Theoretical astrophysics. Current work: Theoretical astrophysics. Office: PO Box 26732 Tucson AZ 85726

DE YOUNG, LAWRENCE MARK, skin biologist; b. N.Y.C., Aug. 8, 1948; s. Norman and Marilyn Lydia (Schneider) De Y. B.S., SUNY, Albany, 1970; Ph.D., Syracuse U., 1976. Teaching asst. Syracuse U., 1973-75; cons. Fight for Sight, Upstate Med. Ctr., 1974; research assoc. McArdle Lab. Cancer Research, U. Wis., 1976-78; staff researcher Syntex Research, Palo Alto, Calif., 1978-81, staff researcher II, 1981—; NSF grad. trainee, 1970-72. Contbr. articles in field to profl. jours. Human support vol. El Camino Hosp., Mountain View, Calif. N.Y. State Regents Scholar, 1966-70; NSF fellow, 1970-74. Mem. Am. Assn. Cancer Research, Soc. Investigative Dermatologists., AAAS, Beta Beta Beta. Subspecialties: Cancer research (medicine); Dermatology. Current work: Epidermal differentiation, chemoprevention of cancer. Office: 3401 Hillview Ave Palo Alto CA 94304

DHALIWAL, RANJIT SINGH, mathematician, educator; b. Bilaspur, India, June 21, 1930; s. Bharpur Singh and Bachan Kaur (Grewal) D.; m. Gurdev Kaur, July 1, 1958; 1 son, Gurminder Singh. M.A., Punjab U., 1955; Ph.D., Indian Inst. Tech., Kharagpur, 1960. Lectr. Indian Inst. Tech., New Delhi, 1961-63, asst. prof., 1963-66; asso.prof. U. Calgary, Alta., Can., 1966-71, prof. math., 1971—; visitor Imperial Coll. Sci. and Tech., London, 1964-65; vis. prof. City U. London, 1971-72. Author book; contbr. articles to profl. jours. Mem. Am. Math. Soc., Soc. Indsl. and Applied Math., Can. Math. Congress, London Math. Soc. Subspecialty: Applied mathematics. Current work: Solid mechanics; elasticity,thermoelasticity, and fracture mechanics. Home: 44 Patterson Dr Calgary AB Canada Office: Dept Math U Calgary Calgary AB Canada

DHANIREDDY, RAMASUBBAREDDY, neonatologist, researcher; b. Gunthachiyyapadu, India, July 8, 1951; came to U.S., 1976; s. Peddaeswarareddy and Veeramma (Kasa) D.; m. Brezeetha Peddireddy, Dec. 15, 1972; children: Shireesha, Kirankumar. M.B., B.S., Kurnool Med. Coll., 1974. Diplomate: Am. Bd. Pediatrics (Neonatal and Perinatal Medicine). Intern Govt. Gen. Hosp., Kurnool, India, 1974-75; jr. resident Jawaharlal Inst. Postgrad. Med. Edn. and Research, Pondicherry, India, 1975-76; pediatrics resident La. State U., Med. Ctr., 1976-78; neonatology fellow Georgetown U. Med. Ctr., Washington, 1978-80; asst. prof. La. State U. Med. Ctr., Shreveport, 1980-82; staff fellow NIH, Bethesda, Md., 1982—; asst. prof. Georgetown U. Med. Ctr., 1982—. Nat. Merit scholar, 1967-69; state spl. scholar, 1969-74. Fellow Am. Acad. Pediatrics, Am. Coll. Nutrition; mem. Am. Fedn. Clin. Research, AMA, So. Soc. Pediatric Research. Hindu. Subspecialties: Neonatology; Pediatrics. Current work: Lipoprotein lipase, triglyceride clearing in premature infants. Role of prolactin in fetal lung development, prolactin receptors, hormonal regulation of glycogen metabolism in hepatocytes. Office: Nat Inst Health Bldg 6 Room 126 Bethesda MD 20205 Home: 2 Anamosa Ct Derwood MD 20855

DHARAN, C.K. HARI, engineering educator; b. Cannanore, India, Dec. 31, 1942; came to U.S., 1964, naturalized, 1978; s. Karthik and Sushila (Kunhiraman) D.; m. Selma Sarah Meyerowitz, June 21, 1970; children: Nila, Anjali. B.Tech., Indian Inst. Tech., Bombay, 1964; M.Engring., U. Calif.-Berkeley, 1965, Ph.D., 1968. Registered profl. engr., Calif. Sr. research scientist, sci. research staff Ford Motor Co., Dearborn, Mich., 1968-75; sr. staff scientist Ford Aerospace & Communications Corp., Palo Alto, Calif., 1975-80; mgr. spacecraft engring. Communications Satellite Corp., Palo Alto, 1980-82; assoc. prof. mech. engring. U. Calif.-Berkeley, 1982—. Contbr. articles to profl. jours. Mem. ASME, AIAA, Soc. for Advancement of Material and Process Engring. Subspecialties: Composite materials; Mechanical engineering. Current work: Mechanical behavior and processing of composite materials; mechanical design of spacecraft structures and components. Office: Dept Mech Engring U Calif Berkeley CA 94720

DIACONIS, PERSI, mathematics educator; b. N.Y.C., Jan. 31, 1945. B.S., CCNY, 1971; M.A., Harvard U., 1973, Ph.D. in Stats, 1974. Asst. prof. dept. stats. Stanford (Calif.) U., 1974-80, assoc. prof., 1980-81. Prof. Stanford (Calif.) U. (1981—). Subspecialty: Statistics. Office: Dept Stats Stanford U Stanford CA 94305

DIACUMAKOS, ELAINE G., cell biologist, consultant; b. Chester, Pa., Aug. 11, 1930; d. Gregoris and Olga (Dezes) D.; m. James Chimonides, Nov. 24, 1958. B.S., U. Md., 1951; M.S., N.Y.U., 1955, Ph.D. in Cell Physiology and Embryology, 1958; postdoctoral fellow, Rockefeller U., 1962-64. Instr. Grad. Sch. Med. Scis., Cornell U. Med. Coll., N.Y.C., 1963-71, research assoc., 1959-63, Sloan-Kettering Inst. Cancer Research, N.Y.C., 1958-60, assoc., 1960-71, assoc. and sect. head, 1965-71; guest investigator biochem. genetics, spl. USPHS fellow Rockefeller U., N.Y.C, 1962-64; sr. research assoc., 1971-75, sr. research assoc. and head cytobiology, 1976—; cons. Nat. Cancer Inst., others. Contbr. articles to sci. jours. Mem. adv. bvd. vis. guest scholars program Fulbright-Hays Act in Met. N.Y., 1980-82; bd. dir. Children's Sch., Rockefeller U., 1981—. Recipient Founder's Day award N.Y.U., 1958. Mem. AAAS, Am. Genetics Assn., Am. Soc. Cell Biology, Genetics Soc. Am., Internat. Cell Cycle Soc., Harvey Soc., N.Y. Acad. Scis., Sigma Xi (exec. com. chpt.). Subspecialties: Cell and tissue culture; Genetics and genetic engineering (medicine). Office: 1230 York Ave New York NY 10021

DIAMANDOPOULOS, GEORGE THEODORE, physician, researcher; b. Herakleion, Crete, Greece, Nov. 21, 1929; came to U.S., 1948, naturalized, 1964; s. Theodore George and Rita Theodore (Mouzenidis) D. B.A., Lawrence Coll., Appleton, Wis., 1951; M.D., U. Vt., Burlington, 1955; A.M. (hon.), Harvard U., 1979. Diplomate: Am. Bd. Pathology. Intern New Eng. Ctr. Hosp., Boston, 1955-56, resident in medicine, 1956-57; resident in pathology Peter Bent Brigham Hosp., Boston, 1957-59, Children's Hosp., 1960, Boston Lying-In Hosp., 1960; mem. faculty Harvard U. Med. Sch., 1965—, prof. pathology, 1980—. Mem. AAAS, Am. Soc. Microbiology, Soc. Exptl. Biology and Medicine, Am. Assn. Acad. Pathologists, Am. Assn. Cancer Research, Am. Soc. Exptl. Pathology, Alpha Omeha Alpha. Subspecialties: Pathology (medicine); Cancer research (medicine). Current work: etiopathology of human leukemia.

DIAMOND, BRUCE I., neuropharmacology educator; b. N.Y.C., June 7, 1945; s. Gerson D. and Gertrude F. (Steiger) D.; m. Ana Hitri, Jan. 20, 1976; 1 child, Sania Vorih. B.A., Bradley U., 1967; M.S., L.I. U., 1969; Ph.D., Chgo. Med. Sch., 1975. Asst. prof. Chgo. Med. Sch., 1976-81, Rush U., Chgo., 1978-81; assoc. prof. Med. Coll. Ga., Augusta, 1981—. Contbr. articles to sci. publs. Grantee Ill Dept. Mental Health, 1977-79, 80-81, David Pratt Found., 1981-83. Mem. AAAS, Soc. Neurosci., Am. Soc. Anesthesiologists, Sigma Xi. Club: Westlake Social (Augusta). Subspecialties: Psychopharmacology; Neuropharmacology. Current work: Neuropharmacology of movement disorders with emphasis on clinical and basic research as well as correlation of psychotropic blood levels in psychotic patients with clinical conditions. Home: 3531 Pebble Beach Dr Martinez GA 30907 Office: Dept Psychiatry Med Coll Ga Augusta GA 30912

DIAMOND, DAVID JOSEPH, nuclear engineer; b. Bklyn., Dec. 31, 1940; s. Harry and Blanche (Benin) D.; m. Ellen Schatz, June 23, 1962; children: Gary, Russell, Brian. B.Eng.Phys., Cornell U., 1962; M.S., U. Ariz., 1963; Ph.D., M.I.T., 1968. Nuclear engr. Westinghouse Astronuclear Lab., Pitts., 1963-64, Brookhaven Nat. Lab., Upton, N.Y., 1968—; mem. Energy Edn. Exponents, Inc., Patchogue, N.Y., 1980—; adj. prof. Poly. Inst. N.Y., 1977-78. AEC spl. fellow, 1967-68. Mem. Am. Nuclear Soc. (sect. chair 1982-83). Club: Mt. Sinai Sailing Assn. Subspecialty: Nuclear engineering. Current work: Light water reactor safety; neutronic and thermal hydraulic computer code development and application. Home: 4 Settlers Path Port Jefferson NY 11777 Office: Brookhaven Nat Lab Upton NY 11973

DIAMOND, JARED MASON, biologist; b. Boston, Sept. 10, 1937; s. Louis K. and Flora K. D. B.A., Harvard U., 1958; Ph.D., Cambridge (Eng.) U., 1961. Jr. fellow Soc. Fellows, Harvard U., 1962-65; asso. in biophysics (Med. Sch.), 1965-66; asso. prof. physiology U. Calif. Med. Center, Los Angeles, 1966-68, prof., 1968—; cons. in conservation and nat. park planning govts., Papua New Guinea, Solomon Islands, Indonesia. Author: Avifauna of the Eastern Highlands of New Guinea, 1972, Ecology and Evolution of Communities, 1975. Recipient Burr medal Nat. Geog. Soc., 1979, Bowditch prize Am. Physiol. Soc., 1976, Disting. Achievement award Am. Gastroent. Assn., 1975. Fellow Am. Acad. Arts and Scis.; mem. Nat. Acad. Scis. Subspecialties: Ecology; Physiology (biology). Research in membrane physiology, ecology. Office: Physiology Dept UCLA Med Center Los Angeles CA 90024

DIANZANI, FERDINANDO, microbiologist, educator; b. Grosseto, Italy, Sept. 12, 1932; s. Edgardo and Inma (Bocelli) D.; m. Giuliana Bini; children: Lorenzo, Caterina. M.D., U. Siena, Italy, 1959, Ph.D. in Microbiology, 1963, 1965. Intern U. Siena, 1959-61, resident dept. microbiology, 1961-63, asst. prof. microbiology, 1963-69; prof. U. Turin, Italy, 1970-76, U. Tex. Med. Br., 1976-80, adj. prof., 1980—; prof. virology U. Rome, 1980—. Contbr. numerous articles to profl. jours. Recipient Richardson-Merrel Pharms. award, 1968; G. Lenghi Internat. award Italian Lincei Acad., 1972. Mem. Am. Soc. Microbiology, Am. Assn. Immunology, Soc. Exptl. Biology and Medicine. Roman Catholic. Subspecialties: Virology (medicine); Immunopharmacology. Current work: Mechanisms of production and action of human interferons; mechanisms of recovery from viral infections. Office: Dept Microbiology U Tex Med Br Galveston TX 77550

DIBARTOLA, STEPHEN PAUL, veterinary scientist; b. Pitts., Oct. 17, 1950; s. Philip Edward and Martha Melinda (Weiman) DiB.; m. Maxey Lee Wellman, May 12, 1979. B.S., Loyola Marymount U., Los Angeles, 1972; D.V.M., U. Calif.-Davis, 1976. Intern small animal medicine and surgery Cornell U., 1976-77; resident in small animal medicine Ohio State U., 1977-79; asst. prof. small animal medicine U. Ill., Urbana, 1979-81, Ohio State U., Columbus, 1981—. Mem. AVMA, Am. Coll. Vet. Internal Medicine, Am. Animal Hosp. Assn., Vet. Urology Soc. Subspecialty: Internal medicine (veterinary medicine). Current work: Renal disease. Home: 210 Irving Way Columbus OH 43214 Office: 1935 Coffey Rd Columbus OH 43210

DIBIANCA, FRANK ANTHONY, physicist, research scientist; b. Atlantic City, N.J., Apr. 8, 1940; s. Vincent Joseph and Sarina Cardamone DiB.; m. Kay Carpenter, Mar. 14, 1970; 1 son, Arthur Nicholas. Ph.D. in Physics, Carnegie-Mellon U., 1970. Research assoc. dept. physics Case Western Res. U., 1970-73; research assoc. dept. physics FermiLab, Batavia, Ill., 1973-76; sr. physicist Gen. Electric Med. Systems, Milw, 1976-81; assoc. prof. radiology and biomedical engring. U. N.C., Chapel Hill, 1981—, dir. biomed. microelectronics program, 1981—. Contbr. articles in field to profl. jours. Served with USNR, 1962-64. Recipient Gen. Electric Co. award for devel. of digital radiography instrument, 1979. Mem. Am. Assn. Physicists in Medicine, Internat. Soc. Optical Engring., Internat. Soc. Hybrid Microelectronics. Subspecialties: Biomedical engineering; Imaging technology. Current work: Medical imaging instrumentation research, biomedical microelectronics research. Office: University of North Carolina 144 Mac Nider Hall Chapel Hill NC 27514

DIBNER, MARK DOUGLAS, neurobiologist, educator; b. N.Y.C., Nov. 7, 1951; s. David Robert and Dorothy Joyce (Siegel) D. B.A., U. Pa., 1973; Ph.D., Cornell U., 1977. Postdoctoral fellow Med. Sch., U. Colo., Denver, 1977-79; research fellow, instr. U. Calif., San Diego, 1979-80; lectr. psychology Nat. U., San Diego, 1980; research neurobiologist E.I. DuPont Glenolden Lab., Glenolden, Pa., 1980—. Treas. Woodside Hills Civic Assn., 1982—. Mem. Am. Soc. Pharmacology and Exptl. Therapeutics, Soc. for Neurosci., AAAS. Subspecialties: Neuropharmacology; Neurobiology. Current work: Researcher in study of receptors on cell surfaces and their pharmacologic regulation. Office: E I DuPont Glenolden Labs Glenolden PA 19036

DICELLO, JOHN FRANCIS, JR., physicist, educator; b. Bradford, Pa., Dec. 18, 1938; s. John Francis and Nicolina Carmilla (Costello) D.; m. Shirley Ann Rodgers, Aug. 25, 1962; children—John Francis III, Paul T. B.S., St. Bonaventure U., 1960; M.S., U. Pitts., 1962; Ph.D., Tex. A&M U., 1968. Instr. St. Bonaventure U., 1962-63; AEC-Assoc. Western Univs. grad fellow Los Alamos Sci. Lab., 1965-67, staff scientist, 1973—; on leave as research asso. Columbia U., N.Y.C., 1967-73; prof. physics Clarkson Coll., Potsdam, N.Y., 1982—. Contbr. articles to profl. jours. Bd. dirs. N.Mex. div. Am. Cancer Soc., 1978-82. Recipient Young Scientist travel award Am. Assn. Physicists in Medicine, 1972. Mem. Am. Assn. Physicists in Medicine, Radiation Research Soc., Am. Phys. Soc. Roman Catholic. Subspecialties: Biophysics (physics); Cancer research (medicine). Current work: Medical physics, dosimetry, microdosimetry, radiation biology, hyperthermia, cancer research integrated circuits, and nuclear physics. Office: Dept Physics Clarkson Coll Potsdam NY 13676

DICHTER, MARC ALLEN, physician, researcher, educator; b. N.Y.C., Dec. 1, 1943; s. Jack and Frances (Kleid) D.; m. Carole Ruth Salz, Dec. 20, 1964; children: Harold, Eric. B.A., Queens Coll., CUNY, 1965; M.D., NYU, 1969, Ph.D. in Neurophysiology, 1969. Diplomate: Am. Bd. Psychiatry and Neurology. Clin. assoc. Nat. Inst. Neurol. and Communications Disorders, 1970-72; resident in neurology Beth Israel Hosp., Children's Hosp., Peter Bent Brigham Hosp., Boston, 1972-75; asst. prof. neurology Harvard Med. Sch., Boston, 1975-78, asso. prof. neurology, 1978—; clin. practice medicine; researcher in field. Contbr. articles in biomed. research to publs. Served to lt. comdr. USPHS, 1970-72. Recipient research career devel. award, 1975-80. Mem. Am. Acad. Neurology, Am. Neurol. Assn., AAAS, Soc. for Neurosci., Phi Beta Kappa, Alpha Omega Alpha. Subspecialties: Neurology; Neurophysiology. Current work: Cellular neurophysiology and neuropharmacology of mammalian cortex. Mechanisms of epilepsy. Office: Dept Neurosc Children's Hosp Med Ct 300 Longwood Ave Boston MA 02115

DICK, CHARLES EDWARD, physicist; b. Ft. Wayne, Ind., Apr. 24, 1937; s. Melvin L. and Virginia Viola (Laemmle) D.; m. Vivian Claire Dick, Aug. 16, 1958; children: Timothy M., Victoria M. B.S., Ill. Benedictine Coll., Lisle, Ill., 1958; Ph.D., U. Notre Dame, 1963. Physicist Nat. Bur. Standards, Washington, 1962—. Contbr. articles to profl. jours. Mem. Am. Phys. Soc., AAAS, Soc. Photo-Optical Instrumentation Engrs., Sigma Xi. Democrat. Roman Catholic. Subspecialties: Atomic and molecular physics; Imaging technology. Current work: Applications of digital techniques to diagnostic imaging, interaction of electrons and photons with matter. Home: 14000 Manorvale Rd Rockville MD 20853 Office: Nat Bur Standards C 216 Bldg 245 Washington DC 20234

DICK, DANIEL EGGLESTON, energy educator, librarian; b. Worcester, Mass., Apr. 22, 1924; s. Richard George and Florence Isabella (Eggleston) D.; m. Marjory Stephenson, June 18, 1949; children: Gary, Carol, Marjory, John, Peter, Joseph, Mary, Katherine, Elizabeth. B.A., Amherst Coll., 1947; M. Forestry, Yale U., 1949; M.L.S., U. R.I., 1970. Cert. librarian, Mass., 1970. In various drafting and layout positions, Boston area, 1949-52; mng. partner Dick Bros., Worcester, 1952-69; reference librarian Worcester State Coll., 1969—, coordinator energy studies, 1977—. Author energy studies curricula, Worcester State Coll., 1977—. Corporator Worcester Natural History Soc., 1971—; mem. Worcester Peace Network, 1981—; treas. Worcester Area Campus Ministry. Served with USN, 1943-46. Worcester State Coll. office Community Services mini-grantee, 1979, 80, 81; Mass. State Coll. System profl. devel. grantee, 1982. Mem. New Eng. Solar Energy Assn., Internat. Solar Energy Soc. Democrat. Subspecialties: Software engineering; Wind power. Current work: Energy conservation and management; architectural design and solar energy; teaching and research in energy studies with orientation towards consumer understanding and application. Home: 41 Iroquois St Worcester MA 01602 Office: 486 Chandler St Worcester MA 01602

DICK, HENRY JONATHAN BIDDLE, marine geologist; b. Portland, Oreg., Aug. 30, 1946; s. Hugh Lenox Hodge and Helene (Biddle) D. B.A., U. Pa., 1969; M.Phil., Yale U., 1971, Ph.D., 1976. Postdoctoral investigator Woods Hole (Mass.) Oceanographic Instn., 1975-76, asst. scientist, 1976-80, assoc. scientist, 1980—. Editor: Magma Genesis, 1977; contbr. articles to profl. jours. Pres. Pinecrest Beach Neighborhood Assn., Falmouth, Mass., 1980—; bd. dirs. Citizens for Protection of Waquoit Bay, Falmouth, 1981—; active Big Bros./Big Sisters of Cape Cod. Mem. Am. Geophys. Union, Geol. Soc. Am. Republican. Episcopalian. Subspecialties: Petrology; Sea floor spreading. Current work: Igneous petrology of the abyssal upper mantle, geology of ophiolites, petrology and tectonics of the ocean crust. Office: Woods Hole Oceanographic Instn Woods Hole MA 02543

DICK, STEVEN JAMES, astronomer; b. Evansville, Ind., Oct. 24, 1949; s. James Edward and Elizabeth Emma (Grieshaber) D.; m. Mary Theresa Milharcic, Mar. 11, 1951; 1 son, Gregory James. B.S., Ind. U., Bloomington, 1971, M.A., 1974, Ph.D., 1977. Asst. editor Adventures in Exptl. Physics, Princeton, N.J., 1976-77; assoc. editor sci. and tech. Arete Pub. Co., Princeton, N.J., 1977-79; astronomer U.S. Naval Obs., Washington, 1979—. Author: Plurality of Worlds: The Origins of the Extraterrestrial Life Debate from Democritus to Kant, 1982. Mem. Am. Astron. Soc., History of Sci. Soc., Soc. for History of Tech. Subspecialties: Optical astronomy; astrometry. Current work: Astrometry, extraterrestrial life, history of American science.

DICK, WILLIAM ALLEN, composite materials engr.; b. Belleville, Ill., June 7, 1956; s. William Allen and Ruth Anne (Racine) D.; m. Kim Mary Smith, Aug. 11, 1978. B.M.E., U. Del., 1979. Instnl. research and fin. planning U. Del., 1975-79; research assoc. I, 1979-80, research assoc. II, 1980-81, asst. dir., 1981-82, research assoc. III, asst. dir. Ctr., 1982—; cons.; guest lectr. Contbr. articles to tech. jours. Recipient K. C. Citizenship award, 1974. Mem. ASME, ASTM. Republican. Christian Scientist. Subspecialties: Composite materials; Theoretical and applied mechanics. Current work: Composite materials; processing science; damage and repair mechanisms in composite structures; mechanical behavior of materials; composites testing and design. Home: 9 Elan Hall Newark DE 19711 Office: 201 Spencer Lab Newark DE 19711

DICKE, ROBERT HENRY, educator, physicist; b. St. Louis, May 6, 1916; s. Oscar H. and Flora (Peterson) D.; m. Annie Henderson Currie, June 6, 1942; children—Nancy Jean (Mrs. John Rapoport), John Robert, James Howard. A.B., Princeton, 1939; Ph.D., U. Rochester, 1941, D.Sc. (hon.), 1981, U. Edinburgh, 1972, Ohio No. U., 1981. Microwave radar devel. Radiation Lab., Mass. Inst. Tech., 1941-46; physics faculty Princeton, 1946—, Cyrus Fogg Brackett prof. physics, 1957-75, Albert Einstein Univ. prof. sci., 1975—, chmn. physics dept., 1967-70; Mem. adv. panel for physics NSF, 1959-61; chmn. adv. com. atomic physics Nat. Bur. Standards, 1961-63; mem. com. on physics NASA, 1963-70, chmn., 1963-66; chmn. physics adv. panel Com. on Internat. Exchange of Persons (Fulbright-Hays Act), 1964-66; chmn. adv. com. on radio astronomy telescopes NSF, 1967, 69; mem. Nat. Sci. Bd., 1970-76; vis. com. Nat. Bur. Standards, 1975-79, chmn., 1979. Author: (with Montgomery, Purcell) Principles of Micro-wave Circuits, 1948, (with J.P. Wittke) An Introduction to Quantum Mechanics, 1960, The Theoretical Significance of Experimental Relativity, 1964, Gravitation and the Universe, 1970. Trustee Asso. Univs. Inc., 1980—. Recipient Nat. Medal Sci., 1970, NASA medal for exceptional sci. achievement, 1973, Cresson medal Franklin Inst., 1974. Mem. Nat. Acad. Scis. (Comstock prize 1973), Am. Philos. Soc., Am. Geophys. Union, Am. Phys. Soc., Am. Astron. Soc., Am. Acad. Arts and Scis. (Rumford medal 1967). Subspecialty: Relativity and gravitation. Home: 321 Prospect Ave Princeton NJ 08540

DICKENS, ELMER DOUGLAS, JR., research physicist; b. Charleston, W.Va., Dec. 26, 1942; s. Elmer D. and Gennevie (Duff) D.; m. Helen Marie Lively, Aug. 3, 1962; children: Tony, Mark. B.S., Morris Harvey Coll., Charleston, 1965; M.S., W.Va. U., 1967, Ph.D., 1970. Research physicist B.F.Goodrich Research and Devel. Center, Brecksville, Ohio, 1970-72, sr. research physicist, 1972-74, group leader new products, 1974-77, sect. leader new ventures, 1977-78, mgr. new ventures, 1978-81, mgr. corp. research, 1981-82, research fellow, 1983—. NASA fellow, 1967-70. Mem. Am. Phys. Soc., Combustion Inst., Sigma Xi, Sigma Pi Sigma. Subspecialties: Polymer physics; Electronic materials. Current work: Polymer combustion, fire testing, fire modeling, low temperature plasmas, conductive polymers, magnetic materials, mathematical modeling. artificial intelligence. Patentee in field. Home: 4160 Maple Dr Richfield OH 44286 Office: BF Goodrich Research and Devel Center 9921 Brecksville Rd Brecksville OH 44141

DICKENS, JUSTIN KIRK, physicist, nuclear engineer; b. Syracuse, N.Y., Nov. 2, 1931; s. Milton Clifford and Jennette Martin (Holmes) D.; m. Marcay Cosette Jordan, Dec. 21, 1957; children: Alan Russell, Leonard Raymond, Steven Kenneth, Michael Loren. A.B., U. So. Calif., 1955, Ph.D. in Physics, 1962; M.S., U. Chgo., 1956. Mem. research staff Oak Ridge Nat. Lab., 1962—. Treas. Oak Ridge High Sch. PTA, 1981-83; bd. dirs. Oak Ridge Community Playhouse, 1975-76. Served with U.S. Army, 1950-52. Mem. Am. Phys. Soc., Am. Nuclear Soc., Sigma Xi. Subspecialties: Nuclear physics; Nuclear fission. Current work: Experimental research; nuclear interaction; safety of nuclear reactors; nuclear fission; computer control of nuclear experiments; nuclear damage in fusion energy; man-to-machine interactions. Office: Oak Ridge Nat Lab PO Box X Oak Ridge TN 37830

DICKERMAN, CHARLES EDWARD, physicist; b. Carbondale, Ill., Mar. 9, 1932; s. Bert Lewis Dickerman and Bessie May (Crowder); m. Barbara Ellen Swartz, Sept. 7, 1954; children: Mary, Joel, John, James. B.A., So. Ill. U., 1951, M.A., 1952; diploma, U. London, 1953; Ph.D., State U. Iowa, 1957. Sect. mgr. Argonne (Ill.) Nat. Lab., 1968—, sr. physicist, 1979—. Chmn. Pub. Sch. Bds. Caucus, Downers Grove, Ill., 1979-80; vice chmn. Coll. DuPage Caucus, Glen Ellyn, Ill., 1981-82. Mem. Am. Phys. Soc., Am. Nuclear Soc., Sigma Xi., Phi Mu Alpha. Presbyterian. Club: R1Music (Downers Grove) (pres. 1982). Lodge: Rotary. Subspecialties: Nuclear fission; Nuclear engineering. Current work: Fission reactor safety; core behavior, energy conversion, radiological source. Home: 4802 Highland Ave Downers Grove IL 60515 Office: Argonne National Laboratory 9700 S Cass Ave Argonne IL 60439

DICKEY, JOHN MILLER, astronomer, educator; b. Wiles-Barre, Pa., July 24, 1950; s. John and Harriet Marcy (Hunt) D.; m. Robin Lee Schiff, June 4, 1978. B.S., Stanford U., 1972; M.S., Cornell U., 1975, Ph.D., 1977. Research asst. Arecibo Obs., P.R., 1975-76; teaching asst. Cornell U., 1976-77; research assoc. Five Coll. Radio Astronomy Obs., Amherst, Mass., 1977-79; astronomer Observatoire de Paris, 1979; assoc. astronomer Nat. Radio Astronomy Obs., Charlottesville, Va., 1979-82; asst. prof. U. Minn., 1982—. Contbr. articles to profl. jours. NSF fellow, 1972-75. Mem. Am. Astron. Soc., Sigma Xi. Quaker. Subspecialties: Radio and microwave astronomy; Theoretical astrophysics. Current work: Observations at radio frequencies of galactic and extra-galactic objects. Office: Dept Astronomy 116 Church St Minneapolis MN 55455

DICKEY, THOMAS EDGAR, software engineer, researcher; b. Washington, Aug. 10, 1951; s. Thomas Oliver and Elizabeth (Parks) D.; m. Helen Kay Cimbala, June 10, 1978; children: Nathan, Ruth. B.S. Engring., Widener Coll., 1973, 1973; M.S.E.E., Carnegie-Mellon U., 1974, Ph.D. in Elec. Engring, 1981. Engring. technician Naval Electronic Systems Command, Washington, summers 1971-73; research asst. Carnegie-Mellon U., 1974-76; sr. engr. Westinghouse Research and Devel. Center, Pitts., 1976-81; sr. mem. tech. staff ITT-Advanced Tech. Ctr., Shelton, Conn., 1981—. Mem. IEEE, Math. Assn. Am., Assn. Computing Machinery. Republican. Baptist. Subspecialties: Computer architecture; Artificial intelligence. Current work: Distributed computing networks, man-machine interfaces, database. Office: ITT Advanced Tech Center 1 Research Dr Shelton CT 06484

DICKINSON, WILLIAM JOSEPH, biologist, educator, researcher; b. Pasadena, Calif., June 17, 1940; s. Bernard Neil and Oleta (Scott) D.; m. Carolyn Louise Huskey, June 9, 1963; children: Jeffrey Alan, Jennifer Ann. B.A. in Zoology, U. Calif., (1963), Berkeley; Ph.D. in Biology, Johns Hopkins U., 1969. Asst. prof. Reed Coll., Portland, Oreg., 1969-72; vis. scientist dept. zoology U. B.C., 1971; asst. prof. dept. biology U. Utah, Salt Lake City, 1972-75, assoc. prof., 1975-81, prof., 1981—; vis. scientist dept. genetics U. Hawaii, Honolulu, 1978-79. Author: Gene Enzyme Systems in Drosophila, 1975; contbr. articles to sci. jours. Served to lt. (j.g.) USN, 1963-65. NSF grantee, 1973-75, 78-81; NIH grantee, 1977—. Mem. AAAS, Genetics Soc. Am., Soc. for Devel. Biology, Am. Soc. Naturalists. Subspecialties: Gene actions; Genome organization. Current work: Developmental gene regulation, evolution of gene regulation, patterns of enzyme expression during development, mechanisms of gene regulation. Home: 1408 Laird Ave Salt Lake City UT 84105 Office: Biology Dept U Utah Salt Lake City UT 84112

DICKINSON, WINIFRED BALL, biologist, educator, researcher; b. Pitts., Sept. 10, 1933; d. Breese Morse and Winifred (Brown) D. B.S., Pa. Coll. for Women, (now Chatham Coll.), 1955; M.S., U. Colo., 1957; Ph.D., U. Pitts., 1971. Lab. asst. Biophysics Research Lab., Eye and Ear Hosp., Pitts., 1952-55, Pa. Coll. for Women, 1951-55; teaching asst. U. Colo., 1955-57; research asst. U. Pitts. Sch. Medicine, 1957-61; adminstrv. asst., head dept. biology St. Paul's Sch., Walla Walla, Wash., 1961-63; instr. Point Park Coll, 1964-65, 78; teaching asst. U. Pitts., 1965-66; teaching fellow, 1965-66, 67-71, adv. to undergrads., 1967-71, instr. univ. community ednl. programs, 1971-72, asst. instr., 1971-72, instr., 1972-73; asst. prof. biology Pa. State U., Beaver Campus, Monaca, 1973-79; assoc. prof. U. Steubenville, 1979—. Contbr. articles, revs. to profl. jours. Am. Cancer Soc. grantee, 1970-71; Pa. State U. grantee, 1974-75, 77-78. Mem. Pa. Acad. Sci., Western Pa. Conservancy, Nature Conservancy, AAAS, Am. Inst. Biol. Sci., Bot. Soc. Am. Republican. Subspecialties: Cell and tissue culture; Developmental biology. Current work: Development Pteridium aquilinum gametophyte. Home: 83 Union Ave Pittsburgh PA 15205 Office: U Steubenville Franciscan Way Steubenville OH 43952

DICKMAN, ROBERT LAURENCE, astrophysicist; b. N.Y.C., May 16, 1947; s. Sidney and Eve D.; m. Albertina Catharina Otter, Sept. 18, 1975; children: Joshua, Ilana. A.B., Columbia U., 1969, M.A., 1972, Ph.D., 1976. Postdoctoral research assoc. physics dept. Rensselaer Poly. Ins., Troy, N.Y., 1975-78; mem. tech. staff Aerospace Corp., Los Angeles, 1978-80; faculty research assoc. U. Mass., Amherst, 1980—, mgr., 1980—. Contbr. articles to profl. jours. Mem. Am. Astron. Soc., Am. Phys. Soc., Sigma Xi. Subspecialty: Radio and microwave astronomy. Current work: Interstellar cloud dynamics, star formation, turbulence. Receiver development for radio astronomy. Office: Radio Astronomy GRC Univ Mass Amherst MA 01003

DICKMAN, STEVEN RICHARD, geology educator; b. Bklyn., June 24, 1950; s. Sidney and Eve D.; m. Barbara Alexander, May 16, 1981. B.A., Columbia U., 1972; M.A., U. Calif.-Berkeley, 1974, Ph.D., 1977. Asst. prof. dept. geol. scis. SUNY-Binghamton, 1977—. Recipient Van Buren prize in math. Columbia U., 1972; NSF fellow, 1973-76; SUNY Found. fellow, 1978; NASA grantee, 1981-83. Mem. Am. Geophys. Union, Math Assn. Am., Sigma Xi. Subspecialty: Geophysics. Current work: Effects of global phenomena (tides, plate motions, etc.) on earth's rotation, especially on Chandler wobble and other polar motion; time series analysis of various geophysical data. Office: Dept Geol Scis SUNY Binghamton NY 13901 Home: 408 Club House Rd Apt 5 Binghamton NY 13903

DICKSON, JAMES FRANCIS, III, asst. Surgeon Gen.; b. Boston, May 4, 1924; s. James Francis and Mary Elizabeth (Rich) D. A.B., Dartmouth Coll., 1944; M.D., Harvard U., 1947. Diplomate: Am. Bd. Surgery. Intern, resident Harvard Surg. Service, Boston City Hosp., 1947-50; thoracic surgeon Harvard Med. Sch., 1950-52; NIH spl. fellow engring. M.I.T., Cambridge, 1961-65; program dir. biomed. engring. Nat. Inst. Gen. Med. Sci., HEW, 1965-74, dep. asst. sec. for health, 1975; now asst. Surgeon Gen., Dept. Health and Human Services.; Cons. Pres.'s Commn. Tech. Automation and Econ. Progress, 1965; dir. health Pres.'s Adv. Council on Mgmt. Improvement, 1970-73. Editor books on sci., tech. and health; contbr. articles to profl. jours. Served as 1st lt. M.C. Aus, 1952-54. Fellow A.C.S.; mem. Inst. Medicine, Nat. Acad. Scis., Biomed. Engring. Soc. (pres. 1972—). Subspecialties: Cardiac surgery; Biomedical engineering. Current work: Biological control systems. Office: Asst Surgeon Gen Office of Asst Sec for Health Dept Health and Human Services 200 Independence Ave SW Washington DC 20201

DICKSON, LEROY DAVID, optical engr.; b. New Brighton, Pa., June 26, 1934; s. David LeRoy and Cleova Catherine (Ortelt) D.; m. Nola Scoggins, Aug. 15, 1954; children: Dana, Kenneth, Alan, Cole. B.E.S., Johns Hopkins U., 1960, M.S.E., 1962, Ph.D., 1968. With IBM, Research Triangle Park, N.C., 1968—, sr. engr. scanner devel., 1981—. Contbr. articles to profl. jours. Served with U.S. Army, 1954-57. Recipient Outstanding Innovation award IBM, 1981, Corp. award, 1982. Mem. Optical Soc. Am., Soc. Photo-optical Instrumentation Engrs., Laser Inst. Am. Republican. Methodist. Subspecialties: Laser research; Optical engineering. Current work: Development of laser scanning systems for reading and writing. Patentee in field. Office: IBM PO Box 12195 Research Triangle Park NC 27709

DICKSON, PAUL WESLEY, JR., physicist; b. Sharon, Pa., Sept. 14, 1931; s. Paul Wesley and Elizabeth Ella (Trevethan) D.; m. Eleanor Ann Dunning, Nov. 17, 1952; children: Gretchen Ann, Heather Elizabeth, Paul Wesley. B.S. in Metall. Engring, U. Ariz., 1954, M.S., 1954; Ph.D. in Physics, N.C. State U., 1962. With Westinghouse Electric Corp., Large, Pa., 1963—, mgr. advanced projects, 1969-72, mgr. reactor analysis and core design, Madison, Pa., 1975-79, tech. dir., Oak Ridge, 1979—; mem. adv. com. on advanced propulsion systems NASA, Washington, 1970-72; mem. adv. com. reactor physics AEC/Dept. Energy, 1974-79; mem. rev. com. applied physics Argonne (Ill.) Nat. Lab., 1978-83, chmn., 1980; mem. rev. com. engring. physics Oak Ridge Nat. Lab., 1982—. Contbr. numerous sci. articles to profl. publs. Served to capt. USAF, 1955-63. Phelps Dodge fellow, 1953-54. Mem. Am. Nuclear Soc., Am. Phys. Soc., N.Y. Acad. Scis., AIME (pres. student chpt. 1953-54), AAAS. Republican. Methodist. Subspecialties: Nuclear fission; Nuclear engineering. Current work: Nuclear breeder reactor development. Office: Westinghouse Elec Corp: PO Box W Oak Ridge TN 37801 Home: 1103 W Outer Dr Oak Ridge TN 37801

DICKSON, PHILIP F., chemical engineer, educator; b. Huron, S.D., Aug. 5, 1936. B.S., S.D. Sch. Mines and Tech., 1958; Ph.D. in Chem. Engring, U. Minn., 1962. Research engr. Esso Research & Engring. Co., 1958-59; research engr. Humble Products Research div. Humble Oil & Refining Co., 1962-63; assoc. prof. chem. engring. Colo. Sch. Mines, Golden, 1963-70, prof., 1970-72, head. chem. and petroleum engring., 1972—. Mem. Am. Inst. Chem. Engrs., Am. Soc. Engring. Edn., Assn. Asphalt Paving Tech. Subspecialties: Chemical engineering; Petroleum engineering. Office: Dept Chem and Petroleum Refining Colo Sch Mines Golden CO 80401

DICKSON, ROBERT CARL, biochemistry educator; b. Coeur d'Alene, Idaho, Apr. 22, 1943; s. William Carl and Jean (Hamilton) D. B.S., U. Redlands, 1965; Ph.D., UCLA, 1970. Cert. molecular biologist. Postdoctoral fellow Calif. Inst. Tech., Pasadena, 1970-72, UCLA, 1972-73, U. Calif.-San Diego, 1973-75; assoc. prof. U. Ky., Lexington, 1975—. Am. Cancer Soc. fellow, 1979-84; USPHS grantee, 1976—. Mem. Am. Soc. Microbiology, Am. Soc. Biol. Chemists, Sigma Xi. Subspecialties: Molecular biology; Genetics and genetic engineering (biology). Office: Dept Biochemistry Coll Medicine U Ky Lexington KY 40536

DI DONATO, ARMIDO RICHARD, mathematician, researcher; b. Pitts., June 8, 1922; s. Aronne and Delia DiD.; m. Annie Jean Breaux,

Nov. 22, 1971; children: Judy, Laurie, Thomas, Robert. B.S., Duquesne U., 1950; S.M., M.I.T., 1951; postgrad., Carnegie-Mellon U., 1964-66, Ph.D., 1972. Mathematician duPont de Nemours, Wilmington, Del., 1951-53, Melpar, Alexandria, Va., 1953-54, Naval Surface Weapons Center, Dahlgren, Va., 1954—. Subspecialties: Numerical analysis; Applied mathematics. Home: PO Box 405 Dahlgren VA 22448 Office: Naval Surface Weapons Center Dahlgren VA 22448

DIEDRICHSEN, LOREN DALE, system engineer, consultant; b. Fremont, Nebr., Feb. 17, 1936; s. Orville and LaVera A. (Hanson) D.; m. Marie F. DeSantis, Sept. 30, 1938; children: John, (dec.), David, William, Mark, Catherine. B.S. in E.E, Iowa State U., 1958; M.S., Stevens Inst. Tech., 1967. Engr. Army Communication/ADP Lab., Ft. Monmouth, N.J., 1958-67; chief systems div. Mallard project, Ft. Monmouth, 1967-71, Joint Tactical Communications Office, Red Bank., N.J., 1971-79; dir. Ctr. System Engring. and Integration U.S. Army Communications Elecgronics Commd., Ft. Monmouth, 1979—. Umpire Amateur Softball Assn., Monmouth County, N.J.; coach Little League Baseball, Eatontown, N.J. Served to 2d lt. U.S. Army, 1959. Recipient Army Decoration for Exceptional Civilian Service, 1975, Army Decoration for Meritorious Civilian Service, 1968-78. Mem. Armed Forces Communications-Electronics Assn. (Meritorious Service award 1982). Subspecialties: Systems engineering; Information systems, storage, and retrieval (computer science). Current work: Command, control, communication system engineering system design, engineering, analysis and optimization. Home: 4 Abis Pl Oakhurst NJ 07755 Office: Director CENSEI DRSEL-SEI Fort Monmouth NJ 07703

DIEHL, ANDREW KEMPER, medical educator; b. San Antonio, Sept. 14, 1946; s. Kemper Wilson and Mary Suzanne (Farnam) D.; m. Nancy Salling, Sept. 19, 1970; children: Marley Suzanne, Audrey Aline. B.A., Yale U., 1968; M.D., Harvard U., 1972; M.Sc., London Sch. Hygiene and Tropical Medicine, 1980. Diplomate: Am. Bd. Internal Medicine. Intern Harbor-UCLA Med. Ctr., Torrance, 1972-73; resident in internal medicine U. Tex. Health Sci. Ctr., San Antonio, 1975-77, instr., 1977-79, asst. prof., 1979—, chief div. gen. medicine, 1981—. Contbr. articles to med. jours. Served with USPHS, 1973-75. Milbank Meml. Fund scholar in epidemiology, 1979-84. Fellow ACP. Subspecialties: Internal medicine; Epidemiology. Current work: Research into causes and prevention of gallbladder cancer. Research in general area of clinical epidemiology. Home: 306 Canterbury Hill San Antonio TX 78209 Office: U Tex Health Sci Ctr 7703 Floyd Curl Dr San Antonio TX 78284

DIELS, JEAN-CLAUDE MARCEL, physics educator, research scientist; b. Brussels, Dec. 12, 1943; s. Albert Fern and Marquerite (Colin) D.; m. Marie-Luise Margarete Schnauck, Dec. 2, 1971; children: Nathalie Genevieve Juliette, Nicoletta Gabriele Tania, Natacha Dominique Claire. M.S., U. Brussels, 1965, Ph.D., 1973. Asst. chercheur Ecole Royale Militaire, Brussels, 1966; with Philips Research Labs., Eindhoven, the Netherlands, 1967-71; postgrad. research physicist U. Calif.-Berkeley, 1971-73; with Max Planck Inst., Gottingen, Germany, 1973-75; research scientist, adj. asst. prof. physics and elec. engring. Ctr. Laser Studies, U. So. Calif., Los Angeles, 1975—, research assoc., prof. physics and elec. engring., 1980-81; prof. physics North Tex. State U., Denton, 1981—. Served with Inf. Belgian Army, 1965-67. Mem. Am. Phys. Soc., IEEE, Optical Soc. Am., Interam. Photochem. Soc. Subspecialties: Laser Optics; Laser-induced chemistry. Current work: Interaction of laser light with matter, generation of the shortest laser light pulses, detection of frequency shifts of coherent radiation propagating through matter. Patentee in field. Home: 2117 Woodbrook Dr Denton TX 76201 Office: Center Applied Quantum Electronics Dept Physics PO Box 5368 Denton TX 76203

DIENER, EDWARD FRANCIS, psychologist; b. Glendale, Calif., July 25, 1946; s. Frank C. and Mary Alice (Ferry) D.; m. Carol I. Merk, Dec. 27, 1966; children: Marissa, Mary Beth, Robert. B.A., Calif. State U.-Fresno, 1968; Ph.D., U. Wash., 1974. Asst. prof. U. Ill., Champaign, 1974-79, assoc. prof. psychology, 1979—. Author: Ethics in Social and Behavioral Research, 1978; Contbr. articles to profl. jours. Mem. Soc. Exptl. Social Psychology, Internat. Soc. for Research on Aggression, Am. Psychol. Assn., Soc. Psychol. Study of Social Issues. Subspecialty: Social psychology. Current work: Mood; positive effect; life satisfaction. Home: 1711 Mayfair Rd Champaign IL 61821 Office: Dept Psychology U Ill 603 E Daniel St Champaign IL 61820

DIENER, THEODOR OTTO, plant pathologist; b. Zurich, Switzerland, Feb. 28, 1921; came to U.S., 1949, naturalized, 1955; s. Theodor Emanuel and Hedwig Rosa (Baumann) D.; m. Sybil Mary Fox, May 11, 1968; children by previous marriage: Theodor W., Robert A., Michael S. Dipl. Sc. Nat. E.T.H., Swiss Fed. Inst. Tech., 1946, D. Sc. Nat., 1948. Asst. Swiss Fed. Inst. Tech., Zurich, 1946-48; plant pathologist Swiss Fed. Exptl. Sta., Waedenswil, 1949-50; asst. prof. plant pathology R.I. State U., Kingston, 1950; asst. plant pathologist Wash. State U., Prosser, 1950-55, asso. plant pathologist, 1955-59; research plant pathologist Agrl. Research Service, USDA, Beltsville, Md., 1959—; lectr. univs. and research insts.; Regents' lectr. U. Calif., Riverside, 1970; Andrew D. White prof.-at-large Cornell U., 1979-81. Author: Viroids and Viroid Diseases, 1979; asso. editor: jour. Virology, 1964-66, 74-76; editor, 1967-71; mem. editorial com.: Ann. Rev. Phytopathology, 1970-74, Annales de Virologie; contbr. articles to sci. publs. Recipient Campbell award Am. Inst. Biol. Scis., 1968; Alexander von Humboldt award, 1975; Superior Service award USDA, 1969; Distinguished Service award, 1977. Fellow Am. Phytopath. Soc. (Ruth Allen award 1976), N.Y. Acad. Scis., Am. Acad. Arts and Scis.; mem. AAAS, Nat. Acad. Scis., Leopoldina, German Acad. Natural Scientists. Subspecialties: Plant pathology; Plant virology. Current work: Fundamental research and research leadership on subviral pathogens of plants and animals. Discoverer novel class of pathogens (viroids), 1971. Home: 4530 Powder Mill Rd PO Box 272 Beltsville MD 20705 Office: Plant Virology Lab Agrl Research Center USDA Beltsville MD 20705

DIENER, URBAN LOWELL, mycotoxicology educator, plant pathology consultants; b. Lima, Ohio, May 26, 1921; s. Urban Edward and Jenny Ethel (Hoverman) D.; m. Mary Jacqulyn Maund, Aug. 11, 1956. B.A., Miami U., Oxford, Ohio, 1943; M.A., Harvard U., 1945; Ph.D., N.C. State U., 1953. Indsl. mycologist Sindar Corp., N.Y.C., 1945-47; asst. plant pathologist Clemson (S.C.) U., 1947-48; grad. research asst. N.C. State U., 1948-51; asst. prof. dept. botany, plant pathology and microbiology Auburn (Ala.) U., 1952-57, assoc. prof., 1957-63, prof., 1963—. Editor: Aflatoxin, 1969; contbr. articles to profl. jours. Bd. dirs. Auburn United Fund, 1968-78; mem. exec. bd. Chattahoochee council Boy Scouts Am., 1968—, dist. fin. chmn., 1968—. Recipient Golden Peanut research award Nat. Peanut Council, 1972. Fellow AAAS; mem. Am. Phytopathol. Soc., Am. Soc. Microbiology, Ala. Acad. Sci., Am. Peanut Research Edn. Soc., Sigma Xi, Phi Kappa Phi. Republican. Methodist. Lodges: Auburn Lions (past treas.); Shriners; Men's Camellia. Subspecialties: Plant pathology; Microbiology. Current work: Aflatoxin in peanuts and corn; physiology of aspergillus flavus and A. parasiticus regarding aflatoxin production. Home: 750 Sherwood Dr Auburn AL 36830 Office: Dept Botany Auburn U Auburn AL 36449

DIENSTAG, JULES LEONARD, physician, educator; b. N.Y.C., Dec. 10, 1946; s. Baruch and Josephine D.; m. Judy Iris Gordon, Feb. 3, 1974; children: Joshua, Jonathan. A.B., Columbia U., 1968, M.D., 1972. Diplomate: Am. Bd. Internal Medicine, Nat. Bd. Med. Examiners. Intern U. Chgo., Billings Hosp, 1972-73, resident, 1973-74; research assoc. Lab. Infectious Diseases, Nat. Inst. Allergy and Infectious Diseases, NIH, 1974-76; research fellow in medicine Harvard Med. Sch., 1976-78, asst. prof. medicine, 1978-82, assoc. prof., 1982—; clin. and research fellow in medicine Mass. Gen. Hosp., Boston, 1976-78, asst. in medicine, 1979-82, asst. physician, 1982—; vis. scientist Lab. Epidemiology, Lindsley F. Kimball Research Inst., N.Y. Blood Center, 1980—. Mem. editorial bd.: Hepatology, 1980—, Jour. Clin. Microbiology, 1977—, Gastroenterology, 1981—, Infectious Disease series, Marcel Dekker Med. Div., 1981—; contbr. articles to profl. jours. Served with USPHS, 1974-76. USPHS grantee, 1978-79, 79-82. Fellow ACP; mem. Am. Soc. Microbiology, AAAS, Am. Fedn. Clin. Research, Am. Assn. Immunologists, Internat. Assn. Study of the Liver, Am. Assn. Study of Liver Diseases, Am. Gastroent. Assn., N.Y. Acad. Sci., Phi Beta Kappa. Subspecialties: Virology (medicine); Immunology (medicine). Current work: Viral hepatitis, liver immunology, medical research. Home: 4 Lincoln Rd Wayland MA 01778 Office: Gastrointestinal Unit Mass Gen Hosp Boston MA 02114

DIERKS, RICHARD ERNEST, veterinarian, ednl. adminstr.; b. Flandreau, S.D., Mar. 11, 1934; s. Martin and Lillian Ester (Benedict) D.; m. Eveline Carol Amundson, July 20, 1956; children—Jeffrey Scott, Steven Eric, Joel Richard. Student, S.D. State U., 1952-55; B.S., U. Minn., 1957, D.V.M., 1959, M.P.H., 1964, Ph.D., 1964. Supervisory microbiologist Communicable Disease Center, Atlanta, 1964-68; prof. coll. veterinary medicine Iowa State U., Ames, 1968-74; head dept. veterinary sci. Mont. State U., Bozeman, 1974-76; dean Coll. Veterinary Medicine, U. Ill., Urbana, 1976—; mem. tng. grant rev. com. Nat. Inst. Allergy and Infectious Diseases, 1973-74. Contbr. articles to profl. jours. Served with USPHS, 1964-67. Career Devel. awardee Nat. Inst. Allergy and Infectious Diseases, 1969-74. Mem. Am., Ill. veterinary medicine assns., Am. Soc. Microbiologists, Am. Coll. Veterinary Microbiologists, Am. Coll. Vet. Preventive Medicine, Am. Assn. Immunologists, Soc. Exptl. Biology and Medicine, Gamma Sigma Delta, Phi Kappa Phi, Phi Zeta. Republican. Lutheran. Club: Rotary. Subspecialties: Virology (veterinary medicine); Microbiology (veterinary medicine). Current work: Research administration; immunological response to viral antigens. Home: 2801 E Holcolm Dr Urbana IL 61801 Office: College of Veterinary Medicine University of Illinois Urbana IL 61801

DIESEM, CHARLES DAVID, vet. anatomist, educator; b. Galion, Ohio, July 5, 1921; s. John Elmer and Mary Florence (Burwell) D.; m. Janet Moore, Jan. 18, 1945; children: Mary Lynn Diesem Peoples, Nancy Sue, Robert C. D.V.M., Ohio State U., 1943, M.Sc., 1949, Ph.D., 1956. Gen. practice vet. medicine, Mt. Gilead, Ohio, 1943-44; instr. dept. vet. anatomy Ohio State U., Columbus, 1947-56, asst. prof., 1956-59, assoc. prof., 1959-61, prof., 1961—. Co-author: Anatomy and Histology of the Eye and Orbit and Orbit in Domestic Animals, 1960, The Rabbit in Eye Research, 1964, The Anatomy of Domestic Animals, 5th edit, 1975. Dep. health commr. City of Upper Arlington, Ohio, 1954-76. Served with Vet. Corps. U.S. Army, 1945-47. Mem. AVMA, Ohio Vet. Med. Assn., Am. Assn. Anatomists, Am. Assn. Vet. Anatomists, World Assn. Vet. Anatomists, Sigma Xi. Lodge: Masons. Subspecialty: Ophthalmology. Current work: Ophthalmic anatomy and peripheral nervous system. Home: 1872 Berkshire Rd Columbus OH 43221 Office: Dept Vet Anatomy Ohio State U 1900 Coffee Rd Columbus OH 43210

DIETENBERGER, MARK ANTHONY, research physicist; b. Reedsburg, Wis., Aug. 10, 1952; s. William Karl and Cleo (Rockweiler) D.; m. Joleen Ann Soper, June 31, 1979; children: Nicole, Angela, Elizabeth. B.S., U. Wis.-Milw., 1974; M.S., U. Dayton, 1978. Research physicist U. Dayton, Ohio, 1977—. Contbr. articles to profl. jours. Mem. AIAA, Am. Phys. Soc. Roman Catholic. Subspecialties: Aerospace engineering and technology; Numerical analysis. Current work: Primary general interest is in computer modeling of physical systems; current research is in aviation safety whereby numerical analysis is applied to atmospheric (winds, rain, ice, frost, fire) impacts on aerospace systems. Office: U Dayton Research Inst Applied Systems Analysis 300 College Park Ave Dayton OH 45469

DIETERICH, ROBERT ARTHUR, veterinary science educator; b. Salinas, Calif., Mar. 22, 1939; s. Louis Gunther and Ruth Mable (Alexander) D., m. Jamie Kay, Aug. 17, 1977, children. Dan, Mark. B.S., U. Calif.-Davis, 1961, D.V.M., 1963. Pvt. vet. practice, 1963-67; research assoc. U. Alaska, 1967-74, prof. vet. sci., 1975—; project leader FAO Wildlife Project, Kenya, 1974-75; judge equine competitive events. Contbr. articles to profl. jours., chpts. in books. Subspecialty: Wildlife diseases. Current work: Wildlife Diseases, reindeer husbandry. Office: Inst of Arctic Biology U Alaska Fairbanks AK 99701

DIETRICH, JOHN WILLIAM, endocrinologist; b. Syracuse, N.Y., June 28, 1946; s. Joseph F. and Elizabeth (Lawler) D.; m. Marilyn Jean Fuller, July 26, 1969; children: Tamera, Brian. B.S., LeMoyne Coll., 1968; M.S., U. Dayton, 1970; Ph.D. in Pharmacology, U. N.C., 1973. Postdoctoral fellow U. Conn. Health Center, Farmington, 1974-76; asst. prof. U. Ill. Sch. Medicine, Peoria, 1976-79; dir. endocrinology Revlon Health Care Group, Tuckahoe, N.Y., 1979—. Contbr. articles to profl. jours. Pharm. Mfrs. research grantee, 1974; NIH grantee, 1975. Mem. AAAS, Am. Soc. Bone and Mineral Research, Endocrine So., Fedn. Am. Soc. Exptl. Biology. Roman Catholic. Subspecialty: Endocrinology. Current work: Endocrinology drug discover, bone and calcium metabolism. Office: Revlon Health Care Center 1 Scarsdale Rd Tuckahoe NY 10707

DIETRICH, W. DALTON, III, neuroanatomist, educator, medical researcher; b. Richmond, Va., May 31, 1952; m. Denise N. Nelms. B.S. in Biology, Va. Poly. Inst. and State U., 1974; Ph.D. in Anatomy, Med. Coll. Va., 1979. Postdoctoral fellow dept. pharmacology Washington U., St. Louis, 1980-81; asst. prof. neurology and anatomy U. Miami (Fla.) Sch. Medicine, 1981—. Mem. Soc. Neurosci., Am. Assn. Anatomists, Electron Microscopic Soc. Am. Subspecialties: Neurobiology; Neurochemistry. Current work: Neurosciences, stroke research. Office: Dept Neurology D4-5 U Miami Sch Medicine 1501 NW 9th Ave Miami FL 33136

DIETSCHY, JOHN MAURICE, internal medicine educator, researcher; b. Alton, Ill., Sept. 23, 1932; s. John C. and Clara A. (Sahner) D.; m. Beverly A. Robertson, Apr. 18, 1959; children: John, Daniel, Michael, Karen. A.B., Washington U., St. Louis, 1954, M.D., 1958. Intern St. Joseph's Hosp., Denver, 1958-59; resident in medicine VA Hosp., Denver, 1959-61; research fellow Boston U., 1961-63, U. Tex. Southwestern Med. Sch., 1963-65, asst. prof. internal medicine, 1965-69, assoc. prof., 1969-71, prof., 1971—, dir. div. gastroenterology, 1979—. Med. research publs.; editor: Gastroenterology Monographs, 1976—, Lipid Metabolism, 1978, Textbooks of Medicine, 1978-83. Markle scholar, 1966-71; NIH grantee, 1964—; recipient Heinrich-Wieland prize, 1983. Mem. Am. Physiol. Soc. (chmn. com. 1977), Am. Gastroent. Assn. (Disting. achievement award 1978), Am. Soc. Clin. Investigation, Assn. Am. Physicians, Am. Soc. Biol. Chemists, Am. Fedn. Clin. Research (pres. So. sect. 1975), So. Soc. Clin. Investigation (pres. 1983). Roman Catholic. Subspecialties: Internal medicine; Gastroenterology. Current work: Regulation cholesterol metabolism, atherosclerosis; intestinal absorption; gastrointestinal infections. Office: Dept Internal Medicine U Tex Health Sci Ctr 5323 Harry Hines Blvd Dallas TX 75235

DIETZ, SHERL M. (SAM), botanist; b. Ames, Iowa, Nov. 29, 1927; s. Sherlock Melvin and Lorraine Annetta (Best) D.; m. Ida Dietz; children: Lorraine Dietz McConnell, David H. Student, Iowa State U., 1945-47; B.S. in Botany, Oreg. State U., 1950; Ph.D. in Plant Pathology, Wash. State U., 1963. Teaching and research asst. Wash. State U., Pullman, 1950-54; biol. technician in plant pathology U.S. Dept. Agr. Cereal Disease Lab., Pullman, 1954-57; plant pathologist Western Regional Plant Introduction Sta., U.S. Dept. Agr., Pullman, 1957-66, research leader and coordinator, 1966—; cons. U.S. AID mission to set up nat. plant germplasm system for Pakistan, 1981. Active Boy Scouts Am. Recipient cert. of merit U.S. Dept. Agr., 1979, 82. Mem. Am. Phytopathol. Soc., Soc. Economic Botany, Sigma Xi, Phi Sigma, Phi Kappa Phi. Club: Whitman County Sportsman's Assn. Office: Wash State U Western Regional Plant Introduction Sta Room 59 Johnson Hall Pullman WA 99164

DIETZ, WILLIAM BRUCE, computer scientist; b. Canonsburg, Pa., Apr. 8, 1951; s. Herman Fredrick and Martha Lucille (McCrory) D.; m. Susan Lynn Martin, Aug. 29, 1970; 1 son, Christopher Martin. B.S. in Physics and Math, Muskingum Coll., 1973; M.S. in Computer Engring, Carnegie-Mellon U., 1974. Research asst. Carnegie-Mellon U., Pitts., 1973-75, computer architecture project supr., 1977-82; cons. for Westinghouse Research, Pitts., 1975-77; sr. tech. staff Tartan Labs., Inc., Pitts., 1982—. Active United Presbyterian Ch. Mem. ACM, IEEE, Sigma Pi Sigma. Subspecialties: Computer architecture; Software engineering. Current work: Computer architecture, validation, optimizing compiler back ends, code generators. Co-designer Nebula computer architecture (current standard 32-bit U.S. Army and U.S. Air Force tactical computer). Office: 477 Melwood Ave Pittsburgh PA 15213

DI FRANCO, ROLAND BARTHOLOMEW, mathematics educator; b. N.Y.C., July 26, 1936; s. Salvatore Philip and Josephine Dorothy (Fiscella) di F.; m. Toni Lee Merrell, June 5, 1965; children: Tamar Lisa, Gianna Rebekah. B.S., Fordham U., 1958; M.S., Rutgers U., 1960; Ph.D., Ind. U., 1965. Asst. prof. Fordham U., N.Y.C., 1965-66, Swarthmore (Pa.) Coll., 1966-72; assoc. prof. math. U. Pacific, Stockton, Calif., 1972-78, prof., 1978—; vis. prof. U. Calif.-Berkeley, 1969-70, 77, Harvey Mudd Coll., Claremont, Calif., 1982. NSF fellow, 1969-70. Mem. Am. Math. Soc., Math. Assn. Am., Soc. Indsl. and Applied Math., Danforth Assocs. (regional liaison officer 1979-82), Sigma Xi. Subspecialty: Applied mathematics. Current work: Reduction of grid-orientation effects in oil reservoir simulation using generalized upstream weighting. Office: U Pacific Stockton CA 95211

DIGGS, CARTER LEE, immunologist, researcher, army officer; b. Deltaville, Va., Dec. 31, 1934; s. Harvey Lee and Jewel (Carter) D.; m. Virginia Mabry, June 5, 1956; children: Carter Lee, Diana, Daniel Christopher. B.S., Randolph-Macon Coll., 1956; M.D., Med. Coll. Va., 1960; Ph.D. (Univ. fellow), Johns Hopkins U., 1968. Intern, Med. Coll. Va., 1960-61, resident in pathology, 1961-62; commd. capt. M.C., U.S. Army, 1962, advanced through grades to col., 1976; research assoc. dept. med. zoology Walter Reed Army Inst. Research, Washington, 1962-64, sr. research assoc. dept. med. zoology, dept. dir. div. communicable diseases and immunology, 1970-73, chief dept. immunology, 1973-80, dir. div. communicable diseases and immunology, 1979—; chmn. dept. parasitology SEATO Med. Research Lab., Bangbok, Thailand, 1968-70; cons. in field. Contbr. numerous articles to profl. jours. Mem. Am. Assn. Immunologists, Am. Soc. Tropical Medicine and Hygiene, Phi Beta Kappa. Subspecialties: Infectious diseases; Microbiology (medicine). Current work: Parasite immunology; malaria, trypanosomiasis; vaccine development. Home: 11202 Landy Ct Kensington MD 20895 Office: Walter Reed Army Inst Research Div Communicable Diseases and Immunology WRAMC Washington DC 20012

DILCHER, DAVID LEONARD, paleobotany researcher, educator; b. Cedar Falls, Iowa, July 10, 1936; s. Leonard George and Hannah Eliza (Short) D.; m. Katherine Rose Swanson, Sept. 10, 1961; children: Peter Corbin, Ann Katherine. B.S., U. Minn., 1958, M.A., 1960; Ph.D., Yale U., 1964. NSF postdoctoral fellow Senckenberg Mus., Frankfurt, W.Ger., 1964-65; instr. Yale U., New Haven, 1965-66; asst. prof. paleobotany Ind. U., Bloomington, 1966-69, assoc. prof., 1969-75, prof., 1975—; Mem. Utility Service Bd., 1975-77. Author books and articles. Cullman fellow, 1964; NSF fellow, 1964-65; Guggenheim fellow, 1972-73; Amax Research grantee, 1979-80; NSF grantee, 1966-84. Fellow Linnean Soc., Ind. Acad. Sci.; mem. Bot. Soc. Am., AAAS, Geol. Soc. Am., Orgn. Tropical Biology. Democrat. Subspecialties: Evolutionary biology; Paleobiology. Current work: Evolution of flowering plants. Office: Dept Biology Ind U Bloomington IN 47405

DILL, KENNETH AUSTIN, pharmaceutical chemistry educator; b. Oklahoma City, Dec. 11, 1947; s. Austin Glenn and Margaret (Blocker) D. S.B., Mass. Inst. Tech., 1971, S.M., 1971; Ph.D., U. Calif.-San Diego, 1978. Damon Runyon-Walter Winchell fellow Stanford (Calif.) U., 1978-81; asst. prof. chemistry U. Fla., Gainesville, 1981-82; asst. prof. pharm. chemistry and pharmacy U. Calif., San Francisco, 1982—. Contbr. numerous sci. articles to profl. publs. Mem. Am. Chem. Soc., Am. Phys. Soc., Biophys. Soc., AAAS. Subspecialties: Biophysical chemistry; Polymer physics. Current work: Statistical mechanical theory of biological molecules, especially surfactants. Holder U.S. patent. Home: 665 Lancaster St Moss Beach CA 94038 Office: Univ Calif Pharm Chemistry Dept San Francisco CA 94143

DILLEY, RICHARD A., biologist, educator; b. South Haven, Mich., Jan. 12, 1936; s. Varnum M. and Marion (Dahlquist) D.; m. Janette G. Fitzsimons, Aug. 13, 1960; children: John, Thomas, David, Neil. B.S., Mich. State U., 1958, M.S., 1959; Ph.D., Purdue U., 1963. Research assoc. C.F. Kettering Lab., Yellow Springs, Ohio, 1963-64, U. Rochester, 1965; staff scientist C.F. Kettering Research Lab., Yellow Springs, Ohio, 1966-70; assoc. prof. Purdue U., West Lafayette, Ind., 1970-75, prof., 1975—; mem. peer. rev. panel NSF, U.S. Dept. Agr. Mem. editorial bd.: Jour. Biol. Chemistry. NSF, NIH, Dept. Agr. grantee; von Humboldt fellow, 1982-83. Mem. Am. Soc. Biol. Chemists, Am. Soc. Plant Physiologists, Am. Soc. Photobiology, AAAS. Roman Catholic. Subspecialties: Photosynthesis; Biochemistry (biology). Current work: Research in photosynthesis, membrane biochemistry and membrane structure. Office: Dept Bicl Scis Purdue U West Lafayette IN 47907

DILLON, DONALD JOSEPH, psychologist; b. N.Y.C., Apr. 26, 1926; s. Will F. and Mae (Hockman) D.; m. Helene R. Cooney, Sept. 3, 1955; children: Linda M., Donald J., James J., Kathleen M., Richard P. B.S., Yale U., 1949; M.A., Fordham U., 1952, Ph.D., 1955. Lic. psychologist, N.Y. Teaching fellow Fordham U., 1953-54; Research psychologist N.Y. Psychiat. Inst., N.Y.C., 1955-57, sr. research psychologist, 1957-74; instr. psychology Columbia U., N.Y.C., 1958-67; adj. asst. prof. Manhattan Coll., Riverdale, N.Y., 1963-67; research assoc. psychology Columbia U., 1967-69; assoc.

research psychologist N.Y. Psychiat. Inst., 1974—; adj. assoc. prof. Manhattan Coll., 1967—; asst. prof. Columbia U., 1969—. Chmn. religious edn. bd. St. William the Abbot Ch., Seaford, N.Y., 1973-76; committeeman Boy Scouts Am., Massapequa, 1970-71. U.S. Navy research asst., 1952-54. Mem. AAAS, Am. Psychol. Assn., Soc. Sigma Xi, Psychonomic Soc., N.Y. Acad. Sci. Republican. Roman Catholic. Subspecialties: Psychobiology; Sensory processes. Current work: Research, diagnosis and treatment anxiety disorders; research into basic psychophysical measures of pain, teaching undergrads. Office: 722 W 168th St New York NY 10032 Home: 77 Chicago Ave Massapqua NY 11758

DILMORE, ROGER H., electrical engineer; b. Rochester, N.Y., Feb. 8, 1946; m. Judy M., 1968; children: Jonathan, Gregory. B.S.E.E., Clarkson Coll. Tech., 1968; postgrad., U. Mich.-Ann Arbor, 1969, SUNY-Binghamton, 1970-82. Registered profl. engr., N.Y. Engr. N.Y. State Electric & Gas Corp., Binghamton, 1968, staff engr., 1968-74, sr. engr., 1974-77, engring. supr., 1977—. Chmn. Town of Nanticoke (N.Y.) Planning Bd., 1980-82; mem. Community Resources Devel., Broome County, N.Y., 1982-83. Mem. Am. Nuclear Soc. Subspecialties: Utility operations; Resource management. Current work: Business/government relations. Home: RD 4 95A Dunham Hill Rd Binghamton NY 13905 Office: NY State Electric & Gas Corp 4500 Vestal Pkwy Binghamton NY 13902

DIMANT, JACOB, physician, medical educator; b. Rehovot, Israel, Apr. 27, 1947; s. Symcha and Ita D.; m. Rose Bea Jearolman, Sept. 11, 1974. M.D., Hebrew U., Jerusalem, 1972. Diplomate: Am. Bd. Internal Medicine, Am. Bd. Rheumatology, Am. Bd. Quality Assurance and Utilization Rev. Physicians. Resident in medicine Maimonides Med. Ctr., Bklyn., 1972-76, asst. dir. med. edn., 1978-80, dir. rheumatology, 1978—; fellow in rheumatology SUNY-Downstate Med. Ctr., Bklyn., 1976-78, asst. prof. medicine, 1978—; med. dir. Prospect Park Nursing Home, Bklyn., 1977—; adj. prof. clin. pharmacy Bklyn. Coll. Pharmacy, 1975-76; hon. prof. Universidad Autonoma de Guadalajara, Mex., 1979-80; hon. police surgeon, N.Y.C., 1982—. Contbr. articles to profl. jours. Research fellow Arthritis Found., 1977-78; recipient research award Maimonides Med. Ctr., 1980. Fellow ACP; mem. Am. Fedn. Clin. Research, Am. Rheumatism Assn., AMA, Am. Geriatric Soc. Subspecialties: Internal medicine; Gerontology. Current work: Rheumatic disease in the elderly; aging research. Office: The Rheumatology Ctr 921 49th St Brooklyn NY 11219

DIMENT, WILLIAM HORACE, research geophysicist, consultant; b. Oswego, N.Y., Oct. 15, 1927; s. James Smith and Priscilla Rose (Faatz) D.; m. Evelyn Virginia East, Nov. 12, 1958; children: Evelyn Patricia Diment Chamberlain, James Howell, William David. A.B., Williams Coll., 1949; A.M., Harvard U., 1951, Ph.D., 1954. Registered geophysicist, Calif. Geophysicist Standard Oil Co. Calif., New Orleans, 1953-56; geophysicist, br. chief U.S. Geol. Survey, Washington, Menlo Park, Calif., Denver, 1956-65, research geophysicist, 1973-83, Golden, Colo., 1973—; prof. geology U. Rochester, N.Y., 1965-73; cons. on reactor siting U.S. AEC, Washington, 1965-69, U.S. Dept. Interior, 1965-69, cons. on radioactive waste disposal, 1965-69, 1971; mem. various panels, coms. NSF, NRC. Served with USNR, 1945-46; served to 1st lt. USAFR, 1949-58. Sr. postdoctoral fellow NSF, Yale U., 1964-65; prin. investigator NSF grants, 1966-73. Fellow AAAS (council 1961-62), Geol. Soc. Am.; mem. Soc. Exploration Geophysicists (rep. AAAS council 1961-62). Republican. Congretationalist. Subspecialties: Geophysics; Tectonics. Current work: Regional geophysics; exploration geophysics, seismicity eastern U.S.; geothermal systems and geothermal energy; physical limnology. Home: 1822 Arapahoe St Golden CO 80401 Office: US Geol Survey 1711 Illinois St Golden CO 80401

DIMICCO, JOSEPH ANTHONY, pharmacology educator, neuroscience researcher; b. New Haven, June 13, 1947; s. Joseph John and Helen (Vergoni) DiM.; m. Deborah Ann Hofman, Oct. 22, 1972 (div.); m. Susan Young, Oct. 2, 1982. B.S. in Biology, Tufts U., 1969; Ph.D. in Pharmacology, Georgetown U., 1978. Staff fellow NIMH, Bethesda, Md., 1978-80; asst. prof. pharmacology Ind. U.-Indpls., 1980—; researcher. Contbr. articles profl. jours. Served with AUS, 1970-71. Recipient Carrie E. Wolff award Ind. affiliate Am. Heart Assn., 1982. Mem. Soc. Neurosci. Subspecialties: Neuropharmacology; Pharmacology. Current work: Central nervous control of autonomic nervous function, role of neurons using inhibitory neurotransmitter GABA in brain pathways regulating autonomic cardiovascular function. Home: 4980 W 59th St Indianapolis IN 46254 Office: 635 Barnhill Dr Indianapolis IN 46202

DIMMICK, DAVID MICHAEL, nuclear engineer; b. Glenn Ridge, N.J., Dec. 10, 1951; s. Charles Henry and Adaline (Wynkoop) D. B.S. in Mech. Engring, Tufts U., 1976. Engr. research and design Fluid Dynamics Co., Cedar Knolls, N.J., 1976-79; regional mgr. Air Maze Co., Cleve., 1979-80; sr. applications engr. Hittman Nuclear & Devel. Co., Columbia, Md., 1980-82, mgr. services and planning, 1982—. Mem. Am. Nuclear Soc., Soc. Plastics Engrs., Soc. Automotive Engrs. Republican. Presbyterian. Subspecialties: Nuclear fission; Nuclear engineering. Current work: Development of safe means of processing, transporting and disposal of low-level radioactive waste. Office: Hittman Nuclear & Devel Co 9151 Rumsey Rd Columbia MD 21045

DINGLE, ALBERT NELSON, atmospheric science educator; b. Bismarck, N.D., May 22, 1916; s. Victor Stanley and Nanna Bergetha (Nelson) D.; m. Eleanor Amelia Nelson, Nov. 20, 1941; children: Karen Louise, Timothy Nelson. B.Sc., U. Minn., 1939; M.Sc., Iowa State Coll., 1940; S.M., MIT, 1945, Sc.D., 1947. Asst. prof. physics Ohio State U., 1947-54; research assoc. U. Mich., Ann Arbor, 1954-56, assoc. prof. atmospheric sci., 1956-63, prof., 1963-81, prof. emeritus, 1982—. Contbr. articles to profl. jours. Mem. Ann Arbor City Council, 1958-60. Grantee NSF, AEC, Dept. Energy, NASA, NIH, 1950-81. Fellow AAAS; mem. Am. Geophys. Union, Sigma Xi. Democrat. Lutheran. Club: Ann Arbor Golf and Outing. Subspecialties: Meteorology; Atmospheric chemistry. Current work: Cloud and precipitation processes; expert witness in litigations involving weather factors. Home: 8140 Huron River Dr Dexter MI 48130 Office: U Mich 200 Research Activities Bldg Ann Arbor MI 48109

DINGLE, RICHARD DOUGLAS HUGH, entomology educator; b. Penang, Malaysia, Nov. 4, 1936; came to U.S., 1942; s. Walter Hugh and Mildred Burns (Porter) D.; m. Geraldine Joyce Palmer, Aug. 29, 1959; children: Jennifer Leigh, Hilary Alison, Tracy Alexandra. B.A., Cornell U., 1958; M.S., U. Mich., 1959, Ph.D., 1962. Asst. prof. U. Iowa, Iowa City, 1964-67, assoc. prof., 1967-73, prof., 1973-82; vis. lectr. U. Nairobi, Kenya, 1969-70; prof. U. Calif.-Davis, 1982—. Editor: Evolution of Insect Migration and Diapause, 1978; co-editor: Insect Life History Patterns, 1981, Evolution and Genetics of Life Histories, 1982. NIH fellow, 1969-70; NSF grantee, 1964-83. Fellow AAAS; mem. Ecol. Soc. Am., Am. Soc. Naturalists, Soc. Study Evolution, Animal Behavior Soc. Democrat. Subspecialties: Genome organization; Behavioral ecology. Current work: Evolution and genetics of insect migration and life histories. Office: U Calif Dept Entomology Davis CA 95616 Home: 1204 Colby Dr Davis CA 95616

DINGLEDINE, RAYMOND JOSEPH, research scientist; b. Celina, Ohio, Dec. 17, 1948; s. Raymond J. and Katherine L. (Roettger) D.; m. Catherine L. VanDeest, Sept. 11, 1971; children: Brian, Roger. B.S. in Biochemistry, Mich. State U., 1971; Ph.D. in Pharmacology, Stanford U., 1975. Postdoctoral fellow Med. Research Council Neurochem. Pharmacology Unit, Cambridge, Eng., 1975-77, U. Oslo, Norway, 1977-78; research assoc. dept. physiology Duke U., Durham, N.C., 1978; asst. prof. dept. pharmacology U. N.C.-Chapel Hill, 1978—. Sloan Found. fellow, 1978-82. Mem. Brit. Pharmacology Soc., European Neurosci. Assn., Soc. for Neurosci., Am. Soc. Pharmacology and Exptl. Therapeutics, Internat. Narcotics Research Conf. Subspecialties: Neuropharmacology; Neurophysiology. Current work: Cellular actions of opioids in hippocampal brain slice; cellular basis for epilepsy. Home: 111 Virginia Dr Chapel Hill NC 27514 Office: Univ NC Dept Pharmacology Bldg 231H Chapel Hill NC 27514

DINNEEN, GERALD PAUL, elec. engr., corp. exec., former govt. ofcl., educator; b. Elmhurst, N.Y., Oct. 23, 1924; s. Walter James and Anna Constance (Costello) D.; m. Mary Purington, June 28, 1947; children—Patricia, Barbara (Mrs. Timothy J. Sehr), Michael. B.S., Queens Coll., 1947; M.S., U. Wis., 1948, Ph.D., 1952. Mathematician Goodyear Aircraft Corp., Akron, Ohio, 1951-53; with Mass. Inst. Tech., Lexington, Mass., 1953-77; dir. Lincoln Lab., 1970-77, prof. elec. engring., 1971-77; asst. sec. def. for communications, command, control and intelligence, 1977-81; corp. v.p. sci. and tech. Honeywell, Inc., Mpls., 1981—; Mem. sci. adv. bd. USAF, 1960-64, 70-77; vice chmn. 1971-75, chmn., 1975-77; cons. Def. Dept., NASA, USN, USAF. Served with AC AUS, 1943-46. Recipient Exceptional Civilian Service award USAF, 1966; Disting. Public Service award Dept. Def., 1981. Mem. Nat. Acad. Engring., Am. Math. Soc., Math. Assn. Am., Sigma Xi. Club: Cosmos (Washington). Subspecialty: Electrical engineering. Home: 6400 Barrie Rd Edina MN 55435 Office: Honeywell Inc Honeywell Plaza Minneapolis MN 55408

DIORIO, MARK LEWIS, mech. engr., metallurgist; b. Norwalk, Conn., Feb. 18, 1957; s. Joseph P. and Susan DiO.; m. Constance Pratt, June 27, 1981. B.S. in Mech. Engring. and Materials Engring, U. Conn., 1979; postgrad., St. Louis U. Sch. Bus., 1980-81. Market devel. engr. Olin Fineweld Tube, Olin Brass div. Olin Corp., East Alton, Ill., 1979-81; market devel. engr. Somers Thinstrip, Olin Brass, Waterbury, Conn., 1981; sales engr. Olin Brass, San Francisco, 1982—. Mem. ASME, Micro Electronic Packaging and Processing Engrs. Subspecialties: Alloys; Mechanical engineering. Current work: Marketing and developing copper alloys for semiconductor applications. Home: 10194 Parwood Dr Cupertino CA 95014 Office: 20430 Town Center Ln 5-I Cupertino CA 95014

DIPAOLO, JOSEPH AMEDEO, geneticist, lavoratory director; b. Bridgeport, Conn., June 13, 1924; s. John Anthony and Nancy (Montagano) DiP.; m. Arleta Mae Schreib, June 14, 1952; children: Nancy, John. A.B., Wesleyan U., 1948; M.S., Western Res. U., 1949; Ph.D., Northwestern U., 1951. Instr. genetics and bacteriology dept. biology Loyola U., Chgo., 1951-53; instr. clin. and exptl. pathology dept. pathology Northwestern U. Sch. Med., Evanston, Ill., 1953-55; sr. cancer research scientist Roswell Park Meml. Inst., Buffalo, N.Y., 1955-63; research pharmacologist, cell biologist biology br. div. cancer cause and prevention Nat. Cancer Inst., Bethesda, Md., 1963-76, chief lab. biology, div. cancer cause and prevention, 1976—; assoc. profl. lectr. anatomy George Washington U., Washington, 1973-76; chmn. U.S.-USSR Mammalian Somatic Cell Genetics Related to Neoplasia, 1973-76, U.S.-Germany Cancer Program Area for Environ. Carcinogenesis, 1979—. Assoc. editor: Jour. Nat. Cancer Inst, 1968-71, Cancer Research, 1970-78, Teratogenesis, Carcinogenesis and Mutagenesis, 1982—; editor: Chemical Carcinogenesis, 1974. Served with USN, 1943-46. Fellow AAAS, N.Y. Acad. Sci.; mem. Am. Assn. Cancer Research (dir. 1983-87), Am. Soc. Human Genetics, Am. Soc. Exptl. Pathology, Genetics Soc. Am., Teratology Soc., Hamster Soc., Tissue Culture Assn., Sigma Xi. Roman Catholic. Subspecialties: Cancer research (medicine); Cell biology. Current work: Modulation of neoplasia, DNA metabolism, cell surface changes, cytogenetics, in vitro transformation. Office: Bldg 37 Room 2A-19 NIH-Nat Cancer Inst 9000 Rockville Pike Bethesda MD 20205

DI PASQUALE, GENE, pharmacologist, researcher; b. N.Y.C., July 17, 1932; s. Emidio and Maria (De Gennaro) Di P.; m. Anita Famiglietti, Sept. 7, 1962; children: Lora, Dean. B.S., Iona Coll., 1954; M.S., L.I. U., 1960; Ph.D., N.Y.U., 1970. From asst. scientist to assoc. dir. pharmacodynamics Warner Lambert Research Inst., Morris Plains, N.J., 1957-77; sect. mgr. immunopharmacology I.C.I., Wilmington, Del., 1977—. Contbr. articles to sci. jours. Served with M.C. U.S. Army, 1954-56. Recipient Founders Day award N.Y.U., 1970. Mem. Soc. Study of Reprodn., Am. Physiol. Soc., Am. Soc. Pharmacology and Exptl. Therapeutics, Endocrine Soc., N.Y. Acad. Scis., AAAS, Sigma Xi, Phi Sigma. Subspecialties: Pharmacology; Biochemistry (medicine). Current work: Anti-arthritic research. Co-discoverer Benisone (Flurobate) and Isoxicam (Maxicam). Office: Murphy and Concord Rd Wilmington DE 19897

DIRKS, LESLIE CHANT, government official; b. New Ulm, Minn., Mar. 7, 1936; s. Emereld Francis and Eva (Gay) D.; m. Eleanor G. McPeake, Feb. 10, 1959; children: Anthony, Jason, Elizabeth. B.S. in Physics, MIT, 1958, Oxford (Eng.) U., 1960. Instr. physics Phillips Acad., Andover, Mass., 1960-61; with CIA, 1961—, dep. dir. sci. and tech., 1976—. Recipient Disting. Intelligence medal CIA, 1977, Nat. Security medal, 1978; ann. award IEEE, 1980. Mem. Nat. Acad. Engring. Subspecialties: Microelectronics; Aerospace engineering and technology. Home: 45 Hancock St Lexington MA 02173 Office: Raytheon Co Lexington MA 02173

DIRKSEN, THOMAS REED, II, dental educator, biochemist, dentist; b. Pekin, Ill., Nov. 5, 1931; s. Thomas Reed and Mildred Roslyn (Neville) D.; m. Jean Kathryn Twietmeyer, Dec. 17, 1955; children: Thomas R., Peter T., John S., James C., Robert S., Kathryn A. B.S., Bradley U., 1953; D.D.S., U. Ill.-Chgo., 1957; M.S., U. Rochester, 1960, Ph.D., 1967. Lic. tchr., Ga.; lic. dental bds. Ill., N.Y. Assoc. prof. Sch. Dentistry, Med. Coll. Ga., Augusta, 1967-70, prof., 1970—, assoc. dean, 1977—; mem. Nat. Adv. Dental Research Council, Nat. Inst. Dental Research, Washington, 1982—. Biomed. Research Support Subcom., HEW, 1979-80. Co-editor: Boucher's Clinical Dental Terminology, 1982; contbr. to books. Bd. dirs. Dirksen Ctr., Pekin, 1979—, Bd. Edn. Adv. Com., Augusta, 1978—; bd. dirs. Richmond County Library, Augusta, 1974-79, chmn., 1977-79. Served to capt. USAF, 1960-62. Research grantee Nat. Inst. Arthritis and Metabolic Diseases, 1968, 71, Juvenile Diabetes Found., 1976, Nat. Inst. Dental Research, 1981. Fellow Am. Coll. Dentists; mem. Internat. Assn. Dental Research, Am. Assn. Dental Research (chmn. nat. affairs com. 1982—), Am. Assn. Dental Schs. (sect. chmn. 1974-75, 82-83), ADA (dental and dental hygiene test const. com. 1978-82). Subspecialties: Oral biology; Biochemistry (medicine). Current work: Lipids of calcified issues, cariology, lipid metabolism of oral structures.

DI SALVO, NICHOLAS ARMAND, dental educator, orthodontist; b. N.Y.C., Nov. 2, 1920; s. Frank and Mary (Ruberto) DiS; m. Pauline Rose Pluta, June 2, 1945; children: Allan, Donald. B.S., CCNY, 1942; D.D.S., Columbia U., 1945, Ph.D. in Physiology, 1952, cert. in orthodontics, 1957. Diplomate: Am. Bd. Orthodontics. Fellow Inst. Dental Research, Columbia U., 1950-52; instr. in physiology Coll. Physicians and Surgeons, Columbia U., 1948-51, asst. prof. physiology, 1952-57, assoc. prof., 1957-58, prof. dentistry, 1958—, dir. orthodontics, 1957—; attending dentist Presbyn. Hosp., N.Y.C., 1975—; cons. N.Y. State Dept. Health, 1970—, VA, N.Y.C., 1975—, Project/HOPE/Egypt, Alexandria and Cairo, 1976, Nat. Def. Med. Center, Taipei, Taiwan, 1982. Contbg. editor book chpts.; contbr. articles to profl. jours. Pres. Hartsdale-Fels Civic Assn., 1960-66. Served to lt. USNR, 1945-50. Recipient Disting. Service award Orthodontic Alumni Soc. Columbia U., 1973; fellow 8th Inst. Advanced Edn. in Dental Research. Mem. Am. Assn. Orthodontists (del. 1970-76), Northeastern Soc. Orthodontists (pres. 1974-75), Angle Soc. of Orthodontists (pres. 1977-79), Internat. Soc. Craniofacial Biology (pres. 1965-66). Republican. Roman Catholic. Subspecialty: Orthodontics. Current work: Growth and development of occlusion. Office: Columbia U Dental Sch 630 W 168th St New York NY 10032

DISHMAN, RODNEY KING, exercise and sport psychologist, psychology educator; b. Springfield, Mo., Feb. 4, 1951; s. Willard King and Virginia Lanette (Potter) D.; m. Sharon Emily Alter, Aug. 17, 1974; children: Jessica E., Amanda Corinne. B.S., Southwest Mo. State U., 1973; M.S., U. Wis., 1975, Ph.D., 1978. Grad. asst. U. Wis.-Madison, 1973-77, research asst., 1976; vis. lectr. N. Tex. State U., 1977-78; assoc. prof. Southwest Mo. State U., 1978—; cons. research psychologist Inst. Aerobics Research, Dallas, 1977-81. Author: Essentials of Fitness, 1980; Contbr. articles to profl. jours. Recipient A. J. McDonald award S.W. Mo. State U., 1973. Fellow Am. Coll. Sports Medicine, AAHPER and Dance (chmn. research div. Mo. chpt. 1981); mem. Am. Psychol. Assn. Methodist. Subspecialties: Execise and sport psychology; Sports medicine. Current work: Mental health; medical compliance psychogenic aids in sports. Office: Southwest Mo State U 901 S National St Springfield MO 65804

DISRAELI, DONALD JAY, ecology educator, coastal resource consultant; b. San Diego, Jan. 15, 1953; s. Richard Israel and Joyce Harriet (Greenbaum) D.; m. Sally Jackson Harper, Dec. 28, 1974 (div. Aug. 1977); m. Randee S. Recht, July 26, 1983. B.S. magna cum laude in Biology and Ecology, Western Wash. U., 1975, M.S. in Biology, 1977; Ph.D. in Botany, U. Mass., Amherst, 1982. Teaching asst., research asst. Huxley Coll. Environ. Scis., Western Wash. U., 1974-77; teaching assoc., teaching assoc., research asst. dept. botany U. Mass., Amherst, 1977-81; vis. asst. prof. depts. environ. studies and geography U. Calif.-Santa Barbara, 1981—; cons. U.S. Army C.E., Vicksburg Environ. Lab., Nat. Park. Cape Cod Nat. Sea Shore. Contbr. articles on ecology to profl. jours. Mem. AAAS, Internat. Soc. Biometeorology, Ecol. Soc. Am., Commn. on Coastal Environments. Subspecialties: Ecology; Micrometeorology. Current work: Coastal plant ecology, modeling coastal ecosystems, formation of vegetated dunes, sediment (sand) transport. Office: Dept Geography U Calif Santa Barbara CA 93106

DISTEFANO, MICHAEL KELLY, JR., psychologist, research coordinator, consultant; b. Roseland, La., Nov. 27, 1929; s. Michael Kelly and Marie (Easley) D.; m. Laura Bea Harvey, June 2, 1957; children: Laura Lynne, Michael Emmette. B.S., Tulane U., 1951; M.A., La. State U., 1957, Ph.D., 1967. Prisoner classification offer La. State Penitentiary, Angola, 1955-57, dir. classification, 1959-62; psychologist Pinecrest State Sch., Pineville, La., 1957-59; research and vocat. psychologist Central La. State Hosp., Pineville, 1962-69, research coordinator, personnel cons., 1969—; cons. personnel, tng.; mem. adj. faculty La. Coll., 1962-81; vocat. cons. Contbr. articles to profl. jours. Served in U.S. Amy, 1951-53; Korea. Mem. Am. Psychol. Assn., La. Psycol. Assn., Southeastern Psychol. Assn. Subspecialty: Industrial and organizational psychology. Current work: Research interests include: personnel selection, valididty and test fairness, performance evaluation, and turnover; program evaluation and related research. Home: 803 Shell Rd Pineville LA 71360 Office: Central La State Hosp PO Box 31 Pineville LA 71360

DISTLER, JACK JOUNIOR, biochemist; b. Pontiac, Mich., Dec. 7, 1928; s. John and Julia (Diedrich) D. B.S., Mich. State U., 1952; M.S., U. Mich., 1954, Ph.D., 1964. Teaching asst. dept. botany U. Mich.-Ann Arbor, 1954-55; research asst. Rackham Arthritis Research, 1956-59, instr. dept. biol. chemistry, 1966-71, research assoc. dept. internal medicine, 1964-74, asst. research scientist, 1974-77, assoc. research scientist, 1977—. Arthritis Found. postdoctoral fellow, 1968-71. Mem. Am. Soc. Biol. Chemistry, AAAS, Sigma Xi. Republican. Lutheran. Subspecialty: Biochemistry (medicine). Current work: Biochemistry of complex carbohydrates. Home: 575 Scio Church Rd Ann Arbor MI 48103 Office: U Mich 4633 Kresge Med Research I Ann Arbor MI 48109

DITTO, FRANK HASELWOOD, applied mathematician, systems engineer; b. LaCrosse, Wis., Nov. 28, 1929; s. Weir Hays and Anna Marie (Haselwood) D.; m. Annemarie Elisabeth Rose, Oct. 29, 1956; children: Niels Detlev, Quinn Everett. B.S., Roosevelt U., Chgo., 1955; M.S., Northeastern U., 1959. Staff engr. M.I.T. Lincoln Lab., Lexington, 1955-59; mathematician FAA Bur. Research and Devel., Atlantic City, 1959-61; engr., scientist fed. systems div. IBM, Bermuda, 1961-62, Bethesda, Md., 1962, Houston, 1963-67, Riverdale, Md., 1967-71, Gaithersburg, Md., 1971—. Served with USAF, 1950-54. Mem. Math. Assn. Am., Ops. Research Soc. Am., Meteorol. Assn. Am., Mensa (treas. 1978-79). Subspecialties: Applied mathematics; Astronautics. Current work: Numerical analysis on micro-processors especially on programmable calculators. Home: 5505 Norbeck Rd Rockville MD 20853 Office: IBM Federal Systems Div 18100 Frederick Pike Gaithersburg MD 20760

DIWAN, JOYCE JOHNSON, biology educator; b. Bklyn., Dec. 25, 1940; d. John Henry and Lillian Freida (Russ) Johnson; m. Romesh Kumar Diwan, Oct. 25, 1970. A.B., Mt. Holyoke Coll., 1962; Ph.D., U. Ill., 1967. Postdoctoral fellow U. Pa., 1966-69; asst. prof. Rensselaer Poly. Inst., 1969-75, assoc. prof., 1975—; vis. fellow U. Warwick, Eng., 1976-77. Contbr. to profl. jours. USPHS fellow, 1966-69; NSF grantee, 1970-72; USPHS grantee, 1974-76, 77-81, 83—. Mem. Am. Assn. Biol. Chemists, Am. Soc. Cell Biology, Biophys. Soc., AAAS, AAUP, Assn. Women in Sci., N.Y. Acad. Sci. Subspecialties: Cell biology; Membrane biology. Current work: Bioenergetics, mitochondrial ion transport and metabolism. Home: 6 Bolivar Ave Troy NY 12180 Office: Dept Biology Rensselaer Polytechnic Inst Troy NY 12181

DIX, ROLLIN CUMMING, mechanical engineering educator; b. N.Y.C., Feb. 8, 1936. B.S., Purdue U., 1957, M.S., 1958, Ph.D. in Mech. Engring, 1963. Sr. engr. Bendix Mishawaka Div., 1962-64; asst. prof. dept. mech. engring. Ill. Inst. Tech., Chgo., 1964-68, assoc. prof., 1964-80, prof., 1980—, assoc. dean computer sci., 1981—. Mem. Am. Soc. Engring. Educators, ASME. Subspecialties: Mechanical engineering; Theoretical computer science. Office: MMAE Dept Ill Inst Tech Chicago IL 60616

DIXEN, JEAN MARIE, psychologist; b. Owatonna, Minn., May 31, 1954; d. Jens Alfred and Mary Ann (Johnson) D. B.A., U. Minn., 1975; M.A., Mankato State U., 1977; Ph.D., U. Ga., 1971. Lic. marriage, family and child counselor, Calif. Clin. psychology intern Palo Alto VA Med. Ctr., Calif., 1979-80; research asst. U. Ga., Athens, 1980-81; practicum supr., 1980-81, marriage and family counselor in pvt. practice, Palo Alto, Calif., 1982—; postdoctoral scholar Stanford U., 1981—; research cons. Gender Dysphoria Program, Palo Alto, 1980—. Contbr. articles to profl. jours. Recipient Sci. award Bausch &

Lomb, 1972; U. Ga. fellow, 1978-79, 80-81. Mem. Am. Psychol. Assn., AAAS, Calif. Assn. Marriage and Family Therapists, Assn. for Advancement Psychology, Mensa. Democrat. Lutheran. Subspecialties: Psychophysiology; Behavioral psychology. Current work: Effects of estrogen replacement therapy on sexual function and depression in postmenopausal women; the relative effects of age and menopause on sexuality in women; evaluation of surgical sex reassignment for gender dysphoria. Home: 145 Carmel St Apt 7 San Francisco CA 94117 Office: Dept Physiolog Stanford Univ Stanford CA 94305

DIXIT, BALWANT NARAYAN, pharmacologist, academic adminstrator; b. Kerawde, Maharashtra State, India, Jan. 7, 1933; came to U.S., 1962, naturalized, 1969; s. Narayan V. and Janakibai N. (Gokhale) D.; m., Dec. 26, 1969; children: Sunil, Sanjay. B.S. in Biology and Chemistry, Fergusson Coll., Poona, India, 1954, 1955; M.S. with honors in Biochemistry, Poona U., 1956, M.S. Univ., Baroda, India, 1961; Ph.D. (Internat. Union Physiol. Scis. fellow), U. Pitts., 1965. Research asst. Faculty of Medicine, M.S. U., Baroda, India, 1956-59, sr. research fellow, 1960-62; asst. prof. U. Pitts., 1965-68, assoc. prof., 1968-73, asst. chmn., 1970-73, acting chmn., 1973-74, prof. pharmacology, 1974—, chmn. dept., 1974—, assoc. dean, 1976—, acting dean, 1976-78. Recipient Disting. Alumnus award U. Pitts. Sch. Pharmacy, 1981. Mem. N.Y. Acad. Scis., AAAS, Am. Assn. Colls. Pharmacy, Soc. Neurosci., Am. Soc. Pharmacology and Exptl. Therapeutics, Sigma Xi, Rho Chi. Subspecialty: Pharmacology. Current work: Biochemical pharmacology, autonomic pharmacology, and drug metabolism.

DIXON, EARL, JR., veterinary educator, writer, consultant; b. Halifax County, Va., Oct. 20, 1937; s. Earl and Sallie (Davis) D.; m. Anna Boykin, June 28, 1983; children by previous marriage: Felicia, Andre, Stacy. B.S., St. Augustine Coll., N.C., 1959; M.S., Atlanta U., 1961; Ph.D., Howard U., 1971. Assoc. prof. Tuskegee (Ala.) Inst., 1971-81, prof., 1981—; cons. NIH, 1976—, NSF, 1980—. Grantee NIH, NSF. Mem. AAAS, Am. Physiol. Soc. Democrat. Baptist. Subspecialties: Animal physiology; Biophysical chemistry. Current work: Red blood cell metabolism. Home: 16 Oslin Dr Tuskegee AL 36083 Office: Tuskegee Inst Tuskegee Institute AL 36088

DIXON, FRANK JAMES, educator, med. scientist; b. St. Paul, Mar. 9, 1920; s. Frank James and Rose Augusta (Kuhfeld) D.; m. Marion Edwards, Mar. 14, 1946; children—Janet Wynne, Frank, Michael. B.S., U. Minn., 1941, M.B., 1943, M.D., 1944. Diplomate: Am. Bd. Pathology. Intern U.S. Naval Hosp., Great Lakes, Ill., 1943-44; research asst. dept. pathology Harvard, 1946-48; instr. dept. pathology Washington U., 1948-50, asst. prof., 1950-51; prof., chmn. dept. pathology U. Pitts. Med. Sch., 1951-60; chmn. dept. exptl. pathology Scripps Clinic and Research Found., La Jolla, Calif., 1961-74, chmn. biomed. research depts., 1970-74, dir. research inst., 1974—; research asso. dept. biology U. Calif. at San Diego, 1961-64, prof. in residence in dept. biology, 1965-68, adj. prof. dept. pathology, 1968—; sci. adviser NIH, Nat. Found., Helen Hay Whitney Found., St. Jude's Med. Center; mem. Christ Hosp. Inst., Cin.; mem. expert adv. panel on immunology WHO; sci. adv. bd. Nat. Kidney Found.; Pahlavi lectr. Ministry of Sci. and Higher Tech., Iran, 1976. Co-editor: Advances in Immunology; Editorial bd.: Excerpta Medica; Contbr. articles to profl. jours. Served with M.C. USNR. Recipient Theobald Smith award, 1952; Parke-Davis award exptl. pathology, 1957; Distinguished Achievement award Modern Medicine, 1961; Martin E. Rehfuss award in internal medicine, 1966; Von Pirquet medal Ann. Forum on Allergy, 1967; Bunim medal Am. Rheumatism Assn., 1968; Gairdner Found. Internat. award, 1969; Mayo Soley award Western Soc. Clin. Research, 1969; Albert Lasker Basic Med. Research award, 1975; Dickson prize, 1975; Homer Smith award N.Y. Heart Assn., 1976; Rous-Whipple award, 1979. Mem. Nat. Acad. Scis., N.Y. Acad. Scis. Western Assn. Physicians, Western Soc. Clin. Research, Soc. Exptl. Biology and Medicine, Transplantation Soc., AAAS, Am. Soc. Clin. Investigation, Am. Acad. Allergists, Interurban Path. Soc., Harvey Soc. (lectr. 1962), Am. Soc. Exptl. Pathology (pres. 1966), Am. Assn. Immunologists (pres. 1972), Am. Assn. for Cancer Research, Assn. Am. Physicians, Am. Acad. Arts and Scis., Sigma Xi (Nu Sigma Nu), Alpha Omega Alpha. Subspecialties: Immunology (medicine); Immunogenetics. Current work: Genetics influencing autoimmunity. Home: 2355 Avenida de La Playa La Jolla CA 92037 Office: 10666 N Torrey Pines Rd La Jolla CA 92037

DIXON, GORDON HENRY, biochemist; b. Durban, South Africa, Mar. 25, 1930; s. Walter James and Ruth (Nightingale) D.; m. Sylvia W. Gillen, Nov. 20, 1954; children: Frances Anne, Walter Timothy, Christopher James, Robin Jonathan. B.A. with honors, U. Cambridge, Eng., 1951; Ph.D., U. Toronto, 1956. Research asso. U. Wash., 1956-58; research asso. U. Oxford, Eng., 1958-59; asst. prof. biochemistry U. Toronto, 1959-61, asso. prof., 1961-63; prof. U. B.C., 1963-72; prof., chmn. dept. biochemistry U. Sussex, Eng., 1972-74; prof. med. biochemistry U. Calgary, Alta., Can., 1974—, chmn., 1983—. Contbr. articles to profl. jours. Recipient Steacie prize, 1966. Fellow Royal Soc. London, Royal Soc. Can. (Flavelle medal 1980); mem. Am. Soc. Biol. Chemists, Can. Biochemistry Soc. (pres. 1982-83, Ayerst award 1966). Subspecialties: Molecular biology; Genetics and genetic engineering (biology). Current work: Mechanism of differential gene expression; organization and expression of sperm-specific genes. Home: 3424 Underwood Pl NW Calgary AB Canada T2N 4G7 Office: Dept Med Biochemistry Health Scis Centre 3330 Hospital Dr NW Univ of Calgary Calgary AB Canada T2N 4N1

DIXON, HELEN ROBERTA, geologist; b. Belvidere, Ill., Aug. 13, 1927; d. Elmer Lauritz and Helen Amanda (Johnson) D.; m.; children: Dalvin, Catherine. B.A., Carleton Coll., Northfield, Minn., 1949; M.A., U. Calif.-Berkeley, 1956; Ph.D., Harvard U., 1968. Geologist U.S. Geol. Survey, Denver, 1954—, Boston, 1960-65; guest instr. San Diego State Coll., 1969. Contbr. articles to profl. jours. Fellow Geol. Soc. Am., Am. Mineral Soc.; mem. Am. Geophys. Union, AAAS, Sigma Xi. Democrat. Subspecialties: Petrology; Tectonics. Current work: Petrology and structure of high grade metamorphic terranes; specially New England and Wyoming. Home: 30111 Rainbow Hills Golden CO 80401 Office: US Geol Survey Denver Fed Center Denver CO 80225

DIXON, JACK E., biochemistry educator; b. Nashville, June 16, 1943; s. Margaret and Jesse D. D.; m. Claudia Kent, July 25, 1981. B.A., UCLA, 1966; Ph.D., U. Calif.-Santa Barbara, 1971. Teaching asst. U. Calif.-Santa Barbara, 1967-68, research asst., 1968-71; asst. prof. Purdue U., West Lafayette, Ind., 1973-78, assoc. prof., 1978-82, prof. biochemistry, 1982—; adj. assoc. prof. Ind. U., Bloomington, 1978—. NSF postdoctoral research fellow, 1971-73; recipient outstanding counselor Purdue U., 1975; travel award Am. Soc. Biol. Chemistry, 1976-81; career devel. award USPHS, 1976-81. Mem. Am. Chem. Soc., AAAS, Am. Assn. Biol. Chemists. Subspecialties: Biochemistry (biology); Molecular biology. Current work: Molecular biology of peptide hormones. Office: Purdue U West Lafayette IN 47907 Home: 3743 Capilano Dr West Lafayette IN 47906

DIXON, PAUL NICHOLS, educational psychologist, educator, evaluation consultant; b. St. Louis, Aug. 17, 1944; s. Roy Nichols Dixon and Dorothea Eloise (Matthews) Rolls; m. Carla Kay Pinson, July 19, 1980. B.A. in Biology, U. Tex.-Austin, 1969, Ph.D., 1973. Asst. prof. U. Tex.-Austin, 1973-74; personnel research scientist Lackland AFB (Tex.) Human Resources Lab., 1974-75; asst. prof. ednl. psychology Tex. Tech. U., 1975-78, assoc. prof., 1978-82, prof., 1982—, chmn. dept., 1982—; cons. in field. Author: (with Jerry Willis and Lamont Johnson) Computers, Teaching and Learning, 1982; contbr. articles, monographs to sci. jours. Recipient Teaching Excellence award Grad. Students Assn. Tex. Tech. U., 1978, Research award Dads Assn. Tex. Tech. U., 1979. Mem. Am. Psychol. Assn., Am. Ednl. Reseasrch Assn., N.Am. Soc. Adlerian Psychologist, Internat. Council Psychologists. Subspecialties: Social psychology; Microcomputers in education. Current work: Research on relationships between several personality variables and learning under a variety of conditions; studies involve computer-managed instruction, humor-induced arousal, locus of control and the effects of abortion. Home: 2606 24th St Lubbock TX 79410 Office: Ednl Psychology Tex Tech U Box 4560 Lubbock TX 79409

DIXON, ROBERT L., pharmacologist, toxicologist; b. Sacramento, Feb. 9, 1936; s. Wilbur Harold and Frances M. (Schafer) D.; m. Marilyn Veva Roth, June 8, 1958; children—Wendy C., Diane F., David R. B.S. in Pharmacy, Idaho State U., 1958; M.S. in Pharmacology, U. Iowa, 1961, Ph.D., 1963. Asst. prof. dept. pharmacology U. Wash., Seattle, 1965-69; sr. investigator Nat. Cancer Inst., Bethesda, Md., 1963-65, chief toxicology lab., 1969-72; chief lab. reproductive and devel. toxicology Nat. Inst. Environ. Health Scis., Research Triangle Park, N.C., 1972—, asst. to dir. for internat. programs, 1979-80; sr. policy analyst Office of Sci. and Tech. Policy, Exec. Office of Pres., Washington, 1977-78. Mem. editorial bd.: Environ. Health Perspectives, 1972—, Fundamental and Applied Toxicology, 1982—, Toxicology and Applied Pharmacology, 1973—, Jour. Toxicology and Environ. Health, 1978—, others; contbr. articles to encys. and profl. jours. Served with U.S. Army, 1959. Recipient Dir.'s award NIH, 1977. Mem. AAAS, Am. Assn. for Cancer Research, Am. Pub. Health Assn., Am. Soc. for Pharmacology and Exptl. Therapeutics, Am. Soc. Andrology, Am. Chem. Soc. (dir. chem. health and safety), Internat. Soc. Study of Xenobiotics, Internat. Union Pharmacology (sect. toxicology), Mt. Desert Island Biol. Lab. Soc., Soc. for Exptl. Biology and Medicine, Soc. for Occupational and Environ. Health, Soc. for Risk Analysis, Soc. Study of Reprodn., Soc. Ecotoxocology and Environ. Safety, Soc. Toxicology (pres. 1982-83, Achievement award 1972), Western Pharmacology Soc., Sigma Xi, Rho Chi. Lutheran. Subspecialties: Toxicology (medicine); Pharmacology. Current work: Reproductive toxicology, developmental toxicology; extrapolation of laboratory data to man; risk analysis. Home: 6208 Winthrop Dr Raleigh NC 127612 Office: PO Box 12233 Research Triangle Park NC 27709

DIXON, WILFRID JOSEPH, statistics educator; b. Portland, Oreg.; m. Glorya Duffy, June 25, 1983; children—Janet Dixon Elashoff, Kathleen Dixon Nebent. B.A., Oreg. State Coll., 1938; M.A., U. Wis., 1939; Ph.D., Princeton, 1944. Asst. prof. math. U. Okla., 1942-44, 45-46; mem. joint Army-Navy Target Group, Washington and, Guam, 1944-45; asso. prof., then prof. math. U. Oreg., 1946-55; prof. preventive medicine UCLA, 1955-67, prof., 1967—, chmn. dept. biomath., 1967-74; pres. BMDP Statis. Software, Inc., 1981—; math. stat. VA Brentwood. Author: (with F.J. Massey) Introduction to Statistical Analysis, 4th edit., 1982; also articles.; Asso. editor: Biometrics, 1955-65, Annals of Math. Statistics, 1955-58. Cons. NIH, 1960—, NRC, 1948—, NSF, 1968—, Calif. Dept. Mental Hygiene and Public Health, 1963—. Fellow AAAS, Royal Statis. Soc.; mem. Inst. Math. Statistics, Internat. Statis. Inst., Am. Statis. Assn. (v.p. 1969-70, 78-81). Subspecialties: Mathematical software; Statistics. Current work: Statistical software with applications to environmental and medical research on analyzing incomplete data. Home: 1909 Pelham Ave Apt 308 Los Angeles CA 90025 Office: Univ Calif Los Angeles CA 90024

DIZER, JOHN THOMAS, JR., indsl. engr., educator, coll. dean, cons.; b. Norwood, Mass., Nov. 7, 1921; s. John Thomas and Eunice Haven (Homer) D.; m. Marie Leerkamp, Dec. 25, 1947; children: John Thomas III, Jane E., William D., Ann E., Mary L. B.S., Northeastern U., 1943; M.S. in Indsl. Engring, Purdue U., 1947, Ph.D., 1969. Registered profl. engr., N.Y. State; cert. mfg. engr. Standards engr. E. I. du Pont de Nemours & Co., Inc., East Chicago, Ill., 1947-50; prodn. engr., supr. Cummins Engine Co., Columbus, Ind., 1952-59; mem. faculty Mohawk Valley Community Coll., Utica, 1959—, head mech. engring. tech. dept., 1968-82, dean tech. and bus., 1982—; NSF cons., India, 1969. Author: Tom Swift & Co, 1982; contbr. numerous articles on engring. edn. to profl. jours.; also writer juvenile lit. Bd. dirs. Oneida Hist. Soc.; dist. advancement chmn. Boy Scouts Am. Served to lt. USN, 1944-46, 50-52; PTO; Korea. Recipient Excellence in Adminstrn. award Mohawk Valley Community Coll., 1982. Mem. Am. Soc. for Engring. Edn., ASME (past chmn. Mohawk Valley chpt., Outstanding Engr. award 1971, Centennial medal 1981), Soc. Mfg. Engrs. (past chmn. Mohawk Valley chpt., Outstanding Engr. award 1973), Inst. Indsl. Engrs. (past pres. Mohawk Valley chpt.), Mohawk Valley Engrs. Exec. Council (past chmn., Outstanding Engr. award 1978), N.Y. State Engring. Tech. Assn. (pres.), Tau Beta Pi. Mem. Ch. of Christ. Lodge: Masons. Subspecialties: Industrial engineering; Mechanical engineering. Current work: Electrochemical machining; physics of metal removal. Home: 10332 Ridgecrest Rd Utica NY 13502 Office: 1101 Sherman Dr Utica NY 13501

DJERASSI, CARL, educator, chemist; b. Vienna, Austria, Oct. 29, 1923; s. Samuel and Alice (Friedmann) D.; m. Norma Lundholm (div. 1976); children: Dale, Pamela (dec.). A.B. summa cum laude, Kenyon Coll., 1942, D.Sc. (hon.), 1958; Ph.D., U. Wis., 1945; D.Sc. (hon.), Nat. U. Mex., 1953, Fed. U., Rio de Janeiro, 1969, Worcester Poly. Inst., 1972, Wayne State U., 1974, Columbia, 1975, Uppsala U., 1977, Coe Coll., 1978, U. Geneva, 1978. Research chemist Ciba Pharm. Products, Inc., Summit, N.J., 1942-43, 45-49; asso. dir. research Syntex, Mexico City, 1949-52, research v.p., 1957-60; v.p. Syntex Labs., Palo Alto, Calif., 1960-62, Syntex Research, 1962-68, pres., 1968-72, Zoecon Corp., 1968—; Prof. chemistry Wayne State U., 1952-59, Stanford, 1959—; dir. Cetus Corp., Ridge Vineyards, Catalytica, Inc., Teknowledge, Inc.; Andrews lectr. U. New South Wales, Australia; Debye lectr. Cornell U.; Reynaud lectr. Mich. State U.; Venable lectr. U. N.C.; Edgar Fahs Smith Meml. lectr. U. Pa.; O.H. Smith lectr. Okla. State U.; Stieglitz lectr. U. Chgo.; Bachman lectr. U. Mich.; Mack lectr. Ohio State U.; Dreyfus lectr. Dartmouth; Fuson lectr U. Nev.; Dreyfus Disting. scholar Duke U.; Gregory Pincus Meml. lectr. Harvard U.; Baker lectr. U. Calif. (Santa Barbara); Osborne lectr. Rockefeller U.; Purves lectr. McGill U.; Redmen lectr. McMaster U.; ann. chemistry lectr. Royal Swedish Acad. Engring.; Scheele lectr. Swedish Pharm. Soc. Mem. editorial bd.: Jour. Organic Chemistry, 1955-59; Editorial bd.: Tetrahedron, 1958—, Steroids, 1963—, Proc. of Nat. Acad. Scis, 1964-70, Jour. Am. Chem. Soc, 1966-75, Organic Mass Spectrometry, 1968—; Author 7 books.; Contbr. numerous articles to profl. jours. Recipient Intrasci. Research Found. award, 1969; Freedman Patent award Am. Inst. Chemists, 1970; Chem. Pioneer award, 1973; Nat. Medal Sci., 1973; Perkin medal, 1975; Wolf prize in chemistry, 1978; John and Samuel Bard award in Sci. and Medicine, 1983; named to Nat. Inventors Hall of Fame, 1978. Mem. Nat. Acad. Scis., Am. Chem. Soc. (award pure chemistry 1958, Baekeland medal 1959, Fritzsche award 1960, award for creative invention 1973, Centenary lectr. 1964), Swiss Chem. Soc., Royal Soc. Chemistry (hon. fellow), Am. Acad. Arts and Scis., German Acad. (Leopoldina), Royal Swedish Acad. Scis. (fgn.), Am. Acad. Pharm. Scis. (hon.), Brazilian Acad. Scis. (fgn.), Mexican Acad. Sci. Investigation, Bulgarian Acad. Scis. (fgn.), Phi Beta Kappa, Sigma Xi, Phi Lambda Upsilon (hon.). Subspecialties: Organic chemistry; Mass spectrometry. Current work: Chemistry of natural products (antibiotics, alkaloids, steroids and terpenoids), medicinal chemistry (oral contraceptives, antiinflamatory agents), applications of physical measurements (optical rotatory dispersion, circular dichroism and mass spectrometry) to organic chemical problems. Office: Dept Chemistry Stanford U Stanford CA 94305

DJURIC, DUSAN, meteorology educator; b. Novi Sad, Yugoslavia, Jan. 17, 1930; came to U.S., 1965; s. Ljubomir and Vasilija (Vukasinovic) D.; m. Jelena Milojkovic, Sept. 24, 1955; children: Zora, Mara. B.S., U. Belgrade, Yugoslavia, 1953, Dr. Sci, 1960. Teaching asst. U. Belgrade, 1954-60, asst. prof., 1960-65; research asst. U. Stockholm, 1955-57; postdoctoral fellow Nat. Ctr. Atmospheric Research, Boulder, Colo., 1965-66; asst. prof. meteorlogy Tex. A&M U., 1966-68, assoc. prof., 1968-82, prof., 1982—. Mem. Am. Meteorol. Soc., Royal Meteorol. Soc., German Meteorol. Soc., Yugoslavian Meteorol. Soc. Subspecialties: Synoptic meteorology; Meteorology. Current work: Low-level jet, description and numerical modeling. Home: 1018 Holt St College Station TX 77840 Office: Dept Meteorology Tex A&M U College Station TX 77843

DLHOPOLSKY, JOSEPH GERALD, psychology educator; b. Bronx, Mar. 4, 1950; s. Joseph Jaraslav and Sophie (Kist) D.; m. Patrice Sweeney, June 2, 1973; children: Heather, Gregory. B.A., St. John's U., 1972; Ph.D., SUNY-Stony Brook, 1978. Asst. prof. dept. psychology St. John's U., S.I., 1978—; cons. Life Sci. Assoc., Bayport, N.Y., 1980-82; cons. software design. Contbr. articles to profl. jours.; author: microcomputer software Grade File System, 1982, Extra-Sensory Perception, 1981, ICON: Visual Sensory Store, 1981, Hemispheric Information Processing, 1982. Mem. N.Y. Acad. Sci., AAAS, Eastern Psychol. Assn., Nat. Space Inst., Danforth Assocs. Subspecialties: Software engineering; Space psychology. Current work: Designing microcomputer software for cognitive research; entering area: Research on space environment as perceptual distortion and human adjustment to same. Home: 27 Wilson St Station NY 11776 Office: Saint Johns Univ 300 Howard Ave Staten Island NY 10301

DO, HIEN DUC, chemist, microbiologist, toxicologist; b. Saigon, Viet Nam, June 30, 1943; s. Luan Duc and Ngo Mau-Thi (Vu) D.; m. Thu Ngoc-Thi Nguyen, Oct. 10, 1974; children: Steven Q.D., Cardine M.A.D., David M.D. M. Biology, Faculty of Pharmacy, Saigon, 1969; Ph.D. in Microbiology, U. Scis., Saigon, 1972. Dir. quality assurance VANCO Pharm. Labs., Saigon, 1970-74; assoc. prof., head pharmacology lab. Minh-Duc Med. Sch., Saigon, 1971-75; supr. quality control Diagnostic Products Corp., Los Angeles, 1976-80, sr. prodn. chemist supr., 1980—. Served to capt. Vietnamese Armed Forces, 1970. Mem. Am. Soc. Quality Control. Republican. Roman Catholic. Subspecialties: Toxicology (medicine); Pharmacology. Current work: Production of 57 Cobalt Vitamin B12 from cultures of Streptomyces species. Office: 5700 W 96th St Los Angeles CA 90045

DOANE, WILLIAM MCKEE, biomaterials research administrator, researcher; b. Covington, Ind., Sept. 26, 1930; s. Earle Edward and Mildred Rowena (McKee) D.; m. Joan Marie, June 6, 1952; children: Diane, Steven, Robert, Karen. B.S., Purdue U., 1952, M.S., 1960, Ph.D., 1962. Cert. tchr., Ind. Exec. trainee Gen. Motors, 1952; tchr. Fountain County (Ind.) Public Schs., 1954-55; grad. asst. Purdue U., 1955-62; with No. Regional Research Center, Dept. Agr., Peoria, Ill., 1962—, research leader, 1970-79; chief Biomaterials Conversion Lab., 1980—; lectr. Bradley U., 1965-80. Contbr. numerous articles, chpts. to profl. publs. Served with U.S. Army, 1952-54. Recipient Indsl. Research IR-100 award Dept. Agr., 1975, 78, Disting. Service award, 1976, Pollution Abatement award, 1978, Superior Service award, 1979; Don Wood award Am. Electroplaters Soc., 1978. Mem. Am. Chem. Soc., AAAS, Controlled Release Soc., Weed Sci. Soc. Subspecialties: Biomaterials; Polymers. Current work: Chemistry and biochemistry of plant materials; research leadership on structure-property realtinships of biomaterials—plant polymers and composites. Patentee in field. Office: 181 Regional Research Center 1815 N University Peoria IL 61604

DOANE, WINIFRED WALSH, zoologist, geneticist, educator, researcher, cons.; b. Bronx, N.Y., Jan. 7, 1929; d. Harold Vandervoort and Helen Harper (Loucks) Walsh; m. Charles Chesley Doane, July 5, 1952; 1 son, Timothy Price. B.A. magna cum laude, Hunter Coll., CUNY, 1950; M.S., U. Wis., 1952; Ph.D. (NSF fellow), Yale U., 1960. E. Seringhaus scholar Woods Hole Marine Biol. Lab., 1950; teaching asst. U. Wis., Madison, 1950-51, research asst., 1951-53; asst. prof. Millsaps Coll., Jackson, Miss., 1954-55; lab. asst. Yale U., New Haven, 1956-58, NIH postdoctoral research trainee in genetics, 1960-62, faculty research assoc., 1962-75, lectr., 1965-75, assoc. prof. biology, 1975-77; prof. zoology Ariz. State U., Tempe, 1977—; prin. investigator research grants NSF, 1965-75, NIH, 1973—, cons. genetics study sect. div. research grants, 1972, 74-77, 79; cons. genetic basis of disease rev. com. Nat. Inst. Gen. Med. Scis., NIH, 1980—; cons. Alan T. Watermand award com. NSF, 1979-81, chair com., 1981. Assoc. editor: Devel. Genetics, 1979-81; contbr. numerous articles to sci. jours. and books. Recipient Hall of Fame award Hunter Coll. Alumni Assn., 1972; NSF grantee, 1969-74; NIH grantee, 1973—. Fellow AAAS; mem. Am. Inst. Biol. Scis., Am. Soc. Cell Biology, Am. Soc. Naturalists, Am. Soc. Zoologist, Genetics Soc. Am., Internat. Soc. Devel. Biology, Soc. for Devel. Biology (sec. 1976-79), Phi Beta Kappa. Democrat. Episcopalian. Subspecialties: Gene actions; Developmental biology. Current work: Developmental, biochemical and molecular genetics of Drosophila; genetic regulatory mechanisms in cellular differentiation using A-amylase gene-enzyme system as a model; current work uses recombinant DNA technology.

DOBBINS, JAMES GREGORY HALL, computer engineer, educator; b. Ashland, Ky., June 30, 1943; s. James Edward and Opal Jeanette (Hall) D.; m. Susan Jeanne Culver, Dec. 19, 1971; children: Julia Christine, Stephen Gregory. B.A., U. Ky., 1965, M.A., 1966, Ph.D., 1969. Asst. prof. Marshall U., Huntington, W.Va., 1969-70; asst. prof. Wheaton (Ill.) Coll., 1970-74, assoc. prof., 1974-76; prof. Mt. Vernon (Ohio) Nazarene Coll., 1976-81; part-time prof. engring. Ohio State U., Columbus, 1978-81; sr. prin. engr. NCR Corp., West Columbia, S.C., 1981—. Contbr. articles to profl. jours. Recipient various research grants. Mem. Am. Math. Soc., Assn. for Computing Machinery, Phi Beta Kappa. Republican. Presbyterian. Subspecialties: Applied mathematics; Statistics. Current work: Fault tolerant computers, highly reliable computers. statistical quality control, semigroups. Home: 743 Trafalgar Columbia SC 29210 Office: NCR Corp 3325 Platt Springs Rd West Columbia SC 29210

DOBBS, GREGORY MELVILLE, chemist, computer scientist, educator; b. Teaneck, N.J., Aug. 19, 1947; s. Melville George and Madeline Veronica (Schlesler) D.; m. Mary Elizabeth Connor, Jan. 15, 1977; children: Katherine Michelle, John Gregory. A.B., Dartmouth Coll., 1969; M.A., Princeton U., 1972, Ph.D. in Chemistry, 1975. Systems programmer Kiewit Computation Center, Hanover, N.H., 1966-69; resident research fellow, central research dept. E.I. DuPont de Nemours & Co., Wilmington, Del., 1969; NSF trainee Princeton U.,

1969-70, asst. in instrn., 1971-74; postdoctoral research assoc. dept. chemistry M.I.T., 1974-76; sr. research scientist, chem. physics United Techs. Research Center, East Hartford, Conn., 1976—; adj. lectr. dept. computer and info. sci. Hartford (Conn.) Grad. Center, 1978—. Contbr. articles to profl. jours. Recipient John G. Kemeny prize in computing Dartmouth Coll., 1969, E.B. Hartshorn medal in chemistry, 1969. Mem. Am. Chem. Soc., Am. Phys. Soc., Combustion Inst., Phi Beta Kappa. Current work: Laser diagnostic spectroscopy; chemical lasers; laser-induced chemistry; systems programming; laboratory data acquisition. Home: 103 Farmstead Ln Glastonbury CT 06033 Office: Mail Stop 90 United Techs Research Center Silver Ln East Hartford CT 06108

DOBELBOWER, RALPH RIDDAL, JR., physician; b. Bellefonte, Pa., Mar. 23, 1940; m. Deborah Downing Lyon, Aug. 29, 1963; children: Ralph Riddal, Michael Christian, Barbara Dee. B.S. in Physics, Pa. State U., 1962, A.B. in Liberal Arts, 1963; M.D., Jefferson Med. Coll., 1967; Ph.D. in Radiation Biology, Thomas Jefferson U., 1975. Diplomate: Am. Bd. Radiology. Intern Balboa Naval Hosp., San Diego, 1967-68; resident in radiation therapy and nuclear medicine Thomas Jefferson U. Hosp., Phila., 1971-74, instr., 1974-75, asst. prof., 1975-79, assoc. prof., 1979-80; asst. to dir. Am. Coll. Radiology, Phila., 1974-79; asst. attending physician Bryn Mawr (Pa.) Hosp., 1975-79; assoc. prof. radiology, dir. radiation oncology Med. Coll. Ohio, Toledo, 1980—. Contbr. chpts. to books, articles to profl. jours. Served to lt. comdr. U.S. Navy, 1964-70. Mem. Radiol. Soc. N.Am., Am. Soc. Therapeutic Radiologists, Am. Radium Soc., Am. Assn. Physicists in Medicine, Am. Coll. Radiology (com. radiotherapy equipment), Toledo Acad. Medicine, Ohio State Med. Assn., Ohio Radiol. Soc., Radiation Therapy Oncology Group, Sigma Xi. Subspecialties: Oncology. Current work: Hyperthermia; radiation therapy for cancer of the pancreas; intraoperative radiation therapy. Home: 2908 River Rd Maumee OH 43537 Office: Med Coll Ohio CS 10008 Toledo OH 43699

DOBKIN, DAVID PAUL, computer science educator; b. Pitts., Feb. 29, 1948; s. Ben G. and Sylvia June (Swartz) D.; m. Kathy Kram, Sept. 6, 1970 (div. Dec. 1976); m. Suzanne Gespass, April 17, 1983. S.B., M.I.T., 1970; M.S., Harvard U., 1971, Ph.D., 1973. Asst. prof. Yale U., 1973-78; assoc. prof. U. Ariz., 1978-81; prof. computer sci. Princeton U., 1981—. Editor: Foundations of Secure Computation, 1978; contbr. articles profl. jours. Mem. Assn. Computing Machinery, Soc. Indsl. and Applied Math., Sigma Xi. Subspecialties: Algorithms; Graphics, image processing, and pattern recognition. Home: 463 Prospect Ave Princeton NJ 08540 Office: EECS Dept Princeton U Princeton NJ 08544

DOBROVOLNY, JERRY STANLEY, engring. educator; b. Chgo., Nov. 2, 1922; s. Stanley and Marie (Barone) D.; m. Joan Gretchen Baker, June 14, 1947; children—James Lawrence, Janet Lee. B.S. in Mech. Engring, U. Ill., 1943, M.S., 1947. Registered profl. engr. Faculty U. Ill., Urbana, 1945—, asso. prof., 1957—, prof., head dept. gen. engring., 1959; Geophys. research engr. Ill. Geol. Survey, summers 1949-52; design and traffic survey engr. Ill. Div. Hwys., summers 1948, 53, 54; cons. soil mechanics, 1955—; Mem. Ill. Adv. Council on Vocational Edn., 1969-72, Nat. Adv. Council on Vocational Edn., 1970-73. Author: (with others) Basic Drawing for Engineering Technology, (with R.P. Hoelscher and C.H. Springer) Graphics for Engineers, 1968. Past pres. Champaign County Young Republican Club; mem. Champaign County Rep. Central Com. Served with C.E. AUS, 1942-44. Fellow A.A.A.S.; mem. Am. Legion, 40 and 8, Soc. for History and Tech., Ill. Acad. Sci., Am. Soc. Engring. Edn. (Arthur Williston award 1971), Am. Soc. C.E., Am. Tech. Edn. Assn. (trustee 1964-67, 69-74, pres. 1967-68), Nat. Soc. Profl. Engrs., Ill. Soc. Profl. Engrs. (pres. Champaign County chpt. 1964-65, state v.p. 1971-73, pres. 1974-75), Champaign County Soc. Profl. Engrs., Newcomen Soc. N.Am., Sigma Xi, Scabbard and Blade, Sigma Iota Epsilon, Tau Nu Tau. Subspecialty: Civil engineering. Home: 1104 S Prospect Ave Champaign IL 61820 Office: Coll Engring U Ill Urbana IL 61801

DOBSON, ALAN, veterinary physiology educator, physiologist; b. London, Dec. 20, 1928; U.S., 1962; s. Albert Percy and Dorothy Blanche D.; m. Marjorie Jean Masson, Mar. 29, 1954; children: Ian, Janet, Graham, Barry. B.A. in Biochemistry with honors, Cambridge (Eng.) U., 1952, M.A., 1970, Sc.D. in Physiology, 1982; Ph.D. in Physiology of Nutrition, Aberdeen (Scotland) U., 1956. Exhibitioner Corpus Christi Coll., Cambridge U., 1949-52; sci. officer Rowett Research Inst., Aberdeen, Scotland, 1952-57, sr. sci. officer, 1957-64, prin. sci. officer, 1964; vis. prof. N.Y. State Vet. Coll., Cornell U., 1961-62, assoc. prof. vet. physiology, 1964-70, prof., 1970—; Wellcome fellow Sch. Vet. Medicine, Cambridge U., 1970-71; vis. worker Physiol. Lab., Cambridge, 1977-78, 79, 80, 82. Contbr. articles to profl. jours. Served to cpl. RAF, 1947-49. Mem. Biochem. Soc. (U.K.), Physiol. Soc. (U.K.), Am. Physiol. Soc., Sigma Xi, Phi Zeta. Subspecialties: Physiology (biology); Physiology (medicine). Current work: Research in control and mechanism of absorption from and blood flow to gastrointestinal tract with special reference to ruminant stomach; physiology and digestion of ruminant. Home: 21 Etna Ln Etna NY 13062 Office: Dept Physiology NY State Coll Vet Medicine Cornell U Ithaca NY 14853

DOBSON, WAYNE LAWRENCE, nuclear fuels engineer; b. Wilkes Barre, Pa., Sept. 21, 1954; s. John W. and Dorothy (Meade) D.; m. Susan S. Smith, Oct. 11, 1975; children: Holley, Corrie. B.S. in Engring. Physics, Cornell U., 1976, M.Nuclear Engring., 1977. Registered profl. engr., Pa. Nuclear engr. Am. Electric Power Co., N.Y.C., 1977-81; sr. nuclear engr. Gilbert/Commonwealth Cos., Reading, Pa., 1981—. Mem. Am. Nuclear Soc. (mem. working group com. design criteria for dry spent fuel storage facilities). Democrat. Subspecialties: Nuclear fission; Nuclear engineering. Current work: Spent nuclear fuel storage and disposal; development of spent fuel rod consolidation and dry spent fuel storage. Office: Gilbert Commonwealth PO Box 1498 Reading PA 19603

DOCHERTY, JOHN JOSEPH, microbiologist, educator; b. Youngstown, Ohio, Dec. 5, 1941; s. John Henry and Viola Jean (Sovak) D.; m. Pamela Ann, Aug. 21, 1965; children: Patricia, Susan. B.A., Youngstown U., 1964; M.S., Miami U., Oxford, Ohio, 1966; Ph.D., U. Ariz., 1970. Postdoctoral fellow Med. Sch., Pa. State U., 1970-72, asst. prof. microbiology, 1972-76, asso. prof., 1976—. Contbr. articles to profl. jours. Mem. Am. Soc. Microbiology, AAAS, Sigma Xi, Phi Sigma. Subspecialties: Virology (biology); Microbiology. Current work: Herpes simplex virus genital infections and relation to cancer induction. Home: 852 Webster Dr State College PA 16801 Office: Pa State U 310 S Frear Bldg University Park PA 16802

DODD, RICHARD ARTHUR, engineering educator; b. Eng., Feb. 11, 1922. B.S., U. London, 1944, M.S., 1947, D.Sc. in Metallurgy, 1974; Ph.D., U. Birmingham, Eng., 1950. Research metallurgist Rolls Royce Ltd., Eng., 1944-47; sr. lectr. U. Witwatersrand, S.Africa, 1950-54; research metallurgist Dept. Mines, Ottawa, 1954-56; asst. prof. metall. engring. U. Pa., Phila., 1956; prof. metall. engring. U. Wis.-Madison, 1956, dept. chmn., 1974—. Fellow Royal Inst. Chemists; mem. AIME, Am. Soc. Metals, Brit. Iron and Steel Inst., Brit.Inst. Metals. Subspecialty: Metallurgical engineering. Office: Dept Mining and Metals U Wis Coll Engring Madison WI 53706

DODSON, CHARLES LEON, JR., chemistry educator; b. Knoxville, Tenn., Mar. 15, 1935; s. Charles Leon and Margaret Glen (Berry) D.; m. Vernell Laura Woodard, Sept. 6, 1958; children: Alyssa, Bronwyn. B.S., Emory and Henry Coll., 1957; M.S., U. Tenn., 1962, Ph.D., 1963. Postdoctoral fellow U. Birmingham, Eng., 1963-64, Nat. Research Council, Ottawa, Ont., Can., 1964-66; mem. faculty U. Ala., Huntsville, 1966-81, assoc. prof. chemistry, 1969-81; applications specialist Beckman Instruments, Inc., 1981—; vis. prof. Oxford U., 1972-73. Contbr. articles to profl. jours. Mem. Am. Chem. Soc., Am. Phys. Soc., Blue Key, Sigma Pi Sigma. Subspecialties: Inorganic chemistry; Physical chemistry. Home: 1403 Appalachee Dr Huntsville AL 35801 Office: PO Box 1247 Huntsville AL 35807

DOEHRMAN, STEVEN R(ALPH), research clinical psychologist, consultant; b. Ft. Wayne, Ind., July 12, 1942; s. Ralph C. and Virginia Rita (Drury) D.; m. Margery Jean Adelson, May 5, 1968 (div. Apr. 1981); children: Eric, David. A.B., U. Mich., 1965, Ph.D., 1971. Supervising clin. psychologist U. Mich., 1974-80, lectr. dept. psychology, 1974-81, research coordinator psychol. clinic, 1976-81; project dir. Inst. Social Research, Ann Arbor, Mich., 1977-81; clin. psychologist, research dir. Orchard Hills Psychiat. Center, Farmington Hilsl, Mich., 1980—; lectr. Univ. Extension Service, Grand Rapids, Mich., 1971; speaker for civic, bus. groups, Ann Arbor area, 1979—, cons. in field. Author: (with J.R. French) Stress, social support and adjustment, 1982. NIMH postdoctoral fellow, 1973-75; Center for Clin. Study Personality grantee, 1977-80. Mem. Am. Psychol. Assn., Soc. Psychotherapy Research, Assn. Advancement Psychology. Clubs: Ski, Parents Without Partners (speaker), Ann Arbor Track (Ann Arbor). Subspecialties: Social psychology; Health Psychology. Current work: Social support as a buffer against deleterious effects of environmental stress upon individual strain; prevention and treatment for anorexia nervosa and bulimia. Home: 555 E William Ann Arbor MI 48104 Office: Orchard Hills Psychiat Center 23800 Orchard Lake Rd Farmington Hills MI 48024

DOERNER, ROBERT CARL, reactor physicist; b. St. Cloud, Minn., Sept. 26, 1926; s. Carl A. and Marcella (Krieger) D.; m. Elizabeth Dalton, Oct. 25, 1954; children: Katherine, Mary, Joanne, David, James. B.S., St. Johns U., Collegeville, Minn., 1949; Ph.D., St. Louis U., 1955. Research assoc. Argonne Nat. Lab., Ill., 1955-56, asst. physicist, 1956-59, assoc. physicist, 1959—. Chmn. Gov.'s Planning Council Developmental Disabilities, 1981-83; bd. dirs. Ill. Advocacy Authority, 1977—. Served with USN, 1944-46; PTO, ETO. Mem. Am. Nuclear Soc. (vice chmn. sect. 1969-70, chmn. 1970-71, Meritorious award 1972). Subspecialties: Nuclear engineering; Nuclear physics. Current work: Water reactor safety analysis, code development and implimentation. Fast-reactor physics experiments. Home: 615 Knollwood Wheaton IL 60187 Office: Argonne Nat Lab 9600 Cass Ave Argonne IL 60439

DOERR, ROBERT DOUGLAS, psychologist; b. Burlington, Vt., Apr. 9, 1944; s. Robert Joseph and Betty Jane (Whitney) Stubbings D. B.A., Rollins Coll., 1966; M.A., San Francisco State U., 1971; Ph.D., Saybrook Inst., 1978. Cert. Biofeedback Cert. Inst. Am. Clinician behavioral medicine, dir. Alameda (Calif.) Biofeedback Center, 1978—; prof. psychology and communication arts Columbia Coll., 1978—; prof. sci. fiction Chabot Coll., Hayward, Calif., 1971—; mem. adj. faculty Internat. Coll., 1981. Editor: Saybrook Review and Humanistic Psychology Inst. Review, 1978—; author: Canto Libre, 1979, The Peace Corps Experience: A Dialogal Analysis, 1981; 7 books of poetry. Mem. Kensington Symphony Orch., 1979—. Fellow Am. Psychol. Assn.; mem. Biofeedback Soc. Am., Assn. Humanistic Psychology, Am. Fedn. Tchrs., Alameda C. of C. Taoist. Subspecialties: Biofeedback; Physiological psychology. Current work: behavioral medicine, the use of biofeedback to aid in self-regulation of vascular and muscular disorders; the phenomenon of mind-body resistance in the above context. Home: 1517 B Saint Charles St Alameda CA 94501

DOGGETT, LEROY ELSWORTH, astronomer, editor; b. Waterloo, Iowa, Oct. 22, 1941; s. Albert Elsworth and Lura (Thompson) D.; m. Rachel Harrington, Aug. 7, 1965. B.S. in Astronomy, U. Mich., 1964, M.A., Georgetown U., 1970; Ph.D. in Engring. Mechanics, N.C. State U., 1981. Astronomer U.S. Naval Obs., Nautical Almanac Office, Washington, 1965—; adj. prof. Union Grad. Sch., Cin.; cons. Editor: Archaeoastronomy; Contbr. articles to profl. jours. Chmn. Friends of Music of Smithsonian Instn. Mem. Am. Astron. Soc. (div. dynamical astronomy, hist. astronomy div.), Inst. of Navigation. Subspecialties: Celestial Mechanics; Numerical analysis. Current work: Development of planetary theories for analysis of high precision observations from space craft; representation of astronomical data in most efficient forms for computer applications; archaeoastronomy, history of astronomy. Developer Almanac for Computers, 1977. Office: Nautical Almanac Office US Naval Obs Washington DC 20390

DOHERTY, LOWELL RALPH, astronomy educator; b. San Diego, Mar. 12, 1930. B.A., UCLA, 1952; M.S., U. Mich., 1954, Ph.D. in Astronomy, 1962. Lectr. astronomy, research fellow Harvard U., Cambridge, Mass., 1957-63; asst. to assoc. prof. astronomy U. Wis.-Madison, 1963-76, prof., 1976—; Chmn. Washburn Obs. Mem. Am. Astron. Soc., Internat. Astron. Union, AAAS, Sigma Xi. Subspecialty: Optical astronomy. Address: 3202 Knollwood Way Madison WI 53713

DOHERTY, MARK FITZGERALD, computer systems analyst; b. Norfolk, Va., Dec. 8, 1953; s. Robert Emmett and Mary Elizabeth (Fitzgerald) D.; m. Linda Jean Noeske, Sept. 6, 1980; children: Jonathan Alexander, Anna Elise. Student, Chaminade U. Honolulu, 1972-74; B.A., Cath. U. Am., 1976; postgrad., U. Md., 1982—. Programmer Son-Chief Electrics, Winstead, Conn., 1974-75; sr. mem. tech. staff ConTel Info. Systems, Bethesda, Md., 1976-82. State chmn. Libertarian party Md., Bethesda, Md., 1979. Mem. Assn. Computing Machinery, Phi Beta Kappa. Libertarian. Subspecialties: Distributed systems and networks; Graphics, image processing, and pattern recognition. Current work: Autonomous navigation, spatial representations of objects. Home: 7324 Willow Ave Takoma Park MD 20912

DOI, KUNIO, med. physicist; b. Tokyo, Japan, Sept. 28, 1939; s. Umekiti and Mitiko (Nakamura) D.; m. Akiko Doi, Feb. 15, 1962; children: Hitoshi, Takeshi. B.Sc., Waseda U., Tokyo, 1962, Ph.D., 1969. Dir. Kurt Rossmann Labs. for Radiologic Image Research dept. radiology U. Chgo., 1976—, prof. radiology, 1976—. Mem. Am. Assn. Physicists in Medicine. Subspecialties: Diagnostic radiology; Imaging technology. Current work: Digital radiography, monte carlo simulation, transfer function analysis, monoenergetic x-ray source, screen film systems. Home: 6415 Lane Ct Hinsdale IL 60521 Office: 950 E 59th St Chicago IL 60637

DOI, ROY HIROSHI, biochemistry educator; b. Sacramento, Calif., Mar. 26, 1933; s. Thomas Toshiteru and Ima (Sato) D.; m. Joyce Takahashi, Aug. 30, 1958; children: Kathryn Ellen, Douglas Alan. B.A. in Physiology, U. Calif.-Berkeley, 1953, 1957, M.S., U. Wis.-Madison, 1958, Ph.D., 1960. Asst. prof. biochemistry Syracuse U., 1963-65; asst. prof. U. Calif.-Davis, 1965-66, assoc. prof., 1966-69, prof., 1969—; chmn. microbial chemistry study sect. NIH, Bethesda, 1978-79; v.p., treas. Internat. Spore Conf., Inc., Boston, 1982—. Served with U.S. Army, 1953-55. NIH fellow, 1960-63; NSF fellow, 1971-72; recipient Sr. U.S. Scientist award Alexander von Humboldt Found., 1978-79. Mem. AAAS, Am. Soc. Biol. Chemists, Am. Soc. Microbiology, Sigma Xi. Democrat. Unitarian. Subspecialties: Molecular biology; Microbiology. Current work: Use of expression-probe plasmids to study promoter structure and function, heterologous gene expression, and gene organization in Bacillus subtilis. Office: Dept Biochemistry and Biophysics U Calif Davis CA 95616 Home: 1520 Lemon Ln Davis CA 95616

DOLAN, JOSEPH FRANCIS, astronomer; b. Rochester, N.Y., Sept. 17, 1939; s. Edwin Joseph and Helen Virginia (McAnnally) D.; m. Susan Marie, Aug. 14, 1971; children: Lee Marie, John Henry. B.S., St. Bonaventure U., Olean, N.Y., 1961; A.M., Harvard U., 1963, Ph.D., 1966. Physicist Smithsonian Astrophys. Obs., Cambridge, Mass., 1964-66; sr. scientist Jet Propulsion Lab., Pasadena, Calif., 1966-68; asst. prof. Warner & Swasey Obs., Case Western Res. U., Cleve., 1968-75; sr. fellow Nat. Acad. Scis., NASA Goddard Space Flight Ctr., Greenbelt, Md., 1975-77, astrophysicist, 1977—; co-investigator high speed photometer experiment Space Telescope Satellite. Contbr. articles to profl. jours. in field. Mem. Internat. Astron. Union, Am. Astron. Soc. Subspecialties: 1-ray high energy astrophysics; Optical astronomy. Current work: Galactic X-ray sources, close binaries, astronomical polarization. Home: 16007 Jerald Rd Laurel MD 20707 Office: Code 681 NASA Goddard Space Flight Ctr Greenbelt MD 29771

DOLAN, LINDA CAPANO, nuclear fuels company executive; b. Cin., Dec. 7, 1953; d. Arnold Frank and Velma Emily (Craig) Capano; m. Jerome Francis Dolan, July 23, 1977; 1 dau.: Jacqueline Michelle. B.S., Xavier U., Cin., 1976, postgrad., 1981—; M.S. in Nuclear Engring, U. Cin., 1977. Instr. Xavier U., Cin., 1976-77; engr. Gen. Electric Co., San Jose, Calif., 1977-79, Evendale, Ohio, 1979-81; chief nuclear safety NLO, Inc., Cin., 1981—; Chmn. Gen. Electric Task Force on Human Factors, Evendale, 1980. U. Cin. grad. fellow, 1976. Mem. Am. Nuclear Soc., Soc. Women Engrs., Sigma Pi Sigma. Republican. Subspecialties: Nuclear engineering; Nuclear fission. Current work: Nuclear criticality safety of fuel elements, particularly in transportation applications. Home: 1961 Augusta Blvd Fairfield OH 45014 Office: NLO Inc PO Box 39158 Cincinnati OH 45239

DOLAN, LINDA SUTLIFF, alternative energy research ofcl., cons.; b. Danville, Pa., Sept. 10, 1951; d. William Bruce and June (Mausteller) Sutliff; m. Roderick Norman Dolan, Nov. 26, 1952. B.A. in Biology and Earth Sci, Clarion (Pa.) State Coll., 1974; M.S. in Forest Mgmt. (music scholar, Weyerhaeuser Found. fellow), Oreg. State U., Corvallis, 1977. Exec. asst. Clarion County Conservation Dist., 1972-74; forest researcher Oreg. State U., 1976-77; dist. mgr. Spokane County Conservation Dist., 1978; chief cons. Dolan & Assocs., Seattle, 1978-79; biomass program mgr. Seattle City Light, 1979—; cons., curriculum adv., program reviewer. Contbr. articles to profl. jours. Assoc. supr. King County Conservation Dist., 1979—; program adv. MetroCenter YMCA, 1981—; mem. Democratic Women's Caucus, 1982. Mem. Soc. Am. Foresters, Biomass Energy Research Assn., Poplar Council Can., Biomass Invisible Coll. Brazil, Bio-Energy Council. Methodist. Subspecialties: Biomass (energy science and technology); Biomass (agriculture). Current work: Cultivating and marketing crops as fuel for energy production biomass, energy farms, renewable resources, wood-fired powerplants, bio-fuels, bio-energy, silviculture. Office: Seattle City Light 1015 3d Ave Seattle WA 98104

DOLENZ, JOHN JOSEPH, psychologist; b. Barton, Wis., July 17, 1931; s. Joseph John and Martha Mary (Kircher) D.; m. Jetta Maxine Parmer, Dec. 27, 1962; 1 son, John Andrew. Cert., Cath. U. Am., 1957; M.A., U. Tulsa, 1962; postgrad., U. Kans., 1961-64; Ph.D., U. Ottowa, Ont., 1970. Lic. psychologist, Tex., Kans. Clin. psychologist Psychol. Services, Kansas City, Mo., 1962-68; dir. services Tarrant County Hosp. Dist., Ft. Worth, Tex., 1970-73; with alcohol unit VAMC, Marion, Ind., 1973-76; dir. mental hygiene clinic VA-Outpatient, Evansville, Ind., 1976-79; coordinator alcohol treatment unit VA Med. Ctr., Big Spring, Tex., 1979—; cons. Tex. Tech Med. Sch. Alcohol Unit, Lubbock, 1983. Editor, co-author: Participant Manual, 1982; contbr. articles to profl. jours. Recipient Superior Performance award VA Med. Ctr., Big Spring, 1980, 82, 83. Mem. Am. Psychol. Assn., Canadian Psychol. Assn., Tex. Psychol. Assn., Kans. Psychol. Assn., Council for Nat. Register of Health Service Providers in Psychology, Psychol. Assn. Greater West Tex. (pres. 1983). Clubs: Sertoma (sec. Evansville 1978-79), Garden (pres. 1982—). Lodge: Rotary (dir. Big Spring 1982-83). Subspecialties: Neuropsychology; Toxicology (medicine). Current work: Cognitive impairment from excessive use mind altering drugs especially alcohol. Home: 2519 E 25th St Big Spring TX 79720 Office: 2400 S Gregg St Big Spring TX 79720

DOLLING, DAVID STANLEY, fluid dynamics researcher, aerospace engineer, educator; b. Bournemouth, Dorset, Eng., Mar. 21, 1950; came to U.S., 1976; s. Stanley Henry and Irene (Lucas) D. B.Sc. in Engring, London U., 1971, Ph.D., 1977; diploma, Von Karman Inst., Brussels, 1974. Aerodynamicist Hawker-Siddeley Dynamics, Hertfordshire, Eng., 1971-73; researcher, lectr. Princeton U., 1976—. Contbr. numerous articles to profl. publs., 1976—. Mem. Royal Aero. Soc. (London) (Undergrad. prize 1971), AIAA (chmn. Princeton sect. 1980-81, sr.), AAAS. Subspecialties: Aeronautical engineering; Aerospace engineering and technology. Current work: Supersonic flows, interactions of shock waves with boundary layers, high frequency instrumentation. Office: Gas Dynamics Lab James Forrestal Campus Princeton U Princeton NJ 08544 Home: 65 S Stanworth Dr Princeton NJ 08540

DOLLINGER, ELWOOD JOHNSON, agronomy educator, researcher; b. Lynchburg, Ohio, Apr. 20, 1920; s. Andrew Leroy and Reba Mable (Johnson) D.; m. Beth Elaine Goodrich, May 9, 1952; children: John, Emily. Ph.D., Columbia U., 1953. NIH postdoctoral fellow Brookhaven Nat. Lab., 1953-55; asst. prof. agronomy Ohio State U. Ohio Agrl. Research and Devel. Ctr., 1955-61, assoc. prof., 1961-62, prof., 1962—. Contbr. numerous articles to profl. publs. Mem. AAAS, Am. Soc. Agronomy, Genetics Soc. Am., Am. Genetics Assn., Council Agrl. Sci. and Tech., Am. Inst. Biol. Sci., Sigma Xi. Presbyterian. Subspecialty: Plant genetics. Current work: Maize genetics and breeding. Office: Dept Agronomy Ohio State U Ohio Agrl Research and Devel Ctr Wooster OH 44691

DOLLINGER, MALIN ROY, physician; b. San Francisco, Oct. 7, 1935; s. Mel King and Marilyn Hinda D.; m. Lenore Carole Levy, June 5, 1960; children: Jeffrey, Marc, Deborah, Cynthia. A.B., Stanford U., 1956; M.D., Yale U., 1960. Diplomate: Am. Bd. Internal Medicine. Spl. fellow Meml. Sloan Kettering Cancer Ctr., 1965-68; asst. clin. prof. radiology U. So. Calif., 1969-70; dir. med. oncology Harbor UCLA Med. Ctr., Torrance, Calif., 1970-72; practice medicine specializing in med. oncology, Torrance, Calif., 1972—; mem. Hematology Oncology Asocs., Torrance, 1972—; assoc. clin. prof medicine U. So. Calif., 1980—; dir. oncology South Bay Hosp., Redondo Beach, Calif., 1974—. Author: Cancer Chemotherapy, 1978; contbr. articles to profl. jours. Pres. br. Am. Cancer Soc., 1975; pres. Los Angeles Theatre Organ Soc., 1976. Served with USNR, 1961-63. Meller vis. scholar Meml. Sloan Kettering Cancer Ctr., 1975; recipient 10 Yr. Service award Am. Cancer Soc., 1982. Mem. Am. Soc. Hematology, Am. Assn. Cancer Research, Am. Soc. Clin. Oncology.

Subspecialties: Chemotherapy; Hematology. Current work: Experimental cancer chemotherapy, cancer medical oncology. Office: 3440 Lomita 252 Torrance CA 90505

DOLPH, GARY EDWARD, biology educator, researcher; b. Binghamton, N.Y., Oct. 17, 1946; s. Allen Nelson and Sophie Alice (Bajkowski) D.; m. Debra Doris Young, Jan. 2, 1982; children: Laura Lee, Brenda Leigh, Ann. A.B., SUNY-Binghamton, 1968; A.M., Ind. U., 1973, Ph.D., 1974. Lectr. Ind. U.-Kokomo, 1972-74, asst. prof., 1974-79, affiliate mem. faculty, 1981—, assoc. prof., 1979—. Contbr. articles to profl. jours.; editor: Internat. Assn. Angiosperm Paleobotanists, 1978-81. Exxon Edn. Found. grantee, 1978; NSF grantee, 1980. Mem. Biometric Soc., Bot. Soc. Am. (chmn. paleobot. sect. 1981), Paleobot. Soc. (editor 1982—), Paleontol Soc., AAAS, Ind. Acad. Scis., Am. Statis. Assn., Torrey Bot. Club, Sigma Xi. Republican. Methodist. Lodge: Kiwanis. Subspecialties: Evolutionary biology; Theoretical ecology. Current work: Paleoecology of the angiosperms, computer simulations in teaching undergraduate students. Home: 2209 Beauvoir Ct Kokomo IN 46902 Office: Ind U at Kokomo 2300 S Washington St Kokomo IN 46902

DOMANGUE, BARBARA B., psychology educator; b. Roselle, N.J., Aug. 21, 1931; d. Edward Woodward and Eleanor Ingrid (Petersen) Buttery; m. Norris J. Domangue, Jr., Apr. 21, 1951; children: Michelle Ann, Norris J., III. Student, Barnard Coll., 1949-51; B.A., West Chester (Pa.) State Coll., 1973; M.A., U. Del., 1976, Ph.D., 1978. Asst. prof. dept. psychiatry and human behavior Jefferson Med. Coll. of Thomas Jefferson U., Phila., 1978—, dir. continuing ed., 1982—. Author manual and articles. Mem. Am. Psychol. Assn., Pa. Psychol. Assn., Phila. Soc. Clin. Psychologists, Soc. for Clin. and Exptl. Hypnosis (assoc.), Sigma Xi, Psi Chi, Pi Gamma Mu. Subspecialties: Neurochemistry; Cognition. Current work: Biochemical correlates of pain. Biochemical correlates of hynoanalgesia, cognitive style, nonverbal communication. Office: 327 Curtis Jefferson Med Coll 1025 Walnut St Philadelphia PA 19107

DOMER, FLOYD RAY, pharmacologist, educator; b. Cedar Rapids, Iowa, July 12, 1931; s. William Ray and Caroline Anne (Zimmer) D.; m. Judith Elaine Kofroth, 1965. B.S., State U. Iowa, 1954, M.S., 1956; Ph.D., Tulane U., 1959. Life Inst. Med. Research Fund postdoctoral fellow Nat. Inst. Med. Research, London, 1959-60; with USAF Research and Devel. Command, Instituto Superior di Sanita, Rome, 1960-61; asst. prof. pharmacology U. Cin., 1961-62, Tulane U., 1963-64, asso. prof., 1965-74, prof., 1974—. Author: Animal Experiments in Pharmacological Analysis, 1971. Recipient award for teaching Owl Club, 1982. Mem. Am. Soc. Pharmacology and Exptl. Therapeutics, Soc. Neurosci., Soc. Exptl. Biology and Medicine. Club: Trojan. Subspecialties: Pharmacology; Neuropharmacology. Current work: The effects of drugs on function of blood-brain barrier, genitourinary tract. Home: 4420 Copernicus St New Orleans LA 70114 Office: 1430 Tulane Ave New Orleans LA 70112

DOMER, JUDITH ELAINE, microbiologist; b. Millersville, Pa., Apr. 9, 1939; d. Richard Harvey and Dorothy Alice (Peters) Kofroth; m. Floyd R. Domer, Apr. 15, 1965. B.A., Tusculum Coll., 1961; Ph.D., Tulane U., 1966. Diplomate: Am. Bd. Med. Microbiology. Instr. dept. biology St. Mary's Dominican Coll., New Orleans, 1966-67, asst. prof., 1967-68; research assoc. dept. microbiology and immunology Tulane U. Med. Sch., New Orleans, 1968-71; research fellow Kennedy Inst. Rheumatology, London, 1971-72; asst. prof. dept. microbiology Tulane U. Med. Sch., 1972-77, assoc. prof., 1977—; mem. bacteriology and mycology study sect. NIH, 1975-79. Mem. editorial bd.: Exptl. Mycology, 1976-79; assoc. editor, 1979-83; mem. editorial bd.: Infection and Immunity, 1981-83; Contbr. articles to profl. jours. NIH grantee; Cancer Assn. Greater New Orleans grantee. Mem. Am. Soc. Microbiology (chmn.-elect med. mycology div. 1982-83, chmn. 1983-84), Med. Mycol. Soc. Ams., Internat. Soc. Human and Animal Mycology, Am. Assn. Immunologists, Am. Acad. Microbiology, Infectious Diseases Soc. Am., Sigma Xi. Democrat. Methodist. Subspecialties: Microbiology; Infectious diseases. Current work: Immunology of fungal diseases. Home: 4420 Copernicus St New Orleans LA 70114 Office: 1430 Tulane Ave New Orleans LA 70112

DOMINO, GEORGE, psychologist; b. Torino, Italy, June 13, 1938; came to U.S., 1949; s. Tommaso and Maria (Oglietti) D.; m. Valerie Gerencser, Aug. 14, 1965; children: Brian, Marisa, Marla. B.S., Loyola U., Los Angeles, 1960; Ph.D., U. Calif.-Berkeley, 1967. Cert. clin. psychologist, Ariz. Instr. U. San Francisco, 1962-65; asst. prof. Fresno State Coll., 1965-66; prof., dir. counseling Fordham U., 1966-75; prof., dir. clin. tng. U. Ariz., Tucson, 1975—; cons., v.p. George W. Fotis, Greenwich, Conn., 1968—. Contbr. articles to profl. jours. USPHS predoctoral fellow, 1963-65; postdoctoral fellow Am. Coll. Testing, 1970. Mem. Am. Psychol. Assn., Western Psychol. Assn., Rocky Mountain Psychol. Assn., Sigma Xi. Roman Catholic. Subspecialty: Clinical psychology. Current work: Psychometrics, creativity, dream content. Office: University of Arizona Tucson AZ 85721

DON, NORMAN STANLEY, psychologist; b. Port Chester, N.Y., Oct. 2, 1934; s. William and Betty (Berson) D.; m. Ruth Stevens Tolman, June 28, 1958; children: Bronson Whitmarsh, Brent Tolman. M.S., U. Chgo., 1960; Ph.D., Union Grad. Sch., 1974. Researcher U. Chgo., 1961-65; cons. in field industry, fed., state govt., 1965-74; research asso. Dept. Psychiatry, U. Chgo., 1974-75; investigator Am. Dental Assn., Chgo., 1976-80; pvt. practice psychology, cons., Chgo., 1975-81; dir., pres. Kairos Found., Chgo., 1981—; lectr. in field; cons. in field. Author: The Transpersonal Crisis, 1983; contbr. articles to profl. jours.; discoverer/prin. investigator: The Canonical Effect, 1974. Epilepsy Found. Am. fellow, 1975. Mem. Biofeedback Soc. Ill. (dir. 1976-79), Am. Psychol. Assn., Biofeedback Soc. Am., AAAS. Subspecialties: Physiological psychology; Consciousness. Current work: Psychophysiology of altered states of consciousness; consciousness research; phenomenology of conscious experience; cybernetic modeling of perceptual control. Address: Kairos Found 35 E Wacker Dr Chicago IL 60601

DONABEDIAN, AVEDIS, physician, public health educator; b. Beirut, Lebanon, Jan. 7, 1919; came to U.S., 1955, naturalized, 1960; s. Samuel and Maritza (Der Hagopian) D.; m. Dorothy Salibian, Sept. 15, 1945; children: Haig, Bairj, Armen. B.A., Am. U. Beirut, 1940, M.D., 1944; M.P.H., Harvard U., 1955. Intern Am. U. of Beirut Hosps., 1943-44; physician, acting supt. English Mission Hosp., Jerusalem, 1945-47; instr. physiology, clin. asst. dermatology and venereology Am. U. Med. Sch., Beirut, 1943-51, univ. physician, dir. univ. health service, 1949-54; med. assoc. United Community Services Met., Boston, 1955-57; asst. prof. to assoc. prof. preventive medicine N.Y. Med. Coll., N.Y.C., 1957-61; mem. faculty U. Mich. Sch. Pub. Health, Ann Arbor, 1961—, prof. med. care orgn., 1964—, Nathan Sinai Disting. prof. pub. health, 1979—. Author: A Guide to Medical Care Administration: Medical Care Appraisal-Quality and Utilization, 1969, Aspects of Medical Care Administration, 1973, Benefits in Medical Care Programs, 1976, The Definition of Quality and Approaches to Its Assessment, 1980, The Criteria and Standards of Quality, 1982, Medical Care Chartbook, 1980. Recipient Dean Conley award Am. Coll. Hosp. Adminstrs., 1969; Norman A. Welch award Nat. Assn. Blue Shield Plans, 1976; Elizur Wright award Am. Risk and Ins. Assn., 1978; Nat. Merit award Delta Omega, 1978. Fellow Am. Coll. Hosp. Adminstrs. (hon.), Am. Pub. Health Assn.; mem. Inst. Medicine, Assn. Tchrs. Preventive Medicine. Subspecialties: Health services research; Health Care Organization and Administration. Current work: Organization; administration; financing and evaluation of personal health care services. Home: 1739 Ivywood Dr Ann Arbor MI 48103 Office: Univ Mich Sch Public Health 109 Observatory St Ann Arbor MI 48109

DONAHUE, FRANCIS MARTIN, chemical engineering educator; b. Phila., May 8, 1934. B.A., LaSalle Coll., 1956; Ph.D. in Engring, UCLA, 1965. Research chemist Tasty Baking Co., 1956-59; group leader corrosion research Betz Labs. Inc., 1959-61; electrochemist Stanford (Calif.) Research Inst., 1961-63; research engr. UCLA, 1963-65; asst. prof. chem. engring. U. Mich., Ann Arbor, 1965-69, assoc. prof., 1969-79; vis. prof. Swiss Fed. Inst. Tech., Zurich, 1972-73. Mem. Electrochem. Soc., Nat. Assn. Corrosion Engrs., Internat. Soc. Electrochemistry. Subspecialty: Electrochemical engineering. Office: Coll Engring U Mich Ann Arbor MI 48109

DONAHUE, MICHAEL JAMES, engineer; b. Barre, Vt., Oct. 23, 1953; s. Howard James and Beatrice Therese (Demers) D. B.S. in Engring, Northeastern U., 1982. Technician Gen. Electric Ordnance Systems, Pittsfield, Mass., 1973-76; programmer/analyst Honeywell Info. Systems, Brighton, Mass., 1976-78; research and devel. engr. communications systems div. GTE Products Corp. Systems Group, Needham Heights, Mass., 1978—. Mem. IEEE, Assn for Computing Machinery. Subspecialties: Computer-aided design; Distributed systems and networks. Current work: The integration of information processing facilities in an industrial environment, inparticular product description data defined by engineering units andutilized by manufacturing units. Home: 148 First Parish Rd Scituate MA 02066 Office: GTE Products Corp SSG/CSD 77 A St Needham Heights MA 02194

DONAHUE, THOMAS MICHAEL, educator; b. Healdton, Okla., May 23, 1921; s. Robert Emmett and Mary (Lyndon) D.; m. Esther Marie McPherson, Jan. 1, 1950; children—Brian M., Kevin E., Neil M. A.B., Rockhurst Coll., 1942, D.Sc., 1981; Ph.D., Johns Hopkins, 1947. Research asso., asst. prof. Johns Hopkins, 1947-51; asst. prof. U. Pitts., 1951-53, assoc. prof., 1953-57, prof., 1957-74, dir., 1966-74, 1966-74; chmn. dept. atmospheric and oceanic sci. and Space Physics Research Lab., U. Mich., Ann Arbor, 1974—; mem. phys. scis. com. NASA, 1972-77, adv. council, 1982—, Mars sci. working group, 1976—, solar system exploration com., 1981—; mem. Arecibo adv. bd. Cornell U., 1971-76; chmn. solar terrestrial relations com., mem. atmospheric scis. com., geophysics research bd., climate bd. Nat. Acad. Scis.; chmn. sci. steering groups Pioneer Venus multi-probe and orbital missions to Venus; Trustee-at-large Upper Atmosphere Research Corp., 1975—; vice-chmn. exec. com., trustee Univ. Corp. for Atmospheric Research, 1978—; chmn. bd. trustees Univs. Space Research Assn., 1978—. Editor: Space Research X, 1969; assoc. editor numerous publs., particularly specializing in atomic physics and properties of planetary atmospheres.; editor: Venus, 1983; assoc. editor: Planetary and Space Sci. Served with AUS, 1944-46. Recipient Public Service award NASA, 1977, 7 achievement awards, Disting. Public Service medal, 1980; Anictowski medal Nat. Acad. Sci., 1981; Fleming medal Am. Geophys. Union, 1981; Guggenheim fellow U. Paris, 1960. Fellow Am. Phys. Soc., Am. Geophys. Union (pres. solar-planetary relations 1972-75, v.p. 1969-72, pres., 1972-75), AAAS.; Mem. Nat. Acad. Scis. Club: Cosmos. Subspecialties: Planetary science; Aeronomy. Current work: Develop theories to explain the origin of planets and evolution of their atmospheres. Observe and explain composition and structure of planetary atmospheres. Participant Voyager mission to outer planets, Galileo mission to Jupiter, Spacelab 1. Home: 1781 Arlington Blvd Ann Arbor MI 48104

DONALDSON, COLEMAN DUPONT, consulting engineer; b. Phila., Sept. 22, 1922; s. John W. and Renee (duPont) D.; m. Barbara Goldsmith, Jan. 17, 1945; children: B. Beirne, Coleman duPont, Evan F., Alexander M., William M. B.S. in Aero. Engring., Rennselaer Poly. Inst., 1943; M.A., Princeton U., 1954, Ph.D., 1957. Staff, NACA, Langley Field, Va., 1943-44, head aerophysics sect., 1946-52; gen. aerodynamics USAC, Wright Field, Ohio, 1945-46; aerodynamic evaluation Bell Aircraft, Niagara Falls, N.Y., 1946; sr. cons., pres. Aero Research Assos. of Princeton, N.J., 1954-79, chmn. bd., 1979—; cons. missile guidance and control Gen. Precision Equipment Corp., 1957-68; cons. magnetohydro-dynamics Thompson Ramo Wooldridge, Inc., 1958-61; cons. aerodynamic heating, gen. aerodynamics Martin Marietta Corp., 1955-72; gen. editor Princeton series on high speed aerodynamics and jet propulsion, 1955-64; cons. boundary layer stability, aerodynamic heating, missile and ordnance systems dept. Gen. Electric Co., 1956-72; cons. Grumman Aerospace Corp., 1959-72; Robert H. Goddard vis. lectr. with rank of prof. Princeton U., 1970-71; mem. research and tech. adv. council panel on research NASA, 1969-76; mem. indsl. profl. adv. com. Pa. State U., 1970-77; mem. Pres.'s Air Quality Adv. Bd., 1973-74; chmn. lab. adv. bd. for air warfare Naval Research Adv. Com., 1972-77; mem. Marine Corps panel Naval Res. Adv. Com., 1972-77; chmn. adv. council dept. aerospace and mech. scis. Princeton U., 1973-78. Author articles on aerodynamics. Fellow AIAA (Dryden Research lecture award 1971, gen. chmn. 13th aerospace scis. meeting 1975), Nat. Acad. Engring., Am. Phys. Soc., Sigma Xi, Delta Phi. Subspecialty: Aeronautical engineering. Home: PO Box 279 Gloucester VA 23061 Office: 1800 Old Meadow Rd Suite 114 McLean VA 22102

DONCHIN, EMANUEL, psychologist, educator; b. Tel Aviv, Apr. 3, 1935; U.S., 1961; s. Michael and Guta D.; m. Rina Greenfarb, June 3, 1955; children: Gill, Opher, Ayala. B.A., Hebrew U., 1961, M.A., 1963; Ph.D., UCLA, 1965. Teaching and research asst. dept. psychology Hebrew U., 1958-61; research asst. dept. psychology UCLA, 1961-63, research psychologist, 1964-65; research asso. div. neurology Stanford U. Med. Sch., 1965-66, asst. prof. in residence, 1966-68; research asso. neurobiology br. NASA, Ames Research Center, Moffett Field, Calif., 1966-68; asso. prof. dept. psychology U. Ill., Urbana-Champaign, 1968-72, prof. psychology and physiology, 1972—, head dept. psychology, 1980—. Author: (with Donald B. Lindsley) Averaged Evoked Potentials, 1969; contbr. articles to profl. jours. Served with Israeli Army, 1952-55. Fellow AAAS, Am. Psychol. Assn.; mem. Soc. Psychophysiol. Research (pres. 1980), Fedn. Behavioral, Cognitive and Psychol. Socs. (v.p. 1981—), Am. EEG Soc., Psychonomic Soc., Soc. Neuroscis., AAAS. Subspecialties: Psychobiology; Neuropsychology. Current work: Biology of cognitive function; event related brain potentials; engineering psychology; complex skills; psychophysiology. Office: Dept Psychology U Ill 603 E Daniel St Champaign IL 61820

DONEGAN, WILLIAM LAURENCE, surgeon, educator; b. Jacksonville, Fla., Nov. 3, 1932; s. William Elton and Mildred Louise (Bullock) D.; m. Judith Higgins, Dec. 21, 1963; children: William David, Elizabeth Kathleen. B.A., Yale U., 1955, M.D., 1959. Intern Barnes Hosp., St. Louis, 1959-60, asst. resident surgery, 1960-63, resident in surgery, 1963-64; USPHS cancer clin. fellow, 1963-64; clin. asst. prof. surgery U. Mo., Columbia, 1964-65; instr. surgery Washington U., St. Louis, 1964-67; asst. prof. surgery U. Mo., Columbia, 1965-69, assoc. prof., 1969-74; prof. surgery Med. Coll. Wis., Milw., 1974—; prof. clin. oncology Am. Cancer Soc., 1975—; surgeon Ellis Fischel State Cancer Hosp., Columbia, 1964-74; dir. clin. research unit Cancer Research Ctr., Columbia, 1965-74, assoc. scientist, 1967-74; attending surgeon Milw. County Med. Complex, 1974—, Columbia Hosp., Milw., 1974—, Mt. Sinai Med. Ctr., 1976—; acting chief surgery, 1982-83, chief surgery, 1983—; cons. VA Med. Ctr., 1975—; sr. attending surgeon Froedtert Meml. Luth. Hosp., Milw., 1981—; cons. in field. Contbr. numerous articles to profl. jours. Sterling-Winthrop Research Inst. grantee; Am. Cancer Soc. grantee, 1975—; Nat. Cancer Inst. grantee, 1982—; others. Fellow ACS; mem. Am. Assn. Cancer Research, Soc. Surgery of Alimentary Tract, Central Surg. Assn., Am. Assn. Cancer Edn., Milw. Acad. Surgery, Soc. of Head and Neck Surgeons, Soc. Surg. Oncology, Milw. Acad. Medicine, Wis. Surg. Soc., Am. Soc. Clin. Oncology, Pan-Pacific Surg. Assn., Collegium Internationale Chirurgiae Digestivae, Am. Radium Soc., Sigma Xi, Phi Beta Kappa. Subspecialties: Surgery; Oncology. Current work: Research in cancer. Home: 9421 N Lake Dr Bayside WI 53217 Office: 950 N 12th St Milwaukee WI 53201

DONGARRA, JACK JOSEPH, computer scientist; b. Chgo., July 18, 1950; s. Joseph and Anne (Danca) D.; m. Susan Sauer, Oct. 11, 1980. B.S., Chgo. State U., 1972; M.S., Ill. Inst. Tech., 1973; Ph.D., U. N.Mex., 1980. Scientist Argonne Nat. Lab., Ill., 1973—. Mem. Soc. Indsl. and Applied Math., Assn. Computing Machinery. Subspecialties: Numerical analysis; Mathematical software. Current work: Numerical linear algebra, mathematical software, high performance computing. Office: Math and Computer Sci Dept Argonne Nat Lab Argonne IL 60439

DONIGER, JAY, molecular biologist; b. Bklyn., Mar. 22, 1944; s. Irving and Ann (Tuchschneider) D.; m. Marcia Sherman, Mar. 26, 1967; children: Elizabeth, Jeremy. B.S., Bklyn. Coll., 1965; Ph.D., Purdue U., 1972. Postdoctoral fellow Brandeis U., Waltham, Mass., 1972-75; asst. biologist Brookhaven Nat. Lab., 1975-77; expert scientist Nat. Cancer Inst., Bethesda, Md., 1977-82, sr. staff fellow, 1982—. Contbr. numerous articles to profl. jours. Mem. Am. Assn. Cancer Research, Biophys. Soc., Am. Assn. Photobiology. Subspecialties: Molecular biology; Gene actions. Current work: Relationship of DNA metabolism to carcinogenesis. Office: Lab Biology Nat Cancer Inst Bethesda MD 20205

DONIKIAN, MARY ADRIENNE, microbiologist, analyst; b. Istanbul, Turkey, May 7, 1923; d. Puzant P. and Eunice V. (Malhasian) D. M.S. in Microbiology, U. Pa., 1948. Research assoc. Charles Pfizer, N.Y.C., 1948-56; microbiologist Smith Kline Beckman, Phila., 1956-67, sr. scientist, 1967-71, sr. clin. data/sci. analyst, 1971—. Mem. N.Y. Acad. Scis., Am. Soc. Microbiology. Subspecialties: Virology (medicine); Microbiology (medicine). Current work: Bacterial and antiviral chemotherapy; interferon; vaccines (viral); adminstrative work, interfacing with various departments; assessing, evaluating and writing clinical reports in process of drug development. Office: 1500 Spring Garden St Philadelphia PA 19101

DONIS, JOSE MARIA, utility company executive; b. Santiago de, Cuba, Dec. 28, 1944; came to U.S., 1961; s. Jose and Maria (Pareira) D.; m. Margarita Alfaro, Dec. 9, 1972; 1 son: Joseph. B.S., U. Miami, 1965; M.S., U. Va.-Chalrottesville, 1971. Registered profl. engr., Fla., N.Y., N.J., Calif. Project engr. E.I. DuPont de Nemours & Co., Inc., Richmond, Va., 1965-67; engr. Babcock & Wilcox, Lynchburg, Va., 1967-71; supervising engr. Burns & Roe Inc., Oradell, N.J., 1971-79, Fla. Power & Light Co., Juno Beach, Fla., 1979—. Mem. ASME, Am. Nuclear Soc., Sigma Xi (assoc.). Subspecialties: Mechanical engineering; Nuclear engineering. Current work: Development of computerized systems to perform piping design and stress analysis; supervision of design liquid processing systems for nuclear power plants. 1C8525 Bonita Isle Dr: Lake Worth FL 33463 Office: Fla Power & Light Co 7900 Universe Blvd Juno Beach FL 33410

DONIVAN, FRANK FORBES, JR., astronomer; b. Inglewood, Calif., Oct. 19, 1943; s. Frank Forbes and Daisy Dean (Lambert) D.; m. Margaret R. Gates; children: Laura, Erin. B.A., UCLA, 1966; Ph.D., U. Fla., Gainesville, 1970. Assoc. prof. astronomy U. Fla., 1970-79; asst. dean (Grad. Sch.), 1971-73; vis. asst. scientist Nat. Radio Obs. Astronomy, Charlottesville, VA., 1976; mem. tech. staff Jet Propulsion Lab., Pasadena, Calif., 1979—. Mem. Am. Astron. Soc., Sigma Xi, Sigma Pi Sigma. Subspecialties: Radio and microwave astronomy; Astronautics. Current work: Very Long Baseline Interferometry (VLBI) applied to interplanetary and earth orbiter navigation. Office: 4800 Oak Grove Dr Pasadena CA 91103

DONN, WILLIAM L., research scientist, consultant; b. Bklyn., Mar. 2, 1918; s. Nathan and Tina D.; m. Renee M. Brilliant, Jan 23, 1960; children: Matthew, Tara. B.A., Bklyn. Coll., 1939; M.A., Columbia U., 1946, Ph.D., 1951. Sr. research scientist Lamont-Doherty Obs., Columbia U., Palisades, N.Y., 1951—; instr. to prof. geology Bklyn. Coll., 1946-63; prof. geology CCNY, N.Y.C., 1963-77; research cons. Woods Hole (Mass.) Oceanographic Instn., 1947-49; heat meteorology sect. U.S. Merchant Marine Acad., 1942-45; aerologist Naval Air Navigation Sch., 1945; geologist N.Atlantic Dist. C.E. U.S. Army, 1941-42, Del. Aqueduct Program, 1941; cons. various indsl. and legal orgns., 1951—; White House cons. Office Sci. and Tech., Washington, 1979-81; mem. com. on long waves ASCE; commn. on microseisms Internat. Union Geodesy and Geophysics; review com. prevention and mitigation of flood losses NRC. Author: textbooks Meteorology - With Marine Applications, 1946, 2d edit., 1951, 3d edit., 1965, 4th edit., 1975, Graphic Methods in Structural Geology, 1958, The Earth, 1973; editor: Glossary of Geology, 4th edit, 1980, International Geophysics Series, 1979—; contbr. research articles to profl. pubs. Trustee Village Bd. Grand View-on-Hudson, N.Y., 1968—, dep. mayor, 1975—, policy commr., 1968—. Served with USCGR, 1942; served to lt. s.g. USNR, 1942-46. Disting. nat. lectr. Am. Assn. Petroleum Geologists, 1960, Soc. Exploration Geophysicists, 1960; NSF sr. postdoctoral fellow, 1959-60; fellow in geology Bklyn. Coll., 1940-41; chief scientist, prin., co-prin., dir. project and grants, 1941—. Fellow AAAS, Geol. Soc. Am., Explorers Club; mem. Seismol. Soc. Am., Am. Geophys. Union, N.Y. Acad. Scis. (council 1950-52, chmn. sect. oceanography and neteorology 1950-52), Am. Meteorol. Soc. (profl., chmn. com. on paleoclimatology 1965), Phi Beta Kappa, Sigma Xi. Subspecialties: Climatology; Atmospheric Infrasound. Current work: Prediction of monthly and seasonal climate with thermodynamic model, study of evolution of climate; application of infrasound as atmospheric probe. Home: 302 River Rd Grand View-on-Hudson NY 10990 Office: Lamont-Doherty Geol Observatory Palisades NY 10960

DONNELLEY, JAMES ELLIS, computer scientist, consultant; b. Palo Alto, Calif., July 5, 1948; s. John Donovan and Rachael Ellis (Millard) D. B.S. in Math, U. Calif.-Davis, 1970, B.A. in Physics, 1970, M.S. in Math, 1972. Technician Hewlett Packard Co., Palo Alto, 1966; hydrologist U.S. Geol. Survey, Menlo Park, Calif., 1968-72; computer scientist Lawrence Livermore Lab., Calif., 1972—; cons. Aerospace Corp., Los Angeles, 1980—, Dietrich, Glasrud & Jones (Law Firm), Fresno, Calif., 1981—. Contbr. articles to profl. jours. Recipient cert. for outstanding achievement in undergrad. math. U. Calif.-Davis, 1970. Mem. Assn. Computing Machinery, IEEE, Planetary Soc., Nat. Space Inst., ACLU. Mem. Libertarian party. Subspecialties: Distributed systems and networks; Computer architecture. Current work: Capability-based network operating systems; asynchronous cellular data flow computers. Home: 5241 Norma Way Apt 191 Livermore CA 94550 Office: PO Box 808 Livermore CA 94550

DONNELLY, RUSSELL JAMES, educator, physicist; b. Hamilton, Ont., Can., Apr. 16, 1930; s. Clifford Ernest and Bessie (Harrison) D.; m. Marian Card, Jan. 21, 1956; 1 son, James. B.Sc., McMaster U., 1951, M.Sc., 1952; M.S., Yale, 1953, Ph.D., 1956. Faculty U. Chgo., 1956-66, prof. physics, 1965-66, U. Oreg., Eugene, 1966—, chmn. dept., 1966-72, 82-83; vis. prof. Niels Bohr Inst., Copenhagen, Denmark, 1972; co-founder Pine Mountain Obs., 1967—; cons. Gen. Motors Co. Research Labs., 1958-68, NSF, 1968-73, 79—, mem. adv. panel for physics, 1970-73, chmn., 1971-72, mem. adv. coms. on materials research, 1979—; cons. Jet Propulsion Lab., Calif. Inst. Tech., Pasadena, 1973—. Contbr. papers to profl. lit.; editor: (with Herman, Prigogine) Non-Equilibrium Thermodynamics Variational Techniques and Stability, 1966, (with Parks, Glaberson) Experimental Superfluidity, 1967; assoc. editor: Physics of Fluids, 1966-68; mem. editorial bd.: Phys. Rev. A, 1978—. Bd. dirs. U. Oreg. Devel. Fund, 1970-72; chmn. Lane County Coop. Mus. Commn., 1975—. Alfred P. Sloan fellow, 1959-63; sr. vis. fellow Sci. Research Council, U.K., 1978. Fellow Am. Phys. Soc. (exec. com. div. fluid dynamics 1966-72, 80—, sec-treas. 1967-70, chmn. 1971-72, 82-83, Otto Laporte Meml. lectr. 1974); mem. AAUP, AAAS, Am. Assn. Physics Tchrs. Episcopalian. Club: Cosmos (Washington). Subspecialties: Low temperature physics; Fluid mechanics. Current work: Superfluidity; coutte and Benard flows. Research on physics fluids, especially hydrodynamic stability and superfluidity. Home: 2175 Olive St Eugene OR 97405 Office: Dept Physics Univ Oreg Eugene OR 97403

DONNELLY, THOMAS EDWARD, JR., pharmacology educator; b. Chelsea, Mass., Sept. 16, 1943; s. Thomas Edward and Catherine S. (Ross) D.; m. Thorkatla Thorkelsdottir, Jan. 31, 1975; children: Karina, Erling. B.S., Mass. Coll. Pharmacy, 1966; M.A., Harvard U., 1968; Ph.D., Yale U., 1972. Postdoctoral fellow U. Copenhagen, 1972-73; biochemist Leo Pharm. Products, Ballerup, Denmark, 1973-74; research assoc. Emory U., Atlanta, 1974; asst. prof. U. Nebr. Med. Ctr., Omaha, 1974-78, assoc. prof. pharmacology, 1978—; mem. research com. Nebr. affiliate Am. Heart Assn. Contbr. articles to sci. jours. NIH grantee, 1976-79, 78-81. Mem. Am. Soc. Pharmacology and Exptl. Therapeutics, N.Y. Acad. Sci., AAAS. Subspecialties: Molecular pharmacology; Biochemistry (medicine). Current work: Role of calcium-dependent and cyclic nucleotide-dependent protein kinases in cellular proliferation; biochemical basis of beta-adrenergic supersensitivity in cardiac tissue. Home: 10713 Valley St Omaha NE 68124 Office: 42nd St and Dewey Ave Omaha NE 68105

DONOGHUE, EDWARD SYLVESTER, mathematics educator, physical chemistry researcher; b. Phila., Apr. 27, 1945; s. Edward Sylvester and Anna (Murphy) D. B.S., St. Joseph's Coll., Phila., 1967; M.S., Cornell U., 1971, Ph.D., 1974. Research assoc. dept. chemistry Brown U., 1974-78, asst. prof. research, 1978-79; vis. asst. prof. dept. math. Amherst (Mass.) Coll., 1979-80, vis. assoc. prof., 1982—; research assoc. polymer sci. and engring. U. Mass., Amherst, 1980-82. Mem. Am. Chem. Soc., Am. Math. Soc., Am. Soc. Indsl. and Applied Math., Sigma Xi. Subspecialties: Applied mathematics; Physical chemistry. Current work: Gelation in polycondensing systems; kinetics of polymer gelation; antigen-antibody reactions; condensation theory; theory of liquid water; thermal analysis. Home: 1-A Merrill House Merrill Pl Amherst MA 01002 Office: PO Box 607 Amherst MA 01004

DONOHO, PAUL LEIGHTON, physicist, researcher; b. Fort Worth, Sept. 7, 1931; s. David Hubert and Martelle (Hicks) D.; m.; children: David, Andrew, Julia. B.A., Rice U., 1952; Ph.D., Calif. Inst. Tech., 1958. Mem. tech. staff Bell Labs., Murray Hill, N.J., 1957-59; prof. physics Rice U., Houston, 1959-79; research scientist U. Tex. Marine Sci. Inst., 1977-81, acting dir., 1981, research scientist, 1981-82; sr. research physicist Chevron Oil Field Research Co., La Habra, Calif., 1982—. Contbr. articles in field to sci. jours. Mem. Am. Phys. Soc., IEEE. Republican. Unitarian. Subspecialties: Offshore technology; Magnetic physics. Current work: Applying physics to solution of problems in petroleum exploration and production. Home: 2512 San Carlos St Fullerton CA 92631 Office: PO Box 446 La Habra CA 90631

DONOVAN, TERRENCE JOHN, geologist, research/test pilot; b. Waterbury, Conn., July 27, 1936; s. Francis Aubry and Jane Anne (Rasinski) Sando; m. Sharon Anne McIlwain, Aug. 11, 1956; children: Kathleen Anne, Terrence John, Michael Patrick. B.S., Midwestern U., Wichita Falls, Tex., 1961; M.A., U. Calif.-Riverside, 1963; Ph.D., UCLA, 1972. Geologist Mobil Oil Corp., Los Angeles, 1963-68; instr. Midwestern U., 1968-70, asst. prof., 1970-72; geologist U.S. Geol. Survey, Denver, 1972-75, geologist/pilot, Flagstaff, Ariz., 1975—; mem. U.S./USSR Energy Coop. Com., 1975-79. Contbr. numerous tech. articles to profl. jours. Served with USAF, 1955-58; Philippines. Fellow Geol. Soc. Am.; mem. Am. Assn. Petroleum Geologists (Leverson Meml. award 1971), AAAS, Soc. Exptl. Test Pilots. Democrat. Subspecialties: Geology; Geophysics. Current work: Petroleum exploration research; research, development, test and evaluation of unconventional airborne geophysical exploration techniques for energy resources. Home: 2696 N Oakmont Dr Flagstaff AZ 86001 Office: US Geol Survey 2255 Gemini Dr Flagstaff AZ 86001

DONOVICK, PETER J., psychologist, educator; b. Champaign, Ill., Jan. 14, 1938; s. Richard and Anne L. (Wolfson) D.; m. Valerie P. Perdue; children: Paul, Roger, Joshua. A.B., Lafayette Coll., 1961; M.A., U. Wis., 1963, Ph.D., 1967. Asst. prof. SUNY-Binghamton, 1966-71, assoc. prof., 1971-76, prof. psychology, 1976—; chmn. dept., 1978-81, assoc. dean Arts and Scis., 1982-83; vis. asst. prof. anatomy Milton S. Hershey Med. Center, 1970-71. Contbr. articles to profl. publs. Grantee NIMH, NSF, Nat. Inst. Child Health and Human Devel., Biomed-Research Support Council. SUNY-Binghamton. Mem. Am. Psychol. Assn. (div. 6), N.Y. Acad. Scis., Soc. Neuroscis., Behavior Genetics Assn. Subspecialties: Neurobiology; Neuropsychology. Current work: Environ. neuropsychology; developmental psychobiolgy; behavorial toxicology. Office: Dept Psychology SUNY Binghamton NY 13901

DOOB, JOSEPH LEO, mathematician, educator; b. Cin., Feb. 27, 1910; s. Leo and Mollie (Doerfler) D.; m. Elsie Haviland Field, June 26, 1931; children—Stephen, Peter, Deborah. B.A., Harvard U., 1930, M.A., 1931, Ph.D., 1932; D.Sc. (hon.), U. Ill., 1981. Faculty U. Ill., Urbana, 1935—, successively asso., asst. prof., asso. prof., 1935-45, prof. math., 1945—, now emeritus prof. Recipient Nat. Medal of Sci., 1979. Mem. Nat. Acad. Scis., Am. Acad. Arts and Scis., Acad. Scis. (Paris) (fgn. asso.). Subspecialties: Probability. Home: 208 W High St Urbana IL 61801

DOOLEY, HARRISON LEROY, research plant pathologist; b. Cherokee, Okla., Feb. 25, 1932; s. Harrison and Opal Irene (Andrews) D.; m. Peggy Diane Pickrel, Sept. 1, 1953; children: Harrison, Diane, Susan, Chris. B.S., Okla. State U., 1954, M.S., 1961. Sr. exptl. aide Prosser Irrigation Expt. Sta., Wash. State U., 1956-58; research asst. dept. botany and plant pathology Okla. State U., Stillwater, 1958-60, Iowa State U., Ames, 1960-61; asst. in plant pathology Oreg. State U., Corvallis, 1961-64; plant pathologist pesticide regulation div. plant biology lab. Agrl. Research Service, U.S. Dept. Agr., Corvallis, 1964-70; plant pathologist div. pesticide regulation EPA, Corvallis, 1970-72, supervisory plant pathologist, plant biology lab. supr., 1972-79, supervisory plant pathologist, plant biology lab. supr., acting sta. chief biol. investigations sta., 1975-78, supervisory plant patnologist benefits and field studies div., 1979-81; research plant pathologist hort. crops research lab. Agrl. Research Service, U.S. Dept. Agr., Corvallis, 1981—. Contbr. articles to profl. jours. Served with U.S. Army, 1954-56. Mem. Am. Phytopathological Soc. Lutheran. Lodge: Elks. Subspecialty: Plant pathology. Current work: Plant rusts of ornamentals and foliar diseases. Office: 3420 Orchard Ave Corvallis OR 97330

DOOLITTLE, DONALD PRESTON, geneticist, educator; b. Torrington, Conn., May 14, 1933; s. Merton Elford and Louva Edna (Mack) D.; m. Maria Zergenyi, Nov. 23, 1957; children: Andrew, Donald, Thomas Robert. B.S., U. Conn., 1954; M.S., Cornell U., 1956, Ph.D., 1959. Postdoctoral fellow Jackson Lab., Bar Harbor, Maine, 1958-60; asst. research prof. biometry Grad. Sch. Public Health, U. Pitts., 1960-65; asst. prof. genetics W.Va. U., Morgantown, 1966-67, assoc. prof., 1967; assoc. prof. animal scis. Purdue U., West Lafayette, Ind., 1967—. Contbr. articles to profl. jours. Mem. West Lafayette Citizens Community Devel. Com., 1979—. Mem. Am. Genetics Assn., Genetics Soc. Am. Lutheran. Subspecialties: Animal genetics; Evolutionary biology. Current work: Researcher in population genetics, quantitative genetics, genetics of the mouse. Office: Dept Animal Scis Purdue Univ West Lafayette IN 47907

DOOLITTLE, ROBERT F., II, physicist, computer systems programmer; b. Chgo., Dec. 21, 1925; s. Arthur K. and Dortha B. D.; m. Mary P., Apr. 30, 1955 (dec.); children: Robert A., Nancy E.; m. Karen K., Dec. 28, 1976. A.B., Oberlin Coll., 1948; M.S., U. Mich., 1950, Ph.D., 1958. Asst. prof. physics San Diego State U., 1958-60; sr. scientist TRW Electronics and Def., Redondo Beach, Calif., 1960-83; sr. systems programmer Ashton-Tate, Culver City, 1983—. Served with USNR, 1944-46; served to lt. comdr., 1952-54. Mem. Am. Astron. Soc., Am. Phys. Soc. Subspecialties: Cosmic ray high energy astrophysics; Gamma ray high energy astrophysics. Current work: NASA space telescope science operation ground system; computer graphics, algorithms and operating systems. Home: 1290 Monument Pacific Palisades CA 90272 Office: TRW 119/2845 One Space Park Redondo Beach CA 90278

DOOLITTLE, RUSSELL F., biochemistry educator; b. New Haven, Jan. 10, 1931. B.A., Wesleyan U., 1952; M.A., Trinity Coll., 1957; Ph.D. in Biochemistry, Harvard U., 1962. Instr. biology Amherst (Mass.) Coll., 1961-62; Nat. Heart Inst. fellow, 1962-64; asst. research biologist U. Calif.-San Diego, La Jolla, 1964-65, asst. prof. to assoc. prof. chemistry, 1967-72, prof. biochemistry, 1972—, chmn. dept. chemistry; mem. staff Marine Biol.Lab., 1962-69; mem. Blood Adv. Com., NIH, 1974-78. USPHS awardee, 1969-74. Mem. Am. Heart Assn. (thrombosis com.), Am. Soc. Biol. Chemists. Subspecialty: Biochemistry (biology). Office: Dept Chemistry U Calif-San Diego La Jolla CA 92093

DORA, EORS ISTVAN, medical educator, researcher; b. Budapest, Hungray, May 1, 1943; emigrated to U.S., 1981; s. Tivadar Tibor and Ilona (Toth) D.; m. Tunde Nagy, Dec. 7, 1974; 1 dau., Melinda. M.D., Semmelweis U. Med. Sch., Budapest, 1968; Ch.D., Hungarian Acad. Scis., 1978. Diplomate: Med. diplomate. Researcher Semmelweis U. Med. Sch., 1968-78, adj., 1978—; research assoc. U. Pa., Phila., 1972-73, assoc. prof., 1981—. Contbr. numerous articles to med. jours. Mem. Hungarian Physiol. Soc., Internat. Soc. for Oxygen Transport to Tissue, Am. Physiol. Soc., Internat. Soc. Cerebral Blood Flow and Metabolism. Subspecialties: Neurochemistry; Neurophysiology. Current work: Regulation of cerebral blood flow and energy metabolism under hypoxia, increased brain activity and autoregulation; mechanism of brain damage during ischemia and shock; mechanism of oxidative phosphorylation of in vivo, in situ mitochondria. Home: Hajanlka U R Budapest Hungary 1221 Office: Cerebrovascular Research Center U Pa Philadelphia PA 19104

DORE-DUFFY, PAULA, neuroimmunologist, researcher; b. Hyannis, Mass., Feb. 23, 1948; d. Paul David and Doris Jeanette (Hochu) Dore; m. Michael Charles Duffy, Dec. 27, 1972. B.S., Simmons Coll., 1970; Ph.D., La. State U., 1976. Postdoctoral fellow U. Conn. Sch. Medicine, 1977-78, asst. prof. medicine, 1978, asst. prof. neurology, 1979-82, assoc. prof. neurology, 1982—; also head div. neuroimmunology dept. neurology, dir. Multiple Sclerosis Ctr. Contbr. articles to profl. jours. Bd. dirs. Conn. River Valley chpt. Nat. Multiple Sclerosis Soc. Multiple Sclerosis Soc. grantee, 1978; fellow, 1978; Kroc Found. grantee, 1978-80; NIH grantee, 1979—. Mem. Am. Assn. Immunologists, Am. Acad. Neurologists, Am. Fedn. Clin. Research, Am. Soc. Microbiology. Subspecialty: Neuroimmunology. Current work: Multiple Sclerosis and autoimmune diseases; role of the prostaglandins in etiology. Office: Dept Neurology U Conn Health Ctr Farmington CT 06032

DORFMAN, MYRON HERBERT, petroleum engineer, educator; b. Shreveport, La., July 3, 1927; s. Samuel Yandell and Rose (Gold) D.; m.; children: Shelley Fonda Dorfman Roberts, Cynthia Renee. B.S., U. Tex., 1950, M.S., 1972, Ph.D., 1975. Registered profl. engr., Tex. Geologist engr. Sklar Oil Co., Shreveport, 1950-56, mgr. prodn. and devel., 1957-59, partner, 1958-59; owner Dorfman Oil Properties, Shreveport, 1950-71, Austin, Tex., 1971—; prof. petroleum engring. U. Tex., Austin, 1976—, H.B. Harkins prof. petroleum engineering, 1980—, dir. Center Energy Studies, 1977—, chmn. dept. petroleum engring., 1978—; dir. Tex. Petroleum Research Commn., Tex. R.R. Commn., 1982—; disting. lectr. Soc. Petroleum Engrs. of AIME, 1978-79, disting. author, 1982—. Contbr. articles to profl. jours. Pres. Shreveport Community Council, 1966; bd. dirs. Gov.'s Com. Employment Handicapped, 1966-68, La. Youth Opportunity Center, Shreveport, 1966-71, ARC, Caddo Parish, La., 1964-71; pres. La. Mental Health Center, Shreveport, 1967. Served with USNR, 1945-46; PTO. Recipient medal State of Israel, 1963. Fellow Geol. Soc. Am.; mem. Am. Geophys. Union, Nat. Acad. Scis., Am. Assn. Petroleum Geologists, Soc. Profl. Well Log Analysts, AIME, Shreveport Geol. Soc., Petroleum Club Shreveport, Shreveport Jewish Fedn. (pres. 1967), Pi Epsilon Tau., Tau Beta Pi. Club: Shreveport Skeet (pres. 1964). Subspecialties: Petroleum engineering; Sedimentology. Current work: Geopressured geothermal energy. director geothermal studies, center for energy studies, university of Texas, facies characterization of gelogic environments by use of well logs, general petroleum engineering studies of enhanced oil and gas recovery. Home: 1918 Cypress Point W Austin TX 78746 Office: Dept Petroleum Engring U Tex Austin TX 78712

DORIA, VICTOR, software engr.; b. Phila., Apr. 3, 1951; s. Anthony Notarnicola and Diane Leah (Joslow) D.; m. Wendy Jean Whitehead, Feb. 15, 1974; 1 son, Seth Joshua. B.A. with distinction, U. R.I., 1978; M.A., U. Chgo., 1979. Systems programmer U. Chgo. Computer Center, 1978-81; software engr. operating subsystems Digital Equipment Corp., Marlboro, Mass., 1981—. Mem. Assn. Computing Machinery, IEEE Computer Soc. Jewish. Subspecialties: Software engineering; Operating systems. Current work: Special interest and ongoing research into philosophical aspects of artificial intelligence and engineering enterprise.

DORKO, ERNEST A(LEXANDER), chemist, educator, cons., researcher; b. Detroit, Sept. 16, 1936; s. John and Julia Anne (Pala) D.; m. Betty Jane, June 18, 1971; 1 son, Thomas. B.Ch.E., U. Detroit, 1959; M.S., U. Chgo., 1961, Ph.D. in Chemistry, 1964. Research chemist Phys. Scis. Lab., Redstone Arsenal, Ala., 1964-67; asst. prof. chemistry, dept. physics Air Force Inst. Tech., Wright-Patterson AFB, Ohio, 1967-70, assoc. prof., 1970-77, prof., 1977—; cons. Air Force labs. Contbr. numerous articles to profl. publs. Trustee Dayton (Ohio) View Triangle Assn. Served to capt. U.S. Army, 1964-66. Decorated Army Commendation medal; recipient Sci. Achievement award USAF, 1972. Mem. Am. Chem. Soc., AAAS, Sigma Xi, Tau Beta Pi. Roman Catholic. Subspecialties: Kinetics; Photochemistry. Current work: Kinetics and spectroscopic analysis in flow tubes and shock tubes, kinetics and mechanism of deradation of laser dyes. Patentee in field. Office: Air Force Inst. Tech./ENP Dept Physics Wright-Patterson AFB OH 45433

DORN, C. RICHARD, vet. scientist, educator; b. London, Ohio, June 13, 1933; m. Barbara J. Monroe, 1964; children. D.V.M., Ohio State U., 1957; M.P.H., Harvard U., 1962. Research specialist Calif. Dept. Pub. Health, Berkeley, 1962-68; vis. lectr. epidemiology Sch. Vet. Medicine, U. Calif., Davis, 1966-68; assoc. prof. Sch. Medicine, U. Mo., Columbia, 1968-75, Coll. Vet. Medicine, 1968-74, prof., 1974-75; vis. scientist, epidemiology br. Nat. Cancer Inst., Bethesda, Md., 1975-76; prof. Ohio Agrl. Research and Devel. Center, Wooster, 1975—; prof. dept. preventive medicine Coll. Medicine, Ohio State U., Columbus, 1975—, chmn., 1975—; cons. Nat. Bd. Vet. Med. Examiners, 1980, U.S. Dept. Agr., 1979, NIH, 1979, others. Contbr. articles to profl. jours. Served to lt. col. U.S. Air Force. Grantee Am. Cyanamid Co., Canine Research Fund, NIH, FDA, EPA, USPHS, Nat. Cancer Inst., others. Mem. AVMA (co-chmn. public health sect. 1970), Ohio Vet. Med. Assn., Central Ohio Vet. Med. Assn., Am. Pub. Health Assn., Mo. Public Health Assn. (exec. bd. 1969-75), Ohio Public Health Assn., Assn. Tchrs. Vet. Public Health and Preventive Medicine (exec. com. 1970—, pres. 1981—), Am. Assn. Vet. Med. Colls. (sec. 1981—), Conf. Pub. Health Veterinarians, Conf. Tchrs. Food Hygiene (chmn. 1972-73), Am. Assn. Food Hygiene Veterinarians (dir. 1977—), Internat. Assn. Vet. Food Hygienists, Vet. Cancer Soc., Conf. Research Workers in Animal Disease, Soc. Environ. Geochemistry and Health, Internat. Assn. for Comparative Research on Leukemia and Related Diseases, Sigma Xi, Phi Zeta, Gamma Sigma Delta. Subspecialty: Preventive medicine (veterinary medicine). Office: Coll Vet Medicine Ohio State U Sisson Hall 1900 Coffey Rd Columbus OH 43210

DORN, GORDON LEE, inst. adminstr., educator; b. Chgo., June 8, 1937; s. Irvin Arleigh and Grace (Jahr) D.; m. Kathie Lee Dorn, Oct. 30, 1969; children: Scott Lee, Kelly Lee. B.Sc. cum laude, Purdue U., 1958, M.Sc., 1960, Ph.D., 1961. Teaching asst. Purdue U., 1958-59; research assoc. Glasgow (Scotland) U., 1961-63; asst. prof. Albert Einstein Coll. Medicine, 1964-67; assoc. prof. microbiology Baylor U., 1968-70; adj. prof. N. Tex. State U., Denton, 1974—; chmn. dept. microbiology Wadley Insts. Molecular Medicine, Dallas, also; dir. clin. microbiology; pres. Dorn Microbiol. Assocs. Dallas Inc. Contbr. articles to profl. jours. NSF grantee, 1964; NIH grantee, 1965-67; Blanche Mary Taxis Found. grantee, 1970-71; recipient IR 100 award, 1982. Mem. Genetics Soc. Am., AAAS, Am. Genetics Assn., N.Y. Acad. Sci., Am. Soc. Microbiology, Am. Soc. Quality Control. Christian Sci. Subspecialties: Microbiology (medicine); Genetics and genetic engineering (medicine). Current work: Major research interests include interferon prodn., diagnostics test in the areas of bacteriology and mycology; spl. interests include blood, sputum, urine, and throat cultures and antibiotic susceptibility testing. Patentee in field. Home: 1232 Lausanne St Dallas TX 75208 Office: 9000 Harry Hines Blvd Suite 311 Dallas TX 75235

DORNBUSH, RHEA L., psychologist, cons.; b. N.Y.C.; d. Barnett and Betty (Shore) D. B.A., Queens Coll., 1962, M.A., 1963; M.P.H., Columbia U., 1981; Ph.D., CUNY, 1967. Lic. psychologist, N.Y. Asst. prof. psychology Rutgers U., New Brunswick, N.J., 1965-68; asst., then assoc. prof. psychiatry N.Y. Med. Coll., N.Y.C., 1968-73, assoc. prof., 1978-80, prof. psychiatry, 1980—; lectr. in med. psychology in psychiatry Washington U. Sch. Medicine, St. Louis, 1976-78; sr. research scientist Reproductive Biology Research Found., St. Louis, 1976-78; dir. Bradford Nat. Corp., Bradford Trust Co. Contbr. articles to profl. jours. Grantee NIMH, Am. Philos. Soc., Sigma Xi, 1966-80. Mem. AAAS, AAUP, Am. Psychol. Assn. Soc., Biol. Psychiatry, N.Y. Acad. Scis., Am. Public Health Assn. Subspecialty: Psychopharmacology. Current work: Behavioral effects of psychoactive agents (brain peptides). Research, consulting, teaching, administration. Office: Dept Psychiatry NY Ned Coll Valhalla NY 10595

DORNFELD, RICHARD LOUIS, mechanical engineer; b. Oak Park, Ill., Feb. 9, 1947; s. Raymond Louis and Elta V. (Overby) D. B.S.M.E., U. Ill., Urbana, 1975. Registered profl. engr., Ill.; registered patent agt. With Chgo. Bridge & Iron Co., 1975-77, Walker Process Corp. subs. Chgo. Bridge & Iron Co., Aurora, Ill., 1977—; now design engr. water, waste water and indsl. waste water treatment equipment. Chgo. Bridge & Iron Co. (Walker Process Corp. subs.). Served with USN, 1969-73; Served as officer Ordnance Corps USAR, 1975. Mem. ASME., Res. Officers' Assn. Subspecialties: Mechanical engineering; Water supply and wastewater treatment. Current work: High efficiency, low-energy water treatment mechanisms; initial design through final installation of electro-mechanical waste water and water treatment devices.

DOROS, MARIA HECZEY, educator; b. Budapest, Hungary, Oct. 25, 1937; came to U.S., 1963, naturalized, 1970; d. Gabriel and Ethel (Tima) D.; m. Ivan M. Heczey, Apr. 6, 1963 (div. 1980). B.A., CUNY, 1973, M.A., 1975, Ph.D., 1977. Biofeedback technician Grad. Ctr., CUNY, 1974-77; instr. Calif. Sch. Profl. Psychology, Los Angeles, 1977-79; program coordinator N.W. Ctr., Devel. Disabled, Chippewa Falls, Wis., 1979-80; vis. scholar Carthage Coll., Kenosha, Wis., 1980-81; asst. prof. Marietta (Ohio) Coll., 1981—; cons., 1981—. Author: Woman to Woman: On the Menstrual Experience, 1978, Deep Relaxation Exercies, 1979, Behavioral Approach to Treatment of Dysmenorrhea, 1979. Recipient Exxon-Marietta Coll. stipend, 1982-83. Mem. Am. Psychol. Assn., Soc. Behavioral Medicine, Biofeedback Soc. N.Y., N.Y. Acad. Sci., Phi Beta Kappa. Republican. Presbyterian. Subspecialties: Psychobiology; Biofeedback. Current work: Applications of biofeedback, autogenics, imagery to athletic performance; preventive stress mgmt. in large groups. Home: 218 Pennsylvania Ave Apt 4 Marietta OH 45750 Office: Marietta Coll Marietta OH 45750

DORREN, JOHN DAVID, physicist, educator; b. Edinburgh, Scotland, Sept. 3, 1944; came to U.S., 1979; s. Robert and Helen M. (Johnstone) D. B.Sc. with 1st class honours, U. Glasgow, 1966; Ph.D. in Theoretical High Energy Physics, Oxford (Eng.) U., 1969. Royal Soc. exchange fellow Weizmann Inst., Rehoboth, Israel, 1969-70, Inst. fellow, 1970-71; postdoctoral fellow dept. applied math. and theoretical physics U. Cambridge, Eng., 1971-72; research assoc. Seminar fur Theoretische Physik Eidgenossische Technische Hochschule, Zurich, Switzerland, 1972-74; assoc. prof. dept. physics and Biruni Obs., Pahlavi U., Shiraz, Iran, 1975-79; lectr. dept. astronomy and astrophysics U. Pa., Phila., 1980—. Contbr. articles to profl. publs. Fellow Royal Astron. Soc.; mem. Am. Astron. Soc. Subspecialty: Optical astronomy. Current work: Photoelectric photometry of close binary stars; observation and analysis of stars with active chromospheres; optical and ultraviolet studies of X-ray binaries. Office: Dept Astronomy and Astrophysics U Pa Philadelphia PA 19104

DORRIS, ROY LEE, pharmacologist, educator; b. Choctaw, Okla., Oct. 2, 1932; s. Walter James and Beulah (Blakeley) D.; m. Rebecca Joy Wachel, Dec. 31, 1956; children: Joseph Edgar, Rebecca Lee, James David. B.S., Bethany Nazarene Coll., 1959; M.A., Peabody Coll., 1960; Ph.D., Vanderbilt U., 1969. Asst. prof. pharmacology Southwestern Med. Sch., Dallas, 1969-74; prof. Baylor Coll. Dentistry, 1974—. Served with USAF, 1951-53. Mem. Am. Soc. Pharmacology and Exptl. Therapeutics, Am. Assn. Dental Schs., Internat. Assn. Dental Research, Sigma Xi. Republican. Nazarene. Subspecialties: Pharmacology; Neuropharmacology. Current work: Drug actions on brain neurotransmitter functions. Home: 724 Carriage Way Duncanville TX 75137 Office: Dept Pharmacology Baylor College Dentistry 3302 Gaston Ave Dallas TX 75246

DORSON, WILLIAM JOHN, JR., engineering educator, consultant; b. Manchester, N.H., May 9, 1936; s. William John and Josephine (Pinska) D.; m. Denise Estelle Larivee; children: Mark John, Peter George. B.Ch.E., Rensselaer Poly. Inst., 1958, M.Ch.E., 1960; Ph.D., U. Cin., 1967. Research engr. Knolls Atomic Power Lab., Gen. Electric Co., 1958-64, research engr., space power and propulsion, 1964-65; instr. U. Cin., 1965-66; assoc. prof. Ariz. State U., Tempe, 1966-71, prof. engring., 1971—; cons. in field. Mem. editorial bd.: Artificial Organs. Rensselaer Poly. Inst. scholar; Doehla Found. scholar. Fellow AAAS; mem. Assn. for Advancement of Med. Instrumentation, Am. Soc. Artificial Internal Organs (mem. exec. com., trustee 1982-83), Biomed. Engring. Soc., Internat. Soc. for Oxygen Transport to Tissue, Am. Inst. Chem. Engrs., ASME, Sigma Xi, Tau Beta Pi, Phi Kappa Phi, Phi Lambda Upsilon. Club: Mesa Country. Subspecialties: Artificial organs; Biomedical engineering. Current work: Biomedical engineering, artificial kidney, hemofiltration, plasmiapheresis, detoxification, hemoperfusion, blood rheology, artificial lungs, mass transfer, mathematical models. Office: Engineering Center Arizona State University Tempe AZ 85287

DOS, SERGE JACQUES, surgeon, physiology researcher; b. Paris, Jan. 24, 1934; came to U.S., 1957; s. Octave Pierre Marie and Fernande Lucienne (Daire) D.; m. Rasma Kupers, Aug. 19, 1966; children: Soshana, Yasmin, Maiya. M.D., U. Paris, 1964; Ph.D. in Physiology, U. Minn., 1965. Lab. instr. physiology U. Minn., Mpls., 1962-65; instr. in surgery Cornell U., N.Y.C., 1971-73; asst. prof. surgery SUNY-Stony Brook, 1973, asst. prof. clin. physiology, 1973-76; surgeon St. John's Episcopal Hosp., Smithtown, N.Y., 1978—; research com. VA Hosp., Northport, N.Y., 1974-76. Contbr. chpt. to book. USPHS trainee., 1962-65; various research grants NIH, Am. Heart Assn., pvt. labs.; Laureate (Silver Medal) Faculty of Medicine U. Paris, 1966. Fellow N.Y. Acad. Scis.; mem. Am. Fedn. Clin. Research, AAAS, Am. Physiol. Soc., Assn. Acad. Surgery. Subspecialties: Surgery; Cardiac surgery. Current work: Physiology, history. Home: 16 Crooked Oak Rd Belle Terre NY 11777 Office: St John's Episcopal Hosp Route 25A Smithtown NY 11787

DOSAMANTES-ALPERSON, ERMA, psychology educator, consultant; b. Mexico City, Jan. 12, 1940; U.S., 1950; d. Francisco and Martha (Badash) Dosamantes; m. Burton L. Alperson, Jan. 12, 1964. B.S., CUNY, 1959, M.A., 1962; Ph.D., Mich. State U., 1967; postgrad., UCLA, 1972-73. Lic. psychologist, registered movement therapist, Calif. Dir. counseling SUNY-Stony Brook, 1968-72; assoc. prof. psychology Calif. State U.-Los Angeles, 1974-81; prof., dir. grad. movement therapy program UCLA, 1977—; cons. Dept. HHS, 1980—, Dept. Edn., 1980—, Am. Psychiat. Assn., 1980—. Editor: Am. Jour. Dance Therapy, 1978-80; contbr. articles to profl. jours. NIMH grantee, 1962-66; USPHS tng. fellow, 1963-66; UCLA research grantee, 1977-82. Mem. Am. Dance Therapy Assn. (pres. 1980-82), Am. Assn. Study of Mental Imagery, Am. Psychol. Assn. Subspecialties: Imagory and movement; Psychotherapy. Current work: Psychotherapy, movement therapy. Home: 2346 Panorama Terr Los Angeles CA 90039 Office: University of California 405 Hilgard Ave Los Angeles CA 90024

DOSS, ROBERT PAUL, plant scientist; b. Madera, Calif., May 25, 1945; s. Jesse Paul and Rhea March (Trethewey) D.; m. Elke F. Mittmann, June 7, 1968; children: Lynn E., Christopher P. A.B., Calif. State U., Fullerton, 1968; Ph.D. (NDEA fellow), U. Calif., Davis, 1974. Research asso. U. Calif., Davis, 1974-76; plant physiologist U.S. Dept. Agr., Puyallup, Wash., 1976—. Contbr. numerous articles to profl. publs. Served with U.S. Army, 1969-71. Mem. AAAS, Am. Soc. Plant Physiologists, Entomol. Soc. Am., Scandanavian Soc. Plant Physiology, Japanese Soc. Plant Physiologists. Subspecialty: Plant physiology (biology). Current work: Plant growth and development, phytochemical ecology; working on phytochemical mechanisms of plant disease and insect resistance and on plant growth and development.

DOSTER, JOSEPH MICHAEL, nuclear engineering educator, consultant; b. Chapel Hill, N.C., Dec. 3, 1954; s. Joseph C. and Ann Howard D.; m. Ellen Winship Rogers, June 6, 1981. B.S., N.C. State U., Raleigh, 1977, Ph.D., 1982. Vis. instr. N.C. State U., Raleigh, 1980-81, instr., 1981-82, asst. prof. nuclear engring., 1982—; cons. Research Triangle Inst., Research Triangle Park, N.C., 1980—. Mem. Am. Nuclear Soc., Sigma Xi. Democrat. Subspecialties: Nuclear fission; Fluid mechanics. Current work: Nuclear reactor systems simulation, thermal-hydraulics, two phase flow, radiation transport, radiation shielding Monte Carlo simulation. Office: 1110 Burlington Engring Labs NC State U Raleigh NC 27650

DOSZKOCS, TAMAS ENDRE, computer specialist, information scientist; b. Banhida, Hungary, Mar. 26, 1942; came to U.S., 1967; s. Kalman and Margit (Lowi) D.; m. Veronica Petrics, June 29, 1963 (div. 1968); 1 son, Tamas; m. Christine Anne Aquino, July 10, 1971; children: Matthew, Christopher, Adam. B.A., Kossuth U., Hungary, 1964; M.L.S., U. Md., 1968, M.S., 1972, Ph.D., 1979. Tchr. Fazekas Gimnazium, Debrecen, Hungary, 1964-65; librarian U. Md., 1968-74; computer specialist Nat. Cancer Inst., Bethesda, Md., 1977-79; adminstrv. librarian Nat. Library of Medicine, Bethesda, 1979-80, supervisory tech. info. specialist, 1980—; lectr. U. Md., 1970-77; cons. Tracor-Jitco, Rockville, Md., 1973-74, Md. State Dept. Edn., 1974-80. Recipient Nat. Library of Medicine Bd. Regents award for Tech. Excellence, 1979. Mem. Assn. Computing Machinery, Am. Soc. Info. Sci. Republican. Roman Catholic. Subspecialties: Information systems, storage, and retrieval (computer science); Automated language processing. Current work: Ergonomics; machine-aided indexing; automatic classification; natural language searching. Home: 9627 Lawndale Dr Silver Spring MD 20901 Office: Nat Library of Medicine 8600 Rockville Pike Bethesda MD 20209

DOTTER, CHARLES THEODORE, radiology educator, researcher; b. Boston, June 14, 1920; s. John Maury and Rosalind (Allin) D.; m. Pamela Beattie, Sept. 30, 1944; children: Barbara Allin, Jeffrey Churchill, Jane Huntington. A.B., Duke U., 1941; M.D., Cornell U., 1944. Diplomate: Am. Bd. Radiology, 1950. Instr. medicine Cornell U. Med. Coll., N.Y.C., 1948-52, instr. radiology, 1948-51, asst. prof. radiology, 1951-52; mem. faculty Oreg. Health Scis. U., Portland, 1952—, now prof. radiology, chmn. dept.; cons. VA Hosp., Portland; mem. rev. bd. Nat. Heart and Lung Inst., 196568; mem. surg. drugs adv. com. FDA, 1974-77, chmn. com., 1977-78. Author: (with Israel Steinberg) Angiocardiography; contbr. over 300 articles to sci. jours.; author/producer 2 med. films. Served to lt. (j.g.) USNR, 1944-45; surgeon USMC, 1945-46; CBI. Recipient Gold medal Chgo. Med. Soc./Chgo. Radiol. Soc., 1981, Radiol. Soc. N.Am., 1981, Am. Coll. Radiology, 1983. Fellow Am. Coll. Angiology, Am. Coll. Radiology; mem. Am. Heart Assn., AMA, Am. Roentgen Ray Soc., Assn. Univ. Radiologists, Czechoslovak Med. Soc. (hon.), Internat. Cardiovascular Soc., Internat. Soc. Angiology, Multnomah County Med. Soc., Oreg. Heart Assn., Oreg. Radiol. Soc., Oreg. Med. Soc., Oreg. Thoracic Soc., Pacific N.W. Radiol. Soc., Radiol. Soc. N.Am., Soc. Chairmen Acad. Radiology Depts., Soc. Cardiovascular Radiology, Western Angiography Soc. Clubs: American Alpine, Mazamas (Portland). Subspecialties: Radiology; Diagnostic radiology. Current work: Medical devices, invention of angiography; nonoperative cateter therapy; interventional radiology; instruction. Office: 3181 SW Sam Jackson Park Rd Portland OR 97201

DOTY, ROBERT WILLIAM, neuroscience educator; b. New Rochelle, N.Y., Jan. 10, 1920; s. Earle Birdsell and Ethel (Mack) D.; m. Elizabeth Natalie Jusewich, Aug. 30, 1941; children: Robert William, Mary E., Cheryl A., Richard M. B.S., U. Chgo., 1948, M.S., 1949, Ph.D., 1950. Fellow U. Ill.-Chgo., 1950-51; asst. prof. U. Utah, Salt Lake City, 1951-56; assoc. prof. U. Mich.-Ann Arbor, 1956-61; prof. U. Rochester, N.Y., 1961—; sci. advisor NIMH, 1975-79; mem. visual sci.-study sect. NIH, 1968-72, exptl. psychol. study sect., 1978-81; advisor Yerkes Primate Ctr., Atlanta, 1975-78. Chief editor: Neurosci. Translations, 1963-70; assoc. editor: jours. including Exptl. Neurology, 1965-75, Jour. Physiologie, Paris, 1971-81, Behavioral Brain Research, 1981—. Served to capt. AUS, 1942-46. Fellow Am. Psychol. Assn. (pres. Div. 6 1983); mem. Soc. for Neurosci. (council 1970-75, pres. 1976), Internat. Brain Research Orgn., Am. Physiol. Soc. Subspecialties: Neuropsychology; Neurophysiology. Current work: Interhemispheric mnemonic transfer; correlation of neuronal with visual responses in man and macaque. Office: Ctr for Brain Research U Rochester Rochester NY 14642

DOUBEK, CLIFFORD JAMES, chem. engr.; b. Chgo., Mar. 27, 1925; s. James Frank and Sylvia (Fara) D.; m. Mary Lillian Vinduska, Nov. 5, 1949; 1 dau., Annette Mary. B.S. in Chem. Engring, Ill. Inst. Tech., 1949. Registered profl. engr., Calif. Research engr., mgr. quality assurance, asst. nat. dir. microbiology Johnson & Johnson Co., Chgo. and New Brunswick, N.J., 1949-69; mgr. quality assurance C.R. Bard Co., Murray Hill, N.J., 1970, Abbott Labs., North Chicago, Ill., 1971-74; dir. quality assurance Hancock Extracorporeal Co., Anaheim, Calif., 1974—. Served with U.S. Army, 1943-46; ETO. Fellow Am. Soc. Quality Control (cert. reliability engr., nat. chmn. biomed. div.), Soc. Advancement Mgmt./Am. Mgmt. Assn. (profl. mgr. 1979, internat. chmn. bd., internat pres.); mem. VFW. Subspecialties: Chemical engineering; Biomedical engineering. Current work: Heart valve development; quality assurance in the manufacture of porcine and pericardial heart valves and cardiovascular devices. Office: 4633 E La Palma Ave Anaheim CA 92807

DOUD, ERIC LEO, architect; b. Glenwood Springs, Colo., July 5, 1950; s. Franklin R. and Marjorie L. (Myers) D. B.Arch., U. Colo., 1974. Lic. architect, Colo. Prin. Glauth Surveying, Woodland Park, Colo., 1970-72; field and office staff Childress Paulin Architects, Denver, 1973; prin. Telluride Designworks, Colo., 1974—; Mem. Telluride Hist. Preservation Commn., 1976-78. Contbr. articles to profl. jours. Mem. Internat. Solar Energy Soc. Subspecialty: Solar energy. Current work: Involved in development of passive solar design features and their integration into building systems. Home: Box 713 401 E Columbia Telluride CO 81435 Office: Box 1248 117 N Willow Telluride CO 81435

DOUGHERTY, GEORGE JOHN, computer sciences company manager; b. Lakewood, N.J., Jan. 30, 1943; s. Edward George and Jean Gilroy (McIntyre) D.; m. Karen Ann O'Hern, Aug. 30, 1969; children: Dawn Ann, Erin Jean. B.Gen. Studies, U. Nebr., 1971; grad, U.S. Army Command and Gen. Staff Coll., 1978; M.S., Boston U., Heidelberg, Ger., 1979. Commd. officer U.S. Army, advanced through grades to maj. Res.; comdr. Hdqrs. Battery 1st Bn. 5th Field Arty., Fort Riley, Kans., 1976-77, intelligence officer, 1977; emergency action officer Hdqrs. U.S Army Europe, Heidelberg, Ger., 1977-80, programmer analyst, 1980-81, Tampa, Fla., 1981-82; sr. mem. tech. staff Computer Scis. Corp., Tampa, 1982-83, work area mgr., 1983—; adj. lectr. City Colls. Chgo., Heidelberg, 1980-81, Boston U., 1980-81, U. Tampa, 1982; adj. instr. Fla. Inst. Tech., St. Petersburg, 1982—. Pres., Butterfield Homeowners Assn., Manhattan, Kans., 1974; bd. dirs. Twelve Oaks Homeowners Assn.; treas. Soc. Pen and Sword, Omaha, 1970; mem. bus. adv. council Abilities Inc. of Clearwater (Fla.). Decorated Silver Star. Mem. Assn. U.S. Army, Inst. Cert. of Computer Profls., Assn. Computing Machinery (v.p. chpt.), Data Processing Mgmt. Assn., Processing Mgmt. Assn. Roman Catholic. Subspecialties: Database systems; Distributed systems and networks. Current work: Distributed database structures for decision support systems. Home: 7517 Mayfair Ct Tampa FL 33614 Office: Computer Scis Corp PO Box 19188 Tampa FL 33686

DOUGHERTY, JOHN ALFRED, JR., psycholpharmacologist; b. Shreveport, La., Jan. 5, 1943; s. John A. and Ruth M. (Kloeck) D. B.A., U. Minn., 1969, Ph.D. in Psychology and Pharmacology, 1973. Staff psychologist VA Med. Ctr., Lexington, Ky.; now assoc. prof. depts. psychiatry and pharmacology Gead. Ctr. Toxicology, U. Ky., Lexington. Contbr. articles on psychopharmacology to profl. jours. Served with USAF, 1961-66. VA grantee, 1973-82; Nat. Inst. Drug Abuse grantee, 1975-81. Mem. Behavioral Pharmacology Soc., Am. Psychol. Assn., Neurobehavioral Toxicology Soc., Soc. Neuroscis. Current work: Behavioral effects of drugs and chemicals, psychology of pain, researcher on behavioral effects of nicotine and agent orange (herbicides), pain management services for inpatients. Office: 116-CDD VA Med Center Lexington KY 40511 Office: Depts Psychiatry and Pharmacology and Toxicology U Ky Lexington KY 40536

DOUGHERTY, JOSEPH CHARLES, nephrologist; b. Troy, N.Y., Feb. 13, 1934; s. William Joseph and Dorothy Dewar (Anker) D.; m. Katherine Irene Barron, June 6, 1959; children: William R., Suzanne V., Timothy J., Laura E. B.S., Manhattan Coll., 1956; M.D., Cornell U., 1960. Diplomate: Am. Bd. Internal Medicine, Intern Albany (N.Y.) Med. Center Hosp., 1960-61. Resident in internal medicine Bellevue Hosp. Center, N.Y.C.; fellow N.Y. Heart Assn., 1964-67; asst. prof. medicine Cornell U. Med. Coll., 1967-68, Albert Einstein Coll. Medicine, 1968-71; assoc. prof. medicine U. Tex. Health Sci. Ctr., San Antonio, 1971-76; assoc. dir. Moses Taylor Hypertension Inst., Scranton, Pa., 1976-80; ptnr. Valley Diagnostic Clinic, Harlingen, Tex., 1980—; exec. com. ESRD Network 11, Dallas, 1980-82; med. dir. Watson W. Wise Dialysis Ctr., Harlingen, 1980—. Author: Concise Textbook on Nephrology, 1977; contbr.: to Hand Book of Nutrition, 1978; contbg. editor: Med. Times Resident and Staff Physician, 1966-78; contbr. articles sci. jours. Med. adv. bd. Kidney Found. So. Tex., 1973-76, Tex. Dept. Pub. Safety, Drivers Lic., Austin, 1973-76, Keystone Heart Assn., 1977-80; dir. Cameron County Heart Assn., 1980-83. Mead Johnson fellow, 1962-63; N.Y. Heart Assn. fellow, 1964-68; sr. investigator, 1969-71. Fellow ACP; mem. Am. Soc. Artificial Internal Organs, Transplantation Soc., Am. Soc. Nephrology. Roman Catholic. Subspecialties: Nephrology; Internal medicine. Current work: Economics of health care especially outpatient services, assessment of nutritional needs in dialysis patients. Office: Valley Diagnostic Clinic 2200 Haine Dr Harlingen TX 78550

DOUGHERTY, ROBERT MALVIN, microbiologist, researcher; b. Long Branch, N.J., May 25, 1929; s. Robert L. and Justine (Mernone) D.; m. Barbara Geran, June 26, 1950; children: Karen, Joyce. B.S., Rutgers U., 1952, M.S., 1954, Ph.D., 1957. Instr. U. Rochester, 1957-60; USPHS Spl. Research fellow Imperial Cancer Research Fund, London, 1960-62; assoc. prof. microbiology SUNY Upstate Med. Center, Syracuse, 1962-69, prof., 1969—. Contbr. numerous articles profl. jours. Served to cpl. AUS, 1946-48. Research grantee Am. Cancer Soc., 1963-68, USPHS, 1965-79, NSF, 1967-72. Mem. Am. Assn. Immunology, AAAS, Am. Soc. Microbiology, Soc. Exptl. Biology and Medicine, Sigma Xi. Subspecialties: Microbiology; Virology (biology). Current work: Biology of oncogenic viruses. Office: 766 Irving Ave Syracuse NY 13210

DOUGHERTY, THOMAS JOHN, radiobiology and photobiology educator; b. Buffalo, Aug. 2, 1933. B.S., Canisius Coll., 1955; Ph.D. in Organic Chemistry, Ohio State U., 1959. Research chemist film dept. Yerkes Research and Devel. Lab., E.I. du Pont de Nemours & Co. Inc., 1959-67, staff scientist, 1967-70; cancer research scientist to assoc. cancer research scientist Roswell Park Meml. Inst., Buffalo, 1970-75, prin. cancer research scientist, 1975—, head div. radiobiology, dept. radiol. medicine, 1976—; assoc. prof. SUNY-Buffalo, 1975—. Recipient Benjamin Franklin award, 1981, James H. Crowdle award, 1982, W.D. Mark award, 1983. Mem. AAAS, Am. Chem. Soc., Am. Assn. Cancer Research, Am. Soc. Photobiology, Sigma Xi. Subspecialty: Laser medicine. Office: Dept Radiation Medicine Roswell Park Meml Inst Buffalo NY 14263

DOUGHMAN, DONALD JAMES, ophthalmologist; b. Des Moines, Sept. 26, 1933; s. Edward Gilmore and Edith Marie (Johnson) D. B.M.E., Drake U., Des Moines, 1955; M.D., State U. Iowa, 1961. Diplomate: Am. Bd. Ophthalmology. Intern Los Angeles County Hosp., 1961-62; physician Permentente Med. Group, Panorama City, Calif., 1962-63; resident in internal medicine, then resident in ophthalmology U. Iowa Hosp., 1963-68; chief ophthalmologist USPHS Hosp., Boston, 1963-70; fellow cornea and external disease Mass. Eye and Ear Infirmary and Retina Found., Boston, 1970-72; mem. faculty U. Minn. Med. Sch., 1972—, prof. ophthaomology, chmn. dept., 1979—. Author papers corneal transplantation, donor corneal storage. Fellow A.C.S.; mem. Am. Acad. Ophthalmology, Am. Ophthal. Assn., AMA, Minn. Med. Assn., Hennepin County Med. Soc., Minn. Acad. Ophthalmology, Assn. Research in Vision and Ophthalmology, Alpha Omega Alpha. Subspecialty: Ophthalmology. Office: Box 493 Mayo 516 Delaware St SE Minneapolis MN 55455

DOUGLAS, BEN HAROLD, physiologist, educator; b. Monticello, Miss., Feb. 20, 1935; s. Ben H. and Nell (Rich) D; m. Jo Ann Rutland, Aug. 16, 1953; children: Ben H., Pamela. B.S., Miss. Coll., 1956; Ph.D., U. Miss. Med. Ctr., 1964. Assoc. prof. medicine U. Miss. Med. Ctr., Jackson, 1968-77, assoc. prof. physiology, 1970—, assoc. prof. anatomy, 1977-78, prof. anatomy, 1978, asst. vice chancellor, 1981—, chmn. grad. council; vis. prof. Med. Research Council Blood Pressure Unit, Glasgow, Scotland, 1973-74. Editorial bd.: Clin. and Exptl. Hypertension, 1979; contbr. chpts. to sci. publs. Am. Heart Assn. established investigator, 1972-76; recipient Silver Disting. Achievement award, 1978. Subspecialties: Toxicology (medicine); Physiology (medicine). Current work: Cardiovascular, hypertension. Office: U. Miss Med Ctr 2500 N State St Jackson MS 39216

DOUGLAS, BRYCE, pharmaceutical company executive; b. Glasgow, Scotland, Jan. 6, 1924; came to U.S., 1958; s. Alexander and Mary (Turner) D.; m. Joyce M. Flynn, Aug. 24, 1955; children: Alan David, Neal Malcolm, Iain Graham. B.Sc. with honors, Glasgow U., 1944; Ph.D. in Organic Chemistry, Edinburgh (Scotland) U., 1948. Chemotherapy researcher, research lab. Royal Coll. Physicians, Edinburgh, 1947-49; research asst. biol. chemistry Aberdeen (Scotland) U., 1949; research fellow in alkaloid chemistry NRC Can., Ottawa, 1949-51; research fellow dept. pharmacology Harvard U., 1952-53; research asso., lectr. Ind. U., Bloomington, 1953-56; vis. research asso. U. Malaya, Singapore, 1956-58; with Smith Kline & French Labs., Phila., 1956—, v.p. research and devel., 1971-80, pres. research and devel., 1980-81, v.p. sci. and tech., 1981—. Contbr. articles to profl. jours. Bd. dirs. Royal Soc. Medicine Found.; bd. overseers U. Pa. Sch. Dental Medicine; bd. mgrs. Franklin Inst. of Phila.; bd. dirs. Franklin Research Ctr., Bartol Found.; bd. dirs. Southeastern Pa. chpt. ARC. Fellow Royal Soc. Chemistry (U.K.), Coll. Physicians Phila.; mem. N.Y. Acad. Scis., AAAS, Am. Chem. Soc. Subspecialty: Medicinal chemistry. Patentee in field. Home: Box 672 Kimberton PA 19442 Office: PO Box 7929 Philadelphia PA 19101

DOUGLAS, DEXTER RICHARD, plant pathologist, consultant, researcher; b. Benton, Ohio, Nov. 14, 1937; s. Richard Loren and Dorthey Ann (Phillabaum) D.; m. Bernadine Rose, May 26, 1962; children: Laura Ann, Cynthia Marie. B.S. in Biology, Kent State U., 1962; M.S. in Plant Pathology, U. Wyo., 1965, Ph.D., U. Minn.-St. Paul, 1968. Research plant pathologist U.S. Dept. Agr., Aberdeen, Idaho, 1968-75; pvt. researcher Chem. Supply Co. Inc., Twin Falls, Idaho, 1976-77; pvt. cons., pres., gen. mgr. Hi-Alta Inc., Moore, Idaho, 1978—; area mgr. Idaho Crop Improvement Assn., Potato Seed Cert., Idaho Falls, Idaho, 1980-82; expert witness on potato diseases. Contbr. articles to profl. jours. Served with U.S. Army, 1962-64. Mem. Am. Phytopathol. Soc., Am. Potato Assn. Lodge: Lions. Subspecialty: Plant pathology. Current work: Potato diseases, control of seed piece decay, seed certification; seed potato prodn. Home: RD 1 Box 394 Arco ID 83213 Office: Box 916 Moore ID 83255

DOUGLAS, J. FIELDING, toxicologist, biochemist; b. Delta, Utah, Jan. 25, 1927; s. Ben and Amelia (Fielding) D.; m. Rose Terrazzino, Sept. 16, 1951; children: David, Pamela, Jason. B.S. with high honors, U. Ill., 1948; M.A., Columbia U., 1950, Ph.D., 1953. Project leader Johnson & Johnson, New Brunswick, N.J., 1952-58; dir. biochemistry Carter-Wallace, Cranbury, N.J., 1958-74; dep. dir. carcinogenesis testing program Nat. Cancer Inst., Bethesda, Md., 1976-80; chief program ops. br. nat. toxicology program Nat. Inst. Environ. Health Scis., Bethesda, 1980—; cons., 1974-80. Contbr. articles to sci. jours. Served in U.S. Army, 1944-46; ATO. Recipient award Richard Neff Soc., 1966; USPHS fellow, 1950-52. Mem. Am. Chem. Soc. (dir. sect.), Soc. Toxicology, Am. Soc. Pharmacology and Exptl. Therapeutics, Soc. Exptl. Biology and Medicine, AAAS, N.Y. Acad. Scis. Subspecialties: Toxicology (medicine); Cancer research (medicine). Current work: Chemical carcinogenesis and toxicology; evaluation of environmental and industrial chemicals for carcinogenicity and toxicology. Patentee in field. Home: Hermitage Farm PO Box 533 Front Royal VA 22630 Office: NIH Landlow Bldg Bethesda MD 20205

DOUGLAS, MICHAEL GILBERT, biochemistry educator; b. Perth, Australia, June 9, 1945; came to U.S., 1948; s. Claude Earl and Margaret Mary (Anderton) D.; m. Joanne Beth Arnall, Nov. 24, 1973; children: Hannah, Peter. B.S., Southwestern U., Georgetown, Tex., 1967; M.S., St. Louis U., 1969, Ph.D., 1974. Postdoctoral fellow U. Tex. Health Sci. Ctr., Dallas, 1974-75; postdoctoral fellow Biozentrum, Basel, Switzerland, 1975-77; asst. prof. U. Tex. Health Sci. Ctr., San

Antonio, 1977-82, assoc. prof., 1982—; cons. New Eng. Nuclear, Newton, Mass., 1982—, Fusion Biotech, Inc., San Antonio, 1982—, Sandoz Pharm., Basel, Switzerland, 1983—. Contbr. articles to sci. jours. Chmn. bldg. com. Prince of Peace Roman Catholic Ch., 1979-83. Served in U.S. Army, 1969-70; Vietnam. NIH grantee, 1978—; Robert A. Welch Found. grantee, 1980—; Burroughs Wellcome Fund grantee, 1982. Mem. Am. Soc. Microbiology, Am. Soc. Biol. Chemists, Am. Soc. Cell Biology, N.Y. Acad. Sci., AAAS. Democrat. Lodge: KC. Subspecialties: Biochemistry (biology); Genetics and genetic engineering (biology). Current work: Molecular genetics and cell biology of subcellular membrane targeting and assembly membrane bound enzyme complexes, control of genes involved in [illegible] biogenesis. Home: 8518 Ridge Stone San Antonio TX 78251 Ofice: 7703 Floyd Curl Dr San Antonio TX 78284

DOUGLAS, TOMMY C., immunologist, educator; b. Durant, Okla., Sept. 29, 1946; s. Harry G. and Florence E. (Teague) D. A.B., Princeton U., 1969; M.S., Calif. Inst. Tech., 1970, Ph.D., 1974. Postdoctoral fellow Calif. Inst. Tech., Pasadena, 1974; mem. Basel (Switzerland) Inst. Immunology, 1974-76; asst. prof. U. Tex. Grad. Sch. Biomed. Sci., Houston, 1977-82, assoc. prof., 1982—; vis. asst. prof. Calif. Inst. Tech., 1980. NIH grantee, 1979-82. Mem. Am. Assn. Immunologists, Sigma Xi. Subspecialties: Genome organization; Immunobiology and immunology. Current work: Immunogenetics, cell surface antigens, immunogenicity, immunodeficiency, gene mapping, isozymes. Office: Med Genetics U Tex Grad Sch Biomed Sci PO Box 20334 Houston TX 77025

DOUGLAS, WILLIAM HUGH, dental educator; b. Belfast, No. Ireland, Aug. 18, 1937; came to U.S., 1977; s. James and Susan (Marno) D.; m. Margaret Elizabeth Cartwright, July 17, 1971; children: Emma Susan, Harriet Irene. B.S., U. Belfast, 1959, M.S., 1961, Ph.D., 1965; B.D.S., U. London, 1970. Licentiate in dental surgery Royal Coll. Surgeons (Eng.). Lectr. dental materials U. Belfast, 1961; Nuffield fellow Guys Hosp., London, 1965-70; lectr. dentistry Welsh Nat. Sch. Medicine, Cardiff, Wales, 1971-78; vis. assoc. prof. Sch. Dentistry, U. Mich., Ann Arbor, 1977; dir. biomaterials program Sch. Dentistry, U. Minn., Mpls., 1978—, assoc. prof., 1983—; cons. Contbr. articles to dental jours. Mem. Internat. Assn. Dental Research, Omicron Kappa. Upsilon. Presbyterian. Subspecialties: Biomaterials; Restorative dentistry. Current work: Composite technology and adhesives; composite, resin, filler, dentin, enamel, acid etch, adhesive, robotics, computer graphics, clinical measurement. Co-patentee in field; developer artifical mouth for testing dental materials. Office: University of Minnesota 16-212 HSUA Minneapolis MN 55455

DOUTHIT, HARRY ANDERSON, JR., microbial physiology educator; b. Raymondville, Tex., June 18, 1935; s. Harry Anderson and Myra May (Miller) D.; m. Karin Mechthild Weishaar, May 2, 1935; children: Timothy, Lorel. B.A., U. Tex., 1961, Ph.D., 1965; Postdoctoral fellow, U. Wis., 1964-67. Asst. prof. U. Mich., 1967-73, assoc. prof., 1973—. Served with U.S. Army, 1956-59. Recipient Dinsting. Service award U. Mich., 1973. Mem. Am. Soc. Microbiology, AAAS. Subspecialties: Microbiology; Developmental biology. Current work: Biochemical and developmental aspects of dormant systems. Office: Div Biology U Mich Ann Arbor MI 48109

DOW, BRUCE MACGREGOR, neurophysiologist, educator; b. Newton, Mass., Oct. 30, 1938; s. Norman MacGregor and Marjorie (Stone) D.; m. Ann Powell, Jan. 21, 1967 (div. Feb. 1981); children: Michael, Sara; m. Soonja Choi, Dec. 4, 1982. B.A., Wesleyan U., 1960; M.D., U. Rochester, 1966. Intern in medicine Balt. City Hosp., 1966-67; research assoc. Nat. Inst. Dental Research, Bethesda, Md., 1967-70; sr. staff fellow Nat. Eye Inst., Bethesda, 1970-75; vis. scientist NIMH, 1975-76; research assoc. prof. physiology SUNY-Buffalo, 1976-78, assoc. prof., 1978—. Served with USPHS, 1967-70. Nat. Eye Inst. grantee, 1978-88; Fulbright fellow, France, 1960-61; Danforth fellow, 1960-62. Mem. Soc. Neurosci., Assn. Research in Vision and Ophthalmology, Phi Beta Kappa. Democrat. Episcopalian. Subspecialty: Neurophysiology. Current work: Neural mechanisms of foveal vision in primates, analysis of single cell responses in trained monkeys. Home: 77 Garnet Rd Buffalo NY 14226 Office: SUNY Neurobiology Div Dept Physiology 4234 Ridge Lea Rd Buffalo NY 14226

DOW, DANIEL GOULD, educator; b. Ann Arbor, Mich., Apr. 26, 1930; s. William Gould and Edna Lois (Sontag) D.; m. Kathleen Mary Bond, June 19, 1954; children—Sarah, Suzanne, Jennifer, Gordon. B.S. in Engring, U. Mich., 1952, M.S., 1953; Ph.D., Stanford U., 1958. Asst. prof. elec. engring. Calif. Inst. Tech., Pasadena, 1958-61; with Varian Assos., Palo Alto, Calif., 1961-68; prof. U. Wash., Seattle, 1968—, chmn. dept. elec. engring., 1968-77; asso. dir. Applied Physics Lab., 1977-79; dir. Washington Energy Research Center, 1979-81; cons. Hughes Aircraft, Malibu, Calif., 1958-61, Varian Assos., 1968-71, Boeing Co., 1973-74, John Fluke Co., 1979—; mem. Adv. Group on Electron Devices, Microwave Working Group, 1965-76, Wash. Energy Policy Council, 1973-74; mem. subpanel on energy research Energy Research Adv. Bd., 1980; mem. panel on measurement services Nat. Acad. Scis.-Nat. Bur. Standards. Served to lt. USAF, 1953-55. Mem. IEEE. (sr. mem.). Subspecialties: Electronics; 3emiconductors. Current work: Microwaves, semiconductor modeling. Home: 9620 NE 31st St Bellevue WA 98004

DOW, LOIS WEYMAN, physician, consultant; b. Cin., Mar. 11, 1942; d. Albert Dames and Else Marion (Krug) W.; m.; children: Elizabeth, Alan. B.A. summa cum laude, Cornell U., 1964; M.D. cum laude, Harvard U., 1968. Diplomate: Am. Bd. Internal Medicine with spltys. hematology, med. oncology. Intern Bronx Mcpl. Hosp. Center, 1968-69; resident Presbyn. Hosp., N.Y.C., 1969-70; fellow in hematology Columbia U., 1970-72; instr. U. Tenn., Memphis, 1972-73, assoc. prof., 1973-74; research assoc. St. Jude's Children's Research Hosp., Memphis, 1974-77, asst. mem., 1977-80, assoc. mem., 1980—; cons. Nat. Cancer Inst., 1978-82. Contbr. articles to profl. jours. NIH grantee. Mem. Am. Fedn. Clin. Research, ACP, Am. Soc. Hematology, Am. Assn. Cancer Research, Internat. Soc. Exptl. Hematology, Am. Soc. Clin. Oncology, Phi Beta Kappa, Phi Kappa Phi, Alpha Lambda Delta, Alpha Epsilon Delta. Subspecialties: Cancer research (medicine); Cell and tissue culture. Current work: Cell culture studies of normal and malignant cells; cell kinetics. Office: Div Hematology/Oncolog St Jude Children's Research Hosp Memphis TN 38101

DOW, MARTHA ANNE, microbiology educator; b. Little Rock, Jan. 3, 1939; d. Clarence E. and Gretchen D. (Eudy); m. Gary E. Dow, Aug. 28, 1960; children: Julie Denise, Kevin A., Jerilyn Kay. B.S., No. Mont. Coll., 1961; M.S., Mont. State U., 1969. Registered sanitarian Nat. Registry Microbiologists; cert. wastewater treatment operator Class I, Mont. Lab. technician Kem Data, Havre, Mont., 1971-76; dairy microbiologist Vita-Rich, Havre, 1976; asst. prof. microbiology No. Mont. Coll., Havre, 1976—, dir. water quality tech., 1976—. Contbr. to, Natural Systems Assessment, 1982. Chmn. Hill County Health Bd., 1981-82; dist. supr. Soil Conservation Service, 1981—. NSF grantee, 1979; East-West Ctr. fellow, Honolulu, 1982. Mem. Mont. Water Pollution Control Fedn. (bd. dirs.), AAUW (pres. 1976), Delta Kappa Gamma. Subspecialties: Water supply and wastewater treatment; Microbiology. Current work: Microbiology, water and wastewater technology, environmental virology. Office: Northern Montana College Havre MT 59501

DOWD, JOHN PETER, physics educator and researcher; b. New Bedford, Mass., Feb. 1, 1938; s. John Henry and Estelle (Fournier) D.; m. Mary Beth Vancini, Feb. 12, 1960; children: Michael, Paul. B.S., MIT, 1959, Ph.D., 1966. Vis. researcher German Electron Synchrotron, Hamburg, W.Ger., 1966-67; asst. prof. physics Southeastern Mass. U., North Dartmouth, 1967-72, assoc. prof., 1972-77, prof., 1978—; guest researcher Cambridge (Mass.) Electron Accelerator, 1967-72, Brookhaven Nat. Lab., Upton, N.Y., 1972—; guest scientist U. Bonn, W.Ger., 1978-79. NSF grantee. Mem. Am. Phys. Soc., Am. Assn. Physics Tchrs., AAAS, Sigma Xi. Subspecialty: Particle physics. Current work: Experiment high energy physics; application of microcomputers to physics laboratory instruction. Home: Box 833 Fairhaven MA 02719 Office: Dept Physics Southeastern Mass U North Dartmouth MA 02747

DOWELL, EARL HUGH, university dean, aerospace and mechanical engineering educator; b. Macomb, Ill., Nov. 16, 1937; s. Earl S. and Edna Bernice (Dean) D.; m. Lynn M. Cary, July 21, 1981; children: Marla Lorraine, Janice Lynelle, Michael Hugh. B.S., U. Ill., 1959; S.M., Mass. Inst. Tech., 1961, Sc.D., 1964. Research engr. Boeing Co., 1962-63; research asst. Mass. Inst. Tech., 1963-64, research engr., 1964, asst. prof., 1964-65; asst. prof. aerospace and mech. engring. Princeton U., 1964-68, asso. prof., 1968-72, prof., 1972-83, assoc. chmn., 1975-77, acting chmn., 1979; dean Sch. Engring. Duke U., Durham, N.C., 1983—; cons. to industry and govt. Author: Aeroelasticity of Plates and Shells, 1974, A Modern Course in Aeroelasticity, 1978; Assoc. editor: AIAA Jour, 1969-72; Contbr. articles to profl. jours. Chmn. N.J. Noise Control Council, 1972-76. Named outstanding young alumnus U. Ill. Sch. Aero. and Astronautical Engring., 1973, disting. alumnus, 1975. Assoc. fellow AIAA (Structures, Structural Dynamics and Material award 1980, v.p. publs. 1981-83); mem. Acoustical Soc. Am., ASME, Am. Acad. Mechs. Subspecialties: Aerospace engineering and technology; Theoretical and applied mechanics. Current work: Nonlinear dynamics, aeroelastilcity, structural dynamics, acoustics. Home: 2207 Chase St Durham NC 27706 Office: School of Engineering Duke University Durham NC The pursuit of excellence is exciting; the achievement is sometimes anti-climatic

DOWLEY, MARK WILLIAM, physicist, business exec.; b. Dundalk, Ireland, Apr. 28, 1934; came to U.S., 1959, naturalized, 1964; s. Arthur Gerard and Sheila Mary (Williams) D.; m. Mary F. Donnelly, Mar. 29, 1967; children: A. David, Aoife, Patrick. B.Sc. with 1st class honours, Univ. Coll., Dublin, 1956; M.A., U. Toronto, 1957, Ph.D., 1959. Postdoctoral research fellow in physics U. Calif., Berkeley, 1959-61; with IBM Research Labs., San Jose, Calif., 1961-67, Spectra Physics, Mountain View, Calif., 1967-68; dir. researcher Coherent Radiation, Palo Alto, Calif., 1968-72; pres. chmn. bd. Liconix, Sunnyvale, Calif., 1972—; Lectr. U. Calif., Berkeley, Calif. State U., San Jose. Contbr. numerous articles to sci. jours. Numerous scholarships and fellowships, 1952-62. Mem. IEEE, Optical Soc. Am., Am. Phys. Soc. Roman Catholic. Clubs: University (Palo Alto); Richmond (Calif.) Yacht. Subspecialties: Laser research; Optical engineering. Current work: Design and physics of helium cadmium lasers; strategic planning and finance in high technology company, laser optics. Patentee laser tech. field. Office: 1390 Borregas Sunnyvale CA 94086

DOWLING, JOHN ELLIOTT, educator; b. Pawtucket, R.I., Aug. 31, 1935; s. Joseph Leo and Ruth W. (Tappan) D.; m.; children by previous marriage: Christopher, Nicholas.; m. Judith Falco, Oct. 18, 1975; 1 dau., Alexandra. A.B., Harvard U., 1957, Ph.D., 1961; M.D. (hon.), U. Lund (Sweden), 1982. Asst. prof. biology Harvard U., 1961-64; prof. Harvard, 1971—; asso. prof. Johns Hopkins Sch. Medicine, 1964-71. Contbr. numerous articles on vision to profl. jours. Recipient ann. award N.E. Ophthal. Soc., 1979; award of merit Retina Research Found., 1981. Fellow Am. Acad. Arts, Scis., AAAS; mem. Assn. Research in Vision and Ophthalmology (Friedenwald medal 1970), Nat. Acad. Sci., Neurosci. Soc., Soc. Gen. Physiologists. Subspecialty: Neurobiology. Current work: Physiology; anatomy and biochemistry of the retina. Home: Master's Lodgings Leverett House 25 De Wolfe St Cambridge MA 02138 Office: Biol Labs Harvard U Cambridge MA 02138

DOWNEY, JAMES MERRITT, physiology educator; b. Wabash, Ind., Nov. 1, 1944; s. Richard Merritt and Janet Mildred (Fisher) D.; m. Patty Ann Froebe, June 18, 1967 (div. 1979); children: Michael, Douglas. B.S., Manchester Coll., 1967; M.S., U. Ill., 1969, Ph.D., 1971. Research fellow Harvard Med. Sch., Boston, 1970-72; asst. prof. physiology U. South Fla., Tampa, 1972-75; assoc. prof. physiology U. South Ala., Mobile, 1975-79, prof., 1979—; cons. scientist Rayne Inst., St. Thomas Hosp., London, 1980—. Co-author: PET Interfacing, 1980, Interfacing Projects for the Sinclair Computers, 1983. Served with U.S. Army, 1962. NIH grantee, 1976—; Am. Heart Assn. grantee, 1982—. Fellow Am. Heart Assn. (Circulation Council), Am. Physiol. Soc. (circulation group); mem. Am. Fedn. Clin. Research. Democrat. Methodist. Subspecialties: Physiology (medicine); Pharmacology. Current work: Laboratory research concerning protection of the heart in the presence of coronary artery disease; basic biophysics and drug evaluation. Home: 5453 Old Shell Rd 264 Mobile AL 36608 Office: Coll Medicine U South Ala MSB 3024 Mobile AL 36688

DOWNEY, RONALD JOSEPH, microbiologist; b. Manitowoc, Wis., Apr. 8, 1933; s. Reginald Joseph and Ellen Mary (Hannaway) D.; m. Mary Ann McAleer, Sept. 14, 1957. B.S., Regis Coll., 1955; M.S., Creigton U., 1958; Ph.D., U. Nebr., 1961. Instr. microbiology U. Nebr., 1959-60; research assoc. U.S. Dept. Agr. Labs., Beltsville, Md., 1961-62; prof. microbiology U. Notre Dame, 1962-72; prof., chmn. dept. zoology and microbiology Ohio U., Athens, Ohio, 1972—; Mem. Ohio Osteo. Med. Adv. Bd., 1977-84. Contbr. articles to profl. jours. USPHS career devel. awardee, 1967-72. Mem. Am. Soc. Microbiology, Am. Soc. Cell 'Biology, AAAS, Sigma Xi, Phi Beta Kappa. Subspecialties: Microbiology; Biochemistry (biology). Current work: Regulations of nitrogen assimilation in eucaryotic cells, form and function of inducible enzymes in fungi, nitrate reductase in Aspergillus nidulans. Home: 7 Pleasantview Dr Athens OH 45701 Office: Depts Zoology and Microbiolog Ohio U Athens OH 45701

DOWNS, ASA CHRIS, psychology educator; b. South Bend, Ind., Nov. 28, 1951; s. Asa Woodrow and Opal Frances (House) D. B.A., Ind. U.-South Bend, 1973; Ph.D., U. Tex., Austin, 1978. Asst. prof. Moorhead (Minn.) State U., 1978-79; asst. prof. U. Houston - Clear Lake City, 1979—. Contbr. articles to profl. jours. Mem. Am. Psychol. Assn., Soc. for Research in Child Devel., Western Psychol. Assn., Southwestern Psychol. Assn., Southeastern Psychol. Assn. Mem. Christian Ch. Subspecialty: Developmental psychology. Current work: Research on sex roles, television and impact of physical attractiveness on young children. Home: 3737 Watonga #31 Houston TX 77009 Office: Univ Houston - Clear Lake City 2700 Bay Area Blvd Houston TX 77058

DOWNS, ROBERT JACK, plant physiologist, educator; b. Sapulpa, Okla., June 25, 1923; s. Lester and Elizabeth (McGhie) D.; m. Rosa Joy Griffin, Oct. 18, 1945; 1 dau., Kathleen Cheryl. B.S., George Washington U., 1950, M.S., 1951, Ph.D., 1954. Phys. scientist Smithsonian Instn., Washington, 1952; plant physiologist Plant Physiology Lab., U.S. Dept. Agr., Beltsville, Md., 1952-65; mem. faculty N.C. State U., Raleigh, 1965—, prof. botany, 1965—, dir., 1965—; cons. in field. Author: Controlled Environments for Plant Research, 1975, Environment and the Experimental Control of Plant Growth, 1975, (with L.B. Smith) Tillandsioideae, 1977; author: Bromelioideae, 1979; contbr. numerous sci. articles to profl. publs. Served with USN, 1941-47. Mem. Am. Soc. Hort. Sci. (Alex Laurie award 1954), N.Y. Bot. Garden (Henry Allan Gleason award 1978), Am. Soc. Agrl. Engring., Am. Soc. Plant Physiology, Bot. Soc. Am., Internat. Soc. Biometerol., Am. Inst. Biol. Sci., Sigma Xi, Phi Epsilon Phi, Pi Alpha Xi. Presbyterian. Subspecialties: Plant physiology (agriculture); Developmental biology. Current work: Environmental physiology. Home: 3605 Octavia St Raleigh NC 27606 Office: NC State U Phytotron Raleigh NC 27607

DOYLE, FRANK LAWRENCE, hydrogeologist, research administrator; b. San Antonio, Oct. 16, 1926; s. William Michael and Elizabeth Lillian (Black) D.; m. Giovanna Maria Scorza, June 9, 1962; 1 son, Michael Joseph. B.S. in Geology, U. Tex.-Austin, 1950; M.S., La. State U., 1955; Ph.D., U. Ill., 1958. Registered geologist, Calif. Instr. St. Mary's U., San Antonio, 1950-53, asst. prof. geology, 1958-60, assoc. prof., 1960-62, chmn. dept. geology, 1961-62; petroleum geologist Seeligson Engring. Com., San Antonio, 1952-53; asst. geologist Ill. Geol. Survey, Urbana, 1956-58, assoc. geologist/research affiliate, 1959-61; geologist U.S. Geol. Survey, Mont., 1955, Colo., Ariz., 1962-63; assoc. prof. U. Conn., 1963-65; cons. hydrogeologist, Panama, Nicaragua, Algeria, 1965-71; regional geologist for North Ala. Geol. Survey Ala., 1971-77; cons. Kenneth E. Johnson Environ. and Energy Ctr., U. Ala.-Huntsville, 1971-77; adj. prof. hydrology, chmn. environ. sci. program Sch. Sci. and Engring., U. Ala.-Huntsville, 1971-77; cons. hydrogeologist, Fla., 1977-78; chief hydrogeologist Metcalf and Eddy, Inc., Boston, 1978-79; sr. hydrogeologist/program mgr. for waste mgmt. research U.S. Nuclear Regulatory Commn., Washington, 1979—. Sr. author: Environmental Geology and Hydrology, Huntsville and Madison County, Alabama, 1975; co-editor: Karst Hydrogeology, 1977. Active Huntsville/Madison County Local Govt. Study Com., Huntsville Solid Waste Mgmt. Com., 1972-75. Served with AUS, 1945-46. U. Ill. fellow, 1954-55. Fellow Geol. Soc. Am.; mem. Am. Assn. Petroleum Geologists (del. 1975-76), Internat. Assn. Hydrogeologists (gen. chmn. 12th Internat. Congress 1975, adv. council 1977-80, sec.-treas. U.S. com. 1980—), Am. Geophys. Union, Am. Inst. Profl. Geologists, Sigma Xi. Subspecialties: Hydrogeology; Geology. Current work: Research management: hydrogeology of high-level nuclear waste management; application of remotely-sensed data to ground water contamination; geomorphology; field geology. Home: 4875 Wheatstone Dr Fairfax VA 22032 Office: US Nuclear Regulatory Commn Mail Stop 1130-SS Washington DC 20555

DOYLE, ROBERT JOSEPH, electronics engineer; b. Somerville, Mass., Sept. 13, 1936; s. Frederick C. and Catherine D. (McTaggart) D.; m. Marilyn Norton, Apr. 23, 1960; children: William F., Stephen M., Christopher R. B.S. in Elec. Engring., Northeastern U., 1959; M.Engring. Adminstrn., George Washington U., 1980. Project engr. Westinghouse, Horseheads, N.Y., 1960-63; sr. project engr. CBS, Inc., Stamford, Conn., 1963-72; engring. mgr. CGR Med. Corp., Balt., 1983—; project engr. Gould DED, Clen Burnie, Md., 1983—. Contbr. articles to profl. jours. Served to 2d lt. Signal Corps, U.S. Army, 1959-60. Mem. Am. Assn. Physicists in Medicine, IEEE (sr.). Democrat. Roman Catholic. Subspecialties: Imaging technology; Electronics. Current work: Developer of computer controlled x-ray systems, microprocessor x-ray generators, digital radiography systems, digital image processing, passive sonar sytstems and x-ray image systems. Patentee in field. Home: 3431 Arcadia Dr Ellicott City MD 21043 Office: Gould DED 6711 Baymeadow Dr Clen Burnie MD 21061

DOYLE, WALTER ARNETT, pedodontist, orthodontist; b. Los Angeles, Aug. 9, 1933; s. Walter and Ruth D.; m. Betty Ann Parrot, Dec. 27, 1957 (div. June 1975); children: Shannon, Elizabeth, Sarah, Walter; m. Elizabeth Lewis, July 17, 1977. D.D.S., Emory U., 1959; M.S.D., Ind. U.-Indpls., 1961; postgrad., Boston U., 1974-76. Diplomate: Am. Bd. Pedodontics, Am. Bd. Orthodontics. Pvt. practice pedodontics, Lexington, Ky., 1962—, pvt. practice orthodontics, Lexington, 1976—; instr. pedodontics Ind. U., 1961-62, U. Ky., 1964-65, guest lectr. dept. community dentistry, 1972-74, asst. field prof., 1972-74; vis. assoc. prof. pedodontics Northwestern U., 1972-74; vis. clin. prof. pedodontics Boston U. Sch. Grad. Dentistry, 1975—; mem. staff Good Samaratin, St. Joseph, Central Baptist, Humana hosps.; cons., contbr. Health Info. Systems, Inc.; dental cons. Medcom, Inc.; bd. dirs. Ky. Dental Service Corp., 1964-69; S. S. White Centennial teaching fellow. Contbr. articles to profl. jours. Trustee Hunter Found., 1972-74; mem. Bluegrass Trust for Historic Preservation, 1968—, Lexington Council for Arts, 1976—; Boston U. Alumni Area Rep. Fellow U. Ky.; Recipient award for leadership in dental progress Thomas Hinman, 1972, 76. Fellow Am. Acad. Pedodontics, Internat. Coll. Dentists, Am. Coll. Dentists.; Mem. Am. Soc. Preventive Dentistry (pres. Ky. unit 1972), Am. Soc. Dentistry for Children (mem. exec. council 1972—, pres. 1976-77), Internat. Assn. Dental Research, Am. Dental Assn., Assn. Pedodontic Diplomates (pres.-elect. 1968), Southeastern Soc. Pedodontics (pres. 1968), Ky. Soc. Dentistry for Children (pres. 1963), Psi Omega, Lexington C. of C. Clubs: Lexington Polo, Keeneland, Sierra, Ind. U. Century. Subspecialties: Dental growth and development; Orthodontics. Current work: Orthodontics; pedodontics. Home: 3284 Paris Pike Lexington KY 40511 Office: 1628 Nicholasville Rd Lexington KY 40503

DRAGO, JOSEPH ROSARIO, urologist; b. Jersey City, N.J., Oct. 28, 1947; s. Rosario P. and Betty L. (Brisgal) D.; m. Diane Mary Lavacca, June 17, 1972; children: Andrea, Daniella, Denise. B.S., U. Ill., 1965; M.D., 1972. Diplomate: Am. Bd. Urology. Intern gen. surgery Milton S. Hershey Med. Ctr., Pa. State U., Hershey, 1972-73, resident in urology, 1973-77, instr. urology, 1976-77, asst. prof. urology, dir. urologic oncology, 1979-80, assoc. prof. surgery, dir. urologic oncology, 1980—; asst. prof. urology, dir. urologic oncology U. Calif., Davis, 1977-79; mem. Nat. Prostatic Cancer Task Force, 1980-83; mem. cancer research coordinating com. U. Calif., 1978-79. Contbr. articles in field to profl. jours. Elsa U. Pardee Found. grantee, 1978; U. Calif. faculty research grantee, 1977. Mem. Am. Urologic Assn., Assn. Acad. Surgery, AMA, Am. Fertility Soc., Am. Soc. Andrology, Crippled Children's Services, Soc. Clin. Trials, Soc. Univ. Urologists, Pan-Pacific Surg. Assn., Western Assn. Transplant Surgeons, Pa. Med. Soc., Dauphin County Med. Soc., Phila. Urologic Soc. Roman Catholic. Subspecialties: Oncology; Urology. Current work: Urologic oncology, tumor model systems, both bladder cancer model and prostrate cancer model, urologic-endocrinologic management carcinoma, chemotherapeutic measurements in animal model systems. Home: 104 Banbury Circle Hummelstown PA 17036 Office: 500 University Dr Div Urology Hershey PA 17033

DRAKE, CHARLES LUM, geology educator; b. Ridgewood, N.J., July 13, 1924; s. Ervin Thayer and Elizabeth (Lum) D.; m. Martha Ann Churchill, June 24, 1950; children—Mary Aiken, Caroline Elizabeth, Susannah Churchill. B.S. in Engring, Princeton, 1948; Ph.D., Columbia, 1958. Research assoc. Lamont Geol. Obs. Columbia U., N.Y.C., 1948-56; sr. scientist Lamont Geol. Obs., 1956-58, acting asst. dir. Lamont Geol. Obs., 1963-65, became mem. faculty univ. Lamont Geol. Obs., 1958, prof. geology, chmn. dept. Lamont Geol.

Obs., 1967-69; prof. dept. geology Dartmouth Coll., Hanover, N.H., 1969—, chmn. dept. geology, 1978-79, dean grad studies, assoc. dean sci. div., 1979—; mem. coms. Nat. Acad. Sci.; cons. NSF, 1964-82; mem. Nat. Adv. Com. on Oceans and Atmosphere, 1971-74; chmn. earth scis. div. NRC, 1973-76, mem. geophys. research bd., 1968-82. Trustee Village S. Nyack, N.Y., 1963-65, 66-69, dep. mayor, 1968-69. Served with AUS, 1943-46. NSF postdoctoral fellow, 1965-66. Mem. Internat. Council Sci. Unions (pres. interunion commn. on geodynamics 1970-75, chmn. U.S. Geodynamics com. 1970-78, chmn. U.S. nat. commn. geology 1979—), Am. Geophys. Union (pres.-elect 1982-84), AAAS, Am. Assn. Petroleum Geologists, Geol. Soc. Am. (pres. 1976-77), Geol. Soc. France (hon.), Seismol. Soc. Am., Royal Astron. Soc., Soc. Exploration Geophysicists, Marine Tech. Soc., Sigma Xi. Club: Cosmos. Subspecialty: Geophysics. Home: RFD 1 East Thetford VT 05043

DRAKE, FRANK DONALD, astronomer; b. Chgo., May 28, 1930; s. Richard Carvel and Winifred Pearl (Thompson) D.; m. Amahl Z. Shakhashiri, Mar. 4, 1978; children: Nadia Meghann, Leila Marlyss; children by previous marriage: Stephen David, Richard Procter, Paul Robert. B. Engring. Physics, Cornell U., 1952; M.A., Harvard, 1956, Ph.D., 1958. Mem. Harvard Radio Astronomy Project, 1955-58; dir. Astron. Research Group, Ewen-Knight Corp., 1958; head telescope operations div. and scil services div., radio studies Venus and Jupiter Nat. Radio Astronomy Obs., 1958-63; chief lunar and planetary scis. sect. Jet Propulsion Lab., 1963-64; assoc. prof. astronomy Cornell U., Ithaca, N.Y., 1964-66, prof., 1966—, Goldwin Smith prof. astronomy, 1976—, chmn. dept., 1968-71, assoc. dir. Center for Radiophysics and Space Research, 1965-74; dir. Arecibo Ionospheric Obs., 1966-68, Nat. Astronomy and Ionosphere Center, 1971-81; mem. NRC, 1969-71; adviser govt. coms. on space research and astronomy. Author: Intelligent Life in Space, 1962, Murmurs of Earth: The Voyager Interstellar Record, 1978; Editorial adv. bd.: World Book Ency. Vice-pres. CETI Found.; bd. dirs. Extrasolar Planetary Found. Mem. Am. Astron. Soc. (councillor, past chmn. div. planetary scis.), AAAS (past v.p., past chmn. astronomy sect.), Nat. Acad. Scis., Am. Acad. Arts Scis., Internat. Astron. Union (v.p. commn. on life in universe, vice chmn. U.S. nat. com.), Internat. Sci. Radio Union, Planetary Soc. (dir.), Sigma Xi, Tau Beta Pi. Club: Explorers. Subspecialty: Radio and microwave astronomy. Organized pioneer search for extra-terrestrial life, project OZMA, 1960. Office: 422 Space Scis Bldg Cornell U Ithaca NY 14853

DRAKE, MICHAEL JULIAN, geochemistry educator; b. Bristol, Eng., July 8, 1946. B.Sc., U. Manchester, Eng., 1967; Ph.D. in Geology, U. Oreg., 1972. Research assoc. lunar sci. Smithsonian Astrophysics Obs., 1972-73; asst. prof. planetary sci. U. Ariz., Tucson, 1973-78, assoc. dir., 1978-80, assoc. prof. planetary sci. dept., 1978—. Assoc. editor: Jour. Lunar Sci. Confs, 1975-78. Mem. Am. Geophys. Union, Meteoritical Soc. Subspecialty: Planetology. Office: Planetary Sci and Lunar and Planetary Lab U Ariz Tucson AZ 85721

DRAKE, MICHAEL LEE, applied research engineer; b. Dayton, Ohio, Feb. 9, 1949; s. Ralph L. and Fannie R. (Britton) D.; m. Rebecca Ann Kerns, Oct. 16, 1971; children: Dawn Michelle, Benjamin Phillip. B.S. in Aerospace Engring, U. Cin., 1972, M.S., 1973. Research engr., group leader vibration analysis and control group U. Dayton, 1973—. Contbr. articles to profl. jours. Mem. ASME, Soc. Exptl. Stress Analysis, Inst. Environ. Scis. Subspecialties: Mechanical engineering; Aerospace engineering and technology. Office: 300 College Park Dr Dayton OH 45469

DRAKONTIDES, ANNA B., anatomy educator; b. N.Y.C., Aug. 21, 1933. Ph.D., Cornell U. Med. Coll., 1971. With Chas. Pfizer & Co., Groton, Conn., 1955-61; with Geigy Chem. Corp., Ardsley, N.Y., 1962-63; instr. Hunter Coll., 1967-69, asst. prof., 1969-73; instr. dept. pharmacology Cornell U. Med. Coll., Ithaca, N.Y., 1974-76; assoc. prof. pharmacology Sch. Nursing, Cornell U., 1974-76; assoc. prof. anatomy N.Y. Med. Coll., 1973—. Co-author: Workbook and Laboratory Manual in Anatomy and Physiology, 1972, 77, Anatomy and Physiology, 17th edit, 1977; Contbr. articles to sci. jours. USPHS predoctoral service grantee, 1968-71; NIH postdoctoral grantee, 1971-72; co-investigator NIH grants, also Whitehall Found.; Pharm. Mfrs. Assn. Found. fellow, 1972-74. Mem. N.Y. Acad. Sci., N.Y. Soc. Electron Microscopy, Am. Assn. Anatomists, Soc. Neurosci. Subspecialties: Microscopy; Regeneration. Current work: Structure-function neuromuscular junction in normal and pathological states, e.g. denervation-regeneration, disease state-diabetes; ultrastructural morphology of synapses related to function of neural area. Office: NY Med Coll Dept Anatomy Valhalla NY 10595

DRANCE, S. M., medical educator; b. Bielsko, Poland, May 22, 1925. M.B., Ch.B., U. Edinburgh, 1948; Surg. diploma, Royal Coll., 1953. Ophthalmology fellow, 1956; research assoc. ophthalmology Oxford(Eng.) U., 1955-57; assoc. prof. ophthalmology and dir. glaucoma clinic U. Saskatoon, Can., 1957-63; assoc. prof. ophthalmology U. B.C., Vancouver, 1963-68, prof., 1968—, head dept., 1974—, dir. glaucoma services, 1963—; mem. sub. com. ophthalmology Can. Dept. Health and Welfare, 1964; lectr. in field U.S., Gt. Britain; grantee Can. Dept. Health and Welfare, 1967—. Editor: Can. Jour. Ophthalmology. Mem. Assn. Research in Vision and Ophthalmology, Can. Med. Assn., Can. Ophthalmology Soc. (pres. 1975), Brit. Med. Assn., Am. Acad. Ophthalmology (v.p. 1983), Ophthalmology Soc. U.K., Internat. Glaucoma Soc. (pres. 1980). Subspecialty: Ophthalmology. Office: Dept Ophthalmology U BC Vancouver BC Can V6T 1W5

DRAPALIK, DONALD JOSEPH, biology educator, plant ecology consultant; b. Chgo., Dec. 10, 1934; s. Frank Anton and Rose Annette (Veverka) D.; m. Janice Jane Moore, June 18, 1960; children: Lisa Bauguss, Debbie Moore, Bruce Moore; m. Jo Ann Horne, Aug. 2, 1975. B.A., So. Ill. U., 1959, M.A., 1962; Ph.D., U. N.C.-Chapel Hill, 1970. Jr. forester, naturalist Cook County Forest Preserve Dist., Ill., 1953-58; grad. teaching and research asst. So. Ill. U., 1959-62, U. N.C., 1962-68; asst. prof. biology Ga. Soc. Coll., 1968-73, assoc. prof., 1973—; cons. Contbr. articles profl. jours. Williams Chambers Coker fellow, 1964, 68. Mem. Assn. Southeastern Biologists, Am. Soc. Plant Taxonomists, Internat. Assn. Plant Taxonomy, Bot. Soc. Am., So. Appalachian Bot. Club, Sigma Xi, Phi Kappa Phi. Democrat. Club: Ga. So. Faculty. Subspecialties: Taxonomy; Ecology. Current work: Vascular plant taxonomy and ecology. Home: 109A Valley Circle Statesboro GA 30458 Office: Dept of Biology GA So Coll Statesboro GA 30460 8042

DRAPER, ERNEST LINN, JR., electric utility executive, nuclear engineer; b. Houston, Feb. 6, 1942; s. Ernest L. and Marcia Lee (Saylor) d.; m. Mary Deborah Doyle, June 9, 1962; children: Susan, Robert, Barbara, David. Student, Williams Coll., 1960-62; B.A., Rice U., 1964, B.S., 1965; Ph.D., Cornell U., 1970. Registered profl. engr., Tex. Asst. prof. U. Tex.-Austin, 1969-72, assoc. prof. mech. engring., 1972-79; tech. asst. to chmn. bd. Gulf States Utilities, Beaumont, Tex., 1979-80, v.p. nuclear tech., 1980-81, sr. v.p. engring. and tech. service, 1981-82, sr. v.p. external affairs, 1982—; cons.; mem. Commn. Radioactive Waste Mgmt., Tex. Energy and Nat. Resources Adv., Austin, 1979—; cons. Congressional Office of Tech. Assessment, Washington, 1982—. Editor: Proc. Tex. Symposium on Tech. Controlled Thermonuclear Fusion Experiments and Engring. Aspects of Fusion Reactors, 1974, Proc. Implications of Nuclear Power for Tex, 1973. NSF fellow, 1965, 66; AEC fellow, 1967, 68; recipient Faculty award U. Tex. Mem. Am. Nuclear Soc. (exec. com., dir.), Am. Phys. Soc. Subspecialties: Nuclear engineering; Nuclear fusion. Current work: Nuclear waste management, public understanding of energy issues. Home: 1190 Dowlen Beaumont TX 77706 Office: Gulf States Utilities PO Box 2951 Beaumont TX 77704

DRAPER, GRENVILLE, geologist, geology educator; b. Leeds, Yorkshire, U.K., Sept. 24, 1950; came to U.S., 1978; s. Frederick and Betty (Beadnell) D. B.A., Cambridge U., 1973; M.Sc., London U., 1974; Ph.D., U. West Indies, Kingston, 1979. Asst. lectr. U. West Indies, Kingston, Jamaica, 1974-75, research asst., 1975-78; asst. prof. geology Fla. Internat. U., Miami, 1978—. Fellow Geol. Soc. London; mem. Am. Geophys. Union, Geol. Soc. Am., Geol. Soc. Jamaica (council mem. 1976-78), Miami Geol Soc. (program chmn. 1981-83, pres. 1983-84). Subspecialties: Geology; Tectonics. Current work: Structural geology and tectonics with particular reference to the Caribbean region, structure and tectonics of blueschist belts, island arc structure and tectonics. Office: Fla Internat U Tamiami Trail Miami FL 33199

DRAPER, THOMAS WILLIAM, social science educator; b. American Fork, Utah, June 19, 1947; s. Terry Parshall and Lillian (Anderson) D.; m. Linda Gordon, Apr. 5, 1972; children: Thomas Gordon, Janine Louise, Robert Anderson. B.S., Brigham Young U., Provo, 1971, M.S., 1973; Ph.D., Emory U., Atlanta, 1976. Research asst. Ednl. Testing Service, Atlanta, 1973-75; instr. dept. psychology Atlanta Univ. Ctr., 1975-76; asst. prof. dept. child devel. and family relations U. N.C., Greensboro, 1976-82; assoc. prof. dept. family scis. Brigham Young U., Provo, 1982—; cons. Butternick Pub., N.Y.C., 1979-80, Head Start, Greensboro, 1980-81. Editor: (with others) See How They Grow, 1980; co-author battery of tests, CIRCUS, 1976. Scoutmaster Boy Scouts Am., Greensboro, 1981. Recipient Research Excellence award U. N.C., Greensboro, 1979. Mem. AAAS, Am. Psychol. Assn., Nat. Council on Family Relations, Nat. Assn. for Edn. of Young Children, Soc. for Research in Child Devel., Sigma Xi. Mormon. Subspecialties: Developmental psychology; Social psychology. Current work: Longitudinal research methods, famological research methods, socialization and personality development in young children. Home: 3100 Bannock Dr Provo UT 84604 Office: Brigham Young Univ Dept Family Scis 1407 SFLC Provo UT 84602

DRAPKIN, ROBERT L., oncologist, clinical investigator; b. Albany, N.Y., Sept. 22, 1944; s. Isadore and Francis D.; m. Renee Mary Kumaraperv, Oct. 24, 1978; children: Julia, Jessica. B.S., Union Coll., 1966; M.S., Rensselaer Poly. Inst., 1967; M.D., Wayne State U., 1971. Intern U. Ill.Hosp., Chgo., 1971-72, resident in internal medicine, 1972-74, instr. in medicine, 1974-75; fellow oncology Meml. Hosp., Sloan Kettering Cancer Ctr., N.Y.C., 1975-78; physician Roswell Meml. Inst., Buffalo, 1978-79; assoc. prof. U. South Fla., Tampa, 1979—; attending physician Morton Plant Hosp., Clearwater, Fla., 1979—, Clearwater Community Hosp., 1979—, Mease Hosp., Dunedin, Fla., 1979—. Contbr. chpts. to book, articles to profl. jours. Fellow ACP; mem. Am. Soc. Clin. Oncology. Subspecialties: Chemotherapy; Oncology. Current work: Physician and clinical investigator.

DRAY, ANDRE, neurobiology educator, researcher; b. Hanover, W. Ger., July 3, 1946; came to U.S., 1977; s. Jack and Maria (Bondar) D.; m. Karen Anne Jackson, July 3, 1978; children: Amalia, Alexis. B.Sc. with honours, U. St. Andrews, 1968; Ph.D. (Med. Research Council Scholar), U. Birmingham, Eng., 1971. Research fellow dept. preclin. pharmacology U. Birmingham, 1971-72; Med. Research Council research fellow dept. pharmacology Sch. Pharmacy, London, 1972-75, lectr., 1975-77; research assoc. dept. physiology Duke U., Durham, N.C., 1977-79; asst. prof. dept. pharmacology Coll. Medicine, U. Ariz., Tucson, 1979-80, assoc. prof., 1981—; mem. sr. research staff Med. Research Council neurochem. pharmacology unit Med. Sch., Cambridge (Eng.) U., 1980-81; cons. NIH, NSF. Editor: Life Scis, 1979-80, Neuropharmacology, 1981—; contbr. numerous articles and abstracts to profl. jours., also chpts. to books. Recipient Royal Soc. travel award, 1977. Mem. Brit. Pharm. Soc., Brit. Physiol. Soc., Soc. for Neurosci., European Neurosci. Soc., Am. Physiol. Soc., Am. Soc. Pharmacology and Exptl. Therapeutics. Subspecialties: Neuropharmacology; Neurophysiology. Current work: Neurobiology of the nervous system; neurobiology and neuropharmacology of central nervous system; single cell electro-recording; chemical communication; drug action; neurological disease. Office: Dept Pharmacology Med Sch U Ariz Tucson AZ 85724

DRAYER, BURTON PAUL, neuroradiologist, neurologist, educator; b. N.Y.C., Mar. 19, 1946; s. Alexander and Marion (Horowitz) D.; m. Michaele Gerri Cohen, June 13, 1968; children: Aron Stuart, Alex Nathan. A.B. in Polit. Sci, U. Pa., 1967; M.D., Chgo. Med. Sch., 1971. Diplomate: Am. Bd. Psychiatry and Neurology, Am. Bd. Radiology. Asst. prof. neurology U. Pitts. Health Center, 1977-79, assoc. prof. radiology, 1978-79; dir. neuroradiology U. Pitts. Health Center Children's Hosp., 1978-79; asso. prof. radiology Duke U. Med. Center, Durham, N.C., 1979—, asst. prof. neurology, 1979—, chief sect. neuroradiology, 1981—; mem. exec. com. stroke council Am. Heart Assn. Contbr. articles to profl. jours. Recipient Cornelius G. Dyke award Am. Soc. Neuroradiology, 1977; Pres.'s award Pitts. Roentgen Ray Soc., 1977; Am. Heart Assn. grantee, 1979-81; Squibb Research Inst. grantee, 1981-82; Nat. Heart, Lung and Blood Inst. grantee, 1983—. Mem. Am. Acad. Neurology, Am. Neurologic Assn., Radiol. Soc. N.Am., Am. Soc. Neuroradiology (Outstanding Paper award 1978, 81), Am. Roentgen Ray Soc. (President's award 1977), Soc. Neurosci., Sigma Xi, Alpha Omega Alpha. Jewish. Subspecialties: Neurology; Diagnostic radiology. Current work: The in vivo characterization of brain anatomy and function using radiographic techniques. Home: 4011 Nottaway Rd Durham NC 27707 Office: Department Radiology Box 3808 Duke University Medical Center Durham NC 27710

DRAYER, JAN IGNATIUS, physician; b. Amsterdam, Netherlands, Jan. 31, 1946; came to U.S., 1980; s. Roelof Pieter and Anna Betsie (DeSwart) D.; m. Thea Jacoba van Kalmthout, July 3, 1971; children: Myke, Joris. H.B.S.-B., St. Joris Coll., 1963; M.D., U. Nijmegen, 1971, Ph.D., 1975. Research asso. Cornell U., N.Y.C., 1975-77; internist U. Nijmegen, Netherlands, 1977-80; asso. chief clin. pharmacology/hypertension VA Med. Center, Long Beach, Calif., 1980—. Contbr. articles to profl. jours.; editor: Introduction to Echocardiography, 1980, Mineralocorticoids in Essential and Secondary Hypertension, 1982. Bd. dirs. Am. Heart Assn., Orange County, 1981—. Grantee Am. Heart Assn., 1982. Fellow Council for High Blood Pressure Research, Am. Coll. Clin. Pharmacology; mem. Am. Soc. Clin. Pharmacology and Therapeutics, Internat. Soc. Hypertension, Endocrine Soc. Subspecialties: Internal medicine; Pharmacology. Current work: Clin. pharmacology of cardiovascular agts.; endrocine aspects of hypertension; non-invasive monitoring and evaluation of hypertensive patients. Home: 5401 Catowba Ln Irvine CA 92715 Office: VA Med Center 5901 E 7th St Long Beach CA 90822

DRAZNIN, BORIS, physician, researcher, educator; b. Kharkov, USSR, Oct. 1, 1945; came to U.S., 1977; s. Nahum and Rosa (Rips) D.; m. Elena Lerman, Dec. 25, 1965; children: Julie, Micky, Ann. M.D., Minsk (USSR) State Med. Inst., 1968; Ph.D., Vilnus (USSR) Inst. Exptl. Medicine, 1972. Diplomate: Am. Bd. Internal Medicine. Intern Molodechno City Hosp., 1968-69; resident in medicine Minsk Regional Hosp., 1969-71, research assoc., 1971-73, staff physician, 1973; research assoc. Tel Aviv Med. Ctr., 1974-77; fellow endocrinology U. Colo. Med. Ctr., 1977-80; asst. prof. medicine U. Colo. Health Sci. Ctr., Denver, 1980—; research assoc. VA Med. Ctr., Denver, 1980-83, clin. investigator, 1983—. Mem. Am. Fedn. Clin. Research, Am. Diabetes Assn., Endocrine Soc., ACP, Am. Physiol. Soc., Am. Soc. Cell Biology. Jewish. Subspecialties: Endocrinology; Cell biology. Current work: Insulin release and action; intracellular movement of proteins and receptors. Office: U Colo Health Sci Ctr 1055 Clermont St Denver CO 80220

DREES, THOMAS CLAYTON, health care company executive; b. Detroit, Feb. 2, 1929; s. Clayton Henry and Mildred (Stevenson) D.; m. Elaine Hnath, Feb. 9, 1952; children: Danette, Clayton, Barry, Nancy. B.A. with honors, Coll. Holy Cross, 1951; M.B.A., Pacific Western U., 1979, Ph.D., 1980. With Spaulding Fibre Co., Inc., 1953-70, sales engr., N.Y.C., 1953-56, br. mgr., Toronto, Ont., Can., 1957-60, asst. to pres., Tonawanda, N.Y., 1960-63, sec., 1961-70, v.p. internat., 1963-66, exec. v.p., 1966-70; also dir., mem. exec. com.; mng. dir. Spauldings, Ltd., London, 1964-70, dir., 1963-70, chmn. bd. dirs., 1964-70; gen. mgr. Spaulding Fibre of Can., Ltd., Toronto, 1957-60, v.p., dir., 1957-70; pres., dir. La Fibre Vulcanisee Spaulding, Paris, 1964-70; v.p., dir. Mycalex Corp. Am., Clifton, N.J., 1967-70, Spaulding Norton, Inc., North Westchester, Conn., 1968-70; group v.p. Ipco Hosp. Supply Corp., 1970-72; pres., vice chmn. IVAC Corp., San Diego, 1972-73; v.p., gen. mgr. Abbott Labs., South Pasadena, Calif., 1973-78; v.p., dir. AMEC, Houston; pres., vice chmn. Alpha Therapeutic Corp., 1978—; dir. Green Cross Corp., Osak, Japan. Chmn. Sch. Bd.; mem. pres. council Holy Cross Coll.; bd. dirs. Hemophilia Found., Am. Blood Commn., Am. Blood Resources Assn., La Jolla Cancer Research Found.; mem. bus. adv. bd. U.S. Senate.; dir. president's adv. bd. Calif. State U.-Los Angeles; mem. exec. forum Calif. Inst. Tech.; trustee Thomas Aquinas Coll.; bd. dirs. Alliance Tech. Fund, Alliance Internat. Health Care Trust. Served from ensign to lt. (j.g.) USNR, 1951-53. Fellow Inst. Dirs.; mem. I.E.E.E., Nat. Sales Execs. Assn., Am. C. of C. Republican. Roman Catholic (bd. advisers). Clubs: Rotary, San Gabriel Country. Subspecialties: Genetics and genetic engineering (medicine); Hematology. Current work: Artificial blood; recombinant DNA; monoclona antibodies. Home: 784 Saint Katherine Dr Flintridge La Canada CA 91001 Office: 220 Pasadena Ave South Pasadena CA 91030

DREESMAN, GORDON RONALD, medical researcher, consultant; b. Grundy County, Iowa, Nov. 28, 1935; s. Edwin Thomas and Mabel H. (Buskohl) D.; m. Dolores J. Wisdom; children: Kimberlyn Lynn, Thomas Lee, Andrea Beth; m. Margaret Carter Neville. B.A. in Chemistry, Central Coll., Pella, Iowa, 1957; M.T., St. Luke's Meth. Hosp., Cedar Rapids, Iowa, 1958; M.A. in Bacteriology, Kans. U., 1963; Ph.D. in Microbiology, U. Hawaii, 1965. NIH predoctoral fellow U. Hawaii, 1964-65; postdoctoral trainee Baylor U., 1965-66; asst. prof. Inst. Molecular Biology, St. Louis U., 1966-69; cons. immunochemist St. John's Mercy Hosp., St. Louis, 1969; asst. prof. virology and epidemiology Baylor U., 1969-73, assoc. prof., 1973-79, prof., 1979—. Contbr. numerous articles profl. jours. Mem. Am. Soc. Microbiology, AAAS, Am. Assn. Immunologists, Sigma Xi, Phi Kappa Phi. Presbyterian. Subspecialties: Immunobiology and immunology; Immunology (medicine). Current work: Research in fields of immunochemistry of hepatitis B virus, association of herpes simplex virus with human cancer. Patentee in field. Home: 6704 Community Dr Houston TX 77005 Office: Dept Virology Baylor U Coll Medicine Houston TX 77030

DREGNE, HAROLD ERNEST, agronomy educator; b. Ladysmith, Wis., Sept. 25, 1916. B.S., Wis. State U., 1938; M.S., U. Wis., 1940; Ph.D. in Soil Chemistry, Oreg. State U., 1942. Asst. prof. Oreg. State U., Corvallis, 1940-42; jr. soil scientist Soil Conservation Service, U.S. Dept. Agr., 1942-43, 46; asst. prof. agronomy, agronomist Exptl. Sta., N.Mex. State U., 1949-50, prof. soils, 1950-69; prof. agronomy, chmn. dept. Tex. Tech. U., Lubbock, 1969-72, chmn, dept. plant and soil sci., 1972-78, Horn prof. soil sci., 1972—; dir. Internat. Ctr. Arid and Semi-Arid Land Studies, 1976—; head Internat. Coll. Exchange Program, Pakistan, 1955-57; mem. U.S. soil salinity del., USSR, 1960; soil expert UN, Chile, 1961; cons. UNESCO, Tunisia, 1967, UN Environ. Program, 1975-80, 79-82. Fellow AAAS, Am. Soc. Agronomists, Am. Soil Sci. Soc. Subspecialty: Soil science. Office: Dept Plant and Soil Sci Tex Tech U Lubbock TX 79409

DREIFUSS, FRITZ EMANUEL, neurologist, educator; b. Dresden, Germany, Jan. 20, 1926; came to U.S., 1958, naturalized, 1964; s. Alfred and Erika (Ballin) D.; m. Daphne Guthrie, Feb. 10, 1954; children: Simone, Donald Alfred. M.B., Ch.B., Otago (N.Z.) U., 1950. Diplomate: Am. Bd. Psychiatry and Neurology. Intern Auckland (N.Z.) Pub. Hosp., 1951-52, resident, 1953; registrar Nat. Hosp. London, 1954-56, resident med. officer, 1956-57; asst. prof. neurology U. Va., 1959-64, assoc. prof., 1964-68, prof., 1968—. Author: books, including Epilepsy Case Studies, 1981, Pediatric Epilepsy, 1983; contbr. numerous articles to profl. jours. Fellow Royal Coll. Physicians (London), Royal Australasian Coll. Physicians; mem. Am. Epilepsy Soc. (pres.), Epilepsy Found. Am. (pres.-elect), Internat. League Against Epilepsy (sec.-gen.), Alpha Omega Alpha. Jewish. Subspecialties: Neurology; Neuropharmacology. Current work: Pediatric neurology, epilepsy.

DREITLEIN, RAYMOND PAUL, alcoholism counselor, consultant; b. Bklyn., Sept. 14, 1943; s. Michael George and Madelyn (Zamitka) D.; m. Carol Ann Lays, Jan. 23; children: Raymond, William, Karen, Scott, Adam, James, Kevin. B.A., St. Francis Coll., Bklyn., 1967; M.A., Seton Hall U., 1969. Cert. alcoholism counselor, N.J. Dept. head counseling and evaluation Mt. Carmel Guild, Newark, 1971-73. Clin. and project dir. A.T.U. Runnels Hosp., Berkeley Heights, N.J., 1973-76; dir., founder SSDC Detoxication Ctr., Elizabeth, N.J., 1977-79; regional counselor Employee Adv. Service, State of N.J., Newark, 1979-82; alcoholism counselor in pvt. practice, Morristown, N.J., 1982—; dir. out patient dept. alcoholism treatment unit Fair Oaks Hosp., Summit, N.J., 1983—; mem. faculty Rutgers U., 1976—; aftercare coordinator Mountainside Hosp., Montclair, N.J., 1980—; police instr. Union County Police Chief Acad., Cranford, N.J., 1974—. Recipient Alcoholism Recognition award Union County Counselors in Alcoholism, 1977. Mem. N.J. Assn. Alcoholism Counselors, N.J. Psychol. Assn. (assoc.), Am. Psychol. Assn. (assoc.), N.J. Assn. Profl. Psychologists (charter). Democrat. Roman Catholic. Subspecialties: Alcoholism rehabilitation counseling; Burnout alcoholism treatment personnel. Current work: Research on alcoholism treatment; professionals and dynamic of burnout and its aftermath on staffs and program, developed treatment mode for burnout dynamic. Home: 726 Prospect St Maplewood NJ 07040 Office: 181 South St Morristown NJ 07960

DREIZEN, SAMUEL, oncology educator, researcher; b. N.Y.C., Sept. 12, 1918; s. Charles and Rose (Schneider) D.; m. Jo Gilley, Aug. 3, 1956; 1 dau., Pamela. B.A., Bklyn. Coll., 1941; D.D.S., Western Res. U., 1945; M.D., Northwestern U., 1958. Research asso. Nutrition Clinic, Hillman Hosp., Birmingham, 1945-47; instr. nutrition and metabolism Northwestern U. Med. Sch., Chgo., 1947-48, asst. prof., 1948-58, asso. prof., 1958-66; prof. pathology U. Tex. Dental Br., Houston, 1966-76, prof. and chmn. dental oncology, 1976—; cons. Dental Assn., M.D. Anderson Hosp., Houston, 1966—. Author: multi media package Mouth in Medicine, 1979; book Experimental Stomatology, 1981; guest editor profl. jours., 1971, 74. Served to capt. U.S. Army, 1953-60. Recipient Propylaea award Bklyn. Coll., 1940. Fellow AAAS; mem. Internat. Assn. Dental Research, Am. Assn. Phys. Anthropologists, Soc. Research in Child Devel., Sigma Xi, Alpha Omega Alpha, Omicron Kappa Upsilon. Subspecialties: Oncology; Nutrition (biology). Current work: Oral complications of cancer therapy; effect of substandard nutrition on human growth; role of nutrition in diseases of the mouth. Home: 5218 Dumfries Dr Houston TX 77096 Office: U Tex Dental Br PO Box 20068 Houston TX 77025

DREN, ANTHONY THOMAS, research pharmacologist; b. Chisholm, Minn., Feb. 15, 1936; s. Anthony F. and Angeline R. (Mehle) D.; m. Catherine Marie Danjanic, Aug. 5, 1961; children: Christina, A. Michael, Mary Beth, Amy. B.S., Duquesne U., 1959, M.S., 1961; Ph.D., U. Mich., 1966. Registered pharmacist, Pa., 1959. Sr. pharmacologist Abbott Labs., North Chicago, Ill., 1966-72, assoc. research fellow, 1972-75, sect. head neuropharmacology, 1975-79; sr. clin. research scientist Burroughs Wellcome Co., Research Triangle Park, N.C., 1979—. Contbr. chpts. to books, articles to profl. jours. Mem. Am. Soc. Pharmacology and Exptl. Therapeutics, AAAS, Am. Soc. Clin. Pharmacoloty and Therapeutics, Soc. Neurosci., N.Y. Acad. Scis., Am. Chem. Soc., Sigma Xi. Subspecialties: Psychopharmacology; Neuropharmacology. Current work: Clinical testing and development of psychotropic drugs. Patentee in field (9). Office: 3030 Cornwallis Rd Research Triangle Park NC 27709

DRESSELHAUS, MILDRED SPIEWAK, engineering educator; b. Bklyn., Nov. 11, 1930; d. Meyer and Ethel (Teichtheil) Spiewak; m. Gene F. Dresselhaus, May 25, 1958; children: Marianne, Carl Eric, Paul David, Eliot Michael. A.B., Hunter Coll., 1951, D.Sc. (hon.), 1982; Fulbright fellow, Cambridge (Eng.) U., 1951-52; A.M., Radcliffe Coll., 1953; Ph.D. in Physics, U. Chgo., 1958; D.Engring. (hon.), Worcester Poly. Inst., 1976, D.Sc., Smith Coll., 1980. NSF postdoctoral fellow Cornell U., 1958-60; mem. staff Lincoln Lab., MIT, 1960-67, prof. elec. engring., 1968—, asso. dept. head elec. engring., 1972-74; Abby Rockefeller Mauzé vis. prof. MIT, 1967-68, Abby Rockefeller Mauzé prof., 1973—, dir., 1977-83; vis. prof. dept. physics U. Campinas (Brazil), summer 1971, Technion, Israel Inst. Tech., Haifa, Israel, summer 1972, Nihon and Aoyama Gakuin Univs., Tokyo, summer 1973, IVIC, Caracas, Venezuela, summer 1977; mem. solid state scis. panel and com. NRC, 1973—; mem. exec. com. assembly of math. and phys. scis. Nat. Acad. Scis., 1975-78; chmn. steering com. of evaluation panels Nat. Bur. Standards, 1978-83. Contbr. articles to profl. jours. Named to Hunter Coll. Hall of Fame, 1972; recipient Alumnae medal Radcliffe Coll., 1973. Fellow Am. Phys. Soc. (chmn. nominating com. 1975, chmn. Buckley Prize com. 1977, v.p. 1982, pres.-elect 1983, pres. 1984), Am. Acad. Arts and Scis., IEEE; mem. Nat. Acad. Engring., Soc. Women Engrs. (Achievement award 1977); corr. mem. Brazilian Acad. Sci. Subspecialties: Condensed matter physics; Materials (engineering). Current work: Condensed matter physics, especially modification of material properties by intercalation and ion implantation. Home: 147 Jason St Arlington MA 02174 Office: Mass Inst Tech Cambridge MA 02139

DRESSER, MILES JOEL, physicist, educator; b. Spokane, Wash., Dec. 19, 1935; s. Lloyd Joel and Stella Christine (Nelson) D.; m. Muriel Louise Hunt, June 7, 1959; children: Don Joel, Marilyn Louise, Laura Jill. B.A., Linfield Coll., McMinnville, Oreg., 1957; Ph.D., Iowa State U., Ames, 1964. Teaching asst. Iowa State U., 1957-60; research asst. Ames Lab., 1960-63; asst. prof. physics Wash. State U., Pullman, 1963-70, assoc. prof., 1970—; physicist Nat. Bur. Standards, 1972. Contbr. articles to profl. jours. Bd. dirs. Pullman United Way, 1972-75. Mem. Am. Assn. Physics Tchrs., Am. Phys. Soc., Am. Vacuum Soc., Sigma Xi. Baptist. Subspecialties: Condensed matter physics; Surface chemistry. Current work: Desorption of ions by electronic tranisitions, vacuum surface phenomena, impact excitation processes, ionization mechanisms, electro luminescence. Office: 2814 Physics Washington State U Pullman WA 99164

DRESSLER, ALAN, astronomer; b. Cin., Mar. 23, 1948; s. Charles and Gay (Stein) D. B.A. in Physics, U. Calif., Berkeley, 1970, Ph.D. in Astronomy and Astrophysics, 1976. Carnegie fellow Hale Obs., Pasadena, Calif., 1976-78, Las Campanas fellow, 1978-81; mem. sci. staff Mt. Wilson and Las Campanas Obs., Carnegie Instn. Washingtin, Pasadena, 1981—; researcher. Contbr. papers to sci. jours. Mem. Am. Astron. Soc. (Pierce prize 1983). Subspecialty: Optical astronomy. Current work: Formation and evolution of galaxies; studies structure, morphology and stellar populations of galaxies as function of environmental and cosmological age. Office: 813 Santa Barbara St Pasadena CA 91106

DREWINKO, BENJAMIN, physician, researcher; b. Buenos Aires, Argentina, Feb. 10, 1940; came to U.S., 1963, naturalized, 1973; s. Aaron Joseph and Aida (Kadecka) D.; m.; children: Andrea P., Henry D., Marla G., Alejandra J. B.Sc. and B.A., Nat. Coll., 1956; M.D., Buenos Aires U., 1961; Ph.D., U. Tex., 1970. Cert. Am. Bd. Pathology. Intern Mt. Sinai Hosp., Chgo., 1963-64; resident Mt. Sinai Hosp. (N.Y.), N.Y.C., 1964-65; fellow M.D. Anderson Hosp. & Tumor Inst., Houston, 1967-70, chief sect. hematology dept. lab. medicine, 1970—, assoc. pathologist, 1973; dir. Schs. Med. Tech. and Blood Banking, 1973-74, prof. pathology, 1979—. Assoc. editor: Cancer Research, 1982—; contbr. chpts. to books and articles to profl. jours. Served to capt. MC U.S. Army, 1964-65. NIH grantee, 1976—; recipient Research Career Devel. award Nat. Cancer Inst., 1969. Mem. Am. Soc. Clin. Pathologists, Internat. Acad. Pathology, Am. Soc. Hematology, Cell Kinetics Soc. (pres. 1980-81), AAAS, Am. Assn. Cancer Research, Am. Soc. Clin. Oncology. Jewish. Subspecialties: Cancer research (medicine); Cellular pharmacology. Current work: Cellular pharmacology, interactions of antitumor drugs and cells, in vitro cell killing mechanisms, cell cycle progression delay, models of cell killing, combination of drugs in vitro; cell kinetics, growth kinetics of tissue cultured cells, exptl. tumors and human (clinical) neoplasms, methods of, model, exptl. therapy based on. Home: 2603 Rice Blvd Houston TX 77005 Office: 6723 Bertner Ave Houston TX 77030

DREXLER, HENRY, microbial geneticist, educator; b. Glendale, Pa., June 24, 1927; s. John Henry and Helena Catherine (Kieffer) D.; m. Susan Jane Schneider, June 15, 1957; children: Patricia Ann, Wesley Mark. B.Sc., Pa. State U., 1954; Ph.D., U. Rochester, 1960. Instr. U. Soc. Calif., Los Angeles, 1960-62; USPHS postdoctoral fellow Karolinska Inst., Stockholm, 1962-64; asst. prof. microbiology Med. Sch., Wake Forest U., Winston-Salem, N.C., 1964-69, assoc. prof., 1969-75, prof., 1975—. Served with U.S. Army, 1945-46. Mem. Am. Soc. Microbiology. Subspecialties: Genetics and genetic engineering (biology); Microbiology. Current work: Packaging of DNA by bacterial viruses, basic recombination, and the genetics of bacteriophage T1. Office: Dept Microbiolog Sch Medicine Wake Forest U Winston-Salem NC 27103

DREYFUS, EDWARD A., clin. psychologist; b. N.Y.C., Mar. 27, 1937; s. Herbert and Estelle (Soussi) D.; m. Estelle Dobbs, June 15, 1958 (div. June 1972); children: David E., Ronald C., Lydia M.; m. Judith K. Jones, Aug. 3, 1980 (div. 1983). B.B.A., CUNY, 1958, M.S., 1960; Ph.D., U. Kans., 1964. Lic. psychologist, lic. marriage, family and child counselor, Calif.; diplomate: Am. Bd. Psychotherapy. Psychologist VA Hosp., Palo Alto, Calif., 1964-65, UCLA, 1965-73; pvt. practice clin. psychology, Santa Monica, Calif., 1965—. Author: Youth: Search for Meaning, 1972, Adolescence, 1976; contbr. articles to profl. jours. Fellow Am. Univ. Sex. Educators and Therapists; mem. Am. Psychol. Assn. Am. Assn. Marriage and Family Therapists, Calif. State Psychol. Assn., Los Angeles County Psychol. Assn., Los Angeles Soc. Clin. Psychologists. Subspecialties: Behavioral psychology; Cognition. Current work: Psychotherapy, psycho-social issues. Office: 1431 Santa Monica Blvd Santa Monica CA 90101

DREYFUSS, JACQUES, drug metabolism researcher; b. St. Gallen, Switzerland, Jan. 20, 1937; came to U.S., 1940, naturalized, 1948; s. Hugo and Marga (Wachtel) D.; children: Julie, Heidi. B.S., Beloit Coll., 1958; Ph.D., Johns Hopkins U., 1963. Postdoctoral fellow Princeton U., 1963-64; sr. research fellow Squibb Inst. Med. Research, Princeton, N.J., 1964—. Contbr. numerous articles in biochemistry and drug metabolism to profl. jours. Mem. Am. Soc. Clin. Pharamacology and Therapeutics, AAAS, Am. Soc. Pharmacology and Exptl. Therapeutics, N.Y. Acad. Scis., Internat. Soc. Study Xenobiotics. Current work: Drug metabolic studies of new drugs developed at Squibb. Office: PO Box 4000 Princeton NJ 08540

DRICKAMER, HARRY GEORGE, chemistry educator; b. Cleve., Nov. 19, 1918; s. George Henry and Louise (Strempel) D.; m. Mae Elizabeth McFillen, Oct. 28, 1942; children: Lee Charles, Lynn Louise, Lowell Kurt, Margaret Ann, Priscilla. B.S., U. Mich., 1941, M.S., 1942, Ph.D., 1946. Chem. engr. Pan Am. Refining Corp., 1942-46; asst. prof. U. Ill. at Urbana, 1946-49, asso. prof., 1949-53, prof. phys. chemistry and chem. engring., 1953—. Recipient Bendix award, 1968; P.W. Bridgman award Internat. Assn. High Pressure Sci. and Tech., 1977; Guggenheim fellow, 1952; Michelson-Morley award Case Western Res. U., 1978. Fellow Am. Phys. Soc. (Buckley Solid State Physics award 1967), Am. Geophys. Union; mem. Nat. Acad. Engring., Am. Chem. Soc. (Ipatieff prize 1956, Langmuir award in chem. physics 1974), Am. Inst. Chemists (Chem. Pioneers award 1983), Am. Inst. Chem. Engrs. (Colburn award 1947, Alpha Chi Sigma award 1967, Walker award 1972), Faraday Soc., Nat. Acad. Scis., Am. Acad. Arts and Sci., Am. Philos. Soc., Center for Advanced Studies. Subspecialties: Physical chemistry; Condensed matter physics. Current work: Use of high pressure to study electronic phenomena in condensed phases. Home: 304 E Pennsylvania St Urbana IL 61801

DRICKAMER, KURT, biochemistry educator; b. Champaign, Ill., Nov. 4, 1952; s. Harry George and Mae Elizabeth (McFillen) D. B.S., Stanford U., 1973; Ph.D., Harvard U., 1978. Fellow Duke U., 1978-79; fellow Cold Spring Harbor Lab., Cold Spring Harbor, N.Y., 1979-81, sr. staff scientist, 1982; asst. prof. biochemistry U. Chgo., 1982—. NSF fellow, 1973; Helen Hay Whitney Found. fellow, 1978. Mem. Am. Soc. Biol. Chemists, Phi Beta Kappa. Subspecialties: Biochemistry (biology); Cell and tissue culture. Current work: Study of protein structure and synthesis as it relates to receptors and endocytosis. Home: 5419 S Dorchester Ave Chicago IL 61615 Office: Department of Biochemistry University of Chicago 920 E 58th St Chicago IL 60637

DRIESSEL, KENNETH R., applied mathematician; b. Milw., 1940; s. Richard H. and Margaret (Otto) D. B.S., U. Chgo., 1962; M.S., Oreg. State U., Corvallis, 1965, Ph.D., 1967. Research scientist Amoco Research, Tulsa, 1971—; asst. prof. U. Colo., Denver, 1967-71. Mem. Am. Math. Soc., Soc. Indsl. and Applied Math., Assn. Computer Machinery, Assn. Symbolic Logic, U.S. Cycling Fedn. (Okla. dist. rep. 1974-81). Club: Tulsa Bicycle (pres. 1974-75). Subspecialties: Applied mathematics; Mathematical software. Current work: Mathematical aspects of exploration seismology. Office: Amoco Research PO Box 591 Tulsa OK 74102 Home: 3734 S Madison Tulsa OK 74105

DRISCOLL, TIMOTHY JOHN, physicist; b. Washington, Oct. 17, 1941; s. Timothy John and Rose Mary (Gallogly) D. B.A., Cath. U. Am., 1964, M.S., 1966, Ph.D., 1977. Physicist U.S. Bur. Mines, Avondale, Md., 1964-69, research physicist, 1969—. Contbr. articles to profl. publs. Mem. Am. Phys. Soc., Am. Vacuum Soc. Subspecialties: Surface physics; Alloys. Current work: Applying techniques of surface physics to study of oxidation and corrosion of metals; studying incentive behavior of light metals in mines. Office: 4900 LaSalle Rd Avondale MD 20782

DRITSCHILO, ANATOLY, radiation oncologist; b. Reigersfeld, Ger., Oct. 10, 1944; s. Peter and Maria (Kardash) D.; m. Joy Ann Ickenroth, Apr. 6, 1968; children: Peter, Andrea, Lisa. B.S., U. Pa., 1967; M.S., Newark Coll. Engring., 1969; M.D., Coll. Medicine N.J., 1973. Cert. in therapeutic radiology. Intern Cin. Gen. Hosp., 1973-74; resident Joint Center Radiation Therapy, Harvard U., 1974-77; chmn. dept. radiation medicine Georgetown U., 1980—; dir. radiation oncology Vincent Lombardi Comprehensive Cancer Center, 1979—. Contbr. articles to profl. jours. NSF fellow, 1967-68; Am. Cancer Soc. clin. fellow, 1978. Mem. Am. Soc. Therapeutic Radiologists, Radiation Research Soc., Am. Assn. Cancer Research, Am. Soc. Clin. Oncologists. Republican. Russian Orthodox. Subspecialties: Radiology; Cell biology. Current work: Response of human cells in tissue culture to radiation. Office: 3800 Reservoir Rd NW Washington DC 20007

DRNEVICH, VINCENT PAUL, civil engineering educator; b. Wilkinsburg, Pa., Aug. 6, 1940. B.S.C.E., U. Notre Dame, 1962, M.S.C.E., 1964; Ph.D., U. Mich., 1967. Research asst. soil dynamics U. Notre Dame, Ind., 1962-64; research asst. soil dynamics U. Mich., Ann Arbor, 1964-67; asst. to assoc. prof. civil engring. U. Ky., Lexington, 1967-77, prof., 1977—, chmn. dept., 1980—. NSF grantee, 1968-69. Mem. Internat. Soil Mechanics and Found. Engrs., ASCE (Huber prize 1980), ASTM (award 1966, Hogentogler award 1979). Subspecialty: Civil engineering. Office: Dept Civil Engring U Ky Lexington KY 40506

DROLSOM, PAUL NEWELL, agronomist; b. Martell, Wis., July 15, 1925; s. Peter and Inga Marie (Qualle) D.; m. Marian Eda Zwerg, Aug. 19, 1950; children: Amy Ann, Ann Marie. B.S., U. Wis., 1949, M.S., 1950, Ph.D., 1953. Research asst. U. Wis., Madison, 1949-52, asst. prof., 1958-61, asso. prof., 1961-66, prof., 1966—; pathologist U.S. Dept. Agr., Oxford, N.C., 1953-58. Contbr. articles in field to profl. jours. Served with U.S. Army, 1943-46. Mem. Am. Soc. Agronomy, AAAS, Am. Forage and Grassland Council. Lutheran. Club: Kiwanis. Subspecialty: Plant genetics. Current work: Researcher in corn breeding, especially for earliness with cold tolerance and disease resistance. Office: University of Wisconsin Dept.Agronomy Madison WI 53706

DROZD, ANDREW LOUIS STEPHAN, electromagnetics engr., software support cons.; b. Vucht, Belgium, Jan. 14, 1956; U.S., 1956, naturalized, 1978; s. Matthew Thomas and Mary (Sandak) D. B.S. in Physics-Math, Syracuse U., Utica Coll., 1978, M.S. in Elec. Engring, 1982. Electronmagnetics/systems engr. IIT Research Inst., Griffiss AFB, N.Y., 1978—. Mem. Am. Inst. Physics, Assn. for Computing Machinery, IEEE. Democrat. Roman Catholic. Club: Optimist. Lodge: KC (Rome, N.Y.). Subspecialties: Systems engineering; Computer engineering. Current work: Software applications to electromagnetics analysis; software/database support, electromagnetic theory; conducting workshops/courses in computer-aided electromagnetics analysis. Home: 7755 Turin Rd Rome NY 13440 Office: RADC/RBCTI Bldg 104 Griffiss AFB NY 13441

DRUCKER, DANIEL CHARLES, engineer, dean; b. N.Y.C., June 3, 1918; s. Moses Abraham and Henrietta (Weinstein) D.; m. Ann Bodin, Aug. 19, 1939; children: R. David, Mady. B.S., Columbia U., 1937, C.E., 1938, Ph.D., 1940; D.Engring. (hon.), Lehigh U., 1976, D.Sc. in Tech., Technion, Israel Inst. Tech., 1983. Instr. Cornell U., 1940-43; supt. Armour Research Found., Chgo., 1943-45; asst. prof. Ill. Tech., 1946-47; assoc. prof. Brown U., Providence, 1947-50, prof., 1950-64, L. Herbert Ballou Univ. prof., 1964-68, chmn. div. engring., 1953-59, chmn. phys. scis. council, 1960-63; dean Coll. Engring., U. Ill., Urbana, 1968—; Marburg lectr. ASTM, 1966; Mem., past chmn. U.S. Nat. Com. on Theoretical and Applied Mechanics; treas. Internat. Union Theoretical and Applied Mechanics, 1972-80, pres., 1980-84; mem. gen. com. Internat. Council Sci. Unions; past chmn. adv. com. for engring. NSF; hon. chmn. 3d SESA Internat. Congress on Exptl. Mechanics. Author: Introduction to Mechanics of Deformable Solids, 1967; Contbr. chpts. in tech. books, also tech. papers to mech. and sci. jours. Guggenheim fellow, 1960-61; NATO Sr. Sci. fellow, 1968; Fulbright travel grantee, 1968; Gustave Trasenster medal U. Liège, Belgium, 1979; Thomas Egleston medal Columbia U. Sch. Engring. and Applied Sci., 1978. Fellow ASME (chmn. applied mechanics div. 1963-64, v.p. policy bd. communications 1969-71, pres. 1973-74, Timoshenko medal 1983), Am. Acad. Mechanics (past pres.), Am. Acad. Arts and Scis. (mem. Midwest Council), AAAS (past chmn. sect. engring., mem. council 1980), Am. Inst. Aero. and Astronautical Scis. (asso. fellow), ASCE (von Karman medal 1966, past pres. New Eng. council, past pres. Providence sect., past chmn. exec. com. engring. mechanics div.); mem. N.R.I., Ill. socs. profl engrs., Soc. Exptl. Stress Analysis (hon.; past pres., W. M. Murray lectr. 1967, M.M. Frocht award 1971), Am. Technion Soc. (past pres. So. N.E. chpt.), Soc. of Rheology, Am. Soc. Engring. Edn. (charter fellow mem., past 1st v.p., past chmn. engring. coll. council, dir., pres. 1981-82, Lamme award 1967), Nat. Acad. Engring. (mem. com. on pub. engring. policy 1972-75), Nat. Acad. Scis. (chmn. subcom. on sci. unions, bd. internat. ops. and programs), Soc. Engring. Sci. (Wiliam Praeger medal 1982), Polish Acad. Scis. (fgn. mem.), Sigma Xi (past pres. Brown U. chpt.), Phi Kappa Phi, Tau Beta Pi, Pi Tau Sigma. Subspecialties: Theoretical and applied mechanics; Materials (engineering). Current work: Stress-strain relations; finite plasticity; stability; fracture and flow on macroscale and microscale. Office: 106 Engring Hall U Ill 1308 W Green St Urbana IL 61801

DRUGER, MARVIN, educator; b. Bklyn., Feb. 21, 1934; s. Harry and Ida (Taks) D.; m. Patricia Sylvia Meyers, June 9, 1957; children: Lauren Jane, Robert Kenneth, James Meyers. B.S., Bklyn. Coll., 1955; M.A., Columbia U., 1957, Ph.D., 1961. NIH postdoctoral fellow Commonwealth Sci. Indsl. Research Orgn., 1961-62; faculty Syracuse U., 1962—, prof. biology and sci. edn., 1971—. Chmn. adv. bd.: Jour. Coll. Sci. Teaching, 1982-84; Contbr. articles to profl. jours. Served with USCGR, 1957-65. Recipient James Howard McGregor prize Columbia U., 1959; Sci. Teaching Achievement Recognition award Nat. Sci. Tchrs. Assn., 1968; Gustav-Ohaus award, 1978; Outstanding Achievement award Syracuse U., 1973; Fulbright lectr. U. Sydney, 1969-70; Western Australian Inst. Tech. vis. fellow, 1981; Danforth assoc., 1980. Mem. AAAS (chmn. edn. sect. 1983-84); Mem. Nat. Sci. Tchrs. Assn. (coll. div. dir. 1982-84), Soc. Coll. Sci. Tchrs. (pres. 1981-82), Assn. for Edn. Tchrs. Sci. (region pres. 1981-82), Nat. Assn. Biology Tchrs., Genetics Soc. Am., Soc. Study of Evolution, AAAS, Nat. Assn. Research in Sci. Teaching, N.Y. Acad. Sci., Am. Inst. Biol. Sci. Subspecialty: Evolutionary biology. Current work: Evolutionary genetics, individualized and audio-tutorial instruction.

DRUM, BRUCE ALAN, visual psychophysicist, consultant; b. Wauseon, Ohio, May 18, 1947; s. Virgil Ward and Clela Laverne (Overly) D.; m. Pamela Joy Neff, June 16, 1973; 1 dau., Rachel Lynne Neff. B.S., Ohio State U., 1969, Ph.D., 1973. Vis. research assoc. Ohio State U., 1973; postdoctoral fellow Johns Hopkins U., Balt., 1973-75; asst. research prof. George Washington U., Washington, 1975-79, research scientist, 1979—; co-founder, sec. Vision Research Assos., Inc., Balt., 1981—. Postdoctoral fellow Seeing Eye, Inc., Balt., 1974; biomed. research support grantee George Washington U., 1978-79; research grantee Nat Eye Inst., NIH, 1975-78, 79-82, 81—. Mem. AAAS, Assn. for Research in Vision and Ophthalmology, Optical Soc. Am., Psychonomic Soc., Internat. Research Group for Color Vision Deficiencies, Internat. Perimetric Soc., Fedn. Am. Scientists. Subspecialties: Psychophysics; Sensory processes. Current work: Correlations between neural processes and visual sensation; color vision; light and dark adaptation; rod-cone interactions; chromatic and achromatic brightness sensations; peripheral vision and perimetry; visual functioning glaucoma. Home: 5503 Calhoun Ave Alexandria VA 22311 Office: Dept Ophthalmology George Washington Univ 2150 Pennsylvania Ave NW Washington DC 20037

DRUM, DONALD A., chemist, educator; b. Warren, Ohio, Apr. 7, 1942; s. Willis A. and Kathryn H. (Meade) D.; m. Pamela A. Lewis, Aug. 8, 1964; children: Deborah, David, Patricia. B.Sc., Ohio State U., 1964; M.Sc., U. Mass., 1968, Ph.D., 1970. With Ross Labs., Columbus, Ohio, 1964-65; mem. faculty Columbia-Greene Community Coll., Hudson, N.Y., 1969—, prof. chemistry, 1977—, chmn. sci. and tech. div., 1979; cons. in field. Contbr. articles to profl. jours. Recipient Mfg. Chemists' Assn. award for excellence in coll. chemistry teaching, 1980; award for excellence in teaching N.Y. State Chancellor, 1976. Mem. Am. Chem. Soc., N.Y. State Coll. Chemistry Tchrs. Assn. (pres. 1977-78), Two-Yr. Coll. Chemistry Assn., Nat. Environ. Tng. Assn., Am. Soc. Engring. Edn., Nat. Council Instructional Adminstrs., Sigma Xi. Subspecialty: Water supply and wastewater treatment. Current work: Incineration of chemical wastes; waste compatibility and handling, incinerator design, BACT for control of emissions, wastewater and water. Home: Box 290 RD 2 Catskill NY 12414 Office: Columbia-Greene Community Coll Box 1000 Hudson NY 12534

DRUMMOND, ROBERT JOHN, psychologist, consultant; b. Newark, Mar. 30, 1929; s. Lester Linwood and Marie (Pester) D.; m. Gloria E. Erickson, Nov. 11, 1968; children: Robin, Heather. A.B., Waynesburg Coll., 1949; A.M., Columbia U., 1952, 56, Ed.D., 1959. Prof., chmn. dept. psychology Waynesburg (Pa.) Coll., 1957-69; vis. prof. U. Pitts., 1966-67; prof., coordinator field research U. Maine, Orono, 1969-81; acting chmn. div. ednl. services and research U. North Fla., Jacksonville, 1981—; cons. Inst. Can. Bankers, Montreal, 1977-82; evaluator dept. Edn. and Cult Services, Augusta, Maine, 1969-81. Contbr. articles to profl. jours. Recipient Impact Computer Guidance Systems award Maine Occupational Info. Coordinating Com., 1980. Fellow Am. Psychol. Assn.; mem. Am. Personnel and Guidance Assn., Am. Edn. Research Assn. Methodist. Subspecialties: Learning; Social psychology. Current work: Computer guidance systems, computer applications in education, computer anxiety, learning style and computer utilization. Home: 3405 Compass Rose Dr Jacksonville FL 32216 Office: U N Fla PO Box 17074 Jacksonville FL 32216

DRUMMOND, WILLIAM ECKEL, physicist, educator; b. Portland, Oreg., Sept. 18, 1927. B.S., Stanford U., 1951, Ph.D. in Physics, 1958.

Physicist Hanford Labs., Gen. Electric Co., 1951-52, Calif. Research & Devel., 1952-54; mem. staff radiation lab. U. Calif., 1954; mem. staff Stanford (Calif.) U. Research Inst., 1955-58; prin. scientist Research Lab., Avco Mfg. Co., 1958-59; physicist Gen. Atomic div. Gen. Dynamics Corp., 1959-65; prof. physics U. Tex.-Austin, 1965—, dir., 1966—; lectr. Stanford U., 1955. Fellow Am. Phys. Soc.; mem. AAAS. Subspecialties: Plasma physics; Theoretical physics. Office: Dept Physics U Tex Austin TX 78712

DRYE, CHARLES EDWIN, plant pathologist, educator; b. Cleveland County, N.C., June 12, 1946; s. Columbus Herman and Myrtle Betty (Tucker) D.; m. Lynne Brunson, July 14, 1979. B.S., Lenoir Rhyne Coll., 1969; M.A., Appalachian State U., 1972; Ph.D., Clemson U., 1976. Secondary sch. tchr., N.C., 1969-72; grad. research asst. Clemson U., 1972-76; asst. prof. plant biology Rutgers U., New Brunswick, N.J., 1976-78; asst. prof. plant pathology and physiology Clemson U., Coop. Extension Service, Blackville, S.C., 1978—, extension plant pathologist, Savannah Valley sect., S.C. Mem. Am. Phytopath. Soc., Am. Peanut Research and Edn. Soc., Southeastern Phytopath. Soc. Methodist. Subspecialties: Plant pathology; Integrated pest management. Current work: Diagnosis and control of crop diseases, education, IPM program development. Home: PO Box 167 Jackson SC 29831 Office: PO Box 247 Edisto Expt Sta Blackville SC 29817

DU, DAVID HUNG-CHANG, computer science educator; b. Kaohsiung, Taiwan, Republic of China, July 1, 1951; came to U.S., 1977; s. Shien-Wen and Min Tsao Du; m. Jasmine Du/Ran, Aug. 23, 1980; 1 son, Albert Jeng. B.S., Nat. Tsing-Hua U., Taiwan, 1974; M.S., U. Wash.-Seattle, 1980, Ph.D., 1981. Asst. prof. U. Minn., Mpls., 1981—. Contbr. articles in field to profl. jours. Mem. Assn. Computing Machinery, IEEE. Subspecialties: Distributed systems and networks; Database systems. Current work: File design and file allocation problems for database systems, packet switching networks, parallel algorithms. Home: 1439 19th Ave NW New Brighton MN 55112 Office: Dept Computer Sci U Minn 207 Church St SE Minneapolis MN 55455

DUARTE, CRISTOBAL G., physician, educator, researcher, scientist; b. Concepcion, Paraguay, July 17, 1929; s. Cristobal Duarte and Emilia Miltos. B.S., Colegio de San Jose, Asuncion, Paraguay, 1947; M.D., Nat. U. Asuncion, 1953. Intern De Goesbriand Meml. Hosp., Burlington, Vt., 1956; resident in medicine Carney Hosp. and St. Elizabeth's Hosp., Boston, 1956-58; fellow in medicine Lahey Clinc, Boston, 1959; fellow hypertension and renal medicine Hahnemann Hosp., Phila., 1960; assoc. in medicine U. Vt. Coll. Medicine, 1962-65; clin. investigator VA, 1966-68, staff physician, 1968-73; dir. Renal Function Lab., Mayo Clinic and Found., Rochester, Minn., 1973-77; asst. prof. lab. medicine Mayo Med. Sch., 1973-77; asso. prof. medicine and physiology Uniformed Services U. Health Scis., Bethesda, Md., 1977—; attending in medicine commd. lt. col. U.S. Army, 1977; Walter Reed Army Med. Ctr., Washington. Editor: Renal Function Tests, 1980; contbr. articles to profl. jours., chpts. to books. Recipient cert. of accomplishment VA, 1969; physician's recognition award AMA, 1981; Cordell Hull Found. fellow, 1958-59. Fellow Am. Coll. Nutrition; mem. Nat. Kidney Found., Latin Am. Soc. Nephrology, Am. Fedn. Clin. Research, Am. Physiol. Soc., Am. Soc. Pharmacology and Exptl. Therapeutics, Midwest Salt and Water Club, Am. Soc. for Clin. Research, Central Soc. for Clin. Research, Am. Soc. Nephrology, Sigma Xi. Roman Catholic. Subspecialty: Nephrology. Current work: Interrelations between potassium and magnesium at renal and tissue levels. Office: 59621 Walter Reed Station Washington DC 20012

DUAX, WILLIAM LEO, biol. researcher; b. Chgo., Apr. 18, 1939; s. William Joseph and Alice B. (Joyce) D.; m. Caroline Townsend Dowell, May 6, 1966; children: Julia, Sarah, William, Stephen. B.A., St. Ambrose Coll., 1961; Ph.D., U. Iowa, 1967. Postdoctoral research fellow Ohio U., Athens, 1967-68; research assoc. Med. Found. Buffalo, 1968-69, head molecular physics dept., 1970—; adj. research prof. SUNY, Buffalo, 1973—, assoc. research prof. dept. biochemistry, 1981—. Contbr. articles to profl. jours. Active Am. Field Service. Served with USAR, 1961-67. NIH grantee, 1972—. Mem. Am. Crystallographic Assn., Am. Chem. Soc., AAAS, Am. Cancer Soc., Biophys. Soc. Democrat. Roman Catholic. Subspecialties: X-ray crystallography; Endocrinology. Current work: X-ray crystallography, biol. active molecules, steroid and peptide hormone action. Office: 73 High St Buffalo NY 14203

DUBAR, JULES RAMON, geogolist; b. Canton, Ohio, June 30, 1923; s. Joseph Adolphe and Inez Ismay (Simlar) DuB.; m. Susan Stokes Davidson, July 29, 1964; children: Nicole Mae, Scott Johnson. B.S., Kent State U., 1949; M.S., Oreg. State U., 1950; Ph.D., U. Kans., 1957. Assoc. prof. geology Duke U., Durham, N.C., 1962-64; sr. research assoc. Esso Prodn. Research Co., Houston, 1964-67; chmn., prof. geosci. Morehead (Ky.) State U., 1967-81; exploration mgr. Internat. Resource Devel. Corp., Pepper Pike, Ohio, 1981-82; research scientist U. Tex. Bur. Econ. Geology, Austin, 1982—; cons. Fla. Geol. Survey, Tallahassee, 1953-58, Internat. Minerals and Chem. Corp., Lakeland, Fla., 1963-64, 76-77, William Bird Sales Co., Charleston, S.C., 1972-74; professorial research appointee U.S. Geol. Survey, 1979-81. Author: Stratigraphy Neogene Stratigraphy Southern Florida, 1958, Neogene Stratigraphy of Carolinas, 1971, Biostratigraphy of Southwestern Florida, 1962; author, editor: Post-Miocene Stratigraphy, 1974. Served with USCG, 1942-46; PTO. Grantee NSF, 1959-65, 68-70, 77-79, 79-81. Fellow Geol. Soc. Am., AAAS, Explorers Club; mem. Am. Assn. Petroleum Geologists. Democrat. Subspecialties: Geology; Paleoecology. Current work: Evaluations of hydrocarbon potentials in Gulf of Pexico Province, Michigan Basin, and Appalachian Basin based on siesmic, geophysical log, and sample analyses. Home: 12600 Esplanade St Austin TX 78758 Office: Bur Econ Geology Univ Tex Univ Sta Box X Austin TX 78712

DUBE, DONALD ARTHUR, nuclear engineer, educator; b. Biddeford, Maine, Mar. 4, 1954; s. Arthur A. and Annette L. (Babineau) D.; m. Marianna Koziol, Sept. 19, 1980; 1 dau., Kasia Koziol. B.S., Cornell U., 1976; S.M., MIT, 1978, Ph.D., 1980. Summer tech. intern U.S. Nuclear Regulatory Commn., Washington, 1976; grad. research asst. Los Alamos Nat. Lab., summer 1977, 78; summer engr. N.E. Utilities Service Co., Hartford, Conn., summer 1979, nuclear engr., 1982—; mem. tech. staff Sandia Nat. Labs., Albuquerque, 1980-82; instr. Hartford State Tech. Coll., 1982—. Author: National Lab Report, 1982. Mem. Com. to Save Maine Yankee, 1980. Sherman R. Knapp fellow, 1979. Mem. Am. Nuclear Soc., Sigma Xi, Tau Beta Pi, Phi Kappa Phi. Democrat. Roman Catholic. Subspecialty: Nuclear fission. Current work: Nuclear reactor safety; fire protection; reactor transient analysis; degraded core analysis. Home: 130 Belmont St New Britain CT 06053 Office: PO Box 270 Hartford CT 06141

DUBE, GEORGE, opticist, optical glass company executive; b. Denver, Dec. 10, 1942; s. Paul Henderson and Levena Maria (Paddock) Dube.; m. Margaret Ann Hollander, Jan. 28, 1971; children: Jean-Paul, Michelle. B.S., Inst. Optics, U. Rochester, 1964, M.S. (NDEA fellow), 1969, Ph.D., 1972. Sr. scientist Owens-Illinois Inc., Toledo, 1971-77, product mgr., 1978-80; mgr. corp. tech. services Schott Optical Glass Inc., Duryea, Pa., 1981—. Contbr. articles onlaser damage and optical techniques to profl. jours. Mem. Scranton (Pa.) C. of C. Served to lt. j.g. USNR, 1964-66; Vietnam. Mem. IEEE, Optical Soc. Am., Sigma Xi, Tau Beta Pi. Subspecialties: Laser fusion; Optical engineering. Current work: Laser glasses, optical materials, solid state lasers. Patentee laser designs. Home: RD 1 Stone Rd Dalton PA 18414 Office: 400 York Ave Duryea PA 18642

DUBE, RAJESH, operations research analyst; b. Indore, India, Dec. 11, 1952; came to U.S., 1977; s. Jayadeo Prasad and Sharad Kumari (Dubey) D. B.E. (Mech.), U. Indore, 1974; M.Tech., Indian Inst. Tech., 1977; M.S. in Indsl. Engring, Rutgers U., 1979; M.Sc. in Computer Sci, Rutgers U., 1980, postgrad., 1980—. Systems analyst RCA, Princeton, N.J., 1980-82, ops. research analyst, 1982—. Mem. Inst. Indsl. Engrs., Assn. Computing Machinery, IEEE Computer Soc. (affiliate), Indsl. Engring. Honors Soc. Subspecialties: Information systems, storage, and retrieval (computer science); Operations research (engineering). Current work: Distributed systems and networks, packet switched networks, software engineering. Home: 36-16 Fox Run Dr Plainsboro NJ 08536 Office: PO Box 2023 Princeton NJ 08540

DUBE, ROGER RAYMOND, exptl. physicist; b. Portland, Maine, Nov. 24, 1949; s. Roger Joseph and Doris Ruth (Roy) D.; m. Marilyn Markman, Dec. 9, 1972; children: Dawn, Danielle, Laura. A.B., Cornell U., 1972; M.A., Princeton U., 1974, Ph.D., 1976. Postdoctoral Kitt Peak Nat. Obs., 1976-77; mem. sr. staff Jet Propulsion Lab., Pasadena, Calif., 1977-78; asst. prof. physics U. Mich., 1978-80, U. Ariz, 1980-82; mem. staff IBM Gen. Products div., Tucson, 1982—; cons. Grantee in field. Mem. AAAS, N.Y. Acad. Sci., Soc. Advancement of Chicanos and Native Americans in Sci., Am. Astron Soc. Subspecialties: Relativity and gravitation; Optical astronomy. Current work: Astronomy instrumentation, computer applications, research and devel., high speed instrumentation, electronic data processing performance evaluation. Office: Department 67E/041-2 IBM Corporation Tucson AZ 85744

DUBES, GEORGE RICHARD, geneticist; b. Sioux City, Iowa, Oct. 12, 1926; s. George W. and Regina E. (Kelleher) D.; m. Margaret J. Tumberger, July 25, 1964; children: George, David, Deanna, Kenneth, Deborah, Keith. B.S., Iowa State U., 1949; Ph.D., Calif. Inst. Tech., 1953. Research assoc. Johns Hopkins U., Balt., 1953-54; successively research assoc., asst. prof., assoc. prof. U. Kans. Sch. Medicine, Kansas City, 1954-64; assoc. prof., then prof. med. microbiology U. Nebr. Coll. medicine, Omaha, 1964—. Contbr. articles and abstracts to sci. lit. Mem. citizen's adv. com. Omaha Pub. Schs., 1977-80. Served with U.S. Army, 1945-46; PTO. NIH grantee, 1966-69. Mem. Am. Assn. Cancer Research, AAAS, Am. Genetic Assn., Am. Inst. Biol. Scis., Am. Soc. Microbiology, Biometric Soc., Genetics Soc. Am., Nebr. Acad. Scis., N.Y. Acad. Scis., Sigma Xi. Subspecialties: Genetics and genetic engineering (biology); Virology (biology). Current work: Methods for transfecting cells; mechanism of copper-mediated inactivation of nucleic acids; opal mutants of viruses; effects of virus tansformation of mammalian cells on their requirements for hormones and other growth factors; effects of oxygen limitation and virus multiplication. Home: 7515 Lawndale St Omaha NE 68134 Office: 42d and Dewey Sts Omaha NE 68105

DUBNER, RONALD, neurobiologist, dental scientist; b. N.Y.C., Oct. 12, 1934; s. Louis and Matilda (Fox) D.; m. Mary Ann P. Pollack, June 22, 1958; children: Susan R., Andrew D., Julia P. A.B., Columbia U., 1955, D.D.S., 1958; Ph.D., U. Mich.-Ann Arbor, 1964. Intern USPHS Hosp., Balt., 1958-59; staff dentist Clin. Center, NIH, Bethesda, Md., 1959-61; research scientist Nat. Inst. Dental Research, Bethesda, 1961-68, chief neural mech. sect., 1968-73, chief neurbiol. and anesthesiology br., 1973—; vis. scientist Univ. Coll., London, 1970-71; vis. assoc. prof. Howard U., Washington, 1968-80; chmn. Subcom. on Research, Interagency Com. on New Therapies for Pain and Discomfort, 1978-80. Author: Neural Basis of Oral and Facial Function, 1978; Editor: Oral-Facial Sensory and Motor Mechanisms, 1971, Oral-Facial Sensory and Motor Functions, 1981. Pres. Glade Spring Assn., Manns Choice, Pa., 1977, bd. dirs., 1982. Recipient Meritorious Service medal USPHS, 1975; Birnberg Research award Columbia U., 1981. Mem. Internat. Assn. Study Pain (v.p. 1981-84), Internat. Assn. Dental Research (pres. neurosci. group 1976-77), Am. Pain Soc. (bd. dirs. 1980-82, chmn. sci. program com. 1983), Soc. Neurosci., Am. Physiol. Soc. Subspecialties: Neurobiology; Oral biology. Current work: Conduct and direct research on somatic sensation, with emphasis on pain mechanisms and methods of pain control. Office: NIH Bldg 30 Room B-18 Bethesda MD 20205

DUBOIS, ANDRE, medical educator, administrator, researcher; b. Liège, Belgium, Mar. 16, 1939; came to U.S., 1971. B.S. magna cum laude, U. Brussels, 1959, M.D., 1963, Ph.D., 1975. Intern University Hosp. St. Pierre, Brussels, 1962-63, resident in surgery, 1963-64, 1965-66, 1967-69; fellow in exptl. surgery U. Brussels, 1966-67, instr. dept. anatomy, 1965-66, staff fellow dept. surgery, 1969-71; NATO fellow NIMH, NIH, Bethesda, Md., 1971-72; vis. assoc. lab. theoretical biology Nat. Cancer Inst., 1972-73; vis. assoc. and vis. scientist sect. gastroenterology, digestive diseases br. Nat. Inst. Arthritis, Metabolism and Digestive Diseases, 1973-75; instr. dept. medicine Uniformed Services U. Health Scis., Bethesda, 1975-76, asst. prof., 1977-78, assoc. prof., 1978—, asst. dir. digestive diseases div., 1980—, assoc. prof. (research) dept. surgery (secondary), 1981—. Recipient van Engelen prize Univ. Hosp., U. Brussels, 1963. Mem. Am. Gastroent. Assn., Am. Physiol. Soc., Am. Fedn. of Clin. Research, Am. Assn. for Acad. Surgery, William Beaumont Soc. Gastroenterology. Subspecialty: Gastroenterology. Current work: Gastric secretion, gastric emptying and motility, effect of physical and psychological stress. Home: 9210 Bardon Rd Bethesda MD 20814 Office: Uniformed Services Univ Health Scis 4301 Jones Bridge Rd Bethesda MD 20814

DUBOW, JOEL BARRY, electrical engineering educator; b. Bklyn., June 13, 1943; s. Harold Louis and Natalie DuB.; m. Donna Kay McGregor; children: Therlla; Tammy. Ph.D., Case Western Res. U.-Cleve., 1972. Assoc. prof. elec. engring. Colo. State U., 1974-81; prof. elec. engring. Poly. Inst. N.Y., Bklyn., 1981-82, chmn. dept., 1981-82; prof. elec. engring. Boston U. Coll. Engring., 1982—, chmn. dept., 1982—; cons. in field. Contbr. over 200 sci. articles to profl. publs. Recipient Cert. of Achievement award NASA, 1968. Mem. Am. Chem. Soc., IEEE, AAAS. Subspecialties: Semiconductors; Oil shale. Current work: Integrated circuits; energy systems for remote areas. Patentee in field. Office: 110 Cummington St Boston MA 02215

DUCHAMP, DAVID JAMES, physical chemist; b. St. Martinville, La., Oct. 15, 1939; s. Clarence Joseph and Marie (Fournet) D.; m. Annette Vivienne Sleigh, 1964; children: James, Vivienne, Joe. B.S., U. Southwestern La., Lafayette, 1961; Ph.D. (NSF fellow), Calif. Inst. Tech., 1965. Research asst. Calif. Inst. Tech., 1965; scientist pharm. research and devel. The Upjohn Co., Kalamazoo, 1965—; mem. U.S. nat. com. on crystallography Nat. Acad. Scis., 1981-83. Contbr. articles to profl. jours. Active PTA, Boy Scouts Am. Woodrow Wilson fellow. Mem. Am. Crystallographic Assn., Am. Chem. Soc., AAAS, Assn. for Computing Machinery, N.Y. Acad. Scis., Kappa Mu Epsilon, Sigma Pi Sigma. Roman Catholic. Subspecialties: Crystallography; Physical chemistry. Current work: Three-dimensional structure of biologically active molecules, molecular mechanics, computer graphics, laboratory automation computing, crystallographic computer programming. Home: 6209 Litchfield Ln Kalamazoo MI 49009 Office: Upjohn Co Kalamazoo MI 49001

DUCHARME, DONALD WALTER, pharmacologist, researcher; b. Saginaw, Mich., June 14, 1937; s. WalterArnold and Marion (Law); m. Doris Barbara Rieck, Aug. 30, 1958; children: Michael, Mark, Daniel. A.B., Central Mich. U., 1959; Ph.D. (USPHS fellow), U. Mich., 1965. Scientist Upjohn Co., Kalamazoo, Mich., 1965-67, research scientist, 1967-70, sr. research scientist, 1970-73, sr. scientist, 1973-78, research head, 1978—; adj. asso. prof. dept. pharmacology Med. Coll. of Ohio. Contbr. articles on pharmacology to profl. jours. Pres. Kalamazoo (Mich.) County Heart Unit Bd., 1971-72. Recipient W. E. Upjohn award, 1977. Fellow Council for High Blood Pressure Research; mem. Am. Heart Assn. (pres. Mich. affiliate 1981-82, chmn. north central region research adv. com. 1981-83, regional heart com., nat. regional research com.), Am. Soc. for Pharmacology and Exptl. Therapeutics, Council for Kidney in Cardiovascular Disease, Mich. Steelhead and Salmon Fisherman's Assn. Subspecialties: Pharmacology; Cardiology. Current work: Cardiovascular pharmacology, etiology of hypertension, research administration, hypertension, neurogenic and humoral control of circulatory system, prostaglandins. Patentee in field. Home: 287 Fineview Kalamazoo MI 49007 Office: Cardiovascular Disease Research The Upjohn Co Kalamazoo MI 49001

DUCHOWNY, MICHAEL SAMUEL, neurologist, neurophysiologist, educator; b. N.Y.C., Nov. 17, 1945; s. Boris M. and Helen (Ledman) D.; m. Bonnie L. (maiden name please) Levin, May 26, 1979; 1 dau., Alexandra. A.B., Cornell U., 1966; M.D., Albert Einstein Coll. Medicine, 1970. Research assoc. NIH, Bethesda, Md., 1972-74; clin. fellow in neurology Harvard U. Med. Sch., Boston, 1974-77, instr. in neurology, 1977-80, asst. prof. neurology, 1980; dir. EEG labs. and seizure unit Miami (Fla.) Children's Hosp., 1980—; clin. asst. prof. neurology and pediatrics U. Miami, 1980-83, clin. assoc. prof., 1983—. Contbr. articles to profl. publs. Served to lt. comdr. USPHS, 1972-74. Manealoff fellow, 1969; Grass fellow, 1978-79. Mem. Am. Acad. Pediatrics, Am. Acad. Neurology, Child Neurology Soc., Soc. Neurosci., Soc. Clin. Neurologists, Am. EEG Soc., Am. Epilepsy Soc. Subspecialty: Neurology. Current work: Clinical neurophysiology and epilepsy. Home: 5420 SW 92d St Miami Fl 33156 Office: Miami Children's Hosp 6125 SW 31 St Miami FL 33155

DUCKLES, SUE PIPER, pharmacologist, educator, researcher; b. Oakland, Calif., Mar. 1, 1946; d. Carl Frank Piper and Joan (Brashares) Robert; m. Lawrence T. Duckles, Mar. 21, 1968; children: Ian Muir, Galen Vincent. B.A., U. Calif., Berkeley, 1968, Ph.D., 1973. Fellow UCLA, 1973-76, asst. prof. in residence, 1976-79; asst. prof. pharmacology U. Ariz., Tucson, 1979—; established investigator Am. Heart Assn., 1982—. Assoc. editor: Life Scis. Jour, 1980—; mem. editorial adv. bd.: Jour. Pharmacology and Exptl. Therapeutics, 1983—. Recipient Faculty Devel. award Pharm. Mfrs. Assn., Inc., 1976-78. Mem. Am. Soc. Pharmacology and exptl. Therapeutics, Soc. Neuroscience, Western Pharmacology Soc., Phi Beta Kappa. Subspecialties: Pharmacology; Neuropharmacology. Current work: Control of vascular smooth muscle, cerebral circulation, autonomic nervous system, peptide neurotransmitters, receptors. Office: Department Pharmacology College Medicine University of Arizona Tucson AZ 85724

DUCKWORTH, DONNA HARDY, microbiologist, educator; b. Balt., Sept. 12, 1935; d. Albert Victor and Grace (Campbell) Hardy; m. Alistair Duckworth, June 13, 1964; children: Alexandra, Edward. B.A., Fla. State U., 1957; Ph.D., Johns Hopkins U., 1966. Asst. prof. U. Va., Charlottesville, 1967-73; asst. prof. U. Fla., Gainesville, 1973-77, assoc. prof., 1977-81, prof. dept. immunology and med. microbiology, 1981—. Pres.'s scholar U. Fla., 1976; Macy Faculty scholar, 1977. Mem. Am. Soc. Microbiology, AAAS, Sigma Xi. Subspecialties: Genetics and genetic engineering (biology); Genetics and genetic engineering (medicine). Current work: Role of bacterial plasmids in disease; interactions between bacteriophage and plasmids, molecular genetics of plasmids. Home: 1720 NW 26 Way Gainesville FL 32605 Office: Box J266 JHM Health Center Gainesville FL 32610

DUCOFFE, ARNOLD L., aerospace engineering educator; b. Montreal, Que., Can., Mar. 22, 1921. B. Aero. Engring., Ga., Tech. U., 1943, M.S., 1947; Ph.D. in Gas Dynamics, U. Mich., 1952. Asst. prof. aerodynamics Ga. Tech. U., Atlanta, 1945-48, assoc. prof., 1951-55, acting dir. aerospace engring., 1963-64, prof. aerospace engring., 1955—, dir. aerospace engring., 1964—; bd. dirs. Universal Co. Ltd., Quebec, Rich's Inc., Ga., Unitron Internat. Systems Inc., Calif. Mem. AIAA. Subspecialty: Aerospace engineering and technology. Office: Aerospace Engring Sch Ga Tech U Atlantia GA 30332

DUDA, RICHARD FRANK, nuclear fuel cycle planning mgr.; b. N.Y.C., Sept. 23, 1923; s. Frank Joseph and Emma (Jazek) D.; m. Wynema Jane Bond, May 3, 1945; children: Wynema J. Duda Duffy, Richard F., Jr., Lesley J. Duda Koluder, Desiree J. B.Chem. Engring., Rensselaer Polytechnic Inst., 1948; cert. meteorology, N.Y. U., 1944. Registered profl. engr., N.Y. Design engr., project mgr. Vitro Engring. Co., N.Y.C., 1948-60, chief process engr., mgr. chem. program, 1960-68; program dir., project mgr. Numec, Apollo, Pa., 1968-71; mgr. design and constrn., design engr. mgr. nuclear fuels Westinghouse Electric Corp., Pitts., 1971-78, mgr. fuel cycle activities, 1978—; project mgr. Unique Extractive Metallurgy, 1956-57. Cubmaster North Bergen Council Boy Scouts Am., Paramus, N.J., 1960, asst. cubmaster, 1962. Served to 1st lt. AC U.S. Army, 1943-46. Mem. Am. Nuclear Soc., Inst. Nuclear Materials Mgmt. (chmn. govt. laison subcom.), ASTM, Phi Lambda Upsilon. Subspecialties: Nuclear fission; Chemical engineering. Current work: Currently investigating commercialization of reprocessing, plutonium conversion and mixed oxide fuel fabrication; providing in-house consultation on nuclear fuel cycle technology. Home: RD 9 Box 535 Greensburg PA 15601 Office: Westinghouse Electric Corporation Advanced Energy Systems Division PO Box 10864 Pittsburgh PA 15236

DUDEK, BRUCE CRAIG, biopsychologist, statis. cons.; b. Lakeland, Fla., Nov. 10, 1951; s. Anton and Maysele Valeria (Ringgenberg) D. B.A., Fla. State U., 1974; M.A., SUNY, Binghamton, 1976, Ph.D., 1978. Asst. prof. psychology SUNY, Albany, 1978—. Author: (with Hahn and Jensen) Development and Evolution of Brain Size: Behavioral Implications, 1979; also articles and book chpts. SUNY Research Found. grantee, 1981. Mem. Behavior Genetics Assn., Internat. Soc. Neurochemistry, Soc. Neurosci., Am. Statis. Assn., Animal Behavior Soc., Internat. Soc. Developmental Psychobiology, Research Soc. on Alcoholism. Democrat. Subspecialties: Neuropharmacology; Behavior genetics. Current work: Genetic control of neural and behavioral response to alcohol; pharmacogenetics; genetic neurobiology; behavioral neurochemistry; monoamines and behavior. Office: Dept Psychology SUNY Albany NY 12222

DUDLEY, ALDEN WOODBURY, JR., pathologist, medical researcher; b. Lynn, Mass., May 15, 1937; s. Alden Woodbury and Dorothy Helen (Newth) D.; m. Mary Elinor Adams, Sept. 12, 1959; children: Raymond Adams, Eric Clark. A.B., Duke U., 1958, M.D., 1962. Cert. in anatomic pathology and neuropathology. Intern Duke U., 1962-63, resident, 1963-65, asst. prof. pathology, 1967-68; asst. prof. U. Wis.-Madison, 1968-71, assoc. prof., 1971-76; prof., chmn. U. South Ala., Mobile, 1976-80; dir. neuropathology Cleve. Clinic Found., 1980—; cons. Nat. Biomed. Research Found., Washington, Life Systems, Cleve. Dir. park devel. Arbor Hills Community Assn.,

Madison, 1970. Served with USPHS, 1963-65. Recipient Golden Forceps award Armed Forces Inst. Pathology, 1974, Tchr. of Yr. award U. South Ala., 1977, 79, Prix de l'Institut Muscle et Nerf U. Marseilles, France, 1982. Fellow Am. Coll. Pathologists; mem. Group for Research in Pathology Edn. (pres. 1975-77), Am. Assn. Neuropathologists, Internat. Soc. Neuropathologists, World Fedn. Neurology. Republican. Subspecialties: Graphics, image processing, and pattern recognition; Neurology. Current work: High computer technology analysis of medical tissues for diagnosis of type and extent of disease and efficacy of treatment; measure and map cells of brain, muscle, bone, etc.; response of each tumor to drugs. Patentee rapid tissue processor. Home: 21925 Parnell Rd Shaker Heights OH 44122 Office: Director of Neuropathology Cleveland Clinic Cleveland OH 44106

DUDNIK, ELLIOTT ELIASAF, architect, educator; b. Tel-Aviv, Israel, July 24, 1943; s. Isaac and Sabina (Frank) D.; m. Laura Susan Mall, June 4, 1972; children: Nina Simone, Sara Arona. B.Arch., Ill. Inst. Tech., 1965, M.S., 1967; Ph.D., Northwestern U., 1983. Registered architect, Ill., Wis. Architect C.F. Murphy & Assoc., Chgo., 1964-66; structural engr., project architect Skidmore, Owings & Merrill, Chgo., 1966-69; prin. Elliott Dudnik & Assocs., Chgo. and Evanston, Ill., 1969—; lectr. Ill. Inst. Tech., 1966-67, U. Ill., Chgo., 1967-70, asst. prof., 1970-73, assoc. prof., 1973-80, prof., 1980—, dir. grad. studies, 1982—; vis. prof. U. Newcastle, Australia, 1976; vis. assoc. prof. U. Sydney, 1976; vis. lectr. Loyola U., Chgo., 1975; cons. archtl. and engring. firms, 1969-76, Rutgers Archeol. Survey Office, 1980, City of Evanston, 1981—; lectr. in field. Contbr. tech. papers to profl. jours. Walter P. Murphy fellow, 1970-71; Sr. Fulbright-Hays lectr.-scholar, 1976; grantee U. Ill. Research Bd., 1972-73, 77, Am. Inst. Steel Constrn., 1978, Nat. Endowment for Arts, 1979. Mem. AIA, ACM, Internat. Solar Energy Soc., Assn. Architects and Engrs. in Israel. Subspecialties: Graphics, image processing, and pattern recognition; Solar energy. Current work: Computer-aided design, energy conscious design, housing value and preferences, building systems optimization. Office: Sch Architecture Univ ILL Chicago IL 60680

DUERINCKX, ANDRE JOZEF, research scientist; b. Leuven, Belgium, Jan. 25, 1952; came to U.S., 1974. Ph.D., Stanford U., 1979. Mem. staff IBM Watson Research Center, Yorktown Heights, N.Y., 1979-80; mem. research staff Philips Ultrasound, Inc., Santa Ana, Calif., 1980-82, research mgr., 1982—; chmn. Picture Archiving and Communications Systems Conf., Newport Beach, Calif., 1982, editor procs., 1982. Mem. IEEE, Am. Assn. Physicists in Medicine, Assn. Computing Machinery. Subspecialties: Imaging technology; Acoustical engineering. Current work: Acoustical imaging, image archiving, image communications, computer animation, filmless radiology. Office: Philips Ultrasound Inc 2722 S Fairview St Santa Ana CA 92704

DUERR, J. STEPHEN, materials consultant, metallurgist; b. Erie, Pa., Apr. 8, 1943; s. John S. and Jodine J. (Sparks) D.; m. Judith M. Duerr, Oct. 17, 1964; children: Karen L., Kristen M., Craig M. S.B., M.I.T., 1965, S.M., 1967, Ph.D., 1971. Registered profl. engr., N.J. Sr. metallurgist Westinghouse Bettis Atomic Power Lab., West Mifflin, Pa., 1971-74; dir. analytical services PhotoMetrics, Inc., Woburn, Mass., 1974-77; tech. dir. Structure Probe, Inc., Metuchen, N.J., 1977—; pres. Metuchen Analytical, Inc., 1978—; course dir. Center Profl. Advancement, East Brunswick, N.J., 1980—. Contbr. articles on electron microscopy, microanalysis and metall. failure analysis to profl. jours. Mem. Microbeam Analysis Soc., Am. Soc. Metals (chmn. N.J. chpt.), ASTM, Internat. Soc. Hybrid Microelectronics, Alpha Tau Omega. Subspecialties: Metallurgy; Metallurgical engineering. Current work: Primary interest is characterization of materials by electron and ion microbeam techniques applied to analytical investigations and their court presentation. Office: 230 Forrest St Metuchen NJ 08840

DUFF, JAMES THOMAS, microbiologist, virologist; b. Sandusky, Ohio, Jan. 23, 1925; s. William John and Winifred Kathryn (Breining) D. B.S., Ohio State U., 1947, M.S., 1949; Ph.D., U. Tex., 1960. Microbiologist, immunology br., med. investigations div. U.S. Army Biol. Lab., Ft. Detrick, Md., 1949-56, 59-65; with Nat. Cancer Inst., Bethesda, Md., 1965-83, chief biol. carcinogenesis br., 1978-83. Contbr. articles to profl. jours. Served with U.S. Navy, 1943-46. Mem. Am. Soc. Microbiology., Tissue Culture Assn., Internat. Assn. Comparative Research on Leukemia and Related Diseases, Sigma Xi. Republican. Subspecialties: Virology (biology); Cancer research (medicine). Current work: Azotobacter bacteriophage, Clostridium botulinum toxins and toxoids, tissue culture, psittacosis vaccines, viral oncology. Home: 1329 Midwood Pl Silver Spring MD 20910

DUFF, RONALD GEORGE, virologist, cell biologist; b. Billings, Mont., Dec. 8, 1936; s. Ross I. and Alda Mable (Markholt) D.; m. Naomi Darlene, Aug. 12, 1962; children: Kelle Amber, Ross Alan. Mus.B., U. Mont., 1959; Ph.D. in Pathology, U. Colo. Med. Sch., 1968. Assoc. prof. microbiology Milton S. Hershey (Pa.) Med. Center, 1969-74; head viral and cell biology Abbott Labs., North Chicago, Ill., 1974-83; prof. microbiology Chgo. Med. Sch., North Chicago, 1978-83; v.p. devel. Damon Biotech, Needham Heights, 1983—. Contbr. numerous articles to profl. publs. Bd. dirs. Deerfield (Ill.) Park Dist. Bands, 1981—. Mem. Am. Soc. Microbiology, AAAS, Am. Assn. Cancer Research, Sigma Xi, Phi Sigma. Subspecialties: Virology (biology); Cell and tissue culture. Current work: Antiviral agents, anti-cancer agents, viral diagnostics, cancer diagnostics, molecular biology of cancer. Home: 11 Shaunee Rd Medfield MA 02052 Office: 119 4th Ave Needham Heights MA 02194

DUFFEY, DICK, nuclear engineering educator, consultant; b. Wasash County, Ind., Aug. 26, 1917; s. Glen and Kate (Parker) D. B.S. in Chem. Engring, Purdue U., 1939, M.S., U. Iowa-Iowa City, 1940; Ph.D. in Nucelar Engring, U. Md.-College Park, 1956. Engr. Union Carbide, Tonawanda, N.Y., 1940-42, AEC, Washington and Richland, Wash., 1946-54; prof. nuclear engring. U. Md., College Park, 1954—; cons. AEC, U.S. Dept. Energy. Served to capt. U.S. Army, 1942-46. Mem. Am. Nuclear Soc., Am. Phys. Soc., Am. Chem. Soc., Am. Inst. Chem. Engrs., Am. Geophys. Union. Subspecialties: Nuclear engineering; Chemical engineering. Current work: Neutron uses, particularly nuclear spectroscopy using Californium 252. Patentee in field. Office: University of Maryland College Park MD 20740

DUFFIE, JOHN ATWATER, chem. engr., educator; b. White Plains, N.Y., Mar. 31, 1925; s. Archibald Duncan and Lulie Adele (Atwater) D.; m. Patricia Ellerton, Nov. 22, 1947; children: Neil A., Judith A. Duffie Schwarzmeier, Susan L. B.Ch.E., Rensselaer Poly. Inst., 1945, M.Ch.E., 1948; Ph.D., U. Wis., 1951. Registered profl. engr., Wis. Instr. chem. engring. Rensselaer Poly. Inst., 1946-49; research asst. U. Wis., 1949-1951; research engr. DuPont, 1951; sci. liaison officer Office Naval Research, 1952-53; mem. faculty dept. chem. engring. U. Wis.-Madison, 1954—, prof., 1957—, dir. solar energy lab., 1956—; Fulbright scholar U. Queensland, Australia, 1964; sr. Fulbright-Hays scholar Commonwealth Sci. and Indsl. Research Orgn., Australia, 1977. Author: (with W.A. Beckman) Solar Energy Thermal Processes, 1974, (with W.A. Beckman, S.A. Klein) Solar Heating Design, 1977, (with W.A. Beckman) Solar Engineering of Thermal Processes, 1980. Served with USN, 1943-46. Recipient Charles G. Abbot award Am. sect. Internat. Solar Energy Soc., 1976. Fellow Am. Inst. Chem. Engrs.; mem. Internat. Solar Energy Soc. (past pres.), AAAS. Subspecialties: Solar energy; Chemical engineering. Current work: Research on solar energy thermal processes and teaching chemical engineering. Home: 5710 Dorsett Dr Madison WI 53711 Office: 1500 Johnson Dr Madison WI 53706

DUFFY, FRANK HOPKINS, neurologist, biomedical engineer, neurophysiologist, educator; b. Honolulu, Jan. 22, 1937; s. Irving Arthur and Frances (Hopkins) D.; m. Tanya Maloff, Oct. 10, 1965; m. Heidelise Als, Jan. 1, 1977; children: Stephen W.H., Lisa G., Victoria C.; m.; 1 stepson, Christopher Rivinus. B.S.E. in Elec. Engring, U. Mich., 1958, 1958; M.D., Harvard U., 1963. Intern Yale New Haven Hosp., 1963-64; resident in neurosurgery and neurology Mass. Gen. Hosp., Boston, 1964-68, Lahey Clinic, 1964-68, Peter Bent Brigham Hosp., 1964-68, Children's Hosp. Med. Ctr., 1964-68; staff neurologist Beth Israel Hosp. and Children's Hosp. Med. Ctr., Boston, 1970—; assoc. prof. neurology Harvard U., 1979—; dir. devel. neurophysiology, co-dir. seizure unit Children's Hosp. Med. Ctr. Research numerous publs. in neurophsyiology and neurology. Served to maj. M.C. U.S. Army, 1968-70. Mem. Eastern Assn. Electroencephalographers (pres. 1981-82), Am. EEG Soc., Am. Neurol. Assn., Soc. Neurosci. Club: Harvard (Boston). Subspecialties: Neurophysiology; Biomedical engineering. Current work: Clinical neurology, clinical and animal neurophysiology, biomedical engineering, pediatric and adult neurology; research in computer analysis of human brainwave activity, animal modeling of environmental deprivation. Patentee brain elec. activity mapping. Home: 49 Harrison Brookline MA 02146 Spring Rd Tunbridge VT 05077 Office: Harvard Med Sch Children's Hosp Med Center 300 Longwood Ave Boston MA 02115

DUFFY, ROBERT ALOYSIUS, aeronautical engineer; b. Buck Run, Pa., Sept. 9, 1921; s. Joseph Albert and Jane Veronica (Archer) D.; m. Elizabeth Reed Orr, Aug. 19, 1945; children: Michael Gordon, Barclay Robert, Marian Orr, Judith Elizabeth, Patricia Archer. B.S. in Aero. Engring, Ga. Inst. Tech., 1951. Commd. 2d lt. U.S. Army, 1942; commd. U.S. Air Force, advanced through grades to brig. gen, 1967; service in C.Z., Morocco, Algeria, Tunisia, Sicily, Italy, Vietnam; vice comdr. USAF Space and Missile Systems Orgn., Los Angeles, 1970-71; ret., 1971; v.p., dir. Draper Lab. div. M.I.T., Cambridge, Mass., 1971-73; pres., dir., chief exec. officer Charles Stark Draper Lab., Inc., 1973—; chmn. USAF-NOAA weather satellite program rev. Dept Def.-NASA, 1972; chmn Fed. Contract Research Center Task Force, Dept. Def., 1975; mem. indsl. and profl. adv. council Pa. State U. Sch. Engring., 1979—. Contbr. articles to profl. jours. Decorated Disting. Service medal, Legion of Merit; recipient Thomas D. White award Nat. Geog. Soc., 1970. Fellow AIAA; mem. Nat. Acad. Engring., Inst. Navigation (Thurlow award 1964, pres. 1976-77), Air Force Assn., U.S. Naval Inst. Clubs: Algonquin (Boston); Concord Country. Subspecialties: Aeronautical engineering; Research Administration. Home: 115 Indian Pipe Ln Concord MA 01742 Office: 555 Technology Sq Cambridge MA 02139

DUFFY, THOMAS EDWARD, neurochemist; b. Balt., July 10, 1940; s. James Henry and Mary Cecilia (Broczykowski) D.; m. Marie Milkowski, Sept. 3, 1966; children: Janet Theresa, Lisa Marie, Michael Thomas. B.S., Loyola Coll., Balt., 1962; Ph.D. (USPHS fellow), U. Md., 1967. NIH postdoctoral fellow, dept. neurobiology U. Goteborg, Sweden, 1967-68; dept. pharmacology Washington U. Med. Sch., St. Louis, 1968-70; asst. prof. biochemistry in neurology Cornell U. Med. Coll., N.Y.C., 1970-75; assoc. prof., 1975-81, prof. biochemistry, 1981—. Contbr. articles to profl. jours. Recipient Tchr. Scientist award Mellon Found., 1974-75; established investigator Am. Heart Assn., 1975-80. Fellow Stroke Council Am. Heart Assn.; mem. AAAS, Am. Soc. Biol. Chemists, Am. Soc. Neurochemistry, Internat. Soc. Cerebral Blood Flow and Metabolism, Internat. Soc. Neurochemistry, Soc. neuroscience. Roman Catholic. Subspecialty: Neurochemistry. Current work: Regulation of cerebral energy metabolism in altered functional states; pathogenesis of hepatic coma; neurochemistry of stroke and perinatal asphyxia; mechanisms of cerebral ammonia toxicity.

DUFILHO, HAROLD LOUIS, electronic engineer; b. Opelousas, La., Sept. 26, 1944; s. Louis Joseph and Celeste Theresa (St. Am) D.; m. Erna Christl Click, Nov. 3, 1965; children: Angela Marie, Jody Lynn, Karl Michael, Brian Stephen, David Wayne. B.S., Fla. State U., 1973. Electronic technician RCA Service Co., Cape Canaveral, Fla., 1969-70; research asst. Fla. State U., Tallahassee, 1971-73; research physicist Naval Coastal Systems Lab., Panama City, Fla., 1973-77, Naval Ocean Research and Devel. Activity, NSTL Station, Miss., 1977-79; engring. sect. head U.S. Naval Oceanographic Office, NSTL Station, Miss., 1979—, adult adv., 1981-82. Served with USAF, 1963-68. Recipient Sustained Superior Performance award Naval Coastal Systems Lab., 1976. Mem. Marine Tech. Soc., Res. Officers Assn. Democrat. Club: NSTL Diving (v.p. 1979-80). Subspecialties: Electronics; Acoustics. Current work: Acoustic measurements; acoustic signal processing; data acquisition instrumentation and automation. Office: Navoceano Code 6212 NSTL Station MS 39522

DUFOUR, REGINALD JAMES, educator; b. Marksville, La., July 29, 1948; s. Benjamin Joseph and Virginia (Deshotelles) D. B.S., La. State U., 1970; M.S., U. Wis., 1971, Ph.D., 1974. NRC postdoctoral research assoc. NASA (Johnson Space Center), Houston, 1974-75; asst. prof. space physics, astronomy Rice U., Houston, 1976-81, assoc. prof., 1981—. Contbr. articles to profl. jours. Research Corp. grantee, 1976-78; NSF grantee, 1978-80, 81; NASA grantee, 1978—. Mem. Houston Astron. Soc. (adv. dir.), Am. Astron. Soc., Internat. Astron. Union, Royal Astron. Soc., Astron. Soc. Pacific, AAAS, Sigma Xi, Phi Kappa Phi. Subspecialties: Optical astronomy; Graphics, image processing, and pattern recognition. Current work: Imagery and spectroscopy of nebulae and galaxies; UV space astronomy; chem. evolution of galaxies. Home: 2800 Jeanetta St Apt 2011 Houston TX 77063 Office: Rice Univ 210 H Space Scis Houston TX 77251

DUGAN, CHARLES HAMMOND, physicist, educator; b. Balt., Apr. 2, 1931; s. Hammond J. and Frances L. (Smith) D.; m. Gwendolyn Finn, Nov. 4, 1954; children: Melanie, Alison, Ann, Frances, John. B.S., U. Ky., 1951; M.A., UCLA, 1954; Ph.D., Harvard U. Staff Smithsonian Astrophys. Obs., 1963-66; asst. prof. physics York U., Downsview, Ont., Can., 1967-74, prof., 1975—, assoc. dean, 1978-79; vis. fellow dept. elec. engring. Cornell U., Ithaca, N.Y., 1981; vis. fellow dept. physics Imperial Coll., London, 1973-74. Contbr. articles to profl. jours. Served with U.S. Army, 1954-56. Grantee NRC Can., Def. Research Bd. Can., Province Ont. Mem. Am. Phys. Soc., AAAS, Can. Assn. Physicists. Subspecialties: Atomic and molecular physics; Spectroscopy. Current work: Research in photodissociation of molecules. Home: 37 Johnson St Thornhill ON Canada L3T 2N9 Office: York U Petrie Bldg Downsview ON Canada M3J 1P3

DUGGIN, MICHAEL JOHN, physics educator; b. Dorking, Surrey, Eng., July 30, 1937; came to U.S., 1979; s. Walter J. and Winnifred L. (Button) D.; m. Maggie Amelia Beveridge, July 16, 1978; children: John Bruce, Blake Michael. B.Sc., Melbourne (Austrlia) U., 1959; Ph.D., Monash U., Melbourne, 1965. Teaching fellow Monash U., Melbourne, Australia, 1962-64; postdoctoral fellow U. Pitts., 1965, asst. prof., 1966; research scientist CSIRO, Sydney, Australia, 1967-71, sr. research scientist, 1971-79; prof. physics, dept. forest engring. Coll. Environ. Sci. and Forestry SUNY-Syracuse, 1979—; cons. in field. Contbr. numerous articles to profl. jours. Assoc. fellow AIAA; fellow Royal Astron. Soc.; mem. Am. Inst. Physics, Am. Soc. Photogrammetry, Internat. Remote Sensing Assn., Soc. of Photo-Optical Instrumentation Engrs. Subspecialties: Satellite studies; Aerospace engineering and technology. Current work: Physics of remote sensing processes; fundamental remote sensing research; spectral reflectance factor measurement; remote sensing data acquisition and analysis optimization. Home: 212 Robinhood Ln Camillus NY 13031 Office: 308 Bray Hall Suny Syracuse NY 13210

[illegible], BRENT KERNS, earth science educator; b. Waverly, N.Y., July 5, 1945; s. Max Frank and Jean Margaret (Kerns) D.; m. Susan Marie Mackie, Sept. 18, 1971; 1 dau., Kristine Alicia. B.S., Syracuse U., 1967; M.S., 1972; Ph.D., U. Hawaii, 1976. Lectr. geology U. Wis.-Oshkosh, 1976-77; econ. geologist W.Va., Geol. Survey, Morgantown, 1977, head econ. geology sect., 1977-80; asst. prof. geology State Univ. Coll., Oneonta, N.Y., 1980—. Served with U.S. Army, 1969-71; Vietnam. Mem. Geol. Soc. Am., Soc. Econ. Paleontologists and Mineralogists, Internat. Oceanographic Found. Subspecialties: Deep-sea biology; Sedimentology. Current work: Description, identification and classification of newly discovered species of deep-sea agglutinating Foraminifera and their roles in deep-sea ecology. Home: RD 1 Box 506 Otego NY 13825 Office: Earth Scis Dept Oneonta State Coll Oneonta NY 13820

DUKE, JAMES ALAN, botanist, musician; b. Birmingham, Ala., Apr. 4, 1929; s. Robert Edwin and Martha (Turrs) D.; m. Carolyn Jean Saylor; m. Peggy-Ann Wetmore Kessler; children: John Carl, Celia Gayle. A.B. in Botany, U. N.C., 1952, M.A., 1955, Ph.D., 1960. Asst. curator Mo. Bot. Garden, St. Louis, 1959-62; botanist New Corps REsearch, Dept. AGr., Beltsville, Md., 1963-65, research botanist, 1971-77, chief, 1977—; research ecologist Battelle Meml. Inst., Columbus, Ohio, 1965-71; tchr., lectr., field workers in field. Composer, guitarist, bass violinist: recs. Ginseng; Author: Isthmian Ethnobotanical Dictionary, 1972, Handbook of Legumes of World Economic Importance, 1981. Served with Chem. Corps U.S. Army, 1956-58. Mem. Soc. Econ. Botany, Internat. Assn. Plant Taxonomists, Internat. Soc. Tropical Ecology, Assn. Tropical Biology, Nat. Assn. Profl. Bureaucrats, Weed Sci. Soc. Am., Internat. Weed Sci. Soc. Baptist. Lodge: Elks. Subspecialties: Taxonomy; Ecosystems analysis. Current work: Economic botany, ethnobotany, nutrition, medicinal plants, germplasm, ecosystematics, climatology, data retrieval. Office: Dept Agr Room 133 B-001 Beltsville MD 20705

DUKE, STANLEY HOUSTEN, plant physiologist, educator; b. Battle Creek, Mich., Oct. 9, 1944; s. Oscar and Azalee (Tallent) D.; m. Cyhtnia Arlene Henson, Sept. 12, 1978. B.S., Henderson State U., 1966; M.S., U. Ark., 1968; Ph.D., U. Minn., 1975. Teaching asst. dept. botany U. Ark., Fayetteville, 1966-68; teaching asst. dept. botany U. Minn., St. Paul, 1970-74, teaching asso., 1975; postdoctoral fellow dept. agronomy U. Wis., Madison, 1976-78, asst. prof., 1978-83, assoc. prof., 1983—. Contbr. numerous articles to profl. jours. Served to 1st lt. Chem. Corps U.S. Army, 1968-70. NSF trainee, summer 1971; U.S. Dept. Agr. grantee, 1978-83; Am. Soybean Found. fellow, 1976-78; McKnight Found. grantee, 1983—. Fellow Linnean Soc. London; mem. Am. Soc. Plant Physiologists, Societas Physiologiae Plantarum Scandinavica, Japanese Soc. Plant Physiologists. Subspecialty: Plant physiology (biology). Current work: Research on plant nitrogen and carbon metabolism, environ. stress, and seed germination. Office: Dept Agronom U Wis Madison WI 53706

DUKE, STEPHEN OSCAR, plant physiologist; b. Battle Creek, Mich., Oct. 9, 1944; s. Oscar and Azalee Rosa (Tallant) D.; m. Barbara Alice Rowe, June 2, 1967; children: Gregory Ivan, Robin Anne. B.S., Henderson State U., 1966; M.S., U. Ark., 1968; Ph.D., Duke U., 1975. Instr. Duke U., 1974-75; NSF research assoc. So. Weed Sci. Lab., Stoneville, Miss., 1975-76, staff scientist, 1976—. Served to 1st lt., M.S.C. U.S. Army, 1968-70. Decorated Bronze Star. Mem. AAAS, Am. Inst. Biol. Sci., Bot. Soc. Am., Am. Soc. Plant Physiology, Am. Soc. Photobiology, Am. Soc. Plant Physiology, Scandinavian Soc. Plant Physiology, Japan Soc. Plant Physiology, Weed Sci. Soc. Am., Sigma Xi. Club: Soccer (Greenville, Miss.). Subspecialties: Plant physiology (biology); Photosynthesis. Current work: Herbicide mechanism of action, plant photobiology, secondary metabolism. Home: 1741 W Azalea St Greenville MS 38701 Office: US Dept Ag Agrl Research Service So Weed Sci Lab PO Box 225 Stoneville MS 38776

DUKE, WINSTON LAVELLE, nuclear utility executive, lawyer, engineer; b. Atlanta, Nov. 8, 1941; s. Herbert Windham and Eva Virginia (Renteria) D.; m. Beverly Ann Conklin, June 21, 1970; children: Grant, Heather, Heidi. B.S. in Physics, Ga. Inst. Tech., 1963; M.S. in Nuclear Engring, 1964; M.B.A., Harvard U., 1970; J.D., Ill. Inst. Tech., 1978. Bar: Bar: Colo., 1979, Ill 1979; registered profl. nuclear engr., Calif. Adminstrv. asst. to congressman, Dallas, 1969-70; nuclear engr. Commonwealth Edison, Chgo., 1970-78, adminstrv. asst. to vice chmn. bd., 1970-71; area mgr. North Shore, Northbrook, Ill., 1975-78, nuclear security adminstr., Chgo., 1981—; corp. counsel Cotter Corp., Denver, 1978-81; sole practice law, Barrington Hills, Ill. Contbr. articles to profl. jours. in field. Served to capt. USAF, 1964-68. Mem. ABA, Chgo. Bar Assn., Am. Nuclear Soc., IEEE, Edison Electric Inst. Atomic Indsl. Forum (mem. rep.), Delta Tau Delta (chpt. pres. 1963). Subspecialties: Nuclear engineering; Nuclear fission. Home: 10 Crawling Stone Rd Barrington Hills IL 60010 Office: Commonwealth Edison C Nuclear Security Adminstrn PO Box 767 Room 1248 E Chicago IL 60690

DUKER, NAHUM JOHANAN, experimental pathologist; b. N.Y.C., Oct. 27, 1942; s. Abraham G. and Lillian (Sandrow); m. Naomi Ruth Maisel, June 4, 1972; children: Eli, Joshua, Jonathan, Ezra. M.D., U. Ill., 1966. Diplomate: Am. Bd. Pathology, 1978. Intern Bellevue Hosp., N.Y.C., 1966-67; resident in pathology N.Y. U. Med. Ctr., 1970-76; instr. pathology N.Y. U. Med. Sch., 1976-77; asst. prof. pathology Fels Research Inst., Temple U. Med. Sch., 1977-82, assoc. prof. pathology, 1982—. Author papers on DNA damage and repair. Served to capt. M.C. USAR, 1967-69. Grantee USPHS, 1978—; USPHS research career devel. award, 1983. Mem. Internat. Acad. Pathology, Am. Assn. Pathologists, Am. Assn. Cancer Research, Am. Soc. Photobiology., Environ. Mutagen Soc. Jewish. Subspecialties: Biochemistry (biology); Molecular biology. Current work: Physical and chemical damage to DNA and its repair. Office: 3400 N Broad St Philadelphia PA 19140

DUKLER, ABRAHAM EMANUEL, chemical engineer; b. Newark, Jan. 5, 1925; s. Louis and Netty (Charles) D.; m.; children—Martin Alan, Ellen Leah, Malcolm Stephen. B.S., Yale U., 1945; M.S., U. Del., 1950, Ph.D., 1951. Devel. engr. Rohm & Haas Co., Phila., 1945-48; research engr. Shell Oil Co., Houston, 1950-52; mem. faculty dept. chem. engring. U. Houston, 1952—, prof., 1963—, chmn. dept., 1967-73, dean engring., 1976-83; dir. State of Tex. Energy Council, 1973-75; cons. U.S. Nuclear Regulatory Commn., Brookhaven Nat. Lab., Shell Devel. Co., Exxon, others. Contbr. chpts. to books, articles to profl. jours. Recipient Research award Alpha Chi Sigma, 1974. Fellow Am. Inst. Chem. Engrs., Nat. Acad. Engring., Am. Soc. Engring. Edn. (research lectureship award 1976); mem. Am. Inst. Chem. Engrs., ASME, AAAS, Am. Chem. Soc., AAUP, Sigma Xi, Tau Beta Pi. Subspecialty: Fluid mechanics. Current work: Theoretical and experimental studies of two phase gas liquid flow and related energy and mass transfer questions. Office: Coll of Engring Univ of Houston Houston TX 77004

DULEY, WALTER WINSTON, physicist, educator; b. Montreal, Que., Can., Oct. 8, 1941; s. Walter Albert and Ella (Harnum) D.; m. Irmgardt Zunker, July 3, 1965; children: Nicholas, Mark. B.S., McGill U., 1963; D.I.C., Imperial Coll., 1966; Ph.D., U. London, Eng., 1966, D.Sc., 1982. Scientist Def. Research Bd., Quebec City, Que., Can., 1966-67; asst. prof. York U., Toronto, 1967-70, assoc. prof., 1970-74, prof., 1974—; assoc. research chemist U. Calif., Berkeley, 1973; vis. prof. Swiss Inst. Tech., Zurich, 1974; vis. prof. theoretical physics Oxford U., 1981; research chemist, Harwell, Eng., 1981; cons. lasers; pres. Powerlasers Ltd., King City, Ont., Can., 1976—. Contbr. numerous articles to profl. jours.; Author: CO_2 Lasers, 1976, Laser Processing, 1982, Interstellar Chemistry, 1983. Recipient Can. SRC Sr. Vis. Research Fellowship award, 1978, 80. Fellow Royal Astron. Soc., Royal Soc. Arts; mem. Internat. Astron. Union. Subspecialties: Spectroscopy; Laboratory astrophysics. Current work: Laser spectroscopy, materials processing with lasers, laboratory studies of interstellar dust, theoretical astrophysics. Home: 136 Dew St King City ON Canada LOG 1KO Office: 4700 Keele St Toronto ON Canada M3J 1P3

DUMAS, HERBERT MONROE, (JR.), physicist, elec. engr., research and devel. lab. exec.; b. Eldorado, Ark., Dec. 16, 1927; s. Herbert Monroe and Emma Villa (Woodard) D.; m. Patricia Ann Johnson, May 9, 1953; 1 son, Herbert Scott. B.A. in Physics, U. Ark., 1954, B.S., 1955, M.S., 1956. With Sandia Nat. Labs., Albuquerque, 1956—, super. div. seismic systems, 1965-76, mgr. dept. space systems, 1976—. Served with USN, 1946-49; to 2d lt. USAR, 1952-53. Recipient Physics Achievement award U. Ark., 1952. Mem. Optical Soc. Am., Kappa Sigma. Democrat. Club: U. Ark. Alumni (Albuquerque) (v.p. and dir. 1965—). Subspecialties: Satellite studies; Optical engineering. Current work: Instrumentation systems for satellite applications; mgmt. research engring. of satellite instrumentation and sensor systems and satellite data processing stas. Home: 1304 Florida NE Albuquerque NM 87110 Office: PO Box 5800 7240 (880/B-42) Albuquerque NM 87165

DUMAS, NEIL STEPHEN, senior scientist, research psychologist; b. N.Y.C., Oct. 22, 1940; s. Isadore and Eva (Horowitz) D.; m. Betsy Ellen Magid, Jan. 21, 1962; children: Leslie S., Meredith L., Abigail L. B.A., CCNY, 1962; Ed.M., U. N.D., 1964; Ph.D., U. Wis., 1967; M.S., 1968. Asst. research prof. U. Fla., 1968-70; research psychologist Nat. Ctr. Health Services Research and Devel., Rockville, Md., 1970-72; study dir. NIH, Bethesda, Md., 1972-74; planning dir. HEW, Dallas, 1974-76; program dir. NSF, Washington, 1976-80; sr. scientist U.S. Army Research Inst., Alexandria, Va., 1980—. Author: The Master's Degree in Science and Engineering, 1981; editor: Decision Maker's Guide to Science, 1971; contbr. articles to profl. jours. HEW research grantee, 1968-72. Mem. Am. Psychol. Assn., Ops. Research Soc. Am. Subspecialties: Industrial psychology; Information systems, storage, and retrieval (computer science). Current work: Design of computer based, decision support systems to fit organizational and human requirements including work restructuring and software development. Home: 7110 Davis Ct McLean VA 22101 Office: U S Army Research Institute for the Behavioral Sciences 5001 Eisenhower Ave Alexandria VA 22333

DUMONTELLE, PAUL BERTRAND, geologist; b. Kankakee, Ill., June 22, 1933; s. Lester Vernon and Helen (McKinstry) DuM.; m. Dollie Louise Bridgewater, June 5, 1955; children: John, Jeffrey, Jo, James, Jay. B.A., DePauw U., 1955; M.S., Lehigh U., 1957. Geologist Lehigh Portland Cement Co., Allentown, Pa., 1956-57, Homestake Mining Co., Lead, S.C., 1957-63; asst. geologist Ill. State Geol. Survey, Champaign, 1963-70, assoc. geologist, 1970-78, geologist, 1979—, coordinator environ. geology, 1975-79, geologist, head engring. geology sect., 1979—. Fellow Geol. Soc. Am.; mem. Assn. Engring. Geologists (chmn. nat. awards com.), Am. Inst. Profl. Geologists, Am. Congress on Surveying and Mapping, Ill. Geol. Soc. Club: Radio Control (Champaign). Lodge: Kiwanis. Subspecialties: Geology. Current work: Engineering geology, including slope stability, soil problems, mine subsidence, and computerization of geological information. Home: 2020 Burlison Dr Urbana IL 61801 Office: Ill State Geol Survey 615 E Peabody St Champaign IL 61801

DUN, NAE J., pharmacologist, educator; b. Shanghai, China, Aug. 12, 1945; s. Meng-chieh and Shiaoshing (Chou) D.; m. Siok Le Tan; children: Erica C., Bryant T. B.Sc., Coll. Pharmacy, U. Ill., 1969; Ph.D., Loyola U. Chgo., 1975. Registered pharmacist, Ill. Asst. prof. Loyola U. Stritch Sch. Medicine, Maywood, Ill., 1976-81, asso. prof. pharmacology, 1981—. Contbr. articles to profl. jours, chpts. to books. Am. Parkinson Disease Found. research grantee, 1978-80; Nat. Inst. Neurol. and Communicative Disorders and Stroke grantee, 1980-82. Mem. Am. Soc. Pharmacology and Exptl. Therapeutics, Soc. Neurosci. Subspecialties: Neuropharmacology; Neurophysiology. Current work: Synaptic transmission in autonomic ganglia and spinal neurons. Office: Dept Pharmacology Loyola Univ Med Center Maywood IL 60153

DUNAYEVSKY, VICTOR ARKADY, applied mathematician, researcher; b. Ashchabad, Turkmenia, USSR, 1942; came to U.S., 1978; s. Arkady Samual and Rozalia (Shklovsky-Tictinsky) D.; m. Evelina Aisenberg, Oct. 1, 1968 (div. 1977). M.S., Civil Engring. Inst., Kharkov, USSR, 1964; Ph.D., Siberian Div. Acadamy of Sci., Novosibirsk, USSR, 1973, Northwestern U., 1980. Assoc. prof. Elec. Engring. Inst., Novosibirsk, USSR, 1974-78; project leader Standard Oil Co. Research Center, Warrensville, Ohio, 1982—; vis. scholar Northwestern U., 1978-82. Mem. Soc. Indsl. and Applied Math., ASME. Subspecialties: Fracture mechanics; Solid mechanics. Current work: Crack propagation in an elastic-plastic medium; fracture mechanics; dynamic fracture; stability of elastic structures. Home: 6805 Mayfield Rd Apt 1004 Mayfield Heights OH 44124 Office: Standard Oil Co Research Center 4440 Warrensville Rd Cleveland OH 44128

DUNBAR, ROBERT COPELAND, chemistry educator and researcher; b. Boston, June 26, 1943; s. William H. and Carolyn (Roorbach) D.; m. Mary Asmundson; children: Geoffrey, William. B.S., Harvard U., 1965; Ph.D., Stanford U., 1970. Prof. chemistry Case Western Res. U., Cleve., 1970—. Contbr. writings to profl. publs. Sloan fellow, 1973-75; Guggenheim fellow, 1978-79. Mem. Am. Chem. Soc., Am. Soc. Mass Spectrometry, Inter-Am. Photochem. Soc., Am. Phys. Soc. Subspecialties: Physical chemistry; Laser-induced chemistry. Current work: Ion-molecule reactions, ion photochemistry, ion cyclotron resonance. Home: 2880 Fairfax Rd Cleveland Heights ON 44118 Office: Dept Chemistry Case Western Res U Cleveland OH 44106

DUNCAN, DORIS GOTTSCHALK, information systems educator; b. Seattle, Nov. 19, 1944; d. Raymond Robert and Marian (Onstad) D.; m. Robert George Gottschalk, Sept. 12, 1971. B.A., U. Wash., Seattle, 1967, M.B.A., 1968; Ph.D., Golden Gate U., 1978. Cert. data processor. Communications cons. Pacific NW Bell Telephone Co., Seattle, 1968-71; mktg. supr. AT&T, San Francisco, 1971-73; sr. cons., project leader Quantum Sci. Corp., Palo Alto, Calif., 1973-75; dir. analysis program Input Inc., Palo Alto, 1975-76; assoc. prof. accounting, information systems Calif. State U., Hayward, 1976—; dir. information sci. dept. Golden Gate U., San Francisco, 1982—; cons. pvt. cos., 1975—. Author: Computers and Remote Computing Services, 1983; contbr. articles to profl. jours. Loaned exec. United Good Neighbors, Seattle, 1969; nat. committeewoman, bd. dirs. Young Republicans, Wash., 1970-71; adv. Jr. Achievement, San Francisco, 1971-72. Mem. Data Processing Mgmt. Assn. (1982, Meritorious Service award, div. San Francisco chpt. 1984-85). Club: Junior (Seattle). Subspecialties: Information systems (information science); Database systems. Current work: curriculum development, professionalism in data processing field, professional certification, industry standards, computer literacy and user education, design of data bases and data banks. Office: Golden Gate Univ 536 Mission St San Francisco CA 94105

DUNCAN, IAN WILLIAM, geneticist, educator; b. Winnipeg, Man., Can., July 11, 1952; s. Morris Robert and Maida Torrance (McKenzie) D. B.Sc., U. B.C., 1974; Ph.D., U. Wash., 1978. Postdoctoral researcher div. biology Calif. Inst. Tech., Pasadena, 1979-82; asst. prof. biology Washington U., St. Louis, 1982—. Mem. Genetics Soc. Am. Subspecialties: Genetics and genetic engineering (biology); Developmental biology. Current work: Developmental genetics of drosophila. Office: Dept Biology Washington U Saint Louis MO 63130

DUNCAN, STARKEY DAVIS, JR., behavioral sciences educator; b. San Antonio, Aug. 24, 1935; s. Starkey Davis and Catherine (Poulson) D.; m. Susan Morton, June 30, 1960; children: Arne, Sarah, Owen. B.A., Vanderbilt U., 1959; Ph.D., U. Chgo., 1965. Postdoctoral fellow U. Chgo., 1965-67, asst. prof. behavioral scis., 1967-74, assoc. prof., 1974-81, prof., 1981—. Author: (with others) Face-to-face Interaction, 1977; contbr. articles profl. jours. Served to lt. (j.g.) USNR, 1957-59. Grantee NSF, 1972, 75, 80, NIMH, 1978. Mem. Am. Psychol. Assn., AAAS, Linguistic Soc. Am. Subspecialties: Cognition. Current work: Structure and strategy of face-to-face interaction, nonverbal communication. Office: Dept of Behavioral Sciences University of Chicago 5848 S University Ave Chicago IL 60637

DUNCAN, WILLIAM R., immunologist, educator; b. Harlingen, Tex., Aug. 14, 1949; s. Willim and Margaret (Marlor) D.; m. Marilyn Ruth Duncan, May 28, 1970; children: Jennifer, Graham. B.A., U. Tex., Austin, 1976; Ph.D., Dallas, 1976. Postdoctoral fellow U. Tex. Health Sci. Ctr., Dallas, 1976, 77, asst. prof. dept. cell biology, 1978—. Mem. Winnetka Heights Neighborhood Assn., 1981. Mem. Am. Genetics Assn., Am. Assn. Immunologists, Transplantation Soc., N.Y. Acad. Scis. Subspecialties: Immunogenetics; Cellular engineering. Current work: Studies concerning cellular mechanisms of allograft rejection and characterization of major histocompatibility genes and immunoglobulin genes of Syrian hamster. Office: 5323 Harry Hines Blvd Dallas TX 75235

DUNDON, JEFFREY MICHAEL, physicist, educator; b. Cambridge, Mass., Apr. 30, 1946; s. John Peter and Janet Le Hentz (Watson) D. A.B. cum laude, Harvard U., 1968; Ph.D. in Physics, U. Calif., San Diego, 1974. Asst. prof. physics U. Nev., Las Vegas, 1974-79, assoc. prof., 1979—. Contbr. articles to profl. jours. U. Nev. Research Council grantee, 1974, 76, 78, 80; Research Corp. scholar, 1979; Asso. Western Univs. scholar, 1982. Mem. Am. Phys. Soc. Subspecialties: Low temperature physics; Condensed matter physics. Current work: Research into magnetic properties of condensed oxygen. Home: Box 70243 Las Vegas NV 89170 Office: Dept Physics U Nev Las Vegas NV 89154

DUNHAM, EARL WAYNE, pharmacologist, educator; b. Peoria, Ill., Sept. 22, 1942; s. Richard Vincent and Dorothy Elizabeth (Eddington) D.; m. Valerie Jean Dunham, Aug. 30, 1964; children: Bradley, Bruce, Elaine. B.S., Drake U., 1965; Ph.D., U. Minn., 1971. Mem. faculty U. Minn. Coll. Pharmacy and Sch. Medicine, Mpls., 1971—, now assoc. prof. pharmacology. Contbr. chpts. to books, articles to sci. jours. USPHS fellow, 1966-71. Mem. Am. Soc. Pharmacology and Exptl. Therapeutics, Am. Assn. Colls. Pharmacy, Am. Heart Assn. Subspecialties: Pharmacology; Physiology (medicine). Current work: Synthesis and vascular effects of renal lipids; teaching of pharmacology, research; cardiovascular and renal pharmacology and physiology and pathophysiology (hypertension). Office: Dept Pharmacolog U Minn Sch Medicine 435 Delaware St SE Minneapolis MN 55455

DUNHAM, PHILIP BIGELOW, biology educator, physiology researcher; b. Columbus, Ohio, Apr. 26, 1937; s. T. Chadbourne and Margaret (Bigelow) D., m. Joyce Enderle, Aug. 20, 1965 (div. Dec. 1969). B.A., Swarthmore Coll., 1958; Ph.D., U. Chgo., 1962. USPHS postdoctoral fellow Carlsberg Found., Copenhagen, 1962-63; asst. prof. zoology Suracuse U., 1963-67, assoc. prof., 1967-71, prof., 1971—; vis. assoc. prof. physiology Yale U. Sch. Medicine, 1968-70; vis. scientist Physiol. Lab., U. Cambridge, Eng., 1969; vis. honors examiner Swarthmore Coll., 1966-67, 73-74, mem. alumni council, 1971-73; mem. exec. com. of bd. trustees Marine Biol. Lab., Woods Hole, Mass., 1972-76. Research numerous publs. in field. Mem. Soc. Gen. Physiologists (council 1967-69), Am. Physiol. Soc., Biophys. Soc. Subspecialties: Physiology (biology); Biophysics (biology). Current work: Mechanism and cellular function of passive and active membrane transport of sodium and potassium in mammalian erythrocytes. Home: 2311 E Genesee St Syracuse NY 13210 Office: Syracuse U 130 College Pl Syracuse NY 13210

DUNHAM, WOLCOTT BALESTIER, cancer researcher; b. Boston, June 15, 1900; s. Theodore and Josephine (Balestier) D.; m. Isabel Bosworth, Oct. 7, 1940; children: Wolcott Balestier, Anne Huntington Dunham Ewart. Student, Harvard Coll., 1920-22; A.B., Columbia U., 1924, M.D., 1928. Diplomate: Am. Bd. Microbiology. Asst. bacteriologist N.Y. Postgrad. Med. Sch. and Hosp., 1936-46; research biologist VA Hosp., Memphis, 1946-61, Asst. dir. profl. services for research, 1956-61, assoc. chief staff, 1961-68; vis. investigator Jackson Lab., Bar Harbor, Maine, 1968-76; assoc. research prof. Linus Pauling Inst., Palo Alto, Calif., 1978—. Author articles. Recipient Civic award VA, 1967. Fellow ACP, N.Y. Acad. Medicine, N.Y. Acad. Scis.; mem. Am. Assn. Immunologists, Am. Soc. for Microbiology, Am. Tissue Culture Assn., Physicians for Social Responsibility. Democrat. Episcopalian. Club: Harvard (N.Y.C.). Subspecialties: Cancer research (medicine); Virology (medicine). Current work: processes involved in carcinogenesis and methods for counteracting these processes. Home: 270 W Floresta Way Menlo Park CA 94025 Office: 440 Page Mill Rd Palo Alto CA 94306

DUNIGAN, PAUL FRANCIS XAVIER, chemical engineer; b. Boston, Mar. 9, 1918; s. John Joseph and Therese Florence (Donoghue) D.; m. Eva Lucile Reckley, July 2, 1942; 1 son, Paul Francis Xavier. B.A., Boston Coll., 1939; M.Ed. in Biology/Chemistry, Mass. State Tchrs. Coll., Boston, 1940. Tchr. sci. St. Rose High Sch., Chelsea, Mass., 1940; with E.I. duPont de Nemours, 1941-46; sr. supr. Hanford (Wash.) Engr. Works, 1944-46; mgr. facilities operation Gen. Electric Co., Hanford, 1946-65; mgr. plant ops. Battelle N.W., Richland, Wash., 1965-70; mgr. facilities utilization Westinghouse Hanford Co., Richland, 1970—. Dist. chmn. Blue Mountain council Boy Scouts Am., Richland, 1967; mem. Lay Com. on Edn., Richland, 1957-58; res. officer Aux. Police, Richland, 1962-63; sec. CeeKay Fed. Credit Union, Richland, 1970. Mem. Am. Chem. Soc. (sec. Richland 1952), Am. Nuclear Soc. (vice chmn. remote systems tech. div. 1961), Am. Inst. Chem. Engrs. (nuclear div.), AAAS. Democrat. Roman Catholic. Subspecialties: Nuclear engineering; Chemical engineering. Current work: Design and utilization of research and development facilities. Home: 1942 Davison Ave Richland WA 99352 Office: Westinghouse Hanford Co Box 1970 Richland WA 99352

DUNIWAY, JOHN M., plant pathologist; b. San Francisco, Nov. 6, 1942; s. Ben C. and Ruth M. D.; m. Catherine C. Cohrs, June 10, 1965; children: Sarah, Michael. B.A. in Biology, Carleton Coll., Northfield, Minn., 1964; Ph.D. in Plant Pathology, U. Wis., 1969. NSF postdoctoral fellow Australian Nat. U., Canberra, 1969-70; mem. faculty dept. plant pathology U. Calif., Davis, 1970—, prof., 1982—. Contbr. articles to profl. jours. Mem. Am. Phytopath. Soc. (CIBA-GEIGY award for research 1982), Am. Soc. Plant Physiologists. Subspecialties: Plant pathology; Plant physiology (agriculture). Current work: Water relations of plants and soil fungi. Office: Dept Plant Pathology U Calif Davis CA 95616

DUNLAP, BRETT IRVING, research physicist; b. Oakland, Calif., Apr. 24, 1947; s. George Robert and Betty Jean (Faust) D.; m. Barbara Ann Dunlap, June 14, 1969; children: Tanya Faust, Corrie Jessica. B.A. in Math. and Physics, U. Iowa, 1969; M.A. in Physics, Johns Hopkins U., 1972, Ph.D., 1976. Postdoctoral assoc. physics U. Fla., 1975-77; postdoctoral assoc. surface sci. Nat. Bur. Standards, 1977-79; postdoctoral asso. chemistry George Washington U., 1979-80; research physicist Chemistry div. Naval Research Lab., Washington, 1980—. Chmn. adminstrv. bd. Hughes United Methodist Ch., Wheaton, Md., 1982. Subspecialties: Theoretical chemistry; Surface chemistry. Current work: Theoretical surface science. Home: 13419 Tamarack Rd Silver Spring MD 20904 Office: Naval Research Lab Code 6171 Washington DC 20375

DUNLAP, R. BRUCE, chemist, educator; b. Elgin, Ill., Oct. 14, 1942; s. Robert J. and Carol C. D.; m.; children: Heather Diane, Edward Joseph. B.S. with honors, Beloit (Wis.) Coll., 1964; Ph.D. in Chemistry, Ind. U., 1968. NIH postdoctoral fellow dept. biochemistry Scripps Clinic and Research Found., LaJolla, Calif., 1968-71; asst. prof. dept. chemistry U. S.C., 1971-74, assoc. prof., 1974-78, prof., 1978—. Contbr. articles to profl. jours. Am. Cancer Soc. awardee, 1976-80. Mem. AAAS, Am. Assn. Cancer Research, Am. Chem. Soc., Am. Soc. Biol. Chemists, Am. Cancer Soc. (dir. S.C. div.). Subspecialties: Biochemistry (biology); Nuclear magnetic resonance (biotechnology). Current work: Enzymology and protein chemistry, selenium biochemistry, mechanism of action of folate enzymes, development of analytical technique of room temperature phosphorescence, application of nuclear magnetic resonance spectroscopy to biochemical problems. Home: 1409 Brookview Columbia SC 29210 Office: Dept Chemistr U SC Columbia SC 29208

DUNLOP, TERRENCE WARD, psychologist; b. San Antonio, Nov. 29, 1943; s. Ward Carl and Laura Louise (Laue) D. B.A., Bradley U., 1972; M.A., U. Conn., 1974, Ph.D., 1976. Pvt. practice psychology, Balt., 1976—; postdoctoral fellow dept. psychiatry Johns Hopkins U., Balt., 1976-78, assoc. dir. Cortical Function Lab., 1978-80; chief psychologist Social Security Adminstrn., Office Disability Program, Balt., 1980—. Mem. profl. service com. United Cerebral Palsy, Annapolis, Md., 1981—. Served with U.S. Army, 1966-69. Recipient Acad. Excellence Citation, Bradley U., 1972; cert. of recognition Psychol. Info., 1981; Johns Hopkins U. neuropsychol. grantee, 1978. Mem. Internat. Neuropsychol. Soc., Am. Psychol. Assn., Nat. Acad. Neuropsychologists, Neurol. Soc. Md. Subspecialties: Neuropsychology; Behavioral psychology. Current work: Neuropsychological evaluation and rehabilitation, determination of disability. Home: 1231-L Gemini Dr Annapolis MD 21403 Office: Social Security Adminstrn Dickinson Bldg Suite 2414 Woodlawn MD 21241

DUNN, ADRIAN JOHN, neurosci. educator; b. London, June 16, 1943; U.S., 1970; s. John Charles and Gwendolyn (Gracie) D.; m. Glenda Bradley, Oct. 6, 1973. B.A., U. Cambridge, 1965, M.A., 1968, Ph.D., 1968. Asst. prof. biochemistry U. N.C., Chapel Hill, 1973; asst. prof. neurosci. U. Fla., Gainesville, 1973-77, assoc. prof., 1977—; mem. neuropsychology research rev. com. NIMH, 1980—. Author: Functional Chemistry of the Brain, 1974, Peptides, Hormones and Behavior, 1983. Mem. Biochem. Soc., Internat. Soc. Neurochemistry, Am. Soc. for Neurochemistry, Soc. for Neurosci., AAAS. Subspecialties: Neurochemistry; Neuroendocrinology. Current work: Mechanisms of ACTH action on the brain, neurochemistry of stress. Office: U Fla PO Box J-244 Gainesville FL 32610

DUNN, ANNE ROBERTS, optical engr.; b. Champaign, Ill., Nov. 23, 1940; d. Howard Creighton and Elizabeth (Clifford) Roberts; m. Karl Lindemann Dunn, June 24, 1967. B.S., Beloit (Wis.) Coll., 1962; M.S., Rensselaer Poly. Inst., 1965, Ph.D., 1969. Research physicist Teledyne Brown Engring., Huntsville, Ala., 1969-76; staff engr. McDonnell Douglas Astronautics Co., Huntsville, 1976-78; sr. scientist Nichols Research Corp., Huntsville, 1978—. Contbr. articles to profl. jours. Mem. Am. Astron. Soc., AAAS. Subspecialties: Optical engineering; Infared technology. Current work: Processing and analysis of data from large infrared detector arrays; military infrared optics; electrooptics; photoelectric mosaic arrays; sensor calibration; data reduction and analysis. Home: 1044 Joe Quick Rd Hazel Green AL 35750 Office: 4040 S Memorial Pkwy Suite A Huntsville AL 35802

DUNN, CHARLETA JESSIE, psychology educator; b. Clarendon, Tex., Jan. 18, 1927; d. James Arthur and Ruby (Burcham) Sisk; m. Roy E. Dunn, Sept. 13, 1947; children: Thomas A., Roy E., Sharleta E. B.S., West Tex. U., 1951, M.Ed., 1954; Ed.D., U. Houston, 1966; postgrad., U. Tex., Galveston, 1970. Lic. psychologist, Tex. Tchr. Dalhart (Tex.) Pub. Schs., 1951-52, Amarillo (Tex.) Pub. Schs., 1952-62; asst. prof. U. Houston, 1966-69; pediatric psychologist U. Tex. Med. Br., Galveston, 1970; dir. pupil appraisal Goose Creek Ind. Sch. Dist., Baytown, Tex., 1971-74; prof. psychology Tex. Women's U., Houston, 1974—; cons. Vidor Ind. Sch. Dist., 1969—, Goose Creek Ind. Sch. Dist., 1974—, Stafford Ind. Sch. Dist., 1982-83. Author: Songs of Sharleta, 1966, World of Work, 1970; contbr. articles to profl. jours. Gasreda fellow U. Houston, 1965; Hogg Found. grantee, 1966-69; Tex. Women's U. research grantee, 1982—; others. Mem. Am. Psychol. Assn., Western Tex. Psychol. Assn., Tex. Psychol. Assn. Prebyterian. Subspecialties: Behavioral psychology; Cognition. Current work: Biofeedback, contingency management, emotional adjustment, psychological appraisal. Office: Tex Women's U 1130 M D Anderson St Houston TX 77030

DUNN, FLOYD, biophysicist, bioengineer, educator; b. Kansas City, Mo., Apr. 14, 1924; s. Louis and Ida (Leibtag) D.; m. Elsa Tanya Levine, June 11, 1950; children: Andrea Susan, Louis Brook. Student, Kansas City Jr. Coll., 1941-42, Tex. A. and M. U., 1943; B.S., U. Ill., Urbana, 1949, M.S., 1951, Ph.D., 1956. Research asso. elec. engring. U. Ill., Urbana, 1954-57, research asst. prof. elec. engring., 1957-61, asso. prof. elec. engring. and biophysics, 1961-65, prof., 1965—, prof. elec. engring., biophysics and bioengring., 1972—, dir. bioacoustics research lab., 1976—, chmn. bioengring. faculty, 1978-82; vis. prof. dept. microbiology Univ. Coll., Cardiff, Wales, 1968-69; vis. sr. scientist Inst. Cancer Research, Sutton, Surrey, Eng., 1975-76, 82-83;

vis. prof. Inst. Chest Diseases and Cancer, Tohoku U., Sendai, Japan; mem. radiation study sect. NIH, 1976-81; steering com. NSF Workshop on Interaction of Ultrasound and Biol. Tissues, 1971-72; chmn. WHO working group on health aspects of exposure to ultrasound radiation, London, 1976; mem. tech. elec. products radiation standards com. FDA, 1974-76. Editorial bd.: Ultrasonic Imaging, others; manuscript reviewer: Jour. Phys. Chemistry, Jour. Acoustical Soc. Am., IEEE Transactions, others; Contbr. articles on biophys. acoustics to profl. jours. Trustee Hensley Twp., Ill., 1980-81. Served with AUS, 1943-46. NIH Spl. Research fellow Univ. Coll., Cardiff, 1968-69; Am. Cancer Soc.-Eleanor Roosevelt-Internat. Cancer fellow, 1975-76, 82–83; Fulbright Fellow, 1982–83; Japan Soc. for Promotion of Sci. Fellow, 1982. Fellow Acoustical Soc. Am. (asso. editor Jour., v.p. 1981-82), Am. Inst. Ultrasound in Medicine, IEEE, Inst. Acoustics (U.K.); mem. Am. Inst. Physics, Biophys. Soc., Nat. Acad. Engring., AAAS, Sigma Xi, Sigma Tau, Eta Kappa Nu, Tau Beta Pi, Pi Mu Epsilon, Phi Sigma. Subspecialties: Biophysics (physics); Biomedical engineering. Current work: Research on all aspects of ulrtasonic propagation in, and interaction with, biological media. Home: Rural Route 3 Box 295 Champaign IL 61820 Office: Bioacoustics Research Lab U Ill 1406 W Green St Urbana IL 61801 Excellent, dedicated and understanding teachers, bright and energetic students, and a single-mindedness to see a problem to solution are the ingredients for a modicum of success.

DUNN, JAMES, electronics and speech communications researcher; b. Harrisburg, Pa., 1932. B.S.E.E., Rutgers U.; M.S.E.E., Ph.D., Columbia U. With Def. Communications div. ITT, 1950's; developed ANDVT and Quintrel voice-processing systems, now tech. dir. div., named exec. scientist ITT, 1983—. Contbr. articles to tech. jours. Subspecialty: Electronics. Patentee in electronic communications fields (6).

DUNN, MICHAEL JAMES, engineering executive; b. Bklyn., July 13, 1947; s. Frank Nelson and Helen Macaluso (Rau) D.; m. Ann Elizabeth Lutz, Feb. 20, 1982; children: Peter James, Steven Christopher, John Andrew. B.S. in Nuclear Engring, U. Fla., 1971, M.Engring., 1972. Registered profl. engr., Calif., Mich. Engr., group leader Bechtel Corp., San Francisco, 1972-75; project engr. Black & Veath, Kansas City, Kans., 1975-77; supr. engring. Bechtel Corp., Ann Arbor, Mich., 1977-82; sect. mgr. FDS Nuclear, Norcross, Ga., 1982—. Author tech. papers in field. Mem. admissions com. U. Fla., 1972. Mem. Am. Nuclear Soc. Republican. Lutheran. Subspecialties: Nuclear engineering; Systems engineering. Current work: Development of technology for radwastertreatment, volume reduction and operations management software. Home: 9060 Martin Rd Roswell GA 30076 Office: FDS Nuclear 333 Technology Park/Atlanta Norcross GA 30092

DUNN, THOMAS M., physical chemistry educator; b. Sydney, Australia, Apr. 25, 1929. B.Sc., U. Sydney, Australia, 1949, M.Sc., 1959; Ph.D. in Phys. Chemistry, U. London, Eng., 1957. Teaching fellow chemistry U. Sydney, Australia, 1950-52; asst. lectr. to lectr. phys. chemistry U. London, 1954-63; prof. phys. chemistry U. Mich., Ann Arbor, 1963—, head dept., 1974. Fellow The Chemistry Soc.; mem. Am. Chem. Soc. Subspecialty: Physical chemistry. Office: Dept Chemistry U Michigan Ann Arbor MI 48109

DUNN, WILLIAM LAWRENCE, principal scientist, consulting psychologist; b. Richmond, Va., Apr. 19, 1924; s. William Lawrence and Emily Chenault (Noble) D.; m. Elisabeth Oleknovitch, June 5, 1948; children: Olga, William Mark, Alexandra Noble, Lawrence Alexis. B.S., Lynchburg (Va.) Coll., 1947; Ph.D., Duke U., 1953. Clin. psychologist VA Hosp., Richmond, 1953-61; assoc. prin. scientist Philip Morris Research Ctr., Richmond, 1961-75, prin. scientist, 1975—; adj. prof. Va. Commonwealth U., 1956-61; cons. psychologist Va. Penitentary for Women, Goochland, 1957—; Mem. Bd. Psychologist Examiners, Richmond, 1965-75; mem. wine adv. com. Va. Alcoholic Beverage Control Bd., Richmond, 1976—. Editor: Smoking Behavior, 1973. Served to lt. USNR, 1942-46. Mem. Va. Psychol. Assn. (pres. 1965-66), Va. Acad. Clin. Psychologists (charter pres. 1976-78), Am. Psychol. Assn. Eastern Orthodox. Subspecialties: Clinical psychology; Sensory processes. Current work: Investigation of the psychodynamics of smoking; methods development in consumer evaluation of products. Home: 4701 New Kent Ave Richmond VA 23225 Office: Philip Morris Research Center PO Box 26583 Richmond VA 23261

DUNN, WILLIAM LEE, radiation and med. physicist; b. Muncie, Ind., May 6, 1944; s. Wilbur Francis and Betty Ruth (Boyd) D.; m. Donna Joy Widstrom, Aug. 16, 1969; children: Colin, Adam. B.A. in Physics, Vanderbilt U., 1969, M.S., 1975. Radiation physicist Vanderbilt U. Hosp., 1975-76, King Faisal Specialist Hosp. and Research Centre, Riyadh, Saudi Arabia, 1976-78; radiation physicist nuclear medicine Mayo Clinic, Rochester, Minn., 1979—. Contbr. articles to profl. jours. Served with USMC, 1962-73. Mem. Am. Assn. Physicists in Medicine, Health Physics Soc., Soc. Nuclear Medicine. Democrat. Unitarian. Subspecialties: Nuclear medicine; Imaging technology. Current work: Nuclear medicine instrumentation and procedure development; quantitation of bone mineral in vivo; health physics. Home: 2665 Riverside Ln Rochester MN 55901 Office: Nuclear Medicine Mayo Clinic C66 Hilton Rochester MN 55901

DUNNE, THOMAS, geology educator; b. Prestbury, Eng., Apr. 21, 1943; came to U.S., 1964, naturalized, 1979; s. Thomas and Monica Mary (Whitter) D. B.A., Cambridge (Eng.) U., 1964; Ph.D., Johns Hopkins U., 1969. Research hydrologist U.S. Geol. Survey, 1968-71; asst. prof. McGill U., Montreal, 1969-73; asst. prof. dept. geol. scis. U. Wash., Seattle, 1973-74, assoc. prof., 1974-79, prof., 1979—, U. Nairobi, 1969-71; cons. Author: (with L. B. Leopold) Water in Environmental Planning, 1978; contbr. articles to sci. publs. Mem. Am. Geophys. Union, AAAS, Geol. Soc. Am. Subspecialties: Geology; Surface water hydrology. Current work: Hillslope erosion processes; river sedimentation. Office: Department of Geological Sciences University of Washington Seattle WA 98195

DUNNICK, NICHOLAS REED, radiologist, educator, researcher; b. Waukegan, Ill., Aug. 23, 1943; s. Paul A. and Marceil (Reed) D.; m. June Kaiser, Aug. 3, 1968; children: Cory, Amanda. B.S., Purdue U., 1965; M.D., Cornell U., 1969. Diplomate: Am. Bd. Radiology, Nat. Bd. Med. Examiners. Intern U. Rochester, N.Y., 1969-70, resident, 1970-71; staff assoc. NIH, Bethesda, Md., 1971-73, staff radiologist, 1976-80; resident Stanford (Calif.) U. Med. Ctr., 1975-76, acting asst. prof., 1976; assoc. prof. radiology Duke U. Med. Sch., Durham, N.C., 1980—. Served to comdr. USPHS, 1976-78. Mem. Radiol. Soc. N.Am., Am. Coll. Radiology, Assn. Univ. Radiologists, Am. Roentgen Ray Soc., Soc. Uroradiology, Soc. Computed Body Tomography, Sigma Xi. Subspecialty: Diagnostic radiology. Office: Dept Radiology Duke U Med Center Durham NC 27710

DUPERON, DONALD FRANCIS, dental educator, pediatric dentist; b. Regina, Sask., Can., Dec. 18, 1937; came to U.S., 1974; s. Francis and Eugenie (Dhuez) D.; m. Donna Joy Hill, Aug. 20, 1960; children: Lori Anne, Mona Lee. Cert., Children's Hosp., Winnipeg, Man., Can., 1968; M.Sc., U. Man., 1970; D.D.S., U. Alta., 1961. Pvt. practice dentistry Weiker & Duperon, Regina, 1961-67; asst. prof. dentistry U. Man., Winnipeg, 1968-70, assoc. prof., chmn., 1970-74; assoc. prof. dentistry UCLA, 1974—, also chmn. grad. program. Mem. Can. Dental Assn., Can. Acad. Pedodontics, Royal Coll. Dentists Can., Man. Dental Assn., Calif. Soc. Pediatric Dentistry. Subspecialties: Pediatric dentistry; Dental growth and development. Current work: Computerized cephalometric analysis, oral problems of bone marrow transplant patients, pulp therapy in pedodontics. Home: 30169 Via Victoria Rancho Palos Verdes CA 90274 Office: UCLA 23-020 CHS Los Angeles CA 90024

DUPONT, HERBERT LANCASHIRE, medical educator, researcher; b. Toledo, Nov. 12, 1938; s. Robert L. and Martha (Lancashire) DuP.; m. Margaret Wright, June 9, 1963; children: Denise Lorraine, Andrew Wright. B.A., Ohio Wesleyan U., 1961; M.D., Emory U., 1965. Diplomate: Am. Bd. Internal Medicine. Infectious Disease fellow U. Md. Sch. Medicine, Balt., 1986-69; intern U. Minn. Med. Ctr., Mpls., 1965-67; mem. faculty program infectious diseases U. Tex. Health Sci. Ctr., Houston, 1973—, prof. dir. program, 1973—; prof. U. Tex. Sch. Pub. Health, Houston, 1975—; adj. prof. microbiology Baylor U. Coll. Medicine, Houston, 1975—. Author: Practical Antimicrobial Therapy, 1978, Infections of the Gastrointestinal Tract, 1980, Travel with Health, 1981, also 200 sci. articles.; Assoc. editor: Am. Jour. Epidemiology, 1978-81. Served to lt. comdr. USPHS, 1967-69. Fellow ACP; mem. Am. Soc. Clin. Invetigation, Infectious Diseases Soc. Am. (counselor, sec.), Am. Soc. Clin. Investigation, Nat. Found. Infectious Diseases (bd. dirs. 1981-84), Alpha Omega Alpha. Subspecialties: Infectious diseases; Internal medicine. Current work: Enteric infectious diseases - etiology, epidemiology, therapy, prevention. Home: 147 Hickory Ridge Houston TX 77024 Office: 6431 Fannin Houston TX 77030

DUPONT, JACQUELINE (LOUISE), food and nutrition educator, university administrator; b. Plant City, Fla., Mar. 4, 1934; d. Albert Pierre and Bessie (Clemons) D. B.S., Fla. State U., Tallahassee, 1955; M.S., Iowa State U., 1959; Ph.D., Fla. State U., 1962. Home economist U.S. Dept. Agr., Beltsville, 1955-57, research nutritionist, 1962-64; asst. prof. biochemistry Howard U., 1964-66; asst. prof. Colo. State U., 1966-69, assoc. prof., 1969-73, prof., 1973-78; prof. food and nutrition Iowa State U., Ames, 1978—, chmn., 1978—; mem. nutrition study sect. NIH, Bethesda, 1972-76; vis. scientist Inst. Nutrition Central Am. and Panama, Guatemala, 1973. Author, editor: Cholesterol Systems in Insects and Animals, 1982; contbr. articles to profl. jours. U.S. Dept. Agr. grantee, 1976-81; NIH grantee, 1968-80; Am. Heart Assn. grantee, 1982; recipient Research Career Devel. award NIH, 1972-77. Fellow Council Arteriosclerosis Am. Heart Assn.; mem. Am. Inst Nutrition, Am. Oil Chemists Soc., Am. Dietetic Assn. Democrat. Subspecialties: Nutrition (medicine); Biochemistry (medicine). Current work: Administration of food and nutrition department; teaching nutrition, research in cholesterol and prostaglandin metabolism. Home: 309 Crane Ave Ames IA 50010 Office: Iowa State University 107 Mackay Hall Ames IA 50011

DUPREE, SAMUEL HARDY, JR., programmer/analyst; b. Phila., Feb. 25, 1953; s. Samuel H. and Louise D. B.S., Pa. State U., 1974, M.S., 1978. Programming cons. Computation Ctr., Pa. State U., 1975-78; instr. physics and computer sci. Rose-Hulman Inst. Tech., 1978-81; programmer/analyst Gen. Electric Space Systems Div., Phila., 1981—. Mem. Am. Astron. Soc., Nat. Soc. Indsl. and Applied Math., Astron. Soc. Pacific, Nat. Computer Graphics Assn., IEEE, IEEE Computer Soc. Baptist. Subspecialty: Scientific Computing. Current work: Computer graphics, software engineering, scientific computing, man-machine interfacing. Office: Gen Electric Space System Div PO Box 8555 Philadelphia PA 19101

DUPUY, DAVID LORRAINE, astronomer, educator; b. Asheville, N.C., Mar. 7, 1941; s. Edward Lorraine and Attawa (Dixon) DuP.; m. Lieselotte Kruska, Nov. 15, 1969. A.B., King Coll., 1963; M.A., Wesleyan U., 1967; Ph.D., U. Toronto, 1972; postgrad., U. Heidelberg, W.Ger., 1963-64. Astronomer US. Naval Obs., Washington, 1963; asst. Kitt Peak Nat. Obs., Tucson, 1967; assoc. prof. dir. obs. St. Mary's U., Halifax, N.S., Can., 1972-82; assoc. prof. physics, dir. obs. and planetarium Va. Mil. Inst., Lexington, 1982—. Mem. Am. Astron. Soc., Astron. Soc. Pacific, Royal Astron. Soc., Internat. Astron. Union. Subspecialty: Optical astronomy. Current work: Period search methods in variable stars; instrumentation and electronic imaging devices for astronomy, especially charge coupled devices. Office: Dept Physics Va Mil Inst Lexington VA 24450

DURACK, DAVID TULLOCH, physician, medical educator; b. Perth, Western Australia, Dec. 18, 1944; U.S., 1974; m. Carmen Elizabeth Prosser, 1970; children: Jeremy Charles, Kimberley David, Sonya Elizabeth, Justin Rooks. M.B., B.S. with honors, U. Western Australia, 1969; D.Phil., Oxford (Eng.) U., 1973. Diplomate: Am. Bd. Internal Medicine. House physician Nuffield dept. medicine Radcliffe Infirmary, Oxford, 1969, house surgeon, 1969-70, sr. house physician, 1973; sr. house physician dept. medicine Hammersmith Hops., London, 1973-74; chief resident in medicine U. Wash. Hosp., Seattle, 1974-75; acting instr. medicine U. Wash., Seattle, 1974-75, acting asst. prof., 1975-77; assoc. prof. medicine Duke U. Med. Ctr., 1977-82, chief div. infectious diseases, 1977—, assoc. prof. microbiology and immunology, 1977-83, prof. microbiology and immunology, 1983—, acting chmn. dept. medicine, 1982—, prof. medicine, 1982—. Contbr. articles to profl. jours. Recipient E. Simpson prize for biochemistry, 1965; M.M. Bergin Meml. prize, 1966; R.P.H. Clin. Staff prize in surgery, 1966; others; Rhodes scholar, 1968; Lilly Internat. fellow, 1974; NIH grantee, 1979—. Fellow ACP, Royal Australasian Coll. Physicians, Royal Coll. Physicians; mem. Am. Fedn. for Clin. Research, Infectious Diseases Soc. Am., Am. Soc. Microbiology, So. Soc. for Clin. Investigation, Am. Rhodes Scholars, Am. Heart Assn. (com. on rheumatic fever 1978-83, exec. com. of council on cardiovascular disease in young 1981-83), Alpha Omega Alpha. Subspecialties: Internal medicine; Infectious diseases. Current work: Experimental pathology, endocarditis and meningitis. Home: 1700 Woodstock Rd Durham NC 27705 Office: Duke U Med Ctr Box 3867 Durham NC 27710

DURBIN, RICHARD DUANE, plant pathologist; b. Santa Ana, Calif., Sept. 6, 1930; s. Frederick H. and Erma Rebecca (Schooley) D.; m. Elizabeth Letts, Apr. 2, 1964; children: Stephen D., Christopher D., Jeffrey A. Ph.D., U. Calif.-Berkeley, 1957. Asst. prof. dept. plant pathology U. Minn., Mpls., 1958-62; mem. faculty dept. plant pathology U. Wis.-Madison, 1962—, prof., 1967—; research leader U.S. Dept. Agr., Madison, 1964—. Contbr. articles to profl. jours. Served with USMC, 1947. Recipient Dir.'s award Agrl. Research Service, U.S. Dept. Agr., 1982; named Scientist of Yr. Agrl. Research Service, U.S. Dept. Agr., 1983. Mem. Am. Phytopath. Soc., Sigma Xi, Alpha Zeta. Subspecialties: Plant physiology (biology); Microbiology. Current work: Synthesis, structure, metabolism and role in disease causation of toxins. Home: Route 3 Cross Plains WI 53528 Office: Dept Plant Pathology U Wis Madison WI 53706

DURELLI, AUGUST JOSEPH, mech. engr.; b. Buenos Aires, Argentina, Apr. 30, 1910; came to U.S., 1939, naturalized, 1960; s. August F. and Jeannette (Natzi) D.; m. Marie-Marthe Baril, Oct. 2, 1943; children: Ana, Monica, Andrew. Civil engr., U. Buenos Aires, 1932; D. Engring., U. Paris, 1937; D. Social Scis., Cath. U. Paris, 1937. Prof. Ill. Inst. Tech., 1956-61; prof. Cath. U., Washington, 1961-75; Dodge prof. engring. Oakland U., Mich., 1975-80; Nabor Carrillo prof. U. Mexico, 1981; prof. mech. engring. U. Md., College Park, 1981—. Author: Essai sur les Mentalites Contemporaines, 1937, Nacionalismo frente al Cristianismo, 1940, Del Universo de la Universidad al Universo del Hombre, 1942, Liberation de la Liberte, 1942, (with Phillips and Tsao) Introduction to the Theoretical and Experimental Analysis of Stress and Strain, 1958, (with Riley) Introduction to Photomechanics, 1965, (with Parks) Moire Analysis of Strain, 1970, Applied Stress Analysis, 1967; also 300 articles. Recipient Sr. research award Am. Soc. Engring. Edn., 1980. Mem. AAUP (pres. Cath. U. chpt. 1973-75), ASME, N.Y. Acad. Sci., Soc. Exptl. Stress Analysis (hon.), Sigma Xi. Democrat. Roman Catholic. Subspecialties: Solid mechanics; Optical engineering. Current work: Optimization of structural components, development of experimental stress analysis methods. Home: PO Box 6 Myersville MD 21773 Office: Mech Engring U Md College Park MD 20742

DURGUN, KANAT, mathematician; b. Istanbul, Turkey, Mar. 30, 1940; s. Resat A. and Feriha (Ulu) D.; m. Lynn K. Hogan, June 12, 1971 (div. Jan. 1979); 1 dau. Sarah K.; m. Ann Mary Moskiewicz, May 13, 1979. Ph.D., Tech. U. Istanbul, 1965; M.S., Syracuse U., 1972, Ph.D., 1976. Research collaborator Brookhaven Nat. Lab., Upton, L.I., N.Y., 1965-67; asst. prof. Tech. U. Istanbul, 1967-69, Utica (N.Y.) Coll., 1976-77, U. South Ala., Mobile, 1977-79; assoc. prof. U. Ark., Little Rock, 1981—; mem. vis. faculty NASA Johnson Space Center, Houston, 1982. Author: General Equation of Heat Conduction in Semi-Infinite and Infinite Mediums. Mem. Soc. Indsl. and Applied Math., Am. Math. Soc., Math. Assn. Am., Sigma Xi. Democrat. Roman Catholic. Subspecialties: Applied mathematics; Civil engineering. Current work: Applied mathematics. Home: 6200 Asher St Apt 272 Little Rock AR 72204 Office: Dept Math Univ Ark Little Rock AR 72204

DURIE, BRIAN GEORGE MARTIN, physician, educator; b. Gullane, Scotland, Dec. 22, 1942; m. Veronica; 1 son, Benjamin. M.B. Ch.B., U. Edinburgh, 1966. Diplomate: Am. Bd. Internal Medicine. Clin. and research fellow in hematology Mayo Clinic, Rochester, Minn., 1970-72; instr. sect. hematology/oncology, dept. internal medicine U. Ariz. Coll. Medicine, Tucson, 1972-73, asst. prof., 1973-76, assoc. prof., 1976-80, prof., 1980—; dir. clin. hematology Ariz. Health Scis. Center; hematologist Regional Hemophilia Center, U. Ariz; dir. clin. cell kinetics service Cancer Center; cons. NIH, Nat. Cancer Inst. Leukemia Soc. Am. scholar, 1976-81. Mem. Am. Soc. Hematology, Am. Soc. Clin. Oncology, Am. Assn. Cancer Research. Subspecialties: Hematology; Oncology. Office: Sect Hematology/Oncology Ariz Health Scis Center Tucson AZ 85724

DURST, TONY, organic chemistry educator; b. San Martin, Romania, Jan. 21, 1938. B.Sc., U. Western Ont., 1961, Ph.D. in Chemistry, 1964. Asst. to assoc. prof. dept. chemistry U. Ottawa, Ont., Can., 1967-77, prof., 1977—. Recipient Merck, Sharp & Dohme lectr. award, 1980. Mem. Am. Chem. Soc., Chem. Inst. Can. Subspecialty: Organic chemistry. Office: Dept Chemistry U Ottawa Ottawa ON Canada

DUS, KARL M(ARIA), biochemistry educator; b. Vienna, Austria, Jan. 2, 1932; came to U.S., 1958; s. Karl Arthur and Johanna (Novak) D.; m. Martha Heath Mahler, June 5, 1971; children: Johanna M., Melinda M. Ph.D., U. Vienna, 1958. Research fellow Mass. Gen. Hosp.-Harvard Med. Sch., Boston, 1958-60; research assoc. grad. dept. biochemistry Brandeis U., 1960-61, U. Calif.-San Diego, 1961-68; asst. prof. biochemistry U. Ill.-Urbana, 1968-73; assoc. prof. St. Louis U., 1974—. Contbr. chpts. to books, articles to profl. jours. NIH grantee, 1962-65, 69—; NSF grantee, 1975—; NASA contract awardee, 1966-68; recipient Maître de recherches in genetics and physiology Centre Nationale de Recherches Scientifiques, Eif-Sur Yvette, France, 1966-68. Mem. Am. Chem. Soc., Am. Soc. Biol. Chemists, AAAS, N.Y. Acad. Sci., Am. Inst. Chemists. Roman Catholic. Subspecialties: Biochemistry (biology); Analytical chemistry. Current work: Molecular oxygen fixation-drug metabolism, detoxification, steroid hormone biosynthesis and regulation in liver and adrenal; enzyme structure-function studies of P-450 hemeproteims, mechanism of action, photoaffinity labeling, affinity labeling and purification techniques. Patentee in field. Home: 411 S Holmes Ave Saint Louis MO 63122 Office: Saint Louis U Med Sch Biochemistry 1402 S Grand Blvd Saint Louis MO 63104

DUSANIC, DONALD GABRIEL, parasitologist, educator; b. Chgo., Dec. 15, 1934; s. Garbriel John and Harriet (Rojewski) D.; m. Roberta Leona Drost, June 22, 1957; children: Belinda Conrad, Donald, Karla Conrad, Allan Conrad, Robert; m. Jane Mitchell Haw, June 11, 1971. B.S., U. Chgo., 1957, M.S., 1959, Ph.D., 1963. Instr. U. Chgo., 1963-64; vis. prof. U. Philippines, 1964; asst. prof. U. Kans., 1964-68, assoc. prof., 1968-71, prof., 1971; vis. prof. Nat. Taiwan U. Sch. Medicine, 1971; prof. dept. life scis. Ind. State U., Terre Haute, 1971—; adj. prof. Ind. U. Sch. Medicine; guest faculty Rockefeller U., 1965; Universidade Catolica de Pelotas, Rio Grande do Sul, Brazil, 1980; cons. NATO, NIH, NSF, U.S. Navy. Contbr. articles to profl. jours. NSF grantee, 1982-84; NIH grantee, 1982-85. Mem. Am. Soc. Tropical Medicine and Hygiene, Am. Soc. Parasitologists, Soc. Protozoologists, Am. Assn. Immunologists, N.Y. Acad. Scis., AAAS, Sigma Xi. Subspecialties: Immunobiology and immunology; Parasitology. Current work: Immunity in rodent and human trypanosomiases; hybridoma technology in the study of trypanosome antigens, their functions during infections and their use as vaccines and serodiagnostic reagents. Home: BOX 176A Route 24 Terre Haute IN 47802 Office: Dept Life Scis Ind State U Terre Haute IN 47809

DUSENBERY, DAVID BROCK, Biophysics educator; b. Portland, Apr. 30, 1942; s. Harris and Evelyn (Shields) D. B.A., Reed Coll., 1964; Ph.D., U. Chgo., 1970. Postdoctoral fellow Calif. Inst. Tech., Pasadena, 1970-73; asst. prof. Ga. Inst. Tech., Atlanta, 1973-79, assoc. prof., 1979—. Contbr. articles to profl. jours. Mem. AAAS, Biophys. Soc., Soc. Nematologists, Genetics Soc. Am. Subspecialties: Biophysics (biology); Neurobiology. Current work: Nematode behavior.

DUSKO, HAROLD GEORGE, military technology branch executive, geography educator; b. Tarentum, Pa., July 1, 1942; s. Harold Richard and Josephine Mary (Goralka) D.; m. Janet Lamonby Craig, Sept. 1, 1963; children: Steven Harold, Christopher David, Jeffrey Craig. B.S. in Edn, Slippery Rock State Coll., 1964; M.B.A., U. Dayton, 1970; postgrad., Southern Ill. U., 1965-66. Cartographer Aero. Chart and Info. Ctr., St. Louis, 1964-66; imagery analyst fgn. tech. div. Wright-Patterson AFB, Fairborn, Ohio, 1966-69, methods officer, 1970-79, br. chief, 1979—; faculty Wright State U., Fairborn, 1970—, adj. prof. geography, 1970—, cartographic cons., 1973-82; U.S. Air Force rep. Mensuration Standards Group, Washington, 1975-80, Digital Image Work Group, 1976-80; cartographic cons. Dayton (Ohio) Family Services, 1979-80. Author: Caesar Creek and Toddsfork maps, 1974, Laboratory Exercises in Cartography, 1975, Map Exercises in Remote Sensing, 1976, Digital Image Processing, 1980. Bd. dirs. New Carlisle (Ohio) Baseball Assn., 1974-78; coach Miami County (Ohio) Area Youth Soccer Assn., 1979; commr. Bethel Baseball Assn., Miami County, 1980; mem. Steering Com. for Remote Sensing, Columbus, Ohio, 1982. Recipient Outstanding Performance awards Air Force Systems Command/Fgn Tech Div, U.S. Air Force, 1975, 78, 79, Sustained Superior Performance award, 1982. Mem. Am. Soc.

Photogrammetry. Democrat. Presbyterian. Subspecialties: Graphics, image processing, and pattern recognition; Remote sensing (geoscience). Current work: Digital image processing using distributed systems and microprocessors; computer-aided design systems and software, low-cost microprocessors for education in earth sciences. Home: 5244 Eastland Dr New Carlisle OH 45344 Office: Imagery Analysis Branch Fgn Tech Div Bldg 856 Wright-Patterson AFB Fairborn OH 45433

DUTCHER, JANICE PHILLIPS, oncology educator; b. Bend, Ore, Nov. 10, 1950; d. Charles Glen and May Belle (Fluit) P. B.A., U. Utah, 1971; M.D., U. Calif.-Davis, 1975. Diplomate: Am. Bd. Internal Medicine. Intern Presbyn. St. Lukes Hosp., Chgo., 1975-76, resident in medicine, 1976-78; clin. assoc. Balt. Cancer Research Ctr., Nat. Cancer Inst. NIH, 1978-81, med. investigator, 1981-82; asst. prof. med. oncology U. Md., Balt., 1982; asst. prof. medicine and oncology Albert Einstein Coll. Medicine, Bronx, 1983—; cons. in field. Bd. advisers Sunshine Found., Balt., 1982—. Served to lt. comdr. USPHS, 1978-82. Recipient Beecham award, 1983. Fellow ACP; mem. NOW, Am. Soc. Clin. Oncology, Am. Soc. Hematology, Am. Assn. Cancer Research, Am. Fedn. Clin. Research, Am. Assn. Blood Banks, Am. Med. Women's Assn., Phi Beta Kappa, Phi Kappa Phi, Alpha Omega Alpha. Democrat. Subspecialty: Oncology. Current work: Transfusion supportive care of cancer and leukemia patients; clin. trials in acute leukemia and solid tumors. Home: 640 W 239th St Apt 6D Riverdale NY 10463 Office: Section of Oncology Albert Einstein Coll Medicine 1300 Morris Park Ave Bronx NY 10411

DUTE, ROLAND ROY, botany educator; b. Amherst, Ohio, Nov. 9, 1947; s. Roy Casper and Donna Yvonne (Fazey) D. B.S., Ohio State U., 1969, M.S., 1972; Ph.D., U. Wis.-Madison, 1976. Research assoc. Lab. Molecular Biology U. Wis., 1976-78; research asso. dept. botany U. Ill., 1978-79; asst. prof. biology St. Ambrose Coll., 1979-82; asst. prof. botany, plant pathology and microbiology Auburn U., 1982—. Contbr. articles to profl. jours. Served with U.S. Army, 1970-72. Mem. AAAS, Bot. Soc. Am., Iowa Acad. Sci. Subspecialties: Cell biology; Developmental biology. Current work: Ultrastructure of food-conducting systems in vascular plants; ultrastructure; sieve elements; phloem. Office: Dept Botany Auburn U Auburn AL 36849

DU TEMPLE, OCTAVE JOSEPH, scientific society executive; b. Hubbell, Mich., Dec. 10, 1920; s. Octave Joseph and Marguerite (Gadoury) Du T.; m. Orpha Jane Servies, June 9, 1943 (div. 1945); 1 son, Michael; m. Susan Margaret Keach, June 9, 1951; children: Lesley Ann, Octave Jr. B.S., Mich. Tech. U., Houghton, 1948, M.S., 1949; M.B.A., Northwestern U., 1955. Chem. engr. Argonne (Ill.) Nat Lab., 1949-58; exec. dir. Am. Nuclear Soc., 1958—. Editor: Prehistoric Copper Mining in Lake Superior Area, 1963; sci. news jour. Nuclear News, 1959-63; contbr. articles to profl. jours. Served with USAAF, 1945-46. Recipient Disting. Service award Am. Nuclear Soc., 1978; Outstanding Mgmt. award Am. Soc. Assn. Execs., 1972. Mem. Am. Assn. Engring. Socs. (dir. 1980—), AAAS (mem. internat. adv. com. 1980—), Council Engring. and Sci. Soc. Execs. (dir. 1974-78), Am. Nuclear Soc., Am. Inst. Chem. Engrs., Am. Chem. Soc., Am. Soc. for Engring. Edn. Republican. Episcopalian. Subspecialties: Nuclear engineering; Nuclear fission. Current work: Periodic reports on nuclear programs overseas. Chief executive officer of American Nuclear Society — Scientific and Educational Society. Home: Route 2 Box 468 Kenosha WI 53142 Office: American Nuclear Soc 555 N Kensington Ave La Grange Park IL 60525

DUTKO, FRANCIS JOSEPH, JR., molecular virologist; b. Trenton, N.J., July 5, 1951; s. Francis Joseph and Helen Antoinette (Zyla) D.; m. Angela Lynn Grossi, Apr. 19, 1975; 1 dau., Rachel. B.S. in Biology, Rensselaer Poly. Inst., 1973, M.S., 1975, Ph.D., 1977. Research fellow dept. immunology Scripps Clinic and Research Found., La Jolla, Calif., 1977-80, research assoc., 1980-83; with dept. microbiology Sterling-Winthrop Research Inst., Rensselaer, N.Y., 1983—. Contbr. articles to profl. jours. NIH fellow, 1977-79; Leukemia Soc. Am. spl. fellow, 1982-84. Mem. Am. Soc. for Microbiology, AAAS. Democrat. Roman Catholic. Subspecialties: Molecular biology; Virology (biology). Current work: Mechanism of persistent and latent viral infections. Office: Sterling-Winthrop Research Inst Rensselaer NY 12204

DUTTA, SISIR KAMAL, educator; b. Bengal, India, Aug. 28, 1928; came to U.S., 1956, naturalized, 1974; s. Krishna K. and Satyabati (Chanda) D.; m. Minati Roy, July 1, 1955; children—Mahasweta, Basabi. M.S., Kans. State U., 1958, Ph.D., 1960. Dir., chief research officer Nat. Pineapple Research Inst., Malaysia, 1961-64; research assoc. Rice U., 1964-65; asst. prof. biology Tex. So. U., Houston, 1965-66; chmn. div. sci. and math., assoc. prof. biology Jarvis Christian Coll., 1966-67; prof. molecular genetics dept. botany Howard U., 1967—; cons. pineapple industries, Formosa, Philippines, Malaysia, various univs.; collaborator Pasteur Inst., Carnegie Instn.; lectr. Univs., U.S. and abroad. Contbr. articles to profl. jours. Vis. scientist Rockefeller U., 1968-69, Pasteur Inst., Paris, 1974-75, NIH, Bethesda, Md., 1974-75. Grantee NSF, Dept. Energy, NIH, Olin Found., EPA, Research Corp. N.Y., Anna Fuller Found., USNR. Mem. AAAS, Indian Sci. Congress, Genetics Soc. Am., AAUP, Am. Mycol. Soc., Am. Soc. Environ. Mutagen, Sigma Xi, Beta Kappa Chi. Subspecialties: Genetics and genetic engineering (biology); Genome organization. Current work: Gene isolation, gene expression, molecular genome organization. Home: 8841 Tuckerman Ln Potomac MD 20854 Office: Dept Botany Howard U Washington DC 20059

DUTTON, JAMES WILLIAM, nuclear utility executive; b. Detroit, Aug. 26, 1948; s. James Munroe and Ruth Ludeen (Emery) D.; m. Julie Ann Fisher, Dec. 14, 1968; children: James Thomas, Jonathan Andrew, Daniel William. B. Profl. Studies in Nuclear Tng. Mgmt, Memphis State U., 1979. Tng. instr. Ind. and Mich. Electric Co., Bridgeman, Mich., 1972-76; tng. supr. Memphis State U., 1976-79; tng. mgr. Wash. Pub. Power Supply System, Richland, 1979-81; asst. dir. nuclear tng. Detroit Edison Co., 1981—. Bd. chmn. United Meth. Ch., Benton City, Wash., 1981. Served with USN, 1966-71. Mem. Am. Nuclear Soc. (exec. com. tech. group for human factor systems), Nat. Rifle Assn., Nat. Muzzle Loading Rifle Assn., Phi Kappa Phi. Subspecialties: Human factors engineering; Nuclear engineering. Current work: Management systems for industrial training. Home: 462 Borgess Ave Monroe MI 48161 Office: Detroit Edison GTOC Fermi 2 6400 N Dixie Hwy Newport MI 48166

DUTTON, JOHN ALTNOW, meteorologist; b. Detroit, Sept. 11, 1936; s. Carl Evans and Velma (Altnow) D.; m. Frances Elizabeth Andrews, Jan. 13, 1962; children—Christopher Evan, John Andrews, Jan Frederik. B.S., U. Wis., 1958; M.S., 1959; Ph.D., 1962. Mem. faculty Pa. State U., University Park, 1965—, assoc. prof. meteorology, 1968-71, prof., 1971—, head dept. meteorology, 1981—; expert aero. system div. USAF, 1965-71; vis. scientist Riso Research Establishment, Roskilde, Denmark, 1971-72, summer 1975, 78-79; vis. prof. Tech. U., Denmark, 1978-79. Author: The Ceaseless Wind: An Introduction to the Theory of Atmospheric Motion, 1976; assoc. editor: Meteorol. Monographs, 1973-79; editor, 1979—; contbr. articles to profl. jours. Trustee Univ. Corp. for Atmospheric Research, 1974-81, sec., 1977, treas., 1978-79, vice-chmn., 1980—; Mem. bd. atmospheric scis. and climate Nat. Acad. Scis., 1982—; mem. space and earth scis. adv. com. NASA, 1982—. Served with USAF, 1962-65. Fellow Am. Meteorol. Soc.; mem. Math. Assn. Am., Soc. Indsl. and Applied Math., Sigma Xi, Phi Kappa Phi, Theta Delta Chi. Subspecialties: Meteorology; Applied mathematics. Current work: Atmospheric dynamics; spectral modeling of atmospheric and hydrodynamic flow, predictability; effects of atmospheric turbulence on structures. Home: 447 Nimitz Ave State College PA 16801 Office: 503 Walker Bldg University Park PA 16802

DUWEZ, POL EDGARD, educator; b. Mons, Belgium, Dec. 11, 1907; s. Arthur and Jeanne (Delcourt) D.; m. Nera Faisse, Sept. 4, 1935; 1 dau., Nadine. Metall.E., Sch. Mines, Mons, 1932; D.Sc., U. Brussels, 1933, Calif. Inst. Tech., 1935. Instr., prof. Sch. Mines, Mons, 1935-40; research engr. Calif. Inst. Tech., Pasadena, 941-45, chief materials sect. jet propulsion lab., 1945-54, asso. prof. materials sci., 1947-52, prof., 1952-78, prof. emeritus, 1978; Campbell Meml. lectr., 1967; mem. sci. adv. bd. to chief of staff USAF, 1945-55. Contbr. articles to profl. jours. Recipient Charles B. Dudley award ASTM, 1951; Francis J. Clamer medal Franklin Inst., 1968; Gov. Cornez prize, Belgium, 1973; Paul Lebeau medal French Soc. for High Temperature, 1974; Heyn medal Deutsche Gesellschaft für Metallkunde, 1981; W. Hume-Rothery award Metall. Soc. of AIME, 1981. Fellow AIME (C.H. Mathewson Gold medal 1964), Am. Soc. Metals (Albert Sauveur Achievement medal 1973); mem. Nat. Acad. Scis., Nat. Acad. Engring., Am. Ceramic Soc., Am. Acad. Arts and Scis., Am. Phys. Soc. (internat. prize for new materials 1980), AAAS, Brit. Inst. Metals, Société Francaise des Ingenieurs Civils, Sigma Xi. Subspecialties: Materials; Amorphous metals. Home: 1535 Oakdale St Pasadena CA 91106

DU WORS, ROBERT JEROME, software engineer; b. Lewisburg, Pa., June 17, 1952; s. Richard Edward and Luella Maude (Manter) DuWors; m. Leona Ion Locklin, July 17, 1982; 1 dau., Alexa Celeste. B.S. in Computer Sci. with distinction, U. Calgary, Alta., Can., 1980. Cert. computer programmer, cert. data processor. Systems programmer Technion-Israel Inst. Tech., 1971-73, Control Data France, 1973-74; systems designer Calgary Bd. Edn., 1978-79; cons. ALTEL, 1975-77; instr. U. Calgary, 1979-81; software engr. Can. Systems Group, 1981-82; sr. tech. computer analyst Petro-Can., Calgary, 1982-83; sofeware engr. radar group Intera, Environ. Cons. Ltd., Calgary, 1983—. Mem. Assn. Computing Machinery, IEEE Computer Soc., Can. Info. Processing Soc., AICCP. Club: Cu-Nim Gliding (Calgary). Subspecialties: Operating systems; Information systems, storage, and retrieval (computer science). Current work: System design, application languages, graphics, configuration analysis, image processing, synthetic operture radar, systems analysis. Home: RR4 Calgary AB Canada T2M 4L4 Office: Intera 1200-510 5th St SW Calgary AB Canada T2P 3S2

DVORCHIK, BARRY HOWARD, pharmacologist; b. Bridgeport, Conn., Feb. 29, 1944; s. Bernard and Esther (Zeidel) D.; m. Susan Louise Garcy, Aug. 7, 1966; children: Keith, Lawrence, Beth. B.S. in Pharmacy, U. Conn., 1966; PhD. in Pharmacology, U. Fla., 1972. Instr. Milton S. Hershey Med. Ctr., Pa. State U., 1972-74, asst. prof., 1974-79, assoc. prof. depts. ob/gyn and pharmacology, 1979-83; sect. head drug disposition dept. drug metabolism McNeil Pharm., Spring House, Pa., 1983—; cons. FDA, Heather Drug Co., Cert. Labs., Inc., Pa. Dept. Health. Contbr. articles to profl. jours. Fellow Am. Coll. Clin. Pharmacology; mem. AAAS, Am. Soc. Clin. Pharmacology and Therapeutics, Am. Soc. Pharmacology and Exptl. Therapeutics, N.Y. Acad. Scis., Soc. Toxicology. Subspecialties: Pharmacology; Pharmacokinetics. Current work: Effect of pregnancy on drug metabolism and pharmacokinetics; fetal/neontal development of drug metabolism; effect of pregnency on maternal pharmacokinetics. Office: McNeil Dept Drug Metabolism Spring House PA 19477

DWINELL, LEW DAVID, plant pathologist; b. Albuquerque, Mar. 24, 1938; m. Patricia Leech, July 1, 1961; children: Kathryn, Erica, Jeanine. B.S., Colo. State U., 1961; M.S., U. Denver, 1963; Ph.D., Cornell U., 1967. Research plant pathologist Southeastern Forest Expt. Sta., U.S. Dept. Agr. Forest Service, Athens, Ga., 1966—. Contbr. articles in field to profl. jours. Mem. Am. Phytopath. Soc., Ga. Assn. Plant Pathologists, Soc. Am. Foresters. Roman Catholic. Club: Athens Photography. Subspecialties: Information systems, storage, and retrieval (computer science); Resource conservation. Current work: Basic and applied research on pitch canker and fusiform rust diseases of southern pines in plantations and seed orchards. Home: 125 Curtis Dr Athens GA 30605 Office: US Dept Agriculture Forest Service Carlton St Athens GA 30602

DWIVEDI, CHANDRADHAR, biochemist; b. Jaunpur, U.P., India, July 1, 1948; s. Abhaya N. and Maharaji (Dubey) D.; m. Prabha Dwivedi, June 21, 1966; children: Sudhanshu, Neeraja, Himanshu. B.Sc., Gorakhpur U., 1964, M.Sc., 1966; Ph.D., Lucknow U., 1972. Clin. chemist Tenn. Dept. Health.; Research fellow K.G. Med. Coll., Lucknow, India, 1969-73; research assoc. Vanderbilt U., Nashville, 1973-76; asst. prof. dir. biochem. lab. Meharry Med. Coll., Nashville, 1976-82, assoc. prof., dir. biochem. lab., 1982—. Contbr. articles to profl. jours. Lady Tata Meml. scholar, 1968-69; Indian Council Med. Research scholar, 1969-73; NSF grantee, 1977-80; NIH grantee, 1981—; Bur. Maternal and Child Health Service Adminstrn. grantee, 1976-82. Mem. Soc. Explt. Biology and Medicine, Soc. Environ. Toxicology and Chemistry, Am. Assn. Clin. Chemistry, Soc. Neurosci., Am. Soc. Human Genetics, Genetics Soc. Am., Internat. Soc. Devel. Neurosci., Am. Physiol. Soc., Internat. Soc. Biochemical Pharmacology, Internat. Union Pharmacology. Hindu. Subspecialties: Cell study oncology; Neurochemistry. Current work: in developmental neurochemistry, environmental biochemical toxicology and oncology.

DYBOWSKI, CECIL RAY, chemistry educator; b. Yorktown, Tex., Sept. 23, 1946; s. Hermin Romana and Ruth Joyce (Geffert) D.; m. Mary Agnes Kaiser, May 12, 1979. B.S. in Chemistry, U. Tex., 1969, Ph.D., 1973. Research fellow Calif. Inst. Tech., Pasadena, 1973-76; asst. prof. chemistry U. Del., Newark, 1976-82, assoc. prof., 1982—; vis. scientist E.I. Dupont de Nemours, Wilmington, Del., 1982-83. Mem. Am. Chem. Soc., Am. Phys. Soc., Am. Inst. Chem. Engrs. (assoc.), Materials Research Soc., Soc. Applied Spectroscopy (chmn. sect. 1982-83). Subspecialties: Physical chemistry; Nuclear magnetic resonance (chemistry). Current work: Polymer NMR; surface spectroscopy; inelastic electron tunneling spectroscopy. Home: 19 Linden St Newark DE 19711 Office: Dept Chemistry U Del Newark DE 19711

DYE, JAMES LOUIS, physical chemistry educator; b. Soudan, Minn., July 18, 1927. A.B., Gustavus Adolphus Coll., 1949; Ph.D. in Chemistry, Iowa State U., 1953. Asst. phys. chemist Inst. Atomic Research and dept. chemistry, Iowa State U., Ames, 1949-53; prof. phys. chemistry Mich. State U., East Lansing, 1953—; NSF sci. faculty fellow Max Planck Inst., Gottingen, W. Ger., 1961-62; vis. scientist Ohio State U., Columbus, 1968-69. Fulbright scholar, 1975-76; Guggenheim fellow, 1975-76. Mem. AAAS, Am. Chem. Soc. Subspecialty: Physical chemistry. Office: Dept Chemistry Mich State U East Lansing MI 48824

DYKHUIZEN, DANIEL EDWARD, research scientist, geneticist, educator; b. Muskegon, Mich., Oct. 31, 1942; s. Harold D. and Lucille J. (Walvoord) D.; m. Phyllis Rosalyn Clark, Sept. 7, 1968; children: Marta Lu, Edward Daniel, Emily Carla. B.S., Stanford U., 1965; Ph.D., U. Chgo., 1970. Research fellow Stanford (Calif.) U., 1970-72, Australian Nat. U., Canberra, 1972-76; asst., then assoc. research scientist Purdue U., West Lafayette, Ind., 1976-81; assoc. research prof. genetics Washington U. Med. Sch., St. Louis, 1982—. Contbr. articles to profl. jours. Bd. dirs. Lafayette Urban Ministry, 1977-81, pres., 1978-79. Mem. Genetics Soc. Am., Am. Soc. Microbiology, Soc. Study Evolution. Presbyterian. Subspecialties: Population biology; Evolutionary biology. Current work: An experimental study of natural selection using bacteria and chemostats. Home: 543 W Glendale Rd Webster Groves MO 63119 Office: Washington U Med Sch Dept Genetics Box 8031 Saint Louis MO 63110

DYKSTRA, CLIFFORD ELLIOT, educator, research chemist; b. Chgo., Oct. 30, 1952; s. Raymond and Vivian (Mishkutz) D. B.S. in Chemistry and Physics, U. Ill., 1973; Ph.D., U. Calif.-Berkeley, 1976. Research assoc. U. Calif., 1976-77; asst. prof. chemistry U. Ill., 1977-83, assoc. prof. chemistry, 1983—; cons. chemistry div. Argonne Nat. Lab., Ill., 1978-80. Contbr. articles to profl. jours. Alfred P. Sloan fellow, 1979-81. Mem. Am. Chem. Soc., Am. Phys. Soc. Subspecialties: Physical chemistry; Theoretical chemistry. Current work: Electron correlation in molecules, hydrogen bonding, molecular structure and stability, interstellar and prebiotic chemistry. Office: Dept Chemistry U Ill 505 S Matthews Urbana IL 61801

DYM, CLIVE LIONEL, engineering educator; b. Leeds, Eng., July 15, 1942; came to U.S., 1949, naturalized, 1954; s. Isaac and Anna (Hochmann) D.; children: Jordana, Miriam. B.C.E., Cooper Union, 1962; M.S., Poly. Inst. Bklyn., 1964; Ph.D., Stanford U., 1967. Asst. prof. SUNY, Buffalo, 1966-69; asso. professorial lectr. George Washington U., Washington, 1969; research staff Inst. Def. Analyses, Arlington, Va., 1969-70; asso. prof. Carnegie-Mellon U., Pitts., 1970-74; vis. asso. prof. TECHNION, Israel, 1971; sr. scientist Bolt Beranek and Newman, Inc., Cambridge, Mass., 1974-77; prof., head civil engring. dept. U. Mass., Amherst, 1977—; vis. sr. research fellow Inst. Sound and Vibration Research, U. Southampton, Eng., 1973; vis. scientist Xerox PARC, 1983; vis. prof. civil engring. Stanford U., 1983, MIT, 1984; cons. Bell Aerospace Co., 1967-69, Dravo Corp., 1970-71, Salem Corp., 1972, Gen. Analytics Inc., 1972, ORI, Inc., 1979, BBN Inc., 1979. Editorial bds.: Jour. of Sound and Vibration; author: (with I.H. Shames) Solid Mechanics: A Variational Approach, 1973, Introduction to the Theory of Shells, 1974, Stability Theory and Its Applications to Structural Mechanics, 1974, (with A. Kalnins) Vibration: Beams, Plates, and Shells, 1977, (With E.S. Ivey) Principles of Mathematical Modeling, 1980, (with I.H. Shames) Energy and Finite Element Methods in Structural Mechanics, 1983; contbr. (with I.H. Shanes) articles and tech. reports to profl. publs. NATO sr. fellow in sci., 1973. Fellow Acoustical Soc. Am., ASME, ASCE (Walter L. Huber research prize 1980); mem. AAAS, Inst. Noise Control Engring. Jewish. Subspecialties: Theoretical and applied mechanics; Civil engineering. Current work: Structural dynamics and stability; vibration and acoustics; expert systems for engineering design. Office: Civil Engineering Dept U Mass Amherst MA 01003

DYMINSKI, JOHN W(LADYSLAW), immunologist, researcher; b. Lindau, Germany, Feb. 13, 1945; came to U.S., 1949, naturalized, 1955; s. Jan and Wanda (Michionek) D. A.B., U. Rochester, 1967; M.S., Syracuse U., 1970, Ph.D., 1972. Postdoctoral fellow U. Fla. Coll. Medicine, 1972-76; asst. prof. Children's Hosp., Cin., 1976-80; sect. leader Bethesda Research Labs., Gaithersburg, Md., 1980-82; research mgr. Mast Immunosystems, Mountain View, Calif., 1982-83, Paragon Diagnostics, Sunnyvale, Calif., 1983—. Contbr. articles to profl. jours. Nat. Inst. Arthritis and Infectious Diseases fellow, 1975-76. Mem. Am. Assn. Immunologists, Transplantation Soc. Democrat. Roman Catholic. Subspecialties: Immunology (medicine); Immunobiology and immunology. Current work: Development of in vitro diagnostic systems in fields of immunobiology, oncology, infectious diseases and allergy. Office: Mast Immunosystems 630 Clyde Ct Mountain View CA 94043

DYMOND, PATRICK WILLIAM, computer scientist, educator; m. Barbara Ann Dymond, Sept. 24, 1969; children: Michael, Paul. Ph.D., U. Toronto, 1980. Asst. prof. computer sci. U. Calif.-San Diego, 1980—; dir. Dymond Inst., 1982—. Mem. Assn. for Computing Machinery, IEEE, Computer Soc. Subspecialties: Foundations of computer science; Theoretical computer science. Office: U Calif San Diego Dept EECS C-014 La Jolla CA 92037

DYSON, FREEMAN JOHN, physicist; b. Crowthorne, Eng., Dec. 15, 1923; s. George and Mildred Lucy (Atkey) D.; m. Verena Haefeli-Huber, Aug. 11, 1950 (div. 1958); children—Esther, George; m. Imme Jung, Nov. 21, 1958; children—Dorothy, Emily, Miriam, Rebecca. B.A., Cambridge U., 1945. Operations research R.A.F. Bomber Command, 1943-45; fellow Trinity Coll., Cambridge U., Eng., 1946-49; Commonwealth fellow Cornell U., Princeton, 1947-49; prof. physics Cornell U., 1951-53; prof. Inst. Advanced Study, Princeton, 1953—. Author: Disturbing the Universe, 1979. Fellow Royal Soc. London; mem. Am. Phys. Soc., Nat. Acad. Scis. Subspecialty: Theoretical physics. Current work: Quantum field theory, statistical mechanics. Home: 105 Battle Rd Circle Princeton NJ 08540

DZIEWONSKI, ADAM MARIAN, geology educator; b. Lwow, Poland, Nov. 15, 1936. M.S., U. Warsaw, 1960; Dr. Tech. Sci. in Applied Geology, Acad. Mining and Metallurgy-Krakow, 1965. Research asst. seismology Inst. Geophys., Polish Acad. Sci., 1961-65, research assoc., 1965, S.W. Center Advanced Studies, 1965-69; asst. prof. geophysics U. Tex.-Dallas, 1969-71; assoc. prof. geophysics, assoc. Ctr. Earth and Planetary Physics, Harvard U., Cambridge, Mass., 1972-76; prof. geology, mem. Center Earth and Planetary Physics, Harvard U., 1976—; chmn. dept. geol. sci. Harvard U., 1982—; Disting. Fairchild Scholar Calif. Inst. Tech., 1983-84; mem. Polish Sci. del., N. Vietnam, 1958-59. Mem. Seismology Soc. Am., Am. Geophys. Union, Soc. Explorative Geophysicists, AAAS. Subspecialty: Geophysics. Office: Dept Geol Scis Harvard U Cambridge MA 02138

DZIUK, HAROLD EDMUND, educator; b. Foley, Minn., Apr. 27, 1930; s. Edmund William and Ellen Mary (Carlin) D.; m. Elizabeth Jane Pagels, Dec. 27, 1952; children—Barbara Jane, Michael Peter, Robert Joseph, David Alan, Stephanie Joan. Student, St. John's U., 1947-48; B.S., U. Minn., 1951, D.V.M., 1954, M.S., 1955, Ph.D., 1960. Instr. U. Minn., St. Paul, 1951-54, 57-59, asst. prof., 1961-64, asso. prof., 1964-69, prof. vet. physiology, 1969—, head dept. vet. physiology and pharmacology, 1972-73, chmn. dept. vet. biology, 1973—. Served with Vet. Corps U.S. Army, 1955-56. Mem. AVMA, Minn. Vet. Med. Assn., Am. Soc. Vet. Physiologists and Pharmacologists, Conf. Research Workers in Animal Diseases, Midwest Assn. Avian Veterinarians, Farm House Frat., Sigma Xi, Gamma Sigma Delta, Alpha Zeta, Phi Zeta. Republican. Home: 2826 Marion St Saint Paul MN 55113 Office: Coll Vet Medicine U Minn Saint Paul MN 55108

EAGLESON, PETER STURGES, educator, hydrologist; b. Phila., Feb. 27, 1928; s. William Boal and Helen (Sturges) E.; m. Marguerite Anne Partridge, May 28, 1949 (div.); children: Helen Marie, Peter Sturges, Jeffrey Partridge; m. Beverly Grossmann Rich, Dec. 27, 1974. B.S. in Civil Engring, Lehigh U., 1949, M.S., 1952; Sc.D., MIT, 1956. Jr. engr. George B. Mebus (cons. engr.), Glenside, Pa., 1950-51;

teaching asst. Lehigh U., 1951-52; research asst. Mass. Inst. Tech., 1952-54; mem. faculty MIT, 1954—, prof. civil engring., 1965—, head dept. civil engring., 1970-75; vis. asso. Calif. Inst. Tech., 1975-76; Fulbright sr. research scholar Commonwealth Sci. and Indsl. Research Orgn., Canberra, Australia, 1966-67. Author: (with others) Estuary and Coastline Hydrodynamics, 1966, Dynamic Hydrology, 1970. Served to 2d lt. C.E. AUS, 1949-50. Recipient Desmond Fitzgerald medal, 1959, Clemens Herschel prize, 1965; both Boston Soc. Civil Engrs.; research prize Am. Soc. C.E., 1963. Fellow Am. Geophys. Union (Robert E. Horton award 1979); mem. Nat. Acad. Engring. Subspecialty: Surface water hydrology. Current work: Interrelationships of climate, soil and vegetation. Office: Dept Civil Engring Room 48-335 Mass Inst Tech Cambridge MA 02139

EAGLETON, LEE CHANDLER, chemical engineer, educator; b. Vallejo, Calif., July 27, 1923; s. William L. and Mary Louise (Chandler) E.; m. Mary E. Stewart, Feb. 21, 1953; children: James C., William L., Elizabeth L. S.B., MIT, 1946, S.M., 1947; D.Eng., Yale U., 1950. Research asso. Columbia U., 1950-51; devel. engr. Rohm & Haas Co., Phila., 1951-56; lectr. Drexel Inst. Tech., 1954, U. Pa., Phila., 1954-55, asso. prof., 1956-65, prof., 1966-69; prof. head dept. chem. engring. Pa. State U., 1970—; cons. Rohm & Haas, 1956-74, Inst. for Def. Analyses, 1961-63, Martin Marietta Co., 1970-72, 74. Served with AUS, 1942-46. Fellow Am. Inst. Chem. Engrs. (dir. 1980-82); mem. Am. Soc. Engring. Edn. (chmn. chem. engring. div. 1970-71), Am. Chem. Soc., AAUP. Subspecialty: Chemical engineering. Home: 445 Cricklewood Dr State College PA 16801 Office: 160 Fenske Lab Pa State U University Park PA 16802

EAGON, ROBERT GARFIELD, microbiology educator; b. Salesville, Ohio, Oct. 29, 1927; m. Margretta Isabel Buchanan, Aug. 30, 1952; 1 dau., Victoria Margretta. B.Sc., Ohio State U., 1951, M.Sc., 1952, Ph.D., 1954. Fulbright scholar Pasteur Inst., Paris, 1954-55; asst. prof. microbiology U. Ga., Athens, 1955-59, assoc. prof., 1959-66, prof., 1966—. Contbr. chpts. to books, articles to profl. jours. Served to col. U.S. Army. Recipient M.G. Michael award U. Ga., 1962. Fellow Am. Acad. Microbiology; mem. Am. Soc. Microbiology (P.R. Edwards award S.E. br. 1976), Soc. Indsl. Microbiology, AAAS. Subspecialties: Microbiology; Biochemistry (biology). Current work: Structure and function of the bacterial cell envelope; mechanisms and bioenergetics of solute transport in the bacterial cell; mechanisms of bacterial antibiotic susceptibility and resistance. Office: Dept Microbiology U Ga Athens GA 30602

EAKER, CHARLES WILLIAM, chemistry educator, research; b. St. Louis, May 25, 1949; s. Charles Mayfield and Mildred Catherine (Staples) E.; m. Mary Alice Eisenmann, July 6, 1974; 1 dau., Stephanie Eisenmann. B.S., Mich. State U., 1971; Ph.D., U. Chgo, 1974. Postdoctoral fellow U. Tex. - Dallas, Richardson, 1974-76; instr. U. Dallas, Irving, 1976-78, asst. prof., 1978-81, assoc. prof. chemistry, 1981—. Arthur Vining Davis Found. devel. grantee, 1980; Robert A. Welch Found. research grantee, 1983-85. Mem. Am. Chem. Soc. (research grantee 1978-80). Subspecialties: Theoretical chemistry; Graphics, image processing, and pattern recognition. Current work: Determination of potential energy surfaces, reaction dynamics calculations, computer graphics. Office: U Dallas U Dallas Sta Irving TX 75061

EAMES, DAVID ROBSON, plasma physicist; b. Detroit, July 15, 1952; s. Robson MacDonald and Jane Bell (Brunton) E. B.S. with honors in Math., U. Mich., 1974, B.S.E. in Physics, 1974, 1974; Ph.D. in Plasma Physics, Princeton U., 1981. Research asst. Princeton Plasma Physics Lab., 1975-80; sr. scientist GA Technologies, San Diego, Calif., 1981—. Recipient NSF grad. fellowship award, 1975-78. Mem. Am. Phys. Soc. Clubs: San Diego Orienteering, San Diego Table Tennis. Subspecialty: Plasma physics. Current work: Plasma physics of tokamaks; laser-induced fluorescence diagnostics on the Doublet III Tokamak. Home: 5710 Ferber St San Diego CA 92122 Office: GA Technologies PO Box 85608 San Diego CA 92138

EARHART, CHARLES FRANKLIN, JR., microbiologist, educator; b. Melrose Park, Ill., Oct. 26, 1941; s. Charles Franklin and Katherine Anne (Laho) E. B.A., Knox Coll., 1962; Ph.D., Purdue U., 1967. Assoc. prof. dept. microbiology U. Tex., Austin, 1977-83, prof. microbiology, 1983—. Editorial bd.: Jour. Bacteriology. Mem. Am. Soc. Microbiology, Genetics Soc. Am., Sigma Xi, Phi Lambda Upsilon. Subspecialties: Microbiology; Membrane biology. Current work: Microbial iron assimilation Escherichia coli, siderophores, outer membrane.

EARL, BOYD LOREL, chemist, educator, researcher; b. Burley, Idaho, Aug. 17, 1944; s. Boyd W. and Alismae (Melton) E.; m. Judy Mathewson Nye, June 1, 1980. B.S. in Chemistry, U. Idaho, 1966; M.S. in Phys. Chemistry (NSF fellow), U. Calif., Berkeley, 1969, Ph.D., 1973. Postdoctoral fellow CUNY, 1973-76; asst. prof. Bklyn. Coll., CUNY, 1973-74, adj. asst. prof., 1974-75; asst. prof. chemistry U. Nev., 1976-81, assoc. prof., 1981—. Contbr. articles to profl. jours. Mem. Am. Chem. Soc., AAAS, Sigma Xi. Subspecialties: Laser photochemistry; Laser-induced chemistry. Current work: Carbon dioxide laser induced chemical reactions. Office: Dept Chemistry U Nev Las Vegas NV 89154

EARLE, SYLVIA ALICE, oceanographer; b. Gibbstown, N.J., Aug. 30, 1935; d. Lewis Reade and Alice Freas (Richie) E. B.S., Fla. State U., 1955; M.A., Duke U., 1956, Ph.D., 1966. Resident dir. Cape Haze Marine Lab., Sarasota, Fla., 1966-67; research scholar Radcliffe Inst., 1967-69; research fellow Farlow Herbarium Harvard U., 1967-75, research assoc., 1975—; research assoc. botany Natural History Mus. Los Angeles County, 1970-75; research biologist, curator Calif. Acad. Scis., San Francisco, 1976—; research assoc. U. Calif., Berkeley, 1969—; v.p., sec.-treas. Deep Ocean Tech., Inc., Oakland, Calif. Author: Exploring the Deep Frontier, 1980; editor: Scientific Results of the Textite II Project, 1972-75; contbr. articles to profl. jours. V.p., sec.-treas. Deep Ocean Engring., Oakland, 19782—; trustee World Wildlife Fund U.S., World Wildlife Fund Internat., Charles A. Lindbergh Fund, Ocean Trust Found.; council mem. Internat. Union Conservation Nature; corp. mem. Woods Hole Oceanographic Inst.; mem. Nat. Adv. Com. Oceans and Atmosphere. Recipient Conservation Service award U.S. Dept. Interior, 1970, Boston Sea Rovers award, 1972, 79, Nogi award Underwater Soc. Am., 1976, Conservation service award Calif. Acad. Sci., 1979, Lowell Thomas award Explorer's Club, 1980, Order of Golden Ark Prince Netherlands, 1980; named Woman of Year Los Angeles Times, 1970, Scientist of Year Calif. Mus. Sci. and Industry, 1981. Mem. Internat. Phycological Soc. (sec. 1974-80), Phycological Soc. Am., Am. Soc. Ichthyologists and Herpetologists, Am. Inst. Biol. Scis., AAAS, Brit. Phycological Soc., Marine Tech. Soc., Ecol. Soc. Am., Internat. Soc. Plant Taxonomists. Subspecialties: Ecology; Deep-sea biology. Current work: Ecology and distribution of marine plants; deepwater ecology; marine mammal behavior; development and use of technology for ocean exploration andwork; conservation of natural resources, management of Rand D and operations companies for ocean technology. Office: Calif Acad Scis Golden Gate Park San Francisco CA 94118

EARLEY, JOSEPH EMMET, chemist, educator; b. Providence, Apr. 6, 1932; s. Daniel McGlynn and Margaret T. (Doran) E.; m. Shirley Ann Titus, June 23, 1956; children: Thomas D., David G., Joseph Emmett. B.S., Providence Coll., 1954; Ph.D., Brown U., 1957. Faculty U. Chgo., 1958; faculty Georgetown U., Washington, 1958—, prof. chemistry, 1969—; vis. faculty Calif. Inst. Tech., 1968-69, Free U. Brussels, 1976; coordinator chemistry research evaluation U.S. Air Force Office Sci. Research. Author articles, chpts. in books, revs. Served to 1st lt. U.S. Army, 1958. Recipient Potter prize Brown U., 1958. Mem. Am. Chem. Soc., AAAS. Democrat. Roman Catholic. Lodge: KC. Subspecialties: Inorganic chemistry; Kinetics. Current work: Solution-phase redox reactions involving ruthenium and titanium compounds. Home: 2348 Greenwich St Falls Church VA 22046 Office: Reiss Sci Ctr Georgetown U Washington DC 20057

EARLY, JAMES MICHAEL, semiconductor company executive; b. Syracuse, N.Y., July 25, 1922; s. Frank J. and Rhoda Gray E.; m. Mary Agnes Valentine, Dec. 28, 1948; children—Mary, Kathleen, Joan Early Farrell, Rhoda Early Alexander, Maureen Early Mathews, James, Margaret Mary. B.S., N.Y. Coll. Forestry, Syracuse, N.Y., 1943; M.S., Ohio State U., 1948, Ph.D., 1951. Instr., research asso. Ohio State U., Columbus, 1946-51; dir. lab. Bell Telephone Labs., Murray Hill, N.J., 1951-64, Allentown, Pa., 1964-69; research and devel. dir. Fairchild Camera and Instrument Corp., Palo Alto, Calif., 1969—. Served with U.S. Army, 1943-45. Fellow IEEE (recipient J.J. Ebers award IEEE Electron Device Soc. 1979); mem. AAAS, Am. Phys. Soc., Electrochem. Soc., Internat. Platform Assn. Roman Catholic. Club: Palo Alto (Calif.) Yacht. Subspecialties: Microchip technology (engineering); Microelectronics. Current work: Direct development of VLSI (very large scale integration)technology. Home: 740 Center Dr Palo Alto CA 94301 Office: 4001 Miranda Ave Palo Alto CA 94304

EAST, JAMES LINDSAY, virologist, educator; b. Senatobia, Miss., Nov. 5, 1936; s. Lindsay Smith and Juanita Brents (Pipkin) E.; m. Judy Allen, May 2, 1963; 1 dau., Allison Anne. B.S., Memphis State U., 1963, M.S., 1967; Ph.D., U. Tenn., 1970. Research technologist, trainee St. Jude Children's Research Hosp., Memphis, 1963-70; asst. prof. virology M.D. Anderson Hosp. and Tumor Inst., Houston, 1970-1979—. Contbr. articles to profl. jours. USPHS fellow, 1967-70. Mem. Am. Soc. Microbiology, Internat. Assn. Comparative Research on Leukemia and Related Diseases. Baptist. Subspecialties: Virology (biology); Cancer research (medicine). Current work: Molecular viral oncology, nucleic acid sequencing. Home: 4915 Droddy St Houston TX 77091 Office: M D Anderson Hosp and Tumor Inst 6723 Bertner Ave Houston TX 77030

EASTER, ROBERT ARNOLD, nutritionist, educator; b. San Antonio, Oct. 10, 1947; s. Edward and Hazel (Boyle) E.; m. Cheryl Kay Williams, Dec. 28, 1972; children: Brian, Johanna, Aaron. A.A., Southwest Tex. Jr., Coll., 1968; B.S., Tex. A&M U., 1970, M.S., 1972; Ph.D., U. Ill., 1976. Grad. fellow Tex. A&M U., 1970-72; grad. research asst. U. Ill., 1972-75; asst. prof., 1976-80, assoc. prof., 1980—; cons. in field. Contbr. articles to profl. jours., chpts. to books. Served to capt. USAR, 1972—. Mem. Am. Soc. Animal Sci., Am. Registry Cert. Animal Scientists. (cert.). So. Baptist. Subspecialties: Nutrition (biology); Animal nutrition. Current work: Nutrition of the reproducing female pig. Office: 318 Mumford Hall 1301 W Gregory Urbana IL 61801

EASTER, STEPHEN SHERMAN, JR., biology educator; b. New Orleans, Feb. 12, 1938; s. Stephen Sherman and Myrtle Olivia (Bekkedahl) E.; m. Janine Eliane Piot, June 4, 1963; children: Michele, Kim. B.S., Yale U., 1960; postgrad., Harvard U., 1961; Ph.D., Johns Hopkins U., 1966. Postdoctoral fellow Cambridge (Eng.) U., 1967, U. Calif., Berkeley, 1968-69; asst. prof. biology U. Mich., Ann Arbor, 1970-74, assoc. prof., 1974-78, prof., 1978—. Editor: Jour. Vision Research, 1978—. Mem. Soc. Neuroscience, Assn. Research in Vision and Ophthalmology. Subspecialties: Neurobiology; Regeneration. Home: 2204 Brockman Blvd Ann Arbor MI 48104 Office: Division Biological Sciences University of Michigan 2109 Natural Science Bldg Ann Arbor MI 48109

EASTERDAY, BERNARD CARLYLE, university administrator, veterinarian; b. Hillsdale, Mich., Sept. 16, 1929. D.V.M., Mich. State U., 1952; M.S., U. Wis., 1958, Ph.D., 1961. Gen. practice vet. medicine, Mich., 1952; veterinarian Ft. Detrick, Md., 1955-56, 58-61; research asst. vet. sci. U. Wis.-Madison, 1956-58, assoc. prof., 1961-66, prof., 1966—, dept. chmn., 1968-74, dean sch. vet. medicine, 1979—; mem. expert panel zoonoses WHO; mem. Com. Animal Health, NRC-Nat. Acad. Scis. Mem. Am. Coll. Vet. Microbiology, Assn. Am. Vet. Med. Colls. (pres. 1975-76), AVMA, Conf. Research Workers Animal Disease. Subspecialty: Virology (veterinary medicine). Office: Sch Vet Medicine U Wis Madison WI 53706

EASTMAN, ALAN RICHARD, biochemist, researcher; b. Ewell, Surrey, Eng., Oct. 16, 1949; came to U.S., 1976, naturalized, 1979; s. Leonard Ernest and Edith Dorcas (Beakhust) E. B.Tech., Brunel U., London, 1972; Ph.D., London U., 1975. Research asst. Chester Beatty Research Inst., London, 1975-76; research asso. dept. biochemistry U. Vt., 1976-78, research asst. prof., 1978—. Contbr. articles profl. jours. Recipient Jr. Faculty Research award Am. Cancer Soc., 1980-83, ROI Nat. Cancer Inst., 1981-84. Mem. Am. Assn. Cancer Research. Subspecialties: Molecular biology; Cancer research (medicine). Current work: Mechanism of action of chem. carcinogens and cancer chemotherapeutic agts., specifically reactions with DNA, mechanisms of resistance of cells to cancer chemotherapeutic agts. Office: Dept Biochemistry U Vt College Medicine Burlington VT 05405

EASTMAN, DEAN ERIC, physicist; b. Oxford, Wis., Jan. 21, 1940; s. Eric and Mildred (Benson) E.; m. Ella Mae Staley, Aug. 18, 1979. B.S.E.E., M.I.T., 1962, M.S.E.E., 1963, Ph.D.E.E., 1965. Research staff mem. IBM, Yorktown Heights, N.Y., 1963-74, IBM fellow, 1974—, mgr. surface physics and photoemission, 1972—. Contbr. numerous articles on solid state physics to profl. jours. Fellow Am. Phys. Soc. (Oliver C. Buckley prize 1980, councilor at large), Nat. Acad. Scis. Subspecialty: Solid state physics. Home: 806 Pines Bridge Rd Ossining NY 10562 Office: IBM T J Watson Research Center Yorktown Heights NY 10598

EASTMAN, MICHAEL PAUL, chemistry educator, researcher; b. Lancaster, Wis., Apr. 14, 1941; s. LeRoy Irons and Virginia Marie (Anderson) E.; m. Frances Rose Barto, Oct. 24, 1963; children: Michael E., Nathanial L.; m. Carol Oden Eastman, Aug. 23, 1980. B.A., Carleton Coll., 1963; Ph.D., Cornell U., 1968. Fellow Los Alamos Nat. Lab., 1968-70; asst. prof. U. Tex., 1970-74, assoc. prof., 1974-80, prof. chemistry, 1980—, asst. dean sci., 1981—; vis. staff mem. U. Calif. Los Alamos Nat. Lab. Contbr. articles to profl. jours. Mem. Am. Chem. Soc. Subspecialties: Physical chemistry; Geochemistry. Current work: Magnetic resonance and low temperature geochemistry. Home: 1308 Madeline St El Paso TX 79902

EASTON, DEXTER MORGAN, biologist, educator; b. Rockport, Mass., Sept. 13, 1921; s. Oscar Wallerius and Catherine Sophia (Bragner) E.; m. Jean Renfrew Mattoon, Dec. 18, 1953; children: Matthew, Andrew, Karl, Sylvia. B.A., Clark U., 1943; M.A., Harvard U., 1944, Ph.D., 1947. From instr. to asst. prof. zoology U. Wash., Seattle, 1947-49, from research asst. prof. to asst. prof. physiology, 1951-55; from asst. prof. to prof. biol. sci. Fla. State U., Tallahassee, 1955—. Author: Mechanisms of Body Functions, 2d edit, 1973; author articles. Fulbright research scholar, Dunedin, N.Z., 1949-50; NIH grantee, NSF grantee, 1954—. Mem. AAAS, Soc. Zoologists, Physiol. Soc., Biophys. Soc., Sierra Club, Sigma Xi. Unitarian-Universalist. Subspecialty: Neurophysiology. Current work: Theory of excitable membrane (nerve) specific ion currents. Home: 2908 Lasswade Dr Tallahassee FL 32312 Office: Dept Biol Sci Fla State U Tallahassee FL 32306

EATON, BARBARA RUTH, biochemist; b. Somerville, Mass., Aug. 14, 1928; d. Laurence Clifford and Josephine H. (Connell) E. A.B., Emmanuel Coll., Boston, 1950; M.S., Cath. U. Am., Washington, 1963; Ph.D., U. Calif., San Diego, 1975. Tchr. high sch. sci., sci. supr., Boston, 1953-66; instr. dept. chemistry Emmanuel Coll., Boston, 1966-76, asst. prof., 1976-80, assoc. prof. biochemistry, 1980—. Mem. Sisters of Notre Dame de Namur, Boston province, 1950—. USPHS intern, 1970-75; NSF fellow, 1960-63. Mem. Am. Chem. Soc., New Eng. Assn. Chemistry Tchrs., Hastings Center, Sigma Xi. Democrat. Roman Catholic. Subspecialties: Biochemistry (medicine); Membrane biology. Current work: Characterization and kinetics of phospholipases. Office: Chemistry Dept Emmanuel Coll Boston MA 02115

EATON, DAVID J., environmental sciences educator; b. Detroit, Dec. 18, 1949; s. Joseph W. and Helen F. E. A.B., Oberlin Coll., 1971; M.P.W., U. Pitts., 1972, M.Sc., 1972; Ph.D., Johns Hopkins U., 1977. Asst. prof. environ. systems LBJ Sch. Pub. Affairs, U. Tex., Austin, 1976-80, assoc. prof., 1980—. Author: A Systems Analysis of Grain Reserves, 1980; contbr. articles to profl. jours. Fulbright fellow, 1981-82; Lady Davis fellow, 1982. Subspecialties: Water supply and wastewater treatment; Operations research (engineering). Current work: Applications of systems analysis to natural resources. Office: LBJ Sch Public Affairs U Tex Austin TX 78712

EATON, GARETH RICHARD, chemist, educator; b. Lockport, N.Y., Nov. 3, 1940; s. Mark Dutcher and Ruth Emma (Ruston) E.; m. Sandra Y. Shaw, Mar. 29, 1969. B.A., Harvard U., 1962; Ph.D., M.I.T., 1972. Asst. prof. chemistry U. Denver, 1972-76, asso. prof., 1976-80, prof., 1980—. Contbr. numerous articles on chemistry to profl. jours. Served to lt. USN, 1962-67. Mem. Am. Chem. Soc., AAAS, Chem. Soc. (London), Internat. Soc. for Magnetic Resonance, Soc. Applied Spectroscopy. Subspecialties: Inorganic chemistry; Physical chemistry. Current work: Synthetic chemistry and magnetic resonance spectroscopy. Office: Dept Chemistry U Denver Denver CO 80208

EATON, JOEL A., astronomer; b. Paducah, Ky., Jan. 2, 1948; s. C. Warren and Molly A. E. B.S. in Physics with highest honors, Auburn U., 1970; M.S., Vanderbilt U., 1971; Ph.D., U. Wis., 1975. Vis. asst. prof. U. Ala., Tuscaloosa, 1975-76; Nat. Acad. Scis.-NRC resident research assoc. NASA-Goddard Space Flight Ctr., Greenbelt, Md., 1976-78; asst. prof. Pa. State. U., State College, 1978-79; teaching assoc. physics Vanderbilt U., Nashville, 1979-80, asst. prof. astronomy, 1980—. Contbr. articles to profl. jours. Recipient Comer medal and outstanding sr. physics student award Auburn U., 1970. Mem. Am. Astron. Soc., Internat. Astron. Union, Soc. Mfg. Engrs., Astron. Soc. Pacific. Subspecialties: Optical astronomy; Ultraviolet high energy astrophysics. Current work: Computer modeling of close binary star systems, dark star spots, chromspheric emission, contact binary stars. Office: PO 1803 Sta B Nashville TN 37235

EATON, JOHN KELLY, mechanical engineering educator, consultant; b. Camden, N.J., May 22, 1954; s. George Warren and Dorothy (Kelly) E.; m. Marilyn Louise Keller, Sept. 10, 1977 (div. July 1982). B.S.M.E., Stanford U., 1976, M.S.M.E., 1977, Ph.D. in mech. Engring, 1980. Student engr. Airesearch Corp., Phoenix, 1974; mech. engr. Hewlett Packard, San Diego, 1975-76; asst. prof. mech. engring. Stanford U., 1980—; founder, pres. GDK Engring.; dir. Bus. Computer Corp., Santa Clara, Calif.; cons. to industry. Contbr. articles to profl. jours. Recipient Silver medal Royal Soc. Arts, 1976; NSF grad. fellow, 1976-79; NSF and NASA research grantee, 1980-82; Wheeler Found. research grantee, 1982. Mem. ASME, AIAA. Democrat. Subspecialty: Fluid mechanics. Current work: Experimental studies of complex turbulent flows and convective heat transfer; particle transport in fluidized beds; computer-aided experimentation. Home: 12 Peter Coutts Circle Stanford CA 94305 Office: Dept Mechanical Engineering Stanford University Stanford CA 94305

EATON, MORRIS LEROY, statistics educator; b. Sacramento, Aug. 10, 1939; s. Franklin LeRoy and Dorothy Evelyn (Ward) E.; m. Marcia Mae Muelder, Aug. 13, 1964; 1 son, Dennis Owen. B.S., U. Wash., 1961; M.S., Stanford U., 1963, Ph.D., 1966. Research asso. Stanford U., 1966; asst. prof. statistics U. Chgo., 1966-70, asso. prof., 1970-71; vis. prof. U. Copenhagen, 1971-72; asso. prof. theoretical stats. U. Minn., 1972-73, prof., 1973—, chmn. dept. theoretical stats., 1978-81. Author: Multivariate Statistical Analysis, 1972, Multivariate Statistics: A Vector Space Approach, 1983; asso. editor: Annals Stats, 1978-83. Population Council grantee, 1969-70; NSF grantee, 1972-79. Fellow Inst. Math. Stats., Am. Stat. Assn. (asso. editor Jour. Am. Stat. Assn. 1970-72). Subspecialties: Probability; Statistics. Current work: Research in multivariate statistical analysis, statistical decision theory and multivariate probabilty ineqalities. Home: 33 Park Ln Minneapolis MN 55416 Office: 206 Church St Vincent Hall 270 Minneapolis MN 55455

EATON, ROBERT CHARLES, biology educator; b. Los Angeles, Aug. 14, 1946; s. Charles Hardman and Ruth Irene (Cook) E.; m. Janet Carol Conner, June 24, 1967 (div. 1975); 1 son, Christopher Charles. B.A. in Zoology, U. Calif.-Riverside, 1968, Ph.D., 1974; M.S. in Biology, U. Oreg., 1970; postgrad., U. Calif.-San Diego, 1974-78. Asst. research neuroscientist U. Calif.-San Diego, 1974-78; asst. prof. biology U. Colo., Boulder, 1978-83, assoc. prof., 1983—. Editor: Neural Mechanisms in Startle Behavior, 1983. Nat. Research Service awardee NIH, 1975; NSF grantee, 1978, 79, 82. Mem. Soc. Neursci., AAAS. Subspecialties: Comparative neurobiology; Neurophysiology. Current work: Neurophysiological basis of behavior with specific emphasis on neuronal networks responsible for triggering behavior patterns. Office: U Colo Box 334 Boulder CO 80309

EAVES, BURCHET CURTIS, educator, researcher; b. Shreveport, La., Nov. 25, 1938; s. Everett and Mary Curtis (King) E.; m.; children: Lesley Allen, Everett Jordon. B.S. in Mech. Engring, Carnegie Tech. Inst., 1961; M.B.A., Tulane U., 1965; M.S., Stanford U., 1969, Ph.D., 1969. Asst. prof. bus. U. Calif.-Berkeley, 1968-70; asst. prof. ops. research Stanford U., 1970-72, assoc. prof. ops. research, 1972-75, prof. ops. research, 1975—; vis. assoc. prof. Yale U., New Haven, 1974-75. Author: A View of Complementary Pivot Theory, Functional Differential Equations and Approximation of Fixed Points, 1979, Constructive Approaches to Mathematical Models, 1979; contbr. articles to profl. jours. Served as 1st lt. U.S. Army, 1961-63. Guggenheim fellow, 1979-80. Mem. Am. Math. Soc., Math. Programming Soc., Math. Assn. Am., Ops. Research Soc. Am., Soc. Indsl. and Applied Math., Inst. Mgmt. Sci., Sigma Xi, Beta Gamma Sigma. Subspecialties: Operations research (mathematics); Operations research (engineering). Office: Dept Ops Research Terman Engring Ctr Stanford U Stanford CA 94305

EBDON, DAVID WILLIAM, chemist, educator, researcher; b. Detroit, Apr. 9, 1939; s. William George and Lillian (Smith) E.; m. Priscilla AnnRagle, June 9, 1967; children: Melanie Ann, Derek William, Deren George. B.S.Chem., U. Mich., 1961; Ph.D. in Phys. Chemistry, U. Md., 1967. Sr. research scientist Nat. Biomed. Research Found., Washington, 1967-68; lectr. U. Md., 1967-68; asst. prof. chemistry Eastern Ill. U., 1968-73, assoc. prof., 1973-80, prof., 1980—, chmn. dept. chemistry, 1977—; vis. prof. U. Okla., summers 1972, 75; research scientist U. Tex., Austin, 1978-79. Mem. Am. Chem. Soc., Royal Soc. Chemistry, Ill. State Acad. Sci. Subspecialties: Physical chemistry; Surface chemistry. Current work: Physical chemistry of electrolyte solutions; solution chemistry of surfactant systems; enhanced oil recovery; surfactant ion-selective electrodes; computer modeling of natural water systems; ion assn. equilibria; chem. oceanography. Home: 2506 Salem Charleston IL 61920 Office: Dept Chemistry Eastern Ill U Charleston IL 61920

EBELING, DOLPH GEORGE, computer consulting firm executive; b. N.Y.C., Aug. 1, 1920; s. Adolph Jacob and Katheryn Gertrude (Phillippi) E.; m. Sylvia Ruth Ryder; children: Dianne, Scott. B.S., Rensselaer Poly. Inst., 1940; M.S., 1948; Ph.D., 1950; postgrad., Carnegie Inst. Tech., 1940-41, Union Coll., 1950-51. Metallurgist U.S. Steel Corp., Pitts., 1940-41; with Gen. Electric Co., Scenectady, 1946-70, mgr. metallurgy turbine div., 1953-60, sr. materials and process cons., mgr. corp. engring. edn., 1960; pres. Ebeling Assocs., Scotia, N.Y., 1970—. Served to lt. comdr. USN, 1941-46. Recipient Cordonnier award Gen. Electric Co., 1963. Mem. ASME, Am. Soc. Metals, AIME, Sigma Xi, Tau Beta Pi, Phi Lambda Upsilon. Subspecialties: Software engineering; Metallurgical engineering. Current work: Development of integrated, management information analysis and computer systems. Office: 1 Glen Ave Scotia NY 12302

EBENEZER, JOB SELVARAYAN, univ. adminstr., cons.; b. Vellore, India, Oct. 10, 1941; came to U.S., 1967; s. Titus and Kanthammal (Selvarayan) E.; m. Marjorie Rasilini Doraiswamy, July 4, 1969; children: Roshini, Suresh, Arul. M.S., U. Madras, India, 1962, Indian Inst. Sci., Bangalore, 1967; Ph.D., Stevens Inst. Tech., 1973. Asst. prof. N.Y. Inst. Tech., 1972-76; vis. asst. prof. U. N.Mex., Albuquerque, 1976-79, dir. div. energy conservation design, 1978-79, asst. dir. tech./vocat. edn., Belen, 1981—; dir. Rural Appropriate Tech. Ctr., Madras, 1980; sr. mgr. tech. application Solar Am., Inc., Albuquerque, 1981; exec. dir. Appropriate Rural Tech. Assn., Inc., Los Lunas, N.Mex.; cons. Appropriate Tech. Internat., Washington. Contbr. articles to profl. publs. Mem. N.Mex. Solar Energy Assn., Am. Vocat. Assn., N.Mex. Vocat. Assn. Lutheran. Club: Optimists (Belen). Subspecialties: Solar energy; Pedal power. Current work: Third world rural technologies. Inventor bicycle conversion attachment for prime mover function; modifications of small-scale agrl. implements.

EBERHARDT, ALLEN CRAIG, engring. educator; b. Cin., Aug. 30, 1950; s. Alfred John and Elfriede Marie (Vollmer) E.; m. Mary Drake, June 9, 1973; 1 son, William Craig. Ph.D., N.C. State U., 1977. Asst. prof. mech. engring. N.C. State U., Raleigh, 1977-81, assoc. prof., 1981—; cons. structural mechanics and acoustics. Mem. ASME, Soc. Automotive Engrs. (Teetor award 1980), Am. Soc. Engring. Educators, Acoustical Soc. Am., Sigma Xi. Subspecialties: Acoustical engineering; Mechanical engineering. Current work: Computer aided design, signal analysis, experimental data acquisition design, acoustics, vibration, stress analysis. Office: NC State U 3211 Broughton Hall Raleigh NC 27650

EBERLE, HELEN I., radiation biology and biophysics educator; b. Oakland, Calif., Mar. 2, 1932; d. Oakland B. and Imogene T. (Musser) Guard; m. Rolf A. Eberle, Feb. 6, 1958; children. B.S., Calif. State Coll.-Los Angeles, 1956; Ph.D., UCLA, 1965. Public health microbiologist Los Angeles County, 1957-60; mem. faculty U. Rochester (N.Y.) Sch. Medicine, 1968—, assoc. prof. radiation biology and biophysics, 1976—; cons. in field. Contbr. numerous sci. articles to profl. publs. Grantee Dept of Energy, 1969—, Cheveron, 1980—, NIH, 1972-75, 81-84. Mem. Am. Chem. Soc. (Faculty Research award 1969-71, grantee 1970-72), AAAS, Biophys. Soc., Am. Soc. Microbiology, Sigma Xi. Subspecialties: Molecular biology; Genetics and genetic engineering (biology). Current work: Regulation of DNA replication; gene expression; genetic engineering. Office: U Rochester Sch Medicine Radiation Biology and Biophysics Dept Rochester NY 14642

EBERT, MARLIN J., business executive, nuclear engineering consultant; b. Bryn Athyn, Pa., Dec. 2, 1938; s. Stanley F. and Sarah Jane (Heilman) E.; m. Linda Brinsley, Jan. 28, 1961 (div. 1970); children: Winfrey, Nina, Benjamin; m. Norma Jean Smith, Ja. 23, 1982. B.S., Pa. State U., 1961, M.Sc., 1963. Registered profl. nuclearengr., Calif. Research engr. Nuclear Materials and Equipment Corp. Atlantic Richfield Corp., Apllo, Pa., 1968-70; sr. mktg. engr. Gen. Elec. Nuclear Corp., Pleasanton, Calif., 1970-76; pres. Life Science Systems, Inc., Livermore, Calif., 1976—; cons. U.S. Navy, 1978-81, U.S. Army, 1972-76. Bd. dirs. Water Mgmt. Agy., 1978-80; mem. Livermore City Council, 1978-82. Served to capt., C.E. U.S. Army, 1961-68; Vietnam. NDEA fellow, 1961-63; decorated Bronze Star, Air medal. Mem. Soc. Nuclear Medicine, Am. Nuclear Soc. Republican. Lodge: Masons. Subspecialties: Nuclear medicine; Nuclear engineering. Current work: Toxic material personal anticontamination; nuclear biological chemical warfare protection; medical and industrial application of radionuclides; wastewater irradiation. Developer toxic material nuclear anti-contamination kit, 1978; co-developer Xenon-133 ventilation study system, 1974. Home: 1609 4th St Livermore CA 94550 Office: 4049 1st St Livermore CA 94550

EBERT, PAUL STOUDT, biochemist; b. Palmerton, Pa., Apr. 13, 1933; s. Paul F. and Ellen S. (Stoudt) E.; m. Linda P. Romig, Nov. 26, 1964; children: Robert, Mindy. B.S., Pa. State U., 1955; Ph.D., Rutgers U., 1962. Chemist Merck & Co., Rahway, N.J., 1955-57; research asst. Rutgers U., New Brunswick, N.J., 1957-61; postdoctoral fellow U. Pa., Phila., 1961-65; research chemist Lab. Molecular Oncology, Div. Cancer Cause and Prevention, Nat. Cancer Inst., NIH, Bethesda, Md., 1965—. Contbr. numerous sci. articles to profl. publs. Deacon 4th Presbyterian Ch., Bethesda, 1978-81. Muscular Dystrophy Assn. grantee, 1964-65. Mem. Am. Assn. Cancer Research, N.Y. Acad. Scis., Sigma Xi. Subspecialties: Cancer research (medicine); Cell study oncology. Current work: Control mechanisms in erythropoiesis; pharmacology of heme pathway inhibitor, succinylacetone; mechanism of photoinactivation of tumor cells with hematoporphyrin; mechanisms of cell transformation. Home: 10004 Sinnott Dr Bethesda MD 20817 Office: Nat Cancer Inst Frederick Cancer Research Facility Frederick MD 21701

EBERT-FLATTAU, PAMELA, science policy analyst; b. Chgo., Dec. 24, 1946; d. Raymond C. and Sylvia Ann (Jones) Ebert; m. Edward S. Flattau, Feb. 1, 1977; 1 son, Jeremy Paul. B.Sc. with honors, Leeds U., 1969; M.S., U. Ga., 1972, Ph.D., 1974. AAAS-Am. Psychol. Assn. Congressional Sci. fellow U.S. Senate, 1974-75; staff officer Nat. Research Council, Washington, 1975-80, sr. staff officer, 1980-81; sci. policy analyst Sci. Indicators unit NSF, Washington, 1981—. Author: A Legislative Guide, 1980. Mem. Am. Psychol. Assn., AAAS, Assn. Advancement Psychology (trustee 1976—), Sigma Xi, Psi Chi. Subspecialty: Behavioral psychology. Current work: Patterns of national support for research and development; and academic science and engineering. Office: National Science Foundation L-611 Washington DC 20550

EBNER, FORD FRANCIS, neuroscience educator; b. Colfax, Wash., Feb. 10, 1934. B.S., Wash. State U., 1954, D.V.M., 1958; Ph.D. in Neuroanatomy, U. Md., 1965; M.S. (hon.), Brown U., 1969. Asst. prof. anatomy and physiology U. Md., College Park, 1965-66; asst. to assoc. prof. biology and medicine Brown U., Providence, 1966-73, prof., co-dir. Ctr. for Neural Sci., 1973—; NIH fellow in physiology Johns Hopkins U., Balt., 1960-63; spl. fellow anatomy U. Md., 1963-65. Served to capt. U.S. Army, 1958-60. Mem. AAAS, Am. Assn. Anatomists, Am. Assn. Neuropathologists, Neurosci. Soc. Subspecialty: Neurophysiology. Office: Neurobiology Sect Div Biology and Medicine Brown U Providence RI 02912

EBY, WILLIAM CLIFFORD, clin. pathologist; b. Portland, Oreg., July 25, 1942; s. Wilder Schell and Helen Lillian (Emmerson) E.; m. Karen Joan White, June 9, 1966; children: Michael William, Mary Ellen, James Wilder, Thomas Eugene. B.S., Walla Walla Coll., 1963; M.D., Loma Linda U., 1967; Ph.D., U. Ill., 1978. Diplomate: Am. Bd. Pathology. Intern Gorgas Hosp., Ancon, C.Z., 1967-68; resident Mayo Clinic, Rochester, Minn., 1975-77; assoc. cons., 1977-78; asst. prof. depts. pathology and microbiology Loma Linda (Calif.) U., 1978—. Contbr. articles to profl. jours. Served with U.S. Army, 1968-70. Decorated Army Commendation medal. Mem. AMA, Am. Soc. Clin Pathologists, Coll. Am. Pathologists, Am. Assn. Clin. Chemistry. Republican. Seventh-day Adventist. Subspecialties: Pathology (medicine); Immunology (medicine). Current work: Clinical laboratory immunology, immunochemistry. Office: Loma Linda U Med Ctr Loma Linda CA 92350

ECK, BERNARD JOHN, product engineer, railway wheel manufacturing company executive researcher; b. Springfield, Ill., May 2, 1928; s. Edward Franz and Pauline Cecelia (Schafer) E.; m. Janice May Carlson, Apr. 7, 1956; children: William, Robert, John, James, Julie. B.S., Mo. Sch. Mines, 1950. Technologist U.S. Steel Corp., Chgo., 1950-51, 53-57; dir. product engring. Griffin Wheel Co., Chgo., 1957—. Contbr. articles to profl. publs. Served to sgt. CIC U.S. Army, 1951-53; Korea. Decorated Bronze Star. Mem. ASME (Rail Transp. Div. Best Paper award 1973), Am. Soc. Metals, ASTM, Am. Ceramic Soc. Roman Catholic. Subspecialties: Mechanical engineering; Metallurgical engineering. Current work: Designing and testing railroad wheels; product engineering. Office: 200 W Monroe S 23d Floor Chicago IL 60606

ECKARDT, MICHAEL JON, psychologist; b. Glendale, Calif., Apr. 3, 1943; s. Ralph Benjamin and Betty June (Davey) Voelker; m. Johneen Pofahl, Aug. 9, 1968; children: Shea Michael, Neil Edward, Rory Lawrence. B.A., Calif. State U.-Northridge, 1966; M.S., U. So. Calif., 1967, U. Mich., 1970; Ph.D., U. Oreg. Helath Sci. Ctr., 1975. Lic. psychologist, Md. Vice pres. J & M Enterprises, Inc., Glendale, Calif., 1963-65; lectr. U. Mich., 1970; psychologist VA Hosp., Sepulveda, Calif., 1975-76; fellow U. Calif.-Irvine, 1976; psychologist Nat. Inst. Alcohol Abuse, Rockville, Md., 1976—. Contbr. articles to profl. jours. Trustee Bridge, A Way Across, Inc., Burbank, Calif., 1976-76; co-chmn. community services com. Cinnamon Woods, Inc., Germantown, Md., 1982—. Recipient Outstanding Biologist award Calif. State U., 1966; Superior Service award USPHS, 1982. Fellow Md. Psychol. Assn.; mem. Am. Psychol. Assn., Soc. Neurosci., AAAS. Subspecialties: Neuropsychology; Medical psychology. Current work: Investigating the acute and chronic effects of alcohol and other addictive substances on various anatomic and physiological systems. Home: 18626 Turmeric Ct Germantown MD 20874 Office: Laboratory of Preclinical Studies National Institute on Alcohol Abuse and Alcoholism 12501 Washington Ave Rockville MD 20852

ECKELS, DAVID DEAN, immunology educator; b. St. Helena, Calif., Dec. 2, 1952; s. Paul Coolidge and Claire Ann (Griffith) E.; m. Julianne Riedel, Aug. 9, 1975; 1 son, Daniel William. B.A., U. Calif.-Santa Barbara, 1975, M.S., 1977, Ph.D., 1979. Instr. Georgetown U., Washington, 1979-82, asst. prof. pediatrics and microbiology, 1982—. Contbr. articles to profl. jours. Calif. State scholar, 1971-75; Georgetown U. Biomed. Research Support grantee, 1980; Nat. Inst. Allergy and Infectious Disease research grantee, 1983-86. Mem. Am. Assn. Clin. Histocompatibility Testing, Am. Assn. Immunologists. Republican. Subspecialties: Immunogenetics; Cellular engineering. Current work: Genetic regulation of immune responsiveness; human immunogenetics; immunomodulation-therapeutic manipulation of human immunity; immune engineering. Office: Georgetown Univ Hosp Vincent T Lombardi Cancer Research Center E Level 3800 Reservoir Rd NW Washington DC 20007

ECKENHOFF, JAMES EDWARD, physician, educator; b. Easton, Md., Apr. 2, 1915; s. George L. and Ada (Ferguson) E.; m. Bonnie Lee Youngerman, June 4, 1938 (div. Jan. 1973); children—Edward Alvin, James Benjamin, Walter Leroy, Roderic George; m. Jane M. Mackey, Sept. 22, 1973. B.S., U. Ky., 1937; M.D., U. Pa., 1941; D.Sc., Transylvania U., 1970. Diplomate: Am. Bd. Anesthesiology (bd. dirs. 1965-73, pres. 1972-73). Intern Good Samaritan Hosp., Lexington, Ky., 1941-42; Harrison fellow anesthesia U. Pa., 1945-48, mem. faculty, 1948-65, prof. anesthesiology, 1955-65; physician anesthetist Hosp. U. Pa., 1948-65; prof. anesthesia Northwestern U. Med. Sch., Chgo., 1966—, chmn. dept., 1966-70, dean, 1970—; pres. McGaw Med. Center, 1980—; Fellow faculty anesthesia, also Hunterian prof. Royal Coll. Surgeons.; chief anesthesia Passavant Meml. Hosp., Chgo., 1966-70; chmn. anesthesia Chgo. Wesley Hosp., 1966-70; cons. VA Research Hosp., Childrens Hosp., Chgo., 1966—; surgeon gen. U.S. Navy, 1964—; Mem. surgery study sect. NIH, 1962-66, anesthesia tng. com., 1966-70; vis. prof. Australian and New Zealand Soc. Anesthetists, 1968, South African Soc. Anesthetists, 1970; dir. Nat. Bd. Med. Examiners, 1975—, treas., 1979—. Author: (with others) Introduction to Anesthesia, 6th edit, 1982, Anesthesia from Colonial Times, 1966, also numerous articles.; Editor: Science and Practice in Anesthesia, 1965, (with J. Beal) Intensive and Recovery Room Care, 1969, Jour. Anesthesiology, 1958-62, Yearbook of Anesthesia, 1970-81, Controversy in Anesthesiology, 1979. Trustee Evanston Hosp., 1972—, Rehab. Inst. Chgo., 1972—, Northwestern Meml. Hosp., 1973—, Children's Meml. Hosp., 1977—. Served to capt. M.C. AUS, 1942-45; ETO. Commonwealth Fund fellow Queen Victoria Hosp., East Grinstead, Eng., 1961-62. Fellow Inst. Medicine Chgo., A.C.P.; mem. Australian, New Zealand, South African socs. anesthesiologists, Soc. Acad. Anesthesia Chairmen (pres. 1967-68), Soc. Med. Consultants to Armed Forces, Am. Soc. Anesthesiologists (Disting. Service award 1981), AMA, Assn. Univ. Anesthetists (pres. 1962), Am. Assn. U. Profs., Ill. Council Med. Deans (pres. 1973-74), Chgo., Ill. med. socs., Am. Physicians Art Assn., Am. Physiol. Soc. Home: 1242 Lake Shore Dr Chicago IL 60610 Office: 303 E Chicago Ave Chicago IL 60611

ECKERT, CHARLES ALAN, chemical engineering educator; b. St. Louis, Dec. 13, 1938; m., 1961; 2 children. S.B., MIT, 1960, S.M., 1961; Ph.D. in Chem. Engring, U. Calif.-Berkeley, 1965. NATO fellow high pressure physics High Pressure Lab., Nat. Ctr. Sci. Research, Bellevue, France, 1964-65; from asst. prof. to assoc. prof. U. Ill., Urbana, 1965-73, prof. chem. engring., 1973—, head dept. chem, engring., 1980—; cons. various cos.; vis. prof. Stanford U., 1971-72. Guggenheim fellow, 1971. Mem. AIME, Am. Chem. Soc. (Ipatieff prize 1977), Am. Inst. Chem. Engrs. (Allan Colburn award 1973), Am. Soc. Engring. Edn., Nat. Acad. Engring., Chem. Soc. London. Subspecialty: Chemical engineering. Office: Dept Chem Engring 113 RAL U Ill Urbana IL 61801

ECKERT, ERNST R. G., emeritus mechanical engineering educator; b. Prague, Czechoslovakia, Sept. 13, 1904; came to U.S., 1945, naturalized, 1955; s. Georg and Margarete (Pfrogner) E.; m. Josefine Binder, Jan. 30, 1931; children: Rosemarie Christa Eckert Koehler, Elke, Karin Eckert Winter, Dieter. Diploma Ing., German Inst. Tech., Prague, 1927, Dr.Ing., 1931; Dr. habil., Inst. Technology, Danzig, 1938; Dozent, Inst. of Technol., Braunschweig, Germany, 1940; hon. doctorates, Inst. Tech., Munich, 1968, Purdue U., 1968, U. Manchester, Eng., 1968, U. Notre Dame, 1970, Poly. Inst. Romania, Jassy, 1973. Registered profl. engr., Minn. Chief engr., lectr. Inst. Technology, Danzig, 1934-38; sect. chief thermodynamics Aero. Research Inst., Braunschweig, 1938-45; prof., dir. Inst. Technology, Prague, 1943-45; cons. USAF, 1945-49, Lewis Flight Propulsion Lab., NASA, 1949-51; prof. mech. engring. dept. U. Minn., 1951-73, dir. thermodynamics and heat transfer and of heat transfer lab., 1955-73, Regents' prof. emeritus mech. engring., 1966-73; former vis. prof. Purdue U.; cons. Gen. Electric Co.; former cons. Trane Co.; U.S. rep. aerodynamics panel Internat. Com. Flame Radiation. Author: Introduction to the Transfer of Heat and Mass, 1950, 2d edit., 1959, Heat and Mass Transfer, (translated by J.F. Gross), 1963; others in German and Russian, (with Goldstein) Measurement Techniques in Heat Transfer, 1970, 2d edit., 1976, (with Drake) Analysis of Heat and Mass Transfer, 1972; Chmn. hon. editorial adv. bd.: Internat. Jour. Heat and Mass Transfer; Editor: Thermal Sciences series, Wadsworth Pub. Co., Belmont, Cal.; editor: Thermo and Fluid Dynamics; co-chmn. adv. editorial bd.: Heat Transfer-Japanese Research; co-editor: Energy Developments in Japan; chmn. hon. editorial adv. bd.: Letters in Heat and Mass Transfer; editorial adv. bd.: Numerical Heat Transfer; Contbr. articles to sci. mags. Mem. Nat. Commn. Fire Prevention and Control, 1970-73. Recipient Max Jacob Meml. award, 1961, Distinguished Teaching award U. Minn., 1965, Western Electric Fund award, 1965, Gold medal French Inst. Energy and Fuel, 1967, Vincent Bendix award, 1972; Alexander von Humboldt U.S. Sr. Scientist award, 1980, A.V. Luikov medal, 1979; research fellow Japan Soc. Promotion Sci., 1982. Fellow N.Y. Acad. Scis., AIAA; mem. Am. Soc. Engring. Edn. (hon.), Wissenschaftliche Gesellschaft für Luft and Raumfahrt, Sigma Xi, Pi Tau Sigma, Tau Beta Pi. Subspecialties: Mechanical engineering. Current work: Heat transfer in energy production and conservation, advanced gas turbines, energy storage. Home: 60 W Wentworth Ave W St Paul MN 55118 Office: U Minn Minneapolis MN 55455

ECKHART, WALTER, molecular biologist, virologist; b. Yonkers, N.Y., May 22, 1938; s. Walter and Jean (Fairnington) E.; m. Karen, June 5, 1965. B.S., Yale U., 1960; postgrad., Cambridge (Eng.) U., 1960-61; Ph.D., U. Calif., Berkeley, 1965. Mem. Salk Inst., La Jolla, Calif., 1970-73; assoc. prof. molecular biology, 1973-79, prof., 1979—, dir., 1976—; adj. prof. U. Calif., San Diego. Contbr. articles on molecular biology, tumor virology to profl. jours. NIH grantee, 1967—. Mem. Am. Soc. Microbiology, AAAS, Phi Beta Kappa. Subspecialties: Molecular biology; Virology (biology). Current work: Mechanisms of cell growth control, viral gene organization and regulation of expression.

ECKROAT, LARRY RAYMOND, biology educator; b. Bloomsburg, Pa., July 18, 1941; s. Raymond E. and Helen M. E.; m. Cozella E. Harvey, Aug. 14, 1971; 1 dau., Andrea. B.S. in Biology, Bloomsburg State Coll., 1964; M.S., Pa. State U., 1966, Ph.D. in Zoology, 1969. Assoc. prof. biology Behrend Coll. of Pa. State U., Erie, 1969—. Mem. Genetics Soc. Am., Am. Genetics Assn. Democrat. Lutheran. Subspecialty: Genetics and genetic engineering (biology). Home: 8564 Knoyle Rd Erie PA 16510 Office: Station Rd Erie PA 16563

ECONOMOU, NICHOLAS PHILIP, microelectronics specialist; b. Bklyn., Oct. 20, 1948; s. George N. and Helen (Mourtsakis) E.; m. Donna Wlodkoski, Apr. 5, 1974. B.A., Dartmouth Coll., 1970; M.A., Harvard U., 1973; Ph.D., 1977. Mem. tech. staff Bell Telephone Labs., Holmdel, N.J., 1977-78; mem. tech. staff M.I.T. Linclon Lab., Lexington, Mass., 1979-80, asst. leader microelectronics group, 1981—. Contbr. articles to profl. jours. Served to 1st lt. USAF, 1973-74. Mem. Optical Soc. Am., IEEE, Am. Phys. Soc. Subspecialties: Microchip technology (engineering); Microelectronics. Current work: Technology for miniaturization of semiconductor devices and integrated circuits; microfabrication technology, high speed electronic devices. Office: MIT Lincoln Lab 244 Wood St Lexington MA 02173

ECONOMOU, STEVEN GEORGE, surgeon, educator; b. Chgo., July 4, 1922; s. George Demetrious and Helen (Kandrevas) E.; m. Kathryn Dotska, Sept. 16, 1950; children: James Steven, Tasia Stephanie Khan, Elena Saclarides. A.A., U. Chgo., 1943; M.D., Hahnemann Med. Coll., 1947. Diplomate: Am. Bd. Surgery, 1955. Intern St. Francis Hosp., Evanston, Ill., 1947-48, resident, 1948-49, Presbyn. Hosp., Chgo., 1950-52,54; prof. surgery Rush Med. Coll., Chgo., 1971—, Jack Fraser Smith prof., 1981; assoc. chmn. dept. gen. surgery Rush-Presbyn.-St. Luke's Med. Ctr., 1973—; chmn. tumor com. Presbyn.-St. Luke's Hosp., 1964—; clin. prof. surgery U. Ill. Coll. Medicine, Chgo., 1968—. Author: (with E.J. Beattie) An Atlas of Advanced Surgical Techniques, 1968; guest editor: Surg. Clinics of N.Am, Oct. 1970, Feb. 1979. Mem. ACS, AMA, Am. Surg. Assn., Central Surg. Assn., Chgo. Surg. Soc., Ill. Med. Soc., Chgo. Med. Soc., Soc. Head and Neck Surgeons, Soc. Surg. Oncology. Greek Orthodox. Subspecialties: Cancer research (medicine); Oncology. Current work: The stability of estrogen and progesterone receptors as an operative advantage; two-dimensional protein electrophoresis and the identification of histologically indeterminate human cancers. Office: 1725 W Harrison St Suite 378 Chicago IL 60612

EDBERG, STEPHEN J., astronomer; b. Pasadena, Calif., Nov. 3, 1952; s. Joseph and Sophie (Pasternak) E.; m. Janet Lynn Greenstein, Dec. 23, 1979. B.A, U. Calif.-Santa Cruz, 1974; postgrad., U. Calif.-San Diego, 1974-75; M.A., UCLA, 1976, 1976-77. Research assoc. San Fernando Obs., Calif., 1978-79; mem. staff Galileo and Space Telescope Projects, Jet Propulsion Lab., Pasadena, 1979-81; coordinator amateur observations Internat. Halley Watch, Jet Propulsion Lab., 1981—; expdn. leader total solar eclipses in, Can., 1979, Kenya, 1980, Hawaii, 1981, Java, 1983. Contbr. articles to sci. jours. U. Calif. Pres.'s undergrad. fellow, 1973; Sigma Xi grantee, 1974; Australian Nat. U. scholar, 1974; UCLA Chancellor's Intern fellow, 1975-77. Mem. Am. Astron. Soc., Astron. Soc. Pacific, Planetary Soc., Polaris Obs. Assn. Subspecialties: Planetary science; Solar physics. Current work: Coordinating the contributions of amateur astronomers to Comet Halley observations in 1985-86; science administration; solar prominences and corona structure and activity; cometary tail activity. Office: Jet Propulsion Lab T 1166 4800 Oak Grove Dr Pasadena CA 91109

EDDS, LOUISE LUCKENBILL, biologist; b. Lebanon, Pa., Nov. 19, 1936; d. Fred E. and Anna M. (Luckenbill) E.; B.A., Oberlin Coll., 1958; postgrad., Washington U., St. Louis, 1958-61; Ph.D., Brown U., 1964. Postdoctoral fellow, research instr. Boston U. Med. Sch., 1964-68; sci. fellow Hubrecht Lab., Utrecht, Netherlands, 1968-69; asst. prof. Smith

Coll., Northampton, Mass., 1969-75; instr. neuropathology Harvard U. Med. Sch., 1975-77; research assoc. Boston Children's Hosp. Med. Ctr., 1975-77; now assoc. prof. zoology and biomed. sci. Coll. Osteo. Medicine, Ohio U.; vis. asst. prof. Harvard U., 1973-74. Mem. Soc. Devel. Biology, Am. Soc. Cell Biology, Soc. Neurosci., AAAS. Subspecialties: Developmental biology; Neurobiology. Current work: Development of neural crest and sympathetic neurons. Office: Ohio U Coll Osteopathic Medicine Irvine Hall Athens OH 45701

EDDY, EDWARD MITCHELL, biomedical researcher; (married); 2 children. B.S., Kans. State U., 1962, M.S., 1964; Ph.D. in Anatomy, U. Tex., 1967. Fellow in anatomy Harvard U. Med. Sch., 1967-69, instr., 1969-70; asst. prof. U. Wash., Seattle, 1970-76, assoc. prof. biol. structure, 1976-82, prof., 1982-83, vice chmn. dept., 1975-78, acting chmn., 1978-81; research biologist NIEHS, NIH, Research Triangle Park, N.C., 1983—. NSF fellow; NIH fellow. Mem. Am. Soc. Cell Biologists, Am. Assn. Anatomists, Soc. Devel. Biology, Am. Soc. Biology, Am. Soc. Zoology, Internat. Soc. Devel. Biologists. Subspecialty: Cell biology. Office: NIH NIEHS LRDT Research Triangle Park NC 27709

EDDY, JOHN ALLEN, astrophysicist; b. Pawnee City, Nebr., Mar. 25, 1931; m.; 4 children. B.S., U.S. Naval Acad., 1953; Ph.D. in Astrogeophysics, U. Colo., 1962. Physicist Nat. Bur. Standards, 1962-63; sr. scientist High Altitude Obs., Boulder, Colo., 1963—; adj. prof. U. Colo., 1963—; research assoc. Harvard-Smithsonian Ctr. Astrophysics, 1967-70, 77-79. Recipient Boulder Scientist award Sci. Research Soc. Am., 1965. Mem. Am. Astron. Soc., AAAS, Am. Geophys. Union, Internat. Astron. Union, Sigma Xi. Subspecialties: Infrared optical astronomy; Archaeoastronomy. Office: High Altitude Obs PO Box 3000 Boulder CO 80307

EDELMAN, ANN LYNN, psychologist, computer programmer, systems analyst, laboratory supervisor; b. N.Y.C., Dec. 17, 1944; d. Eugene and Sarah Dorothy (Paris) E.; m. Alfred Henry Letourneau, Aug. 6, 1972. B.A., Case Western Res. U., 1967; M.A., Boston U., 1969, Ph.D., 1974. Cert and lic. psychologist, Mass. Psychologist Emmanuel Coll., Boston, 1971-76; chief psychologist Brockton (Mass.) Multi-Service Ctr., 1976-78; clin. dir. TRIAD, Watertown, Mass., 1970—; sr. lab. coordinator Cambridge Inst., Boston, 1982—; cons. Youth Community Action, Arlington, Mass., 1980-82. NIMH fellow, 1969, 70; Rehab. Services Adminstrn. fellow, 1968; VA fellow, 1967. Mem. Am. Psychol. Assn., Mass. Psychol. Assn., Nat. Register Health Services Providers in Psychology. Jewish. Subspecialties: Social psychology; Information systems (information science). Current work: Systems psychology—the integration of computer science and clinical psychology to study human factors in computer-related fields. Home: 80 Thorndike St Arlington MA 02174 Office: Cambridge Inst Computer Programming 480 Boylston St Boston MA 02116

EDELMAN, GERALD MAURICE, biochemist; b. N.Y.C., July 1, 1929; s. Edward and Anna (Freedman) E.; m. Maxine Morrison, June 11, 1950; children: Eric, David, Judith. B.S., Ursinus Coll., 1950, Sc.D., 1974; M.D., U. Pa., 1954, D.Sc., 1973; Ph.D., Rockefeller U., 1960; M.D. (hon.), U. Siena, Italy, 1974; D.Sc., Gustavus Adolphus Coll., 1975; Sc.D., Williams Coll., 1976. Med. house officer Mass. Gen. Hosp., 1954-55; asst. physician hosp. of Rockefeller U., 1957-60, mem. faculty, 1960—, assoc. dean grad. studies, 1963-66, prof., 1966-74, Vincent Astor Distinguished prof., 1974—; Mem. biophysics and biophys. chemistry study sect. NIH, 1964-67; mem. Sci. Council, Center for Theoretical Studies, 1970-72; assoc., sci. chmn. Neurosciences Research Program, 1980—, dir. Neurosci. Inst., 1981—; mem. adv. bd. Basel Inst. Immunology, 1970-77, chmn., 1975-77; non-resident fellow, trustee Salk Inst.; bd. overseers Faculty Arts and Scis., U. Pa., 1976-83; trustee, mem. adv. com. Carnegie Inst., Washington. Bd. govs. Weizmann Inst. Sci.; trustee Rockefeller Bros. Fund., 1972-82. Served to capt. M.C. AUS, 1955-57. Recipient Spencer Morris award U. Pa., 1954; Ann. Alumni award Ursinus Coll., 1969; Nobel prize for physiology or medicine, 1972; Albert Einstein Commemorative award Yeshiva U., 1974; Buchman Meml. award Calif. Inst. Tech., 1975; Rabbi Shai Shacknai meml. prize Hebrew U.-Hadassah Med. Sch., Jerusalem, 1977. Fellow N.Y. Acad. Scis., N.Y. Acad. Medicine; mem. Am. Philos. Soc., Am. Soc. Biol. Chemists, Am. Assn. Immunologists, Genetics Soc. Am., Harvey Soc. (pres. 1975-76, Am. Chem. Soc., Eli Lilly award biol. chemistry 1965), AAAS, Am. Acad. Arts and Scis., Nat. Acad. Sci., Am. Soc. Cell Biology, Acad. Scis. of Inst. France (fgn.), Japanese Biochem. Soc. (hon.), Pharm. Soc. Japan (hon.), Soc. Developmental Biology, Council Fgn. Relations, Sigma Xi, Alpha Omega Alpha. Subspecialties: Developmental biology; Molecular biology. Research structure of antibodies, molecular and devel. biology.

EDELMAN, ISIDORE SAMUEL, educator, scientist; b. N.Y.C., July 24, 1920; s. Abraham and Fannie (Thaler) E.; m.; children: Arthur, Susan, Joseph, Ann. B.A., Ind. U., 1941, M.D., 1944. Intern, Greenpoint Hosp., Bklyn., 1944-45; resident Montefiore Hosp., Bronx, N.Y., 1947-49; postdoctoral research fellow Harvard Med. Sch., 1949-52; prof. medicine and physiology U. Calif.-San Francisco, 1960-67, prof. biophysics, 1969-78, Samuel Neider Research prof. medicine, 1967-78; Robert Wood Johnson, Jr. prof., chmn. biochemistry Coll. Physicians and Surgeons Columbia U., 1978—; Harry T. Dozor vis. prof. biochemistry Ben-Gurion U., Beer Sheva, Israel, 1980; mem. research career awards com. Nat. Inst. Gen. Med. Sci., NIH, 1969-72; bd. sci. counselors Nat. Heart, Lung and Blood Inst., 1978-82; mem. U.S. nat. com. Internat. Union Pure and Applied Biophysics, 1971-73. Editor: Ann. Revs. Physiology; editorial bd.: Current Topics in Membranes and Transport, Jour. Membrane Biology. Served to capt. AUS, 1945-47. Mem. Assn. Am. Physicians, Am. Acad. Arts and Scis., Am. Physiol. Soc., Inst. Medicine of Nat. Acad. Scis., Am. Soc. Clin. Investigation, Biophys. Soc. (council 1974-77), Soc. Gen. Physiology, Western Soc. Clin. Research, Am. Soc. Biol. Chemistry, Endocrine Soc. (publs. com. 1974-77, council 1979—), Western Assn. Physicians, Harvey Soc. Subspecialty: Biochemistry (medicine). Research transport solutes and water across cell membranes; molecular mechanisms in actions of adrenal, posterior pituitary and thyroid hormones. Dept Biochem Coll Physicians & Surgeons Columbia U 630 W 168th St New York NY 10032

EDELSON, MARTIN CHARLES, research chemist; b. N.Y.C., Nov. 18, 1943; s. Samuel and Rebecca (Stamler) E.; m. Wendy Alice Lipton, July 9, 1967; 1 son, Steven Jonathan. B.S., CCNY, 1964, M.A., 1967; Ph.D., U. Oreg., 1973. Postdoctoral teaching fellow U. B.C., Vancouver, 1973-77; postdoctoral fellow Ames Lab., Iowa State U., 1978-79, assoc. chemist, 1979-82, chemist, 1982—. Contbr. articles to sci. jours. Pres. Prairie Ridge Homeowners Assn., 1981—. N.Y. State Regents scholar, 1960-64. Mem. Soc. Applied Spectroscopy, Optical Soc. Am., ASTM. Jewish. Subspecialties: Analytical chemistry; Spectroscopy. Current work: Analytical spectroscopy applied to nuclear safeguards analysis; laser-basd spectroscopy; multi-photon spectroscopy. Home: RR4 Prairie Ridge Ames IA 50010 Office: Ames Lab Iowa State U Ames IA 50011

EDELSON, PAUL JEFFREY, physician, educator; b. Newport News, Va., Dec. 5, 1943; s. Harry and Ruth (Levine) E.; m. Ingrid Rosner, Jan. 11, 1981; 1 son, Jonathan M.R.; m.; children by previous marriage: Christopher Peter, Nicholas James. A.B., U. Rochester, 1964; M.D., SUNY-Downstate Med. Ctr., 1969. Diplomate: Am. Bd. Pediatrics. Intern, asst. resident Yale-New Haven Hosp., 1969-71; fellow dept. medicine U. Calif.-San Francisco, 1971-72; postdoctoral fellow Rockefeller U., 1973-75, asst. prof., 1975-77, Harvard U., 1977-82; assoc. prof., dir. div. pediatric infectious diseases and immunology Cornell U., 1982—; tutor in field. Attending physician Camp for Displaced Persons, Thailand, 1980; adv. bd. Cambodian Community Mass., 1980-81. Co-editor: Methods for The Study of Mononuclear Phagocytes, 1982; contbr. articles to profl. jours. Kerb fellow Oxford (Eng.) U., 1974; Research Career Devel. award NIH, 1978-83. Mem. Am. Assn. Immunologists, Am. Soc. Cell Biology, Soc. Pediatric Research, Mass. Audubon Soc., Physicians for Social Responsibility. Subspecialties: Infectious diseases; Cell biology. Current work: Cell biology of macrophanges. Home: 104 St Marks Ave Brooklyn NY 11217 Office: 1300 York Ave New York NY 10021

EDEN, FRANCINE CLAIRE, molecular biologist, researcher; b. Oakland, Calif., Sept. 6, 1945; d. Lorin August and Clara Elizabeth (Decker) E.; m. Thomas Frank McCutchan, Nov. 28, 1980; 1 son, Micah Thomas. B.A., Occidental Coll., 1967; Ph.D., U. Wash., 1973. Predoctoral fellow dept. microbiology U. Wash., Seattle, 1969-73; research assoc. Calif. Inst. Tech., Pasadena, 1974-75; dept. zoology Ind. U., Bloomington, 1976; staff fellow Nat. Cancer Inst., NIH, Bethesda, Md., 1977-83, cancer expert, 1983—. Mem. Am. Soc. Biol. Chemists. Presbyterian. Subspecialties: Molecular biology; Biochemistry (biology). Current work: Eucaryotic genome organization, repeated DNA, DNA methylation, gene cloning, DNA sequencing. Office: NIH Bldg 37 Room 3C19 Bethesda MD 20910

EDEN, WILLIAM MURPHEY, physicist; b. Macon, Ga., Sept. 26, 1928; s. John F. and Sula M. (Wommack) E.; m. Clara May Edwards, July 7, 1961; 1 son, Andrew Mark. A.B., Mercer U., 1955; M.S., U. Miami, Coral Gables, Fla., 1964. Surg. technician Macon Hosp., 1954-56; sanitarian Jones County Health Dept., Gray, Ga., 1956-59; public health sanitarian Volusia County Health Dept., Daytona Beach, Fla., 1959-63; public health physicist Fla. Dept. Health and Rehab. Services, Tallahassee, 1964—. Contbr. articles to profl. jours. Served with USN, 1948-52. Mem. Health Physics Soc., Internat. Radiation Protection Assn., Am. Assn. Physicists in Medicine, Am. Conf. Govt. Indsl. Hygienists, Soc. Photooptical Instrumentation Engrs. Democrat. Baptist. Subspecialty: Health Physics. Home: 2812 Duffton Lopp Tallahassee FL 32303 Office: 1317 Winewood Blvd Tallahassee FL 32301

EDENBERG, HOWARD JOSEPH, biochemist; b. N.Y.C., Jan. 29, 1948; s. Benjamin and Frances Rose (Kasper) E.; m. Susan Ann Grow, June 4, 1978; 1 dau., Elizabeth. B.A., CUNY, 1968; M.A., Stanford U., 1970, Ph.D., 1973. Fellow MIT, 1973-76, Harvard U., 1976-77; asst. prof. biochemistry Ind. U., 1977-82, assoc. prof., 1982—. Contbr. articles to profl. jours. Woodrow Wilson fellow, 1968; NSF fellow, 1968; Damon Runyon fellow, 1973-75; NIH fellow, 1975-77; NIH grantee, 1978-81. Mem. Am. Soc. Biol. Chemists, Am. Soc. Microbiology, AAAS, Biophys. Soc., N.Y. Acad. Sci. Subspecialties: Molecular biology; Genetics and genetic engineering (biology). Current work: DNA repair and replication in mammalian cells and simian virus 40; structure and regulation of mammalian genes. Home: 5960 Lieber Rd Indianapolis IN 46208 Office: Department Biochemistry Indiana University School Medicine Indianapolis IN 46223

EDGAR, ARLAN LEE, biology educator; b. Gratiot County, Mich., June 3, 1925; s. Sherman J. and Letha (Perdew) E.; m. Bonnie Jean Anderson, Mar. 30, 1952; children: Rosemary, Amy, Andrew. B.A., Alma Coll., 1949; M.A., U. Mich., 1950, M.S., 1957, Ph.D., 1960. Instr. Alma (Mich.) Coll., 1950-51, asst. prof., 1953-58, assoc. prof., 1958-65, prof., 1965-76, Charles A. Dana prof. biology, 1976—; vis. prof. zoology U. Mich. Biol. Sta., Pellston, 1965-75; vis. prof. U. Los Andes, Merida, Venezuela, 1971, 74. Contbr. chpts. to books. Served with U.S. Army, 1951-53. Recipient citation for scholarly achievement Mich. Acad. Scis., Arts, and Letters, 1973. Mem. Am. Inst. Biol. Scis., Am. Soc. Zoologists (sec. ecology div. 1976-78), Sigma Xi (past pres. chpt.). Presbyterian. Subspecialties: Morphology; Ecology. Current work: Physiological ecology and taxonomy of Opiliones; biology of invertebrates; human health and behavior. Home: 602 Woodworth St Alma MI 48801 Office: Dept Biology Alma Coll Alma MI 48801

EDGERTON, HAROLD EUGENE, educator, elec. engr.; b. Fremont, Nebr., Apr. 6, 1903; s. Frank Eugene and Mary Nettie (Coe) C.; m. Esther May Garrett, Feb. 25, 1928; children—Mary Louise, William Eugene, Robert Frank. B.S., U. Nebr., 1925, Dr.Engring. (hon.), 1948; M.Sc., Mass. Inst. Tech., 1927, D.Sc., 1931; LL.D. (hon.), Doane Coll., 1969, U. S.C., 1969. Elec. engr. Nebr. Light & Power Co., 1920-25, Gen. Electric Co., 1925-26; Inst. prof. emeritus Mass. Inst. Tech. Author: (with James R. Killian, Jr.) Moments of Vision, 1979, Electronic Flash, Strobe, 1979, also numerous tech. articles. Recipient medal Royal Photog. Soc.; Gold medal Nat. Geog. Soc.; Modern Pioneer award; Potts medal Franklin Inst.; Albert A. Michelson medal, 1969. Fellow I.E.E.E., Am. Inst. Elec. Engrs., Soc. Motion Pictures and TV Engrs., Royal Soc. Gt. Britain; mem. Nat. Acad. Scis., Nat. Acad. Engrs., Marine Tech. Soc., Sigma Xi, Eta Kappa Nu, Sigma Tau. Republican. Conglist. Club: Mason. Subspecialty: Electrical engineering. Inventor of stroboscopic high-speed motion and still photography apparatus; designer underwater camera and high-resolution sonar equipment. Home: 100 Memorial Dr Cambridge MA 02142 Office: MIT Room 4-405 Cambridge MA 02139

EDGERTON, MARY ELIZABETH, kinetic researcher, consultant; b. Austin, Tex., Jan. 5, 1956; d. George Headley and Sally Sue (Williams) E. B.S. with honors in Physics, U. Tex.-Austin, 1976; Ph.D. in Biophysics, U. East Anglia, 1979. Research asst. SUNY, Stony Brook, 1980-81; project analyst Exxon Research and Engring. Co., Linden, N.J., 1981—. Contbr. articles to profl. jours. Mem. first aid team Exxon Research and Engring. Co. Cardio-Vascular Research Inst. fellow U. Calif.-San Francisco, 1979-80; Marshall scholar Marshall Commn., U.K., 1976-79. Mem. Brit. Biochem. Soc., Biophys. Soc., Soc. Indsl. and Applied Math. Subspecialties: Biophysics (biology); Mathematical Modelling. Current work: Mathematical modelling and computer simulation of complex systems. Office: Exxon Research and Engring Co Computing Tech and Services Div Linden NJ 07036

EDGREN, RICHARD ARTHUR, biologist, sci. adminstr.; b. Chgo., May 28, 1925; s. Richard Arthur and Helga D. (Corydon) E.; m. Margery Edith Kelly, June 7, 1952; children: Susan Ann, Jean Elizabeth. B.S., Northwestern U., 1949, M.S., 1951, Ph.D., 1952. Sr. investigator G.D. Searle Co., Chgo., 1952-60; mgr. endocrinology Wyeth Labs., Phila., 1960-71; dir. endocrinology Warner-Lambert/Parke-Davis, Morristown, N.J., 1971-75, Ann Arbor, Mich., 1975-78; dir. sci. affairs Syntex Labs., Palo Alto, Calif., 1978—; cons. in field. Editorial bd., assoc. editor: Internat. Jour. Fertility; contbr. numerous articles to profl. jours., chpts. to books. Bd. dirs., treas. U.S. Internat. Found. Studies in Reprodn. Served with U.S. Army, 1943-46; ETO. Decorated Purple Heart. Mem. Endocrine Soc., Am. Fertility Soc., Soc. Exptl. Biology and Medicine, Soc. Study of Reprodn., Am. Soc. Pharmacology and Exptl. Therapeutics, Ecol. Soc. Am., Am. Soc. Ichthyologists and Herpetologists, Royal Soc. Medicine, Sigma Xi, Internat. Soc. Reproductive Medicine (dir.). Subspecialties: Reproductive biology; Endocrinology. Current work: Safety of oral contraceptives; development of new contraceptive modalities, particularly analogues of LHRH. Patentee in field of contraceptives. Office: Syntex Labs 3401 Hillview Ave Polo Alto CA 94304

EDLUND, MILTON CARL, physicist; b. Jamestown, N.Y., Dec. 13, 1924. B.S., M.S., U. Mich., 1948, Ph.D., 1966. Physicist reactor physics, gaseous diffusion plant, 1948-49, Oak Ridge Nat. Lab., 1949-50; physicist, lectr. Sch. Reactor Tech., 1950-51, sr. physicist and sect. chief, 1953-55; mgr. devel. dept. Babcock & Wilcox Co., 1955-65, asst. mgr. atomic energy div., 1965-66; prof. U. Mich., 1966-67; planning cons. AEC, 1967-68; exec. v.p. Nuclear Assurance Corp., Atlanta, 1968-70; chmn. nuclear engring. Va. Poly. Inst. and State U., Blacksburg, 1970-74; dir. Center for Energy Research, 1974-78, prof. nuclear engring., 1978—; Vis. lectr. Swedish Atomic Energy Com., 1953. Author: (with S. Glasstone) Elements of Nuclear Reactor Theory, 1952, (with J. Fried) Desalting Technology, 1971. Recipient Ernest Orlando Lawrence award, 1965. Fellow Am. Nuclear Soc.; mem. Nat. Acad. Engring. Subspecialty: Nuclear fission. Current work: Development of new breeder reactors. Spl. research neuron diffusion, nuclear reactor design, energy policy analysis. Address: 302 Neil St Blacksburg VA 24060

EDMOND, JOHN MARMION, research geochemist, educator; b. Glasgow, Scotland, Apr. 27, 1943; s. Andrew John Sheilds and Christina (Marmion) E.; m. Mssoudeh Vafaei, Jan. 26, 1947; 1 son, Kazem Vafaei. B.Sc. with 1st class honors in Pure Chemistry, U. Glasgow, 1965; Ph.D. in Chem. Oceanograph, Scripps Instn. Oceanography, U. Calif.-San Diego, 1970. Asst. prof. to prof. marine geochemistry dept. earth and planetary scis. MIT, 1970—. Research, numerous publs. in field. Recipient Macelwane award Am. Geophys. Union, 1979. Mem. AAAS, Geochem. Soc. Subspecialties: Geochemistry; Oceanography. Current work: Processes controlling chemical composition of natural waters; geochemical cycle; geochemical evolution of environment of surface of earth. Office: MIT 334-201 Cambridge MA 02139

EDMONDS, HARVEY LEE, JR., neuropharmacologist, educator, cons. toxicologist; b. Leavenworth, Kans., Sept. 23, 1942; s. Harvey Lee and Esther Jane E.; m. Jeanne Carolyn Ford, July 10, 1970; 1 son: Harvey Lee III. B.A., U. Kans., 1964, B.S. in Pharmacy, 1967; Ph.D., U. Calif., Davis, 1974. Registered pharmacist, Kans., 1967. Grad. asst. U. Calif., Davis, 1971-74; asst. prof. pharmacology Wash. State U., Pullman, 1974-77; asso. prof., dir. research dept. anesthesiology U. Louisville, 1977-82, prof., dir. research, 1982—; cons. toxicologist, 1977—. Mem. bus. adv. bd. U.S. Senate, 1980—. Served to capt. U.S. Army, 1968-70. Decorated Army Commendation medal.; Grantee Epilepsy Found. Am., 1973, G.D. Searle & Co., 1974, Ky. Heart Assn., 1979, Distilled Spirits Council, 1980; Upjohn Pharms., 1982. Mem. Am. Epilepsy Soc., Am. Soc. Pharmacology and Exptl. Therapeutics, Internat. Union Pharmacology, Research Soc. Alcoholism, Soc. Neuroscis. Republican. Subspecialty: Neuropharmacology. Current work: Epilepsy and cerebral trauma; direct a series of basic science and clinical studies concerning the causes and treatment of epilepsy and brain damage associated with head trauma. Office: Dept Anesthesiology U Louisville Sch Medicine Louisville KY 40292

EDMONDS, MARY PATRICIA, biochemistry educator; b. Racine, Wis., May 7, 1922; s. Millard Samuel and Sarah (Gibbons) E. B.A., Milw. Downer Coll., 1943; M.A., Wellesley Coll., 1945; Ph.D., U. Pa., 1951. Research assoc. Montefiore Hosp. Research, Pitts., 1955-65; asst. prof. biochemistry U. Pitts., 1965-71, assoc. prof., 1971-76; prof., 1976—; mem. molecular biology study sect. NIH, 1973-77, mem. devel. therapeutics study sect., 1978-81. Contbr. articles to profl. jours. NIH grantee, 1962—. Mem. Am. Soc. Biol. Chemists, Am. Assn. Cancer Research. Subspecialties: Molecular biology; Biochemistry (biology). Current work: Nucleic acid metabolism and structure. Office: University Pittsburgh 527 Lansley Hall Pittsburgh PA 15260

EDMONDS, ROBERT LESLIE, forest microbiology educator; b. Sydney, Australia, May 6, 1943; s. Harold Melville and Elizabeth (Osborne) E.; m. Victory Corliss Lesher, Apr. 25, 1969; children: Nicole Thais, Stephen Robert. B.S., U. Sydney, Australia, 1964; M.S., U. Wash., 1968, Ph.D., 1971. Research scientist Forest Research Inst., Canberra, Australia, 1964-65, Australian Nat. U., Canberra, 1965-66; research asst. Coll. Forest Resources, U. Wash., Seattle, 1966-71, asst. prof. to prof., 1973-82, prof. forest microbiology, 1982—. Contbr. over 50 sci. articles to profl. publs. Mem. Ecol. Soc. Am., Am. Phytopath. Soc., Am. Meterol. Soc., Sigma Xi, Sigma Pi. Episcopalian. Subspecialties: Plant pathology; Microbiology. Current work: Diseases of forest trees; soil microbiology; aerobiology. Home: 4517 48th St NE Seattle WA 98105 Office: Coll Forest Resources U Wash Seattle WA 98195

EDMONDSON, DALE EDWARD, biochemistry educator; b. Morris, Ill., Oct. 13, 1942; s. Allen Dean and Evelyn (Johnson) E.; m. Joan Warren, Apr. 23, 1972. B.S., No. Ill. U., 1964; Ph.D., U. Ariz., 1970. Asst. research biochemist U. Mich.-Ann Arbor, 1970-72; asst./assoc. research biochemist U. Calif.-San Francisco, 1972-80; assoc. prof. Emory U., Atlanta, 1980—; mem. study sect. NIH, Bethesda, Md., 1980-84. Contbr. articles in field to profl. jours. NIH predoctoral fellow U. Ariz., 1967-70; NIH postdoctoral fellow U. Mich., U. Calif., 1971-74; NIH, NSF grantee U. Calif., Emory U., 1974—. Mem. Am. Chem. Soc., Am. Soc. Biol. Chemists, Sigma Xi. Subspecialty: Biophysical chemistry. Current work: Structure-funciton studies of oxidation-reduction enzymes. Home: 2866 Woodland Park Dr Atlanta GA 30345 Office: Emory U Dept Biochemistry Atlanta GA 30322

EDMONDSON, W(ALLACE) THOMAS, limnologist, educator; b. Milw., Apr. 24, 1916; s. Clarence Edward and Marie (Kelley) E.; m. Yvette Hardman, Sept. 26, 1941. B.S., Yale U., 1938, Ph.D., 1942; postgrad., U. Wis., 1938-39. Research asso. Am. Mus. Natural History, 1942-43, Woods Hole Oceanographic Instn., 1943-46; lectr. biology Harvard U., 1946-49; faculty U. Wash., Seattle, 1949—, prof., 1957—. Editor: Freshwater Biology (Ward and Whipple), 2d edit, 1959; contbr. articles to profl. jours. NSF sr. postdoctoral fellow, Italy, Eng. and Sweden, 1959-60; recipient Einar Naumann-August Thienemann Medal Internat. Assn. Theoretical and Applied Limnology, 1980. Mem. Nat. Acad. Scis., Am. Micros. Soc., Nat. Acad. Scis. (Cottrell award 1973), AAAS, Am. Soc. Limnology and Oceanography, Internat. Assn. Limnology, Am. Soc. Naturalists, Phycol. Soc. Am., Ecol. Soc. Am. Subspecialties: Zooplankton limnology. Current work: Mechanism of control of population productivity and abundance in lakes; eutrophication. Office: Dept Zoology U Wash Seattle WA 98195

EDWARDS, CARL NORMAND, psychologist, institute administrator, educator; b. Norwood, Mass., Jan. 22, 1943; s. Wilfred Carl and Cecile Marie-Anne (Pepin) E.; m. Mary Louise Buyse, Jan. 22, 1982. Student, Bridgewater (Mass.) State Coll., 1960-63; M.Ed., Suffolk U., 1969; postgrad., Harvard U., 1964-71, MIT, 1977-80. Cert. Mass. Bd. Registration of Psychologists. Cons. Harvard U., 1966-69, research fellow, 1969-71, lectr. social relations, 1971-72; asst. clin. prof. psychiatry Tufts U. Sch. Medicine, 1971-78, assoc. clin. prof., 1978—; dir. Four Oaks Inst., Norfolk, Mass., 1974—; cons. research analyst Cambridge Computer Assocs., 1966—; mem. field faculty Goddard Coll., Plainfield, Vt., 1972—; sr. assoc. Justice Resource Inst., Boston, 1972-74; chmn. Info. Industry Assn. Task Force on Edn. and Human Resource Devel., Washington, 1982—. Author: Drug Dependence, 1974; contbr. articles to profl. jours. Chmn. permanent bldg. com.

Town of Norfolk, Mass., 1981-83. Served with N.G. U.S. Army, 1960-65. Mem. Am. Psychol. Assn., Am. Soc. Info. Sci., Info. Industry Assn., Am. Statis. Assn. Clubs: Harvard of Boston, Appalachian Mt. Subspecialties: Information systems (information science); Health services research. Current work: Application of information science and technology to information and decision support systems in the health sciences and education; conceptual and organizational development of innovative programs incorporating applied information technology, and design and programming of university, corporate and government buildings to house and facilitate such programs. Home: 61 Winthrop St West Newton MA 02165

EDWARDS, DAVID JOEL, educator; b. Durham, N.C., Dec. 5, 1943; s. Joseph Philip and Rebecca Josephine (Kornblut) E.; m. Mary Dawn Herring, Oct. 8, 1972; children: Leah, Rachel. B.A., Duke U., 1966; Ph.D., U. N.C.-Chapel Hill, 1971. Postdoctoral fellow U. Pitts., 1971-73, asst. prof., 1973-78, 1978-81, assoc. prof., 1981—. Mem. AAAS, Soc. Neurosci., Am. Soc. Pharmacology and Exptl. Therapeutics, Sigma Xi. Subspecialties: Neurochemistry; Neuropharmacology. Current work: Amine metabolism in the brain. Home: 6652 Ridgeville St Pittsburgh PA 15217 Office: 546 Salk Hall U Pitts Pittsburgh PA 15261

EDWARDS, DONALD MERVIN, educator; b. Tracy, Minn., Apr. 16, 1938; s. Mervin B. and Helen L. (Halstenrud) E.; m. Judith Lee Wilson, Aug. 8, 1964; children: John, Joel, Jeffrey, Mary. B.S., S.D. State U., 1960, M.S., 1961; Ph.D. in Agrl. Engring, Purdue U., 1966. Registered profl. engr., Nebr. With Soil Conservation Service, U.S. Dept. Agr., Marshall, Minn., 1957-62; teaching, research asst. S.D. State U. and Purdue U., 1960-66; asso. prof. agrl. engring. U. Nebr. at Lincoln, 1966-71, prof., 1971-80; asst. dean Coll. Engring. and Tech., 1970-73; asso. dean, dir. Engring. Research Center, 1973-80; dir. Energy Research and Devel. Center, 1976-80; prof. and chmn. dept. agrl. engring Mich. State U., East Lansing, 1980—; collaborator, cons. to numerous industries and agys., 1966—; mem. Engring Accreditation Commn. of Accreditation Bd for Engring. and Tech. Contbr. numerous articles on irrigation, water pollution, remote sensing, energy, engring. edn. to profl. jours. Active Boy Scouts Am., Am. Field Service, 4-H; bd. dirs. Nat. Safety Council; mem. adv. bd. local sch.; past chmn. bd. dirs. Lincoln Transp. System.; mem. Christian edn. com. East Lansing Trinity Ch. Mem. Profl. Engrs. Nebr. (v.p. 1976-77), Mich. Soc. Profl. Engrs. (nat. dir.), Nat. Soc. Profl. Engrs., AAAS, Am. Soc. Agrl. Engrs. (nat. dir. profl. dept. 1977-79), Nat. Assn. Coll. Tchrs. Agr., Internat. Water Resources Assn., Sigma Xi, Alpha Gamma Rho. Clubs: Farmhouse, Triangle. Subspecialty: Agricultural engineering. Current work: Professor and chairman of programs in architectural engineering, agricultural engineering technology, building construction, power equipment technology, electrical technology. Home: 4557 Arrow Head Rd Okemos MI 48864

EDWARDS, GORDON STUART, toxicology cons., cancer researcher; b. Greenwich, Conn., Feb. 11, 1938; s. Alfred Conway and Eleanor Angela (Turnbull) E.; m.; children: Alexis, Margot. B.A., Amherst (Mass.) Coll., 1959; M.A., Harvard U., 1963; Sc.D., M.I.T., 1970. Diplomate: Am. Bd. Toxicology. Postdoctoral fellow Rockefeller U., N.Y.C., 1970-72; assoc. prof. pharmacology and genetics George Washington U., Washington, 1972-77, vice chmn. genetics program, 1974-77; chief biology sect. New Eng. Inst. Life Scis., Waltham, Mass., 1977-80; pres. ToxiCon Assocs., Natick, Mass., 1981—. Contbr. articles sci. jours. Damon Runyon fellow, 1970-72. Mem. Mutagenesis Assn. New Eng. (steering com., past pres.), Soc. Toxicology, Am. Assn. Cancer Research, Environ. Mutagen Soc., AAAS. Subspecialties: Environmental toxicology; Cancer research (medicine). Current work: Chem. carcinogens, environ. and occupational toxicology, cons. toxicologist, nitroso compounds, mutagenesis, carcinogenesis, occupational toxicology. Office: 34 Everett St Natick MA 01760

EDWARDS, JOHN ROBERT, psychology educator; b. Southampton, Eng., Dec. 3, 1947; emigrated to Can., 1954; s. Robert A. and Marjorie (Cass) E.; m. Suzanne De Lariceliere, Jan. 5, 1974; children: Colin, Emily, Katherine. B.A., U. Western Ont., 1969; M.A., McGill U., 1970, Ph.D., 1974. Research officer Que. Govt., Montreal, 1971; evaluator U.S. Govt., Island Pond, Vt., 1972-73; research fellow St. Patrick's Coll., Dublin, Ireland, 1974-77; asst. prof. psychology St. Francis Xavier U., Antigonish, N.S., Can., 1977-80, assoc. prof., 1980—; mem. adv. bd. Inst. Modern Langs., Silver Spring, Md., 1980—. Author: Language and Disadvantage, 1979, The Irish Language, 1983; editor: The Social Psychology of Reading, 1981, Language Minorities and Cultural Pluralism, 1983; rev. editor: Jour. of Language and Social Psychology, Bristol, Eng., 1982—. Mem. Can. Psychol. Assn., Brit. Psychol. Soc., Am. Assn. Applied Linguistics. Subspecialty: Sociolinguistics. Current work: Sociolinguistics in education; ethnicity and identity; bilingual education. Office: Dept Psychology St Francis Xavier U Antigonish NS Canada

EDWARDS, JOHN STUART, zoology educator, researcher; b. Auckland, N.Z., Nov. 25, 1931; came to U.S., 1962; s. Charles Stuart Marten and Mavis Margaret (Wells) E.; m. Ola Margery Shreeves, June 21, 1957; children: Richard Charles, Duncan Roy, Marten John, Andrew Zachary. B.Sc., U. Auckland, 1954, M.Sc. with 1st class honors, 1956; Ph.D., U. Cambridge, Eng., 1960. Asst. prof. biology Western Res. U., 1963-67, assoc. prof., 1967; assoc. prof. zoology U. Wash., Seattle, 1967-70, prof., 1970—, dir. biology program, 1982—. Guggenheim fellow, 1972-73; recipient Alexander von Humboldt award, 1981. Fellow Royal Entomol. Soc., AAAS; mem. Soc. Neurosci., Am. Soc. Zoologists, Western Apicultural Soc. (v.p. 1983-). Subspecialties: Neurobiology; Developmental biology. Current work: Developmental neurobiology, neuroembryology, regeneration neural specificity; alpine ecosystems snowfield fauna. Home: 5747 60th Av Seattle WA 98095 Office: Zoology Dept U Wash Seattle WA 98195

EDWARDS, LOIS ADELE, science educator; b. Kansas City, Mo., Oct. 14, 1940; d. Arch Paul and Phyllis Elmo (Enos) E. B.A., N.W. Nazarene Coll., 1962; Ph.D., U. Mich., 1979. Asst. prof. biology and physics Spring Arbor (Mich.) Coll., 1970-80, assoc. prof., 1980—; vis. assoc. prof. biochemistry Mich. State U., 1981-82. Subspecialty: Biochemistry (biology). Office: Spring Arbor College Spring Arbor MI 49283

EDWARDS, RAY CONWAY, physicist, engineering company executive; b. 1b. Belleville, Ont., Can., Sept. 1, 1913; came to U.S., 1915, naturalized, 1938; s. Ernest Alfred and Augusta Ann (Fee) E.; m. Marjorie Baisch, Dec. 17, 1951; children: David, Douglas, Diane, Ruth, Robert (dec.), Helen. B.A., UCLA, 1935. Registered profl. engr., N.Y., N.J., Va., Pa. Engr. Carrier Corp., Syracuse, N.Y., 1935-42; physicist U.S. Rubber Co., Passaic, N.J., 1943-46; founder, pres., chmn. bd. Edwards Engring. Corp., Pompton Plains, N.J., 1946—. Contb. chpts. to books. Mem. ASHRAE (life), Theta Delta Chi. Republican. Club: Smoke Rise. Subspecialties: Heat transfer; Environmental engineering. Current work: Heat transfer, hydrocarbon vapor condensation, noise control, air conditioning, refrigeration, automatic temperature controls, air pollution control, mechanical engineering, chemical engineering and physics shale oil. Patentee in heat transfer field. Home: 396 Ski Trail Kinnelon NJ 07405 Office: Edwards Engring Corp 101 Ancker Ave Pompton Plains NJ 07444

EDWARDS, SUZAN, astronomer, educator; b. Columbia, Mo., June 15, 1951; d. William Clark and Rose (Szywriel) E.; m. Duncan MacKinnon Chesley. B.A. in Physics, Dartmouth Coll., 1973; Ph.D. in Astronomy, U. Hawaii, 1980. Asst. prof. astronomy Smith Coll., Northampton, Mass., 1980—. Mem. Am. Astron. Soc., Astron. Soc. Pacific, Phi Beta Kappa, Sigma Xi. Subspecialties: Infrared optical astronomy; Radio and microwave astronomy. Current work: Star formation, stellar winds. Office: Dept Astronomy Smith Coll Northampton MA 01063

EDWARDS, TERRY WINSLOW, astrophysicist, educator; b. Sheboygan, Wis., Nov. 2, 1935; s. Albert C. and Helen L. E.; m. Alice J., Feb. 8, 1958; 1 son, Kent D. B.S. in Elec. Engring, U. Wis., 1958, M.S., 1961, Ph.D., 1968. Instr. physics U. Mo., Columbia, 1966-67, asst. prof., 1967-71, assoc. prof. physics and astronomy, 1971—; vis. assoc. prof. U. Rochester, N.Y., 1981-82. Contbr. articles to profl. jours. Mem. Am. Astron. Soc., Internat. Astron. Union, Astron. Soc. of Pacific, Tau Beta Pi, Eta Kappa Nu. Subspecialties: Theoretical astrophysics; Statistical physics. Current work: Teaching and research in stellar astrophysics, stellar structure and dense plasma theory. Office: U Mo 420 Physics Bldg Columbia MO 65201

EDWARDS, WILLIAM CHARLES, educator, naturalist; b. Waukegan, Ill., May 17, 1934; s. Henry Charles and Lillian E. (Yockey) E.; m. Nancy Beal, June 10, 1961; children: Jon, Ben. B.A., Carleton Coll., 1956; M.S., U. Wyo., 1958; Ph.D., U. Nebr., 1966. Cert. tchr., Wyo. Ranger-naturalist Grand Teton Nat. Park, summers 1959, 60, 61, Rocky Mountain Nat. Park, summer 1962; biology tchr. Central High Sch., Cheyenne, Wyo., 1958-63; asso. prof. Mankato (Minn.) Sate U., 1966-70; prof. biology, ecology and energy Laramie County Community Coll., Cheyenne, 1971—; dir. Wyo. Postsecondary Energy Edn. Consortium, 1980-82; cons. Author articles and editorials. Mem. Wyo. Ho. of Reps., 1974—. Nat. Endowment Humanities grantee, 1973—. Mem. Audubon Soc. (past pres.), Phi Delta Kappa, Sigma Xi. Democrat. Episcopalian. Lodge: Kiwanis. Subspecialties: Ecology. Current work: Translating scientific discoveries into lay language; updating state legislation dealing with science; research on alternative energy.

EDWARDSON, JOHN RICHARD, agronomist; b. Kansas City, Mo., Apr. 17, 1923; s. George Edward and Louise Marie (Sundstrom) E.; m. Mickie Newbill, Dec. 26, 1969; children: George, Elizabeth, Sarah. B.S., Tex. A. and M. U., 1948, M.S., 1949; Ph.D., Harvard U., 1954. Asst. agronomist Fla. Agrl. Expt. Sta., Gainesville, 1953-60, assoc. agronomist, 1960-66, agronomist, 1966—. Served with U.S. Army, 1942-45. Mem. AAAS, Genetics Soc. Am., Am. Phytopath. Soc. Subspecialties: Plant genetics; Plant virology. Current work: Cytoplasmic inheritance, cytology of virus induced inclusions, research on cytoplasmic male sterility in plants and on morphology, structure and location of virus inclusions. Home: 2721 SW 3d Pl Gainesville FL 32607 Office: U Fla 2559 HS and PP Bldg Gainesville FL 32611

EGAN, JOHN THOMAS, mathematician, computer scientist, consultant; b. Troy, N.Y., Mar. 20, 1937. B.S., St. Louis U., 1965; M.S., SUNY-Buffalo, 1969, Ph.D., 1976. Physicist Xerox Corp., Rochester, N.Y., 1965-66; computer scientist Bell Aerospace Corp., Niagara Falls, N.Y., 1967-70; asst. prof. SUNY-Buffalo, 1970-80; researcher NASA-Ames Research Ctr., Moffet Field, Calif., 1978-79; mathematician Naval Research Lab., Washington, 1980—; adj. prof. U. Santa Clara, 1978-80, George Mason U., 1982—; cons. in field. Recipient Outstanding Performance award Naval Research Lab., 1981; NASA-Ames Research Center intergovtl. exchange grantee, 1978-79; Stanford U. Am. Soc. Engring. Edn. summer faculty fellow, 1977, 80. Mem. AAAS, Assn. Computer Machinery, Soc. Cert. Data Processors. Democrat. Roman Catholic. Subspecialties: Graphics, image processing, and pattern recognition; Computer architecture. Current work: Video disc applications; enhancement of underwater imagery; special applications of computer graphics to chemical modeling; applying very large scale integrated circuit technology to passive sonar systems. Home: 4600 S Four Mile Run Dr Apt 731 Arlington VA 22204 Office: Naval Research Lab 4555 Overlook Ave Washington DC 20375

EGAN, MARIANNE LOUISE, immunologist; b. Jersey City, June 9, 1942; d. Joseph Lawrence and Thecla (Roesch) E.; m. David G. Pritchard, Dec. 27, 1975; 1 dau., Barbara Lynn. A.B. with high honors, Coll. St. Elizabeth, Convent Station, N.J., 1964; Ph.D., Jefferson Med. Coll., Phila., 1969. Instr. in biochemistry Jefferson Med. Coll., Phila., 1969-70; asst. research scientist City of Hope Nat. Med. Center, Duarte, Calif., 1970-72, assoc. research scientist, 1972-76; assoc. scientist Comprehensive Cancer Center U. Ala.-Birmingham, 1977—, assoc. scientist, 1979—, research asst. prof. immunobiology and immunology, 1976—; mem. med. adv. bd. Nat. Multiple Sclerosis Soc., Central Ala., 1977—; co-chmn. diagnosis and therapy working group Breast Cancer Task Force Com., Nat. Cancer Inst., Bethesda, Md., 1980—. Reviewer: Cancer Research Jour. of Nat. Cancer Inst, 1977—, Arthritis and Rheumatism, 1977-81; adv. editor: Molecular Immunology, 1971-74; contbr. numerous sci. articles to profl. publs. Recipient McClung award Beta Beta Beta, 1964; Paul Pinchunk award Jefferson Med. Sch., 1969; Nat. Multiple Sclerosis Soc. grantee, 1980-83; NIH grantee, 1976—. Mem. Am. Assn. Immunologists, Am. Assn. Cancer Research, Sigma Xi. Subspecialties: Immunobiology and immunology; Cell and tissue culture. Current work: Control of cellular interactions in autoimmune diseases; the production of murine and human hybridomas structure and function of human DR antigens. Office: Univ Ala 450 LHR Dept Microbiology Birmingham AL 35294

EGEL, LAWRENCE, psychologist, researcher; b. Chgo., June 11, 1940; s. Milton and Dorothy (Levin) E.; m. Maria Gutsmiedl, Jan. 12, 1967; children: Barbara Natasha, Lisa Jeanne. B.A., Roosevelt U., 1965; M.S., Ill. Inst. Tech., 1971, Ph.D., 1975. Lic. sch. and clin. psychologist, Ill. With Chgo.-Read Mental Health Ctr., 1968-72, asst. unit chief, 1971-72, research supr., 1972; crisis worker Ravenswood Hosp., Chgo., 1972; gen. psychologist Will County Mental Health Clinic, Joliet, Ill., 1973-76, chief psychologist, 1976-78; intern in sch. psychology So. Will County Coop. for Spl. Edn., Joliet, 1978-79; pvt. practice forensic, psychodiagnostic, neuropsychodiagnostic and therapeutic psychology, Joliet, 1976—; instr. in exptl. psychology and physiol. psychology Coll. St. Francis, 1974-76; adj. prof. psychology/counseling div. Govs. State U., 1978—; adj. prof. Loyola U., 1982—; v.p. HEM, Inc., Joliet, 1980—. Mem. Am. Psychol. Assn., Midwest Psychol. Assn., Ill. Psychol. Assn., Internat. Neuropsychology Soc., AAAS, Assn. Symbolic Logic, Nat. Acad. Neuropsychology, Neurosci. Soc., Nat. Assn. Sch. Psychologists, Ill. Sch. Psychologists Assn., Psy-Law, Amnesty Internat. Jewish. Subspecialties: Neuropsychology; Neurochemistry. Current work: Co-researcher on relationship of cell associated water to neuronal membrane ion transport. Co-developer Psy-DX, an electronic Halstead Neuropsychol. Test Battery with integrated computer for correlation of computerized tests for comprehensive neuropsychol diagnosis, 1980.

EGGER, ERICK LOWELL, veterinary surgeon, educator; b. Washington, Nov. 14, 1951; s. Henry Albert and Dollie E. B.S. with high distinction, Colo. State U.-Ft. Collins, 1973, D.V.M., 1975. Diplomate: Am. Coll. Vet. Surgeons. Intern Purdue U., 1975-76; resident in vet. surgery U. Mo., 1976-78; asst. prof. surgery Iowa State U., 1978-82, Colo. State U., 1982—. Mem. AVMA, Vet. Orthopedic Soc. Subspecialty: Surgery (veterinary medicine). Current work: External skeletal fixation for fracture repair, cruciate ligament injury, growth deformity correction. Office: Veterinary Hospital Colorado State University Fort Collins CO 80523

EGGER, MAURICE DAVID, neurobiologist; b. Bakersfield, Calif., June 21, 1936; d. Henry and Ida (Hoffman) E.; m. Ellen M., Sept. 4, 1958; children: Daniel, Rachel, Gideon. B.S. in Physics, Stanford U., 1958; M.S., Yale U., 1960, Ph.D., 1962. Instr. Sch. Medicine, Yale U., 1965-66, asst. prof., 1966-69, assoc. prof., 1969-74, Rutgers U. Med. Sch., 1974-78, prof., 1978—; research scientist devel. rev. group NIMH, HEW, 1975-79; neurobiology rev. group NIH, Dept. HHS, 1982—. Fellow AAAS; mem. Soc. Neurosci., Internat. Brain Research Orgn., Am. Psychol. Assn., Sigma Xi. Democrat. Jewish. Subspecialties: Neurophysiology; Neurobiology. Current work: Relationships between structure and function in the central nervous system. Patentee scanning optical microscope. Office: Anatomy Rutgers Medical School Piscataway NJ 08854

EGGERT, FRANK MICHAEL, dental educator, immunology researcher, dentist; b. Hamburg, W.Ger., Apr. 24, 1945; emigrated to Can., 1954, naturalized, 1960; s. Frank Paul and Suse (Schilling) E.; m. Susan Louise Denny, June 19, 1976; 1 son, Frank Matthew Arthur. D.D.S., U. Toronto, Ont., Can., 1969, M.Sc., 1971; Ph.D., U. Cambridge, Eng., 1978. Lectr. London Hosp. Med. Coll. Dental Sch., 1979-81; assoc. prof. stomatology Faculty Dentistry, U. Alta. (Can.), Edmonton, 1981-83, prof., 1983—. Contbr. articles on secretory immunity, histochemistry to profl. jours. Royal Coll. Surgeons Eng. research fellow, 1976; Alta. Found. Med. Research establishment grantee, 1982-85. Fellow Royal Soc. Medicine; mem. Royal Coll. Dentists Can., Biochem. Soc., Brit. Soc. Immunology, Internat. Assn. Dental Research. Club: Royal Can. Yacht (Tornto). Subspecialties: Immunology (medicine); Periodontics. Current work: Immunochemistry of secretory glycoproteins that aggregate bacteria; experimental pathology and histochemistry of periodontal disease. Office: U Alta Faculty Dentistry Edmonton AB Canada T6G 2N8

EGGLER, DAVID H(EWITT), geochemistry educator; b. Ashland, Wis., May 15, 1940; s. Willis A and Dorothy (Smith) E.; m. Betsey A. Eggler, Feb. 9, 1974; children: Aimee, Willis. A.B., Oberlin Coll., 1962; Ph.D., U. Colo., 1967. Research assoc. Pa. State U., 1967-70; asst. prof. Tex. A&M U., 1970-72; mem. staff Geophys. Lab., Washington, 1972-77; assoc. prof. geochemistry Pa. State U., 1977—; cons. NSF, Washington, 1982—. Assoc. editor: Am. Mineralogist, 1978—. Recipient L.R. Wager prize Internat. Assn. Volcanic Geochemistry of Earth's Interior, 1979. Fellow Mineral Soc. Am., Geol. Soc. Am.; mem. Am. Geophys. Union, Geochem. Soc., Geol. Soc. Washington. Subspecialties: Geochemistry; Petrology. Current work: Experimental petrology on multisystems related to mineralogy, geochemistry, and magmatic production from planetary interiors. Home: RD 1 Box SW 24 Belleville PA 17004 Office: Pa State U 211 Deike Bldg University Park PA 16802

EGORIN, MERRILL JON, oncologist; b. Balt., May 25, 1948; s. Nathan A. and Toba Rose (Rombro) E.; m. Karen Deborah Kantor, Aug. 6, 1969; children: Melanie Ann, Noah Michael. B.A., Johns Hopkins U., Balt., 1969, M.D., 1973. Diplomate: Am. Bd. Internal Medicine. Intern Johns Hopkins Hosp., 1973-74, asst. resident in medicine, 1974-75; clin. assoc. Balt. Cancer Research Center, 1975-78, sci. expert, 1978-81; staff physician U. Md. Hosp., 1981-82; assoc. prof. oncology, head div. devel therapeutics U. Md. Cancer Center, Balt., 1982—. Served to sr. asst. surgeon USPHS, 1975-78. Fellow ACP; mem. Am. Assn. Cancer Research, Am. Soc. Clin. Pharmacology, Am. Soc. Pharmacology and Exptl. Therapeutics. Jewish. Subspecialties: Cancer research (medicine); Pharmacology. Current work: Pharmacology of antineoplastic chemotherapeutic agts.

EHLERS, ERNEST GEORGE, geology educator, petrologic researcher; b. N.Y.C., Jan. 17, 1927; s. Ernest Frederick and Elsie Frieda (Buchenroth) E.; m. Diane Wiersema, June 17, 1950; children: Karen Alice Ehlers Chipman, Ernest George. M.S., U. Chgo., 1950, Ph.D., 1952. Geologist New Jersey Zinc Co., Sweetwater, Tenn., 1952-54; asst. prof. geology Ohio State U., 1954-57, assoc. prof., 1957-65, prof., 1965—; sr. lectr. U. Utrecht, Netherlands, 1965-66, U. Athens and Greek Geol. Survey, 1970-71. Author: The Interpretation of Geological Phase Diagrams, 1972, Petrology, 1982; contbr. articles to profl. jours. Served with USN, 1945. Recipient Fulbright Hays award, 1965-66, 70-71. Fellow Geol. Soc. Am., Mineral. Soc. Am.; mem. Sigma Xi. Subspecialties: Petrology; Geochemistry. Current work: Igneous and metamorphic petrology; optical crystallography; mineral synthesis. Office: Dept Geology and Mineralogy Ohio State U 104 W 19 Ave Columbus OH 43210

EHLERS, KENNETH WARREN, physicist, consultant; b. Dix, Nebr., Aug. 3, 1922; s. Walter Richard and Clara (Sievers) E.; m. Marion W., Mar. 4, 1947; 1 son, Gary Walter. Student, U. Colo., 1940-42; B.S., Okla. A&M U., 1944; postgrad., M.I.T., 1945; Ph.D., U. Calif., 1967. Head electronic aids dept. Landing Aids Expt. Sta., Arcata, Calif., 1946-50; staff sr. physicist Lawrence Berkeley Lab., U. Calif., 1950—; cons. Brobeck Industries, Avco Corp., Cyclotron Corp., TRW. Editorial bd.: Rev. Sci. Instruments; contbr. articles to profl. jours. Served with USN, 1942-46. Ford Found. grantee U. Chile, Santiago, 1970-74. Fellow Am. Phys. Soc.; mem. IEEE, Am. Vacuum Soc. (Sr., exec. bd. fusion div.), AAAS. Subspecialties: Plasma physics; Particle physics. Current work: Developing neutral beam injectors for heating controlled fusion reactor plasmas. Patentee. Home: 3129 Via Larga Alamo CA 94507 Office: Lawrence Berkeley Lab Bldg 4 Berkeley CA 94720

EHRENFELD, ELVERA, biochemist, virologist, cons.; b. Phila., Mar. 1, 1942; s. Henry and Eughenia (Frantz) E.; m. Donald F. Summers; 1 dau., Cynthia. B.A., Brandeis U., 1962; Ph.D., U. Fla., 1967. Asst. prof. cell biology Albert Einstein Coll. Medicine, 1969-74, assoc. prof., 1974; assoc. prof. biochemistry and cell, viral, molecular biology U. Utah Coll. Medicine, 1974-79, prof., 1979—. Recipient Career Devel. award USPHS, 1971-76, Tchr.-Scholar award Dreyfus Found., 1975-80, Faculty award Merck, 1971; NIH grantee, 1974—; NSF grantee, 1970-80. Mem. Am. Soc. Biol. Chemists, Am. Soc. Microbiology. Subspecialties: Virology (biology); Biochemistry (biology). Current work: Replication of RNA viruses. Office: Dept Microbiology U Utah Med Center Salt Lake City UT 84132

EHRENPREIS, SEYMOUR, pharmacologist, educator; b. N.Y.C., June 20, 1927; s. William and Ethel (Balk) E.; m. Bella R. Goodman, June 30, 1954; children: Mark, Eli, Ira. B.S., CCNY, 1949; Ph.D., N.Y. U., 1953. Research assoc. U. Pitts., 1953-55; instr. chemistry Cornell U., Ithaca, N.Y., 1955-57; asst. prof. biochemistry and neurology Coll. Physicians and Surgeons, Columbia U., N.Y.C., 1957-61; assoc. prof. pharmacology Georgetown U., Washington, 1961-69; head neuropharmacology Inst. Med. Research and Studies, N.Y.C., 1969-70; head pharmacology N.Y. State Research Inst. Neurochemistry and Drug Addiction, Ward's Island, N.Y., 1971-76; adj. prof. pharmacology Columbia Coll. Pharm. Sci., N.Y.C., 1972-76; prof. chmn. dept. pharmacology Univ. Health Sci. (Chgo. Med. Sch.), North Chgo., 1976—; vis. prof. Keio and Tokyo univs., Japan, summer 1974. Contbr. numerous articles on pharmacology to profl. jours.; editor: Cholinergic Mechanisms, 1966, Neurosci. Research, 1967-71,

Neurosci. Revs, 1974-76, (with A. Neidle) Methods in Narcotics Research, 1974. Served with USN, 1945-46. Recipient Meritorious Service award Coll. Pharm. Sci., Columbia U., 1976; Morris L. Parker award Univ. Health Sci./Chgo. Med. Sch., 1981; NIH grantee, 1961-82; NSF grantee, 1963-68; Hoffman-LaRoche grantee, 1976-78. Fellow AAAS, Am. Inst. Chemists; mem. Am. Soc. Biol. Chemistry, Am. Soc. Pharmacology and Exptl. Therapeutics, Soc. Neurosci., Sigma Xi. Subspecialties: Neuropharmacology; Neurobiology. Current work: Analgesic mechanisms, functional role of endorphins, enkephalinase inhibitors. Office: 3333 Green Bay Rd North Chicago IL 60064

EHRLICH, CLARENCE EUGENE, gynecologist, educator; b. Rosenberg, Tex., Oct. 19, 1938; m.; children: Bradley, Tracey, Suzanne. B.A., U. Tex., 1961; M.D., Baylor U., 1965. Intern Phila. Gen. Hosp., 1965-66; resident in ob-gyn Tulane U., 1966-69; fellow in gynecol. oncology M.D. Anderson & Tumor Inst., Houston, 1971-73; mem. faculty Ind. U. Med. Center, Indpls., 1973—, prof. ob-gyn, 1981—, chmn. ob-gyn dept., 1982—, Coleman prof., 1982—, dir. gynecol. oncology, 1973-82. Contbr. chpts. to med. textbooks, articles to profl. jours. Served to maj. USAF, 1969-72. Gynecologic Oncology Group grantee, 1980—. Fellow ACS; mem. AMA, Am. Assn. Cancer Research, Radium Soc., Conrad G. Collins Soc., Am. Soc. Clin. Oncology, Assn. Profs. Ob&Gyn, Central Assn. Obstetrics and Gynecology, Gynecol. Oncology Group, Sigma Xi. Subspecialties: Obstetrics and gynecology; Chemotherapy. Office: 926 W Michigan Indianapolis IN 46223

EHRLICH, GERT, educator; b. Vienna, Austria, June 22, 1926; came to U.S., 1939, naturalized, 1945; s. Leopold and Paula Marie (Kucera) E.; m. Anne Vodges Alger, Apr. 27, 1957. A.B., Columbia U.; A.M., Harvard U., 1950, Ph.D., 1952. Research asso. dept. physics U. Mich., Ann Arbor, 1952-53; mem. research staff Gen. Electric Research Lab., Schenectady, 1953-68; prof. phys. metallurgy U. Ill., Urbana-Champaign, 1968—, research prof. Contbr. articles to profl. jours. Served with U.S. Army, 1945-47. Fellow Am. Phys. Soc., N.Y Acad. Sci.; mem. Am. Chem. Soc. (Kendall award 1982), Am. Vacuum Soc. (Medard W. Welch award 1979), Phi Beta Kappa, Sigma Xi. Subspecialties: Surface Physics; Surface chemistry. Current work: Direct observation of atomic phenomena on solid surfaces; surface reactions and surface diffusion. Office: Coordinated Science Lab University of Illinois 1101 W Springfield Ave Urbana IL 61801

EHRLICH, KENNETH C., chemist; b. N.Y.C., Sept. 23, 1943; s. Louis H. and Sylvia (Stark) E.; m. Melanis Ellis, June 6, 1966; children: Emily Myung-Hee, Anilin. B.A., Columbia Coll., 1965; Ph.D., SUNY-Stony Brook, 1969. Research assoc. Columbia U., N.Y.C., 1969-72, La. State U. Sch. Medicine, New Orleans, 1972-75; sr. chemist Gulf South Research Inst., New Orleans, 1975-80; research chemist So. Regional Research Ctr., U.S. Dept. Agr., New Orleans, 1981—; adj. assoc. prof. Tulane Sch. Medicine, New Orleans, 1975—. Pres. New Orleans Suzuki Parents Forum, 1981-82. Mem. Am. Chem. Soc., N.Y. Acad. Scis., Assn. Ofcl. Analytical Chemists, Sigma Xi. Democrat. Jewish. Subspecialties: Organic chemistry; Biochemistry (biology). Current work: Mycotoxins, proteoglycan biochemistry. Home: 1450 Crescent New Orleans LA 71022 Office: USDA So Regional Research Ctr PO Box 19687 New Orleans LA 70179

EHRLICH, LOUIS WILLIAM, mathematician; b. Balt., Oct. 4, 1927; s. Isaac and Rae (Arge) E.; m. Barbara Joyce Belkin, Feb. 8, 1959; children: Tamietta Lynn, Kimberly Ann, Todd Aaron. B.S. in Chem. Engring, U. Md., 1951, M.S. in Math, 1956, Ph.D., U. Tex., Austin, 1963. Project engr. Hercules Powder Co., Cumberland, Md., 1951-54; mathematician TRW-STL, Los Angeles, 1956-59; programmer analyst U. Tex.-Austin, 1959-62; mathematician, numerical analyst Johns Hopkins Applied Physics Lab., Laurel, Md., 1962—; instr. Evening Coll., 1963—. Contbr. articles to profl. jours. Bd. dirs. Montessori Soc. Central Md., 1975-81. Served with AUS, 1945-47. Mem. Am. Math. Soc., Assn. Computing Machinery (bd. dirs. Signum 1977-80), Soc. Indsl. and Applied Math. Democrat. Jewish. Subspecialties: Applied mathematics; Numerical analysis. Current work: Numerical solution of partial differential equations; numerical linear algebra. Home: 3 Falshire Ct Randallstown MD 21133 Office: Johns Hopkins Applied Physics Lab Johns Hopkins Rd Laurel MD 20707

EHRLICH, PAUL RALPH, biology educator; b. Phila., May 29, 1932; s. William and Ruth (Rosenberg) E.; m. Anne Fitzhugh Howland, Dec. 18, 1955; 1 dau., Lisa Marie. A.B., U. Pa., 1953; A.M., U. Kans., 1955, Ph.D., 1957. Research assoc. U. Kans., Lawrence, 1958-59; asst. prof. biol. scis. Stanford, 1959-62, asso. prof., 1962-66, prof., 1966—, Bing prof. population studies, 1976—, dir. grad. study dept. biol. scis., 1966-69, 1974-76; cons. Behavioral Research Labs., 1963-67; cons. biology, editor in population biology McGraw Hill Book Co., N.Y.C., 1964—. Author: How to Know the Butterflies, 1961, Process of Evolution, 1963, Principles of Modern Biology, 1968, Population Bomb, 1968, 2d edit., 1971, Population, Resources, Environment: Issues In Human Ecology, 1970, 2d edit., 1972, How to Be a Survivor, 1971, Global Ecology: Readings Toward a Rational Strategy for Man, 1971, Man and the Ecosphere, 1971, Introductory Biology, 1973, Human Ecology: Problems and Solutions, 1973, Ark II: Social Response to Environmental Imperatives, 1974, The End of Affluence: A Blueprint for the Future, 1974, Biology and Society, 1976, Race Bomb, 1977, Ecoscience: Population, Resources, Environment, 1977, The Golden Door: International Migration, Mexico, and the U.S, 1979, Extinction: The Causes and Consequences of the Disappearance of Species, 1981; contbr. articles to profl. jours. Fellow Calif. Acad. Scis., Am. Acad. Arts and Scis.; mem. Soc. for Study Evolution, Soc. Systematic Zoology, Am. Soc. Naturalists, Lepidopterists Soc., Am. Mus. Natural History (hon. life mem.). Subspecialties: Population biology; Ecology. Current work: Evolution and ecology of natural population; plant-herbivore coevolution; policy research on human population/resources/environment. Address: Biological Scis Stanford U Stanford CA 94305

EHRLICH, ROBERT STARK, biochemist; b. N.Y.C., Aug. 30, 1940; s. Louis Herman and Sylvia Ray (Stark) E.; m. Marion Faith Stern, May 28, 1966; 1 dau., Heather. A.B., Columbia U., 1962; Ph.D., Rutgers U., 1969. Asst. prof. Muskingum Coll., New Concord, Ohio, 1969-70; instr. CCNY, 1970-71; postdoctoral fellow Rutgers Med. Sch., Piscataway, N.J., 1971-73; research assoc. U. Del., Newark, 1973-83, assoc. scientist, 1983—. Contbr. articles to profl. jours. Chmn. conservation com. Audubon Soc., New Castle County, Del., 1982—; adv. com. Biden, Wilmington, 1981—. N.Y. State Regents Sci. fellow, 1958. Mem. Am. Soc. Biol. Chemists, Am. Phys. Soc. Democrat. Subspecialty: Biophysical chemistry. Current work: Elucidation of structure-function relationships in enzymes; techniques used are binding studies, NMR, chemical modification and unfolding and refolding of proteins. Home: 1424 Carson Rd Wilmington DE 19803 Office: Chemistry Dept U Del Newark DE 19711

EHRLICH, WALTER, psychiatry researcher, educator; b. Bosicany, Bohemia, Czechoslovakia, Sept. 22, 1915; came to U.S., 1967; s. Karl and Irma (Stein) E.; m. Helli Egerer, Jan. 10, 1940; children: Eva, Karel, Sonia. B.S., Gymnasium, Carlsbad, 1934; M.D., Charles U., Prague, 1947; Cand. Sci., Acad. Sci., Prague, 1961. Asst. in pathology Charles U., Prague, Czechoslovakia, 1947-48, asst. in internal medicine, 1948-51; chief research group Inst. Cardiovascular Research, Prague, 1951-63, Inst. Hygiene, 1963-66; assoc. prof. psychiatry Johns Hopkins U., Balt., 1967—, assoc. prof. environ. health, 1967—. Author: Pharmacodynamische Analyse, 1962. Served with Allied Armies, World War II. NIH research grantee, 1967—. Mem. Am. Physiol. Soc. (fellow cardiovascular sect.), Am. Heart Assn. Democrat. Jewish. Subspecialty: Physiology (medicine). Current work: Cardiovascular and respiratory changes in awake, active animals, effects of respiration on circulation, peripheral circulation. Home: 6104 Northwood Dr Baltimore MD 21212 Office: Johns Hopkins Sch Hygiene 615 N Wolfe St Baltimore MD 21205

EHRLICH, YIGAL H., neurochemist, research, educator; b. Tel Aviv, Oct. 9, 1943; U.S., 1972, naturalized, 1982; s. Arthur and Regina (Eisenberg) E.; m. Elizabeth H. Kornecki, Mar. 18, 1983. M.Sc. in Microbiology, Tel Aviv U., 1968; Ph.D. in Biochemistry, Weizmann Inst. Sci., Rehovot, Israel, 1972. Postdoctoral fellow Cresap Neurosci. Lab., Northwestern U., Evanston, Ill., 1972-73, vis. asst. prof. dept. psychology and dept. physiology, 1973-75; research scientist Mo. Inst. Psychiatry, St. Louis, 1975-79; research assoc. prof. depts. psychiatry and biochemistry U. Vt. Coll. Medicine, Burlington, 1980—. Sr. editor: Modulators, Mediators and Specifiers in Brain Function, 1979; contbr. articles to sci. jours. Served with Israeli Def. Army. Recipient award Epilepsy Found. Am., 1977; NSF grantee, 1975, 79, 82; NIH grantee, 1979-82. Mem. Am. Soc. for Neurochemistry, Soc. for Neurosci., Internat. Soc. for Neurochemistry, N.Y. Acad. Scis., Sigma Xi. Subspecialties: Neurochemistry; Psychobiology. Current work: Research on molecular mechanisms underlying neuronal adaptation and synaptic plasticity, focusing on the role of protein phosphorylation in the regulation of neural receptors and long-lasting alterations induced in their function by hormonal and pharmacological stimulations. Home: 16 Brookwood Dr South Burlington VT 05401 Office: Dept Psychiatry U Vt Coll Medicine Burlington VT 05405

EHRREICH, STEWART JOEL, pharmacologist, government official; b. Bklyn., Mar. 24, 1936; s. Harry and Mollie Frances (Teller) E.; m. Rhoda Elaine Ehrreich, Aug. 13, 1960; children: William, Steven. B.S., CUNY, 1957; M.S., SUNY–Bklyn., 1961, Ph.D., 1963. Sr. investigator Smith Kline & French Labs., Phila., 1965-69; sect. leader Geigy Chem. Corp., Arosley, N.Y., 1969-71, Schering Corp., Bloomfield, N.J., 1971-77; pharmacologist FDA, Rockville, Md., 1977-79, dep. dir. cardiorenal drug products div., 1979—. Contbr. numerous sci. articles to profl. publs. USPHS fellow Cornell U. Med. Coll., 1963-65. Mem. Am. Soc. Pharmacology and Exptl. Therapeutics, Acad. Pharm. Scis. Club: Montgomery Amateur Radio (Rockville, Md.). Subspecialty: Pharmacology. Current work: Pharmacology of vasodilator and cardiac stimulating agents. Patentee antihypertensive agents.

EHST, DAVID ALAN, physicist; b. Bryn Mawr, Pa., May 2, 1948; s. Donald D. and Ruth Newman (Mann) E.; m. Barbara Ann Gramley, Jan. 30, 1949; children: Benjamin, Michael. B.A., M.S., U. Pa., 1970; Sc.D., MIT, 1976. Nuclear Engr. Argonne (Ill.) Nat. Lab., 1977—; lectr. in field; conducts seminars in field. Contbr. articles to profl. jours. Mem. Am. Phys. Soc., Phi Beta Kappa. Republican. Methodist. Club: Soc. for Preservation and Encouragement of Barber Shop Quartet Singing in Am. Subspecialties: Nuclear fusion; Plasma. Current work: Magnetic fusion reactor research/design systems studies. Office: 205 Argonne Nat Lab Argonne IL 60439 Home: 6028 Woodward Ave Downers Grove IL 60516

EIBER, ROBERT JAMES, civil engr.; b. Cleve., July 7, 1933; s. Harry E. and Grace A. E.; m. Carol A. Rankin, May 28, 1960; children: Jeffrey, Jill. B.S. in Civil Engring, Case Western Res. U., 1955, M.S. in Structural Engring, 1958. Instr. Case Western Res. U., Cleve., 1955-59; researcher Battelle Columbus (Ohio) Labs., 1959-75, mgr. fracture sect., 1975-80, mgr. stress analysis and fracture sect., 1980—. Contbr. articles to profl. jours. Mem. ASME, ASCE, ASTM, Sigma Xi, Theta Tau, Tau Beta Pi. Lutheran. Club: Columbus Yacht. Subspecialties: Materials (engineering); Fracture mechanics. Current work: Fracture initiation, propagation and arrest of fractures in piping systems. Fracture control, piping failure analyses, pipelines, railroad tank cars, nuclear piping fractures. Patentee in field. Home: 4279 Camborne Rd Columbus OH 43220 Office: 505 King Ave Columbus OH 43201

EICHHOLZ, GEOFFREY G., nuclear engineering educator; b. Hamburg, W. Ger., June 29, 1920; s. Max and Adele Daisy (Elias) E. B.S. in Physics, U. Leeds, Eng., 1942, PhD., 1948, D.Sc., 1979. Exptl. officer Brit. Admiralty, Whitley, Surrey, Eng., 1942-46; asst. prof. physics U. B.C. (Can.), Vancouver, 1947-51; head physics and radiotracer subdiv. Can. Bur. Mines, Ottawa, Ont., 1951-63; prof. nuclear engring. Ga. Inst. Tech., Atlanta, 1963—; nuclear cons., cons. archtl. acoustics, 1965—. Author: Environmental Aspects of Nuclear Power, 1977; author: Nuclear Radiation Detection, 1979; editor: Radioisotope Engineering, 1972. Recipient Outstanding Tchr. award Ga. Inst. Tech., 1973. Fellow Am. Nuclear Soc. (chmn. isotopes and radiation div. 1967-68); mem. Am. Phys. Soc., Health Physics Soc., Can. Assn. Physicists, Inst. Physics. Subspecialties: Nuclear engineering; Radiation Protection. Current work: Migration of radioactive wastes; radiation detectors, applied radiation technology, natural radiation background, architectural acoustics. Office: Ga Inst Tech Atlanta GA 30332 Home: 1784 Noble Dr N Atlanta GA 30306

EICHHORN, GUNTHER LOUIS, chemist; b. Frankfurt am Main, Ger., Feb. 8, 1927; s. Fritz David and Else Regina (Weiss) E.; m. Lotti Neuhaus, June 25, 1964; children—David Mark, Sharon Julie. A.B. in Chemistry, U. Louisville, 1947; M.S., U. Ill., 1948, Ph.D., 1950. Asst. prof., then asso. prof. chemistry La. State U., 1950-57; commd. officer USPHS, 1954-57; asso. prof. chemistry Georgetown U., 1957-58; guest scientist Naval Med. Research Inst., 1957-58; chief sect. molecular biology Gerontology Research Center, Nat. Inst. Aging, NIH, Balt., 1958-78, chief lab. cellular and molecular biology and head sect. inorganic biochemistry, 1978—; pres. Nat. Inst. Child Health and Human Devel. Assembly Scientists, 1972-73; mem. panel nickel NRC, 1974; distinguished lectr. Mich. State U., 1972; condr. seminars, lectr. in field. Editor: Inorganic Biochemistry, 1973; co-editor: Advances in Inorganic Biochemistry, 1978—; mem. editorial bds. profl. jours.; Author papers in field. Gen. Aniline and Film Co. grantee, 1949; postdoctoral fellow Ohio State U., summers 1951, 52; recipient Woodcock medal U. Louisville, 1947; Md. Chemist award, 1978; NIH Dir.'s award, 1979. Fellow AAAS, Am. Inst. Chemists, Gerontol. Soc. (fin. com. 1980); mem. Am. Chem. Soc., N.Y. Acad. Scis., Am. Inst. Biol. Chemists, Biophys. Soc. Subspecialty: Molecular biology. Home: 6703 97th Ave Seabrook MD 20801 Office: Gerontology Research Center NIH Baltimore City Hosps Baltimore MD 21224

EIFRIG, DAVID ERIC, ophthalmology educator; b. Oak Park, Ill., Jan. 4, 1935; m.; 4 children. B.A., Carleton Coll., 1956; M.D., Johns Hopkins U., 1960. Diplomate: Am. Bd. Ophthalmology. Asst. prof. ophthalmology U. Ky., 1968-70; assoc. prof. U. Minn., Mpls., 1970-77; prof., chmn. dept. ophthalmology U. N.C., Chapel Hill, 1977—. Mem. Retina Soc., Am. Acad. Ophthalmology, ACS. Subspecialty: Ophthalmology. Office: Dept Ophthalmology U NC Chapel Hill NC 27514

EIGEN, DARYL JAY, psychologist, engineering administrator; b. Milw., July 29, 1947; s. David J. and Pearl (Rice) E.; m. Carol A. Kois, Mar. 30, 1972; children: Tony, Molly. B.A. in Psychology, U. Wis.-Milw., 1972, M.S. in Elec. Engring, 1973; postgrad., Rutgers U., 1974-75; Ph.D. in Indsl. Engring, Northwestern U., 1981. Teaching asst. U. Wis., Milw., 1971-72, research asst., 1972-73; mem. tech. staff Bell Labs., Piscataway, N.J., 1973-75, Naperville, Ill., 1975-81, supr. tech. staff, 1981—, reviewer, Naperville, 1980—; session chmn. Automatic Control Symposium, Milw., 1976; organizer Workplace in the Info. Age, Murray Hill, N.J., 1982. Contbr. articles in areas of pattern, recognition, human factors and methodologies exptl. to profl. jours. Adviser Explorer Scouts, Naperville, 1982; coordinator Affirmative Action. Served with USMC, 1966-68; Vietnam. Recipient outstanding personal contbns. and commitment award Bell Labs., Naperville, 1982; Chancellor's Office scholar U. Wis.-Milw., 1970; U.Wis.-Milw. grantee, 1971; NASA Grantee, 1972; Bell Labs. grantee, 1980. Mem. IEEE (sec., treas. Computer Soc. Chgo. 1982, v.p., reviewer transactions Piscatawa 1973, Naperville, 1977), Am. Psychol. Assn., Human Factors Soc., AAAS, Tau Beta Pi. Subspecialties: Human factors engineering; Graphics, image processing, and pattern recognition. Current work: Human-machine dialogues, quasi-experimental designs, computer based tools for human factors research, controlled preservice testing of new telephone services, system analysis and performance analysis. Designer phone service charge-a-call, 1976, calling card service, 1978, teleconferencing, 1982. Home: 1541 Fender Rd Naperville IL 60540 Office: Bell Labs Naperville-Wheaton Rd Naperville IL 60566

EIL, CHARLES, endocrinologist, researcher; b. Mpls., Dec. 15, 1946; s. Harry Meyer and Lois Helen (Latts) E.; m. Adele Ruth Geffen, July 9, 1978; children: Andrew, Matthew. B.A., U. Rochester, 1968; Ph.D., 1972; M.D., 1974. Intern U. Mich. Med. Ctr., 1974-75, resident, 1975-76; fellow in endocrinology NIH, Bethesda, Md., 1976-80; commd. lt., comdr U.S. Navy, 1980; staff endocrinologist Nat. Naval Med. Ctr., Bethesda, 1980-82, asst. br. chief, 1982—; cons. endocrinology NIH, 1980—; asst. prof. medicine Uniformed Services U. Health Scis., 1980—. Mem. Endocrine Soc., Am. Fedn. Clin. Research, Am. Soc. Bone and Mineral Research. Jewish. Subspecialties: Endocrinology; Cell and tissue culture. Current work: Molecular defects in diseases of hormone resistance, such as testicular feminization, vitamin D-dependent rickets. Home: 2940 Northampton St NW Washington DC 20015 Office: PO Box 396 Nat Naval Med Ctr 8901 Rockville Pike Bethesda MD 20814

EILERT, JEFFRIES HARVEY, chemistry educator, consultant; b. Aurora, Ill., Nov. 16, 1940; s. William G. and Geraldine L. (Divekey) E.; m. Mary I. Breckenridge, Dec. 31, 1971; 1 dau., Eloise B. B.S., So. Meth. U., 1962; Ph.D., U. Tex.-Austin, 1971. Instr. chemistry Bee County Coll., Beeville, Tex., 1971-73, George Williams Coll., Downers Grove, Ill., 1973; chief adminstr. St. Charles Med. Ctr., Aurora, 1974-78; prof., chmn. dept. chemistry Aurora Coll., 1978—. Contbr. articles to sci. jours. Bd. dirs. Vis. Nurse Assn., 1976—, Sr. Citizens Planning Services, 1975-77, Aurora YMCA Swim Team, 1978-81; vestry Trinity Episcopal Ch., 1982. NSF fellow, 1968-71. Mem. Am. Chem. Soc. (acad. standards com. Chgo. sect.), Assoc. Colls. Chgo. Subspecialties: Organic chemistry; Biochemistry (biology). Current work: Investigation of chemical reaction mechanisms. Office: 347 S Gladstone Aurora IL 60507

EINHELLIG, FRANK ARNOLD, plant physiologist, educator, researcher, clergyman; b. Independence, Mo., July 7, 1938; s. Robert Frank and Bernice Louise (Landsberg) E.; m. Gertrude Inez Norris, Apr. 1, 1961; children: Robert Frank, Richard Ray. A.A., Graceland Coll., 1957; B.S. in Agr. (Fribourg scholar), Kans. State U., 1960, U. Kans., Lawrence, 1961; M.N.S., U. Okla., 1964; Ph.D. in Botany, U. Okla., 1969. Tchr. sci. Shawnee Mission (Kans.) High Sch. Dist., 1961-67; mem. faculty U. S.D., 1969—, asst. prof. biology, 1969-73, assoc. prof., 1973-78, prof., 1978—; dir. various grants; ordained to ministry Reorganized Ch. of Jesus Christ of Latter Day Saints, 1959; asst. pastor congregation, Sioux City, Iowa, 1979, 82-83. Contbr. articles in allelopathy to profl. jours. Served to sgt. USNG, 1956-64. NSF fellow, summers 1962-64, 65-67, 69; NDEA fellow, 1967-69; NSF grantee, 1974-80, 81-82; Dept. Energy grantee, 1979; Dept. Interior Office Water Resources grantee, 1979-82. Mem. Am. Soc. Plant Physiologists, Plant Growth Regulators Soc., Weed Sci. Soc. Am., S.D. Acad. Sci., Sigma Xi, Phi Sigma, Phi Delta Kappa. Democrat. Subspecialty: Plant physiology (biology). Current work: Research on mechanisms of plant interaction through biochemical transfer (allelopathy). Home: 1111 Ridgecrest Vermillion SD 57069 Office: Dept Biology Churchill-Haines Lab U SD Vermillion SD 57069

EINHORN, DANIEL, physician; b. Tel Aviv, Israel, Mar. 1, 1951; came to U.S., 1953, naturalized, 1953; s. Marcel and Lori (Haller) E. B.A., Yale U., 1973; M.D., Tufts U., 1977. Diplomate: Am. Bd. Internal Medicine, Am. Bd. Endocrinology. Asst. chief service Douglas Hosp., Montreal, 1966-67; intern in medicine Beth Israel Hosp., Boston, 1977-78, resident in psychiatry, 1978-79, resident in medicine, 1979-81, staff, 1981—; clin. fellow Harvard Med. Sch., Boston, 1977-81, research fellow, 1981—; staff physician Jewish Meml. Hosp., Boston, 1981—, Southwood Hosp., Norfolk, Mass., 1982—; courtesy staff Malden (Mass.) Hosp., 1983. Mem. AAAS, Am. Heart Assn., Am. Fedn. Clin. Research, ACP, Phi Beta Kappa, Alpha Omega Alpha. Subspecialties: Neuroendocrinology; Internal medicine. Current work: Metabolism, obesity, hypertension, clinical endocrinology, opiates, sympathetic nervous system, adrenal, catecholamines, steroids. Home: 105 Trowbridge St Cambridge MA 02138 Office: Harvard Med Sch 330 Brookline Ave Boston MA 02215

EINZIGER, ROBERT EMANUEL, research scientist, physicist; b. Asbury Park, N.J., Feb. 19, 1945. B.S. in Physics, Ga. Tech., 1967, M.S., Rensselaer Poly. Inst., 1973, Ph.D., 1973. Postdoctoral scientist Argonne (Ill.) Nat. Lab., 1974-76, asst. scientist, Idaho Falls, 1976-79; sr. scientist Westinghouse Hanford Co., Richland, Wash., 1979—. Mem. Am. Nuclear Soc., Am. Phys. Soc., Sigma Xi, Sigma Pi Sigma, Tau Beta Pi. Subspecialties: Metallurgical engineering; Nuclear fission. Current work: Behavior of spent nuclear fuel during dry storage or disposal. Home: 2363 Davison Ave Richland WA 99352 Office: Westinghouse Hanford Co PO Box 1970 Mail Stop W/A-40 Richland WA 99352

EISEN, EUGENE J., animal science and genetics educator; b. N.Y.C., May 14, 1938; s. Abraham and Fay (Hantman) E.; m. Jacqueline Serxner, Aug. 27, 1960; children: Arri, Avram, Andrea. A.A.S., SUNY-Farmingdale, 1957; B.S.A., U. Ga., 1959; M.S., Purdue U., 1962, Ph.D., 1965. Research asst. Purdue U., West Lafayette, Ind., 1959-62, instr. genetics, 1962-64; asst. prof. animal sci. and genetics N.C. State U., 1964-67, assoc. prof., 1967-73, prof., 1973—; vis. assoc. prof. U. Calif.-Davis, 1970-71; vis. prof. Inst. Animal Genetics, U. Edinburgh, Scotland, 1979-80. Contbr. numerous articles to profl. jours. Mem. Genetics Soc. Am., Am. Soc. Animal Sci., AAAS, Sigma Xi (N.C. State U. chpt. Young Researchers award 1972), Phi Kappa Phi, Gamma Sigma Delta. Subspecialties: Animal genetics; Animal breeding. Current work: Genetics of maternal effects; selection; genetics of obesity. Home: 308 Northfield Dr Raleigh NC 27609 Office: Dept Animal Sci NC State U Raleigh NC 27650

EISENBARTH, GEORGE STEPHEN, physician, educator, researcher; b. N.Y.C., Sept. 17, 1947; s. John and Esther (Davidowitz) E.; m. Frieda Sauer, July 12, 1969; children: Stephanie, Stephan. B.A., Columbia U., 1969; Ph.D., Duke U., 1974, M.D., 1975. Successively intern, resident, endocrine fellow dept. medicine Duke U., Durham,

N.C., 1975-77, asst. prof. dept. medicine, 1979-82; research assoc. Nat. Heart, Lung, Blood Inst., NIH, Bethesda, Md., 1977-79; assoc. prof. dept. medicine Brigham Hosp.-Harvard Med. Sch., Boston, 1982—; sr. investigator immunology sect. Joslin Diabetes Ctr., Boston, 1982—. Editor: Monoclonal Antibody Endocrine Research, 1982, Monoclonal Antibodies: Autoimmunity, 1983. Recipient Career Devel. award Juvenile Diabetes Found., 1982; Kroc Found. grantee, 1982-84; NIH grantee, 1979-86. Mem. Am. Soc. Clin. Investigation, Am. Fedn. Clin. Research, Am. Diabetes Assn. Democrat. Subspecialties: Neuroendocrinology; Immunology (medicine). Current work: Immunology of type I diabetes; monoclonal antibodies; cell culture. Office: Joslin Diabetes Ctr 1 Joslin Pl Boston MA 02215

EISENBERG, LAWRENCE, electronic engineer, educator, researcher; b. N.Y.C., Dec. 21, 1919; m.; 2 children. B.S., CCNY, 1940, B.E.E., 1944; M.E.E., Poly. Inst. Bklyn., 1952, Ph.D. in Elec. Engring, 1966. Sr. instr. electronics Sch. Indsl. Tech., 1950-52; project engr. Poly. Research & Devel. Corp., 1952-56; sr. logician Digitronics Corp., L.I., 1956-58; lectr. elec. engring. CCNY, 1958; research assoc. in electronics Rockefeller U., 1958-66, asst. prof., 1966—, co-head depts. electronics and computer sci. and sr. research assoc., 1970—; instr. in charge grad. dept. elec. engring. Poly. Inst. Bklyn., 1956—. Mem. IEEE. Subspecialties: Biomedical engineering; Microelectronics. Office: Depts Electronics and Computer Sci Rockefeller U New York NY 10021

EISENBERG, LEON, child psychiatrist; b. Phila., Aug. 8, 1922; s. Morris and and Elizabeth (Sabreen) E.; m. Ruth Harriet Bleier, June 11, 1948 (div. 1967); children: Mark Philip, Kathy Bleier; m. Carola Blitzman Guttmacher, Aug. 31, 1967; children: Laurence, Alan. A.B., U. Pa., 1944, M.D., 1946; M.A. (hon.), Harvard, 1967, D.Sc., U. Manchester, Eng., 1973. Diplomate: in child psychiatry and psychiatry Am. Bd. Psychiatry and Neurology. Intern Mt. Sinai Hosp., N.Y.C., 1946-47; instr. physiology U. Pa., 1947-48; resident psychiatry Sheppard-Pratt Hosp., Towson, Md., 1950-52; with Johns Hopkins, 1952-67; prof. child psychiatry Med. Sch., 1961-67; psychiatrist-in-charge children's psychiat. service Harriet Lane Home, 1958-67; prof. psychiatry Harvard Med. Sch., 1967—, Maude and Lillian Presley prof. psychiatry, 1975-80, chmn. exec. com. dept. psychiatry, 1973-80, Maude and Lillian Presley prof. social medicine and chmn. dept. social medicine and health policy, 1980—; psychiatrist-in-chief Mass. Gen. Hosp., 1967-74, mem. bd. consultation, 1974—; sr. asso. in psychiatry Children's Hosp., Boston, 1974—; psychiat. cons. Crownsville (Md.) State Hosp., 1954-58, Rosewood State Tng. Sch., Owings Mills, Md., 1957-60, Balt. City Hosp., 1959-62, Children's Guild, Balt., 1954-61; cons. Sinai Hosp., Balt., 1963-67; Mapother-Lewis ann. lectr. Maudsley Hosp., London, 1977; Baan Meml. lectr. Netherlands Psychiat. Soc., Amsterdam, 1978; Royal Soc. Medicine vis. prof., London, 1983; Mem. subcom. psychiat. nomenclature, com. vital statistics USPHS; chmn. WHO Conf. Developmental Regulation, 1964-67; mem. Joint Commn. Mental Health of Children; cons. Office Mental Health, World Health Assn., 1974—; mem. adv. com. to dir. NIH, 1977-80. Editor: Am. Jour. Orthopsychiatry, 1963-73; editorial bd.: Medicine and Psychiatry. Served to capt. M.C.; Served to capt. AUS, 1948-50. Theobald Smith award Albany Med. Coll., 1979; Recipient Orton award Orton Soc., 1980. Fellow Am. Psychiat. Assn. (trustee 1973-76), Am. Orthopsychiat. Assn., A.A.A.S., Soc. Research Child Devel.; mem. Inst. Medicine of Nat. Acad. Scis. (council 1975-77, program and membership coms. 1979-82), AAUP (past pres. Johns Hopkins), Am. Acad. Pediatrics (Aldrich award 1980), Am. Pediatric Soc., Assn. Research Nervous and Mental Disease, Am. Psychopath. Assn., Md. Psychiat. Soc. (past pres.), Am. Acad. Arts and Scis., Psychiat. Research Soc. (past pres.), Soc. Neurosci., Mass. Med. Soc., Greek Soc. Neurology and Psychiatry (hon.), Johns Hopkins Soc. Scholars, Phi Beta Kappa (chpt. pres.), Sigma Xi, Alpha Omega Alpha. Subspecialties: Psychiatry; Epidemiology. Current work: Effects of social class and ethnicity on illness and on care-seeking behavior. Home: 9 Clement Circle Cambridge MA 02138 Office: Dept Social Medicine and Health Policy Harvard Med Sch Boston MA 02115

EISENBERG, MORRIS, electrochem. exec., scientist; b. Poland, Aug. 26, 1921; came to U.S., 1947, naturalized, 1952; s. Solomon and Haya Nenna (Troppe) E.; m. Edith Weiser Boxer, Aug. 26, 1981; children: Zachary, David. Student, Tech. Hochschule Munich, 1946-47; B.S. in Chem. Engring, U. Calif., Berkeley, 1950, M.S., 1952, Ph.D. in Chem. Engring, 1953. Registered profl. engr., Calif. Mgr. electrochemistry Stanford Research Inst., 1953-56; lectr. Stanford U., 1953-56; v.p. research and devel. Thermomaterials, Inc., 1956-57; dir. electrochem. lab. Lockheed Missile Systems, Sunnyvale, Calif., 1957-62; pres., chmn. bd. Electrochimica Corp., Mountain View, Calif., 1962 ; chmn. bd. Battery Systems, Inc.; dir. Elca Battery Co. Contbr. articles profl. jours. Mem. Sigma Xi. Jewish. Lodge: B'nai B'rith. Subspecialties: Physical chemistry. Current work: Electrochemistry, batteries, fuel cells, energy storage, electric vehicles. Patentee field of batteries.

EISENBERG, M(YRON) MICHAEL, surgery educator, academic administrator; b. N.Y.C., Jan. 27, 1931; s. George H. and Dorothy E.; m. (div.); children: Elysa Debra, Ellen Beth, Andrea Carla. B.A., NYU, 1952; M.D., Harvard U., 1956. Diplomate: Am. Bd. Surgery. Intern Peter Bent Brigham Hosp., Boston; resident Yale-New Haven Med. Ctr.; instr. surgery U. Fla., 1962-63, asst. prof. surgery, 1963-67, assoc. prof., 1967-68; prof. U. Minn.-Mpls., 1968-81, head gastrointestinal surgery, 1968-81; prof. SUNY Downstate Med Ctr., Bklyn., 1981—; chief surgery Mt. Sinai Hosp., Mpls., 1968-75; Chief surgery L.I. Coll. Hosp., 1981—. Author: Ulcers, 1978; contbr. articles to profl. publs., chpts. to books. Served as capt. M.C. U.S. Army, 1958-60. Sr. investigator NIH, 1963-80. Mem. Internat. Soc. for Surgery of Digestive Tract, Phi Beta Kappa, Alpha Omega Alpha. Subspecialties: Surgery; Psychophysiology. Current work: Physiology and surgery of gastrointestinal tract.

EISENBERG, PHILLIP, consulting engineer; b. Detroit, Nov. 6, 1919; s. Morris and Ida (Blaizovsky) E.; m. Edith S. Rosenbaum, Nov. 21, 1942; children: Elyse, Jean. B.S., Wayne State U., 1941; postgrad., U. Iowa, 1942; C.E., Calif. Inst. Tech., 1948. Instr. U. Iowa, Iowa City, 1941-42; head research br. David Taylor Model Basin, Navy Dept., Carderock, Md., 1942-44, 46-53; pres. Hydronautics, Inc., Laurel, Md., 1959-74, 78-82, chmn. exec. com., Washington, 1974-82, also dir.; pvt. cons., 1982—; Mem. bd. cons. Iowa Inst. Hydraulic Research, 1954-56, 59-61; mem. Md. Ocean Sci. Com., 1969-70; mem. sea grant adv. panel NOAA, 1969-76; Mem. vis. com. ocean engring. MIT, 1974—; mem. Marine Bd., Maritime Transp. Research Bd., NRC; bd. dirs. Am. Bur. Shipping. Contbr. articles to publs. in field. Served to lt. (j.g.) USNR, 1944-47. Recipient Meritorious Civilian award U.S. Navy, 1944, Distinguished Alumni award Wayne State U., 1958; tech. achievement award ASME, 1959; Gold medal Nat. Acad. Scis., 1974. Fellow Royal Inst. Naval Architects, ASME, Soc. Naval Architects and Marine Engrs. (pres. 1973-74, hon. mem. gold medal 1972), Marine Tech. Soc. (pres. 1976, recipient Lockheed award for ocean sci. and engring. 1980); mem. Nat. Acad. Engring., Am. Inst. Aeros. and Astronautics, Am. Phys. Soc., Am. Inst. Physics, Acoust. Soc. Am. Club: Cosmos (Washington). Subspecialties: Fluid mechanics; Ocean engineering. Current work: Retired but still interested in cavitation phenomena and ocean development. Patentee in field. Home: 6402 Tulsa Ln Bethesda MD 20034

EISENBERG, RONALD LEE, radiology educator; b. Phila., July 11, 1945; s. Milton and Betty Ruth (Klein) E.; m. Zina Schiff, Sept. 19, 1970; children: Avlana, Cherina. A.B. in Chemistry, U. Pa., 1965, M.D., 1969. Diplomate: Am. Bd. Radiology. Intern Mt. Zion Med. Ctr., San Francisco, 1969-70; resident Mass. Gen. Hosp., Boston, 1970-71, U. Calif., San Francisco, 1973-75, staff radiologist, 1975-80; asst. prof. radiology La. State U. Med. Ctr., Shreveport, 1980—, chmn. dept., 1980—; radiology cons. Berlex Labs., Hannover, N.J., 1983—. Author: Gastrointestinal Radiology, 1983; author: Atlas of Signs in Radiology, 1984; editor: Critical Diagnostic Pathways in Radiology, 1981. Served to maj. M.C. U.S. Army, 1971-73. VA grantee, 1977; NIH grantee, 1979. Mem. Radiol. Soc. N.Am., Am. Coll. Radiology, Am. Roentgen Ray Soc., Soc. Gastrointestinal Radiology, Assn. Univ. Radiologists, Am. Physicians Fellowship for Medicine in Israel, Phi Beta Kappa, Alpha Omega Alpha. Subspecialty: Diagnostic radiology. Current work: Referral criteria, optimization of projects, cost-containment in radiology; development of imaging consultancy program in radiology residency; algorithmic approach to sequencing of radiographic imaging procedures. Office: Dept Radiology La State U Med Center PO Box 33932 Shreveport LA 71130

EISENBUD, MERRIL, environmental scientist; b. N.Y.C., Mar. 18, 1915; s. Kalman and Leonora (Kopaloff) E.; m. Irma Onish, Jan. 22, 1939; children—Elliott, Michael, Fredrick. B.S. in Elec. Engring, N.Y. U., 1936; Sc.D. (hon.), Fairleigh Dickinson U., 1960; D.H.C., Catholic U., Rio de Janiero. Diplomate: Am. Acad. Environ. Engrs. Indsl. hygienist Liberty Mut. Ins. Co., 1936-47; asso. prof. indsl. medicine Sch. Medicine, N.Y. U., 1945-55, adj. prof., 1956-59, prof. environ. medicine, dir. lab. environ. studies, 1959—; adminstr. N.Y.C. EPA, 1968-70; dir. health and safety lab. AEC, 1947-59; mem. Nat. Commn. on Radiation Protection and Measurements, 1965—, dir., 1971-76; mem. expert panel on radiation hazards WHO, 1956—; mem. N.Y. State Health Adv. Council, 1975-80. Author: Environmental Radioactivity, 2d edit, 1973, Environment, Technology, and Health, 1979. Bd. dirs. Blue Cross-Blue Shield Greater N.Y., 1968-75; bd. mgrs. State Community Aid Assn.; mem. adv. council Electric Power Research Inst. Recipient Gold medal AEC, 1974; Hermann Biggs medal N.Y. State Pub. Health Assn.; Arthur Holly Compton award Am. Nuclear Soc.; Power-Life award Am. Inst. Elec. and Electronic Engrs. Fellow AAAS, N.Y. Acad. Scis. (hon. life mem., gov., v.p. 1979-80), N.Y. Acad. Medicine; mem. Nat. Acad. Engring., Health Physics Soc. (pres. 1965-66), Am. Indsl. Hygiene Assn., Radiation Research Soc., Am. Bd. Health Physics. Clubs: Cosmos (Washington); Explorers, Century (N.Y.C.). Subspecialty: Environmental toxicology. Current work: Environmental effects of technology development. Home: PO Box 837 Tuxedo NY 10987 Office: NYU Med Center Tuxedo NY 10987

EISERLING, FREDERICK ALLEN, microbiology educator; b. San Diego, Calif., May 8, 1938; s. Allen Frederick and Nancy Lucille (Simpson) E.; m. Monica Eiserling, 1963 (div.); children: Erik Robert, Ingrid Monica. B.A., UCLA, 1959, Ph.D. in Microbiology, 1964. USPHS postdoctoral fellow in biophysics, Geneva, 1964-66; asst. prof. to prof. microbiology UCLA, 1966-74, prof., 1974—, chmn., 1981—. Contbr. articles to profl. jours. NIH and NSF research grantee. Mem. AAAS, Am. Soc. Microbiology, Am. Soc. Virology, Electron Microscope Soc. Am. Subspecialties: Microbiology; Molecular biology. Current work: Virus structure and assembly, strucuture and assembly of photosynthetic light-harvesting organelles. Office: Dept Microbiology U Calif Los Angeles CA 90024

EISGRUBER, LUDWIG MARIA, educator; b. Mallersdorf, Germany, Dec. 12, 1931; came to U.S., 1955, naturalized, 1964; s. Ludwig and Maria E.; m. Eva Renate Eisgruber, Dec. 17, 1960; children—Christopher, Karen, Michelle, Ingrid. Diploma in Agr, Technische Hochschule Munchen, W. Ger., 1955; M.S., Purdue, 1957, Ph.D., 1959. Asst. prof. Purdue U., Lafayette, Ind., 1959-62, asso. prof., 1962-65, prof., 1965-73, asst. head dept. agrl. econs., 1969-73, asst. dean, 1970-73, program leader applications requirements, 1970-73; prof., head dept. agrl. and resource econs Oreg. State U., Corvallis, 1973-81, asso. dean, dir. internat. agr., 1981—; cons. in field. Author: (with E.M. Babb) Management Games for Teaching and Research, 1966, (with J.L. Hesselbach) Betriebliche Entscheidungen Mittels Simulations, 1967. IBM travel grantee, 1972; Sears, Roebuck & Co. grantee, 1958. Mem. Am. Agrl. Econs. Assn., Internat. Assn. Agrl. Economists, Am. Econs. Assn., Western Agrl. Econs. Assn. (pres. 1978-79), Sigma Xi (sec.-treas. Purdue chpt. 1970-72), Gamma Sigma Delta. Subspecialty: Agricultural economics. Current work: Agricultural policy and analysis of New Eastern countries. Economic feasibility analyses of agricultural development projects. Office: Extension Hall Oreg State U Corvallis OR 97331

EISINGER, JOSEF, biophysics researcher, medicine historian; b. Vienna, Austria, Mar. 19, 1924; s. Rudolf and Grete (Lindner) E.; m. Styra Jean Avins, 1962; children: Alison A., Simon E. B.A., U. Toronto, 1947, M.A., 1948; Ph.D., MIT, 1951. Research asst. MIT, 1951-52; research assoc. NRC Can., 1952-53, Rice U., Houston, 1953-54; biophysics researcher Bell Labs., Murray Hill, N.J., 1954—; adj. prof. dept. physics NYU, 1968-71. Contbr. articles to profl. jours. Served with Canadian Army, 1944-45. John Simon Guggenheim fellow, 1963, 77. Fellow Am. Soc. Exptl. Biology; mem. Biophys. Soc., Am. Soc. Photobiology, Am. Soc. Hematology. Subspecialties: Biophysics (physics); Environmental toxicology. Current work: Membrane biology, fluorescence spectroscopy of biological molecules, resonance energy transfer, lead toxicology, red blood cells, history of environmental medicine. Office: Bell Lab 600 Mountain Ave Murray Hill NJ 07974

EISNER, DONALD ALAN, psychologist; b. Cleve., Nov. 16, 1943; s. Leon and Ruth E. B.A., Ohio U., 1965; M.A., W.Va. U., 1968, Ph.D., 1970. Lic. psychologist, Calif., diplomate: Internat. Acad. Profl. Counseling and Psychotherapy. Dir. geriatrics unit Lincoln Hosp., Bronx, N.Y., 1974-75; clin. psychologist Manhattan State Hosp., Wards Island, N.Y., 1975-77, Peninsula Gen. Hosp., Queens, N.Y., 1977; dir. psychol. services Orange County Assn. Help of Retarded Children, Montgomery, N.Y., 1978-79; psychologist Wright Psychiat. Med. Ctr., La Puente, Calif., 1979-81, Hillview Mental Health Ctr., Lake View Terrace, Calif., 1981—; dir. Eisner Psychol. Assocs., Los Angeles, 1982—; cons. Claremont Counseling Ctr., 1979-81, Hope Community Mental Health Ctr., Lake View Terrace, Calif., 1981; instr. U. So. Calif., 1971; asst. prof. psychology William Patterson Coll. N.J., 1971-74; adj. faculty CUNY, 1974, Mercy Coll., White Plains, N.Y., 1978, Cerritos (Calif.) Coll., 1980, Calif. State U.-Long Beach, 1980, Antioch U. West, 1980. Contbr. articles to sci. jours. Mem. Am. Psychol. Assn., Calif. Psychol. Assn. Current work: Gerontology, forensic psychology, clinical psychology. Home: 1260 Veteran Ave 316 Los Angeles CA 90024 Office: 12304 Santa Monica Blvd Los Angeles CA 90025

EISNER, THOMAS, biologist, educator; b. Berlin, June 25, 1929; s. Hans Edouard and Margarete (Heil) E.; m. Maria Lobell, June 10, 1952; children: Yvonne, Vivian, Christina. B.A., Harvard U., 1951, Ph.D., 1955; D.Sc. hon., U. Wurzburg, W. Ger., 1982, U. Zurich, Switzerland, 1983. Postdoctoral fellow Harvard U., 1955-57; asst. prof. biology Cornell U., Ithaca, N.Y., 1957-62, assoc. prof., 1962-65, prof., 1965-76, Jacob Gould Schurman prof. biology, 1976—; vis. scientist dept. entomology Sch. Agr., Wageningen, Netherlands, 1964-65; vis. scientist Smithsonian Tropical Research Lab., Barro Colorado Island, C.Z., 1968; sr. vis. scientist Max Planck Inst. für Verhaltensphysiologie, Seewisen, W. Ger., 1971, Div. Entomology, Canberra, Australia, 1972-73; Rand fellow Marine Biol. Labs., Woods Hole, Mass., 1974; vis. research prof. U. Fla., Gainesville, 1977-78; vis. prof. Stanford U., 1979-80, U. Zurich, 1980-81. Co-author: Animal Adaptation, 1964, Life on Earth, 1973, and 3 other books.; Mem. editorial bd.: Sci, 1970-71, Am. Naturalist, 1970-71, Jour. Comparative Physiology, 1974-80, Chem. Ecology, 1974—, Cornell Rev, 1976-77, Behavioral Ecology and Sociobiology, 1976—, Sci. Yr. World Books, 1979—; contbr. articles to profl. jours. Guggenheim fellow, 1964-65, 72-73; Recipient Newcomb Cleveland prize AAAS, 1967; Founder's Meml. award Entomol. Soc. Am., 1969. Fellow Explorers Club, AAAS (chmn. sect. biology 1979—, mem. com. for sci. freedom and responsibility 1980—), Am. Acad. Arts and Scis., Royal Soc. Arts; mem. Nat. Acad. Sci., Zero Population Growth (dir. 1969-70), Nat. Audubon Soc. (dir. 1970-75), Nature Conservancy (nat. council 1969-74), Fedn. Am. Scientists (mem. council 1977-81). Subspecialties: Ecology; Ethology. Current work: Behavioral and chemical ecology of insects. Office: Dept Neurobiology and Behavior W347 Mudd Hall Cornell U Ithaca NY 14853 I am a naturalist, interested primarily in field exploration and discovery. My research deals with the behavior and ecology of insects, and with the photographic and cinematographic documentation of little-known aspects of the life of these animals. My chief goal in life is to relate my findings to the cause of wildlife and wilderness preservation, to which I am fiercely devoted.

EISON, MICHAEL STEVEN, psychopharmacologist, researcher; b. N.Y.C., Jan. 21, 1953; s. Fred and Freda (Berg) E.; m. Arlene Stark, Aug. 19, 1979. B.A., SUNY-Buffalo, 1975; M.A., UCLA, 1976, Ph.D., 1980. Postdoctoral fellow NSF-NATO, Cambridge, Eng., 1980-81; scientist Mead Johnson Pharms, Evansville, Ind., 1981-82; sr. scientist Bristol-Myers Co., Evansville, 1982—. Recipient Feldman-Cohan award SUNY-Buffalo, 1975. Mem. Soc. for Neurosci., N.Y. Acad. Sci., Am. Psychol. Assn., AAAS, Sigma Xi, Phi Beta Kappa. Subspecialties: Neuropharmacology; Physiological psychology. Current work: Biological basis of human psychopathology. Office: Pharm and Research and Devel Div Bristol-Myers Co Evansville IN 47721

EISS, NORMAN SMITH, JR., mech. engr., educator; b. Buffalo, Mar. 13, 1931; s. Norman Smith and Elizabeth Charlotte (Hengerer) E.; m. Nancy Jean Siegrist, Mar. 27, 1975; children: Martin E., Christine C., Jennifer L. B.S. in M.E, Rensselaer Poly. Inst., 1953; M.S., Cornell U., 1959, Ph.D., 1961. With textile fibers div. duPont, Tonawanda, N.Y., 1953-54; with Cornell Aero. Lab., Cheektowaga, N.Y., 1956-58, 61-66, Va. Poly. Inst. and State U., Blacksburg, 1966—, prof. mech. engring., 1977—; cons. tribology, stress analysis, surface topography characterization. Contbr. articles to profl. jours. Served with USAF, 1954-56. NSF Sci. Faculty fellow, 1970-71. Mem. ASME, Am. Soc. Lubrication Engrs., Am. Soc. Engring. Edn., ASTM, Sigma Xi, Pi Tau Sigma, Tau Beta Pi, Phi Kappa Phi. Subspecialties: Mechanical engineering. Current work: Abrasive and fatigue wear of polymers. Office: Mech Engring Dep Va Poly Inst and State U Blacksburg VA 24061

EKANADHAM, KATTAMURI, computer scientist, educator; b. India; s. Kattamuri V. and Kattamuri Kanaka (Bhramaramba) Nanmadha Rao; m. Visalakshi K. Ekanadham, 1982. B.S.E.E., Rec, Warangal, India, 1969; M.S. in Computer Sci, Indian Inst. Tech., Kanpur., 1972; Ph.D., SUNY-Stonybrook, 1976. Asst. prof. computer sci. SUNY-Stony Brook, 1977-80; research staff mem. IBM Research Ctr., Yorktown Heights, N.Y., 1981—; cons. DVI Communications, 1979-80; mem. faculty univs. and colls. Mem. IEEE, Assn. Computing Machinery. Subspecialties: Operating systems; Computer architecture. Current work: Parallel architectures, data flow principles, resource synchronization. Office: IBM Research Center PO Box 218 Yorktown Heights NY 10598

EKBERG, DONALD ROY, fisheries administrator; b. Hinsdale, Ill., Dec. 23, 1928; s. Roy Harley and Evelyn Beatrice (Newman) E.; m. Anneliese G. Nattermann, May 27, 1961; children: Kenneth, Dale. B.S., U. Ill., 1950, Ph.D., 1957; M.S., U. Chgo., 1952. Physiologist Gen. Electric Co., Phila., 1958-59; postdoctoral fellow USPHS, 1959-60, physiologist and mgr. bioscis., Phila. and Bay St. Louis, Miss., 1960-76; biologist Nat. Marine Fisheries Service, St. Petersburg, Fla., 1976-77, chief environ. and tech. service div., 1977—. Served to col. USAFR, 1952—. Mem. Soc. Gen. Physiologists, Am. Soc. Zoologists, Am. Physiol. Soc., Sigma Xi. Subspecialties: Resource management; Deep-sea biology. Current work: Fish ecosystem alteration, technology transfer to fisheries. Office: Nat Marine Fisheries Service NOAA 9450 Koger Blvd Duval Bldg St Petersburg FL 33702

EKDALE, ALLAN ANTON, geology educator; b. Burlington, Iowa, Aug. 30, 1946; s. Warren E. and Marian L. (Nielsen) E.; m. Susan Faust Rostberg, July 5, 1969; children: Joan Diane, Eric Gregory. A.B., Augustana Coll., 1968; M.A., Rice U., 1973, Ph.D., 1974. Asst. prof. geology U. Utah, Salt Lake City, 1974-78, assoc. prof., 1978—. Served with U.S. Army, 1969-70; Vietnam. Fellow AAAS; mem. Soc. Econ. Paleontologists and Mineralogists, Paleontol. Soc., Geol. Soc. Am., Nat. Assn. Geology Tchrs. Subspecialties: Paleoecology; Sedimentology. Current work: Ichnology, trace fossils, bioturbation of sediments, animal-sediment interrelationships, structure of fossil communities. Office: Dept Geology and Geophysics University of Utah Salt Lake City UT 84112 1183

EKERS, RONALD DAVID, astronomer; b. Victor Harbour, South Australia, Sept. 18, 1941; came to U.S., 1967; s. Laurence and Elsie (Plaisted) E.; m. Jennifer A. Brooks, June 1, 1940; children: Brook, Daen, Erik. B.Sc., U. Adelaide, 1962, 1963; Ph.D., Australian Nat. U., 1967. Postdoctoral fellow Calif. Inst. Tech., Pasadena, 1967-70, Inst. Theoretical Astronomy, Cambridge, Eng., 1970-71; research worker Kapeteyn Lab., Groningen, Netherlands, 1971-80, prof., 1977—; asst. dir. very large array ops. Nat. Radio Astronomy Obs., Socorro, N.Mex., 1980—; bd. dirs. Netherlands Found. Radio Astronomy, 1975-80; mem. vis. com. Meudon (France) Radio Obs., 1975-77; vis. scientist Commonwealth Sci. and Indsl. Research Orgn., 1977-78; vis. fellow Australian Nat. U., 1979; mem. sci. adv. com. Inst. Millimeter Astronomy, 1979-80; adj. prof. U. N.Mex., 1981—; mem. telecommunication and data acquisition adv. council Jet. Propulsion Lab., 1982—. Contbr. articles to sci. jours. Mem. Internat. Astron. Union, Astron. Soc. Australia, Am. Astron. Soc., Royal Astron. Soc. Subspecialty: Radio and microwave astronomy. Office: NRAO PO Box 0 Socorro NM 87801

ELACHI, CHARLES, scientist, educator; b. Lebanon, Apr. 18, 1947; s. Rokos and Yvonne (Obeid) E. B.S., U. Grenoble, France, 1968; M.S., Calif. Inst. Tech., 1969, Ph.D., 1971; M.B.A., U. So. Calif., 1978. Research fellow Calif. Inst. Tech., 1971-74; sr. scientist Jet Propulsion Lab., Pasadena, Calif., 1971-74, team leader, 1975—, sr. research scientist, program mgr., 1981—; lectr. Calif. Inst. Tech., 1982—. Contbr. articles to profl. jours. Recipient Autometric award Am. Soc. Photogrammetry, 1981. Mem. AAAS, Am. Phys. Soc., Am. Geophys. Union, IEEE, Electromagnetic Soc., Sigma Xi. Subspecialties: Remote sensing (geoscience); Planetary science. Current work: Use of remote sensors in earth and planetary sciences; principal investigator on shuttle imaging radar. Patentee in field.

EL-ACKAD, TAREK M., cardiovascular clinical research pharmacologist, human physiology educator; b. Alexandria, Egypt, Sept. 25, 1941; came to U.S., 1964; s. Mohamed Soliman and Fatma (Fahmy) El-A. B.Sc., Alexandria U., 1962; M.S., N.C. State U., 1967, Rutgers U., 1970, Ph.D., 1972. Pharmacology research fellow U. Iowa Coll. Medicine, 1972-75; head cardiovascular pharmacology Pennwalt Pharms., Rochester, N.Y., 1975-77; sr. clin. research assoc. Abbott Internat., Ltd., Abbott Park, Ill., 1977-79, Abbott Hosp. Products, Abbott Park, 1979-81; sr. med. research assoc. Schering-Plough Corp., Kenilworth, N.J., 1981—; adj. assoc. prof. Fairleigh-Dickinson U. Sch. Dentistry, Hackensack, N.J., 1981—. Contbr. articles to profl. jours. Chmn. social com. Rutgers U. Grad. Student Assn., 1970-72; mem. curriculum com. Rutgers U. Physiology Program, 1970-72. Bio-Med. Telemetry scholar, 1970; NSF research asst., 1968-71; research fellow Am. and Iowa Heart Assns., 1973-74. Mem. N.Y. Acad. Scis., AAAS, Am. Heart Assn., Am. Fedn. Clin. Research, Histamine Research Soc. N.Am., Am. Soc. Parenteral and Enteral Nutrition, Schering Classical Music Lovers Club (pres.), Sigma Xi. Club: Schering Classical Music Lovers (pres.). Subspecialties: Pharmacology; Gastroenterology. Current work: Clinical pharmacology research in cardiovascular antihypertensive drugs; academic teaching in gastroenterology. Office: Schering Plough Corp Galloping Hill Rd Kenilworth NJ 07033

ELANDER, RICHARD PAUL, microbiologist, educator; b. Worcester, Mass., Sept. 17, 1932; s. Arthur Waldemar and Edith Alma-Louise (Engstr) E.; m. Barbara Ann Sudz, Feb. 8, 1958; children: Tracy, Richard T., Ronald P. B.S. cum laude, U. Detroit, 1955, M.S., 1956; Ph.D. in Botany and Bacteriology, U. Wis., 1960; postdoctoral, U. Minn., 1967. Research scientist Lilly Research Labs., Indpls., 1960-68; assoc. dir. antibiotic mfg. and devel. Wyeth Labs., West Chester, Pa., 1968-72; assoc. dir. microbiology research Smith Kline & French Labs., Phila., 1972-75; dir. fermentation devel. Bristol-Myers Co., Syracuse, N.Y., 1976-80, sr. dir., 1980-81, v.p. biotech., 1982—; research prof. biology Syracuse U. Contbr. articles to tech. jours. Recipient biology award U. Detroit, 1960. Fellow Am. Acad. Microbiology; mem. Am. Soc. Microbiology, Am. Chem. Soc., Soc. Indsl. Microbiology (pres. 1973-74), AAAS, N.Y. Acad. Scis., Sigma Xi, Phi Sigma, Gamma Alpha. Republican. Subspecialties: Microbiology; Genetics and genetic engineering (biology). Current work: Antibiotic and enzyme fermentation technology and development, microbial fermentations, scale-up of genetically-engineered microorganisms. Patentee in field. Home: 8376 Vassar Dr Manlius NY 13104 Office: Industrial Div Bristol Myers Co Syracuse NY 13201

EL-ASHRY, MOHAMED TAHA, geologist, federal agency administrator; b. Cairo, Jan. 21, 1940; arrived U.S., 1961; s. Taha; ; s. 5yka and (Fadda) El-A.; m. Patricia R. Murphy, Oct. 12, 1962; children: Dorraya, Sumaya. B.S., Cairo U., 1959; M.S., U. Ill-Urbana, 1963, Ph.D., 1966. Asst. prof. Cairo U., 1966-69; asst. prof. Wilkes Coll., Wilkes-Barre, Pa., 1969-71, assoc. prof., 1971-75, chmn. dept., 1973-75; staff scientist Environ. Def. Fund, Denver, 1975-79; dir. environ. quality TVA, Knoxville, 1979—; cons. Pan-Am.-UAR Oil Co., Cairo, 1966-69, various coal cos., Wilkes-Barre, 1969-75. Co-author, editor: Air Photography and Coastal Problems, 1977. Recognized as top geology grad. Govt. of Egypt, 1959; doctoral studies grantee, 1961. Fellow Geol. Soc. Am., AAAS; mem. Am. Assn. Petroleum Geologists, Am. Quaternary Assn., Am. Shore and Beach Preservation Assn. (bd. dirs. 1978-81), Sigma Xi. Moslem. Subspecialties: Geology; Environmental management. Current work: Environmental geology, energy/environment, water resources, coastal geomorphology, remote sensing. Home: 4604 Tahoe Ln Knoxville TN 37918 Office: Tenn Valley Authority Knoxville TN 37902

EL-AWADY, ABBAS ABBAS, chemistry educator; b. Damas, Egypt, Jan. 2, 1939; came to U.S., 1959; s. Abbas M. El-A. B.S., U. Cairo, 1958; Ph.D., U. Minn., 1965. Postdoctoral research assoc. UCLA, 1965-66; with EPA, Central Regional Lab., Chgo., 1974-75; vis. scholar, vis. research prof. SUNY-Buffalo, 1979-80; prof. and head inorganic chemistry div. Faculty of Sci., King Abdulaziz, Jeddah, Saudi Arabia, 1980-82; prof. chemistry Western Ill. U., Macomb, 1976—. Contbr. articles on chemistry to profl. jours. Mem. Am. Chem. Soc., Sigma Xi. Muslim. Subspecialties: Inorganic chemistry; Kinetics. Current work: Physical-inorganic chemistry, kinetics and thermodynamics of complex formation, mechanisms of inorganic reactions in solution, photochemistry of coordination compounds, envrionmental chemistry. Office: Dept Chemistry Western Ill U Macomb IL 61455

ELBAUM, CHARLES, physicist, educator, researcher; b. May 15, 1926; m.; 3 children. M.A.Sc., U. Toronto, Ont., Can., 1952, Ph.D. in Applied Sci, 1954; M.A. (hon.), Brown U., 1961. Research fellow in metal physics U. Toronto, 1954-57, Harvard U., 1957-59; asst. prof applied physics Brown U., Providence, 1959-61, assoc. prof. physics, 1961-63, prof. physics, 1963—, chmn. dept. physics, 1980—; cons. to industry. Fellow Am. Phys. Soc.; mem. AIME, Soc. Neurosci., AAAS. Subspecialties: Solid State Physics; Biophysics (physics). Office: Dept Physics Brown U Providence RI 02912

EL-BAZ, FAROUK, corporate executive; b. Zagazig, Egypt, Jan. 2, 1938; came to U.S., 1960, naturalized, 1970; s. El-Sayed Mohammed and Zahia Abul-Ata (Hammouda) El-B.; m. Catherine Patricia O'Leary, 1963; children—Monira, Soraya, Karima, Fairouz. B.Sc., Ain Shams U., 1958; M.S., U. Mo., 1961; Ph.D., U. Mo. and Mass. Inst. Tech., 1964. Demonstrator geology dept. Assiut U., Egypt, 1958-60; lectr. Mineralogy-Petrography Inst., U. Heidelberg, Ger., 1964-65; geologist exploration dept. Pan Am.-UAR Oil Co., Egypt, 1966; supr. lunar exploration Bellcomm and Bell Telephone Labs., Washington, 1967-72; research dir. Center for Earth and Planetary Studies, Nat. Air and Space Mus., Smithsonian Instn., Washington, 1973-82; v.p. internat. devel. Itek Optical Systems, Litton Industries, Lexington, Mass., 1982—; cons. geology; prof. geology and geophysics U. Utah, 1975-77; prof. geology Ain Shams U., Egypt, 1976-81; sci. adviser Pres. Anwar Sadat of Egypt, 1978—. Author or co-author: Say It in Arabic, 1968, Coprolites: An Annotated Bibliography, 1968, Glossory of Mining Geology, 1970, The Moon as Viewed by Lunar Orbiter, 1970, Astronaut Observations from the Apollo-Soyuz Mission, 1977, Apollo Over the Moon: A View from Orbit, 1978, Egypt As Seen by Landsat, 1979, Apollo-Soyuz Test Project Summary Science Report: Earth Observations and Photography, 1979; Author: Desert Landforms of Southwest Egypt: A basis for Comparison with Mars, 1982; Author or co-author also articles. Decorated Order of Merit 1st class, Egypt; recipient certificate merit U.S. Bur. Mines, 1961, Exceptional Sci. Achievement medal NASA, 1971, Alumni Achievement award U. Mo., 1972, Honor citation Assn. Arab-Am. U. Grads., 1973. Fellow Royal Astron. Soc., Geol. Soc. Am. (certificate commendation 1973); mem. AAAS, Sigma Xi. Clubs: Explorers., University. Subspecialties: Planetary science; Satellite studies. Current work: Photographic systems and photointerpretation of planetary surface features and application of space technology to high resolution photography for mapping and reconnaissance. Office: Itek Optical Systems 10 Maquire Rd Lexington MA 02173

ELBEIN, ALAN DAVID, biochemistry educator; b. Lynn, Mass., Mar. 20, 1933; s. Gersh and Golda (Stryer) E.; m. Elaine J. Brooks, June 13, 1953; children: Stever C., Bradley M., Richard C. A.B., Clark U., 1954; M.S., U. Ariz., 1956; Ph.D., Purdue U., 1960. Postdoctoral fellow U. Mich., Ann Arbor, 1960-63; research assoc. U. Calif.-Berkeley, 1963-64; asst. prof. Rice U., Houston, 1964-67, assoc. prof., 1967-69; prof. dept. biochemistry U. Tex. Health Sci. Ctr., San Antonio, 1970—; vis. prof. Ciba Geigy, Basel, Switzerland, 1980-81; cons. NSF, NIH, VA. Editorial bd.: Plant Physiology, Jour. Bacteriology, Arch. Biochem Biophys.; contbr. chpts. and revs. to books, articles to publs. in field. USPHS fellow, 1958-63; career devel. awardee, 1967-69; NIH fellow, 1957-60, 60-63; career devel. awardee, 1967-69. Fellow Am. Acad. Microbiology; mem. AAAS, Am. Soc. Microbiology, Am. Soc. Biol. Chemistry, Am. Soc. Plant Physiology, Am. Chem. Soc. Subspecialties: Biochemistry (biology); Microbiology. Current work: Biosynthesis of glycoproteins, use of inhibitors for biosynthetic studies, mechanism of antibiotic action, plant carbohydrate biochemistry. Home: 6106 Sun Dial San Antonio TX 78238 Office: Dept Biochemistry U Tex Health Sci Ctr San Antonio TX 78284

EL DAREER, SALAH MOHAMMED, pharmacologist, researcher; b. Tanta, Egypt, Nov. 4, 1926; s. Mohammed Elsayed and Nabiha Mohammed (El Aroussy) El D.; m. Beulah Mae Campbell, Aug. 17, 1951; children: Nadia, Suzanne, Tanya-Nabiha. D.V.M., Cairo U., 1947; M.S., Mich. State U., 1951. Past mem. faculty Cairo U.; mem. faculty U. Mich., Ann Arbor; now sr. pharmacologist So. Research Inst., Birmingham, Ala. Contbr. articles to profl. jours. Pres. Com. for Better Am. Relations in the Middle East. Mem. Am. Assn. Cancer Research, Am. Soc. Pharmacology and Exptl. Therapeutics, AAAS, Sigma Xi. Muslim. Subspecialties: Animal genetics; Preventive medicine (veterinary medicine). Current work: Anti cancer compounds, xenobiotics. Office: So Research Inst 2000 9th Ave S Birmingham AL 35255

ELDEFRAWI, AMIRA TOPPOZADA, pharmacologist, toxicologist, neurobiologist, researcher, educator; b. Giza, Egypt, Feb. 10, 1937; came to U.S., 1968, naturalized, 1974; d. Hussein K. and Fadila I. (Aref) Toppozada; m. Mohyee E. Eldefrawi, July 18, 1957; children: Mohsen, Mona, Mohab. B.Sc. in Agr, U. Alexandria, Egypt, 1957; Ph.D., U. Calif., Berkeley, 1960. Asst. prof. U. Alexandria, 1960-68; research asso., then sr. research asso. sect neurobiology and behavior Cornell U., Ithaca, N.Y., 1968-76; research prof. pharmacology U. Md. Sch. Medicine, Balt., 1976—; cons. in field. Co-editor: Myasthenia Gravis, Molecular and Clinical Aspects, 1982; contbr. over 100 articles to sci. jours. NIH grantee, 1975—. Mem. Am. Soc. Pharmacology and Exptl. Therapeutics, Soc. Neurosci., Entomol. Soc. Am. Republican. Moslem. Subspecialties: Molecular pharmacology; Neuropharmacology. Current work: Neuropharmacology, neurotransmitter receptors and ionic channels. Home: 8403 Topping Rd Pikesville MD 21208 Office: Dept Pharmacology and Exptl Therapeutics U Md Sch Medicine Baltimore MD 21201

ELDER, CURTIS HAROLD, geologist, researcher; b. Laramie, Wyo., Mar. 30, 1921; s. Cecil and Agnes Christine (Miller) E.; m. Wiese Wild, Jan. 2, 1948; children: George W., Christian N., Robin T., Melody C. Student, U. Mo., Columbia, 1939-43, B.S. in Geology, 1950, postgrad., 1950-52. Research geologist Am. Oil Co., Tulsa, Salt Lake City and Denver, 1952-63; pvt. practice cons. geologist, Denver, 1963-65; research geologist U.S. Bur. Mines, Pitts., 1965-82, Office Surface Mining, U.S. Dept. Interior, 1982—; assoc. dir. Allegheny County Conservation Dist., Pitts., 1965-81; pvt. practice cons. geology, Pitts. Author reports in field. Mem. Allegheny Trails Council, Boy Scouts Am., Pitts., 1964—. Served with U.S. Army, 1943-46. Recipient Silver Beaver award Boy Scouts Am., 1979. Fellow Geol. Soc. Am.; mem. Pitts. Geol. Soc., Soc. Econ. Paleontologists and Mineralogists, Kappa Sigma. Republican. Presbyterian. Subspecialties: Geology; Tectonics. Current work: Research in methane gas control and drainage relative to mining; subsidence reclamation related to old abandoned mines; application of remote sensing imagery to geologic and mining problems. Home: 410 Summit Dr Pittsburgh PA 15228 Office: US Dept Interior Office Surface Mining Eastern Tech Ctr 10 Parkway Center Pittsburgh PA 15220

ELDER, REX ALFRED, civil engr.; b. Pa., Oct. 4, 1917; s. George Alfred and Harriet Jane (White) E.; m. Janet Stevens Alger, Aug. 10, 1940; children—John A., Carol S., Susan A., William P. B.S. in Civil Engring, Carnegie Inst. Tech., 1940; M.S., Oreg. State Coll., 1942. Hydraulic engr. TVA, Norris, Tenn., 1942-48, dir. hydraulic lab., 1948-61, dir. engring. lab., 1961-73; engring. mgr. Bechtel Inc., San Francisco, 1973—. Contbr. numerous articles on hydraulic structures, reservoir stratification and water quality, hydraulic research and hydraulic machinery to profl. jours. Served with USN, 1945-46. Fellow ASCE (James Laurie prize 1949); mem. Nat. Acad. Engring., ASME, Internat. Assn. Hydraulic Research, Permanent Internat. Assn. Nav. Congresses. Subspecialties: Civil engineering; Fluid mechanics. Current work: Application of fluid mechanics to civil engineering structures, aqueous waste disposal and major industrial problems. Home: 2180 Vistazo E Tiburon CA 94920 Office: PO Box 3965 San Francisco CA 94119

ELDRED, EARL, anatomy educator; b. Tacoma, Feb. 27, 1919; s. Stephen and Amalia Virginia (Goellnor) E.; m. Bergliot Marie Stockland, Mar. 4, 1944; children: Stig Michael, Daniel Bent, Heidi Ann. B.S., U. Wash., 1940; M.S., Northwestern U., 1950, M.D., 1950. Intern Virginia Mason Hosp., Seattle, 1950-51; instr. dept. anatomy UCLA, 1951-53, asst. prof. anatomy, 1953-57, assoc. prof., 1957-63, prof., 1963—, vice chmn. dept., 1979—; fellow Nobel Inst. of Neurophysiology, Karolinska Inst., Stockholm, Sweden, 1952-53. Served to 1st lt. coast arty. and inf. U.S. Army, 1941-46; ETO. Markle fellow, 1951-56. Mem. World Brain Orgn., Am. Assn. Anatomists, Soc. Neurosci. Subspecialty: Neurophysiology. Current work: Muscle receptors; motor control; neuromuscular trophic interrelations. Home: 16840 Oakview Dr Encino CA 91436 Office: Dept Anatomy UCLA Los Angeles CA 90024

ELDRED, KENNETH MCKECHNIE, acoustical consultant; b. Springfield, Mass., Nov. 25, 1929; s. Robert Mosley and Jean McKechnie (Ashton) E.; m. Helene Barbara Koerting Fischer, May 31, 1957; 1 dau., Heidi Jean. B.S., MIT, 1950, postgrad., 1951-53; postgrad., UCLA, 1960-63. Engr. in charge vibration and sound lab. Boston Naval Shipyard, 1951-54; supervisory physicist, chief phys. acoustics sect. U.S. Air Force, Wright Field, Ohio, 1956-57; v.p., cons. acoustics Western Electro-Acoustics Labs., Los Angeles, 1957-63; v.p., tech. dir. sci. services and systems group Wyle Labs., El Segundo, Calif., 1963-73; v.p., dir. div. environ. and noise control tech. Bolt Beranek and Newman Inc., Cambridge, Mass., 1973-77, prin. cons., 1977-81; dir. Ken Eldred Engring.; mem. exec. standards council Am. Nat. Standards Inst., 1979—, vice-chmn., 1981—; mem., past chmn. Acoustical Standards Bd.; mem. com. hearing, bioacoustics and biomedics NRC, 1963—. Served with USAF, 1954-56. Fellow Acoustical Soc. Am. (chmn. coordinating com. environ. acoustics), Nat. Acad. Engring., Inst. Noise Control Engring. (pres. 1976), Inst. Environ. Scis. (chmn. tech. com. on acoustics 1963-71), Soc. Automotive Engrs., Soc. Naval Architects and Marine Engrs., U.S. Yacht Racing Union. Subspecialties: Acoustical engineering; Environmental engineering. Current work: Development of better methods to control aircraft operations at airports to minimize noise impact on airport neighbors; analysis of distribution of worker noise exposure in industry and costs for its control. Home: 722 Annursnac Hill Rd Concord MA 01742 Office: PO Box 1037 Concord MA 01742

ELDRED, WILLIAM DOUGHTY, III, anatomist, educator; b. Denver, June 16, 1950; s. William Doughty and Margaret Helen (Hargrove) E.; m. Felicitas Barbara Pointner, May 22, 1976. Ph.D. in Anatomy, U. Colo. Health Sci. Ctr., 1979. Postdoctoral fellow dept. anatomy SUNY-Stony Brook, 1979-80, postdoctoral fellow dept. neurobiology, 1980-81, research asst prof. dept. neurobiology, 1981-82; asst. prof. dept. biology Boston U., 1982—. Recipient New Research Investigator award Nat. Eye Inst., 1981. Mem. Soc. for Neurosci. Subspecialties: Neurobiology; Immunocytochemistry. Current work: Neurotransmitters and peptides in visual systems. Retina amacrine cells, neuropeptides, neurotransmitters, immunocytochemistry, ultrastructure, inner plexiform layer. Office: 2 Cummington St Boston MA 02215

ELDREDGE, NILES, paleobiologist, museum curator; (married). A.B., Columbia U., 1965, Ph.D. in Geology, 1969. Asst. curator Am. Mus. Natural History., N.Y.C., 1969-74, assoc. curator invertebrate paleontology, 1974-79, curator invertebrates, 1979—; adj. prof. biology CUNY, 1972-80; adj. assoc. prof. geology Columbia U., 1975—. Co-editor: Systematic Zool, 1974-77. Mem. Paleontology Soc. (Schuchert award 1979), Brit. Palaeontology Assn., Soc. Study Evolution, Soc. Systematic Zoology. Subspecialty: Paleobiology. Office: Dept Invertebrates Am Mus Natural History New York NY 10024

ELDRIDGE, FREDERIC LOUIS, medical educator, physiology researcher; b. Kansas City, Mo., July 8, 1924; s. Charles Judson and Dorothea (Hackbusch) E.; m. Mary Frances Hill, Feb. 11, 1951; children: Charles Frederic, Karen Marie. A.B., Stanford U, 1945, M.D., 1948. Instr. to assoc. prof and prof. medicine Stanford U., 1954-73; prof. medicine and physiology U. N.C.-Chapel Hill, 1973—; cons. NIH, 1975-79, VA, 1979-82, U.S. Army, 1981—. Served with AUS, 1942-46; to capt. USAF, 1951-53. John and Mary Markle Found. scholar, 1956-61; NIH research grantee, 1957-87. Mem. Am. Physiol. Soc., Soc. Neurosci., Am. Fedn. Clin. Research, Am. Soc. Clin. Investigation. Subspecialties: Physiology (medicine); Neurophysiology. Current work: Neural control of respiration.

ELDRIDGE, JOHN CHARLES, medical educator; b. Chgo., June 7, 1942; s. John Godfrey and Carol Spier (Boedeker) E. B.A., N.Central Coll., Naperville, Ill., 1965; M.S., No. Ill. U., 1967; Ph.D., Med. Coll. Ga., 1971. Instr. dept. biol. and health scis. Orange County Community Coll., Middletown, N.Y., 1967-68; attaché de recherche INSERM, Bordeux, France, 1971-72; research assoc. Med. Coll. Ga., Augusta, 1973; assoc. dept. lab. medicine Med. U. S.C., Charleston, 1973-76, asst. prof., 1976-78; asst. prof. dept. physiology/pharmacology Bowman Gray Sch. Medicine, Winston-Salem, N.C., 1978—; cons. in field. Contbr. papers to profl. confs. and articles to profl. jours. Mem. Endocrine Soc., AAAS, Soc. Study Reprodn., Am. Fertility Soc., Am. Soc. Andrology. Presbyterian. Club: Twin City (Winston-Salem). Lodge: Masons. Subspecialties: Neuroendocrinology; Reproductive endocrinology. Current work: Mechanisms controlling pituitary hormone secretion; role of adrenals in reproductive function; techniques of analyzing hormones and hormone receptor interactions. Home: 2458 Tantelon Pl Winston-Salem NC 27107 Office: Bowman Gray Sch Medicine 300 S Hawthorne Rd Winston-Salem NC 27103

ELEUTERIUS, LIONEL NUMA, botanist, laboratory administarator, researcher; b. Biloxi, Miss., Dec. 25, 1936; s. Lionel Adam and Martha Elizabeth (Tiblier) E.; m. Kathryn Sarah Poole, Dec. 25, 1969; children: Chris L., Lee L. A.S., Perkinston Jr. Coll., 1958; B.S. in Botany, U. So. Miss., 1966, M.S., 1968, Ph.D., Miss. State U., 1974. Biol. technician U.S. Dept. Agr. Forest Service, Plant Pathology lab., Gulfport, Miss., 1961-65; lab. instr., research asst. U. So. Miss., Hattiesburg, 1965-68; biology and botany instr. Resident Center, U. So. Miss., Keesler AFB, Biloxi, 1968-69, genetics and gen. biology instr., 1969-72; prof. botany, head botany sect. Gulf Coast Research Lab., Ocean Springs, Miss., 1968—, instr. salt marsh ecology, summers 1976—, coastal vegetation, summers 1982—; adj. assoc. prof botany Miss. State U., 1976—; adj. asst. prof. biology U. Miss., 1977—; adj. prof. biology U. So. Miss., 1981—; mem. Deer Island Study Com. (adv.), 1978-79, Nat. Wetlands Tech. Council, Washington, Miss. Gov.'s Conf. on Coastal Zone Mgmt., 1974. Author: Tidal Marsh Plants, 1980; contbr. numerous articles on botany to profl. jours. Served to maj. Army N.G., 1959-80. Recipient grants in field. Mem. Am. Bot. Soc., Am. Soc. Plant Taxonomists, Ecol. Soc. Am., So. Appalachian Bot. Club (v.p. 1979-80), Miss. Acad. Sci. (chmn. botany sect. 1970-71), Phi Theta Kappa, Beta Beta Beta. Republican. Episcopalian. Subspecialties: Taxonomy; Ecology. Current work: Plant growth, plant morphology, tide-plant relations, plant genetics, plant populations, plant taxonomy. Home: 123A Red Bluff Circle Ocean Springs MS 39564 Office: E Beach Dr Ocean Springs MS 39564

ELFSTROM, GARY MACDONALD, aeronautical engineer, consultant; b. Vancouver, C., Can., Aug. 14, 1944; s. Roy Harold and Vera Marie (Macdonald) E.; m. Carol Ann Skelton, Sept. 25, 1969; children: David Roy, Julie Ann. B.A.Sc., U. B.C., Vancouver, 1968; Ph.D., Imperial Coll., London, 1971. Registered profl. engr., Ont. Research fellow U. Tenn., Tullahoma, 1971-73; research officer Nat. Research Council, Ottawa, Ont., 1973-81; sr. engr. DSMA Internat., Toronto, Ont., 1981—. Athlone fellow U.K. Bd. Trade, London, 1968. Fellow Can. Aeronautics and Space Inst. (assoc., councillor 1978, 81); mem. AIAA. Subspecialty: Aeronautical engineering. Current work: Research and development of aerodynamics test facilities. Home: 1032 Deer Run Mississauga ON Canada L5C 3N4 Office: DSMA Internat Inc 10 Park Lawn Rd Toronto ON Canada M8Y 3H8

ELGERT, KLAUS DIETER, immunologist, researcher, educator; b. Schwarmstedt, Germany, Mar. 5, 1948; s. Adam and Irmgard Eva (Christowzik) E.; m. Kathleen Carol McCall, Aug. 9, 1969; children: Heather Kathleen, Colleen Elizabeth. B.S., Evangel. Coll., Springfield, Mo., 1970; Ph.D., U. Mo., 1973. Postdoctoral fellow U. Mo., 1974; asst. prof. microbiology Va. Poly. Inst. and State U., 1974-80, assoc. prof., 1980—. Contbr. articles to profl. jours. NSF grantee NATO Advanced Study Inst., Greece, 1977; NIH grantee, Israel, 1978; recipient Teaching Excellence award Va. Poly. Inst. and State U., 1979; Disting. Alumnus award Evangel. Coll., 1981; NIH-Nat. Cancer Inst. grantee, 1981—; Anna Fuller, Elsa U. Pardee, Whitehall, Lane Found. grantee, 1975-84. Mem. Am. Assn. Immunologists, Reticuloendothelial Soc., Am. Soc. Microbiology, AAAS, N.Y. Acad. Sci., Am. Soc. Microbiology, Va. Acad. Sci., Internat. Leucocyte Culture Conf., Southeastern Immunology Conf., Sigma Xi. Subspecialties: Immunobiology and immunology; Cancer research (medicine). Current work: Compromising of T lymphocyte- and macrophage-mediated immuno-regulation by tumor growth. Office: Dept Biology Microbiology Sect Va Poly Inst and State U Blacksburg VA 24061

EL-GEWELY, MOHAMED RAAFAT, molecular geneticist; b. Damanhour, Egypt, June 2, 1942; s. Ahmed Abdelsalam and Aziza Khatab (Makey) El-G.; m. Sara, Nov. 16, 1981. B.S., Alexandria (Egypt) U., 1963; Ph.D., U. Alta., Can., 1971. Postdoctoral fellow McGill U., Montreal, Que., Can., 1971-73; asst. prof. Cairo (Egypt) U., 1973-77; assoc. prof. Alexandria U. Sci. Ctr., 1980; research scientist U. Mich., Ann Arbor, 1977—. Contbr. articles in field to profl. publs. UNESCO fellow, 1977. Mem. Am. Soc. Microbiology, Genetics Soc. Am., Genetics Soc. Can., Egyptian Soc. Genetics, N.Y. Acad. Scis. Moslem. Subspecialty: Genetics and genetic engineering (biology). Current work: Gene organization and regulation in eucaryotes and eycaryotic cell organelles using trecombinant DMA methodology. Office: Dept Biol Chemistry Med Sci I U Mich Ann Arbor MI 48109

EL GHATIT, ZEINAB MOHAMMED, psychologist; b. Cairo, June 11, 1936; emigrated to Can., 1974, naturalized, 1977; d. Mohammed Ali and Shahwer El G.; m. Mustafa Kamel A. Rostom, June 4, 1965; children: Alaa, Wael. M.A. in Clin. Psychology, Ohio State U., 1960. Pvt. practice psychology, Cairo, 1961-74; cons. psychologist Family Guidance, Dokki, Cairo, 1965-74, Mil. Hosp. and El Nile Hosp., 1965-74; lectr. psychology Am. U. Cairo, 1967-74; psychometrist II Prince Edward Heights, Picton, Ont., 1974-76, unit dir., 1976-79; program supr. mental retardation services Children's Services, Ministry of Community & Social Services, Kingston Office, Ont., 1979, 1980—; prof. spl. edn. Ottawa U., 1981-83. Author: Adaptive Behavior, 1975; contbr. articles to profl. jours. Ford Found. grantee, 1973. Mem. Am. Psychol. Assn., Am. Assn. Mental Deficiency, Canadian Psychol. Assn., Que. Assn. Profl. Psychologists, Brit. Psychol. Soc., Egyptian Psychol. Assn., Am. Assn. Edn. Severely/Profoundly Handicapped, Internat. Assn. Applied Psychology, Council of Exceptional Children. Subspecialties: Developmental psychology; clinical psychology. Current work: Applied deinstitutionalization. Home: #227 1695 Playfair Dr Ottawa ON Canada K1H 8J6 Office: 10 Rideau St Ottawa ON Canada K1H 7X3 Home: 27 1695 Playfair Dr Ottawa ON Canada K1H 8J6

ELGHOBASHI, SAID (ELSAYED), mechanical engineering educator, consultant; b. Cairo, Egypt, Aug. 17, 1941; came to U.S., 1968, naturalized, 1978; s. Elsayed A. and Zenab (Elshinawy) E.; m. Mary Ellen Brooks, July 5, 1969; children: Karim, Nadia, Magda. B.S. in Engring, Cairo U., 1962, M.S., U. So. Calif., 1971, Ph.D., Imperial Coll., U. London, Eng., 1974, diploma (hon.), 1975. Group leader CHAM Ltd., London, 1974-76; sr. cons. CHAM. N.Am., Huntsville, Ala., 1976-78; asst. prof. U. Calif.-Irvine, 1978-82, assoc. prof. mech. engring., 1982—; cons. S.A.I. Inc., Los Angeles, 1978—, Jet Propulsion Lab., Pasadena, Calif., 1978—. Mem. ASME, Combustion Inst., AIAA. Subspecialties: Fluid mechanics; Combustion processes. Current work: Research in mathematical modeling of turbulence, two-phase flows, combustion and numerical methods. Office: Dept Mech Engring Univ Calif Irvine Irvine CA 92717 Home: 35 Cool Brook Irvine CA 92715

ELGIN, SARAH CARLISLE ROBERTS, biology educator; b. Washington, July 16, 1945; d. Carlisle Bishop and Lorene (West) Roberts; m. Robert Lawrence Elgin, June 9, 1967; 1 son, Benjamin Carlisle. B.A. magna cum laude in Chemistry, Pomona Coll., 1967; Ph.D. in Biochemistry, Calif. Inst. Tech., 1971. Research fellow in biology Calif. Inst. Tech., Pasadena, 1971-73; asst. prof. biochemistry and molecular biology Harvard U., Cambridge, Mass., 1973-77, assoc. prof. biochemistry and molecular biology, 1977-81; assoc. prof. biology Washington U., St. Louis, 1981—. Editorial bd.: Jour. Cell Biology, 1980-82; contbr. articles to sci. jours. Nat. Merit scholar, 1963-67; Pomona Coll. scholar, 1963-67; recpient Wilson prize, 1967; NSF fellow, 1967-71; Jane Coffin Childs Meml. Fund fellow, 1971-73; NIH Research Career Devel. awardee, 1976-81. Mem. Am. Chem. Soc., Am. Soc. Biol. Chemists, AAAS, Genetics Soc. Am., Am. Soc. Cell Biology (council 1983—), Phi Beta Kappa, Mortar Bd. Subspecialties: Gene actions; Genome organization. Current work: Structure of chromatin and control of gene expression in Drosophila. Home: 7261 Kingsbury St University City MO 63130 Office: Washington U Box 1137 Saint Louis MO 63130

ELGOMAYEL, JOSEPH IBRAHIM, engineering educator, consultant; b. Maghagha, Egypt, Apr. 18, 1928; came to U.S., 1959, naturalized, 1978; s. Ibrahim D. and Aziza E. (ElKhoury) ElG.; m. Theresa Shoukry, Oct. 10, 1955; children: Dina, Ramzy. M.S., Purdue U., 1961, Ph.D., 1963. Instr. Cairo U., 1950-59; dep. dir. Mgmt. Cons. Ctr., Cairo, 1964-68; asst. prof. Purdue U., West Lafayette, Ind., 1969-75, assoc. prof. indsl. engring., 1975—; cons. in field, 1975—. Contbr. tech. articles to profl. publs. Mem. ASME, Soc. Mfg. Engrs., Soc. Carbide and Tool Engrs. (Service award 1980), Am. Inst. Indsl. Engrs. Democrat. Roman Catholic. Lodge: KC. Subspecialties: Industrial engineering; Robotics. Current work: Computer aided manufacturing, manufacturing science teaching. Home: 855 Ashland St West Lafayette IN 47906 Office: MGL Dept Indsl Engring Purdue U West Lafayette IN 47907

EL-HAWARY, FERIAL MOHAMED, engineering educator; b. Mansoura, Egypt, Mar. 20, 1943; emigrated to Can., 1968; s. Mohamed El-Sayed and Waheba (El-Kadaah) El-Bebani; m. Mohamed El-Aref, Sept. 10, 1966; children: Bahaa el-Deen (Bobby), Rany, Elizabeth Sarah. B.Sc., Alexandria (Egypt) U., 1967; M.Sc., Alta. U., Edmonton, 1971; Ph.D., Meml. U. Nfld., St. John's, 1981. Elec. engr. Msr. Beida Dyers, Alexandria, 1967-68; grad. asst. U. Alta., 1969-71; teaching and research asst. Meml. U., Nfld., 1975-80, postdoctoral fellow, 1981; research asst. prof. Tech. U. N.S., Halifax, 1981—. Mem. IEEE, Marine Tech. Soc., Sigma Xi. Subspecialties: Sea floor spreading; Ocean engineering. Current work: Modeling and signal processing for identifying features of the ocean subsurface layered media on the basis of acoustic reflection data. Also estimation theory and parameters identification. Home: 23 Bayview Rd Halifax 9 NS Canada B3M 1N8 Office: Tech Univ Nova Scotia PO Box 1000 Halifax NS Canada B3J 2X4

ELIAS, HANS GEORG, research institute executive, consultant; b. Bochum Westfalia, Germany, Mar. 29, 1928; came to U.S., 1971; s. Hermann Ludwig Georg and Elisabeth Charlotte (Rowlin) E.; m. Maria Hanke, Mar. 22, 1956; children: Peter Cornelius, Rainer Martin. Diploma in Chemistry, Tech. U. Hannover, W.Ger., 1954; Dr. rer. nat., Tech. U. Munich, W.Ger., 1957; Habilitation, Swiss. Fed. Inst. Tech., Zurich, 1961. Sci. asst. Tech. Univ. Munich, 1956-59; sci. head asst. Swiss Fed. Inst. Tech., 1959-63, privatdozent, 1961-63, asst. prof., 1963-71; pres. Mich. Molecular Inst., Midland, 1971-83; cons. firms, U.S., W.Ger., Switzerland, 1956—. Author: Macromolecules, German, 4 edits., 1971-81, English, 1977; contbr. chpts. to books, articles to profl. jours. Hon. adv. bd. Midland Symphony Orch., 1975—; mem. Midland Beautification Adv. Com., 1979-83. Served with German Armed Forces, 1944-45. Fellow Sigma Xi (paper award 1982); mem. Am. Chem. Soc. (chmn. polymer internat. membership com. 1983), Am. Phys. Soc., German Chem. Soc., Swiss Chemists Assn., Bunsen Soc. for Phys. Chemistry. Club: Torch (dir 1982—). Subspecialties: Polymer chemistry; Polymer engineering. Current work: Stereocontrol in polymerizations; association in solution; polymers from renewable resources. Patentee in field. Home and Office: 4009 Linden Dr Midland MI 48640

ELIAS, THOMAS ITTAN, mechanical engineering educator; b. South Piramadom, Kerala, India, Mar. 16, 1947; came to U.S., 1975, naturalized, 1980; s. Varkey and Sara (Paul) Ittan; m. Gracy Thomas, Sept. 3, 1979; 1 son, Joe. B.S., U. Kerala, 1970, M.S., 1972; M.S., U. Cin., 1977; Ph.D., U. Minn., 1981. Devel. engr. Vikram Sarabhai Space Ctr., Trivandrum, India, 1972-75; research asst. U. Cin., 1976-77; teaching assoc., research asst. U. Minn., Mpls., 1977-80; asst. prof. Ind. Inst. Tech., Ft. Wayne, 1980-82; asst. prof. mech. engring. Youngstown (Ohio) State U., 1982—. Contbr. research papers to profl. publs. Merit scholar U. Kerala, 1966-72. Mem. ASME, Nat. Soc. Profl. Engrs., AIAA. Subspecialties: Mechanical engineering; Fluid mechanics. Current work: Research in fluid mechanics, thermal sciences, other related areas of mechanical and aerospace engineering. Home: 435 Garver Dr Youngstown OH 44512 Office: Mech Engring Dept Youngstown State U 410 Wick Av Youngstown OH 44555

ELIASON, MORTON ALBERT, chemist; b. Fargo, N.D.; s. Albert L. and Frances Marguerite (Martinson) E.; m. Dorothy Bartlett, Aug. 18, 1956; children: Ellen, Eric, Kent, Karl. B.A., Concordia Coll., Moorhead, Minn., 1954; Ph.D., U. Wis.-Madison, 1959. Mem. faculty Augustana Coll., Rock Island, Ill., 1958—, assoc. prof. chemistry, 1963-69, prof., 1969—, head dept., 1980—. Mem. Am. Chem. Soc., Sigma Xi. Lutheran. Subspecialties: Physical chemistry; Theoretical chemistry. Current work: Chemical kinetics; quantum theory; oscillating reactions; physical properties of corn syrups. Home: 4026 26th Ave Rock Island IL 61201 Office: Dept Chemistry Augustana Coll Rock Island IL 62101

ELIEL, ERNEST LUDWIG, educator, chemist; b. Cologne, Germany, Dec. 28, 1921; came to U.S., 1946, naturalized, 1951; s. Oskar and Luise (Tietz) E.; m. Eva Schwarz, Dec. 23, 1949; children—Ruth Louise, Carol Susan. Student, U. Edinburgh, (1939-40), Scotland; Dr.Phys.-Chem. Sci., U. Havana, Cuba, 1946; Ph.D., U. Ill., 1948. Mem. faculty U. Notre Dame, South Bend, Ind., 1948-72, prof. chemistry, 1960-72, head dept., 1964-66; W.R. Kenan Jr. prof. chemistry U. N.C., Chapel Hill, 1972—; Le Bel Centennial lectr., Paris, 1974; Benjamin Rush lectr. U. Pa., Phila., 1978; Sir C.V. Raman vis. prof. U. Madras, India, 1981. Author: Stereochemistry of Carbon Compounds, 1962, Conformational Analysis, 1965, Elements of Stereochemistry, 1969; Co-editor: Topics in Stereochemistry, Vols. I-XII, 1967-80. Pres. Internat. Relations Council, St. Joseph Valley, 1961-63. NSF sr. research fellow Harvard, 1958; Calif. Inst. Tech., 1958-59; E.T.H., Zurich, Switzerland, 1967-68; Recipient Coll. Tchrs. award Mfg. Chemists Assn., 1965; Morley medal Cleve. sect. Am. Chem. Soc., 1965; Laurent Lavoisier medal French Chem. Soc., 1968; Guggenheim fellow Stanford, Princeton U., 1975-76. Mem. Nat. Acad. Scis., Am. Acad. Arts and Scis., Am. Chem. Soc. (chmn. St. Joseph Valley sect. 1960, councillor 1965-73, 75—, chmn. com. publs. 1972, 76-78), AAAS, Chem. Soc. London, AAUP (chpt. pres. 1971-72, 78-79), Sigma Xi (chpt. pres. 1968-69), Phi Lambda Upsilon, Phi Kappa Phi. Subspecialties: Organic chemistry; Nuclear magnetic resonance (chemistry). Current work: Stereochemistry, conformational analysis, asymmetric synthesis, carbanions, organosulfur chemistry. Home: 725 Kenmore Rd Chapel Hill NC 27514

ELIEZER, ISAAC, chemistry educator, educational administrator; b. Sofia, Bulgaria, Jan. 19, 1934; came to U.S., 1965, naturalized, 1982; m. Naomi Stein, June 30, 1960; children: Eran, David, Ken. M.Sc., Hebrew U., 1956, Ph.D., 1960. Vis. prof. Weizmann Inst., Rehovoth, Israel, 1972-75; prof. chemistry, mgr. energy research Mont. State U., 1975-79; prof. Oakland U., Rochester, Mich., 1979—, assoc. dean, 1979—. Editor: Israel Jour. Chemistry, 1972-75; contbr. numerous articles to profl. jours. Grantee in field. Fellow Royal Inst. Chemistry, Chem. Soc. London, N.Y. Acad. Scis.; mem. Am. Chem. Soc., AAAS. Subspecialties: Physical chemistry; Inorganic chemistry. Current work: Thermodynamics, high-temperature reactions and equilibria, solutions, donor-acceptor complexes, computer simulations, energy and the environment, information storage and retrieval. Office: Dept Chemistry Oakland U Rochester MI 48063

ELIOT, JOHN, educator; b. Washington, Oct. 28, 1933; s. Charles William and Regina (Dodge) E.; m. Sylvia Ashley Hewitt, July 3, 1959; children: Mary, Catherine. A.B., Harvard U., 1956, A.M.T., 1958; Ed.D., Stanford U., 1966. Asst. prof. edn. Northwestern U., Evanston, Ill., 1966-69; mem. faculty dept. edn. U. Md., College Park, 1969—, prof. edn., 1977—. Author 5 books in field; contbr. articles to profl. jours. Mem. Soc. Neurosci., Am., Psychol. Assn., Soc. for Research in Child Devel., Brit. Psychol. Soc. Subspecialties: Cognition; Spatial ability. Home: 2705 Silverdale Silver Springs MD 20906 Office: Inst Child Study University of Maryland College Park MD

ELITZUR, MOSHE, astronomer, educator; b. Poland, Apr. 29, 1944; came to U.S., 1977; s. Yechieal and Sophia E.; m. Shlomit Yoskowitz; children: Ofer, Haggai, Ben. B.Sc., Hebrew U., Jerusalem, 1964; M.Sc., Weizmann Inst., Rehovot, Israel, 1966, Ph.D., 1970. Research assoc. Rockefeller U., N.Y.C., 1970-72; research fellow Calif. Inst. Tech., Pasadena, 1972-74; scientist Weizmann Inst., 1974-75, Sr. scientist, 1975-80; vis. research asst. prof. U. Ill., Urbana, 1977-80; assoc. prof. physics and astronomy U. Ky., Lexintgon, 1980—. Contbr. articles to profl. jours. Served to lt. Israel Def. Forces, 1966-69. Fulbright awardee, 1970-72; NSF grantee, 1981. Mem. Am. Astron. Soc., Internat. Astron. Union. Subspecialties: Radio and microwave astronomy; Theoretical astrophysics. Current work: Research on problems in theoretical astrophysics, in particular problems related to interstellar medium and molecular emission. Office: Dept Physics U Ky Lexington KY 40506

ELIZAN, TERESITA S., neurologist, neuroscientist, educator; b. Naga City, Philippines, Dec. 12, 1931; d. Paulo and Nicolasa Rosales (Siguenza) E. M.D., U. Philippines, Manila, 1955. Diplomate: Am. Bd. Psychiatry and Neurology. Resident in neurology Yale U. Sch. Medicine, 1956-58, Montreal Neurol. Inst., McGill U., 1959-60; Dazian Found. fellow in neurology Mt. Sinai Hosp., N.Y.C., 1960-62, attending neurologist, head lab. neurovirology, 1977—; vis. scientist neurology and neurovirology NIH, Bethesda, Md., 1963-68; asst. prof. neurology Mt. Sinai Sch. Medicine, 1968-71, asso. prof., 1971-77, prof., 1977—. Contbr. articles to profl. jours. Fellow Am. Acad. Neurology; mem. Am. Neurol. Assn., Am. Soc. Virology, Am. Assn. Neuropathology, Am. Soc. Microbiology, Soc. Neurosci., AAAS. Subspecialties: Neurology; Neurovirology. Current work: Role of viruses in etiology and pathogenesis of certain degenerative brain diseases; dementing illnesses the aging nervous system. Home: 245 E 63d St New York NY 10021 Office: 1200 Fifth Ave New York NY 10029

ELKAYAM, URI, Cardiologist; b. Petach Tikva, Israel, Mar. 11, 1945; came to U.S., 1976, naturalized, 1978; s. Mordechai and Dvora (Shapira) E.; m. Yael Batia Nachum, Aug. 17, 1975; children by previous marriages: Ifaat, Yehonatan, Danielle. M.D., Sackler Med. Sch., Israel, 1973. Intern Ichilov Med. Ctr., Tel Aviv, Israel, 1973-74, resident, 1974-76; dir. coronary care unit U. Calif.-Irvine, Orange, 1979-81; dir. inpatient cardiology U. So. Calif., Los Angeles, 1981—. Editor: Cardiac Problems in Pregnancy, 1982; contbr. articles to med. jours. Fellow Am. Coll. Chest Physicians; mem. ACP, Am. Fedn. Clin. Research, Am. Heart Assn., Am. Coll. Cardiology. Subspecialty: Cardiology. Current work: Congestive heart failure; ischemic heart disease; cardiovascular pharmacology. Home: 2956 Queensbury Dr Los Angeles CA 90064 Office: U So Calif Med Sch Section of Cardiology 2025 Zonal Ave Los Angeles CA 90033

ELKIN, ROBERT GLENN, nutritional biochemist, educator; b. Passaic, N.J., May 7, 1953; s. Abe and Jean Estaire (Edelman) E.; m. Emily Jill Furumoto, May 23, 1981. B.S., Pa. State U., 1975; M.S., Purdue U., 1977, Ph.D., 1981. Asst. prof. animal scis. Purdue U., West Lafayette, Ind., 1981—. Contbr. articles to profl. jours. Ralston Purina Co. research fellow, 1979-80; Monsanto Co. grantee, 1982. Mem. Poultry Sci. Assn., World's Poultry Sci. Assn., AAAS, Sigma Xi, Gamma Sigma Delta. Jewish. Subspecialties: Animal nutrition; Analytical chemistry. Current work: Avian amino acid nutrition and metabolism; amino acid analysis by high performance liquid chromatography. Home: 3031 Courthouse Dr Apt 2C West Lafayette IN 47906 Office: Dept Animal Scis Purdue U West Lafayette IN 47907

ELKIND-HIRSCH, KAREN ELIZABETH, scientific researcher; b. N.Y.C., Apr. 28, 1954; d. Maurice Paul and Lilyan Gloria (Shuman) Elkind; m. Gary Stephen Hirsch, June 17, 1978; 1 son, Jordan Elkind. B.A., Skidmore Coll., 1974; M.S., Tulane U., 1977, Ph.D., 1980. Research asst. Endocrine and Polypeptide Lab., Tulane U., New Orleans, 1977-78; research asst. dept. physiology Harvard Med. Sch., Boston, 1978-80, Mellon postdoctoral research fellow dept. ob-gyn, 1980—. Author articles. NSF fellow, 1974-78. Mem. AAUP, Soc. for Neurosci., Phi Beta Kappa, Sigma Xi. Club: Houston Metropolitan Racquet. Subspecialty: Neuroendocrinology. Current work: Neuroendocrine control of gonadotropin secretion. Office: Reproductive Research Lab Baylor Coll Medicine Houston TX 77030

ELKINS, GARY RAY, psychologist; b. Hot Springs, Ark., Oct. 5, 1952; s. Billy Ray and Jewel Dean (Edwards) E.; m. Dorothy J. Sutton, May 31, 1975. B.A., Henderson State U., 1975; M.A., East Tex. State U., 1976; Ph.D., Tex. A&M U., 1980. Sr. staff psychologist Scott and White Clinic, Temple, Tex., 1982—, program dir., 1982—; instr. dept. psychiatry Tex. A&M Coll. Medicine, College Station, 1982—. Contbr. articles to profl. jours. Served to capt. USAF, 1979-82. Mem. Am. Psychol. Assn., Tex. Psychol. Assn., Internat. Assn. for Study of Pain, Am. Soc. for Clin. Hypnosis. Current work: Behavioral medicine, psychology of pain, hypnosis, psychotherapy research. Home: 3409 Forest Trail Temple TX 76501 Office: Dept Psychiatry Scott and White Clinic 2401 S 31st St Temple TX 76508

EL KOUNI, MAHMOUD HAMDI, medical scientist, educator; b. Cairo, May 30, 1942; s. Mustapha Mahmoud and Zeinab A. El K.; m. Fardos Naguib Mohammed Naguib, May 30, 1966; children: Mustapha, Sarah. B.Sci. with honors, U. Alexandria, Egypt, 1964, M.Sc., 1968; Ph.D., U. Alta., 1977. Teaching asst. Cairo U., 1964-66, U. Alexandria, 1966-70; teaching, then research asst. U. Alta., 1970-77; research assoc. Syracuse U., 1977-79, Brown U., Providence, 1979-81, asst. prof. med. scis., 1981—. Contbr. articles to profl. jours. Grantee Am. Cancer Soc., 1980-82, WHO, 1981-83. Mem. AAAS, Genetics Soc. Am. Moslem. Subspecialties: Biochemistry (medicine); Cancer research (medicine). Current work: Biochemistry and regulation of nucleotide metabolism. Mechanisms of action of antimetabolites in treatment of cancer and parasites. Office: Bio-Med Scis Brown U Providence RI 02912

ELLEMAN, THOMAS SMITH, nuclear engineering educator, utility company executive; b. Dayton, Ohio, June 19, 1931; m.; 3 children. B.S., Denison U., 1953; Ph.D. in Chemistry, Iowa State Coll., 1957. Chemist Inst. Atomic Research, Ames Lab., 1953-57; radiochemist Battelle Meml. Inst., 1957-64; assoc. prof. N.C. State U., Raleigh, 1964-67, prof. nuclear engring., 1967—, head dept., 1974-79; v.p. nuclear safety and research dept. Carolina Power & Light Co., 1979—. Mem. Am. Chem. Soc., Am. Nuclear Soc. Subspecialty: Nuclear engineering. Office: Dept Nuclear Engring NC State U Raleigh NC 27607

ELLENBOGEN, LEON, biochemist, nutritionist; b. Bklyn., May 3, 1927; s. Martin and Bella (Zalesnick) E.; m. Roslyn Barban, June 30, 1951; children: Kenneth, Richard, Cheryl. B.S., CCNY, 1949; M.S., N.Y.U., 1951; Ph.D., Ind. U., 1954. Research technician Columbia U., 1949-51; teaching asst. Ind. U., 1951-53; research biochemist Lederle Labs., Pearl River, N.Y., 1953-59, sr. research biochemist and group leader, 1959-77, chief nutritional sci., asso. dir. profl. pharm. services, 1977—; adj. prof. nutrition in medicine Cornell U., 1978—; adj. prof. nutrition N.Y. Med. Coll., 1982—. Contbr. articles to profl. jours. Served with USN, 1945-47. Fellow N.Y. Acad. Scis.; mem. Am. Heart Assn., Am. Soc. Hematology, Am. Inst. Nutrition, Am. Soc. Clin. Nutrition, Am. Soc. Biol. Chemistry, Am. Soc. Pharmacology and Exptl. Therapeutics, Am. Chem. Soc., Soc. Exptl. Sci., Sigma Xi, Phi Lambda Upsilon. Subspecialties: Nutrition (medicine); Biochemistry (medicine). Home: 16 Morris Dr New City NY 10956 Office: Lederle Labs Pearl River NY 10965

ELLENSON, JAMES L., biophysicist, plant researcher; b. Ft. Dodge, Iowa, Apr. 25, 1946; m.; 1 child. B.A., Oberlin Coll., 1968; Ph.D. in Chemistry, U. Calif.-Berkeley, 1973. Research fellow Harvard U., 1973-79; research assoc. Boyce Thompson Inst. Plant Research, Ithaca, N.Y., 1979—. Mem. Biophys. Soc., AAAS. Subspecialty: Plant physiology (agriculture). Office: Boyce Thompson Inst Plant Research Tower Rd Ithaca NY 14853

ELLINGBOE, JAMES, biochemist; b. Wilmington, Del., June 10, 1937; s. Ellsworth Knowlton and Helen Adele (Jones) E.; m. Karin Ester Sofia Westlin, Sept. 16, 1967; children: Randi, Christopher. A.B., Oberlin Coll., 1959; postgrad., U. Geneva, 1957-58; Ph.D., Harvard U., 1966. NIH postdoctoral fellow and amanuens Royal Caroline Inst., Stockholm, 1966-68; asst. research biochemist Sch. Medicine, U. Calif.-San Diego, 1968-70; assoc. in biochemistry Harvard U. Med. Sch., Boston, 1970-72, prin. research assoc. in psychiatry, 1972—; chief drug surveillance and biochemistry lab. Boston City Hosp., 1970-73; chief biochemistry lab., alcohol and drug abuse research ctr. McLean Hosp., Belmont, Mass., 1973—. Contbr. articles to sci. jours. Chmn. Littleton (Mass.) Conservation Commn., 1980—; trustee Lake Matawanakee Assn., 1975—. USPHS fellow and grantee, 1959—. Mem. Am. Chem. Soc., Endocrine Soc., Am. Soc. Pharmacology and Exptl. Therapeutics, Soc. Neurosci., Internat. Soc. Psychoneuroendocrinology, Internat. Soc. Biomed. Research on Alcoholism, N.Y. Acad. Scis., Fedn. Am. Scientists, AAAS, New Eng. Hist. Genealogic Soc., Sierra Club, Mass. Audubon Soc. Mem. Soc. of Friends. Clubs: Valdres Samband, Valdres Historielag, Svenska Folkdansens Vanner. Subspecialties: Pharmacology; Neuroendocrinology. Current work: Studies of the effects and mechanisms of action of alcohol and drugs of abuse on neuroendocrine regulatory systems. Patentee materials and process inventions. Home: 63 Matawanakee Trail Littleton MA 01460 Office: 115 Mill St Belmont MA 02178

ELLINGWOOD, BRUCE RUSSELL, structural engineering researcher; b. Evanston, Ill., Oct. 11, 1944; s. Robert W. and Carolyn L. (Ehmen) E.; m. Lois J. Drager, June 7, 1969; 1 son, Geoffrey D. B.S.C.E., U. Ill., Urbana, 1968, M.S.C.E., 1969, Ph.D., 1972. Structural engr. Naval Ship Research and Devel. Ctr., Bethesda, Md., 1972-75; research structural engr. Ctr. Bldg. Tech., Nat. Bur. Standards, Washington, 1975—; lectr., cons. Contbr. articles to profl. jours. Recipient Dural Research prize U. Ill., 1968; Nat. Capital award for Engring. Achievement D.C. Joint Council Engring. and Archtl. Socs., 1980; Walter L. Huber prize ASCE, 1980; Silver medal U.S. Dept. Commerce, 1980. Mem. ASCE., ASTM, Am. Nat. Standards Inst., Am. Inst. Steel Constrn., Sigma Xi, Chi Epsilon. Presbyterian. Subspecialties: Civil engineering; Structural reliability. Current work: Application of probability and statistics to structural engineering, specifically in developing structural loading and strength criteria for use in standards and other regulatory documents. Office: Nat Bur Standards Washington DC 20234

ELLINWOOD, EVERETT HEWS, JR., psychiatrist, pharmacologist, educator; b. Wilmington, N.C., June 27, 1934; s. Everett Hews and Hulda Eggleston (Holloman) E.; m. Cornelia Dawn Strain, Mar. 28, 1963; children: Everett Hews III, Susan Bradley. B.S., U. N.C., 1956, M.D., 1959. Diplomate: Am. Bd. Psychiatry and Neurology. Intern Watts Hosp., Durham, N.C., 1959-60; resident N.C. Meml. Hosp., Chapel Hill, 1960-63; asst. prof. psychiatry Duke U. Med. Ctr., Durham, 1967-70, assoc. prof., 1970-72, prof., 1972—, asst. prof. pharmacology, 1973-80, prof., 1980—; chmn rev. com. Nat. Inst. Drug Abuse, 1976-77; chmn. drug abuse rev. com. FDA, 1979-81; pres. Biomed. Inst. Inc., developer drug dispensing devices for internal implantation, 1978—. Editor 2 books; contbr. over 140 articles on neuropsychopharmacology to sci. jours. Served to lt. comdr. USPHS, 1963-65. Numerous research grants. Fellow Am. Psychiat. Assn. (research council), Am. Coll. Neuropsychopharmacology; mem. Soc. Biol. Psychiatry (sec.-treas.). Presbyterian. Subspecialties: Neuropharmacology; Psychopharmacology. Current work: General area of neuropharmacology; neuropsychopharmacology of psychosis; psychopharmacology of drug induced psychoses; implanted drug dispensing devices. Patentee self-powered implanted programable medication system and methods, also others. Office: Dept Psychiatry Duke U Med Center PO Box 3870 Durham NC 27710 Home: 3519 Tonbridge Way Durham NC 27706

ELLIOT, JAMES LUDLOW, astronomer; b. Columbus, Ohio, June 17, 1943; s. James Ludlow and Doris Belle (Eckfeld) E.; m. Elaine Kasparian, Nov. 24, 1967; children—Lyn, Martha. S.B., M.I.T., 1965, S.M., 1965; A.M., Harvard U., 1967, Ph.D., 1972. Research asso. Cornell U., Ithaca, N.Y., 1972-74, sr. research asso., 1974-77, asst. prof. astronomy, 1977-78; asso. prof. astronomy M.I.T., Cambridge, 1978—; dir. George R. Wallace Jr. Astrophys. Obs., 1978—. Recipient medal for exceptional sci. achievement NASA, 1977; NSF fellow, 1965-71. Mem. Am. Astron. Soc., Internat. Internat. Astron. Union. Subspecialties: Optical astronomy; Planetary science. Discovered rings of Uranus, 1977. Home: 27 Forest St Wellesley MA 02181 Office: Bldg 54-422A MIT Cambridge MA 02139

ELLIOTT, CHARLES H., clinical pediatric psychologist, researcher; b. Kansas City, Mo., Dec. 30, 1948; s. Joe Bond and Suzanne (Weider) E.; m. Barbara Lynn Johnson, Apr. 24, 1971; 1 son, Brian D. B.A., U. Kans., 1971, M.A., 1974, Ph.D., 1976. Asst. prof. East Central U., Ada, Okla., 1976-79; Okla. U. Health Sci. Ctr., Oklahoma City, 1979—; clin. cons., staff psychologist Mental Health Services So. Okla., Ada, 1977-79; psychologist pediatric psychology service Okla. CHildren's Meml. Hosp., Oklahoma City, 1979-80, dir, cons.-liaison div. mental health service, 1980—; cons. Nat. Cancer Inst., Los Angeles, 1982-85. NIH grantee, 1982-85; recipient Research and Devel. Com. award East Central U., 1977, 78; U. Okla. Health Sci. Ctr. award, 1981. Mem. Okla. Psychol. Assn., S.W. Psychol. Assn., Am. Psychol. Assn., Assn. Advancement Behavior Therapy, Soc. Behavioral Medicine, Central Okla. Pediatric Soc., AAAS, Soc. Pediatric Psychology, Assn. Advancement Psychology, Psi Chi. Subspecialties: Behavioral psychology; Pediatric psychology. Current work: Consultation-liaison psychology, pediatric psychology, and the assessment of the interaction between subject variables; also, pain management, asthma, smoking, and treatment of depression. Office: Okla U Health Sci Ctr Dept Psychiatry and Behavioral Sci PO Box 26901 South Pavilion Oklahoma City OK 73190

ELLIOTT, DANA RAY, biology educator, entomology consultant; b. Grain Valley, Mo., Feb. 7, 1945; s. Franklin Ellwood and Edna Mae (Row) E.; m. Cheryl Jeanne Boyd, July 18, 1970 (div.); 1 dau., Rebecca Leigh; m. Harriet Margaret Thompson, Mar. 17, 1978; 1 son, Daniel Paul. B.A., William Jewell Coll., 1967; M.S., Central Mo. State U., 1971; Ph.D., U. Mo.-Columbia, 1981. Tchr. sci., Raytown, Mo., 1967-68, Kansas City, Mo., 1971-72, Liberty, Mo., 1972-74; assoc. prof. biology Central Meth. Coll., Fayette, Mo., 1974—; cons. in entomology, Fayette, 1981—. Served as sgt. U.S. Army, 1968-70; Vietnam. Mem. Ecol. Soc. Am., Entomol. Soc. Am., Mo. Acad. Sci., Archaeol. Soc. Mo. Clubs: Fayette Jaycees (v.p. 1974-77), Fayette Optimist.). Subspecialties: Theoretical ecology; Ecosystems analysis. Current work: Analysis and correlation of insects and vascular plants in old field succession. Home: 103 Lucky St Fayette MO 65248 Home: 103 Lucky St Fayette MO 65248 Office: Dept Biology Central Methodist College Fayette MO 65248

ELLIOTT, JOHN FRANK, engineering educator; b. St. Paul, July 31, 1920; s. Stowe E. and Helen (Grube) E.; m. Frances Pendleton, May 4, 1946; children: William S., Dorothy E. Sempolinski. B.S., U. Minn., 1942; Sc.D., MIT, 1949. Phys. chemist Fundamental Research Lab. U.S. Steel Corp., Kearny, N.J., 1949-51; research metallurgist Inland Steel Co., East Chicago, Ind., 1951-54, asst. supt. quality control, 1954-55; asso. prof. dept. metallurgy MIT, Cambridge, 1955-60, prof. metallurgy dept. materials sci. and engring., 1960—; now AISI Disting. prof.; dir. MIT (Mining and Mineral Resources Research Inst.), 1978—. Author: Thermochemistry for Steelmaking, vol. I, 1960, vol. II 1963, Steelmaking: The Chipman Conference, 1965; editor: The Physical Chemistry of Steelmaking, 1958; contbr. articles to profl. jours. Served to lt. comdr. USNR, 1942-46. Guggenheim fellow, 1965; Disting. mem. Iron and Steel Soc., 1976. Fellow Metall. Soc., AIME (hon. mem. 1982;, Douglas Gold medal 1976, Howe Meml. lectr. 1963, extractive metallurgy lectr. 1975), Am. Soc. Metals (White disting. teaching award 1971), Am. Inst. Chem. Engrs.; mem. Nat. Acad. Engring., Am. Acad. Arts and Scis., Metals Soc. (Gt. Britain), Iron and Steel Inst. Japan (hon.), Can. Inst. Mining and Metallurgy, Venezuelan Soc. Mining and Metall. Engrs. (hon.), Société Française de Métallurgie (hon.), AAAS, Sigma Xi, Tau Beta Pi. Subspecialties: Metallurgical engineering; High-temperature materials. Current work: Physical chemistry of high temperature inorganic materials process and extractive metallurgy. Home: 118 Arlington St Winchester MA 01890 Office: 77 Massachusetts Ave Cambridge MA 02139

ELLIOTT, JOHN OTHNIEL, electronics engineer; b. Eau Claire, Wis., Feb. 7, 1920; s. Othniel Alonzo and Flora (McDonald) E.; m. Sarah Fox Matthews, Oct. 4, 1949; children: Sarah M., Flora M., Jane F., John Othniel. B.S.E.E., Purdue U., 1941. Engr. Zenith Radio Corp., Chgo., 1941-44, Bell and Howell Corp., 1944-45; engr. Fairchild Pilotless Plant, Farmingdale, N.Y., 1947-58, Ramo-Wooldrich, Patrick AFB, Fla., 1958-60, Aerospace Corp., Cape Canaveral Air Force Sta., Fla., 1960-63, TRW Def. and Space Systems, Cape Canaveral, 1963—. Pres. governing bd. Cocoa Beach (Fla.) Community Ch., 1980-83; audio supr. Cocoa Beach Pub. Library, 1960-83. Served to sgt. U.S. Army, 1945-47. Mem. IEEE, AIAA. Mem. United Ch. of Christ. Clubs: Indian River Amateur Radio (Brevard Repeater Assn) (pres. 1972-74); (Cocoa Beach)). Subspecialties: Aerospace engineering and technology; Unmanned vehicle launches. Current work: Aerosapce systems engineering; primarily concerned with communications and telemetry. Home: PO Box 41 Cocoa Beach FL 32531 Office: TRW Def and Space Systems PO Box 903 Cape Canaveral FL 32920

ELLIOTT, LARRY PHILLIP, microbiologist, educator; b. Fleming, Mo., Sept. 27, 1938; s. Melvin J. and Margaret (Hedrick) E.; m. Wilma Lee Grove, Jan. 5, 1938; children: Kerrie Lynn, Kimberly Ann, Kelly Jo. B.A., William Jewell Coll., 1960; M.S., U. Wis., 1962, Ph.D., 1965. Teaching asst. U. Wis.-Madison, 1960-61, research asst., 1961-65; asst. biology; Western Ky. U., Bowling Green, 1965-68, assoc. prof., 1968-79, prof., 1979—. NSF grantee, 1972; Council on Higher Edn. grantee, 1977-83. Mem. Ky. Acad. Sci., Sigma Xi, Beta Beta Beta. Republican. Baptist. Lodge: Odd Fellows. Subspecialty: Microbiology. Current work: Medical technology coordinator; isolation and identification of microscopic organisms from various habitats. Office: Dept Biology Western Ky U Bowling Green KY 42101 Home: 706 Wedgewood Way Bowling Green KY 42101

ELLIOTT, RICHARD AMOS, physicist; b. Lowbanks, Ont., Can., May 17, 1937; came to U.S., 1966, naturalized, 1983; s. Richard Irwin and Flora Hope (Rittenhouse) E.; m. Iva Joy Clark, June 21, 1958; children: Richard Clark, Jennifer Amelia. B.A., Queen's U., 1960, B.Sc., 1961, M.Sc., 1963; Ph.D., U. B.C., 1966. Research assoc. Inst. Fundamental Studies, U. Rochester, 71966-67, research assoc., instr., 1967-69; asst. prof. dept. applied physics Oreg. Grad. Ctr., 1969-77, assoc. prof. dept. applied physics, 1977—. Contbr. numerous articles to tech. jours. Mem. Can. Assn. Physicists, Optical Soc. Am. Subspecialties: Atmospheric Optics; Remote sensing (atmospheric science). Current work: Remote sensing; optical scattering and wave propagation in turbid and turbulent media; picosecond electro-optic phenomena and devices; nonlinear optics; lasers devices. Patentee in field. Home: 15585 NW Barkton St Beaverton OR 97006 Office: 19600 NW Walker Rd Beaverton OR 97006

ELLIS, DONALD GRIFFITH, bioengr., design cons.; b. Colorado Springs, Colo., Aug. 10, 1940; s. William Eugene and Lucile (Mathews) E.; m. Merle Elizabeth, May 21, 1977. B.S. in Mech. Engring, U. Colo., Boulder, 1962; postgrad. in metallurgy, U. Denver, 1962-63; M.S., U. Mich., 1964; Ph.D. in Bioengring, U. Mich., 1970. Research engr. U. Mich. Hosp., 1964-68, research assoc., 1968-70, 72-75; postdoctoral fellow Webb-Waring Inst., 1970-72; bioengr. U. Colo. Med. Center, Denver, 1973; research assoc. U. Colo., Boulder, 1975-78, U. Colo. Health Scis. Center, Denver, 1978—; cons. design. Contbr. articles on connective tissue mechanics and lab. instrumentation and equipment to profl. jours. Mem. ACLU, Inst. Food and Devel. Policy, Town Forum., ASME, Forth Interest Group, Assn. Humanistic Psychology, AAAS. Unitarian. Subspecialties: Biomedical engineering; Bioinstrumentation. Current work: Biomedical research instrumentation and equipment. Patentee inflatable insulating apparatus. Home: Geneva Park Boulder CO 80302 Office: CVP Research B1338471 U Colo Health Sci Center Denver CO 80262

ELLIS, HENRY CARLTON, psychology educator; b. New Bern, N.C., Oct. 23, 1927; s. Henry Alford and Frances Lee (Mays) E.; m. Florence Pettyjohn, Aug. 24, 1957; children: Joan, Diane Elizabeth, John Weldon. B.S., Coll. of William and Mary, 1951; M.A., Emory U., 1952; Ph.D., Washington U., St. Louis, 1958. Asst. prof. psychology U. N. Mex., Albuquerque, 1957-62, assoc. prof., 1962-67, prof., 1967—, chmn. dept., 1975—, ann. research lectr., 1978; disting. vis. prof. U.S. Air Force Med. Ctr., San Antonio, 1978; vis. prof. U. Calif.-Berkeley, 1971, U. Hawaii, Honolulu, 1977; cons. Am. Psychol. Assn., Washington, 1978—. Author: Human Learning and Cognition, 1972, Human Learning, Memory and Cognition, 1978, Psychology of Learning and Memory, 1979, Human Memory and Cognition, 1983. Served with USAAF, 1946-47. Van Blarcom fellow, 1956-67. Fellow Am. Psychol. Assn. (mem. and chmn. edn. and tng. bd. 1981-83, council of reps. 1983—, sec.-treas. exec. com. div. exptl. psychology 1982—), AAAS; mem. Rocky Mountain Psychol. Assn. (pres. 1967-68, disting. service award 1983), Southwestern Psychol. Assn. (pres. 1978-79), Council of Grad. Depts. Psychology (chmn. 1977-79), Psychonomic Soc., Phi Kappa Phi. Methodist. Clubs: Tennis, Twenty-One (Albuquerque). Subspecialties: Cognition; Learning. Current work: Human memory and cognitive psychology, emotional mood states and memory, organization and memory individual differences in memory and cognition, perceptual memory, eyewitness identification. Office: Dept Psychology U NMex Albuquerque NM 87131 Home: 1905 Amherst Dr NE Albuquerque NM 87106

ELLIS, KEITH OSBORNE, pharmacologist; b. Albany, N.Y., Oct. 18, 1941; s. Osborne Winfield and Helen (Scrafford) E.; m. Bonnie Jean, Aug. 17, 1964; children: Randall Keith, Darci Lynn. B.A., Heidelberg Coll., 1963; Ph.D., U. Cin., 1969. With Norwich (N.Y.) Pharm., 1964—, mgr. licensing and acquisitions, 1982—. Republican. Subspecialties: Pharmacology; Physiology (medicine). Current work: Researcher in skeletal muscle pharmacology. Home: 23 Aurora Hills Dr Norwich NY 13815 Office: Norwich Pharmaceuticals 17 Eaton Ave Norwich NY 13815

ELLIS, PAUL, JOHN, physics educator; b. Northampton, Eng., May 25, 1941; came to U.S., 1966; s. Alfred Sidney and Olive Edith (Meacock) E.; m. Alicia Irena Dudek, Jan. 24, 1973; 1 son, Aleksander. B.Sc., U. Bristol, Eng., 1962; Ph.D., U. Manchester, Eng., 1966. Research assoc. U. Mich., Ann Arbor, 1966-68, Rutgers U., New Brunswick, N.J., 1968-70; research officer Oxford (Eng.) U., 1970-73; asst. prof. physics U. Minn., Mpls., 1978-77, assoc. prof., 1977-82, prof., 1982—. Contbr. articles to profl. jours. Recipient Outstanding Tchr. award Inst. Tech., U. Minn., 1980-81. Mem. Am. Phys. Soc. Subspecialties: Nuclear physics; Theoretical physics. Current work: The theory of nuclear structure and nuclear reactions. Office: Sch Physics and Astronomy U Minn Minneapolis MN 55455

ELLIS, WILLIAM R., fusion scientist; b. Greenville, S.C., Jan. 22, 1940; s. William Rufus and Mary Louise (Rogers) E.; m. Gail Maxcine Gladden, Aug. 13, 1966; children: Benjamin Brian, Jaman Nathaniel. B.S., Clemson U., 1962; M.A., Princeton U., 1965, Ph.D., 1967. Vis. scientist A.E.R.E. Culham Lab., Abingdon, Berks, Eng., 1967-69; mem. staff Los Alamos Nat. Lab., 1970-73, assoc. group leader, 1974-75; chief open systems br. Office of Fusion Energy, Dept. Energy, Washington, 1976-78, dir. mirror systems div., 1979—. Author articles and newspaper columns. Chmn. U.S.-Japan Joint Coordinating Com. for Tandem Mirror and Bumper Taxes Fusion Programs, 1982—; vice chmn. Los Alamos County Council, 1975-76. NSF fellow, 1961; NASA grantee, 1962-66. Mem. Am. Phys. Soc. (div. plasma physics, exec. com. 1980), Am. Nuclear Soc. Subspecialties: Fusion; Plasma physics. Current work: U.S. Fusion program technical management and oversight. Patentee in field. Home: 1613 Auburn Ave Rockville MD 20850 Office: Office of Fusion Energy US Dept Energy ER-56/MS G234 Washington DC 20545

ELLISON, ALFRED HARRIS, government research administrator, environmental scientist; b. Quincy, Mass., Dec. 23, 1923; m.; 5 children. B.S., Boston Coll., 1950; M.S., Tufts U., 1951; Ph.D. in Surface Chemistry, Georgetown U., 1956. Chemist U.S. Naval Research Lab., 1951-56; research chemist Texaco Research Ctr., 1956-65, Gillette Research Inst., 1965-69; dep. dir. Environ. Sci. Research Lab., Office Research and Devel., EPA, Research Triangle Park, N.C., 1969-79, dir., 1979—. Mem. AAAS, Air Pollution Control Assn., Am. Chem. Soc. Subspecialty: Environmental Science Research Administration. Office: Environ Sci Research Lab Room N 317 Tech Ctr Research Triangle Park NC 27711

ELLISON, CAROL RINKLEIB, psychologist; b. Santa Barbara, Calif., Sept. 7, 1938; d. Edwin Henry and Margaret Bovard (Skinner) E.; m.; children: Randy, Teresa, Karen Monica. A.B., U. Chgo., 1960; M.S., U. Calif.-Davis, 1970, Ph.D., 1975. Lic. psychologist, Calif. Human relations rep. Coop. Ext./U. Calif., Davis, 1971-75; psychology instr. Sacramento City Coll., 1973-78, Calif. Sch. Profl. Psychology, Berkeley, 1979-80; coordinator didactic tng. Human Sexuality Program, U. Calif.-San Francisco, 1978-79, assoc. staff, 1977-81; pvt. practice psychology, Oakland, Calif., 1976—; lectr. in field. Author: (with others) Understanding Sexual Interaction, 1977, 2d edit., 1981, Understanding Human Sexuality, 1980. Founder, 1st pres. Atascadero Presch. Assn., Calif., 1962-68; vol. counselor Planned Parenthood, Sacramento, 1972-73; coordinator Family Life Edn. Council, Sacramento, 1973-75, counseling sect. chmn. Western Council Family Relations, 1976-77. Mem. Am. Psychol. Assn., Soc. Sci. Study of Sex (treas. chpt. 1980—), Western Psychol. Assn. Subspecialties: Psychology of human sexuality; Psychobiology. Current work: Female sexuality, orgasm, arousal; sex therapy; medical psychology; mind-body interaction; psychosomatic issues; hypnotherapy; individual and couple therapy. Address: 39 Park Way Piedmont CA 94611

ELLISON, MARLON LOUIS, biology educator; b. Woodbine, Iowa, Dec. 18, 1916; s. Russel and Adeline (Mefferd) E.; m. Johnetta Perry, Apr. 30, 1949. B.S., Iowa State U., 1940; M. Biology, Trinity U., 1961; Ph.D., U. Kans., 1964. Commd. 2d lt. U.S. Army, 1940, advanced through grades to lt. col.; ret., 1960; prof. biology Trinity U., 1961; mem. faculty Stephen Austin Coll., 1964; prof. biology U. Tampa, Fla., 1964—. Mem. Am. Inst. Biol. Scis., Am. Bryology Assn. Democrat. Methodist. Subspecialty: Taxonomy. Current work: Marine botany, botany of Tampa Bay, bryologist. Office: U Tampa Tampa FL 33606

ELLISON, WILLIAM THEODORE, acoustics scientist; b. Wilmington, N.C., Nov. 30, 1941; s. Robert Jay and Marie Catherine (Robinson) E.; m. Jean Elizabeth Salter, Dec. 21, 1968; children: Britt Kirsten, Hans Salter. B.S., U.S. Naval Acad., 1963; M.S.M.E., MIT, 1968, Nav.E., 1968; Ph.D., M.I.T., 1970. Vice-pres., sr. scientist Cambridge Acoustical Assocs., Inc., Mass., 1974-83; pres., chief scientist Marine Acoustics, Cotuit, Mass., 1983—; founding mem. sci. adv. com. Alaska Eskimo Whaling Commn., Barrow, 1980—; witness on Arctic ecology Nat. Energy Bd. Can., Ottawa, Ont., 1982—. Mem. Acton (Mass.) Conservation Commn., 1976-81. Served to lt. comdr. USN, 1963-74. Mem. Acoustical Soc. Am., Marine Tech. Soc., Am. Polar Soc., Sigma Xi, Tau Beta Pi. Episcopalian. Subspecialties: Acoustical engineering; Behavioral ecology. Current work: Impact of industrial activity on the Arctic environment including field studies on underwater noise and animal behavior, principally endangered species. Office: Marine Acoustics 116 Trout Brook Rd Cotuit MA 02635

ELLMAN, GEORGE, biochemistry educator; b. Chgo., Dec. 27, 1923; m. Phyllis Butcher, Aug. 20, 1948; 1 dau., Judith. B.A., U. Ill.-Urbana, 1948; M.S. in Biochemistry, Wash. State Coll., 1949; Ph.D. in Organic Biochemistry, Calif. Inst. Tech., 1952. Mem. staff Biochem. Research Labs., Dow Chem. Co., Midland, Mich., 1952-58; research specialist Langley Porter Neuropsychiat. Inst., San Francisco, 1959-62, chief research biochemist, 1962-63, research specialist IV, 1968-73; mem. faculty U. Calif.-San Francisco, 1958—, assoc. prof. biochemistry, 1973—. Contbr. numerous sci. articles to profl. publs. Commr. Met. Transp. Commn. Parks and Recreation Dept., 1967-71; mem. City Council Tiburon, Calif., 1972-76; mayor City of Tiburon, 1975. Served with USAF, 1943-46. Recipient Chancellor's Pub. Service award U. Calif.-San Francisco, 1978; McCallum fellow Calif. Inst. Tech. Grad. Sch., Pasadena, 1950-51. Mem. Am. Soc. Pharmacology and Exptl. Therapeutics, Western Pharmacology Soc., Am. Soc. Neurochemistry, Phi Lambda Upsilon. Subspecialties: Analytical chemistry; Neurobiology. Current work: Lipofuscin in aging, assay methods, characterization; glutathione transfease-an adaptive enzyme that is significant in detoxification of xenobiotics; norepinephrine in brains of rats raised under enriched and impoverished conditions. Office: Brain Behavior Research Ctr Sonoma State Hosp Eldridge CA 95431

ELLSTRAND, NORMAN CARL, geneticist, educator; b. Elmhurst, Ill., Jan. 1, 1952; s. Edwin A. and Beverly (Singer) E. B.S., U. Ill., 1974; Ph.D., U. Tex., 1978. Postdoctoral research assoc. Duke U., Durham, N.C., 1978-79; asst. prof. plant ecology, dept. botany and plant scis. U. Calif., Riverside, 1979—. Contbr. articles to sci. jours. Regents Jr. Faculty Fellow, 1981; U.S. Dept. Agr. grantee, 1980-84; NSF grantee, 1983-85. Mem. Soc. Study Evolution, Ecol. Soc. Am., Am. Soc. Plant Taxonomists, Internat. Soc. Plant Population Biologists, Nat. Audubon Soc., Calif. Native Plant Soc., Riverside Art Center, Phi Beta Kappa, Sigma Xi, Phi Kappa Phi. Club: University (Riverside). Subspecialties: Evolutionary biology; Population biology. Current work: Genetic structure of populations; plant breeding systems. Office: Dept Botany and Plant Scis U Calif Riverside CA 92521

ELLSWORTH, ROBERT KING, biochemist, educator; b. Plattsburgh, N.Y., Nov. 22, 1941; s. Francis K. and Virginia E. (LaValley) E.; m. Nancy B. Forgette, Sept. 24, 1963; children: Lisa, Tonia. B.S., SUNY-Plattsburgh, 1963, M.S., 1966; Ph.D., Iowa State U., 1968. Asst. prof. SUNY-Plattsburgh, 1968-70, assoc. prof., 1970-73, prof., 1973—. Contbr. numerous articles to sci. jours. NIH grantee, 1969-71, 74-76; NSF grantee, 1971-73, 73-75, 75-78. Mem. Am. Chem. Soc., Am. Soc. Biol. Chemists, Am. Soc. Plant Physiologists, AAAS. Subspecialties: Biochemistry (biology); Photosynthesis. Current work: Biosynthesis of photosynthetic pigments and the roll the pigments play in chloroplast development and in photosynthesis. Home: 31 Addoms St Plattsburgh NY 12901 Office: Dept Chemistry State University Coll of Arts and Scis Plattsburgh NY 12901

ELLWOOD, BROOKS BERESFORD, paleomagnetist, researcher; b. Chgo., July 18, 1942; s. John Fiske and Doris (Hammill) E. Ph.D., U. R.I.-Narragansett, 1976. Research assoc. Ohio State U., Columbus, 1976-77; asst. prof. U. Ga., Athens, 1977-80, assoc. prof. geology, 1980-83; assoc. prof. U. Tex.-Arlington, 1983—; vis. assoc. prof. U. Tex.-Arlington, 1983. Contbr. articles to profl. jours. Served with U.S. Army, 1964-66. Subspecialties: Geophysics; Geology. Current work: Magnetic properties of rocks and sediments, terrigenous and marine. Home: 522 Waggoner Dr Arlington TX 76013 Office: Dept Geology U Tex Arlington TX 76019 Home: 522 Waggoner Dr Arlington TX 76013

ELMEGREEN, BRUCE GORDON, astronomy educator; b. Milw., Feb. 24, 1950; s. George Lester and Jean (White) E.; m. Debra Meloy, Aug. 21, 1976. B.A., U. Wis.-Madison, 1971; Ph.D., Princeton U., 1975. Jr. fellow Harvard U., 1975-78; vis. assoc. prof. U. Calif.-Berkeley, 1981; asst. prof. Columbia U., 1978—; vis. scientist U. Sussex, Eng., 1981, Cambridge (Eng.) U., 1981. Co-editor: Jour Fundamentals of Cosmic Physics, 1982—; contbr. articles to profl. jours. NSF fellow, 1971-74, 79, 80. Mem. Math. Assn. Am., Am. Astron. Soc., Royal Astron. Soc., Internat. Astron. Union, Phi Beta Kappa, Phi Kappa Phi, Phi Eta Sigma. Subspecialties: Theoretical astrophysics; Radio and microwave astronomy. Office: Dept Astronom Pupin Labs Columbia U New York NY 10027

ELMEGREEN, DEBRA ANNE MELOY, astronomer; b. South Bend, Ind., Nov. 23, 1952; d. Thurston George and Anne Elizabeth (Clubb) Meloy; m. Bruce Gordon Elmegreen, Aug. 21, 1976. A.B., Princeton U., 1975; postgrad., U. Calif.-Santa Cruz, 1975-76; M.A., Harvard U.,

1977, Ph.D., 1979. Teaching fellow Harvard U., 1977; Carnegie fellow Mt. Wilson and Las Campanas Obs., Pasadena, Calif., 1979-81; vis. astronomer Royal Greenwich Obs., Sussex, Eng., 1981, Inst. Astronomy, Cambridge, Eng., 1981; vis. scientist IBM T.J. Watson Research Center, 1982—. Contbr. articles to profl. jours. Westinghouse scholar, 1971-75; Amelia Earhart fellow Zonta Internat., 1976-79. Mem. Am. Astron. Soc., Royal Astron Soc., Internat. Astron. Union, Sigma Xi. Subspecialties: Optical astronomy; Radio and microwave astronomy. Current work: Spiral structure of galaxies; star formation. Office: IBM TJ Watson Research Center PO Box 218 Yorktown Heights NY 10598

ELSENBAUMER, RONALD LEE, chemist; b. Allentown, Pa., July 18, 1951. B.S., Purdue U., 1973; Ph.D. in Organic Chemistry, Stanford U., 1977. Research chemist Allied Chem. Co., 1977—. Subspecialty: Organic chemistry. Current work: Research in conducting polymers, batteries. Officer: Corp Materials Research Ctr Allied Chem Co PO Box 1021R Morristown NJ 07960

ELSON, LEE STEPHEN, atmospheric physicist; b. Chgo., June 26, 1947; m. Anne Barton, Sept. 23, 1972; children: Andrea, Marieka. B.A., U. Calif., Berkeley, 1969; M.S., U. Wash., 1971, Ph.D., 1975. Cons. Jet Propulsion Lab., Calif. Inst. Tech., Pasadena, 1975-76, resident research assoc., 1976-78, mem. tech. staff, 1978—. Contbr. articles to profl. jours. Mem. Am. Astron. Soc., Am. Meteorol. Soc. Subspecialties: Planetary atmospheres; Remote sensing (atmospheric science). Current work: Researcher in atmospheric physics and meteorology. Office: Calif Inst Tech Jet Propulsion Lab 4800 Oak Grove Dr Pasadena CA 91109

ELSTON, WOLFGANG EUGENE, geology educator, consultant; b. Berlin, Germany, Aug. 13, 1928; came to U.S., 1945, naturalized, 1950; s. Frederick G. and Anny (Halpert) E.; m. Lorraine Hind, Dec. 26, 1952; children: Stephen, Richard. B.S., CUNY, 1949; M.A., Columbia U., 1953, Ph.D., 1953. Cert. profl. geol. scientist. Geologist N.Mex. Bur. Mines and Mineral Resources, Socorro, summers 1950-64; asst. prof. Tex. Technol. U., Lubbock, 1955-57; faculty mem. U. N.Mex., Albuquerque, 1957—, prof. geology, 1967—, acting dept. chmn., 1982; cons. to numerous mining and petroleum cos. Los Alamos and Sandia nat. labs., 1957—. Co-editor, author: Cenozoic Volcanism in Southwestern New Mexico, 1976, Cauldrons and Ore Deposits of Southwestern New Mexico, 1978; co-editor: Ash-flow Tuffs, 1979. Served with AUS, 1953-55. Research grantee NASA, NSF, U.S. Geol. Survey, N.Mex. Energy Research and Devel. Inst., others, 1964—. Fellow Geol. Soc. Am., AAAS; mem. Am. Inst. Profl. Geologists (pres. N.Mex. sect. 1983), Nat. Assn. Geology Tchrs. (pres. Southwest sect. 1974-75), N.Mex. Geol. Soc. (pres. 1963-64). Subspecialties: Volcanology; Economic geology. Current work: Volcanology, tectonics and mineral deposits of the basin and range province, especially N.Mex.; application of volcanology to planetology. Home: 1023 Columbia Dr NE Albuquerque NM 87106 Office: Dept Geology U NMex Albuquerque NM 87131

EL-TAHAN, MONA SALAH, project engineer; b. Cairo, Egypt, Aug. 27, 1950; d. Salah and Nawal (Sherif) Shahwan; m. Hussein Wabha El-Tahan, Dec. 28, 1975; 1 child, Tahmir. B.Sc. in Engring, Cairo U., 1975; M.Eng., Meml. U., Nfld., 1980. Lic. profl. engr., Nfld. Research asst. Meml. U., St. John's, Nfld., 1976-80; ocean engr. Fenco Nfld. Ltd., St. John's, 1980-82, project engr., 1982—. Mem. Assn. Profl. Engrs. Nfld., Marine Tech. Soc. Muslim. Subspecialty: Ocean engineering. Current work: Iceberg drift precition - iceberg deterioration. Developer, Iceberg Drift Prediction Model, 1980; patentee in field. Office: P O Box 8246 Kenmount Rd Saint Johns NF Canada A1B 3N4 Home: 70 Blothuck St Apt 206 Saint Johns NF Canada A1B 4C6

ELTIMSAHY, ADEL H., electrical engineer, educator; b. Damanhoor, Egypt, June 10, 1936; s. Hassan H. and Hamida A. (Elzohairy) E.; m. Kathleen H. Hoag, July 1, 1967; children: Ann Mirrette, Todd Tarek. B.S.E.E., Cairo U., 1958; M.S.E.E., U. Mich., 1961, Ph.D., 1967. Instr. Cairo U., 1958-59; teaching fellow U. Mich., 1966-67; asst. prof. U. Tenn., 1967-68; asst. prof. elec. engring. U. Toledo, 1968-73, assoc. prof., 1973-78, prof., 1978—, chmn. dept. elec. engring., 1980—. Contbr. articles to profl. jours. Named Elec. Engr. of Yr. Toledo sect. IEEE, 1979-80; Dept. Energy grantee; Libbey-Owens Ford Co. grantee. Mem. IEEE, Internat. Solar Energy Soc., Am. Soc. Engring. Edn., Simulation Council. Subspecialties: Electrical engineering; Solar energy. Current work: Optimal control of solar energy systems. Home: 4903 S Arvilla Toledo OH 43623 Office: Dept Elec Engring U Toledo Toledo OH 43606

ELY, BERTEN E., III, educator; b. Newark, Nov. 26, 1948; s. Berten E. and Ruth Dorothy (Bloy) E.; m. Tracey Allison Ward, May 30, 1970; children: Marc, Gregory. B.S., Tufts U., 1969; Ph.D., Johns Hopkins U., 1973. Asst. prof. dept. biology U. S.C., Columbia, 1973-79, assoc. prof., 1979—; asst. prof. microbiology/immunology Sch. Medicine, U. S.C., 1975-79, assoc. prof., 1979-82. Contbr. articles to profl. jours. Mem. Am. Soc. Microbiology. Mem. Christian Ch. Subspecialties: Gene actions; Microbiology. Current work: Constructed a genetic map for Canulobacter crescentus cloning C. crescentus genes; analyzing gene organization and regulation. Office: Dept Biology Univ SC Columbia SC 29208

ELZERMAN, ALAN WILLIAM, environmental systems engineering educator, researcher; b. Ann Arbor, Mich., Apr. 2, 1949; s. Alvah William and Yvonne G. (Hearn) E.; m. Eleanor A. Mitchell, Aug. 16, 1970; children: Sara, Ashley. B.A., Williams Coll., 1971; Ph.D., U. Wis.-Madison, 1976. Postdoctoral fellow Woods Hole (Mass.) Oceanographic Inst., 1976-77, NATO, Harwell, Eng., 1978; asst. prof. environ. systems engring. Clemson U., 1978-81, assoc. prof., 1981—. Contbr. articles to profl. publs. Chmn. Alpha Montessori Sch. Clemson, 1981-82. Mem. Am. Chem. Soc. (co-editor newsletter Environ. Chemistry Div. 1981—), Am. Soc. Limnology and Oceanography, Water Pollution Control Fedn., Assn. Environ. Engring. Profs., Am. Geophys. Union. Subspecialties: Environmental engineering; Analytical chemistry. Current work: Environmental chemistry, fate and distribution of chemicals in environment, acid deposition/ rain. Office: Environ Systems Engring Clemson U Clemson SC 29631

ELZINGA, DONALD JACK, engineering educator; b. Coupeville, Wash., Jan. 16, 1939; s. Martin Jay and Phyllis M. (Dickson) E.; m. Marley Ann Plomer, July 18, 1962; children: Erik, Bruce, Mark; m. Virginia Collins, Aug. 14, 1981. B.E., U. Wash., 1960; M.S., Northwestern U., 1965, Ph.D., 1968. Engr. Shell Devel. Co., Emeryville, Calif., 1960-61; vol. Peace Corps, 1961-63; teaching asst. Northwestern U., Evanston, Ill., 1963-65; asst. prof. chem. engring. dept. Johns Hopkins U., Balt., 1967-68, asst. prof., 1968-72, asst. prof. math. sci. dept., 1972-73, assoc. prof., 1973-78, research scientist and assoc. prof., 1978-79; mathematician Bur. Health Manpower, 1976-77; prof., chmn. dept. indsl. and systems engring. U. Fla., Gainesville, 1979—; cons. in field. Contbr. articles to profl. jours. Standard Oil Co. of Calif. fellow, 1965-67. Mem. Ops. Research Soc. Am., Inst. Indsl. Engrs., Math Programming Soc., Sigma Xi, Tau Beta Pi. Subspecialties: Operations research (mathematics); Industrial engineering. Current work: Mathematical programming, operations research. Office: 303 Weil Hall U Fla Gainesville FL 32611

EMELE, JANE FRANCES, consumer products company executive; b. Phillipsburg, N.J., Nov. 14, 1925; d. Karl A. and Mary E. (Shafer) E. B.S., Upsala Coll., 1947; M.S., U. Ill., 1949; Ph.D., Yale U., 1954. Asst. scientist dept. pharmacology Schering Corp., 1947-48; lab. asst. U. Ill., 1948-49; chief sect. pharmacodynamics Eaton Labs., Norwich Pharmacal Co., N.Y., 1954-55; sr. scientist Warner Lambert Research Inst., Warner-Lambert Pharm. Co., Morris Plains, N.J., 1955-61, sr. research assoc., 1961-65, mgr. dept. proprietary pharmacology and toxicology, 1965-66, dir. dept. pharmacology and toxicology consumer products research div., 1966-70; assoc. dir. biol. research consumer products research div., 1970-73, dir. consumer products group, 1973-74, dir. biol. research, 1975-77, dir. biol. and clin. affairs, 1977-80, dir. proprietary clin. research and toxicology, 1979—; vis. scientist Rutgers U., 1964-67. Contbr. numerous articles to profl. jours. Bd. dirs. Morris County Assn. Health and Welfare Agys., 1964-66; mem. Morris County Bd. Mental Health, 1959; bd. dirs. N.J. Assn. Mental Health, 1962-64; mem. investigational rev. com. Morristown Meml. Hosp., 1978—; trustee Upsala Coll., 1978—. Mem. Am. Soc. Pharmacology and Exptl. Therapeutics, Fedn. Am. Socs. Exptl. Biology, Am. Pharm. Assn., Acad. Pharm. Scis., Am. Soc. Clin. Pharmacology and Teherapeutics, Am. Coll. Toxicology, Inst. Food Tech., Toxcology Forum, Assos. Clin, Pharmacology, AAAS, N.Y. Acad. Scis., Internat. Soc. Biochem. Pharmacology, W.T. Salter Soc., Yale U. Alumni Assn., U. Ill. Alumni Assn., Upsala Coll. Alumni Assn. (coll. Pres.s' Forum 1973—), Sigma Xi, Phi Sigma, Sigma Delta Epsilon. Subspecialties: Toxicology (medicine); Pharmacology.

EMENER, WILLIAM GEORGE, psychology educator; b. N.Y.C., June 10, 1943; s. William George and Rose (Donner) E.; m. Rae Dorothy Torgesen, June 25, 1965; children: Karen, Barbara, Scott. B.A., Trenton State Coll., 1965; M.A., N.Y. U., 1968; Ph.D., U. Ga., 1971. Lic. psychologist, Fla. Tchr. English N.E. Burlington (N.J.) County High Sch., 1965-66; rehab. counselor E.R.J. Tng. & Research Ctr., Bordentown, N.J., 1976-96; asst. prof. Murray (Ky.) State U., 1971-74; assoc. prof. Fla. State U., Tallahassee, 1974-78; prof./ program dir. U. Ky., Lexington, 1978-80; prof. rehab. edn. U. So. Fla., Tampa, 1980—; pvt. practice psychology, 1971—. Author/editor: Rehabilitation Administration and Supervision, 1981; contbr. articles to profl. jours.; co-editor: Jour. Applied Rehab. Counseling, 1977-82. Coach youth soccer Little League, youth softball, 1978—. Recipient Nat. Research award Am. Rehab. Counseling Assn., 1980; Disting. Researcher award Nat. Rehab. Adminstrn. Assn., 1982. Mem. Nat. Rehab. Adminstrn. Assn. (pres.-elect), others. Subspecialties: Counseling Psychology; Rehabilitation Psychology. Current work: Rehabilitation counseling, administration, psychology. Home: 16404 Shagbark Pl Tampa FL 33618 Office: Univ S Fla Tampa FL 33620

EMERSON, KENNETH, chemistry educator; b. Pasadena, Calif., Nov. 9, 1931; s. Robert and Claire (Garrison) E.; m. Helen Margaret Walker, June 14, 1956; children: Bruce Lloyd, Brian Gordon, Lee Parrish. B.A., Harvard U., 1953; M.A., U. Oreg., 1958; Ph.D., U. Minn., 1961. With dept. chemistry Mont. State U., 1962—, prof., 1970—; Fulbright fellow U. Canterbury, Christchurch, N.Z., 1968-69; vis. prof. U. Los Andes, Merida, Venezuela, 1972-73, U. Canterbury, 1979; vis. lectr., Republic of China, 1981. Contbr. articles to profl. jours. Served with U.S. Army, 1953-55. Noyes fellow, 1961-62. Mem. Am. Chem. Soc., Am. Crystallographic Assn., Nat. Sci. Tchrs. Assn., Mont. Acad. Sci., Sigma Xi. Subspecialties: Inorganic chemistry; Solid state chemistry. Current work: Synthetic inorganic chemistry, crystal structure, magnetic properties of solids. Office: Dept Chemistry Montana State U Bozeman MT 59717

EMERSON, ROBERT CHARLES, neuroscientist, research center administrator; b. Detroit, July 17, 1939; s. Kenneth Harwood and Charlotte Magdalena (Steiner) E.; m. Marie Rose Huston, Aug. 1, 1964; children: Robert Charles, David Kenneth. B.S. in Elec. Engring, Lehigh U., 1961, 1962; M.S. in Biology, Adelphi U., 1966; Ph.D. in Physiology, U. Pa., 1973. Engr., physicist Biomed. Engring. Lab., Airborne Instruments Lab., Melville, NY., 1962-66; hardware and software cons., Phila., 1967-72; assoc. dir. Ctr. for Visual Sci., U. Rochester, N.Y., 1974—; mem. spl. study sects. NIH, 1977, 79, 80; Mem Finger Lakes Trail Conf. N.Y.; trustee Am. Diabetes Assn. (Rochester regional affiliate), 1983. Contbr. articles in field to profl. jours. USPHS trainee Inst. Neurol. Scis., U. Pa., 1966-71; Nat. Eye Inst. research grantee, 1974-81; recipient Research Career Devel. award, 1977-82, conf. grantee, 1978. Mem. Assn. Research in Vision and Ophthalmology, Soc. Neurosci, Eta Kappa Nu, Sigma Xi. Subspecialties: Neurophysiology; Neuroanatomy. Current work: Neural coding, computers, nonlinear interactions, morphology-injected neurons. Home: 10 Squire Ln Pittsford NY 14534 Office: Ctr Visual Sci U Rochester Rochester NY 14627

EMERSON, VICTOR F., research psychologist, educator, consultant; b. Ottawa, Ont., Can., Jan. 17, 1948. B.S., McGill U., 1971; Ph.D., Queen's U., 1976. NIH postdoctoral fellow Physiol. Lab., Cambridge, Eng., 1977-79; lectr. Concordia U., Montreal, Que., Can., 1979-80; postdoctoral fellow Queen's U., Kingston, Ont., Can., 1980-82, Nat. Scis. and Engring. Research Council univ. research fellow, asst. prof. dept. psychology, 1982-83; mem. sci. staff Bell-No. Research, Ottawa, Ont, Can., 1983—. Bd. dirs. Nat. Film Theatre, Kingston, 1982—. Nuffield travelling fellow Nuffield Found., Cambridge, 1977. Mem. Can. Psychol. Assn., Am. Psychol. Assn., Assn. for Research in Vision and Ophthalmology. Subspecialties: Sensory processes; Comparative neurobiology. Current work: Experimental design and statistics, pattern vision, comparative aspects of visual development and visual behavior. Office: Bell No Research 2733 Carling PO Box 3511 Station C Ottawa ON Canada K1Y 4H7

EMERY, KEITH ALLEN, research scientist; b. Lansing, Mich., Oct. 31, 1954; s. Roy S. and Ruth W. (Simons) E.; m. Patricia Withey, July 17, 1982. A.S., Lansing Community Coll., 1973; B.S. in Physics, Mich. State U., 1976; M.S. in Elec. Engring, 1978. Research assoc. Mich. State U., East Lansing, 1975-78; research assoc. photovoltaics lab. Colo. State U., Ft. Collins, 1978-80, research assoc. laser lab., 1983—; staff scientist Solar Energy Research Inst., Golden, Colo., 1980-83. Contbr. articles to profl. jours. Mem. Optical Soc. Am., IEEE, Am. Phys. Soc., Laser Inst. Am., Electrochem. Soc. Subspecialties: Solar energy; 3emiconductors. Current work: Electro-optical characterization of thin film solar cells with emphasis on understanding surfaces and interfaces and device physics; photon induced chemical vapor deposition of thin film solar cells. Office: Engring Research Ctr Colo State U Fort Collins CO 80523

EMERY, KENNETH ORRIS, marine geologist; b. Swift Current, Sask., Can., June 6, 1914; s. Clifford Almon and Agnes (Baird) E.; m. Caroline Roberta Alexander, Oct. 3, 1941; children—Barbara Kathryn Emery Alvarado, Charlet Adelia Emery Shave. Student, N. Tex. Agrl. Coll., 1933-35; B.S., U. Ill., 1937, Ph.D., 1941. Staff Ill. State Geol. Survey, Urbana, 1941-43; staff div. war research U. Calif., San Diego, 1943-45; asso. marine geologist, prof. geology U. So. Calif., Los Angeles, 1945-62; marine geologist Woods Hole (Mass.) Oceanographic Inst., 1962-75, Henry Bryant Bigelow oceanographer, 1975-79, emeritus, 1979—; mem. U.S. Geol. Survey, Los Angeles, 1945-58. Author: books including Sea Off Southern California, 1960, Oceanography in a Coastal Pond, 1967, (with E. Uchupi) Western North Atlantic, 1972; contbr. numerous articles to profl. jours. Guggenheim fellow, 1959; recipient Shepard prize for marine geology Soc. Econ. Paleontologists and Mineralogists, 1969; Outstanding Alumnus award U. Tex. at Arlington, 1969; Prince Albert de Monaco medal U. Paris, 1971; Compass Distinguished Achievement award Marine Tech. Soc., 1974; Rosenstiel-AAAS award in oceanographic sci., 1974; Illini Achievement award U. Ill., 1977. Fellow Am. Geophys. Union; mem. Am. Assn. Petroleum Geologists, Geol. Soc. Am., Soc. Econ. Paleontologists and Mineralogists, Nat. Acad. Scis., Am. Acad. Arts and Scis., China Acad. Sci., Swedish Royal Acad. Sci. Subspecialty: Sea floor spreading. Current work: Synthesis of knowledge of ocean floor. Home: 74 Ransom Rd Falmouth MA 02540 Office: Woods Hole Oceanographic Inst Woods Hole MA 02543

EMERY, PHILIP ANTHONY, hydrologist; b. Neodesha, Kans., Oct. 20, 1934; s. Vincent Anthony and Whilomena Birtha (Kempker) E.; m. Janet Louise Bayne, Apr. 13, 1960; 1 son, David A. B.S. in Geology, U. Kans.-Lawrence, 1960, M.S., 1962. Geologist Kans. Geol. Survey, Lawrence, 1961-62; geologist water resources div. U.S. Geol. Survey, Lincoln, Nebr., 1962-66, hydrologist, Alamosa, and Pueblo, Colo., 1966-74, dist. ground-water specialist, Menlo Park, Calif., 1974-76, dist. chief, Louisville, 1976-81, Anchorage, 1981—. Contbr. articles to profl. jours. Served to sgt. U.S. Army, 1954-57; Germany. Fellow Geol. Soc. Am.; mem. AAAS, Internat. Assn. Hydrogeologists, Am. Water Resources Assn., Sigma Xi. Subspecialties: Hydrology; Ground water hydrology. Current work: Scientific-technical management, arctic hydrology, climatology. Office: US Geol Survey WRD 1515 E 13th Ave Anchorage AK 99501

EMINO, EVERETT RAYMOND, horticulture educator; b. Milford, Mass., Feb. 8, 1942; s. Gerald C. and Dorothy H. (Taft) E.; m. Sarah Jean Tripp, Dec. 16, 1967; children: James, Kathryn. A.A.S., Stockbridge Sch. Agr., 1962; B.S., U. Mass., 1965; M.S., Mich. State U., 1967, Ph.D., 1972. Instr. Mich. State U., East Lansing, 1967-72; asst. prof. U. Mass., Amherst, 1972-75; asso. prof. Tex. A&M U., College Station, 1975-82; prof. U. Conn., Storrs, 1983—. Mem. Am. Soc. Hort. Sci., Bot. Soc. Am., Internat. Plant Propagators Soc., Am. Soc. Agronomy, Crop Sci. Soc. Am. Subspecialties: Morphology; Plant physiology (agriculture). Current work: Leaf phytotoxicity from foliar applied nutrients. Office: Dept Plant Sci U Conn Storrs CT 06268

EMMATTY, DAVY ALLEASU, food mfg. ofcl.; b. Trichur, Kerala, India, Sept. 29, 1941; s. Alleasu and Eliakutty (Chammanam) E.; m. Gracy Joseph, Aug. 1, 1944; children: Anil, Lisa. B.Sc., Kerala (India) U., 1961; M.S., Purdue U., 1966, Ph.D., 1967. Sr. research pathologist Heinz USA, Bowling Green, Ohio, 1968-77, assoc. mgr. agrl. research dept., 1977-81, mgr., 1981—. Mem. Am. Phytopath. Soc. Episcopalian. Subspecialties: Plant pathology; Plant genetics. Current work: Develop tomato varieties suitable for processing under Midwest conditions, mechanized harvesting, rot-free tomatoes. Home: 1067 Revere Dr Bowling Green OH 43402 Office: 13737 Middleton Ave Bowling Green OH 43402

EMMERICH, WERNER SIGMUND, physicist; b. Duesseldorf, Germany, June 3, 1921; s. Adolph and Julia (Frank) E.; m. Eva G. Pauson, June 13, 1953; children: Fay Lillian, Ralph Austin, Bertram Frank. B.S., Ohio State U., 1949, M.S., 1950, Ph.D., 1953. Research physicist Westinghouse Research and Devel. Center, Pitts., 1954-57, adv. physicist, 1957-64, mgr. arc and plasma research, 1964-73, dir. applied physics, 1973-75, dir. corp. research, 1975-79, dir. power systems, 1979-83, dir. corp. and comml. research, 1983—. Author: Fast Neutron Physics, 1963. Served with AUS, 1942-46; ETO. Fellow Am. Phys. Soc.; mem. AAAS (life), Am. Nuclear Soc., Combustion Inst., N.Y. Acad. Scis., Am. Mgmt. Assn., Sigma Xi, Phi Beta Kappa, Zeta Beta Tau. Subspecialties: Nuclear physics; Atomic and molecular physics. Current work: Fast neutronphysics, magnetohydrodynamics, circuit interrupters, research planning and administration. Patentee in field. Home: 1883 Beulah Rd Pittsburgh PA 15235 Office: Westinghouse Research and Devel Center 1310 Beulah Rd Pittsburgh PA 15235

EMMERS, RAIMOND, physiologist, researcher, educator; b. Liepaja, Latvia, Apr. 19, 1924; U.S., 1951, naturalized, 1956; s. Fricis and Anna Maria (Ozol) E.; m. Amy Elizabeth Puett, June 14, 1971. B.A., East Tex. Baptist Coll., 1953; M.A., U. N.C., 1955; Ph.D., Syracuse U., 1958. Instr. Syracuse U., 1958-59; postdoctoral fellow U. Wis., 1959-61; asst. prof. physiology Coll. Physicians and Surgeons, Columbia U., 1961-71, assoc. prof., 1971—. Author: (with K. Akert) A Stereotaxic Atlas of the Squirrel Monkey Brain, 1963, (with R. R. Tasker) The Human Somesthetic Thalamus, 1975, Pain, 1981. Nat. Inst. Neurol. and Communicable Diseases and Stroke grantee, 1961—. Mem. Am. Physiol. Soc., Soc. Neurosci., Soc. Exptl. Biology and Medicine, Am. Assn. Anatomists. Subspecialties: Physiology (medicine); Neurophysiology. Current work: Neurophysiology of somesthetic sensibilities; involvement of taste in food and water intake; orgn. of sensory nerve impulses in brain. Office: Dept Physiology Coll Physicians and Surgeons Columbia U 630 W 168th St New York NY 10032

EMMERSON, JOHN L., toxicology research administrator; b. Princeton, Ind., Nov. 21, 1933; s. John Carter and Marjorie (Woods) E.; m. Karen N. Nethery, Sept. 7, 1957; children: Michael, James. B.S in Pharmacy, Purdue U., 1958, M.S in Pharmacology, 1959, Ph.D., 1962. Dir. toxicology studies Eli Lilly Research Labs., Greenfield, Ind., 1978—. Subspecialty: Toxicology (medicine). Office: PO Box 708 Greenfield IN 46140

EMMONS, SCOTT WILSON, molecular biology educator; b. Boston, July 14, 1945; s. Howard W. and Dorothy A. (Allen) E. A.B., Harvard U., 1967; Ph.D., Stanford U., 1974. Postdoctoral fellow Carnegie Instn. Washington, Balt., 1974-76, U. Colo., Boulder, 1976-79; asst. prof. molecular biology Albert Einstein Coll. Medicine, N.Y.C., 1979—. NIH research grantee, 1981; Am. Cancer Soc. faculty research awardee, 1981. Subspecialties: Genome organization; Developmental biology. Current work: Molecular and developmental biology, eukaryotic genome organization. Home: 320 Riverside Dr Apt 2H New York NY 10025 Office: Dept Molecular Biology Albert Einstein Coll Medicine 1300 Morris Park Ave Bronx NY 10461

EMPTAGE, MICHAEL R(OLLINS), chemistry educator, researcher; b. Jersey City, June 10, 1939; s. Edward R. and Dorothy (B(ierman)) E.; m. Cathy Simdars, Ag. 30, 1969; children: Dorothea Clare, Nicholas Paul. A.B., Middlebury Coll., 1960; PhD., Harvard U., 1965. Research assoc. Brown U., Providence, 1965-66; asst. prof. U. Md., Balt., 1966-68, So. Ill. U, Carbondale, 1968—. Woodrow Wilson fellow, 1960-61; NSF fellow, 1962-64; NATO fellow, Brussels, 1964-65. Mem. Am. Phys. Soc., Am. Statis. Assn., Phi Beta Kappa. Democrat. Roman Catholic. Subspecialties: Statistical mechanics; Statistics. Current work: Kinetic equations for spin systems; helix-coil transition in polypeptides; bootstrapping and other resampling plans in mathematical statistics. Home: RR 2 Box 254 Murphysboro IL 62966 Office: So Ill U Carbondale IL 62901

EMRICH, GROVER HARRY, geotechnical/engineering consultant; b. Englewood, N.J., Apr. 9, 1929; s. Grover A. and Florence L. (Olson) E.; m. Charlotte Peterson, Aug. 12, 1972; children: Charles, John, Craig. B.S., Franklin and Marshall Coll., 1952; M.S., Fla. State U., 1957; Ph.D., U. Ill., 1962. Teaching asst. geology dept. Fla. State U., Tallahassee, 1954-56; research asst./asst. geologist groundwater and geophys. research sect. Ill. State Geol. Survey (stratigraphy and areal geology sect.), 1956-63; groundwater geologist/acting dir. div. mine drainage Bur. San. Engring., Pa. Dept. Health, Harrisburg, 1963-71; with SMC Martin Inc., Valley Forge, Pa., 1971—; vis. scientist Am. Geol. Inst., 1971. Contbr. articles to profl. jours. Fellow Geol. Soc. Am.; mem. Assn. Inst. Profl. Geologists, Water Pollution Control Fedn., Nat. Water Well Assn., Assn. Engring. Geologists, Internat. Assn. Water Pollution Research, Pa. Acad. Scis., Ill. Acad. Scis., Harrisburg Area Geological Soc. (1st chmn., co-founder), Phila. Geol. Soc. (pres.), Wissahickon Valley Watershed Assn. (dir. 1980-83), Sigma Xi. Republican. Methodist. Subspecialties: Ground water hydrology; Environmental engineering. Current work: Cost effective land management of liquid and hazardous wastes. Office: SMC Martin PO Box 859 Valley Forge PA 19482

EMSLIE, A. GORDON, educator; b. Hamilton, Scotland, Sept. 6, 1956; came to U.S., 1976, naturalized, 1979; s. Norman and Isabel Marion (Cowie) E.; m. Buff Day Watson, Oct. 21, 1978. B.Sc., U. Glasgow, 1976, Ph.D., 1979. Sr. research asst. Harvard Coll. Obs., 1978-79; research assoc. Inst. Plasma Research, Stanford U., 1979-81; Von Braun prof. space physics U. Ala., Huntsville, 1981—. Contbr. numerous articles to profl. jours. Mem. Am. Astron. Soc. Subspecialties: Solar physics; Cosmology. Current work: Physics of energy transport in solar flares; study of early evolution of universe. Office: U Ala Dept Physics SB 211 Huntsville AL 35899

ENDLER, NORMAN SOLOMON, psychology educator; b. Montreal, Que., Can., May 2, 1931; s. Beatrice Kerdman and June 26, 1955; m.; children: Mark, Marla. B.Sc., McGill U., 1953, M.Sc., 1954; Ph.D., U. Ill., 1958. Registered psychologist, Ont., Can. Psychologist Pa. State U., 1958-60; lectr. psychology York U., Downsview, Ont., 1960-62, asst. prof., 1962-65, assoc. prof., 1965-68, prof., 1968—; cons. Toronto East Gen. Hosp., 1964—; mem. adv. bd. Addiction Research Found.; cons. Clarke Inst., Toronto, 1972. Author: Holiday of Darkness, 1982; co-author: (with E.J. Shipton, F.D. Kemper) Maturing in a Changing World, 1971, (with D. Magnusson) Interactional Psychology and Personality, 1976, (with L.R. Boulter, H. Osser) Contemporary Issues in Developmental Psychology, 1976, (with D. Magnusson) Personality at the Crossroads: Current Issues in Interactional Psychology, 1977. Recipient Can. Silver Jubilee medal, 1978; Ont. Mental Health grantee, 1968-74; Can. Council grantee, 1969-78; Social Scis. and Humanities Research Council grantee, 1979-80. Fellow Am. Psychol. Assn., Can. Psychol. Assn. Subspecialties: Social psychology; Personality. Current work: Interactional psychology of personality; anxiety. Home: 52 Sawley Dr Willowdale Ont Canada M2K 2J5 Office: Psychology Dept York U 4700 Keele St Downsview Ont Canada M3J 1P3

ENDRIZZI, JOHN E., geneticist, educator; b. Wilberton, Okla., July 28, 1923; m. Yvonne V. Barbot, June 6, 1955; children: Colette, George, Regina, Carisa, Karena. B.S., Tex. A&M U., 1949, M.S., 1951; postgrad., U. Va., 1951-52; Ph.D., U. Md., 1955. Asst. Prof. Tex. A&M U., 1955-63; prof., head plant breeding dept. U. Ariz., Tucson, 1963-71, prof. plant gentics, 1971—. Contbr. articles to sci. jours. Served in U.S. Army, 1943-46. Nicholson fellow, 1949-50; Dupont fellow, 1951-52; Nat. Cotton Council grantee, 1958-61; recipient Cotton Genetics Research award, 1969. Mem. Genetics Soc. Am., Am. Genetics Assn., Genetics Soc. Can., Am. Inst. Biol. Scis., AAAS, Ariz.-Nev. Acad. Sci., Sigma Xi, Gamma Sigma Delta. Democrat. Roman Catholic. Subspecialties: Plant genetics; Genome organization. Current work: Genetics and cytology. Home: 2335 E 9th St Tucson AZ 85719 Office: U Ariz Tucson AZ 85721

ENG, NORMAN, engineering administrator, consultant; b. Chgo., Dec. 21, 1952; s. Shang Hon (Eugene) and Hop Yee (Wong) E.; m. Candice (Wei June) Chiang, June 26, 1982. B.S. in Civil Engring. (scholar), U. Calif., Berkeley, 1974. Prin. engr. EDS Nuclear Inc., San Francisco, 1975-78; advanced engr. Westinghouse Hanford, Richland, Wash., 1978-79; lead engr. Duke Power Co., Charlotte, N.C., 1979-80; project mgr. URS/John A. Blume & Assocs., Engrs., San Francisco, 1980-81; project engr. NUTECH Engrs., Inc., San Jose, Calif., 1981—. Dir. East Bay Asians Community Action, Oakland, Calif., 1973-74; tchr. Chinese Community Adult Sch., Oakland, 1975-77; sec. U.S. Jaycees, Berkeley, 1979-82. Mem. Am. Nuclear Soc., Am. Welding Soc., ASME, AAAS, Am. Concrete Inst., Am. Soc. Engring. Edn., N.Y. Acad. Sci., Nat. Soc. Profl. Engrs. Subspecialties: Civil engineering; Nuclear engineering. Current work: Nuclear power piping and pipe support technology, fast breeder reactor technology, intergranular stress corrosion cracking, induction heating stress improvement. Home: 3285 Padilla Way San Jose CA 95148 Office: NUTECH Engrs 6835 Via del Oro San Jos CA 95119

ENGEL, BERNARD THEODORE, psychologist; b. Chgo., Apr. 18, 1928; s. Marvin I. and Hannah (Hollander) E.; m. Rae Goldberg, Mar. 10, 1951; children: Sandra E., Jeffrey P., Lauren C. B.A., UCLA, 1954, Ph.D., 1956. Cert. biofeedback, 1981. Jr. research psychologist UCLA, 1956; research psychologist Inst. Psychosomatic and Psychiatric Research and Tng., Michael Reese Hosp., Chgo., 1957-58; lectr. med. psychology, mem. sr. staff Cardiovascular Research Inst., U. Calif. Sch. Medicine, San Francisco, 1959-67; chief lab. behavioral scis., chief psychophysiology sect. Gerontology Research Center, Nat. Inst. Aging, NIH, Balt., 1967—; assoc. prof. behavioral biology Johns Hopkins Sch. Medicine, Balt., 1970-82, prof., 1982—. Contbr. 100 articles to sci. jours. Served in U.S. Army, 1950-52. Recipient award Pavlovian Soc., 1979. Fellow AAAS, Gerontol. Sci.; mem. Soc. Psychophysiol. Research (pres. 1970-71), Biofeedback Soc. (pres. 1981-82), Am. Psychosomatic Soc. (sec.-treas. 1981—), Pavlovian Soc., Internat. Coll. Psychosomatic Medicine, Soc. Behavioral Medicine, Acad. Behavioral Medicine Research, Sigma Xi. Subspecialties: Psychophysiology; Biofeedback. Current work: Behavioral medicine, application of methods and principles of behavioral sciences to the assessment and treatment of patients with medical disorders; analysis of behavioral and physiological mechanisms underlying biological adaptiveness of circulation. Home: 106 Welford Rd Lutherville MD 21093 Office: Balt City Hosp Gerontology Research Ctr Baltimore MD 21224

ENGEL, GEORGE LIBMAN, psychiatrist, educator; b. N.Y.C., Dec. 10, 1913; m.; 2 children. B.A., Dartmouth Coll., 1934; M.D., Johns Hopkins U., 1938, U. Bern, Switzerland. Fellow in medicine Harvard U., 1941-42; from instr. to asst. prof. medicine and psychiatry U. Cin., 1942-46; from asst. prof. to assoc. prof. medicine U. Rochester, 1946-57, prof., 1957—; psychiatrist, physician Strong Meml. Hosp., Rochester, 1957—; clinician Med. Service, Cin. Gen. Hosp., 1942-44, asst. attending psychiatrist, 1943-44; cons. Office Surgeon Gen. Recipient Career Research award USPHS, 1962. Mem. AAAS, Am. Soc. Clin. Investigation, Am. Psychosomatic Soc., Am. Psychiat. Assn., Am. Psychoanalytic Assn., Inst. Medicine (sr.). Subspecialty: Psychiatry. Office: Strong Meml Hosp 260 Crittenden Blvd Rochester NY 14642

ENGEL, JOANNE BOYER, psychology educator and administrator, consultant, researcher; b. Meadville, Pa., Mar. 15, 1944; d. Edward Charles and Wanda Ann (Chasco) Boyer; m. Richard E. Hammer, Aug. 12, 1965 (dec. 1969); m. Harold N. Engel, Jr., Mar. 12, 1971; children: Cynthia, Keith. B.S., Pa. State U., 1965; M.Ed., U. Sydney, Australia, 1972; M.S., Iowa State U., 1979, Ph.D., 1979. Tchr. Pub. Schs. Broomall Wallingford, Pa., 1965-69; dir. clin. research lab. Aurubn (Ala.) U., 1971-75; instr. Iowa State U., Ames, 1975-79; asst. prof. Oreg. State U., Corvallis, 1979-80; assoc. prof. psychology, dept. head Willamette U., Salem, Oreg., 1980—, chmn. edn. dept., 1981—; adj. grad. faculty Oreg. State U., Salem, 1980—; cons. various ednl.-psychol. groups, Ala., Iowa, Oreg., 1970—. Author: Kalediscope of Women: Women and Technology, 1982, Women: The Way We Work, 1981, Cognition and Pre School, 1975, Teaching Mathematics, 1974. Active Oreg. Women's Polit. Caucus, Corvallis. HEW grantee, 1974; Atkinson research grantee, 1981-82; Oreg. State U. research grantee, 1980. Mem. Am. Psychol. Assn., Am. Ednl. Research Assn., Oreg. Psychol. Assn. (officer 1981-82), Phi Delta Kappa (v.p. 1982—), Kappa Delta Pi (chpt. advisor 1980—). Presbyterian. Club: Music Soc. (Corvallis). Subspecialties: Cognition; Learning. Current work: Learning research. Home: 2855 NW Skyline Dr Corvallis OR 97330 Office: Willamette Univ 900 State St Salem OR 97301

ENGEL, JOHN WILLIAM, educator, researcher; b. Effingham, Ill., Mar. 17, 1946; s. Edwin V. and Anastasia (Fischenich) E.; m. Barbara Schiltgen, Nov. 22, 1975; children: Theresa Nani, Patricia Ann. B.S., St. John's U., Minn., 1968; Ph.D., U. Minn., 1978. Cert. sex educator and counselor, Am. Assn. Sex Educators, Counselors and Therapists. Family therapist E.C. Family Service Ctr., St. Paul, 1975-78, exec. dir., 1978-79; instr. U. Minn., Mpls., 1974-79; asst. prof. U. Hawaii, Honolulu, 1979—; cons. U.S. Navy, 1979-81, various community groups, Honolulu, 1979—; marriage and family therapist, Honolulu, 1979—. Editor: Dynamics of Marriage, 1974; author: Human Sexual Behavior, 1975, Women's Employment, 1978, Family Life in China, 1982. NDEA fellow, 1968; Rehab. Services Adminstrn. trainee, 1969; Hawaii Inst. Tropical Agr. grantee, 1981; U.S. Dept. Agr. grantee, 1982. Mem. Hawaii Assn. Marriage and Family Therapists (pres. 1982—), Nat. Council Family Relations, Am. Assn. Sex Educators, Counselors and Therapists, Am. Psychol. Assn., Hawaii Home Econs. Assn. (dir.). Subspecialties: Behavioral ecology; Social psychology. Current work: Cross-cultural family life; evaluation of family life and sex education; work-family interface: unemployment, work ethics, maternal employment, etc. Home: 1201 Wilder Ave #3003 Honolulu HI 96822 Office: U Hawaii Honolulu HI 96822

ENGEL, LARS NORLICK, applied mathematicians; b. Portland, Oreg., Nov. 2, 1934; s. Ernest Herman and Joyce (Graham) E.; m. Emily Jo Flachmeier, Dec. 28, 1957; children: Jan Kristin, Karen Gale. B.S. with honors, U. Tex., 1961, M.S.E.E, 1964. Research engr. Elec. Engring. Research Lab., U. Tex., Austin, 1961-63; microwave engr. Electro-Mechanics Co., Austin, 1963-64; engr. Westinghouse Electric., Balt., 1964-65; staff mem. Los Alamos Nat. Lab., N.Mex., 1965—. Pres. and dir. No. N.Mex. Meml. and Funeral Soc., Los Alamos, 1969-71; trustee Dad's Assn. Tex. Tech. U., Lubbock, 1982-83. Served to staff sgt. USAF, 1954-58. Mem. IEEE, Am. Math Soc., Soc. Indsl. and Applied Math. Subspecialties: Mathematical software; Nuclear physics. Current work: Application of non-linear optimization techniques to nuclear physics problems. Home: 1210 Myrtle Los Alamos NM 87544 Office: Los Alamos Nat Lab PO Box 1663 Los Alamos NM 87545

ENGEL, PETER ANDRAS, research mechanical engineer; b. Kassa, Hungary, July 10, 1935; s. Geza and Herta (Reisz) E.; m.; children: Gregory, David. B.S., Vanderbilt U., 1958; M.S., Lehigh U., 1960; Ph.D., Cornell U., 1968. Registered profl. engr., N.Y. Structural analyst Praeger Kavanagh Waterbury Engrs., N.Y.C., 1959-62; research engr. Boeing Co., New Orleans, 1962-65; staff engr. IBM, Endicott, N.Y., 1968-71, adv. engr., 1971—; sr. engr. Endicott Lab., 1975—; mem. faculty SUNY-Binghamton. Author: Impact Wear of Materials, 1976; contbr. chpts., numerous articles to profl. publs. Recipient A. J. Dyer Meml. prize Vanderbilt U., 1958, Charles Russ Richards Meml. award ASME, 1983. Mem. ASME. Subspecialties: Theoretical and applied mechanics; Mechanical engineering. Current work: Impact, wear, packaging, structural mechanics, adhesion science, tribology. Office: Endicott Lab IBM PO Box 6 Endicott NY 13760

ENGEL, THOMAS WALTER, chemistry educator, researcher; b. Yokohama, Japan, Apr. 2, 1942. B.A., Johns Hopkins. U., 1963; M.A., 1964; Ph.D. in Chemistry, U. Chgo., 1969. Research assoc. in surface sci. Clausthal Tech. U., 1969-75, U. Munich, W.Ger., 1975-78; research assoc. in surface sci. IBM Research Lab., Zurich, 1978-80; assoc. prof. chemistry U. Wash., Seattle, 1980—. Subspecialty: Surface chemistry. Office: Dept Chemistry U Wash Seattle WA 98195

ENGEL, TOBY ROSS, physician, medical educator and researcher; b. N.Y.C., Mar. 6, 1942; s. Fred and Pauline (Bienstock) E.; m. Lorraine Barbara Rodney, Aug.15, 1965; children: Joshua, Jeffrey, Benjamin. B.A., N.Y.U., 1962, M.D., 1966. Diplomate: Am. Bd. Internal Medicine (subcert. in cardiovascular disease). Resident U. Pa., Phila., 1966-68; fellow, instr. Ohio State U., Columbus, 1970-73; asst. prof. Med. Coll. Pa., Phila., 1973-75, assoc. prof., 1976-79, prof. medicine, 1980—. Assoc. editor: Annals of Internal Medicine, 1977—. Served to capt. U.S. Army, 1968-70; Korea. Recipient awards Am. Heart Assn., NIH, others. Fellow Am. Coll. Cardiology, ACP, Am. Coll. Clin. Pharmacology, Am. Heart Assn. Council on Clin. Cardiology, others. Jewish. Subspecialties: Cardiology; Pharmacology. Current work: Electrophysiology, electrocardiography, pharmacologic and surgical and pace maker treatment of arrhythmia. Home: 30 Crestline Rd Wayne PA 19087 Office: Med Coll Pa 3300 Henry Ave Philadelphia PA 19125

ENGELBERGER, JOSEPH FREDERICK, robotics mfg. co. exec.; b. N.Y.C., July 26, 1925; s. Joseph H. and Irene E. E.; m. Margaret B. Thomas, May 24, 1954; children—Gay, Jeff. B.S., Columbia U., 1946, M.S. in Physics, 1949. With Manning Maxwell & Moore, Stanford, Conn., 1946-57, chief engr., 1953-56, div. gen. mgr., 1956-57; founder, gen. mgr. Consol. Controls Corp. (sold to Condec Corp. 1964), Old Greenwich, Conn., 1957-77, chmn. bd., 1977; also v.p., mem. exec. com. Condec Corp., 1965—; also dir.; founder, pres. Unimation Inc., Danbury, Conn., 1962—; dir. Cooper Thermometer, Anderson Labs., State Nat. Bank; founder, past chmn. Conn. Product Devel. Corp., 1973, chmn., 1973-79; mem. Pres.'s Commn. on Indsl. Innovation. Contbr. numerous articles on robotics to profl. jours. Vice pres. Fairfield County council Boy Scouts Am. Recipient Nyselius award Die Casting Inst., 1978; Progress award Soc. Mfg. Engrs., 1979. Mem. Chief Execs. Forum, World Bus. Council, Robot Inst. Am. (founder 1973, pres. 1974, established Joseph F. Engelberger Ann. award 1977), Tau Beta Pi. Subspecialty: Robotics. Current work: Applications of robotics throughout the factory and information service area; provide robots with sensory perception. Patentee in field. Office: Unimation Inc Shelter Rock Ln Danbury CT 06810 also Durant Ave Bethel CT 06810

ENGELBRECHT, RICHARD STEVENS, environmental engineering educator; b. Ft. Wayne, Ind., Mar. 11, 1926; s. William C. and Mary Elizabeth (Stevens) E.; m. Mary Condrey, Aug. 21, 1948; children: William, Timothy. A.B., Ind. U., 1948; M.S., M.I.T., 1952, Sc.D., 1954. Teaching asst. Ind. U. Sch. Medicine, Indpls., 1949-50; research asst. M.I.T., Cambridge, 1950-52, instr., 1952-54; asst. prof. U. Ill., Urbana-Champaign, 1954-57, assoc. prof., 1957-59, prof. environ. engring., 1959—; dir. Advanced Environ. Control Tech. Research Center, 1979—; cons. Ill. EPA, U.S. EPA, WHO; mem. Ohio River Valley Water Sanitation Commn., chmn., 1980-82. Named Ernest Victor Balsom Commemoration Lectr., 1978; recipient Eric H. Vick award Inst. Public Health Engrs., U.K., 1979. Mem. Internat. Assn. Water Pollution Research and Control (pres. 1980—), Am. Water Works Assn. (George W. Fuller award 1974, Publ. award 1975), Water Pollution Control Fedn. (Eddy medal 1966, Arthur Sidney Bedell award 1973, pres. 1978), Nat. Acad. Engring., AAAS, Am. Soc. Microbiology, N. Am. Benthological Soc., Ill. Soc. Microbiology., Abwasser-technische Vereini-gung (hon.). Subspecialty: Water supply and wastewater treatment. Current work: Microbiological (bacteria, viruses) problems associated with water quality management, including water and wastewater treatment. Home: 2012 Silver Ct W Urbana IL 61801 Office: 3230 Newmark Civil Engring Lab 208 N Romine St Urbana IL 61801

ENGELHARD, ARTHUR WILLIAM, plant pathologist, educator; b. Dayton, Ohio, Apr. 9, 1928; s. Paul George and Louise Emma (Stroh) E.; m.; children: Eric, Lisa, Arthur William. B.S., Ohio U., 1950; M.S., Yale U., 1952; Ph.D., Iowa State U., 1955. Grad. asst. Iowa State U., 1952-55; asst. plant pathologist Ill. State Natural History Survey, Urbana, 1955-56; research biologist E. I. duPont, Wilmington, Del., 1956-64, sr. research biologist, Bradenton, Fla., 1964-66; assoc. prof. plant pathology U. Fla.-Bradenton, 1966-78, prof., 1978—; cons. in field. Contbr. numerous articles to profl. jours. Recipient ann. research award for outstanding research and service to Fla. growers Fla. Ornamental Growers Assn., 1981, Medal of Merit for outstanding contbns. in plant pathology Ohio U. Nat. Alumni Bd., 1983. Mem. Am. Phytopath Soc., Fla. State Hort. Soc. (Outstanding Paper Award 1970, 74, 80, 81), Internat. Soc. Plant Pathology, Sigma Xi, Gamma Sigma Delta, Phi Kappa Phi, Theta Chi. Subspecialties: Integrated pest management; Plant pathology. Current work: Etiology and control of diseases of ornamental plants; integration of chemical-cultural-nutritional systems of disease control. Patentee methods to control arachnids. Home: 5306 7th Ave Dr W Bradenton FL 33529 Office: 5007 60th St E Bradenton FL 33508

ENGELHARDT, DEAN LEE, biotechnology company executive; b. Oak Park, Ill., Jan. 15, 1940; s. Bruce Haas and Doris Pricilla (O'Grady) Nixon; m. Sara Hilary Lawrence, June 20, 1970; children: Barbara Elizabeth, Margaret Ann. B.A., Amherst Coll., 1961, M.A., 1963; Ph.D., Rockefeller U., 1967. Postdoctoral fellow Philip I. Marcus, Albert Einstein Coll. Medicine, Bronx, N.Y., 1969; asst. prof. U. Conn.-Storrs, 1969-73; asst. prof. dept. microbiology Columbia U. Coll. Physicians and Surgeons, N.Y.C., 1973-79, assoc. prof., 1980-81, adj. assoc. prof., 1981—; v.p. research Enzo Biochem, Inc., N.Y.C., 1981—. Contbr. articles to profl. jours. Subspecialties: Genetics and genetic engineering (biology); Developmental biology. Current work: Cloning pathognemonic DNA sequences and making monoclonal antibodies for diagnosis. Patentee in field. Home: 173 Riverside Dr New York NY 10024 Office: Enzo Biochem Inc 325 Hudson St New York NY 10013

ENGELHARDT, JOHN KERCH, neurosci. educator; b. Camden, N.J., Sept. 15, 1946; s. David LeRoy and Mildred Marian (Henisee) E.; m. Rita Faye Studdard, Dec. 22, 1968. B.A., Trinity U., San Antonio, 1968; Ph.D., UCLA, 1973. Postdoctoral fellow City of Hope Med. Center, Duarte, Calif., 1974-77; research scholar U. So. Calif. Sch. Medicine, Los Angeles, 1977-79, asst. research prof. dept. neurology, 1979—. Contbr. articles to profl. jours. USPHS awardee, 1975. Mem. Biophys. Soc., Soc. Neurosci., AAAS, Philosophy of Sci. Assn. Subspecialty: Neurophysiology. Current work: Electrophysiology of nerve and muscle cells in tissue culture. Office: Dept Neurology U So Calif 2025 Zonal Ave Los Angeles CA 90033

ENGELKING, HENRY MARK, virologist; b. Burbank, Calif., May 3, 1949; s. Henry Christian and Lorraine Katherine (Miehl) E.; m. Judy Ann Hagner, Sept. 15, 1979. B.A., U. Calif., San Diego, 1971; M.S., Oreg. State U., 1974. Research asst. dept. biochemistry-biophysics Oreg. State U., 1974-78, sr. research asst. dept. microbiology, 1978—; cons. in field. Contbr. articles to profl. jours. Mem. Am. Soc. Microbiology, Sigma Chi, Phi Kappa Phi. Subspecialties: Virology (biology); Molecular biology. Current work: Genetic engineering, sub unit vaccines, persistent viral infections, viral diseases of fish, rhabdovirus cloning, rapid diagnosis of IHN virus in salmonids. Home: 234 NW 29th St Corvallis OR 97330 Office: Dept Microbiology Oregon State U Corvallis OR 97331

ENGELKING, PAUL CRAIG, physical chemist; b. Glendale, Calif., May 11, 1948; s. Fred Carl and Gladys (Nicol) E.; m. Patricia Donaldson, Aug. 2, 1975; children: Kirstin, Gwynne. B.S., Calif. Inst. Tech., 1971, M.Phil., Yale U., 1974, Ph.D., 1976. Research asst. Joint Inst. Lab. Astrophysics, Boulder, Colo., 1976-78; asst. prof. chemistry U. Oreg., Eugene, 1978—. Alfred P. Sloan fellow, 1982. Mem. Am. Chem. Soc., Am. Phys. Soc. Subspecialties: Physical chemistry; Spectroscopy. Current work: Laser induced fluorescence of radicals and ions, photodetachment. Office: Dept Chemistry U Oreg Eugene OR 97403 Home: 32 N Alder St Lowell OR 97452

ENGELS, WILLIAM ROBERT, geneticist; b. Mineral Point, Wis., Nov. 19, 1950; s. James Edward and Betty (Marquardt) E. Ph.D., U. Wis., 1978. Research scientist genetics U. Wis., Madison, 1978—, instr., 1979—. Contbr. articles to profl. jours. NSF fellow, 1980; NIH fellow, 1982; NSF grantee, 1981; NIH grantee, 1982. Mem. Genetics Soc. Am. Subspecialties: Genome organization; Statistics. Current work: Researcher in transposable genetics elements in Drosophila, the P factor and hybrid dysgenesis; population genetics. Home: 1214 Spring St #25 Madison WI 53715 Office: Dept Genetics U Wis Madison WI 53706

ENGEN, EUGENE PAUL, clin. psychologist, mental health center adminstr.; b. Yankton, S.D., Mar. 4, 1931; s. Oscar Leonard and Cena Caroline (Paulson) E.; m. Eunice Lorraine Erickson, June 1, 1957; children: Paul Douglas, Brendan Clark. B.A., Yankton Coll., 1952; M.A., Mills Coll., 1955; Ph.D., La. State U., 1959. Diplomate: in clin. psychology Am. Bd. Profl. Psychology. Psychologist Yankton State Hosp., 1954-56, chief psychologist, 1959-62, 64-67, dir. adolescent treatment program, 1967-71; scientist USPHS Hosps., Ft. Worth, also Lexington, Ky., 1963; exec. dir. Lewis and Clark Mental Health Center, Yankton, 1970—; Chmn. Mental Health Interface Com., Yankton, 1979-81; chmn. S.D. Mental Health Adv. Com., Yankton, 1977-80; mem. Gov.'s Com. on Mental Health, Drugs and Alcohol, 1980. Served with USPHS, 1963. USPHS fellow, 1957; recipient fellowships and grants. Mem. Am. Psychol. Assn., S.D. Psychol. Assn. (pres. 1961-62, 72-74), Assn. S.D. Mental Health Centers (pres. 1981—). Lodge: Rotary. Subspecialties: Behavioral psychology. Current work: Development of short-term, intensive psychotherapies. Home: Box 146 RR 1 Lewis and Clark Lake Rd Yankton SD 57078 Office: Lewis and Clark Mental Health Center Inc 1028 Walnut St Yankton SD 57078

ENGLAND, TALMADGE RAY, physicist; b. Bonham, Tex., Dec. 22, 1929; s. Bascomb Curtis and Dora May (Hobbs) E.; m. Carol Elizabeth Odell, Mar. 26, 1949; children: Cheryl, Ana, Rebecca, Rhonda. B.S., Lincoln Meml. U., 1956; M.S., U. Pitts., 1962; Ph.D., U. Wis.-Madison, 1969. Chief engr. Sta. WJMA, Orange, Va., 1949-59, Sta. WMIK, Middlesboro, Ky., 1951-56; jr. scientist Bettis Atomic Power Lab., Pitts., 1956-57, scientist, 1957-60, sr. scientist, 1960-72;

staff mem. physics Los Alamos Nat. Lab., 1972—; chmn. cross sect. evaluation working group yields com. Brookhaven Nat. Lab., Upton, N.Y., 1975—. Mem. Am. Nuclear Soc. (chmn. yields standard 1975—, sec. decay power standard 1973—), IEEE, AAAS. Democrat. Unitarian. Subspecialties: Nuclear physics; Theoretical physics. Current work: Nuclear data for national data files and fission yield standards. Home: 613 Meadow Ln Los Alamos NM 87544 Office: Los Alamos Nat Lab PO Box 1666 Los Alamos NM 87545

ENGLE, RANDALL WAYNE, educator; b. Falling Rock, W.Va., Dec. 2, 1946; s. Harold Eugene and Maxine Luverne (Walker) E.; m. Eileen Jo White, July 7, 1967; children: Holly, Matthew. B.A., W.Va. State Coll., 1968; M.A., Ohio State U., 1969, Ph.D., 1973. Asst. prof. King Coll., Bristol, Tenn., 1972-74; asst. prof. to prof. psychology U. S.C., Columbia, 1974—. Mem. Am. Psychol. Assn., Psychonomic Soc. Democrat. Subspecialties: Cognition; Developmental psychology. Current work: Characteristics of auditory sensory memory, development of sensory memory, development of memory strategies. Address: Dept Psychology Univ SC Columbia SC 29208

ENGLEMAN, EDGAR GEORGE, pathology educator; b. Palm Springs, Calif., July 11, 1945; s. Ephraim Phillip and Jean (Sinton) E.; m. Judith, Dec. 21, 1967; children: Eric, Jason. B.A. magna cum laude, Harvard U., 1967; M.D., Columbia U., 1971. Diplomate: Am. Bd. Internal Medicine. Intern, resident in medicine U. Calif. Hosps., San Francisco, 1971-73; research assoc. Heart and Lung Inst., NIH, Bethesda, Md., 1973-76; postdoctoral fellow in immunogenetics and rheumatology Stanford (Calif.) U., 1976-78, asst. prof. pathology and medicine, 1978—; dir. Stanford Med. Sch. Blood Ctr., 1979—. Contbr. articles to profl. jours. Served to lt. comdr. USPHS, 1973-76. Am. Cancer Soc. fellow, 1976-78. Mem. Am. Soc. Clin. Investigation, Am. Fedn. Clin. Research, Am. Assn. Immunologists, Am. Rheumatism Assn. Jewish. Subspecialty: Immunology (medicine). Current work: Control of immunity in man. T lymphocytes; T lymphocyte subsets; monoclonal antibodies; hybridomas; immunoregulation; autoimmunity. Office: Dept Pathology L23 Stanford U Med Sch Stanford CA 94305

ENGLER, MARK J, medical physicist; b. N.Y.C., Dec. 25, 1945; m. Joyce M. B.S. in Chemistry cum laude, CCNY, 1966; Ph.D. in Phys. Chemistry, M.I.T., 1969. Cert. in therapeutic radiol. physics Am. Bd. Radiology. Instr., founding mem. exptl. studies group M.I.T., 1969-70; radiation physicist, asst. prof. Boston U. Med. Center, 1975-79, Duke U. Med. Ctr. and Duke Comprehensive Cancer Ctr., Durham, N.C., 1979—; cons. Radiation Physics Inc., Newton, Mass., 1975-79. Self-employed musician, 1971-74; Contbr. articles to profl. jours. Mem. AAAS, Am. Assn. Phys. Medicine, Am. Coll. Radiology, Am. Phys. Soc., N.Y. Acad. Scis., Radiation Research Soc., Radiol. Soc. N.Am., N.Am. Hyperthermia Group, South East Cancer Study Group, Am. Soc. Therapeutic Radiology and Oncology, Am. Radium Soc., Health Physics Soc., Phi Beta Kappa. Subspecialties: Biophysics (physics); Cancer research (medicine). Current work: Therapeutic hyperthermia; synchronous field shaping of teletherapy beams.

ENGLISH, JOSEPH T., physician, med. adminstr.; b. Phila., May 21, 1933; m. Ann Carr Sanger, Dec. 20, 1969; 3 children. A.B., St. Joseph's Coll., 1954; M.D., Jefferson Med. Coll., 1958. Intern Jefferson Med. Coll. Hosp., Phila., 1958-59; resident in psychiatry Inst. of Pa. Hosp., Phila., 1959-61, NIMH, Bethesda, Md., 1961-62; practice psychiatry, 1962—; psychiatrist Office of Dir., NIMH, 1964-65, asst. chief policy and program co-ordination, 1965-66, dept. chief office interagy. liaison, 1966; chief psychiatrist med. program div. Peace Corps, Washington, 1962-66; dep. asst. dir. health affairs OEO, Washington, 1966, acting asst. dir., 1966-68, asst. dir., 1969; adminstr. Health Services and Mental Health Adminstrn., HEW, 1969-70; pres. N.Y.C. Health and Hosps. Corp., 1970—; dir. dept. psychiatry St. Vincent's Hosp. and Med. Center, N.Y.C., 1973—; also dean; prof. psychiatry N.Y. Med. Coll., 1979—; adj. prof. psychiatry Cornell U.; Chmn. interagy. task force emergency food and med. program for U.S. OEO-HEW, U.S. Dept. Agr., 1968-69; chmn. Alaska Subcom. Fed. Health Programs Pres.'s Rev. Commn. Alaska, 1969—; chmn. adv. com. on accessible environments for disabled Bldg. Research Adv. Bd., Washington, 1974—; chmn. exec. com. of com. on mental health services Greater N.Y. Hosp. Assn., 1974—; exec. coordinator panels on mental health services delivery Pres.'s Commn. on Mental Health, 1977; mem. Health Adv. Council Gov. State N.Y., 1981. Author spl. reports on Peace Corps, other govtl. programs.; Contbr. articles to profl. jours. Served to capt. USAF Res., 1958-63; sr. surgeon USPHS, 1963-66. Named One of Outstanding Young Men of Year U.S. Jr. C. of C., 1964; recipient John XXIII medal Coll. New Rochelle, N.Y., 1966; Meritorious award for exemplary achievement pub. adminstrn. William A. Jump Meml. Found., 1966; Flemming award, also personal commendation Pres. of U.S., 1968. Fellow Am. Psychiat. Assn. (chmn. council financing psychiat. sers. 1981), N.Y. Acad. Medicine, Am. Coll. Psychiatrists, Inst. Medicine of Nat. Acad. Scis.; mem. AMA (com. mental health services to poor 1965-66), Insts. Religion and Health (profl. adv. bd. 1966—), Am. Public Health Assn., Group Advancement Psychiatry, Pa. Med. Soc., Soc. of Jefferson for Research (charter), Am. Coll. Mental Health Adminstrs., Am. Hosp. Assn. (chmn. bd. govs. Center Mental Health and Psychiatry), Arnold Air Soc., Alpha Omega Alpha, Kappa Beta Phi, Alpha Sigma Nu. Subspecialty: Psychiatry. Home: 7 Valley Rd Bronxville NY 10708 Office: St Vincent's Hosp and Med Center 203 W 12th St New York NY 10011

ENGLUND, CARL ERNEST, research psychologist, educator; b. Cleve., Aug. 11, 1936; s. Ernest Carl and Anne Delice (Olson) E.; m. Sandra Brooks, Jan. 20, 1955 (div. Jan. 1968); children: Eric Carl, Kristin Delice, Svea Pilar; m. Linda Sue Trumbull, Dec. 16, 1978; 1 stepson, Timothy Scott. B.A., San Diego State U., 1960; M.A., U.S. Internat. U., 1973, Ph.D., 1980; cert. human factors engring, U. Mich., 1964. Research engr. Gen. Dynamics Corp., SAn Diego, 1955-61; sr. human factors specialist System Devel. Corp., Lexington, Mass., 1961-62; project engr. Philco-Ford Corp., Palo Alto, Calif., 1962-69; cons. behavioral sci., San Diego, 1969-72; research psychologist Decision Sci. Inc., SAn Diego, 1972-74, USN, 1974—; pres. Applied Research and Cons. Assocs., SAn Diego, 1970-72; prof. psychology Nat. U. and Mesa Coll., 1974—. Contbr. articles to profl. jours. Mem. Internat. Soc. Chronobiology, Am. Psychol. Assn., Human Factors. Soc. Democrat. Subspecialties: Human factors engineering; Applied experimental psychology. Current work: Determining the relationships between mental efficiency and physical work exertion and fitness; environmental effects upon circadian rhythms. Home: 14162 Recuerdo Dr Del Mar CA 92014 Office: Naval Health Research Center PO Box 85122 San Diego CA 92138

ENGSTROM, DAVID RALPH, clinical psychologist, researcher; b. Chgo., Nov. 14, 1942; s. Ralph H. and Roberta E. (Coleman) E.; m. Bonnie B. Buhl, Aug. 20, 1965; children: Loren, Brian, Dana. A.B., George Washington U., 1964; Ph.D., U. So. Calif., 1970. Diplomate: Am. Bd. Profl. Psychologists; lic. psychologist, Calif. Asst. clin. prof. psychiatry U. Calif.,-Irvine, 1970—, dir. behavioral medicine, 1975—; pvt. practice clin. psychologist, Newport Beach, Calif., 1971—; dir. biofeedback Hoag Meml. Hosp., Newport Beach, 1974—; v.p., dir. psychol. service Advanced Health System, Irvine, 1975—. Author: Consciousness and Self-regulation, 1976, Biofeedback and Self-Control, 1979; contbr. articles to profl. jours. Fellow Soc. Clin. and Exptl. Hypnosis; mem. Am. Psychol. Assn., Orange County Psychol. Assn. (pres. 1978). Subspecialties: Behavioral psychology; Cognition. Current work: Cognitive-behavioral assessment and treatment; pain control in humans; relationship between behavior change and attitude change. Office: Student Health Service Univ Calif Irvine CA 92717

ENK, GORDON A., natural resources economist; b. Milw., June 24, 1940; s. Benedict W. and Irene G. (Kopp) E.; m. Elise Werner, Aug. 25, 1962; children: Terrence, Christopher. B.S., Ripon Coll., 1962; M.F., Yale U., 1967, M. Philosophy, 1970, Ph.D., 1975. Researcher forest products mktg. U.S. Forest Service, 1967; with State of Wash. Dept. Natural Resources, 1969; dir. econ. and environ. studies Inst. Man and Sci., Rensselaerville, N.Y., 1970-81; pres. Gordon A. Enk & Assos., Inc., Medusa, N.Y., 1981—, Research and Decision Ctr., Medusa, 1981—. Contbr. articles to profl. jours. Served to 1st lt. U.S. Army, 1962-65. Mem. AAAS, Am. Econs. Assn., Soc. Am. Foresters, ASCE (com. impact analysis for water resources planning and mgmt. div.), Sigma Xi. Subspecialties: Resource management; Resource conservation. Current work: Organizational decisionmaking concerned with economic and environmental quality, energy and toxic substances. Office: Makely House Medusa NY 12120

ENNA, SALVATORE JOSEPH, pharmacologist, cons.; b. Kansas City, Mo., Dec. 19, 1944; s. Veto Anthony and Fannie Silvia (Bonello) E.; m. Colleen Anne Nestor, July 26, 1969; children: Anne, Matthew, Katherine. B.A., Rockhurst Coll., 1965; M.S., U. Mo., Kansas City, 1967, Ph.D., 1970. Postdoctoral fellow U. Tex. Med. Sch., Dallas, 1970-72; Roche fellow F. Hoffmann-LaRoche & Co., Basel, Switzerland, 1973-74; research fellow dept. pharmacology Johns Hopkins U., 1974-76; asst. prof U. Tex. Med. Sch., Houston, 1976-77, assoc. prof., 1977-80, prof., 1980—; cons. in field. Contbr. numerous articles to profl. jours. Bd. dirs. Houston Area Parkinson Soc., 1980—. Trustee fellow U. Mo., 1967-70; NIMH fellow, 1974-76; NIH Research Career Devel. award, 1978—; Basic Sci. Teaching award U. Tex. Med. Sch., 1980-81; John Jacob Abel award, 1980. Mem. Soc. Neurosci., Am. Soc. Neurochemistry, Am. Soc. Pharmacology and Exptl. Therapy, AAAS, Am. Chem. Soc. Subspecialties: Molecular pharmacology; Neuropharmacology. Current work: Neurotransmitter biochemistry and pharmacology. Home: 6227 Cheena Houston TX 77096 Office: Dept Pharmacology U Tex Med Sch Box 20708 Houston TX 77025

ENNIS, FRANCIS A., physician, educator; b. Boston, May 14, 1938; s. Lewis and Veronica (Pittman) E.; m. Anne M. Cavanagh, Aug. 10, 1963. A.B., Boston Coll., 1960; M.D., Tufts U., 1964. Diplomate: Am. Bd. Internal Medicine, 1971. NIH research assoc., Bethesda, Md., 1966-68; resident in medicine Cornell U., 1968-70; assoc. prof. medicine Boston U. Sch. Medicine, 1970-73; dir. div. virology Bur. Biologics, FDA, Bethesda, Md., 1973-81; prof. medicine and molecular genetics U. Mass. Med. Ctr., Worcester, 1981—. Contbr. articles to profl. jours. Served with USPHS, 1966-68, 74-81. Recipient award of Merit FDA, 1977. Mem. Am. Soc. Clin. Investigation, Am. Assn. Immunologists, Am. Soc. Virology. Subspecialties: Infectious diseases; Immunology (medicine). Current work: Immune responses to viruses, interferons, lymphocytes. Office: 55 Lake Ave N Worcester MA 01605

ENNIS, HERBERT L., biochemist; b. Bklyn., Jan. 6, 1932; s. Rudolph and Fannie (Stringer) E.; m. Judith A. Wolper, June 5, 1960; children: Ronald D., Ethan W. B.S., Bklyn. Coll., 1953; M.S., Northwestern U., 1954, Ph.D., 1957. Postdoctoral fellow Northwestern U., 1957-58, Harvard Med. Sch., 1958-59; Am. Cancer Soc. research fellow Brandeis U., 1959-60; instr. dept. pharmacology Harvard Med. Sch., 1960-64; mem. St. Jude Children's Research Hosp.-U. Tenn. Med. Sch., 1964-69, Roche Inst. Molecular Biology, Nutley, N.J., 1969—. Contbr. articles to profl. jours.; editor: Antimicrobial Agents and Chemotherapy, 1977—. Mem. AAAS, Am. Soc. Biol. Chemistry, N.Y. Acad. Sci., Am. Soc. Microbiology, Sigma Xi. Subspecialties: Molecular biology; Developmental biology. Current work: Protein and nucleic acid synthesis; messenger RNA decay; cloning of developmentally regulated genes; mechanism of antibiotic action. Office: Roche Inst Molecular Biology Nutley NJ 07110

ENOCH, JAY MARTIN, visual scientist, educator; b. N.Y.C., Apr. 20, 1929; s. Jerome Dee and Stella Sarah (Nathan) E.; m. Rebekah Ann Feiss, June 24, 1951; children—Harold Owen, Barbara Diane, Ann Allison. B.S. in Optics and Optometry, Columbia U., 1950; postgrad., Inst. Optics U. Rochester, 1953; Ph.D. in Physiol. Optics, Ohio State U., 1956. Asst. prof. physiol. optics Ohio State U., Columbus, 1956-58, assoc. prof., 1957-58; fellow Nat. Phys. Lab., Teddington, Eng., 1959-60; research instr. dept. ophthalmology Washington U. Sch. Medicine, St. Louis, 1958-59, research asst. prof., 1959-64, research asso. prof., 1965-70, research prof., 1970-74; fellow Barnes Hosp., St. Louis, 1960-64, cons. ophthalmology, 1964-74; research prof. dept. psychology Washington U., St. Louis, 1970-74; grad. research prof. ophthalmology and psychology U. Fla. Coll. Medicine, Gainesville, 1974-80, grad. research prof. physics, 1979-80; dir. Center for Sensory Studies, 1976-80; dean Sch. Optometry, prof. physiol. optics and optometry U. Calif., Berkeley, 1980—, prof. physiol. optics in ophthalmology, San Francisco, 1980—; chmn. subcom. contact lens Standards Am. Nat. Standards Inst., 1970-77; mem. nat. advisory eye council Nat. Eye Inst., NIH, 1975-77, 80—; exec. com., com. on vision NAS-NRC, 1973-76; mem. U.S. Nat. Com. Internat. Commn. Optics, 1976-79. Author numerous chpts. and articles on visual sci., receptor optics, perimetry, contact lenses and infant vision to sci. jours.; contbr. chpts. in field to med. books; hon. editorial bd.: Vision Research, 1974-80; editorial bd.: Internat. Ophthalmology, 1977—; asso. editor: Investigative Ophthalmology, 1965-75, Sight-Saving Rev, 1974—, Sensory Processes, 1974—; editorial bd. optical scis.: Springer-Verlag, Heidelberg, 1978—. Mem. nat. sci. advisory bd. Retinitis Pigmentosa Found., 1977—; U.S. rep. Internat. Perimetric Soc., 1974—; also exec. com., chmn. Research Group Standards.; Bd. dirs. Friends of Eye Research, 1977—; trustee Illuminating Engring. Research Inst., 1977-81. Served to 2d lt. U.S. Army, 1951-52. Recipient Career Devel. award NIH, 1963-73. Fellow AAAS, Am. Acad. Optometry (Glenn A. Fry award 1972, Charles F. Prentice medal award 1974), Optical Soc. Am. (chmn. vision tech. sect. 1974-76), Am. Acad. Ophthalmology Otolaryngology (asso.); mem. Assn. for Research in Vision and Ophthalmology (trustee 1967-73, pres. 1972-73, Francis I. Proctor medal 1977), Internat. Strabismological Assn., Internat. Soc. for Clin. Electro-retinography, Biophys. Soc., Psychonomic Soc., Am. Soc. for Photobiology, AAUP, Am. Psychol. Assn. (sect. 3), Contact Lens Soc. of U.K., Sigma Xi. Subspecialty: Optometry. Home: 54 Shuey Dr Moraga CA 94556 Office: Sch Optometry U Calif Berkeley CA 94720

ENQUIST, LYNN WILLIAM, molecular biologist, biotechnology co. research administrator, educator; b. Denver, Oct. 23, 1945; s. Clarence Andrew and Doris Alice (Hajenga) E.; m. Kathleen Marie, Aug. 10, 1968; 1 son, Brian Joseph. B.S. (Woodrow Wilson fellow), S.D. State U., 1967; Ph.D., Med. Coll. Va., 1971. Postdoctoral fellow Roche Inst. Molecular Biology, Nutley, N.J., 1971-73; staff fellow NIH, Bethesda, Md., 1973-77, staff scientist, 1977-81; research dir. Molecular Genetics, Inc., Minnetonka, Minn., 1981—; instr. advanced bacterial genetics Cold Spring Harbor (N.Y.) Labs., 1981, 82, 83. Contbr. numerous articles to profl. jours.; mem. editorial bd.: Jour. Virology, 1979-81. Served to comd. USPHS, 1973-81. Mem. Am. Soc. Microbiology, AAAS, Sigma Xi, Phi Kappa Phi. Subspecialties: Genetics and genetic engineering (biology); Molecular biology. Current work: Organization and expression of genes; gene manipulation. Office: 10320 Bren Rd E Minnetonka MN 55343

ENRIGHT, ROBERT DAVID, educational psychology educator; b. Holyoke, Mass., Jan. 9, 1951; s. William F. and Margaret T. (Redding) E.; m. Nancy I. Harms, Sept. 7, 1974. B.A., Westfield State Coll., 1973; Ph.D., U.Minn.-Mpls., 1976. Asst. prof. U. New Orleans, 1977-78; asst. prof. ednl. psychology U. Wis.-Madison, 1978-81, assoc. prof., 1981—; presenter, lectr. in field. Author Minn. Dept. Edn. monographs and articles for profl. educators; author mans.; contbr. numerous articles, chpts. to profl. publs. Title III ESEA grantee, 1974, 75; Minn. Dept. Edn. Pupil Personnel Div. grantee, 1978; Wis. Alumni Research Found. grantee, 1978, 79, 80; Spencer Found. grantee, 1980, 81. Mem. Am. Psychol. Assn., Am. Ednl. Research Assn. Subspecialty: Developmental psychology. Current work: Child and adolescent developmental psychology; social development; social cognitive development. Home: 801 Glenview Dr Madison Wi 53716 Office: Dept Ednl Psychology U Wis 1025 W Johnston St Madison WI 53706

ENZER, NORBERT BEVERLY, psychiatry educator, college dean; b. Milw., Nov. 26, 1930; m.; 3 children. B.A., Yale U., 1952; M.D., McGill U., Montreal, Que., Can., 1956. Intern in pediatrics Med. Ctr., Duke U., 1956-57, resident in pediatrics, 1957-58, 60-61, resident in psychiatry, 1961-64, asst. prof. psychiatry and assoc. pediatrician, 1965-68; from assoc. prof. to prof. psychiatry and pediatrics U. New Orleans, 1968-73, head dept. psychiatry and biobehavioral sci., 1971-73; prof. psychiatry Mich. State U., East Lansing, 1973—, chmn. dept. psychiatry, 1973-81, assoc. dean acad. affairs, 1981—; fellow in child psychiatry Duke U., 1963-65. Mem. Am. Acad. Pediatrics, Am. Psychiat. Assn., Am. Acad. Child Psychiatry, Soc. Research in Child Devel. Subspecialties: Psychiatry; Pediatrics. Office: A118 East Fee Hall Mich State U Coll Human Medicine East Lansing MI 48824

EPP, EDWARD RUDOLPH, medical physicist, educator, hospital administrator; b. Saskatoon, Sask., Can., July 21, 1929; m. B.A., U. Sask., 1950, M.A., 1952; Ph.D. in Physics, McGill U., Montreal. Que., Can., 1955. Asst. in physics NRC Can., 1952-53; physicist radiation physics, dept. radiology Montreal Gen. Hosp., 1955-57; asst. in biophysics Sloan-Kettering div. Cornell U. Med. Coll., 1957-58, assoc., 1958-60, from asst. prof. to prof. biophysics, 1960-74, chmn. dept., 1966-74; asst. Sloan-Kettering Inst. Cancer Research, 1957-60, assoc., 1960-64, assoc. mem., 1964-68, mem. and chief div. physics biology, 1968-74; head div. radiation biophysics, dept. radiation medicine Mass. Gen. Hosp., Boston, 1974—; prof. radiation therapy Harvard U. Med. Sch., Boston, 1974—; cons. to hosps., 1955-57; mem. task group Internat. Commn. Radiol. Units and Measurements, 1965-70; assoc. attending physicist, dept. med. physics Meml. Hosp. for Cancer and Allied Diseases, 1967-74; mem. radiation study sect. NIH, 1971-75; mem. ad hoc com. hot particles Adv. Com. Biol. Effects Ionizing Radiations, Nat. Acad. Sci., 1974-76, mem. com. rev. use ionizing radiations for treatment benign diseases, 1975-77; mem. Clin. Cancer Program Project Rev. Com., Nat. Cancer Inst., 1977-81, Com. Dept. Energy Research Health Effects Ionizing Radiation, Nat. Acad. Sci., 1978-79, Com. Fed. Research Biol. and Health Effects Ionizing Radiation, 1979—. Assoc. editor: Internat. Jour. Radiation Oncology Biol. Physics, 1979—. Mem. AAAS, Am. Phys. Soc., Health Physics Soc., Radiation Research Soc. (sec.-treas. 1981—), Am. Assn. Physicists in Medicine. Subspecialties: Radiology; Medical Physics. Current work: Radiobiology, especially cellular radiobiology; biophysics; health physics; effects of ionizing radiation of ultra-high intensity on living cells. Office: Mass Gen Hosp Div Radiation Biophysics Boston MA 02114

EPP, MELVIN DAVID, plant geneticist, tissue culture specialist; b. Newton, Kans., June 16, 1942; s. John, Jr. and Marie (Harder) E.; m. Sylvia K. Rieger, June 26, 1964; children: David S., J. Terry. B.S., Wheaton (Ill.) Coll., 1964; M.S., U. Conn., 1967; Ph.D., Cornell U., 1971. NIH genetics trainee Cornell U., Ithaca, N.Y., 1967; Hort. trainee Pan Am. Seed Co., Paonia, Colo., 1964-65; Damon Runyon fellow Brookhaven Nat. Lab., Upton, N.Y., 1972-74; sr. research biologist Monsanto Co., St. Louis, 1974-77; research supt. Philippine Packing Corp. subs. Del Monte Corp., Manila, 1977-82; mgr. plant propagation and tissue culture research parent co., San Leandro, Calif., 1982—. Contbr. articles to sci. jours. Mem. Genetics Soc. Am., Bot. Soc. Am., AAAS. Subspecialties: Plant cell and tissue culture; Genetics and genetic engineering (agriculture). Current work: The application of genetic selection and genetic engineering together with plant cell and tissue culture to develop superior varieties of temperate and tropical fruits and vegetables. Office: 850 Thornton St San Leandro CA 94577

EPPERLY, WILLIAM ROBERT, energy company executive; b. Christiansburg, Va., Mar. 17, 1935; s. William Rangeley and Myrtle Claire (Vest) E.; m. Sarah Ann Owen, June 9, 1957; children: William Robert, Jennifer Ann, Thomas. B.S., Va. Poly. Inst., 1956, M.S., 1958. With Exxon Research & Engring. Co., and parent co., 1957—; mgr. Baytown (Tex.) reseach research and devel. div., 1973-76, mgr. project devel. and planning, Florham Park, N.J., 1976-77, gen. mgr. liquefaction, 1977-79, gen. mgr. synthetic fuels dept., 1980-83, sr. program mgr., 1983—; mem. air pollution research adv. com. Coordinating Research Council, 1969-71; mem. fossil energy program adv. com. Oak Ridge Nat. Lab., 1978-81; mem. com. synthetic fuels safety NRC, 1982, mem. com. on coop. govt. industry research, 1983. Author. Mem. Am. Inst. Chem. Engrs. (award for chem. engring. practice 1983), Am. Petroleum Inst., AAAS. Methodist. Subspecialties: Fuels; Chemical engineering. Current work: Liquid and gaseous fuels from coal and oil shale; modern management systems for complex projects; enhancement of productivity and creativity. Patentee in synthetic fuels, automotive emissions/gasoline composition, iron ore reduction, fuel cells, others. Home: 18 Gloucester Rd Summit NJ 07901 Office: PO Box 101 Florham Park NJ 07932

EPPERSON, JAMES FELTS, mathematician, educator; b. Detroit, July 13, 1953; s. John Wallace Walker E. and Anne (Felts) Forest. B.S., U. Mich., Ann Arbor, 1975; M.S., Carnegie-Mellon U., Pitts., 1978, Ph.D., 1980. Asst. prof. U. Ga., Athens, 1980—. Actor, technician, Town and Gown Theatre, Athens, 1980—. Mem. Soc. Indsl. and Applied Math., Am. Math Soc. Subspecialties: Numerical analysis; Applied mathematics. Current work: Numerical solution of partial differential equations. Office: Dept Math U Ga Athens GA 30602

EPPSTEIN, DEBORAH ANNE, biochemical researcher; b. Kalamazoo, Mich., Oct. 16, 1948; s. Samuel Hillel and Dorothy Jean (Dodd) E.; m. Jim D. Allen, Dec. 31, 1975. A.B. with honors, Grinnell Coll., 1970; Ph.D. in Biochemistry, U. Ark., 1975. Research assoc. plant pathology U. Ark., 1974-75; research assoc. dept. biol. sci. U. Calif.-Santa Barbara, 1976-78; sect. leader biochemistry Inst. Bio-Organic Chemistry Syntex, Palo Alto, Calif., 1978—. Contbr. in field. NIH fellow, 1976-78. Mem. Am. Chem. Soc., Am. Soc. Microbiology, AAAS, N.Y. Acad. Sci. Subspecialties: Cell and tissue culture; Virology (biology). Current work: Biochemical mechanism of action studies (antiviral, (anticancer) , drug delivery systems. Office: 3401 Hillview Ave Palo Alto CA 94304

EPSTEIN, ALAN LEE, cancer biologist, educator, consultant; b. Bklyn., Aug. 14, 1949; s. Arthur Victor and Shirley (Blatt) E.; m. Lindsay Diane Mount, Dec. 19, 1977; children: Aaron Jacob, Seth David. B.A., Wesleyan U., Middletown, Conn., 1971; M.D., Stanford U., 1978, Ph.D., 1978. Postdoctoral fellow Eleanor Roosevelt Inst. Cancer Research, Denver, 1978-80; asst. prof. medicine Northwestern U., 1980—; cons. Techniclone Internat., Santa Ana, Calif. Hubert H. Humphrey fellow of Damon Runyon-Walter Winchell Fellowship Fund, 1979-80; Leukemia Research Found. grantee, 1980-81; Milheim Found. grantee, 1979-80; Nat. Cancer Inst. grantee, 1980—; recipient Jr. Faculty Research award Am. Cancer Soc., 1980-83; Searle scholar, 1982-83. Mem. Am. Soc. Cell Biology, Am. Soc. Hematology, Am. Assn. Cancer Research, Am. Soc. Clin. Oncology, N.Y. Acad. Sci. Democrat. Jewish. Subspecialties: Cancer research (medicine); Cell study oncology. Current work: The study of the biology of the human malignant lymphomas and leukemias; in particular, initiation of tumor cell lines and production of monoclonal antibodies to tumor-associated antigens. Home: 812 Colfax St Evanston IL 60201 Office: Medical Oncology Section Northwestern U 303 E Chicago Ave Chicago IL 60611

EPSTEIN, EDWARD S., meteorologist; b. N.Y.C., Apr. 29, 1931; s. Herman and Julia E.; m. Alice Katzenstein, June 6, 1954; children: Debra, Harry, Nancy, William. A.B., Harvard U., 1951; M.B.A., Columbia U., 1953; M.S., Pa. State U., 1954, Ph.D., 1960. Lectr. U. Mich., 1959-61, asst. prof., 1961-63, asso. prof., 1964-68, prof., 1969-73, chmn. dept. atmospheric and oceanic sci., 1971-73; asso. adminstr. for environ. monitoring and predictions NOAA, 1973-77, acting asst. adminstr. for research and devel., 1977-78, dir. Nat Climate Program Office, Rockville, Md., 1978-81, chief Climate and Earth Scis. Lab., 1981—, acting dir. research and applications Nat. Environ. Satellite, Data and Info. Services, 1982—; bd. dirs. Univ. Corp. for Atmospheric Research, 1969-73. Editor: Jour. Applied Meteorology, 1971-73; contbr. articles to profl. jours. Served with USAF, 1953-57. Fellow Am. Meteorol. Soc. (councillor 1974-77), AAAS (chmn. sect. hydrospheric scis. 1980); mem. Am. Geophys. Union. Jewish. Subspecialty: Meteorology. Home: 8216 Inverness Hollow Terr Potomac MD 20854 Office: Fed Office Bldg 4 Suitland MD 20233

EPSTEIN, EMANUEL, plant physiologist; b. Duisburg, Germany, Nov. 5, 1916; came to U.S., 1938, naturalized, 1946; s. Harry and Bertha (Lowe) E.; m. Hazel M. Leask, Nov. 26, 1943; children: Jared H. (dec.), Jonathan H. B.S., U. Calif.-Davis, 1940, M.S., 1941; Ph.D., U. Calif.-Berkeley, 1950. Plant physiologist Dept. Agr., Beltsville, Md., 1950-58; lectr., assoc. plant physiologist U. Calif., Davis, 1958-65, prof. plant nutrition, plant physiologist, 1965—, prof. botany, 1974—; cons. to govt. and pvt. agys. Author: Mineral Nutrition of Plants: Principles and Perspectives, 1972; editorial bd.: Plant Physiology, 1962-71, 76—, CRC Handbook Series in Nutrition and Food, 1975—, The Biosaline Concept: An Approach to the Utilization of Underexploited Resources, 1978—, Plant Sci. Letters, 1981—, Advances in Plant Nutrition, 1981—. Served with U.S. Army, 1943-46. Recipient Gold medal Pisa (Italy) U., 1962; Guggenheim fellow, 1958; Fulbright sr. research scholar, 1965-66, 74-75. Fellow AAAS; mem. Nat. Acad. Scis., Am. Soc. Plant Physiologists, Scandinavian Soc. Plant Physiology, Australian Soc. Plant Physiologists, Am. Inst. Biol. Scis., Crop Sci. Soc. Am., Am. Soc. Agronomy, Common Cause, Save-the-Redwoods League, Sierra Club, Sigma Xi. Club: U. Calif. at Davis Faculty. Subspecialties: Plant physiology (agriculture); Plant genetics. Current work: Mineral nutrition of plants, salt relatives of plants, salt tolerant crops. Research, publs. on ion transport in plants, mineral nutrition and salt relations of plants, salt tolerant crops. Office: Land Air and Water Resources U Calif Davis CA 95616

EPSTEIN, HENRY FREDRIC, molecular biologist, educator; b. N.Y.C., Oct. 13, 1944; s. Kenneth Akiba and Esther (Glassner) E.; m. Maxine L. Rosenberg; children: Shana Ellen, Elizabeth Anne, Daniel Zachary, Adam Sherman. A.B., Columbia U., 1964; M.D., Stanford U., 1968. Fellow Stanford U., 1968-69, asst. prof., 1973-78; staff assoc. NIH, Bethesda, Md., 1969-71; fellow Nat. Research Ctr. Lab. Molecular Biology, Cambridge, Eng., 1971-73; assoc. prof. Baylor Coll. Medicine, Houston, 1978-81, prof., 1981—; dir. Jerry Lewis Neuromuscular Dystrophy Research Ctr.; mem. rev. com. Nat. Inst. Aging, 1981; sci. adv. com. Muscular Dystrophy Assn., 1974-83. Editor: Genetic Analysis of the X Chromosome, 1982, Muscle Development: A Molecular and Cellular Approach, 1983. Recipient alumni award Stanford Med. Sch., 1968; Borden award, 1969. Mem. AAAS, Soc. Neurosci., Am. Soc. Biol. Chemists, Am. Soc. Cell Biology, Am. Chem. Soc. Jewish. Subspecialties: Cell biology (medicine); Genetics and genetic engineering (medicine). Current work: Genetic and molecular analysis of muscle development and myofibrillar organization. Home: 5426 Ariel Houston TX 77096 Office: Baylor Coll Medicine 1 Baylor Plaza Houston TX 77030

EPSTEIN, IRVING ROBERT, chemistry educator; b. Bklyn., Aug. 9, 1945; s. Milton and Marion (Hillsberg) E.; m. Ellen Fisher, Oct. 31, 1971; children: David, Peter. A.B. (Merit scholar), Harvard U., 1966, M.A. (Woodrow Wilson fellow), 1968, Ph.D., 1971; diploma (Marshall scholar), Oxford U., 1967. NATO postdoctoral fellow Cambridge U., Eng., 1971; Asst. prof. chemistry Brandeis U., Waltham, Mass., 1971-75, assoc. prof., 1975-81, prof., 1981—, chmn. dept. chemistry, 1983—. Contbr. numerous articles on chemistry to profl. jours. NSF fellow Max Planck Inst., Gottingen, W.Ger., 1977-78; Guggenheim fellow, 1977; Humboldt fellow, 1977; recipient Dreyfus award, 1973. Mem. Am. Chem. Soc. (Liebmann award 1962), Phi Beta Kappa. Subspecialties: Kinetics; Biophysical chemistry. Current work: Design and analysis of oscillating chemical reactions and dynamic instabilities in chemical systems, mathematical modeling of biochemical kinetic processes. Home: 28 Otis St Newton MA 02160 Office: Dept Chemistry Brandeis U Waltham MA 02254

EPSTEIN, JOSHUA, cancer research scientist; b. Petah-Tikvah, Israel, May 10, 1940; came to U.S., 1972, naturalized, 1980; s. Walter and Dina (Zeldowicz) E.; m. Anna L. Marks; children: Yifat, Ayelet, Karen H. B.Sc., Bar-Ilan U., Israel, 1967, M.Sc., 1969; D.Sc., Technion, Haifa, Israel, 1972. Research fellow pediatrics Mass. Gen. Hosp., Boston, 1972-75; instr., asst. prof. pediatrics Johns Hopkins U. Sch. Medicine, Balt., 1974-77; asst. prof. Ind. U. Sch. Medicine, Indpls., 1977-78; cancer research scientist IV, leukemia service, dept. med. oncology Roswell Park Meml. Inst., Buffalo, 1978—. Contbr. articles to profl. jours. Served with Israeli Army, 1959-61; Res., to 1980. Recipient Encouragement prize Israeli Gerontol. Soc., 1972; NIH research fellow, 1974-76; Am. Cancer Soc. research grantee, 1979-81. Mem. Am. Assn. Cancer Research, Am. Soc. Clin. Oncology. Jewish. Subspecialties: Cell study oncology; Cell and tissue culture. Current work: Developing predictive assays for response and cure of malignant diseases. Office: 666 Elm St Buffalo NY 14263

EPSTEIN, LOIS BARTH, physician, cancer researcher, pediatric immunologist; b. Cambridge, Mass., Dec. 29, 1933; d. Benjamin and Mary Frances (Perlmutter) Barth; m. Charles Joseph Epstein, June 10, 1956; children: David Alexander, Jonathan Akiba, Paul Michael, Joanna Marguerite. A.B. cum laude, Radcliffe Coll., 1955; M.D., Harvard U., 1959. Resident in pathology Peter Bent Brigham Hosp., Boston, 1959-60; intern in medicine New Eng. Center Hosp., Boston, 1960-61; research med. officer Nat. Inst. Arthritis and Metabolic Diseases, 1962-63, Nat. Inst. Allergy, Immunology and Infectious Diseases, 1966-67; spl. NIH fellow, 1964-66; mem. faculty U. Calif. Med. Center, San Francisco, 1969—, asso. dir., 1972-74, prof. pediatrics, 1980—; cons. in field, mem. Study Sects. NIH, 1972-81. Author numerous articles in field.; Mem. editorial bd.: Jour. Interferon Research, Jour. Clin. Immunology, Jour. Soviet Oncology. Bd. dirs. Marin Symphony Assn., 1979—, Marin Dance Assn., 1980—; fundraising chmn. Marin Youth Orch., Vienna tour, 1979; mem. Israel tour com. Bd. Jewish Edn., 1976-79. Mem. Am. Soc. Clin. Investigation, Am. Assn. Immunologists, Soc. Pediatric Research, Am. Assn. Cancer Research, Am. Soc. Hematology, Assn. Women in Sci., Tissue Culture Assn., Western Assn. Physicians, Phi Beta Kappa. Subspecialties: Cancer research (medicine); Immunology (medicine). Current work: Research on prodn., structure, antitumnor, immunomodulatory and antiviral actions and genetic control in interferons, clin. investigations primary and secondary immunodeficiency diseases, role interferon in their pathogenesis and/or correction. Home: 19 Noche Vista Ln Tuburon CA 94920 Office: Cancer Research Inst U Calif Med Center San Francisco CA

EPSTEIN, MURRAY, medical educator, researcher; b. Tel Aviv, Aug. 11, 1937; U.S., 1940; s. Louis and Susanna (Mendelowitz) E.; m. Anna Kristina Nilsson, June 25, 1978; children: David, Susanna. B.A., Columbia U., 1959, M.D., 1963. Diplomate: Am. Bd. Internal Medicine, Am. Bd. Nephrology. Intern U. Wis., 1963-64, resident in internal medicine, 1964-65, Cleve. Met. Gen. Hosp., 1965-66; med. investigator USAF Sch. Aerospace Medicine, San Antonio, 1968-70; asst. prof. medicine U. Miami (Fla.), 1970-74; assoc. prof. medicine (Fla.), 1974-78, prof. medicine, 1978—; cons. Nat. Acad. Scis., Washington, 1974—; NASA Johnson Space Ctr., Houston, 1972—. Author: Hypertension, A Practical Approach, 1983; Editor: The Kidney in Liver Disease, 1978, 83; editorial bd.: Am. Jour. Kidney Diseases. Served as maj. USAF, 1968-70. Fellow ACP; mem. Am. Soc. Nephrology, Am. Soc. Clin. Investigation, So. Soc. Clin. Investigation, Am. Physiol. Soc., Soc. Exptl. Biology and Medicine. Subspecialties: Nephrology; Internal medicine. Office: Nephrology Sect VA Med Ctr 1201 NW 16th St Miami FL 33125

EPSTEIN, SEYMOUR, psychology educator; b. Bklyn., July 15, 1924; s. Jacob and Dina (Bugorad) E.; m. Alice H. Hopper, Nov. 23, 1950; children: Lisa B., Martha L. B.A., Bklyn. Coll., 1948; M.A., U. Wis.-Madison, 1951, Ph.D., 1953. Diplomate: Am. Bd. Clin. Psychology. Trainee in clin. psychology VA, Madison, Wis., 1948-52; pvt. practice psychology, Amherst, Mass., 1953-69; asst. prof. U. Mass., Amherst, 1953-58, assoc. prof., 1958-62, prof. psychology, 1962—, head personality program, 1969—; research cons., bd. mem. NIMH, NSF, Can. Research Council, Australian Research Council, 1965—. Editorial bd.: Jour. Cons. and Clin. Psychology, 1973-77, Psychophysiology, 1972-75, Jour. Personality, 1975—, Jour. Exptl. Research in Personality, 1976—; contbr. articles to profl. jours. Served with U.S. Army, 1943-46; ETO. Recipient Sigma Xi award U. Mass., 1962; Clarke Lecky Meml. award Clarke Assocs., 1981; Chancelor's lectr. U. Mass., 1981; NIMH research grantee, 1955—. Fellow Am. Psychol. Assn.; mem. Eastern Psychol. Assn., Phi Kappa Phi. Democrat. Subspecialty: Personality. Current work: The development of unified theory of personality that integrates the views of other theories, the improvement of research methodology in personality. Home: 37 Bay Rd Amherst MA 01002 Office: Psychology Dept U Mass Amherst MA 01003

EPSTEIN, SHELDON LEE, electrical engineer; b. Chgo., July 16, 1938; s. Morris Louis and Mae (Levin) E.; m. Suzanne Buch Latt, June 18, 1964; children: Samuel Latt, Elizabeth Anne. M.B.A., U. Chgo., 1978; J.D., Columbia U., 1964; B.S.E.E., M.I.T., 1961, 1961. Assoc. Silverman & Cass, Chgo., 1964-66; project mgr. Booz Allen, Chgo., 1967-68; new bus. evaluation Brunswick Corp., Chgo., 1968-72, sr. licensing atty., 1972-78; owner Epstein Assocs., Northbrook, Ill., 1978—. Mem. IEEE, Assn. Computing Machinery, Instrument Soc. Am., Soc. Computer Simulation, Soc. Info. Display. Republican. Jewish. Subspecialties: Computer engineering; Graphics, image processing, and pattern recognition. Current work: Microprocessor based custom design and service engr.; particularly process control, instruments, data transmission and computer graphics. Home: PO Box 400 Wilmette IL 60091 Office: 3657 Woodhead Dr Northbrook IL 60062

EPSTEIN, WILLIAM, experimental psychologist; b. N.Y.C., Nov. 23, 1931; s. Jacob and Sarah (Kaplan) E.; m. Arlene Rita Cohen, Mar. 25, 1956; children: Sarah Ann, Edith Lynn. B.A., NYU, 1955; M.A., New Sch. Social Research, 1957, Ph.D., 1959. Asst. prof. psychology U. Kans., 1959-68, asso. prof., 1962-65, prof., 1965-68; prof. psychology U. Wis.-Madison, 1968—, chmn. dept., 1975-79; vis. prof. Cambridge (Eng.) U., 1972-73; Fulbright research fellow and vis. prof. Delhi (India) U., 1981-82. Author: Varieties of Perceptual Learning, 1967, (with F.C. Shontz) Psychology in Progress, 1971, Stability and Constancy in Visual Perception, 1977; Cons. editor: Perception and Psychophysics, 1971-82; editor: Jour. Exptl. Psychology: Human Perception and Performance, 1982—. NSF sr. postdoctoral fellow U. Uppsala, Sweden, 1966-67; NIMH grantee, 1959—. Mem. Am. Psychol. Assn., AAAS, Psychonomic Soc., Sigma Xi. Office: Psychology Bldg Univ of Wis Madison WI 53706

EPSTEIN, WOLFGANG, biochemist, educator; b. Breslau, Germany, May 7, 1931; came to U.S., 1936, naturalized, 1943; s. Stephan and Elsbeth (Lauinger) E.; m. Edna Selan, June 12, 1961; children: Matthew, Ezra, Tanya. B.A. with high honors, Swarthmore Coll., 1951; M.D., U. Minn., 1955. Postdoctoral fellow in physiology U. Minn., Mpls., 1959-60; postdoctoral fellow Pasteur Inst., Paris., 1963-65; postdoctoral fellow in biophysics Harvard Med. Sch., 1961-63, research asso., then asso. in biophysics, 1965-67; asst. prof. biochemistry U. Chgo., 1967-73, asso. prof., 1973-79, prof., 1979—, chmn. com. on genetics, 1976—. Served with M.C. U.S. Army, 1957-59. Mem. Am. Soc. Biol. Chemists, Am. Soc. for Microbiology, Biophys. Soc., AAAS. Subspecialties: Membrane biology; Genetics and genetic engineering (biology). Current work: Genetic and biochemical investigation of: bacterialion transport systems; regulatory mechanisms. Home: 1120 E 50th St Chicago IL 60615 Office: 920 E 58th St Chicago IL 60637

ERB, HOLLIS NANCY, veterinarian, educator; b. Oakland, Calif., July 2, 1948; d. Russel Garis and Arline May (Coe) E. B.S., U. Calif.-Davis, 1972, D.V.M., 1974; M.S., U. Guelph, Ont., Can., 1976, Ph.D., 1979. Asst. prof. animal health epidemiology Cornell U., Ithaca, N.Y., 1979—. Mem. Am. Dairy Sci. Assn., AVMA, Assn. Tchrs. Vet. Pub. Health and Preventive Medicine, Soc. for Epidemiol. Research, Soc. for Med. Decision Making. Subspecialties: Preventive medicine (veterinary medicine); Integrated systems modelling and engineering. Current work: Epidemiology and economics of dairy cattle heath; epidemiology, cost benefit analysis, herd health, preventive medicine, medical decision making, dairy cattle. Home: 118 Snyder Hill Rd Ithaca NY 14850 Office: NY State Coll Veterinary Medicine Cornell U Ithaca NY 14853

ERBER, JOAN T., psychology researcher and educator; b. Rochester, N.Y., June 21, 1943; s. Milton and Harriet (Frank) Tatelbaum; children: Stephanie, Melanie. B.A., Washington U., St. Louis, 1965; M.S., St. Louis U., 1969, Ph.D. in Exptl./Devel. Psychology, 1971. Research asst. dept. vital stats. St. Louis County (Mo.) Health Dept., 1966-67; research assoc. Social Sci. Inst., Washington U., 1971, postdoctoral fellow in aging and devel., 1972-74, research assoc. dept. psychology aging and devel. program, 1974-75, 77-82; lectr. in psychology Univ. Coll., 1974-77; health care edn. specialist, geriatric research Edn. and Clin. Center, VA Med. Center, St. Louis, 1982; assoc. prof. psychology Fla. Internat. U., 1982—; vis. asst. prof. psychology U. Mo.-St. Louis, 1971-72, adj. asst. prof., spring 1977; research proposal reviewer NSF Social and Devel. Program, Div. Behavioral and Neural Scis., 1977. Book reviewer, C. V. Mosby Co., 1979; book reviewer, Sci. Research Assocs., Inc.; Contbr. articles, revs. to profl. publs., papers to meetings. NDEA Title IV fellow, 1967-70. Fellow Gerontol. Soc. Am.; mem. Am. Psychol. Assn., Phi Beta Kappa, Sigma Xi, Psi Chi. Subspecialties: Cognition; Developmental psychology. Home: 7887 SW 105 Pl Miami FL 33173 Office: Fla Internat U Tamiami Campus Miami FL 33199

ERDOGAN, HAYDAR, environmental engineering educator; b. Tunceli, Turkey, Mar. 18, 1951; came to U.S., 1972, naturalized, 1980; s. Hasan and Cicek E. B.S., U. Ankara, 1972; M.S., Mich. State U., 1976; Ph.D., U. Pitts., 1981. Researcher Pitts. Energy and Environ. Systems Inc., 1977; teaching, research fellow U. Pitts., 1978-81, cons. to model bioreactors for activated sludge treatment, 1981-82; asst. prof. environ. engring. Rutgers U., Piscataway, N.J., 1982—; cons. Nat. Steel Co., Pitts., 1980. Contbr. articles to profl. lit. Turkish Ministry Edn. scholar, 1964-72; State Scholarship Found. Turkey scholar, 1972-77. Mem. Am. Inst. Chem. Engring., AAUP, Assn. Environ. Engring. Profs., Turkish Inst. Chem. Engring., Can. Inst. Chem. Engring. Subspecialties: Water supply and wastewater treatment; Chemical engineering. Current work: Modeling of mass transfer through porous media, disposal of solid liquid hazardous waste, detoxification of contaminated soils, physical-chemical and biological treatment of water and waste water, energy production from biomass. Office: Dept Engring Rutgers U PO Box 909 Piscataway NJ 08854

ERECINSKA, MARIA, pharmacology educator, biochemist, biophysicist; b. Warsaw, Poland, Aug. 17, 1939; came to U.S., 1969; d. Kazimierz and Matylda (Branczewska) E. M.D. summa cum laude, Med. Sch., Gdansk, Poland, 1961; Ph.D., Polish. Acad. Sci., Warsaw, 1967. Research assoc. Inst. Biochemistry Biophysics, Polish Acad. Sci., Warsaw, 1964-67, asst. prof., 1967-69; postdoctoral asst. Johnson Research Found., U. Pa., Phila., 1969-71, asst. prof. dept. biochemistry, 1971-78, assoc. prof. dept. pharmacology, 1978—. Editor: 02 Transport to Tissue, 1978, Inhibitors of Mitochondrial Function, 1981. CNRS travel award, France, 1968; PennPlan scholar, 1970; recipient Merck award, 1971; Am. Heart Assn. investigator, 1979. Mem. Am. Soc. Biol. Chemistry. Subspecialties: Biochemistry (medicine); Neurochemistry. Current work: Regulation of cellular energy production, structure and function of mitochondrial respiratory chain, molecular basis of hypoxic and ischemic brain damage, molecular basis of ion transport. Office: U Pa Dept Pharmacology Philadelphia PA 19104

ERF, ROBERT K., physicist; b. Bellevue, Ohio, Oct. 29, 1931; s. Herbert A. and Frances E. (Knapp) E.; m. Mary Elva Congleton, July 3, 1954; children: Keith, Karen, Kate, Frank. B.S.E., U. Mich., 1953; M.S. in Applied Physics, Harvard U., 1954. With United Technologies Corp., East Hartford, Conn., 1954—, chief optics and acoustics, 1975—. Author: Holographic Nondestructive Testing, 1974, Speckle Metrology, 1978. Bd. mgrs. Glastonbury YMCA, 1976—, chmn., 1980-82. Mem. Optical Soc. Am., Soc. Photo-optical Instrumentation Engrs. Republican. Mem. United Ch. of Christ. Subspecialties: Holography; Acoustics. Current work: Laser applications for nondestructive testing, measurement, machining, holography, and speckle; ultrasonics, acoustic emission. Patentee in field. Home: 127 Carriage Dr Glastonbury CT 06033 Office: Silver Ln MS 86 East Hartford CT 06108

ERICKSEN, GEORGE EDWARD, geologist; b. Butte, Mont., Mar. 17, 1920; s. Edward Martin and Ruby Clara (Johnson) E.; m. Mary Frances Kelly, Dec. 21, 1948. A.B., U. Mont., 1946; M.A., Ind. U., 1949; Ph.D., Columbia U., 1954. Geologist U.S. Geol. Survey, Missoula, Mont., 1942-45; instr. Ind. U., Bloomington, 1947-49; Geologist U.S. Geol. Survey, Lima, Peru, 1949-52, Santiago, Chile, 1954-62, Washington, 1962-73, Reston, Va., 1973—. Contbr. articles to profl. jours.; editor: Mineral Resources of the Appalachian Region, 1968. Recipient Meritorious Service award Dept. Interior, 1973, Disting. Service award Govt. Peru, 1982, Bernardo O'Higgins medal Govt. Chile, 1983. Fellow Geol. Soc. Am., Mineral. Soc. Am.; mem. Soc. Econ. Geologists, Geochem. Soc., Am. Geophys. Union, Soc. Mining Engrs., Geol. Soc. Peru, Inst. Mining Engrs. Chile. Club: Cosmos (Washington). Subspecialties: Geochemistry; Mineralogy. Current work: Investigation of saline deposits of the central Andes; metalliferous deposits of western Mont., tin deposits of Bolivia; study of origin and mineralogy of Chilean nitrate deposits. Office: US Geol Survey Nat Center MS 954 Reston VA 22092 Home: 11903-PH3 Winterthur Ln Reston VA 22091

ERICKSEN, JERALD LAVERNE, educator, physicist; b. Portland, Oreg., Dec. 20, 1924; s. Adolph and Ethel Rebecca (Correy) E.; m. Marion Ella Pook, Feb. 24, 1946; children: Lynn Christine, Randolph Peder. B.S., U. Wash., 1947; M.A., Oreg. State Coll., 1949; Ph.D., Ind. U., 1951. Mathematician, solid state physicist U.S. Naval Research Lab., 1951-57; faculty Johns Hopkins U., 1957-83, prof. theoretical mechanics, 1960-83, U. Minn., Mpls., 1983—. Editorial adv. bd.: Internat. Jour. Solids and Structures; editorial bd.: Jour. Elasticity. Served with USNR, 1943-46. Recipient Bingham medal, 1968, Timoshenko medal, 1979. Mem. Soc. Rheology, Soc. Natural Philosophy, Soc. Interaction Mechanics and Math. Subspecialties: Theoretical and applied mechanics; Theories of crystals. Current work: Transition phenomena in crystals and liquid crystals. Home: 10 Poplar Ln North Oaks MN 55110 Office: Mechanics Dept U Minn Minneapolis MN 55455

ERICKSON, CARL JOHN, psychology educator, researcher; b. Fitchburg, Mass., Aug. 25, 1937; s. Carl Hjalmar and Doris Esther (Kilpatrick) E.; m. Bonnie Elise Daulton, Aug. 25, 1962; 1 dau., Christine. A.B., Clark U., 1959; postgrad., Harvard U., 1960-61; Ph.D., Rutgers U., 1961-65, U. Groningen, Netherlands, 1965-66. Asst. prof. psychology Duke U., 1966-70, assoc. prof., 1970-76, prof., 1976—; chmn. small grants com. NIMH, 1977-80. Recipient Outstanding Tchr. award Students of Duke U., 1969. Mem. AAAS, Animal Behavior Soc., Sigma Xi. Democrat. Subspecialty: Psychobiology. Current work: Role of hormones in animal social behavior; teaching courses in biological bases of behavior; research concerns effects of social interaction on hormone secretion. Home: 5224 Inverness Dr Durham NC 27712 Office: Dept Psychology Duke U Durham NC 27706

ERICKSON, ERIC HERMAN, JR., research entomologist, educator; b. Denver, Apr. 26, 1940; s. Eric Herman and Emma Rocelia (Quist) E.; m. Ruth Charlene Ashford, Sept. 1, 1962 (div. Sept. 1982); children: Eric Herman, III, Jeffrey Paul; m. Barbara J. Hanny, Apr. 13, 1983. B.S., Colo. State U., 1963, M.S., 1965; Ph.D., U. Ariz., 1970. Registered profl. entomologist. Research entomologist U.S. Dept. Agr. Agrl. Research Service, North Central Region Bee Research Unit U. Wis., Madison, 1970-78; supervisory research entomologist, 1978—.

Served to 1st lt. U.S. Army, 1965-67. Mem. Entomol. Soc. Am., Internat. Bee Research Assn. (U.S. Dept. Agr. rep. to governing council 1972—), Am. Soc. Agronomy, Internat. Commn. for Bee Botany, Eastern Apicultural Soc., Western Apicultural Soc., Sigma Xi, Gamma Sigma Delta. Subspecialties: Apiculture; Crop Pollination. Current work: Honey bee biology and behavior, crop pollination, floral biology. Home: 2206 McKenna Blvd Madison WI 53711 Office: USDA/ARS/NCR Bee Research Unit University of Wisconsin 436 Russell Lab Entomology Madison WI 53706

ERICKSON, HOWARD HUGH, veterinary medicine educator, researcher; b. Wahoo, Nebr., Mar. 16, 1936; s. Conrad R.N. and Laurene (Swanson) E.; m. Ann E. Nicolay, June 6, 1959; children: James, David. B.S., Kans. State U., Manhattan, 1959, D.V.M., 1959; Ph.D., Iowa State U.-Ames, 1966. Commd. lt. U.S. Air Force, 1959, advanced through grades to col., 1979; spl. asst. to dir. (Directorate of Research and Devel.), Brooks AFB, Tex., 1975-76, chief tech. plans and analysis div., 1976-79, chief mech. forces div., 1979-81; prof. physiology dept. anatomy and physiology Coll. Vet. Medicine, Kans. State U., Manhattan, 1981—; clin. asst. prof. lab. animal medicine U. Tex. Health Sci. Ctr., San Antonio, 1972-81; USAF liaison rep. study sect. applied physiology NIH, Bethesda, Md., 1974-81. Decorated Meritorious Service medal; recipient Midland Luth. Coll. Alumni Achievement award, 1977. Fellow AAAS, Aerospace Med. Assn., Royal Soc. Health; mem. AVMA (chmn. research council 1983-84), Am. Physiol. Soc. Republican. Lutheran. Subspecialties: Biomedical engineering; Physiology (medicine). Current work: Physiology of exercise, stress and the cardiopulmonary system in domestic animals. Home: 2017 Arthur Dr Manhattan KS 66502 Office: Coll Vet Medicine Kansas State U Manhattan KS 66506 Home: 2017 Arthur Dr Manhattan KS 66502

ERICKSON, JOHN (ELMER), biology educator; b. Sioux City, Iowa, June 17, 1923; s. Elmer Eric and Edith Pearl June (Snyder) E.; m. Betty Jean Davenport, July 26, 1946; children: Steven J., Kae A. Erickson Wingate, David E.; m. Roberta Blanche Craft, Feb. 28, 1973. Student, Yankton Coll., 1941-43; B.A., Omaha U., (now U. Nebr.-Omaha), 1948; M.A., Ind. U., 1951; Ph.D., U. Oreg., 1964. Instr. McCook Coll., 1954-59; instr. U. Oreg., 1959-62; asst. prof. biology Western Wash. State Coll. (now Western Wash. U.), 1964-66, assoc. prof., 1966—. Contbr. articles to profl. jours. Served to cpl. A.C. U.S. Army, 1943-46. Recipient Danforth Tchr. award Danforth Found, 1962-63; NIH fellow, 1963-64; NSF grantee, 1965-71. Mem. Genetics Soc. Am., Am. Genetic Assn., Rocky Mountain Biol. Lab., AAAS, Am. Biology Tchrs., Physicians for Social Responsibility, Nuclear Arms Freeze, Sigma Xi (pres. chpt. 1981-82). Presbyterian. Clubs: Soc. Preservation and Encouragement of Barbershop Quartet Singing in Am., Birchwood Garden. Lodge: Elks (Bellingham, Wash.). Subspecialties: Genetics and genetic engineering (biology); Genome organization. Current work: Meiotic drive; chromosome behavior at meiosis in Drosophila; writing genetics text under contract. Home: 2028 Huron Bellingham WA 98226 Office: Dept Biology Western Wash U Bellingham WA 98225

ERICKSON, LEONARD CHARLES, cell biologist; b. Waltham, Mass., Dec. 17, 1946; s. Charles Leonard and Priscilla Faith (Wheeler) E.; m. Pamella Prothro, Dec. 15, 1973. B.A., Ottawa (Kans.) U., 1968; M.S., Fla. State U., 1972, Ph.D. in Biology, 1974. Staff fellow Nat. Cancer Inst., NIH, Bethesda, Md., 1974-77, sr. staff fellow, 1977-79, cancer expert, 1979-83, sr. investigator, 1983—. Contbr. articles to profl. jours. Served to lt. (j.g.) USN, 1968-69. Mem. Am. Assn. Cancer Research, Am. Soc. for Cell Biology, Sigma Xi. Clubs: Old Red Rugby Football, Potomac Soc. Rugby Football Referees (Washington). Subspecialties: Cell and tissue culture; Cancer research (medicine). Current work: Mechanisms of action of anti-tumor agents. DNA damage and repair. The origins of differential cytotoxicity of anti-tumor agents to normal and tumor human cells. Office: Bldg 37 Room 5D17 NIH Bethesda MD 20205

ERICKSON, ROBERT ARLEN, engr.; b. Brainerd, Minn., Jan. 12, 1932; s. Benjamin Bernhart and Ruth Amelia (Linn) E.; m. Naomi Marie Hepburn, May 18, 1956; children—Catherine, John, Benjamin. Student, Brainerd Jr. Coll.; B.A. in Physics, U. Minn., 1958. Physicist Sperry Univac Def. Systems div., St. Paul, Minn., 1958-62, sr. physicist, 1962-64, supr. engr., 1964-65, mgr., 1965-66, group mgr., 1966-70, dir. research and devel., 1970-74, dir. engring., 1974-78, v.p. product engring., 1978-80, v.p., gen. mgr. Semicondr. div., 1980—. Contbr. articles to profl. jours. Served with USAF, 1951-55. Mem. Am. Legion, various mgmt. and tech. groups, several conservation orgns. Republican. Subspecialty: Semiconductors. Patentee in field. Home: 2182 Garnet Point St Paul MN 55122 Office: 333 Pilot Knob Rd St Paul MN 55165

ERICKSON, ROBERT PORTER, psychologist, educator; b. South Bend, Ind., Feb. 13, 1930; s. Carl G. and Elinor (Porter) E.; children: Lars, Nils, David. Sc.B., Northwestern U., 1951; Ph.D., Brown U., 1958. Faculty Duke U., Durham, N.C., 1961—, assoc. prof. physiology, 1972—, prof. psychology, 1974—. Contbr. articles to profl. publs. Served to lt. (j.g.) USN, 1951-54; Korea, Mediterranean. Mem. Am. Physiol. Soc., Soc. Neurosci., Sigma Xi. Subspecialties: Neuropsychology; Neurophysiology. Current work: Neural bases of behavior. Home: 564 Rockwood Rd Route 1 Hillsborough NC 27278 Office: Psychology Dept Duke U Durham NC 27706

ERIKSON, RAYMOND LEO, cellular and developmental biology educator, researcher; married. B.S., U. Wis., 1958, M.S., 1961, Ph.D. in Molecular Biology, 1963. USPHS fellow U. Colo., Denver, 1963-65, from asst. prof. to assoc. prof., 1965-72, prof. pathology, 1972-82; prof. cellular and devel. biology Harvard U., 1982—. Am. Cancer Soc. scholar, 1972-73. Mem. Am. Soc. Biol. Chemists, Am. Soc. Microbiology. Subspecialties: Developmental biology; Cell study oncology. Office: Dept Cellular and Devel Biology Harvard U Cambridge MA 02138

ERLENMEYER-KIMLING, L., psychiatric and behavior genetics researcher, educator; b. Princeton, N.J.; d. Floyd M. and Dorothy F. (Dirst) Erlenmeyer; m. Carl F. E. Kimling. B.S. magna cum laude, Columbia U., 1957, Ph.D., 1961. Sr. research scientist N.Y. State Psychiat. Inst., N.Y.C., 1960-69, assoc. research scientist, 1969-75, prin. research scientist, 1975-78, dir. div. devel. behavior studies, 1978—; asst. in psychiatry Columbia U., 1962-66, research assoc., 1966-70, asst. prof., 1970-74, assoc. prof., psychiatry and human genetics, 1974-78, prof., 1978—; vis. prof. psychology New Sch. Social Research, 1971—; mem. peer rev. group NIH, 1976-80; mem. work group on guidance and counseling Congressional Commn. on Huntington's Disease, 1976-77; mem. task force on intervention Pres.'s Commn. on Mental Health, 1977-78; mem. initial rev. group NIMH, 1981-85. Editor: Life-span Research in Psychopathology, 1983; issue editor: Differential Reprodn, 1971, Genetics and Mental Disorders, 1972; editorial bd.: Social Biology, 1970-79, Schizophrenia Bull, 1978—, Jour. Preventive Psychiatry, 1980—. NIMH grantee, 1966-69, 71-85; Scottish Rite Com. on Schizophrenia grantee, 1970-74; W. T. Grant Found. grantee, 1978-84; MacArthur Found. grantee, 1981. Fellow Am. Psychol. Assn.; mem. Am. Soc. Human Genetics, Am. Psychopath. Assn., World Psychiat. Assn. (com. epidemiology and community psychiatry), Behavior Genetics Assn. (mem.-at-large 1972-74), Soc. Study Social Biology (dir. 1969—, sec. 1972-75, pres. 1975-78), Scientists Center for Animal Welfare, Scientists Group for Reform of Animal Experimentation, Phi Beta Kappa, Sigma Xi. Subspecialties: Psychobiology; Psychiatry. Current work: Study of risk and protective factors in mental disorders. Office: NY State Psychiat Inst 722 W 168th St New York NY 10032

ERNST, WALLACE GARY, geology educator; b. St. Louis, Dec. 14, 1931; s. Fredrick A. and Helen Grace (Mahaffey) E.; m. Charlotte Elsa Pfau, Sept. 7, 1956; children: Susan, Warren, Alan, Kevin. B.A., Carleton Coll., 1953; M.S., U. Minn., 1955; Ph.D., Johns Hopkins U., 1959. Geologist U.S. Geol. Survey, Washington, 1955-56; fellow (Geophys. Lab.), Washington, 1956-59; mem. faculty UCLA, 1960—, prof. geology and geophysics, 1968—, chmn. geology dept. (now earth and space scis. dept.), 1970-74, 78-82. Author: Amphiboles, 1968, Earth Materials, 1969, Metamorphism and Plate Tectonic Regimes, 1975, Subduction Zone Metamorphism, 1975, Petrologic Phase Equilibria, 1976, The Geotectonic Development of California, 1981, The Environment of the Deep Sea, 1982. Mem. Nat. Acad. Sci. (chmn. geology sect. 1979-82), AAAS, Am. Geophys. Union, Am. Geol. Inst., Geol. Soc. Am., Geochem. Soc., Mineral. Soc. Am. (recipient award 1969, pres. 1979-80), Mineral. Soc. London. Subspecialties: Tectonics; Geochemistry. Home: 16939 Livorno Dr Pacific Palisades CA 90272 Office: Dept Earth and Space Scis U Calif Los Angeles CA 90024

ERON, LEONARD DAVID, psychologist, educator; b. Newark, N.J., Apr. 22, 1920; s. Joseph I. and Sarah (Hileman) E.; m. Madeline G. Marcus, May 21, 1950; children: Joan Hobson, Don, Barbara. B.S., CCNY, 1941; M.A., Columbia U., 1946; Ph.D., U. Wis.-Madison, 1949. Diplomate: Am. Bd. Profl. Psychology. Asst. prof. psychology Yale U., New Haven, 1948-55; dir. research Rip Van Winkle Found., Hudson, N.Y., 1955-62; prof. psychology U. Iowa, Iowa City, 1962-69; prof. psychology, research prof. social scis. U. Ill., Chgo., 1969—; bd. sci. advisors Psychsystems, Balt., 1980—; mem. psychology edn. rev. com. NIMH, Rockville, Md., 1968—; mem. commn. social scis. and humanities Am. Council Learned Socs. and Polish Acad. Scis., N.Y.C., 1982—. Author: Learning of Aggression, 1971, Growing Up To Be Violent, 1977, Experimental Approach to Projective Techniques, 1965; editor: Jour. Abnormal Psychology, 1973-80. Served to 1st lt. AUS, 1942-45; ETO. Fulbright lectr., Netherlands, 1967-68; recipient Sr. Scientist award Fulbright Found., Australia, 1977. Fellow Am. Psychol. Assn. (Disting. Contbn. to Knowledge award 1980), AAAS, Am. Orthopsychiat. Assn. Subspecialty: Developmental psychology. Current work: Development of aggression in children; solialization of behavior and effect of media. Home: 4250 Marine Dr Chicago IL 60613 Office: Dept Psychology Univ Illinois at Chgo Box 4348 Chicago IL 60680

ERPINO, MICHAEL JAMES, biology educator; b. Schenectady, May 30, 1939; s. Michael and Marian (McCarty) E.; m. Alicia Zawadzki, Dec. 16, 1961; children: Michael, Ann, Susan. B.S., Pa. State U., 1962; M.S., U. Wyo., 1964, Ph.D., 1967. Postdoctoral fellow Cornell U., Ithaca, N.Y., 1967-68; mem. faculty dept. biology Calif. State U.-Chico, 1968—, prof., 1976—. NDEA fellow, 1962-65; NSF grantee, 1971-72. Fellow Internat. Soc. Research on Aggression; mem. AAAS, Am. Ornithologists Union, Am. Soc. Zoologists, Sigma Xi. Democrat. Subspecialties: Reproductive biology; Evolutionary biology. Office: Dept Biol Sci Calif State U Chico CA 95929

ERTURK, ERDOGAN, pathologist, educator; b. Havran/Balikesir, Turkey, Sept. 25, 1930; s. Musa Kazim and Ayse E.; m. Gulten Alp Erturk, Aug. 10, 1953. D.V.M., U. Ankara, 1955, M.S., 1960, Ph.D. in Pathology, 1965. Assoc. prof. dept. pathol. anatomy U. Ankara (Turkey) Sch. Vet. Medicine, 1972-78; adj. prof. Clin. Sci. Ctr., U. Wis. Med. Sch., 1978—. Served with Turkish Army, 1955-57. Mem. Am. Assn. Cancer Research, AAAS, N.Y. Acad. Scis. Subspecialties: Animal pathology; Oncology. Current work: Chemical carcinogenesis and histogenesis of several malignancies including renal, liver and breast cancers. Office: 600 Highland Ave Madison WI 53792

ERULKAR, SOLOMON DAVID, pharmacology educator, researcher; b. Calcutta, India, Aug. 18, 1924; came to U.S., 1958, naturalized, 1966; s. David Solomon and Esther Flora (Killekar) E.; m. Catherine Joyce Bear, Sept. 2, 1950; children: Matthew David, Benjamin. B.A., U. Toronto, Ont., Can., 1948, M.A., 1949; Ph.D., Johns Hopkins U., 1952; D.Phil., Oxford (Eng.) U., 1957. Fellow in physiology Johns Hopkins U., 1951-52, USPHS postdoctoral fellow, 1952-54; dept. demonstrator Oxford U., 1955-58; asst. prof. Temple U., 1959-60; asst. prof. pharmacology U. Pa., 1960-62, assoc. prof., 1962-66, prof., 1966—; cons. to govt., 1973-75; mem. neurology B study sect. Nat. Inst. Neurol. and Communicative Diseases and Stroke, Washington, 1974-78, mem. program project neurology B, 1982—. Recipient Personal award Brit. Med. Research Council, 1955-58; Guggenheim Meml. Found. fellow, Jerusalem, 1974-75; Josiah Mary Found. sr. fellow, Paris, 1981-82. Mem. Am. Physiology Soc., AAAS, N.Y. Acad. Scis., Soc. Neurosci., Internat. Brain Research Orgn., Physiol. Soc. Gt. Britain (assoc.). Democrat. Subspecialties: Neurobiology; Neurophysiology. Current work: Synaptic transmission; hormonal effects on neurons; membrane characteristics; ion channels; spinal cord physiology. Office: Dept Pharmacology U Pa Med Sch 36th and Hamilton Walk Philadelphia PA 19104 Home: 318 Kent Rd Bala Cynwyd PA 19004

ERWIN, TERRY LEE, research entomologist, research administrator; b. St. Helena, Calif., Dec. 1, 1940; s. June L. and (Gebhardt) E. B.A., San Jose State U., 1963, M.A., 1966; Ph.D., U. Alta., Can., 1969. Fellow Harvard U., 1969-70, Lund (Sweden) U., 1970-71; curator, research entomologist Smithsonian Instn., Washington, 1970—; dep. dir. research Nat. Mus. Natural History, 1980—. Author: (with others) Carabid Beetles: Their Evolution, Natural History and Classification, 1979; contbr. articles to profl. jours. Recipient Outstanding Service award Nat. Mus. Natural History, 1976. Fellow Calif. Acad. Sci., Royal Entomol. Soc.; mem. Entomol. Soc. Am., Am. Inst. Biol. Scis. (governing bd.), Am. Soc. Naturalists. Subspecialties: Systematics; Ecology. Current work: Systematics and biogeography of carabid ground beetles and ecology of tropical forest canopies. Office: Smithsonian Institution Washington DC 20560

ESAKI, LEO, physicist; b. Osaka, Japan, Mar. 12, 1925; came to U.S., 1960; s. Soichiro and Niyoko (Ito) E.; m. Masako Araki, Nov. 21, 1959; children—Nina Yvonne, Anna Eileen, Eugene Leo. B.S., U. Tokyo, 1947, Ph.D., 1959. With Sony Corp., Japan, 1956-60; with Thomas J. Watson Research Center, IBM, Yorktown Heights, N.Y., 1960—, IBM fellow, 1967—, mgr. device research, 1965—; dir. IBM-Japan. Recipient Morris N. Liebmann Meml. prize I.E.E.E., 1961; Stuart Ballantine medal Franklin Inst., 1961; Japan Acad. award, 1965; Nobel Prize in physics, 1973; decorated Order of Culture Govt. of Japan, 1974. Fellow Am. Phys. Soc. (councillor-at-large 1971-74), I.E.E.E., Japan Phys. Soc., Am. Vacuum Soc. (dir. 1973-74); mem. Am. Acad. Arts and Scis., Nat. Acad. Scis. (fgn. asso.), Nat. Acad. Engring. (fgn. asso.), Academia Nacional de Ingenieria Mex. (corr.), Japan Acad. Subspecialty: Solid-state physics. Inventor tunnel diode, 1957. Home: 16 Shady Ln Chappaqua NY 10514 Office: Watson Research Center IBM PO Box 218 Yorktown Heights NY 10598

ESCHENBACH, ARTHUR EDWIN, behavioral scientist, human factors engineer, educator; b. N.Y.C., Jan. 31, 1923; s. Karl Godfried and Magdelen (Rupert) E.; m. Patricia Lucille Flowers, Apr. 29, 1942 (dec. 1965); children: Mary Patricia, Karl; m. Marie Louise Perez, June 22, 1968; children: Deborah, Charmion, Roxanne. A.B., Cornell U., 1947; M.A., U. Fla., 1949, Ph.D., 1955. Lic. psychologist, Fla. Commd. 2d lt. U.S. Air Force, 1943, advanced through grades to lt. col., 1966; behavioral scientist and human engr. USAF, NASA, U.S. and Korea, 1951-66; ret., 1967; assoc. prof. psychology Jacksonville U., 1967—, head dept. psychology, 1982—. Contbr. numerous articles to profl. jours. Mem. Am. Psychol. Assn., Fla. Psychol. Assn., Southeastern Psychol. Assn., Psychology Dept. Heads. Subspecialties: Human factors engineering; Physiological psychology. Current work: Human factors engineering in missile, aviation and space programs; teaching and research in experimental psychology. Home: 11440 Starboard Dr Jacksonville FL 32225 Office: Jacksonville U Jacksonville FL 32211

ESEN, ASIM, botany educator; b. Fethiye, Mugla, Turkey, Nov. 10, 1938; s. Hasan and Hasibe E.; m. Gulbun Esen, Feb. 5, 1966; children: S. Evren, Filiz. Diploma, U. Ankara, Turkey, 1965; M.S., U. Calif.-Riverside, 1968, Ph.D., 1971. Research asst. U. Calif.-Riverside, 1969-71, postdoctoral fellow, 1971-75; asst. prof. botany Va. Poly. Inst., Blacksburg, 1975-81, assoc. prof., 1981—. Mem. Am. Chem. Soc., Am. Genetics Assn., AAAS. Subspecialties: Plant genetics; Genetics and genetic engineering (biology). Current work: Biochemistry and genetics of seed storage proteins; purification and characterization of maize prolamins and mRNAs and genes coding for maize prolamins; immunochemistry of zeins. Office: Dept Biology Va Poly Inst Blacksburg VA 24601

ESHELMAN, FRED NEVILLE, clinical research administrator; b. High Point, N.C., July 20, 1948; s. John Alfred and Lossie (Neville) E. B.S., U. N.C., 1972; Pharm.D., U. Cin., 1974. Registered pharmacist, Ohio, Ill., Tenn. Clin. asst. prof. U. Ill. Coll. Pharm., Chgo., 1974-76; asst. dir. Bio Basics, N.Y.C., 1976-77; dir. profl. relations and assoc. dir. clin. research Beecham Lab., Bristol, Tenn., 1977-79; assoc. dir. clin. research Glaxo, Inc., Research Triangle Park, N.C., 1979—. Contbr. articles to profl. jours. Mem. Am. Soc. Clin. Pharmacology and Therapeutics, Am. Fedn. Clin. Research, Am. Coll. Clin. Pharmacology, Am. Coll. Clin. Pharmacy. Republican. Methodist. Subspecialties: Pharmacology; Pharmacokinetics. Current work: New drug development; clinical research. Home: 1 Brumley Pl Chapel Hill NC 27514 Office: Glaxo Inc 3306 Hwy 54 Research Triangle Park NC 27709

ESHER, HENRY JEMIL, microbiologist, laboratory consultant; b. Lebanon, Aug. 28, 1938; s. J. and F. (Khoury) Esber; m. Tina R. Nicassio, July 29, 1967; children: Sonya A., Tanya M. Lab. technologist Leigh Meml. Hosp., Norfolk, Va., 1958-62; bacteriologist-in-charge Portsmouth (Va.) Gen. Hosp., 1963-64; immunologist EG&G Mason Research Inst., Worcester, Mass., 1967-75, sect. chief immunology and microbiology, 1975-76, dir. Lab Immunology and Clin. Services, 1979—, sr. scientist, 1976-78; affiliate Grad. Sch., Anna Maria Coll., Paxton, Mass., 1975—; instr. clin. immunology and med. microbiology Hahnemann Hosp., Worcester, 1969—, Worcester City Hosp., 1979—; instr. immunology Clark U., Worcester, 1969. Contbr. articles to sci. jours. Nat. Cancer Inst. grantee. Mem. Am. Assn. Cancer Research, Am. Assn. Immunologists, Am. Soc. Microbiology, Am. Pub. Health Assn., AAAS, Nat. Registry Microbiologists, Environ. Mutagen Soc., Mutagenesis Assn. New Eng., Am. Assn. Clin. Chemistry. Greek Orthodox. Subspecialties: Microbiology (medicine); Transplantation. Current work: Cancer chemotherapy, immunotherapy, immunosuppression, immune enhancement adjuvant activity, radioimmunoassay, clinical assay development, immunoperoxidase, hybridoma development and characterization. Office: 57 Union St Worcester MA 01608

ESHLEMAN, VON RUSSEL, radar astronomer; b. Darke County, Ohio, Sept. 17, 1924; s. Earl Ellsworth and Lydia Mae (Kneisly) E.; m. Patricia May Middleton, Mar. 6, 1947; children—Mary Angela, Kathleen Carol, Eric Earl, David Middleton. Student, Ohio State U., 1946-47; B.E.E., George Washington U., 1949; M.S., Stanford U., 1950, Ph.D., 1952. Research asso. electronics labs. Stanford U., 1957-57, mem. faculty, 1957—, prof. elec. engring., 1961—, co-dir. center radar astronomy, 1961—, dir. radiosci. lab., 1974—; dir. Watkins-Johnson Co.; cons. NASA, Nat. Acad. Scis., SRI Internat., N.Am. Rockwell, Inc., Nat. Oceanographic and Atmospheric Adminstrn.; dep. dir. Office of Technology Policy and Space Affairs, Dept. State Bur. Internat. Sci. and Technol. Affairs, Washington, 1973-74. Contbr. articles to profl. jours. Served with USNR, 1943-46. Fellow IEEE, AAAS, Royal Astron. Soc. (Britain); mem. Nat. Acad. Engring., Internat. Sci. Radio Union, Am. Astron. Soc., Am. Inst. Aeros. and Astronautics, Internat. Astron. Union, Internat. Aero. Congress, Am. Geophys. Union, Sigma Xi, Sigma Tau. Subspecialties: Planetary science; Remote sensing (geoscience). Pioneer radar astronomy and radio links to spacecraft as techniques for studying moon, sun, planets, meteors, astron. space, man-made satellites; attained 1st radar echoes from sun, 1959. Home: 576 Gerona Rd Stanford CA 94305

ESLINGER, PAUL JOSEPH, neuropsychologist; b. N.Y.C., Feb. 3, 1952; s. Frank Joseph and Rita (Quinn) E. B.S. in Psychology, Philosophy and Chemistry, Fordham U., 1974, Ph.D., Tex. Christian U., 1980. Cert. psychologist, Iowa. Postdoctoral fellow in behavioral neurology U. Iowa Hosps. and Clinics, Iowa City, 1979-82, adj. asst. prof. neurology, assoc. research scientist, 1983—. Contbr. articles to profl. jours. Andrus Found. grantee, 1980-83; Nat. Insts. Neurol. and Communicative Disorders and Stroke grantee, 1981-83. Mem. Am. Psychol. Assn., Internat. Neuropsychol. Soc., Soc. for Neurosci. Subspecialty: Neuropsychology. Current work: The effects of brain lesions on human behavior; psychological and neurological changes in aging; the human olfactory system. Office: Division of Behavioral Neurology University of Iowa Iowa City IA 52242

ESLYN, WALLACE EUGENE, plant pathologist; b. Lawrenceville, Ill., Nov. 13, 1924; s. Nicholas John and Beatrice Marie (Browé) Eiselein; m. Nola Sampayo, Mar. 1, 1947; children: Cynthia, Cole, Wayne, Christine, Lance. B.S., U. Mont., 1950, M.S., 1953; Ph.D., Iowa State U., 1956. With Forest Products Lab., U.S. Forestry Service, Madison, Wis., 1956, 60—, supervisory research plant pathologist, 1969—; research plant pathologist Rocky Mountain Forest & Range Exptl. Sta., Albuquerque, 1956-60; lectr. U. Wis., 1969-83. Contbr. articles in field to profl. jours. Served with USMC, 1943-45, 50-51. Decorated Air medal. Mem. Am. Phytopath. Soc., Mycol. Soc. Am., Internat. Research Group on Wood Preservation, Sigma Xi. Subspecialties: Plant pathology; Microbiology. Current work: Researcher in biodeterioration of wood products. Home: 525 S Segoe Rd Madison WI 53711 Office: Forest Products Lab US Forest Service PO Box 5130 Madison WI 53705

ESPINOZA, LUIS ROLAN, physician, educator; b. Pisco, Peru, July 3, 1943; came to U.S., 1969, naturalized, 1983; s. Luis R. and Luz L. (Bernales) E.; m. Carmen G. Gonzalez, June 14, 1969; children: Luis M., Gabriela M. M.D., Cayetano Heredia U., Lima, Peru, 1969. Diplomate: Am. Bd. Internal Medicine. Intern Jersey City Med. Ctr., 1969-70; resident in medicine Barnes Hosp., St. Louis, 1970-72; fellow in rheumatology Barnes Hosp.-Washington U., St. Louis, 1972-73, McGill U., Montreal, Que., Can., 1973-74, asst. prof. medicine, 1976-78; research fellow Rockefeller U., 1974-76; prof. medicine U. South Fla., 1978—, dir., 1978—; cons. rheumatology James Haley VA Hosp.,

Tampa, Fla., 1978—. Author: Circulating Immune Complexes: Their Clinical Significance, 1983; contbr. numerous articles to profl. publs. Nat. Inst. Arthritis Metabolic Diseases fellow, 1972-73; Can. Arthritis Soc. fellow, 1973-74; USPHS fellow, 1974-75; Arthritis Soc. assoc., 1976-78. Fellow ACP; mem. Am. Rheumatism Assn., Am. Assn. Immunologists, So. Soc. Clin. Investigation. Roman Catholic. Subspecialties: Immunology (medicine); Rheumatology. Current work: Immunogenetic factors in rheumatic disorders; patient care; teaching medical students and house staff, including fellows in rheumatology; and clinical investigation.

ESPOSITO, LARRY WAYNE, planetary scientist; b. Schenectady, Apr. 15, 1951; s. Albert and Beverly Jane (De La Mater) E.; m. Diane Marie McKnight, July 26, 1975; 1 dau., Rhea Marie McKnight. S.B. in Math, M.I.T., 1973; Ph.D. in Astronomy, U. Mass., 1977. Research assoc. Lab. for Atmospheric and Space Physics, U. Colo., Boulder, 1977—, univ. lectr. dept. astro-geophysics, 1979—. Contbr. articles to profl. jours. Mem. Am. Astron. Soc. (div. for planetary sci.), Am. Geophys. Union, Internat. Astron. Union. Methodist. Club: Boulder Go. Subspecialty: Planetary science. Current work: Studies of planetary atmospheres and planetary rings. Spacecraft studies of Venus, Jupiter, Saturn, Uranus. Office: LASP Campus Box 392 Boulder CO 80309

ESPOSITO, RALPH UMBERTO, neuroscientist; b. N.Y.C., Nov. 10, 1948; s. Ralph James and Mary Jane E.; m. Cynthia Pillars, Oct. 31, 1981. Ph.D., Boston U., 1978. Research fellow div. psychiatry Boston U., 1977-79, asst. research prof. psychiatry, 1980-82; research psychologist Lab. of Psychology and Psychopathology NIMH, Bethesda, Md., 1981-82, sr. staff fellow, 1982—. Contbr. articles to profl. jours. NIMH grantee, 1977-79, 78-81; Nat. Inst. Drug Abuse, 1979-82. Mem. Soc. Neuroscience, AAAS, Brit. Brain Research Assn., European Brain and Behavior Soc., Union Concerned Scientists, Sigma Xi. Subspecialties: Neuropharmacology; Neuropsychology. Current work: Neuropsychological studies of reward and attention; relationship of basic brain mechanisms to normal cognition and psychopathology in man. Home: 11065 Seven Hill Ln Potomac MD 20854 Office: NIMH Lab Psychology and Psychopathology Bldg 31 Room 4C-35 9000 Rockville Pike Bethesda MD 20205

ESPOSITO, ROCHELLE EASTON, geneticist, educator; b. N.Y.C., June 28, 1941. B.S., Bklyn. Coll., 1962; Ph.D., U. Wash., 1967. NIH postdoctoral fellow U. Wis.-Madison, 1967-69, asst. prof. collegiate div. biology, 1969-70; asst. prof. dept. biology U. Chgo., 1970-75, assoc. prof., 1975-82, prof. dept. biology, 1982—. Mem. Genetics Soc. Am., Am. Soc. Microbiology., AAAS. Subspecialties: Genome organization; Gene actions. Current work: Genetic recombination, chromosome segregation, gene regulation. Office: Dept Biology U Chicago Chicago IL 60637

ESSENBERG, MARGARET KOTTKE, biochemistry educator; b. Troy, N.Y., Apr. 21, 1943; d. Frank Joseph and Esther Crissey (Hendee) Kottke; m. Richard Charles Essenberg, July 17, 1967; children: Gavin Richard, Carla Jean. A.B., Oberlin Coll., 1965; Ph.D., Brandeis U., 1971. NSF fellow Leicester (U.K.) U., 1971, NIH fellow, 1972-73; research asso. Okla. State U., Stillwater, 1973-75, asst. prof. biochemistry, 1975-81, asso. prof., 1981—. Contbr. articles in field to profl. jours. Herman Frasch Found. grantee, 1977-82; U.S. Dept. Agr. grantee, 1979-81; NSF grantee, 1982—. Mem. Am. Soc. Biol. Chemists, Am. Phytopath. Soc., AAAS, Sigma Xi. Democrat. Roman Catholic. Subspecialties: Biochemistry (medicine); Plant pathology. Current work: Researcher in biochemistry of interactions between plants and plant pathogenic bacteria. Office: Oklahoma State University Dept Biochemistry Stillwater OK 74075

ESSENWANGER, OSKAR MAXIMILIAN KARL, supervisory research physicist, educator; b. Munich, Bavaria, Germany, Aug. 25, 1920; came to U.S., 1956, naturalized, 1969; s. Oskar and Anna E.; m. Katharina D. Dorfer, June 17, 1947. B.S., U. Danzig, Germany, 1941; M.S., U. Vienna, Austria, 1943; Ph.D., U. Wurzburg, Germany, 1950. Instr., meteorologist German Air Force, 1944-45; research meteorologist German Weather Service, 1946-57; project assoc. dept. meteorology U. Wis., 1956; prin. investigator Nat. Weather Records Center, Asheville, N.C., 1957-60; supervisory research physicist, research dir. U.S. Army Missile Command, Huntsville, Ala., 1961—; adj. prof. earth and environ. sci. U. Ala.-Huntsville, 1970—. Author: Applied Statistics in Atmospheric Science, 1976; contbg. author, editor: International Compendium World Survey of Climatology, vol. I, 1982; contbr. numerous articles to profl. jours. Recipient Sci. and Engring. Achievement award Missile Command, Redstone Arsenal, Ala., 1965. Fellow AIAA (assoc., Ala.-Miss. div. Hermann Oberth award 1981); mem. Am. Soc. Quality Control (sr.), Am. Meterol. Soc. (profl.), Ala. Acad. Sci (v.p. 1973), Sigma Xi (v.p. club 1976-77, pres. chpt. 1977-82, Outstanding Researcher 1977). Subspecialties: Climatology; Statistics. Current work: Environmental design criteria for rockets, statistical analysis in climatology, solar energy. Home: 610 Mountain Gap Dr Huntsville AL 35803

ESSER, ALFRED F., scientific educator; b. Lauf/Pegnitz, Germany, Feb. 11, 1940; came to U.S., 1969; s. Wilhelm J. and Gretel A. (Mayer) E.; m. Karin E. Lueben, Feb. 9, 1970. Diploma in Chemistry, J.W. Goethe U., Frankfurt am Main, W.Ger., 1966, Ph.D., 1969. Postdoctoral fellow U. Calif.-Santa Barbara, 1969-71; research assoc. NASA-Ames Research Ctr., Moffett Field, Calif., 1971-73; asst. prof. Calif. State U.-Fullerton, 1973-75; asst. mem. Scripps Clinic and Research Found., La Jolla, Calif., 1975-80, assoc. mem., 1980-81; prof. dept. comparative and exptl. pathology J. Hellis Miller Health Ctr., U. Fla., Gainesville, 1981—, assoc. chmn. for research dept. comparative and exptl. pathology, 1982—, head lab. for structural biology, 1981—. Established investigator Am. Heart Assn., 1976-81. Mem. Am. Soc. Biol. Chemists, Am. Assn. Immunologists, Biophys. Soc. Subspecialties: Membrane biology; Biochemistry (biology). Current work: Biochemistry and biophysics of cell membranes, biochemistry of complement system. Office: J Hellis Miller Health Ctr U Fla Box J145 Gainesville FL 32610

ESSEX, MYRON ELMER, microbiologist, educator; b. Coventry, R.I., Aug. 17, 1939; s. Ruth and (Knight) E.; m. Elizabeth Katherine Jordan, June 19, 1966; children: Holly Anne, Carrie Lisa. B.S., U. R.I., 1962; D.V.M., Mich. State U., 1967, M.S., 1967; Ph.D., U. Calif., 1970; M.S. (hon.), Harvard U., 1978. Vis. scientist Karolinska Inst., Stockholm, Sweden, 1970-72; asst. prof. microbiology Sch. Pub. Health, Harvard U., 1972-76, assoc. prof., 1976-78, prof., chmn. dept. microbiology, 1979—, lectr. in pathology, 1976—; adj. prof. Cornell U., 1980—. Editor: Viruses in Naturally Occurring Cancer, 1980, Feline Leukemia Virus, 1980; Contbr. articles to profl. jours. Trustee Leukemia Soc. Am., 1979—; mem. research com. Mass. chpt. Am. Cancer Soc., 1975—. Recipient Bronze medal Am. Cancer Soc., 1980; Leukemia Soc. Am. scholar, 1972-77; research grantee Nat. Cancer Inst., Am. Cancer Soc. Mem. Am. Assn. Cancer Research, Am. Assn. Immunologists, Am. Soc. Microbiology, Reticuloendothelial Soc., Internat. Assn. for Research on Leukemia, AVMA. Subspecialties: Microbiology; Cancer research (medicine). Current work: Role of viruses in naturally occurring cancers especially feline and human retroviruses and hepatitis B virus. Office: 665 Huntington Ave Boston MA 02115

ESSIG, HENRY WERNER, animal nutrition scientist, educator, researcher, conultant; b. Paragould, Ark., Dec. 9, 1930; s. George and Emma Elisabeth (Rudi) E.; m. Alice H., June 7, 1953; children: Stephen W., Rebecca A. B.S.A., U. Ark., 1953, M.S., 1956; Ph.D., U. Ill., 1959. Mem. faculty dept. animal sci. Miss. State U., 1959—, asst. prof. animal sci., 1959-61, assoc. prof., 1961-66, prof., 1966—; cons. nutrition. Contbr. numerous articles to profl. jours., popular publs. Served to 1st lt. inf. U.S. Army, 1953-55. Recipient Research award Miss. State U. chpt. Gamma Sigma Delta, 1973, faculty achievement award for research Miss. State U. Alumni Assn., 1976. Mem. Am. Soc. Animal Sci., Am. Soc. Dairy Sci., Animal Nutrition Research Council, Am. Forage and Grassland Council (merit cert. 1978, dir. 1982—), Alpha Zeta. Lutheran. Subspecialty: Animal nutrition. Current work: Ruminant nutrition, forage evaluation. Office: Miss State U Dept Animal Sci Box 5228 Mississippi Station MS 39762

ESTABROOK, RONALD WINFIELD, chemistry educator; b. Albany, N.Y., Jan. 3, 1926; s. George Arthur and Lillian Florence (Childs) E.; m. June Elizabeth Templeton, Aug. 23, 1947; children: Linda Estabrook Gilbert, Laura Estabrook Verinder, Jill Estabrook, David Estabrook. B.S., Rensselaer Poly. Inst., 1950; Ph.D., U. Rochester, 1954, D.Sc. (hon.), 1980, M.D., Karolinska Inst., Stockholm, 1981. Johnson Research Found. fellow U. Pa. Sch. Medicine, 1955-58; research asso., 1958-59, asst. prof. phys. biochemistry, 1959-62, asso. prof., 1961-65, prof., 1965-68; prof. biochemistry U. Tex. Health Sci. Center, Dallas, 1968-82, dean, 1973-76; chmn. basic sci. rev. com. VA, 1972-74; cons. in field. Bd. sci. advisers St. Judes Hosp., Memphis, 1978-81; chmn. bd. toxicology and environ. health Nat. Acad. Sci., 1980—; mem. Atlantic Richfield Sci. Adv. Council, 1981—. Exec. editor: Archives of Biochemistry and Biophysics, 1966-73, Cancer Research, 1980—; editor: Jour. Pharmacology and Exptl. Therapeutics, 1969-74, Xenobiotica, 1970—, Life Scis, 1973—; Contbr. articles to profl. jours. Served with USNR, 1943-46. Recipient Disting. Scientist award Fedn. Am. Socs. Exptl. Biologist, 1977; Claude Bernard medal U. Montreal, 1969. Mem. Inst. Medicine, Nat. Acad. Scis., Pan Am. Assn., Biochem. Socs. (sec.-gen. 1972-75), Am. Assn. Med. Schs. (adminstrv. bd. council acad. socs.; task force cost med. edn. 1971-72, liaison com. med. edn. 1975-80), Am. Soc. Biol. Chemists, Am. Soc. Pharmacology and Exptl. Therapeutics, Sigma Xi. Subspecialty: Biochemistry (medicine). Home: 5208 Preston Haven Dallas TX 75229 Office: U Tex Health Sci Center 5323 Harry Hines Blvd Dallas TX 75235

ESTEBAN, MARIANO, biochemist educator, microbiologist; b. Villalon, Spain, July 26, 1944; came to U.S., 1974, naturalized, 1981; s. Victorino and Victoria (Rodriguez) E.; m. Victoria Jimenez, Dec. 27, 1979. M.S., U. Santiago, Spain, 1967, 1971, Ph.D., 1970. Vis. scientist Nat. Inst. Med. Research, London, 1970-74; instr. Rutgers U. Med. Sch., 1974-77; vis. prof. Molecular Biology Inst., Gent, Belgium, 1978; asst. prof. biochemistry SUNY-Bklyn., 1979-81, assoc. prof. biochemistry, 1981—. NIH grantee, 1980—; Health Research Council N.Y. prin. investigator, 1980-81. Fellow European Molecular Biology Orgn.; mem. Spanish Soc. Microbiology, Brit. Soc. Microbiology, Am. Soc. Microbiology and Virology, Harvey Soc., N.Y. Acad. Scis., Sigma Xi. Subspecialties: Animal virology; Genetics and genetic engineering (agriculture). Current work: Mode of action of antiviral and antitumor drugs; interferon, prostaglandins, vaccinia virus genome organization, biomedical engineering, cancer research. Home: 304 E 20th St Apt 6G New York NY 10003 Office: Downstate Med Ctr SUNY 450 Clarkson Ave Brooklyn NY 11203

ESTES, EDWARD HARVEY, JR., medical educator; b. Gay, Ga., May 1, 1925; s. Edward Harvey and Veola (Jarrell) E.; m. Jean Anderson, Oct. 15, 1948; children: Sara Estes Brown, Susan Estes Jones III, Rebecca Estes Dunn, John, Elizabeth Estes Smith. B.S., Emory U., 1944, M.D., 1947. House officer, research fellow Grady Meml. Hosp., Atlanta, 1947-50; mem. faculty dept. medicine Duke U., 1952—, prof. chmn. dept. community and family medicine, 1966—. Author: (with R.P. Grant) Spatial Vector Electrocardiography, 1950. Mem. Inst. Medicine, Am. Soc. Internal Medicine (named Distinguished Internist of Year 1975), AMA, ACP, N.C. Med. Soc. (pres. 1977-78). Subspecialties: Internal medicine; Preventive medicine. Current work: Cost and cost effectiveness of medical services, medical manpower needs, payment system. Home: 3542 Hamstead Ct Durham NC 27707

ESTES, JAMES RUSSELL, botanist, curator herbarium; b. Burkburnett, Tex., Aug. 28, 1937; s. Dow Worley and Bessie B. Seidlitz (Sidlet) E.; m. Nancy Elizabeth Arnold, Dec. 21, 1962; children: Jennifer Lynn, Susan Elizabeth. B.S. in Biology, Midwestern State U., 1959; Ph.D. in Plant Systematics, Oreg. State U., Corvallis, 1967. Asst. prof. botany dept. botany and microbiology U. Okla., Norman, 1967-70, assoc. prof. botany, 1970-82, prof. botany, 1982—, curator, 1979—; dir. Okla. Natural Heritage Program, 1981-82; ecol. cons. Contbr. numerous articles to profl. jours. Mem. adv. bd. Sutton Urban Wilderness Park, Norman, Okla. Served to 1st lt. U.S. Army, 1960-62. Recipient Ortenburger award Phi Sigma, 1975; recipient Baldwin award Okla. Alumni Found., 1976; NSF grantee, 1968-70, 80, 1982—; others. Mem. Bot. Soc. Am. (sec., program chmn. for systematic sect.), Am. Soc. Plant Taxonomists (sec., program chmn.), Southwestern Assn. Naturalists (bot. editor, bd. govs.). Democrat. Presbyterian. Subspecialties: Systematics; Taxonomy. Current work: Plant systematics, cytotaxonomy, pollination ecology, systematics of the grasses, systematics and comparative pollination ecology of the composites, taxonomy of the grasses of the U.S., especially Oklahoma, S.E. and S.W. U.S. Home: 1906 Burnt Oak Norman OK 73071 Office: 770 Van Vleet Oval Norman OK 73019

ESTES, WILLIAM KAYE, psychologist, educator; b. Mpls., June 17, 1919; s. George D. and Mona; m. Katherine Walker, Sept. 26, 1942; children: George E., Gregory W. Mem. faculty Ind. U., 1946-62, prof. psychology, 1955-60, research prof. psychology, 1960-62; faculty research fellow Social Sci. Research Council, 1952-55; lectr. psychology U. Wis., summer 1949; vis. prof. Northwestern U., spring 1959; fellow Center Advanced Study Behavioral Scis., 1955-56; spl. univ. lectr. U. London, Eng., 1961; prof. psychology, mem. Inst. Math. Studies Social Scis., Stanford, 1962-68; prof. Rockefeller U., 1968-79, Harvard U., 1979—. Author: An Experimental Study of Punishment, 1944, Learning Theory and Mental Development, 1970, Models of Learning, Memory, and Choice, 1982; co-author: Modern Learning Theory, 1954, Models of Learning, Memory, and Choice, 1982; also numerous articles.; Editor: Handbook of Learning and Cognitive Processes, 1975, Jour. Comparative and Physiol. Psychology, 1962-68, Psychol. Rev., 1977-82; asso. editor: Jour. Exptl. Psychology, 1958-62. Served with AUS, 1944-46. Fellow Am. Psychol. Assn. (pres. div. exptl. psychology 1958-59, Distinguished Sci. Contbn. award 1962), AAAS, Am. Acad. Arts and Scis.; mem. Nat. Acad. Scis., N.Y. Acad. Sci. (hon. life), Soc. Exptl. Psychologists (Warren medal 1963), Psychometric Soc., Midwestern Psychol. Assn. (pres. 1956-57). Subspecialty: Learning. Home: 95 Irving St Cambridge MA 02138 Office: 620 W James Hall 33 Kirkland Cambridge MA 02138

ESTRADA, GALINDO IGNACIO, scientific researcher, educator; b. Mexico City, Jan. 22, 1934; s. Luis Vazquez Galindo and Carmen Nares E.; m. Lara Beatriz, Dec. 9, 1965; children: Cludia, Guillermo, Andres; m. Johanna Speicher, Nov. 20, 1974; 1 dau., Madeleine. M.Sc. (USPHS fellow), U. Chgo., 1966; M.D., U. Mex., 1967; Ph.D., U. Basel, Switzerland, 1972. Mem. faculty Instituto de Geofisica, 1966-68; mem. faculty Universidad Nacional Autonoma de Mexico, Mexico City, 1977—, prof., dir. Inst. Geophysics, 1977—; cons. World Meteorol. Orgn., others. Contbr. numerous articles to sci. jours. Mem. Internat. Soc. Biometeorology, Internat. Radiation Commn., Acad. Scis. Mex., Asociaion Ibero Latino Americana de Geofisica, Asociacion Latinoamericana de Energia Solar, Asociacion Nacional de Energia Solar (Mex.). Subspecialty: Climatology. Current work: Atmospheric aerosols, evaluation of potential energy sources, solar radiation, solar energy, atmospheric pollution.

ESTRIN, GERALD, computer scientist, engineering educator, academic administrator; b. N.Y.C., Sept. 9, 1921; m.; 3 children. B.S., U. Wis., 1948, M.S., 1949, Ph.D. in Elec. Engring, 1951. Research engr. Inst. Advanced Study, Princeton U., 1950-53, 55-56; dir. electronic computing project Weizmann Inst. Sci., Israel, 1953-55; assoc. prof. engring. UCLA, 1956-58, prof., 1958—, chmn. dept. computer sci., 1979—; mem. adv. bd. applied math. div. Argonne Nat. Lab., 1966-68, mem. assoc. univs. rev.com. for chmn., 1976-77, mem. applied math. div., 1974-80; dir. Computer Communications, Inc., 1966-67, Systems Engring. Labs., 1977—; mem. internat. program com. Internat. Fedn. Info. Processing Congress, 1968; internat. program chmn. Jerusalem Conf. Info. Tech., 1971; mem. math. and computer sci. research adv. com. AEC; mem. sci. com. Gould, Inc., Rolling Meadows, Ill., Nitron, Inc., Cupertino, Calif., 1981—. Bd. govs. Weizmann Inst. Sci., 1971. Lipsky fellow, 1954; Guggenheim fellow, 1963, 67. Fellow IEEE (Disting. Speaker); mem. Assn. Computing Machinery (lectr.), Am. Soc. Engring. Edn., N.Y. Acad. Sci. Subspecialty: Digital computer systems. Office: Dept Computer Sci UCLA Los Angeles CA 90024

ETGEN, ANNE MARIE, research scientist, biologist, educator; b. Ft. Atkinson, Wis., Oct. 1, 1953; d. William Mathias and Jesslyn Joan (Skeen) E. B.S., Coll. William and Mary, 1975; Ph.D., U. Calif., Irvine, 1979. Teaching asst. U. Calif., Irvine, 1975-79, NSF fellow, 1976-79; research scientist IV L.I. Research Inst., SUNY, Stony Brook, 1979; asst. prof. dept. biol. sci. Rutgers U., New Brunswick, N.J., 1979—; vis. scholar Columbia U., 1979-80. Contbr. articles to profl. jours. NIMH grantee, 1980-81, 82-85. Mem. Soc. Neurosci., Hamster Research Soc.; mem. European Soc. Comparative Physiology and Biochemistry; Mem. Phi Beta Kappa. Subspecialties: Neuroendocrinology; Neurobiology. Current work: Cellular and molecular mechanisms of hormone action in central nervous system. Office: Dept Biol Sci Rutgers U Livingston Campus New Brunswick NJ 08903

ETTENBERG, AARON, psychopharmacologist, research consultant; b. Montreal, Que., Can., Oct. 12, 1953; came to U.S., 1980; s. Bernard and Rosa (Shaffer) E.; m. Ina Dale Greenspon, Aug. 26, 1978. B.A., Concordia U., Montreal, 1975; M.A., McGill U., Montreal, 1977, Ph.D., 1980. Research technician Center for Research on Drug Abuse and Dependence, 1973-75; lectr. McGill U., Montreal, 1975-80; research assoc. Salk Inst., San Diego, 1980-82; asst. prof. U. Calif., Santa Barbara, 1982—. Editorial cons. numerous jours.; contbr. chpts. to books, articles to profl. jours. Med. Research Council fellow, 1980, 81, 82. Mem. Am. Psychol. Assn., Can. Psychol. Assn., Soc. Neurosci., AAAS. Subspecialties: Neuropharmacology; Psychobiology. Current work: Drug reinforcement; neurobiology of learning and memory; behavioral effects of psychoactive drugs; neuropeptides and behavior; brain stimulation reinforcement; neurochemistry of motivation. Office: Dept Psychology Univ Calif Santa Barbara CA 93106

ETTENSOHN, FRANK ROBERT, geology educator; b. Cin., Feb. 6, 1947; s. Robert Frank and Aileen Frances (Keman) E.; m. Beth Ann Mosher, June 3, 1978; 1 dau., Clare Marie. B.S., U. Cin., 1969, M.S., 1970; Ph.D., U. Ill.-Urbana, 1975. Instr. math. Greenhills-Forest Park City Sch. Dist., Cin., 1971; teaching and research asst. U. Ill., Urbana, 1971-75; asst. prof. U. Ky., Lexington, 1975-81, assoc. prof. geology, 1981—; cons. geology, Lexington, 1976—; prin. investigator U.S. Dept. Energy, 1976-80. Contbr. articles, guidebooks, maps and charts to profl. lit. Asst. scoutmaster Boy Scouts Am., Cin. and Lexington, 1964—. Served as 2d lt. AUS, 1970. Fellow Geol. Soc. Am.; mem. Paleontol. Soc., Paleontol. Assn., Paleontol. Research Instn., Internat. Paleontol. Assn., AAAS, Ky. Acad. Sci., Sigma Xi. Republican. Roman Catholic. Subspecialties: Paleoecology; Geology of gas shales. Current work: Geology of black gas shales, paleontology and paleontology of carbonate rocks. Home: Route 7 Old Richmond Rd Lexington KY 40511 Office: Dept Geology U Ky Lexington KY 40506

ETTER, PAUL COURTNEY, oceanographer; b. Phila., Oct. 27, 1947; s. Richard T. and Ellen M. (Cunliffe) E.; m. Alice D. Eblighatian, June 21, 1969; children: Gregory, Andrew. B.S., Tex. A. and M. U., 1969, M.S., 1975. Technician Technitrol, Inc., Phila., 1969; research asst. Tex. A. and M. U., College Station, Tex., 1973-76; sr. engr. MAR, Inc., Rockville, Md., 1976-82; tech dir. ODSI Defense Systems, Inc., Rockville, 1982—; pvt. cons., Gaithersburg, Md., 1980-82; reviewer NSF, Washington, 1981. Served to lt. USN, 1969-73. Mem. Am. Meterol. Soc., Am. Geophys. Union, Acoustical Soc. Am., Marine Tech. Soc. Democrat. Subspecialties: Oceanography; Acoustics. Current work: Apply principles of physical oceanography and underwater acoustics to the development of undersea acoustic sensing technology for the U.S. Navy. Home: 39 Honey Brook Ln Gaithersburg MD 20878 Office: ODSI Defense Systems Inc 6110 Executive Blvd Rockville MD 20852

ETTINGER, ANNA MARIE, anatomist; b. Janesville, Wis., Nov. 4, 1925; d. Martin and Anna (Dawson) Conway; m. Ralph Ettinger, Apr. 26, 1969. B.S., U. Wis., 1946, M.S., 1950; Ph.D., U. Ill. Chgo. Med. Ctr., 1967. Tchr. elem. and high schs., Barrington, Ill., 1946-49; tchr. Joliet (Ill.) High Sch., Joliet Jr. Coll., 1950-55; teaching fellow in anatomy U. Wis., 1955-57; instr. dept. anatomy St. Louis U. Med. Sch., 1957-63; asst. prof., chmn. sect. anat. scis. basic sci. dept. U. Detroit Sch. Dentistry, 1967-69, assoc. prof., chmn., 1969-76, prof., chmn. dept. anatomy, 1976—. Contbr. articles to profl. jours. NSF grantee, 1965-66; NIH grantee, 1969-72. Mem. Am. Assn. Immunologists, Am. Assn. Anatomists, Assn. Anatomy (chmn.), Sigma Xi, Omicron Kappa Upsilon. Subspecialties: Anatomy and embryology; Immunobiology and immunology. Current work: The study of cross reaction of natrual agglutinins, natural, antibody, chicks, human milk. Office: U Detroit Sch Dentistry 2985 E Jefferson Detroit MI 48207

ETTINGER, DAVID SEYMOUR, physician; b. Bklyn., Mar. 16, 1942; s. Harry and Frieda (Rose) E.; m. Phyllis Evellen Katz, June 4, 1964; children: Laura, Daniel, Kathryn. B.A., Yeshiva U., 1963; M.D., U. Louisville, 1967. Diplomate: Am. Bd Internal Medicine and Med. Oncology. Intern Albany (N.Y.) Med. Center, 1967-68; fellow in internal medicine Mayo Clinic, Rochester, Minn., 1968-71; fellow in oncology Johns Hopkins U. Sch. Medicine, Balt., 1973-75, instr. oncology and medicine, 1976-77, asst. prof., 1977-81, assoc. prof., 1981—; Mem. patient service com. Md. div. Am. Cancer Soc., 1977—. Contbr. articles to profl. jours. Served to maj. M.C. U.S. Army, 1971-73. Decorated Army Commendation medal. Fellow ACP; mem. Eastern Coop. Oncology Group, Am. Soc. Clin. Oncology, Am. Assn. Cancer Research. Democrat. Jewish. Subspecialties: Oncology; Internal medicine. Current work: Development and utilization of new cancer chemotherapeutic agents, drug combinations and concomitant chemotherapy with other modalities in the treatment of malignant tumor. Home: 2511 Lawnside Rd Timonium MD 21093 Office: Johns Hopkins Oncology Center Baltimore MD 21205

ETTINGER, HELEN CLARICE, veterinary surgeon; b. Long Beach, Calif., Sept. 14, 1946; d. Charles Anthony and Marina Ellen (Thomas) Meeks; m. Daniel C. Ettinger, July 1, 1975. B.S., U. Nebr., 1967; D.V.M., Ohio State U., 1972. Diplomate: Am. Coll. Vet. Surgeons. Surg. resident Hosp. for Spl. Surgery, N.Y.C., 1973-75; asst. prof. surgery N.Y. State Vet. Coll., Ithaca, 1975-78; research fellow in surgery Cornell U. Med. Coll., N.Y.C., 1978-79; assoc. prof. vet. surgery U. R.I., Kingston, 1980—; assoc. research scientist Animal Med. Ctr., N.Y.C., 1978-80. Named Outstanding Alumnus Ohio State U., 1982. Mem. AVMA, Orthopedic Research Soc., Assn. Women Veterinarians. Subspecialty: Surgery (veterinary medicine). Home: Werik Ct 1 Salvati Way Providence RI 02909

ETTINGER, PHILIP OWEN, cardiologist, internist; b. N.Y.C., Oct. 3, 1936; s. Samuel and Charlotte (Adler) E.; m. Roxanne M. Miller, May 31, 1964; children: Alyssa Anne, Jonathan Seth. A.B., Swarthmore Coll., 1956; M.D., N.Y.U., 1960. Diplomate: Am. Bd. Internal Medicine(subspecialty cardiovascular disease). Intern Lenox Hill Hosp., N.Y.C., 1960-61; resident Bronx (N.Y.) VA Hosp., 1961-63, N.Y.U. Hosp., 1963-64; clin. assoc. prof. medicine N.J. Med. Sch., Newark, 1966—; pvt. practice cardiology, Teaneck, N.J., 1969—; attending cardiologist Englewood (N.J.) Hosp., 1969—. Served to capt. USAF, 1964-66. N.Y. State med. scholarship, 1956; NIH cardiology grantee, 1975. Fellow ACP, Am. Coll. Cardiology; mem. Am. Physiol. Soc., Phi Beta Kappa. Subspecialties: Cardiology; Internal medicine. Current work: Cardiology practice, research involving myocardial metabolism and function, cardiac rhythm disturbances. Office: 185 Cedar Ln Teaneck NJ 07666

ETU, PAUL DAVID, psychologist; b. Glens Falls, N.Y., Sept. 26, 1954; s. James Lawrence and Cecelia Joan (Jordan) E.; m. Marcia Edna Mears, Aug. 13, 1977; children: Eric Mears, Joshua Jordan, Nathan Patrick. A.A.S., SUNY-Adirondack, 1975; B.A., SUNY-Oswego, 1975; M.S., Marquette U., 1980. Asst. mgr. Louis Wolf, Inc., Rochester, N.Y., 1979-80; grad. asst., dir. Ctr. Psychol. Services, Marquette U., Milw., 1978-80; psychologist, program coordinator Community Workshop Inc., Glens Falls, N.Y., 1980-82; pvt. practice psychology, Glens Falls, 1982—; program coordinator Rehab. Support Services, Albany, N.Y., 1983—. Contbr. articles to profl. jours. Bd. dirs. Monument Sq. Day Ctr., Glens Falls, 1980-81, Warren County Post Am. Legion Baseball, 1982—. Mem. Am. Psychol. Assn., Am. Assn. Mental Deficiency, Saratoga County Assn. Psychologists, Tri-County Assn. Psychologists, Psi Chi, Pi Gamma Mu. Democrat. Roman Catholic. Subspecialties: Behavioral psychology; Developmental psychology. Current work: Infant nutrition, allergy and mental illness; reading; deinstitutionalization of handicapped; childbirth. Address: Box 54 RD4 Glens Falls NY 12801

ETZEL, KENNETH RAYMOND, physiologist, researcher; b. Milw., Apr. 6, 1948; s. Raymond Joseph and Vera Catherine (Burdette) E.; m. Virginia Dennise Heider, July 26, 1975; children: Andrew Kenneth, Matthew Jared. B.A., St. Thomas Coll., 1970; M.S., U. Wis., 1972; Ph.D., Creighton U., 1978. Postdoctoral research asst. Rutgers U., New Brunswick, N.J., 1978-81; asst. prof. U. Tex. Health Sci. Ctr. Dental Sch., San Antonio, 1981—. U. Tex. Health Sci. Ctr. grantee, 1982. Mem. Am. Assn. Dental Research, Sigma Xi. Roman Catholic. Subspecialties: Psychophysiology; Nutrition (medicine). Current work: Trace mineral metabolism and nutrition, periodontal disease and nutritional interactions. Home: 5415 Princess Donna San Antonio TX 78229 Office: Dept Periodontics U Tex Health Sci Ctr Dental Br 7703 Floyd Curl Dr San Antonio TX 78284

ETZLER, MARILYNN EDITH, biochemistry educator; b. Detroit, Oct. 30, 1940; d. Elmer Ellsworth and Doris (Tegge) E. B.S. and B.A., Otterbein Coll., 1962; Ph.D., Washington U., 1967. Postdoctoral fellow Coll. Physicians and Surgeons, Columbia U., N.Y.C., 1967-69; asst. prof. biochemistry U. Calif.-Davis, 1969-75, assoc. prof., 1975-79, prof. biochemistry, 1979—. Contbr. articles to profl. jours. NIH grantee, 1970—; U.S. Dept. Agr. grantee, 1979—. Mem. Am. Soc. Biol. Chemists, Am. Soc. Cell Biology, Am. Soc. Plant Physiology, Soc. Complex Carbohydrates, AAAS. Subspecialties: Biochemistry (biology); Cell and tissue culture. Current work: Structure and function of lectins. Office: Dept Biochemistry and Biophysics U Calif Davis CA 95616 Home: 1112 Drake Dr Davis CA 95616

EUBANK, RANDALL LESTER, statistics educator; b. Dallas, Jan. 17, 1952; s. Lester and Mary Katherine (Kelley) E.; m. Elisa Marie Cordova, Dec. 20, 1979. B.S. in Agr., N. Mex. State U., 1974, M.S. in Agrl. Econs, 1975, Tex. A&M U., 1976, Ph.D., 1979. Research asst. dept. agrl. econs. N.Mex. State U., Las Cruces, 1974-75; teaching asst. Inst. Stats., Tex. A&M U., College Station, 1975-78, lectr., 1978-79; asst. prof. dept. math. Ariz. State U., Phoenix, 1979-80; asst. prof. dept. stats. So. Meth. U., Dallas, 1980—. Contbr. articles to profl. publs.; referee: Am. Statistician; speaker profl. and endl. confs. Mem. Am. Statis. Assn., Inst. Math. Stats., Soc. Indsl. and Applied Math., Signa Xi. Subspecialties: Statistics; Probability. Current work: Research relating to spline approximation in statistics. Home: 3719 Casa del Sol Dallas TX 75228 Office: Dept Statistics Southern Meth Univ Dallas TX 75257

EULER, ARTHUR RAY, pediatric gastroenterologist; b. Hammond, Ind., Oct. 20, 1942; s. John Stanley and June Alice Biestek; m. Becky Suzanne Brashares, Feb. 6, 1966 (div. 1980); children: Elizabeth, Katherine; m. Dana Mary Pederson, May 22, 1982. B.S., Purdue U., 1965; M.D., Ind. U., 1969. Intern Ind. U. Med. Ctr., Indpls., 1969-70, resident in Pediatrics, 1970, Harbor Gen. Hosp.-UCLA Med. Ctr., Torrance, Calif., 1973-74; fellow in cystic fibrosis-gastroenterology UCLA, 1974-76, asst. prof. pediatrics, 1976-77; head div. pediatric gastroenterology Ark. Children's Hosp., Little Rock, 1977-81; research physician gastrointestinal prostaglandin program Upjohn Co., Kalamazoo, 1981—; vis. prof. Universidad Autonoma de Guadalajara, Mex., 1980, 81, 82. Contbr. numerous articles to profl. jours. Served to lt. comdr. USN, 1970-73; Vietnam. Fellow Am. Acad. Pediatrics, Am. Gastroenterology Assn., Am. Coll. Gastroenterology, Soc. Pediatric Research; mem. N.Am. Soc. Pediatric Gastroenterology, So. Calif. Soc. Gastrointestinal Endoscopy, Western Gastroenterology Assn., So. Calif. Soc. Gastroenterology, Am. Soc. Parenteral and Enteral Nutrition, Central Ark. Pediatric Soc., So. Soc. Pediatric Research, Am. Fedn. Clin. Research, AAAS, N.Y. Acad. Scis., Am. Motility Soc., Soc. Clin. Trials. Lutheran. Subspecialties: Gastroenterology; Pediatrics. Current work: Gastroesophageal reflux and its complications, gastric acid secretion physiology, gastrin physiology in infants and children, therapeutic use of prostaglandins in upper and lower gastrointestinal diseases. Home: 2101 Bronson Blvd Kalamazoo MI 49007 Office: Upjohn Co Kalamazoo MI 49001

EURICH, ALVIN CHRISTIAN, psychologist; b. Bay City, Mich., June 14, 1902; s. Christian H. and Hulda (Steinke) E.; m. Nell Plopper, Mar. 15, 1953; children: Juliet Ann, Donald Alan. A.B., North Central Coll., 1924; M.A., U. Maine, 1926; Ph.D., U. Minn., 1929. Asst. to pres. U. Minn., Mpls., 1935-36; v.p. Stanford (Calif.) U., 1944-47, acting pres., 1948; first chancellor SUNY-Buffalo, 1949-51; exec. dir. Ford Found. Edn. Program, N.Y.C., 1951-64; pres. Aspen Inst. for Humanistic Studies, Colo., 1963-67, spl. adviser, 1972—, hon. trustee, 1982—; pres. Acad. for Ednl. Devel., Inc., N.Y.C., 1961—. Bd. dirs. Ctr. Pub. Resources; trustee Lovelace Med. Ctr. and Found. Served to comdr. USNR, 1942-44. Recipient Disting. Achievement award U. Minn., 1951; 4th Ann. award Times Sq. Club, 1953; Annual award N.Y. Acad. Pub. Edn., 19. Fellow AAAS, Am. Psychol. Assn.; mem. Am. Ednl. Research Assn. (pres. 1944), Internat. Council Ednl. Devel. (dir.), Sigma Xi. Clubs: Cosmos (Washington); University Century, Coffee House (N.Y.C.). Subspecialties: Learning; Behavioral psychology. Current work: Learning, education and aptitude measurement. Office: Acad for Ednl Devel 680 5th Ave New York NY 10019

EVANS, ALBERT E., nuclear physicist; b. Tarrytown, N.Y., Apr. 21, 1930; s. Albert Edwin and Gerardine (Ogilby) E.; m. Patricia Corbin Flynn, May 12, 1956; children: Leslie L. Evans Smith, Keith, Andrea, Hilary. B.S., Yale U., 1952; M.S., Ohio State U., 1953; Ph.D., U. Md., 1965. Research assoc. dept. biophysics Yale U., White Plains, N.Y., 1957; reactor test engr. Martin Co., Balt., 1957-58; exptl. physicist U.S. Naval Ordnance Lab., Silver Springs, Md., 1958-67; staff Los Alamos Nat. Lab., 1967—. Search and rescue coordinator CD; dist. com. Boy Scouts Am. Served as 1st lt. USAF, 1952-56. Mem. Am. Phys. Soc., Am. Nuclear Soc., Sigma Xi, Sigma Pi Sigma; affiliate IEEE. Republican. Episcopalian. Subspecialties: Nuclear fission; Nuclear physics. Current work: Neutron and gamma-ray spectrometry, material assay by nuclear techniques, properties of delayed neutrons from fission, nuclear critical experiments, small accelerator facility construction, operation, and research. Office: Los Alamos Nat Lab Mail Stop J-562 Los Alamos NM 87545 Home: 349 Venado Los Alamos NM 87544

EVANS, BILLY JOE, chemist, educator, researcher, cons.; b. Macon, Ga., Aug. 18, 1942; s. Will and Mildred (Owens) E.; m. Adye Bel Sampson, Aug. 31, 1963; children: William Joseph, Carole Elizabeth and Jesse Niles (twins). B.Sc. summa cum laude in math with honors (Merrill scholar), Morehouse Coll., 1963; Ph.D., U. Chgo., 1968. Research assoc. U. Chgo, 1968; NRC Can. postdoctoral fellow in physics U. Man. (Can.), Winnipeg, 1968-69; asst. prof. chemistry Howard U., 1969-70; asst. prof. mineralogy/crystal physics U. Mich., 1970-73, assoc. prof. mineralogy/crystal physics, 1973-74, assoc. prof. inorganic chemistry, 1974-78, prof. inorganic chemistry, 1978—; cons. Nat. Bur. Standards, Gaithersburg, Md., 1970-78; guest prof. U. Marburg, W.Ger., 1977-78; cons. U.S. Geol. Survey, Reston, Va., 1979—; research scientist dept. chemistry Sci. Research Labs., Ford Motor Co., 1977. Contbr. chpt., numerous articles to profl. publs. Hon. Woodrow Wilson fellow, 1963; Alfred P. Sloan fellow, 1972; Alexander Von Humboldt Found. spl. fellow, 1977-78; Danforth Found. assoc., 1977-86. Mem. Am. Chem. Soc., Am. Mineral. Assn., Am. Phys. Soc., Sigma Xi. Subspecialties: Solid state chemistry; Electronic materials. Current work: Synthesis of megnetic electronic materials; structure property relationships for non-metallic materials; chemistry of iron; inorganic phases in fossil fuels. Home: 810 Oxford Rd Ann Arbor MI 48104 Office: Dept Chemistry U Mich Ann Arbor MI 48109

EVANS, BOB OVERTON, electronics executive; b. Grand Island, Nebr., Aug. 19, 1927; s. Walter Bernard and Lillian (Overton) E.; m. Maria Bowman, Nov. 19, 1949; children: Cathleen L., Robert W., David D., Douglas B. B.E.E., Iowa State U., 1949. Electric operating engr. No. Ind. Pub. Service Co., Hammond, 1949-51; with IBM, 1951—; v.p. devel. Data Systems div., 1962-64; pres. Fed. Systems div., 1965-69, Systems Devel. div., 1970-74, Systems Communication div., 1975-77; v.p. IBM engring., programming and tech., 1977—; mem. Stark Draper Labs., Inc.; cons. govt. agys.; area bd. mem. Md. Nat. Bank; mem. Def. Sci. Bd. Mem. exec. bd. Nat. Capital Area council Boy Scouts Am.; trustee Rensselaer Poly. Inst., N.Y. Pub. library; mem. elec. engring. vis. com. MIT. Served with USNR, 1945-46. Recipient Disting. Pub. Service award NASA; Disting. Alumni citation Iowa State U. Fellow IEEE (chmn. computer group conf. 1970); mem. Nat. Acad. Engring., Profl. Group Electronic Computers, Nat. Security Indsl. Assn. (trustee), Armed Forces Communications and Electronics Assn. (trustee), Aerospace Industries Assn. (exec. bd.). Presbyterian. (elder). Subspecialty: Computer engineering. Designed and developed large digital electric computers. Home: Ivanhoe Ln Greenwich CT 06830 Office: Old Orchard Rd Armonk NY 10504

EVANS, CHARLES HAWES, med. scientist, physician; b. Orange, N.J., Apr. 16, 1940; s. Charles Hawes and Jean Marie (Robinson) E.; m. Nancy Engel, Aug. 21, 1965; 1 dau., Heather Leigh. B.S., Union Coll., 1962; M.D., U. Va., 1969, Ph.D., 1969. Diplomate: Nat. Bd. Med. Examiners. Intern pediatrics U. Va. Hosp., Charlottesville, 1969-70, resident, 1970-71; research assoc. Nat. Cancer Inst., NIH, Bethesda, Md., 1971-73, sr. scientist, 1973-76, chief tumor biology sect., 1976—. Assoc. editor: Jour. Nat. Cancer Inst, 1981—; contbr. articles in field to med. sci. jours. Served with USPHS, 1971—. Recipient John Horsley Meml. Prize U. Va., 1982. Mem. Am. Assn. Cancer Research, Am. Assn. Immunologists, Internat. Soc. Immunopharmacology, Assn. Mil. Surgeons U.S., AAAS. Presbyterian. Lodge: Rotary. Subspecialties: Cancer research; Immunology, immunopharmacology. Current work: Director biomedical research laboratory, investigating immunological and other physiological mechanisms for their potential to prevent or control cancer. Home: 9233 Farnsworth Dr Potomac MD 20854 Office: NIH Room 2A17 Bldg 37 Bethesda MD 20205

EVANS, DAVID C., computer company executive; b. Salt Lake City, Feb. 24, 1924; s. David W. and Beatrice (Cannon) E.; m. Joy Frewin, Mar. 21, 1947; children: Gayle Evans Scheidel, Susan Evans Foote, David F., Ann Evans Brown, Peter F., Douglas F., Katherine E. B.S., U. Utah, 1949, Ph.D. in Physics, 1953. Dir. engring. computer div. Bendix Corp., Los Angles, 1953-62; prof. elec. engring. and computer sci. U. Calif.-Berkeley, 1962-66; prof. elec. engring. and computer sci., chmn. dept. U. Utah, Salt Lake City, 1965-73; chmn. bd., pres. Evans & Sutherland Computer Corp., Salt Lake City, 1968—. Served in U.S. Army, 1942-45. Recipient Silver Beaver award Boy Scouts Am.; named to Computer Hall of Fame. Fellow IEEE; mem. Nat. Acad. Engring. Republican. Mem. Ch. of Jesus Christ of Latter-day Saints. Subspecialties: Graphics, image processing, and pattern recognition; Computer architecture. Current work: General mangement. Home: 1393 E South Temple Salt Lake City UT 84102 Office: PO Box 8700 Salt Lake City UT 84108

EVANS, DAVID STANLEY, astronomer, educator; b. Cardiff, Wales, Jan. 28, 1916; came to U.S., 1968; s. Arthur Cyril and Kate (Priest) E.; m. Betty Hall Hart, Mar. 8, 1949; children: Jonathan Gareth Weston, Barnaby Huw Weston. B.A., Cambridge U., 1937, M.A., Ph.D., ScD., 1941. Research asst. Univ. Obs., Oxford, Eng., 1938-46; 2d asst. Radcliffe Obs., Pretoria, South Africa, 1946-51; chief asst., sr. prin. sci. officer Royal Obs., Cape of Good Hope, South Africa, 1951-68; prof. astronomy U. Tex., Austin, 1968—; Sr. vis. scientist NSF fellow, 1965-66. Author or editor 8 books; contbr. over 200 articles to sci. and hist. publs. Fellow Royal Astron. Soc. (London), Inst. Physics (London), Royal Soc. South Africa; mem. Astron. Soc. So. Africa (hon. mem.; past v.p., McIntyre award for astron. history 1972), Am. Astron. Soc., Internat. Astron. Union (past pres. commn.). Clubs: Owl, West Province Sports (Cape Town, South Africa); Town and Gown (Austin). Subspecialties: Optical astronomy; History of Astronomy. Current work: Lunar occultations flare stars, multiple stars. Home: 6001 Mountainclimb Dr Austin TX 78731 Office: Dept Astronomy U Tex Austin TX 78712

EVANS, EDWIN VICTOR, nutrition educator, researcher; b. Toronto, Ont., Can., Mar. 30, 1914; m.; 2 children. B.A., U. Western Ont., 1936, M.A., 1937. From asst. prof. to assoc. prof. animal nutrition Ont. Agrl. Coll., 1941-48, assoc. prof. nutrition, 1951-68; biochem. dir. W. R. Drynan Nutrition Lab., Hamilton, Ont., 1948-50; assoc. prof. and acting head dept. biochemistry nutrition and food sci. U. Ghana, 1965-66, prof., head dept. biochemistry, nutrition and food sci., 1968-71; assoc. prof. nutrition Coll. Biol. Sci., U. Guelph, Ont., Can., 1971—. Mem. AAAS, Animal Nutrition Research Council, Am. Soc. Animal Sci., Can. Physiol. Soc., Nutrition Soc. Can. (sec. 1960-65). Subspecialty: Nutrition (biology). Office: Dept Nutrition U Guelph Guelph ON Canada

EVANS, ERSEL ARTHUR, manufacturing company executive; b. Trenton, Nebr., July 17, 1922; s. Arthur E. and Mattie Agnes (Perkins) E.; m. Patricia A. Powers, Oct. 11, 1945; children: Debra Lynn (dec.), Paul Arthur. B.A., Reed Coll., Portland, Oreg., 1947; Ph.D., Oreg. State U., 1950. Registered profl. engr., Calif. With Gen. Electric Co., 1951-67, supr. ceramics research and devel., Hanford, Wash., 1961-64; mgr. plutonium devel. Vallecitos Lab., Pleasanton, Calif., 1964-67; mgr. fuels and materials dept. Battelle Meml. Inst., Richland, Wash., 1967-70; with Westinghouse Electric Corp., 1970—; v.p. Westinghouse Hanford Co., Richland, 1972—; tech. dir. Hanford Engring. Devel. Lab., 1976—; mem. vis. com. Coll. Engring., U. Wash., Seattle, 1974—. Author. Bd. dirs. Mid-Columbia Mental Health Center, Richland, 1975—. Served with USNR, 1943-45. DuPont fellow, 1950-51; grantee Research Corp. Am., 1949-50; recipient Westinghouse Order of Merit. Fellow Am. Nuclear Soc. (Spl. Merit award 1964, Performance award 1980), Am. Inst. Chemists, Am. Soc. Metals, Am. Ceramic Soc.; mem. Nat. Acad. Engring., Phi Kappa Phi. Subspecialty: Nuclear fission. Patentee in field. Home: 2033 Weiskopf Ct Richland WA 99352 Office: PO Box 1970 Richland WA 99352 Inspiration and guidance for my career have often been provided by Justice Oliver Wendell Holmes, "certainty generally is illusion, and repose is not the destiny of man." (Harvard Law Review 1897)

EVANS, FREDERICK E., supervisory research chemist; b. Springfield, Mass., Nov. 11, 1948; s. Edward E. and Hedwig J. (Zeletzky) E.; m. Huey-Ing Tseng, Aug. 20, 1978. B.S., U. Mass., 1970; Ph.D., SUNY-Albany, 1974. Postgrad. research chemist U. Calif.-San Diego, 1975-78; chief spectroscopic techniques br., research chemist Nat. Ctr. Toxicol. Research, FDA, Jefferson, Ark., 1978—; adj. assoc. prof. U. Ark., Little Rock, 1980—. Contbr. articles to profl. jours. Mem. Am. Assn. Cancer Research, Soc. Magnetic Resonance in Medicine, Am. Chem. Soc. Subspecialties: Nuclear magnetic resonance (chemistry); Analytical chemistry. Current work: Responsible for the operation and supervision of a high level nuclear magnetic resonance spoectroscopy laboratory. Application of nuclear magnetic resonance techniques to analytical chemistry and chemical carcinogenesis. Office: Dept Chemistry Nat Center Toxicol Research FDA Jefferson AR 72079

EVANS, GARY WILLIAM, social ecology educator; b. Summit, N.J., Nov. 22, 1948. A.B. with high honors in Psychology, Colgate U., 1971; M.S., U. Mass., 1973, Ph.D. in Psychology, 1975; postgrad., U. Calif.-Irvine, 1978. Research asst. in psychology U. Mass., Amherst, 1971-73, instr. psychology and Bklyn. Career Opportunities Program, 1973-75; asst. prof. social ecology U. Calif.-Irvine, 1975-80, assoc. prof. social ecology, 1980—, assoc. dir. undergrad. affairs, 1979-81. Contbr. articles to profl. publs.; Editorial rev. bd.: Rep. Research in Social Psychology, 1973-75, Man-Environ. Systems, 1975—, Environ. Psychology and Nonverbal Behavior, 1976-79, Jour. Population and Environ, 1981—; ad hoc reviewer, NSF, NIH, various jours. Grantee in field.; Univ. scholar Colgate U., 1966-71; George Cobb fellow, 1970; Phil R. Miller psychology prize, 1971; NSF dissertation yr. fellow, 1974-75; Regents' jr. faculty fellow U. Calif., 1977; Fulbright award Council Internat. Exchange of Scholars, U. Poona, India, 1981-82. Subspecialties: Environmental psychology. Office: Program in Social Ecology Univ CaliF Irvine Irvine CA 92717

EVANS, HAROLD J., plant physiologist, biochemist, educator; b. Franklin, Ky., Feb. 19, 1921; s. James H. and Allie (Uhls) E.; m. Elizabeth Dunn, Dec. 14, 1946; children: Heather Mary, Pamela. B.S., U. Ky., 1946, M.S., 1948; Ph.D. (Cook-Vorhees fellow), Rutgers U., 1950. Asst. prof. botany N.C. State U., 1952-54, asso. prof., 1954-57, prof., 1957 61; postdoctoral fellow Johns Hopkins U., Balt., 1952, prof. plant physiology Oreg. State U., Corvallis, 1961—; dir. Lab. for Nitrogen Fixation, 1978—; vis. prof. U. Sussex, Eng., 1967; George A. Miller vis. prof. U. Ill., Urbana, 1973; mem. panel for metabolic biology NSF, 1964-68; mem. U.S.-Japan Coop. Sci. Program, 1976. Contbr. articles to profl. jours. Recipient Hoblitzelle Nat. award Tex. Research Found., 1964; Basic Research award Oreg. State U., 1965; N.W. Sci. award Gov. Oreg., 1967; named Disting. Alumnus U. Ky., 1975; recipient George G. Ferguson Disting. Prof. award and Milton Harris research award Ohio State U., 1983. Mem. Am. Soc. Plant Physiologists (pres. 1971, trustee 1977—), Biochem. Soc. (U.K.), Am. Soc. Biol. Chemists, U.S. Nat. Acad. Scis., Sigma Xi (award 1968), Phi Kappa Phi. Democrat. Subspecialties: Nitrogen fixation; Enzyme technology. Current work: Cloning and transfer of hydrogenase gene in Rhizobium with goal of increasing efficiency of nitrogen fixation in legumes. Home: 2939 Mulkey St Corvallis OR 97330 Office: Lab for Nitrogen Fixation Research Oreg State U Corvallis OR 97331

EVANS, JAMES THOMAS, surgical oncologist, researcher; b. Niagara Falls, N.Y., Apr. 12, 1942; s. James Edward and Kathleen (Walsh) E.; m. Joyce Gray, May 9, 1981. B.S., Tulane U., 1963; M.D., La. State U., 1967. Diplomate: Am. Bd. Surgery. Intern Charity Hosp., New Orleans, 1967-69; oncology fellow Roswell Park Meml. Inst., Buffalo, New Orleans, 1970-71, cancer research surgeon, 1973-80; resident in surgery La. State U., dept. surgery, New Orleans, 1972-73; assoc. prof. SUNY, Buffalo, 1980—. Am. Cancer Soc. fellow, 1978-81. Mem. Am. Cancer Soc. (chmn. colorectal com. 1979—, dir. N.Y. state 1980—, pres. Erie County 1981—), Am. Assn. Cancer Research, ACS (sec.-treas. 1979—), Am. Soc. Clin. Oncology, Soc. Surgery Alimentary Tract, Soc. Surgical Oncology. Mem. Christ. Ch. Subspecialties: Laser medicine; Cancer research (medicine). Current work: Research on treatment of solid tumors with laser; research on genetic variability of chemical carcinogenesis. Home: 33 Crosby Blvd Eggertsville NY 14226 Office: SUNY Dept Surgery 1462 Grider St Buffalo NY 14215

EVANS, JAMES WARREN, physiologist, equine cons.; b. Edna, Tex., Oct. 31, 1938; s. Calvin Cecil and Thelma Waley (Williamson) E.; m. Benita M. Evans, Sept. 2, 1959; 1 son, Scott Allen. Student, Tex. A&M U., 1958-59; B.S., Colo. State U., 1964; Ph.D., U. Calif., 1968. Prof. dept. animal sci. U. Calif., 1968—, assoc. dean, 1982—. Author: Horses, 1981, (with Borton, Hintz & VanVleck) The Horse, 1977; contbr. numerous articles in field to profl. jours. Mem. Am. Physiol. Soc., Am. Soc. Animal Sci., Endocrine Soc., Equine Nutriton and Physiology Soc. Club: Cross Court (Woodland, Calif.). Subspecialties: Reproductive biology; Chronobiology. Current work: Interaction of hormones at cellular level to control reproduction in mare and stallion. Office: Dept Animal Sci U Calif Davis CA 95616

EVANS, JOHN VAUGHAN, physicist; b. Manchester, Eng., July 5, 1933; s. Gyril John and Gertrude Veronica (Bayliss) E.; m. Maureen Vervain Patrick, Oct. 19, 1958; children: Carol, David, Lesley. B.Sc. in Physics with honors, Manchester U., 1954, Ph.D., 1957. Leverhulme research fellow Jodrell Bank Exptl. Sta., U. Manchester, 1957-60; mem. staff Lincoln Lab. MIT, Lexington, 1960-66, 67-70; G.A. Miller vis. prof. U. Ill.-Urbana, 1966-67; assoc. group leader surveillance techniques group Lincoln Lab. MIT, 1970-72, group leader, 1972-74, assoc. div. head aerospace div., 1974-77, asst. dir lab., 1977-83; dir. Haystack Obs., MIT, 1980-83; also prof. meteorology; dir. research COMSAT Labs., Gaithersburg, Md., 1983—. Editor: (with T. Hagfors) Radar Astronomy, 1968; contbr. numerous articles to sci. jours. Served with Royal Brit. Army, 1951-57. Recipient Appleton prize Royal Soc., London, 1954. Fellow IEEE; mem. Am. Geophys. Union, AAAS, Internat. Astron. Union, Internat. Union Radio Scis., Sigma Xi. Unitarian. Club: Cosmos (Washington). Subspecialties: Aeronomy; Remote sensing (atmospheric science). Current work: Studies by radar of the earth's upper atmosphere, meteors, the moon and terrestrial planets. Office: COMSAT Labs Gaithersburg MD 20877

EVANS, LANCE SAYLOR, research plant biologist, agronomist; b. Phila.; s. Edward C. and Elsie (Saylor) E.; m. Patricia May Watson, Sept. 11, 1965; children: Stephanie, Gail, Matthew. B.S., Calif. State Poly. U., 1967; Ph.D., U. Calif.-Riverside, 1970. With Brookhaven Nat. Lab., 1972-75, cons., 1975—; with Manhattan Coll., Bronx, N.Y., 1975—, assoc. prof. biology, 1981—. Pres. Glenville Taxpayers Assn., 1980—. Mem. Air Pollution Control Assn., Air Pollution Workshop, AAAS, Am. Phytopath. Soc., Am. Soc. Plant Physiologists, Bot. Soc. Am., N.Y. Acad. Scis., Sigma Xi, Beta Beta Beta. Subspecialties: Plant physiology (biology); Plant physiology (agriculture). Current work: Cell cycle functions; plant roots; effects of air pollutants on crops. Home: 14 Spring St Tarrytown NY 10591 Office: Plant Morphogene Lab Manhattan Coll Bronx NY 10471

EVANS, LATIMER RICHARD, chemist, educator; b. Washington, Nov. 4, 1918; s. Clifford V. and Ruth (Latimer) E.; m. Eloise Swick, Aug. 24, 1942; children: Carol Sandoval, Beth Sells, Marget Hensley, Scott. B.S., Am. U., 1941; Ph.D., Purdue U., 1945. Postdoctoral fellow Purdue U., 1945; research chemist E.I. DuPont Co., Wilmington, Del., 1946-50; asst., then assoc. prof. N.Mex. State U., Las Cruces, 1950-60, prof. chemistry, 1960—, sometime acting dept. chmn., also asst. chmn.; N.Mex. state chemist, 1950-58. Author lab. manual, tech. publs. Mem. Am. Chem. Soc., Sigma Xi, Phi Lambda Upsilon, Omicron Delta Kappa. Democrat. Unitarian. Subspecialties: Organic chemistry; Synthetic chemistry. Current work: Undergraduate education in chemistry, advising chairman for prehealth professions, director regional science fair. Patentee in field. Home: Box 3592 University Park NM 88003 Office: Box 3-C Las Cruces NM 88003

EVANS, MARY JO, cancer research scientist, educator; b. Maysville, Mo., Nov. 28, 1935; d. William Lloyd and Lillian Berle (Reeves) Smith; m. Jack R. Olds, Feb. 3, 1957; m. Richard T. Evans, Apr. 5, 1968; children: David Todd, Douglas Alden. B.A., William Jewell Coll., 1957; M.S., U. Mo., 1965; Ph.D., U. Tenn., 1968. Research assoc. U. Mo., 1957-58, 64-65; teaching fellow U. Tenn., 1965-66; trainee St. Jude Children's Research Hosp., Memphis, 1966-68; cancer research scientist Roswell Park Meml. Inst., 1968-69, sr. cancer research scientist, 1969-78, cancer research scientist IV, 1978—; assoc. research prof. microbiology SUNY, Buffalo. Contbr. articles in field to profl. jours. NIH grantee, 1975—. Mem. Am. Soc. Microbiology. Methodist. Subspecialties: Molecular biology; Cancer research (medicine). Current work: Regulation of DNA replication, oncogenes, DNA polymerase. Home: 38 Fiddler's Green East Amherst NY 14051 Office: 666 Elm St Buffalo NY 14263

EVANS, NANCY JEAN, information science educator; b. Homestead, Pa., Sept. 3, 1953; d. Jack W. and Elizabeth (Pudleiner) E. B.S., Pa. State U., 1975; Ph.D., Duke U., 1979. Vis. instr. psychology Colgate U. Hamilton, N.Y., 1979-80; asst. prof. psychology Mt. Holyoke Coll., South Hadley, Mass., 1980-81; asst. prof. info. sci. U. Pitts., 1981—. Contbr. articles to profl. jours. U. Pitts. research devel. grantee, 1982. Mem. Am. Psychol. Assn., Eastern Psychol. Assn., Am. Soc. Info. Sci., Assn. Computing Machinery. Subspecialty: Cognition. Current work: Human memory, visual perception, human info. processing, social cognition. Office: Univ Pitts 727 L I S Pittsburgh PA 15260

EVANS, NANCY REMAGE, astronomer; b. Taunton, Mass., May 19, 1944; d. Russell and Esther (Swaffield) Remage; m. Martin Griffith Evans, Aug. 3, 1968; children: Lisa Remage, Katherine Griffith. B.A., Wellesley Coll., 1966; M.S., U. Toronto, Ont. Can., 1969, Ph.D., 1974. Postdoctoral fellow U. Toronto, 1975-77, research asso., 1977-82, asst. prof., 1982-83; RDAF spectropist IVE Obs. Computer Scis. Corp., Greenbelt, Md., 1983—. Contbr. articles to profl. jours. Grantee in field. Mem. Am. Astron. Soc., Can. Astron. Soc., Internat. Astron. Union, Sigma Xi. Subspecialty: Optical astronomy. Current work: Variable stars, classical cepheids, radii, ultraviolet spectroscopy masses. Home: 7333 New Hampshire Ave Apt 1102 Hyattsville MD 20783 Office: Computer Scis Corp Code 685.9 Goddard Space Flight Center Greenbelt MD 20771

EVANS, NEAL JOHN, II, astrophysicist, educator; b. San Antonio, Sept. 22, 1946; s. Neal John and Lucie D. (Barbour) E.; m. Carol Sue Kirschenbaum, Dec. 19, 1973; 1 son, Daniel Spencer. B.A. in Physics, U. Calif.-Berkeley, Ph.D., 1973. Research fellow Calif. Inst. Tech., 1973-75; asst. prof. U. Tex., Austin, 1975-80, assoc. prof. astronomy, 1980—; also writer, cons. Contbr. articles to profl. jours. Recipient 1st prize Griffith Observer essay contest, 1974. Mem. Am. Astron. Soc., Internat. Astron. Union, Internat. Union Radio Sci. Democrat. Subspecialties: Radio and microwave astronomy; Infrared optical astronomy. Current work: Studies of molecular clouds and star formation using millimeter and infrared astronomy. Home: 5400 Shoalwood St Austin TX 78756 Office: Dept Astronom U Tex Austin TX 78712

EVANS, ROBERT, research scientist; b. Cardiff, South Wales, Feb. 2, 1940; m. Penelope Susan Butler, Sept. 24, 1966; children: Timothy Robert, Alexa Catharine. B.A., Cambridge (Eng.), U., 1962; Ph.D., London U., 1967, D.Sc., 1979. Research asst. London Hosp. Med. Coll., 1962-67; postdoctoral fellow Sloan Kettering Inst., Rye, N.Y., 1967; research assoc. Pacific N.W. Research Found., Seattle, 1967-68; staff scientist Chester Beatty Research Inst., Sutton, Eng., 1968-78; sr. staff scientist Jackson Lab., Bar Harbor, Maine, 1978—; mem. exptl. immunology rev. group Nat. Cancer Inst., NIH. Contbr. articles to profl. jours. Mem. Am. Assn. Immunologists, Am. Assn. Cancer Research, Reticuloendothelial Soc. (pres. 1980-81), N.Y. Acad. Sci. Episcopalian. Subspecialties: Cancer research (medicine); Immunology (medicine). Current work: Tumor immunobiology; the role of macrophages during tumor progression and drug induced regression. Home: Route 308 Mount Desert ME 04660 Office: Jackson Lab Bar Harbor ME 04609

EVANS, WILLIAM EDWARD, pharmacokineticist, clinical pharmacist, educator; b. Clarksville, Tenn., June 22, 1950; s. Buford Joe and Wanda (Wilson) E.; m. Dianne Dewit Miller, Sept. 2, 1972; children: Leslie Rhea, Kelli Nicol. Student, Austin Peay State U., 1968-70; B.Sc., U. Tenn., 1973, Pharm.D., 1974. Diplomate: Am. Bd. Bioanalysis. Research assoc. St. Jude Children's Research Hosp., Memphis, 1976-78, asst. mem., 1978-80, dir. clin. pharmacokinetics-pharmacodynamics sect., 1981—; asst. prof. pharmaceutics U. Tenn. Ctr. Health Scis., Memphis, 1974-76, assoc. prof., 1981-83, prof., 1983—; mem. adv. com. hematologic and neoplastic diseases U.S. Pharmacopeaia, 1980—. Editor: Applied Pharmacokinetics, 1980; assoc. editor: Pharmacotherapy jour; editorial adv. bd.: Jour. Investigation Anticancer Drugs; contbr. articles to profl. jours. Recipient Young Investigator award NIH, 1978-80; NIH grantee. Mem. AAAS, Am. Soc. Cancer Research, Am. Soc. Clin. Oncology, Am. Soc. Clin. Pharmacology and Therapeutics, Am. Coll. Clin. Pharmacology (pres.'s award 1982, chmn. bd. trustees Research Inst. 1982), Am. Assn. Colls. Pharmacy (research and grad. affairs com.). Methodist. Subspecialties: Pharmacokinetics; Cancer research (medicine). Current work: Metabolism and disposition of anticancer drugs in children. Pharmacokinetics, clinical pharmacology of antineoplastic drugs. Home: 2425 Willinghurst Germantown TN 38138 Office: St Jude Children's Research Hosp 332 N Lauderdale Memphis TN 38105

EVARTS, EDWARD VAUGHAN, neurophysiologist; b. N.Y.C., Mar. 28, 1926; m., 1950, ; 2d, 1971; 3 children. M.D., Harvard U., 1948. Med. house officer Peter Bent Brigham Hosp., 1948-49; research assoc. Yerkes Labs. Primate Biology, 1949-50; Moseley traveling fellow in neurology Nat. Hosp., London, 1950-51; asst. research psychiatrist Payne Whitney Clinic, 1951-53; chief physiology sect. Lab. Clin. Sci., NIMH, Bethesda, Md., 1953-71, chief, 1971-77. Mem. Nat. Acad. Sci., Am. Physiology Soc., Internat. Brain Research Soc., Psychiat. Research Soc., Soc. Neuroscience, Inst. Medicine. Subspecialty: Neurophysiology. Office: Bldg 36 Room 2D10 NIMH Bethesda MD 20014

EVENS, RONALD GENE, physician; b. St. Louis, Sept. 24, 1939; s. Robert and Dorothy (Lupkey) E.; m. Hanna Blunk, Sept. 3, 1960; children—Ronald Gene, Christine, Amanda. B.A., Washington U., St. Louis, 1960, M.D., 1964, postgrad. in Bus. and Edn, 1970-71. Intern Barnes Hosp., St. Louis, 1964-65; resident Mallinckrodt Inst. Radiology, 1965-66, 68-70; research asso. Nat. Heart Inst., 1966-68; asst. prof. radiology, v.p. Washington U. Med. Sch., 1970-71; prof., head dept. radiology, dir. Mallinckrodt Inst. Radiology, 1971-72; Elizabeth Mallinckrodt prof., head radiology dept., dir. Mallinckrodt Inst., 1972—; radiologist in chief Barnes and Children's Hosp., St. Louis, 1971—; chmn. bd. Med. Care Group, St. Louis, 1980—; mem. bd. Washington U. Med. Center, 1980—; mem. adv. com. on splty. and geog. distbn. of physicians Inst. Medicine, Nat. Acad. Scis., 1974-76; dir. City Bank St. Louis. Contbr. over 120 articles to profl. jours. Lodge adviser Order Arrow, Boy Scouts Am., 1975—; elder Glendale and Kirkwood Presbyn. Ch., 1971-74; bd. dirs. St. Louis Comprehensive Neighborhood Health Center, OEO, 1970-74. Served with USPHS, 1966-68. James Picker Found. advanced acad. fellow, 1970; Hickey lectr., 1976; recipient Disting. Service award St. Louis C. of C., 1972. Fellow Am. Coll. Radiology; mem. Mo. Radiol. Soc. (pres. 1977-78), Soc. Nuclear Medicine (trustee 1971-75), AMA, St. Louis Med. Soc., Mo. Med. Assn., Soc. Chairmen Acad. Radiology Depts. (pres. 1979), Radiol. Soc. N.Am., Assn. Univ. Radiologists, Am. Roentgen Ray Soc. (v.p. 1982), Phi Beta Kappa, Alpha Omega Alpha (Sheard-Sanford award). Subspecialties: Radiology; Nuclear medicine. Office: 510 S Kingshighway Saint Louis MO 63110

EVENSON, EDWARD BERNARD, geology educator, researcher; b. Milw., Dec. 30, 1942; s. Bernard John and Loraine (Willich) E.; m.; 1 son, Mark John. B.S., U. Wis.-Milw., 1965, M.S., 1969; Ph.D., U. Mich., 1972. Sr. research geologist Exxon Research Lab., Houston, 1972-73; assoc. prof. geology Lehigh U., Bethlehem, Pa., 1973—, dir., 1973—. Editor: Tills and Related Deposits, 1983. NSF trainee U. Mich., 1970-73. Fellow Geol. Soc. Am.; mem. Nat. Assn. Geology Tchrs., Sigma Xi (treas. 1975-82). Republican. Subspecialty: Geology. Current work: Glacial geology and geomorphology, mineral exploration in glaciated areas, glacial stratigraphy. Home: 18 E Goepp St Bethlehem PA 18018 Office: Lehigh Univ Williams Hall Bethlehem PA 18015

EVENSON, WILLIAM EDWIN, ecol. physicist, cons., educator, researcher; b. Martinez, Calif., Oct. 12, 1941; s. Raymond Fox and Berta (Woolley) E.; m. Nancy Ann Woffinden, Dec. 21, 1964; children: Brian, Elizabeth, Joann, Andrew, Bengte. B.S. in Physics, Brigham Young U., 1965; Ph.D. in Theoretical Solid State Physics (Woodrow Wilson fellow, NSF fellow, Danforth fellow), Iowa State U., 1968. Research asso. U. Pa., Phila., 1968-70; NSF postdoctoral fellow, 1968-69; asst. prof. Brigham Young U., Provo, Utah, 1970-73, assoc. prof., 1973-79, prof. physics, 1979—, asso. dir. gen. edn., 1980-81, dir., 1981-82, dean, 1982—; vis. colleague in botany U. Hawaii at Manoa, Honolulu, 1977-78; cons. Eyring Research Inst., Provo, 1980-82; cons. computer programs for ecol. analysis, 1975—. Contbr. articles on ecol. physics to profl. jours. Bishop 108th ward Ch. Jesus Christ of Latter-day Saints, 1971-74, high council 1st stake, 1970-71, 1974-75, mem. Mission to France, Paris, 1961-63; chmn. Utah County Democratic Party, 1981-82, vice chmn., 1979-81; mem. Utah State Dem. Party Central Com., 1979-82. Named Brigham Young U. Prof. of Month, Feb. 1979. Mem. Am. Assn. Physics Tchrs., Am. Bot. Soc., Hawaii Bot. Soc., Hawaii Audubon Soc., Nat. Audubon Soc., Sigma Xi. Subspecialties: Theoretical ecology; Theoretical physics. Current work: Use of computers in ecological analysis, applications of physics to solving problems in theoretical ecology, computer programming, ecological analysis, reproductive energetics. Office: Dept Physics and Astronomy Brigham Young U Provo UT 84602

EVERETT, ARDELL GORDON, geochemist/geologist; b. Cambridge, Mass., July 27, 1937; s. Ardell Tillman and Iva Lorene (Ferrier) E.; m. Natalie Randle Eubank, Dec. 27, 1960; children: Elizabeth Randle, Virginia Monroe, William Gordon. A.B., Cornell U., 1959; M.S., U. Okla., 1962; Ph.D., U. Tex., Austin, 1968; student, Rice Inst., 1955-56; postgrad., U. Tex. Law Sch., 1965-67. Instr. to asst. prof. dept. geology Ohio State U., Columbus, 1967-68, 68-69; spl. asst. to asst. sec. U.S. Dept. Interior, Washington, 1969-70, dep. asst. sec. applied scis., 1970; dir. Office Tech. Analysis, EPA, Washington, 1971-74; dir. regulatory litigation Am. Petroleum Inst., Washington, 1974-77; pres., owner Everett & Assocs., Rockville, Md., 1978-82; v.p., sec., dir. Ardell T. Everett Inc., Arlington Heights, Ill., 1973—; cons. Dept. Minerals and Energy, Papua New Guinea, 1979—. Sr. author: Western Wyoming. . . Development of Sour Gas Resources, 1981, Withdrawal of Public Lands from Access to Minerals and Fuels, 1980, Estimate of Geochemical Processes in Fly River Drainage, Papua New Guinea, 1980. Republican precinct chmn. Travis County (Tex.), 1964. Served to 1st. lt. USAR, 1959-67. Recipient cert. of Merit Am. Inst. Profl. Geologists, 1982; U. Tex. fellow, 1963-66; U. Okla. scholar, 1961. Fellow Geol. Soc. Am.; mem. Am. Assn. Petroleum Geologists Geochem. Soc., Am. Inst. Profl. Geologists, Soc. Mining Engrs. of AIME. Club: Cosmos (Washington). Subspecialties: Geochemistry; Geology. Current work: Application of multidisciplinary research to minerals, fuels and environ. problems; application of technical knowledge to litigation and regulatory issues. Home: 203 Dale Dr Rockville MD 20850 Office: Suite 300 416 Hungerford Dr Rockville MD 20850

EVERETT, ELWOOD DALE, physician, educator; b. Ft. Gibson, Okla., Aug. 18, 1939; s. Elwood Humphrey and Lola Mae (Tennison) E.; m. Billie Jean Cain, June 17, 1962; children: Kevin, Brett, jennifer, Kathryn. Student, Northeastern State U., Tahlequah, Okla., 1956-59; M.D., U. Okla., 1963. Commd. capt. M.C. U.S. Army, advanced through grades to lt. col., 1971; chief resident (Fitzsimmon Army Hosp.), Denver, 1968-69, asst. chief medicine, 1970-73, chief internal medicine, Ft. Ord, Calif., 1969-70, chief infectious diseases, San Antonio, 1975-77, ret., 1977; dir. infectious diseases U. Mo., Columbia, 1977—, interim. chmn. dept. medicine, 1982—. Assoc. editor: Jour. Lab. and Clin. Medicine, 1981. Named Outstanding Tchr. Dept. Medicine, U. Mo., 1981. Fellow ACP, Infectious Diseases Soc. Am.; mem. Am. Soc. Microbiology, Central Soc. Clin. Research, Am. Fedn. Clin. Research. Republican. Mem. Ch. of Christ. Subspecialties: Internal medicine; Infectious diseases. Current work: Peritonitis; fungal diseases.

EVERETT, ROBERT RIVERS, manufacturing company executive; b. Yonkers, N.Y., June 26, 1921; s. Chester McKenzie and Ruth (Melius) E.; m. Helen Burns, Oct. 21, 1944 (div. 1972); children—Robert F., Bruce M., Douglas F., Theodore J., Michael B.; m. Jean M. McGrath, Nov. 4, 1972 (dec. Nov. 1980); m. Ann T. Russell, Mar. 26, 1982. B.S., Duke, 1942; M.S., Mass. Inst. Tech., 1943. With Servomechanisms Lab. of Mass. Inst. Tech., 1942-51, asso. dir., 1951; asso. div. head Lincoln Lab., 1951-56, div. head, 1956-58; tech. dir. The Mitre Corp., Bedford, Mass., 1958-59, v.p. tech. operations, 1959-69, exec. v.p., 1969, pres., 1969—; mem. sci. adv. bd. USAF; mem. sci. adv. group Def. Communications Agy.; trustee No. Energy Corp.; cons. Def. Sci. Bd.; cons. div. adv. group Electronic Systems div. Air Force Systems Command.; Mem. sci. adv. bd. U.S. Air Force. Contbr. articles to tech. jours. Fellow IEEE; mem. Assn. Computing Machinery, AAAS, Nat. Acad. Engring., Phi Beta Kappa, Sigma Xi, Tau Beta Pi. Club: Cosmos (Washington). Subspecialties: Information systems, storage, and retrieval (computer science); Digital computer design. Patentee digital computers. Home: 80 Rollingwood Ln Concord MA 01742 Office: PO Box 208 Bedford MA 01730

EVERHART, DONALD LEE, immunology educator; b. Erie, Pa., Jan. 27, 1932; s. Watson H. and Ruth (Swarner) E.; m. Barbara Spiess, Aug. 27, 1955. B.S., Grove City Coll., 1954; M.S., Boston U., 1957, Ph.D., 1961. Research fellow U. Tenn., Knoxville, 1961-63; research assoc. Ill. Inst. Tech. Research Center, Chgo., 1962-66; asst. prof. Med. Coll. Va., Richmond, 1966-71; assoc. prof., chmn. N.Y. U. Coll. Dentistry, N.Y.C., 1971-78, prof. immunology, chmn., 1978—. Served with U.S. Army, 1956-57. Mem. Am. Assn. Dental Research (sec. 1975—), Am. Chem. Soc., N.Y. Acad. Sci. Subspecialties: Immunocytochemistry; Microbiology. Current work: Cures vaccine and immunology of periodontal disease. Home: 63 Ridgeview St New Providence NJ 07974 Office: NY U Coll Dentistry 345 E 24th St New York NY 10010

EVERHART, DONALD LOUGH, geologist; b. Troy, Ohio, July 18, 1917; s. William Alfred and Mary Elder (Lough) E.; m. Dorothy Alice Lindaman, June 27, 1942; children: Mary Ellen, Lawrence, Gregory, Douglas. A.B., Denison U., 1939; M.A., Harvard U., 1942, Ph.D., 1953. Geologist U.S. Geol. Survey, Washington, 1942-48; chief geol. br. U.S. AEC, N.Y.C., 1949-54, geol. adv., Denver, 1954-59; div. v.p. Internat. Minerals & Chem. Corp., Libertyville, Ill., 1959-77; mgr. office U.S. Dept. Energy, Grand Junction, Colo., 1977-81; cons. geologist, Grand Junction, 1981—; U.S. del. UN Conf., Geneva, 1958, ECAFE Conf., Bangkok, 1963. Author, editor: Uranium in the USA, 1980; author: (with E.H. Bailey) Quicksilver-New Almaden, California, 1964. Mem. adv. bd. Salvation Army, Grand Junction, 1980—; moderator United Ch. of Christ Congregational Ch., Grand Junction, 1982; pres. Community Concerts, Arlington Heights, Ill., 1974-75; mem. Cultural Commn. Arlington Heights, 1974-77. Recipient Disting. Alumni award Denison U., 1972. Fellow Geol. Soc. Am.; mem. Soc. Econ. Geologists (council 1969-72), Am. Inst. Mining Engrs., Phi Beta Kappa. Republican. Clubs: Bookcliff Country, Harvard (Colo.). Subspecialties: Geology; Fuels. Current work: Innovative mineral resource exploration and assessment methods; data base application to geologic models. Home: 2700 G Rd Unit 10-B Grand Junction CO 81501 Office: 715 Horizon Dr Suite 401 Grand Junction CO 81501

EVERHART, WATSON HARRY, natural resources educator, academic administrator, fishery biologist; b. Connellsville, Pa., June 5, 1918; m.; 3 children. B.S., Westminster Coll., Pa., 1940; M.S., U. Pitts., 1942; Ph.D. in Fishery Biology, Cornell U., 1948. Asst. in embryology and anatomy U. Pitts., 1940-42; asst. fishery biologist Cornell U., 1945, asst. biologist, 1947-48, chmn. dept. natural resources, 1972—, prof. natural resources, 1980—; state fishery biologist, State of Conn., 1947; asst. prof. fishery biology and ichthyology U. Maine, 1948-53, from assoc. prof. to prof. zoology, 1953-67; fishery biologist Maine Inland Fisheries and Game, 1948-50, chief fishery researcher and mgr., 1950-67; prof. biology, chmn. fishery maj. Colo. State U., 1967-72; mem. Atlantic Sea Run Salmon Commn., 1953-67; cons. Colo. Game, Fish and Parks Dept., 1967—. Mem. Am. Fisheries Soc., Am. Soc. Ichthyologists and Herpetologists, Am. Soc. Limnology and Oceanography. Subspecialties: Resource management; Fishery Biology. Office: Dept Natural Resources Cornell U Ithaca NY 14853

EVERLY, GEORGE STOTELMYER, JR., psychophysiologist; b. Balt., May 31, 1950; s. George Stotelmyer and Kathleen (Webster) E.; m. Gayle May Schabdach., Apr. 27, 1975; 1 dau., Marideth Rose. B.S., U. Md., 1972, M.A., 1974, Ph.D., 1978. Instr. U. Md., 1975-80, assoc, prof., 1980-82; assoc. prof. Loyola Coll. Md., Balt., 1980—, dir. behavioral medicine lab., 1982—; dir. behavioral medicine service (Psychol. Scis. Inst.), Balt., 1982—. Author: Controlling Stress and Tension, 1979, Natural Treatment of the Stress Response, 1981, Occupational Health Promotion, 1984; Developer: Everly Stress and Coping Inventory, 1984. Mem. Am. Psychol. Assn., Acad. Psychosomatic Medicine, Biofeedback Soc. Am., Am. Acad. Behavioral Medicine, Soc. Behavioral Medicine. Subspecialties: Physiological psychology; Psychobiology. Current work: Cognitive influences in psychosomatic medicine; clinical applications for biofeedback therapy. Home: 204 Glenmore Ave Catonsville MD 21888 Office: Dept Psychology Loyola Coll Md Baltimore MD 21210

EVERNDEN, JACK FOORD, research geophysicist; b. Okeechohee, Fla., Mar. 12, 1922; s. Hans Foord and Rose (Wagner) E.; m. Roberta Katherine Smith, Dec. 31, 1965. B.S., U. Calif.-Berkeley, 1948, Ph.D., 1951. Asst. prof. to prof. dept. geology and geophysics U. Calif.-Berkeley, 1953-65; researcher, program mgr. Air Force Tech. Applications Ctr. and Advance Research Projects Agy., Dept. Def., Washington, 1965-71; research scientist U.S. ACDA, Washington, 1971-73; program mgr., research geophysicist Nat. Ctr. for Earthquake Research, U.S. Geol. Survey, Menlo Park, Calif., 1973—. Contbr. articles to profl. jours. Served with USAAF, 1943-46. Recipient Newcomb-Cleveland prize AAAS, 1962; Disting. Civilian Service award U.S. Air Force, 1967; Meritorious Service medal U.S. Dept. Interior, 1983. Mem. Seismological Soc. Am. Subspecialties: Geophysics; Applied mathematics. Current work: Earthquake analysis and predictability; research to determine or modify understanding of earthquakes as they relate to the structure of the earth, premonitory phenomena, and explosions. Office: US Geol Survey 345 Middlefield Rd Menlo Park CA

EVERSMEYER, HAROLD EDWIN, biology educator; b. Randolph, Kans., July 7, 1927; s. Gideon F. and Susie Elizabeth (Kintigh) E.; m. Ruth Josephine Stinson, Oct. 18, 1953; children: Clair, Elaine, Kent, Denise. B.S. in Agrl. Edn, Kans. State U., 1951, Ph.D. in Botany and Plant Pathology, 1965. 4-H club agt. Johnson County (Kans.) Extension Service, 1951-54, Lyon County (Kans.) Extension Service, 1956-60; mem. faculty dept. biol. scis. Murray (Ky.) State U., 1964—, asst. prof. biology, 1964-67, assoc. prof., 1967-69, prof., 1969—. Served with U.S. Army, 1954-56. Mem. Soc. Nematologists, Ky. Acad. Scis. (dir.), Helminthological Soc. Washington, Sigma Xi, Alpha Zeta. Methodist. Current work: Currentwork: Aeromycology. Home: 820 N 19th St Murray KY 42071 Office: Dept Biol Scis Murray State U Murray KY 42071

EVES, EVA MAE, geneticist, educator; b. Danville, Pa., Mar. 4, 1949; d. E. Eugene and Laura Irene (Metz) E.; m. Jeffrey B. Schamis, Oct. 22, 1977. B.A. in Biology, U. Chgo., 1970, Ph.D. in Genetics, 1979. Postdoctoral fellow U. Chgo., 1979-81, research assoc., 1981-82; research geneticist VA Westside Hosp., 1982—; instr. Ctr. for Genetics U. Ill.-Chgo., 1982—. Contbr. articles on genetics to profl. jours. Am. Cancer Soc. fellow, 1979-81. Mem. Genetics Soc. Am., Am. Soc. of Cell Biology. Club: East Bank (Chgo.). Subspecialties: Gene actions; Molecular biology. Current work: Molecular biology, mechanisms of gene expression. Office: 820 S Damen Chicago IL 60612

EVIAN, CYRIL IAN, periodontics researcher, educator, periodontist; b. Johannesburg, Transvaal, South Africa, july 31, 1948; came to U.S., 1977; s. Solly and Sonia (Dembo) E.; m. Cheryl Adrienne Freedman, Dec. 12, 1971; 1 son, Allon; m. Andrea Michelle Goldin, Febv. 3, 1976; children: Samantha, Tracy, Debbie. B.D.S., U. Witwatersrand, Johannesburg, 1971, Higher Dental Diploma, 1974; cert. periodontics, U. Pa., 71979, D.M.D., 1981. Clin. instr. U. Witwatersand, 1971, 75, lectr. in preventive dentistry, 1975-76; adj. asst. prof. periodontics U. Pa., 1979-80; research assoc. in periodontics Center for Oral Health Research, 1979-80, asst. prof. periodontics, 1980—; dir. grad. periodontics center for Oral Health Research, 1983—. Contbr. articles, abstracts to profl. jours. Mem. South African Soc. Periodontology, South African Soc. Endodontics, Internat. Assn. Dental Research, ADA, Pa. Dental Soc., Am. Acad. Periodontology, Phila. Dental County Soc., South African Dental Assn. (chmn. com. preventive dentistry 1975), Brit. Dental Soc. Subspecialty: Periodontics. Current work: Periodontal research. Home: 1124 Woodbine Ave Narberth PA 19072 Office: U Pa 4001 Spruce St Philadelphia PA 19104

EWAN, RICHARD COLIN, animal scientist; b. Cuba, Ill., Sept. 10, 1934; s. John Grafton and Zelma (Shoop) E.; m. Arlene Francis Ewan, June 23, 1956; children: William, Richard, Daniel, Christopher. B.S., U. Ill., 1956, M.S., 1957; Ph.D., U. Wis., 1966. Research asst. U. Ill., 1956-57, U. Wis., 1962-66; prof. animal sci. Iowa State U., 1966—. Contbr. in field. Active PTA, Ames, Iowa, 1967-78; active Mid-Iowa council Boy Scouts Am., 1966—. Served to 1st lt. USAF, 1957-62. Calcium Carbonate travel fellow, 1976; recipient Boy Scouts Am. Dist. award of merit, 1974, Silver Beaver award, 1980. Mem. Am. Inst. Nutrition, Am. Soc. Animal Sci. Subspecialty: Animal nutrition. Current work: Energy metabolism of pigs; vitamin E and selenium metabolism of pigs nutrition, swine, energy, vitamins and minerals.

EWING, BENJAMIN BAUGH, engineering educator; b. Donna, Tex., Apr. 4, 1924; s. Joshua Fulkerson and Bula Betty (Baugh) E.; m. Elizabeth Malone, Apr. 3, 1947; children: Melissa, Douglas Malone, Frederick Joshua. B.S., U. Tex., Austin, 1944, M.S., 1949; Ph.D., U. Calif. at Berkeley, 1959. Diplomate: Am. Acad. Environ. Engrs. Instr., asst. prof. U. Tex., Austin, 1947-55; asso. in civil engring., asst. research engr. U. Calif. at Berkeley, 1955-58; asso. prof., prof. U. Ill., Urbana, 1958—, dir., 1966-73, 1972—; Cons. engr., 1959—. Trustee Urbana and Champaign San. Dist., 1974-80; public mem. Ill. Water Resources Commn., 1975—. Served to lt. (j.g.) CEC. USNR, 1943-46. Recipient Epstein award dept. civil engring. U. Ill., 1961, Harrison Prescott Eddy award for noteworthy research, 1968. Fellow ASCE; mem. Am. Water Works Assn. (life), Am. Geophys. Union, Water Pollution Control Fedn., AAAS, Assn. Environ. Engring. Profs. Club: Rotarian. Subspecialties: Water supply and wastewater treatment; Civil engineering. Current work: Water quality management and pollution control, water treatment, wastewater treatment, water resources management. Home: 2212 Cottage Grove Urbana IL 61801 Office: 408 S Goodwin St Urbana IL 61801

EWING, DAVID LEON, radiation biology researcher; b. Shreveport, La., Aug. 20, 1941; s. Arlington Burleson and Mary Lenore (Bryant) E.; m. Mary Dessagene Crawford, Feb. 6, 1965; children: Kelley Michelle, Scott Emlyn. B.S. in Physics magna cum laude, Centenary Coll., 1963; M.S. in Environ. Health, U. Calif.-Berkeley, 1965; Ph.D. in Zoology, U. Tex., 1969. Radiobiologist NASA, Johnson City, Tex., 1965; NIH postdoctoral fellow, Sutton, Surrey, Eng., 1970-72; research scientist, asst. prof. U. Tex., Austin, 1972-76; assoc. prof. Med. and Grad. Schs., Hahnemann U., Phila., 1976-82, prof., 1982—, carcinogenic safety officer, 1982—. Contbr. articles to profl. jours. Mem. Hist. Soc. Phoenixville (Pa.) Area, 1979—, Phila. Mus. Art., 1980—, Brandywine Art Mus. and Conservatory, Chadds Ford, Pa., 1980-82; active Boy Scouts Am. Recipient Sr. Sci. award Bell Telephone Co., 1959; Magale Found. fellow, 1960-63; Nat. Cancer Inst.-NIH grantee, 1976—. Mem. N.Y. Acad. Sci., Radiation Research Soc., AAAS, Environ. Mutagen Soc., Alpha Chi, Alpha Sigma Pi, others. Subspecialties: Cancer research (medicine); Radiation Biology. Current work: Chemical mechanisms for cellular damage from ionizing radiation; mechanisms of action for agents which modify radiation sensitivity. Office: Hahnemann U Health Scis 230 N Broad St Philadelphia PA 19102

EWING, JUNE SWIFT, scientific program administrator, electron microscopist; b. Fayetteville, Ark., July 19, 1938; d. Albert Duane and Anice Gertrude (Carlisle) Swift; m. George Edward Ewing, Feb. 18, 1961; 1 dau., Alice Adair; m. Thomas Delaney Wilkerson, Jan. 1, 1978. Student, Cornell Coll., Mt. Vernon, Iowa, 1955-57; B.S. cum laude in Chemistry, U. Wis., 1959; M.S. in Phys. Chemistry, U. Calif.-Berkeley, 1961; M.P.A., U. Colo., 1976. Staff cons. CHEM study project Harvey Mudd Coll., Claremont, Calif., 1961-63; electron microscopist, dept. chemistry Ind. U., Bloomington, 1963-66; instr. dept. chemistry Rutgers U., New Brunswick, N.J., 1969-70; lab. mgr. dept. molecular, cellular and devel. biology U. Colo., Boulder, 1971-74; program mgr. Univs. Space Research Assn., Houston, 1975-77; cons. Sci. Applications, Inc., McLean, Va., 1977-79; staff officer, biomed. programs Nat. Acad. Scis., Washington, 1979—; v.p. Environ. Sci. Communications, Inc.; sec. Applied Sci. Tech., Inc. Author: (with M. A. Bonneville) Laboratory Exercises, 1973. Pres. Unitarian Ch., Boulder, 1974-76; bd. dirs. End World Hunger Benefit Com., Washington, 1981—. Mem. AAAS. Subspecialties: Neurophysiology. Current work: Electron microscopy of brain tissue; administration of biomedical studies; funding of neuroscience research programs; funding of international scientific travel, planning and funding of international scientific meetings. Office: Nat Acad Scis 2101 Constitution Ave NW Washington DC 20416

EWING, MARTIN SIPPLE, radio astronomer, research engr.; b. Albany, N.Y., May 4, 1945; s. Galen W. and Alice C. (Sipple) E.; m. Eva R., June 11, 1966; children: Margaret, Robert, Eric. B.A., Swarthmore Coll., 1966; Ph.D., M.I.T., 1971. Research fellow Calif. Inst. Tech., Pasadena, 1971-73, mem. profl. staff, 1973—. Contbr. articles to profl. jours. Mem. Internat. Union Radio Sci., Internat. Astron. Union, Am. Astron. Soc., IEEE. Subspecialties: Radio and microwave astronomy; Computer engineering. Current work: Researcher in digital systems for radio astronomy, instrumentation design and data analysis facility mgmt. Office: Calif Inst Tech MS 105-24 Pasadena CA 91125

EWING, RICHARD EDWARD, educator, oil company researcher; b. Kingsville, Tex., Nov. 24, 1946; s. Floyd Ford and Olivia Clara (Henrichson) E.; m. Rita Louise Williams, Aug. 8, 1970; children: John Edward, Lawrence Alan. B.A., U. Tex.-Austin, 1969, M.A., 1972, Ph.D., 1974. NSF postdoctoral fellow U. Chgo., 1976-77; vis. prof. Math. Research Center, U. Wis.-Madison, 1978-79; assoc. prof. Ohio State U., Columbus, 1977-80; sr. research mathematician Mobil Research and Devel., Dallas., 1980-82, assoc., 1982—; J.E. Warren Disting. prof. energy and environ. U. Wyo., Laramie, 1982—; mem. research adv. council Center for Computational Studies in Petroleum, Fort Collins, Colo., 1981-82; cons. Mobil Field Research Lab., Dallas, 1982—. Contbr. articles to profl. jours. Active Boy Scouts Am., Dallas. U.S. Army Research grantee, 1978-82, 83—; NSF grantee, 1979-83. Mem. Am. Math. Soc., Soc. Indsl. and Applied Math. (adv. council Tex.-Okla. sect. 1981-82), Soc. Petroleum Engrs., Internat. Assn. Math. and Computers in Simulation, Sigma Xi. Presbyterian. Subspecialties: Composite materials; Petroleum engineering. Current work: Applied mathematics, numerical analysis, partial differential equations, inverse problems in geophysics, petroleum reservoir simulation. Home: 4403 Grays Gables Rd Laramie WY 82070 Office: Univ Wyoming Dept Math and Dept Petroleum Engring Laramie WY 82071

EWING, RODNEY CHARLES, geology educator; b. Abilene, Tex., Sept. 20, 1946; s. Charles Thomas and Mary Louise (Cobos) E.; m. Jerrilyn Ann Harris, June 17, 1973; 1 son, Travis Russell. B.S., Tex. Christian U., 1968; M.S., Stanford U., 1972, Ph.D., 1974. Assoc. prof. U. N.Mex., Albuquerque, 1978—, asst. prof., 1974-78, chmn. dept. geology, 1978—; guest scientist Hahn-Meitner Inst., Berlin, 1979-82; vis. scientist Oak Ridge Nat. Lab., 1980-82; cons. Sandia Nat. Lab., Albuquerque, 1979-81, Prentice Hall, Englewood Cliffs, N.J., 1979-81. Served with U.S. Army, 1969-70. NSF grad. fellow Stanford U., 1970-74; U. Queensland travel grantee, 1982. Fellow Geol. Soc. Am.; mem. Mineral Soc. Am. (assoc. editor 1979-81), Materials Research Soc, councilor 1982-85), Mineral. Soc. Can., Geochem. Soc. Subspecialties: Mineralogy; Materials. Current work: Radiation effects in crystalline materials, genesis of pegmatites and rare earth mineralogy. Home: 821 Solano St NE Albuquerque NM 87110 Office: U N Mex Dept Geology Albuquerque NM 87131

EWING, RONALD IRA, physicist; b. Dallas, July 13, 1935; s. Ira D. and F. Barbara (Oswald) E.; m. Carol Nasby, June 7, 1957 (div. 1963); children: Anne, Amy; m. Judy Williams, Mar. 20, 1966; children: Megan, Guy. B.A., Rice U., 1956, M.A., 1957, Ph.D., 1959. Mem. tech. staff Sandia Nat. Lab., Albuquerque, 1959—. Mem. Am. Phys. Soc., Phi Beta Kappa, Sigma Xi. Subspecialty: Lightning. Current work: Lightning phenomenology, experimental and descriptive. Home: 1320 Dakota SE Albuquerque NM 87108 Office: 7554 Sandia Nat Lab PO Box 5800 Albuquerque NM 87185

EWING, SOLON ALEXANDER, animal scientist, university administrator; b. Headrick, Okla., July 21, 1930; m.; 2 children. B.S., Okla. State U., 1952, M.S., 1956, Ph.D. in Animal Nutrition, 1958. Instr. in animal sci. Okla. State U., 1956-58; from asst. prof. to assoc. prof. Iowa State U., Ames, 1958-64, head dept. animal sci., 1973—; prof. Okla. State U., 1964-68; asst. dir. Iowa Agrl. and Home Econs. Expt. Sta., 1968-73. Mem. Am. Soc. Animal Sci. Subspecialty: Animal nutrition. Office: 101 Kildee Hall Iowa State U Ames IA 50010

EXTEIN, IRL LAWRENCE, psychiatrist, researcher; b. St. Louis, Mar. 3, 1948; s. Alvin M. and Leadora S. E.; m. Barbara Sundheimer, June 2, 1974; children: Melissa, Jason. B.A., U. Chgo., 1970; M.D., Yale U., 1974. Diplomate: Am. Bd. Psychiatry and Neurology. Resident in psychiatry Yale U., 1975-77; clin. assoc. clin. psychobiology br. NIMH, 1977-79; co-dir. neuropsychiat. evaluation unit Fair Oaks Hosp., Summit, N.J., 1979—, dir. research and tng., 1980—. Contbr. articles to profl. publs. Served assurgeon USPHS, 1977-79. Mem. AMA, Am. Psychiat. Assn., Soc. Neurosci., Soc. Biol. Psychiatry, Soc. Psychoneuroendocrinology, Am. Acad. Clin. Psychiatrists. Jewish. Subspecialties: Psychiatry; Psychopharmacology. Current work: Psychopharmacology, neuropsychiatric evaluation, mood disorders, neuroendocrine diagnostic tests, drug abuse. Office: 19 Prospect St Summit NJ 07901

EYDEN, BERNARD, mechanical engineer, engineering, consultant; b. Rugby, Eng., Oct. 18, 1919; s. John and Violet Cecilia (Clay) E.; m. Lona Constance Webster, Nov. 11, 1921; children: Allan John, Janice Anne. Cert. in Mech. Engring, Southall Tech. Coll., London. Registered profl. engr., Eng., Ont., Mass., Colo. Design and procurement engr. Brazilian Light and Power Co. Ltd., Toronto, Ont., Can., 1947-55; cons. Stone & Webster Engring. Corp., Boston, 1957-76; chief mech. engr. Tippetts-Abbett-McCarthy-Stratton, Belo Horizonte, Brazil, 1976-79; engring. cons. Brown & Root, Inc., Houston, 1979—. Contbr. papers and tech. articles to profl. publs. Served to capt. Royal Elec. and Mech. Engrs., 1942-47. Recipient Whitworth Soc. prize, 1938; Instn. Mech. Engrs. prize, 1940. Mem. ASME, ASCE, Instn. Mech. Engrs. Subspecialties: Mechanical engineering; Hydroelectric Power. Current work: Hydraulic turbines, pumps and related equipment for hydroelectric power stations and water pumping stations. Home: 9411 Westheimer Rd Apt 116-B Houston TX 77063 Office: PO Box 3 Houston TX 77001

EYLER, JOHN ROBERT, chemist, educator; b. Wilmington, Del., May 29, 1945; s. Robert Wilson and Doris Leota (Robinson) E.; m. Fonda Page Davis, June 24, 1967; children: Lisa Todd, Jason Nathaniel. B.S. in Chemistry with honors, Calif. Inst. Tech., 1967; Ph.D. in Chem. Physics, Stanford U., 1972. NRC-Nat. Bur. Standards postdoctoral assoc. Nat. Bur. Standards, Gaithersburg, Md., 1972-74; asst. prof. U. Fla., 1974-79, assoc. prof., 1979—. Contbr. articles to profl. jours. Recipient award of Merit Chem. and Engring. News, 1967, Tchr.-Scholar award Camille and Henry Dreyfus Found., 1978. Mem. Am. Chem. Soc., Inter Am. Photochem. Soc., Am. Soc. Mass Spectrometry. Subspecialties: Physical chemistry; Laser-induced chemistry. Current work: Ion-molecule reactions; laser irradiation of ions, surfaces, and neutrals near surfaces; developed Fourier transform mass spectrometry. Office: Dept Chemistry U Fla Gainesville FL 32611

EZRIN, ALAN MARK, biology researcher; b. Phila., Mar. 21, 1952; s. Sidney and Edie (Hassman) E.; m. Nancy Salz, Aug. 12, 1973; children: Jodi Elyse, Kara Rochelle. B.S., U. Miami, 1973, M.S., 1977; Ph.D., Sch. Medicine, 1980. Research assoc. dept. pharmacology and pediatrics U. Miami Sch. Medicine, 1973-82; sr. research biologist Sterling Winthrop Research Inst., Rensselaer, N.Y., 1982—, antiarrhythmic project leader, 1982—. Fla. Heart Assn. grantee, 1980-82; Am. Heart Assn. grantee, 1982. Mem. Am. Physiol. Soc., Teratology Soc., Am. Heart Assn. Subspecialties: Cardiology; Cellular pharmacology. Current work: Antiarrhythmic drug development, cardiac electrophysiology, cardiovascular pharmacology, neonatal physiology. Office: Sterling Winthrop Research Inst 81 Columbia Turnpike Rensselaer NY 12144 Home: 17 Alpine Dr Latham NY 12110

FAABORG, JOHN RAYNOR, ecology educator; b. Hampton, Iowa, Jan. 23, 1949; s. Rolfe Folmer and Darlene Margaret (Jurgensen) F.; m. Janice Elaine Winters, June 7, 1980; children from previous marriage: Jason, Jodine. B.S., Iowa State U., 1971; Ph.D., Princeton U., 1975. Asst. prof. ecology U. Mo.-Columbia, 1975-81, assoc. prof., 1981—. Contbr. articles to profl. jours. Mem. Am. Ornithologists Union, Wilson Ornithol. Soc/., Cooper Ornithol. Soc. Subspecialties: Ecology; Behavioral ecology. Current work: Factors limiting distribution and abundance of birds, especially species interactions, island biogeography, applied biogeography and behavioral ecology. Office: Div Biol Sci U Mo 110 Tucker Hall Columbia MO 65211 Home: Route 12 13 Bearfield Ct Columbia MO 65201

FABER, DONALD STUART, physiologist, educator; b. Buffalo, Mar. 3, 1943; s. Gilbert and Mildred (Brothman) F.; m. Jo W. Welch, Dec. 26, 1964; children: Eve S., Amy E. S.B., M.I.T., 1964; Ph.D., SUNY-Buffalo, 1968. Grad. asst. dept. physiology SUNY-Buffalo, 1964-68, postdoctoral research fellow in neurobiology, 1968-70, research assoc. prof. dept. physiology, 1975-78, assoc. prof., 1978-81, prof., 1981—, dir. div. neurobiology, 1978—; vis. research assoc. neurobiology div. Max Planck Inst. for Brain Research, Frankfurt, W.Ger., 1970-72; vis. research assoc. Lab. Physiology, U. Paris, 1972,73,76,79,81,82; asst. prof. physiology U. Cin., 1972-74. Mem. editorial bd.: Neurosci; contbr. articles to sci. jours. Grass Found. fellow Marine Biol. Lab., Woods Hole, Mass., summer 1969. Mem. AAAS, Soc. for Neurosci., Am. Physiol. Soc., Internat. Brain Research Orgn., Assn. Neurosci. Depts. and Program. Subspecialties: Neurobiology; Neurophysiology. Current work: Excitability and synaptic interactions in the vertebrate CNS. Office: SUNY Div Neurobiology 313 Cary Hall Buffalo NY 14214

FABER, SANDRA MOORE, astronomer; b. Boston, Dec. 28, 1944; d. Donald Edwin and Elizabeth (Borwick) Moore; m. Andrew Leigh Faber, June 9, 1967; children: Robin Leigh, Holly Ilena. B.A., Swarthmore Coll., 1966; Ph.D., Harvard U., 1972. Asst. prof. Lick Obs., U. Calif.-Santa Cruz, 1972-77, assoc. prof., 1977-79, prof./ astronomer, 1979—. Contbr. articles to profl. jours. NSF fellow, 1966-71; Woodrow Wilson grad. fellow, 1966-71; Alfred P. Sloan Found. fellow, 1977-81; recipient Bart J. Bok prize Harvard U., 1978. Mem. Am. Astron. Soc., Internat. Astron. Union, NOW, ACLU. Subspecialties: Optical astronomy; Cosmology. Current work: Extraglactic astronomy, cosmology, formation and evolution of galaxies. Office: Lick Obs U Calif Santa Cruz CA 95064

FABER, VANCE, mathematician; b. Buffalo, Dec. 1, 1944; s. Norman and Selma (Greenberg) F.; m. Karen M. Peterson, Aug. 16, 1967 (div. 1972); m. Noni K. Soldo, Aug. 31, 1980. A.B., Washington U., St. Louis, 1966, A.M., 1969, Ph.D., 1971. Mem. faculty U. Colo., Denver, 1970-80; mem. staff Los Alamos Nat. Lab., 1980—. Subspecialties: Theoretical computer science; Numerical analysis. Current work: Linear algebra, integral equations, mathematical physics, parallel processors, algorithms, numerical analysis, graph theory, artifical intelligence. Home: PO Box 345 Los Alamos NM 87544 Office: Los Alamos Nat Lab PO Box 1663 Los Alamos NM 87545

FABIATO, ALEXANDRE, biophysicist, physiologist, educator; b. Paris, Nov. 7, 1937; U.S., 1970; s. Nicolas and Edith (Laisne) F; m. Francoise Loulergue, Apr. 6, 1968; children: Nicolas, Francois, Denys, Helene. B.S. in Math, Lycee Saint-Louis, 1954; M.D., Université de Paris, 1969, Ph.D., 1970. Chef de Clinique Faculte de Medecine de Paris, 1969-71; instr. Harvard Med. Sch., Boston, 1971-72, asst. prof., 1972-75; prof. Med. Coll. Va., Richmond, Va., 1975—. Mem. editorial bd.: Circulation Research. Recipient Va. Commonwealth U. 1st Disting. Prof. award for research, 1982. Fellow AAAS; mem. Am. Physiol. Soc., Biophys. Soc., Soc. Gen Physiologists Assn. des Physiologistes de Langue Francaise. Roman Catholic. Subspecialties: Physiology (biology); Biophysics (biology). Current work: Subcellular biophysics of cardiac muscle cells; skinned cardiac cells, sarcoplasmic reticulum, microscopy, micromanipulations, application of microcomputers and microprocessors to physiology. Home: 1404 Westridge Rd Richmond VA 23229 Office: Medical Coll V Dept Physiology Box 55 Richmond VA 23298

FABRICANT, CATHERINE GRENCI, microbiologist, researcher; b. Davoli, Calabria, Italy, Sept. 24, 1919; came to U.S., 1921, naturalized, 1927; d. Frank and Maria Antonia (Sinopoli) Grenci; m. Julius Fabricant, Dec. 8, 1946; children: Barbara Louise, Daniel Grenci B.S., Cornell U., 1942, M.S., 1948. Chief technician, infirmary Cornell U., Ithaca, N.Y., 1942-44, teaching asst., 1945-49, 59-60, research assoc., 1960-62, 63-73, sr. research assoc. dept. vet. microbiology, 1973—, acting asst. prof., 1962-63; cons. NIH Site Visit, 1980, 81. Recipient award Am. Animal Hosp. Assn., 1976; Ralston Purina grantee, 1975-82; NIH Nat. Inst. Heart, Lung and Blood grantee, 1976-77, 77-83. Fellow Morris Animal Found.; mem. Am. Soc. Microbiology, Am. Soc. Pathology, Am. Soc. Virology, Sigma Xi, Sigma Delta Epsilon (sec.-treas. chpt. 1960-61), Phi Zeta (hon.). Subspecialties: Microbiology; Virology (biology). Current work: Pathogenesis of herpesvirus-induced atherosclerosis. Office: Dept Microbiology NY State Coll Vet Medicine Cornell U Ithaca NY 14853

FABRIKANT, VALERY ISAAC, mechanics and applied mathematics researcher, consultant; b. Minsk, USSR, Jan. 28, 1940; emigrated to Can., 1979, naturalized, 1983; s. Isaac Haim and Pesya Yudel (Turetskaya) F.; m. Maya Tyker, Dec. 23, 1981; 1 son, Isaac. B.Sc., Power Inst., Ivanovo, USSR, 1962, Ph.D., 1966. Asst. prof. Aiation Technol. Inst., Rybinsk, USSR, 1967-69; prof. engring. mechanics Poly. Inst., Ulyanovsk, USSR, 1970-73; sr. researcher Automation and Control Research Inst., Ivanovo, 1973-78; research asst. Concordia U., Montreal, Que., Can., 1979-80, research assoc., 1980-82, research prof., 1982—; cons. Paper Inst., Montreal, 1981—. Contbr. articles to profl. jours. Ministry Edn. USSR spl. scholar, 1958-62. Mem. Internat. Union Theoretical and Applied Mechanics, Soc. Engring. Sci., Soc. Indsl. and Applied Math. Subspecialties: Solid mechanics; Applied mathematics. Current work: Elasticity theory, integral equations, potential theory, numerical methods, contact problems. Home: 2170 Lincoln Ave Apt 505 Montreal PQ Canada H3H 2N5 Office: Concordia U 1455 De Maissonneuve W Montreal PQ Canada H3G 1M8

FABRO, SERGIO EDIGIO, physician, educator, consultant, researcher; b. Trieste, Italy, Sept. 3, 1932; came to U.S., 1967, naturalized, 1979; m. Susan Sieber, July 31, 1971. M.D. summa cum laude, U. Milan, 1956; Ph.D. in Biol. Chemistry, U. Rome, 1966, U. Rome, 1968, U. London, 1967. Diplomate: Am. Bd. Ob-Gyn, Sub-Bd. Maternal-Fetal Medicine. Rotating intern U. Milan, 1956-57; instr. dept. internal medicine Milan U. Hosp., 1956-58; asst. prof. dept. gen. pathology U. Modena, Italy, 1958-59, asst. prof. dept. biochemistry, 1960-62; research asst. dept. biochemistry U. London, St. Mary's Hosp., 1963-67; Brit. Council fellow in biochemistry St. Mary's Hosp. Med. Sch., 1964-66; assoc. research prof. dept. pharmacology George Washington U., Washington, 1967-71, research prof. dept. pharmacology, resident in ob-gyn, 1971-74, 1971-81, prof. dept. ob-gyn, 1974-78; prof. dept. pharmacology and ob-gyn, 1981—; dir. maternal-fetal medicine div. George Washington U. Hosp., 1978-81;

med. dir. ambulatory care center, dir. maternal-fetal medicine div. Columbia Hosp. for Women, Washington, 1981—, dir., 1981—; cons. Nat. Cancer Inst., 1966-70, Nat. Found., 1977, Office Sci. and Tech. Policy, Exec. Office of Pres., 1979-80, Council Environ. Quality, Exec. Office of Pres., 1980-81, Nat. Inst. Environ. Health Scis., 1969-80, 82, sr. cons. investigator devel. and reproductive toxicology br., 1975—; founder, cons. Environ. Teratology Info. Center, 1978—; mem. U.S.A.-USSR Environ. Health Coop. Program, 1977-80; mem. task force project on safety of chems. for human progeny WHO, 1981—; mem. maternal and child health research com. Nat. Inst. Child Health and Human Devel., 1980-82. Mem. editorial bd.: Teratogenesis, Carcinogenesis and Mutagenesis, Pediatric Pharmacology; ad hoc reviewer: Sci, Teratology, Ob-Gyn, Toxicology and Applied Pharmacology, Pharmacology and Exptl. Therapeutics; contbr. numerous articles to sci. jours. Recipient Biochemistry prize Accademia Nazionale dei Lincei, 1965, best tchr. award ob-gyn resident tng. program Georgetown U., 1981, 82. Mem. Am. Coll. Obstetricians and Gynecologists, Soc. Gynecol. Investigation, Perinatal Research Soc., Soc. Perinatal Obstetricians, Med. Soc. D.C., Washington Gynecol. Soc., Am. Soc. Pharmacology and Exptl. Therapeutics, Soc. Toxicology, Teratology Soc., European Teratology Soc., Biochem. Soc. (U.K.), Soc. Occupational and Environ. Health. Subspecialties: Maternal and fetal medicine; Reproductive biology (medicine). Office: Columbia Hosp for Women 2425 L St NW Washington DC 20037 Home: 9621 Annlee Terr Bethesda MD 20817

FACCINI, ERNEST CARLO, mech. engr., cons.; b. Livo, Trento, Italy, May 28, 1949; s. Carlo Emmanuel and Elena Agnes (Pancheri) F.; m. Sharon Louise, July 23. A.A., Western Wyo. Community Coll., 1969; B.S., U. Wyo., 1972, M.S., 1976. Registered profl. engr., Wyo., 1981. Mech. engr. dept. research and devel. Naval Explosive Ordinance Disposal Tech. Ctr., Indian Head, Md., 1976-79; mech. engr. ops. dept. Laramie Energy Tech. Ctr., Wyo., 1979-80; sr. engr. tech. info. dept. Naval Explosive Ordinance Disposal Tech. Ctr., Indian Head, 1980—. Contbr. articles to profl. jours. Recipient Letter of Commendation USAF, 1978; Outstanding Performance award USN, 1978; Spl. Achievement award, 1981; Letter of Commendation Project Sand Dollar, Naval Explosive Ordinance Disposal Tech. Ctr., 1982. Mem. ASME, AAAS, Am. Soc. Metals, Am. Phys. Soc. Roman Catholic. Subspecialties: Mechanical engineering; Explosives phenomena. Current work: Explosive effects—non-nuclear and nuclear containment; working on various methods to disable and/or contain explosives and their effects (military, non-military, nuclear and conventional). Patentee, inventor in field. Home: PO Box 201 Marbury MD 20658 Office: Naval Explosive Ordinance Disposal Tech. Center Indian Head MD 20640

FACTOR, STEPHEN MICHAEL, anatomic pathologist, educator, researcher; b. Far Rockaway, N.Y., Oct. 28, 1942; m. Sandra Basner, Aug. 17, 1967; children: Jason Robert, Rachel Elizabeth. B.A., Queens Coll, 1964; M.D., Albert Einstein Coll. Medicine, 1968; cert. anat. and clin. pathology, 1975. Diplomate: Am. Bd. Pathology. Intern and resident in surgery U. Mich., Ann Arbor, 1968-70; resident in pathology Albert Einstein Coll. Medicine, Bronx, N.Y., 1970-71, resident/chief resident pathology, 1973-75, asst. prof., 1975-80, assoc. prof., 1980—. Served to maj. M.C. U.S. Army, 1971-73. Spector fellow, 1980—. Fellow Am. Coll. Cardiology; Mem. Am. Assn. Pathologists, Am. Heart Assn., N.Y. Heart Assn., Pathologists Club, N.Y. Path. Soc., Alpha Omega Alpha. Subspecialties: Pathology (medicine); Cardiology. Current work: Morphology and pathogenesis of cardiomyopathy; pathogenesis of ischemic heart disease;anatomy and pathophysiology of cardiac microcirculation. Office: 1300 Morris Park Ave Bronx NY 10461

FADEN, ALAN IRA, neurobiologist, neurologist; b. Phila., Jan. 11, 1945; s. Leon Lable and Rebecca (Balter) F. B.A., U. Pa., 1966; postgrad., Ind. U., 1966-67; M.D., U. Chgo., 1971. Diplomate: Am. Bd. Neurology. Intern, resident U. Calif.-San Francisco, 1972-75; assoc. prof. neurology and medicine Uniformed Services U. Health Service, Bethesda, Md., 1978-81, prof. neurology, 1981-83, prof. physiology, 1983—, vice chmn. dept. neurology, 1980-82, dir. research, 1980-82, chief neurobiology research, 1982—; dir. neurology consultation Bethesda Naval Hosp., 1980-81; guest scientist Naval Med. Research Inst., Bethesda, 1981—; cons. clin. neurosci. NIH, 1982—. Served to maj. M.C. U.S. Army, 1975-80. Recipient Newman award San Francisco Neurol. Soc., 1975. Fellow Am. Acad. Neurology, ACP; mem. Am. Soc. Clin. Investigation, Am. Neurol. Assn., Am. Physiol. Soc., Soc. Neuroscis. Subspecialties: Neurobiology; Neurology. Current work: Neuropeptides in shock, spinal cord injury and stroke; leukotrienes in shock. Office: Neurobiology Research Unit Uniformed Services U 4301 Jones Bridge Rd Bethesda MD 20814

FADULU, SUNDAY O., microbiology educator; b. Ibadan, Nigeria, Nov. 11, 1940; s. William Cornelius and Ruth (Olumade) F.; m. Jacqueline F. Counter, Oct. 26, 1968; children: Sunday, Tony, Jeannie. B.S., Okla. Bapt. U., 1964; M.S., U. Okla., 1965, Ph.D., 1969. Lectr. U. Ife, Nigeria, 1969-70; research assoc. U. Okla. Med. Ctr., Oklahoma City, 1970-72; prof. microbiology Tex. So. U., Houston, 1972—; adj. prof. U. Houston, 1981—; v.p. Adoxy Corp., Houston, 1982—. Bd. dirs. Nigerian Found., Houston, 1982—, Econ. of African Nations Aid, Inc. Mem. N.Y. Acad Scis., Am. Soc. Microbiology (planning com. Tex. br. 1983—), Beta Beta Beta. Baptist. Subspecialties: Infectious diseases; Hematology. Current work: Microbial pathogenesis; isolation of fungal toxins in immunological, hematological, cardiovascular activities; sickle cell anemia; isolation of natural products as potential antisickling agents. Patentee in field. Home: 20115 Wickham Ct Katy TX 77450 Office: Tex So U 3200 Wheeler Ave Houston TX 77004

FADUM, RALPH EIGIL, university dean; b. Pitts., July 19, 1912; s. Torgeir Bleken and Mimi (Knudsen) F.; m. Nancy Isabelle Fields, July 19, 1939; 1 dau., Jane Fields. B.S. in Civil Engring, U. Ill., 1935; M.S., Harvard, 1937, S.D., 1941; D.Eng., Purdue U., 1963. Registered profl. engr., N.C. Parttime asst. civil engring. Harvard, 1935-37, instr., 1937-41, faculty instr., 1941- 43; asst. prof. soil mechanics Purdue U., 1943-45, asso. prof., 1945-47, prof., 1947-49; head of civil engring. dept. and prof. of civil engring. N.C. State U., Raleigh, 1949-62, dean of engring., 1962-78; cons. Dept. Def., U.S. Corps Engrs.; Mem. Army Sci. Bd. Dept. Army, 1959-81; mem. research adv. com. Fed. Hwy. Adminstrn., 1963-70; adv. bd. Ford Found., 1963-69; vice chmn. Army Sci. Adv. Panel, Dept. Army, 1966-70; chmn. adv. group to comdr. gen. Tank Automotive Command, 1967-70. Contbr. articles to profl. jours. Chmn. N.C. Water Control Adv. Council; bd. dirs. Nat. Driving Center, 1973-77; commr. Raleigh Housing Authority, 1962-72; pres. Atlantic Coast Conf., 1966-67, 71-72; v.p. Nat. Collegiate Athletic Assn., 1972-76; Chmn. bd. dirs. N.C. Water Resources Research Inst., U. N.C. Recipient Patriotic Civilian Service award Dept. Army, 1967, Meritorious Civilian Service medal, 1967, Outstanding Civilian Service medal, 1973, 77; Distinguished Civil Engring. Alumnus award U. Ill., 1969. Mem. ASCE (hon. mem., Outstanding Civil Engr. N.C. award 1971); mem. Nat. Acad. Engring., U.S. Nat. Council Soil Mechanics and Found. Engring., Nat. Soc. Profl. Engrs., N.C. Soc. Engrs. (Outstanding Engring. Achievement award 1971), Raleigh Engrs. Club (Outstanding Engr. award), Am. Soc. Engring. Edn. (hon. mem.; v.p., mem. exec. com. 1973-74, dir.), Sigma Xi, Tau Beta Pi, Chi Epsilon (nat. honor mem.), Phi Kappa Phi, Delta Upsilon. Clubs: Rotary (Raleigh); Carolina Country. Subspecialty: Civil engineering. Current work: Geotechnical Engineering. Address: 408 Mann Hall NC State U Raleigh NC 27650

FAGAN, RAYMOND, epidemiologist, educator; b. Bklyn., Dec. 27, 1914; s. Louis and Bertha (Fagan) F.; m. Esther Fried, Sept. 20, 1936; children: Susan Barbara, Kathleen Ellen, Deborah Jill. B.A., N.Y. U., 1935; D.V.M., Cornell U., 1939; M.P.H., Harvard U., 1949. Jr. veterinarian Dept. Agr., 1939-42; sr. scientist USPHS, 1946-54; assoc. prof. U. Pa., 1954-56; sr. virologist Wyeth Labs., Radnor, Pa., 1956-67; prin. scientist Philip Morris, Inc., Richmond, Va., 1967—; adj. assoc. prof. toxicology Med. Coll. Va.; adj. prof. epidemiology Drexel U., 1964-67; cons. WHO, 1963, 66, 70. Mem. Richmond Air Pollution Control Bd., 1975-81; chmn. adv. council environ. health Phila. Dept. Health, 1963-67; chmn. pub. health com. Chester County (Pa.) Health and Welfare Council, 1957-60. Served to capt. U.S. Army, 1942-46. Mem. Am. Pub. Health Assn., AVMA, Soc. Microbiology, N.Y. Acad. Scis., AAAS, Sigma Xi. Subspecialty: Epidemiology.

FAGERSTROM, JOHN A., geologist, educator; b. Ypsilanti, Mich., Jan. 4, 1930; s. Simon Emanuel and Lorena (Dowlin) F.; m. Marilyn Landis, Oct. 3, 1953; children: Linda, Christine, Eric. A.B., Oberlin Coll., 1952; M.S., U. Tenn., 1953; Ph.D., U. Mich., 1959. From asst. prof. to prof. geology U. Nebr., Lincoln, 1958—. Contbr. articles to profl. jours. Served to cpl. U.S. Army, 1954-56. Fellow Geol. Soc. Am.; mem. Paleontol. Soc. (treas. 1975-81), Internat. Paleontol. Union, Sierra Club (chmn. Lincoln br. 1975-77), Sigma Xi. Subspecialties: Paleoecology; Sedimentology. Current work: Devonian history of the Michigan basin. Home: 7321 York Ln Lincoln NE 68505 Office: U Nebr Dept Geolog Morrill Hall Lincoln NE 68506

FAGET, MAXIME A(LLAN), government aeronautical administrator, aeronautical engineer; b. Stann Creek, Brit. Honduras, Aug. 26, 1921; m.; 4 children. B.S., La. State U., 1943, D.Eng. (hon.), 1972, U. Pitts., 1966. Aero. research scientist Nat. Adv. Com. Aeros., 1946-58; chief Flight Systems Div., NASA, 1958-62; asst. dir. engring. and devel. Manned Spacecraft Ctr., Houston, 1962-66, dir. engring. and devel., 1966—. Recipient Arthur S. Flemming award, 1959, Golden Plate award, 1962; medal Outstanding Leadership NASA, 1963; Disting. Service medal, 1969; medal Exceptional Service, 1969; award for outstanding accomplishment IEEE, 1971; Harry Diamond award, 1976; Space Flight award, 1976; Daniel and Florence Guggenheim Internat. Astronaut award, 1973; Gold medal ASME, 1975; Albert F. Sperry medal Instrument Soc. Am., 1976. Fellow AIEE (Spacecraft Design award 1970), Am. Astronautical Soc. (William Randolph Lovelace II award 1971); mem. Nat. Acad. Engring., Internat. Acad. Astronautics. Subspecialty: Aerospace engineering and technology. Office: Engring and Devel NASA Manned Spacecraft Ctr Houston TX 77058

FAGIN, RONALD, computer scientist; b. Oklahoma City, May 1, 1945; s. George J. and Maxine (Appleman) F.; 1 son, Joshua Harris. A.B., Dartmouth Coll., 1967; Ph.D., U. Calif.-Berkeley, 1973. Research staff IBM Research Ctr., Yorktown Heights, N.Y., 1973-75; research staff mem. IBM Research Lab., San Jose, Calif., 1975-79, mgr. founds. computer sci., 1979—. NSF grad. fellow, 1967-72; recipient Outstanding Innovation awards IBM, 1981. Mem. Assn. Computing Machinery. Democrat. Jewish. Subspecialties: Foundations of computer science; Database systems. Current work: Research in applications of mathematical logic to theoretical computer science, especially the theory of relational databases. Home: 162 Loma Alta Ave Los Gatos CA 95030 Office: IBM Research Lab K51/281 5600 Cottle Rd San Jose CA 95193

FAGOT, ROBERT FREDERICK, psychologist, educator; b. Cape Gracias, Nicaragua, July 4, 1921; s. Fred Clark and Ruby Mary (Howorka) F.; m. Beverly I. Fields, Apr. 1, 1961; children: Brian Kevin, Clark Albert. B.S., MIT, 1946; Ph.D., Stanford U., 1956. Asst. prof. psychology U. Oreg., Eugene, 1956-61, assoc. prof., 1961-66, prof., 1966—, head dept. psychology, 1968-80. Contbr. articles to profl. jours. Served to lt. USNR, 1942-46, 54-55. USPHS spl. fellow, 1962-63; Netherlands Inst. Advanced Study fellow, 1972-73. Mem. Am. Psychol. Assn., Am. Statis. Assn., Psychonomic Soc., Psychometric Soc. Subspecialties: Measurement in psychology; Statistics. Current work: Research in measurement theory, psychophysics, judgement models. Home: 680 W 35 Pl Eugene OR 97405 Office: Dept Psychology U Oreg Straub Hall Eugene OR 97403

FAHEY, JOHN LESLIE, immunologist, university research center administrator; b. Cleve., Sept. 8, 1924; m. (married); 3 children. M.S., Wayne State U., 1949; M.D., Harvard U., 1951. Intern in medicine Presbyterian Hosp., N.Y.C., 1951-52, asst. resident, 1952-53; clin. assoc. Nat. Cancer Inst., NIH, 1953-54, sr. investigator metabolism, 1954-63, chief immunology br., 1964-71; prof. microbiology and immunology, chmn. dept. Sch. Medicine, UCLA, 1971-81, dir., 1978—. Mem. Am. Physiol. Soc., Soc. Exptl. Biology and Medicine, Am. Assn. Cancer Research, Am. Fedn. Clin. Research, Am. Soc. Clin. Investigation. Subspecialty: Immunology (medicine). Office: Center Interdisciplinary Research Immunological Diseases UCLA Los Angeles CA 90024

FAHIM, MOSTAFA SAFWAT, reproductive biologist, cons.; b. Cairo, Egypt., Oct. 7, 1931; came to U.S., 1966; s. Mohamed and Amna (Hussin) F.; m. Zuhal Fahim, Feb. 23, 1959; 1 child, Ayshe. B.S. in Chemistry, U. Cairo, 1953; M.S., U. Mo., 1958, Ph.D., 1961. Research asso. U. Mo. Health Scis. Center, Columbia, 1966-68, asst. prof., 1968-71, assoc. prof., 1971-75, prof., 1981—, prof., chief reproductive biology, 1975—; cons. in field. Contbr. articles to profl. jours. Mem. Am. Public Health Assn., Mo. Public Health Assn., Nutrition Today Soc., Internat. Andrology Soc., Internat. Toxicology Soc., Am. Coll. Clin. Pharmacology, Am. Soc. Pharmacology and Exptl. Therapeutics, Internat. Fertility Soc., Am. Fertility Soc., Fedn. Am. Socs. Exptl. Biology, N.Y. Acad. Scis., Soc. Environ. Geochemistry and Health, Soc. Study Reprodn., AAAS, Sigma Xi, Gamma Alpha. Current work: Contraception, electronics and reprodn. Patentee in field. Office: Dept Obstetrics and Gynecology U Mo Health Scis Center Columbia MO 65212

FAHN, STANLEY, neurologist; b. Sacramento, Nov. 6, 1933; s. Ernest and Sylvia (Schumer) F.; m. Charlotte Zmora, June 21, 1958; children: Paul N., James D. B.A., U. Calif.-, Berkeley, 1955, M.D., 1958. Diplomate: Am. Bd. Neurology. Resident in neurology Neurol. Inst. N.Y., 1959-62; research asso. NIH, 1962-65; faculty Columbia, 1965-68, prof., 1973-78, H. Houston Merritt prof., 1978—; dir. (Dystonia Research Center), 1981—; faculty U. Pa., 1968-73. Served with USPHS, 1962-65. NIH grantee, 1974-77, 80-82. Subspecialties: Neurology; Neuropharmacology. Current work: Clin. and basic sci. movement disorders, including pharmacology of neurotransmitters. Home: 155 Edgars Ln Hastings NY 10706 Office: 710 W 168th St New York NY 10032

FAINBERG, ANTHONY, physicist; b. London, Jan. 14, 1944; U.S., 1947; s. Benjamin and Elizabeth (Martelli) F.; m. Louise Vasvari, Jan. 23, 1964. A.B., NY U, 1964; Ph.D., U. Calif.-Berkeley, 1969. Research asst. Lawrence Radiation Lab., Berkeley, 1969; physicist U. Turin, Italy; researcher European Orgn. for Nuclear Research (CERN), Geneva, 1970-72; research asst. prof. Syracuse (N.Y.) U., 1972-77; physicist Brookhaven Nat. Lab., Upton, N.Y., 1977—. NSF fellow, 1964-65; Am. Phys. Soc. Congl. fellow, 1983-84. Mem. Am. Phys. Soc., Inst. for Nuclear Materials Mgmt., Fedn. Am. Scientists, Am. Nuclear Soc. Subspecialties: Nuclear fission; Particle physics. Current work: Nuclear safeguards, instrumentation, particle detectors. Office: DNE Brookhaven Nat Lab 197C Upton NY 11973

FAINGOLD, CARL LAWRENCE, pharmacologist, educator, researcher; b. Chgo., Feb. 1, 1943; s. Charles and Ann (Glassman) F.; m. Carol Ann Baskin, June 21, 1964; children: Scott, Charles, Robert. B.S. in Pharmacy, U. Ill., Chgo., 1965; Ph.D. in Pharmacology, Northwestern U., 1970. Postdoctoral fellow Inst. Psychiatry, U. Mo., St. Louis, 1970-72; asst. prof. pharmacology Sch. Medicine, So. Ill. U., Springfield, 1972-76, asso. prof., 1976—, acting chmn. dept., 1981-83. Contbr. articles to profl. jours. Nat. Inst. Neurol and Communicative Disorders grantee, 1979-84. Mem. Am. Soc. Pharmacology and Exptl. Therapeutics, Soc. for Neurosci., AAAS, Am. Epilepsy Soc., N.Y. Acad. Sci., Sigma Xi. Subspecialties: Neuropharmacology; Neurophysiology. Current work: Neurophysiology and neuropharmacology of epilepsy, hearing and neuronal correlates of behavior. Home: 60 Danbury Dr Springfield IL 62704 Office: PO Box 3926 Springfield IL 62708

FAIR, JAMES RUTHERFORD, JR., chemical engineering educator, consultant; b. Charleston, Mo., Oct. 14, 1920; s. James Rutherford and Georgia Irene (Case) F.; m. Merle Innis, Jan. 14, 1950; children: James Rutherford III, Elizabeth, Richard Innis. Student, The Citadel, 1938-40; B.S., Ga. Inst. Tech., 1942; M.S., U. Mich., 1949; Ph.D., U. Tex., 1955; D.Sc. (hon.), Wash. U., 1977. Research engr. Shell Devel. Co., Emeryville, Calif., 1954-56; with Monsanto Co., 1942-52, 56-79, engring. dir. corp. engring. dept., St. Louis, 1969-79; Cockrell prof. chem. engring. U. Tex., Austin, 1979—; dir., v.p. Fractionation Research, Inc., Bartlesville, Okla., 1969-79. Author: North Arkansas Line, 1969, Distillation, 1971; Contbr. numerous articles to profl. publs. Bd. dirs. Nat. Mus. Transport. Recipient profl. achievement award Chemical Engineering mag., 1968. Fellow Am. Inst. Chem. Engrs. (bd. dirs. 1965-67, Walker award 1973, Practice award 1975, Founders award 1977, Inst. lectr. 1979); mem. Am. Chem. Soc., Nat. Acad. Engring., Am. Soc. Engring. Edn., Nat. Soc. Profl. Engrs., Sigma Nu. Republican. Presbyterian (elder). Clubs: Faculty (U. Tex.); Headliners (Austin). Subspecialty: Chemical engineering. Current work: Head separations research program at University of Texas. Home: 2804 Northwood Rd Austin TX 78703 Office: Dept Chem Engring U Tex Austin TX 78712

FAIRBANK, WILLIAM MARTIN, physicist, educator; b. Mpls., Feb. 24, 1917; s. Samuel Ballantine and Helen Leslie (Martin) F.; m. Jane Davenport, Aug. 16, 1941; children—William Martin, Robert Harold, Richard Dana. A.B., Whitman Coll., Walla Walla, Wash., 1939, D.Sc. (hon.), 1965; postgrad. fellow, U. Wash., 1940-42; M.S., Yale, 1947, Ph.D. (Sheffield fellow), 1948; D.Sc., Duke U., 1969, Amherst Coll., 1972. Mem. staff radiation lab. Mass. Inst. Tech., 1942-45; asst. prof. physics Amherst Coll., 1947-52; asso. prof. Duke, 1952-58, prof., 1958-59; prof. physics Stanford, 1959—. Bd. overseers Whitman Coll. Named Calif. Scientist of Year Calif. Museum Sci. and Industry, 1961; recipient Fritz London award, 1968; Wilbur Lucius Cross medal Yale U., 1968; Guggenheim fellow, 1976-77. Fellow Am. Phys. Soc. (Oliver E. Buckley Solid State Physics prize 1963, Research Corp. award 1965); mem. AAAS (chmn. physics sect. 1980-81), Nat. Acad. Scis., Am. Acad. Arts and Scis., Am. Philos. Soc. Subspecialties: Low temperature physics; Superconductors. Spl. research microwave radar systems, microwave propagation, cryogenics, quantized flux in superconductors, properties liquid helium II, He3, liquid helium bubble chambers, superconducting electron accelerators, quarks, exptl. gravitations. Home: 141 E Floresta Way Menlo Park CA 94025 Office: Physics Dept Stanford Univ Stanford CA 94305

FAIRBANK, WILLIAM MARTIN, JR., physicist; b. New Haven, Jan. 7, 1946; s. William Martin and Jane (Davenport) F.; m. Donna Lorraine Witter, Aug. 30, 1975; children: William Henry, Mary Helen. B A., Pomona Coll., Claremont, Calif., 1968; M.S., Stanford U., 1969, Ph.D., 1974. Research assoc. U. Ariz., 1974-75; asst. prof. physics Colo. State U., 1975-78, assoc. prof., 1978-83, prof., 1983—; cons. in field. Contbr. articles in field to profl. jours. Alfred P. Sloan fellow, 1976-78; NSF fellow, 1968-71. Mem. Am. Phys. Soc., Optical Soc. Am., Sigma XI. Subspecialties: Spectroscopy; Atomic and molecular physics. Current work: Frontier research in applications of dye laser spectroscopy and single atom detection, particularly to elementary particle physics. Home: 1712 Clearview Ct Fort Collins Co 80521 Office: Atom Science PO Box 138 Oak Ridge TN 37830

FAIRCHILD, RALPH GRANDISON, radiol. physicist; b. Trenton, N.J., Sept. 24, 1935; s. Ralph Grandison and Sara Gertrude (Edgerton) F.; m. Frances Woods, June 14, 1958; children: David, James, Stefanie, Jovi. B.S., St. Lawrence U., Canton, N.Y., 1958; M.S., Cornell U., 1961; Ph.D., Adelphi U., 1976. Physicist Lawrence Radiation Lab., Livermore, Calif., 1959, Savannah River Lab., Aiken, S.C., 1960; radiol. physicist Brookhaven Nat. Lab., Upton, N.Y., 1961—; asso. prof. SUNY, Stony Brook, 1980—; radiol. physicist VA Hosp., Northport, N.Y., 1975—. Contbr. articles to profl. jours. NIH grantee. Mem. Am. Assn. Physicists in Medicine, Soc. Nuclear Medicine, Sigma Xi. Quaker. Subspecialties: Cancer research (medicine); Biophysics (physics). Current work: Radiological physics; development of particle beams for radiotherapy; mixed field dosimetry; neutron capture therapy. Home: 6 Huckleberry Ln Setauket NY 11733 Office: Med Dept Brookhaven Nat Lab Upton NY 11973

FAIRHURST, CHARLES, civil engineering educator; b. Widnes, Lancashire, Eng., Aug. 5, 1929; came to U.S., 1956, naturalized, 1967; s. Richard Lowe and Josephine (Starkey) F.; m. Margaret Ann Lloyd, Sept. 7, 1957; children: Anne Elizabeth, David Lloyd, Charles Edward, Catherine Mary, Hugh Richard, John Peter, Margaret Mary. B.Eng., U. Sheffield, Eng., 1952, Ph.D., 1955. Mining engr. trainee Nat. Coal Bd., St. Helens, Eng., 1949-56; research assoc. U. Minn., Mpls., 1956-67, prof., 1967-70, head Sch. Mineral and Metall. Engring., 1969—, prof. dept. civil and mineral engring., 1970—, head dept., 1972—, E.P. Pfleider prof. mining engring. and rock mechanics, 1983; cons. U.S. Army C.E., Petrobras, Brazil.; Chmn. U.S. nat. commn. rock mechanics Nat. Acad. Scis., 1971-74. Mem. AIME, S. African Inst. Mining and Metallurgy, ASCE (chmn. rock mechanics com. 1978-80), Internat. Soc. Rock Mechanics (past dir.), Am. Underground Space Assn. (pres. 1976-77), Royal Swedish Acad. Engring. Scis. (fgn.), Sigma Xi. Roman Catholic. Subspecialties: Mining engineering; Civil engineering. Current work: Mining engineering; rock mechanics; underground space and construction. Home: 417 5th Ave N South Saint Paul MN 55075 Office: Dept Civil and Mineral Engring U Minn Minneapolis MN 55455

FAITH, ROBERT EARL, JR., lab. animal veterinarian, immunotoxicologist, cons.; b. El Paso, Mar 7, 1942; s. Robert Earl and Mittie Lane (Condon) F.; m. Marsha Kae Nelson, Apr. 17, 1967; 1 son, Richard Trent; m.; stepchildren—Trechon Laurice Duncan, Christopher Scott Duncan; m. Stacia Luanne Behrens, Oct. 24, 1980. B.S., Tex. Western Coll., 1965, Tex. A&M U., 1966, D.V.M., 1968; M.S. in Med. Microbiology, U. Fla., 1971; Ph.D. in Immunology, U. Fla., 1979. Postdoctoral fellow div. comparative medicine U. Fla.,

Gainesville, 1968-71, USPHS postdoctoral fellow dept. immunology and med. microbiology, 1971-74; staff fellow Nat. Inst. Environ. Health Scis., Research Triangle Park, N.C., 1974-76, sr. staff fellow research and animal colony adminstrn., 1976-77; dir. Biomed. Research Ctr., Oral Roberts U., Tulsa, 1978—. Contbr. articles and abstracts to profl. jours. Mem. Am. Coll. Lab. Animal Medicine (cert.; profl. standards subcom.), AVMA, Tex. Vet. Med. Assn., Am. Assn. Lab. Animal Sci. (pres. Research Triangle br. 1977-78), Am. Assn. Immunologists, AAAS. Republican. Presbyterian. Subspecialties: Laboratory animal medicine; Immunotoxicology. Current work: Effects of chemicals of environmental concern on developing immune system; research animal colony management and research on the effects of chemicals on environmental concern on developing immune system. Office: 7777 S Lewis Ave Tulsa OK 74171

FAJANS, STEFAN STANISLAUS, physician; b. Munich, Ger., Mar. 15, 1918; came to U.S., 1936, naturalized, 1942; s. Kasimir M. and Salomea (Kaplan) F.; m. Ruth Stine, Sept. 6, 1947; children—Peter S., John S. B.S., U. Mich., Ann Arbor, 1938, M.D., 1942. Intern Mount Sanai Hosp., N.Y.C., 1942-43; resident U. Mich., 1947-49, research fellow, 1946-47, 49-51; mem. faculty U. Mich. Med. Sch., 1950—, prof. internal medicine, 1961—, head div. endocrinology and metabolism, also dir. metabolism research unit, 1973—; dir. Mich. Diabetes Research and Tng. Center, 1977—; mem. endocrinology study sect. NIH, 1958-62, mem. diabetes and metabolism tng. grants com., 1966-70; chmn. Am. zone internat. sci. adv. com. Congresses Internat. Diabetes Fedn., 1977-79; Banting meml. lectr., 1978. Contbr. articles med. publns. Served as officer M.C. AUS, 1943-46. Research fellow medicine A.C.P., 1949-50; fellow Life Ins. Med. Inst., 1950-51. Mem. Am. Diabetes Assn. (pres. 1971-72, Banting medal 1972, Banting Meml. award 1978), Endocrine Soc. (council 1967-71, 78-81, v.p. 1970-71), Am. Fedn. Clin. Research, Am. Soc. Clin. Investigation, Assn. Am. Physicians, Central Soc. Clin. Research, Sigma Xi, Alpha Omega Alpha. Subspecialty: Endocrinology. Current work: Genetics; natural history classification and diagnosis of diabetes; diagnosis and treatment of organic hypothemia. Home: 2485 Devonshire Rd Ann Arbor MI 48104 Office: Univ Mich Hosp Ann Arbor MI 48109

FAKHARZADEH, ALI M., software engineer, researcher; b. Tehran, Iran, June 14, 1947; came to U.S., 1966; s. Ali-Asghar and Fatemeh (Daeian) F.; m. Marya Montazeri, Sept. 26, 1976; children: Ali-Reza, Bahman Benjamin. B.S.E.E., U. Utah, 1970, M.S., 1977. Sr. software engr. Applicon, Inc., Burlington, Mass., 1977-82; prin. software engr. Adage, Inc., Billerica, Mass., 1982—. Contbr. articles to profl. jours. Mem. IEEE, Assn. for Computing Machinery, Eta Kappa Nu. Subspecialties: Graphics, image processing, and pattern recognition; Software engineering. Current work: Computer graphics, solids modeling, interactive man-machine communication, robotics, developing high performance graphics firmware. Home: 556 Concord Ave Lexington MA 02173 Office: Adage Inc 1 Fortune Dr Billerica MA 01821

FALCON, WALTER PHILLIP, economist; b. Iowa, Sept. 28, 1936; s. Norman F.; m. Laura Hann; children—Lesley, Phillip, Andrew. B.A., Iowa State U., 1958; Ph.D., Harvard U., 1962. Asst. prof. dept. econs. Harvard U., 1963-66, lectr., devel. adviser, 1966-72; gen. econ. and agrl. econ. adviser Pakistan Planning Commn., Harvard U. Devel. Adv. Service, 1964-65; cons. agrl. econs. BAPPENAS, Govt. of Indonesia, 1968-72, dep. dir., 1970-72, dir. research, 1966-70; prof. econs., dir. Food Research Inst., Stanford U., 1972—; cons. AID, Dept. of State, 1963-66, Ford Found., World Bank; mem. Presdl. Commn. on World Hunger, 1978—. Editor: (with Gustav Papanek) Development Policy—Theory and Practice, 1968, Jour. Food Research Inst. Studies, 1977. Trustee Agrl. Devel. Council. Mem. Am. Agrl. Econs. Assn. (Best Article award 1971), Western Agrl. Econs. Assn. Subspecialty: Agricultural economics. Home: 415 Gerona Rd Stanford CA 94305 Office: Food Research Inst Stanford U Stanford CA 94305

FALES, HENRY MARSHALL, organic chemist, mass spectrometrist; b. N.Y.C.; s. Henry Marshall and Cecile Marie (Vatet) F.; m. b. Caroline Eleanor McCullagh, Dec. 19, 1947; children: Marsha Kent Mazz, Suzanne kent Palmer, Henry Richard. B.S., Rutgers U., 1948, Ph.D., 1953. Research asst. Rutgers U., New Brunswick, N.J., 1952, instr., 1952; chemist Nat. Heart Inst., NIH, Bethesda, Md., 1953—; chief lab. chemistry Nat. Heart, Lung and Blood Inst., 1966—; mem. adv. panel NSF; instr. Found. for Advanced Edn. in Scis. Contbr. numerous articles on organic chemistry to profl. jours.; mem. editorial bds.: Biomed. Mass Spectrometry, Analytical Chemistry, Drug Metabolism and Disposition. Served with USN, 1944-46. Mem. Am. Chem. Soc., Am. Soc. for Mass Spectroscopy. Subspecialties: Organic chemistry; Analytical chemistry. Current work: Structural analysis of organic compounds of biological interest using mass spectrometry x-ray crystallography, nuclear magnetic resonance. Office: NIH Bldg 10 N318 Bethesda MD 20205

FALK, HANS LUDWIG, biochemist; b. Breslau, Germany, Sept. 15, 1919; came to U.S., 1947, naturalized, 1953; s. Hermann and Gertrude (Raphaelsohn) F.; m. Gabrielle Clara Freund, June 30, 1950; children: Raymond Walter, Donald Herman, Stephen Thomas. Student, U. London, 1938-40; B.Sc., McGill U., 1944, Ph.D., 1947. Instr. U. Chgo., 1947-52; adj. asst. to assoc. prof. U. So. Calif., Los Angeles, 1952-62; assoc. sci. dir. Nat. Cancer Inst., NIH, Bethesda, Md., 1962-68; assoc. dir. program/assoc. dir. health hazard assessment Nat. Inst. Environ. Health Scis., Res. Triangle Park, N.C., 1968—; cons. in field. Author: Chemical Mutagens - Environmental Effects on Biolological Systems, 1970; assoc. editor, co-author: Handbook on Environmental Physiology, 1977; contbr. articles to profl. jours. Maj. Hiram Mills scholar, 1943; recipient USPHS Superior Service award, 1968. Mem. AAAS, Am. Assn. Cancer Research, Am. Soc. Cell Biology, Am. Soc. Exptl. Pathology, Soc. Toxicology, Royal Soc. Chemists, N.Y. Acad Sci., Sigma Xi. Subspecialties: Toxicology (agriculture); Cancer research (medicine). Current work: Health hazard assessment, carcinogenesis, mutagenesis, DNA repair mechanism and inhibition, interaction chems. causing synergism or antagonism. Home: 4508 Pitt St Raleigh NC 27609 Office: PO Box 12233 South Bldg 101 A2 05 Research Triangle Park NC 27709

FALK, JAMES ROBERT, data processing cons.; b. N.Y.C., May 31, 1952; s. Arthur Eugene and Ruth Amelia (Hall) F.; m. Frances Anne Hines, Feb. 25, 1978. B.S., Fordham U., 1974, M.A., 1976, Ph.D., 1980. Electronics cons. NY U Med. Center-Millhauser Labs., N.Y.C., 1971-74; research asst. Fordham U., 1975-77, N.Y.State Poly. Inst./Columbia Presbyn. Hosp., N.Y.C., 1977; human factors engr. Gould, Inc., Simulation Systems, Melville, N.Y., 1979-80; research analyst/statistician Cornell U. Med. Coll., White Plains, N.Y., 1980-83; dir., research cons. ComStat, Croton-on-Hudson, N.Y., 1979—; adj. instr. Marymount Coll., Tarrytown, N.Y., 1975-79, Mercy Coll., Dobbs Ferry, N.Y., 1978—, Fordham U., Bronx, 1978-79. Contbr. articles to profl. jours. Loyola fellow, 1977. Mem. N.Y. Acad. Sci., AAAS, Am. Psychol. Assn., Eastern Psychol. Assn., Sigma Xi. Roman Catholic. Subspecialties: Statistics; Software engineering. Current work: Computer utilization in science, computer/sensory information processing.

FALK, LAWRENCE ADDNESS, JR., virologist; b. c. Houston, May 5, 1938; s. Lawrence A. and Lorraine O. (Wilson) F. B.A., Centenary Coll., 1962; M.S., U. Houston, 1966; Ph.D., U. Ark., 1969. With Rush-Presbyn.-St. Luke's Med. Center, Chgo., 1969-78; chmn. div. microbiology New Eng. Regional Primate Research Center; virologist Harvard Med. Sch., 1978—, also asso. prof. microbiology; vis. scientist Karolinska Inst., Stockholm, 1976-77. Contbr. numerous articles to profl. jours. Scholar Leukemia Soc. Am., 1975-80; grantee in field. Mem. Am. Assn. Immunologists, AAAS, Tissue Culture Assn., Am. Soc. Microbiology, Am. Assn. Cancer Research, Sigma Xi. Subspecialties: Virology (medicine); Virology (veterinary medicine). Current work: Lymphotropic herpes viruses of humans and primates. Office: 1 Pine Hill Dr Southborough MA 01772

FALK, SANDOR A., renal physiologist; b. Paterson, N.J., Dec. 20, 1949; s. Malvin Donald and Lois Helen (Slotkin) F.; m. Barbara Carol Rudenstein, June 14, 1972. Student, The Citadel, 1967-69; B.A., Franklin Pierce Coll., 1971. Research technologist U. Pa. Hosp., Phila., 1973-74; biol. researcher VA Hosp., Denver, 1974-75; researcher U. Colo. Health Sci. Center, Denver, 1975—; cons. glomerular dynamics VA Hosp., Denver, 1981—. Contbr. articles to profl. jours. Mem. Am. Fedn. for Clin. Research. Subspecialties: Physiology (medicine); Nephrology. Current work: Using micropuncture techniques to study glomerular dynamics of acute renal failure. Home: 9591 Warhawk Rd Conifer CO 80433 Office: U Colo Health Scis Center 4200 E 9th Ave Denver CO 80262

FALKLER, WILLIAM ALEXANDER, JR., microbiologist, educator, researcher, adminstr.; b. York, Pa., Sept. 9, 1944; s. William Alexander and Kathryn (Grove) F.; m. Patricia Eileen Bonville, Dec. 23, 1967; children: Rachel Beth, Laurie Marie. B.A., Western Md. Coll., 1966; M.S., U. Md., 1969, Ph.D., 1971; postgrad., U. Hawaii Sch. Medicine, 1971-73. Instr. U. Hawaii Sch. Medicine, Honolulu, 1971-73; asst. prof. U. Md. Dental Sch., Balt., 1973-77, assoc. prof., 1977—, chmn. dept. microbiology, 1981—; cons. Nat. Inst. Dental Research, Bethesda, Md., 1975-77, U.S. Army, Research and Devel. Command, Washington, 1981—. Contbr. numerous articles on microbiology to profl. jours. Pres. Stewartstown Jaycees, 1979, 80; councilman Stewartstown Borough Council, 1980—. Mem. Am. Soc. Microbiology (J. Howard Brown award Md. chpt. 1971, v.p. 1976-77, pres. 1977-78), Internat. Assn. Dental Research, Am. Assn. Dental Schs., Sigma Xi, Phi Kappa Phi. Subspecialties: Microbiology (medicine); Immunology of periodontal diseases. Current work: Immunology of periodontal diseases, immune response to anaerobic bacteria, colonization mechanisms of bacteria. Home: 67 W Pennsylvania Ave Stewartstown PA 17363 Office: Dept Microbiology U Md Dental Sch 666 W Baltimore St Baltimore MD 21201

FALKOW, STANLEY, microbiologist, educator; b. Albany, N.Y., Jan. 24, 1934; s. Jacob and Mollie (Gingold) F.; m. Rhoda Mae Falkow, Jan. 18, 1958; children: Lynn Beth, Jill Stuart.; m. Lucy Stuart Thompkins, Dec. 3, 1983. B.S. in Bacteriology cum laude, U. Maine, 1955, D.Sc. (hon.), 1979; M.S. in Biology, Brown U., 1960, Ph.D., 1961. Asst. chief dept. bacterial immunity Walter Reed Army Inst. Research, Washington, 1963-66; prof. microbiology Med. Sch. Georgetown U., 1966-72; prof. microbiology and medicine U. Wash., Seattle, 1972-80; prof., chmn. dept. med. microbiology Stanford (Calif.) U., 1981—; Karl H. Beyer vis. prof. U. Wis., 1978-79; Sommer lectr. U. Oreg. Sch. Medicine, 1979; Kinyoun lectr. NIH, 1980; Rubbro orator Australian Soc. Microbiology, 1981; Stanhope Bayne-Jones lectr. Johns Hopkins U., 1982; Mem. Recombinant DNA Molecule Com.; mem. task force on antibiotics in animal feeds FDA; mem. microbiology test com. Nat. Bd. Med. Examiners. Author: Infectious Multiple Drug Resistance, 1975; Editor: Jour. Bacteriology, Jour. Infection and Immunity, Jour. Infectious Diseases. Recipient Ehrlich prize, 1981. Fellow Am. Acad. Microbiology; mem. Infectious Disease Soc. Am. (Squibb award 1979), Am. Soc. Microbiology, Genetics Soc. Am., AAAS, Sigma Xi. Subspecialty: Microbiology (medicine). Home: 87 Peter Coutts Circle Stanford CA 94305 Office: Dept Microbiology Stanford U Stanford CA 94305

FALLAW, WALLACE CRAFT, geology educator; b. Durham, N.C., Apr. 15, 1936; s. Walter Robert and Amy Wilson (Childs) F.; m. Sarah Steppe Howle, Mar. 25, 1966; children: Ben, James. B.S., Duke U., 1958; M.S., U. N.C., 1963, Ph.D., 1965. Geologist Chevron Oil Co., Lafayette, La., 1965-70; prof. dept. geology Furman U., Greenville, S.C., 1970—. Contbr. articles to profl. jours. Fellow Geol. Soc. Am.; mem. Am. Assn. Petroleum Geologists, AAAS. Subspecialties: Paleobiology; Paleoecology. Current work: Effect of plate tectonics on the distribution of fossil organisms. Home: Rt 9 Starsdale Circle Greenville SC 29609 Office: Dept Geology Furman U Greenville SC 29613

FALLON, ANN MARIE, microbiologist, educator; b. Westerly, R.I., Nov. 10, 1949; d. Richard Francis and Theresa Frances (Bessette) F. Student, Annhurst Coll., Woodstock, Conn., 1967-69; B.A. in Biology summa cum laude (Lt. Drotch Meml. scholar), U. Conn., Storrs, 1972; M.S. in Biology (univ. fellow), Yale U., 1974; Ph.D (R. Samuel McLaughlin fellow, Ont. fellow), Queen's U., Kingston, Ont., Can., 1976. Am. Cancer Soc. fellow Inst. Enzyme Research, U. Wis., Madison, 1976-78; NSF fellow dept. entomology Tex. A&M U., College Station, 1978-79; NIH fellow Rutgers U. Med. Sch., Piscataway, N.J., 1979-81, instr. microbiology, 1981-82; now adj. asst. prof.; asst. prof. N.J. Sch. Osteo. Medicine, 1982—. Contbr. articles to sci. jours. Recipient Edmond Schmidt Research award U. Conn., 1972; Woods Hole Physiology Course scholar, 1973. Mem. Am. Soc. Microbiologists, Am. Soc. Zoologists, AAAS, Phi Beta Kappa, Phi Kappa Phi. Subspecialties: Cell biology; Molecular biology. Current work: Biochemistry and genetics of insect cells in culture, with particular emphasis on gene transfer tech. and regulation of ribosomal protein synthesis. Office: Dept Microbiology UMDNJ-NJSOM Rutgers Med Sch Piscataway NJ 08854

FAMILY, FEREYDOON, physicist; b. Tehran, Iran, Sept. 18, 1945; came to U.S., 1964; s. Hassan and Aghdas (Keramaty) F.; m. Soheila Family, Apr. 20, 1966; 1 son, Afsheen. B.S., Worcester Poly. Inst., 1968; M.S., Tufts U., 1970; Ph.D., Clark U., 1974. Asst. prof. Central New Eng. Coll., Worcester, Mass., 1971-74; research assoc. MIT, 1974-75; head solid state div. Atomic Energy Orgn. of Iran, Tehran, 1975-79; research assoc. Boston U., 1979-81; vis. asst. prof. Worcester Poly. Inst., 1980-81; asst. prof. physics Emory U., 1981—; vis. research physicist Inst. Theoretical Physics, U. Calif.-Santa Barbara, 1982. Contbr. articles; referee sci. jours. Recipient Lawton-Plimpton prize Worcester Poly. Inst., 1968, award for teaching excellence Central New Eng. Coll., 1971; NSF grantee, 1983; Research Corp. grantee, 1982; NIH biocsis. grantee, 1982. Mem. Am. Phys. Soc., Am. Assn. Physics Tchrs., Sigma Xi, Sigma Pi Sigma. Subspecialties: Polymer physics; Condensed matter physics. Current work: Condensed matter physics, theoretical polymer physics, critical phenomena, many-body problem. Office: Dept Physics Emory U Atlanta GA 30322

FAN, HUNG Y, virology educator, researcher, consultant; b. Beijing, China, Oct. 30, 1947; came to U.S., 1949, naturalized, 1958; s. Hsu Yun and Li Nien (Bien) F. B.S., Purdue U., 1967; Ph.D., MIT, 1971. Postdoctoral fellow MIT, Cambridge, 1971-73; asst. prof. Salk Inst., San Diego, 1973-81; asst. prof. virology U. Calif.-Irvine, 1981-83, assoc. prof., 1983—. Contbr. articles on virology to profl. jours. Nat. sec. Gay Acad. Union, 1982-84. Woodrow Wilson fellow, 1967; Helen Hay Whitney fellow, 1971-73; NIH grantee, 1974—. Mem. Am. Soc. for Microbiology, Am. Soc. Virology. Democrat. Subspecialties: Virology (biology); Molecular biology. Current work: Molecular biology of murine leukemia virus, studies of viral genetic structure and chromatin conformation, use of the virus as a recombinant DNA vector. Office: Dept Molecular Biology and Biochemistr U Calif Irvine CA 92717

FAN, LIANG-TSENG (L.T. FAN), chemical engineering educator; b. Yang-Mei, Tao-Yuan, Taiwan, Aug. 7, 1929; came to U.S., 1952, naturalized, 1970; s. Chung-chan and Chien-mei (Huang) F.; m. Eva Cheung, June 2, 1958; children: Tso Yee, Judith Tso-ling. B.S. in Chem. Engring, Nat. Taiwan U., Taipei, 1951, M.S., Kans. State U., 1954, Ph.D., W.Va. U., 1957, M.S. in Math, 1958. Mem. faculty Kans. State U., Manhattan, 1958—, prof. chem. engring., 1963—, head chem. engring. dept., 1968—. Author: The Discrete Maximum Principle, 1964, The Continuous Maximum Principle, 1966, Flow Models for Chemical Reactors, 1975, Environmental Systems Engineering, 1977; U.S. editor: Biotech. Series, 1981—; editor: Particle Sci. and Tech, 1981—; mem. editorial com.: Applied Transport Phenomena Series, 1978—. Recipient Disting. Grad. Faculty award Kans. State U., 1972-73. Mem. Soc. Engring. Scis. (founding mem.), Am. Chem. Soc., Am. Inst. Chem. Engring., Am. Soc. Engring. Edn., Sigma Xi. Subspecialties: Chemical engineering; Fuels and sources. Current work: Systems engineering; energy resources conversion; solids mixing; fluidization; transport phenomena; chemical process dynamics; chemical reactor analysis and design; environmental pollution control; biochemical engineering and biotechnology. Patentee method for wastewater treatment in fluidized bed biol. reactors, 1982. Office: Kans State Univ Chem Engring Dept Manhattan KS 66506 Home: 830 Lee St Manhattan KS 66502

FAN, L.T. *See also* **FAN, LIANG-TSENG**

FANG, SHU-CHERNG, research engineer; b. Nantou, Taiwan, Republic of China, June 14, 1952; s. Shaw-Han F.; m. Chi-Hsin Chao, Aug. 5, 1982. B.S., Nat. Tsing Hua U., Hsinchu, Taiwan, 1974; M.S., Johns Hopkins U., 1977; Ph.D., Northwestern U., 1979. Asst. prof. ops. research U. Md.-Balt., 1979-80; mem. research staff Western Electric Co., Princeton, N.J., 1980-82, sr. mem. research staff, 1982—. Author tech. papers and research reports. Johns Hopkins U. fellow, 1976; Murphy fellow Northwestern U., 1977; research grantee U. Md., summer 1980. Mem. Am. Soc. Indsl. Engring., Soc. Indsl. and Applied Math., Ops. Research Soc. Am., Nat. Engring. Honor Soc. Subspecialties: Fiber optics; Operations research (engineering). Current work: Lightwave system design, optical fiber manufacturing, operations research in computer-aided manufacturing. Office: Western Electric Engineering Research Ctr PO Box 900 Princeton NJ 08540

FANNING, GEORGE RICHARD, biochemist; b. Princeton, W.Va., Sept. 2, 1936; s. George Emory and Catherine Annalee (Raney) F.; m. Nancy Lawrence, June 29, 1962; children: Nancy Catherine, David Lawrence. B.S., Concord Coll., 1958; postgrad., Purdue U., 1958-59; M.S., George Washington U., 1963. Chemistry instr. Kimball Union Acad., Meriden, N.H., 1959-60; biochemist NIH, Bethesda, Md., 1960-63; biochemist div. biochemistry Walter Reed Army Inst. Research, Washington, 1963—. Mem. Am. Soc. Microbiology, Blue Key. Democrat. Presbyterian. Subspecialty: Molecular biology. Current work: Bacteriological relatedness, DNA hybridization of bacteria and their plasmids. Home: 13200 Wilton Oaks Dr Silver Spring MD 20906 Office: Div Biochemistry Walter Reed Army Inst Research Washington DC 20012

FANO, ROBERT MARIO, educator; b. Torino, Italy, Nov. 11, 1917; came to U.S., 1939, naturalized, 1947; s. Gino and Rosetta (Cassin) F.; m. Jacqueline M. Crandall, Mar. 26, 1949; children—Paola C., Linda, Carl. B.S. in Elec. Engring, Mass. Inst. Tech., 1941, Sc.D., 1947. Teaching asst. elec. engring. dept. Mass. Inst. Tech., 1941-43, instr., 1943-44, staff mem., 1944-46, research asso. elec. engring. dept. and research lab. electronics, 1946-47, asst. prof. elec. engring dept., 1947-51, group leader, 1950-53, asso. prof. elec. engring. dept., 1951-56, prof., 1956-62, Ford prof., 1962—; dir. Project MAC, 1963-68, asso. head for computer sci. and engring. elec. engring. dept., 1971-74; cons. to indsl. labs. Author: (with R. B. Adler and L. J. Chu) Electromagnetic Fields, Energy and Forces; Electromagnetic Energy Transmission and Radiation, 1960, Transmission of Information, 1961. Fellow IEEE (Edl. medal 1977), Nat. Acad. Scis., Nat. Acad. Engring., Am. Acad. Arts and Scis.; mem. Assn. Computing Machinery, Sigma Xi, Eta Kappa Nu. Subspecialties: Computer science education; Electrical engineering. Current work: Developing program of lifelong cooperative engineering education. Home: 9 Edmonds Rd Concord MA 01742 Office: Mass Inst Tech Cambridge MA 02139

FANO, UGO, physicist, educator; b. Turin, Italy, July 28, 1912; came to U.S., 1939, naturalized, 1945; s. Gino and Rosa (Cassin) F.; m. Camilla V. Lattes, Feb. 8, 1939; children: Mary, Virginia. Sc.D., U. Turin, 1934; D.Sc. (hon.), Queen's U., Belfast, No. Ireland, 1978, U. Pierre and Marie Curie, Paris, 1979. Lectr., U. Rome, 1937-38; fellow, resident investigator Carnegie Instn., Washington, 1940-46; cons. ballistition U.S. Army Ordnance, 1944-45; physicist X-ray sect. Nat. Bur. Standards, Washington, 1946-49, chief radiation theory sect., 1949-60, sr. research fellow, 1960-66; prof. physics James Franck Inst., U. Chgo., 1966-82, prof. emeritus, 1982—, chmn. dept., 1972-74; lectr. George Washington U., 1946-47; vis. prof. U. Calif., Berkeley, summer 1958, 68, Cath. U., Washington, 1963-64. Author: (with G. Racah) Irreducible Tensorial Sets, 1959, (with L. Fano) Basic Physics of Atoms and Molecules, 1959, Physics of Atoms and Molecules, 1972; also articles. Recipient Rockefeller Pub. Service award, 1956, Exceptional Service award Dept. Commerce, 1957; Stratton award Nat. Bur. Standards, 1963; Davisson-Germer prize Am. Phys. Soc., 1976. Mem. Nat. Acad. Scis., Am. Acad. Arts and Scis., Am. Phys. Soc., Radiation Research Soc. Subspecialties: Atomic and molecular physics; Theoretical physics. Current work: Research describing the mechanisms of transformations of atomic and molecular structures. Home: 5801 S Dorchester Ave Chicago IL 60637

FANSELOW, MICHAEL SCOTT, psychology educator; b. Bklyn., May 2, 1954; s. Hyman and Rhoda Lois (Blatt) F.; m. Mary T. Zatloukal, July 20, 1979. B.S., Bklyn. Coll., 1976; Ph.D., U. Wash., 1980. Asst. prof. Rensselaer Poly. Inst., Troy, N.Y., 1980-81; asst. prof. Dartmouth Coll., Hanover, N.H., 1981—. Contbr. articles to profl. jours. Recipient Edwin Newman award for excellence in research Am. Psychol. Assn./Psi Chi, 1980. Mem. Psychonomic Soc., Am. Psychol. Assn. (D.O. Mebb Young Sicentist award 1983), Eastern Psychol. Assn., AAAS, Sigma Xi, Psi Chi. Subspecialties: Learning; Neuropsychology. Current work: Endogenous opiates, aversively motivated behavior. Home: Rural Route 1 Box 386 Norwich VT 05055 Office: Dept Psychology Dartmouth Coll Hanover NH 03755

FANSLER, KEVIN SPAIN, research physicist; b. Thomas, Okla., Jan. 13, 1938; s. Ralph Kibbee and Alma Ruth (Spain) F.; m. Sherry Rulana Hall, Feb. 22, 1969; children: Zoya, Kira. B.S., Okla. State U., 1960; M.S., U. Hawaii, 1964; Ph.D., U. Del., 1974. Research physicist Naval Ordnance Lab., Silver Spring, Md., 1960-62; chem. research lab. Edgewood Arsenal, 1965-67; ballistic research lab. Aberdeen Proving Ground, 1967—. Contbr. articles to profl. jours. Recipient Special Act award U.S. Army, 1980; Quality Step Increase award, 1981. Mem. AIAA. Subspecialty: Aerospace engineering and technology. Current work: Fluid flow about projectiles and other objects. Home: 501 Jamestown Ct Edgewood MD 21040 Office: Launch & Flight Div US Army Ballistics Research Lab Aberdeen Proving Grounds MD 21005

FARAH, BADIE NAIEM, educator, consultant; b. Nazareth, Palestine, Jan. 15, 1946; came to U.S., 1970, naturalized, 1983; s. Naim R. and Afifi (Takla) F. B.S., Damascus U., 1967, M.A., 1968; M.S., Wayne State U., 1973; M.S.I.E., Ohio State U., 1976, Ph.D., 1977. Teaching asst. Wayne State U., Detroit, 1971-73; research assoc. Ohio State U., Columbus, 1973-77; sr. systems analyst Gen. Motors Co., Detroit, 1977-78; asst. prof. Oakland U., Rochester, Mich., 1978-82; asst. prof. info. systems, ops. research Eastern Mich. U., Ypsilanti, 1982—; advisor to bd. dirs. S & G Grocer Co., Detroit, 1979-81, vis. gen. mgr., 1980-81. Mem. Am. Inst. Indsl. Engrs., Assn. Computing Machinery, Ops. Research Soc. Am., Inst. Mgmt. Scis., Mich. Acad. Sci., Arts and Letters, Alpha Pi Mu. Syrian Orthodox. Subspecialties: Distributed systems and networks; Software engineering. Current work: Data communications and networks of computers, decision support systems for microcomputers, management information systems. Home: 37 Foxboro Dr Rochester MI 48063 Office: Eastern Mich U 511 Pray-Harrold Ypsilanti MI 48197

FARAH, FUAD SALIM, dermatologist, immunologist, educator; b. Haifa, Palestine, Apr. 5, 1929; came to U.S., 1976; s. Salim and Nada (Fuleihan) F.; m. Mona Haddad, June 25, 1955; children: Richard-Salim, Ronald-Samir, Joyce-Bahia, Ramsay-Sami. B.A., Am. U., Beirut, 1950, M.D., 1954. Diplomate: Am. Bd. Dermatology. Asst. prof. medicine Am. U., Beirut, 1959-60, assoc. prof., 1960-66, prof., 1966-76; dir. WHO Immunology Research and Tng. Ctr., Beirut, 1972-76; prof. medicine, chief sect. dermatology SUNY Upstate Med. Ctr., Syracuse, 1976—. Contbr. numerous articles in immunology and dermatology to profl. jours. Decorated knight and officer Order of Cedar, Lebanon). Mem. Soc. Investigative Dermatology, Am. Assn. Immunologists, Soc. Tropical Dermatology, Reticuloendothelial Soc., AAAS. Subspecialties: Immunology (agriculture); Genetics and genetic engineering (agriculture). Home: 113 Victoria Park Dr Liverpool NY 13088 Office: 750 E Adams St Syracuse NY 13210

FARBER, JOSEPH, solar energy company executive, consultant, lecturer, solar energy researcher; b. Newark, June 1, 1924; s. Samuel and Anna (Sielunchik) F.; m. Margaret A. Farber, Dec. 28, 1951; children: Jennie Lee, Steven Eric. B.S. in Chemistry and Physics, CCNY, 1945; Ph.D. in Phys. Chemistry, U. Wis.-Madison, 1951. Research engr., sr. thermal engr. Convair, Gen. Dynamics, San Diego, 1951-55; mgr. real gas engring., mgr. aero. physics Gen. Electric Co., Valley Forge, Pa., 1955-65, mgr. advanced system engring., 1965-67; chief engr. Ford Space and Re-entry Systems, Newport Beach, Calif., 1967-70; pres. KMS Tech. Center, Irvine, Calif., 1970-73; pres., cons. Solar Research Systems, Santa Ana, Calif., 1973—; lectr. U. Calif.-Irvine, Orange Coast Coll.; cons. to govt., nonprofit orgns., industry; bd. dirs. Solar Age Mag., 1977—. Contbr. articles to profl. jours., chpts. to books. Chmn. Solar Coalition of Orange County, 1980-81. Mem. Am. Chem. Soc., Am. Phys. Soc., AIAA, Am. Solar Energy Soc. (dir.), Internat. Solar Energy Soc. Democrat. Unitarian. Subspecialties: Solar energy; Systems engineering. Current work: Solar technology, thermal, photovoltaic, plastics systems development, solar and aerospace, energy power generation. Patentee in field. Home: 1605 Sherington Pl Y212 Newport Beach CA 92663 Office: 2116 S Yale St Santa Ana CA 92704

FARBER, PHILLIP ANDREW, clinical cytogeneticist, educator; b. Wilkes-Barre, Pa., Sept. 19, 1934; s. Phillip Henry and Josephine Mary (Penkala) F.; m. Larice May Krebs, Oct. 11, 1974; children: Michael, Steven, Phillip. B.S., King's Coll., Wilkes-Barre, 1956; M.S., Boston Coll., 1958; postgrad., Cath. U. Am., 1963. Instr. dept. biology Georgetown U., Washington, 1962-63; research biologist Lab. Perinatal Physiology NIH, 1963-64; research instr. Inst. Phys. Medicine and Rehab., N.Y.U. Med. Center, 1964-66; prof. dept. biol. and allied health scis. U. Pa., Bloomsburg, 1966—; cons. clin. cytogenetics, dept. lab. medicine, cytogenetic lab. Geisinger Med. Center, Danville, Pa., 1967—. Contbr. articles to profl. jours. Served with USPHS, 1960. NSF grantee, 1962; NIH grantee, 1965. Mem. Am. Soc. Human Genetics, Assn. Cytogenetics Technologists, Nat. Soc. Histotech., Teratology Soc., Tissue Culture Assn., AAAS, N.Y. Acad. Scis., Pa. Acad. Scis., AAUP, Am. Fedn. Tchrs., Sigma Xi. Subspecialties: Cytogenetics; Cytology and histology. Current work: Human genetics, cytology and cytogenetics, histology, histological and histochemical techniques, developmental biology. Home: PO Box 92 Mifflinville PA 18631 Office: Dept Biol and Allied Health Scis Hartline Sci Center U Pa Bloomsburg PA 17815

FARHI, LEON ELIE, physiology educator; b. Cairo, Oct. 9, 1923; U.S., 1958; s. Elie and Victoria (Anzarut) F.; m., July 12, 1949; children: Nitza Farhi Ellis, Eli Ralph. M.D., St. Joseph U., Beirut, 1947. Asst. prof. physiology SUNY-Buffalo, 1958-62, assoc. prof., 1962-66, prof., 1966—; cons. NIH, Washington. Contbr. articles to profl. jours. Mem. Am. Physiol. Soc., Am. Thoracic Soc., Biomed. Engring. Soc. Jewish. Subspecialties: Psychophysiology; Information systems, storage, and retrieval (computer science). Current work: Cardiorespiratory physiology, environmental physiology, computer-assisted instruction. Home: 158 North Dr Eggertsville Buffalo NY 14226 Office: Department of Physiology SUNY Buffalo NY 14214

FARINOLA, ANTHONY LARRY, physicist, electrical engineer; b. Springfield Gardens, N.Y., Sept. 1, 1948; s. Antonio and Loretta Henrietta (Helwig) F.; m. Nancy Jane, Apr. 11, 1982; 1 dau., Alicia Marie. B.S. in Physics, U. Scranton, 1970, M.S., St. Bonaventure U., Olean, N.Y., 1977. Lab. engr. Schott Optical Glass, Inc., Duryea, Pa., 1972-80; sr. engr. Electro-Optical Products div. ITT, Roanoke, Va., 1980—; speaker 22d Ann. Tech. Symposium, San Diego, 1978. Mem. Am. Phys. Soc., Optical Soc. Am., Soc. Photo-Optical Instrumentation Engrs., Triumph Sports Owners Assn., Am. Motorcycle Assn., Mensa, Sigma Pi Sigma. Republican. Lutheran. Subspecialties: Fiber optics; Systems engineering. Current work: Real-time applications of computer automation specializing in fiber-optical measurements. Home: Route 2 Box 591 Troutville VA 24175 Office: 7635 Plantation Rd Roanoke VA 24019

FARLEY, DONALD THORN, JR., electrical engineering educator; b. N.Y.C., Oct. 26, 1933; s. Donald Thorn and Rebecca (Hamlin) F.; m. Jennie Tiffany Towle, June 16, 1956; children: Claire, Anne, Peter. B.E., Cornell U., 1956, Ph.D., 1960. NATO postdoctoral fellow Cambridge U., Eng., 1956-60; docent Chalmers U. Tech., Gothenburg, Sweden, 1960-61; physicist Jicamarca Radio Obs., Lima, Peru, 1961-64, dir., 1964-67; prof. elec. engring. Cornell U., Ithaca, N.Y., 1967—. Assoc. editor: Revs. Geophysics and Space Physics, 1963-69, Jour. Geophys. Research, 1975-77, Radio Sci, 1976-79; contbr. numerous articles to sci. jours. Recipient U.S. Dept. Commerce Gold medal, 1967, Disting. Authorship award, 1963, 64. Mem. Am. Geophys. Union, IEEE, Internat. Sci. Radio Union, AAAS, Sigma Xi, Tau Beta Pi, Phi Kappa Phi. Current work: Radar probing of the upper atmosphere; radar technology; plasma instabilities in the ionosphere. Home: 711 Triphammer Rd Ithaca NY 14850 Office: Phillips Hall Cornell U Ithaca NY 14853

FARLEY, JOHN RANDOLPH, biochemist, educator; b. Stockton, Calif., Oct. 12, 1948; s. John and Dawn (Krueger) F.; m. Sally Marie Gibson, June 26, 1971; 1 son, Patrick J. B.S., U. Calif.-Davis, 1972, Ph.D., 1977. Postgrad. research assoc. U. Calif.-Davis, 1977-78; research fellow U. Wash., Seattle, 1978-80, acting research instr. medicine, 1979-80, research asst. prof., 1980-81; asst. prof. biochemistry/medicine Loma Linda (Calif.) U., 1981—. Mem. AAAS, Am. Soc. for Bone and Mineral Research, Am. Fedn. for Clin. Research, Soc. for Exptl. Biology and Medicine. Democrat. Subspecialties: Biochemistry (biology); Cell and tissue culture. Current work: Enzymology and biochemistry of skeletal metabolism; regulation of bone volume by systemic and local effectors—biochemical mechanisms of regulation; role of alkaline phosphatase in bone formation. Office: Research Pettis VA Hosp 11201 Benton St Loma Linda CA 92357

FARLEY, JOHN WILLIAM, physicist, educator; b. N.Y.C., Feb. 7, 1948; s. John and Eileen Gertrude (Gray) F. B.A., Harvard Coll., 1970; M.A., Columbia U., 1974, M. Ph., 1974, Ph.D., 1977. Postdoctoral research assoc. U. Ariz.-Tucson, 1976-80, research assoc. prof. physics, 1980-81; asst. prof. physics U. Oreg., Eugene, 1981—. Contbr. articles to sci. publs. in field. NSF nat. needs postdoctoral fellow, 1978-79. Mem. Am. Phys. Soc., Optical Soc. Am. Subspecialties: Atomic and molecular physics; Spectroscopy. Current work: Lasers and other quantum-electronic areas; nonlinear optics, studies of simple atomic and molecular systems; molecular ions; computers and automated instrumentation. Home: 744 E 21st Ave Eugene OR 97405 Office: Physics Dept U Oreg Eugene OR 97403

FARMAN, ALLAN GEORGE, dental educator, academic administrator; b. Birmingham, Eng., July 26, 1949; came to U.S., 1980; s. George and Lily (Hewitt) F.; m. Francoise Jeanne Lemaire, July 1, 1972; children: Julie Melinda, Wendy Claire. B.D.S., Birmingham U., 1971; L.D.S., Royal Coll. Surgeons, London, Eng., 1972; Ph.D. in Oral Pathology, Stellenbosch U., South Africa, 1977. Lectr. U. of Witwatersrand and South African Inst. Med. Research, Johannesburg, South Africa, 1972-74; sr. lectr. U. Stellenbosch, Parowvallei, South Africa, 1974-77, U. Western Cape, Bellville, South Africa, 1974-77; head oral pathology U. Riyadh, Saudi Arabia, 1978-79; assoc. prof. oral diagnosis U. Louisville, 1980—, dir. radiology, 1982—; vis. prof. dental diagnostic scis. U. Tex.-San Antonio, 1981—. Numerous sci. presentations.; Contbr. numerous articles to internat. profl. jours., 1975—. Instr. CPR ARC, Louisville, 1982—. Recipient Harry Crossley Found. award, South Africa, 1974, 75; South African Med. Research Council grantee, 1975-80. Mem. Royal Soc. South Africa, Internat. Assn. Oral Pathologists (charter), Am. Acad. Dental Radiology, Am. Acad. Oral Medicine, Internat. Assn. Dental Research, Am. Assn. Dental Research. Subspecialties: Oral pathology. Current work: Epidemiology of oral cancer and related conditions; correlation of histopathologic and diagnostic imaging findings for oral disease; dental radiology education in United States. Home: 12517 Farmbrook Dr Louisville KY 40243 Office: U Louisville Sch Dentistry 501 Preston St Louisville KY 40292

FARMER, JAMES LEE, geneticist; b. South Gate, Calif., Aug. 8, 1938; s. James Ira and Eliza Ellen (Sheeks) F.; m. Gladys Clark, Jan. 27, 1967; children: Sarah Lynn, James Clark, Rachel Lee, Jared Randall, Deborah Ann. B.S., Calif. Inst. Techn., 1960; postgrad., Brigham Young U., 1960-61; Ph.D., Brown U., 1966. Instr. biophysics U. Colo. Med. Ctr., Denver, 1966-68; asst. prof. zoology Brigham Young U., Provo, Utah, 1969-78, assoc. prof., 1978—. Editorial bd.: Dialogue: A Jour. of Mormon Thought, 1975-82; contbr. articles to profl. jours. Mem. AAAS, Fedn. Am. Scientists, Genetics Soc. Am. Mem. Ch. Jesus Christ of Latter-day Saints. Subspecialties: Gene actions; Genome organization. Current work: Researcher in gene structure of Drosophila melanogaster, recombinant DNA, gene interaction. Home: 150 E 4200 N Provo UT 84604 Office: Dept Zoology Brigham Young U Provo UT 84602

FARMER, JOSEPH CLARENCE, JR., physician, educator; b. Fayetteville, N.C., Oct. 14, 1937; s. Joseph Clarence and Bettie (Eatman) F.; m. Margery Jean Newton, Aug. 19, 1957; 1 son, Thomas Hackney Richardson. M.D., Duke U., 1962. Diplomate: Am. Bd. Otolaryngology. Intern, resident Duke U., Durham, N.C., 1962-65, 67-70, asst. prof. dept. surgery, 1970-75, assoc. prof., 1975—; clin. assoc. surgery br. Nat. Cancer Inst., Bethesda, Md., 1965-67; mem. core faculty F.G. Hall Environ. Research Lab., Duke U. Med. Ctr., 1970—. Contbr. articles to profl. jours. Fellow ACS, Am. Laryingological, Rhinological and Otological Soc., Am. Acad. Otolaryngology Head and Neck Surgery; mem. AMA, N.C. Med. Soc., Durham-Orange County Med. Soc., Alpha Omega Alpha. Democrat. Episcopalian. Subspecialties: Otorhinolaryngology; Surgery. Current work: Academic otolaryngology; effects of altered pressures on hearing and balance; otoneurology; diving medicine-otology. Office: Div Otolaryngology Dept Surgery Duke U Med Ctr Durham NC 27710

FARNELL, GERALD WILLIAM, educator; b. Toronto, Ont., Can., Aug. 31, 1925; s. Jack and Alice (Turner) F.; m. Norma Catherine McRae, Sept. 14, 1948; children—Sandra Rae (Mrs. Moisl), Gerald Douglas. B.A. Sc., U. Toronto, 1948; M.S., Mass. Inst. Tech., 1950; Ph.D., McGill U., 1957. Research asst. Mass. Inst. Tech., 1948-50; asst. prof. McGill U., Montreal, Que., 1950-57, asso. prof., 1957-61, prof. engring. physics, 1962—, chmn. dept. elec. engring., 1967-74, dean faculty of engring., 1974—; Nuffield fellow Clarendon Lab., Oxford, 1960-61. Served with Canadian Army, 1943-45. Fellow IEEE; mem. Order Engrs. Que., Am. Soc. Engring. Edn., Sigma Xi. Subspecialties: Electrical engineering; Acoustical engineering. Current work: Acoustic propogation, acoustic surface waves, acoustic microscopy. Home: 1509 Sherbrooke St Montreal PQ Canada H3G 1M1

FARNER, DONALD SANKEY, biologist; b. Waumandee, Wis., May 2, 1915; s. John and Lillian O. (Sankey) F.; m. Dorothy S. Copps, Dec. 21, 1940; children—Carla M., Donald C. B.S., Hamline U., 1937, D.Sc. (hon.), 1962; M.A., U. Wis., 1939, Ph.D., 1941. Instr. zoology U. Wis., 1941-43; asst. prof. zoology U. Kans., 1946-47; faculty Wash. State U., 1947-65, prof. zoophysiology, 1952-65, dean, 1960-64; prof. zoophysiology U. Wash., Seattle, 1965—, chmn. dept. zoology, 1966-81; Fulbright research scholar, hon. lectr. zoology U. Otago, N.Z., 1953-54; Guggenheim fellow U. Western Australia, 1958-59; chmn. div. biology and agr. Nat. Acad. Sci.-NRC, 1969-73; sr. U.S. scientist Alexander von Humboldt Stiftung, 1978; pres. XVII Internat. Ornithol. Congress, 1978. Served to lt. USNR, 1943-46; capt. Res. ret. Fellow AAAS (council 1964—), Am. Ornithologists Union (Brewster award 1960, pres. 1973-75); mem. Am. Physiol. Soc., Am. Soc. Zoologists, Am. Inst. Biol. Scis., Am. Chem. Soc., Internat. Union Biol. Scis. (pres. 1967-73, chmn. div. zoology 1973—), Soc. Systematic Zoology, Am. Soc. Naturalists, Cooper Ornithol. Soc. (hon. mem.; bd. govs. 1965-71), Deutsche Ornithologen-Gesell. (hon.), Ornitologiska Foreningen: Finland (hon.), Soc. for Endocrinology, Explorer's Club, Phi Beta Kappa, Sigma Xi, Phi Kappa Phi, Phi Sigma (hon. pres. 1973-80), Gamma Alpha, Omicron Delta Kappa. Methodist. Club: Cosmos (Washington). Subspecialties: Animal physiology; Reproductive biology. Current work: Reproductive biology, protoperiodism in birds, avian chrono biology, control of annual cycles. Research and publs. in avian biology and physiology. Home: 4533 W Laurel Dr Seattle WA 98105 Office: Dept Zoology U Washington Seattle WA 98195

FARQUHAR, JOHN WILLIAM, physician, educator; b. Winnipeg, Man., Can., June 12, 1927; came to U.S., 1934, naturalized, 1950; s. John Giles and Marjorie Victoria (Roberts) F.; m. Christine Louise Johnson, July 14, 1968; children: Margaret F., John C.M.; children by previous marriage: Bruce E., Douglas G. A.B., U. Calif., Berkeley, 1949, M.D., 1952. Intern U. Calif. Hosp., San Francisco, 1952-53; resident, 1953-54, 57-58, postdoctoral fellow, 1955-57; resident U. Minn., Mpls., 1954-55; research asso. Rockefeller U., N.Y.C., 1958-62; asst. prof. medicine Stanford (Calif.) U., 1962-66, asso. prof., 1966-73, prof., 1973—; dir. Stanford Heart Disease Prevention Program, 1973—; mem. staff Stanford U. Hosp.; adviser, cons. Inst. of Medicine of Nat. Acad. Scis. Author: The American Way of Life Need Not Be Hazardous to Your Health, 1978; contbr. articles to profl. jours. Served with U.S. Army, 1945-46. Recipient James D. Bruce award ACP, 1983. Mem. Am. Soc. Clin. Investigation, Acad. Behavioral Medicine, Harvey Soc., Gold Headed Cane Soc., Sigma Xi, Alpha Omega Alpha. Episcopalian. Subspecialty: Preventive medicine. Office: Sch Medicine Stanford U Stanford CA 94305

FARR, WILLIAM ROGERS, nuclear engineer, computer consultant; b. Memphis, Feb. 21, 1949; s. William Glenn and Miriam Virginia (Rogers) F.; m. Carolyn Sue Bryant, May 27, 1972. B.S., Ga. Inst. Tech., 1972, M.S., 1973; postgrad., N.C. State U., 1979—. Registered profl. engr., N.C.; Registered profl. engr., N.C. Coop. student/tng. analyst NASA/Marshall Space Flight Ctr., Huntsville, Ala., 1968-72; project engr. Carolina Power & Light Co., Raleigh, N.C., 1972—; com. mem. Electric Power Research Inst., Palo Alto, Calif., 1979-81; computer cons. Univ. Computer Co., Dallas, 1982, Precision Computer Systems, Havelock, N.C., 1981-82. Contbr. articles in field to profl. jours. Dist. commr. Boy Scouts Am., Raleigh, N.C., 1975-76, Eagle Scout, 1966. AEC fellow, 1972-73. Mem. Am. Nuclear Soc., Assn. Computing Machinery, Soc. Indsl. and Applied Math., Ops. Research Soc. Baptist. Subspecialties: Operations research (engineering); Mathematical software. Current work: Application of mathematical programming to electric utility decision problems in transportation, scheduling and nuclear fuel design engineering. Home: 406 Morrison Ave Raleigh NC 27608 Office: Carolina Power & Light Co 411 Fayetteville St Raleigh NC 27602

FARRAR, JAMES MARTIN, chemistry educator; b. Pitts., June 15, 1948; s. Martin W. and Lorraine (Williams) F.; m. Kathy June Meyer, Mar. 20, 1971; children: Stacey Elizabeth, Andrew Martin. A.B., Washington U., 1970; M.S., U. Chgo., 1972, Ph.D., 1974. Research asst. U. Chgo., 1971-74; research assoc. U. Calif.-Berkeley, 1974-76; asst. prof. U. Rochester, N.Y., 1976-82, assoc. prof. chemistry, 1982—. NSF fellow, 1970-73; Sloan Found. fellow, 1981-83. Mem. Am. Chem. Soc., Am. Phys. Soc. Subspecialties: Physical chemistry; Spectroscopy. Current work: Molecular beam and laser spectroscopic studies of chemical reaction dynamics. Home: 440 Wellington Ave Rochester NY 14619 Office: Dept Chemistry U Rochester Rochester NY 14627

FARRER, DONALD NATHANAEL, research executive, behavior toxicology researcher; b. Dardanelle, Ark., Mar. 18, 1935; s. Hye and Nina (Adney) F.; m. Claire Rafferty, Feb. 2, 1957 (div. 1973); 1 dau., Suzanne; m. Colene Harrington, May 6, 1977. B.A., San Jose (Calif.) State Coll., 1957, M.S., 1958; Ph.D., Wash. State U., Pullman, 1962. Lic. psychologist, N.Mex. Research psychologist Aeromed. Research Lab., Holloman AFB, N.Mex., 1961-65, supervisory research psychologist, 1965-69, research dir. behavioral toxicology, 1969-71, USAF Sch. Aerospace Med., Brooks AFB, Tex., 1971—. Contbr. articles to profl. jours. Subspecialties: Neuropsychology; Behavioral psychology. Current work: Behavior toxicology, vision, radiation bioeffects. Home: 5810 Gomer Pyle San Antonio TX 78240 Office: USAF Sch Aerospace Medicine Brooks AFB TX 78235

FARRIS, PAUL LEONARD, agricultural economist; b. Vincennes, Ind., Nov. 10, 1919; s. James David and Fairy Julia (Kahre) F.; m. Rachel Joyce Rutherford, Aug. 16, 1953; children: Nancy, Paul, John, Carl. B.S., Purdue U., 1949; M.S., U. Ill., 1950; Ph.D., Harvard U., 1954. Asst. prof. agrl. econs. Purdue U., West Lafayette, Ind., 1952-56, asso. prof., 1956-59, prof., 1959—, head dept. agrl. econs., 1973-82; agrl. economist Dept. Agr., Washington, 1962; project leader for meat and poultry Nat. Commn. Food Mktg., Washington, 1965-66. Editor: Market Structure Research, 1964; contbr. articles to profl. jours. Served with AUS and USAAF, 1941-46. Mem. Am. Agrl. Econs. Assn., Am. Econ. Assn. Subspecialty: Agricultural economics. Current work: Agricultural marketing policy; agricultural futures markets. Home: 1510 Woodland Ave West Lafayette IN 47906 Office: Dept Agrl Econs Purdue U West Lafayette IN 47907

FARSHIDI, ARDESHIR B., cardiologist, educator, cardiac electrophysiologist, researcher; b. Kerman, Iran, June 13, 1945; came to U.S., 1972, naturalized, 1977; s. Jamshid and Farangis F.; m. Katayoon Kavoussi, Jan. 2, 1982. M.D., Tehran U., 1969. Diplomate: Am. Bd. Internal Medicine, Am. Bd. Cardiovascular Disease. Intern, Washington, 1972-73; resident U. Pa., Phila., 1973-75, resident in cardiology, 1975-77, electrophysiologist, 1977-78; asst. prof. medicine U. Conn., Farmington, 1978-79, dir. electrophysiology, 1982—, assoc. prof., 1982—, attending cardiologist Hosp., 1982—, dir. electrophysiology, 1982—; co-dir. electrophysiology Yale U., 1979-82, asst. prof. medicine, 1979-82, attending cardiology Hosp., 1979-82, co-dir. electrophysiology, 1979-82; chief cardiology sect. VA Hosp., Newington, Conn., 1982—. Mem. editorial bd.: Jour. Am. Coll. Cardiology, 1983; contbr. articles in field to profl. jours. Served to lt. Iran Army, 1969-72. Am. Heart Assn. researcher, 1981. Fellow ACP, Am. Coll. Cardiology, Am. Heart Assn.; mem. Am. Fedn. Clin. Research, Am. Electrophysiologic Soc. Zoroastrian. Subspecialties: Cardiology; Internal medicine. Current work: Clinical cardiac electrophysiology and arrhythmia. Office: University of Connecticut Medical Center 263 Farmington Ave Farmington CT 06032

FARWELL, SHERRY OWEN, analytical chemistry educator, research consultant; b. Miles City, Mont., June 1, 1944; s. Sherrill and Clara (Bazil) F.; m. Judy Ann Lovec, June 24, 1963; children: Gary Owen, Jodey Shawn. B.S., S.D. Sch. Mines and Tech., Rapid City, 1966, M.S., 1969; Ph.D., Mont. State U., 1973; postgrad., Wash. State U., Pullman, 1973-74. Chemist John Deere, Dubuque, Iowa, 1966-67; analytical chemist Tektronix, Beaverton, Oreg., 1969; asst. prof. analytical chemistry Wash. State U., Pullman, 1974-77; asst. prof. U. Idaho, 1977-80, assoc. prof., 1981—; Mem. exec. com. Moscow (Idaho) Baseball Assn.; cons. Hewlett-Packard, Palo Alto, Calif., 1978—. Contbr. articles to profl. jours. NDEA fellow, 1969-72. Mem. Am. Chem. Soc. (pres. Wash.-Idaho sect. 1981-83), Air Pollution Control Assn., ASTM. Democrat. Subspecialties: Analytical chemistry; Atmospheric chemistry. Current work: Development and validation of new, analytical methods for sulfur compounds, halogen compounds and precious metals. Home: 820 Courtney St Moscow ID 83843 Office: Dept Chemistry U Idaho Moscow ID 83843

FASS, BARRY, neuroscientist; b. Bridgeport, Conn., Jan. 29, 1953; s. Elias N. and Rosalind B. (Bernstein) F. B.A., Clark U., 1975, M.A., 1977, Ph.D., 1979. Postdoctoral fellow UCLA, 1979-80; postdoctoral fellow U. Va. Sch. Medicine, 1980-81, research assoc., 1981—. Contbr. articles to profl. jours. Campaign worker Central Va. Citizens Party. NIMH fellows, 1975-76; trainee, 1979-80. Mem. AAAS, Internat. Soc. Research Aggression, Soc. Neurosci., Phi Beta Kappa. Unitarian-Universalist. Club: U. Va. Raquetball. Subspecialties: Psychobiology;

Regeneration. Current work: Neuroplasticity and behavioral recovery after brain lesions; feeding behavior; agonistic behavior. Home: 128 Georgetown Apt 3 Charlottesville VA 22901 Office: Dept Neurosurgery Box 420 U Va Sch Medicine Charlottesville VA 29908

FASSETT, JAMES ERNEST, geologist, stratigrapher; b. Dearborn, Mich., May 1, 1933; s. Ernest Elihu and Mary (Sleiva) F.; m. Mary Margaret Mlynarowich, Nov. 22, 1959 (div. 1969); children: Melissa Jo, Douglas Paul, Leslie Ann; m. Sarah Lynn Sheafe, June 13, 1970; step-children: Susan Lynne Badsgard, Tracy Anne Badsgard. B.S., Wayne State U., 1959, M.S., 1964. Cert. profl. geologist. Geologist U.S. Geol. Survey, Washington, 1960-61, dist. geologist, Farmington, N.Mex., 1961-81; dep. minerals mgr. Minerals Mgmt. Service, Albuquerque, 1981-83, chief br., Reston, Va., 1983—; cons. Time-Life Books, 1973, Harper & Row, 1977. Editor: Canyonlands Country, 1975, San Juan Basin III, 1977, Oil and Gas Fields of the Four Corners Area (2 vols.), 1978; contbr. articles to profl. jours. Chmn. San Juan County Human Rights Com., Farmington, 1976; mem. adv. bd. Legal Aid of Farmington, 1978; v.p. Citizens for Preservation Pub. Lands, 1978. Served with AUS, 1953-55. Recipient Superior Performance awards U.S. Geol. Survey, 1963, 81. Fellow Geol. Soc. Am.; mem. Am. Assn. Petroleum Geologists (A.I. Levorsen award 1969, del. 1972-79), Am. Inst. Profl. Geologists (del. 1968), Four Corners Geol. Soc. (hon. life mem., pres. 1971-72), N.Mex. Geol. Soc. (pres. 1976). Democrat. Club: Farmington Roadrunners. Subspecialties: Stratigraphy; Geology. Current work: Research on the stratigraphy, environments of deposition and geologic history of Cretaceous rocks of southern Colorado Plateau, especially as related to the origin and resource evaluation of cretaceious oil and gas and coal deposits; studies related to dinosaur extinction and the definition of the cretaceous-tertiary boundary. Office: Br Eastern Tech Reports US Geol Survey Mail Stop 904 Reston VA 22093

FAST, EDWIN, physicist, safety engineer; b. Major County, Okla., July 2, 1914; s. Daniel Peter and Elizabeth Ida (Martens) F.; m. Evelyn Marie McElmurry, June 4, 1943; children: Margaret Louise Fast Whiting, Larry Gleason, Genita Gail Fast Angel, Cecilia Junell Fast Glenn. A.A., Tabor Coll., Hillsboro, Kans., 1936; A.B., Friends U., 1939; M.S., U. Okla., 1941, Ph.D., 1946. Instr. physics U. Okla., Norman, 1943-45; research physicist Phillips Petroleum Co., Bartlesville, Okla., 1945-51; nuclear energy researcher, Idaho Falls, 1951-66, Idaho Nuclear, 1966-69; cons. nuclear physics, 1969-79; criticality safety engr. Exxon Nuclear Idaho, Idaho Falls, 1980—; assoc. prof. physics and math. Mo. Bapt. Coll., St. Louis, 1973-74. Mem. Am. Phys. Soc., Am. Nuclear Soc., Am. Sci. Affiliation. Baptist. Subspecialties: Nuclear fission; Nuclear engineering. Home: 1549 Beverly Rd Idaho Falls ID 83402 Office: Exxon Nuclear Idaho Idaho Falls ID 83401

FATELEY, WILLIAM GENE, scientist, educator, adminstr.; b. Franklin, Ind., May 17, 1929; s. Nolan William and Georgia (Scott) F.; m. Wanda Lee Glover, Sept. 1, 1953; children—Leslie Kaye, W. Scott, Kevin L., Jonathan H., Robin L. A.B., Franklin Coll., 1951, D.Sc. (hon.), 1965; postgrad., Northwestern U., 1951-53, U. Minn., 1956-57; Ph.D., Kans. State U., 1956. Head phys. measurement Dow Chem. Co., Williamsburg, Va., 1958-60; fellow Mellon Inst., Pitts., 1960-62, head sci. relations, 1962-64, asst. to pres., 1964-67, sr. fellow in ind. research, 1965-72; asst. to v.p. for research, 1967-72; prof. chemistry Carnegie-Mellon U., 1970-72; prof., head dept. chemistry Kans. State U., 1972-79; vis. prof. chem. dept. U. Tokyo, 1973, 81; pres. D.O.M. Assos., Internat., 1979—; dir. Pitts. Conf. on Analytical Chemistry and Applied Spectroscopy, 1964-65, pres., 1970-71; editor Jour. Applied Spectroscopy, Raman Newsletter, also finance chmn., steering com. for interferometry. Author: Infrared and Raman Selection Rules, 1973, Characteristic Raman Frequencies, 1974, also numerous sci. papers.; Contbr. articles to profl. jours. Recipient Coblentz award for outstanding contbn. to molecular spectroscopy, 1965; Spectroscopy award Pitts. Conf. Analytical Chemistry and Applied Spectroscopy, 1976; named 1st outstanding grad. chemistry Kans. State U., 1964; H.H. King award, 1979. Fellow Optical Soc. Am.; mem. Am. Chem. Soc. (pres. phys.-inorganic sect. Pitts. 1969-70), Phi Beta Kappa (hon.), Sigma Xi, Sigma Alpha Epsilon, Phi Lambda Epsilon, Pi Mu Epsilon. Subspecialties: Analytical chemistry; Physical chemistry. Current work: Application of Fourier transform infrared spectroscopy to analytical chemistry and to time-resolved kinetic studies. Home: 1928 Leavenworth Manhattan KS 66502 Be nice to young people on their way up. Students are our greatest natural resources.

FATER, DENNIS CARROLL, research physiologist; b. Dearborn, Mich., Oct. 7, 1953; s. Paul Ervin and Mildred Lucille (Wilkins) F.; m. Vonnie Kay Stanford, Aug. 9, 1974. B.S., U. Mich., 1975; Ph.D., U. Kans. Med. Ctr.-Kansas City, 1981. Teaching asst. U. Kans. Med. Ctr., 1976-81; postdoctoral fellow U. Fla., Gainesville, 1981—. Am. Heart Assn. fellow, 1982-83. Mem. Am. Heart Assn., Am. Physiol. Soc., Sigma Xi. Subspecialty: Physiology (medicine). Current work: Influence of neutral amino acids on development of hypertension. Office: Dept Physiology U Fla Box J 274 JHMHC Gainesville FL 32610 Home: PO Box 395 Williston FL 32696

FATHMAN, CHARLES GARRISON, physician, immunologist; b. Clarksville, Mo., Aug. 30, 1942; s. Alfred Stewart and Rachel Erritt (Gillum) F.; m. Ann Kohlmoos, June 16, 1968; children: Christopher Erritt, Carrie Ann, John Warner. B.A., U. Ky., 1964; M.D., Washington U., 1969. Intern, resident in medicine Dartmouth Affiliated Hosps., Hanover, N.H., 1969-71; postdoctoral fellow in immunology Stanford (Calif.) U., 1971-73; postdoctoral fellow immunology br. Nat. Cancer Inst. NIH, Bethesda, Md., 1973-75; mem. Basel Inst. Immunology, Switzerland, 1975-77; assoc. prof. Mayo Med. Sch., Mayo Clinic, Rochester, Minn., 1977-81; assoc. prof. medicine div. immunology Stanford U., 1981—; sci. advisor Cetus Corp.; acad. assoc. Nichols Inst. Contbr. articles to profl. jours. Served as lt. comdr. USPHS, 1971-73. Mem. Am. Rheumatism Assn., Am. Assn. Immunologists, Transplantation Soc., Am. Soc. Clin. Investigation. Democrat. Subspecialties: Immunogenetics; Internal medicine. Current work: Isolation and characterization of immunoregulatory lymphocyte subsets as well as cell-free products involved in normal immunoregulation. Office: Div Immunology S102A Stanford Med Ctr Stanford CA 94305

FATIADI, ALEXANDER JOHANN, research chemist; b. Kharkov, Ukraine, Oct. 22, 1923; came to U.S., 1952, naturalized, 1957; s. Johann George and Maria Ivan (Goncharenko) F.; m. Irina Ivan Matussevich, July 20, 1952; children: Elena Fatiadi Zahirpour, Irina Fatiadi Weiss, Tamara Fatiadi Stoner, Julia. Dr. Nat. Sci. in Organic Chemistry, Inst. Tech. Regensburg (Germany) and U. Mainz (Germany), 1950; M.S. in Phys. Organic Chemistry, George Washington U., 1959. Research assoc. dept. chemistry George Washington U., 1954-59; research chemist in organic chemistry Nat. Bur. Standards, Washington, 1959—. Contbr. numerous articles on organic chemistry, synthetic methods, oxidation reactions, semicondrs., revs. to profl. jours.; author: Reaction of Potassium Metal with Carbon Monoxide, 1959, Reactions of Cyano Compounds, 1983. Recipient cert. commendation Nat. Bur. Standards, 1965, 68, 73, 82, Hillebrand award Chem. Soc. Washington, 1982. Mem. Am. Chem. Soc., Chem. Soc. (London), German Chem. Soc. Republican. Russian Orthodox. Subspecialties: Organic chemistry; Synthetic chemistry. Current work: Research on electrochemical properties of transition metal complexes of oxo-and pseudo-oxocarbons, biomedical technology (protein analysis). Patentee synthesis inositol hexasulfate; introduced periodic acid for oxidation for polycyclic aromatic hydrocarbons; proved structure of active manganese dioxide; discovered semiconducting oxocarbons. Office: Nat Bur Standards Washington DC 20234

FATT, IRVING, optometry and bioengineering educator; b. Chgo., Sept. 16, 1920; s. David and Annie Lily (Arkin) F.; m.; 1 dau., Lois Fatt White. B.S. in Chemistry, UCLA, 1947, M.S., 1948, Ph.D., U. So., Calif., 1955. Sr. research chemist Standard Oil Co., La Habra, Calif., 1948-52, group supr., 1952-57; mem. faculty U. Calif.-Berkeley, 1957—, prof. physiol. optics and engrng. sci., 1962-63, Miller Research prof. engrng., 1962-67, asst. dean, 1964-69, acting dean, 1975-78. Served to 1st lt. USAF, 1942-46. Petroleum Fund Research Career grantee, 1957. Mem. Biomed. Engrng. Soc., Am. Acad. Optometry, U.K. Biol. Engrng. Soc. Clubs: Berkeley Faculty, Berkeley Yacht. Subspecialties: Bioinstrumentation; Optometry. Current work: Development of blood chemistry sensors for use in operating rooms; also research and development of new contact lenses and instruments for measuring properties of new contact lenses. Patentee in field. Home: 406 Boynton Ave Berkeley CA 94707 Office: U Calif Minor Hall Berkeley CA 94720

FAULKNER, JAMES RANDALL, design engineer; b. Dallas, Apr. 3, 1955; s. Leslie E. and Eloise F.; m. Vickie Lynn Hibbitts, Jan. 8, 1977; 1 dau., Malia. B.S. Mech. Engrng, Tex. A&M U., 1977, 1977. Devel. engr. Arco Oil & Gas Co., Houston, 1977-78, design engr., Dallas, 1978-81, sr. design engr., 1981—, design engr., 1979-81, lead engr., 1980-81, 1981-82, 1981-82. Mem. Marine Tech. Soc., ASME, Phi Tau Sigma. Republican. Mem. Ch. of Christ. Subspecialties: Offshore technology; Ocean engineering. Current work: Development of deep water and unique ocean engineering concepts for purposes of petroleum exploration and production in offshore areas. Office: Arco Oil & Gas Co PO Box 2819 Dallas TX 75221 Home: 2630 Chatham St Grand Prairie TX 75052

FAUNCE, FRANK ROLAND, pediatric dentist; b. Richmond, Ind., Sept. 5, 1938; s. Frank Harry and Leota Parilee (Elliot) F. A.B., Ind. U., 1960; D.D.S., 1964; postgrad., U. Tex., 1974. Asst. prof. Med. Coll. Ga., Augusta, 1974-75; asst. prof., grad. dir. U. Tex., Houston, 1975-79; assoc. prof., undergrad. dir. U. Miss., Jackson, 1979-82; cons. Plimark Lab., Muncie, Ind., 1982—. Author: Bonded Aesthetic Dentistry, 1982; contbr. articles in field to profl. jours. Fellow Am. Acad. Pedodontics; mem. ADA, Houston Acad. Pedodontics, Am. Assn. Dental Research, Am. Assn. Dental Schs. Presbyterian. Subspecialties: Biomaterials; Preventive dentistry. Current work: Micro structured ceramic materials and fluoridated phosphate compounds and their anti-microbial activities. Developer: laminate veneers, fluoridated phosphates. Home: 2020 Enterprise Dr Muncie IN 47302 Office: Plimark Labs. Inc PO Box 191 Muncie IN 47305

FAUST, MIKLOS, horticultural scientist, government research administrator; b. Nagyberény, Hungary, Dec. 25, 1927; m. (married); 1 child. B.S., Agrl. U. Budapest, Hungary, 1952; M.S., Rutgers U., 1960; Ph.D. in Pomology, Cornell U., 1965. Mgr. Csaszartoltes State Farm, Hungary, 1952-54; regional supr. Ministry State Farms, 1955-57; research assoc. Rutgers U., 1958-60; research horticulturist United Fruit Co., N.Y., 1960-62; research assoc. Cornell U., 1963-65, N.Y. State Agrl. Expt. Sta., 1965-66, Beltsville (Md.) Agrl. Research Ctr., Agrl. Research Service, Dept. Agr., 1966-69, leader pomological fruit investigations, 1969-72, chief fruit lab., 1973—. Mem. Am. Soc. Hort. Sci., Internat. Soc. Hort. Sci. Subspecialty: Pomology. Office: Agrl Research Service Dept Agr Beltsville MD 20705

FAW, FREDERICK LEE, radiological physicist; b. Yakima, Wash., July 5, 1936; s. Wendell Enoch and Gladys Irene (Moberley) F.; m. Mae Turner, June 9, 1965; children: Lisa Diane, Kathy Lee. B.S., U. Oreg., 1960; M.S., Columbia U., 1966; postgrad., Cath. U. Am., 1970-82. Diplomate: in X-Ray and Gamma Physics, Am. Bd. Radiology; cert. radiation equipment safety officer, N.Y. Med. physicist Nat. Cancer Inst., Bethesda, Md., 1962-74; med. physicist Arnot-Ogden Meml. Hosp., Elmira, N.Y., 1974—. Contbr. articles to tech. jours. Chmn. bldg. com. Maranatha Bible Chapel, 1979-83; chmn. bd. dirs. Twin Tiers Youth for Christ, 1981—. Mem. Am. Assn. Physicist in Medicine, Health Physics Soc., British Inst. Radiology, Mumps Users Group. Mem. Christian Missionary Alliance Ch. Current work: Radiation applied to medicine; therapeutic and diagnostic use, radiological physics; quality assurance-diagnostic radiology; radiation safety. Office: Arnot-Ogden Meml Hosp Elmira NY 14901

FAWCETT, DON WAYNE, anatomist; b. Springdale, Iowa, Mar. 14, 1917; s. Carlos J. and Mabel (Kennedy) F.; m. Dorothy Marie Secrest, 1941; children: Robert S., Mary Elaine, Donna, Joseph. A.B. cum laude, Harvard, 1938, M.D., 1942; D.Sc. (hon.), U. Siena, Italy, 1974, N.Y. Med. Coll., 1975, U. Chgo., 1977, U. Cordoba, Argentina, 1978, M.D., U. Heidelberg, Germany, 1977, D.V.M., Justus Liebig U., Giessen-Lahn, Germany, 1977. Intern surgery Mass. Gen Hosp., Boston, 1942-43; instr. anatomy Harvard Med. Sch., 1946-48, asso. anatomy, 1948-51, asst. prof. anatomy, 1951-55, Hersey prof. anatomy, 1958—, James Stillman prof. comparative anatomy, 1962—, sr. asso. dean preclin. affairs, 1975-77; prof. anatomy Cornell Med. Coll., 1955-58; scientist Internat. Lab. Research on Animal Diseases, Nairobi, Kenya, 1980—. Author: The Cell, 1966, 2d edit., 1981, Textbook of Histology, 1968, 10th edit., 1975. Served as capt. M.C. AUS, 1943-46; bn. surgeon A.A.A. John and Mary Markle scholar med. sci., 1949-54; recipient Lederle Med. Faculty award, 1954. Fellow Am. Acad. Arts and Sci., Nat. Acad. Sci., Royal Microscopical Soc. (hon.); mem. AAAS, N.Y. Acad. Sci., Am. Assn. Anatomists (pres. 1964- 65, Henry Gray award 1983), N.Y. Soc. Electron Microscopists (pres. 1957-58), Histochem. Soc., Tissue Culture Assn. (v.p. 1954-55), Soc. Exptl. Biology and Medicine, Assn. Anatomy Chairmen (pres. 1973-74), Am. Soc. Zoologists, Am. Soc. Mammalogists, Electron Microscope Soc. Am., Soc. Study Devel. and Growth, Harvey Soc., Am. Soc. Cell Biology (pres. 1961-62), Argentine Nat. Acad. Sci., Anat. Soc. So. Africa (hon.), Japanese Anat. Soc. (hon.), Anat. Soc. Australia and N.Z. (hon.), Japanese Electron Microscope Soc., Internat. Fedn. Soc. Electron Microscopy (pres. 1976-78), Am. Soc. Andrology (pres. 1977-78), Soc. Study Reprodn., Mexican (hon.), Canadian (hon.) assns. anatomists. Subspecialties: Cell and tissue culture; Parasitology. Current work: Mechanisms of entry of oblilgate intracellular parasites into cells of arthropod vectors and mammallian hosts. Office: Internat Lab Research in Animal Diseases PO Box 30709 Nairobi Kenya

FAWCETT, JAMES JEFFREY, geology educator, researcher, college dean; b. Blyth, Eng., July 6, 1936; m. (married); 2 children. B.Sc., U. Manchester, Eng., 1957, Ph.D. in Geology, 1961. Asst. geologist U. Manchester, 1960-61; fellow Carnegie Inst. Geophys. Lab., 1961-64; prof. geology U. Toronto, Ont., Can., 1964—, assoc. chmn. dept., 1970-75, assoc. dean, 1977-80; assoc. dean sci. Erindale Coll., 1980—. Mem. Am. Geophys. Union, Mineral. Soc. Am., Geol. Assn. Can., Mineral. Soc. Gt. Brit. and Ireland, Mineral. Assn. Can. Subspecialties: Geology; Petrology. Office: Dept Geology U Toronto Toronto ON Canada M5S 1A1

FAWCETT, NEWTON CREIG, chemistry educator, researcher; b. Fargo, N.D., Nov. 28, 1941; s. Newton W. and Vera (Strong) F.; m. Karen Scrivner, Dec. 28, 1964; children: Dennis, Jennifer. B.S., Denver U., 1964; M.S., U. N.Mex., 1970, Ph.D., 1973. Staff asst. Sandia Lab., Albuquerque, 1965-68; staff scientist Los Alamos Sci. Lab., Albuquerque, 1972-75; Robert A. Welch fellow S.W. Tex. State U., San Marcos, 1975-76, vis. scientist, 1979; asst. prof. U. So. Miss., Hattiesburg, 1976-81, assoc. prof. chemistry, 1981—; cons. in field. Mem. Am. Chem. Soc., Phi Kappa Phi. Subspecialties: Analytical chemistry; Polymer chemistry. Current work: Photogalvanic cells for solar energy conversion. Office: U So Miss Box 5043 Hattiesburg MS 39406

FAWWAZ, RASHID ADIB, physician, educator; b. Sao Paulo, Brazil, May 19, 1935; came to U.S., 1961, naturalized, 1973; s. Adib Rashid and Salwa Asma (Sabra) F.; m. Marcia Mary Cosse, Jan. 30, 1966; children: Marc, Michael. M.D., Am. U. Beirut, 1961; Ph.D., U. Calif., Berkeley, 1968. Diplomate: Am. Bd. Nuclear Medicine. Intern Am. U. Beirut, 1961, resident, 1962-63; research scientist Lawrence Berkeley Lab., 1966-74; asst. prof. radiology Columbia U., N.Y.C., 1976-78, assoc. prof., 1979—; guest scientist Brookhaven Nat. Lab., Upton, N.Y., 1981—. Contbr. articles to profl. jours. Internat. Atomic Energy fellow, 1963; Donner fellow, 1964. Mem. Am. Assn. Cancer Research, Am. Soc. Nuclear Medicine, Transplantation Soc. Democrat. Presbyterian. Subspecialties: Nuclear medicine; Transplantation. Current work: Immunotherapy. Office: 622 W 168th St New York NY 10032

FAXON, DAVID PARKER, cardiologist, researcher, educator; b. Manchester, N.H., Nov. 22, 1944; s. William Otis and Francis Mary (Parker) F.; m. Monica Kersten, Dec. 29, 1968; children: Kimberly, Nathaniel. B.A., Hamilton Coll, 1967; M.D., Boston U., 1971. Diplomate: Am. Bd. Internal Medicine. Intern Mary Hitchcock Hosp., Hanover, N.H., 1971-72, resident, 1972-74; fellow Univ. Hosp., Boston, 1974-76, instr., 1976-77, asst. prof. cardiology, 1977-82, dir. Cardiac Catherization Lab., 1976-83; mem. exec. com. Mass. affiliate Am. Heart Assn., Boston, 1981-83. Fellow Am. Coll. Cardiology, Am. Heart Assn., Soc. Cardiac Angiography; mem. Am. Fedn. Clin. Research, ACP. Subspecialty: Cardiology. Current work: Renin-engiotensin system; lasers; coronary angioplasty. Office: Univ Hosp 75 E Newton St Boston MA 02118

FAXVOG, FREDERICK R., elec. engr.; b. Mpls., Jan. 7, 1943; s. Ivar and Bessie (Roggeman) F.; m. Julie M. Klug, Aug. 31, 1965; children: Mark, Todd. B.S. in Elec. Engring, U. Minn., 1965, M.S., 1968, Ph.D., 1971. Scientist, group leader Gen. Motors Research Labs., Warren, Mich., 1971-80; sect. chief active devices Honeywell Systems and Research Ctr., Mpls., 1980—. Contbr. articles to profl. jours. Mem. Am. Phys. Soc., Am. Optical Soc., Sigma Xi. Subspecialties: Laser research; Infrared spectroscopy. Current work: Co-2- lasers for optical radars, guidance and remote sensing, fiber optics, and solid state accelerometers. Patentee in field. Office: 2600 Ridgway Minneapolis MN 55413 Home: 615 Brockton Ln Plymouth MN 55447

FAY, JAMES ALAN, engineering educator; b. Southold, N.Y., Nov. 1, 1923; s. William Joseph, Jr. and Margaret (Keenan) F.; m. Agatha Marie Kelly, Jan. 12, 1946; children: David Anthony, Mark Bernard, Colin Michael, Jamie Martin, Peter Robert, Michele Marie. B.S., Webb Inst. Naval Architecture, 1944; M.S., MIT, 1947; Ph.D., Cornell U., 1951. Research engr. Lima-Hamilton Corp., 1947-49; asst. prof. engrng. mechanics Cornell U., 1951-55; mem. faculty Mass. Inst. Tech., 1955, prof. mech. engring., 1960—; dir. SCA Services, Inc.; cons. to govt. and industry; mem. NRC Environ. Studies Bd., 1973-78, 80—, . Author: Molecular Thermodynamics, 1965; also articles. Chmn. Boston Air Pollution Commn., 1969-72, Mass. Port Authority, 1972-77; bd. dirs. Union Concerned Scientists, 1978—. Served with USNR, 1942-46. Fellow Am. Acad. Arts and Scis., Am. Phys. Soc. (exec. com. div. fluid dynamics 1964-67), AAAS, AIAA (chmn. plasmadynamics com. 1966-68); mem. ASME, Air Pollution Control Assn., AAUP, Mass. Audubon Soc. (dir. 1978-82), Sigma Xi. Subspecialties: Fluid mechanics; Environmental engineering. Current work: Dispersion of dense gases; long distance transport of air pollutants; renewable energy systems. Home: 36 Spruce Hill Rd Weston MA 02193 Office: Mass Inst Tech Cambridge MA 02139

FAY, JOSEPH WAYNE, hematologist, oncologist; b. Barberton, Ohio, May 4, 1946; s. Joseph N. and Rosalind P. (Starling) F.; m. Joanne Marie Marinich, Aug. 24, 1968; children: Nathan Wayne, Lauren Marie. B.A., Coll. of Wooster, 1968; M.D., Ohio State U., 1972. Diplomate: Am. Bd. Internal Medicine, Nat. Bd. Med. Examiners. House officer Duke U., Durham, N.C., 1972-74, asst. prof., 1976-81; clin. assoc. NIH, Bethesda, Md., 1974-76; dir. marrow transplantation Baylor U. Med. Ctr., Dallas, 1982—. Served to lt. comdr. USPHS, 1974-76. Mem. Am. Soc. Clin. Oncology, Am. Soc. Hematology, Alpha Omega Alpha. Christian. Subspecialties: Marrow transplant; Cancer research (medicine). Current work: Clinical research in marrow transplantation, hemopoiesis in culture, transplantation immunology. Office: Baylor U Med Ctr 3500 Gaston Ave Dallas TX 75426

FAY, THEODORE DENIS, JR., astrophysicist, electro-optical engr.; b. Franklin, La., Dec. 7, 1940; s. Theodore Denis and Beatrice Marie (Moresi) F.; m. Ann Katherine Wolfe, Jan. 26, 1968; children: Theodore Denis III, Aaron Wolfe. B.S., La. state U., 1962; M.S., Ind. U., 1964, Ph.D., 1968. Postdoctoral fellow Ind. U., 1970-73; postdoctoral fellow, vis. asst. prof. U. Ala., Tuscaloosa, 1973-75, La. State U., 1973-75; sr. postdoctoral fellow NRC, NASA, Huntsville, Ala., 1975-77; systems analyst Teledyne Brown Engring., Huntsville, 1977-81; tech. specialist McDonnell Douglas, Huntsville, 1981—. Mem. AAAS, Am. Astron. Soc., Internat. Astron. Union, Sigma Xi. Roman Catholic. Club: Huntsville Swim. Subspecialties: Infrared optical astronomy; Optical image processing. Current work: Optical image and signal processing for infrared sensors. Patentee in field.

FAZIO, MICHAEL VINCENT, electrical engineer; b. Houston, Feb. 22, 1952; s. Louis B. and Isabelle (Matassa) F. B.S. in Elec. Engring, Rice U., 1974, M.E.E., 1975, Ph.D., 1978. Mem. staff Los Alamos Nat Lab., 1978-80, sect. leader, 1980—, assoc. group leader, 1983—. Contbr. articles on elec. engring. to profl. jours. Mem. Nat. Ski Patrol.; active Sierra Club. Subspecialty: Electrical engineering. Current work: Advancing state of the art in the particle accelerator technology, research and development in the field of accelerator technology and high power RF amplifier systems.

FEARN, RICHARD LEE, engineering sciences educator, research engineer; b. Mobile, Ala., Mar. 24, 1937; s. Lee Syson and Miriam (Clinton) F.; m. Barbara Ann Hubert, June 21, 1969. B.S. in Physics, Auburn U., 1960, M.S., 1961, Ph.D., U. Fla., 1965. Engr. ARO, Inc., Arnold Engring Devel. Ctr., Tullahoma, Tenn., 1967; asst. prof. engring. sci. U. Fla., 1965-74, assoc. prof., 1974-79, prof., 1979—; NASA-Am. Soc. Engring. Edn. faculty fellow Langley Research Ctr., Hampton, Va., 1969-70. Author NASA reports. NASA grantee, 1969-77, 78—. Mem. AIAA. Democrat. Subspecialties: Aeronautical engineering; Fluid mechanics. Current work: Teaching low-speed aerodynamics; research in V/STOL aerodynamics, jet-in-crossflow, vortex wakes, wing and airfoil theory. Home: 2001 NW 12th Rd

Gainesville FL 32605 Office: Dept Engring Sci U Fla Gainesville FL 32611

FEARNOT, NEAL EDWARD, electrical engineering educator, researcher; b. Salem, Mass., June 7, 1953; s. Charles D. and Muriel (Thomas) F.; m. Sharon Lee Tagliaferro, Aug. 23, 1975; 1 dau., Michelle Marie. B.S.E.E., Purdue U., 1975, M.S.E.E., 1978, Ph.D., 1980. Research and design engr. Am. Vet. Supply Co., Erie, Pa., 1975-76; cons., research engr. Cook Group Co., Bloomington, Ind., 1981—; Vis. asst. prof. elec. engring. Purdue U., West Lafayette, Ind., 1981—, vis. asst. prof. mech. engring., 1981—; assoc. research scholar Biomed. Engring Ctr., 1981—. Contbr. articles and videotapes in ultrasonic imaging and cardiac pacemaking to profl. jours. and orgns. Youth leader Covenant Presbyterian Ch., West Lafayette; CPR instr. Am. Heart Assn. Cook Meml. fellow, 1980; CTS Fdn. fellow, 1979-80; recipient Harold Lamport Young Investigator award, 1982. Mem. IEEE, Biomed. Engring. Soc. Presbyterian. Subspecialties: Computer engineering; Biomedical engineering. Current work: Research and development of cardiac related electronic equipment employing new physiological principles. Patentee med. devices. Home: 832 Ashland St West Lafayette IN 47906 Office: Purdue U 204 Potter Bldg West Lafayette IN 47907

FEASBY, WILFRID HAROLD, dental educator, pedodontist; b. St. Thomas, Ont., Can., Sept. 30, 1920; s. Harold George and Vera E. (Rowley) F.; m. Jean I. Erskine, Sept. 21, 1943; children: Thomas E., James H., Anne E. D.D.S., U. Toronto, 1943; M.Sc., U. Man., 1971. Pvt. practice dentistry, 1943-61; assoc. prof. U. Man., Winnipeg, Can., 1961-67; chief dentistry Winnipeg Children's Hosp., 1961-67; prof., chmn. pediatric and community dentistry U. Western Ont., London, 1967-81, prof. pediatric dentistry, 1967—; chief dentistry Victoria Hosp., London, 1968-79; cons. St. Joseph Hosp., London, 1970—, Univ. Hosp., 1973—. Served to capt. Royal Canadian Dental Corps, 1943-46. Fed. Dept. Health and Welfare grantee, 1969-70. Fellow Am. Acad. Pedodontics, Royal Coll. Dentists Can. (pres. 1971); mem. Canadian Dental Assn. (chmn., council edn. 1975-79). Subspecialty: Pedodontics. Current work: Dentistry for handicapped children, radiographic study of dental eruption. Office: U Western Ont Faculty Dentistry London ON Canada N6A 5C1

FEDAN, JEFFREY STEPHEN, pharmacologist, educator; b. Bklyn., May 31, 1944; s. Stephen Anthony and Sophie (Chanas) F.; m. Judith Roche, June 14, 1969; children: David, Ashley, Scott. B.A., Case Western Res. U., 1967; Ph.D. in Pharmacology, U. Ala.-Birmingham, 1974. Postdoctoral fellow W. Va. U., 1974-75, Am. Heart Assn., 1975-76; NIH, 1976-77; adj. asst. prof. dept. pharmacology and toxicology W. Va. U., Morgantown, 1977-82, adj. assoc. prof., 1982—; mem. planning com. Med. Sch. W. Va. U., Morgantown, 1978-79, mem. curriculum com., 1980—, mem. vis. speakers com., 1980—; research pharmacologist Nat. Inst. Occupational Safety and Health, Morgantown. Contbr. chpts. to books, articles to profl. jours. Bd. dirs. W.Va. affiliate Am. Heart Assn., 1979—, mem research policy and allocations com., 1978-81, chmn. research policy and allocations com., 1981-82, mem. exec. com., 1982—, mem. program planning and review com. Mem. Am. Soc. Pharmacology and Exptl. Therapeutics, AAAS, N.Y. Acad. Scis., Am. Heart Assn., Research Discussion Group (pres. 1979-81, sec.-treas. 1981—), Sigma Xi. Episcopalian. Subspecialties: Pharmacology; Molecular pharmacology. Current work: Autonomic nervous system; Ca^{2+}-metabolism; photoaffinity labels; muscle Ca^{2+}-metabolism; smooth muscle; receptors. Home: 137 Bryan Dr Morgantown WV 26505 Office: Physiology Sect Nat Inst Occupational Safety and Health 944 Chestnut Ridge Rd Morgantown WV 26505

FEDERICO, PAT-ANTHONY, research psychologist; b. Newark, Mar. 4, 1942; s. Pasquale and Vincenza (Caramanna) Frederico; m. Suzanne Marie Boudreaux, Nov. 24, 1967. B.A., U. St. Thomas, Houston, 1965; M.S., Tulane U., 1967, Ph.D., 1969. Sr. research psychologist Navy Personnel Research and Devel Ctr., San Diego, 1972—; mem. hon. faculty U. Colo.-Denver, 1969-71; lectr. San Diego State U., 1972-73, 77. Sr. author: Management Information Systems and Organizational Behavior, 1980; co-editor: Aptitude, Learning, and Instruction: Vol. 1, Cognitive Process Analyses of Aptitude, 1980, Vol. 2, Cognitive Process Analyses of Learning and Problem Solving, 1980. Served to capt. USAF, 1969-72. NDEA fellow, 1966-69; NSF fellow, 1972-73. Mem. Human Factors Soc. (sec.-treas. San Diego chpt. 1980-81, pres. 1981-82, exec. dir. 1982-83), Cognitive Sci. Soc., Psychonomic Soc., Am. Ednl. Research Assn. Subspecialties: Cognition; Learning. Current work: Human information processing; computer-based instruction and simulation; management information systems and organizational behavior. Office: Navy Personnel Research and Devel Center San Diego CA 92152

FEDERMAN, STEVEN ROBERT, physicist, researcher; b. Queens, N.Y., Nov. 19, 1949; s. Joseph Meyer and Adele Louise (Strome) F. B.S., Poly. Inst. N.Y., 1971; M.S., NYU, 1976, Ph.D., 1979. Sr. research technician Hosp. for Spl. Surgery, N.Y.C., 1972-75; teaching fellow NYU, 1975-78, research asst., 1978-79, postdoctoral fellow, 1979, U. Tex., Austin, 1979-83; research assoc. NRC Jet Propulsion Lab, Pasadena, Calif., 1983. Mem. Am. Astron. Soc., Internat. Astron. Union., Am. Phys. Soc. Subspecialties: Optical astronomy; Theoretical astrophysics. Current work: Studies of the interstellar medium at the atomic and molecular level. Home: 500 S Los Robles #319 Pasadena CA 91101 Office: Jet Propulsion Lab MS 183-601 4800 Oak Grove Dr Pasadena CA 91109

FEDNER, MARK LEE, psychologist, researcher; b. Phila., Dec. 10, 1947; s. Walter and Gloria (Rodin) F.; m. Sherry Lynn Weiner, Sept. 15, 1974; children: Randall Lawrance, Matthew Brant. B.S. in Psychology, Pa. State U., 1969; Ed.M. in Ednl. Psychology, Temple U., 1972; cert. in sch. psychology, Temple U., 1974; Ph.D. in Sch. Psychology, Temple U., 1975. Lic. psychologist, Pa.; cert. sch. psychologist, Pa., N.J. Grad. asst. asst. supr. dept. sch. psychology Temple U., 1973, 74; sch. psychologist Montgomery County Intermediate Unit, Norristown, Pa., 1975-77; child and youth service dir. N.E. Mental Health Ctr., Phila., 1977-78, clin. dir., 1979-81; psychologist in pvt. practice, Phila., 1981—; presenter, trainer profl. orgns. and agencies; cons. to schs. and social service agencies; presenter, staff trainer N.E. Community Ctr., Phila., 1977-81; participating specialist Health Maintenance Orgn. of Pa.; allied health staff Montgomery Hosp., Norristown, Pa. Mem. Nat. Rifle Assn., Pa. State Alumni Assn., Sports Range, Inc.; Fellow Pa. Psychol. Assn.; mem. Am. Psychol. Assn., Phila. Soc. Clin. Psychologists, Nat. Register Health Service Providers in Psychology, Data Processing Mgmt. Assn. Phila. Current work: Survey of univariate and multivariate statistical techniques, psychoneurological and psychobiological advances in science for psychological evaluation and assessments. Home: 639 Arbor Rd Cheltenham PA 19012

FEDOROWICZ, JANE, information systems educator; b. Ansonia, Conn., Mar. 5, 1955; d. Joseph S. and Dorothy Helen (Daiuto) F. B.S. in Health Systems, U. Conn.-Storrs, 1976; M.S. in Systems Sci, Carnegie-Mellon U., 1978, Ph.D., 1981. Asst. prof. info. systems Northwestern U., Evanston, Ill., 1980—; researcher Center for Health Services and Policy Research, Evanston, 1980—; lectr., cons. in field. Merit fellow Carnegie-Mellon U., 1976-78; IBM fellow, 1978. Mem. Am. Acctg. Assn., Am. Inst. Decision Scis., Assn. Computing Machinery (reviewer 1979—), Inst. Mgmt. Scis., Soc. Info. Mgmt. Roman Catholic. Club: Orgn. Women Faculty (treas. 1982—). Subspecialties: Information systems (information science); Database systems. Current work: Logic and deduction in a decision support system model base; the effects of FASB 13 on the computer industry as lessors; hospital information systems; decision support system design. Home: 623 Sheridan Rd Apt 1 Evanston IL 60202 Office: Northwestern Univ J L Kellogg Grad Sch Mgmt Evanston IL 60201

FEEMSTER, JOHN RONALD, high technology company executive, educator, consultant; b. Hartshorne, Okla., Aug. 14, 1934; s. Floyd Raymond and Eleanor Elizabeth (Ingram) F.; m. Janet Hepburn, Nov. 7, 1956; m. Phyllis Dale, Nov. 20, 1967; children: Edward, Randal, Margaret, Karen. B.S.M.E., U. Mo., 1956; M.S.M.E., UCLA, 1962; M.B.A., U. Santa Clara, 1973; Ph.D., Golden Gate U., 1980. Cert. life coll. instr. bus. and engring., Calif. Project engr. McCormick Selph Assocs., 1961-62; mgr. engring. United Tech., 1962-74; research/ program mgr. Gen. Electric, San Jose, Calif., 1974-82; programs dir. Star Struck Inc., Redwood City, Calif., 1982—; staff instr. West Valley Coll., 1974—, chmn. energy reduction, 1979; pres. Ave' Cons., 1976—. Author classified publs. on rocket propulsion research. Served to lt. comdr. USN, 1956-60. Recipient Superior Research Accomplishment award U.S. Navy, 1958; Spl. Accomplishment award Gen. Electric, 1980; United Tech. Ctr. mgmt. scholar, 1970. Mem. AIAA, ASME, GE 400. Republican. Episcopalian. Subspecialties: Aerospace engineering and technology; Nuclear fission. Current work: Energy management; solar, high technology business management; propulsion; system integration; communication; sea operations. Patentee aerospace apparatus. Home: 18800 Ten Acres Rd Saratoga CA 95070 Office: 837 2d Ave Redwood City CA 94063

FEESER, LARRY JAMES, civil engineering educator, researcher; b. Hanover, Pa., Feb. 23, 1937; s. Cyrus Myers and Arelia Cecilia (Stonesifer) F.; m. Patricia Marianne Reinhold, Aug. 19, 1961; children: Anne Elizabeth, David John. B.S. in Civil Engring., Lehigh U., 1958, M.S., U. Colo., 1961; Ph.D., Carnegie-Mellon U., 1965. Registered profl. engr., Colo., N.Y. Instr. to prof. civil engring. U. Colo., Boulder, 1958-74; prof. and chmn. dept. civil engring. Rensselaer Poly. Inst., Troy, N.Y., 1974-82, assoc. dean engring., 1982—; cons. Jorgensen & Hendrickson Engrs., Denver. Contbr. articles on civil engring. to profl. jours. NSF fellow, 1961-62; Named to Those Who Made Marks in 1981, Engring. News Record, 1982. Fellow Am. Concrete Inst., ASCE (nat. dir. 1979-82); mem. Nat. Soc. Profl Engrs., Am. Soc. Engring. Edn. Subspecialties: Civil engineering; Graphics, image processing, and pattern recognition. Current work: Computer graphics, computer aided design, engineering administrator, engineering researcher. Office: Sch Engring Rensselaer Poly Inst Troy NY 12181

FEFFERMAN, CHARLES LOUIS, educator, mathematician; b. Washington, Apr. 18, 1949; s. Arthur Stanely and Liselott Ruth (Stern) F.; m. Julie Anne Albert, Feb. 1975; children: Nina Heidi, Elaine Marie. B.S., U. Md., 1966; hon. doctorate, 1979; Ph.D., Princeton U., 1969; hon. doctorate, Knox Coll., 1981. Instr. math. Princeton U., 1969-70, prof. math., 1974—; mem. faculty U. Chgo., 1970-74, prof. math., 1971-74. Author research papers. Recipient Salem prize for outstanding work in fourier analysis by young mathematician, Alan T. Waterman award, 1976, Fields medal Internat. Congress Mathematicians, 1978. Mem. Nat. Acad. Scis., Am. Math. Soc., Am. Acad. Arts and Scis. Subsepcialty: Mathematical analysis. Home: 234 Clover Ln Princeton NJ 08540 Office: Fine Hall Princeton Univ Princeton NJ 08540

FEHER, LESLIE, psychologist; b. N.Y.C., Mar. 6, 1944; d. Alexander and Elizabeth (Geller) F.; m. Ralph R. Ferney, Nov. 3, 1967; 1 dau., Vanessa. B.A., Pace U., 1967; M.A., New Sch. for Social Research, 1970. Lic. therapist, N.Y. Asst. dir. Analytic Inst. for Motivational Edn., N.Y.C., 1968-74; dir., founder The E.F. Natal Therapy Inst., N.Y.C., 1974—; exec. dir., founder, pres. Assn. for Birth Psychology, N.Y.C., 1978—; pvt. practice psychology, N.Y.C., 1974—; guest lectr. in field. Author: The Psychology of Birth: The Roots of Human Personality, 1980; editor-in-chief: Birth Psychology Bull, 1979—; contbr. articles to profl. jours. Mem. Am. Psychol. Assn., Am. Orthopsychiat. Assn., Eastern Psychol. Assn., Assn. Women in Sci., Authors League. Democrat. Unitarian. Subspecialty: Perinatal psychology. Current work: Research into perinatal theory and effect of birth on later personality. Home: 444 E 82d St New York NY 10028 Office: New York NY 10028

FEIFEL, HERMAN, psychologist, researcher; b. N.Y.C., Nov. 4, 1915; s. Jacob and Rebecca (Katz) F. B.A., CCNY, 1935; M.A., Columbia U., 1939, Ph.D., 1948. Diplomate: Am. Bd. Profl. Psychology. Research psychologist Adj. Gen. Office War Dept., Washington, 1946-49; supervisory clin. psychologist Winter VA Hosp., Topeka, 1950-54; vis. sr. scientist Research Ctr. for Mental Health, N.Y.U., N.Y.C., 1960-61; vis. prof. psychology U. SO. Calif., Los Angeles, 1966-67; clin. prof. psychiatry and behavioral scis. U. So. Calif. Sch. Medicine, Los Angeles, 1965—; chief psychologist VA Outpatient CLinic, Los Angeles, 1960—; adv. mem. NIMH, Bethesda, Md., 1967-70. Editor: The Meaning of Death (Feifel), 1959, 65, New Meaning of Death (Feifel), 1977 (Book of Yr. Am. Jour. Nursing 1977). Bd. govs. acad. affairs U. Judaism, Los Angeles, 1968—; bd. mem. Nat. Adv. Council Hospice, Washington, 1975-78. Served to capt. U.S. Army, 1942-46. Recipient Disting. Sci. Achievement award Calif. State Psychol. Assn., 1974; Disting. Human Services award Yeshiva U., 1979; Disting. Practioner of Psychology Nat. Acads. Practice, 1982. Fellow Am. Psychol. Assn. (bd. profl. affairs 1974-77, Harold M. Hildreth disting. award 1978, cert. of recognition 1981), AAAS, Gerontol. Soc., Soc. Sci. Study Religion. Jewish. Club: Athletic (Los Angeles). Subspecialties: Clinical psychology; Thanatology. Current work: Personality theory and assessment, psychology of dying death and bereavement, adult development and aging, religion and mental health, psychotherapy. Home: 360 S Burnside Ave Los Angeles CA 90036 Office: VA Outpatient Clinic 425 S Hill St Los Angeles CA 90013

FEIGELSON, ERIC DENNIS, astrophysicist, educator; b. Springfield, Ohio, Apr. 23, 1953; s. Philip and Muriel F. (Horowitz) F. B.A., Haverford (Pa.) Coll., 1975; M.A., Harvard U., 1978, Ph.D., 1980. Teaching fellow Harvard U., Cambridge, Mass., 1976-77; research assoc. Smithsonian Astrophys. Obs., Cambridge, 1977-80; staff research scientist Center for Space Research and dept. physics M.I.T., Cambridge, 1980-82; asst. prof. dept. astronomy Pa. State U., 1982—; vis. faculty Sci. Coll., Concordia U., Montreal, 1982. Contbr. articles to profl. jours. Smithsonian fellow, 1975-77; John Parker fellow, 1978; recipient NASA Group Achievement award, 1980. Mem. Am. Astron. Soc., N.Y. Sci., Fedn. Am. Scientists, Union Concerned Scientists, Phi Beta Kappa, Sigma Xi. Subspecialties: 1-ray high energy astrophysics; Radio and microwave astronomy. Current work: Observational x-ray and radio astronomy; active galactic nuclei, radio galaxies, young stars. Office: PA State U Dept Astronomy State College PA 16801

FEIGEN, LARRY PHILIP, physiologist, researcher, educator; b. Everett, Mass., Mar. 27, 1942; s. Robert and Celia (Beecher) F.; m. Judy Lee Segel, Dec. 26, 1965; children: Scott, Sharon. B.A. in Physics, Northeastern U., 1964, M.S., 1966; Ph.D. in Physiology and Biophysics, U. Health Scis., Chgo. Med. Sch., 1974. Asst. prof. Tulane U., New Orleans, 1974-81, asso. prof. physiology, 1981—. Contbr. articles, book chpts. to profl. lit. Recipient Edward G. Schlieder Found. award, 1974-78; Am. Heart Assn. grantee, 1975-77, 79-81; NIH grantee, 1980—. Mem. AAAS, AAUP, Am. Physiol. Soc., Am. Soc. Pharmacology and Exptl. Therapeutics, N.Y. Acad. Scis., Am. Heart Assn., La. Heart Assn. Jewish. Subspecialties: Physiology (medicine); Pharmacology. Current work: Hormonal control of the circulation and mechanisms which regulate these hormones with special emphasis on kinins, prostaglandins. and leukotrienes. Home: 2157 LaSalle Ave Gretna LA 70053 Office: 1430 Tulane Ave New Orleans LA 70112

FEIGENBAUM, EDWARD ALBERT, computer science educator; b. Weehawken, N.J., Jan. 20, 1936; s. Fred J. and Sara Rachman; m. H. Penny Nii, 1975; children: Janet Denise, Carol Leonora, Sheri Bryant, Karin Bryant. B.S. in Elec. Engring., Carnegie Inst. Tech., 1956, Ph.D. in Indsl. Adminstrn., 1960. Asst., then assoc. prof. bus. adminstrn. U. Calif. at Berkeley, 1960-64; assoc. prof. computer sci., then prof. Stanford U., 1965—; prin. investigator heuristic programming project, 1965—; dir. Computation Center Stanford U., 1965-68, chmn. dept. computer sci., 1976-81; pres. Intelli Genetics Inc., 1980-81, mem. tech. adv. bd., 1983—; cons. to industry, 1957—; Mem. computer and biomath. scis. study sect. NIH, 1968-72, mem. adv. com. on artificial intelligence in medicine, 1974—; mem. adv. com. Health Care Tech. Center, U. Mo., Columbia; mem. Math. Social Sci. Bd., 1975-78; computer sci. adv. com. NSF, 1977-80; mem. Internat. Joint Council on Artificial Intelligence, 1973—. Author: (with others) Information Processing Language V Manual, 1961, (with P. McCorduck) The Fifth Generation; author: (with R. Lindsay, B. Buchanan, J. Lederberg) Applications of Artificial Intelligence to Organic Chemistry: the Dendral Program; Editor: (with J. Feldman) Computers and Thought, 1963, (with A. Barr and P. Cohen) Handbook of Artificial Intelligence, 1981, 82; Mem. editorial bd.: Jour. Artificial Intelligence, 1969—. Fulbright scholar Gt. Britain, 1959-60. Fellow AAAS; Mem. Assn. Computing Machinery (nat. council 1966-68, chmn. spl. interest group on biol. applications 1973-76), Am. Assn. Artificial Intelligence (pres. 1980-81), Cognitive Sci. Soc. (council 1979-81), Am. Psychol. Assn., AAAS, Sigma Xi, Tau Beta Pi, Eta Kappa Nu, Pi Delta Epsilon. Subspecialties: Artificial intelligence; Computer architecture. Current work: Artificial intelligence, heuristic programming, expert systems, knowledge engineering, advanced computer architectures. Home: 1017 Cathcart Way Stanford CA 94305 Office: Computer Sci Dept Stanford U Stanford CA 94305

FEIGENBAUM, MITCHELL JAY, theoretical physicist; b. Phila., Dec. 19, 1944; s. Abraham Joseph and Mildred (Sugar) F. B.E.E., CCNY, 1964; Ph.D. in Physics, MIT, 1970. Research assoc., instr. dept. physics Cornell U., Ithaca, N.Y., 1970-72, prof. physics, 1982—; research assoc. dept. physics Va. Poly. Inst., Blacksburg, 1972-74; mem. staff theory div. Los Alamos Nat. Lab., 1974-81, lab. fellow, 1981-82. Contbr. articles to profl. jours. Recipient Disting. Performance award Los Alamos Nat. Lab., 1980; E.O. Lawrence Meml. award Dept. Energy, 1983. Mem. Am. Phys. Soc., N.Y. Acad. Scis., Sigma Xi. Subspecialties: Chaotic Phenomena; Theoretical physics. Current work: Development of physical and mathematical techniques to describe the onset of turbulence (or more generally chaos) in complex dynamical systems. Developer of universal scaling theory for the onset of chaos; prin. developer of the subdiscipline of chaotic phenomena. Office: Cornell U 538 Clark Hall Ithaca NY 14853

FEIN, MICHAEL E., laster engr., gas discharge device engr. B.A., Harvard U., 1963; M.S.E.E., U. Ill., 1966, Ph.D. in Elec. E.E, 1969. Engr. Zenith Radio Corp. Research Div., Chgo., 1963-64, Owens-Illinois, Inc., Toledo, 1970-79; engring. sect. mgr. Spectra-Physics, Inc., Mountain View and San Jose, Calif., 1979-82; with KLA Instruments Corp., Santa Clara, Calif., 1982—. Contbr. papers to profl. jours. Mem. IEEE, Am. Phys. Soc., Soc. Photo-optical Instrumentation Engrs. Subspecialties: Optical engineering; Laser design. Current work: Designing laser-based instruments. Patentee U.S. and abroad. Home: 1909 Lime Tree Ln Mountain View CA 94040 Office: 2051 Mission College Blvd Santa Clara CA 95054

FEINBERG, GERALD, educator, physicist; b. N.Y.C., May 27, 1933; s. Leon and Florence (Weingarten) F.; m. Barbara J. Silberdick, Aug. 9, 1968; children—Jeremy Russell, Douglas Loren. B.A., Columbia U., 1953, M.A., 1954, Ph.D., 1957. Mem. Inst. Advanced Study, Princeton, N.J., 1956-57; research assoc. Brookhaven Nat. Lab., Upton, N.Y., 1957-59, cons., 1960-74; prof. physics dept. Columbia U., N.Y.C., 1959—, chmn., 1980-82. Author: The Prometheus Project, 1969, What is the World Made Of?, 1977, Consequences of Growth, 1977, Life Beyond Earth, 1980, The Future of Science, 1984; div. assoc. editor: Phys. Rev. Letter, 1983—; Contbr. articles to profl. jours. Sloan Found. fellow, 1960-64; Overseas fellow Churchill Coll., Cambridge, Eng., 1963-64; Guggenheim fellow, 1973-74. Fellow Am. Phys. Soc.; mem. Sigma Xi. Subspecialties: Theoretical physics; Extraterrestrial life. Current work: Theory of subatomic particles; quantum field theory; conditions for, and types of, extraterrestrial life. Home: 535 E 86th St New York NY 10028

FEINBERG, ROBERT JACOB, radiation physicist, scientific administrator; b. Chelsea, Mass., Apr. 6, 1931; s. Charles S. and Mary A. (Melamed) F.; m. Carole I. Young, May 31, 1964; children: Curt Michael, Mark William. B.S. in Physics, Boston Coll., 1953, M.S., 1954; M.S. in Radiation Biology, U. Rochester, 1955; grad. in nuclear engring, Oak Ridge Sch. Reactor Tech., 1956. Physicist Picatinny Arsenal, Dover, N.J., 1951, Nat. Bur. Standards, Washington, 1952; astrophysicist Air Force Cambridge (Mass.) Research Center, 1953; instr. in physics Boston Coll., Chestnut Hill, Mass., 1954; health physicist Brookhaven Nat. Lab., Upton, N.Y., 1955; nuclear engr. Oak Ridge Nat. Lab., 1955-56; physicist-radiol. devel. Gen. Electric Co., Knolls Atomic Power Lab., Schenectady, 1956-58, supr. nuclear and radiol. safety, 1958-66, mgr. health physics and nuclear safety, 1966-76, project dir. radiol. controls, 1976—. AEC fellow, 1954-55. Mem. Internat. Health Physics Soc. (chmn. admissions com. 1969-72, chmn. membership com. 1972-75, dir. 1982—), Health Physics Soc. (pres. Northeastern N.Y. chpt. 1957, 69), Am. Nuclear Soc. Jewish. Subspecialties: Biophysics (physics); Nuclear fission. Current work: Radiological physics and engineering; reactor safeguards and nuclear safety; radiation dosimetry and radiation control; environmental hazards analysis; radiation effects; waste management. Inventor beta radiation dosimeter, 1958. Home: 1223 Godfrey Ln Schenectady NY 12309 Office: Gen Electric Co Knolls Atomic Power Lab River Rd Schenectady NY 12301

FEINERMAN, BURTON, physician, cancer researcher; b. N.Y.C., July 2, 1929; s. David and Pauline (Mendelsohn) F.; m. Judith Einhorn, Oct. 23, 1960; children: Steven, Gregg. B.A., N.Y.U., 1950; M.D., N.Y. Med. Coll., 1954. Diplomate: Am. Bd. Pediatrics. Intern L.I. Coll. Hosp., 1954-55; resident in pediatrics Flower & Fifth Ave. Hosp., N.Y.C., 1955-56, Mayo Clinic, Rochester, Minn., 1956-57; chief pediatrics Cloverleaf Hosp., North Miami Beach, Fla., 1962-64, North Miami Gen. Hosp., 1972-74, Internat. Hosp., Miami, 1982—; adj. prof. microbiology also asst. clin. prof. pediatrics Southeastern Coll. Medicine, 1981—; research assoc. Papanicolaou Cancer Research Inst., 1982—; dir. research Cancer Tech., Inc., 1981—. Author articles in field. Served to capt. M.C. AUS, 1957-59. Fellow Am. Soc.

Hematology; mem. Am. Soc. Clin. Oncology, Am. Acad. Pediatrics, Pediatric Soc. Hematology-Oncology, Am. Coll. Allergy, Am. Assn. Clin. Immunology and Allergy, Am. Soc. Microbiology. Subspecialties: Immunopharmacology; Cancer research (medicine). Current work: Control cancer metastasis with biological response modifers, immunology of malignancy, immunodefiencies, brain tumors and blood brain barrier. Office: 640 NW 18301 St Miami FL 33169

FEINLEIB, MARY ELLA (HARMAN), educator, academic administrator, photobiologist; b. Italy, May 21, 1938. A.B., Cornell U., 1959; A.M., Radcliffe Coll., 1961; Ph.D. in Biology, Harvard U., 1966. From instr. to asst. prof. Tufts U., 1965-72, assoc. prof. biology, 1972-82, prof. biology, 1982—, chmn. dept., 1976-82, dean Colls. Liberal Arts and Jackson Coll., 1982—. Mem. AAAS, Am. Soc. Photobiology. Subspecialty: Plant physiology (biology). Office: Dept Biology Tufts U Medford MA 02155

FEINSTEIN, ALVAN RICHARD, physician; b. Phila., Dec. 4, 1925; s. Joel B. and Bella (Ukasz) F.; m. Linda Louise Marean, Oct. 20, 1968; children: Miriam Anne, Daniel Joel Bennett. B.S., U. Chgo., 1947, M.S. in Math, 1948, M.D., 1952; M.A. (hon.), Yale U., 1969. Intern, then resident Yale-New Haven Hosp., 1952-54; research fellow Rockefeller Inst., 1954-55; resident Columbia-Presbyn. Med. Center, N.Y.C., 1955-56; clin. dir. Irvington House, N.Y.C., 1956-62; instr., then asst. prof. N.Y. U. Sch. Medicine, 1956-62; chief clin. pharmacology VA Hosp., West Haven, Conn., 1962-64, chief clin. biostatistics, 1964-74; mem. faculty Sch. Medicine, Yale U., 1962—, prof. medicine and epidemiology, 1969—, dir. clin. scholar program, 1974—; chief Eastern Research Support Center, VA, 1967-74; Pres. New Haven area chpt. Assn. Computing Machinery, 1968-69. Author: Clinical Judgment, 1967, Clinical Biostatistics, 1977; editor: Jour. Chronic Diseases; also articles. Served with AUS, 1944-46. Recipient Francis G. Blake award for outstanding teaching Yale Med. Sch., 1969. Mem. Assn. Am. Physicians, Am. Soc. Clin. Investigation, Am. Epidemiol. Soc., A.C.P., Inst. Medicine, Am. Fedn. Clin. Research, Am. Soc. Clin. Pharmacology Therapeutics, Am. Statis. Assn., Assn. Computing Machinery, Biometric Soc., Am. Heart Assn., Am. Assn. History Medicine, Alpha Omega Alpha. Subspecialties: Internal medicine; Epidemiology. Current work: Clinical epidemiology for evaluation of diagnostic, prognostic, and therapeutic strategies in medicine; clinical biostatistics. Home: 45 Edgehill Rd New Haven CT 06511 Office: 333 Cedar St New Haven CT 06510

FEINSTEIN, JOSEPH, electronics engineer, educator; b. N.Y.C., July 8, 1925; s. David and Edith (Morgenstern) F.; m. Elaine Cantor, Mar. 2, 1952; children: Susan, David, Jonathan. B.E.E., Cooper Union, 1944; M.A. in Physics, Columbia U., 1947, Ph.D., NYU, 1951. Mem. staff Nat. Bur. Standards, Washington, 1949-54; mem. tech. staff, dept. head Bell Telephone Labs., Murray Hill, N.J., 1954-59; dir. research S-F-D Labs., Union, N.J., 1959-64, exec. v.p., 1960-64; v.p. research Varian Assocs., Palo Alto, Calif., 1964-79; dir. electronics and phys. scis. Dept. Def., 1980-83; vis. prof., cons. in electronics Stanford U. (Calif.), 1983—. Editor, contbg. author: Crossed Field Microwave Devices, 1968, Electronics Engineers Handbook, 1975, 81. Fellow IEEE; mem. Nat. Acad. Engring. Subspecialties: Laser power generation; Microelectronics. Current work: Free electron laser for optical power generation; E-beam and x-ray lithography for microelectronics. Patentee in communications, microwaves, electron tubes. Home: 2398 Branner Dr Menlo Park CA 94025 Office: Stanford U Dept Elec Engring Stanford CA

FEIR, DOROTHY JEAN, insect physiologist, educator, researcher; b. St. Louis, Jan. 29, 1929; d. Alex R. and Lillian (Smith) F. B.S., U. Mich., 1950; M.S., U. Wyo, 1956; Ph.D., U. Wis.-Madison, 1960. Instr. biology U. Buffalo, 1960-61; asst. prof. biology St. Louis U., 1961-64, assoc. prof., 1964-67, prof., 1967—; cons. in field; mem. study sect. NIH, 1980-84. Editor: Environ. Entomology, 1977—. Mem. Entomol. Soc. Am. (governing bd. 1982-85), Am. Physiol. Soc., AAAS, Mo. Acad. Sci. (program chmn. 1983), Sigma Xi (pres. St. Louis U. chpt. 1977-78). Subspecialty: Physiology (biology). Current work: Insect hormones; air pollutants; invertebrate immunology. Office: Dept Biology Saint Louis U Saint Louis MO 63103

FEIT, WALTER, mathematics educator, researcher; b. Vienna, Austria, Oct. 26, 1930; m., 1957; 2 children. B.A., M.S., U. Chgo., 1951; Ph.D. in Math, U. Mich., 1954. Instr. in math. Cornell U., 1953-55, from asst. prof. to assoc. prof., 1956-64; prof. math. Yale U., 1964—. NSF fellow Inst. Advanced Study, 1958-59. Mem. Nat. Acad. Sci., Am. Math. Soc. (Cole prize 1965), Math. Assn. Am. Office: Dept Math Yale U New Haven CT 06520

FELD, MICHAEL STEPHEN, educator; b. N.Y.C., Nov. 11, 1940; s. Albert and Lillian R. Norwalk; m. Frances Aschheim, Mar. 2, 1980; children—David A., Jonathan R. S.B. in Humanities and Sci, M.I.T., 1963, S.M. in Physics, 1963, Ph.D., 1967. Postdoctoral fellow M.I.T., Cambridge, 1967-68, asst. prof., 1968-73, asso. prof., 1973-79, prof. physics, 1979—, dir. spectroscopy lab., 1976—, dir. regional laser center, 1979—. Editorial bd.: Laser Focus Mag; co-editor: Fundamental and Applied Laser Physics, 1973, Coherent Nonlinear Optics, 1980. Alfred P. Sloan research fellow, 1973; recipient Disting. Service award M.I.T. Minority Community, 1980. Fellow Am. Optical Soc., Sigma Xi; mem. Am. Phys. Soc. Subspecialty: Atomic and molecular physics. Current work: Lasers, coherent processes, biomeidical applilcations. Condr. research laser saturation spectroscopy. First exptl. demonstration of superradiance. Home: 56 Hinckley Rd Waban MA 02168 Office: 77 Massachusetts Ave Cambridge MA 02139

FELDMAN, ALBERT, physicist; b. Jersey City, May 20, 1936; s. Harry and Rebecca (Kosowsky) F.; m. Rosalind Sylvia Brodofsky, Aug. 16, 1959; children: Gail, June, Jonathan. B.S., CCNY, 1959; M.S., U. Chgo., 1960, Ph.D., 1966. Physicist Nat. Bur. Standards, Washington, 1966—. Contbr. artiles to profl. jours. Recipient Bronze medal U.S. Dept. Commerce, 1980. Mem. ASTM (editor subcom. 1972—), Am. Phys. Soc., Optical Soc. Am., ASTM, Phi Beta Kappa, Sigma Xi. Subspecialty: Condensed matter physics. Current work: Optical materials; optical properties of thin films. Office: Nat Bur Standards B328 Mat Bldg Washington DC 20234

FELDMAN, ALBERT WILLIAM, educator; b. Gardner, Ill., Aug. 6, 1918; s. Joseph M. and Ann (Miller) F.; m. Helen Taylor, July 21, 1944; children: Michael Ann Feldman Smmerso, William Taylor. A.B., U. Ill., 1942; M.S., N.C. State U., 1944; Ph.D., U. Minn., 1947. Asst. prof. plant pathology U.R.I., Kingston, 1947-51; mgr. research and devel. Uniroyal Chem. Co., Naugatuck, Conn., 1951-58; faculty U. Fla., Lake Alfred, 1958—, prof. plant pathology, 1958—. Contbr. articles in field to profl. jours. Mem. Internat. Orgn. Citrus Virologists, Am. Phytopath. Soc., AAAS, Am. Soc. Plant Physiologists, Soil and Crop Sci. Soc. Fla., Sigma Xi. Lodge: Elks. Subspecialties: Plant pathology; Plant physiology (agriculture). Current work: Researcher in disease physiology and virology. Patentee in field. Office: University of Florida Lake Alfred FL 33850

FELDMAN, GARY JAY, physicist, educator; b. Cheyenne, Wyo., Mar. 22, 1942; m. (married); 2 children. B.S., U. Chgo., 1964; A.M., Harvard U., 1965, Ph.D. in Physics, 1971. Research assoc. in physics Stanford Lineaer Accelerator Ctr. Stanford U., 1971-74; staff physicist Stanford Lineaer Accelerator Center Stanford U., 1974-79, assoc. prof., 1979-83; prof. Stanford Linear Accelerator Center Stanford U., 1983—; sci. assoc. European Orgn. Nuclear (Research (CERN)), Switzerland, 1982-83. Mem. Am. Phys. Soc. Subspecialty: Particle physics. Office: Stanford Linear Accelerator Ctr Stanford U Stanford CA 94305

FELDMAN, HENRY R., physicist; b. N.Y.C., June 28, 1932; s. Morton and Dorothy (Roman) F.; m. Dorothy C. Feldman, Dec. 22, 1956; 1 dau., Fern. A.B., Harvard U., 1953; A.M., Columbia U., 1958, Ph.D., 1963. Research assoc. Columbia U., 1963; research asst. prof. U. Wash., 1963-65, physicist, 1966, sr. physicist, 1966—. Contbr. articles to profl. jours. Served as cpl. AUS, 1953-55. Mem. Acoustical Soc. Am., Am. Phys. Soc., Fedn. Am. Scientists, Sigma Xi. Subspecialties: Acoustics; Atomic and molecular physics. Current work: Underwater acoustics, acoustic lenses. Home: 4823 NE 42d St Seattle WA 98105 Office: U Wash HN 10 Seattle WA 98105

FELDMAN, JACK MICHAEL, psychologist, educator; b. Chgo., Sept. 4, 1944; s. Daniel Roy and Lauretta G. (Zaslaw) F.; m. Susan Jane Pfeifer, Aug. 26, 1967; children: Joshua David, Zachary William. B.S., U. Ill.-Urbana, 1966, M.A., 1968, Ph.D., 1972. Asst. prof. mgmt. U. Fla., 1972-75, assoc. prof., 1975—. Contbr. articles to profl. jours. Fellow Am. Psychol. Assn.; mem. Acad. Mgmt., Am. Inst. Decision Scis. Jewish. Subspecialties: Social psychology. Current work: Application of theory and data of social and cognitive psychology to problems of human behavior in organizational settings.

FELDMAN, JOSEPH DAVID, pathologist, researcher; b. Hartford, Conn., Dec. 13, 1916; m., 1949; 3 children. B.A., Yale U., 1937; M.D., L.I. Coll. Medicine, 1941. From assoc. prof. to prof. pathology Sch. Medicine, U. Pitts., 1954-61; mem. staff Scripps Clinic and Research Found., La Jolla, Calif., 1961—, chmn. dept. immunopathology, 1976-79; lectr. Hadassah Med. Sch., Hebrew U., Jerusalem, 1950-54; cons. USPHS, 1967—; chmn. Pathology B Study Sect., 1967-70; adj. prof. pathology U. Calif.-San Diego, 1968—; cons. Nat. Cancer Inst. Virus Cancer Program Sci. Editor-in-chief: Jour. Immunology, 1971—. Rev. Com. and Sci. Adv. Bd. Council Tobacco Research, 1974. Mem. Internat. Acad. Pathology, Histochem. Soc., Endocrine Soc., Electron Microscopy Soc. Am., Assn. Pathology. Subspecialty: Pathology (medicine). Office: Jour Immunology 3770 Tansy St San Diego CA 92121

FELDMAN, MARTIN, physicist, utility co. research executive; b. Bklyn., July 13, 1935; s. Herman and Mollie (Nemsure) F.; m. Ellen S. Feldman, Aug. 21, 1961; children: Jerald, Richard, Nina. B.S. in Physics, Rensselaer Poly. Inst., 1957; Ph.D in Exptl. Physics, Cornell U., 1962. Research asst. Cornell U., 1962-63; asst. prof. U. Pa., 1963-68; mem. staff Bell Labs., Murray Hill, N.J., 1968—, supr., 1973—. Subspecialties: Optics research management; Optical engineering. Current work: Applications of optics to integrated circuit manufacture; inspection systems; alignment systems. Home: 141 Murray Hill Blvd Murray Hill NJ 07974 Office: Bell Labs Room 2A 221 Murray Hill NJ 07974

FELDMAN, ROY SAMUEL, dentist, periodontist, researcher; b. Winthrop, Mass., Apr. 9, 1947; s. Henry and Marilyn (Lundy) F. B.A., Columbia U., 1969, D.D.S., 1973; D.M.Sc., Harvard U., 1978. Cert. in periodontology and oral medicine, 1977. Assoc. investigator VA, Boston, 1978-80, research assoc., 1982—; instr. Harvard Sch. Dental Medicine, Boston, 1978-81; asst. prof., 1982—; cons. in periodontics VA, Brockton, Mass., 1978-80, Bedford, Mass., 1980—. Contbr. articles on bone destruction in lab. animal research and human disease and aging. Mem. Am. Assn. for Dental Research (v.p. Boston chpt. 1981-82, pres. 1982-83), Internat. assn. for Dental Research, Am. Acad. Periodontology, N.Y. Acad. Scis. Club: Old North Bridge Hounds (Concord, Mass.). Subspecialties: Periodontics; Oral biology. Current work: Bone resorption processes; longitudinal studies of human aging and disease; radiographic analysis of bone destruction. Office: Harvard U Sch Dental Medicine 188 Longwood Ave Boston MA 02115

FELDMAN, SUSAN C., anatomy educator; b. Bklyn., Oct. 1, 1953; d. Saul and Anne (Richman) F. B.A., Hofstra U., 1963; M.S., Rutgers U., 1967; Ph.D., CUNY, 1976. Teaching asst. CUNY, 1967-75; postdoctoral fellow Albert Einstein Coll. Medicine, Bronx, N.Y., 1975-77, Coll. Physicians and Surgeons, Columbia U., 1977-79; asst. prof. dept. anatomy N.J. Med. Sch., Newark, 1979—. Contbr. articles to sci. jours. Mem. Am. Soc. Cell Biology, AAAS, N.Y. Acad. Scis., Soc. Neurosci., Am. Assn. Anatomists, Union Concerned Scientists. Subspecialties: Neurobiology; Neuroendocrinology. Current work: Anatomy and evolution of peptides in the nervous system function and distribution of calcium-binding proteins; light microscopic immunocythochemistry in combination with pathway tracing techniques. Office: NJ Med Sch Dept Anatomy 100 Bergen St Newark NJ 07103

FELLOWS, W(ALTER) SCOTT, JR., nuclear engineer; b. Columbus, Ohio, Sept. 26, 1918; s. Walter Scott and Nellie Bonner (Rinesmith) F.; m. Julia Ann Grassbaugh, Feb. 14, 1948 (dec. 1979); children: Marsha, Nancy, Joanne. B.Mech.Engring., Ohio State U., 1940; M.S., Purdue U., 1950; Ph.D., Sch. Reactor Tech., Oak Ridge, 1954. Registered prof. engr., Ohio, Ala. Commd. 2d lt. U.S. Army Air Force, 1940, advanced through grades to col., 1958, ret., 1965; chief direct cycle office AEC, Washington, 1954-57, asst. mgr. for ops., Cin., 1957-60; project mgr. Marshall Space Flight Ctr., NASA, Huntsville, Ala., 1960-65, dir. ops. mgmt. office, 1965-70; sr. research engr. So. States Energy Bd., Atlanta, 1970—. Decorated D.F.C. with 2 oak leaf clusters, Legion of Merit. Fellow AIAA (assoc.); mem. Am. Nuclear Soc. Subspecialties: Nuclear engineering; Astronautics. Current work: Nuclear propulsion for spacecraft. Home: 9450 Coleman Rd Roswell GA 30075 Office: So States Energy Bd 2300 Peachford Rd Atlanta GA 30338

FELNER, ROBERT DAVID, psychology educator; b. Norwich, Conn., June 3, 1950; s. Joseph and Roslyn (Aptaker) F. B.A., U. Conn., 1972; M.A., U. Rochester, 1975, Ph.D., 1977. Lic. psychologist, Ala. Research asst. U. Conn., Storrs, 1971-72; clin. psychology trainee U. Rochester, N.Y., 1972-73, U. Rochester Med. Center/Convalescent Hosp., 1973-76; asst. prof. psychology Yale U., New Haven, 1976-81; assoc. prof., dir. clin. psychology program Auburn (Ala.) U., 1981—; cons. Conn. Bar Assn., 1978-81, City of New Haven Schs., 1977-82, NIMH, 1978—; mem. grant com. NIH. Author: Community and Preventive Psychology, 1984; editor: Preventive Psychology, 1983; contbr. chpts. to books, articles to profl. jours. NIMH grantee, 1976-77; Edward W. Hazen Found. grantee, 1978-81; NSF grantee, 1978—. Mem. Am. Psychol. Assn., Am. Orthopsychiat. Assn., Council Clin. Program Dirs., Council Community Psychology Program Dirs. Democrat. Jewish. Current work: Primary prevention of emotional problems of children; life transitions; pediatric psychology; family disruption; child custody. Home: 830 Moores Hill Rd Auburn AL 36830 Office: Dept Psychology Auburn U Haley Center Auburn AL 36849

FELPEL, LESLIE P., pharmacology educator; b. West Ghent, N.Y., Aug. 17, 1939; s. Pierce L. and Mildred E. (Leggett) F.; m. Vivian P. Snook, Apr. 6, 1968; children: Kurt, Daryl, Heidi. B.S., Albany Coll. Pharmacy, 1961; M.S., Purdue U., 1964, Ph.D., 1967. Registered pharmacist, N.Y. Sr. staff mem. Ayerst Labs., 1967-68; postdoctoral fellow Rockefeller U., 1968-71; asst. prof. Cornell U. Med. Sch., N.Y.C., 1971-72; asst. prof. pharmacology U. Tex. Health Sci. Ctr., San Antonio, 1972-77, assoc. prof., 1977—. Contbr. chpts. to books. Mem. AAAS, Soc. Neurosci., Am. Soc. Pharmacology and Exptl. Therapeutics, Am. Assn. Dental Schs., Am. Assn. Dental Research, Internat. Assn. Dental Research, Omicron Kappa Upsilon. Subspecialty: Neuropharmacology. Current work: Central nervous system neurotransmitters; alterations in neurotransmitters following drug treatment or after injury. Home: 3320 Rock Creek Run San Antonio TX 78230 Office: U Tex Health Sci Ctr 7703 Floyd Curl Dr San Antonio TX 78284

FELSENFELD, GARY, govt. ofcl.; b. N.Y.C., Nov. 18, 1929; m. (married), 1956; 3 children. A.B., Harvard, 1951; Ph.D. in Chemistry (NSF fellow), Calif. Inst. Tech., 1955; postgrad. (NSF fellow), Oxford (Eng.) U., 1954-55. Officer NIH, USPHS, HEW, 1955-58; chief phys. chemistry sect., lab. molecular biology Nat. Inst. Arthritis and Metabolic Diseases, 1961—; asst. prof. biophysics U. Pitts., 1958-61; cons. NIMH, 1958-61; vis. prof. Harvard, 1963; Merck disting. lectr. Rutgers U., 1977. Mem. editorial bd.: Jour. Biol. Chemistry, 1965-70, Jour. Molecular Biology, 1967-73, Biophys. Chemistry, 1973-76, Biopolymers, 1975-77, Ann. Rev. of Biochemistry, 1975-80, Quar. Rev. of Biophysics, 1975—, Cell, 1979—, Jour. Molecular and Cellular Biology, 1980—. Fellow AAAS; mem. Am. Chem. Soc., Am. Biophys. Soc., Am. Assn. Biol. Chemists, Nat. Acad. Scis., Am. Acad. Arts and Scis. Subspecialties: Molecular biology; Biophysical chemistry. Current work: Nucelic and physical chemistry; protein-nucleic acid interaction; chromatin structure and function. Office: Lab Molecular Biology Nat Inst Arthritis Diabetes, Digestive, and Kidney Diseases Bethesda MD 20205

FELTON, JAMES STEVEN, genetic toxicologist; b. San Francisco, Jan. 31, 1945; s. Leland R. and Jeanne D. (Sichel) F.; m. Bette Borden, Aug. 18, 1968; children: Alisa, Diana. A.B., U. Calif. -Berkeley, 1967; Ph.D., SUNY-Buffalo, 1973. Staff fellow NIH, Bethesda, Md., 1973-76; sr. biomed. scientist Lawrence Livermore (Calif.) Nat. Lab., 1976—; adj. assoc. prof. genetic toxicology San Jose State U., 1982—. NIH fellow, 1968-72. Mem. AAAS, Am. Cancer Research Assn., Environ. Mutagen Soc., Genetic and Environ. Toxicology Assn. Subspecialties: Genetics and genetic engineering (medicine); Cancer research (medicine). Current work: Genetic toxicology and environmental carcinogenesis isolation and characterization of mutagenic compounds formed during the cooking of foods; metabolism and genetic damage associated with those compounds. Office: PO Box 5507 Livermore CA 94550

FELTS, WILLIAM ROBERT, JR., physician; b. Judsonia, Ark., Apr. 24, 1923; s. Wylie Robert and Willie Etidorpha (Lewis) F.; m. Jeanne E. Kennedy, Feb. 17, 1954 (div. 1971); children—William R. III, Thomas Wylie, Samuel Clay, Melissa Jeanne. B.S., U. Ark., 1944, M.D., 1946. Intern Garfield Meml. Hosp., Washington, 1946-47; resident in medicine Gallinger Mcpl. Hosp., Washington, 1949-51, George Washington U. Hosp., 1951-53, trainee in rehab., 1955-57; asst. chief arthritis research unit VA Hosp., Washington, 1953-54, adj. asst. chief, 1954-58, chief, 1958-62; cons. in rheumatology U.S. Naval Hosp., Bethesda, Md., 1957-70; mem. faculty dept. medicine George Washington U., 1962—, asso. prof., 1962-80, prof., 1980—, dir. div. rheumatology, 1970-79; mem. Nat. Commn. on Arthritis and Related Musculoskeletal Diseases, 1975-76, Nat. Arthritis Adv. Bd., 1977-80; mem. nat. com. on health policy Project Hope, 1977; cons. health affairs and mem. profl. adv. bd. Control Data Corp., 1976—; mem. D.C. Health Planning Adv. Com., 1969-72; chmn. med. adv. com. D.C. chpt. Arthritis Found., 1963—. Author articles in field, especially med. socioecons.; Mem. editorial adv. bd., cons. internal medicine: Current Procedural Terminology, 3d edit, 1972-73; chmn. editorial cons. panel, CPT-TV, 1980—; editorial adv. bd.: Internal Medicine News, 1976—. Bd. dirs. Nat. Capital Med. Found., 1979—, pres., 1980-81. Served with AUS, 1943-46, 47-49. Mem. Am. Soc. Internal Medicine (dir. 1969-78, pres. 1976-77), AMA, Am. Fedn. Clin. Research, Am. Rheumatism Assn., Inst. Medicine of Nat. Acad. Scis., D.C. Med. Soc. (chmn. legis. com. 1972-76), D.C. Soc. Internal Medicine (exec. council 1975-78), N.Y. Acad. Scis., So. Med. Assn. (sec. sect. internal medicine 1978-79, vice-chmn. 1979-80, chmn. 1980-81, asso. councilor 1979-81), Rheumatism Soc. D.C. (pres. 1963-64), Alpha Epsilon Delta, Phi Chi, Kappa Sigma. Republican. Baptist. Clubs: Masons, George Wever. Subspecialties: Internal medicine. Current work: Rheumatology Home: 4827 N 27th Pl Arlington VA 22207 Office: 2150 Pennsylvania Ave NW Washington DC 20037

FENDERSON, CONSTANTINE LLEWELLYN, animal scientist, educator; b. St. Thomas, Jamaica, Feb. 7, 1937; s. Cleveland L. and Gladys (McWhinney) F.; m. Una Mildred Johnston, Sept. 21 1963; 1 son, John Constantine Fitzgerald. Diploma, Jamaica Sch. Agr., 1960; B.S., Tuskegee Inst., 1969; M.S., Mich. State U., 1972, Ph.D., 1974. Agrl. asst. Ministry Agr. and Lands, Jamaica, 1960-61; sr. livestock officer Reynolds Metals Co., Jamaica, 1961-67; grad. research asst. Mich. State U., 1969-74, postdoctoral fellow, 1974-75; asso. prof. Tenn. State U., 1975—, also acting head dept. animal sci.; cons. AID. Contbr. articles to profl. jours. Mem. AAAS, Am. Soc. Animal Sci., Sigma Xi. Baptist. Subspecialty: Animal nutrition. Office: Tenn State U 3500 John Merritt Blvd Nashville TN 37203

FENDLER, JANOS HUGO, educator; b. Budapest, Hungary, Aug. 12, 1937; came to U.S., 1964, naturalized, 1982; s. Janos and Vilma (Csiky) F.; m. Anne-Marie Martiuele, Feb. 14, 1976; children: Peter, Monika; m.; children by previous marriage: Michael, Lisa. B.Sc., U. Leicester, Eng., 1960; Dipl., Radiochemistry Leicester Coll., 1961; Ph.D., U. London, 1964, D.Sc., 1978. Postdoctoral fellow U. Calif.-Santa Barbara, 1964-66; fellow Mellon Inst., Pitts., 1966-70; asso. prof. chemistry Tex. A.&M. U., College Station, 1970-75, prof., 1975-81; prof. chemistry Clarkson Coll. Tech., Potsdam, N.Y., 1982—. Editorial bd.: Jour. Organic Chemistry, 1978—, Jour. Colloid & Interface Sci, 1981—; author: Catalysis in Micellar and Macromolecular Systems, 1975, Membrane Mimetic Chemistry, 1982. Recipient Kendall award Am. Chem. Soc., 1982. Mem. Am. Chem. Soc., AAAS, Royal Soc., Internat. Assn. Colloid and Interface Sci., Soc. Applied Spectroscopy. Subspecialties: Analytical chemistry; Laser photochemistry. Current work: Membrane mimetic chemistry; characterization and utilization for synthesis, energy conversion, recognition and drug delivery; stereochemistry in excited state; photophysical investigation of chiral discrimination and enantiomeric recognition; picosecond spectroscopy.

FENDRICH, JOHN WILLIAM, computer science educator, consultant; b. Sioux Falls, S.D., Mar. 20, 1935; s. John Anton and Rose Marie (McConnell) F.; m. Lu Anne Frances Young, Feb. 6, 1958; children: Anne Marie, John Anton, Lisa Rene, Julie Christine, Christopher Edward, Andrew Kurt. B.S., U. Colo., 1962, M.A., 1964; A.M., U. Ill., 1966, Ph.D., 1971. Asst. prof. Butler U., Indpls., 1970-71, S.C. State Coll., Orangeburg, 1971-73; assoc. prof. Southwest State U., Marshall, Minn., 1973-76; asst. prof. St. Lawrence U., Canton, N.Y., 1976-80; assoc. prof. computer sci Bradley U., Peoria, Ill., 1980—; pvt. practice cons., Peoria, 1982—. Served to capt. USAF, 1956-51. Summer fellow NASA/ASEE, Stanford, 1980-81, Dept. Energy, 1982.

Mem. Assm. Computing Machinery, IEEE (assoc.), Math. Assn. Am., Am. Math. Soc., Futurists Soc. Subspecialties: Software engineering; Foundations of computer science. Current work: Systems and software engineering, distributed architectures and network vertification and validation. Office: Computer Sci Dept Bradley U Peoria IL 61625 Home: 3613 N Sterling Peoria IL 61604

FENG, SUNG YEN, biology educator, marine biologist; b. Shanghai, China, Oct. 1, 1929; m., 1963; 1 child. B.S., U. Taiwan, 1954; M.A., Coll. William and Mary, 1958; Ph.D. in Parasitology, Rutgers U., 1962. Research asst. dept. zoology and N.J. Oyster Research Lab., Rutgers U., 1960-62; research assoc. N.J. Oyster Research Lab., 1962-66; asst. prof. systs. and environ. biology Marine Sci. Inst. and biol. sci. group U. Conn., Storrs, 1966-68, assoc. prof. biology, 1968-74, prof. biol. sci., 1974—, prof. marine sci., 1979—; asst. dir. Marine Sci. Inst., 1972-77, dir., 1977—, acting head dept. marine scis., 1979—. Mem. Am. Soc. Parasitology, Am. Soc. Protozoology, Nat. Shellfisheries Assn., Soc. Invertebrate Pathology, Am. Soc. Zoology. Subspecialty: Marine Biology. Office: Marine Sci Inst U Conn Groton CT 06340

FENNEMA, OWEN RICHARD, food science educator, researcher; b. Hinsdale, Ill., Jan. 23, 1929; m., 1948; 3 children. B.S., Kans. State U., 1950; M.S., U. Wis., 1951, Ph.D. in Food Sci, 1960. Project leader food process research Pillsbury Co., Minn., 1954-57; from asst. prof. to assoc. prof. U. Wis., Madison, 1960-69, prof. food sci., 1969—, chmn. dept. food sci., 1977-81. Mem. Inst. Food Tech., Am. Chem. Soc., Am. Dairy Sci. Assn., Soc. Cryobiology. Subspecialty: Food science and technology. Office: U Wis Dept Food Sci Babcock Hall Madison WI 53706

FENNESSY, JOHN JAMES, radiologist; b. Clonmel, Ireland, Mar. 8, 1933; s. John and Ann (McCarthy) F.; m. Ann M. O'Sullivan, Aug. 20, 1960; children—Deirdre, Conor, Sean, Emer, Rona, Nial, Ruairi. M.B., B.Ch., BAO, Univ. Coll., Dublin, Ireland, 1958. Asso. prof. U. Chgo., 1971-74, prof., 1974—, chief chest and gastrointestinal radiology, 1971-73, acting chief diagnostic radiology, 1973-74, chmn. dept. radiology, 1974—. Hon. fellow Royal Coll. Surgeons Ireland; mem. Am. Assn. Univ. Radiologists, Soc. Gastrointestinal Radiology, Chgo. Radiol. Soc. (trustee), AMA, Am. Coll. Radiology, AAUP, Am. Gastroent. Soc., Fleischner Soc., Soc. Chmn. Academic Radiology Depts., Sigma Xi, Alpha Omega Alpha. Democrat. Roman Catholic. Subspecialty: Diagnostic radiology. Office: Dept Radiology U Chgo 950 E 59th St Chicago IL 60637

FENOGLIO, CECILIA METTLER, pathologist, researcher, educator; b. N.Y.C.; d. Frederick Albert and Cecilia Charlotte (Asper) Metteler; m. John Fenoglio, Jr., May 27, 1967 (div. 1977); 1 son, Timothy John. M.D., Georgetown U., 1969. Diplomate: Am. Bd. Pathology. Intern Presbyn. Hosp., N.Y.C., 1969-70; dir. Central Tissue Facility, Columbia-Presbyn. Med. Ctr., N.Y.C., 1976—; co-dir. div. surg. pathology Presbyn. Hosp., N.Y.C., 1978-82, dir., 1982—, Electron Microscop. Lab., Internat. Inst. for Human Reprodn., N.Y.C., 1978—; assoc. prof. pathology Coll. Physicians and Surgeons, Columbia U., N.Y.C., 1981-82, prof., 1982—, attending pathologist, 1982—, vice chmn. dept. pathology, 1982—. Author: General Pathology, 1983; editor: Advances in Pathobiology. Cell Membranes, 1975, Advances in Patholbiology: Aging and Neoplasia, 1976, Progress in Surgical Pathology, Vols. I-IV (also author Vol. IV), 1980-82. NIH grantee, 1973, 79-82; Cancer Research Ctr. grantee, 1975—; Population Council grantee, 1977—; Nat. Ileitis and Colitis Found. grantee, 1979-80. Mem. Internat. Acad. Pathology (edn. com. 1980—), AAAS (life), Am. Assn. Pathologists, N.Y. Acad. Sci., N.Y. Acad. Medicine, Fedn. Am. Scientists for Exptl. Biology, Gastrointestinal Pathologist Group (founding mem.). Subspecialties: Pathology (medicine); Cell study oncology. Current work: Primary interst is in the field of oncology especially as it relates to gastrointestinal or gynecologis disorders; am also interested in interaction of viruses and human cancers; a final area of interest is in the expression of tumor markers. Office: Columbia U Coll Physicians and Surgeons 630 W 168th St New York NY 10032

FENSTERHEIM, HERBERT, psychologist; b. N.Y.C., July 22, 1921; s. Harry and Mollie (Feder) F.; m. Jean Baer, June 20, 1968. B.A., N.Y.U., 1941; M.A., Columbia U., 1942; Ph.D., N.Y.U., 1958. Assoc. prof. N.Y. Med. Coll., N.Y.C., 1964-72; clin. assoc. prof. Cornell U. Med. Coll., N.Y.C., 1972—; sport psychologist U.S. Olympic Fencing Team, 1982—. Author: Don't Say Yes When You Want to Say No, 1975; contbr. articles to profl. jours. Fellow Behavior Therapy and Research Soc. (charter); mem. Am. Psychol. Assn., Assn. for Advancement Behavior Therapy (dir. 1971-73), Am. Group Psychotherapy Assn. Jewish. Subspecialties: Behavioral psychology; Psychiatry. Current work: Clinical application of behavioral concepts to psychol. treatment. Home: 151 E 37th St New York NY 10016

FENTON, NOEL JOHN, computer systems manufacturing company executive; b. New Haven, May 24, 1938; s. Arnold Alexander and Carla (Mathiasen) F.; m. Sarah Jane Hamilton, Aug. 14, 1965; children: Wendy, Devon, Peter, Lance. B.S., Cornell U., 1959; M.B.A., Stanford U., 1963. Research asst. Stanford (Calif.) U., 1963-64; v.p. Mail Systems Corp., Redwood City, Calif., 1964-66; v.p., gen. mgr. products div. Acurex Corp., Mountain View, Calif., 1966-72, pres., chief exec. officer, dir., 1972-83, Covalent Systems Corp., Santa Clara, Calif., 1983—; dir. Elpac Electronics, Inc., Micro Mask Inc., RPC Industries. Chmn. adv. council Resource Center for Women; mem. San Jose Econ. Devel. Task Force, 1983, Pres. Reagan's Bus. Adv. Panel. Served to lt. (j.g.) USN, 1959-61. Mem. Am. Electronics Assn. (chmn. 1978-79, dir. 1976-80), Young Pres.'s Orgn., Santa Clara County Mfrs. Group (dir.), Stanford Bus. Sch. Alumni Assn. (pres. 1976-77, dir. 1971-76). Republican. Episcopalian. Subspecialty: Electronics company management. Home: 60 Hayfields Rd Portola Valley CA 94025 Office: Acurex Corp 485 Clyde Ave Mountain View CA 94042

FENVES, STEVEN JOSEPH, civil engr.; b. Subotica, Yugoslavia, June 6, 1931; came to U.S., 1950, naturalized, 1953; s. Louis and Clara (Gereb) F.; m. Norma Jean Horwitz, July 3, 1955; children—Gregory L., Carol E., Peter D., Laura R. B.C.E., U. Ill., 1957, M.S., 1958, Ph.D., 1961. Prof. U. Ill., 1957-71; prof., head dept. civil engring. Carnegie-Mellon U., Pitts., 1972-75, Univ. prof., 1975—; vis. prof. M.I.T., 1962-63, Nat. U. Mex., 1965, 70, Cornell U., 1970-71; cons. to pvt. corps., govt. agys. Author: Stress Programming System, 1964, Computer Methods in Civil Engineering, 1967; contbr. numerous articles to tech. jours. Served with U.S. Army, 1952-53. Mem. Nat. Acad. Engring., ASCE (Research prize 1965), Assn. Computing Machinery, Sigma Xi, Tau Beta Pi, Chi Epsilon. Subspecialties: Civil engineering; Software engineering. Current work: Computer-aided design in civil engineering, use of expert systems, database methods, computer graphics and network theory in design. Home: 1125 Folkstone Dr Pittsburgh PA 15243

FERCHAU, HUGO ALFRED, plant ecologist, educator; b. Mineola, N.Y., July 22, 1929; s. Hugo and Melita (Roller) F.; m. Mary Ellen Shea, Dec. 5, 1952; children: Hugo A., Andrea Marie, Erich Marshall, Michele Suzanne. B.S., Coll. of William and Mary, 1951; postgrad., State Coll. of Wash., 1953; Ph.D., Duke U., 1959. Assoc. prof. biology Wofford Coll., Spartanburg, S.C., 1958-62; vis. assoc. prof. biology Duke U., Durham, N.C., summer 1962; asst. prof. botany Western State Coll., Gunnison, Colo., 1962-66, assoc. prof., 1966-69, prof., 1969—; vis. prof. environ. scis. Colo. Sch. Mines, Golden, 1981-82; cons. Contbr. articles on plant ecology to profl. jours. Mem. Bot. Soc. Am., Ecol. Soc. Am., AAAS, Am. Inst. Biol. Scis., S.W. Naturalist Soc., Can. Reclamation Assn., Soc. Range Mgmt., Sigma Xi. Republican. Episcopalian. Subspecialties: Ecology; Microbiology. Current work: Undergraduate and graduate education development of reclamation programs for industry (mining), supervision of reclamation monitoring, soil microbiology research. Home: 819 N Pine Gunnison CO 81230 Office: Biology Dept Western State Coll Gunison CO 82130

FERCHEK, GARY RANDALL, operations research manager; b. Detroit, Jan. 28, 1947; s. Matthew and Stephanie Barbara (Glowacki) F.; m. Mary Ann Pounders, Feb. 2, 1970; 1 dau., Tracie Ann. B.S., U.S. Mil. Acad., 1969; M.S. in Engring, U. Mich., 1976. Commd. 2d lt. U.S. Army, 1969, advanced through grades to maj., 1980; mem. staff and faculty U.S. Army Personnel and Adminstrn. Ctr., Ft. Benjamin Harrison, Ind., 1977-80; resigned, 1980; mgr. ops. research ISACOMM, Atlanta, 1980—. Mem. Ops. Research Soc. Am., Am. Inst. Indsl. Engrs., Mensa. Subspecialty: Operations research (engineering). Current work: Design of an intercity satellite based network utilizing state of the art digital switching and providing voice communications, teleconferencing and digital termination service. Home: 1868 Remington Rd Atlanta GA 30341 Office: ISACOMM 1815 Century Blvd Suite 500 Atlanta GA 30345

FERER, KENNETH MICHAEL, oceanographer; b. Pitts., Dec. 12, 1937; s. Michael Kenneth and Bertha (Fonos) F.; m. Joan Byrne; children: Michael Kenneth, David Craig. Student, U. Md., 1968-70; B.S., George Washington U., 1972; student, Catholic U. Am., 1973; M.S., U. So. Miss., 1982. Nuclear reactor operator Westinghouse Testing Reactor, Waltz Mills, Pa., 1960-62; test engr. Westinghouse Astronuclear Lab., Large, Pa., 1962; reactor operator Naval Research Lab., Washington, 1963-66, test and oceanographic engr., 1966-75; project engr. naval ocean research and devel. Nat. Sci. Testing Lab. Sta., Miss., 1975-79, project mgr., 1979—; naval rep. Nuclear Power Com., Washington, 1972-75. Contbr. articles to sci. jours. Bd. dirs. 4th Ward Assn., 1981. Served with USN, 1956-58. Mem. Marine Tech. Soc. (med. edn. com.), Am. Geophys. Union. Republican. Roman Catholic. Club: Pontchartrain Yacht. Subspecialties: Petroleum engineering; Ocean engineering. Current work: Development and use of new and unique oceanographic instrumentation in order to study the physical, chemical, optical, biological and magnetic properties of the ocean. Patentee cable elongation measurement; aramid fiber cable. Home: 108 Country Club Dr Covington LA 70433 Office: Naval Research and Devel Code 542 NSTA Sta MS 39529

FERGENSON, P. EVERETT, marketing educator, consultant; b. Bklyn., Feb. 15, 1940; m. Laraine R. Leberfeld, Aug. 5, 1967; children: Jon, David, Hilary. B.A., L.I.U., 1962; M.S., U. Mass., 1966, Ph.D., 1969. Sr. dynamics engr. Sikorsky Aircraft, Stamford, Conn., 1966-67; prof. Stevens Inst. Tech., Hoboken, N.J., 1967-77; assoc. mgr. strategic research Gen. Foods Corp., White Plains, N.Y., 1977-79; research group head Batton, Barton, Durstine & Osborn, N.Y.C., 1979-81; assoc. dir. research Ally & Gargano, N.Y.C., 1981-82; prof. mktg. Manhattan Coll., Riverdale, N.Y., 1982—. Contbr. articles in field to profl jours. Mem. Am. Psychol. Assn., Am. Mktg. Assn., Human Factors Soc. Subspecialties: Consumer behavior; Human factors engineering. Current work: Marketing research and consumer behavior, new product development, human-information processing, sensory evaluation as it relates to product design. Home: 9 Marcotte Ln Tenafly NJ 07670 Office: Manhattan Coll Manhattan Coll Pkwy Riverdale NY 10471

FERGUSON, ELDON EARL, physicist; b. Rawlins, Wyo., Apr. 23, 1926; s. George Earl and Bess (Pierce) F. B.S., Okla. U., 1949, M.S., 1950, Ph.D., 1953. Physicist U.S. Naval Research Lab., Washington, 1954-57; prof. physics U. Tex., Austin, 1957-62; dir. aeronomy lab. NOAA, Dept. Commerce, Boulder, Colo., 1962—. Served with U.S. Army, 1944-45. Guggenheim Found. fellow, 1960; Humboldt fellow, 1979-80. Mem. Am. Phys. Soc., Am. Chem. Soc., Am. Geophys. Union. Subspecialty: Aeronomy. Office: 325 Broadway Boulder CO 80302

FERGUSON, JAMES MALCOLM, physicist; b. Chgo., June 6, 1931; s. Robert H. and Anne (Cox) F.; m. Elizabeth S. Gee, May 29, 1970. B.S., Antioch Coll., 1953; Ph.D., MIT, 1957. Physicist U.S. Naval. Radiol. Def. Lab., San Francisco, 1957-69, Lawrence Livermore (Calif.) Lab., 1969—. Recipient Gold medal for research U.S. Naval. Radiol. Def. Lab., 1969. Fellow Am. Phys. Soc.; mem. Am. Nuclear Soc. Subspecialties: Nuclear physics; Nuclear fission. Current work: Theoretical and computational research on neutron transport in matter; theory of neutron-nuclear and atomic interactions. Office: Lawrence Livermore Lab PO Box 808 Livermore CA 94550 Home: 1856 Grand View Dr Oakland CA 94618

FERGUSON, JOHN CARRUTHERS, biology educator; b. Tuscaloosa, Ala., Mar. 2, 1937; s. John Howard and Rosalind Vera (Carruthers) F.; m. Rebecca Alretta Folsom, July 15, 1961; children: Katherine, Joellyn, John. B.A., Duke U., 1958; M.A., Cornell U., 1961, Ph.D., 1961. Asst. prof. biology Fla. Presbyn. Coll., St. Petersburg, 1963-67, assoc. prof., 1967-72; prof. biology Eckerd Coll., St. Petersburg, 1972—; vis. investigator Marine Biol. Lab., Woods Hole, Mass., 1962-66, Friday Harbor (Wash.) Lab., 1970. Author: Chemical Zoology, 1969, Echinoderm Nutrition, 1983. Mem. St. Petersburg Environ. Devel. Commn., 1973-75; mem. Coastal Zone Mgmt. Citizens Adv. Com., 1976-81; bd. dirs. Pinellas Point Civic Assn., 1976-80. NSF fellow, 1962; NSF grantee; Bird Found.grantee. Mem. AAAS, Am. Soc. Zoologists, Am. Microscopical Soc. Republican. Episcopalian. Subspecialties: Physiology (biology); Nutrition (biology). Current work: Echinoderm nutrition; nutrient translocations; use of dissolved free nutrients. Home: 2127 Inner Circle S Saint Petersburg FL 33712 Office: Eckerd Coll Box 12560 Saint Petersburg FL 33733

FERGUSON, KINGLSEY GEORGE, psychologist, educator; b. Newcastle-on-Tyne, Eng., Apr. 13, 1921; emigrated to Can., 1927; s. William George and Isobel (Finnegan) F. B.A. in English and French, U. Western Ont., 1943; M.A. in Psychology, U. Toronto, 1951, Ph.D., 1956. Diplomate: Am. Bd. Profl. Psychology. Staff psychologist Sunnybrook Vets. Hosp., Toronto, Ont., Can., 1949-50; chief psychologist Westminster Vets. Hosp., London, Ont., Can., 1950-61; chief psychologist Montreal (Que., Can.) Gen. Hosp., 1961-68; psychologist-in-chief Clarke Inst. Psychiatry, Toronto, 1968—; mem.-at-large Nat. Sci. and Planning Council (Can. Mental Health Assn.), nat. office), Toronto, 1970—; chmn. Ont. Bd. Examiners in Psychology, Toronto, 1972-77. Served to lt. Can. Navy, 1942-45. Fellow Can. Psychol. Assn., Behavior Therapy and Research Soc. (clin.); mem. Am. Psychol. Assn., Ont. Psychol. Assn. (pres. 1959-60). Subspecialties: Behavioral psychology; Primary prevention in mental health. Current work: Head of clinical/research department. Therapeutic orientation is cognitive behavioral. Volunteer consultant to national mental health body. Office: Clarke Inst Psychiatry 250 College St Toronto ON Canada M5T 1R8

FERGUSON, RONALD MAX, educator; b. South Beaver Twp., Pa., Sept. 29, 1936; s. Myrle Delbert and Nellie Madeline (Boyer) F.; m. Dorothy Smith, Mar. 15, 1964; children: Clifford Scott, Laura Gail. B.S., Clarion State Coll., 1958; M.S., Temple U., 1962, U. Conn., 1982. Techr. physics and chemistry, Crafton, Pa., 1958-61; tchr. physics Moses Brown, Providence, 1962-65; prof. chemistry Eastern Conn. State Coll., Willimantic, 1965—, assoc. prof. earth and phys. scis. dept., 1982—. Mem. Am. Chem. Soc., New Eng. Chemistry Tchrs. Assn. Subspecialties: Analytical chemistry; Atmospheric chemistry. Current work: Atmospheric chemistry, acid rain, trace pollutants research in chem. oceanography. Home: 91 Northfield Rd Coventry CT 06238 Office: Eastern Conn State Coll Goddard Hall Willimantic CT 06226

FERKO, ANDREW PAUL, pharmacologist; b. b, Trenton, N.J., Aug. 19, 1942; s. Andrew and Margaret (Rogaczewski) F.; m. Joan M., Apr. 24, 1971; children: Carolyn, Karen, Katharine. B.S., Phila. Coll. Pharmacy and Scis., 1965; Ph.D., Hahnemann Med. Coll. and Hosp., 1969. Jr. instr. Hahnemann Med. Coll., Phila., 1967-69, sr. instr., 1969-71, asst. prof., 1971-81, asso. prof., 1981—. Contbg. editor: Basic Pharmacology in Medicine, 1976, 81; contbr. articles to profl. jours. NIH grantee, 1965-69; recipient Lindback Found. teaching award, 1981. Mem. Am. Soc. Pharmacology and Exptl. Therapeutics, Soc. Toxicology, AAAS, N.Y. Acad. Sci., Mid-Atlantic Soc. Toxicology, Phila. Physiol. Soc., Phila. Neurosci Soc., Internat. Soc. Study Xenobiotics, Rho Chi. Subspecialties: Pharmacology; Neuropharmacology. Current work: Effects of ethanol on central nervous system. Office: Hahnemann Univ Broad and Vine Sts Philadelphia PA 19102

FERL, ROBERT JOSEPH, geneticist, educator; b. Conneaut, Ohio, Jan. 19, 1954; s. Joseph Edward and Joanne Marie (Rositer) F.; m. Mary Blythe Ferl, May 10, 1980. B.A., Hiram Coll., 1976; Ph.D., Ind. U., Bloomington, 1980. Vis. scientist div. plant industry Commonwealth Sci. and Indsl. Research Orgn., Canberra, Australia, 1981; asst. prof. botany U. Fla., 1980—. Mem. Genetics Soc. Am., Bot. Soc. Am., Sigma Xi. Subspecialties: Genetics and genetic engineering (biology); Molecular biology. Current work: Gene structure and function in plants; gene structure and function; recombinant DNA technology. Office: Department of Botany University of Florida Gainesville FL 32611

FERLAND, GARY JOSEPH, astrophysicist, educator; b. Washington, May 10, 1951; s. Andrew Joseph and Ida Maria (Schneemann) F.; m. Ann Elizabeth. B.S. in Physics, U. Tex., 1973; Ph.D. in Astrophysics, 1978. Research assoc. Cambridge (Eng.) U., 1978-80; asst. prof. astrophysics U. Ky., Lexington, 1980—; cons. NASA. Contbr. chpt. to book, articles to profl. jours. Welch Found. fellow, 1977-78; NASA grantee, 1982; NSF grantee, 1980-82. Mem. Am. Astron. Soc., Astron. Soc. Pacific, Internat. Astron. Union, Royal Astron. Soc. Subspecialties: Ultraviolet high energy astrophysics; Theoretical astrophysics. Current work: Active galactic nuclei, novae, emission line formation; formation of emission lines in photoionized environments, computer modeling of gaseous nebulae. Home: 3514 Brookview Dr Lexington KY 40503 Office: Physics Dept U Ky Lexington KY 40506

FERNALD, RUSSELL DAWSON, biology educator; b. Chuquicamata, Chile, Nov. 20, 1941; s. Russell G. and Catherine (Graf) F.; m. Anne Jones, May 25, 1969; children: Lia, Anya. B.S., Swarthmore Coll., 1963; Ph.D., U. Pa., 1968. Postdoctoral fellow Max Planck Inst., Munich, W.Ger., 1969-71, staff scientist, Seewiesen, W.Ger., 1971-77; assoc. prof. biology U. Oreg., Eugene, 1977—. Med. Research Council Fogarty fellow, London, 1982-83. Mem. Soc. Neuroscience. Subspecialties: Neurobiology; Ethology. Current work: Visual pattern recognition; development of the nervous system. Office: Dept Biology U Oreg Eugene OR 97403

FERNANDES, DANIEL JAMES, biochemist; b. Fall River, Mass., June 22, 1948; s. Casey Louis and Gilda (Carreiro) F.; m. Kay Galliher, Mar. 13, 1982; 1 son from previous marriage, Duane. B.S., Providence Coll., 1970; Ph.D., George Washington U., 1978. Teaching fellow dept. pharmacology George Washington U., Washington, 1972-77; postdoctoral research fellow Yale U. Sch. Medicine, 1977-80; asst. prof. dept. biochemistry Bowman Gray Sch. Medicine, Wake Forest U., Winston-Salem, N.C., 1980—; pharmacology cons. Flavor and Extract Mfrs. Assn. U.S., 1976. Contbr. articles to profl. jours. USPHS fellow, 1977-80. Mem. Am. Assn. Cancer Research, Southeastern Cancer Research Assn. Subspecialties: Cancer research (medicine); Molecular biology. Current work: Genetic regulation of enzymes in drug-resistant cancer cells; design and development of cancer chemotherapeutic agents. Office: 300 S Hawthorne Rd Winston-Salem NC 27103

FERNÁNDEZ, EDUARDO BUGLIONI, computer engineering educator; b. Concepción, Chile, Sept. 13, 1936; came to U.S., 1970, naturalized, 1976; s. Plácido and Elena (Buglioni) F.; m. Stacy Amber Lundgren, July 3, 1971. Ingeniero Electricista, U. F. Santa María, Valparaíso, Chile, 1960; M.S.E.E., Purdue U., 1963; Ph.D. in Computer Sci, UCLA, 1972. Investigator, prof. U. Chile, Santiago, 1963-69; sci. specialist IBM, Los Angeles and New Haven, 1973-81; prof. elec. and computer engring. U. Miami, Coral Gables, Fla., 1981—; cons., lectr. IBM, Boca Raton, 1982. Author: Database Security and Integrity, 1981; contbr. 30 articles to profl. jours. Mem. Assn. for Computing Machinery, IEEE. Subspecialties: Computer architecture; Database systems. Current work: Research on database security and performance, high level computer architecture. Home: 9015 SW 78 Ct Miami FL 33156 Office: U Miami PO Box 248294 Coral Gables FL 33124

FERNÁNDEZ, FERNANDO LAWRENCE, applied research co. executive; b. N.Y.C., Dec. 31, 1938; s. Fernando and Luz Esther F.; m. Carmen Dorothy Mays, Aug. 26, 1962; children: Lisa Marie, Christopher John. M.S., Stevens Inst. Tech., 1960, 1961; Ph.D. in Aeros, Calif. Inst. Tech., 1969. Thermodynamicist Lockheed Missiles, Sunnyvale, Calif., 1961-63; group dir. Aerospace Corp., San Bernadino, Calif., 1963-72; program mgr. R & D Assocs., Santa Monica, Calif., 1972-75; v.p. Physical Dynamics Inc., La Jolla, Calif., 1975-76; pres. Arete Assocs., Encino, Calif., 1976—. Mem. AIAA, N.Y. Acad. Sci. Subspecialties: Aeronautical engineering; Oceanography. Home: 4404 Conchita Way Tarzana CA 91356 Office: Arete Assocs 5445 Balboa Blvd PO Box 350 Encino CA 91356

FERNÁNDEZ, SALVADOR M., physiology educator; b. Guantanamo, Cuba, Mar. 14, 1943; came to U.S., 1961, naturalized, 1983; s. Salvador Fernández-Álvarez and Onelia Mola de F. B.A., Wayne State U., 1965; M.S., U. Conn., 1968, Ph.D., 1975. Asst. prof. physiology U. Conn. Health Ctr., Farmington, 1975—; cons. NIH, Bethesda, Md., 1981—, Conn. Research Found., Storrs, 1980-83. Editor: Fast Methods in Physical Biochemistry, 1983; contbr. articles to profl. jours. Mem. Am. Biophys. Soc., AAAS, Sigma Pi Sigma. Subspecialties: Spectroscopy; Biophysics (biology). Current work: Time-resolved fluorescence spectroscopy of living cells and biomolecules employing mode-locked lasers. Office: U Conn Health Center Farmington Ave Farmington CT 06032

FERNANDEZ-CRUZ, EDUARDO P., medical researcher, physician; b. Santiago de Compostela, Spain, July 16, 1946; came to U.S., 1978, naturalized, 1982; s. Arturo L. Fernandez-C. and Filis Maria S.

(Perez); m. Pilar C. Sarrate, Sept. 22, 1973; children: Eduardo, Arturo, Laura. B.Scis., Loyola Coll., Spain, 1957-63; student, Sch. Medicine, U. Barcelona, Spain, 1963-66, Sch. Medicine, U. Madrid, 1968-70. Intern, resident Clinica Puerta de Hierro, Autonomous U. Madrid, 1971-76; asst. mem. dept. immunopathology, 1977-82; research fellow, research assoc. dept. immunopathaology Scripps Clinic and Research Found., La Jolla, Calif., 1978-82, sci. assoc., 1982—; research fellow dept. immunology Middlesex Hosp. Med. Sch., London, 1973-74; sci. cons. Med. Biology Inst., 1982—. Contbr. articles to profl. jours. Named Outstanding Med. Student, 1971; Juan March Found. fellow, 1974; others. Mem. Brit. Soc. Immunology, Brit. Soc. Allergy and Clin. Immunology, Spanish Soc. Immunology, Spanish Soc. Hepatology, Am. Assn. Immunologists. Roman Catholic. Subspecialty: Immunobiology and immunology. Current work: Cellular mechanisms involved in the host's immune response to tumors of different origin. Experimental models of immunotherapy of cancer. Home: 5120 Bothe Ave San Diego CA 92122 Office: Scripps Clinic and Research Found 10666 N Torrey Pines La Jolla CA 92037

FERNANDEZ-MORAN, HUMBERTO, biophysicist; b. b, Maracaibo, Venezuela, Feb. 18, 1924; s. Luis and Elena (Villalobos) Fernandez-M.; m. Anna Browallius, Dec. 30, 1953; children—Brigida Elena, Veronica. M.D., U. Munich, Germany, 1944, U. Caracas, Venezuela, 1945; M.S., U. Stockholm, Sweden, 1951, Ph.D., 1952. Fellow neurology, neuropath. George Washington U., 1945-46; intern George Washington U. Hosp., 1945-46; resident Serafimerlasarettet, Stockholm, 1946-58; fgn. asst. Neurosurg. Clinic, Stockholm, 1946-48; research fellow Nobel Inst. Physics, Stockholm, 1947-49, Inst. Cell Research & Genetics, Karolinska Institutet, 1948-51, asst. prof., 1952; prof., chmn. dept. biophysics U. Caracas, 1951-58; dir. Venezuelan Inst. Neurology and Brain Research, Caracas, 1954-58; asso. biophysicist neurosurg. service Mass. Gen. Hosp., Boston, 1958-62; vis. lectr. dept. biology Mass. Inst. Tech., 1958-62; research asso. neuropath. Harvard, 1958-62; prof. biophysics U. Chgo., 1962—, now A.N.; Pritzker prof. biophysics. Sci. and cultural attaché to Venezuelan legations, Sweden, Norway, Denmark, 1947-54; head Venezuelan commn. Atomic Energy Conf., Geneva, 1955; chmn. Venezuelan commn. 1st Inter-Am.-Symposium on Nuclear Energy, Brookhaven, N.Y., 1957; minister of edn., Venezuela, 1958; mem. Orgn. Am. States adv. commn. on sci. devel. in Latin Am., Nat. Acad. Scis., 1958; mem. U.S. Nat. Com. UNESCO, 1957. Author: The Submicroscopic Organization of Vertebrate Nerve Fibres, 1952, The Submicroscopic Organization of the Internode Portion of Vertebrate Myelinated Nerve Fibers, 1953, Cryoelectronmicroscopy; Superconductivity; Diamond Knife Ultramicrotomy, 1955-76; author series publs. in fields molecular biology, nerve ultrastructure, electron and cryo-electron microscopy, electron and x-ray diffraction, cell ultrastructure, neurobiology, superconducting lenses, superconductivity, others.; Editorial bd., Jour. of Cell Biology, 1961. Decorated Knight of Polar Star, Sweden; Claude Bernard medal, Canada; Medalla Andres Bello, Venezuela, 1973; Recipient Gold medal City Maracaibo, 1968, John Scott award for invention of diamond knife, 1967; medal Bolivarian Soc. U.S., 1973. Fellow Am. Acad. Arts and Sci.; mem. Venezuelan Acad. Medicine (hon.), Academia Ciencias Fisicas y Matematicas (Caracas), Am. Acad. Neurology (corr. mem.), Internat. Soc. Cell Biology, Buenos Aires, Santiago, Lima, socs. Neurology, Buenos Aires, Santiago, Lima, Porto Alegre societies surgery, Electron Microscopy Soc. Am. (spl. citation), Am. Nuclear Soc., Pan Am. Med. Assn., Sociedad Bolivárianade Arquitectos (Venezuela) (hon.), Pan Am. Assn. of Anatomy (hon.). Home: Apartado 362 Maracaibo Venezuela Office: Research Insts U Chgo 5640 S Ellis Ave Chicago IL 60637

FERNSTROM, MADELYN HIRSCH, neuroscientist, psychiatric educator; b. N.Y.C., Oct. 4, 1952; d. Emanuel M. and Marylin J. (Schechter) Hirsch; m. John Dickson Fernstrom, Jan. 8, 1978; 1 son, Aaron David. A.B. in Biology, Boston U., 1974; Ph.D., MIT, 1978. Research fellow Harvard U. Med. Sch., Boston, 1978-80; research assoc. lab. brain and metabolism M.I.T., Cambridge, 1980-82; asst. prof. psychiatry U. Pitts., 1982—; instr. physiology Emmanuel Coll., Boston, 1979-80; instr. endocrinology Boston U., 1980-81. Contbr. articles to profl. jours. NIMH fellow, 1976-78; Juvenile Diabetes Found. fellow, 1978-80. Mem. Endocrine Soc., Soc. Neurosci., Soc. Nutrition Edn., Sigma Xi. Subspecialties: Neuroendocrinology; Neuropharmacology. Current work: Effects of diet, drugs and disease on neurotransmitters and neuropeptides in brain and retina. Office: Dept Psychiatry U Pitts Sch Medicine Western Psychiat Inst and Clinic 38110 Hara St Pittsburgh PA 35213

FERNSTRROM, JOHN DICKSON, neuroscientist, cons.; b. N.Y.C., July 9, 1947; s. Karl D. and Dorothy W. (Bond) F.; m. Madelyn Hirsch, Jan. 8, 1978; 1 son, Aaron D. S.B., M.I.T., 1969, Ph.D., 1972. Postdoctoral fellow Roche Inst. Molecular Biology, Hoffmann-LaRoche, Nutley, N.J., 1972-73; asst. prof. physiology M.I.T., Cambridge, Mass., 1973-77; asso. prof. neuroendocrinology, 1977-82; asso. prof. psychiatry and pharmacology U. Pitts., 1982—; mem. life scis. com. NASA, 1980—; mem. NINCDS program project rev. com. NIH, 1978-81, chmn., 1981-82. Contbr. articles to sci. jours. Recipient NIMH research scientist devel. award, 1979-82; Alfred P. Sloan fellow in neurochemistry, 1974-76. Mem. Am. Soc. Pharmacology and Exptl. Therapeutics, Am. Physiol. Soc., Soc. Neurosci., Internat. Soc. Neurochemistry, Internat. Soc. Neuroendocrinology. Subspecialties: Neuroendocrinology; Neuropharmacology. Current work: Control of neurotransmitter and neuropeptide synthesis and release. Regulation of somatostatin synthesis in hypothalamus; regulation of serotonin and catecholamine synthesis and release in brain and retina. Office: Dept Psychiatry U Pitts Sch Medicine 3811 O'Hara St Pittsburgh PA 15213

FERONE, ROBERT, microbiologist; b. Mt. Vernon, N.Y., Nov. 8, 1936; s. Nicholas and Anne (Zucker) F.; m. Diane F. Pasciolla, Sept. 10, 1960; children: Thomas, Steven, Michael, Elizabeth. B.A., NYU, 1958, M.S., 1963. Research technician Wellcome Research Labs., Tuckahoe, N.Y., 1958-64, jr. research microbiologist, 1964-66, sr. research microbiologist, Research Triangle Park, N.C., 1969-79, group leader, 1979—. Mem. Am. Soc. Microbiology, Am. Soc. Protozoologists, Am. Soc. Biol. Chemists. Democrat. Subspecialties: Microbiology; Biochemistry (biology). Current work: Folate biosynthesis in microorganisms, folate analogs, chemotherapy, biochemical parasitology, folate metabolism in malaria. Home: 5500 Knollwood Dr Raleigh NC 27609 Office: Wellcome Research Labs Research Triangle Park NC 27709

FERRAIO, NICHOLAS LAVERNE, psychotherapist, consultant; b. Rochester, N.Y., Sept. 9, 1946; s. LaVerne Leslie and Blanche Rose (Yates) F.; m. Carroll Ann Doolittle, June 4, 1971; children: Amy Lynn, Shannon Alicia. B.S. in Psychology and Sociologywith distinction, U. Rochester, 1976, M.S. in Community Service, 1980. Research technician U. Rochester Environ. Health Sci. Center, 1968-80, cons., 1980—; psychotherapist assoc. Joseph A. Dipoala, M.D., Rochester, N.Y., 1977. Mem. Am. Psychol. Assn., Assn. Behavior Analysis (affiliate), Nat. Psychiat. Assn., Acad. Holistic Medicine, Am. Council Counselors, Educators and Therapists, Biofeedback Soc. Am., Genessee Valley Psychol. Assn., Am. Guild Hypnotherapists (registered hypnotherapist), Rochester Acad. Pain Mgmt., Acad. Sci. Hypnotherapy, Am. Assn. Profl. Hypnotherapists. Republican. Unitarian. Subspecialty: Clinical psychology. Current work: Holistic psychology, holistic health and medicine, psychotherapy, relaxation training, biofeedback, hypnosis. Home: 410 C Audino Ln Rochester NY 14624 Office: 2128 E Henrietta Rd Rochester NY 14623

FERRARI, DOMENICO, electrical engineering educator; b. Gragnano, Piacenza, Italy, Aug. 31, 1940; came to U.S., 1970; s. Giacomo and Erina (Fracchioni) F.; m. Alessandra Ferrari Cella-Malugani, Apr. 16, 1966; children: Giuliarachele, Ludovica. Dr.Ing., Politecnico di Milan, Italy, 1963. Asst. Politecnico di Milan, 1964-67, asst. prof., 1967-70, prof., 1976-77; asst. prof. U. Calif.-Berkeley, 1970-75, assoc. prof., 1975-79, prof., dept. elec. engring. and computer sci., 1979—, dep. vice chmn., 1977-79; cons. in field. Author: Computer Systems Performance Evaluation, 1978, (with Serazzi and Zeigner) Measurement and Tuning of Computer Systems, 1983; editor: Performance of Computer Installations, 1978, Experimental Computer Performance Evaluation, 1981, Theory and Practice of Software Technology, 1983; contbr. articles to profl. jours. Recipient Libera Docenza Italian Govt., 1969; O. Bonazzi award AEI, 1970; NSF grantee, 1974—; U. Calif. grantee, 1982—. Mem. IEEE, Computer Measurement Group, Assn. for Computing Machinery. Clubs: Berkeley City, Croara Country (Italy). Subspecialties: Operating systems; Distributed systems and networks. Current work: Research in performance evaluation of computer systems, especially distributed systems. Office: Computer Sci Div EECS Dept Univ Calif Berkeley CA 94720

FERRARO, DOUGLAS PETER, psychologist; b. White Plains, N.Y., Nov. 27, 1939; s. Peter Mario and Edith Isabella (Lewendon) F.; m. Sandra Jean Odell, Jan. 5, 1980; children: Craig Alan, Kim Elizabeth. A.B., Columbia U., 1961, M.A., 1963, Ph.D., 1965. Cert. psychologist, N. Mex. Asst. prof. U. N. Mex., 1965-69, asso. prof., 1969-72, prof., 1973—; disting. prof. Universidad del Noreste, Mexico, 1977; adj. prof. Union Experimenting Colls. and Univs., 1980—; vis. prof. Nat. U. Mexico, 1981—; cons. in field. Author: The Pharmacology of Marihuana, 1976, Systematic Analysis of Learning and Motiviation, 1978, Psychology: Contemporary Concepts, 1981; contbr. articles to profl. jours. Mem. Nat. Inst. Drug Abuse Public Advisory Com., 1978-81. Fellow Am. Psychol. Assn.; mem. Am. Soc. Pharmacology and Exptl. Therapeutics, Soc. Behavioral Medicine, Soc. Pyschologists in Substance Abuse, Psychonomic Soc. Subspecialties: Behavioral Pharmacology; Behavioral Medicine. Current work: Behavioral approaches to prevention, diagnosis, treatment and rehab. of phys. illness. Home: One Pool NW La Luz Albuquerque NM 87120 Office: Dept Psychology U N Mexico Albuquerque NM 87131

FERRENDELLI, JAMES ANTHONY, neurologist, educator; b. Trinidad, Colo., Dec. 5, 1936; s. Alex and Edna F.; m. Judith A. Chavers, June 8, 1957; children: Elisabeth, Cynthia, Michael. A.B. Cum laude in Chemistry, U. Colo., 1958, M.D., 1962. Diplomate: Am. Bd. Psychiatry and Neurology. Intern U. Ky. Med. Center, 1962-63; resident in neurology Cleve. Met. Gen. Hosp., 1965-68; research fellow in neurochemistry Sch. Medicine, Washington U., St. Louis, 1968-70, asst. prof. neurology and pharamacology, 1970-74, asso. prof., 1974-77, prof., 1977-, Seay prof. clin. neuropharmacology in neurology, 1977—. Contbr. numerous articles to profl. jours. Served to capt., M.C. U.S. Army, 1963-1965. Recipient Research Career Devel. award USPHS, 1971-76; Founders Day award Washington U., 1981; NIH grantee, 1971—. Mem. Am. Acad. Neurology, Am. Neurol. Assn., Am. Soc. for Pharmacology and Exptl. Therapeutics (Epilepsy award 1981), Am. Epilepsy Soc. Subspecialties: Neurology; Neuropharmacology. Current work: Molecular mechanisms of neurologic diseases and neurotropic drugs; investigations of mechanisms of action of antiepileptic drugs, of pathophysiological mechanisms of seizure disorders, and molecular mechanisms of retinal function. Home: 87 Lake Forest Saint Louis MO 63117 Office: Dept Neurology Washington U Med Sch 660 S Euclid Ave Saint Louis MO 63110

FERRETTI, JAMES ALFRED, Research chemist; b. Sacramento, Aug. 1, 1939; s. Emilio Alfred and Thelma (Selmi) F.; m. Martine Polin, July 10, 1970; children: Dominique, Sophie. B.S., San Jose (Calif.) State U., 1961; Ph.D., U. Calif.-Berkeley, 1965. Lectr. in chemistry U. Naples, Italy, 1965-66; research chemist Nat. Heart, Lung and Blood Inst., NIH, Bethesda, Md., 1966—. Contbr. articles to profl. jours. Soc. Magnetic Resonance in Medicine grantee, 1982. Mem. Am. Chem. Soc., Am. Phys. Soc., Biophys. Soc., Sigma Xi. Roman Catholic. Subspecialties: Polymer chemistry; Biophysics (physics). Current work: Nuclear magnetic resonance spectroscopy; studies in conformational and dynamic properties of biologically important molecules. Office: NIH Bldg 10 Room 7N316 Bethesda MD 20205

FERRETTI, JOSEPH JEROME, microbiology educator; b. Chgo., Dec. 23, 1937; s. John and G. Marie (Nemeth) F.; m. Martha J. Bang, Nov. 27, 1965; children: Joseph, John, Anne-Marie. B.S., Loyola U., Chgo., 1960; M.S., U. Minn., 1965, Ph.D., 1967. Postdoctoral fellow Johns Hopkins U., Balt., 1967-69; asst. prof. microbiology U. Okla. Health Sci. Ctr., Oklahoma City, 1969-72, assoc. prof., 1972-78, prof., 1978—, prof. and head dept. microbiology, 1983—; Bd. dirs. Headlands Indian Health Careers, Mackinaw City, Mich, Career Opportunities in Health Scis., Oklahoma City, 1970-80. Author: Basic Bacteriology and Genetics, 1978; contbr. articles to profl. jours. Bd. dirs. Omniplex Sci. Mus., Oklahoma City, 1981-83. NIH grantee, 1973; NSF grantee, 1973; Am. Heart Assn. grantee, 1971-77; vis. scientist Nat. Acad. Sci., 1981, NSF, Paris, 1982. Mem. Am. Soc. Microbiology, Am. Acad. Microbiology, AAAS, Johns Hopkins Soc. Scholars, Sigma Xi. Subspecialties: Microbiology (medicine); Genetics and genetic engineering (medicine). Office: Dept Microbiology and Immunology U Okla Oklahoma City OK 73190 Home: 1115 NE 55th St Oklahoma City OK 73111

FERRIANS, OSCAR JOHN, JR., geologist; b. Touchet, Wash., Mar. 9, 1928; m., 1953. B.S., State Coll. Wash., 1952, M.S., 1958. Geologist U.S. Geol. Survey, 1953—, now in, Menlo Park, Calif. Mem. AAAS, Geol. Soc. Am., Arctic Inst. N.Am., Am. Quaternary Assn. Subspecialties: Remote sensing (geoscience); Geology. Office: US Geol Survey 345 Middlefield Rd Menlo Park CA 94025

FERRIER, JACK MORELAND, biophysicist, dentistry educator; b. Cleve., Aug. 8, 1943; went to Can., 1975, naturalized, 1981; s. Jack Howard Moreland and Ruth Marie (Routa) F.; m. Sue Ellen Lamb, Dec. 28, 1966; 1 dau., Megan Elizabeth; m. Laura Elizabeth Watson, Mar. 1, 1981. B.Sc., Ohio State U., Columbus, 1966, M.Sc., 1968, Ph.D., 1973. Lectr. physics Ohio State U., Columbus, 1973-75; research assoc. botany U. Toronto, 1975-79, asst. prof. dentistry, 1979—. Mem. Am. Phys. Soc. New Democrat. Buddhist. Subspecialty: Biophysics (physics). Current work: Electrophysiology of small mammalian cells, intercellular communication via ion concentration waves or electrical signals, viscoelastic properties of biological tissues. Office: U Toronto 1 Kings College Circle Toronto ON Canada M5S 1A8

FERRIS, DEAM HUNTER, microbiologist, epidemiologist, parasitologist; b. Mankato, Minn., July 8, 1912; s. Joseph Alexander and Ruby (Dawson) F.; m. Merle Wildey Ferris; children: Sara Josephine Ferris Decker, Tary Jeane Ferris Tobin, Deborah, Karen. A.S., Mo. Western Coll., 1932; A.B. in Zoology, Drake U., 1934; M.A. in Parasitology, Drake U., 1938; Ph.D. in Veterinary Sci, U. Wis., 1953. Prof. zoology Graceland Coll., Lamoni, Iowa, 1948-57, dir. audio-visual program, 1948-57; asso. prof. pathobiology Coll. Veterinary Medicine U. Ill., 1957-72; microbiologist Plum Island Animal Disease Center Agrl. Research Service U.S. Dept. Agr., Greenport, N.Y., 1972—; asso. team leader, dir. Arbovirus Lab. FAO-UN Near East Animal Health Inst., Cairo, Egypt, 1964-66. Contbr. articles in field to profl. jours. Served with U.S. Army, 1942-46; Served with USAR, 1946-66. NIH grantee, 1958-64. Fellow Royal Soc. Health, AAAS; mem. Am. Veterinary Med. Assn., Am. Soc. Virology, Am. Soc. Microbiology, Am. Soc. Parasitologists, Am. Soc. Tropical Medicine and Hygiene, Conf. Research Workers in Animal Disease, Conf. Public Health Veterinarian, Wildlife Disease Assn., Sigma XI, Phi Sigma Soc., Sigma Tau Delta, Phi Zeta. Subspecialties: Microbiology (veterinary medicine); Parasitology. Current work: Foreign animal diseases, diagnosis and control, African swine fever, malignant catarrhal fever, trypanosomiasis, devel. of diagonostic tests for these and other foreign animal disease. Home: 843 Main St Greenport NY 11944 Office: PO Box 848 Greenport NY 11944

FERRIS, STEVEN HOWARD, psychologist, neuroscientist, gerontologist, educator; b. N.Y.C., June 27, 1943; s. Jack and Dora (German) F.; m. Arlene Carol Toubin, Jan. 21, 1968; children: David, Marc. B.S., Rensselaer Poly. Inst., 1965; Ph.D., CUNY, 1970. Cert. psychologist, N.Y. Postdoctoral research assoc. Naval Submarine Med. Research Lab., Groton, Conn., 1970-73; research psychologist NYU Med. Ctr., 1973-79, assoc. prof. dept. psychiatry, 1981—, exec. dir., 1975—. Contbr. articles to profl. jours. Mem. Am. Coll. Neuropsychopharmacology, Am. Gerontol. Soc., Soc. Neurosci., Am. Psychol. Assn. Subspecialties: Neuropharmacology; Cognition. Current work: Gerontology, neurobiology of aging and senile dementia; neuropsychobiology and neuropsychopharmacology of aging, senie dementia and Alzheimer's disease. Office: Dept Psychiatry NYU Med Ctr 550 First Ave New York NY 10016

FERRY, JASON HUGHES, mathematics educator; b. Pensacola, Fla., Nov. 29, 1933; s. Irving M. and Thelma Jane (Hughes) F.; m. Lucinda Kaye Boyle, Jan. 8, 1961; children: Dale, Carla, Mark. B.S. in Math., Northwestern U., 1957, M.A., 1964; Ph.D. in Engring. Sci., U. Ala., 1968. Research assoc. NRC, 1968-70; mathematician NASA, Huntsville, Ala., 1970-75; assoc. prof. math. U. Ala., Huntsville, 1975-78, Stanford U., (Calif.), 1978-82, prof. math., 1982—; cons. NASA, 1978—; vis. prof. math. Tex. A&M U., 1980. Contbr. articles to profl. jours. Served to 1st lt. U.S. Army, 1957-60. Grantee Nat. Acad. Sci.-Inst. Math., 1981-82. Mem. Am. Math. Soc., Math. Assn. Am., AIAA, Soc. for Indsl. and Applied Math., Pi Mu Epsilon. Club: University (San Francisco). Subspecialties: Applied mathematics; Mathematical software. Office: Werik Ctr 24 California St Rm 312 San Francisco CA 94111

FERRY, JOHN DOUGLASS, chemist; b. Dawson, Can., May 4, 1912; s. Douglass Hewitt and Eudora (Bundy) F.; m. Barbara Norton Mott, Mar. 25, 1944; children—Phyllis Leigh, John Mott. A.B., Stanford U., 1932, Ph.D., 1935; student, U. London, 1932-34. Pvt. asst. Hopkins Marine Sta., Stanford, 1935-36; instr. biochem. scis. Harvard, 1936-38; mem. Soc. Fellows, 1938-41; asso. chemist Woods Hole Oceanographic Inst., 1941-45; research asso. Harvard U., 1942-45; asst. prof. chemistry U. Wis., 1946, asso. prof., 1946-47, prof., 1947—, Farrington Daniels research prof., 1973—, chmn. dept., 1959-67; chmn. Internat. Com. on Rheology, 1963-68; vis. lectr. Kyoto (Japan) U., 1968, Ecole d'Eté, U. Grenoble, France, 1973. Author: Viscoelastic Properties of Polymers, 1961, 2d edit., 1970, 3d edit., 1980; co-editor: Fortschritte der Hochpolymeren Forschung. Recipient Eli Lilly award Am. Chem. Soc., 1946, Bingham medal Soc. Rheology, 1953; Kendall Co. award Am. Chem. Soc., 1960; Witco award, 1974; Colwyn medal Instn. Rubber Industry, U.K., 1972; Tech. award Internat. Inst. Synthetic Rubber Producers, 1977. Fellow Am. Phys. Soc. (high polymer physics prize 1966), Am. Acad. Arts and Scis.; mem. Nat. Acad. Sci., Am. Chem. Soc. (Goodyear medal Rubber div. 1981), Am. Soc. Biol. Chemists, Soc. Rheology (pres. 1961-63), Internat. Soc. Hematology, d'Honneur Groupe Français Rhéologie, Phi Beta Kappa, Sigma Xi, Phi Lambda Upsilon, Alpha Chi Sigma. Club: Rotary. Subspecialties: Polymer chemistry; Polymers. Current work: Rheology of polymers; viscoelastic and other physical properties of biological macromolecules. Home: 137 N Prospect Ave Madison WI 53705

FESHBACH, HERMAN, physicist, educator; b. N.Y.C., Feb. 2, 1917; s. David and Ida (Lapiner) F.; m. Sylvia Harris, Jan. 28, 1940; children—Carolyn Barbara, Theodore Philip, Mark Frederick. B.S., CCNY, 1937; Ph.D., MIT, 1942; D.Sci., Lowell Tech. Inst., 1975. Tutor CCNY, 1937-38; instr. MIT, 1941-45, asst. prof., 1945-47, asso. prof., 1947-55, prof., 1955—, Cecil and Ida Green prof. physics, 1976-83, Inst. prof., 1983—, dir., 1967-73, head dept. physics, 1973-83; cons. AEC; chmn. nuclear sci. adv. com. of Dept. Energy and NSF, 1979-82. Author: (with P.M. Morse) Methods of Theoretical Physics, 1953, (with A. deShalit) Theoretical Nuclear Physics, 1974; also sci. articles tech. jours.; Editor: Annals of Physics. Recipient Harris medal Coll. City N.Y., 1977; John Simon Guggenheim Meml. Found. fellow, 1954-55; Ford fellow CERN, Geneva, Switzerland, 1962-63. Mem. Am. Phys. Soc. (chmn. div. nuclear physics 1970-71, divisional councillor 1974-78, exec. com. 1974-78, chmn. panel on pub. affairs 1976-78, v.p. 1979-80, pres. 1980-81, Bonner prize 1973), Nat. Acad. Scis., NRC, Am. Acad. Arts and Scis. (v.p. Class I 1973-76, pres. 1982-85). Subspecialties: Theoretical physics; Nuclear physics. Current work: Theoretical Nuclear Physics. Home: 5 Sedgwick Rd Cambridge MA 02138

FESHBACH, SEYMOUR, educator; b. N.Y.C., June 21, 1925; s. Joseph and Fannie (Katzman) F.; m. Norma Deitch, Aug. 16, 1947; children—Jonathan, Laura, Andrew. B.S., Coll. City N.Y., 1947; M.A., Yale U., 1948, Ph.D., 1951. Project dir. Army Attitude Assessment Br., 1951-52; from asst. prof. to asso. prof. U. Pa., Phila., 1952-63; prof. U. Colo., Boulder, 1963-64; prof. psychology U. Calif., Los Angeles, 1964—, chmn. dept., 1977—; dir. Fernald Sch., 1964-73; cons. CBS, Ednl. TV, 1972; vis. fellow Wolfson Coll., Oxford (Eng.) U., 1980-81. Author: Television and Aggression, 1970, Psychology, An Introduction, 1977, others; editor: Aggression and Behavior Change: Biological and Social Processes, 1979; cons. editor: Jour. Abnormal Psychology, 1973—; Contbr. chpts. to books, articles to profl. jours. Served to 1st lt., inf. AUS, 1943-46; PTO. Recipient Ward medal Coll. City N.Y., 1947, Townsend Harris medal, Distinguished Alumnus award, 1972, Fellowship award Found. Fund Advancement of Psychiatry, 1980-81; NIMH grantee; NSF grantee. Fellow Am. Psychol. Assn., mem., Western Psychol. Assn. (pres. 1976-77), AAAS, Soc. for Study of Social Issues, Soc. for Research in Child Devel., Internat. Soc. for Applied Psychology, Internat. Soc. for Study of Aggression, Internat. Soc. for Study of Behavior Devel., ACLU, Phi Beta Kappa. Democrat. Jewish. Subspecialties: Developmental psychology; Personality. Home: 743 Hanley Ave Los Angeles CA 90049 Office: Dept Psychology U Calif 405 Hilgard Ave Los Angeles CA 90024

FETT, WILLIAM FREDERICK, plant pathologist; b. Hinsdale, Ill., Oct. 30, 1952; s. Walter W. and Francis H. F.; m. Maria I. DaSilva, Nov. 24, 1977; children: Andrew, Melanie. B.S., U. Ill., 1974; M.S., U. Wis.-Madison, 1977, Ph.D in Plant Pathology, 1979. NRC postdoctoral research assoc. Eastern Regional Research Ctr., U.S. Dept. Agr., Phila., 1979-80; research plant pathologist, Phila., 1980—.

Contbr. articles to profl. jours. Mem. Am. Soc. Plant Physiologists, Am. Phytopathological Soc., Am. Soc. Microbiology, Sigma Xi, Phi Kappa Phi, Gamma Sigma Delta. Subspecialties: Plant pathology; Plant physiology (agriculture). Current work: Determine the physiological basis for plant resistance to phytopathogenic bacteria. Home: 128 Reiffs Mill Rd Ambler PA 19002 Office: 600 E Mermaid Ln Philadelphia PA 19118

FETTERMAN, HAROLD RALPH, electrical engineer, educator, researcher; b. Jamaica, N.Y, Jan. 17, 1941; s. Maurice and Gloria (carchnowitz) F.; m. Susan P. Rauchway, Aug. 15, 1965; children: David, Matthew B.A., Brandeis U., 1962; Ph.D. in Exptl. Physics, Cornell U., 1968. Asst. prof. in residence physics dept. UCLA, 1967-69, prof. elec. engring., 1982—, chmn. dept. elec. engring., 1984—; staff physicist Lincoln Lab., M.I.T., Lexington, 1969-82; cons. on millimeter wave systems; dir. Millitech Corp. Patentee: Pasers, modulators, quasi-Optical systems. Recipient IR-100 award Indsl. Research Mag., 1979. Fellow Optical Soc. Am.; mem. IEEE (sr.), Sigma Xi. Subspecialties: Microelectronics; Atomic and molecular physics. Current work: Millimeter and submillimeter detectors and sources: solid state sources for millimeter and submillimeter wave generation; antenna arrays for imaging applications. Patentee lasers, modulators, quasi-optical systems. Office: Elec Engring Dept UCLA Los Angeles CA 90024

FEUERSTEIN, MICHAEL, clinical psychology educator; b. Buffalo, July 26, 1950; s. Irving Seymor and Shirley (Lapides) F.; m. Michele D. Kaplan, Dec. 26, 1971; children: Sara Elizabeth, Andrew Scott. A.B., Boston U., 1972; M.S., U. Ga., 1975, Ph.D., 1977. Lic. psychologist, N.Y., Fla. Health psychologist Stanford Research Inst., Menlo Park, Calif., 1977-78; clin. instr. Stanford U. Sch. Medicine, 1977-78; asst. prof. McGill U., Montreal, 1978-82; co-dir. behavior therapy unit Allan Meml. Inst., Montreal, 1979-82; assoc. prof. clin. psychology U. Fla., Gainesville, 1982—; cons. Med. Comp. Health Systems Ltd., Montreal, 1979—. Author: Mastering Pain, 1979, Health Psychology: A Psychobiological Perspective, 1983; editor: Readings in Behavioral Medicine, 1979. Recipient M.H. Erickson award of scientific excellence Am. Soc. Clin. Hypnosis, 1978; Med. Research Council Can. grantee, 1981; Health and Welfare Can. grantee, 1981; Zimmer scholar, 1977. Mem. Am. Psychol. Assn., Soc. Behavioral Medicine, Soc. Psychophysiol. Research, Assn. Advancement Behavior Therapy, Sigma Xi. Jewish. Subspecialties: Clinical psychology; Psychophysiology. Current work: Psychophysiological mechanism of chronic pain, stress management, medical psychology, corporate health promotion. Home: 2925 NW 22 Pl Gainesville FL 32605 Office: Dept Clin Psychology U Fla Box J165 JHMHC Gainesville FL 32610

FEYNMAN, RICHARD PHILIPPS, physicist; b. N.Y.C., May 11, 1918; s. Melville Arthur and Lucille (Phillips) F. B.S., Mass. Inst. Tech., 1939; Ph.D., Princeton, 1942. Staff atomic bomb project Princeton, 1942-43, Los Alamos, 1943-45; asso. prof. theoretical physics Cornell U., 1945-50; prof. theoretical physics Calif. Inst. Tech., 1950—. Contbr. theory of quantum electrodynamics, beta decay and liquid helium. Recipient Einstein award, 1954; Nobel prize in physics, 1965; Oersted medal, 1972; Niels Bohr Internat. Gold medal, 1973. Mem. Am. Phys. Soc., AAAS, Royal Soc. (fgn. mem.), Pi Lambda Phi. Subspecialty: Theoretical physics. Address: Physics Dept California Institute of Technology Pasadena CA 91125

FICHTEL, CARL EDWIN, physicist; b. St. Louis, July 13, 1933; s. Edwin Blanke and Eleanora Alice (Gutsch) F. B.S., Washington U., St. Louis, 1955, Ph.D., 1960. Teaching and research asst. Washington U., 1956-59; physicist NASA Goddard Space Flight Center, Greenbelt, Md., 1959-60, sect. head, 1960-68, br. head, 1968—, sr. scientist, 1975—; vis. lectr. U. Md., 1963-80, adj. prof., 1980—; mem. com. on space astronomy and astrophysics, mem. Astronomy Survey for the 1980's, Nat. Acad. Sci. Editor: (with others) High Energy Particles and Quanta in Astrophysics, 1974; author: Gamma Ray Astrophysics, New Insight into the Universe, 1981; contbr. numerous articles profl. jours. Recipient John C. Lindsay Meml. award NASA Goddard Space Flight Center, 1968, spl. Achievement award, 1978; Exceptional Sci. Achievement medal NASA, 1971. Fellow Am. Phys. Soc. (sec.-treas. cosmic physics div. 1974-76, vice chmn. 1982); mem. Am. Astron. Soc. (chmn. high energy div. 1979-81), Internat. Astron. Union, Sigma Xi. Subspecialties: High energy astrophysics; Gamma ray high energy astrophysics. Current work: High energy astrophysics, especially gamma ray astrophysics, diffuse galactic and extragalactic radiation, active galaxies, cosmic ray physics and gamma ray instruments. Office: NAS Goddard Space Flight Cente Code 660 Greenbelt MD 20771

FIDLER, JOHN MICHAEL, medical research scientist; b. Balt., Aug. 16, 1947; s. Glenn Leroy and Mary Kathryn (Weyandt) F. B.A. with honors in Biology, Johns Hopkins U., 1969; Ph.D. in Immunology, Purdue U., 1972. Postdoctoral research assoc. dept. biol. sci. Purdue U., 1973-74; Am. Cancer Soc. postdoctoral fellow in biophysics and biochemistry Walter and Eliza Hall Inst. Med. Research, Melbourne, Australia, 1974-76; asst. mem. dept. immunopathology Scripps Clinic and Research Found., La Jolla, Calif., 1976-82; vis. prof. dept. immunology Biomed. Research Inst., U. Mex., Mexico City, 1982; group leader, head cellular and molecular immunoregulation research unit, cellular immunology, inflammation and immunology research dept. metabolic disease research sect. Lederle Labs., Pearl River, N.Y., 1982—; adj. prof. dept. clin. immunology Johns Hopkins U. Sch. Medicine, Balt., 1982—; adj. prof. dept. microbiology N.Y. Med. Coll., Valhalla, 1983—. Recipient Md. Legis. Scholarship award Johns Hopkins U., 1965-69, David Ross predoctoral fellow Purdue U., 1971-72; Am. Cancer Soc. postdoctoral research fellow, 1974-76. Mem. Internat. Soc. Exptl. Hematology, AAAS, Am. Assn. Immunologists, Soc. Analytical Cytology. Democrat. Subspecialties: Immunobiology and immunology; Immunopharmacology. Current work: Immunoregulation, T lymphocyte subsets, lymphocyte differentiation and activation, immunological tolerance, immunobiological research. Office: Lederle Labs 60B-305 Pearl River NY 10965

FIDONE, SALVATORE JOSEPH, neuroscientist, researcher; b. N.Y.C., June 10, 1939; s. Salvatore and Grace Gladys (Fuchs) F.; m. Patricia Aldridge O'Brien, Aug. 11, 1962; children: Keith Hammond, Kristofer William. B.S., Georgetown U., 1962; Ph.D., SUNY-Syracuse, 1967. Asst. prof. U. Utah Sch. Medicine, Salt Lake City, 1969-76, assoc. prof., 1976-81, prof. physiology, 1981—; mem. neurology A study sect. NIH, Washington, 1981-85. Author: Physiology of the Nervous System, 1975. NIH research grantee, 1976-86. Mem. Am. Physiol. Soc., Soc. Neurosci. Roman Catholic. Subspecialties: Neurophysiology; Neurochemistry. Current work: Research on the chemical transmission in arterial chemoreceptors. Home: 3031 Cascade Way Salt Lake City UT 84109 Office: U Utah Sch Medicine 50 N Medical Dr Salt Lake City UT 84112

FIEDLER, FRED EDWARD, psychology, management and organization educator, consultant on management; b. Vienna, Austria, July 3, 1922; came to U.S., 1938; s. Victor and Hilda (Schallinger) F.; m. Judith Joseph, Apr. 4, 1946; children: Phyllis Decky, Ellen Victoria, Robert Joseph, Carol Ann. A.M., U. Chgo., 1947, Ph.D., 1949. Cer. Psychologist, Wash. Clin. psychol. trainee US VA, Chgo., 1947-50; research assoc., instr. U. Chgo., 1950-51; asst. prof. psychology U. Ill.-Urbana, 1951-69; prof. psychology U. Wash.-Seattle, 1969—; vis. prof. U. Amsterdam, Netherlands, 1958-59; guest prof. U. Louvain, Belgium, 1963-64; cons. State of Wash., 1981—, King County, Wash., 1970-80; cons. various govt., mil., pvt. orgns., U.S., Europe, 1953—. Author: Boards, Management and Company Success, 1959, A Theory of Leadership Effectiveness, 1967, Improving Leadership Effectiveness, 1976, Leadership and Effective Management, 1974; contbr. numerous articles to profl. jours. Mem. Wash. Gov.'s Transition Team, 1980; co-chmn. Task Force on Tech. Transfer, State of Wash., 1980-81; pub. mem. State Med. Disciplinary Bd., 1981—. Served with M.C. U.S. Army, 1942-45. Fellow Am. Psychol. Assn.; mem. Western Psychol. Assn., Internat. Assn. Applied Psychology, Soc. Organizational Behavior, Soc. Indsl. Orgns. Subspecialties: Organizational psychology; Social psychology. Current work: Leadership and organizational effectiveness. Office: Dept Psychology U Wash NI 25 Seattle WA 98195

FIELD, CYRUS WEST, geology educator; b. Duluth, Minn., May 5, 1933; s. Thorold Farrar and Katherine (Van Vleck) F.; m. Rebecca Hooker Paine, June 28, 1959; children: Cyrus W., Thorold F., Frederic P., Edward H. B.A., Dartmouth Coll., 1956; M.S., Yale U., 1957, Ph.D., 1961. Research asst. Yale U., New Haven, Conn., 1958-60; research geologist Kennecott Copper, Salt Lake City, 1960-63; asst. prof. geology Oreg. State U., Corvallis, 1963-68, assoc. prof., 1968-76, prof., 1976—. Fellow Geol. Soc. Am.; mem. Soc. Econ. Geologists, Soc. Mining Engrs., Geochem. Soc. Republican. Episcopalian. Current work: Genesis, mineralogy, petrology and geochemistry of metallic mineral deposits, and applications to exploration. Home: 121 Peterson Rd Philhomath OR 97370 Office: Dept Geology Oreg State U Corvallis OR 97331

FIELD, DAVID ANTHONY, researcher, consultant; b. Brunswick, Maine, Aug. 9, 1943; s. Myron Bradford and Cecile Adeline (Ouellette) F.; m. Maureen Bell, Dec. 27, 1969; children: Rebecca, Brendan, Adrienne. A.B., Bowdoin Coll., 1965; M.S., Oakland U., 1966; Ph.D., U. Colo., 1970. Prof. Holy Cross Coll., Worcester, Mass., 1970-78; scientist Gen. Motors Research Labs., Warren, Mich., 1978—. Ford Found. research fellow, Eng., 1975-76. Mem. Soc. Indsl. and Applied Math., Math. Assn. Am., Am. Math. Soc., Internat. Assn. for Math. and Computers in Simulation, Sigma Xi. Roman Catholic. Subspecialties: Numerical analysis; Applied mathematics. Current work: Computer graphics, mathematical software, numerical analysis, applied mathematics, applied mechanics. Office: Gen Motors Research Labs Tech Center Warren MI 48009

FIELD, FRANK HENRY, chemistry educator; b. Keansburg, N.J., Feb. 27, 1922; s. Frank Aretus and Mary (Fleischmann) F.; m. Elma Louise Randall, June 22, 1944; m. Doris Baugh Bodenheimer, Oct. 18, 1959; m. Carolyn Bryan Wilson, Jan. 8 1977; children: Elaine Bodenhelimer, Jonathan Randall, Christopher Randall. B.S., Duke U., 1943, M.A., 1944, Ph.D., 1948. Instr. to asst. prof. dept. chemistry U. Tex., Austin, 1947-52; research chemist to research assoc., sect. head Humble Oil & Refining Co., Baytown, Tex., 1952-66; group leader, research assoc., sr. research assoc. Esso Research & Engring. Co., Linden, N.J., 1966-70; prof. Rockefeller U., N.Y.C., 1970—. Author: (with Franklin) Electron Impact Phenomena, 1957; contbr. articles to profl. jours. Guggenheim fellow Leeds (Eng.) U., 1963-64. Mem. Am. Chem. Soc., AAAS., Am. Soc. Mass Spectrometry (pres. 1972-74), Phi Beta Kappa, Sigma Xi, Phi Lambda Upsilon. Unitarian. Subspecialty: Mass spectrometry. Home: Apt 12 H 430 E 63d St New York NY 10021 Office: Rockefeller U New York NY 10021

FIELD, RICHARD JEFFREY, physical chemistry educator; b. Attleboro, Mass., Oct. 26, 1941; s. Jeffrey Hazzard and Edna Catherine (Hawkins) F.; m. Judith Ann Lauchaire, Sept. 5, 1966; children: Elijah, Sara. B.S., U. Mass.-Amherst, 1963; M.S., Holy Cross Coll., 1964; Ph.D., U. R.I.-Kingston, 1968. Research assoc. U. Oreg., Eugene, 1968-74, vis. asst. prof., 1970-73; sr. research chemist Carnegie-Mellon U., Pitts., 1974-75; asst. prof. U. Mont., Missoula, 1975-78, assoc. prof., 1978-83, prof. phys. chemistry, 1983—. Editor: Oscillations and Traveling Waves in Chemical Systems, 1984. NSF research grantee, 1978, 81. Mem. Am. Chem. Soc. (chmn. Mont. sect. 1978-80), Sigma Xi. Roman Catholic. Subspecialties: Kinetics; Physical chemistry. Current work: Basic research into the theory and characterization of oscillating chemical reactions. Home: 317 Livingston Ave Missoula MT 59801 Office: U Mont Dept Chemistry Missoula MT 59812

FIELDING, STUART, psychopharmacologist; b. Bronx, N.Y., Oct. 31, 1939; s. Harry and Ethel (Weisberg) Feinblatt; m. Maralyn Lowy, Aug. 26, 1962; children: Kimberly Ellen, Bradford Scott. B.A., Monmouth Coll., 1962; M.S., Howard U., 1964; Ph.D., U. Del., 1968. Lic. clin. psychologist, Md.; cert. Biofeedback Inst. Am. Mgr. psychopharmacology research Ciba-Geigy Corp., Summit, N.J., 1967-75; asst. dir. pharmacology Hoechst-Roussel Pharms., Inc., Somserville, N.J., 1975-76, asso. dir. biol. sci./mgr. pharmacology, 1977—; adj. asso. prof. Fairleigh Dickinson U.; adj. prof. U. RI. Editor: Drug Devel. Research; co-editor: Neuroleptics, 1974, Antidepressants, 1975, Anxiolytics, 1979, New Frontiers in Psychotropic Drug Research, 1979, GABA Neurotransmission, 1980, Psychopharmacology of Clonidine, 1981; contbr. articles to profl. jours. Mem. AAAS, Am. Psychol. Assn., Am. Soc. Pharmacology and Exptl. Therapeutics, Biofeedback Soc. N.J., Eastern Psychol. Assn., Soc. Neurosci., N.Y. Acad. Sci. Subspecialties: Pharmacology; Learning. Current work: Psychopharmacology; aging; cardiovascular disease. Home: 16 Bromleigh Way Morris Plains NJ 07950 Office: Route 202-260 N Somerville NJ 08869

FIELDS, CARL CLARENCE, nuclear engring. educator; b. Detroit, Dec. 30, 1944; s. Ralph and Janette (Golosoff) F.; m. Donna Jean Smith, June 5, 1971. B.S. in Physics, Mich. State U., 1967; M.S., Carnegie-Mellon U., 1973. Sr. scientist Westinghouse-Bettis Atomic Power Lab., West Mifflin, Pa., 1967-82; instr. Pa. State U., Altoona, 1983—. Mem. Am. Nuclear Soc., Am. Soc. for Engring. Edn. Subspecialty: Nuclear engineering. Current work: Fission core design, fuel management. Improving productivity in engineering work. Patentee improved nuclear fuel system. Office: Pa State U 6 Ivy Hall Altoona PA 16603

FIELDS, PATRICK F., mus. preparator, paleobotanist, editor; b. Palo Alto, Calif., May 30, 1954; s. Earl F. and Elizabeth R. (Reay) F. B.S. in Palo-Plan Ecology, U. Calif., Davis, 1977, postrad. in paleontology, 1982—. Sr. mus. preparator paleobotany U. Calif., Berkeley, 1978—; researcher macroscopic plant fossils. Contbr. articles on paleobotany to profl. jours.; editor: Paleontologic Jour, 1981—. Mem. Internat. Assn. for Angiosperm Paleobotany, Bot. Soc. Am., Paleobot. Sect. Bot. Soc. Am., Calif. Bot. Soc., Calif. Native Plant Soc., Calif. Hort. Soc. (Golden Gate Park, San Francisco). Subspecialties: Paleoecology; Paleobotany. Current work: Tertiary megafloral paleobotany and paleoecology, paleobotany, macrofossils, tertiary, paleoecology, curation, biostratigraphy. Home: 2839 Ashby Ave Berkeley CA 94705 Office: Mus Paleontology U Calif Berkeley CA 94720

FIENBERG, STEPHEN ELLIOTT, mathematics educator; b. Toronto, Ont., Can., Nov. 27, 1942; came to U.S., 1964. B.S., U. Toronto, 1964; A.M., Harvard U., 1965, Ph.D., 1968. Asst. prof. dept. stats. and theretical biology U. Chgo., 1968-72; assoc. prof. dept. applied stats. U. Minn., St. Paul, 1972-76, prof., 1976-80, chmn. dept., 1972-78; prof. dept. statistics and social sci. Carnegie-Mellon U., Pitts., 1980—, head dept. stats., 1981—; chmn. com. nat. statistics NRC, 1981—. Author: Analysis of Cross-Classified Categorical Data, 1977, 2d edit., 1980; co-author: (with others) Discrete Multivariate Analysis: Theory and Practice, 1975; editor: (with A. Zellner) Studies in Bayesian Econometrics and Statistics, 1975, (with D.V. Hinkley) R.A. Fisher: An Appreciation, 1980, (with A.J. Reiss Jr.) Indicators of Crime and Criminal Justice: Quantitative Studies, 1980. Fellow AAAS, Am. Statis. Assn., Inst. Statisticians, Inst. Math Stats., Royal Statis. Soc.; mem. Biometric Soc., Council on Social Graphics, Internat. Statis. Inst., Psychometric Soc. Subspecialty: Statistics. Office: Dept Stats Carnegie-Mellon U Pittsburgh PA 15213

FIENE, RICHARD JOHN, psychologist; b. Bklyn., Nov. 11, 1949; s. Richard Ludwig and Vera Ann (Garone) F.; m. Judith Theresa Gibson, July 18, 1970; children: Corinne Leah, Christopher Alan. B.A., SUNY-Stony Brook, 1971, M.A., 1973; Ph.D., Newport U., 1978. Research asst. SUNY-Stony Brook, 1970-73; instr. U. N.C., Greensboro, 1973-75; asst. prof. Guilford Tech. Inst., High Point, N.C., 1974-75; research psychologist Gov.'s Office, Harrisburg, Pa., 1975-77; adj. prof. Pa. State U., 1977-82; dir. info. systems Children and Youth, Harrisburg, 1977—; info. systems cons. Health Edn. and Welfare, 1975-77; research assoc. Millersville State U., 1977-78; panel reviewer Child Care Quar., Mpls., 1978-82. Author: Instrument Based Program Monitoring System, 1981; Editor: In the Best Interest of Children, 1977. Bd. dirs. Mulberry House Delinquents, Harrisburg, 1975-76. Nat. Child Service grantee, 1980; named Boss of the Yr. Am. Bus. Woman's Assn., 1981; N.Y. State Regents scholar, 1971; scholar incentive, 1973. Mem. Pa. Psychol. Assn. (pres. 1980-81), Am. Psychol. Assn., Evaluation Research Soc., Pa. Children's Consortium (chmn. 1977-80). Roman Catholic. Subspecialties: Information systems (information science); Developmental psychology. Current work: Cognitive and information processing theory in young children; compliance benefit theory in information systems science/computer science. Home: 1800 Pineford Dr Middletown PA 17057 Office: Information Systems 1514 N 2d St Harrisburg PA 17102

FIENUP, JAMES RAY, physicist, researcher; b. St. Louis, Apr. 17, 1948; s. Wilbur G. and Helen (Rozanski) F.; m. Patricia Ann, Dec. 19, 1970; children: Jonathan, Matthew, Daniel, David. A.B. in Physics and Math. magna cum laude, Holy Cross Coll., 1970; M.S. in Applied Physics (NSF fellow), Stanford, U., 1972, Ph.D, 1975. Research asst. Stanford Electronics Labs., Stanford U., 1972-75; research physicist Environ. Research Inst. Mich., Ann Arbor, 1975—. Editor 1 book; Assoc. editor: Optics Letters; Contbr. articles to profl. jours. Mem. Optical Soc. Am., Soc. Photo-optical Instrumentation Engrs. (Rudolf Kingslake medal and prize 1979), Sigma Xi, Sigma Pi Sigma, Alpha Sigma Nu. Subspecialties: Optical image processing; Algorithms. Current work: Research in optical and digital information processing; holography; coherent optics. Patentee in field. Home: 3951 Waldenwood Dr Ann Arbor MI 48105 Office: PO Box 8618 Ann Arbor MI 48107

FIERER, JOSHUA A(LLAN), pathology educator, physician; b. N.Y.C., Nov. 25, 1937; s. Norman and Evelyn (Bolstein) F.; m. Mary Ellen Bailey, June 14, 1959; children: Pamela, Robin, Jonathan, Lisa. B.A., Alfred U., 1949; M.D., SUNY-Downstate Med. Ctr., 1963. Diplomate: Am. Bd. Pathology. Intern Strong Meml. Hosp., Rochester, N.Y., 1963-64; resident Columbia-Presbyn. Med. Ctr., N.Y.C., 1967-69; asst. in surgery U. Rochester, N.Y., 1964-65; instr. in pathology Columbia U., N.Y.C., 1969-70, asst. prof. pathology, 1970-75; guest investigator Rockefeller U., N.Y.C., 1969-70; prof., dir. div. anat. pathology Creighton U., Omaha, 1975-78; prof., chmn. dept. pathology U. Ill. Coll. Medicine, Peoria, 1978—, dir. labs., 1978—; dir. microbiology and immunopathology Francis Delafield Hosp., N.Y.C., 1970-75; asst. attending pathologist Presbyn. Hosp., N.Y.C., 1972-75; cons. Ill. Cancer Council, Chgo., 1981—. Contbr. chpts. to books, articles to profl. pubs. Bd. dirs. Ill. div. Am. Cancer Soc., Chgo., 1981—, pres. Peoria unit, 1982—; med. adv. bd. Heart of Ill. region ARC, Peoria, 1981—; bd. dirs. Peoria Civic Opera Co., 1978-81. Served to capt. USAF, 1965-67. Fellow Am. Cancer Soc., 1968, NIH, 1969; recipient Gold award Ednl. Class, Am. Soc. Clin. Pathology/Coll. Am. Pathologists, 1980. Fellow Coll. Am. Pathologists; mem. Am. Assn. Pathologists, Am. Assn. Immunologists, Soc. Exptl. Biology and Medicine, Central Ill. Pathology Soc. (pres. 1981-82), Sigma Xi. Club: Ill. Valley Yacht (Peoria). Subspecialties: Pathology (medicine); Immunology (medicine). Current work: Lung connective tissue, pulmonary emphysema, modulation of immune system, immunopathology, cellular aspects of aging, asbestos related disease, initiation and promotion of cancer. Home: 565 N Minnesota Ave Morton IL 61550 Office: Univ Ill Coll Medicine One Illini Dr PO Box 1649 Peoria IL 61656

FIGLEY, CHARLES RAY, psychology educator, psychotherapist; b. Chgo., Oct. 6, 1944; s. John and Geneva (Bartley) F.; m. Marilyn Gaye Reeves, 1983; 1 step dau., Jessica Reeves Burns. B.S., U. Hawaii, 1970; M.S., Pa. State U., 1971, Ph.D., 1974. Cert. marriage and family therapist; cert. sex therapist; lic. psychologist, Ind. Instr. Bowling Green State U., 1970-71; instr., teaching asst. Pa. State U., 1971-74; asst. prof. Purdue U., 1974-77, assoc. prof., 1977-83, prof., 1983—; dir. Child and Family Research Inst., 1978—; psychotherapist in pvt. practice, West Lafayette, Ind., 1976—, Chgo., 1983—; research cons.; founder, dir. Consortium on Vet. Studies, 1975—; pres. Groves Conf. on Marriage and the Family, 1981-84. Author, editor: Stress Disorders Among Vietnam Veterans, 1978, (with S. Leventman) Strangers at Home: Vietnam Veterans Since the War, 1980, (with H. McCubbin) Stress and the Family, Vols. I and II, 1983; contbr. articles to sci. jours. Bd. dirs. Lafayette ARC, 1980—; bd. dirs., pres. Planned Parenthood, 1975-79. Served with USMC, 1963-67; Vietnam. Recipient Instrnl. Innovation award Purdue Parents Assn., 1974, Sanger Founder award Planned Parenthood, 1979, Tchr. of Yr. award Sigma Delta Chi, Purdue U., 1979, Disting. Scholar award Forgotten Warriors Project, Cleve., 1970. Fellow Am. Assn. Marriage and Family Therapists, Am. Orthopsychiat. Assn., Am. Psychol. Assn. Subspecialties: Social psychology; Psychiatry. Current work: Immediate and long-term effects of highly stressful and noxious situations on psychosocial and emotional parameters in search of effective mitigating intervention methods. Office: Purdue University 525 Russell St West Lafayette IN 47906

FIGLEY, MELVIN MORGAN, physician, educator; b. Toledo, Dec. 5, 1920; s. Karl Dean and Margaret (Morgan) F.; m. Margaret Jane Harris, Mar. 16, 1946; children: Karl Porter, Joseph Dean, Mark Thompson. Student, Dartmouth, 1938-41; M.D. magna cum laude (John Harvard fellow), Harvard, 1944. Diplomate: Am. Bd. Radiology (trustee 1967-72). Intern, resident internal medicine Western Res. U., 1944-46; resident radiology U. Mich., 1948-51, instr., asst. prof., asso. prof. radiology, 1950-58; practice medicine, specializing in radiology, Seattle, 1958—; prof. radiology, chmn. dept. U. Wash., 1958-78, prof. radiology and medicine, 1979—; mem. radiation study sect. NIH, 1963-67; mem. com. on radiology Nat. Acad. Scis.-NRC, 1964-69, chmn., 1968-69. Editor: Am. Jour. Roentgenology, 1976—; Contbr. articles profl. jours. Bd. dirs. James Picker Found., 1970—. Served to capt. M.C. AUS, 1946-48. John and Mary R. Markle scholar, 1952-57. Fellow Am. Coll. Radiology; hon. fellow Royal Coll. Radiologists (London), Royal Australian Coll. Radiologists; hon. mem. Royal Soc. Medicine; mem. Assn. Univ. Radiologists (pres. 1966, Gold medal 1983), Am. Roentgen Ray Soc. (exec. council 1970—), N. Am. Soc.

Cardiac Radiology (pres. 1974), Fleischer Soc., Radiol. Soc. N.Am., AMA, Boylston Med. Soc., Wash. Heart Assn. (past trustee), Soc. Chmn. Acad. Radiology Depts. (exec. council 1969-71), Phi Beta Kappa, Sigma Xi, Alpha Omega Alpha, Sigma Alpha Epsilon. Episcopalian. Home: 7010 51st Ave NE Seattle WA 98115 Office: Univ Hosp Dept Radiology Seattle WA 98105

FIGUERES, MAURICE CHRISTIAN, data processing executive; b. Lyon, France, July 24, 1932; came to U.S., 1961, naturalized, 1968; s. Pierre and Antoinette Jeanne (Garde) F. Student, U. Lyon, France, 1952-53, McGill U., 1958-59; B.A., CUNY, 1971; M.B.A., NYU, 1978. Internal auditor Bankers Trust Co., N.Y.C., 1969-72; sr. EDP auditor Am. Express Co., N.Y.C., 1972-80; EDP auditor Home Life Ins. Co., N.Y.C., 1980-81; EDP audit mgr. Rockefeller Center, Inc., N.Y.C., 1981—. Reviewer: Computing Reviews, 1982. Treas. French Vets. Assn., Algerian War, N.Y.C., 1981—. Mem. IEEE (mem. tech. coms. computer soc. 1978), Assn. Computing Machinery (mem. spl. interest groups 1976—), Soc. Mgmt. Info. Systems, EDP Auditors Assn., Inst. Internal Auditors, N.Y. Acad. Scis. Republican. Roman Catholic. Subspecialties: Software engineering; Cryptography and data security. Current work: Verification and validation, simulation, data communication security. Patentee in field. Office: Rockefeller Center Inc 1230 Ave of the Americas New York NY 10020

FIGURSKI, DAVID HENRY, microbiologist, educator; b. Erie, Pa., Apr. 12, 1947; s. Henry J. and Lydia P. (Rodgers) F.; m. Donna Marie O'Donnell, Aug. 9, 1969; children: Kiersten Ann, Jared David. B.S., U. Pitts., 1969; Ph.D., U. Rochester, 1974. USPHS fellow biology dept. U. Calif.-San Diego, 1974-78; asst. prof. microbiology Coll. of Physicians and Surgeons, Columbia U., N.Y.C., 1978—. NSF grantee, 1978-80; NIH grantee, 1979—. Mem. Am. Soc. Microbiology. Subspecialties: Molecular biology; Gene actions. Current work: Molecular basis of plasmid replication control and host range, gene regulation and expression, molecular genetics, plasmids, recombinant DNA. Office: Microbiology Dept Coll Physicians and Surgeons Columbia U 701 W 168th St New York NY 10032

FILBY, ROYSTON HERBERT, chemistry educator; b. London, Feb. 16, 1934; m., 1965; 2 children. B.Sc., U. London, 1955; M.Sc., McMaster U., 1957; Ph.D. in Chemistry, Wash. State U., 1971. Research fellow in geochemistry U. Oslo, 1961-64; head dept. chemistry U. El Salvador, 1964-67; chemist Wash. State U., 1967-70, asst. prof. chemistry, 1970-71, assoc. prof., 1971-74, prof., 1974—; asst. dir. Nuclear Radiation Ctr., 1970-74, assoc. dir., 1974-76, dir., 1976—; Orgn. European Econ. Cooperation sr. vis. fellow European Atomic Energy Comm., Mol, Belgium, 1962; guest worker Nat. Bur. Standards, Washington, 1975-76. Fellow Am. Inst. Chemistry; mem. AAAS, Geochem. Soc., Am. Chem. Soc., Geol. Soc. Finland. Subspecialties: Geochemistry; Nuclear chemistry. Office: Nuclear Radiation Ctr Wash State U Pullman WA 99163

FILIPESCU, NICOLAE, chemistry educator, physician, researcher; b. Predeal, Romania, July 30, 1935; came to U.S., 1960, naturalized, 1965; s. Nicolae and Lucretia F.; m. Louise Thelma Filipescu, Sept. 14, 1963; children: Sanda, Patricia, Laura Caroline, Christina Louise. M.S. in Chem. Engring, Poly. Inst., Bucharest, 1957; Ph.D. in Chemistry, George Washington U., 1964, M.D., 1975. Diplomate: Am. Bd. Ob-Gyn. Resident in ob-gyn George Washington U. Hosp., 1975-78; sr. scientist Melpar Inc., 1960-63; prof. chemistry, ob-gyn George Washington U., Washington, 1963—; cons. NASA. Contbr. articles in field to profl. jours. Recipient Hillebrand award of Chem. Soc. Washington, 1971. Mem. AMA, Washington Acad. Sci., N.Y. Acad. Sci., Am. Chem. Soc., Am. Fertility Soc., Am. soc. Gynecology Laparoscopists, Chem. Soc. London. Subspecialties: Organic chemistry; Obstetrics and gynecology. Current work: Organic photochemistry, endocrinology, reproduction.

FINCH, CLEMENT A., physician; b. Broadalbin, N.Y., July 4, 1915; s. Percy Henry and Marion Elizabeth F.; m. Eugenia C. English, 1966; children—Clifton A., Carin A., Lisa, Derel. B.A., Union Coll., Schenectady, 1936; M.D., U. Rochester, 1941. Diplomate: Am. Bd. Internal Medicine. Fellow in pathology U. Rochester, 1938-39; med. intern Peter Bent Brigham Hosp., Boston, 1941-42, asst. resident in medicine, 1942-43, resident in medicine, 1944-46, jr. asso. in medicine, 1946-48; fellow in hematology Boston U., 1943-44; instr. medicine Harvard U., Boston, 1946-48, asso. in medicine, 1948-49; asso. prof. medicine U. Wash., 1949-55, prof., 1955—, head div. hematology, 1949—; attending physician Univ. Hosp., Seattle; cons. in field; mem. coms. Nat. Acad. Scis., NRC.; mem. adv. com. Nat. ARC; mem. adv. com. for biology and medicine AEC; mem. study sect. and council USPHS; chmn. blood diseases and resources adv. com. NIH. Contbr. numerous articles to profl. jours.; editorial positions: Haematologia. Recipient Goldberger award AMA, 1973, Wahle award in hematology SUNY, Buffalo, 1973. Mem. A.C.P., Am. Fedn. Clin. Research, Am. Soc. Clin. Investigation (pres. 1961), Am. Soc. Clin. Nutrition, Am. Physiol. Soc., Am. Soc. Hematology (pres. 1966, adv. com.), Assn. Am. Physicians, Internat. Soc. Hematology, King County Med. Soc., Nat. Acad. Arts and Scis., Nat. Acad. Scis., Soc. Exptl. Biology and Medicine, Societas Haematologica Helvetica (hon.), Wash. State Med. Assn., Western Assn. Physicians (pres. 1960-61), Western Soc. Clin. Research (Mayo Soley award 1972), Sigma Xi, Alpha Omega Alpha. Subspecialty: Internal medicine. Current work: Iron Metabolism. Office: U Wash Med Sch Seattle WA 98105

FINCH, STUART CECIL, physician; b. Broadalbin, N.Y., Aug. 6, 1921; s. Cecil Clement and Olga Ulrika (Lofgren) F.; m. Patricia O'Brien, June 15, 1946; children: James, Ellen, Sheldon, Polly. Student, Dartmouth Coll., 1941; M.D., U. Rochester, 1944. Intern in surgery, resident in pathology Balt. City Hosps., 1944-46; fellow in hematology, resident in medicine Peter Bent Brigham Hosp., Boston, 1948-50; fellow in hematology Mass. Meml. Hosp., Boston, 1950-53; asst. prof. to prof. Yale U., 1953-77; chief research, permanent dir. Radiation Effects Research Found., Hiroshima, Japan, 1975-79; prof. medicine Rutgers Med. Sch., 1979—; chief medicine Cooper Hosp./ Unit. Med. Center, 1979—. Contbr. articles to profl. jours. Served to capt. AUS, 1946-48. Recipient Humanitarian award Nat. Hemophilia Found. Mem. Am. Soc. Clin. Investigation, Am. Assn. Physicians, AMA, Interurban Clin. Club. Subspecialties: Internal medicine; Hematology. Current work: Radiation effects, agranulocytosis, lymphocyte migration. Office: One Cooper Plaza Camden NJ 08103

FINCK, ELMER JOHN, behavioral ecologist, ornithologist, mammalogist; b. Mandan, N.D., June 23, 1947; s. Glenn and Monica (Boehm) F.; m. LaVonne Catherine Altenburg, Nov. 9, 1968; children: Eric, Aaron. A.S., Coll. of Lake County, 1972; B.S., U.N.D., 1974; M.S. (HEW fellow), 1979; Ph.D., Kans. State U., 1982. Research asst. Kans. State U., Konza Prairie Research Natural Area, Manhattan, 1981—. Served with USN, 1965-68. Mem. Ecology Soc. Am., Am. Ornithology Union, Cooper's Ornithology Soc., Wilson Ornithol. Soc., Am. Soc. Mammalogists, Animal Behavioral Soc. Roman Catholic. Subspecialty: Behavioral ecology. Current work: Behavioral ecology of birds and mammals of the Great Plains, behavioral ecology of Dickcissel and Neotoma floridana.

FINDLAY, JOHN WILLIAM ADDISON, biochemist; b. Buckie, Banffshire, Scotland, Mar. 27, 1945; s. Alexander and Jessie Ross (Addison) F.; m. Jean Marjorie Hey, July 11, 1970; children: Kathleen Moira, Fiona Jean. B.Sc., U. Aberdeen, 1966, Ph.D., 1969. Research assoc. pharm. chemistry U. Va., Charlottesville, 1969-72; research biochemist dept. pathology Norfolk Gen. Hosp., Norfolk, 1972-73; research scientist Radiochem. Centre, Amersham, Eng., 1973-75; sr. research scientist Wellcome Research Labs., Research Triangle Park, N.C., 1975-81, group leader, 1981—; cons. Tobacco Research Council, 1979. Contbr. articles to profl. jours. Mem. Am. Soc. Pharmacology and Exptl. Therapeutics. Presbyterian. Subspecialties: Pharmacokinetics; Pharmacology. Current work: Development of antisera and immunoassays: application of these to studies of drug disposition and pharmacology in animals and man. Office: Burroughs Wellcome Co 3030 Cornwallis Rd Research Triangle Park NC 27709

FINE, BERNARD J., psychologist; b. Hartford, Conn., Nov. 2, 1926; s. William and Ida (Kase) F.; m. Eleanor Walker, Apr. 6, 1957; children: Rachel Walker Fine, Jonathan Walker Fine. B.S., Am. Internat. Coll., 1951; M.S., Boston U., 1953, M.A., 1954, Ph.D., 1956. Lic. psychologist, Mass. Study dir. Boston U., 1951-56; research psychologist U.S. Army QM Research & Engring. Ctr., Natick, Mass., 1956-59, chief psychophysiology, 1959-61, research psychologist, 1961—; prof. Mass. Bay Community Coll., Watertown, 1970-71, 76, Mt. Wachusetts Community Coll, Gardner, Mass., 1973-75, Boston State Coll., 1972-73. Trustee Harvard Pub. Library, 1980—; pres. Harvard Unitarian Ch., 1979-81; bd. dirs. Concerts-at-the Common, 1975—. Served with U.S. Army, 1945-46. Fellow AAAS, Am. Psychol. Assn., Inter-Univ. Seminar on Armed Forced and Soc.; mem. Fedn. Am. Scientists, Union Concerned Scientists, Sigma Xi, Alpha Chi. Subspecialty: Individual differences and environmental stress. Current work: Basic and applied research on individual differences in cognitive, perceputal and sensory function as affected by heat, cold, altitude and sustained physical and mental work. Home: Woodside Rd Harvard MA 01451 Office: U S Army Research Inst Environ Medicine Natick MA 01760

FINE, DONALD LEE, microbiologist, research administrator; b. Nanticoke, Pa., Jan. 14, 1943; s. Donald Harmon and Doris (Searles) F.; m. Judith Ann, June 12, 1965; children: Dawn Ann, Heather, Matthew. Ph.D. in Microbiology, Pa. State U., 1968. Chief dept. virology and cell biolgoy Frederick (Md.) Cancer Research Ctr., 1970-72, head devel. research sect., 1972-76, head immunobiology of type B retroviruses sect., 1976-78; mgr. NIH intramural research support program, 1978—; cons. in field. Author: Biological Markers of Neoplasia, 1978, Breast Cancer: New Concepts in Etiology and Control, 1980, others; contbr. articles to profl. publs. Recipient Outstanding Achievement award Dept. Army, 1969. Mem. Am. Soc. Microbiology, N.Y. Acad. Sci., AAAS, Am. Assn. Cancer Research, Soc. Exptl. Cell Biology and Medicine, Tissue Culture Assn., Internat. Leukemia Soc., Sigma Xi (chpt. program chmn. 1981), Iota Mu Pi. Subspecialties: Cancer research (medicine); Microbiology (medicine). Office: Frederick Cancer Research Facility Box B Frederick MD 21701

FINE, MICHAEL LAWRENCE, biologist, educator; b. Bklyn., Feb. 10, 1946; s. Herbert A. and Helen R. F.; m. Judith R. Slavsky, Aug. 7, 1977; 1 son, Benjamin A. B.S., U. Md., 1967; M.A., Coll. William and Mary, 1970; Ph.D., U. R.I., 1976. Oceanographer U.S. Naval Oceanographic Office, summers 1966-67; postdoctoral assoc. Cornell U., Ithaca, N.Y., 1976-79; physiologist Tunison Lab. Fish Nutrition, Fish and Wildlife Service, Cortland, N.Y., summer 1979; asst. prof. biology Va. Commonwealth U., Richmond, 1979—. Contbr. articles to profl. jours. Mem. Soc. Neurosci., AAAS, Animal Behavior Soc., Am. Soc. Ichthyologists and Herpetologists. Jewish. Subspecialties: Ethology; Comparative neurobiology. Current work: Acoustic communication in fishes, toadfish, neural substrates of behavior, sexual dimorphism of brain, hormones and behavior, auditory neurophysiology. Home: 9419 Broad Meadows Rd Glen Allen VA 23060 Office: Dept Biology Va Commonwealth U Richmond VA 23284

FINE, MORRIS EUGENE, materials engineer, educator; b. Jamestown, N.D., Apr. 12, 1918; s. Louis and Sophie (Berrington) F.; m. Mildred Eleanor Glazer, Aug. 13, 1950; children: Susan Elaine, Amy Lynn. B.Metall. Engring. with distinction, U. Minn., 1940, M.S., 1942, Ph.D., 1943. Instr. U. Minn., 1942-46; mem. tech. staff Bell Telephone Labs., Murray Hill, N.J., 1946-54; prof., chmn. dept. metallurgy Tech. Inst., Northwestern U., Evanston, Ill., 1955-57, chmn. dept. materials sci., 1958-60, prof. and chmn. materials research center, 1960-64, Walter P. Murphy prof. materials sci., 1963—, asso. dean grad. studies and research, 1973—; vis. prof. dept. materials sci. Stanford U., 1967-68; JSPS vis. scholar, Japan, 1979; asso. engr. Manhattan Project, U. Chgo., also, Los Alamos, World War II; mem. materials adv. bd. Nat. Acad. Sci., 1963-68; mem. com. geol. and materials scis. NRC, 1979-82; co-chmn. Engring. Socs. Conf. on Fatigue Crack Initiation, 1980; chmn. adv. bd., program on modular methods for teaching materials Pa. State U., 1973-77; chmn. vis. com. metallurgy and materials Sci. and Materials Research Center, Lehigh U., 1965-75; vis. com. Lawrence Berkeley Lab., 1978-81, chmn., 1981, vis. com. Ames Dept. Energy Lab., 1976-80. Author numerous tech. and sci. articles on mech. properties of metals and ceramics, fatigue of metals, phase transformations and other subjects.; author: Introduction to Phase Transformation in Condensed Systems. Named Chicagoan of Year in Sci., 1961. Fellow Am. Phys. Soc., AAAS, Am. Soc. Metals (chpt. chmn. 1963, Campbell lectr. 1979), Metall. Soc. of AIME (chmn. inst. metals div. 1966-68, dir. 1968-71, dir. inst. 1972-75, Mathewson gold medal for research 1981, James Douglas Gold medal), Am. Ceramic Soc. (keynote lectr. electronic materials div. 1972); mem. Nat. Acad. Engring. (mem. astronautics space engring. bd. 1973-77, membership com. 1974-79, chmn. 1977-78), Am. Assn. Engring Edn., ASTM, AAUP, Fedn. Am. Scientists, Engring. Com. for Profl. Devel. (accreditation panel for metallurgy and materials), Sigma Xi, Tau Beta Pi, Alpha Sigma Mu, Sigma Alpha Sigma. Subspecialty: Metallurgy. Home: 1101 Manor Dr Wilmette IL 60091 Office: Northwestern U Evanston IL 60201

FINGER, DENNIS ROBERT, psychologist; b. Bklyn., May 10, 1949; s. Irving I. and June (Rose) F. B.A., CUNY, 1971; M.S., L.I.U., 1974; M.S.Ed., Pace U., 1979; postgrad., Rutgers U., 1979—. Cert. sch. psychologist, N.Y., N.J.; cert. tchr., N.Y.; cert. guidance counselor, N.J. Tchr. spl. edn. N.Y.C. Bd. Edn., Bklyn., 1972-79; coordinator emotionally handicapped program Canarsie High Sch., Bklyn., 1976-79; trainer/cons. Mitchell & Assocs., Caldwell, N.J., 1976—; adj. asst. prof. psychology Pace U., N.Y.C., 1976—; psychologist Jersey City Bd. Edn., 1980—; research cons. Ontologic Edn. Inc., Monmouth, N.J., 1981—. Pace U. faculty scholar, 1979. Mem. Am. Psychol. Assn., Div. 16 Psychol. Assn., N.J. Psychol. Assn., Nat. Assn. Sch. Psychologists, N.J. Assn. Sch. Psychologists, NEA, Psi Chi. Subspecialty: Psychology. Current work: Measurement of emotional expressiveness (in men and women); in clinical applications of a measure of emotional expressiveness. Office: Jersey Bd Edn 241 Erie St Jersey City NJ 07302

FINGERMAN, MILTON, biologist, educator, academic administrator; b. Boston, May 21, 1928; m. (married); 2 children. B.S., Boston Coll., 1948; M.S., Northwestern U., 1949, Ph.D. in Biology, 1952. Asst. Northwestern U., 1949-51; from instr. to assoc. prof. zoology Tulane U., 1954-63, prof. biology, 1963—; chmn. dept. biology; mem. adv. panel regulatory biology NSF, 1966-69, 80; mem. supply dept. com. Marine Biol. Lab., Woods Hole, 1970-73, chmn. com., 1971-73; mem. com. animal models biomed. research invertebrate Inst., Lab. Animal Resources, NRC, 1972-73; mem. com. marine invertebrates Gulf Univ. Research Consortium, 1976-81, mem. environ. sci. program planning council, 1977—78. Mem. editorial bd.: Physiol. Zoology, 1976—; mem. editorial bd.: Jour. Crustacean Biology, 1980—; mng. editor: Am. Zoologist, 1981—. Mem. AAAS, Am. Inst. Biol. Sci., Internat. Soc. Chronobiology, Am. Soc. Zoologists (chmn. nominating com. Div. Comparative Endocrinology 1972, program officer 1977-78); mem. Crustacean Soc. Subspecialty: Physiology (biology). Office: Dept Biology Tulane U New Orleans LA 70118

FINK, ANTHONY LAWRENCE, biochemist, educator; b. Hertford, Eng., Jan. 25, 1943; came to U.S., 1968, naturalized, 1971; s. Lawrence and Mimi (Wagner) F.; m. Shirley Ann Mullen, Aug. 27, 1966; 1 dau., Christa. B.Sc., Queen's U., Kingston, Ont., Can., 1964, Ph.D., 1968. NRC Can. postdoctoral fellow Northwestern U., 1968-69; asst. prof. chemistry U. Calif., Santa Cruz, 1969-75, asso. prof., 1976-82, prof., 1982—; vis. fellow All Souls Coll., Oxford (Eng.) U., 1981. NIH grantee, 1972—; NSF grantee, 1972—. Mem. AAAS, Am. Chem. Soc., Am. Soc. Biol. Chemists, Biochem. Soc. Subspecialties: Biochemistry (biology); Enzyme technology. Current work: Mechanisms of enzyme action, protein folding, cryoenzymology, beta-lactamases, enzyme immobilization. Office: Natural Scis II U Calif Santa Cruz CA 95064

FINK, GERALD RALPH, genetics educator; b. Bklyn., July 1, 1940; s. Benjamin and Rebecca F.; m. Rosalie P. Lewis, June 15, 1961; children: Julia, Jennifer. B.A., Amherst Coll., 1962; M.S., Yale U., 1964, Ph.D., 1965; D.Sc. (hon.), Amherst Coll., 1982. Asst. prof. genetics Cornell U., Ithaca, N.Y., 1967-71, asso. prof., 1971-76, prof., 1976-79, prof. biochemistry, 1979-82, Am. Cancer Soc. Life prof., 1981; prof. genetics Whitehead Inst. MIT, Cambridge, 1982—. Asso. editor: Genetics, 1970-74, Jour. Bacteriology, 1973-78, Molecular and Gen. Genetics, 1980—; editor: Gene, 1978—. Chmn. Am. Cancer Soc., 1976-77. Recipient U.S. Steel award in molecular biology Nat. Acad. Scis., 1981; Guggenheim Found. fellow, 1974-75. Mem. Genetics Soc. Am. (award 1982), Am. Soc. Microbiologists. Subspecialty: Genetics and genetic engineering (biology). Office: Dept Biology MIT Cambridge MA 02139

FINK, JOANNE KRUPEY, physicist, chemical engineer; b. Greensburg, Pa., Feb. 19, 1945; d. Wasyl and Mildred (Zemlansky) Krupey; m. Charles Lloyd Fink, July 15, 1967. B.S. summa cum laude in Physics, U. Pitts., 1966, Ph.D. in Physics, 1972. Presdl. internship fellow Argonne (Ill.) Nat. Lab., 1972-73, research assoc., 1974-75, asst. physicist, 1975-79, chem. engr., 1979—; vis. scientist Los Alamos Sci. Lab., 1973-74, postdoctoral fellow, 1974; prof. physics U. N.Mex., Los Alamos, 1974. Contbr. articles and tech. reports to profl. jours. Mem. Am. Phys. Soc., Am. Nuclear Soc., Phi Beta Kappa. Roman Catholic. Subspecialties: High-temperature materials; Numerical analysis. Current work: Thermophysical properties of materials, numerical analysis of data, computer modeling, heat transfer, reactor safety research. Home: 2022 Coach Dr Naperville IL 60565 Office: Argonne Nat Lab 9700 S Cass Ave Argonne IL 0439

FINK, LOUIS MAIER, pathologist, educator; b. Bklyn., Mar. 28, 1942; s. Jacob and Pauline (Kurtz) F. B.A., Boston U., 1961; M.D., Albany Med. Coll., 1965. Diplomate: Am. Bd. Pathology. Intern U. Colo. Med. Center, 1965-66, resident, 1966-67; resident in pathology Coll. Physicians and Surgeons, Columbia-Presbyn. Hosp., Columbia U., N.Y.C., 1969-70, asst. prof., 1970, U. Colo. Med. Sch., 1972-77, assoc. prof., 1977-1981; prof. pathology Vanderbilt U., Nashville, 1981—. Mem. AAAS, Am. Assn. Cell Biology, Am. Assn. Pathologists, Coll. Am. Pathologists, Harvey Soc., Internat. Acad. Pathologists. Subspecialties: Pathology (medicine); Oncology. Current work: Membrane structure. Office: Dept Pathology U Colo Med Sch 4200 E 9th Ave Denver CO 80262

FINK, MARTIN RONALD, aerodynamicist, aeroacoustics researcher; b. N.Y.C., Apr. 27, 1931; s. David Peter and Etta Alice (Checker) F.; m. Jacqueline Fay Klein, Aug. 24, 1952; children: Howard Jeffrey, Andrew Charles, Douglas Reuben. B.S. in Aero. Engring, MIT, 1952, M.S., 1953. Research asst. MIT, Cambridge, 1952-53; research engr. United Aircraft Research Labs., East Hartford, Conn., 1953-58, supr. missile aeros., 1959-63, supr. aerodynamics, 1964-67, sr. cons. engr. aerodynamics, 1967-80; chief aerodynamics Norden Systems, Norwalk, Conn., 1980—. Assoc. fellow AIAA; mem. Sigma Xi (pres. Hartford chpt. 1980-81), Tau Beta Pi (pres. Cen. Conn. chpt. 1972-73). Jewish. Subspecialties: Aeronautical engineering; Acoustical engineering. Current work: Development of analytical methods for predicting aerodynamic pressures, forces and moments on missiles and projectiles. Home: 183 Woody Ln Fairfield CT 06430 Office: Norden Systems PO Box 5300 Norwalk CT 06856

FINK, MARY ALEXANDER, health scientist adminstr.; b. Camden, Tenn., Oct. 18, 1919; d. Mitchell Trotter and Alice (Blackwell) Alexander; m. Charles Dennis Fink, Dec. 2, 1950. B.S., Okla. State U., 1939; M.S., U. Mich., 1945; Ph.D., George Washington U., 1949. Research assoc. Jackson Lab., Bar Harbor, Maine, 1949-51; mem. facuty U. Colo. Med. Sch., 1951-58; research microbiologist Nat. Cancer Inst., NIH, 1959-70, health scientist adminstr., 1970—. Contbr. articles to profl. jours. Mem. Am. Assn. Cancer Research, Am. Assn. Immunologists, Nat. Acad. Microbiology, Soc. Exptl. Biology and Medicine, Brit. Soc. Immunology., Sigma Xi. Subspecialties: Oncology; Immunology (medicine). Current work: Health science grants administration. Home: 9414 Locust Hill Rd Bethesda MD 20814 Office: Westwood 425 NIH Bethesda MD 20205

FINKE, RONALD ALAN, psychology educator; b. Columbus, Ohio, June 10, 1950; s. Phillipp William and Sally (Watson) F. B.S., U. Tex.-Austin, 1972, B.A., 1974; M.A., U. N.H.-Durham, 1976; Ph.D., M.I.T., 1979. Inst. staff scholar M.I.T., Cambridge, 1976-77, Spencer Found. predoctoral fellow, 1978-79; NSF postdoctoral fellow Cornell U., Ithaca, N.Y., 1979-80; fellow in cognitive sci. Stanford U., Palo Alto, Calif., 1980-81; asst. prof. U. Calif., Davis, 1981-83, SUNY, Stony Brook, 1983—. Contbr. articles in field to profl. jours. Mem. AAAS, Am. Psychol. Assn., Psychonomic Soc., Western Psychol. Assn. Subspecialties: Cognition; Psychophysics. Current work: Nature of mental imagery, visual attention, visual-motor processes, neural mechanisms in vision, spatial representation. Home: 27 Atterbury Dr Smithtown NY 11787 Office: Dept Psychology SUNY Stony Brook NY 11794

FINKEL, MADELON LUBIN, public health educator; b. Mt. Vernon, N.Y., Oct. 11, 1949; d. Ralph H. and Lorraine (Alper) Lubin; m. David J. Finkel, Mar. 25, 1973; 1 dau., Rebecca Anne. B.A., NYU, 1971, M.P.A., 1973, Ph.D., 1980. Research asst. NYU, N.Y.C., 1972-74; staff assoc. Columbia U. Sch. Pub. Health, N.Y.C., 1974-75; cons. N.Y.C. Dept. Health, 1975-77; asst. prof. pub. health Cornell U. Med. Coll., N.Y.C., 1977—; cons. Amalgamated Meat Cutters, Queens, N.Y., 1981—, Mobil Oil Corp., N.Y.C., 1980-81; commentator on health Nat. Pub. Radio, 1982. Author: Fundamentals of Second Opinion Programs, 1980, Second Opinion Elective Surgery, 1981. Mem. Am. Fedn. for Clin. Research, Am. Pub. Health Assn., Internat. Epidemiol. Assn., Assn. for Health Services Research, Population Assn. Am. Subspecialties: Health services research; Epidemiology.

Current work: Adolescent sexual behavior; medical care cost containment; health services research—second opinion surgery programs. Office: Cornell U Med Coll 411 E 69th S New York NY 10021

FINKELSTEIN, JAMES DAVID, internist, researcher, educator; b. N.Y.C., Oct. 16, 1933; s. Harry and Sylvia (Bernstein) F.; m. Barbara Joan Eisenberg, Dec. 12, 1959; children: Donna I., Laura H. A.B. summa cum laude, Harvard Coll., 1954; M.D., Columbia U., 1958. Diplomate: Am. Bd. Internal Medicine. Intern Presbyn. Hosp., N.Y.C., 1958-59; resident in medicine Presbyterian Hosp., N.Y.C., 1959-61; trainee in gastroenterology Columbia U., 1961-63; clin. investigator VA Med. Center, Washington, 1965-68, chief biochemistry research, 1965—, chief gastroenterology, 1968-79, med. investigator, 1970-75, assoc. chief of staff for research and devel., 1975-79, chief med. service, 1979—; asst. prof. to prof. medicine George Washington U., from 1974; cons. Children's Hosp., Washington, 1968—; mem. nutrition study sect. NIH, 1972-78; clin. prof. Georgetown U., 1981—. Served to surgeon USPHS, 1963-65. Recipient Arthur S. Flemming award D.C.C. of C., 1970. Mem. Am. Soc. Clin. Investigation, Am. Inst. Nutrition, Am. Soc. Clin. Nutrition, Am. Gastroent. Assn., Am. Assn. Study Liver Diseases, Phi Beta Kappa, Alpha Omega Alpha. Jewish. Club: Harvard (Washington). Subspecialties: Internal medicine; Biochemistry (medicine). Current work: Metabolism of methionine and other sulfur amino acids in mammals; teaching of internal medicine and gastroenterology (nutrition). Office: VA Med Ctr 50 Irving St NW Washington DC 20422

FINKELSTEIN, RICHARD ALAN, microbiologist; b. N.Y.C., Mar. 5, 1930; s. Frank and Sylvia (Lemkin) F.; m. Helen Rosenberg, Nov. 30, 1952; children: Sheri, Mark, Laurie; m. Mary Boesman, June 20, 1976; 1 dau., Sarina Nicole. B.S., U. Okla., 1950; M.A., U. Tex., Austin, 1952, Ph.D., 1955. Teaching fellow, research scientist U. Tex., Austin, 1950-55; fellow, instr. U. Tex. Southwestern Med. Sch., Dallas, 1955-58; chief bioassay sect. Walter Reed Army Inst. Research, Washington, 1958-64; dep. chief, chief dept. bacteriology and mycology U.S. Army Med. Component, SEATO Med. Research Lab., Bangkok, Thailand, 1964-67; asso. prof. dept. microbiology U. Tex. Southwestern Med. Sch., Dallas, 1967-73, prof., 1973-79; prof., chmn. dept. microbiology Sch. Medicine U. Mo.-Columbia, 1979—; mem. Nat. Com. for Coordination of Cholera Research, Ministry of Public Health, Bangkok, 1965-67; cons. WHO, 1970—; cons. to comdg. gen. U.S. Army Med. Research and Devel. Command, 1975-79; cons. Schwarz-Mann Labs., 1974-79; vis. asso. prof. U. Med. Scis., Bangkok, 1965-67; vis. prof. U. Chgo. Med. Sch., 1977; vis. scientist Japanese Sci. Council, 1976; Ciba-Geigy lectr. Waksman Inst., Rutgers U., 1975. Contbr. articles on cholera, enterotoxins, gonorrhea, role of iron in host parasite interactions to profl. jours. Recipient Robert Koch prize, Bonn, W. Ger., 1976. Fellow Am. Acad. Microbiology, Infectious Diseases Soc. Am.; mem. Am. Soc. Microbiology (div. councilor, chmn. program com. 1979-82, pres. Tex. br. 1974-75), Am. Assn. Immunologists, Soc. Gen. Microbiology, Pathol. Soc. Gt. Britain and Ireland. Subspecialties: Microbiology; Immunobiology and immunology. Current work: Pathogenesis and immunology of cholera and related diarrheal diseases; enterotoxins; critical role of iron in host-parasite interactions. Home: 3207 Honeysuckle Dr Columbia MO 65201 Office: Dept Microbiology Sch Medicine Univ of Mo-Columbia Columbia MO 65212

FINKIN, EUGENE FELIX, mech. engr.; b. N.Y.C., Dec. 24, 1940; s. Michael C. and Dorothy (Korman) F.; m. Lillian Weiss, Apr. 19, 1970; children: Nathaniel, Suzanne. B.S., M.I.T., 1962; M.S., Rensselaer Poly. Inst., 1962, Ph.D., 1966. Tech. dir. D.A.B. Industries, Inc., Troy, Mich., 1972-74; v.p., dir. research and devel. Amsted Industries, Geneva, Ill, 1974-77; v.p. Devco, Allegheny Internat. Inc., Pitts., 1977-81; sr. cons. Westinghouse Electric Corp., Pitts., 1981—. Contbr. articles to profl. jours. Mem. ASME, Soc. Automotive Engrs., ASTM, Licensing Execs. Soc., N.Am. Soc. Corp. Planning, Am. Soc. Lubrication Engrs. Subspecialties: Mechanical engineering; Materials (engineering). Current work: Thin films, coatings, amorphous metals, powder metallurgy, metal working, productivity, technology management, strategic planning, ventures, transportation, aerospace, automotive, steel and metals. Patentee in field. Home: 409 Gaywood Circle Pittsburgh PA 15241 Office: Westinghouse Bldg Gateway Ctr Pittsburgh PA 15222

FINKLE, BERNARD J., plant biochemist, cryobiologist; b. Chgo., Mar. 17, 1921; s. Nathan Ephraim and Lena F.; m. Evelyn Cohen, Dec. 9, 1944; children: Wayne Nathaniel, Claudia. B.S., U. Chgo., 1942; Ph.D., UCLA, 1950. Research assoc. U. Calif., Berkeley, 1951-53; lab. dir. Atomic Research Lab., Los Angeles, 1953-54; research instr. U. Utah, Salt Lake City, 1954-57; research biochemist Western Regional Research Center, Berkeley, 1957—; adj. assoc. prof. food sci. U. Calif., Berkeley, 1970-72; cons. cryobiology, plant and food enzymology, Berkeley, 1983—; lectr. Symposium Academia Sinica, Peking, China, 1982. Editor: Phenolic Compounds and Metabolic Regulation, 1967; contbr. chpts. to books. Nat. Cancer Inst. fellow UCLA, 1947-49; AEC fellow U. Cambridge, Eng., 1949-50; U.S.-Japan Coop. Sci. Found. grantee U. Kyoto, 1966-67; Japan Sci. Promotion grantee U. Hokkaido, 1974. Mem. Am. Soc. Plant Physiologists, Soc. Cryobiology, Internat. Assn. Plant Tissue Culture, Am. Soc. Biol. Chemists, Phytochem. Soc. N.Am. (v.p. 1964-65, pres. 1965-66), Orgn. Profl. Employees Dept. Agr. (chpt. pres. 1968-69, Nat. Honor Award 1970), Nat. Council Gene Resources (mem. adv. bd. 1980—), Internat. Bd. Plant Genetics Resources (mem. working group on tissue culture 1982—). Subspecialties: Plant physiology (agriculture); Resource conservation. Current work: Plant tissue culture and biochemistry; cryobiology; freezing damage and the actions of cryoprotective agents; enzyme technology. Patentee treatment to prevent browning of plant tissue, 1964; prevention of freeze damage in food, 1973. Home: 21 Kingston Rd Berkeley CA 94707 Office: Western Regional Research Center US Dept Agr 800 Buchanan St Berkeley CA 94710

FINKS, ROBERT MELVIN, paleontology educator; b. Portland, Maine, May 12, 1927; s. Abraham Joseph and Sarah (Bendette) F. B.S. in Biology magna cum laude, Queens Coll., 1947; M.A. in Geology, Columbia U., 1954, Ph.D., 1959. Lectr. geology Blyn. Coll., 1955-58, instr., 1959-61; lectr. geology Queens Coll., Flushing, N.Y., 1961-62, asst. prof., 1962-65, assoc. prof., 1966-70, prof. earth and environ. scis., 1971—, acting chmn. dept. geology and geography, 1963-64; geologist U.S. Geol. Survey, Washington, 1952-54, 63—; research assoc. Smithsonian Inst., Washington, 1968—; doctoral faculty CUNY; research assoc. Am. Mus. Natural History, N.Y.C., 1961-77. Author: Late Paleozoic Sponge Faunas of the Texas Region, 1960; editor: Guidebook to Field Excursions, 1968; contbr. articles to profl. jours. Queens Coll. scholar, 1947. Fellow AAAS, Geol. Soc. Am., Explorers Club; mem. Paleontol. Soc. (chmn. N.E. sect. 1977-78), Internat. Paleontol. Assn., Paleontol. Assn. Great Britain, Planetary Soc., Phi Beta Kappa, Sigma Xi (exec. sec. chpt. 1982—). Subspecialties: Paleobiology; Paleoecology. Current work: Fossil sponges, taxonomy and evolution; paleoecology of reefs; life-span and growth studies of fossil populations; natural selection in fossil populations. Home: 166-40 Powells Cove Blvd Beechhurst NY 11367

FINLAND, MAXWELL, physician; b. Russia, Mar. 15, 1902; came to U.S., 1906, naturalized, 1925; s. Frank and Rebecca (Povza) F. Ed., Wendell Phillips Sch., Boston, English High Sch.; B.S., Harvard U., 1922, M.D., 1926, D.Sc. (hon.), 1982, Western Res. U., 1964, D.H.L., Thomas Jefferson U., 1978. Asst. resident Boston Sanatorium, 1926-27; med. house officer Boston City Hosp., 1927-28, resident physician for pneumonia and med. service, 1928-29, became jr. vis. physician, 1938; chief IV Harvard) Med. Service, 1939-62, dir. II and IV, 1963-68, head dept. medicine, 1963-68, hon. physician, 1972—; asst. resident Thorndike Meml. Lab., 1929-32, asst. physician, 1932-41, asso. physician, 1941-50, asso. dir. lab., 1950-63, dir., 1963-68; epidemiologist Boston City Hosp., 1968-72; vis. physician Pondville Hosp., 1933-69; successively Charles Follen Folson teaching fellow in hygiene Harvard Med. Sch., 1928-29, asst., 1929-32, Francis Weld Peabody fellow, 1932-37, instr., 1935-37, asso., 1937-40; asst. prof. 1940-46, asso. prof., 1946-62, prof. medicine, 1962-63, George Richards Minot prof. medicine, 1963-68, George Richards Minot Prof. emeritus, 1968—; mem. subcom. infectious diseases NRC, 1946-54, chmn., 1955-59; mem. advisory com. on influenza research USPHS, 1959-63; mem. bacteriol. and mycol. study sect. NIH, 1958-63; asso. mem. commn. acute respiratory diseases Armed Forces Epidemological Bd., 1950-67, mem., 1967-72; mem. drug research bd. Nat. Acad. Scis.-NRC, 1964-71; cons. VA, 1945-73, mem. clin. investigators com. dept. medicine and surgery, 1955-69, chmn., 1964-69, distinguished physician, 1973—; chmn. Com. for Lederle Med. Faculty awards, 1952-68. Contbr. articles, editorials and revs. to sci. jours. and med. text books.; Editor, co-editor numerous monographs on infectious diseases.; Mem. editorial bd.: N.E. Jour. Medicine, 1945-68, Applied Microbiology, 1964-74, Antimicrobiol. Agents and Chemotherapy, 1960-71, Jour. Infectious Diseases, 1969-72, Jour. AMA, 1973—, Jour. Clin. Microbiology, 1974—. Recipient Charles V. Chapin award City of Providence, 1960; Bristol award Infectious Diseases Soc. Am., 1966; Modern Medicine award, 1969; John Phillips Meml. award ACP, 1971; Oscar B. Hunter Meml. award Am. Soc. Clin. Pharmacology and Therapuetics, 1971; Sheen award AMA, 1971; Outstanding contbns. award in field of antibiotic research Bristol Meyers Co., Internat. Div., 1981; named hon. citizen City of Panama, 1970. Master ACP; fellow AAAS; mem. Assn. Am. Physicians (emeritus, Kober medalist 1978), Mass. Med. Soc., Am. Med. Assn., Soc. Exptl. Biology and Medicine, Am. Soc. Clin. Investigation (councillor 1942-45, v.p. 1947-48), Am. Bd. Internal Medicine, Infectious Diseases Soc. Am. (pres. 1963-64), Am. Assn. Immunologists, Soc. Am. Bacteriologists, Am. Acad. Arts and Scis., N.Y. Acad. Scis., Am. Epidemiol. Soc. (councillor 1957-60, v.p. 1961-62), Nat. Acad. Scis., Harvard Med. Alumni Assn. (pres. 1971-72), Assn. Clin. Pathologists (London) (corr.), Sigma Xi, Alpha Omega Alpha. Subspecialties: Immunology (medicine); Pharmacokinetics. Current work: Clinical infectious diseases; infectious control in hospital. Address: Boston City Hospital Boston MA 02118

FINLAYSON, BRUCE ALAN, chemical engineering educator; b. Waterloo, Iowa, July 18, 1939; s. Rodney Alan and Donna Elizabeth (Gilbert) F.; m. Patricia Lynn Hills, June 9, 1961; children: Mark, Catherine, Christine. B.A., Rice U., 1961, M.S., 1963; Ph.D., U. Minn., 1965. Asst. prof. to assoc. prof. U. Wash., Seattle, 1967-77, prof. dept. chem. engring. and applied math., 1977—, co-holder Rehnberg Chair, 1983—; vis. prof. Univ. Coll., Swansea, Wales, U.K., 1975-76, Denmark Tekniske Hójskole, Lyngby, 1976, Universidad Nacional del Sur, Bahia Blanca, Argentina, 1980; trustee Computer Aids to Chem. Engring. Edn., Salt Lake City, 1980—. Mem. editorial bd.: Internat. Jour. Numerical Methods Engring, Swansea, 1974—; mem. editorial bd.: Internat. Jour. Numerical Methods in Fluids, Swansea, 1980—; Author: The Method of Weighted Residuals and Variational Principles, 1972, Nonlinear Analysis in Chemical Engineering, 1980. Served to lt. USN, 1965-67. Mem. Am. Inst. Chem. Engrs. (div. vice chmn. 1981—, William H. Walker award 1983), Am. Chem. Soc., Soc. Petroleum Engrs., Soc. Indsl. and Applied Math., Soc. Rheology. Subspecialties: Chemical engineering; Numerical analysis. Current work: Fluid mechanics, finite element methods, variational principles, collocation methods, chemical reactor modeling. Office: Dept Chem Engring U Wash Benson Hall 105 BF-10 Seattle WA 98195 Home: 6315 22nd Ave NE Seattle WA 98115

FINLEY, GORDON ELLIS, psychologist, educator; b. Evanston, Ill, July 30, 1939; s. Malcolm Hedges and Marguerite (Rheinert) F. B.A., Antioch Coll., Yellow Springs, Ohio, 1962; M.A., Harvard U., 1965, Ph.D., 1968. Asst. prof. U. B.C., Vancouver, Can., 1967-69; U. Toronto, 1969-71; vis. asst. prof. U. Calif.-Berkeley, 1971-72; assoc. prof. Fla. Internat. U., Miami, 1972-76, prof., 1976—. Editor: Avances en Psicología Contemporánea, 1979, (with others) Review of Human Development, 1982; contbr. articles to profl. jours. Mem. Interam. Soc. Psychology (editor 1977-82), Internat. Assn. Cross-Cultural Research (treas. 1971-72), Internat. Assn. Cross-Cultural Psychology (cons. editor 1974—), Am. Psychol. Assn., Gerontol. Soc., Assn. for Gerontology and Anthropology. Subspecialties: Developmental psychology; Gerontology. Current work: Aging; cross-cultural comparisons, cognition and memory. Office: Dept Psychology Fla Internat U Miami FL 33199

FINN, PATRICIA ANN, group leader, researcher; b. Oak Park, Ill.; d. LeRoy and Anne (Pavlinic) F. B.S., Mundelein Coll., 1967; Ph.D., U. Calif.-Berkeley, 1971; postdoctoral, Iowa State U., 1972-73, Argonne (Ill.) Nat. Lab., 1973-75. Asst. chemist chem. tech. div. Argonne Nat. Lab., 1975-80, chemist, 1980-82, group leader, 1982—. Contbr. numerous articles to profl. jours. Mem. Am. Ceramics Soc., Am. Chem. Soc., Am. Nuclear Soc., Assn. Women in Sci., Sigma Xi. Subspecialties: Fusion; Materials (engineering). Current work: System design (tritium and/or blanket system for fusion). Home: 339 S Park Westmont IL 60559 Office: Argonne Nat Lab 9700 S Cass Argonne IL 60439

FINN, RONALD DENNET, radiochemistry educator; b. Weymouth, Mass., Aug. 15, 1944; m., 1969. B.S., Worcester Poly. Inst., 1966; Ph.D. in Nuclear Chemistry and Radiochemistry, Va. Poly. Inst., 1971. Research assoc. chemistry Brookhaven Nat. Lab., 1971-72, research assoc. medicine, 1972-73, assoc. chemist solid state physics, 1973-74, collaborator dept. chemistry, 1974—; radiochemist Mt. Sinai Med. Ctr., Miami Beach, Fla., 1974-75, tech. dir. radiochemistry and radiopharmacology, 1975-79, dir. cyclotron facility, 1979—; asst. research prof. Med. Sch., U. Miami, 1974-77, asst. prof., 1978—. Mem. Am. Chem. Soc., Soc. Nuclear Medicine, N.Y. Acad. Sci., AAAS, Sigma Xi. Subspecialty: Radiochemistry. Office: Cyclotron Facility Mt Sinai Med Ctr 4300 Alton Rd Miami Beach FL 33140

FINNERTY, WILLIAM ROBERT, microbiologist, educator, cons.; b. Keokuk, Iowa, May 2, 1929; s. William T. and Harriet E. (Vandervort) F.; m. Margaret V. Gallagher, Aug. 14, 1952; children: Deborah, Steven, Carrie, Beth, Todd, Stan. B.S. Gen. Sci. and Math, U. Iowa, 1955, Ph.D. in Organic Chemistry and Microbiology, 1961. Instr. U. Iowa, 1959-60; USPHS postdoctoral fellow Oak Ridge Nat. Lab., 1960-62; asst. prof. microbiology Ind. U. Med. Center, 1962-65, assoc. prof., 1965-68, U. Ga., 1968-75, prof., 1975—, head dept. microbiology, 1977—; cons. to industry; vis. prof. U. Gottingen, W.Ger., 1973. Contbr. numerous articles, chpts., revs. on microbiology or biochemistry to profl. publs. Served to sgt. USAF, 1948-52. AEC grantee, 1963-66; NSF grantee, 1965-74; NIH grantee, 1980-83; Dept. Energy grantee, 1975-83. Mem. Am. Soc. Microbiology (P.R. Edwards award S.E. sect. 1980), Am. Chem. Soc., Am. Soc. Biol. Chemists, AAAS, Sigma Xi. Subspecialties: Microbiology; Genetics and genetic engineering (biology). Current work: Hydrocarbon and petroleum microbiology; microbial physiology, genetics and biochemistry; microbial enhanced oil recovery and production.

FINNEY, JOSEPH J., geology educator, researcher; b. N.Y.C., Mar. 11, 1927; m., 1961; 3 children. B.S., U.S. Mcht. Marine Acad., 1950; M.S., U. N. Mex., 1959; Ph.D. in Structural Mineralogy, U. Wis., 1962. Prof. geology, head dept. geology Colo. Sch. Mines, 1962—. Colo. Sch. Mines Found. grantee, 1962—; Research Corp. grantee, 1963. Mem. Mineral. Soc. Am., Mineral. Assn. Can., Mineral. Soc. Gt. Britain and Ireland, Am. Crystallography Assn. Subspecialty: Mineralogy. Office: Dept Geology Colo Sch Mines Golden CO 80401

FINNEY, ROY PELHAM, urology surgeon; b. Gaffney, S.C., Dec. 7, 1924; s. Roy P. Finney and Mary Frances (Cannon) Woodard; m. Kay Harkness, Apr. 5, 1962; children: Wright C., James L., Joella R., Gray, Kevin. M.D., Med. Univ. S.C., 1952. Diplomate: Am. Bd. Urology. Resident in urology Johns Hopkins U., Balt., 1952-57, instr. in urology, 1957; prof. surg. urology U. South Fla., Tampa, 1972—, dir. div. urology, 1972-82. Fellow ACS; mem. Am. Urology Assn., Soc., Internationale D'Urologie, Internat. Continence Soc., Urodynamic Soc. Republican. Club: Tampa Yacht. Subspecialties: Urology; Biomedical engineering. Current work: Design implantable medical prostheses for impotence, incontinence, of the genito-urinary tract, originate new surgical procedures. Designer and inventor implantable prosthese, 1973, penile prostheses, incontinence device treatment urinary incontinence; developer new surg. proceedings treatment impotence, 1977. Home: 92 Adriatic Ave Tampa FL 33606 Office: U South Fla PO Box 16 12901 N 30th St Tampa FL 33612

FINSTON, HARMON LEO, chemistry educator, researcher; b. Chgo., Feb. 16, 1922; s. Leo M. and Minnie (Fogel) Finkelstein; m. Edythe Heller, Jan. 8, 1950; children: Martin, David, Mira, Leo. B.S.A.S., Ill. Inst. Tech., Chgo, 1943; Ph.D., Ohio State U., 1950. Jr. chemist Manhattan Project, Chgo., 1943-45; chemist Brookhaven Nat. Lab., Upton, N.Y., 1950-63; prof. chemistry Bklyn. Coll. CUNY, 1963—. Author: Acid-Base Theory, 1982; contbr. chpts. to books and articles to profl. jours. Israel AEC fellow, 1965; Japan Soc. Promotion of Sci. fellow, 1972; I.M. Kolthoff fellow, 1973. Mem. Am. Chem. Soc. Democrat. Jewish. Subspecialties: Analytical chemistry; Inorganic chemistry. Current work: Solvent extraction; applications of nuclear and radiochemical techniques to chemical analysis. Office: Dept Chemistry Brooklyn Coll CUNY Ave H and Bedford Ave Brooklyn NY 11210

FIORE, NICHOLAS FRANCIS, chemical and metals company executive, physical metallurgy researcher, consultant; b. Pitts., Sept. 24, 1939; s. William Henry and Margaret Angeline (Scinto) F.; m. Sylvia Marie Chinque, Aug. 13, 1960; children: Maria, Nicholas Francis, Kristin, Anthony. B.S. in Metall. Engring., Carnegie-Mellon U., 1960, M.S., 1963, Ph.D., 1964. Asst. prof. metall. engring. and materials sci. U. Notre Dame, 1966-68, assoc. prof., 1968-70, prof., 1970-81, chmn. dept. metall. engring. and materials sci., 1969-81; v.p., corp. dir. tech. Cabot Corp., Boston, 1982—; vis. scientist Argonne Nat. Labs., 1974-75; cons., 1966-81. Contbr. numerous articles on phys. metallurgy, wear, embrittlement, internal friction to profl. jours. Chmn. bd. dirs. Primary Day Sch., Inc., South Bend, Ind. Served to 1st lt. Signal Corps U.S. Army, 1964-66. Recipient Adams Meml. award Am. Welding Soc., 1972. Fellow Am. Soc. Metals; mem. AIME, AAUP. Roman Catholic. Club: Downtown (Boston). Subspecialties: Metallurgical engineering; Materials (engineering). Current work: Degradation of materials by wear, corrosion, embrittlement; management and long-range planning of technology and research. Home: 15 Rockport Rd Weston MA 02193 Office: 125 High St Boston MA 02110

FIREMAN, EDWARD LEONARD, physicist; b. Pitts., Mar. 23, 1922; s. Nathan and Anna (Caplan) F.; m. Rita Hinda, Sept. 9, 1947; children: Bruce, Ellen, Gary. B.S., Carnegie Inst. Tech., 1943; Ph.D., Princeton U., 1949. Instr. physics Carnegie Inst. Tech., 1943-45; postdoctoral fellow Princeton U., 1949-50; research physicist (Brookhaven Nat. Lab.), 1950-56; physicist Smithsonian Astro-phys. Observatory, Cambridge, Mass., 1956—. Mem. Am. Phys. Soc., Am. Chem. Soc., Am. Geophys. Union, AAAS, Meteorol. Soc. Democrat. Subspecialties: Cosmic ray high energy astrophysics; Nuclear physics. Current work: Meteorites, cosmic rays. Home: 57 Clifton Rd Newton MA 02159 Office: 60 Garden St Cambridge MA 02138

FIRESTER, ARTHUR HERBERT, research physicist; b. N.Y.C., 1942; s. Jacob and Dorothy F.; m. Lynne Balber, 1964; children: Jonathan, Alana. B.A., Brandeis U., 1962; Ph.D. in Physics, Princeton U., 1967. Mem. staff RCA Labs., Princeton U., 1967—, heat photovoltaic process and applications research, 1978—. Contbr. numerous articles to profl. publs. Mem. Am. Phys. Soc., Optical Soc. Am., IEEE. Subspecialties: Photovoltaics; Laser applications. Current work: Amorphous silicon photovoltaics, laser applications, optics and holography. Patentee in field. Office: RCA Labs Princeton NJ 08540

FIRESTONE, PHILIP, psychologist, educator; b. Russia, Dec. 28, 1945. B.A., Carleton U., Can., 1969, M.A., 1971; Ph.D., McGill U., Can., 1974. Adj. prof. psychology U. Ottawa, Ont., Can., 1976-79, asst. prof., 1979-80, assoc. prof., 1980—, coordinator internship program, 1981-82; dir. Child Study Centre, 1981—; adj. prof. psychology Carleton U., 1975—. Contbr. articles to profl. jours. Mem. Am. Psychol. Assn., Can. Psychol. Assn., Soc. Behavioral Medicine, Assn. Advancement of Behavior Therapy, Soc. Pediatric Psychology. Subspecialties: Behavioral psychology; Developmental psychology. Current work: Behavioral medicine; evaluation of psychotherapy. Canada

FIRESTONE, RICHARD FRANCIS, chemistry educator; b. Canton, Ohio, June 18, 1926; s. Lester Ellis and Elizabeth Mary (Corkran) F.; m. Olwen Margaret Huskins, Aug. 21, 1954; children: William, Mark, Robert. A.B., Oberlin Coll., 1950; Ph.D., U. Wis., 1954. Resident research assoc. Argonne Nat. Lab., Ill., 1954-56; asst. prof. chemistry Western Res. U., Cleve., 1956-60; assoc. prof. chemistry Ohio State U., 1961-66, prof. chemistry, 1967—. Served with USNR, 1944-46. Fellow AAAS; mem. Am. Chem. Soc., Am. Phys. Soc. Subspecialties: Physical chemistry; Kinetics. Current work: Fast chemical reactions; kinetics and mechanism of reactions of excited atoms; rare gas excimer formation and decay mechanisms; energy transfer and relaxation mechanisms. Office: 140 W 18th Ave Columbus OH 43210

FIRESTONE, ROGER MORRIS, marketing manager, computer scientist; b. Washington, Aug. 23, 1945; s. Linn Jacob and Regina Caroline (Steiner) F. A.B., Brown U., 1967, Sc.M., 1967; M.S., NYU, 1969, Ph.D., 1971; M.B.A., Coll. St. Thomas, 1976. Prin. programmer Sperry Univac, St. Paul, 1971-72, supervising programmer, 1972-77, mgr./cons., Blue Bell, Pa., 1977-79, sr. product mgr., 1979-82, prin. mktg. specialist, 1982-83, mktg. mgr., 1983—. Contbr. articles in field to profl. jours. Mem. AAAS, Soc. Indsl. and Applied Math., Assn. for Computing Machinery; mem. Am. Assn. Artificial Intelligence; Mem. IEEE Computer Soc.; mem. Soc. Computer Simulation. Jewish. Clubs: St. Paul Athletic, N.Y.U., Scottish Rite. Subspecialties: Programming languages; Artificial intelligence. Current work: Primarily involved with marketing of commercially viable products in scientific computing, including numerical methods, mathematical modelling, and artificial intelligence. Home: 1000 Wick Ln Meadow Wick Norristown PA 19401 Office: Sperry Corp Mail Sta B216M PO Box 500 Blue Bell PA 19424

FIRK, FRANK WILLIAM KENNETH, physics educator, university administrator, nuclear physicist; b. London, Nov. 2, 1930; m., 1952; 3 children. B.Sc., U. London, 1956, M.Sc., 1965, Ph.D. in Physics, 1967. Asst. experiment officer for physics Atomic Energy Research Establishment, Eng., 1952-56, expt. officer, 1956-59, sr. sci. officer, 1959-62, prin. sci. officer, 1962-65; sr. research assoc. Yale U., 1965-68, assoc. prof. physics, 1968-77, prof., 1977—, dir. electron accelerator lab., 1976—, chmn. dept. physics, 1980—; vis. scientist Oak Ridge Nat. Lab., 1960-61. Mem. Am. Inst. Physics, Brit. Inst. Physics and Phys. Soc. Subspecialty: Nuclear physics. Office: Dept Physics Yale U New Haven CT 06520

FISCHBARG, JORGE, physiology educator, biomedical researcher; b. Buenos Aires, Argentina, Aug. 14, 1935; came to U.S., 1964, naturalized, 1975; s. Julio and Dora Beatriz (Hadis) F.; m. Zulema Fridman, Jan. 9, 1964; children: Gabriel Julian, Victor Ernest. M.D., Universidad de Buenos Aires, 1962; Ph.D. in Physiology, U. Chgo., 1971. Postdoctoral fellow U. Buenos Aires, 1962-63, U. Louisville, 1964-65, U. Chgo., 1965-70; asst. prof. physiology/ophthalmology Columbia U., 1970-76, assoc. prof., 1976—; cons. Nat. Eye Inst., NIH, Bethesda, 1975; mem. VISA study sect. div. research grants, 1980—; vis. scientist Centre Energie Nucleaire, Saclay, France, 1974, 78. Served with Argentine Army, 1956-57. Nat. Eye Inst. grantee, 1970—; Fellow Commoner Cambridge U., 1976-77. Mem. Am. Physiol. Soc., Biophys. Soc., Assn. Research in Vision and Ophthalmology (physiol. program com. 1980-83), N.Y. Acad. Sci. Democrat. Jewish. Subspecialties: Physiology (medicine); Membrane biology. Current work: Epithelial transport of water and electrolytes; epithelial electrophysiology; theoretical biophysics; corneal physiology. Home: 175 E 62d St New York NY 10021 Office: Columbia U Dept Physiology 630 W 168th St New York NY 10032

FISCHELL, DAVID ROSS, physicist, electrical engineer; b. Washington, Dec. 4, 1953; s. Robert Elentuch and Marian (Standard) F.; m. Sarah Thole, June 11, 1977. B.S. in Engring. Physics, Cornell U., 1975, M.S. in Applied Physics, 1978, Ph.D., 1980. Teaching asst., then research asst. Cornell U., 1975-79; mem. tech. staff Bell Labs., Holmdel, N.J., 1979-82, supr. performance objectives studies, 1982—. Mem. Optical Soc. Am., IEEE. Democrat. Jewish. Subspecialties: Graphics, image processing, and pattern recognition; Acoustical engineering. Current work: Performance evaluation of AT&T communication's new audiographic teleconferencing service. Home: 62 Stratford Rd Tinton Falls NJ 07724 Office: 3 E 505 Bell Labs Holmdel NJ 07733

FISCHER, (ALBERT) ALAN, medical educator, physician; b. Indpls., June 30, 1928; m.; 4 children. M.D., Ind. U., 1952. Diplomate: Am. Bd. Family Practice. Intern St. Vincent Hosp., Indpls., 1952-53, dir. family practice residency program, 1969-75; practice medicine specializing in family medicine, 1953-70; prof. family medicine, chmn. dept. family medicine Ind. U.-Indpls., 1974—; med. dir. Lakeview Convalescent Ctr.; mem. nat. joint practice commn. Nat. Acad. Sci. Mem. Inst. Medicine Nat. Acad. Sci., Am. Acad. Family Physicians (v.p. 1971-72), Internat. Acad. Family Physicians (pres. 1964-66), AMA, Sigma Xi. Subspecialty: Family practice. Office: Ind U Sch Medicine 1100 W Michigan St Indianapolis IN 46202

FISCHER, EDMOND H., biochemistry educator, researcher; b. Shanghai, China, Apr. 6, 1920; m., 1948; 2 children. Mat. Fed., State Coll. Geneva, 1939; Lic. es Sc, U. Geneva, 1943, diploma, 1944, Ph.D. in Chemistry, 1947. Asst. organic chemistry labs. U. Geneva, 1946-48, Swiss Found. research fellow in chemistry, 1948, privat-docent, 1950, Rockefeller Found. research fellow, 1950-53; from asst. prof. to assoc. prof. biochemistry U. Wash., 1953-61, prof., 1961—. Recipient Lederle Med. Faculty award, 1956-69, Jaubert prize, 1968. Mem. Nat. Acad. Sci., Am. Acad. Arts and Sci., Am. Chem. Soc., Am. Soc. Biol. Chemists, Swiss Chem. Soc. (Warner medal 1952). Subspecialty: Biochemistry (biology). Office: 5540 NE Windermere Rd Seattle WA 98105

FISCHER, GEORGE J., metallurgical engineering educator, researcher; b. Bronx, N.Y., Mar. 30, 1925; m., 1948; 2 children. B. Met.E., Poly. Inst. Bklyn., 1949, M. Met.E., 1953. Instr. metallurgy Poly Inst. Bklyn., 1948-50, from asst. prof. to assoc. prof., 1955-65, prof. metall. engring., 1965—, adminstrv. officer div. metall. engring., 1961-71, head dept., 1971-76, dean student services, 1976-80; plant metallurgist Western Electric Co., 1950-53; dept. head Sam Tour & Co., 1953-55. Internat. Nickel Co. grantee, 1960-61; Curtiss-Wright Corp. grantee, 1962-65; NSF grantee, 1963-65. Mem. Am. Soc. Metals, AIME, ASTM. Subspecialty: Metallurgical engineering. Office: Dept Metall Engring Poly Inst Bklyn 333 Jay St Brooklyn NY 11201

FISCHER, HARRY W., radiologist, educator; b. St. Louis, 1921; s. Harry William and Amy Babette (Gieselman) F.; m. Kay Fischer, 1943; 5 children. B.S., U. Chgo., 1943, M.D., 1945. Diplomate: Am. Bd. Radiology. Asst. prof., then asso. prof. radiology U. Ia. Med. Sch., 1956-63, prof., head sect. diagnostic radiology, 1963-66; prof. radiology U. Mich. Med. Sch., 1966-71; dir. dept. radiology Wayne County Gen. Hosp., 1966-71; prof. radiology, chmn. dept. U. Rochester (N.Y.) Sch. Medicine and Dentistry, 1971—. Editorial bd.: Investigative Radiology, 1966—, Radiology, 1971—. Served to lt. (j.g.) M.C. USNR, 1946-48. Fellow Am. Coll. Radiology; mem. Radiol. Soc. N.Am., Assn. Univ. Radiologists (Gold medal), Am. Roetgen Ray Soc., Uroradiology Soc., U. Chgo. Med. Sch. Alumni Assn., Sigma Xi. Subspecialty: Diagnostic radiology. Current work: Contrast media, development and toxicity; radiology department planning and administration. Home: 3565 Elmwood Ave Rochester NY 14610 Office: 601 Elmwood Ave Rochester NY 14642

FISCHER, TRAUGOTT ERWIN, physicist; b. Aarau, Switzerland, Jan. 21, 1932; came to U.S., 1963; s. Traugott J. and Josephine A. (Kuhn) F.; m. Marie-Claude Blanc, Sept. 7, 1958; children: Pierre F., Jacques F., Anne C. B.A., St. Michel Coll., Fribourg, Switzerland, 1952; diploma in physics, Fed. Inst. Tech., Zurich, 1956, Dr. Sc. Nat., 1963. Mem. tech. staff Bell Telephone Labs., Murray Hill, N.J., 1963-66; assoc. prof. Yale U., New Haven, 1966-72; with Exxon Corp. Research Labs., Linden, N.J., 1972—, sr. research assoc., 1977—; cons. CBS Labs., 1968-72. Contbr. articles to profl. jours. Served with Swiss Air Force. Recipient Kern prize Swiss Phys. Soc., 1962. Fellow Am. Phys. Soc.; mem. Am. Vacuum Soc., AIME, Am. Soc. Metals, Am. Soc. Lubrication Engrs. Roman Catholic. Club: Metals Sci. of N.Y. Subspecialties: Condensed matter physics; Mechanical engineering. Current work: Catalysis from viewpoint of electronic structure. Surface chemistry aspects of metallurgy. Interdisciplinary approach to friction, lubrication and wear. Home: 107 Passaic Ave Summit NJ 07901 Office: Exxon Research & Engring Co PO Box 45 Linden NJ 07036

FISCHETTE, CHRISTINE THERESA, neurobiologist, researcher; b. Jersey City, Apr. 3, 1951; s. Carmen and Concetta F.; m. Philip Anthony Femano, Jan. 18, 1981. B.A., Rutgers U., 1973; Ph.D., U. Medicine and Dentistry N.J., 1979. Postdoctoral fellow Rockefeller U., N.Y.C., 1979-82, research assoc., 1982-83; sr. scientist Hoffmann-La Roche, Inc., Nutley, N.J., 1983—. Contbr. articles to profl. jours. USPHS fellow Rockefeller U., 1979-82; Internat. Physiol. Soc. travel grantee, 1983. Mem. Am. Physiol. Soc., Soc. Neurosci., AAAS, European Soc. Neurosci., N.Y. Acad. Scis. Roman Catholic. Subspecialties: Neuroendocrinology; Neurochemistry. Current work: Sexual differentiation of the brain, male and female sex differences, how hormones effect the brain function. Office: Hoffman-La Roche Inc Nutley NJ 07110

FISCHL, MYRON ARTHUR, psychologist; b. N.Y.C., Oct. 13, 1929; s. Louis Frank and Sally Maeth F.; m. Suzanne Dona Borowsky, June 21, 1958; children: Jeffrey, Sally, Amy. B.A., NYU, 1950; M.A., Hofstra U., 1954; Ph.D., Purdue U., 1956. Diplomate: Am. Bd. Profl. Psychology. Sr. research assoc. Courtney & Co., Phila., 1957-63; human factors scientist Gen. Electric Co., Phila., 1958-59; dir. personnel research Nat. Analysts Corp., Phila., 1964-65; research assoc. Applied Psychol. Services, Wayne, Pa., 1965-71; chief selection and classification research U.S. Army Research Inst., Alexandria, Va., 1971-80, chief officer selection and classification research, 1980—; lectr. U. Md., 1972-81, Villanova U., 1966-70, Temple U., 1964. Contbr. articles to profl. jours. Pres. Tantallon S. Homeowners Assn., Ft. Washington, Md., 1981, 82; bd. dirs. Brotherhood of Temple Beth El, Alexandria, 1980—, program chmn., 1981, 82. Served as cpl. U.S. Army, 1951-52. Fellow Am. Psychol. Assn.; mem. Am. Statis. Assn., Eastern Psychol. Assn., Sigma Xi, Psi Chi. Subspecialty: Industrial and organizational psychology. Home: 317 Rexburg Ave Fort Washington MD 20744 Office: US Army Research Inst 5001 Eisenhower Ave Alexandria VA 22333

FISCHMAN, STUART LEE, dentistry educator, dental researcher; b. Buffalo, Nov. 29, 1935; s. Ben C. and Lilliam R. (Friedl) F.; m. Jane Vogel, June 25, 1960; children: Lisa, Everett. Student, Cornell U., 1953-56; D.M.D. cum laude, Harvard U., 1960; hon. prof., Nat. U., Asunción (Paraguay), 1976. Diplomate: Am. Bd. Oral Pathology. Dental intern VA Hosp., Boston, 1960-61; faculty SUNY-Buffalo, 1961; prof. oral medicine Sch. Dentistry, 1972—; dir. dentistry Eric County Med. Ctr., Buffalo, 1974—; cons. Lever Bros. Co., Edgewater, N.J., 1974—, Pan. Am. Health Orgn., Washington, 1968-70, ADA, Chgo., 1975—, Am. Assn. Dental Schs., Washington, 1979-80. Contbr. articles to publs. in field. Vice pres. Temple Beth-El, Buffalo, 1980—; trustee Kadimah Sch., Buffalo, 1978-81; asst. chmn. alumni fund Harvard Sch. Dental Medicine, Boston, 1979—; adv. bd. Hemophilia Ctr., Buffalo, 1975—. Research grantee Lever Bros. Co., 1962, Nat. Inst. Dental Research, 1972, USPHS, 1977. Fellow Am. Acad. Oral Pathology, Am. Coll. Dentists, Internat. Coll. Dentists. Jewish. Subspecialties: Oral pathology; Forensic dentistry. Current work: Clinical research, oral ulcers, dental caries, periodontal disease, epidemiology of oral cancer, forensic dentistry. Home: 255 Louvaine Dr Buffalo NY 14223 Office: Dept Oral Medicine Sch Dentistry SUNY-Buffalo Buffalo NY 14214

FISH, ANDREW JOSEPH, JR., electrical engineering educator, researcher; b. New Haven, Aug. 15, 1944; s. Andrew Joseph and Katherine Pauline (Frey) F. B.S.E.E., Worcester Poly. Inst., 1966; M.S.E.E., U. Iowa, 1973, St. Mary U., San Antonio, 1974; Ph.D., U. Conn.-Storrs. Asst. prof. elec. engring. U. Hartford, West Hartford, Conn., 1979—; co-chmn. nonlinear systems group Am. Control Conf., 1980. Contbr. articles to profl. publs. Gate keeper West Suffield Grange, 1982; Fellow Yale U., 1972-73. Mem. IEEE, Soc. Indsl. and Applied Math., ASME. Subspecialties: Systems engineering; Applied mathematics. Current work: Modeling, analysis, control of nonlinear systems, particularly Hybrid Analog-Digital Systems.

FISHBEIN, MICHAEL CLAUDE, physician, pathologist; b. Brussels, May 25, 1946; U.S., 1949, naturalized, 1957; s. Fred F. and Celia S. (Feldman) F.; m. Astrid Lorette duMortier, Aug. 11, 1974; children: Danielle, Gregory. B.S., U. Ill.-Urbana, 1967; M.D., U. Ill.-Chgo., 1971. Diplomate: Am. Bd. Pathology. Intern Harbor Gen. Hosp./UCLA, Torrance, 1971-72, resident, 1972-75; assoc. pathologist Peter Bent Brigham Hosp., Boston, 1975-78; asst. prof. pathology Harvard Med. Sch., Boston, 1975-78; assoc. pathologist Cedars-Sinai Med. Ctr., Los Angeles, 1978—; assoc. prof. pathology UCLA Med. Sch., 1978—. Mem. Phi Beta Kappa, Alpha Omega Alpha. Subspecialties: Pathology (medicine); Cardiology. Current work: Cardiovascular pathology. Office: Cedars-Sinai Med Center 8700 Beverly Blvd Los Angeles CA 90048

FISHER, CHARLES HAROLD, research adminstr.; b. Hiawatha, W.Va., Nov. 20, 1906; s. Lawrence D. and Mary (Akers) F.; m. Elizabeth Dye, Nov. 4, 1933 (dec. 1967); m. Lois Carlin, July 1968. B.S., Roanoke Coll., 1928; M.S., U. Ill., 1929, Ph.D., 1932; D.Sc. (hon.), Tulane U., 1953, Sc.D., Roanoke Coll., 1963. Teaching asst. chemistry U. Ill., 1928-32; instr. Harvard, 1932-35; asso. organic chemist U.S. Bur. Mines, Pitts., 1935-40; head carbohydrate div. E. Regional Research Lab., U.S. Dept. Agr., 1946-50, dir., New Orleans, 1950-72; adj. research prof. Roanoke Coll., 1972—. Pres. New Orleans Sci. Fair, 1967-69. Recipient So. Chemists award, 1956, Herty medal, 1959; Chem. Pioneer award Am. Inst. Chemists, 1966. Mem. Am. Inst. Chemists (hon., pres. 1962-63, chmn. bd. dirs.), Sci. Research Soc. Am., Oil Chem. Soc., Am. Chem. Soc. (dir. region IV), Chemurgic Council (dir.), Am. Assn. Textile Chemists and Colorists, Sigma Xi, Alpha Chi Sigma, Gamma Alpha, Phi Lambda Upsilon. Club: Cosmos (Washington). Subspecialties: Organic chemistry; Polymer chemistry. Current work: Polymer additives; relations between physical and thermocynamic properties and molecular structure. Office: Chemistry Dept Roanoke College Salem VA 24153 I have worked hard as a physical scientist and research administrator because research is fun and offers the best way of benefiting mankind.

FISHER, DONALD BOYD, plant physiologist, educator; b. Phila., Oct. 14, 1935; s. Alexander M. and Catherine (Boyd) F.; m. Rita Campbell, June 8, 1963; children: Jennifer, Gregory. B.S. in Botany, U. Wash., 1957; M.S., U. Wis., 1961; Ph.D. in Biochemistry and Biophysics, Iowa State U., 1965. NIH postdoctoral fellow dept. botany U. Calif.-Berkeley, 1965-67, research assoc., 1967-48; from asst. prof. to assoc. prof. botany U. Ga., Athens, 1968-78; prof. botany Wash. State U.-Pullman, 1978—. Contbr. chpts. to books, articles to profl. jours. Served to lt. (j.g.) USN, 1957-59. NIH fellow, 1963-67; NSF grantee, 1970—; U.S. Dept. Agr. grantee, 1979—. Mem. AAAS, Am. Inst. Biol. Scis., Am. Soc. Plant Physiologists. Subspecialty: Plant physiology (agriculture). Current work: Intercellular transport of organic materials in plants, with emphasis on phloem transport. Office: Wash State U Pullman WA 99164

FISHER, EDWARD RICHARD, chemical engineering educator, university research administrator, researcher; b. Detroit, Mar. 24, 1938; m., 1973; 5 children. B.Sc., U. Calif.-Berkeley, 1961; Ph.D. in Chem. Engring. Sci, Johns Hopkins U., 1965. Research chemist Lawrence Radiation Lab., 1961; asst. prof. chemistry U. Copenhagen, 1965-66, lectr., 1966; phys. chemist space sci. lab. Gen. Electric Co., Pa., 1966-68; assoc. prof. chem. engring. Wayne State U., 1968-74, prof., 1974—, dir., 1977—; lectr. Gen. Electric grad. program Rensselaer Poly. Inst., 1967. Mem. Am. Inst. Physics, Am. Inst. Chem. Engrs. Subspecialties: Chemical engineering; Chemical physics. Office: Wayne State U Research Inst 220 Engring Bldg Detroit MI 48202

FISHER, GEORGE WESCOTT, university dean, geology educator; b. New Haven, May 16, 1937; s. Irving Norton and Virginia (Hays) F.; m. Frances Louisa Gilbert, Dec. 26, 1959; children: Catherine Anne, Lynn Ellen, Cynthia Lee. A.B., Dartmouth Coll., 1959; Ph.D. (Woodrow Wilson fellow, NSF fellow), Johns Hopkins U., 1963. Postdoctoral fellow Geophys. Lab., Carnegie Instn., Washington, 1964-66; asst. prof. geology Johns Hopkins U., Balt., 1966-71, asso. prof., 1971-74, prof., 1974—, chmn. dept. earth and planetary scis., 1978-83, dean Div. Arts and Scis., 1983—; geologist U.S. Geol. Survey, Beltsville, Md., 1967-72. Editor: Studies in Appalachian Geology: Central and Southern, 1970. Served with Signal Corps U.S. Army, 1962-64. NSF grantee, 1971—. Mem. Geol. Soc. Am., Mineral. Soc. Am. (treas. 1974-76), Geol. Soc. Washington, Geochem. Soc., AAAS, Phi Beta Kappa, Sigma Xi. Subspecialties: Petrology; Tectonics. Current work: The dynamics of metamorphic processes; relation between metamorphism and rock deformation. Research in Appalachian geology and metamorphic petrology. Home: 936 Cromwell Bridge Rd Towson MD 21204 Office: Div Arts and Sciences Johns Hopkins U Baltimore MD 21218

FISHER, H. LEONARD, information scientist; b. Bronx, N.Y., Nov. 7, 1936; s. Harold Leonard and Betty (Kahn) F.; m. Yvonne Derrick, Aug. 2, 1968. B.S. in Physics, St. John's U., 1957, M.S., 1961. Research physicist United Aircraft Research Labs., East Hartford, Conn., 1960-63; physicist Lawrence Livermore (Calif.) Nat. Lab., 1963-69, tech. info. specialist, 1969-79, head research info. group, 1981—; mem. faculty Golden Gate U., San Francisco, 1980—. Mem. Am. Soc. Info. Sci., Am. Phys. Soc., Sigma Xi, Sigma Pi Sigma. Subspecialties: Information systems, storage, and retrieval (computer science); Societal effects of information technology. Current work: Online retrieval, literature searching, videotex, societal effects of information technology.

FISHER, HAROLD WALLACE, chemical engineer; b. Rutland, Vt., Oct. 27, 1904; s. Dean Wallace and Grace Minot (Cheney) F.; m. Hope Elisabeth Case, Sept. 29, 1930; 1 son, Dean Wallace. S.B. in Chem. Engring, Mass. Inst. Tech., 1927; D.Sc., Clarkson Coll. Tech., Potsdam, N.Y., 1960. With Exxon Corp. (and affiliates), 1927-69; joint mng. dir. Iraq Petroleum Co., 1957-59; dir. Exxon Corp., 1959-69, v.p., 1962-69; mem. adv. bd. energy lab. Mass. Inst. Tech., 1974-80, mem. corp. devel. commn., 1975—; mem. marine bd. Nat. Acad. Engring., 1969-74. Co-author: The Process of Technological Innovation, 1969; also articles. Chmn. exec. com. Community Blood Council Greater N.Y., 1969-71; trustee Sloan-Kettering Inst. Cancer Research, N.Y.C., 1964—, chmn., 1970-74. Recipient Chem. Industry medal Am. sect. Soc. Chem. Industry, 1968; Bronze Beaver award Mass. Inst. Tech. Alumni Assn., 1970. Fellow AAAS; mem. Nat. Acad. Engring., Am. Chem. Soc., Am. Inst. Chem. Engrs., Pilgrims of U.S., Kappa Sigma, Alpha Chi Sigma, Tau Beta Pi. Republican. Clubs: University (N.Y.C.); Duxbury Yacht (Mass.); Community Men's; American (London). Subspecialty: Chemical engineering. Patentee petroleum processing, petrochem. mfr. Home: 68 Goose Point Ln PO Box 1792 Duxbury MA 02332

FISHER, JAMES HAROLD, geology educator, petroleum cons.; b. Mayfield, Ky., Nov. 8, 1919; s. Clyde and Lillian (Smithson) F.; m. Anne Brown, June 6, 1943 (div.); children: James Michael, John Allan, Jeanne Ann; m. Normalee Velma, June 12, 1982. A.B., U. Ill.-Urbana, 1943, B.S., 1947, M.S., 1949, Ph.D., 1953. Geologist McCurtain Limestone Co., Idabel, Okla., 1948-49, Pure Oil Co., Casper, Wyo., 1949-51; asst. prof. U. Ill.-Urbana, 1952-55, U. Nebr.-Lincoln, 1955-57; prof. geology Mich. State U., East Lansing, 1957—; cons. Tenneco Oil Co., Houston, 1960-67; exploration mgr. McClure Oil Co., Alma, Mich., 1968-69. Served to 1st lt. U.S. Army, 1943-46; ETO. Shell Oil Co. fellow, 1951. Fellow Geol. Soc. Am.; mem. Am. Assn. Petroleum Geologists (pres. ea. sect. 1974-75), Mich. Basin Geol. Soc. (pres. 1961-61). Republican. Unitarian. Subspecialties: Petroleum engineering; Sedimentology. Current work: Petroleum occurrences in the Michigan basin, tectonic analysis of intercratonic basins. Home: 1175-F Arbor Dr East Lansing MI 48823 Office: Dept Geology Mich State East Lansing MI 48824

FISHER, JAMES W., pharmacologist, educator; b. Startex, S.C., May 22, 1925; s. Earnest Amaziah and Mamie Viola (Turner) F.; m. Carol Brodarick, June 5, 1947; children: Candis, Patricia, Richard, William, John, Elaine. B.S., U.S.C., 1947; Ph.D. in Pharmacology, U. Louisville Med. Sch., 1958. Pharmacologist, Armour Pharm. Research Labs., Chgo., 1950-53; Pharmacolgist, Lloyd Bros., Inc. Pharm. Research Labs., Cin., 1954-56; instr. dept. pharmacology U. Tenn. Med. Units, Memphis, 1958-60, asst. prof., 1960-62, assoc. prof., 1962-66, prof., 1966-68; prof., chmn. dept. pharmacology Tulane U. Sch. Medicine, New Orleans, 1968—; cons. Pharmacia Labs., Piscataway, N.J., 1973-75, Schering Corp., Bloomfield, N.J., 1976—, Upjohn Co., Kalamazoo, 1977—, Biogen, Inc., Cambridge, Mass., 1983—. Editorial bd.: Proc. Soc. Exptl. Biology and Medicine, 1971-74, 74-77; editor: Kidney Hormones, vols. I and II, 1977; mem. various editorial rev. bds.; contbr. articles to profl. jours. Served to lt. (j.g.) USNR, 1943-46; PTO. USPHS fellow, 1956-58. Mem. Am. Soc. Pharmacology and Exptl. Therapeutics, Soc. Exptl. Biology and Medicine, Am. Soc. Nephrology, Am. Soc. Hematology, Assn. Med. Sch. Pharmacology, AAAS, Internat. Soc. Nephrology, AAUP, N.Y. Acad. Scis., Am. Fedn. Clin. Research, Sigma Xi. Subspecialties: Cellular pharmacology; Hematology. Current work: Hematopharmacology; endocrine pharmacology; renal pharmacology; drug effects on renal blood flow and erythropoietin; hormones and erythropoietin; anemia of chronic renal failure; prostaglandins and erythropoiesis. Home: 4025 Pin Oak Ave New Orleans LA 70114 Office: 1430 Tulane Ave New Orleans LA 70112

FISHER, JEFFREY DAVID, social psychologist, educator, writer, consultant; b. Bklyn., Apr. 23, 1949; s. Silvan and Helen (Friedberg) F.; m. Sherry L. Blank, June 10, 1972; children: Andrew Jay, Aaron Michael. B.A., U. Wis.-Madison, 1971; M.S., Purdue U., 1973, Ph.D., 1975. Asst. prof. U. Conn.-Storrs, 1975-79, assoc. prof. dept. psychology, 1979—; vis. scholar Stanford (Calif.) U., 1981. Cons. editor: Jour. Personality and Social Psychology, 1979-83, Population: Behavioral, Social, and Environ. Issues, 1977—, Environ. Psychology and Nonverbal Behavior, 1977-79, Jour. Social and Clin. Psychology, 1982—. Grantee U.S.-Israel Binat. Sci. Found., Jerusalem, 1979, U. Conn. Research Found., 1975-82. Mem. Am. Psychol. Assn., Eastern Psychol. Assn., Soc. Exptl. Social Psychology, Midwestern Psychol. Assn., Western Psychol. Assn. Subspecialty: Social psychology. Current work: Prosocial behavior; especially recipient reactions to aid; environmental psychology. Home: 98 Timber Dr Storrs CT 06268 Office: Univ Conn U-20 Storrs CT 06268

FISHER, JERID MARTIN, clinical neuropsychologist, health promotion consultant; b. Houston, July 12, 1953; s. Seymour and Rhoda (Lee) F. B.S. in Psychology, Duke U., 1975; Ph.D. in Clin. Psychology, U. Rochester, 1981. Lic. psychologist, Calif., N.Y. Dir neuropsychology service U. Rochester Med. Ctr., 1981—; ptnr. Am. Data Systems, Rochester, 1982—; pub. speaker health promotion, 1980—. Author pamphlet: Stress-Don't let it distress you, 1982. Mem. Am. Psychol. Assn., Internat. Neuropsychol. Soc., N.Y. Neuropsychol. Group, Phi Beta Kappa. Subspecialties: Neuropsychology; Clinical psychology. Current work: Differential diagnosis of dementing disorders; treatment and rehabilitation of the brain injured; life extension; health promotion and modification of the American lifestyle. Home: 248 Crittenden Way Apt 5 Rochester NY 14623 Office: U Rochester Med Center 300 Crittenden Blvd Rochester NY 14642

FISHER, JOHN COURTNEY, surgical physicist, consultant; b. Wilkinsburg, Pa., Apr. 19, 1922; s. Edwin Henry and Elizabeth (Walden) F.; m. Patricia Kingsbury, Nov. 26, 1942; children: Carolyn Fisher Ellis, John Courtney, Stephen Kingsbury; m. Jane Clauss, July 7, 1976. B.S., Harvard U., 1942, M.S., 1947, Sc.D., 1952. Teaching fellow Harvard U., 1942-52; sonar engr. Submarine Signal Co., Boston, 1945-46; dir. electromech. engring. Calidyne Co., Winchester, Mass., 1952-55; dir. devel. privately funded research project, Maynard, Mass., 1955-60; chmn., treas. Am. Dynamics Corp., Cambridge, Mass., 1960-68; northeastern regional sales mgr. Princeton Applied Research Corp., N.J., 1968-72; cons. in sci. instrumentation, Weston, Mass., 1972—; dir. med. devel. Cavitron Lasersonics div., Stamford, Conn., 1976-81; mem. surg. staff St. Barnabas Med. Center, Livingston, N.J., 1980—. Contbr. articles to profl. jours. Mem. Weston Planning Bd., 1974-80. Fellow Am. Soc. Laser Medicine and Surgery; mem. Internat. Soc. Laser Surgery, ASME, IEEE, AAAS, Gynecologic Laser Soc., Midwest Biolaser Inst., N.Y. Acad. Scis. Subspecialties: Laser medicine; Surgery. Current work: Applications of lasers and other light sources to medicine, biology, surgery, and therapy, development of sophisticated instrumentation for diagnosis, therapy, and surgery. Patentee in field. Home: 417 Palmtree Dr Wildewood Springs Bradenton FL 33507

FISHER, LAWRENCE, psychologist, researcher; b. Phila., Oct. 18, 1941; s. Marvin and Louise Greenblatt; m.; children: David, Gabrielle. B.S., Pa. State U., 1963; M.A., Temple U., 1965; Ph.D., U. Cin., 1968. Diplomate: Am. Bd. Profl. Psychology. Postdoctoral fellow U. Colo. Med. Sch., Denver, 1968-69; asst. prof. U. Rochester, N.Y., 1969-79; assoc. prof. U. Calif.-San Francisco, Fresno, 1979; chief. psychol. service VA Med. Ctr., Fresno, 1979—. Contbr. articles to profl. jours. Mem. Am. Psychol. Assn., Am. Orthopsychiat. Assn., Am. Assn. Marital and Family Therapy. Democrat. Jewish. Subspecialty: Clinical research. Current work: Research and teaching in family issues in health outcome, risk to psychopathology, vulnerability to maladaptation. Home: 1635 W Celeste St Fresno CA 93711 Office: U Calif San Francisco 2615 E Clinton Ave Fresno CA 93703

FISHER, MICHAEL ELLIS, mathematical physicist, chemist; b. Trinidad, W.I., Sept. 3, 1931; s. Harold Wolf and Jeanne Marie (Halter) F.; m. Sorrel Castillejo, Dec. 12, 1954; children: Caricia J., Daniel S., Martin J., Matthew P.A. B.S. with 1st class honors in Physics, King's Coll., London, 1951, Ph.D., 1957. Lectr. math. RAF, 1952-53; lectr. theoretical physics King's Coll., 1958-62, reader physics, 1962-64; prof. physics U. London, 1965-66; prof. chemistry and math. Cornell U., 1966-73, Horace White prof. chemistry, physics and math., 1973—, chmn. dept. chemistry, 1975-78; guest investigator Rockefeller Inst., 1963-64; vis. prof. applied physics Stanford U., 1970-71; Buhl lectr. theoretical physics Carnegie-Mellon U., 1971; Richtmyer Meml. lectr. Am. Assn. Physics Tchrs., 1973; S.H. Klosk lectr. N.Y. U., 1975; 17th F. London Meml. lectr. Duke U., 1975; Walker-Ames prof. U. Wash., Seattle, 1977; Loeb lectr. physics Harvard U., 1979; vis. prof. physics M.I.T., 1979; Welsh Found. lectr. in physics U. Toronto, Ont., Can., 1979; 21st Alpheas Smith lectr. Ohio State U., 1982. Author: (with D.M. MacKay) Analogue Computing at Ultra-High Speed, 1962, The Nature of Critical Points, 1964, The Theory of Equilibrium Critical Phenomena, 1967; assoc. editor: Jour. Math. Physics, 1963-68, 72-74; adv. bd.: Jour. Theoretical Biology, 1969-82, Chem. Physics, 1972—, Discrete Math, 1971-78, Jour. Statis. Physics, 1978-81; contbr. 200 articles to profl. jours. Recipient award in phys. and math. scis. N.Y. Acad. Scis., 1978; Guthrie medal and prize Inst. Physics, London, 1980; Wolf prize in physics, 1980; Michelson-Morely award Case Western Res. U., 1982; Guggenheim fellow, 1970-71, 78-79. Fellow Am. Acad. Arts and Scis., Royal Soc. (London) (Bakerian lectr. 1979), Phys. Soc. London, Am. Phys. Soc. (Langmuir prize chem. physics 1970), Kings Coll. London; mem. Am. Chem. Soc., Soc. Indsl. and Applied Math., Math. Assn. Am., Nat. Acad. Scis. (Fgn. assoc., James Murray Lack award 1983), N.Y. Acad. Scis. Subspecialties: Statistical physics; Statistical mechanics. Office: Baker Lab Cornell U Ithaca NY 14853

FISHER, PAUL B., cell biologist, virologist, oncologist; b. Bklyn., July 5, 1945; s. George and Ray M. (Fisher) Cherkis; m. Marlene J. Weintraub, May 16, 1976; 1 dau., Danielle Leah. B.A., Hunter Coll., 1968; M.A., H. Lehman Coll., CUNY, 1971; M.Phil., Rutgers U., 1973, Ph.D., 1974. Busch postdoctoral fellow Waksman Inst. Microbiology Rutgers U., 1974-75; research assoc. dept. microbiology and immunology Albert Einstein Coll. Medicine, N.Y.C., 1975-76; staff assoc. Cancer Center-Inst. Cancer Research Coll. Physicians and Surgeons, Columbia U., N.Y.C., 1976-80, research assoc. prof. dept. microbiology, 1980—; asso. prof. Bronx Community Coll. of CUNY, 1979—; cons. EPA, Nat. Cancer Inst., NSF. Contbr. abstracts and articles to profl. jours. Mem. Am. Assn. Cancer Research, AAAS, Am. Soc. Microbiology, N.Y. Acad. Scis., Harvey Soc., Internat. Pigment Cell Soc., Tissue Culture Assn., Nat. Found. Infectious Diseases, Am. Soc. Cell Biology, Am. Inst. Biol. Sci. Subspecialties: Cell study oncology; Virology (medicine). Current work: Molecular basis of chemical and viral carcinogenesis; gene structure and function; mechanisms of tumor promotion and progression; cancer chemotherapy; chemical and viral transformation; membrane structure and function; cellular differentiation. Home: 15 Gordon Pl Scarsdale NY 10583 Office: Dept Microbiology Cancer Ctr Inst Cancer Research Columbia U Coll Physicians and Surgeons New York NY 10032

FISHER, RAY W., mechanical engineering educator, university administrator, researcher; b. Anamosa, Iowa, Nov. 27, 1921; m., 1945; 4 children. B.S., Iowa State U., 1948. Asst. in atomic research and phys. chemistry Iowa State U., 1943-48, adminstrv. aide, 1949-61, asst. prof. mech. engring., 1961-63, assoc. prof., 1963—, dir. fossil energy program, 1975—; dir. Mining and Mineral Resources Research Inst., 1981—; research assoc. Ames Lab., 1949-57, assoc. engr., 1957-62, engr., 1962-64, assoc. prof. nuclear engring., 1963-71, head bldg. and engring. services, 1964-69, plant mgr., 1969-77; prin. lectr. coal preparation research Royal Swedish Acad. Engring. Sci., Stockholm, 1979. Mem. Soc. Mining Engrs. Subspecialties: Mining engineering; Coal. Office: Iowa State U Spedding Hall Room 320 Ames IA 50011

FISHER, RICHARD ROYAL, astrophysicist; b. Wichita, Kans., June 2, 1941; s. James B. and Winifred R. (Royal) F.; m. Patricia A. Pierce, Jan. 30, 1962; children: Alson, Matthew Pierce. B.A., Grinnell Coll., 1961; Ph.D., U. Colo., 1965. Asst. astronomer U. Hawaii, 1965-70, assoc. astronomer, 1970-71, asst. fiscal officer, 1971; staff scientist Sacramento Peak Obs., 1971-76; research assoc. Harvard Coll. Obs., 1973-76; scientist Coronal Dynamics Project, High Altitude Obs., 1976-78, program mgr., 1978-79, staff scientist III, 1979-82; sr. scientist Nat. Ctr. Atmospheric Research, Boulder, Colo., 1982—; telescope scientist NASA Solar Optical Telescpe Project, 1981—; leader Joint U.S.A.-USSR Eclipse Expdn., 1981. Contbr. articles to sci. and popular jours. Pres. Conservation Council of Hawaii, Maui, 1969; active Boy Scouts Am. Recipient Lynn Smith prize in math., 1961; High Altitude Obs. fellow, 1961. Mem. Optical Soc. Am., Am. Astron. Soc., AAAS, Internat. Astron. Union. Club: Big Island Road Runners (Hilo, Hawaii). Subspecialties: Optical astronomy; Solar physics. Current work: Design and implementation of astronomical instrumentation systems. Coronal physics; telescope design; solar optical telescope (SOT-1.25M, shuttle payload). Office: PO Box 3000 Boulder CO 80307

FISHER, ROBERT ALAN, physicist; b. Berkeley, Calif., Apr. 19, 1943; s. Leon Harold and Phyllis (Kahn) F.; m. Andrea, Mar. 18, 1967; children: Andrew, Derek. A.B. with honors, U. Calif.-Berkeley, 1965, M.A., 1967, Ph.D., 1971. Lab. asst., programmer Lockheed Missiles & Space Co., Palo Alto, Calif., summers, 1962, 63; NSF grantee Stanford (Calif.) U. Genetics Dept., summer, 1964; programmer Stanford Linear Accelerator Center, summer 1965; with Granger Assoc., Palo Alto, 1966 (summer); teaching asst., research asst., reader U. Calif.-Berkeley, 1965-71; instr. Dept. Applied Sci., U. Calif.-Davis, 1972-74; physicist Lawrence Livermore Lab., 1971-74, Los Alamos (N.Mex.) Nat. Lab., 1974—; presider OSA Symposium on Optical Phase Conjugation, 1981; program com. Gordon Conf. on Nonlinear Optics, 1981, Internat. Quantum Electronics Conf., 1982, ann. meeting Optical Soc. Am., 1981. Contbr. articles to profl. jours.; editor spl. issue on optical phase conjugation: Jour. Optical Soc. Am., 1983. Vol., cons. N.Mex. No. Schs. Chess League, 1980-82; coach N.Mex. elem. state champion chess team, 1982, 83; mem. adv. com. to provost U. N.Mex., 1983—; mem. Air Force Red Team Study Panel, 1983. Mem. Am. Phys. Soc., Optical Soc. Am., IEEE (sr.), Soc. Photo-Optical Instrumentation Engrs. Subspecialties: Nonlinear laser optics; Optical phase conjugation. Current work: Research in new lasers, optical phase conjugation and wavefront reversal, nonlinear laser optics, high energy lasers, molecular spectroscopy, ultraviolet laser tech. Patentee in field. Home: Jacona Plaza Route 5 Box 23 Santa Fe NM 8750 Office: Los Alamos Nat Lab Mail Stop E535 Los Alamos NM 87545

FISHER, RONALD RICHARD, chemistry educator; b. Peoria, Ill., Oct. 3, 1941; s. Richard William and Vivienne (Allen) F.; m. Kira Virginia Baldini, Nov. 24, 1969. B.A., Ariz. State U., 1963; Ph.D., Cornell U., 1970. Asst. prof. chemistry U. S.C., Columbia, 1971-75, assoc. prof., 1975-79, prof., 1979—, chmn. dept. chemistry, 1976-82; mem. phys. biochem. study sect. NIH, 1982—. NIH fellow, 1966-70. Mem. Am. Chem. Soc., Am. Soc. Biol. Chemists. Subspecialties: Biochemistry (biology); Biophysical chemistry. Current work: Membrane biochemistry, structure and function of biological ion transport, mechanism of enzyme action. Office: Dept Chemistry U SC Columbia SC 29208

FISHER, STEVEN KAY, biology educator; b. Rochester, Ind., July 18, 1942; s. Stewart King and Hazel Madeline (Howell) F.; m. Dinah Dawn Marschall, May 2, 1971; children: Jenni Dawn Ward, Brian Andrew. B.S., Purdue U., 1964, M.S., 1966, Ph.D., 1969. Predoctoral fellow Purdue U., Lafayette, Ind., 1967-69; postdoctoral fellow Johns Hopkins U., Balt., 1969-71; asst. prof. U. Calif.-Santa Barbara, 1971-77, assoc. prof., 1977-82, prof. biology, 1982—. Grantee Nat. Eye Inst., 1972—; Research Career Devel. award, 1980-84. Mem. Assn. Research in Vision and Ophthalmology (program chmn. anatomy and pathology 1980), AAAS. Democrat. Subspecialties: Neurobiology; Cell biology. Current work: Cell biology and ultrastructure of the vertebrate retina with special emphasis on the renewal mechanisms of cone outer segments, retinal detachment, and metabolism and synaptic connectivity of retinal cells. Home: 6890 Sabado Tarde Isla Vista CA 93117 Office: U Calif Dept Biol Scis Santa Barbara CA 93106

FISHER, STUART GORDON, ecology educator; b. Elmhurst, Ill., Mar. 1, 1943; s. Gordon Burt and Jessie (Paige) F.; m. Nancy Beth Grimm, Dec. 19, 1981; 1 dau., Caitlin Paige. B.S., Wake Forest Coll., 1965, M.A., 1967; Ph.D., Dartmouth Coll., 1967. Asst. prof. Amherst (Mass.) Coll., 1970-76; asst. prof. ecology Ariz. State U., Tempe, 1976-78, assoc. prof., 1978—; adv. subcom. for ecology NSF, 1982, ecosystem studies, 1981—, long term ecol. research, 1981. Contbr. articles to profl. jours. NSF grantee, 1973-75, 77—; EPA grantee, 1974-76; Water Resources & Tech. grantee, 1978-79. Mem. Ecol. Soc. Am., Southwestern Assn. Naturalists, Desert Fishes Council, Internat. Soc. Limnology, Am. Soc. Limnology and Oceanography. Subspecialties: Ecosystems analysis; Ecology. Current work: Limnology, stream ecology; ecology of natural and managed waters in arid regions; biol. productivity. Home: 1714 Kings Ranch Apache Junction AZ 85220 Office: Ariz State Univ Dept Zoology Tempe AZ 85287

FISHMAN, ALFRED PAUL, physician; b. N.Y.C., Sept. 24, 1918; s. Isaac and Anne (Tiner) F.; m. Florence Howitz, Aug. 23, 1948; children—Mark, Jay. A.B., U. Mich., 1938, M.S., 1939; M.D., U. Louisville, 1943; M.A. (hon.), U. Pa., 1971. Diplomate: Nat. Bd. Examiners, Am. Bd. Internal Medicine. Intern Jewish Hosp., Bklyn., 1943-44; Dazian Found. fellow pathology Mount Sinai Hosp., N.Y.C., 1946-47, asst. resident, resident medicine, 1947-48; Dazian Found. fellow cardiovascular physiology Michael Reese Hosp., Chgo., 1948-49; Am. Heart Assn. research fellow Bellevue Hosp., N.Y.C., 1949-50, established investigator cardiopulmonary lab., 1951-55; Am. Heart Assn. research fellow physiology Harvard U., Boston, 1950-51; instr. physiology N.Y. U., 1951-53; asso. in medicine Columbia Coll. Physicians and Surgeons, N.Y.C., 1953-55, asst. prof., 1955-58, asso. prof., 1958-66; prof. medicine U. Chgo., 1966-69; dir. Cardiovascular Inst., Chgo., 1966-69; dir. div. cardiovascular disease Michael Reese Hosp., Chgo., 1966-69; prof. medicine U. Pa., 1969—, William Maul Measey prof., 1972—, asso. dean, 1969-75; dir. cardiovascular-pulmonary div., dir. Robinette Found., Clin. Cardiovascular Research Center, U. Pa. Med. Center, 1969—; dir. Specialized Center of Research (Lung), 1973—; attending physician Hosp. U. Pa., 1969—; sr. attending physician Phila. Gen. Hosp., 1970—; physician Mass. Gen. Hosp., 1979; cons. to chancellor U. Mo., Kansas City, 1973-78; vis. prof. Harvard U., 1970, Oxford (Eng.) U., 1972, Washington U., St. Louis, 1973, Johns Hopkins U., 1974, Ben Gurion U., 1975, Emory U., Atlanta, 1976, U. Porto Alegra, Brazilia, Brazil, 1976, U. Zurich, Switzerland, 1978, Fu Wai Hosp., Peking, China, 1980; cons. Exec. Office Pres., 1961-69, U. Athens, Greece, 1980; mem. WHO Expert Panel, Geneva, 1973—, Nat. Adv. Heart and Lung Council, NIH, 1968-71; chmn. Gov.'s Com. for Research on Respiratory Diseases in Coal Miners, 1974, Internat. Conf. on Lung, Titisee, Germany, Florence, Italy, 1976; mem. Inst. of Medicine, Nat. Acad. Sci., 1980—. Editor: (with D.W. Richards) Men and Ideas, 1964, (with H.H. Hecht) The Pulmonary Circulation and Interstitial Space, 1969, Handbooks of Respiratory Physiology, Am. Physiol. Soc., 1967-72, 79—, Physiology in Medicine, New Eng. Jour. Medicine, 1969-79, Jour. Applied Physiology, 1981—; editorial bd.: Merck Manual, 1972—, Ann. Rev. Physiology, 1977—, Heart Failure, 1979, (with E. M. Renkin) Pulmonary Edema, 1979, Pulmonary Diseases and Disorders, 1979; contbr. articles to profl. jours. Bd. dirs. Polachek Found., Phila. Zool. Soc. Served to capt. M.C. U.S. Army, 1944-46. Fellow Am. Coll. Chest Physicians (hon.), Royal Coll. Physicians, A.C.P.; mem. Am. Physiol. Soc. (chmn. publs. bd. 1974-81), Am. Soc. Clin. Investigation, AAAS, Royal Soc. Medicine (London), Assn. Am. Physicians, Am. Heart Assn. (dir. 1973—, chmn. council on cardiopulmonary disease 1972-74, research council 1974—, Disting. Achievement award 1980), Am. Coll. Cardiology (hon.), Interurban Clin. Club, N.Y. County Med. Soc., Phila. Coll. Physicians, Heart Assn. Southeastern Pa. (bd. dirs.), Alpha Omega Alpha. Subspecialties: Animal physiology; Pulmonary medicine. Current work: Cardiopulmonary physiology; integrative physiology; comparative physiology. Home: 2401 Pennsylvania Ave Apt 20-A7 Philadelphia PA 19130 Office: Hospital U Pennsylvania 3400 Spruce St Philadelphia PA 19104

FISHMAN, GERALD JAY, astrophysicist; b. St. Louis, Feb. 10, 1943; s. Irwin M. and Minnie (Kalish) F.; m. Nancy Dale Neyman, June 18, 1967; children: Lisa R., Jodi L. B.S. with honors in Physics, U. Mo., Columbia, 1965; M.S., Rice U., Houston, 1968, Ph.D., 1969. Research asst. Rice U., 1965-69, research assoc., 1969; sr. research physicist Teledyne Brown Engring. Co., Huntsville, Ala., 1969-74; part-time instr. U. Ala., Huntsville, 1972-74; with NASA (George C. Marshall Space Flight Center), Ala., 1974—, on detail, Washington, 1977-78, prin. investigator, 1978—, astrophysicist, 1974—. Contbr. numerous articles to sci. jours. Mem. Am. Astron. Soc., Am. Phys. Soc., AAAS, Internat. Astron. Union. Subspecialty: Gamma ray high energy astrophysics. Current work: Gamma-ray astronomy (balloon-borne and satellites), gamma ray astrophysics, nuclear astrophysics, nuclear radiation monitors and detectors. Patentee radiation detectors using multiple scintillation crystal pieces. Office: ES62 George C Marshall Space Flight Center NASA Space Flight Center AL 35812

FISHMAN, JACK, research scientist, educator; b. St. Louis, Apr. 11, 1948; s. Irwin M. and Minnie (Kalish) R.; m. Sue A. Gluck, June 24, 1972; children: David, Jason, Melissa. B.A., U. Mo., 1971; M.S., St. Louis U., 1974, Ph.D., 1977. Research assoc. Colo. State U., Ft. Collins, 1976-79; vis. scientist Max Planck Inst., Mainz, W.Ger., 1979; research scientist NASA-Langley Research Ctr., Hampton, Va., 1979—; cons. U.S. EPA, Research Triangle Park, N.C., 1977-79. O.M. Stewart scholar, 1967. Mem. Am. Meteorol. Soc., Am. Geophys. Union, AIAA. Subspecialty: Atmospheric chemistry. Current work: Theoretical study of global tropospheric geochemical cycles. Home: 8 Woodhaven Dr Poquoson VA 23662 Office: NASA-Langley Research Ctr Mail Stop 401 B Hampton VA 23665

FISHMAN, WILLIAM H(AROLD), cancer research foundation executive, biochemist; b. Winnipeg, Man., Can., Mar. 2, 1914; came to U.S., 1940, naturalized, 1946; s. Abraham and Goldie (Chmelnitsky) F.; m. Lillian Waterman, Aug. 6, 1939; children: Joel, Nina, Daniel. B.S., U. Sask.(Can.), Saskatoon, 1935; Ph.D., U. Toronto, Ont., Can., 1939. Dir. cancer research New Eng. Med. Ctr. Hosp., Boston, 1958-72; research prof. pathology Tufts U. Sch. Medicine, 1961-70, prof. pathology, 1970-77, dir., 1972-76; pres. La Jolla (Calif.) Cancer Research Found., 1976—, dir., 1981—; mem. basic sci. problems merit rev. bd. com. VA, 1971-75; mem. pathobiol. chemistry sect. NIH, Bethesda, Md., 1977-81. Author in field. Recipient Research Career award NIH, 1962-77; Royal Soc. Can. research fellow, 1939, 40; 17th Internat. Physiol. Congress-U.K. Fedn. fellow, 1947. Mem. Am. Assn. Cancer Research, Am. Soc. Biol. Chemists, Am. Soc. Cell Biology, Am. Soc. Exptl. Biology, Histochem. Soc. (pres. 1983-84). Jewish. Club: Univ. (San Diego). Subspecialties: Biochemistry (biology); Cancer research (medicine). Current work: Basic research on expression of placental genes by cancer cells; monoclonal antibodies; oncodevelopmental markers; immunocytochemistry. Office: La Jolla Cancer Research Found 10901 N Torrey Pines Rd La Jolla CA 92037

FISK, LANNY HERBERT, geology educator, petroleum consultant; b. Edmore, Mich., Feb. 24, 1944; s. Paul James and Mildred Pauline (Courser) F.; m. Carol M. McDowell, Jan. 29, 1967. B.A. with honors in Biology, Andrews U., 1971; Ph.D. in Biology (NSF fellow, Univ. fellow), Loma Linda U., 1976; doctoral candidate in Geology, Mich. State U., 1978—. Asst. then assoc. prof. paleobiology Walla Walla Coll., College Place, 1974-79; assoc. prof. geology, chmn. dept. geol. scis. Loma Linda U., Riverside, Calif., 1979—; petroleum cons. Amoco, Davis Oil Co., Unioil, Barrick Exploration, Valero Energy. Contbr. numerous articles to profl. jours. Served with U.S. Army, 1967-69. Mem. AAAS, Am. Assn. Petroleum Geologists, Am. Assn. Stratigraphic Palynologists, Geol. Soc. Am., Internat. Orgn. Palaeobotany, Nat. Assn. Geology Tchrs., Paleontol. Soc. Seventh-Day Adventist. Subspecialties: Geology; Paleoecology. Current work: Petroleum and coal exploration geology including palynostratigraphy, source and reservoir rock analyses, paleoenvironmental interpretations, structural analyses and economic potential. Home: 4620 Ambs Dr Riverside CA 92505 Office: Dept Geol Scis Loma Linda U Riverside CA 92515

FISS, HARRY, psychology educator; b. Vienna, Apr. 15, 1926; U.S., 1939, naturalized, 1944; s. Emil and Gertrude (Roemer) F.; m. Gerda May, Oct. 20, 1962; children: Karen, Naomi. B.A., N.Y. U., 1949, Ph.D., 1961. Diplomate: Dipolmate Am. Bd. Profl. Psychology. Instr. Albert Einstein Coll. Medicine, 1960-63, dir. psychology tng. program, 1969-71; asst. prof. N.Y. U., 1963-69; prof. psychology L.I. U., 1971-73; prof. U. Conn., Farmington, 1973—; vis. prof. Israel Inst. Tech., Haifa, 1982. Contbr. chpts. to books and articles to profl. jours. Mem. adv. bd. Whiting Forensic Inst., Middletown, Conn., 1982. Served with AUS, 1944-46; Served with USAR, 1949-54. Recipient Founders Day award N.Y. U., 1961; NIH grantee, 1967; U. Conn. Research Found. grantee, 1974. Fellow Soc. Personaliy Assessment; mem. Am. Psychol. Assn., Sleep Research Soc., Assn. Sleep Disorders Ctrs., Phi Beta Kappa. Democrat. Jewish. Subspecialties: Psychology education; Psychobiology. Current work: Research contributions in sleep, dreaming, sleep disorders and substance abuse. Home: 75 Westmont St West Hartford CT 06117 Office: U Conn Sch Medicine Farmington CT 06032

FISTEDIS, STANLEY H., nuclear and mechanical engineer; m., 1953; 2 children. B.S., Robert Coll., Istanbul, 1947; M.S., Mont. State U., 1949; Ph.D. in Engring, Mechanics, U. Mo., 1953; M.B.A., U. Chgo., 1965. Designer Babcock & Wilcox Co., 1948-49; instr. in engring. mechanics U. Mo., 1949-52; structural engr. Western Knapp Engring. Co., 1952-53, Johnson & Johnson Co., 1953, Allen & Garcia Co., 1953-54; spl. assignments engr. Girdler Co., 1954-57; assoc. engr. Argonne Nat. Lab., 1957-63, group head engring. mechanics, 1963-66, mgr., 1966-71, program mgr., 1971—; sci. and gen. chmn. 7th Internat. Conf. Structural Mech. Reactor Tech., Chgo., 1983; chmn. Internat. Seminars Containment Nuclear Reactors, San Francisco, 1977, Berlin, 1979, Ispra, Italy, 1981, Chgo., 1983. Prin. editor: Internat. Jour. Nuclear Engring. and Design, 1980—. Fellow ASCE; mem. Am. Nuclear Soc., Nat. Soc. Profl. Engrs., ASME, Internat. Assn. Structural Mech. Reactor Tech. (pres. 1981-83). Subspecialties: Nuclear engineering; Theoretical and applied mechanics. Address: 500 N Parkwood Park Ridge IL 60068

FITCH, FRANK WESLEY, pathologist, educator; b. Bushnell, Ill., May 30, 1929; s. Harold Wayne and Mary Gladys (Frank) F.; m. Shirley Dobbins, Dec. 23, 1951; children—Mary Margaret, Mark Howard. M.D., U. Chgo., 1953, S.M., 1957, Ph.D., 1960. USPHS postdoctoral research fellow, 1954-55, 57-58; faculty U. Chgo., 1957—, prof. pathology, 1967—, Albert D. Lasker prof. med. sci., 1976—, asso. dean med. and grad. edn., 1976—; vis. prof. Swiss Inst. Exptl. Cancer Research, Lausanne, Switzerland, 1974-75. Contbr. chpts. to books, articles profl. jours. Recipient Borden Undergrad. Research award, 1953, Lederle Med. Faculty award, 1958-61; Markle Found. scholar, 1961-66; Commonwealth Fund fellow U. Lausanne (Switzerland) Institut de Biochimie, 1965-66; Guggenheim fellow, 1974-75. Mem. Am. Assn. Immunologists, Am. Assn. Pathologists, Chgo. Path. Soc., Radiation Research Soc., Reticuloendothelial Soc., Sigma Xi, Alpha Omega Alpha. Subspecialties: Immunobiology and immunology;

Pathology (medicine). Current work: Cellular immunology, T cell clones, antibody-producing hybridomas. Home: 5449 Kenwood Ave Chicago IL 60615

FITCH, VAL LOGSDON, physicist; b. Merriman, Nebr., Mar. 10, 1923; s. Fred B. and Frances Marion (Logsdon) F.; m. Elise Cunningham, June 11, 1949 (dec. 1972); children—John Craig, Alan Peter; m. Daisy Harper Sharp, Aug. 14, 1976. B.Eng., McGill U., 1948; Ph.D., Columbia U., 1954. Instr. Columbia, 1953; instr. physics Princeton, 1954-56, asst. prof., 1956-59, 1959-60, prof., 1960—, Class 1909 prof. physics, 1968-76, Cyrus Fogg Bracket prof. physics, 1976—; Mem. Pres.'s Sci. Adv. Com., 1970-73. Trustee Asso. Univ., Inc., 1961-67. Served with AUS, 1943-46. Recipient Research Corp. award, 1967; E.O. Lawrence award, 1968; Wetherill medal Franklin Inst., 1976; Nobel prize in physics, 1980; Sloan fellow, 1960. Fellow Am. Phys. Soc., Am. Acad. Arts and Sci., A.A.A.S.; mem. Nat. Acad. Sci. Subspecialty: Particle physics. Home: 292 Hartley Ave Princeton NJ 08540

FITCH, WALTER MONROE, physiological chemistry educator; b. San Diego, May 21, 1929; s. Chloral Harrison Monroe and Evelyn Charlotte (Halliday) F.; m. Eleanor E. McLean, Sept. 1, 1951; children: Karen Allyn, Kathleen Leslie, Kenton Monroe. A.B., U. Calif.-Berkeley, 1953, Ph.D., 1958. USPHS postdoctoral fellow U. Calif.-Berkeley, 1958-59, Stanford U., Palo Alto, Calif., 1959-61; lectr. Univ. Coll., London, 1961-62; asst. prof. U. Wis., Madison, 1962-67, assoc. prof., 1967-72, prof. physiol. chemistry, 1972—; Fulbright vis. lectr., London, 1961-62; vis. prof. NIH, Hawaii, 1973-74, Macy Found., Los Angeles, 1981-82. Editor: Classification Literature, 1975-80; editor in chief: Molecular Biology and Evolution; assoc. editor jour.: Molecular Evolution, 1976-82; contbr. articles to profl. publs.; reviewer various jours. Chmn. Reapportionment, Madison, 1979-81; mem. Dane County Regional Planning, Madison, 1968-73, Madison Planning Commn., 1965-68, Cupertino Planning Commn., 1960-61. Research grantee NIH, NSF, 1963-83. Mem. Am. Soc. Biol. Chemists, AAAS, Am. Chem. Soc., Genetics Soc. Am., Am. Soc. Naturalists, Soc. for Study Evolution, Biochem. Soc. (Brit.)., Soc. Systematic Zoology. Subspecialty: Evolutionary biology. Home: 1210 Salisbury Pl Madison WI 53711 Office: Univ Wisconsin 1215 Linden Dr Madison WI 53706

FITE, KATHERINE VIRGINIA, neuroscientist; b. Tallahassee, Fla., Oct. 12, 1941; s. Van Roy and Elah Katherine (Porter) F. B.Sc., Fla. State U., 1963; M.Sc., Brown U., 1967, Ph.D. (NIH fellow), 1969. NIH postdoctoral fellow Brown U., 1968-70; asst. prof. psychology U. Mass., Amherst, 1970-74, assoc. prof., 1975-80, prof., 1981—; mem. neuropsychology study sect. NIMH, 1975-79. Editor: The Amphibian Visual System: A Multidisciplinary Approach, 1976; contbr. articles to profl. jours. Recipient NIMH Career Devel. research awards, 1977-82. Mem. Soc. Neurosci., Assn. Research in Vision and Ophthalmology, Am. Assn. Anatomists, AAAS, Sigma Xi. Subspecialties: Neurobiology; Comparative neurobiology. Current work: Comparative anatomy and physiology of vertebrate vision, biomedial studies of inherited retinal degeneration. Office: Div Neuroscience and Behavior U Mass Amherst MA 01003

FITZGERALD, EDWARD ALOYSIUS, microbiologist; b. Washington, Feb. 11, 1942; s. Thomas J. and Cecilia M. (O'Connor) F.; m. Joan L. Markey, May 13, 1967; children: Brian T., Laura M., Janet L., Michael C., Jeffrey P. B.S. in Zoology, Georgetown U., 1963; Ph.D. in Microbiology, Cath. U. Am., 1970. Contract ops. officer div. biologics standards NIH, 1967-68, microbiologhist lab. control activites, 1969-72; microbiologist div. control activities Bur. Biologics, FDA, 1972-74, dep. dir., 1974-80, dep. dir. product quality control, 1980-82; dep. dir. div. product quality control Office of Biologics, Nat. Ctr. Drugs and Biologics, FDA, 1983—. Contbr. articles profl. jours. Served with USPHS, 1967—. Mem. Am. Soc. Microbiology, AAAS, Internat. Assn. Biol. Standardization, Sigma Xi. Subspecialties: Microbiology; Virology (biology). Current work: Quality control of vaccines and other biological products, potency testing of rabies vaccine and rabies immune globulin; research to improve testing of biological products. Office: 8800 Rockville Pike Bethesda MD 20205

FITZGERALD, EDWIN ROGER, educator, physicist; b. Oshkosh, Wis., July 14, 1923; s. James C. and Edwina (Brown) F.; m. Carolyn H. Johnson, Aug. 30, 1946; children: Lucia Edwina, Margaret Mary, William Maurice, Alice Ann, Roger Edwin, Douglas Brendan, Thomas Michael, Jane Carolyn. B.S. in Elec. Engring, U. Wis., 1944, M.S. in Physics, 1950, Ph.D., 1951. Registered profl. engr., Md. Physicist Phys. Research Lab., B.F. Goodrich Co., 1944-46; Project asso. chemistry U. Wis., 1951-52; mem. faculty Pa. State U., 1953-61, prof. physics, 1959-61; prof. dept. mechanics Johns Hopkins U., 1961—. Author: Particle Waves and Deformation in Crystalline Solids, 1966; Contbr. numerous tech. articles to profl. jours., sects. in books. Fellow Am. Phys. Soc. (exec. com., chmn. high polymer Physics 1958-59); mem. Acoustical Soc. Am., Materials Research Soc., Phi Beta Kappa, Sigma Xi, Eta Kappa Nu, Tau Beta Pi. Subspecialties: Polymer physics; Theoretical and applied mechanics. Current work: Wave mechanical explanation of deformation fracture (theoretical); dynamic mechanical measurement apparatus (experimental). Spl. research on mech. and dielectric properties solids. Home: 2445 Tracey's Store Rd Parkton MD 21120

FITZGERALD, JOSEPH JAMES, radiopharmaceutical executive; b. Boston, Mar. 3, 1919; s. Edward J. and Mary J. (Murphy) F.; m. Claire E. Whelan, Aug. 4, 1946; children: Claire Marie, Joseph Francis, Joanne Jacqueline, Edward Gerard, Frances Xaviar, Kevin James. B.S. in Physics, Boston Coll., 1949, M.S., 1950; AEC fellow, U. Rochester, 1950-51. Cert. health physicist, Health Physics Soc., 1956. Research scientist G.E. Knolls Atomic Power Lab., Niskayuna, N.Y., 1951-58; prof. environ. physics Harvard U., 1958-61; radiopharm. exec. Cambridge Nuclear Corp., Billerica, Mass., 1961—; cons. Los Alamos Sci. Lab., 1960-70; mem. Mass. Atomic Energy Commn., Boston, 1959-64; cons. HEW, 1960-73, AEC, 1964-67. Author: Mathematical Theory of Radiation Dosimetry, 1967, Applied Radia, Protection & Control, 1969; contbr. articles to profl. jours. Served with USAAF, 1942-45. Recipient Cross & Crown Soc. award Boston Coll., 1948-49. Mem. Am. Nuclear Soc., Soc. Nuclear Medicine, Clin. Radioassay Soc. Subspecialty: Nuclear medicine. Patentee. Office: Cambridge Nuclear Corp 575 Middlesex Turnpike Billerica MA 01865

FITZGIBBON, WILLIAM EDWARD, III, mathematics educator, consultant; b. Cambridge, Mass., July 21, 1945; s. William Edward, Jr. and Florence (Steuteville) F.; m.; 1 son, William Edward. B.A., Vanderbilt U., 1968, Ph.D., 1972. NASA trainee Vanderbilt U., Nashville, 1968-72; asst. prof. U. Houston, 1972, assoc. prof., 1972-80, prof. dept. math., 1981—; vis. assoc. prof. U. Calif.-San Diego, 1979-81; summer visitor Argonne (Ill.) Nat. Lab., 1980. Editor: Nonlinear Diffusion, 1976. Recipient various fed. and univ. grants. Mem. Am. Math. Soc., Soc. Indsl. and Applied Math., Sigma Xi. Subspecialty: Applied mathematics. Current work: Qualitative theory of nonlinear partial differential equations and dynamical systems. Home: 5401 Chimney Rock 798 Houston TX 77081 Office: Dept Math Univ Houston Houston TX 77081

FIVOZINSKY, SHERMAN PAUL, physicist; b. Hartford, Conn., Aug. 2, 1938; s. Irving and Betty P. (Pivnick) F.; m. Marilyn Leibowitz, June 18, 1961; children: Karen Beth, Laurie Sue. B.A., U. Conn., 1961, M.S., 1963, Ph.D., 1971. Nuclear structure researcher Nat. Bur. Standards, Washington, 1966-76; asst. to chief Office of Standard Reference Data, 1976—; adj. prof. physics Montgomery Coll., Rockville, Md., 1973—. Author: Measurements for the Safe Use of Radiation, 1976, Medical Physics Data Book, 1982; also articles. Mem. Am. Phys. Soc., Am. Assn. Physicists in Medicine, AAAS, Sigma Xi. Subspecialties: Nuclear physics; Radiology. Home: 5503 Manorfield Rd Rockville MD 20853 Office: Physics Bldg A323 Washington DC 20234

FIX, GEORGE JOSEPH, mathematics educator, consultant; b. Dallas, May 10, 1939; s. George Joseph and Francis (Barlett) F.; m. Linda Mitchell, June 30, 1962; children: Paige, Blake. B.S., Tex. A&M U., 1963; M.S., Rice U., 1965; Ph.D., Harvard U., 1968. Engr. Tex. Instruments, Dallas, 1963-64; assoc. prof. U. Md., 1972-73, U. Mich., 1973-75; prof., head dept. math. Carnegie-Mellon U., Pitts., 1975—; cons. Author 2 books, numerous articles. Served in USMC, 1958-62. NASA fellow, 1981-83; grantee Office Naval Research, Army Research Office, NSF. Mem. Am. Math. Soc., Soc. Indsl. and Applied Math., AAAS, Sigma Xi, Tau Beta Pi. Subspecialties: Numerical analysis; Applied mathematics. Current work: Numerical analysis and scientific computing. Office: Carnegie-Mellon U Pittsburgh PA 15213

FIXMAN, MARSHALL, educator, chemist; b. St. Louis, Sept. 21, 1930; s. Benjamin and Dorothy (Finkel) F.; m. Marian Ruth Beatman, July 5, 1959 (dec. Sept. 1969); children—Laura Beth, Susan Ilene, Andrew Richard; m. Branka Ladanyi, Dec. 7, 1974. A.B., Washington U., St. Louis, 1950; Ph.D., Mass. Inst. Tech., 1954. Jewett postdoctoral fellow chemistry Yale, 1953-54; instr. chemistry Harvard, 1956-59; sr. fellow Mellon Inst., Pitts., 1959-61; prof. chemistry, dir. Inst. Theoretical Sci., U. Oreg., 1961-64, prof. chemistry, research asso. inst., 1964-65; prof. chemistry Yale, New Haven, 1965-79; prof. chemistry and physics Colo. State U., Ft. Collins, 1979—. Asso. editor: Jour. Chem. Physics, 1962-64, Jour. Phys. Chemistry, 1970-74, Macromolecules, 1970-74. Served with AUS, 1954-56. Fellow Alfred P. Sloan Found., 1961-63; recipient Governor's award Oreg. Mus. Sci. and Industry, 1964. Mem. Nat. Acad. Scis., Am. Acad. Arts and Scis., Am. Chem. Soc. (award pure chemistry 1964), Am. Phys. Soc. (high polymer physics award 1980), Fedn. Am. Scientists. Subspecialties: Statistical mechanics; Theoretical chemistry. Current work: Theory of liquids and polymers. Address: Dept of Chemistry Colo State U Fort Collins CO 80523

FLACK, RONALD DUMONT, JR., mechanical engineer, educator, researcher, consultant; b. South Bend, Ind., Dec. 24, 1947; s. Ronald Dumont and Alpha Jeanette F.; m. Nancy Lee Slauson, Aug. 30, 1969; children: Melissa Beth, Todd Alan. B.S.M.E., Purdue U., 1970, M.S.M.E., 1973, Ph.D. in Mech. Engring, 1975. Registered profl. engr., Va. Analytical design engr. Pratt & Whitney Aircraft Co., West Palm Beach, Fla., 1970-71; research and teaching asst. Purdue U., West Lafayette, Ind., 1971-75; asst. prof. mech. engring. U. Va., Charlottesville, 1976-81, assoc. prof., 1981—; cons. on flows on pumps, compressors and turbines, devel. lab. equipment, thermal analysis of electronic equipment. Contbr. articles on bearing lubrication, laser velocimetry, heat transfer, rotor dynamics to profl. jours. Asst. dir. Y-Indian Guides, Charlottesville, 1982. Research grantee NIH, ERDA/Dept. Energy, NASA, NSF, Dept. Agro. Mem. ASME, Am. Soc. Lubrication Engring., Am. Soc. Engring. Edn. (Outstanding Young Faculty award 1980), Sigma Xi, Pi Tau Sigma, Tau Beta Pi, Phi Beta Phi. Subspecialties: Mechanical engineering; Fluid mechanics. Current work: Fluid mechanics on turbomachines, lubrication, laser velocimetry. Home: 4265 Viewmont Rd Earlysville VA 22936 Office: Dept Mech and Aero Engring Thornton Hal U Va Charlottesville VA 22901

FLAGG, RAYMOND OSBOURN, biologist, scientific supply company executive; b. Martinsburg, W. Va., Jan. 31, 1933; s. Dorsey Slemons and Dorothy (Hobbs) F.; m. Ann Birmingham, May 19, 1956; children: Richard M., Elizabeth Flagg Laseau, Catherine G. B.A. with honors, Shepherd Coll., W.Va., 1957; Ph.D. in Biology, U. Va., 1961. Math. tchr. Boonsboro (Md.) High Sch., 1957; research asst. Blandy Exptl. Farm, Boyce, Va., 1957-61; research assoc. U. Va., Charlottesville, 1961-62; dir. botany Carolina Biol. Supply Co., Burlington, N.C., 1962-80, v.p., 1980—; mem. N.C. Plant Conservation Bd., 1980—. Contbr. articles to sci. jours. Chmn. Burlington Beautification Commn., 1976-80, Burlington Hist. Commn., 1981-82; pres. Williams High Sch. PTA, 1979-80. Served to cpl. U.S. Army Security Agy., 1952-55. Recipient Progressive Community Leadership award No. Piedmont Area Devel. Assn., 1977. Mem. Assn. Southeastern Biologists (pres. 1978-79), N.C. Acad. Sci. (pres. 1983-84), AAAS, Bot. Soc. Am., So. Appalachian Bot. Club, Va. Acad. Sci., Sigma Xi. Presbyterian (elder). Subspecialties: Systematics; Genetics and genetic engineering (biology). Current work: Improved lab instruction in the sciences; administrative, endangered species, science education, lab supplies, taxonomy, genetics, plants, animals, Zephyranthes, Drosophila. Home: 712 W. Davis St Burlington NC 27215 Office: Carolina Biol. Supply Co Burlington NC 27215

FLAHERTY, LORRAINE AMELIA, medical institute administrator; b. Rochester, N.Y., Jan. 8, 1946; d. John Thomas and Amelia Grace (Molinari) Plati; m. Joseph Edward Flaherty, June 17, 1967. B.S., Tufts U., 1967; Ph.D., Cornell U. Med. Sch., 1973. Research scientist N.Y. State Dept. Health, Albany, 1973—; dir. Kidney Disease Inst.; adj. asst. prof. Duke U., Durham, N.C., 1977—; mem. immunobiology study sect. NIH, 1978-82. Mem. Am. Assn. Immunologists, Sigma Xi. Subspecialties: Immunobiology and immunology; Immunogenetics. Current work: Immunogenetics of major histocompatibility complex; lymphocyte differentiation; tumor biology; genetics and biology of cell surface antigens. Office: NY State Dept Health Empire State Plaza Albany NY 12201

FLAIM, STEPHEN FREDERICK, cardiovascular physiologist; b. San Jose, Calif., May 28, 1948; s. Francis Richard and Cecilia Martha F.; m. Kathryn Erskine, Aug. 16, 1975; children: Bryna Kathryn, Celia Elizabeth. B.S., U. Santa Clara, Calif., 1970; Ph.D., U. Calif.-Davis, 1975. Asst. prof. med. and physiol. scis. Pa. State U. Coll. Medicine, Hershey, 1978-82; prin. scientist McNeil Pharm., Spring House, Pa., 1982—. Editor: Calcium Blockers: mechanisms of Action and Clinical Applications, 1982; author numerous sci. abstracts and book chpts.; editorial referee numerous sci. jours. Edwin J. Brown fellow U. Santa Clara, 1971; NIH research fellow, 1976-78; NIH Young Investigator awardee, 1980-82. Fellow Am. Heart Assn. (research fellow 1975-76), Am. Coll. Clin. Pharmacology; mem. Am. Physiol. Soc., AAAS, Soc. Exptl. Biology and Medicine. Subspecialties: Physiology (medicine); Pharmacology. Current work: Cardiovascular physiology and pharmacology—study of cardiocirculatory dynamics and cardiac output distribution; study of control of calcium ion flux and contraction in vascular smooth muscle; calcium antagonists. Home: 900 Cross Ln Blue Bell PA 19422 Office: Dept Biol Research McNeil Pharmaceutical Spring House PA 19477

FLANAGAN, CHARLES ALLEN, engineering administrator; b. Aultman, Pa., Oct. 14, 1931; s. Paul John and Mabel Celia (Bloomquist) F.; m. Jane Hoyt Pell, Nov. 27, 1959; children: Catherine Anne, William Patrick. B.S. in Physics, Lafayette Coll., 1953; M.B.A., U. Pitts., 1972. Jr. scientist Westinghouse-Bettis, West Mifflin, Pa., 1956-57, assoc. scientist, 1957-58, scientist, 1958-62, sr. scientist, 1962-66, supr., mgr., 1966-75, mgr., Madison, Pa., 1975-80; dep. mgr. fusion engring. design ctr., Oak Ridge, 1980—; fusion reactor studies participant IAEA, Vienna, 1980-82. Served as 1st lt. U.S. Army, 1953-55. Recipient cert. of appreciation Dept. Energy, 1981. Mem. Am. Nuclear Soc. (chmn. fusion energy div., program com. 1977-78, vice-chmn. fusion div. 1982, chmn. div. 1983). Unitarian. Subspecialties: Nuclear fusion; Nuclear fission. Current work: Design and analysis of next generation fusion devices. Home: 106 Dana Dr Oak Ridge TN 37830 Office: Fusion Engring Design Ctr Westinghouse Oak Ridge Nat Lab PO Box Y Oak Ridge TN 37830

FLANAGAN, JAMES LOTON, electrical engineer. B.S. in Elec. Engring, Miss. State U., 1948, S.M., M.I.T., 1950, Sc.D., 1955. Mem. elec. engring. faculty Miss. State U., 1950-52; mem. tech. staff Bell Labs., Murray Hill, N.J., 1957-61, head dept. speech and auditory research, 1961-67, head dept. acoustics research, 1967—. Author: Speech Analysis, Synthesis and Perception, 1972; contbr. numerous articles to profl. jours. Mem. evaluation panel Nat. Bur. Standards/NRC, 1972-77; mem. adv. panel on White House tapes U.S. Dist. Ct. for D.C., 1973-74; bd. govs. Am. Inst. Physics, 1974-77; mem. sci. adv. bd. Callier Center, U. Tex., Dallas, 1974-76; mem. sci. adv. panel on voice communications Nat. Security Agy., 1975-77; mem. sci. adv. bd. div. communications research Inst. Def. Analyses, 1975-77. Recipient Disting. Service award in sci. Am. Speech and Hearing Assn., 1977. Fellow IEEE (mem. fellow selection com. 1979-81), Acoustical Soc. Am. (asso. editor Speech Communication 1959-62, exec. council 1970-73, v.p. 1976-77, pres. 1978-79); mem. Acoustics, Speech and Signal Processing Soc. (v.p. 1967-68, pres. 1969-70, Achievement award 1970, Soc. award 1976), Nat. Acad. Engring. of Nat. Acad. Scis. Subspecialty: Electrical engineering. Current work: Digital communications. U.S. and fgn. patentee in field. Office: 600 Mountain Ave Murray Hill NJ 07974

FLANEGAN, JAMES BERT, microbiologist, educator, researcher; b. Tampa, Fla., Aug. 2, 1946; s. James H. and Julia L. F.; m. Dawn Elaine, July 12, 1980; children: Ryan, Jennifer, Amy. B.S., Fla. State U., 1968, M.S., 1969; Ph.D., U. Mich., 1975. Postdoctoral fellow Center for Cancer Research, M.I.T., 1975-78; assoc. prof. immunology and med. microbiology Coll. Medicine, U. Fla., 1978—. Contbr. articles to profl. jours. Served to lt. (s.g.) USNR, 1970-72. NIH postdoctoral fellow, 1975-77; NIH grantee, 1979—; NSF grantee, 1979—. Mem. Am. Soc. Microbiology, Am. Soc. Virology, Sigma Xi (Fla. chpt. Faculty Research award 1981). Subspecialties: Virology (biology); Molecular biology. Current work: Molecular biology of RNA viruses. Office: Dept Immunology and Med Microbiology Coll Medicine Box J-266 U Fla Gainesville FL 32610

FLANNERY, BRIAN PAUL, applied mathematician; b. Utica, N.Y., July 30, 1948; s. Richard F. and Estelle R. F.; m. Sharon A. Flannery, May 23, 1970; children: Colleen C., Paul E. B.A., Princeton U., 1970; Ph.D. in Astrophysics, U. Calif.-Santa Cruz, 1974. Research assoc. Inst. Advanced Study, Princeton, N.J., 1974-76; asst. prof. astronomy Harvard U., 1976-80, assoc. prof., 1980; sr. staff physicist Exxon Research & Engring., Clinton, N.J., 1982—, research assoc., 1982—. Contbr. articles to profl. jours. Mem. Am. Astron. Soc., Internat. Astron. Union. Subspecialties: Theoretical astrophysics; Theoretical physics. Current work: Climate modeling, fluid flow modeling, tomography, astrophysics. Home: 24 River Bend Rd Clinton NJ 08809 Office: Exxon Research and Engineering Rt 22 E Clinton Twp Annandale NJ 08801

FLASAR, F. MICHAEL, physicist; b. East St. Louis, Ill., Feb. 20, 1946; s. Frank A. and Kathryn C. (Zedolek) F.; m. Diane L. Hallen, May 31, 1969; 1 son, Mark. B.S. in Physics, M.I.T., 1967, Ph.D., 1972. Research fellow Ctr. for Earth and Planetary Physics, Harvard U., 1971-75; NRC resident research assoc. NASA Goddard Space Flight Ctr., 1975-77, staff scientist, 1977—. Contbr. articles to profl. jours. NSF predoctoral fellow, 1967-71; NRC research grantee, 1975-77. Mem. Am. Astron. Soc. (div. planetary scis.), Am. Geophys. Union. Subspecialties: Planetary science; Planetary atmospheres. Current work: Dynamics of planetary atmospheres. Office: NASA/Goddard Space Flight Ctr Code 693 Greenbelt MD 20771

FLASCHEN, STEWARD SAMUEL, corporation executive; b. Berwyn, Ill., May 28, 1926; s. Hyman Herman and Ethel (Leviton) F.; m. Joyce Davies, Apr. 21, 1949; children: John, Sheryl, David, Evan. B.S. in Chemistry, U. Ill., 1947; M.S., Miami U., Oxford, Ohio, 1948; Ph.D. in Geochemistry, Pa. State U., 1953. Supr. research dept. Bell Telephone Labs., Murray Hill, N.J., 1952-59; dir. phys. scis., research and devel., semiconductor products div. Motorola, Inc., Phoenix, 1959-64; sr. v.p., gen. tech. dir. ITT Corp., N.Y.C., 1964—; lectr. Pace U. Grad. Sch. Bus. Author: Search and Research, 1965; also articles. Mem. Phoenix Bd. Edn., 1962-64. Served with USNR, 1944-46. Fellow Am. Inst. Chemists, IEEE; mem. Electromech. Soc. Am., Am. Ceramic Soc., AAAS, Indsl. Research Inst., N.Y. Acad. Scis. Subspecialty: Research and development management. Patentee in field. Home: 592 Weed St New Canaan CT 06840 Office: 320 Park Ave New York NY 10022 I was fortunate in my education to have had broad exposure to philosophy, the natural sciences and English. This early training in the reduction to basics of problems, concepts, and decisions, and in the skill of communicating effectively, has been of the utmost value to me professionally and personally.

FLAX, ALEXANDER HENRY, administrator, aeronautical engineer; b. Bklyn., Jan. 18, 1921; s. David and Etta (Schenker) F.; m. Ida Leane Warren, Aug. 25, 1951; 1 dau., Laurel Elizabeth. B.Aero. Engring., N.Y. U., 1940; Ph.D. in Physics, U. Buffalo, 1958. Structure, vibration engr. airplane div. Curtiss-Wright Corp., 1940-44; chief aerodynamics and structures Piasecki Helicopter Corp., 1944-46; asst. head aeromechanics dept. Cornell Aero. Lab., 1946-49, head aerodynamics dept., 1949-55, asst. dir., 1955-56, v.p., tech. dir., 1956-59, 61-63; chief scientist USAF, 1959-61; asst. sec. Air Force for research and devel., 1963-69; v.p. for research Inst. Def. Analyses, Arlington, Va., 1969, pres., 1969—; mem. com. aerodynamics NACA, 1952-54, subcom. highspeed aerodynamics, 1954-58; adv. com. aircraft aerodynamics NASA, 1958-62; mem. sci. com. nat. reps. SHAPE Tech. Center, The Hague, Netherlands, 1963-69, chmn., 1965-67; U.S. del. adv. com. aero. research and devel. NATO, 1969—; mem. bd. direction Von Karman Inst., Brussels, 1969—; mem. adv. council Stanford U. Sch. Engring., 1981—. Contbr. sect. to book, numerous articles to profl. jours. Recipient Air Force Exceptional Civilian Service awards, 1961, 69; NASA Distinguished Service medal, 1968; Civilian Service medal Def. Intelligence Agy., 1974; Von Karman medal NATO Adv. Group for Aerospace Research and Devel., 1978; Medal for Disting. Pub. Service Dept. Def., 1983. Hon. fellow AIAA (Lawrence Sperry award 1949, Wright Bros. lectr. 1959); fellow Royal Aero. Soc. (Wright Bros. Meml. Lectr. 1974); mem. Nat. Acad. Engring., Sigma Xi. Club: Cosmos (Washington). Subspecialty: Aeronautical engineering. Home: 9007 Belmart Rd Potomac MD 20854 Office: 1801 N Beauregard St Alexandria VA 22311

FLAY, BRIAN RICHARD, university researcher, educator; b. Hamilton, N.Z., Feb. 1, 1947; came to U.S., 1976. Student, Canterbury U., 1967-68; B.Social Sci., Waikato U., New Zealand, 1972, M.Social Sci., 1973; postgrad., London Sch. Econs., 1973-74; Ph.D., Waikato U., 1976, (Fulbright-Hays fellow), Northwestern U., 1976-78. Instr. stats. Sarawak Agr. Dept., Malaysia, 1966; instr. Waikato Tech. Inst.,

1970-75; lectr. dept. psychology Waikato U., 1972-76; vis. asst. prof. Northwestern U., 1977-78; asst. prof. health studies dept. U. Waterloo, Ont., Can., 1978-80, assoc. chmn. grad. affairs, 1979-80, adj. asst. prof., 1980—; asst. prof. health behavior research U. So. Calif. Sch. Pharmacy, 1980—, asst. dir. Health Behavior Research Inst., 1980—, adj. asst. prof. Annenberg Sch. Communication, 1982—; cons. in field. Contbr. articles to profl. jours. HEW grantee, 1976-77; Rockefeller Found. grantee, 1977-78; Ont. Lung Assn. grantee, 1979-80; Ont. Ministry Health grantee, 1980; Can. Cancer Soc. grantee, 1980; Nat. Inst. Drug Abuse grantee, 1982—; Am. Lung Assn. grantee, 1981—; Nat. Cancer Inst. grantee, 1983—. Mem. AAAS, Am. Psychol. Assn., Am. Pub. Health Assn., Evaluation Research Soc., Internat. Union Health Edn., Soc. Psychol. Study of Social Issues, Soc. Behavioral Medicine. Subspecialties: Preventive medicine; Social psychology. Current work: Prevention of cigarette smoking and drug abuse; use of news media for health promotion; evaluation research. Office: Health Behavior Research Inst Sch Pharmac U So Calif 1985 Zonal Ave Los Angeles CA 90033

FLECK, DAVID CHARLES, elec. engr.; b. Findlay, Ohio, Aug. 13, 1954; s. Charles Edward and Mary Catherine (Thiry) F. B.S.E.E., U. Cin., 1978. Engr. EPA, Cin., Annapolis, Md., 1976-80; pres. Capella Cons., Laurel, Md., 1980—; sr. engr. GTCO Corp., Rockville, Md., 1982—. Mem. Assn. Computing Machinery, IEEE, AAAS. Democrat. Roman Catholic. Subspecialties: Computer engineering; Computer architecture. Current work: Computer graphics, image processing, pattern recognition; microcomputer applications, computer architectures. Home: 16127 Kenny Rd Laurel MD 20707 Office: 1055 1st St Rockville MD 20850

FLECK, GEORGE MORRISON, chemist, educator; b. Warren, Ind., May 13, 1934; s. Ford Bloom and Deloris Magdalene (Morrison) F.; m. Margaret Dyer Reynolds, June 27, 1959; children: Margaret Morrison, Louise Elizabeth. B.S., Yale U., 1956; Ph.D., U. Wis., 1961. Asst. prof. Smith Coll., Northampton, Mass., 1961-67, assoc. prof., 1967-76, prof. chemistry, 1976—. Author: Equilibria in Solution, 1966, Chemical Reaction Mechanisms, 1971, Chemistry: Molecules That Matter, 1974, Carboxylic Acid Equilibria, 1973, Patterns of Symmetry, 1974; Contbr. articles to profl. jours. Griffin scholar, 1952-56; DuPont fellow, 1960-61; Danforth fellow, 1956-61; Danforth assoc., 1981—; grantee NSF, NIH, U.S. Office Edn., Am. Philos. Soc. Subspecialties: Physical chemistry; Analytical chemistry. Current work: Kinetics and equilibria of reactions in aqueous solution; history of physical chemistry (especially kinetics) in the nineteenth century. Home: Williamsburg MA 01096 Office: Clark Science Center Northampton MA 01063

FLECK, ROBERT DAVIS, chem. engr.; b. Woodbury, N.J., Nov. 17, 1922; s. William Chalmer and Ethel Marie (Davis) F.; m. Mollie Reeves, Nov. 6, 1943; children: Robert Davis, E. Thomas, David M. B.S.Ch.E., Drexel U., 1947. Process engr. Budd Polychem, Newark, Del., 1947-51, quality control mgr., 1951-62; research engr. Spaulding Fibre, Tonawanda, N.Y., 1962-75; dir. research Griffin Wheel Anchor Div., West Chicago, Ill., 1975—. Served as 1st lt. USAAC, 1942-45. Decorated Air medal. Mem. ASME, Chgo. Rubber Group, Soc. Plastics Engrs., Central Air Brake Club, Air Brake Assn. Subspecialties: Chemical engineering; Composite materials. Current work: Railroad composition brake shoes. Development of improved brake shoe materials to use in stopping freight cars, locomotives and transit cars. Home: 849 Woodland Hills Rd Batavia IL 60510 Office: 1920 Downs Dr West Chicago IL 60185

FLEISCH, JEROME HERBERT, pharmacologist; b. Bronx, N.Y., June 6, 1941; s. Wolf and Miriam (Glaser) F.; m. Marlene L. Cohen, Aug. 8, 1976; children: Abby Faye, Sheryl Brynne. B.S., Columbia U., 1963; Ph.D., Georgetown U., 1967. Research fellow Harvard U. Med. Sch., Boston, 1967-68; research asso./sr. staff fellow Nat. Heart and Lung Inst., NIH, Bethesda, Md., 1968-74; research asso. Lilly Research Labs., Eli Lilly & Co., Indpls., 1976—. Contbr. articles to sci. jours. Served to lt. comdr. USPHS, 1968-70. Mem. Am. Soc. Pharmacology and Exptl. Therapeutics, Soc. Exptl. Biology and Medicine; mem. Am. Acad. Allergy and Immunology; Mem. Collegium Internationale Allergologicum. Subspecialties: Pharmacology; Allergy. Current work: Mechanisms involved in antigen-induced release of mediators of anaphylaxis - leukotriene receptors. Office: Lilly Research Labs MC905 Indianapolis IN 46285

FLEISCHER, ROBERT, research management center executive, science administration consultant; b. Flushing, N.Y., Aug. 20, 1918; m. Avis Collins, June 14, 1942; children: Martha, Warren, Stephen; m. Marle Moorefield, May 10, 1980. B.S., Harvard U., 1940, fellow, 1942, M.A., 1947, Ph.D., 1949; Steward Obs. fellow, U. Ariz., 1940-41. Asst. in astronomy Harvard U., 1941-42; instr. physics, astronomy Rensselaer Poly. Inst., 1946-49, asst. prof., 1941-55, assoc. prof., 1955-57, prof. astronomy, head obs., 1958-62; program dir. solar-terrestrial research, coordinator Internat. Years of Quiet Sun, NSF, Washington, 1962-66; dep. head Office Internat. Sci. Activities, Washington, 1966-68, head astronomy sect., 1968-75; program dir. spl. activities, exec. sec. interagy. coordinating com. on astronomy Fed. Council for Sci. and Tech., 1975-76; dir. The Greylock Center, Washington, 1976—; cons. Mem. Internat. Sci. Radio Union, Internat. Astron. Union, Am. Astron. Soc., Astron. Soc. Pacific, AAUP, Am. Mgmt. Assn. Soc. Research Adminstrs. Democrat. Episcopalian. Club: Cosmos. Subspecialties: Optical astronomy; Radio and microwave astronomy. Current work: Management and Support of research. Home: 1733 Church St NW Washington DC 20036 Office: The Greylock Center 1346 Connecticut Ave Suite 925 Washington DC 20036

FLEISCHMANN, HANS HERMANN, physicist, educator; b. Munich, Germany, June 2, 1933; came to U.S., 1963, naturalized, 1978; s. Paul and Gertrud (Jaenicke) F. Dipl. Phys., Tech. U. Munich, 1959, Dr.rer.nat., 1962. Research assoc. Tech. U. Munich, 1962-63; cons. Rohde & Schwarz, Messgeraetebau, Munich, 1962-63; mem. staff Gen. Atomic, San Diego, 1963-67; prof. applied physics Sch. Applied and Engring. Physics, Cornell U., Ithaca, N.Y., 1967—; cons. to govt. and industry. Contbr. articles to profl. jours. Fellow Am. Phys. Soc. (div. plasma physics, past mem. exec. com.), IEEE, Plasma Sci. Soc. (exec. com., adminstrv.com.); mem. Am. Nuclear Soc. (past mem. exec. com.). Subspecialties: Plasma physics; Fusion. Current work: Plasma physics and controlled fusion, in particular in area of compact toroids, EBT, and intense electron and ion beams, atomic collisions. Office: Sch Applied and Engring Physics Cornell U Ithaca NY 14853

FLEISHMAN, MORTON ROBERT, nuclear engineer; b. N.Y.C., Jan. 21, 1933; s. Nathan and Frieda (Ginsburg) F.; m. Rhea Carol Watnick, June 18, 1961; children: Kevin Marc, Shari Lynn. B.S.E., U. Mich.-Ann Arbor, 1954; M.S.E., 1955; diploma reactor tech., Oak Ridge Sch. Reactor Tech., 1956; postgrad. physics, N.Y. U., 1956-59. Power plant engr. Martin Co., Balt., 1954; project engr. U. Mich. Engring. Research Inst., Ann Arbor, 1955; sr. scientist United Nuclear Corp., White Plains, N.Y., 1956-63; nuclear physicist U.S. AEC Space Nuclear Propulsion, Cleve., 1963-72, Space Nuclear Systems, Washington, 1972-73; sr. nuclear engr. U.S. Nuclear Regulatory Commn., Washington, 1973—. Henry Earle Riggs fellow U. Mich., 1954; recipient Spl. Achievement award U.S. Nuclear Regulatory Commn., 1982. Mem. AIAA, Am. Nuclear Soc., Aircraft Owner and Pilots Assn., Sigma Xi, Tau Beta Pi. Club: West River Sailing (Galesville, MD.). Subspecialties: Nuclear fission; Nuclear engineering. Current work: Responsible for writing rules and regulations covering the safety of nuclear power reactors. Home: 10623 Great Arbor Dr Potomac MD 20854 Office: US Nuclear Regulatory Commn Office Research 20555 DC

FLEISIG, ROSS, aeronautical engineer, engineering manager; b. Montreal, Que., Can., Oct. 12, 1921; came to U.S., 1922, naturalized, 1922; s. Samuel and Ethel (Levy) F.; m. Majorie H. Hall, June 6, 1943; children: Ann, Dale. B.S. in Aero. Engring, Poly. Inst. Bklyn., 1942, M.S., 1955. Sr. aero-dynamicist Chance Vought Aircraft Corp., Stratford, Conn., from 1942, Dallas, to 1950; engring. sect. head Sperry Gyroscope Co., Great Neck, N.Y., 1950-61; project mgr. Grumman Aerospace Corp., Bethpage, N.Y., 1961—. Editor: Lunar Flight Programs, 1964, Lunar Exploration and Spacecraft Systems, 1962; contbr. articles to tech. jours. Fellow Am. Astronautical Soc. (pres. 1957-58, dir. 1958-68), AIAA, AAAS; corr. mem. Internat. Acad. Astronautics. Clubs: Garden City Casino, University of L.I. (dir. 1979-81). Subspecialties: Aeronautical engineering; Theoretical and applied mechanics. Current work: Aerospace technology management; aerospace guidance and control systems. Home: 58 Kilburn Rd Garden City NY 11530

FLEMING, JAMES STUART, JR., pharmacologist; b. Buffalo, Sept. 1, 1936; s. James Stuart and Pauline (McClurg) F.; m. Marilyn Joyce Bartsch, June 7, 1960; children: Lois Vernette, James Stuart III. B.A., Northwestern U., 1958; M.A., U. Buffalo, 1962; Ph.D., Ohio State U., 1965; M.B.A., Syracuse U., 1983. Teaching asst. U. Buffalo, 1961-62; research asst. Ohio State U., Columbus, 1962-65; dir. cardiovascular biology (Bristol-Myers Sci. and Tech div.), Syracuse, N.Y., 1965—. Contbr. articles to profl publs. NIH tng. grantee, 1958-61. Mem. Am. Soc. Pharmacology and Exptl. Therapeutics, Am. Heart Assn., Council on Thrombosis, Microcirculatory Soc., Internat. Soc. on Oxygen Transport to Tissue.; mem. Beta Gamma Sigma. Subspecialties: Pharmacology; Hematology. Current work: Supervision and development of a comprehensive research program for identification and development of new drugs for the prevention and treatment of occlusive vascular disease, including planning of early stages of clinical testing, general cardiovascular profiling of pharmacodynamic agents. Patentee pharmacologic agts in prevention of thrombosis. Office: Dept Cardiovascular Biology Bristol-Myers Thompson Rd Syracuse NY 13201

FLEMING, KARL NEILL, nuclear scientist/engineer, consultant; b. Uniontown, Pa., July 18. 1947; s. John Nesbit and Gladys Ruth (Helmick) F.; m. Suzanne Watson, July 14, 1980. B.S. in Physics, Pa. State U., 1969; M.S. in Nuclear Sci. and Engring, Carnegie-Mellon U., 1974. Engr. Gen. Atomic Co. (name now GA Techs.), San Diego, 1974-75, sr. engr., 1975-77, staff engr., 1977-80, br. mgr., 1980-81; dir. probabilistic risk assessment methods Pickard, Lowe & Garrick, Irvine, Calif., 1981-82; project mgr. Seabrook Sta. Probabilistic Safety Assessment, 1982—; chmn. working group on probabilistic risk assessment U.S. Dept. Energy, 1982—. Author: IEEE/Am. Nuclear Soc. Probabilistic Risk Assessment Procedures Guide, 1981. Dudley Meml. scholar, 1968-69. Mem. Am. Nuclear Soc., AAAS, Sigma Pi Sigma, Pi Mu Epsilon, Phi Kappa Phi. Current work: Made major contributions to early research and development in the new technology of probabilistic risk (safety) assessment as it is applied to assessment of nuclear reactor safety; refined the state-of-the-art in systems reliability engineering in the analysis of dependent (common cause) failures. Developer model for reliability analysis of systems called Beta Factor Method. Office: Pickard Lowe & Garrick 17840 Sky Park Irvine CA 92714

FLEMING, REX JAMES, scientist; b. Omaha, Apr. 25, 1940; s. Robert Leonard and Doris Mae (Burrows) F.; m. Kathleen Joyce Ferry, Sept. 3, 1969; children: Thane, Manon, Mark, Noel. B.S., Creighton U., 1963; M.S., U. Mich., 1968, Ph.D., 1970. Commd. lt. U.S. Air Force, 1963, resigned commn. as capt., 1972; research scientist, Offutt AFB, Nebr., 1963-67; sci. liaison to Nat. Weather Service for Air Weather Service, Suitland, Md., 1970-72; ret., 1972; mgr. applications mktg. advanced sci. computer Tex. Instruments, Inc., Austin, 1972-75; dir. U.S. Project Office for Global Weather Expt., NOAA, Rockville, Md., 1975-80, Spl. Research Projects Office, 1980-82, Office of Climate and Atmospheric Research, 1983—. Contbr. articles to profl. jours. Recipient Gold Medal award Dept. Commerce, 1980. Mem. Am. Meteorol. Soc. (chmn. probability and statistics com. 1976-77), Am. Soc. Photogrammetry, Am. Geophys. Union, AAAS, The Ocean Soc. Republican. Subspecialty: Government program administration. Home: 9200 Bells Mill Rd Potomac MD 20854 Office: 6010 Executive Blvd Rockville MD 20852 One need only be inspired by its spring-morning freshness, stimulated by its magnificent variety of color and form, and humbled by the power of its ever-present energy, to be driven to unveil the secrets of our life-sustaining atmosphere.

FLEMINGS, MERTON CORSON, metallurgy educator, university administrator, researcher; b. Syracuse, N.Y., Sept. 20, 1929; m.; 3 children. S.B., MIT, 1951, S.M., 1952, SC.D. in Metallurgy, 1954. Metallurgist Am. Brake Shoe Research Lab., 1954-56; from asst. prof. to assoc. prof. metallurgy MIT, 1956-69, Abex prof., 1970-75, Ford prof., 1975-81, prof. metallurgy, 1969—, Toyota prof. materials processing, 1981—; assoc. dir. Ctr. Materials Sci. and Engring., 1973-77; dir. Materials Processing Ctr., 1979—; cons. to govt. labs., industry, 1970—. Recipient Albert Sauveur Achievement award, 1978; overseas fellow Churchill Coll., Eng., 1970-71. Fellow Am. Soc. Metals (Henry Marion Howe medal 1973); mem. Nat. Acad. Engring., Am. Foundrymen's Soc. (Simpson Gold medal 1961), Inst. Metals London, AIME (Mathewson Gold medal 1969). Subspecialty: Materials (engineering). Office: MIT Room 8-407 Cambridge MA 102139

FLESSA, HERBERT CHRISTIAN, physician, educator; b. Cin., Mar. 21, 1926; s. Herbert L. and Eleanor (Vockell) F.; m. Jeannine McCue, May 31, 1952; children: Thomas, John, James, Joseph. B.S., U. Cin., 1950, M.D., 1952. Cert. Bd. Internal Medicine, Hematology, Oncology. Intern Cin. Gen. Hosp., 1952-53, resident in internal medicine, 1961-62; gen. practice medicine, cin., 1954-61, cons. hematology/oncology, 1964—; asst. prof. medicine U. Cin., 1964-67, assoc. prof., 1967-72, prof., 1972—. Bd. dirs. Ohio chpt. Leukemia Soc. Am. Served with USNR, 1944-46. Mem. ACP, Cin. Soc. Internal medicine (pres. 1978), Am. Soc. Hematology, Am. Soc. Oncology, Clin. Research Soc. Clubs: Faculty, Cin. Tennis. Subspecialties: Hematology; Oncology. Current work: Clinical research leukemia, coagulation disorders, selected cancers. Office: U Cin Coll Medicine 231 Bethesda Ave Cincinnati OH 45267 Home: 44 Rawson Woods Cresent Cincinnati OH 45220

FLETCHER, EDWARD ABRAHAM, educator; b. Detroit, July 30, 1924; s. Morris and Lillian (Protes) F.; m. Roslyn Silber, June 15, 1948; children—Judith Ellen, Deborah Gail, Carolyn Ruth. B.S., Wayne State U., 1948; Ph.D. (DuPont fellow, AEC fellow), Purdue U., 1952. Head propellant chemistry and flame mechanics sects. NASA, Cleve., 1952-59; asso. prof. U. Minn., Mpls., 1959-60, prof., 1960—, dir. grad. studies in mech. engring., 1965—; vis. scientist Byelloryssian Acad. Scis., 1964; vis. Fulbright prof. U. Poitiers, 1968; cons. U.S. Dept. Commerce Study Waste Heat Mgmt., Minn. Energy Agy., No. States Power Co., Public Systems Research Corp.; co-chmn. com. on fire resistant hydraulic fluids NRC-Nat. Acad. Scis. Nat. Materials Adv. Bd., 1977-78; Participant adv. group for aero. research and devel. NATO Confs., 1960, 61. Editor: Isotopes, 1958-59. Bd. dirs. Minn. Com. for Technion., New Friends of Chamber Music. Served with USNR, 1943-46. Recipient NASA Spl. award, 1961; Outstanding Ski Patrolman of Western Region award Nat. Ski Patrol, 1969-70. Mem. Combustion Inst. (bd. advisers, sec. Central States sect. 1967-78, vice chmn. 1978-79, chmn. 1979—), Am. Chem. Soc., AAAS, Sigma Xi, Tau Beta Pi, Pi Tau Sigma, Phi Lambda Upsilon. Subspecialties: Solar energy; Thermodynamics. Current work: High temperature solar thermochemistry electrolysis and applied thermodynamics and combustion. Home: 3909 Beard Ave S Minneapolis MN 55410

FLETCHER, HARVEY, mathematician; b. N.Y.C., Apr. 9, 1923; m., 1953; 6 children. B.S., MIT, 1944; M.S., Calif. Inst. Tech., 1948; Ph.D. in Math, U. Utah, 1954. Instr. in physics U. Utah, 1953; from instr. to prof. math. Brigham Young U., 1954-63, chmn. dept. math., 1958-61, 62-63, prof., 1964-74, mem. staff, 1980—; mem. tech. staff Bellcom, 1963-64; prof. S.I. Community Coll., 1974-75; mem. staff Eyring Research Inst., 1975-80; mem. tech. staff Bell Labs., 1961-62, 73-74; sr. tech. specialist Hercules Inc., 1967-68. Mem. Math. Assn. Am. Subspecialty: Applied mathematics. Address: 1175 Locust Circle Provo UT 84601

FLETCHER, JAMES CHIPMAN, cons. engr.; b. Millburn, N.J., June 5, 1919; s. Harvey and Lorena (Chipman) F.; m. Fay Lee, Nov. 2, 1946; children—Virginia Lee, Mary Susan, James Stephen, Barbara Jo. A.B., Columbia U., 1940; Ph.D., Calif. Inst. Tech., 1948; D.Sc. (hon.), U. Utah, 1971, Brigham Young U., 1977; LL.D., Lehigh U., 1978. Research physicist bur. ordnance Dept. Navy, 1940-41; spl. research asso. Cruft Lab., Harvard U., 1941-42; instr. Princeton U., 1942-45; teaching fellow Calif. Inst. Tech., 1945-48; instr. U. Calif. at Los Angeles, 1948-50; dir. theory and analysis lab. Hughes Aircraft Co., 1948-54; asso. dir. guided missile lab., dir. electronics guided missile research div., later in space tech. labs. Ramo-Wooldridge Corp., 1954-58; organizer, pres. Space Electronics Corp., 1958-60, Space-Gen. Corp. subs. Aerojet-Gen. Corp., 1960-62, chmn. bd. subs., 1962-64; v.p. systems Aerojet-Gen. Corp., 1962-64; pres. U. Utah, 1964-71; adminstr. NASA, Washington, 1971-77; Whiteford prof. U. Pitts., 1977—; cons. engr., McLean, Va., 1977—; mem. subcom. on stability and control NACA, 1950-54; cons. Office Sec. Def., 1959-64; asst. sec. USAF, 1961-64, to ACDA, 1962-63, Aerojet-Gen. and Space-Gen. Corps., 1964-71; cons., then mem. Pres.'s Sci. Adv. Com., 1958-70; chmn. com. rev. Minuteman Command and Control System, 1961; mem. Air Force Sci. Adv. Bd., 1962-67; chmn. physics panel rev. com. NIH, 1962-64; mem. strategic weapons panel, 1959-61, mil. aircraft panel, 1964-67, chmn. naval warfare panel, 1967-73; mem. Pres.'s Nat. Crime Commn., 1966; mem. tech. assessment adv. council Office of Tech. Assessment; vice chmn. energy resource adv. bd. Dept. Energy; governing bd. NRC, 1978—; v.p. Nat. Space Inst. Author classified papers, sci. papers, chpts. in books; bd. editors, Addison-Wesley Pub. Co., 1958-64. Trustee Aerospace Edn. Found., 1966, Internat. Astronautical Fedn.; bd. regents Nat. Library Medicine, 1971—; bd. visitors Def. Intelligence Sch., 1970—. Recipient Disting. Service medal NASA; Exceptional Civilian Service award USAF; John Jay award Columbia U. Fellow IEEE, Am. Acad. Arts and Sci., AIAA (hon.), Am. Astronautical Soc.; mem. Am. Phys. Soc., Nat. Space Club (bd. govs.), Nat. Acad. Engring. (council, rep. to governing bd. NRC, governing bd. 1978-81), Am. Ordnance Assn., Air Force Assn., Sigma Xi. Club: Cosmos. Subspecialties: Systems engineering; Astronautics. Home: 7721 Falstaff Rd McLean VA 22102

FLETCHER, JOHN DEXTER, research psychologist; b. Providence, Dec. 9, 1940; s. John Dexter and Agnes (McClell) F.; m. Sheila Gates, June 17, 1968; children: Scott Dexter, Jeffrey Quinn, Brian Whitford. B.A., U. Ariz., 1965; M.S. in Computer Sci, Stanford U., 1973; Ph.D. in Ednl. Psychology, Stanford U., 1973. Research assoc. Stanford U., 1969-73; asst. prof. psychology and computer sci. U. Ill., Chgo., 1973-74; supervisory research psychologist U.S. Navy Personnel Research and Devel. Ctr., San Diego, 1974-78; program mgr. Def. Advanced Research Projects Agy., Arlington, Va., 1978-80; coordinator tri-service tng. research and devel. U.S. Army Research Inst., Alexandria, Va., 1980-81; dir. WICAT Edn. Inst., Provo, Utah, 1981—; cons. computer-assisted instrn. Contbr. articles to profl. jours. Mem. bd. advisors Apple Edn. Found., Cupertino, Calif., 1979—; mem. Def. Sci. Bd. Panel on Tng. and Tng. Tech., Washington, 1982—; Webelos leader Cub Scouts Am., 1980-81; active Boy Scouts Am., 1981-82. Served with U.S. Army, 1960-63. Recipient Sustained Superior Performance award Dept. Def., 1979. Mem. AAAS, Am. Ednl. Research Assn., Am. Psychol. Assn., Assn. Computing Machinery, Assn. Devel. of Computer-Based Instructional Systems. Subspecialties: Cognition; Computer-assisted instruction. Current work: Research and development on computer-assisted education, training, and human performance technology. Home: 586 W 440 S Orem UT 94057 Office: WICAT Edn Inst 931 E 300 N Provo UT 84601

FLETCHER, JOHN EDWARD, research mathematician, research laboratory administrator; b. Banner Elk, N.C., June 12, 1937; s. James Clair and Goldie Erin (Critcher) F.; m. Carol Ann Serra, July 18, 1964; children: Leah Renée, Craig Alan. B.S.M.E.A., N.C. State U., 1959, M.S.A.M., 1961; Ph.D. in Math., U. Md., 1972. Research asst. N.C. State U., 1959-61; sr. analyst Lockheed Ga. Co., Marietta, 1964-66; mathematician NIH, Bethesda, Md., 1966-72, research mathematician div. computer research and tech., 1972—; chmn. math. and computer sci. Found. Advanced Edn. Sci., Bethesda, 1972—. Contbr. numerous articles, chpts. to profl. publs. Bd. dirs. Manor Boys Baseball League, Rockville, Md., 1978-82. Served to capt. USAF, 1961-64. Recipient Merit award NIH, 1972, Dir.'s award NIH, 1980; NIH in-service fellow, 1968. Mem. Soc. Indsl. and Applied Math., Internat. Soc. Oxygen Transport to Tissue. Subspecialties: Applied mathematics; Numerical analysis. Current work: Mathematical modelling in the biological and medical sciences; reaction-diffusion models; mathematical models of the micro-circulation; simulation of physiological processes. Home: 4211 Norbeck Rd Rockville MD 20853 Office: 12A 2041 DCRT NIH 9000 Rockville Pike Bethesda MD 20205

FLETCHER, LEROY S(TEVENSON), college dean, mechanical and aerospace engineer; b. San Antonio, Oct. 10, 1936; m., 1966; 2 children. B.S., Tex. A&M U., 1958; M.S., Stanford U., 1963, Engr., 1964; Ph.D. in Mech. Engring., Ariz. State U., 1968. Aero. engr. Ames Aero. Lab., Nat. Adv. Com. Aeros., 1958-60; aerospace engr. Ames Research Ctr., NASA, 1961-63; asst. heat transfer dept. mech. engring. Stanford U., 1962-63; asst. thermodynamics, 1963-64; asst. mech. engr. Ariz. State U., 1964-65, instr., 1965-68; from asst. prof. to prof. aerospace engring. Rutgers U., 1968-75, acting assoc. dean, 1974-75; prof. mech. engring., chmn. dept. U. Va., 1975-77, prof. mech. and aero. engring., chmn. dept., 1977-80, mem. pres.'s adv. com. energy conservation, 1975-80; assoc. dean Coll. Engring., Tex. A&M U., 1980—; mem. Va. Gov.'s Com. Conservation Energy Resources, 1976-80. Recipient Ralph R. Teeter award Soc. Automotive Engrs., 1970. Fellow ASME (Charles Russ Richards award 1982, Centennial medallion 1980), AAAS, AIAA (assoc.); mem. Am. Soc. Engring. Edn. (Ralph Coats Roe award 1983, George Westinghouse award 1981), AAAS. Subspecialties: Mechanical engineering; Aerospace engineering and technology. Office: Coll Engring Tex A&M U College Station TX 77843

FLETCHER, ROBERT H., medical educator; b. Abington, Pa., Mar. 26, 1940; s. Stevenson Whitcomb and Wanda (Moss) F.; m. Suzanne Wright, June 15, 1963; children: John Wright, Grant Selmer. B.A., Wesleyan U., Middletown, Conn., 1962; M.D., Harvard U., 1966; M.Sc., Johns Hopkins U., 1973. Diplomate: Am. Bd. Internal Medicine. Asst. prof. Faculty of Medicine, McGill U., Montreal, Que., Can., 1973-78; assoc. prof. medicine Sch. Medicine, U. N.C.-Chapel Hill, 1978-83, prof. medicine, clin. prof. epidemiology, 1983—, dir., Robert Wood Johnson Clin. Scholars Program, 1978—. Sr. author: Clinical Epidemiology, The Essentials, 1982. Served to maj. M.C., U.S. Army, 1968-71. Fellow ACP; mem. Am. Fedn. Clin. Research, Am. Pub. Health Assn., Soc. Clin. Trials, Soc. Research and Edn. in Primary Care Internal Medicine, Phi Beta Kappa, Sigma Xi. Democrat. Quaker. Subspecialties: Internal medicine; Epidemiology. Current work: Clinical epidemiology, health services research. Home: 208 Boulder Bluff Chapel Hill NC 27514 Office: Robert Wood Johnson Program U NC Sch Medicine Chapel Hill NC 27514

FLETCHER, RONALD DARLING, microbiologist, educator, research systems coordinator; b. Foxboro, Mass., Jan. 18, 1933; s. Howard Wendell and Ada (Darling) F.; m. Barbara Gundersen, Jan. 30, 1954; children: Deborah, Mark Ronald, Christopher Gary. B.S., U. Conn., 1954, M.S., 1959, Ph.D., 1963; postdoctoral fellow, U. Zurich, Switzerland, 1963-64. Cert. spl. and registered microbiologist Nat. Registry of Microbiology. Research virologist Am. Cyanamid Co., Pearl River, N.Y., 1964-67; dir. microbiology McKeesport (Pa.) Hosp., 1971-79; prof. microbiology, assoc. chmn. dept. U. Pitts., 1967—; instr. U. Conn., 1959-63; lectr. Contbr. articles and abstracts to profl. jours. Mem. Republican Nat. Com., 1971—, U.S. Senatorial Club. Served to col. MSC AUS, 1954-57, 78-79. Recipient cert. of achievement U.S. Army Surgeon Gen., 1973; grantee Am. Cancer Soc., 1968, United Fund, 1970, U.S. Army, 1969-72, NIDR, 1977-79, U. Pitts., 1980-81. Fellow Am. Acad. Microbiology, AAAS; mem. Internat. Assn. Dental Research (pres. Pitts. chpt. 1979-80), Am. Soc. Microbiologists, Mil. Surgeons U.S., Res. Officers Assn. (sec. W. Pa. council 1973-74, sec. Pitts. chpt. 1974-76), Am. Soc. Cell Biology, VFW, Am. Legion. Republican. Presbyterian. Subspecialties: Biomedical engineering; Microbiology (medicine). Current work: Quantitation of mammalian cell adhesion to high-energy surfaces, Herpes Simplex viruses effect on attached cells and inhibitors. Office: University of Pittsburgh 645 Salk Hall Pittsburgh PA 15261

FLEXNER, LOUIS BARKHOUSE, scientist, educator; b. Louisville, Jan. 7, 1902; s. Washington and Ida (Barkhouse) F.; m. Josefa Barba Gosé, Aug. 23, 1937. B.S., U. Chgo., 1923; M.D., Johns Hopkins, 1927; LL.D., U. Pa. Fellow medicine Johns Hopkins Hosp., 1928-29; resident physician U. Chgo. Clinics, 1929-30; instr. and asso. anatomy Johns Hopkins Med. Sch., 1930-39; with dept. physiology Cambridge (Eng.) U., 1933-34; staff mem. dept. embryology Carnegie Instn., Washington, 1939-51, research asso., 1951—; prof. anatomy Sch. Med. U. Pa., 1951—, chmn. dept., 1951-67; dir. Inst. Neurol. Scis., 1953-66. Contbr. articles to profl. jours. Sci. adv. bds. USPHS, United Cerebral Palsy, Nat. Council to Combat Blindness, Nat. Paraplegic Soc., NRC, Nat. Found. Mem. Am. Assn. Anatomists, Nat. Acad. Scis., Am. Physiol. Soc., Am. Soc. Biol. Chemists, Am. Acad. Arts and Scis., Am. Philos. Soc. Subspecialty: Neurobiology. Current work: Memory: mechanism of cerebral spread of an engram. Home: 4631 Pine St Philadelphia PA 19143

FLIEGEL, HENRY FREDERICK, astronomer; b. Ridley Park, Pa., Apr. 25, 1936; s. Henry Frederick and Mary Alice (Reed) F. Ph.D. in Astronomy, U. Pa., Phila., 1963. Mem. tech. staff Jet Propulsion Lab., Pasadena, Calif., 1969-81; project engr. Aerospace Corp., El Segundo, Calif., 1982—. Mem. Am. Astron. Soc., Am. Geophys. Union. Democrat. Lutheran. Subspecialties: Optical astronomy; Tectonics. Current work: Geodesy, astrometry, celestial mechanics. Home: 3640 5th Ave La Crescenta CA 91214 Office: 2350 E El Segundo Blvd El Segundo CA 90245

FLIKKE, ARNOLD MAURICE, educator; b. Viroqua, Wis., July 8, 1919; s. Arthur P. and Mabel Christine (Hermanson) F.; m. Bethel Irene Christie, June 11, 1942; children—Gary, Craig, Karen. B.S. in Agrl. Engring, U. Wis., 1941; M.S., U. Minn., 1943; Ph.D., Auburn U., 1972. Instr. agrl. engring. U. Minn., 1946-49, asst. prof., 1949-56, asso. prof., 1956-63, prof., 1963-72, prof., head dept. agrl. engring., 1972—; cons. in field. Served with USNR, 1944-46. NSF faculty fellow, 1966. Mem. Am. Soc. Agrl. Engrs. (chmn. Minn. sect. 1962, chmn. North Central region 1972, nat. dir. 1974), Am. Soc. for Engring. Edn., IEEE, AAAS, Sigma Xi, Phi Kappa Phi, Alpha Zeta, Gamma Sigma Delta, Alpha Gamma Rho. Lutheran. Club: Mason. Subspecialties: Biomass (energy science and technology); Resource conservation. Research and publs. on application of electricity for grain drying, farm power. Home: 3409 Downers Dr NE Minneapolis MN 55418 Office: Dept of Agrl Engring U Minn St Paul MN 55108

FLINN, EDWARD AMBROSE, III, government research official; b. Oklahoma City, Aug. 27, 1931; s. Edward Ambrose and Marion Catalina (Prater) F.; m. Jane Margaret Bott, Dec. 29, 1962; 1 dau., Susan Katherine. B.S. in Geophysics (William Barton Rogers scholar), Mass. Inst. Tech., 1953; Ph.D. in Geophysics (NSF fellow 1953-54), Calif. Inst. Tech., 1960; postgrad. (Fulbright scholar), Australian Nat. U., 1958-59; certificat, Le Cordon Bleu Ecole de Cuisine et de Patesserie, 1971. Seismologist United ElectroDynamics Inc., Pasadena, Calif., 1960-62; chief seismologist, lab. seismic data Teledyne Geotech, Alexandria, Va., 1962-64, dir. research, 1964-68; asso. dir. Alexandria labs., 1968-74; dir. lunar programs office space sci. NASA, Washington, 1975, dep. dir., chief scientist lunar and planetary programs, 1976-77, chief scientist earth and ocean programs, 1977, chief scientist geodynamics program, 1978—; mem. joint research panel AEC-U.K. Atomic Energy Authority, 1963-74; vis. research asso. Calif. Inst. Tech., 1969, 78; vis. asso. prof. geophysics Brown U., 1970; cons. subcom. planetology steering com. space sci. and applications NASA, 1969-70, Nat. Swedish Inst. Bldg. Research, Stockholm, 1970; participant sci. exchange Nat. Acad. Sci.-Acad. Scis. USSR, 1970; mem. com. lunar and planetary exploration Nat. Acad. Scis., 1973-75, mem. com. on seismology, 1973-75, mem. com. on internat. geology, 1982—; mem. adv. com. earthquake studies U.S. Geol. Survey, 1975—. Trans., editor two books; asso. editor for gen. seismology: Geophysics, 1965-67; editor sect. earth and planetary surfaces and interiors: Jour. Geophys. Research, 1973-78. Mem. adv. bd. No. Va. br. Urban League, 1970-71; mem. Alexandria Council on Human Relations, 1970—; mem. traffic bd., Alexandria, 1972-74; mem. Alexandria Democratic Com., 1970-74, exec. bd., 1971-73. Mem. AAAS, Am. Astron. Soc., Royal Astron. Soc. (editorial bd. Geophys. Jour. 1969-74), Am. Geophys. Union (sec. sect. seismology 1970-74), Assn. Earth Sci. Editors, Internat. Union Geodesy and Geophysics (chmn. commn. planetary scis. 1976—, sec. Commn. on Internat. Coordination of Space Techniques for Geodesy and Geodynamics), Seismol. Soc. Am., Soc. Exploration Geophysicists, Inter-Assn. Com. Math. Geophysics (sec. 1971-75), Internat. Council Sci. Unions (sec.-gen. Inter-Union Commn. on Lithosphere, mem. com. on publs. and communications 1981—), 89ers Soc., Sigma Xi, Beta Theta Pi. Club: Cosmos. Subspecialties: Tectonics; Geophysics. Current work: Application of space technology to geodesy, geophysics and geodynamics. Research, numerous publs. on seismology, geophysics, applied maths., computer sci. Home: 3605 Tupelo Pl Alexandria VA 22304 Office: NASA Hqrs Code EE-8 Washington DC 20546

FLINN, RICHARD ALOYSIUS, metall. engr., cons., educator; b. N.Y.C.; s. Richard A. and Anna M. F.; m. Edwina R. Flinn, 1944; children: Ellen, John, Paul, Mark, Brian. B.S.Ch.E., CUNY, 1936; M.S. in Metallurgy, M.I.T., 1937, Sc.D., 1941. Research in metallurgy Internat. Nickel Co., Bayonne, N.J., 1937-39; asst. chief metallurgist ABEX, Mahwah, N.J., 1941-51; prof. metall. engring. U. Mich., Ann Arbor, 1951—; cons. legal work in failure analysis. Author: Fundamentals of Metal Casting, 1964, Engineering Materials and Their Applications, 1978, 2d edit., 1980; contbr. numerous articles profl. jours. Trustee Barton Hills Village, 1979—. Fellow Am. Soc. Metals (Howe medal 1944, 61); mem. Am. Foundrymen's Soc. (Simpson Gold medal 1947, hon. life mem.). Republican. Subspecialties: Materials processing; Alloys. Current work: Failure analysis, alloy devel., metal casting. Patentee in field. Office: U Mich 4305 E Engineering Ann Arbor MI 48109 Home: 140 Underdown Dr Ann Arbor MI 48105

FLIPPEN, LUTHER DANIEL, JR., fluid mechanics educator, researcher; b. Richmond, Va., Oct. 20, 1955; s. Luther Daniel and Norma Jean (Snead) F.; m. Meredith Lee Trella, May 7, 1979. B.S., U. Va., 1978; M.S., Yale U., 1979; Ph.D., Duke U., 1982. Asst. prof. mech. and nuclear engring. Miss. State U., Starkville, 1982—. Mem. Am. Nuclear Soc., Pi Tau Sigma, Alpha Nu Sigma, Tau Beta Pi, Pi Kappa Phi. Methodist. Subspecialties: Fluid mechanics; Nuclear engineering. Current work: Theoretical electrohydrodynamics. Office: Miss State U Starkville MS 39762

FLODIN, NESTOR WINSTON, biochemistry educator; b. Chgo., Jan. 30, 1915; s. Eric Anton and Anna Beata (Nygren) F.; m. Betty Shea, Sept. 18, 1941; children: Andrea, Mark. B.S., U. Chgo., 1935, Ph.D., 1938. Research chemist Elchem dept. E. I. duPont De Nemours & Co., Inc., Wilmington, Del., 1940-52, tech. mgr. lysine div., 1953-60, planning specialist biochems. dept., 1961-66, product mgr. pharms. div., 1967-69, market research mgr., 1969-73; prof. biochemistry U. South Ala., Mobile, 1974—. Author: Vitamin/Trace Mineral/Protein Interactions, vols. 1-4, 1979-81. Fellow Am. Coll. Nutrition; mem. Am. Inst. Nutrition, N.Y. Acad. Sci., AAAS, Sigma Xi. Subspecialties: Nutrition (medicine); Nutrition (biology). Current work: Clinical nutrition and nutritional biochemistry with special emphasis on vitamins and trace elements. Home: 16 Town Crier Ct Mobile AL 36608 Office: Dept Biochemistry Univ South Ala Coll Medicine Mobile AL 36688

FLOM, TERRENCE EDSEL, elec. engr.; b. Moose Lake, Minn, Sept. 21, 1940; s. Andrew Felix and Olga (Koskey) F. B.E.E., Ga. Inst. Tech., 1964. Electronic engr. Control Data Corp., Mpls., 1964-67; electronic engr. Xerox, Pasadena, Calif., 1967-68, ITT, Van Nuys, Calif., 1968-75; supr. laser communications GTE, Mountain View, Calif., 1975-77, mgr. dept. systems engring., 1977-81, mgr. tech. devel., 1982—. Contbr. articles to profl. jours. Mem. Optical Soc. Am., Laser Inst. Am. Subspecialties: Laser communications; Laser radar. Current work: Laser communications and laser radar systems; research and devel. of new communication and radar systems using lasers. Home: 873 Balboa Ln Foster City CA 94404 Office: PO Box 7188 Mountain View CA 94039

FLOOD, DOROTHY GARNETT, neurobiologist; b. Sayre, Pa., Oct. 7, 1951; d. James Murlin and Dorothy Garnett (Dietrich) F. Student, U. Ill., 1972-73; B.A. cum laude, Lawrence U., Appleton, Wis., 1973; M.S., U. Rochester, N.Y., 1980, Ph.D., 1980. Sr. instr. U. Rochester, 1980—. Contbr. articles to profl. jours. Mem. AAAS, Soc. Neurosci. Subspecialties: Anatomy and embryology; Neurobiology. Current work: Development and aging of nervous system; plasticity. Office: Dept Anatomy Box 603 U Rochester Med Center 601 Elmwood Ave Rochester NY 14642

FLORES, IVAN, computer science educator, computer consultant; b. N.Y.C., Jan. 3, 1923; s. Angel and Ruth (Blumauer) F.; m. Helen Hubert, Mar. 5, 1955 (div.); children: Pamm, Glenn. With various comml. and indsl. firms, 1950-60; founder, pres. Flores Assocs., Bklyn., 1960—; assoc. prof. elec. engring. Poly. Inst. Bklyn., 1958-65, Stevens Inst., Hoboken, N.J., 1965-67; prof. computer sci. Baruch Coll., CUNY, 1967—; lectr. at profl. meetings, univs. Author: books including Data Base Architecture, 1981, Microcomputer Systems, 1982, Word Processing Handbook, 1982; editor: Jour. Computer Lang; contbr. articles to profl. jours. and encys. Subspecialties: Operating systems; Database systems. Current work: Small business computer systems, operating systems, database, microcomputers, graphics office automation. Office: Flores Associates 108 8th Ave Brooklyn NY 11215

FLORY, PAUL JOHN, chemist; b. Sterling, Ill., June 19, 1910; s. Ezra and Martha (Brumbaugh) F.; m. Emily Catharine Tabor, Mar. 7, 1936; children—Susan, Melinda, Paul J. B.S., Manchester Coll., 1931, Sc.D. (hon.), 1950; M.S., Ohio State U., 1931, Ph.D., 1934, Sc.D., 1970. Engaged in research on synthetic fibers, synthetic rubber and other polymeric substances Dupont Exptl. Sta., Wilmington, Del., 1934-38, U. Cin., 1938-40, Standard Oil Devel. Co., Elizabeth, N.J., 1940-43; dir. fundamental research Goodyear Tire & Rubber Co., Akron, Ohio, 1943-48; prof. chemistry Cornell U., 1948-57; exec. dir. research Mellon Inst., Pitts., 1956-61; J.G. Jackson-C.J.Wood prof. chemistry Stanford, 1961—. Author: Principles of Polymer Chemistry and of Statistical Mechanics of Chain Molecules; Contbr. to sci. publs. Recipient Sullivant medal Ohio State U., 1945; Baekeland award Am. Chem. Soc., 1947; George Fisher Baker non-resident lectureship in chemistry Cornell U., 1948; Peter Debye award in phys. chemistry Am. Chem. Soc., 1968; Gibbs medal, 1973; Priestley medal, 1974; Cresson medal Franklin Inst., 1971; Nobel prize for chemistry, 1974; Nat. medal of sci., 1974. Fellow AAAS; mem. Am. Chem. Soc., Nat. Acad. Scis., Am. Acad. Arts and Scis., Am. Phys. Soc., Am. Philos. Soc. Subspecialties: Polymer chemistry; Polymer physics. Current work: Theory of rubber elasticity; configurational statistics of macromolecules; theory of liquid crystalline state; interphases; morphology of crystalline polymers. Pioneered research on constitution and properties of substances composed of giant molecules (rubbers, plastics, fibers, films, proteins, etc). Home: 210 Golden Oak Dr Portola Valley CA 94025 Office: Stanford U Stanford CA 94305

FLORY, RANDALL KEAN, psychology educator; b. Elgin, Ill., Sept. 13, 1942; s. James Robert and Libbie Emma (Roney) F.; m. Patricia Lou King, Mar. 25, 1967; children: Mark Randall, Sarah Louanne. B.A. in Psychology, Ill. Wesleyan U., 1964, Ph.D., Ariz. State U., 1969. Lic. psychologist, Va. Asst. prof. psychology Hollins Coll., 1969-75, assoc. prof., 1976—; cons. Center for Human Devel./Assn. Retarded Citizens, Roanoke, Va., 1980—. Author: (with Sherman) Student Laboratory Experiments in Operant Conditioning, 1974; research publs. in field. Danforth Found. fellow, 1980—; NIMH grantee, 1971, 72; NSF grantee, 1970, 74. Mem. Am. Psychol. Assn., Psychonomic Soc., Southeastern Psychol. Assn., Eastern Psychol. Assn., Va. Psychol. Assn., Sigma Xi. Congregationalist. Subspecialties: Behavioral psychology; Learning. Current work: Research on aggressive behavior, adjunctive or schedule-induced behavior, behavior modification. Home: 866 Peyton St NW Roanoke VA 24019 Office: Dept Psychology Hollins Coll Roanoke VA 24020

FLORY, WILLIAM EVANS SHERLOCK, govt. ofcl.; b. Canton, Ohio, Apr. 25, 1914; s. Wilson Reese and Frances (Sherlock) F.; m. Anne Randolph Putney, June 4, 1938; children—William, Anne. A.B., Coll. Wooster, 1935; A.M., Duke, 1938, Ph.D., 1941. Various teaching positions, Ohio and Ga., 1935-39; analyst Princeton U. govt. surveys, 1940; dir. research N.J. Municipal Aid Adminstrn., Trenton, 1940-42; analyst N.Y. Joint Legis. Economy Commn. Albany, 1942-43, Bur. Budget, Washington, 1943-44; dep. asst. to under-sec. State, Washington, 1944-50; econ. policy adviser to sec. Interior, Washington, 1950-53; staff economist Bur. Mines and Office Minerals and Solid Fuels, Washington, 1953-61, dir., 1961-69; asso. and adj. prof. mktg. and transp. Am. U. Sch. Bus. Adminstrn., Washington, 1969-79. Author: Prisoners of War, 1941, Restoration of Historic Bel Air Plantation; Contbr. to govt. publs. Bd. dirs. Prince William County Indsl. Devel. Authority, Manassas, Va.; curator Weems-Botts Mus., Dumfries, Va. Democrat. Episcopalian. Subspecialties: Organic chemistry; Biochemistry (biology). Current work: Biosynthesis of natural products, fermentations, steocochemistry of biological reactions. Home: Bel Air Plantation 14313 Minnieville Rd Woodbridge VA 22193

FLOURNOY, DAYL JEAN, clinical microbiologist; b. San Antonio, Dec. 17, 1944; s. Dayl J. and Bonnie (Allen) Floyrnoy; m. Mary Virginia Patrick, June 2, 1967; children: David D., Michael P., Michelle S. B.S., S.W. Tex. State U., 1965; A.S., San Antonio Coll., 1966; M.A., Incarnate Word Coll., 1968; Ph.D., U. Houston, 1973. Med. technologist Santa Rosa Med. Ctr., San Antonio, 1966-69; teaching fellow U. Houston, 1969-72; med. technologist St. Luke's Episcopal Hosp., Houston, 1972-73, microbiologist, postdoctoral fellow, 1974-75; dir. microbiology VA Med. Ctr., Oklahoma City, 1975—. Contbr. articles to profl. jours. Soccer coach Tri-City Athletic Assn., Bethany, Okla., 1978—. Mem. Am. Soc. Microbiology, Okla. Acad. Sci., Southwestern Assn. Clin. Microbiology, Sigma Xi. Subspecialties: Microbiology; Pathology (medicine). Current work: Antimicrobials, infectious diseases, clinical pathology (microbiology). Office: VA Med Ctr 921 NE 13th St Oklahoma City OK 73104

FLOWER, PHILLIP JOHN, astronomer, educator; b. Toledo, Feb. 4, 1948; s. Chester R. and Ursula (Malikowski) F. B.S. in Physics, U. Toledo, 1970; Ph.D. in Astronomy, U. Wash., Seattle, 1976. Research assoc. Inst. Astronomy, Polish Acad. Sci., Warsaw, 1974-75; postdoctoral research assoc. Joint Inst. Lab. Astrophysics, Boulder, Colo., 1976-78; asst. prof. physics and astronomy Clemson U., 1978-83, assoc. prof., 1983—. Mem. Am. Astron. Soc., AAAS, Am. Phys. Scis., Astron. Soc. Pacific, S.C. Acad. Scis. Subspecialties: Theoretical astrophysics; Optical astronomy. Current work: Stellar evolution, stellar interiors, Magellanic clouds, star clusters, dwarf galaxies. Office: Dept Physics and Astronomy Clemson U Clemson SC 29631

FLOWERS, JOHN HAWKINS, psychology educator; b. Pittsfield, Mass., Sept. 10, 1946; s. John Wilson and Edmay (Vienneau) F.; m., June 14, 1970; children: Andrew Tucker, Michael Wilson. B.A., Wesleyan U., Middletown, Conn., 1968; Ph.D., Yale U., 1972. Asst. prof. U. Nebr.-Lincoln, 1972-76, assoc. prof. psychology, 1977-83, prof., 1983—. Editor: Nebraska Symposium on Motivation, 1981. NSF grantee, 1979-81, 81-83. Mem. Psychonomic Soc., Am. Psychol. Assn., Midwestern Psychol. Assn., AAAS, Sigma Xi. Democrat. Subspecialty: Cognition. Current work: Human attention and performance, visual information processing. Home: 1221 Piper Way Rt 13 Lincoln NE 68527 Office: Univ Nebr Psychology Dept 209 Burnett Lincoln NE 68588

FLOYD, ROBERT A., biochemist; b. Yosemite, Ky., Oct. 7, 1940; s. Aaron and Clarice (Williams) F.; m. Marlene Gale Rohner, Aug. 21, 1965; children: Matthew Christopher, Patrick Aaron. B.S., U. Ky., Lexington, 1963, M.S., 1965; Ph.D., Purdue U., Lafayette, Ind., 1969. Postdoctoral work U. Calif., Davis, 1968-69, U. Pa., Phila., 1969-71; research assoc. Washington U., St. Louis, 1971-74; biochemist Okla. Med. Research Found., Oklahoma City, 1974—. Contbr. sci. articles to profl. publs. NIH grantee, 1975-82. Mem. Am. Soc. Biol. Chemists, Biophys. Soc., Am. Chem. Soc., Am. Soc. Photobiology, Sigma Xi. Subspecialties: Biochemistry (biology); Toxicology (agriculture). Current work: Free radicals in biological systems, oxygen free radicals, oxidation-reduction reactions in biochemistry, bioenergetics. Home: 207 NW 18th St Oklahoma City OK 73103 Office: 825 NE 13th St Oklahoma City OK 73104

FLUCK, MICHELE MARGUERITE, microbiology educator, researcher; b. Geneva, Aug. 5, 1940; U.S., 1972; d. Wilhelm and Henriette (Delaloye) F. B.S., U. Geneva, 1964, M.S., 1966, Ph.D., 1972. Postdoctoral fellow N.Y. Pub. Health Inst., N.Y.C., 1972-73; instr. Harvard Med. Sch., Boston, 1973-78, asst. prof., 1978-79; assoc. prof. microbiology Mich. State U., East Lansing, 1979—. Damon Runyon Found. fellow, 1972; Leukemia Soc. Am. spl. fellow, 1976; Young Investigator award Nat. Cancer Inst., 1978; Leukemia Soc. Am. scholar, 1979; Nat. Cancer Inst. grantee, 1980. Mem. Assn. Women in Sci., Am. Soc. Virology, AAAS, Sci. for People, Union of Concerned Scientists, NOW. Subspecialties: Molecular biology; Cancer research (medicine). Current work: Molecular and cellular biology of cells transformed to a neoplastic state by DNA tumor virus; state of viral genome in such cells; induction of cellular genes (incl. oncogenes) in such cells. Office: Dept Microbiology Mich State Univ East Lansing MI 48823

FLUEGGE, RONALD MARVIN, nuclear engineer; b. Cape Girardeau, Mo., Nov. 22, 1948; s. Marvin Alvin and Maxine Louise (Hamilton) F.; m. Vicki Sue Oldham, Aug. 9, 1969; children: Terasa Dawn, Jennifer Beth. B.S., U. Mo.-Rolla, 1970. Registered profl. engr., Mo., Md. Engr. Balt. Gas & Electric Co., 1970-74; reactor engr. Nuclear Regulatory Commn., Bethesda, Md., 1974-76; med. physicist Shoss Radiology Group, Cape Girardeau, Mo., 1976-78; pres. Diagnostic Services Unltd., Jackson, Mo., 1978-79; dir. Mo. Pub. Service Commn., Jefferson City, 1979-83; nuclear cons. analyst Univ. Computing Co., Dallas, 1983—; curriculum adv. U. Mo., Rolla., 1980—; econs. adv. Atomic Indsl. Forum, Washington, 1982—; tech. adv. Gov's. Office, Jefferson City, Mo., 1979-83; radiation cons. Oliver, Oliver, Waltz & Cook/Geo-Log, 1978. Vol. Mo. Nuclear Emergency Team, Jefferson City, 1979-83. Served with Md. Nat. Guard, 1970-71. Mem. Am. Nuclear Soc., ASME, Nat. Soc. Profl. Engrs., Tex. Soc. Profl. Engrs. Republican. Methodist. Subspecialties: Nuclear engineering; Nuclear fission. Current work: Nuclear engineering, computer code development regarding nuclear fuel management physics and economics, and simulations of nuclear power plant incident emergency response. Home: 5633 N Colony Rd The Colony TX 75056 Office: Mo Pub Service Commn 100 E Capitol St Jefferson City MO 65101

FLYNN, GEORGE WILLIAM, chemistry educator; b. Hartford, Conn., July 11, 1938; s. George William and Rose Margaret (Tummillo) F.; m. Jean Van Alystine Pieri, Oct. 3, 1970; children: David, Suzanne. B.S., Yale U., 1960; A.M., Harvard U., 1962, Ph.D., 1965. Postdoctoral fellow M.I.T., Cambridge, 1964-66; asst. prof. Columbia U., N.Y.C., 1967-72, assoc. prof., 1972-76, prof. chemistry, 1976—; vis. scientist M.I.T., Cambridge, 1975; dir. Columbia

Radiation Lab., N.Y.C., 1979—; research collaborator Brookhaven Nat. Lab., Upton, N.Y., 1969—. Sloan Found. fellow, 1968-70; Guggenheim Found. fellow, 1974-75; NSF fellow, 1960-64, 64-65. Mem. Am. Phys. Soc., Am. Chem. Soc., N.Y. Acad. Scis., Sigma Xi. Roman Catholic. Subspecialties: Physical chemistry; Laser photochemistry. Current work: Application of lasers to the study of vibrational relaxation, chemical reactions and photofragmentation processes. Office: Columbia U 116th St New York NY 10027 Home: 382 Summit Ave Leonia NJ 07605

FLYNN, JOHN THOMAS, physiologist, medical educator; b. Chester, Pa., Mar. 14, 1948; s. Deward Belmont and Pauline (Dolski) F.; m. Harriet Yvonne Medwid, July 18, 1970; children: Susan Michelle, Mark Brian. B.S., Widener U., 1970; Ph.D., Hahnemann Med. Coll., 1974. Postdoctoral fellow Thomas Jefferson U., 1974-76; asst. prof. physiology Jefferson Med. Coll., 1976-82, assoc. prof., 1982—. Contbr. numerous articles, abstracts to profl. publs. Am. Lung Assn. young investigator, 1979-81; NIH peer rev. grantee, 1980-83. Mem. Am. Physiol. Soc., N.Y. Acad. Scis., Phila. Physiol. Soc., Fedn. Am. Scientists, Circulatory Shock Soc. Subspecialties: Physiology (medicine); Pharmacology. Current work: Medical research related to role of prostaglandin-like materials in pathophysiology of cellular injury as related to pulmonary edema, circulatory shock, septicemia, burn injury, myocardial infarction and metabolic poisoning; regulation of arachidonic acid cascade during pathophysiologic conditions. Office: Thomas Jefferson U Med Coll Dept Physiology 1020 Locust St Philadelphia PA 19107

FLYNN, KEVIN FRANCIS, nuclear chemist, health physicist; b. Chgo., Oct. 28, 1927; s. Edward Joseph and Anna Mary (McDonnell) F.; m. Norman Jean Williams, May 20, 1950; children: Karen, Nance, James, Mary. B.S. in Chem. Engring, Ill. Inst. Tech., 1950, M.S., 1952. Nuclear chemist, health physicist Argonne Nat. Lab., Ill., 1951—. Contbr. numerous articles to profl. jours. Bd. dirs. Human Relations Council, Chgo., 1957-82, Advocacy for Deaf/Blind, Chgo., 1975-82; mem. Ind. Voters Ill, 1965-82. Served with USN, 1945-47. Mem. Am. Chem. Soc., Am. Nuclear Soc., Health Physics Soc., Sigma Xi, Phi Lambda Psi. Democrat. Roman Catholic. Subspecialties: Nuclear fission; Environmental effects of nuclear technology. Current work: Nuclear chemistry (research), nuclear waste management. Home: 10057 S Longwood Dr Chicago IL 60643 Office: Argonne Nat Lab 9700 S Cass Ave Argonne IL 60439

FODOR, GABOR BELA, chemistry educator, foundation researcher; b. Budapest, Hungary, Dec. 5, 1915; came to U.S., 1969, naturalized, 1976; s. Domokos Victor and Paula Maria (Bayer) F. B.S., Poly. Inst. Graz, Austria, 1934; Ph.D., U. Szeged, Hungary, 1937; D.Sc., Hungarian Acad. Sci., Budapest, 1952. Univ. demonstrator Lab. Organic Chemistry, Szeged, Hungary, 1935-38; research chemist Chinoin Pharm., Ujpest, Hungary, 1938-45; assoc. prof. chemistry U. Szeged, 1945-49, prof. organic chemistry, 1949-57; head Lab. Stereochemistry Hungarian Acad. Sci., 1958-65; prof. Laval U., Que., 1965-69; Centennial prof. W.Va. U., 1969—; cons., regional dir. Nat. Found. Cancer Research, Bethesda, Md., 1977—. Author: Organic Chemistry, 1960, Organishche Chemie, 1966; contbr. numerous articles to profl. jours. Recipient Kossuth medal, Hungary, 1950, 54, Silver medal U. Helsinki, 1958; fellow Churchill Coll., Cambridge, Eng., 1961. Mem. Am. Chem. Soc., Chem. Soc. London, Swiss Chem. Soc., Hungarian Acad. Sci., Canadian Inst. Chemistry. Republican. Calvinist. Club: Lakeview Country (Morgantown). Subspecialties: Organic chemistry; Synthetic chemistry. Current work: Synthetic organic chemistry, determination of the three dimensional structure of alkaloids; sphingosine, sugars; new reactions of vitamin C related cancer research. Patentee in field. Home: 829 Augusta Ave Morgantown WV 26505 Office: West Virginia University Department of Chemistry Morgantown WV 26506

FODOR, MAGDA MARIA, aerospace engineer. Civil squad leader Am. Cyanamid Co., Wayne, N.J., 1974-76; sr. engr. Merck & Co., Rahway, N.J., 1976-77; project engr. Exxon Research and Engring., Florham Park, N.J., 1977-79; mem. tech. staff TRW, Redondo Beach, Calif., 1980-81; project engr. research and devel. Tood Pacific Shipyards Corp., San Pedro, Calif., 1981-83; aerospace engr. U.S. Air Force, Los Angeles, 1983—. Mem. AIAA, Am Soc. Naval Engrs. Subspecialties: Robotics; Aerospace engineering and technology. Current work: Worked for DOD-STS program at Vandenberg; also MX-program; research and development of robotics for U.S. Navy; Air Force B-1 B program. Office: 710 N Fron St San Pedro CA 90733

FOEGE, WILLIAM HERBERT, pub. health adminstr.; b. Decorah, Iowa, Mar. 12, 1936; s. William August and Anne Erika (Ermisch) F.; m. Paula S. Ristad, Dec. 23, 1958; children—David, Michael, Robert. B.A., Pacific Luth. U., 1957; M.D., U. Wash., 1961; M.P.H., Harvard U., 1965. Intern USPHS Hosp., S.I., N.Y., 1961-62; epidemic intelligence service officer Communicable Disease Center, Atlanta, 1962-64; med. officer Immanuel Med. Center, Yahe, Eastern Nigeria, 1965-66; epidemiologist smallpox eradication/measles control program, Eastern Nigeria, 1969-70; dir. smallpox eradication program Center Disease Control, Atlanta, 1970-73, dir., 1977—; med. epidemiologist assigned to SE Asia Regional Office smallpox program WHO, New Delhi, 1973-75; WHO cons., Bangkok, Thailand, 1967, Kinshasha, Zaire, 1968. Dep. field coordinator Internat. Red Cross Joint Relief Action, Nigeria. Subspecialty: Public health agency administration. Office: Director Centers for Disease Control 1600 Clifton Rd NE Room 2000 Bldg 1 Atlanta GA 30333

FOGEL, MAX LEONARD, psychologist, educator; b. Des Moines, Mar. 19, 1934; s. Louis and Peggy (Hoff) F.; m. Joy S., June 6, 1957; children: Max, Celia, Carol, Alex. B.A., U. Iowa, Iowa City, 1956, M.A., 1959, Ph.D., 1960. Lic. profl. psychologist, Pa. Research assoc. U. Iowa, 1960-62; chief psychologist VA Hosp., Iowa City, 1962-63; sr. research scientist Eastern Pa. Psychiat. Inst., Phila., 1963-80; prof. U. Pa., Pitts., 1965—; pres. Personnel Profiles, Gwynedd, Pa., 1976—; research investigator Children's Hosp. of Phila., 1977-81; psychol. cons., 1963—. Contbr. numerous psychol. articles to profl. publs. Mem. bd. govs. Internat. Sch. for Future, N.Y.C., 1980—; mem. adv. bd. Parent Resource Assn., Glenside, Pa., 1979—. Mensa Edn. and Research Found. grantee, 1968-73; Nat. Inst. Environ. Health Scis. grantee, 1976-82. Fellow Inst. Rational-Emotive Therapy (tng. supr.); mem. Internat. Neuropsychol. Soc. (mem. exec. bd., program chmn.), Mensa (dir. sci. and edn. Bklyn. chpt. 1967-81, exec. v.p. Edn. and Research Found. 1972—, chief editor Research Jour. 1971—), Am. Assn. Behavior Therapy, Am. Assn. Marriage and Family Therapy, Am. Assn. Sex Educators, Counselors and Therapists, Am. Personnel and Guidance Assn., Am. Psychol. Assn., Assoc. Rational Thinkers, Eastern Psychol. Assn., Nat. Assn. Gifted Children, Pa. Psychol. Assn., Phila. Soc. Clinical Psychologists, Psychonomic Soc., Soc. Neurosci., Southeastern Psychol. Assn., Sigma Xi. Subspecialties: Neuropsychology; Cognition. Home: Box 276 Gwynedd PA 19436-0276 Office: Personnel Profiles Box 276 Gwynedd PA 19436-0276

FOGLER, HUGH SCOTT, chem. engr., educator, cons.; b. Normal, Ill., Oct. 28, 1939; s. Ralph Waldo and Ann (Scott) F.; m. Janet Meadors, July 1, 1962; children: Peter, Robert, Kirstin. B.S., U. Ill., 1962; M.S., U. Colo., 1963, Ph.D., 1965. Registered profl. engr., Calif. Vis. scientist Nat. Center Atmospheric Research, 1965; mem. faculty U. Mich., 1965—, asst. prof. chem. engring., 1968-71, assoc. prof., 1971-75, prof., 1975—; research scientist Jet Propulsion Lab., Pasadena, Calif., 1966-68; cons. chem., petroleum engring. to industry. Author: The Elements of Chemical Kinetics and Reactor Calculations, 1974; Contbr. numerous articles to profl. jours. Served as capt. U.S. Army, 1966-68. Mem. Am. Inst. Chem. Engrs. (Detroit chpt. Chem. Engr. of Yr. 1980), Am. Chem. Soc., Am. Soc. Engring. Edn. Methodist. Subspecialties: Chemical engineering; Surface chemistry. Current work: Emulsion stability, dissolution catalysis, petroleum engineering. Office: Dept. Chem. Engring U Mich Ann Arbor MI 48109

FOLAND, KENNETH AUSTIN, geoscience educator; b. Frederick, Md., May 25, 1945; s. Austin Franklin and P. Lillian F.; m. Ellen Spero, June 19, 1968. B.Sc., Bucknell U., 1967; M.Sc., Brown U., 1969, Ph.D., 1972. Postdoctoral fellow U. Pa., Phila., 1972-73; asst. prof. dept. geology, 1973-80, assoc. prof., 1980; assoc. prof. dept. geology and mineralogy Ohio State U., Columbus, 1980—; cons. Lawrence Livermore Nat. Lab., 1982—. Subspecialties: Geochemistry; Petrology. Current work: Isotope geochemistry of natural and artificial materials. Office: Dept Geology and Mineralogy Ohio State U Columbus OH 43210

FOLDS, JAMES DONALD, immunologist; b. Augusta, Ga., Sept. 26, 1940; s. James Earl and Kathleen Louise (Smith) F.; m. Alice Carolyn Cadle, Dec. 27, 1962; children: James Donald, John David. B.S., U. Ga., 1962; Ph.D., Med. Coll. Ga., 1967. Diplomate: Am. Bd. Med. Lab. Immunology. NIH postdoctoral fellow Case Western Res. U., Cleve., 1967-69; instr. Sch. Medicine U. N.C., Chapel Hill, 1969-70, asst. prof., 1970-76, assoc. prof., 1976-82, prof. bacteriology and immunology, 1982—; dir. clin. microbiology immunology labs. N.C. Meml. Hosp. Mem. Am. Soc. Microbiology, Am. Assn. Immunologists, AAAS, Soc. Exptl. Biology and Medicine. Episcopalian. Subspecialties: Infectious diseases; Cellular engineering. Current work: Immunology of T. Pallidum infection. Office: Dept Microbiology and Immunology U NC Sch Medicine 605 FLOB Chapel Hill NC 27514

FOLEY, CRAY LYMAN, aerospace corporation executive; b. Tulsa, Apr. 15, 1927; s. Lyndon Lyman and Margaret (Cray) F.; m. Paula Ann Vincent, Nov. 30, 1957 (div. 1980); children: Kelly Ann, Jill, Cray, Seth. Student, U. Tulsa, 1945-46, 47-48; B.S. in Mech. Engring, Okla. State U., 1951, M.S., 1953. Registered profl. engr., Okla. Jr. research engr. Lockheed Aircraft Co., Burbank, Calif., 1951-52; engr. Sperry Gyroscope Co., Great Neck, N.Y., 1953-57; adv. systems staff engr. Lockheed Missile & Space Co., Inc., Sunnyvale, Calif., 1957—. Pres. Homeowners Assn., San Jose, Calif., 1964-65; com. chmn. Cub Scouts of Santa Clara County, San Jose, 1971-72; mem. Republican Nat. Com., Washington, 1971—. Mem. Soc. Automotive Engrs., Nat. Soc. Profl. Engrs., Am. Def. Preparedness Assn., AIAA, Okla. Soc. Profl. Engrs., Cousteau Soc. (founding). Subspecialties: Mechanical engineering; Aerospace engineering and technology. Current work: Studies and new business proposals in support of the company's aerospace programs, both technical aspects and cost. Home: 7090 Galli Dr San Jose CA 95129 Office: Lockheed Missile & Space Co Inc 1111 Lockheed Way Sunnyvale CA 94086

FOLEY, DANIEL PATRICK, psychology educator; b. Cin., Oct. 15, 1920; s. Daniel Patrick and Mildred Dowell (Lamborn) F. B.Litt. in Greek, Xavier U., Cin., 1945: Ph.L. in Philosophy, Loyola U., Chgo., 1948; M.A. in Exptl. Psychology, Loyola U., Chgo., 1951; S.T.L. in Theology, Loyola U., Chgo., 1955; Ph.D. in Clin. Psychology, U. Ottowa, Ont., Can., 1962. Prof. psychology Loyola U., Chgo., 1948-50; dir. pub. relations St. Xavier High Sch., Cin., 1958-59; prof. psychology Xavier U., Cin., 1962—. Contbr. articles to profl. jours. Roman Catholic. Subspecialty: Social psychology. Current work: Social attitudes and personality dynamics. Home and Office: Dept Psychology Xavier Univ Cincinnati OH 45207

FOLGER, DAVID WINSLOW, marine geologist; b. Woburn, Mass., Nov. 21, 1931; s. Joseph Butler and Marion (Allen) (Folger); m. Joan Carol Throckmorton, June 30, 1956 (dec. Sept. 1980); children: Susan W., Peter F., John B.; m. Carolyn Gail Merrihew, Aug. 14, 1982. B.A., Dartmouth Coll., 1953; M.A., Columbia U., 1958, Ph.D., 1968. Petroleum geologist Chevron Oil Co., Jackson, Miss., 1958-63; postdoctoral investigator Woods Hole (Mass.) Oceanographic Inst., 1968-69; prof. Middlebury (Vt.) Coll., 1969-75; coordinator environ. studies, br. chief U.S. Geol. Survey, Woods Hole, 1975-82, marine geologist, 1982—. Served to lt. (j.g.) U.S. Navy, 1953-56. Fellow Geol. Soc. Am.; mem. Am. Geophys. Union, AAAS, Soc. Exploration Paleontologists and Mineralogists, Sigma Xi. Subspecialties: Geology; Oceanography. Current work: Geologic evaluation of continental margins and ocean basins. Home: 99 Nursery Rd Falmouth MA 02540 Office: Br of Atlantic Geology US Geol Survey Quissett Campus Woods Hole MA 02543

FOLK, GEORGE EDGAR, JR., environmental physiology educator; b. Natick, Mass., Nov. 14, 1914; s. George Edgar and Minnie May (Davis) F. A.B., Harvard U., 1937, M.A., 1940, Ph.D., 1947. Instr. New Eng. secondary schs., 1937-39, 40-42; asst. prof. Bowdoin Coll., Brunswick, Maine, 1947-52; prof. environ. physiology U. Iowa, Iowa City, 1952—. Author: Textbook of Environmental Physiology, 1965, 3d edit., 1984; contbr. over 140 articles to sci. jours., chpts. to books. Mem. Internat. Hibernation Soc., Internat. Soc. Biometeorology, Internat. Soc. Biotelemetry, Bear Biology Assn., Am. Physiol. Soc., Am. Soc. Zoologists, Am. Soc. Mammalogy, Am. Meteorol. Soc., Am. Soc. Circumpolar Health, Am. Soc. Chronobiology, AAAS, AAUP, Ecol. Soc. Am., Polar Soc., Arctic Inst. N.Am., Nat. Soc. Med. Research, Am. Inst. Biol. Sci., Explorers Club, Iowa Acad. Sci., Sigma Xi. Subspecialties: Mammalian environmental physiology; Comparative physiology. Current work: Effects of environments on man and other mammals: cold, heat, altitude, high pressure, hibernation, temperature regulation, light and darkness. Office: Dept Physiology and Biophysics Coll Medicine U Iowa Iowa City IA 52242

FOLKERS, KARL AUGUST, chemist; b. Decatur, Ill., Sept. 1, 1906; s. August William and Laura Susan (Black) F.; m. Selma Leona Johnson, July 30, 1932; children—Cynthia Carol, Richard Karl. B.S., U. Ill., 1928, D.Sc., 1973; Ph.D., U. Wis., 1931; postdoctoral research, Yale, 1931-34; D.Sc., Phila. Coll. Pharmacy and Sci., 1962, U. Wis., U. Uppsala (Sweden), 1969. With Merck & Co., Inc., Rahway, N.J., summer 1933, 34-63, asst. dir. research, 1938-45, dir. organic and biochem. research, 1945-63, asso. dir. research and devel., 1951-53, dir. organic and biol. chem. research, 1953-56, exec. dir. fundamental research, 1956-62, v.p. exploratory research, 1962-63; pres., chief exec. officer Stanford Research Inst., Menlo Park, Calif., 1963-68, mem. council bd. trustees, 1971-74; courtesy prof. chemistry Stanford, 1963-68; Ashbel Smith prof., dir. Inst. Biomed. Research, U. Tex., 1968—; Baker non-resident lectr. in chemistry Cornell U., 1953; Regents lectr. UCLA, 1960; lectr. vitamin chemistry U. Calif., Berkeley, 1963; F.F. Nord lectr. biochemistry Fordham U., 1971; mem. sci. adv. com. Inst. Microbiology, Rutgers; chmn. symposium chmn. 3d Internat. Congress Pure and Applied Chemistry, Boston, 1971; adv. council dept. chemistry Princeton, 1958-64; Walter Hartung lectr. U. N.C., Chapel Hill; chmn., lectr. sect. 1 Isolation, Chemistry and Radioimmunoassay Nobel Symposium on Substance P, Stockholm, 1976; lectr. Tohoku U. Sch. Medicine, Japan, 1981, Assn. Advancement Med. Instrumentation, Washington, 1981, U. Athens, 1981; plenary lectr., chmn. sect. on chemistry of hypothalamic hormones 2d European Colloqium on Hypothalamic Hormones U. Tübingen, Germany, 1976; chmn., lectr. Internat. Symposium on Coenzyme Q, Japan, 1976; Burger lectr. U. Va., 1977; Plenary lectr. 500th Anniversary U. Uppsala, 1977; organizer, chmn. Gordon Research Conf. on Chemotherapy of Exptl. and Clin. Cancer, 1978; lectr. Ferring Symposium, Munich, 1979, Internat. Brain Research Symposium, Zurich, 1979; co-chmn., lectr. Internat. Symposium on Coenzyme Q, Tokyo, 1979; Dreyfus Disting. scholar Reed Coll., Portland, Oreg., 1981. Mem. editorial bd.: Jour. Molecular Medicine; Contbr. sci. jours. on organic chemistry. Trustee Gordon Research Confs., 1971-77. Recipient Am. Chem. Soc. award in pure chemistry, 1941, Spencer award, 1959; Julius Sturmer Lecture award, 1957; Perkin medal Soc. Chem. Industry, 1960; Nichols medal N.Y. sect. Am. Chem. Soc., 1967; Robert A. Welch Internat. award and medal for research on life processes, 1972; award in pharm. and medicinal chemistry Am. Pharm. Assn. Found. and Acad. Pharm. Scis., 1974, Alexander von Humboldt-Stiftung, 1977; 2d S.W. Sci. Forum award, 1979; co-recipient Van Meter prize Am. Thyroid Assn., 1969. Mem. Nat. Acad. Sci., Am. Chem. Soc. (pres. 1962), Am. Soc. Biol. Chemistry, Am. Inst. Nutrition, Soc. Exptl. Biology, N.Y. Acad. Sci., Am. Soc. Biol. Chemistry, Am. Inst. Nutrition, Soc. Exptl. Biology and Medicine, A.A.A.S., Am. Inst. Chemists, Royal Swedish Acad. Engring. Scis. (fgn. mem.), Societa Italiana di Scienze Farmaceutiche (hon.), Sigma Xi (Phi Lambda Upsilon, hon.), Alpha Chi Sigma, Rho Chi. Methodist. Subspecialty: Bromedical research administration. Home: 6406 Mesa Dr Austin TX 78731

FOLLINGSTAD, HENRY GEORGE, mathematics educator, consultant; b. Wanamingo, Minn., Jan. 6, 1922; s. Henry A. and Lottie R. (Johnson) F.; m. Helen Jane Chrislock, May 26, 1945; children: Nancy Ellen, Daniel Mark, Karen Joy, Sharon Ruth, Carl Martin. B.E.E., U. Minn., 1947, M.S., 1971. Mem. tech. staff Bell Telephone Labs., Inc., Murray Hill, N.J., 1948-62; instr. Augsburg Coll., Mpls., 1962-66; sci. research cons. Honeywell, Mpls. and St. Paul, 1964-81; asst. prof. Augsburg Coll., 1966-78, assoc. prof., 1978—; electronics research cons. North Star Research and Devel. Inst., Mpls., 1965-66. Contbr. numerous articles to profl. jours. Trustee Luther Coll. Bible and Liberal Arts, Teaneck, N.J., 1960-63; Bible lectr. Augsburg Coll., 1983, Central Luth. Ch., Mpls., 1969-71, 81-82. Served in USAAF, 1943-46. Mem. IEEE (sr.), Math. Assn. Am., Tau Beta Pi, Sigma Pi Sigma. Lutheran. Subspecialties: Applied mathematics; Relativity and gravitation. Current work: Mathematical-logical-physical absurdities, experimental interpretation errors, and inadequacies in Einstein special and general relativity, with space-age alternatives; abuses of mathematics in scientific-academic-social-theological areas, with space-age alternatives; math-model analyses of complex physical systems. Home: 3506 Garfield Ave S Minneapolis MN 55408 Office: 731 21st Ave S Minneapolis MN 55454

FOMON, SAMUEL JOSEPH, physician, educator, researcher; b. Chgo., Mar. 9, 1923; s. Samuel and Isabel (Sherman) F.; m. Betty Lorraine Freeman, Aug. 20, 1948 (div. Apr. 1978); children: Elizabeth Ann Fomon Seiberling, Kathleen Lenore Fomon Anderson, David Bruce, Christopher, Mary Susan. A.B. cum laude, Harvard U., 1945; M.D., U. Pa., 1947; Doctor Honora Causis, Universidad Catolica de Cordoba, Argentina, 1974. Cert. Am. Bd. Pediatrics, Am. Bd. Nutrition. Intern Queen's Gen. Hosp., Jamaica, N.Y., 1947-48; resident Children's Hosp., Phila., 1948-50; research fellow Cin. Children's Hosp. Research Found., 1950-52; asst. prof. pediatrics U. Iowa, Iowa City, 1954-58, assoc. prof., 1958-61, prof., 1961—, dir. program in numan nutrition, 1980—; mem. rev. com. child health and human devel. program project NIH, 1966-69, nutrition study sect., 1978-81; mem. Select Com. GRAS substances Life Sci. Research Office, 1974-80. Author: Infant Nutrition, 1967, 2d edit., 1974. Recipient Career Devel. award NIH, 1962-67, Rosen von Rosenstein award Swedish Pediatric Soc., 1975, McCollum award Am. Soc. Clin. Nutrition, 1979; F. Cuenca Villoro Found. award, 1981. Mem. Am. Acad. Pediatrics (chmn. com. nutrition 1960-63, recipient Borden award 1956), Am. Soc. Clin. Nutrition (pres. 1981-82), Fedn. Am. Socs. Exptl. Biology, Midwest Soc. Pediatric Research (pres. 1963-64). Subspecialties: Nutrition (medicine); Pediatrics. Current work: Research in growth, body composition and nutrition in infancy; teaching in human nutrition. Office: Dept Pediatrics University Hospitals Iowa City IA 52242

FONCK, RAYMOND JOHN, physicist; b. Joliet, Ill., Nov. 1, 1951; s. Joseph H. and Rosalie M. (Ochs) F.; m. Rosalie Ann Migas, Aug. 22, 1977. B.A. with honors, U. Wis., 1973; postgrad., Princeton U., 1973-74; Ph.D., U. Wis., 1978. Research assoc. Plasma Physics Lab, Princeton U., 1978-80, research staff II, 1980-83; research physicist Plasma Physic Lab. Princeton U., 1983—. Contbr. articles to profl. publs. NSF grad. fellow, 1974. Mem. Am. Phys. Soc., Optical Soc. Am., AAAS, Union Concerned Scientists, Fedn. Am. Scientists, ACLU, Phi Beta Kappa, Phi Kappa Phi. Subspecialties: Nuclear fusion; Atomic and molecular physics. Current work: Plasma spectroscopy; optical instrumentation; tokamak diagnostics; plasma confinement Development of spectroscopic diagnostics for fusion plasmas. Atomic and plasma physics research in high temperature plasmas. Office: Princeton Plasma Physics Lab PO Box 451 Princeton NJ 08544

FONTAINE, GILLES JOSEPH, physics educator; b. Levis, Que., Can., Aug. 13, 1948; s. Emilien Joseph and Marie Louise (Roy) F; m. Francine Fortin, July 5, 1969; children: Marc-Andre, Julie. B.Sc., U. Laval, 1969; Ph.D., U. Rochester, 1973. Postdoctoral fellow U. Montreal, Que., Can., 1973-75, research assoc., 1975-76, asst. prof. physics, 1977-80, assoc. prof., 1980—; research assoc. U. Western Ont., London, 1976-77. Contbr. articles to profl. jours. Mem. Internat. Astron. Union, Am. Astron. Soc., Canadian Assn. Physicists, Canadian Astron. Soc. Subspecialties: Theoretical astrophysics; Optical astronomy. Current work: Stellar structure and evolution, white dwarf stars, pulsating white dwarfs, dense matter, diffusion theory.

FONTANA, MARS GUY, engr.; b. Iron Mountain, Mich., Apr. 6, 1910; s. Dominic and Rosalie (Amico) F.; m. Elizabeth Frances Carley, Aug. 21, 1937; children—Martha Jane, Mary Elizabeth, David Carley, Thomas Edward. B.S., U. Mich., 1931, M.S., 1932, Ph.D., 1935, D.Eng. (hon.), 1975. Research asst., dept. engring. research U. Mich., 1929-34; metall. engr., group supervisor engring. dept. duPont Co., Wilmington, Del., 1934-45; prof., chmn. dept. metall. engring. Ohio State U., 1945-75, prof. emeritus, 1976, Regents prof., 1967—, Duriron prof., 1970-75; dir. Corrosion Center; supr. metall. research; dir. Worthington Industries, 1973—, mem. audit com., 1975—; research NASA, USN, USAF, Nat. Sci. Found., Alloy Casting Inst.; cons. engr. several pvt. and govtl. orgns. Author: Corrosion: A Compilation, 1957, Corrosion Engineering, 1967, 2d edit., 1978; contbr.: column Indsl. and Engring. Chemistry, 1947-56; also other tech. publs. Recipient distinguished alumnus citation U. Mich., 1953, Sesquicentennial award, 1967; Frank Newman Speller award in corrosion engring. Nat. Assn. Corrosion Engrs., 1956; Native Son award Iron Mountain (Mich.) Rotary Club, 1969; Neil Armstrong award Ohio Soc. Profl. Engrs., 1973; MacQuigg Teaching award Coll. Engring., Ohio State U., 1973; Mars G. Fontana Labs. at Ohio State U. named in his honor, 1981. Fellow Am. Soc. Metals (hon.; Gold medal 1979), Am. Inst. Mining, Metall. and Petroleum Engrs., Am. Inst. Chem. Engrs.; mem.

Nat. Assn. Corrosion Engrs. (pres. 1952, editor Jour. Corrosion 1962-74), Electrochem. Soc., Materials Tech. Inst. of Chem. Process Industries (exec. dir. 1977—), Nat. Acad. Engring., Nat. Soc. Profl. Engrs., Am. Soc. Engring. Edn. (award for excellence in engring. instruction 1969), Sigma Xi, Tau Beta Pi, Alpha Chi Sigma, Iota Alpha, Phi Eta Sigma, Phi Lambda Upsilon, Sphinx, Texnikoi. Clubs: Port au Villa (Naples, Fla.) (pres. 1967-70); Faculty, Univ. Golf. Subspecialty: Materials. Current work: Solving corrosion problems primarily in chemical industry. Patentee on corrosion testing and recording devices, also iron ore reduction and corrosion resistant alloys. Home: 2086 Elgin Rd Columbus OH 43221

FONTANA, PETER ROBERT, physicist, educator, cons.; b. Bern, Switzerland, Apr. 20, 1935; came to U.S., 1956, naturalized, 1970; divorced. M.S., Miami U., Oxford, Ohio, 1958; Ph.D., Yale U., 1960. Research assoc. U. Chgo., 1960-62; asst. prof. U. Mich., Ann Arbor, 1962-67; assoc. prof. Oreg. State U., Corvallis, 1967-74, prof. physics, 1974—; vis. prof. physics Swiss Inst. Tech., Lausanne, 1974-77, U. Tubingen, W. Ger., 1982. Author: Atomic Radiative Processes, 1982; also articles on atomic physics. Mem. Am. Phys. Soc. Subspecialties: Theoretical physics; Atomic and molecular physics. Current work: Atomic radiative processes; quantum optics; inter-atomic forces; atomic fine and hyperfine structure. Patentee multi-detector intensity interferometer.

FONTANELLA, JOHN JOSEPH, physics educator, researcher; b. Rochester, Pa., Dec. 20, 1945; s. John P. and Anna (Mikita) F.; m. Mary Catherine Wintersgill, 1978; 1 son; John Michael Thomas. B.S., Westminster Coll., 1967; M.S., Case Western Reserve U., 1969; Ph.D., Case Inst. Tech., 1971. Asst. prof. U.S. Naval Acad., 1974-77, assoc. prof. physics, 1978—. Contbr. numerous articles to profl. jours.; Asst. editor: IEEE Transactions on Electrical Insulation Jour, 1978—. Served to lt. (j.g.) USNR, 1971-74. Mem. Am. Assn. Physics Tchrs., Am. Phys. Soc., IEEE. Subspecialties: Condensed matter physics; Polymer physics. Current work: Electrical properties of materials. Home: 106 Cresston Rd Arnold MD 21012 Office: US Naval Acad Physics Dept Annapolis MD 21402

FONTENELLE, DON HARRIS, child psychologist, author; b. New Orleans, Mar. 11, 1946; s. Irvin Joseph and Olga (Thoele) F.; m. Carla M. Leto, aug. 26, 1967; children: Jason, Alan. Student, Southeastern La. U., 1963-64; B.A., La. State U.-New Orleans, 1967; M.S., Northwestern State U. La., 1969; postgrad., U. So. Miss., 1969; Ph.D., Okla. State U., 1972, U. Ark. Med. Ctr., 1971-72. Lic. clin. psychologist, cert. sch. psychologist, La. Psychol. trainee Central La. State Hosp., Pineville, 1969; grad. teaching asst. Okla. State U., 1970-71; mem. staff Payne County Mental Health Ctr., Stillwater, Okla., 1970-71, Psychol. Guidance Ctr., Stillwater, 1970-71, Bi-State Mental Health Ctr., 1971; intern in clin. psychology U. Ark. Med. Ctr., Little Rock, 1971-72; instr. La. State U. Med. Sch., 1972; instr. dept. psychiatry and neurology Tulane U. Sch. Medicine, 1972-73, guest lectr. dept. speech and hearing, also; pvt. practice psychology Orleans Clinics of Psychology, New Orleans, 1973-78; instr. St. Bernard Community Coll., 1975-76; dir. St. Bernard Devel. Ctr., Chalmette, La., 1979—; cons. in field; mem. adv. bd. St. Bernard Mental Health Ctr., Chalmette, St. Louise De Marillac Sch., Arabi, La.; condr. parent tng. workshops/programs. Author: How To Live With Your Children: A Guide for Parents Using a Positive Approach to Child Behavior, 1981, (M. M. Collins) Changing Student Behaviors: A Positive Approach, 1982, Understanding and Managing Overactive Children: A Guide for Parents and Teachers, 1983. Mem. Am. Psychol. Assn., Nat. Assn. Sch. Psychologists, Southeastern Psychol. Assn., Southwestern Psychol. Assn., La. Psychol. Assn., La. Assn. Sch. Psychologists, New Orleans Assn. Children with Learning Disabilities, Phi Kappa Phi, Psi Chi. Subspecialty: Child psychology. Current work: Parent training in techniques of child management, working with children (ages two years to 14 years) with a variety of behavioral/emotional and/or academic difficulties. Home: 2009 Aycock St Arabi LA 70032 Office: Saint Bernard Devel Center 3114 Paris Rd Chalmette LA 70043

FONTENOT, JOSEPH PAUL, animal science educator, researcher; b. Mamou, La., May 11, 1927; married, 1946; 6 children. B.S., Southwestern La. U., 1951; M.S., Okla. State U., 1953, Ph.D. in Animal Nutrition, 1954. Instr. in physiology and pharmacology Okla. State U., 1954-55; asst. prof. animal husbandry Miss. State U., 1955-56; assoc. prof. Va. Poly. Inst. and State U., 1956-63, prof. animal sci., 1963—. Recipient Nutrition Research award Am. Feed Mfrs. Assn. Mem. AAAS, Am. Soc. Animal Soc. (Morrison award), Am. Inst. Nutrition, Animal Nutrition Research Council. Subspecialty: Animal nutrition. Office: Dept Animal Sci Va Poly Inst and State U Blacksburg VA 24061

FORAL, RALPH FRANCIS, mechanical engineer; b. Omaha, June 18, 1934; s. Ralph Adolph and Alice Marie (Hakel) F.; m. Kathryn Margaret Gaffney, May 4, 1957; children: R. David, Michael J., Mary E., Jeanne M., James E. Student, Creighton U., 1952-53; B.S.M.E., U. Nebr., 1956; M.S., U. Colo., Boulder, 1958, Ph.D., 1963. Assoc. engr. to research specialist Martin Marietta Corp., Denver, 1956-64; assoc. prof. U. Nebr., Lincoln, 1964-72, prof. engring. mechanics, 1972—; cons. Brunswick Corp., Lincoln, 1965—. Recipient Skylab Achievement award NASA, 1974, 1st Shuttle Flight Achievement award, 1981. Mem. ASTM, ASME, Am. Soc. Engring. Edn., Am. Acad. Mechanics. Roman Catholic. Subspecialties: Solid mechanics; Composite materials. Current work: Mechanics of composite materials, mechanical behavior characterization of fiber/epoxy composite materials, filament wound composite pressure vessels. Patentee in field. Home: 1836 Morningside Dr Lincoln NE 68506 Office: U Nebr Lincoln NE 68588

FORBES, GEORGE FRANKLIN, cons. mathematician; b. Boston, June 26, 1915; s. Mark A. and Leah (Boudreau) F.; m. Mary Eleanor Kimball, Mar. 9, 1943; children: Marie, Dorothy, Carolyn, Kenneth, Barbara, Marjorie, Marilyn, Christopher, William, Patricia. B.S., Northeastern U., 1939; postgrad., St. Louis U., Harvard U., U. N.H., M.I.T. Physicist Edwards AFB, Lancaster, Calif., 1951-52; computer programmer Lockheed Aircraft Co., Burbank, Calif., 1952-56; sci. specialist Litton Industries, Los Angeles, 1956-73; contract engr. Singer Simulation Products, Silver Spring, Md., 1974, Nat. Cash Register Co., Cambridge, Ohio, 1974, Naval Weapons Ctr., China Lake, Calif., 1974-76, N.Am. Rockwell, Downey, Calif., 1977, Boeing Aircraft, Wichita, 1977-79, LTV, Grand Prairie, Tex., 1979-80, AIL-Eaton at Edwards AFB, 1980, Boeing, Seattle, 1981-82, Gen. Electric Co., Lynn, Mass., 1982-83; pres. Sperry-Univac, Salt Lake City. Author: Digital Differential Analyzer, 1957; contbr. articles to profl. jours. Served to lt. comdr. USNR, 1942-47. Fellow Brit. Interplanetary Soc., AIAA (assoc.); mem. Assn. Computing Machinery, Naval Inst., Nat. Speleological Soc., Mensa, Soc. Indsl. and Applied Math. Roman Catholic. Lodge: KC. Subspecialties: Mathematical software; Applied mathematics. Current work: Digital differential analyzers; differential equations; computer applications. Home: 9813 Monogram Ave Sepulveda CA 91343

FORBES, JUDITH, project engineer; b. Fullerton, Calif., Sept. 27, 1942; d. James Franklin and Lois Virginia (Couse) F.; m. Edward John Resha, Jr., Aug. 2, 1966; children: Laurel, James, Edward John. B.A. in Physics, Calif. State U.-Fullerton, 1974, M.S. in Engring., 1979. Engr., Electromech. div. Northrop Electronics, Anaheim, Calif., 1975-79, project engr. Electronics div., Hawthorne, Calif., 1980—; mem. tech. staff TRW, San Bernardino, Calif., 1979-80; cons. Calif. State U., Fullerton, 1980-82. Contbr. articles and papers to sci. jours. Mem. council Canyon High Sch., Anaheim, 1980-81. Served with USN, 1960. Calif. State U.-Fullerton Found. grantee, 1974. Mem. Soc. Women Engrs. (sr.; pres. 1980-81, regional dir. 1983-84), AIAA (sr.; vice chmn. membership services 1981-82, treas. 1983-84), Orange County Engring. Council (del. 1980-81). Subspecialties: Systems engineering; Materials (engineering). Current work: Gyroscope and accelerometers for ballistic missiles; integration of interial instrumentation and testing into guidance systems. Home: 7066 Country Club Anaheim CA 92807 Office: Northrop Electronics 5320/N4 2301 W 120th St Hawthorne CA 90250

FORCE, RONALD C(LARENCE), psychologist, researcher; b. Toledo, Apr. 10, 1917; s. Rockwell Clarence and Anna Elizabeth (Briner) F.; m. Winifred Amelia Schnatz, Dec. 15, 1941; children: Eric R., Hugh B., Bryan P., Gregory M. B.A., Heidelberg Coll., Tiffin, Ohio, 1940; M.A., Miami U., Ohio, 1941; postgrad., U.Calif.-Berkeley, 1948-50, U. Tex.-Austin, 1952. Lic. psychologist, Kans.; cert. health services provider in psychology Nat. Register Health Service Providers in Psychology. Commd. 2d lt. U.S. Air Force, 1947, advanced through grades to lt. col., 19; clin. and research psychologist (3320th Retng. Group), Amarillo AFB, Tex., 1951-56, dir. clin. services, 1961-64, psychologist, Lackland AFB, Tex., 1957-60, ret., 1964; clin. coordinator St. Francis Boys Homes, Salina, Kans., 1966-81, cons., dir. research, 1982—; dir. Test Systems, Inc., Wichita, Kans., 1972-82; cons. U.S. Army Rehab. Ctr., Ft. Riley, Kans., 1973—, Episcopal Ch. Commn. on Ministry, Kans., 1976—, Presbyterian Ch., 1982—. Author, developer: Biographical and Personal Inventory, 1971; co-author, developer: Multi-Instrument Screen, 1959; contbr. articles to profl. jours. Bd. dirs., officer Salina Symphony Orch., 1977-82. Decorated Air Force Commendation Medal, 1954-64. Fellow AAAS, Am. Orthopsychiat. Assn.; mem. Am. Psychol. Assn., Fedn. Am. Scientists, Union Concerned Scientist. Democrat. Presbyterian. Subspecialties: Behavioral psychology; Corrections psychology. Current work: Development of factorially-derived description/predictive instruments for early intervention, differential placement and treatment of behavior-disordered youth and young adults, feedback directed treatment modification for optimum results. Home: 2811 Melanie Ln Salina KS 67401 Office: Saint Francis Boys Homes Inc PO Box 1348, 509 E Elm St Salina KS 67401

FORD, ALBERT LEWIS, JR., physics educator; b. Ft. Worth, May 12, 1946; s. Albert Lewis and Mary Louise (Garrison) F.; m. Linda Margarette Alexander, Aug. 25, 1968; children: Benjamin, Jason. B.A. in Chemistry, Rice U., Houston, 1968; Ph.D., U. Tex., Austin, 1972. Research assoc. Harvard Coll. Obs., Cambridge, Mass., 1972-73; asst. prof. dept. physics Tex. A&M U., 1973-79, assoc. prof., 1979—. Author: Solutions Guide to accompany College Physics, 1981, Solutions Guide to accompany University Physics, 1982. Recipient Faculty Disting. Achievement award in teaching Tex. A&M Assn. Former Students, 1981. Mem. Am. Phys. Soc. Episcopalian. Subspecialty: Atomic and molecular physics. Current work: Ab Initio calculation of properties and structure of few-electron diatomic molecules; excitation, ionization and charge transfer in ion-atom collisions. Home: 3807 Tanglewood St Bryan TX 77801 Office: Dept Physics Tex A&M U College Station TX 77843

FORD, BYRON MILTON, computer systems consultant; b. Hayden, Colo., Feb. 24, 1939; s. William Howard and Myrtle Oretta (Christian) F.; m. Shirley Ann Edwards, Sept. 4, 1959; children: Gregory Scott, Barry Matthew. B.S., Engring., Colo. U., 1964; M.S., Johns Hopkins U., 1971. Sr. mathematician Johns Hopkins U. Applied Physics Lab., Laurel, Md., 1964-79; cons. computer systems, Laurel, 1979—. Mem. Ops. Research Soc. Am., Nat. Assn. Profl. Cons., Nat. Assn. Self-Employed. Subspecialties: Mathematical software; Operations research (mathematics). Current work: Structured programming languages; data communications. Address: 6909 Remiles Rd Laurel MD 20707

FORD, DONALD HERBERT, neuroanatomy research consultant; b. Kansas City, Mo., Aug. 18, 1921; s. Horace G. and Gladys E. (Newell) F.; m. Dorothy Anne Glander, Aug. 5, 1944; 1 dau., Linda Anne. B.A., Wesleyan U., Middletown, Conn., 1947, M.A., 1949; Ph.D., U. Kans., 1952. Faculty mem. Downstate Med. Ctr.-SUNY, N.Y.C., 1952-77, assoc. prof., 1959-68, prof. neurosci., 1968-77; assoc. research dir. Council Tobacco Research, N.Y.C., 1977—; cons. neuroanatomy Bklyn. VA Hosp., 1954-77, L.I. Coll. Hosp., 1954-77, neuroanat. research cons. N.Y. State Research Inst. Neurochemistry and Drug Addiction, 1968-71. Author: books including The Brain Barrier Systems, 1968, Basic Neurology, 2d edit, 1973, Tissue Responses to Addictive Drugs, 1978, Primary Intracranial Neoplasms, 1979, Atlas of the Human Brain, 1978; contbr. numerous articles in field to profl. jours. Served with M.C. AUS, 1942-45. Fellow Woodshole (Mass.) Marine Lab., 1957. Mem. Am. Assn. Anatomists, Am. Soc. Neurochemistry, Internat. Soc. Neurochemistry, Internat. Soc. Psychoneuroendocrinology, Internat. Soc. Devel. Neurosci., Am. Physiol. Soc., Endocrine Soc., Sigma Xi, Phi Sigma. Republican. Subspecialties: Neurology; Neuroendocrinology. Current work: Role of thyroid hormone on neuronal function and development; influence of opiates on neuroendocrinology and neuron protein synthesis. Office: Council for Tobacco Research 110 E 59th St New York NY 10022

FORD, LARRY HOWARD, psychologist, statistician; b. Orlando, Fla., May 20, 1953; s. William Curtis and Etha (Howard) F.; m. Diane Marie Klepper, Nov. 19, 1973. Student, Eckerd Coll., 1971-73; B.A., No. Mich. U., 1976; M.A., U. Mich., 1979. Asst. study dir. Inst. Social Research, Ann Arbor, Mich., 1978-79; psychologist Tng. Analysis and Evaluation Group, Dept Navy, Orlando, 1980; teaching fellow U. Mich., 1977-79; cons., Ann Arbor, 1979. NIMH trainee (3), 1977-79. Mem. Acad. Mgmt., Am. Psychol. Assn. (assoc.), Phi Kappa Phi. Democrat. Subspecialties: Organizational and educational psychology; Statistics. Current work: Training systems, exploratory analysis, social system evaluation. 32813

FORD, LESTER R., JR., mathematician, computer scientist; b. Houston, Sept. 23, 1927; s. Lester R. and Marguerite Eleanor (John) F.; m. Janet Lux, Apr. 20, 1950 (div. 1967); children: Diana, Barbara, Pamela, Andrea, Lester, Ilisa, Melinda, Robert, Francis; m. Naoma Gower, Aug. 3, 1968. Ph.B., U. Chgo., 1949, S.M., 1950; Ph.D., U. Ill., 1953. Research instr. Duke U., Durham, N.C., 1953-54; analyst RAND Corp., Santa Monica, Calif., 1954-57; project dir. Gen. Analysis Corp., Sierra Vista, Ariz., 1957-60, C.E.I.R., Beverly Hills, Calif., 1960-63; sr. scientist Gen. Research Corp., Santa Barbara, Calif., 1963—. Author: Flows in Networks, 1961, Calculus, 1963. Served with AUS, 1946-47. Mem. Math. Assn. Am., Soc. Indsl. and Applied Math., Ops. Research Soc. Am. Subspecialties: Algorithms; Operations research (mathematics). Current work: Application of computer simulation techniques to the prediction and evaluation of large-scale system performance. Office: Gen Research Corp PO Box 6770 5383 Hollister Ave Santa Barbara CA 93111 Home: 1140 Mission Ridge Rd Santa Barbara CA 93103

FORD, LINCOLN EDMOND, cardiologist, educator; b. Boston, May 14, 1938; s. John B. and Mary Margaret (Clark) F.; m. Erica Winifred Roy, Feb. 24, 1973; children: Catherine Leigh, Gretchen Anne, Vanessa Erica, Emily Winifred. A.B., Harvard U., 1960; M.D., U. Rochester, 1965. Intern Bassett Hosp., Cooperstown, N.Y., 1965-66; staff assoc. NIH, 1966-68; research fellow-research assoc. Peter Bent Brigham Hosp., Boston, 1970-71; NIH Hon. Research fellow U. Coll. London, 1971-74; assoc. prof. medicine U. Chgo., 1974—; adj. prof. biomed. engring. Northwestern U., Chgo., 1980—. Editorial bd.: Jour. Muscle Research and Cell Mobility, 1979—, Am. Jour. Physiology: Heart and Circulatory, 1981—, Circulation Research, 1983—. Chmn. research council Chgo. Heart Assn., 1978-79; established investigator Am. Heart Assn., 1975-80. Served with USPHS, 1966-68. Mem. Am. Physiol. Soc., Biophys. Soc., Soc. Gen. Physiologists. Clubs: Harvard (Boston); Quadrangle (Chgo.). Subspecialties: Physiology (medicine); Biofeedback. Current work: Muscle physiology and cardiovascular hemodynamics. Office: U Chgo Box 249 950 E 59th St Chicago IL 60637 Home: 4949 S Woodlawn Ave Chicago IL 60615

FORD, MARY ELIZABETH, clinical psychologist; b. Burlington, Iowa, May 4, 1949; d. John Calvin and Iva Elizabeth (Ernest) Ferguson. A.A., Southeastern Iowa Community Coll., Burlington, 1971; B.A., U. Iowa, 1973, Ph.D., 1980. Lic. psychologist, Ga., Iowa. Grad. asst. psychology U. Iowa, Iowa City, 1974-77, research asst., 1977; staff psychologist St. Luke's Methodist Hosp., Cedar Rapids, Iowa, 1977-79, dir. clin. psychology, 1979-81; clin. psychologist Community Mental Health Ctr. Linn County, Cedar Rapids, 1981-82, N. DeKalb Mental Health Ctr., Atlanta, 1982—; cons. psychologist Linn County Orthopedists, Cedar Rapids, 1981-82. Explorer post leader Hawkeye Area council Boy Scouts Am., Cedar Rapids, 1980-81. Mem. Am. Psychol. Assn., Acad. Psychologists in Marital Sex and Family Therapy, Ga. Psychol. Assn., Pi Lambda Theta. Subspecialty: Behavioral psychology. Current work: Psychotherapy with adults and children, psychotherapy outcome, family therapy. Office: North DeKalb Mental Health Center 3007 Hermance Dr NE Atlanta GA

FORD, PETER CAMPBELL, chemist, educator; b. Salinas, Calif., July 10, 1941; s. Clifford and Thelma Mae (Martin) F.; m. Katherine Ellen Howland, June 20, 1963; children: Vincent Howland, Jonathan Martin. B.S. with honors, Calif. Inst. Tech., 1962; M.S., Yale U., 1963, Ph.D. (NIH fellow), 1966. NSF postdoctoral fellow Stanford U., 1966-67; asst. prof. chemistry U. Calif., Santa Barbara, 1967-72, assoc. prof., 1972-77, prof., 1977—; fellow Australian Nat. U., 1974; guest prof. U. Copenhagen, 1981. Mem. Am. Chem. Soc., Royal Soc. Chemistry (U.K.). Subspecialties: Inorganic chemistry; Photochemistry. Office: Dept Chemistry U Calif Santa Barbara CA 93106

FORD, RICHARD EARL, plant virologist, educator; b. Des Moines, May 25, 1933; s. Victor S. and Gertrude (Headlee) F.; m. Roberta Jean Essig, June 20, 1954; children—Nina Diane, Linda Marie, Kent Richard (dec.), Steven Earl. B.S., Iowa State U., 1956; M.S., Cornell U., 1959, Ph.D., 1961. Undergrad. research technician Iowa State U., 1952-55, undergrad. tchr. botany, 1956; grad. research asst. plant pathology Cornell U., 1956-59, grad. research asst. virology tchr., 1959-61; research plant pathologist, asst. prof. U.S. Dept. Agr. and Oreg. State U., 1961-65; prof. plant virology, virus research Iowa State U., Ames, 1965-72; prof., head dept. plant pathology U. Ill., Urbana, 1972—, cons. agrl. improvement, Europe, Asia, S.Am., others. Editor: Jour. Phytopathology; contbr. articles to profl. jours. Mem. Am. Phytopath. Soc. (sec., councilor, pres. N.C. div., nat. pres. and v.p.), Assn. Dept. Heads of Plant Pathology in U.S. (chmn.), Sigma Xi (v.p., treas. U. Ill. chpt.), Phi Kappa Phi, Phi Sigma Gamma, Gamma Sigma Delta, Gamma Gamma, FarmHouse Frat. (nat. pres., Master Builder of Men award, tres. FarmHouse Found.). Clubs: Lions, Toastmasters. Subspecialties: Plant pathology; Plant virology. Current work: Diagnosis of plant virus diseases, identification, purification, characterization and serology of plant viruses; virus-virus/virus-host interactions; physiology and biochemistry of disease; virus, fungal and bacterial interractions in disease. Home: 11 Persimmon Circle Urbana IL 61801 Principles guiding my life are honesty, integrity and complete openness and candor. Hard work and long, often tedious, hours on the farm taught me the value of these attributes in all jobs. My goal is to provide an atmosphere conducive for all associates to attain their pinnacle of success by honest, intelligent use of time with hard work.

FORD, STEPHEN PAUL, reproductive physiologist, educator; b. Palo Alto, Calif., Oct. 11, 1948; s. Frank George and Rosemary (Bonnot) F.; m. Marsha Ann Ford, Sept. 12, 1970; children: Tamara Lynn, Joanna Christine, Jessica Gale. B.S., Oreg. State U., 1971, Ph.D., 1977; M.S., W. Va. U., 1973. Grad. research asst. dept. animal and vet. sci. W. Va. U., 1971-73; grad. research asst. dept. animal sci. Oreg. State U., 1974-77; research physiologist Roman L. Hruska U.S. Meat Animal Research Center, Clay Center, Nebr., 1977-79; assoc. prof. animal sci. Iowa State U., Ames, 1979—. Contbr. to profl. jours. The Upjohn Co. grantee, 1980-81; NIH grantee, 1982-85. Mem. Am. Soc. Animal Sci., Soc. Study Reprodn., Sigma Xi, Gamma Sigma Delta. Subspecialties: Reproductive biology; Animal physiology. Current work: Role of the uterus and ovaries in potentiating conceptus survival in domestic animals; factors controlling uterine and ovarian blood-flow during the estrous cycle and early pregnancy in the ewe, cow, and sow. Home: 3317 Jewel Dr Ames IA 50010 Office: 11 Kildee Hall Iowa State U Ames IA 50011

FORD, WARREN THOMAS, chemistry educator; b. Kalamazoo, Mar. 22, 1942; m. Sharon R. Tenney, 1967; children: Sarah, Emily. A.B., Wabash Coll., 1963; Ph.D., UCLA, 1967. Postdoctoral fellow Harvard U., Cambridge, Mass., 1967-68; asst. prof. chemistry U. Ill., Urbana, 1968-75, Okla. State U., Stillwater, 1978-80, assoc. prof., 1980-83, prof., 1983—; sr. scientist Rohm & Haas Co., Spring House, Pa., 1975-78. Contbr. articles to profl. jours. Mem. Am. Chem. Soc., AAAS, Phi Beta Kappa, Sigma Xi, Phi Lambda Upsilon. Subspecialties: Organic chemistry; Polymer chemistry. Current work: Polymer-supported reagents and catalysts, polymer synthesis, polymer spectroscopy. Home: 217 S Ridge Rd Stillwater OK 74074 Office: Dept Chemistry Okla State U Stillwater OK 74078

FOREMAN, KENNETH M(ARTIN), aeronautical research engineer; b. N.Y.C., July 30, 1925; s. Louis and Jennie (Gordon) F.; m. Shirley Eleanor Teiger, June 24, 1954; children: Elissa, Michael. B.Aero. Engring., N.Y. U., 1950, M.Aero. Engring., 1953; M.S. in Mgmt, Poly. Inst. N.Y., 1981, Grad. Sch. Cert. Energy Policy and Tech., 1976. Research engr. Bendix Corp., Teterboro, N.J., 1951-52; sect. head ramjet devel. Wright Aero. Div., Woodridge, N.J., 1952-56; research engr. Fairchild Engine Div., Deer Park, N.Y., 1956-59; sci. research engr. Republic Aviation Corp., Farmingdale, N.Y., 1959-65; chief project engr. EDO Corp., College Point, N.Y., 1965-66; head advanced fluid concepts lab. Grumman Aerospace Corp., Bethpage, N.Y., 1966—. Contbr. writings to profl. jours.; Patentee acoustic emission flow measuring system, diffuser augmented wind turbine. Vice pres., historian Hist. Soc. Bellmores, N.Y., 1980-83; v.p. edn. Temple Beth-El of Bellmore, 1970-72; merit badge counselor, committeeman Merogue and Peguot councils Boy Scouts Am., Nassau County, N.Y., 1974—. Recipient Cervantes medal Inst. de los Espanas, N.Y.C., 1942. Assoc. fellow AIAA (chmn. L.I. sect. 1979-80, Basil Staros Meml. award 1982, tech. com. atmospheric environ. 1974-76, L.I. sect. award 1978); mem. ASME (chmn. aerospace engring. div. 1972-73, industry dept. policy bd. 1971-74), AAAS, Tau Beta Pi. Jewish. Subspecialties: Aerospace engineering and technology; Wind power. Current work: Fluid mechanics; aircraft propulsion; solar energy; planetary atmospheres; design of experiments; instrumentation; history of aviation; economic evaluation of scientific invention, engineering. Patentee acoustic emission flow measuring system, diffuser augmented wind turbine. Home: 32 Stratford Ct North Bellmore NY 11710 Office: Grumman Aerospace Corp Mail Stop A08-35 Bethpage NY 11714

FOREMAN, MARK MORTENSEN, physiologist, pharmacologist, researcher; b. Stillwater, Okla., May 8, 1949; s. Paul Breck and Lydia Marie (Mortensen) F.; m. Lynne Dalglish, Dec. 14, 1974; children: Lisa Anne, Scott David. B.S. in Zoology, Pa. State U., 1972, M.S. in Physiology, 1973, Ph.D., U. Tex.-Dallas, 1977. Grad. teaching asst. U. Tex. Health Sci. Ctr.-Dallas, 1973-77, postdoctoral fellow, 1977-80; sr. physiologist dept. endocrine research Eli Lilly Research Labs., Eli Lilly & Co., Indpls., 1980—. NIH fellow, 1977-80; Cecil and Ida Green fellow, 1977-80. Mem. Neurosci. Soc. Subspecialties: Neuroendocrinology; Neuropharmacology. Current work: Neuroendocrine and neuropharmacological control of neurotransmitter release, sexual behavior, thermoregulation. Office: Dept Endocrine and CNS Research Eli Lilly Research Labs Eli Lilly and Co Indianapolis IN 46285

FOREST, CHARLENE LYNN, biology educator, researcher; b. Bklyn., Feb. 27, 1947; d. Harold Matthew and Sadie (Biller) Friedman; m. Richard Mark Forest, June 29, 1969. B.S., Cornell U., 1968; M.S., Adelphi U., 1972; Ph.D., Ind. U., 1976. Post-doctoral fellow Harvard U., 1976-79; asst. prof. biology Bklyn. Coll., 1979—. Cabot Found. fellow, 1976-79; NIH grantee, 1980-83. Mem. Am. Soc. Cell Biology, Genetics Soc. Am, AAAS, Assn. Women in Sci., N.Y. Acad. Sci., Sigma Xi. Subspecialties: Gene actions; Membrane biology. Current work: I am studying the interactions between the membranes of the single-celled alga, Chlamydomonas; I am looking at interactons of flagella, that lead to signaling between cells and interactions between mating structure membranes that lead to adhesion and fusion. Office: Dept Biology Brooklyn College Bedford Ave and Ave H Brooklyn NY 11210

FOREYT, JOHN PAUL, psychologist, educator; b. Manitowoc, Wis., Apr. 6, 1943; s. John O. and Ann R. (Hynek) F. B.S., U. Wis.-Madison, 1965; M.S., Fla. State U., 1967, Ph.D., 1969. Psychol. cons. Fla. Div. Vocat. Rehab., Tallahassee, 1969-71; asst. prof. dept. psychology Fla. State U., Tallahassee, 1971-74; asst. prof. dept. medicine Baylor Coll. Medicine, Houston, 1974-80, assoc. prof., 1980—; clin. assoc. prof. Tex. Woman's U., Houston, 1980—, U. Houston, 1980—; cons. Galena Park Ind. Sch. Dist., Houston, 1979—, Harris County Ind. Sch. Dist., 1977—. Editor: Behavioral Treatments of Obesity, 1977, (with D. Rathjen) Cognitive Behavior Therapy: Research and Application, 1978, Social Competence: Interventions for Children and Adults, 1980; author: (with others) The Living Heart Diet, 1983. Mem. Am. Psychol. Assn., Acad. Behavioral Medicine Research, Assn. Advancement of Behavior Therapy, Behavior Therapy and Research Soc. Subspecialty: Behavioral psychology. Current work: Behavioral treatments of obesity research; research on adherence and compliance to medical regimens; research on cardiovascular risk reduction. Office: Baylor Coll Medicine 6535 Fannin MS F700 Houston TX 77030 Home: 7708 Nairn St Houston TX 77074

FORGHANI, BAGHER, virologist; b. Bandar Anzali, Iran, Mar. 10, 1936; came to U.S., 1969; s. Baba and Jahan (Rahimi) F.; m. Nikoo Alavi, June 12, 1969; children: Niki, Nikta. Ph.D., Justus Liebig U., Giessen, W.Ger., 1965. Postdoctoral fellow Utah State U., 1965-67; asst. prof. Nat. U. Iran, Tehran, 1967-69; postdoctoral trainee Calif. Dept. Health Services, Berkeley, 1970-72, research specialist, 1972—. Contbr. chpt., articles to profl. publs. Mem. Am. Soc. Microbiology. Moslem. Subspecialties: Virology (medicine); Infectious diseases. Current work: Production and characterization of monoclonal antibodies to human immunoglobulin, several viruses, and their application to diagnosis of viral infections. Home: 134 Lombardy Ln Orinda CA 94563 Office: Calif Dept Health Virus Lab 2151 Berkeley Way Berkeley CA 94704

FORISHA-KOVACH, BARBARA ELLEN, psychology educator, consultant, writer; b. Ann Arbor, Mich., Dec. 28, 1941; d. Harry A. and Margaret Mayne (Buell) Lusk; m. Craig Randall Duncan, Dec. 28, 1963; m. Bill Edward Forisha, July 19, 1973; m. Randy Louis Kovach, May 3, 1981; children: Deborah Louise, Mark Randall. B.A., Stanford U., 1963, M.A., 1964; Ph.D., U. Md., 1973. Asst. prof. U. Mich.-Dearborn, 1973-77, assoc. prof., 1977-83, prof., 1983—, chairperson dept. psychology, 1980-83; pres. Human Systems Analysis, Inc., Ann Arbor, 1980—; cons. Mich. Bell Telephone Co., 1981, Ford Motor Co., 1982, Gen. Motors Co., 1982-83. Author: Sex Roles and Personal Awareness, 1978, Power and Love, 1982, The Experience of Adolescence, 1983, Organizational Synchrony, 1983, Pyramids and Circles, 1983; editor: Outsiders on the Inside, 1981. Daniel E. Prescott fellow, 1972; recipient Faculty Recognition award U. Mich., 1980, Susan B. Anthony award, 1981. Mem. Am. Psychol. Assn., Phi Beta Kappa, Pi Lambda Theta. Republican. Episcopalian. Subspecialties: Cognition; Developmental psychology. Current work: Major research area in the relationship of mental imagery to creativity and application of cognitive processes to organizational behavior. Home: 2606 Traver Rd Ann Arbor MI 48105 Office: U Mich Dearborn MI 48128

FORKEL, CURT EMIL, mechanical engineer; b. Cameron, Tex., Oct. 26, 1922; s. Curt Emil and Mary Louise (Baade) F.; m. Chloe Tidwell, Oct. 30, 1943; children: Kaye (Mrs. Keith Lawton Bentzen), Mary Candace (Mrs. Thomas Brame Wilkinson IV), Ruth Ann. B.S., U. Tex., 1944, postgrad., 1947; postgrad., U. Denver, 1946; M.S., U. Idaho, 1969. Registered profl. engr., Okla. Draftsman N.Am. Aviation, Dallas, 1943-44; instr. U. Tex., Austin, 1944-45, U. N.Mex., 1945-46, U. Denver, 1946-47, asst. prof., 1947-48; metall. engr. Phillips Petroleum Engring., Bartlesville, Okla., 1948-60, design group supr., Borger, Tex., 1960-63, Idaho Falls, Idaho, 1963-69, Idaho Nuclear Corp., Idaho Falls, 1969-71, Aerojet Nuclear Co., Idaho Falls, Idaho, 1971-76; sr. project engr. EG & G Idaho Inc., Idaho Falls, 1976-77, br. mgr. design rev. and cost estimating, 1977, design sect. supr., 1977-78, br. mgr. engring. support for nuclear energy programs, 1978-80, supr. design, 1980-81, sr. program specialist, 1981—. Republican precinct committeeman, 1963-64. Mem. ASME, Pi Tau Sigma, Phi Eta Sigma. Republican. Baptist. Clubs: Idaho Nuclear Sportsmen's, Acad. Model Aeronautics, Red Baron RC Modelers Ltd. Subspecialties: Mechanical engineering; Nuclear engineering. Current work: Flow induced heat exchanger tube vibration; high temperature gas cooled nuclear reactor systems and components. Patentee pebble heater heat transfer apparatus. Home: 2306 Koro Ave Idaho Falls ID 83401 Office: PO Box 1625 Idaho Falls ID 83415

FORMAL, SAMUEL BERNARD, microbiologist, researcher; b. Providence, Aug. 28, 1923; s. Jacob and Inez (Simons) F.; m. Rosamond Anne Martin, Oct. 27, 1951; children: Christopher S., David J., James M. A.B., Brown U., 1945, Sc.M., 1949; Ph.D., Boston U., 1952. Microbiologist Walter Reed Army Inst. Research, Washington, 1952-56, chief dept. applied immunology, 1956-76, chief dept. bacterial diseases, 1976—; professorial lectr. Georgetown U., 1972—. Served to lt. j.g. USN, 1943-46. Mem. Am. Soc. Microbiology, Infectious Disease Soc. Am., Assn. Am. Immunologists. Subspecialty: Microbiology (medicine). Current work: Pathogenesis and immunity in enteric infections.

FORMAN, DAVID SHOLEM, neurobiologist; b. Detroit, Dec. 10, 1942; s. Louis B. and Gertrude (Baroff) F. B.A. summa cum laude, Harvard U., 1964; postgrad (Knox fellow), Cambridge U., 1963-65; Ph.D., Rockefeller U., 1971. Postdoctoral fellow Lab. Neuropharmacology NIMH, Washington, 1971-75, Sloan fellow, 1974-75; research physiologist Naval Med. Research Inst., Bethesda, Md., 1975-81; asst. prof anatomy Uniformed Services U. of Health Scis., Bethesda, Md., 1981—. Contbr. articles to profl. jours. Recipient Golden Eagle award Council Internat. Nontheatrical Events, 1976, Spl. Merit award Naval Med. Research Inst., 1978, Fed. Sustained Superior Performance award, 1980, 81. Mem. Soc. Neurosci., Am. Soc. Cell Biology, Am. Soc. Neurochemistry, Am. Assn. Anatomists, Internat. Brain Research Orgn. Jewish. Subspecialties: Cell biology; Neurobiology. Current work: Neurobiology: axonal transport, nerve regeneration, video microscopy, cell motility. Home: 4301 Mass Ave NW#1012 Washington DC 20016 Office: Dept Anatomy Uniformed Services U Health Scis 4301 Jones Bridge Rd Bethesda MD 20814

FORMAN, HENRY JAY, biochemistry researcher, educator; b. N.Y.C., Apr. 20, 1947; s. Albert and Lillian (Klein) F. B.A., Queens Coll., 1967; Ph.D., Columbia U., 1971. Research assoc., postdoctoral fellow Duke U., Durham, N.C., 1971-73; research assoc., asst. prof. U. Kans. Med. Sch., Kansas City, 1973-77, VA Hosp., Kansas City, Mo., 1973-77; asst. prof., assoc. prof. biochemistry U. Pa., Phila., 1978—; mem. Nat. Research Rev. Bd., Med. Coll. Pa., Phila., 1982—. Mem. Am. Soc. Biol. Chemists, John Morgan Soc. (U. Pa.). Jewish. Subspecialties: Biochemistry (biology); Physiology (biology). Current work: Mechanism of free radical production in cells in relation to oxygen and paraquat toxicity and the respiratory burst of phagocytic cells. Office: Univ Pa Richards Bldg G4 Philadelphia PA 19174

FORNES, RAYMOND EARL, SR., physics educator; b. Pitt County, N.C., Jan. 16, 1943; s. Henry Loyd and Lillian (Tucker) F.; m. Geraldine Hudson, June 26, 1966; children: Raymond Earl, Gregory Todd, Timothy Dean. B.A., East Carolina U., 1965; Ph.D., N.C. State U., 1970. Asst. prof. Sch. Textiles, N.C. State U., Raleigh, 1970-74, assoc. prof., 1974-79, prof., 1979—, assoc. mem. physics dept., 1970—, assoc. dean, 1983—. Named Outstanding Tchr. N.C. State U., 1974; recipient Young Scientist Research award Sigma Xi, N.C. State U., 1975, Disting. Research medal Fiber Soc., 1982. Mem. Am. Phys. Soc., Am. Chem. Soc., AAAS. Democrat. Club: Torch (Raleigh, N.C.). Subspecialties: Polymers; Polymer physics. Current work: Magnetic resonance, x-ray diffraction, spectroscopy, composites, fiber and polymer physics, characterization of cotton dusts, polymer blends, liquid crystal solutions. Home: 1008 Maple Ave Apex NC 27502 Office: NC State U Sch Textiles PO Box 5006 Raleigh NC 27502

FORNEY, LEROY S., laboratory executive, device development engineer; b. Lancaster, Pa., Aug. 11, 1938; s. Roy S. and Mary R. (Snyder) F.; children: Matthew, Ellen. B.S., Juniata Coll., Huntingdon, Pa., 1960; Ph.D., Syracuse U., 1965. Research project leader Mobil Chem. Co., Edison, N.J., 1965-71; mem. adj. faculty Rutgers U., New Brunswick, N.J., 1971-72; chem. supply specialist Mobil Oil Corp., N.Y.C., 1971-74; sr. devel. scientist Surgikos (a Johnson & Johnson Co.), Piscataway, N.J., 1974-79; mgr. quality assurance engrng. and systems Extracorporeal (a Johnson & Johnson Co.), King of Prussia, Pa., 1979-81; pres. Qarex Labs., Phila., 1981—. Contbr. articles to profl. jours. Chmn. Unitarian Soc. New Brunswick, 1976-78. Mem. Am. Soc. Artificial Internal Organs, Am. Chem. Soc., AAAS, Sigma Xi. Subspecialties: Biomedical engineering; Biomaterials. Current work: Medical device development and commercialization. Patentee in field. Home and Office: 3530 Henry Ave Philadelphia PA 19129

FOROULIS, Z. ANDREW A., metallurgical engineer; b. Volos, Greece, Dec. 6, 1926; s. Andreas and Athena (Pontikis) F.; m. Diana Kane, Mar. 9, 1962. B.S., Nat. U. Athens, 1954; M.S.Ch.E., MIT, 1956, Met. Eng., 1960, D.Sc. in Metall. Engring, 1961. Registered profl. engr., Calif. Chem. and metall. engring. positions W.R. Grace & Co., Cambridge, Mass., 1956-57; with Exxon Research & Engring. Co., Florham Park, N.J., 1961—; now engring. adv. materials tech.; part-time faculty MIT, 1956, NYU, 1968-73; cons. Aramco. Contbr. numerous articles to profl. jours.; editor/co-editor 5 books in field. Served in Greek Army, 1949-51. MIT fellow, 1958-61. Mem. Electrochem. Soc. (Merit award 1979), Am. Inst. Chem. Engrs., AIME, ASME, Am. Soc. Metals, Am. Chem. Soc., Nat. Assn. Corrosion Engrs., Sigma Xi, Water Pollution Control Fedn., Am. Concrete Inst. Subspecialties: Metallurgical engineering; Chemical engineering. Current work: Engineering consultant in materials technology. Patentee (18).

FORREST, HUGH SOMMERVILLE, zoology educator; b. Glasgow, Scotland, Apr. 28, 1924; came to U.S., 1951; s. Archibald and Margaret Watson (Peden) F.; m. Rosamond Scott Baker, June 12, 1953; children: Eleanor Scott, Anne Sommerville, Hugh Watson. B.S. with honors, U. Glasgow, 1944; Ph.D., U. London, 1947; D.Sc., 1970; Ph.D., U. Cambridge, 1951. Research scientist Med. Research Council Gt. Britain, 1947-51; research fellow Calif. Inst. Tech., Pasadena, 1951-54, sr. research fellow, 1954-55; research scientist U. Tex., Austin, 1955-56, asso. prof., 1956-63, prof. zoology, 1963—, chmn. dept., 1974-78. Editor: Biochemical Genetics, 1971—; Contbr. articles to profl. jours. Carnegie scholar, 1944-45; Gt. Britain Dept. Sci. and Indsl. Research fellow, 1948-51; USPHS research fellow, 1951-53; spl. research fellow, 1973; numerous research grants NSF, USPHS, Robert A. Welch Found. Fellow Royal Soc. Edinburgh, Royal Chem. Soc.; mem. Am. Chem. Soc., Soc. Gen. Physiologists, Soc. Biol. Chemists. Subspecialties: Biochemistry (biology); Gene actions. Current work: Biochemistry of new coenzyme, methoxatin, from methane-oxidizing baceteria; biochemistry of development; insect pigments. Home: 3302 River Rd Austin TX 78703

FORRESTER, THOMAS, physiologist, educator; b. Glasgow, Scotland, Nov. 22, 1936; s. Thomas and Mary (Kirk) F.; m. Norma Carless, July 11, 1962; children: Linda, Mary Claire, Robert. M.B.Ch.B., U. Glasgow, 1962, Ph.D., 1967, M.D. (with honors), 1982. Resident physician So. Gen. Hosp., Glasgow, 1962-63; resident surgeon Glasgow Royal Infirmary, 1963; asst. lectr. physiology U. Glasgow, 1963-65, lectr. physiology, 1965-71, sr. lectr. physiology, 1971-75; assoc. prof. physiology St. Louis U., 1975—, asst. prof. internal medicine, 1983—. Contbr. numerous articles to nat. and internat. sci. jours. Pres., St. Andrews Soc. Greater St. Louis, 1979-81. Mem. U.K. Physiol. Soc., Am. Physiol. Soc., Sci. Research Soc. N.Am. Presbyterian. Subspecialty: Physiology (medicine). Current work: control of blood flow through skeletal and cardiac musculature transmission of the nerve impulse at the neuromuscular junction. Office: St Louis U Med Center 1402 S Grand Blvd Saint Louis MO 63130

FORSEE, AYLESA, author; b. Kirksville, Mo.; d. Edward W. and Lena (Moore) F. B.S., S.D. State U.; Mus.B., MacPhail Coll. Music, Mpls., 1938; M.A., U. Colo., 1939. Instr. history and music, Rochester, Minn., 1939-45; tchr. history and music U. Iowa, 1945-46, U. Denver, 1946-49; mem. staff writers conf. Temple Buell Coll., 1967, 68; Mem. adv. bd. Nat. Writers Club. Author: The Whirly Bird, 1955, Miracle for Mingo, 1956, Too Much Dog, 1957, American Women Who Scored Firsts, 1958, Louis Agassiz: Pied Piper of Science, 1958, Frank Lloyd Wright: Rebel in Concrete, 1959, Women Who Reached for Tomorrow, 1960, My Love and I Together, 1961, Beneath Land and Sea, 1962, Albert Einstein, 1963, William Henry Jackson, 1964, Pablo Casals: Cellist for Freedom, 1965, Men of Modern Architecture, 1966, Headliners, 1967, Famous Photographers, 1968, Artur Rubinstein: King of the Keyboard, 1969, They Trusted God, 1980; also articles in adult and juvenile publs. Recipient Helen Fish award, 1955. Mem. Colo. Author's League. (Tophand award 1966, 69). Christian Scientist (practitioner). Subspecialty: Systems engineering. Current work: System dynamics: simulation modeling of managerial, economic, urban and financial systems. Analysis of how alternative policies alter social and economic behavior. Address: 1845 Bluebell Ave Boulder CO 80302

FORSHAM, PETER HUGH, physician; b. New Orleans, Nov. 15, 1915; s. John S. and Augusta (Kahnweiler) F.; m. Constance Campbell, Aug. 2, 1947; children—Barbara Forsham Cheetham, Elizabeth, Ann. B.A., Cambridge U., 1937, M.A., 1941; M.D. cum laude, Harvard, 1943. Intern, resident research fellow Peter Bent Brigham Hosp., Boston, 1944-51; also instr. Harvard, 1950-51; asso. prof. medicine and pediatrics U. Calif., 1952-57, prof., 1957—, also dir. metabolic research unit, 1952—, chief endocrinology dept. medicine, 1957—; dir. Gen. Clin. Research Center of U. Calif. Hosps., 1963—; mem. nat. adv. council on health research socs. NIH; cons. Oak Knoll, Naval, San Francisco VA hosps. Mem. editorial bd. of: Diabetes, 1961-71, Metabolism, 1962-69; Contbr. articles profl. jours., chpts. in books. Fellow N.Y. Acad. Sci.; mem. A.M.A., Mass., Calif., San Francisco med. assns., Endocrine Soc. (council), Am. Diabetes Assn., San Francisco Diabetes Assn. (past pres.), No. Calif. (past pres.), Western Assn. Physicians, Calif. Soc. Internal Medicine, Assn. Am. Physicians, Am. Soc. Clin. Investigation, Am. Fedn. Clin. Research (pres. San Francisco br. 1956), Western Soc. Clin. Research, A.A.A.S., Soc. Exptl. Biology and Medicine, Sigma Xi. Subspecialties: Internal medicine; Pulmonary medicine. Current work: Clinical endocrine research in adrenals; pituitary and diabetes. Home: 267 Hillside Ave Mill Valley CA 94941 Office: University of California Hospitals San Francisco CA 94143

FORSLUND, DAVID WALLACE, physicist; b. Ukiah, Calif., Feb. 18, 1944; s. Dero Bradford and Myrtle Ruth (Conner) F.; m. Jean Carolyn Monson, Aug. 17, 1968; children: Daniel, Luke. B.S., U. Santa Clara, 1964; M.A., Princeton U., 1967, Ph.D., 1969. Postdoctoral fellow Los Alamos Sci. Lab., 1969-71, mem. staff, 1971-77, assoc. group leader, 1977-78, alt. group leader, 1978-80, fellow, 1981—. Contbr. articles to profl. jours; assoc. editor: Physics of Fluids, 1982-84. Ruling elder Presbyn. Ch. in Am., 1975—. Recipient Disting. Performance award Los Alamos Nat. Lab., 1982. Fellow Am. Phys. Soc; mem. Am. Geophys. Union, Am. Astron. Soc., Astron. Soc. of Pacific. Subspecialties: Plasma physics; Theoretical physics. Current work: Inertial confinement fusion, laser interaction with matter, plasma simulation methods, space plasma physics. Home: 309 Aragon Ave Los Alamos NM 87544 Office: Los Alamos Nat Lab MS E531 Los Alamos NM 87545

FORSTER, DENIS, chemist; b. Newcastle, Eng., Feb. 28, 1941; s. Reginald Thomas and Margaret (Dobson) F.; m. Hazel Frances Onions, Apr. 18, 1964; children: Juliet, Rachel. B.Sc., Imperial Coll., London, 1962, Ph.D., 1965. Postdoctoral fellow Princeton (N.J.) U., 1965-66; research chemist Monsanto Co., St. Louis, 1966, sci. fellow, 1974, sr. fellow, 1980—. Contbr. articles to profl. jours. Mem. Am. Chem. Soc. (Ipatieff prize 1980), Am. Inst. Chemists (Chem. Pioneer award 1980). Club: Racquet (St. Louis). Subspecialties: Catalysis chemistry; Inorganic chemistry. Current work: Homogeneous and heterogeneous catalysis. Patentee in field. Address: 800 N Lindbergh Blvd Saint Louis MO 6316

FORSTER, ROBERT E., II, physiology and surgery educator, physician; b. St. Davids, Pa., Dec. 23, 1919; m., 1947; 4 children. B.S., Yale U., 1941; M.D., U. Pa., 1943. Life Ins. Med. Research Fund fellow in physiology Harvard U., 1948-50; from asst. prof. to assoc. prof. physiology Grad. Sch. Medicine, U. Pa., 1950-58, prof. physiology, 1958—, Isaac Ott prof., 1959—, prof. physiology and surgery, 1961—, chmn. dept. physiology, 1959-70, 70—; mem. cardiovascular study sect. NIH, 1960-64, mem. gen. clin. research ctr. com., 1964-71; mem. Nat. Adv. Heart Council, 1967-71. Lowell M. Palmer sr. fellow Grad. Sch. Medicine, U. Pa., 1954-60. Mem. Nat. Acad. Sci., Am. Physiol. Soc. (pres. 1966-67), Am. Soc. Clin. Investigation, Biophys. Soc., Soc. Gen. Physiology. Subspecialty: Physiology (medicine). Office: U Pa Dept Physiology A-201 Richards Bldg G-4 Philadelphia PA 19174

FORSTER, WILLIAM OWEN, research ecologist, government official; b. Dearborn, Mich., July 2, 1927; s. Clarence William and Florence Veda (Spencer) F.; m. Ruth Lynn Austin, Sept. 4, 1948; children: Vicki Lynn, Suzette Mari, Janis Kaye. B.S., Mich. State U., 1951, M.S., 1952; Ph.D., U. Hawaii, 1966. Asst. prof. oceanography dept. Oreg. State U., 1966-69; dir. marine biology P.R. Nuclear Ctr., Mayaguez, 1969-72; head oceanography research AEC-ERDA, Germantown, Md., 1972-77; head waste mgmt. IAEA, Vienna, Austria, 1977-81; asst. dir. environ. U.S. Dept. Energy, Germantown, Md., 1981—. Served with USNR, 1945-47. NSF fellow, 1959. Mem. Am. Chem. Soc., Am. Soc. Lymnology and Oceanography, Am. Geophys. Union, Soc. Applied Spectroscopy, Nuclear Sci. and Engring. Assn., Sigma Xi. Methodist (lay leader 1982-83). Subspecialties: Oceanography; Chemical oceanography. Current work: Energy-related materials in the marine environment: movment of these materials through marine ecosystems; information on research in marine sciences. Home: 20145 Darlington Dr Gaithersburg MD 20879 Office: Dept Energy ER-75 Washington DC 20585

FORSYTH, DALE M., nutritionist, animal scientist; b. Charles City, Iowa, Feb. 4, 1945; s. Wilbur B. and Mabel J. (Winters) F.; m. Judi H. Reynolds, May 28, 1967. B.S., Iowa State U., 1967; Ph.D., Cornell U., 1971. Mem. faculty dept. animal scis. Purdue U., West Lafayette, Ind., 1972—, assoc. prof., 1978—. Mem. Am. Soc. Animal Sci., AAAS, Nutrition Today Soc., Am. Inst. Nutrition, Fedn. Am. Socs. for Exptl. Biology, Sigma Xi, Phi Kappa Phi, Gamma Sigma Delta, Phi Eta Sigma. Subspecialty: Animal nutrition. Current work: Nutrition of swine, especially nutrition disease interactions, Fe metabolism, needs of the young pig. Office: Purdue U Dept ANSC Lilly Hall West Lafayette IN 47907

FORSYTH, JANE LOUISE, geology educator; b. Hanover, N.H., Nov. 9, 1921; s. Chester Hume and Louise Ann (James) F. A.B., Smith Coll., 1943; M.A., U. Cin., 1946; Ph.D., Ohio State U., 1956. Pleistocene geologist Ohio Geol. Survey, Columbus, 1955-65; asst. prof. geology Bowling Green (Ohio) State U., 1965-68, assoc. prof., 1968-74, prof., 1974—; mem. Ohio Natural Areas Council, Columbus, 1975—. Author: Physical Features for the Toledo Regional Area, 1968, Geology of the Lake Erie Islands and Adjacent Shores, 1971; contbr. articles to profl. jours. Recipient Citation award Ohioana Library, 1976. Fellow Geol. Soc. Am., Ohio Acad. Sci. (v.p. geology 1961-62, 82-84, editor in chief jour. 1964-74); mem. Nat. Assn. Geology Tchrs.

(councilor-at-large 1981—), Ecol. Soc. Am. Episcopalian. Subspecialties: Geology. Current work: Glacial geology of Ohio, geologic history of Lake Erie, distribution of plant and animal species relative to geology. Office: Dept Geology Bowling Green State U Bowling Green OH 43403

FORTIER, CLAUDE, medical scientist, physiology educator; b. Montreal, Que., Can., June 11, 1921; s. Carolus and Flore-Edith (Lanctôt) F.; m. Elise Gouin, Sept. 8, 1953; children: Anne, Michele, Nicole, Nathalie. B.A., U. Montreal, 1941, M.A. in Polit. Sci, 1941, M.D., 1948, Ph.D., 1952, D.U. (hon.), 1981, LL.D., Dalhousie U., 1977; D.U., Ottawa U., 1981. Research asst. U. Montreal Inst. Exptl. Medicine and Surgery, 1948-51; research cons. U. Lausanne, Switzerland, 1952-53; research asso. dept. neuroendocrinology U. London, 1953-55; asso. prof. physiology, dir. neuroendocrine research lab. Baylor U. Coll. Medicine, 1955-60; dir. endocrine lab. Laval U., Quebec, Que., Can., 1960—, prof. exptl. physiology, 1961—, chmn. dept. physiology, 1964—; mem. Med. Research Council Can., 1963-68, 70-72, vice chmn., 1965-67; chmn. study group on med. research, Govt. Que., 1968-70; vice chmn. med. research coordinating, adv. coms. Def. Research Bd. Can., 1967-70; chmn. Can. nat. com. Internat. Union Physiol. Scis., 1969-72; cons. physician Laval U. Med. Center, 1968—; mem. Killam com. Can. Council, 1967-72, Que. Med. Research Council, 1963-70; chmn. bd. Canadian Fedn. Biol. Socs., 1973-74; mem. neuroendocrinology panel IBRO, UNESCO, 1958—; advisory council Order of Can., 1974-75; chmn. Sci. Council Can., 1978-81, vice chmn., 1975-78, chmn. task force on research in Can., 1976-78; asso. Humanities and Social Scis. Research Council Can., 1981—. Trustee Inst. Research in Pub. Policy, 1974-75, 78-81. Decorated companion Order of Can., 1970; recipient Archambault Research award French Canadian Assn. Advancement Sci., 1972; Sci. award Govt. Que., 1972; Wightman award Gairdner Found., 1979; Marie-Victorin sci. award Govt. of Que., 1980; sci. achievement award French Can. Med. Assn., 1982. Fellow Royal Soc. Can. (pres. 1974-75), Royal Coll. Physicians Can.; mem. Can. Physiol. Soc. (pres. 1966-67), Am. Physiol. Soc., Endocrine Soc., Am. Thyroid Soc., AAAS, N.Y. Acad. Sci., Soc. Exptl. Biology and Medicine, Assn. Am. Physicians, Biomed. Engring. Soc., Peripatetic Club, Can. Soc. Clin. Investigation, Internat. Soc. Neuroendocrinology, Can. Soc. Endocrinology and Metabolism, Assn. Sci., Engring. and Tech. Communities Can. (hon.), Can. Assn. Club of Rome. Subspecialties: Endocrinology; Neuroendocrinology. Current work: Science policy; medical education; consultant in endocrinological research (endocrine bio-controls). Research, publs. neurohumoral control adenohypophysial functions, pituitary-thyroid-adrenocortical interactions, biostatistics, bio-control systems, role plasma steroid-binding proteins, boundaries of knowledge, science policy. Home: 1014 DeGrenoble St Quebec PQ Canada G1V 2Z9 Faculty Medicine Laval U Quebec PQ G1K 7P4 Canada

FORTNA, JOHN DAVID EDWARD, physicist, educator, cons.; b. Boston, Nov. 15, 1928; s. Clyde Bishop and Vera Millicent (Jackson) F.; m. Viola Manville Shields, Feb. 21, 1959; children: Clyde Bishop, Frances Millicent, Culver Shields. Student, U. Richmond, Va., 1945-48; B.A., U. Va., Charlottesville, 1950; Ph.D., U. Md., College Park, 1969. Instr. Miller Sch., Albemarle, Va., 1950-51; ordnance engr. U.S. Naval Surface Weapons Center, Dahlgren, Va., 1951-53; electronic engr. U.S. Naval Research Lab., Washington, 1957-61, physicist, 1961-71; assoc. prof. physics U. D.C., Washington, 1971—. Author ednl. modules.; Contbr. articles to profl. jours. Active YMCA, Boy Scouts Am., Arlington Citizens Adv. Com. Served with MC AUS, 1953-55. Recipient 1st place comml. exhibit demonstration Nat. Sci. Tchrs. Assn., 1977; NSF indsl. research fellow, 1981. Mem. Am. Phys. Soc., Soc. Applied Spectroscopy, Am. Assn. Physics Tchrs., Fusion Power Assocs. Club: Wilson (Alexandria, Va.). Subspecialties: Plasma engineering; Information systems, storage, and retrieval (computer science). Current work: High temperature plasma diagnostics, computer-assisted mgmt. of instrn. Designer electronic ednl. hardware. Home: 2104 N Troy St Arlington VA 22201 Office: 3121 H St NW Washington DC 20005

FORTNEY, LLOYD RAY, physics educator, university administrator, researcher; b. Enid, Okla., June 22, 1936; m., 1962; 2 children. B.S., N. Mex. State U., 1958; Ph.D. in Physics, U. Wis., 1962. Research assoc. in high energy physics U. Wis., 1962-63; research assoc. Duke U., 1963-64, asst. prof., 1964-70, assoc. prof. physics, 1970—, dir. undergrad. studies, 1980—. Mem. Am. Phys. Soc. Subspecialty: High energy physics. Office: Dept Physic Duke U Durham NC 27706

FORTSON, EDWARD NORVAL, physicist, educator; b. Atlanta, June 16, 1936; s. Charles Wellborn and Virginia (Norval) F.; m. Alix Madge Hawkins, Apr. 3, 1960; children: Edward Norval, Lucy Frear, Amy Lewis. B.S., Duke U., 1957; Ph.D., Harvard U., 1963. Research fellow U. Bonn, W. Ger., 1965-66; research asst. prof. physics U. Wash., Seattle, 1963-65, asst. prof., 1966-69, assoc. prof., 1969-74, prof., 1974—. Fulbright travel grantee, 1965-66; Nat. Research Council fellow Oxford (Eng.) U., 1977; Guggenheim fellow, 1980-81. Fellow Am. Phys. Soc. Subspecialties: Atomic and molecular physics; Spectroscopy. Current work: Atomic properties, including parity and time-reversal symetries, studied by radio-frequency and laser-optical techniques. Office: Dept Physics FM-15 U Wash Seattle WA 98195

FOSS, DONALD JOHN, research psychologist, psychology educator; b. Mpls., Mar. 28, 1940; s. Bernard J. and Elizabeth (Cody) F.; m. Patricia R. Diamond, Sept. 18, 1965; children: Melissa, Lara. B.A., U. Minn., 1962, Ph.D., 1966. Postdoctoral fellow Harvard U., 1966-67; asst. prof. psychology U. Tex.-Austin, 1967-71, assoc. prof., 1971-75, prof. psychology, 1975—. Co-author: Psycholinguistics, 1978; Editor: Contemporary Psychology jour, 1980—; Contbr. articles to profl. jours. Army Research Inst. grantee, 1982-85; NIMH grantee, 1976-81; NSF grantee, 1969-72. Fellow Am. Psychol. Assn.; mem. Human Factors Soc., Psychonomics Soc., Soc. Engring. Psychologists, AAAS. Clubs: Chancellor's Council, President's. Subspecialties: Cognition; Software engineering. Current work: Research on the processes of comprehension, including comprehension of computer-based systems, manuals, etc. Home: 3710 Gilbert St Austin TX 78703 Office: Dept Psychology U Tex Austin TX 78712

FOSTER, CHARLES STEPHEN, physician, surgeon, educator; b. Charleston, W.Va., May 19, 1942; s. Carson and Martha F.; 1 son, Marc David. B.S., Duke U., 1965, M.D., 1969. Diplomate: Am. Bd. Ophthalmology. Intern Duke U., 1969-70; resident in opthalmology Washington U., St. Louis, 1972-75; fellow in cornea and immunology Harvard U., 1975-77, instr. ophthalmology, 1976-79, asst. prof., 1979-82, assoc. prof., 1982—; dir. immunology and uveitis unit Harvard U.-Mass. Eye and Ear Infirmary, 1981—; dir. immunopathology lab. Eye Research Inst. Contbr. articles to profl. jours. Served with USPHS, 1970-72. Fellow ACS; mem. AMA, Am. Acad. Ophthalmology, Mass. Med. Soc., New Eng. Ophthal. Soc., Am. Soc. Immunology, Royal Soc. Medicine, Phi Beta Kappa, Sigma Xi, Alpha Omega Alpha. Subspecialties: Ophthalmology; Immunology (medicine). Current work: Ocular immunology; immunoregulation in the eye, corneal transplantation; diagnosis and therapy of autoimmune diseases affecting the eyes; ocular microsurgery. Home: 348 Glen Rd Weston MA 02193 Office: 243 Charles St Boston MA 02114

FOSTER, EDWIN MICHAEL, microbiologist, educator, consultant; b. Alba, Tex., Jan. 1, 1917; s. Edward Lee and Katherine (Ryan) F.; m. Winona Lively, Apr. 25, 1941; 1 son, Michael Stewart. B.A., North Tex. State Tchrs. Coll., Denton, 1936, M.A., 1937; Ph.D., U. Wis., 1940. Instr. microbiology U. Wis., Madison, 1940-41; asst. prof., 1945-46, assoc. prof., 1946-52, prof., 1952—; instr. U. Tex., Austin, 1941-42; cons. in field. Author: with others Dairy Microbiology, 1957; contbr. numerous articles to profl. jours. Served to capt. Chem. Corps U.S. Army, 1942-45. Recipient Pasteur award Ill. Soc. Microbiology, 1969; W.O. Atwater lectr. U.S. Dept. Agr., 1982. Fellow Am. Acad. Microbiology (pres. 1965-66), Inst. Food Technologists (Nicholas Appert award 1969); mem. Am. Soc. Microbiology (pres. 1969-70), Internat. Assn. Milk and Food Sanitation, Toxicology Forum (dir. 1980-82). Club: Cosmos. Subspecialties: Microbiology; Food science and technology. Current work: I am director of a research institute that is totally involved in food safety investigations; with a faculty of nine we have major research projects in all the principal areas of food borne disease. Home: 1111 Amherst Dr Madison WI 53705 Office: University of Wisconsin 1925 Willow Dr Madison WI 53706

FOSTER, HENRY LOUIS, veterinarian, laboratory executive; b. Boston, Apr. 6, 1925; s. Louis and Clara Friedman; m. Lois Ann Foster, June 1948; children: James C., John S., Neal R. D.V.M., Middlesex Coll., 1946. Diplomate: Am. Coll. Lab. Animal Medicine, 1961. Cons. veterinarian UNRRA, 1946-47; founder, pres. Charles river Breeding Labs., Inc., Wilmington, Mass., 1947—; dir. Century Bank & Trust Co.; Chmn. vis. com. Tufts Vet. Sch.; chmn. bd. trustees Brandeis U. Contbr. numerous articles to sci. jours.; sr. editor: The Mouse in Biomedical Research, 4 vols. Trustee Boston Mus. Fine Arts, Tufts U.; mem. surgery vis. com. Mass. Gen. Hosp. Paul Harris fellow, 1972. Mem. AVMA, N.Y. Acad. Sci., Am. Assn. Lab. Animal Sci. (Charles A. Griffin award 1976), Am. Coll. Lab. Animal Medicine, Am. Inst. Biol. Scis. Subspecialty: Animal breeding and embryo transplants. Current work: Commercial production of caesarean-originated, barrier sustained and gnotobiotic rats, rabbits, hamsters, guinea pigs and non-human primates, production methods, elimination of specific diseases, environmental control. Home: 11 Drumlin Rd Newton MA 02159 Office: 251 Ballardvale St Wilmington MA 01887

FOSTER, HENRY WENDELL, obstetrician, gynecologist, educator; b. Pine Bluff, Ark., Sept. 8, 1933; m., 1960; 2 children. B.S., Morehouse Coll., 1954; M.D., U. Ark., 1958. Diplomate: Am. Bd. Ob-Gyn (examiner 1976—). Intern Receiving Hosp., Detroit, 1958-59; resident Malden Hosp., Mass., 1961-62, Hubbard Hosp., Hashville, 1962-65; asst., prof. Meharry Med. Coll., 1967-77, prof. ob-gyn., 1977—, chmn. dept., 1973; chief ob-gyn John A. Andrew Meml. Hosp., 1965—; attending physician Macon County Hosp.; mem. cons. staff VA Hosp., Tuskegee, Ala.; dir. maternity and infant care project 556 John A. Andrew Meml. Hosp., 1970-80. Bd. dirs. Planned Parenthood Fedn. Am., 1975—; mem. Ethics Adv. Bd., 1977—. Fellow Am. Coll. Obstetricians and Gynecologists; mem. Inst. Medicine of Nat. Acad. Sci., AMA, Nat. Med. Assn. Subspecialties: Obstetrics and gynecology. Office: Dept Ob-Gyn 1005 18th Ave N Nashville TN 37208

FOSTER, JOHN STUART, JR., mfg. co. exec.; b. New Haven, Sept. 18, 1922; s. John Stuart and Flora (Curtis) F.; m. Frances Schnell, Dec. 28, 1978; children—Susan, Cathy, Bruce, Scott, John. B.S., McGill U., 1948; Ph.D. in Physics, U. Calif., Berkeley, 1952; D.Sc. (hon.), U. Mo., 1979. Dir. Lawrence Livermore (Calif.) Lab., 1952-65; dir. def. research and engring. Dept. Def., Washington, 1965-73; v.p. TRW Energy Systems Group, Redondo Beach, Calif., 1973-79; v.p. sci. and tech. TRW Inc., Cleve., 1979—; mem. energy adv. bd. Calif. Inst. Tech.; mem. nat. adv. bd. Am. Security Council. Decorated knight Comdr.'s Cross, Badge and Star of Order of Merit, W. Ger.; comdr. Legion of Honor, France; recipient Ernst Orlando Lawrence Meml. award AEC, 1960; Disting. Public Service medal Dept. Def., 1969, 73. Mem. Am. Def. Preparedness Assn, Nat. Petroleum Council, Nat. Security Indsl. Assn., AIAA. Office: 23555 Euclid Ave Cleveland OH 44117

FOSTER, ROBERT JOHN, geology educator, author; b. Cambridge, Mass., Apr. 19, 1929; s. John T. and Margaret (Loftus) F.; m. Joan Callin, Sept. 10, 1951; children: Helen L., Ralph C. S.B., M.I.T., 1951; M.S., U. Wash., 1955, Ph.D., 1957. Seismologist Geophys. Service Inc., Dallas, 1951-53; engr. Boeing Airplance Co., Seattle, 1953-55; asst. prof. Mont. State U., 1957-61; prof. geology San Jose State U., 1961—; cons. geology, geology texts; active in amateur radio. Author: texts, including General Geology, 4th edit, 1983; editor, contbg. author: numerous texts, including Physical Geology, 1982; editor phys. sci. series. Served to 1st lt. U.S. Army, 1957-58. Fellow Geol. Soc. Am., AAAS (life); mem. Am. Assn. Petroleum Geologists. Subspecialties: Geophysics; Tectonics. Current work: Volcanic and metamorphic petrology; geophysics as applied to tectonics. Home: 3605 Hilltop Rd Soquel CA 95073 Office: San Jose State U Dept Geology San Jose CA 95192

FOSTER, WILLIAM SAMUEL, chem. engr., co. exec.; b. Franklin, N.J.; m. Elaine, Apr. 9, 1949; children: Howard Elliot, Meg Robin. B.S. in Chem. Engring, N.J. Inst. Tech., 1948, M.S., 1952. Registered profl. engr., N.J.; pub. sewage lic., N.J. Vice pres., plant mgr. Thermo-Nat. Industries, South Plainfield, N.J., 1977—; tchr., cons. in field. Served with U.S. Army, 1945-46. Mem. Am. Chem. Soc., ASME, Am. Electroplater's Soc. Jewish. Lodges: Masons; Shriners. Subspecialties: Materials processing; Corrosion. Current work: Management in materials applications.

FOTOPOULOS, SOPHIA STRATHOPOULOS, life science executive; b. Kansas City, Mo., Nov. 6, 1936; C. Marinos G. and Stauroula F. Stathopoulos. B.A., U. Kans., Lawrence, 1958, M.A., 1964, Ph.D., 1970. Research asst. U. Kans. Med. Center, Kansas City, 1958-61; NIH research fellow U. Kans., Lawrence, 1962-64; research assoc. Inst. Community Studies, Kansas City, Mo., 1965-66; lectr. U. Kans., Lawrence, 1969-70, adj. prof., 1975—; dir. Psychophysiology-Psychopharmacology Lab., Greater Kansas City Mental Health Found., 1970-73; staff assoc. neuropsychophysiology Midwest Research Inst., Kansas City, Mo., 1974-75; head psychophysiology lab., 1975-77, assoc. dir. chem. scis. div., 1977-79, dir. life scis. dept., 1979—; mem. spl. rev. com. Nat. Cancer Inst., Kansas City, Mo., 1978—; mem. nat. adv. com. Am. Cancer Soc., Kansas City, 1982—; lectr. U. Mo.-Kansas City Sch. Medicine, 1970—; NIH research fellow, 1963-64, Health and Human Services research fellow, 1965-69. Contbr. articles in field to profl. jours. Recipient Creative Scientist award Am. Inst. Research, 1971. Mem. Claude Bernard Soc., AAAS, N.Y. Acad. Scis., Sigma Xi. Club: Zonta (Kansas City, Mo) (pres. 1983—). Current work: Medical and health research related to in vitro and in vivo metabolism studies, carcinogenesis and mutogenesis, psychophysiology; human factors and performance, effects of drugs on pharmacological and biochemical mechanisms, biochemical psychophysiological and neuroimmunologic aspects of disease onset and treatment, biodetection and toxicology of toxic substances, cell biology, microbial biochemistry and neurosciences. Office: Midwest Research Inst 425 Volker Blvd Kansas City MO 64110

FOTOS, PETER GEORGE, dentist, microbiology researcher; b. Maysville, Ky., Oct. 5, 1952; s. George Pete and Bessie (Nicholas) F.; m. Angelica Petredis, June 26, 1980; 1 dau., Stephanie Irene. B.A., Marshall U., 1974; D.D.S., W.Va. U., 1978, M.S., 1980. Cert. gen. practice residency Miami Valley Hosp., Dayton, Ohio, 1980. Research fellow W.Va. U., 1978-79, asst. prof. oral diagnosis, 1980—, research asso. in microbiology, 1981—. Contbr. articles, abstracts to profl. jours. Packmaster Boy Scouts Am., 1981—. Mem. Acad. Gen. Dentistry, Internat. Assn. Dental Research, Hellenic Am. Dental Soc., Am. Assn. Dental Schs., Am. Assn. Dental Research (sec. W.Va. sect., officer). Republican. Orthodox Christian. Subspecialties: Oral biology; Cell and tissue culture. Current work: Microbiology and immunology of oral disease: pulp cell tissue culture classification and investigation of oral Actinomycetes. Home: Route 6 Box 3 Morgantown WV 26505 Office: Dept Oral Diagnosis W Va U Med Center Morgantown WV 26506

FOUGERE, PAUL FRANCIS, physicist, consultant; b. Cambridge, Mass., Feb. 29, 1932; s. Louis N. and Helen V. (Weissbach) F.; m. Marguerite M. Burwell, Dec. 27, 1953; children: Paul J., Peter L., Gregory J., Mark J., Daniel E., Gabrielle. Physicist Naval Research Lab., Washington, 1951-53; tech. engr. Gen. Electric Co., Lynn, Mass., 1954; physicist Air Force Geophysics Lab., Hanscom AFB, Mass., 1955—; math. cons. Contbr. articles to profl. jours. Recipient Sci. Achievement award Air Force Geophysics Lab., 1970, 78. Mem. Am. Geophys. Union, Internat. Assn. Geomagnetism and Aeronomy, IEEE. Democrat. Roman Catholic. Subspecialties: Geophysics; Graphics, image processing, and pattern recognition. Current work: Ionospheric scintillation; maximum-entropy power spectral estimation; space physics; geophysical data analysis; computer graphics for science. Office: Air Force Geophysics Laboratory Hanscom Air Force Base MA 01731

FOULKE, JUDITH DIANE, radiobiologist; b. Bucyrus, Ohio, Nov. 22, 1945; d. Lawrence Kern and Alberta (Durnwald) F.; m. Mark Allen Elrod, July 17, 1981. B.A., St. Mary of the Springs, Columbus, Ohio, 1967; M.S., U. Mich., 1969; Ph.D., Purdue U., 1973. Health physicist NASA/Goddard Space Flight Center, Greenbelt, Md., 1969-71, U.S. AEC, Washington, 1973-77; radiobiologist U.S. Nuclear Regulatory Commn., Washington, 1977—. Mem. Federally Employed Women of Suburban Md., 1979—. Mem. AAAS, Health Physics Soc., Am. Nuclear Soc., Sigma Xi, Delta Epsilon Sigma, Iota Sigma Pi, Kappa Gamma Pi, Phi Kappa Phi, Rho Chi. Democrat. Roman Catholic. Club: Toastmasters (sec.-treas. 1980). Subspecialty: Radiobiology. Current work: Health effects of exposure to ionizing radiation. Home: 9644 Shadow Oak Dr Gaithersburg MD 20879 Office: Office Nuclear Regulatory Research US Nuclear Regulatory Commn Washington DC 20555

FOULKES, DAVID, research psychologist, educator; b. East Orange, N.J., May 29, 1935; s. Paul Bergen and Alice (Hinson) F.; m. Nancy Helen Kerr, Apr. 19, 1978; children: Alice, Anne, William, Thomas, Sarah. B.A., Swarthmore Coll., 1957; Ph.D., U. Chgo., 1960. Instr. Lawrence Coll., Appleton, Wis., 1960-63; research assoc. U. Chgo., 1963-64; from asst. prof. to prof. psychology U. Wyo., Laramie, 1964-77; prof. psychiatry Emory U., Atlanta, 1977—, adj. prof. psychology, 1977-83; dir. Cognition Research Lab., Ga. Mental Health Inst., Atlanta, 1977—. Author: The Psychology of Sleep, 1966, Grammar of Dreams, 1978, Children's Dreams, 1982. NSF grantee, 1965-70; NIMH grantee, 1970-74, 78-81; Nat. Inst. Child Health and Human Devel. grantee, 1970-75. Fellow Ctr. for Advanced Study in Behavioral Scis, AAAS; mem. Assn. for Psychophysiol. Study of Sleep (exec. sec. 1966-69), Midwest Psychol. Assn., Psychonomic Soc., Internat. Neuropsychol. Soc. Subspecialty: Cognition. Current work: Information-processing models of dreaming; dream ontogeny. Home: 1420 Berkeley Ln Atlanta GA 30329 Office: 1256 Briarcliff Rd Atlanta GA 30306

FOULKES, ERNEST C(HARLES), science educator, research administrator; b. Karlsruhe, Germany, Aug. 20, 1924; came to U.S., 1952. B.S., U. Sydney, Australia, 1946, M.S., 1947; Ph.D., Oxford (Eng.) U., 1952. Investigator, Med. Research Council, Sydney, New South Wales, Australia, 1946-49; assoc. May Inst., Cin., 1952-65; asst. prof., assoc. prof., prof. depts. environ. health and physiology U. Cin., 1965—; established investigator Am. Heart Assn., Cin., 1956-61; mem. various adv. coms. fed. govt. Contbg. author, editor publs. in field. Active various civic orgns. including Human Relations Commn., Sch. Found., Neighborhood Orgn. Mem. Am. Nephrology Soc., Am. Physiol. Soc., Biochem. Soc., Am. Soc. Biol. Chemistry, Soc. Exptl. Biology and Medicine, Biophys. Soc., Soc. Toxicology, AAAS, AAUP. Subspecialties: Physiology (medicine); Toxicology (medicine). Current work: Renal physiology and toxicology; intestinal absorption of heavy metals. Home: 4082 Rose Hill Ave Cincinnati OH 45229 Office: Univ Cincinnati Coll Medicine ML 56 Cincinnati OH 45267

FOUNDS, HENRY WILLIAM, JR., microbiologist, virologist, laboratory manager; b. Drexel Hill, Pa., Mar. 6, 1942; s. Henry W. and Elaine V. (Smargis) F.; m. Ferne A. Brunner, Nov. 6, 1971; children: Steven, Jeffrey, Jennifer. B.S., Villanova U., 1964; M.S., U. Notre Dame, 1968; Ph.D., Rutgers U., 1981. Instr. Bayley-Ellard High Sch., Madison, N.J., 1964-68; mem. faculty County Coll. Morris, Randolph Twp., N.J., 1968-83, asst. prof., microbiology, 1971-77, assoc. prof., 1978-83; mgr. Biotech. Products Group research and devel. Ventrex Labs., Inc., Portland, Maine, 1983—; cons. in field; research asso. Rutgers U. Mem. Am. Soc. Microbiology, Theobald Smith Soc., N.Y. Acad. Sci. Subspecialties: Microbiology; Virology (biology). Current work: Study of early interactions of virus with host cells, of RNA virus-host cell interactions, cell growth research and development. Home: 13 Hemlock Circle Scarboroughŧ ME 04074

FOURNEY, M(ICHAEL) E(UGENE), aeronautic and mechanical engineer, educator researcher; b. Blue Jay, W.Va., Jan. 30, 1936. B.S., W.Va. U., 1958; M.S., Calif. Inst. Tech., 1959, Ph.D. in Aeros, 1963. Aerospace engr. Boelkow Entwicklungen KG, Germany, 1963-64; from research asst. prof. to research assoc. prof. aeros. and astronautics U. Wash., 1964-72; assoc. prof. UCLA, 1972-75, prof., 1975—, chmn. dept. mechanics and structures, 1979—; cons. in field. Mem. Soc. Exptl. Stress Analysis (pres. 1980-81), Optical Soc. Am. Subspecialties: Mechanical engineering; Aerospace engineering and technology. Office: Dept Mechanics and Structures UCLA Los Angeles CA 90024

FOURNIER, RAYMOND EMILE KEITH, genetics educator; b. Attleboro, Mass., July 26, 1949; s. Roland E. and Marguerite Claire (Fortier) F. B.S., Providence Coll., 1971; Ph.D., Princeton U., 1974. Postdoctoral fellow Yale U., New Haven, 1975-78; asst. prof. U. So. Calif., 1978—; mem. adj. faculty W. Alton Jones Cell Sci. Ctr., Lake Placid, N.Y., 1978-82. Contbr. articles to sci. jours. Am. Cancer Soc. grantee, 1979-82; NIH grantee, 1979—; Leukemia Soc. Am. fellow, 1975-77. Mem. AAAS, Sigma Xi. Subspecialties: Gene actions; Molecular biology. Current work: Genetic and molecular mechanisms that control gene expression in eukaryotic cells. Office: USC Comprehensive Cancer Center 1441 Eastlake Ave Los Angeles CA 90031

FOUTS, GREGORY TAYLOR, psychology educator, media consultant; b. Logansport, Ind., May 22, 1943; emigrated to arrived Canada, 1974; s. Frederic Wagner and Mae (Vodicka) F.; m. Geraldine Janet Mattea, Aug. 19, 1967; children: Benjamin, Courtney. B.A., Ind. U., 1965; M.A., U. Iowa, 1968, Ph.D., 1970. Tchr. math.

Paul Hadley Jr. High Sch., Mooresville, Ind., 1965-66; asst. prof. psychology U. Denver, 1969-74; assoc. prof. psychology U. Calgary (Alta., Can.), 1974-80; prof. U. Calgary, 1980—; pvt. practice psychology, Calgary, 1976—; cons. Royal Commn. on Violence in the Communications Industry, Province of Ont., 1976-77, Nat. Film Bd., Ottawa, 1978; bd. dirs. Can. Jour. Communications, Sask., 1982—. Host TV prodn.: Children and Television, 1980; contbr. numerous articles to profl. jours. Royal Commn. on Violence in the Communications Industry grantee, 1976-77; Alta. Environ. Research Trust grantee, 1977. Mem. Soc. Research in Child Devel., Am. Psychol. Assn., AAAS, Internat. Communication Assn., Internat. Assn. Applied Psychology. Subspecialties: Developmental psychology; Learning. Current work: Media technologies; media and education; communications science; media effects; impact of media technologies on society; socialization of children; emotional development. Home: RR 1 Site 3 Box 10 Cochrane AB Canada T0L 0W0 Office: adrt Psychology U Calgary 2500 University Dr Calgary AB Canada T2N 1N4

FOUTS, JAMES RALPH, pharmacologist, educator; b. Macomb, Ill., Aug. 8, 1929; s. Ralph Butler and Mary May (Lingenfelter) F.; m. Joan Laverne Van Dyke, June 20, 1964; children: Mary, Jeffrey, Carolyn. B.S. with highest honors (Merit Scholar), Northwestern U., 1951, Ph.D., 1954. Instr. biochemistry Northwestern U. Med. Sch., Chgo., 1952-54; postdoctoral fellow Lab. Chem. Pharmacology, Nat. Heart Inst., NIH, Bethesda, Md., 1954-56; sr. research biochemist Wellcome Research Labs., Burroughs Wellcome & Co., Tuckahoe, N.Y., 1956-57; asst. prof., assoc. prof., prof. dept. pharmacology Coll. Medicine, U. Iowa, Iowa City, 1957-70, dir., 1968-70; Claude Bernard prof. U. Montreal, (Que., Can.) Inst. of Medicine, 1970; chief pharmacology and toxicology br. Nat. Inst. Environ. Health Scis., Research Triangle Park, N.C., 1970-76, sci. dir., 1976-78, chief lab. pharmacology, 1978-81, sr. exec. service, 1981—; cons. pharmacology Cos.; adj. prof. pharmacology U. N.C. Sch. Medicine, 1970—; adj. prof. toxicology N.C. State U., 1971—; cons. coms. NIH, FDA, NIDA, EPA; chmn. Gordon Conf. on Drug Metabolism, 1977-78; vis. prof. Swiss Fed. Inst. Tech. and U. Zurich, Switzerland, 1978. Contbr. numerous articles on drug metabolism to profl. jours. Served to lt. USPHS, 1954-56. Recipient Marple-Schweitzer award Northwestern U., 1950; recipient Superior Achievement award NIH, 1975; Spl. award U. N.C. Sch. Nursing, 1978. Mem. Am. Soc. Pharmacology and Exptl. Therapeutics (Abel award 1964), Am. Assn. for Cancer Research, Soc. of Toxicology, Mt. Desert Island Biol. Lab.; mem. Phi Beta Kappa; Mem. Sigma Xi., Phi Lambda Upsilon. Democrat. Episcopalian. Subspecialties: Pharmacology; Environmental toxicology. Current work: Metabolism of chemicals/drugs; role of metabolism in effect/toxicity of chemical/drug, species differences in metabolism/toxicity, perinatal drug metabolism, metabolism of drugs by isolated organs, tissues, cells, effects of pollutants on metabolizing systems, drug/chemical interactions at metabolizing systems. Home: 212 Ridge Trail Chapel Hill NC 27514 Office: NIEHS PO Box 12233 Research Triangle Park NC 27709

FOWLER, BRUCE ANDREW, toxicologist, researcher; b. Seattle, Dec. 28, 1945; s. Andrew and Dolores (Ivey) F.; m. Mary Glenn Oler, June 9, 1968; children: Glenn Andrew, Randall Bruce. B.S. in Fisheries, U. Wash., 1968; Ph.D. in Pathology, U. Oreg., 1972. Staff fellow Nat. Inst. Environ. Health Sci., NIH, Research Triangle Park, N.C., 1972-74, sr. staff fellow, 1974-77, research biologist, 1977—; temporary adviser WHO, Geneva, 1978-79; mem. working group Internat. Agy. for Research in Cancer, Lyon, France, 1978-79. Editor: The Biological and Environmental Effects of Arsenic, 1983. Mem. Am. Soc. Cell Biology, Am. Assn. Pathologists, Am. Soc. Pharmacology and Exptl. Therapeutics, Soc. Toxicology (councilor mechanisms sect. 1981-83). Subspecialties: Toxicology (medicine); Biochemistry (medicine). Current work: Mechanisms of toxic cell injury from trace metals; role of metal-binding proteins in regulating intracellular bioavailability of metals; comparative biochemistry of metal-binding proteins in marine invertebrates. Office: PO Box 12233 Research Triangle Park NC 27709

FOWLER, CHARLES ALBERT, electronics engineer; b. Centralia, Ill., Dec. 17, 1920; s. Clarence J. and Bess (Maxwell) F.; m. Kathryn Elizabeth Grimes, Oct. 23, 1943; children: Patricia Ann, Mary Catherine. B.S. in Engring. Physics, U. Ill., 1942. Mem. staff radiation lab. MIT, 1942-45; head radar systems dept. Airborne Instruments Lab., Deer Park, N.Y., 1946-66; dep. dir. (tactical warfare) def. research and engring. Dept. Def., 1966-70; v.p., mgr. equipment devel. labs. Raytheon Co., Sudbury, Mass., 1970-76; v.p., gen. mgr. Bedford (Mass.) ops. Mitre Corp., 1976—; mem. sci. adv. com. Def. Intelligence Agy., 1971—, chmn. sci. adv. com., 1976-82; mem. Air Force Sci. Adv. Bd., 1971-77, Def. Sci. Bd., 1972—. Contbr. articles in field. Mem. East Norwich Sch. Bd., 1955-61, East Norwich Library Bd., 1956-62. Fellow IEEE, AAAS; assoc. fellow AAAS. Subspecialty: Electronics. Home: 15 Woodberry Rd Sudbury MA 01776 Office: Mitre Corp Bedford MA 01730

FOWLER, ELIZABETH, immunologist, educator; b. Schenectady, Apr. 18, 1943; d. Francis Raynor and Julia Rachel (Merchant) F. A.B., Cornell U., 1965; Ph.D., Harvard U., 1972. Postdoctoral fellow U. Wis.-Madison, 1972-76, asst. scientist, 1976-77; asst. prof. bacteriology and immunology U. N.C., Chapel Hill, 1977-83; assoc. prof. U. South Ala., Mobile, 1983—. Contbr. articles to profl. jours. NSF fellow, 1968-71; NIH fellow, 1972-74; Am. Cancer Soc. awardee, 1980-83. Mem. Am. Assn. Immunologists, Am. Soc. Cell Biology, AAAS, Am. Women in Sci., Phi Beta Kappa. Subspecialties: Gene actions; Molecular biology. Current work: Isolation of receptor for Epstein Barr virus; role of chromosomal proteins in control of gene action. Home: 5562 Vanderbilt Ct Mobile AL 36608 Office: U South Ala Dept Microbiology MSB Mobile AL 36688

FOWLER, HOWLAND AUCHINCLOSS, scientific administrator, researcher; b. N.Y.C., Jan. 25, 1930; s. Robert H. and Caroline (Auchincloss) F.; m. Shirley Boers, May 5, 1962; children: Joanna L., Amy A. A.B., Princeton U., 1952; M.S., Brown U., 1955, Ph.D. in Physics, 1957. With Nat. Bur. Standards, Washington, 1957—, successively physicist, project leader atomic physics div. and electricity div., 1957-71; sci. asst. to dir. Inst. Basic Standards, 1971-77, 1977—; research in computer-graphics simulation of non-linear equations. Contbr. articles to profl. publs. Mem. bd. ushers Presbyterian Ch., 1965—. Mem. Am. Phys. Soc., Soc. Indsl. and Applied Math., IEEE, Sigma Xi. Republican. Club: Princeton of N.Y. Subspecialties: Condensed matter physics; Applied mathematics. Current work: Nonlinear systems, Josephson effect; applied mathematics, theoretical solid-state physics, general technical writing.

FOWLER, WILLIAM ALFRED, physics educator; b. Pitts., Aug. 9, 1911; s. John McLeod and Jennie Summers (Watson) F.; m. Ardiane Foy Olmstead, Aug. 24, 1940; children: Mary Emily, Martha Summers Fowler Schoenemann. B.Eng., Ohio State U., 1933, D.Sc. (hon.), 1978; Ph.D., Calif. Inst. Tech., 1936; D.Sc. (hon.), U. Chgo., 1976. Research fellow Calif. Inst. Tech., Pasadena, 1936-39, asst. prof. physics, 1939-42, assoc. prof., 1942-46, prof. physics, 1946-70, Inst. prof. physics, 1970—; asst. dir. research NDRC, 1941-45; tech. observer OSRD, South Pacific Theatre, 1944; sci. dir. project VISTA, Dept. Def., 1951-52; Fulbright lectr. Cavendish Lab., U. Cambridge, 1954-55; Guggenheim fellow dept. applied math. and theoretical physics St. John's Coll., U. Cambridge, 1961-62; vis. fellow Inst. Theoretical Astronomy, 1967-72; mem. nat. sci. bd. NSF, 1968-74; mem. space bd. Nat. Acad. Scis., 1970-73, 77-80, chmn., 1981—; mem. space program adv. council NASA, 1971-73; mem. nuclear sci. adv. com. Dept. Energy/NSF, 1977-80; cons. and lectr. in field. Contbr. articles to profl. jours. Bd. dirs. Am. Friends, Cambridge U., 1970-78. Recipient Naval Ordnance Devel. award USN, 1945, medal of merit, 1948; Lammé medal Ohio State U., 1952; U. Liège medal, 1955; Calif. Co-scientist of Yr. award, 1958; Barnard medal for sci. Columbia U., 1965; Apollo Acheivement award NASA, 1969; Vetlesen prize, 1973; Nat. Medal of Sci., 1974, Nobel prize, 1983; Benjamin Franklin fellow Royal Soc. Arts, Bruce gold medal Astron. Soc. Pacific, 1979. Fellow Am. Phys. Soc. (Bonner prize 1970, pres. 1976), Am. Acad. Arts and Scis., Royal Astron. Soc. (assoc., Eddington medal 1978); mem. Nat. Acad. Scis. (council 1974-77), AAAS, Am. Astron. Soc., Am. Inst. Physics (governing bd. 1974-80), AAUP, Am. Philos. Soc., Soc. Royal Sci. Liège (corr. mem.), Brit. Assn. Advancement Sci., Mark Twain Soc. (hon.), Phi Beta Kappa (vis. scholar 1980-81), Sigma Xi, Tau Beta Pi, Tau Kappa Epsilon. Democrat. Clubs: Athenaeum (Pasadena); Cosmos (Washington). Subspecialty: Nuclear physics. Current work: Research on nuclear forces and reaction rates, nuclear spectroscopy, structure of light nuclei, thermonuclear sources of stellar energy and element synthesis in stars and supernovae; study of general relativistic effects in quasar and pulsar models. Office: Calif Inst Tech Pasadena CA 91125

FOX, C. FRED, microbiologist, educator, consultant; b. Springfield, Ohio, Aug. 19, 1937; s. Charles L. and Geneva F. A. B. cum laude, Wittenberg U., 1960, D.Sc. (hon.), 1974; M.S., Ohio State U., 1961; Ph.D. (NSF predoctoral fellow), U. Chgo., 1964. Teaching asst. Ohio State U., 1959-61; research assoc. U. Chgo., 1964; postdoctoral fellow NSF, Harvard Med. Sch., 1964-66; asst. prof. biochemistry U. Chgo., 1966-70, assoc. prof., 1970-71; assoc. prof. microbiology UCLA, 1970-72, prof., 1972—, chmn. dept., 1976-81; dir., cons. INGENE, Inc., Santa Monica, Calif. Editor: more than 40 books including Membrane Research, 1972, (with A. Keith) Membrane Molecular Biology, 1972, Biochemistry of Cell Walls and Membranes, 1975, (with D. Oxender) Molecular Aspects of Membrane Transport, 1978, (with Cunningham, Watson and Goldwasser) Control of Cellular Division and Development, 1980; series editor: UCLA Symposia on Molecular and Cellular Biology, 1972; editor: Jour. Cell Biochemistry, 1972; contbr. numerous articles to sci. jours. Recipient Research Career Devel. award USPHS, 1969-70, 71-76. Mem. Am. Soc. Microbiology, Am. Soc. Biol. Chemists, Am. Chem. Soc. (Eli Lilly award 1973), Endocrine Soc. Subspecialties: Cell biology; Membrane biology. Current work: Hormone action in cell culture. Office: Molecular Biology Inst UCLA Los Angeles CA 90024

FOX, EUGENE NOAH, microbiologist, cons.; b. Chgo., Dec. 9, 1927; s. Noah and Beatrice (Sender) F.; m. Eloise Block, June 14, 1964; children: Mara, Daniel. B.S., U. Ill., 1949, M.S., 1950; Ph.D., Case-Western Res. U., 1955. Successively asst. prof., assoc. prof., prof. microbiology U. Chgo., 1960-77; dir. research and devel. Cutter Labs., Berkeley, Calif., 1977-81; prin. Kensington Cons., Calif., 1981—. Subspecialties: Infectious diseases; Microbiology. Current work: Consultant to biotechnology companies and venture capital institutions. Address: 235 Willamette Ave Kensington CA 94708

FOX, GEORGE EDWARD, biochemist, educator; b. Syracuse, N.Y., Dec. 17, 1945; s. Charles Dainer and Henrietta L. (Carpentier) F.; m. Carolyne Ann Tordiglione, Sept. 1, 1973; children: Brian Trevor, Kevin William. B.S., Syracuse U., 1967, Ph.D., 1974. Research assoc. dept. genetics and devel. U. Ill., Urbana-Champaign, 1973-77; asst. prof. dept. biophys. scis. U. Houston, 1977-82, assoc. prof. dept. biochem. and phys. scis., 1982—. Mem. editorial bd.: Jour. Molecular Evolution, 1980—; contbr. articles in field to profl. jours. NASA grantee, 1978—; Robert A. Welch Found. grantee, 1979—. Mem. AAAS, Am. Chem. Soc., Am. Inst. Chem. Engrs., Am. Soc. Microbiology, Sigma Xi, Theta Tau. Subspecialties: Molecular biology; Systematics. Current work: Archaebacteria, 5S rRNA structure and function, molecular evolution. Office: Dept Biochem Sci U Houston Houston TX 77004

FOX, HAROLD EDWARD, obstetrician, gynecologist, educator; b. East Orange, N.J., Feb. 19, 1945; s. Willis E. and Elizabeth E. (Strathearn) F.; m. Rhea L. Keller, June 18, 1966; children: Harold H., Andrhea A. B.A., U. Rochester, 1963-67, M.Sc. in Physiology, 1972, M.D. with honor and distinction, 1972. Diplomate: Am. Bd. Obstetrics-Gynecology, Nat. Bd. Med. Examiners. Resident in ob.-gyn. Strong Meml. Hosp., Rochester, N.Y., 1972-75, fellow in maternal and fetal medicine, 1975-77; asst. prof. ob.-gyn. pediatrics U. Rochester Sch. Medicine, 1975-79, chief div. maternal and fetal medicine, 1978-79; dir. regional perinatal program, Rochester, 1975-79; assoc. prof. clin. ob.-gyn. Coll. Physicians and Surgeons, Columbia U., N.Y.C., 1979—; med. dir. Western and Upper Manhattan Regional Perinatal Network, 1979—. Contbr. articles to profl. jours. Grantee in field. Fellow Am. Coll. Ob.-Gyn.; mem. Am. Inst. Ultrasound in Medicine, Soc. Gynecol. Investigation, Sigma Xi, Alpha Omega Alpha. Subspecialties: Maternal and fetal medicine; Perinatal diagnosis and therapy. Current work: Research in ultrasound, fetal activity studies and medical care. Home: 6 Whippoorwill Lake Rd Chappaqua NY 10514 Office: Div Ob-Gyn Coll Physicians & Surgeons Columbia U 622 W 168th St New York NY 10032

FOX, IRVING HARVEY, biochemistry educator; b. Montreal, Que., Can., Dec. 7, 1943; came to U.S., 1976; s. Nathan and Phyllis (Maron) F.; m. Gloria Phyllis Godine, June 22, 1966; children: Caroline, Sharon, Joanna. B.S., McGill U., 1965, M.D., 1967. Intern Royal Victoria Hosp., Montreal, 1967-68; resident, 1968-69, Duke U. Med. Ctr., Durham, N.C., 1969-72; asst. prof. U. Toronto, Ont., Can., 1972-76; assoc. prof. medicine U. Mich., Ann Arbor, 1976-78, asst. prof. biochemistry, 1976-80, prof. medicine, 1978—, assoc. prof. biochemistry, 1980—, program dir. clin. research ctr., 1977—; cons. Warner and Lambert/Parke Davis, 1980, Procter and Gamble, Cin., 1982. Contbr. numerous articles to med. jours. Pres. bd. trustees Temple Beth Emeth, Ann Arbor, 1981-83. Mem. Am. Fedn. Clin. Research (sec.-treas. region 1979-82, chmn. 1982-83), Am. Soc. Clin. Investigation, Am. Soc. Biol. Chemists. Subspecialties: Internal medicine; Biochemistry (medicine). Current work: Human prime metabolism focused upon disorders of purine nucleotide degradation. Home: 3425 Andover Dr Ann Arbor MI 48105 Office: U Mich Univ Hosp Ann Arbor MI 48109

FOX, JAMES GAHAN, veterinarian, educator, researcher; b. Reno, Nev., Mar. 8, 1943; m., 1970; 1 child. D.V.M., Colo. State U., 1968; M.S., Stanford U., 1972. Resident veterinarian Biol. Lab. Animal Div., U.S. Army Vet. Corp., Ft. Detrick, 1968-70; asst. prof., staff veterinarian Med Ctr., U. Colo., 1973-74; inst. veterinarian, dir. animal care facility MIT, 1974-75, inst. vet., assoc. prof., dir. div. comparative medicine, 1975-82, dir., prof. comparative medicine, 1983—; faculty affiliate dept. clinics and surgery Colo. State U., 1973-74; prin. investigator NIH Animal Research Diagnostic Lab. grant, 1975-85, Nat. Cancer Inst. Animal Research Ctr. Grant, 1977-80; cons., 1976-78. NIH fellow in lab. animal medicine and med. microbiology Stanford U., 1970-72. Mem. Am. Assn. Accreditation Lab. Animal Care, AVMA, Am. Coll. Toxicology, Am. Assn. Lab. Animal Sci., Am. Coll. Lab. Animal Medicine. Subspecialty: Laboratory animal medicine. Office: MIT Div Comparative Medicine Cambridge MA 02139

FOX, JOHN PERRIGO, medical educator, physician; b. Chgo., Nov. 10, 1908; s. John S. and Myrtle (Perrige) F.; m. Helen Duffell, July 14, 1934 (dec. July 1973); children: Judith M., John D., Haigh P., Joanne M.; m. Eleanor Bell Arterton, Dec. 11, 1974. B.S., Haverford Coll., 1929, LL.D., 1977; M.D., Ph.D., U. Chgo., 1936; M.P.H., Columbia U., 1948. Diplomate: Am. Bd. Preventive Medicine. With No. Trust Co., Chgo., 1929-31; fellow, later asst. dept. pathology U. Chgo., 1933-36; intern Evanston (Ill.) Hosp., 1937-38; staff internat. health div. Rockefeller Found., 1938-49; prof. epidemiology Sch. Medicine, Tulane U., 1949-58, William Hamilton Watkins prof. epidemiology, dir. div. grad. pub. health, 1958-60, vis. prof. epidemiology, 1960-70; chief dept. epidemiology, mem. Pub. Health Research Inst., City N.Y., Inc., 1960-1965; adj. prof. epidemiology Columbia Sch. Pub. Health, 1960-65, NYU Sch. Medicine, 1960-65; prof. preventive medicine Med. Sch., U. Wash., 1965-70, prof. epidemiology and internat. health, 1970-76, prof. emeritus, 1976—; assoc. dean Sch. Pub. Health and Community Medicine, 1970-72; cons. Communicable Disease Ctr., USPHS; mem. rickettsial disease commn. Armed Forces Epidemiological Bd., 1955-72; mem. virus and rickettsial study sect. NIH, 1958-64, chmn., 1962-64, mem. bd. sci. counselors div. standards, 1962-64, chmn., 1964, mem. com. epidemiology and biometry tng. grants, 1965-69; mem. bd. sci. counselors Nat. Inst. Allergy and Infectious Disease, 1969-73, chmn., 1973; mem. viral and rickettsial vaccine rev. panel Bur. Biologies, FDA, 1973-79. Author: Epidemiology: Man and Disease, 1970, Viruses in Families, 1980; assoc. editor: Am. Jour. Epidemiology, 1964-69; asst. mng. editor, 1969-71; editor, 1971—; contbr. articles to sci., med. jours. Recipient Howard Taylor Ricketts prize U. Chgo., 1936. Fellow Am. Pub. Health Assn. (governing council 1958-62), AAAS, Royal Belgian Soc. Tropical Medicine (hon.), N.Y. Acad. Sci.; mem. Am. Soc. Tropical Medicine and Hygiene, Am. Soc. Bacteriologists, Harvey Soc., Am. Coll. Preventive Medicine, Internat. Epidemiol. Soc., Assn. Tchrs. Preventive Medicine (exec. com. 1957-60), Soc. Exptl. Biology and Medicine, Am. Assn. Immunologists, Am. Epidemiol. Soc. (pres. 1969), Am. Acad. Microbiology. Club: Cosmos. Subspecialties: Epidemiology; Virology (medicine). Current work: Expert in vaccine liability litigation. Home: 900 University St Apt 17Q Seattle WA 98101

FOX, KENNETH, physicist, astronomer; b. Highland Park, Mich., Aug. 16, 1935; s. Abraham and Jennie (Krakowski) F. B.S., Wayne State U., 1957; M.S., U. Mich., Ann Arbor, 1958, Ph.D., 1962. Research physicist Nuclear Physics Inst., Netherlands, 1962-63; vis. asst. prof. Vanderbilt U., 1963-64; prof. physics and astronomy U. Tenn., Knoxville, 1964—; cons. Oak Ridge Nat. Lab., Jet Propulsion Lab., Los Alamos Nat. Lab. Editor NASA Procs., 1982; Contbr. articles in field to profl. jours and books. Nat. Acad. Sci. fellow, 1967-69, 77-78; grantee in field; Fulbright scholar, 1974-75. Mem. Am. Astron. Soc., Am. Phys. Soc., N.Y. Acad. Sci., Tenn. Acad. Sci., Internat. Astron. Union, Phi Beta Kappa, Sigma Xi. Club: Explorers (N.Y.C.). Subspecialties: Atomic and molecular physics; Cosmology. Current work: Interstellar molecules, planetary atmospheres, terrestrial environ.; laser isotope separation; theoretical physics and chem. physics; ultra high-resolution spectroscopy. Office: Department of Physics and Astronomy University of Tennessee Knoxville TN 37996

FOX, MICHAEL HENRY, researcher, radiation biology educator; b. Great Bend, Kans., Mar. 19, 1946; s. Robert Loren and Wilma Mae (Ulrich) F.; m. Mary Ann Ryser, Oct. 8, 1946; children: Nathan Michael, Jennifer Marie. B.S., McPherson Coll., 1968; M.S., Kans. State U., 1972, Ph.D., 1977; postgrad., Colo. State U., 1979. Vol. Peace Corps, Universidad Mayor de San Andres, La Paz, Bolivia, 1968-70; asst. prof. McPherson (Kans.) Coll., 1972-73; grad. research asst. Kans. State U., Manhattan, 1973-77; postdoctoral fellow Colo. State U., Fort Collins, 1977-79, asst. prof. radiation biology, 1979—. Exec. com. Larimer County Democratic party, Fort Collins, Colo., 1980-83; mem. ch. bd. Foothills Unitarian Ch., Fort Collins, 1980-83; Spanish tchr. Amigos de las Americas, Fort Collins, 1980-81; fireman Wellington Vol. Fire Dept., Wellington, Colo., 1981-83. Mem. Soc. Analytical Cytology, Radiation Research Soc. Democrat. Unitarian. Subspecialties: Cell and tissue culture. Current work: Studying the response of cells to hyperthermia utilizing multiparameter flow cytometry and cell sorting techniques to measure cell cycle, membrane fluidity, and enzyme kinetics. Home: 901 Eggleston St Fort Collins CO 80524 Office: Dept Radiation Biology Colo State U Fort Collins CO 80523

FOX, ROBERT ALAN, educator; b. Marshfield, Wis., Nov. 10, 1951; s. Lawrence George and Barbara Jane (Dern) F.; m. Theresa Ann Blaskowski, Aug. 2, 1975; children: Jennifer Mary, Thomas Lawrence. B.S., U. Wis., 1973, M.A., 1975, Ph.D., 1978. Lic. psychologist, Ill., Wis. Asst. prof. psychology Western Ill. U., Macomb, 1978-80, Ohio State U., Columbus, 1980-82; staff psychologist Nisonger Ctr., Columbus, 1980-82; asst. prof. edn. Marquette U., Milw., 1982—. Author: (with A.F. Rotatori) Behavioral Weight Reduction Programs for Mentally Handicapped Persons: A Self-Control Approach, 1981; contbr. chpts. to books, articles to profl. jours. Recipient Presdl. Merit award for teaching, research and service excellence Western Ill. U., 1980. Mem. Am. Psychol. Assn. (newsletter editor 1978—), Am. Assn. Mental Deficiency. Roman Catholic. Subspecialties: Behavioral psychology; Developmental psychology. Current work: Mental retardation; parent training; behavior therapy. Home: 5810 N Ames Terr Glendale WI 53209 Office: Dept Edn Marquette U Schroeder Complex Milwaukee WI 53233

FOXX, RICHARD MICHAEL, psychologist; b. Denver, Oct. 28, 1944; s. James Martin and Marie Louise (Harris) F.; m. Carolyn Lee Crodutt, Apr. 4, 1966; children: Carrie, Christopher. B.A., U. Calif., Riverside, 1967; M.A., Calif. State Coll., 1970; Ph.D., So. Ill. U., 1971. Research scientist IV State of Ill., Anna, 1970-74; asst. prof. pediatrics U. Md. Med. Sch., Balt., 1974-80, asso. prof. psychology, 1976-80; prof. rehab. So. Ill. U., Carbondale, 1981—; behavior modification coordinator State of Ill., Anna, 1980—; pres. HELP Services, Inc., Murphysboro, Ill., 1976—; mem. AABT Peer Review Com., N.Y.C., 1982—. Bd. editors various psychol. jours.; adv. bd., Assn. for Retarded Citizens, Arlington, Tex., 1982—; Author: Increasing Behavior, 1982, Decreasing Behavior, 1982, Toilet Training in Less Than a Day, 1974, Toilet Training the Retarded, 1973. Recipient Distinguished Alumnus award Calif. State U., Fullerton, 1981; Commendation Am. Film Festival, 1981; Research award Am. Assn. Mental Deficiency, 1979; First place Internat. Rehab. Film Fetival, 1980. Fellow Am. Psychol. Assn., Am. Assn. Mental Deficiency, Behavior Therapy and Research Soc.; mem. AAAS, Psi Chi. Subspecialties: Behavioral psychology; Learning. Current work: Application of Learning based theories to clin. treatment of normal, retarded, autistic, emotionally disturbed and deviant individuals - behavioral treatments. Office: Anna Mental Health & Devel Center 1000 North Main St Anna IL

FRADKIN, DAVID MILTON, physics educator; b. Los Angeles, Apr. 20, 1931; s. Aaron and Annie (Gordon) F.; m. Dorothea Fairweather, Nov. 25, 1959; children: Lee, Mark, Steven. B.S., U. Calif.-Berkeley, 1954; Ph.S., Iowa State U., 1963. Exploitation engr. Shell Oil Co., Los

Angeles, 1954-56; research assoc. Iowa State U., 1963-64; NATO fellow U. Rome, 1964-65; asst. prof. physics Wayne State U., 1965-69, assoc. prof., 1969-75, prof., 1975—, chmn. dept., 1981—; sr. postdoctoral fellow U. Edinburgh, 1977-78. Contbr. articles to profl. jours. Vice chmn. local sch. adv. bd. Detroit Pub. Schs., 1972-73. Recipient Probus Club award, 1973; Presdl. citation Wayne State U., 1977. Mem. Am. Phys. Soc., Sigma Xi. Subspecialty: Theoretical physics. Current work: Electron-laser interactions, electron dynamics. Office: Dept Physics Wayne State U Detroit MI 48202 Home: 18506 Parkside Detroit MI 48221

FRAENKEL-CONRAT, HEINZ, research biochemist; b. Breslau, Germany, July 29, 1910; came to U.S., 1936, naturalized, 1941; s. Ludwig Fraenkel and Lili Conrat; m. Jane Opermann, July 14, 1939 (div.); children—Richard, Charles; m. Bea A. Singer, 1964. Student univs., Breslau, Munich, Vienna, Geneva; Dr. Med., U. Breslau, 1933; Ph.D. in Biochemistry, U. Edinburgh, 1936. Research in enzymatic peptide synthesis Rockefeller Inst., N.Y.C., 1936-37; crystallization of rattlesnake venom neurotoxin Inst. Butantan, Sao Paulo, Brazil, 1937-38; chemistry and biology pituitary hormones Inst. Exptl. Biology, U. Calif. at Berkeley, 1938-42; research methods protein modification Western Regional Research Lab., Dept. Agr., Albany, Calif., 1942-49; Rockefeller fellow, Eng., Denmark, 1950; staff virus lab., prof. virology U. Calif. at Berkeley, 1951—, also prof. molecular biology. Author: Design and Function at the Threshold of Life: the Viruses, 1962, the Chemistry and Biology of Viruses, 1969, Comprehensive Virology, Vol. 1, 1974, Virology, 1981. Recipient (with Schramm and Hershey) Lasker award, 1958; named Calif. Scientist of Year, 1958. Mem. Am. Soc. Biol. Chemists, Am. Chem. Soc., Nat. Acad. Scis., Am. Acad. Arts and Scis. Subspecialty: Molecular biology. Also research degradation and reconstitution tobacco mosaic virus, chem. research protein and nucleic acid of viruses, viral and nonviral plant enzymes, mechanism of snake venom neurotoxicity. Home: 870 Grizzly Peak Blvd Berkeley CA 94708

FRAHM, RICHARD RAY, geneticist, educator; b. Scottsbluff, Nebr., Nov. 17, 1939; s. Robert L. and Murial E. (Nolke) F.; m. Joyce L. Frahm, Aug. 27, 1961; children: Lorinda S., Kathy R. B.S. in Animal Sci, U. Nebr., 1961; M.S. in Genetics, N.C. State U., 1961; Ph.D. in Genetics and Stats, N.C. State U., 1965. Cert. animal scientist. Asst. prof. dept. animal sci. Okla. State U., 1967-71, assoc. prof., 1971-76, prof., 1976—. Contbr. articles to profl. jours. Served to capt. U.S. Army, 1965-67. NSF fellow, 1961-65; grantee in field. Mem. Am. Soc. Animal Sci., Genetics Soc. Am., Okla. Cattlemen's Assn., Sigma Xi. Subspecialties: Animal breeding and embryo transplants; Animal genetics. Current work: Breed evaluation and development of effective crossbreeding systems to improve efficiency of beef production. Home: 2106 Warren Dr Stillwater OK 74074 Office: Animal Sci Dept Okla State U Stillwater OK 74078

FRALEY, ELWIN EUGENE, med. educator; b. Sayre, Pa., May 3, 1934; s. George L. and Alene J. (Callear) F.; m. Jeanne Meikle, Nov. 26, 1954; children—George, William, Elwin, Karen, Christopher, Andrew. Grad., Phillips Exeter Acad., 1953; A.B. cum laude, Princeton, 1957, M.D., Harvard, 1961. Diplomate: Am. Bd. Medicine, Am. Bd. Urology. Intern, then resident Mass. Gen. Hosp., Boston, 1961-63, resident urology, 1963-66; sr. investigator, staff urologist NIH, Bethesda, Md., 1967-69; prof., chmn. dept. urol. surgery U. Minn. Med. Sch., Mpls., 1969—; Researcher John Hartford Found., 1966-68. Contbr. articles to research jours. Recipient Clin. Research prize Am. Urol. Assn., 1964, 1st prize Clinic Research Essay Contest, 1967, 1st prize Lab. Research Essay contest, 1968, 1st prize for movies Ann. meeting, 1969, 1st prize Lab. Research Essay Contest, 1969, Ann. Motion Pictures award, 1969; Clin. Research fellow Am. Cancer Soc., 1964-66. Fellow Am. Coll. Surgeons; mem. Am. Nephrology Soc., Soc. U. Urologists, AMA, Hennepin County Med. Soc., AAAS, Minn. State Med. Assn., Am. Assn. Cancer Research, Mpls. Pediatric Soc., Soc. U. Surgeons, Minn. Acad. Gen. Practice, A.C.S., Internat. Urol. Soc., Surg. Biology Club II, Sigma Xi, Alpha Omicron Alpha. Clubs: Mason., Princeton (N.Y.C.); Interlachen Country (Edina, Minn.). Home: 19355 Cedarhurst Wayzata MN 55391

FRAME, ROGER EVERETT, psychologist; b. Lansing, Mich., Feb. 15, 1949; s. James Sutherl and Emily (Boyce) F.; m. Marsha Wiggins, Dec. 18, 1982. B.S., Denison U., 1971; M.A., Western Mich. U., 1973; Ph.D., Mich. State U., 1979. Children's therapist North Central Mich. Mental Health Services Bd., Cadillac, 1973-76; vis. asst. prof., project dir. Rehab. Inst., So. Ill. U., Carbondale, 1979-81; sch. psychologist Collier County Pub. Schs., Naples, Fla., 1981—; mental health counselor in pvt. practice, Naples, 1983—. Active Boy Scouts Am., Golden Gate United Meth. Ch., Naples. Title XX grantee, 1980, 81. Mem. Am. Psychol. Assn., Assn. Advancement of Behavior Therapy, Nat. Assn. Sch. Psychologists, Fla. Assn. Sch. Psychologists, Fla. Assn. Behavior Analysis. Subspecialties: School psychology; Behavioral psychology. Current work: Diagnostic decision making. Behavioral intervention strategies. Home: 4215 23d Pl SW Naples FL 33999 Office: Collier County Public Schs 3710 Estey Ave Naples FL 33942

FRANCE, OLIN KENNETH, JR., psychologist, educator; b. Miami Beach, Fla., Feb. 22, 1949; s. Olin Kenneth and Eve (Center) F.; m. Mary Duncan, Aug. 16, 1969; 1 son, Micah. Student, Davidson (N.C.) Coll., 1967-69; B.A., Wake Forest U., 1971; M.S., Fla. State U., 1973, Ph.D., 1975. Lic. psychologist. Intern in psychology U. Fla., Gainesville, 1974-75; asst. prof. Francis Marion Coll., Florence, S.C., 1975-78; asst. prof., assoc. prof. dept. psychology Shippensburg (Pa.) U., 1978—; pvt. practice psychology, S.C. and Pa., 1975—. Author: Crisis Intervention, 1982; author: Excercise, Thinking and Your Heart, 1984; contbr. articles to profl. jours. Fellow Pa. Psychol. Assn.; mem. Am. Psychol. Assn. Subspecialty: Cognition. Current work: Health psychology - stress management applied to exercise; relationship between exercise and cardiovascular disease. Home: 30 Faith Circle Carlisle PA 17013 Office: Dept Psychology Shippensburg Univ Shippensburg PA 17257

FRANCIOSA, JOSEPH ANTHONY, cardiologist, researcher; b. Easton, Pa., Apr. 24, 1936; s. Joseph and Letizia Beatrice (Cascioli) F.; m. Antonietta Battistoni, Feb. 8, 1964 (div. 1972); m. Barbara Ann Neilan, Aug. 3, 1973; 1 son, Christopher David. B.A., U. Pa., 1958; M.D., U. Rome, Italy, 1963. Diplomate: Am. Bd. Internal Medicine. Intern USPHS Hosp., S.I., N.Y., 1964-65; resident Washington Hosp. Ctr., 1967-69; cardiology fellow VA Hosp.-Georgetown U., Washington, 1969-71; chief ICU VA Hosp., Washington, 1971-73; assoc. dir. cardiac research Georgetown U., 1973-74; dir. CCU VA Hosp., Mpls., 1974-76, cardiac research, 1976-79; chief cardiology VA Hosp., Phila., 1979-82; dir. cardiology div. U. Ark., Little Rock, 1982—; asst. prof. medicine Georgetown U., 1973-74, U. Minn., Mpls., 1974-77, assoc. prof., 1977-79, U. Pa., Phila., 1979-82; prof. U. Ark., 1982—. Contbr. numerous articles to med. jours. Mem. med. research com. Am. Heart Assn., Mpls., 1976-79 Mem. med. research com. Am. Heart Assn., Phila., 1981-82. Served to lt. comdr. USPHS, 1964-67. VA grantee, 1974-84; U. Ark. grantee, 1982-83. Fellow ACP, Am. Coll. Cardiology, Am. Coll. Chest Physicians (chmn. hypertension com. 1981—), Am. Heart Assn. (circulation council 1978—, council high blood pressure research 1982—); mem. Am. Soc. Clin. Pharmacology and Therapeutics (vice chmn cardiopulmonary com. 1981—). Roman Catholic. Subspecialties: Cardiology; Internal medicine. Current work: Cardiovascular physiology, hemodynamics, heart failure, exercise pathophysiology, hypertension, cardiovascular clinical pharmacology. Home: 3027 Painted Valley Dr Little Rock AR 72212 Office: Univ Ark Med Scis 4301 W Markham St Little Rock AR 72205

FRANCIS, JOHN ELBERT, aerospace, mech. and nuclear engr., educator, researcher; b. Kingfisher, Okla., Mar. 14, 1937; s. John Amos and Virginia Lorain (Mitchell) F.; m. Susan Ruth Bentley, June 2, 1962; children: John Carl, Steven Michael. B.S. in Mech. Engring, U. Okla., 1960, M.S., 1963, Ph.D. in Engring. Sci, 1965. Registered profl. engr., Okla. Engr. Allis-Chalmers Co., Springfield, Ill., 1960; grad. asst. in mech. engring. U. Okla., Norman, 1960-61, asst. prof. aerospace and mech. engring., 1966-69, assoc. prof., 1969-74, prof. aerospace, mech. and nuclear engring., 1974—, asst. dean, 1968-71, now assoc. dean; asst. research scientist Continental Oil Co., Ponca City, Okla., summer 1963; asst. prof. mech. engring. U. Mo., Rolla, 1964-66; Phys. scientist Nat. Bur. Standards, summers 1978-79. Contbr. articles to profl. jours. Recipient Ralph Teetor award Soc. Automotive Engrs., 1969; Wonders of Engring. award Okla. Soc. Profl. Engrs., 1974. Mem. AIAA (chmn. thermophysics tech. com. 1978-81, gen. chmn. 15th Thermophysics Conf. 1980), ASME, Am. Soc. for Engring. Edn., Optical Soc. Am., N.Y. Acad. Scis., Sigma Xi, Pi Tau Sigma. Democrat. Roman Catholic. Lodge: Norman Rotary. Subspecialties: Mechanical engineering; Biomedical engineering. Current work: Heat transfer, thermography, solar energy; research on combined conductive and radiative heat transfer in insulating materials; administrative duties in academic programs for Engineering College. Home: 1406 Greenbriar Dr Norman OK 73069 Office: U Okla 22 W Boyd Room 107 Norman OK 73019

FRANCIS, KENNON THOMPSON, physiologist, educator; b. Camp LeJune, N.C., July 8, 1945; s. James Ballard and Dorothy (Thompson) F.; m. Sheryl East, Dec. 29, 1966; children: Connie Jill, Wendy Kay. B.S. in Zoology, Auburn U., 1967, M.S. in Biochemistry, 1969, Ph.D., 1972. Asst. prof. dept. sci. Troy State U., Montgomery, Ala., 1973-74; prof. dept. allied health U. Ala. in Birmingham, 1974—; mem. nat. adv. com for guidance of Ph.D. Program in Phys. Therapy U. Calif., 1977; mem. Nat. Adv. Council in Phys. Therapy Edn., 1980-83. Assoc. editor Jour. Ala. Acad. Sci.; Contbr. articles to sci. jours. Served to capt. Signal Corps. U.S. Army, 1970-72. Faculty grantee U. Ala., 1975-77, 81-82; Linn Henley Charitable Trust grantee, 1977; also rehab. and medicine grants. Mem. Am. Physiol. Soc., Sigma Xi, Gamma Sigma Delta, Alpha Eta. Baptist. Subspecialties: Physiology (medicine); Physiology (biology). Current work: Neuroendocrine aspects of exercise in relation to psychological stress and heart disease.

FRANCIS, MARION DAVID, research chemist; b. Campbell River, B.C., Can., May 9, 1923; came to U.S., 1949; s. George Henry and Marian (Flanagan) F.; m. Emily Liane Williams, Aug. 27, 1949; children: William Randall, Patricia Ann. B.A., U. B.C., Vancouver, 1946, M.A., 1949; Ph.D., U. Iowa, 1953. Cert. profl. chemist, Am. Inst. Chemistry. Instr. U. B.C. (Can.), Vancouver, 1946-49; chemist Can. Fishing Co., Vancouver, 1946; research asst. U. Iowa, Iowa City, 1949-51; research chemist Procter & Gamble Co., Cin., 1952-76, sr. scientist, 1976—; chmn. Gordon Research Conf., N.H., 1968, 79. Contbr. articles to sci. jours. Dist. chmn. United Appeal, Cin., 1956-60. Recipient Profl. Accomplishment award Tech. and Sci. Socs. Cin., 1979. Fellow Am. Inst. Chemists, AAAS; mem. Soc. Nuclear Medicine, Am. Assn. Dental Research, Internat. Assn. Dental Research, Am. Pharm. Assn., Am. Chem. Soc. (program chmn. central regional meeting 1983, Cin. Chemist of Yr. 1977), N.Y. Acad. Scis., Ohio Acad. Sci. Republican. Roman Catholic. Club: Dance (pres. 1972-73). Subspecialties: Nuclear medicine; Medicinal chemistry. Current work: Chemistry and physiology of calcium phosphates of bones and teeth; bone and gastrointestinal imaging with technetium-99m; diphosphonate physiology and pharmacology. Patentee. Office: Procter & Gamble Co 2N142 Miami Valley Labs Cincinnati OH 45247

FRANCIS, MICHAEL SKOK, air force officer, research scientist; b. Sheboygan, Wis., May 29, 1947; s. Louis I. and Stanza Skok (Moegenburg) F.; m. Donna S. Weber, Jan. 3, 1982. B.S., U. Colo., 1969, M.S., 1970, Ph.D., 1976. Grad. research asst. Los Alamos Sci. Lab., 1969, 70; teaching asst. U. Colo., 1970-71, research asst., 1971-73; commd. U.S. Air Force, 1969, advanced through grades to maj., 1983—; research assoc. F.J. Seiler Research Lab., U.S. Air Force Acad., Colo., 1974-75, div. chief, 1975-80; program mgr. Air Force Office of Sci. Research, Bolling AFB, Washington, 1980—. Contbr. articles to profl. jours. NSF trainee, 1972-73; L. Brown Found. fellow, 1973; recipient USAF Research and Devel. award, 1980. Mem. AIAA (chmn. fluid dynamics tech. com. 1983—), ASME, Am. Phys. Soc. Subspecialties: Fluid mechanics; Aeronautical engineering. Current work: Manager of Air Force basic research program in fluid mechanics; research in unsteady fluid dynamics. Office: Air Force Office Sci Research AFOSR/NA Bolling AFB Washington DC 20332

FRANCIS, ROBERT DORL, microbiologist, educator, researcher; b. West Liberty, Ohio, Sept. 28, 1920; s. Joseph C. and Nellie Viola (Wartenbe) F.; m. Lisbeth Ann Innis, Feb. 20, 1943. A.B., Franklin Coll., 1942; M.S., U. Chgo., 1945; Ph.D., U. Mich., 1955. Chemist Swift & Co., Chgo., 1942-44; research asst. U. Chgo., 1945-47; bacteriologist Emulsol Corp., Chgo., 1947-49; instr. U. Mich., 1950-52; assoc. prof. microbiology U. Ala., Birmingham, 1956—; cons. virology Ala. Health Dept., 1966-77. Served with USPHS, 1953-56. NIH grantee, 1957-74. Mem. AAAS, Am. Soc. Microbiology (E.O. King award Southeastern Br. 1976), Sigma Xi. Presbyterian. Subspecialties: Microbiology (medicine); Virology (medicine). Current work: Comparative and molecular virology; chlamydial agents. Research, publs. virology. Office: Univ Sta CHSB 445 U Ala in Birmingham Birmingham AL 35294

FRANCO, JOHN VINCENT, computer science educator; b. N.Y.C., Mar. 27, 1947; s. Vincent Joseph and Concetta (Scandura) F.; m. Myra Suzanne Rowe, Feb. 16, 1975; children: Veronica, Vanessa. B.E.E., CCNY, 1969; M.S., Columbia U., 1971, Rutgers U., 1978, Ph.D., 1981. Mem. tech. staff Bell Telephone Labs., Holmdel, N.J., 1969-75; prof. computer sci. Case Western Res. U., Cleve., 1980—. Vice-pres. Ludlow Community Assn., Cleve., 1982; dir. radio broadcasting Ch. of the Saviour, Cleveland Heights, Ohio, 1982. Recipient Alumni award CCNY Engring. Alumni, 1969; Rutgers U. fellow, 1977-80. Mem. Assn. Computing Machinery, Soc. Indsl. and Applied Math., Eta Kappa Nu. Methodist. Subspecialties: Algorithms; Theoretical computer science. Current work: Probabilistic analysis of algorithms for NP hard problems, theoretical computer science, complexity classes, approximation algorithms, duality. Inventor multiple digital FSK demodulator, 1975. Home: 3164 Ludlow Rd Shaker Heights OH 44120 Office: Dept Computer Engring and Sci Case Western Res U Cleveland OH 44106

FRANCO, VICTOR, physicist, educator; b. N.Y.C., Dec. 15, 1937; s. Isaac and Regina (Ferezy) F. B.S., N.Y.U., 1958; M.A., Harvard U., 1959, Ph.D, 1963. Research asso. M.I.T., 1963-65, Lawrence Radiation Lab., Berkeley, Calif., 1965-67, Los Alamos (N. Mex.) Nat. Lab., 1967-69; asso. prof. physics Bklyn. Coll., 1969-72, prof., 1973—. Contbr. articles to profl. jours. Grantee NASA, NSF, others. Fellow Am. Phys. Soc.; mem. Sigma Xi. Subspecialties: Theoretical physics; Nuclear physics. Current work: Theoretical studies of nuclear and atomic scattering at medium and high energies. Office: Dept Physics Brooklyn Coll Brooklyn NY 11210

FRANK, ELLEN RYAN, management scientist; b. N.Y.C., July 10, 1954; d. Joseph Thomas and Cleone Irene (Mekos) Ryan; m. Stephen Z. Frank, Aug. 1, 1976. B.S., C.W. Post Coll., 1976; M.A., Hofstra U., 1978, Ph.D., 1981. Research mgr. L.I. Cancer Council, Melville, N.Y., 1978-79; research asso. Inst. for Research/Eval., 1979-80; mgmt. scientist Pfizer, Inc., N.Y.C., 1980-81; mgr. ops. research Roerig/Pfizer, Inc., N.Y.C., 1981—; adj. faculty/doctoral supr. Hofstra U., Hempstead, N.Y., 1981—; lectr., cons. in field, 1980—; lectr. Micro Computer confs., Washington, 1982—. Hofstra U. grad. fellow, 1976-80. Mem. Am. Psychol. Assn., Ops. Research Soc. Am., Metro. N.Y. Assn. for Applied Psychology, Eastern Psychol. Assn., Phi Eta, Psi Chi. Subspecialties: Operations research (engineering); Algorithms. Current work: Industrial psychology, decision support systems, multivariate modeling. Home: 235 E 42d St New York NY 10017 Office: Roerig Div Pfizer Inc 235 E 42 St New York City NY 10017

FRANK, IRWIN NORMAN, urology educator, urologist; b. Rochester, N.Y., Mar. 24, 1927; s. Harry and Bess (Smalline) F.; m. Marilyn Ellowitch, June 13, 1954; children: Gary, Steven, Lawrence. B.A., U. Rochester, 1950, M.D., 1954. Diplomate: Am. Bd. Urology. Intern Strong Meml. Hosp., Rochester, 1954-55, asst. resident, 1955-58, Nat. Cancer Inst. trainee, 1957-59, chief resident urology, 1958-59; asst. prof. urology U. Rochester, 1959-67, assoc. prof., 1967-74, prof., 1974—, acting chmn. dept. urology, 1967-69. Contbr. chpts., numerous articles to profl. publs. Med. adv. bd. Kidney Found. Upstate N.Y., Rochester. Served with USN, 1944-46; PIO. Am. Cancer Soc. fellow, 1951. Fellow ACS, Am. Acad. Pediatrics; mem. Am. Urol. Assn. (pres. Northeastern Sect.), N.Y. State Urol. Soc. (pres.). Club: Irondequoit Country (Rochester) (v.p. 1980-82). Subspecialties: Urology; Cancer research (medicine). Current work: Medical education, cancer research (bladder and kidney cancer), cytologic diagnosis of urologic cancer, kidney stone disease, practice of urology (patient care), ultrasonic and laser treatment of kidney stones and urologic cancer. Home: 221 Monteroy Rd Rochester NY 14618 Office: U Rochester Med Sch 601 Elmwood Ave Rochester NY 14642

FRANK, LOUIS ALBERT, physicist, astronomer, educator, researcher; b. Chgo., Aug. 30, 1938; m., 1960; 2 children. B.A., U. Iowa, 1960, M.S., 1961, Ph.D. in Space Sci, 1964. From asst. prof. to assoc. prof. U. Iowa, 1964-71, prof. physics and astronomy, 1971—. Mem. Am. Geophys. Union. Subspecialty: High energy astrophysics. Office: Dept Physics and Astronomy U Iowa Iowa City IA 52242

FRANK, MICHAEL M., physician; b. Bklyn., Feb. 28, 1937; s. Robert and Helen (Prakin) F.; m. Ruth Sybil Pudolsky, Nov. 5, 1961; children—Robert E., Abigail B., Brice S.H. A.B., U. Wis., 1956; M.D., Harvard U., 1960. Intern Boston City Hosp., 1960-61; resident in pediatrics Johns Hopkins Hosp., 1961-62, 64-65; vis. scientist Nat. Inst. Med. Research, London, 1965-66; with NIH, 1967—; chief lab. of clin. investigation, clin. dir. Nat. Inst. Allergy and Infectious Diseases, Bethesda, Md., 1977—. Editor: Blood. Mem. Assn. Am. Physicians, Am. Soc. Clin. Investigation, Soc. Pediatric Research, Infectious Diseases Soc., Am. Acad. Allergy, A.C.P., Soc. Hematology. Subspecialties: Allergy; Infectious diseases. Current work: Role of complement and immune complexes in host defense and in production of human disease. Office: Room 11N232 Clin Center NIH Bethesda MD 20205

FRANK, NEIL LAVERNE, meteorologist, meteorol. center adminstr.; b. Kans., Sept. 11, 1931; s. Clarence E. and Mary Violet F.; m. Velma L. Becker, Sept. 12, 1952; children—Pamela, Debra, Ron. B.A., Southwestern Coll., 1953; M.S., Fla. State U., 1959, Ph.D., 1967. Meteorologist Nat. Hurricane Center, Miami, Fla., 1961-73, dept. dir., 1973-74, dir., 1974—. Contbr. articles to profl. jours. Bd. dirs. Dade County Citizens Safety Council, Dade County ARC. Served with USAF, 1953-57. Mem. Am. Meteorology Soc. Methodist. Subspecialty: Meteorology. Office: 1320 S Dixie Miami FL 33146

FRANK, ROBERT GEORGE, psychology educator; b. Paris, Mar. 25, 1952; s. Fred James and Dorothea (Plaut) F.; m. Carol Ann Cross, Aug. 17, 1974. B.S., U. N.Mex., 1974, M.A., 1977, Ph.D., 1979. Lic. psychologist, Mo. Asst. prof. psychology dept. phys. medicine and rehab. U. Mo. Sch. Medicine, Columbia, 1979—; grad. research asst. dept. psychology U. N.Mex., Albuquerque, 1974-78; psychology intern dept. psychiatry U. Wash., Seattle, 1978-79; asst. prof. dept. psychiatry Sch. Medicine, U. Mo., 1979—. Contbr. articles to profl. jours. Coach Columbia Master's Swim Club, 1980—. NIMH trainee, 1977. Mem. Am. Psychol. Assn., Central States Ctr. A.K. Rice Inst., Soc. Behavioral Medicine, Am. Soc. Clin. Hypnosis, Assn. Advancement of Psychology. Roman Catholic. Subspecialties: Health psychology; Clinical psychology. Current work: Psychogical response and coping mechanisms after catastrophic injury, hypnosis and management of chronic pain; smoking cessation, affective disorders following catastrophic illness. Home: 4508 Georgetown Dr Columbia MO 65201 Office: Dept Phys Medicine and Rehab 501 Rusk Sch Medicine 1 Hospital Dr Columbia MO 65212

FRANK, SIDNEY RAYMOND, meteorologist; b. Mpls., Mar. 16, 1919; m., 1950; 1 child. B.A., U. Minn., 1940; M.A., UCLA, 1941. Lab. asst. UCLA, 1941; forecaster Trans World Airlines, Calif., 1941-45, researcher and instr. meteorology, Kansas City, Mo., 1945-52; project dir. Aerophys. Research Found., Calif., 1952-56; v.p. Aerometric Research Inc., 1956-68; pres. Aerometric Research Found., Calif., 1952-68, bd. dirs., 1956-68; pres. Sidney R. Frank Group, Goleta, Calif., 1968—, SRF Research Inst., 1968—; lectr. U. Kansas City, 1949; v.p. N. Am. Weather Cons., 1956-68; mem. tech. rev. com. USPHS, 1963-65; lectr., research assoc. U. Calif.-Riverside, 1960-65, U. Calif.-Santa Barbara, 1965—; meteorol. coordinator Air Pollution Control Inst., U. So. Calif., 1967-70. Editor: Jour. Aero. Meteorology, 1945-47; assoc. editor: Jour. Applied Meteorology, 1962-67. Recipient Air Transp. Assn. awards, 1943-49. Mem. AAAS, Air Pollution Control Assn., Am. Meterol. Soc., Solar Energy Soc., Am. Geophys. Union. Subspecialty: Synoptic meteorology. Office: Sidney R Frank Group 444 David Love Pl Goleta CA 93117

FRANKE, RICHARD HERBERT, mgmt. educator; b. Washington, Dec. 19, 1937; s. Herbert A. and Edna L. (Mathisen) F.; m. Elke Körner, Dec. 23, 1963; children: Martin Richard, Andrea Elke, Erik Nicolas. B.Ch.E., Cornell U., 1960; postgrad., U. Hamburg, W.Ger., 1963-64; M.B.A., U. Pitts., 1965; Ph.D., U. Rochester, 1974. Prodn. engr. Union Carbide Internat. Co., Tex., 1960-63, Italy, 1960-63; research and mktg. research engr. St. Joseph Lead Co., Pitts., 1966-67; research asst. Mgmt. Research Center, U. Rochester, N.Y., 1968-72; research fellow Nat. Acad. Scis., Belgrade, Yugoslavia, 1972-73; asst. prof. U. Wis.-Milw., 1973-78; assoc. prof. mgmt. Worcester (Mass.) Poly. Inst., 1978-83; prof. mgmt. Loyola Coll., Balt., 1983—. Co-author, editor: The Science of Productivity, 1983; contbr. articles to profl. jours. NDEA fellow, 1968, 69; Japanese Found. grantee, 1979; NSF grantee, 1979; others. Mem. Am. Psychol. Assn., Acad. Mgmt., AAAS, Am. Sociol. Assn., Am. Chem. Soc. Subspecialties: Behavioral psychology; Chemical engineering. Current work: Research on productivity, economic growth, and on various maladaptations to industrial society (homicide, suicide, pollution- and consumer goods-related death rates, etc.), integrating evidence from psychology, economics, and statistics. Office: Sch Bus and Mgmt Loyola Coll 4501 N Charles St Baltimore MD 21210

FRANKE, RICHARD HOMER, mathematics educator, researcher; b. Herndon, Kans., April 11., 1937; s. Claude E. and Beulah E. (Tannehill) F.; m. J. Amelia Franklin, July 6, 1963; children: Evan, Tanna, Hailey. B.S., Ft. Hays (Kans.) State Coll., 1959; M.S., U. Utah, 1961, Ph.D., 1970. Research engr. Boeing Co., Wichita, Huntsville and New Orleans, 1961-64; research scientist Kaman Nuclear, Colorado Springs, Colo., 1964-66; teaching asst. U. Utah, Salt Lake City, 1966-70; asst. prof. math. Naval Postgrad. Sch., Monterey, Calif., 1970-76, assoc. prof., 1976—. NASA trainee, 1969. Mem. Soc. Indsl. and Applied Math., Math. Assn. Am., Calcutta Math. Soc., Sigma Xi. Democrat. Subspecialties: Numerical analysis; Mathematical software. Current work: Applied approximation theory, scattered data interpolation and approximation multivariate approximation. Home: 877 Jefferson St Monterey CA 93940 Office: Dept Math Naval Postgrad Sch Monterey CA 93940

FRANKEL, JOSEPH, zoology educator, biologist; b. Vienna, Austria, July 30, 1935; came to U.S., 1940, naturalized, 1947; s. Arie and Ruth (Wirth) F.; m. Anne Wijtske Koopmans, Dec. 21, 1961; children: Rebecca Eva, Martin David. B.A. in Zoology, Cornell U., 1956, Ph.D., Yale U., 1960. Asst. prof. zoology U. Iowa, 1962-65, assoc. prof., 1965-71, prof., 1971—. Research numerous publs. in field. NIH grantee, 1974-83. Mem. AAAS, Am. Inst. Biol. Scis., Am. Soc. Zoologists, Soc. Devel. Biologists, Genetics Soc. Am., Soc. Protozoologists. Subspecialties: Developmental biology; Gene actions. Current work: Analysis of cell-surface patterns in ciliated protozoa: pattern formation, cell surface development, developmental genetics, ciliate development, Tetrahymena thermophila development. Office: Dept Zoology U Iowa Iowa City IA 52242

FRANKEL, RICHARD BARRY, physicist, chemist, researcher; b. St. Paul, June 24, 1939; m., 1960; 2 children. B.S., U. Mo., 1961; Ph.D. in Chemistry, U. Calif.-Berkeley, 1965. Research asst. Lawrence Radiation Lab., U. Calif., 1962-65; mem. research staff MIT, 1965-79; sr. research scientist Nat. Magnet Lab., 1979—. NATO fellow Munich (W.Ger.) Tech., 1967-68. Fellow Am. Phys. Soc.; mem. AAAS, Bioelectromagnetic Soc., Explorers Club. Subspecialty: Biophysical chemistry. Office: Nat Magnet Lab Bldg Nw 14 MIT Cambridge MA 02139

FRANKEL, VICTOR HIRSCH, orthopaedic surgeon; b. Wilmington, Del., May 14, 1925; s. Harry and Estelle (Hillersohn) F.; m. Elna Ruth Olsen, Feb. 15, 1958; children—Victor Hirsch, Dana G., Lars-Erik, Carl S., Paul A. B.A., Swarthmore Coll., 1946; M.D., U. Pa., 1951; D.Sc., U. Upsala, Sweden, 1960. Intern Grad. Hosp. U. Pa., 1951-52; resident Charlotte (N.C.) Meml. Hosp., 1954-55, Hosp. Joint Diseases, N.Y.C., 1955-58, attending orthopaedic surgeon, 1960-66; prof. orthopaedic surgery and bioengring. Case Western Res. U., 1966-75; prof., chmn. dept. orthopaedic surgery U. Wash. Sch. Medicine, Seattle, 1976-81; dir. orthopaedic surgery, surgeon-in-chief Hosp. Joint Diseases, Orthopaedic Inst., N.Y.C., 1981—; prof. orthopaedic surgery Mount Sinai Med. Sch., N.Y.C.; chmn. orthopaedic panel FDA Bur. Med. Devices, 1972-75. Author: Orthopaedic Biomechanics, 1970, Basic Biomechanics of Skeletal System, 1980. Served with M.C. U.S. Army, 1952-54. Recipient Klinkicht award Am. Orthopaedic Foot Soc., 1972; award of merit U.S. Ski Assn., 1972; Citation award Am. Coll. Sports Medicine, 1974; award of Merit ASTM, 1978; citation Outstanding Performance Boeing Comml. Airplane Co., 1978; Clemson award Internat. Biomaterials Symposium, 1980; Nat. Found. fellow, 1958-60; Frauenthal fellow Hosp. Joint Diseases, 1958-60; Am. Orthopaedic Assn. exchange fellow, 1965. Mem. Internat. Soc. Orthopaedic Surgery and Traumatology, Internat. Soc. Study Lumbar Spine, AMA, Am. Acad. Orthopaedic Surgeons, Can. Orthopaedic Assn., Am. Orthopaedic Assn., Hip Soc., Am. Orthopaedic Soc. Sports Medicine. Subspecialty: Orthopedics. Office: 301 E 17th St New York NY 10003

FRANKEN, ROBERT EARL, psychology educator; b. Stickney, S.D., May 19, 1939; emigrated to Can., 1965; s. Henry E. and Harriet (Dykshorn) F.; m. Helen Stirling, Aug. 17, 1969; 1 son, Ryan Stirling, 1 dau., Renee Joline. B.A., Hope Coll., 1960; Ph.D., Claremont Grad. Sch., 1965. Cert. psychologist, Alta. Asst. prof. psychology U. Calgary, Alta., 1965-68, assoc. prof., 1968-78, prof., 1978—. Author: Human Motivation, 1982; contbr. articles to profl. jours. Mem. Am. Psychol. Assn., Can. Psychol. Assn., Western Psychol. Assn., Midwestern Psychol. Assn., Sigma Xi. Subspecialties: Behavioral psychology; Cognition. Current work: Human motivation (thinking styles of risk takers and sensation seekers). Home: 5512 Dalwood Way NW Calgary AB Canada T3A 1S7 Office: Univ Calgary 2500 University Dr Calgary AB Canada T2N 1N4

FRANKL, WILLIAM STEWART, physician; b. Phila., July 15, 1928; s. Louis and Vera (Simkin) F.; m. Razelle Sherr, June 17, 1951; children—Victor S. (dec.), Brian A. B.A. in Biology, Temple U., 1951, M.D., 1955, M.S. in Medicine, 1961. Diplomate: Am. Bd. Internal Medicine, Am. Bd. Cardiovascular Disease. Intern Buffalo Gen. Hosp., 1955-56; resident in medicine Temple U., Phila., 1956-57, 59-61, mem. faculty, 1962-68, dir. EKG sect. dept. cardiology, 1966-68, dir. cardiac care unit, 1967-68; research fellow U. Pa., Phila., 1961-62; prof. medicine, dir. div. cardiology Med. Coll. Pa., Phila., 1970-79; prof. medicine, asso. dir. cardiology div. Thomas Jefferson U., Phila., 1979—; physician-in-chief Springfield (Mass.) Hosp., 1968-70; practice medicine specializing in cardiology, Phila., 1962-68, 70—; cons. cardiology Phila. Va Hosp., 1970-79; Fogarty Sr. Internat. fellow Cardiothoracic Inst., U. London, 1978-79; Vice-pres., Pa. affiliate Am. Heart Assn., 1979-80. Contbr. articles to profl. jours. Served with M.C. U.S. Army, 1957-59. Recipient Golden Apple award Temple U. Sch. Medicine, 1967; award Med. Coll. Pa., 1972; Lindback award for distinguished teaching, 1975. Fellow A.C.P., Am. Coll. Cardiology, Phila. Coll. Physicians, Am. Coll. Clin. Pharmacology (regent), Council Clin. Cardiology, Am. Heart Assn. (council on arteriosclerosis); mem. N.Y. Acad. Scis., Am. Fedn. Clin. Research, AAUP, AAAS, Assn. Am. Med. Colls., Am. Heart Assn. (bd. govs. S.E. Pa. chpt. 1972—, pres. 1976), Am. Soc. Clin. Pharmacology and Therapeutic Therapeutics. Subspecialties: Cardiology; Pharmacology. Current work: Cardiovascular pharmacology; electrophysiology; cardiovascular hemodynamiscmyocardial metaolism; education of medical students, house staff, fellows and practicing physicians; administration; research. Home: 536 Moreno Rd Wynnewood PA 19096 Office: 111 S 11 St Philadelphia PA 19107 The essence of humanity and being human is caring. When one cares, life takes on a new dimension and provides one the ability to transcend the thin veneer which separates human and animal.

FRANKLIN, PAULA ANNE, psychologist; b. Wheaton, Ill., Feb. 2, 1928; d. Paul Spangler and Ella (Daniels) Fowler; m. Richard Franklin, Aug. 13, 1950; children: Jan Patience Franklin BenDor, Timothy Vicery, Edward Lee. B.Sc., Northwestern U., 1949, postgrad., 1975; M.A., W. Va. U., 1970; Ph.D., Union of Experimenting Colls. and Univs., Cin., 1980. Lic. psychologist, Md. Group social worker Girl Scouts U.S.A., El Paso, Tex., 1949-51, Las Cruces, N.M., 1949-51; staff mem. Placement Office, Columbia U., N.Y.C., 1952-54; adult educator, writer, Kan., Ill., W.Va., 1954-69; dir. Franklinc Applied Behavioral Cons., W. Va., 1969-73, Balt., 1969-73; faculty applied behavior sci. Johns Hopkins Evening Coll., Balt., 1972—; social sci. research analyst Social Security Adminstrn., Balt., 1973—. Author: (with Richard Franklin) Tomorrow's Track, 1976, Urban Decisionmaking, 1967; contbr. articles to profl. jours. Active Girl Scouts U.S.A., Boy Scouts Am. Recipient Spl. Achievement award Social Security Adminstrn., 1982; Certificate of Recognition of Psychol. Service in Pub. Interest Md. Psychol. Assn., 1981. Mem. Am. Psychol. Assn. (mem. pub. policy com. 1981—, mem. research support network 1980—), Md. Psychol. Assn., AAAS, Assn. Women in Sci., League Women Voters, Evaluation Research Soc. Democrat. Unitarian. Club: Toastmasters. Subspecialties: Social psychology. Current work: Math. modelling to estimate costs of various policy and legislative alternatives for chronically ill. Involved labor force participation, utilization of health care facilities, estimation of health care costs. Also implemented social experimentation in disability program to estimate more accurately impact of future changes from more precisely targetted data bases. Home: 3946 Cloverhill Rd Baltimore MD 21218 Office: Div Disability Studies Office Research and Statistics Office of Policy Social Security Administration Security Blvd Baltimore MD 21235

FRANKLIN, RANDALL MORROW, director regional forensic unit, clinical psychologist; b. Bloomington, Ind., May 13, 1947; s. Owen Ellsworth and Dicy Lou (Morrow) F.; m. Mary Ann La Cava, Dec. 18, 1971 (div. 1975). B.A., U. Va., 1969; M.A., Temple U., 1976, Ph.D., 1979. Lic. psychologist, Pa. Clinic asst. Temple U. Psychol. Services Center, Phila., 1974-75; ward psychologist Phila. State Hosp., 1975-78, forensic ward adminstr. and ward psychologist, 1978-80, chief psychology services and forensic ward adminstr., 1980-81, chief forensic psychologist, 1981, dir. regional forensic unit, 1981—; pvt. practice clin. psychology, Phila., 1981—; clin. asst. prof. Hahnemann Med. Center, Phila., 1980—; cons. Fairmount Inst., Phila., 1981—. Mem. Pa. Psychology Polit. Action Com., 1980—. Served with USNR, 1969-71. Temple U. fellow, 1972; Pa. Psychiatry Dept. fellow, 1973. Mem. Nat. Register Health Service Providers in Psychology, Internat. Transactional Analysis Assn., Am. Psychol. Assn., Pa. Psychol. Assn., Am. Psychology-Law Soc., Assn. Advancement Psychology, Phi Beta Kappa, Phi Sigma, Alpha Phi Omega. Current work: Research: Competency to receive/refuse treatment. Treatment and research issues in forensic psychology and law. Home: 589 Forest Rd Wayne PA 19087 Office: Regional Forensic Unit Phila State Hosp 14000 Roosevelt Blvd Philadelphia PA 19114

FRANKLIN, ROBERT FRASER, JR., research scientist, consultant; b. Phoenix, Sept. 18, 1943; s. Robert Fraser and Marjorie Elizabeth (Bolen) F. B.A., San Francisco State U., 1972; Ph.D., U. Kans., 1979. Sr. programmer Computing Center U. Kans., Lawrence, 1975-77, research assoc. dept. entomology, 1977-79; NIH postdoctoral fellow in neurophysiology dept of biology U. Oreg., Eugene, 1979-82, asst. prof., computer and info. scis., 1982-83; research scientist Environ. Research Inst. of Mich., Ann Arbor, 1983—. Contbr. articles to profl. jours. Recipient Ramon Y. Cajal award for Excellence in Neurohistology Soc. for Neuroscience, 1980. Mem. AAAS, Am. Soc. Zoology, Robotics Internat. Soc. for Exptl. Biology. Subspecialties: Ethology; Robotics. Current work: Control of movement in animals and machines; studies of natural and artificial intelligence with respect to the generation and control of movement. Office: Environ Research Inst of Mich PO Box 8618 Ann Arbor MI 48107

FRANKO, BERNARD VINCENT, pharmacologist; b. West Brownsville, Pa., June 9, 1922; m. Marie Burke, June 25, 1945; 9 children. B.S. in Pharmacy, W.Va. U., 1954, M.S. in Pharmacology, 1955, Ph.D., Med. Coll. Va., 1958. With A. H. Robins Co., Richmond, Va., 1958—, assoc. dir. pharmacology, 1968-71, 73-77, dir. pharmacologic research, 1971-73, dir. good lab. practices dept., 1978-79; mgr. research coordination and tng. sect., 1981—; asst. prof., adj. asst. prof. pharmacology Med. Coll. Va., Va. Commonwealth U., Richmond, 1961—. Contbr. numerous sci. articles to profl. publs. Fellow AAAS; mem. Am. Soc. Pharmacology and Exptl. Therapeutics, Soc. Exptl. Biology and Medicine, Internat. Soc. Biochem. Pharmacology, Va. Acad. Scis. Subspecialties: Pharmacology; Toxicology (medicine). Current work: Autonomic, cardiovascular and renal pharmacology; toxicology. Patentee in field. Home: 4012 Patterson Ave Richmond VA 23221 Office: 1211 Sherwood Ave Richmond VA 23220

FRANKS, JOHN JULIAN, physician, educator, researcher; b. Pueblo, Colo., Apr. 9, 1929; s. Frank Alec and Lila Ethelda (Ownbey) F.; m. Kathryne Jean Sammon, Dec. 27, 1951; children: John A., William T., Margaret L., Elizabeth E. B.A., U. Colo., 1951, M.D., 1954; research fellow, Harvard Med. Sch., 1963-64. Med. intern Bellevue Hosp., N.Y.C., 1954-55; med. resident Colo. Gen. Hosp., Denver, 1955-58; assoc. dir., dir. Clin. Research Ctr., Denver, 1968-80; assoc. chief of staff (research) Denver VA Med. Ctr., 1968-82, chief of hemotology, 1982—; asst. prof., assoc. prof. U. Colo., Denver, 1963—; vis. prof. U. Cape Town, South Africa, 1975-76. Served to maj. USAF, 1958-63, 68; to col. Air N.G., 1976—. NIH Research Career Devel. awardee, 1963-69; research grantee, 1964-69, VA, 1969—. Fellow AAAS; mem. Am. Physiool Soc., Central Soc. Research, Am. Gastroent. Assn., Am. Fedn. Research. Democrat. Subspecialties: Internal medicine; Hematology. Current work: Coagulation research, mathematical models of plasma protein metabolism, clinical research in hematology. Office: Denver VA Med Center 1055 Clermont St Denver CO 80220 Home: 1035 Oneida St Denver CO 80220

FRANSON, RICHARD CARL, biochemistry educator; b. Woburn, Mass., Dec. 14, 1943; s. Robert Carl and Elizabeth Anne (Fitzpatrick) F.; m. Teresa Anne Joseph, July 13, 1967; children: Kristen, Kate, Kevin. B.S., U. Mass., 1965; M.S., Bowman Gray Sch. Medicine, 1970, Ph.D., 1972. Postdoctoral fellow N.Y. U. Med. Center, 1972-74, instr., 1974-75; asst. prof. biophysics and biochemistry Med. Coll. Va., 1975-79, assoc. prof. biochemistry, 1979—. Served with USN, 1966-68. Recipient Young Investigator award NIH, 1976-79, Research Career Devel. award, 1980-85; NIH tng. grantee, 1977-87; research grantee, 1979-83; Am. Heart Assn. grantee, 1979-82. Mem. N.Y. Acad. Sci., Harvey Soc., Am. Soc. Biol. Chemistry, Biophys. Soc., AAAS, Sigma Xi. Subspecialties: Biochemistry (medicine); Biochemistry (biology). Current work: Phospholipid metabolism, phospholipases, regulation of lipolytic activites. Home: 3010 Comet Rd Richmond VA 23229 Office: Dept Biochemistry Med Coll Va Box 614 Richmond VA 23298

FRANTI, CHARLES ELMER, biostatistician, consultant, researcher; b. Ewen, Mich., Apr. 29, 1933; s. John Elmer and Ida Mary (Nykanen) F.; m. Carole Joan Wisti, Aug. 25, 1956; children: Rebecca Lee, Sara Lynne, Daniel John, Michael Jacob, Matthew Benjamin. B.S., U. Mich., 1955, A.M., 1960; M.S., Mich. State U., 1960; Ph.D., U. Calif.-Berkeley, 1967. Dir. athletics, instr. Suomi Coll., Hancock, Mich., 1955-59; instr. math. Mich. Tech. U., Houghton, 1956-59, 60-62; asst. prof. U. Calif.-Davis, 1966-70, assoc. prof., 1970-76, prof. biostatistics, 1976—; cons. U. Alta., Edmonton, 1981, Royal Vet. and Agrl. Coll., Copenhagen, 1981, U.S. Dept. Agr., Sacramento, 1972-76, Calif. Dept. Mental Health, 1968. Author: (with others) Epidemiology in Veterinary Practice, 1977. Pres. Upper Peninsula Intercollegiate Athletic Conf., Mich., Wis., 1956-57; v.p. Little League Baseball, Davis, Calif., 1971-73; officer, coach Davis Softball League, Calif., 1971-76; bd. dirs. Davis Youth Baseball, 1978-80, 82—. Martin Luther fellow Luth. Brotherhood Ins., Mich. State U., 1959-60; named Outstanding Instr. Sch. Vet. Medicine, U. Calif.-Davis, 1967-68. Mem. Math. Assn. Am. Biometric Soc., Am. Statis. Assn., Inst. Math. Stats. Soc. Epidemiological Research, Soc. Indsl. and Applied Math., Nat. Council Tchrs. Math., Am. Pub. Health Assn., Statis. Soc. Can., Sigma Xi, Delta Omega, Zeta Psi. Lutheran. Current work: Applications of statistical methods in biomedical research, epidemiologic methods and applications in medicine, zoonotic diseases. Office: Dept Community Health U Calif Sch Medicine Davis CA 95616

FRANTZ, FREDERICK STRASSNER, JR., nuclear engineer; b. Lebanon, Pa., Jan. 21, 1922; s. Frederick Strassner and Charlotte Elizabeth (Toor) F.; m. Emma Louise Keller, June 26, 1946; children: Charlotte L., Rachel L., Marjorie E., Frederick K. B.S., Lebanon Valley Coll., 1943; M.S., U. Pa.-Phila., 1949. Physicist Nat. Bur. Standards, Washington, 1949-55; nuclear engr. Westinghouse Electric Co., Pitts., 1955—. Served to lt. (j.g.) USNR, 1943-46. Mem. Am. Nuclear Soc. Republican. Lutheran. Subspecialty: Nuclear fission. Current work: Project engineer for the nuclear reactor subsystem of the Clinch River breeder reactor plant, responsible for cost and schedule management, integration of sub-systems, operating procedures. Home: 325 McClellan Dr Pittsburgh PA 15236

FRANZ, DONALD NORBERT, pharmacologist, researcher, author; b. Indpls., Sept. 23, 1932; s. Norbert John and Henrietta Pauline (Bluemel) F.; m. Barbara Lorraine Stiver, Sept. 14, 1958; children: Diane V., Beth L. B.S., Butler U., 1954, M.S., 1962; Ph.D., U. Utah, 1966. Research assoc. U. Edinburgh, Scotland, 1966-68; asst. prof. U. Utah Sch. Medicine, 1968-75, assoc. prof., 1975—. Contbg. author: The Pharmacological Basis of Therapeutics, 5th edit, 1975, 6th edit., 1980; contbr. research articles to profl. jours. Served with AUS, 1954-56. Grantee Utah Heart Assn., 1976-78, Mont. Heart Assn., 1977-78, Am. Parkinson Disease Assn., 1979-81; Nat. Heart, Lung and Blood Inst. NIH grantee, 1979—; Gardner Faculty fellow U. Utah, 1981. Mem. Am. Soc. Pharmacology and Exptl. Therapeutics, Soc. Neurosci., Am. Heart Assn., AAAS. Lutheran. Subspecialties: Neuropharmacology; Neurophysiology. Current work: Regulation of blood pressure and the autonomic nervous system by the central nervous system; essential hypertension; pharmacology of central monoaminergic systems and opiate addiction. Office: Dept Pharmacology U Utah Sch Medicine Salt Lake City UT 84132

FRANZ, ELDON HENRY, ecologist; b. Omaha, Dec. 22, 1943; s. Henry Alfred and Virginia Grace (Christensen) F.; m. Kristi Rennebohm, June 17, 1967; children: Wendy Elizabeth, Benjamin Robert, Matthew Zachary. A.B., Grinnell Coll., 1966; M.S., U. Ill., Urbana-Champaign, 1970, Ph.D., 1971. Teaching asst. dept. botany U. Ill., 1966-68, instr. dept. landscape architecture, 1968-70, grad. research asst. dept. botany, 1971-77; asst. prof. dept. botany and Inst. Ecology U. Ga., Athens, 1971-77; asst. prof., asst. environ. scientist Wash. State U., Pullman, 1977-81, assoc. prof., assoc. environ. scientist, 1981—; vis. research asst. Brookhaven Nat. Lab., summer, 1967; naturalist tchr. City of Omaha Dept. Parks and Recreation, Hummel Park Natural Sci. Day Camp, summer, 1966; cons. in field; trustee The Inst. of Ecology, 1980-82. Contbg. author: Growth without Ecodisasters?, 1980, Stress Effects on Natural Ecosystems, 1981; contbr. articles on ecology to profl. jours. Mem. Fed. Interagy. Com. on Edn., Subcom. on Envirn. Edn., 1977, Whitman County Regional Planning Council, Agrl. Lands Preservation Tech. Adv. Com., 1978-80. NSF grantee; EPA grantee; Nat. Park Service grantee; Dept. Energy grantee. Mem. AAAS, Bot. Soc. Am., Ecol. Soc. Am. (chmn. edn. com. 1980-83, council 1980-83), Internat. Union for the Conservation of Nature and Natural Resources, Assn. of Research Profs. (Wash. State U.), Sigma Xi. Subspecialties: Ecology; Resource management. Current work: Energy from biomass, plant/insect interactions, biomass, nitrogen flux and biogeochemical cybernetics, disturbance and stress ecology, environmental impact assessment, modelling. Home: SE 865 Green Hill Rd Pullman WA 99163 Office: Environmental Research Center Wash State U Pullman WA 99164

FRANZBLAU, DANIEL ERIC, computer graphics software design engineer; b. Chgo., Jan. 29, 1955; s. Sanford Asher and Eugenia (Wysat) F. B.S. in Art and Design, MIT, 1977, M.S., 1979. Computer graphics programmer MIT, 1980—; computer graphics programmer Capital Children's Museum, Washington, 1980-81; programmer, cons. Graphic Devices, Somerville, Mass., 1981—; cons. Mass. Coll. Art, Boston, 1982—, Most Media, Somerville, 1981-82. Vol. digital arts project Boston Film and Video Found., 1982—. Mem. IEEE, Assn. for Computing Machinery. Subspecialties: Graphics, image processing, and pattern recognition; Software engineering. Current work: Interactive computer graphics applied to page composition and makeup, video paint systems, igh quality font generation, education, training. Development of computer graphic tools for artistic creation of images, computer as darkroom and postprocessor for images. Creator computer video art work. Office: Graphic Devices 129 Highland Ave Somerville MA 02143

FRASCH, CARL EDWARD, microbiologist; b. Albuquerque, Sept. 7, 1943; s. Eldon Dell and Laura Kay (Mahon) F.; m. Charlotte Ann Miller, Oct. 24, 1963; children: Douglas, Mark. B.S., U. Wash., 1967; Ph.D., U. Minn., 1972. Postdoctoral fellow Rockefeller U., N.Y.C., 1972-75; research microbiologist Office Biologics, FDA, Bethesda, Md., 1975—; expert meningococcal disease WHO. Assoc. editor: Jour. Immunology, 1978-80; contbr. numerous articles to sci. publs. Asst. scoutmaster Nat. Capital Area council Boy Scouts Am., 1980-82. Mem. Am. Soc. Microbiology, Am. Assn. Immunologists, Harvey Soc. Methodist. Subspecialties: Microbiology; Immunobiology and immunology. Current work: Studies on bacterial outer membrane structure with reference to Neisseria development of protein and polysaccharide vaccines for prevention of bacterial meningitis; mechanisms by which the meningococcus invades and causes meningitis. Office: Office of Biologics 8800 Rockville Pike Bethesda MD 20205 Home: 13100 Burlwood Dr Rockville MD 20853

FRASER, THOMAS HUNTER, biotechnology company executive; b. Dansville, N.Y., Mar. 19, 1948; s. H. Hunter and Harriette Louise (Bryant) F.; m. Janis Koehler, Aug. 19, 1972; children: Whitney, Hunter. B.A., U. Rochester, 1970; Ph.D., MIT, 1975. Instr. MIT, Cambridge, Mass., 1975-76; postdoctoral fellow U. Colo., Boulder, 1976-77; research scientist Upjohn Co., Kalamazoo, Mich., 1977-81; v.p. research and devel. Repligen Corp., Cambridge, 1981-83, pres., 1983—, also dir. Contbr. articles to profl. jours. Subspecialties: Enzyme technology; Genetics and genetic engineering (biology). Current work: Manage the development and application of new technologies resulting in unique products addressing important market needs. Patentee in field. Office: 101 Binney St Cambridge MA 02142

FRAUENFELDER, HANS, physicist, educator; b. Neuhausen, Switzerland, July 28, 1922; came to U.S., 1952, naturalized, 1958; s. Otto and Emma (Ziegler) F.; m. Verena Anna Hassler, May 16, 1950; children: Ulrich Hans, Kätterli Anne, Anne Verena. Diploma, Swiss Fed. Inst. Tech., 1947, Ph.D. in Physics, 1950. Asst. Swiss Fed. Inst. Tech., 1946-52; asst. prof. physics U. Ill. at Urbana, 1952-56, asso. prof., 1956-58, prof., 1958—; Cons. Los Alamos (N.M.) Sci. Lab.; Guggenheim fellow, 1958-59, 73; vis. scientist CERN, Geneva, Switzerland, 1958-59, 63, 73. Author: The Mössbauer Effect, 1962, (with E.M. Henley) Subatomic Physics, 1974, Nuclear and Particle Physics, 1975; Contbr. articles to profl. publs. Fellow Am. Phys. Soc.,

N.Y. Acad. Sci., AAAS; mem. Nat. Acad. Scis., Am. Acad. Arts and Sci., Acad. Leopoldina. Subspecialties: Biophysics (physics); Nuclear physics. Current work: Dynamics of biomolecules; nuclear and particle symmetries. Home: 8 Hagan Blvd Urbana IL 61801 Office: Dept Physics 1110 West Green Street U Ill Urbana IL 61801

FRAVEL, DEBORAH R., botanist; b. Morgantown, W.Va., May 11, 1950; d. Harvey Nixon and Janyce Gertrude (Ballengee) Rexroad; m. Frederic Dean Fravel, Dec. 30, 1972. B.A., Duke U., 1972, M.S., 1976; Ph.D., N.C. State U., 1981. Research asst. dept. plant pathology U.N.C., Raleigh, 1974-76, teaching asst., 1976-81; research asso. dept. botany U. Md., College Park, 1981—. Mem. Am. Phytopath. Soc., Mycological Soc. Am., Sigma Xi, Gamma Sigma Delta. Subspecialties: Plant pathology; Biocontrol. Current work: Researcher in biological control of plant pathogens. Office: Plant Protection Institute Soilborne Diseases Lab Room 281 Bldg 011A BARC-West Beltsville MD 20705

FRAZER, DAVID GEORGE, JR., bioengineer, physiology educator; b. East Brady, Pa., Feb. 25, 1941; s. David George and Margaret (Simpson) F.; m. Nancy Ethel Nehrig, Aug. 19, 1966; children: Jennifer, Jason. B.S., Pa. State U., 1962, M.S., 1964, Ph.D., 1970; Ph.D. in Physiology, W. Va. U.-Morgantown, 1974. Test set design engr. Western Electric Co., Allentown, Pa., 1965-66; asst. prof. physiology W. Va. U., Morgantown, 1974-76; research physiologist Nat. Inst. Occupational Safety and Health, Morgantown, 1976—. Mem. IEEE (sec.-treas. 1982-83, chmn. engring. in med. biology Pitts sect. 1981-83), Am. Physiol. Soc. Subspecialties: Biomedical engineering; Physiology (medicine). Current work: Application of bioengineering techniques to studying lung function. Home: Route 4 Box 219 Fairmont WV 26554 Office: NIOSH 944 Chestnut Ridge Rd Morgantown WV 26505

FRAZIER, DONALD THA, physiology educator, researcher; b. Martin, Ky., Sept. 26, 1935; s. Thaoe M. and Ethel (Smith) F.; m. Sara Houston, Aug. 28, 1957 (div.); children: Jane, Donald Tha, James. B.S., U. Ky., 1958, Ph.D., 1964. Asst. prof. physiology U. N.Mex., 1964-68, asso. prof., 1968-74; prof. physiology U. Ky., Lexington, 1974—, chmn. dept., 1980—; dir. Grass Found. Fellowships, 1969-79; vis. prof., 1969, 70, 74, mem. rev. panels, NIH. Contbr. over 100 articles to sci. jours., chpts. to books. Active leadership devel. program United Way, 1979-82; active Vol. Action Center, 1981—. NIH fellow, 1959-61; NIH grantee, 1966-70, 74-79. Mem. Soc. for Neurosci., Am. Physiol. Soc., Marine Biol. Lab. (Woods Hole). Democrat. Mem. Christian Ch. (Disciples of Christ). Subspecialties: Neurophysiology; Physiology (medicine). Current work: Neural control of respiration; neuropharmacology administration, research, teaching.

FRAZIER, HOWARD STANLEY, physician; b. Oak Park, Ill., Jan. 16, 1926; s. Cecil Austin and Harriet DeGolyer (Greenleaf) F.; m. Lenore Callahan, June 10, 1950; children—Mark C., Reid J., Anne K., Peter B. Ph.B., U. Chgo., 1949; M.D., Harvard U., 1953. Intern, then resident in medicine Mass. Gen. Hosp., Boston, 1953-55; postdoctoral fellow Harvard U. Med. Sch., 1955-56, Cambridge U., 1956-57, Case Western Res. U. Med. Sch., 1957-58; mem. faculty Harvard U. Med. Sch., 1958—, prof. medicine, 1978—, dir. center analysis health practices, 1975—; cons. NIH, Nat. Center Health Care Tech. Author papers in field. Served with USNR, 1943-46. Mem. Am. Soc. Clin. Investigation, Am. Physiol. Soc., Am. Soc. Nephrology, Inst. Medicine. Subspecialties: Health services research; Internal medicine. Current work: Medical technology evaluation, health practices research, research administration. Office: 677 Huntington Ave Boston MA 02115

FRAZIER, LOY WILLIAM, JR., physiologist, educator; b. Ft. Smith, Ark., Aug. 14, 1938; s. Loy William and Louise (Allen) F.; m. Mary-Jane White, Aug. 16, 1961; children: Loy William, Jennifer Jane. B.S., U. Tex.-Arlington, 1968; Ph.D., U. Tex.-Dallas, 1972. Asst. prof. physiology Coll. Dentistry, Dallas, 1972-75, assoc. prof., 1975-78, prof., 1978—, chmn., 1982—; prof. Sch. Nursing, 1975—. Author: Laboratory Exercises in Physiology, 1981; contbr. articles to profl. jours. Served with USAF, 1957-60. Tex. Med. Found. fellow, 1966-68; NIH grantee, 1976—. Mem. Am. Physiol. Soc., Soc. Exptl. Biology and Medicine, AAAS. Lodge: Lions. Subspecialties: Physiology (medicine); Membrane biology. Current work: H , NH4 , Na , K transport across epithelial tissues. Office: Baylor College Dentistry 3302 Gaston Ave Dallas TX 75246 Home: 1413 Austrian Rd Grand Prairie TX 75050

FRAZIER, OSCAR HOWARD, thoracic and cardiovascular surgeon, researcher; b. Stephenville, Tex., Apr. 16, 1940; s. Oscar H. and Adelle (Jones) F.; m. Rachel Merriman, Feb. 15, 1964; children: Todd, Allison. B.A., U. Tex., 1963; M.D., Baylor U., 1967. Diplomate: Am. Bd. Surgery, 1975, Am. Bd. Thoracic Surgery, 1977. Intern Baylor Affiliated Hosps., Houston, 1967-68, resident in surgery, 1970-74; resident in thoracic and cardiovascular surgery Tex. Heart Inst. of St. Luke's Episcopal Hosp., Tex. Children's Hosp., Houston, 1974-76; assoc. surgeon Tex. Heart Inst., 1976—; staff M.D. Anderson Hsp., Hermann Hosp., both Houston, 1976—; dir. cardiovascular research St. Luke's Hosp., Houston, 1981—, also mem. courtesy staff; assoc. prof. surgery U. Tex. Med. Sch., Houston, 1976—, U. Tex.-M.D. Anderson Hosp. and Tumor Inst., 1976—; guest examiner Am. Bd. Surgery, 1982. Assoc. editor: Tex. Heart Inst. Jour, 1981—; contbr. numerous articles to sci. and med. jours. Bd. dirs. Nat. Soc. Prevention of Child Abuse, 1977—; mem. Fun Football Bd., 1978-80, team physician, 1978-80; mem. adminstrv. council St. Luke's United Methodist Ch., 1978. Served as capt. U.S. Army, 1968-70; Vietnam. Recipient Outstanding Surgery Student award Baylor U., 1967. Fellow ACS, Am. Coll. Cardiology; mem. Soc. Thoracic Surgeons, AMA, Am. Coll. Chest Physicians, Am. Soc. Clin. Oncology, Tex. Med. Assn., Tex. Surg. Assn., Denton A. Cooley Cardiovascular Surg. Soc., Internat. Soc. Study of Lung Cancer, AAAS, Houston Surg. Soc., Houston Cardiology Soc., N.Y. Acad. Scis., Internat. Soc. Heart Transplantation, N.Am. Hyperthermia Group, Am. Soc. Artificial Internal Organs, N.Am. Soc. Pacing and Electrophysiology, Radiation Research Soc., Sigma Chi. Subspecialties: Cardiac surgery; Transplant surgery. Current work: Development of an implantable left heart assist system; study of systemic hyperthermia by extracorporeal circulation; role of lasers in vascular surgery; development of a new heart valve. Home: 2241 Chilton Houston TX 77019 Office: PO Box 20345 Houston TX 77025

FRAZZETTA, THOMAS H., evolutionary biologist, functional morphologist, educator; b. Rochester, N.Y., May 12, 1934; s. Joseph H. and Louise V. (Cross) F. B.S., Cornell U., 1957; Ph.D., U. Wash., 1964. Instr. in zoology U. Wash., Seattle, 1963-64; assoc. in herpetology Harvard U., Cambridge, Mass., 1964-65; asst. prof. U. Ill., Urbana, 1965-71, assoc. prof., 1971-76, prof. dept. ecology, ethology, evolution, 1976—. Author: Complex Adaptations in Evolving Populations, 1975; contbr. articles to jours. Active ACLU, World Wildlife Fedn., Nat. Abortion Rights Action League, Zero Population Growth, Amnesty Internat. NIH Postdoctoral fellow, 1964; NSF research grantee, 1969, 77. Mem. AAAS, Am. Soc. Naturalists, Soc. Study of Evolution. Democrat. Subspecialties: Evolutionary biology; Morphology. Current work: Evolution of complex adaptive systems; development and population aspects as related to macroevolution; functional morphology of jaws; shark and reptile biology. Office: Dept Ecology Ethology and Evolution Univ Ill 515 Morrill Hall Urbana IL

FREDERICKSON, ROBERT, nuclear engineer; b. N.Y.C., Feb. 7, 1945; s. Robert and Helena (Sands) F.; m. Susan Chambers, Jan. 3, 1976; 1 son, Robert III. B.S., Stanford, 1966; M.S., M.I.t., 1969. Registered prof. nuclear engr., Calif. Reactor engr. AEC, Washington, 1966-67; vol. Peace Corps, Patna, India, 1969-71; sci. master Ascham Sch., Sydney, New South Wales, Australia, 1972; sr. engr. Bechtel Corp., San Francisco, 1973-75; program mgr. Electro-Nucleonics, Inc., Fairfield, N.J., 1975—. Mem. Am. Nuclear Soc., Am. Vacuum Soc., Phi Beta Kappa. Subspecialties: Nuclear fission; Mechanical engineering. Current work: Isotope separation technology; development of diagnostic, functional test and inspection equipment; optical engineering. Home: 80-22 New Rd Parsippany NJ 07054 Office: Electro-Nucleonics Inc 368 Passaic Ave Fairfield NJ 07006

FREDERICKSON, ROBERT C.A., neuropharmcologist, educator, consultant; b. Winnipeg, Man., Can., June 29, 1941; s. Robert Harold and Lorna E. G. (Johnson) F.; m.; children: Robert Murray, Karen Elizabeth, Vicki Jane. Student in architecture, U. Man., 1961-62, in engring. physics, 1959-61; B.S. in Physics with honors, 1966; Ph.D. in Pharmacology, 1971; postdoctoral in neurophysiology, 1971-72. Sr. pharmacologist Eli Lilly & Co., Indpl.s., 1972-76, research scientist, 1976-80, research asso., 1980—; asst. prof. psychiatry Ind. U., 1976—; cons. Pain Clnic. Contbr. articles to sci. jours. Isbister scholar, 1959-60; INCO scholar, 1959-61; E. M. Brydon Meml. scholar, 1960; NRC Can. scholar, 1966-67; Med. Research Council scholar, 1968-71. Mem. Soc. Neurosci., Am. Soc. Pharmacology and Exptl. Therapeutics, Pharmacol. Soc. Can., Internat. Narcotics Research Conf., N.Y. Acad. Scis. Subspecialties: Neuropharmacology; Neurophysiology. Current work: Opioid peptides, devlopment of therapeutic utility; peptides and amino acids as neurotransmitters, mechanisms of action and therapeutic possibilities. Patentee in field. Office: Eli Lilly & Co Indianapolis IN 46285

FREDRICK, LAURENCE WILLIAM, astronomer, educator; b. Stroudsburg, Pa., Aug. 27, 1927; s. Ishmeal T. and Grace (Slider) F.; m. Frances I. Schwenk, Feb. 5, 1949; children—Laura Grace, Theodore David, Rebecca Lyn. B.A, Swarthmore Coll., 1952, M.A., 1954; Ph.D., U. Pa., 1959. Research asst. Sproul Obs., Swarthmore, Pa., 1952-56; research asso. Flower and Cook Obs., Malvern, Pa., 1957-59; astronomer Lowell Obs., Flagstaff, Ariz., 1959-63; mem. faculty U. Va., Charlottesville, 1963—, prof. astronomy, 1965—; chmn. dept., dir. Leander McCormick Obs., 1963-79; cons. in field.; Fulbright-Hays exchange lectr., Austria, 1972-73; prof. U. Vienna, 1972-73. Co-author: An Introduction to Astronomy, 9th edit, 1980, Astronomy, 10th edit, 1976, Descriptive Astronomy, 1978. Served with USN, 1945-48. Named Alumnus of Year Milton Hershey Sch., 1961. Mem. Am. Astron. Soc. (sec. 1969-80), Internat. Astron. Union (sec. U.S. nat. com. 1970-80), Am. Inst. Physics (bd. govs. 1969-79), Univs. for Space Research Assn. (trustee), Sigma Xi. Subspecialty: Optical astronomy. Current work: Binary stars; stellar masses; planerary nebulae; identification of field galaxies. Home: 2602 Bennington Rd Charlottesville VA 22901

FREE, CHARLES ALFRED, biochemist, pharmacologist; b. Cleve., Apr. 19, 1936; s. Alfred H. and Dorothy (Hoffmeister) F.; m. Thora C. Meade, Oct. 21, 1961; children: Charles M., Maia E. B.S. in Chemistry, Purdue U., 1957; Ph.D. in Physiol. Chemistry, UCLA, 1962. USPHS postdoctoral research fellow Sloan-Kettering Inst. for Cancer Research, N.Y.C., 1962-65; research investigator Squibb Inst. for Med. Research, New Brunswick, N.J., 1965-69, sr. research investigator, Princeton, N.J., 1969-82, research fellow, 1982—; adj. asso. prof. chemistry Rider Coll., Lawrenceville, N.J., 1980—. Contbr. articles to biochem. and pharmacologic publs. Mem. AAAS, Am. Chem. Soc., Am. Soc. for Pharmacology and Exptl. Therapeutics, Sigma Xi. Subspecialties: Biochemistry (medicine); Molecular pharmacology. Current work: Steroid hormones, receptors and mechanisms of action; enzyme inhibitors; therapy of inflammatory and hypertensive diseases. Home: 103 Bradford Ln Pennington NJ 08534 Office: Squibb Inst for Med Research PO Box 4000 Princeton NJ 08540

FREEBERG, CARLIN HENRY, consulting educational psychologist, research specialist; b. Bellingham, Wash., Aug. 13, 1931; s. Albin Johan and Inga Mathilda (Hammer) F.; m. Mary Ann Ohrt, Aug. 15, 1954; children: Kristofer Norman, Morgan Karlsson. B.A., Western Wash. State Coll., 1954, M.Ed., 1958; Ph.D., Ariz. State U., 1978. Cert. gen. edn. specialist, tchr., Wash. Sch. psychologist Edmonds Sch. Dist., Lynnwood, Wash., 1961-67; teaching asst., research asst. Ariz. State U., 1968-72; lectr. in ednl. psychology Western Wash. U., 1972-73; sch. psychologist Edn. Service Dist., Whatcom County, Wash., 1974-78; cons. ednl. psychologist, Bellingham, Wash., 1978—; rehab. research specialist Whatcom County Juvenile Ct., Bellingham, 1981—; mem., com. chmn. Whatcom County Mental Health Bd., 1979-82; external mem. Sch. Psychology Curriculum Bd., Western Wash. U., 1981; grant author, coordinator Whatcom County Juvenile Ct., Bellingham, 1981—. Served in USAF, 1949-52. Ariz. State U. scholar, 1969-71. Mem. Am. Ednl. Research Assn., Nat. Council Measurement in Edn., Am. Psychol. Assn., Wash. Psychol. Assn., Nat. Assn. Sch. Psychologists. Republican. Club: Men's. Subspecialties: Cognition; Behavioral psychology. Current work: Modifying undesirable behavior; developing strategies for isolating dependent and independent behavioral variables; designing change strategies, measuring strategy effects. Developed retinal image fade tachistoscope, 1977. Home: 825 Liberty St Bellingham WA 98225 Office: Whatcom County Juvenile Ct Grand Ave Bellingham WA 98225

FREED, CURT RICHARD, medical educator, physician; b. Seattle, Jan. 14, 1943. A.B., Harvard U., 1965, M.D., 1969. Resident in medicine Los Angeles County Harbor Gen. Hosp., 1969-71; resident in psychiatry Mass. Gen. Hosp., Boston, 1971-72; research fellow in clin.pharmacology U. Calif.-San Francisco, 1972-75; asst.prof. medicine and pharmacology U. Colo. Health Scis. Ctr., Denver, 1975-81, assoc. prof., 1981—; chmn. research com. Colo. Heart Assn., 1981-83. Contbr. articles to profl. jours. Mem. Am. Fedn. Clin. Research, Soc. Neurosci., Am. Soc. Pharmacology and Exptl. Therapeutics, Western Soc. Clin. Investigation, Sigma Xi. Subspecialties: Pharmacology; Neurochemistry. Current work: Dopamine in movement disorders, brain regulation of blood pressure. Office: U Colo Med Ctr Box C237 Denver CO 80262

FREED, JOHN HOWARD, immunologist, educator; b. New Brighton, Pa., Mar. 10, 1943; s. Howard Frederick and Harriet Frazier (MacKinney) F.; m. Debora Dow, May 30, 1945; children: Peter, Emily. S.B. in Chemistry, M.I.T., 1965; Ph.D. in Organic Chemistry, Stanford U., 1971. Postdoctoral fellow Stanford U. Sch. Medicine, 1970-72; postdoctoral fellow Albert Einstein Coll. Medicine, Bronx, N.Y., 1973-76; asst. prof. dept. biophysics Johns Hopkins U. Sch. Medicine, Balt., 1976—. Contbr. articles to profl. jours. NSF fellow, 1966-69; Damon Runyon-Walter Winchell postdoctoral fellow, 1970-72; Arthritis Found. fellow, 1973-76. Mem. Am. Assn. Immunology, AAAS. Subspecialties: Immunobiology and immunology; Immunogenetics. Current work: Immunochemistry, transplanation biochemistry, membrane biochemistry. Office: Dept Biophysic Johns Hopkins U Sch Medicine 725 N Wolfe St Baltimore MD 21205

FREED, KARL FREDERICK, chemist, educator; b. Bklyn., Sept. 25, 1942; s. Nathan and Pauline (Wolodarsky) F.; m. Gina P. Goldstein, June 14, 1964; children: Nicole Yvette, Michele Suzanne. B.S., Columbia U., 1963; A.M., Harvard U., 1965, Ph.D., 1967. NATO postdoctoral fellow U. Manchester, Eng., 1967-68; asst., then assoc. prof. chemistry James Franck Inst., U. Chgo., 1968-76, prof., 1976—. Adv. editor: Chem. Revs; assoc. editor: Jour Chem. Physics; Contbr. articles on chem. physics to profl. jours. Fellow Am. Phys. Soc.; mem. Royal Soc. Chemistry (London). Subspecialties: Theoretical chemistry; Atomic and molecular physics. Current work: Radiationless processes and photochemistry; statis. mechanics of polymer systems; many-body theory; theoretical founds. of semi-empirical quantum chemistry. Home: 921 E 54th Pl Chicago IL 60615 Office: 5640 S Ellis Ave Chicago IL 60637

FREED, VIRGIL HAVEN, toxicology educator; b. Mendota, Ill., Nov. 18, 1919; s. Jay R. and Olive Rebecca (Edgell) F.; m. Anna May Carlson, Jan. 30, 1944; children: Kathleen, John, Linda, (dec.), David. B.S., M.S., Oreg. State U.; Ph.D., U. Oreg. Asst. prof. agrl. chemistry and farm crops Oreg. State U., Corvallis, 1944-48, assoc. prof., 1948-54, assoc. prof. chemistry, 1954-61, prof., 1960-61, head agrl. chemistry, 1961—, dir., 1967-81. Co-editor: Environ. Dynamics of Pesticides, 1975, Agromedical Practices in Pesticide Management, 1982. Active Concerned Citizens of Oreg. Recipient Gov.'s N.W. Sci. award Oreg. Mus. Sci. and Industry, 1971. Fellow AAAS, Weed Sci. Soc. (hon.); mem. Am. Weed Sci. Soc., Oreg. Acad. Scis., Jamaican Agromed. Assn., Soc. Toxicology, Phi Lambda Upsilon. Republican. Baptist. Club: Triads (Corvallis). Lodge: Kiwanis. Subspecialties: Toxicology (agriculture); Environmental toxicology. Current work: Biochemistry of pesticides, environmental chemodynamics, behavior of pesticides in soils and in biological systems. Office: Dept Agrl Chemistry Oreg State U Corvallis OR 97331 Home: 11 NW Edgewood Way Corvallis OR 97330

FREEDLAND, RICHARD A., biochemical nutritionist, educator, metabolic researcher; b. Pitts., May 9, 1931; s. Milton and Gertrude (Davis) F.; m. Beverly Jane Packefsky, June 22, 1958; children: Howard M., Judith L., Stephen J. B.S., U. Pitts., 1953; M.S., U. Ill., 1955; Ph.D., U. Wis., 1958. Research assoc. U. Wis.-Madison, 1958-60; lectr. U. Calif., Davis, 1960-62, asst. prof., 1962-65, assoc. prof., 1965-69, prof., 1969—, chmn. dept. physiol. scis., 1974—. Author: A Biochemical Approach to Nutrition, 1977; mem. editoral bd.: Archives of Biochemistry and Biophysics; contbr. articles to profl. jours. NIH postdoctoral awardee, 1973-74; recipient Magnar Ronning award for teaching excellence, 1981. Fellow AAAS; mem. Am. Soc. Biol. Chemists, Am. Inst. Nutrition, Am. Physiol. Soc., Biochem. Soc., Soc. Exptl. Biology and Medicine. Subspecialties: Biochemistry (medicine); Nutrition (biology). Current work: Nutritional and hormonal control of metabolism in higher animals. Home: 1401 Notre Dame Dr Davis CA 95616 Office: Dept Physiol Sci Sch Vet Medicine U Calif Davis CA 95616

FREEDLE, ROY OMER, research psychologist, editor; b. Chgo., Aug. 8, 1933; s. Edwin Lee and Goldie (Dolance) F. B.S., Roosevelt U., 1957; M.S., Columbia U., 1959, Ph.D., 1964. Research scientist Am. Insts. for Research, Washington, 1964-67; research psychologist Ednl. Testing Service, Princeton, N.J., 1967—, sr. research psychologist, 1971—. Editor-in-chief: Discourse Processes, 1978; series editor: Advances in Discourse, 1977—; author, editor: Discourse Production and Comprehension, 1977; author: Culture and Language, 1975. Mem. Am. Psychol. Assn., Am. Statis. Assn., Psychometric Soc. Republican. Subspecialties: Cognition; Learning. Current work: Integrate and evaluate language performance as presented by specialists in the fields of psychology, computer science, linguistics and anthropology. Office: Educational Testing Service Rosedale Rd Princeton NJ 08541

FREEDMAN, DANIEL X., psychiatrist, educator; b. Lafayette, Ind., Aug. 17, 1921; s. Harry and Sophia (Feinstein) F.; m. Mary C. Neidigh, Mar. 20, 1945. B.A., Harvard U., 1947; M.D., Yale U., 1951; grad., Western New Eng. Inst. Psychoanalysis, 1966; D.Sc. (hon.), Wabash Coll., 1974, Indiana U., 1982. Intern pediatrics Yale Hosp., 1951-52, resident psychiatry, 1952-55; from instr. to prof. psychiatry Yale U., 1955-66; chmn. dept. U. Chgo., 1966-83, Louis Block prof. biol. scis., 1969-83; Judson Braun prof. psychiatry and pharmacology UCLA, 1983—; career investigator USPHS, 1957-66; dir. psychiatry and biol. sci. tng. program Yale U., 1960-66; cons. NIMH, 1960—, U.S. Army Chem. Center, Edgewood, Md., 1965-66; chmn. panel psychiat. drug efficacy study Nat. Acad. Sci.-NRC, 1966; chmn. Nat. Acad. Scis.-NRC, 1977—; mem. adv. com. FDA, 1967—; rep. to div. med. scis. NRC, 1971—, mem. com. on brain scis., 1971-73, mem. com. on problems of drug dependence, 1971—; mem. com. problems drug dependence Nat. Inst. Medicine, 1971-76, com. substance abuse, and habitual behavior, 1976—; advisor Pres.'s Biomed. Research Panel, 1975-76; mem. selection com., coordinator research task panel Pres.'s Commn. Mental Health, 1977-78; mem. Jt. Commn. Prescription Drug Use, Inc., 1977—. Author: (with N.J. Giarman) Biochemical Pharmacology of Psychotomimetic Drugs, 1965, What Is Drug Abuse?, 1970, (with F.C. Redlich) The Theory and Practice of Psychiatry, 1966, (with D. Offer) Modern Psychiatry and Clinical Research, 1972; editor: (with J. Dyrud) American Handbook of Psychiatry, Vol. V, 1975, The Biology of the Major Psychoses: A Comparative Analysis, 1975; chief editor: Archives Gen. Psychiatry, 1970—. Bd. dirs. Founds. Fund for Research in Psychiatry, 1969-72, Drug Abuse Council, 1972-80; vice chmn. Drug Abuse Council Ill., 1972—. Served with AUS, 1942-46. Recipient Distinguished Achievement award Modern Medicine, 1973; William C. Menninger award ACP, 1975; McAlpin medal for research achievment, 1979; Vestermark award for edn., 1981. Fellow Am. Acad. Arts and Scis., Am. Psychiat. Assn. (chmn. commn. on drug abuse 1971—), Am. Coll. Neuropsychopharmacology (pres. 1970—); mem. Inst. Medicine Nat. Acad. Scis., A.C.P. (William C. Menninger award 1975), Ill. Psychiat. Soc. (pres. 1971-72), Social Sci. Research Council (dir. 1968-74), Chgo. Psychoanalytic Soc., Western New Eng. Psychoanalytic Inst., Am. Soc. Pharmacology and Exptl. Therapeutics, AAAS, Am. Assn. Chairmen Depts. Psychiatry (pres. 1972-73), Am. Psychiat. Assn. (v.p. 1975-77, pres.-elect 1980-81, pres. 1981-82), Group Advancement Psychiatry, Psychiat. Research Soc., Am. Psychosomatic Soc. (councillor 1970-73), Assn. Research in Nervous and Mental Disease (pres. 1974), Soc. Biol. Psychiatry, Sigma Xi, Alpha Omega Alpha. Subspecialties: Psychopharmacology; Neuropharmacology. Current work: Biological basis of behavioral disorders pursued by use of drugs as tools to understand brain function. Home: 806 Leonard Rd Los Angeles CA 90049 Office: 760 Westwood Plaza Los Angeles CA 90024

FREEDMAN, HERBERT ALLEN, educator, research scientist; b. N.Y.C., Jan. 22, 1944; s. Abraham and Pearl (Maloff) F. B.A., N.Y. U., 1965, M.S., 1969, Ph.D., 1972. Research fellow Inst. Muscle Diseases, N.Y.C., 1967-68, Inst. Med. Research and Studies, 1968-71; adj. lectr. CCNY, 1972; postdoctoral research fellow in genetics Albert Einstein Coll. Medicine, 1972-74, assoc. in genetics, 1975-76; asst. prof. pathology SUNY-Downstate Med. Center, Bklyn., 1977—. Contbr. articles in field to profl. jours. Leukemia Soc. Am. fellow, 1973-75; Spl. fellow, 1975-77; Am. Cancer Soc. grantee, 1978-82; NIH grantee, 1981-83; recipient N.Y. U. Founders Day Scholars award, 1972, Dupont-Sorvall Bios award, 1981. Mem. Am. Assn. Cancer Research, Am. Assn. Immunologists, N.Y. Acad. Sci., AAAS. Subspecialties:

Cancer research (medicine); Immunogenetics. Current work: Basic research on the mechanisms of genetic control of malignancy and target cell specificity of carcinogens. Office: 450 Clarkson Ave PO Box 25 Brooklyn NY 11203

FREEDMAN, ROBERT RUSSELL, psychologist, researcher; b. Phila., Apr. 30, 1947; s. Bernard and Sarah (Lichtenstein) F.; m. Mary Ann Morris, July 12, 1980. B.A., U. Chgo., 1969; Ph.D., U. Mich., 1975. Lic. psychologist, Mich. Dir. Behavioral Medicine Lab., Lafayette Clinic, Detroit, 1975—; adj. asst. prof. psychology Wayne State U., 1976—; cons. Nat. Inst. Drug Abuse, Washington, 1976-77. USPHS fellow, 1975-76; NSF grantee, 1967; Nat. Heart Lung and Blood Inst. grantee, 1980-83. Mem. Acad. Behavioral Medicine Research, Assn. Psychophysiol. Study of Sleep, Biofeedback Soc. Am., Soc. Behavioral Medicine, Soc. Psychophysiol. Research. Subspecialties: Biofeedback; Behavioral psychology. Current work: Etiology and behavioral treatment of psychophysiological disorders, Raynaud's disease and insomnia; generalization of laboratory behavioral procedures to real life situations. Office: Lafayette Clinic 951 E Lafayette St Detroit MI 48207 Home: 605 Kellogg St Ann Arbor MI 48105

FREEHILL, PETER EUGENE, nuclear engineer; b. Fort Wayne, Ind., July 19, 1941; s. Clarence Edward and Rosemary (Brady) F.; m. Sandra Kay Ward, Apr. 3, 1976; children: Peter P., Timothy B., Christopher A. Student, U. Ala. Mgr. reactor repair project Gen. Electric Co., Fukushima, Japan, 1977, mgr. ops., 1977-79; supt. startup and test Gulf States Utilities, River Bend Sta., St. Francisville, La., 1980—; shift supr. Gen. Electric Co., Fayetteville, Ark., 1967-71, ops. analyst, 1971-72, startup engr., San Jose, N.C., 1972-75, ops. supt., Southport, N.C., 1975-77. Editor, contbg. author: Nuclear Startup Manual, 1981. Served with USN, 1959-67. Mem. Am. Nuclear Soc., La. Nuclear Soc. Republican. Roman Catholic. Subspecialties: Nuclear engineering; Electronics. Current work: Develop and manage the river bend nuclear station startup test program, approve all test procedures and test results. Office: Gulf State Utilities Startup and Test Box 220 St Francisville LA 70811 Home: 10817 Shoe Creek Dr Baton Rouge LA 70811

FREEMAN, AARON ELIOT, bacteriologist; b. Buena Vista, Va., Mar. 5, 1928; s. Jacob A. and Sadie (Schulman) F.; m. Fredda D. Lipps, Mar. 30, 1952; children: Jeffrey, Joan. A.B., George Washington U., 1951, M.S., 1954; Ph.D., Cath. U. Am., 1964. Bacteriologist NIAID, NIH, Bethesda, Md., 1954-62; dir. quality control and prin. investigator Microbiol. Assos., Inc., Bethesda, 1962-72; dir. exptl. pathology Childrens Hosp., Akron, Ohio, 1972-76; dir. research Microbiol. Assos., Inc., Torrey Pines Research Center, LaJolla, Calif., 1976-78; scientist, prin. investigator LaJolla (Calif.) Cancer Research Found., 1978-81; head, toxicology and carcinogenesis dept. Center for Neurol. Study, San Diego, 1981—; adj. prof. pathology U. Calif., Davis, 1982—. Contbr. articles in field to profl. jours. Mem. Tissue Culture Assn. Subspecialties: Cell study oncology; Toxicology (medicine). Current work: Researcher in selective response of function skin, liver and lung non-fibroblast cells to carcinogenic and otherwise toxic agents. Home: 1156 Sidonia Ct Leucadia CA 92024 Office: Univ Calif Davis Sacramento Med Ctr Rm J FOLB II Sacramento CA 95817

FREEMAN, GORDON RUSSEL, chemistry educator; b. Hoffer, Sask., Can., Aug. 27, 1930; s. Winston Spencer Churchill and Aquila Maud (Chapman) F.; m. Phyllis Joan Elson, July 9, 1927; children: Mark Russel, Michele Leslie. B.A., U. Sask., 1952, M.A., 1953; Ph.D., McGill U., 1957; D.Phil., Oxford (Eng.) U., 1957. Postdoctoral fellow Centre D'Etudes Nucleaires, Saclay, France, 1957-58; asst. prof., then assoc. prof. chemistry U. Alta. (Can.), Edmonton, 1958-65, prof., 1965—, chmn. div. phys. and theoretical chemistry, 1965-75; exec. Chem. Inst. Can., 1974-80, chmn. phys. chemistry div., 1976-78, councillor, 1978-80. Contbr. articles to jours., chpts. to books. Research grantee Nat. Research Council Can., 1959-78, Natural Scis. and Engring. Research Council Can., 1978—, Def. Research Bd. Can., 1965-72. Mem. Chem. Inst. Can., Am. Phys. Soc., Radiation Research Soc., Can. Assn. Physicists. Subspecialties: Physical chemistry. Current work: Electron and ion behavior in gases, liquids and solids; radiation chemistry; optical and transport properties of electrons in disordered materials. Office: Chemistry Dept U Alta Edmonton AB Canada T6G 2G2

FREEMAN, HARVEY ALLEN, engineering company executive; b. Phila., Mar. 4, 1944. B.S.E.E., U. Pa., 1966; M.S.E.E., U. Ill., 1968, Ph.D., 1970. Mem. engring. staff RCA Corp., Camden, N.J., 1966-73; prin. systems design engr. Def. Systems div. Sperry Univac, Eagan, Minn., 1973-77, systems mgr., Roseville, Minn., 1977-81; v.p. engring. Architecture Tech. Corp., Mpls., 1981—; adj. prof. computer sci. U. Minn., Mpls., 1979-80. Author: Data Base Computers, 1979. Mem. IEEE, Assn. for Computing Machinery, IEEE Computer Soc. (governing bd. 1983—). Subspecialties: Distributed systems and networks; Database systems. Current work: All aspects of consulting and research in local networks, office automation and data base management systems. Office: PO Box 24344 Minneapolis MN 55424

FREEMAN, JOHN WRIGHT, JR., space physics and astronomy educator, space systems analyst; b. Chgo., July 12, 1935; s. John Wright and Evelyn (Laier) F.; m. Phyllis Palmer, June 13, 1935; children: Laura, David. B.A., Beloit Coll., 1957; M.S., U. Iowa, 1961, Ph.D., 1963. Staff scientist NASA, Washington, 1963-64; mem. faculty Rice U., Houston, 1964—, prof. space physics and astonomy, 1972—; systems analyst NASA/Goddard Space Flight Ctr., Greenbelt, MD., 1982—. Editor: Space Solar Power Rev, 1979—; contbr. numerous sci. articles to profl. publs. Recipient Disting. Service citation Beloit Coll., 1972; Apollo Achievement award NASA, 1970; Exceptional Sci. Achievement medal, 1972. Mem. AAAS, AIAA. Subspecialties: Satellite studies; Space agriculture. Current work: Magnetospheric physics and solar wind studies.

FREEMAN, LEONARD MURRAY, radiologist, educator; b. N.Y.C., Apr. 20, 1937; s. Joseph and Tillie (Krautman) F.; m. Marlene Carolyn Held, Apr. 28, 1967; children: Eric Lawrence, David Robert, Joy Esther. B.A., N.Y. U., 1957; M.D., Chgo. Med. Sch., 1961. Diplomate: Am. Bd. Radiology, Am. Bd. Nuclear Medicine. Intern Beth Israel Hosp. and Med. Center, N.Y.C., 1961-62; resident in radiology Bronx Municipal Hosp. Center, 1962-65; mem. staff Bronx Municipal Hosp. Center and Hosp. of Albert Einstein Coll. Medicine, N.Y.C., 1965—, co-dir. div. nuclear medicine, 1965-83; dir. nuclear medicine Montefiore Hosp. and Med. Center, N.Y.C., 1976—, attending radiologist, 1977—; cons. nuclear medicine USPHS Hosp., S.I., N.Y., 1967—, St. Barnabas Hosp., Bronx, 1967—, Beth Israel Hosp. and Med. Center, 1974—, Maimonides Hosp. and Med. Center, 1974—; asst. instr. radiology Albert Einstein Coll. Medicine, Bronx, 1964-65, instr., 1965-67, asst. prof., 1967-72, asso. prof., 1972-77, prof., 1977—, prof. nuclear medicine, 1983—; mem. adv. com. nuclear medicine program Brookhaven Nat. Labs., Upton, N.Y., 1972—; examiner nuclear medicine Am. Bd. Radiology. Author: Clinical Scintillation Scanning, 1969, Clinical Scintillation Imaging, 1975, Clinical Radionuclide Imaging, 1984; co-editor: Seminars in Nuclear Medicine, 1970—, Physicians Desk Reference for Radiology and Nuclear Medicine, 1971-80; reviewer: Jour. Nuclear Medicine, 1972—; editor: Nuclear Medicine Ann, 1977; editorial bd.: European Jour. Nuclear Medicine, 1979—; contbr. numerous articles to jours., also book chpts. Fellow Am. Coll. Radiology; fellow Am. Coll. Nuclear Physicians; mem. Soc. Nuclear Medicine (gov. local chpt. 1973—, nat. trustee 1973-77, nat. v.p. 1977-78, nat. pres. 1979-80, pub. relations com. 1981—, correlative imaging council 1982-84, chmn. awards com. 1983—), Assn. Univ. Radiologists, Am. Roentgen Ray Soc., Radiol. Soc. N.Am., N.Y. Roentgen Soc., L.I. Radiol. Soc., Soc. Gastrointestinal Radiologists, N.Y. State Med. Soc., Nassau County Med. Soc., Pan Am. Med. Assn. (hon. life), Gissellschaft für Nuklearmedizin (hon. corr.), L.I. Soc. Nuclear Med. Technologists (hon. life). Subspecialties: Nuclear medicine; Nuclear medicine. Current work: Introduction of new and diagnostic imaging techniques using radionuclides and comparison of their usefulness with existing radiologic techiques. Home: 65 Oak Dr East Hills NY 11576 Office: 111 E 210th St Bronx NY 10467

FREEMAN, LOUIS BARTON, electric equipment manufacturing company nuclear engineer; b. N.Y.C., May 12, 1935. B.A., Colgate U., 1955; M.A., Harvard U., 1957; Ph.D., U. Pitts., 1965. With Westinghouse Corp. Bettis Atomic Power Lab., West Mifflin, Pa., 1958—, adv. scientist, 1976—. Mem. Am. Nuclear Soc. Subspecialties: Nuclear engineering; Nuclear fission. Current work: Reactor physics methods, nuclear core analysis, nuclear design. Office: Westinghouse Corp Bettis Atomic Power Lab Box 79 West Mifflin PA 15122

FREEMAN, MARC EDWARD, biology educator; b. Phila., Feb. 18, 1944; s. Harry and Dorothea (Glickman) F.; m. Ada Louise Anderson, June 7, 1970. B.S., Moravian Coll., 1965; M.S., W.Va. U., 1967, Ph.D., 1970. Research assoc. Emory U., Atlanta, 1970-72; asst. prof. Fla. State U., Tallahassee, 1972-77, assoc. prof., 1977-82, prof. biol. sci., 1982—, assoc. chmn. dept., 1983—. Contbr. articles to profl. jours.; editorial bd.: Endocrinology, 1982-86. Mem. Common Cause. NIH career devel. awardee, 1978-83; Fla. State U. developing scholar, 1980. Mem. Endocrine Soc. (membership com. 1982-84), Am. Physiol. Soc. (membership com. 1981-84), Soc. Study of Reproduction (membership com. 1977-79), Soc. Study of Fertility, AAAS. Subspecialties: Neuroendocrinology; Reproductive biology. Current work: Teaching and research in reproduction neuroendocrinology. Office: Fla State Univ Dept Biol Sci Tallahasse FL 32306

FREEMAN, MARK PHILLIPS, physical chemist, researcher; b. Palemhang, Sumatra, June 9, 1928; s. Mark and Gwen (Jones) F.; m. Helen Millin, June 17, 1949; children: Mark, Aaron, Frances, Ion, Charles. B.S. in Physics magna cum laude, U. Wash., 1953, Ph.D. in Phys. Chemistry, 1956. Instr. chemistry U. Calif., Berkeley, 1956-58; from research scientist to research assoc. Am. Cyanamid Co., Stamford, Conn., 1958-72; vis. prof. chem. engring. MIT, 1967-68; sr. cons. Humphries Corp., Bow, N.H., 1972-73; sr. scientist Dorr-Oliver Inc., Stamford, 1973—; dir. Engring. Socs. Library, 1978—. Contbr. articles to sci. jours. Served with USAAF, 1946-48; with USAF, 1949, 50-51. Mem. Am. Chem. Soc., Am. Inst. Chem. Engrs., Am. Phys. Soc., Phi Beta Kappa, Sigma Xi. Democrat. Presbyterian. Club: Chemists (N.Y.C.). Subspecialties: Physical chemistry; Surface chemistry. Current work: Presently interested in the properties of aqueous colloidal systems, their stabilization and control. Patentee in field; inventor electrofilter; developer incineration system for wet sludges. Office: Dorr-Oliver Co Stamford CT 06904

FREEMAN, MARSHA GAIL, scientific magazine editor; b. N.Y.C., June 3, 1947; d. Joseph and Nettie (Bomze) Osofsky; m. Richard Edward Freeman, July 14, 1970 (div. 1973); m. Robert Laurence Gallagher, Mar. 26, 1982. B.A., Queens Coll.; M.A., Columbia U. Tchr., Detroit pub. schs., 1970-72; dir. indsl. research Fusion Energy Found., N.Y.C., 1975—, Washington editor, 1976—. Sci. and tech. editor and writer: Executive Intelligence Rev. Mem. AIAA, AAAS, Nat. Assn. Sci. Writers, Nat. Space Club. Democrat. Jewish. Subspecialties: Astronautics; Fusion. Current work: Civilian space program science and technology; advanced energy research including fusion. Office: Fusion Energy Found 304 W 58th St New York NY 10019

FREEMAN, PETER KENT, chemist; b. Modesto, Calif., Nov. 25, 1931; s. Russell Arthur and Helen Aleth (Surryhne) F.; m. Marilyn Taber, June 16, 1955; children: Diane, Irene, Theodore, Michael. B.S., U. Calif.-Berkeley, 1953; Ph.D., U. Colo., 1958. Chemist Shell Oil Co., Martinez, Calif., 1958; postdoctoral fellow Pa. State U., 1958-59; asst. prof. U. Idaho, 1959-62, assoc. prof., 1962-65, prof., 1965-68, Oreg. State U., Corvallis, 1968—. Contbr. articles to profl. jours. Mem. Am. Chem. Soc., Sigma Xi, Phi Lambda Upsilon, Phi Kappa Phi. Subspecialty: Organic chemistry. Current work: Organic reaction mechanisms; environmental chemistry. Home: 3335 Chintimini Ave Corvallis OR 97333 Office: Dept Chemistry Oreg State Univ Corvallis OR 97331

FREEMAN, RICHARD BENTON, physician; b. Allentown, Pa., July 24, 1931; s. Benton C. and Evelyn B. (Boyer) F.; m. Margaret C. McGuire, July 24, 1954; children—Richard, Charles C., Patricia Anne. B.S., Franklin and Marshall Coll., Lancaster, Pa., 1953; M.D., Thomas Jefferson U., 1957. Intern Pa. Hosp., Phila., 1957-58, resident, 1958-61; fellow in nephrology Georgetown U., 1960; clin. instr. medicine U. Pa., 1958-61, U. Calif., San Francisco, 1963; instr. medicine Georgetown U., 1964-67; asst. prof. U. Rochester, N.Y., 1967-69, asso. prof., 1969—, head nephrology unit, 1974—; cons. Contbr. articles to profl. jours. Served with USPHS, 1961-67. Fellow A.C.P.; mem. Am. Fedn. Clin. Investigation, Am. Heart Assn., Am. Soc. Nephrology, Internat. Soc. Nephrology, Am. Soc. Artificial Internal Organs, Renal Physicians Assn. (dir. 1977, v.p. 1979, pres. 1981—), AAAS, Nat. Kidney Found., Kidney Found. Upstate N.Y., Internat. Soc. Internal Organs. Subspecialty: Nephrology. Home: 25 Hearthstone Rd Pittsford NY 14534 Office: 601 Elmwood Ave Rochester NY 14642

FREEMAN, ROBERT GLEN, pathologist, dermatologist, educator; b. Kerrville, Tex., Feb. 3, 1927; s. Clifford and Lucy Mae (Garnett) F.; m. Lilaree Crawford, Sept. 9, 1950; children: David, Mark, Angie, Sara. Student, Tarleton State Coll., 1943-44, U. Tex.-Austin, 1944-45; M.D., Baylor Coll. Medicine, 1949. Diplomate: Am. Bd. Pathology, Am. Bd. Dermatology. Instr. anatomy Baylor Coll. Medicine, Houston, 1950-52; asst. prof. anatomy U. Tenn. Med. Br., Memphis, 1954-55; resident pathology Baylor Affiliated Residency, Houston, 1956-59; asst. prof. dermatology and pathology Baylor Coll. Medicine, Houston, 1959-64, assoc. prof., 1964-68, prof., 1968-70; prof. pathology and dermatology U. Tex. Health Sci. Ctr., Dallas, 1970-73, clin. prof. pathology and dermatology, 1973—; dermatol. cons. Brooke Army Med. Ctr., Fort Sam Houston, Tex., 1968-83; cons. pathology VA Hosps., Dallas and Temple, Tex., 1973-83, Baylor U. Med. Ctr., Dallas, 1973-83; mem. test com. dermatopathology Am. Bd. Pathology, Tampa, Fla., 1978-83. Author: Treatment of Skin Cancer, 1967; contbr. articles to profl. jours. Served to 1st lt. U.S. Army, 1952-54. Robert Freeman hon. lectr. U. Tex. Health Scis. Ctr. at San Antonio, 1982; Geoffrey Hunter lectr. Austrian Soc. Dermpathology, Melbourne, 1981. Fellow Am. Soc. Dermatopathology (pres. 1979); mem. Am. Dermatol. Assn., Am. Acad. Dermatology (Gold exhibit 1968), So. Med. Assn. (chmn. dermatology sect. 1979). Subspecialties: Pathology (medicine); Dermatology. Current work: Dermatopathology, skin cancer, photobiology, computer use in dermatopathology. Office: U Tex Health Scis Center 5323 Harry Hines St Dallas TX 75235

FREEMAN, WADE AUSTIN, chemist, educator; b. Evanston, Ill., Nov. 20, 1940; s. Raymond Barrett and Eleanor (Knaus) F.; m. Sally Fairchild Miller, Mar. 22, 1975; children: Andrew Taylor, Neil Barrett. B.S., U. Ill., 1962; M.S., U. Mich., 1964, Ph.D., 1967. Mem. faculty U. Ill., Chgo., 1967—, asst. dean, 1969-79, dir. gen. chemistry, 1979—. Mem. Am. Chem. Soc., AAAS. Subspecialties: X-ray crystallography; Crystallography. Current work: Structures of transition metal complexes and relationship to their spectra. Office: U Ill Dept Chemistry Box 4348 Chicago IL 60680

FREI, EMIL, III, internist, institute administrator, educator, researcher; b. St. Louis, Feb. 21, 1924; m. Elizabeth Frei, 1948; 3 children. M.D., Yale U., 1948. Intern Univ. Hosp., St. Louis, 1948-49, resident in pathology, 1952-53, resident in internal medicine, 1953-55; head chemotherapy service, chief med. br. and assoc. sci. dir. Nat. Cancer Inst., 1955-65; assoc. dir. U. Tex. M.D. Anderson Hosp. and Tumor Inst., 1967-53; dir., physician-in-chief Sidney Farber Cancer Inst., Boston, 1973—; prof. medicine Harvard U., Boston, 1980—. Recipient Lasker award, 1972, Man of Yr. award Am. Cancer Soc., 1981, Kettering prize Gen. Motors Corp., 1983. Mem. Inst. Medicine of Nat. Acad. Sci., Am. Soc. Clin. Investigation, Am. Assn. Cancer Research (pres. 1971-72), Assn. Am. Physicians. Subspecialties: Chemotherapy; Internal medicine. Office: Dana-Farber Cancer Inst 44 Binney St Boston MA 02115

FREIBERGER, WALTER FREDERICK, mathematics educator; b. Vienna, Austria, Feb. 20, 1924; came to U.S., 1955, naturalized, 1962; s. Felix and Irene (Tagany) F.; m. Christine Mildred Holmberg, Oct. 6, 1956; children: Christopher Allan, Andrew James, Nils Henry. B.A., U. Melbourne, 1947, M.A., 1949; Ph.D., U. Cambridge, Eng., 1953. Research officer Aero. Research Lab. Australian Dept. Supply, 1947-49, sr. sci. research officer, 1953-55; tutor U. Melbourne, 1947-49, 53-55; asst. prof. div. applied math. Brown U., 1956-58, asso. prof., 1958-64, prof., 1964—, dir. Computing Center, 1963-69; dir. Center for Computer and Info. Scis., 1969-76, chmn. div. applied math., 1976—; mem. fellowship selection panel NSF. Author: (with U. Grenander) A Short Course in Computational Probability and Statistics, 1971; Editor: The International Dictionary of Applied Mathematics, 1960, (with others) Applications of Digital Computers, 1963, Advances in Computers, Volume 10, 1970, Statistical Computer Performance Evaluation, 1972; Mng. editor: Quarterly of Applied Mathematics, 1965—; Contbr. numerous articles to profl. jours. Served with Australian Army, 1943-45. Fulbright fellow, 1955-56; Guggenheim fellow, 1962-63; NSF Office Naval Research grantee in field. Mem. Am. Math. Soc. (asso. editor Math. Reviews 1957-62), Soc. for Indsl. and Applied Math., Inst. Math. Statistics, Assn. Computing Machinery. Republican. Episcopalian. Club: Univ. (Providence). Subspecialties: Applied mathematics; Statistics. Current work: Computational probability and statistics, pattern theory and applications, computer science. Home: 24 Alumni Ave Providence RI 02906 Office: 182 George St Providence RI 02912

FREINKEL, NORBERT, physician, educator, researcher; b. Mannheim, Germany, Jan. 26, 1926; s. Adolf and Veronika (Kahn) F.; m. Ruth Kimmelstiel, June 19, 1955; children: Susan Elizabeth, Andrew Jonathan, Lisa Ann. A.B., Princeton U., 1947; M.D., NYU, 1949, Uppsala (Sweden) U., 1981. Postdoctoral tng. in medicine, endocrinology and metabolism Bellevue Hosp., N.Y.C., Boston City Hosp., Thorndike Meml. Lab., Harvard U., ARC Inst. Animal Physiology, Cambridge, Eng., 1949-56; from research fellow asst. prof. medicine Harvard Med. Sch. and Thorndike Meml. Lab., Boston City Hosp., 1952-66; chief metabolism div. Thorndike Meml. Lab., 1957-66; Kettering prof. medicine, chief sect. endocrinology, metabolism and nutrition, dir. Endocrine Clinics, Northwestern U. Med. Sch., 1966—, prof. biochemistry, 1969—; dir. Center for Endocrinology, Metabolism and Nutrition, 1973—; Mem. metabolism study sect. NIH, 1967-69, chmn. designate, 1970; mem. adv. com. on alcoholism NIMH, 1967-70; mem. subcom. on diabetes Fogarty Internat. Center, NIH, 1972—; mem. com. on renal and metabolic effects Space Flight Space Sci. Bd., Nat. Acad. Sci., 1973-74; cons. surg. gen. U.S. Army, 1962-79; mem. endocrinology and metabolism adv. com. Bur. Drugs, FDA, 1973-76, cons., 1976—; mem. career devel. com. VA, Washington, 1975-77; mem. sci. adv. com. Solomon A. Berson Fund for Med. Research, Inc., 1976—; mem. spl. study sect. DRTC NIAMDD, 1976-77; mem. nutrition coordinating com. NIH, 1978-80; dir. BioTechnica Internat. Inc., 1981—. Co-editor: Handbook of Physiology Series, Am. Phys. Soc.; Editorial bd.: Endocrinology, Jour. Developmental Physiology, Ann. Rev. Medicine, Jour. Clin. Investigation, Jour. Clin. Endocrinology, Jour. Lab. Clin. Medicine, Bull. Internat. Diabetes Fedn.; editor-in-chief: The Year in Metabolism, 1975-79, Contemporary Metabolism, 1979—; Contbr. articles to profl. jours., chpts. in textbooks. Served with USNR, 1943-45; Served with USNR AUS, 1950-52. Recipient Lilly award and medal Am. Diabetes Assn., 1966; Woodyatt award No. Ill. Diabetes Assn., 1976; Mosenthal award N.Y. Diabetes Assn., 1978; Banting Meml. medal Am. Diabetes Assn., 1978, 80; Joslin medal New Eng. Diabetes Assn., 1978; Kellion medal Australian Diabetes Assn., 1981; Am. Cancer Soc. fellow, 1953-55; Nat. Found. fellow, 1955-56. Fellow A.C.P., AAAS, Diabetes Assn. of India (hon.); mem. Assn. Am. Physicians, Am. Soc. Clin. Investigation (editorial com. 1971-76), Am. Physiol. Soc., World Med. Assn. (mem. med. bd. advisers 1975—), Endocrine Soc. (council 1969-72, chmn. meetings com. 1980—), Am. Thyroid Assn. (chmn. Van Meter award com. 1977-78), Am. Diabetes Assn. (dir. 1968-79, chmn. com. sci. programs 1971-75, v.p. profl. sect. 1975-76, pres. 1977-78, exec. com. 1975-79), Soc. Exptl. Biology and Medicine, Alpha Omega Alpha, Phi Beta Kappa, Sigma Xi; hon. mem. High Table, King's Coll., Cambridge, Eng. Subspecialty: Endocrinology. Current work: Intermediary metabolism regulation; fuel homeostasis in pregnancy and perinatal life; stimulus-secretion coupling for insulin and islet physiology; metabolic regulation in embryogenesis, organogenesis and teratogenesis. Home: 938 Edgemere Ct Evanston IL 60202 Office: Northwestern U Med Sch 303 E Chicago Ave Chicago IL 60611

FREIWALD, DAVID ALLEN, engineering consulting executive; b. Cleve., June 4, 1941; s. Harry Herman and Arline (Woehrman) F.; m. Joyce Darlene Gross, Apr. 3, 1976; children by previous marriage: Wesley, Todd, Christopher. B.S. in Mech. Engring, Northwestern U., 1963, Ph.D., 1968. Staff mem. Sandia Nat. Labs., Albuquerque, 1967-72; staff mem., mgmt. dir. Office of Dir., Los Alamos Nat. Labs., 1972-81; sr. scientist SEA, Inc., McLean, Va., 1981-82; cons. MRJ, Inc., Fairfax, Va., 1982—. Author: booklet Introduction to Laser Fusion, 1975. Mem. N.Mex. Gov.'s Land Use Legis. Com., 1971; N.Mex. rep. environ. com. Western Power Systems Coordinating Com., 1972-73; mem. Gov.'s Energy Task Force, 1973-74; adv. council State N.Mex. Energy Inst., 1978-81. Mem. Am. Phys. Soc., Internat. Assn. H-2 Energy, Am. Nuclear Soc., N.Mex. Acad. Sci. (pres. 1981), Tau Beta Pi, Pi Tau Sigma, Sigma Xi. Republican. Methodist. Subspecialties: Fuels and sources; Fluid mechanics. Current work: Secure, survivable, sustainable energy supplies/logistics for U.S. Armed Forces, including advanced concepts and alternative fuels for present and projected force structures for deterrence and war scenarios. Home: 10401 Lloyd Rd Potomac MD 20854 Office: MRJ Inc Suite 300 10400 Eaton Pl Fairfax VA 22030

FRELINGER, JEFFREY ALLEN, geneticist, educator; b. Bklyn., July 16, 1948; s. John Edgar and Alice (Andersen) F.; m. Joy Vaughan; 1

son, Jacob Jeffrey. B.A., U. Calif. at San Diego, 1969; Ph.D., Calif. Inst. Tech., 1973. Asst. prof. to prof. U. So. Calif., 1975-82; prof. microbiology U. N.C., 1983—; Mem. NIH Mammalian Genetics Study Sect. Contbr. numerous articles to profl. jours. James Coffin Childs fellow; NIH grantee. Mem. AAAS, Am. Assn. Immunologists, Sigma Xi. Subspecialties: Immunogenetics; Gene actions. Current work: Use of molecular biology and hybridoma tech. to understand genetic orgn. and function of the maj. histocompatability complex. Home: 1721 Allard Rd Chapel Hill NC 27514 Office: Dept Microbiology and Immunology U NC Chapel Hill NC 27514

FREMOUNT, HENRY NEIL, biologist, educator; b. Easton, Pa., Sept. 29, 1933; s. Henry Anthony and Lucille (DeRenzis) F.; m. Rosalyn Arlene Malone, Sept. 7, 1957. B.S., East Stroudsburg State Coll., 1956, M.Ed. in Biology, 1964; M.S. in Parasitology, Columbia U., 1966, D.P.H., 1970. Tchr. sci. jr. high sch. Delaware Valley Joint Sch. System, Milford, Pa., 1956-58; tchr. sci. Belvidere (N.J.) High Sch., 1958-65; assoc. prof. biology East Stroudsburg (Pa.) State Coll., 1966-71, prof., 1971—, chmn. dept., 1974-80; instr. parasitic diseases, lectr. histology Columbia U. Coll. Physicians and Surgeons, N.Y.C., 1965-71, NIH trainee, 1965-66, 68-70; NSF grantee Pa. State U., 1963-64; NIH trainee U. Mich. Biol. Sta., 1965; internat. fellow in tropical medicine La. State U. Faculty Medicine, 1970; vis. scientist Gorgas Meml. Lab., Panama, 1970; adj. assoc. prof. pathobiology Inst. for Pathobiology, Lehigh U., Bethlehem, Pa., 1971-78, sr. mem., 1973-78; assoc. sci. staff parasitic diseases Sacred Heart Hosp., Allentown, Pa., 1974—, mem. assoc. sci. staff dept. pathology, 1978—. Contbr. articles to sci. jours. Recipient cert. of excellence in teaching East Stroudsburg State Coll., 1977; NIH grantee, 1970; Commonwealth teaching fellow, 1977. Fellow Royal Soc. Tropical Medicine and Hygiene; mem. Am. Soc. Tropical Medicine and Hydiene, Am. Soc. Parasitologists, Am. Inst. Biol. Scis., AAAS, Am. Soc. Clin. Pathologist, N.Y. Soc. Tropical Medicine, N.Y. Zool. Soc., Pa. Acad. Sci. (exec. com., editorial com.), Sigma Xi. Roman Catholic. Subspecialties: Microscopy; Immunology (medicine). Current work: Ultrastructure and pathophysiology of malaria. Home: RD 4 Box 4441 Stroudsburg PA 18360 Office: Dept Biology East Stroudsburg Univ East Stroudsburg PA 18301

FREMOUW, EDWARD JOSEPH, ionospheric physicist, research exec.; b. Northfield, Minn., Feb. 23, 1934; s. Fred J. and Marion Elizabeth (Drozda) F.; m. Rita Lorraine Johnson, June 26, 1960; children: Thane Edrik, Sean Fredrik. B.S.E.E., Stanford U., 1957; M.S. in Physics, U. Alaska, Fairbanks, 1963; Ph.D. in Geophysics, U. Alaska, Fairbanks, 1966. Engr. Boeing Co., Seattle, 1957; chief auroral obs. U.S. Antarctic Research Program, South Pole, 1958-59; grad. teaching and research asst. Geophys. Inst., U. Alaska, Fairbanks, 1960-66, asst. prof. geophysics, 1966; physicist Stanford Research Inst., Menlo Park, Calif., 1967-70, sr. physicist, 1970-76, program mgr., 1976-77; v.p., dir. Phys. Dynamics, Inc., Bellevue, Wash., 1977—; assoc. La Jolla Inst.; conf. chmn., presenter. Contbr. articles prof. jours. Mem. student learning objectives com. Mercer Island Sch. Dist. Recipient Antarctic Service Medal, 1959. Mem. AAAS, Am. Geophys. Union, IEEE, Internat. Radio Sci. Union. Unitarian. Club: Seattle Mountaineers. Subspecialties: Aeronomy; Satellite studies. Current work: Morphology and dynamics of plasma-density irregularities in ionosphere and their effects on transionospheric communication systems. Patentee in field. Home: 8232 E Mercer Way Mercer Island WA 98040 Office: 300 120th Ave NE Bldg 7220 Bellevue WA 98005

FRENCH, A. JAMES, physician; b. Van Houten, N.Mex., Sept. 3, 1912; s. A. P. and Elizabeth (Williams) F.; m. Genevieve Fetter, July 19, 1937; 1 dau., Patricia Sue. A.B., U. Colo., 1933, M.A., 1936, M.D., 1936. Diplomate: Am. Bd. Pathology (trustee 1962-74, sec.-treas. 1964-74, exec. dir. 1974-78, cons. 1979-80). Intern Kansas City (Mo.) Gen. Hosp., 1936-37; resident pediatrics Children's Hosp., Denver, 1937-38; resident pathology St. Louis City Hosp., 1938-40; resident, instr. pathology U. Mich. Hosp., 1940-41, chief clin. labs., 1952—; asst. prof. pathology. Mich. Med. Sch., 1946-47, asso. prof., 1947-53, prof., 1953-80, chmn. dept., 1956-80; also editor Med. Bull., 1955-57; pathologist Surgeon Gen.'s Office, Washington and, Far East, 1941-46, cons., 1947-50; cons., mem. pathology adv. council VA Hosp., Ann Arbor, also; Wayne County Gen. Hosp., 1959—; Dir. Mich. Maternal Tissue Registry, 1957—; mem. sci. adv. bd. Armed Forces Inst. Pathology, 1965-70, chmn., 1968-70; mem. etiology com. Am. Cancer Soc., 1962-65. Contbr. to med. jours. Col. AUS Res. Fellow A.C.P.; mem. Mich. Pathol. Soc. (pres. 1953, 73), Internat. Acad. Pathology (council 1957-60, pres. 1966), Am. Soc. Clin. Pathology, Am. Assn. Pathologists and Bacteriologists (mem. council 1970, sec.-treas. 1971-74, pres. 1975), Coll. Am. Pathologists (chmn. acad. sect. 1960-61, gov. 1964-70, sec.-treas. 1969-70), Am. Acad. Oral Pathology (hon. mem.; vice chmn. 1974-76, chmn. 1976—), AMA (sec. sect. council on pathology 1972—), Frederick A. Coller Surg. Soc. (hon.). Subspecialty: Pathology (medicine). Home: 356 Ausable Pl Ann Arbor MI 48104 Office: 1335 E Catherine Dr Ann Arbor MI 48109

FRENCH, ALEXANDER MURDOCH, plant pathologist; b. New Bedford, Pa., Apr. 23, 1920; s. William McClell and May Huey (Murdoch) F.; m. Naomi Margaret Schoeller, June 14, 1947; children: Thomas Richard, Patricia French Grimm. B.S., Muskingum Coll., 1942; Ph.D., Cornell U., 1950. Phys. chemist Manhattan Project, N.Y.C., 1943-46; assoc. plant pathologist Calif. Dept. Agr., Riverside, 1950-59; program supr. Calif. Dept. Food and Agr., Sacramento, 1959-71, prin. plant pathologist, nematologist, 1971—. Mem. Am. Phytopath. Soc., Soc. Nematologists. Subspecialty: Plant pathology. Current work: Quarantine and regulatory work involving plant disease organisms.

FRENCH, CHARLES STACY, scientist; b. Lowell, Mass., Dec. 13, 1907; s. Charles Ephraim and Helena (Stacy) F.; m. Margaret Wendell Coolidge, Dec. 10, 1938; children—Helena Stacy, Charles Ephraim. Student, Loomis Inst., Windsor, Conn., 1921-25; B.S., Harvard, 1930, A.M., 1932, Ph.D., 1934; Ph.D. (hon.), U. Göteborg, Sweden, 1974. Asst. in gen. physiology Harvard, 1930-33; research fellow Calif. Inst. Tech., 1934-35; guest worker with Otto Warburg Kaiser Wilhelm Inst., Berlin-Dahlem, Germany, 1935-36; Austin teaching fellow in biochemistry Harvard Med. Sch., 1936-38; instr. (research) in chemistry with James Franck U. Chgo., 1938-41; asst. prof. dept. botany U. Minn., 1941-45, asso. prof., 1945-47; dir. div. plant biology Carnegie Instn. of Washington at Stanford U., 1947-73, dir. emeritus, 1973—; prof. (by courtesy) Stanford. Contbr. articles on plant physiology to sci. jours. Bd. dirs. Hidden Villa Inc., 1979—. Mem. Am. Soc. Plant Physiologists (chmn. Western Sect. 1954, Charles Ried Barnes life mem.), Bot. Soc. Am. (award of merit 1973), Nat. Acad. Scis., Am. Acad. Arts and Scis., Am. Soc. Biol. Chemists, Soc. Gen. Physiologists (v.p. 1954, pres. 1955-56), AAAS, Biophys. Soc., Deutsche Akademie der Naturforscher Leopoldina, Friends of Hidden Villa (pres. 1977-79). Clubs: Am. Alpine, Harvard of Peninsula (pres. 1973-75), Explorers.). Subspecialties: Photosynthesis; Plant physiology (agriculture). Current work: Spectroscopy of chlorophyll-protein complexes. Photo-oxidation by chloroplast components. Home: 11970 Rhus Ridge Rd Los Altos Hills CA 94022 Office: Carnegie Institution Stanford CA 94305

FRENCH, DAVID NICHOLS, metallurgist; b. Newton, Mass., Jan. 24, 1936; s. Sydney Perkins and Donalda (Roy) F.; m. Louise Murray French, June 25, 1960; children: Katherine, Andrew, Stephen, Jonathan. B.S., MIT, 1958, M.S., 1959, Sc.D., 1962. Research scientist Linde Co., Indpls., 1962-63; mem. tech. staff Ingersoll-Rand Co., Princeton, N.J., 1963-68; phys. metallurgist Abex Corp., Mahwah, N.J., 1968-72; mem. tech. staff P.R. Mallory Co., Burlington, Mass., 1972-73; dir. corp. quality assurance Riley Stoker Corp., Worcester, Mass., 1973-82; v.p. Leighton Industries, Inc., Phoenixville, Pa., 1982-83; chief metallurgist D.G. Peterson & Assocs., Greenfield, Mass., 1983—. Author: Metallurgical Failures in Fossil Fired Boilers, 1983; contbr. articles to profl. jours. Mem. ASME, Nat. Assn. Corrosion Engrs., Am. Soc. Metals, AIME, Sigma Xi. Subspecialties: Metallurgy; Corrosion. Current work: Boiler corrosion, metallurgy, failure analysis, metall. engring., boilers and fuel burning equipment, quality assurance. Home: 1 Lancaster Rd Northborough MA 01532 Office: 13 Newall Ct Greenfield MA 01302

FRENKEL, EUGENE PHILLIP, physician; b. Detroit, Aug. 27, 1929; s. David Eugene and Eva (Antin) F.; m. Rhoda Beth Smilay, Dec. 21, 1958; children: Lisa Michelle, Peter Alan. B.S., Wayne State U., 1949; M.D., U. Mich., 1953. Diplomate: Am. Bd. Internal Medicine (hematology, med. oncology; bd. govs. 1980—, chmn. subspecialty com. hematology 1980—). Intern Wayne County Gen. Hosp., Eloise, Mich., 1953-54; resident in internal medicine Boston City Hosp., 1954-55; resident in internal medicine, then instr. U. Mich. Med. Center, 1957-62; mem. faculty U. Tex. Health Sci. Center, Dallas, 1962—, prof. internal medicine and radiology, 1969—, chief div. hematology-oncology, 1962—; chief nuclear medicine, cons. hematology-oncology VA Med. Center, Dallas, 1962—; cons. com. evaluation research hematology Nat. Inst. Arthritis and Metabolic Diseases. Author numerous research papers in field. Served as officer M.C. USAF, 1955-57. Fellow A.C.P., Internat. Soc. Hematology; mem. Am. Soc. Hematology (treas. 1976—), Am. Soc. Clin. Oncology (chmn. membership com. 1982—), Am. Cancer Soc. (pres. Dallas unit 1970-71, dir. Tex. div. 1978—, sci. adv. com. on clin. investigations II—chemotherapy and hematology 1978—, Emma Freeman prof. 1981, nat. clin. fellowship com. 1978—), Assn. Am. Physicians, Am. Assn. Cancer Research, Am. Assn. Cancer Edn., Am. Soc. Biol. Chemists, Am. Soc. Clin. Investigation, So. Soc. Clin. Investigation, Soc. Nuclear Medicine, Am. Fedn. Clin. Research, Internat. Assn. Study Lung Cancer, Alpha Omega Alpha. Subspecialties: Hematology; Oncology. Office: 5323 Harry Hines Blvd Dallas TX 75235

FRESCO, JACQUES ROBERT, biochemist, educator; b. N.Y.C., May 30, 1928; s. Robert and Lucie (Asseo) F.; m. Rosalie Sarah Bernstein, Dec. 22, 1957; children—Lucille Deborah, Suzette Josie, Linda Hannah. B.A., N.Y. U., 1947, M.S., 1949, Ph.D., 1953; M.D. (hon.), U. Göteborg, Sweden, 1979. Postdoctoral fellow Sloan Kettering Inst. for Cancer Research, N.Y.C., 1952-54; instr. biochemistry N.Y. U. Coll. Medicine, 1953-54, instr. pharmacology, 1954-56; research fellow dept. chemistry Harvard, 1956-60, tutor biochem. scis., 1957-60; vis. fellow Cavendish Lab., Cambridge, Eng. and; Institut de Biologie Physico-Chimique, Paris, France, 1957; asst. prof. dept. chemistry Princeton, 1960-62, asso. prof., 1962-65, prof., 1965—, acting chmn. biochem. scis., 1965-66, prof. dept. biochem. scis., 1970—, chmn. dept., 1974-80; vis. prof. Hebrew U. of Jerusalem, 1973; dir. Nat. Cancer Inst. Basic Sci. Cancer Center, 1976—, Pfeiffer prof. life scis., 1977—; mem. adv. bd. Biopolymers, 1963-70; cons. sci. adv. com. Helen Hay Whitney Found.; vis. scientist MRC Lab. Molecular Biology, Cambridge, Eng., 1969-70. Mem. editorial bd.: Jour. Phys. Chemistry, 1963-70, Analytical Biochemistry, 1969-81. Recipient Am. Scientist Writing award AAAS, 1962; NIH fellow, 1952-54; Lalor Found. fellow, 1957; established investigator; Am. Heart Assn., 1958-63; Guggenheim fellow, 1969-70. Mem. Am. Chem. Soc., Am. Soc. Biol. Chemists, Sigma Xi. Subspecialties: Biochemistry (biology); Molecular biology. Current work: Nucleic acids; base pairing; mutagenesis; carcino genesis. Home: 282 Hartley Ave Princeton NJ 08540

FREUDENHEIM, MILTON B., journalist; b. New Rochelle, N.Y., Mar. 4, 1927; s. Milton Benjamin and Lenore Patricia (Kroh) F.; m. Elizabeth Ege, Mar. 7, 1951; children: Jo Louise, Susan Patricia, John Milton Otto, Tom Henry. A.B., U. Mich., 1948. Reporter Louisville (Ky.) Courier-Jour., 1948-49; reporter Akron (Ohio) Beacon Jour., 1949-52, Washington corr., 1953-56; UN corr. Chgo. Daily News, 1956-66, nat. and fgn. editor, 1966-69, Paris corr., 1969-77; dir. public affairs for Region V HEW, Chgo., 1978-79; copy editor, writer N.Y. Times Week in Rev., 1979—; adv. U.S. del. UNESCO Gen. Conf., 1978; Pres. UN Corrs. Assn., 1966, Anglo-Am. Press Assn., 1975. Mem. Phi Beta Kappa, Sigma Delta Chi. Subspecialty: Mechanical engineering. Current work: Mechanisms, kinematics, dynamics and mechanical design. Home: 123 W 74th St New York NY 10023 Office: 229 W 43d St New York NY 10036

FREUND, LAMBERT BEN, engineering educator, researcher, consultant; b. McHenry, Ill., Nov. 23, 1942; s. Bernard and Anita (Schaeffer) F.; m. Colleen Jean Hehl, Aug. 21, 1965; children: Jonathan Ben, Jeffrey Alan, Stephen Neil. B.S., U. Ill., 1964, M.S., 1965; Ph.D., Northwestern U., 1967. Postdoctoral fellow Brown U., Providence, 1967-69, asst. prof., 1969-73, assoc. prof., 1973-75, prof. engring., 1975—, chmn. div., 1979—; vis. prof. Stanford (Calif.) U., 1974-75; Cons. Aberdeen Proving Ground, U.S. Steel Corp. Editor-in-chief: ASME Jour. Applied Mechanics, 1983—; contbr. articles to tech. jours. NSF trainee, 1964-67; grantee NSF, Office Naval Research, Army Research Office, Nat. Bur. Standards. Fellow ASME (Henry Hess award 1974); mem. Am. Geophys. Union, Am. Acad. Mechanics. Subspecialties: Solid mechanics; Materials. Current work: Mechanics of solids and structures; fracture mechanics; stress waves in solids; theoretical seismology. Home: 3 Palisade Ln Barrington RI 02806 Office: Brown U Providence RI 02912

FREUND, ROBERT STANLEY, physicist; b. Newark, Jan. 26, 1939; s. Herman and Evelyn (Osterweil) F.; m. Joan Kesselman, June 17, 1962; children: Kevin, Wendy. B.A., Wesleyan U., 1960; M.A., Harvard U., 1962, Ph.D., 1965. Mem. tech. staff Bell Labs., Murray Hill, N.J., 1966-76, head dept. environ. chemistry, 1976-79; head dept. chem. kinetics research Bell Labs, 1979—; mem. com. atomic and molecular scis. NRC, 1976-82; mem. N.J. Gov's Sci. Adv. Com., 1979-83. Treas., N. Jersey chpt. Juvenile Diabetes Found., 1979-81, v.p., 1982—. Fellow Am. Phys. Soc.; mem. Am. Chem. Soc., Am. Vacuum Soc. Subspecialties: Atomic and molecular physics; Physical chemistry. Current work: Electron-molecule collision experiment, spectroscopy, high-Rydberg molecules. Office: AT&T Bell Labs Murray Hil NJ 07974

FREY, ELMER J(ACOB), systems engineering consultant; b. Buffalo, Jan. 3, 1918; s. Joseph and Nettie (Thieberge) F.; m. Barbara R. Dunnan, Sept. 20, 1945 (dec. 1962); 1 son, Eric D.; m. (m. 2d), Sept. 1, 1963; 1 dau., Maria. Student, Faculty Scis., U. Paris, 1935-36; B.S., CCNY, 1937; M.S., NYU, 1940; Ph.D., MIT, 1949. Lectr. dept. aeronautics and astronautics, assoc. dir. Instrumentation Lab., MIT, Cambridge, 1946-72; mgr. satellite systems engring. Fairchild Space and Electronics Co., Germantown, Md., 1972-75; cons. Frey Assoc., Inc., Amherst, N.H., 1975-79, Efrey Internat., Inc., Amherst, 1979—; vis. prof. Ecole Speciale de Travaux Aeronautiques, Paris, 1963-64; dir. ATAC (S.A.), Meudon, France, 1974-77; lectr. course U. Naples (Italy), U. Louvain (Belgium), Nat. Aero. Sch., Paris, 1969-70; cons. USAF Sci. Adv. Bd., 1962-65, Adv. Group Aero. Research and Devel. NATO, Paris, 1964, 69. Contbr. articles to profl. jours. and encys. Served to capt. U.S. Army, 1941-46. Naumburg scholar CCNY, 1935. Mem. Am. Geophys. Union, Soc. Indsl. and Applied Math., Phi Beta Kappa, Sigma Xi. Subspecialties: Systems engineering; Petroleum engineering. Current work: Development of well survey equipment, inertial instrumentation, control systems engineering, dynamic gravimetry, transportation safety studies, operations research. Developer various inertial guidance systems for USAF including Titan. Office: Efrey Internat Inc Chestnut Hill Rd Amherst NH 03031

FREY, HENRY EDWIN, JR., mechanical engineer; b. Tiffin, Ohio, Aug. 4, 1936; s. Henry Edwin and Mildred Louise (Gardner) F.; m. Carol Eileen Blair, Feb. 26, 1936; children: Bradly George, Pamela Sue. B.S., Case Inst. Tech., 1958. Lic. Profl. Engr., Ohio. With Nat. Machinery Co., Tiffin, Ohio, 1953—, group leader, 1960-79, chief engr. cold headers, 1979—. Mem. ASME. Mem. United Ch. Christ. Lodges: Masons; Shriners. Subspecialties: Mechanical engineering; Industrial engineering. Current work: Work in design of cold forming machinery responsible for machinery and tooling. Home: 88 Pearl St Tiffin OH 44883 Office: Nat Machinery Co Greenfield St Tiffin OH 44883

FREY, WILLIAM HOWARD II, research biochemist, research neurochemist; b. Atlanta, Nov. 19, 1947; s. William Howard and Brena (Feldman) F.; m. Barbara Katherine Hursh, Nov. 3, 1977; children: Brandl Laurie, William Howard III. B.A., Washington U., St. Louis, 1969; Ph.D., Case Western Res. U., 1974. Research specialist U. Minn., Mpls., 1974-77, asst. prof. psychiatry dept., 1977—; dir. Psychiatry Research Labs., St. Paul-Ramsey Med. Center, 1980—; adviser Geriatric Research, Edn. and Clin. Center, VA Med. Center, Mpls., 1980—. Grantee NIH, 1978, 82. Mem. Am. Soc. Biol. Chemists. Subspecialties: Neurochemistry; Behavioral psychology. Current work: Research in human brain biochemistry and the neurochemistry of Alzheimer's disease, research on human emotional crying and emotional tears. Home: 1100 W Montana Ave St Paul MN 55108 Office: Psychiatry Research Labs St Paul-Ramsey Med Center 640 Jackson St St Paul MN 55101

FRIAS, JAIME LUIS, medical eduator, pediatrician; b. Concepcion, Chile, Mar. 20, 1933; s. Humberto and Olga (Fernandez) F.; m. Jacqueline M. Steel, Apr. 8, 1961; children: Jaime A., Juan Pablo, Maria Josefina, Patricio Andres. M.D., U. Concepcion, 1959. Diplomate: Am. Bd. Pediatrics, Am. Bd. Med. Genetics. Asst. prof. pediatrics U. Concepcion, 1965-69, prof., 1969-70; asst. prof. pediatrics U. Fla., Gainesville, 1970-73, assoc. prof., 1973-76, prof., 1976—. Contbr. articles to profl. jours. Fellow Am. Acad. Pediatrics; mem. Am. Soc. Human Genetics, Teratology Soc., AAAS, Sigma Xi. Roman Catholic. Subspecialties: Pediatrics; Genetics and genetic engineering (medicine). Current work: Genetics and dysmorphology. Office: Dept Pediatrics U Fla Gainesville FL 32610

FRIBERG, EMIL EDWARDS, cons. engr.; b. Wichita Falls, Tex., Apr. 11, 1935; s. John Walter and Anne (Crumpton) F.; m. Jo Ann Rutta, Jan. 26, 1957; children: Emil Edwards, Vicki Lynn, Joe Alan. B.S. in Mech. Engring, U. Tex., 1958. Registered profl. engr., Tex., Okla., La., Miss. With Tex. Electric Service Co., Wichita Falls, 1958-64, engring. cons., Ft. Worth, 1964-69; prin. Friberg Alexander Maloney Gipson Weir, Inc., Ft. Worth, 1969—, pres., 1973—. Chmn. Energy Conservation Adv. Com., City of Ft. Worth, 1981. Served to 1st lt. C.E. U.S. Army, 1958-66. Recipient Young Engr. of Yr. award Ft. Worth chpt. Tex. Soc. Profl. Engrs., 1968; award of merit Cons. Engrs. Council Tex., 1976; award of excellence, 1977; Disting. Service award ASHRAE, 1979. Mem. Cons. Engrs. Council Tex. (pres.), ASHRAE (past dir.), Nat. Soc. Profl. Engrs., Tex. Soc. Profl. Engrs., ASME, AIME. Baptist. Clubs: Ft. Worth, Sierra. Lodge: Rotary. Subspecialties: Mechanical engineering; Electrical engineering. Current work: Energy conservation in building heating, air conditioning and lighting systems; management of counsulting engineering firm with emphasis on energy conservation. Home: 3406 Woodford Arlington TX 76013 Office: PO Box 2080 Fort Worth TX 76113

FRIDOVICH, IRWIN, biochemistry educator; b. N.Y.C., Aug. 2, 1929; s. Louis and Sylvia (Appelbaum) F.; m. Mollie Finkel; children: Sharon E., Judith L. B.S., CCNY, 1951; postgrad., Cornell U. Med. Coll., 1951-52; Ph.D., Duke U., 1955; hon. doctorate, U. Rene Descartes, Paris, 1980. Instr. biochemistry Duke U., Durham, N.C., 1956-58, assoc., 1958—; vis. research assoc. Harvard U., Cambridge, Mass., 1961-62; asst. prof. biochemistry Duke U., 1961-66, assoc. prof., 1966-71, prof., 1971-76, James B. Duke prof., 1976—; mem. study sect. Am. Cancer Soc., mem. adv. com. biochemistry and chem. carcinogenesis. Editorial bd.: Jour. Biol. Chemistry; contbr. articles to sci. jours. Recipient Founders' award Chem. Industry Inst. Toxicology, 1980; Herty award Ga. sect. Am. Chem. Soc., 1980; Research Career Devel. award NIH, 1959-69. Mem. Nat. Acad. Scis., Am. Acad. Arts and Scis., Am. Soc. Biol. Chemists (pres. 1982), N.C. Acad. Scis., Phi Beta Kappa, Sigma Xi. Subspecialty: Biochemistry (medicine). Current work: Enzymology, metabolism of oxygen, oxygen free radicals, superoxide dismutase. Home: 3517 Courtland Dr Durham NC 27707 Office: Duke U Med Center Durham NC 27710

FRIED, BERNARD, biology educator; b. N.Y.C., Aug. 17, 1933; s. Harry and Anna (Bergstein) F.; m. Janet Avery, Aug. 29, 1959 (div. 1967); 1 son, Neil; m. Grace J. Evans, Jan. 29, 1969; 1 stepson, David. A.B., N.Y. U., 1954; M.S., U. N.H., 1956; Ph.D., U. Conn., 1961; postdoctoral, Emory U., 1961-63. Asst. prof. biology Lafayette Coll., Easton, Pa., 1963-69, assoc. prof. biology, 1969-75, Kreider prof. biology, 1975—. Author: Thin Layer Chromatography, 1982, In Vitro Cultivation, 1978; contbr. articles to jours. Grantee NIH, 1981-84, NSF, 1981, Research Corps., 1977-81; recipient Darbaker prize in Microsc. Biology, 1966. Mem. Am. Soc. Parasitologists (exec. council 1981-82), Pa. Acad. Sci. (pres. 1970-72), N.J. Soc. Parasitologists (pres. 1971-72), Am. Inst. Biol. Sci., N.Y. Soc. Tropical Medicine, Helm. Soc. Washington, Sigma Xi. Subspecialties: Parasitology; Analytical chemistry. Current work: Biology, physiology and biochemistry of trematodes. Applications of thin layer chromatography to biology. Office: Lafayette Coll High St Easton PA 18042

FRIED, GEORGE HERBERT, biology educator; b. N.Y.C., Apr. 16, 1926; s. Alexander and Celia Sandra (Goldberg) F.; m. Lillian Huchital, Dec. 15, 1954; children: Susan, Heidi, Amy. A.B., Bklyn. Coll., 1947; M.S., U. Tenn., 1949, Ph.D., 1952. Postdoctoral assoc. NYU, 1953-54; adj. prof. biology, 1961—; research assoc. Beth Israel Med. Ctr., N.Y.C., 1954-65; asst. prof. biology Bklyn. Coll., CUNY, 1963-67, assoc. prof., 1967-70, prof., 1970—, chmn. dept., 1983; cons. N.Y. U. Dental Coll. Contbr. numerous articles to sci. jours. NIH Research grantee, 1963-66; NSF Edn. grantee, 1970-71; Sci. and Disabled grantee, 1979-82. Fellow AAAS; mem. Am. Phys. Soc., Soc. Exptl. Biology and Medicine, Am. Soc. Zoologists. Democrat. Jewish. Subspecialties: Physiology (biology); Nutrition (medicine). Current work: Obesity, comparative metabolism, enzymes, oxygen consumption. Home: 2985 Bedford Ave Brooklyn NY 11210 Office: Biology Dep Brooklyn Coll Brooklyn NY 11210

FRIED, JERROLD, biomedical researcher; b. N.Y.C., Mar. 3, 1937; s. Max and Irene (Shapiro) F.; m. Shirley M. Rozen, Dec. 26, 1965; children: Daniel Alan, Andrew Mark. B.S., Calif. Inst. Tech., 1958; M.S., Stanford U., 1960, Ph.D., 1964. Postdoctoral fellow Hunter Coll., N.Y.C., 1965; research assoc. Sloan-Kettering Inst., N.Y.C.,

1965-67, assoc., 1968-76, asso. mem., 1976—; asst. prof. biophysics Cornell U. Grad. Sch. Med. Scis., Sloan-Kettering div., 1968-78, assoc. prof., 1978—. Nat. Cancer Inst. research grantee, 1974-82. Mem. Biophys. Soc., Am. Assn. Cancer Research, Cell Kinetics Soc., Soc. Analytical Cytology, Sigma Xi. Democrat. Jewish. Subspecialties: Cell and tissue culture; Information systems, storage, and retrieval (computer science). Current work: Cell kinetics, effects of cytotoxic drugs on cell survival and kinetics, application to leukemia, flow cytometry, computer simulation models. Home: 11 Winslow Pl Larchmont NY 10538 Office: 1275 York Ave New York NY 10021

FRIED, JOSEF, chemist, educator; b. Przemysl, Poland, July 21, 1914; came to U.S., 1938, naturalized, 1944; s. Abraham and Frieda (Fried) F.; m. Erna Werner, Sept. 18, 1939; 1 dau., Carol Frances. Student, U. Leipzig, 1934-37, U. Zurich, 1937-38; Ph.D., Columbia U., 1941. Eli Lilly fellow Columbia U., 1941-43; research chemist Givaudan, N.Y., 1943; head dept. antibiotics and steroids Squibb Inst. Med. Research, New Brunswick, N.J., 1944-59, dir. sect. organic chemistry, 1959-63; prof. chemistry, biochemistry and Ben May Lab. Cancer Research, U. Chgo., 1963—, Louis Block prof., 1973—, chmn. dept. chemistry, 1977-79; mem. med. chem. study sect. NIH, 1963-67, 68-72, chmn., 1971; mem. com. arrangements Laurentian Hormone Conf., 1964-71; Knapp Meml. lectr. U. Wis., 1958. Mem. bd. editors: Jour. Organic Chemistry, 1964-69, Steroids, 1966—, Jour. Biol. Chemistry, 1975-81, 83—; contbr. articles to profl. jours. Recipient N.J. Patent award, 1968. Fellow AAAS, N.Y. Acad. Scis.; mem. Am. Chem. Soc. (award in medicinal chemistry 1974), Nat. Acad. Scis., Am. Acad. Arts and Scis., Am. Soc. Biol. Chemists, Brit., Swiss chem. socs., Sigma Xi. Subspecialties: Organic chemistry; Biophysical chemistry. Patentee in field. Home: 5715 S Kenwood Ave Chicago IL 60637

FRIED, PETER MARC, physicist; b. N.Y.C., Apr. 1, 1948; s. Ernest and Eleanor (Lustig) F.; m. Wendy Ann Zisfein, Aug. 26, 1971; children: Lisa Jean, David Michael, Stephen Ernest. A.B., Cornell U., 1969; M.S., U. Wis., Madison, 1971, Ph.D., 1978. Mem. tech. staff Bell Telephone Labs., Whippany, N.J., 1978—. Served to lt. USNR, 1971-74. Mem. Am. Astron. Soc., Sigma Xi. Subspecialties: Acoustics; 1-ray high energy astrophysics. Current work: Exptl. studies in underwater acoustics, diffuse X-ray emission in interstellar medium. Office: Bell Telephone Labs Whippany NJ 07981

FRIED, VOJTECH, chemist, educator; b. Czechoslovakia, Aug. 27, 1921; came to U.S., 1965, naturalized, 1970; s. Maurice and Helene (Lefkovits) F.; m. Katarina Altmann, July 10, 1951; 1 son, Thomas. Student, Charles U., Prague, Czechoslovakia; M.A. in Chem. Engring. with honors, Chem. Tech. Inst., Prague, 1951, Dr.T.Sci., 1953, Dr. Chem. Sci., 1957, Dr. Phys. Chem., 1963. Instr. Chem. Tech. U., Prague, 1950-53, successively asst. prof., assoc. prof., dozent, 1953-65; prof. chemistry Bklyn. Coll. of CUNY, 1965—, Disting. prof. chemistry, 1974—; disting. vis. prof. Arya Mehr U. Tech., Tehran, Iran, 1974; M. Jorizane chair vis. prof. U. Hiroshima, Japan, 1975. Author articles and books in field. Recipient Excellence in teaching awards Chem. Tech. Inst., 1959, Bklyn. Coll., 1974; Czechoslovak State Prize in Sci., 1963; grantee Petroleum Research Fund, Iranian Govt.; others. Fellow Japan Soc. Promotion Sci.; mem. Am. Chem. Soc., Czechoslovak Chem. Soc. Lodge: B'nai B'rith. Subspecialties: Physical chemistry; Statistical mechanics. Current work: Equations of state; theories of liquid solutions experimental and theoretical studies of PVT behavior of gases and the behavior of nonelectrolyte solutions. Patentee. Home: 1655 Flatbush Ave Apt C-1602 Brooklyn NY 11210 Office: Bedford Ave and Ave H Brooklyn NY 11210

FRIEDBERG, ARTHUR LEROY, educator, ceramic engr.; b. River Forest, Ill., Mar. 25, 1919; s. Oscar and Fannie (Blumenthal) F.; m. Marian Davis, Feb. 4, 1944; children—Richard Charles, Anne. B.S. in Ceramic Engring, U. Ill., 1941, M.S., 1947, Ph.D., 1952; postgrad., U. Chgo., 1943-44. Mem. faculty U. Ill., Champaign-Urbana, 1946-79, prof. ceramic engring., 1957-79, prof. emeritus, 1979—, head dept., 1963-79; adj. prof. ceramic engring. Ohio State U., 1979—. Trustee Edward Orton, Jr. Ceramic Found., 1979—. Served to lt. (s.g.) USNR, 1943-46. Mem. Am. Ceramic Soc. (exec. dir. 1979—), Nat. Inst. Ceramic Engrs. (exec. dir. 1979—). Subspecialties: Ceramic engineering; Materials (engineering). Home: 1375 Kingsgate Rd Columbus OH 43221

FRIEDBERG, CARL E., computer systems company executive; b. Pitts., May 2, 1942. B.A., Harvard U., 1964; M.A., Princeton U., 1969, Ph.D., 1969. Instr. Princeton (N.J.) U., 1969-71; research fellow U. Calif., Lawrence Berkeley Lab., 1971-76; asst. prof. Harvard U., Boston, 1976-79; asst. physicist Meml. Sloan Kettering Cancer Center, N.Y.C., 1979-81; pres. In House Systems, N.Y.C., 1979—; chmn. VAX Systems, Maynard, Mass., 1980; mem. steering com. N.Y. Met. Local Users Group, N.Y.C., 1979. Mem. Am. Assn. Physicists Medicine, Am. Inst. Physics, Assn. Computing Machinery, N.Y. Acad. Scis., Sigma Xi. Subspecialties: Operating systems; Particle physics. Current work: Systems studies, integration, automated software development systems. Office: In House Systems 165 William St New York NY 10038

FRIEDBERG, ERROL CLIVE, molecular biologist; b. Johannesburg, South Africa, Oct. 2, 1937; came to U.S., 1965, naturalized, 1974; s. Edward and Rena F.; m.; children: Malcolm Bradley, Andrew Seth. B.Sc., U. Witwatersrand, South Africa, 1958, M.B., 1961, B.Ch., 1961. Intern King Edward VIII Hosp., Durban, South Africa, 1961; resident in pathology Cleve. Met. Gen. Hosp., 1965; postdoctoral fellow in biochemistry Case Western Reserve U., 1966-68; maj. Walter Reed Army Inst. Research, 1969-71; asso. prof. pathology Stanford U., 1971—, dir. program in cancer biology, 1983—. Contbr. numerous articles to profl. jours. Served to maj. AUS, 1969-71. USPHS grantee, 1971—; DOE grantee, 1971—; Josiah Macy Jr. Faculty Scholar award, 1978-79. Mem. Am. Soc. Microbiologists, Biophys. Soc., Am. Soc. Pathologists. Subspecialty: Molecular biology. Current work: DNA damage and repair.

FRIEDENSON, BERNARD ALLEN, biological chemist, educator; b. Duluth, Minn., July 6, 1943; s. Fred and Sarah (Berger) F.; m. Louise Pinsky, Nov. 10, 1968; children: Rachel, Jennifer, Joel. B.A., U. Minn., 1965; Ph.D., 1970. Cancer research scientist Roswell Park Meml. Inst., Buffalo, 1971-72, sr. cancer research scientist, 1973; asst. prof. biol. chemistry U. Ill. Med. Center, Chgo., 1973-78, assoc. prof., 1978—. Contbr. articles to profl. jours. USPHS grantee. Mem. Am. Assn. Immunologists. Subspecialties: Immunology (medicine); Biochemistry (medicine). Current work: Study of mechanisms of regulation of the immune system. Home: 9348 Home Circle Des Plaines IL 60016 Office: 1853 W Polk Chicago IL 60612

FRIEDHOFF, ARNOLD, medical scientist; b. Johnstown, Pa., Dec. 26, 1923; s. Abraham M. and Stella (Beerman) F.; m. Frances Wolfe, Feb. 24, 1946; children: Lawrence, Nancy, Richard. B.A., U. Pa., 1944, M.D., 1947. Diplomate: Am. Bd. Psychiatry and Neurology. Intern Western Pa. Hosp., 1947-48; resident psychiatry U.S. Army, 1952-53, Bellevue Hosp., N.Y.C., 1953-55; instr., prof. psychiatry N.Y. U. Sch. Medicine, 1956—, head psychopharmacology research unit, 1956-63, co-dir., 1963-69, dir., 1970—, 1970—; mem. clin. projects research rev. com. NIMH, 1970-74, chmn. treatment devel. and assessment research rev. com., 1979-81; Mem. Mayors Com. on Prescription Drugs, N.Y.C. Co-editor: Yearbook of Psychiatry and Applied Mental Health, 1968-80; mem. adv. bd.: Biological Psychiatry, 1969—; Contbr. numerous reports on biochem. psychiatry, psychopharmacology. Served to 1st lt. M.C. U.S. Army, 1951-53. Recipient Research Scientist award NIMH, 1967—. Fellow Am. Coll. Neuropsychopharmacology (past councillor and past pres. 1978-79), Am. Psychiat. Assn., Am. Soc. Clin. Pharmacology and Therapeutics, Royal Coll. Psychiatrists (Gt. Britain); mem. Am. Chem. Soc., Internat. Soc. Neurochemistry, Assn. for Research in Nervous and Mental Diseases (past asst. sec.-treas.), Am. Psychopath. Assn. (past pres., Samuel B. Hamilton award), Soc. Biol. Psychiatry (past pres.). Subspecialties: Neurobiology; Neuropharmacology. Current work: Neuropharmacology, psychobiology, biological substrate of behavior, particularly adaptive biological processes in the brain. Office: 550 1st Ave New York NY 10016

FRIEDLAND, AARON J., chemical engineer; b. Spring Lake, N.J., Dec. 1, 1929; s. Moses and Eva (Naidich) F.; m. Sandra M. Gothardt, Jan. 30, 1960; children: Gregory J., Andrew J. B.Ch.E., CCNY, 1951; M.S., Columbia U., 1954; Dr.Engring. Sci., 1960. Chem. engr. Picatinny Arsenal, Dover, N.J., 1951-52, Brookhaven Nat. Lab., Upton, N.Y., 1952-53; head engring. analysis Atomic Power Devel. Assocs., Detroit, 1960-73; fellow engr. Advanced Reactors div. Westinghouse Electric Corp., Waltz Mill, Pa., 1973-81, Monroeville, Pa., 1981—. Contbg. author, editor: Fast Reactor Technology, 1966; contbr. articles to profl. jours. Served to cpl. U.S. Army, 1954-56. Mem. ASME, Am. Nuclear Soc. Subspecialty: Nuclear engineering. Current work: Thermal and hydraulic design and analysis of nuclear power reactors. Inventor fuel subassembly for nuclear reactor, 1972. Home: 107 Briar Crest Dr Monroeville PA 15146 Office: Westinghouse Electric Corp NFD PO Box 3912 Pittsburgh PA 15230

FRIEDLANDER, GERHART, chemist; b. Munich, Germany, July 28, 1916; came to U.S., 1936, naturalized, 1943; s. Max O. and Bella (Forchheimer) F.; m. Gertrude Maas, Feb. 6, 1941 (dec. 1966); children: Ruth Ann F. Huart, Joan Claire F. Hurley; m. Barbara Strongin, 1983. B.S., U. Calif. - Berkeley, 1939, Ph.D., 1942. Instr. U. Idaho, Moscow, 1942-43; staff Los Alamos Sci. Lab., 1943-46; research asso. Gen. Electric Co. Research Lab., Schenectady, 1946-48; vis. lectr. Washington U., St. Louis, 1948; chemist Brookhaven Nat. Lab., Upton, N.Y., 1948-52, sr. chemist, 1952-81, chmn. chemistry dept., 1968-77; Chmn. Gordon Research Conf. on Nuclear Chemistry, 1954; mem. adv. com. for chemistry Oak Ridge Nat. Lab., 1966-70; mem. program adv. com. Los Alamos Meson Physics Facility, 1971-75; chmn. vis. com. nuclear chemistry Lawrence Berkeley Lab., 1974-75; exec. sec. basic energy scis. lab. program panel Dept. Energy, 1976-80; mem. adv. com. for nuclear chemistry div. Lawrence Livermore Lab., 1977—. Author: (with J.W. Kennedy) Introduction to Radiochemistry, 1949, Nuclear and Radiochemistry, 1955, (with J.M. Miller) Nuclear and Radiochemistry, 1964, (with E.S. Macias) Nuclear and Radiochemistry, 1981; also articles.; assoc. editor: Ann. Rev. Nuclear Sci., 1958-67; editor: Radiochimica Acta, 1972-73. Recipient Alexander von Humboldt award Institut für Kernchemie, Mainz, W. Ger., 1978-79. Fellow Am. Phys. Soc.; mem. Nat. Acad. Scis. (mem. assembly math. and phys. scis. 1981—, mem. commn. phys. scis., math. and resources 1982—, chmn. ad hoc panel on future nuclear sci. 1975-76, chmn. com. on recommendations for U.S. Army basic sci. research Nat. Acad. Scis.-NRC 1977-81, mem. com. on postdoctoral and doctoral research staff NRC 1977-80), Am. Acad. Arts and Scis., Am. Chem. Soc. (chmn. div. nuclear chemistry and tech. 1967, award for nuclear applications in chemistry 1967), AAAS. Subspecialties: Physical chemistry; Nuclear chemistry. Current work: Solar neutrino detection; interaction of high-energy particles with complex nuclei. Research chem. effects of nuclear transformations, properties of radioactive isotopes, mechanisms of nuclear reactions, especially those induced by protons of very high energies, solar neutrino detection. Home: 5 Lorraine Ct Smithtown NY 11787 Office: Brookhaven Nat Lab Upton NY 11973

FRIEDLANDER, SHELDON KAY, educator; b. Bronx, N.Y., Nov. 17, 1927; s. Irving and Rose (Katzewitz) F.; m. Marjorie Ellen Robbins, Aug. 17, 1958; children—Eva Kay, Amelie Elise, Antonia Zoe, Josiah. B.S., Columbia, 1949; M.S., Mass. Inst. Tech., 1951; Ph.D., U. Ill., 1954. Asst. prof. chem. engring. Columbia, 1954-57; asst. prof. then asso. prof. Johns Hopkins, 1957-62, prof. chem. engring., 1962-64; prof. chem. engring. and environ. engring. Calif. Inst. Tech., 1964-78; prof. engring. and applied sci., vice-chmn. dept. chem. engring. UCLA, 1978—; dir. Intermedia Transport Research Center, 1980—; mem. environ. sci. and engring. study sect. B USPHS, 1965-68; chmn. panel on particulate emissions NRC-Nat. Acad. Engring., 1970—; chmn. panel on photochem. oxidants and ozone NRC, 1973-76; cons. Los Angeles Air Pollution Control Dist., 1973-75; mem. environ. studies bd. NRC, 1977-80; chmn. clean air sci. adv. com. EPA, 1978—. Served with AUS, 1946-47. Fulbright scholar, 1960-61; Guggenheim fellow, 1969-70. Mem. Nat. Acad. Engring., Am. Inst. Chem. Engrs. (Colburn award 1959, Alpha Chi Sigma award 1974, Walker award 1979), Am. Chem. Soc., Sigma Xi, Tau Beta Pi. Subspecialties: Environmental engineering; Chemical engineering. Originator theory self-preserving size distbns., aerosol dynamics, filtration theory, turbulent deposition, relation of air quality to emission sources for particulate pollution, facilitated transport

FRIEDLANDER, SUSAN JEAN, mathematics educator, researcher; b. London, Jan. 26, 1946; U.S., 1967; d. Richard McLaughlin and Kate Poate; m. Eric M. Friedlander, July 20, 1968. B.Sc., London U., 1967; M.S., MIT, 1970; Ph.D., Princeton U., 1972. Vis. mem. Courant Inst. NYU., 1972-74; instr. Princeton U., 1974-75; asst. prof. U. Ill., Chgo., 1975-82, assoc. prof., 1982—; cons. Goddard Inst. Space Studies, N.Y., 1974-75. Author: Introduction to the Mathematical Theory of Geophysical Fluid Dynamics, 1980. Vis. fellow Sonderforschungsbereich, Bonn, Germany, 1983; vis. lectr. Oxford (Eng.) U., 1977; assoc. prof. U. Paris, 1982. Mem. Am. Math. Soc. (council 1983—), Soc. Indsl. and Applied Math., Assn. Women Math. Subspecialties: Applied mathematics; Fluid mechanics. Current work: Applications of mathematical techniques to problems arising from geophysical fluid dynamics. Home: 107 Garrison Ave Wilmette IL 60091 Office: Dept Math U Ill Chicago IL 60680

FRIEDLER, GLADYS, developmental psychopharmacologist; b. Lewiston, Maine, Sept. 7, 1926; s. Max Herman and Anna Diana (Feld) F. B.A., U. Maine; M.A., U. Pa.; Ph.D., Boston U. NIH postdoctoral fellow U. Calif. San Francisco Med. Ctr.; research fellow Harvard Med. Sch., 1970-72; research fellow anesthesiology Lying-In div. Boston Hosp. for Women, 1970-72; asst. prof. psychiatry and pharmacology Boston U. Med. Sch., 1973-81, assoc. prof., 1981—. Contbr. articles to profl. jours. NIH fellow, 1964-68; NIMH trainee, 1968-70; Nat. Inst. Drug Abuse grantee, 1975-82. Mem. Behavioral Teratology Soc., Internat. Soc. Devel. Psychobiology, Am. Soc. Pharmacology and Exptl. Therapeutics, AAAS, Phi Beta Kappa, Phi Kappa Phi. Current work: Long term and cross generational alterations in growth, behavioral and neuroendocrine development of progeny which follow parental exposure to pharmacologic agents and other environmental contaminants; exploration of comparative roles of genetic and environmental factors in etiology of the imprint. Office: 80 E Concord St Boston MA 02118

FRIEDMAN, DAVID PAUL, neuroscientist, educator; b. Bklyn., May 17, 1946; s. Herbert Meyer and Sylvia (Cohen) F.; m. Susan Barbara Gelber, June 25, 1972. B.S., U. Pitts., 1968; M.S., N.Y. Med. Coll., 1977, Ph.D., 1978. Instr. physiology N.Y. Med. Coll., Valhalla, 1975-77; postdoctoral fellow Washington U. Sch. Medicine, St. Louis, 1977-80; staff fellow NIMH, Bethesda, Md., 1980-82, sr. staff fellow, 1982—. Contbr. articles to profl. jours. Recipient Nat. Research Service award USPHS, 1977-80. Mem. AAAS, Soc. for Neurosci. Subspecialties: Comparative neurobiology; Neuroanatomy. Current work: Mapping of brain regions devoted to specific functions. Localization of neurotransmitters. Functional neuroanatomy, receptor localization. Office: Lab Neuropsychology NIMH Bethesda MD 20205

FRIEDMAN, EDWARD ALAN, college dean, educator; b. Bayonne, N.J., Sept. 29, 1935; s. Philip Arthur and Esther (Weinstein) F.; m. Arline Joan Lederman, Jan. 13, 1963; children: Millard Timur. B.S., MIT, 1957; postgrad., Stanford U., 1957-58; Ph.D., Columbia U., 1963. Philip Kerim. Asst. prof., asso. prof. physics Stevens Inst. Tech., Hoboken, N.J., 1963-69, dean of coll., 1973—, prof., 1980—, chmn. computer tng. com., 1982; v.p. Mentor Systems, Inc., 1980—; vis. prof. Kabul U., 1965-67, dir. engring. coll. devel. program, Afghanistan, 1970-73; Cons. Hudson Inst., 1962-63, Doubleday Book Co., 1969; Chmn. Civic Affairs Com., Hoboken, 1968-69; vice chmn. bd. edn. Am. Internat. Sch. of Kabul, 1971-72; chmn. bd. dirs. N.Y. Scientists Com. for Pub. Info., 1977-78; chmn. Council for Understanding of Tech. in Human Affairs, 1979-82; sr. v.p. Afghanistan Relief Com., 1980—; mem. N.J. Commn. on Grad. Tchr. Edn., 1982-83. Co-founder, co-editor: Machine-Mediated Learning Jour., 1983. Trustee Hudson Higher Edn. Consortium, 1974; bd. dirs. Assn. Ind. Colls. and Univs. N.J., 1978-82. Recipient (with R.D. Andrews) Ottens research award Stevens Inst. Tech., 1970, 1st Class Edn. medal Govt. of Afghanistan, 1973; NSF grantee to direct elementary sch. sci. program for Orange, N.J., 01969. Mem. Am. Phys. Soc., AAAS, Am. Assn. Engring. Edn., Computer Soc., Am. Soc. Tng. and Devel. Subspecialty: Technological Education. Home: Colonial House Castle Point Hoboken NJ 07030

FRIEDMAN, GARY DAVID, epidemiologist; b. Cleve., Mar. 8, 1934; s. Howard N. and Cema (Cort) F.; m. Ruth Helen Schleien, June 22, 1958; children: Emily, Justin, Richard. Student, Antioch Coll., 1951-53; B.S., U. Chgo., 1956, M.D., 1959; S.M. in Hygiene, Harvard U., 1965. Diplomate: Am. Bd. Internal Medicine. Intern, resident in internal medicine Harvard Med. Services, Boston City Hosp., 1959-61; resident in medicine U. Hosps. of Cleve., 1961-62; research fellow Harvard Med. Sch., 1962-64, research assoc. in preventive medicine, 1964-66; commd. sr. asst. surgeon USPHS, 1962, advanced through grades to sr. surgeon; med. officer heart disease epidemiology study, Framingham, Mass., 1962-66; chief epidemiology unit, epidemiology field and tng. sta. Heart Disease Control Program, San Francisco, 1966-68, resigned, 1968; sr. epidemiologist dept. med. methods research Kaiser-Permanente Med. Care Program, Oakland, Calif., 1968-76, asst. dir. for epidemiology and biosts., dept. med. methods research, 1976—; asst. clin. prof. dept. family and community medicine U. Calif.-San Francisco, 1967-75, assoc. clin. prof., 1975—; lectr. epidemiology U. Calif.-Berkeley, 1968—; mem. com. on epidemiology and vets. follow-up studies NRC; mem. epidemiology and disease control study sect. NIH. Author: Primer of Epidemiology, 1974, 2d edit., 1980. Fellow Council on Epidemiology Am. Heart Assn., ACP; mem. Am. Epidemiol. Soc., Soc. for Epidemiologic Research, Am. Pub. Health Assn., Internat. Epidemiol. Assn., Am. Soc. for Preventive Oncology, Internat. Soc. for Twin Studies, Phi Beta Kappa, Alpha Omega Alpha, Delta Omega. Jewish. Club: Vintage Chevrolet Am. Subspecialties: Epidemiology; Internal medicine. Current work: Epidemiology of Chronic diseases and effects of smoking, alcohol and medicinal drugs; epidemiologic research. Office: Kaiser-Permanente Medical Care Program 3451 Piedmont Ave Oakland CA 94611

FRIEDMAN, GERALD MANFRED, geology educator, geologist; b. Berlin, July 23, 1921; U.S., 1946, naturalized, 1950; s. Martin and Frieda (Cohn) F.; m. Sue Tyler, June 27, 1948; children: Judith Fay Friedman Rosen, Sharon Mira Friedman Azaria, Devorah Paula Friedman Zweibach, Eva Jane Friedman Scholle, Wendy Tamar. Student, U. Cambridge, Eng., 1938-39; B.Sc., U. London, 1945, D.Sc., 1977; M.A., Columbia U., 1950, Ph.D., 1952. Lectr. Chelsea Coll., London, 1944-45; analytical chemist E. R. Squibb & Sons, New Brunswick, N.J.; also J. Lyons & Co., London, 1945-48; asst. geology Columbia U., 1950; temporary geologist N.Y. State Geol. Survey, 1950; instr., asst. prof. geology U. Cin., 1950-54; cons. geologist, Sault Ste. Marie, Ont., Can., 1954-56; mem. research dept. Amoco Prodn. Co., Tulsa, 1956-64, sr. research scientist, 1956-60, research assoc., 1960-62, supr. sedimentology research, 1962-64; Fulbright vis. prof. geology Hebrew U., Jerusalem, 1964; prof. geology Rensselaer Poly. Inst., Troy, N.Y., 1964—, adv. Judo Club, 1964—; research scientist Hudson Labs., Columbia U., 1965, 66-69, research assoc. dept. geology, 1968-72; vis. prof. U. Heidelberg, Germany, 1967; cons. scientist Inst. Petroleum Research and Geophysics, Israel, 1967-71; vis. scientist Geol. Survey Israel, 1970, 78. Author: (with J.E. Sanders) Principles of Sedimentology, 1978, Exploration for Carbonate Petroleum Reservoirs, 1982, (with K. G. Johnson) Exercises in Sedimentology; editor: Jour. Sedimentary Petrology, 1964-70 (Best Paper award 1961), Northeastern Geology, 1979—, Earth Scis. History, 1982—; pub.: Northeastern Environ. Sci, 1982—; sect. editor: Chem. Abstracts, 1962-69; editorial bd.: Sedimentary Geology, 1967—, Israel Jour. Earth Scis, 1971-76, Jour. Geology, 1977—, GeoJour, 1977—; co-editor, contbr.: Carbonate Sedimentology in Central Europe, 1968, Modern Carbonate Enviroments, 1983; editor, contbr.: Depositional Environments in Carbonate Rocks, 1969; contbr. articles in field to profl. jours. Mem. phys. edn. com. Tulsa YMCA, 1958-63; bd. dirs. Troy Jewish Community Council, 1966-72, 74-77; pres. Northeastern Sci. Found., 1979—. Fellow Mineral. Soc. Am. (nominating com. for fellows 1967-69, awards com. 1976-77), Geol. Soc. Am. (mem. com. on publs. 1980-82), Geol. Assn. Can., Geol. Soc. Gt. Britain, AAAS (chmn. sect. geology and geography 1978-79, council mem. 1979-80); mem. Am. Chem. Soc. (group leader 1962-63), Am. Assn. Petroleum Geologists (disting. lectr. 1972-73, chmn. carbonate rock com. 1965-69, mem. research com. 1965-71, 75-82, lectr. continuing edn. program 1967—, adv. council 1974-75, disting. lectr. com. 1975-78, membership com. 1981—, ho. of dels. 1977-79, treas. eastern sect. 1980-81, v.p. eastern sect. 1981-82, pres. 1982-83), Soc. Econ. Paleontologists and Mineralogists (nat. pres.

1974-75, sect. pres. 1967-68, Best Paper award Gulf Coast sect. 1974), Assn. Geology Tchrs. (pres. Okla. 1962-63, pres. Eastern sect. 1983—), Am. Geol. Inst. (governing bd. 1971-72, 74-75), Assn. Earth Sci. Editors (pres. 1971-72), Internat. Assn. Sedimentologists (hon., pres. 1975-78), N.Y. State Geol. Assn. (pres. 1978-79), Geol. Soc. Israel, Geol. Vereinigung, U.S. Judo Fedn., Sigma Gamma Epsilon (nat. pres. 1981—). Subspecialty: Sedimentology. Current work: Research in carbonate and clastic sedimentology, especially depositional environments and diagenetic overprint. Home: 32 24th St Troy NY 12180 Office: Rensselaer Poly Inst Troy NY 12181 also Office: Northeastern Sci Found Rensselaer Ctr Applied Geology PO Box746 Troy NY 12181

FRIEDMAN, HAROLD IRA, surgeon; b. N.Y.C., Oct. 22, 1946; s. Joseph and Dorothy (Asnin) F.; m. Clarke Emmons, Nov. 24, 1976. B.Sc., Hobart Coll., 1967; Ph.D., U. Va., 1972, M.D., 1974. Diplomate: Nat. Bd. Med. Examiners. Intern surgery U. Va. Med. Center, Charlottesville, 1974-75; cell biologist gen. surgeon Letterman Army Inst. Research, San Francisco, 1975-78; resident gen. surgery U. Ariz. Med. Center, Tucson, 1978-82; resident in plastic surgery U. Va. Med. Center, Charlottesville, 1982—. Contbr. articles to profl. jours. Served to maj., M.C. U.S. Army, 1975-78. Recipient Van Winkle award U. Ariz., 1982; Upjohn Achievement award, 1982; decorated Army Commendation medal; recipient James Kembrough award for outstanding urologic research, 1976. Mem. Am. Inst. Nutrition, Am. Assn. Clin. Nutrition, Am. Soc. Exptl. Biology, Am. Soc. Parenteral and Enteral Nutrition, AMA, Assn. Mil. Surgeons. Subspecialties: Cell biology (medicine); Surgery. Current work: Electron microscopic investigations of nutrition, prostate cancer, resuscitative fluids, obesity, blood storage products, others. Home: 136 Bennington Rd Charlottesville VA 22901 Office: Dept Plastic Surgery Univ Va Med Center Charlottesville VA 22901

FRIEDMAN, HARVEY MICHAEL, physician, virologist, researcher, educator; b. Montreal, Que., Can., May 29, 1944; came to U.S., 1971; s. Sidney and Sybil (Garfinkle) F.; m. Cynthia D. Mickey, Apr. 12, 1980; 1 dau., Julie. M.D., McGill U., Montreal, 1969. Intern Jewish Gen. Hosp., Montreal, 1969-70, resident, 1970-71; fellow in virology Wistar Inst., Phila., 1971-73; fellow in infectious disease Hosp. U. Pa., 1973-75; dir. Diagnostic Virology Lab., Children's Hosp., Phila., 1975—; assoc. prof. medicine U. Pa., 1981—; cons. infectious diseases. Contbr. numerous articles to profl. jours. NIH grantee, 1981-84. Fellow Infectious Disease Soc. Am.; mem. AAAS, Am. Soc. Microbiology. Subspecialties: Infectious diseases; Virology (medicine). Current work: Mechanisms of viral-induced vessel wall injury. Office: Children's Hospl Phila 34th and Civic Center Blvd Philadelphia PA 19146

FRIEDMAN, HERBERT, physicist; b. N.Y.C., June 21, 1916; s. Samuel and Rebecca (Seligson) F.; m. Gertrude Miller, 1940; children—Paul, Jon. B.A., Bklyn. Coll., 1936; Ph.D. in Physics, Johns Hopkins U., 1940; D.Sc. (hon.), U. Tubingen, W. Ger., 1977, U. Mich., 1979. With U.S. Naval Research Lab., 1940—, supt. atmosphere and astrophysics div., 1958-63, supt. space sci. div., 1963-80; chief scientist E.O. Hulburt Center Space Research, 1963—; part-time prof. physics U. Md., adj. prof., 1960—, U. Pa., 1974—; vis. prof. Yale, 1966-68; Chmn. COSPAR working group II, Internat. Quiet Sun Year; v.p. COSPAR, 1970-75; mem. Gen. Adv. Com. on Atomic Energy, 1968-74; pres. Interunion Com. on Solar-Terrestrial Physics, 1967-74; chmn. com. on solar-terrestrial research Nat. Acad. Scis.-NRC, 1968-71; mem. geophysics research bd. Nat. Acad. Scis., 1969-71, chmn., 1976-79, mem. adv. com. int. orgns. and programs, 1969-77, mem. com. sci. and pub. policy, 1967-71; mem. space sci. bd. Nat. Acad. Scis.-NRC, 1962-75; mem. Pres.' Sci. Adv. Com., 1970-73. Recipient Disting. Service award Navy Dept., 1945, 80, R.D. Conrad medal Navy Dept., 1964, Distinguished Achievement in Sci. award, 1962; medal Soc. Applied Spectroscopy, 1957; Distinguished Civilian Service award Dept. Def., 1959; Janssen medal French Photog. Soc., 1962; Eddington medal Royal Astron. Soc., 1964; Presdl. medal for distinguished fed. service, 1964; Space Sci. award Am. Inst. Aeros. and Astronautics, 1963, also; Dryden Research award, 1973; Rockefeller Pub. Service award, 1967; Nat. medal Sci., 1969; NASA medal for exceptional sci. achievement, 1970, 78; Michelson medal Franklin Inst., 1972; Lovelace award Am. Astronautical Soc., 1973. Fellow Am. Phys. Soc., Am. Optical Soc., Am. Geophys. Union (pres. sect. on solar-planetary relationships 1967-70, Bowie medal 1981), Am. Astronautical Soc., Am. Inst. Aeros. and Astronautics; mem. AAAS (v.p. 1972), Nat. Acad. Scis. (council 1979—, chmn. Assembly of Math. and Phys. Scis. 1980—), Am. Acad. Arts and Scis., Internat. Acad. Astronautics, Am. Philos. Soc. Club: Cosmos. Subspecialties: High energy astrophysics; Solar physics. Current work: X-ray/gamma ray astronomy; solar terrestrial physics. Spl. research V-2 rocket, satellite launchings, solar cycle variations X-ray and ultra-violet radiations from sun, X-ray astronomy; produced 1st X-ray and ultra-violet photographs of sun, also discovered hydrogen geocorona, measured ultraviolet fluxes of early-type stars. Home: 2643 N Upshur St Arlington VA 22207 Office: Code 7100 Naval Research Lab Washington DC 20375

FRIEDMAN, HERMAN, educator; b. Phila., Sept. 22, 1931; s. Max and Rose (Snechtman) F.; m. Ilona Gottfried, Dec. 27, 1958; children: Franklin, Michelle, Suzanne, Andrea. A.B., Temple U., 1953, M.A., 1955; Ph.D., Hahnemann Med. Coll., 1957. Diplomate: Am. Bd. Med. Lab. Microbiology, Am. Bd. Immunology. Instr. Hahnemann Med. Coll., Phila., 1957-58; chief allergy research lab. Pitts. VA Hosp., 1958-59; head Dept. Microbiology, Albert Einstein Med. Center, Phila., 1959-78; asst. prof. to prof. microbiology Temple U., Phila., 1959-78; prof., chmn. dept. microbiology U.S. Fla., Tampa, 1958—; cons. in field; review com. mem. NIH, Washington, 1975—. Contbr. articles to profl. jours.; author: numerous books including Manual of Clinical Immunology, 1976, 80, Macrophages of Lymphocytes, 1980. Fellow Am. Acad. Microbiology; mem. Reticuloendotholocial Soc. (pres. 1977-78), Eastern Pa. Soc. Microbiology (pres. 1970-71), Fla. Soc. Microbiology (pres. 1980-83), Am. Soc. Microbiology. Democrat. Jewish. Current work: Infectious diseases and role of immune response in resistance to bacterial and viral infection. Address: Univ South Fla Coll Medicine 12901 N 30th St Tampa FL 33612

FRIEDMAN, JEROME ISAAC, physics educator, researcher; b. Chgo., Mar. 28, 1930; m., 1956; 3 children. A.B., U. Chgo., 1950, M.S., 1953, Ph.D. in Physics, 1956. Research assoc. in physics U. Chgo., 1956-57, Stanford U., 1957-60; from asst. prof. to assoc. prof. MIT, 1960-67, prof. physics, 1967—, dir. lab. nuclear sci., 1980-83, head dept. physics, 1983—. Fellow Am. Phys. Soc., Am. Acad. Arts and Scis. Subspecialty: Particle physics. Office: Dept Physics MIT Cambridge MA 02139

FRIEDMAN, JULES DANIEL, geologist; b. Poughkeepsie, N.Y., Oct. 24, 1928; s. Jack and Sophie (Seltzer) F.; m.; children: Susanne Kerstin, Jack Arne, Lisa Kari. A.B., Cornell U., 1950; M.S., Yale U., 1952, Ph.D., 1958. Asst. in instrn., dept. geology Yale U., 1950-53; geologist, research supr. mil. geology U.S. Geol. Survey, Washington, 1953-65, geologist, br. theoretical geophysics, 1965-67, staff geologist Office Regional Geology, 1967-69, geologist, br. regional geophysics, 1969-75, br. petrophysics and remote sensing, Denver, 1975-82, chief remote sensing sect., br. geophysics, 1983—; cons. in field. Contbr. articles to profl. jours.; cons. editor: Am. Edit. of Earth Sci. Dokl Akad Nauk, 1962—; contbg. author: 1980 Eruptions of Mount St. Helens, 1982, Manual of Remote Sensing, 1983, Skylab Views the Earth, 1975, Volcanic Land Forms and Surface Features, 1971, Glossary of Geology, 1972. Recipient Group Achievement award NASA, 1975; Quality of Sci. Work award U.S. Geol. Survey, 1979; others. Fellow Geol. Soc. Am; mem. Am. Geophys. Union (sec.), Internat. Assn. Volcanology. Subspecialties: Geology; Remote sensing (geoscience). Current work: Infrared aerial surveys in geology; application of geology and remote sensing techniques in natural hazard analyses,volcanology, heat flow, geothermal resources, geologic and geophysical aspects of national energy, natural resource and strategic foreign policy considerations. Home: 658 S Reed CT NE 31 Lakewood CO 80226 Office: US Geol Survey Mail Stop 964 Denver Federal Center Denver CO 80226

FRIEDMAN, LOUIS DILL, planetary scientist; b. Kingston, N.Y., July 7, 1940; s. Benjamin Alexander and Letha (Dill) F.; m. Connie Lou, Aug. 16, 1963; children: Michael, Christopher, Rachel. B.S., U. Wis., 1961; M.S., Cornell U., 1963; Ph.D., M.I.T., 1970. Mem. tech. staff space systems div. Avco, Wilmington, Mass., 1963-69; program mgr. Jet Propulsion Lab., Calif. Inst. Tech., Pasadena, 1970-81; congressional fellow U.S. Senate Com. on Commerce, Sci. and Transp., Washington, 1978-79; exec. dir. Planetary Soc., Pasadena, Calif., 1980—. Contbr. articles to profl. jours. Bd. dirs. Pasadena Jewish Temple and Ctr., 1973-82. Mem. AIAA, AAAS, Am. Astron. Soc., Sigma Xi. Subspecialties: Aerospace engineering and technology; Astronautics. Current work: Space scientist in celestial mechanics, aerospace engineering, space policy and program management. Office: The Planetary Society 110 S Euclid Ave Suite 103 Pasadena CA 91101

FRIEDMAN, MELVIN, geologist, educator; b. Orange, N.J., Nov. 14, 1930; s. Leonard and Hannah L. (Sholk) F.; m. Deborah Friedman, June 13, 1954; children: Barry D., Cheryl A. B.S., Rutgers U., 1952, M.S., 1954; Ph.D., Rice U., 1961. Geologist, research assoc., sect. leader Shell Devel. Co., 1954-67; assoc. prof. geology Tex. A&M U., College Station, 1967-69, prof., 1969—, dir. Ctr. for Tectonophysics, 1979-82, assoc. dean Coll. Geoscis., 1982-83, dean Coll. Geoscis., 1983—. Co-editor-in-chief: Jour. Tectonophysics, 1980—; contbr. articles to profl. jours. Recipient research award Internat. Soc. Commn. U.S. Rock Mechanics, 1969; Disting. Faculty Achievement award Tex. A&M U., 1975. Fellow Geol. Soc. Am., AAAS; mem. Am. Geophys. Union. Jewish. Subspecialties: Tectonics; High-temperature materials. Current work: Observational studies of deformation mechanisms in experimentally and naturally deformed rocks; application of same to problems in petroleum exploration and production, engineering rock mechanics, faulting and folding. Office: Ctr Tectonophysics Tex A&M U College Station TX 77843

FRIEDMAN, NATHAN BARUCH, pathologist; b. N.Y.C., Jan. 30, 1911; s. Emanuel David and Rose (Borgenicht) F.; m. Helen Eugenie McFrancis; m. Isabel Williams Speers, Jan. 3, 1960; children: Sharon, Janifer, Robert, Marylou, Emily, Charles. B.S., Harvard U., 1930; M.D., Cornell U., 1934. Diplomate: Am. Bd. Pathology. Intern Montefiore Hosp., N.Y.C., 1935-36; resident U. Chgo. Clinics, 1937-38; asst. in pathology U. Chgo., 1938-39; Littauer fellow Harvard U., 1939-40; instr. in pathology Stanford U., 1941-42; dir. labs. Cedars of Lebanon Hosp., Los Angeles, 1948-70; sr. cons. in pathology Cedars Sinai Med. Ctr., Los Angeles, 1970—; clin. prof. pathology U. So. Calif. Author: Tumors of Bladder, 1959; contbr. articles on cancer, radiation, endocrines, urology to profl. jours. Served to maj. M.C. AUS, 1942-46. Mem. Am. Assn. Pathologists, Am. Soc. Clin. Pathology, Am. Assn. Cancer Research, Endocrine Soc. Subspecialties: Pathology (medicine); Oncology. Current work: Germ cell tumors, breast cancer, experimental immunology. Home: 15150 Mulholland Dr Los Angeles CA 90077 Office: 8700 Beverly Blvd Los Angeles CA 90048

FRIEDMAN, ORRIE MAX, biotechnology company executive, b. Grenfell, Sask., Can., June 6, 1915; came to U.S., 1946, naturalized, 1962; s. Jack and Gertrude (Shulman) F.; m. Laurel E. Leeder, Jan. 2, 1959; children: Mark David, Gertrude Jane, Hugh Robert. B.Sc., U. Man., 1935, McGill U., 1941, Ph.D., 1944. Research chemist NRCCan., Ottawa, Ont., 1944-46; research fellow chemistry Harvard U., 1946-49, research assoc., 1949-51; asst. prof. Harvard Med. Sch., 1951-53, Brandeis U., 1953-55, assoc. prof., 1955-58, prof. chemistry, 1958-62; pres., sci. dir. Collaborative Research, Inc., Waltham, Mass., 1962-82, chmn. bd., 1982—; cons. in field; dir. Nat. Assn. Life Sci. Industries, 1977-79; dir. Canada Fibre Can Co. Ltd., Montreal. Contbr. articles to profl. jours. Bd. dirs. Boston chpt. Am. Technion Soc., also nat. bd. dirs.; trustee Beth Israel Hosp., Boston, Worcester Found. Exptl. Biology; mem. corp. Sidney Farber Cancer Inst., Mus. Sci., Boston.; bd. govs. Technion, Israel Inst. Tech. Recipient numerous grants from various govt. agys. Fellow AAAS, Chem. Soc. London; mem. Am. Chem. Soc., Am. Assn. Cancer Research, Radiation Soc., N.Y. Acad. Sci., Sigma Xi. Subspecialties: Genetics and genetic engineering (biology); Genetics and genetic engineering (medicine). Current work: Developer, manufacturer products having commercial potential in medicine, agricultural and industry and research. Home: 49 Warren St Brooklin MA 02146 Office: 128 Spring St Lexington MA 02173

FRIEDMAN, SARAH L(ANDAU), research psychologist; b. Tel Aviv, Nov. 18, 1943; U.S., 1969, naturalized, 1974; d. Faivel and Esther Haya (Vloska) Landau; m. Moshe Friedman, June 11, 1967; 1 dau., Daphne Ruth. B.A., Hebrew U., Jerusalem, 1969; M.A., Cornell U., 1971; Ph.D., George Washington U., 1975. Research asst. Cornell U., 1969-70, teaching asst., 1970-71, George Washington U., 1971-73; guest worker Nat. Inst. Child Health and Human Devel., NIH, Bethesda, Md., 1972-73; research fellow NIMH, Bethesda, 1974-76; research psychologist Lab. Devel. Psychology, 1976-79, 83-; assoc. Nat. Inst. Edn., Washington, 1980-83. Contbr. chpts., articles to profl. publs., conf. presentations; editor: (with M. Sigman) Preterm Birth and Psychological Development, 1981, (with K. A. Klivington and R. W. Peterson) The brain, cognition and education, in press, (with E. K. Scholnick and R.R. Cocking) The growth of cognitive planning skills, in press; editorial bd.: Child Devel, 1981—, Jour. Applied Devel. Psychology, 1983—; reviewer profl. jours., conf. papers. Recipient Tuition award George Washington U., 1972-73; NIMH postdoctoral fellow, 1974-76; Sigma Xi grantee, 1973-74. Mem. AAAS, Am. Ednl. Research Assn., Am. Psychol. Assn., N.Y. Acad. Sci., Soc. Research in Child Devel., Sigma Xi, Psi Chi (Nat. Research award 1974), Phi Kappa Phi, Phi Lambda Theta. Jewish. Subspecialty: Developmental psychology. Current work: Environmental influences on cognitive skills; risk (medical, environmental) and psychological development. Home: 4511 Yuma St NW Washington DC 20016 Office: NIMH 9000 Rockville Pike Bethesda MD 20205

FRIEDMAN, SELWYN MARVIN, microbiologist, educator; b. N.Y.C., May 17, 1929; s. Louis and Leah (Weinstein) F.; m. Rivka Teitelbaum, May 28, 1972. Student, CCNY, 1946-48; B.S., U. Mich., 1951; M.S., Purdue U., 1953, Ph.D., 1961. Postdoctoral fellow dept. biochemistry Western Res. U., 1961-62; dept. cell biology Albert Einstein Coll. Medicine, 1962-63; research fellow Columbia U. Coll. Physicians and Surgeons, 1962-66; vis. scientist dept. chemistry Columbia U., 1973; asst. prof. Hunter Coll., CUNY, 1966-69, assoc. prof. biology, 1969—; U.S. coordinator joint U.S.-Japan seminar in Biochemistry of Thermophily U.S.-Japan Coop. Sci. Program, Honolulu, 1977. Research publs. in field; editor: Biochemistry of Thermophily, 1978. Served with U.S. Army, 1954-56. USPHS grantee, 1967-70, 75-78; CUNY Faculty Research Award Program grantee, 1975-79, 81-83. Mem. AAAS, Am. Soc. Microbiology, Harvey Soc. Democrat. Jewish. Subspecialties: Microbiology; Genetics and genetic engineering (biology). Current work: Molecular aspects of growth at elevated temperatures; genetic exchange in thermophilic bacteria. Home: 340 E 64th St Apt 14-L New York NY 10021 Office: 695 Park Ave New York NY 10021

FRIEDMANN, E(MERICH) IMRE, biology educator; b. Budapest, Hungary, Dec. 20, 1921; s. Hugo and Gisella (Singer) F.; m. Naomi Kraus, Sept. 16, 1953; m. Roseli Ocampo, July 22, 1974; 1 dau., Daphna. Ph.D., U. Vienna, 1951. Instr., lectr., assoc. prof. Hebrew U., Jerusalem, Israel, 1952-66; vis. prof. Fla. State U., Tallahassee, 1966-67, assoc. prof., 1968-76, prof. dept. biol. scis., 1976—; assoc. prof. Queens U., Kingston, Ont., Can., 1967-68; vis. prof. U. Vienna, 1975. Contbr. articles to profl. jours. Recipient Congl. Antarctic Service medal NSF, 1979. Fellow Linnean Soc. London; mem. Am. Soc. Microbiology, Internat. Soc. for Study of Origins of Life, Internat. Phycological Soc., Am. Phycological Soc., Bri. Phycological Soc., French Phycological Soc., Indian Phycological Soc. Jewish. Subspecialties: Microbiology; Ecology. Current work: Microbial ecology of extreme environments; endolithic (rock-inhabiting) microorganisms in Antarctica and in deserts; exobiology and origins of life; taxonomy of cyanobacteria. Discoverer of microorganisms (cryptoendolithic lichens) living in Antarctic rocks, 1976. Office: Department of Biological Sciences Florida State University Tallahassee FL 32306

FRIEDMANN, NAOMI KRAUS, biochemistry educator; b. Budapest, Hungary, July 4, 1933; came to U.S., 1965; d. Jacob and Vilma Kraus; 1 dau., Daphna. M.Sc., Hebrew U., Jerusalem, 1960, Ph.D., 1965. Postdoctoral fellow Columbia U., 1965-66, Vanderbilt U., 1966-68; instr. U. Pa., 1968-74; asst. prof. biochemistry U. Tex.-Houston, 1974-76, assoc. prof., 1976—. Mem. Am. Physiol. Soc., Am. Soc. Biol. Chemists, Assn. Women in Sci. (pres. Gulf Coast chpt. 1975-76). Subspecialties: Biochemistry (medicine); Endocrinology. Current work: Hormonal regulation of gluconeogenesis, hormonal effects on calcium sequestering.

FRIEDRICH, DONALD MARTIN, chemist, educator; b. Bay City, Mich., Jan. 3, 1944. B.Sc, U. Mich., 1966; Ph.D., Cornell U., 1973. Research assoc. Wayne State U., Detroit, 1973-75; asst. prof. chemistry Hope Coll., Holland, Mich., 1975-81, assoc, prof., 1982—. Contbr. articles to profl. jours. NDEA fellow, 1966-70. Mem. Optical Soc. Am., Am. Chem. Soc., AAAS. Subspecialties: Physical chemistry; Atomic and molecular physics. Current work: Chemical and physical properties of molecules in excited states. Office: Dept Chemistry Hope Coll Holland MI 49423

FRIEDRICH, OTTO MARTIN, JR., elec. engr., researcher, educator, cons.; b. Austin, Tex., Jan. 29, 1939; s. Otto Martin and Lillie Louise (Stark) F. B.S. in Elec. Engring, U. Tex., 1961, M.S., 1962, Ph.D., 1965. Registered profl. engr., Tex. Research engr. U. Tex., Austin, 1961-65, mem. faculty dept. elec. engring., 1965—, faculty research engr., 1969—, asst. dir., 1971-75; cons. profl. engring. to industry, fed., state and local govt. Contbr. articles on electro-optics, lasers, plasma diagnostics and instrumentation to publs. of, IEEE, Instrument Soc. Am. Mem. IEEE (sr.), Instrument Soc. Am. (sr., Dr. Charles Stark Draper award 1972), Am. Phys. Soc., AAAS, AIAA, Am. Geophys. Soc., Optical Soc. Am., Nat. Soc. Profl. Engrs., Tex. Soc. Profl. Engrs. Lutheran. Subspecialties: Electronics; Laser research. Current work: Laser instrumentation and applications, including holography; electro-optics, lasers, holography, plasmas, instrumentation. Office: U Tex Austin TX 78712

FRIEDRICHS, KURT OTTO, mathematics educator, researcher; b. Kiel, Germany, Sept. 28, 1901; m., 1937; 5 children. Ph.D. in Math, U. Gottingen, 1925, Aachen Tech. U., 1970, Uppsala U., 1974, Nat. Med. Sci., Washington, 1976. Asst. U. Gottingen, 1925-27, privat-dozent, 1929-30; asst., privat-dozen Aachen Tech. U., 1927-29; prof. math. Brunswick Tech. U., 1930-37; vis. prof. applied math. Courant Inst. Math. Sci., NYU, 1937-39, from assoc. prof. to prof., 1939-74, disting. prof. math., 1974—, from assoc. dir. to dir. ins., 1953-67; mem. staff Office Sci. Research and Devel., U.S. Navy. Fellow Am. Acad. Arts and Sci.; mem. Nat. Acad. Sci. (Applied Math. and Numerical Analysis award 1972), Am. Math. Soc. Subspecialty: Applied mathematics. Office: Courant Inst Math Sci NYU New York NY 10012

FRIEMAN, EDWARD ALLAN, physicist, educator; b. N.Y.C., Jan. 19, 1926; s. Joseph and Belle (Davidson) F.; m. Ruth Rodman, June 19, 1949 (dec. May 1966); children—Jonathan Paul and Michael Rodman (twins), Linda, Wendy, Joshua Adam; m. Joy Fields, Sept. 17, 1967; children—Jonathan Paul and Michael Rodman (twins), Linda, Wendy, Joshua Adam. B.S., Columbia, 1945; M.S., Poly. Inst., Bklyn., 1948, Ph.D., 1951. Research asso. Poly Inst., Bklyn., 1947-49, instr., 1949-51; research asso. Project Matterhorn (B), Princeton, 1952-53, head theoretical div. of project, 1953-64, dir. plasma physics program, 1959-62, lectr. physics, 1959-60, prof. astronomy, 1961—; dep. dir. Plasma Physics Lab., 1964-79; dir. energy research U.S. Dept. Energy, 1980—; exec. v.p. Sci. Applications, Inc., 1981—; cons. Inst. Def. Analyses, Los Alamos Sci. Lab.; cons., bd. dirs. Aero. Research Assos. Princeton; sci. adv. com. Gen. Precision Instruments Co. Deep-sea diving officer Bikini tests, 1945-46; mem. research adv. com. nuclear processes NASA, 1959-60; mem. Watson com. Weapons Systems Evaluation Group, 1961—; mem. physics adv. com. NSF, 1976-79; mem. Naval Research Adv. Com., 1978-79; mem. vis. com. dept. nuclear energy Brookhaven Nat. Labs.; cons. laser fusion div. Dept. Energy.; Dir. Trans-East Airlines. Author articles plasma physics, hydromagnetics, statis. mechanics.; Bd. editors: Physics of Fluids, 1964-66, Am. Phys. Soc, 1965, Phys. Rev, 1966-67. Trustee Jersey City State Coll., 1968-75. John Simon Guggenheim Meml. Found. fellow.; Recipient Service award Dept. Energy. Fellow Am. Phys. Soc.; mem. Nat. Acad. Scis., Am. Astron. Soc., AAUP, AAAS, Sigma Xi. Subspecialty: Theoretical physics Office: 1200 Prospect St La Jolla CA 92037

FRIEND, CHARLOTTE, medical microbiologist, medical school administrator, educator; b. N.Y.C., Mar. 11, 1921. B.A., Hunter Coll., 1944; Ph.D. in Bacteriology, Yale U., 1950. Assoc. mem. Sloan-Kettering Inst., 1949-66; assoc. prof. microbiology Sloan-Kettering div. Med. Coll. Cornell U., 1952-66; prof., dir. Ctr. Exptl. Cell Biology, Mt. Sinai Sch. Medicine, 1966—. Recipient Alfred P. Sloan award, 1954, 57, 62, Am. Cancer Soc. award, 1962, Presdl. medal and Centennial award Hunter Coll., 1970, Virus-Cancer program award NIH, 1974, prix Griffuel, 1979. Mem. Nat. Acad. Sci., Am. Assn. Cancer Research, Am. Assn. Immunology, Am. Soc. Hematology, Tissue Culture Assn. Subspecialty: Cell biology. Office: Ctr Exptl Cell Biology 100th St and Fifth Ave New York NY 10029

FRIEND, JAMES PHILIP, chemistry educator; b. Hartford, Conn., Nov. 30, 1929; s. Edward Stanley and Estelle (Magaram) F.; m. Etta Levy Zuckerman, Mar. 27, 1955. S.B., MIT, 1951; M.A., Columbia U., 1953, Ph.D., 1956. Project engr. Perkin-Elmer Corp., Norwalk, Conn., 1956-57; sr. research scientist Isotopes, Inc., Westwood, N.J., 1957-67; assoc. prof. atmospheric chemistry NYU, N.Y.C., 1967-72, prof., 1972-73; prof. atmospheric chemistry Drexel U., Phila., 1973-74, R.S. Hanson prof., 1974—; cons.; mem. coms. Nat. Acad. Scis. Served with Chem. Corps. U.S. Army, 1953-55. NSF grantee, also NASA grantee, 1969—. Mem. Am Meteorol. Soc., Am. Geophy. Soc., Am. Chem. Soc., AAAS, Sigma Xi. Subspecialties: Atmospheric chemistry; Geophysics. Current work: Atmospheric particle formation from photochemical reactions of sulfur dioxide; stratospheric aerosols; volcanic emissions in the atmosphere; nuclear warfare effects. Office: Department of Chemistry Drexel University Philadelphia PA 19104

FRIESEN, LARRY JAY, software engineer; b. Salina, Kans., Sept. 7, 1946; s. Eric Jacob and Dorothy Marie (LeClair) F. B.A., U. Kans., 1967; M.S., Rice U., 1972, Ph.D., 1974. Postdoctoral research asso. U. Ga., Athens, 1974-75; engr. guidance and control mechanics McDonnell Douglas, Houston, 1976-80, sr. engr., 1980—. Served with U.S. Army, 1969-70. Mem. AAAS, Nat. Space Inst., Am. Astron. Soc., World Space Found., L-5 Soc., Gulf Coast Sci. Fiction Soc. Republican. Methodist. Club: Planetary Soc. Subspecialties: Software engineering; Aerospace engineering and technology. Current work: Lunar and planetary studies, space exploration, Space Shuttle mission planning software, Flight Design System. Copyrighted board game: Quadrants. Office: Johnson Space Ctr Mail Code DF7 Houston TX 77058

FRIESEN, WOLFGANG OTTO, biology educator; b. Elbing, Germany, Oct. 31, 1942; came to U.S., 1950, naturalized, 1960; s. Helmuth and Trude F. B.S., Bethel Coll., Newton, Kans., 1964; M.A. in Physics, U. Calif.-Berkeley, 1966; Ph.D. in Neurosci, U. Calif.-San Diego, 1974. Research physicist Cardiovascular Research Inst., U. Calif.-San Francisco, 1969-70; research asso., research fellow dept. molecular biology U. Calif.-Berkeley, 1974-77; asst. prof. biology U. Va., Charlottesville, 1977-82, assoc. prof., 1982—. Contbr. articles on pulmonary physiology, neurophysiology, neuroanatomy, math. modeling and instrumentation to sci. jours. NIH fellow, 1975-77; NIH research grantee, 1978-82; NSF research grantee, 1982—. Mem. AAAS, Soc. for Neurosci. Subspecialties: Neurobiology; Bioinstrumentation. Current work: Biology, neurobiology. Inventor infant respiratory monitor and a sonic level. Home: 2821 Northfields Rd Charlottesville VA 22901 Office: Dept Biology U Va Charlottesville VA 22901

FRISCH, JOSEPH, mech. engr., educator, cons.; b. Vienna, Austria, Apr. 21, 1921; came to U.S., 1940, naturalized, 1946; s. Abraham and Rachel (Lieberman) F.; m. Joan S., May 26, 1962; children: Nora Theresa, Erich Martin, Jonathan David. B.S.M.E., Duke U., 1946; M.S., U. Calif., 1950. Registered profl. engr., Calif. Mem. faculty U. Calif., Berkeley, 1947—, asst. prof. mech. engring., 1951-57, asso. prof., 1957-63, prof., 1963—, asst. dir., 1961-63, chmn. div. mech. design, 1966-70, asso. dean, 1972-75; cons. to indsl. and govtl. labs. Contbr. numerous articles to profl. jours. Fellow ASME; mem. Phi Beta Kappa, Sigma Xi, Tau Beta Pi, Pi Tau Sigma. Clubs: U. Calif. at Berkeley Faculty, Berkeley City. Subspecialties: Mechanical engineering; Computer-aided design. Current work: Computer-aided design and manufacturing; computer-aided systems analysis. Office: Dept Mech Engring U Calif Berkeley CA 94720

FRISCH, NORMAN W(ILLIAM), chemical engineer, energy company executive; b. N.Y.C., Dec. 8, 1923; m., 1949; 3 children. B.Ch.E., CCNY, 1948; D.Eng. in Chem. Engring, Yale U., 1954. Research chem. engr., research labs. Rohm & Haas Co., 1952-60; research assoc. Princeton Chem. Research Inc., 1960-64, dir. catalyst research, 1964-67; mgr. chem. process devel. Cottrell Environ. Systems div. Research-Cottrell, Inc., 1968-70, dir. research, 1970-74, sr. sci. cons., 1974-80; v.p. research and devel. Affiliated Energy and Environ. Techs., Inc., 1980-83; pres. W.W. Frisch Assoc. Inc., 1983—. Mem. Air Pollution Control Assn., AAAS, Am. Inst. Chem. Engrs., Am. Chem. Soc. Subspecialty: Chemical engineering. Address: 145 Ridgeview Circle Princeton NJ 08540

FRISELL, WILHELM RICHARD, biochemist, educator; b. Two Harbors, Minn., Apr. 27, 1920; s. Olof Wilhelm and Thyra Magnina (Falk) F.; m. Margaret Jane Fleagle, Mar. 6, 1948; children William Richard, Robert Benjamin. B.A., St. Olaf Coll., Minn., 1942; M.A., Johns Hopkins U., 1943, Ph.D., 1946, postdoctoral, 1946-49. Instr. physiol. chemistry Johns Hopkins U. and Sch. Medicine, Balt., 1950-51; asst. prof. biochemistry U. Colo. Sch. Medicine, Denver, 1951-58, asso. prof., 1958-64, prof., 1964-69; prof., chmn. dept. biochemistry N.J. Med. Sch., Newark, 1969-76; acting dean Grad. Sch. Biomed. Sci., Coll. Medicine and Dentistry N.J., Newark, 1971-73; prof., chmn. dept. biochemistry East Carolina U. Sch. Medicine, also asst. dean grad. affairs, Greenville, N.C., 1976—; mem. fellowships com. Fogarty Internat. Center, NIH, 1962-66, 67-71, 81—, chmn., 1968-71, 81, chmn. sr. fellowship com., 1981—. Author: Acid-Base ehemistry in Medicine, 1968, Human Biochemistry, 1982; also articles and revs. NDRC fellow, 1943-44; Am. Scandinavian Found. fellow, Uppsala, Sweden, 1949-50. Fellow AAAS; mem. Am. Chem. Soc., Am. Soc. Biol. Chemists, Am. Soc. Microbiology, Soc. Exptl. Biology and Medicine, N.Y. Acad. Scis., Harvey Soc., Phi Beta Kappa, Sigma Xi (pres. Colo. chpt. 1968-69). Subspecialties: Biochemistry (medicine); Microbiology. Current work: Biosynthesis and reaction mechanisms of flavin and folate-coenzymes; metabolic pathways in prokaryores and eukaryotic cells. Home: 209 Fairlane Rd Greenville NC 27834 Office: Sch Medicine E Carolina U Greenville NC 27834 As a first generation American who has been given so many opportunities, I can never forget the sacrifices of my immigrant parents and their compatriots. The likes of them may never pass this way again.

FRISHMAN, WILLIAM HOWARD, physician, researcher; b. Bronx, N.Y., Nov. 9, 1946; s. Aaron and Frances (Fishel) F.; m. Esther Rose Sandowsky, Mar. 11, 1971; children: Sheryl Renee, Amy Helene, Michael Aaron. B.A., Boston U., M.D., 1969. Diplomate: Am. Bd. Internal Medicine. Intern Montefiore Hosp., Bronx, 1969-70, resident in medicine, 1970-71, Bronx Mcpl. Hosp., 1971-72; fellow in cardiology Cornell-N.Y. Hosp., N.Y.C., 1972-74; instr. in medicine Cornell U. Med. Sch., N.Y.C., 1974-76; asst. prof. medicine, dir. cardiac labs. Albert Einstein Coll. Medicine, 1976-80, assoc. prof., chief clin. cardiology service, 1980-82, chief dept. medicine coll. hosp., 1982—; cons. cardiology Martin Luther King Health Center, Bronx. Author: Clinical Pharmacology of the Beta Adronoceptor Blocking Drugs, 1980, 2d edit., 1983; contbr. numerous articles, rev. articles, chpts. on cardiovascular pharmacology and other cardiovascular topics to profl. publs.; editor: Clinical Pharmacology of the Calcium-Entry Blocking Drugs. Served with USAR, 1969—. Decorated Army Commendation medal; recipient Teaching Scholar award Am. Heart Assn., 1979, Preventive Cardiology Acad. award NIH Heart, Lung and Blood Inst., 1980. Fellow ACP, Am. Coll. Cardiology, Council Clin. Cardiology of Am. Heart Assn., Am. Coll. Chest Physicians; mem. Am. Fedn. Clin. Research, Assn. Program Dirs. in Internal Medicine. Democrat. Jewish. Subspecialties: Cardiology; Pharmacology. Current work: Cardiovascular pharmacology, thrombosis, psychomic heart disease, hypertension, medical education.

FRITSCH, ARNOLD RUDOLPH, nuclear science administrator; b. Passaic, N.J., Mar. 28, 1932; s. Arno Rudolph and Martha Emma (Werner) F.; m. Betsey Anne Rorapaugh, June 13, 1953; children: Kristin, Kerry A. Read, Paul. B.S. in Chemistry, U. Rochester, 1953, Ph.D., U. Calif.-Berkeley, 1957. Sr. scientist atomic power dept. Wetinghouse Corp., Pitts., 1957-61; exec. asst. to chmn. AEC, Washington, 1961-68; coordinator Allied Gulf Nuclear Services Gen. Atomic Co., La Jolla, Calif., 1968-71; pres. Gulf United Nuclear Fuels Corp., Elmsford, N.Y., 1971-74; coordinator nuclear affairs Gulf Oil Corp., Pitts., 1974—. Trustee Episcopal Diocese, Pitts., 1976-79; bd. dirs. Bach Choir of Pitts., 1975-78. Recipient Arthur S. Flemming award Washington Jaycees, 1966; AEC fellow, 1955. Fellow Am. Inst. Chemists; mem. Am. Nuclear Soc., Am. Chem. Soc., Am. Phys. Soc. Club: Pitts. Athletic. Subspecialties: Nuclear fission; Nuclear engineering.

FRITZ, IRVING BAMDAS, scientist; b. Rocky Mount, N.C., Feb. 11, 1927; s. Henry Norman and Rose (Bamdas) F.; m. Helen Bridgman, Aug. 20, 1950 (dec. 1971); m. Angela McCourt, Oct. 21, 1972; children: David Bamdas, Jonathan Bridgman, Winston Romaine, Rachel Bamdas, Zoë McCourt, Daniel William. D.D.S., Med. Coll. Va., 1948; Ph.D., U. Chgo., 1951. Instr., Harvard U., 1951; asst. dir. metabolism and endocrinology Michael Reese Hosp., Chgo., 1954-56; asst. prof. physiology U. Mich., Ann Arbor, 1956-60, asso. prof., 1960-64, prof., 1964-68, Banting and Best Dept. Med. Research, U. Toronto, 1968—, chmn. dept., 1968-78; vis. scholar dept. biochemistry, U. Wash., 1963-64; vis. prof. U. B.C., 1970. Editor: Insulin Action, 1971; editorial bd.: Jour. Lipid Research, 1962-67, Can. Jour. Biochemistry, 1969-72, Molecular and Cellular Endocrinology, 1974-82, Am. Jour. Physiology, 1976-81; contbr. articles to profl. jours. Served with U.S. Army, 1951-53. Recipient Gairdner award, 1980; USPHS fellow U. Copenhagen, 1953-55; Guggenheim fellow, 1978-79. Mem. Am. Physiol. Soc., Am. Biochem. Soc., Am. Soc. Cell Biology, Can. Biochem. Soc., Endocrinology Soc., Soc. Reproductive Biology, Can. Med. Research Council (chmn. metabolism grants com. 1969-72). Jewish. Subspecialties: Reproductive biology; Developmental biology. Current work: Controls of spermatogenesis; cell-cell interactions; hormone actions; gonadal somatic cell-germinal cell interactions. Office: 112 College St Toronto ON M5G 1L6 Canada

FRITZ, ROBERT BARTLETT, immunologist; b. Milw., Nov. 16, 1937; s. William Hardy and Jane Ann (Bartlett) F.; m. Onalee Grindle, Feb. 22, 1964; children: Michael, Wendy. A.B., Bowdoin Coll., 1959; M.S., U. Maine, 1964; Ph.D., Duke U., 1967. Postodoctoral research assoc. Duke U., 1967-69; asst. prof. Emory U., Atlanta, 1969-75, assoc. prof., 1975—; guest investigator NIH, 1980-81; Jackson Lab., Bar Harbor, Maine, 1982. Contbr. articles to profl. jours. Served with USAR, 1959-65. NIH grantee, 1970—; Nat. Multiple Sclerosis Soc. grantee, 1982—; Damon Runyon Fund grantee, 1970-72. Mem. Am. Assn. Immunologists, Sigma Xi. Subspecialties: Immunology (medicine); Immunobiology and immunology. Current work: Experimental autoimmune disease, immunology, cellular immunology, neuroimmunology. Home: 1618 Emory Rd NE Atlanta GA 30306 Office: Dept Microbiology Emory U Atlanta GA 30322

FRITZE, KAREN J., biology educator; b. Ft. Meade, Md., July 15, 1954; d. Bernard H. and Florence E. (Eckelman) F. A.A., Broward Community Coll.; B.S., M.S., Fla. State U. Grad. asst. Fla. State U., Tallahassee, 1977-79, research asst., 1980; instr. biology, lab. coordinator U. Southwestern La., Lafayette, 1980—. Contbr. articles to profl. jours. Sigma Xi grantee, 1978, 79; U. Southwestern La. grantee, 1981. Mem. Bot. Soc. Am. Democrat. Lutheran. Subspecialties: Evolutionary biology; Systematics. Current work: SEM studies of pollen Gesneriaceae of New World tropics—relation to classification. Office: Dept Biology U Southwestern La Lafayette LA 70504

FRITZSCHE, HELLMUT, educator, physicist; b. Berlin, Germany, Feb. 20, 1927; came to U.S., 1952, naturalized, 1966; s. Carl Hellmut and Anna (Jordan) F.; m. Sybille Charlotte Lauffer, July 5, 1952; children—Peter Andreas, Thomas Alexander, Susanne Charlotte, Katharina Sabine. Diploma, U. Göttingen, Germany, 1952; Ph.D., Purdue U., 1954. Instr., then asst. prof. physics Purdue, 1954-56; mem. faculty U. Chgo., 1957—, prof. physics, 1963—, dir. materials research lab., 1973-77, chmn. dept. physics, 1977—; v.p., dir. Energy Conversion Devices, Inc., 1968—; mem. adv. bd. Ency. Brit., 1968—, Office Naval Research, 1976-79; mem. materials scis. panel Nat. Acad. Scis., 1975-80; bd. dirs. Bull. Atomic Scientists, 1981—. Editorial bd.: Jour. Applied Physics, 1975-80. Fellow Am. Phys. Soc. (chmn. div. condensed matter physics 1979-80); mem. N.Y. Acad. Scis., Am. Arbitration Assn. Subspecialty: Condensed matter physics. Home: 5801 Blackstone Ave Chicago IL 60637

FROEMSDORF, DONALD H., college dean, chemist; b. Cape Girardeau, Mo., Mar. 4, 1934; s. Rudolph F. and Marie A. (Mammon) F.; m. Joy L. Kasten, May 29, 1954; 1 dau., Dawn Elaine. B.S., S.E. Mo. State U., 1955; Ph.D. in Organic Chemistry, Iowa State U., 1959. Asst. project chemist Standard Oil Co., (Ind.), Whiting, 1959-60; assoc. prof. chemistry S.E. Mo. State U., Cape Girardeau, 1960-66, prof. chemistry, 1966—, chmn. div. sci. and math., 1970-76, dean, 1976—. Contbr. articles to profl. jours. Citizen mem. S.E. Mo. Regional Council, Mo. Council Criminal Justice, 1974-76; mem. del. council Mo. Area V Health Services Agy., 1972—. DuPont research fellow, 1958. Mem. Am. Chem. Soc., Chem. Soc. London, AAAS, Mo. Acad. Sci. Lutheran. Subspecialties: Organic chemistry; Nuclear magnetic resonance (chemistry). Current work: Organic reaction mechanisms, especially beta-elimination reactions. Home: 600 Highland St Cape Girardeau MO 63701 Office: Coll Scis SE Mo State U Cape Girardeau MO 63701

FROGEL, JAY ALBERT, astronomer; b. N.Y.C., Apr. 28, 1944; s. Herman and Dorothy (Schaiman) F. B.A., Harvard U., 1966; Ph.D., Calif. Inst. Tech., 1971. Research fellow, lectr. Harvard U., 1971-75; staff astronomer Cerro Tololo Interam. Observatory, La Serena, Chile, 1975—. Contbr. articles to profl. jours. Hon. Woodrow Willson fellow. Mem. Am. Astron. Soc., Internat. Astron. Union, Phi Beta Kappa. Jewish. Club: Sierra. Subspecialties: Infrared optical astronomy; Optical astronomy. Current work: Stellar populations in star clusters and galaxies. Home: Cerro Tololo Interam Observatory Casilla 603 La Serena Chile Office: Cerro Tololo Interam Observatory PO Box 26732 Tucson AZ 85726

FROHLICH, EDWARD DAVID, physician; b. N.Y.C., Sept. 10, 1931; s. William and May (Zneimer) F.; m. Sherry Linda Fine, Nov. 1, 1959; children—Marjorie, Bruce, Lora. B.A., Washington and Jefferson Coll., 1952; M.D., U. Md., 1956; M.S., Northwestern U., 1963. Diplomate: Am. Bd. Internal Medicine. Intern, resident D.C. Gen. Hosp., 1956-58; fellow in cardiovascular research Georgetown U. Hosp., Washington, 1958-59, resident in internal medicine, 1959-60; clin. investigator VA Research Hosp., Chgo., 1962-64; asso. in medicine Northwestern U., 1963-64; staff mem. research div. Cleve. Clinic, 1964-69; faculty medicine, physiology, Biophysics, dir. div. hypertensive diseases U. Okla., Oklahoma City, 1969-76, George Lynn Cross research prof., 1975-76; v.p. edn., research Alton Ochsner Med. Found., New Orleans, 1976—; staff mem. div. hypertensive diseases Ochsner Clinic, 1976—; prof. medicine, physiology La. State U., 1976—; prof. medicine, adj. prof. pharmacology Tulane U., 1976—; cons. FDA, 1971-74, VA, 1972—, NIH, 1972—, WHO, 1975-82, U.S. Pharmacopeia, 1975—. Contbr. many chpts. to books, numerous articles in field to profl. jours.; Editor: Pathophysiology - Altered Regulatory Mechanisms in Disease, 1972, 76, Rypins' Medical Licensure Examinations, 1981; editor.: Jour. Lab. and Clin. Medicine, 1974-76, Am. Jour. Cardiology, 1977—, Archives of Internal Medicine, 1978—, Modern Medicine, 1980—, Hypertension, 1980—. Served to capt. M.C. AUS, 1960-62. Recipient Honors Achievement award Angiology Research Found., 1964, Disting. Faculty award U. Okla., 1970, So. Med. Assn. Ann. award, 1971. Fellow A.C.P., Am. Coll. Cardiology, AAAS; mem. Am. Soc. for Clin. Investigation, Am. Soc. for Clin. Pharmacology and Therapeutics (pres. 1973-74), Internat. Soc. Hypertension (sci. council 1974—, treas. 1980—), Am. Heart Assn. (dir. La. 1979—), Am. Physiol. Soc., Am. Soc. Nephrology, Central Soc. for Clin. Research, Chi Epsilon Mu, Phi Sigma, Alpha Kappa Alpha. Jewish. Subspecialties: Internal medicine; Cardiology. Current work: Cardiovascular, physiological and pharmacological aspects of experimental and clinical hypertension. Home: 5353 Maria Ave New Orleans LA 70124 Office: Alton Ochsner Med Found 1516 Jefferson Hwy New Orleans LA 70121

FRÖLICHER, FRANZ, geologist, consultant, educator; b. Ridgewood, N.J., Jan. 11, 1936; s. Victor and Helene (Stehli) F.; m. Margrit Grundmann, Jan. 23, 1976; 1 dau., Britta. B.A., Don Bosco Coll., 1959, Alaska Methodist U., 1970, Alaska Methodist U., 1972; Ph.D., U. Edinburgh, Scotland, 1977. Researcher U. Tubingen, W.Ger., 1974-79; asst. prof. geology U. So. Miss., 1979—; dir. Swiss Field Camp, Solothurn, 1980—, Geo Kartier Kurs, 1979—; bd. dirs. Geo Dirs. Union, Tubingen, 1978—. Bd. dirs. West Lake Property Owners Assn., Hattiesburg, Miss., 1980—. Served to 1st lt. USAF, 1959-62. Mem. Geol. Soc. Am. Subspecialties: Paleoecology; Coal. Current work: Paleoecology of estuarine system; modern coalforming environments; carboniferous paleoecology. Office: U So Miss SS Box 9364 Hattiesburg MS 39406

FROMHOLD, ALBERT THOMAS, JR., physics educator; b. Birmingham, Ala., Nov. 25, 1935; s. Albert Thomas and Mary Lillian (Rutherford) F.; m. Regina Emilie Giordano, July 27, 1960; children: Thomas William, Matthew Albert. A.S., St. Bernard Jr. Coll., Cullman, Ala., 1955; B. Engring. Physics, Auburn U., 1957, M. Nuclear Sci., 1958; Ph.D., Cornell U., 1961. Research asst. Oak Ridge Nat. Lab., 1957, Cornell U., Ithaca, N.Y., 1959-61; mem. staff Sandia Labs., Albuquerque, 1961-65; vis. scientist Nat. Bur. Standards, Gaithersburg, Md., 1969-70; prof. physics Auburn (Ala.) U., 1965—. Author: Theory of Metal Oxidation, vol. 1, 1976, vol. 2, 1980, Quantum Mechanics for Applied Physics and Engineering, 1981. NSF awardee, 1980; Japan Soc. for Promotion of Sci. fellow, 1979. Mem. Am. Phys. Soc. (Jesse Beams research award 1981), Electrochem. Soc., Am. Assn. Physics Tchrs., Ala. Acad. Sci. Roman Catholic. Club: Toastmasters (pres.). Lodges: Elks; K.C. Subspecialties: Materials; Condensed matter physics. Current work: Theory of oxide film formation, solid-state electronics; optical properties of metals; transidnet nuclear magnetic resonance; nuclear relaxation in metals and magnetic insulators; solid state transport. Home: 2111 Robin Dr Auburn AL 36830 Office: Auburn U 313 Allison Lab Auburn AL 36849

FROMING, WILLIAM JOHN, psychology educator; b. New Holstein, Wis., July 3, 1950; s. George Henry and Cecilia Marie (Helf) F. B.A., U. Wis., 1972; Ph.D., U. Tex., 1977. Teaching asst. U. Tex., 1974-77; research assoc. Research and Devel. Ctr. for Tchr. Edn., 1977; asst. prof. psychology U. Fla., 1977-82, assoc. prof., 1982—; cons. reviewer NSF, Washington, 1979—, NIMH, 1979—; reviewer numerous profl. jours. Author: Computer Applications of Statistical Techniques, 1983; contbr. articles to profl. jours. NSF grantee, 1981—; Office of Atty. Gen. Fla. grantee, 1982. Mem. Am. Psychol. Assn., Soc. Research in Child Devel. Subspecialties: Social psychology; Developmental psychology. Current work: Impact of self-focused attention on behavior; the development of norms governing altruistic behavior; the relationship between moral reasoning and moral behavior. Office: Dept Psycholog Psychology Bldg U Fla Gainesville FL 32611 Home: 5907 NW 52d Terr Gainesville FL 32606

FROMM, GERHARD HERMANN, neurologist, educator; b. Konigsberg, Germany, Sept. 7, 1931; s. Fritz Wilhelm and Ilse (Pflaum) F.; m. Antionette McKenna, May 26, 1973; children: Allison, Devin. B.S., U. P.R., Rio Piedras, 1949; M.D., Jefferson Med. Coll., Phila., 1953. Diplomate: Am. Bd. Psychiatry and Neurology. Instr. neurology Tulane U. Sch. Medicine, New Orleans, 1961-62, asst. prof., 1962-66, asso. prof., 1966-68; asso. prof. neurology U. Pitts. Sch. Medicine, 1968-81, prof., 1981—; attending physician Presbyn.-Univ. Hosp., Pitts. Contbr. articles to med. jours. Served to lt., M.C. USNR, 1956-58. NIH fellow, 1959-61; career devel. awardee, 1962-68. Mem. Am. Acad. Neurology, Am. Epilepsy Soc., Soc. Neurosci., Am. EEG Soc., Eastern Assn. EEG, Internat. Assn. for Study of Pain, Am. Pain Soc., Am. Neurol. Assn. Subspecialties: Neurology; Neuropharmacology. Current work: Neuropharmacology of anticonvulsant drugs; epilepsy; trigeminal neuralgia; pain mechanisms. Home: 1401 N Negley Ave Pittsburgh PA 15206 Office: University of Pittsburgh 322 Scaife Hall Pittsburgh PA 15261

FROMM, PAUL OLIVER, physiology educator, researcher environmental toxicology; b. Ramsey, Ill., Dec. 2, 1923; s. August Mottke and Edith Marie (Wollerman) F.; m. Mary Magdalene Shaw, June 15, 1947; children: David, Emily. B.S., U. Ill., 1949, M.S., 1951, Ph.D., 1954. Instr. dept. physiology Mich. State U., East Lansing, 1954-58, asst. prof., 1958-62, assoc. prof., 1962-65, prof., 1965—; cons. U.S.-Can. Great Lakes Commn., Windsor, Ont., 1981, Nat. Research Council Can., Ottawa, 1983. Contbr. articles to profl. jours. Served with USMC, 1943-46. Mem. Midwest Benthological Soc. (pres. 1958), Am. Soc. Zoologists, Am. Physiol. Soc., Soc. Exptl. Biology and Medicine. Subspecialties: Comparative physiology; Environmental toxicology. Current work: Effect of pollutants on aquatic animals; mainly freshwater teleost fish; comparative animal physiology. Home: 1238 Ivanhoe Dr East Lansing MI 48823 Office: Michigan State University Department Physiology East Lansing MI 48824

FROMMER, J. PEDRO, physician, medical educator; b. Santiago, Chile, Nov. 24, 1949; came to U.S., 1978; s. Guillermo and Eva (Holota) F. M.D., U. Chile, 1974. Diplomate: Am. Bd. Internal Medicine. Intern U. Toronto, Ont., Can., 1974-75, resident in internal medicine, 1975-78; fellow in nephrology Baylor U. Coll. Medicine, Houston, 1978-81; instr. medicine U. Ill. Med. Sch., Chgo., 1981-82, asst. prof., 1982-83; asst. prof. medicine Baylor Coll. Medicine, Houston, 1983—; chief renal sect. VA Med. Ctr., Houston, 1983—. Contbr. numerous articles to med. jours. Nat. Kidney Found. fellow, 1980. Fellow ACP, Royal Coll. Physicians and Surgeons Can., Chgo. Council Fgn. Affairs. Subspecialties: Internal medicine; Nephrology. Current work: Research interest in renal physiology, salt and water metabolism and acid base metabolism. Office: Renal Sect VA Med Ctr Research Bldg 211 (151-B) Houston TX 77211

FRONING, GLENN WESLEY, food scientist, educator; b. Gray Summit, Mo., Sept. 8, 1930; s. Gilbert Charles and Marie Juanita (Buchanan) F.; m. Lynne S., Aug. 4, 1962; children: Teri Ann, Sharon Marie. B.S. in Poultry Sci, U. Mo., 1953, M.S. in Poultry Products, 1957; Ph.D. in Food Tech, U. Minn., 1961. Asst. prof. food sci. Rutgers U., 1961-63; asst. prof. poultry sci. U. Conn., 1963-66; assoc. prof. poultry sci. U. Nebr., Lincoln, 1966-70, prof. poultry sci., 1970-72, prof. poultry and wildlife sci., chmn. dept. poultry and wildlife sci., 1972-77, prof. animal sci., 1977—. Contbr. numerous articles to profl. jours. Served to 1st lt. U.S. Army, 1953-55. Recipient Research award Nat. Turkey Fedn., 1972; named Nebr. Poultry Man of Yr. Nebr. Poultry Industries, 1982; Dept. Agr. grantee, 1967; Am. Egg Bd. grantee, 1979, 82; also various industry grants. Mem. Am. Chem. Soc., Poultry Sci. Assn. (sec.-treas.), Inst. Food Technologists, AAAS, Nat. Assn. Colls. and Tchrs. Agr., World's Poultry Sci. Assn., Sigma Xi, Phi Tau Sigma, Gamma Sigma Delta. Methodist. Lodge: Kiwanis. Subspecialty: Food science and technology. Current work: Functional and chem. properties of mechanically deboned poultry meat, functional properties of eggs, color chemistry of poultry meat, study of poultry meat and egg proteins as related to functionality. Office: Dept Animal Sci U Nebr 122 Mussehl Hall Lincoln NE 68583

FROST, DOUGLAS OWEN, neuroscientist, educator; b. N.Y.C., June 1, 1948; s. Seymour and Bernyce J. F. B.S., M.I.T., 1971, M.S., 1971, Ph.D., 1975. Premier asst. Inst. Anatomy, Faculty Medicine, U. Lausanne, Switzerland, 1975-78, maitre asst., 1979-80; asst. prof. sect. neuroanatomy Yale U. Med. Sch., 1980—. Contbr. articles to sci. publs. Woodrow Wilson fellow, 1969-70; grantee NIH, Fonds National Suisse. Fellow Swiss Am. Found. Sci. Exchange; mem. Soc. Neurosci., Internat. Brain Research Orgn., Sigma Xi, Eta Kappa Nu. Subspecialties: Neurobiology; Nervous system development and plasticity. Current work: Development of nervous system, plasticity of nervous system, regeneration, neuroanatomy, neurophysiology, neuropsychology. Office: Sect Neuroanatomy Yale Med Sch 333 Cedar St New Haven CT 06510

FROST, JOHN ELLIOTT, minerals company executive; b. Winchester, Mass., May 20, 1924; s. Elliott Putnam and Hazel Lavera (Carley) F.; m. Carolyn Catlin, July 12, 1945 (div. 1969); children: John Crocker, Jeffrey Putnam, Teresa Baird, Virginia Nicholl; m. Martha Hicks, June 6, 1969. B.S., Stanford U., 1949, M.S., 1950, Ph.D., 1965. Geologist Asarco, Salt Lake City, 1951-54; chief geologist, surface mines supt. Philippine Iron Mines Inc., Larap, Camarines Norte, 1954-60; chief geologist Duval Corp. (Pennzoil Corp.), Tucson, 1961-67; minerals exploration mgr. Exxon Corp., Houston, 1967-71; minerals mgr. Esso Eastern Inc. div., 1971-80; sr. v.p. Exxon Minerals Co. div., N.Y.C., 1980—; bd. dirs., mem. real estate com. United Engring. Trustees, N.Y.C., 1982—. Mem. adv. bd. Sch. Earth Scis., Stanford (Calif.) U., 1983—. Served to 1st lt. USAAF, 1943-45; PTO. Fellow Geol. Soc. Am.; mem. Soc. Econ. Geologists (councilor 1982—, program com., nominating com. chmn. 1982), AIME (chmn. edn. com. Soc. Mining Engrs. 1971), Australian Inst. Mining and Metallurgy, Sigma Xi. Republican. Congregationalist. Clubs: Mining (N.Y.C.); Mining of Southwest (Tucson). Subspecialties: Geoscience management; Geochemistry. Current work: Management responsibilities for minerals subsidiary including overall management of minerals exploration and minerals research programs. Originator porphyry copper zoning concept, 1958-60. Home: 12 Stonehenge Dr New Canaan CT 06840 Office: Exxon Minerals Company 1251 Ave of the Americas New York NY 10020

FROST, MELVIN JESSE, geography educator; b. Salt Lake City, Aug. 14, 1920; s. Clarence Alfred and Seraphine (Smith) F.; m. Dorothy Skousan, Mar. 16, 1951; children: Susan, Luisa, Melvin S., Robert S., Rita, Paul S. B.S., Ariz. State U.-Tempe, 1959; M.S., Brigham Young U., 1961; Ph.D., U. Fla., 1964. Gen. mgr. C.A. Frost Co., Monticello, Utah, 1945-47; owner, mgr. Frost Implement Co., Monticello, 1948-52; instr. phys. sci. U. Fla., 1962-64; assoc. prof. U. So. Miss., Hattiesburg, 1964-65; dir. geography student tchrs. Ariz. State U., Tempe, 1969-74, asst. prof. geography, 1965—. Author: A Decimal System for Calculating Angles, Time & Location, 1961. Pres. Mesa Hist. and Archeol. Soc., Mesa, 1973-77; chmn. Farm Museum, Mesa, 1982. Pan Am. Research Found. grantee, 1961; Caribbean Research Found. grantee, 1962; Ariz. State U. grantee, 1966, 68. Mem. Ariz. Acad. Sci. (sec. 1969-71), Latin Am. Studies Assn., Nat. Council Geographic Edn. (state coordinator 1966-68). Republican. Mormon. Subspecialties: Ecology; Resource management. Current work: Teaching geography of Latin America and environmental studies; research in solid waste recycling and management. Patentee device for showing time and dates. Home: 1748 N Lindsay St Mesa AZ 85203 Office: Department of Geography Arizona State University Tempe AZ 85287

FROYD, JAMES DONALD, plant pathologist; b. N.Y.C., May 25, 1939; s. Milton Carl and Mildred Grace (Nordine) F.; m. Elizabeth, Dec. 28, 1963; children: Allison, Jamie, Karl. B.S., Denison U., 1961; M.S., U. Minn., 1964, Ph.D., 1967. Instr., extension plant pathologist U. Minn., St. Paul, 1964-67; research plant pathologist Eli Lilly & Co., Greenfield, Ind., 1967—; instr. microbiology Butler U., Indpls., 1969-70. Mem. Am. Phytopath. Soc., Soc. Nematologists, Sigma Xi. Subspecialties: Plant pathology; Integrated pest management. Current work: Research in discovery and devel. of new chems. for control of plant diseases. Office: Eli Lilly & Co Greenfield IN 46140

FRUEH, ALFRED JOSEPH, crystallography educator; b. Passaic, N.J., Sept. 2, 1919; s. Alfred Joseph and Giuliette (Fanciulli) F.; m. Anne Torrey, Dec. 18, 1943; children: Jonathan, Carol, Timothy. B.S., MIT, 1942, M.S., 1947, Ph.D., 1949. Teaching asst. MIT, Cambridge, 1946-49; asst. prof. U. Chgo., 1949-58; assoc. prof. McGill U., Montreal, Que., Can., 1959-66, prof., 1966-69; dept. head, prof. dept. geology U. Conn., Storrs, 1969—; vis. prof. U. Oslo, 1958-59; vis. research assoc. Univ. Coll., U. London, 1976; mem. com. teaching crystallography Internat. Crystallographic Union, Chester, Eng., 1960-66; mem. com. on teaching sci. Internat. Council Sci. Unions, Paris, 1963-66. Editor: Structures Reports for 1956, 1963, Structure Reports for 1967, 1975; contbr. articles to profl. jours. Served to lt. USNR, 1942-46; PTO. Fellow Mineral. Soc. Am. (assoc. editor 1963-65); mem. Mineral. Soc. Can. (exec. com. 1962-65). Subspecialties: Mineralogy; X-ray crystallography. Current work: Stability of sulfide minerals, crystal structure and crystal chemistry, x-ray diffraction and non-bragg diffraction, order disorder and modulated structures, group theory in crystallography. Home: 23 Bundy Ln Storrs CT 06268 Office: Dept Geology and Geophysics U Conn Storrs CT 06268

FRUEND, GERHARD, medical educator; b. Frankfurt/Main, Germany, Apr. 21, 1926; came to U.S., 1951, naturalized, 1960; s. Adolf and Martha (Neuhaus) F.; m. Marion Healy, Sept. 24, 1955; children: Anne, Michael. M.D., Goethe U., Germany, 1951; M.S., McGill U., Can., 1957. Diplomate: Am. Bd. Internal Medicine. Assoc. prof. medicine U. Fla. Coll. Medicine, Gainesville, 1970-75, prof. medicine, 1975—, prof. neurosci., 1976—; mem. Ctr. for Neurobiol. Sci., 1967—; chief endocrinology VA Med. Ctr., Gainesville, 1970—. Fellow A.C.P.; mem. Soc. for Neurosci., Soc. for Biol. Psychiatry, Endocrine Soc., Research Soc. on Alcoholism (exec. com. 1981—). Subspecialties: Internal medicine; Neuropharmacology. Current work: Effects of aging and chronic exposure to alcohol on brain synaptic receptor functions, membranes and behavior in animals and man. Office: U Fla/VA Med Center Archer Rd Gainesville FL 32602 Home: 2031 NW 14th Ave Gainesville FL 32605

FRUMERMAN, ROBERT, chemical engineer; b. Rochester, Pa., Aug. 1924; m. Marcia; children: Bruce, Julie. B.S. in Chem. Engring, U. Pa.; M.S. in Chemistry, Carnegie Mellon U. Registered engr., Pa., N.Y., N.J., Fla. Process engr. Elliott Co., Blaw-Knox Chem. Plants Div.; project engr. Koppers Co. E & C Div.; mgr. chem. processing NUMEC; founder, pres. Frumerman Assocs Inc., Pitts., 1962—; cons. devel. of processes and systems for chem., metall., and coal conversion, and nuclear applications. Author publs. in field. Fellow Am. Inst. Chem. Engrs. Subspecialty: Chemical engineering. Patentee. Office: Frumerman Assocs 218 S Trenton Ave Pittsburgh PA 15221

FRY, FRANCIS J(OHN), research center administrator, electrical engineer, medical educator; b. Johnstown, Pa., Apr. 2, 1920; m., 1946; 9 children. B.S., Pa. State U., 1940; M.S., U. Pitts., 1946. Design engr. Westinghouse Electric Corp., 1940-46; research assoc. in elec. engring. U. Ill., Urbana, 1946-50, research asst. prof., 1950-57, research assoc. prof., 1957-68, assoc. prof., 1968-72, vis. assoc. prof. elec. engring., 1972—; assoc. dir. ultrasound research labs. Indpls. Ctr. Advanced Research, 1972-78, dir. ultrasound research labs., 1978—; assoc. prof. surgery Ind. Sch. Medicine, 1972—; v.p. Intersci. Research Inst., 1957-68, pres., 1968-70, sec., chmn. bd. dirs., 1970-72. Mem. Am. Inst. Ultrasound Medicine (Pioneer award 1981), Am. Soc. Artificial Internal Organs, Neurosci. Soc., Biomed. Engring. Soc., Acoustical Soc. Am. Subspecialties: Electrical engineering; Ultrasound. Office: Fortune-Fry Ultrasound Research Labs IUPUI 410 Beauty Ave Indianapolis IN 46202

FRY, JAMES LESLIE, chemistry educator; b. Fostoria, Ohio, May 24, 1941; m. Sara Ann Reiss; 1 son, Christopher Wellington. B.S. in Chemistry, Bowling Green State U., 1963; Ph.D. in Organic Chemistry, Petroleum Research Fund fellow, Mich. State U., 1967. NIH postdoctoral fellow Princeton U., N.J., 1967-69; asst. prof. chemistry U. Toledo, 1969-73, assoc. prof., 1973-78, prof., 1978—; vis. scholar Université des Sciences et Techniques du Languedoc, Montpellier, France, 1982-83; tech. cons. Contbr. articles to profl. jours. Grantee NSF, Research Corp., Petroleum Research Fund. Mem. Am. Chem. Soc., Royal Soc. Chemistry (London), Sigma Xi. Subspecialty: Organic chemistry. Current work: Organic and physical organic chemistry; organosilicon chemistry; deuterium kinetic isotope effects. Patentee in organic chemistry. Office: Dept Chemistry Univ Toledo 2801 W Bancroft St Toledo OH 43606 Home: 2527 Gradwohl Rd Toledo OH 43617

FRY, JAMES N., astrophysicist; b. Phila., Aug. 6, 1952; s. C. Herbert and Barbara (McGuire) F.; m. Kristin Schleich, Mar. 11, 1982. B.S., Cornell U., 1974; M.A., Princeton U., 1976, Ph.D., 1979. Robert McCormick postdoctoral fellow U. Chgo., 1979-81, research asso., 1981-82, sr. research asso., 1982—; vis. scholar Inst. Astronomy, Cambridge, Eng., 1982. Contbr. articles in field to profl. jours. Mem. Am. Astron. Soc., Am. Phys. Soc. Subspecialties: Theoretical astrophysics; Cosmology. Office: Astronomy and Astrophysics Center 5640 S Ellis Ave Chicago IL 60639

FRY, JAMES PALMER, research scientist, university research group director; b. Detroit, May 2, 1939; s. Palmer Edmund and Anne Joanna (Olson) F.; m. Rosemary S. King, June 24, 1961; children: Mary Ann, James Palmer Jr., Leif Christian, Erika Diane, Benjamin Jotham, Joshua. B.S., U. Mich., 1962, M.S. in Engring, 1963, 1974. Area engr. E.I. DuPont de Nemours & Co., Montague, Mich., 1962-63; assoc. research engr. Boeing Co., Renton, Wash., 1964-65; mem. tech. staff and project leader MITRE Corp., McLean, Va., 1965-69; mem. faculty U. Mich., Ann Arbor, 1970—, dir. info. systems research group, 1977—. Author: Structured Databases, 1981; contbr. articles to Design sci. publs. Mem. Assn. Computing Machinery (dir. 1973—), Conf. on Data Systems Lang. (chmn. stored-data definition task group 1969-74), Share Inc. (chmn. database project 1968-70). Lutheran. Subspecialties: Distributed systems and networks; Information systems, storage, and retrieval (computer science). Home: 6130 Munger Rd Ypsilanti MI 48197 Office: Univ Mic Grad Sch Bus Adminstrn Ann Arbor MI 48109

FRY, RICHARD JEREMY MICHAEL, physiologist, radiobiologist, physician; b. Dublin, Ireland, July 8, 1925; m., 1956; 3 children. B.A., U. Dublin, 1946, M.B., B.Ch. and B.A.O., 1949, M.D., 1962. Lectr. in physiology U. Dublin, 1952-59, 61-63; resident research assoc. in radiobiology Argonne Nat. Lab., 1959-61, assoc. scientist, 1963-70, sr. scientist in radiobiology, 1970-77; sect. head, cancer and toxicology sect., biol. div. Oak Ridge Nat. Lab., 1977—; prof. radiology U. Chgo., 1970-77. Mem. Radiation Research Soc., Am. Assn. Cancer Research, Am. Soc. Photobiology. Cell Kinetics Soc. Subspecialty: Radiobiology. Office: Biol Div Oak Ridge Nat Lab PO Box Y Oak Ridge TN 37830

FRYE, CECIL LEONARD, chemist, researcher; b. Dearborn, Mich., Apr. 3, 1928; m., 1952; 4 children. B.S., U. Mich., 1950, M.S., 1951; Ph.D. in Organic Chemistry, Pa. State U., 1960. Research chemist Dow Corning Corp., Midland, Mich., 1951-56, project leader research dept., 1960-63, lab. supr., 1963-70, mgr. sealants research unit, 1970-72, mgr. bio-sci. chem. research, 1973-74, scientist, corp. research, 1975-80, scientist dept. health and environ. sci., 1980—. Mem. Am. Chem. Soc., Sigma Xi. Subspecialties: Silicon chemistry; Environmental chemistry. Office: Dow Corning Corp PO Box 176 2200 W Salzburg Rd Midland MI 48640

FRYE, GERALD DALTON, biomedical researcher, educator; b. Winchester, Va., Aug. 27, 1950; s. Eldred Dalton and Rhoda Mae (Fletcher) F.; m. Janet Pierce, Sept. 11, 1971. B.S., Va. Poly. Inst. and State U., 1972; Ph.D. in Pharmacology, U. N.C., 1977. Postdoctoral fellow Biol. Scis. Research Ctr., U. N.C., Chapel Hill, 1977-79, research asst. prof. dept. psychiatry, 1979-83; asst. prof. dept. med. pharmacology and toxicology Tex. A&M U., 1983—. Recipient Nat. Research Service award Nat. Inst. Alcohol Abuse and Alcoholism, 1977-79, grantee, 1982—. Mem. Am. Soc. for Pharmacology and Exptl. Therapeutics, Soc. for Neuroscis., AAAS, Research Soc. on Alcoholism. Baptist. Subspecialties: Neuropharmacology; Neurochemistry. Current work: The role of the neurochemical gamma aminobutyric acid in drug-induced intoxication and physical dependence and in epilepsy. Office: Dept Med Pharmacology and Toxicology Tex A&M U College Station TX

FRYE, HERSCHEL GORDON, chemist, educator; b. Long Beach, Calif., Apr. 6, 1920; s. Herschel Guy and Neva Jane (Yancey) F.; m. Grace Helen Tener, Sept. 26, 1940; children: Wayne Herschel, Gordon Evans. B.A., Coll. of Pacific, 1947, M.A., 1949; Ph.D., U. Oreg., 1956. Asst. prof. chemistry U. of Pacific, Stockton, Calif., 1956-59, assoc. prof., 1959-62, prof., 1962—; partner FM Analytical Cons., Stockton, 1968—. Contbr. articles to profl. jours. Cons. to dist. atty. and pub. defender, San Joaquin County (Calif.), 1970—. Served with U.S. Army, 1944-46. Amn. Philos. Soc. grantee; Research Corp. grantee, 1958-68. Mem. Am. Chem. Soc., Am. Acad. Forensic Scis., AAAS, Cons. Chemists Assn., Calif. Assn. Toxicologists, N.Y. Acad. Scis. Subspecialties: Analytical chemistry; Forensic chemistry. Current work: Analytical chemistry of restricted drugs in biofluids; synthetic inorganic chemistry of platinum group metals. Office: 1231 W Robinhood Dr Suite D-3 Stockton CA 95207

FRYER, JOHN L., microbiology educator; b. Ft. Worth, July 4, 1929; s. George G. and LoRene W. (Sonnendriker) F.; m. Shirley White, June 9, 1952; 1 child, Shelley; m. Mary, July 3, 1966; children: Heather, Holy, Stephanie. B.A., Oreg. State U., 1956, M.S., 1957, Ph.D., 1964. Fisheries biologist dept. fisheries and wildlife Oreg. State U., 1957-58; biologist Oreg. Fish Commn., 1958-60, state fisheries pathologist, 1960-63; research assoc., asst. prof. dept. food sci. and tech. Oreg. State U., Corvallis, 1963-64, from asst. prof. to prof. dept. microbiology, 1964—, acting chmn. dept. microbiology, 1976-77, chmn., 1977—. Contbr. articles to profl. jours. Served with USMC, 1950-52. Decorated Purple Heart; recipient Elizabeth P. Ritchie Disting. Prof. award Oreg. State U., 1972, Carter award, 1972. Mem. Am. Soc. Microbiology (Carski Disting. Teaching award 1982), Am. Fisheries Soc. (S.F. Snieszko Disting. Service award 1982), Am. Acad. Microbiology, Am. Soc. Virology, European Assn. Fish Pathologists, Sigma Xi, Phi Kappa Phi, Gamma Sigma Delta. Subspecialties: Microbiology; Microbiology (veterinary medicine). Current work: Infectious diseases of fishes; tissue culture and viral diseases of fish; pathogenic microbiology and immunology of cold blooded animals. Office: Dept Microbiology Oreg State U Corvallis OR 97331

FRYXELL, PAUL ARNOLD, research botanist, educator; b. Moline, Ill., Feb. 2, 1927; s. Hjalmar Edward and Hulda Eunice (Peterson) F.; m. Greta A. Fryxell, Aug. 23, 1940; children: Karl J., Joan E., Glen E. B.A., Augustana Coll., Rock Island, Ill., 1949; M.S., Iowa State U., 1951, Ph.D., 1955. Asst. agronomist N.Mex State U., Las Cruces, 1952-55; asst. prof. botany U. Wichita, 1955-57; research geneticist U.S. Dept. Agr., Tempe, Ariz., 1957-65; research botanist, College Station, Tex., 1965—; mem. grad. faculty Tex. A&M U., 1965—, U. Tex., 1980—. Author: Natural History of the Cotton Tribe, 1979; contbr. over 100 articles to sci. jours. Pres. local chpt. ACLU, 1968. Served in USAAF, 1945-46. Mem. Am. Soc. Plant Taxonomists (pres. 1983-84), Bot. Soc. Am., AAAS, Internat. Assn. Plant Taxonomists, Am. Soc. Naturalists, Soc. Study of Evolution. Unitarian-Universalist. Subspecialties: Taxonomy; Evolutionary biology. Current work: Taxonomy of the family Malvaceae; revision of selected genera of the family; floristic studies. Office: PO Drawer DN College Station TX 77840

FU, KAREN KING-WAH, radiation oncologist; b. Shanghai, China, Oct. 15, 1940; came to U.S., 1959, naturalized, 1975; d. Ping Sen and Lein Sun (Ho) F. Student, Ind. U., 1959-61; A.B., Barnard Coll., Columbia U., 1963, M.D., 1967. Cert. radiation oncologist. Intern Montreal (Que., Can.) Gen. Hosp., 1967-68; resident Princess Margaret Hosp., Toronto, Ont., Can., 1968-69, Stanford U. Hosp., 1969-71; instr. U. Utah, 1971-72; clin. instr. U. Calif.-San Francisco, 1972-73, asst. prof., 1973-76, assoc. prof., 1976-82, prof., 1982—, research assoc. Cancer Research Inst., 1973—. Contbr. articles to profl. jours. Mem. San Francisco Opera Guild, San Francisco Symphony Assn., San Francisco Ballet, Calif. Acad. Sci., De Young Mus. Am. Cancer Soc. grantee, 1982; NIH grantee, 1982. Mem. Am. Soc. Therapeutic Radiologists, Am. Med. Women's Assn., Am. Coll. Radiology, Calif. Radiation Therapy Assn., Calif. Radiol. Soc., No. Calif. Acad. Clin. Oncology, Radiation Research Soc., Am. Soc. Clin. Oncologists, Cell Kinetics Soc., Assn. Women in Sci. Subspecialties: Oncology; Cancer research (medicine). Current work: Cancer research, patient care, teaching in medical school. Office: U Calif San Francisco Dept Radiation Oncology L-75 San Francisco CA 94143

FU, KING-SUN, electrical engineer; b. China, Oct. 2, 1930; s. Tzao-jen and Tzao-wen (Hsiang) F.; m. Viola Ou, Apr. 7, 1958; children: Francis, Thomas, June. B.S., Nat. Taiwan U., 1953; M.A.Sc., U. Toronto, 1955; Ph.D., U. Ill., 1959. Research engr. Boeing Airplane Co., 1959-60; mem. faculty Purdue U., 1960—, prof. elec. engring., 1966—, Goss disting. prof., 1975—. Author: Sequential Methods in Pattern Recognition and Machine Learning, 1968, Syntactic Methods in Pattern Recognition, 1974, Statistical Pattern Classification using Contextual Information, 1980, Syntactic Pattern Recognition and Applications, 1982. Guggenheim fellow, 1972. Fellow IEEE; mem. Nat. Acad. Engring., Academia Sinica, Am. Soc. Engring. Edn., Assn. Computing Machinery. Subspecialties: Graphics, image processing, and pattern recognition; Artificial intelligence. Current work: Pattern recognition and image processing; image database management; expert systems; special computer architectures. Home: 132 Rockland Dr West Lafayette IN 47906 Office: Sch Elec Engring Purdue U West Lafayette IN 47907

FU, SHOU-CHENG JOSEPH, biomedicine educator; b. Peking, China, Mar. 19, 1924; came to U.S., 1946; s. W.C. Joseph and W.C. (Tsai) F.; m. Susan B. Guthrie, June 21, 1951; children: Robert W.G., Joseph H.G., James B.G. B.S., M.S., Catholic U., Peking, 1944; Ph.D., Johns Hopkins U., 1949. Gustav Bissing fellow Johns Hopkins U., at Univ. Coll., London, 1955-56; chief Enzyme and Bioorganic Chemistry Lab. Children's Cancer Research Found., 1956-67; research assoc. Harvard U. Med. Sch., Boston, 1956-67; prof., chmn. bd. chemistry Chinese U., Hong Kong, 1967-70, dean sci. faculty, 1967-69; vis. prof. Coll. Physicians and Surgeons, Columbia U., N.Y.C., 1970-71; prof. biochemistry U. Medicine & Dentistry, Newark, 1971—, asst. dean, 1975-77, acting dean, 1977-78. Contbr. articles to profl. jours. Served to lt. commdr. USPHS Res., 1959—. Fellow AAAS, Royal Soc. Chemistry (London); mem. Sigma Xi. Club: Royal Hong Kong Jockey: American (Hong Kong). Subspecialties: Biochemistry (medicine); Biophysical chemistry. Current work: Amino acids, peptides and proteins; folate and anti-folate agents in cancer research; mammalian lens protein in cataractogenesis. Home: 693 Prospect St Maplewood NJ 07040 Office: 100 Bergen St Newark NJ 07103

FU, SHUMAN, research scientist; b. Canton, China, July 29, 1942; s. Tak-Yu and Lai-Sim (Tang) F.; m. Felicia Gaskin, Jan. 29, 1969;

children: Kai Ming, Kai Mei. B.S., Dickinson Coll., 1965; M.D., Stanford U., 1970; Ph.D., Rockefeller U., 1975. Diplomate: Am. Bd. Internal Medicine. Intern, resident in medicine Stanford U. Med. Center, 1970-71, N.Y. Hosp., 1975-76; grad. fellow Rockefeller U., 1971-75, asst. prof., then asso. prof., 1975-82; prof. U. Okla., 1982—; mem. Okla. Med. Research Found., 1982—. Scholar Leukemia Soc. Am., 1976-81. Mem. Am. Soc. Clin. Investigation, Am. Assn. Immunolgists, AAAS, Am. Rheumatism Assn., Phi Beta Kappa. Subspecialty: Immunology (medicine). Current work: Human immunology with emphasis on B cell proliferation and differentaition in normal and disease states. Office: 825 NE 13th St Oklahoma City OK 73104

FU, WEI-NING, genetics educator; b. Huang Hsien, China, Feb. 24, 1925; s. Dung-Chen and Shih (Yu) F.; m. Lucia Yueh, Feb. 6, 1955; children: Jean, Aileen, Daniel. B.S., Nat. Honan U., 1949; M.S., Okla. State U., 1960, Ph.D., 1963. Tchr. biology Wen-Shan High Sch., Taipei, Taiwan, 1949-52; assoc. agronomist Taiwan Tobacco Research Inst., Taichung, 1952-57; grad. asst. Okla. State U., Stillwater, 1960-63; mem. faculty dept. biology Central Conn. State U., New Britain, 1963—, prof. genetics, 1963—. NSF grantee, 1967. Mem. AAAS, Am. Soc. Human Genetics, Am. Hort. Soc., Sigma Xi. Subspecialties: Animal genetics; Plant genetics. Current work: Interest in study of human chromosome, diagnosis, banding and population cytogenetics in the Chinese population. Office: Dept Biological Sciences Central Conn State U New Britain CT 06050

FUCCILLO, DAVID ANTHONY, virologist, immunologist; b. Chelsea, Mass., Aug. 20, 1930; s. Anthony and Marion (Catalano) F.; m. Norma Ellen Regis, Apr. 19, 1955; children: Joanne, David Regis. B.S. in Bacteriology, U. Mass., Amherst, 1953, M.S. in Microbiology, 1955, Ph.D., Purdue U., 1960. Commd. 2d lt. U.S. Air Force, 1955, advanced through grades to capt., 1965; chief lab. services USAF Hosp., Williams AFB, 1955-57, officer charge microbiology dept. and blood bank transfusion service, Lackland AFB, 1960-62, officer in charge microbiology dept. and clin. lab., Weisbaden, W.Ger., 1962-65; ret., 1965; asst. head sect. infectious diseases Perinatal Research br. Nat. Inst. Neurol. Diseases and Stroke, NIH, 1965-71, asst. chief infectious diseases br. collaborative and field research, 1971-76; dir. advanced testing and devel. labs. Biomed. Research div. Litton Bionetics, Inc., 1976-80, mgr. biol. lab. products, Kensington, Md., 1978-80; dir. research and devel. M.A. Bioproducts, Walkersville, Md., 1980—; chmn. subcom. human herpesvirus Nat. Com. Lab. Standards; advisor grad. students Hood Coll. Contbr. articles to profl. jours. Recipient USPHS commendation medal for outstanding achievement, 1976. Mem. Am. Assn. Immunologists, Am. Soc. Microbiology, Soc. Exptl. Biology and Medicine, Sigma Xi. Subspecialties: Virology (biology); Immunobiology and immunology. Current work: Development of immunological assays, development of enzyme-linked immunosorbent assays for detection of antibodies and antigens for various infections. Home: 11801 Seven Locks Rd Rockville MD 20854 Office: Bldg 100 Biggs Ford Rd Walkersville MD 21793

FUDENBERG, H. HUGH, immunologist, university administrator; b. N.Y.C., Oct. 24, 1928; m., 1955; 4 children. A.B., UCLA, 1949; M.D., U. Chgo., 1953; M.A., Boston U., 1957; Dr. h.c., U. Kuopio (Finland), 1982. Intern in medicine U. Utah, 1953-54; fellow in hematology Sch. Medicine, Tufts U., 1954-56; asst. resident in medicine Mt. Sinai Hosp., N.Y.C., 1956-57, Peter Bent Brigham Hosp., Boston, 1957-58; research assoc. in immunology Rockefeller Inst., 1958-60; prof. bacteriology and immunology U. Calif.-Berkeley, 1966-75; from asst. prof. to prof. medicine U. Calif.-San Francisco, 1966-75, chief hematology unit, 1962-75; chmn. dept. basic and clin. immunology Med. U.S.C., 1975—; mem. expert com. immunology WHO, 1962-77; participant Alfred Nobel 150th Birthday Celebration, 1983; mem. sci. adv. bd. Pasteur Inst., UNESCO Research Inst. Chief editor: Jour. Clin. Immunology. Recipient Pasteur medal Inst. Pasteur, Paris, 1962, Robert A. Cooke Meml. medal Am. Acad. Allergy, 1967. Fellow AAAS; mem. Am. Assn. Immunologists, Am. Soc. Human Genetics, Am. Soc. Clin. Investigation, Genetics Soc. Am. Subspecialties: Immunology (medicine); Immunogenetics. Office: Dept Basic and Clin Immunology Med U SC Charleston SC 29425

FUERSTENAU, DOUGLAS WINSTON, materials science educator; b. Hazel, S.D., Dec. 6, 1928; s. Erwin Arnold and Hazel Pauline (Karterud) F.; m. Margaret Ann Pellett, Aug. 29, 1953; children: Lucy, Sarah, Stephen. B.S., S.D. Sch. Mines and Tech., 1949; M.S., Mont. Sch. Mines, 1950; Sc.D., MIT, 1953; Mineral Engr., Mont. Coll. Mineral Sci. and Tech., 1968. Asst. prof. mineral engring. MIT, 1953-56; sect. leader, metals research lab. Union Carbide Metals Co., Niagara Falls, N.Y., 1956-58; mgr. mineral engring. lab Kaiser Aluminum & Chem. Corp., Permanente, Calif., 1958-59; asso. prof. metallurgy U. Calif. at Berkeley, 1959-62, prof. metallurgy, 1962—, Miller research prof., 1969-70, chmn. dept. materials sci. and mineral engring., 1970—; dir. Homestake Mining Co.; Chmn. Engring. Found. Research Conf. on Comminution, 1963; Mem. adv. bd. Sch. Earth Scis., Stanford, 1970-73; mem. Nat. Mineral Bd., 1975—; Am. rep. Internat. Mineral Processing Congress Com., 1978—. Editor: Froth Flotation-50th Anniversary Vol, 1962; co-editor-in-chief: Internat. Jour. of Mineral Processing, 1972—; Mem. editorial adv. bd.: Jour. of Colloid and Interface Sci, 1968-72, Colloids and Surfaces, 1980—; Contbr. articles to profl. jours. Recipient Distinguished Teaching award U. Calif., 1974; Fellow Instn. Mining and Metallurgy, London. Mem. Nat. Acad. Engring., Am. Inst. Mining and Metall. Engrs. (chmn. mineral processing div. 1967, Robert Lansing Hardy Gold medal 1957, Rossiter W. Raymond award 1961, RobertH. Richards award 1975, Antoine M. Gaudin award 1978, Mineral Industry Edn. award 1983), Soc. Mining Engrs. (dir. 1968-71, Distinguished mem.), Am. Chem. Soc., Am. Inst. Chem. Engrs., Sigma Xi, Theta Tau. Congregationalist. Subspecialties: Metallurgical engineering; Surface chemistry. Current work: Extractive metallurgy, mineral processing, surface and collid chemistry, particle science and technology, marine resources, coal benefication. Home: 1440 LeRoy Ave Berkeley CA 94708

FUHR, JEFFREY ROBERT, physicist; b. Balt., Oct. 27, 1946; s. Irvin and Ruby Lillian (Rubenstein) F.; m. Barbara Ann, Sept. 17, 1972; children: Lisa, Daniel. A.B., Earlham Coll, 1968; M.S. in Physics, Purdue U., 1970. Physicist Nat. Bur. Standards, Washington, 1970—. Contbr. writings to profl. publs. in field. Pres. Relda Sq. Homeowners Assn., 1981—. Mem. Am. Phys. Soc. Club: Saturday in Style Square Dance (pres.). Subspecialty: Atomic and molecular physics. Current work: Compile and publish critically evaluated data on atomic transition probabilities; cons. availability and reliability atomic data. Home: 9 Pavilion Dr Gaithersburg MD 20878 Office: Nat Bur Standards Physics Bldg Room A267 Washington DC 20234

FUJIMOTO, GEORGE I., biochemistry researcher, educator; b. Seattle, July 1, 1920; s. Richard R. and Hisano (Okuda) F.; m. Mary Yano, June 13, 1949; children: Tara, Gary, Brad. B.A., Harvard U., 1942; M.S., U. Mich.-Ann Arbor, 1945, Ph.D., 1947. Research fellow Calif. Inst. Tech., 1947-49; research asst. prof. U. Utah, Salt Lake City, 1949-55; asst. prof. biochemistry Albert Einstein Coll. Medicine, Bronx, N.Y., 1955-58, assoc. prof., 1958—. Mem. Am. Soc. Biol. Chemists, Endocrine Soc., AAAS, Soc. Study Reprodn. Subspecialties: Neuroendocrinology; Reproductive biology. Current work: Reproductive endocrinology. Home: 101 Carthage Rd Scarsdale NY 10583 Office: Albert Einstein College of Medicine 1300 Morris Park Ave Bronx NY 10461

FUJIMURA, OSAMU, physicist, researcher; b. Tokyo, Aug. 29, 1927; U.S., 1973; s. Susumu and Sei (Tsuneyoshi) F.; m. Yoko Sugai, Dec. 15, 1957; children: Akira, Makoto. B.S., U. Tokyo, 1952, D.Sc., 1962. Guest researcher Royal Inst. Tech., Stockholm, 1963-65; staff research mem. M.I.T., Cambridge, 1958-61; prof. Research Inst. Logopedics and Phoniatrics, U. Tokyo, 1965-73; head dept. Bell Labs., Murray Hill, N.J., 1973—; mem. sci. adv. bed. Voice Found., N.Y.C.; mem. vis. com. for linguistics and philosophy M.I.T.; mem. vis. com. for linguistics Harvard U. Mem. editorial bd.: Sprachwissenschaft und Kommunikationsforschung, Phonetica; contbr. over 300 articles to sci. jours. Fellow Acoustical Soc. Am., N.Y. Acad. Scis.; mem. IEEE, Linguistic Soc. Am. (exec. com.), Soc. for Studies Japanese Lang., Phys. Soc. Japan, AAAS, Am. Assn. Physicists in Medicine, Internat. Soc. Phonetic Scis., Internat. Coll. Exptl. Phonology, Am. Speech and Hearing Assn. Clubs: Internat. House of Japan (Tokyo); Gakusikai (U. Tokyo). Subspecialties: Linguistics and speech science; Acoustics. Current work: Speech and language research. Patentee in U.S. and Japan. Office: 600 Mountain Ave Room 2D-545 Murray Hill NJ 07974

FUJIMURA, ROBERT KANJI, biochemist, researcher, lecturer; b. Seattle, July 28, 1933; s. Tatsuo and Tamiko Ruth F.; m. Shigeko Ichikawa, Dec. 1, 1962; children: Dan, Tomi, Kei. B.S., U. Wash., 1956; M.S., U. Wis.-Madison, 1959, Ph.D., 1961. Postdoctoral fellow Osaka (Japan) U., 1961-62, U. Wis.-Madison, 1963; sr. research staff Oak Ridge Nat. Lab., 1963—; lectr. U. Tenn., Oak Ridge, 1971—; chmn. Biology Div. Research Conf., Gatlinburg, Tenn., 1980. Research fellow Japan Soc. Promotion Sci., 1981. Mem. Am. Soc. Biol. Chemists, AAAS, Sigma Xi. Methodist. Subspecialties: Biochemistry (biology); Molecular biology. Current work: Control of DNA replication in vitro; interactions of DNA polymerase and proteins involved in DNA replication with each other and with DNA; correlation of structure and function of these proteins. Home: 109 Emerson Circle Oak Ridge TN 37830 Office: Biology Div Oak Ridge Nat Lab Oak Ridge TN 37830

FUJINAMI, ROBERT SHIN, immunovirologist, educator; b. Salt Lake City, Dec. 8, 1949; s. Mitsuru and Taiko (Kawaguchi) F.; m. Christine Duder, June 23, 1973. B.S. with honors, U. Utah, 1972; Ph.D., Northwestern U., 1977. Instr. dept. microbiology-immunology Northwestern U., Chgo., 1973-76; research fellow dept. immunopathology Scripps Clinic and Research Found., La Jolla, Calif., 1977-80, research assoc., 1980-81, asst. mem., 1981—; research immunopathologist dept. pathology U. Calif., San Diego, 1980—. Contbr. articles to profl. jours. Recipient Harry M. Weaver award Nat. Multiple Sclerosis Soc., 1982; NIH fellow, 1977-1979; NIH grantee, 1981-84. Mem. AAAS, Am. Soc. Microbiology, Am. Assn. Immunologists. Subspecialties: Neuroimmunology; Virology (medicine). Current work: Viral Persistence, neuroimmunology, autoimmunity. Office: Scripps Clinic and Research Found Dept Immunology 10666 N Torrey Pines Rd La Jolla CA 92037

FUJISHIRO, KATAKAZU KENNETH, regional and environmental planner; b. Cambridge, Mass., Sept. 25, 1932; s. Shinji and Yasu (Matsudaira) F.; m. Jane Foster Eubanks, Nov. 22, 1973; 1 stepdau., Joni Eubanks. B.S.C.E., U. S.C., 1964; cert., Civil Def. Staff Coll., Battle Creek, Mich., 1967, Rensselaer Poly. Inst., Troy, N.Y., 1969, Ga. Inst. Tech., Atlanta, 1970, Mich. Tech. U., Houghton, 1972. Casualty underwriter Am. Internat. Underwriters, Tokyo and N.Y.C., 1948-57; engr. Charles J. Craig Constrn. Co., Columbia, S.C., 1958-65; engr., planner LBC&W Assocs., Columbia, 1965-73; planner dir. BCD Regional Planning Council, Charleston, S.C., 1973-76; prin. planner Metro Washington Council of Govt., 1976-79; sr. program engr. Advanced Technology, Inc., McLean, Va., 1979—. Author devel. plans, environ. assessments, cons. reports on facilities engring., devel. and environment. Commr. Charleston County (S.C.) Parks, Recreation and Tourist Commn., 1973-76; mem. Charleston Am. Revolution Bicentennial Sterring Com., 1975; mem. environment com. Nat. Assn. Regional Councils, 1975; mem. S.C. Coastal Zones Planning Tech. Com., Columbia, 1973-76, Trident Transp. Planning Com., Charleston, 1974-76. Served with U.S. Army, 1954-56. Mem. Am. Inst. Cert. Planners (charter), Am. Inst. Planners (nat. bd. examiners 1972-78, v.p. S.C. chpt. 1975-76, Service award 1976), Am. Planning Assn. (v.p. Nat. Capital Area chpt. 1978-82, Service award 1979), Soc. Am. Mil. Engrs., Am. Nuclear Soc. Episcopalian. Subspecialties: Civil engineering; Environmental engineering. Current work: Urban, regional and environmental planning including environmental impact statements, areawide water quality management plans, nuclear defense and multi-diaster design curricula development and teaching; local and regional government planning. Home: 5804-81 Merton Ct Alexandria VA 22311 Office: Advanced Technology Inc 7923 Jones Branch Dr McLean VA 22102

FUKS, JOACHIM ZBIGNIEW, physician, educator clinical researcher; b. Poland, Mar. 18. 1948; came to U.S., 1971; s. Herszlik and Bajla (Slawna) F.; m. Gail Katz, Mar. 3, 1978; 1 dau., Sara Esther. M.D., Med. U. Barcelona, 1975. Diplomate: Am. Bd. Internal Medicine. Intern Mt. Sinai Hosp., N.Y., 1975-76, resident in internal medicine, 1976-78; clin. assoc. BCRP U. Md., Balt., 1978-81, sr. investigator, 1981-82, staff physician, 1982—, asst. prof. medicine, oncology, 1982—; staff physician VA Balt. Hosp., 1982—. Served with USPHS, 1980-81. Recipient Clin. research award Polish Student's Assn., 1967, Polish Student's Sci. Orgn., 1968. Mem. Am. Soc. Clin. Oncology, Am. Fedn. Clin. Research, Am. Coll. Physicians, N.Y. Acad. Scis. Democrat. Jewish. Subspecialties: Chemotherapy; Cancer research (medicine). Current work: Currwnt work: New drug development, pharmacokinetics, assessment of new drugs, and clinical trials in solid tumors. Home: 6708 Bonnie Ridge Dr Baltimore MD 21209 Office: UMCC U Md 22 S Green St Baltimore MD 21201

FUKUDA, MINORU, cancer research scientist; b. Hiroshima, Japan, July 6, 1945; came to U.S., 1975, naturalized, 1980; s. Iwao and Sueko (Fujiwara) F.; m. Michiko Nishida, Apr. 8, 1970; children: Ko, Shun. B.S. in Biochemistry, U. Tokyo, 1968, M.S., 1970, Ph.D., 1973. Research assoc. U. Tokyo, 1973-75; postdoctoral assoc. Yale U., 1975-77; assoc. Hutchinson Cancer Research Ctr., Seattle, 1977-81; asst. prof. U. Wash., Seattle, 1980-81; staff scientist La Jolla (Calif.) Cancer Research Found., 1982—. Author: Biology of Glycoproteins, 1983. Nat. Cancer Inst. grantee, 1981, 83. Mem. Am. Soc. Cell Biology, Am. Soc. Biol. Chemists. Subspecialties: Biochemistry (biology); Cancer research (medicine). Current work: Biochemistry of glycoproteins in differentiation and oncogenesis; now directing a laboratory composed of 6 researchers, of which the objective is to study membrane differentiation in normal and leukemic human blood cells. Home: 2818 Passy Ave San Diego CA 92122 Office: La Jolla Cancer Research Foundation 10901 N Torrey Pines Rd La Jolla CA 92037

FULBRIGHT, DENNIS WAYNE, plant pathology educator; b. Lynwood, Calif., Aug. 20, 1952; s. Ross Edward and Beulah Mae (Hollingshead) F.; m. Joanne Shutt, May 30, 1953; 1 dau., Kimberly Ann. B.A., Whittier Coll., 1974; Ph.D., U. Calif.-Riverside, 1979. Asst. prof. botany and plant pathology Mich. State U., East Lansing, 1979—. Mem. Sigma Xi. Subspecialties: Plant pathology; Microbiology. Current work: Wheat pathology, genetics of pathogenicity of pathogens, mycoviruses in chestnut blight pathogen Endothia parasitica and its hypovirulent nature. Office: Dept Botany and Plant Patholog Mich State U East Lansing MI 48824

FULFORD, PHILLIP JAMES, nuclear engineer; b. Winnipeg, Man., Can., Oct. 18, 1935; came to U.S., 1964; s. Philip J. and Christina (Kirton) F.; m. Patricia Lee Parker, Oct. 6, 1962; childen: Christy, John, Peter. B.Sc., U. Man., 1957; M.Sc., U. Birmingham, Eng., 1958; Ph.D., Purdue U., 1968. Registered profl. engr., Ind. Engr. Atomic Energy Can., Toronto, Ont., 1960-63, Dilworth Secord Meagher, Toronto, 1963-64; assoc. prof. Purdue U., West Lafyette, Ind., 1968-80; mgr. safety analysis NUS Corp., Gaithersburg, Md., 1980—; cons. Recipient Gold medal U. Man., 1957; Athlone fellow Govt. U.K., 1957-59; ASEE/Ford Found. fellow, 1971. Mem. Am. Nuclear Soc., Sigma Xi. Subspecialties: Nuclear engineering; Nuclear fission. Current work: Degraded core phenomenology—applications to quantify risk of nuclear power. Home: 12001 Devilwood Dr Potomac MD 20854 Office: 910 Clopper Rd Gaithersburg MD 20878

FULKERSON, SAMUEL COLE, psychology educator, consultant; b. Cedar Rapids, Iowa, Jan. 5, 1923; s. Samuel and Ruth Amanda (DeSilva) F.; m. Katharine Alger James, May 6, 1947; children: Gregory, Heidi. B.A., U. Iowa, 1948, M.A., 1951; Ph.D., U. Tex.-Austin, 1955. Intern Worchester (Mass.) State Hosp., 1952-53; psychologist U.S. Air Force Sch. Aviation Medicine, San Antonio, 1954-58; assoc. prof. psychology U. Pitts., 1958-61; prof. U. Louisville, 1961—; cons. in field. Contbr. articles to profl. jours., 1955—. Served to cpl. USAAF, 1943-45; CBI. Recipient award for journalistic excellence Sigma Delta Chi, U. Iowa, 1948. Fellow Am. Psychol. Assn.; mem. Psychonomic Assn., U.S. Chess Fedn. (life), Am. Mycologyical Assn., Sierra Club, Sigma Xi. Democrat. Subspecialty: Cognition. Current work: Cognitive analysis of intelligence tasks and personality items. Home: 1021 Watterson Trail Louisville KY 40299 Office: Dept Psychology U Louisville Louisville KY 40292

FULLAGAR, PAUL DAVID, geology educator; b. Ft. Edward, N.Y., Dec. 19, 1938; s. William Alfred and Evelyn Louise (Hoyt) F.; m. Patricia Ann Kelley, June 6, 1959; children: Scott David, Eric Cgraig. B.A., Columbia U., 1960; Ph.D., U. Ill., 1963. Asst. prof. geology Old Dominion Coll., Norfolk, Va., 1963-67; asst. prof. geology U. N.C., Chapel Hill, 1967-69, assoc. prof. geology, 1969-73, prof. geology, 1973—; analytical geochemist Goddard Space Flight Center, Greenbelt, Md., 1964-68. NSF research grantee, 1964-67, 67-70, 70-72, 72-75, 75—. Fellow Geol. Soc. Am.; mem. Elisha Mitchell Sci. Soc. (sec.-treas. 1970-72), AAUP, AAAS, Am. Geophys. Union, Nat. Assn. Geology Tchrs., Geochemistry Soc., Carolina Geol. Soc., internat. Assn. Geochemistry and Cosmochemistry, Sigma Xi. Home: 354 Wesley Dr Chapel Hill NC 27514 Office: Dept Geology U N C Chapel Hill NC 27514

FULLER, JAMES LESLIE, nuclear engineering scientist; b. Cleve., Aug. 18, 1948; s. Norman Judson and A. Bernadine (Adkins) F.; m. Dona Christine Wilfong, Aug. 22, 1970; children: Christopher James, Calvin Lewis. B.S., Purdue U., 1970; M.Sc., U. Fla., 1972, Ph.D., 1975. Sr. engr. Westinghouse Hanford Co., Richland, Wash., 1975—; lectr. dept. nuclear engring. Joint Center for Grad. Study, U. Wash., 1978—. Contbr. articles to profl. jours. Pres. Clipper Ridge Homeowners Assn., Richland, 1976-78. Served to USAR, 1972-80. Mem. Am. Nuclear Soc., Am. Phys. Soc. Republican. Presbyterian. Subspecialties: Nuclear fusion. Current work: Cognizant engineer fusion materials irradiation test facility instrumentation systems. Co-developer nuclear-pumped laser. Home: 257 Saint Ct Richland WA 99352 Office: Westinghouse Hanford Co PO Box 1970 Richland WA 99352

FULLER, MILTON EUGENE, chemistry educator; b. Mesa, Ariz., Aug. 27, 1926; s. Horace Ralph and Hortense (McClellan) F.; m. Evelyn L. Palmer, Dec. 29, 1952; children: Allison, Randal, Tami Lynne. B.S., Ariz. State U., 1948; Ph.D., Northwestern U., 1956. Chemist Shell Devel. Co., Everyville, Calif., 1956-60, IBM Corp., Poughkeepsie, N.Y., 1960-61; prof. U. Pacific, Stockton, Calif., 1961-64; prof. chemistry Calif. State U., Hayward, 1964—; vis. scientist Lawrence Livermore Nat. Lab., 1968, 69, 78-79, cons., 1979-80; Fulbright lectr., Pakistan, 1971-72. Served with USAF, 1944-5, 1951-52. Mem. Am. Chem. Soc., Calif. Assn. Chemistry Tchrs. Club: Flying Particles (Livermore, Calif.). Subspecialties: Physical chemistry; Foundations of computer science. Current work: Computer assisted instruction. Patentee in field. Home: 55 Provo Ln Danville CA 94526 Office: Dept Chemistry Calif State U Hayward CA 94542

FULLER, RAY WARD, pharmacologist; b. Dongola, Ill., Dec. 16, 1935; s. Lloyd M. and Wanda (Keller) F.; m. Sue Brown, Dec. 22, 1956; children: Ray W. II, Angela Lea. B.A., So. Ill. U., 1957, M.A., 1958; Ph.D., Purdue U., 1961. Dir. Biol. Research Lab., Ft. Wayne (Ind.) State Hosp., 1961-63; sr. pharmacologist Eli Lilly & Co., Indpls., 1963-67, research scientist, 1967-68, head dept. metabolic research, 1968-71, research asso., 1971-75, research advisor, 1976—; vis. lectr. M.I.T., 1976-81; asso. prof. biochemistry Ind. U., 1974—; cons. NIMH, 1980—, Pharm. Mfrs. Assn. Found., 1978—. Contbr. articles to profl. jours., chpts. in books. Mem. Am. Soc. Pharmacology and Exptl. Therapeutics, Am. Soc. Biol. Chemistry, AAAS, Am. Soc. Neurochemistry, Soc. Neurosci., Endocrine Soc. Subspecialties: Pharmacology; Neuropharmacology. Current work: Biochem. pharmacology of brain and peripheral nervous system. Patentee field of medicinal chemistry. Office: 307 E McCarty St Indianapolis IN 46285

FULLER, RENEE NUNI, psychologist, publisher; b. Mannheim, Ger., Apr. 14, 1929; came to U.S., 1938; d. Eric Woldemar and Fridel Gronau (Henning) Stoetzner; m. George Ripley Fuller, Dec. 21, 1949 (dec. 1955). B.A., Swarthmore Coll., Pa., 1951, 1951; M.A., Columbia U., 1953; Ph.D., NYU, 1963. Cert. psychologist, Md. Research scientist Letchworth Village, Thiells, N.Y., 1960-66; project dir. S.I. Soc. Mental Health, S.I., N.Y., 1966-67; chief psychol. services Rosewood Hosp., Owings Mills., Md., 1967-75; pres. Ball-Stick-Bird Publs., Inc., Stony Brook, N.Y., 1975—. Author: In Search of the IQ Correlation, 1977. Mem. Am. Psychol. Assn., Am. Psychopath. Assn., Ednl. Research Assn., Behavior Genetics Assn. Subspecialties: Cognition; Developmental psychology. Current work: Theretical and applied technologies for enhancing human cognition. Developed Ball-Stick-Bird learning systems, 1974, 75. Office: Ball-Stick-Bird Publications Inc Box 592 Stony Brook NY 11790

FULLMER, HAROLD MILTON, biomedical research administrator and researcher; b. Gary, Ind., July 9, 1918; s. Howard and Rachel Eva (Tiedge) F.; m. Marjorie Lucile Engel, Dec. 31, 1942; children: Angela Sue, Pamela Rose. B.S., Ind. U., 1942, D.D.S., 1944; D. (hon.), U. Athens, Greece, 1981. Diplomate: Am. Bd. Oral Pathology. Intern Charity Hosp., New Orleans, 1946-47, resident, 1947-48, vis. dental surgeon, 1948-53; instr. Loyola U., New Orleans, 1948-49, asst. prof., 1949-50, assoc. prof. gen. and oral pathology, 1949-53; cons. pathology VA Hosps., Biloxi and Gulfport, Miss., 1950-53; asst. dental surgeon Nat. Inst. Dental Research, NIH, Bethesda, Md., 1953-54, dental surgeon, 1954-56, sr. dental surgeon, 1956-60, dental dir., 1960-70; chief sect. histochemistry Nat. Inst. Dental Research, 1967-70, chief exptl. pathology br., 1969-70, cons. to dir., 1971-72; mem. dental caries program adv. com. HEW, 1975-79, chmn., 1976-79; dir. Inst. Dental

Research, prof. pathology, prof. dentistry, assoc. dean Sch. Dentistry U. Ala. Med. Center, Birmingham, 1970—, sr. scientist cancer research and tng. program, sci. adv. com., 1977—; mem. med. research career devel. com. VA, 1977—. Co-editor: Histopathologic Technic and Practical Histochemistry, 1976; cons. editor: Oral Surgery, Oral Medicine, Oral Pathology, 1970; editor, founder: Jour. Oral Pathology, 1972—; assoc. editor: Jour. Cutaneous Pathology, 1973—; editorial bd.: Tissue Reactions, 1976—. Served to capt. U.S. Army, 1944-46. Recipient Isaac Schour award for outstanding research and teaching in anat. scis. Internat. Assn. Dental Research, 1973; Fulbright grantee, 1962; Disting. Alumnus of Year Ind. U. Sch. Dentistry, 1978; Disting. Alumnus award Ind. U., 1981. Fellow Am. Coll. Dentists, Am. Acad. Oral Pathology, AAAS (chmn. sect. 1976-78, sec. sect. 1979—); mem. ADA (cons. Council Dental Research 1973-74), Internat. Assn. Dental Research (v.p. 1974-75, pres. 1976-77), Am. Assn. Dental Research (pres. 1976-77), Internat. Assn. Pathologists, Internat. Assn. Oral Pathologists (co-founder, 1st pres. 1976, editor), Histochem. Soc., Nat. Soc. Med. Research (dir. 1977-79), Biol. Stain Commn. (trustee 1977—), Commd. Officers Assn. Republican. Presbyterian. Club: Exchange (pres. New Orleans 1952-53). Subspecialties: Pathology (medicine); Oral pathology. Current work: Biology and pathology of connective tissues; histochemical methods. Home: 3514 Bethune Dr Birmingham AL 25223 Office: Univ Ala Birmingham Birmingham AL 35294

FULTON, CHANDLER MONTGOMERY, cell biologist; b. Cleve., Apr. 17, 1934; s. Ralph Eugene and Margaret E. (Montgomery) F.; m. Margaretta Lyon, June 11, 1955; children: Thomas, Margaretta, William; m. Elaine Lai, June 27, 1981. A.B. in Biology, Brown U., 1956; Ph.D., Rockefeller U., 1960. From instr. to assoc. prof. Brandeis U., Waltham, Mass., 1960-76, prof. biology, 1976—, chmn. dept., 1981—; mem. adv. panel for devel. biology NSF, 1974-77. Contbr. articles to profl. jours.; author: (with A. O. Klein) Explorations in Developmental Biology, 1976. NSF grantee, 1960—; NIH grantee, 1978—. Mem. Soc. for Devel. Biology, Am. Soc. Cell Biology, Corp., Marine Biol. Lab., Sigma Xi, Phi Beta Kappa. Subspecialties: Developmental biology; Molecular biology. Current work: Analysis of molelcular and cellular events of cell differentiation and cell organelle morphogenesis using the amebo-flagellate Naegleria gruberi. Office: Dept Biology Brandeis U Waltham MA 02254

FULTZ, DAVE, meteorology educator, researcher; b. Chgo., Aug. 12, 1921; m., 1946; 3 children. S.B., U. Chgo., 1941, Ph.D. in Meteorology, 1947. Asst. U.S. Weather Bur., Chgo. Sta., 1942; research assoc. U. Chgo. and PR, 1942-44, U. Chgo., 1946-47, instr. in meteorology, 1947-48, from asst. prof. to assoc. prof., 1948-60, prof. meteorology, 1960—, in charge hydrodyn lab., 1946—; ops. analyst U.S. Army Air Force, 1945; mem. sci. adv. bd. U.S. Air Force, 1959-63; mem. Nat. Com. Fluid Mechanics Films, 1962-71; mem. research grants adv. com. Air Pollution Control Office, 1970-71. Recipient Rossby Research medal, 1967; Guggenheim fellow, 1950-51; NSF sr. fellow, 1957-58. Fellow Am. Meteorol. Soc. (Meisinger award 1951), Am. Geophys. Union; mem. AAAS, Am. Astron. Soc., Nat. Acad. Sci. Subspecialty: Meteorology. Office: Dept Geophys. Sci U Chgo Chicago IL 60637

FUNCH, PAUL GERARD, neurobiologist, bioengr., educator; b. Bklyn., June 19, 1950; s. Willard and Florence (Buyser) F.; m. Donna Patricia Funch, June 8, 1974. B.S., Cornell U., Ithaca, N.Y., 1970; postgrad., U. Wyo., 1974-75, La. State U. Med. Center, 1975-77; Ph.D., SUNY-Buffalo, 1980. Postdoctoral fellow dept. neurol. surgery U. Wash., Seattle, 1980-81; research asst. prof. div. neurobiology, dept. physiology SUNY, Buffalo, 1982—. Mem. Am. Wildlife Fedn., Soc. Neurosci., Am. Physiol. Soc., AAAS, Biophys. Soc., IEEE. Subspecialties: Neurobiology; Neurophysiology. Current work: Neurobiology of myelinated axons, axonal-glial relationships and neurotoxicology related to axonal conduction. Home: 4194 Harlem Rd Snyder NY 14226 Office: SUNY 313 Cary Hall Buffalo NY 14214

FUNDER, DAVID CHARLES, psychologist, educator; b. Long Beach, Calif., Feb. 15, 1953; s. Elvin Lawrence and Edna Marie (Steele) F. B.A., U. Calif.-Berkeley, 1975; Ph.D., Stanford U., 1979. Asst. prof. psychology Harvey Mudd Coll., Claremont, Calif., 1979-82, Harvard U., Cambridge, Mass., 1982—. Cons. editor: Jour. Personality and Social Psychology, Washington, 1980—; Contbr. articles to profl. jours. NSF fellow, 1976-79. Mem. Am. Psychol. Assn., Phi Beta Kappa. Democrat. Subspecialties: Social psychology; Personality. Current work: Personality psychology and research into processes of person perception with special attention to accuracy. Home: 60 Saint Germain St Boston MA 02115 Office: Dept Psychology and Social Relations Harvard U 33 Kirkland St Cambridge MA 02138

FUNDERBURK, NOEL ROGER, microbiologist, educator; b. San Angelo, Tex., Mar. 16, 1940; s. Paul Dennis and Viva Gladys (Copple) F.; m. Sandra Kay Eldred, June 13, 1964; children: Carol, Jennifer. B.S., Baylor U., 1963; M.S., North Tex. State U., 1969, Ph.D., 1971. Diplomate: Am. Bd. Med. Microbiology. Chief microbiology, instr. Hillcrest Bapt. Hosp., Waco, Tex., 1963-67; instr. N. Tex. State U., 1967-71; supr. clin. labs. NASA Manned Space Ctr., Houston, 1971-74; instr. pathology U. Tex. Health Sci. Ctr., 1974-76; chief microbiology Severance & Assocs., San Antonio, 1976-78; clin. microbiologist dept. pathology Tex. Coll. Osteo. Medicine, Ft. Worth, 1978—; cons. in field. Artist bronze western and wildlife sculpture. Founder Fellowship of Christian Microbiologists, 1978. Mem. Tex. Soc. Clin. Microbiologists (pres. 1983—), Am. Soc. Microbiology, Tex. Soc. Microbiology, Southwest Assn. Clin. Microbiologists, Christian Med. Soc. Subspecialties: Microbiology (medicine); Infectious diseases. Current work: Antimicrobial therapy and rapid diagnostic techniques in infectious diseases. Office: Texas College of Osteopathic Medicine Department of Pathology Camp Bowie at Montgomery Fort Worth TX 76107

FUNG, BING-MAN, chemistry educator; b. Hong Kong, Aug. 15, 1939; came to U.S., 1963, naturalized, 1973; s. Hon-Wah and Fung-Ngor (Lam) F.; m. Mildred Wai-Yum Mah, Aug. 26, 1967; 1 son: Archon. Diploma with honors, Chung Chi Coll., Hong Kong, 1963; Ph.D., Calif. Inst. Tech., 1967. Asst. prof. Tufts U., Medford, Mass., 1966-72; assoc. prof. U. Okla., Norman, 1972-76, prof. chemistry, 1976—. Contbr. articles to profl. jours. Recipient Research Career Devel. award USPHS, 1975-80. Mem. Am. Chem. Soc., AAAS, Sigma Xi (faculty research award 1974). Subspecialty: Physical chemistry. Current work: Nuclear magnetic resonance. Home: 1528 Homeland St Norman OK 73069 Office: Dept Chemistry U Okla Norman OK 73019

FUNG, LESLIE WO-MEI, chemistry educator, biophysics researcher; b. Nanking, China, Sept. 23, 1946. B.Sc., U. Calif.-Berkeley, 1968; Ph.D. in Phys. Chemistry, MIT, 1971. Research fellow in phys. chemistry Rice U., 1971-72; lectr. in chemistry Chinese U., Hong Kong, 1972-73; research assoc. in biophysics U. Pitts., 1973-77; asst. prof. chemistry Wayne State U., beginning 1977; now assoc. prof. chemistry Loyola U., Chgo. Recipient Nat. Research Service award NIH, 1975-77, Research Career Devel. award NIH, 1982—. Mem. AAAS, Am. Biophys. Soc., Am. Chem. Soc., Assn. Women Scientists, Sigma Xi. Subspecialty: Biophysical chemistry. Office: Loyola U Dept Chemistry 6526 N Sheridan Rd Chicago Il 60626

FUNG, YUAN-CHENG BERTRAM, bioengineering, author; b. Yuhong, Changchow, Kiangsu, China, Sept. 15, 1919; came to U.S., 1945, naturalized, 1957; s. Chung-Kwang and Lien (Hu) F.; m. Luna Hsien-Shih Yu, Dec. 22, 1949; children: Conrad Antung, Brenda Pingsi. B.S., Nat. Central U., Chungking, China, 1941, M.S., 1943; Ph.D., Calif. Inst. Tech., 1948. Research fellow Bur. Aero. Research China, 1943-45; research asst., then research fellow Cal. Inst. Tech., 1946-51, mem. faculty, 1951-66; prof. aeros. Calif. Inst. Tech., 1959-66; prof. bioengring. and applied mechanics U. Calif.-San Diego, 1966—; cons. aerospace indsl. firms, 1949—. Author: The Theory of Aeroelasticity, 1956, Foundations of Solid Mechanics, 1965, A First Course in Continuum Mechanics, 1969, 77, Biomechanics, 1972, Biomechanics: Mechanical Properties of Living Tissues, 1980; also papers.; Editor: Jour. Biorheology, Jour. Biomech. Engring. Recipient Achievement award Chinese Inst. Engrs., 1965, 68; Landis award Microcirculatory Soc., 1975; von Karman medal ASCE, 1976; Guggenheim fellow, 1958-59. Fellow AIAA, ASME (Lissner award 1978); mem. Nat. Acad. Engring, Soc. Engring. Sci., Microcirculatory Soc., Am. Physiol. Soc., Nat. Heart Assn., Basic Sci. Council, Sigma Xi. Subspecialty: Biomechanics. Current work: Biomechanics, pulmonary circulation, international cooperation in science. Home: 2660 Greentree Ln La Jolla CA 92037

FUNK, GLENN ALBERT, virologist, educator, consultant; b. Highland Park, Mich., Feb. 28, 1942; s. John Louis and Dorothy Emily (Fritz) F.; m. Heidi Anne Reutiman, July 16, 1966; children: Kevin Hunter, Colin Jeffrey. A.B. in Bacteriology, U. Calif.-Berkeley, 1963; M.S. in Med. Microbiol. Immunology, UCLA Med. Sch., 1965; Ph.D. in Med. Microbiology, Stanford U. Sch. Medicine, 1975. Postdoctoral fellow Baylor U. Coll. Medicine, 1974-76; asst. prof. microbiology San Jose State U., 1976-81, assoc. prof., 1981-83; mgr. research Internat. Diagnostic Tech., Santa Clara, 1981-82; expt. support scientist for spacelab GE-Mgmt. and Tech. Services Co., 1982-83, sr. scientist for spacelab, 1983—; cons. to industry on virus and tissue culture products and prodn. Contbr. articles to profl. publs. Served to lt. USNR, 1966-70; Vietnam. Mem. Am. Soc. Microbiology, Calif. Acad. Sci., Taiwan Internat. Med. Soc., Soc. Wine Educators. Subspecialties: Virology (medicine); Animal virology. Current work: Management and technical service to NASA's spacelab missions in the area of microbiology and science management and operations. Home: 1664 Fairlawn Ave San Jose CA 95125 Office: Mailstop 240A-4 NASA-Ames Research Center Moffett Field CA 94035

FUNK, VICKI ANN, research scientist, herbarium curator; b. Owensboro, Ky., Nov. 26, 1947; divorced. B.S., Murray State U., 1969, M.S. in Biology, 1975; Ph.D., Ohio State U., 1980. Grad. asst. Murray State U., 1974-75; teaching assoc. Ohio State U., Columbus, 1975-76, 77-78, research assoc., 1978-79; asst. curator Herbarium, 1976-77; research fellow, 1979-80; mus. intern N.Y Bot. Garden, 1980-81; assoc. curator U.S. Nat. Herbarium, Smithsonian Instn., Washington, 1981—; curator in charge (Dynamics of Evolution Hall), Nat. Mus. Natural History. Contbr. articles to profl. jours. NSF grantee, 1977-80; Sigma Xi grantee, 1979-80; WHO grantee, 1980; Smithsonian Instn. grantee, 1981-83; NATO travel grantee, 1982. Mem. AAAS, Am. Soc. Plant Taxonomists, Bot. Soc. Am., Calif. Bot. Soc., Internat. Assn. for Plant Taxonomy, Sociedad Botanica de Mexico, Soc. Systematic Zoology, Sigma Xi, others. Subspecialties: Systematics; Evolutionary biology. Current work: Origin and evolution of Andean flora, especially the family Asteraceae; Cladistics; biogeography; computer assisted taxonomy. Office: US Nat Herbarium Smithsonian Instn NHB 166 Washington DC 20560

FUNT, RICHARD CLAIR, horticulturist, fruit grower, educator, researcher; b. Gettysburg, Pa., Feb. 13, 1946; s. Sterling Samuel and Dorothy M. (Guise) F.; m. Shirley May Fox., Sept. 6, 1969; 1 dau., Elizabeth Anne. B.S., Del. Valley Coll., 1968; M.S., Pa. State U., 1971, Ph.D., 1974. Research assoc. Pa. State U.-University Park, 1970-74; asst. prof. U. Md.-College Pk., 1974-78; assoc. prof. Ohio State U., Columbus, 1978-83, student adviser, 1979-83; adv. Md. Hort. Club, College Park, 1975-78. Contbr. chpts. to hort. books. Bd. dirs. Methodist Ch., Columbus, 1982-83. Served to sgt. U.S. Army, 1969-70; Vietnam. Decorated Bronze Star, Army Commendation medal, Vietnam. Mem. Am. Soc. Hort. Sci., Am. Agr. Econ. Soc., N. Am. Strawberry Grower Assn. Methodist. Subspecialties: Plant growth; Plant physiology (agriculture). Current work: Apple, peach growth regulators, rootstock cultivar evaluations, strawberry, bramble-herbicides; tricke irrigation production economics. Home: 1877 Stockwell Drive Columbus OH 43220 Office: Ohio State Univ 2001 Fyffe Ct Columbus OH 43210

FURIE, BRUCE, hematologist, medical educator; b. Coral Gables, Fla., Apr. 17, 1944; s. J. Leon and Bernice (Wolfe) F.; m. Barbara Cantor, May 5, 1966; children: Eric, Gregg. A.B., Princeton U., 1966; M.D., U. Pa., 1970. Diplomate: Am. Bd. Internal Medicine (subsplty. hematology). Intern Hosp. of U. Pa., Phila., 1970-71, resident in medicine, 1971-72; fellow in hematology Upstate Med. Ctr., SUNY, Syracuse, 1974-75; chief coagulation unit New Eng. Med. Ctr., Boston, 1975—; assoc. prof. medicine Tufts U. Sch. Medicine, Boston, 1975—, dir. med. scientist tng. program, 1979-83; Chmn. clin. scis. 4 study sect. NIH, Bethesda, Md., 1981-85; established investigator Am. Heart Assn., Boston, 1976-81. Contbr. articles to profl. publs. Served with USPHS, 1972-74. Recipient Creskoff prize U. Pa., Phila., 1970; NIH grantee, 1977—. Mem. Am. Soc. Clin. Investigation, Am. Soc. Biol. Chemists. Subspecialties: Hematology; Biochemistry (medicine). Current work: Biochemistry of blood coagulation; vitamin K; immunochemistry. Home: 175 Oakland St Wellesley MA 02181 Office: Tufts-New England Med Center 171 Harrison Ave Boston MA 02111

FURMAN, SEYMOUR, surgery educator, researcher; b. N.Y.C., July 12, 1931; m., 1957; 3 children. M.D., SUNY, 1955. Intern in medicine Montefiore Hosp. and Med. Ctr., N.Y., 1955-56, resident in surgery, 1956-60, adj. attending surgeon, 1963, clin. assoc. surgeon, 1964-66, assoc. surgeon, 1966-67, assoc. attending surgeon, 1970-78, attending surgeon, 1978—; resident in thoracic surgery Baylor U., 1962-63, instr. in surgery, 1963; from asst. prof. to assoc. prof. Albert Einstein Coll. Medicine, 1968-77, prof. surgery, 1977—; assoc. attending surgeon Polyclin. Med. Sch. and Hosp., 1966-67. Fellow ACS; mem. AMA, Am. Soc. Artificial Internal Organs, Assn. Advancement Med. Instrumentation, Am. Heart Assn. Subspecialty: Cardiac surgery. Office: 111 E 210th St Bronx NY 10467

FURSHPAN, EDWIN JEAN, neurobiology educator, researcher; b. Hartford, Conn., Apr. 18, 1928; m., 1957; 3 children. B.A., U. Conn., 1950; Ph.D. in Animal Physiology, Calif. Inst. Tech., 1955; A.M. (hon.), Harvard U. Med. Sch., 1967. Fellow and hon. asst. Univ. Coll., U. London, 1955-58; instr. in neurophysiology Med. Sch., Johns Hopkins U., 1958-59, assoc. in neurophysiology and neuropharmacology, 1959-62; from asst. prof. to assoc. prof. Harvard U. Med. Sch., 1962-69, prof. neurobiology, 1969—. Mem. Soc. Neurosci., Nat. Acad. Scis., Am. Acad. Arts and Sci., Harvey Soc. (hon.). Subspecialties: Neurophysiology; Neurobiology. Office: Dept Neurobiology Harvard U Med Sch Boston MA 02115

FURTH, HAROLD PAUL, physicist, educator; b. Vienna, Austria, Jan. 1930; came to U.S., 1941, naturalized, 1947; s. Otto and Gertrude (Harteck) F.; m. Alice May Lander, June 19, 1959 (div. Dec. 1977); 1 son, John Frederick. Grad., Hill Sch., 1947; A.B., Harvard U., 1951, Ph.D., 1960; postgrad., Cornell U., 1951-52. Physicist U., Calif. Lawrence Radiation Lab., Livermore, 1956-65, group leader, 1965-67; prof. astrophys. scis. Princeton U., 1967—; dir. Plasma Physics Lab., 1981—. Bd. editors: Physics of Fluids, 1965-67, Nuclear Fusion, 1964—, Revs. Modern Physics, 1975-80; Contbr. articles to profl. jours. Recipient E.O. Lawrence award U.S. AEC, 1974. Fellow Am. Phys. Soc.; mem. Nat. Acad. Scis. Subspecialties: Fusion; Nuclear fusion. Patentee in field. Home: 36 Lake Ln Princeton NJ 08540

FURTH, JOHN JACOB, molecular biologist, pathologist; b. Phila., Jan. 25, 1929; s. Jacob and Olga (Berthauer) F.; m. Mary Autry, June 24, 1959; children: Karen, Susan, Robin. B.A., Cornell U., 1950; student, Yale Law Sch., 1950-51; M.D., Duke U., 1958; M.A. (hon.), U. Pa., 1972. Intern Bellevue Hosp., N.Y.C., 1958-59; resident in pathology N.Y.U. Sch. Medicine, N.Y.C., 1959-60, postdoctoral fellow dept. microbiology, 1960-62; mem. faculty dept. pathology U. Pa. Med. Sch., Phila., 1962—, prof., 1978—. Contbr. articles to profl. jours. Pres. Darby Creek Valley Assn., 1972—. Served to 2d lt. Q.M.C. U.S. Army, 1951-53. Recipient Hoffman LaRoche award, 1958; Eleanor Roosevelt fellow, 1977-78. Mem. Am. Soc. Biol. Chemists, AAAS, Am. Assn. Cancer Research, Am. Assn. Pathologists. Democrat. Quaker. Subspecialties: Molecular biology; Pathology (medicine). Current work: Gene regulation in normal and malignant cells. Home: 43 Roselawn St Lansdowne PA 19050 Office: Dept Pathology G3 U Pa Sch Medicine Philadelphia PA 19104

FUSSELL, CATHARINE PUGH, biology educator; b. Phila., July 13, 1919; d. Milton H. and Isabel R. P. A.B., Colby Coll., 1941; M.S., Cornell U., 1958; Ph.D., Columbia U., 1966. Adminstrv. asst. Am. Friends Service Com., 1947-55; research asst. Brookhaven Nat. Lab., Upton, N.Y., 1957-60; research assoc. Inst. Cancer Research, Phila., 1966-67; NIH postdoctoral fellow Fels Research Instn., Phila., 1967-68; asst. prof. Pa. State U., McKeesport, 1968-74, assoc. prof., 1974-77, assoc. prof. biology, Abington, Pa., 1977—; vis. scientist Clin. and Population Cytogenetics Unit, Med. Research Council, Edinburgh, Scotland, 1975-76. Contbr. articles in field to profl. jours. USPHS predoctoral fellow Columbia U., 1960-66; NIH postdoctoral fellow Fels Research Instn., 1967-68. Mem. AAAS, Am. Soc. Cell Biology, Bot. Soc. Am., Genetics Soc. Am., LWV, Sigma Xi, Phi Kappa Phi. Subspecialties: Cell biology; Developmental biology. Current work: The properties and functions of interphase chromosome arrangements; cytological, cytochemical and biosynthetic changes of nuclei during development and differentiation. Office: Ogontz Campus Pa State U Abington PA 19001

FUTUYMA, DOUGLAS JOEL, ecology and evolution educator; b. N.Y.C., Apr. 24, 1942; s. Joseph J. and Eleanor (Haessler) F. B.S., Cornell U., 1963; M.S. in Zoology, U. Mich., 1966, Ph.D., 1969. Asst. prof. ecology and evolution SUNY-Stony Brook, 1969-76, assoc. prof., 1976-83, prof., 1983—; occasional faculty mem. Orgn. Tropical Studies, Costa Rica; book cons. various pubs. Author: Evolutionary Biology, 1979, Science on Trial: The Case for Evolution, 1983; contbr. numerous articles to profl. jours.; editor: (with M. Slatkin) Coevolution, 1983, Evolution, 1981-83. Subspecialties: Evolutionary biology; Ecology. Current work: Population biology and ecological genetics of insect-plant interactions; adaptations of insects to host plants. Office: Dept Ecology and Evolution SUNY Stony Brook NY 11794

FYSTROM, DELL ORREN, educator, physicist, cons.; b. Mpls., Aug. 29, 1937; s. Orren Alvin and Florence (Benson) F.; m. Linda Mae Conger, Sept. 7, 1962 (div.); children: Eileen Teresa, Heather Lee, Michael Dell. B.A., St. Olaf Coll., Northfield, Minn., 1959; Ph.D. in Physics, U. Colo., Boulder, 1969. Asst. prof. U. Wis., LaCrosse, 1969-72, assoc. prof., 1972—, chmn. dept. physics, 1971-73, 82—; chmn. dept. physics Nat. U., Addis Ababa, Ethiopia, 1974-75. Chmn. Mayor's Ad Hoc Energy Commn. of LaCrosse. Mem. Am. Assn. Physics Tchrs. Lodge: Rotary. Subspecialty: Atomic and molecular physics. Current work: Magnetic moment of proton in nuclear magnetons (exptl). Home: 2918 N Marion Rd LaCrosse WI 54601 Office: Cowley Hall University of Wisconsin 1775 Pine St LaCrosse WI 54601

GABBE, STEVEN GLENN, medical educator, physician; b. Newark, Dec. 1, 1944; s. Charles P. and Marcia (Abrams) G.; m. Patricia C. Temple, July 26, 1981; children: Adam, Erica, Amanda, Daniel. B.A. Magna cum laude, Princeton U., 1965; M.D., Cornell U., 1969. Diplomate: Am. Bd. Obstetrics and Gynecology; lic. physician, Calif., Colo., Pa. Med. intern N.Y. Hosp., N.Y.C., 1969-70; research fellow in reproductive medicine Boston Hosp. for Women, 1970-71, resident in obstetrics and gynecology, 1972-74; research fellow in biol. chemistry Harvard Med. Sch., Boston, 1970-71, clin. fellow in obstetrics and gynecology, 1972-74; asst. prof. dept. obstetrics and gynecology U. So. Calif., Los Angeles, 1975-77; assoc. prof. dept. obstetrics and gynecology U. Colo. Sch. Medicine, Denver, 1977-78; assoc. prof. dept. obstetrics and gynecology, dept. pediatrics U. Pa. Sch. Medicine, Phila., 1978-82, prof., 1982—, assoc. prof., 1981—; dir. Jerrold R. Golding div. maternal-fetal medicine, dept. obstetrics and gynecology Hosp. of U. Pa., Phila., 1978—; dir. maternal and infant care program City of Phila. Dept. Pub. Health Disease Prevention and Health Promotion, 1982—; mem. adv. bd. Children's Diabetes Found. of Denver, 1978—; mem. sci. adv. bd. Phila. chpt. Juvenile Diabetes Found., 1980—; chmn. adv. subcom. Nat. Diabetes Research Interchange, 1981—. Mem. editorial bd.: Perinatology, Neonatology, Pediatric Nutrition Currents, Ross Labs., 1980—; mem. bd. editorial advisors: Advances in Reproductive Medicine, Med. Econs. Co., 1982—; contbr. articles to med. jours. Fellow Am. Coll. Obstetricians and Gynecologists; mem. Perinatal Research Soc., Soc. for Gynecologic Investigation, Soc. for Perinatal Obstetricians, Phila. Neonatal Soc., Obstet. Soc. Phila., Phila. Perinatal Soc. (pres. 1982), Phi Beta Kappa, Alpha Omega Alpha. Subspecialty: Maternal and fetal medicine. Current work: Diabetes mellitus in pregnancy. Office: Department of Obstetrics and Gynecology Hospital of the University of Pennsylvania 3400 Spruce St Philadelphia PA 19104

GABEL, J. RUSSEL, biology educator; b. Pottstown, Pa., Aug. 21, 1918; s. Charles Franklin and Emma Minerva (Kessinger) G.; m. Sonia Garnet Venger, June 25, 1944; 1 son, Jon Michael. B.S. in Edn, Lock Haven (Pa.) State Coll., 1947; Ph.D., U. Pa., 1953. Asst. prof. biology Fisk U., 1953-5; assoc. prof. biology West Liberty State Coll., 1956-59; prof. biology San Francisco State U., 1959—. Served to 2d lt. USAAF, 1942-46. Mem. Am. Arachnol. Soc., Calif. Acad. Scis., Am. Nature Study Soc., Pan Pacific Entomol. Soc., Western Soc. Naturalists. Subspecialties: Behaviorism; Morphology. Current work: Life history and behavior of tarantulas, protozoan populations in gut of marmots. Home: 440 Monticello St San Francisco CA 94127 Office: Dept Biology San Francisco State U 1600 Holloway Ave San Francisco CA 94132

GABELMAN, JOHN WARREN, geologist, engineer; b. Manila, Philippines, May 18, 1921; s. Charles Grover and Cyprienna (Turcotte) G.; m. Olive Alexander Thompson, Sept. 22, 1945; children: Barbara Grace, Joan Lynn. B.S., Colo. Sch. Mines, 1943, M.S. in Geol. Engring, 1948, Sc.D., 1949. Registered profl. engr., Colo.; cert. profl. geologist. Jr. engr. N.J. Zinc Co., Gilman, Colo., 1943-44; geologist Colo. Fuel & Iron Corp., Pueblo, 1949-51, Am. Smelting & Refining Co., Salt Lake City, 1951-54; dist. geologist, br.

chief AEC, Western U.S., S.Am., Washington, 1954-73, program mgr. geothermal, Washington, 1973-75; mgr. exploration research Utah Internat. Inc., San Francisco, 1975—; v.p. Geosat Com., Inc., San Francisco, 1981—; adv. AEC, Latin-Am. rep., 1958-61. Author: Migration of Thorium and Uranium, 1977; Contbr. numerous articles to profl. jours. Served with USNR, 1944-46. Fellow Geol. Soc. Am.; mem. Am. Assn. Petroleum Geologists, Soc. Econ. Geologists, Am. Inst. Mining Engrs., Am. Geophys. Union. Republican. Roman Catholic. Subspecialties: Geology; Remote sensing (geoscience). Current work: Research in metallic ore deposits, particularly uranium, spectral remote sensing in geology; tectonics related to ore deposits; geochemistry of mineralization processes. Home: 23 Portland Ct Danville CA 94526 Office: Mineral Exploration and Devel Div Utah International Inc 550 California St San Francisco CA 94104

GABLE, RALPH KIRKLAND, psychology educator; b. Canton, Ohio, Mar. 21, 1934; s. Harry C. and Mary (Blackburn) Schwitzgebel; m. Colleen Ryan, Dec. 28, 1963; children: Eric, Sandra. B.S., Ohio State U., 1956; Ed.D., Harvard U., 1962, J.D., 1970. Diplomate: Am. Bd. Forensic Psychology. Lectr. Harvard U., 1968-71, Northeastern U., 1969-73, Harvard Med. Sch., 1970-74, asst. prof., 1975-78; asst. prof. psychology Calif. Luth. Coll., Thousand Oaks, 1975-77, assoc. prof., 1978-83, prof., 1983—; mem. Harvard Civil Rights-Civil Liberties Law Rev., Cambridge, 1968-70; chmn. Crime and Delinquency Rev. Com. NIMH, 1974-76; mem. Inventors Workshop Internat., Ventura, Calif., 1976—. Author: Streetcorner Research, 1964, Law and Psychological Practice, 1980, Changing Human Behavior, 1974; mem. editorial bd.: Jour. Law and Human Behavior, 1974-82; asst. editor: Internat. Jour. Psychiatry, 1963-64. Mem. Am. Psychol. Assn. (chairperson ethics com. 1974-76), Assn. Advancement Behavior Therapy (dir. 1968-80). Democrat. Subspecialties: Social psychology; Bioinstrumentation. Current work: Behavioral electronics; product development law; regualtion of ethanol tolerance; interpersonal communications technology. Patentee behavioral superivison system, 1969, antomobile camera system. Office: California Lutheran College 60 Olsen Rd Thousand Oaks CA 91360 Home: 515 Fargo St Thousand Oaks CA 91360

GABLEHOUSE, REUBEN HAROLD, aerospace co. exec.; b. Berthoud, Colo., Aug. 31, 1923; s. Daniel Henry and Mollie Emma (Henry) G.; m. Genevieve Margaret Willburn, June 19, 1949; children: Timothy, R. Daniel, Nancy, Kelly. B.S.E.E., U. Colo., 1951; M.S.E.E., U. N.Mex., 1956. Electronics engr. instrumentation design Chance-Vought, 1951-52; staff mem. instrumentation devel. Sandia Corp., Albuquerque, 1952-60; with Ball Aerospace Systems Div., Boulder, Colo., 1960—, pres., 1980—. Bd. dirs. Fiske Planetarium, 1982—; vice chmn. United Way Campaign, Boulder, 1983. Served to 2d lt. U.S. Army, 1943-46. Mem. AIAA, Am. Astron. Soc. Methodist. Subspecialty: Electronics. Home: 1840 Forest Ave Boulder CO 80302 Office: 1600 Commerce St Boulder CO 80303

GABOUREL, JOHN DUSTAN, pharmacologist, educator, researcher; b. San Francisco, Oct. 16, 1928; s. John Richard and Mary Josephine (Mourterot) G.; m. Carol E. Roberts, Aug. 5, 1951; children: Michaele A., Linda S., Allison J., Candice M., John R. B.S., U. Calif., Berkeley, 1950; M.S., U. San Francisco, 1951; Ph.D., U. Rochester, 1957. Instr., asst. prof. Stanford U. Sch. Medicine, 1957-64; asso. prof. pharmacology Oreg. Health Scis. U., Portlant, 1964-70, prof., 1970—; vis. scientist Walter and Eliza Hall Inst., Melbourne, Australia, 1973-74; mem. research grants com. Oreg. Heart Assn. Contbr. numerous articles on cellular, molecular and immunopharmacology to profl. jours. Served with Chem. Corps. U.S. Army, 1953-55. Recipient Allen J. Hillaaward for teaching excellence Oreg. Health Scis. U., 1968; Carl J. and Alma Johnson Trust travel award, 1973. Mem. Am. Assn. for Pharmacology and Exptl. Therapeutics. Subspecialties: Pharmacology; Immunopharmacology. Current work: Drug effects on the growth, metabolism and function of lymphocytes, leukocytes and other mammalian cells. Drug induced immunosuppression. Drug induced antoimmunity. Home: 6825 SW Raleighwood Way Portland OR 97225 Office: Dept Pharmacology Oreg Health Scis U Portland OR 97201

GABRENYA, WILLIAM KARL, JR., psychologist; b. Cleve., Jan. 20, 1952; s. William Karl and Rose J. (Lupo) G.; m. Yue-Eng Wang, May 24, 1980; 1 son, William Karl. A.B., Ohio U., 1974; M.A., U. Mo., Columbia, 1977, Ph.D., 1979. Postdoctoral research asst. Ohio State U., Columbus, 1979-81; asst. prof. psychology Fla. Inst. Tech., Melbourne, 1981—. Contbr. articles to profl. jours. Mem. Am. Psychol. Assn., Internat. Assn. for Cross-cultural Psychology, Soc. for Psychol. Study of Social Issues. Subspecialties: Social psychology; Psychological anthropology. Current work: Labeling theory and social cognition; cross-cultural cognition; cross-cultural group productivity. Office: Fla Inst Tech Sch of Psychology Melbourne FL 32901

GABRIEL, OTHMAR, biochemist, educator, researcher; b. Vienna, Austria, Jan. 10, 1925; came to U.S., 1958; s. Othmar and Rosa (Fellner) G.; m. Elisabeth Rehurek, May 28, 1949 (div.); children: Harriet, Annamaria; m. Rachelle Rothenberg, Aug. 5, 1965. Ph.D., U. Vienna, 1954. Asst. prof. U. Vienna, 1954-58; research assoc. Columbia U., N.Y.C., 1958-60; research chemist NIH, Bethesda, Md., 1960-65; assoc. prof. Georgetown U., Washington, 1965-70, prof., 1970—. Contbr. chpts. to books, articles to profl. jours. Mem. Am. Soc. Biol. Chemists, Am. Chem. Soc., Am. Soc. Microbiology, Soc. for Exptl. Biology and Medicine, Sigma Xi. Subspecialty: Biochemistry (medicine). Current work: Enzymology of carbohydrate metabolism, role of carbohydrates in cellular recognition. Home: 6508 Wilmett Rd Bethesda Md 20817 Office: Georgetown Univ 3900 Reservoir Rd Washington DC 20007

GADBERRY, JOSEPH LAFAYETTE, microbiologist; b. Fargo, N.D., June 8, 1940; s. Joseph Lafayette and Helen A. (Ydstie) G.; m. Betty Helen Eisenbeis, Nov. 25, 1966; 1 son, Brett Jason. B.A., Concordia Coll., 1962; M.S., N.D. State U., 1966; Ph.D., U. Nebr. Med. Ctr., 1975. Registered microbiologist, Mo. Asst. instr. U. Nebr. Med. Ctr., Omaha, 1970-76; asst. prof. Miami U., Oxford, Ohio, 1977-80, U. Health Sci., Kansas City, Mo, 1980—; cons. Nat. Bd. Examiners Pub. Health, Am. Osteopathic Assn., 1980—. Contbr. articles to profl. jours. Precinct election worker Douglas County Bd. Electons, Omaha, 1975; prison rehab. worker Minimum Security Nebr. Penitentiary, Lincoln, 1976-77; exec. com. Christian Coop. Presch., Oxford, Ohio, 1978-79; com. sec. Cub Scouts, Kansas City Council Boy Scouts Am., 1981-82, com. chmn., 1982—. Analytab Products Inc. grantee, 1978-79; Remel Labs. grantee, 1980-82. Mem. Am. Soc. Microbiology, Assn. Practitioners in Infection Control (reviewer 1982), Soc. Hosp. Epidemiologists of Am., Infectious Disease Soc. Kansas City, Assn. Microbiologists of Greater Kansas City, Phi Sigma Upsilon, Beta Beta Beta. Republican. Lutheran. Lodge: Lions. Subspecialties: Microbiology (medicine); Neuroimmunology. Current work: Identify rapid methods of isolating and determines bacterial agents from clinical specimens, especially Pasteurella multicida and Group B Streptococci. Office: Univ Health Scis 2105 Independence Blvd Kansas City MO 66212

GAFFAR, ABDUL, microbiologist; b. Borman, Burma, Dec. 10, 1940; came to U.S., 1964; s. Ismail and Khatija (Mohamed) Darji; m. Maria C. Solis, May 23, 1970; 1 son, Yousof A. B.S., U. Karachi, 1963; M.S., Brigham Young U., 1965; Ph.D., Ohio State U., 1967. Research chemist PSIR, Karachi, Pakistan, 1963-64; research asst. Brigham Young U., Provo, Utah, 1964-65, Ohio State U., Columbus, 1965-67; research immunologist Colgate-Palmolive Research Center, Piscataway, N.J., 1968-70, sr. research immunologist, 1970-73, research assoc., 1973-75, sr. research assoc., 1975-77, sr. scientist, 1977-80; research fellow, assoc. dir., 1980—. Contbr. articles to sci. publs. Mem. Am. Soc. for Microbiology, AAAS, Internat. Assn. Dental Research. Muslim. Subspecialties: Immunocytochemistry; Microbiology. Current work: Calcium phosphate chemistry, bone, teeth; periodontal diseases; antimicrobial agents and vaccines; molecular biology of connective tissues. Patentee on caries vaccine, caries control, periodontal disease and terminization. Home: 30 Macafee Rd Somerset NJ 08873

GAFFNEY, PAUL GOLDEN, II, naval officer, oceanographer; b. Attleboro, Mass., May 30, 1946; s. Paul G. and Elfrieda L. (Piepenstock) G.; m. Linda L. Myers, Sept. 7, 1974; 1 dau., Crista Lee. B.S., U.S. Naval Acad., 1968; M.S.E., Cath. U. Am., 1969; grad. with highest distinction, Naval War Coll, 1979. Commd. ensign USN, 1968, advanced through grades to comdr., 1982; grad. research asst. Cath. U. Am., Washington, 1968-69; hydrographic advisor Vietnamese Navy, Saigon, 1971-72; oceanographic officer FleetWeather Central, Rota, Spain, 1972-75; exec. asst. Oceanographer of the Navy, Alexandria, Va., 1975-78; research fellow Naval War Coll., Newport, R.I., 1978-79; comdg. officer Ocean Unit 4, Hydrographic Survey, Indonesia, 1979-80; acting dir. arctic and earth sci. Office Naval Research, Washington, 1980-81; exec. asst. Asst. Sec. Def., Washington, 1981-83; exec. officer Naval Oceanography Command Facility, Jacksonville, Fla., 1983—. Decorated Bronze Star medal, Def. Superior Service medal. Mem. Hydrographic Soc., Marine Tech. Soc., Explorers Club, U.S. Naval Inst. Republican. Roman Catholic. Subspecialties: Oceanography; Synoptic meteorology. Current work: Field hydrography, research and development management in ocean environment, ocean policy, synoptic meterology. Home: 2757 Pebblendge Ct Orange Park FL 32073 Office: Naval Oceanography Command Facility Jacksonville FL 32212

GAGNE, ELLEN DALTON, psychology educator; b. New London, Conn., Jan. 15, 1947; d. Robert Mills and Harriet (Towle) G.; m. William Wiant Davis, Jan. 2, 1982. A.B., Oberlin Coll., 1968; M.S.Ed., Bucknell U., 1972; Ph.D., U. Wis., 1974. Asst. prof. U. Ga., Athens, 1974-79, assoc. prof. ednl. psychology, 1979—; Tchr. Peace Corps, Sierra Leone, 1968-70. Cons. editor: The Ednl. Psychologist, 1979—; contbr. articles in field to profl. jours. NSF postdoctoral fellow, 1978; Nat. Inst. Edn. research grantee, 1978-81. Mem. Am. Psychol. Assn., Am. Ednl. Research Assn., Sigma Xi. Democrat. Subspecialties: Learning; Cognition. Current work: Research on training cognitive strategies, learning processes, cognition and instruction. Home: 430 Gran Ellen Dr Athens GA 30606 Office: U Ga Athens GA 30602

GAGNEBIN, ALBERT PAUL, mining executive; b. Torrington, Conn., Jan. 23, 1909; s. Charles A. and Marguerite E. (Huguenin) G.; m. Genevieve Hope, Oct. 26, 1935; children: Anne Hope Gagnebin Coffin, Joan DeVere Gagnebin Wicks. B.S. in Mech. Engring., Yale U., 1930, M.S., 1932. With Internat. Nickel Co., Inc., 1932-74, successively staff research lab., research ferrous metallurgy, devel. ductile iron, staff devel. and research div., 1932-55, mgr. nickel sales dept., 1956-61, v.p., 1958-64, exec. v.p., 1964-67, pres., 1967-72, also dir., mem. exec. com.; v.p. Internat. Nickel Co. Can., Ltd., 1960-64, exec. v.p., 1964-67, pres., 1967-72, chmn. bd., 1972-74, ret., 1974, mem. exec. com.; trustee Atlantic Mut. Ins. Co.; mem. adv. bd. Inco Ltd.; dir. Abex Corp., Centennial Ins. Co., Am.-Swiss Assos., Inc.; dir. emeritus Ill. Central Industries; former dir. Ingersoll-Rand Co., Schering-Plough Corp., Bank of N.Y.; mem. N.Am. adv. bd. Swissair; cons. Stauffer Chem. Co. Author: The Fundamentals of Iron and Steel Castings. Decorated Ordre National du Merite, France; recipient ann. award Ductile Iron Soc., 1965; Grande Medaille d'Honneur L'Association Technique de Fonderie, 1967. Mem. ASME (Charles F. Rand Meml. Gold medal 1977), AIME, Am. Soc. Metals, Am. Foundrymen's Soc. (hon. life, Peter L. Simpson gold medal award 1952), Nat. Acad. Engring., Mining and Metall. Soc. Am., Yale Engring. Assn. (dir.), Sigma Xi. Clubs: Seabright (N.J.) Beach; Down Town Assn., Univ., Yale (N.Y.C.); Rumson Country. Subspecialty: Metallurgy. Co-inventor ductile iron. Home: 143 Grange Ave Fair Haven NJ 07701 Office: 1 New York Plaza New York NY 10004

GAINER, JOSEPH HENRY, virologist, researcher; b. Atlanta, Ill., Oct. 24, 1924; s. Henry Albert and Erma Irene (Hoose) G.; m. Bridget Ginty, Jan. 2, 1954; children: Karen, Lisa, Patricia, Kelly, David, Erin. D.V.M., Ohio State U, 1946, M.S., 1947; M.S. in Virology, U. Mich., 1958. Diplomate: Am. Coll. Veterinary Microbiologists.; lic. veterinarian, Ohio, Ark., Mich. Instr. Ohio State U., 1947-49; research asst. Mayo Clinic, Rochester, Minn., 1949-50; research assoc. U. Chgo., 1950-51; asst. prof. U. Ark., 1951-53; virologist Fla. Dept. Agr., Kissimmee, 1959-66; research virologist NIH, 1967-73; veterinary med. officer FDA, Beltsville, Md., 1973—; adj. prof. N.C. State U., 1967-73. Contbr. articles in field to profl. jours. Pres. elect Kissimmee High Sch. PTA, 1965-66. Served to capt. U.S. Army, 1953-55. NIH grantee, 1962-65. Mem. Am. Veterinary Med. Assn., Am. Soc. Microbiology, Internat. Acad. Pathology, Am. Soc. Virology, Mayo Clinic Alumni Assn., AAAS, Sigma Xi, Phi Eta Sigma, Phi Zeta. Roman Catholic. Lodge: Rotary. Subspecialties: Virology (veterinary medicine); Immunotoxicology. Current work: Adverse/beneficial effects of chemicals on immune and interferon . Home: 12408 Willow Green Ct Potomac MD 20854 Office: FDA Bldg 328 A Beltsville MD 20705

GAINES, ALBERT LOWERY, mechanical engineer; b. Selma, Ala., Feb. 28, 1920; s. Drummond Fletcher and Rubye Lee (Cocke) G.; m. Dorothy Lea Conley, Apr. 25, 1942; children: Albert Lowery, John Bruce, Richard Andrew, William Stuart. B.M.E., Auburn U., 1946; M.M.E., U. Mo.-Columbia, 1950. Registered profl. engr., Tenn. Mech. engr. Humble Oil & Refining, Baytown, Tex., 1948-49; asst. instr. U. Mo.-Columbia, 1949-50; engr. Union Carbide Nuclear, Oak Ridge, 1950-56; mgr. spl. products Combusion Engring., Chattanooga, 1956-65; project mgr. Combustion Engring., Windsor, Conn., 1965—; cons. MIT, Cambridge, 1981, Gen. Atomics, San Diego, 1982. Served to 1st lt., arty. U.S. Army, 1942-46. Mem. ASME, Am. Nuclear Soc., Am. Inst. Chem. Engrs. Episcopalian. Subspecialties: Mechanical engineering; Nuclear engineering. Current work: Mechanical and system engineering of advanced power technologies, project management of PWR power plant and studies in fusion and liquid metal breeder reactors. Patentee in field. Office: Combustion Engring Inc 1000 Prospect Hill Rd Windsor CT 06095

GAITHER, WILLIAM SAMUEL, college dean; b. Lafayette, Ind., Dec. 3, 1932; s. William Marcius and Susan Frances (Kirkpatrick) G.; m. Robin Cornwall McGraw, Aug. 1, 1959; 1 dau., Sarah Curwen. Student, Purdue U., 1950-51; B.S. in Civil Engring, Rose Poly. Inst., 1956; M. Sci. Engring., Princeton, 1962; M.A. (Arthur Le Grand Doty fellow), Princeton, 1963; Ph.D. (Ford Found. fellow), Princeton, 1964. Engr. Dravo Corp. (marine constrn.), Pitts., 1956-60; supt. Myer Corp., Neenah, Wis., 1960-61; supervising engr., chief engr. port and coastal devel., pipeline div. Bechtel Corp., San Francisco, 1965-67; asso. prof. coastal engring. dept. U. Fla. at Gainesville, 1964-65; mem. faculty U. Del. at Newark, 1967—, asso. prof. civil engring., 1967-70, prof. civil engring., 1970, prof., dean, 1970—, also dir. sea grant coll. program; dir. Roy F. Weston Inc.; mem. marine bd. NRC, 1975-81. Chmn. Gov.'s Oil Transp. Study Com., 1971-73; mem. Gov.'s Task Force Marine and Coastal Affairs, 1970-72, Gov.'s Council Sci. and Tech., Del., 1970-72; mem. ocean affairs adv. com. U.S. Dept. State; chmn. adv. council dept. civil engring. Princeton U., 1973—. Served as pvt. C.E. AUS, 1953. Recipient Distinguished Achievement award Rose Poly. Inst., 1975. Fellow ASCE (chmn. offshore policy com. 1979—); mem. Del. Acad. Sci. (pres. 1971-72), Soc. Naval Architects and Marine Engrs., Marine Tech. Soc., Am. Geophys. Union, Assn. Sea Grant Program Instns. (pres. 1973-74). Club: Cosmos (Washington). Subspecialties: Civil engineering; Systems engineering. Home: 240 Beverly Rd Newark DE 19711

GAJDUSEK, DANIEL CARLETON, pediatrician, research virologist; b. Yonkers, N.Y., Sept. 9, 1923; s. Karl A. and Ottilia D. (Dobroczki) G.; m.; children: Ivan Mbagintao, Josede Figirliyong, Jesus Raglmar, Jesus Mororui, Mathias Maradol, Jesus Tamel, Jesus Salalu, John Paul Runman, Yavine Borima, Arthur Yolwa, Joe Yongorimah Kintoki, Thomas Youmog, Toni Wanevi, Toname Ikabala, Magame Prima, Senavayo Anua, Igitava Yariga, Luwi Ivavara, Irvam'bin'ai Undae'mai, Steven Malruio. B.S., U. Rochester, 1943; M.D., Harvard U., 1946; NRC fellow, Calif. Inst. Tech., 1948-49; D.Sc. (hon.), U. Rochester, 1977, Med. Coll. Ohio, 1977, Washington & Jefferson Coll., 1980, Harvard U., 1982, Hahnemann U., 1983; D.H.L., Hamilton Coll., 1977, U. Aix-Marseille, France, 1977; LL.D. (hon.), U. Aberdeen, Scotland, 1980. Diplomate: Am. Bd. Pediatrics. Intern, resident Babies Hosp., Columbia Presbyn. Med. Center, N.Y.C., 1946-47; resident pediatrics Children's Hosp., Cin., 1947-48; pediatric med. mission, Germany, 1948; resident, clin. and research fellow Childrens Hosp., Boston, 1949-51; research fellow pediatrics and infectious diseases Harvard U., 1949-52; with Walter Reed Army Inst. Research, Washington, 1952-53, Institut Pasteur, Teheran, Iran and dept. medicine U. Md., 1954-55; vis. investigator Nat. Found. Infantile Paralysis, Walter and Eliza Hall Inst. Med. Research, Melbourne, Australia, 1955-57; dir. program for study child growth and devel. and disease patterns in primitive cultures and lab. slow, latent and temperate virus infections Nat. Inst. Neurol. and Communicative Disorders and Stroke, NIH, Bethesda, Md., 1958—; chief Central Nervous System Studies Lab., 1970—; chief scientist research vessel Alpha Helix expdn. to, Banks and Torres Islands, New Hebrides, South Solomon Islands, 1972. Author: Hemorrhagic Fevers and Mycotoxicoses in the USSR, 1951, Journals, 35 vols., 1954-82, Hemorrhagic Fevers and Mycotoxicoses, 1959, Slow Latent and Temperate Virus Infections, 1965, Correspondence on the Discovery of Kuru, 1976, (with Judith Farquhar) Kuru, 1980. Recipient E. Meade Johnson award Am. Acad. Pediatrics, 1963, Superior Service award NIH, HEW, 1970, Disting. Service award HEW, 1975, Prof. Lucian Dautrebande prize in pathophysiology, Belgium, 1976, Nobel prize in physiology and medicine, 1976; GudaKunst lectr. U. Mich., 1973; Dyer lectr. NIH, 1974; Heath Clark lectr. U. London, 1974; B.K. Rachford lectr. Children's Hosp. Research Found., Cin., 1975; Langmuir lectr. Center for Disease Control, Atlanta, 1975; Withering lectr. U. Birmingham, Eng., 1976; Cannon Elie lectr. Boston Children's Med. Center, 1976; Zale lectr. U. Tex., Dallas, 1976; Bayne-Jones lectr. Johns Hopkins Med. Sch., Balt., 1976; Harvey lectr. N.Y. Acad. Medicine, 1977; J.E. Smadel lectr. Infectious Disease Soc. Am., 1977; Burnet lectr. Australasian Soc. Infectious Disease, 1978; Mapother lectr. U. London, 1978; Disting. lectr. in medicine Mayo Clinic, 1978; Kaiser Meml. lectr. U. Hawaii, 1979; Eli Lilly lectr. U. Toronto, 1979; Payne lectr. Children's Hosp. D.C., 1981; Ray C. Moon lectr. Angelo State U., Tex., 1981; Silliman lectr. Yale U. Sch. Medicine, 1981; Blackfan lectr. Children's Hosp. Med. Ctr., Boston, 1981; Hitchcock Meml. lectr. U. Calif.-Berkeley, 1982; Nelson lectr. U. Calif.-Davis, 1982; Derick-MacKerres lectr. Queensland Inst. Med. Research, 1982; Bicentennial lectr. Harvard U. Sch. Medicine, 1982; Cartwright lectr. Columbia U., 1982; lectr. Chinese Acad. Med. Sci., 1983. Mem. Nat. Acad. Scis., Am. Acad. Arts and Scis., Am. Philos. Soc., Soc. Pediatric Research, Am. Pediatric Soc., Am. Soc. Human Genetics, Am. Acad. Neurology (Cotzias prize 1979), Soc. Neurosci., Am. Epidemiol. Soc., Infectious Diseases Soc. Am., Société des Oceanistes, Paris, Papua and New Guinea Sci. Soc., Micronesian Acad. Sci., Slovak Acad. Scis., Academia Nacional de Medicina de Mexico, Phi Beta Kappa, Sigma Xi. Subspecialty: Virology (medicine). Home: Prospect Hill 6552 Jefferson Pike Frederick MD 21701 Office: NIH Bethesda MD 20205

GAJEWSKI, WALTER MICHAEL, nuclear reactor equipment engineer; b. Hartford, Conn., Apr. 4, 1923; s. Michael Walter and Mary Ann (Boron) G.; m. Mary Christine Maglieri, Dec. 27, 1949; children: Lisa Marie, Stephen Walter, Paul Michael. B.S.E.E., U. Conn., 1949, M.S.E.E., 1951. Registered profl. engr., Calif. Engr., mfg. engring. Bettis Atomic Power Lab., Pitts., 1950-60; mgr. reactor plant Naval Reactors Facility, Arco, Idaho, 1960-68; asst. project mgr. Bettis Atomic Power Lab., 1968-70; engring. mgr. Westinghouse Hanford Co., Richland, Wash., 1970—. Served with AUS, 1943-46; ETO. Mem. IEEE (sr.), ASME (vice chmn. advanced reactors com./nuclear engring div. 1982—), Am. Nuclear Soc. (chmn. Pitts. chpt. 1954-56). Lodge: Elks. Subspecialties: Nuclear engineering; Mechanical engineering. Current work: Design of nuclear reactors; reactor plant equipment; fuel and waste handling; surveillance inspection—robotics. Inventor unique pool water reactor, various nuclear components. Home: 3103 S Everett Pl Kennewick WA 99336 Office: Westinghouse Hanford Co Richland WA 99352

GAJSKI, DANIEL DANKO, computer scientist, educator; b. Zagreb, Yugoslavia, Oct. 10, 1938; came to U.S., 1969; s. Dako and Nada (Mirol) G.; m. Ana Veselic, Mar. 2, 1963. B.S.E.E., U. Zagreb, 1963, M.S., 1967; Ph.D., U. Pa., 1974. Engr. Inst. Tesla, Zagreb, Yugoslavia, 1963-69; vis. asst. prof. U. Ill, Urbana-Champaign, 1977-78, asst. prof., 1978-79, assoc. prof., 1980—; project engr. Burroughs Corp., Paoli, Pa., 1974-77; mgr. semicondr. structures, 1979-80. Contbr. articles in field to profl. jours. Mem. IEEE, Assn. Computing Machinery. Subspecialties: Computer architecture; Algorithms. Current work: Numerical algorithms, architecture of supercomputers, hardware and silicon compilers, design methodology and expert systems. Patentee in field. Office: 1304 W Springfield Ave Urbana IL 61801

GAKENHEIMER, DAVID CHARLES, mechanical engineer; b. Balt., Aug. 3, 1943; s. Albert Christian and Doris (Bader) G. B.Eng. Sci., Johns Hopkins U., 1965; M.S., Calif. Inst. Tech., 1966, Ph.D., 1968. Engr. Rand Corp., Santa Monica, Calif., 1968-72, R&D Assocs., Marina Del Rey, Calif., 1972—, program mgr., 1974—, div. dir., 1983—; instr. U. So. Calif., 1970-71. Contbr. articles on stress wave propagation and laser effects to profl. jours. Recipient Hamilton Watch award, 1965; NASA traineeship, 1965-68. Mem. ASME (Henry Hess award 1972), Tau Beta Pi, Pi Tau Sigma. Subspecialty: Mechanical engineering. Current work: High power laser effects. Home: PO Box 1782 Santa Monica CA 90406 Office: PO Box 9695 Marina del Rey CA 90291

GALABURDA, ALBERT MARK, neurologist, researcher, educator; b. Santiago, Chile, July 20, 1948; came to U.S., 1963, naturalized, 1968; s. John and Eva (Drinberg) G.; m. Margaret Okun, July 21, 1969; children: Adam, Daniel, Laura, Julia. B.A. cum laude, Boston U., 1971, M.D., 1971. Diplomate: Am. Bd. Internal Medicine, Am. Bd. Psychiatry and Neurology, Nat. Bd. Med. Examiners. Med. intern Boston City Hosp., 1971-72, asst. resident in medicine, 1972-73, resident in neurology, 1973-75, chief resident, 1975-76, asst. vis. physician, 1976-77, asso. dir. neurol. unit, 1977-82; teaching asso. Boston U. Sch. Medicine, 1972-73; clin. fellow in neurology Harvard U. Med. Sch., Boston, 1973-76, instr., 1976-80, asst. prof., 1980—; asst. neurologist Beth Israel Hosp., Boston, 1976—, asso. dir., 1976-81, dir., 1980—; dir. neurology Hebrew Rehab. Center for Aged, Roslindale, Mass., 1976—; cons. Sabin and Mark CAT Scan Lab., Boston, 1976-80; guest lectr., Stockholm, 1980, Frankfurt, W.Ger., 1980, Paris, 1980, Monastir, Tunisia, 1981. Contbr. articles to sci. jours. Served with U.S. Army Rev., 1971-77. Mem. Am. Acad. Neurology, Boston Soc. Psychiatry and Neurology, AAAS, N.Y. Acad. Scis., Pan Am. Med. Assn., Mass. Med. Soc., Soc. for Neurosci., Am. Assn. Anatomists, Orton Soc. (cons. 1980—). Subspecialty: Neurology. Current work: Anatomical aspects of cerebral lateralization and language; research on neurol. basis of learning disabilities. Office: 330 Brookline Ave Room K-420 Boston MA 02215

GALAMBOS, JAMES ANDREW, cognitive scientific researcher; b. Cleve., Feb. 26, 1951; s. Henry J. and Jeannette (Turvey); m. Sylvia Joseph, Dec. 29, 1974. B.A., U. Pa.Phila., 1974; Ph.D., U. Chgo., 1981. Research assoc. Yale U., New Haven, 1981-83; exec. v.p. Compu-Tech, Inc., New Haven, 1983—. Co-editor: Yale Cognitive Science: New Approaches to the Study of Cognition. USPHS grantee, 1975-80; IBM grantee, 1982. Mem. Am. Psychol. Assn., Assn. for Computing Machinery, Cognitive Sci. Soc. (assoc.), Psychonomic Soc. (assoc.). Subspecialties: Cognition; Artificial intelligence. Current work: Investigating the representation and utilization of knowledge about common activities, in human and computer memory, including design of man-machine interfaces. Office: Cognitive Sci Project Yale U POBox 2158 Yale Sta New Haven CT 06520

GALAS, DAVID JOHN, biology educator; b. St. Petersburg, Fla., Feb. 25, 1944; s. David Emanuel and Catherine Elizabeth (Filan) G.; m. Linda Elaine Hubbard, July 1, 1967; children: David John, John Ryan. B.A., U. Calif.-Berkeley, 1967; M.S., U. Calif.-Davis, 1968, Ph.D., 1972. Sr. scientist U. Calif. Lawrence Livermore Lab., 1974-77; charge de recerche U. Geneva, 1977-81; asst. prof. molecular biology U. So. Calif., Los Angeles, 1981-83, assoc. prof., 1983—. Served to capt. USAF, 1972-74. Mem. Am. Phys. Soc., AAAS, Sigma Xi. Subspecialties: Molecular biology; Genetics and genetic engineering (biology). Current work: Molecular biology of DNA recombination transposition and DNA-protein interactions. Office: Dept Biol Scis U So Calif University Park Los Angeles CA 90089

GALASSO, GEORGE JOHN, health scientist, administrator; b. N.Y.C., June 3, 1932; s. Giorgio and Lucia (Surico) G.; m. Joan Catherine Walsh, June 7, 1958; children: Catherine J., G. John, George J. B.S., Manhattan Coll., 1954; Ph.D., U. N.C., 1960. Research asst. prof. U.N.C., Chapel Hill, 1963-64; assoc. prof. U. Va., Charlottesville, 1964-68; grants assoc. Div. Research Grants, NIH, Bethesda, Md., 1968-69; antiviral substances program officer Nat. Inst. Allergy and Infectious Diseases, NIH, Bethesda, 1969-76, chief infectious disease br., 1973-77, chief devel. application br., 1977-83; assoc. dir. extramural affairs NIH, 1983—; dir. WHO Collaborating Ctr. for Interferon, 1982—, mem. adv. panel infectious diseases. Editor: Antiviral Agents and Viral Diseases of Man, 1979, Biology of the Interferon System, 1981, Jour. of Med. Virology, 1978—, Jour. Antiviral Research, 1983—. Served with AUS, 1954-56. Named Boss of Yr. Am. Bus. Women's Assn., 1978; recipient USPHS Superior Service award, 1978, Spl. Achievement award, 1981, Exceptional Achievement award Asst. Sec. Health, 1983. Fellow Infectious Diseases Soc.; mem. Inter-Am. Soc. Chemotherapy (virus sect. co.-chmn.), Am. Acad. Microbiology (ethics com. 1979-82), Am. Soc. Microbiology. Subspecialties: Infectious diseases; Virology (medicine). Current work: Help direct extramural research for the NIH through direct involvement with the various institutes and policy direction in general. Involved in interferon and antiviral research development.
Office: Assoc dir Extramural Affairs NIH Bldg 1 Room 111 Bethesda MD 20205

GALBIATI, LOUIS JOSEPH, elec. engr., educator; b. Vineland, N.J., Feb. 17, 1925; s. Louis Anthony and Mary Theresa G.; m. Jane Horan, Mar. 30, 1952; children: Louis, James, Philip, Andrew, Susan. B.E. in Elec. Engring, Johns Hopkins U., 1951; M.S., Cornell U., 1956, Ph.D., 1960; M.Ed. Adminstrn., Northeastern U., 1968. Registered profl. engr., Ohio. Instr. Cornell U., Ithaca, N.Y., 1954-60; prof. Merrimack Coll., Andover, Mass., 1960-62; engr. MITRE, Bedford, Mass., 1962-68; computer mgr. Service Tech Corp., Boston, 1968-71; engring leader RCA, Morestown, N.J., 1971-73; acad. dean, asst. pres. Hartford (N.Y.) State Coll., 1973-75; prof. U. Ark., Little Rock, 1975-77; dean SUNY Coll. Tech, Utica, 1977—, exec. dir. 1982—. Mem. Mohawk Valley Engring. Exec. Council; mem. Council Engring. Tech. in N.Y. State. Served with Signal Corps U.S. Army, 1943-47. Rockefeller Found. grantee, 1976-78. Mem. IEEE, Soc. Mfg. Engrs. (grantee 1980-82, chmn. Robotics Internat. 1982—), Mohawk Valley Engring. Exec. Soc., Sigma Xi, Tau Alpha Pi. Subspecialties: Electronics; Software engineering. Current work: Robotics, CAD/CAM. Office: SUNY 811 Court St Utica NY 13502

GALBRETH, TIMOTHY MICHAEL, nuclear engr., health physicist; b. Ft. Wayne, Ind., July 8, 1946; s. John Edward and Jeanne Marie (Junk) G.; m. Marilyn Kay Ulrich, July 16, 1966 (dec. Nov. 1966); m. Betty Margaret Veers, July 4, 1969; children: Jennifer, Michael. B.S. in Engring. Physics, U. Toledo, 1968; M.S. in Nuclear Engring, Ga. Inst. Tech., 1971. Registered profl. engr., Tenn; cert. health physicist. Nuclear engr. TVA, Knoxville and Chattanooga, 1972—. Served with U.S. Army, 1969-70. Mem. Am. Nuclear Soc., Health Physics Soc., Tau Beta Pi. Republican. Lutheran. Subspecialty: Nuclear engineering. Current work: Safety review of nuclear power plant operations. Home: 221 Stratford Way Signal Mountain TN 37377 Office: TVA 701 Chestnut St Chattanooga TN 37614

GALE, ROBERT PETER, physician/scientist; b. N.Y.C., Oct. 11, 1945; m. Tamar Tishler; children: Tal Britt, Shir Jessica, Elan Adam. A.B. with high honors, Hobart Coll., 1966; M.D., SUNY, Buffalo, 1970; Ph.D. in Microbiology and Immunology, UCLA, 1978. Diplomate: Am. Bd. Internal Medicine, Am. Bd. Med. Oncology. Postdoctoral scholar in immunology UCLA, 1972-76, asst. prof. medicine div. hematology-oncology, 1974-79, sr. mem. immunobiology group, 1974—, dir., 1977—, assoc. prof.medicine div. hematology-oncology, 1979—; chmn. Internat. Bone Marrow Transplant Registry, 1982—; Bogert fellow Leukemia Soc. Am., 1974-76; Leukemia Soc. scholar, 1976-81, SUNY Research Found. fellow, 1967, Excerpta Medica Found. traveling prof., 1979-81; Meyerhoff vis. scientist Weizmann Inst. Sci., 1982-83. Contbr. articles to sci. jours. Mem. AAAS, Am. Assn. Immunologists, ACP, Am. Fedn. Cancer Research, Am. Fedn. Clin. Research, Am. Soc. Clin. Oncology, Am. Soc. Hematology, Internat. Soc. Exptl. Hematology, N.Y. Acad. Sci., Soc. Cryobiology, Soc. Exptl. Biology and Medicine, Transplantation Soc., Western Soc. Clin. Research, Sigma Xi, Epsilon Pi Sigma. Subspecialties: Cell biology; Cancer research (medicine). Office: 14-121 UCLA Center Health Sciences Los Angeles CA 90024

GALEENER, FRANK LEE, physicist; b. Long Beach, Calif., July 31, 1936; s. Floras Frank and Daisy Elizabeth (Lee) G.; m. Janet Louise Trask, June 7, 1959; children: Keith Lee, Matthew Lee. S.B. in Physics, MIT, 1958, S.M. in Physics (Woodrow Wilson fellow), 1962, Ph.D., Purdue U., 1970. Physicist MIT Lincoln Lab., 1959-61, Nat. Magnet Lab., 1961-64; scientist Xerox Palo Alto Research Ctr., Calif., 1970-73, mgr. semicondr. research, 1973-77, prin. scientist, 1977—; mem. com. on recommendations U.S. Army Basic Sci. Research, 1976-79; mem. adv. panel on solid state physics Office Naval Research, 1980; co-organizer adv. panel on amorphous materials Dept. Energy, 1980; vis. scientist dept. theoretical physics Oxford U., fall 1982; vis. scientist Cavendish Lab. Cambridge U., 1983-84. Contbr. numerous articles to physics jours.; editor: (with others) Structure and Excitations of Amorphous Solids, 1976, The Physics of MOS Insulators, 1980. Mem. Am. Phys. Soc., Am. Ceramic Soc., Optical Soc. Am., Am. Vacuum Soc., Sigma Xi, Sigma Pi Sigma. Subspecialties: Condensed matter physics; Materials. Current work: Structure and excitations of disordered solids; raman spectroscopy; fundamental studies of condensed matter; experiment and phenomenological theory; electromagnetic theory. Home: 4035 Orme St Palo Alto CA 94306 Office: 3333 Coyote Hill Rd Palo Alto CA 94304

GALEY, JOHN TAYLOR, natural gas operator, cons. geologist; b. Beaver, Pa., Aug. 30, 1907; s. George Banks and Vera (Taylor) G.; m. Blanche Georgene Fishback, Nov. 19, 1938; children: Margaret Elizabeth, John Taylor. B.A., princeton U., 1932; postgrad., U. Pitts., 1932-33. Certified profl. geologist, Pa. Natural gas operator John T. Galey, Pitts., 1934-66, Somerset, Pa., 1966—; cons. in field. Contbr. articles to profl. jours. Trustees Carnegie Mus. Natural History, Pitts., 1974—; trustee Mt. Lebanon Presbyn. Ch., 1947-57, pres., 1948-50, 52-54. Recipient Distinguished Service award Am. Assn. Petroleum Geologists, Ben H. Parker Medal Am. Inst. Profl. Geologists, 1978. Mem. Am. Assn. Petroleum Geologists (hon.), Geol. Soc. Am. (fellow), Pitts. Geol. Soc. Republican. Episcopalian. Clubs: Duquesne, Harvard-Yale-Princeton, Rolling Rock, Cottage. Subspecialty: Geology. Current work: Investigation deeper hitherto unexplored natural gas/petroleum possibilities of Appalchians. Home and Office: Galecrest RD 4 Somerset PA 15501

GALIL, KHADRY AHMED, health science educator, researcher, dentist, oral surgeon; b. Aswan, Egypt, Sept. 1, 1942; emigrated to Can., 1969; s. Ahmed and Amna G.; m. Kathleen McCully, Aug. 1974; 1 child, Ramzy. B.D.S., U. Alexandria, Egypt, 1964, D.Oral Surgery, 1967; Ph.D., U. Western Ont., 1973. Asst. prof., lectr. Alexandria (Egypt) U. Oral Surgery, 1964-69; co-dir. Mich. Alexandria X-Ray Pharoah Project, 1967-69; vis. prof. U. Mich.; fellow U. Sask., U. Western Ont., 1969-73; assoc. prof. dept. anatomy, dept. oral and maxillofacial surgery Health Scis. Centre, U. Western Ont., London, Can., 1976—; sec./treas. Midwest Anatomist Assn., U.S. and Can., 1975. Contbr. chpt. to book in field. Judge, referee Can. Amateur Boxing Assn., Ont., 1976—. Recipient awards in oral diagnosis Alexandria U., 1964, awards in pharmacology, 1962. Fellow Acad. Gen. Dentistry; mem. Royal Coll. Dentists and Surgeons Ont., Internat. Assn. Dental Research (pres. Ont. chpt. 1973, 82-83, Am. Assn. Anatomy.). Subspecialty: Biomaterials. Current work: Tissue adhesives, surgical glue; use of surgical glue as substitute for sutures, experimental and clinical trials, scanning electron microscopy. Inventor Galil, Wright & Way device (acid etch space maintainer). Office: Dept Anatomy Health Scis Centre Univ Western Ont London ON Canada N6A 5C1

GALINAT, WALTON CLARENCE, genetics educator; b. Manchester, Conn., Dec. 9, 1923; s. Clarence Clayton and Edith Hargraves (Gillette) G.; m. Elizabeth Ruth Warren, Nov. 28, 1945; children: David W., Alice R. B.S. with distinction, U. Conn, 1949; M.S., U. Wis., 1951, Ph.D., 1953. Asst. in genetics Conn. Agr. Expt. Sta., 1946-50; asst. in agronomy Wis. Agr. Expt. Sta., 1950-53; research fellow, research assoc. Bussey Inst., Harvard U., 1953-54; assoc. prof. Waltham Field Sta. U. Mass., 1964-68; prof. genetics Suburban Expt. Sta., 1968—. Contbr. numerous articles in field to profl. jours. Served with USCG, 1943-46. Mem. AAAS, Am. Genetic Assn., Am. Soc. Naturalists, Bot. Soc. Am., Genetics Soc. Am., Soc. Econ. Botany. Republican. Subspecialties: Plant genetics; Genome organization. Current work: Origin, evolution and improvement of maize, sweet corn breeding, the origin and comparative morphology of the maize cob. Home: 33 Mossfield Rd Waban MA 02168 Office: 240 Beaver St Waltham MA 02154

GALINSKY, RAYMOND ETHAN, pharmacy educator; b. Hartford, Conn., Jan. 27, 1948; s. Max and Cecille Marie (Smith) G. Student, San Francisco State Coll., 1966-68; B.A. in Biology, U. Calif.-Berkeley, 1970; Pharm.D., U. Calif.-San Francisco, 1975. Registered pharmacist, Calif. Resident in hops. pharmacy U. Calif.-San Francisco Hosp., 1975-76; pharmacist Summit Pharmacy, Oakland, Calif., 1976-77; instr. Phila. Coll. Pharmacy and Sci., 1977-78; postdoctoral fellow Sch. Pharmacy, SUNY-Amherst, 1978-80, research asst. prof. pharmaceutics, 1980—; guest faculty mem. Eli Lilly & Co., Indpls., 1982—; mem. speakers bur. Syva Co., Palo Alto, Calif., 1982—. Mem. Am. Fedn. Clin. Research, Am. Coll. Clin. Pharmacy, N.Y. Acad. Scis., AAAS, Federation Internationale Pharmaceutique (assoc.). Jewish. Subspecialty: Pharmacokinetics. Current work: Teaching graduate and undergraduate pharmaceutics; research in drug metabolism, drug disposition, pharmacokinetics, toxicokinetics. Home: 273 Lisbon Ave Buffalo NY 14215 Office: Dept Pharmaceutics Sch Pharmac SUNY Amherst NY 14260

GALL, JOSEPH GRAFTON, biologist, researcher; b. Washington, Apr. 14, 1928; s. John Christian and Elsie (Rosenberger) G.; m. Dolores Marie Hogge, Sept. 17, 1955 (div. 1982); children: Lawrence, Barbara.; m. Diane Marie Dwyer, July 17, 1982. B.S., Yale, 1949, Ph.D., 1952. Faculty U. Minn., 1952-63, prof., 1963; prof. biology and molecular biophysics Yale, 1963-83; staff dept. embryology Carnegie Instn., Balt., 1983—; Mem. cell biology study sect. NIH, 1963-67, chmn., 1972-74. Contbr. articles profl. jours. Mem. AAAS, Am. Soc. Cell Biology (past pres.), Genetics Soc. Am., Am. Soc. Zoologists, Nat. Acad. Scis., Am. Acad. Arts and Scis. Subspecialty: Cell biology. Home: 81 North Lake Dr Hamden CT 06517 Office: Dept Embryology Carnegie Instn 115 W University Pkwy Baltimore MD 21210

GALLAGER, HARRY STEPHEN, surgical pathologist, educator, researcher; b. Chester, Pa., June 5, 1922; s. Harry and Josephine Emily (Johnson) G.; m. Bette A. Wehrung, Apr. 1, 1967. B.A., Temple U., 1943, M.D., 1946. Diplomate: Am. Bd. Pathology. Intern Chester Hosp, 1946-48; resident Germantown Dispensary, Phila., 1949-53; asst. pathologist Germantown Hosp., Phila., 1953-54; asst. prof. pathology U. Tex. System Cancer Ctr., Houston, 1956-62, assoc. prof., 1962-74, prof., 1974—. Editor: Early Breast Cancer, 1975; author, editor: The Breast, 1978; mem. editorial bd.: Breast, 1975—, Breast Cancer Research & Treatment, 1980, Clinics in Oncology, 1980—. Served to capt. U.S. Army, 1954-56. Fellow Coll. Am. Pathologists, Am. Soc. Clin. Pathologists, Am. Coll. Radiology (hon.); mem. Am. Assn. Pathologists, Soc. Surg. Oncology, AMA. Democrat. Episcopalian. Club: U. Tex. Faculty (Houston). Subspecialties: Pathology (medicine); Cancer research (medicine). Current work: Pathology of breast diseases; pathology of gynecologic neoplasms. Home: 10220 Memorial Dr #15 Houston TX 77024 Office: University of Texas System Cancer Ctr MD Anderson Hospital and Tumor Inst 6723 Bertner St Houston TX 77030

GALLAGHER, BRIAN BORU, neurologist; b. Chgo., Sept. 2, 1934; s. Thomas Francis and Beatrice (Sheehan) G.; m. Suzanne G., June 17, 1978; children: Ellen Elizabeth, Eileen Catherine, Ian Boru. B.S., U. Notre Dame, 1956; Ph.D. (NIH fellow), U. Chgo., 1960, M.D., 1963. Intern U. Chgo., 1963-64; resident in neurology Yale U., 1964-67, instr. in neurology, 1967-68, asst. prof. neurology, 1968-72, asso. prof., 1972; asso. prof. pharmacology and neurology Georgetown U., 1972-77; prof. neurology Med. Coll. Ga., 1977—; cons. to govt. agys. Contbr. numerous research articles, chpts. and revs. to profl. publs. Recipient Research Career Devel. award NIH, 1972; NIH grantee, 1967—. Mem. Am. Acad. Neurology, Am. Neurochemistry Soc., AAAS, N.Y. Acad. Sci., Siginati, Am. Soc. Pharmacology and Exptl. Therapeutics, Am. Epilepsy Soc. Subspecialties: Neurology; Neuropharmacology. Current work: Research in neuropharmacology and neuroendocrinology of epilepsy. Office: Dept Neurology Med Coll Ga Augusta GA 30912

GALLAGHER, JACK *See also* **GALLAGHER, JOHN JOSEPH, JR.**

GALLAGHER, JOAN SHODDER, immunologist, educator; b. Camden, N.J., June 19, 1941; d. Emanuel S. and Marie T. (Danella) Shodder; m. Joseph P. Gallagher, Jan. 30, 1965; children: Joseph, Timothy. B.S., Drexel U., 1964; M.S., U. Ill., 1967, Ph.D., 1970. Research asst. Wistar Inst., Bockus Inst. U. Pa., Phila., 1960-64; research asst. to asst. prof. depts. microbiology and biochemistry U.Ill., Urbana-Champaign, 1965-72; asst. prof. exptl. medicine dept. internat. medicine-immunology U. Cin., 1974—. Contbr. articles on immunology to profl. jours.; Editor: 2d edit. Bellbrook History Book for Bellbrook Hist. Soc, 1982. Mem. Am. Assn. Immunologists, Am. Acad. Allergy and Immunology, Sigma Xi. Democrat. Roman Catholic. Club: Bellbrook Athletic Boosters. Subspecialties: Immunology (medicine); Allergy. Current work: General allergic disorders, insect allergens, industrial allergens, sperm allergen and food allergens. Home: 3340 Ferry Rd Bellbrook OH 45305 Office: U Cin Med Ctr 231 Bethesda Av Immunology Div Cincinnati OH 45267

GALLAGHER, JOHN JOSEPH, JR. (JACK GALLAGHER), energy company manager, researcher; b. Boston, Oct. 7, 1940; s. John Joseph and Margaret Theresa (Leahy) G.; m. Judy Gail Kilgore, Mar. 2, 1968; children: John Joseph, Michael Patrick, Judy Gail, Mary Elizabeth. B.S., Boston Coll., 1962; M.A., U. Mo., Columbia, 1965; Ph.D., Tex. A&M U., 1971. Cert. petroleum geologist, Am. Assn. Petroleum Geologists; cert. profl. geologist Am. Inst. Profl. Geologists. Asst. field party chief Que. Dept. Natural Resources, Que., Can., 1964; exploration geologist Am. Metal Climax of Can., Noranda, Que., 1965; research assoc. Cities Service Exploration and Prodn. Research Lab., Tulsa, 1971-78; sr. geol. assoc. Cities Service Internat. New Ventures, Houston, 1978-79, Cities Services Internat. Tech. Services, 1979-80; basin study mgr. Cities Service Co. Exploration Resources, Tulsa, 1980—; trustee, vice chmn. for confs. Basement Tectonics Com. Inc., Denver and Cairo, Egypt, 1982—; session convenor 2d World Oil and Gas Show and Conf., Dallas, 1982-83; assoc. editor tectonics Am. Geophys. Union and European Geophys. Soc., Washington and Paris, France, 1982—; session chmn. Basin Analysis Research Conf. Am. Assn. Petroleum Geology, Tucson, 1982, 83. Contbr. chpt. to book in field; convenor exploration symposium. Served to capt. C.E. U.S. Army, 1966-68. Recipient M.T. Halbouty Award Tex. A&M U., 1969. Fellow Geol. Soc. Am.; mem. Am. Assn. Petroleum Geologists, Am. Inst. Profl. Geologists, Am. Geophys. Union (supporting), Sigma Xi (ife). Republican. Roman Catholic. Subspecialties: Tectonics; Geophysics. Current work: Evaluation of hydrocarbonn possibilities in sedimentary basins worldwide; global tectonics; physics of basin development; rock mechanics: cataclastic flow; fractography; lithosphers mechanics. Regional Geology: China, Arctic, Africa, India, Western North America, South America, Southeast Asia-Australia; research management; remote sensing. Home: 6219 E 99th St Tulsa OK 74136 Office: Cities Service Tech Center Box 3908 Tulsa OK 74102

GALLAGHER, JOSEPH PATRICK, research structural engr.; b. Phila., July 20, 1941; s. Joseph Michael and Margaret Regina (Balson) G.; m. Joan Mary Shodder, Jan. 30, 1965; children: Joseph Patrick, Timothy Adrian. B.S.C.E., Drexel U., 1964; M.S. (NSF trainee), U. Ill., Urbana, 1965, Ph.D., 1968. Grad. teaching asst. dept. theoretical and applied mechanics U. Ill., 1964-68, asst. prof. theoretical and applied mechanics, 1968-72; aerospace engr. Air Force Flight Dynamics Lab., Dayton, Ohio, 1972-77; group leader, service life mgmt. U. Dayton Research Inst., 1977—; adj. prof. materials engring., grad. program U. Dayton, 1980—; short course lectr. on fracture mechanics, 1974—; research engr. metall. div. U.S. Naval Research Lab., summer 1969, cons. Rockwell-Internat., 1982. Editor: (with T.A. Cruse) Fatigue Life Technology, 1977, (with T.W. Crooker) Structural Integrity Technology, 1979. Recipient Systems Command Tech. Achievement award USAF, 1973, 77, 2d Pl. Engring. award Air Force Assn., 1973, Gen. Foulois award Air Force Flight Dynamics Lab., 1975. Mem. ASME (exec. com. materials div. 1976-81), ASTM (Sam Tour award 1975, chmn. subcom. on subcritical crack growth), Am. Soc. Metals, Sigma Xi. Subspecialties: Fracture mechanics; Materials (engineering). Current work: Management of research and development projects that lead to improved crack damage control technologies, to enhanced structural maintenance plans or to solutions of fracture problems. Research, numerous publs. on applying fracture mechanics techniques to solution of subcritical crack growth problems; patentee crack growth gage. Home: 3340 Ferry Rd Bellbrook OH 45305 Office: U Dayton Research Inst Dayton OH 45469

GALLAGHER, KIM PATRICK, research physiologist, educator; b. Linz, Austria, Sept. 4, 1950; s. William Bernard and Mary Ellen (Myers) G.; m. Mary Alice Collins, July 30, 1977; children: Patrick, Kathryn. B.S., U. Wis.-Madison, 1973, M.S., 1975, Ph.D., 1977. Postdoctoral physiologist U. Calif.-San Diego, 1978-81, asst. research physiologist, 1981-82; asst. prof. depts. physiology and surgery U. Mich., Ann Arbor, 1982—; dir. Thoracic Surgery Research Lab., 1982—. Contbr. articles to profl. jours. NIH fellow, 1979-81; Am. Heart Assn. grantee, 1981—; recipient New Investigator award NIH, 1981—. Mem. Am. Physiol. Soc., Council Circulation Council Basic Sci. Am. Heart Assn. Subspecialty: Physiology (biology). Current work: Coronary blood flow, ventricular performance, relationship between myocardial blood flow and regional myocardial contration. Office: University of Michigan Thoracic Surgery Research Laboratory R3484 Kresge I Ann Arbor MI 48109

GALLAGHER, MICHAEL TERRENCE, immunologist; b. Gadsden, Ala., Nov. 26, 1943; s. Patrick Paul and Lola G. (Carter) G.; m. Linda Kay Barnard, Feb. 14, 1974; children: Kelly Margaret, Lisa Kathleen. B.S., U. Houston, 1966; M.S., Northwestern U., 1969; Ph.D., Baylor Coll., 1974. Instr. div. exptl. biology Baylor Coll. Medicine, 1974-75, asst. prof., 1975-80; asst. research scientist div. clin. pathology City of Hope Med. Center, Duarte, Calif., 1980-81, asst. research scientist dept. clin. and exptl. immunology, 1981-82, asso. research scientist, 1982—. Contbr. numerous articles to profl. jours. Mem. Am. Assn. Cancer Research, Am. Assn. Immunologists, Internat. Soc. Exptl. Hematology, Transplantation Soc. Roman Catholic. Subspecialty: Immunology (agriculture). Current work: Transplantation immunology with emphasis on bone marrow transplantation and graft vs. host disease.

GALLAGHER, PATRICK XIMENES, mathematics educator, researcher; b. Elizabeth, N.J., Jan. 2, 1935; m., 1961; 2 children. A.B., Harvard U., 1956; Ph.D. in Math, Princeton U., 1959. Asst. in math. Princeton U., 1957-59; instr. MIT, 1959-61; asst. prof. Columbia U., 1962-64; from assoc. prof. to prof. math. Barnard Coll., 1965-72; prof. Columbia U., 1972—; mem. Inst. Advanced Study, 1964-65. Mem. Am. Math. Soc. Office: Dept Math Columbia U New York NY 10027

GALLAGHER, RICHARD HUGO, university dean, engineer; b. N.Y.C., Nov. 17, 1927; s. Richard Anthony and Anna (Langer) G.; m. Therese Marylyn Doyle, May 17, 1952; children—Marylee, Richard, William, Dennis, John. B.C.E., N.Y. U., 1950, M.C.E., 1955; Ph.D., SUNY, Buffalo, 1966. Field engr. CAA, Dept. Commerce, Jamaica, N.Y., 1950-52; structural designer Texaco, N.Y.C., 1952-55; asst. chief engr. Bell Aerospace Co., Buffalo, 1955-67; prof. civil engring. Cornell U., 1967-78, chmn. dept. structural engring., 1969-78; dean Coll. Engring., U. Ariz., 1978—; cons. in field. Author: Finite Element Analysis, 1975, Matrix Structural Analysis, 1979; editor: Internat. Jour. Numerical Methods in Engring, 1969—. Served with USNR, 1945-47. Fulbright fellow, Australia, 1973; Sci. Research Council fellow U. Wales, 1974. Fellow ASCE, ASME (chmn. pressure vessel piping com. 1981—); mem. AIAA, Am. Soc. Engring. Edn., Nat. Acad. Engring., Soc. Exptl. Stress Analysis, Sigma Xi, Chi Epsilon, Tau Beta Pi. Roman Catholic. Club: Rotary (Tucson). Subspecialties: Solid mechanics; Civil engineering. Current work: Finite element analysis of plates and shells fo geometrically nonlinear behavior; complementary formulations of finite element analysis. Home: 6735 Camino Padre Isidoro Tucson AZ 85718 Office: Bldg 72 U Ariz Tucson AZ 85721

GALLAGHER, THOMAS FRANCIS, physicist; b. Bronxville, N.Y., Nov. 19, 1944; s. Thomas Francis and Margaret Ann (Sheekey) G.; m. Betty Barbara Cassiman, Sept. 21, 1974; 1 son Thomas Francis. A.B., Williams Coll., 1966; M.A., Harvard U., 1968, Ph.D., 1971. Research assoc. Harvard U., 1971; research assoc. U. Utah, 1971-72; postdoctoral physicist Stanford Research Inst., Menlo Park, Calif., 1972-74, physicist, 1974-79, sr. physicist, 1979—. Contbr. articles in field to profl. jours. Fellow Am. Phys. Soc. Roman Catholic. Subspecialty: Atomic and molecular physics. Current work: Collisions and spectroscopy of highly excited atoms. Patentee in field. Home: 1011 Fulton Ave Palo Alto CA 94301 Office: 333 Ravenswood St Menlo Park CA 94025

GALLANT, NANETTE, medical clinic executive, consultant; b. Salt Lake City, Mar. 14, 1956; d. Richard Austin and Kathryn Amelia (Grogan) G. Student, U. Mont., 1973-74; cert., Utah Tech. Coll., 1976; B.S., U. Utah, 1980, postgrad., 1980—. Cert. surg. technologist. Cardiac surgery specialist St. Mark's Hosp., Salt Lake City, 1976-80; ops. engr. KUED-TV, Salt Lake City, 1978-80; remote ops. engr. Skaggs Telecommunications, Salt Lake City, 1981-82; chief of research Western Spinal Clinic, Salt Lake City, 1982—; cons. NGP Co., 1974—; cons. engring. to physicians, 1981-82, 82—, Zimmer-Jackson Assocs., 1982—. Mem. Assn. Surg. Technologists (pres. 1977, regional rep. 1978, Nat. Technologist of Month 1978), Soc. Photo-Optical Instrumentation Engrs., Soc. Broadcast Engrs., IEEE; assoc. mem. Soc. Motion Picture and TV Engrs., Am. Nuclear Soc. Democrat. Roman Catholic. Subspecialties: Microsurgery; Bioinstrumentation. Current work: Application of broadcast equipment to microsurgical situation to provide accurate imaging; design and application bioinstruments to microsurgery. Home: 1408 Military Way Salt Lake City UT 84103

GALLARDO-CARPENTIER, ADRIANA, pharmacologist; b. Valparaiso, Chile, Sept. 13, 1930; d. Gustavo Rene and Lily Rebeca (Saavedra) Gallardo-C.; m. Robert Georges Carpentier, Feb. 23, 1967. D.D.S., Dental Sch. U. Chile, 1955. Pvt. practice dentistry, Santiago, Chile, 1955-73; instr. Coll. Dentisty U. Chile, 1964-67, asst. prof., 1966-74; asst. research pharmacologist Sch. Medicine UCLA, 1966-67; vis. fellow dept physiology SUNY Downstate Med. Center, Bklyn., 1970-72; asso. prof. exptl. medicine Med. Coll. U. Chile, 1973-76; research asso. dept. pharmacology Med. Coll. Howard U., Washington, 1977, asst. prof., 1977—. Contbr. articles to profl. jours. Mem. Am. Soc. Pharmacology and Exptl. Therapeutics, N.Y. Acad. Scis. Subspecialties: Pharmacology; Neuropharmacology. Current work: Alcohol and cardiac fiber, CNS. Office: 520 W St NW Washington DC 20059

GALLATIN, JUDITH ESTELLE, author, research psychologist; b. Grand Rapids, Mich., Feb. 15, 1942; d. Marcus and Marilyn (Gallatin) Laniado; m. Charles Helppie, Jan. 2, 1975; step-children: Charles, Bruce, Kathleen. B.A., U. Mich., 1962, M.A., 1963, Ph.D., 1967. Ltd. lic. psychologist, Mich. Teaching fellow U. Mich., 1966-67; asst. prof. psychology Eastern Mich. U., 1967-72, assoc. prof., 1972-77, prof., 1977; research assoc. U. Strathclyde, 1968; cons. Ford Found., 1975; reviewer NSF, Washington, 1979-82. Author: Adolescence and Individuality, 1975; author: Abnormal Psychology: Concepts, Issues, Trends, 1982; contbg. author: Understanding Adolescence, 1976, 80, Handbook of Adolescent Psychology, 1980, Personality Theory, Moral Development and Criminal Behavior, 1983. Mem. Portland (Oreg.) Com. on Fgn. Relations, 1978-82. USPHS research fellow, 1963-66. Mem. Am. Psychol. Assn., Soc. Research in Child Devel., Am. Orthopsychiat. Assn., Phi Beta Kappa. Subspecialties: Behavioral psychology; Cognition. Current work: Author of general books and textbooks in developmental and abnormal psychology; research into virtually all areas of psychology. Home: 436 Briarlane NE Grand Rapids MI 49503

GALLETTI, PIERRE MARIE, university executive, medical science educator; b. Monthey, Switzerland, June 11, 1927; s. Henri and Yvonne (Chamorel) G.; m. Sonia Aiden, Dec. 31, 1959; 1 son, Marc-Henri. B.A. in Classics, St. Maurice Coll., Switzerland, 1945; M.D., U. Lausanne, Switzerland, 1951, Ph.D. in Physiology and Biophysics, 1954. Asst. prof. physiology Emory U., 1958-62, assoc. prof., 1962-66, prof., 1966-67, vis. prof., 1967-68; prof. med. sci. Brown U., 1967—, chmn. div. biol. sci., 1968-72, acting dean of medicine, 1980-81, v.p. biology and medicine, 1972—; mem. acad. rev. bd. Exxon; mem. polymer adv. panel Johnson and Johnson; chmn. sci. adv. council pulmonary SCOR in adult respiratory failure Mass. Gen. Hosp.; chmn. Consensus Devel. Conf., NIH; chmn. devices and tech. br. task force NIH; plenary lectr. World Biomaterials Conf., 1980; Hastings lectr. NIH, 1979; McNeil Pharm. Spring Sci. lectr., 1982. Author: Heart-Lung Bypass: Principles and Techniques of Extracorporeal Circulation, 1962; contbr. numerous chpts., abstracts, articles to profl. publs. Recipient John H. Gibbon award Am. Soc. Extracorporeal Technology, 1980; NIH grantee, 1969—. Fellow Am. Coll. Cardiology; mem. AAAS, Biomed. Engring. Soc., Am. Physiol. Soc. Subspecialties: Artificial organs; Biomaterials. Current work: Implantable artificial lung; hybrid artificial pancreas; insulin delivery systems; piezoelectric polymers; hybrid articial liver. Office: Box G 97 Waterman St Providence RI 02912

GALLIE, THOMAS MUIR, JR., computer scientist; b. N.Y.C., Aug. 25, 1925; s. Thomas Muir and Mavis (Chubb) G.; m. Mary Frances Cordts, Nov. 22, 1945; children: Thomas Muir III, Charles Cordts, Ann Walston, Margaret Elizabeth. A.B., Harvard, 1947; M.A., U. Tex., 1949, Rice U., 1952, Ph.D., 1954. Research engr. Humble Oil Co., Houston, 1954, 55-56; mem. faculty Duke U., Durham, N.C., 1954, 56—, prof. math, 1967-71, prof. computer sci., 1971—, v.p., 1980-81, chmn. dept. computer sci., 1982—; sect. head NSF, 1968-69; vis. prof. Eidgenossische Technische Hochschule, Zurich, 1962-63, 73-74, Universidad Central de Venezuela, Caracas, 1965; chmn. bd. dirs. Triangle Univs. Computation Ctr., 1965-67, mem., 1969-73, 79—. Author: Computer Science, 1976; contbr. articles to profl. jours. Mem. Assn. Computing Machinery (nat. lectr. 1965-66), Math. Assn. Am., Sigma Xi. Subspecialty: Numerical analysis. Current work: Medical applications of computers. Office: Dept Computer Sci Duke U Durham NC 27706 Home: 21 Glenmore Dr Durham NC 27707

GALLIN, JOHN I., physician, researcher; b. N.Y.C., March 25, 1943; s. Nathaniel M. and Helen R. (Cohen) G.; m. Elaine B. Klimerman, June 23, 1966; children: Alice, Michael. B.A. cum laude, Amherst Coll., 1965; M.D., Cornell Med. Coll., 1969. Intern, resident, then chief resident in medicine Bellevue Hosp., NYU Med. Ctr., N.Y.C., 1969-71, 74-75; clin. assoc. Nat. Inst. Allergy and Infectious Diseases, NIH, Bethesda, Md., 1971-74, sr. investigator, 1974-76, chief bacterial diseases sect. Lab. Clin. Investigation, 1976—. Editor: Advanced in Host Defense Mechanisms, 1978; assoc. editor: Jour. Immunology, 1979-83; mem. editorial bd.: Infection and Immunity; contbr. over 100 articles to profl. jours. Recipient Commendation medal USPHS, 1980. Mem. Am. Soc. Clin. Investigation, Assn. Am. Physicians, Am. Fedn. Clin. Research, Am. Assn. Immunologists, Am. Soc. Cell Biology. Subspecialties: Infectious diseases; Internal medicine. Current work: Infectious diseases, phagocyte function, inflammation; leukocyte chemotaxis and secretion. Office: Bldg 10 Nat Inst Allergy and Infectious Diseases NIH Bethesda MD 20205

GALLO, CHARLES FRANCIS, JR., physicist; b. Mt. Vernon, N.Y., July 22, 1935; s. Charles F. and Ann Marie (Del Vecchio) G.; m. Marilyn R. Gallo, Oct. 5, 1957; 1 dau., Bonnie. B.S. in Physics (scholar), Rensselaer Poly. Inst., 1957. Engr. Gen. Electric Co.,-Rensselaer Poly. Inst. Coop. Program, Schenectady and Cleve., 1955-57; scientist Westinghouse Research, Pitts., 1957-64, Xerox Corp., Webster, N.Y., 1965-81, 3M Corp., St. Paul, 1981—; cons. sci. Rochester (N.Y.) Inner City Schs.; lectr. univs.; conf. presenter. Contbr. articles to sci. and engring. jours. Active Jobs for Inner City Students, Rochester. Recipient hon. mention for best publ. IEEE/Indsl. Applications Soc., 1972-73, 75-76; fgn. fellow Japan Inst. Electrostatics, 1977—; recipient Bausch & Lomb hon. sci. award, 1954. Mem. Am. Phys. Soc., Electrostatics Soc. Am., Japan Inst. Electrostatics. Clubs: Carlton Racquet, Plymouth Racquet, Penfield Racquet, Tartan (St. Paul). Subspecialties: Reprographics; Optical engineering. Current work: Reprographics, optical engring., gas discharge physics, electrostatics, atomic and molecular physics, data storage and reprodn., elec. engring., electronics. Patentee in field. Home: 2440 Lisbon Ave Lake Elmo MN 55042 Office: 3M Corp 3M Center Bldg 201 PO Box 33221 Saint Paul MN 55133

GALLO, MARIO MARTIN, clinical psychologist, neuropsychologist; b. Chgo., Sept. 13, 1947; s. Mike Vito and Christina Mary (Serritella) G. B.S., Loyola U., Chgo., 1969; M.A., Roosevelt U., 1972; Ph.D., Miami U., Oxford, Ohio, 1977. Lic. psychologist, Ill., Ohio, Ariz., Calif. Intern Greater Lawn Mental Health Ct., Chgo., 1970-71; trainee Miami U. Psychol. Clinic, Oxford, Ohio, 1972-75; psychiat. ward specialist U.S. Air Force Res., Chgo., 1972-76; intern Good Samaritan Hosp., Dayton, Ohio, 1974-75; resident Northwestern U., 1975-77; clin. dir. Kevin Coleman Mental Health Ctr., Kent, 1978—; pvt. practice psychology, Kent, Ohio, 1978—; adj. asst. prof. psychology Kent State U., 1978—. Served with USMC, 1969-70. Mem. Am. Psychol. Assn., Am. Acad. Psychotherapists, Am. Soc. Clin. Hypnosis, Internat. Neuropsychol. Soc., Nat. Acad. Neuropsychologists. Subspecialties: Neuropsychology; Clinical psychology. Current work: Neuropsychological assessments and rehabilitation; psychotherapy, hypnotherapy and biofeedback. Home: 5215 Cline Rd Apt A Kent OH 44240 Office: Kevin Coleman Mental Health Center 275 Martinel Dr Kent OH 44240

GALLO, ROBERT CHARLES, research scientist; b. Waterbury, Conn., Mar. 23, 1937; s. Francis Anton and Louise Mary (Ciancuilli) G.; m. Mary Jane Hayes, July 1, 1961; children—Robert Charles, Marcus. B.A., Providence Coll., 1959, D.Sc. (hon.), 1974; M.D., Jefferson Med. Coll., 1963. Clin. asso. med. br. Nat. Cancer Inst., NIH, Bethesda, Md., 1965-68, sr. investigator human tumor cell biology br., 1968-69, head sect. cellular control mechanisms, 1969-72, chief lab. tumor cell biology, 1972—; adj. prof. genetics George Washington U.; adj. prof. microbiology Cornell U.; cons. M.D. Anderson Hosp. and; Tumor Inst., Roswell Park Meml. Inst., U. S.C., Georgetown U. Cancer Center, Internat. Inst. Genetics, Naples, Italy, Hahnemann Med. Sch. Cancer Center; U.S. rep. to world com. Internat. Comparative Leukemia and Lymphoma Assn., 1981—. Served with USPHS, 1965-68. Recipient Dameshek award Am. Hematol. Soc., 1974, CIBA-GEIGY award in biomed. sci., 1977, USPHS Superior Service award, 1979, F. Stohlman lecture award, 1979. Mem. Internat. Soc. Hematology, Am. Soc. Clin. Investigation, Am. Soc. Biol. Chemists, Am. Microbiology Soc., Am. Soc. Pharmacology and Explt. Therapeutics, Biochem. Soc., Am. Assn. Cancer Research, Am. Fedn. Clin. Research, Fedn. for Advanced Edn. in Scis., Alpha Omega Alpha. Subspecialty: Cancer research (medicine). Research on viruses, biochemistry and leukemia. Home: 8513 Thornden Terr Bethesda MD 20034 Office: Nat Cancer Inst 9000 Rockville Pike Bethesda MD 20014

GALT, JOHN ALEXANDER, physicist, radio astronomer; b. Toronto, Ont., Can., Mar. 8, 1925; s. Ian and Mabel Claire (Johnston) G.; m. Rena Mabel Smith, Aug. 6, 1955; children: Sheila Jean, David Richard. B.A., U.Toronto, 1949, M.A., 1952, PH.D., 1955. Physicist DuPont of Can., Kingston, Ont., 1956-57; radio astronomer Nat. Research Council Can., Penticton, B.C., Can., 1957—; guest prof. Chalmers Tekniska Hogskola, Goteborg, Sweden, 1972-73. Serviced with Royal Can. Navy, 1944-45. Subspecialties: Radio and microwave astronomy; Atomic and molecular physics. Office: Box 248 Penticton BC Canada V2A 6K3

GALTON, PETER MALCOLM, biology educator; b. London, Mar. 14, 1942; U.S., 1967; s. Sidney and Ena (Childs) G.; m. Natalie S. Rawls, Feb. 14, 1975. B.Sc., U. London, 1964, Ph.D., 1967, D. Sc., 1983. Curatorial assoc. Peabody Mus. Natural History, Yale U., 1967-70, research staff dept. geology, 1967-70; asst. prof. biology U. Bridgeport, Conn., 1970-74, assoc. prof., 1974-78, prof., 1978—. Contbr. articles to profl. jours. Mem. Soc. Vertebrate Paleontology, Am. Soc. Zoologists, Sigma Xi. Subspecialty: Paleobiology. Current work: Vertebrate paleontology, functional morphology, systematics, relationships, evolution and zoogeography of dinosaurs. Office: Dept Biology U Bridgeport Bridgeport CT 06601

GALTON, VALERIE ANNE, endocrinology educator; b. Louth, Eng., May 6, 1934; came to U.S., 1959; s. Wilfrid and Eileen (Watson) Hamilton; m. Michael Galton, Aug. 26, 1956 (dec. 1968); children: Ian Andrew, Kenneth Anthony; m. Reed Detar, Dec. 28, 1976; stepchildren: James, Elizabeth, Stephen, Susan. B.Sc. with honors, U. London, 1955, Ph.D., 1958. Research assoc. Nat. Inst. for Med. Research, Mill Hill, London, 1955-58, Harvard Med. Sch., Boston, 1959-61; instr., asst. prof. Dartmouth Med. Sch., Hanover, N.H., 1961-66, assoc. prof., 1968-75, prof., 1975—; cons. NIH, Bethesda, Md., 1973—. Contbr. articles to profl. jours.; mem. editorial bd.: Endocrinology, 1982-86, Am. Jour. Physiology, 1982—. NIH grantee, 1962—. Mem. Am. Thyroid Assn., Endocrine Soc. Subspecialties: Endocrinology; Receptors. Current work: Mechanism of action of thyroid hormones. Office: Dartmouth Med. Sch Hanover NH 03756 Home: 20 Rayton Rd Hanover NH 03755

GAMBERT, STEVEN ROSS, medical educator; b. N.Y.C., Aug. 22, 1949; s. Lawrence and Mildred G.; m.; children: Christopher, Iselin. B.A. with honors, NYU, 1971; M.D., Columbia U., 1975. Intern Darmouth Med. Sch., Hanover, N.H., 1975-76, resident, 1976-77; fellow in endocrinology and geriatrics Harvard U. Med. Sch., Boston, 1977-79; asst. prof. medicine Med. Coll. Wis., Milw., 1979-81, assoc. prof. medicine, 1979-81, assoc. prof. medicine and physiology and chief geriatics and gerontology, 1981—; cons. Wis. Bur. Health, Madison, 1982—. TV host: To Better Health for the Elderly, Milw., 1982; writer health column: Sr. Citizen News, Milw., 1982; Editor: Contemporary Geriatric Medicine, 1983; contbr. numerous articles to med. jours.; mem. editorial bds.: Jour. Gerontology, 1981—, Psychit. Medicine, 1982—. Nat. Inst. Aging grantee, 1982-87; recipient Goldberger Research award AMA, 1974. Fellow ACP; mem. Endocrine Soc., Gerontol. Soc. Am. (sec.-treas. 1981-82), Soc. for Exptl. Biology and Medicine, Am. Geriatrics Soc., Am. Aging Assn., Am. Fedn. Clin. Research (councilor MW sect. 1982—), State of Wis. Med. Soc. (chmn. aging com. 1982—), Milw. County Med. Soc. (aging com.). Subspecialties: Gerontology; Neuroendocrinology. Current work: Aging, endocrine, thyroid and aging, neuropeptide, health service delivery to elderly, geriatrics, gerontology, health education. Office: Med Coll Wis Care Wood VA Hosp 5000 National Ave Milwaukee WI 53193 Home: 9043 N Tennyson Dr Bayside WI 53217

GAMBERTOGLIO, JOHN GINO, pharmacy educator; b. San Francisco, Sept. 5, 1947; s. John and Olympia (Benedetti) G.; m. Mary Madeline Scolini, June 25, 1972; children: Lisa, Emily, Catherine, John, Gina. Student, U. San Francisco, 1965-68; Pharm.D., U. Calif., San Francisco, 1972. Lic. pharmacist, Calif. Clin. instr. pharmacy U. Calif., San Francisco, 1972-73, asst. clin. prof., 1973-81, assoc. adj. prof., 1981—. Author: Clinical Use of Drugs in Renal Failure, 1976; contbr. articles to profl. jours. Chmn. instl. rev. bd. Calif. Dept. Corrections, 1981—. NIH grantee, 1980—. Mem. Am. Fedn. Clin. Research, Am. Soc. Nephrology, Am. Soc. Clin. Pharmacology and Therapeutics, Am. Pharm. Assn., Acad. Pharm. Scis. Democrat. Roman Catholic. Subspecialties: Nephrology; Pharmacokinetics. Current work: The pharmacokinetics and pharmacodynamics of prednisone and antibiotics in kidney transplant patients and renal failure and dialysis patients. Home: 1309 Bernal Ave Burlingame CA 94010 Office: U Calif Parnassus Ave San Francisco CA 94143

GAMBORG, OLUF LIND, plant genetic engineer, consultant; b. Denmark, Nov. 9, 1924; s. Peter Andersen and Anna Marie (Antonson) G.; m. Gertrude Katherine Christensen, July 11, 1953; children: Brian, Cheryl, Linda. B.Sc., U. Alta., 1956, M.Sc., 1958; Ph.D., U. Sask., 1962. Assoc. and sr. research scientist NRC of Can. Lab., Saskatoon, Sask., 1958-79; sci. research dir. plant cell biology Internat. Plant Research Inst., San Carlos, Calif., 1979-81; now sr. research fellow, sr. research scientist, 1981-83, cons., speaker; internat. organizer, instr. courses in plant biotech. for developing countries sponsored by UNESCO. Mng. editor: Plant Cell Report, 1980; adv. bd.: Internat. Rev. Cytology, 1979; contbr. numerous articles profl. jours., chpts. in books. Served with Danish Air Force, 1946-47. Mem. Am. Soc. Plant Physiologists, Can. Soc. Plant Physiologists (Gold medal 1977), Genetics Soc. Can., Internat. Assn. Plant Tissue Culture, Internat. Soc. Plant Molecular Biology. Lutheran. Subspecialties: Plant cell and tissue culture; Genetics and genetic engineering (biology). Current work: Plant cell genetic manipulation; genetic engineering and plant cell and tissue culture biotechnology in food crop plant improvement. Home: 1404 Solana Dr Belmont CA 94002

GAME, JOHN CHARLES, research scientist, geneticist; b. Tonbridge, Kent, eng., Dec. 14, 1946; s. Philip Malcolm and Vera Isobel (Blackburn) G. B.A. with 2d class honors, Oxford (Eng.) U., 1968, D.Phil. in Botany/Genetics, 1971. NATO fellow York U., Toronto, Ont., Can., 1971-73, U. Calif., Berkeley, 1971-73; staff scientist Nat. Inst. Med. Research, London, 1975-78; researcher dept. genetics U. Calif., Berkeley, 1978-82; research fellow Lawrence Berkeley Lab., 1982—. Contbr. articles to sci. jours. Mem. Genetic Soc. U.K., Genetic Soc. Am., Calif. Native Plant Soc., World Wildlife Fund, Nature Conservancy, Defenders of Wildlife, Oceanic Soc., Friends of the Earth. Subspecialties: Gene actions; Molecular biology. Current work: Genetics of yeast with emphasis on recombination, meiosis and radiation-repair mechanisms. Home: 2414 Parker St #16 Berkeley CA 9704 Office: Donner Lab Bldg 1 Lawrence Berkeley Lab Berkeley CA 94720

GAMMON, JAMES ROBERT, zoology educator; b. Sparta, Wis., Apr. 24, 1930; s. Abner James and LeVerne Marie (Robertson) G.; m. Carolyn Patricia O'Beirne, June 16, 1952 (div. Nov. 1977); children: David M., Clifford W., Kathleen L., Robert J.; m. Sharon Ann Sanders Garner, Mar. 22, 1980; children: Shannon, Bradley. B.S., Wis. State U., Whitewater, 1956; M.S., U. Wis., 1956, Ph.D., 1961. Research asst. U. Wis., Madison, 1957-61; asst. prof. dept. zoology DePauw U., Greencastle, Ind., 1961-66, assoc. prof., 1966-73, prof., 1973—; mem. panel biol. cons. ORSANCO, Cin., 1978—; mem. Indsl. Pesticide Rev. Bd., 1983—; cons. to bus. and govt. agys. Contbr. articles to profl. jours. Served with USN, 1948-52. Danforth fellow, 1957-61. Fellow Ind. Acad. Sci. (sec. 1967-69); mem. Am. Fisheries Soc., Am. Soc. Limnology and Oceanography. Methodist. Subspecialties: Ecology; Ecosystems analysis. Current work: Development of methods focusing aquatic biotic communities as an index of water and environmental quality. Office: Dept Zoology DePauw Greencastle IN 46135

GAMOTA, GEORGE, physics educator, researcher, university administrator; b. Lviv, Western Ukraine, May 6, 1939; m., 1961; 3 children. B.Physics, U. Minn., 1961, M.S., 1963; Ph.D. in Physics, U. Mich., 1966. Research asst. and teaching asst. U. Minn., 1959-63; research asst. U. Mich., 1963-66, research assoc., lectr., 1966-67; prof. physics, dir. Inst. Sci. and Tech., 1981—; mem. tech. staff Bell Labs., Murray Hill, N.J., 1967-74; research specialist Office Under Sec. Def., Research and Engring., Dept. Def., 1976-78, dir. research, 1978-81; mem. adv. Council Research, 1973-75, N.J. Gov's Commn. to Evaluate Capital Needs N.J., 1975; chmn, founder Sci. and Tech. Council for Congressman M. Rinaldo, 1974-75; exec. sec. Def. Sci. Bd. Study Fundamental Research in Univs., 1976; mem. Pres. Sci. Adv. Fed. Coordinating Council on Sci. and Engring. Tech., 1976-77; exec. sec. Def. Shale Oil Task Group, 1978; chmn Def. Com. research, 1978; mem. space system and tech. adv. com. NASA, 1979—, mem. adv. subcom. for electronics, 1983—, mem. research designee adv. com., 1983—; bd. dirs. Mich. Tech. Council, 1981—. Recipient Meritorious Civilian Service award U.S. Sec. Def., 1981. Fellow AAAS; mem. Am. Phys. Soc., Ukrainian Am. Engrs. Soc., IEEE, Sigma xi. Office: Inst Sci and Tech U Mich 2200 Bonisteel Blvd Ann Arbor MI 48109

GAMZU, ELKAN RAPHAEL, research psychopharmacologist; b. Liverpool, Eng., Jan. 27, 1943; came to U.S., 1967; s. Avrom Zeev and Rachel Mary (Susman) G.; m. S. Zelda Opwald, Aug. 10, 1965; children: Yiphat Pnina, Amir. B.A., Hebrew U., Jerusalem, 1967; M.S. in Psychology (Univ. fellow), U. Pa., 1968; Ph.D. in Psychology (Harrison fellow), U. Pa., 1971. Research and teaching asst. Hebrew

U., 1965-67, U. Pa., Phila., 1968-71; sr. pharmacologist Hoffmann-La Roche Inc., Nutley, N.J., 1971-77, asst. group chief, 1978-81, research group chief, 1981—; adj. asso. prof. dept psychology Hunter Coll., CUNY, N.Y.C., 1977-79. Bd. editors: Jour. Exptl. Analysis of Behavior, 1975-78, Behavioral Analyst, 1980-82; guest reviewer various jours.; contbr. articles to sci. jours. Treas. Temple Emmanuel, 1978-80; pres. Men's Club, 1983-84; v.p. Temple Emmanuel, 1980—, Men's Club, 1980-82; pres. Men's Club, 1983-84; alt. mem. bd. dirs. Neighborhood Resources of Passiac, 1978-79; treas. Hill Home Owners Assn., 1979-81. Fellow Am. Psychol. Assn. (sec.-treas. div. 25 1977-79); mem. N.Y. Acad. Scis. (conf. com. 1982—, fin. affairs com.), Soc. Neurosci., Am. Soc. Pharmacology and Exptl. Therapeutics. Subspecialties: Psychopharmacology; Learning. Current work: Preclinical psychopharmacology with specific emphasis on discovery and development of drugs to improve learning and memory and to counteract senility. Office: Dept Pharmacology Hoffmann-La Roche Inc Kingsland Ave Nutley NJ 07110

GANAPATHY, RAMACHANDRAN, scientist, researcher; b. Tellicherry, India, Jan. 16, 1939; s. Ramachandran and Annapoorni (Ammal) G.; m. Radha Narayanan; 1 son, Hari. B.Sc., U. Madras, 1959; Ph.D., U. Ark., 1968. Research assoc. Enrico Fermi Inst., U. Chgo., 1967-70, sr. research assoc., 1970-74; scientist J.T. Baker Chem. Co., Phillipsburg, N.J., 1976—. Contbr. articles to profl. jours. Mem. AAAS, Geochem. Soc. Subspecialties: Geochemistry; Analytical chemistry. Current work: Meteorites, trace element chemistry, impacts and extinction. Patentee in field. Office: Research Lab JT Baker Chem Co Phillipsburg NJ 08865

GANAPOL, BARRY DOUGLAS, nuclear and energy engineering educator, consultant; b. San Francisco, May 15, 1944; s. Manny Marvin and Miriam (Comar) G.; m. Susan Partnow, Sept. 16, 1967 (div. Feb. 1980); m. Starr Storms, Nov. 6, 1982. B.S., U. Calif.-Berkeley, 1966, Ph.D., 1971; M.S., Columbia U., 1967. Engr. Swiss Fed. Inst. for Reactor Research, Wurenlingen, 1971-72, Ctr. Nuclear Studies, Saclay, France, 1972-74, Argonne (Ill.) Nat. Lab., 1974-76; assoc. prof. nuclear and energy engring. U. Ariz., 1976—; vis. prof. U. Bologna, Italy, winters 1979-81; cons. in field. Mem. Am. Nuclear Soc., Soc. Indsl. and Applied Math. Subspecialties: Nuclear engineering; Applied mathematics. Current work: Mathematical methods in transport theory. Home: 4012 Calle Chica Tucson AZ 85711 Office: U Ariz Dept Nuclear and Energy Engring Room 106 Engring Bldg Tucso AZ 85721

GANDA, OM P., endocrinology educator, researcher; b. West Punjab, India, June 17, 1944; came to U.S., 1971; s. Udho D. and Ishwar D. (Dhingra) G.; m. Kanchan M. Ganda, Dec. 21, 1975; children: Anjali, Kiran. M.B.B.S., S.M.S. Med. Sch., Jaipur, India, 1966; M.D., All India Inst. Med. Scis., 1970. Diplomate: Am. Bd. Internal Medicine. Intern Tufts U. Med. Sch., 1971-72; resident Boston VA Hosp., 1972-73; fellow Joslin Diabetes Ctr., 1973-76; staff physician Joslin Clinic, Boston, 1976—; investigator Joslin Research Lab., 1976—; mem. staff New Eng. Deaconess Hosp., Boston, 1977—; asst. prof. medicine Harvard U. Med. Sch., Boston, 1979—. Author: Clinical Diabetes Mellitus, 1982, Diabetes Mellitus, 1983, Joslin's Diabetes Mellitus, 1984; tech. reviewer: New Eng. Jour. Medicine, 1979—. Fellow ACP; mem. Endocrine Soc., Am. Fedn. Clin. Research, Am. Diabetes Assn. Subspecialties: Endocrinology; Internal medicine. Current work: Diabetes-related research focused on etiology and epidemiology. Home: 187 Varick Rd Newton MA 02168

GANDHI, SHIRISH MANILAL, consulting engineering company executive; b. Jinja, Uganda, Nov. 17, 1931; came to U.S., 1967, naturalized, 1975; s. Manilal Tricumlal and Savitri M. (Shah) G.; m. Judith Aileen Hardy, Jan. 18, 1964; children: Sunita, Rohini, Ranjana, Suschiel. B.E., Poona U., 1956, 1957; M.Sc., London U., 1961. Elec. engr. Central Electricity Generating Bd., London, 1962-67; elec. and mech. engr. Westinghouse Elec. Corp., Pitts., 1967-70; elec. engr. Bechtel Corp., San Francisco, 1970-73, Kaiser Engrs., Oakland, Calif., 1974-75; cons., dir. U. Calif.-Berkeley (Hope Cons. Group), San Francisco, 1977-81; pres. SMG Cons. Engrs., Walnut Creek, Calif., 1975—, also dir. Author: Digital Computer in Power, 1962. Gen. sec., treas. Assn. Indian Engrs. in U.K., London, 1965-67; organizor ceremonies Boy Scouts Am., Walnut Creek, Calif., 1981-83. Mem. IEEE, Am. Nuclear Soc. (legislative panel 1977-78), Instrument Soc. Am., Chem. Engring. Product (research panel 1982-83). Clubs: Royal Soc. Vis. Scientis (commonwealth rep. 1959-63), Goats Internat. (London) (sr. rep. 1960-67). Subspecialties: Electrical engineering; Mechanical engineering. Current work: Advancement of design of electrical and mechanical systems and machines with computer automation at frontiers of science and technology for processes in nuclear fission and fusion power programs, reactor development and other process facilities, including use of laser and fiber optics for power and communication purposes. Home: 2595 Chippewa Ct Walnut Creek CA 94598 Office: SMG Cons Engrs 2595 Chinook Dr Walnut Creek CA 94598

GANDOLFI, ALLEN JAY, anesthesiology researcher; b. San Mateo, Calif., Dec. 11, 1946; s. Frank Joseph and Lorine (Wittmeier) G.; m. Judith Anne Monks, July 20, 1968; children: Christopher, Matthew, Jason. B.A., U. Calif.-Davis, 1968; Ph.D., Oreg. State U., 1972. USPHS fellow Oreg. State U., 1968-72; Mayo fellow Mayo Clinic, Rochester, Minn., 1972-75; sr. research scientist Battelle Meml. Inst., Richland, Wash., 1975-78; research asst. prof. anesthesiology and pharmacology U. Ariz., 1978—. Mem. Soc. Toxicology, Am. Soc. Pharmacology and Exptl. Therapeutics, Am. Soc. Anesthesiologists. Subspecialties: Toxicology (medicine); Anesthesiology. Current work: Toxicity and disposition of anesthetics, drugs and xenobiotics. Office: U Ariz Dept Anesthesiology Tucson AZ 85724 Home: 3231 Camino Suerte Tucson AZ 85715

GANDOUR, RICHARD DAVID, bioorganic chemist; b. Sistersville, W.Va., Feb. 12, 1945; s. Jackson Thomas and Mary Frances (Valent) G.; m. Ruth Frances Wells, Dec. 26, 1971 (dec.); 1 dau., Rochelle Marie. B.S., Wheeling Coll., 1967; Ph.D., Rice U., 1972. Teaching assoc. Kans. U., 1971-72, research assoc., 1973-1975; asst. prof. chemistry La. State U., 1975-80, assoc. prof., 1980—. Contbr. articles to profl. jours.; author 2 undergrad. teaching supplements; editor: (with R.L. Schowen) Transition States of Biochemical Processes, 1978. Served with U.S. Army, 1969. Recipient Faculty Service award La. State U., 1979; Petroleum Research Fund grantee, 1976-79; Research Corp. grantee, 1977, 80; Am. Heart Assn.-La. grantee, 1979, 80; Dow Chem. Co. grantee, 1980, 81; Exxon Chem. Co. grantee, 1981; Nat. Inst. Gen. Med. Scis. grantee, 1981-84. Mem. Am. Chem. Soc., AAAS. Roman Catholic. Subspecialty: Organic chemistry. Current work: Design and synthesis of chemical models of enzymatic catalysis; reaction rates, isotope effects and mechanism of organic reactions. Home: 4659 Bennett Dr Baton Rouge LA 70808 Office: Dept Chemistry La State U Baton Rouge LA 70803

GANESAN, ADAYAPALAM TYAGARAJAN, geneticist, educator, researcher; b. Madras, India, May 15, 1932; s. Adayapalam Vasudeva Tyagarajan and Savitri; m. Ann Katherine Cook, Aug. 3, 1963. B.S., Annamalai U., Madras State, India, 1952, M.A., 1953; Ph.D. (NIH predoctoral fellow, 1959-63), Stanford U., 1963. Research fellow Indian Inst. Sci., Bangalore, 1953-55; research asst. Indian Agrl. Research Inst., New Delhi, 1955-57; research fellow Carlsberg Lab., Copenhagen, Denmark, 1957-59; successively grad. student, research assoc., asst. prof. assoc. prof. Stanford (Calif.) U., 1959-76, prof. genetics, 1977—. Contbr. writings to profl. publs. Charter mem. World Wildlife Fund, U.S.A. Fellow Indian Inst. Sci., 1953-55, Rask-Orsted Found. of Denmark, 1957-59; research career devel. awardee NIH, 1970-75; recipient Calif. Higher Med. Edn. award Am. Lung Assn., 1975-76; sabbatical yr. dept. biochemistry Oxford (Eng.) U. Mem. AAAS, Am. Soc. Microbiology, Genetics Soc. Am. Subspecialties: Biochemistry (biology); Molecular biology. Current work: Genetic and biochemical basis of DNA replication and recombination in prokaryotes and eukaryotes. Office: Dept Genetics Stanford Univ Stanford CA 94305

GANGAROS, LOUIS PAUL, SR., dentist, pharmacologist; b. Rochester, N.Y., June 8, 1929; s. Biagio and Carmella (Bellassi) G.; m. Clara Amalfi, Sept. 4, 1950; children: Michael, Louis Paul, Maria, Alyssa. B. A. with high distinction, U. Rochester, 1952, M.S., 1961, Ph.D. in Pharmacology, 1965; D.D.S., U. Buffalo, 1955. Lic. dentist, Ga., N.Y. Pvt. practice dentistry, Rochester, 1958-61; asst. prof. U. Rochester, 1965-68; asso. prof., coordinator pharmacology Med. Coll. Ga., Augusta, 1968-71, prof. oral biology, coordinator pharmacology, 1971—, prof. pharmacology, 1971—; cons. on iontophoresis ALZA Pharms Inc., Palo Alto, Calif., Motion Control Inc., Salt Lake City, Del Commerce Lab., Farmingdale, N.Y.; grant reviewer NIH, Nat. Inst. Dental Research. Author: (with Ciarlone and Jeske) Pharmacotherapeutics in Dentistry, 1983; also articles, monographs and abstracts.; Reviewer: Jour. ADA, 1965—, Jour. Acad. Gen. Dentistry, 1982; Contbg. editor: Methods and Findings in Exptl. and Clin. Pharmacology, 1979—. Served to capt. USAF, 1954-58. Recipient C.V. Mosby Co. award, 1955; Nat. Inst. Dental Research grantee, 1966-75, 82-84; also indsl. grantee. Mem. Am. Soc. Pharmacology and Exptl. Therapeutics, Internat. Assn. Dental Research, ADA, Soc. Exptl. Biology and Medicine, Am. Assn. Dental Schs., Am. Coll. Dentists, Am. Assn. Dental Research. Methodist. Subspecialties: Pharmacology; Oral biology. Current work: Iontophoresis; application of drugs for treatment of caries, periodontal disease, viral ulcers and other conditions. Home: 380 Folkstone Circle Augusta GA 30907 Office: Med Coll Ga Sch Dentistry Dept Oral Biology-Pharmacology Augusta GA 30901-9990

GANGULI, MUKUL CHANDRA, nutritional physiologist, researcher; b. Comilla, Bengal, India, Feb. 28, 1938; came to U.S., 1964; s. Sushil C. and Kamala (Chatterjee) G.; m. Aparna Banerjee, May 29, 1970; children: Suman, Suvranu, Ina. B.S. with distinction, U. Calcutta, India, 1956, B.V.Sc. and A.H., 1961; M.Vet. Sci., Agra U., 1963; Ph.D., U. Minn., 1968. Tech. asst. Nat. Dairy Research Inst., Karnal, India, 1963-64; research asst. U. Minn., Mpls., 1964-68; research assoc., 1970—; postdoctoral assoc. Iowa State U., Ames, 1968-70. Author: Renal Papilla and Hypertension, 1980; contbr. articles to profl. jours. Mem. Am. Physiol. Soc., Sigma Xi, Gamma Sigma Delta. Subspecialties: Nutrition (medicine); Physiology (medicine). Current work: Sodium, prostaglandins and their relationship to kidney and hypertension. Home: 1901 Malvern St Saint Paul MN 55113 Office: Univ Minn Box 78 Maye Bldg Minneapolis MN 55455

GANGULY, SUNILENDU NARAYAN, cardiologist, educator; b. Calcutta, West Bengal, India, Dec. 27, 1936; came to U.S., 1971; s. Ramesh Chandra and Sailabasini G.; m. Kuniko Tanii, Sept. 24, 1939; children: Joya Anne, Maya Lynne. M.D., Med. Coll., Calcutta, 1960. Intern Hutzel Hosp., Detroit, 1963; resident in medicine Wayne State U., 1964-66, fellow in cardiology, 1966; asst. prof. medicine, 1972-75, assoc. prof., 1975—; fellow in cardiology Vancouver (B.C., Can.) Gen. Hosp.; chief cardiology Hutzel Hosp., 1974—. Fellow A.C.P., Am. Coll. Cardiology, Am. Coll. Angiology; mem. AMA, N.Y. Acad. Sci., Bichitra (pres. 1973), Physicians India (pres. Detroit 1979). Subspecialties: Cardiology; Internal medicine. Current work: Cardiovascular hemodynamics; streptoinase therapy and percutaneous transluminal coronary angioplasty in acute myocardial infarction. Home: 3630 Estates Dr Troy MI 48084 Office: 4727 Saint Antoine Detroit MI 48201

GANNON, PATRICK THOMAS, meteorology educator; b. Ft. Bragg, N.C., May 4, 1930; s. Michael Vincent and Mary Lee (Ayers) G.; m. Donna Merle Neilson, Aug. 15, 1953; children: Patrick Thomas, Mary, Jane, Megan, Stephen, Elizabeth, Emily. B.S., U. Fla., 1956; postgrad., Fla. State U., 1956-57, MIT, 1965-67; M.S., U. Chgo., 1961; Ph.D., U. Miami, 1977. Weather officer USAF Air Weather Service (various locations), 1956-72; meteorology researcher NOAA/Environ. Research Labs., Coral Gables, Fla., 1975-80, Boulder, Colo., 1980-81; asst. prof. meteorology Lyndon State Coll., Lyndonville, Vt., 1981—. Served with U.S. Army, 1948-52. Decorated Bronze Star, Purple Heart; NOAA grantee, 1982-83. Mem. Am. Meteorol Soc. (chmn. Miami chpt. 1979-80), Am. Water Resources Assn., Fla. Acad. Scis., Phi Kappa Phi, Epsilon Tau Lambda. Republican. Roman Catholic. Subspecialties: Synoptic meteorology. Current work: Numerical modeling of mesoscale processes. Office: Dept Meteorology Lyndon State Coll Vail Hill Lyndonville VT 05851

GANS, CARL, zoologist; b. Hamburg, Ger., Sept. 7, 1923; came to U.S., 1939, naturalized, 1945; s. Samuel S. and Else Hubertine (Leeser) G.; m. Kyoko Andow, Nov. 18, 1961. B.M.E., N.Y. U., 1944; M.S., Columbia U., 1950; Ph.D. in Biology, Harvard U., 1957. Contract and service engr. Babcock & Wilcox Co., 1947-55; from asst. prof. to prof. biology, chmn. dept. biology State U. N.Y., Buffalo, 1958-71; prof. zoology U. Mich., Ann Arbor, 1971—, chmn. dept., 1971-75; research scientist Mus. Zoology, 1971—; research asso. Carnegie Mus., 1953—, Am. Mus. Natural History, 1958—; sec., bd. dirs. Zool. Soc. Buffalo, 1961-71; med. adv. council Detroit Zool. Park, 1973—; cons. in field, vis. prof. univs. and colls. Author: Biomechanics, 1974, Reptiles of the World, 1975; co-author: Photographic Atlas of Shark Anatomy, 1964; Gen. editor: Biology of the Reptilia, 13 vols., 1969-82; mng. editor: Jour. Morphology, 1968—. Served with AUS, 1944-47. Guggenheim fellow, 1953, 77; NSF predoctoral fellow, 1956-57; postdoctoral fellow U. Fla., Gainesville, 1957-58; grantee NSF, NIH, others. Fellow N.Y. Zool. Soc., Zool. Soc. London, AAAS, Zool. Soc. India, Acad. Zoology India; mem. Am. Soc. Zoologists (pres. 1977), ASME, Soc. Study Evolution (v.p. 1971), Am. Soc. Ichthyology and Herpetology (gov. 1961, 70, 76, pres. 1979), Am. Inst. Biol. Scis. (gov. bd. 1975-78), Soc. Study Amphibians and Reptiles (pres. 1983), Am. Assn. Anatomists, Soc. Exptl. Biology, Am. Physiol. Soc., Senckenberg. Naturforsch. Gesellschaft (corr.). Subspecialties: Evolutionary biology; Morphology. Current work: Functional morphology, herpetology, electromyography. Home: 2811 Park Ridge Dr Ann Arbor MI 48103 Office: 2127 Natural Scis Bldg Univ Mich Ann Arbor MI 48109

GANTT, DAVID GRAHAM, dental researcher, physical anthropologist; b. Alliance, Ohio, Sept. 18, 1943; m. Phyllis Arlene Pollard, Aug. 14, 1963; children: David Graham Jr., Tasha, Bryan. B.A., U. Wash., Seattle, 1971; M.A., Washington U., St. Louis, 1976, Ph.D., 1977. Instr., research asst. Washington U., 1974-77; asst. prof. anthropology Fla. State U., Tallahassee, 1977-81; research fellow, adj. asst. prof. U. Ala., Birmingham, 1981—, cons., 1982—, co-dir. electron microscopy core facility, 1981—. Served with U.S. Army, 1961-64. Alexander von Humboldt-Stiftung Research fellow, Frankfurt, Germany, 1979; recipient Young Investigator award Nat. Inst. Dental Research, Washington, 1978; others. Mem. Internat. Assn. for Dental Research, Am. Assn. Phys. Anthropologists, Internat. Primatology Assn. Club: Mid-State Soccer (exec. bd. St. Louis 1982—). Subspecialties: Cariology; Oral biology. Current work: SEM and X-ray microanalysis of teeth, especially enamel and dentin structures. Evolutionary history of calcified dental tissues in mammals, primates, and cariology. Home: 5927 Creekwood Rd Birmingham AL 35210 Office: Inst Dental Research Sch Dentistry Univ Alabama Birmingham AL 35294

GAPOSCHKIN, EDWARD MICHAEL, geophysicist, educator-researcher; b. Boston, May 29, 1935; m., 1959; 3 children. B.S., Tufts U., 1957; Diploma, Cambridge (Eng.) U., 1959; Ph.D. in Geophysics, Harvard U., 1969. Prin. scientist Satellite Geophysics Group, Smithsonian Astrophys. Obs. Harvard U., 1959—, research assoc., 1969—, mem., 1972—, lectr. in astronomy, 1975—; cons. adv. com. geodesy and cartography NASA, 1968-70; mem. Com. Space Research, 1968; lectr. Inst. Geodetic Sci. Ohio State U., 1969; dir. research Research Group Spatial Geodesy, Mardon Obs. U. Paris., 1972-73. Mem. Am. Geophys. Union, Royal Astron. Soc., Internat. Astron. Union, Internat. Assn. Geodesy (chmn. spl. study group 1975). Subspecialties: Satellite studies; Geodesy. Office: Ctr Astrophysics 60 Garden St Cambridge MA 02138

GAPOSCHKIN, PETER JOHN ARTHUR, physicist, educator; b. Bston, Apr. 5, 1940; s. Sergei and Cecilia (Payne) G. B.Sc., MIT, 1961; M.A., U. Calif.-Berkeley, 1965, 1966, Ph.D., 1971. Research asst. Lawrence Berkeley Lab., 1965-70; physicist Naval Plant Rep. Office, Sunnyvale, Calif., 1974-75; computer programmer Fleet Numerical Oceanography Center, Monterey, Calif., 1975-79; sr. analyst Informatics, Palo Alto, Calif., 1979-80; Instr. community colls., 1980-82; mem. job acquisition com. Experience Unltd., Oakland, Calif., 1982—. Mem. University Ave. Ctr. Council of Consumers Coop. of Berkeley, 1978-82. Mem. Am. Astron. Soc., Math. Assn. Am., Assn. Computing Machinery. Democrat. Unitarian Universalist. Club: Toastmaster. Subspecialties: Relativity and gravitation; Cosmology. Current work: General relativity; cosmology, number theory, group theory. Home: 711 Harrison Berkeley CA 94710 Office: 1442-A Walnut St 371 Berkeley CA 94709

GARA, ROBERT IMRE, forest entomology educator, researcher; b. Santiago, Chile, Dec. 16, 1931; came to U.S., 1941; s. Emery Imre and Marina G. (Briones) G.; m. Ola O. Alexander, Jan. 31, 1958 (div. Jan. 1979); children: Jennifer Gail, Katheryn A.; m. Marcela B. García-Huidobro, Sept. 1, 1979; 1 son, Robert I. B.S., Utah State U., 1953; M.S., Oreg. State U., 1962, Ph.D., 1964. Forester Kirby Lumber Corp., Houston, 1957-60; sr. scientist Boyce Thompson Inst., Yonkers, N.Y., 1961-64, project leader, 1964-66; assoc. prof. SUNY Coll. Forestry-Sryacuse, 1966-68; assoc. prof. to prof. Coll. Forest Resources, U. Wash., 1968—. Served to capt. USAF, 1953-57. UN grantee, Costa Rica, 1970, Chile, 1978; NSF grantee, Chile, 1977. Mem. Entomol. Soc. Am., Entomol. Soc. Can., Soc. Am. Foresters. Roman Catholic. Subspecialties: Ecology; Behaviorism. Current work: Study host-insect interactions that lead to integrated pest management schemes of forest insects; study host selection behavior of forest insects; tropical forest entomology. Home: 15742 Burke Ave N Seattle WA 98133 Office: Coll Forest Resources U Wash Seattle WA 98195

GARABEDIAN, PAUL ROESEL, mathematics educator, researcher; b. Cin., Aug. 2, 1927. A.B., Brown U., 1946, Harvard U., 1947; Ph.D. in Math, Harvard U., 1948. NRC fellow, 1948-49; asst. prof. math. U. Calif., 1949-50; from asst. prof to prof. Stanford U., 1950-59; prof. math. N.Y.U., 1959—. Recipient Birkhoff prize, 1983; Sloan fellow, 1960-62; Guggenheim fellow, 1966-67; Sherman Fairchild fellow Calif. Inst. Tech., 1975. Mem. Nat. Acad. Sci., Soc. Indsl. and Applied Math., Am. Math Soc., Am. Acad. Arts and Sci. Subspecialty: Applied mathematics. Office: Dept Math NYU New York NY 10012

GARB, FORREST ALLAN, petroleum engineer; b. San Antonio, Dec. 15, 1929; s. Julius and Sada K. (Pullen) G.; m. Janelda D. Duke, Feb. 7, 1959; children: David, Kara Lee. B.S. in Petroleum Engring., Tex. A&M U., 1951, profl. degrees in petroleum engring., 63. Registered profl. engr., Tex. Petroleum engr. Magnolia Petroleum Co., 1951-52, Socony Mobile of Venezuela, 1955-57; reservoir engr. H.J. Gruy and Assocs., Inc., Irving, Tex., 1957-59, v.p., 1959-63, exec. v.p., 1963-73, pres., chief operating officer, 1973—. Contbr. articles to profl. jours. Dist. chmn. Boy Scouts Am., 1982-83. Served to 1st lt. USAF, 1952-53. Recipient Silver Beaver award Boy Scouts Am., 1982. Mem. Soc. Petroleum Engrs., Assn. Computing Machinery, Dallas Geol. Soc., Am. Assn. Petroleum Geologists. Jewish. Clubs: Engrs. of Dallas, Dallas Corinthian Yacht, Los Colinas Country. Subspecialties: Petroleum engineering; Geology. Current work: Petroleum reservoir engring. and computer applications. Home: 2973 Sunbeck Circle Farmers Branch TX 75234 Office: The Gruy Cos 150 W John Carpenter Freeway Irving TX 75062

GARBARINI, EDGAR JOSEPH, civil engineer, engineering company executive; b. Jackson, Calif., Aug. 1, 1910; s. Henry Casamero and Elvira (Gardella) G.; m. Lillian Rosemarie Arata, Nov. 14, 1936; children—Paul Henry, Ann Elisabeth. B.S., U. Calif. at Berkeley, 1933. Registered profl. engr., several states. Jr. research engr. U. Cal. at Berkeley, 1933-34, research engr., 1934-38; field engr. W.A. Bechtel & Six Cos. Calif., San Francisco, 1934; civil engr. Calif. Commn., Golden Gate Internat. Expn., 1938-39, Dewell & Earl (cons. engrs.), San Francisco, 1939, Pacific Gas & Electric Co., 1939-40; with Bechtel Group of Cos., San Francisco, 1940—, now sr. exec. cons. Fellow ASCE; mem. Nat. Acad. Engring., Structural Engrs. Assn. No. Calif., Mining and Metall. Soc. Am., Order of Golden Bear, U. Calif. Alumni Assn., Sigma Xi, Tau Beta Pi, Chi Epsilon. Clubs: Family, World Trade, Pacific Union (San Francisco). Subspecialty: Civil engineering. Current work: Improving large organizations' ability to produce large governmentand industrial facilities. Office: PO Box 3221 San Francisco CA 94119

GARCIA, JULIO M., physician; b. Camaguey, Cuba, Sept. 29, 1937; s. Julio C. and Hortensia C. (Sedeno) G.; m. Marta E. Ismail, May 20, 1962; children: Marta M., Teresita M. B.S., Champagnat Sch., Cuba, 1954; M.D., Harvard U., 1963. Intern Riverside Hosp., Newport News, Va., 1969-70; resident in internal medicine Mount Sinai Hosp., Miami Beach, Fla., 1970-72; med. oncology fellow U. Miami, 1972-73; practice medicine, Miami, 1973—; cons. Liga Contra el Cancer, Miami. Chmn. Ballet Concerto, Miami. Mem. Dade County Med. Assn., Fla. Med. Assn., AMA, Am. Soc. Clin. Oncology, Fla. Soc. Clin. Oncology, Am. Cancer Soc. Republican. Roman Catholic. Clubs: Big Five (Miami); Grove Isle (Coconut Grove). Subspecialties: Chemotherapy; Internal medicine. Current work: Five year follow-up in cancer of breast with adjunct chemotherapy. Office: 3661 S Miami Ave Suite 203 Miami FL 33133

GARDIN, JULIUS MARKUS, cardiologist; b. Detroit, Jan. 14, 1949; s. Abram and Fania Toba (Garden) G.; m. Susan Deanne Kelemen, Dec. 19, 1982. B.S. with high distinction, U. Mich., 1968, M.D. cum laude, 1972. Diplomate: Am. Bd. Internal Medicine. Resident in medicine U. Mich., Ann Arbor, 1972-75; fellow in cardiology Georgetown U., Washington, 1975-77; dir. cardiology noninvasive lab., staff cardiologist Lakeside VA Med. Ctr., Chgo., 1977-79; staff cardiologist, asst. prof. Northwestern U. Med. Sch., Chgo., 1977-79; dir. cardiology noninvasive lab. U. Calif.-Irvine Med. Ctr., Orange,

1979—; acting chief cardiology Long Beach (Calif.) VA Med. Ctr., 1982—. Co-editor: Textbook of Two-Dimensional Echocardiography, 1983; editor: Update on Cardiovascular Diagnostics, 1982; corr. editor: Archives of Internal Medicine and Chest, 1978—; contbr. articles to profl. jours. Served to maj. M.C. USAR. Am. Heart Assn. grantee, 1980-82, 83-84. Fellow ACP, Am. Coll. Cardiology, Am. Coll. Chest Physicians, Am. Heart Assn. (council clin. cardiology); mem. U. Mich. Med. Center Alumni Assn. (bd. govs. 1979-81), Phi Beta Kappa, Alpha Omega Alpha, Phi Delta Epsilon. Jewish. Subspecialty: Cardiology. Current work: Doppler echocardiography. Office: U Calif-Irvine Med Center 101 City Dr S Orange CA 92668

GARDIN, T. HERSHEL, epidemiologist, research and evaluation consultant; b. Detroit, July 21, 1947; s. Abraham and Ruth (Miedzwinski) G.; m. Joy Beth Lewis, Oct. 10, 1972; children: Naftali M., Dov E., Miriam S., Yehudis K. B.A., Wayne State U., 1969, M.A., 1971, Ph.D., 1983. Instr. Wayne State U., 1971-75; dir. psychol. services Alexandrine House, Inc., Detroit, 1975-77; planner Wayne County Substance Abuse, Detroit, 1977-79; assoc. MFA and Assocs., Bingham Farms, Mich., 1979-81; sr. research analyst Comprehensive Health Services Detroit, 1981—; grant investigator, contract reviewer, research cons.; mem., contbr. Tex. Instruments, Inc., Profl. Program Exchange, 1980—. Contbr. articles to profl. jours., chpts. to books. Mem. Am. Psychol. Assn., Am. Pub. Health Assn., Soc. Psychologists in Substance Abuse Soc., Psychol. Study of Social Issues, Psi Chi. Jewish. Subspecialties: Health services research; Information systems, storage, and retrieval (computer science). Current work: Health care delivery systems, experimental design and data analysis. Home: 1DOak Park Oak Park MIOffice: Comprehensive Health Services of Detroit 6500 John C Lodge Expressway Detroit MI 48237

GARDNER, CHARLES OLDA, agronomy educator, consultant; b. Tecumseh, Nebr., Mar. 15, 1919; s. Olda Cecil and Frances E (Stover) G.; m. Wanda Marie Steinkamp, June 9, 1947; children: Charles Olda, Lynda, Thomas, Richard. B.Sc. with high distinction, U. Nebr., 1941, M.S., 1948; M.B.A., Harvard U., 1943; Ph.D., N.C. State U., 1951. Asst. extension agronomist U. Nebr.-Lincoln, 1946-48, assoc. prof. agronomy, 1952-57, prof., 1957-70, Found. prof. (Regents Disting. prof.) agronomy, 1970—, chmn., 1957-68; asst. statistician N.C. State U., 1951-52; vis. prof. genetics U. Wis.-Madison, 1962-63; internat. lectr. and cons. in plant breeding, quantitative genetics and applied stats.; mem. Nat. Plant Genetics Resources Bd., 1975-81; mem. Dept. Agr. Competitive Grants Program Rev. Panel, 1978-80, Nat. Com. to Develop Title XII Internat. Maize Planning Grant Proposal, 1978-80, Nat. Corn Research Coordinating Com., 1980-82. Contbr. chpts., numerous articles to profl. publs.; assoc. editor: Crop Sci, 1964-66, Agronomy Jour, 1971-73; mem. adv. bd.: Egyptian Jour. Genetics and Cytology, 1976—, Brazilian Jour. Genetics, 1978—, Agricultural Handbook, Plant Sci. Series, 1977—. Elder, local ch. Served to capt. Q.M.C. U.S. Army, 1943-46; New Guinea, Philippines. Recipient award for outstanding research and creative activity U. Nebr., 1981. Fellow Am. Soc. Agronomy (editorial bd. 1970-73, bd. dirs. 1973-76, 81-83, chmn. bd. dirs. 1982, mem. exec. com. 1973-76, 81-83, chmn. exec. com. 1982, pres. 1982), AAAS; mem. Crop Sci. Soc. Am. (editorial bd. 1963-66, mem. exec. com. 1973-76, chmn. exec. com. 1975, bd. dirs. 1973-76, chmn. bd. dirs. 1975, pres. 1975, Crop Sci. award 1978), Biometric Soc., Genetic Soc. Am., Am. Genetic Assn., Sigma Xi, Gamma Sigma Delta (Internat. award 1977), Phi Kappa Phi. Independent Republican. Presbyterian. Subspecialties: Plant genetics; Plant breeding. Current work: Plant quantitative genetics; plant breeding systems; population improvement methods; germplasm collection, maintenance and utilization; isoenzyme studies in maize populations; corn and sorghum breeding. Office: Dept Agronomy U Nebr Lincoln NE 68583

GARDNER, DANIEL, neurophysiologist, educator; b. N.Y.C., Jan. 23, 1945; s. Harold and Beatrice (Hornick) G.; m. Esther Polinsky; children: Benjamin, Deborah. A.B., Columbia U., 1966; Ph.D., N.Y.U., 1971; postdoctoral, U. Wash., 1971-73. Computer programmer, lectr. Inst. Space Studies, N.Y.C., 1962-66; sr. fellow dept. physiology U. Wash., Seattle, 1971-73; asst. prof. physiology Cornell U. Med. Coll., N.Y.C., 1973-79, assoc. prof., 1979—; prin. investigator NIH. Contbr. sci. papers to profl. lit. Mem. Am. Physiol. Soc., Soc. Neurosci., Biophys. Soc., N.Y. Acad. Scis., AAAS. Subspecialties: Neurophysiology; Biophysics (biology). Current work: Research in synaptic biophysics. Office: 1300 York Ave Suite C-508 New York NY 10021

GARDNER, ELDON JOHN, genetics educator; b. Logan, Utah, June 5, 1909; s. John and Cynthia Evelyn (Hill) G.; m. Helen Richards, Aug. 21, 1939; children: Patricia Mahrt, Donald E., Betty Morrison, Cynthia Pulley, Alice, Mary Jane Neville. B.S., Utah State U., 1934, M.S., 1935, D.Sc. (hon.), 1980; Ph.D., U. Calif.-Berkeley, 1939. Instr., dean Salinas (Calif.) Jr. Coll., 1939-46; faculty U. Utah, Salt Lake City, 1946-49, Utah State U., 1949—, research prof. emeritus, 1974—, dean, 1962-67, 1967-74; research prof. Coll. Medicine, U. Utah, Salt Lake City, 1977—; cons. in field; dir. Utah Div. Am. Cancer Soc., 1981-82. Author: Principles of Genetics, 6th edit, 1981, History of Biology, 3rd edit, 1972, Genetics Laboratory Investigations, 7th edit, 1980; contbr. articles to profl. jours. Mem. Logan City Sch. Bd., 1962-64. Recipient Disting. Service award Utah Acad. Sci., Arts and Letters, 1957; Willard Gardner Sci. award, 1975; Sci. Achievement award Brigham Young U., 1979. Mem. Genetics Soc. Am., Am. Soc. Human Genetics, Am. Soc. Naturalists, Am. Inst. Biol. Scis., AAAS, Sigma Xi, Phi Kappa Phi. Lodge: Rotary. Subspecialties: Genetics and genetic engineering (biology); Cell and tissue culture. Current work: Mechanism, prevention and cure of hereditary precancers; Gardner syndrome. Home: 369 N 5th E St Logan UT 84321 Office: Utah State Univ 137 NRB Bldg Logan UT 84322

GARDNER, HOWARD EARL, research psychologist, educator; b. Scranton, Pa., July 11, 1943; s. Ralph and Hilde (Weilheimer) G. A.B., Harvard U., 1965, Ph.D., 1971. Co-dir. Harvard Project Zero, 1971—; research psychologist VA Med. Ctr., Boston, 1974—; assoc. prof. neurology Boston U. Sch. Medicine, 1979—. Author: The Quest for Mind, 1973, The Shattered Mind, 1975, Artful Scribbles, 1980, Art, Mind and Brain, 1982. Bd. dirs. Social Sci. Research Council, 1982. Grantee NIH, NSF, Spencer Found., Carnegie Corp., Sloan Found., Markle Found., Van Leer Found.; MacArthur fellow, 1981. Fellow Am. Psychol. Assn.; mem. Internat. Neuropsychology Symposium, Acad. Aphasia, Soc. for Research in Child Devel. Subspecialties: Developmental psychology; Neuropsychology. Current work: Conduct full-time research on the development and breakdown of symbol-using capacities. Home: 15 Lancaster St Cambridge MA 02140 Office: Harvard Project Zero Longfellow 326 Cambridge MA 02138

GARDNER, JERRY DAVID, digestive diseases researcher; b. Chanute, Kans., Jan. 15, 1941; s. Maurice K. and Ruth N. (Hiatt) G.; m. Marsha C. Wertzberger, Aug. 25, 1962; children: Brett Elizabeth, Carolyn Paige. A.B., U. Kansas, 1962; M.D., U. Pa., 1966. Resident in medicine Hosp. U. Pa., Phila., 1966-68; clin. assoc. NIH, Bethesda, Md., 1968-70, sr. staff fellow, 1970-73, chief gastrointestinal sect., 1973—, chief digestive diseases br., 1976—. Contbr. articles to profl. jours. Coach basketball teams, Potomac Valley, 1976—. Served with USPHS, 1968—. Recipient H. Marvin Pollard award U. Mich., 1980. Mem. Am. Gastroenterol. Assn., Am. Fedn. Clin. Research, Am. Soc. Clin. Investigation, AAAS. Democrat. Subspecialties: Gastroenterology; Cell biology (medicine). Current work: Biochemical basis of action of gastrointestinal hormones. Cellular function of gastrointestinal tissues. Home: 4716 Waverly Ave Garrett Park MD 20896 Office: NIH Bldg 10 Room 9C-103 Bethesda MD 20205

GARDNER, JOHN ARVY, JR., physics educator; b. Winona, Miss., Nov. 5, 1939; s. John A. and Ruth Elizabeth (Bryson) G.; m. Carolyn Kerchner, Aug. 28, 1965; children: Daniel Taylor, Jeffrey Alan. B.A., Rice U., 1961; M.S., U. Ill.-Urbana, 1963, Ph.D., 1966. Research assoc. Tech. U., Munich, W. Ger., 1966-67; asst. prof. U. Pa., Phila., 1967-73; assoc. prof. Oreg. State U., Corvallis, 1973-81, prof., 1981—; vis. research fellow U. Warwick, Eng., 1979-80; sponsor, sec. Oreg. Materials Sci. Symposium, Corvallis, 1976—. Contbr. articles to physics jours. NATO postdoctoral fellow NSF, 1966; Sci. Research Council fellow, 1979-80; NSF, ONR research grantee, 1975—. Mem. Am. Phys. Soc., AAAS. Subspecialty: Condensed matter physics. Current work: liquid and amorphons semiconductors, organic conductors, ceramics, nuclear magnetic resonance, hyperfine interactions in solids. Home: 3223 NW McKinley St Corvallis OR 97330 Office: Physics Dep Oreg State U Corvallis OR 97331

GARDNER, JOHN C., research agronomist; b. Kansas City, Mo., Jan. 28, 1958; s. John Jess and Christine (Liddy) G.; m. Julie Rae Abare, Aug. 7, 1976; 1 dau., Kate Christine. B.S., Kans. State U., 1978, M.S., 1980. Asst. agronomist N.D. State U., Carrington, 1980—. Recipient Agrl. Research award Gamma Sigma Delta, 1977. Mem. Am. Soc. Agronomy, Crop Sci. Soc. Am., Republican. Roman Catholic. Subspecialties: Plant physiology (agriculture); Integrated systems modelling and engineering. Current work: Improvement of sunflower, corn, wheat and sorghum productivity through the study of crop physiology as affected by climatic and management variables. Office: ND State U Carrington Br Agrl Experiment Sta Box 219 Carrington ND 58421

GARDNER, JOHN WILLARD, physician, med. educator; b. Boston, May 10, 1948; s. John Hale and Olga Helen (Dotson) G.; m. Kathryn Joyce Burton, Dec. 17, 1971; children: Lisa Kathryn, Kirsten Marie, John Burton, Michael Grant. B.A., Brigham Young U., 1972, M.S., 1973; M.D., U. Utah, 1976; M.P.H., Harvard U., 1979, D.P.H., 1981. Diplomate: Nat. Bd. Med. Examiners, Am. Bd. Preventive Medicine. Intern U. Ariz. Affiliated Hosps., Tucson, 1976-77, resident in pediatrics, 1977-78; asst. prof. family and community medicine U. Utah, 1980—, adj. asst. prof. pediatrics, 1981—; bd. dirs. Utah div. Am. Cancer Soc., 1981—. Served to maj. MC USAR, 1978—. Am. Cancer Soc. Research grantee, 1981; Nat. Cancer Inst. Research grantee, 1982. Republican. Mormon. Subspecialties: Epidemiology; Preventive medicine. Current work: Research and teaching in the causes of cancer, chronic disease, and maternal and childhood diseases; epidemiologic studies in human populations. Home: 2957 Ida Circle Salt Lake City UT 84106 Office: Dept Family and Community Medicine U Utah Sch Medicine 50 N Medical D Salt Lake City UT 84132

GARDNER, MARJORIE HYER, chemistry educator, consultant; b. Logan, Utah, Apr. 25, 1923; d. Saul Edward and Gladys (Christenson) Hyer. B.S., Utah State U., 1947, Ed.D. (hon.), 1975; M.A., Ohio State U., 1968, Ph.D., 1960; cert. ednl. mgmt., Harvard U., 1975. Asst. exec. sec. Nat. Sci. Tchrs. Assn., Washington, 1961-64; asst. prof. to prof. chemistry U. Md., after 1964; Fulbright prof. Australian Nat. Commn., Sydney, Canberra, 1973-74, Nigerian Nat. Commn., Port Harcourt, 1982-83; UNESCO cons. UNESCO, France, China, Thailand, Qatar, 1976—; dir. SERI div. NSF, Washington, 1979-81. Author: Chemistry in Space Age, 1965; dir.: monograph series Vistas of Science, 1963; texts and tchrs.' guides Interdisciplinary Approaches to chemistry, 1973-79. Recipient O'Haus award Nat. Sci. Tchrs. Assn., 1972, Carleton award, 1974; Catalyst award Chem Mfrs. Assn., 1982. Fellow AAAS (sect. chairperson 1975-78, mem. council 1980-83), Am. Inst. Chemists; mem. Am. Chem. Soc. (Western Conn. Vis. Scientist award 1972, div. chairperson 1982-85), Fulbright Assn. Alumni (dir. 1972-82, pres. 1980-81), Internat. Union Pure and Applied Chemistry (exec. b. com. teaching 1982—), Internat. Orgn. for Chem. Scis. in Devel. (edn., bd. 1982—), Kappa Delta (pres. 1946-47, province officer 1948-52). Subspecialties: Chemical Education; Learning. Current work: Chemical education, a rapidly developing frontier research area; curriculum development; learning theory; assessment. Office: U Md Dept Chemistry College Park MD 20742

GARDNER, WAYNE SCOTT, plant pathologist, educator; b. Clifton, Colo., Jan. 11, 1920; s. Dora Levi and Violette Ruth (Scott) G.; m. Leona M. Oberly, Nov. 22, 1944; children: Susan E. Larock, Barbara J. Gee, Janet L., Kathrine A. A.S., Mesa Coll., 1948; B.S., Utah State U., 1950, M.S., 1951; Ph.D., U. Calif.-Davis, 1967. Instr. Mesa Coll., Grand Junction, Colo., 1951; plant pathologist U.S. Army Chem. Corps, Dugway, Utah, 1951-54; agrl. technician U.S. Steel Corp., Provo, Utah, 1954-63; research asst., technician U. Calif.-Davis, 1963-67; prof. dept. plant sci. S.D. State U., Brookings, 1967—; cons. air pollution, electron microscopist. Contbr. articles to profl. jours. Active Brookings Art Council. Served with USAAF, 1941-45; PTO. Recipient Electron Microscopy award U. Calif-Davis., 1968. Mem. Am. Phytopath. Soc., S.D. Acad. Sci., Sigma Xi, Gamma Sigma Delta. Subspecialties: Plant pathology; Plant virology. Current work: Virus and abiotic plant diseases; electron microscopy; photography; research and teaching in areas of virus diseases, non-parasitic diseases, electron microscopy and photography of diseased plants. Home: 417 Dakota Ave Brookings SD 57006 Office: Dept Plant Sci South Dakota State Univ Brookings SD 57007

GARDNER, WILLARD HALE, computer systems analyst; b. Logan, Utah, Dec. 22, 1925; s. Willard and Rebecca Viola (Hale) G.; m. Lillian DeAnn Rich, Aug. 28, 1956; children: Julie, Bonnie, Wendy, Scott, Craig, Paul. B.S., Utah State Agrl. Coll. dir. computer services Brigham Young U., Provo, Utah, also assoc. prof. computer sci. Subspecialties: Information systems (information science); Software engineering. Current work: Development of software and interfaces for information networks in distributed information systems. Home: 1495 Oak Ln Provo UT 84604 Office: Brigham Young U Provo UT 84602

GAREY, MICHAEL RANDOLPH, mathematician, researcher; b. Manitowoc, Wis., Nov. 19, 1945; m., 1965. B.S., U. Wis., 1967, M.S., 1969; Ph.D. in Computer Sci, 1970. Mem. tech. staff Math. Research Ctr. Bell Labs, Murray Hill, N.J., 1970; assoc. editor Jour. Assn. Computing Machinery, 1975-79, editor in chief, 1979-82. Mem. Math. Assn. Am., Assn. Computing Machinery, Soc. Indsl. and Applied Math., Ops. Research Soc. Am. (Lancaster prize 1979). Subspecialties: Algorithms; Operations research (mathematics). Office: Math Research Ctr Bell Labs Murray Hill NJ 07974

GARFIELD, ALAN J., art and computer graphics educator; b. Utica, N.Y., Jan. 17, 1950; s. Robert and Charlotte Hirsch G.; m. Phyllis H. Rafferty, Sept. 2, 1979. B.A., U. Iowa-Iowa CIty, 1971; M.A., SUNY-Binghamton, 1974. Technician Univ. Art Gallery, SUNY-Binghamton, 1972-74; instr. art Converse Coll., Spartanburg, S.C., 1974-76; asst. prof. art S.W. Mo. State U., Springfield, 1976-77; asst. prof. art dept. Creighton U., Omaha, 1977-79; dir. Krasl Art Center, St. Joseph, Mich., 1979-80; chmn. art and computer graphics Marycrest Coll., Davenport, Iowa, 1980—; cons., lectr. in field. Author: Drawings by Isabel Bishop, 1976, Pastels by Pierre Prins, 1976, Alexander Beary Gavalas, 1982. Mem. Spl. Interest Group in Graphics, Assn. Computing Machinery Spl. Interest Group in Personal Computers, Coll. Art Assn., Am. Assn. Museums. Jewish. Subspecialty: Graphics, image processing, and pattern recognition. Current work: Interactive computer graphic software; visible line algorithms. Home: 34 Oak Ln Davenport IA 52804 Office: Computer Graphics Marycrest Coll 1607 W 12th St Davenport IA 52804

GARFIELD, EUGENE, information science educator; b. N.Y.C., Sept. 16, 1925; s. Ernest and Edith (Wolf) Garofano; m. Faye Byron, 1945 (div.); 1 son, Stefan; m. Winifred Koziolek, 1955 (div.); children: Laura, Joshua, Thea. B.S., Columbia U., 1949, M.S., 1954; Ph.D., U. Pa., 1961. Research chemist Evans Research & Devel. Corp., 1949-50; research chemist Columbia U., N.Y.C., 1950-51; mem. staff machine index project Johns Hopkins U., Balt., 1951-53; pres. Eugene Garfield Assocs., Phila., 1954-60; pres., founder Inst. Sci. Info., Phila., 1960—; weekly columnist Current Comments in Current Contents, 1956—; adj. prof. info. sci. U. Pa., Phila., 1974—. Author: Essays of an Information Scientist, 1977 (Book of Yr., Am. Soc. Info. Sci.), Citation Indexing: Its Theory and Application in Science, 1979; editor-in-chief: Scientometrics; mem. editorial bd.: Progress in Info. Sci. and Tech; contbr. articles to profl. jours. Served with AUS, 1943-45. First Grolier Soc. fellow, 1953-54. Fellow AAAS (chmn. sect. T), Inst. Info. Scientists London; mem. IEEE (sr.), Info. Industry Assn. (corp. mem., past chmn. bd., past pres., Hall of Fame award), Spl. Libraries Assn., ACM, Authors League Am., Med. Library Assn., Am. Soc. Info. Sci. (award of merit 1975, past pres. Delaware Valley chpt.), Am. Chem. Soc. (div. award 1977), Drug. Info. Assn., Fedn. Am. Scientists. Developer Sci. Citation Index, 1961—, Index Chemicus, 1960—; developer sci. info. service; patentee in field. Office: ISI 3501 Science Ctr Philadelphia PA 19104

GARFINKEL, BORIS, astronomer; b. Rjev, USSR, Nov. 18, 1904; came to U.S., 1921, naturalized, 1928; s. Myron and Fanny (Kaplan) G. B.S., CCNY, 1927; M.A., Columbia U., 1926; Ph.D., Yale U., 1943. Instr. Yale U., New Haven, 1943-46; sr. research scientist Ballistic Research Lab., Aberdeen Proving Ground, Md., 1946-66; sr. research astronomer Yale U., 1966-73, emeritus, 1973—, faculty, 1966—. Contbr. articles to sci. jours. Recipient Cert. of Achievement Dept. of Army, 1957; R.H. Kent award Ballistic Research Labs., 1959. Mem. Am. Astron. Soc. (Dirk Brouwer award 1980); Royal Astron. Soc. Internat. Astron. Union. Subspecialties: Celestial mechanics; Optical astronomy. Current work: Celestial mechanics; artificial satellite theory, ideal resonance problem, theory of Trojan asteroids. Home: 738 Whitney Ave Apt 6 New Haven CT 06511 Office: Dept Astronomy Yale U New Haven CT 06511

GARFINKEL, DAVID, researcher, educator; b. N.Y.C., May 18, 1930; s. Louis and Leah (Markosfeldt) G.; m. Lillian Magid, June 26, 1960; children: Susan Laura, Beth Diane. A.B., U. Calif.-Berkeley, 1951; Ph.D., Harvard U., 1955; M.A. (hon.), U. Pa., 1972. Postdoctoral fellow in biophysics U. Pa., Phila., 1955-58, research assoc., 1961-63, asst. prof., 1963-65, assoc. prof. biophysics, 1965-72, assoc. prof. computer sci., 1972-77, prof. computer sci., 1977—; research biochemist N.Y. State Psychiat. Inst., Columbia U., N.Y.C., 1958-60. Mem. editorial bds.: Computers and Biomed. Research; mem. editorial bds.: Jour. Theoretical Biology; Contbr. articles to profl. jours. NIH research grantee, 1961—. Fellow IEEE; mem. Soc. Computer Simulation (sr.), Biomed. Engring. Soc. (bd. dirs. 1982—), Am. Soc. Biol. Chemists, Am. Physio. Soc. Subspecialties: Biomedical engineering; Biochemistry (medicine). Current work: Computer simulation of biological systems. Office: Moore Sch Elec Engring Univ Pennsylvania Philadelphia PA 19104

GARFINKLE, BARRY DAVID, pharmaceutical company executive; b. Newark, Oct. 3, 1946; s. Philip and Adella (Rauchwager) G.; m. Sherry Ellen Frank, Aug. 11, 1968; children: Jack, Stacey. B.S. in Bacteriology, Kans. State U., 1968; M.S. (NSF fellow), Pa. State U., 1970, Ph.D., 1972. Postdoctoral fellow Roche Inst. Molecular Biology, Nutley, N.J., 1972-74; sr. project microbiologist Merck Sharp & Dohme, West Point, Pa., 1975-77, mgr. biol. quality control tech. services, 1977-81, mgr. process validation, 1981—; tchr. course on process validation. Contbr. articles to profl. jours. NIH fellow, summer 1971. Mem. Am. Soc. Microbiology, Parenteral Drug Assn. Jewish. Subspecialties: Microbiology; Virology (biology). Current work: Sterilization validation and process development for sterile pharmaceuticals and biologicals. Home: 1274 Georgia Ln Hatfield PA 19440 Office: Merck Sharp & Dohme Sumneytown Pike Bldg 29-M West Point PA 19486

GARG, DEVENDRA PRAKASH, mech. engr., educator; b. Roorkee, India, Mar. 22, 1934; came to U.S., 1965; s. Chandra Gopal and Godawari (Devi) G.; m. Prabha Govil, Nov. 19, 1961; children—Nisha, Seema. B.Sc., Agra (India) U., 1954; B.S. in Mech. Engring, U. Roorkee, 1957; M.S. (Tech. Coop. Mission Merit scholar), U. Wis.-Madison, 1960; Ph.D., N.Y. U., 1969. Lectr. mech. engring. U. Roorkee, 1957-62, reader, 1962-65, vis. prof., 1978; instr. N.Y. U., 1965-69; asst. prof. Mass. Inst. Tech., 1969-71, assoc. prof., 1971-72, chmn. engring. projects lab., 1971-72, lectr., 1972-75; prof. Duke U., 1972—, dir. undergrad. studies dept. mech. engring. and materials sci., 1977—; cons. in field. Author: An Introduction to the Theory and Use of the Analog Computer, 1963, A Textbook of Descriptive Geometry, 1964; asso. editor: Jour. Interdisciplinary Modeling and Simulation, 1978—; contbr. numerous articles to profl. jours. Recipient Founder's Day award N.Y. U., 1969. Mem. IEEE (reviewer), Instrument Soc. Am. (reviewer), ASME (reviewer, co-guest editor spl. issues on ground transp. 1974, also socioecon. and ecol. systems 1976, sec. dynamic systems and control div. 1980—), Sigma Xi (sec. chpt. 1970-72). Subspecialties: Mechanical engineering; Systems engineering. Current work: Technological forecasting and assessment; advanced transportation systems; analysis and design of linear and nonlinear feedback control systems. Home: 2815 DeKalb St Durham NC 27705 Office: Sch Engring Duke Univ Durham NC 27706

GARG, LAL CHAND, pharmacology educator; b. India, Jan. 22, 1933; s. Amar Nath and Kalawati (Jindal) G.; m. Shakuntala Goyal, Jan. 1, 1940; children: Arun, Lovey. B.S., Panjab U., 1954, M.S., 1956; Ph.D., U. Fla., 1969. Instr. Vet. Coll., Mhow, India, 1956-61, asst. prof., 1961-63; asst. prof. pharmacy Panjab U., 1963-65; grad. asst. U. Fla., Gainesville, 1965-69, instr., 1969-71, asst. prof., 1971-76, assoc. prof., 1976—. Mem. Am. Soc. Pharmacology and Exptl. Therapeutics, Am. Soc. Nephrology, Internat. Soc. Nephrology, N.Y. Acad. Scis., Sigma Xi. Subspecialties: Pharmacology; Renal Pharmacology. Current work: Electrolyte transport, acid-base balance, hormonal regulation of transport enzymes. Office: Box J-267 JHMHC Gainesville FL 32610

GARING, JOHN SEYMOUR, physicist, research executive; b. Toledo, Ohio, Nov. 6, 1930; s. Harry Raymond and Catherine Marie (Gomer) G.; m. Ione Davis, Apr. 26, 1952; children: John Davis, Susan Carolyn. B.S. summa cum laude, Ohio State U., 1951, M.S., 1954, Ph.D. in Physics, 1958. Physicist Air Force Geophysics Lab., Hanscom AFB, Mass., 1958-61, chief infared physics br., 1961-63, dir. optical physics div., 1973—. Contbr. articles to profl. jours. Served to 1st lt. USAF, 1954-57. Fellow Optical Soc. Am., AAAS; mem. AIAA, Sigma Xi, Phi Beta Kappa, Phi Eta Sigma, Pi Mu Epsilon.

Subspecialties: Atomic and molecular physics; Infrared spectroscopy. Current work: Infrared phsyics, molecular spectroscopy, atmospheric optics and transmission, research management. Home: 157 Cedar St Lexington MA 02173 Office: Air Force Geophysics Lab Hanscom AFB MA 01731

GARLAND, HOWARD, psychologist, educator; b. Bklyn., June 22, 1946; s. Murray and Norma (Luft) Garlitsky; m. Eileen Mary Cohen, Aug. 21, 1968; children: Eric Lee, Adam Marc. B.A., Bklyn. Coll., 1968; M.S., Cornell U., 1971, Ph.D., 1972. Asst. prof. psychology Upsala Coll., East Orange, N.J., 1972-74; asst. prof. psychology and mgmt. U. Tex., Arlington, 1974-75, assoc. prof. psychology and mgmt., 1976-78, prof. psychology and mgmt., 1979—. Mem. editorial bd.: Group and Orgn. Studies; contbr. articles to profl. jours. Army Research Inst. grantee, 1978-80. Mem. Am. Psychol. Assn., Soc. for Personality and Social Psychology, Soc. of Indsl. and Organizational Psychology, Soc. for Advancement of Social Psychology. Subspecialties: Organizational psychology; Social psychology. Current work: Work in the area of goals levels and task performance. In particular, research centers on looking at the facilitative effects of impossible goals on performance. Home: 4706 Basswood Ct Arlington TX 76016 Office: Univ Texas Dept Mgmt Box 19467 Arlington TX 76010

GARLAND, JAMES C., physics educator, university research laboratory administrator; b. Columbia, Mo., Aug. 11, 1942; m., 1965; 2 children. A.B., Princeton U., 1964; Ph.D. in Physics, Cornell U., 1969. NSF fellow in physics Cambridge (Eng.) U., 1969-70; asst. prof. Ohio State U., 1970-75, assoc. prof. physics, 1975—, now dir. materials research lab. Mem. Am. Phys. Soc. Subspecialty: Low temperature physics. Office: Dept Physics Smith Lab Ohio State U Columbus OH 43210

GARMIRE, GORDON PAUL, astronomer, educator; b. Portland, Oreg., Oct. 3, 1937; s. Paul W. and Ethel V. (Alsen) G.; m. Audrey B. Cook, Feb. 14, 1976; children: Geoffrey, Lisa, Marla, Chris, Rosemary, David. A.B. cum laude, Harvard U., 1959; Ph.D., M.I.T., 1962. Staff M.I.T., Cambridge, 1962-64, asst., then assoc. prof., 1964-68; sr. research fellow Calif. Inst. Tech., 1966-68, assoc. prof., 1968-72, prof. physics, 1972-81; prof. astronomy Pa. State U., State College, 1980—; cons. TRW, Los Alamos Lab., NASA. Sr Hays Fulbright fellow, 1973-74; Guggenheim fellow, 1973-74; NASA Exceptional Sci. Achievement awardee, 1978. Mem. Am. Astron. Soc., Internat. Astron. Union, Sigma Xi. Subspecialties: 1-ray high energy astrophysics; High energy astrophysics. Current work: Supernova remnants, hot interstellar medium, x-ray sources, radiation detectors, x-ray telescopes. Co-discoverer cosmic high energy Gamma radiation; pioneer 1st diamond turned x-ray telescope; co-prin. investigator high energy astron. obs.; pioneer 1st all-sky soft x-ray survey. Home: Route 2 Box 256 Huntingdon PA 16652 Office: 504 Davey Lab University Park PA 16802

GARNER, ANDREW MORRIS, aerospace engineer, consultant; b. Palestine, Tex., Oct. 7, 1932; s. Andrew Morris and Essie (LaRue) G.; m. Mary Frances Kerns, Sept. 19, 1955; children: David Andrew, Karen Frances. B.A. in Math, Tex. A&M U., 1955. Registered profl. engr., Fla. Engr. Anderson, Greenwood, Houston, 1957-63; engring. specialist Rocketdyne, McGregor, Tex., 1963-66, Martin Marietta, Orlando, Fla., 1966-69; program mgr. Transp. Tech., Irving, Tex., 1969-71; engring. mgr. Martin Marietta, Orlando, 1971—; cons. in field. Pres. River Oaks Civic Orgn., Edgewood, Fla., 1973; mem. Edgewood Zoning Bd., 1976. Served to 1st lt. AUS, 1955-57. Recipient award for meritorious service Martin Marietta, 1978. Mem. AIAA, Nat. Soc. Profl. Engrs., Fla. Engring. Soc. Republican. Subspecialties: Aerospace engineering and technology; Mechanical engineering. Current work: Mechanical engineering manager battlefield interdiction systems, technical director supersonic target system; consultant in propulsion, fluid mechanics, materials and structures. Inventor modular propulsion, low radar cross section supersonic inlet, variable supersonic inlet. Home: 5141 The Oaks Circle Orlando FL 32809 Office: Martin Marietta Sand Lake Rd Orlando FL 32855

GARNER, HAROLD E., physiologist, educator; b. Eldorado, Kans., May 8, 1935; s. Carl W. and Viola M. G.; m. Patsy McClenahan, Aug. 4, 1957; children: Dustin, Genell, Gayla, Tony. B.S. in Animal Sci, Kans. State U., 1957, D.V.M., 1962, M.S. in Vet. Pathology, 1964; Ph.D. in Cardiovascular Physiology and Biomed. Physics, Baylor Coll. Medicine, 1971. Pvt. practice vet. medicine, Eureka, Kans., 1962-65; instr. dept. clin. medicine U. Ill., Urbana, 1965-66, asst. prof., 1966-68; instr., spl. research fellow, asst. prof. Baylor Coll. Medicine, Houston, 1968-71; prof. vet. medicine and surgery, interim assoc. dean for research and grad. studies U. Mo., Columbia, 1971—. Contbr. articles to profl. jours. Chmn. 4-H Beef Com., 1980-82; active Boy Scouts Am., 1976-79; trustee Community United Meth. Ch., Columbia, 1980—. NIH spl. research fellow, 1968-71; Mo. Heart Assn. grantee, 1982-83. Mem. AVMA, Am. Assn. Equine Practitioners, Mo. Heart Assn., Am. Physiol. Soc., Am. Soc. Vet. Physiology and Pharmacology, N.Y. Acad. Sci., Am. Soc. Vet. Anesthesiology, Alpha Zeta, Phi Zeta, Omega Tau Sigma. Subspecialty: Comparative physiology. Current work: Ischemic heart disease. Equine laminitis hypertension. Home: Route 5 Box 222C Columbia MO 65201 Office: Coll Vet Medicin U Mo W203 Vet Medicine Bldg Columbi MO 65211

GARNER, JAMES KIRKLAND, fusion engineer; b. Amityville, N.Y., Jan. 7, 1951; s. Ray C. and Virginia L. (Kirkl) G.; m. Margaret Roark, July 6, 1946; children: Sean, Kirk. B.S.M.E., U. Calif.-Santa Barbara, 1980. Fusion engr. TRW Corp., Redondo Beach, Calif., 1980—. Mem. Am. Nuclear Soc., Soaring Soc. Am. Subspecialties: Nuclear fusion; Mechanical engineering. Current work: Fusion first wall blanket/shield design; fusion synfuel and power plant design; fusion-fission hybrid blanket design. Home: 1133 Nowita Pl Venice CA 90291 Office: TRW Corp One Space Park Redondo Beach CA 90278

GARNER, LAFORREST DEAN, orthodontist, educator; b. Muskogee, Okla., Aug. 20, 1933; s. Sanford and Fannie (Thompson) G.; m. Alfreida Thomas, July 18, 1964; children: Deana Yvette, Thomas LaForrest, Sanford Ernest. D.D.S., Ind. U.-Indpls., 1957, M.S., 1959. Diplomate: Am. Bd. Orthodontists. Program chmn. craniofacial biology sect. Internat. Assn. Dental Research, Washington, 1982—; chmn. dept. orthodontics Ind. U. Sch. Dentistry, Indpls., 1969—. Contbr. articles to sci. jours., chpts. to books. Bd. dirs. Ind. Boys Clubs, 1975—, Park-Tudor Sch., Indpls., 1978—, Ind. Health Careers, 1974—, Vis. Nurses Assn., 1973-76. Fellow Am. Coll. Dentists; mem. Am. Assn. Orthodontists, Edward H. Angle Soc. Orthodontists, Internat. Assn. Dental Research, Am. Cleft Palate Assn. (bd. dirs. 1974-78), Ind. Dental Assn., Ind. Soc. Orthodontists (pres. 1972-73), Omicron Kappa Upsilon. Democrat. Presbyterian. Club: Boule (Indpls.). Subspecialties: Orthodontics; Dental growth and development. Current work: Cleft lip and palate rehabilitation, growth and development, mechaniotherapy and instrumentation for delivery of services, student direction in growth and development. Home: 6245 Riverview Dr Indianapolis IN 46260 Office: Indiana University School of Dentistry 1121 W Michigan St Indianapolis IN 46202

GARNER, WENDELL RICHARD, psychology educator; b. Buffalo, Jan. 21, 1921; s. Richard Charles and Lena Belle (Cole) G.; m. Barbara Chipman Ward, Feb. 18, 1944; children: Deborah Ann, Peter Ward, Elinor. A.B., Franklin and Marshall Coll., 1942, D.Sc., 1979; A.M., Harvard U., 1943, Ph.D., 1946; D.H.L., Johns Hopkins U., 1983. Teaching fellow Harvard U., 1942-43, research assoc., 1943-46; instr. Johns Hopkins U., 1946; asst. prof. Johns Hopkins, 1947-51; assoc. prof. Johns Hopkins U., 1951-55, prof., 1955-67; dir. Psychol. Lab. Inst. Coop. Research, 1949-55, chmn. dept. psychology, 1954-64; James Rowland Angell prof. psychology Yale U., 1967—, dir. social scis., 1972-73, 81—, chmn. dept. psychology, 1974-77, dean, 1978-79; Paul M. Fitts Meml. lectr. U. Mich., 1973. Author: Uncertainty and Structure as Psychological Concepts, 1962, Processing of Information and Structure, 1974; editor: Ability Testing, 1982. Recipient alumni citation and award Franklin and Marshall Coll., 1975. Fellow Am. Psychol. Assn. (Distinguished Sci. Contbn. award 1964, pres. div. exptl. psychology 1974), AAAS (v.p. psychology 1967—), Acoustical Soc. Am.; mem. Soc. Exptl. Psychologists (chmn. 1959, 75, Warren medal 1976), AAUP, Md. Psychol. Assn. (pres. 1961-62), Eastern Psychol. Assn., Nat. Acad. Scis., Sigma Xi. Subspecialties: Cognition; Psychophysics. Current work: Visual information processing, especially perceptical structure and dimensional interactions of stimuli. Home: 48 Yowago Ave Branford CT 06405 Office: Yale U New Haven CT 06520

GARNICK, MARC BENNETT, physician, educator; b. Lawrence, Mass., Jan. 8, 1947; s. Philip and Ruth Rose (Sodnowsky) G.; m. Barbara Kates, Apr. 1, 1973; children: Alexander, Nathaniel. A.B., Bowdoin Coll., 1968; M.D., U. Pa., 1972. Diplomate: Am. Bd. Internal Medicine, 1976. Intern Hosp. of U. Pa., Phila., 1972-73, resident in medicine, 1973-74; clin. assoc. Div. Arthritis, Metabolism and Digestive Diseases, NIH, Phoenix and Bethesda, Md., 1974-76; clin. fellow in medicine Harvard Med. Sch., Boston, 1976-78; clin. fellow in med. oncology Sidney Farber Cancer Inst., Boston, 1976-78; asst. in medicine Peter Bent Brigham Hosp., Boston, 1976-78; instr. medicine Harvard Med. Sch., 1978-79; asst. prof. medicine Sidney Farber Cancer Inst., Harvard Med. Sch., 1979—, asst. physician, 1978—; jr. assoc. medicine Brigham & Women's Hosp., 1978—; cons. physician Nantucket (Mass.) Cottage Hosp., 1981—; med. coordinator new drug liaison Nat. Cancer Inst.-NIH, 1980—. Contbr. chpts. to books, articles to profl. jours. Served to lt. comdr. USPHS, 1974-76. Fellow ACP; mem. Am. Assn. Cancer Research, Am. Soc. Clin. Oncology, Am. Fedn. Clin. Research, Am. Urologic Assn., Phi Beta Kappa, Alpha Omega Alpha. Subspecialties: Cancer research (medicine); Chemotherapy. Current work: Research in protocol design, new chemotherapeutic agents, genitourinary cancer treatment. Office: 44 Binney St Boston MA 02115

GARNSEY, STEPHEN MICHAEL, plant pathologist; b. Oceanside, Calif., Aug. 3, 1937; s. Felix R. and Theodora Louise (Mueller) G.; m. Rosalee Zepik, June 28, 1958; children: Lisa Ann, Paul Michael. B.A., U. Calif.-Riverside, 1958; Ph.D., U. Calif.-Davis, 1964. Research plant pathologist U.S. Hort. Research Lab., Orlando, Fla., 1963—. Contbr. articles to profl. jours. Recipient cert. of merit U.S. Dept. Agr., 1969, 80; Binat. Sci. Found. grantee, 1978; BARD grantee, 1981. Mem. AAAS, Am. Phytopathol. Soc. (Lee M. Hutchins award 1981), Internat. Orgn. Citrus Virologists, Fla. State Hort. Soc., Sigma Xi. Republican. Lutheran. Subspecialties: Plant pathology; Plant virology. Current work: Identification, characterization and host interactions of citrus viruses. Transmission of virus and virus-like pathogens. Home: 2313 Sherbrooke Rd Winter Park FL 32792 Office: 2120 Camden Rd Orlando FL 32803

GAROVOY, MARVIN R., physician, researcher; b. N.Y.C., June 14, 1943; s. Nathan and Ann (Braffman) G.; m. Seena Fischer, June 7, 1969; children: Natara, Jocelyn. B.A., NYU, 1964; M.D., SUNY, Downstate Med. Ctr., 1969. Diplomate: Am. Bd. Internal Medicine, 1974. Research assoc., fellow in nephrology Harvard Med. Sch., 1971-73; asst. prof. medicine, assoc. dir. tissue typing lab. Harvard Med. Sch. and Peter Bent Brigham Hosp., 1975-81; assoc. prof. surgery and medicine U. Calif.-San Francisco, 1981—, dir. immunogenetics and transplantation lab., 1981—. Contbr. articles to profl. jours. Served to maj. USAF, 1973-75. Mem. Am. Assn. Immunologists, Am. Soc. Nephrology, Am. Assn. Clin. Histocompatibility Testing, Am. Fedn. Clin. Research. Subspecialties: Transplantation; Immunogenetics. Current work: Mechanisms of kidney graft rejection; induction of graft tolerance, controlling mechanisms of antibody formation, flow cytometry. Office: U Calif Med Ctr 3d and Parnassus Sts San Francisco CA 94143

GARRELS, ROBERT MINARD, geology educator; b. Detroit, Aug. 24, 1916; s. John C. and Margaret A. (Gibney) G.; m. Jane M. Tinen, Dec. 21, 1940 (div. 1969); children: Joan F., James C., Katherine G.; m. Cynthia A. Hunt, 1970. B.S., U. Mich., 1937, Sc.D. (hon.), 1980; M.S., Northwestern U., 1939, Ph.D., 1941; M.A. (hon.), Harvard U., 1955, Sc.D., U. Brussels, 1969, U. Louis Pasteur, Strasbourg, France, 1976. From instr. to assoc. prof. geology Northwestern U., Evanston, Ill., 1941-52, prof. geology, 1965-69, 72-80, Scripps Instn. Oceanography, 1969-71; prof. U. South Fla., 1980—; geologist U.S. Geol. Survey, 1952-55; assoc. prof. geology Harvard U., 1955-57, prof., 1957-65, chmn. dept. geol. scis., 1963-65; Henri Speciael prof. sci. U. Brussels, Belgium, 1962-63; Capt. James Cook prof. oceanography U. Hawaii, Honolulu, 1972-74. Author: Textbook of Geology, 1951, Mineral Equilibria, 1959, (with C.L. Christ) Solutions, Minerals and Equilibria, 1965, (with F.T. Mackenzie) Evolution of Sedimentary Rocks, 1971, (with C.A. Hunt) Water, The Web of Life, 1972, (with F.T. Mackenzie, C. Hunt) Chemical Cycles and the Global Environment, 1975. Recipient Wollaston medal Geol. Soc. London, 1981. Fellow AAAS, Geol. Soc. Am. (Arthur L. Day medal 1966, Penrose medal 1978), Mineral. Soc. Am.; mem. Geochem. Soc. (pres. 1962, V.M. Goldschmidt award 1973), Nat. Acad. Scis., Soc. Econ. Geologists, Am. Acad. Arts and Sci., Am. Chem. Soc., Sigma Xi. Subspecialties: Geochemistry; Geology. Current work: Geochemistry of low-temperature and pressure mineral-water systems; mathematical modeling of environments of geologic past. Office: Dept Marine Sci U South Fla Saint Petersburg FL 33701

GARRETT, REGINALD HOOKER, biology educator, researcher; b. Roanoke, Va., Sept. 24, 1939; s. William Walker and Lelia Elizabeth (Blankenship) G.; m. Linda Joan Harrison, Mar. 15, 1958; children: Jeffrey David, Randal Harrison, Robert Martin. B.S., Johns Hopkins U., 1964, Ph.D., 1968. Asst. prof. biology U. Va., 1968-73, assoc. prof., 1973-82, prof., 1982—; sci. cons. Internat. Plant Research Inst., 1981—. Contbr. articles in field to profl. jours. NIH fellow, 1964-68; Fulbright Hays fellow, 1975-76; Thomas Jefferson vis. fellow, 1983; grantee NIH, NSF. Mem. Am. Soc. Biol. Chemists, Am. Soc. Microbiology, Am. Soc. Plant Physiology, Soc. Gen. Physiology, Sigma Xi, Phi Lambda Upsilon, Phi Sigma. Subspecialties: Biochemistry (biology); Molecular biology. Current work: The enzymology, genetics and regulation of nitrate assimilation, mechanism of action of electron transfer enzymes, gene expression in eucaryotic cells. Office: Department Biology Gilmer Hall University of Va Charlottesville VA 22901

GARRETT, WILLIAM N., animal science educator; b. Cresson, Pa., June 8, 1926; s. William M. and Hilda M. (Wiel) G.; m. Ida T. Terkildsen, June 26, 1954; children: Joan E., W. Karsen. B.S., Pa. State U., 1950, M.S., 1951; Ph.D., U. Calif.-Davis, 1958. Prof. animal sci. U. Calif.-Davis, 1958—. Subspecialties: Animal nutrition; Nutrition (biology). Current work: Ruminant nutrition; energy metabolism. Home: 1300 W 8th St Davis CA 95616 Office: Dept Animal Sci U Calif Davis CA 95616

GARRETTSON, LORNE KEITH, physician, educator, researcher; b. Pasadena, Calif., Mar. 2, 1934; s. George Webbert and Grace Loa (Ingham) G.; m. Elizabeth Stabler Miller, Apr. 20, 1963; children: Elizabeth Brook, Linda Janney, Mariana. B.A., Pomona Coll., 1955; M.D., John Hopkins U., 1959. Intern John Hopkins U., 1959-60; resident, 1960-61, Children's Hosp., Boston, 1961-62; assoc. in pediatrics Emory U., 1967-68; asst. prof. pediatrics SUNY-Buffalo, 1968-73; asso. prof. pediatrics and pharmacy/pharmaceutics Va. Commonwealth U., Richmond, 1973—; dir. Central Va. Poison Center, 1975—. Contbr. articles to profl. jours. Served to sr. asst. surgeon USPHS, 1962-64. Fellow Am. Acad. Pediatrics; mem. Am. Soc. Pharmacology and Exptl. Therapeutics, Soc. Pediatric Research, Am. Assn. Poison Control Centers, League Am. Wheelmen. Democrat. Quaker. Subspecialties: Pediatrics; Pharmacology. Current work: Drug metabolism in children; clinical toxicology. Office: Va Commonwealth U Box 522 MCV Station Richmond VA 23298

GARRICK, B. JOHN, engineering consulting firm executive; b. Eureka, Utah, Mar. 5, 1930; s. Morrison H. and Zelma (Hoffman) G.; m. Amelia Madson, Sept. 18, 1952; children; Robert Stephen, John Morrison, Ann. B.S., Brigham Young U., 1952; M.S., UCLA, 1962, Ph.D., 1968. Physicist Phillips Petroleum Co., Idaho Falls, Idaho, 1952-54, U.S. AEC, Washington, 1955-57; pres. nuclear and systems scis. group Holmes & Narver, Inc., Anaheim, Calif., 1957-75; prin. Pickard, Lowe and Garrick, Inc., Irvine, Calif., 1975—; adj. prof. UCLA, 1976-83; pres. Los Angeles Maintainability Assn., 1973-74; U.S. rep. Internat. Conf. on Nuclear Relia Germany, 1971. Contbr. articles to profl. jours. Fellow Inst. Advancement Engring.; mem. Am. Nuclear Soc., Atomic Indsl. Forum, Pacific Coast Elec. Assn., N.Y. Acad. Scientists. Subspecialties: Nuclear fission; Nuclear engineering. Current work: Quantitative risk assessment of nuclear power plants and other high technology facilities plant performance analysis, management consulting, and teaching at graduate level. Office: Pickard Lowe and Garrick Inc 17840 Skypark Blvd Irvine CA 92714

GARRIE, STUART ALLEN, physician; b. Chgo., June 20, 1943; s. Leon J. and Pauline (Pancoe) G.; m. Erica Vera Ohlbaum, June 11, 1967; children: Daniel Benjamin, Nicole Leya, Michael. Student, Pasadena City Coll., 1961-62, U. Calif.-Berkeley, 1962-63; M.D., U. So. Calif., 1967. Intern Los Angeles County - U. So. Calif. Med. Ctr., 1967-68, resident in dermatology, 1968-71; resident in psychiatry U. Utah Med. Ctr., 1974-77; research fellow U. Copenhagen, Denmark, 1971; pvt. practice medicine, specializing in dermatology, dermatopathology and psychiatry, Seattle; assoc. clin. prof. U. Wash., Seattle, 1979—. Contbr. articles to profl. jours. Mem. Am. Acad. Dermatology, Soc. Investigative Dermatology, Salt Lake County Med. Soc., AMA, Wash. Dermatol. Soc., Wash. Psychiat. Assn. Subspecialties: Dermatology; Pathology (medicine). Current work: Skin window technique, anxiety and atopic dermathology. Address: 14434 Ambaum Blvd SW Seattle WA 98166

GARRISON, ROBERT EDWARD, geological educator, researcher; b. Dallas, Oct. 25, 1932; s. Robert Edward II and Marlen Heloise (Simpson) G.; m. Jan Erin Dodge, June 8, 1963; 1 son, James Edward. B.S., Stanford U., 1955, M.S., 1958; postgrad., Innsbruck U., 1958-59; Ph.D., Princeton U., 1964. Geologist, Sunray-DX Oil Co., Casper, Wyo., 1959-61; asst. U. Calif.-Santa Barbara, 1965-66, U. B.C. (Can.), Vancouver, 1966-68; assoc. prof. U. Calif.-Santa Cruz, 1968-73, prof., 1973—; cons. U.S. Geol. Survey, Menlo Park, Calif., 1974-83. Co-author: Electron Micrographs of Limestones, 1967. Served as 1st lt. USAF, 1955-57. Guggenheim fellow, Oxford, Eng., 1972-73; Fulbright scholar, Austria, 1958-59. Fellow Geol. Soc. Am., AAAS; mem. Am. Assn. petroleum Geologists, Soc. Econ. Paleontologists and Mineralogists. (pres. Pacific sect.). Subspecialty: Sedimentology. Current work: Research on the origin and diagenesis of deep water sedimentary rocks. Office: Earth Scis Bd U Calif Santa Cruz CA 95064

GARRISON, ROBERT FREDERICK, astronomer, educator; b. Aurora, Ill., May 9, 1936; s. Robert W. and Dorothy I. (Rydquist) G.; m. Ada V. Mighell, June 7, 1957 (div. 1978); children: Forest L., Alexandra, David C. B.A. in Math, Earlham Coll., 1960; postgrad., U. Wis., 1961-62; Ph.D. in Astronomy and Astrophysics, U. Chgo., 1966. Research assoc. Mt. Wilson and Palomar Obs., Pasadena, Calif., 1966-68; asst. prof. U. Toronto, Ont., Can., 1968-74, assoc. prof., 1974-78, prof. astronomy, 1978—, also research, assoc. dir. Chile ops. Contbr. articles to profl. jours. Bd. dirs. Bruce Trail Assn., 1975-76. Served with USMC, 1954-56. Mem. Can. Astron. Soc., Am. Astron. Soc., Astron. Soc. of Pacific, Am. Assn. Variable Star Observers, Royal Astron. Soc. Can., Internat. Astron. Union. Club: U. Chgo. of Can. (v.p. schs. 1982—). Subspecialty: Optical astronomy. Current work: Stellar spectra, spectral classification, galactic structure, morphology of galaxies, instrumentation. Office: David Dunlap Obs Box 360 Richmond Hill ON Canada L4C 4Y6

GARRITSON, GRANT RICHARD, naval officer; b. Kokomo, Ind., Oct. 18, 1939; s. Ulysses Grant and Geraldine Irene (Gerhart) G.; m. Joan Mary Synave, June 8, 1961; children: Richard, John. B.S., U.S. Naval Acad., 1961; Ph.D., Notre-Dame U., 1968. Commd. ensign U.S. Navy, 1961, advanced through grades to capt., 1982; served on various submarines, New London, Conn., 1969-73, exec. officer, Charleston, S.C., 1973-76; asst. repair officer Portsmouth (N.H.) Naval Shipyard, 1976-79; OIC Naval Engring., asst. prof. mech. engring. Naval Postgrad. Sch., Monterey, Calif., 1979-82; asst. project. mgr. Trident Submarine; UK trident program mgr. NAVSEA, Washington, 1982—. Mem. Am. Nuclear Soc., Am. Soc. Naval Engrs., U.S. Naval Inst. Subspecialties: Nuclear engineering; Nuclear fission. Home: 11897 Antietam Rd Woodridge VA 22192 Office: Naval Sea Systems Command PMS396 Washington DC 20370

GARTE, SEYMOUR JAY, research chemist; b. Bklyn., Oct. 31, 1947; s. Samuel M. and Esther (Goober) G.; m. Michal R. Soltes, June 15, 1969; children: Rebecca, Cora. B.S. in Chemistry, CCNY, 1970; Ph.D. in Biochemistry, CUNY, 1976. Lectr. chemistry CCNY, 1970-75; postdoctoral fellow environ. medicine N.Y.U. Med. Center, 1975-76, asso. research scientist, 1976-79, asst. prof., 1979—. Author papers in field. Recipient Young Investigator award Nat. Cancer Inst., 1979-82. Mem. Am. Assn. Cancer Research, Am. Chem. Soc., Mid-Atlantic Soc. Toxicology. Current work: Research into biochem. mechanics of chem. carcinogenesis using receptor studies, gene transfer, cell culture, in vivo systems. Office: Dept Environ Medicine NYU Med Sch 550 1st Ave New York NY

GARTLAND, WILLIAM JOSEPH, scientific research administrator, biochemist; b. N.Y.C., Apr. 15, 1941; s. William Joseph and Mary Elizabeth (Klik) G.; m. Margaret Louise Wenstadt, June 20, 1981. B.S. in Chemistry, Holy Cross Coll., 1962; M.A., Princeton U., 1964, Ph.D. in Biochemistry, 1967. Asst. research scientist dept. biochemistry N.Y.U. Med. Center, 1967-69; postgrad. research biologist dept. biology U. Calif.-San Diego, La Jolla, 1969-70; grants assoc. div. research grants NIH, Bethesda, Md., 1970-71, program adminstr. genetics program Nat. Inst. Gen. Med. Scis., 1971-76, dir. Office Recombinant DNA Activities, 1976—, dir., 1979—; exec. sec. Recombinant DNA Adv. Com. NIH, 1975—; U.S. head U.S.-Japan

Coop. Program for Recombinant DNA Research, 1980—. Contbr. articles on genetics and biochemistry to profl. jours. Recipient NIH Dir.'s award, 1978. Mem. Am. Soc. for Human Genetics, AAAS; mem. Am. Soc. Microbiology. Subspecialties: Genetics and genetic engineering (biology); Molecular biology. Current work: Oversight of recombinant DNA research, administration of NIH guidelines, risk assessment studies. Office: NIH 3B10 Bldg 31 Bethesda MD 20205

GARTLER, STANLEY MICHAEL, genetics educator, geneticist; b. Los Angeles, June 9, 1923; s. George David and Delvira (Cupferberg) G.; m. Marion Ruth Mitchelson, Nov. 7, 1948. B.S., UCLA, 1948; Ph.D., U. Calif.—Berkeley, 1952. Research assoc. Columbia U., N.Y.C., 1952-57; research asst. prof. U. Wash., Seattle, 1957-60, assoc. prof., 1960-64, prof. genetics, 1964—; dir. NATO meeting on mosaicism, Venice, Italy, 1972. Author: (with R. E. Cole) Inactivation Sexual Differentiation, 1978. NIH and NSF research grantee, 1956—. Mem. Am. Soc. Human Genetics (dir. 1970), Genetics Soc. Am., Soc. Cell Biology, Am. Soc. Naturalists. Subspecialty: Genetics and genetic engineering (medicine). Current work: Mammalian somatic cell genetics with emphasis on x-chromosome inactivation. Home: 9009 42d St NE Seattle WA 98115 Office: Dept Genetics U Wash Seattle WA 98195

GARTNER, LAWRENCE MITCHEL, pediatrician, med. coll. adminstr.; b. Bklyn., Apr. 24, 1933; s. Samuel and Bertha (Brimberg) G.; m. Carol Sue Blicker, Aug. 12, 1956; children—Alex David, Madeline Hallie. A.B., Columbia U., 1954; M.D., Johns Hopkins U., 1958. Intern pediatrics Johns Hopkins Hosp., 1958-59; resident pediatrics Albert Einstein Coll. Medicine, 1959-60, chief resident, 1960-61, instr. pediatrics, 1962-64, asst. prof., 1964-69, asso. prof., 1969-74, prof., 1974-80, dir. div. neonatology, 1967-80, dir. div. pediatric hepatology, 1967-80; dir. clin. research unit Rose F. Kennedy Center, 1972-80; attending physician Hosp. Albert Einstein Coll. Medicine; prof., chmn. dept. pediatrics U. Chgo. Pritzker Sch. Medicine, 1980—; dir. Wyler Children's Hosp., U. Chgo. Med. Center, 1980—; med. dir. Gail I. Zuckerman Found., 1967—. Contbr. articles to med. jours. and textbooks. Mem. adv. bd. Children's Liver Found.; Trustee Home for Destitute Crippled Children, La Rabida Children's Hosp. Recipient award NIH, 1967-74; Appleton Century Crofts prize, 1956; Mosby book award, 1958; NIH grantee, 1967—. Mem. Am. Pediatric Soc., Soc. Pediatric Research, Perinatal Research Soc., Am. Assn. Study Liver Disease, N.Y. Pediatric Soc., Am. Acad. Pediatrics, Harvey Soc., N.Y. Acad. Sci., AAAS, N.Am. Soc. Pediatric Gastroenterology (pres. 1974- 75), LaLeche League Internat., Chgo. Pediatric Soc., Phi Beta Kappa, Alpha Omega Alpha. Subspecialties: Pediatrics; Neonatology. Current work: Jaundice in newborn infants and studies of bilirubin metabolism in fetus and newborn, including animal studies; studies of liver disease in children. Office: U Chgo Pritzker Sch Medicine 950 E 59th St Chicago IL 60637

GARTNER, LESLIE PAUL, anatomy educator, researcher, consultant; b. Szolnok, Hungary, Mar. 18, 1943; came to U.S., 1956, naturalized, 1962; s. Janos Gartner and Mary (Schwartz) Karmer; m. Roseann C. Kollar, June 13, 1971; 1 dau., Jennifer. B.A., Rutgers U.-Newark, 1965; M.S., Rutgers U.-New Brunswick, 1968, Ph.D., 1970. Instr. U. Md. Dental Sch., Balt., 1970-71, asst. prof. anatomy, 1971-75, assoc. prof., 1975—; cons. U.S. Army Inst. Dental Research at Walter Reed Hosp., Washington, 1981—. Author: Textbook of Head and Neck Anatomy, 1981, Essentials of Oral Histology and Embryology, 1982; biol. scis. editor: Jour. Balt. Coll. Dental Surgery, 1981—. Mem. horseshow com. McDonogh (Md.) Sch. USPHS fellow, 1965-70. Mem. Am. Assn. Anatomists, Internat. Assn. Dental Research, Pan Am. Assn. Anatomists, Chesapeake Electron Microscopy Assn., PTA. Subspecialties: Anatomy and embryology; Teratology. Current work: Main research interest involves; effects of teratologic agents on craniofacial development and epithelio-mesenchymal interactions; subcellular alterations as a function of age and environmental insults; histochemistry. Home: 119 Cherry Valley Rd Reisterstown MD 21136 Office: U Md Dental Sch 666 W Baltimore St Baltimore MD 21201

GARTRELL, CHARLES FREDERICK, space scientist, engineer, researcher; b. Balt., Nov. 4, 1951; s. Charles Collins and Viviane Jeanne (Brown) Cole; m. Vanessa Lynn vanManen, May 19, 1975; 1 son, Charles Michael. B.A., U. Md.-Balt., 1973; postgrad., Johns Hopkins U., 1974-75, 79-80, Rutgers U., 1976-77. Task mgr./analyst Computer Scis. Corp., Silver Spring, Md., 1973-75; systems analyst RCA Am. Communications, Princeton, N.J., 1975-78; sr. engr. Gen. Research Corp., McLean, Va., 1978—. Author/editor: Military Space Systems Technology Model, 1982, 83, NASA Space Systems Technology Model, 1980, 81, 82, 83. Mem. Optical Soc. Am., AIAA. Republican. Baptist. Clubs: Nat. Model R.R. Assn. (Indpls.); B & O R.R. Hist. Soc. Subspecialties: Aerospace engineering and technology; Satellite studies. Current work: Research and planning for advanced civilian and military space systems technology, covering next 20-30 years; design/studies of advanced earth observation, radar and communication satellites, orbital transfer vehicles and spacecraft control. Home: 10332 Ridgeline Dr Gaithersburg MD 20879 Office: Gen Research Corp 7655 Old Springhouse Rd McLean VA 22102

GARVIN, EVERETT ARTHUR, psychologist; b. Worcester, Mass., July 30, 1922; s. Arthur Leo and Rose (Berthiame) G.; m. Norma J. Ibbotson, Sept. 5, 1955; 4 children. A.B., Antioch Coll., 1950; M.S., Tulane U., 1953; Ph.D., Washington U., 1962. Lic. psychologist, Mass. Tech. staff The Mitre Corp., Bedford, Mass., 1962-64; research dir. N. Central Mental Health Ctr., Fitch, Mass., 1964-67; prof. psychology Fitch (Mass.) State Coll., 1967—; cons. Mass. Rehab. Commn., Boston, 1967—; project dir. Mass. State Coll. System, Fed. Grant Alcohol Tng., 1978. Contbr. articles to profl. jours. Served with USAAF, 1940-45. Fellow Mass. Psychol. Assn.; mem. Am. Psychol. Assn. Subspecialties: Behavioral psychology; Human factors engineering. Current work: Memory; biofeedback; cognitive development; behavioral changes under stress. Home: Common St Groton MA 01450 Office: Fitchburg State Coll Highland Ave Fitchburg MA 01420

GARVIN, PAUL JOSEPH, JR., toxicologist; b. Toledo, Nov. 16, 1928; s. Paul Joseph and Laura Mary (Blanchet) G.; m. Priscilla Ann Haines, Nov. 24, 1929; children: Peter, Thomas, Paul, Peggy, Priscilla, Polly. B.A., St. John's U., 1950; M.S., U. Minn., 1959. Research assoc. Sterling-Winthrop Research Inst., Rensselaer, N.Y., 1954-58; sr. research pharmacologist Baxter-Travenol Labs., Inc., Morton Grove, Ill., 1958-72, mgr. safety evaluation, 1972-77; dir. toxicology Standard Oil Co. (Ind.), Chgo., 1977—. Mem. Soc. Toxicology, Am. Soc. Pharmacology and Exptl. Therapeutics, Am. Indsl. Hygiene Assn., AAAS, European Soc. Toxicology, N.Y. Acad. Sci. Subspecialties: Toxicology (medicine); Cancer research (medicine). Current work: Toxicology of petroleum and petrochemical products and processes. Home: 309 N Wille St Mount Prospect IL 60056 Office: Standard Oil Co 200 E Randolph Dr MC 4901 Chicago IL 60601

GARWIN, RICHARD LAWRENCE, physicist; b. Cleve., Apr. 19, 1928; s. Robert and Leona (Schwartz) G.; m. Lois Levy, Apr. 20, 1947; children: Jeffrey L., Thomas M., Laura J. B.S. in Physics, Case Inst. Tech., 1947, Ph.D., U. Chgo., 1949. Mem. faculty U. Chgo., 1949-52; with IBM, 1952—; dir. applied research IBM Thomas J. Watson Research Center, Yorktown Heights, N.Y., 1965-66; dir. IBM Watson Lab., Columbia U., 1966-67; fellow Thomas J. Watson Research Center, 1967—, mem. corp. tech. com., 1970-71; prof. public policy Kennedy Sch. Govt., Harvard U., 1979-81; Andrew D. White prof.-at-large Cornell U., 1983—; adj. prof. Columbia U.; adj. research fellow Harvard U., 1982—; congl. witness on nat. security, transp., energy policy, tech; cons. to industry and govt.; mem. Pres.'s Sci. Adv. Com., 1962-65, 69-72; mem. Def. Sci. Bd., 1966-69. Co-author: Nuclear Weapons and World Politics, 1977, Nuclear Power Issues and Choices, 1977, Securing the Seas, 1979, Energy: The Next Twenty Years, 1979, The Dangers of Nuclear War, 1979, Nuclear Energy and Nuclear Weapons Proliferation, 1979, The Genesis of New Weapons, 1980, Science Advice to the President, 1980; contbr. numerous articles to profl. publs. Ford Found. fellow CERN, 1959-60. Fellow Am. Phys. Soc. (chmn. panel on public affairs 1978-79), Am. Acad. Arts and Scis.; mem. Nat. Acad. Scis. (council 1983-85), Inst. Medicine, Nat. Acad. Engring., Council Fgn. Relations, Inst. for Strategic Studies (London) (council), Am. Philos. Soc. Jewish. Subspecialty: Applied Physics. Patentee in field. Office: PO Box 218 Yorktown Heights NY 10598

GARWOOD, VERNON ABINGTON, geneticist, animal breeder, educator; b. Carroll, Nebr., Oct. 29, 1924; s. Rodney Vernon and Helen Marie (Abington) G.; m. Elaine Lawton, Jan. 19, 1953 (div. 1976); children: Douglas, Laurie, Bradley, Jennifer, Blaine, Amy; m. Alma Margaret Linton, July 20, 1979. B.Sc., U. Nebr., 1950, M.Sc., 1954, Ph.D., 1956. Asst. animal husbandman U. Nebr., Lincoln, 1954-56; asst. prof. animal husbandry Purdue U., West Lafayette, Ind., 1956-59, assoc. prof., 1959-62, 64-70, Rural Estado de Minas Gerais, Brazil, 1962-64; animal geneticist Agrl. Research Service, U.S. Dept. Agr., West Lafayette, 1970-83, Poultry Research Lab., Agr., Georgetown, Del., 1983—; cons. in swine breeding. Served with U.S. Navy, 1943-45. Mem. Poultry Sci. Assn., Animal Sci. Soc., Genetics Soc. Am. Subspecialties: Animal genetics; Animal breeding and embryo transplants. Current work: Poultry breeding, fat metabolism.

GARZIA, MARIO RICARDO, mathematician; b. Buenos Aires, Mar. 6, 1955; US, 1967; s. Ricardo Francisco and Julia Elisa (Berrud) G.; m. Marjorie Ann Wilson, Oct. 3, 1975; 1 son, Daniel Ricardo. B.S. in Math, U. Akron, 1975; M.S., 1977; Ph.D. in Systems Engring, Case Western Reserve U., 1982. Grad. teaching asst. dept. math. sci. U. Akron, 1975-76; sci. programmer Babcock & Wilcox Co., Barberton, Ohio, 1976-82; mem. tech. staff Bell Labs., Holmdel, N.J., 1982—. Contbr. articles in field to profl. jours. Mem. IEEE, Soc. Indsl. & Applied Math., Am. Math. Soc., Pi Mu Epsilon. Subspecialties: Applied mathematics; Systems engineering. Current work: Computer simulation, mathematical modeling and geometric systems theory. Home: 126-1000 Oaks Dr Atlantic Highlands NJ 07716 Office: Bell Labs WB1K207 Holmdel NJ 07733

GARZIA, RICARDO FRANCISCO, computer company executive; b. Lomas de Zamora, Argentina, Sept. 19, 1926; came to U.S., 1967; s. Mario Francisco and Zulema (Alvarez) G.; m. Julia E. Berrud, Oct. 2, 1948; children: Liliana Julia, Silvia Cristina, Mario Ricardo, Fernando Marcelo. B.S.EE., Otto Krause Sch., Argentina, 1945; M.S.EE., La Plata (Argentina) U., 1950. Prof. Nat. Indsl. Sch., Buenos Aires, 1950-53; prof. electricity Nat. Tech. U., Buenos Aires, 1954-67, chmn. dept., dir. computer ctr., 1954-67; leader computer applications Gen. Dynamics, Rochester, N.Y., 1967-69; sr. computer scientist Computer Sci. Corp., Huntsville, Ala., 1969-71; mgr. tech. applications Babcock & Wilcox Co., Barberton, Ohio, 1971-83, sr. engring. cons., 1983—. Author: Transformada Z, 1966, Introduccion a la Computacion Digital, 1968; contbr. chpts. to books in field. Mem. IEEE (chmn. tech. com. 1983—), Instrument Soc. Am. (editor jour. 1975-78), Ops. Research Soc. Roman Catholic. Subspecialties: Algorithms; Mathematical software. Current work: Mathematical modeling and simulation, continuous and discrete, control systems. Home: 509 Vosello Ave Akron OH 44313 Office: Babcock & Wilcox Co 4282 Strausser St NW North Canton OH 44720

GASH, DON MARSHALL, neuroscientist; b. Harrodsburg, Ky., Aug. 27, 1945; s. Donald and Adeline (Phillips) G.; m. Sharon Lee Whisler, Aug. 17, 1967; children: Sheryl, Clelland. B.S., U. Ky., 1967; Ph.D., Dartmouth Coll., 1975. Postdoctoral fellow U. So. Calif., Los Angeles, 1975-77; asst. prof. U. Rochester, N.Y., 1977-81, asso. prof. dept. anatomy, 1981—. Contbr. articles to sci. jours. Served to capt. USAF, 1968-71. Recipient John W. Oswald award U. Ky., 1967. Mem. Am. Assn. Anatomists, AAAS, Soc. Neurosci. Democrat. Baptist. Subspecialties: Neurobiology; Neuroendocrinology. Current work: Developmental neurobiology; development, structure and function of transplanted neurons; Vasopressin and hypothalamic neurosecretory system.

GASPARRINI, WILLIAM GERARD, clinical psychologist; b. Greenwich, Conn., Oct. 25, 1951; s. Joseph William and Rose Marie (Cianci) G.; m. Martha Sue Dodd, May 24, 1981. B.A., Lehigh U., 1972, M.A., 1973; Ph.D., U. Fla., 1977. Lic. psychologist, Miss. Psychology trainee VA Hosp., Gainesville, Fla., 1974-76; teaching asst. U. Fla., Gainesville, 1973-74; psychology intern VA Hosp., New Orleans, 1976-77, staff psychologist, Gulfport, Miss., 1978-81; adj. faculty U. So. Miss., Long Beach, 1978-83; pvt. practice clin. psychology, Gulfport, 1983—. Contbr. articles to profl. jours. Mem. Am. Psychol. Assn., Southeastern Psychol. Assn., Miss. Psychol. Assn., Internat. Neuropsychol. Assn., Internat. Council Psychologists, Phi Beta Kappa. Subspecialties: Neuropsychology; Behavioral psychology. Current work: Neuropsychology, biofeedback, behavioral medicine, personality assessment, behavior therpay, children's temperament, oncology. Home: 153 Markham Dr Gulfport MS 39501

GASS, GEORGE HIRAM, academic administrator, medical educator; b. Sunbury, Pa., Sept. 23, 1924; s. George Calvin and Lillian May (Stahl) G.; m. Dorothy Kinsey, June 12, 1948; children: Nancy E., George David, Patricia Ann. B.S., Bucknell U., 1948; M.S., U. N. Mex., 1952; Ph.D., Ohio State U., 1955. Research biologist Lederle Labs., Pearl River, N.Y., 1948-50; pharmacologist FDA Washington, 1955-59, br. chief div. pharmacology, 1957-59; assoc. prof. physiology So. Ill. U., Carbondale, 1959-62, prof. pharmacology, 1962-83, dir. endocrinologic pharm. research, 1960-79; prof. basic scis., chmn. dept. basic scis. Okla. Coll. Osteo. Medicine, 1979—. Editor, contbg. author: Handbook of Endocrinology, 1983. Alexander von Humboldt Found. sr. research fellow, Munich, W.Ger., 1967-68, 79; Fulbright fellow, Germany, Denmark, 1972. Fellow AAAS; mem. Ill. State Acad. Sci. (v.p. 1978-79, editorial bd. 1983—), Endocrine Soc., Am. Physiologist Soc., Soc. Study Reprodn. Lodge: Eagles. Subspecialties: Endocrinology; Cancer research (medicine). Current work: Oncogenic agents. Inventor animal care products, including metabolism cages, 1968-78. Home: 602 W 32d St Sand Springs OK 74063 Office: Okla Coll Osteo Medicine 1111 W 17th St Tulsa OK 74107

GASSMAN, MERRILL LOREN, biological sciences educator; b. Chgo., Feb. 10, 1943; s. Alfred E. and Elvina (Chessen) G.; m. Beverly Sue Sacks, Sept. 3, 1967; children: Debra, Sharyl, Aaron. B.S. in Biology, U. Chgo., 1964, M.S. in Botany, 1965, Ph.D., 1967. Guest investigator Rockefeller U., N.Y.C., 1967-68; research plant physiologist Internat. Minerals and Chems. Co., Libertyville, Ill., 1968-69; asst. prof. U. Ill., Chgo., 1969-75, assoc. prof., 1975-82, prof. biol. scis., 1982—; vis. assoc. research botanist U. Calif.-Davis, 1976; vis. prof. botany and microbiology Ariz. State U.-Tempe, 1983; content cons. Coronet Films, Chgo., 1973-80; ad hoc reviewer NSF, Washington, 1973—. Contbr. articles on plant physiology and biochemistry to profl. jours.; mem. editorial bd.: Am. Soc. Plant Physiologists, 1983. Pres. Edgebrook PTA, Chgo., 1981—. NSF grantee, 1972, 75, 79, 82. Mem. AAAS, Am. Soc. for Photobiology (charter mem.), Am. Soc. Plant Physiologists, Am. Soc. Biol. Chemists, Sigma Xi. Subspecialties: Plant physiology (biology); Biochemistry (biology). Current work: Biochemical studies on the formation of chlorophyll in plants. Home: 5504 W Lunt Ave Chicago IL 60646 Office: U Ill at Chgo PO Box 4348 Chicago IL 60680

GASSMAN, PAUL GEORGE, chemistry educator; b. Alden, N.Y., June 22, 1935; s. Joseph Martin and Florence Marie (Rautenstrauch) G.; m. Gerda Ann Rozler, Aug. 17, 1957; children: Deborah, Michael, Vicki, Nancy, Amy, Kimberly, Eric. B.S., Canisius Coll., 1957; Ph.D., Cornell U., 1960. Asst. prof. Ohio State U., Columbus, 1961-66, assoc. prof., 1966-69, prof., 1969-74; prof. dept. chemistry U. Minn., Mpls., 1974—, chmn. dept., 1975-79. Editor books in field; contbr. articles to profl. jours.; editor bd.: Reviews of Chem. Intermediates, 1978—. Recipient James R. Crowdle Disting. Alumni award Canisius Coll., 1971; Alfred P. Sloan Found fellow, 1967-69; Japan Soc. for Promotion of Sci. fellow, 1981. Fellow AAAS; mem. Am. Chem. Soc. (award in petroleum chemistry 1972, chmn. Columbus sect. 1970, nat. councilor 1971-74, chmn. organic div. 1980), Chem. Soc. (London), Am. Inst. Chemists. Roman Catholic. Subspecialties: Organic chemistry; Catalysis chemistry. Current work: Organic chemistry: catalysis chemistry, photochemistry, organoelectrochemistry, and synthetic chemistry. Patentee. Office: University of Minnesota 207 Pleasant St SE Minneapolis MN 55455

GASTORF, JOHN WAYNE, psychologist, consultant; b. St. Louis, Mar. 27, 1944; s. Jack Charles and Geraldine Nina (Taylor) G.; m. Janice L. Burton, 1967; children: Jeffrey, Jason, Jennifer. B.A., U. Mo.-St. Louis, 1974; M.A., So. Ill. U., 1976; Ph.D., SUNY-Albany, 1979. Dir. rehab. medicine St. Vincent's Hosp., St. Louis, 1969-76; statis. cons. N.Y. State Edn. Dept., Albany, 1977-78; instr. SUNY-Albany, 1979; asst. prof. Coll. Community Health Services U. Ala., 1979-81, asst. prof., 1981; dir. behavioral sci. Tulsa Med. Coll., 1981—; cons. Jenks (Okla.) Med. Ctr., 1982—. Author: Family Violence: Social Learning Theory Explanation, 1980; contbr. articles to profl. jours. Treas. bd. dirs. Spouse Abuse Shelter, Tuscaloosa, Ala., 1979-82; mem. Mayor's Adv. Council on Alcohol, Tulsa, 1981—. Served with U.S. Army, 1966-69. USPHS grantee, 1979, 80. Mem. Am. Psychol. Assn., Am. Acad. Behavioral Medicine, Soc. Tchrs. Family Medicine, Soc. Personality and Social Psychology, Assn. Behavioral Scis. and Med. Educators, Assn. Med. Rehab. Dirs. and Coordinators. Subspecialties: Social psychology; Health psychology. Current work: Type a coronary prone behavior pattern; domestic violence. Home: 8817 E 134th St Bixby OK 74008 Office: Tulsa Medical College 2808 S Sheridan St Tulsa OK 74129

GAT, URI, nuclear researcher; b. Jerusalem, June 28, 1936; U.S., 1969; s. Werner and Jenny Lore Kastor (Goldschmidt) Hagelberg; m. Ruth Tasse, July 24, 1961; children: Erann, Irit. B.Sc., Israel Inst. Tech., Haifa, 1963; D. Ing., Tech. U., Aachen, Germany, 1969. Registered profl. engr., Ky. Research assoc. KFA, Juelich, Germany, 1963-69; asst. prof. U. Ky., Lexington, 1963-74; mgr. gas-cooled fast reactor Oak Ridge Nat. Lab., 1974-80, mgr. fast breeder reactor, 1980-82, group leader research and devel., 1982—. Contbr. articles to profl. jours. Vice pres. Oak Ridge chpt. ACLU, 1981-82. Mem. Am. Nuclear Soc., U.S. Metric Assn., Am. Soc. Engring. Mgmt., Sigma Xi. Subspecialties: Nuclear fission; Mechanical engineering. Current work: Development and safety of nuclear reactors. Breeder reactors and energy supply. Patentee in field. Home: 238 Gum Hollow Rd Oak Ridge TN 37830 Office: Oak Ridge Nat Lab PO Box Y 9108 MS-2 Oak Ridge TN 37830

GATELY, MAURICE KENT, research immunologist; b. Omaha, Feb. 3, 1946; s. Harold Stephen and Alya Marie (Witt) G.; m. Celia Lin, July 8, 1972; children: Lynn Christine, Mark Stephen. Student, Creighton U., 1964-66; B.A., Johns Hopkins U., 1968, Ph.D. in Microbiology, 1974, M.D., 1975. Resident in pediatrics St. Louis Children's Hosp., 1975-76; research fellow in pathology Harvard U. Med. Sch., Boston, 1976-79; sr. staff fellow NIH, Bethesda, Md., 1979—; chief cellular immunology unit surg. neurology br. Nat. Inst. Neurol. and Communicative Disorders and Stroke, 1980—. Contbr. articles to profl. jours. Helen Hay Whitney Found. fellow, 1976-79. Mem. Am. Assn. Immunologists, AAAS, N.Y. Acad. Scis. Presbyterian. Subspecialty: Immunobiology and immunology. Current work: Cellular immune response to brain tumors; immunotherapy of brain tumors; mechanisms of lymphocyte-mediated tumor cell lysis; T-lymphocyte cloning and growth in vitro; lymphokines. Office: NIH Bldg 9 Rm 1W121 Bethesda MD 20205

GATES, DAVID MURRAY, botany educator; b. Manhattan, Kans., May 27, 1921; s. Frank Caleb and Margaret Henry (Thompson) G.; m. Marian Francis Penley, June 4, 1944; children: Murray Penley, Julie Mary, Heather Margaret, Marilyn Joan. B.S., U. Mich., 1942, M.S., 1944, Ph.D., 1948. Faculty, U. Denver, 1947-57; sci. dir. Office Naval Research, Am. embassy, London, 1955-57; cons. to dir., asst. chief upper atmosphere and space physics div. Boulder Labs., Nat. Bur. Standards, Colo., 1957-64; prof. natural history U. Colo., 1964-65; prof. biology Washington U., dir. Mo. Bot. Garden, St. Louis, 1965-71; prof. botany U. Mich.; dir. Biol. Sta., Ann Arbor, 1971—; dir. Detroit Edison Corp.; mem. nat. sci. bd. NSF, 1970-76; chmn. environ. studies bd. Nat. Acad. Scis. and Nat. Acad. Engring., 1970-73; mem. panel sci. and tech. Com. Sci. and Astronautics, U.S. Ho. of Reps., 1970-74. Author: Energy Exchange in the Biosphere, 1962, Atlas of Energy Exchange for Plant Leaves, 1971, Man and His Environment: Climate, 1972, Perspectives of Biophysical Ecology, 1975, Biophysical Ecology, 1980; contbr. numerous articles to profl. jours. Bd. dirs. Conservation Found., Washington, 1970—, Nat. Audubon Soc., 1972-78, Cranbrook Inst. Sci. Recipient Gold Seal award Nat. Council State Garden Clubs, 1971; Disting. Faculty Achievement award U. Mich., 1982. Mem. Am. Inst. Biol. Scis. (dir. 1970-76, pres. 1975, Outstanding Achievement in Biolclimatology award), Am. Meteorol. Soc. Club: Cosmos (Washington). Subspecialty: Ecology. Home: 442 Huntington Pl Ann Arbor MI 48104

GATES, FREDERICK TAYLOR, III, research biochemist; b. Boston, Apr. 5, 1949; s. Frederick Taylor and Helen Elizabeth (Swift) G.; m. Susan Julian, Dec. 6, 1975; children: John Taylor, Alan Russell. B.S., Yale U., 1971; Ph.D., U. Calif.-Berkeley, 1976. Postdoctoral fellow Rockefeller U., 1976-77; staff fellow NIH, Bethesda, 1977-80; sr. staff fellow Office Biologics, FDA, Bethesda, 1980-83, research chemist, 1983—. Mem. Harvey Soc., Am. Assn. Immunologists. Subspecialties: Genetics and genetic engineering (biology); Immunobiology and immunology. Current work: Genetic basis for antibody diversity; structure and genetics of histocompatability antigens; gene regulation; cloning and characterization of genes for oncofetal antigens. Home: 2915 McComas Ave Kensington MD 20895 Office: Office of Biologics FDA Bldg 29 8800 Rockville Pike Bethesda MD 20215

GATES, JOSEPH SPENCER, geologist; b. Des Moines, Jan. 18, 1935; s. Leslie Dean and Ella Wheeler (Hicks) G.; m. Constance Maree Clemens, May 11, 1964; children: Jennifer A., Karin M. Geol. E., Colo. Sch. Mines, 1956; M.S., U. Utah, 1960; Ph.D., U. Ariz., 1972.

Engr. U.S. Geol. Survey, Denver, 1956-58; geologist, Salt Lake City, 1958-64; tech. adv. AID, Cairo, 1965-67; hydrologist, project chief U.S. Geol. Survey, El Paso, 1971-77, chief investigation sect., Salt Lake City, 1977—; chmn. 11th Rocky Mountain Ground Water Conf., Salt Lake City, 1982. Treas. South Valley Unitarian Soc., Salt Lake City, 1981—. Served to capt. USAR, 1956-67. Fellow Geol. Soc. Am.; mem. Am. Geophys. Union, Utah Geol. Assn. (program chmn. 1980), AAAS. Subspecialties: Ground water hydrology; Hydrogeology. Current work: Ground water hydrology of the Great Basin, modeling of ground water basins, use of geophysics in ground water studies. Home: 2560 Cavalier Dr Salt Lake City UT 84121 Office: US Geol Survey 1745 W 17th South St Salt Lake City UT 84104

GATES, MARSHALL DEMOTTE, JR., chemistry educator emeritus; b. Boyne City, Mich., Sept. 25, 1915; s. Marshall DeMotte and Virginia (Orton) G.; m. Martha Louise Meyer, Sept. 9, 1941; children—Christopher David, Catharine Louise, Marshall DeMotte III, Virginia Alice. B.S., Rice Inst., 1936, M.S., 1938; Ph.D., Harvard, 1941; D.Sc. (hon.), MacMurray Coll., 1963. Asst. prof. chemistry Bryn Mawr Coll., 1941-43; vis. prof. Harvard, 1946; asso. prof., 1947-49, Max Tishler lectr., 1953; tech. aid NDRC, 1943-46; lectr. chemistry U. Rochester, 1949-52, part-time prof., 1952-60, prof., 1960-68, Charles Frederick Houghton prof. chemistry, 1968-81, prof. emeritus, 1981—; Welch Found. lectr., 1960; adv. bd. Chem. Abstracts Services, 1974-76; vis. prof. Dartmouth Coll., 1982. Mem. com. on drug addiction and narcotics, div. med. scis. NRC, 1956-70, also com. on organic nomenclature div. of chemistry; mem. Pres.'s Com. on Nat. Medal of Sci., 1968-70. Recipient Edward Peck Curtis award for excellence in undergrad. teaching, 1967; Armed Services certificate Appreciation, 1946. Fellow Am. Acad. Arts and Scis., N.Y. Acad. Scis.; mem. Am. Chem. Soc. (editor Jour. 1963-69), Nat. Acad. Scis. Subspecialties: Organic chemistry; Synthetic chemistry. Current work: Synthesis of narcotics and narcotic antagonist; chemistry of opium alkaloids. Home: 41 W Brook Rd Pittsford NY 14534 Office: U Rochester Rochester NY 14627

GATES, WILLIAM LAWRENCE, meteorologist, educator; b. South Pasadena, Calif., Sept. 14, 1928; m., 1951; 3 children. S.B., M.I.T., 1950, S.M., 1951, Sc.D., 1955. Asst. M.I.T., 1950-53; research meteorologist Air Force Cambridge Research Ctr., 1953-57; asst. prof. meteorology UCLA, 1957-59, assoc. prof., 1959-66; research scientist Rand Corp., 1966-76; prof. atmospheric sci., chmn. dept. atmospheric sci. Oreg. State U., 1976—. Mem. Am. Meteorol. Soc., Am. Geophys. Union, Royal Meteorol. Soc. Subspecialties: Meteorology; Climatology. Office: Dept Atmospheric Sci Oreg State U Corvallis OR 97331

GATEWOOD, GEORGE DAVID, astronomer; b. St. Petersburg, Fla., May 10, 1940; s. George Harry and Virginia V. G.; m. Carolyn Virginia Scott, Mar. 10, 1959; children: Sara, Ann. B.A., U. South Fla., 1965, M.A., 1968; Ph.D., U. Pitts., 1972. Asst. prof. U. Pitts., 1972-77, assoc. prof., 1977—; dir. Allegheny Obs., 1977—; cons. NASA. Contbr. articles to profl. jours. NSF grantee, 1974—; NASA grantee, 1975—. Mem. Am. Astron. Soc., Internat. Astron. Union. Subspecialties: Astrometry; Computer interfacing. Current work: Determination of stellar distances and masses. Office: Allegheny Obs Pittsburgh PA 15214

GATEWOOD, LAEL CRANMER, health computer scientist, biometrist, university administrator, educator; b. Cleve., Nov. 16, 1938; m., 1961; 2 children. B.A., Rockford Coll., 1959; M.S., U. Minn., 1966, Ph.D. in Biometry, 1971. Technician biochem. research Mayo Clinic, 1959-61, technician in biophysics, 1962-67; scientist health computer sci. and biometry U. Minn., Mpls., 1967-68, asst. prof. lab. medicine and biometry, asst. dir. div. health computer sci., 1971-74, sr. assoc. dir. div. health computer sci., 1974-79, dir. div. health computer sci., 1979—, assoc. prof. lab. med. pathology and biometry, 1974—. Mem. Am. Pub. Health Assn., AAAS, N.Y. Acad. Sci., Am. Assn. Med. Systems Informatics, Assn. Computing Machinery. Subspecialty: Biomedical computation. Address: 4932 Stevens Minneapolis MN 55409

GATHERUM, GORDON ELWOOD, forestry educator; b. Salt Lake City, Oct. 22, 1923; s. James Elwood and Jessie Margaret (Robertson) G.; m. Patricia Jeanne Brandley, July 31, 1947; children: Laurie Patricia, Mark Gordon, Kristin Lee. B.S. in Forest Mgmt., U. Wash., 1949; M.S. in Range Mgmt., Utah State U., 1951; Ph.D., Iowa State U., 1959. Mem. faculty dept. forestry Iowa State U., 1953-69; chmn. dept. forestry Ohio State U., also Ohio Agrl. Research and Devel. Center, 1969-75; dir. Sch. Natural Resources, asso. dean Coll. Agr., Ohio State U. and Ohio Agrl. Research and Devel. Center, 1975—; AID cons., Brazil. Contbr. articles to profl. jours. NSF grantee. Mem. Nat. Assn. Profl. Forestry Schs. and Colls. (regional chmn.), Soc. Am. Foresters, Sigma Xi, Phi Eta Sigma, Xi Sigma Pi. Subspecialties: Plant physiology (agriculture); Resource management. Current work: The interactions of genetic and enviromental factors on metabolic processes underlying forest tree production. Home: 5710 Strathmore Ln Dublin OH 43017

GATROUSIS, CHRISTOPHER, chemist; b. Norwich, Conn., Oct. 8, 1928; s. George John and Irene (Romeliotou) G.; m. Patricia O'Brien, May 16, 1951; 1 son, John F. B.S., De Paul U., 1956; M.S., U. Chgo., 1960; Ph.D., Clark U., 1965. Research assoc. Argonne (Ill.) Nat. Lab., 1956-61; asst. scientist Woods Hole (Mass.) Oceanographic Inst., 1964-66; chemist Lawrence Radiation Lab., Livermore, Calif., 1966-71, Lawrence Livermore Lab., Livermore, 1971-72, asst. div. leader, 1972-73, assoc. div. leader, 1973-74, dep. div. leader, 1974-77, leader nuclear chemistry div., 1977—. Author: (with R.R. Heinrich and C.E. Crouthamel) Progress in Nuclear Energy, 1961. Chmn. Livermore Beautification Com., 1969-71, Livermore Design Rev. Study Com., 1970-71. Served with USMCR, 1950-52. AEC grantee, 1972-73. Mem. Am. Phys. Soc., Am. Chem. Soc. Subspecialties: Nuclear physics; Cosmology. Current work: Heavy-ion reactions; isotope geochemistry and cosmochronology; director basic and applied research. Office: Lawrence Livermore National Laboratory PO Box 808 Livermore CA 94550

GATTO, LOUIS ALBERT, biological sciences educator. M.S., Fordham U., 1974, Ph.D., 1978. Assoc. prof. biol. scis. SUNY-Cortland, 1978—. Contbr. articles to profl. jours. NSF grantee; SUNY Found. grantee. Mem. Am. Physiol. Soc., Am. Soc. Zoologists, Am. Micros. Soc., AAAS. Subspecialty: Physiology (biology). Current work: Physiology of mucociliary function in respiratory airways. Office: Dept Biol Scis SUNY Corland NY 13045

GATZ, EDWARD ERWIN, anesthesiologist, clinical pharmacologist; b. O'Neill, Nebr., Apr. 19, 1937; s. Edward Erwin and Mable Helen (Morton) G.; m. Helen Jeanne Gurnett, June 23, 1962; 1 son, Bart Gerard. B.S., Creighton U., Omaha, 1961; M.S., U. Colo., 1964; Ph.D., U. Nebr., 1968, M.D., 1974. Intern U. Nebr. Med. Center, 1974, resident in anesthesiology, 1974-77; research asst. U. Colo., 1962-64, research assoc., 1964-65; asst. inst. U. Nebr. Coll. Medicine, Omaha, 1965-68, instr., 1968-69, asst. prof., 1969-71; cons. FDA, USPHS, Dept. HEW, 1972-76; mem. staff Bergan Mercy Hosp., Omaha, 1977—. Contbr. articles to profl. jours. Asst. scout master Mid-Am. council Boy Scouts Am., Omaha, 1981—. Served with U.S. Army, 1955-57. McNeil Labs., Inc. research grantee, 1968-71; Ayerst Labs. research grantee, 1970-75. Mem. Internat. Anesthesia Research Soc., Biochem. Soc., Am. Fedn. Clin. Research, AMA, Nebr. Med. Assn., N.Y. Acad. Scis., Am. Coll. Clin. Pharmacology, Am. Soc. Anesthesiology, Nebr. Soc. Anesthesiology, Greater Omaha Met. Med. Soc. (mem. exec. com.), Sigma Xi, Phi Chi. Roman Catholic. Subspecialties: Anesthesiology; Pharmacology. Current work: Malignant hyperthermia; human stress syndrome; oxidative phosphorylation; clinical anesthesiology. Home: 10029 Frederick St Omaha NE 68124 Office: Bergan Mercy Hosp 7500 Mercy Rd Omaha NE 68124

GAULT, CHARLES S., research physicist; b. Harrisburg., Pa., July 18, 1943; s. Sheldon B. and Hellen I. (Pick) G.; m. Marcia Herring, June 19, 1965; children: Brian, Emily. A.B., Gettysburg (Pa.) Coll., 1965; M.S., U. Md., 1969; Ph.D., Am. U., Washington, 1973. Physicist SPARCOM, Alexandria, Va., 1972-74; research physicist Nat. Security Agy., Ft. Meade, Md., 1974-82; physicist Ford Aerospace and Communication Corp., Hanover, Md., 1982—. Contbr. articles to profl. jours. Mem. Optical Soc. Am. Lutheran. Subspecialties: Systems engineering; Optical signal processing. Current work: Fiber optics applications, systems integration engineering. Office: 7235 Standard Dr Hanover MD 21076

GAULT, N.L., JR., physician, educator; b. Austin, Tex., Aug. 22, 1920; s. N.L. and Pauline (Johnson) G.; m. Sarah Jane Dickie, June 28, 1947; children—Elizabeth Jean, John Dickie, Paul Alan. Student, U. Tex., 1938-42, Baylor U. Med. Sch., 1946-48; B.A., U. Tex., 1950; M.B., U. Minn., 1950, M.D., 1951, student Grad. Sch., 1951-54. Intern Mpls. Gen. Hosp., 1950-51; resident internal medicine Mpls. VA Hosp., 1951-52; chief resident internal medicine Ancker Hosp., St. Paul, 1952, U. Minn. Hosp., 1953-54; faculty U. Minn. Med. Sch., Mpls., 1953-67, 72—, asso. prof. internal medicine, asso. dean, 1962-67, prof. medicine, dean, 1972—; prof. medicine U. Hawaii, asso. dean, 1967-72; chief adviser Seoul (Korea) Nat. U. Coll. Medicine, 1959-61; med. edn. cons. China Med. Board, N.Y.C., 1963, 71, AID, 1964-68; dir. postgrad. med. edn. program for, Ryukyu Islands, 1967-69; cons. Mpls. VA Hosp., 1956-67. Sec.-treas. Minn. Med. Found., 1956-67. Served to capt. AUS, USAAF, 1942-46. Decorated Commendation medal.; Recipient Supreme award Japan Med. Assn., 1969. Mem. AMA, Assn. Am. Med. Colls. (exec. council 1974-80, chmn. council deans MidWest-Gt. Plains region 1974-76), Minn. Med. Assn., Hennepin County Med. Soc. Subspecialties: Internal medicine; Medical Education. Address: Medical Sch U Minn Minneapolis MN 55455

GAUT, ZANE NOEL, physician, medical research administrator; b. Nauvoo, Ala, Aug. 29, 1929; s. Noel and Gladys (Odom) G.; m. Laura Tarence, June 25, 1955; children: Douglas, Julie, David. B.S., Birmingham-So. Coll., Ala., 1950; M.D., Tulane U., 1954; Ph.D. in Biochemistry, 1964; postgrad., Oak Ridge (Tenn.) Inst. Nuclear studies, 1966. Intern Vanderbilt U. affiliated St. Thomas Hosp., Nashville, 1955; flight surgeon and examiner Gen. Dynamics Corp., Fort Worth, 1958-60; asst. prof. depts. biochemistry and medicine Tulane U, New Orleans, 1961-66; mem. attending and teaching staff dept. medicine Martland Med. Center, Newark, 1966-71; staff physician, 1966-71; assoc. dept. clin. pharmacology Hoffmann-La Roche Inc., Nutley, N.J., 1966-71, sect. head dept. biochem. nutrition, 1971-76, asst. dir. dept. clin. pharmacology, 1976-78, dir. clin. research/endocrinology-metabolism, dept. med. research, 1978—; mem. attending staff Newark Presbyn. and Newark Beth Israel Hosps., 1966-69, Newark Beth-Israel Hosp., 1971—; asst. attending staff St. Luke's Hosp., N.Y.C., 1972—; mem. attending staff East Orange VA Hosp., N.J., 1969—; clin. asst. prof. N.J. Coll. Medicine and Dentistry, Newark, 1967-72; asst. clin. prof. Columbia U. Coll. Physicians and Surgeons, 1972—. Contbr. numerous articles on pharmacology to profl. jours. Served to lt. USN, 1955-58. Fellow Am. Coll. Clin. Pharmacology, Sci. Council of Internat. Coll. Angiology; mem. AAAS, Am. Diabetes Assn., Am. Fedn. for Clin. Research, Am. Inst. Nutrition, AMA, Am. Soc. Clin. Nutrition, Am. Soc. for Clin. Pharmacology and Therapeutics, Essex County Med. Soc., Med. Soc. of N.J., N.J. acad. Medicine, N.Y. Acad. Scis., Royal Soc. Health, Am. Med. Writers Assn., Am. Chem. Soc., Sigma Xi. Subspecialties: Endocrinology; Nutrition (medicine). Office: Hoffmann-La Roche Inc 340 Kingsland St Nutley NJ 07110

GAUTIERI, RONALD FRANCIS, pharmacologist, educator; b. Providence, Oct. 10, 1933; s. Emilio and Frances (Amalfitano) G.; m. Bernadette Howell, Jan. 20, 1962; 1 dau., Anne Marie. B.S., R.I. Coll. Pharmacy, 1955; M.S., Temple U., 1957, Ph.D., 1960. Asst. prof. pharmacology Temple U. Sch. Pharmacy, Phila., 1960-66, asso. prof., 1966-70, prof., 1970—, chmn. dept., 1971—. Contbr. articles to sci. jours. Co-chmn. Cheltenham Democratic party, 1981-82. NIH grantee. Mem. Am. Pharmac. Assn., Am. Soc. Pharmacology and Exptl. Therapeutics, Sigma Xi, Rho Chi. Roman Catholic. Subspecialties: Pharmacology; Teratology. Current work: Birth defects; animal models; hypertensive screening; placental perfusion. Home: 418 Bolton Rd Glenside PA 19038 Office: 3307 N Broad St Philadelphia PA 19140

GAUTREAUX, MARCELIAN FRANCIS, JR., chem. co. exec.; b. Nashville, Jan. 17, 1930; s. Marcelian Francis and Mary Eunice (Terrebonne) G.; m. Mignon Alice Thomas, Apr. 26, 1952; children—Marcelian, Marian, Kevin, Andrée. B.S.Ch.E. magna cum laude, La. State U., 1950, M.S.Ch.E., 1951, Ph.D. in Chem. Engring, 1958. With Ethyl Corp., Baton Rouge, 1951-55, 58—, gen. mgr. dept. research and devel., 1968-69, v.p., 1969-74, sr. v.p., 1974—, advisor to exec. com., 1981—, also dir.; instr. La. State U., 1955-56, asst. prof. chem. engring., 1956-58. Bd. dirs. Baton Rouge Community Concerts Assn., 1974—, pres., 1981—; trustee La. Arts and Sci. Center, Baton Rouge, 1974-77; mem. La. State U. Found.; chmn. adv. com. dept. engring. La. State U. Recipient (charter) Personal Achievement in Chem. Engring. award Chem. Engring. Mag., 1968; Charles E. Coates Meml. award Am. Chem. Soc./Am. Inst. Chem. Engrs., 1976; Ann. Meml. award Chem. Mktg. Research Assn., 1978; Best Paper award, 1980; charter mem. La. State U. Engring. Hall of Distinction, 1979. Fellow Am. Inst. Chem. Engrs. (Best Presented Paper award 1952); mem. Nat. Acad. Engring., Soc. Chem. Industry, Soc. Engring. Sci. (dir.). Roman Catholic. Clubs: Baton Rouge Country, Baton Rouge City, Baton Rouge Camelot. Subspecialty: Chemical engineering. Current work: Chemical processing, biotechnology, photovoltaics. Patentee and author in field. Home: 1662 Pollard Pkwy Baton Rouge LA 70808 Office: 451 Florida Blvd Baton Rouge LA 70801 Any successes I have had are no more or less than the composite result of a supportive and loving wife and children, professional associates who have never let me down, a corporation whose ethics are the highest, a religious heritage from my parents and early schooling, and some God-given talents for chemistry and engineering.

GAUTSCH, JAMES WILLARD, biologist, researcher; b. Rockford, Ill., Oct. 13, 1941; s. Joseph Arthur and Lucina Martha (Williams) G.; m. Elizabeth Sanford Clark, Aug. 12, 1968; children: Pollie Alisa, Anna Nell. B.A., Denver U., 1963; M.S., U. Wyo., 1968; Ph.D., U. Calif., Irvine, 1973. Postdoctoral fellow Jackson Lab., Bar Harbor, Maine, 1974-76; asst. mem. Research Inst. of Scripps Clinic, La Jolla, Calif., 1976—. Served with U.S. Army, 1966-68. Subspecialties: Molecular biology; Genetics and genetic engineering (biology). Current work: Mechanisms governing gene expression; recombinant DNA technology, teratocarcinoma stem cells, retrovirus, regulation of gene expression in preimplantation embryo cells. Office: 10666 N Torrey Pines Rd MB7 La Jolla CA 92037

GAUTSCHI, WALTER, mathematics educator; b. Basel, Switzerland, Dec. 11, 1927; came to U.S., 1955, naturalized, 1961; s. Hans and Margrit Eugster G.; m. Erika Wuest, Apr. 8, 1960; children: Thomas, Theresa, Doris, Caroline. Ph.D., U. Basel, 1953. Research mathematician Nat. Bur. Standards, Washington, 1956-69; mathematician Oak Ridge Nat. Lab., 1959-63; vis. prof. Tech. U. Munich, Germany, 1970-71, Math. Research Ctr., U. Wis.-Madison, 1976-77; prof. math. and computer scis. Purdue U., Lafayette, Ind., 1963—; cons. Argonne Nat. Lab., Ill., 1967-77. Assoc. editor: Math. of Computation, 1966—, Soc. Indsl. and Applied Math. Math. Analysis, 1970-73, Numerische Mathematik, 1971—, Calcolo, 1975—. Fulbright research scholar, Munich, 1970-71. Mem. Am. Math. Soc., Math. Assn. Am., Soc. Indsl. and Applied Math., Assn. Computing Machinery, Schweizerische Mathematische Gesellschaft. Subspecialties: Numerical analysis; Mathematical software. Current work: Applied orthogonal polynomials, gaussian quadrature, continued fractions, Padé approximation, special functions, numerical linear algebra, theory of condition. Office: Dept Computer Scis Purdue U Lafayette IN 47907

GAVELIS, JONAS RIMVYDAS, dentist, educator; b. Boston, Jan. 1, 1950; s. Mykolas and Janina (Povydis) G. B.S., U. Mass., Amherst, 1971; D.M.D., U. Conn., 1975. Resident in dentistry Cabrini Health Care Center, N.Y.C., 1975-76; fellow in prosthetic dentistry Harvard U. Sch. Dental Medicine, Boston, 1976-78, instr., 1978-79; asst. prof. U. Conn. Sch. Dental Medicine, Farmington, 1979-82; practice dentistry specializing in prosthodontics Harvard Community Health Plan, Boston, 1982—; asst. prof. Harvard Sch. Dental Medicine, 1982—. Contbr. articles on prosthetic dentistry to profl. jours. Fellow Acad. Gen. Dentistry (Vernon S. Johnson award 1981); mem. ADA, Northeast Prosthodontic Soc., Internat. Assn. Dental Research, Am. Assn. Dental Schs. Roman Catholic. Clubs: Scuba Shack Dive (Rocky Hill, Conn.) (treas. 1980-82); Southboro (Mass.); Rod and Gun; New Eng. Aquarium Dive (Boston). Subspecialties: Prosthodontics; Dental materials. Current work: Clinical application and biocompatibility of ceramics, gold-palladium based and base metal alloys. Home: 30 Hilltop Dr Millis MA Office: Harvard Community Health Plan 2 Fenway Plaza Boston MA 02215

GAVIN, JOSEPH GLEASON, JR., aerospace corporation executive, aeronautical engineer; b. Somerville, Mass., Sept. 20, 1920; s. Joseph Gleason and Elizabeth (Tay) G.; m. Dorothy Dunklee, Sept. 27, 1943; children: Joseph Gleason, III, Tay Gavin Erickson, Donald Lewis. S.B., MIT, S.M., 1942. With Propeller div. Curtiss-Wright Corp., Clifton, N.J., 1941; design engr. Grumman Aircraft Engring. Corp., Bethpage, N.Y., 1946-48, with preliminary design group, 1948-50, project engr., 1950-56, chief exptl. projects engr., 1956-57, chief missile and space engr., 1957-62; pres. Grumman Corp., 1962—; mem. Charles Stark Draper Lab., Inc.; dir. European Am. Bancorp. Mem. MIT Corp. vis. com. for dept. aero. and astron.; trustee Huntington (N.Y.) Hosp., Poly. Inst. N.Y. Served with U.S. Navy, 1942-46. Recipient Aerospace Ednl. Council Man of the Yr. award, 1968; C.W. Post Coll. Leadership award for mgmt., 1969; NASA Disting. Pub. Service medal, 1971; Poly. Inst. N.Y.-L.I. Tech. Leadership award, 1979. Fellow Am. Astron. Soc., AIAA; mem. Am. Assn. Engring. Scos., Nat. Acad. Engring. Subspecialties: Aeronautical engineering; Astronautics. Patentee in field. Home: 6 Endicott Dr Huntington LI NY 11743 Office: 1111 Stewart Ave Bethpage LI NY 11714

GAVINI, MURAL B., testing company executive, researcher; b. Guntur, A.P., India, July 1, 1947; came to U.S., 1973, naturalized, 1982; s. Suryanarayana and Mangamamba G.; m. Anuradha Ginjupalli, Aug. 15, 1977; children: Anita, Deepa, Rekha. B.Sc., A.C. Coll., Guntur, A.P., India, 1970; M.Sc., Andhra U., Waltair, A.P., India, 1972; Ph.D., U. Ark.-Fayetteville, 1976. Research asst. U. Ark., Fayetteville, 1973-76; postdoctoral investigator Woods Hole Oceanographic Instn., Mass., 1976-79; mgr. Radiation Mgmt. Corp., Phila., 1980; tech. contract adminstr. U.S. Testing Co., Inc., Richland, Wash., 1981, v.p., tech. dir., 1982-83, v.p., gen. mgr., 1983—. Mem. Am. Chem. Soc., Am. Nuclear Soc., Health Physics Soc. Subspecialties: Analytical chemistry; Ecology. Current work: Behavior of natural and artificially produced radionuclides in the environment, radioanalytical methods development, laboratory management, business adminstration, marketing. Home: 2356 Ferndale Richland WA 99352 Office: US Testing Co Inc 2800 George Washington Way Richland WA 99352

GAY, CHARLES FRANCIS, physical chemist, solar engineer, research and development executive; b. Redlands, Calif., Oct. 2, 1946. B.S., U. Calif.-Riverside, 1968, Ph.D. in Chemistry, 1978. Engr. Spectrolab, 1975-78; dir. research Arco Solar, Chatsworth, Calif., 1978-80, v.p. research and devel., 1980—. Mem. Electrochem. Soc., Sigma Xi. Subspecialty: Solar energy. Office: Arco Solar 21011 Warner Center Ln Woodland Hills CA 91367

GAY, TIMOTHY JAMES, physics educator; b. Ashtabula, Ohio, Mar. 23, 1953; s. William Coddington and Anne Elizabeth (McClelland) G.; m. Anna Christine Nothstine, Sept. 6, 1975. B.S., Calif. Inst. Tech., 1975; M.S., U. Chgo., 1976, Ph.D., 1980. Research asso., instr. Yale U., New Haven, 1980-82; asst. prof. U. Mo.-Rolla, 1983—. Mem. Am. Phys. Soc. Subspecialty: Atomic and molecular physics. Current work: Electron-Atomic collisions, ion-atom collisions.

GAYLORD, NORMAN GRANT, chemist, research laboratory executive; b. Bklyn., Feb. 16, 1923; s. Irving M. and Tillie (Horowitz) G.; m. Marilyn Einhorn, June 24, 1945; children: Lori Gaylord Wright, Kathy Gaylord Fleegler, Richard, Cory Gaylord-Ross. B.S., CCNY, 1943; M.S., Poly. Inst., Bklyn., 1949, Ph.D., 1950. Chemist Elko Chem. Works, Pittstown, N.J., 1943-44, Pa. Salt Mfg. Co., Pittstown, 1945, Merck & Co., Rahway, N.J., 1946-48; research chemist E.I. duPont de Nemours & Co., Buffalo, 1950-54; group leader Interchem. Corp., N.Y.C., 1955-56, asst. dept. dir., 1957-59; v.p. Western Petrochem. Corp., Newark, 1959-61; pres., research dir. Gaylord Research Inst., Whippany, N.J., 1961—; adj. prof. polymer chemistry Canisius Coll., Buffalo, 1951-54, Poly. Inst. Bklyn., 1955-62, U. Lowell, Mass., 1981—. Author: Reduction with Complex Metal Hydrides, 1956, Linear and Stereoregular Addition Polymers, 1959, Polyaklylene Sulfides and Other Polythioethers, 1962, Polyalkylene Oxides and Other Polyethers, 1963; mem. editorial adv. bd.: Jour. Macromolecular Sci.-Chemistry, 1968—, Macromolecular Syntheses, 1963—, Jour. Polymer Sci, 1959—, Jour. Applied Polymer Sci, 1959-74, Ency. of Polymer Sci. and Tech, 1964-72, Soc. Plastics Engrs. Transactions, 1963-64, Polymer Engring. and Sci, 1965-66, Revs. in Macromolecular Chemistry, 1968-73; contbr. articles to profl. publs. Served with USAAF, 1945-46. Mem. Am. Chem. Soc., Soc. Plastics Engrs., TAPPI, Chem. Soc. Japan, Polymer Soc. Japan, Sigma Xi. Subspecialties: Polymer chemistry; Photochemistry. Current work: Charge transfer polymerization; graft copolymerization; polymer modification. Patentee in field. Home: 28 Newcomb Dr New Providence NJ 07974 Office: 156 Algonquin Pkwy Whippany NJ 07981

GAYLORD, RICHARD J, physics educator; b. Plainfield, N.J., Dec. 20, 1947; s. Norman G. and Marilyn G.; m. Carole S., Feb. 20, 1970. B.S. summa cum laude, Poly. Inst. Brklyn., 1969; Ph.D., SUNY-Syracuse, 1972. Research assoc. Polymer Research Inst., U. Mass., Amherst, 1973-74; asst. prof. U. Ill.-Urbana, 1974-80, assoc. prof. dept. metallurgy, 1980—. Mem. Am. Phys. Soc. Subspecialty: Polymer physics. Current work: Theoretical polymer physics, rubber elasticity, colloidal stabilization. Home: 11 Hawthorn Ln Mahomet IL 61853 Office: Dept Metallurgy U Ill 1304 W Green St Urbana IL 61801

GAYLORD, THOMAS KEITH, electrical engineering educator; b. Casper, Wyo., Sept. 22, 1943; s. Earl F. and Vesta (Kinsley) G.; m. Janice L. Smith, June 5, 1966; 1 dau., Grace May. B.S., U. Mo.-Rolla, 1965, M.S.E.E., 1967; Ph.D., Rice U., 1970. Prof. elec. engring. Ga. Inst. Tech., 1972—. Contbr. numerous articles to tech. jours. Recipient Curtis W. McGraw Research award, 1979; Sigma Xi research awardee, 1977-80. Fellow IEEE, Optical Soc. Am.; mem. AAAS, AAUP, Am. Soc. Engring. Edn., Soc. Photo-optical Instr. Engrs., Sigma Xi, Omicron Delta Kappa. Subspecialties: Optical signal processing; Electrical engineering. Current work: Holographic optical digital parallel processing. Home: 3180 Verdun Dr NW Atlanta GA 30305 Office: Sch Elec Engring Ga Inst Tech Atlanta GA 30332

GAYNOR, HAROLD MARVIN, dentist, consultant; b. Phila., Apr. 25, 1930; s. David J. and Ida (Kremens) G.; m. Sandra L. Woodoff, June 25, 1955; 1 son, Eric Reid. B.S. in Pharmacy cum laude, Phila. Coll. Pharmacy and Scis., 1953; D.D.S., Temple Dental Sch., 1957; postgrad., Yale Sch. Pub. Health, 1973-75. Registered pharmacist. Dir. U. Conn. Sch. Dental Medicine, Farmington, 1975-76, asst. dean, 1976-77, assoc. dean, 1977-82, univ. dir., 1982-83; pvt. practice dentistry, Branford, Conn., 1983—; cons. Bowman Gray Med. Sch., Winston-Salem, N.C., 1982—, ADA Council on Dental Therapeutics. Dir. Bd. Health Branford, 1974-76; bd. dirs. East Shore Health Dist., Branford, 1974-76. Served to capt. U.S. Army, 1957-59. Fellow Am. Coll. Dentistry, Acad. Gen. Dentistry (mastership 1982); mem. Conn. State Dental Assn., Pierre Fauchard Acad., New Eng. Dental Soc. Club: Cosmopolitan (marshall 1960-61). Subspecialties: Oral biology; Periodontics. Current work: Research directed to clinical application and clinical research into pharmacology of drugs applicable to dentistry. Home: 35 Oxbow Ln Woodbridge CT 06525

GEALT, MICHAEL ALAN, molecular biologist, educator; b. Phila., Nov. 27, 1948; s. Edward Leonard and Lillian Rose (Brenner) G.; m. Antonia Malandrucco, May 12, 1967 (div.); children: Lillian, Marina, Benjamin; m. Maryjanet McNamara, Jan. 2, 1981. B.A., Temple U., 1970; Ph.D., Rutgers U., 1974. Postdoctoral fellow Rutgers Med. Sch., Piscataway, N.J., 1974-76, Fox Chase Cancer Ctr., Phila., 1976-78; asst. prof. dept. biol. scis. Drexel U., Phila., 1978—. Contbr. articles to profl. jours. NSF grantee, 1980. Mem. AAAS, Am. Soc. Cell Biology, Am. Soc. Microbiology, Sigma Xi. Subspecialties: Cell biology; Genetics and genetic engineering (biology). Current work: Regulation of gene expression; plasmid transfer in wastewater. Secondary metabolism/fungal gene regulation; bacterial conjugation/waste treatment microbiology. Office: 32d and Chestnut Sts Philadelphia PA 19104

GEAR, CHARLES WILLIAM, computer science educator; b. London, Feb. 1, 1935; U.S., 1962, naturalized, 1977; s. Charles William and Margaret (Dumbleton) G.; m. Sharon Sue Smith, Jan.25, 1958 (div. Oct. 1970); children: Kathlyn Jo, Christopher William Gilpin; m. Ann Lee Morgan, Nov. 19, 1976. B.A., Cambridge U., 1956, M.A., 1960; M.S., U. Ill.-Urbana, 1957, Ph.D., 1960. Engr. IBM, Hursley, Eng., 1960-62; prof. dept. computer sci. U. Ill.-Urbana, 1962—; vis. prof. Stanford (Calif.) U., 1969-70, Yale U., New Haven, 1976. Author: Computer Organization and Programming, 1969, 74, 80, Numerical Initial Value Problems, 1971, Introduction to Computer Science, 1973, Introduction to Computers, Program Application, 1978. Recipient Fulbright award, 1956; Forsythe award Spl. Interest Group for Numerical Analysis, 1979. Mem. IEEE, Soc. Indsl. and Applied Math. (council 1980—), Assn. Computing Machinery (council 1976-78). Subspecialties: Mathematical software; Numerical analysis. Current work: Automatic solution of differential equations, numerical methods of computer graphics. Home: 3302 Lakeshore Dr Champaign IL 61820 Office: U Ill Dept Computer Sci 1304 W Springfield St Urbana IL 61801

GEBALLE, THEODORE HENRY, educator; b. San Francisco, Jan. 20, 1920; s. Oscar and Alice (Glaser) G.; m. Frances C. Koshland, Oct. 19, 1941; children—Gordon, Alison, Adam, Monica Ruth, Jennifer, Ernest. B.S. in Chemistry, U. Calif. at Berkeley, 1941, Ph.D., 1950. Research asso. Low Temperature Lab., U. Calif. at Berkeley, 1949-51; mem. staff Bell Telephone Lab., Murray Hill, N.J., 1952—, head low temperature physics dept., 1958-67, research cons., 1967—; prof. applied physics Stanford, 1967—. Served to capt. AUS, 1941-46. Guggenheim fellow, 1974-75. Fellow Am. Phys. Soc. (Oliver E. Buckley solid state physics prize 1970); mem. Nat. Acad. Scis., Am. Acad. Arts and Scis., Am. Chem. Soc., Phi Beta Kappa, Sigma Xi. Subspecialty: Low temperature physics. Home: 259 Kings Mountain Rd Woodside CA 94062 Office: Dept Applied Physics Stanford Univ Stanford CA 94305

GEBALLE, THOMAS RONALD, astronomer, educator; b. Seattle, Nov. 16, 1944; s. Ronald and Marjorie L. (Cohn) G.; m. Carole Gem, June 11, 1967; children: Anneke Lee, Matthew Thomas. Student, U. Wash., 1962-64; student, U. Amsterdam, Netherlands, 1964-65; B.A. in Physics, U. Calif., Berkeley, 1967, Ph.D., 1974. Research fellow dept. physics U. Calif., Berkeley, 1974-75, Leiden (Netherlands) U., 1975-77; sr. research fellow Mount Wilson and Las Campanas Obs., Carnegie Instn. of Washington, Pasadena, Calif., 1977-81; mem. research faculty dept astronomy Groningen (Netherlands) U. assigned to U.K. Infrared Telescope, Hilo, Hawaii, 1981—. Contbr. articles to profl. jours. James Monroe MacDonald scholar U. Calif., Berkeley, 1966. Mem. Am. Astron. Soc., Phi Beta Kappa. Subspecialty: Infrared optical astronomy. Current work: Development of infrared astronomical instrumentation; infrarfed spectroscopy of solar system, regions of star formation, young stars, evolved stars, ionized nebulae and galactic nuclei. Home: 1645 Oneawa Pl Hilo HI 96720 Office: 900 Leilani St Hilo HI 96720

GEBER, WILLIAM FREDERICK, pharmacologist, educator; b. Rahway, N.J., Oct. 26, 1923; s. William F. and M. Ruth (Babel) G.; m. Joan M. Rezny, June 29, 1946; 1 dau., Sharron Ruth. A.B., Dartmouth Coll., 1947; M.S., Ind. U.-Bloomington, 1950, Ph.D., 1954. Research assoc. U. Minn.-Mpls., 1954; asst. prof. physiology St. Louis U. Med. Sch., 1954-58; assoc. prof. U. S.D. Med. Sch., 1958-65; prof. pharmacology Med. Coll. Ga., 1965—; cons., expert witness in field. Contbr. numerous articles to profl. jours. Lectr. on drug abuse to community groups. Served with USN, 1942-46. Mem. Am. Soc. Pharmacology and Exptl. Therapeutics, Am. Physiol. Soc., Teratology Soc., Soc. Toxicology. Subspecialties: Pharmacology; Teratology. Current work: Causes and prevention of birth defects; relationships between cancer, birth defects, mutations. Home: 1319 Martinique Dr Augusta GA 30909 Office: Med Coll Ga Augusta GA 30912

GEBHART, GERALD FRANCIS, pharmacologist; b. Chgo., June 23, 1943; s. Frank L. and Anne (Muschong) G.; m. Valda A. Zalums, Jan. 22, 1944; 1 son, David J. B.S., U. Ill., 1967; M.S. in Pharmacology, U. Iowa, 1969, Ph.D., 1971. Fellow dept. physiology U. Montreal, Que., Can., 1971-73; asst. to prof. dept. pharmacology U. Iowa, Iowa City, 1973—; vis. scientist dept. physiology U. Heidelberg, W.Ger., 1981-82. Contbr. articles on pharmacology to profl. jours. Med. Research Council of Can. fellow, 1971-73; Alexander von Humboldt Stiftung fellow, 1981-82; NIH grantee, 1973—. Mem. Am. Soc. for Pharmacology and Exptl. Therapeutics, Soc. for Neurosci., Internat. Assn. for Study of Pain. Subspecialties: Neuropharmacology; Neurophysiology. Current work: Sites and mechanisms of opioid analgetics, pain research neurophysiologic and neuropharmacologic research. Office: Dept Pharmacology U Iowa Iowa City IA 52242

GEDDES, LESLIE ALEXANDER, bioengr., physiologist, educator; b. Scotland, May 24, 1921; s. Alexander and Helen (Humphrey) G.; m. La Nelle E. Nerger, Aug. 3, 1962; 1 son, James Alexander. B.E.E., M.S., Sc.D. (hon.), McGill U.; Ph.D. in Physiology, Baylor U. Demonstrator in elec. engring. McGill U., 1945, research asst. dept. neurology, 1945-52; cons. elec. engring. to various indsl. firms, Que., Can.; biophysicist dept. physiology Baylor Med. Coll., Houston, asst. prof. physiology, 1956-61, asso. prof., 1961-65, prof., 1965—; dir. Lab. of Biophysics, Tex. Inst. Rehab. and Research, Houston, 1961-65; prof. physiology Coll. Vet. Medicine, Tex. A. and M. U., College Station, 1965—, prof. biomed. engring., 1969—; Showalter Distinguished prof. bioengring. Purdue U., West Lafayette, Ind., 1974—; cons. NASA Manned Spacecraft Center, Houston, 1962-64, USAF, Sch. Aerospace Medicine, Brooks AFB, 1958-65. Author: (with H.E. Hoff) Experimental Physiology, 1967, (with others), 5 books, also numerous articles on bioengring.; Cons. editor: Med. and Biol. Engring, 1969—, Med. Research Engring, 1964—, Med. Electronics and Data, 1969—; editorial bd.: Jour. Electrocardiology, 1968—. Served with Canadian Army. Fellow Am. Coll. Cardiology, Australasian Coll. Physicists in Biology and Medicine; mem. IEEE, Tex., Nat. socs. profl. engrs., Houston Engring. and Sci. Soc., Biomed. Engring. Soc., Am. Inst. Biol. Scis., Assn. for Advancement Med. Instrumentation, Am. Physiol. Soc., N.Y. Acad. Scis., Neuroelectric Soc., Sigma Xi, Tau Beta Pi. Subspecialties: Biomedical engineering; Electronics. Home: 400 N River Rd Apt 1724 West Lafayette IN 47906 Office: APEC Bldg Purdue Univ West Lafayette IN 47907

GEDNEY, LARRY DANIEL, seismologist, geophysicist; b. Salt Lake City, Jan. 22, 1938; s. Roy Jay and Vinita (Manley) G.; m. Jean Marie Talarski, Mar. 28, 1957; children: Gregory Monroe, Jeffrey Jay, Douglas Kevin, Laura Anne. B.S., U. Nev.-Reno, 1960, M.S., 1966. Grad. Asst. U. Nev., Reno, 1964-66; asst. geophysicist Geophys. Inst., U. Alaska, Fairbanks, 1966-70; geophysicist U.S. Geol. Survey, San Francisco, 1970-71; assoc. prof. Geophys. Inst., U. Alaska, Fairbanks, 1971—. Contbr. articles in field to profl. jours. Soc. Pres. Ester (Alaska) Community Assn., 1972-74; bd. Dirs., co-fire chief Ester Vol. Fire Dept., 1976-78. Served to capt. U.S. Army, 1960-64. Mem. Seismol. Soc. Am., Am. Geophys. Union, AAAS, Alaska Geol. Subspecialty: Geophysics. Current work: Ongoing tectonic deformation of Alaska. Home: Willow Ct AK 99725 Office: Geophys. Inst U Alaska Fairbanks AK 99701

GEE, ADRIAN PHILIP, immunologist, educator; b. Whitehead, Antrim, No. Ireland, Apr. 30, 1952; s. Gordon Stanley and Sally Allen. B.S., U. Birmingham, U.K., 1973; Ph.D., U. Edinburgh, U.K., 1977. Research scientist Western Gen. Hosp., U. Edinburgh, Scotland, U.K., 1977-78; vis. fellow Nat. Cancer Inst., Bethesda, Md., 1978-81; research fellow U. Toronto, Ont., Can., 1981-82; asst. prof. U. Fla., Gainesville, 1982—. Contbr. articles to sic. publs. U. Edinburgh Hastilow research scholar, 1973; Lady Tata Meml. fund fellow, 1981; recipient Internat. Union Against Cancer Icrett award, 1977. Mem. Am. Assn. Immunologists, Brit. Soc. Immunology, Can. Soc. Immunology, N.Y. Acad. Scis., Brit. Inst. Biology. Episcopalian. Subspecialties: Immunology (medicine); Marrow transplant. Current work: Complement research, granulocyte immunology bone marrow transplantation. Home: 3215 NW 51st Pl Gainesville FL 32605 Office: Dept Pediatrics U Florida Box J-296 Gainesville FL 32610

GEELHOED, GLENN WILLIAM, surgery educator, researcher, author; b. Grand Rapids, Mich., Jan. 19, 1942; s. William and Alice (Stuk) G.; m. Sally Elaine Ryden, Sept. 11, 1965 (div. 1973); children: Donald William, Michael Alan. A.B. cum laude, Calvin Coll., 1964, B.S., 1965; M.D. cum laude, U. Mich., 1968. Diplomate: Am. Bd. Surgery. Assoc. in surgery Harvard Med. Sch., Boston, 1968-70; clin. assoc. NIH, Bethesda, Md., 1970-72; sr. investigator Nat. Cancer Inst., 1972-73; clin. scholar Robert Wood Johnson Found., Washington, 1975-77; successively instr., asst. prof., assoc. prof. George Washington U., Washington, 1973-75, prof. surgery, 1975—; cons. Nat. Cancer Inst., NIH, 1973—, Walter Reed Army Hosp., 1974—, VA, 1975—, World Bank, WHO, 1977—. Author: Problem Management of the Endocrine Surgical Patient, 1982, Correlative Surgical Endocrinology, Endocrine and Metabolic Care of the Critically Ill, 1983; contbr. articles to profl. jours. Chmn. med. adv. com. ARC, Washington, 1980—, chmn. blood services com., 1978—; chmn. blood com. S.C. Med. Soc., 1975—. Served to comdr. USPHS, 1970-73. Life sci. scholar Life Ins. Med. Research Found., Boston, 1968; clin. scholar Robert Wood Johnson Found., Princeton, N.J., 1975; recipient Borden award U. Mich., 1967; James B. Welch award AMA, 1966; Conrad Tobst gold medal Southeastern Surg. Congress, 1974. Fellow , ACS; mem. Soc. Univ. Surgeons, Halstad Soc., Am. Assn. Endocrine Surgeons, Internat. Assn. Endocrine Surgeons, Sociéé Internationale de Chirurgie, other profl. orgns. Republican. Mem. Christian Reformed Church. Subspecialties: Surgery; Transplant surgery. Current work: International medical education, surgical research, natural history and ecology, cancer, physiology, immunology, endocrinology. Office: George Washington U Med Ctr 2150 Pennsylvania Ave NW Washington DC 20037

GEER, JAMES FRANCIS, mathematics educator, consultant; b. Syracuse, N.Y., Oct. 3, 1940; s. Francis Bion and Dorothy E. (Wilder) G.; m. Linda Rae Sundquist, Sept. 18, 1965; children: Jim, Bill, Wendy, Jennifer. B.A., Harpur Coll., SUNY-Binghamton, 1962; M.A., U. Va., 1964; Ph.D., NYU, 1967. Mathematician IBM, Endicott, N.Y., 1967-69; asst. prof. SUNY-Binghamton, 1969-71, assoc. prof., 1971-82, prof. math., 1982—; cons. IBM, Gen. Electric, NASA Langley. Contbr. research papers to publs. Research grantee NSF, 1979-81, Gen. Electric, 1979-81, Research Found. SUNY, 1970, 71, 73. Mem. Soc. Indsl. and Applied Math., Am. Acad. Mechanics. Methodist. Subspecialties: Applied mathematics; Numerical analysis. Current work: Perturbation, asymptotic, numerical methods to solve partial differential equations; use of sumbolic computation in applied mathematics. Home: RD 3 Box 328 Endicott NY 13760 Office: SUNY Binghamton NY 13901

GEER, RONALD LAMAR, mech. engr., oil co. exec.; b. West Palm Beach, Fla., Sept. 2, 1926; s. Marion Wood and Bertha (Lightfoot) G.; m. Geneva Yvonne Chappell, Dec. 24, 1951; children—Ronald Lamar, Mark Randall. B.M.E., Ga. Inst. Tech., 1951. With Shell Oil Co., 1951—, sr. staff mech. engr., head office, Houston, 1969-71, cons. mech. engr., 1971—; mem. various govt., univ. adv. coms. Contbr. articles on petroleum drilling and prodn. to profl. jours. Mem. Nat. Acad. Engring., NRC (marine bd.), Nat. Security Indsl. Assn. (petroleum panel, research and engring. adv. com.), ASME, Marine Tech. Soc., Am. Petroleum Inst., Model-A Ford Club Am., Classic T-Bird Club Internat., Thistle Class Assn., Pi Tau Sigma. Republican. Subspecialty: Petroleum engineering. Patentee petroleum drilling and prodn. equipment; mem. Shell Oil Co. team recognized in Offshore Tech. Conf. Disting. Achievement award to co., 1971. Home: 14723 Oak Bend Dr Houston TX 77079 Office: One Shell Plaza Houston TX 77001

GEHA, ALEXANDER SALIM, Cardiothoracic surgeon, educator; b. Beirut, Lebanon, June 18, 1936; came to U.S., 1963; s. Salim M. and Alice I. (Hayek) G.; m. Diane L. Redalen, Nov. 25, 1967; children: Samia, Rula, Nada. B.S. in Biology, Am. U. Beirut, 1955, M.D., 1959; M.S. in Surgery and Physiology, U. Minn.-Rochester, 1967, Yale U., 1978. Asst. prof. U. Vt., Burlington, 1967-69; asst. prof. Washington U., St. Louis, 1969-70, assoc. prof., 1970-75, Yale U., New Haven, 1975-78, prof., chief cardiothoracic surgery, 1978—; cons. VA Hosp., West Haven, Conn., 1975—, Waterbury Hosp., 1976—, Sharon Hosp., 1981—; mem. study sect. Nat. Heart, Lung and Blood Inst., 1981-85. Editor: Thoracic and Cardio-vascular Surgery, 1983; editor: Basic Surgery, 1984. Bd. dirs. New Haven Heart Assn., 1981—. Mem. Assn. Clin. Cardiac Surgery (chmn. membership com. 1979-80, sec.-treas. 1980-83), Am. Heart Assn. (bd. dirs. 1981-85, councilon cardiovascular surgery), Am. Coll. Chest Physicians (steering com. 1980-84), Am. Assn. Thoracic Surgery, Am. Coll. Cardiology, ACS (coordinating com. on edn. in thoracic surgery), AMA, Am. Physiol. Soc., Am. Surg. Assn., Am. Thoracic Soc., Assn. for Acad. Surgery, Central Surg. Assn., Internat. Sco. Cardiovascular Surgery, Lebanese Order Physicians, New Eng. Surg. Soc., Pan Am. Med. Assn., Halsted Soc., Soc. Thoracic Surgery (govt. relations com.), Soc. Univ. Surgeons, Soc. Vascular Surgery, Societa di Richerche in Chirugia, Sigma Xi, Alpha Omega Alpha. Subspecialties: Cardiac surgery; Transplant surgery. Current work: Clinical cardiothoracic surgery and cardiovascular research into cardiac hypertrophy and its physiological effects and characteristics; cardiac transplantation and cardiac preservation, short term (hours) and long term (days). Home: 345 Ridge Rd Hamden CT 06517 Office: Yale U Sch Medicine 333 Cedar St New Haven CT 06510

GEHRELS, NEIL, astrophysicist; b. Lake Geneva, Wis., Oct. 3, 1952; s. Tom and Aleida (de Stoppelaar) G.; m. Ellen D. Williams, Apr. 5, 1980. Mus.B., U. Ariz., 1976; B.S. in physics, 1976; Ph.D., Calif. Inst. Tech., 1981. Research asst. Calif. Inst. Tech., Pasadena, 1977-81; NRC research assoc. Goddard Space Flight Center, Greenbelt, Md., 1981-83, astrophysicist, 1983—. Contbr. articles to profl. jours. Mem. Am. Phys. Soc., Am. Geophys. Union, Am. Astron. Union, Phi Beta Kappa. Subspecialty: Gamma ray high energy astrophysics. Current work: Researcher in gamma ray line spectroscopy. Office: Code 661 Goddard Space Flight Center Greenbelt MD 20771

GEHRELS, TOM, astronomer. Ph.D., U. Chgo., 1956. Research assoc. Ind. U., 1956-61; assoc. prof. U. Ariz., Tucson, 1961-67, prof., 1967—; fellow Phys. Research Lab, Ahmedabad, India, 1978—. Author articles on astron. polarimetry, surveying of asteroids and comets, environ. issues. Served with Spl. Services, 1944-47. Recipient NASA medal for exceptional sci. achievement, 1974. Mem. Am. Astron. Soc., Internat. Astron. Union. Subspecialty: Planetary science. Current work: Surveying of the solar system. Office: Lunar Lab U Ariz Tucson AZ 85721

GEHRING, FREDERICK WILLIAM, mathematician, educator; b. Ann Arbor, Mich., Aug. 7, 1925; s. Carl E. and Hester McNeal (Reed) G.; m. Lois Caroline Bigger, Aug. 29, 1953; children: Kalle Burgess, Peter Motz. B.S.E. in Math, U. Mich., 1946, M.A., 1949; Ph.D. (Fulbright fellow) in Math, Cambridge U., Eng., 1952, Sc.D., 1976; Ph.D. (hon.), U. Helsinki (Finland), 1977. Benjamin Peirce instr. Harvard, Cambridge, Mass., 1952-55; instr. math. U. Mich., Ann Arbor, 1955-56, asst. prof., 1956-59, asso. prof., 1959-62, prof., 1962—, chmn. dept. math., 1973-75, 77—; vis. prof. Harvard U., 1964-65, Stanford U., 1964, U. Minn., 1971, Inst. Mittag-Leffler, Sweden, 1972. Editor: Duke Math. Jour, 1963-80, D. Van Nostrand Pub. Co., 1963-69, North Holland Pub. Co., 1970—, Springer-Verlag, 1974—; editorial bd., Procs. Am. Math. Soc., 1962-65, Ind. U. Math. Jour., 1967-75, Math. Revs., 1969-75, Bull. Am. Math. Soc., 1979—, Complex Variables, 1981—; contbr. numerous articles on research in pure math. to sci. jours. Served with USNR, 1943-46. NSF fellow, 1959-60; Fulbright fellow, 1958-59; Guggenheim fellow, 1958-59; Sci. Research Council sr. fellow, 1981; Humboldt fellow, 1981. Mem. Assn. Women in Math., Math. Assn. Am., Am. Math Soc. (council 1980-83, trustee 1983—), Swiss Math Soc., Swedish Math. Soc., Finnish Math Soc., London Math. Soc., Finnish Acad. Sci. Subspecialty: Complex Analysis. Home: 2139 Melrose Ave Ann Arbor MI 48104

GEHRZ, ROBERT DOUGLAS, physicist, astronomer, educator; b. Evanston, Ill., Dec. 28, 1944; s. Robert Gustave and Mary Gilbert (Laubscher) G; m. Susan Lucille Laurel, May 1, 1970; children: Alexander Robert, Andre Laurel. B.A. in Physics, U. Minn., 1967, Ph.D., 1971. Research asst. U. Minn., 1968-71, research assoc., 1971-72; asst. prof. physics and astronomy U. Wyo., 1972-78, asso. prof. physics and astronomy, 1978-83, prof. physics and astronomy, 1983—; dir. at large AURA, Inc., 1976-79; Chmn. NSF Optical-Infrared Subcom., 1978-80, Astronomy Adv. com., 1977-80; chmn. AURA Obs. Vis. Com., 1982—; mem. U. Calif. Ten Meter Telescope Infrared Working Group, 1981—. Contbr. numerous articles to profl. jours. Chmn. Laramie (Wyo.) Parks and Recreation Bd., 1981—. Mem. Am. Astron. Soc. (mem. astronomy survey com. 1978—, chmn. galactic astronomy working group 1978—), Royal Astron. Soc., Internat. Astron. Union, Sigma Xi. Episcopalian. Clubs: Explorers Club, Laramie Adult Hockey (sec.-treas.). Subspecialties: Infrared optical astronomy; Low temperature physics. Current work: Observational astrophysics, experimental instrumentation and development; teaching, research, consulting. Home: 1719 Mill St Laramie WY 82070 Office: Dept Physics and Astronomy Univ Wyo Laramie WY 82071

GEIBEL, VALERIE HENKEN, research psychologist, statistical consultant; b. Long Branch, N.J., Mar. 2, 1954; d. August E. and Margaret M. (Gabriel) Henkensiefken; m. Philip L. Geibel, May 18, 1980; 1 dau., Michaela Anne. B.A. summa cum laude, Montclair State Coll., 1975; M.A., N.Y.U., 1978, Ph.D., 1980. Jr. statistician Acad. Computing Center, N.Y.U., N.Y.C., 1977-78; clin. instr. Coll. Health Related Professions, SUNY Downstate Med. Center, Bklyn., 1979-81, sr. statis. cons., 1978-81, research assoc., statistician dept. obstetrics and gynecology, 1983—; research psychologist. sr. statistician, postdoctoral fellow in psychiatry N.Y. U.Sch. Medicne, 1982. Contbr. articles to profl. jours. NIMH trainee, 1975-77. Mem. Am. Psychol. Assn., Soc. for Personality and Social Psychology, Health Psychology Soc., Am. Statis. Assn., Psi Chi. Subspecialties: Social psychology; Statistics. Current work: Psychology, statistics research methods, biomedical and neuropsychological applications. Home: 24 Crescent Dr Whippany NJ 07981

GEIDEL, GWENDELYN, geochemical research scientist, consultant; b. Lowville, N.Y., May 4, 1953; d. Henry Augustus and Elizabeth (Wait) G., Jr.; m. Frank T. Caruccio, Oct. 23, 1976. B.S., U. S.C., 1974, M.S., 1976, Ph.D., 1982. Research scientist geology dept. U. S.C., Columbia, 1976—; sec. W.Va. Acid Mine Drainage Tech. Adv. Com., 1980-82. Co-author: Paleoenvironment of Coal, 1977. EPA grantee, 1977; U.S. Bur. Mines grantee, 1980; W.Va. Dept. Natural Resources grantee, 1982. Mem. Soc. Environ. Geochemistry and Health, Am. Council for Reclamation Research, Can. Land Reclamation Assn.,

AAAS, Phi Beta Kappa, Sigma Xi. Subspecialties: Geochemistry; Hydrology. Current work: Prediction and prevention of acid drainage from coal, copper, gold, lead, and uranium mines; determination of mining hydrology using stable isotopes and chemical tracers. Patentee method to predict coal mine drainage quality. Office: Dept Geology U SC Columbia SC 29208

GEIDUSCHEK, E(RNEST) PETER, biophysicist, educator; b. Vienna, Austria, Apr. 11, 1928; came to U.S., 1945, naturalized, 1946; s. Sigmund and Frieda (Tauber) G.; m. Joyce Barbara Brous; 2 children. B.A., Columbia, 1948; A.M., Harvard, 1950, Ph.D., 1952. Instr. chemistry Yale, 1952-53, 55-57; asst. prof. chemistry U. Mich., 1957-59; asst. prof. biophysics U. Chgo., 1959-62, asso. prof., 1962-64, prof., 1964-70; prof. biology U. Calif. at San Diego, LaJolla, 1970—, chmn. dept., 1981—; Cons. USPHS, 1963-69. Editorial bd.: Biophys. Jour, 1967-69, Ann. Revs. Biophysics and Bioengring, 1971-74, Virology, 1972—, Science, 1977-81. Served with AUS, 1953-55. Recipient Research award Am. Postgrad. Med. Assn., 1962, Research Career Devel. award USPHS, 1962; Guggenheim Found. fellow, 1964-65. Fellow Am. Acad. Arts and Scis.; mem. Nat. Acad. Scis., Am. Soc. Biol. Chemists, Biophys. Soc. (council 1964-66), AAAS, Am. Soc. for Microbiology. Subspecialties: Molecular biology; Virology (biology). Current work: Research in biophysical chemistry, molecular biology, virology, genetic regulation, enzymology of transcription. Research in biophys. chemistry, molecular biology, virology. Home: 8460 Cliffridge Ln La Jolla CA 92037

GEIGER, JON ROSS, bacterial geneticist; b. Phila., Nov. 28, 1943; s. Gordon R. and Violet M. (Anderson) G.; m. Margaret Jeffrey, June 25, 1966; children: Michael Gil, Douglas Jeffrey. B.S., Pa. State U., 1965; Ph.D., U. Conn., 1976. Asst. prof. Smith Coll., Northampton, Mass., 1976-80; vis. scientist biology dept. M.I.T., Cambridge, Mass., 1980-81; sr. research biologist Olin Research Center, New Haven, 1981—. Contbr. articles to sci. lit. Served to lt. USN, 1965-70; Vietnam. NSF, Commonwealth Fund grantee. Mem. AAAS, Am. Soc. Microbioloy, Environ. Mutagen Soc., Genetics Soc. Am., Inst., Soc., Ethics and Life Scis., Sigma Xi. Unitarian. Subspecialties: Genetics and genetic engineering (biology); Microbiology. Current work: Genetics and regulation of biochemical pathways in bacteria; mutagenesis in bacteria; applications of biotech. in chem. industry. Office: Olin Research Center PO Box 30-275 New Haven CT 06511

GEIST, JACOB MYER, chemical engineer; b. Bridgeport, Conn., Feb. 2, 1921; s. David and Anne Rose (Steinschreiber) G.; m. Sandra Levy, Nov. 17, 1972; children by previous marriage—Eric D., Ellen A., David C. B.S. in Chem. Engring, Purdue U., 1940; M.S., Pa. State U., 1942; Ph.D., U. Mich., 1951. Instr. Pa. State U., 1943-44; teaching fellow, part-time instr. U. Mich., 1946-48; instr., then asst. prof. Mass. Inst. Tech., 1950-52; sr. lectr. Technion, 1952-55; with Air Products and Chems., Inc., Allentown, Pa., 1955-82, assoc. dir. research and devel., 1961-63, assoc. chief engr., 1963-69, chief engr., 1969-82; pres. Geist Tech., 1982—; lectr., adj. prof. Lehigh U., Bethlehem, Pa., 1960—. Author. Served to 2d lt. AUS, 1944-46. Hon. fellow Indian Cryogenic Soc., 1975. Fellow Am. Inst. Chem. Engrs. (award chem. engring. practice 1976); mem. Nat. Acad. Engring., Am. Chem. Soc., AAAS, Internat. Inst. Refrigeration (v.p.), Cryogenic Engring. Conf. (dir.), Nat. Soc. Profl. Engrs., Sigma Xi, Tau Beta Pi, Phi Lambda Upsilon. Subspecialties: Cryogenics. Current work: Cryogenics, separations, process engineering. Patentee in field. Home and Office: 2720 Highland St Allentown PA 18104

GELB, ALVIN MEYER, gastroenterologist; b. Stamford, Conn., May 4, 1930; s. Jacob and Rose (Rodansky) G.; m. Ronda Ann Shainmark, Mar. 7, 1954; children: Janet, Daniel, Michael. A.B., NYU, 1950; M.D., SUNY, 1954. Diplomate: Am. Bd. Internal Medicine, Am. Bd. Gastroenterology. Intern Cin. Gen. Hosp., 1954-55; resident in internal medicine and gastroenterology West Haven (Conn.) VA Hosp., 1955-56, Mt. Sinai Hosp., N.Y.C., 1958-61, clin. instr., 1961-70; dir. dept. medicine French Polyclinic Med. Ctr., N.Y.C., 1970-75; chief div. gastroenterology Beth Israel Med. Ctr., N.Y.C., 1975—. Author: Geriatrics 2, 1983. Served as capt. U.S. Army, 1956-58. Fellow N.Y. Acad. Gastroenterology (pres. 1981-82), Am. Coll. Gastroenterology (gov. 1982-83), N.Y. Acad. Medicine, ACP; mem. Am. Gastroent. Assn., Phi Beta Kappa, Alpha Omega Alpha. Democrat. Jewish. Subspecialties: Internal medicine; Gastroenterology. Current work: Colon cancer, liver disease in alcoholics and drug abusers; intestinal absorbtion. Office: Beth Israel Med Ctr 10 Perlman Pl New York NY 10003

GELB, ARTHUR FRANKLIN, physician, educator; b. Bklyn., Apr. 19, 1942. B.A., Vanderbilt U., 1963; M.D., St. Louis U., 1967. Diplomate: Am. Bd. Internal Medicine (pulmonary diseases). Intern, resident Kings County Hosp-Downstate Med. Ctr., Bklyn; resident U. Calif.-San Francisco, Cardiovascular Research Inst.; attending Drs. Hosp., Lakewood, Calif., Los Altos Hosp., Long Beach, Calif., Meml. Hosp., Long Beach; assoc. clin. prof. medicine UCLA Sch. Medicine, 1973—; practice medicine, specializing in pulmonary medicine, Lakewood, Calif., 1973—. Served to maj. M.C. U.S. Army, 1971-73. Fellow ACP, Am. Coll. Chest Physicians, Am. Thoracic Soc.; mem. AMA, Am. Fedn. Clin. Research. Subspecialties: Laser medicine; Pulmonary medicine. Office: 3650 E South St Suite 308 Lakewood CA 90712

GELBOIN, HARRY VICTOR, biochemist; b. Chgo., Dec. 21, 1929; s. Herman and Eva (Jurkowsky) G.; m. Marlena Maisels, Apr. 1, 1962; children: Michele Ida, Lisa Rebecca, Sharon Anna, Tamara Rachel. A.B., U. Ill., 1951; M.S., U. Wis., 1956, Ph.D., 1958. Devel. chemist U.S. Rubber Co., 1952-54; research asst. McArdle Meml. Lab. Cancer Research, U. Wis., 1954-58; biochemist NIMH, 1958-61; supervisory biochemist Nat. Cancer Inst., 1962-64, head chemistry sect., 1964-66; chief Lab. Molecular Carcinogens div. cancer cause and prevention, 1966—; adj. prof. Mt. Vernon Jr. Coll., Washington, Georgetown U., 1978; cons. Am. Cancer Soc., EPA, Nat. Acad. Sci., Fedn. Am. Soc. Exptl. Biology, others.; Predoctoral fellow Nat. Cancer Inst., 1957-58; keynote speaker Gordon Research Conf. Cancer, 1965; Franz Bielschowsky Meml. lectr., Dunedin, New Zealand, 1966, Smith, Kline and French hon. lectr., 1974, 76. Author articles, chpts. in books.; Asso. editor: Chem.-Biol. Interactions; assoc. editor: Cancer Research, 1983-86; editor: Environmental Health Sciences, 1976. Recipient Claude Bernard award U. Montreal, 1970; New Horizons award lectr. Radiol. Soc. N.Am., 1970; Superior Service award NIH, 1970. Mem. AAAS, Am. Assn. Cancer Research Am., Soc. Biol. Chemists, Am. Soc. Pharmacology and Exptl. Therapeutics. Jewish religion (trustee congregation 1972—). Club: Mem. B'nai B'rith. Subspecialty: Biochemistry (biology). Home: 2806 Abilene Dr Chevy Chase MD 20015 Office: Nat Insts Health Bethesda MD 20014

GELBWACHS, JERRY AVRON, laser scientist; b. N.Y.C. B.S., CCYN, 1965; M.S. (NSF grad. fellow, 1965-70), Stanford U., 1966, Ph.S., 1970. Research scientist, sect. mgr. Aerospace Corp., El Segundo, Calif., 1970—. Contbr. articles to sci. publs. Mem. Am. Phys. Soc., Optical Soc. Am., IEEE. Subspecialty: Spectroscopy. Current work: Applications of laser spectroscopy to trace atomic and molecular detection, to novel methods of IR detection, to energy transfer, to diagnosis of submicron-size semiconductors. Office: PO Box 92957 Los Angeles CA 90009

GELEHRTER, THOMAS DAVID, medical and genetics educator, physician; b. Liberec, Czechoslovakia, Mar. 11, 1936; m., 1959; 2 children. B.A., Oberlin Coll., 1957; M.A., U. Oxford, Eng., 1959; M.D., Harvard U., 1963. Intern, then asst. resident in internal medicine Mass. Gen. Hosp., Boston, 1963-65; research assoc. in molecular biology Nat. Inst. Arthritis, Metabolic and Digestive Diseases, NIH, Bethesda, Md., 1965-69; fellow in med. genetics U. Wash., 1969-70; asst. prof. human genetics, internal medicine and pediatrics Sch. Medicine, Yale U., 1970-73, assoc. prof., 1973-74, U. Mich., 1974-76, prof. internal medicine and human genetics, 1976—, dir. div. med. genetics, 1977—; trustee Oberlin Coll., 1970-75; mem. cell biology ad hoc rev. study sect. NIH, 1978; vis. scientist Imperial Cancer Research Fund Labs., London, 1979-80. Mem. Am. Soc. Human Genetics, Am. Soc. Clin. Investigation, Am. Soc. Biol. Chemists. Subspecialties: Genetics and genetic engineering (medicine); Cell biology. Office: Dept Human Genetics Box 015 Med Sch U Mich 1137 E Catherine St Ann Arbor MI 48109

GELFAND, DAVID H., biologist, genetic research adminstr.; b. N.Y.C., June 9, 1944; s. Sidney J. and Gigi P. (Levinson) G.; m. Ellen Daniell, Dec. 29, 1980; 1 dau., Duskie Lynn. A.B., Brandeis U., 1966; Ph.D. in Biology, U. Calif., San Diego, 1970. Research asst. biochemistry Brandeis U., Waltham, Mass., 1965; research asso. biology U. Calif.-San Diego, 1966; NIH trainee molecular genetics U. Calif., 1966-70, staff research biochemist, 1970-72; asst. research prof. U. Calif.-San Francisco, asst. research biochemist and lab. mgr., 1972-76; dir. recombinant molecular research Cetus Corp., Berkeley, Calif., 1977-81, v.p. sci. affairs, 1979—; mem. sci. adv. council NSF, 1980—. Contbr. articles on molecular research in genetics to sci. jours. Mem. Am. Soc. Biol. Chemists, Am. Soc. for Microbiology, AAAS. Subspecialties: Molecular biology. Office: 1400 53rd st Emeryville CA 94608

GELIEBTER, ALLAN, psychology educator, researcher; b. Frankfurt, Germany, Jan. 22, 1947; came to U.S., 1950; s. Leo and Bella G. B.S., CCNY, 1968; M.A. in Biology, Columbia U., 1970; M.Phil. in Psychology, Columbia U., 1973, Ph.D., 1976. Lic. psychologist, N.Y. Adj. lectr. Lehman Coll., 1974-76, adj. asst. prof., 1976-78; asst. prof. psychology Touro Coll., 1976-81, assoc. prof., 1981—, chmn. dept. psychology, 1978—; research assoc. Obesity Research Ctr., St. Luke's Hosp., N.Y.C., 1978—. Contbr. chpt., articles on food intake and obesity to profl. publs. Mem. Eastern Psychol. Assn., Am. Psychol. Assn., N. Am. Assn. Study Obesity. Jewish. Subspecialties: Physiological psychology; Nutrition (medicine). Current work: Control of food intake; treatments for obesity. Home: 165 Bennett Ave Apt 6A New York NY 10040 Office: Obesity Research Ctr St Luke's Hosp 114th and Amsterdam Ave New York NY 10025

GELINAS, ROBERT JOSEPH, scientist; b. Muskegon, Mich., Sept. 25, 1937; s. Robert Joseph and Vivian (Van Westen) G.; m. Mary R. Halfpap, June 18, 1960; 1 son, Robert H. Student, Muskegon (Mich.) Community Coll., 1955-57; B.S. in Chem. Engring, U. Mich., Ann Arbor, 1960; M.S. in Nuclear Engring, U. Mich., Ann Arbor, 1961, Ph.D., 1965. With Gen. Atomic, San Diego, 1960; student research asso. U. Mich., Ann Arbor, 1963-65; group leader, theoretical computational physics Lawrence Livermore (Calif.) Lab., 1966-75; sr. scientist, mgr. basic scis. div. Sci. Applications, Inc., Pleasanton, Calif., 1975—. Contbr. articles to sci. jours. Mem. Am. Nuclear Soc. (Mark Mills award 1965), Am. Phys. Soc., AAAS, Am. Chem. Soc., Combustion Inst., Internat. Assn. Math., Sigma Xi, Phi Kappa Phi. Subspecialties: Non-equilibrium fluid dynamics; Combustion processes. Current work: Non-equilibrium fluid dynamics; combustion, atmospheric photo-chem. modeling; free electron laser modeling; shock phenomena in gases and solids; transport of radiation and matter; continuum mechanics. Home: 580 E Angela St Pleasanton CA 94566 Office: 1811 Santa Rita Rd Suite 104 Pleasanton CA 94566

GELLER, HERBERT MILES, neurobiologist, educator; b. N.Y.C., Feb. 20, 1945; s. Edward and Sylvia (Tannenbaum) G.; m. Nancy Lorch, Sept. 3, 1967. B.E., CCNY, 1965; Ph.D., Case-Western Res. U., 1970. Asst. prof. pharmacology U. Medicine and Dentistry N.J.-Rutgers Med. Sch., Piscataway, 1972-78, assoc. prof., 1978—; cons., site visitor NIMH; mem. neurobiology adv. com. NSF, 1980-83. Contbr. articles to profl. jours. Mem. Soc. for Neurosci., Am. Physiol. Soc., Am. Soc. Pharmacology and Exptl. Therapeutics. Subspecialties: Neuropharmacology; Neurophysiology. Current work: Neurotransmission. Office: Dept. Pharmacology UMDNJ-Rutgers Med Sch Piscataway NJ 08854

GELLES, RUBIN, metrological instrumentation company executive; b. N.Y.C., Feb. 16, 1926; s. Louis and Celia (Jordan) G.; m. Dorothy Gelles, Jan. 31, 1927; 1 son, Adam. B.A., Cooper Union, 1948; B.S., CCNY, 1958; postgrad., NYU, 1958-60, Baruch Coll. Sch. Bus., 1979-81. Optical engr. Servo Corp. Am., Hicksville, N.Y., 1957-60; sr. optical engr. Kollsman Instrument Co., Syosset, N.Y., 1960-68, exec. engr., 1972-74, cons., 1974-77; v.p. engring. Spl. Optics, Little Falls, N.J., 1977-79; dir. research and devel. MKG Optical Instrument Corp., Woodside, N.Y., 1968-72; ops. mgr. Keuffel & Esser Co., Morristown, N.J., 1979—. Contbr. articles on optical engring., lens design, geometrical optics to profl. jours. Mem. Optical Soc. Am., Soc. Photo-Optical Instrumentation Engrs., Am. Mgmt. Assn. Subspecialties: Optical engineering; Robotics. Current work: Development of integrated metrology systems for automated factory and robotics control.

GELL-MANN, MURRAY, theoretical physicist; b. N.Y.C., Sept. 15, 1929; s. Arthur and Pauline (Reichstein) Gell-M.; m. J. Margaret Dow, Apr. 19, 1955 (dec. 1981); children: Elizabeth, Nicholas. B.S., Yale, 1948; Ph.D., Mass. Inst. Tech., 1951. Mem. Inst. for Advanced Study, 1951; instr. U. Chgo., 1952-53, asst. prof., 1953-54, asso. prof., 1954, Calif. Inst. Tech., Pasadena, 1955-56, prof., 1956—, now R.A. Millikan prof. physics.; Mem. Pres.'s Sci. Adv. Com., 1969-72. Author: (with Y. Ne'eman) Eightfold Way. Regent Smithsonian Instn., 1974—; bd. dirs. J.D. and C.T. MacArthur Found., 1979—. NSF post doctoral fellow, vis. prof. Coll. de France and U. Paris, 1959-60; Recipient Dannie Heineman prize Am. Phys. Soc., 1959; E.O. Lawrence Meml. award AEC, 1966; Franklin medal, 1967; Carty medal Nat. Acad. Scis., 1968; Research Corp. award, 1969; Nobel prize in physics, 1969. Fellow Am. Phys. Soc.; mem. Nat. Acad. Scis., Royal Soc. (fgn.), Am. Acad. Arts and Scis. Club: Cosmos. Subspecialty: Theoretical physics. Research theory of weak interactions, research dispersion relations. Developed eightfold way theory and Quark scheme developed strangeness theory. Office: Dept Physics Calif Inst Tech Pasadena CA 91125

GELMAN, DONALD, physics educator; b. Bklyn., Sept. 13, 1938; s. Samuel and Etta (Silverman) G.; m. Doraine Sills, Aug. 11, 1962; children: Sharon Beth, Deborah Lauren. B.S., Bklyn. Coll., 1959; M.S., NYU, 1964, Ph.D., 1969. Research physicist Kollsman Instrument Corp., Elmhurst, N.Y., 1961-64; assoc. prof. physics L.I. U., Greenville, N.Y., 1964—. Subspecialties: Theoretical physics; Theoretical astrophysics. Current work: Path integrals as applied to scattering theory. Home: 11 Fielding Ave Dix Hills NY 11746 Office: CW Post College Greenvale NY 11548

GEMMELL, DONALD STEWART, government research administrator, physicist; b. Adelaide, Australia, Nov. 6, 1934; m. Anne Gemmell, 1958; 3 children. B.Sc., U. Adelaide, 1955; Ph.D. in Physics, Australian Nat. U., 1958. Research fellow in physics Atomic Energy Research Establishment, Harwell, Eng., 1959-62; mem. staff Argonne (Ill.) Nat. Lab., 1962—, dir. physics div., 1982—; vis. fellow in physics Max Planck Inst. Nuclear Physics, Heidelberg, W. Ger., 1967-68; guest prof. U. Munich, W. Ger., 1974-75. Subspecialty: Physics research administration. Office: Argonne Nat Lab 9700 S Cass Ave Argonne IL 60439

GENCO, ROBERT JOSEPH, scientist; b. Silver Creek, N.Y., Oct. 31, 1938; s. Joseph and Santa (Barone) G.; m. Sandra C. Genco, Sept. 14, 1957; children: Deborah, Robert, Julie. Student (N.Y. State Regents scholar), Canisius Coll., Buffalo, 1956-59; D.D.S. cum laude (N.Y. State Regents scholar), SUNY, Buffalo, 1963; Ph.D. in Microbiology and Immunology (USPHS fellow), U. Pa., 1967. Cert. periodontics, 1967. Asst. prof. oral biology and periodontology SUNY Dental Sch., Buffalo, 1967-69, assoc. prof., 1969-74, prof., 1974—, chmn. dept. oral biology, 1977—, dir. grad. periodontology, 1968—, dir. fellowship program in immunology and periodontology, 1974—; dir. Periodontal Disease Clin. Research Center, 1978—; mem. dental caries program adv. com. Nat. Inst. Dental Research, NIH, 1975-78. Cons.: Infection and Immunity, 1972-77; editorial bd., 1977—; adv. editor: Immunochemistry, 1973-77; adv. editorial bd.: Jour. Dental Research, 1976—; editorial bd.: Jour. Periodontal Research, 1976-80; contbr. numerous articles to sci. jours. Recipient George Thorn award, 1977. Mem. Internat. Assn. Dental Researchers (Basic Research in Oral Sci. award 1975, Research in Periodontal Disease award 1981), Am. Assn. Dental Research (v.p. 1983-84), Am. Acad. Periodontology (chmn. Orban Prize com. 1975, chmn. Periodontology Pathology Center 1976, chmn. research com. 1977-78, 81-82, Gies award 1983), AAAS (chmn. dental sect. 1980), Am. Assn. Immunologists, Am. Assn. Microbiologists, ADA, N.Y. Acad. Scis., Sigma Xi, Omicron Kappa Upsilon, Alpha Omega. Roman Catholic. Subspecialties: Immunobiology and immunology; Periodontics. Current work: Role of immune phenomena in oral infections.

GENES, ANDREW NICHOLAS, geology educator, researcher; b. Boston, Oct. 28, 1932; s. Constantine and Antigone (Gavostes) G.; m. Aagot Marie Berg (div.); children: Thalia, Karl; m. Anne Frances Marr, May 13, 1978. A.B., M.A., Boston U.; Ph.D., U. Bergen, Norway, Syracuse U. Field editor McGraw-Hill Book Co., N.Y.C., 1960-66; assoc. prof. geology Boston State Coll., 1966-82, U. Mass.-Boston, 1982—; field mapper Maine Geol. Survey, Augusta, 1974—; cons. staff geologist Hydell/Eyster Assocs., Weymouth, Mass., 1979—. Contbr. articles to profl. jours. Mem. Hull (Mass.) Conservation Commn., 1982. NSF fellow, 1968-71. Mem. Norwegian Geol. Soc., Friends of Pleistocene. Greek Orthodox. Subspecialty: Geology. Current work: Mapping surficial geology of northern Maine wilderness areas; cirque features. Home: 3 C St Hull MA 02045 Office: Dept Geography and Earth Sci U Mass Boston MA 02125

GENEST, JACQUES, physician, research administrator, educator; b. Montreal, Que., Can., May 29, 1919; s. Rosario and Annette (Girouard) G.; m. Estelle Deschamps, Oct. 3, 1953; children: Paul, Suzanne, Jacques, Marie, Helene. B.A., Coll. Jean de Brebeuf, Montreal, 1937; M.D., U. Montreal, 1942; LL.D. (hon.), Queen's U., 1966, U. Toronto, Ont., Can., 1970, U. Ottawa, Ont., Can., 1983, D.Sc., Laval U., Quebec, Que., 1973, Sherbrooke U., 1974, Meml. U. Nfld., 1978, St. Xavier U., 1983. Intern Harvard Med. Sch., Boston, 1938, 39; resident Hôtel-Dieu Hosp., Montreal, 1942-45, Johns Hopkins Hosp., Balt., 1945-48, Harvard Sch. Chemistry, Boston, 1948, Rockefeller Hosp. Med. Research, N.Y.C., 1948-51; cons. practice medicine specializing in nephrology, endocrinology and internal medicine, dir. clin. research physician Hôtel-Dieu Hosp., Montreal, 1952—; prof. medicine U. Montreal, 1965—; prof. exptl. medicine McGill U., Montreal, 1960—; sci. dir. Clin. Research Inst. Montreal, 1965—; dir. Merck & Co., Rahway, N.J., Montreal Trust Co. Editor: (with Erich Koiw) Hypertension, 1972, (with Erich Koiw and Otto Kuchel) Hypertension: Physiopathology and Treatment, 1977, 2d edit., 1983. Decorated companion Order of Can.; recipient award Gairdner Found., 1963; Archambault medal Can. Assn. Advancement of Sci., 1965; Stouffer prize Am. Heart Assn., Cleve., 1969; Marie-Victorin Sci. prize Govt. Que., Que., 1977; Royal Bank award, 1980; Sims Commonwealth travelling prof., 1970. Fellow Royal Coll. Physicians and Surgeons Can., A.C.P., Royal Soc. Can. (Flavelle medal and award 1968); mem. Peripatetic Club U.S.A., Assn. Am. Physicians, Am. Clin. and Climatol. Assn., Am. Heart Assn. Roman Catholic. Clubs: St.-Denis (Montreal); Century Assn. and Club (N.Y.C.). Subspecialties: Nephrology; Endocrinology. Current work: Mechanisms and management of arterial hypertension. Home: 1171 Mont-Royal Blvd Montreal Canada PQ H2V 2H6 Office: 110 Pine Ave W Montreal Canada PQ H2W 1R7

GENET, RUSSELL MERLE, astronomer; b. Bell, Calif., Aug. 7, 1940; s. Gilbert R. and Elizabeth (Gillis) G.; m. Anna Marie Bailey; children: David, Michael, Karen, Shirley, Russell. B.S.E.E., U. Okla., 1964; M.S., Air Force Inst. Tech., 1980. Research engr., ops. research analyst USAF, 1975—, dir., 1979—, chief acquisition logistics research, 1979—. Author: Photoelectric Photometry of Variable Star, 1981, RealTime Control with Microcomputers, 1982; contbr. numerous articles to profl. jours. Served to capt. USAF, 1957-64. Mem. Am. Astron. Soc., Astron. Soc. Pacific, Am. Assn. Variable Star Observers, Internat. Amateur-Profl. Phooelectrc Photometry Assn. Subspecialties: Optical astronomy; Computer engineering. Current work: Devel. and application of microcomputer-controlled photoelectric photometers for optical astronomy. Home: 1247 Folk Rd Fairborn OH 45324 Office: Fairborn Obs 1247 Folk Rd Fairborn OH 45324

GENIGEORGIS, CONSTANTIN, veterinary public health educator, researcher; b. Drama, Greece, June 26, 1933; came to U.S., 1960; s. Angelos and Chrysoula (Hatzitaki) G.; m. Maudeen Boger, Jan. 29, 1966; children: Alexander, Kristina. D.V.M., Aristotle U., Thessaloniki, Greece, 1957; M.S., U. Calif.-Davis, 1963, Ph.D., 1966. Pvt. practice vet. medicine, Greece, 1959-60; research U. Calif.-Davis, 1964-70, asst. prof. vet. food safety, 1970-72, assoc. prof., 1972-78, prof., 1978—. Contbr. chpts., numerous articles to profl. publs. Served to 2d lt. Vet. Corps Greek Army, 1957-59. Named Outstanding Tchr. Vet. Food Hygiene Am. Assn. Food Hygiene Veterinarians, 1982; Greek State fellow, 1960-63; grantee NIH, 1964-65, 66-72, 72-74, WHO, 1973-75, U.S. Dept. Commerce, 1982-84, others. Fellow Royal Soc. Health; mem. Inst. Food Technologists, Am. Soc. Microbiology, Am. Meat Sci. Assn., AVMA, Soc. Applied Bacteriology, Hellenic Vet. Med. Assn. (hon.). Democrat. Greek Orthodox. Subspecialties: Preventive medicine (veterinary medicine); Food science and technology. Current work: Teaching and research in food safety, foodborne diseases, food processing and hygiene. Home: 4044 Vista Way Davis CA 95616 Office: Dept Epidemiology and Preventive Medicine Sch Vet Medicine U Calif Davis CA 95616

GENNIS, ROBERT BENNETT, chemist, biochemist, educator; b. N.Y.C., Oct. 7, 1944; s. Joseph and Sylvia (Levine) G.; m. Louise Slade, June 26, 1971 (div. 1977). B.S., U. Chgo., 1966; Ph.D., Columbia U., 1971. Helen Hay Whitney fellow Harvard U., 1971-73; asst. prof. U. Ill., Champaign-Urbana, 1973-80, assoc. prof. chemistry and biochemistry, 1980—. Bd. editors: Jour. Bacteriology, 1980—. Sloan fellow, 1978-80; NIH grantee, 1973—; Dept. Energy grantee,

1980—. Mem. Am. Soc. Microbiology, Biophys. Soc., Fedn. Am. Scientists for Exptl. Biology. Subspecialties: Biophysical chemistry; Biochemistry (biology). Current work: Membrane biochemistry; bioenergetics. Home: 507 W Illinois St Urbana IL 61801 Office: Noyes Lab U Ill 505 S Mathews 1 Room 49 Urbana IL 61801

GENSLER, HELEN LYNCH, molecular biologist; b. Stamford, Conn., Jan. 12, 1934; d. Joseph Martin and Lauretta (Murphy) Lynch; m. William George Gensler, July 8, 1961. B.A., Albertus Magnus Coll., 1955; M.S., U. Rochester, 1958; Ph.D., U. Ariz., 1979. Grad. fellow Sch. Medicine, U. Rochester, N.Y., 1955-58; research asst. Arthur D. Little Co., Cambridge, Mass, 1958-59; instr. physiology Coll. Nursing, U. Buffalo, 1961-62; cancer research scientist to sr. cancer research scientist Roswell Pk. Meml. Inst., Buffalo, 1959-63; research assoc. U. Pitts., 1967-68; research fellow dept. microbiology U. Ariz., Tucson, 1974-79, research assoc. dept. radiology/div. radiation oncology, 1981—; research assoc. Ariz. Center Occupational Safety and Health, Tucson, 1980—. Contbr. articles to profl. jours. USPHS fellow, 1978-79; NIH grantee, 1980-81; Cancer Biology grantee, 1981-82. Mem. Genetics Soc. Am., Am. Soc. Microbiology. Subspecialties: Genetics and genetic engineering (biology); Oncology. Current work: Genetic toxicology, DNA repair, oncology, gerontology. Office: Dept Radiology Div Radiation Oncology Univ Ariz Health Sci Center Tucson AZ 85724

GENTILE, ARTHUR CHRISTOPHER, university dean; b. N.Y.C., Nov. 24, 1926; s. Leo and Grace (Leone) G.; m. Gloria Lenore Ennevor, Jan. 22, 1949; 1 dau., Flora. B.S., CCNY, 1948; M.S., Brown U., 1951; Ph.D., U. Chgo., 1953. Plant physiologist U.S. Forest Service, Lake City, Fla., 1955-56; mem. faculty U. Mass., Amherst, 1956-72, prof. botany, 1964-72, asso. dean, 1965-72; dean Grad. Coll., v.p. research adminstrn. U. Okla., 1972-74; v.p. for acad. affairs U. Nev., Las Vegas, 1974-79; exec. dir. Am. Inst. Biol. Scis., 1979-83; dean acad. affairs Ind. U.-Kokomo, 1983—; Bd. dirs. Midcontinent Environ. Council Assn., 1972-74; mem. Sci. Manpower Commn., 1980-83; trustee Am. Type Culture Collection, 1980-83, Bioscis. Info. Service, 1981—. Author: Plant Growth, 1971. Served with AUS, 1945-46. USPHS research fellow Nat. Cancer Inst., 1954-55. Mem. Am. Soc. Plant Physiologists, AAUP, AAAS, Sigma Xi, Alpha Epsilon Delta. Subspecialties: Plant physiology (biology); Cell and tissue culture. Office: Ind U-Kokomo 2300 S Washington St Kokomo IN 46902

GENTILE, JAMES MICHAEL, biologist, researcher, educator; b. Chgo., Aug. 31, 1946; s. Michael A. and Theresa M. (Velluntini) G.; m. Glenda Jean, Dec. 29, 1973; 1 son, Michael Everett. B.A., St. Mary's Coll., Winona, Minn., 1968; M.S., Ill. State U., 1970, Ph.D., 1974. NDEA fellow Ill. State U., Normal, 1970-74, teaching asst., 1968-70, 71-74; postdoctoral research asso. dept. human genetics Sch. Medicine, Yale U., New Haven, 1974-76; asst. prof. biology Hope Coll., Holland, Mich., 1976-82, assoc. prof., 1982—; vis. asst. prof. genetics Inst. Environ. Studies, U. Ill., Urbana, 1977-78, 79-80, 82-83; cons. Lear Siegler Corp., Zeeland, Mich., 1977-78, Shering-Plough Corp., N.J., 1978-79, Squirtpak, Holland, Mich., 1977—, City of Muskegon, Mich., 1980, Kamatsu Am. Corp., Carol Stream, Ill., 1980. Contbr. chpts. to books, articles, abstracts to publs. in field; reviewer: U.S. EPA contracts reports Jour. Agriculture and Food Chemistry; apptd. reviewer for research grants, NSF, NIH. Research grantee Beta Lambda chpt. Phi Sigma Soc., 1974, Wilson Found., Hope Coll., 1979-80, 80-81; Mellon research grantee Hope Coll., 1977-78, 79-80; Research Corp. research grantee, 1978-79; Shering-Plough research awardee, 1978-79; Nat. Inst. Environ. Health Scis. research grantee, 1980; EPA research contracts, 1977-80, 80-82; research grants cons. NIH, 1981; Fed. Water Pollution Assn. trainee, 1966-68. Fellow Environ. Mutagen Soc.; mem. Genetics Soc. Am., Am. Soc. Microbiology, AAAS, Sigma Xi. Subspecialties: Genetics and genetic engineering (biology); Cancer research (medicine). Current work: Interaction of chemical and physical agents with genome of pro- and eukaryotes-environmental mutagenesis. Office: Dept Biolog Hope Coll Holland MI 49423

GENTRY, R. THOMAS, physiological psychologist; b. Coral Gables, Fla., Jan. 15, 1944; s. Roy Thomas and Margaret Louise (Tyler) G.; m. Donna Marie Bisesi, July 8, 1972; children: Russell Thomas, Sarah Victoria. B.S., B.A., Fla. Atlantic U., 1969; M.S., U. Mass., 1974, Ph.D., 1976. Research asst. neurosurgery Albany Med. Coll., 1970-72; instr. psychology U. Mass., 1973-74; Amherst Coll., 1972-76; postdoctoral fellow dept. of psychiatry and Inst. for Steroid Research, Albert Einstein Coll. Medicine, 1976-79; asst. prof. physiol. psychology Rockefeller U., N.Y.C., 1979—. Mem. AAAS, Soc. Neurosci., N.Y. Acad. Scis., Sigma Xi. Subspecialties: Physiological psychology; Neuropharmacology. Current work: Animal models of alcoholism, pharmacokinetics of ethanol, neuropharmacology of ethanol, voluntary consumption of ethanol, plasma concentrations of ethanol. Office: 1230 York Ave New York NY 10021

GENZEL, REINHARD LUDWIG, astrophysicist; b. Bad Homburg, W.Ger., Mar. 24, 1952; s. Ludwig and Eva Maria (Schlueter) G.; m. Orsolya Boroviczeny, Jan. 19, 1976; 1 dau., Daria. Diplom, U. Bonn, W.Ger., 1975, Dr. Rer. Nat., 1978. Postdoctoral fellow Center for Astrophysics, Cambridge, Mass., 1978-80; Miller fellow dept. physics U. Calif.-Berkeley, 1980-82, assoc. prof. physics, assoc. research astronomer, 1981—. Recipient Otto Hahn medal Max-Planck Soc., 1978. Mem. Am. Astron. Soc., Internat. Astron. Union, Am. Phys. Soc. Current work: Infrared and microwave spectroscopy, interferometry. Office: Dept Physics U Calif 567 Birge Hall Berkeley CA 94720

GEORGE, AJAX ELIS, neuroradiologist; b. Athens, Greece, Jan. 16, 1941; came to U.S., 1947; s. Elis P. and Helen E. (Palaiogiannidoy) G.; m. Bia Lapiotis, Aug. 22, 1982. M.D., Nat. U., Athens, Greece, 1966. Diplomate: Am. Bd. Diagnostic Radiology. Intern St. Joseph's Hosp., Paterson, N.J., 1966-67; resident in radiology NYU Med. Ctr., 1968-69, neuroradiology fellow, 1970-71, dir. computed tomography, 1979—, prof. radiology (neuroradiology), 1981—; dir. neuroradiology Bellevue Hosp., N.Y.C., 1972-78, attending radiologist, 1972—; med. adv. bd. Gen. Electric Co., Milw., 1983. Author: Neuroradiology Case Studies, 1977; con. reviewer: Am. Journal Neuroradiology, 1980—. NIH grantee, 1982—. Mem. Am. Soc. Neuroradiology, Radiologic Soc. N.Am., N.Y. Med. Soc. Greek Orthodox. Subspecialties: Diagnostic radiology; CAT scan. Current work: Research activities; aging and cognitive dysfunctions of aging, computed tomography, cat scanning, positron tomography, pet scanning, nuclear magnetic resonance. Home: 245 E. 40th St Apt 33E New York NY 10016 Office: Dept Radiolog NYU Sch Medicine 550 1st Ave New York NY 10016

GEORGE, ARNOLD, chemistry educator, researcher; b. Haverhill, Mass., May 15, 1941; s. Oliver and Louise (Zani) G.; m. Ismini Poliou, July 31, 1965; 1 child, Thanasi. B.S., U. Mass.-Amherst, 1964, Ph.D., 1969. Successively asst. prof., assoc. prof., prof. chemistry Mansfield (Pa.) State Coll., 1968—, chmn. chemistry dept., 1977—. NSF grantee, 1980, 81; recipient excellence in teaching award Pa. Dept. Edn., 1981. Mem. Am. Chem. Soc., Pa. Acad. Sci., Sigma Zeta. Subspecialty: Inorganic chemistry. Current work: Computer-assisted video instruction, chemistry for children, environmental and practical applications of chemistry. Office: Chemistry Dept Mansfield State Coll Mansfield PA 16933

GEORGE, CLAY EDWIN, psychology educator, consultant; b. Phoenix, June 1, 1926; s. Clay Heenan and Ruby (Hyde) G.; m. Mary Elizabeth Cook, May 29, 1958; children: Elizabeth Lynn, John Clay, Edgar Earle. B.S., Ariz. State U., 1949; M.A., U. ARiz., 1953; Ph.D., U. Houston, 1962. Instr. Tex. A&M U., 1953-56; lectr. U. Houston, 1956-58; asst. prof. Ariz. State U., 1958-61; sr. scientist Human Resources Research Office, Alexandria, Va., 1961-67; prof. psychology Tex. Tech. U., 1967—; cons. in field. Contbr. articles to profl. jours. Served in USMC, 1944-46, 50-52; PTO, China. Grantee Office Naval Research, 1955-61, Army Research Office, 1961-67, Dept. Def., 1968-74; Tex. State Organized Research grantee, 1978-81. Mem. Am. Psychol. Assn. Subspecialties: Social psychology; Behavioral psychology. Current work: Theoretical integration of small group behavior with organizational, environmental and community psychology. Home: 5423 16th Pl Lubbock TX 79416 Office: Dept Psychology Tex Tech U Box 4100 Lubbock TX 79409

GEORGE, JAMES, research physicist; b. Lynn, Mass., Dec. 29, 1922; s. John Jeanto and Thalia Anne (Pappas) Zorzy; m. Winifred Rose Nazzaro, Mar. 8, 1952; children: James Kevin, Carolyn Anne. B.S., Northeastern U., 1948; postgrad., Georgetown U., 1951-55. Physicist high energy radiation physics div. NIH, Nuclear Radiation Physics Lab., Bethesda, Md., 1948-53; physicist Mass Spectrometry Lab., U.S. Naval Med. Research Inst., Bethesda, 1953-55; sr. physicist dept. physics Cesium atomic beams Nat. Co., Melrose, Mass., 1955-66; dir. atomic research Frequency Electronics Inc., New Hyde Park, N.Y., 1973-78; research physicist JG Research & Tech., Salem, Mass., 1969—; dir. Electronautics Corp., Maynard, Mass., 1964-66; pres., treas. Frequency Control Corp., Topsfield, Mass., 1966-69. Contbr. articles to profl. jours. Mem. Am. Phys. Soc. Subspecialties: Atomic and molecular physics; Statistical physics. Current work: Investigations in the hyperfine energy transitions of cesium, surface ionization of atoms, atomic collisions with surfaces, atomic and ionic impact phenomena on metal surfaces; secondary electron emissions. Patentee in field. Home: 15 Oakledge Rd Swampscott MA 01907 Office: 89 Washington St Melrose MA 02176

GEORGE, NICHOLAS, optics educator, researcher; b. Council Bluffs, Iowa, Oct. 29, 1927; s. Nicholas and Marguerite (Hunsinger) G.; m. Carol Neufeld, June 18, 1966. B.S. with highest honors, U. Calif.-Berkeley, 1949; M.S., U. Md., 1956; Ph.D., Calif. Inst. Tech., 1959. Sect. chief Nat. Bur. Standards, 1951-53; sr. staff physicist Hughes Aircraft Co., Culver City, Calif., 1956-60; prof. elec. engring. and applied physics Calif. Inst. Tech., 1960-77; prof. optics U. Rochester, N.Y., 1977—, dir., 1977-81. Contbr. articles to profl. jours. Served with U.S. Army, 1946-47. Howard Hughes fellow, 1956-59. Fellow Optical Soc. Am., Soc. Photo-Optical Instrumentation Engrs.; mem. Am. Phys. Soc., IEEE, AAUP, Sigma Xi, Phi Beta Kappa, Tau Beta Pi, Phi Kappa Phi. Subspecialties: Optical image processing; Holography. Current work: Opto-electronic systems; hybrid computers for automatic image assessment; speckle; statistical optics; x-ray diffraction, interference and holography. Patentee in field. Office: Inst of Optics U Rochester Rochester NY 14627

GEORGE, STEPHEN ANTHONY, neurobiologist; b. Seattle, May 31, 1943; s. S. Anthony and Doris Lorene G.; m. Kathleen Ann Spalding, Sept. 6, 1969; children: Madeleine, Jennifer. B.Sc., U. B.C., 1964; Ph.D., John Hopkins U., 1970. Asst. prof. biol. scis. U. Md., Baltimore County, 1970-73; asst. prof. biology Amherst (Mass.) Coll., 1973-77, assoc. prof., 1977—. Contbr. articles to profl. jours. NIH fellow, 1966-70; grantee, 1975-81. Mem. AAAS, Soc. Neurosci., Assn. Research in Vision and Ophthalmology. Subspecialties: Neurobiology; Physiology (biology). Current work: Development and physiology of visual systems. Home: 23 Dana St Amherst MA 01002 Office: Dept. Biology Amherst Coll Amherst MA 01002

GEORGE, THOMAS CARL, nuclear systems test engineer; b. Lancaster, Ohio, May 25, 1956; s. Richard Harman and Sara Ann (Bertshin) G.; m. Barbara Ann Dirr, Apr. 12, 1980. B.S., U. Cin., 1979. Registered profl. engr., Ohio. Coop. student engr. Toledo Edison Co., 1975-78; nuclear systems test engr. Cleve. Elec. Illuminating Co., 1979—. Mem. Am. Nuclear Soc., Nat. Soc. Profl. Engrs. Republican. Roman Catholic. Subspecialties: Nuclear engineering; Nuclear fission. Current work: Nuclear supply systems. Home: 10242 Lola Ct Concord OH 44077 Office:: Cleve Electric Illuminating Co 55 Public Square Cleveland OH 44077

GEORGE, THYCODAM VARKKEY, government engineering executive, physicist; b. Pulthuppally, India, Aug. 19, 1938; came to U.S., 1956, naturalized, 1971; s. T. K. and Mariamma Varkkey; m. Achamma Mathew, June 14, 1964; children: Asha, Shobha, Sageer. B.Sc., U. Madras, India, 1956; Ph.D., U. Ill., 1964. Asst. prof. elec. engring. U. Ill.-Urbana, 1963-66; sr. engr. Westinghouse Research and Devel. Ctr., Pitts., 1966-75; with Office Fusion Energy, Dept. Energy, 1975—, program mgr., 1975-82, program mgr. microwave (plasma) heating tech., gyrotron devel. program, Washington, 1982—. Mem. Am. Phys. Soc., Sigma Xi. Democrat. Subspecialties: Electrical engineering; Fusion. Current work: Research in Thomson scattering, Rayleigh scattering, measurement of plasma properties with far-infrared absorption techniques. Management of research and development efforts related to fusion programs in industry and universities.

GEORGI, HOWARD MASON, III, physicist, educator; b. San Bernardino, Calif., Jan. 6, 1947; s. Howard Mason and Mary Alice (Mack) G.; m. Ann Rutledge Blake, June 14, 1969; children: Geoffrey Barnes, Justin Avery. B.A., Harvard U., 1967; Ph.D., Yale U., 1971. Mem. faculty Harvard U., Cambridge, Mass., 1976—, prof. physics, 1980—. Contbr. articles on theoretical particle physics to profl. jours.; author Lie Algebras in Particle Physics. NSF postdoctoral fellow, 1971-73; Sloan Found. fellow, 1976-80. Mem. Am. Phys. Soc., Harvard Soc. of Fellows.; fellow Am. Acad. Arts and Scis. Episcopalian. Subspecialties: Particle physics; Theoretical physics. Current work: Grand unified theories; properties of quarks and leptons. Office: Physics Dept Harvard U Cambridge MA 02138

GEORGIADES, JERZY ALEXANDER, biological research and development company executive; b. Drohobycz, Lwow, Poland, Apr. 18, 1928; came to U.S., 1963, naturalized, 1982; s. Jerzy and Elizabeth G.; m. Izolda Schroeder, Sept. 20, 1960; 1 son, George B. Ph.D., Med. Acad. Krakow, Poland, 1963; M.D., Sch. Medicine, Gdansk, Poland, 1960; postgrad., Sch. Chemistry, Szczecin, Poland, 1943-48. Asst. prof. State Inst. Marine and Tropical Medicine, Gdansk, 1953-60; assoc. prof. dept. microbiology Krakow Coll. Medicine, 1960-63, prof., 1964-67, 69-70; vis. prof. Inst. Med. Pathology, U. Naples, Italy, 1967-69, 70-71; asst. prof. virology, virologist, dept. virology U. Tex. System, Cancer Ctr., M.D. Anderson Hosp. and Tumor Inst., Houston, 1971-75; vis. prof. K.U. Leuven, Belgium, 1976-77; research assoc., dept. microbiology U. Tex. Med. Br., Galveston, 1977-78, faculty assoc., dept. microbiology, 1978-80; v.p., sci. dir. Immuno Modulators Labs., Inc., Stafford, Tex., 1980—. Fellow WHO, London, 1958, State Inst. Microbiology, Epidemiology and Hygiene, Prague, Czechoslovakia, 1956. Subspecialties: Virology (medicine); Microbiology (medicine). Current work: Research and development of interferon and related products. Office: Immuno Modulators Labs Inc 10511 Corporate Dr Stafford TX 77477

GERAN, RUTH I., biologist in cancer research; b. Middletown, Ohio, Nov. 27, 1922; d. Harry Clayton and Ethel Lacy (Kinney) G. B.A., Am. U., 1944; M.S., George Washington U., 1954, M.Phil., 1969, Ph.D., 1971. Biologist Nat. Cancer Inst., Bethesda, Md., 1958—. Contbr. articles to profl. jours. Mem. Am. Assn. for Cancer Research, Am. Inst. Biol. Scis., Am. Soc. Zoologists, Mortar Board, Kappa Delta, Beta Beta Beta. Methodist. Club: Potomac Appalachian Trail. Subspecialties: Cancer research (medicine); Chemotherapy. Current work: Searching for new or improved compounds for the treatment of cancer. Office: Nat Cancer Inst Room 415 Blair Bldg Bethesda MD 20205

GERBER, GWENDOLYN L., psychologist, educator; b. Calgary, Alta., Can.; came to U.S., 1958; d. Ernest and Alma G. A.B., UCLA, 1961, M.A., 1964, Ph.D., 1967. Lic. psychologist, N.Y. Research psychologist Bronx (N.Y.) State Hosp., 1969-70; clin. psychologist Hillside Hosp., Glen Oaks, N.Y., 1970-73; asst. prof. psychology John Jay Coll. Criminal Justice, CUNY, 1973-77, assoc. prof., 1977—; pvt. practice psychoanalysis, N.Y.C., 1969—. Contbr. articles to profl. jours. Mem. Am. PSychol. Assn., Eastern Psychol. Assn., Am. Women in Sci., Phi Beta Kappa, Chi Delta Pi. Subspecialties: Clinical psychology; Social psychology. Current work: Research into sex-role stereotypes. Office: John Jay Coll Criminal Justice 445 W 59th St New York NY 10012

GERBER, JOSEPH CHARLES, III, medical researcher, pharmacologist; b. Camden, N.J., June 20, 1952; s. Joseph Charles and Anna Jeanette (Buohl) G.; m. Frances Theresa Martin, June 1, 1974; children: Christopher, Erin, Kristin. Ph.D. in Pharmacology, Thomas Jefferson U. Postdoctoral fellow in neuropsychopharmacology U. Pa., Phila. Contbr. articles to profl. jours. Mem. AAAS, N.Y. Acad. Sci., Soc. for Neurosci., Am. Pharm. Assn., Sigma Xi. Subspecialties: Neurobiology; PET scan. Current work: Investigation of the actions and effects of neuroleptic drugs. Examination of and prediction of susceptability to malignant hyperthermia, actions of anesthetic agents. Office: Hahneman Med Coll and Hosp Dept Anesthesiology Philadelphia PA 19102

GERBLICH, ADI ABRAHAM, physician, educator; b. Borislaw, Poland, Jan. 2, 1945; s. Jacob and Bracha (Neuman) G.; m. Judith Chassis, July 14, 1970; children: Michal, Ilanit. M.D., Tel-Aviv U., 1973. Diplomate: Am. Bd. Internal Medicine. Intern Shorok Med. Ctr., Beer-Sheva, Israel, 1972-73; resident in internal medicine Mt. Sinai Med. Ctr., Cleve., 1973-75; fellow in pulmonary medicine Case Western Res. U., Cleve., 1975-77, postdoctoral fellow in microbiology, 1977-79, asst. prof. medicine, 1979—; staff physician VA Med. Ctr., Cleve., 1979—. Contbr. articles to med. jours. E.L. Trudeau fellow, VA research assoc., 1981. Mem. Am. Thoracic Soc., Ohio Thoracic Soc. Jewish. Subspecialties: Internal medicine; Immunology (medicine). Current work: Lung interstitial inflammation, asthma. Office: VA Med Ctr 10701 East Blvd Cleveland OH 44106

GEREN, COLLIS ROSS, biochemist; b. Miami, Ohio, Mar. 28, 1945; s. Elmer Ross and Margaret Ann (Halteman) G.; m. Lois Marie Kountz, Mar. 31, 1967; 1 dau., Tanya. B.S. in Sci. Edn, Northeastern Okla. State U., 1967; M.S. in Chemistry, Kans. State Coll. at Pittsburg, 1972; Ph.D. in Biochemistry, Okla. State U., 1974. Cert. tchr., Okla. Tchr. high sch., Picher, Okla., 1967-70; research asso. U. Kans. Med. Center, Kansas City, 1974-76; asst. prof. chemistry U. Ark., Fayetteville, 1976-79, asso. prof., 1979—; HEW-NIH-GM grantee, 1977-80, 1980-83, Research Corp. grantee, 1977-79. Contbr. articles on biochemistry to profl. jours. Recipient HEW NIH Research Career Devel. award, 1979-84. Mem. Am. Soc. Biol. Chemists, Am. Chem. Soc., AAUP, Sigma Xi. Subspecialties: Biochemistry (medicine); Biochemistry (biology). Current work: Purification and characterization of toxic and/or potentially useful molecules from N. Am. spider and snake venoms. Office: Chemistry Bldg U Ark Fayetteville AR 72701

GERETY, ROBERT JOHN, research physician; b. Jersey City, Oct. 16, 1939; s. James Leo and Helen (Beck) G.; m. Joan Imelda Grant, Feb. 3, 1967; children: Andrew Lawrence, Kathleen Suzanne, Nancy Grant. B.A., Rutgers U., 1962; M.A., Stanford U., 1966, Ph.D., 1971; M.D., George Washington U., 1970. Research assoc. Stanford U., 1970, intern, resident in pediatrics, 1971, 74-75; Commd. Capt. USPHS, 1971; advanced to med. dir.; clin. assoc. NIH, 1971-73; chief hepatitis br. Office of Biologics, FDA, Bethesda, Md., 1972—; cons. NIH, WHO; prof. Found. for Advanced Edn. in the Scis., NIH, 1973-81. Editor: Non-A, Non-B Hepatitis, 1981; contbr. articles to profl. jours. USPHS fellow, 1962-66; recipient Commendation medal USPHS, 1979, Outstanding Service medal, 1982. Mem. Am. Assn. Immunologists, William Beaumont Soc., Sigma Xi. Subspecialties: Health services research; Infectious diseases. Current work: Vaccine development, in vitro diagnostics virology, immune globulins; research hepatitis, vaccine development and regulation of biologics, specifically those related to monoclonal antibodies and gene splicing technology. Patentee in field. Home: 8807 Maxwell Dr Potomac MD 20854 Office: 8800 Rockville Pike Bethesda MD 20205

GERGIS, SAMIR DANIEL, anesthesia educator; b. Beni-Suef, Egypt, Sept. 24, 1933; came to U.S., 1968; s. Daniel and Hekmat (Assaad) G.; m. Dorothy K. Auen, June 16, 1973 (div. 1983); 1 son, Michael. M.B., Ch.B., Cairo U., 1954, D.A., 1957, D.M., 1958, M.D. in Anesthesia (Ph.D.), 1962; D.A., U. Copenhagen, 1963. Intern Cairo U. Hosp., 1955-56, resident, 1957-59; instr. dept. anesthesia Coll. Medicine U. Iowa, Iowa City, 1968-69, asst. prof., 1969-72, assoc. prof., 1972-76, prof. anesthesia, 1976—. Fellow Am. Coll. Anesthesiology; mem. Am. Soc. Anesthesiologists, Internat. Anesthesia Research Soc., N.Y. Acad. Scis., Am. Soc Pharmacology and Exptl. Therapeutics, Soc. Exptl. Biology and Medicine, AAAS, Nat. Soc. Med. Research, Soc. Neurosurg. Anesthesia and Neurologic Supportive Care. Coptic Orthodox Christian. Subspecialty: Anesthesiology. Current work: Neuromuscular physiology and pharmacology. Research in anesthesia for neurosurgery. Home: 1019 Sunset St Iowa City IA 52240 Office: Dept Anesthesia Coll Medicine U Iowa Iowa City IA 52242

GERHARDT, LESTER A., electrical and systems engineering educator; b. Bronx, N.Y., Jan. 28, 1940; m., 1961; 2 children. B.E.E., CCNY, 1961; M.S., SUNY-Buffalo, 1964, Ph.D. in Communication Systems, 1969. Sr. elec. engr. avionics systems Bell Aerospace Corp., 1961-64, sect. head signal and info. processing research, 1964-69, asst. to dir. advanced research and cons., 1969-75; assoc. prof. systems div. Rensselaer Poly. Inst., Troy, N.Y., 1969-74, prof. elec. and systems engring., 1974—, chmn. dept. elec. and systems engring., 1975—; cons. U.S. Air Force-Rome Air Devel. Ctr., 1972—, Gen. Elec. Co., 1978—; mem. panel Adv. Group Aerospace Research and Devel., NATO. Fellow IEEE; mem. Sigma Xi. Subspecialties: Systems engineering; Electrical engineering. Office: Dept Elec and Systems Engring Rensselaer Poly Inst Troy NY 12181

GERHARDT, PHILIPP, microbiologist, educator; b. Milw., Dec. 30, 1921; s. Philipp W. and Agnes (Daigh) G.; m. Vera Mary Armstrong, Feb. 24, 1945; children: Ellen Daigh, Stephen Philipp, Doris Mary. Ph.B. with honors, U. Wis., 1943, M.S., 1947, Ph.D., 1949. Diplomate: Am. Bd. Med. Microbiology. Faculty microbiology Oreg. State U., 1949-51, U. Mich. Med. Sch., 1953-65; prof., chmn. dept. microbiology and pub. health Colls. Natural Sci., Human Medicine, Osteo.

Medicine, Vet. Medicine and Agr. Expt. Sta., Mich. State U., 1965-75; prof., asso. dean for research and grad. study Coll. Osteo. Medicine, Mich. State U., 1975—; cons. various univs. and corps.; mem. U.S. nat. com. Internat. Union Biol. Scis. Editor: Manual of Methods for General Bacteriology, 1981. Served with AUS, 1943-46, 51-52. Wis. Alumni Research Found. fellow, 1946-47; NIH research fellow, 1947-49. Fellow AAAS; mem. Am. Soc. for Microbiology (sec. 1961-67, v.p. 1973-74, pres. 1974-75), Am. Acad. Microbiology (charter fellow, bd. govs. 1970-76), Brit. Soc. Gen. Microbiology, Internat. Union Microbiol. Socs. (pres. 1982—), Phi Beta Kappa, Sigma Xi. Subspecialties: Microbiology; Microbiology (medicine). Research and publs. on microbial endospores, permeability, fermentations. Home: 529 Woodland Dr East Lansing MI 48823

GERKEN, LOUIS CHARLES, research and development company executive; b. Lakewood, N.J., Nov. 27, 1925; s. Louis Charles and Meta G.; m. Carmen Hermenia Rios Mellado, Apr. 12, 1950; children: Louis, Valerie, Sherry. B.S.E.E., Port Arthur (Tex.) Coll., 1948; M.B.A., So. Meth. U., 1972. Cert. profl. logistician. With Pan Am. Airways, 1948-51; electronics cons. U.S. Air Force, 1951-54; electronics and anti-submarine warfare cons. U.S. Navy, 1954-70; mgr. H-G & Assocs., National City, Calif., 1970—; pres., gen. mgr. Am. Sci. Corp., National City, 1972—. Served to col. U.S. Army Res. Mem. Am. Soc. Naval Engrs., Soc. Logistics Engrs., Assn. Old Crows, Assn. U.S. Army, Res. Officers Assn. Subspecialties: Electronics; Oceanography. Current work: Research and development of electronics and electro-mechanical products. Home: 3250 Holly Way Chula Vista CA 92010 Office: 205 W 35th St National City CA 92050

GERMAN, JAMES LAFAYETTE, III, research physician, human geneticist; b. Grayson County, Tex., Jan. 2, 1926; s. George J. and Mary (Davis) G.; m. Margaret (Fohring), Jan. 14, 1956; children: James Lafayette IV, Ann Elizabeth. B.S., La. Poly. Inst., 1945; postgrad., U. Tex., Galveston, 1945-47; M.D., Southwestern Med. Coll., 1949. Diplomate: Am. Bd. Internal Medicine. Intern Cook County Hosp., Chgo., 1949-51; resident in medicine VA Hosp., Dallas, 1952-55; clin. assoc. NIH, Bethesda, Md., 1956-58; from research assoc. to asst. prof. Rockefeller Inst., N.Y.C., 1958-63; from assoc. prof. to prof. Cornell U. Med. Ctr., N.Y.C., 1963—; investigator N.Y. Blood Ctr., N.Y.C., 1968, now sr. investigator, dir., 1968—. Editor-in-chief Chromosomes series, 1974; author articles. Served with USN, 1943-46; Served with USPHS, 1955-58. Mem. Am. Soc. Human Genetics, Am. Soc. Clin. Investigation, Soc. Colonial Wars, S.R. Episcopalian. Subspecialty: Genetics and genetic engineering (medicine). Current work: The study of genetic and cytogenetic factors of importance in human health and disease, employing clinical observation and laboratory techniques. Home: 270 Riverside Dr Apt 11C New York NY 10035 Office: 310 E 67th St New York NY 10021

GERMAN, RANDALL M(ICHAEL), materials engineering, educator, consultant; b. Bainbridge, Md., Nov. 12, 1946; s. Eugene Knox and Helen (Schrufer) G.; m. Carol Jean Hosmer, Dec. 21, 1968; children: Eric James, Garth Thomas. B.S., San Jose State U., 1968; M.S., Ohio State U., 1971; Ph.D. U. Calif.-Davis, 1975. Materials scientist Battelle Labs., Columbus, Ohio, 1968-69; mem. tech. staff Sandia Labs., Livermore, Calif., 1969-77; dir. research Mott Metall. Co., Farmington, Conn., 1977-78, J.M. Ney Co., Bloomfield, Conn., 1978-80; prof. materials engring. Rensselear Poly. Inst., 1980—; dir. Newmet Products Inc., Terryville, Conn.; cons. in field. Contbr. numerous articles to profl. jours.; editor: Processing of Metal and Ceramic Powders, 1982. Recipient Electron Microscopy award Am. Soc. Metals and Internat. Microstructural Soc., 1972; diploma Internat. Inst. Sci. Sintering, 1980. Mem. Am. Soc. Metals (Alfred Geisler award 1981), Metall. Soc. (com. chmn. 1981-83), Internat. Precious Metals Inst., Am. Powder Metallurgy Inst., Am. Assn. Dental Research. Club: TI 99/4 (Albany, N.Y.). Subspecialties: Metallurgical engineering; Materials processing. Current work: Powder metallurgy processing for net shape forming of engineering components with emphasis on the sintering process. Patentee dental alloy. Office: Rensselaer Poly Inst Materials Research Troy NY 12181

GERMAN, RICHARD BARRY, engineering company executive; b. Halifax, N.S., Can., Nov. 19, 1950; s. Andrew Barry Crawford and Sage Janet (Ley) G.; m. Debra June Pepler, Sept. 9, 1978; 1 son, Andrew. B.S., Queen's U., Can., 1973. Registered profl. engr., Ont. Oceanological engr. Geocon Ltd., Toronto, Ont., Can., 1972-74, sr. project engr., joint venture, Lagos, Nigeria, 1974-75, mgr. field services, 1977-82; mgr. Geocon Offshore, Toronto, 1976—; mgr. bus. devel. Geocon, Inc., Toronto, 1982—; mem. task group on marine geotech. engring. NRC of Can., 1978-82. Richardson scholar Queen's U., 1969. Mem. Marine Tech. Soc., Arctic Inst. N.Am., Canadian Inst. Mining and Metallurgy, Prospectors and Developers Assn., Cons. Engrs. of Ont. Subspecialties: Ocean engineering; Mining engineering. Current work: Marine geotechnical engineering, offshore drilling and sampling, ocean mining, systems development, offshore technology. Home: 147 Old Orchard Grove Toronto ON Canada M5M 2E1 Office: Geocon Offshore 33 Yonge St Toronto ON Canada

GERMANO, DON JOSEPH, chemist; b. Chgo., July 18, 1953; s. Joseph Michael and Eleanor Louise (Herbst) G. B.A., Benedictine Coll., 1975; Ph.D., U. Pitts., 1980. Sr. research chemist Dow Chem. U.S.A., Freeport, Tex., 1980-82, project leader, 1982—. Actor Brazosport Little Theatre, Lake Jackson, Tex. Mem. Am. Phys. Soc., Am. Chem. Soc., Am. Soc. Testing and Materials, N.Am. Thermal Analysis Soc., Phi Lambda Upsilon. Roman Catholic. Subspecialties: Polymers; Thermodynamics. Current work: Direct a research group which utilizes dynamic mechanical testing, thermal analysis, microscopy and x-ray techniques to studypolymers; hazard evaluation. Home: 420 Garland Dr Apt 513 Lake Jackson TX 77566 Office: Dow Chem USA Tex Div B-1219 Freeport TX 77541

GERNANT, ROBERT EVERETT, geology educator; b. Geneseo, Ill., Dec. 3, 1941; s. Everett Adam and Dorothy (Frank) G.; m. Virginia Marie Kramer, July 1, 1967; children: Timothy Robert, Daniel Everett. B.S., U. Ill., 1963; postgrad., U. Tex., 1964; M.S., U. Mich., 1965, Ph.D., 1969. Geologist Shell Oil Co., Houston, 1963-65, Humble Oil and Refining Co., Lafayette, La., 1965-68; mem. faculty dept. geology U. Wis.-Milw., 1968—, prof., 1972—, chmn. dept., 1976-80; curator Greene Meml. Mus., Milw., 1983; editorial cons. Kendall/Hunt Pub. Co., Dubuque, Iowa, 1976—. Author: Glossary of Geology, 1972; contbr. articles to profl. jours. Mem. Water Planning Bd., Mequon, Wis., 1981—. Mem. Paleontol. Soc., Friends of Bivalves (chmn. 1970), Friends of Ostracodes, Sigma Xi, Phi Kappa Phi. Presbyterian. Subspecialties: Paleoecology; Paleobiology. Current work: Paleoecology of marine, invertebrate fossils, especially microfossils; evolution of fossil ostracoda; quantitative morphology of fossil ostracoda; geologic analysis of Nineteenth century photography of Western United States. Home: 10017 N Greenview Dr Mequon WI 53092 Office: Dept Geol and Geophys Scis U Wis Milwaukee WI 53201

GEROLA, HUMBERTO C., physicist; b. Buenos Aires, Argentina, Feb. 7, 1943; came to U.S., 1972; s. Umberto and Ada (Pepe) G.; m. Marina O. Pothe, Nov. 16, 1966; children: Wolfgang H., Werner A. Ph.D. in Physics, U. Rome, Italy, 1969. Asst. prof. U. Buenos Aires, 1969-72; research asso. U. Colo., 1972-79, N.Y. U., 1974-76; mem. research staff IBM T. J. Watson Research Center, Yorktown Heights, N.Y., 1976-82; tech. planning staff IBM Research Div., 1982-83; mgr. IBM San Jose Research Lab., 1983—; vis. prof. 1st di Astrofisico Spaziale CNR, Frascati, Italy, 1981; dir. Nat. Center Cosmic Radiation, Argentina, 1970-71. Contbr. numerous articles to sci. jours. Recipient IBM Outstanding Invention award, 1978. Mem. Internat. Astron. Union, Am. Astron. Soc., N.Y. Acad. Scis. Subspecialty: Theoretical astrophysics. Current work: Manager physics of storage.

GERONE, PETER JOHN, microbiologist, research institute administrator; b. Oakfield, N.Y., Apr. 11, 1928; s. John and Rose (Cipolla) G.; m. Anne Lindberg, Aug. 25, 1951; children: Roseanne, John, Carl, Susan, Paul, Mary Ellen, Laura Beth. B.A., U. Buffalo, 1949, M.A., 1951; Sc.D., Johns Hopkins U., 1954. Am. Bd. Med. Microbiology. Supervisory microbiologist U.S. Army, Frederick, Md., 1954-71; dir. Delta Primate Ctr. Tulane U., Covington, La., 1971—; adj. prof. microbiology Tulane U. Med. Sch., New Orleans, 1973—, adj. prof. tropical medicine, 1977—; dir. Gulf So. Research Inst., Baton Rouge, 1978—. Mem. Am. Soc. Microbiology, AAAS, Am. Soc. Primatologists, Internat. Soc. Primatology. Democrat. Roman Catholic. Subspecialties: Virology (medicine); Microbiology (medicine). Current work: Infectious diseases, virology, research administration. Home: 150 Cheron Dr Mandeville LA 70448 Office: Delta Regional Primate Research Center Tulane U Three Rivers Rd Covington LA 70433

GERSHBEIN, LEON LEE, biochemist, research inst. exec., clin. chemist; b. Chgo., Dec. 22, 1917; s. Meyer and Ida (Shutman) G.; m. Ruth, Sept. 30 1956; children: Joel Dan, Marcia Renee, Carla Ann. S.B. (Oscar Blumenthal Meml. Scholar), U. Chgo., 1938, M.S. in Chemistry, 1939, Ph.D., Northwestern U., 1944. Research asso. in chemistry, Nat. Def. Research Council and Pitts. Plate Glass postdoctoral fellow Northwestern U., 1944-47; asst. prof. biochemistry U. Ill. Coll. Medicine, Chgo., 1947-53; asso. prof. biology Ill. Inst. Tech., 1953-57, adj. prof., 1957—; pres. and dir. Northwest Inst. for Med. Research, Chgo., 1957—; dir. labs. Northwest Hosp., Chgo., 1957—. Research numerous publs. in clin. chemistry and biochemistry, neurobiology, sebaceous lipids, liver regeneration, nutrition and endocrine metabolism. Recipient merit award in chromatography Chgo. Gas Chromatography Discussion Group, 1978, citation Ill. State Acad. Sci., 1975-79. Mem. Am. Chem. Soc., Am. Inst. Chemists, Am. Oil Chemists' Soc., AAAS, Ill. State Acad. Scis., N.Y. Acad. Sci., Soc. Exptl. Biology and Medicine, Soc. Applied Spectroscopy, Soc. Cosmetic Chemists, Am. Fedn. Clin. Research, Am. Assn. Cancer Research, Sigma Xi, Phi Lambda Upsilon. Subspecialties: Biochemistry (medicine); Clinical chemistry. Current work: Tumorigenesis; liver regeneration; tissue enzymes and isoenenzymes, lipid metabolism with emphasis on sebaceous types. Home: 2836 Birchwood Ave Wilmette IL 60091 Office: 5659 W. Addison St Chicago IL 60634

GERSHBERG, HERBERT, physician, educator, researcher; b. N.Y.C., Dec. 1, 1917; s. Abraham and Rose (Goldstein) G.; m. Herta L. Stadelmann, Aug. 30, 1947 (dec. Aug. 1964); m. Beatrice G. Gottler, Oct. 11, 1968; children: Viviane, Denise, Danielle. B.S., CCNY, 1936; M.S., U. Md., 1937; M.D., Med. Coll. Va., 1941. Intern Jewish Hosp., Bklyn., 1941-42; resident U.S. Army Evacuation Hosp., 1942-45; fellow physiol. chemistry Yale U., 1945-48, fellow internal medicine, 1948; asst. prof. medicine and physiology NYU, 1950-56, assoc. prof., 1966—; dir. diabetes and endocrine clinics Bellevue Hosp., 1960-66; consulting endocrinologist N.Y. Infirmary, Beekman Downtown Hosp., N.Y.C.; attending physician Bellevue, Univ., Doctors' hosps.; vis. prof. endocrinology U. Miami Sch. Medicine, 1972; cons. WHO, 1964. Contbr. reviews and articles to profl. jours. Mem. malpractice mediation panel N.Y. Supreme Ct. Served to maj. AUS, 1942-45. Decorated Bronze Star. Mem. Harvey Soc., Soc. Exptl. Biology and Medicine, Am. Physiol. Soc., Am. Fedn. Clin. Research, Am. Diabetes Assn., Endocrine Soc. Jewish. Subspecialties: Endocrinology; Internal medicine. Current work: Pituitary, parathyroid, diabetes, calcium metabolism, growth hormone. Home: 34 Skyridge Rd Greenwich CT 06830 Office: 614 2d Ave New York NY 10016

GERSHEN, JAY ALAN, dental educator, evaluator dental-care programs; b. Bronx, N.Y., Apr. 9, 1946; s. Julius and Ver (Sherman) G.; m. Arleen Yvonne Corman, July 3, 1971. B.A. in Psychology, SUNY-Buffalo, 1968; D.D.S., U. Md., 1972; Ph.D. in Edn, UCLA, 1976. Prof. pediatric dentistry, pub. health and preventive dentistry Sch. Dentistry, UCLA, 1976—; cons. in field; dir. UCLA Sch. Dentistry Mobile Dental Clinic, 1976—. Contbr articles to profl. publs.; author instructional videotapes, 1980. Robert Wood Johnson Health Policy fellow Inst. Medicine, Nat. Acad. Scis., Wshington, 1982-83. Mem. Am. Assn. Dental Schs. (chmn. sect. behavioral scis. 1982-83), Am. Pub. Health Assn., Internat. Assn. Dental Research, Behavioral Scientists in Dental Research, Am. Ednl. Research Assn. Subspecialty: Behavioral sciences in dentistry. Current work: Behavioral science curriculum development; evaluator dental-care programs and quality assurance systems. Office: Sch Dentistry CHS 23-019 UCLA Los Angeles CA 90024

GERSHON, ELLIOT SHELDON, psychiatrist; s. David and Ann (Pohorille) G.; m. Faye Deborah Saltman, Nov. 4, 1967; children: Ari Andrew, Ethan Daniel. A.B., Harvard U., 1961, M.D., 1965. Intern Mt. Sinai Hosp., N.Y.C., 1965-66; teaching fellow psychiatry Harvard U. Med. Sch.; also resident in psychiatry Mass. Mental Health Center, Boston, 1966-69; cons. Peter Bent Brigham Hosp., Boston, 1968-69, Prince George's County (Md.) Health Dept., 1969-70; clin. asso. lab. clin. sci. NIMH, Bethesda, Md., 1969-71, unit chief, sect. psychogenetics, biol. psychiatry br., 1974—, sect. chief, 1978—, mem. faculty staff coll., 1977—; dir. research Jerusalem (Israel) Mental Health Center, 1971-74; mem. faculty Washington Sch. Psychiatry, 1976—; sci. adv. bd. Israel Center Psychobiology, 1972-74; sr. surgeon USPHS, 1969-71, 75-80, med. dir., 1980—; Mem. sci. adv. bd. Am. Friends of Jerusalem Mental Health Center, 1978—. Author: Impact of Biology on Modern Psychiatry, 1977, Genetic Research Strategies for Psychobiology and Psychiatry, 1981; also articles; editorial bd.: Jour. Affective Disorders, 1978—, Psychiatry Research: An Internat. Jour. for Rapid Communications, 1978—, Psychiat. Devels., 1982—, Jour. Psychiat. Research, 1983. Recipient Anna Monika Found. prize, 1979; USPHS Commd. Officer's Commendation medal, 1979. Mem. AAAS, Am. Psychiat. Assn., Am. Psychopath. Assn., Psychiat. Research Soc., Am. Coll. Neuropsychopharmacology, Soc. for Neurosci. Jewish. Subspecialty: Psychopharmacology. Current work: Clinical and biological inherited factors in major psychiatric disorders. Pschopharmacology and pharmacogenetics. Human population genetics, biochemical genetics, chromosomal linkage. Address: Bldg 10 Room 3N218 NIMH 9000 Rockville Pike Bethesda MD 20205

GERSHON, MICHAEL DAVID, anatomist, educator; b. N.Y.C., Mar. 3, 1938; s. Murray Huda and Juliette (Levinson) G.; m. Elda Anne Angen, June 10, 1961; children: Perry, Timothy, Dana. B.A., Cornell U., 1958, M.D., 1963. Fellow, instr. Cornell U. Med. Coll., N.Y.C., 1963-65, asst. prof., 1965-69, assoc. prof. anatomy, 1969-74, prof., 1974-75; research assoc. Oxford (Eng.) U., 1965-66; prof. anatomy and cell biology, chmn. dept. Columbia U. Coll. Phys. and Surgeons, 1975—; mem. neurol. disorders program project rev. com. NIH, 1972-75, Neurology A study sect., 1980—. Contbr. med. jours.; Editorial bd.: Neurochemistry Internat., Jour. Histochemistry, Anat. Record, Jour. Comparative Neurology and Anatomy and Embryology. Recipient Borden Undergrad. research prize, 1963; N.Y.C. Health Research Council Career Scientist award, 1971; Markle Found. scholar acad. medicine, 1968; grantee NIH. Mem. Am. Assn. Anatomists, Am. Gastroenterol. Assn., Am. Soc. Cell Biology, Am. Physiol. Soc., AAAS, Endocrine Soc., Soc. Neurosci., N.Y. Soc. Electron Microscopists (pres. 1977-78), N.Y. Acad. Sci., Internat. Soc. Devel. Neurosci., Phi Beta Kappa, Sigma Xi, Alpha Epsilon Delta, Phi Kappa Phi. Club: Cajal. Subspecialty: Neurobiology. Office: 630 W 168th St New York NY 10032

GERSTEN, JOEL IRWIN, physicist, educator; b. N.Y.C., Mar. 18, 1942; s. Lester and Esther (Berman) G.; m. Harriet Riva Rubenstein, June 7, 1964; children: Samuel, Bonnie, Sarah, Eli. B.S., CCNY, 1962; Ph.D., Columbia U., 1968. Mem. Tech. staff Bell Telephone Labs., Murray Hill, N.J., 1968-70; prof. physics CUNY, 1970—; exec. officer doctoral program, 1982—; researcher, cons. Contbr. articles to sci. jours. Fellow Am. Phys. Soc.; Mem. Phi Beta Kappa. Subspecialties: Condensed matter physics; Theoretical physics. Current work: Surface physics, optics, atomic and molecular physics, computer applications. Office: Dept Physics CUNY New York NY 10031

GERSTENHABER, MURRAY, mathematician, educator; b. Bklyn., May 6, 1927; s. Joseph and Pauline (Rosenzweig) G.; m. Ruth P. Zager, June 3, 1956; children: Jeremy, David, Rachel. B.S., Yale U., 1948; Ph.D., U. Chgo., 1951; J.D., U. Pa., 1973. Asst. to assoc. prof. math. U. Pa., Phila., 1953-60, prof., 1961—; lectr. Law Sch., 1973—, cons. on statis. questions in law. Editor: Some Mathematical Problems in Biology, 1968; Contbr. articles to profl. jours. Bd. dirs. Solomon Schecter Day Sch., Merion, Pa., 1982. NSF grantee. Fellow AAAS; mem. Am. Math. Soc., Math. Assn. Am., Soc. for Indsl. and Applied Math., ABA. Current work: Deformation of rings and algebras; statistical questions in law. Office: Dept Math U Pa Philadelphia PA 19104

GERSTL, SIEGFRIED ADOLF WILHELM, physicist, researcher; b. Neuhausen/Filder, Wuerttemberg, Germany, Aug. 5, 1939; came to U.S., 1968, naturalized, 1974; s. Franz Xaver and Pauline (Haug) G.; m. Andrea Haug, Oct. 10, 1970; children: Susanne, Stephan. Dipl. Phys., U. Stuttgart, Germany, 1964; Dr. rer. nat., U. Karlsruhe, Germany, 1967. Scientist U. Stuttgart, 1964-65, U. Karlsruhe, 1965-68; sr. engr. advanced reactors div. Westinghouse Corp., Pitts., 1968-71; staff physicist Argonne (Ill.) Nat. Lab., 1971-74; mem. staff Los Alamos Nat. Lab., 1974—; cons to govt. Contbr. articles to profl. jours.; inventor fusion reactor blanket and shield, 1978; author computer codes. Mem. Am. Nuclear Soc. (chmn. radiation protection and shielding div. 1979, Spl. Services award 1978, 79), Am. Geophys. Union. Subspecialties: Climatology; Theoretical physics. Current work: Transfer of solar radiation through atmosphere in ultra violet, visible and infra red regions; remote sensing by satellites; laser propagation and multiple scattering in atmosphere; effects of aerosols and ozone depletion on atmospheric radiation balance; modelling of light scattering and reflection from vegetative plant stands; fusion reactor neutronics; climatology. Home: 15 El Nido Los Alamos NM 87544 Office: Theoretical Div Lso Alamos Nat Lab T-DOT MS B279 Los Alamos NM 87545

GERSTLE, FRANK P., JR., materials scientist, research adminstr.; b. Louisville, June 23, 1942; s. Frank P. and Marie E. (Marrone) G.; m. Louise T., Aug. 23, 1967; children: Ron, David. B.A. in Math.Physics, St. Joseph's Coll., 1965; S.B. and S.M. in Mech. Engring, M.I.T., 1966; Ph.D. in Mech. Engring. and Materials Sci, Duke U., 1972. Mem. staff Sandia Labs., Albuquerque, 1966-68, mem. tech. staff div. composite materials, 1972-73, div. mechanics of materials, 1973-74, supr. div. composites and polymer mechanics, 1974—; grad. asst. Duke U., 1968-72; mem. com. on organic matrix composites Nat. Materials Adv. Bd., 1978-80, mem. com. on lightweight mil. vehicles, 1981-82; chmn. Gordon Research Conf. on Composites, 1981. Contbr. articles to profl. jours. Bd. dirs. Soc. Devel. Gifted and Talented Students, Albuquerque, 1980-81; chmn. Albuquerque Mountain Rescue Council, 1982. Mem. ASME, Soc. Advancement of Materials Processing, Soc. Engring. Sci. Club: N.Mex. Mountain. Subspecialties: Composite materials; Theoretical and applied mechanics. Current work: Mech. behavior of materials; design of composite structures; supervision activities in composite materials, textiles, and failure in polymers; research in creep rupture in composites. Office: Sandia Nat Labs Div 1814 Albuquerque NM 87185

GERWICK, BEN CLIFFORD, JR., construction engineer, educator; b. Berkeley, Calif., Feb. 22, 1919; s. Ben Clifford and Bernice (Coultrap) G.; m. Martelle Louise Beverly, July 28, 1941; children Beverly (Mrs. Robert A. Brian), Virginia (Mrs. Roy Wallace), Ben Clifford III, William. B.S., U. Calif., 1940. With Ben C. Gerwick, Inc., San Francisco, 1939-70, j1pres, 1952-70; exec. v.p. Santa Fe-Pomeroy, Inc., 1968-71; prof. civil engring. U. Calif. at Berkeley, 1971—; sponsoring mgr. Richmond-San Rafael Bridge substructure, 1953-56, San Mateo-Hayward bridge, 1964-66; lectr. constrn. engring. Stanford U., 1962-68; cons. major bridge and marine constrn. projects; cons. constrn. engr. for ocean structures, also concrete offshore structures in, North Sea, Arctic Sea, Japan, Australia, Indonesia, Arabian Gulf; mem. exec. com., past chmn. marine bd., mem. polar research bd. NRC; mem. Nat. Acad. Engring., Norwegian Acad. Tech. Scis. Author: (with Peter V. Peters) Russian-English Dictionary of Prestressed Concrete and Concrete Construction, 1966, Construction of Prestressed Concrete Structures, 1971, (with John C. Woolery) Construction and Engineering Marketing for Major Project Services; Contbr. articles to profl. jours. Served with USNR, 1940-46; comdr. Res. ret. Recipient Lockheed award Marine Tech. Soc., 1977. Fellow ASCE (Karp award 1976), Am. Concrete Inst. (dir. 1960, Turner award 1974, Corbetta award 1981); mem. Federation Internationale de la Precontrainte (pres. 1974-78, now hon. pres., Freyssinet medal 1982), Prestressed Concrete Inst. (pres. 1957-58, hon.), Deutscher Beton Verein (hon., Emil Mörsch medal 1979), Concrete Soc. U.K. (hon.), Association Francaise du Beton (hon.), Moles, Soc. Naval Architects and Marine Engrs. (Blakely Smith award 1981), Beavers (Engring. award 1975), Phi Beta Kappa, Tau Beta Pi, Sigma Xi, Chi Epsilon, Kappa Sigma. Conglist. Clubs: Bohemian (San Francisco); Claremont Country (Oakland). Subspecialties: Offshore technology; Ocean engineering. Current work: Offshore construction with emphasis on Artic and sub-arctic marine environment; prestressed concrete application to marine construction. Home: 5874 Margarido Dr Oakland CA 94618 Office: 217 McLaughlin Hall U Calif Berkeley CA 94720 also 500 Sansome St San Francisco CA 94111

GERWIN, BRENDA ISEN, biochemist, researcher; b. Boston, May 2, 1939; d. Maurice M. and Jeannette Hershon Isen; m. Robert David Gerwin, Dec. 18, 1960; children: David, Daniel, Joel. B.A. cum laude in Biochemistry, Radcliffe Coll., 1960; Ph.D. in Biochemistry, U. Chgo., 1964. Instr. biochemistry Rockefeller U., 1964-66, Case Western Res. U., 1966-69; biochemist molecular anatomy program Oak Ridge Nat. Lab., 1969-70; sr. staff fellow viral leukemia and lymphoma br. Nat. Cancer Inst., Bethesda, Md., 1971-73, chemist viral biochemistry sect., 1973-81, research chemist, 1981—. Contbr. chpts. and articles to profl. publs. Mem. Am. Soc. Biol. Chemists, Am. Soc. Microbiology, Am. Chem. Soc., AAAS, Sigma Xi. Subspecialties: Biochemistry (biology); Genome organization. Current work: Sequencing DNA of mutant RNA tumor viruses to determine which sequences encode or control important functions; examining intermediates in copying of RNA to DNA by reverse transcriptase. Office: Nat Cancer Inst Bldg 41 Suite 100 Bethesda MD 20205

GESCHEIDER, GEORGE ALBERT, psychologist, educator; b. Steubenville, Ohio, Aug. 28, 1936; s. George Floto and Mary Fieta (Clauss) G.; m. Margaret Elizabeth Nicola, Sept. 6, 1958 (div. 1981); children: Mary, Margaret. B.S., Denison U., 1958; M.S., Tulane U., 1961; Ph.D., U. Va., 1964. Faculty dept. psychology Hamilton Coll., Clinton, N.Y., 1964—, prof. psychology, 1974—. Author: Psychophysics: Method and Theory, 1976; Contbr. articles to profl. jours. Mem. Soc. Neurosci., Sigma Xi. Subspecialties: Psychophysics; Sensory processes. Current work: Tactile sensitivity. Home: Fanchel House Hamilton Coll Clinton NY 13323 Office: Dept Psychology Hamilton Coll Clinton NY 13323

GESCHWIND, NORMAN, physician, educator; b. N.Y.C., Jan. 8, 1926; s. Morris and Anna (Blau) G.; m. Patricia Dougan, Sept. 8, 1956; children: Naomi, David, Claudia. B.A., Harvard U., 1946, M.D., 1951; LL.D. (hon.), Northwestern U., 1981; D.h.c., U. Lyon, 1981. Moseley Travelling fellow Nat. Hosp., London, 1952-53, USPHS research fellow, 1953-55; resident in neurology Boston City Hosp., 1955-56; research fellow MIT, 1956-58; staff neurologist Boston VA Hosp., 1958-62, chief neurology service, 1962-66; asso. prof. neurology Boston U., 1962-66, prof., chmn. dept., 1966-68; James Jackson Putnam prof. neurology Harvard Med. Sch., 1969—. Fellow Am. Acad. Neurology; mem. Am. Acad. Arts and Scis., Acad. Aphasia, Am. Neurol. Assn., Royal Belgian Acad. Medicine (corr.). Subspecialties: Neurology; Neuropsychology. Current work: Anatomical bases of behavior, syndromes of cortical connections, biological foundations of cerebral dominance. Research, publs. on anat. basis higher functions nervous system, biological foundations of cerebral dominance, brain and emotion. Office: 330 Brookline Ave Boston MA 02215

GESLICKI, MARK LOUIS, engring. adminstr.; b. Rome, N.Y., Nov. 24, 1949; s. Paul and Elaine (Legendre) G. B.S. in Physics, Rochester Inst. Tech., 1972. Assoc. engr. optical design Xerox Corp., Rochester, N.Y., 1972-74, sr. engr. optical design, Dallas, 1974-79, engring. mgr. design laser scanning systems, El Segundo, Calif., 1980-81, imaging systems devel. mgr., 1981—. Mem. Optical Soc. Am. Republican. Subspecialties: Laser scanning systems; Laser optics design. Current work: Program management of raster scanning systems for digital printing, using lasers as light sources, system integration and architecture of laser scanners in host printing systems. Office: 701 S Aviation Blvd El Segundo CA 90245

GESSOW, ALFRED, aerospace engineer, educator; b. Jersey City, Oct. 1922; s. Morris Samuel and Emma (Levovsky) G.; m. Elaine E. Silverman, Nov. 23, 1947; children—Laura Gessow Goldman, Lisa Gessow Michelson, Miles Jory, Andrew Jody. B.C.E., CCNY, 1943; M.Aero. Engring., N.Y. U., 1944. Aero. research scientist, nat. adv. com. on aeros. Langley Research Center, Va., 1944-59; chief fluid physics research NASA, Washington, 1959-66, asst. dir. research div., 1966-70, dir. for aerodynamics, 1970-79; prof., chmn. dept. aerospace engring. U. Md., College Park, 1979—; lectr. U. Va., Va. Poly. Inst.; adj. prof. N.Y. U., Cath. U. Am.; vis. prof. Korean Inst. Advanced Sci.; cons. NATO; chmn. aerospace div. adv. council Pa. State U.; mem. Nat. Acad. Sci/NRC bd. Army Sci. and Tech., 1982—. Sr. author: Aerodynamics of the Helicopter, 3d edit, 1967; contbr. articles to encys. and profl. jours. Recipient medal for exceptional service NASA, 1974. Fellow Am. Helicopter Soc., AIAA. Jewish. Subspecialties: Aeronautical engineering; Aerospace engineering and technology. Current work: Aerospace education; research in rotocraft aerodynamics and flight dynamics. Home: 7308 Durbin Terr Bethesda MD 20817 Office: U Md College Park MD 20742 The human psychology is such as to require and respond to an awards system which is made up of many factors—challenge, variety, peer recognition and economic benefits. The young generally place more importance on the first two, while the last two are more significant in later years. The chances of achieving such rewards are enhanced by starting early to develop expertise in a specialty area, to be interested and knowledgeable in one or more allied areas, and to take on challenging tasks which are not required but are self motivated. If these steps are followed, success and satisfaction will follow.

GETZELS, JACOB WARREN, psychologist, educator; b. Bialystok, Poland, Feb. 7, 1912; came to U.S., 1921, naturalized, 1933; es. Hirsch and Frieda (Solon) G.; m. Judith Nelson, Dec. 24, 1949; children: Katharine, Peter, Julia. B.A., Bklyn. Coll., 1936; M.A., Columbia U., 1939; Ph.D., Harvard U., 1951. Instr. ednl. psychology U. Chgo., 1951, asst. prof. ednl. psychology, 1952-54, asso. prof., 1955-57, prof., 1957—, now R. Wendell Harrison Disting. Service prof. edn. and behavioral scis.; vis. prof. psychology U. P.R., summer, 1962, Stanford U., summer, 1963; mem. U.S. Office Edn. Mission to Soviet Russia, 1960, mem. research adv. council, 1964-70; mem. council of scholars Library of Congress, 1980—. Author: (with A. Coladarci) The Use of Theory in Educational Administration, 1955, (with P.W. Jackson) Creativity and Intelligence: Explorations with Gifted Students, 1962, (with J.M. Lipham and R.F. Campbell) Educational Administration as a Social Process, 1968, (with I. Taylor) Perspectives in Creativity, 1975, (with M. Csikszentmihalyi) The Creative Vision: A Longitudinal Study of Problem Finding in Art, 1976; contbr. articles to profl. jours. Mem. bd. visitors Learning Research and Devel. Center, U. Pitts., 1973-79; bd. dirs. Spencer Found., 1971—. Recipient Research award Am. Personnel and Guidance Assn., 1959; Tchrs. Coll. medal, 1977; Nicholas Murray Butler medal for theory or philosophy of edn. Columbia U., 1980; Center for Advanced Study in Behavioral Scis. fellow, 1960-61; Center for Policy Study (U. Chgo.) fellow, 1967-75. Mem. Am. Psychol. Assn., Am. Sociol. Assn., Nat. Acad. Edn. (1st. v.p. 1972-76), Nat. Soc. for Study Edn. (dir. 1975-77), Am. Ednl. Research Assn. Subspecialties: Cognition; Social psychology. Current work: Longitudinal studies of development of artists; research on creative thought processes of children and artists; the social context of creativity. Office: 5835 S Kimbark Ave Chicago IL 60637

GEULA, GIBORI, endocrinologist, educator; b. Beirut, Aug. 8, 1945; U.S., 1973; s. Aslan and Sophie (Cheka) Derzie; m. Shimon Gibori, Mar. 1, 1970; children: Gil, Ilan, Ron. B.S. in Biology, Lebanese U., Beirut, 1967; M.S. in Reproductive Biology, Sorbonne U., Paris, 1968; Ph.D. in Physiology, Tel Aviv U., 1973. Fulbright fellow Case Western Res. U., Cleve., 1973-75; Ford Found. fellow U. Mich., Ann Arbor, 1975-76; mem. faculty U. Ill., Chgo., 1976—, assoc. prof. physiology and biophysics, 1980—. Editorial bd.: Biology of Reproduction, 1982—; reviewer: Endocrinology; contbr. chpts. to books, sci. articles to profl. publs. Centre National de Recherche Scientifique fellow, 1967-68; NIH grantee, 1977—; NSF grantee, 1982—. Mem. Soc. Study of Fertility, Soc. Study of Reproduction (chmn. edn. com. 1980-81), Endocrine Soc., AAAS, Am. Physiology Soc. Subspecialties: Endocrinology; Reproductive biology. Current work: Ovarian and placental function. Home: 8932 Pottowattomi St Skokie IL 60076 Office: Univ Ill Physiology and Biophysics Dept PO Box 6998 Chicago IL 60680

GEYER, LYNETTE ARNASON, zoologist; b. Salt Lake City, Nov. 12, 1938; d. Julius Bradford and Fern Mildred (Farrer) Arnason; m.; children: Bradford Lee, Melora. B.S., Brigham Young U., 1958; M.A. in Ednl. Psychology, U. Calif., Berkeley, 1964; Ph.D. in Zoology, Rutgers U., 1976. Fellow Monell Chem. Senses Center, Phila., 1976-78, asst. mem., 1978-81; sr. research scientist Am. Cyanamid Consumer Product Research Div., Clifton, N.J., 1981-82; sr. scientist Rocky Mountain Biol. Lab., 1978-81; lectr. U. Pa., 1981. Contbr. articles in field to profl. jours. Mem. Animal Behavior Soc., Assn. for Chemoreception Scis., Soc. Neuroscience, Am. Soc. Mammalogists, Eastern Psychol. Assn. Subspecialties: Psychobiology; Integrated pest management. Current work: Behavioral or physiological responses of humans to odors; biol. approaches to pest mgmt.; behavioral profiles in dog breeds.

GEYER, WAYNE ALLAN, forest scientist; b. Oak Park, Ill., Nov. 21, 1933; s. Herman M. and Alice J. (Miller) G.; m. Patricia Joyce Wheeler, Aug. 20, 1960; children: Keith, Kevin. B.S., Iowa State U., 1955; M.S.F., Purdue U., 1962; Ph.D., U. Minn., 1971. Field forester Ga. Kraft Paper Co., Macon, 1957-59; research assoc. forestry U. Ill., Simpson, 1962-65; forest scientist Kans. State U., Manhattan, 1966—, prof. forestry, 1982—. Contbr. articles to profl. jours. Served with USNR, 1955-57. Mem. Soc. Am. Foresters. Subspecialty: Biomass (agriculture). Current work: Reclamation of mined areas; fuelwood forestry plantations; herbicides; soil-site studies. Office: Forestry Dept Kans State U Manhattan KS 66506

GEZARI, DANIEL YSA, astrophysicist; b. N.Y.C., Nov. 6, 1942; s. Zvi and Temima G.; m. Pirjo Kortelainen, July 27, 1974; 2 children. A.B., Cornell U., 1966; M.S., N.Y.U., 1969; Ph.D., Stony Brook U., 1973. Research asst. Woods Hole Oceanographic Inst., 1963, Kitt Peak Nat. Obs., 1969; research fellow Calif. Inst. Tech., Pasadena, 1973-76; resident research asso. NRC, Washington, 1976-78; astrophysicist NASA/Goddard Spare Flight Center, Greenbelt, Md., 1978—; Robert A. Millikan research fellow in astrophysics Calif. Inst. Tech., 1973. Contbr. 30 articles in field to profl. jours. Mem. Internat. Astron. Union, Am. Astron. Soc. Subspecialties: Infrared optical astronomy; Biomedical engineering. Current work: Infrared observational astrophysics star formation regions, speckle interferometry, cryogenic instrumentation, space physics and technology; infared array detectors. Patentee in field. Office: NASA/Goddard Space Flight Center Greenbelt MD 20771

GHAFFAR, ABDUL, immunologist, educator; b. India, July 6, 1942; s. Abdul and Faheema Sattar; m. Anita Kristensen, Oct. 4, 1966; children: Tariq, Omar, Yasmeen. B. Sc., U. Karachi, Pakistan, 1962; M. Phil., U. London, 1970; Ph.D., U. Edinburgh, Scotland, 1973. Research fellow U. Edinburgh, 1973-75; research asst. prof. U. Miami, Fla., 1975-78; assoc. prof. immunology U. S.C. Med. Sch., Columbia, 1978—. Fellow Am. Acad. Microbiology; Mem. British Soc. Immunology, Am. Assn. Immunologists, Am. Soc. Microbiology, Reticuloendothelial Soc., Royal Coll. Pathology. Muslim. Subspecialties: Immunobiology and immunology; Immunotoxicology. Current work: Macrophage functions, immunosuppression, immunopotentiation; macrophage subpopulations, lymphocyte proteinases, immunosuppression, immunopotentiation. Home: 808 Coldbranch Dr Columbia SC 29204 Office: U SC Sch Medicine Columbia SC 29208

GHANGAS, GURDEV S., molecular biologist; b. Ghangas, Punjab, India, Nov. 15, 1942; s. Gurbachan S. and Dalip Kaur (Virk) G.; m. Surinder P. Kaur, Aug. 1972; children: Imroz, Param, Roop. B.Sc. (Merit scholar), Punjab Agrl. U., 1963, M.S., 1966; Ph.D., Syracuse U., 1971. Lectr. Punjab Agrl. U., 1966; Carver fellow Tuskegee Inst., Ala., 1966-67; research asso. Cornell U., Ithaca, N.Y., 1971-74; postgrad. research biochemist U. Calif., Berkeley, 1974-76; asso. research scientist N.Y. U. Med. Center, N.Y.C., 1977-78; vis. asst. prof. Cornell U., 1979-80; vis. asso. prof. N.Y. Med. Coll., 1981; mgr. cell biology Biol. Energy Corp., Valley Forge, Pa., 1982—; cons. Contbr. articles to profl. jours. NIH fellow, 1977-79. Mem. Am. Chem. Soc., Am. Assn. Immunologists, Genetics Soc. Am. Subspecialties: Genetics and genetic engineering (agriculture); Molecular biology. Current work: Gene structure and function in procaryotes, eucaryotes, viruses and plasmids, biotechnology. Inventor. Home: 61 Spring Run Rd Freeville NY 13068 Office: PO Box 766 Valley Forge PA 19482

GHANTA, BABU MADHUR, computer science educator; b. Guntur, Andhra Pradesh, India, Mar. 5, 1941; s. Rattaiah and Venkata Subbamma (Gogineni) G.; m. Vijaya Madhur Vemulappalli, June 24, 1970. B.S., Andhra U., 1959; B.Tech. with honors, Indian Inst. Tech., 1963; M.S., MIT, 1969; Ph.D., N.Y. U., 1980. Indsl. engr. Guest Keen Nettlefolds, Bombay, India, 1963-64; tool engr. Godrej & Boyce, Inc., Bombay, 1964-65; mfg. engr. IBM, Bombay, 1965-67; sr. systems cons. Equitable Life Assurance Co., N.Y.C., 1971-77; sr. staff analyst Arthur D. Little, Cambridge, Mass., 1979-80; prof. mgmt. sci. Manhattan Coll., 1980—. Mem. Assn. Computing Machinery, Am. Mktg. Assn. Subspecialties: Software engineering; Graphics, image processing, and pattern recognition. Current work: Human factors aspects of software engring.; automated factories. Home: 89 Oakview Terr Short Hills NJ 07078 Office: Manhattan Coll Riverdale NY 07041

GHAVAMIKIA, HAMID, research engineer; b. Tehran, Iran, Mar. 23, 1951; s. Zia and Masoumeh (Majidi) G.; m. Sheila Farrokhalaee, Sept. 25, 1972; children: Mike, Kevin. B.S., Arya-Mehr U., Tehran, 1973; M.S., Imperial Coll., London, 1976; Ph.D., Cambridge U., Eng., 1979. Instr. Arya-Mehr U., 1973-75; asst. prof. Isfahan (Iran) U. Tech., 1979-81; research assoc. Case Western Res. U., 1982; sr. research engr. Gen. Motors Research Ctr., Warren, Mich., 1982—. Arya-Mehr U. scholar, 1969; Isfahan U. Tech. scholar, 1975. Mem. Am. Chem. Soc., Am. Phys. Soc., Am. Inst. Chem. Engrs. Subspecialties: Chemical engineering; Polymers. Current work: Physics of polymers, in particular: polymer-solvent interactions, polymer crystallization, composites interfaces. Home: 38104 Lorostown Dr Sterling Heights MI 48077 Office: Gen Motors Research Center Warren MI 48090

GHEBREHIWET, BERHANE, immunologist, educator; b. Asmara, Ethiopia, Sept. 28, 1946. D.V.M., Sch. Vet. Medicine, Warsaw, Poland, 1971; M.V.Sc., Ecole Nationale Veterinaire D'Alfort, France, 1973; D.Sc., U. Paris, 1974. Research assoc. dept. molecular immunology Scripps Clinic and Research Found., La Jolla, Calif., 1974-79; asst. prof. medicine SUNY, Stony Brook, 1979—, asst. prof. pathology, 1983—. Contbr. articles to profl. jours. Mem. Am. Assn. Immunology, Am. Fedn. Clin. Research, N.Y. Acad. Sci., Am. Chem. Soc., AAAS, Am. Assn. Vet. Immunology. Coptic Orthodox. Subspecialties: Immunology (medicine); Biochemistry (medicine). Office: SUNY at Stony Brook HSC T-16 Room 040 NY 11794

GHEORGHIADE, MIHAI, cardiologist, physician, educator; b. Bucarest, Rumania, Feb. 19, 1946; came to U.S., 1972, naturalized, 1978; s. Cosma and Matilda (Sneer) G.; m. Anborg Reidun Lindblom, Sept. 8, 1968. Maturita, Matei Basarab, Bucharest, 1963; M.D., U. Rome, 1972. Diplomate: Am. Bd. Internal Medicine, Am. Bd. Cardiology. Intern Miriam Hosp. Brown U., Providence, 1972-73, resident, 1973-76, chief resident, 1976-77, fellow in cardiology, 1977-79; chief cardiology sect. VA Med. Ctr., Salem, Va., 1979—; asst prof. medicine U. Va., Charlottesville, 1979-83, assoc. prof., 1983—. Contbr. articles to profl. jours. Recipient Outstanding Attending award U. Va., 1981. Fellow Am. Coll. Cardiology; mem. ACP, Roanoke Acad. Medicine, Am. Fedn. Clin. Research, Am. Heart Assn. (dir. 1979—), Porsche Club Am. Subspecialties: Internal medicine; Cardiology. Current work: Left ventricular functions and cardiac pharmacology (inotropic agents). Home: 5260 Crossbow Circle Roanoke VA 24014 Office: VA Medical Center University of Virginia Salem VA 24011

GHER, MARLIN EUGENE, JR., periodontist, naval officer, educator; b. Harrisburg, Pa., Jan. 8, 1947; s. Marlin E. and Joan E. (Ewing) G.; m. Sharon R. Sumler, May 23, 1970; children: Steven Todd, Jaime Malia. D.D.S., U. Mo.-Kansas City, 1971; M.Ed., George Washington U., 1981. Diplomate: Am. Bd. Oral Medicine, Am. Bd. Periodontics; cert. in periodontics Nat. Naval Dental Center, Bethesda, Md., 1979. Commd. lt. U.S. Navy Dental Corps, 1971, advanced through grades to comdr., 1979; asst. dental officer USN, Twentynine Palms, Calif., 1971-73, Kaneohe, Hawaii, 1973-77, resident in periodontics, Bethesda, 1977-79; mem. staff periodontal dept. Nat. Naval Dental Ctr., Bethesda, 1979-81; head periodontal dept., regional cons. Naval Dental Center, San Diego, 1981—; assoc. prof. U. Md., 1979-81; guest lectr. U. So. Calif., 1982; cons. Uniform Services U. of Health Scis., 1979-81; Research in immunogenicity of freeze-dried skin allografts, 1977-81, inventor collapsible image viewing device. Mem. ADA, Am. Acad. Periodontology, Am. Acad. Oral Medicine, Internat. Assn. Dental Research, Am. Assn. Dental Research. Republican. Episcopalian. Subspecialty: Periodontics. Current work: Immunology and clinical application of freeze-dried skin used as . Home: 2804 Atadero Ct Carlsbad CA 92008 Office: NRDC Naval Sta Box 147 San Diego CA 92136

GHETTI, BERNARDINO, neuropathologist, neurobiology researcher; b. Pisa, Italy, Mar. 28, 1941; s. Getulio and Iris (Mugnetti) G.; m. Caterina Genovese, Oct. 8, 1966; children: Chiara, Simone. M.D. cum laude, U. Pisa, 1966, specialist in mental and nervous diseases, 1969. Lic. physician, Italy; cert. Edn. Council for Fgn. Med. Grads., diplomate: Am. Bd. Pathology. Postdoctoral fellow U. Pisa, 1966-70; research fellow in neuropathology Albert Einstein Coll. Medicine, Bronx, N.Y., 1970-73, resident, clin. fellow in pathology, 1973-75, resident in neuropathology, 1975-76; asst. prof. pathology Ind. U., Indpls., 1976-77, asst. prof. pathology and psychiatry, 1977-78, assoc. prof. pathology and psychiatry, 1978-83, prof., 1983—. Contbr. articles and abstracts to profl. jours. Mem. Am. Assn. Neuropathologists, Soc. Neurosci., Assn. Research in Nervous and Mental Diseases, Am. Soc. Cell Biology, Italian Soc. Psychiatry, Italian Soc. Neurology, Sigma Xi. Roman Catholic. Subspecialties: Pathology (medicine); Neurobiology. Current work: Pathologic reactions of neuronal cytoskeleton to toxic compounds, experimental models induced by natural agents, neurobiology and neuropathology of genetically determined cerebellar degenerations in mutant mice. Home: 1124 Frederick S Dr Indianapolis IN 46260 Office: 635 Barnhill Dr Room 157 Indianapolis IN 46223

GHIL, MICHAEL, research scientist; b. Budapest, Hungary, June 10, 1944; came to U.S., 1971; s. Louis and Ilona V. (Dobo) Cernat; m. ELiza M. Cristescu, Feb. 28, 1968 (div. July 1981); m. Michele R. J. Denizot, July 8, 1982. B.Sc., Technion-Israel Inst. Tech., 1966, M.Sc., 1971; M.S., NYU, 1973, Ph.D., 1975. Registered profl. engr., Israel. Research asst., instr. Technion-Israel Inst. Tech., Haifa, 1966-71; research scientist Courant Inst., NYU, N.Y.C., 1971-75; Nat. Acad. Sci./NRC research assoc. NASA Goddard Inst. Space Studies, N.Y.C., 1975-76; research asst. prof. math. Courant Inst., 1976-79, research assoc. prof. atmospheric sci., 1979-82, research prof., 1982—; faculty research assoc. NASA Goddard Lab. Atmospheric Sci., Greenbelt, Md., 1977—; cons. U.S. Nat. Meteorol. Ctr., Washington, 1982—. Editor: Dynamic Meteorology: Data Assimilation Methods, 1981; author: Topics in Geophysical Fluid Dynamics, 1984; editor: Atmospheric Sci. and Applied Math, 1981—. Served to lt. Israeli Navy, 1967-71. Mem. Am. Meteorol. Soc., Am. Geophys. Union, Am. Math. Soc., Soc. Indsl. and Applied Math. Jewish. Subspecialties: Meteorology; Applied mathematics. Current work: Applied mathematics, climate dynamics, dynamic meteorology, estimation theory, geophysical fluid dynamics, nonlinear phenomena, numerical methods, remote sensing and applications. Office: Courant Inst NYU 251 Mercer St New York NY 10012

GHISELLI, WILLIAM BARRON, psychology educator; b. Berkeley, Calif., June 30, 1940; s. Edwin E. and Louisa (Hickox) G.; m. Bettina Bost, Apr. 5, 1966; children: Michelle Bost, Peter von Arx. B.A., San Francisco State Coll., 1966; M.S., U. Pitts., 1970, Ph.D., 1972. Research scientist E.R. Johnstone Tng. and Research Ctr., Bordentown, N.J., 1972-74; asst. prof. U. Mo., Kansas City, 1974-80, assoc. prof. psychology, 1980—; research and program evaluator Osawatomie (Kans.) State Hosp., 1975—; cons. in field. Contbr. articles to profl. jours. Sec. Ctr. for Developmentally Disabled, Kansas City, Mo., 1978—; bd. dirs. Kansas City Adv. Bd. Mental Retardation, 1979. Served with U.S. Army, 1960-63. Mem. Common Cause, Psychonomic Soc., Am. Psychol. Assn., Internat. Soc. Research on Aggression, Midwestern Psychol. Assn., Southwestern Psychol. Assn. (Mo. rep. 1981-82). Subspecialties: Behavioral psychology; Learning. Current work: Animal behavior, aggression, developmental disabilities. Office: Dept Psychology Univ Mo 5301 Holmes St Kansas City MO 64110

GHONIEM, NASR MOSTAFA, engineering educator, consultant; b. Alexandria, Egypt, Mar. 5, 1948; came to U.S., 1974, naturalized, 1980; s. Mostafa H. and Etmad (Elzarda) G.; m. Virginia L. Schink, June 19, 1976; children: Amira Anne, Adam Tarek. B.S., Alexandria U., 1971; M.Eng., McMaster U., 1974; M.S., U. Wis., 1975, Ph.D., 1977. Research engr. Whiteshell Can. Nuclear Research Establishment, Man., 1974; research asst. fusion design group U. Wis.-Madison, 1974-77; asst. prof. engring. UCLA, 1977-81, assoc. prof., 1982—; vis. research scientist Oak Ridge Nat. Lab., 1981, cons., 1981—; cons. space and def. systems TRW, Inc., Los Angeles, 1980—. Contbr. articles to tech. publs., also reviewer articles. Bd. dirs. Egyptian-Am. Scholars Assn., 1981—. Egyptian Govt. fellow, 1967-71. Mem. Am. Nuclear Soc., Atomic Indsl. Forum. Democrat. Moslem. Subspecialties: Nuclear engineering; Nuclear fusion. Current work: Radiation damage and effects, material selection for energy systems, fusion engineering and design, structural design, ion-solid interactions, mechanical metallurgy and microstructure-property relationships. Home: 10415 Danube Ave Granada Hills CA 91344 Office: 6291 Boelter Hall UCLA Los Angeles CA 90024

GHORAI, SUSANTA K., physics educator; b. Kandarparur, W. Bengal, India, Aug. 16, 1937; came to U.S., 1964; s. Debendra Nath and Late Pankajini (Maity) G.; m. Mamata, Aug. 14, 1964; 1 child, Sujoy K. B.Sc. with honors, Calcutta (India) U., 1958, M.Sc., 1960; Ph.D., Auburn U., 1971. Lectr. physics Contai (India) P.K. Coll., 1961-64; grad. teaching asst. dept. physics Auburn U., 1964-68; asst. prof. physics Ala. State U., Montgomery, 1968-71, assoc. prof., 1973—, dir. physics grants, 1975-80; asst. prof. physics Lincoln Meml. U., Harrogate, Tenn., 1971-73; cons. in field. Author: Modern College Physics Experiments, 1982. Coach soccer YMCA, Montgomery, 1979—. Office Edn. grantee, 1975-80; NSF research grantee, 1978; NSF LOCI grantee, 1978-81. Mem. Am. Phys. Soc., Am. Assn. Physics Tchrs., Sigma Xi. Democrat. Hindu. Subspecialty: Nuclear physics. Current work: Research in partial neutron cross sections; acid rain and asbestos pollution analysis by the PIXE method. Home: 1442 Greenway Pl Montgomery AL 36117 Office: Ala State U 915 S Jackson St Montgomery AL 36195

GHOSHAL, NANI GOPAL, veterinary anatomist, educator; b. Dacca, Bangladesh, Dec. 1, 1934; came to U.S., 1963; s. Priya Kanta and Kiron Bala (Thakurata) G.; m. Chhanda Banerjee, Jan. 24, 1971; 1 dau., Nupur. G.V.Sc., B.V.C., India, 1955; D.T.V.M., U. Edinburgh, 1961; Dr. med. vet., Tieraerztliche Hochschule Hannover, W. Ger., 1962; Ph.D., Iowa State U., 1966. Vet. asst. surgeon West Bengal (India) State Govt., 1955-56; instr. Bengal Vet. Coll., U. Calcutta, 1955-56; research asst. M.P. Govt. Coll. Vet. Sci. and Animal Husbandry, Mhow, India, 1956-59; research officer FAO, India, 1963; instr. Iowa State U., Ames, 1963-66, asst. prof., 1967-70, assoc. prof., 1970-74, prof. vet. gross anatomy, 1974—; chmn. Internat. Vet. Medicine Com., 1967-79. Co-author, editor: Getty's Anatomy of Domestic Animals, 1975; author: (with Tankred Koch, Peter Popesko) Venous Drainage of Domestic Animals, 1981; contbr. chpts. to books, articles to profl. jours. Recipient various scholarships and grants. Fellow Royal Zool. Soc. Scotland (life); mem. World Assn. Vet. Anatomists, Am. Assn. Vet. Anatomists, AAAS, Am. Assn. Anatomists, Pan Am. Assn. Anatomy, N.Y. Acad. Scis., Iowa Vet. Med. Assn., Sigma Xi, Phi Zeta, Gamma Sigma Delta, Phi Kappa Phi. Subspecialties: Morphology; Biomedical engineering. Current work: Morphology of various organ systems, brain temperature regulation, audio-visual education, developmental anatomy. Home: 1310 Glendale Ave Ames IA 50010 Office: Coll Vet Medicine Iowa State U 1070 Vet Anatomy Ames IA 50011

GIACCONI, RICCARDO, astrophysicist; b. Genoa, Italy, Oct. 6, 1931; came to U.S., 1956, naturalized, 1967; s. Antonio and Elsa (Canni) G.; m. Mirella Manaira, Feb. 15, 1957; children—Guia Giaconni Chmiel, Anna Lee, Marc A. Ph.D., U. Milan, Italy, 1954. Asst. prof. physics U. Milan, 1954-56; research asso. Ind. U., 1956-58, Princeton U., 1958-59; exec. v.p., dir. Am. Sci. & Engring. Co., Cambridge, Mass., 1959-73; prof. astronomy Harvard U.; also asso. dir. high energy astrophysics div. Center Astrophysics, Smithsonian Astrophys. Obs./Harvard Coll. Obs., Cambridge, 1973-81; dir. Space Telescope Sci. Inst., Balt., 1981—; mem. space sci. adv. com. NASA, 1978-79, mem. adv. com. innovation study, 1979—, astronomy adv. com., 1979—; mem. high energy astronomy survey panel Nat. Acad. Scis., 1979-80. Co-editor: X-ray Astronomy, 1974; Author numerous articles, papers in field. Fulbright fellow, 1956-58; recipient Röntgen prize astrophysics Physikalish-Medizinische Gesellschaft, Wurzburg, Germany, 1971; Exceptional Sci. Achievement medal NASA, 1971, 80; Disting. Public Service award, 1972; Space Sci. award AIAA, 1976; Elliott Cresson medal Franklin Inst., 1980; also Bruce medal; Heinneman award; Russell lectr. Mem. Am. Astron. Soc. (Helen B. Warner award 1966, chmn. high energy astrophysics div. 1976-77, councilor 1979-82), Italian Phys. Soc. (Como prize 1967), AAAS, Internat. Astron. Union (Nat. Acad. Scis. astron. rep. 1979-82), Nat. Acad. Scis., Am. Acad. Arts and Scis. Club: Cosmos (Washington). Subspecialty: 1-ray high energy astrophysics. Inventor x-ray telescope, discovered x-ray stars. Home: 4205 Underwood Rd Baltimore MD 21218 Office: Space Telescope Sci Inst Homewood Campus Baltimore MD 21218

GIAEVER, IVAR, physicist; b. Bergen, Norway, Apr. 5, 1929; came to U.S., 1957, naturalized, 1963; s. John A. and Gudrun (Skaarud) G.; m. Inger Skramstad, Nov. 8, 1952; children—John, Anne Kari, Guri, Trine. Siv. Ing., Norwegian Inst. Tech., Trondheim, 1952; Ph.D., Rensselaer Poly. Inst., 1964. Patent examiner Norweigian Patent Office, Oslo, 1953-54; mech. engr. Can. Gen. Electric Co., Peterborough, Ont., 1954-56; applied mathematician Gen. Electric Co., Schenectady, 1956-58, physicist, 1958—. Served with Norwegian Army, 1952-53. Recipient Nobel Prize for Physics, 1973; Guggenheim fellow, 1970. Fellow Am. Phys. Soc. (Oliver E. Buckley prize 1965); mem. IEEE, Norwegian Profl. Engrs., Nat. Acad. Sci., Nat. Acad. Engring. (V.K. Zworykin award 1974), Am. Acad. Arts and Scis., Norwegian Acad. Sci., Norwegian Acad. Tech. Subspecialties: Tissue culture; Condensed matter physics. Current work: Immunology, tissue culture, biophysics. Spl. research in pioneer tunneling into superconductors, 1962. Home: 2080 Van Antwerp Rd Schenectady NY 12309 Office: Gen Electric Co Research and Devel Center PO Box 1088 Schenectady NY 12301

GIAMALVA, MIKE JOHN, agronomist, educator; b. Independence, La., Oct. 21, 1924; s. Natale and Josephine N. G.; m. Louise Markham, June 25, 1952; children: John, David, Bruce, Benjamin, Michael, Amy. B.S. in Horticulture, Southeastern La. U., 1949; M.S., La. State U., 1950, Ph.D. in Horticulture-Plant Pathology, 1960. Asst. prof. horticulture La. State U., Hammond, from 1950; agronomist in, Somalia, 1965-67; mem. faculty La. State U., Baton Rouge, 1969—, now prof. plant pathology, head. Chmn. bd. Baton Rouge Internat. Hospitality Found.; bd. dirs. Southside Civic Assn. Served with USAAC, 1943-46. Mem. Am. Phytopath. Soc., Internat. Soc. Sugarcane Technologists. Democrat. Lodges: Kiwanis; KC. Subspecialties: Genetics and genetic engineering (agriculture); Biomass (agriculture). Current work: Breeding sugarcane for sugar; disease resistance and biomass. Office: 143 Agr Center La State U Baton Rouge La 70803

GIAMEI, ANTHONY FRANCIS, metallurgist, materials scientist; b. Corning, N.Y., Oct. 14, 1940; m., 1962; 2 children. B.E., Yale U., 1962; Ph.D. in Materials Sci, Northwestern U., 1967. Research assoc. alloy studies Advanced Materials Research and Devel. Lab., Pratt & Whitney Aircraft div. United Techs. Research Ctr., Hartford, Conn., 1966-68, sr. research assoc., 1968-69, group leader, 1969-71, group leader alloy research, 1971-77, sr. staff scientist, 1977-81, sr. cons. scientist, 1981—. Recipient George Mead Gold medal, 1981. Mem. Am. Soc. Metals, AIME, Sigma Xi. Subspecialties: Materials; Alloys. Office: United Techs Research Ctr Silver Ln East Hartford CT 06108

GIAMMATTEO, MICHAEL CHARLES, psychotherapist; b. Pitts., Dec. 5, 1931; s. Charles and Venera (Cicco) G.; m. Dolores L. Minoli, Oct. 3, 1953; children: Michele Rae, Michael John. B.S., Slippery Rock State U., 1953; M.S., U. Pitts., 1957, Ph.D., 1963; Ed.D., NIMH, 1964. Faculty Penn Hills Sch. Dist., Pa., 1953-59; biomed. research prof. Temple U., Phila., 1959-64; br. chief U.S. Office of Edn., Washington, 1964-65; human/med. researcher Regional Lab., Oreg. Med. Program, Portland, 1965-70; clin. dir. Sylvan Inst., Vancouver, Wash., 1970—; cons. AMA, NASA, VA, others. Contbr. articles to profl. jours.; author: Executive Well Being, 1980, Forces on Leadership, 1981; others; author/presentor: 13 1/2 hr. TV shows Family in Crises, 1979-80. Organizer Parent-Child Centers, Portland, 1966-68; facilitator Model Cities, various locations, 1965-70; cons. White House Conf. on Children and Youth, 1970, others. Served with U.S. Army, 1953-55. U.S. and Trust Ters. grantee, 1965-74. Mem. Am. Assn. Marriage and Family Therapists (supr.), Am. Soc. Clin. and Exptl. Hypnosis (pres. 1972-73), Am. Psychol. Assn. Roman Catholic. Subspecialties: Cognition; Sensory processes. Current work: Use of imagery in self-healing and to speed up recovery; futuristic concepts using integrative knowledge. Home: 509 NW 80th St Vancouver WA 98665 Office: 7104 NE Hazel Dell Vancouver WA 98665

GIANNETTI, RONALD A., clin. psychologist, educator; b. Chgo., May 21, 1946; s. Armando Edward and Olga (Santarelli) G.; m. Carolyn Jean Openshaw, Nov. 23, 1975; 1 son, Anthony Michael. A.B., U. Calif., Berkeley, 1967, Ph.D., 1973. Lic. clin. psychologist, Va. Evaluation coordinator VA Hosp., Salt Lake City, 1975-76, chief psychol. assessment, 1976-78; asst. prof. psychology Eastern Va. Med. Sch., Norfolk, 1978-79, assoc. prof., 1979—; clin. psychologist in pvt. practice, Salt Lake City, 1974-78, Norfolk, 1978—; Chmn. Va. Consortium for Profl. Psychology, Norfolk, 1979—; dir. psychology intern tng., Norfolk, 1978-81; cons. VA Hosp., Hampton, Va., 1982—. Found. for Applied Communication Tech. grantee, 1974-75. Mem. Am. Psychol. Assn., Nat. Conf. on Use of On-Line Computers in Psychology, Soc. Personality Assessment, Biofeedback Soc. Am. Subspecialties: Clinical psychology; Software engineering. Current work: Applications of on-line computer technology to the evaluation of mental health patients and services. Home: 713 Redgate Ave Norfolk VA 23507 Office: Dept Psychiatry and Behavioral Scis Eastern Va Med Sch PO Box 1980 Norfolk VA 23501

GIANNINI, A. JAMES, psychiatrist, educator; b. Youngstown, Ohio, June 11, 1947; s. Matthew and Grace Carla (Nistri) G.; m. Judith Ludvik, Apr. 26, 1975; 1 dau., Juliette Nicole. B.S., Youngstown State U., Ohio, 1970; M.D., U. Pitts., 1974; postgrad., Yale U., 1974-78. Diplomate: Nat. Bd. Med. Eximiners. Intern St. Elizabeth Med. Ctr., Youngstown, 1974; resident dept. psychiatry Yale U., New Haven, 1975-78, chief resident, 1977-78; assoc. psychiatrist Elmcrest Psychiat. Inst., Portland, Conn., 1976-78; acting ward chief Conn. Mental Health Center, New Haven, 1977; assoc. dir. family medicine, psychiatry St. Elizabeth Med. Ctr., Youngstown, 1978-80; assoc. prof. dept. psychiatry Northeast Ohio Med. Coll., 1978—, program dir., 1980—; sr. cons. Fair Oaks Hosp., Summit, N.J., 1980—, Regent Hosp., N.Y.C., 1981—. Authur: (with Henry Black) Psychiatric, Psychogenic, Sumatopsychic Disorders, 1978, (with Robert Gilliland) Neurologic, Neurogenic and Neuropsychiatric Disorders, 1982; contbr. numerous articles to profl. jours. Vice chmn. Mahoning County (Ohio) Mental Health Bd.; councilor Nat. Italian Am. Found. Recipient James Earley award U. Pitts., 1974, Upjohn Research prize Upjohn Co., 1974; recipient Fair Oaks Research award Fair Oaks Hosp., 1979, Bronze award Brit. Med. Assn. Fellow N.J. Acad. Medicine, Am. Coll. Clin. Pharmacology; mem. Soc. Neurosci. Brit. Brain Soc., Brit. Brain Soc., European Neurosci. Soc., Am. Psychiat. Assn., Acad. Clin. Psychiatry. Roman Catholic. Club: Youngstown. Subspecialties: Psychopharmacology; Neurophysiology. Current work: Central effects of phencyclidine; effect of ambient ions upon mood, mood-modulating effects of clonidine. Office: PO Box 2169 Youngstown OH 44504

GIANUTSOS, ROSAMOND R., psychology educator; b. Lawrence, Mass., Apr. 1, 1945; d. S. Forbes and Ursula (Ingalls) Rockwell; m. John G. Gianutsos, Sept. 11, 1965; children: Gerasimos John, Matthew Nicholas. B.A., Barnard Coll., 1966; Ph.D., NYU, 1970. Lic. psychologist, N.Y. Asst. prof. Adelphi U., Garden City, N.Y., 1970-74, assoc. prof., 1974-83; sr. psychologist NYU Med. Center, N.Y.C., 1978-83, adj. assoc. prof. neurology, 1979—; dir. Cognitive Rehab. Services, Sunnyside, N.Y., 1983—. Author computer software package; contbr. articles to profl. jours. Fellow N.Y. Acad. Scis. (chmn. sect. linguistics 1974-75); mem. Am. Psychol. Assn., Am. Congress Rehab. Medicine, Psychonomic Soc., Eastern Psychol. Assn. Subspecialties: Cognition; Neuropsychology. Current work: Rehabilitation of cognitive dysfunction in brain-injured adults. Clinical neuropsychology; personal computing for injured persons. Home: 38-25 52d St Long Island City NY 11104 Office: Cognitive Rehab Services 38-25 52d St Sunnyside NY 11104

GIARDINA, PAUL ANTHONY, environmental hazard administrator; b. Rockville Centre, N.Y., June 7, 1949; s. Anthony J. and Carolyn A. (Tamburello) G.; m. Jane Cranston, May 13, 1972. B.S., U. Mich., 1971; M.S., N.Y.U., 1973. Engr. Consolidated Edison, N.Y.C., 1971-74, Ebasco Services, 1974-75; chief radiation EPA, N.Y.C., 1975-79; dir. N.J. DEP, Trenton, 1979—; adv. council Environ. Hazard Mgmt. Inst., Portsmouth, N.H., 1979—; dir. chem. control clean-up, 1980-81. Author: govt. report Summary Report on the Low-Level Radioactive Waste Burial Site, West Valley, 1977. Named Outstanding Young Man of Am. Jaycees, 1979; USPHS fellow, 1971; U.S. Army scholar, 1967. Mem. AAAS, Am. Nuclear Soc., Health Physics Soc., N.Y. Acad. Scis., Planetary Soc. Republican. Current work: The control of hazardous materials and toxic substances, including radioactive materials, specializing in developing new clean-up and disposal technology. Home: 152 Three Mile Harbor Rd East Hampton NY 11937

GIBBONS, MICHAEL FRANCIS, JR., anthropology educator, anatomy researcher; b. Laconia, N.H., Mar. 20, 1941; s. Michael Francis and Mary Jo (Leso) G. B.A., Yale U., 1963, M.Phil., 1970, Ph.D., 1974. Teaching fellow anatomy Yale U., New Haven, 1970-72; instr. anthropology U. Mass., Boston, 1972-74, asst. prof., 1974-78, assoc. prof., 1979—, vis. prof. U. Alaska-S.E., Juneau, 1977-78; cons. Children's Hosp., Boston, 1974-76; vis. scientist Mass. Audubon Soc., Wenham, 1978-79. Author: Dissection Guide to Higher Primates, 1979; monograph Phyletics of Oreopithecus, 1963 (Dean's prize). Sec. The Conservation Agy., Jamestown, R.I., 1981—. Served to lt. USN, 1963-68. Yale Univ. fellow, 1968, 70; NIH fellow, 1969; NSF grantee, 1973. Fellow Am. Anthrop. Assn.; mem. Am. Assn. Anatomists, N.Y. Acad. Scis., Systematics Assn. (London), Conservation Agy. Clubs: Elizabethan; Yale (New Haven). Subspecialties: Anatomy and embryology; Acoustics. Current work: Mechanisms of speech sound generation; anatomy, embryology and evolution of the human head and neck; evolution of speech. Home: 125 Chestnut St Wakefield MA 01880 Office: U Mass Harbor Campus Boston MA 02125

GIBBONS, WALTER RAY, physiology and biophysics educator, researcher; b. Ironton, Mo., Oct. 10, 1938; s. Walter L. and Dorothy Marie (McNabb) G.; m. Martha Jane Mowe, Mar. 23, 1963; children: James Lawrence, Anne Elizabeth. B.S., Washington U., St. Louis, 1961, Ph.D., 1967. Research fellow U. Chgo., 1967-71, instr., 1969-71; asst. prof. U. Vt., Burlington, 1971-75, assoc. prof., 1975-79, prof. physiology and biophysics, 1979—; Mem. editorial bd. Circulation Research, 1977-82. Assoc. editor: Am. Jour. Physiology: Heart and Circulation, 1981—; contbr. articles in field to profl. jours. Mem. Essex Junction Planning Commn., 1976, chmn., 1977. NIH research grantee, 1971—. Mem. Cardiac Muscle Soc., Am. Heart Assn. (Louis N. Katz basic sci. research prize, council basic sci. 1970). Subspecialties: Physiology (medicine); Biophysics (biology). Current work: Membrane properties responsible for electrical activity in heart muscle, investigations of mechanisms that couple electrical activity and contraction in heart muscle. Home: RD 1 Box 123 Charlotte VT 05445 Offic: Dept Physiology and Biophysic Coll Medicine U Vt Burlington VT 05405

GIBBS, ANN, chemical manufacturing corporation nuclear chemist; b. Corpus Christi, Tex., May 19, 1941; d. Frank James and Elizabeth Ann (Setzer) G. B.A., U. Tex.-Austin, 1962; M.S., U. Ark., 1964. Research asst. U. Tex. Marine Sci. Inst., Port Aransas, 1959-64; research asst. U. Ark., Fayetteville, 1962-66; staff chemist E.I. DuPont de Nemours & Co., Inc., Aiken, S.C., 1966—. Vol. Election Commn., Aiken, 1976—. AEC fellow, 1962-66. Mem. Am. Chem. Soc. (chmn. sect. 1980-81), Am. Phys. Soc., Am. Nuclear Soc., AAAS, Inst. Nuclear Materials Mgmt. Presbyterian. Subspecialties: Nuclear chemistry; Nuclear fission. Current work: Nuclear spectroscopy of Pu and others for analytical purposes; moderator chemistry studies. Home: PO Box 6624 North Augusta SC 29841 Office: E I DuPont de Nemours & Co Inc Savannah River Lab Aiken SC 29808

GIBBS, CLARENCE JOSEPH, JR., virologist, research administrator, educator; b. Washington, Dec. 10, 1924. A.B., Catholic U. Am., 1950, M.S., 1952, Ph.D., 1962, D.Sc. (hon.); M.D., U. Mass. Med. bacteriologist clin. pathology, div. vet. medicine Walter Reed Army Inst. Research, 1952-55; virologist dept. Hazardous Ops., Div. Communicable Diseases, 1955-59; virologist arbovirus sect. Lab. Tropical Virology, Nat. Inst. Allergy and Infectious Diseases, NIH, 1959-63; dep. chief Lab. Central Nervous System Studies, chief Lab. Slow, Latent and Temperate Virus Infections, Nat. Inst. Neurol. and Communicable Diseases and Stroke, NIH, Bethesda, Md., 1963—, neurovirological coordinator, intramural research, 1974—; assoc. prof. epidemiology Sch. Pub. Health and Hygiene and assoc. prof. neurology Sch. Medicine Johns Hopkins U.; mem. fellow comm. Nat. Soc. Med. Sci.; mem. com. virology Nat. Comm. Med. Sci. Recipient Outstanding Service award Nat. Inst. Allergy and Infectious Diseases, 1962; Disting. Service award HEW, 1970. Mem. Am. Assn. Immunology, World Fedn. Neurology, AAAS, Am. Soc. Tropical Medicine and Hygiene, Am. Assn. Pathology. Subspecialty: Virology (medicine). Office: Nat Inst Neurol Communicable Diseases and Strok NI Room 4-A15 Bldg 36 Bethesd MD 20014

GIBBS, HYATT MCDONALD, physicist, educator; b. Hendersonville, N.C., Aug. 6, 1938; s. Robert Shuford and Isabella Frances (Gamble) G.; m. Lethia Elizabeth Archer, June 3, 1960; children: Alexander Robert, Vanetta Lea. A.A., Mars Hill Coll., 1958; B.S.E.E., B.S.Engring. Physics, N.C. State U., Raleigh, 1960; Ph.D., U. Calif., Berkeley, 1965. Acting asst. prof. U. Calif., Berkeley, 1965-67; exchange scientist Philips Research Labs., Eindhoven, 1975-76; mem. tech. staff Bell Labs., Murray Hill, N.J., 1967-80; prof. optical scis. U. Ariz., Tucson, 1980—; vis. lectr. Princeton U., 1978-79. Mem. editorial bd.: Phys. Review A, 1982-83; mem. editorial bd.: Optics Letters, 1983—. Mem. Warren Twp. Planning Bd., 1973, 74, Bd. Edn., 1978-80. Recipient Michelson medal Franklin Inst., 1984. Fellow Am. Phys. Soc., Optical Soc. Am., AAAS; mem. IEEE, N.Y. Acad. Sci., Sigma Xi, Sigma Pi Sigma, Phi Kappa Phi, Eta Kappa Nu, Tau Beta Pi. Presbyterian. Subspecialties: Condensed matter physics; Optical signal processing. Current work: Optical bistability: basic experiments and development practical all-optical logic and switching devices, nonlinear optics, coherent pulse propagation, optical chaos, fundamental tests. Patentee in field. Home: 4900 E Calle Barril Tucson AZ 85718 Office: Optical Scis Center Tucson AZ 85721

GIBBS, MARTIN, educator, biologist; b. Phila., Nov. 11, 1922; s. Samuel and Rose (Sugarman) G.; m. Svanhild Karen Kvale, Oct. 11, 1950; children—Janet Helene, Laura Jean, Steven Joseph, Michael Seland, Robert Kvale. B.S., Phila. Coll. Pharmacy, 1943; Ph.D., U. Ill., 1947. Scientist Brookhaven Nat. Lab., 1947-56; prof. biochemistry Cornell U., 1957-64; prof. biology, chmn. dept. Brandeis U., Waltham, Mass., 1965—; Cons. NSF, 1961-64, 69-72, NIH, 1966-69; mem. corp. Marine Biol. Lab., Woods Hole, Mass., 1970—; RESA lectr., 1969; NATO cons. fellowship bd., 1968-70; mem. Council Internat. Exchange of Scholars, 1976—; chmn. adv. com. selection Fulbright scholars for Eastern Europe; adj. prof. Bot. Inst., U. Münster, W.Ger. Author: Structure and Function of Chloroplasts; Editor-in-chief: Plant Physiology, 1963—; asso. editor: Physiologie Vegetale, 1966—, Ann. Rev. Plant Physiology, 1966-71. Mem. AAUP, Am., Japanese socs. plant physiologists, Am. Acad. Arts and Scis., Am. Soc. Biol. Chemists, Council Biology Editors, Nat. Acad. Scis., Sigma Xi. Subspecialty: Plant physiology (biology). Current work: Plant: carbohydrate metabolism, carbon flow in photosynthesis. Home: 32 Slocum Rd Lexington MA 02173 Office: Brandeis Univ Waltham MA 02154

GIBBS, SAMUEL JULIAN, radiologist/radiobiologist, educator; b. Amory, Miss., Apr. 1, 1932; s. Samuel John and Inez (McGarity) G.; m. Emily Jane Starnes, Feb. 16, 1958; children: Samuel Phillip, Stephen Julian, Julie Ann. Student, U. Ala., 1950-52; D.D.S., Emory U., 1956; Ph.D., U. Rochester, 1969. Diplomate: Am. Bd. Oral and Maxillofacial Radiology. Pvt. practice dentistry, Vernon, Ala., 1959-63; asst. prof. U. Rochester, N.Y., 1968-70; asst. prof. radiology Vanderbilt U., Nashville, 1970-76, assoc. prof., 1976—; cons. ADA, Chgo., 1971—, VA Med. Center, Nashville, 1970—. Co-author: The Physical Basis of Medical Imaging, 1981, Dental Radiology: Principles and Interpretation, 1982, Digital Radiography: A Focus on Clinical Utility, 1982. Active Boy Scouts Am. Served to capt. USAF, 1955-59. NIH grantee, 1982; Bur. Radiol. Health grantee, 1974. Fellow Am. Acad. Dental Radiology (pres. 1979-80); mem. Am. Assn. Dental Schs. (chmn. council on advanced edn. 1973-74), ADA, Tenn. Dental Assn., Nashville Dental Assn., Radiol. Soc. N.Am., Internat. Assn. for Dental Research. Methodist. Subspecialty: Radiology. Current work: Radiation carcinogenesis; risks of low dose radiation. Computer simulation, using Monte Carlo methods, of photon transport to generate radiation dose distribution to patients from diagnostic (especially dental) radiography. Office: Dept Radiology and Radiol Scis Vanderbilt U Med Center Nashville TN 37232 Home: 784 Greeley Dr Nashville TN 37205

GIBIAN, MORTON J., chemistry educator; b. N.Y.C., Mar. 15, 1939; s. Walter N. and Cele J. G.; m. Maxine F., July. 2, 1961; children: Jocelyn, Craig. A.B., Columbia Coll., 1960; Ph.D., Columbia U., 1965. NIH postdoctoral fellow Northwestern U., Evanston, Ill., 1965-66; asst. prof. U. Calif.-Riverside, 1966-70, assoc. prof., 1970-76, prof., 1976-79; assoc. prof. CUNY, Queens Coll., N.Y.C., 1971-72; USPHS-Nat. Research Service research fellow Harvard U., Cambridge, Mass., 1975-76; prof., chmn. chemistry Seton Hall U., South Orange, N.J., 1979—. Contbr. articles in field to profl. jours. USPHS/NIH grantee, 1966-70, 73-78; NSF grantee, 1971-75, 76-79; Research Corp. grantee, 1966-71. Mem. Am. Chem. Soc., Am. Assn. Biol. Chemists, Sigma Xi. Subspecialties: Biochemistry (biology); Organic chemistry. Current work: Enzymology, mechanisms of enzymes, especially oxygenases, bioorganic chemistry, catalysis, active sites. Office: Dept Chemistry Seton Hall U South Orange NJ 07079

GIBLETT, ELOISE ROSALIE, hematology educator; b. Tacoma, Wash., Jan. 17, 1921; d. William Richard and Rose (Godfrey) G. B.S., U. Wash., 1942, M.S., 1947, M.D. with honors, 1951. Mem. faculty U. Wash. Sch. Medicine, 1957—, research prof., 1967—; asso. dir., head immunogenetics Puget Sound Blood Center, 1955-79, exec. dir., 1979—; former mem. several research coms. NIH. Author: Genetic Markers in Human Blood, 1969; Editorial bd.: Transfusion; Contbr. over 180 articles to profl. jours. Recipient fellowships, grants, Emily Cooley, Karl Landsteiner, Philip Levine and Alexander Wiener immunohematology awards. Fellow AAAS; Mem. Nat. Acad. Scis., Am. Soc. Human Genetics (pres. 1973), Am. Soc. Hematology, Am. Assn. Immunologists, Brit. Soc. Immunology, Internat. Soc. Hematologists, Am. Fedn. Clin. Research, Western Assn. Physicians, Assn. Am. Physicians, Sigma Xi, Alpha Omega Alpha. Subspecialties: Genetics and genetic engineering (medicine); Hematology. Current work: Administration, teaching, research in blood transfusion science, immunohematology, medical genetics, genetic markers in human blood, biochemistry of immunodeficiency disease, enzyme polymorphisms. Home: 6533 53d St NE Seattle WA 98115 Office: Puget Sound Blood Center Terry and Madison Sts Seattle WA 98104

GIBSON, ELEANOR JACK (MRS. JAMES J. GIBSON), psychology educator; b. Peoria, Ill., Dec. 7, 1910; d. William A. and Isabel (Grier) Jack; m. James J. Gibson, Sept. 17, 1932; children:

James J., Jean Grier. B.A., Smith Coll., 1931, M.A., 1933, D.Sc., 1972; Ph.D., Yale U., 1938; D.Sc., Rutgers U., 1973, Trinity Coll., 1982. Asst., instr., asst. prof. Smith Coll., 1931-49; research asso. psychology Cornell U., Ithaca, N.Y., 1949-66, prof., 1972—, Susan Linm Sage prof. psychology, 1972—; fellow Inst. for Advanced Study, Princeton, 1959-60, Inst. for Advanced Study in Behavioral Scis., Stanford, Calif., 1963-64; vis. prof. Mass. Inst. Tech., 1973, Inst. Child Devel., U. Minn., 1980; vis. disting. prof. U. Calif., Davis, 1978; vis. scientist Salk Inst., La Jolla, Calif., 1979. Author: Principles of Perceptual Learning and Development, 1967 (Century award), (with H. Levin) The Psychology of Reading, 1975. Recipient Wilbur Cross medal Yale U., 1973; Howard Crosby Warren medal, 1977; medal for disting. service Tchrs. Coll., Columbia U., 1983; Guggenheim fellow, 1972-73. Fellow AAAS (div. chairperson 1983), Am. Psychol. Assn. (Distinguished Scientist award 1968, G. Stanley Hall award 1970, pres. div. 3 1977); mem. Eastern Psychol. Assn. (pres. 1968), Soc. Exptl. Psychologists, Nat. Acad. Edn., Psychonomic Soc., Soc. Research in Child Devel. (Disting. Sci. Contbn. award 1981), Nat. Acad. Sci., Am. Acad. Arts and Scis., Brit. Psychol. Soc. (hon.), Phi Beta Kappa, Sigma Xi. Subspecialties: Developmental psychology; Cognition. Current work: Perceptual development. Home: 111 Oak Hill Rd Ithaca NY 14850

GIBSON, GEORGE WILLIAM, metallurgical engineer; b. Los Angeles, Nov. 5, 1921; s. George Imboden and Florence (Huse) G.; m. Aleene Neely, Sept. 11, 1948; children: George N., David H., Howard E., Richard B. B.S., U. Ariz., 1943; postgrad., Yale U., 1946; M.S., U. Idaho, 1966. Metall. engr. ASARCO, Los Angeles, 1946-55, Braden Copper Co., Caletones, Chile, 1955-58, Idaho Nat. Engring. Lab., Idaho Falls, 1958-71, So. Peru Copper Co., Ilo, 1971-73, Idaho Nat. Engring. Lab., Idaho Falls, 1973—. Served to 1st lt. USAF, 1943-46. Mem. AIME, Am. Nuclear Soc. Republican. Subspecialties: Metallurgical engineering; Nuclear engineering. Current work: Development of processes for disposal of nuclear waste. Home: PO Box 923 Hailey ID 83333 Office: EG&G Idaho PO Box 1625 Idaho Falls ID 83415

GIBSON, JAMES EDWIN, toxicologist; b. Des Moines, Aug. 22, 1941; s. Donald Edwin and Lorene Jane (Faris) G.; m. Karen Rae Hasselquist; children: Debra Rae, Bradley James, Mark Alan. B.A., Drake U., Des Moines, 1964; M.S., U. Iowa, Iowa City, 1967, Ph.D., 1969. Diplomate: Acad. Toxicological Scis. Prof. dept. pharmacology Mich. State U., East Lansing, 1969-76; vis. prof. Pharmakologisches Instit, Universitat Mainz, Ger., 1975-76; v.p., dir. research Chem. Industry Inst. Toxicology, Research Triangle Park, N.C., 1976—; adj. prof. curriculum toxicology U. N.C., Chapel Hill, 1981—. Recipient Alexander von Humboldt sr. U.S. scientist award, 1975-76. Mem. Soc. Toxicology (Achievement award 1977), Am. Soc. Pharmacology and Exptl. Therapeutics, Teratology Soc., Soc. Developmental Biology, Internat. Soc. Study Xenobiotics. Subspecialties: Toxicology (medicine); Teratology. Current work: Mechanisms of toxic chemical injury; chemical risk analysis. Home: 8605 Caswell Pl Raleigh NC 27612 Office: PO Box 12137 Research Triangle Park NC 27709

GIBSON, JEAN M., radiation physicist; b. Columbia, Utah, Dec. 1, 1930; d. George J. and Mary (Pizzuto) G. B.A., Coll. St. Benedict, 1956; M.S., U. Iowa, 1966, Ph.D., 1969. Tchr. high sch., Cold Spring, Minn., 1958-63; asst. prof. physics Coll. St. Benedict, St. Joseph, Minn., 1969-75; physicist St. Benedict's Hosp., Ogden, Utah, 1975—. Contbr. articles to profl. jours. NASA trainee, 1968-69. Mem. Am. Assn. Physicists in Medicine, Am. Registry Radiol. Technologists. Roman Catholic. Subspecialty: Medical physics. Current work: Computer software evaluation for radiation therapy, calibration of linear accelerators and other x-ray machines; computer treatment planning for radiation therapy. Home: 319 E 5450 S Ogden UT 84403 Office: 5475 S 500 E Ogden UT 84403

GIBSON, JOHN EGAN, educator, univ. dean; b. Providence, June 11, 1926; s. Arthur and Judith Agnes (Egan) G.; m. Nancy Gertrude McGuinness, Sept. 7, 1950; children—William Francis, John Egan, Robert Alan, Nancy Regina. B.S., U. R.I., 1950; M.E., Yale U., 1952, Ph.D., 1956. Asst. prof. elec. engring. Yale, 1956; asso. prof. elec. engring. Purdue U., 1957-60, prof., 1960-65; founding dean Sch. Engring., prof. Oakland U., 1965-73, John Dodge prof. engring., 1972-73, dean, U. Va., Charlottesville, 1973—; leader U.S. team to U.S. Republic of China Conf. on Urban Systems, 1972; cons. on urban system design methodology Battelle Inst., 1971-73. Author: Control System Components, 1958 (transl. into Japanese, Polish), Nonlinear Automatic Control (transl. into Japanese, Polish, Romanian), 1963, Introduction to Engineering Design, 1968, Designing the New City, 1977. Served with AUS, 1944-46. Mem. IEEE, ASME, Am. Soc. Engring. Edn. (dir. nat. electronics conf. 1965-67), Soc. Automotive Engrs., Sigma Xi, Tau Beta Pi. Roman Catholic. Subspecialties: Graphics, image processing, and pattern recognition; Operations research (engineering).

GIBSON, MRS. JAMES J. *See also* **GIBSON, ELEANOR JACK**

GIBSON, SAM THOMPSON, physician; b. Covington, Ga., Jan. 1, 1916; s. Count Dillon and Julia (Thompson) G.; m. Alice Chase Gibson, Oct. 31, 1942 (dec. Dec. 1971); children: Lena S., Stephen C., Judith Gibson Hammer, Lucy F. B.S. in Chemistry, Ga. Inst. Tech., 1936; M.D., Emory U., 1940. Diplomate: Am. Bd. Internal Medicine. Med. house officer Peter Bent Brigham Hosp., Boston, 1940-41, asst. resident in medicine, 1946-47; research fellow in medicine Harvard Med. Sch., 1941-43, Milton research fellow, 1947-49; asst. to assoc. dir. ARC Blood Program, Washington, 1949-56, nat. dir., 1956-66, sr. med. officer, 1957-67; clin. asst. prof. medicine George Washington U. Med. Sch., 1963—, Uniformed Services U. Health Scis., 1981—; asst. dir. Div. Biol. Standards, NIH, Bethesda, Md., 1967-72; asst. to dir. Bur. Biologics, FDA, Bethesda, 1972-77, dir. div. biologics evaluation, 1977—; cons. Care-Medico, 1970—. Editorial bd.: Jour. Vox Sanguinis, 1956-76. Served to capt. USNR, 1941-46. Recipient service award FDA, 1977, merit award, 1978. Mem. Internat. Soc. Blood Transfusion (dir. 1962-66), Internat. Soc. Hematology, Am. Soc. Hematology, N.Y. Acad. Scis., AMA. Club: George Washington U. Subspecialty: Regulatory medicine. Current work: Regulation of biologics production. Office: Office Biologics Div Biologics Evaluation 8800 Rockville Pike Bethesda MD 20205 Home: 5801 Rossmore Dr Bethesda MD 20814

GIBSON, THOMAS PATRICK, renal pharmacologist, nephrologist; b. Cin., July 9, 1942; s. James George and Viola Catherine (Blanke Meyer) G.; m. Joyce Ann Goefft, Aug. 5, 1967; children: Caroline, Patrick, Brian. B.S., Xavier U., 1964; M.D., U. Cin., 1968. Med. intern, then resident U. Iowa, 1968-71, renal fellow, 1972, Walter Reed Med. Center, 1973; research fellow Walter Reed Army Inst. Research, 1974, staff nephrologist, 1974-76; asst. prof. medicine Northwestern U. Med. Center, Chgo., 1976-78, asso. prof., 1978-81, prof., 1981—; staff physician VA Lakeside Med. Center, Chgo., 1976. Contbr. articles to profl. jours. Served with M.C. U.S. Army, 1972-76. Mem. Am. Soc. Pharmacology and Exptl. Therapeutics, Am. Soc. Clin. Pharmacology and Therapeutics. Republican. Roman Catholic. Subspecialties: Pharmacology; Nephrology. Current work: Drug metabolism and disposition in renal failure, renal digoxin elimination. Office: 333 F Huron St Chicago IL 60611

GIDDA, JASWANT SINGH, physiologist, educator; b. Panjab, India, Oct. 1, 1946; s. Ram R.S. and Shakuntla (Devi) G.; m. Rajinder K., May 2, 1948; children: Archana R., Vipul J. B.S. with honors, Panjab U., 1967, M.S., 1968, Ph.D., 1973. Research scientist U. Tex.-Dallas, 1973-78; instr. U. Tex. Health Sci. Center-San Antonio, 1978-81; instr. physiology Harvard Med. Sch., 1981—; Jr. research fellow Council Sci. and Indsl. Research, India, 1968-72. Contbr. articles to sci. jours. Mem. AAAS, Neurosci. Soc. Subspecialties: Gastroenterology; Neurophysiology. Current work: Neuromuscular control of gut motility. Neurophysiology, motility, smooth muscle physiology, neuropharmacology. Home: 41 Country Ln Westwood MA 02090 Office: 330 Brookline Ave Boston MA 02215

GIDDINGS, JOHN CALVIN, educator, researcher; b. American Fork, Utah, Sept. 26, 1930; s. Luther W. and Berniece (Crandall) G.; m.; children: Steven B., Michael C. B.S., Brigham Young U., 1952; Ph.D., U. Utah, 1954. Asst. prof. chemistry U. Utah, 1957-59, assoc. prof., 1959-62, research prof., 1962-66, prof., 1966—; mem. adv. bd. Negative Population Growth, Inc. Author: Dynamics of Chromatography, 1965, Chemistry, Man and Environmental Change, 1973; editor: Our Chemical Environment, 1972, Advances in Chromatography, Vols. 1-19; exec. editor: Separation & Technology, 1966—; mem. editorial bd.: Jour. Liquid Chromatography, 1978—; contbr. articles to profl. jours. Fulbright grantee, 1974; recipient Tswett medal in chromatography, 1978, Stephen Dal Nogare Chromatography award, 1979, U. Utah Disting. Research award, 1979, Russian Sci. Council Chromatography award, 1980. Mem. Am. Chem. Soc. (award in chromatography and electrophoresis 1967, award in analytical chemistry 1980). Club: Wasatch Mt. (Salt Lake City). Subspecialties: Analytical chemistry; Ecosystems analysis. Current work: Chromatography and field flow fractionation; active in research relating chromatographic separability to the underlying molecular processes and research on general separation theory and methods; new separation methodology, macromolecular separations, devel. of techniques for diffusion coefficient measurements, theory of diffusion and chemical kinetics. Organizer expedition for exploration and descent of the upper canyons of the Apurimac River in Peru, 1975. Home: 3978 Emigration Canyon Salt Lake City UT 84108 Office: University of Utah Department of Chemistry Salt Lake City UT 84112

GIDDINGS, LUTHER VAL, genetics researcher, educator; b. Salt Lake City, Apr. 21, 1953; s. Crandall Bland and Elizabeth Ann (Kiefer) G. B.S., Brigham Young U., 1975; M.S., U. Hawaii, 1977, Ph.D. in Genetics, 1980. Grad. research asst. Hawaiian Drosophila Research project U. Hawaii, Honolulu, 1975-80; cons., liaison Open U./BBC, 1979—; NIH postdoctoral research assoc. in biology Washington U., St. Louis, 1980—, lectr. biology, 1983. Contbr. articles to profl. jours. NSF grantee; NIH grantee. Mem. Genetics Soc. Am., Soc. Study Evolution, AAAS, Am. Mus. Natural History, Sierra Club, Nature Conservancy, Arnold White Water Soc., Sigma Xi. Subspecialties: Evolutionary biology; Genetics and genetic engineering (biology). Current work: Evolutionary genetics and biology of Drosophila with particular emphasis on processes of speciation. Ancillary studies with other non-insect species. Applying genetic engineering techniques. Home: 2011 McCausland St Saint Louis MO 63143 Office: Washington U Dept Biology Campus Box 1137 Saint Louis MO 63130

GIEN, TRAN TRONG, physicist, educator, researcher; b. Quan Yen, Vietnam, Oct. 31, 1937; m. Lan Xuan Nguyen, July 23, 1966; children: Daniel, Lilian, Aileen. Licensees-Sciences, U. Saigon, 1959; M.Sc., Ohio U., 1962, Ph.D., 1965. Maitre de conferences U. Bordeaux, France, 1965-66; assoc. prof. dept. physics Meml. U. Nfld. (Can.), St. Johns, 1966-74, prof., 1974—. Subspecialties: Atomic and molecular physics; Theoretical physics. Current work: Atomic and molecular physics; theoretical physics. Home: 11 Parliament St Saint Johns NF Canada A1A 2Y6 Office: Dept Physics Meml U Nfld Saint Johns NF Canada A1B 3X7

GIESSEN, BILL CORMANN, chemistry educator, cons.; b. Pitts., June 8, 1932; s. Ernst Aloys and Gustel (Cormann) G.; m. Mary Burns; 1 dau., Nora. Dr.Sci.Nat., U. Göttingen, W.Ger., 1958. Research assoc. MIT, Cambridge, 1959-68; assoc. prof. chemistry Northeastern U., Boston, 1968-72, prof., 1972—, assoc. dir., 1973—; dir. Marko Materials, Inc., North Billerica, Mass., Energy Materials Corp., South Lancaster, Mass., Cambridge Analytical Assocs., Inc., Watertown, Mass., Pilgrim Materials Corp., South Windsor, Conn. Editor: Structural Chemistry of Alloy Phases, 1969, (with others) Rapidly Quenched Metals, 1976, Rapidly Solidified Amorphous and Crystalline Alloys, 1982, Alloy Phase Diagrams, 1983. Mem. AIME, Materials Research Soc. (sec. 1980-83), Am. Chem. Soc., Am. Crystallographic Assn. Subspecialties: Amorphous metals; Solid state chemistry. Current work: Synthesis and structural, thermal, mechanical and electronic characterization of new metallic glasses and metastable alloys. Semiconductor studies. Office: Northeastern U Boston MA 02115

GIFFORD, ERNEST MILTON, JR., botanist, educator; b. Riverside, Calif., Jan. 17, 1920; s. Ernest and Mildred (Campbell) G.; m. Jean Duncan, July 15, 1942; 1 dau., Jeanette. A.B., U. Calif., Berkeley, 1942, Ph.D., 1950. Asst. prof. botany U. Calif., Davis, 1950-56, asso. prof., 1957-61, prof., 1962—, chmn. dept., 1963-68, 73-78, asst. botanist agrl. expt. sta., 1950-56, asso. botanist, 1957-61, botanist, 1962—. Editor-in-chief: Am. Jour. Botany, 1975-79; co-author: Comparative Morphology of Vascular Plants, 2d edit, 1974; co-editor: Mechanisms and Control of Cell Division, 1977; contbr. over 85 articles to sci. jours. Served with U.S. Army, 1942-46; to col. Res., 1948-73. Decorated Bronze Star; Merck sr. postdoctoral fellow Harvard U., 1956; Guggenheim fellow, 1966; Fulbright scholar, France, 1966; NATO sr. sci. fellow, 1973; NSF grantee, 1958-66, 79-81. Mem. Bot. Soc. Am. (cert. of merit 1981, pres. 1982), Internat. Soc. Plant Morphologists (v.p. 1980—), Am. Inst. Biol. Scis. Democrat. Subspecialties: Developmental biology; Plant Anatomy. Current work: Morphology and anatomy of vascular plants; morphogenesis and ultrastructure, especially of meristems; spermatogenesis. Home: 1023 Ovejas Ave Davis CA 95616 Office: U Calif Davis CA 95616

GIFFORD, GEORGE EDWIN, microbiologist, researcher; b. Mpls., Dec. 6, 1924; s. Ernest Wilbur and Hulda Victoria (Widen) G.; m. June Marie Pirila, Dec. 27, 1955 (dec.); children: Charles Stephen, Sheryl Byrne. B.S. cum laude, U. Minn., Mpls., 1949, M.S., 1953, Ph.D., 1955. Grad. asst. U. Minn., 1950-55, instr., 1955-56; asst. prof. U. Fla. Coll. Medicine, Gainesville, 1957-64, assoc. prof., 1964-68, prof., 1968—, co-prof. microbiology and cell sci., 1969—, mem. ctr. macromolecular sci., 1969—. Contbr. numerous research articles to publs. USPHS fellow, London, 1962; USPHS and Am. Cancer Soc. grantee, 1957—. Fellow Am. Acad. Microbiology, AAAS; mem. Am. Soc. Microbiology, Am. Assn. Immunologists, Am. Assn. Virologists, Union Concerned Scientists, Sigma Xi (pres. Fla. chpt. 1981). Subspecialties: Immunology (medicine); Cancer research (medicine). Current work: Natural resistance to viral disease at the cellular level; nature, production and action interferon; production and action of tumor necrosis factor and oncolytic factors. Home: 1013 NW 91st Terr Gainesville FL 32606 Office: Dept Immunology and Med Microbiology U Fla Coll Medicine Box J-266 Gainesville FL 32610

GIL, JOAN, medical educator, physician; b. Barcelona, Catalonia, Spain, June 26, 1940; came to U.S., 1976; s. Antonio and Petra (Camino) G.; m. m Brigitte A. Sollereder, Aug. 12, 1970; children: Daniel J., Isadora B. M.D., U. Barcelona, 1964, Dr. Medicine, 1968. Mem. faculty U. Berne, Switzerland, 1966-76; assoc. prof. medicine U. Miami, 1976-77, U. Pa., 1977—; visitor Webb-Waring Lung Inst., U. Colo. Med. Ctr., Denver, 1971-72, Pulmonary div. VA Hosp., Washington, 1972. Mem. Internat. Stereology Soc., Am. Physiol. Soc., Am. Thoracic Soc., So. Soc. Clin. Investigation. Subspecialties: Cytology and histology; Graphics, image processing, and pattern recognition. Current work: Structure-function relationships in the lung; quantitative histology (morphometry); computer-based reconstruction of serial sections; lung physiology; lung injury. Home: 518 General Lafayette Rd Merion Station PA 19066 Office: Dept Medicine Hosp U Pa Gl 3400 Spruce St Philadelphia PA 19104

GILBERT, CHARLES D., neurobiologist, educator; b. N.Y.C., Jan. 15, 1949; s. Gustave M. and Matilda S. (Safran) G. B.A. summa cum laude, Amherst Coll., 1971; M.D., Ph.D., Harvard U., 1977. Research fellow in neurobiology Harvard U., 1977-79, prin. research assoc. in neurobiology, 1979-81; asst. prof., 1981-83; asst. prof. neurobiology Rockefeller U., N.Y.C., 1983—. Mem. AAAS, Soc. Neurosci. Subspecialty: Neurobiology. Current work: Study of visual cortex, brain processing of sensory information, relationship between neuronal circuitry and brain function. Office: Rockefeller U 1230 York Ave New York NY 10021

GILBERT, DANIEL LEE, physiologist; b. Bklyn., July 2, 1925; s. Louis and Blanche (Lutz) G.; m. Claire Gilbert, July 26, 1964; 1 son: Raymond Louis. A.B., Drew U., 1948; M.S., State U. Iowa, 1950; Ph.D., U. Rochester, 1955. Instr. U. Rochester, 1955-56; instr. Albany Med. Coll., 1956-59, asst. prof. physiology, 1959-60; asst. prof. Jefferson Med. Coll., 1960-62, assoc. prof., 1962-63; research physiologist Nat. Inst. Neurol. and Communicative Diseases and Stroke, NIH, Bethesda, Md., 1962—, head sect. cellular biophysics, 1963-7l; Bowditch lectr. Am. Physiol. Soc., 1964. Contbr. articles to profl. books, jours.; editor: Oxygen and Living Processes, 1981. Served with U.S. Army, 1943-45. Decorated Purple Heart. Fellow AAAS; mem. Am. Chem. Soc., Am. Inst. Biol. Sci., Am. Physiol. Soc., Am. Soc. Pharm. and Exptl. Therapeutics, Biophys. Soc., Corp. of Marine Biol. Lab., Internat. Soc. Study Origin of Life, N.Y. Acad. Sci., Soc. Exptl. Biology and Medicine, Soc. Neurosci., Soc. Gen. Physiology, Undersea Med. Soc., Sigma Xi. Subspecialties: Physiology (biology); Membrane biology. Current work: Research in membrane properties, oxygen evolution, oxygen toxicity, history of respiration physiology. Home: 10324 Dickens Ave Bethesda MD 20814 Office: Lab Biophysics Nat Inst Neurol and Communicative Diseases and Stroke NIH Bldg 36 Room 2A29 Bethesda MD 20205

GILBERT, DONALD GEORGE, biologist; b. Oswego, N.Y., Apr. 24, 1953; s. Harold George and Alice Mabel (Fuller) G.; m. Betty Ruth Bernard, Dec. 1, 1975 (div. 1980). B.S., U. Ill., 1975; Ph.D., Ind. U., 1981. Instr. U. Ill., Champaign, 1974; instr. Ind. U., Bloomington, 1977-79, research asst., 1980-81; research assoc. Syracuse (N.Y.) U., 1981—. Contbr. articles to profl. jours. Recipient fellowships and grants. Mem. Soc. Study of Evolution, Genetics Soc. Am., Ecol. Soc. Am., Soc. Am. Naturalists, AAAS, Nature Conservancy, Nat. Audubon Soc., Sierra Club, Nat. Wildlife Fedn., Phi Beta Kappa. Subspecialties: Evolutionary biology; Population biology. Current work: Evolutionary biology of Drosophila species; drosophila-yeast coevolution; population genetics.

GILBERT, FRED IVAN, JR., med. research dir., educator; b. Newark, Mar. 5, 1920; s. Fred Ivan and Gertrude Olga (Lund) G.; m. Gayle H. Yamashiro, Sept. 16, 1978; children—Rondi, Kristin, Galen, Gerald, Fred Ivan, Lisa, Heidi, Cara, John. Jr. cert., U. Hawaii, Honolulu, 1940; B.S., U. Calif., Berkeley, 1942; M.D., Stanford U., Palo Alto, 1945; postgrad., Neurol. Inst., U. London Nat. Hosp., 1960-61. Diplomate: Am. Bd. Internal Medicine. Intern Stanford Hosp., San Francisco, 1945-46; resident Ft. Miley VA Hosp., San Francisco, 1948-50; asst. clin. prof. medicine Stanford U., Palo Alto, Calif., 1948-51; internist Straub Clinic, Honolulu, 1951—; med. dir. Pacific Health Research Inst., Honolulu, 1961—; prof. public health and preventive medicine Sch. Public Health and Medicine, U. Hawaii, Honolulu, 1959—. Contbr. to books and jours. in field. Pres. Hawaii Heart Assn., 1959. Served with U.S. Army, 1943-48. Recipient award Nat. Acad. Sci. Inst. Medicine, 1978. Fellow A.C.P., Royal Coll. Health (Eng.); mem. AMA, Hawaii Med. Assn., Honolulu County Med. Assn., AAAS, Soc. Nuclear Medicine (pres. Hawaii 1980-82), Hawaii Assn. Med. Clinics (pres. 1973-74), Internat. Health Evaluation Assn. (pres. 1971-72). Club: Sons of Norway. Subspecialties: Internal medicine; Nuclear medicine. Current work: Clinical nuclear medicine; health service research. Home: 2112 Mott-Smith Honolulu HI 96822 Office: 800 S King St Suite 200 Honolulu HI 96813

GILBERT, JEFFREY MORTON, neurochemist, psychiatrist; b. Portland, Oreg., Apr. 26, 1941; s. Samuel G. and Ethel (Gross) G.; m. Marilyn Louise Minyard, Mar. 22, 1973; children: Jeremy Aaron, Michael John, Meredith Anne. B.A., Reed Coll., 1962; M.D., Washington U., St. Louis, 1966. Diplomate: Am. Bd. Psychiatry and Neurology. Intern Montefiore Hosp., N.Y.C., 1967-68; resident in medicine, 1967-68; research assoc. Nat. Heart, Lung and Blood Inst., NIH, Bethesda, Md., 1968-70; resident in psychiatry Mass. Gen. Hosp., Boston, 1970-73; instr. psychiatry Harvard Med. Sch., Boston, 1973, asst. prof. psychiatry, 1973-82, assoc. prof., 1982—; chief neurochemistry lab. Labs. Psychiat. Research, Mailman Research Ctr., McLean Hosp., Belmont, Mass., 1977—. Served with USPHS, 1968-70. King Trust fellow Mass. Gen. Hosp., 1974-75. Mem. Am. Psychiat. Assn., AAAS, Soc. Neuroscience, Am. Soc. Neurochemistry. Subspecialties: Neurochemistry; Psychopharmacology. Current work: The characterization and metabolism of brain proteins.

GILBERT, ROBERT PERTSCH, mathematician, educator; b. N.Y.C., Jan. 8, 1932; s. Ralph H. and Ruth (Pertsch) G.; m. E. Eileen Manton, Oct. 28, 1955 (div. Jan. 1975); m. Ursula Murach, June 27, 1975 (div. Mar. 1979); m. Elizabeth Page Cogswell, Aug. 12, 1979. B.S., Bklyn. Coll., 1952; M.S. in Physics, Carnegie-Mellon U., 1955, Carnegie-Mellon U., 1955, Ph.D., 1958. Faculty U. Pitts., 1957-60, Mich. State U., 1960-63; research asst. prof. Inst. for Fluid Dynamics and Applied Math., U. Md., 1961-64, research asso. prof., 1964-65; prof. dept. math. Georgetown U., Washington, 1965-66; prof. math. Ind. U., Bloomington, 1966-75, dir., 1973-75; Unidel prof. math. U. Del., 1975—, also dir., 1975—; Cons. spl. coal research div. U.S. Bur. Mines, 1958-60, Naval Ordnance Lab., 1961-64; vis. Unidel prof. U. Del., 1972-73; guest prof. U. Glasgow, 1972, U. Dortmund, Germany, 1972, Hahn Meitner Inst. Nuclear Physics, Berlin, 1974, Free U. Berlin, 1974-75; vis. prof. Tech. U. Denmark, 1979, U. Karlsruhe, 1980, Oxford U., 1981-82. Author: Function Theoretic Methods in Partial Differential Equations, 1969, Constructive Methods for Elliptic Equations, 1973; co-author: Foundations of Applied Mathematics, First Order Elliptic Systems, 1983; Co-editor: Analytic Methods in Mathematical Physics, 1970; editor-in-chief: an internat. jour. Applicable Analysis; assoc. editor: Jour. Nonlinear Analysis; adv. editor: Math. Method in Applied Scis; cons. editor, Pitman Press, London. Recipient von Humboldt Sr. Scientist award, 1975. Mem. Am. Math. Soc. (mem. council), Soc. for Indsl. and Applied Math. (asso. editor jour.), Washington Acad. Scis., Sigma Xi, Pi Mu Epsilon.

Subspecialty: Applied mathematics. Current work: Moving boundary problems; porous media problems. Research and publs. on analysis, especially harmonic functions, boundary value problems, math. physics, partial differential equations, numerical analysis. Home: 112 Briar Ln Newark DE 19711

GILBERT, WALTER, scientist, molecular biologist; b. Boston, Mar. 21, 1932; s. Richard V. and Emma (Cohen) G.; m. Celia Stone, Dec. 29, 1953; children—John Richard, Kate. B.A., Harvard U., 1953, A.M., 1954; D.Phil., Cambridge U., Eng., 1957; D.Sc. (hon.), U. Chgo., 1978, Columbia U., 1978, U. Rochester, 1979. NSF postdoctoral fellow Harvard, 1957-58, lectr. physics, 1958-59, asst. prof. physics, 1959-64, asso. prof. biophysics, 1964-68, prof. molecular biology, 1968—, Am. Cancer Soc. prof. molecular biology, 1972—. Recipient U.S. Steel Found. award Nat. Acad. Sci., 1968; Ledlie prize Harvard, 1969; Guggenheim fellow, 1968-69; V.D. Mattia lectr. Roche Inst. Molecular Biology, 1976; Warren triennial prize Mass. Gen. Hosp., 1977; Louis and Bert Freedman Found. award N.Y. Acad. Scis., 1977; Prix Charles-Leopold Mayer Academie des Scis., Inst. de France, 1977; Nobel prize in chemistry, 1980; co-winner Louisa Gross Horwitz prize Columbia U., 1979; Gairdner prize, 1979; Albert Lasker Basic Sci. award, 1979. Mem. Nat. Acad. Scis., Am. Phys. Soc., Am. Soc. Biol. Chemists, Am. Acad. Arts and Scis. Subspecialty: Molecular biology

GILBERTSON, JOHN ROBERT, physiologist, educator; b. Bemidji, Minn., Oct. 17, 1928; s. John Leonard and Helen Margaret G.; m. Barbara Irene, Oct. 12, 1953; children: Lesley, David, Hollis. A.B., U. Minn., 1951, M.S., 1956, Ph.D., 1960. Instr. U. Minn., 1959-60; research asso. Harvard U. Med. Sch., 1960-62; asst. research prof. biochemistry and nutrition U. Pitts., 1963-69, asso. prof., 1969-76, prof., 1976—, adj. prof. Contbr. articles to profl. jours. Served with USNR, 1951-53. Nat. Inst. Neurol. Diseases and Blindness fellow. Mem. Am. Chem. Soc., Biochem. Soc., Fedn. Am. Soc. Exptl. Biology, Pharmacology Soc., Am. Cancer Soc., Sigma Xi. Subspecialties: Biochemistry (biology); Membrane biology. Office: U Pitts 559 Salk Hall Pittsburgh PA 15261

GILBOA, ELI, biochemistry educator, researcher; b. Timisoara, Romania, July 6, 1947; came to U.S., 1977; s. Tibor and Irena Böhm. B.Sc., Hebrew U., Jerusalem, 1971; Ph.D., Weizmann Inst., Rehovot, Israel, 1977. Research assoc. MIT, Cambridge, 1977-81; asst. prof. biochemistry Princeton (N.J.) U., 1981—. Subspecialties: Molecular biology; Gene actions. Current work: Mechanism of gene expression in mammalian cells; gene therapy - correction of genetic disorders in individual patients. Home: 67 Stanworth Ln Princeton NJ 08540 Office: Princeton U Princeton NJ 08544

GILES, NORMAN HENRY, educator, geneticist; b. Atlanta, Aug. 6, 1915; s. Norman Henry and Alice (Guerard) G.; m. Dorothy Lunsford, Aug. 26, 1939 (dec. Jan. 1967); children—Annette Guerard, David Lunsford; m. Doris Vos Weaver, Aug. 1, 1969; stepchildren—Gayle Weaver (dec.), Alix Weaver. A.B., Emory U., 1937, Sc.D. (hon.), 1980; M.A., Harvard U., 1938, Ph.D., 1940; M.A. (hon.), Yale, 1951. Instr. botany Yale, 1941-45, asst. prof., 1945-46, asso. prof., 1946-51, prof., 1951-61, Eugene Higgins prof. genetics, 1961-72; Fuller E. Callaway prof. genetics U. Ga., 1972—; prin. biologist Oak Ridge Nat. Lab., 1947-50; cons. AEC, 1954-64; Mem. genetics study sect. NIH, 1960-64, mem. genetics tng. com., 1966-70; ednl. adv. bd. John Simon Guggenheim Meml. Found., 1977—. Editorial bd.: Radiation Research, 1953-58, Am. Naturalist, 1961-64, Devel. Genetics, 1979—. Bd. dirs. U. Ga. Research Found., 1979—. Parker fellow Harvard U., 1940-41; Fulbright and Guggenheim fellow U. Genetics Inst., Copenhagen, 1959-60; Guggenheim fellow Australian Nat. U., Canberra, 1966. Fellow Am. Acad. Arts and Scis., AAAS; mem. Nat. Acad. Scis. (chmn. genetics sect. 1976-79), Genetics Soc. Am. (treas. 1954-56, pres. 1970), Bot. Am., Am. Soc. Naturalists (pres. 1977), Am. Inst. Biol. Scis., Genetics Soc. Japan (hon.), Am. Ornithologists Union, Phi Beta Kappa, Sigma Xi. Subspecialties: Gene actions; Genetics and genetic engineering (biology). Current work: Gene organization and regulation in lower eukaryotes. Home: 289 Hanover Dr Bogart GA 30622 Office: Dept Molecular and Population Genetics Univ Ga Athens GA 30602

GILES, RALPH EDSON, pharmacologist; b. Rahway, N.J., Mar. 26, 1941; s. Edson Remmick and Lena Margaret (Tram) G.; m. Trudy Anne Moran, Dec. 28, 1963; children: Thomas, Deborah, Monica, Mary Ellen, Patrick. B.S. in Pharmacy, Fordham U., 1962; Ph.D. in Pharmacology, U. Minn., 1966. Asst. prof. pharmacology Fordham U., Bronx, N.Y., 1966-67; scientist Warner-Lambert, 1967-69, sr. scientist, 1969-72, sr. research assoc., 1972-75; mgr. pharmacology Stuart Pharms. div. ICI-Ams., Wilmington, Del., 1975-79, dir. pharmacology, 1979—. Contbr. articles and abstracts to sci. jours. Mem. Soc. Exptl. Biology and Medicine, Royal Soc. Medicine, Western Pharm. Soc., Am. Pharm. Assn., Acad. Pharm. Scis., N.Y. Acad. Scis., Am. Chem. Soc. (div. med. chemistry), Sigma Xi, Rho Chi., Am. Soc. Pharmacology and Exptl. Therapeutics. Roman Catholic. Subspecialties: Pharmacology; Biochemistry (medicine). Current work: Pulmonary, CNS, GI, renal, cardiovascular, anti-arthritic pharmacology administration; director of pharmacology department. Home: 1303 Grayson Rd Wilmington DE 19803 Office: Stuart Pharms Div ICI-Americas Murphy Rd and Concord Pike Wilmington DE 19897

GILL, D. MICHAEL, molecular biology and microbiology, educator; b. London, Sept. 17, 1940; U.S., 1967; m. Gillian, Catherine (Scobie), 1964; children: Christopher, Catherine. M.A., Cambridge (Eng.) U., 1964, Ph.D., 1966. Asst. prof. biology Harvard U., 1969-74, assoc. prof., 1974-77; vis. assoc. prof. Yale U., 1978; assoc. prof. molecular biology and microbiology Tufts U., 1979-81, prof., 1981—. NIH grantee, 1973—. Subspecialties: Biochemistry (biology); Microbiology. Current work: Mode of action of bacterial toxins: cholera and other enterotoxins, toxic shock toxin, Clostridial toxins, and others; NAD metabolism: poly ADP-ribose and its role in DNA repair, NAD glycohydrolase. Home: 31 Oakland St Lexington MA 02173 Office: Tufts U Med Sch 136 Harrison Ave Boston MA 02111

GILL, DANIEL EMMETT, structural analyst; b. Celina, Ohio, Sept. 1, 1946; s. Homer Robert and Thelma A. G.; m. Leslie Jennifer Davis, June 21, 1975; 1 son, Patrick Danielson. B.S.M.E. with honors, Purdue U., 1973, M.S.M..E., 1975. Structural engr. David W. Taylor Naval Ship Research and Devel. Center, Bethesda, Md., 1975-79; analytical engr. Gen. Electric Aircraft Engine Group, Cin., 1979-81; sr. engr. Cummins Engine Co., Columbus, Ind., 1981—. Served with USN, 1966-69. Mem. ASME. Lutheran. Subspecialties: Mechanical engineering; Theoretical and applied mechanics. Current work: Computational methods to predict service life of engine components under complex load/temperature cycles. Office: Cummins Engine Co Mail Code 50181 Box 3005 Columbus IN 47201

GILL, DENNIS HOWARD, electrical engineer; b. Monticello, Akr., Jan. 5, 1939; s. William Howard and Dorothy Lauree G.; m. Opal Lee Clark, Sept. 3, 1939; children: John Howard, David Dennis. B.A., Rice U., 1961, B.S.E.E., 1962; M.S.E.E., U. Tex., 1963, Ph.D., 1966. With Los Alamos Nat. Lab., 1966—, group leader tunable laser research group, 1980-83, assoc. div. leader chemistry div., 1983—. Contbr. articles to profl. jours. Mem. Los Alamos Bd. Edn., 1979—, v.p., 1981-83, pres., 1983—; pres. adv. bd. Los Alamos br. U. N.Mex., 1981—. Mem. IEEE (sr.), Optical Soc. Am. Republican. Methodist. Subspecialties: Laser-induced chemistry; Optical engineering. Current work: Technical management of laser system research and development, especially for applications related to photochemistry, spectroscopy, remote detection and laser fusion. Home: 2403 Club Rd Los Alamos NM 87544 Office: Los Alamos Nat Lab MS J563 Los Alamo NM 87545

GILL, GERALD CLIFFORD, atmospheric science educator, instrument manufacturer; b. Mt. Elgin, Ont., Can., Apr. 5, 1911; s. Clarence C. and Dora M. (Parsons) G.; m.; children: Susan S., Thomas G. B.A. in Math. and Physics, U. Western Ont., London, 1934, M.A. in Physics, 1936. Field engr. Meteorol. Service of Can., 1936-49; research engr. M.I.T., 1949-55; lectr., assoc. prof. to prof. meteorol. instruments, 1955-75, ret., 1975; v.p. R.M. Young Co., Traverse City, Mich., 1975—. Contbr. articles sci. jours. Fellow Am. Meteorol. Soc. Congregationalist. Lodge: Ann Arbor Rotary. Subspecialty: Meteorologic instrumentation. Current work: Developing and testing of improved radiation shields for temperature measuring sensors. Patentee in field. Home: 1303 W Madison St Ann Arbor MI 48103 Office: U Mich 200 Research Activities Bldg Ann Arbor MI 48109

GILL, PIARA SINGH, chemistry educator; b. Bassawal, Panjab, India, Feb. 15, 1940; came to U.S., 1962; s. Gurdial S. and Bhagwan K. G.; m. Gurjatinder Bains, Dec. 25, 1967; children: Sukhmani, Brijesh. B.Sc., Panjab U., 1961; M.S., Kans. State U., 1965, Ph.D., 1967. Research assoc. Wright-Patterson AFB, Ohio, 1968-69; asst. prof. Tuskegee (Ala.) Inst., 1969-72, assoc. prof., 1972-77, prof., 1977—. Contbr. articles to profl. jours. Mem. Am. Chem. Soc., Sigma Xi. Subspecialty: Kinetics. Current work: Atmospheric chemistry. Home: 3132 Fitzgerald Rd Montgomery AL 36106 Office: Dept Chemistry Tuskegee Inst Tuskegee AL 36088

GILL, THOMAS JAMES, III, physician, educator; b. Malden, Mass., July 2, 1932; s. Thomas James and Marguerite (Capobianco) G.; m. Faith Libbie Etoll, July 8, 1961; children—Elizabeth Ruth, Thomas James IV. A.B. summa cum laude, Harvard U., 1953, M.A. in Chemistry, 1957, M.D., 1957. Asst. pathology Peter Bent Brigham Hosp., Boston, 1957-58; intern N.Y. Hosp.-Cornell Med. Center, 1958-59; jr. fellow Soc. Fellows Harvard U., 1959-62; mem. faculty Harvard Med. Sch., 1962-71, asso. prof. pathology, 1970-71; prof. pathology, chmn. dept. U. Pitts. Med. Sch., 1971—; pathologist-in-chief Univ. Health Center Pitts., 1971—; cons. to govt. and industry; mem. sci. adv. bd. St. Jude Children's Research Hosp., Memphis, 1969-77, chmn., 1974-76; mem. allergy and immunology research com. Nat. Inst. Allergy and Infectious Diseases, 1973-76; mem. med. research service merit rev. bd. in immunology VA, 1976-79, chmn., 1977-79; mem. sci. adv. com. Damon Runyon-Walter Winchell Cancer Fund, 1978—; mem. com. on animal models and genetic stocks NRC, 1978—, mem. com. on rabbit genetic resources, 1979-80; mem. surgery, anesthesiology and trauma study sect. NIH, 1983—; mem. Armed Forces Epidemiol. Bd., 1966-72. Editorial bds. several sci. and med. jours.; contbr. articles to profl. jours. Bd. dirs. Easter Seal Soc., Allegheny County, 1972-77, Univs. Asso. for Research and Edn. in Pathology, 1979—; trustee Am. Bd. Pathology, 1981—. Recipient Lederle med. faculty award, 1962-65, research career devel. award NIH, 1965-71; certificate appreciation for patriotic civilian service Dept. Army, 1973. Fellow Am. Soc. Clin. Pathologists, Am. Acad. Allergy, Assn. Pathology Chairmen (pres. 1978); mem. Am. Assn. Immunologists, Am. Assn. Pathologists, Am. Soc. Biol. Chemists, Internat. Acad. Pathology, Transplantation Soc. (v.p. 1982-84), Am. Chem. Soc., Am. Soc. Cell Biology, Genetics Soc. Am., AMA. Clubs: Harvard (Western Pa.); (Boston); Fox Chapel Racquet (Pitts.); Pitts. Athletic Assn., Harvard Varsity. Subspecialties: Immunogenetics; Transplantation. Current work: Major interest is in basic immunogenetics of the rat and application of this knowledge to problems of transplantation and reproduction. Home: 117 Crofton Dr Pittsburgh PA 15238

GILLASPIE, ATHEY GRAVES, JR., research plant pathologist; b. Asheville, N.C., July 30, 1938; s. Athey Graves and Virginia (Shackelford) G.; m. Margaret Ellen Fleming, Aug. 14, 1965; children: Timothy G., Jonathan T. B.A. in Botany, Miami U., Oxford, Ohio, 1960; M.S., Purdue U., 1962, Ph.D. in Plant Pathology, 1965. Research plant pathologist U.S. Dept. Agr., Houma, La., 1965-72, Beltsville, Md., 1972—, chief plant pathology lab., 1980—. Co-editor: Diseases of Sugarcane, 1983; editor: Sugarcane Pathologists' Newsletter, 1981-83; contbr. articles to profl. jours. Mem. Am. Phytopath. Soc., Internat. Soc. Sugar Cane Tech., Am. Soc. Sugar Cane Tech. Baptist. Subspecialties: Plant pathology; Plant virology. Current work: Pathogens of sugarcane especially viruses and bacteria nature, mode of action and control of ratoon stunting disease bacterium and sugarcane mosaic virus. Office: Beltsville Agrl Research Ctr Beltsville MD 20705

GILLASPY, JAMES EDWARD, biology educator; b. Bartlett, Tex., Oct. 15, 1917; s. Arthur Porter and Mae Lou (Mitchusson) G.; m. Dorene Lannen, Aug. 2, 1948; children: Zena Jolaine, Kyna Dorene. B.S., Tex. A&M U., 1940; postgrad., Ohio State U., 1941, U. Tex.-Austin, 1951-52; Ph.D., U. Calif.-Berkeley, 1954. Registered profl. entomologist, Tex. Tchr. sci. West Covina High Sch., Calif., 1958-59; asst. prof. biology Tex. Luth. Coll., Seguin, 1959-60, Sul Ross State Coll., Alpine, Tex., 1960-61; research assoc. Mus. Comparative Zoology, Cambridge, Mass., 1961-63; asst. prof. biology Mankato (Minn.) State Coll., 1963-66; prof. biology Tex. A&I U., Kingsville, 1966—. Author: Revisionary Study of the Genus Steniolia, 1954. Served to 1st lt. USAAF, 1942-46. Disting. Research award Tex. A&I U. Alumni Assn., 1981. Mem. AAAS, Entomol. Soc. Am., Southwestern Entomol. Soc., Am. Inst. Biol. Scis., Kans. Entomol. Soc. Democrat. Presbyterian. Subspecialties: Integrated pest management; Allergy. Current work: Use of predatory wasps for caterpillar control; extraction of paper wasp (polistes) venom for allergy research; insect faunistics, U.S. and Mexico. Home: 1414 W Santa Gertrudis Kingsville TX 78363 Office: Tex A&I U West Santa Gertrudis Ave Kingsville TX 78363

GILLELAND, JOHN ROGERS, research scientist; b. Gadsden, Ala., Jan. 12, 1941; s. Earl Rogers and Margaret (Kilpatrick) G.; m. Ruth Magnuson, Nov. 14, 1964; children: John Rogers Jr., David Reid. B.S. in Physics, Yale U., 1963, M.S., U. Mich., Ann Arbor, 1964, Ph.D., 1969. Research scientist U. Mich., 1964-69; assoc., sr. to staff scientist Gen. Atomic Co. (name now GA Techs. Inc.), San Diego, 1970-73, project mgr., 1973-78, mgr. fusion projects, 1978-80, dir., 1980—; mem. U.S. del. to UN Internat. Tokanak Reactor design team, 1979; exec. dir. Fusion Engring. Device Tech. Mgmt. Bd., Washington, 1980-81; mem., panel chmn. Magnetic Fusion Adv. Com., Washington, 1982—; participant Nat. Research Council on Fusion, Washington; founding mem. Atomic Indsl. Forum Com. on Fusion; mem. exec. adv. bd. U.S. Internat. Tokanak Reactor Study, 1980-82. Mng. editor, author: The Fusion Engineering Device, vols. L-IV, 1981. Recipient Young Mem. Achievement award Am. Nuclear Soc., 1980; Cert. of Appreciation U.S. Dept. Energy, 1981. Mem. Am. Phys. Soc., Am. Nuclear Soc., Yale Sci. and Engring. Assn., Fusion Power Assocs., Sigma Xi. Subspecialties: Fusion; Plasma. Current work: Director of largest operating magnetic fusion experiment in the world. Member of magnetic fusion advisory committee, reports through Director of Research to Secretary of Energy. Office: GA Techs Inc 10955 John Jay Hopkins Dr San Diego CA 92121

GILLESPIE, ELIZABETH, biochemical pharmacologist; b. Montreal, Que., Can., May 7, 1936; came to U.S., 1966; d. George and Madeline Maud (Deller) G. B.Sc., McGill U., 1957, Ph.D., 1966. Info. scientist Explosives and Ammunition divs. Can. Industries Ltd., Montreal, 1957-59; sr. research technician U. Mich., Ann Arbor, 1959-62; postdoctoral fellow dept. pharmacology Yale U. Sch. Medicine, New Haven, 1966-68, research asso., 1968-74; instr. clin. immunology div. dept. medicine Johns Hopkins U. Sch. Medicine, Balt., 1971-73, asst. prof. medicine, 1973-81; with Wellcome Research Labs., Beckenham, Kent, Eng., 1978-79; sect. mgr. Respiratory Pharm. div. Mead Johnson Pharm. div. Bristol-Myers Co., Evansville, Ind., 1981—. Contbr. articles and abstracts to sci. jours., also chpts. to books. NIH grantee, 1974-81. Mem. Am. Assn. Immunology, Am. Soc. Pharmacology and Exptl. Therapeutics. Subspecialties: Immunopharmacology; Allergy. Current work: Drug development. Home: 619 College Hwy Evansville IN 47714 Office: Mead Johnson Pharm Div Bristol Myers Co 2400 Pennsylvania Ave Evansville IN 47721

GILLESPIE, GEORGE HUBERT, physicist; b. Dallas, Sept. 9, 1945; s. Hubert W. and Frieda G.; m. Susan Gilbert, Sept. 7, 1968; children: James S., Colin H., Ian G. B.A., Rice U., 1968, M.E.E., 1968, M.S. in Physics, 1969; Ph.D., U. Calif., San Diego, 1974. Engr., IBM, Lexington, Ky., summer 1967; research asst. Los Alamos Sci. Lab., summer 1968, U. Calif., San Diego, 1968-74; assoc. LaJolla (Calif.) Inst., 1976—; staff scientist Physical Dynamics, Inc., LaJolla, 1975—. Contbr. articles to profl. jours. Served to capt. USAR, 1968-76. Mem. Am. Phys. Soc., AAAS. Subspecialties: Theoretical physics; Atomic and molecular physics. Current work: Theoretical applied physics; primary interests in scattering theory, quantum mechanics, atomic physics and plasma physics; with applications in magnetic confinement fusion, ion-beam fusion, and particle accelerators. Home: 364 Hillcrest Dr Leucadia CA 92024 Office: Physical Dynamics Inc PO Box 1883 La Jolla CA 92038

GILLESPIE, JAMES HOWARD, veterinary microbiologist, educator; b. Bethlehem, Pa., Nov. 26, 1917; s. John C. and Mary (Kidd) G.; m. Virginia Arnts, Dec. 19, 1941; children: H. Thomas, Janet M., Ian Scott. D.V.M., U. Pa., 1939. Diplomate: Am. Coll. Vet. Microbiologists. Asst. poultry pathologist U. N.H., Durham, 1939-40; asst. prof. poultry diseases N.Y. State Coll. Vet. Medicine, Cornell U., Ithaca, 1946-48, asst. prof. vet. bacteriology, 1948-52, assoc. prof. vet. microbiology, 1952-56, asst. dir., 1962-67, prof. vet. microbiology dept., 1972-81, dir., 1981—. Author: Hagan and Bruner's Infectious Diseases of Domestic Animals, 7th edit, 1980; contbr. over 100 sci. articles to profl. publs. Served to lt. col. U.S. Army, 1940-45. NIH spl. fellow, 1953-54, 60-61; Nat. Cancer Inst. spl. fellow, 1968-69. Mem. Am. Soc. Microbiologists, AVMA (Gaines award 1971, Merit award 1971), N.Y. State Vet. Med. Soc., AAAS, Am. Coll. Vet. Microbiologists, Soc. Am. Microbiologists, Sigma Xi. Club: Ithaca Yacht. Subspecialties: Virology (veterinary medicine); Preventive medicine (veterinary medicine). Current work: Vertical transmission of bovine viruses via embryo transplants; equine viral infectious diseases; genetics and genetic engineering as it applies to preventive medicine. Office: Vet Microbiology Dept NY State Coll Vet Medicine Room 215 Vet Research Tower Ithaca NY 14853

GILLESPIE, JESSE SAMUEL, JR., research institute administrator, chemistry educator; b. Lynchburg, Va., Dec. 20, 1921; m., 1950; 4 children. B.S., Va. Mil. Inst., 1943; Ph.D. in Chemistry, U. Va., 1949. Asst. prof. chemistry U. Richmond, 1949-51; sr. chemist Va.-Carolina Chem. Corp., 1951-53, group leader, 1953-54, asst. div. mgr., 1954-56, mgr. organic and agrl. chemistry, 195-58; ptnr. Cox & Gillespie Chemists & Chem. Engrs., 1958-62; sr. chemist Va. Inst. Sci. Research, U. Richmond, 1962-68, acting dir. inst., 1968-69, dir., 1969—, prof. chemistry, 1972—; dir. sponsored programs U. Richmond, 1972—. Mem. Am. Chem. Soc., Sigma Xi. Subspecialty: Organic chemistry. Office: Va Inst Sci Researc U Richmond Sci Ct Richmond VA 23173

GILLESPIE, SHERRY JACQUELINE, microcircuit engineer; b. N.Y.C., June 2, 1943; d. Rubin and Anne (Sloter) Victor; m. (div.); children: Neil Gillespie, Tristan Gillespie. A.B., Vassar Coll., 1965; M.S., U. Pa., 1966; Ph.D., Temple U., 1975. Tchr. physics Abington (Pa.) High Sch., 1967-68; engr. IBM, Burlington, Vt., 1975—. Contbr. articles to physics, engring. jours. NSF trainee, summer 1971. Mem. Am. Phys. Soc. Subspecialties: Microchip technology (engineering); 3emiconductors. Current work: Manager of process development for electron beam technology. Home: 6 Nahma Ave Essex Junction VT 05452 Office: IBM Corp Essex Junction VT 05452

GILLETTE, DEAN, mathematician, consultant; b. Chgo., Aug. 11, 1925; s. Frank Kenneth and Ruth (Whitmore) G.; m. Helen Klamt, Dec. 19, 1949; 1 son, Troy. B.S. in Chemistry, Oreg. State U., 1948; M.A. in Math, U. Calif.-Berkeley, 1950, Ph.D., 1953. Mem. tech. staff Bell Labs., Holmdel, N.J., 1953—, exec. dir. labs., 1966—; mem. various coms. Dept. Def.; mem. com. communications Office Tech. Assessment; mem. indsl. liaison com. U. Calif.-Berkeley. Author papers on math., systems engring., mgmt.; editorial com. Networks, Telcom Policy. Mem. Sch. Bd. Mendham Twp., N.J., 1966. Served with USN, 1944-46; PTO. Mem. IEEE (sr.; com. communications and info. policy, chmn. subcom. indsl. productivity), Am. Math. Soc., Soc. Indsl. and Applied Math. Republican. Subspecialties: Systems engineering; Information systems (information science). Current work: Application of information systems to industry and management. Office: Bell Labs Holmdel NJ 07733

GILLHAM, NICHOLAS WRIGHT, geneticist, educator; b. N.Y.C., May 14, 1932; s. Robert Marty and Elizabeth (Enright) G.; m. Carol Lenore Collins, June 2, 1956. A.B., Harvard, 1954, A.M., 1955, Ph.D. (USPHS fellow), 1962. From instr. to asst. prof. Harvard, 1963-68; asso. prof. zoology Duke, 1968-72, prof., 1973-82, James B. Duke prof. zoology, 1982—; Mem. biochemistry, molecular genetics and cell biology interdisciplinary cluster Pres.'s Biomed. Research Panel, 1975; mem. study sect. in genetics NIH, 1976-80. Author: (with R. Kreuger and J. Coggin) Introduction to Microbiology, 1973, Organelle Heredity, 1978; Mem. editorial bd.: Genetics, 1975-78, Jour. Cell Biology, 1977-79; sr. editor: Plasmid, 1977—. Served to 1st lt. Med. Service Corps USAF, 1955-58. Postdoctoral fellow, 1962-63; Spl. fellow, 1967-68; Research Career Devel. Award grantee, 1972-77; all USPHS). Mem. Am. Soc. Naturalists, Genetics Soc. Am., Soc. for Cell Biology, Sigma Xi. Subspecialties: Plant genetics; Molecular biology. Current work: Genetics and biogenesis of chloroplasts and mitochondria; the chloroplast genome—what it does, its genetics and how it interacts with the nuclear genome. Home: 1211 Woodburn Rd Durham NC 27705 Office: Dept Zoology Duke U Durham NC 27706

GILLIGAN, JOHN GERARD, nuclear engineering educator; b. Indpls., Jan. 17, 1949; s. John Bernard and Mary Ann (Prieshoff) G.; m. Barbara Ann Bertolami, Dec. 8, 1979; 1 dau.: Theresa Marie. B.S., Purdue U., 1971; M.S., U. Mich., 1973, Ph.D., 1977. Research assoc. Princeton (N.J.) Plasma Physics. Lab., 1974-77; asst. prof. dept. nuclear engring. U. Ill., Champaign, 1977-83, affiliate asst. prof. mech. engring., 1982-83; assoc. prof. dept. nuclear engring. N.C. State U., 1983—; cons. SAI, La Jolla, Calif., 1980—. Argonne Nat. Lab. summer faculty fellow, 1982, 83. Mem. Am. Nuclear Soc. (fusion exec.

com. 1980-83), IEEE, Univ. Fusion Assn., Fusion Power Assocs. Roman Catholic. Subspecialties: Nuclear fusion; Nuclear fission. Current work: Technology of fusion reactors for electrical production, plasma physics, particle and energy transport. Home: 1028 Sturdivant Dr Cary NC 27511 Office: Dept Nuclear Engring 214 NEL 103 S Goodwin St Urbana IL 61801

GILLILAND, RONALD LYNN, astrophysicist; b. Emporia, Kans., July 16, 1952; s. Rodney and Reta Frances (Jones) G. B.A., U. Kans., 1974; Ph.D., U. Calif., Santa Cruz, 1979. Postdoctoral fellow Advanced Study Program, Nat. Ctr. for Atmospheric Research, Boulder, Colo., 1979-81; staff scientist High Altitude Obs., Boulder. Mem. Am. Astron. Soc., Pi Mu Epsilon, Sigma Pi Sigma. Subspecialties: Theoretical astrophysics; Solar physics. Current work: Solar and stellar variability. Office: PO Box 3000 Boulder CO 80307

GILLIN, JOHN CHRISTIAN, research phychiatrist; b. Columbus, Ohio, Apr. 28, 1938; s. John Philip and Helen (Norgord) G.; m. Frances Davis, May 29, 1966; children: John Lorin, Peter Daniel. B.A. magna cum laude, Harvard U., 1961; M.D., Western Res. U., 1966. Diplomate: Am. Bd. Phychiatry and Neurology. Intern, Cleve. Met. Gen. Hosp., 1966-67; resident in phychiatry Stanford (Calif.) U., 1967-69; commd. USPHS, 1969; clin. assoc. NIMH, Bethesda, Md., 1969-71, research phychiatrist, chief unit on sleep studies, 1971-82, dep. chief adult psychiatry br., 1982; prof. psychiatry U. Calif., San Diego, 1982—; professorial lectr. George Washington Sch. Medicine, 1982; adj. prof. pharmacology Uniformed Services U. Health Scis. Sch. Medicine, 1982. Contbr. over 200 articles to sci. publs. Recipient Commendation award USPHS, 1978, Outstanding Service Medal, 1982. Mem. Am. Psychiat. Assn., Assn. Psychophysiol. Study of Sleep, Psychiat. Research Soc., Am. Coll. Neuropsychopharmacology, Soc. Neurosci., So. Sleep Soc., Am. Assn. Geriatric Psychiatry. Unitarian. Subspecialties: Psychiatry; Psychopharmacology. Current work: Sleep research; biological psychiatry; pharmacology; chronobiology. Home: 2110 Calle Guaymas La Jolla CA 92037 Office: Va Med Center U Calif Psychiatry 116A San Diego CA 92161

GILLIOM, RICHARD D., chemist, educator, researcher, consultant; b. Bluffton, Ind., June 25, 1934; s. Andrew and Roberta Aldine (Dowty) G.; m. Patricia Ann Hastings, Mar. 7, 1958; children: Laura Rhea, Andrea Lee, Bruce Hastings. Student, Centenary Coll., 1952-54; B.S., Southwestern at Memphis, 1956; Ph.D. (Sun Oil Co. fellow), M.I.T., 1960. Research chemist Esso Research Labs., Baton Rouge, 1960-61; asst. prof. chemistry Southwestern at Memphis, 1961-64, assoc. prof., 1964-71, prof., 1971—; cons. in field; undergrad. research participation dir. NSF, 1966-68, 71, 75, 81. Author: Introduction to Physical Organic Chemistry, 1970; contbr. articles to profl. jours. Fulbright Hays lectr., 1968-69; NSF grantee, 1970-73; Research Corp. grantee, 1962-63. Mem. Am. Chem. Soc. Subspecialties: Organic chemistry; Theoretical chemistry. Current work: Mechanisms of hydrogen abstraction reactions, linear free energy relationships, drug design, computational chemistry. Patentee in field. Home: 3017 Dumbarton Rd Memphis TN 38128 Office: 2000 N Parkway Memphis TN 38112

GILLIS, STEVEN, immunologist; b. Phila., Apr. 25, 1953; s. Herbert and Rosalie Henrietta (Segal) G.; m. Anne Cynthia Edgar, June 26, 1976. B.A., Williams Coll., 1975; Ph.D., Dartmouth Coll., 1978. Research fellow Norris Cotton Cancer Ctr., Hanover, N.H., 1977-78; research assoc. Dartmouth Med. Sch., Hanover, 1978-79; vis. assoc. Sloan Kettering Inst., N.Y.C., 1979-80; asst. mem. Hutchinson Cancer Research Ctr., Seattle, 1980-82; also asst. prof. U. Wash.; exec. v.p. Immunex Corp., Seattle, 1982—; affiliate Hutchinson Cancer Ctr., 1982—. Contbr. articles to profl. jours. Leukemia Soc. spl. fellow, 1978-82. Mem. Am. Assn. Immunologists, N.Y. Acad. Sci., Phi Beta Kappa, Sigma Xi. Subspecialties: Cellular engineering; Immunopharmacology. Current work: Biochemical and molecular characterization of lymphokines, hormones which control the immune response. Office: 51 Univ Bldg Suite 600 Seattle WA 98101

GILLMAN, CLIFFORD BRIAN, computer scientist; b. Montreal, Que., Can., Jan. 30, 1941; s. Hyman and Fanny (Izenberg) G.; m. Joan Mira Freidlander, June 18, 1967; children: Gail, Joshua. A.B., Stanford U., 1962; M.A., Calif. State U., 1964; Ph.D., Ind. U., 1970. Asst. prof. dept. psychology U. Wis.-Madison, 1968-73; assoc. scientist Waisman Ctr., 1973-79, sr. scientist, 1979—; cons. in field. Contbr. articles to profl. jours. NIMH grantee, 1969-72; Nat. Inst. Child Health and Human Devel. grantee, 1973—. Mem. Am. Psychol. Assn., Harris Users Exchange (pres. 1979-80), Assn. Computing Machinery, Psychometric Soc., AAAS, Soc. Math. Psychology, IEEE Computer Soc. Club: Four Lakes Amateur Radio. Subspecialties: Distributed systems and networks; Sensory processes. Current work: Basic research in acoustic phonetics, infant vocalizations; also research/ applications in man-machine interaction, distributed computing, graphics. Home: 26 S Owen Dr Madison WI 53705 Office: 1500 Highland Ave Madison WI 53706

GILMAN, DONALD LAWRENCE, meteorologist; b. Hartford, Conn., Oct. 15, 1931; s. Ralph Lawrence and Ruby Gertrude (Weaver) G.; m. Barbara Anne Burrows Feys, Dec. 21, 1961 (div. 1977); children—Jennifer Barbara, Christopher James, Valerie Doe. A.B. cum laude, Harvard U., 1952; S.M., Mass. Inst. Tech., 1954, Ph.D., 1957. Research asst. dept. meteorology Mass. Inst. Tech., Cambridge, 1952-54, 56-57, research asso., 1957-58; mem. staff Nat. Meteorol. Center, Nat. Weather Service, NOAA, Dept. Commerce, 1958—, chief devel. and testing sect., extended forecast div., 1965-72, chief long range prediction group, Washington, 1972-78, 1978—, chief prediction br., 1979—. Asso. editor: Jour. Applied Meteorology, 1968-72. Recipient Pub. Service award NOAA, 1977. Fellow Am. Meteorol. Soc.; mem. Am. Geophys. Union, AAAS, Sigma Xi. Subspecialty: Meteorology. Home: 8127 Pinelake Ct Alexandria VA 22309 Office: Nat Meteorological Center NOAA Nat Weather Service Washington DC 20233

GILMAN, HENRY, organic chemist, educator; b. Boston, May 9, 1893; s. David and Jane (Gordon) G.; m. Ruth V. Shaw, July 20, 1929; children—Jane Gordon, Henry Shaw. B.S., Harvard, 1915, A.M., 1917, Ph.D., 1918; postgrad., Zurich Polytechnikum, Oxford, 1916. Instr. chemistry Harvard, 1918-19; asso. in chemistry U. Ill., 1919; prof. organic chemistry Iowa State U., 1919—; Cons. AEC; Air Force research dir.; plenary lectr. Internat. Symposia Organometallic and Organometalloidal Chemistry; invited lectr. internat. confs., Bordeaux, 1970, Moscow, 1971, Madison, 1972. Author: (with C.J. West) Organomagnesium Compounds in Synthetic Chemistry, 1922, (with R.J. Jones) Organo-lithium Compounds in Organic Reactions, (with J.W. Morton Jr.) Metalation in Organic Reactions, (with D. Wittenberg) Silylmetallic Compounds, (with F.K. Cartledge) Characterization of Organometallics, 1968, More Than One-half Century of Organometallic Chemistry, 1968; Editor: Organic Chemistry (2 vols.), 1943, Vols. III, IV Organic Chemistry: Organometallic Compounds in Encyclopedia of Chem. Tech, (with Benkeser), (with R.K. Ingham) Organopolymers of Silicon, Germanium, Tin, and Lead, (with W.H. Atwell and F.K. Cartledge) Catenated Organic Compounds of Group IV-B, (with G.L. Schwebke) Organic Substituted Cyclosilanes, (with H.J.S. Winkler) Organosilylmetallic Chemistry; Syntheses of Some Perfluoro-organometallic Types, 1975; Mem. editorial bd.: Current Contents-Chemical Sciences; Contbg. editor ann.: Survey of Am. Chemistry, 1928, 1929-30, Organometallic Syntheses, Sci. Citation Index, Organometallic Reactions; asso. editor: Chem. Revs, 1936, Jour. Organometallic Chemistry, Organic Preparations and Procedures Internat, Acta Chimica Inorganica; editorial bd., exec. com.: Jour. Organic Chemistry; asso. editor: Jour. Am. Chem. Soc; Contbr. to ency.; also; co-author monographs; contbr. articles in field to sci. periodicals, with current publs. particularly in area of polyhalo-organometallic chemistry and perfluoroorganic chemistry. Trustee Carver Research Found. Served with CWS, World War I; Nat. Def. research work (Manhattan Project). Holder of various, lectureships and mem. awards coms.; recipient Mid-West Gold medal, Iowa Am. Chem. Soc. Medal award; Frederick Stanley Kipping award Am. Chem. Soc., also; Priestly medal; First Firestone Internat. Lectures award in organometallic chemistry; Distinguished Prof. in Sci. and Humanities, 1962; Iowa State U. chemistry bldg. named in his honor, 1973; Recipient initial Merit award Iowa Acad. Scis., 1973, 100th Anniv. Distinguished Fellow award, 1975. Fellow AAAS (v.p., chem. sect. chmn. 1931), Chem. Soc. London (hon.), Am. Chem. Soc. (councillor at large 1939-41, 42-44, chmn. organic div.), Nat. Acad. Scis. (ofcl. del., lectr. in Soviet Union 1963), Royal Soc. (London); mem. N.Y. Acad. Scis. (hon. life mem.), Phi Beta Kappa, Sigma Xi, Phi Kappa Phi, Phi Lambda Upsilon (hon.). Subspecialty: Organometallic chemistry.. Home: 111 Lynn Ave Apt 902 Ames IA 50010

GILMAN, SID, neurology educator, neuroscientist; b. Los Angeles, Oct. 19, 1932; s. Morris and Sarah Rose (Cooper) G. B.A., UCLA, 1954, M.D., 1957. Diplomate: Am. Bd. Psychiatry and Neurology. Research assoc. NIH, Bethesda, Md., 1958-60; resident in neurology Harvard Med. Sch., Boston, 1960-63, instr. to assoc., 1965-68; from asst. prof. to prof. neurology Columbia U., N.Y.C., 1968-76, H.H. Merritt prof., 1976-77; prof., chmn. dept. neurology U. Mich., Ann Arbor, 1977—; mem. sci. programs adv. com. Nat. Inst. Neurol. Communicative Diseases and Stroke, NIH, 1982—; chmn. sci. adv. council Nat. ALS Found., N.Y.C., 1981—; mem. profl. adv. bd. Epilepsy Found. Am., 1976—; mem. research adv. council United Cerebral Palsy, 1973—. Author: Disorders of the Cerebellum, 1981, Essentials of Clinical Neuroanatomy and Neurophysiology, 1982; contr. numerous articles to sci. jours. Served as sr. asst. surgeon USPHS, 1958-60. Recipient Weinstein-Goldenson award United Cerebral Palsy, 1981, Lucy G. Moses prize Columbia U., 1973, NIH Career devel. award, 1966; NIH fellow, 1962. Fellow Am. Acad. Neurology; mem. Am. Neurol. Assn. (v.p. 1980-81), Am. Soc. Clin. Investigation, Am. Physiol. Soc., Soc. Neurosci. Subspecialties: Neurology; Neurophysiology. Current work: Cerebellar neurophysiology; mechanisms underlying spasticity and movement disorders; studies of cerebral metabolism with positron emission tomography; regenerative activity in the nervous system after injury. Home: 3411 Geddes Rd Ann Arbor MI 48105 Office: 1405 E Ann St Ann Arbor MI 48109

GILMARTIN, AMY JEAN, botanist, herbarium director, educator; b. Red Bluff, Calif., Oct. 15, 1932; d. Ruy H. and Margaret Helena (Harvey) Finch; m.; children: Malvern, Dale Moana, Sheila Ann, Ian Harvey. B.A., Pomona Coll., 1954; M.S., U. Hawaii, 1956, Ph.D (AAUW fellow), 1968. Research asst. Hawaii Agrl. Exptl. Sta., 1954-56; curatorial asst. U. B.C. (Can.), Vancouver, 1956-58; prof. contratada U. Guayaquil, Ecuador, 1962-64; asso. researcher Smithsonian Instn., Washington, 1969-70; instr. biology Monterey (Calif.) Peninsula Coll., 1970-75; dir. Ownbey Herbarium, Pullman, Wash., 1975—; asst. prof. Wash. State U., Pullman, 1975-79, assoc. prof., 1979-83, prof., 1983—; cons. environ. surveys. Author: Bromeliads of Ecuador, 1972; contbr. articles on botany to profl. jours.; editorial bd., Calif. Bot. Soc., Internat. Assn. Plant Taxonomy. NSF grantee, 1971-83; sr. Fulbright prof., Bogota, Colombia, 1980; AAUW postdoctoral fellow, 1982-83. Mem. Bromeliad Soc. (dir.), Soc. Study of Evolution, Am. Soc. Plant Taxonomists, Soc. Systematic Zoology, Internat. Assn. Plant Taxonomy, N.W. Sci. Assn., AAAS. Democrat. Subspecialties: Evolutionary biology; Numerical analysis. Current work: Evolution and phylogeny of flowering plants, numerical taxonomy, phylogenetic inference, speciation, bromeliads, umbellifers. Office: Ownbey Herbarium Wash State U Pullman WA 99164

GILMORE, EARL HOWARD, college energy management program administrator, research educator; b. Turkey, Tex., July 9, 1923; s. Joseph Lee and Nora Edna (Cowart) G.; m. Carmen Earline Thomson, Aug. 23, 1946; children: Carmen Dianne Gilmore Harris, Steven Lee. B.S. in Chem. Engring, Tex. Tech. U., 1943, M.S. in Chemistry, 1947, Ph.D., U. Calif.-Berkeley, 1951. Physical chemist U.S. Naval Ordnance Test Sta., China Lake, Calif., 1950-53; assoc. prof. chemistry Okla. State U., 1953-58; assoc. prof. math. Tex. Tech. U., 1958-68; physicist Helium Research Ctr., U.S. Bur. Mines, Amarillo, Tex., 1968-70; prof. phys. sci. Amarillo Coll., 1970—; dir. energy mgmt. program, 1981—; research prof. Alternative Energy Inst., West Tex. State U.; cons. wind energy, alternative energy techs. Author publs. in chem. analysis, photochemistry and photometry, 1948-60, publs. in wind energy tech., 1973-81. Served to lt. j.g. USNR, 1943-46. Du Pont fellow, 1949-50. Mem. Am. Phys. Soc., Am. Chem. Soc., Tex. Solar Energy Soc. Methodist. Current work: Development and direction of an energy management program at Amarillo College; educational programs, test projects, and performance analyses of wind energy conversion systems.

GILMORE, RICHARD ALLEN, university administrator, educator; b. Indpls., July 27, 1933; s. Thomas Byard and Lena Irvine (Gregory) G.; m. Janine Raymonde, Jan. 3, 1959; children: Steven E., Sheryl L. Cochran, Janet M. A.B., Hanover Coll., 1955; Ph.D., U. Calif.-Berkeley, 1966. Radiation biophysicist USPHS, Washington, 1956-64, Las Vegas, Nev., 1956-64, Berkeley, Calif., 1956-64; asst. prof. microbiology So. Ill. U., Carbondale, 1968-75, Office Research and Projects grantee, 1971-72; assoc. prof. biochemistry Marshall U. Sch. Medicine, Huntington, W.Va., 1975-81; assoc. dean Sch. Medicine, Oral Roberts U., Tulsa, 1981—; mem. adv. com. nuclear medicine Doctor's Hosp., Carbondale, Ill., 1974-75. Contbr. articles to profl. jours. Chmn. Ind. Elks scholarship selection com., 1972; active Boy Scouts Am., 1971-75; v.p. Carbondale elem. sch. PTA, 1971-72, elem. sch. dist. 95 behavioral code com., 1972-75. AEC fellow, 1955-56; USPHS postdoctoral fellow, 1966-68; NIH grantee, 1964-66; Danforth Found. Assoc., 1971—. Mem. Genetics Soc. Am., Health Physics Soc., AAAS, N.Y. Acad. Scis., Sigma Xi. Methodist. Subspecialties: Biochemistry (medicine); Genetics and genetic engineering (medicine). Current work: Molecular genetics and environmental effects. Office: Oral Roberts Univ 7777 S Lewis St Tulsa OK 74171

GILMOUR, WILLIAM ALEXANDER, computer based energy conservation systems co. exec.; b. Elizabeth, N.J., May 26, 1934; s. Alexander W. and Kathleen (Sullivan) G.; m. Margaret E. Gillich, Oct. 6, 1954; m. Madeleine C. Clark, Dec. 7, 1973; children: William J., Joyce M., Lynn A., Margaret E., Barbara Anne. Student, Union Coll., Cranford, N.J., 1957; A.S., U. Md., 1958. Research chemist Exxon Research & Devel., Linden, N.J., 1957-61; asst. div. mgr. Sargent Welch Sci. Co., Skokie, Ill., 1961-70; v.p. eastern ops. Environ. Data Corp., Monrovia, Calif., 1970-79; founder, pres., chmn. bd. Princeton Sensors Inc., N.J., 1979—. Mem. Presdl. Task Force; trustee Covenent Christian Sch., Cranford, N.J. Served with U.S. Army, 1952-54. Mem. Instrument Soc. Am. (sr., author tech. publs. 1979-80), TAPPI. Republican. Subspecialties: Infrared spectroscopy; Combustion processes. Current work: Development of optical sensors, computer based, for non-compliance monitoring of gases, specifically as related to improving efficiency of fossil fuel fired combustion processes. Patentee in field; developed microprocessor-based stack gas monitor, 1979. Office: Princeton Sensors Inc Research Park 1101-B Princeton NJ 08540

GILROY, BEVERLY ANN, veterinarian; b. Framingham, Mass., Feb. 14, 1949; d. George E. and Mary Jane (McGlynn) G. B.S., Mich. State U., 1970, D.V.M., 1971; M.A. in Edn, Chapman Coll., 1976. Diplomate: Am. Coll. Vet. Anesthesiologists. Staff veterinarian Allen Animal Hosp., Livonia, Mich., 1972, Weston (Mass.) Vet. Clinic, 1972-74; asst. base veterinarian March AFB, Calif., 1974-76; resident in anesthesiology U. Calif.-Davis Sch. Vet. Medicine, 1976-78; vis. lectr. Nat. Zoo, 1978; postdoctoral fellow U. Calif.-San Diego Sch. Medicine, 1978-79; asst. prof. Sch. Vet. Medicine, U. Mo., Columbia, 1979-82; assoc. prof. Sch. Vet. Medicine, N.C. State U., Raleigh, 1982—, chief of anesthesia, dir.intensive care unit, 1982—. Contbr. articles to profl. jours. Served as capt. USAF, 1974-76. Decorated Air Force Commendation medal; recipient Mich. State Vet. Faculty award, 1970; Catherine Patton Physiology award Mich. State U., 1970; Upjohn Small Animal Medicine award, 1971. Mem. AVMA, Mich. Vet. Med. Assn., Assn. for Women Veterinarians, Vet. Critical Care Soc., Am. Soc. Anesthesiologists. Subspecialties: Anesthesiology; Critical care. Current work: Neuroresuscitation, cardiopulmonary resuscitation, exotic animal anesthesia. Home: 1502 Edgeside Ct Raleigh NC 27609 Office: Dept Anatomy Physiol Scis and Radiology Sch Vet Medicine NC State U Raleigh NC 27606

GILRUTH, ROBERT ROWE, aerospace consultant; b. Nashwauk, Minn., Oct. 8, 1913; s. Henry Augustus and Frances Marion (Rowe) G.; m. Jean Barnhill, Apr. 24, 1937 (dec. 1972); 1 dau., Barbara Jean (Mrs. John Wyatt); m. Georgene Hubbard Evans, July 14, 1973. B.S. in Aero. Engring, U. Minn., 1935, M.S., 1936, D.Sc., 1962; D.Sc., George Washington U., 1962, Ind. Inst. Tech., 1962; D.Eng., Mich. Tech. U., 1963; LL.D., N.Mex. State U., 1970. Flight research engr. Langley Aero. Lab., NACA, Langley Field, Va., 1937-45, chief pilotless aircraft research div., 1945-50, asst. dir., 1950-58; dir. NASA Project Mercury, 1958-61, NASA Manned Spacecraft Center, Houston, 1961-72, dir. key personnel devel., 1972-73, ret., 1973; cons. to adminstr. NASA, 1974—; dir. Bunker Ramo Corp. Ind. experimenter and cons. hydrofoil craft, 1938-58; advisor on guided missiles, aeros. and structures, high temperature facilities U.S. Dept. Def., 1947-58; mem. com. space systems NASA Space Adv. Council, 1972—; chmn. mgmt. devel. edn. panel NASA, 1972-73; mem. ad hoc com. fire safety aspects of polymeric materials Nat. Materials Adv. Bd., 1973-74. Recipient Outstanding Achievement award U. Minn., 1954, Great Living Am. award U.S. C. of C., 1962, Distinguished Fed. Civilian Service award Pres. U.S., 1962, Americanism award CBI Vets. Assn., 1965, Spirit of St. Louis medal, 1965, Internat. Astronautics award Daniel and Florence Guggenheim, 1966, Distinguished Service medal NASA, spring 1969, fall 1969, Pub. Service at Large award Rockefeller Found., 1969, ASME medal, 1970, James Watt Internat. medal, 1971; Achievement award Nat. Aviation Club, 1971; Robert J. Collier trophy with Nat Aero. Assn., 1972; Space Transp. award Louis W. Hill; Distinguished Service medal NASA; medal of honor N.Y.C.; Robert H. Goddard Meml. trophy Nat. Rocket Club; named to Nat. Space Hall of Fame, 1969, Internat. Space Hall of Fame, 1976. Mem. Nat. Acad. Engring. (aeros. and space bd. 1974—), Nat. Acad. Scis. Subspecialties: Aerospace engineering and technology; Astronautics. Home: 5128 Park Ave Dickinson TX 77539

GIMARC, BENJAMIN MAURICE, chemist, educator, researcher; b. Nogales, Ariz., Dec. 5, 1934; s. John and Virgie Dill (Ringo) G.; m. Jerry Dell, Nov. 4, 1936. B.A. in Chemistry, Rice U., 1956, Ph.D., Northwestern U., 1963. Instr. Northwestern U., Evanston, Ill., 1961-62; USPHS fellow, lectr. Johns Hopkins U., Balt., 1962-64; asst. prof. Ga. Inst. Tech., Atlanta, 1964-66; asst. prof. chemistry U. S.C., Columbia, 1966-72, assoc. prof., 1972-78, prof., 1978—, dept. head, 1973-76, chmn. dept. chemistry, 1982—; Nat. Acad. Scis. exchange scientist, Zagreb, Yugoslavia, 1978, 82. Author: Molecular Structure and Bonding, 1978; contbr. numberous articles to profl. jours. Served to lt. (j.g.) USN, 1956-58. Mem. Am. Chem. Soc., Am. Phys. Soc., AAAS. Subspecialty: Theoretical chemistry. Current work: The development and application of qualitative theories of chemical valence. Home: 316 Wateree Ave Columbia SC 29205 Office: Dept. Chemistry U SC Columbia SC 29208

GINER-SOROLLA, ALFRED, chemist; b. Vinaros, Spain, Sept. 23, 1919; came to U.S., 1959, naturalized, 1969; s. Sebastian and Agustina Sorolla-Domenech; m. Hilary Harry, Mar. 21, 1964; 1 son: Roger. B.S., Tchrs. Coll., Castellon, 1942; M.S., U. Valencia, 1946, U. Madrid, 1946, D.Pharmacy, 1954; Ph.D. in Biochemistry, Cornell U., 1958. Instr. U. Madrid, 1944-46; chemist Andreu Pharms., Barcelona, 1947-52, Farbwerke Hoechst (Pharms.), 1952-54; mem. staff Sloan-Kettering Inst., N.Y.C., 1954—, head chem. lab., assoc. mem., 1963—; vis. scientist Cambridge (Eng.) U., 1960; mem. faculty Cornell U. Med. Sch. (Sloan-Kettering div.); cons. in field. Contbr. numerous articles to profl. publs. Recipient Valencia Regional Sci. award, 1975. Mem. Am. Assn. Cancer Research, Am. Chem. Soc., AAAS, Cambridge U. Chem. Soc. Subspecialties: Medicinal chemistry; Chemotherapy. Current work: Synthesis of immunostimultants, Anticancer and antiviral agents chemical carcinogenesis and its prevention. Patentee in field. Office: 145 Boston Post Rd Rye NY 10580 Home: 9 Chapel Ln Riverside CT 06878

GINGERICH, PHILIP DEAN, geology educator, curator; b. Goshen, Ind., Mar. 23, 1946; s. Orie J. and Miriam (Derstine) G.; m. B. Holly Smith. A.B., Princeton U., 1968; Ph.D., Yale U., 1974. Asst. prof. geol. scis. U. Mich., 1974-79, assoc. prof., 1979-83, prof., 1983—; dir. Mus. Paleontology, 1981—. Recipient Russel award U. Mich., 1980. Mem. Geol. Soc. Am., Paleontol. Soc. (councillor 1979-81, Schuchert award 1981), Soc. Study Evolution, Soc. Systematic Zoology (councillor 1983—), Soc. Vertebrate Paleontology. Subspecialties: Evolutionary biology; Paleobiology. Current work: Vertebrate paleontology, especially evolution of early Cenozoic mammals; evolutionary process; evolution of primates; origin of whales. Office: Museum Paleontology U Mich Ann Arbor MI 48109

GINGERY, ROY EVANS, research agricultural chemist; b. Lodi, Ohio, June 3, 1942; s. Orley Fred and Beatrice Ruth (Evans) G.; m. Judith Ann; children: Julie, Stephen. B.S., Carnegie-Mellon U., 1964; M.S., U. Wis.-Madison, Ph.D., 1968. Asst. prof. to prof. Ohio Agrl. Research and Devel. Ctr., Ohio State U., 1968—; research chemist Agrl. Research Service-U.S. Dept. Agr., Wooster, 1969—. Mem. Am. Phytopath. Soc. Subspecialties: Virology (biology); Biochemistry (biology). Current work: Characterization of viruses; virology, maize viruses, characterization. Home: 2769 Tanglewood St Wooster OH 44691 Office: Dept Plant Pathology Ohio Agrl Research and Devel Ctr- Ohio State U Agrl Research Service-U.S. Dept Agr Wooster OH 44691

GINGOLD, WILLIAM, psychologist; b. Warsaw, Poland, Sept. 20, 1939; came to U.S., 1952; s. David and Lillian (Weintal) G.; m. Phyllis K. Gingold, Aug. 22, 1965; children: Steven, Shara, Tamara, Jason. B.S., U. Wis., 1963, M.S., 1966, Ph.D., 1971. Spl. edn. tchr. Madison & Milw. pub. schs., 1963-66; chmn. Dept. Spl. Edn. Moorhead State U., Minn., 1971-72; exec. dir. spl. edn. Mental Health & Retardation Ctr., Fargo, N.D., 1972-81; exec. dir. Developmental Services Ctr., Champaign, IL, 1981—; faculty U. Minn., Mpls., 1972-74, U. N.D., Fargo, 1974-79; nat. program cons. U. N.C., 1971—; reviewer for state and fed. grant giving agencies, 1975—. Author: Geriatric Curriculum: Persistent Life Functions, 1981, Cost Accounting and Accountability, 1973, Magic Kingdom: A Preschool Screening Program, 1973. U. Wis. honor scholar, 1963; RSA/HEW fellow, 1965; U. Minn. Bush Found. fellow, 1981. Fellow Am. Assn. on Mental Deficiency; mem. Am. Psychol. Assn., Am. Ednl. Research Assn., Council for Exceptional Children (chpt. pres. 1967-69), Nat. Council Community Mental Health (treas. 1979-81). Jewish. Lodge: Kiwanis. Subspecialty: Behavioral psychology. Current work: Pediatric neuropsychology; orgnl. growth and crisis. Home: 307 Pond Ridge Ln Urbana IL 61801 Office: 1304 W Bradley Ave Champaign IL 61821

GINOS, JAMES ZISSIS, research scientist; b. Hillsboro, Ill., Feb. 1, 1923; s. Zissis and Nicoletta M. (Sakellaris) G.; m. Chrisilla Paul Katsas, June 13, 1947; children: Geoffrey, Milton. B.A., Columbia U., 1954; M.S. in Chem. Engring., Stevens Inst. Tech., 1962; Ph.D. in Organic Chemistry, Stevens Inst. Tech., 1964. Chemist, Colgate Palmolive Co., Jersey City, 1953-57; chief chemist Diamond Shamrock Corp., Newark, 1957-58; project coordinator Nopco Chem. Co., Harrison, N.J., 1959-64; asst. scientist Brookhaven Nat. Labs., Upton, N.Y., 1964-68; research asst. prof. Mt. Sinai Sch. Medicine, N.Y.C., 1968-70; assoc. scientist Brookhaven Nat. Labs., 1970-74, scientist, 1974-75; research asso. prof. Cornell U., 1975-80; sr. research asso. neuro-oncology Lab. Meml. Sloan-Ketting Cancer Center, N.Y.C., 1980—. Contbr. articles to profl. jours. Mem. Am. Chem. Soc., AAAS, Harvey Soc., Am. Soc. Pharmacology and Exptl. Therapeutics, N.Y. Acad. Sci. Subspecialties: Nuclear medicine; Synthetic chemistry. Current work: synthesis of radiopharmaceuticals labelled with short-lived positron emitting radioisotopes to be used in positron emission tomography. Patentee in field. Home: 200 Winston Dr Cliffside Park NJ 07010 Office: 1275 York Ave New York NY 10021

GINSBERG, HAROLD SAMUEL, virologist, educator; b. Daytona Beach, Fla., May 27, 1917; s. Jacob and Anne (Kalb) G.; m. Marion Reibstein, Aug. 4, 1949; children: Benjamin Langer, Peter Robert, Ann Meredith, Jane Elizabeth. A.B., Duke, 1937; M.D., Tulane U., 1941. Resident Mallory Inst. Pathology, Boston, 1941-42; intern, asst. resident Boston City Hosp., 4th Med. Service, 1942-43; resident physician, asso. Rockefeller Inst., 1946-51; asso. prof. preventive medicine Western Res. U. Sch. Medicine, 1951-60; prof. microbiology, chmn. dept. U. Pa. Sch. Medicine, 1960-73, Coll. Phys. and Surg. Columbia, 1973—; Mem. commn. acute respiratory diseases Armed Forces Epidemiological Bd., 1959-73; cons. NIH, 1959-72, 75—, Army Chem. Corps, 1962-64, NASA, 1969—, Am. Cancer Soc., 1969-73, mem. council on research and personnel, 1976-80; v.p. Internat. Com. on Nomenclature of Viruses, 1966-75; mem. space sci. bd., chmn. panel microbiology Nat. Acad. Sci., 1973-74; chmn. microbiology exam. com. Nat. Bd. Med. Examiners, 1974-79; mem. microbiology and infectious disease com. Nat. Inst. Allergy and Infectious Disease, NIH, 1976-81, chmn., 1979-81. Contbr. textbooks.; Co-author: Microbiology, 1967, 3d edit., 1980; mem. editorial bd.: Jour. Infectious Diseases; editor: Jour. Virology, 1979—, Cancer Research, 1978—. Served to maj. M.C. AUS, 1943-46. Decorated Legion of Merit. Mem. Nat. Acad. Scis.; Mem. Inst. Medicine of Nat. Acad. Scis., Assn. Am. Physicians, Am. Acad. Microbiologists (chmn. bd. govs. 1971-72), Am. Soc. Clin. Investigation (councillor 1958-60), Am. Assn. Immunologists, Am. Soc. Microbiology (chmn. virology div. 1961-62, councilor div. 1977—), Soc. Exptl. Biology and Medicine, Assn. Med. Sch. Microbiology Chairmen (pres. 1972-73), Harvey Soc. (pres.-elect 1983), Central Soc. Clin. Research, Am. Soc. Biol. Chemists, Am. Soc. Virology (pres. 1983), Alpha Omega Alpha. Subspecialties: Virology (biology); Molecular biology. Current work: Study of viral gene regulation and function of viral gene products using DNA containing animal viruses. Home: 450 Riverside Dr New York NY 10027 Office: Dept Microbiology Columbia U Coll Physicians and Surgeons 701 W 168th St New York NY 10032

GINSBERG-FELLNER, FREDDA VITA, pediatric endocrinologist, researcher; b. N.Y.C., Apr. 21, 1937; d. Nathaniel and Bertha S. (Jagendorf) Ginsberg; m. Michael J. Fellner, Aug. 27, 1961; children: Jonathan R., Melinda B. A.B., Cornell U., 1957; M.D., N.Y.U., 1961. Diplomate: Am. Bd. Pediatrics and Pedatric Endocrinology. Resident in pediatrics Bronx Mcpl. Hosp. Medicine, N.Y.C., 1961-64; fellow in pediatrics Albert Einstein Coll., N.Y.C., 1964-67; assoc. in pediatrics Mt. Sinai Sch. Medicine, N.Y.C., 1967-69, asst. prof. pediatrics, 1969-75, assoc. prof., 1975-81, prof., 1981—; cons. Brookhaven Natl Lab., Upton, N.Y., 1969-75. Contbr. articles to med. jours. Chmn. Camp NYDA for diabetic children, Burlingham, N.Y., 1978—. NIH grantee, 1978—; Am. Diabetes Assn. grantee, 1978; N.Y. Diabetes Assn. grantee, 1972, 74; Juvenile Diabetes Found., 1982; recipient Paul E. Lacy award Nat. Diabetes Research Interchange, 1982. Fellow Am. Acad. Pediatrics; mem. Ambulatory Pediatric Assn., Am. Diabetes, N.Y. Diabetes Assn. (N.Y. state coordinator 1980—, diabetes in youth chmn. soc. 1980), Soc. for Pediatrics Research. Club: Cornell (N.Y.C.). Subspecialties: Endocrinology; Pediatrics. Current work: Insulin-dependent diabetes mellitus in children and adolescents, etiologic factors, methods for improving blood glucose control, obesity in childhood viral and immune factor in diabetes mellitus, growth disorders in children. Office: Mt Sinai Sch Medicine One Gustave L Levy Pl New York NY 10029

GINTER, MARSHALL LLOYD, physicist, educator; b. Oroville, Calif., Aug. 24, 1935; s. Lloyd J. and Lois (Chapman) G.; m. Dorothy Spencer, Sept. 1, 1957; children: Karl Lloyd, Gretchen E. A.B., Chico State U., 1957; Ph.D., Vanderbilt U., 1961. Research assoc. Vanderbilt U., Nashville, 1961, acting dir. Spectroscopy Lab., 1961-62; research assoc. U. Chgo. Lab. Molecular Structure and Spectra, 1962-66; vis. research assoc. Physikalische Anstalt der Universitat Basel, Switzerland, 1966; asst. prof. Inst. for Phys. Sci. and Tech., U. Md., College Park, 1966-69, assoc. prof., 1969-74, prof., 1974—; cons. Naval Research Lab. Contbr. articles to profl. jours. Grantee NSF, NASA, Air Force Office of Sci. Research. Mem. Am. Phys. Soc., Am. Chem. Soc., Optical Soc. Am., Sigma Xi. Subspecialty: Atomic and molecular physics. Current work: Atomic and molecular electronic structure, high resolution electronic spectroscopy, high vacuum optical systems. Home: 12240 Valerie Ln Laurel MD 20708 Office: Inst. Phys Sci and Tech U Md College Park MD 20742

GINZBURG, LEV, ecology and evolution educator, consultant; b. Moscow, USSR, Jan. 11, 1945; came to U.S., 1976; s. Ruvim and Raisa (Matzahova) G.; m. Tatyana Porotova, Oct. 1, 1970; children: Maria, Alexander. M.S., U. Leningrad, USSR, 1967; Ph.D., Agrophys. Inst., Leningrad, 1970. Research assoc. Agrophys. Inst., Leningrad, USSR, 1967-69, sr. research assoc., 1969-70, lead theoretical ecology group, 1970-75; asst. prof. Northeastern U., Boston, 1976-77; assoc. prof. SUNY-Stony Brook, 1977-83, prof. dept. ecology and evolution, 1983—. Author: The Dynamic Theory of Biological Populations (in Russian), 1974, Theory of Natural Selection and Population Growth, 1983, Benjamin Cummings, California Lectures in Theoretical Population Biology, 1983. Subspecialties: Theoretical ecology; Population biology. Office: Ecology and Evolution Dept SUNY Stony Brook Stony Brook NY 11794

GINZTON, EDWARD LEONARD, engineering corporation executive; b. Dnepropetrovsk, Ukraine, Dec. 27, 1915; came to U.S., 1929; s. Leonard Louis and Natalie P. (Philipova) G.; m. Artemas A. McCann; children: Anne, Leonard, Nancy, David. B.S., U. Calif., 1936, M.S., 1937; E.E., Stanford, 1938, Ph.D., 1940. Research engr. Sperry Gyroscope Co., N.Y.C., 1940-46; asst. prof. applied physics and elec. engring. Stanford, 1946-47, assoc. prof., 1947-50, prof., 1951-68; dir. Microwave Lab., 1949-59, Varian Assocs., 1948—, chmn. bd., 1959—, chief exec. officer, 1959-72; pres. Varian Assos., 1964-68; dir. Stanford Bank, 1967-71, Stanford Project M, Stanford Linear Accelerator Center, 1957-60; Mem. commn. 1 U.S. nat. com. Internat. Sci. Radio Union, 1958-68; mem. Lawrence Berkeley Lab. Sci. and Adv. Com., 1972-79; chmn. adv. bd. Sch. Engring., Stanford, 1968-70; co-chmn. Stanford Mid-Peninsula Urban Coalition, 1968-72, mem. exec. com., 1968-74, mem. bd., 1968-74. Author: Microwave Measurements, 1957; Contbr. articles to tech. jours. Bd. dirs. Mid-Peninsula Housing Devel. Corp., 1970—, Stanford Hosp., 1975-80; trustee Stanford U., 1977—. Recipient Morris Liebmann Meml. prize I.R.E., 1958, Calif. Manufacturer of Yr. award, 1974. Fellow IEEE (bd. dirs. 1971-72, chmn. awards bd. 1971-72, medal of honor 1969); mem. Nat. Acad. Scis. (chmn. com. on motor vehicle emissions 1971-74, co-chmn. com. nuclear energy study 1975-80), Am. Acad. Arts and Scis., Nat. Acad. Engring. (mem. council 1974-80), Sigma Xi, Eta Kappa Nu, Tau Beta Pi. Subspecialty: Electronics. Current work: Microwave tubes; linear electron acceleration. Patentee in field. Home: 28014 Natoma Rd Los Altos Hills CA 94022 Office: 611 Hansen Way Palo Alto CA 94303

GIOVANELLI, RICCARDO, astronomer; b. Reggio Emilia, Italy, Aug. 30, 1946; came to U.S., 1969; s. Enzo and Franca Dolia (Gatti) G.; m. Ilca Baroni, July 8, 1971 (dec.); m. Martha Patricia Haynes, Dec. 5, 1977. Laurea di dottorato, U. Bologna, Italy, 1968; M.A., Ind. U., Bloomington, 1971, Ph.D., 1976. Jr. research assoc. Nat. Radio Astronomy Obs., Charlottesville, Va., 1972; vis. prof. Nat. U. El Salvador, San Salvador, 1973-75; astronomer U. Bologna, Italy, 1976-77; sr. staff scientist Nat. Astronomy and Ionosphere Center, Arecibo, P.R., 1978—; mem. users com. Nat. Radio Astronomy Obs., 1982-85; Karcher vis. fellow U. Okla., Norman, fall 1977, adj. assoc. prof. dept. physics. Contbr. articles to profl. jours. Mem. Internat. Astron. Union, Am. Astron. Soc., Societa Astronomica Italiana. Subspecialties: Radio and microwave astronomy; Optical astronomy. Current work: Radio spectroscopy, structure clusters and superclusters galaxies, gas distribution in galaxies, interstellar medium our galaxy, high velocity clouds neutral hydrogen. Office: Nat Astronomy and Ionosphere Center PO Box 995 Arecibo PR 00613

GIPSON, GARY STEVEN, civil engineering educator; b. Jackson, Miss., Jan. 4, 1952; s. James Luther and Johnnie Lavera (Ferguson) G.; m. Cheryl Martinez, Feb. 24, 1953. B.S., La. State U., 1975, M.S., 1978, Ph.D., 1982. Instr. dept. civil engring. La. State U., 1977-80; research asso. Inst. Environ. Studies, 1980-82, asst. prof. dept. civil engring, 1982—. Contbr. articles to profl. jours. Mem. Am. Acad. Mechanics, ASME, Soc. Engring. Sci., Am. Phys. Soc., Nat. Soc. Profl. Engrs., La. Engring. Soc., Mensa, Sigma Xi. Subspecialties: Theoretical and applied mechanics; Applied mathematics. Current work: Numerical modelling, applied physics, theoretical mechanics; boundary integral equations applied to mechanics. Home: 17735 Creek Hollow Rd Baton Rouge LA 70816 Office: Dept Civil Engring La State U Baton Rouge LA 70803

GIRDEN, ELLEN ROBINSON, psychology educator; b. Bklyn., May 14, 1936; d. Robert and Sarah (Bellinoff) R.; m. Edward Girden, Sept. 8, 1977. B.A., CUNY-Bklyn., 1956; M.A., 1958; Ph.D., Northwestern U., 1962. Instr. Northwestern U., 1961-62; asst. prof. Hobart and William Smith Coll., 1961-63; assoc. prof. Yeshiva U., 1963-77, Fla. Sch. Profl. Psychology, 1978-81, Nova U., 1981—. Contbr. articles to profl. jours. NIH Research grantee, 1963; CUNY fellow, 1956-58. Mem. AAAS, Am. Psychol. Assn. Democrat. Jewish. Subspecialties: Research design; Statistics. Current work: Amygdala and motivation; time estimation; litter size and crowding effects on learning and retention by white rats. Home: 2851 NE 183d St Apt 1204 North Miami Beach FL 33160 Office: Sch Profl Psychology Nova U 3301 College Ave Fort Lauderdale FL 33314

GIRI, SHRI N., pharmacology and toxicology educator; b. Meghipur, India, Jan. 30, 1934; s. Prasidh N. and Sitwanti N. Giri G.; m. Donna L. Clark, Jan. 27, 1966; children: Patrick E., Eric C. B.Vet. Sci., U. Agra, 1959; M.S., Mich. State U., 1961; Ph.D., U. Calif.-Davis, 1965. Diplomate: Am. Bd. Toxicology. Research assoc. dept. pharmacology Stanford U., Palo Alto, Calif., 1967-68; asst. prof. physiol. scis. U. Calif.-Davis, 1968-74, assoc. prof., 1974-80, prof. pharmacology and toxicology, 1980—. Contbr. numerous pharmacology and toxicology articles to profl. publs., 1971. Mem. Western Pharmacology Soc., Am. Soc. Pharmacology and Exptl. Therapeutics, Am. Soc. Toxicology, Calif. Thoracic Soc., AAAS. Subspecialties: Pharmacology; Toxicology (agriculture). Current work: Evaluation of anti-fibrotic drugs; cardoimyopathy fibrosis, biochemical mechanism of lung injury, prostaglandins. Home: 513 Eisenhower St Davis CA 95616 Office: Vet Pharmacology and Toxicology Dept Sch Vet Medicine U Calif Davis CA 95616

GIRSE, ROBERT DONALD, mathematics educator; b. St. Louis, Nov. 4, 1948; s. Russell Charles and Dorothy Eileen (Speichinger) G.; m. Linda Rose Podorski, Aug. 16, 1974; children: Robert Donald II, Derek. B.S., Southern Ill. U., 1972, M.S., 1974; Ph.D., Kans. State U., 1979. Asst. prof. Northern State Coll., Aberdeen, S.D., 1979-81, Idaho State U., Pocatello, 1981—; bd. dirs. Rocky Mountain Math. Consortium, Tempe, Ariz., 1981—. Contbr. articles in field to profl. jours. Mem. camping com. Tendoy council Boy Scouts Am., Pocatello, 1981—; dist. commr. Sioux council, Aberdeen, 1979-81. Mem. Am. Math. Soc., Soc. for Indsl. and Applied Math., Math. Assn. Am., London Math. Soc., Fibonacci Assn. Roman Catholic. Current work: Classical combinatorial problems and graph theory and their uses in computer science and numerical analysis. Office: Dept Math Idaho State U Pocatello ID 83209 Home: 154 Ranch Dr Pocatello ID 83204

GIRVIN, JOHN PATTERSON, neurosurgeon; b. Detroit, Feb. 5, 1934; s. Patterson and Sally (Hawkins) G.; m. Bettye Ruth Parker, Sept. 13, 1959; children: Douglas, Michael, Jane. M.D., U. Western Ont., 1958; Ph.D. in Neurophysiology, McGill U., 1965. Intern Montreal Gen. Hosp., 1958-59, resident in gen. surgery, 1963-64; resident in neurosurgery Montreal Neurol. Inst., 1964-65, Glasgow, Scotland, 1965-66, Victoria Hosp., U. Western Ont., 1967-68; resident in neuropathology Case-Western Res U., Cleve., 1966-67; lectr. physiology McGill U., 1962-63; asst. prof. neurosurgery and physiology U. Western Ont., London, 1968-74, assoc. prof., 1974—. Vice pres. Physicians Services Inc. Found.; bd. dirs. Victoria Hosp. Corp., London Theatre. Med. Research Council Can. grantee, 1968-74; Physicians Services Inc. Found. grantee, 1981. Fellow Royal Coll. Physicians and Surgeons (Can.); mem. Am. Assn. Neurol. Surgeons, Congress Neurol. Surgeons, Research Soc. Neurol. Surgeons, Am. Epilepsy Soc., Soc. for Neurosci., Can. Neurosurg. Soc., London Assn. Surgeons. Subspecialties: Neurosurgery; Neurophysiology. Current work: Research in epilepsy, neurophysiology, neuroprostheses; clinical research in neurosurgery. Office: Univ Hospital 339 Windermere Rd London ON Canada N6A 5A5

GISLASON, ERIC ARNI, chemist, educator; b. Oak Park, Ill., Sept. 9, 1940; s. Raymond Spencer and Jane Ann (Clifford) G.; m. Nancy Davis Brown, Sept. 11, 1962; children: Kristina Elizabeth, John Harrison. B.A. summa cum laude in Chemistry, Oberlin Coll., 1962; Ph.D. in Chem. Physics, Harvard U., 1967. Postdoctoral fellow Nat. Center for Air Pollution Control, U. Calif., Berkeley, 1967-69; asst. prof. chemistry U. Ill. at Chgo., 1969-73, assoc. prof., 1973-77, prof., 1977—; vis. scientist FOM-Inst. for Atomic and Molecular Physics, Amsterdam, Netherlands, 1977-78. Contbr. articles on chemistry to profl. jours. Recipient Silver Circle award U. Ill. at Chgo., 1982. Mem. Am. Chem. Soc., Am. Phys. Soc., Sigma Xi, Phi Beta Kappa. Congregationalist. Subspecialties: Atomic and molecular physics; Theoretical chemistry. Current work: Molecular beam studies of molecular interactions, theoretical studies of molecular collisions, theory of nonadiabatic processes in atom-molecule collisions. Home: 7227 Oak Ave River Forest IL 60305 Office: Chemistry Dept PO Box 4348 Chicago IL 60680

GIULIANO, ARMANDO E., surgeon, ocologist, educator; b. N.Y.C., Oct. 2, 1947; s. Antonio and Victoria G.; m. Cheryl Jane Fallon, June 21, 1970. B.A., Fordham U., 1969; M.D., U. Chgo., 1973. Diplomate: Am. Bd. Surgery. Resident in surgery U. Calif.-San Francisco, 1973-80; fellow in surg. oncology UCLA, 1976-78, asst. prof. surgery, 1980—, dir. breast service, 1980—; med. dir. John Wayne Cancer Clinic, 1981—; attending surgeon VA Hosp., Sepulveda, Calif., 1980—, UCLA, 1980—. Contbr. chpts. to books, articles to jours. Nat. Research Service awardee NIH, 1976-77; tng. grantee for clin. investigator in surgery, 1977-78; recipient James Ewing Soc. resident's award for cancer research, 1978; Golden Scalpel award for teaching excellence, 1982. Mem. Am. Soc. Clin. Oncologists, Am. Assn. Cancer Research, Soc. Univ. Surgeons. Subspecialties: Surgery; Oncology. Current work: Clinical and basic aspects of surgical oncology. 90024

GIVEN, PETER HARVEY, fuel science educator, researcher; b. Swansea, Wales, Jan. 29, 1918; m., 1944; 1 child. B.A., Oxford (Eng.) U., 1940, M.A., 1943, D.Phil. in Chemistry, 1944. Sci. officer Brit. Coal Utilization Research Assn., 1944-60, head organic chemistry sect., 1950-60; assoc. prof. fuel sci. Pa. State U., University Park, 1961-62, prof., 1962—, head dept. materials sci., 1965-69. Mem. Am. Chem. Soc., Geochem. Soc., Royal Soc. Chemists, Geol. Soc. Am. Subspecialty: Coal chemistry. Office: Fuel Sci Sect Dept Materials Sci Pa State U University Park PA 16802

GLACKIN, DAVID LANGDON, astronomer; b. Jamestown, N.Y., July 28, 1952; s. Franklin Pennell and Joann Elizabeth (Langdon) G.; m. Carlotta Allen, June 21, 1975. B.S., Calif. Inst. Tech., 1974; M.S., U. Colo., 1977. Research asst. Calif. Inst. Tech. Big Bear Solar Observatory, summers 1971, 72, 73, undergrad. teaching asst. dept. physics, 1974; research astronomer Sacramento Peak Observatory, Sunspot, N. Mex., 1974; research asst. Joint Inst. for Lab. Astrophysics, U. Colo., Boulder, 1974-77; profl. research asst. Goddard Space Flight Center, Greenbelt, Md., 1978; sr. scientist Jet Propulsion Lab., Pasadena, Calif., 1978—. Contbr. articles to profl. jours. Recipient Calif. Inst. Tech. chpt. Sigma Xi award for Outstanding Research, 1974; Calif. Inst. Tech. President's Fund grantee, 1979,80. Mem. Am. Astronom. Soc., Am. Inst. Physics. Subspecialties: Solar physics; Graphics, image processing, and pattern recognition. Current work: Solar atmosphere, earth's atmosphere, art conservation; exploration of sun with interplanetary spacecraft; physics and remote sensing of earth; image processing of art objects. Office: Jet Propulsion Lab 168-427 4800 Oak Grove Dr Pasadena CA 91103

GLADFELTER, WILBERT EUGENE, physiologist, educator; b. York, Pa., Apr. 29, 1928; s. Paul John and Marea Bernadette (Miller) G.; m. Ruth Ballantyne, Jan. 26, 1952; children: James W., Charles D., Mary A. A.B. magna cum laude, Gettysburg Coll., 1952; Ph.D. (NSF fellow), U. Pa., 1960. Asst. instr. physiology U. Pa., Phila., 1954-56, 58-59, NIH trainee fellow, 1958-59; instr. physiology W. Va. U. Med. Center, Morgantown, 1959-61, asst. prof., 1961-69, asso. prof., 1969—. Contbr. articles to profl. jours. Treas. Monongalia County Heart Assn., 1976—. Served with USN, 1946-48. Mem. Am. Physiol. Soc., Soc. for Neurosci., Am. Soc. Zoologists, N.Y. Acad. Scis., Phi Beta Kappa, Sigma Xi, Beta Beta Beta. Lutheran. Subspecialties: Neurophysiology; Physiology (medicine). Current work: Hypothalamic control of the excitability of the motor system; teaching medical, dental and graduate students; research. Home: Route 7 Box 528 Morgantown WV 26505 Office: Dept. Physiology WVa U Med Center Morgantown WV 26506

GLANZ, WILLIAM EDWARD, zoology educator; b. Ypsilanti, Mich., Jan. 27, 1949; s. Eldin H. and Lillian E. (Mandeville) G.; m. Lori Sue Weiss, Aug. 17, 1980. A.B., Dartmouth Coll., 1970; Ph.D., U. Calif.-Berkeley, 1977. Postdoctoral fellow Smithsonian Tropical Research Inst., Balboa, Panama, 1977-78; instr. environ. studies U. Calif.-Santa Cruz, 1978-79; vis. instr. biology UCLA, 1979; asst. prof. zoology U. Maine, Orono, 1979—. Am. Philos. Soc. grantee, 1981-82. Mem. Am. Soc. Mammalogists, Ecol. Soc. Am., Am. Ornithol. Union, Phi Beta Kappa. Subspecialties: Ecology; Evolutionary biology. Current work: Community ecology and evolution of mammals, emphasizing neotropical species and rodents; behavioral ecology of terrestrial vertebrates; seed predation and fruit use by birds and mammals. Office: Dept Zoology U Maine Orono ME 04469

GLASCOCK, MICHAEL DEAN, nuclear physicist; b. Hannibal, Mo., June 27, 1949; s. Malcolm Dimmitt and Leta Mae (Ruhl) G. B.S., U. Mo.-Rolla, 1971; Ph.D. in Nuclear Physics, Iowa State U., 1975. Research asst. nuclear physics Ames Lab., AEC, Ames, Iowa, 1973-75; research assoc. U. Md. Cyclotron Lab., College Park, 1975-78; sr. research scientist U. Mo.-Columbia, 1979—. Served to 1st lt. U.S.

Army, 1972-75. NATO grantee, Scotland, 1977. Mem. Am. Phys. Soc., Am. Chem. Soc., Am. Nuclear Soc., Sigma Xi. Methodist. Subspecialties: Nuclear physics; Analytical chemistry. Current work: Application of gamma-ray spectroscopy techniques to studies in nuclear physics and neutron activation to perform elemental analysis. Home: 35 Vickie Dr Columbia MO 65202 Office: U Mo Research Reactor Facility Columbia MO 65211

GLASEL, JAY ARTHUR, biochemistry educator, researcher; b. N.Y.C., Apr. 30, 1934; m. Jean Muriel Stewardson, Oct. 2, 1962. B.S., Calif. Tech. Inst., 1955; Ph.D., U. Chgo., 1959. Postdoctoral fellow U. Calif.-San Diego, 1960-61, Imperial Coll. London, 1961-62; asst. prof. Columbia U., N.Y.C., 1964-70; prof. U. Conn. Health Ctr., Farmington, 1970—. Contbr. articles on biochemistry to profl. jours. Served to 1st lt. USAF, 1962-64. Mem. Am. Phys. Soc., Am. Chem. Soc., Soc. of Exptl. Biologists, Sigma Xi. Subspecialties: Biochemistry (biology); Neurochemistry. Current work: Molecular biology of narcotic analgesic reception. Office: Dept Biochemistry U Conn Health Ctr Farmington CT 06032

GLASER, DONALD A(RTHUR), physicist; b. Cleve., Sept. 21, 1926; s. William Joseph Glaser. B.S., Case Inst. Tech., 1946, Sc.D., 1959; Ph.D., Cal. Inst. Tech., 1949. Prof. physics U. Mich., 1949-59; prof. physics U. Calif. at Berkeley, 1959—, prof. physics and molecular biology, 1964—. Recipient Henry Russel award U. Mich., 1955; Charles V. Boys prize Phys. Soc., London, 1958; Nobel prize in physics, 1960; NSF fellow, 1961; Guggenheim fellow, 1961-62. Fellow Am. Physics Soc. (prize 1959); mem. Nat. Acad. Scis., Sigma Xi, Tau Kappa Alpha, Theta Tau. Subspecialties: Psychophysics; Neurobiology. Current work: Understanding human vision through psychophysical experiments and theoretical models. Research on cosmic rays, bacterial evolution, control of biol. cell growth and division, automatic identification of bacterial species; automation of cell biology to study growth of somatic cells and effects of carcinogens, mutagens and teratogens. Constructed 1st bubble chamber for visual demonstrations of movements of high-energy atomic particles. Office: Molecular Biology Dept U Calif Berkeley CA 94720

GLASER, GILBERT HERBERT, educator, physician, neuroscientist; b. N.Y.C., Nov. 10, 1920; s. Burnard Richard and Sidelle (Rogers) G.; m. Morfydd Mai Pugh, Mar. 17, 1946; children—Gareth Evan, Sara Elizabeth. A.B., Columbia, 1940, M.D., 1943, Med. Sc.D., 1951; M.A. (hon.), Yale, 1963. Diplomate: Am. Bd. Psychiatry and Neurology. Intern Mt. Sinai Hosp., N.Y.C., 1943-44; resident neurology N.Y. Neurol. Inst., 1944-46; from research asst. to asso. neurology Columbia Coll. Physicians and Surgeons, 1948-52; research scientist N.Y. Psychiat. Inst., 1948-50; head. sect. neurology Yale Sch. Medicine, 1952-71, chmn. dept. neurology, 1971—, asst. prof. neurology, 1952-55, asso. prof., 1955-63, prof. neurology, 1963—; Commonwealth Fund vis. prof. neurology U. London, Eng., 1965-66; cons. West Haven (Conn.) VA Hosp., 1955—; vis. prof. neurology Nat. Hosp., London, 1972, Park Hosp., Oxford, 1973—; Fulbright disting. prof., 1981; mem. neurology research adv. com. USPHS, 1956-60, 68-72, spl. cons., 1973, epilepsy adv. com., 1974-77, chmn. basic sci. subcom., 1977—; mem. neurobiology rev. com. VA, 1975—, chmn., 1977-78. Author: EEG and Behavior, 1963; Editor: Epilepsia, 1958-76; adv. editor, 1976—; editor: Recent Advances in Clinical Neurology, 1978, 81, 84, Antiepileptic Drugs: Mechanisms of Action, 1980; mem. editorial bd.: Jour. Nervous and Mental Diseases; Contbr. articles to profl. jours. Served as capt. M.C. AUS, 1946-48. Recipient Janeway prize Columbia U., 1943, Bicentennial medal award, 1968, Book award Commonwealth Fund, 1975. Fellow Royal Soc. Medicine, A.C.P.; mem. Am. Neurol. Assn. (1st v.p. 1977-78), Am. Acad. Neurology (pres. 1973-75), Am. Epilepsy Soc. (pres. 1963), Am. Electroencephalographic Soc. (council 1958-61, bd. qualifications), Eastern Assn. Electroencephalographers (pres. 1958), EEG Soc. (Gt. Britain), Assn. Brit. Neurologists, Soc. for Neurosci., Epilepsy Found. Am. (med. adv. bd.), Myasthenia Gravis Foundation (med. adv. bd. chmn. 1964-65), Multiple Sclerosis Soc. (chmn. research programs com. 1973-74, med. adv. bd.). Club: Athenaeum (London). Subspecialty: Neurology. Home: 205 Millbrook Rd Hamden CT 06518 Office: 333 Cedar St New Haven CT 06510

GLASER, MICHAEL, biochemistry educator; b. Cleve., 1945. B.S., UCLA, 1966; Ph.D., U. Calif.-San Diego, 1971. Postdoctoral fellow Washington U. Med. Sch., St. Louis, 1971-74; asst. prof. U. Ill.-Urbana, 1974-79, assoc. prof., 1979-83, prof., 1983—. Subspecialties: Biochemistry (medicine); Biophysical chemistry. Current work: Membrane structure and function and lipid metabolism in cultured mammalian cells and bacteria. Office: Dept Biochemistry U Ill 321 RAL 1209 W Calif St Urbana IL 61801

GLASER, PETER EDWARD, mechanical engineer; b. Zatec, Czechoslovakia, Sept. 5, 1923; came to U.S., 1948; s. Hugo and Helen (Weiss) G.; m. Eva F. Graf, Oct. 16, 1955; children: David, Steven, Susan. Diploma, Coll. Tech., Leeds, Eng., 1943; state exam., Tech. U., Prague, 1948; M.S., Columbia U., 1951, Ph.D., 1955. Head design dept. Werner Mgmt., N.Y.C., 1948-53; v.p. Arthur D. Little, Inc., Cambridge, Mass., 1955—; cons. NASA, 1969-71, Office Tech. Assessment, Washington, 1980-81. Editor: Thermal Imaging, 1964, Aerodynamically Heated Structures, 1962, Solar Energy, 1973—. Trustee Combined Jewish Philanthropies, 1976—. Served in Free Czechoslovak Army, 1943-45. Recipient Carl F. Kayan medal Columbia U., 1974. Mem. Internat. Solar Energy Soc. (pres. 1968-70, Farrington Daniels award 1983), Am. Astronautical Soc. (dir. 1976—), Internat. Astronautical Fedn. (com. chmn. 1982), ASME (com. chmn. 1958-60), Internat. Inst. Refrigeration (v.p. 1969-73). Club: Cosmos (Washington). Subspecialties: Solar energy; Aerospace engineering and technology. Current work: solar energy applications, solar power satellite, space shuttle experiments, space station. Invented concept of solar power satellite, 1973. Home: 62 Turning Mill Rd Lexington MA 02173 Office: Arthur D Little Inc 20 Acorn Park Cambridge MA 02140

GLASER, RONALD, microbiology educator; b. N.Y.C., Feb. 27, 1939; s. Irving and Pauline G.; m. Janice Kiecolt, Jan. 17, 1980; children: Andrew, Eric. B.A., U. Bridgeport, 1962; M.S., U. R.I., 1964; Ph.D., U. Conn., 1968; postgrad., Baylor Coll. Medicine, 1968-69. Instr. biology Eastern Conn. State Coll., Willimantic, 1966-67, asst. prof., 1967-68; asst. prof. virology Ind. State U., Terre Haute, 1969-70; asst. prof. microbiology Pa. State U., Hershey, 1970-73, asso. prof., 1973-77, prof., 1977-78; prof. chmn. dept. med. microbiology and immunology Ohio State U., Columbus, 1978—; ad hoc reviewer NIH study sects. Author: (with T. Gottlieb-Stematsky) Human Herpes Virus Infections: Clinical Aspects, 1982. NIH fellow, 1968-69; Franco-Am. exchange Program; Fogarty Internat. Center; NIH and INSRM fellow, 1975, 77; Leukemia Soc. Am. scholar, 1974-79. Mem. Electron Microscopy Soc. Am., Am. Soc. Microbiology, AAAS, Am. Assn. Cancer Research, Soc. Exptl. Biology and Medicine. Subspecialties: Virology (medicine); Cell study oncology. Current work: Human oncogenic herpesviruses and cancer. Office: 333 W 10th Ave Columbus OH 43210

GLASGOW, GLENN PATRICK, med. physicist; b. Lebanon, Ky., Feb. 8, 1944; s. Glenn Shirley and Frances (Tutt) G.; m. Lyndia Rouse Glasgow, June 2, 1968. B.S., Western Ky. U., 1965; M.S., U. Ky., 1969, Ph.D., 1974. Diplomate: Am. Bd. Radiology, Am. Bd. Health Physics. Fellow Washington U., St. Louis, 1974-75, instr. radiation physic in radiology, 1975-78, asst. prof., 1978—. Contbr. numerous articles to profl. jours. Mem. Am. Phys. Soc., Health Physic Soc., Am. Soc. Therapeutic Radiologists, Am. Coll. Radiology, Southeastern Cancer Study Group, Am. Assn. Physicists in Medicine (Mo. River Valley chpt. sec. treas. 1977-78, pres. elect 1979, pres. 1980). Democrat. Methodist. Subspecialties: Cancer research (medicine); Nuclear physics. Current work: Medical uses of radiation. Home: 4205 Flora Pl Saint Louis MO 63110 Office: 510 S Kingshighway Blvd Saint Louis MO 63110

GLASHOW, SHELDON LEE, physicist, educator; b. N.Y.C., Dec. 5, 1932; s. Lewis and Bella (Rubin) G.; m. Joan Glashow; children: Jason David, Jordan, Brian Lewis, Rebecca Lee. A.B., Cornell U., 1954; A.M., Harvard U., 1955, Ph.D., 1958; D.Sc. (hon.), Yeshiva U., 1978, U. Marseille, 1982. NSF fellow U. Copenhagen, Denmark, 1958-60; research fellow Calif. Inst. Tech., 1960-61; asst. prof. Stanford U., 1961-62; asst. prof., asso. prof. U. Calif. at Berkeley, 1962-66; faculty Harvard U., 1966—, prof. physics, 1967—, Higgins prof. physics, 1979—; cons. Brookhaven Nat. Lab., 1966-73, 75—; mem. sci. policy com. CERN, 1979—; vis. prof. U. Marseille, 1971, M.I.T., 1974, 80, Boston U., 1983; affiliated sr. scientist U. Houston, 1983—. Contbr. articles to profl. jours. and popular mags. Pres. Sakharov Internat. Com., 1980—. Recipient J.R. Oppenheimer Meml. prize, 1977; George Ledlie prize, 1978; Nobel prize in physics, 1979; NSF fellow, 1955-60; Sloan fellow, 1962-66; CERN vis. fellow, 1968. Fellow Am. Phys. Soc., AAAS; mem. Am. Acad. Arts and Scis., Nat. Acad. Scis., Sigma Xi. Subspecialty: Theoretical physics. Home: 30 Prescott St Brookline MA 02146

GLASKY, ALVIN JERALD, pharmaceutical company executive, educator; b. Chgo., June 16, 1933; s. Oscar and Bessie (Akwa) G.; m. Rosalie Hanfling, Aug. 25, 1957; children: Michelle, Karen, Mark, Ira. B.S., U. Ill.-Chgo., 1954, Ph.D., 1958. Dir. Biochem. Research Labs., Inst. Psychosomatic and Psychiat. Research and Tng., Michael Reese Hosp., Chgo., 1959-62; group leader drug enzymology Abbott Labs., North Chicago, Ill., 1962-66; dir. research ICN-Nucleic Acid Research Inst., Internat. Chem. and Nuclear Corp., Irvine, Calif., 1966-68; pres., chmn. bd. dirs. Newport Pharm. Internat., Inc., Newport Beach, Calif., 1968—; assoc. prof. pharmacology Chgo. Med. Sch., 1965-70; asst. prof. biochemistry U.Ill. Coll. Medicine, Chgo., 1960-65. NSF fellow U. Lund, Sweden, 1958-59; Wenner-Grens Inst. for Exptl. Biology, fellow U. Stockholm, 1959; U. Ill. fellow, 1954-56; USPHS fellow, 1956-58. Mem. N.Y. Acad. Scis., AAAS, Am. Soc. Microbiology, Am. Chem. Soc., Orange County Chem. Soc., Am. Pharm. Soc., Calif. Pharm. Assn., Orange County Pharm. Assn., Internat. Soc. Immunopharmacology, Sigma Xi, Rho Chi. Republican. Jewish. Subspecialties: Immunopharmacology; Infectious diseases. Current work: Development of drugs to treat immunodeficiency diseases such as viral infections and cancer. Office: 897 W 16th St Newport Beach CA 92660

GLASS, BILL PRICE, geology educator; b. Memphis, Sept. 9, 1940; s. Floyd Martin and Leona (Bryan) G.; m. Judith Ann Niggl, June 11, 1966; children: Jeffrey Alan, Kelly Lynne. B.S., U. Tenn., Knoxville, 1963; Ph.D., Columbia U., 1968. Instr. Hunter Coll., N.Y.C., 1967; NRC postdoctoral fellow Goddard Space Flight Center, Greenbelt, Md., 1970; asst. prof. geology U. Del., Newark, 1970-75, assoc. prof., 1975—, acting chmn. dept., 1980-82. Author: Introduction to Planetary Geology, 1982. Served to capt. C.E. AUS, 1968-70. Recipient Nininger Meteorite award Ariz. State U., Tempe, 1968. Fellow Geol. Soc. Am.; mem. AAAS, Am. Geophys. Union, Internat. Astron. Union, Meteoritical Soc. (pres. Del. chpt. 1980-81), Nat. Assn. Geology Tchrs., Sigma Xi. Current work: Deep-sea sedimentation, tektites and impact glasses, impact cratering, paleomagnetics. Home: 387 Hobart Dr Newark DE 19713 Office: Dept Geology U Del Newark DE 19711

GLASS, RICHARD MCLEAN, medical educator, psychiatrist; b. Phoenix, Sept. 25, 1943; s. Richard Kirkpatrick and Harriet Margaret (Bradshaw) G.; m. Rita M. Denk, Mar. 4, 1967; children: Kathryn, Brendan. B.A., Northwestern U., 1964, M.D., 1968. Diplomate: Am. Bd. Psychiatry and Neurology. Intern, resident in internal medicine U. Chgo., 1968-70, resident in psychiatry, 1972-75, asst. prof. psychiatry, 1975-82, assoc. prof., 1982—. Asst. to chief editor: Archives of Gen. Psychiatry, 1980—; contbr. articles to profl. jours. Served to maj. M.C. U.S. Army, 1970-72. Decorated Army Commendation medal. Mem. Am. Psychiat. Assn., AAAS. Subspecialties: Psychiatry; Psychopharmacology. Current work: Diagnosis and treatment of affective disorders and anxiety disorders. Office: Department of Psychiatry University of Chicago 950 E 59th St Box 411 Chicago IL 60637

GLASSLEY, WILLIAM EDWARD, geochemist, educator; b. Ventura, Calif., July 1, 1947; s. Firmer Morris and Marguerite (Hickok) G.; m. Lucia Ann Milburn, June 26, 1976; children: Jeffrey, Nina. B.A., U. Calif.-San Diego, 1969; M.Sc., U. Wash., 1971, Ph.D., 1973. Postdoctoral fellow Mineral. Mus., Oslo, 1973-74; research asst. prof. U. Wash., 1974-76; asst. prof. geology Middlebury Coll., 1976-83, assoc. prof., 1983—. G. Unger Vetleson Found. postdoctoral fellow, Oslo, 1973. Mem. Am. Geophys. Union, AAAS. Subspecialties: Geochemistry; Petrology. Current work: Formation and evolution of continental crust; role of fluids in deep crustal processes; mass transport and geochemistry of granulite facies rocks. Office: Dept Geology Middlebury Coll Middlebury VT 05753 Home: RD 2 Middlebury VT 05753

GLASSMAN, IRVIN, engineering educator, consultant; b. Balt., Sept. 19, 1923; s. Abraham and Bessie (Snyder) G.; m. Beverly Wolfe, June 17, 1951; children: Shari Powell, Diane Geinger, Barbara Ann. B.E., Johns Hopkins U., 1943, D.Eng., 1950. Research asst. Manhattan Project, Columbia U., N.Y.C., 1943-46; mem. faculty Princeton (N.J.) U., 1950—, prof. mech. and aero. engring., 1964—; cons. to industry. Author: (with R.F. Sawyer) Performance of Chemical Propellants, 1971, Combustion, 1977; editor: 3 books Combustion; contbr.: articles to tech. jours. Combustion. Served with U.S. Army, 1944-46. NSF fellow, 1966-67. Mem. Combustion Inst. (Sir Alfred Edgerton Gold medal 1982), Am. Chem. Soc., AAUP, Tau Beta Pi. Subspecialties: Combustion processes; Fuels. Current work: Combustion of multicomponent fuels, soot formation, hydrocarbon oxidation kinetics, fire safety. Patentee rocket propellants (2). Home: PO Box 14 Princeton NJ 08540 Office: Princeton U Princeton NJ 08544

GLASSMAN, JEROME MARTIN, clinical pharmacologist, educator; b. Phila., Mar. 2, 1919; s. Martin K. and Dorothea (Largeman) G.; m. Justine Helena Rizinsky, June 15, 1952; children: Martin J., Lorna R., Gary J. A.B., U. Pa., 1939, M.A., 1942; Ph.D., Yale U., 1950. Research asso. lab. applied physiology Yale U., New Haven, 1950-51; head dept. pharmacology Wyeth Labs., Phila., 1951-62; dir. biol. research USV/Revlon, Yonkers, N.Y., 1962-69; dir. clin. research and pharmacology Wampole Labs., Stamford, Conn., 1969-75; asso. dir. clin. investigation Wallace Labs., Cranbury, N.J., 1975—; adj. asso. prof. pharmacology N.Y. Med. Coll., 1973—. Contbr. articles to profl. jours. Scoutmaster Boy Scouts Am., 1957-62, chmn. troop com., 1976-81. Fellow AAAS, N.Y. Acad. Scis., Am. Coll. Clin. Pharmacology, Am. Coll. Clin. Pharmacology and Chemotherapy; mem. Biometric Soc., Am. Soc. Pharmacology and Exptl. Therapeutics, Soc. Exptl. Biol. Medicine, Soc. Toxicology, Sigma Xi. Subspecialties: Pharmacology; Toxicology (medicine). Current work: Pharmacology and toxicology of therapeutic agts. in clinical medicine. Patentee in field. Home: 280 Sleepy Hollow Rd Briarcliff Manor NY 10510 Office: Wallace Labs Half Acre Rd Cranbury NY 08512

GLAZ, HARLAND MITCHELL, applied mathematician; b. Cleve., Dec. 15, 1949; s. Arnold J. and Paula R. (Friedlander) G. B.A., U. Pa., 1971; M.A., U. Calif.-Berkeley, 1975, Ph.D., 1977. Mathematician Lawrence Berkeley Lab., 1975-77, Naval Surface Weapons Ctr., Silver Spring, Md., 1977—. Contbr.: articles to profl. jours. Mem. Am. Math. Soc., Soc. Indsl and Applied Math., AIAA. Subspecialties: Applied mathematics; Numerical analysis. Current work: Analysis and construction of algorithms for conservation laws and related partial differential equations. Office: Naval Surface Weapons Center White Oak St Silver Spring MD 20910 Home: 1101 Fairview Ct Silver Spring MD 20910

GLAZE, NORMAN CLINE, plant physiologist, researcher; b. Washington, Jan. 16, 1934; s. Francis Warner and Ruthanna (Stringer) G.; m. Joan Vivian Stark, July 31, 1960. B.S., U. Md., 1957, M.S., 1963; Ph.D., U. Fla., 1966. Plant physiologist Coastal Plain Expt. Sta., U.S. Dept. Agr., Tifton, Ga., 1966—. Active Am. Cancer Soc. Served in U.S. Army, 1957-59. Mem. Am. Soc. Hort. Sci., So. Weed Sci. Soc., Weed Sci. Soc. Am., Ga. Weed Control Soc., Ga. Vegetable Growers Assn., Sigma Xi. Presbyterian. Club: Exchange (Tifton). Subspecialties: Plant physiology (agriculture); Integrated pest management. Current work: Weed science in horticultural crops; herbicides; weed control; integrated pest management; protectants.

GLAZER, HOWARD IRWIN, psychologist; b. Toronto, Ont., Can., May 8, 1946; s. Max and Leah (Osolky) G.; m. Roberta Lynn Hoff, May 8, 1969 (div. Dec. 1982). B.A. with honors, U. Toronto, 1969; Ph.D., U. Tex., 1972. Lic. psychologist, N.Y. Asst. prof. U. Tex., Austin, 1972-73, Rockefeller U., N.Y.C., 1977-80; clin. asst. prof. Cornell U. Med. Coll., N.Y.C., 1977—; pres. Corp. Psychol. Examiners, N.Y.C., 1978—, Corp. Stress Control Services Inc., 1981—; asst. prof. Loyola U. of Montreal, Que., Can., 1975-77; sr. scientist Postgrad. Center for Mental Health, N.Y.C., 1980-81; asst. attending psychologist N.Y. Hosp., N.Y.C., 1977—. Editor, author: Directive and Behavioral Intervention Strategies in Depression, 1981; editor: Behavioral Psychotherapy, 1983; contbr. articles to profl. jours. Woodrow Wilson fellow, 1969-71; Foundations Fund for Research in Psychiatry fellow, 1973-75. Mem. Am. Psychol. Assn., N.Y. State Psychol. Assn., N.Y. Acad. Scis., Nat. Registers of Health Service Providers in Psychology, Sigma Xi. Subspecialties: Clinical/ Organizational Psychology; Psychobiology. Current work: The effects of psychological stressors on brain chemistry and behavioral deficits. Development of clinical service programs for major U.S. corporations. Office: Corporate Stress Control Services Inc 320 E 65th St 117 New York NY 10021 Home: 127 W 92d St 2A New York NY 10025

GLAZER, ROBERT IRWIN, pharmacologist, exptl. oncologist; b. Bronx, N.Y., May 14, 1942; s. Harry Ernest and Rose Dorothy (Entner) G.; m. Roxane Wasserman, May 29, 1965; children: Michele, Jeffrey. B.S., Columbia U., 1965, M.S., 1967; Ph.D., Ind. U., 1970. Postdoctoral fellow dept. pharmacology Yale U. Sch. Medicine, New Haven, 1970-72; asst. prof. dept. pharmacology Emory U., Atlanta, 1972-76; chief applied pharmacology sect. Nat. Cancer Inst., Bethesda, Md., 1977—. Mem. editorial bd.: Molecular Pharmacology; contbr. numerous articles to sci. jours. Recipient faculty devel. award Pharm. Mfrs. Assn. Found., 1973-75; Nat. Cancer Inst. grantee, 1973-76. Mem. AAAS, Am. Soc. Pharmacology and Exptl. Therapeutics, Am. Assn. for Cancer Research, Am. Soc. Biol. Chemists. Jewish. Subspecialties: Cancer research (medicine); Molecular pharmacology. Current work: Mechanism of action of anticancer drugs. Home: 15209 Apricot Ln Gaithersburg MD 20878 Office: Nat Cancer Inst Bldg 37 Room 6D28 Bethesda MD 20205

GLAZER-WALDMAN, HILDA RUTH, health educator; b. Balt., Aug. 14, 1947; d. Maurice and Louise (Merfeld) Glazer; m. David Joel Waldman, May 30, 1976. B.A., Beaver Coll., Glenside, Pa., 1969; Ed.M., Rutgers U., 1972, Ed.D., 1976. Lic. psychologist, Mo. Staff specialist for curriculum assessment Balt. City Schs., 1974-76; adj. asst. prof. U. Mo.-St. Louis, 1977-80; research assoc. Evaluative Research Assocs., St. Louis, 1977-78; asst. prof. psychology Fontboone Coll., St. Louis, 1978-79; research asst. prof. U. Tex. Health Sci. Ctr., Dallas, 1980-82, asst. prof. allied health edn., 1982—; Mem. adv. bd. Home Health Home Care II, Dallas, 1982—. Co-editor: The Aged Patient, 1982; contbr. chpts. to books, articles to profl. jours. Named Outstanding Young Leader in Allied Health in Southwest Am. Soc. Allied Health Professionals, 1983. Mem. Am. Psychol. Assn., Am. Ednl. Research Assn., Eastern Psychol. Assn., Southwest Ednl. Research Assn., Tex. Soc. Allied Health Professionals (program com. 1982, mem., dir-at-large); Mem. Tex. Soc. Hosp. Educators. Subspecialties: Learning; Behavioral psychology. Current work: Research on human learning, patient education, psychological effects of pediatric cancer on siblings. Office: U Tex Health Sci Ctr 5323 Harry Hines Blvd Dallas TX 75235

GLEASON, ANDREW MATTEI, educator; b. Fresno, Calif., Nov. 4, 1921; s. Henry Allan and Eleanor Theodalinda (Mattei) G.; m. Jean Berko, Jan. 26, 1959; children—Katherine Anne, Pamela, Cynthia. B.S., Yale, 1942; jr. fellow, Soc. Fellows, Harvard, 1946-50, M.A. (hon.), 1953. Asst. prof. math. Harvard, 1950-53, asso. prof., 1953-57, prof., 1957—, Hollis prof. math. and natural philosophy, 1969—. Author: Fundamentals of Abstract Analysis, 1966. Served from ensign to lt. (s.g.) USNR, 1942-46; lt. comdr., 1950-52. Recipient Newcomb Cleveland prize AAAS, 1952. Mem. Am. Math. Soc. (pres. 1981-82), Math. Assn. Am., Am. Philos. Soc., Societe Mathematique de France, Am. Acad. Arts and Scis., Nat. Acad. Scis. Club: Cosmos (Washington). Subspecialties: Probability; Abstract Analysis. Home: 110 Larchwood Dr Cambridge MA 02138

GLEASON, LARRY NEIL, biology educator; b. Independence, Oreg., June 3, 1939; s. Merlin Doyle and Ida Bell (Weigel) G.; m. Wendy Alton Shuey, Aug. 26, 1963 (div. 1976); children: Kelly, Scott. A.B., Chico State Coll., 1964; M.S., U. N.C., 1965, Ph.D., 1969. Research assoc. U. Fla., Gainesville, 1969-70; asst. prof. biology Western Ky. U., Bowling Green, 1970-74, assoc. prof., 1974-81, prof., 1981—. Contbr. articles to profl. jours. Mem. Am. Soc. Parasitologists, Helmithological Soc. Washington, Ky., Acad. Sci. Republican. Subspecialty: Parasitology. Current work: Helminths of fish, ecology and life cycles. Home: 1205 Chestnut St Bowling Green KY 42101 Office: Biology Dept Western Ky U Bowling Green KY 42101

GLEICHER, NORBERT, obstetrician-gynecologist, medical administrator and educator; b. Cracow, Poland, Aug. 20, 1948; came to U.S., 1975; s. Arnold and Eugenie (Volk) G. Student, Sch. Medicine Vienna U., 1966-70; M.D., Tel-Aviv U., 1973. Diplomate: Am. Bd. Ob-Gyn. Rotating intern Ihilov Mcpl. Hosp., Tel-Aviv, Israel, 1973-74; research fellow dept. ob-gyn Mt. Sinai Med. Ctr., N.Y.C., 1975, resident in ob-gyn, 1975-78, chief resident, 1978-79; asst. prof. dept. ob-gyn Mt. Sinai Sch. Medicine of CUNY, N.Y.C., 1979-81, dir. lab. reproductive immunology, course dir. reproduction course, dir. med. education dept. ob-gyn, 1979-81, dir. div. reproductive immunology, 1979-81, chmn. med. edn. com., 1980-81; course dir. reproduction sect.

pathophysiology course CUNY, 1981—; chmn. dept. ob-gyn Mt. Sinai Med. Ctr., Chgo., 1981—; prof. ob-gyn, asst. prof. immunology and microbiology Rush Med. Coll., Chgo.; mem. Internat. Coordination Com. for Immunology of Reproduction, 1981-82; bd. dirs. profl. edn. com., chmn. physicians' subcom. Am. Cancer Soc., Chgo. Unit, 1982—; program dir., program chmn., session chmn., panelist profl. confs., seminars in field. Editor: Reproductive Immunology, vol. 70, 1981; co-editor: Cardiac Problems in Pregnancy, Diagnosis and Management of Maternal and Fetal Disease, 1982; editorial bd.: Mt. Sinai Jour. Medicine, 1979—; editor-in-chief: Am. Jour. Reproductive Immunology, 1980—; contbr. articles to profl. publs., papers to profl. confs., U.S., Bulgaria. Recipient Dr. Solomon Silver award in Clin. Medicine Richard and Hilda Rosenthal Found., 1979; Lalor Found. fellow, 1979-80. Mem. AMA, Am. Coll. Ob-Gyn, Am. Assn. Gynecologic Laparoscopists, Am. Soc. for Immunology of Reproduction (v.p. 1980—), Am. Soc. Abdominal Surgeons, Am. Med. Soc. of Vienna, Assn. Profs. of Gyn and Ob, N.Y. Acad. Scis., Am. Fedn. Clin. Research, Internat. Soc. for Study of Hypertension in Pregnancy, Central Assn. Obstetricians and Gynecologists, Israel Med. Soc., Chgo. Gynecol. Soc. Subspecialties: Microbiology; Immunobiology and immunology. Current work: Co-dir. in vitro fertilization program; research in reproductive immunology, immunology of lipids, red cell immunology, cancer immunology, microbiology of pelvic inflammatory disease, cardiac and other medical diseases in pregnancy, cesarean section indications and rates. Office: Mt Sinai Hosp Med Center Dept Ob Gyn California Ave at 15th St Chicago IL 60608 Home: 1411 N State Pkwy Chicago IL 60611

GLENBERG, ARTHUR MITCHELL, psychology educator; b. Cleve., May 16, 1948; s. Leo and Freieda G.; m. Donna C. Winnick, June 11, 1968; children: Rebecca, Jonathan. B.A., Miami U., Ohio, 1970; Ph.D., U. Mich., 1974. Asst. prof. U. Wis.-Madison, 1974-80, assoc. prof. psychology, 1980—; editorial bd. Memory & Cognition, Child Devel. Reviewer profl. jours.; author sci. articles. Grntee NIMH U. Wis. Grad. Sch., Spencer Found., Nat. Inst. Edn. Mem. Am. Psychol. Assn., Midwestern Psychol. Assn., Psychonomic Soc. Subspecialties: Cognition; Learning. Current work: Human memory, comprehension processes. Office: Dept Psychology U Wis Madison WI 53706

GLENDENNING, KAREN KIRCHER, neuroanatomy educator; b. Niagara Falls, N.Y., Mar. 29, 1949; d. Morton Sumner and Hildegaarde Esa (Murrer) Kircher. B.S., U. Wash., 1967; M.S., Ohio State U., 1969, Ph.D., 1971. NIMH predoctoral fellow Ohio State U., Columbus, 1968-71; postdoctoral fellow Duke U., 1971-74; asst. prof. neuroanatomy Fla. State U., Tallahassee, 1974-80, assoc. prof., 1980—. Contbr. articles to profl. jours. NIMH grantee, 1979—, 82—. Mem. Am. Anat. Soc., Neurosci. Soc., Sierra Club. Clubs: Canoe, Toastmasters. Subspecialties: Neurobiology; Neuropsychology. Current work: Tracing pathways in the brain that allow us to localize sounds. Home: 4718 Charles Samuel Dr Tallahassee FL 32308 Office: Dept Psychology Fla State U Tallahassee FL 32306

GLENN, ALAN HOLTON, mechanical engineer; b. Brigham City, Utah, May 9, 1950; s. Merrill H. and Miriam (Holton) G.; m. Deborah Elaine Smith, Dec. 21, 1974; children: Adam Merrill,Miriam Elaine. B.S.M.E., Brigham Young U., 1975. Registered profl. engr., Utah. Teaching asst. Brigham Young U., Provo, Utah, 1973-74; with Valtek Inc., Springville, Utah, 1975—, project mgr., 1982—. Mem. ASME. Mormon. Subspecialties: Mechanical engineering. Current work: Instrumentation, stress analysis, seismic analysis, noise research, vibration analysis, design, pressure vessels. Home: PO Box 577 Salem UT 84653 Office: PO Box 2200 Springville UT 94663

GLENN, LOYD LEROY, neurophysiologist, educator; b. Camp Roberts, Calif., Dec. 10, 1952; s. Ora Lee and Theresia (Fechter) G.; m. Paula Renee Minnis, July 8, 1978; 1 dau., Christina Leigh. A.B., U.Calif.-Santa Cruz, 1974; Ph.D., Stanford U., 1979. NSF postdoctoral fellow Laval U., Quebec, Que., Can., 1979-80; postdoctoral fellow NIH, Bethesda, Md., 1980-82; asst. prof. Ohio Coll. Podiatric Medicine, Cleve., 1982—. Contbr. articles to profl. jours. Mem. Soc. for Neurosci., Assn. for Psychophysiol. Study of Sleep, Am. Physiol. Soc. Republican. Methodist. Subspecialties: Neurophysiology; Chronobiology. Current work: Spinal cord anatomy and behavioral physiology in sleep-wake. Spinal cord, motoneuron, sleep, muscle spindles, cable theory, motor systems. Home: 230 Virginia Way Barstow CA 92311 Office: Ohio Coll Podiatric Medicine Cleveland OH 44106

GLENNER, GEORGE GEIGER, pathologist; b. Bklyn., Sept. 17, 1927; s. Francis Richard and Jennie (Geiger) G.; m. Joy Arlene Sharp; children: Sheldra, Jonathan, Amanda, Sarah. B.A., Johns Hopkins U., 1949, M.D., 1953. Diplomate: Am. Bd. Anat. Pathology. Intern Mt. Sinai Hosp., N.Y.C., 1953-54; resident in pathology Mallory Inst. Pathology, Boston, 1954-55; chief sect. molecular pathology NIADDK, NIH, Bethesda, Md., 1955-82; prof. pathology U. Calif.-San Diego Sch. Medicine, La Jolla, 1982. Author: Tumors of the Extra Adrenal Paraganglion System, 1974; editor: Amyloid and Amyloidosis, 1980; editorial bd.: Annales de Pathologie, 1982. Pres. Alzheimer's Family Ctr., Inc., San Diego, 1982; bd. dirs. Nat. Alzherimer's Disease Brain Bank. Served with USPHS, 1955. Recipient Meritorious Service medal USPHS, 1971; Weingart Found. grantee, 1982. Fellow Coll. Am. Pathologists; mem. Histochem. Soc. (pres. 1972), Am. Assn. Pathologists, Am. Soc. Biol. Chemistry, Am. Fedn. Clin. Research, Alzheimer's Disease and Related Disorders Assn. (dir. 1980—, med. advisor 1980—), German Nat. Acad. Sci. (life), Phi Beta Kappa. Subspecialties: Pathology (medicine); Biochemistry (medicine). Current work: Study of nature, origin and pathogenesis of amyloidosis and the cerebral lesions of Alzheimer's disease. Home: 3108 Morning Way La Jolla CA 92037 Office: U Calif San Diego Gilman Dr and La Jolla Village Dr (M-012) La Jolla CA 92093

GLICK, J. LESLIE, biotechnology company executive; b. N.Y.C., Mar. 2, 1940; s. Arthur Harvey and Hilda Lillian (Lichtenfeld) G.; m.; children: Geoffrey Michael, Jessica Michele. A.B., Columbia U., 1961, Ph.D., 1964. Nat. Cancer Inst. postdoctoral fellow Princeton U., 1964-65; sr., then asso. cancer research scientist Roswell Park Meml. Inst., Buffalo, 1965-69; asso. research prof. physiology, acting chmn. Roswell Park div. SUNY, Buffalo, 1968-70; from exec. v.p. to chmn. bd. Asso. Biomedic Systems, Inc., Buffalo, 1969-77; pres. Inst. Sci. and Social Accountability, Washington, 1975-79; pres., chief exec. officer Genex Corp., Rockville, Md., 1977—; chmn. HTI Corp., Buffalo, 1972-75; pres. Indsl. Biotech. Assn., Rockville, Md., 1981—; dir. Nat. Assn. Life Sci. Industries, 1975-77; research prof. biology Niagara (N.Y.) U., Canisius Coll., Buffalo, 1968-70; exec. com. SUNY Grad. Sch., Buffalo, 1968-70; vis. lectr. NATO Advanced Study Inst., Brussels, 1970. Author: Fundamentals of Human Lymphoid Cell Culture, 1980; also articles. Mem. Am. Assn. Cancer Research, Am. Physiol. Soc., Tissue Culture Assn., N.Y. Acad. Scis., Sigma Xi. Subspecialties: Cell and tissue culture; Genetics and genetic engineering (biology). Current work: Corporate management; technology assessment; product development. Home: 10899 Deborah Dr Potomac MD 20854 Office: Genex Corp 6110 Executive Blvd Rockville MD 20852

GLICK, ZVI, nutritionist; b. Tel-Aviv, Israel, Nov. 20, 1934; s. Abraham and Gruna (Miller) G.; m. Diane Naomi Schott, Jan. 31, 1965; children: Tamar, Ron, Noam. B.Sc., Hebrew U., 1959; M.S., U. Calif.-Davis, 1962; Ph.D., U. Calif.-Berkeley, 1967. Research assoc. Harvard U., Boston, 1968-69; sr. scientist Heller Inst., Tel-Aviv, Israel, 1969-74; lectr. nutrition Hebrew U., Jerusalem, 1974-78; research assoc. Harbor-UCLA Med. Center, Torrance, Calif., 1978-82, assoc. prof. medicine, 1982; nutrition cons. Israeli Army, 1969-74. Served to Maj, M.C. U.S. Army. Recipient Fellowship awards Israeli Acad. Sci.-British Royal Soc., 1977, Wellcome Trust, 1978; NIH research grantee, 1980—. Mem. Am. Inst. Nutrition, AAAS. Subspecialties: Nutrition (medicine); Comparative physiology. Current work: Regulation of energy balance, control mechanism of food intake, obesity. Home: 2320 Via Rivera Palos Verdes Estates CA 90274 Office: Harbor-UCLA Med Center 1000 W Carson St Torrance CA 90509

GLICKSMAN, MARTIN EDEN, educator; b. N.Y.C., Apr. 4, 1937; s. Nathan Henry and Ruth Elaine (Rosensaft) G.; m. Lucinda Jeanette Mulder, May 7, 1967. B.Met.Engring., Rensselaer Poly. Inst., 1957, Ph.D., 1961. With Engring. div. Procter & Gamble Co., Cin., 1957; research staff Naval Research Lab., Washington, 1963-75, sect. head, 1967-69, br. head, 1969-75, assoc. supt. materials sci. div., 1974-75; prof., chmn. dept. materials engring. Rensselaer Poly. Inst., Troy, N.Y., 1975—; cons. in field. Contbr. articles to profl. jours. Recipient RESA Award in Sci., 1968; A.S. Fleming award Washington Jr. C. of C., 1968; M.E. Grossmann award Am. Soc. Metals, 1971; NASA award for tech. excellence, 1980. Fellow Am. Soc. Metals; mem. AIME, AAAS, AIAA, Am. Assn. Crystal Growth. Jewish. Subspecialties: Metallurgy; Materials. Current work: Solidification, crystal growth, phase transformations; transport phenomena during solidification. Home: 22 Schuyler Hills Rd NY 12211 Office: MRC 102 Troy NY 12181

GLIDEWELL, JOHN CALVIN, psychology educator; b. Okolona, Miss., Nov. 5, 1919; s. Henry Clay and Jessie Kate (Jones) G.; m. Frances Lee Reed, Sept. 12, 1941; children: Pamela Lee, Janis Lynn. M.A., U. Chgo., 1949, Ph.D., 1953. Diplomate: Am. Bd. Profl. Psychology. Dir. psychol. services Meridian (Miss.) Pub. Schs., 1949-50; dir. research/devel. St. Louis County Health Dept., Clayton, Mo., 1953-67; assoc. prof. psychology Washington U., St. Louis, 1953-67; prof. psychology U. Chgo., 1967-81, Vanderbilt U., Nashville, 1981—. Editor: Parental Attitudes and Child Behavior, 1961, Social Context Development and Learning, 1975; author: Choice Points, 1970, (with others) Nurses, Patients, Social Systems, 1968. Served to maj. USAF, 1942-46, 50-52. Recipient award for disting. contbn. Div. 27, Am. Psychol. Assn., 1975. Fellow Am. Psychol. Assn., Am. Sociol. Assn.; mem. Sigma Xi. Subspecialties: Organizational psychology; Social psychology. Current work: Development and change of social norms and social structures in orgns. Home: 101 Longwood Pl Nashville TN 37215 Office: Vanderbilt U Peabody Box 321 Nashville TN 37203

GLIMM, JAMES GILBERT, mathematician; b. Peoria, Ill., Mar.24, 1934; s. William Frederick and Barbara Gilbert (Hooper) G.; m. Adele Strauss, June 30, 1957; 1 dau., Alison. A.B., Columbia U., 1956, A.M., 1957, Ph.D., 1959. From asst. prof. to prof. math. M.I.T., 1960-69; prof. Courant Inst., N.Y. U., 1969-74; prof. math. Rockefeller U., N.Y.C., 1974-82; prof. Courant Inst., NYU, N.Y.C., 1982—. Co-author: Quantum Physics, 1981; mem. editorial bds. profl. jours.; contbr. articles to sci. publs. Recipient Dannie Heineman prize in math. physics, 1980; Guggenheim fellow, 1963, 65. Mem. Internat. Assn. Math. Physicists, Am. Math. Soc., Soc. Indsl. and Applied Math., Am. Acad. Arts and Scis., Soc. Petroleum Engrs., N.Y. Acad. Scis. (award in phys. and math. scis. 1979). Subspecialties: Numerical analysis; Petroleum engineering. Office: Courant Inst 251 Mercer St New York NY 10012

GLISSON, SILAS NEASE, III, biomed. researcher, research cons.; b. Springfield, Ill., May 8, 1941; s. Silas nease and Dorothy Lucille (Reed) G., Jr.; m. Mary Louise Mosimann, Feb. 10, 1968; children: Silas Nease IV, Andrew Edward. Student, Ill. Coll., Jacksonville, 1959-61; B.A., So. Ill. U., 1964; Ph.D. in Pharmacology (NIH fellow), Loyola U., Chgo., 1972. Research asso. Thudichum Research Labs., Galesburg, Ill., 1964-67; instr. dept. pharmacology U. Conn. Med. Center, Farmington, 1971-72; asst. prof. depts. anesthesiology and pharmacology Stritch Sch. Medicine, Loyola U. of Chgo., Maywood, Ill., 1972-78, asso. prof., 1978—, dir., 1972—; research cons.; guest lectr. in field. Contbr. over 70 articles on biomed. research to nat. and internat. publs. Numerous research grants from NIH and pharm. cos., 1972—. Mem. Am. Soc. Anesthesiologists, Am. Soc. Pharmacology and Exptl. Therapeutics, Internat. Anesthesia Research Soc., Soc. Cardiovascular Anesthesiologists, Ill. Soc. Anesthesiologists. Subspecialties: Neuropharmacology. Current work: Interested in mechanisms by which stress responses and their related chemical mediators alter cardiovascular performance. The influence of anesthetic and related drugs on the stress responses and the control of stress responses by pharmacological means. Home: 546 N Stewart Ave Lombard IL 60148 Office: 2160 S 1st Ave Maywood IL 60153

GLITZ, DOHN GEORGE, biochemistry educator; b. Buffalo, Sept. 28, 1936; s. Arthur Theodore and Viola Theophia (Raven) G.; m. Beryl Davey, Jan. 29, 1966; 1 dau.: Rachel. B.S., U. Ill., 1958; M.S., U. Wis.-Madison, 1960; Ph.D., U. Calif.-Berkeley, 1963. Postdoctoral fellow Virus Research unit, Cambridge, Eng., 1964-66; postdoctoral fellow Virus Lab., U. Calif.-Berkeley, 1966-67; asst. prof. dept. biol. chemistry UCLA, 1967-71, assoc. prof., 1971-77, prof., 1977—, vice-chmn. dept., 1979—. Contbr. articles to profl. jours. Guggenheim fellow, 1974-75. Mem. Am. Soc. Biol. Chemists, Am. Soc. Microbiology. Democrat. Unitarian. Subspecialties: Biochemistry (medicine); Molecular biology. Current work: Ribosome structure and function; ribonuclease biochemistry; ribonucleases in cancer. Office: Dept Biol Chemistry UCLA Sch Medicine Los Angeles CA 90024

GLODOWSKI, ROBERT JOHN, metallurgist, researcher; b. Williston, N.D., Jan. 4, 1946; s. Daniel Louis and Katheryne Josephine (Polley) G.; m. Amy Marie Burgess, Nov. 7, 1964; children: Camille, Ronald, Gregory, Teresa. B.S. in Metallurgy, S.D. Sch. Mines and Tech., 1967. Metallurgist Armco Inc., Middletown, Ohio, 1967-75, sr. metallurgist, 1975-81, sr. staff metallurgist carbon and alloy products, 1981—. Contbr. articles on stress relaxation testing to profl. publs.; editor: Through Thickness Tension Testing of Steel, 1982. Mem. ASME, ASTM, Am. Soc. Metals, Pressure Vessel Research Council, Internat. Standards Orgn. Subspecialties: Metallurgical engineering; Metallurgy. Current work: Steel alloy development; steel processing improvements; rod and wire products; grinding media. Invented steel heat treatment. Office: Research and Tech Armco Inc Middletown OH 45043

GLOGOVSKY, ROBERT LOUIS, chemistry educator; b. North Chicago, Ill., May 29, 1936; s. Louis and Mary Jennie (Knotek) G.; m. Carol M. Wintsch, Sept. 2, 1961; children: Cheryl, Dawn, Bob, Susan. B.S., No. Ill. U., 1959; Ph.D., U. Colo., 1964. Teaching/research asst. U. Colo., Boulder, 1959-64; Kettering postdoctoral intern Kalamazoo (Mich.) Coll., 1964-65; prof. chemistry Elmhurst (Ill.) Coll., 1965—, chmn. dept., 1973—; cons. in field. Contbr. articles to profl. jours. Alderman City of Elmhurst, 1981—; mem. citizens adv. com. Park Bd., 1970—, chmn., 1979. Danforth Found. assoc., 1967—; NSF grantee, 1969; Dr. Scholle Found. grantee, 1974, 75; Lilly Found. grantee, 1976. Mem. Am. Chem. Soc., Ill. Acad. Sci., Associated Colls. of Chgo. Area. Subspecialties: Physical chemistry; Solar energy. Current work: Thermodynamics of muscles, mechanisms of chemoreception, solar energy systems, trace metal analysis.

GLOVER, GEORGE I., enzymologist, research manager; b. Oakland, Calif., Mar. 8, 1940; s. Howard and Margaret G.; m. Beverly J. Porter, June 20, 1959; children: Susan L., Ruth A. B.S., U. Calif.-Berkeley, 1962, Ph.D., 1966. Sr. research chemist Air Products Inc., Allentown, Pa., 1966-67; research collaborator Brookhaven Nat. Lab., Upton, N.Y., 1967-70; assoc. prof. Tex. A&M, Coll. Station, 1970-80; research mgr. Monsanto, St. Louis, 1980—. Mem. Am. Chem. Soc., Am. Soc. Biol. Chemistry. Democrat. So. Baptist. Subspecialties: Enzyme technology; Organic chemistry. Current work: Biocatalysis, enzymology, organic chemistry. Office: Monsanto Corp Research Labs 800 N Lindbergh Blvd St Louis MO 63167

GLOVSKY, M. MICHAEL, allergist and clinical immunologist, educator; b. Boston, Aug. 15, 1936; s. Hyman Saul and Sylvia Hazel (Silber) G.; m.; children: Steven Staci, Scott, Romy, Adam. B.S. in Chemistry and Biology magna cum laude, Tufts U., 1958, M.D., 1962. Diplomate: Nat. Bd. Med. Examiners, Am. Bd. Allergy and Immunology. Intern Balt. City Hosp., 1962-63; asst. med. resident New Eng. Med. Ctr., Boston, 1965-66; NIH fellow Walter Reed Army Inst. Research, Washington, 1966-68, U. Calif.-San Francisco, 1968-69; staff physician So. Calif. Permanente Med. Group, Los Angeles, 1969-72, dir. allergy and immunology lab., 1970—; chief, dept. allergy and clin. immunology So. Calif. Permanente Med. Group, also co-dir. resident in allergy and clin. immunology, 1974-83; assoc. clin. prof. medicine UCLA, 1976-83; vis. assoc. in chemistry Calif. Inst. Tech., Pasadena, 1977-83; prof. medicine, head allergy and immunology div. pulmonary medicine U. So. Calif. Sch. Medicine, Los Angeles, 1984—. Contbr. numerous articles to med. jours. Head health and safety com. local council Boy Scouts Am., 1963-65. Served with USPHS, 1963-65. Recipient Order of Arrow award Boy Scouts Am., 1949. Fellow Am. Acad. Allergy; mem. AAAS, Reticuloendothelial Soc., Am. Assn. Immunologists, Los Angeles Soc. Allergy and Clin. Immunology (pres. 1980), Am. Thoracic Soc. Subspecialties: Allergy; Immunobiology and immunology. Current work: Mechanisms of the allergic response. Home: 750 Malcolm Ave Los Angeles CA 90024

GLOWER, DONALD DUANE, university dean, mechanical engineer; b. Shelby, Ohio, July 29, 1926; s. Raymond W.W. and Irva (Scheerer) G.; m. Betty Stahl, June 18, 1953; children: Donald, Michel, Leilani, Jacob. B.S., U.S. Mcht. Marine Acad., 1946, Antioch Coll., 1953; M.S., Iowa State U., 1958; Ph.D. (NSF fellow), Iowa State U., 1960. Engring. officer Grace Lines, Inc., San Francisco, 1947-49; research engr. Battelle Meml. Inst., Columbus, Ohio, 1953-54; asst. prof. Coll. Engring., Iowa State U., 1954-58, 60-61; mem. research staff Sandia Corp., Albuquerque, 1961-63; head radiation effects dept. Gen. Motors Corp., Milw., 1963-64; prof., chmn. dept. mech. and nuclear engring. Ohio State U., 1964-76, dean, 1976—; also dir. Engring. Expt. Sta.; dir. Tech. Exporters Inc.; bd. dirs. Indsl. Tech. Enterprise Bd. Ohio; cons. to industry, 1964—. Author: Graphical Theory and Application, 1957, Basic Drawing and Projection, 1957, Working Drawings and Applied Graphics, 1957, Experimental Reactor Analysis and Radiation Measurements, 1965. Bd. dirs. Ohio Transp. Research Center, Orton Found., OSU Devel. Fund, Nat. Regulatory Research Inst. Recipient Outstanding Bus. Achievement award U.S. Mcht. Marine Acad., 1961; Outstanding Profl. Achievement award Iowa State U., 1979. Fellow Am. Nuclear Soc.; mem. Am. Soc. Engring. Edn., ASME, Ohio Acad. Sci., Argonne Univs. Assn., Ohio Energy Task Force, Sigma Xi, Tau Beta Pi, Texnikoi. Subspecialties: Mechanical engineering; Nuclear engineering. Home: 2338 Kensington Dr Columbus OH 43221 Office: Hitchcock Hall 2070 Neil Ave Columbus OH 43210

GLOYNA, EARNEST FREDERICK, environmental engineer, educator, dean; b. Vernon, Tex., June 30, 1921; s. Herman Ernst and Johanna Bertha (Reithmayer) G.; m. Agnes Mary Lehman, Feb. 17, 1946; children: David Frederick, Lisa Anna (Mrs. Jack Grosskopf). B.S. in Civil Engring, Tex. Technol. U., 1946, M.S., U. Tex., 1949; Dr. Engring., Johns Hopkins U., 1952. Registered profl. engr. Engr. Tex. Hwy. Dept., 1945-46; office engr. Magnolia Petroleum Co., 1946-47; instr. civil engring. U. Tex., Austin, 1947-49, asst. prof., 1949-53, asso. prof., 1953-59, prof., 1959—, Joe J. King prof. engring., 1970-82, Bellie Margaret Smith prof. environ. engring., 1982—, dir. Environ. Health Engring. Labs., 1953-70, dir. Center for Research in Water Resources, 1963-76, dean Coll. Engring., 1970—, dir. Bur. Engring. Research, 1970—; Cons on water and wastewater treatment and water resources, 1947—; dir. Parker Drilling Co.; cons. numerous industries WHO, World Bank, U.S. Air Force, U.S. Army, U.S. Senate, fgn. cities and govts., UN, 1952—; chmn. sci. adv. bd. EPA. Author: Waste Stabilization Ponds, 1971 (also French and Spanish edits), (with Joe O. Ledbetter) Principles of Radiological Health, 1969; Editor: (with W. Wesley Eckenfelder, Jr.) Advances in Water Quality Improvement, 1968, Water Quality Improvement by Physical and Chemical Processes, 1970, (with William S. Butcher) Conflicts in Water Resources Planning, 1972, (with Woodson and Drew) Water Management by Electric Power Industry, 1975, (with Malina and Davis) Ponds as a Wastewater Treatment Alternative, 1976; Contbr. numerous articles to profl. jours. Served with Corps Engrs. AUS, 1942-46; ETO. Recipient Harrison Prescott Eddy medal Water Pollution Control Fedn., 1959, Gordon Maskew Fair medal, 1979, Hon. Mem. award, 1980; Water Resources Div. award Am. Water Works Assn., 1959; named Distinguished Engr. Grad. Tex. Tech. U., 1971, Distinguished Alumnus, 1973, Disting. Engring. Grad. U. Tex., Austin, 1982; recipient Joe J. King award U. Tex., Austin, 1982, EPA regional environ. educator award, 1977, Nat. Environ. Devel. award, 1983. Fellow ASCE (Meritorious Paper award Tex. sect. 1968); mem. Nat. Acad. Engring. (council mem.), Am. Inst. Chem. Engrs., Assn. Environ. Engring. Profs. (past pres.), Am. Soc. for Engring. Edn., Am. Water Works Assn., Am. Acad. Environ. Engrs. (diplomate, past pres., Gorden Maskew Fair award 1981), Water Pollution Control Fedn. (pres.), Tex. Soc. Profl. Engrs. (Engr. of Year award Travis chpt. 1972), Southwestern Soc. Nuclear Medicine (hon.), Nat. Acad. Engring. Mex. (fgn. corr. mem.), Nat. Acad. Scis. Venezuela (fgn. corr.), Sociedad Mexicana de Aguas (Jack Huppert award), Sigma Xi, Tau Beta Pi, Chi Epsilon, Phi Kappa Phi, Pi Epsilon Tau (hon.), Omicron Delta Kappa. Clubs: Cosmos (Washington); Headliners, Faculty Center (Austin). Lodge: Rotary. Subspecialty: Water supply and wastewater treatment. Current work: Environmental, water supply and wastewater treatment. Home: 3317 River Rd Austin TX 78703

GLUSKER, JENNY PICKWORTH, chemist; b. Birmingham, Eng., June 28, 1931; came to U.S., 1955, naturalized, 1977; d. Frederick Alfred and Jane Wylie (Stocks) Pickworth; m. Donald Leonard Glusker, Dec. 18, 1955; children: Ann, Mark John, Katharine. B. A. in Chemistry, Oxford (Eng.) U., 1953, M. A., 1957, D.Phil., 1957. Postdoctoral research fellow Calif. Inst. Tech., Pasadena, 1955-56; research fellow Inst. for Cancer Research, Fox Chase Cancer Center, Phila., 1956, research assoc., 1957-67, asst. mem., 1967, assoc. mem., 1967-79, sr. mem., 1979—; mem. faculty U. Pa., Phila., 1969—, adj. prof. biochemistry and biophysics, 1980—; mem. U.S.A. Nat. Com. for Crystallography, 1974-76, 80, sec.-treas., 1977-79, chmn., 1982—, del. to internat. meeting, Warsaw, 1978, Ottawa, Ont., Can., 1981. Author: (with K.N. Trueblood) Crystal Structure Analysis: A Primer,

1972 (transl. into Russian and Polish); editor: Structural Crystallography in Chemistry and Biology, 1981, (with Dodson and Sayre) Structure of Molecules of Biological Interest, 1981, (with McLachlan) Crystallography in North America, 1982; mem. editorial bd.: Biophys. Jour, 1981—; mem. editorial adv. bd.: Accounts of Chem. Research, 1982—; Contbr. articles to profl. sci. jours. Mem. Am. Chem. Soc. (Phila. sect. award 1978, Garvan medal 1979), Am. Crystallographic Assn. (pres. 1979), Am. Inst. Physics (governing bd. 1980-83), AAAS, Biophys. Soc., Am. Soc. Biol. Chemists, Chem. Soc., Am. Assn. for Cancer Research, Am. Phys. Soc., N.Y. Acad. Scis., Sigma Xi. Subspecialties: Biophysical chemistry; X-ray crystallography. Current work: Research on enzyme mechanisms and chemical carcinogenesis; basic research.

GLUSMAN, SILVIO, neurobiologist, researcher; b. Buenos Aires, Argentina, Apr. 12, 1941; s. Pascual and Fanny G.; m. Marta Susana Killner, Sept. 5, 1964; children: Mariana, Alejandro, Marcela, Andres. M.D., U.Buenos Aires, 1964; D.Sci. in Physiology and Biophysics, Nat. Poly. Inst. Mex., 1977. Asst. prof. dept. physiology Nat. Poly. Inst. Mex., 1972-76, assoc. prof., 1976-77; research assoc. dept. neurobiology Harvard Med. Sch., Boston, 1977-81, prin. research assoc., 1981—. Contbr. articles to profl. jours. Mex. Nat. Inst. Cardiology fellow, 1967; UNESCO fellow, 1974. Mem. Soc. Physiology (Mex.), Soc. Neurosci., Am. Soc. Neurochemistry. Jewish. Subspecialties: Neurobiology; Physiology (medicine). Current work: Modulation of synaptic transmission in peripheral and central nervous system; long lasting phenomena associated with changes in synaptic transmission in the central nervous system. Home: 68 Davis Ave Newton MA 02165 Office: 25 Shattuck St Boston MA 02115

GNANADESIKAN, RAMANATHAN, statistician, researcher; b. Madras, India, Nov. 2, 1932; d. Ambalavanan and Jegathambal (Singaram) Ramanathan; m. Mrudulla R. Gnanadesikan, Feb. 18, 1965; children; Anand, Mukund. B.Sc. with honors, U. Madras, 1952, M.A., 1953; Ph.D., U.N.C., 1957. Sr. research statistician Procter & Gamble Co., Cin., 1957-59; mem. tech. staff Bell Labs., Murray Hill, N.J., 1959—, dept. head, 1968—; vis. prof. U. Cin., 1958-59, Courant Inst. Math. Scis., NYU, 1960-62, Imperial Coll. Sci. and Tech., London, 1968-69, Princeton U., 1971; mem. U.S. Census Adv. Com., 1965-68; mem. com. applied and theoretical stats. Nat. Acad. Sci./NRC, 1978—; cons. Author 2 books, numerous articles. Co-chmn. Asian Indians in Am., Inc., 1970-71. Fellow AAAS, Am. Statis. Assn., Inst. Math. Stats., Royal Statis. Soc.; mem. Internat. Statis. Inst., Biometric Soc., Math. Assn. Am., Internat. Assn. Stetis. Computing, Bernoulli Soc., Order Golden Fleece. Subspecialties: Statistics; Graphics, image processing, and pattern recognition. Current work: Statistical data analysis methodology and applications, including large data sets, graphics and multivariate methods.

GOAD, WALTER BENSON, JR., biophysicist; b. Marlowe, Ga., Sept. 5, 1925; m., 1952; 3 children. B.S., Union Coll., 1945; Ph.D. in Physics, Duke U., 1954. Mem. staff T-div Los Alamos Nat. Lab., 1950—; vis. prof. U. Colo., 1968—; vis. scientist Med. Research Council Lab. Molecular Biology, Cambridge (Eng.) U., 1970-71; Sr. fellow in biophysics U. Colo., 1964-65; scholar Eleanor Roosevelt Inst. Cancer Research, 1977—. Fellow AAAS, Am. Phys. Soc.; mem. Biophys. Soc., N.Y. Acad. Sci. Subspecialty: Biophysics (physics). Office: Los Alamos Nat Lab MS 465 Los Alamos NM 87545

GOAN, HUGH CHARLES, agriculture extension specialist, educator; b. Bulls Gap, Tenn., Apr. 21, 1943; s. Hugh C. and Bessie Louise (Pearson) G.; m. Barbara Kay Wegmeyer, June 17, 1966; children: Scott, David. B.S., U. Tenn.-Knoxville, 1965, M.S., 1968; Ph.D., Mich. State U., 1971. Asst. prof. Agr. Extension Service, U. Tenn., Knoxville, 1971-76, assoc. prof., leader, 1976—. Contbr. writings to bulletins, newsletters, jours. in field, newspapers. U.S. Dept. Agr. grantee, 1982. Mem. Poultry Sci. Assn., Tenn. Agrl. Agts. and Specialists Assn., Tenn. Assn. Extension 4-H Workers, Gamma Sigma Delta, Epsilon Sigma Phi. Lutheran. Lodge: Kiwanis. Subspecialty: Animal nutrition. Current work: Management stresses that affect laying hen and broiler performance; work with egg producers and broiler growers for more efficiency in production. Home: 7608 Sheffield Knoxville TN 37919 Office: PO Box 1071 Knoxville TN 37901

GOATES, JAMES REX, chemistry educator; b. Lehi, Utah, Aug. 14, 1920; s. Stanley and Leona (Lunt) G.; m. Marcia Anderson, June 24, 1948; children: Steven Rex, Liane, Jeffrey James. B.S., Brigham Young U., 1948; Ph.D., U. Wis., 1947. Asst. prof. chemistry Brigham Young U., Provo, Utah, 1947-52, assoc. prof., 1952-54, prof., 1954—, chmn. dept. chemistry, 1965-68, dean, 1981—. Author 2 books; contbr. articles to profl. jours. Served with U.S. Army, 1943-45. Decorated Bronze Star.; Recipient Profl. of Yr. award Brigham Young U., 1960; Karl G. Maeser Excellence in Research award, 1970. Mem. Am. Chem. Soc. (Utah award 1975), Sigma Xi, Phi Kappa Phi. Mormon. Subspecialties: Thermodynamics; Physical chemistry. Current work: Research in the thermodynamics of nonelectroylyte solutions. Home: 1375 Maple Ln Provo UT 84604 Office: Brigham Young U Provo UT 84602

GOBBEL, JOHN RANDALL, systems programmer; b. Durham, N.C., Oct. 17, 1952; s. John Temple and Mary Groover (Bardin) G. B.A. in Communications, Antioch Coll., 1974. Systems programmer Resource One, Inc., San Francisco, 1974-76; systems analyst programmer Tymnet Network Devel., Tymshare, Inc., Cupertino, Calif., 1976-78, sr. systems programmer, 1978-79; sr. mem. programming staff operating systems Xerox Corp., Palo Alto, Calif., 1979-81, sr. mem. programming staff electronic mail, 1981-82, sr. mem. programming staff, 1978—. Mem. Assn. Computing Machinery. Subspecialties: Distributed systems and networks; Operating systems. Current work: Distributed systems, local area networks, personal computers, electronic mail, office automation. Office: Xerox SDD 3333 Coyote Hill Rd Palo Alto CA 94304

GOBEL, STEPHEN, neuroanatomist, dentist; b. N.Y.C., Dec. 27, 1938; s. Benjamin and Jean (Friedman) G.; m. Sharon A Gobel; children: Jennifer, Amy. D.D.S., N.Y. U., 1963. Postdoctoral fellow N.Y. U. Coll. Dentistry, 1963-66; staff scientist Nat. Inst. Dental Research, NIH, Bethesda, Md., 1966-75, chief neurocytology and exptl. anatomy sect., 1975—. Served with USPHS, 1966—. Recipient Founders Day award N.Y. U., 1963. Mem. Internat. Assn. Study Pain, Am. Pain Soc., Am. Assn. Anatomists, Soc. Neurosci., AAAS, Internat. Assn. Dental Research, Omicron Kappa Upsilon. Subspecialties: Neurobiology; Anatomy and embryology. Current work: Conduct light and electron microscopical studies which address basic pain mechanisms and which examine synaptic organization and neural circuitry subserving oro-facial and spinal somatosensory pathways. Home: 5917 Empire Way Rockville MD 20852 Office: 9000 Rockville Pike Bethesda MD 20205

GODFREY, HENRY PHILIP, immunologist; b. Poughkeepsie, N.Y., Aug. 7, 1941; s. Joseph and Mildred (Hoffman) G.; m. Ginger Schnaper, May 6, 1977; 2 sons, Thomas, David. A.B., Harvard U., 1961, M.D., 1965; Ph.D., U. Birmingham, (Eng.), 1980. Intern Barnes Hosp., St. Louis, 1965-66; Mosley travelling fellow Harvard U., 1970-72; hon. research fellow, dept. exptl. pathology U. Birmingham, Eng., 1970-78; asst. prof. Inst. Exptl. Immunology U. Copenhagen, 1972; assoc. prof., 1972-75; asst. prof. pathology SUNY-Stony Brook, 1975-82; assoc. prof. pathology N.Y. Med. Coll., Valhalla, 1982—. Contbr. articles to profl. jours. Served with USPHS, 1966-70. Mem. N.Y. Acad. Scis., Am. Assn. Immunologists, Brit. Soc. Immunology, Harvey Soc., Am. Soc. Microbiology, AAAS, Sigma Xi. Subspecialties: Immunopharmacology; Biochemistry (biology). Current work: Biochemical mechanisms underlying delayed hypersensitivity and cellular immunity; isolation of lymphokines employing monoclonal antibody technology. Office: Dept Pathology NY Med Col Valhalla NY 10595

GODLEY, WILLIE CECIL, animal scientist, educator; b. Miley, S.C., Oct. 3, 1922; m., 1944; 3 children. B.S., Clemson Coll., 1943, M.S., N.C. State Coll., 1949, Ph.D., 1955. From instr. to asst. prof. animal husbandry Clemson U., 1946-47, assoc. prof., 1952-57, prof. animal sci., 1957—; assoc. animal husbandman S.C. Agrl. Expt. Sta., 1954-57, animal husbandman and geneticist, 1957-64, assoc. dir., 1973-75, assoc. dean and dir., 1975—. Mem. Am. Soc. Animal Sci., Am. Genetic Assn. Subspecialty: Animal genetics. Office: 103 Lewis Rd Clemson SC 29631

GODSON, G. NIGEL, biochemistry educator; b. London, June 20, 1936; U.S., 1969; s. Godfrey E. and Elsie G.; m. Barbara Cohen, Aug. 1969; children: Rebecca Charlotte, Vanessa Alexandra. B.Sc., London U., 1957, M.Sc., 1958, Ph.D., 1961. Biochemist/sci. staff Chester Beatty Inst., London, 1961-64; postdoctoral fellow Calif. Inst. Tec., Pasadena, 1964-66; research assoc. Yale U., New Haven, 1966-67, asst. prof., 1969-74, assoc. prof., 1974-80; prof., chmn., dept. biochemistry NYU Md. Ctr., N.Y.C., 1980—; biochemist/sci. staff Nat. Inst. Med. Research, London, 1967-69; mem. exec. com. Sackler Inst. Grad. Biomed. Scis., NYU, 1981—; mem. com. Med. Scientist Tng. Program, NYU, 1981—. Contbr. articles and revs. to sci. jours. Mem. Am. Soc. Exptl. Biology, N.Y. Acad. Scis., Am. Soc. Biol. Chemists, Harvey Soc. Subspecialties: Gene actions; Genetics and genetic engineering (agriculture). Current work: Study of gene expression in E. coli and plasmodium; development of an anti-malarial vaccine using recombinant DNA techniques. Office: 550 1st Ave New York NY 10016

GODWIN, JOHN THOMAS, pathologist; b. Social Circle, Ga., Dec. 2, 1917; s. Hubert Olie and Georgie Ann (Adams) G.; m.; children: Elizabeth, Thomas Adams, Patricia Ann. B.S., Emory U., 1938, M.D., 1941. Diplomate: Am. Bd. Pathology, Am. Bd. Nuclear Medicine. Resident in pathology Touro Infirmary, New Orleans, 1941-42, 47-48; intern U.S. Naval Hosp., Pensacola, Fla., 1942-43, resident, Phila., 1945-46; Am. Cancer Soc. fellow in pathology. Meml. Hosp. Sloan Kettering Inst., N.Y.C., 1948-49, chief fellow, 1949-50; pathologist Oschner Found. Hosp., New Orleans, 1949-50; asst. attending pathologist meml. Sloan Keffering Inst. and James Ewing Hosp., 1951-55; chmn. exptl. pathology Brookhaven Nat. Lab., 1951-55; dir. pathology and lab. medicine St. Joseph's Infirmary, Atlanta, 1955-78, Hughes Spalding Hosp., 1962—; chmn. dept. pathology and lab. medicine King Faisal Specialist Hosp. and Research Ctr., Riyadh, Saudi Arabia, 1980—; prof. pathology Emory U., Atlanta, 1959; lectr. Tulane U., New Orleans, 1950-51; adj. prof. allied health Ga. State U., 1966—; chmn. fund drive James Ewing chair Meml. Sloan Kettering, N.Y.C., 1981—; research scientist Ga. Inst. Tech., Atlanta, 1961; dir. Med. Diagnostics and Research Lab., Atlanta, 1961-78. Contbr. articles to profl. jours.; mem. editorial bd.: King Faisal Hosp. Jour, 1980—; abstract editor for cancer: Jour. Am. Cancer Soc, 1954-55. Mem. Metro. Found., Atlanta, 1966-77; bd. dirs. Ga. div. Am. Cancer Soc., 1956; mem. Gov.'s Med. Adv. Com., Atlanta, 1960-69, Ga. Sci. and Tech. Com., Atlanta, 1969-75; cons. to asst. sec. HEW, Washington, 1966-68. Served to lt. USN, 1942-47. Recipient silver medal for exhibit Am. Soc. Clin. Pathology, 1952, gold medal, 1953; bronze award for film Internat. Film and TV Festival, 1960. Fellow Am. Soc. Clin. Pathology (councilor), Coll. Am. Pathology (del.); mem. James Ewing Soc. (treas. 1974-78), Alumni Soc. Meml. Hosp. (pres. 1980-81), Southeastern Ctr. for Internat. Studies, Fulton County Med. Soc. (pres. 1966). Methodist. Clubs: Phoenix Soc. (dir. 1970-76), Commerce, Cherokee Town and Country (dir. 1970-76). Subspecialties: Pathology (medicine); Nuclear medicine. Current work: Laboratory development and management. Development of national blood banking laboratory and tumor registry standards for Saudi Arabia, medical technology education, clinical-pathological cancer research. Home: 4691 Sentinel Post Rd NW Atlanta GA 30327 Office: King Faisal Specialist Hosp and Research Centre Riyadh Saudi Arabia 11211

GOEBEL, EDWIN MARK, microbiology educator; b. Youngstown, Ohio, Aug. 10, 1951; s. Mark Edwin and Rita Jeanne (Wilkinson) G. B.S., Pa. State U., 1973; M.S., U. Ill., 1975; Ph.D., Va. Poly. Inst. and State U., 1980. Grad. teaching asst. U. Ill., Urbana, 1973-75, Va. Poly Inst. and State U., Blacksburg, 1976-80; asst. prof. biology Bates Coll., Lewiston, Maine, 1980-81, Ind.-Purdue U., Ft. Wayne, 1981—. Contbr. articles to profl. jours. Recipient award for teaching excellence Va. Poly. Inst. and State U., 1979, 80. Mem. AAAS, Am. Soc. Microbiology, Ind. Acad. Sci., South Central Assn. Clin. Microbiology. Lutheran. Subspecialty: Microbiology. Current work: Research on carbohydrate and organic acid metabolism of the nitrogen-fixing genus Azospirillum. Office: Ind Purdue U at Fort Wayne 2101 Coliseum Blvd E Fort Wayne IN 46805

GOEBEL, WALTHER FREDERICK, biochemist; b. Palo Alto, Calif., Dec. 24, 1899; s. Julius and Kathryn (Vreel) G.; m. Cornelia Van Rensselaer Robb, Oct. 23, 1930 (dec. Oct. 1974); children—Cornelia Van Rensselaer Bronson, Anne Kathryn Barkman; m. Alice Lawrence Behn, Nov. 12, 1976. A.B., U. Ill., 1920, A.M., 1921, Ph.D., 1923, scholar in chemistry, 1920-21, fellow, 1921-23; postgrad., U. Munich, Germany, 1923-24; D.Sc. (hon.), Middlebury (Vt.) Coll., 1959, Rockefeller U., 1978. Research asst. Rockefeller U., 1924-27, asso., 1927-34, asso. mem., 1934-44, mem., 1944-57, prof., 1957-70, prof. emeritus, 1970—. Contbr. monographs, reports and articles on chem. and immunological subjects sci. jours. Mem. Nat. Acad. Scis., Am. Chem. Soc., Am. Soc. Biol. Chemists, Harvey Soc., Am. Assn. Immunologists, Am. Soc. Microbiology, Conn. Acad. Sci. and Engring., Gesellschaft für Immunologie (Avery-Landsteiner award 1973), Phi Beta Kappa, Sigma Xi, Phi Lambda Upsilon, Phi Eta. Subspecialties: Immunocytochemistry; Immunology (medicine). Current work: Immunologically active carbohydrates; synthetic antigens. Home: 15 Lyon Farm Dr E Greenwich CT 06830 Office: Rockefeller U New York City NY 10021

GOEDDEL, DAVID VAN NORMAN, biochemist; b. Pasadena, Calif., May 3, 1951; s. Walter Van Norman and Barbara Lou (Haist) G.; m. Carol Lynn Smiley, Jan. 13, 1973. B.A. in Chemistry, U. Calif. at San Diego, 1972; Ph.D. in Biochemistry, U. Colo., 1977. Dir. Stanley Andrews Mountaineering Sch. and Guide Service, San Diego, 1970-73; teaching asst., research asst. dept. chemistry U. Colo., Boulder, 1973-77; postdoctoral fellow Stanford Research Inst., Menlo Park, Calif., 1977-78; research scientist Genentech, Inc., South San Francisco, Calif., 1978—. Mem. Am. Chem. Soc., AAAS. Subspecialty: Molecular biology. Co-inventor first bacterial production of human insulin, 1978, of human growth hormone, 1979, of human interferon, 1980. Office: 460 Point San Bruno Blvd South San Francisco CA 94080

GOEL, NARENDRA SWARUP, systems science educator; b. Muzaffarnagar, UP, India, June 12, 1941; came to U.S., 1962, naturalized, 1978; s. Shiva Charan Das and Prem Vati G.; m. Sudha Prakash, May 27, 1967; children: Vindu, Namni. B.S., Agra U., 1957; M.S. in Physics, Delhi U., 1959, Poona U., 1962; Ph.D., U. Md., College Park, 1965. Research assoc., instr. U. Md., College Park, 1962-66; asst. prof., assoc. U. Rochester, 1966-72; mgr., prin. scientist Xerox Corp., Webster, N.Y., 1972-76; dir. tech. vitality SUNY, Binghamton, 1977-81, prof. dept. systems sci., 1976—; cons. Xerox Corp., Webster, N.Y., 1979—, NASA, Houston, 1981—, Lockheed Corp., 1982—. Author: (with N. Richter-Dyn) Stochastic Models in Biology, 1974, (with others) Non-linear models of Populations, 1972; editorial bd.: Jour. Theoretical Biology, Worcester, Mass., 1973—. Jacobs scholar, 1966; U.S.-USSR Acad. Scis. exchange scholar, 1981. Mem. Bioengring. Soc., Biophys. Soc., Soc. Indsl. and Applied Math., Soc. for Origin of Life, N.Y. Acad. Scis. Subspecialties: Space agriculture; Systems engineering. Current work: Remote sensing of vegetation, reliability of complex electro-mech. machines, failures in digital systems, modeling and analysis of bioligical and engineering systems, aging, cellular aggregation, folding of proteins. Home: 4245 Marietta Dr Binghamton NY 13903 Office: Dept Systems Sci SUNY Binghamton NY 13901

GOEPP, ROBERT AUGUST, dental educator, researcher; b. Chgo., Nov. 3, 1930; s. Charles August and Ernestine Josephine (Mertz) G.; m. Iraida Pineiro, July 9, 1960; children: Robert C., Heidi M., Myra J. B.S., Loyola U., Chgo., 1953, D.D.S., 1957; M.S., U. Chgo., 1961, Ph.D., 1967. Diplomate: Am. Bd. Oral Pathology, Am. Bd. Oral and Maxillofacial Radiology.; Lic. dentist. Instr. Pritzker Sch. Medicine, U. Chgo., 1961-64, asst. prof., 1964-70, assoc. prof., 1970-75; prof. U. Chgo., 1975—. Contbr. to textbooks; author articles to sci. jours. USPHS career devel. awardee, 1970; grantee NIH, Am. Cancer Soc., 1962-80. Fellow Am. Coll. Dentists; mem. ADA (chmn. Council on Dental Research 1981-82), Am. Acad. Dental Radiology (pres. 1974), Soc. Oral Pathologists Ill. (pres. 1977-78), Med. Radiation Adv. Council, Nat. Council on Radiation Protection and Measurements. Roman Catholic. Club: Quadrangle (U. Chgo.). Subspecialties: Oral biology; Radiology. Current work: Biologic effects of ionizing radiation, epithelial cell propulati. Co-inventor custom cervical cap. Home: 5928 N Kilbourn Chicago IL 60646 Office: Zoller Dental Clinic Box 418 950 E 59th St Chicago IL 60637

GOETSCH, CARL ALLEN, researcher, physician; b. San Francisco, June 24, 1949; s. Carl and Anne (Tompkins) G. A.B., Washington U., 1972; M.S., U. Calif.-Berkeley, 1974; M.D. with distinction in research, U. Rochester, 1978. Diplomate: Am. Bd. Internal Medicine. Resident in internal medicine U. Va. Hosp., Charlottesville, 1978-81; fellow gastroenterology U. Calif.-San Francisco, 1981—. Mem. Am. Fedn. Clin. Research, ACP (assoc.). Subspecialties: Gastroenterology; Nutrition (medicine). Current work: Currently involved in research work in folic acid metabolism and small intestinal transport, porphyrin metabolism in small intestine, enterotoxigenic diarrheal disease. Home: 38 Vicente Rd Berkeley CA 94705 Office: U Calif Med Ctr San Francisco CA 94122

GOETZ, ALEXANDER FRANKLIN HERMANN, research scientist and manager, consultant; b. Pasadena, Calif., Oct. 14, 1938; s. Alexander and Sylvia (Scott) G.; m. Rosamaria Cyrus, Aug. 21, 1982. B.S., Calif. Inst. Tech., 1961, M.S. in Geology, 1962, Ph.D. in Planetary Sci, 1967. Mem. tech. staff Bell Telephone Labs., Washington, 1967-70; supr. Jet Propulsion Lab., Pasadena, Calif., 1973-76, mgr. planetology and oceanography sect., 1976-78, mgr. geol. applications research, 1978-82, sr. research scientist, mgr. terrestrial remote sensing research, 1981—; vis. prof. UCLA, 1983; cons.; pres. GeoImages, Inc., 1974-78. Contbr. articles profl. jours. Recipient Charles E. Ives award Soc. Photog. Scientists and Engrs., 1973, Exceptional Sci. Achievement medal NASA, 1982, William T. Pecora award NASA/Dept. of Interior, 1982. Mem. AAAS, Am. Geophys. Union, Am. Soc. Photogrammetry (Autometric award 1981). Club: Los Angeles Yacht. Subspecialty: Remote sensing (geoscience). Current work: Research, development and testing of remote sensing techniques and instruments for earth mapping from space. Patentee in field. Office: Jet Propulsion Lab 183-501 Pasadena CA 91109

GOGUEN, JOSEPH AMADEE, computer scientist, linguist, researcher; b. Pittsfield, Mass., June 28, 1941; s. Joseph Amadee and Helen Elmira (Stratton) G.; m. Kathleen G. Morrow, June 20, 1980; children: Heather H., Healfdene H. B.A., Harvard U., 1963; M.A., U. Calif.-Berkeley, 1966, Ph.D., 1968. Asst. prof. U. Chgo., 1968-73; research fellow T. J. Watson Research Center, IBM, 1972; prof. dept. computer sci. UCLA, 1974-81; mng. dir., gen. ptnr. Structural Sematics, Palo Alto, Calif., 1978—; sr. computer scientist SRI Internat., Menlo Park, Calif., 1979—; sr. vis. fellow dept. artificial intelligence U. Edinburgh, 1976, 77. Contbr. articles to profl. jours. Mem. Assn. Computing Machinery, IEEE (tech. com. on founds. of computing computer soc.), Am. Math., Soc., Math. Assn. Am. Subspecialties: Software engineering; Programming languages. Current work: Visual programming, abstract data types, ultra high level programming language design, contect sensitive graphics editor, super computer architecture. Office: 333 Ravenswood Ave Suite EL 372 Menlo Park CA 94025

GOH, DAVID SHUH JEN, psychology educator; b. Kiangsu, China, July 9, 1941; came to U.S., 1967, naturalized, 1977; s. Ying-Hwa and Pei-Shuo (Chao) G.; m. Jane M. Chen, Sept. 1, 1973; children: Alice, Nancy. M.S, Ill. State U., 1969; Ph.D., U. Wis., 1973. Lic. psychologist, Mich., Wis., Ill. Psychologist Lincoln (Ill.) Devel. Ctr., 1969-70; asst. prof. U. Wis.-LaCrosse, 1973-75; asst. prof., then assoc. prof. Central Mich. U., Mt. Pleasant, 1975-80; assoc. prof., then prof. So. Ill. U., Carbondale, 1980—, also staff psychologist, 1980—. Contbr. articles to profl. jours., chpts. to books.; Editorial bd.: Psychology in the Schs. Jour, 1980—; reviewer various other jours., 1977—. Mem. Am. Psychol. Assn., Nat. Assn. Sch. Psychologists, Soc. Personality Assessment, Am. Assn. Asian Psychologists, Nat. Council for Measurement and Evaluation. Subspecialties: Behavioral psychology; Cognition. Current work: Research/teaching in areas of individual differences in cognitive processes and personality, psychological and behavioral assessment and instrumentation, consultation. Home: 4 View Valley Dr Heritage Hills Carbondale IL 62901 Office: So Ill Univ Carbondale IL 62901

GOH, EDWARD HUA SENG, pharmacologist, educator, researcher; b. Sarawak, Malaysia, Jan. 20, 1942; came to U.S., 1964, naturalized, 1974; s. Chee Kuan and Chai Ting (Lim) G.; m. Sharon Ratliff, June 17, 1949; children: Melisa Hua-Linn, Andrew Hua-Seng. A.A., Warren Wilson Coll., Swannonoa, N.C., 1966; B.A., Berea Coll., 1968; Ph.D., Vanderbilt U., 1974; postgrad., U. Mo., 1974-75. Instr. U. Mo., Columbia, 1975-77; asst. prof. pharmacology Ind. U., Bloomington, 1977-82, asso. prof., 1982—. Contbr. articles on metabolism of blood cholesterol to sci. jours. NSF grantee, 1979-82; Am. Heart Assn. grantee, 1978-81; Pharm. Mfrs. Assn. Found. Inc. grantee, 1977-79; Am. Diabetes Assn. grantee, 1982. Mem. Am. Soc. Pharmacology and Therapeutics, Am. Heart Assn., Internat. Soc. for Study Xenobiotics. Subspecialties: Pharmacology; Biochemistry (medicine). Current work: Current research activities involve the metabolism of lipoprotein cholesterol and the developmental testing of new drugs, in relation to their metabolism and toxicology, used to regulate blood cholesterol levels. Office: Ind U Med Scis Program Myers Hall 306 Bloomington IN 47405

GOHAR, MOHAMED YOUSRY AHMED, nuclear engineer; b. Kom-Hamada, Behira, Egypt, Feb. 3, 1947; s. Ahmed Ahmed and Doria A. (Gharib) G.; m. Iman Esmat O, El-Dib, July 23, 1978. B.S. in Nuclear Engring, Alexandria (Egypt) U., 1967, M.S., 1970, Ph.D., 1974. Asst. prof. Atomic Energy Establishment of Egypt, Cairo, 1967-74; research assoc. nuclear engring. dept. U. Wis., Madison, 1974-77; nuclear engr. Argonne (Ill.) Nat. Lab., 1977—; researcher Fusion Engring. Design Center, Oak Ridge, 1981—. Recipient longevity service award Argonne Nat. Lab., 1982. Mem. Am. Nuclear Soc. Subspecialties: Nuclear engineering; Nuclear fusion. Current work: Nuclear fission and fusion reactor design, computational methods in nuclear engineering. nuclear data, applied numerical methods, mathematical software. Office: Argonne National Laboratory 9700 S Cass Ave Argonne IL 60439

GOINS, WILLIAM (DORIS), III, govt. research adminstr.; b. Cleve., Apr. 6, 1943; s. William D. and Lillian L. (Palmer) G.; m. Diane Johnston, Sept. 5, 1964; children: William D. IV, Clay Paul. B.S. in Nuclear Engring, U. Tenn., 1966, M.S. in Metall. Engring, 1969. Supr. Welding lab Combustion Engring., Chattanooga, 1970-77, lab. Welding engring., 1970-77; with TVA, Chattanooga, 1977—, lab. supr., project mgr., 1977-79, program mgr. nuclear research, 1979—; mem. Edison Electric Inst. Metallurgy, Piping, Welding, and Corrosion Task Force. Mem. ASME (chmn. working group on repair welding Boiler and Pressure Vessel Code Sect. XI), Am. Welding Soc., Am. Nuclear Soc., Welding Research Council (chmn. utilities adv. com), Am. Soc. Metals. Baptist. Subspecialties: Materials (engineering); Metallurgical engineering. Current work: Research projects include light water reactor safety; materials; corrosion; radioactive waste; gamma thermometers; nuclear fuel dry storage; decontamination and advanced systems; fusion; high temperature gas-cooled reactors; liquid metal fast breeder reactors. Home: 2506 Big Cedar Rd Soddy TN 37379 Office: TVA 1050 Chestnut St Tower II Chattanooga TN 37401

GOITEIN, MICHAEL, physicist; b. Worcestershire, Eng., Nov. 14, 1939; came to U.S., 1961; s. Hugh and Freda (Goodman) G.; m. Marcia Elizabeth Angell, May 31, 1968; children: Lara, Elizabeth. B.A., Oxford U., 1961; Ph.D., Harvard U., 1968. Diplomate: Am. Bd. Radiology. Research fellow Harvard U., Cambridge, Mass., 1968-69, 1979—; staff physicist Lawrence Berkeley Lab., Berkeley, Calif., 1969-72; with Mass. Gen. Hosp., Boston, 1972—, assoc. radiation biophysicist, 1975—. Brackenbury scholar, 1958-61; Fulbright travel fellow, 1961-65; Research Career Devel. awardee, 1976-81. Mem. Am. Phys. Soc., Am. Assn. Physicists in Medicine, Am. Soc. Therapeutic Radiologists, Radiation Research Soc. Subspecialties: Biophysics (physics); Imaging technology. Current work: Development and use of diagnostic imaging techniques in cancer therapy. Office: Department Radiation Medicine Massachusetts General Hospital Boston MA 02114

GOJOBORI, TAKASHI, population geneticist; b. Fukuoka-City, Japan, Oct. 24, 1951; s. Sukeyuki and Fukumi (Inoue) G.; m. Mariko Veno, Dec. 5, 1976; children: Jun, Yoko. B.S., Kyushu U., 1974, M.S., 1976, Ph.D., 1979. Lectr. in physics Tokai U., 1979-80; sr. research asso. in population genetics U. Tex. at Houston, 1980—. Contbr. articles to profl. jours. Mem. Genetics Soc. Am., AAAS, Soc. Study Evolution, N.Y. Acad. Sci. Subspecialties: Population biology; Evolutionary biology. Current work: Population genetics, DNA sequences, molecular evolution, gene expression, genome orgn., gene action, computer analysis. Office: PO Box 20334 Houston TX 77025 Home: 2524 Sheridan Houston TX 77030

GOLAND, MARTIN, research institute executive; b. N.Y.C., July 12, 1919; s. Herman and Josephine (Bloch) G.; m. Charlotte Nelson, Oct. 16, 1948; children—Claudia, Lawrence, Nelson. M.E., Cornell U., 1940; LL.D. (hon.), St. Mary's U., San Antonio. Instr. mech. engring. Cornell U., 1940-42; sect. head structures dept. research lab., airplane div. Curtiss-Wright Corp., Buffalo, 1942-46; chmn. div. engring. Midwest Research Inst., Kansas City, Mo., 1946-50, dir. for engring. scis., 1950-55; v.p. Southwest Research Inst., San Antonio, 1955-57, dir., 1957-59, pres., 1959—, S.W. Found. Research and Edn., San Antonio, 1972-82; dir. Nat. Bancshares Corp. Tex.; Chmn. subcom. vibration and flutter NACA, 1952-60; chmn. research adv. com. on aircraft structures NASA, 1960-68, chmn. materials and structures group, aeros. adv. com., 1979-82; sci. adv. com. Harry Diamond Labs., U.S. Amry Materiel Command, 1955-75; adv. panel com. sci. and astronautics Ho. of Reps., 1960-73; mem. high speed ground transp. panel Dept. Commerce, 1966-67, nat. inventors council, 1966-67, mem. state tech. services evaluation com., 1967-69; mem. adv. bd. on undersea warfare Dept. Navy, 1968-70, chmn., 1970-73; mem. spl. aviation fire reduction com. FAA, 1979-80; sci. adv. panel Dept. Army, 1966-77; chmn. U.S. Army Weapons Command Adv. Group, 1966-72; mem. materiels adv. bd. NRC, 1969-74; vice-chmn. Naval Research Adv. Com., 1974-77, chmn., 1977, mem., 1978—; dir. Nat. Bank Commerce, San Antonio.; Dir. Engrs. Joint Council, 1966-69; mem. adv. group U.S. Armament Command, 1972-76; mem. sci. adv. com. Gen. Motors, 1971-81; mem. Nat. Commn. on Libraries and Info. Scis., 1971-78, Nat. Bd. on Grad. Edn., 1972-75; mem. adv. bd. on mil. personnel supplies Nat. Acad. Sci., 1973-76; chmn. NRC Bd. Army Sci. and Tech., 1982—, Army Missile Command ROLAND Blue Ribbon Panel, 1983—. Editor: Applied Mechanics Review, 1952-59; editorial adviser, 1959—. Bd. govs. St. Mary's U., San Antonio, 1970-76; research adv. com. coordinating bd. Tex. Coll. and Univ. System, 1966-68; pres. San Antonio Symphony, 1968-70, chmn. bd., 1970-71; bd. dirs. So. Meth. U. Found. Sci. and Engring., Dallas, 1979—; trustee Univs. Research Assos., Inc., 1979—. Recipient Spirit of St. Louis jr. award ASME, 1945, jr. award, 1946, Alfred E. Nobel prize ASCE, 1947. Fellow A.A.A.S., Am. Inst. Aeros. and Astronautics (pres. 1971); hon. mem. ASME (dir., mem. bd. tech., mem. tech. devel. com., v.p. communications); mem. C. of C. (dir.), Nat. Acad. Engring., Research Soc. Am., Sigma Xi, Tau Beta Pi. Subspecialties: Aeronautical engineering; Theoretical and applied mechanics. Current work: President research institute. Home: 306 Country Ln San Antonio TX 78209 Office: 6220 Culebra Rd San Antonio TX 78284

GOLAY, MICHAEL WARREN, engineering educator; b. Pitts., Sept. 5, 1942; s. George Warren and Patricia (Van Buskirk) G.; m. Michal-Ann Barrett, Apr. 25, 1964; children: Barret, Geddes. B.Mech.Engring., U. Fla., 1964; Ph.D., Cornell U., 1969. Research assoc. Rensselaer Poly. Inst., Troy, N.Y., 1969-71; asst. prof. MIT, Cambridge, Mass., 1971-75, assoc. prof., 1975—; pres. Charles River Research, Inc., Lexington, Mass., 1981—; cons. in field. Mem. Am. Nuclear Soc. (chmn. environ. scis. div. 1979-80), ASME. Subspecialties: Nuclear engineering; Environmental engineering. Current work: Nuclear reactor safety, nuclear regulation, cooling towers, atmospheric plumes. Office: MIT 77 Massachusetts Ave Cambridge MA 02139

GOLBERG, LEON, physician, educator, consultant; b. Limassol, Cyprus, Aug. 22, 1915; came to U.S., 1967, naturalized, 1974; s. Aron and Bertha (Lurie) G.; m. Bertha Klempman, July 4, 1944; children: Michael Gregory, Estelle Laura, Aron Anthony. B.Sc., U. Witwatersrand, 1934, M.Sc., 1936, D.Sc., 1946; D.Phil., Oxford U., 1939; M.A., M.B. B.Chir. (sr. scholar), Cambridge U., 1951; D.Sc. (hon.), Phila. Coll. Pharmacy and Scis., 1983. Registered Gen. Med. Council, U.K. Sr. lectr. chem. pathology U. Manchester, Eng., 1951-55; med. research dir. Fisons Pharms. Ltd., Holmes Chapel, Eng., 1955-61; dir. BIBRA, Carshalton, Surrey, Eng., 1961-67; research prof. pathology Albany (N.Y.) Med. Coll., 1967-76; pres. Chem. Industry Inst. Toxicology, Research Triangle Park, N.C., 1976-81; prof. community/occupational medicine Duke U. Med. Center, Durham, N.C., 1981—; Milroy lectr. Royal Coll. Physicians London, 1967; adj. prof. pharmacology, pathology, toxicology. Author: Toxicology: Has a New Era Dawned?, 1980, The Revolution in Toxicology: Real or Imaginary, 1981, The Hierarchy of Hazard Evaluation, 1982; contbr. articles to sci. jours. Recipient George Scott Meml. award Toxicology Forum, 1983. Fellow Royal Soc. Chemistry, Royal Coll. Pathologists (founder fellow), Soc. Toxicology (pres. 1978-79, ambassador of toxicology 1981, Disting. fellow 1981, John Barnes lectr. 1983); mem. Am. Chem. Soc., Environ. Mutagen Soc., Japanese Cosmetic Sci. Soc. (hon.). Subspecialties: Toxicology (medicine); Cancer research (medicine). Current work: Promoting agents in experimental neoplasia; occupational medicine; toxicology; consulting; teaching; cancer research; biotechnology. Home: 2109 Nancy Ann Dr Raleigh NC 27607 Office: Duke U Med Center Div Community/Occupational Medicine Box 2914 Durham NC 27710

GOLBUS, MITCHELL SHERWIN, physician, educator; b. Chgo., Apr. 6, 1939; s. Leonard and Rose (Klein) G.; m. Antoinette Buiano, Jan. 13, 1967; children: Aaron, Matthew, Aliyah. B.S. in Psychology, Ill. Inst. Tech., 1959; M.D., U. Ill.-Chgo., 1963. Diplomate: Am. Bd. Obstetrics & Gynecology, Am. Bd. Med. Genetics. Intern Los Angeles County Gen. Hosp., 1963-64; resident dept. obstetrics, gynecology and reproductive scis. U. Calif.-San Francisco, 1964-68, asst. research geneticist, research fellow med. genetics dept. pediatrics, 1971-73, clin. instr. dept. obstetrics, gynecology & reproductive scis., 1972-73, asst. prof., 1973-77, assoc. prof., 1977-81, prof., 1981—; medical adviser to numerous orgns. Contbr. articles to profl. jours. and abstracts to profl. confs. Served to major U.S. Army MC, 1968-71. Fellow Am. Coll. Obstetricians and Gynecologists; mem. Am. Soc. Human Genetics, San Francisco Gynecol. Soc., Soc. Gynecologic Investigation, AAAS. Jewish. Subspecialties: Perinatal diagnosis and therapy; Fetal surgery. Current work: Reproductive genetics, prenatal diagnosis and therapy of genetic defects.

GOLD, BARRY I(RA), pharmacologist, neurochemist, educator; b. Everett, Mass., Apr. 30, 1946; s. David Bernard and Harriet Sybil (Wolfson) G.; m. Olga Genevive, July 4, 1971; children: Jessica Frances, Gregory Samuel. B.S., U. Cin., 1968; Ph.D., Boston U., 1976. Research assoc. Squibb Inst. for Med. Research, New Brunswick, N.J., 1968-71; postdoctoral fellow Yale U. Sch. Medicine, 1975-78; asst. prof. pharmacology Uniformed Services U., Bethesda, Md., 1978—. Contbr. articles to profl. publs. Recipient Sandoz award Sandoz Pharm. Co., 1975, Research Service award USPHS, 1975, 76, 77. Mem. AAAS, Soc. Neurosci. Subspecialties: Neuropharmacology; Neurochemistry. Current work: Regulation of amino acid neurotransmitter metabolism in the central nervous system. Office: Dept. Pharmacology Uniformed Services U Bethesda MD 20814

GOLD, JONATHAN W.M., physician, microbiologist; b. N.Y.C., July 6, 1944; s. Abraham D. and Tanya L. G.; m. Christy E. Joyce, Aug. 30, 1979. B.A., Columbia U., 1965, M.D., 1971. Diplomate: Am. Bd. Internal Medicine. Assoc. dir. Spl. Microbiology Lab., Meml. Sloan-Kettering Cancer Ctr., N.Y.C., 1982—, asst. attending physician, 1979—. Served with USPHS, 1972-74. Mem. AAAS, Am. Fedn. for Clin. Research, Am. Soc. Microbiology, Infectious Disease Soc. Subspecialties: Infectious diseases; Internal medicine. Current work: Infections in immunocompromised hosts. Office: Meml Sloan-Kettering Cancer Ctr 1275 York Ave New York NY 10021

GOLD, JUDITH HAMMERLING, psychiatrist; b. N.Y.C., June 24, 1941; d. James S. and Anne (Linder) Hammerling; m. Edgar Gold, June 27, 1965. M.D., Dalhousie U., 1965. Intern Victoria Gen. Hosp., Halifax, N.S., Can., 1964-65; resident Dalhousie U., Halifax, 1967-71; practice medicine specializing in psychiatry, Halifax, 1971—; staff psychiatrist Dalhousie U. Student Health Clinic, 1971-73; vis. colleague U. Wales Med. Sch., 1973-75; asst. prof. dept. psychiatry Dalhousie U., Halifax, 1975-78, assoc. prof., 1978-80, part-time, 1980—. Contbr. articles to profl. jours. Bd. govs. Mt. St. Vincent U. Med. Research Council Can. fellow, 1973-75; Health and Welfare Bd. Can. grantee, 1976-78. Mem. Can. Psychiat. Assn. (pres. 1981-82), Am. Psychiat. Assn., Am. Coll. Psychiatrists, Alpha Omega Alpha. Subspecialty: Psychiatry. Office: 5991 Spring Garden Rd #1 020 Halifax NS Canada B3H 1Y6

GOLD, MARK STEPHEN, psychopharmacologist, physician; b. N.Y.C., May 6, 1949; s. Meyer M. and Helene (Levy) G.; m. Janice Finn, June 19, 1971; children: Steven, Kimberly. B.A., Washington U., 1967; M.D., U. Fla., 1975. Neurobehavior fellow Yale U. Sch. Medicine, 1975-78, lectr., 1978-80; v.p. basic research Psychiat. Inst. Am., Summit, N.J., 1978-80; dir. research Fair Oaks Hosp., Summit, N.J., 1978-82, Psychiat. Diagnostic Labs. Am., Summit, 1979—; cons. substance abuse unit Yale U. Sch. Medicine, 1979—. Contbr. articles to profl. jours. Recipient Seymour F. Lustman award for research Yale U. Sch. Medicine, 1978; Founds. Fund prize for research in psychiatry Am. Psychiat. Assn. Found., 1981; Presdl. award for Disting. Leadership in Psychiat. Research Nat. Assn. Pvt. Psychiat. Hosps., 1982; NIMH grantee. Mem. Soc. Neurosci., Am. Psychiat. Assn., Endocrine I Soc., Com. Impaired Physicians, Internat. Soc. Psychoneuroendocrinology, Med. Soc. N.J., Nat. Assn. Pvt. Psychiat. Hosps., AAAS. Subspecialties: Psychopharmacology; Medicinal chemistry. Office: 1 Prospect St Summit NJ 07901

GOLD, STEVEN IRA, periodontist, dentistry educator; b. N.Y.C., Sept. 27, 1941; s. Jack Martin and Sarah (Rives) G.; m. Marion Susan Marx, Sept. 2, 1962; children: Geoffrey Adam, Lauren Jill. Student, Colgate U., 1959-62; D.D.S., N.Y.U., 1966; cert. periodontology, Columbia U., 1972. Diplomate: Am. Bd. Periodontology. Asst. clin. prof. dentistry Columbia U., N.Y.C., 1972-79, assoc. clin. prof., 1979—; cons. N.E. Periodontology Soc. Case Report Jour., 1982. Vice-pres. Congregation Beth Jacob, Newburgh, N.Y., 1979-80. Served to capt. U.S. Army, 1966-68. Mem. Internat. Assn. Dental Research, Am. Acad. Periodontology, ADA, Am. Acad. History Dentistry. Subspecialty: Periodontics. Current work: Research - clinical wound healing, effects of cultural background on periodontal disease severity, history of periodontology. Home: Osborne Hill Rd Fishkill NY 12524 Office: So Dutchess Profl Park Route 52 Fishkill NY 12524

GOLD, THOMAS, educator, astronomer; b. Vienna, Austria, May 22, 1920; s. Max and Josefine (Martin) G.; m. Merle Eleanor Tuberg, June 21, 1947; children—Linda, Lucy, Tanya; m. Carvel Lee Beyer, Dec. 27, 1972; 1 dau., Lauren. B.A., Cambridge (Eng.) U., 1942, M.A., 1945, Sc.D., 1969; fellow, Trinity Coll., Cambridge, 1947; M.A. (hon.), Harvard, 1957. Lectr. physics Cambridge (Eng.) U., 1948-52; chief asst. to Astronomer Royal, Gt. Britain, 1952-56; prof. astronomy Harvard, 1958, Robert Wheeler Willson prof., 1958-59; prof. astronomy, dir. Center Radiophysics and Space Research Cornell U., 1959-81, chmn. dept., 1959-68, asst. v.p. for research, 1970-71, John L. Wetherill prof., 1971—. Contbr. articles to profl. jours. Fellow Royal Soc. London; mem. U.S. Nat. Acad. Sci., Am. Philos. Soc., Am. Acad. Arts and Scis., Royal Astron. Soc. (past councillor), Am. Astron. Soc., Am. Geophys. Union. Subspecialties: Geophysics; Radio and microwave astronomy. Address: Space Scis Bldg Cornell U Ithaca NY 14853

GOLDBAUM, MICHAEL HENRY, ophthalmology educator, researcher; b. Bklyn., Apr. 17, 1939; s. Samuel Zolomon and Sarah (Kramer) G.; m. Brenda Scott Leggio, July 7, 1964; children: David, Stephen, Rachel. B.A., Syracuse U., 1961; M.D., Tulane U., 1965. Diplomate: Am. Bd. Ophthalmology. Retina fellow Cornell U. Med. Sch., N.Y.C., 1973, instr., fellow ophthalmology, 1972-73; asst. prof. ophthalmology U. Ill. Eye and Ear Infirmary, Chgo., 1973-77; assoc. prof. ophthalmology U. Calif.-San Diego, 1977—. Contbr. articles to profl. jours. Served to lt. comdr. USN, 1967-72. Fellow Am. Acad. Ophthalmology, San Diego Acad. Ophthalmology; mem. AMA, San Diego Assn. Retarded Citizens. Subspecialties: Ophthalmology; Laser medicine. Current work: Neovascularization and angiogenesis in the eye, macular degeneration. Home: 8435 Cliffridge Ln La Jolla CA 92037 Office: U Calif 225 Dickinson St San Diego CA 92103

GOLDBERG, ALAN MARVIN, toxicologist, educator; b. Bklyn., Nov. 20, 1939; s. William and Celia Ida (Tudman) G.; m. Helene S. Schwenbach, Aug. 14, 1960; children: Michael David, Naomi Jill. B.S., Bklyn Coll. Pharmacy, 1961; Ph.D. in Pharmacology, U. Minn., 1962. Research asst. U. Wis., 1961-62, U. Minn., 1962-66; research assoc. Inst. Psychiat. Research Ind. U., 1966-67; asst. prof. dept. pharmacology Ind. U., 1967-69; asst. prof. environ. medicine Johns Hopkins U., 1969-71, assoc. prof., 1971-78; prof. dept. environ. health scis., 1978—, assoc. chmn. dept., 1978-80, acting dir. div. toxicology, 1979-80, dir. div. toxicology, 1980-82, dir., 1980—; prin. research scientist Chesapeake Bay Inst., 1979—; mem. Health Hazard Evaluation Team of Chem. Wastes Dumps, State of Tenn., EPA Rev. Panel, 1980-82; cons. Wrightsville Marine Biomed. Labs., Astra Pharm. Co.; mem. sect. on neurobiology NSF; environ. health scis. NIH. Mem. editorial bd.: Jour. Am. Coll. Toxicology; Contbr. articles to profl. jours. Recipient award Ind. Neurol. Soc., 1967. Mem. AAAS, Am. Soc. Pharmacology and Exptl. Therapeutics, Soc. Neurosci. (pres. Balt. chpt. 1971-73), Am. Soc. Neurochemistry, Am. Epilepsy Soc., Internat. Soc. Neurochemistry, Soc. Toxicology, Internat. Study Group on Memory Disorders, Internat. Union Pharmacology. Subspecialties: Environmental toxicology; Neurobiology. Current work: Neurotoxicology, in vitro methodology development. Home: 2231 Crest Rd Baltimore MD 21209 Office: 615 N Wolfe St Baltimore MD 21205

GOLDBERG, ALLEN FRED, oral surgeon; b. Chgo., Sept. 3, 1933; s. Irving Harry and Frances (Genson) G.; m. Norma Boone, Feb. 15, 1960; children: Nancy, David. B.S., U. Ill., 1956, D.D.S., 1958; M.S., Loyola U., Chgo., 1965. Diplomate: Am. Bd. Oral and Maxillofacial Surgeons. Clin. prof. surgery U. Ill., Chgo., 1971-81; oral and maxillofacial surgeon VA Hosp., Hines, Ill., 1964-71, chief dental service, Chgo., 1971-73, oral and maxillofacial surgeon, North Chicago, Ill., 1973—. Contbr. articles to profl. jours. Served to capt. U.S. Army, 1960-61; ETO. Fellow Am. Assn. Oral and Maxillofacial Surgeons, Am. Coll. Oral and Maxillofacial Surgeons; mem. Internat. Assn. Dental Research, ADA, AAAS, Am. Legion; fellow Am. Coll. Dentists. Subspecialty: Oral and maxillofacial surgery. Home: 8500 W Carol St Niles IL 60648 Office: VA Med Center North Chicago IL 60064

GOLDBERG, EDWARD BLEIER, molecular biology and microbiology educator; b. Bronx, N.Y., July 19, 1935; s. Samuel B. and Sylvia (Bleier) G.; m. Ariella Deem, Feb. 8, 1935; children: Donna Faea, Abigail Miriam. A.B., Columbia U., 1956; Ph.D., Johns Hopkins U., 1961. Research fellow Carnegie Inst. of Washington, Cold Springs Harbor, N.Y., 1961-65; prof. molecular biology and microbiology Tufts U., 1965—. Contbr. in field. NIH fellow, 1957-59; grantee, 1965—; NSF fellow, 1959-61; grantee, 1967—; Med. Found. fellow, 1961-63. Mem. AAAS, Am. Soc. Microbiology, Am. Soc. Virology, Am. Soc. Biol. Chemistry, Sigma Xi. Jewish. Subspecialties: Genetics and genetic engineering (biology); Virology (biology). Current work: Host viral recognition and penetration of viral nuclic acid into host. Office: 136 Harrison Ave Boston MA 02111

GOLDBERG, EDWARD DAVID, geochemist, educator; b. Sacramento, Aug. 2, 1921; s. Edward Davidow and Lillian (Rothholz) G.; m. Kathe Bertine, Dec. 26, 1973; children—David Wilkes, Wendy Jean, Kathi Kiri, Beck Bertine. B.S., U. Calif. at Berkeley, 1942; Ph.D., U. Chgo., 1949. Mem. faculty Scripps Instn. Oceanography, La Jolla, Calif., 1949—, prof. chemistry, 1960—; provost Revelle Coll., U. Calif. at San Diego, 1965-66. Author: (with J. Geiss) Earth Sciences and Meteorites, 1964, Guide to Marine Pollution, 1972, North Sea Science, 1973, The Sea: Marine Chemistry, Vol. V, 1974, The Health of the Oceans, 1976; Contbr. numerous articles to profl. jours. Guggenheim fellow, 1961; NATO fellow, 1970. Mem. Am. Geophys. Union, AAAS, Geochem. Soc., U.S. Acad. Scis., Sigma Xi. Subspecialty: Geochemistry. Research, publs. primarily on marine pollution, chem. composition sea water, sediments, marine organisms, environmental mgmt.; radioactive dating techniques in marine environment and glaciers. Home: 750 Val Sereno Dr Encinitas CA 92024

GOLDBERG, ISADORE, psychology educator; b. Buffalo, Aug. 12, 1928; s. Benjamin Nathan and Norma (Mendelson) G.; m. Adelle Gladstone, Sept. 12, 1954; 1 dau., Lynne Ellen. B.A., Miami U., Ohio, 1955; postgrad., Carnegie Inst. Tech., 1956; M.A., U. Md., 1957, Ph.D., 1959. Lic. psychologist, Md. Sr. research scientist Am. Inst. Research, Washington, 1958-64; dir. ednl. devel. U.S. Industries, Inc., Silver Spring, Md., 1964-66; mgr. edn. Computer Applications, Inc., Silver Spring, 1966-69; mgr. Volt Info. Scis., Washington, 1969-71; prof. U. D.C., Washington, 1971—; cons. Edn. Commn. of States, Denver, 1970-75, Nat. Inst. Edn., Washington, 1975, AT&T, 1976-78, U.S. Army, 1982—. Author: (with others) Design for Study of American Youth, 1962; contbr. articles to profl. jours. Active PTA, 1963-80. Served with U.S. Army, 1950-52. Recipient Community Service award D.C. Psychol. Assn., 1979; U. D.C. grantee, 1980-81; FAA grantee, 1979; New Century Edn. Corp. grantee, 1976. Mem. Am. Psychol. Assn., Indsl. Relations Research Assn., Univ. Coll. Labor Edn. Assn., Am. Ednl. Research Assn., D.C. Psychol. Assn. Democrat. Jewish. Subspecialties: Behavioral psychology; Human resource development. Current work: Research, consultant and practice in human resource development, education, and training, labor relations, community development; applications of computers and media to above. Home: 3 Hearthstone Ct Potomac MD 20854 Office: Labor Studies Center 724 9th St NW Washington DC 20005

GOLDBERG, IVAN, nuclear engineer; b. Bklyn., June 5, 1930; s. Nat B. and Sadie (Weiss) G.; m. Bernice G. Rind, Jan. 27, 1951; children: Jacqueline, Roy, Nancy. B.S., CUNY, 1962. Engr. Bettis Lab., West Mifflin, Pa., 1962-70, mgr. fuel design, 1970-76, mgr. core performance, 1976-82, mgr. reactor engring., 1982—. Judge elections Allegheny County Dept. Elections, Pitts., 1969—. Mem. Am. Nuclear Soc. Subspecialties: Fuels; Nuclear fission. Home: 5719 Solway St Pittsburgh PA 15217 Office: Westinghouse Bettis Lab PO Box 109 West Mifflin PA 15122-0109

GOLDBERG, IVAN D., microbiologist, educator; b. Phila., May 13, 1934; s. Max and Frances G.; m. Noveta McCracken, July 27, 1979;

children: Micki, Judy, Lisa; m.; stepchildren: Nick, Vikki Russell. A.B., U. Pa., 1956; Ph.D., U. Ill., 1961. Postdoctoral fellow Rutgers U., 1961-62, Oreg. State U., 1962-63; NRC postdoctoral research assoc. U.S. Army Biol. Labs., Frederick, Md., 1963-65, microbial geneticist, 1965-71; assoc. prof. dept. microbiology U. Kans. Sch. Medicine, 1971-77, prof. microbiology, 1977—. Contbr. articles to sci. jours. NIH grantee, 1972-80; recipient Leroy D. Fothergill Sci. award, 1970. Mem. Am. Soc. Microbiology (pres. Mo. Valley br. 1974), Sigma Xi. Democrat. Jewish. Subspecialties: Genetics and genetic engineering (biology); Microbiology. Current work: Genetics of Neisseria gonorrhea, Bacillus sp. and Legionella pneumophila. Utilization of recombinant DNA technology to elucidate the genetic bases of thermophily; organisms utilized are B.stearothermophilus, B. coagulans and B. subtilis. Home: 14409 W. 90th Terrace Lenexa KS 66215 Office: 39th and Rainbow Blvd Kansas City KS 66103

GOLDBERG, JACK, medical educator; b. Ulm, Germany, Feb. 7, 1948; came to U.S., 1950, naturalized, 1968; s. Isaac and Mary (Selitzka) G.; m. Doreen A. Sandler, June 28, 1970. B.A., Boston U., 1969; M.D., SUNY-Upstate Med. Ctr., 1973. Diplomate: Am. Bd. Internal Medicine. Intern Univ. Hosp., Boston, 1973-74; jr. asst. resident, 1974-75; fellow in hematology SUNY Upstate Med. Ctr., Syracuse, 1975-77, asst. prof. medicine, 1977-81, assoc. prof., 1981, assoc. prof. pathology, 1982—; mem. Leukemia Intergroup, Syracuse, 1981—; mem. med. adv. com. Syracuse ARC, 1979—. Contbr. articles to med. jours. NIH grantee, 1980-85. Mem. Am. Fedn. Clin. Reseach, Am. Soc. Hematology, Internat. Soc. Exptl. Hematology, Alpha Omega Alpha. Subspecialty: Hematology. Current work: Acute leukemia determinants of response, in vitro studies of granulopoiesis. Office: Upstate Med Ctr 750 E Adams St Syracuse NY 13210

GOLDBERG, LEE DRESDEN, physician, med. educator; b. Point Pleasant, N.J., July 29, 1937; s. Milton J. and Maude (Dresden) G.; m. Lana Ditchek, July 23, 1967; children: Marissa Julie, Sara Amy, Rachel Sherry. B.S. summa cum laude, Yale U., 1959, M.D., 1963. Diplomate: Nat. Bd. Med. Examiners, Am. Bd. Internal Medicine (Endocrinology). Intern Mt. Sinai Hosp., N.Y.C., 1963-64; resident in medicine Montefiore Hosp., Bronx, N.Y., 1964, 66-68; clin. research fellow in endocrinology Albert Einstein Coll. Medicine, Bronx, 1968-69; fellow in endocrinology Bellevue Hosp-N.Y. U., N.Y.C., 1969-70; from clin. instr. to clin. asst. prof. medicine U. Miami (Fla.) Sch. Medicine, 1970-80, clin. assoc. prof., 1980—; chief internal medicine South Shore Hosp., Miami Beach, 1975-79; assoc. chmn. med. services St. Francis Hosp., Miami Beach, 1977-78; co-chief endocrinology Mt. Sinai Hosp., Miami Beach, 1974—. Contbr. articles to profl. jours. Mem. Am. Physicians Fellowship for Medicine in Israel, Boston, 1978—; bd. dirs. Hebrew Acad. Greater Miami, Miami Beach, 1978—; chmn. youth com. Beth Israel Congregation, Miami Beach, 1980—. Served to lt., M.C. U.S. Navy, 1964-66. Recipient Performance award VA, 1971; Physicians Recognition award AMA, 1968, 72. Fellow ACP; mem. Am. Fedn. Clin. Research, Am. Diabetes Assn. (dir. Miami chpt. 1981—), Endocrine Soc., Phi Beta Kappa, Sigma Xi. Jewish. Club: Yale (Miami). Subspecialty: Endocrinology. Current work: Thyroid and parathyroid disorders; public education about medicine. Office: 1674 Meridian Ave Miami Beach FL 33139

GOLDBERG, LEO, astronomer, educator; b. Bklyn., Jan. 26, 1913; s. Harry and Rose (Ambush) G.; m. Charlotte B. Wyman, July 9, 1943; children: Suzanne, David Henry, Edward Wyman. S.B., Harvard U., 1934, A.M., 1937; Ph.D., 1938; Sc.D. (hon.), U. Mass., 1970, U. Mich., 1974, U. Ariz., 1977. Asst. dept. astronomy Harvard U., Cambridge, Mass., 1934-37, Agassiz research fellow, 1937-38, spl. research fellow, 1938-41, research assoc., 1941; Higgins prof. astronomy, 1960-73, Higgins prof. emeritus, 1973—, chmn. dept. astronomy, 1966-71, dir., 1966-71; research assoc. McMath-Hulbert Obs., U. Mich., Ann Arbor, 1941-46, asst. prof. astronomy, 1945-46, assoc. prof., chmn. dept. astronomy, dir. obs., 1946-60, prof., 1948-60; dir. Kitt Peak Nat. Obs., Tucson, 1971-77, dir. emeritus, disting. research scientist, 1977—; chmn. Astronomy Missions Bd., NASA, 1967-70; mem. U.S. nat. com. Internat. Astron. Union, 1954-66, 73—, chmn., 1956-61, U.S. del. to Gen. Assembley, Moscow, 1958; mem. sci. adv. bd. USAF, 1959-62; mem. defense sci. bd. Dept. Def., 1962-64; mem. solar physics subcom. NASA, 1962-65, mem. sci. and tech. adv. com. manned space flight, 1964-71; chmn. adv. group on sci. programs NSF, 1974-76; mem. panel on health sci. and tech. enterprise Office Tech. Assessment, U.S. Congress, 1976—. Author: (with L.H. Aller) Atoms, Stars and Nebulae, 1943; contbr. articles to profl. jours.; collaborating jour. editor: Astrophys. Jour, 1949-51; chmn. editorial bd., 1954; editor: Annual Reviews Astronomy and Astrophysics, 1961-73. Trustee Assoc. Univs. Inc., 1957-66; mem. bd. overseers Harvard U., 1976—. Recipient USN service award, 1946; Disting. Pub. Service medal NASA, 1973; George Darwin lectr. Royal Astron. Soc., London, 1978. Fellow Am. Acad. Arts and Scis.; mem. Nat. Acad. Scis. (space sci. bd. 1958-64, chmn. astronomy sect. 1977—), AAAS, Internat. Astron. Union (v.p. 1958-64, pres. 1973-76), Am. Philos. Soc., Am. Astron. Soc. (pres. 1964-66), Assn. Univs. for Research in Astronomy Inc. (dir. 1966-71), Royal Astron Soc. Can. (hon.), Societe Royale des Sciences de Liege. Subspecialty: Theoretical astrophysics. Home: 6846 E Via Dorado Tucson AZ 85715 Office: 950 N Cherry Ave PO Box 26732 Tucson AZ 85726

GOLDBERG, LEON ISADORE, pharmacologist; b. Charleston, S.C., Sept. 26, 1926; s. Harry and Goldy (Cohen) G.; m. Faye Joan Girsh, Feb. 2, 1958 (div.); children—Mark, Claudia. B.S. Pharmacy, Med. U. S.C., 1946, M.S., 1951, Ph.D., 1952; D.H.L.; M.D. cum laude, Harvard U., 1956. Intern Mass. Gen. Hosp., Boston, 1956-57, intern, asst. resident in medicine, 1957-58, research fellow anesthesia, 1954-56; research asst. in pharmacology Med. Coll. S.C., 1949-52, research asso., 1952-54; clin. asso. Exptl. Therapeutics Br., NIH, 1958-61; prof. medicine, pharmacology, dir. clin. pharmacology program Emory U., Atlanta, 1961-74; prof. medicine, pharmacology, chmn. com. on clin. pharmacology U. Chgo., 1974—; cons. NIH, VA, FDA, Nat. Acad. Scis. Editor: Jour. Cardiovascular Pharmacology; mem. editorial bds. several pharmacology jours., 1961—; contbr. chpts. to books, numerous articles on clin. pharmacology, cardiovascular pharmacology to profl. jours. Served as surgeon USPHS, 1958-61. Burroughs Welcome Fund scholar in clin. pharmacology, 1961-66. Fellow Am. Coll. Cardiology; mem. Am. Soc. Clin. Investigation, Am. Soc. for Pharmacology and Exptl. Therapeutics, Soc. for Exptl. Biology and Medicine, Am. Soc. for Clin. Pharmacology and Therapeutics, Am. Heart Assn., Council for High Blood Pressure Research, Assn. Am. Physicians. Subspecialties: Pharmacology; Internal medicine. Current work: Fundamentals and clinical studies of cardiovascular drugs. Home: 5000 S Cornell Apt 21B Chicago IL 60615 Office: Dept Pharmacology Sch Medicine U Chgo 947 E 58th St Chicago IL 60637

GOLDBERG, LOUIS J., dentist, neuroscientist, educator; b. Middletown, N.Y., July 20, 1936; m., 1963. B.A., Bklyn. Coll., 1956; D.D.S., NYU, 1960; Ph.D. in Anatomy, UCLA, 1968. From asst. prof. to assoc. prof. oral biology UCLA, 1968-72, from asst. prof. to assoc. prof. anatomy, 1969-72, dean research, 1972-79, prof. oral biology, 1977—, prof. anatomy, 1977—. Recipient Career Devel. award NIH, 1968; Nat. Inst. Dental Research fellow NYU, 1961-62, Sch. Dentistry UCLA, 1964-68; USPHS grantee, 1968—. Subspecialty: Oral biology. Office: Sch Dentistry U Calif Ctr Health Sci Los Angeles CA 90024

GOLDBERG, MICHAEL E., neuroscientist, neurologist; b. N.Y.C., Aug. 10, 1941; s. Samuel and Irma (Mandell) G.; m. Deborah L. Baron, July 31, 1966; children: Joshua, Jonathan. A. B., Harvard U., 1963, M.D., 1968. Diplomate: Am. Bd. Psychiatry and Neurology. Med. house officer Peter Bent Brigham Hosp., Boston, 1968-69; staff asso. NIMH, Bethesda, Md., 1969-72; resident in neurology Children's Hosp. Med. Center, Boston, 1972-75; research neurologist Armed Forces Radiobiology Research Inst., Bethesda, Md., 1975-78; research med. officer Lab. of Sensorimotor Research, Nat. Eye Inst., NIH, Bethesda, 1978-81, chief neuro-ophthalmologic mechanism sect., 1981—; clin. assoc., prof. neurology Georgetown U., Washington, 1976—; mem. adv. panel on integrative and motor processes NSF, 1981—. Contbr. articles to profl. jours. Served with USPHS, 1969-72. Recipient S. Weir Mitchell award Am. Acad. Neurology, 1972. Mem. Soc. Neurosci. (com. continuing med. edn. 1980—), Am. Acad. Neurology, Assn. Research in Vision and Ophthalmology, Am. Neurol. Assn., Internat. Neuropsycholo. Symposium, Phi Beta Kappa, Alpha Omega Alpha. Subspecialties: Neurophysiology; Neurology. Current work: Neurophysiology of visuomotor behavior in the monkey; development of computer systems for behavioral neurophysiology. Office: Laboratory Sensorimotor Research National Eye Institute Building 10 Bethesda MD 20205

GOLDBERG, RICHARD ARAN, physicist, researcher; b. Boston, Jan. 6, 1936; s. Samuel and Ida (Breger) G.; m. Paula Rachel Erlick, Dec. 26, 1965; 1 dau., Lisa Rebecca. B.S. in Physics, Rensselaer Poly. Inst., 1957, Ph.D., Pa. State U., 1963. Research assoc. Nat. Acad. Scis., Washington, 1963-64; research scientist NASA Goddard Space Flight Center, Greenbelt, Md., 1964—, chmn. internat. working group, 1981—. Author: (with others) Sun, Weather and Climate, 1978; Contbr. articles to profl. jours. Eastman Kodak fellow, 1961; recipient various Spl. Achievement awards NASA. Mem. Am. Phys. Soc., Am. Geophys. Union, Am. Meteorol. Soc., AAAS, Internat. Sci. Radio Union, Sigma Xi, Sigma Pi Sigma. Clubs: NASA Flying (past pres.), TSS Flying (past pres.). Subspecialties: Atmospheric Electrodynamics; Aeronomy. Current work: Coupling processes between the upper and lower atmosphere, sounding rocket experiments and project scientist, solar-terrestrial relationships, auroral physics. Office: NASA Goddard Space Flight Center Code 961 Greenbelt MD 20771

GOLDBERG, STEPHEN, neurobiologist, physician; b. Bklyn., Dec. 20, 1942; s. Harry and Dinah (Veroba) G.; m. Harriet Esther Sabinson, Nov. 26, 1967; children: Shaani Sari, Rebecca Ann, Marc Joseph, Michael. B.A., Yeshiva Coll., 1963; M.D., Albert Einstein Coll. Medicine, 1967. Diplomate: Am. Bd. Ophthalmology, Am. Bd. Family Practice. Intern Montefiore Hosp., Bronx, N.Y., 1967-68; surgeon S.I. USPHS Hosp., 1968-70; guest researcher Nat. Eye Inst., Bethesda, Md., 1970-71; resident in ophthalmology N.Y. Med. Coll., 1971-75; asst. prof. U. Miami Sch. Medicine, 1975-79, asso. prof., 1979—; med. dir. Nat. Parkinson Found., Miami, 1981—. Author: Clinical Neuroanatomy Made Ridiculously Simple, 1979, Ophthalmology Made Ridiculously Simple, 1982, Clinical Anatomy Made Ridiculously Simple, 1984. Served with USPHS, 1968-70. Nat. Eye Inst. grantee, 1977-82. Mem. Assn. Research in Vision and Ophthalmology, Am. Assn. Anatomists, Soc. Neurosci. Subspecialties: Developmental biology; Comparative neurobiology. Current work: Neuronal development and regeneration, visual system, techniques in medical education. Home: 17500 NE 9th Ave North Miami Beach FL 33162 Office: Dept Anatomy R124 U Miami Sch Medicine PO Box 016960 Miami FL 33101

GOLDBERGER, MARVIN L., educator, physicist, institute technology president; b. Chgo., Oct. 22, 1922; s. Joseph and Mildred (Sedwitz) G.; m. Mildred Ginsburg, Nov. 25, 1945; children: Samuel M., Joel S. B.S., Carnegie Inst. Tech., 1943; Ph.D., U. Chgo., 1948. Research asso. Radiation Lab., U. Calif., 1948-49; research asso. Mass. Inst. Tech., 1949-50; asst.-asso. prof. U. Chgo., 1950-55, prof., 1955-57; Higgins prof. physics Princeton U., 1953-54, 57-78, chmn. dept., 1970-76, Joseph Henry prof. physics, 1977-78; pres. Calif. Inst. Tech., Pasadena, 1978—; Cons. Los Alamos Sci. Lab., Brookhaven Nat. Lab.; Mem. President's Sci. Adv. Com., 1965-69; Chmn. Fedn. Am. Scientists, 1971-73; dir. Gen. Motors Corp., Haskel, Inc. Fellow Am. Phys. Soc., Am. Acad. Arts and Scis.; mem. Nat. Acad. Scis., Am. Philos. Soc., Council on Fgn. Relations. Club: Princeton (N.Y.C.). Subspecialties: Theoretical physics; Particle physics. Home: 415 S Hill Ave Pasadena CA 91106

GOLDE, DAVID WILLIAM, educator, physician, scientist; b. N.Y.C., Oct. 23, 1940; s. Harvey and Esther (Bobrove) Goldstein; m.; children: Daniel, Michael. B.S. in Chemistry, Fairleigh Dickinson U., 1962; M.D., McGill U., Montreal, Que., Can., 1966. Diplomate: Am. Bd. Internal Medicine, 1972, Nat. Bd. Med. Examiners. Intern U. Calif. Hosps., San Francisco, 1966-67; resident in clin. pathology Clin. Center, NIH, Bethesda, Md., 1968-69, hematology fellow and resident in clin. pathology, 1969-70; resident in medicine U. Calif. Hosps., San Francisco, 1970-71; fellow Cancer Research Inst., 1971-72, resident in medicine, 1971-72; staff cons. continuing edn. and tng. br. div. regional med. programs NIH, 1967-68; instr. medicine U. Calif., San Francisco, 1972, asst. prof., 1973, UCLA, 1974, assoc. prof., 1975-79, prof., 1979—, chief div. hematology/oncology, 1981—, co-dir., 1974—; mem. (VA hematology study sect.), 1978-80, 1979-81; mem. cancer coordinating com. U. Calif., 1982—. Contbr. numerous articles on hematology to profl. jours.; mem. editorial bds.: Leukemia Research, 1977, Blood, 1978, Peptides, 1979, Stem Cells, 1980, Leukemia Revs, 1980. Served to lt. comdr. USPHS, 1966-70. Recipient J. Francis Williams prize McGill U. Dept. Medicine, 1966. Fellow ACP; mem. Am. Assn. for Cancer Research, AAAS, Am. Fedn. Clin. Research, Am. Soc. Clin. Investigation, Am. Soc. Hematology, Internat. Assn. Comparative Research on Leukemia and Related Diseases, Internat. Soc. Exptl. Hematology, Reticuloendothelial Soc., Soc. Exptl. Biology and Medicine, Western Soc. Clin. Research, Alpha Omega Alpha, Phi Omega Epsilon. Subspecialties: Hematology; Oncology. Office: UCLA Sch Medicine Los Angeles CA 90024

GOLDEN, DAVID EDWARD, physicist, educator, researcher, cons.; b. N.Y.C., May 27, 1932; s. Barnet Dade and Rose (Rosenbaum) G.; m. Paula Englander, July 18, 1962; children: Jeffrey Bertram, Leila Justine. B.A., N.Y.U., 1954, Ph.D. in Physics, 1960. Asst. prof. physics dept. N.Y.U., 1960-61; asst. prof. Adelphi U., 1961-62; engring. specialist Gen. Telephone & Electronics Lab., Palo Alto, Calif., 1962-63; staff scientist Lockheed Research Lab., Palo Alto, 1962-67; vis. prof. in physics U. Bari, 1967-70; assoc. prof. physics U. Nebr., Lincoln, 1970-72, prof., 1972-75; prof., chmn. dept. physics and astronomy U. Okla., Norman, 1975—; hon. lectr. Mid Am. State Univ. Assn., 1982-83; cons. Autometric Corp., 1961-62, ARIS, 1970-72, Tracor, 1972-74, Lawrence Radiation Lab., 1975-80; gen. com. Internat. Conf. on Physics of Electron and Atomic Collisions, 1979—; com. on atomic and molecular sci NRC, 1982—. Contbr. articles to profl. publs. Grantee NSF, 1982—, Dept. Energy, 1982—. Fellow Am. Phys. Soc. (exec. com. div. electron and atomic physics 1981—). Subspecialty: Atomic and molecular physics. Current work: Atomic scattering, electron spectroscopy, lasers, electron-photon angular correlation, scattering with spin polarized electrons. Patentee. Home: 8 Rustic Hills Norman OK 73069 Office: Dept Physics and Astronomy Univ Okla 440 W Brooks Norman OK 73019

GOLDENBERG, HERBERT JAY, testing and research laboratory executive, consultant failure analysis; b. Bklyn., Oct. 2, 1935; s. Phillip G. and Sadie (Schulman) G.; m. Susan C. Feuer, Aug. 31, 1961; children: Lauren Beth, David Gary, Aileen Janice. A.A.Sci., N.Y. Tech. Coll., 1959; B.S. in Mech. Engring, Cooper Union Sch. Engring., 1968. Registered profl. engr., N.Y., N.J. With City Testing & Research Labs., Inc., Rahway, N.J., 1959—, chief engr., 1972—; failure analysis cons., 1967—. Served with U.S. Air Force Res., 1961-67. Mem. Am. Soc. for Metals (Pres.'s award 1975), ASME (pres. N.Y. chpt. 1978-79), ASTM, Nat. Assn. Corrosion Engrs. Jewish. Clubs: B'nai B'rith (exec. bd. Temple Beth Ohr 1976-83, sec. 1982-83. Subspecialties: Materials (engineering); Metallurgy. Current work: Failure analysis. Office: City Testing & Research Labs Inc 967 E Hazelwood Ave Rahway NJ 07065

GOLDENBERG, MARVIN M., immunopharmacologist, scientific administrator; b. N.Y.C., July 7, 1935; s. Jacob and Sarah (Thaler) G.; m. Esther Kay Gelman, Sept. 8, 1957; children: Sol Jeffrey, Lisa Shari. B.S. in Pharmacy, Bklyn. Coll. Pharmacy, 1957; M.S. in Pharmacology, Temple U., 1959; Ph.D., Woman's Med. Coll. Pa., 1965. Sr. research pharmacologist Norwich-Eaton Pharms., 1965-73, research assoc., 1973-79; dir. immunopharmacology Merck, Sharpe & Dohme Research Labs., Rahway, N.J., 1980—; cons. Contbr. articles profl. jours. Pres. Norwich Jewish Ctr., 1976. Mem. Am. Soc. Pharmacology, Am. Gastroenterology Assn., Internat. Soc. Immunopharmacology, Inflammation Research Assn., N.Y. Acad. Scis. Lodge: Masons. Subspecialties: Immunobiology and immunology; Immunopharmacology. Current work: Inflammation, immunology, gastroenterology, inflammation relative to new drug discovery in inflammatory diseases. Patentee in field. Office: Box 2000 Rahway NJ 07060

GOLDFARB, NORMAN MARC, plant genetic engineering company executive; b. San Francisco, Sept. 1, 1953; s. Eli M. and Marilyn R. (Gillis) G.; m. Melanie Yelton, Feb. 13, 1982. B.A. in Econs., Yale U., 1974; M.B.A., Stanford U., 1976. Mgr. inventory control Fairchild Consumer Products Inc., Palo Alto, Calif., 1976-77; mgr. central planning mfg. Am. Microsystems, Santa Clara, Calif., 1977-78; mgr. Philippine plant Pacific Reliability Corp., 1978-79; mgr. prodn. planning Intel Corp., Santa Clara, 1979-80; pres. Calgene, Inc., Davis, Calif., 1980—. Subspecialties: Genetics and genetic engineering (agriculture); Plant cell and tissue culture. Current work: Manage leading plant genetic engineering company. Office: 1920 5th S Davis CA 95616

GOLDFARB, ROY DAVID, educator; b. N.Y.C., Sept. 23, 1947; s. Walter and Mae (Marcus) G.; m. Arlene Simon, June 10, 1973; children: Sarah, Anna, Rachel. A.B., Colgate U., 1968; M.Sc., Hahnemann Med. Coll., 1971, Ph.D., 1973. Instr. U. South Ala., Mobile, 1973-75; asst. prof. dept. physiology Albany (N.Y.) Med. Coll., 1975-79, asso. prof., 1979—. Mem. AAAS, Shock Soc., Am. Physiol. Soc., N.Y. Acad. Sci. Club: Porsche of Am. (pres. 1983). Subspecialty: Physiology (medicine). Current work: Cardiac physiology. Address: Dept Physiology Albany Med Coll 43 New Scotland Ave Albany NY 12208

GOLDFARB, STANLEY, physician, physiology educator; b. N.Y.C., Dec. 18, 1943; s. Robert Melvin and Mary Ann (Siegel) G.; m. Rayna Block, Aug. 31, 1970; children: Rachael Fay, Michael Louis. A.B. cum laude, Princeton U., 1965; M.D., U. Rochester, 1969. Diplomate: Am. Bd. Internal Medicine, Am. Bd. Internal Medicine in Nephrology. Intern Hosp. U. Pa., 1969-70, resident, 1970-72; clin. instr. U. Pa. Sch. Medicine, 1974-75, asst. prof. medicine, 1975-81, assoc. prof., 1981—. N.Y. State Regents Scholar, 1961-65; NIH Clin. Investigator awardee, 1978-81; grantee, 1983—. Fellow Am. Coll. Physicians; mem. Am. Fedn. Clin. Research, Am. Soc. Nephrology, Internat. Soc. Nephrology, N.Y. Acad. Scis. Jewish. Club: Rolling Green Golf. Subspecialties: Nephrology; Physiology (medicine). Current work: Renal regulation of divalent ion metabolism; nephrolithiasis; thiophosphate pharmacology. Home: 34 Rosedale Rd Overbrook Hills PA 19151 Office: Renal-Electrolyte Sect Hosp U Pa 860 Gates 3400 Spruce St Philadelphia PA 19104

GOLDFEDER, ANNA, cancer and radiobiol. researcher; b. Lubin, Poland, July 25, 1897; came to U.S., 1931, naturalized, 1940; d. Chaim and Tauba (Friedman) G. D.Sc., Karl U., Prague, 1923; M.U.C., Masaryk U., Brno, Czechoslovakia, 1931. Fellow Harvard U. Med. Sch., 1931-33; dir. Cancer Research Lab. City of N.Y., 1934-77; dir. cancer and radiobiological research lab. N.Y.U., 1977—. Contbr. articles to profl. jours. Mem. Radiol. Soc. N. Am. (recipient award 1940), N.Y. Acad. Sci. (recipient Gold medal 1981). Subspecialties: Cell and tissue culture; Membrane biology. Current work: Cell Biology, cell kinetics, cell radiosensitizers. Home: 2 Washington Square Village 91 New York NY 10012 Office: 100 Washington Square New York NY 10003

GOLDFINE, HOWARD, microbiology and biochemistry educator, researcher; b. Bklyn., May 29, 1932; s. Samuel and Ida (Cohen) G.; m. Norah C. Johnston, Jan. 25, 1963; children: Cynthia A., Sarah C. B.S., CCNY, 1953; Ph.D., U. Chgo., 1957; M.S. (hon.), U. Pa., 1970. Instr. Harvard U. Med. Sch., Boston, 1962-63, assoc., 1963-66, asst. prof., 1966-68; assoc. prof. U. Pa., Phila., 1968-76, prof., 1976—; vis. scientist Microbial Genetics Research Unit, London, 1967-68; mem. physiol. chemistry study sect. NIH, 1969-73; cons. Phila. VA Hosp., 1975—. Mem. editorial bd.: Jour. Bacteriology, 1970-75, Jour. Biol. Chemistry, 1974-80, 82—; mem. editorial bd.: Jour. Lipid Research, 1972-82; assoc. editor, 1983—. Am. Cancer Soc. scholar, 1960-63; NIH grantee, 1962—; Macy Found. scholar, 1976-77. Mem. Am. Soc. Biol. Chemists, Am. Soc. Microbiology (chmn. div. K 1981-82), AAAS, Am. Oil Chemists Soc. Subspecialties: Biochemistry (biology); Membrane biology. Current work: The structures, biosynthesis, assembly and functions of the lipids of biological membranes, regulation and coordination of membrane lipid synthesis. Home: 416 Sycamore Ave Merion PA 19066 19104

GOLDFRIED, MARVIN R., psychology educator, psychotherapist; b. Bklyn., Jan. 24, 1936; s. Samuel and Anna (Ozer) G.; m. Anita Powers, Dec. 23, 1967; children: Daniel, Michael. B.A. cum laude, Bklyn. Coll., 1957; Ph.D., SUNY-Buffalo, 1961. Cert. psychologist, N.Y., diplomate: Am. Bd. Profl. Psychology. Research asst., teaching fellow SUNY-Buffalo, 1957-60, instr., 1960-61; asst. prof. U. Rochester, N.Y., 1961-64; prof. psychology and psychiatry SUNY-Stony Brook, 1964—; vis. assoc. prof. Bar-Ilan U., Ramat-Gan, Israel, 1970-71; vis. scholar U. Calif.-Berkeley, 1977-78; pvt. practice, N.Y.C., 1964—; adv. bd. Inst. Behavior Therapy, N.Y.C., 1974—. Author: Rorschach Handbook of Clinical and Research Applications, 1971, Behavior Change Through Self-Control, 1973, Clinical Behavior Therapy, 1976, Converging Themes in Psychotherapy, 1982; editor: Behavior Change Through Self-Control, 1973; assoc. editor: Cognitive Therapy and Research, 1977-82. Bd. dirs. Human Resources, Roosevelt Island, N.Y.; mem. Men's Ctr., L.I. NIMH grantee, 1966-68, 67-71, 73-84, 74. Fellow Am. Psychol. Assn.; mem. Assn. Advancement of Behavior Therapy, N.Y. Acad. Sci. Democrat. Jewish. Subspecialties: Behavioral psychology; Cognition. Current work: Psychological assessment procedures and methods for implementing psychotherapy change. Office: Dept Psychology SUNY Stony Brook NY 11794

GOLDHABER, JACOB KOPEL, mathematician, educator; b. Bklyn., Apr. 12, 1924; s. Joseph and Shirley (Heller) G.; m. Ruth Last, Dec. 25, 1951; children—Doreet, David, Aviva. B.A., Bklyn. Coll., 1944; M.A., Harvard, 1945; Ph.D., U. Wis., 1950. Instr. U. Conn., Storrs, 1950-53; instr. Cornell U., Ithaca, N.Y., 1953- 54; asst. prof. Washington U., St. Louis, 1954-59, asso. prof., 1959- 61, U. Md., College Park, 1961-62, prof., 1962—, chmn. math. dept., 1968-77; exec. sec. Office Math. Scis., NRC, 1975—; Vis. research asso. (NSF Sci. Faculty fellow) U. London (Eng.), 1966-67. Author: (with Gertrude Ehrlich) Algebra, 1970; Contbr. papers to profl. jours. Mem. AAAS, Am. Math. Soc., Math. Assn. Am., Sigma Xi. Subspecialty: Algebra. Home: 5517 39th St NW Washington DC 20015 Office: Dept Math U Md College Park MD 20742

GOLDHABER, MAURICE, physicist; b. Lemberg, Austria, Apr. 18, 1911; came to U.S., 1938, naturalized, 1944; s. Charles and Ethel (Frisch) G.; m. Gertrude Scharff, May 24, 1939; children: Alfred S., Michael H. Ph.D., Cambridge U., Eng., 1936, Tel-Aviv U., Israel, 1974, Dr., U. Louvain-La-Neuve, Belgium, 1982, D.Sc., SUNY, Stony Brook, 1983. Bye fellow Magdalene Coll., Cambridge, 1936-38; asst. prof. physics U. Ill., 1938-43, assoc. prof., 1943-45, prof., 1945-50; sr. sci. Brookhaven Nat. Lab., 1950-60, chmn. dept. physics, 1960-61, dir., 1961-73, AUI distinguished Scientist, 1973—; cons. labs. AEC; Morris Loeb lectr Harvard U., 1955; adj. prof. physics SUNY, Stony Brook, 1965—; Mem. nuclear sci. com. NRC. Assoc. editor: Phys. Rev, 1951-53; Contbr. articles on nuclear physics to sci. jours. Mem. bd. govs. Weizmann Inst. Sci., Rehovoth, Israel, Tel Aviv U.; trustee Univs. Research Assn. Recipient citation for meritorious contbns. U.S. AEC, 1973, J. Robert Oppenheimer meml. prize, 1982. Fellow Am. Phys. Soc. (pres. 1982), Am. Acad. Arts and Scis., AAAS; mem. Nat. Acad. Sci., Am. Philos. Soc. (Tom W. Bonner prize in nuclear physics 1971). Subspecialty: Nuclear physics. Home: 91 S Gillette Ave Bayport NY 11705 Office: Brookhaven Nat Laboratory Upton NY 11973

GOLDIN, STANLEY MICHAEL, neuroscientist, educator; b. N.Y.C., Dec. 17, 1948; s. Abraham and Ethel (Feinstein) G.; m. Kathleen Sweadner, Jan. 5, 1975. B.S. in Chem. Engring, M.I.T., 1970; M.S. in Chem. Engring., MIT, 1970; Ph.D. in Biochemistry, Harvard U., 1977. Jr. fellow Harvard U., Cambridge, Mass., 1976-78, asst. prof. pharmacology Med. Sch., Boston, 1979-83, assoc. prof., 1983—. Contbr. articles to profl. jours. Recipient Alfred P. Sloan award in neurosci., 1979, McKnight Scholar's award, 1981, Searle Scholar's award, 1981. Mem. Soc. Neurosci., Sigma Xi, Phi Lambda Upsilon. Subspecialties: Membrane biology; Neurobiology. Current work: Molecular basis of regulation of neuronal electrical activity; ion gates and pumps; nerve cell membranes; membrane reconstitution; immunocytochemical localization. Patentee biologically active membrane material. Office: 250 Longwood Ave Boston MA 02115

GOLDMAN, ALEXANDER, psychologist; b. Bklyn., Apr. 3, 1910; s. Harry and Bertha (Bloch) G.; m. Florence Halperin, Jan. 19, 1944; children: Harriet Elen; m.; 1 son by previous marriage, Thomas J. B.S., L.I. U., 1931; M.A., Columbia U., 1947; Ph.D., NYU, 1959. Lic. psychologist, N.Y. Research psychologist U.S. Navy-O.N.R., Port Washington, N.Y., 1951-59; head human factors Republic Aviation, Farmingdale, N.Y., 1959-64; chief psychologist phys. medicine Meadowbrook Hosp., East Meadow, N.Y., 1964-66; sr. life scientist Grumman Aerospace, Osyter Bay, N.Y., 1968-72; psychologist U.S. VA, Bklyn., 1972-79; pvt. practice psychology, N.Y.C., 1979—. Contbr. articles to profl. jours. Trustee Inst. for Transp. Safety, Franklin Square, N.Y., 1980—. Served to 1st lt. U.S. Army, 1942-47. Recipient Founders Day award NYU, 1960; commendation U.S. VA, 1978. Mem. Human Factors Soc., Am. Psychol. Assn., Product Safety Assn. (bd. dirs. Los Angeles 1981—), System Safety Soc. (pres. 1977-78). Subspecialties: Human factors engineering; Neuropsychology. Current work: Application of behavioral sciences to design and safety in use of the products of technology to protect health, life, welfare of individuals. Address: 18-65 211 St Bayside NY 11360

GOLDMAN, ALLAN LARRY, physician, educator; b. Mpls., June 3, 1943; s. Oscar and Ruth G.; m. Barbara Elaine Francisco, Aug. 17, 1969; children: Lisa, Carrie, Jennifer, Lindsey. B.A., B.S., U. Minn., 1964; M.D., 1968. Prof., dir. div. pulmonary and critical care medicine U. South Fla. Coll. Medicine, Tampa, 1974—. Contbr. articles to profl. jours. Bd. dirs. Fla. Lung Assn., Jacksonville, 1974—. Served to maj. U.S. Army, 1969-74. Fellow ACP, Am. Coll. Chest Physicians; mem. Am. Thoracic Soc., N.Y. Acad. Scis., Phi Beta Kappa, Alpha Omega Alpha. Jewish. Subspecialties: Pulmonary medicine; Critical care. Current work: Lung cancer, asthma, smoking, pulmonary function testing. Home: 4915 Lyford Cay Rd Tampa FL 33609 Office. 13000 N 30th St Tampa FL 33612

GOLDMAN, ALLEN MARSHALL, physics educator, researcher; b. N.Y.C., Oct. 18, 1937; m., 1960; 3 children. A.B., Harvard U., 1958; Ph.D. in Physics, Stanford U., 1965. From asst. prof. to assoc. prof. U. Minn., Mpls., 1965-75, prof. physics, 1975—, now also co-dir. superconductivity lab. Alfred P. Sloan Found. fellow, 1966-70. Mem. Am. Phys. Soc., AAAS, Sigma Xi. Subspecialty: Condensed matter physics. Office: Sch Physics and Astronom U Minn Minneapoli MN 55455

GOLDMAN, ARMOND SAMUEL, physician, educator, researcher; b. San Angelo, Tex., May 26, 1930; s. David and Rose (Gottesfeld) G.; m. Barbara Jean Goldman, July 31, 1950; children: Lynn, David, Daniel, Paul, Robert. M.D., U. Tex., Galveston, 1953. Diplomate: Am. Bd. Pediatrics, Am. Bd. Allergy and Immunology. Instr. dept. pediatrics U. Tex. Med. Br., Galveston, 1959-60, asst. prof., 1960-67, assoc. prof., 1967-72, prof. dept. human biol. chemistry and genetics, 1972—, dir. div. immunology, 1960—. Contbr. articles to profl. jours. Served to capt. M.C. U.S. Army, 1955-57. Mem. Am. Acad. Pediatrics, Soc. Pediatric Research, Am. Pediatric Soc., Am. Assn. Immunologists, Reticuloendothelial Soc., Sigma Xi. Subspecialties: Pediatrics; Immunology (medicine). Current work: Immunologic system in human milk; pathogenesis of immunodeficiency; ontogeny of immunity. Home: 3801 Pine Manor Dickinson TX 77539 Office: U Tex Med Br Galveston TX 77550

GOLDMAN, IRA STEVEN, gastroenterologist, educator, researcher; b. N.Y.C., May 19, 1951; s. George David and Belle (Hans) G.; m. Niki E Kantrowitz, Jan. 20, 1980. B.A., U. Rochester, N.Y., 1973; M.D., Columbia U., 1977. Diplomate: Am. Bd. Internal Medicine. Intern Columbia Presbyn. Hosp., N.Y.C., 1977-78, resident, 1978-80; fellow U. Calif.-San Francisco, 1980-83, asst. prof. medicine, 1983—. Contbr. articles to med. jours. Eisenhower fellow, 1972; Am. Liver Found. fellow, 1982. Mem. ACP, Am. Fedn. Clin. Research, Am. Gastroent. Assn., Am. Liver Found., AMA. Subspecialties: Gastroenterology; Internal medicine. Current work: Protein and vesicular transport systems in the liver; role of the cytoskeleton in these processes. Home: 400 Upper Terr San Francisco CA 94117 Office: U Calif 1120 H SW San Francisco CA 94143

GOLDMAN, JOSEPH ILYA, theoretical physicist; b. Moscow, July 26, 1926; U.S., 1979; s. Ilya Joseph and Hannah Leibovna (Cherniak) G.; m.; 1 son: Vladimir. M.S. with honors, Moscow State U., 1948; Ph.D. in theoretical physics, Kurchatov Atomic Energy Inst., Moscow, 1952. Lectr. Moscow State U., 1952-54, U. Yerevan, USSR, 1954-65; researcher, sr. researcher Theoretical dept. Inst. Cosmic Ray Studies, Yerevan, 1952-63; sr. research scientist ARUS Physics Inst., Yerevan, 1963-79; vis. research assoc. dept. physics Johns Hopkins U., Balt., 1980; adj. prof. dept. physics Am. U., Washington, 1980—, sr. research scientists, 1980—; cons. Astrophys. Obs. Armenian Acad. Scis., Bjurakan, USSR, 1952-65. Author: (with V. Krivchenkov) Problems in Quantum Mechanics, 1957, 3d English edit., 1975; Contbr. articles to profl. jours. Recipient Moscow State U. prize, 1955; USSR Acad. Sci. prize, 1965. Mem. Am. Phys. Soc., European Phys. Soc., Israel Phys. Soc. Jewish. Subspecialties: Theoretical physics; Particle physics. Current work: High energy nuclear collision, possible nuclear quark phase transitions, commutation relations in QED and non-Abelian quantum field theory. Home: 7308 Meadow Ln Chevy Chase MD 20815 Office: American University Department of Physics Washington DC 20016

GOLDMAN, JOSEPH L., scientist-technical director, consultant; b. San Francisco, Aug. 25, 1932; s. Samuel and Charlotte (Malamud) G.; m. Sylvia Shiffman, 1955 (div. 1977); children: Rachel Ann, Charles Israel, Michelle Sandra. Student, UCLA, 1955-56; B.S., Tex. A&M U., 1958, M.S., 1960; postgrad., U. Chgo., 1960-62; Ph.D., U. Okla., 1971. Cert. consulting meteorologist. Mathematician West Coast Research Lab., Los Angeles, 1955-56; research asst. to scientist Tex. A&M U., 1956-60; research meteorologist U. Chgo., 1960-65; prin. investigator Nat. Engring. Sci. Co., Houston, 1965-66; exec. v.p. Inst. Storm Research, Houston, 1966-75; tech. dir. Internat. Ctr. for the Solution of Environ. Problems, Houston, 1976—; mng. dir. Sea & Storm Service Specialists Ltd., London, 1973-79; assoc. prof. U. St. Thomas, Houston, 1967-76. Contbr. chpts. to books and numerous articles to profl. jours. Bd. dirs. ARC, Houston, 1976-82; gen. chmn. Sci. & Engring. Fair, Houston, 1976—. Served with USAF, 1951-55. NSF grantee, 1971-72. Fellow Royal Meteorol. Soc.; mem. Am. Meteorol. Soc., Am. Geophysical Union, N.Y. Acad. Sci., AAAS, Engrs. Council Houston (v.p., exec. bd. 1975-82), Marine Tech. Soc., Sigma Xi (dir. chpt. 1982). Lodge: B'nai B'rith. Subspecialties: Meteorology. Current work: Risk analysis, projections of urban growth and related problems-health, hazards, transportation, waste utilization, structural integrity, water, socio-economics, energy, climate. Originator Sculpture of Storm motion, 1967. Office: International Center for the Solution of Enviromental Problems 3818 Graustark St Houston TX 77006

GOLDMAN, LAWRENCE, physiologist, educator; b. Boston, May 6, 1936; s. Theodore T. and Sophye (Altshuler) G.; m. Faith Kordis, July 11, 1968 (div.); 1 dau. Ann Kordis. B.S. summa cum laude, Tufts U., 1958; Ph.D., UCLA, 1964. NIH postdoctoral trainee dept. neurology Columbia U. Coll. Physicians and Surgeons, 1964-65; asst. prof. dept. zoology U. Md., 1965-67, asst. prof. dept. physiology, 1967-70, assoc. prof., 1970-77, prof., 1977—, prof. biophysics, 1977—; NATO sr. fellow in sci. Queen Mary Coll., U. London, 1970; NIH spl. research fellow Cambridge (Eng.) U., 1973; cons. NIH, NSF. Contbr. numerous articles to sci. jours. Grantee NIH, NATO, NSF, 1966—. Fellow AAAS; mem. Am. Physiol. Soc., Biophys. Soc., Soc. Gen. Physiologists (rep. to Nat. Soc. Med. Research Council), Soc. Neurosci., N.Y. Acad. Sci., Am. Inst. Biol. Sci., Sigma Xi, Phi Beta Kappa. Subspecialty: Biophysics (biology). Current work: Membrane biophysics, study of generation and conduction of nervous impulse. Home: 3502 Newland Rd Baltimore MD 21218 Office: Dept Physiology Sch Medicine U Md Baltimore MD 21201

GOLDMAN, LEE, physician, educator, researcher; b. Phila., Jan, 6, 1948; s. Marvin and Kathryn (Schwartz) G.; m. Jill Steinhardt, Mar. 21, 1971; children: Jeff, Daniel, Robyn Sue. B.A., Yale U., 1969, M.D., 1973, M.P.H., 1973. Diplomate: Am. Bd. Internal Medicine. Intern U. Calif.-San Francisco, 1973-74, resident in med., 1974-75, Mass. Gen. Hosp., Boston, 1975-76; fellow in cardiology Yale-New Haven Hosp., 1976-78; asst. prof. medicine Harvard U. Med. Sch., 1978-83, assoc, prof., 1983—; asst. physician-in-chief dept. med. Brigham and Women's Hosp., Boston, 1983. Contrbr. numerous articles to profl. jours. Bd. dirs. Temple Shir Tikva, Wayland, Mass., 1982—. ACP teaching and research scholar, 1980-83; Henry J. Kaiser FamilyFound. scholar, 1982—. Fellow ACP and Am. Coll. Cardiology; mem. Am. Fedn. Clin. Research, Soc. Med. Decision Making, Soc. Research and Edn. Primary Care Internal Medicine. Democrat. Jewish. Subspecialties: Internal medicine; Cardiology. Current work: Application of multivariate analytic techniques to problems in medical decision making and medical diagnosis. Office: Dept of Medicine Brigham and Women's Hosp 75 Francis St Boston MA 02115

GOLDMAN, LEE WILLIAM, med. physicist; b. Cape May, N.J., Feb. 27, 1951; s. Barnet and Rita G.; m. Cynthia Ann Weiner, Jan. 12, 1975; 1 son: Jacob. B.S. in Physics, U. Md., 1973; M.S. in E.E, George Washington U., 1979. Radiol. physicist George Washington U., 1973-75; med. physicist Bur. Radiol. Health, FDA, 1975-82; med. physicist in research and devel. Elscint Corp., Columbia, Md., 1982—. Contbr. articles to profl. jours. Mem. Am. Assn. Physicists in Medicine, IEEE. Jewish. Subspecialties: Imaging technology; Software engineering. Current work: Medical imaging quality control, computerized radiological treatment planning, x-ray quality control, computerized radiation treatment planning, medical image processing. Home: 603 Twinbrook Pkwy Rockville MD 20851 Office: 8919 McGaw Ct Columbia MD 21045

GOLDMAN, LEON, hospital laser center and laboratory director; b. Cin., Dec. 7, 1905; m. Belle Hurwitz, Aug. 23, 1933; children: John, Steven, Carol. M.D., U. Cin., 1929. Intern. Univ. Hosp., Cin., 1929-30, resident in dermatology, 1930-36; mem. faculty U. Cin., 1937—, prof. dermatology, 1947-76, prof. emeritus, 1976—, dir., 1961-76, Laser Treatment Ctr., Jewish Hosp., Cin., 1980—, 1980—. Contbr. over 200 articles to med. publs. Served to lt. Med. Res., 1940. Recipient Father of Laser Medicine award, Germany, 1980, Japan, 1981; Devel. of Laser Medicine award Laser Inst. Am., 1980. Mem. Am. Soc. Laser Medicine and Surgery (dir. 1981—, Meml. award 1981), Am. Soc. Dermatologic Surgery (historian 1982-83), Internat. Fedn. Laser Medicine (adv. council 1981—). Subspecialties: Dermatology; Laser medicine. Current work: Laser research and development. Home: 2324 Madison Rd Apt 1807 Cincinnati OH 45208 Office: 711 Carew Tower Cincinnati OH 45202

GOLDMAN, MARVIN, biologist, educator; b. N.Y.C., May 2, 1928; s. Sidney Albert and Mary (Wind) G.; m. Joyce Weiss, Aug. 30, 1953; children: David L., Beth L., Robert S. A.B., Adelphi U., 1949; M.S., U. Md., 1951; Ph.D., U. Rochester, 1957. Biologist NIH, Bethesda, Md., 1951-52; scientist U. Rochester, 1957-58; research radiobiologist U. Calif., Davis, 1958—, lectr., 1959-65, prof. radiobiology, 1972—; dir. Lab. for Energy-Related Health Research, 1973—; cons. to Nuclear Regulatory Commn. Contbr. numerous articles to profl. jours. Recipient E.O. Lawrence award AEC, 1972, citation ERDA, 1977. Subspecialties: Nuclear fission; Environmental toxicology. Current work: Health risk assessment from energy-related agents, metabolic, dosimetric, pathologic effects of exposure to ionizing radiations and fossil fuel combustion products, cancer risk. Patentee in field. Office: Laboratory for Energy Related Health Research University of California Davis CA 95616

GOLDMAN, STEPHEN SHEPARD, medical educator, researcher; b. Brockton, Mass., Nov. 24, 1941; s. Maurice and Ethel (Kramer) G.; m. Karen Mary Tack, May 27, 1977. B.A., Northeastern U, 1964; M.S., U. Ill, 1966; Ph.D., U. Ill., 1970. NIH staff fellow NIH, Bethesda, Md., 1970-78; asst. prof. dept. ophthalmology NYU Sch. Medicine, N.Y.C., 1978—. Fellow Stroke Council of Am. Heart Assn.; mem. Am. Physiol. Soc., Am. Soc. Neurochemistry, N.Y. Acad. Scis., AAAS. Subspecialties: Neurochemistry; PET scan. Current work: Cerebrovascular disease; spinal cord injury and brain tumor diagnosis. Office: 550 1st Ave New York NY 10016

GOLDNER, RONALD B., applied physicist, educator, cons., researcher; b. N.Y.C., Mar. 24, 1935; s. Jack and Ethel (Creditor) G.; m. Judith Olef, June 21, 1959; children: Eric Lee, Rachel Ann, Mark Allen. B. S., M.I.T., 1957, M.S., 1957, E.E., 1959; Ph.D., Purdue U., 1962. Teaching asst., research assoc. M.I.T., Cambridge, 1957-59, asst. prof., Ford Found. postdoctoral fellow, 1962-64, vis. asst. prof., 1965, 66, vis. assoc. prof., 1968, cons., 1979—; instr. Purdue U., 1959-62; asst. prof. elec. engring. Tufts U., Medford, Mass., 1964-67, assoc. prof., 1967-75, prof., 1975—; cons. New Eng. Instrument Co., 1963—, Geometrics, Inc., 1965-67, Raytheon Co., 1969-71, Exxon Research, 1974, Tyco Labs., 1974, G.T.E. Labs., 1975, U. Calif.-Los Alamos Nat. Lab., 1981-82. Book reviewer.; Contbr. articles to profl. jours; letters reviewer: Applied Physics. Served to lt., Signal Corps U.S. Army, 1959-60. Mem. Am. Phys. Soc., IEEE, Solar Energy Soc., Electrochem. Soc., Optical Soc. Am., Soc. Photo-Optical Instrumentation Engrs. Subspecialties: Electronic materials; Optoelectronics materials/devices/systems. Current work: Optoelectronic materials and applications, including solar energy conversion. Patentee magnetic shift register. Office: Dept Elec Engring Tufts U Medford MA 02155

GOLDREICH, PETER MARTIN, educator, astrophysicist; b. N.Y.C., July 14, 1939; s. Paul and Edith (Rosenfield) G.; m. Susan Kroll, June 14, 1960; children—Eric, Daniel. B.Engring. Physics, Cornell U., 1960, Ph.D. in Physics, 1963. Instr. Cornell U., 1963; postdoctoral fellow Cambridge (Eng.) U., 1963-64; asst. prof. astronomy and geophysics U. Calif. at Los Angeles, 1964-66; asso. prof. astronomy and planetary sci. Calif. Inst. Tech., 1966-69, prof., 1969—, Lee A. Dubridge prof. astrophysics and planetary physics, 1981—. Mem. Nat. Acad. Sci., Am. Acad. Arts and Scis. Subspecialty: Planetary science. Home: 999 San Pasqual Apt 7 Pasadena CA 91106 Office: 1201 E California Blvd Pasadena CA 91109

GOLDRICH, STANLEY GILBERT, optometrist, educator, experimental psychologist; b. N.Y.C., Sept. 22, 1937; s. Joseph and Doris (Stelzner) G. B.A., Queens Coll., 1959, M.A., 1965; Ph.D., CUNY, 1966; O.D., Mass. Coll. Optometry, 1974. Lic. optometrist, N.Y., Calif., Mass. Lectr. Queens Coll., 1964-65; research assoc. U. Wis.-Madison, 1965-67; asst. prof. Ohio State U., 1967-72; asst. prof. psychology SUNY-State Coll Optometry, N.Y.C., 1974—; cons. vision care AFL/CIO, N.Y.C., 1981—. NSF grantee, 1968-71; Optometric Extention Program Found. grantee, 1976; Optometric Ctr. Found. grantee, 1979-81. Fellow Am. Acad. Optometry; mem. Am. Psychol. Assn., Am. Optometric Assn., N.Y. State Optometric Assn. Jewish. Subspecialties: Optometry; Physiological psychology. Current work: Therapy for eye movement disorders using electronic and optical devices invented by myself; investigating responsiveness of Strabismus and Nystagmus eye disorders to new form of biofeedback treatment; investigating role of eye movement in learning and reading disabilities and dyslexia. Inventor orthotone eye control biofeedback, Goldrich contour rotator, emergent textural control; originator oculomotor biofeedback therapy technique, 1977. Home: 150 Lexington Ave New York NY 10016 Office: State Coll Optometry SUNY 100 E 24th St New York NY 10010

GOLDSBERRY, FRED LYNN, oil exploration exec.; b. San Antonio, Jan. 3, 1947; s. William Donald and Freda Mae (Case) G.; m. Sharon Faye Williams, Dec. 20, 1969; children: Jeanette Louise, Dennis Hamilton. B.S.M.E. with honors, Tex. A&M U., 1968, M.S.M.E., 1969, Ph.D., 1971. Registered profl. engr., Tex. Research technologist Tex. Petroleum Research Co., College Station, Tex., 1970-71; research engr. ENSERCH Corp., Dallas, 1971-73; supervising engr. TUCO, Inc., Amarillo, Tex., 1973-75; spl. projects engr. J. M. Huber Corp., Borger, Tex., 1975-76; program mgr. Hanford (Wash.) site ERDA, 1976-79, dir., Houston, 1976—; adv. to Common Market, 1982. Contbr. articles to profl. jours. Bd. dirs. Epilepsy Found., Richland, Wash., 1977-79. Recipient award U.S. Dept. Energy, 1979. Mem. Tau Beta Pi, Pi Tau Sigma, Phi Eta Sigma, Phi Kappa Phi. Baptist. Subspecialties: Petroleum engineering; Geothermal power. Current work: Reservoir and production engineering. Patentee in field.

GOLDSCHMIDT, MILLICENT EDNA, microbiology educator, researcher; b. Erie, Pa., June 11, 1926; d. Isaac Jerry and Mary Tillie (Semuel) Cohen; m. Eugene P. Goldschmidt, Apr. 10, 1949 (dec. May 1980); children: Richard B., Carol Goldschmidt Warley. B.A., Western Res. U., 1947; M.S., Purdue U., 1950, Ph.D., 1952. Cert. specialist in med. microbiology and pub. health Nat. Registry Microbiologists, Am. Soc. Microbiology. Reseach instr. Baylor U., 1963-67; acting chief clin. microbiology, asst. prof. U. Tex. System Cancer Ctr., Houston, 1967-71; assoc. prof. U. Tex. Health Sci. Ctr., Houston, 1969-76, 79—; assoc. prof. U. Tex. Health Sci. Ctr., Med. Sch., 1973-76; research fellow NIH Dental Br., 1977-79; mem. sci. rev. bd. Sci. and Engring. Fair, Houston, 1977—; cons. to industry; ofcl. cons. news media Am. Soc. Microbiology, 1976—. Author: Microbiology for allied Health Students, 1983; sr. author: Rapid Methods for Microbiology, 1983; contbr. articles sci. jours. Bd. dirs. Houston LWV, 1963-66; co-leader Sci. Explorer post Boy Scouts Am., 1969-71; vol. Am. Cancer Soc., 1973-76. Fellow Am. Inst. Chemists, Am. Acad. Microbiology; mem. Am. Soc. Microbiology (pres. Tex. br. 1980-81, councillor 1983—, Disting. Service award 1980, Appreciation plaque 1975-81), Houston Assn. Med. Microbiologists (pres. 1982—), Assn. Women in Sci., Sierra Club, Audubon Soc., Sigma Xi, Sigma Delta Epsilon. Jewish. Clubs: Brandeis Women's Orgn., Hadassah. Current work: Instrumentation and other rapid methods for detection and characterization of microorganisms; dental, medical and biochemical and biophysical aspects of microbiology; teaching microbial physiology, medical bacteriology and medical mycology. Office: Tex Health Science Center Dental Br Dental Science Inst PO Box 20068 Houston TX 77225

GOLDSCHMIDT, RAUL MAX, microbial molecular geneticist, educator; b. Santiago, Chile, Sept. 23, 1941; s. Franz and Lore (Vasen) G. Licenciado in Physics, Universidad de Chile, 1964; M.Sc. in Biology, Mass. Inst. Tech., 1967, Ph.D., Columbia U., 1970. Assoc. prof. microbiology Facultad de Ciencias, Universidad de Chile, Santiago, 1971-73; research assoc. microbiology dept. Med. Ctr., U. Ala., Birmingham, 1974, vis. prof. microbiology, 1974—. Contbr. numerous sci. articles to profl. publs. FDA grantee, 1979-82. Mem. Am. Soc. Microbiology. Jewish. Subspecialties: Genetics and genetic engineering (biology); Molecular biology. Current work: Genetic and molecular characterization of bacterial plasmids that specify resistance to antibiotics in enterobacteria and in vivo expression of genes contained on bacterial plasmids. Office: U Ala Med Ct Dept Microbiology Birmingham AL 35294

GOLDSCHMIED, FABIO RENZO, consulting engineer; b. Trieste, Venezia Giulia, Italy, Oct. 25, 1919; came to U.S., 1939, naturalized, 1943; s. Rodolfo and Ada (Frankel) G.; m. Marie P. Perfumo, Mar. 14, 1942; 1 dau., Wanda Ada. Student, Swiss Fed. Poly., Zurich, 1937-39, M.I.T., 1943; B.Sc., Columbia U., 1947, M.Sc., 1948. Registered profl.

engr., Pa., Utah, Mass. Adv. scientist Westinghouse Electric Corp., Pitts., 1968-81; research prof. U. Utah, Salt Lake City, 1965-68; research aerodynamicist Sperry Rand Corp., Blue Bell, Pa. and Salt Lake City, 1962-65; mgr. advanced devel. engring. Westinghouse Electric Corp., Boston, 1954-61; cons. engr. F. R. Goldschmied P.E., Monroeville, Pa., 1981—; cons. M.I.T., Electric Power Research Inst., David W. Taylor Naval Ship Research and Devel. Ctr. Contbr. articles to profl. jours. Served with U.S. Army, 1942-46; ETO. Assoc. fellow AIAA (assoc. editor Jour. Hydronautics 1980-82); mem. ASME, Soc. Naval Architects Marine Engrs. Democrat. Jewish. Subspecialties: Fluid mechanics; Systems engineering. Current work: Optimum integration of hull design, boundary-layer control, propulsion and stability for submerged vehicles such as submarines, torpedoes, and airships. Patentee in field. Home: 1782 McClure Rd Monroeville PA 15146

GOLDSMITH, DOUGLAS HOWARD, oral and maxillofacial surgeon; b. Los Angeles, May 26, 1949; s. Leon R. and Edith (Fish) G.; m. Catherine L. Snerling, Aug. 19, 1973; children: Lisa, Sarah. D.D.S. cum laude, UCLA, 1974. Diplomate: Am. Bd. Oral and Maxillofacial Surgery. Resident in oral and maxillofacial surgery Montefiore Hosp. and Med. Ctr., Bronx, N.Y., 1974-76, chief resident in oral and maxillofacial surgery, 1976-77, adj. attending oral and maxillofacial surgeon, 1977—, oral and maxillofacial surgeon, 1977—; pvt. practice dentistry, specializing in oral and maxillofacial surgery, Scarsdale, N.Y., 1980—; instr. gross anatomy course Albert Einstein Coll. Medicine, 1977-78, asst. prof. surgery, 1977-80, assoc. clin. prof., 1981, Columbia Presbyn. Sch. Dentistry; mem. staff, adj. attending oral and maxillofacial surgeon North Central Bronx Hosp.; oral and maxillofacial surgery cons. Norwalk (Conn.) Hosp.; lectr. dept. oral and maxillofacial surgery NYU; mem. staff Beth Abraham Hosp., Bronx Coll. Hosp., Albert Einstein Coll. Medicine, White Plains (N.Y.) Hosp. and Med. Ctr., Columbia Presbyn. Hosp. and Med. Ctr., N.Y.C.; lectr., presenter in field; guest lectr. and surgeon, various hosps., 1977-80. Contbr. articles, chpts. to profl. publs. Recipient 2d place award Am. Acad. History Dentistry Writing Contest, 1973; Am. Inst. Oral Biology fellow, 1973. Mem. Student ADA, So. Calif. Acad. Oral Pathology, Am. Dental Soc. Anesthesiology, Am. Inst. Oral Biology, Internat. Assn. Dental Research, First Dist. Dental Soc. (N.Y.), Am. Cleft Palate Soc., Am. Soc. Oral and Maxillofacial Surgery, N.Y. State Soc. Oral and Maxillofacial Surgery, Sigma Xi, Omicron Kappa Upsilon, Phi Eta Sigma, Alpha Omega. Subspecialties: Oral and maxillofacial surgery; Dental growth and development. Home: 1215 The Colony Hartsdale NY 10530 Office: Dr. Douglas H Goldsmith DDS PC 495 Central Park Ave Scarsdale NY 10583

GOLDSMITH, MICHAEL ALLEN, oncologist, cancer researcher; b. Bronx, N.Y., Jan. 28, 1946; s. Walter and Bertha (Tannenberg) G.; m. Judith Harriet Plaut, June 6,1971; children: Sharon, Esther, Eva, Steven. B.A., Yeshiva U., 1967; M.D., Albert Einstein Coll. Medicine, 1971. Cert. Am. Bd. Internal Medicine (subcert. in med. oncology). Intern Bronx Mcpl. Hosp. Ctr., 1971-72; staff assoc. Nat. Cancer Inst., Bethesda, 1972-74; resident Mt. Sinai Hosp., N.Y.C., 1974-75, neoplastic disease fellow, 1975-77, asst. clin. prof. medicine and neoplastic diseases, 1977—; practice medicine specializing in oncology, N.Y.C., 1977—; cons. Bronx VA Hosp., 1977—. Vice-pres. Congregation Orach Chaim, N.Y.C., 1978—. Served with USPHS, 1972-74. Recipient Upjohn Achievement award Upjohn Co., 1971. Fellow ACP; mem. Am. Soc. Clin. Oncology, Am. Assn. Cancer Research, N.Y. Acad. Scis., AMA. Subspecialties: Oncology; Chemotherapy. Current work: Clinical cancer chemotherapy, clinical trials of new anticancer agents. Office: Oncology Cons PC 1045 Fifth Ave New York NY 10028

GOLDSMITH, STANLEY JOSEPH, nuclear medicine physician, educator; b. Bklyn., Aug. 17, 1937; s. Jack and May (Greenzweig) G.; m. Miriam Schulman, June 6, 1969; children: Ira, Arthur, Beth, Mark. B.A., Columbia U., 1958; M.D., SUNY-Downstate Med. Ctr., 1962. Diplomate: Am. Bd. Internal Medicine, Am. Bd. Nuclear Medicine. Intern SUNY-Kings County Med. Center, Bklyn., 1962-63, resident, 1965-66, chief resident, 1966-67; fellow in endocrinology Mt. Sinai Hosp., N.Y.C, 1967-68, dir., 1973—; research assoc. radioisotope service Bronx (N.Y.) VA Hosp., 1968-69; dir. nuclear medicine, asst. dir. endocrine dept. Nassau County Med. Ctr., East Meadow, N.Y., 1969-73; asst. prof. medicine radiology SUNY-Stony Brook Health Sci. Ctr., 1971-73; asst. prof. medicine Mt. Sinai Sch. Medicine, 1973-76, assoc. prof., 1976—; research collborator Brookhaven Nat. Labs., Upton, N.Y., 1971-75; cons. nuclear medicine; cons. dept. health State of N.Y., 1973-77, Health Services Adminstrn., N.Y.C., 1976. Editor-in-chief: Newsline; mem. editorial bds.: Am. Jour. Cardiology, 1978-82, Picker Jour. Nuclear Medicine Instrumentation, 1980; reviewer: Israeli Jour. Med. Scis, 1979, Jour. AMA, 1983—. Served to capt. U.S. Army, 1963-65. Fellow Am. Coll. Cardiology, ACP, Am. Coll. Nuclear Physicians (chmn. nuclear med. tech. affairs); mem. AAAS, Am. Fedn. Clin. Research, Endocrine Soc., N.Y. Acad. Scis., Radiol. Soc. N.Am., Soc. Nuclear Medicine (sec. Greater N.Y. chpt. 1975-78, pres. 1979-80, mem. nuclear med. tech. cert. bd. 1979-81). Subspecialties: Nuclear medicine; Internal medicine. Current work: Radionuclidic and NMR quantitation of organ function; quantitative analysis of medical diagnosis and management. Home: 72 Ivy Way Port Washington NY 11050 Office: Mt Sinai Med Center One Gustave L Levy Pl New York NY 10029

GOLDSMITH, WILLIAM ALEE, environmental engineer; b. Memphis, Nov. 5, 1941; s. Jack Gene and Louise Elizabeth (Alston) G.; m. LaVance Davis, June 1, 1965; children: Jack Gregory, William Vance, Lara Ellen. B.S., Miss. State U., 1964, M.S., 1966; Ph.D., U. Fla., 1968. San. engr. U.S. EPA, Dallas, 1971-73; asst. prof. U. So. Miss., Hattiesburg, 1973-74; staff mem. Los Alamos Sci. Lab., 1974-75; research staff mem. Oak Ridge Nat. Lab., 1975-81, remedial action survey and certification activities program mgr., 1981-82; engring. specialist Bechtel Nat., Inc., Oak Ridge, 1982—. Treas. Covenant Presbyn. Ch., Oak Ridge, 1978; stewardship chmn. First United Meth. Ch., Oak Ridge, 1981—. Served to capt. AUS, 1969-71. Mem. Am. Chem. Soc., Am. Nuclear Soc., AAAS, Health Physics Soc. (councilman East Tenn. chpt. 1979-82), N.Y. Acad. Scis., Sigma Xi, Tau Beta Pi, Phi Kappa Phi. Subspecialties: Nuclear fission; Environmental engineering. Current work: Development of environmental health physics monitoring programs and techniques, development of monitoring techniques and instrumentation for radiation dosimetry. Patentee technique for Ac Measurement, cryogenic sampler, mobile gamma-ray van. Office: Bechtel Nat Inc 800 Oak Ridge Turnpike Oak Ridge TN 37830

GOLDSTEIN, ALBERT, med. physicist, educator; b. N.Y.C., May 26, 1938; s. Frank and Belle (Altman) G.; m. Anita Mammano, Sept. 8, 1968; children: Jesse, Eva. B.S., CCNY, 1960; Ph.D., M.I.T., 1965. Researcher L'Ecole Normale Superievre, Paris, 1965-67; staff mem. IBM, Yorktown Heights, N.Y., 1967-71; lectr. physics CCNY, 1971-72; asst. prof. radiology U. Kans., Kansas City, 1972-76; head, div. med. physics Henry Ford Hosp., Detroit, 1976—; adj. assoc. prof. radiology Wayne State U., 1980—. Contbr. articles to profl. jours.; Mem. editorial bd.: Jour. Ultrasound in Medicine, 1982—. Mem. Am. Assn. Physicists in Medicine, Am. Inst. Ultrasound Medicine, Am. Coll. Radiology, IEEE, Mich. Radiol. Soc., Radiol. Soc. N. Am. Subspecialties: Imaging technology; Acoustics. Current work: Physicist in quality assurance of diagnostic ultrasound. Office: Henry Ford Hospital 2799 W Grand Blvd Detroit MI 48202

GOLDSTEIN, BYRON BERNARD, research scientist; b. N.Y.C., Nov. 24, 1939; s. David and Mary (Posnak) G. B.S., CCNY, 1961; Ph.D., N.Y. U., 1967. Assoc. prof. physics Fairleigh Dickinson U., Teaneck, N.J., 1968-75, prof., 1975-77; mem. staff Los Alamos (N.Mex.) Nat. Lab., 1977—. NIH fellow, 1969; grantee, 1980. Mem. AAAS, N.Y. Acad. Sci., Am. Phys. Soc., Am. Assn. Immunologists, Biophys. Soc. Subspecialties: Biophysics (physics); Cell biology. Current work: Applications of mathematics to problems in cellular and molecular biology. Home: 105 Calle Golondrina Santa Fe NM 87501 Office: Los Alamos Nat Lab T-10 Los Alamos NM 87545

GOLDSTEIN, CHARLES IRWIN, applied mathematician, researcher; b. N.Y.C., Nov. 21, 1940; s. Emil and Fay (Greenberg) G.; m. Joyce Kluback, May 31, 1975; children: Andrew, Roger. B.S., CCNY, 1962; M.S., Courant Inst. Math. Scis., NYU, 1964, Ph.D., 1967. Instr. in math. NYU, 1966-67; research mathematician Brookhaven Nat. Lab., Upton, N.Y., 1967—, leader math. research group, dept. applied math., 1975—; vis. researcher in math. Math. Research Ctr., U. Wis.-Madison, 1971, Icase-NASA Langley Research Ctr., Hampton, VA., 1978, dept. math. U. Calif.-Berkeley, 1979, Naval Research Lab., Washington, 1983. N.Y. State Regents scholar, 1958-62. Mem. Am. Math. Soc., Soc. Indsl and Applied Math. Subspecialty: Applied mathematics. Current work: Research in partial differential equations, scattering theory, numerical analysis (finite element methods), wave propagation (acoustic and electromagnetic). Home: 49 Empress Pines Dr Lake Ronkonkoma NY 11779 Office: Brookhaven Nat Lab 61 Brookhaven Ave Upton NY 11973

GOLDSTEIN, DAVID JOEL, med. geneticist, educator; b. N.Y.C., June 25, 1947; s. Milton S. and Thelma (Weinman) G.; m. Pat Brimer, June 3, 1973; children: Benjamin Avry Brimer, Philip Jon. B.A., Franklin and Marshall Coll., 1969; M.D., U. Tenn., 1973, Ph.D., 1975. Resident in pediatrics Mayo Grad. Sch. Medicine, Rochester, Minn., 1975-78; fellow in human genetics U. Pa., Phila., 1978-81; asst. prof. med. genetics Ind. U. Med. Sch., Indpls., 1981—, also research. Contbr. articles to profl. jours. Recipient Quigley award in physiology U. Tenn. Med. Sch., 1973; NIH fellow in genetics, 1978. Mem. Am. Soc. Human Genetics, Soc. for Neurosci., AAAS. Subspecialties: Genetics and genetic engineering (medicine); Pediatrics. Current work: Biochemical genetics of essential hypertension, monoclonal antibodies. Office: Dept Med Genetics Ind U Med Sch Indianapolis IN 46223

GOLDSTEIN, DORA BENEDICT, pharmacology educator; b. Milton, Mass., Apr. 25, 1922; d. George W. and Marjory (Pierce) Benedict; m. Avram Goldstein, Aug. 29, 1947; children: Margaret Wallace, Daniel, Joshua, Michael. Student, Bryn Mawr Coll., 1940-42; M.D., Harvard U., 1949. Research assoc. Stanford U. Sch. Medicine, 1955-70, adj. prof., 1970-78, prof., 1978—. Author: Pharmacology of Alcohol, 1983, also articles; editorial bd.: Psychopharmacology. Recipient Sci. Excellence award Research Soc. on Alcoholism, 1980. Mem. Am. Soc. Pharmacology and Exptl. Therapeutics, Am. Soc. Biol. Chemists, Research Soc. on Alcoholism (pres. 1979-81). Subspecialties: Pharmacology; Biophysics (biology). Current work: Effects of drugs on biomembranes; alcohol; alcoholism; drug addiction; membrane lipids; electron spin resonance. Office: Stanford Med Ct Dept Pharmacology Stanfor CA 94305

GOLDSTEIN, ELLIOTT S., zoology educator; b. Bklyn., July 7, 1942; s. William and Lottie (Roth) G.; m. Suzanne G. Kussner, July 7, 1963; children: Andrew Richard, Hyla Lynn. Nat. Cancer Inst. fellow M.I.T., Boston, 1972-74; asst. prof. Ariz. State U., Tempe, 1974-78, assoc. prof. genetics, 1978—. NIH research grantee, 1975-82. Mem. Genetics Soc. Am., Soc. Developmental Biology, Sigma Xi. Subspecialties: Gene actions; Molecular biology. Current work: Analysis of genetic control of development by use of cloned genes from Drosophila melanogaster. Home: 2342 E Alameda Tempe AZ 85282 Office: Zoology Dept Ariz State U Tempe AZ 85287

GOLDSTEIN, IRVING SOLOMON, wood chemistry educator; b. Bronx, N.Y., Aug. 20, 1921; s. Jacob and Jennie (Rathsprecher) G.; m. Helen Haft, Dec. 16, 1945; children: Ardath Ann Goldstein Weaver, Darra Jane Goldstein Crawford, Jared Haft. B.S., Renssalear Poly. Inst., 1941; M.S., Ill. Inst. Tech., 1944; Ph.D., Harvard U., 1948. Research chemist N.Am. Rayon Corp., 1948-51; sr. research chemist, mgr. wood chemistry research Koppers Co., Inc., Pitts., 1951-63; sr. research scientist Nalco Chem. Co., Chgo., 1963-66; mgr. paper research Continental Can Co., Chgo., 1966-68; prof. forest sci. Tex. A&M U., 1968-71; prof. wood and paper sci. N.C. State U., Raleigh, 1971—, head dept., 1971—. Author: Wood Technology: Chemical Aspects, 1977, Organic Chemicals from Biomass, 1981; contbr. numerous articles to profl. jours. Served to lt. USNR, 1942-46. Mem. Am. Chem. Soc., AAAS, TAPPI, Forest Products Research Soc., Soc. Wood Sci. and Technology, Sigma Xi, Zeta Beta Tau. Jewish. Subspecialties: Wood chemistry; Biomass (energy science and technology). Current work: Teaching, research, and consulting. Patentee in field. Home: 209 Glasgow Rd Cary NC 27511 Office: Dept Wood and Paper Sci NC State U Raleigh NC 27650

GOLDSTEIN, JACK, biochemist, educator; b. Phila., June 24, 1930; s. Max and Sara (Granet) G.; m. Ellen Krueger, July 7, 1967; children: Marcus, Andrea, Elissa. B.S., Bklyn. Coll. Pharmacy, 1952; M.N.S., Cornell U., 1957, Ph.D., 1959. Vis. investigator USPHS research assoc., asst. prof. Rockefeller U., N.Y.C., 1959-66; investigator, head cell biochemistry lab. L.F. Kimball Research Inst., N.Y. Blood Ctr., N.Y.C., 1966—; vis. assoc. prof. Yale U., New Haven, 1966-67; assoc. prof. biochemistry Cornell U. Med. Coll., N.Y.C., 1968—. Contbr. articles to profl. jours. Served with M.C. U.S. Army, 1953-55. Nat. Cancer Inst. fellow, 1959-61; NIH grantee, 1966-77; Office Naval Research grantee, 1979—; Greater N.Y. Blood Program grantee, 1978—. Mem. Am. Soc. Biol. Chemists, Am. Soc. Cell Biology, Harvey Soc., Sigma Xi. Subspecialties: Biochemistry (medicine); Hematology. Current work: Enzymatic conversion of type A and B Erythrocytes to type O; isolation and characterizations of biological response modifiers. Patentee enzymatic conversion of red cells for transfusion. Office: New York Blood Center 310 E 67th St New York NY 10021

GOLDSTEIN, JEFFREY MARC, psychopharmacologist; b. Bronx, N.Y., May 9, 1947; s. Joseph and Shirley (Scher) G; m. Robin; children: Kevin Allen, Neal David. B.S., Colo. State U., 1970; M.S., Seton Hall U., 1973; Ph.D., U. Del., 1980. Assoc. scientist dept. pharmacology Schering Corp., Bloomfield, N.J., 1970-76; research pharmacologist, biomed, research dept. ICI Americas, Inc., Wilmington, Del., 1976—. Contbr. articles to profl. jours. Mem. AAAS, N.Y. Acad. Sci., Am. Soc. Pharmacology and Exptl. Therapeutics, Soc. Neuroscience, Sigma Xi. Subspecialties: Neuropharmacology; Psychopharmacology. Current work: Antidepressant drugs. Home: 4 Curry Ct Wilmington DE 19810 Office: Stuart Pharms Div ICI Americas Inc Wilmington DE 19897

GOLDSTEIN, JEROME ARTHUR, math. educator; b. Pitts., Aug. 5, 1941; s. Morris and Henrietta (Vogel) G.; m. Elizabeth Zakucia, June 20, 1964; children: David Jonathan, Devra. B.S., Carnegie-Mellon U., 1963, M.S., 1964, Ph.D., 1967; S.M.D., Internat. Boswell Inst., Loyola U., New Orleans, 1973. Mem. Inst. Advanced Study, Princeton, N.J., 1967-68; prof. math. Tulane U., New Orleans, 1968—; vis. prof. U. Brasilia, Brazil, 1975, Fed. U. Rio de Janeiro, 1975, U. London, 1980-81. Editor: Partial Differential Equations and Related Topics, 1975. Recipient Research award Sigma Xi, 1972. Mem. Math. Assn. Am., Am. Math. Soc., Soc. Indsl. and Applied Math. (vis. lectr. 1981—), London Math. Soc., Math. Soc. Brazil. Subspecialties: Applied mathematics; Mathematical analysis. Current work: Partial differential equations, quantum theory, operator theory. Home: 1624 Fern St New Orleans LA 70118 Office: Math Dept Tulane U New Orleans LA 70118

GOLDSTEIN, JOEL WILLIAM, health science administrator, psychologist; b. Chgo., June 5, 1939; s. Harold and Adeline (Laskin) G.; m. Marcia Ruth Gray, June 4, 1963; children: Lauren Beth, Daniel Gray. B.A., Grinnell (Iowa) Coll., 1961; M.S., U. Kans., Lawrence, 1963, Ph.D., 1966. Asst. prof. Carnegie-Mellon U., Pitts., 1966-74; social sci. program specialist NIMH, Rockville, Md., 1974-75, exec. sec. basic sociocultural research review com., 1975-82, acting chief basic research review br., 1982, exec. sec. research scientist research review com., 1982—; panelist U.S. Office Personnel Mgmt., Bethesda, Md., 1977—; cons. applied social psychology, drug use, drug edn., 1968-75; reviewer NSF, Washington, 1974-75. Contbr. articles to profl. jours. Reviewer United Mental Services of Allegheny County, 1972-74. Recipient Outstanding Community Service award WQED-TV, Pitts., 1970; NIMH grantee, 1968; Falk Med. Fund grantee, 1968, 70; Nat. Inst. Drug Abuse grantee, 1973. Mem. Am. Psychol. Assn., Soc. Exptl. Social Psychology, Soc. Personality and Social Psychology, AAAS, Fed. Exec. and Profl. Assn. Subspecialties: Psychology research program planning; Neuroscience. Current work: Administer a peer review committee for development and support of mental health research scientists in psychiatry, psychology, neuroscience, epidemiology, social science, biochemistry, etc. Home: 11610 Gowrie Ct Potomac MD 20854 Office: NIMH 5600 Fishers Lane Rockville MD 20857

GOLDSTEIN, JOSEPH LEONARD, physician, genetics educator; b. Sumter, S.C., Apr. 18, 1940; s. Isadore E. and Fannie A. G. B.S., Washington and Lee U., Lexington, Va., 1962; M.D., U. Tex., Dallas, 1966; D.Sc. (hon.), U. Chgo., 1982, Rensselaer Poly. Inst., 1982. Intern, then resident in medicine Mass. Gen. Hosp., Boston, 1966-68; clin. assoc. NIH, 1968-70; postdoctoral fellow U. Wash., Seattle, 1970-72; mem. faculty U. Tex. Health Scis. Center, Dallas, 1972—, Paul J. Thomas prof. medicine, chmn. dept. molecular genetics, 1977—; Harvey Soc. lectr., 1977; mem. sci. rev. bd. Howard Hughes Med. Inst., 1978—. Co-author: The Metabolic Basis of Inherited Disease, 5th edit, 1983. Recipient Heinrich-Wieland prize, 1974, Pfizer award in enzyme chemistry Am. Chem. Soc., 1976; Passano award Johns Hopkins U., 1978; Gairdner Found. award, 1981; award in biol. and med. scis. N.Y. Acad. Scis., 1981; Lita Annenberg Hazen award, 1982. Mem. Nat. Acad. Scis. (Lounsbery award 1979), Assn. Am. Physicians, Am. Soc. Clin. Investigation, Am. Soc. Human Genetics, Am. Soc. Biol. Chemists, A.C.P., Am. Fedn. Clin. Research, Phi Beta Kappa, Alpha Omega Alpha. Subspecialty: Genetics and genetic engineering (medicine). Home: 3730 Holland Ave Apt H Dallas TX 75219 Office: 5323 Harry Hines Blvd Dallas TX 75235

GOLDSTEIN, JOYCE ALLENE, pharmacologist; b. Whittier, Calif., Mar. 8, 1941; d. Clarence and Herma May (Cotton) Winfrey; m. Robert Jay Goldstein, Aug. 10, 1963. B.S., S.W. Mo. State U., 1962; Ph.D., U. Tex. Med. Sch., 1967. Pharmacologist FDA/EPA, Atlanta, 1968-73; sect. chief metabolic effects EPA, Research Triangle Park, N.C., 1973-77; pharmacologist Nat. Inst. Environ. Health Scis., Research Triangle Park, 1977—. Contbr. articles to profl. publs. U. Tex. Med. Sch. Pub. Health Services predoctoral fellow, 1963-67. Mem. Am. Soc. Pharmacology and Exptl. Therapeutics, Soc. Toxicology. Subspecialties: Molecular pharmacology; Toxicology (medicine). Current work: Effects of environmental chemicals on liver cytochrome P-450, metabolism of drugs and heme synthesis; structure-activity relationships of environmental chemicals such as polychlorinated biphenyls (PCBs) and mechanism of action. Home: 4818 N Hills Raleigh NC 27612 Office: Nat Inst Environ Health Scis PO Box 12233 Research Triangle Park NC 17709

GOLDSTEIN, LAWRENCE HOWARD, computer scientist; b. N.Y.C., Jan. 7, 1952; s. Jacob and Regina Gross G. B.E., Cooper Union U., 1973; M.S.E., Princeton U., 1974, M.A., 1975, Ph.D., 1976. Mem. tech. staff computer-aided design div. Sandia Labs., Albuquerque, 1976-80; mgr. sci. computing dept. Inmos Corp., Colorado Springs, Colo., 1980-83; dir. tech. planning United Technologies Microelectronics Ctr., Colorado Springs, Colo., 1983—. Guest editor: IEEE Transactions Computers, 1981; editor: Sigda Newsletter, 1979-81. NSF grad. fellow, 1973-76. Mem. IEEE (Browder J. Thompson award 1981), Assn. Computing Machinery, Soc. Indsl. and Applied Math. Subspecialties: Computer-aided design; Integrated circuits. Current work: Technical and strategic planning for semicustom and custom VLSI company. Home: 3107-D Broadmoor Valley Rd Colorado Springs CO 80906 Office: United Technologies Microelectronics Ctr 1365 Garden of the Gods Rd Colorado Springs CO 80907

GOLDSTEIN, LESTER, biology educator; b. Bklyn., June 28, 1924; s. Charles and Gussie (Silverman) G.; m.; children: Natasha, Nina. A.B., Bklyn. Coll., 1948; Ph.D., U. Pa., 1953. Instr. biology Queens Coll., 1950; teaching asst. zoology U. Pa., Phila., 1950-51, univ. fellow, 1951-52, USPHS predoctoral fellow, 1952-53; USPHS postdoctoral fellow U. Calif.-Berkeley, 1953-55; Damon Runyon fellow U. Calif.-San Francisco, 1955-56; research assoc. U. Calif. Med. Ctr., San Francisco, 1955-59; assoc. prof. biology U. Pa., Phila., 1959-64, prof., 1964-67, Inst. Devel. Biology, U. Colo., Boulder, 1967-68, prof. cell biology, 1968-82; ACS research scholar dept. genetics U. Wash., Seattle, 1977-78; prof., dir. Thomas Hunt Morgan Sch. Biol. Scis., U. Ky., Lexington, 1982—. Cons. editor, McGraw Hill, 1970-80; editor: Acad. Press, 1975-80, Springer Verlag, 1981—; contbr. articles to profl. jours. Served with U.S. Army, 1943-46. Lalor Found. fellow Haverford Coll., summer 1959; UPSPHS spl. fellow, U. Colo. Med. Ctr., Denver, 1965-66; recipient Disting. Alumnus award Bklyn. Coll., 1975; Am. Cancer Soc. scholar, 1977; U. Colo. faculty fellow, 1977. Fellow AAAS; mem. Genetics Soc. Am., Am. Soc. Cell Biology (council 1970-72), AAAS. Subspecialty: Cell biology. Office: Morgan Sch Biol Scis Univ Ky Lexington KY 40506

GOLDSTEIN, MARK KINGSTON LEVIN, high technology executive, researcher; b. Burlington, Vt., Aug. 22, 1941; s. Harold Meyer Levin and Roberta (Butterfield) G. B.S. in Chemistry, U. Vt., 1964, Ph.D., U. Miami-Coral Gables, 1971. Pres. IBR, Inc., Coral Gables, Fla., 1970-74; group leader Brookhaven Nat. Lab., Upton, N.Y., 1974-77; sr. researcher East-West Ctr., Honolulu, 1977-79; sr. tech. advisor JGC Corp., Tokyo, 1979-81; pres., chmn. bd. Quantum Group, Inc., La Jolla, Calif., 1981—; exec. dir. Magnatek, Inc., Brotas, Brazil, 1982—. Contbr. articles to profl. jours.; contbr. poetry to mag. NSF fellow, 1964, 65. Mem. Am. Nuclear Soc., Am. Chem. Soc., AAAS. Club: Hawaii Yacht (Honolulu). Subspecialties: Physical chemistry; Nuclear fission. Current work: Health effects of toxic materials, nuclear waste management, forensic science and engineering, fire protection, energy, toxic gas detection and purification. Patentee in field. Office: Quantum Group Inc 1250 Prospect St Suite B-23 La Jolla CA 92037

GOLDSTEIN, MARVIN E(MANUEL), scientist, consultant; b. Cambridge, Mass., Oct. 11, 1938; s. David and Evelyn (Wilner) G.; m. Priscilla Ann Beresh, Nov. 12, 1965; children: Deborah, Judy. B.S., Northeastern U., 1961; M.S., M.I.T., 1962; Ph.D., U. Mich., 1965. Research assoc. M.I.T., Cambridge, 1965-67; research engr. Lewis Research Center, Cleve., 1967-79, chief scientist, 1979—, sci. cons. to dir. sci. and tech. Author: Aeroacoustics, 1976; contbr. articles to profl. publs. Recipient NASA Outstanding Sci. Achievement award, 1979. Fellow AIAA (chmn. aeroacoustics tech. com., jour. editor, Pendray award 1983, Aeroacoustics award); mem. Am. Phys. Soc. Jewish. Subspecialties: Fluid mechanics; Acoustical engineering. Home: 928 Bennett Dr Elyria OH 44035 Office: NASA Lewis Research Center M S 5-3 21000 Brookpark Rd Cleveland OH 44135

GOLDSTEIN, MENEK, neurochemistry educator; b. Poland, Apr. 8, 1924; s. Jacob and Ceylia (Hirsch) G. Ph.D., U. Berne, Switzerland, 1955; D.Medicine (hon.), Karolinska Inst., 1982. Research asst. in biochemistry U. Berne, 1953-56; research staff mem. Worcester Found. for Exptl. Biology, 1956-57; biochemistry NYU Med. Ctr., 1957-58, instr., 1958-59, asst. prof., 1960-63, assoc. prof., 1963-69, prof. neurochemistry, physiology and biophysics, 1969—. Contbr. articles to profl. jours. Recipient Hellenic Geriatric Soc. medal, 1980. Mem. Am. Soc. Biol. Chemists, Am. Soc. Pharmacology and Exptl. Therapeutics, Soc. Neurosci., Am. Coll. Neuropsychopharmacology. Subspecialties: Biochemistry (biology); Neurochemistry. Current work: Neurotransmitters and mental neurological disorders. Home: 333 E 30th St Apt 9A New York NY 10016 Office: NYU Med Ctr 550 1st Ave New York NY 10016

GOLDSTEIN, MURRAY, osteopathic physician, government official; b. N.Y.C., Oct. 13, 1925; s. Israel and Yetta (Zeigen) G.; m. Mary Susan Michael, June 13, 1957; children—Patricia Sue, Barbara Jean. A.B. in Biology, N.Y. U., 1947; D.O., Des Moines Still Coll. Osteo Medicine, 1950; M.P.H. in Epidemiology, U. Calif., 1959; D.Sc. (hon.), Kirksville Coll. Osteo. Medicine, 1966, D.D.L., N.Y. Inst. Tech., 1982. Intern Des Moines Still Coll. Osteo Medicine Hosp., 1950-51, resident in internal medicine, 1951-53; commd. sr. asst. surgeon USPHS, 1953, advanced through ranks to asst. surgeon gen., 1980; asst. chief grants and tng. br. Nat. Heart Inst., NIH, Bethesda, Md., 1953-58; asst. chief research grants rev. br., dir. epidemiology and biometry tng. grant program div. research grants NIH, Bethesda, 1956-60; acting sect. chief Bur. Acute Communicable Disease, Calif. Dept. Public Health, Berkeley, 1958; chief spl. projects br. Nat. Inst. Neurol. Diseases and Blindness NIH, Bethesda, 1960-61; dir. extramural programs Nat. Inst. Neurol. and Communicative Disorders and Stroke, NIH, Bethesda, 1961-76, dir. stroke and trauma program, 1976-78; dep. dir. NINCDS-NIH, 1978-82, dir., 1982—; vis. scientist Mayo Clinic and Grad. Sch., Rochester, Minn., 1967-68; v.p. Eisenhower Inst. for Stroke Research; cons. WHO; clin. prof. medicine N.Y. Coll. Osteo Medicine, N.Y. Inst. Tech.; trustee Am. Osteo. Coll. Public Health and Preventive Medicine; mem. med. adv. bd. Am. Parkinson's Disease Assn.; bd. dirs. United Cerebral Palsy Research and Edn. Found. Editorial bd.: Osteo. Annals, 1973—, Internat. Jour. Neurology, 1980—, Jour. Neuroepidemiology. Served with U.S. Army, 1943-45. Decorated Silver Star, Purple Heart.; Recipient Meritorious Service medal USPHS, 1971, Disting. Service medal USPHS, 1983. Fellow Am. Public Health Assn., Am. Heart Assn. (liaison mem. exec. com. Council on Stroke, asso. editor Stroke, Jour. Cerebral Circulation 1976—), Am. Coll. Osteo. Internists (hon.), Am. Acad. Neurology; mem. Am. Neurol. Assn. (2d v.p., cert. merit), Am. Osteo. Assn., AAAS, Assn. Research in Nervous and Mental Disease, Soc. Neurosci., Am. Osteo. Coll. Public Health and Preventive Medicine, World Fedn. Neurology. Subspecialties: Neurology; Epidemiology. Office: 9000 Rockville Pike Bethesda MD 20205

GOLDSTEIN, NORMAN, dermatologist; b. Bklyn., July 14, 1934; s. Joseph Harry and Bertha (Doctoroff) G.; m. Ramsay Goldstein, Feb. 14, 1980; children: Richard David, Heidi Lee. B.A., Columbia U., 1955; M.D., SUNY, Bklyn., 1959. Diplomate: Am. Bd.Dermatology. Intern Maimonides Hosp., N.Y.C., 1959-60; dermatology resident skin and cancer unit N.Y.U. Med. Center, 1960-61, Belluvue Hosp., 1961-62; partner Honolulu Med. Group (pvt. practice dermatology), 1972—; assoc. clin. prof. dermatology U. Hawaii, 1973—; bd. dirs. Honolulu Med. Group Research/Edn. Found., 1970—; task force mem., adv. council Am. Acad. Dermatology Biomed. Communications Network, 1979—; tech. adv. com. Cancer Center Hawaii; exec. bd. Physicians Exchange; adv. Nat. Skin Cancer Found.; trustee Dermatology Found., 1979-82. Contbr. articles to profl. jours. Active Hawaii Council on Culture and Arts; trustee Hawaii Jewish Fedn., 1976-79, 82—, Historic Hawaii Found.; mem. Hawaii Visitors Bur., Downtown Improvement Assn.; pres. Friends of the Alexander Young Bldg., 1980. Served with U.S. Army, 1963-64. Decorated Army Commendation medal. Mem. AMA, Soc. Investigative Dermatology, Internat. Soc. Tropical Dermatology, Assn. Mil. Dermatologists, Hawaii Dermatol. Soc. (pres. 1970-72), Honolulu County Med. Soc. (bd. govs. 1978-80), Hawaii Med. Assn., Micronesian Med. Assn., Hawaii Pub. Health Assn., AAAS, Am. Soc. Photobiology, Environ. Health and Light Research Inst., Am. Assn. Clin. Oncology, Pacific Dermatol. Assn., Hawaii Assn. for Physicians for Idemnification (dir. 1977), Hawaii Planned Parenthood, Pacific Health Research Inst., Biologic Photographic Assn., Internat. Solar Energy Soc., Photographic Soc.Am., Health Scis. Communications Assn. Clubs: Outrigger Canoe, Waikiki, Plaza, Honolulu, Honolulu Club. Lodge: Rotary. Subspecialties: Dermatology; Medical photography.

GOLDSTEIN, PAUL, marine engineering company executive, consultant; b. N.Y.C., Dec. 2, 1934; s. Joseph and Gertrude (Myers) G.; m. Susan Jennings Hough (div.); 3 daus. B.S. in Marine Engring., U.S. Mcht. Marine Acad., 1956. Test engr. Gen. Electric Co., Schenectady, 1956-57; marine engr. various cos., 1959; research project engr. Foster Wheeler Corp., N.Y.C., 1959-65; research project mgr. CE Inc., Windsor, Conn., 1965-68; group v.p., gen. mgr., project mgr. Superfund div. NUS corp., Pitts., Arlington, Va., 1968—; cons. environ. engr. Contbr. articles to profl. jours.; editorial bd.: ASME Handbook on Water in Thermal Power Systems. Chmn. twp. planning commn., cognizant exec. minority scholarship program, Pitts. Served to lt. (j.g.) USNR, 1957-59. Recipient Am. Seamen's Friend Soc. award, 1956. Mem. ASME (Prime Movers award 1968), TAPPI, ASTM, Engring. Soc. Western Pa. (meml. scholarship com., internat. water conf. award of merit 1980), Water Pollution Control Fedn. Subspecialties: Environmental engineering; Chemical engineering. Current work: Hazardous waste management, water pollution control, thermal power system water technology. Patentee in field. Home: RD #2 Prosperity PA 15329 Office: NUS Corp Park West 2 Cliffmine Rd Pittsburgh PA 15275

GOLDSTEIN, PAUL, geneticist, educator; b. Bklyn., Apr. 24, 1951; s. Isidore and Ida (Weitzman) G.; m. Mira Segal, Aug. 20, 1972; children: Marcie, Alyson. B.S., SUNY-Albany, 1973; M.S., Ohio U., 1975; Ph.D., York U., 1977. Research assoc. NC. State U., Raleigh, 1977-79; asst. prof. dept. biology U. N.C., Charlotte, 1979—. Mem. Am. Soc. Cell Biology. Subspecialties: Genome organization; Genetics and genetic engineering (medicine). Current work: Computer assisted graphics analysis of chromosomal structure and function; correlation of changes in DNA and chromosome structure and the aging process. Address: Dept Biology U NC Charlotte NC 28223

GOLDSTEIN, RICHARD JAY, engineer, educator; b. N.Y.C., Mar. 27, 1928; s. Henry and Rose (Steierman) G.; m. Anita Nancy Klein, Sept. 5, 1963; children: Arthur Sander, Jonathan Jacob, Benjamin Samuel, Naomi Sarith. B.M.E., Cornell U., 1948; M.S.M.E., U. Minn., 1950, 1951, Ph.D., 1959. Instr. U. Minn., 1948-51, instr., research fellow, 1956-58, mem. faculty, 1961—, prof. mech. engring., 1965—, head dept., 1977—; research engr. Oak Ridge Nat. Lab., 1951-54; asst. prof. Brown U., 1959-61; cons. in field, 1956—; NSF sr. postdoctoral fellow, vis. prof. Cambridge (Eng.) U., 1971-72. Served to 1st lt. AUS, 1954-55. NATO fellow, Paris, 1960-61; Lady Davis fellow Technion, Israel, 1976. Fellow ASME (Heat Transfer Meml. award 1978, Centennial medallion 1980); mem. AAAS, Am. Phys. Soc., Sigma Xi, Tau Beta Pi, Pi Tau Sigma. Subspecialty: Mechanical engineering. Research, publs. in thermodynamics, fluid mechanics, heat transfer, optical measuring techniques. Home: 520 Janalyn Circle Golden Valley MN 55416 Office: Dept Mech Engring U Minn Minneapolis MN 55455

GOLDSTEIN, RONALD E(RWIN), dentist, lecturer, writer; b. Atlanta, Nov. 1, 1933; s. Irving H. and Helen G.; m. Judy Salzberg, Aug. 26 , 1956; children: Cary, Cathy, Richard, Kenneth. Student, U. Mich., 1951-53, U. Ga., summer 1952, Oglethorpe U., summer 1953; D.D.S., Emory U., 1957. Mem. periodontal staff Ben Massell Dental Clinic, Atlanta, 1959-67, vol. in charge cosmetic dental services, 1967—, mem. staff com., 1968—; mem. continuing edn. faculty Sch. Dental Medicine, U. Pitts., 1972-75; spl. lectr. in periodontology Emory U., 1972-79, spl. lectr. in esthetic dentistry, 1980—; assoc. clin. prof. continuing edn. Boston U., 1980—; dental cons. P.M. Magazine, 1980-83; numerous lectures, presentations in field.; Lectr. nationally on mental health Nat. Assn. Mental Health, 1962—. Author: Esthetics in Dentistry, 1976, Esthetic Principles for Ceramo Metal Restorations, 1977, You Can Change Your Smile, 1983; contbr. articles to profl. publs.; contbg. editor: The Wonderful World of Modern Dentistry, 1972; bus. mgr.: The Scribbler, 1950-51; mem. editorial staff: Mich. Daily, 1951-52, The Gargoyle, 1951-52; dental editor: Emory Wheel, 1954-56, 1957, Emory Annual, 1957; editor: The Arouser, 1964-66, Thomas P. Hinman Dental Meeting Program Book, 1966-68, Alpha Oemgan, Atlanta, 1966-68, Preview-Rev, 1966-67; internat. editor: Alpha Oemgan, 1971-74. Bd. dirs. Jewish Children's Services, Atlanta, 1969—. Served as capt. Dental Corps U.S. Army, 1957-59. Fellow Am. Coll. Dentists (nat. winner First Ann. Writing award), Internat. Coll. Dentists; mem. Am. Acad. Esthetic Dentistry (co-founder, pres. 1977-78), Am Acad. Cranio-Mandibular Disorders, Am. Acad. Periodontology (assoc.), ADA, Ga. Dental Assn. (hon. fellow), No. Dist. Dental Soc. (founder and first chmn. emergency dental service), Fifth Dist. Dental Soc. (pres. 1974-75, parliamentarian 1976), Internat. Acad. Gnathology, Am. Acad. History of Dentistry, Emory U. Alumni, Am. Jewish Com., Alpha Omega (pres. Alpha Delta chpt. 1956-57, pres. Atlanta Alumni 1968-69, internat. regent 1970-71, assoc. editor 1965-71, internat. pres. 1976, chief justice 1979). Jewish. Subspecialty: Cosmetic dentistry. Inventor various dental instruments and devices. Office: 1218 W. Paces Ferry Rd NW Atlanta GA 30327

GOLDSTEIN, SIDNEY, pharmacologist; b. Phila., Mar. 27, 1932; s. Israel and Gertrude (Stein) G.; m. Janice Levy, June 19, 1955; children: Rhonda, David, Nina. B.Sc., Phila. Coll. Pharmacy, 1954, M.Sc., 1955, D.Sc., 1958. Registered pharmacist. Head antiinflammatory unit Eaton Labs., 1958-59; head cardiovascular unit Lederle Labs., 1959-61; sect. head Nat. Drug Co., 1961-64, dir. pharmacology, 1964-67, dir. devel., 1967-70; sect. head Merrell Dow Pharms. Inc., 1970-72, assoc. group dir. clin. pharmacology, 1972-74, exec. asst. to v.p. research, 1974-76, dir. pharm. scis., 1976-82, dir. product devel., 1982—; lectr. Rutgers State U., 1963-66, Phila. Coll. Pharmacy, 1967-70. Contbr. articles to profl. jours. Vice-pres., sr. adv. B'nai B'rith Youth Orgn., 1978; bd. dirs. Glen Manor Home for the Aged, 1983—. Recipient Bausch and Lomb Sci. award, 1950. Mem. Am. Soc. Pharmacology and Exptl. Therapeutics, Soc. Exptl. Biology and Medicine, Am. Pharm. Assn., Acad. Pharm. Sci., Am. Soc. Clin. Pharmacology and Therapeutics. Jewish. Club: Cresthill Country (Cin.). Lodge: B'nai B'rith. Current work: Pharmaceutics product devel. Home: 259 Compton Rdge Dr Cincinnati OH 45215 Office: 2110 E Galbraith Rd Cincinnati OH 45215

GOLDSTEIN, STEVEN ALAN, biomechanics educator; b. Reading, Pa., Sept. 15, 1954; m. Nancy Ellen Gehr, Aug. 22, 1976; 1 son, Aaron. B.S. in Mech. Engring, Tufts U., 1976; M.S. in Bioengring, U. Mich., 1981, Ph.D., 1981. Research assoc. Biomed. Engring. Ctr., Tufts New Eng. Med. Ctr., Boston, 1974-76; research assoc. Ctr. for Ergonomics, U. Mich., Ann Arbor, 1978-81, asst. prof. surgery, sect. orthopaedic surgery, co-dir. biomechanics, trauma and sports medicine lab., 1981—. Contbr. articles to profl. jours. Nat. Inst. Occupational Health and Safety trainee, 1979-81. Mem. Am. Soc. Biomechanics, Biomed. Engring. Soc., Orthopaedic Research Soc. Subspecialties: Biomedical engineering; Mechanical engineering. Current work: Development of total joint prostheses; mechanical properties of bone and soft tissues; orthopaedic biomechanics; invention of surgical instrumentation. Inventor bone reamer, total joint arthroplasty system. Office: U Mich Orthopaedic Surgery C4002 Outpatient Bldg Ann Arbor MI 48109

GOLDSTEIN, STEVEN B., biologist, geneticist; b; b. Buffalo, Feb. 20, 1956; s. Frederick Paul and Audrey Sheila (Baron) G. Student, Marymount Coll., Palos Verdes, Calif., 1972-73, Moorpark (Calif.) Coll., 1973-74; B.A. in Biology, U. Calif.-San Diego, 1976; Ph.D. in Genetics, U. Wash., 1980. Lab. asst. dept. biology U. Calif.-San Diego, 1974-76; grad. research asst. U. Wash., Seattle, 1976-80; research assoc. dept. molecular, cellular and developmental biology U. Colo., Boulder, 1980—. Contbr. articles to profl. jours. Mem. Genetics Soc. Am. Club: Boulder Road Runners. Subspecialties: Genetics and genetic engineering (biology); Molecular biology. Current work: Genetic and molecular genetic analysis of milotic spindle structure and function research. Office: Dept Molecular Cellular and Developmental Biology U Colo Boulder CO 80309

GOLDSTEIN, STEVEN EDWARD, psychologist; b. N.Y.C., Nov. 25, 1948; s. Maurice and Matilda (Weiss) G. B.S., CCNY, 1970, M.S., 1971; Ed.D., U. No. Colo., 1976. Registered sch. psychologist, N.Y. Sch. psychologist Denver pub. schs., 1975; teaching asst. U. No. Colo., Greeley, 1975-76; asst. prof. psychology Northeastern Okla. State U., Tahlequah, 1976-78; psychologist/dir. inpatient Winnemucca Mental Health Ctr., Nev., 1978-80; psychologist/dir. tng. Desert Devel. Ctr., Las Vegas, 1980-82; sr. psychologist Las Vegas Mental Health Ctr., 1982—; sec. grad. council CCNY, 1970-71, chmn., 1970-71; tchr./counselor N.Y.C. pub. schs., 1971-74; cons. U.S. Dept. Edn., Region IX, 1981-82. NSF scholar, 1975-76; others. Mem. Am. Psychol. Assn., Biofeedback Soc. Am., Am. Soc. Tng. and Devel., Biofeedback Soc. Nev., Nat. Assn. Sch. Psychologists. Jewish. Subspecialties: Behavioral psychology; Stress management. Current work: Stress management, biofeedback, values clarification, management training, training the trainer, family/couples counseling. Office: Las Vegas Mental Health Center 6161 W Charleston Blvd Las Vegas NV 89158

GOLDSTEIN, YONKEL NOAH, psychologist; b. N.Y.C., Sept. 8, 1949; s. Abraham Joseph and Rose (Fishbein) G.; m. Mary K. Goldstein, June 24, 1979; 1 dau., Keira Anne. B.A., SUNY-Binghamton, 1970; M.A., U. Tex.-Austin, 1972; Ph.D., Mich. State U., 1976. Lic. psychologist, Calif., N.C. Assoc. dir. behavioral sci. Duke/FAHEC, Fayetteville, N.C., 1976-80; dir. behavioral sci. Nat. Med. Ctr., Salinas, Calif., 1980—; clin. assoc. Duke U. Med. Ctr., 1976-80; asst. clin. prof. U. Calif.-San Francisco, 1980—. Contbr. book chpts., articles to profl. jours. HEW grantee, 1980. Mem. Am. Psychol. Assn., Calif. Psychol. Assn. Democrat. Jewish. Subspecialties: Health psychology; Hypnosis. Current work: Teaching psychology to medical personnel, control of overeating, psychological medicine. Office: Natividad Med Ctr PO Box 1611 Salinas CA 93902

GOLDSTINE, HERMAN HEINE, mathematician; b. Chgo., Sept. 13, 1913; s. Isaac Oscar and Bessie (Lipsey) G.; m. Adele Katz, Sept. 15, 1941 (dec. 1964); children—Madlen, Jonathan; m. Ellen Watson, Jan. 8, 1966. B.S., U. Chgo., 1933, M.S., 1934, Ph.D., 1936; Ph.D. honoris causa, U. Lund, Sweden, 1974; D.Sc., Adelphi U., 1978, Amherst Coll., 1978. Engaged as research asst. U. Chgo., 1936-37, instr., 1937-39, U. Mich., 1939-42, asst. prof., 1942-45; asst. project dir., electronic computer project Inst. for Advanced Study, Princeton, 1946-55, acting project dir., 1954-57, permanent mem., 1952—; dir. math. scis. dept. IBM Research; dir. sci. devel. IBM Data Processing Hdqrs., White Plains, N.Y.; cons. to dir. research IBM; now IBM fellow; cons. various govt., mil. agys.; mem. report rev. com. Nat. Acad. Scis.; Mem. vis. com., phys. sci. div. U. Chgo. Author: A History of the Calculus of Variations from the 17th through the 19th Century. Served as lt. col. AUS, World War II. Recipient book award in sci. Phi Beta Kappa, 1973, U. Chgo. Alumni Achievement award, 1974. Mem. Am. Math. Soc., Am. Philos. Soc., Nat. Acad. Sci. Math. Assn. Am., Phi Beta Kappa, Sigma Xi. Subspecialties: Theoretical computer science; Applied mathematics. Current work: Mathematics of computer science and work of John and James Bernoulli in the calculus of variations. Home: 175 Fairway Dr Princeton NJ 08540 Office: IBM TJ Watson Research Center PO Box 218 Yorktown Heights NY 10598 Inst Advanced Study Princeton NJ 08540

GOLDTHWAITE, DUNCAN, petroleum geologist; b. N.Y.C., Mar. 31, 1927; s. George Edgar and Emily Jack (Duncan) G.; m. Margaret Temple, Feb. 4, 1956; children: Madelyn, Virginia, Mary, Martha. B.A., Oberlin (Ohio) Coll., 1950; M.A., Harvard U., 1952. Glacial geologist S.D. State Survey, summer 1950; coal geologist U.S. Geol. Survey, S.E. Mont., summer 1951; petroleum geologist, mineral geologist Chevron U.S.A., Bismarck, N.D., Jackson, Mass., Pensacola, Fla., New Orleans, 1952-82, staff petroleum geologist, New Orleans, 1982—. Mem. City Council, Gulf Breeze, Fla., 1962-64. Served with USNR, 1945-46. Fellow Geol. Soc. Am.; mem. Am. Assn. Petroleum Geologists, New Orleans Geol. Soc. (pres. 1982-83, past editor, past dir.). Republican. Episcopalian. Subspecialties: Geology; Geophysics. Current work: Regional geology and oil and gas prospecting. Home: 4608 James Dr Metairie LA 70003 Office: Chevron USA 935 Gravier St New Orleans LA 70112

GOLDWASSER, EUGENE, biochemist, educator; b. N.Y.C., Oct. 14, 1922; s. Herman and Anna (Ackerman) G.; m. Florence Cohen, Dec. 22, 1949; children—Thomas Alan, Matthew Laurence, James Herman. B.S., U. Chgo., 1943, Ph.D., 1950. Am. Cancer Soc. fellow U. Copenhagen, Denmark, 1950-52; research asso. U. Chgo., 1952-61, mem. faculty, 1962—, prof. biochemistry, 1963—, chmn. com. on developmental biology. Served with AUS, 1944-46. Guggenheim fellow U. Oxford (Eng.), 1966-67. Mem. Am. Soc. Biol. Chemists, Biochem. Soc., AAAS, Internat. Soc. Developmental Biologists, Internat. Soc. Exptl. Hematology, Endocrine Soc., Sigma Xi. Subspecialty: Developmental biology. Research biochemistry red blood cell formation. Home: 5727 Dorchester Ave Chicago IL 60637 Office: Dept Biochemistry U Chgo Chicago IL 60637

GOLDYNE, MARC ELLIS, dermatology educator, researcher; b. San Francisco, Oct. 15, 1944; s. Alfred Josef and Helen Sandra (Newman) G.; m. Gail Anne Sokolow, Sept. 5, 1971; children: Serena, Avram. A.B., U. Calif.-Berkeley, 1966; M.D., U. Calif.-San Francisco, 1970; Ph.D., U. Minn.-Rochester, 1975. Intern, French Hosp. Med. Center, San Francisco, 1970-71; resident in dermatology Mayo Clinic, Rochester, Minn., 1971-75; fellow in dermatology Mayo Grad. Sch. Medicine, 1971-75; instr. Mayo Med. Sch., 1974-75; research fellow Karolinska Inst., Stockholm, 1975-78; asst. prof. dermatology and medicine U. Calif.-San Francisco, 1978—; head dermatol. research San Francisco Gen. Hosp., 1978—. Fellow Am. Acad. Dermatology; mem. Soc. Investigative Dermatology (dir. 1972-74), Western Regional Soc. Investigative Dermatology (pres. 1983-84), Am. Fedn. Clin. Research, Soc. Exptl. Biology Medicine. Subspecialties: Dermatology; Biochemistry (medicine). Current work: Prostaglandin and related eicosanoid metabolism by human monocytes in relation to immune response and host tumor defenses. Office: University of California 501 Parnassus Ave San Francisco CA 94143

GOLIKE, ANN ELIZABETH, computer analyst; b. Buffalo, Aug. 26, 1958; d. Ralph Crosby and Marcelaine Ann (Jarvi) G. B.A., U. Va., 1980; M.S., Johns Hopkins U., 1983. Reasearch asst. dept. astronomy U. Va., Charlottesville, Va., 1977-80; programmer, analyst Computer Scis. Corp., Silver Spring, Md., 1980, mission support, Greenbelt, Md., 1980—. Recipient Shuttle Achievement award NASA, 1981, Silver Snoopy award, 1982. Mem. AIAA (officer student br. 1979-80), Planetary Soc. Subspecialty: Computers in satellite tracking systems. Current work: Use of large-scale computer system in order to provide real-time support for NASA's satellite tracking system. Home: 9821 Good Luck Rd Apt 6 Lanham-Seabrook MD 20706

GOLL, DARREL EUGENE, biochemist, educator; b. Garner, Iowa, Apr. 19, 1936; s. Leon Oscar and Maire Eleanor (Nonnweiller) G.; m. Rosalie Elaine Bullock, Jan. 11, 1958; children: Laurene Elaine, Jeffrey Eugene, Kathleen Kay. B.S., Iowa State U., 1957, M.S., 1959; Ph.D., U. Wis., 1962. Asst. prof. depts. animal sci. and food tech. Iowa State U., Ames, 1962-65, assoc. prof. depts. animal sci., food tech., and biochemistry and biophysics, 1965-70, prof., 1970-76; prof. depts. biochemistry and nutrition and food sci., head dept. nutrition and food sci. U. Ariz., Tucson, 1976—; mem. com. on animal products, adv. bd. on mil. personnel supplies NRC, 1975-78; ad hoc mem. molecular cytology study sect. NIH, 1979—; mem. tech. adv. peer panel for grants in food quality and safety U.S. Dept. Agr., 1982. Contbr. chpts. to books, articles to profl. jours. NIH spl. fellow UCLA, 1966-67, Oxford (Eng.) U., 1972; NSF scholar, 1974. Mem. AAAS, Am. Chem. Soc., Am. Meat Sci. Assn. (Disting. Meats Research award 1972), Am. Soc. for Animal Sci. (Meats Research award 1974), Am. Soc. Biol. Chemists, Biophys. Soc., Inst. Food Technologists (Samuel Cate Prescott award 1971), Nutrition Today Soc., Phi Kappa Phi. Presbyterian. Subspecialties: Biochemistry (biology); Biophysics (biology). Current work: Biochemistry and metabolic turnover of contractile proteins, muscle growth, interactions of contractile proteins on their functions in motility and development, properties of proteolytic enzymes involved in metabolic turnover of contractile proteins. Office: Muscle Biology Group U Ariz Tucson AZ 85721

GOLLAHALLI, SUBRAMANYAM RAMAPPA, mechanical engineer; b. Sadali, India, Nov. 26, 1942; s. Bagepalli and Nagalakshamma (Rao) Ramappa; m. Rangamani Gollahalli, Dec. 25, 1967; children: Suma, Anil. B.E., Mysore U., 1963; M.E., Indian Inst.

Sci., 1965; M.A.Sc., U. Waterloo, Ont., Can., 1970, Ph.D., 1973. Registered profl. engr., Okla. Lectr. Indian Inst. Sci., 1965-68; research and devel. engr. John Fowler's Ltd., Bangalore, India, 1968; research asst. prof., asst. prof. U. Waterloo, 1973-76; asst. prof. U. Okla., Norman, 1976-80, assoc. prof., 1980—. Contbr. articles to profl. jours. Recipient Robert Angus medal Can. Soc. Mech. Engrs., 1978. Mem. ASME, AIAA, Soc. Auto. Engrs. (Ralph Teetor award 1978), Combustion Inst., Okla. Acad. Sci., Sigma Xi, Pi Tau Sigma. Subspecialties: Combustion processes; Fuels. Current work: Combustion of low BTU gases, soot production in synthetic fuel flames, emulsified fuel combustion, sulfur oxide production in flames. Home: 4105 Morrison Ct Norman OK 73069 Office: Sch Aerospace Mech Nuclear Engring U Okla Norman OK 73019

GOLLAND, JEFFREY HARRIS, psychologist, educator; b. Bklyn., Apr. 28, 1941; s. Gerald Edward and Rose Alice (Finkelstein) G.; m. Patricia Elaine Yeager, July 14, 1969; children: David Hamilton, Richard Morris. A.B. cum laude, Brandeis U., 1961; M.A., NYU, 1962, Ph.D., 1966. Lic. psychologist, N.Y. Lab. and teaching asst. N.Y.C. Bd. Edn., 1961-63; psychology intern VA, Bronx, 1963-66; psychologist to chief psychology Brooke Gen. Hosp., Ft. Sam Houston, 1966-68; instr. psychiatry NYU, Bellevue Med. Ctr., N.Y.C., 1968-70; pvt. practice clin. psychology, N.Y.C., 1968—; asst. to assoc. prof. edn. Baruch Coll., CUNY, 1970—; field supr. grad. psychology Rutgers U., Piscataway, N.J., 1975—; instr. Am. Inst. Psychotherapy and Psychoanalysis, N.Y.C., 1976-83. Contbr. articles to profl. jours. Bd. dirs. 145 4th Ave. Tenants Assn., 1972—; asst. cubmaster Boy Scouts Am., N.Y.C., 1981-82; bd. dirs., sec. Brandeis U. Alumni Assn., Waltham, Mass., 1977—. Served to capt. U.S. Army, 1966-68. Recipient Founders Day award NYU, 1967. Mem. N.Y. Freudian Soc. (dir. 1982—), Am. Psychol Assn., Council Psychoanalytic Psychotherapists. Democrat. Jewish. Subspecialties: Psychoanalysis; Developmental psychology. Home: 145 4th Ave New York NY 10003 Office: Baruch Coll CUNY 17 Lexington Ave New York NY 10010

GOLLEY, FRANK BENJAMIN, research ecologist, administrator; b. Chgo., Sept. 24, 1930. B.S., Purdue U., 1952; M.S., Wash. State U., 1954; Ph.D., Mich. State U., 1958. Faculty U. Ga., 1958—, dir., 1962-67, exec. dir., 1967-79, research prof., 1981—; dir. environ. biology NSF, 1979-81. Contbr. chpts. to books and articles to profl. jours. Mem. Internat. Assn. Ecology, Ecol. Soc. Am. Subspecialties: Ecology; Theoretical ecology. Current work: Biogeochemical cycling, systems ecology, landscape ecology. Office: Institute of Ecology University of Georgia Athens GA 30601

GOLLOB, HARRY FRANK, psychology educator; b. Newark, June 7, 1939; s. Joseph S. and Doris C. (Miller) G.; m. Maureen M. Morris, Sept. 4, 1959; children: Steven P., David J., Kenneth J. B.A. cum laude, U. Denver, 1960; M.S., Yale U., 1962, Ph.D., 1965. Lic. psychologist, Colo. Asst. to assoc. prof. psychology U. Mich., Ann Arbor, 1965-69; asst. to assoc. research psychologist Mental Health Research Inst., 1965-69; assoc. prof. psychology U. Denver, 1969-73, prof. psychology, 1973—, chmn. dept., 1978-79; statis., psychometric and social psychology cons. Contbr. articles to profl. jours.; editorial bd.: Jour. Exptl. Social Psychology, 1974-79, Multivariate Behavioral Research, 1971—. NIMH fellow, 1964; NIMH grantee, 1969-72; NSF grantee, 1972-81. Fellow Am. Psychol. Assn.; mem. Am. Statis. Assn., Psychometric Soc., AAAS, Ednl. Statisticians, Soc. Multivariate Exptl. Psychologists, Soc. Exptl. Social Psychology. Democrat. Subspecialties: Social psychology; Statistics. Current work: Statistics, quantitative methods; race and sex discrimination, social judgement of traits, feelings and behavior, computer simulation of dyadic interaction. Home: 2558 E Cresthill Ave Littleton CO 80121 Office: Univ Denver Dept Psychology Denver CO 80208

GOLLOBIN, LEONARD PAUL, chemical engineer, research/engineering company executive; b. N.Y.C., July 2, 1928; s. Morris and Jennie (Levine) G.; m. Charlotte Weissman, Jan. 21, 1951; children: Michael L., Susan D. B.Chem. Engring., CUNY, 1951; M.S., Kans. State U.-Manhattan, 1952; SCMP, Harvard U., 1975. Design engr. Foster Wheeler Corp., N.Y.C., 1952-55; mfg. engr. Gen. Electric Co., Waterford, N.Y., 1955-58; program dir. ORI, Inc., Silver Spring, Md., 1958-63; pres. Presearch, Inc., Arlington, Va., 1963—; sec., dir. Petroleum Ops. Support Services Inc., Houston, 1981—. Bd. dirs. Cultural Alliance Greater Washington, 1980—. Mem. Nat. Security Indsl. Assn. (trustee 1980—, chmn. antisubmarine warfare com. 1982—), Naval Undersea Warfare Found. Mus. (v.p. 1982—). Club: Kenwood (Bethesda, Md.). Subspecialties: Systems engineering; Operations research (mathematics). Current work: Major system planning and resources allocation. Home: 6710 Bradley Blvd Bethesda MD 20817 Office: Presearch Inc 2361 S Jefferson Davis Hwy Arlington VA 22202

GOLLUB, EDITH GOLDBERG, microbiologist; b. Bucharest, Romania, June 10, 1928; d. Manuel and Dina (Candeub) Goldberg; m. Seymour Gollub, Aug. 19, 1951; children: Erica Laurel, Marc Jeffrey. B.Sc., Phila. Coll. Pharmacy and Sci., 1948; M.S., U. Pa., 1956, Ph.D., 1961. Research assoc. U. Pa., 1958-63; research assoc. biochemistry dept. Columbia U., N.Y.C., 1964-67, asst. prof., 1967-79; research scientist dept. biol. sci. Barnard Coll., 1980-83; guest investigator Rockefeller U., N.Y.C., 1983—. Contbr. articles to sci. jours. NIH grantee, 1980-83. Mem. AAAS, Am. Soc. Microbiology. Subspecialties: gene regulation; genetics. Current work: Effect of heme on heme-protein synthesis; regulation of heme biosynthesis in yeast in mammals; gene control at molecular level. Office: Rockefeller U 66th St and York Ave New York NY 10021

GOLLUB, JERRY PAUL, physics educator, researcher; b. St. Louis, Sept. 9, 1944. A.B., Oberlin Coll., 1966; A.M., Harvard U., 1967, Ph.D. in Physics, 1971. Asst. prof. Haverford (Pa.) Coll., 1970-76, assoc. prof., 1976-80, prof. physics, 1980—; vis. assoc. prof. Sch. Applied and Engring. Physics, Cornell U., 1978; mem. sci. adv. bd. Environ. Cancer Prevention Ctr., Pub. Interest Law Ctr. Pa., 1980-81; adj. prof. physics U. Pa., 1981—; cons. to numerous environ. agys. Mem. Am. Phys. Soc., Fedn. Am. Scientists, AAAS, Sigma Xi. Subspecialty: Condensed matter physics. Office: Dept Physics Haverford Coll Haverford PA 19041

GOLLUB, LEWIS RAPHAEL, psychology educator; b. Phila., June 9, 1934; s. Benjamin and Edith (Teplitzky) G; m. Sylvia Jean Lipsitz, Oct. 27, 1963; children: Lara, Melissa, Gabrielle. A.B., U. Pa., 1955; Ph.D., Harvard U., 1958. Research fellow Harvard U., Cambridge, Mass., 1958-60; asst. prof. U. Md., College Park, 1960-65, assoc. prof., 1965-70, prof., 1970—; spl. research fellow Royal Coll. Surgeons, London, 1966-67; vis. scientist Central Inst. Exptl. Animals, Kawasaki, Japan, 1973-74, Cambridge (Eng.) U., 1980-81. Contbr. articles to profl. jours.; assoc. editor: Jour. Exptl. Analysis of Behavior, 1981-84. NSF fellow, 1955-58; NIH fellow, 1966-67; USPHS-NIMH grantee, 1962-1972; Nat. Inst. Drug Abuse grantee, 1973-78. Fellow AAAS, Am. Psychol. Assn. (exec. com. div. psychopharmacology 1980-83); mem. Soc. Exptl. Analysis of Behavior (dir. 1982-89). Subspecialties: Psychobiology; Behavioral psychology. Current work: Experimental research on the effects of drugs and toxic agents on behavior of animals, analysis of learned behavior. Home: 2810 Woodstock Ave Silver Spring MD 20910 Office: Dept Psychology U Md College Park MD 20742

GOLTZ, ROBERT WILLIAM, physician, educator; b. St. Paul, Sept. 21, 1923; s. Edward Victor and Clare (O'Neill) G.; m. Patricia Ann Sweeney, Sept. 27, 1945; children: Leni, Paul Robert. B.S., U. Minn., 1943, M.D., 1945. Diplomate: Am. Bd. Dermatology (pres. 1975-76). Intern Ancker Hosp., St. Paul, 1944-45; resident in dermatology Mpls. Gen. Hosp., 1945-46, 48-49, U. Minn. Hosp., 1949-50; practice medicine specializing in dermatology, Mpls., 1950-65; clin. instr. U. Minn. Grad. Sch., 1950-58, clin. asst. prof., 1958-60, clin. asso. prof., 1960-65, prof., head dept. dermatology, 1971—; prof. dermatology, head div. dermatology U. Colo. Med. Sch., Denver, 1965-71. Editorial bd.: Archives of Dermatology; editor: Dermatology Digest. Served from 1st lt. to capt., M.C. U.S. Army, 1946-48. Mem. Am. Dermatol. Assn. (dir. 1976—), Am. Soc. Dermatopathologists (pres. 1981), Am. Dermatology Soc. Allergy and Immunology (pres. 1981), AMA (chmn. sect. on dermatology 1973-75), Dermatology Found. (past dir.), Minn. Dermatol. Soc., Soc. Investigative Dermatology (pres. 1972-73), Histochem. Soc., Am. Acad. Dermatology (pres. 1978-79, past dir.), Colombian Dermatol. Soc. (corr. mem.), Can. Dermatol. Soc. (hon. mem.), Pacific Dermatol. Soc. (hon.-mem.), S. African Dermatol. Soc. (hon. mem.), Rocky Mountain Dermatol. Soc., Chgo. Dermatol. Soc., Assn. Profs. Dermatology (sec.-treas. 1970-72, pres. 1973-74). Subspecialties: Dermatology; Pathology (medicine). Current work: Dermatopathology, immunopathology of skin disorders. Home: 2234 Lee Ave N Minneapolis MN 55422 Office: U Minn Med Sch Dept Dermatology Minneapolis MN 55455

GOLUB, GENE H., computer science educator, mathematician; b. Chgo., Feb. 29, 1932. B.S., U. Ill., 1953; M.A., 1954, Ph.D. in Math, 1959. NSF fellow Math. Lab., U. Cambridge, Eng., 1959-60; mem. staff Lawrence Radiation Lab., U. Calif., 1960-61; mem. tech. staff Space Tech. Labs., Inc., TRW, Inc., 1961-62; vis. asst. prof. computer sci. Stanford U., 1962-64, from asst. prof. to assoc. prof., 1964-70, prof. computer sci., 1970—. Assoc. editor: Jour. Computer and Systems Sci, 1969—, Linear Algebra and Its Application, 1972—, Applied Math. and Optimization, 1974—, Soc. Indsl. and Applied Math. Rev, 1974, Jour. Am. Statis. Assn, 1976—. Mem. Soc. Indsl. and Applied Math. (council 1975-78). Subspecialty: Numerical analysis. Office: Dept Computer Sci Stanford U Stanford CA 94305

GOLUB, LORNE MALCOLM, oral biology and pathology/dentistry educator, researcher; b. Winnipeg, Man., Can., Jan. 13, 1941; came to U.S., 1973; s. Sydney Alexander and Edith (Simovitch) G.; m. Bonny Louise Moss, Aug. 2, 1964; children: Marlo Frances, Michael Benjamin. D.M.D., U. Man., Winnipeg, 1963, M.Sc., 1965; cert. in periodontology, Harvard U., 1968. Postdoctoral fellow Harvard U. Sch. Dental Medicine, Boston, 1965-68; assoc. prof. oral biology U. Man., 1968-70, assoc. prof. periodontics, 1970-73; assoc. prof. oral biology and pathology SUNY-Stony Brook, 1973-77, prof. oral biology, 1977—; cons. in field; mem. sci. adv. com. Center for Oral Health Research, U. Pa., 1982—. Contbr. numerous articles to profl. jours.; editorial rev. bd.: Jour. Dental Research, 1977—. Mem. Hebrew sch. com. Temple Beth Sholom, Smithtown, N.Y., 1982—. Med. Research Council Can. postdoctoral fellow, 1966-68; NIH grantee, 1974—; N.Y. Diabetes Assn. grantee, 1976-78; Kroc Found. grantee, 1981-84. Mem. Am. Acad. Periodontology, Internat. Assn. Dental Research, Internat. Group Periodontal Research, AAAS, Omega Kappa Upsilon. Jewish. Subspecialties: Periodontics; Oral biology. Current work: Gingival collagen metabolism; reduction of excessive collagenolytic enzyme activity by chemo theraphy; leucocyte dysfunctions in chemotaxis and neutral protease activity during diabetes mellitus; gingival crevicular fluid; relationship of these to periodontal disease. Patentee in field. Home: 29 Whitney Gate Smithtown NY 11787 Office: Sch Dental Medicine SUNY Stony Brook NY 11794

GOMER, ROBERT, scientist; b. Vienna, Austria, Mar. 24, 1924; m. Anne Olah, 1955; children: Richard, Maria. B.A., Pomona Coll., 1944; Ph.D. in Chemistry, U. Rochester, 1949; AEC fellow chemistry, Harvard, 1949-50. Instr. dept. chemistry and James Franck Inst. U. Chgo., 1950-51, asst. prof., 1951-54, asso. prof., 1954-58, prof., 1958—; dir. James Franck Inst. U. Chgo., 1977-83. Bd. dirs.: Bull. Atomic Scientists. Served with AUS, 1944-46. Recipient Kendall award in surface chemistry Am. Chem. Soc., 1975; Davisson-Germer prize Am. Phys. Soc., 1981; Sloan fellow, 1958-62; Guggenheim fellow, 1969-70; Bourke lectr., Eng., 1959. Mem. Leopoldina Acad. Scis., Nat. Acad. Scis., Am. Acad. Arts and Sci. Subspecialties: Surface chemistry; Condensed matter physics. Current work: Chemisorption; surface diffusion; electron stimulated desorption. Home: 4824 Kimbark Ave Chicago IL 60615 Office: 5640 Ellis Ave Chicago IL 60637

GOMOLL, ALLEN WARREN, cardiovascular pharmacologist; b. Chgo., July 10, 1933; s. Herbert F. and Sara E. (Cowan) G.; m. Elaine L. Kirkpatrick, Sept. 17, 1955; children: Gary A., Lisa E. B.S., Coll. Pharmacy, U. Ill., 1955; M.S., Coll. Medicine, 1958, Ph.D., 1961. Registered pharmacist, Ill. Asst. prof. pharmacology Coll. Medicine, U. Ill., 1961-66; group leader cardiovascular pharmacology Mead Johnson, Evansville, Ind., 1966-69, sect. leader cardiovascular pharmacology, 1969-76, prin. research assoc. biol. research, 1976-80, sect. mgr. biol. research, 1980-82; prin. research scientist cardiovascular pharmacology Bristol-Myers Research and Devel. Div., Evansville, 1982—; lectr. U. Evansville Coll. Nursing, 1977-81. Contbr. articles to profl. jours. Fellow Councils of Circulation and Basic Sci., Am. Heart Assn., Am. Coll. Cardiology; mem. Am. Soc. Pharmacology and Exptl. Therapeutics, AAAS, Soc. Exptl. Biol. Medicine, N.Y. Acad. Scis., Am. Chem. Soc. Presbyterian. Subspecialties: Pharmacology; Cardiology. Current work: Antiarrhythmic/cardiac dysfunction drug devel., myocardial ischemic models, renal pharmacology. Office: 2404 Pennsylvania Ave Evansville IN 47721

GOMORY, RALPH EDWARD, mathematician, business machines manufacturing company executive; b. Brooklyn Heights, N.Y., May 7, 1929; s. Andrew L. and Marian (Schellenberg) G.; m. Laura Dumper, 1954 (div. 1968); children: Andrew C., Susan S., Stephen H. B.A., Williams Coll., 1950, Sc.D. (hon.), 1973; student, Kings Coll., Cambridge (Eng.) U., 1950-51; Ph.D., Princeton U., 1954. Research assoc. Princeton U., 1951-54, asst. prof. math., Higgins lectr., 1957-59; with IBM, Yorktown Heights, N.Y., 1959—, dir. math. scis., research div., 1968-70, mem. corp. tech. com., 1970, dir. research, 1970—, v.p., 1973—, also mem. corp. mgmt. bd.; Andrew D. White prof.-at-large Cornell U., 1970-76; dir. IBM World Trade Ams./Far East Corp., Bank of N.Y.; Mem. vis. com. Sloan Sch. Mgmt., M.I.T., 1971-77; chmn. council Grad. Sch. Bus., U. Chgo., 1971-80; mem. adv. council dept. math. Princeton, 1974—; mem. adv. council Sch. Engring., Stanford U., 1978—. Trustee Hampshire Coll., 1977—; mem. governing bd. Nat. Research Council., 1980-83. Served with USN, 1954-57. Recipient Lanchester prize Ops. Research Soc. Am., 1963; IBM fellow. Fellow Econometric Soc., Am. Acad. Arts and Scis.; mem. Nat. Acad. Scis. (council 1977-78, 80-83), Nat. Acad. Engring. Subspecialty: Applied mathematics. Research integer and linear programming, non-linear differential equations. Home: 260 Douglas Rd Chappaqua NY 10514 Office: IBM Thomas J Watson Research Center Box 218 Yorktown Heights NY 10598

GOMPF, REBECCA ELAINE, veterinary educator; b. Findlay, Ohio, June 27, 1950; d. John Lawrence and Nettie E. (Sebastian) G. B.S., Ohio State U., 1971, D.V.M., 1975, M.S., 1975. Diplomate: Am. Coll. Vet. Internal Medicine (cardiology). Intern The Animal Med. Ctr., N.Y.C., 1975-76; resident U. Calif.-Davis Coll. Vet. Medicine, 1976-78; asst. prof., then assoc. prof. U. Tenn. Coll. Vet. Medicine, Knoxville, 1978—. Active Little League, 1977-81; mem. choir local Methodist ch., 1976—, sec. ch. fin. com., 1981—; vol. Adult Reading Program. Ohio State Coll. Agr. research scholar, 1970; AVMA Wives' Aux. awardee, 1975; recipient Robert Candon award Alpha Psi, 1975. Mem. Acad. Vet. Cardiology, AVMA, Am. Animal Hosp. Assn., Knoxville Acad. Vet. Medicine. Subspecialties: Internal medicine (veterinary medicine); Veterinary cardiology. Current work: Cardiology. Home: 2409 Alberta Dr Knoxville TN 37920 Office: U Tenn Box 1071 Knoxville TN 37901

GONA, AMOS GNANAPRAKASHAM, anatomy educator, researcher; b. Nandyal, Andhra Pradesh, India, July 16, 1933; came to U.S., 1963, naturalized, 1967; s. Elias and Sarah (Mesa) G.; m. Ophelia D. DeLaine, Apr. 6, 1962; children: Shantha K., Raj P. B.S. with honors, Andrhra U., 1954; M.A., CCNY, 1965; Ph.D., Albert Einstein Coll. Medicine, 1967. Lectr. zoology Andhra Christian Coll., Guntur, India, 1954-55, U. Rangoon, Burma, 1955-57; sci. master Anglo-Chinese Sch., Ipoh, Malaysia, 1957-60; sr. sci. master Opoky Ware Sch., Kumasi, Ghana, 1960-63; asst. prof. U. Medicine and Dentistry of N.J., Newark, 1968-72, assoc. prof., 1972-77, prof. anatomy, 1977. Contbr. articles on anatomy to profl. jours. NSF grantee, 1969-73; NIH grantee, 1973-80; N.Y. State Dept. Health, 1982-85; co-investigator other grants VA, NIH. Mem. Am. Assn. Anatomists, Am. Soc. for Cell Biology, Electron Microscopy Soc. Am., N.Y. Acad. Scis., N.Y. Soc. Electron Microscopists, Bioelectromagnetic Soc., Soc. for Devel. Biology, Soc. for Neuroscis. Democrat. Subspecialties: Neurobiology; Morphology. Current work: Development of the brain, especially cerebellar maturation, and the effects of thyroid hormone on brain development; effects of 60HZ electromagnetic fields on brain development. Office: Anatomy Dept 100 Bergen St Newark NJ 07103

GONCHER, RICHARD SIDNEY, software engineer; b. Bklyn., Apr. 18, 1946; s. Stanley and Sue (Benson) G.; m. Andrea Sue Rosenbaum, Sept. 3, 1972; children: Marc, Daniel, Brett. B.S., Bklyn. Coll., 1966; M.S., Johns Hopkins U., 1968; M.B.A., U. Md., 1981. Data systems analyst Dept. Def., Ft. Meade, Md., 1966-82; assoc. Booz-Allen & Hamilton, Inc., Bethesda, Md., 1982; sr. software engr. Ford Aerospace & Communications Corp., Hanover, Md., 1982—. Mem. Assn. Computing Machinery. Democrat. Jewish. Subspecialties: Information systems, storage, and retrieval (computer science); Software engineering. Current work: Software systems development using state-of-the-art software engring. techniques. Office: 7235 Standard Dr Hanover MD 21076

GONCZY, STEPHEN THOMAS, research scientist, cons.; b. Tuebingen, Germany, Oct. 8, 1947; s. Stephen I. and Doris E. (Eberhardt) G.; m. Anne Marie Laporte, Apr. 21, 1979; 1 dau., Teresa Elizabeth. B.S.M.E., Marquette U., 1969; Ph.D., Northwestern U., 1978. Research materials scientist UOP Corp. Research Ctr., Des Plaines, Ill., 1978—. Served to maj. U.S. Army, 1969—. NSF grad fellow, 1969; Gen. Electric fellow, 1976. Mem. Am. Soc. for Metals, Am. Ceramic Soc., ASME. Subspecialties: Materials; Ceramics. Current work: New materials for advanced ceramics and advanced heat transfer technology. Patentee in field.

GONZALEZ, LUIS FRANCISCO, dental educator; b. San Juan, P.R., July 3, 1947; s. Francisco and Maria E. (Ramos) G.; m. Miriam C. Portela, Dec. 18, 1971; children: Diana Margarita, Eduardo Francisco. B.S., U. P.R.-Rio Piedras, 1968; D.M.D., U. P.R.-San Juan, 1972; D.M.Sc., Harvard U., 1976, cert. in periodontics, 1976. Lic. dentist, P.R., Mass., N.C. Research fellow Harvard U., 1972-76; asst. prof. dentistry U. N.C.-Chapel Hill, 1976-78, U. P.R., 1978-80, assoc. prof., 1980—; asst. dean acad. affairs U. P.R. Sch. Dentistry, San Juan, 1978-81; program dir. (Health Career Opportunity Program), 1978-81; lectr. continuing edn. U. P.R. Contbr. articles sci. jours.; participant oral health programs pub. TV. U. P.R. Outstanding Scholar, 1969, 71, 72; USPHS research scholar, 1972-76; U. N.C. Dental Research Ctr. grantee, 1976-78; Health Career Opportunity Program grantee HEW, 1979-81. Mem. ADA, Am. Assn. Dental Schs., Am. Assn. Dental Research, Am. Acad. Periodontics, Omicron Kappa Upsilon. Clubs: GEO Study (Humacao, P.R.); Cangrejos Yacht (San Juan). Subspecialties: Oral biology; Periodontics. Current work: Ultrastructure of periodontal tissues, dental education, curriculum development, clinical periodontics, applications of computer technology in dentistry. Office: U PR GPO Box 5067 San Juan PR 00936 Home: B10 16 11th St Alturas de Torrimar Guaynabo PR 00657

GONZALEZ-LIMA, FRANCISCO M., neuroanatomist, educator, researcher; b. Havana, Cuba, Dec. 7, 1955; s. Francisco and Jacinta (Lima) Gonzalez-L.; m. Erika V. Musiol, 1981. B.S. in Biology, Tulane U., 1976, B.A. in Psychology, 1977; Ph.D. in Anatomy, U. P.R., 1980, Am. Western U., 1981. Postdoctoral research fellow Lab. Neurophysiology U. P.R., San Juan, 1981; asst. prof. Ponce Sch. Medicine, 1980—; Humboldt research fellow Inst. Zoology, Tech. U., Darmstadt, W.Ger., 1982-83; cons. in field. Contbr. articles to profl. jours. NSF grantee, 1981-83; NIH grantee, ctl2—. Mem. Soc. Neurosci., Nat. Soc. Med. Research, Animal Behavior Soc. Subspecialties: Neurophysiology; Psychobiology. Current work: Definition of neural processes underlying motivational and sensory interactions, structures responsible for normal integration of these processes, alteration of these activities in the brain. Office: Ponce Sch Medicine PO Box 7004 Ponce PR 00732 Home: Cond el Mirador 19F Ponce PR 00731

GOOD, MARY LOWE (MRS. BILLY JEWEL GOOD), business executive, chemist; b. Grapevine, Tex., June 20, 1931; d. John W. and Winnie (Mercer) Lowe; m. Billy Jewel Good, May 17, 1952; children: Billy, James. B.S., Ark. State Tchrs. Coll., 1950; M.S., U. Ark., 1953, Ph.D., 1955, LL.D. (hon.), 1979, D.Sc., U. Ill., Chgo., 1983. Instr. Ark. State Tchrs. Coll., Conway, summer 1949; instr. La. State U., Baton Rouge, 1954-56, asst. prof., 1956-58, asso. prof., New Orleans, 1958-63, prof., 1963-80, Boyd prof. materials sci., div. engring. research, Baton Rouge, 1979-80; v.p., dir. research UOP, Inc., Des Plaines, Ill., 1980—; chmn. Pres.'s Com. for Nat. Medal Sci., 1979-82; mem. Nat. Sci. Bd., 1980-86. Contbr. articles to profl. jours. Bd. dirs. Oak Ridge Asso. Univs. Recipient Agnes Faye Morgan research award, 1969; Distinguished Alumni citation U. Ark., 1973; Scientist of Yr. award Indsl. R & D Mag., 1982; AEC tng. grantee, 1967; NSF internat. travel grantee, 1968; NSF research grantee, 1969-80. Fellow Am. Inst. Chemistry (Gold medal 1983), Chem. Soc. London; mem. Am. Chem. Soc. (1st woman dir. 1971-74, regional dir. 1972-80, chmn. bd. 1978, 80, Garvan medal 1973, Herty medal 1975, award Fla. sect. 1979), Phi Beta Kappa, Sigma Xi, Iota Sigma Pi (regional dir. 1967—, hon. mem. 1983). Clubs: Zonta (past pres. New Orleans club, chmn. dist. status of women com. and nominating com., chmn. internat. Amelia Earhart scholarship com. Subspecialties: Catalysis chemistry; Surface chemistry. Current work: Heterogeneous catalysis, surface science, materials. Home: 295 Park Dr Palatine IL 60067 Office: Corp Research Center UOP Inc Ten UOP Plaza Des Plaines IL 60016

GOOD, MRS. BILLY JEWEL *See also* **GOOD, MARY LOWE**

GOOD, ROBERT ALAN, physician, educator; b. Crosby, Minn., May 21, 1922; s. Roy Homer and Ethel Gay (Whitcomb) G.; m. Joanne

Finstad, May 21, 1967; children: Robert Michael, Mark Thomas, Alan Maclyn, Margaret Eugenia, Mary Elizabeth. B.A., U. Minn., 1944, M.B., 1946, Ph.D., 1947, M.D., 1947; M.D. (hon.), U. Uppsala, Sweden, 1966, D.Sc., N.Y. Med. Coll., 1973, Med. Coll. Ohio, 1973, Coll. Medicine and Dentistry N.J., 1974, Hahnemann Med. Coll., 1974, U. Chgo., 1974, St. John's U., 1977, U. Health Scis., Chgo. Med. Sch., 1978. Teaching asst. dept. anatomy U. Minn., Mpls., 1944-45, instr. pediatrics, 1950-51, asst. prof., 1951-53, asso. prof., 1953-54, Am. Legion Meml. research prof. pediatrics, 1954-73, prof. microbiology, 1962-72, Regents prof. pediatrics and microbiology, 1969-73, prof., head dept. pathology, 1970-72; intern U. Minn. Hosps., 1947, asst. resident pediatrics, 1948-49; pres., dir. Sloan-Kettering Inst. for Cancer Research, 1973-80, mem., 1973-81; prof. pathology Sloan-Kettering div. Grad. Sch. Med. Scis. Cornell U., 1973-81, dir., 1973-80; adj. prof., vis. physician Rockefeller U., 1973-81; prof. medicine and pediatrics Cornell U. Med. Coll., 1973-81; dir. research Meml. Sloan-Kettering Cancer Ctr., v.p., 1980-81; dir. research Meml. Hosp. for Cancer and Allied Diseases, 1973-80, also attending physician depts. medicine and pediatrics; attending pediatrician N.Y. Hosp., 1973-81; mem., head cancer research program Okla. Med. Research Found., 1982—; prof. pediatrics, research prof. medicine, prof. microbiology and immunology U. Okla. Health Scis. Ctr., 1982—; attending physician, head div. immunology Okla. Children's Meml. Hosp., 1982—; vis. investigator Rockefeller Inst. for Med. Research, N.Y.C., 1949-50, asst. physician to Hosp., 1949-50; attending pediatrician Hennepin County Gen. Hosp., 1950-73, cons., 1960-73; Mem. Unitarian Service Commn. Med. Exchange Team to, France, Germany, Switzerland and Czechoslovakia, 1958; cons. VA Hosp., Mpls., 1959-60; cons., sci. adviser Nat. Jewish Hosp., Denver and Childrens Asthma Research Inst. and Hosp., Denver, 1964-69; mem. study sects. USPHS, 1952-69; mem. expert adv. panel on immunology WHO, 1967—; cons. Merck & Co., N.J., 1968—, Nat. Cancer Inst., 1973-74; mem. ad hoc com. President's Sci. Adv. Council on Biol. and Med. Sci., 1970, Pres.'s Cancer Panel, 1972; mem. Lyndon B. Johnson Found. awards com., 1972; mem. adv. com. Bone Marrow Transplant Registry, 1973—; fgn. adv. Acad. Med. Scis., People's Republic of China, 1980—. Author, editor numerous books.; Contbr. articles to profl. jours. Mem. adv. council Childrens Hosp. Research Found., Cin., 1954-58; bd. dirs. Allergy Found. Am., 1973; bd. sci. advisers Jane Coffin Childs Meml. Fund Med. Research, 1972-74, Merck Inst. Therapeutic Research, 1972-76; chmn. Internat. Bone Marrow Registry, 1977-79. Recipient Borden Undergrad. Research award U. Minn. Med. Sch., 1946; E. Mead Johnson First award, 1955; Theobald Smith award, 1955; Parke-Davis 6th Ann. award, 1962; Rectors medal U. Helsinki, 1963-64; Pemberton Lectureship award, 1966; Gordon Wilson Gold medal, 1967; R.E. Dyer Lectureship award, 1967; Clemens Von Pirquet Gold medal 9th Ann. Forum on Allergy, 1968; Presidents medal U. Padua, Italy, 1968; Robert A. Cooke Gold medal Am. Acad. Allergy, 1968; John Stewart Meml. award Dalhousie U., 1969; Borden award Assn. Am. Med. Colls., 1970; Howard Taylor Ricketts award U. Chgo., 1970; Gairdner Found. award, 1970; City of Hope award, 1970; Am. Acad. Achievement golden plate award, 1970; Albert Lasker award for clin. and med. research, 1970; A.C.P. award, 1972; Am. Coll. Chest Physicians award, 1974; Lila Gruber award Am. Acad. Dermatology, 1974; award in cancer immunology Cancer Research Inst. N.Y., 1975; Outstanding Achievement award U. Minn., 1978; award Am. Dermatological Soc. Allergy and Immunology, 1978; 1st Sarasota Med. award, 1979; sect. on mil. pediatrics award Am. Acad. Pediatrics, 1980; recipient Univ. medal Hacettepe U., Ankara, Turkey, 1982; numerous others.; Fellow Nat. Found. for Infantile Paralysis, 1947; Helen Hay Whitney Found. fellow, 1948-50; Markle Found. scholar, 1950-55. Fellow Acad. Multidisciplinary Research, AAAS, N.Y. Acad. Sci., Am. Acad. Arts and Scis.; mem. Am. Assn. History of Medicine, Am. Fedn. Clin. Research, Am. Assn. Anatomists, Am. Assn. Immunologists (past pres.), AAUP, Am., Mpls., Northwestern pediatric socs., Am. Rheumatism Assn., Am. Soc. Clin. Investigation (past pres.), Am. Soc. Exptl. Pathology (past pres.), Am. Soc. Microbiology, Assn. Am. Physicians, Central Soc. Clin. Research (past pres.), Harvey Soc., Infectious Disease Soc. Am. (Squibb award 1968), Internat. Soc. Nephrology, Internat. Acad. Pathology, Internat. Soc. for Transplantation Biology, Minn. State Med. Assn., Nat. Acad. Sci., Nat. Acad. Sci. Inst. Medicine (charter), Reticuloendothelial Soc. (past pres.), Soc. for Exptl. Biology and Medicine, Soc. for Pediatric Research, Am. Clin. and Climatol. Assn. (Gordon Wilson Gold medal 1967), Detroit Surg. Assn. (McGraw medal 1969), Internat. Soc. Blood, Transfusion, Practitioners' Soc., Am. Assn. Pathologists, Internat. Soc. Exptl. Hematology, Transplant Soc., Western Assn. Immunologists, Internat. Soc. Immunopharmacology (founding mem.), Phi Beta Kappa, Sigma Xi, Alpha Omega Alpha. Subspecialties: Immunopharmacology; Pediatrics. Office: Oklahoma Medical Research Foundation 825 NE 13th St Oklahoma City OK 73104

GOOD, ROLAND HAMILTON, JR., educator, theoretical physicist; b. Toronto, Ont., Can., Oct. 22, 1923; came to U.S., 1948, naturalized, 1950; s. Roland Hamilton and Marie (Smith) G.; m. Ferol Hendrickson, May 7, 1944; children—Roland Hamilton III, Patricia Gail, Sue Marie. B.M.E., Lawrence Inst. Tech., 1944; M.A.E., Chrysler Inst. Engring., 1946; M.S., U. Mich., 1948, Ph.D., 1951. Engr. Chrysler Corp., Windsor, Ont. and Highland Park, Mich., 1942-47; instr. U. Calif., Berkeley, 1951-53; asst. prof., asso. prof. Pa. State U., 1953-56, prof., 1972—, head physics dept., 1972-81; asso. prof., prof. physics Iowa State U., Ames, 1956-72, distinguished prof. scis. and humanities, 1970—; physicist, sr. physicist Ames Lab. of U.S. AEC, 1956-72; vis. lectr. U. Colo., summer 1958; NSF sr. postdoctoral fellow Inst. Advanced Study, Princeton, 1960-61; vis. prof. Inst. Math. Sci., Madras, India, 1968, Seoul Nat. U., 1979; guest Stanford Linear Accelerator Center, 1968-69. Author: (with T.J. Nelson) Classical Theory of Electric and Magnetic Fields, 1971. Fellow Am. Phys. Soc. Subspecialties: Theoretical physics; Particle physics. Current work: High energy scattering, field emission processes. Research, publs. theoretical physics, especially relativistic wave equations, polarization of elementary particles, metallic binding, electron emission from metals and spectroscopy of rare earths. Home: 24 S Barkway Ln State College PA 16801

GOODE, DAVID JOHN, physician, educator, researcher; b. Charlotte, N.C., Jan. 16, 1941; s. Thomas David and Florence G.; m. Lehoma Bain, June 10, 1962; children: Christopher Thomas, Elizabeth Daniels. B.A., U. N.C., 1962; M.D., Bowman Gray Sch. Medicine, 1966. Diplomate: Am. Bd. Psychiatry and Neurology. Intern U. Chgo., 1966-67, resident in psychiatry, 1969-72, Found. Fund for Research in Psychiatry fellow, 1972-73, asst. prof., 1974-77; assoc. prof. Bowman Gray Sch. Medicine, 1977—; dir. research Ill. State Psychiat. Inst., 1974-77; dir. clin. research Broughton Hosp., Morgantown, N.C., 1977—. Contbr. articles in field. Served with USPHS, 1967-69. William Reynolds scholar, 1962-66. Mem. Am. Psychol. Assn., Soc. Biol. Psychiatry, Neurosci. Soc., N.C. Med. Soc., N.C. Neuropsychiat. Assn., AAAS. Moravian. Subspecialties: Psychiatry; Psychopharmacology. Current work: Areas of human spinal reflexes as related to central dopamine metabolism in schizophrenia and cortical laterality as related to mental illness. Office: Department of Psychiatry Bowman Gray School of Medicine Winston Salem NC 27103

GOODE, GLENN AMOS, JR., nuclear engineer; b. Yuma, Ariz., May 5, 1946; s. Glenn Amos and Virginia Lee (Altstatt Schilliger) G.; m. Sandra Rose Mackensen, June 17, 1972 (div. Apr. 1982). A.S., Ariz. Western Coll., 1974; B.S., U. Ariz., 1976; M.B.A., Kent State U., 1982. Registered profl engr., Ohio. Electronics technician U.S. Navy, 1965-67; nuclear reactor operator 1967-71; startup engr. Gen. Electric Co., San Jose, Calif., 1976-82, lead startup engr., 1982-83, ops. mgr., 1983—; cons. Hatch Nuclear Plant, Ga., 1983—. Served with USN, 1965-71. Mem. Am. Nuclear Soc., Nat. Soc. Profl. Engrs., Am. Welding Soc., Ohio Soc. Profl. Engrs. Republican. Lutheran. Subspecialties: Nuclear engineering; Nuclear fission. Current work: Nuclear steam supply systems. Office: Gen Electric Co 175 Curtner Ave San Jose CA 95125 Home: 112 Bloomfield Rd Vidalia GA 30474

GOODE, MELVYN DENNIS, cell biologist; b. New Albany, Ind., Feb. 18, 1940; s. Delmar T. and Anita (May) G.; m. Judith Fincham, Aug. 4, 1968. B.S., U. Kans., 1963; Ph.D., Iowa State U., 1967. Research assoc. anatomy U. Pa., 1967-68; asst. prof. U. Md., College Park, 1968-73, assoc. prof., 1973—, dir., 1973—. Contbr. articles to sci. jours. Recipient Sci. Photography Excellence award Nikon, 1979. Mem. Am. Soc. Cell Biology, Chesapeake Soc. Electron Microscopy (pres.), Internat. Soc. Evolutionary Protistology (pres. elect). Subspecialties: Cell biology; Biophysics (biology). Current work: Cell motility, especially the role of microtubules in mitosis and secretion. Home: 1901 Dana Dr Adelphi MD 20783 Office: Dept Zoology U Md College Park MD 20742

GOODE, MONROE JACK, plant pathologist, educator; b. Whitney, Ala., Feb. 15, 1928; s. Paul and Lila Mae (Puckett) G.; m. Ethel Lorene Carter, May 27, 1950; children: Dana Faye Goode Bassi, Paula Carol, Monroe Carter. Diploma, Meridian Jr. Coll., 1950; B.S., Miss. State U., 1952, M.S., 1954; Ph.D., N.C. State U., 1957. Asst. prof. dept. plant pathology Sch. Agr. and Agrl. Expt. Sta., U. Ark., Fayetteville, 1957-61, assoc. prof., 1961-66, prof., 1966—. Contbr. articles to profl. jours. Served with U.S.Army, 1946-49. Mem. Am. Phytopath. Soc., Sigma Xi, Phi Kappa Phi, Beta Beta Beta; Gamma Sigma Delta Alpha Zeta. Subspecialties: Plant pathology; Plant genetics. Current work: Etiology, epidemiology and control of diseases of vegetable crops, developing disease resistant vegetable crops; phenotype breeding for disease resistance; pathogenic variation in microorganisms. Office: Dept Plant Pathology U Ark Fayetteville AR 72701

GOODE, PHILIP RANSON, physicist, educator; b. San Francisco, Jan. 4, 1943; s. Philip Carl and Ruth Starr (Gifford) G.; m. Donna Carol Lundy, Aug. 15, 1964; m. Maureen Anne Mortell, May 21, 1983. A.B., U. Calif.-Berkeley, 1964; Ph.D., Rutgers U., 1969. Research assoc. Rutgers U., New Brunswick, N.J., 1969, asst. prof., 1971-77; mem. tech. staff Bell Telephone Labs., Murray Hill, N.J., 1977-80; assoc. research prof. U. Ariz., Tucson, 1980—; research assoc. U. Rochester, N.Y., 1969-71; cons. in field. Contbr. articles to profl. publs. Recipient numerous research grants from fed. agys. Mem. Am. Phys. Soc. Club: Big C Soc. (Berkeley, Calif.). Subspecialties: Theoretical astrophysics; Theoretical physics. Home: 3027 N Sparkman Blvd Tucson AZ 85716 Offic: U Ari Dept Physics Tucson AZ 85721

GOODELL, HORACE GRANT, environmental sciences educator, consultant; b. Decatur, Ill., Oct. 12, 1925; s. Horace Holbrook and Frieda May (Smith) G.; m. Mona Lee Cluck, Sept. 3, 1956; children: Laurie, Katherine, Zachary. B.A., So. Methodist U., 1955; Ph.D., Northwestern U., 1957. Asst. prof. Fla. State U., 1956-60, assoc. prof., 1960-68, prof., 1968-70; prof. environ. scis. U. Va., 1970—, chmn. dept. environ. scis., 1971-78. Served to capt. USNR, 1943-49, 51-52. Decorated Air medal with 2 gold stars, Purple Heart. Fellow Geol. Soc. Am.; mem. Am. Geophys. Union, Am. Assn. Petroleum Geologists, Soc. Econ. Paleontologists and Mineralogists, Sigma Xi. Subspecialties: Hydrogeology; Geochemistry. Current work: Hydrogeology, marine goelogy—teaching/research. Office: U Va Clark Hall Charlottesville VA 22903

GOODELL, JOHN BOYDEN, optical physicist; b. Montclair, N.J., Feb. 20, 1927; s. John Boyden and Viola Joan (Henry) G.; m. Annette Carla Schirokauer, June 17, 1960. B.A., Johns Hopkins U., 1956, M.A., 1960. Research asst. Johns Hopkins U., Balt., 1958-60, research assoc., Lab. Astrophysics, 1961-62; sr. engr. Westinghouse Electric Corp., Def. and Space Ctr., Balt., 1960-61, fellow, engr. systems devel. div., 1962—; research assoc. Lab. Astrophysics, Johns Hopkins U. Mem. Optical Soc., IEEE, Delta Phi Alpha. Democrat. Subspecialties: Aerospace engineering and technology; Optical astronomy. Current work: Johns Hopkins ultra-violet telescope to fly on space shuttle, ultra-violet optical design, optical components for bulk acousto-optics processors, acousto-optic and acoustic transducer design. Patentee in field. Office: Box 1521 MS 3714 Baltimore MD 21203

GOODENOUGH, SAMUEL HENRY, biomedical engineer, consultant; b. Oakland, Calif., Sept. 8, 1919; s. Samuel Henry and Katherine Mary (Amet) G.; m. Frances Carolyn, Oct. 12, 1940; m. Kathleen Eula, Sept. 15, 1962; 1 son, John Samuel. Student, U. Calif., 1943-51. Registered profl. mech. engr., Ill. Chief tool engr. Warwick Mfg. Corp., Chgo., 1959-61; sr. research engr. Cutter Biomed. Inc., San Diego and Berkeley, Calif., 1963-77; sr. engr. Shiley Labs., Irvine, Calif., 1979-82; v.p. engring. and mfg. Mitral Med. Internat. Inc., Irvine, 1979-82; cons. mech. engring., biomed. engring, Carlsbad, Calif., 1982—; cons. and expert witness in product liability litigation. Mem. Assn. Advancement Med. Instrumentation, Nat. Soc. Profl. Engrs., ASME. Subspecialties: Biomedical engineering; Mechanical engineering. Current work: Cardiac valve prostheses; manufacturing processes for metal working. Patentee cardiac valve prostheses. Home and Office: 5040 Tierra Del Oro Carlsbad CA 92008

GOODFRIEND, PAUL LOUIS, chemist, educator; b. Dallas, Aug. 10, 1930; s. Isidore and Anna (Berger) G.; m. Beverly Lebar, June 28, 1953; children: Benedict, Jason. B.S., U. Va., 1952; Ph.D., Ga. Inst. Tech., 1957. Postdoctoral fellow U. Rochester, N.Y., 1956-58; asst. prof. chemistry Coll. William and Mary, 1958-61; sr. research chemist Texaco Export Inc., Richmond, Va., 1961-63; assoc. prof. chemistry U. Maine, Orono, 1966-70, prof., 1970—. Contbr. articles to sci. jours. Mem. Am. Chem. Soc., Sigma Xi. Democrat. Jewish. Subspecialties: Theoretical chemistry; Physical chemistry. Current work: Molecular quantum mechanics, spectroscopy, kinetics. Home: 118 Howard St Bangor ME 04401 Office: Dept Chemistry U Maine Orono ME 04469

GOODHUE, WILLIAM LEHR, research scientist, educator; b. N.Y.C., June 15, 1946; s. William Stemm and Irma (Lehr) G.; m. Heather P. Thiel, Aug. 6, 1972; children: Elizabeth, Jonathan. B.S., Poly. Inst. Bklyn., 1967, M.S., 1967; Ph.D., Courant Inst., NYU, 1971. Asst. prof. U. Notre Dame, 1971-77; mem. tech. staff GTE Labs., Waltham, Mass., 1977-79; mgr. software tools Xerox Corp., Rochester, N.Y., 1979-82; research scientist Comsat Labs., Rockville, Md., 1982—; instr. Boston U., 1979, U. Md., 1982—. Mem. Am. Math. Soc., Soc. Indsl. and Applied Math., Assn. Computing Machinery, IEEE (affiliate). Subspecialties: Software engineering; Systems engineering. Current work: Engineering development of computer networks and telecommunication systems; software development environments for microprocessor-based systems. Home: 9516 Reach Rd Potomac MD 20854 Office: Comsat Labs 2230 Comsat Dr Clarksburg MD 20871

GOODIN, JOE RAY, plant physiologist, adminstr.; b. Claude, Tex., July 28, 1934; s. Emery Lee and Leta Elizabeth (Culver) G. B.S., Tex. Tech. U., 1955; M.S., Mich. State U., 1958; Ph.D., UCLA, 1963. Asst. prof., asst. agronomist U. Calif., Riverside, 1964-70; assoc. prof., prof. biology Tex. Tech. U., Lubbock, 1970—, dep. dir.; cons. Nat. Acad. Scis., AAAS, U.S. Congress. Contbr. numerous articles on plant physiology to profl. jours. Served to capt. USAF, 1955-57. Grantee Pacific Sci. Congress, Tokyo, 1967; Internat. Grasslands Congress, Sydney, Australia, 1970; Smithsonian Instn., Egypt, 1976. Mem. Am. Soc. Plant Physiologists, Bot. Soc. Am., AAAS, Soc. Developmental Biology. Methodist. Subspecialties: Plant physiology (biology); Tissue culture. Current work: Physiology and biochemistry of water and salt stress in plants, environmental stresses and crop productivity, biomass for energy production. Office: ICASALS Tex Tech U Lubbock TX 79409

GOODING, CHARLES THOMAS, coll. adminstr.; b. Tampa, Fla., Nov. 18, 1931; s. Charles Thomas and Gladys Violet (Bingman) G.; m. Shirley Ann Puckett, June 7, 1953; children: Steven Thomas, Carol Ann, David Lee, Mark Charles. B.A., U. Fla., 1954; postgrad., U. Tampa, 1957-58; M.Ed., U. Fla., 1962, Ed.D., 1964. Tchr. Meml. Sch., Tampa, Fla., 1957-58; tchr., prin., St. Mary's Sch., Tampa, 1958-62; instr. U. Fla., Gainesville, 1963-64; assoc. prof. to prof. SUNY-Oswego, 1964—, dean grad. studies, 1982—; vis. prof. U. Liverpool, Eng., 1979-80; grad. fellow U. Fla., 1962-63; cons. Sodus (N.Y.) Pub. Schs., 1970, Norfolk (Va.) Pub. Schs., 1973, Jamesville-Dewitt (N.Y.) Pub. Schs., 1982, Liverpool (N.Y.) Pub. Schs., 1982. Author: (with Combs) Florida Studies in Helping Professions, 1969, (with Pittenger), Learning Theories in Educational Practice, 1971; contbr. articles to profl. jours. Bd. dirs. Oswego County unit Am. Cancer Soc., 1972-73, 82—. Served to 1st lt. U.S. Army, 1954-56. SUNY Research Found. grantee, 1966, 69, 70; N.Y. Dept. Edn. grantee, 1971-72; vis. scholar grantee U. Liverpool, 1979-80; NSF grantee, 1980-82. Mem. AAAS, AAUP, Am. Ednl. Research Assn., Am. Psychol. Assn., Brit. Ednl. Research Assn., Eastern Ednl. Research Assn. (v.p. 1979-81, council 1982—). Democrat. Episcopalian. Clubs: Classic Jaguar Assn. Internat., Pathfinder Antique Auto (pres. 1980-82). Subspecialties: Learning; Cognition. Current work: Educational psychology and human learning and cognition; evaluation of human interaction and learning. Descriptors: Discussion, wait time and learning interaction. Home: 4169 W River Rd PO Box 231 Minetto NY 13115 Office: Office Grad Studies Culkin Hall SUNY Oswego NY 13126

GOODISMAN, JERRY, chemist, educator, researcher; b. Bklyn., Mar. 22, 1939; s. Abraham and Ida (Machover) G.; m. Mireille Eifermann, June 28, 1963; children: Natalie, Michael. B.A., Columbia U., 1959; M.A., Harvard U., 1960, Ph.D., 1963. Instr. U. Ill., 1963-65, asst. prof. chemistry, 1965-69; assoc. prof. Syracuse U., 1969-71, prof., 1971—. Mem. Am. Chem. Soc., Am. Phys. Soc., AAAS. Subspecialties: Physical chemistry; Surface chemistry. Current work: Metal surfaces, small-angle x-ray scattering. Home: 129 Edgemont Dr Syracuse NY 13214 Office: Syracuse U 308E Bowne Hall Syracuse NY 13210

GOODMAN, ARDEN PATRICIA, chemist, breeder Arabian horses; b. Flushing, N.Y., Aug. 16, 1949; d. Sheldon Stuart and Elizabeth Lillian (Weiss) G.; m. Joseph Theodore Gacsi, Dec. 26, 1976; children: Ted, Vickie. B.S. in Chemistry/Life Sci. (scholar), U. Ill., 1971. Quality assurance chemist Ferro Corp., Huntington Beach, Calif., 1971-72, Beckman Instruments Co., Fullerton, Calif., 1972-73; research chemist Edwards Labs. div. Am. Hosp. Supply Corp., Irvine, Calif., 1974-80, mgr. quality assurance, 1980-81, mgr. quality assurance engring., 1981—. Contbr. articles to sci. jours. Mem. Am. Chem. Soc., Am. Soc. for Quality Control, Am. Horse Show Assn., Internat. Arabian Horse Assn.. Subspecialties: Ophthalmology; Prosthetics. Current work: New product review prior to market introduction of Class II and Class III medical devices, including implants, for use by ophthalmic surgeons; ensure that all good manufacturing practice requirements for product release have been met. Office: 1402 E Alton Ave Irvine CA 92714

GOODMAN, COREY SCOTT, biological sciences educator; b. Chgo., June 29, 1951; s. Arnold Harold and Florence (Friedman) G. B.S., Stanford U., 1972; Ph.D., U. Calif.-Berkeley, 1977. Postdoctoral fellow U. Calif.-San Diego, 1977-79; asst. prof. dept. biol. scis. Stanford (Calif.) U., 1979-82, assoc. prof., 1982—. Contbr. articles to profl. jours. Recipient 2d Internat. award in Neurosci. Demuth Swiss Med. Research Found, 1983, Alan T. Waterman award Nat. Sci. Bd., 1983; Grantee NIH, NSF, McKnight Found., March of Dimes Found.; Alfred P. Sloan Research fellow, 1980-83. Subspecialties: Developmental biology; Neurobiology. Current work: Developmental neurobiology, developmental biology. Home: 768 Arnold Way Menlo Park CA 94025 Office: Dept Biol Scis Stanford U Stanford CA 94305

GOODMAN, DAVID BARRY POLIAKOFF, biochem. endocrinologist, clin. pathologist; b. Lynn, Mass., June 1, 1942; s. Nathan and Eva (Poliakoff) G.; m. Constance Cutler, Aug. 22, 1965; children: Derek, Alex. A.B., Harvard U., 1964; M.D., U. Pa., 1968, Ph.D. in Biochemistry, 1972. Lic. physician, Pa., Conn. Research asst. prof. pediatrics U. Pa. Med. Sch., Phila., 1972-76; from asst. prof. to asso. prof. medicine Yale U. Sch. Medicine, New Haven, 1976-80; asso. prof. dept. pathology and lab. medicine U. Pa. Med. Sch., 1980-82, prof., 1982—, dir. div. lab. medicine, 1980—; cons. NIH, NSF, VA. Mgn. editor: Metabolic Bone Disease and Related Research, 1978—; Contbr. over 100 articles to sci. jours. Recipient Achievement award Upjohn Co., 1968; established investigator Am. Heart Assn., 1977. Mem. N.Y. Acad. Scis. (Laport award 1981), AAAS, Am. Fedn. Clin. Research, Acad. Clin. Lab. Physicians and Scientists, Am. Assn. Pathologists, Am. Heart Assn., Am. Physiol. Soc., Am. Soc. Bone and Mineral Research, Soc. Developmental Biology, Soc. Neurosci. Subspecialties: Membrane biology; Endocrinology. Current work: Role of lipids and hormones in control of membrane permeability; diagnostic biochemistry. Office: 3400 Spruce St Philadelphia PA 19104

GOODMAN, DAVID WAYNE, chemist; b. Glen Allen, Miss., Dec. 14, 1945; s. Henry Grady and Anniebelle (McDonald) G.; m. Sandra Faye Hewitt; 1 son, Jac. B.S. in Chemistry, Miss. Coll., 1968; Ph.D. in Phys. Chemistry, U. Tex., 1974. NATO postdoctoral fellow Technische Hochschule, Darmstadt, W. Ger., 1974-75; NRC postdoctoral fellow Nat. Bur. Standards, Washington, 1976-78, mem. staff, 1978-80; research scientist Sandia Nat. Labs., Albuquerque, 1980—. Contbr. articles to profl. jours. Mem. Am. Chem. Soc. (Ipatieff prize 1983, div. treas.), Am. Vacuum Soc. (mem. div. exec. council), Catalysis Soc. Subspecialty: Catalysis chemistry. Current work: Heterogeneous catalysis using modern surface techniques.

GOODMAN, HOWARD M., biochemistry educator; b. Bklyn., Nov. 29, 1938; s. Samuel G.; children: Sylena, William. Ed., Williams Coll., Mass. Inst. Tech., Cambridge (Eng.) U. Asst. prof. biochemistry U. Geneva, 1969-70; asst. prof. U. Calif., San Francisco, 1970-71, assoc. prof., 1971-76, prof., 1976-81; investigator Howard Hughes Med. Inst., 1978-81; chief dept. molecular biology Mass. Gen. Hosp., Boston, 1981—; prof. genetics Harvard Med. Sch., 1981—. Contbr. articles to profl. jours. Helen Hay Whitney Found. fellow; Am. Cancer Soc. fellow; Josiah Macy Faculty scholar. Mem. Am. Soc. Biol. Chemists. Subspecialty: Molecular biology. Office: Dept Molecular Biology Mass Gen Hosp Boston MA 02114

GOODMAN, JOEL WARREN, research scientist; b. N.Y.C., Jan. 2, 1933; s. Herman Joseph and Ann (Citron) G.; m. Janet Carol Petersen, July 18, 1964; children: Mark Craig, Clifford Scott. B.A., Bklyn. Coll., 1953; Ph.D., Columbia U., 1959. Asst. prof. U. Calif., San Francisco, 1960-65, assoc. prof., 1965-70, prof. microbiology and immunology, 1970—. Research grantee NIH, NSF, ACS. Mem. Am. Assn. Immunologists, AAAS, Planetary Soc. Subspecialties: Immunobiology and immunology; Molecular biology. Current work: Molecular basis of lymphocyte activation in the immune response.

GOODMAN, JULIUS, nuclear engineer, consultant, researcher; b. Odessa, USSR, July 19, 1935; came to U.S., 1979; s. Isaac and Eugenia (Lusher) Guttman; m. Rachel Bezpalko, July 4, 1959; 1 dau., Marina. M.S. in Theoretical Physics, State U., Odessa, 1958, Ph.D., Inst. Nuclear Physics, Tashkent, USSR, 1962, Inst. Tech. Odessa, 1965. Sr. researcher Inst. Nuclear Physics, Tashkent, Acad. Sci., USSR, 1958-63; prof. Inst. Tech., Odessa, 1963-70, Poly U., 1970-76; sr. engr. Bechtel Power Corp., Norwalk, Calif., 1980—. Author: Professional Education, 1975, (with P.U. Arifov) Positron Diagnostic, 1978; contbr. numerous articles to profl. jours. Pres. Hatchiya (Revival), Orange County, Calif., 1982—. Mem. Am. Nuclear Soc. Club: Taostmaster (Fullerton, Calif.). Lodge: B'nai B'rith. Subspecialties: Nuclear engineering; Theoretical physics. Current work: Probabilistic risk assessment; general theory, external hazards-earthquakes, tornado, transportation of toxic and explosive chemicals; general relativity, atomic and nuclear physics, physics of nuclear reactors. Patentee nuclear reactor with UF-6. Home: 1630 Via Linda Fullerton CA 92633 Office: Bechtel Power Corp 12400 E Imperial Hwy Norwalk CA 90650

GOODMAN, MAJOR M., geneticist, educator; b. Des Moines, Sept. 13, 1938; s. Jarrett W. and Mabel O. (Michael) G.; m. Sheila D. Balfour, Aug. 8, 1970; stepchildren: Sean Balfour Dail, Andrew Scot Dail. B.S., Iowa State U., 1960; M.S., N.C. State U., 1963, Ph.D., 1965. NSF postdoctoral fellow Instituto de Genetica, Escola Superior de Agricultura Luiz de Quiroz, Piracicaba, Sao Paulo, Brazil, 1965-67; vis. asst. prof., then asst. prof., then assoc. N.C. State U., Raleigh, from 1967, now prof. crop sci., statistics, genetics and botany; chmn. Maize Crop Adv. Com., U.S. Dept. Agr. Author: Races of Maize in Brazil and Adjacent Areas, 1977; also articles. Recipient research awards, scholarships and fellowships. Mem. Crop Sci. Soc., Genetics Soc. Am., Soc. Econ. Botany, Systematic Zoology Soc., Soc. Am. Naturalists, Am. Bot. Soc., Internat. Soc. Plant Taxonomy, Classification Soc., Sigma Xi. Subspecialties: Plant genetics; Evolutionary biology. Current work: Classification and Utilization of races of maize; maize isozyme genetics. Office: Crop Science Dept NC State U Box 5155 Raleigh NC 27650

GOODMAN, MICHAEL GORDON, immunologist, physician; b. Denver, July 4, 1946; s. Nelson and Florence Rhea G.; m. Jacquelyn Goodman, Feb. 18, 1978; children: Devin, Brielle. B.A., Yale U., 1968; M.D., U. Calif.-San Francisco, 1972. Diplomate: Am. Bd. Internal Medicine. Med. intern U. Miami (Fla.) Affiliated Hosps., 1972-73; resident in medicine Thomas Jefferson U. Hosp., Phila., 1973-74; Wadsworth VA Hosp./UCLA Hosps., 1974-75; postdoctoral fellow in immunology Scripps Clinic and Research Found., La Jolla, Calif., 1975-78, asst. mem. I, 1978-80, asst. mem. II, 1980—; clin. instr. dept. medicine U. Calif-San Diego Sch. Medicine, 1982—; vis. physician San Diego VA Hosp., 1982—. Contbr. articles to profl. jours. NIH fellow, 1975-78; career devel. awardee, 1980—; Arthritis Found. fellow, 1978-80; USPHS grantee, 1978-81, 81—. Mem. ACP, Am. Soc. Internal Medicine, Am. Assn. Immunologists, N.Y. Acad. Scis. Subspecialties: Immunology (medicine); Oncology. Current work: Mechanisms of lymphocyte activation and immune regulation. Immunology, lymphocyte activation, nucleosides, anaphylatoxin, lymphokines, arachidonic acid metabolism, oncology. Office: 10666 N Torrey Pines Rd La Jolla CA 92037

GOODMAN, MORRIS, anatomy educator; b. Milw., Jan. 12, 1925; s. Benjamin and Sara (Bratt) G.; m. Selma Kessler, Apr. 5, 1946; children: Louise, Julia, David. B.S., U. Wis., 1948, M.S., 1949, Ph.D., 1951. NIH postdoctoral fellow Calif. Inst. Tech., Pasadena, 1951-52; research assoc. U. Ill. Coll. Medicine, Chgo., 1952-54, Detroit Inst. Cancer Research, 1954-58; sr. investigator in immunology Lafayette Clinic, Detroit, 1958-65; research assoc. prof. anatomy Wayne State U., Detroit, 1960-66, prof., 1966—; dir. research Plymouth (Mich.) State Home, 1966-72. Editor: Molecular Anthropology, 1976, Macromolecular Sequences and Systematic and Evolutionary Biology, 1982; co-editor: Jour. Human Evolution, 1971-77; mem. editorial adv. bd.: Human Biology, 1974—, Jour. Molecular Evolution, 1975—, Progress in Sci. Culture, 1975—, Advances in Primatology, 1975—, Jour. Human Evolution, 1977—; contbr. over 190 articles to sci. publs. Mem. research com. Mich. Assn. Retarded Citizens, Lansing, 1981—; bd. dirs. Voice of Reason, N.Y.C., 1982—. Serviced with USAF, 1943-45. NSF grantee, 1963—. Mem. Soc. Systematic Zoology, Am. Assn. Anatomists, Soc. for Study of Evolution, Am. Naturalist Soc. Subspecialty: Evolutionary biology. Current work: Use of recombinant DNA and nucleotide sequencing technologies to investigate human origins, mammalian phylogeny, and processes of molecular evolution. Home: 24211 Oneida St Oak Park MI 48205 Office: 540 E Canfield Ave Detroit MI 48201

GOODMAN, RICHARD H., mathematics educator; b. Bklyn., Sept. 8, 1942; s. Joseph L. and Lillian G. A.B. magna cum laude, Harvard U., 1963, A.M., 1964, Ph.D., 1971. Assoc. prof. math and computer sci. U. Miami, Coral Gables, Fla., 1970—. Mem. Sigma Xi, Phi Beta Kappa. Subspecialty: Numerical analysis. Current work: Floating point computer arithmetic, numerical solution of differential equations. Home: 610 Tibidabo Ave Coral Gables FL 33143 Office: Dept Math U Miami Coral Gables FL 33124

GOODMAN, ROBERT L., physician, educator; b. Port Chester, N.Y., Dec. 19, 1940; s. Joseph and May (Jeruss) G.; m. Paula Edelman, June 25, 1961; children—Debra, Ellen. A.B., Dartmouth Coll., 1962; M.D., Coll. of Physicians and Surgeons, Columbia U., 1966; M.A. (hon.), U. Pa., 1977. Diplomate: Am. Bd. Internal Medicine, Am. Bd. Radiology. Intern in medicine Beth Israel Hosp., Boston, 1966-67, resident in medicine, 1967-68, chief resident in medicine, 1969-70; fellow in hematology Presbyn. Hosp., N.Y.C., 1968-69; fellow in radiation therapy Harvard Joint Center for Radiation Therapy, Boston, 1972-74; research fellow in medicine Coll. of Physicians and Surgeons, Columbia U., N.Y.C., 1968-69; instr. medicine Harvard Med. Sch., Boston, 1969-70, instr. radiation therapy, 1974-75, asst. prof. radiation therapy, 1975-77; asst. in medicine Beth Israel Hosp., Boston, 1974-77, oncologist, 1974-77, radiation therapist, 1974-77, Children's Hosp., 1974-77, Peter Bent Brigham Hosp., 1974-77, New Eng. Deaconess Hosp., 1974-77, Boston Hosp., 1974-77; chmn. dept. radiation therapy U. Pa. Sch. Medicine, Phila., 1977—, asso. prof. radiation therapy, 1977-79, prof. radiation therapy, 1979—; dir. dept. radiation therapy Fox Chase Cancer Center, Phila., 1979—; chmn. glioma-misonidazole study Radiation Therapy Oncology Group, 1978-79; mem. cancer clin. investigation rev. com. NIH, 1977-80. Contbr. articles on radiation therapy and oncology to med. jours.; asso. editor: Internat. Jour. of Radiation Oncology, 1977—. Served to maj. M.C. U.S. Army, 1970-72. Mem. Am. Coll. Radiology, Am. Soc. Therapeutic Radiologists, Am. Radium Soc., Am. Soc. Hematology, Am. Soc. Clin. Oncology (mem. program com. 1979—), Keystone Area Soc. Radiation Oncologists (chmn. program com. 1980—), Pa. Oncologic Soc., Soc. for Study of Breast Disease, Pa. Med. Soc., Phila. County Med. Soc., John Morgan Soc., AAAS, N.Y. Acad. Scis., AAUP, Phila. Roentgen Ray Soc., Soc. Chairmen of Acad. Radiation Oncology Programs, Phi Beta Kappa. Subspecialty: Radiology. Home: 1710 Martins Ln Gladwyne PA 19035 Office: 3400 Spruce St Philadelphia PA 19104

GOODMAN, ROBERT MERWIN, virologist, research scientist; b. Ithaca, N.Y., Dec. 30, 1945; s. Robert Browning and Janet Edith (Pond) G.; m. Nancy Aileen Rick, June 22, 1968; m. Linda Joyce Magrum, June 11, 1974. Student, Johns Hopkins U., 1963-65; B.Sc., Cornell U., 1967, Ph.D., 1973. Vis. research fellow John Innes Inst., Norwich, Eng., 1973-74; asst. prof. U. Ill., Urbana, 1974-78, assoc. prof., 1978-81, prof. plant virology and internat. agr., 1981—; v.p. research and devel. Calgene, Inc., Davis, Calif., 1982—. Assoc. editor: Virology, 1976—; editor: Expanding the Use of Soybeans, 1976. Pres. Channing-Murray Found., Urbana, 1976-78. NSF-NATO postdoctoral fellow, 1973. Mem. AAAS, Genetics Soc. Am., Am. Phytopath. Soc., Am. Soc. Virology, N.Y. Acad. Scis. Unitarian. Subspecialties: Genetics and genetic engineering (agriculture); Plant virology. Current work: Molecular biology of plant DNA viruses, gene structure and expression in plants. Home: 209 W California Ave Urbana IL 61801 Office: 1910 5th St Suite F Davis CA 95616

GOODMAN, STEVEN RICHARD, cell biologist; b. N.Y.C., Dec. 29, 1949; s. Martin Jay and Natalie (Hochberg) G.; m. Amy Lou Bobula, Dec. 20, 1980; m. Karen Sue Jacobs, June 20, 1971 (div.); 1 dau., Laela Beth. B.S., SUNY-Stony Brook, 1971; Ph.D., St. Louis U., 1976. Postdoctoral fellow Harvard Med. Sch., Boston, 1976-77; NIH research fellow Harvard U., Cambridge, Mass., 1977-79; asst. prof. Pa. State U., Hershey, 1979—; dir. Multidiscipline Labs. Milton S. Hershey Med. Ctr., Hershey, Pa., 1979—; cons. program project Nat. Heart, Lung and Blood Inst., Nat. Inst. Arthritis, Metabolism and Digestive Diseases, NIH, 1981—; established investigator Am. Heart Assn., 1982—. Recipient Faculty Devel. award Alcoa Found., 1982—; NIH research grantee, 1980—. Mem. Am. Soc. Cell Biology, Biophys. Soc., Am. Soc. Hematology. Jewish. Club: Red Cell. Subspecialties: Membrane biology; Cell biology (medicine). Current work: Research involves the structure and function of the membrane associated cytoskeletal proteins of erythrocytes as well as nonerythroid cells, determination of genetic defects in some of these proteins in pathophysiological erythrocytes. Home: 2405 Rudy Rd Harrisburg PA 17104 Office: Pa State U Milton S Hershey Med Ctr PO Box 850 Hershey PA 17033

GOODRICH, JAMES TAIT, neuroscientist, neurosurgeon; b. Portland, Ore., Apr. 16, 1946; s. Richard and Gail (Josselyn) G.; m. Judy Loudin, Dec. 27, 1970. Student, Golden West Coll., 1971-72; A.A., Orange Coast Coll., 1972; B.S. cum laude, U. Calif., Irvine, 1974; M.Phil., Columbia U. Grad. Sch. Arts and Scis., 1979, Ph.D., 1970; M.D., Coll. Physicians & Surgeons, 1980. Neuroscientist N.Y. Neurol. Inst., N.Y.C., 1981—. Contbr. articles to profl. jours. Recipient Roche Labs. award in neuroscis., 1978, Mead-Johnson award, 1978, Bronze medal Alumni Assn. Coll. Physicians and Surgeons, 1980, Sandoz award for outstanding research, 1980; Willamette Industries scholar; NIH grantee. Mem. N.Y. Acad. Medicine (Melicow award 1980), Am. Assn. History of Medicine (Sir William Osler medal 1977-78), AMA, Brit. Brain Research Assn., European Brain Research Assn., Friends of Columbia U. Libraries, Friends of Osler Library of McGill U., N.Y. Acad. Scis., Med. History Soc. N.J., ISIS History of Sci. Soc., Soc. for Bibliography of Natural History (London), Columbia Presbyterian Med. Soc., U. Calif. Alumni Assn., Soc. Ancient Medicine, AAAS, Am. Osler Soc., Les Amis du Vin, South Coast Wine Explorers Club (past chmn.), Friends of Bacchus Wine Club (past chmn.), Dionysius Council of Presbyn. Hosp. of N.Y.C., Sigma Xi, Alpha Gamma Sigma. Subspecialties: Anatomy and embryology; Regeneration. Current work: Research in neuronal regeneration and brain reconstruction. Home: 214 Everett Pl Englewood NJ 07631 Office: Dept Neurosurgery NY Neurol Inst New York NY 10032

GOODRIDGE, ALAN GARDNER, pharmacology educator; b. Peabody, Mass., Apr. 2, 1937; s. Lester E. and Gertrude E. (Gardner) G.; m. R. Ann Funderburk, Aug. 19, 1960; children: Alan G., Bryant C. B.S. in Biology, Tufts U., 1958; M.S. in Zoology, U. Mich., 1963, Ph.D., 1964. Univ. fellow in biochemistry Harvard Med. Sch., Boston, 1964-66; asst. prof. physiology U. Kans. Med. Ctr., Kansas City, 1966-68; assoc. prof. med. research U. Toronto, Ont., Can., 1968-76, prof., 1976-77; prof. pharmacology and biochemistry Case Western Res. U. Sch. Medicine, Cleve., 1977—; mem. biochemistry study sect. NIH, 1979-83. Editorial bd.: Jour. Biol. Chemistry, 1979—. Served to lt. (j.g.) USN, 1958-61. Macy faculty scholar, Paris, 1975-76. Mem. Am. Soc. Biol. Chemists, Am. Physiol. Soc., AAAS, Sigma Xi. Subspecialties: Biochemistry (biology); Molecular biology. Current work: Molecular basis for hormonal and nutritional regulation of gene expression; regulation of metabolic pathways. Home: 2932 Broxton Rd Shaker Heights OH 44120 Office: 2119 Abington Rd Cleveland OH 44106

GOODSON, RAYMOND EUGENE, systems and control engineer, corporate executive; b. Canton, N.C., Apr. 22, 1935; m., 1957; 2 children. A.B., Duke U., 1957, B.S.M.E., 1959; M.S.M.E., Purdue U., 1961; Ph.D. in Fluids, Automatic Control, 1963. From asst. prof. to assoc. prof. Purdue U., 1963-70, prof. mech. engring., 1970-81, dir., 1976—, 1977-81; group v.p. automotive products group Hoover Universal, Inc., Saline, Mich., 1981—; cons. to industry, govt.; vis. prof. Weizmann Inst. Sci., Rehovot, Israel, 1972; chief scientist Dept. Transp., 1973-75; mem. Fed. Interagy. Task Force Motor Vehicle Goals Beyond, 1980, 1975-76; mem. adv. com. guidance and control NASA, 1973-74; chmn. transp. subcom., energy storage com. Nat. Acad. Sci., 1975; dir. opportunity risk analysis project ERDA, 1977—; assoc. dir. Engring. and Exptl. Sta., 1978—; rep. US/USSR Environ. Agreement, Transp. Source Air Pollution Control Tech., USSR, 1976; mem. com. advanced energy storage systems NRC-Nat. Acad. Engring., 1976; tech. rep. UN Motor Vehicle Conf., Internat. Metalworkers Fedn., Paris, 1976; mem. com. on nuclear and alternative energy systems study and chmn. transp. resource group Demand-Conservation Panel Nat. Acad. Sci./Nat. Acad. Engring., 1976-77; mem. energy engring. bd. NRC Assembly Engrs.; chmn. Office Tech. Assessment, Automobile Adv. Panel. Recipient Arch T. Colwell Merit award Soc. Automotive Engrs., 1973; Meritorious Achievement award Sec. Dept. Transp., 1975; Edward J. Speno Automotive Safety award 5th Internat. Congress Automotive Safety, Cambridge, 1977; Japanese Soc. Promotion Sci. traveling scholar, 1972; grantee NSF, NASA, Office Naval Research. Mem. ASME, Instrument Soc. Am., IEEE. Subspecialties: Systems engineering; Transportation engineering. Office: Automotive Products Group Hoover Univ Inc 135 E Bennett St West Lafayette IN 47907

GOODWIN, BRUCE KESSELI, geology educator; b. Providence, Oct. 14, 1931; s. Thomas William and Lizetta Christina (Kesseli) G.; m. Joan Marilyn Horton, June 9, 1956; children: Stephen Bruce, Susan Joan, Jennifer Ann. A.B., U. Pa., 1953; M.S., Lehigh U., 1957, Ph.D., 1959. Instr. geology U. Pa., Phila., 1959-63; asst. prof. geology Coll. of William and Mary, Williamsburg, Va., 1963-66, assoc. prof., 1966-71, prof., 1971—; geologist Vt. Geol. Survey, Burlington, part-time, summers 1956-59, Pa. Geol. Survey, 1960-63, Va. Div. Mineral Resources, 1965—; chmn. Va. Bd. Geology, Richmond, 1982—. Contbr. articles to profl. jours. Pres. Lafayette Ednl. Fund, Williamsburg, 1973-78, PTA, Williamsburg, 1966-80. Served with U.S. Army, 1953-55. Recipient Thomas Jefferson Teaching award Coll. William and Mary, 1971. Fellow Geol. Soc. Am.; mem. Va. Acad. Sci., Am. Inst. Profl. Geologists, Nat. Assn. Geology Tchrs. (pres. eastern sect. 1982-83), AAAS, Sigma Xi. Republican. Presbyterian. Lodge: Kiwanis. Subspecialties: Tectonics; Petrology. Current work: Structure and stratigraphy of eastern Piedmont of Virginia. Office: Coll William and Mary Williamsburg VA 23185 Home: 103 Wake Robin Rd Williamsburg VA 23185

GOODWIN, RICHARD CLARKE, air force officer, physicist, educator; b. Hancock, Mich., Mar. 24, 1949; s. Robert Clement and Jean (Gibson) G.; m. Linda Wells, Oct. 30, 1971. B.S., U.S. Mil. Acad., 1971; M.S. in Systems Mgmt, U. So. Calif., 1976; M.S. in Nuclear Engring, Air Force Inst. Tech., Wright-Patterson AFB, Ohio, 1978. Commd. 2d lt. U.S. Air Force, 1971, advanced through grades to maj., 1983; navigator, Mather AFB, Calif., 1971-76; asst. prof. physics U.S. Mil. Acad., 1978-81; radar navigator, K.I. Sawyer AFB, Mich., 1981—. Mem. Soc. Am. Mil. Engrs. (pres. 1979-80), Am. Nuclear Soc., Air Force Assn., Alfa Romeo Owners Club, BMW Owners Club (area gov. 1977). Subspecialties: Nuclear engineering; Laser physics. Current work: Involved in computer-aided instruction; mini-computer networking and word processing; computer aided experimentation. Home: 254 Canberra KI Sawyer AFB MI 49843 Office: 644 BMS KI Sawyer AFB MI 49843

GOODWIN, TOMMY LEE, food scientist; b. Little Rock, Apr. 6, 1936; s. Ray N. and Mary Ellen (Mode) G.; m. Alice Jean Featherston, June 20, 1959; children: Greg Alan, Kelli Lynn. B.S. in Agr, U. Ark., 1958, M.S., 1960; Ph.D., Purdue U., 1962. Asst. prof. dept. animal sci. U. Nebr., 1962-64; asst. prof. U. Ark., Fayetteville, 1964-65, assoc. prof., 1965-71, prof., 1971—. Contbr. articles in field to profl. jours. Active Methodist Ch., Fayetteville. Served with USNR, 1954-62. Recipient PEIA Research award, 1975. Mem. Inst. Food Technologists, Poultry Sci. Assn., World Poultry Sci. Assn., Sigma Xi, Gamma Sigma Delta, Alpha Zeta. Subspecialty: Food science and technology. Current work: Food science poultry products, processing, product development, quality improvement. Home: Route 9 Box 82-27 Fayetteville AR 72701 Office: Dept Animal Sci U Ark Fayetteville AR 72701

GOODY, RICHARD MEAD, geophysicist; b. Welwyn-Garden-City, Eng., June 19, 1921; came to U.S., 1958, naturalized, 1966; s. Harold Earnest and Lilian (Rankine) G.; m. Elfriede Koch, Sept. 11, 1946; 1 dau., Brigid. Ph.D., Cambridge U., 1949; M.A. (hon.), Harvard U., 1958. With Brit. Civil Service, 1942-46; fellow St. John's Coll., Cambridge, 1950-53; reader London U., 1953-58; prof. div. applied scis. Harvard U., 1958—; dir. Blue Hill Obs., 1958-70, Center for Earth and Planetary Physics, 1970-71. Author: Physics of the Stratosphere, 1947, Atmospheric Radiation, 1964, Atmospheres, 1974. Mem. Royal Meteorol. Soc. (Buchan prize 1955), Am. Meteorol. Soc. (50th Anniversary medal 1970, Cleveland Abbé award 1977), Am. Acad. Arts and Scis., Nat. Acad. Scis. Club: Cosmos (Washington). Subspecialty: Geophysics. Home: Box 430 Falmouth MA 02541 Office: Pierce Hall 29 Oxford St Cambridge MA 02138

GOORHA, RAKESH MOHAN, virologist, educator; b. Gwallior, M.P., India, Nov. 6, 1940; came to U.S., 1960, naturalized, 1969; s. Shabhu Nath and Jwala (Devi) G.; m. (married), Aug. 5, 1949; children: Salil, Aarti. B.V.Sc. and A.H., Vikram U., India, 1961; M.S. in Microbiology, Agra U., India, 1963; Ph.D., U. Fla., Gainesville, 1969. Postdoctoral fellow Roche Inst. Molecular Biology, Nutley, N.J., 1970-72; asst. mem. St. Jude Children's Research Hosp., Memphis, 1972-75, assoc. mem., 1975—; assoc. prof. microbiology U. Tenn. Med. Center, 1980—. Contbr. articles to profl. jours. Mem. Am. Soc. Microbiology, Am. Soc. Virology. Subspecialties: Molecular biology; Virology (biology). Current work: Structure and replication of DNA; cloning and sequencing of viral genes. Office: Div Virology St Jude Children's Research Hosp PO Box 318 Memphis TN 38101

GOORVITCH, DAVID, research scientist; b. San Pedro, Calif., July 16,1941; s. Philip and Helen (Isbitz) G.; m. Norma Ruth Rubenstein, June 3, 1966; children: Stephen, Laura. B.A., U. Calif.-Berkeley, 1963, Ph.D., 1967. With NASA, Ames Research Center, Moffett Field, Calif., 1967—, research scientist lab. molecular spectroscopy, 1972—. Contbr. articles to physics and astronomy jours. Mem. Am. Phys. Soc., Optical Soc. Am., Am. Astron. Assn., Sigma Xi. Subspecialties: Atomic and molecular physics; Infrared optical astronomy. Current work: Molecular spectroscopy, infrared observations of the planets. Office: N245-6 Moffett Field CA 94035

GOOTMAN, PHYLLIS MYRNA, physiologist, edcator; b. N.Y.C., June 8, 1938; d. Albert S. and Ida (Krieger) Adler; m. Norman Gootman, June 1, 1958; children: Sharon Hillary, Craig Seth. B.A., Barnard Coll., Columbia U., 1959; Ph.D., Albert Einstein Coll. Medicine, Yeshiva U., 1967. Assoc. prof. sch. grad. studies Downstate Med. Center SUNY, 1975-81, prof., 1981—, assoc. prof. dept. physiology, 1975-81, prof., 1981—; cons. pediatrics dept. L.I. Jewish-Hillside Med. Center, New Hyde Park, 1976—; lectr. in field; mem. cardiovascular and renal study sect. NIH, 1981-85. Contbr. numerous articles to profl. jours. John Miles Davidson fellow, 1973. Mem. AAAS, Soc. Neuroscis., Biophys. Soc., Am. Physiol. Soc., Am. Heart Assn., Am. Inst. Biol. Scis., Microcirculatory Soc., Soc. Exptl. Biology and Medicine, Internat. Soc. Devel. Neurosci., Am. Assn. Lab. Animal Sci., Sigma Xi. Subspecialties: Comparative physiology; Neurophysiology. Current work: Research in development of autonomic nervous system, development of autonomic regulation of cardiovascular function. Office: Dept Physiology Downstate Medical Center 450 Clarkson Ave Brooklyn NY 11203

GOPALAKRISHNAN, CHENNAT, agricultural economics educator, researcher, consultant; b. Elankunnapuzha, Kerala, India, Oct. 9, 1936; came to U.S., 1963, naturalized, 1976; d. Palliyil Narayana and Chennat (Sarada) Menon; m. Malini Gopalakrishnan, Sept. 15, 1962; 1 dau., Shalini. B.A., Kerala U., 1955, M.A., 1957; Ph.D., Mont. State U., 1967. Researcher Nat. Council Applied Econ. Research, New Delhi, 1958-61; sr. researcher The Econ. Times newspaper, Bombay, India, 1961-63; grad. asst. Mont. State U., 1963-66, asst. prof., 1967-69; assoc. prof. agrl., econs. U. Hawaii, 1969-74; prof., 1974—; vis. prof. U. So Calif., 1976-77, U. Wyo., 1982-83; Law of the Sea Inst. prof. social scis., Honolulu, 1978. Author: Natural Resources and Energy, 1980; editor: The Emerging Marine Economy of the Pacific, 1983. Named outstanding Researcher Gamma Sigma Delta, 1980; recipient Valuable Service award Marine Tech. Soc., 1981; Inst. Internat. Edn. fellow, 1965-66. Mem. Am. Agrl. Econ. Assn., Internat. Agrl. Econ. Assn., Internat. Assn. Energy Economists, Assn. Evolutionary Econs., Assn. Environ. and Resource Economists. Subspecialties: Agricultural economics; Biomass (agriculture). Current work: Economics of biomass energy; energy demand and supply forecasts; energy and rural development; marine and water resources-role in economic development; technology transfer; energy intensity and factor substitution. Office: U Hawaii 2545 The Mall Bilger 210 Honolulu HI 96822

GORBATY, MARTIN LEO, laboratory director; b. Bklyn., Nov. 17, 1942; s. Julius and Florence (Birnbach) G.; m. Dianne Morse, June 30, 1968; children: H. Mark, Matthew J., Lisa R. B.S., CCNY, 1964; Ph.D., Purdue U., 1969. With Exxon Research & Engring. Co., Linden, N.J., 1969—, sr. research chemist, 1972, group head coal sci., 1975, lab. dir., 1978—. Contbr. articles to profl. jours.; editor: Refining of Synthetic Crudes, 1979, Coal Structure, 1981, Coal Science, Vol. 1, 1982. Mem. Am. Chem. Soc. (chmn. elect petroleum div.), AAAS, N.Y. Acad. Sci., Sigma Xi, Phi Lambda Upsilon. Subspecialties: Coal; Oil shale. Current work: Fundamental chemical and physical structures of refractory hydrocarbons including heavy oils, coal and oil shale, relationships between structure and reactivity and conversion to clean usable fuels and chemicals. Patentee in field. Office: Clinton Twp Route 22 East Annandale NJ 08801 Home: 204 Twin Oaks Terr Westfield NJ 07090

GORBMAN, AUBREY, biologist, educator; b. Detroit, Dec. 13, 1914; s. David and Esther (Korenblit) G.; m. Genevieve D. Tapperman, Dec. 25, 1938; children—Beryl Ann, Leila Harriet, Claudia Louise, Eric Jay. A.B., Wayne State U., 1935, M.S., 1936; Ph.D., U. Calif., 1940. Research asso. U. Calif., 1940-41; instr. zoology Wayne U., 1941-44; Jane Coffin Childs fellow in anatomy Yale, 1944-46; asst. prof. zoology Barnard Coll., Columbia, 1946-49, asso. prof., 1949-53, prof., 1953-63, exec. officer dept. zoology, 1952-55; prof. zoology U. Wash., Seattle, 1963—, chmn. zoology dept., 1963-66; biologist Brookhaven Nat. Lab., 1952-58; Fulbright scholar College de France, Paris, 1951-52; Guggenheim fellow U. Hawaii, 1955-56; vis. prof. biochemistry Nagoya U. Japan, 1956, Tokyo U., 1960. Editor: Comparative Endocrinology; editorial bd.: Endocrinology, 1957-61; editor-in-chief: Gen. and Comparative Endocrinology. Fellow A.A.A.S., N.Y. Zool. Soc., N.Y. Acad. Sci.; mem. Endocrine Soc., Am. Inst. Biol. Scis. (governing bd. 1969-75), Am. Soc. Zoologists (pres. 1976), Soc. Exptl. Biology and Medicine, Phi Beta Kappa. Subspecialties: Animal physiology; Neuroendocrinology. Current work: Comparative endocrinology; textbook writing and editing in endocrinology. Home: 4218 55th Ave NE Seattle WA 98105

GORDON, DAVID HUGH, physician; b. Phila., June 30, 1945; s. Meyer Michael and Syliva Sonia (Robinson) G.; m. Jayne Illene Eisenberg, Aug. 26, 1967; children: Megan Stephanie, Michael Matthew. B.S., Trinity Coll., 1966; M.D., Case Western Res. Med. Sch., 1971. Diplomate: Am. Bd. Internal Medicine., Am. Bd. Ned. Oncology, Am. Bd. Hematology. Oncologist/hematologist S.W. Oncology Assn (P.A.), San Antonio, 1977-79, San Antonio Tumor and Blood Clinic (P.A.), 1979—; clinc. assoc. prof. dept. medicine U. Tex.-San Antonio, 1983—. Am. Cancer Soc. Clin. fellow, 1975-76. Mem. Am. Soc. Clin. Oncology, ACP, Tex. Med. Assn., Bexar County Med. Soc., Am. Soc. Hematology, Phi Delta Epsilon. Subspecialties: Chemotherapy; Hematology. Current work: Clinical chemotherapy studies in treatment of cancer, hyperthermia in treatment of cancer. Home: 2815 Whisper Fawn San Antonio TX 78230 Office: San Antonio Tumor and Blood Clinic 8527 Village Dr San Antonio TX 78217

GORDON, EDWARD BARRY, systems programmer; b. Boston, Sept. 10, 1952; s. Louis and Marion (Shuman) G. B.S. with distinction in Computer Sci, Worcester Poly. Inst., 1974; postgrad., Wayne State U., 1974-77, Villanova U., 1979-81. Engring. programmer Burroughs Corp., Detroit, 1974-77; tech. adv. Fed. Res. Bank, Boston, 1977-78; engring. programmer Racal-Milgo, Inc., Miami, Fla., 1978-79, Decision Data Computer Corp., Horsham, Pa., 1979-81; prin. systems programmer Racal-Milgo, Inc., Ft. Lauderdale, Fla., 1981-82; sr. engring. programmer Autech Data Systems, Pompano Beach, Fla., 1982—. Mem. Assn. Computing Machinery, IEEE, Mensa. Jewish. Subspecialties: Software engineering; Computer architecture. Current work: Specification, design and development of computers and computer peripherals and process controls. Home: 7931 NW 41st Ct Sunrise FL 33321 Office: 1301 W Copans Rd Pompano Beach FL 33604

GORDON, EUGENE IRVING, telecommunications company executive; b. N.Y.C., Sept. 14, 1930; s. Sol and Gertrude (Lassen) G.; m. Barbara Young, Aug. 19, 1956; children—Laurence Mark, Peter Eliot. B.S., CCNY, 1952; Ph.D., Mass. Inst. Tech., 1957. Research asso. Mass. Inst. Tech., Cambridge, 1957; mem. tech. staff Bell Labs., Murray Hill, N.J., 1957-59, supr., 1959-63, dept. head, 1963-68, dir., 1968-83; cons., 1983—. Patentee in field.; Editor: EDS Trans, 1963-64, Jour. Quantum Electronics, 1964-76; Contbr. articles to profl. jours. Fellow IEEE (Zworykin award 1975); mem. Nat. Acad. Engring., Am. Phys. Soc. Democrat. Jewish. Club: B'nai Brith. Subspecialties: Electrical engineering; Electronics. Current work: Development of devices for fiber optic. Communication, display devices. Home: 14 Braidburn Way Convent NJ 07961 Office: Bell Laboratories 600 Mountain Ave Murray Hill NJ 07974 The philosophy underlying my approach to life has been "Better light a candle than curse the dark". As a physicist I know that it is virtually impossible to produce light without some heat, but it always has come as a surprise to me how little heat human beings will tolerate even in the presence of abundant illumination.

GORDON, JAMES ARTHUR, info. systems project manager; b. Midland, Mich., Aug. 19, 1949; s. Arthur Franklin and Adeline Caroline (Okuley) G.; m. Kathleen Marie Bader, Aug. 16, 1975. B.S. in Computer Sci, N.C. State U., 1975. Cert. data processor. Programmer analyst Central Devel. Co., San Francisco, 1975-76; sr. programmer analyst Del Monte Corp., San Francisco, 1976-79, Dow Corning Corp., Midland, Mich., 1979, office systems analyst, 1979-80, project analyst systems and info. mgmt., 1981. Served with USAF, 1969-73. Mem. Assn. Computing Machinery. Subspecialties: Information systems, storage, and retrieval (computer science); Information systems (information science). Current work: Project management, systems design, small business computers, personal computers, process control and instrumentation. Home: 5809 Lamplighter Ln Midland MI 48640 Office: 2200 W Salzburg Rd Midland MI 48640

GORDON, MARK AITKEN, astronomer, observatory official; b. Springfield, Mass., Oct. 13, 1937; s. Alexander Dorward and Josephine Pease (Aitken) G.; m. Mary Abigail Anderson, Aug. 19, 1961; children: Paige Abigail, Sarah Aitken; m. Julia Brown Perry, Aug. 12, 1980. B.A., Yale U., 1959; Ph.D., U. Colo., 1966. Mem. U.S. Antarctic Research Program, 1959-62; staff mem. M.I.T. Lincoln Lab., 1966-69; asst. scientist Nat. Radio Astronomy Obs., Charlottesville, Va., 1969-72, assoc. scientist, Tucson, 1972-77, scientist, 1977—, asst. dir. for Ariz. Ops., 1973—. Contbr. articles to sci. publs. Mem. Am. Astron. Soc., Internat. Astron. Union, U.S. Nat. Com. for Radio Sci. Subspecialty: Radio and microwave astronomy. Current work: Interstellar medium, galactic structure. Office: Suite 100 2010 N Forbes Blvd Tucson AZ 85745

GORDON, RICHARD, developmental biologist; b. N.Y.C., Nov. 6, 1943; s. Jack and Diana (Lazaroff) G.; m. Leslie Joan Biberman, July 22, 1973; children: Leland Terrell, Bryson Philip, Chason Abel. B.Sc. in Math, U. Chgo., 1963; Ph.D. in Chem. Physics, U. Oreg., 1967. Postdoctoral fellow in biophysics U. Colo. Med. Center, Denver, 1967-68; postdoctoral fellow in polymer sci. and engring. U. Mass., Amherst, 1968; research assoc. biol. sci. Columbia U., N.Y.C., 1968-69; research assoc. Ctr. for Theoretical Biology SUNY, Buffalo, 1969-72; staff fellow math. research br. NIH, Bethesda, Md., 1972-75, expert image processing unit, 1975-78; assoc. research prof. radiology George Washington Sch. Medicine, Washington, 1974-78; assoc. prof. pathology, radiology, and elec. engring. U. Man., Winnipeg, 1978—, dir. computer dept. for health scis., 1978-81. Contbr. articles to profl. jours. Can. Med. Research Council grantee, 1979; Children's Hosp. Winnipeg Found. grantee, 1980-83; Natural Sci. and Engring. Research Council Can. grantee, 1979; Man. Med. Services Found. grantee, 1982. Mem. AAAS, Am. Assn. Physicists in Medicine, Am. Inst. Biol. Scis., Assn. Computing Machinery, Soc. Can. Image Processing and Pattern Recognition, Can. Soc. Cell Biology, Internat. Pigment Cell Soc., Internat. Soc. Stereology, Micro. Soc. Can., Soc. Analytical Cytology, Soc. Developmental Biology, Com. Concerned Scientists, Sci. for Peace. Jewish. Subspecialties: Developmental biology; Imaging technology. Current work: Computer simulation of brain morphogenesis; rotating microscope for landsat imagery of vertebrate embyros; diatom shell mophogenesis; pattern formation by melanocytes and migration of melanoma cells; computed tomography via teleradiology; positron tomography; skin scanner for detecting early melanoma. Office: Univ Manitoba Winnipeg MB Canada R3E 0W3

GORDON, ROBERT BOYD, geophysics educator; b. East Orange, N.J., Dec. 25, 1929; s. Myron Boyd and Catherine (Rote) G.; m. Joan Parke Ruttiger, Sept. 13, 1952; children: Penelope, Margaret. B.S., Yale U., 1952, D.Eng., 1955. Asst. prof. Sch. Mines, Columbia U., 1955-57; mem. faculty Yale U., 1957—, assoc. prof. applied sci., 1960-68, prof. geophysics and applied mechanics, 1968—, chmn. dept. geology and geophysics, 1979-82. Author: Physics of the Earth, 1972, (with R.M. Brick and A.W. Pense) Structure and Properties of Engineering Materials, 1977; Contbr. articles to profl. jours. Mem. Am. Phys. Soc., Am. Geophys. Union, Am. Inst. Mining and Metall. Engrs., Phi Beta Kappa, Sigma Xi. Subspecialties: Geophysics; Metallurgy. Current work: Rock mechanics, particularly fracture and its application to surficial processes. Archaeometallurgy and industrial archaeology. Home: 239 Everit New Haven CT 06511 Office: Kline Geol Lab Box 6666 New Haven CT 06511

GORDON, ROBERT THOMAS, surgeon, educator; b. Chgo., Feb. 13, 1950; s. David and Eunice (Wienshienk) G. B.S. in Medicine with highest distinction, Northwestern U., M.D., 1972. Diplomate: Am. Bd. Surgery, Am. Bd. Thoracic Surgery. Resident in gen. surgery Northwestern U. Hosp., Chgo., 1972-77, resident in cardiac surgery, 1977-79; clin. asst. prof. surgery U. Ill., Chgo., 1979—; mem. attending staff, chief dept. cardiac surgery Lutheran Gen. Hosp., Park Ridge, Ill., 1979—; also chief cardiac surgery resident and nurses tng. program; mem. attending staff Highland Park (Ill.) Hosp., Edgewater Hosp., Chgo., Mt. Sinai Hosp., Holy Family Hosp., Des Plaines, Ill., North Suburban Med. Center, Schaumburg, Ill., Good Shepherd Hosp.; instr. gen. and thoracic surgery Northwestern U.; staff assoc. HEW, Bethesda, Md. Contbr. articles to profl. jours. Chmn. Ill. Commn. Conservation, 1966-67. Recipient awards Hoffmann-LaRoche, NASA, Am. Chem. Soc.; G.D. Searle fellow; Macy Found. Research fellow. Mem. Internat. Bio-Electro Magnetic Soc., Internat. Coll. Surgeons, Royal Soc. Medicine, Flying Physicians Assn., Internat. Platform Assn., AMA, Chgo. Thoracic Soc., Am. Coll. Chest Physicians, Chgo. Med. Assn., Ill. State Med. Assn., Soc. Contemporary Medicine and Surgery, Ill. Jr. Acad. Scis. (pres., hon. life mem.), Alpha Omega Alpha, Phi Beta Pi, Phi Eta Sigma. Subspecialties: Cancer research (medicine); Cardiac surgery. Current work: Cardiac surgery and related research; also research and innovations in biophysics (cancer diagnosis, treatment and prevention; also heart disease and other diseases); genetic engineering and instrumentation. Patentee in field. Office: 1775 Dempster St Park Ridge IL 60068

GORDON, ROY GERALD, educator; b. Akron, Ohio, Jan. 11, 1940; s. Nathan Gold and Frances (Teitel) G.; m. Myra Shela Miller, Dec. 24, 1961; children—Avra Karen, Emily Francine, Steven Eric. A.B. summa cum laude, Harvard, 1961, A.M. in Physics, 1962, Ph.D. in Chem. Physics, 1964. Jr. fellow Soc. of Fellows, Harvard, 1964-66, mem. faculty, 1966—, prof., 1969—. Sloan Found. fellow, 1966-69. Fellow Am. Phys. Soc.; mem. Am. Chem. Soc. (award in pure chemistry 1972, Baekeland award 1979), Faraday Soc., Union of Concerned Scientists, Nat. Acad. Scis., Am. Acad. Arts and Scis., Phi Beta Kappa, Sigma Xi. Subspecialties: Physical chemistry; Electronic materials. Theoretical research discovering forms of forces between molecules, the way molecules collide with each other, motion of molecules in liquids and solids. Inventions in energy conservation and solar energy. Office: Harvard U Cambridge MA 02138

GORDON, WAYNE ALAN, research psychologist, educator; b. N.Y.C., Feb. 2, 1946; s. Roy and Caroline (Doheiser) G.; m. Margaret Brown, Nov. 9, 1979. B.A., N.Y. U., 1966; postgrad., New Sch. Social Research, 1966-67; Ph.D., Yeshiva U., 1972. Lab. asst. Inst. Rehab. Medicine, N.Y. U. Med. Ctr., N.Y.C., 1965-67, research asst., 1967-70, clin. research psychologist, 1970-72, sr. clin. research psychologist, 1972-75, supr. research in behavioral scis., 1975—; asst. clin. prof. rehab. medicine Med. Ctr., 1975—; mem. study sect. NIH, 1976-80; cons. United Cerebral Palsy N.Y. State, 1978—. Contbr. articles to profl. jours. W.T. Grant Found. grantee, 1981-84; Nat. Inst. Handicapped Research grantee, 1977-81. Mem. Am. Congress Rehab. Medicine, Am. Psychol. Assn., Internat. Neuropsychol. Soc., Acad. Behavioral Medicine Research, Soc. Behavioral Medicine. Democrat. Jewish. Subspecialties: Behavioral psychology; Neuropsychology. Current work: Research in rehabilitation medicine; behavioral medicine; psychotherapy; health promotion. Home: 420 West End Ave New York NY 10024 Office: 400 E 34th St New York NY 10016

GORDON, WILLIAM BERNARD, mathematician; b. Washington, Nov. 16, 1935; s. Myer W. and Sylvia F. (Feldman) G.; m. Esther Eve Bronstein, Jan. 2, 1939; children: Robert I., David J. B.S., George Washington U., 1959, M.S., 1960; Ph.D., Johns Hopkins U., 1968. Instr. Johns Hopkins U., Balt., 1968-69; cons. radar div. Naval Research Lab., Washington, 1969—. Contbr. articles to in field to profl. jours.; patentee antenna processing. Served with AUS, 1954-56. Mem. Am. Math. Soc., Math. Assn. Am., Soc. for Indsl. and Applied Math., Izaak Walton League, Sigma Xi. Subspecialties: Differential geometry and dynamical systems; Statistics. Current work: Dynamical systems; statistical estimation theory and spectral analysis; radar technology. Home: 14013 Manorvale Rd Rockville MD 20853 Office: Naval Research Lab Code 5308 Washington DC 20375

GORDY, WALTER, emeritus physics educator; b. Miss., Apr. 20, 1909; s. Walter Kalin and Gertrude (Jones) G.; m. Vida Brown Miller, June 19, 1935; children: Eileen, Walter Terrell. A.B., Miss. Coll., 1932, LL.D., 1959; M.A., U. N.C., 1933, Ph.D., 1935; Dr. honoris causa, U. Lille, France, 1955; D.Sc. hon., Emory U., 1983. Assoc. prof. math. and physics Mary Hardin-Baylor Coll., 1935-41; NRC fellow Calif. Inst. Tech., 1941-42; staff radiation lab. Mass. Inst. Tech., 1942-46; assoc. prof. physics Duke, Durham, N.C., 1946-48, prof., 1948-79, James B. Duke prof., 1958-79, James B. Duke prof. emeritus, 1979—; Vis. prof. U. Tex., 1958; Mem. NRC, 1954-57, 68-74. Author: (with W.V. Smith, R.F. Trambarulo) Microwave Spectroscopy, 1953, (with Robert L. Cook) Microwave Molecular Spectra, 1970, Theory and Applications of Electron Spin Resonance, 1980; Assoc. editor: Jour. Chem. Physics, 1954-58, Spectrochimia Acta, 1957-60; editorial bd.: Radiation Research, 1969-72. Recipient Sci. research award Oak Ridge Inst. Nuclear Studies, 1949, Disting. Alumnus award U. N.C., 1976, N.C. award for sci., 1979; 50th Anniversary award Miss. Acad. Scis., 1980. Fellow Am. Phys. Soc. (chmn. S.E. sect. 1953-54, mem. council 1967-71, 73-77, recipient Jessie W. Beams award Southeastern sect. 1974, Earle K. Plyler prize 1980), AAAS (council 1955); mem. Radiation Research Soc. (mem. council 1961-64), Nat. Acad. Scis., Sigma Xi. Subspecialties: Atomic and molecular physics; Biophysics (physics). Current work: Microwave spectroscopy and electron spin resonance spectra. Home: 2521 Perkins Rd Durham NC 27706

GORENSTEIN, PAUL, physicist; b. N.Y.C., Aug. 15, 1934; s. Isidore and Bess (Evans) G. B.Engring. Physics, Cornell U., 1957; Ph.D. in Physics, M.I.T., 1962. Fulbright postdoctoral fellow, cons. Italian Nuclear Energy Com., 1963-65; sr. scientist Am. Sci. and Engring., Cambridge, Mass., 1965-73; sr. astrophysicist Smithsonian Astrophy. Obs., Cambridge, 1973—. Contbr. articles to profl. jours. Recipient contbns. to new tech. awards NASA, medal for outstanding sci. achievement, 1973. Mem. Am. Phys. Soc., Am. Astron. Soc., AAAS. Subspecialties: 1-ray high energy astrophysics; Nuclear physics. Current work: Supernova remanants, clusters of galaxies, instrumentation for x-ray astronomy. Office: 60 Garden St B-434 Cambridge MA 02138

GORGES, HEINZ A., engineering company executive; b. Stettin, Ger., July 22, 1913; came to U.S., 1959; s. Gustav and Marga (Benda) G.; m. Sapienza T. Coco, Sept. 2, 1957. M.S. Mech. Engring, Tech. U. Dresden, Ger., 1938; Ph.D., Tech. U. Hanover, Ger., 1946. Cert. profl. engr., D.C. Group leader Aero. Research Establishment, Braunschweig, Ger., 1940-45; sci. Royal Aircraft Establishment, Farnborough, Eng., 1946-49; prin. sci. officer Weapons Research Establishment, Adelaide, Australia, 1949-59; sci. asst. Marshall Space Flight Ctr., NASA, Huntsville, Ala., 1959-61; dir. advanced projects Cook Tech. Ctr., Morton Grove, Ill., 1961-62; sci. adv. Inst. Tech. Research, Chgo., 1962-66; asst. v.p. environ. and phys. scis. Tracor Inc., Austin, Tex., 1966-72; v.p., Rockville, Md., 1972-75; pres. Vineta Inc., Falls Church, Va., 1975—; prof. U. Ala.-Redstone Extension, 1960-61. Contbr. articles to profl. jours. Fellow AIAA (assoc.); mem. ASME, Acoustical Soc. Am., N.Y. Acad. Scis. Clubs: Cosmos, Palaver, Nat. Press (Washington). Subspecialties: Mechanical engineering; Thermodynamics. Current work: Cogeneration; waste to energy conversion; economics; optimization techniques. Office: Vineta Inc 3705 Sleepy Hollow Rd Falls Church VA 22041

GORLICK, DENNIS LESTER, biological scientist; b. Detroit, Dec. 17, 1944; s. Samuel and Rose Helen (Weiser) G.; m. Sallye Ann Davis, June 21, 1980. B.Sc., Wayne State U., 1967, M.S., 1971; Ph.D., U. Hawaii, 1978. Served with Peace Corps, 1967-69; acting asst. prof. gen. sci. dept. U. Hawaii, Honolulu, 1977, vis. asst. prof. dept. zoology, 1978; vis. research fellow Princeton (N.J.) U., 1980-82; vis. research fellow dept. biol. scis. Columbia U., N.Y.C., 1981-82. Contbr. numerous articles to sci. jours. NIMH postdoctoral tng. fellow, 1980-81; postdoctoral research fellow, 1981—. Mem. Animal Behavior Soc., Soc. for Neurosci., AAAS, Hawaii Road Race Assn. Subspecialties: Ethology; Neurobiology. Current work: Neuroanatomical basis of species-specific aggressive and reproductive behavioral patterns. Office: Biol Scis Dept Fairchild Center Columbia U New York NY 10027

GORLIN, ROBERT JAMES, medical educator; b. Hudson, N.Y., Jan. 11, 1923; s. James A. and Gladys Gretchen (Hallenbeck) G.; m. Marilyn Rhona Alpern, Aug. 24, 1952; children: Cathy Ellen, Jed Baron. A.B., Columbia U., 1943; D.D.S., Washington U., St. Louis, 1947; M.S., U. Iowa, 1956; D.Sc., U. Athens, 1982. Diplomate: Am. Bd., Oral Pathology; charter mem. Am. Bd. Med. Genetics. Research fellow in pathology Columbia U., N.Y.C., 1947-50; oral pathologist Bronx VA Hosp., N.Y.C., 1950-51; instr. dentistry Columbia U., N.Y.C., 1950-51; dental dir., pathologist Operation Blue Jay, Thule, Greenland, 1951-52; assoc. prof. div. oral pathology Sch. Dentistry, U. Minn., Mpls., 1956-58, prof., 1958-79, prof. pathology, dermatology, 1971—, prof. pediatrics, dermatology, ob-gyn and otolaryngology, 1973—, Regents' prof., 1979—, chmn. div. oral pathology, 1956-79, chmn. dept. oral pathology and genetics, 1979—; Fulbright exchange prof., Guggenheim fellow Royal Dental Coll., Copenhagen, Denmark, 1961; 1st Lingamfelter lectr. dermatology U. Va., 1971; 1st Boyle lectr. Case Western Res. U. Med. Ctr., Cleve., 1972; Chase Meml. lectr., 1983; Alpha Omega Alpha vis. prof. Mt. Sinai Hosp., N.Y.C., 1981; vis. prof. Sch. Medicine, Tel Aviv U., 1981; Spies Meml. lectr., Kansas City, 1982, Frank Hoopes lectr., 1977; vis. prof. UCLA-Harbor Gen. Hosp., 1972; asst. chief dental service Glenwood Hills Med. Ctr., Calif., 1959-61, chief, 1962-64, cons., 1969-73; cons. pediatrics and oral pathology U. Minn. Hosps., Mpls., 1956—; cons. oral pathology VA Hosp., Mpls., 1958—, Mt. Sinai Hosp., 1958—; cons. pediatrics Hennepin County Gen. Hosp., 1963—, St. Paul's Children's Hosp., Ramsey County Gen. Hosp., 1968—, Mpls. Children's Hosp., 1972—, Gilette State Hosp., 1979—. Author: (with J. Pindborg) Syndromes of the Head and Neck, 1964, 76, (with R. Goodman) The Face in Genetic Disorders, 1970, (with B. Konigsmark) Genetic and Metabolic Deafness, 1976, (with R. Goodman) Atlas of the Face in Genetic Disorders, 1977, (with H. Sedano and J. Sauk) Oral Manifestations of Genetic Disorders, 1977, (with R. Goodman) The Malformed Infant and Child, 1983; co-contbr.: Computer Assisted Diagnosis in Pediatrics, 2d edit., 1971; editor: (with H. Goldman) Thoma's Oral Pathology, 1970, (with J. Cervenka and B. Koulischer) Chromosomes and Human Cancer, 1972; editorial cons.: Jour. Dental Research, 1962—, Geriatrics, 1962—, Archives of Oral Biology, 1962—, Jour. Pediatrics, 1963—, Pediatrics, 1964—, Am. Jour. Diseases of Children, 1964—, Syndrome Identification, 1973, Radiology, 1976—; editor oral pathology: Oral Surgery, Oral Medicine, Oral Pathology, 1963—; assoc. editor: Am. Jour. Human Genetics, 1970-73, Jour. Oral Pathology, 1972—, Jour. Maxillo Facial Surgery, 1973—, Cleft Palate Jour, 1976—; editorial bd.: Nat. Found. Birth Defects Compendium, 1970—, Am. Jour. Med. Genetics, 1977; mem. bd.: Excerpta Medica, 1976—; cons. editor sect. dentistry: Stedman's Medical Dictionary, 1959; contbr. numerous articles to profl. jours. Mem. adv. com. Human Genetics, Minn., 1959-73; mem. U.S. Congl. Liaison Com. for Dentistry, Minn., 1963—, Ctr. Histologic Nomenclature and Classification of Odontogenic Tumors and Allied Lesions, WHO, 1966—; mem. adv. com. peridontal disease and soft tissue study NIH, 1967—. Served with AUS, 1943-44; to lt. USNR, 1953-55. Fellow Am. Acad. Oral Pathology (v.p. 1957-58, 64-65, sec. 1958-64, pres. 1966-67); mem. ADA (cons. council dental edn. 1967—), Am. Bd. Oral Pathology (dir. 1970-76, v.p. 1974-75, pres. 1975-76), Internat. Assn. Dental Research (sec. Minn. div. 1958-59, pres. 1959-60), Minn. Soc. Pathologists, Am. Soc. Human Genetics, Internat. Skeletal Soc., Internat. Soc. Cranofacial Biology (dir. 1966-67, v.p. 1967-68, pres. 1969-70), Internat. Acad. Oral Pathology (assoc.), Hollywood Acad. Medicine (hon.), Western Study Club Combined Therapy (hon.), Sigma Xi, Omicron Kappa Upsilon. Subspecialties: Oral pathology; Genetics and genetic engineering (medicine). Current work: Craniofacial syndromes, chromosomal disorders, syndromes of deafn. Home: 4600 Chatelain Terr Golden Valley MN 55422 Office: U Minn 16-206 HS Unit A 515 Delaware SE Minneapolis MN 55455

GORMAN, BERNARD SAMUEL, psychologist; b. Far Rockaway, N.Y., Jan. 1, 1943; s. Adrian Ezra and Charlotte J. (Grossberg) G.; m. Dale Wendy Rogoff, Dec. 25, 1967; children: Betsy Rose, Leanne Michelle. B.A., Queens Coll., CUNY, 1964, M.A., 1967; Ph.D.,

CUNY, 1971. Lic. psychologist, N.Y. Instr. to assoc. prof. Queens Coll., Flushing, N.Y., 1966—; instr. to asst. prof. Nassau Community Coll., Garden City, N.Y., 1967-71; asst. prof. CUNY, 1971-76; asst. prof. to assoc. prof. Nassau Community Coll., 1976—; adj. assoc. prof. Hofstra U., Hempstead, N.Y., 1978—; mem. profl. adv. bd. N.Y. State Epilepsy Assn., 1980—; cons. N.Y. State Dept. Mental Hygiene, Albany, 1982—. Author: Developmental Psychology, 1980; author, editor: Personal Experience of Time, 1977; contbr. articles to profl. jours. Advisor N.Y. State Senate, 1976—. Mem. Am. Psychol. Assn., Am. Statis. Assn., Eastern Psychol. Assn., Psychometric Soc., Inst. Rational Psychotherapy, Phi Beta Kappa, Sigma Xi, Psi Chi. Jewish. Subspecialties: Developmental psychology; Statistics. Current work: Multivariate analysis of psychological processes in development; subjective temporal experience; cognitive approaches to personality and developmental psychology. Home: 42 Bonaire Dr Dix Hills NY 11746 Office: Nassau Community Coll Stewart Ave Garden City NY 11530

GORMLEY, PAUL EDWARD, research oncologist; b. Balt., Aug. 30, 1946; s. John Francis and Dorothy Emma (Fabian) G.; m. Barbara Roosen, June 28, 1969; children: Maureen, Patrick, Michael, Megan, Timothy, Anne. B.S., Johns Hopkins U., 1968, M.D., 1972. Diplomate: Am. Bd. Internal Medicine. Intern, resident in internal medicine Pa. State U., Hershey Med. Ctr., 1972-74; clin. assoc., med. oncology fellow NIH, 1974-77; med. officer Lab. Chem. Pharmacology Nat. Cancer Inst., Bethesda, Md., 1977—; Sr. surgeon USPHS, 1974—. Contbr. articles to profl. jours. Mem. Am. Assn. Cancer Researchers. Roman Catholic. Subspecialties: Oncology; Molecular pharmacology. Current work: Pharmacology of antineoplastic drugs; studies on the molecular basis for antineoplastic activity of antimetabolites and DNA binding drugs. Home: 4607 Roland Ave Baltimore MD 21210 Office: NIH Bldg 37 Bethesda MD 20205

GORMUS, BOBBY JOE, research immunologist, biochemist; b. Richmond, Va., Nov. 7, 1941; s. Perkins Alfred and Nelle Catherine (Campbell) G.; m. Ruby Lee Gayle, Aug. 6, 1960 (div. 1969); 1 son: Joseph Henry; m. Molly Jean Obsorne, June 10, 1971. B.S., U. Richmond, 1964; Ph.D., Duke U., 1971. Postdoctoral fellow U. Fla.-Gainesville, 1971-75; research chemist VA Hosp., Mpls., 1975-79; asst. prof. dept. microbiology U. Minn., 1976-79; research scientist Delta Regional Primate Research Ctr., Tulane U., Covington, La., 1979—. Contbr. numerous articles to sci. jours., papers presented. NIH grantee, 1982—. Mem. Am. Assn. Immunologists, Fedn. Am. Socs. for Exptl. Biology. Subspecialties: Biochemistry (medicine); Microbiology (medicine). Current work: Immunology of leprosy; immunology of patients with lymphoma or leukemia; cellular immunology; cell membrane structure/function. Home: 117 Robinhood Rd Covington LA 70433 Office: Delta Regional Primate Research Ctr Tulane U Three Rivers Rd Covington LA 70433

GORSKE, STANLEY FRANCIS, weed scientist, educator; b. Indpls., Feb. 26, 1949; s. Stanley F. and Ruth A. (Fulmer) G.; m. Mary Jo Dowling, Apr. 18, 1970. B.S., Purdue U., 1973; M.S., U. Ill., 1975, Ph.D., 1978. Grad. research asst. U. Ill., Urbana, 1974-77; asst. prof. soils and crops Rutgers U., 1978-79; asst. prof. horticulture Ohio State U., 1979—. Contbr. numerous articles to profl. jours. Grantee in field. Mem. Am. Soc. Hort. Sci., Weed Sci. Soc. Am., Northeastern Weed Sci. Soc., North Central Weed Sci. Soc., Nat. Agrl. Plastics Assn., Gamma Sigma Delta. Roman Catholic. Subspecialties: Weed science; Plant physiology (agriculture). Current work: Conduct research and teach in the area of weed science; secondly I conduct research and teach in the area of vegetable crops physiology. Home: 960 Lambeth Dr Columbus OH 43220 Office: Ohio State University Department of Horticulture 2001 Fyffe Ct Columbus OH 43210

GORZYNSKI, EUGENE ARTHUR, microbiologist, researcher; b. Buffalo, Oct. 8, 1919; s. Charles Stanley and Helen Marie (Prolejko) G.; m. Ruth Repp, Jan. 10, 1946; children: David, Timothy, Kathleen. B.A., U. Buffalo, 1949, M.A., 1953; Ph.D., SUNY-Buffalo, 1968. Cert. Nat. Registry Microbiologists. Clin. and research microbiologist Children's Hosp., Buffalo, 1947-65; sr. cancer research scientist Roswell Park Meml. Inst., Buffalo, 1968-69; asst. dir. Erie County Pub. Health Lab., Buffalo, 1969-75; chief microbiology lab. VA Med. Ctr., Buffalo, 1975—; prof. microbiology Med. Sch. SUNY-Buffalo, 1969—; comdg. officer U.S. Army Res. Hosp., Buffalo, 1966-69. Contbr. chpts. to books in field. Violinist Amherst Symphony Orch., Buffalo, Orchard Park (N.Y.) Symphony Orch. Served to lt. U.S. Army, 1941-46. Recipient meritorious service medal Dept. Def., 1974. Fellow Infectious Diseases Soc. Am., Am. Acad. Microbiology, AAAS; mem. Am. Assn. Immunologists. Democrat. Roman Catholic. Subspecialties: Microbiology; Infectious diseases. Current work: Infectious diseases, immunodiagnosis, endotoxins, antimicrobial therapy. Office: Dept Microbiology SUNY Med Sch 203 Sherman Hall Buffalo NY 14214 Home: S 5557 Oakridge Dr Hamburg NY 14075

GOSINK, JOAN P., geophysics educator, researcher; b. Jamaica, N.Y., Mar. 26, 1941; m., 1961; 4 children. B.S., MIT, 1962; M.S., Old Dominion U., 1973; Ph.D. in Mech. Engring, U. Calif.-Berkeley, 1979. Teaching asst. in thermodynamics and fluid dynamics U. Calif.-Berkeley, 1975-76, research asst., 1976-78; fellow Geophys. Inst., U. Alaska, Fairbanks, 1979-81, asst. prof. geophysics, 1981—; mem. glaciological com. NRC, 1981—, Polar Research Bd., 1981—. NASA fellow, 1972-73; Fulbright-Hayes Found. fellow, 1974-75. Mem. ASCE, ASME, Am. Meteorol. Soc., Internat. Glaciological Soc. Subspecialty: Geophysics. Office: Geophys Inst U Alaska Fairbanks AK 99701

GOSLING, JOHN THOMAS, physicist; b. Akron, Ohio, July 10, 1938; s. Arthur Warrington and Wilhelmena (Bell) G.; m. Marie Ann Turner, Dec. 21, 1963; children: Mark Raymond, Steven Arthur. B.S. in Physics, Ohio U., 1960, Ph.D., U. Calif., Berkeley, 1965. Mem. Sci. staff Nat. Ctr. Atmospheric Research, Boulder, Colo., 1967-75; staff mem. Los Alamos Sci. Lab., 1965-67, 75—. Asso. editor: Jour. Geophys. Research, 1980—; contbr. articles to profl. jours. Recipient Tech. Achievement award Nat. Ctr. Atmospheric Research, 1974. Mem. Am. Geophys. Union, Am. Astron. Soc., Internat. Astron. Union, AAAS. Subspecialties: Space plasma physics; Solar physics. Current work: Solar-terrestrial physics, solar wind physics; collisionless shocks, particle acceleration mechanisms, reconnection, experimental plasma physics and magnetohydrodynamics. Home: 1420 45th St Los Alamos NM 87544 Office: Los Alamos Nat Lab Los Alamos NM 87545

GOSS, RICHARD JOHNSON, biology educator, researcher, university dean; b. Marblehead, Mass., July 19, 1925; m., 1951; 2 children. A.B., Harvard U., 1948, A.M., 1951, Ph.D. in Biology, 1952. From instr. to assoc. prof. biology Brown U., 1952-64, prof., 1964—, chmn. sect. devel. biology, div. biol. and med. sci., 1972-77, dean biol. sci., 1977—; trustee Mt. Desert Island Biol. Lab., 1960-64; dir. Roger Williams Park Zoo, Providence, 1975—. Carnegie Inst. fellow, 1960. Mem. AAAS (sec. sect. biol. sci. 1970-74), Soc. Devel. Biology, Am. Soc. Zoology, Am. Assn. Anatomy, Internat. Soc. Cell Biology. Subspecialty: Developmental biology. Office: Div Biol and MED Sci Brown U Providence RI 02912

GOSSLING, JENNIFER, microbiologist, educator; b. Welwyn Garden City, Hertfordshire, Eng., July 25, 1934; came to U.S., 1962; d. Richard Sidney and Millicent Eveline (Hodson) Sayers; m. William Frank Gossling, Nov. 3, 1956. B.A., Cambridge (Eng.) U., 1955; Ph.D., W.Va. U., 1973. Asst. Ont. (Can.) Vet. Coll., Guelph, 1959-62; dept. vet. pathology and hygiene U. Ill.-Urbana, 1962-66; asst. dept microbiology W.Va. U., from 1969, instr., to 1973; postdoctoral scholar Dental Research Inst., U. Mich., 1979-80; microbiologist Jewish Hosp. of St. Louis, 1980—; asst. prof. microbiology Sch. Dental Medicine, Washington U., St. Louis, 1981—. Nat. Inst. Dental Research grantee, 1982. Mem. AAAS, Am. Soc. Microbiology, N.Y. Acad. Scis., Internat. Assn. Dental Research. Episcopalian. Subspecialties: Microbiology (medicine); Oral biology. Current work: Practice and teaching of clinical bacteriology, research and teaching in oral bacteriology. Office: Washington U Sch Dental Medicine 4559 Scott Ave Saint Louis MO 63110

GOTAY, CAROLYN COOK, psychology educator; b. New Brunswick, N.J., Feb. 12, 1951; d. Richard Cairns and Winifred (Imhof) Cook; m. Mark Joseph Gotay, May 14, 1973. B.A., Duke U., 1973; M.A., U. Md., 1975, Ph.D., 1977. Grad. asst. U. Md., College Park, 1973-77; asst. prof. Gettysburg (Pa.) Coll., 1977-79; research assoc. U. Calgary, Alta., Can., 1979-80, asst. prof. community health scis., 1980—. Contbr. articles to profl. jours. Alta. Heritage Applied Research Cancer Research scholar, 1981; others. Mem. Am. Psychol. Assn., Am. Pub. Health Assn., Can. Assn. Tchrs. Social and Preventive Medicine (sec.). Subspecialties: Cancer research (medicine); Social psychology. Current work: Coping in cancer patients and their families, hospice care for dying patients and families,mental health of elderly. Home: 610 26 Ave NW Calgary AB Canada T2M 2E5 Office: Dept Community Health Sci Faculty of Medicine U Calgary Calgary AB Canada T2N 4N1

GOTCHER, JACK EVERETT, JR., oral and maxillofacial surgeon, researcher; b. Wichita Falls, Tex., May 11, 1949; s. Jack Everett and Josephine Caroline (Kruh) G.; m. Kathyanne Mary King, Dec. 30, 1972; children: Elizabeth Gayle, Jeffrey Everett. B.S., Midwestern U., 1971; D.M.D., Harvard U., 1975; Ph.D., U. Utah, 1979. Postdoctoral fellow U. Utah, Salt Lake City, 1975-78; resident U. Tenn. Meml. Hosp., Knoxville, 1978-82; asst. prof. dept. oral and maxillofacial surgery Sch. Dentistry Emory U., Atlanta, 1982-83; assoc. prof. dept. oral and maxillofacial surgery U. Tenn., Knoxville, 1983—; cons. oral surgery VA Hosp., Atlanta, 1982-83; cons. bone research Proctor & Gamble, Cin., 1977-79. Contbr. articles to profl. jours. in field. First prize resident research award U. Tenn., 1982; Am. Cancer Soc. grantee, 1978; NIH grantee, 1975-78; Johnson & Johnson travel grantee, 1975. Fellow Am. Assn. Oral and Maxillofacial Surgery; mem. ADA, Internat. Assn. Dental Research. Democrat. Presbyterian. Subspecialties: Oral and maxillofacial surgery; Anatomy and embryology. Current work: Mineralized tissues; metabolic inflammatory and nutritional effects on bone, diphosphonate effects on bone; morphometry in bone research; heterotransplantation of human neoplasia. Home: 7620 Hawthorne Dr Knoxville TN 37919 Office: 1928 Alcoa Hwy Suite 305 Knoxville TN 37920

GOTT, JOHN RICHARD, III, astronomer; b. Louisville, Feb. 8, 1947; s. John Richard and Marjorie (Crosby) G.; m. Lucy Pollard-Gott, June 10, 1978. A.B., Harvard U., 1969; Ph.D., Princeton U., 1972. Fellow Calif. Inst. Tech., 1973-74; vis. fellow Cambridge (Eng.) U., 1975; asst. prof. astronomy Princeton U., 1976-80, assoc. prof., 1980—, dir. grad. studies dept. astrophys. scis., 1977—. Contbr. articles to profl. jours. Sloan fellow, 1977; recipient R. J. Trumpler award Astron. Soc. Pacific, 1975. Mem. Internat. Astron. Union. Subspecialties: Cosmology; General relativity. Current work: Cosmology, general relativity, galaxy formation, galaxy clustering. Office: Princeton University Peyton Hall Princeton NJ 08544

GOTT, VINCENT LYNN, physician; b. Wichita, Kans., Apr. 14, 1927; s. Henry Vivian and Helen (Lynn) G.; m. Iveagh Foreman, Sept. 4, 1954; children—Deborah Lynn, Kevin Douglas, Cameron Bradley. B.A., Wichita U., 1951; M.D., Yale, 1953. Intern U. Minn. Hosp., 1953-54; resident surgery U. Minn. Hosps., 1954-60; asst. prof. surgery U. Wis., 1960-65; asso. prof. surgery Johns Hopkins, 1965-68, prof., 1968—; cardiac surgeon in charge Johns Hopkins Hosp., 1965—. Contbr. articles to profl. jours. Served with USNR, 1945-46. Recipient Hektoen gold medal A.M.A., 1957; John and Mary R. Markle scholar, 1962. Fellow A.C.S.; mem. Am. Surg. Assn., Soc. Univ. Surgeons, Am. Assn. Thoracic Surgery, Soc. Thoracic Surgeons, Soc. Vascular Surgeons, Am. Heart Assn. Subspecialty: Cardiac surgery. Co-developer Gott-Daggett artificial heart valve, 1963; developer graphite-benzalkonium-heparin coating for plastic surfaces. Home: 203 Kemble Rd Baltimore MD 21218

GOTTESMAN, IRVING ISADORE, psychiatric genetics educator, consultant; b. Cleve., Dec. 29, 1930; s. Bernard and Virginia (Weitzner) G.; m. Carol Applen, Dec. 22, 1970; children: Adam M., David. B. B.S., Ill. Inst. Tech., 1953; Ph.D., U. Minn., 1960. Diplomate: in clin. psychology; lic. psychologist, Minn., Calif., Mo. Intern clin. psychology VA Hosp., Mpls., 1959-60; lectr. dept. social relations Harvard U., 1960-63; fellow psychiat. genetics Inst. Psychiatry, London, 1963-64; assoc. prof. psychiat. genetics, dept. psychiatry U. N.C., 1964-66; prof. dept. psychology U. Minn., 1966-80; prof. dept. psychiatry Washington U., St. Louis, 1980—; cons. NIMH, Washington, 1975-79; mem. Pres.'s Comm. on Huntington Disease, 1977; tng. cons. VA, Washington, 1968—. Author: Schizophrenia and Genetics, 1972 (Hofheimer prize), Schizophrenia-The Epigenetic Puzzle, 1982; editor: Man, Mind and Heredity, 1971. Served with USN, 1953-56. Guggenheim fellow, 1972; recipient R. Thornton Wilson prize Eastern Psychiat. Research Assn., 1965; David C. Wilson lectr. U. Va. Sch. Medicine, 1967; Parker lectr. Ohio State U. Sch. Medicine, 1983. Fellow Am. Psychol. Assn.; mem. Minn. Human Genetics League (v.p. 1969-71), Soc. Study of Social Biology (v.p. 1976-80), Behavior Genetics Assn. (pres. 1976-77), Royal Coll. Psychiatrists, Am. Psychopath. Assn., Am. Assn. Human Genetics (editorial bd. 1967-72). Subspecialty: Genetics and genetic engineering (medicine). Current work: Genetic aspects of mental illness, personality and intelligence via twin, family and adoption strategies; genetic epidemiology; multifactorial models. Home: 308 N Brentwood Blvd Clayton MO 63105 Office: Washington U Sch Medicine 4940 Audubon Ave Saint Louis MO 63110

GOTTESMAN, STEPHEN THANCY, astronomer, educator; b. N.Y.C., Feb. 23, 1939; s. Jacob Frank and Edna Beatrice (Goldner) G.; m. Celia Frances Docherty, Feb. 24, 1947; children: Lorna Rachel, Ian Kenneth Jacob. B.A. magna cum laude, Colgate U., 1960; Ph.D., Victoria U., Manchester, Eng., 1967. Lectr. in physics and astronomy U. Keele, Eng., 1969; research assoc. Nat. Radio Astron. Obs., Charlottesville, Va., 1969-71; research fellow Calif. Inst. Tech., Pasadena, 1971; asst., then assoc. prof. astronomy U. Fla., Gainesville, 1972-81, prof. astronomy, 1981—; project cons. Nat. Endowment Humanities; cons. astron. textbooks. Contbr. numerous articles to sci. publs. Fulbright scholar, 1960-61; Leverhulme fellow, 1961-64; grantee Internat. Astron. Union, 1972, CNRS, France, 1974, NSF, 1974, 78, 79, 80, 82, So. Regional Edn. Bd., 1975-77, Sigma Xi, 1976, NRC/Nat. Acad. Scis., 1979, 82. Mem. Am. Astron. Soc., Royal Astron. Soc., Internat. Union Radio Sci. (comm. J), Internat. Astron. Union, Phi Beta Kappa. Democrat. Subspecialty: Radio and microwave astronomy. Current work: Structure and kinematics of Milky Way and of extra-galactic nebulae. Atomic and molecular emisions from interstellar medium from our own and other galaxier. Office: Dept Astronomy U Fla Gainesville FL 32611

GOTTESMAN, SUSAN, molecular geneticist, microbiologist; b. N.Y.C., May 19, 1945; d. Robert and Dorothy (Altman) Kemelhor; m. Michael Marc Gottesman, Feb. 5, 1966; children: Daniel, Rebecca. B.A., Radcliffe Coll., 1967; Ph.D., Harvard U., 1971. Postdoctoral fellow NIH, Bethesda, Md., 1971-73; research assoc. MIT, Cambridge, 1973-75; sr. investigator Nat. Cancer Inst., Bethesda, 1975—; mem. recombinant adv. com. NIH, 1977-81, 83-87, chmn. working group on major revisions of guidelines, 1981. Mem. Am. Soc. Microbiology, AAAS, Genetics Soc. Am. Subspecialties: Gene actions; Molecular biology. Current work: Bacterial cell growth regulation. Office: Lab Molecular Biology Nat Cancer Inst Bethesda MD 20205

GOTTFREDSON, GARY DON, psychologist, educator; b. Sonora, Calif., Sept. 4, 1947; s. Don Martin and Betty Jane (Hunt) G.; m. Linda Suzanne Howarth, Apr. 19, 1967 (div. Dec. 1979); m. Denise Claire Ruff, Dec. 31, 1979. B.A., U. Calif.-Berkeley, 1969; M.A., Johns Hopkins U., 1975, Ph.D., 1976. Lic. psychologist, Md. Assoc. adminstrv. officer Am. Psychol. Assn., Washington, 1976-77; asst. prof. psychology and social relations Johns Hopkins U., Balt., 1977-81, assoc. prof., 1981—, research scientist, 1977—, dir. delinquency program, 1978—; sr. partner Decision Research Assocs., Balt., 1979—. Author: Dictionary of Holland Occupational Codes, 1982; editorial bd.: Jour. Vocat. Behavior, 1977—. Mem. com. occupational classification and analysis NRC, Washington, 1978-81, mem. panel research on rehab. techniques, 1978-81. Nat. Inst. Juvenile Justice and Delinquency Prevention grantee, 1977—; Nat. Inst. Edn. grantee, 1977—. Mem. Am. Psychol. Assn. (com. employment and human resources 1982—, task force victims of crime 1982—), Am. Soc. Criminology, Am. Sociol. Assn., Nat. Council Crime and Delinquency. Democrat. Club: Johns Hopkins (Balt.). Subspecialty: Social psychology. Current work: The design, evaluation and improvement of organized activities to improve the effectiveness of organizations; etiology and prevention of crime; vocational behavior. Home: 725 Stoney Spring Dr Baltimore MD 21210 Office: Johns Hopkins U 3505 N Charles St Baltimore MD 21218

GOTTLIEB, ALLAN JOSEPH, computer science educator; b. N.Y.C., Aug. 2, 1945; s. Irving Arthur and Frances (Caggiano) G.; m. Alice Eve Bendix, Jan. 7, 1972; 1 son: David Bendix. B.S., M.I.T., 1967; M.A., Brandeis U., 1968, Ph.D., 1973. Acting instr. U. Calif.-Santa Cruz, 1971-72; instr. State Coll. Mass.-North Adams, 1972-73; asst. prof. math. CUNY, 1973-78, assoc. prof., 1979-81; assoc. research prof. computer sci. N.Y. U., 1981—. Editor puzzle column: Technology Rev, Cambridge, Mass., 1964—; Contbr. articles to profl. jours. Mem. Am. Math. Soc., Assn. Computing Machinery, N.Y. Acad. Scis., IEEE, Sigma Xi. Subspecialties: Computer architecture; Distributed systems and networks. Current work: Manager of NYU Ultracomputer Project, large government funded research project in parallel processing. Office: New York 251 Mercer St New York NY 10012 Home: 500 E 63d St New York NY 10021

GOTTLIEB, ARLAN JAY, medical educator; b. N.Y.C., July 22, 1933; s. Sol and Diana (Begun) G.; m. Ann Catherine Burton, May 24, 1964; children: Daphne, Danielle, Jonathan. B.A., Columbia U., 1954, M.D., 1958. Diplomate: Am. Bd. Internal Medicine. Resident in medicine Mt. Sinai Hosp., N.Y.C., 1958-62; research assoc. NIAMD, Bethesda, Md., 1962-67; clin. asst. prof. medicine George Washington U. Med. Sch., Washington, 1966-67; asst. prof. medicine U. Pa., Phila., 1967-71; assoc. prof. medicine, chief hematology SUNY Upstate Med. Center, Syracuse, 1971-76, prof., chief hematology, 1976—, chmn. Lymphoma CALLGB, 1978—; vice chmn. Leukemia Intergroup (Buffalo), 1982—. Author: The Whole Internist's Catalog, 1980; contbr. chpts. to textbooks, articles to profl. jours. Served to comdr. USPHS, 1962-67. NIH grantee. Fellow ACP; mem. Am. Soc. Hematology, Am. Soc. Clin. Oncology, Dirs. of Hematology/Oncology Programs, Am. Fedn. Clin. Research, Alpha Omega Alpha. Subspecialties: Hematology; Cancer research (medicine). Current work: Controlled, therapeutic trials in cancer, intermediary metabolism of the red cell. Home: Owahgena Rd Manlius NY 13104 Office: SUNY Upstate Med Center 750 E Adams St Syracuse NY 13210

GOTTLIEB, BARBARA WEINTRAUB, psychologist; b. N.Y.C., Jan. 6, 1949; d. Sam and Elaine (Jacobs) Weintraub; m. Jay Gottlieb, Aug. 10, 1975. B.A., U. Pitts., 1971; M.Ed., Lesley Coll., Cambridge Mass., 1974; Ed.D., No. Ill. U., DeKalb, 1980. Dir. Ctr. for Ednl. Research, Larchmont, N.Y., 1980-81; cons. N.Y.C. Bd. Edn., Bklyn., 1980—; asst. prof. spl. edn. Herbert H. Lehman Coll., CUNY, Bronx, N.Y., 1981—; cons. Assn. for Learning Diabilities, White Plains, N.Y., 1980—. Contbr. articles to profl. jours. Mem. Westchester Assn. for Learning Disabilities, Am. Psychol. Assn., Assn. for Advancement Behavior Therapy, Assn. for Severly Handicapped, Council on Exceptional Children. Subspecialties: Learning; Developmental psychology. Current work: Social facilitation influences on behavior. Psychometric evaluation of developmentally delayed persons. Office: Herbert H Lehman Coll CUNY Bronx NY 10468

GOTTLIEB, LEONARD SOLOMON, pathology educator; b. Boston, May 26, 1927; s. Julius and Jeanette (Miller) G.; m. Dorothy Helen Apt, Mar. 23, 1952; children: Julie Ann Gottlieb Texeira, William Apt, Andrew Richard. A.B., Bowdoin Coll., 1946; M.D., Tufts U., 1950; M.P.H., Harvard U., 1969. Diplomate: Am. Bd. Anatomic Pathology. Intern and resident in pathology Boston City Hosp., 1950-55; asst. chief pathology U.S. Naval Hosp., Chelsea, Mass., 1955-57; assoc. pathologist Mallory Inst. Pathology, Boston, 1957-66, assoc. dir., 1966-72, dir., 1972—; prof. pathology Boston U. Sch. Medicine, 1970—, chmn. dept., 1980—; dir. Mallory Inst. Pathology Found., 1980—; lectr. Harvard U., 1963—. Contbr. over 100 articles on exptl. and human gastrointestinal and liver diseases. Served to lt. M.C. USNR, 1955-57; to lt. comdr., 1960-63. James Bowdoin scholar, 1945. Mem. Am. Soc. Exptl. Pathology, Am. Assn. Study of Liver Disease, Internat. Acad. Pathology, Am. Soc. Cell Biology, Am. Gastroent. Assn., New Eng. Soc. Pathologists, Am. Soc. Clin. Pathology, Mass. Soc. Pathology, Coll. Am. Pathologists, Am. Inst. Nutrition. Subspecialties: Pathology (medicine); Gastroenterology. Current work: Tumor antigens in gastrointestinal malignancy; role of colonic polyps in pathogenesis of malignancy; alcoholic liver disease. Home: 120 Willard Rd Brookline MA 02146 Office: 784 Massachusetts Ave Boston MA 02118

GOTTLIEB, MELVIN BURT, astrophysics educator, university administrator; b. Chgo., May 25, 1917; m., 1948; 2 children. B.S., U. Chgo., 1940, Ph.D., 1950. Research assoc. Harvard U., 1943-45; instr. in phys. sci. U. Chgo., 1945-46, asst. in physics, 1946-50; asst. prof. U. Iowa, 1950-54; assoc. dir. Project Matterhorn, Princeton U., 1954-61, prof. astrophys. sci., dir. plasma physics lab., 1961—, assoc. chmn. astrophys. sci., 1974-80. Mem. Am. Phys. Soc., Sigma Xi. Subspecialty: Plasma physics. Office: Dept Astrophys Sci Princeton U PO Box 451 Princeton NJ 08540

GOTTSCHALK, CARL WILLIAM, physician, educator; b. Salem, Va., Apr. 28, 1922; s. Carl and Lula (Helbig) G.; m. Helen Marie Scott, Nov. 22, 1947; children—Carl S., Walter P., Karen E. B.S., Roanoke Coll., 1942, Sc.D., 1966; M.D., U. Va., 1945. Intern, asst.

resident, resident in medicine Mass. Gen. Hosp., Boston, 1945-52; research fellow physiology Harvard, 1948-50; fellow U. N.C. Med. Sch., Chapel Hill, 1952-53, faculty, 1953—, Kenan prof. medicine and physiology, 1969—; established investigator Am. Heart Assn., 1957-61, career investigator, 1961; Bowditch lectr., 1960, Harvey lectr., 1962; Mem. physiology study sect. NIH, 1961-65; mem. research career award com. Nat. Inst. Gen. Med. Scis., 1965-69, mem. physiology tng. com., 1970-73, mem. med. scientist tng. com., 1973; chmn. com. chronic kidney disease Bur. Budget, 1966-67; adv. com. biol. and med. scis. NSF, 1967-69, vice chmn., 1968, chmn., 1969; mem. Inst. Medicine of Nat. Acad. Scis., Nat. Adv. Gen. Med. Scis. Council, 1977-80, Nat. Arthritis, Diabetes and Digestive and Kidney Diseases Adv. Council, 1982—. Author books and papers on physiology of kidney. Mem. adv. com. Burroughs Wellcome Fund for Clin. Pharmacology, 1980—; Pres. Children's Theatre N.C., 1967-68. Served to capt., M.C. AUS, 1946-48. Recipient N.C. award, 1967, Modern Medicine Distinguished Achievement award, 1966, Horsley Meml. prize U. Va., 1956, Homer W. Smith award N.Y. Heart Assn., 1970, David Hume award Nat. Kidney Found., 1976, O. Max Gardner award U. N.C., 1978. Mem. Assn. Am. Physicians, Am. Physiol. Soc., Am. Soc. Clin. Investigation, Am. Clin. and Climatol. Assn., Soc. Exptl. Biology and Medicine, A.C.P., AAUP (council 1970-73), Nat. Acad. Scis., Am. Soc. Nephrology (council 1971-77, pres. 1975-76), Am. Acad. Arts and Scis., Phi Beta Kappa, Sigma Xi. Subspecialties: Physiology (medicine). Current work: Micropuncture studies of mammalian renal physiology; neural control of renal function. Home: 1300 Mason Farm Rd Chapel Hill NC 27514

GOTTWALD, RICHARD LANDOLIN, psychology educator; b. Detroit, Jan. 24, 1941; s. Henry Landolin and Hannah Paulina (Rauch) G.; m. Judith Ann LaFortune, June 7, 1965; children: Katrina, Jennifer, Melissa. Student, M.I.T., 1959-61, Henry Ford Community Coll., 1961-62; B.A., U. Mich., 1963; M.A., Johns Hopkins U., 1966, Ph.D., 1968. Research staff psychologist, lectr. Yale U., New Haven, 1968-70, research assoc. in psychology, 1970-71; asst. prof. psychology Ind. U., South Bend, 1971-75, assoc. prof., 1975—, acting chmn. dept., 1972-73, chmn., 1980—. Mem. Am. Psychol. Assn., Psychonomic Soc., Eastern Psychol. Assn., Medwestern Psychol. Assn., AAAS, Sigma Xi. Subspecialty: Cognition. Current work: Classification, concept attainment, pattern perception, generally the interaction between cognitive and perceptual processes. Home: 1515 E Madison St South Bend IN 46617 Office: Ind U at South Bend 1700 Mishawaka Ave South Bend IN 46634

GOTTWALD, TIM R., plant pathologist, epidemiologist; b. Lynwood, Calif., Feb. 14, 1953; s. Ross G. and Lorraine E. (Dowell) G.; m. Karen Schlegel, July 17, 1979. B.S. in Botany, Long Beach State U., 1975; Ph.D. in Plant Pathology, Oreg. State U., 1979. Research asst. dept. botany Oreg. State U., Corvallis, 1975-79; research plant pathologist Agrl. Research Service, U.S. Dept. Agr., Byron, Ga., 1979—. Contbr. articles to profl. jours. Mem. Am. Phytopathol. Soc., Mycological Soc. Am. Subspecialties: Plant pathology; Integrated pest management. Current work: Investigation of basic biology and epidemiology of fruit and nut trees, especially diseases of pecan and related nuts. Office: USDA-ARS Southeastern Fruit and Nut Research Lab Byron GA 31008

GOUGH, DENIS IAN, physicist, educator; b. Port Elizabeth, South Africa, June 20, 1922; m., 1945; 2 children. B.Sc., Rhodes U. Coll., South Africa, 1943, M.Sc., 1947; Ph.D. in Geophysics, U. Witwatersrand, 1953. Research officer geophysics South African Nat. Phys. Research Lab., 1947-55, sr. research officer, 1955-58; lectr. Univ. Coll., Rhodesian and Nyasaland, 1958-60, sr. lectr., 1961-63; asso. prof. geophysics S.W. Center Advanced Studies, Dallas, 1964-66; prof. U. Alta. (Can.), Edmonton, 1966-68, prof. physics, 1968—; also dir. Inst. Earth and Planetary Physics.; Mem. geodesy and geophysics Central African Fedn., 1960-63. Served with South African Armed Forces, 1943-45. Fellow Royal Soc. Can., Royal Astron. Soc.; mem. Am. Geophys. Union, European Assn. Exploration Geophysics, Can. Geophys. Union (chmn. 1975-77). Subspecialties: Geophysics; Geothermal power. Current work: Electromagnetic induction studies of structures in crust and mantle of tectonic significance, and lithosphere stress. Address: Dept Physics U Alberta Edmonton Canada AB T6G 2J1

GOUGH, FRANCIS JACOB, plant pathologist; b. Grafton, W.Va., Apr. 9, 1928; s. Claude Ernest and Cordie Melissa (Weaver) G.; m. Ruby G. Nestor, Sept. 10, 1950; children: Rodney K., Jennifer Lynn. B.S., W.Va. U., 1951, M.S., 1953, Ph.D., 1957. Research plant pathologist Dept. Agr., Fargo, N.D., 1957-67, College Station, Tex., 1967-74, Stillwater, Okla., 1974—. Contbr. numerous articles to profl. jours., 1956—. Served as cpl. USAAF, then USAF, 1946-49. Mem. Am. Phytopathol. Soc., Internat. Soc. Plant Pathologists. Democrat. Lodge: Masons. Subspecialties: Plant pathology; Plant genetics. Current work: Genetics of host parasite interactions; epidemiology.

GOUGH, PATRICIA MARIE, immunochemistry educator; b. Eagle River, Wis., Jan. 13, 1937; d. Frank C. and Eleanor M. (Johnson) G. B.S., U. Wis.-Madison, 1958; M.S., U. Minn., 1961, Ph.D., 1966. Research chemist Nat. Ctr. Disease Control, USPHS, Atlanta, 1967-68; mem. faculty Vet. Med. Research Inst., Iowa State U., Ames, 1968—, prof. immunochemistry, 1976—. Subspecialties: Infectious diseases; Immunology (medicine). Current work: Immunochemical studies of viral diseases; diagnosis, vaccine development, viral chemical analysis; identification of immune responses. Office: Vet Med Research Inst Iowa State U Ames IA 50011 Home: Rural Route 5 Boone IA 50036

GOULD, CHARLES LAVERNE, aerospace engineer; b. Winston, Mo., Oct. 26, 1933; s. Clem and Nora (Harris) G.; m.; children: Anita, Katherine. B.S.M.E., Iowa State U., 1956; cert. bus. mgmt., UCLA, 1966; Ph.D. in Bus. Adminstrn, Calif. Western U., 1978. Cert. profl. mgr., Calif. Aerospace engr. USAF, Dayton, Ohio, 1959-62; project engr. Rockwell Internat., Downey, Calif., 1962-66, asst. chief engr. space sta., 1968-72, mgr. tech. programs, 1972-76, program mgr. space industrialization, 1976-80, shuttle utilization, 1980—. Author tech. reports; contbr. articles to profl. jours. Chmn. bd. Methodist Ch., Anaheim, Calif., 1967. Served to lt USAF, 1956-59. Recipient Sustained Superior Service award USAF, 1962; Assoc. fellow AIAA. Republican. Methodist. Subspecialties: Aerospace engineering and technology; Satellite studies. Current work: management of high technology development related to utilization of space; marketing of shuttle launch services; identification and development of the global use of space, particularly communications, observation and materials processing. Home: 1832 Sunset Ln Fullerton CA 92633 Office: Rockwell International 12214 Lakewood Blvd Downey CA

GOULD, GORDON, physicist, optical communications executive; b. N.Y.C., July 17, 1920; s. Kenneth Miller and Helen Vaughn (Rue) G. B.S. in Physics, Union Coll., 1941, D.Sc., 1978; M.S. in Physics, Yale U., 1943, Columbia U., 1952. Physicist Western Electric Co., Kearny, N.J., 1941; instr. Yale U., 1941-43; physicist Manhattan Project, 1943-45; engr. Semon Bache Co., N.Y.C., 1945-50; instr. CCNY, 1947-54; research asst. Columbia U., 1954-57; research dir. TRG, Inc./Control Data Corp., Melville, N.Y., 1958-67; prof. electrophysics Bklyn. Poly. Inst., 1967-74; v.p. engring./mktg., dir. Optelecom, Inc., Gaithersburg, Md., 1974—. Contbr. sci. articles to profl. jours. Recipient 63 research grants and contracts, 1958—; named Inventor of Year for laser amplifier Patent Office Soc., 1978. Mem. Am. Inst. Physics, Optical Soc. Am., IEEE, AAAS, Fiber Optic Communications Soc., Laser Inst. Am. (pres. 1971-73, dir. 1971—). Subspecialties: Optical engineering; Fiber optics. Current work: Development of optical communication systems, optical instrumentation; fiber optic cables. Patentee in field. Home: 9101 Deer Park Rd Great Falls VA 22066

GOULD, LAWRENCE, telecommunications executive; b. 1931; m. Anna Gould. Ph.D. in Physics, MIT, 1954. With Microwave Assocs. (now M/A-COM Inc.), Burlington, Mass., chief exec. officer, 1975—, now also chmn. Subspecialty: Telecommunications Management. Office: M/A-COM Inc NW Indsl Park Burlington MA 01803

GOULD, LAWRENCE A., cardiologist; b. Bklyn., Dec. 15, 1930; s. Raymond and Lola (Misel) G.; m. Roberta Berkley, Mar. 31, 1957; children: Julie Ellen, Bruce David. B.A., Bklyn. Coll., 1952; M.D., NYU, 1956. Intern Kings County Hosp., Bklyn., 1956-67; resident in cardiology Bronx VA Hosp., 1959-62, asst. chief cardiology, 1965-67, chief cardiology, 1967-69; chief cardiac catheter lab. Misericordia Hosp., Bronx, 1969-74; chief cardiology Meth. Hosp. Bklyn., 1974—. Editor: Phentolamine in Heart Failure, 1976, Correlative Atlas of Vectorcardiograms and Electrocardiograms, 1977, Vasodilator Therapy for Cardiac Disorder, 1979, Drug Treatment of Cardiac Arrhythmias, 1983. Served as capt. U.S. Army, 1957-59. Fellow N.Y. Cardiovascular Soc. (pres. 1980-81), Am. Coll. Cardiology, Am. Coll. Chest Physicians, Am. Coll. Angiology, Am. Heart Assn. Subspecialty: Cardiology. Home: 4 Effron Pl Great Neck NY 11020 Office: Meth Hosp 506 6th St Brooklyn NY

GOULD, PHILLIP LOUIS, civil engineer, educator; b. Chgo., May 24, 1937; s. David J. and Belle (Blair) G.; m. Deborah Paula Rothholtz, Feb. 5, 1961; children: Elizabeth, Nathan, Rebecca, Joshua. B.S., U. Ill., 1959, M.S., 1960; Ph.D., Northwestern U., 1966. Structural designer Skidmore, Owings & Merrill, Chgo., 1960-63; structural engr. Westenhoff & Novick, Chgo., 1963-64; NASA trainee Northwestern U., Evanston, Ill., 1964-66; Harold D. Jolley prof., chmn. dept. civil engring. Washington U., St. Louis, 1966—; cons. various pvt. firms, utilities U.S. Nat. Bur. Standards, 1968—. Author: Static Analysis of Shells: A Unified Development of Surface Structures, 1977, (with others) Dynamic Response of Structures, 1980, Introduction to Linear Elasticity, 1983; editor: Environmental Forces on Engineering Structures, 1979, (with others) Engring. Structures; contbr. over 100 articles to profl. jours. Served to 1st lt. U.S. Army, 1959-60, 62. Recipient Sr. U.S. Scientist award Alex V. Humboldt Found., W.Ger., 1974. Fellow ASCE; mem. Internat. Assn. Shell Structures, Am. Soc. Engring. Edn., Am. Acad. Mechanics, Sigma Xi. Subspecialties: Civil engineering; Solid mechanics. Current work: Analysis and design of thin-shell structures including dynamic and interaction effects. Finite element analysis of rotational shells. Home: 102 Lake Forest Richmond Heights MO 63117 Office: Dept Civil Engring Washington Univ Campus Box 1130 St Louis MO 63130

GOULD, RICHARD BRUCE, research psychologist, educator; b. Juneau, Alaska, Dec. 16, 1939; s. Donnell Hunting and Emma (Usher) G.; m. Sharry Kay Croffard, July 14, 1960; 1 son, Richard Bruce. A.A., San Antonio Coll., 1960; B.A., St. Mary's U., 1964; M.S., Trinity U., 1965; Ph.D., U. Tex.-Austin, 1978. Computer lab. instr. St. Mary's U., 1963-64; claims adjuster Dunn & Bradstreet, San Antonio, 1964-65; grad. teaching asst. Trinity U., 1965; research psychologist Air Force Human Resources Lab., Brooks AFB, Tex., 1965-79, supervisory psychologist, 1979—; instr. psychology St. Mary's U., 1976—; indsl. cons., San Antonio, 1980—. Contbr. articles to profl. jours. Served with AUS, 1960-62. Mem. Am. Psychol. Assn., Mil. Testing Assn., SAR, SCV. Republican. Roman Catholic. Clubs: Antonian Men's (treas. 1979-80), Bexar County Rep. Men's.). Subspecialty: Industrial/organizational psychology. Current work: Selection and classification research; test development; job redesign; software development; performance measurement research. Home: 7243 N Vandiver St San Antonio TX 78209 Office: Air Force Human Resources Lab AFHRL/MOD Brooks AFB TX 78235

GOULD, ROBERT GEORGE, physicist, educator; b. Plattsburg, N.Y., June 12, 1947; s. Robert Henderson and Sibyl (Renaud) G.; m. Carol Anne Opotow, Oct. 1, 1977. B.A., Coll. of Wooster, 1969; M.S.E.E., U. Pa., 1971; Sc.D., Harvard U., 1978. Hosp. physicist Beth Israel Hosp., Boston, 1973-77; assoc. prof. in residence dept. radiology U. Calif., San Francisco, 1977—. Contbr. articles to profl. jours. Mem. Am. Assn. Physicists in Medicine, Soc. Photo-Optical Instrumentation Engrs. Democrat. Subspecialties: Imaging technology; CAT scan. Current work: Digital radiography, x-ray imaging science, CAT scanning. Office: Dept Radiology Univ Calif San Francisco CA 94143

GOULD, ROY WALTER, engineer, physicist, educator; b. Los Angeles, Apr. 25, 1927; s. Roy Walter and Rosamonde (Stokes) G.; m. Ethel Savage Stratton, Aug. 23, 1952; children: Diana Stratton, Robert Clarke. B.S., Calif. Inst. Tech., 1949, Ph.D., 1956; M.S., Stanford, 1950. Mem. faculty Calif. Inst. Tech., Pasadena, 1955—, exec. officer for applied physics, 1973-79, chmn. div. engring. and applied sci., 1979—, Simon Ramo prof. engring., 1980—; dir. div. controlled thermonuclear research AEC, 1970-72. Served with USNR, 1945-46. Fellow Am. Phys. Soc., IEEE; mem. Nat. Acad. Scis., Nat. Acad. Arts and Scis., Nat. Acad. Engring. Subspecialties: Nuclear fusion; Plasma physics. Current work: Plasma confinement and heating. Home: 808 Linda Vista Ave Pasadena CA 91103

GOULD, STEPHEN JAY, paleontologist, educator; b. N.Y.C., Sept. 10, 1941; s. Leonard and Eleanor (Rosenberg) G.; A.B., Antioch Coll., Yellow Springs, Ohio, 1963; Ph.D., Columbia U., 1967; m. Deborah Ann Lee, Oct. 3, 1965; children—Jesse, Ethan. Mem. faculty Harvard U., 1967—, prof. geology, 1973—. Recipient Nat. Mag. award for essays and criticism, 1980; Nat. Book award in sci., 1981; McArthur Found. prize fellow, 1981—; grantee NSF. Fellow AAAS; mem. Paleontological Soc. (Schuchert award 1975), Soc. Study Evolution, Soc. Systematic Zoology, Am. Soc. Naturalists, Sigma Xi. Author: Ontogeny and Phylogeny, 1977; Ever Since Darwin, 1977; The Panda's Thumb, 1980; A View of Life, 1981; The Mismeasure of Man, 1981; also numerous articles, monthly column This View of Life in Natural History mag. Subspecialties: Evolutionary biology; Paleobiology. Address: Museum Comparative Zoology Harvard Univ Cambridge MA 02138

GOULIANOS, KONSTANTIN, physics educator; b. Salonica, Greece, Nov. 9, 1935; came to U.S., 1958, naturalized, 1967; s. Achilles and Olga (Nakopoulou) G. Student, U. Salonica, 1953-58; Ph.D., Columbia U., 1963. Research assoc. Columbia U., N.Y.C., 1963-64; instr. Princeton U., 1964-67, asst. prof., 1967-71; assoc. prof. Rockefeller U., N.Y.C., 1971-81, prof. physics, 1981—. Fulbright scholar, 1958-59. Subspecialty: Particle physics. Current work: Research in experimental high energy physics. Patentee electronic device for analysis of radioactively labeled gel electrophoretograms.

GOUNARD, BEVERLY ELAINE, psychologist, consultant; b. Hamilton, Ont., Can., Nov. 24, 1942; came to U.S., 1971, naturalized, 1978; d. Cyril and Lily Gladys (Rickard) Roberts; m. Jean-Francois Gounard, Aug. 19, 1967; children: Anne-Marie Christine, Emilie Sarah Marie. B.A., McMaster U., 1964; M.A., Queen's U., 1968; Ph.D., U. Waterloo, 1971. Lic. psychologist , N.Y. State. Asst. prof. psychology SUNY-Buffalo, 1971-78, research cons., 1978-81; pvt. practice psychology, cons., Williamsville, N.Y., 1981—. Contbr. articles to profl. jours. and books. Ont. Mental Health Research Found. grantee, 1970-71; SUNY-Buffalo grantee, 1972-73, 77-78; SUNY-Buffalo Multidisciplinary Center for Study Aging grantee, 1976. Mem. Am. Physchol. Assn., Can. Psychol. Assn., Psychol. Assn. Western N.Y., Gerontol. Soc., AAUW, State Univ. Coll. Buffalo Assn. Women (2d v.p. 1982-83), Sigma Xi. Subspecialties: Developmental psychology; Learning. Current work: Age-related changes in learning and memory abilities; improving learning and memory performance in individuals of any age. Home: 38 Idlewood Dr Tonawanda NY 14150 Office: 5144 Sheridan Dr Williamsville NY 14221

GOUSE, S. WILLIAM, JR., scientist; b. Utica, N.Y., Dec. 15, 1931; s. S. William and Charlotte Virginia (Parzych) G.; m. Jacqueline Ann McLaughlin, Aug. 6, 1955; children: Linda Ellen, S. William III. S.B., S.M., Mass. Inst. Tech., 1954, Sc.D., 1958. Instr. mech. engring. Mass. Inst. Tech., 1956-57, asst. prof., 1957-61, 62-65, asso. prof., 1965-67, lectr., 1967-68; prof. mech. engring., prin. research engr. Transportation Research Inst. of Carnegie-Mellon U., 1967-69; staff mem. Office Sci. and Tech. of Exec. Office of the Pres., Washington, 1969-70; asso. dean Carnegie Inst. Tech. and Sch. Urban and Pub. Affairs of Carnegie-Mellon U., 1971-73; dir. Office Research and Devel. U.S. Dept. Interior, 1973-75; acting dir. Office Coal Research, 1974-75; dep. asst. adminstr. fossil energy ERDA, 1974-77; chief scientist Mitre Corp., 1977-79, v.p., 1979-80, v.p., gen. mgr., 1980—; cons. to industry. Contbr. articles to profl. jours. Served with ordnance AUS, 1961-62. Visking Corp. fellow, 1954-55; Gen. Electric Co. W. Rice Jr. fellow, 1955-56; recipient Ralph Teetor award Soc. Automotive Engrs., 1966. Clubs: Cosmos, Explorers. Subspecialties: Mechanical engineering; Systems engineering. Home: 8410 Martingale Dr McLean VA 22102

GOUST, JEAN-MICHEL CHRISTIAN, immunologist, educator; b. St. Mande, France, Jan. 21, 1941; s. Francois Joseph and Marie Odile (Mirabail) G.; m. Marie Francoise Montassut, Sept. 5, 1963; children: Olivier, Thierry. B.S., U. Paris, 1958, M.D., 1971. Intern Hopitaux de Paris, 1963-66, resident, 1968-72; instr. CHU Pitie Salpetriere, Paris, 1968-72, clin. chief, asst. physician exptl. pathology and medicine, 1972-75; asst. prof. immunology Med. U. S.C., Charleston, 1975-80, asst. prof. medicine, 1976-80, assoc. prof. medicine, 1980-81, assoc. prof. immunology and neurology, 1981—; chmn. med. adv. com. Trident chpt. Multiple Sclerosis Soc., 1980—. Mem. editorial bd.: Clin. Immunology and Immunopathology; contbr. articles to profl. jours. Nat. Multiple Sclerosis Soc. grantee; NIH grantee; Owen-Ceatham Found for Research in Multiple Sclerosis grantee. Mem. Am. Fedn. Clin. Research, Am. Assn. Immunologists, Am. Soc. Microbiology, Soc. Exptl., Biology and Medicine. Roman Catholic. Subspecialties: Immunobiology and immunology; Neurology. Current work: Research in multiple sclerosis and other neurological diseases with immunological abnormalities. Patentee in field. Home: 29 27th Ave Isle of Palms SC 29451 Office: Dept Basic and Clin Immunology and Microbiology Med U SC 171 Ashley Av Charlesto SC 29425

GOUTERMAN, MARTIN PAUL, educator; b. Phila., Dec. 26, 1931; s. Bernard and Melba (Buxbaum) G. B.A., Central High Sch., Phila., 1949, U. Chgo., 1951, M.Sc., 1955, Ph.D. in Physics (NSF Predoctoral fellow), 1958. Faculty Harvard U., Cambridge, Mass., 1958-66; successively postdoctoral fellow, instr., asst. prof. chemistry dept.; faculty U. Wash., Seattle, 1966—, prof. chemistry, 1968—. Fellow Am. Inst. Physics; mem. Am. Chem. Soc., Sigma Xi. Subspecialties: Physical chemistry; Theoretical chemistry. Current work: Electronic spectra and structure of porphyrins and related molecules; solid state electronic properties of porphyrin films. Research and publs. in spectroscopy, quantum chemistry, and solid state electronic properties of porphyrins. Office: Dept Chemistry U Wash Seattle WA 98195

GOVE, HARRY EDMUND, nuclear physicist, educator; b. Niagara Falls, Can., May 22, 1922; came to U.S., 1963, naturalized, 1969; s. Harry Golden and Lucia (Olmsted) G.; m. Elizabeth Alice dePencier, Aug. 20, 1945; children: Pauline Lucia, Diana Elizabeth. B.Sc., Queen's U., Kingston, Ont., 1944; Ph.D., MIT, 1950. Research asst. Nat. Research Council, Chalk River, Can., 1945-46, Mass. Inst. Tech., 1946-50, research asso., 1950-52; asso. research officer Atomic Energy of Can., Ltd., Chalk River, 1952-59, br. head nuclear physics, 1956-63, sr. research officer, 1959-63; on leave with Niels Bohr Inst., Copenhagen, 1961-62; prof. physics, dir. nuclear structure research lab. U. Rochester, 1963—, chmn. dept. physics and astronomy, 1977-80; on leave with Lab. de Physique Nucléaire et d'Instrumentation Nucléaire, C.R.N., Strasbourg, 1971-72; Mem. vis. com. Mass. Inst. Tech. Lab. Nuclear Research, 1966-68, Argonne Nat. Lab. physics div., 1966-69, Queen's U. Coll. Engring., 1968-70; mem. adv. com. physics div. NSF, 1969-71; mem. grant selection com. nuclear physics Nat. Sci. and Engring. Research Council Can., 1979-81; mem. ad hoc panel on meson factories Office Sci. and Tech., 1963; mem. selection panel NSF Postdoctoral Fellows, 1967, 69, 70; mem. vis. com. physics and accelerator depts. Brookhaven Nat. Lab., 1979—; mem. nat. heavy-ion lab. policy com. Oak Ridge Nat. Lab., 1975-76; chmn. panel on basic nuclear data compilations, NRC-Nat. Acad. Scis, 1975-80, mem. panel on future of nuclear sci., 1975-77; chmn. vis. com. Cyclotron Lab., U. Md., 1977-79. Divisional asso. editor: Phys. Rev. Letters, 1975-79; asso. editor: Ann. Rev. Nuclear Sci; Contbr. articles to profl. jours. Pres. Metro Act of Rochester, Inc., 1970-71; trustee Associated Univs., Inc., 1978-83. Served from sublt. to lt. Royal Canadian Navy, 1944-45. Recipient Pergamon Press Jari award, 1980. Fellow Am. Phys. Soc.; mem. Canadian Assn. Physicists. Democrat. Episcopalian. Club: Cosmos (Washington). Subspecialty: Nuclear physics. Current work: Detection of cosmogenic radioisotopes using an accelerator based mass spectrometry technique. Home: 113 Burrows Hill Dr Rochester NY 14625

GOWANS, CHARLES SHIELDS, educator; b. Salt Lake City, Sept. 17, 1923; s. George Henry and Frances Ruth (Shields) G.; m. Ann Midgley, Mar. 30, 1950; children: Kathleen Gowans DeVries, Margaret Gowans Blinn, Susan. A.B. in Zoology, U. Utah, 1949; Ph.D., Stanford U., 1957. USPHS fellow Ind. U., Bloomington, 1956-57; asst. prof. biol. scis. U. Mo., Columbia, 1957-63, assoc. prof., 1963-68, prof., 1968—. Contbr. articles to profl. jours. Served with U.S. Army, 1942-45. NSF grantee, 1961-69; NIH grantee, 1961-63; Ocean Genetics Inc. grantee, 1982—. Mem. AAAS, Am. Inst. Biol. Sci., Internat. Soc. Phycology, Genetics Soc. Am., Phycological Soc. Am., Am. Soc. Microbiology, Soc. Protozoologists. Subspecialties: Genetics and genetic engineering (biology); Microbiology. Current work: Investigation of and engring. prodn. of organic materials by algae. Office: U Mo Div Biological Science Columbia MO 65211

GOWDA, BYRE VENKATARAMANA, research and devel. engr.; b. Bangarapet, India, Mar. 3, 1940; s. Venkataramana and Muniyamma G.; m. Usha Gowda; children: Jayanth, Jeevan, Arathi. M.S. with distinction, Inst. Sci., India, 1965; Ph.D., U. Waterloo, Can., 1968. Registered profl. engr., Pa. Asst. prof. U. Waterloo, 1969-74; analyst Babcock & Wilcox Can., Ltd., 1974; sr. engr. Westinghouse Electric Corp., Pitts., 1974-76, prin. engr., 1976—. Contbr. articles to profl. jours. Exec. com. Hindu Temple. Recipient 3 Gold medals U. Mysore,

India; research grantee U.S. Air Force, 1972-74, NRC, 1969-74. Mem. ASME (chmn. Westmoreland sect.), Acad. Mechanics, Am. Soc. Metals, ASTM. Subspecialties: Nuclear fission; High-temperature materials. Current work: Strength and stability of structural systems, materials applications in energy systems, dynamics of structures degradations mechanisms like fatigue, fracture, corrosion and wear. Patentee in field.

GOWEN, PATRICIA ELIZABETH, forest pathologist; b. Riverside, Calif., Mar. 18, 1952. B.A., Hofstra U., 1974; M.S. in Forest Pathology, SUNY-Syracuse, 1977, Ph.D., 1980. Greenhouse caretaker Hofstra U., Hempstead, N.Y., 1971-74; teaching and research asst. SUNY Coll. Environ. Sci. and Forestry, Syracuse, 1974-80; forest pathologist Mo. Dept. Conservation, Jefferson City, 1980—. N.Y.State Regents scholar, 1950-74; recipient Hofstra U. Disting. Acad. award, 1970-74. Mem. Am. Phytopath. Soc. Subspecialty: Plant pathology. Current work: Investigating the host range of Naemacyclus needlecast of pines in Missouri; diagnosing and recommending controls for the various urban and forest disease problems in the state. Home: 555C Senate Ct Jefferson City MO 65101 Office: 2901 N Ten Mile Dr Jefferson City MO 65101

GOY, ROBERT WILLIAM, psychologist, research administrator; b. Detroit, Jan. 25, 1924; s. George Frederick and Charlotte Elizabeth (McDowell) G.; m. Barbara Elaine Perry, Nov. 13, 1948; children: Michael Frederick, Peter William, Elizabeth Ruth. B.S., U. Mich., 1947; Ph.D., U. Chgo., 1953. Assoc. prof. anatomy U. Kans. Med. Sch., 1954-63; prof. med. psychology U. Oreg. Med. Sch.; chmn. dept. reproductive physiology and behavior Oreg. Regional Primate Research Center, 1963-71; prof. psychology U. Wis.; dir. Wis. Regional Primate Research Center, 1971—. Author: (with B.S. McEwen) Sexual Differentiation of the Brain, 1980; assoc. editor: Archives of Sexual Behavior, 1968—; editor: Hormones and Behavior, 1976—. Served with U.S. Army, 1943-46. USPHS postdoctoral fellow, 1954-56; USPHS grantee, 1954-56; NIMH grantee, 1963—. Mem. Am. Assn. Anatomists, Am. Psychol. Assn., Soc. Study Fertility, Soc. Study Reprodn., Endocrine Soc., Internat. Acad. Sex Research. Subspecialty: Physiological psychology. Home: 1845 Summit Madison WI 53705 Office: 1220 Capitol Ct Madison WI 53706

GRACE, DONALD J., electrical engineer, research administrator; b. Oklahoma City, Feb. 21, 1927; m., 1949; 2 children. B.S.E.E., Ohio State U., 1948, M.S.E.E., 1949; Ph.D., Stanford U., 1962. Lectr. Ohio State U., 1948-49; research engr. Airborne Instruments Lab., 1959-61; research assoc. Stanford U., 1962-63; sr. research assoc. Systems and Techniques Lab., 1963-66, assoc. prof. elec. engring., 1963-67, dir., 1966-67, assoc. dean engring., 1967-69; dir. research Kentron Hawaii, Ltd., 1969-73; dir. Ctr. Engring. Research, U. Hawaii, 1973-76; dir. engring. expt. sta. Ga. Inst. Tech., 1976—; mem. FORECAST Panel U.S. Air Force, 1963, mem. reconnaissance adv. bd., 1964-65; tech. advisor U.S. Army Security Agy., 1965-66; dir. Stanford U. Instructional TV Network, 1967-69; asst. sec. Ga. Tech. Research Inst., 1977; rep. Pub. Service Sattelite Consortium, 1977—; mem. univ. adv. panel Nat. Solar Energy Research Inst., 1978-79; mem. forum Nat. Security Affairs, Pentagon, 1980. Mem. IEEE, AAAS, Sigma Xi. Subspecialty: Engineering research administration. Office: Engring Expt St Ga Inst Tech Atlanta GA 30332

GRACE, RICHARD EDWARD, engineering educator; b. Chgo., June 26, 1930; s. Richard Edward and Louise (Koko) G.; m. Consuela Cummings Fotos, Jan. 29, 1955; children: Virginia Louise, Richard Cummings (dec.). B.S. in Metall. Engring., Purdue U., 1951; Ph.D., Carnegie Inst. Tech., 1954. Registered profl. engr., N.J. Asst. prof. Purdue U., West Lafayette, Ind., 1954-58, asso. prof., 1958-62, prof., 1962—, head sch. materials sci. and metall. engring., 1965-72, head div. interdisciplinary engring. studies, 1969-82, head freshman engring. dept., asst. dean engring., 1981—; cons. to Midwest industries. Contbr. articles to profl. jours. Past dir. and officer engring. edn. and accreditation com. Engrs. Council for Profl. Devel. Mem. Am. Soc. Metals (tchr. award 1962, fellow award 1974), AIME, Am. Soc. Engring. Edn., AAUP, Sigma Xi, Tau Beta Pi, Omicron Delta Kappa, Phi Gamma Delta. Clubs: Rotary, Elks, Lafayette Country. Subspecialties: Materials (engineering); Metallurgical engineering. Current work: Corrsion, transport processes. Home: 2175 Tecumseh Park Ln West Lafayette IN 47906

GRACE, THOMAS PETER, computer scientist, educator; b. Evergreen Park, Ill., Jan. 30, 1955; s. Thomas George and Norma Fay (Rawls) G. B.S., U. Ill., Chgo., 1976, M.S., 1979, Ph.D., 1982. Programmer, analyst N.E. Ill. Planning Commn., Chgo., 1974-76; resident student assoc. Argonne Nat. Lab., Ill., 1976-78; programmer, analyst Speakeasy Computing, Chgo., 1978-80; teaching and research asst. U. Ill., 1977-82; asst. prof. computer sci. Ill. Inst. Tech., Chgo., 1982—; cons. Nat. Opinion Research Ctr., U. Chgo., 1981, Chgo. Area Geog. Info. Survey, 1978. Contbr. articles to profl. jours. Mem. Assn. Computing Machinery, Math. Assn. Am., Am. Math. Soc. Roman Catholic. Subspecialties: Mathematical software; Graphics, image processing, and pattern recognition. Current work: Combinatorial algorithms, graph-theoretic algorithms, computer graphics, image representation and display, general graph theory, error-correcting codes. Home: 650 River Dr Calumet City IL 60409 Office: Dept Computer Sci Ill Inst Tech Chicago IL 60616

GRAD, HAROLD, applied mathematician; b. N.Y.C., Jan. 14, 1923; s. Herman and Helen (Selinger) G.; m. Betty Jane Miller, Jan. 23, 1949; children: Hilary Lynn Grad Goldberg, Michael Jonathan. B.E.E., Cooper Union, 1943; M.S., N.Y. U., 1945, Ph.D., 1948. Research asst. N.Y. U., 1944-48, mem. faculty, 1948—; prof. math. Courant Inst., 1957—, founder, 1956, dir. magneto-fluid dynamics div., 1956-80; adv. com. fusion energy Oak Ridge Nat. Lab., 1964-67, 73-76; dir. Space Scis., Inc., 1966-71; vis. disting. prof. Faculty Sci., Nagoya U. (Japan), 1981; cons. to industry and U.S. govt. Author papers, monograph on kinetic theory gases, statis. mechanics, magneto-fluid dynamics, plasma physics, fusion energy; editorial bd.: Physics of Fluids, 1968-71, Jour. Statis. Physics, 1969-75, Internat. Jour. Engring. Sci, 1963—. Recipient Eringem medal Soc. Engring. Scis., 1982; Guggenheim fellow, 1981-82. Fellow AAAS, Am. Phys. Soc. (chmn. fluid dynamics div. 1963, chmn. plasma physics div. 1968); mem. Nat. Acad. Scis., Soc. Engring. Scis. (dir. 1963-74), N.Y. Acad. Scis. (Pregel award 1970, bd. govs. 1979—), Am. Math. Soc., Soc. Indsl. and Applied Math., Soc. Natural Philosophy. Subspecialties: Applied mathematics; Plasma physics. Current work: Applied mathematics, mathematical models, Kinetical theory, Boltzmann equation, transport, statistical mechanics, fluid dynamics, magneto-fluid dynamics, plasma physics, fusion energy. Home: 248 Overlook Rd New Rochelle NY 10804 Office: 251 Mercer St New York NY 10012

GRADIE, JONATHAN CAREY, research scientist; b. Putnam, Conn., June 20, 1951; s. Robert R. and Avis L. (Gregg) G.; m. Nancy H. Adams, Oct. 22, 1977. Ph.D., U. Ariz., 1978. Research assoc. Center for Radiophysics and Space Research, Cornell U., Ithaca, N.Y., 1978—. Contbr. articles to profl. jours. Mem. AAAS, Am. Astron. Assn., Am. Geophys. Union. Subspecialties: Planetary science; Planetology. Current work: Physical studies of small bodies of the solar system: asteroids, satellites, meteorites, etc. Office: Space Scis Bldg Cornell Univ Ithaca NY 14853

GRADIN, LAWRENCE PAUL, electrical engineer, consultant; b. Bklyn., June 2, 1945; s. Milton and Ann (Miller) G.; m. Helene Ann Fortunah, Aug. 7, 1971; children: Michael, Jennifer, Kevin. B.S. magna cum laude, N.Y. Inst. Tech., 1969. Registered profl. engr., N.Y., Calif. Jr. engr. Met. Transit Authority, N.Y.C., 1968-69; lead engr. Ebasco Services, N.Y.C., 1969-77, supr. tech. tng., 1979-80, supervising engr., 1980-81, equipment qualification program mgr., 1981-82; project mgr. elec. instrumentation and control Nuclear Power Services, Inc., Secaucus, N.J., 1983—; asst. sect. mgr. Burns & Roe, Oradell, N.J., 1977-79; lectr. tng. program U. S. Nuclear Regulatory Commn. personnel, 1979. Lead author: Electrical Technology for Nuclear Plant Safety Systems, 1979, 80. N.Y. State Regents scholar, 1963. Mem. IEEE (sr.), Instrument Soc. Am. (sr.), Am. Nuclear Soc., Atomic Indsl. Forum, Am. Soc. Indsl. Security. Subspecialties: Nuclear fission; Nuclear power. Current work: Electrical, instrumental, control and equipment qualification technology engineering management; technology transfer and scientific and engineering data management. Home: 20 Phillips Rd Edison NJ 08817

GRAEME, MARY LEE, pharmacologist; b. Valentines, Va., Aug. 20, 1922; d. Atwell Joseph and Maben (Huff) Clary. B.A., Westhampton Coll., 1944; postgrad., U. N.C., Chapel Hill, 1944-45. Technician U. N.C., Chapel Hill, 1945-46, U. Mich., Ann Arbor, 1946-47; asst. CIBA Pharm. Co., Summit, N.J., 1947-51; research asst. Johnson and Johnson (and subs.), 1951-62; sr. staff scientist Geigy Pharm., Ardsley, N.Y., 1962—. Contbr. articles to profl. jours. Mem. Dobbs Ferry (N.Y.) Civic Assn. Mem. Fedn. Am. Soc. Pharmacology and Exptl. Therapeutics, Inflammation Research Assn, Sigma Xi. Republican. Episcopalian. Subspecialties: Pharmacokinetics; Pharmacology. Current work: Arthritis; screening and method of development of antiinflammatory and analgesic agents in animals. Office: Saw Mill River Rd Ardsley NY 10502

GRAFF, SAMUEL M., mathematics educator; b. N.Y.C., May 22, 1945; s. Harry and Yetta (Auslander) G. B.S., Rensselaer Poly. Inst., 1966; M.S., N.Y. U., 1971, Ph.D., 1971. Asst. prof. math. John Jay Coll. Criminal Justice, CUNY, N.Y.C., 1971-75, assoc. prof. math., 1976-80, prof. math., 1981—. Mem. Am. Math. Soc., Soc. Indsl. and Applied Math., Math. Assn. Am., Sigma Xi. Subspecialties: Applied mathematics; Operations research (engineering). Current work: I am presently engaged in the modelling of chemical systems as well as rail transportation networks. Office: John Jay Coll Criminal Justice CUNY 445 W 59th St New York NY 10019

GRAFF, WILLIAM JOHN, engring. educator, mech. and civil engring. cons.; b. Marshall., Tex., May 10, 1923; s. William John and Ethel (Kearns) G.; m. Ina Jean Westmoreland, Sept. 7, 1944 (dec.); children: Rebecca Lynn, Judith Ann, Cynthia Jean; m. Ruby Mae Brock, Aug. 17, 1977. Student, Oak Ridge Sch. Reactor Tech., 1952-53; B.S.M.E., Tex. A&M U., 1947, M.S.M.E., 1948; Ph.D., Purdue U., 1951. Registered profl. engr., Tex. Instr. Tex. A&M U., 1947-48; instr. mech. engring. Purdue U., 1948-51; sr. propulsion engr. Gen. Dynamics Corp., Ft. Worth, 1951-54, nuclear group engr., 1954-56; prof., chmn. mech. engring. So. Meth. U., Dallas, 1956-61; dean of instrn. Tex. A&M U., College Station, 1961-65, dean acad. administrn., 1965-66; prof. civil engring. U. Houston, 1966—; lectr.; Fulbright-Hays scholar, vis. prof. structural engring. Aalborg (Denmark) U. Centre, 1977-78; engring. cons. to corps. Author: Introduction to Offshore Structures—Design, Fabrication and Installation, 1981; contbr. articles on tubular structures and offshore platforms to profl. jours. Campaign dir. College Station United Chest, 1962, pres., 1963. Served to 1st lt. USAAF, 1943-45; mem. USAFR, 1945-51. Mem. ASME (past chmn. S. Tex. Sect.), ASCE, Sigma Xi, Phi Kappa Phi, Tau Beta Pi, Chi Epsilon, Pi Tau Sigma. Republican. Presbyterian. Lodges: Masons; Shriners. Subspecialties: Civil engineering; Mechanical engineering. Current work: Ocean engring., specifically structural design and analysis of structures to be placed in the ocean, properties of engring. materials and pipeline behavior. Home: 7815 Windswept St Houston TX 77063 Office: Civil Engring Dept U Houston Houston TX 77004

GRAFMAN, JORDAN HENRY, neuropsychologist, research scientist; b. Chgo., Dec. 21, 1950; s. Joseph and Phyllis (Terkel) G. B.A., Sonoma State U., Rohnert, Park, Calif., 1974; Ph.D., U. Wis.-Madison, 1981. Neuropsychology chief Vietnam Head Injury Study Walter Reed Army Med. Ctr., Washington, 1980-82; guest scientist NIMH, Bethesda, Md., 1982—. Served to capt. USAF, 1981—. Social Sci. Research Council fellow, 1978; recipient U. Wis.-Madison travel award, 1980. Mem. Am. Psychol. Assn., Soc. Neurosci., Internat. Neuropsychol. Soc. Subspecialties: Neuropsychology; Cognition. Current work: Human brain behavior relationships, neuropsychological assessment, memory research, cognitive remediation, cognitive psychology; reading and writing disorders, calculation disorders. Office: Vietnam Head Injury Study WRAMC DCI HSWP-QCR Washington DC 20307

GRAFTON, ROBERT BRUCE, computer scientist; b. Rochester, N.Y., May 15, 1935; s. Corydon M. and Beatrice (Hawes) G.; m. Carolyn P. Kolb, July 8, 1967; children: Geoffrey, Benjamin. Sc.B., Brown U., 1957, Ph.D., 1967. Asst. prof. math. U. Mo., Columbia, 1967-71; vis. lectr. Leicester (Eng.) U., 1969-70; asst. prof.math. Trinity Coll., Hartford, Conn., 1971-75; mathematician Office Naval Research, N.Y.C., 1975-78, computer scientist, Arlington, Va., 1978—. Served to lt. USN, 1958-62. Mem. Math. Assn. Am., IEEE Computer Soc., Assn. Computing Machinery, Soc. Indsl. and Applied Math. Unitarian. Subspecialties: Software engineering; Theoretical computer science. Current work: Manage basic research program in software, emphasizing automation of the programming process, especially the use of graphics, visual aids, very high level languages, and reusable designs. Office: Office Naval Research Code 433 800 N Quincy St Arlington VA 22217 Home: 5131 Portsmouth Rd Fairfax VA 22022

GRAGOUDAS, EVANGELOS STELIOS, ophthalmologist, educator; b. Lesbos, Greece, 1941. M.D., Athens U., 1965. Diplomate: Am. Bd. Ophthalmology. Rotating intern Waltham (Mass.) Hosp., 1969-70; resident in ophthalmology Boston U., 1970-73; research fellow Joslin Research Lab., 1970; retina fellow Mass. Eye and Ear Infirmary, Boston, 1973-75, asst. in ophthalmology, 1975—, dir., 1975—; assoc. dir. retina service; assoc. scientist Eye Research Inst. of Retina Found., 1978—; asst. prof. ophthalmology Harvard U., Boston, 1978-81, assoc. prof., 1981—. Served to 2d lt., M.C. Greek Army, 1966-68. Fellow Am. Acad. Ophthalmology; mem. Assn. Research in Vision and Ophthalmology, Retina Soc. Subspecialty: Ophthalmology. Current work: Proton-beam irradiation of intraocular melanomas. Office: Harvard U Med Sch Boston MA

GRAHAM, DAVID TREDWAY, medical educator, physician; b. Mason City, Iowa, June 20, 1917; s. Evarts Ambrose and Helen (Tredway) G.; m. Frances Jeanette Keesler, June 14, 1941; children: Norma VanSurdam, Andrew Tredway, Polly Brewster. B.A., Princeton U., 1938; M.A., Yale U., 1941; M.D., Washington U., St. Louis, 1943. Intern Barnes Hosp., St. Louis, 1944, asst. resident medicine, 1944-45, 47-48; research fellow medicine Cornell U. Med. Coll., 1948-51; asst. prof. medicine Washington U. Med. Sch., 1951-57, asst. prof. psychiatry, 1956-57; assoc. prof. medicine U. Wis. Med. Sch., 1957-63, prof. medicine, 1963—, assoc. chmn. dept., 1969-71, chmn., 1971-80, asst. dean and/or chmn. med. sch. admissions, 1964-69; vis. prof. psychiatry U. Va. Sch. Medicine, 1960. Research editor: Clin. Research Proc, 1954-59. Alt. del. Democratic Nat. Conv., 1968. Served to capt., M.C. AUS, 1945-47. Mem. State Med. Soc. Wis., Am. Fedn. Clin. Research, Am. Psychosomatic Soc. (council 1952-55, 64-67, pres. 1978-79), Soc. Psychophysiol. Research (bd. dirs. 1964-67, pres. 1969-70), Central Soc. Clin. Research. Subspecialties: Internal medicine; Psychophysiology. Current work: Psychomatic medicine. Home: 2927 Harvard Dr Madison WI 53705

GRAHAM, DONALD LEE, chemical engineer; b. Hymera, Ind., May 1, 1931; s. Ross Raymond and Hazel Mae (McClanahan) G.; m. Phyllis Ann Seymour, Sept. 2, 1950; children: Stephen Lee, Cynthia Ann Graham Bruzewski, Diane Kay Graham McBeath, Robert Bruce. B.S. in Chem. Engring., Purdue U., 1953. With Dow Chem. Co., 1953—, plastics devel. engr., mem. staff plastics tech. service and devel., sect. head fabricated constrn. materials plastics devel. and service, Midland, Mich., 1964-68, tech. mgr. constrn. materials research and devel., 1968-73, dir. research and devel. functional products and systems, 1973-76, dir. research and devel. western div., Pittsburg, Calif., 1976-82, dir. western applied sci. and tech., 1982—; dir. Cynara Co.; trustee Product Research Com., 1975-79. Mem. pres.'s council Purdue U., 1979—. Mem. AAAS, Soc. Chem. Industry, Am. Chem. Soc., Am. Inst. Chem Engrs., Walnut Creek C. of C. (v.p. 1977-78). Republican. Methodist. Subspecialties: Chemical engineering; Organic chemistry. Current work: Directing fundamental and applied research in the chemical, physical and engineering sciences, energy, inorganic, organic, pharmaceutical, plastics and agricultural, chemical engineering unit operation and process and product development. Patentee in field. Home: 7951 SW 167 St Miami Fl 33157 Office: Dow Chem Co Research Ctr 2800 Mitchell Dr Walnut Creek CA 94598

GRAHAM, FREDERICK MITCHELL, educator; b. Des Moines, Feb. 7, 1921; s. Fred and Anna Mae (Mitchell) G.; m. Lillian L. Miller, Aug. 29, 1948; children: Frederick M., Stephen, Anita. B.S., Iowa State U., 1948, M.S., 1950, Ph.D., 1966. Prof., dept. head Prairie View (Tex.) A&M U., 1950-59; prof. engring. sci. and mechanics Iowa State U., Ames, 1962—; cons. McDonnell Douglas, 1972-73, Meredith Pub. Co., 1968-70, VanGorp Corp, 1970—, Sundstrand Corp., 1979. Contbr. articles to profl. jours. Served with USAAF, 1943-46. NSF fellow, 1959-61; named Superior Engring. Tchrs. of the Yr. Iowa State U., 1978. Mem. Am. Soc. Engring. Edn., Nat. Soc. Profl. Engrs. (pres. 1983-84), Tau Beta Pi. Democrat. Episcopalian. Subspecialties: Civil engineering; Theoretical and applied mechanics. Current work: Experimentation with innovative structural components; investigation of complex industrial failures. Home: 134 S Franklin St Ames IA 50010 Office: Iowa State Univ 209 Laboratory of Mechanics Ames IA 50011

GRAHAM, HARRY MORGAN, entomologist, researcher, educator; b. Whittier, Calif., June 18, 1929; s. Harry R. and May A. (Morgan) G.; m. Dorothy D. Denson, Sept. 22, 1962; children: Nancy J., Robert M. A.A., Fullerton Jr. Coll., 1949; B.S. with highest honors, U. Calif.-Berkeley, 1951, M.S., 1953, Ph.D., 1959. Research entomologist Dept. Agr. Agrl. Research Service, Brownsville, Tex., 1958-77, Tucson, 1977—; adj. prof. entomology U. Ariz. Contbr. numerous articles to profl. publs. Served to cpl. U.S. Army, 1953-55. Mem. AAAS, Entomol. Soc. Am., S.W. Entomol. Soc., Ecol. Soc. Am., Internat. Biol. Control Orgn., Ariz.-Nev. Acad. Sci. Presbyterian. Subspecialties: Integrated pest management; Ecology. Current work: Insect ecology, biological control of insects, biological control of insects pests of field crops.

GRAHAM, JAMES ALEXANDER, biomedical researcher; b. Niagara Falls., N.Y., Sept. 9, 1941; s. Archibald and Thelma Ona (Norkum) G; m. Sheryl Lynne, Jan. 20, 1973; children: Ian Scott, Marc Reid. B.S. in Mech. Engring, U. Buffalo, 1963; M.S.M.E., U. R.I., 1967. Mgr. quality assurance Drucker Co., Datambdix Inc.; mgr. quality assurance adminstrn. Modular Computer Systems; dir. quality assurance and regulatory affairs Sontek Corp., Coral Gables, Fla. Served with U.S. Navy, 1959-63. Mem. Am. Soc. Quality Control, Am. Mgmt. Assn. Jewish. Subspecialties: Biomedical engineering; Biomedical engineering. Current work: Research in biomedical plastics. Quality processes related to the biomedical manufacturing field. Office: 4075 Laguna St Coral Cables FL 33146

GRAHAM, JOHN BORDEN, medical educator; b. Goldsboro, N.C., Jan. 26, 1918; s. Ernest Heap and Mary (Borden) G.; m. Ruby Barrett, Mar. 23, 1943; children: Charles Barrett, Virginia Borden, Thomas Wentworth. B.S., Davidson Coll., 1938; M.D., Cornell U., 1942. Asst. Cornell U., 1943-44; mem. faculty U. N.C., Chapel Hill, 1946—, Alumni Distinguished prof. pathology, 1966—, chmn. genetics curriculum, 1963—, asso. dean medicine for basic scis., 1968-70, coordinator interdisciplinary grad. programs in biology, 1968—, dir. hemostasis program, 1974—; vis. prof. haematology St. Thomas's Hosp. Med. Sch., London, 1972; vis. prof. Teikyo U. Med. Sch., Tokyo, 1976; mem. selection com. NIH research career awards, 1959-62; genetics tng. com. USPHS, 1962-66, chmn., 1967-71; mem. genetic basis of disease com. Nat. Inst. Gen. Med. Scis., 1977-80; mem. pathology test com. Nat. Bd. Med. Examiners, 1963-67; mem. research adv. com. U. Colo. Inst. for Behavioral Genetics, 1967-71; mem. Internat. Com. Haemostasis and Thrombosis, 1963-67; chmn. bd. U. N.C. Population Program, 1964-67; sec. policy bd. Carolina Population Center, 1972-78; cons. Environ. Health Center, USPHS, WHO, Bolt, Beranek & Newman, Inc.; mem. med. and sci. adv. council Nat. Hemophilia Found., 1972-76; hon. cons. in genetics Margaret Pyke Centre, London, 1972—. Mem. editorial bd.: N.C. Med. Jour, 1949-66, Am. Jour. Human Genetics, 1958-61, Soc. Exptl. Biology and Medicine, 1959-62, Human Genetics Abstracts, 1962-72, Haemostasis, 1975-80, Christian Scholar, 1958-60. Markle scholar in med. sci., 1949-54; Recipient O. Max Gardner award U. N.C., 1968. Mem. AMA, AAAS, Elisha Mitchell Sci. Soc. (pres. 1963), AAUP, Soc. Exptl. Biology and Medicine, Am. Soc. Exptl. Pathology, Assn. U. Pathologists, Am. Assn. Pathologists and Bacteriologists, Am. Soc. Human Genetics (sec. 1964-67, pres. 1972), Genetics Soc. Am., Internat. Soc. Hematology, Am. Inst. Biol. Sci., Royal Soc. Medicine (London), Med. Soc. N.C., Mayflower Soc., Sigma Xi. Democrat. Presbyn. Club: Cosmos (Washington). Subspecialties: Genetics and genetic engineering (biology); Genetics and genetic engineering (medicine). Current work: Director of graduate curriculum in genetics; program director for research on thrombosis and hemostasis utilizing monoclonal antibodies and recombinant DNA. Publs. on blood clotting, inherited diseases in humans, human population dynamics; co-discoverer blood coagulant Factor X (Stuart factor). Home: 108 Glendale Dr Chapel Hill NC 27514

GRAHAM, KENNETH ROBERT, psychology educator; b. Phila., June 5, 1943; s. Edgar and Margit (Leafgreen) G.; m. Michele C. Monroe, Aug. 10, 1968; children: Mark A., Richard A. B.A., U. Pa.-Phila., 1964; Ph.D., Stanford U., 1969. Research psychologist Inst. Pa. Hosp., Phila., 1969-70; asst. prof. Muhlenberg Coll., Allentown, Pa., 1970-76, assoc. prof. psychology, 1976-83, prof., 1983—. Author: Psychological Research, 1977; editorial bd.: Am. Jour. Clin. Hypnosis, 1974—. Bd. pres. Lehigh Valley Child Care, Inc., Allentown, Pa., 1981—. Mem. Am. Psychol. Assn. (council reps. 1982-85), Am. Soc. Clin. Hypnosis, Soc. Clin. and Exptl. Hypnosis, Internat. Soc. Hypnosis, Eastern Psychol. Assn., Sigma Xi. Democrat. Lutheran.

Lodge: Kiwanis. Current work: Relation between susceptibility to hypnosis and the persuasive effects of the mass media, especially television. Office: Muhlenberg Coll Allentown PA 18104 Home: 2949 Tilghman St Allentown PA 18104

GRAHAM, LESLIE STEPHEN, med. physicist; b. Frankfort, Ind., Jan. 3, 1933; s. Forest P. and Iva E. (Davis) G.; m. Marianne E. Graham, Jan. 25, 1959; children: Michael S., Daryl T. B.A., Pasadena Coll., 1955; B.D., Talbot Sem., La Mirada, Calif., 1959; M.A., Calif. State U., Long Beach, 1962; Ph.D., UCLA, 1971. Diplomate: Am. Bd. Radiology. Med. physicist UCLA, 1971-77, assoc. prof., 1971—; med. physicist VA Med. Center, Sepulveda, Calif., 1977—; cons. in field. Contbr chpts. to books and articles to profl. jours. USPHS fellow, 1968-69. Mem. Am. Assn. Physicists in Medicine, Soc. Nuclear Medicine, IEEE. Subspecialties: Nuclear medicine; Diagnostic radiology. Current work: Computer applications in nuclear medicine; computer modeling, dosimetry, computer analysis of instrumentation function. Office: 16111 Plummer St Sepulveda CA 91343

GRAHAM, ROBERT MONTROSE, computer scientist, educator; b. St. Johns, Mich., Sept. 26, 1929. B.A., U. Mich., 1956, M.A., 1957. Mem. staff U. Mich., 1957-61, research assoc., 1961-63; program coordinator Project Mac, MIT, 1963-66, staff mem. dept. elec. engring., 1965-67, assoc. prof. computer sci., 1967-72; assoc. prof. CCNY, 1972-75; prof. computer sci. U. Mass., 1975—, chmn. dept. computer and info. sci., 1975-81; vis. assoc. prof. U. Calif.-Berkeley, 1970-72; mem. Computer Sci. Bd., 1980—. Mem. Assn. Computing Machinery. Subspecialties: Software engineering; Programming languages. Office: Coins Grad Research Ctr U Mass Amherst MA 01003

GRAHAM, SUSAN LOIS, computer science educator; b. Cleve., Nov. 16, 1942; m., 1971. A.B., Harvard U., 1924; M.S., Stanford U., 1966, Ph.D. in Computer Sci, 1971. Assoc. research scientist, adj. asst. prof. computer sci. Courant Inst. Math. Sci., NYU, 1969-71; asst. prof. computer sci. U. Calif.-Berkeley, 1971-76, assoc. prof., 1976-81, prof., 1981—. Co-editor: Communications, 1975-79; editor-in-chief: Transactions on Programming Language and Systems, 1978—. NSF grantee, 1974—. Mem. Assn Computing Machinery, Assn. Women in Computer Sci., IEEE. Subspecialty: Programming languages. Office: Computer Sci Div EECS U Calif Berkeley CA 94720

GRAHAM, TAD LAURY, computer systems analyst; b. Wichita, Kans., May 12, 1943; s. Robert Virgil and Jeana Ardeth (Black) G.; m. Judith Anne Hathaway, Sept. 1, 1972; 1 son, Geoffrey Todd; m.; children by previous marriage: Carol Lynn, Peter Mark. B.A., U. Ariz., Tucson, 1969; M.S., U. Ill., Urbana, 1974; B.S., Chapman Coll., Orange, Calif., 1981. Dir. info. services Intrec, Inc., Santa Monica, Calif., 1974-77; mem. tech. staff Computer Scis Corp., San Diego, 1977-79, sr. mem. tech. staff, 1979-81, sect. mgr., 1981—. Served with USN, 1960-64. Mem. Nat. Mgmt. Assn. (dir. chpt. 1981—), IEEE, Assn. Computing Machinery. Democrat. Episcopalian. Subspecialties: Database systems; Information systems, storage, and retrieval (computer science). Current work: Data management and management information system development for military surveillance projects. Office: Computer Scis Corp 4045 Hancock St San Diego CA 92110

GRAHAM, TERRENCE LEE, biochemist; b. Corning, N.Y., Sept. 24, 1947; s. Paul N. W. and Grace (Link) G.; m. Lian-Mei Yang, June 7, 1975. B.S., Pa. State U., 1969; Ph.D., Purdue U., 1975. NIH postdoctoral trainee in pathology U. Wis., Madison, 1975-77; project leader plant pathogen project Monsanto Co., St. Louis, 1977-80, sr. group leader host-modification program, 1980-82, sci. fellow, sr. group leader plant biochemistry program, 1982-83, sci. fellow, research mgr., biol. control program, 1983—. Contbr. articles to profl. jours. Charles Gerth scholar., 1967-68; John White fellow, 1969; Purdue U. fellow, 1969. Mem. Am. Chem. Soc., Am. Phytopathol. Soc., Am. Soc. Microbiology, Am. Soc. Plant Physiologists. Subspecialties: Plant physiology (agriculture); Biochemistry (biology). Current work: Biochemical components of the plant cell; cell membrane and cell wall as they relate to cascade regulation of gene expression in whole plants and tissue culture; plant developmental and stress metabolism and the response of plants to chemical messengers, biological control of agricultural pests through genetic engineered microbes. Home: 1559 Meadowside Dr Creve Coeur MO 63141 Office: Monsanto Chem Co 800 N Lindbergh Blvd Saint Louis MO 63166

GRAHN, DOUGLAS, radiobiologist; b. Newark, Apr. 25, 1923; s. Viktor Frederick and Greta Amelia (Franzen) G.; m. Sally Linn Smythe, Dec. 21, 1945; children: Frederick S., Catherine L., Alice A. Grahn Michaels; m. Ann Wagoner, May 19, 1973. B.S., Rutgers U., 1948; M.S., Iowa State U., 1950, Ph.D., 1952. Assoc. biologist Argonne Nat. Lab., Ill., 1953-58; geneticist AEC, 1958-61; assoc. div. dir. div. biol. and med. research Argonne Nat. Lab., 1962-66, sr. biologist, 1966—, div. dir., 1978-81; adj. prof. No. Ill. U., 1971—; mem. Nat. Council Radiation Protection, 1978—; cons. NASA, 1964-70; mem. space sci. bd., radiobiology panel Nat. Acad. Sci./NRC, 1964-74. Contbr. articles to sci. jours. Served in U.S. Army, 1943-45; ETO. AEC fellow, 1951-52. Mem. AAAS, Genetics Soc. Am., Radiation Research Soc., Environ. Mutagen Soc., Soc. Am. Naturalists. Subspecialties: Animal genetics; Radiobiology. Current work: Genetic and somatic effects of low level radiation exposure with emphasis on the effects of neutrons. Home: 606 S Catherine Ave LaGrange IL 60525 Office: Argonne Nat Lab Argonne IL 60439

GRALLO, RICHARD MARTIN, educational psychologist, social science research consultant; b. Winthrop, Mass., Feb. 15, 1947; s. Frederick Michael and Jennie Antonia (Ferrario) G. A.B., Boston Coll., 1969; M.S., MIT, 1972; M.A., NYU, 1976. Research asst. NYU, 1977-79; mem. faculty New Sch. for Social Research, 1978; research assoc. Inst. Developmental Studies, N.Y.C., 1979—; research cons., 1980—; prof. Coll. Human Services, N.Y.C., 1980—. John A. Lyons fellow, 1969-72. Mem. Am. Psychol. Assn. (assoc.), Am. Philos. Assn., N.Y. Acad. Scis. (adv. com. Psychology Sect. 1980—), Eastern Ednl. Research Assn., Phi Delta Kappa. Subspecialties: Developmental psychology; Learning. Current work: Teaching in psycho-educational measurement; research design and evaluation; conducting psychoeducational research; social science research consulting. Home: 202 Ave of Americas New York NY 10013 Office: Inst Devel Studies NY U 32 Washington Pl New York NY 10003

GRAMS, GERALD WILLIAM, atmospheric sciences educator; b. Mankato, Minn., Dec. 7, 1938; s. William Leo and Evelyn Augusta (Leifermann) G.; m. Rita Darlene Carlson, June 9, 1962; children: Theresa Marie, Rebecca Ann. B.S. in Math. and Physics, Mankato State Coll., 1960; Ph.D., M.I.T., 1966. Cert. cons. meteorologist Am. Meteorol. Soc. Tchr., St. Peter (Minn.) High Sch., 1960-61; Ford Found. fellow dept. meteorology MIT, Cambridge, Mass., 1961-64, research asst., 1964-66, research assoc. dept geology and geophysics, 1966-67; research affiliate Research Lab. Electronics, 1967-70; atmospheric physicist meteorology lab. Air Force Cambridge Research Lab., L.G. Hanscomb Field, Bedford, Mass., summer 1962; aerospace technologist in aeronomy NASA Electronics Research Ctr., Cambridge, 1967-70; lectr. in math. Univ. Coll. Northeastern U., Boston, 1967-70; scientist Nat. Ctr. for Atmospheric Research, Boulder, Colo., 1970-77; prof. Sch. Geophys. Scis. Ga. Inst. Tech., Atlanta, 1977—. Contbr. articles to profl. jours. Recipient outstanding publ. award Nat. Ctr. for Atmospheric Research, 1975. Fellow Optical Soc. Am.; mem. AAAS, Am. Geophys. Union, Am. Meteorol. Soc., Soc. Photo-Optical Instrumentation Engrs. Subspecialties: Aeronomy; Meteorologic instrumentation. Current work: Laser atmospheric measurements, atmospheric radiation and optics, atmospheric aerosol particles. Home: 2696 Cosmos Dr Atlanta GA 30345 Office: Sch Geophys Scis Ga Inst Tech Atlanta GA 30332

GRAND, DIANA LEIGH, software engr.; b. Detroit, Dec. 15, 1945; d. Salman and Evelyn (Patt) G.; m. Richard M. Karp, Aug. 12, 1979. B.A., U. Mich., 1966; M.S. in Computer Sci, U. Calif.-Berkeley, 1976. Devel. engr. Hewlett Packard, Cupertino, Calif., 1975; computer scientist Lawrence Berkeley Lab., Berkeley, Calif., 1976—; software engr. Varian Assocs., Walnut Creek, Calif., 1982—. Mem. Assn. for Computing Machinery. Subspecialties: Software engineering; Distributed systems and networks. Current work: Application of structured methodologies to development of real time control systems. Office: 2700 Mitchell Dr Walnut Creek CA 94598

GRANDISON, LINDSEY JAMES, neuroendocrinologist; b. Buffalo, June 17, 1946; s. William Irving and Genevieve (Marchlewski) G.; m. Margaret Anne Burke, Apr. 2, 1946; children: Timothy David, Brian Daniel. B.A., Johns Hopkins U., 1968; M.S., Mich. State U., 1973, Ph.D., 1976. Staff fellow Lab. Preclin. Pharmacology, NIMH, 1976-78; asst. prof. dept. physiology and biophysics U. Med. and Dentistry N.J.-Rutgers, Piscataway, N.J., 1978—; researcher. Contbr. articles to profl. jours. Served as sgt. AUS, 1968-70. Mem. Am. Physiol. Soc., Endocrine Soc., Internat. Neuroendocrine Soc., Soc. Neurosci., N.Y. Acad. Scis. Democrat. Subspecialties: Neuroendocrinology; Neuropharmacology. Current work: Control of hormone secretion by endogenous opiates, brain neurotransmitters and endogenous valium like substance. Home: 96 Lawrence Ave Highland Park NJ 08904 Office: Dept of Physiology and Biophysics U Medicine and Dentistry NJ-Rutgers Med Sch PO Box 101 Piscataway NJ 08854

GRANDON, GARY MICHAEL, psychologist; b. Detroit, Nov. 24, 1948; s. Samuel Francis and Sylvia (Chase) G.; m. Jane Ray Rosen, June 2, 1974; children: Jessica Rose Rosen, Benjamin Seth Rosen. B.S., U. Mich., 1970; M.Ed., Wayne State U., 1972; Ph.D., U. Conn., 1978. Asso. dir. Roper Center Inc., Storrs, Conn., 1977-81, Inst. Social Inquiry, Storrs, 1976-81; asst. prof. U. Conn., Storrs, 1978-81; asso. dir. Computer Research Center, Tampa, Fla., 1981—; asst. prof. U.S. Fla., Tampa, 1981—. Author: Statistical Package A-State 79, 1979-83, Optimizing High School Curriculum Assignments, 1979. Mem. Am. Ednl. Research Assn., Am. Statis. Assn., Assn. Computing Machinery, Am. Psychol. Assn., Assn. Ednl. Data Systems, Sigma Xi, Phi Delta Kappa. Jewish. Subspecialties: Statistics; Educational measurement. Current work: Development of microcomputer based statisical systems; investigations into the causation of human intelligence. Home: 7807 Whittier St Tampa FL 33617 Office: SVC 409 Univ S Fla Tampa FL 33620

GRANGER, CLARK ALLEN, state entomologist; b. Burlington, Vt., Nov. 13, 1941; s. Ralph Hawthorne and Doris Arlene (Hartwell) G.; m. Rosemarie Rowell, Aug. 31, 1965; children: Kimberly, Gregory. B.S., U. N.H., 1963, M.S., 1965; Ph.D., U. Maine, 1968. Registered forester, Maine. Asst. prof. U. Maine, Augusta, Auburn, 1968-69; asst. state entomologist Maine Dept. Forestry, Augusta, 1969-78, state entomologist, 1981—; dir. div. community forestry Maine Dept. Conservation, Augusta, 1978-81; pres. Northwoods Evergreens Corp., Bath, Maine, 1979—; mem. State of Maine Arborist Examining Bd., 1981—. Contbr. articles to profl. jours. Exec. dir. Pine Tree State Arboretum, 1982. Mem. Am. Rhododendron Soc., No. Nut Growers Assn., Maine Christmas Tree Assn. (bd. dirs.), Nat. Christmas Tree Growers Assn. Lodge: Elks. Subspecialty: Resource conservation. Current work: Integrated pest management, soil-plant relationships, herbicides. Home: RFD 3 Wiscasset ME 04578 Office: Maine Forest Service State House Sta 22 Augusta ME 04333

GRANLUND, DAVID JOHN, immunologist, research company executive; b. Duluth, Minn., Dec. 18, 1946; s. Edward Francis G.; m. Margaret Kathrine Swanson, Nov. 21, 1970. B.S., U. Minn.-St. Paul, 1968; M.S., Iowa State U., 1971; Ph.D., U. Mo.-Columbia, 1973. Virologist/immunologist Litton Bionatics, Kensington, Md., 1976-79; sr. scientist Biotech Research Labs., Inc., Rockville, Md., 1979—, dir. products div., 1982—; vis. scientist Karolinska Inst., Stockholm, 1977-78. WHO fellow, Stockholm, 1977-78. Mem. Am. Assn. Immunologists, Am. Soc. Microbiology, Sigma Xi. Subspecialties: Immunobiology and immunology; Virology (biology). Current work: Monoclonal antibodies, clinical in vitro diagnostics. Patentee cryogenic apparatus. Office: Biotech Research Labs Inc 1600 E Gude Dr Rockville MD 20850

GRANT, BRYDON JOHN BRUCE, medical educator; b. Wanstead, Essex, Eng., Apr. 24, 1945; came to U.S., 1977; s. Hope Burce and Emily Hamilton (Robson) G.; m. Meimanat Samimi, Dec. 30, 1944; children: Tannaz Melanie, Gavin Bayan, Nazaneen Nicola. M.B., B.S., Charing Cross Hosp. Med. Sch., London, 1968; M.D., U. London, 1977. Intern New Charing Cross Hosp., London, 1968-69; resident London Chest Hosp., 1969-70, Whittington Hosp., London, 1970; research fellow Royal Postgrad. Med. Sch., London, 1970-72; registrar Middlesex Hosp., London, 1972-74, lectr. medicine, 1974-77; research fellow U. Calif.-San Diego., 1977-79; asst. prof. internal medicine U. Mich., Ann Arbor, 1979—. Contbr. articles to profl. jours. Treas. Spiritual Assembly of the Baha'is, Ann Arbor, 1980-83. Dorothy Temple Cross fellow Med. Research Council, London, 1977-78. Mem. Med. Research Soc., Am. Physiol. Soc., Am. Thoracic Soc., Am. Coll. Chest Physicians. Subspecialties: Physiology (medicine); Internal medicine. Current work: Major interest in pulmonary physiology particularly in pulmonary gas exchange, pulmonary circulation and control of breathing. Home: 3735 Charter Pl Ann Arbor MI 48105 Office: Univ Hosp Box 055 1405 E Ann St Ann Arbor MI 48109

GRANT, DAVID MILLER, molecular biologist; b. Houston, Aug. 11, 1949; s. Roscoe Miller and Elizabeth Irene (Jacobs) G. B.S., SUNY, Stony Brook, 1971; Ph.D., U. Chgo., 1977. Fellow Duke U., 1977-80; trainee dept. biochemistry St. Louis U., 1980-81, asst. research prof., 1983—; project mgr. Pioneer Hi-Bred Internat., Johnston, Iowa, 1983—. Contbr. articles in field. Anna Fuller Fund fellow, 1977-78; NIH fellow, 1978-80. Mem. Am. Soc. Cell Biology, Genetics Soc. Am., Plant Molecular Biology Assn., Sigma Xi. Subspecialties: Genetics and genetic engineering (biology); Molecular biology. Current work: Nucleo-cytoplasmic genome interactions, regulation of gene expression. Home: 4332 NW Country Club Urbandale IA 50322 Office: 7300 NW 62d Ave Box 38 Johston IA 50131

GRANT, DONALD ANDREW, educator; b. Cherryfield, Maine, Jan. 3, 1936; s. Morton Andrew and Eunice Violet (Harrington) G.; m. June Elaine Farren, Aug. 29, 1957; children: Judith Dawn, Jeffrey Donald. B.S. in M.E, U. Maine, 1956, M.S., 1963; Ph.D., U. R.I., 1969. Registered profl. engr., Maine. Flight test engr. NATC, Md., 1956; instr. U. Maine, Orono, 1956-63, asst. prof. mech. engring. dept., 1963-68, assoc. prof., 1968-76, prof., 1976—; cons. Contbr. articles to profl. jours. Mem. ASME. Republican. Baptist. Lodge: Masons. Subspecialties: Mechanical engineering; Theoretical and applied mechanics. Current work: Vibrations of timoshenko beams and vibrations of time dependent boundary conditions. Office: 246 Boardman Hall Orono ME 04469

GRANT, MICHAEL CLARENCE, biologist, statis. cons.; b. Louisville, Oct. 20, 1945; s. Clarence W. and Mary E. (Reed) G.; m. Karen Bolton, Aug. 15, 1961; 1 son, Shane. B.A., Tex. Tech. U., 1969, M.S., 1970; Ph.D., Duke U., 1974. Dir. U. Colo. Mountain Research Sta., Boulder, 1974-76, asst. prof. EPO biology, 1976-80, assoc. prof., 1981—, chmn. dept., 1982—. Contbr. articles on biology to profl. jours. Served with USAF, 1963-67. Grantee in field. Mem. Soc. Study of Evolution, AAAS, Bot. Soc. Am., Phycological Soc. Am., Soc. Systematic Zoologists, Sigma Xi. Subspecialties: Evolutionary biology; Ecology. Current work: Population biology of natural plant communities, nutrient cycling, evolution, ecological genetics, acid rain, charophyte phylogeny. Office: Dept EPO Biology U Colo Boulder CO 80309

GRANT, NICHOLAS JOHN, metallurgy educator; b. South River, N.J., Oct. 21, 1915; s. John and Mary (Sudnik) G.; m. Anne T. Phillips, Sept. 12, 1942 (dec. Apr. 1957); children—Anne P., William D., Nicholas P.; m. Susan Mary Cooper, Aug. 1963; children—Johnathan, Katharine. S.B., Carnegie Inst. Tech., 1938; Sc.D., MIT, 1944. Metallurgist Bethlehem Steel Co., 1938-40; mem. faculty Mass. Inst. Tech., 1942—, prof. metallurgy, 1955—; dir. Center Materials Sci. and Engring., 1968-77, ABEX prof. advanced materials, 1975—; pres., dir. N.E. Materials Lab., Inc., 1954-66; tech. dir. Investment Castings Inst., 1954—; cons. industry, 1947—; Dir. Loomis-Sayles Mut. Fund, Capital Devel. Fund, Interpace Corp., Kimball Physics Corp., Inc., Instron Corp., Indsl. Materials Tech., Inc., Electronic Instrument & Splty. Corp.; mem. materials com. NASA, 1958-67, research com., 1970-76; chmn. working group electrometallurgy and materials U.S.-USSR Joint Sci. and Tech. Agreement. Contbr. articles profl. jours., chpts. in books. Recipient distinguished service award Investment Castings Inst., 1956; Merit award Carnegie Mellon U.; J. Wallenberg award Royal Acad. Engring. Scis., Sweden, 1978. Fellow Am. Soc. Metals, Am. Inst. Mining, Metall. and Petroluem Engrs.; mem. Nat. Acad. Engring., Inst. Metals (London), ASTM, Am. Acad. Arts and Scis., Sigma Xi, Tau Beta Pi, Theta Tau, Alpha Xi Epsilon. Subspecialty: Metallurgy. Home: 10 Leslie Rd Winchester MA 01890 Office: Massachusetts Inst Technology Cambridge MA 02139

GRANT, PATRICK MICHAEL, chemistry researcher, consultant; b. Oakland, Calif., Sept. 20, 1944; s. Dudley-Francis Patrick and Mary Grace G.; m. Joni Marie Bryan, June 28, 1969; children: Lori Lyn, Crystal-Lyn Patricia. B.S., U. Calif.-Santa Barbara, 1967, Ph.D., 1973. Lic. nuclear reactor sr. operator AEC. Tng. and research asst. U. Calif.-Irvine, 1967-73; research radiochemist U. N.Mex. Med. Sch., Albuquerque, 1973-74, adj. asst. prof. chemistry, 1977-81; staff mem. Los Alamos Sci. Lab., 1974-80, assoc. group leader, 1980-81, cons, 1981—; research chemist Chevron Research Co., Richmond, Calif., 1981-83; nuclear and radiochemist Lawrence Livermore Nat. Lab., Livermore, Calif., 1983—. Contbr. in field. Recipient Louis B. Silverman Meml. award in radiobiology Health Physics Soc., 1972. Mem. Am. Chem. Soc., Am. Phys. Soc., Am. Nuclear Soc., AAAS. Democrat. Current work: Radiochemistry, nuclear chemistry, nuclear medicine, analytical chemistry, inorganic chemistry, activation analysis, hot atom chemistry. Patentee radioisotope generator. Office: Lawrence Livermore Nat Lab Nuclear Chemistry Div Livermore CA 94550

GRANT, PHILIP ROBERT, JR., energy exploration company executive, geological consultant; b. Evanston, Ill., Apr. 23, 1930; s. Philip Robert and Nell (Cook) G.; m. Min Lively, Feb. 11, 1956; children: Cynthia, Christie, Tracy. B.Sc., U. N.Mex., Albuquerque, 1951, postgrad., 1951-52. Jr. geologist Sinclair Oil & Gas Co., Albuquerque, 1951-55, intermediate geologist, Denver, 1955-56, sr. petroleum geologist, Billings, Mont., 1956-58, Casper, Wyo., 1958-62, regional exploration geologist, Albuquerque, 1962-65; pres., cons. geologist Energy Resources Exploration, Inc., Albuquerque, 1965—; vice-chmn., mem. exec. com. S.W. Regional Energy Council, Dallas, 1975-78; mem. scis. and tech. com. Nat. Conf. State Legislatures, Washington, 1975-78; bd. dirs. Albuquerque Indsl. Devel. Service, 1978-81; chmn. State N.Mex. Energy Research and Devel. Rev. Com., Santa Fe, 1979-81; mem. adv. council U. N.Mex. Coll. Engring./Geology dept., Albuquerque, 1980—. Author: ENERGYtic New Mexico, 1st edit, 1975, 2d edit., 1977. Chmn. bd. trustees Bernalillo County Mental Health/Mental Retardation Ctr., Albuquerque, 1969-73; mem. State N.Mex. Ho. of Reps., 1972-78; co-chmn. Pres. Ford Com., N.Mex., 1976. Served to cpl. U.S. Army, 1953-54. Fellow Geol. Soc. Am.; mem. Am. Assn. Petroleum Geologists (pres. Rocky Mountain sect. 1980-81, mem. acad. liason com. 1972—), N.Mex. Geol. Soc., Albuquerque Geol. Soc. Republican. Lodge: Rotary. Subspecialties: Geothermal power; Fuels. Current work: Research on exploration techniques for the discovery of geothermal resources and analysis of constraints and institutional barriers to its development and applications. 87112 Home: 1404 Lester Dr NE Albuquerque NM 87112

GRANT, RICHARD EVANS, research paleontologist, museum curator; b. St. Paul, June 18, 1927; s. Charles L. and Gladys N. (Evans) G.; m. Lucy Lee Speaker, July 19,1958; children: Charles Lewis, Evan Richard, Lauren Philip. B.A., U. Minn.-Mpls., 1949, M.S., 1953; Ph.D., U. Tex.-Austin, 1958. Instr. U. Tex.-Austin, 1953-54; research asst. Smithsonian Instn., Washington, 1957-61, paleobiologist, curator, 1972—; geologist U.S. Geol. Survey, Washington, 1961-72; mem. U.S. Nat. Com. on Geology, Nat. Acad. Sci. and Dept. Interior, 1979-83. Author: The Stenoscismatacea, 1965, (with G. A. Cooper) Permian Brachiopods of West Texas, 6 vols., 1972-77, Permian Brachiopods of South Thailand, 1978; contbr. (with G. A. Cooper) articles to profl. jours. Served with USN, 1945-46. Recipient hon. mention Jour. Paleontology, 1968, 72; Dirs. award Nat. Mus. Natural History, Washington, 1977; Daniel Giraud Elliot medal Nat. Acad. Sci., 1979. Fellow Geol. Soc. Am., AAAS; mem. Paleontol. Soc. (pres. 1978-79), Soc. Econ. Paleontologists and Mineralogists, Am. Assn. Petroleum Geologists, Internat. Paleontol. Assn. (treas. 1980—). Subspecialties: Paleobiology; Morphology. Current work: Late Paleozoic brachiopods and stratigraphy; biogeography of Tethyan Seaway (the southern margin of Eurasia, Spain to Indochina). Office: Smithsonian Instn E-206 Natural History Bldg Washington DC 20560

GRANT, VERNE EDWIN, biology educator; b. San Francisco, Oct. 17, 1917; S. Edwin and Bessie (Swallow) G.; m. Alva Day, June 12, 1946 (div. Aug. 1959); children: Joyce Grant Mixon, Brian, Brenda Grant Aley; m. Karen Alt, Nov. 3, 1960. A.B., U. Calif.-Berkeley, 1940, Ph.D., 1949. Teaching asst. botany U. Calif.-Berkeley, 1946-49; NRC fellow Carnegie Inst., Stanford, Calif., 1949-50; geneticist Rancho Santa Ana Bot. Garden, Claremont, Calif., 1950-67; asst. prof. Claremont Grad. Sch., 1951-53, asso. prof., 1953-57, prof., 1957-67; prof. biology Inst. Life Sci., Tex. A&M U., College Station, 1967-68; prof., dir. Boyce Thompson Southwestern Arboretum, U. Ariz.-Superior, 1968-70; prof. botany U. Tex.-Austin, 1970—. Author: Natural History of the Phlox Family, 1959, The Origin of Adaptations, 1963, The Architecture of the Germplasm, 1964, (with Karen Grant) Flower Pollination in the Phlox Family, 1965, Hummingbirds and Their Flowers, 1968, Plant Speciation, 1971, 2d edit., 1981, Genetics of Flowering Plants, 1975, Organismic Evolution, 1977; editorial bd.:

Ency. Americana, 1955-64, Brittonia, 1957-62, Evolution, 1960-62, Am. Naturalist, 1964-67, Biologisches Zentralblatt, 1974—; contbr. numerous articles to profl. jours. Recipient Sci. award Phi Beta Kappa, 1964. Fellow Am. Acad. Arts and Scis.; mem. Nat. Acad. Scis., Am. Soc. Naturalists, Soc. for Study of Evolution (pres. 1968), Bot. Soc. Am. (cert. of merit 1971), Internat. Soc. Plant Taxonomists, Am. Soc. Plant Taxonomists, Southwestern Assn. Naturalists, Soc. Systematic Zoology. Subspecialties: Evolutionary biology; Population biology. Current work: Population biology; teaching, writing, research. Home: 2811 Fresco St Austin TX 78731 Office: Dept Botany U Tex Austin TX 78712

GRANTHAM, JARED JAMES, physician, medical educator; b. Dodge City, Kans., May 19, 1936; s. Jimmie Harrison and Ista Tola (Taylor) G.; m. Carol E. Gabbert, June 15, 1958; children: Janeane, Jared, James, Joel. A.B., Baker U., Baldwin City, Kans., 1958; M.D., U. Kans., 1962. Intern U. Kans. Med. Ctr., Kansas City, 1962-63, resident in medicine, 1963-64; staff investigator NIH, Bethesda, Md., 1966-69; mem. faculty U. Kans., Kansas City, 1969—, prof. medicine, 1975—; pres. Kidney and Urology Research Ctr., 1982—; v.p. Polycystic Kidney Research Found., Kansas City, 1982—. Co-author: Physiology of the Kidney, 1982. NIH grantee, 1969-83; May Found. scholar, 1977-78; Fellow ACP. Mem. Am. Soc. Clin. Investigation, Am. Assn. Physicians, Am. Heart Assn. (chmn. council on kidney 1975-77), Am. Physiol. Soc. Republican. Methodist. Subspecialties: Nephrology; Physiology (medicine). Current work: Transport of anions and cations by kidney tubules; regulation of cell volume; pathogenesis of polycystic kidney disease. Office: U Kans Med Ctr 39th and Rainbow Blvd Kansas City KS 66103

GRANTZ, ARTHUR, geologist; b. N.Y.C., Nov. 9, 1927; s. William and Adele (Glotzer) G.; m. Willene Hatcher, Apr. 1, 1951; children: David Arthur, Eric, Carol Nina, Sara Lynn. A.B., Cornell U., 1949; M.S., Stanford U., 1961, Ph.D., 1966. Registered geologist; registered engring. geologist. Geologist U.S. Geol. Survey, Washington, 1949-51, San Francisco, 1952-53, Menlo Park, Calif., 1954—. Contbr. articles to sci. jours., chpts. to books. Vice chmn. Calif. State Mining and Geology Bd., Sacramento, 1976-79. Fellow Geol. Soc. Am.; mem. Am. Assn. Petroleum Geologists, Am. Geophys. Union. Democrat. Subspecialties: Geology; Tectonics. Current work: Regional geology and tectonics of Alaska and the Arctic Ocean; petroleum, geology; active faults. Office: US Geol Survey 345 Middlefield Rd Menlo Park CA 94025

GRASDALEN, GARY LARS, astronomer; b. Albert Lea, Minn., Oct. 7, 1945; s. Lars G. and Lillie (Olsen) G. A.B., Harvard U., 1967; M.S., U. Calif., Berkeley, 1970, Ph.D., 1972. Research asst. dept. astron. U. Calif., Berkeley, 1970-72; asst. astronomer Kitt Peak Nat. Obs., Tucson, 1972-75, assoc. astronomer, 1975-77; asst. prof. dept. physics and astronomy U. Wyo., Laramie, 1977-80, assoc. prof., 1980—. Contbr. articles to profl. jours. Mem. Am. Astron. Soc., Internat. Astron. Union, AAAS, Astron. Soc. Pacific. Subspecialty: Infrared optical astronomy. Current work: Star formation, H II regions, galaxies; image processing, process control, infrared detectors. Office: Dept Physics and Astronomy P O Box 3905 Univ Sta Laramie WY 82071

GRASSLE, JOHN FREDERICK, biological oceanographer; b. Cleve., July 14, 1939; s. John Kendall and Norah Iris (Fleck) G.; m. Judith Helen Payne, NOv. 21, 1964; 1 son, John Thomas. B.S., Yale U., 1961; Ph.D., Duke U., 1967. Postdoctoral scholar U. Queensland (Australia), Brisbane, 1967-69; asst. scientist Woods Hole (Mass.) Oceanographic Inst., 1969-73, assoc. scientist, 1973-83, sr. scientist, 1983—. Mem. Am. Soc. Naturalists, Am. Soc. Limnology and Oceanography, Soc. Study Evolution, Internat. Union Conservation and Natural Resources (Commn. on Ecology), Ecol. Soc. Am., Gt. Barrier Reef Com., Explorers Club. Subspecialty: Deep-sea biology. Home: PO Box 507 Woods Hole MA 02543 Office: Woods Hole Oceanographic Inst Woods Hole MA 02543

GRATT, BARTON MICHAEL, oral radiology educator, dentist; b. Chgo., Aug. 23, 1945; m. Karren Hakanson, June 3, 1979. D.D.S., UCLA, 1971. Diplomate: Am. Bd. Oral-Maxillofacial Radiology. Asst. prof. U. Calif.-San Francisco, 1975-79; assoc. prof. UCLA, 1979—; cons. dental xeroradiography Xerox Corp., Pasadena, Calif., 1977—; cons. panoramic radiography Gen. Electric Corp., Milw., 1980—. NIH grantee, 1978; Am. Fund Dental Health grantee, 1980. Fellow Am. Acad. Dental Radiology. Subspecialties: Diagnostic radiology; Imaging technology. Current work: Dental xeroradiography, dental radiology quality control, image analysis. Inventor dental xeroradiography, 1976. Home: 2619 Federal Ave Los Angeles CA 90064 Office: UCLA Sch Dentistry Los Angelos CA 90024

GRATT, LAWRENCE BARRY, engineering research company executive; b. Chgo., Sept. 20, 1940; s. Jack J. and Bette Vivian (Goldbloom) G.; m. Dona Jean Janecek, Aug. 21, 1963; children: Robyn An, Gambyl Gale, Alexis Kathleen, Natalee Joyce. B.S., UCLA, 1962, M.S., 1964, Ph.D., 1969. Mem. Tech. staff Hughes Aircraft Co., Culver City, Calif., 1962-65; sect. head TRW Systems, Redondo Beach, Calif., 1965-73; asst. v.p. Sci. Applications, Inc., LaJolla, Calif., 1973-79; pres. IWG Corp., San Diego, 1979—; instr. reactor design UCLA, Westwood, Calif., 1970; instr. systems engring. West Coast U., 1969-72. Author tech. papers and reports. AEC fellow, 1966-69. Mem. Am. Nuclear Soc., AIAA, Geothermal Resources Council, Soc. for Risk Analysis, Am. Inst. Mining Engrs. Subspecialties: Risk analysis; Oil shale. Current work: Risk analysis and assessments of new technological developments. Home: 5463 Coral Reef Ave La Jolla CA 92037 Office: IWG Corp 975 Hornblend St San Diego CA 92109

GRAUPE, DANIEL, electrical engineering educator, researcher; b. Jerusalem, July 31, 1934; m., 1968; 3 children. B.S.M.E., Israel Inst. Tech., 1958, B.S.E.E., 1959, Dipl. Ing., 1960; Ph.D. in Elec. Engring, U. Liverpool, 1963. Engr. automatic control Israel Govt. Industries, Tel Aviv, 1959-60; lectr. in elec. engring. U. Liverpool, 1963-67; sr. lectr. mech. engring. Israel Inst. Tech., 1967-70; from assoc. prof. to prof. elec. engring. Colo. State U., 1970-78; prof. elec. engring. Ill. Inst. Tech., 1978—; vis. prof. U. Notre Dame, 1976; Russell Spring vis. chair and prof. mech. engring. U. Calif.-Berkeley, 1977; v.p. Biocommun Research Corp., Chevy Chase, Md., 1978—. Mem. IEEE, N.Y. Acad. Sci. Subspecialties: Systems engineering; Biomedical engineering. Office: Dept Elec Engring Ill Inst Tech Chicago IL 60616

GRAVER, JACK EDWARD, mathematics educator; b. Cin., Apr. 13, 1935; s. Harold John and Rose L. (Miller) G.; m. Yana Regina Hanus, June 3, 1961; children: Juliet Rose, Yana-Maria, Paul Christopher. B.A., Miami U., Oxford, Ohio, 1958; M.A., Ind. U.-Bloomington, 1961, Ph.D., 1964. John Wesley Young research instr. Dartmouth Coll., Hanover, N.H., 1964-66; asst. prof. Syracuse (N.Y.) U., 1966-69, assoc. prof., 1969-76; vis. prof. U. Nottingham, Eng., 1971-72; chmn. Syracuse (N.Y.) U., 1979-82, prof. math., 1976—. Author: (with M. E. Watkins) Combinatorics with Emphasis on Graph Theory, 1977, (with J. Bagliuo) Incidence and Symmetry in Design and Architecture, 1983; contbr. articles in field to profl. jours. Served with USN, 1953-55. Mem. AAUP, Am. Math. Soc., Math. Assn. Am., Soc. Indsl. and Applied Math., Sigma Xi. Subspecialty: Combinatorics. Current work: Ramsey theory, matroid theory, design theory, graphyfactorization, integer programming, rigidity theory. Home: 871 Livingston Ave Syracuse NY 13210 Office: Dept Math Syracuse U Syracuse NY 13210

GRAVES, HARVEY WILBUR, JR., consulting engineer, educator; b. Rochester, N.Y., June 18, 1927; s. Harvey Wilbur and Margaret (Molloy) G.; m. Nancy Ray Copp, Aug. 6, 1960; children: John H., Thomas C., Rebecca Ann. A.B., Dartmouth Coll., 1950, M.S., 1951; Ph.D., U. Mich., 1973. Registered profl. engr., Pa., Md. Engring. mgr. Westinghouse Electric, Pitts., 1951-68; vis. assoc. prof. dept. nuclear engring. U. Mich., Ann Arbor, 1968-73; cons. engr. Pickard Lowe Assoc., Washington, 1973-75; sole practice cons. engr., Washington, 1975—; adj. prof. U. Md., College Park, 1981—; mem. AEC Com. on reactor physics, 1967-69. Author: Nuclear Fuel Management, 1979. Served with USN, 1945-46. Fellow Am. Nuclear Soc. (pres. 1973). Episcopalian. Subspecialties: Nuclear engineering; Software engineering. Current work: Nuclear fuel management and computer software applications. Home: 7723 Curtis St Chevy Chase MD 20815

GRAVITZ, SIDNEY I., aerospace engineer; b. Balt., June 28, 1932; s. Philip B. and Sophie (Korim) G.; m., Phyllis Bilgrad, ; June 14, 1964; children: Deborah A., Elizabeth E. B.S., MIT, 1953, M.S., 1954, Aero.E., 1957. Research engr. MIT, 1952-57; project engr. Wright Air Devel. Ctr., Dayton, Ohio, 1954-56; dynamics group engr. N.Am. Aviation, Columbus, Ohio, 1957-60; with The Boeing Co., Seattle, 1960—, mgr. surface transp. tech., 1970-72, mgr. space systems product devel. engring., 1972, mgr. inertial upper stage flight ops., 1976-80, mgr. 767 flight mgmt. system analysis and evaluation, 1980, mgr. digital 747 software devel., 1981-82, mgr. energy control systems engring., 1983—; mem. NASA-Industry Space Shuttle Design Criteria Working Group. Contbr. articles to profl. jours. Mem. MIT Ednl. Council. Served as lt. USAF, 1954-56. MIT scholar, 1949; grad. fellow, 1956. Assoc. fellow AIAA; mem. ASME, AAAS, Sigma Xi, Sigma Gamma Tau. Subspecialties: Aerospace engineering and technology; Systems engineering. Current work: Program development, systems analysis and evaluation. Home: 8428 SE 62d St Mercer Island WA 98040 Office: The Boeing Co Seattle WA 98124

GRAY, EOIN WEDDERBURN, physicist, cons.; b. Larne, No. Ireland, May 7, 1942; s. Charles and Irene (Mason) G.; m.; children: Liam Charles, Michael Eoin. B.Sc. (hons.), Queens U., Belfast, No. Ireland, 1964, Ph.D., 1967. Demonstrator in physics Queens U., 1964-67; postdoctoral research U. B.C., Can., 1967-69; mem. tech. staff Bell Labs., Columbus, Ohio, 1969—. Contbr numerous articles to profl. jours. Recipient prize Holm Conf. Elec. Contacts, 1977. Fellow Inst. Physics; mem. IEEE (sr.), Can. Inst. Chemistry (sr.), European Phys. Soc., Am. Inst. Physics, Am. Phys. Soc., Can. Assn. Physicists. Anglican. Subspecialties: Plasma physics; Electronics. Current work: plasma-surface interactions, arc physics, electrical insulation. Patentee in field. Office: Bell Labs 6200 E Broad St Columbus OH 43213

GRAY, FESTUS GAIL, electrical engineering educator, researcher; b. Moundsville, W.Va., Aug. 16, 1943; s. Festus P. and Elsie V. (Rine) G.; m. Caryl Evelyn Anderson, Aug. 24, 1968; children: David, Andrew, Daniel. B.S.E.E., W.Va. U., 1965, M.S.E.E., 1967; Ph.D., U. Mich., 1971. Instr. W.Va. U., Morgantown, 1966-67; teaching fellow U. Mich., 1967-70; asst. prof. Va. Poly. Inst. and State U., Blacksburg, 1971-77, assoc. prof., 1977-82, prof., 1983—; faculty fellow NASA, 1975; cons. Inland Motors, Radford, Va., 1980; researcher Rome Air Devel. Ctr., N.Y., 1980-81, Naval Surface Weapons Ctr., Dahlgren, Va., 1982-83. Contbr. articles to sci. jours. Bd. deacons Northside Presbyterian Ch., Blacksburg, 1980-83; coach S.W. Va. Soccer Assn., Blacksburg, 1980-83. Grantee NSF, Office Naval Research, NASA. Mem. IEEE (chpt. chmn. 1979-80), Assn. Computing Machinery, Heath Users Group, Sigma Xi. Democrat. Subspecialties: Computer engineering; Computer architecture. Current work: Investigation of fault tolerance, diagnosis, testing, and reliability issues for VLSI; distributed and multiprocessor computer architectures. Home: 304 Fincastle Dr Blacksburg VA 24060 Office: Va Poly Inst and State U Blacksburg VA 24061

GRAY, GEORGE AMELUNG, mathematician; b. Chgo., Mar. 7, 1942; s. Joseph Burnham and Susan (Kemp) G.; m. Sara John, Feb. 21, 1979; children: Samsara Willow, Zoalantha Elizabeth. A.B., Franklin and Marshall Coll., 1964; M.S., Rensselaer Poly. Inst., 1966; M.A., U. Del., 1972, Ph.D., 1978. Computer programmer Chrysler Corp., Highland Park, Mich., 1978-79; mathematician Naval Surface Weapons Ctr., Silver Spring, Md., 1979—. Served with USN, 1966-70. Recipient Kershner award Franklin and Marshall Coll., 1963. Mem. Am. Math. Soc., Soc. Indsl. and Applied Math. Subspecialties: Applied mathematics; Mathematical software. Current work: Electromagnetic countermeasures, wave propagation, mathematical modelling. Office: Naval Surface Weapons Ctr Silver Spring MD 20910

GRAY, GRACE WARNER, pharmacology educator, researcher; b. Chgo., Nov. 20, 1924; d. William Scott and Beatrice Warner (Jardine) G. B.A. in Zoology, Mt. Holyoke Coll., 1945; Ph.D. in Pharmacology, U. Mich., 1951. Jr. pharmacologist Merrell Co., Cin., 1945-47; pharmacologist Bristol Labs., Syracuse, N.Y., 1951-54; asst. prof. pharmacology Marquette U. Med. Sch., Milw., 1954-63, Woman's Med. Coll., Phila., 1964-66; research assoc. U. Pa. Med. Sch., Phila., 1966; assoc. prof. vet. pharmacology U. Minn., St. Paul, 1967—. Contbr. numerous articles in pharmacology to profl. jours.; editorial reviewer for sci. jours. NIH grantee, 1957-63; U.S. Air Force grantee, 1969-72; USPHS fellow, 1963-64. Mem. Am. Soc. for Pharmacology and Exptl. Therapeutics, AAAS, AAUP, LWV, Phi Beta Kappa. Unitarian. Subspecialty: Pharmacology. Current work: Research in veterinary pharmacology. Home: 12 Sunset Ln St Paul MN 55110 Office: 1988 Fitch Ave Ave St Paul MN 55108

GRAY, HARRY BARKUS, chemistry educator; b. Woodburn, Ky., Nov. 14, 1935; s. Barkus and Ruby (Hopper) G.; m. Shirley Barnes, June 2, 1957; children: Victoria Lynn, Andrew Thomas, Noah Harry Barkus. B.S., Western Ky. U., 1957; Ph.D., Northwestern U., 1960. Postdoctoral fellow U. Copenhagen, 1960-61; faculty Columbia U., 1961-66, prof., 1965-66; prof. chemistry Calif. Inst. Tech., Pasadena, 1966—, now Arnold O. Beckman prof. chemistry, chmn. div.; vis. prof. Rockefeller U., Harvard U., U. Iowa, Pa. State U., Yeshiva U., U. Copenhagen, U. Witwatersrand, Johannesburg, South Africa, U. Canterbury, Christchurch, New Zealand; cons. govt., industry. Author: Electrons and Chemical Bonding, 1965, Molecular Orbital Theory, 1965, Ligand Substitution Processes, 1966, Basic Principles of Chemistry, 1967, Chemical Dynamics, 1968, Chemical Principles, 1970, Models in Chemical Science, 1971, Chemical Bonds, 1973, Chemical Structure and Bonding, 1980, Molecular Electronic Structures, 1980. Recipient Franklin Meml. award, 1967; Fresenius award, 1970; Shoemaker award, 1970; Harrison Howe award, 1972; award for excellence in teaching Mfg. Chemists Assn., 1972; Remsen Meml. award, 1979; Tolman medal, 1979; Guggenheim fellow, 1972-73; Phi Beta Kappa scholar, 1973-74. Mem. Nat. Acad. Scis., Am. Chem. Soc. (award pure chemistry 1970, award inorganic chemistry 1978, award for disting. service in advancement of inorganic chemistry 1984), Royal Danish Acad. Scis. and Letters, Alpha Chi Sigma, Phi Lambda Upsilon. Subspecialties: Inorganic chemistry; Photochemistry. Current work: Bioinorganic chemistry; inorganic photochemistry; oxidation-reduction reactions. Home: 1415 E California Blvd Pasadena CA 91106

GRAY, JOE WILLIAM, biomed. scientist; b. Hobbs, N.Nex., Apr. 26, 1946; s. Frank Omer and Rosalin Revae (Bauman) G; m. Jane Ellin Madison, Jan. 28, 1967; 1 son, Gerald Todd. Ph.D., Kans. State U., 1972. Mineral engr. Colo. Sch. Mines, 1968; Biomed. scientist Lawrence Livermore Nat. Lab., Calif., 1972—, sect. leader cytophysics, 1982—; asst. adj. prof. U. Calif., Davis, 1976—. Contbr. numerous articles to profl. jours. Served to capt. AUS, 1972. NIH grantee. Mem. Am. Assn. Cancer Research, Cell Kinetics Soc. (pres.-elect), Soc. Analytical Cytology, Am. Phys. Soc., AAAS. Rebublican. Club: Am. Sportsman's. Subspecialties: Cancer research (medicine); Genome organization. Current work: Cell cycle analysis, quantitative cytogenetics, flow cytometry. Office: Biomed Scis Div PO Box 5507 Livermore CA 94550

GRAY, KENNETH EUGENE, petroleum engineering educator, researcher, university administrator; b. Herrin, Ill., Jan. 11, 1930; m., 1955; 3 children. B.S., U. Tulsa, 1956, M.S., 1957; Ph.D. in Petroleum Engring, U. Tex., 1963. Drilling engr. Calif. Co., 1957-59; reservoir engr. Sohio Petroleum Co., 1959-60; from asst. prof. to assoc. prof. petroleum engring. U. Tex., Austin, 1962-68, Halliburton prof. petroleum engring., 1968—, chmn. dept. petroleum engring., 1966-74, dir., 1968—; cons. research dept. Continental Oil Co., 1963—; mem. U.S. Nat. Com. Rock Mechanics. Petroleum Research Fund, Am. Chem. Soc. grantee, 1963—; Tex. Petroleum Research Com. grantee, 1963—; Am. Petroleum Inst. grantee, 1964—; Gulf Research & Devel. Co. grantee, 1964—. Fellow Am. Inst. Chemists; mem. N.Y. Acad. Sci., Am. Acad. Mechanics. Subspecialty: Petroleum engineering. Office: Dept Petroleum Engring U Tex Austin TX 78712

GRAY, PAUL EDWARD, university president; b. Newark, Feb. 7, 1932; s. Kenneth Frank and Florence (Gilleo) G.; m. Priscilla Wilson King, June 18, 1955; children: Virginia Wilson, Amy Brewer, Andrew King, Louise Meyer. S.B., Mass. Inst. Tech., 1954, S.M, 1955, Sc.D., 1960. Mem. faculty Mass. Inst. Tech., 1960-71, Class of 1922 prof. elec. engring., 1968-71, dean, 1970-71, chancellor, mem. corp., 1971-80, pres., 1980—; dir. Shawmut Bank, Boston, New Eng. Mut. Life Ins. Co., A.D. Little Inc., Cambridge, Cabot Corp., Boston. Trustee, mem. corp. Mus. of Sci., Boston, Woods Hole Oceanographic Inst.; chmn. bd. trustees Wheaton Coll., Mass., 1976—; mem. White House Sci. Council, 1982—. Served to 1st lt. AUS, 1955-57. Recipient C.E. Tucker award teaching Mass. Inst. Tech. Fellow Am. Acad. Arts and Scis., IEEE (publs. bd. 1969-70); mem. Nat. Acad. Engring., Mex. Nat. Acad. Engring., AAAS, Sigma Xi, Eta Kappa Nu, Tau Beta Pi, Phi Sigma Kappa. Mem. United Ch. Christ (deacon 1969-72, moderator 1973-77). Subspecialty: Electrical engineering. Current work: Electrical engineering. Home: 111 Memorial Dr Cambridge MA 02142 Office: 77 Massachusetts Ave Cambridge MA 02139

GRAY, PHILIP HOWARD, psychologist, educator; b. Cape Rosier, Maine, July 4, 1926; s. Asa and Bernice (Lawrence) G.; m. Iris McKinney, Dec. 31, 1954; children—Cindelyn, Howard. M.A., U. Chgo., 1958; Ph.D., U. Wash., 1960. Asst. prof. dept. psychology Mont. State U., Bozeman, 1960-65, assoc. prof., 1965-75, prof., 1975—; vis. prof. U. Man., Winnipeg, Can., 1968-70; pres. Mont. Psychol. Assn., 1968-70; chmn. Mont. Bd. Psychologist Examiners, 1972-74; speaker sci. and geneal. meetings on ancestry of U.S. presidents. Organized exhbns. folk art in Mont. and Maine, 1972-79; Author: The Comparative Analysis of Behavior, 1966, (with F.L. Ruch and N. Warren) Working with Psychology, 1963, A Directory of Eskimo Artists in Sculpture and Prints, 1974; contbr. numerous articles on behavior to psychol. jours., poetry to lit. jours. Served with U.S. Army, 1944-46. Recipient Am. and Can. research grants. Fellow Am. Psychol. Assn., AAAS, Internat. Soc. Research on Aggression; mem. History of Sci. Soc., Nat. Geneal. Soc., New Eng. Hist. Geneal. Soc., Deer Isle-Stonington Hist. Soc., Psychonomic Soc., Descs. of Illegitimate Sons and Daus. of Kings of Britain, Piscataqua Pioneers, Animal Behavior Soc., Sigma Xi. Subspecialties: Behavioral psychology; Ethology. Current work: Computer analysis of poetry and humor; quantification of hereditary genius of Ameican presidents; imprinting and murderous behavior; Darwinian evolution of mind; ethnic origins of science in conquests of Europe. Home: 1207 S Black Ave Bozeman MT 59715 Office: Dept Psychology Montana State U Bozeman MT 59717 We are human to the extent that we have bondings and the more bondings we have the more human we are. These attachments include familial bonding (imprinting), friendship bonding, marital bonding, ethnic-religious bonding, possession and goal bondings, and bonding to the land and ocean. My life's work is the study of these bondings and I am thereby more firmly connected to the human race.

GRAY, ROBERT DEE, biochemistry educator; b. Evansville, Ind., May 7, 1941; s. Robert Roy and Bettye Jane (Kane) G.; m. Karen S. Roessler, June 6, 1964; children: Robin Lynn, Michael Robert. B.A., DePauw U., 1963; Ph.D., Fla. State U., 1968. Postdoctoral research fellow Cornell U., Ithaca, N.Y., 1968-71; asst. prof. biochemistry U. Louisville, 1971-76, assoc. prof., 1976—. Mem. Am. Soc. Biol. Chemists, Am. Chem. Soc., AAAS. Subspecialty: Biochemistry (medicine). Office: U Louisville Health Scis Ctr Louisville KY 40292

GRAY, SAMUEL HUTCHISON, mathematician; b. West Chester, Pa., Oct. 7, 1948; s. Samuel Hutchison and Frances Borgia (Connelly) G.; m. Julia Ann Wetz, June 12, 1976; 1 son, Christopher. B.S., Georgetown U., 1970; M.A., U. Denver, 1975, Ph.D., 1978. Mathematician Naval Research Lab., Washington, 1978-79; asst. prof. math Gen. Motors Inst., Flint, Mich., 1979-82; research scientist Amoco Prodn. Co., Tulsa, 1982—. Contbr. articles to profl. jours. Mem. Am. Math. Soc., Soc. Indsl. and Applied Math, Soc. Exploration Geophysicists (assoc.), Sigma Xi. Subspecialties: Applied mathematics; Geophysics. Current work: Wave propagation in geophysics; determining earth's structure from reflected sound waves. Office: Research Ctr Amoco Prodn Co PO Box 591 Tulsa OK 74102

GRAY, TIMOTHY KENNEY, physician, researcher; b. Balt., Oct. 4, 1939; s. Francis Joseph and Eleanor (Kenney) G.; m. Mary Rita Trovato, July 20, 1963; children: Timothy Jr., Kathleen, Maura. B.S., Loyola Coll.-Balt., 1961; M.D., U. Md., 1965. Diplomate: Am. Bd. Internal Medicine. Intern U. Md. Hosp, Balt., 1965-66, resident in internal medicine, 1966-68; asst. prof., medicine U. N.C., Chapel Hill, 1971-74, assoc. prof., 1974-79, dir. clin. research unit, 1976-79, prof., 1979—, chief div. endocrinology, 1982—. Jefferson-Pilot fellow U. N.C., 1973-77. Mem. Am. Soc. Clin. Investigation, Am. Soc. Bone and Mineral Research, Am. Fedn. Clin. Research, The Endocrine Soc. Democrat. Roman Catholic. Subspecialties: Internal medicine; Endocrinology. Current work: Endocrinology and internal medicine. Home: Route 5 Box 325 Chapel Hill NC 27514 Office: Sch Medicine U NC Chapel Hill NC 27514

GRAYBEAL, JACK DANIEL, chemist, educator; b. Detroit, May 16, 1930; s. Paul Herman and Polly Dale (McClintic) G.; m. Evelyn Alice Nicolai, June 13, 1954; children: Daniel L., David E., Dale K. B.S., W.Va. U., 1951; M.S., U. Wis., 1953, Ph.D., 1955. Mem. tech. staff Bell Telephone Labs., 1955-57; asst. prof. chemistry W.Va. U., 1957-62, assoc. prof., 1962-68; assoc. prof. chemistry Va. Poly. Inst. and State U., Blacksburg, 1968-69, prof., 1969—, assoc. head dept., 1975—. Contbr. articles on microwave spectroscopy, nuclear quadrupole spectroscopy, molecular structure to sci. jours. Mem. Am. Chem. Soc., Am. Phys. Soc., Sigma Xi, Phi Lambda Upsilon (nat. editor Register).

Subspecialty: Physical chemistry. Current work: Microwave spectroscopy, nuclear quadropole resonance, molecular structure determination. Home: 312 Apperson Dr Blacksburg VA 24060 Office: Davidson Hall Va Poly Inst and State U Blacksburg VA 24061

GRAYBILL, BRUCE MYRON, chemistry educator; b. Council Bluffs, Iowa, Oct. 2, 1931; s. Amos Dewey and Ada Viola (Larsen) G.; m. Doris M. Green, Sept. 7, 1952; children: David B., Diana L. Graybill McAlister, Steven D. A.A., Graceland Coll., 1952; B.S., Iowa State U., 1955; Ph.D., Fla. State U., 1959. Research chemist Rohm & Haas Co., Huntsville, Ala., 1958-61; chemistry faculty Graceland Coll., Lamoni, Iowa, 1961—. Contbr. research papers to profl. publs. and confs. Pres. Lamoni Sch. Bd., 1979-83; bd. dirs. South Central Iowa Community Action Program, Leon, 1975-80. Recipient sr. chemistry award Am. Inst. Chemists, 1955; Phi Kappa Phi acad. award, 1955; Eastman Kodak fellow Fla State U., 1958. Mem. Am. Chem. Soc., Midwest Assn. Chemistry Tchrs. from Liberal Arts Colls., Sigma Xi, Phi Lambda Upsilon, Pi Mu Epsilon. Democrat. Club: Lamoni Golf (bd. 1975-81). Subspecialty: Organic chemistry. Current work: Organic synthesis and reaction mechanisms. Home: 208 N Elm St Lamoni IA 50140 Office: Graceland Coll Lamoni IA 50140

GRAYHACK, JOHN THOMAS, urologist, educator; b. Kankakee, Ill., Aug. 21, 1923; s. John and Marie (Keckich) G.; m. Elizabeth Houlehin, June 3, 1950; children: Elizabeth, Anne Marie, Linda Jean, John, William. B.S., U. Chgo., 1945, M.D., 1947. Diplomate Am. Bd. Urology. Intern medicine Billings Hosp., Chgo., 1947; intern gen. surgery Johns Hopkins Hosp., 1947-48, asst. resident, 1948-49, fellow urology, 1949-50, asst. resident, 1950-52; resident urology, 1952-53; dir. Kretschmer Lab., Northwestern U. Med. Sch., 1956—, prof. urology, 1963—, chmn. dept.; Cons. VA Research Hosp. Editor: Year Book of Urology, 1963-78; mem. editorial bd.: Surgery, Gynecology and Obstetrics; assoc. editor: Jour. Urology. Served to capt. USAF, 1954-56. Recipient Outstanding Achievement award USAF.; Fellow Am. Cancer Soc., 1949-50, Damon Runyon Fund, 1953-54. Mem. AMA, Ill., Chgo. med. socs., Am. Assn. Genitourinary Surgeons (Barringer medal), Am. Urology Assn. (Hugh H. Young award), Chgo. Urology Soc., Endocrine Soc., Clin. Soc. Genitourinary Surgeons, Am. Surg. Assn., Soc. Univ. Urologists, Nephrology Soc., Phi Beta Kappa, Alpha Omega Alpha. Subspecialty: Urology. Current work: Prostate, normal, benign hyper-plasma; clinical urology. Home: 95 N Park Rd LaGrange IL 60525 Office: 303 E Chicago Ave Chicago IL 60611

GRAYSON, GEORGE WELTON, biology educator, researcher; b. Dixons Mill, Ala., Nov. 1, 1938; s. Aaron and Martha (Harper) G.; m. Lucille Lampkin, Dec. 20, 1963; children; Anthony, Reginald, Dierdre. B.S., M.S., Ala. A & M U.; Ph.D., Vanderbilt U., 1976. Tchr. sci. Tenn. Valley High Sch., Hillsboro, Ala., 1966-67, guidance counselor, asst prin., 1967-68; instr. biology Ala. A&M U., 1968-71, chmn. dept., 1977—. Author: (with Jones) Bio-Learning Guide, 1978; contbr. in field. Founder Huntsville (Ala.) Coalition for Community Action; mem. Madison County (Ala.) Democratic Exec. Com. Served with U.S. Army, 1957-60. Dept. HHS grantee, 1980, 81, 82; TVA grantee, 1981; EPA-TVA grantee, 1981; recipient Disting. Alumnus award, 1980. Mem. NEA, Ala. Edn. Assn., Genetics Soc. Am., Am. Soc. Allied Health Professions, AAUP, Am. Inst. Biol. Scis., Beta Beta Beta., Phi Beta Sigma. Current work: Carcinogenic inhibitors; investigation of meiogenic initated processes with applications to carcinogenic cells and their physiologic properties. Office: PO Box 610 Normal AL 35762

GRAYSON, HENRY, psychoanalyst; b. Atmore, Ala., Oct. 25, 1935; s. Henry T. and Ethel (Sagasen) G.; m. Elizabeth Cauthen, Apr. 1, 1959; children: Regin, Douglas. A.B., Asbury Coll., 1957; S.T.M., Boston U., 1963, Ph.D., 1967; postdoctoral cert., Postgrad. Ctr. for Mental Health, N.Y.C., 1971. Lic. psychologist, N.Y. Instr. dept. psychology Mt. Ida Jr. Coll., Newton, Mass., 1963-67; from asst. to assoc. prof. CUNY, Bklyn., 1967-78; sole practice psychology, N.Y.C., 1967—; founder, exec. dir. Nat. Inst. Psychotherapies, N.Y.C., 1970-82, chmn. bd., 1970—; dir. Counseling and Family Therapy Assocs., Mahopac, N.Y., 1981—; bd. dirs. Ctr. Marital and Family Therapy, N.Y.C., 1981—; pres. F.A.T. Seminars, N.Y.C., 1982—. Author: Three Psychotherapies, 1975; author, editor: Short-term Approaches to the Psychotherapies, 1979, Changing Approaches to the Psychotherapies, 1978. Fellow Am. Group Psychotherapy Assn.; mem. Eastern Group Psychotherapy Soc. (treas. 1978-79), Am. Psychol. Assn., Am. Acad. Psychotherapists, N.Y. State Psychol. Assn. Current work: Eating disorders, mind-body relationships. Office: 330 W 58th St New York NY 10019

GRAYSTON, J. THOMAS, medical educator; b. Wichita, Kans., Sept. 6, 1924; s. Jesse T. and Luzia B. (Thomas) G.; children: Susan, Jesse, David; m. M. Nan Bryant, June 7, 1980. Student, Carleton Coll., 1942-43; B.S., U. Chgo., 1947, M.D., 1948, M.S., 1952. Diplomate: Am. Bd. Internal Medicine, Am. Bd. Preventive Medicine. Intern Albany (N.Y.) Med. Sch., 1948-49; Seymour Coman fellow preventive medicine U. Chgo., 1949-50, asst. resident medicine, 1950-51; epidemiologist epidemic intelligence service USPHS, U. Kans. Med. Center, 1951-53; chief resident medicine U. Chgo., 1953-54, instr. medicine, 1953-55; fellow Nat. Found. Infantile Paralysis, 1954-56; asst. prof. medicine U. Chgo., 1955-60, asso. prof., 1960; chief div. microbiology and epidemiology U.S. Naval Med. Research Unit 2, Taipei, Taiwan, 1957-60, cons., 1960-79; prof. preventive medicine, chmn. dept. Sch. Medicine, U. Wash., 1960-70, dean Sch. Pub. Health and Community Medicine, 1970-71, v.p. for health scis., 1971-83, prof. dept. epidemiology, 1970—, adj. prof. pathobiology, 1982—; mem. exec. com. Regional Primate Research Center, 1964-70, research affiliate, 1967-70; attending physician medicine Univ. Hosp., Seattle, 1960-70; asso. mem. commn. acute respiratory diseases Armed Forces Epidemiol. Bd., 1962-65, mem., 1965-73; mem. research and engring. adv. panel biology and medicine Dept. Def., 1963-67; sci. group trachoma research WHO, 1963; virology and rickettsiology study sect. NIH, 1963-67; mem. internat. centers com. Nat. Inst. Allergy and Infectious Diseases, 1967-71; mem. expert adv. panel on Trachoma, WHO, 1970—; chmn. exec. com., mem. nat. adv. council on health professions edn. NIH, 1972-75. Contbr. numerous articles to profl. jours. Fellow Am. Coll. Preventive Medicine (v.p. gen. preventive medicine 1970-71, regent 1971-74), Am. Pub. Health Assn. (governing bd. 1978-80); mem. Am. Assn. Immunologists, Am. Assn. Physicians, Am. Epidemiol. Soc. (pres. 1982-83), Am. Fedn. Clin. Research, Am. Soc. Clin. Investigation, Am. Soc. Tropical Medicine and Hygiene, Assn. Acad. Health Centers (dir. 1975-80, pres. 1978-79), Assn. Tchrs. Preventive Medicine, Infectious Diseases Soc., Internat. Epidemiol. Assn., Soc. Exptl. Biology and Medicine, Inst. Medicine of Nat. Acad. Scis., Western Assn. Physicians, Western Soc. Clin. Research, Sigma Xi. Subspecialties: Epidemiology; Infectious diseases. Current work: Epidemiology and immunology of infectious diseases. Office: Dept Epidemiology SC-36 U Wash Seattle WA 98195

GRAZIANO, KENNETH DONALD, microbiologist; b. Dunkirk, N.Y., Apr. 10, 1942; s. Russell James and Antoinette Joan (Valvo) G. B.A., Colgate U., 1963; M.S., Syracuse U., 1965; Ph.D., Johns Hopkins U., 1970. Fellow Johns Hopkins U. Sch. Medicine, Balt., 1969-72; staff fellow Nat. Cancer Inst., Balt., 1972-74; instr. Johns Hopkins U., 1974-75; research microbiologist FDA, Bethesda, Md., 1975-78, rev. microbiologist, 1978—. Contbr. articles to profl. jours. Mem. Mt. Washington Community Assn. Mem. Am. Assn. Immunologists, Am. Soc. Microbiology, AAAS, N.Y. Acad. Sci., Johns Hopkins U. Med. and Surg. Assn., Johns Hopkins U. Immunology Council, Balt. Zool. Soc., Nat. Aquarium. Roman Catholic. Clubs: Vintage Japanese Motorcycle, Courts Royal Racquetball., Greenspring Valley Racquet. Subspecialties: Immunobiology and immunology; Cellular engineering. Current work: Use of biological products to alter immune responses. Scientific and regulatory review of data submitted with regard to new biological drugs. Office: 8800 Rockville Pike Bethesda MD 20205

GREASER, MARION LEWIS, science educator, researcher; b. Vinton, Iowa, Feb. 10, 1942; s. Lewis Levi and Elisabeth (Sage) G.; m. Marilyn Sue Pfister, June 12, 1965; children: Suzanne, Scott. B.S., Iowa State U., 1964; M.S., U. Wis.-Madison, 1967, Ph.D., 1969. Postdoctoral fellow Boston Biomed. Research Inst., 1968-71; asst. prof. U. Wis.-Madison, 1971-73, assoc. prof., 1973-77, prof., 1977—. Contbr. research articles to profl. publs. Mem. Am. Soc. Biol. Chemists, Biophys. Soc., Inst. Food Technologists, Am. Meat Sci. Assn. (Disting. Research award 1981), AAAS. Subspecialties: Biochemistry (biology); Cell biology (medicine). Current work: Mechanism of muscle contraction in both heart and skeletal muscle. Home: 2374 Branch St Middleton WI 53562 Office: Muscle Biology Lab Univ Wis 1805 Linden Dr Madison WI 53706

GREBENAU, MARK DAVID, immunologist; b. Newport News, Va., Mar. 26, 1951; s. Franz Herbert and Gisela (Glueck) G.; m. Ruth Cyrille Goodman, Aug. 11, 1974; children: Maurice J., Julie E. A.A., Yeshiva U., 1972, B.A., 1972; M.S., NYU, 1976, M.D., 1978, Ph.D., 1979. Diplomate: Nat. Bd. Med. Examiners. Intern King's County Hosp. Ctr./Downstate Med. Center, Bklyn., 1978-79, resident, 1979-81; practice medicine specializing in immunology, N.Y.C., 1981—; research assoc., assoc. physician Rockefeller U., N.Y.C., 1981—; fellow in allergy N.Y. Hosp., 1983—; mem. staff Rockefeller V. Hosp., N.Y. Hosp. Lita Annenberg Hazen fellow, 1980. Mem. ACP, N.Y. Acad. Scis. Jewish. Subspecialties: Immunology (medicine); Internal medicine. Current work: Clinical immunology and allergy; characterization of polyclonal rheumatoid factors, hybridoma techology. Office: 1230 York Ave Founder's 271 New York NY 10021

GRECO, RICHARD JAMES, software engr.; b. Portland, Oreg., July 6, 1952; s. I. James and Clara Ann (Ferrante) G. B.T.E.E., Oreg. Inst. Tech., 1975. Systems programmer Lewis and Clark Coll., Portland, 1975-79; software engr. Tektronix, Wilsonville, Oreg., 1979—; cons. engr. Technigraph, Portland, 1979—. Active Portland Art Assn., 1981. Mem. Assn. for Computing Machinery, IEEE Computer Soc., Portland Internat. Gormet Soc. Daoist. Subspecialties: Computer architecture; Graphics, image processing, and pattern recognition. Current work: Human interfaces to computers utilizing pictures as a communication media. Home: 8022 SE 60th St Portland OR 97206 Office: PO Box 1000 Wilsonville OR 97070

GREELEY, RONALD, geology educator; b. Columbus, Ohio, Aug. 25, 1939; m. Cynthia Ray Moody, Aug. 28, 1960; children: Randall R., Vanessa L. B.S., Miss. State U.-State College, 1962, M.S., 1963; Ph.D., U. Mo.-Rolla, 1966. Instr. U. Mo.-Rolla, 1965-66; geologist Chevron Oil Co., Lafayette, La., 1966-67; research scientist NASA-Ames Research Center, Moffett Field, Calif., 1967-69; postdoctoral research assoc. NRC-NASA Ames Research Ctr., 1969-71; resarch assoc. U. Santa Clara (Calif.)-Ames Research Ctr., Calif., 1971-77; prof. geology Ariz. State U., Tempe, 1977—; cons. Jet Propulsion Lab., Pasadena, Calif., 1975—, Planetary Sci. Inst., Houston, 1972—. Author: Earthlike Planets, 1981, Geology on the Moon, 1976, Volcanic Features of Hawaii, 1981. Served to capt. U.S. Army, 1967-69. Recipient Pub. Service Award NASA, 1977. Fellow Geol. Soc. Am.; mem. Am. Geophy. Union, AAAS. Subspecialties: Remote sensing (geoscience); Planetology. Current work: Geology of the moon, planets and satellites, volcanology wind-related phenomena. Office: Ariz State U Dept Geology Tempe AZ 85287

GREEN, CORDELL, computer scientist; b. Ft. Worth, Dec. 26, 1941; s. William L. and Rebecca L. G.; m. Christine louise Ochs, June 21, 1979; children: Jeffrey, Laura. B.A.-B.S., Rice U., 1964; M.S. in Elec. Engring, Stanford U., 1965, Ph.D., 1969. Research mathematician artificial intelligence group Stanford Research Inst., 1966-69; research and devel. program mgr. for artificial intelligence Advanced Research Projects Agy., Info. Processing Techniques Office, 1970-71; pub. mem. steering com. speech understanding easst. prof. computer sci. Stanford U., 1971-78; chief scientist computer sci. dept. Systems Control, Inc., Palo Alto, Calif., 1978-81; dir., chief scientist Kestrel Inst., Palo Alto, 1981—; mem. program com. U.S.-Japan Computer Conf., 1974-75. Editor: artificial intelligence area Jour. Assn. for Computing Machinery, 1972-79; mem. editorial bd.: Jour. Cognitive Sci, 1975. Served to capt. U.S. Army, 1969-71. Subspecialty: Artificial intelligence. Current work: Program synthesis, knowledge-based programming, expert systems. Office: 1801 Page Mill Rd Palo Alto CA 94304

GREEN, DANIEL G., physiological optics, psychology and electrical engineering educator; b. N.Y.C., Sept. 3, 1937; m., 1957; 2 children. B.S.E.E., U. Ill., 1959; M.S. and Ph.D. in Elec. Engring, Northwestern U., 1964. Asst. elec. engr. and bioengr. North-western U., 1959-64; NSF fellow physiol. lab. Cambridge U., 1964-65, vis. scientist, 1981; USPHS fellow Nobel Insts. Neurophysiology, Stockholm, 1965-66; asst. prof. physiol. optics U. Mich., 1966-70, assoc. prof. physiol. optics, psychology and elec. engring., 1971-76, prof., 1976—; also now dir. neuroscis. vision research lab.; vis. scientist biol. labs. Harvard U., 1972-73; vis. prof. ophthalmology U. Calif.-San Francisco, 1980-81; vis. scientist physiol. lab. Cambridge U., 1981. Mem. AAAS, Am. Physiol. Soc., Assn. Research in Vision and Ophthalmology, Optical Soc. Am., Soc. Neurosci. Subspecialties: Biomedical engineering; Physiological optics. Office: U Mich Neurosci Bldg 1103 E Huron Ann Arbor MI 48109

GREEN, DANIEL MICHAEL, pediatrician; b. Seattle, May 30, 1946; s. Daniel Marie and Margaret Ann (Johnson) G. B.S. in Elec. Engring, MIT, 1969; M.D., St. Louis U., 1973. Diplomate: Am. Bd. Pediatrics. Intern, in pediatrics Boston City Hosp., 1973-74, resident in pediatrics, 1974-75; fellow in hematology/oncology Children's Hosp., Boston, 1975-78; assoc. prof. pediatrics SUNY, Buffalo, 1983—. Fellow Am. Acad. Pediatrics; mem. Internat. Soc.Pediatric Oncology, Am. Soc.Hematology, Am. Soc. Clin. Oncology, Am. Assn. Cancer Research. Subspecialties: Pediatrics; Chemotherapy. Current work: Treatment of pediatric solid tumors, evaluation of long term complications of cancer treatment. Home: 111 Wallace Ave Buffalo NY 14214 Office: Roswell Park Meml Inst 666 Elm St Buffalo NY 14263

GREEN, DAVID EZRA, educator; b. N.Y.C., Aug. 5, 1910; s. Herman and Jennie (Marrow) G.; m. Doris Cribb, Apr. 15, 1935; children—Rowena (Mrs. Larry Matthews), Pamela (Mrs. Joseph Baldwin, Jr.). B.A., N. Y. U., 1930, M.A., 1932; Ph.D., Cambridge (Eng.) U., 1934. Beit Meml. Research fellow Cambridge, Eng., 1934-40; fellow Harvard, 1940-41; with enzyme lab. Coll. Phys. and Surg., Columbia, 1941-48; asso. prof. biochemistry Columbia, 1947; co-dir. Inst. Enzyme Research, U. Wis., 1948—, prof. enzyme chemistry, 1948—. Author: Mechanisms of Biological Oxidations, 1939, Molecular Insights into the Living Process, 1967. Recipient Paul-Lewis Labs. award enzyme chemistry, 1946. Fellow Am. Acad. Arts and Scis.; fgn. fellow Royal Flemish Acad. Arts and Scis.; mem. Am. Soc. Biol. Chemists, Nat. Acad. Scis., Am. Chem. Soc., Harvey Soc., Biochem. Soc., Am. Soc. Cell Biology, Phi Beta Kappa, Sigma Xi. Subspecialty: Biochemistry (biology). Current work: Biochemistry. Home: 5339 Brody Dr Madison WI 53705

GREEN, DAVID MARVIN, psychologist, educator; b. Jackson, Mich., June 7, 1932; s. George Elmer and Carrie Ruth (Crawford) G.; m. Clara Loftstrom, Jan. 2, 1953 (dec. July 1978); children: Allan, Phillip, Katherine, George; m. Marian Heinzmann, June 7, 1980. B.A., U. Chgo., 1952, U. Mich., 1954, M.A., 1955, Ph.D., 1958; NSF predoctoral fellow, M.I.T., 1956-57. Grad. research asso. Electronic Def. Group, U. Mich., 1954-56, 57-58; asst. prof. psychology M.I.T., 1958-63; cons. Bolt, Beranek & Newman, Inc., Cambridge, Mass., 1958—; asso. prof. psychology U. Pa., 1963-66, vice chmn. dept. psychology, 1964-66; prof. psychology U. Calif., San Diego, 1966-73; prof. psychophysics Harvard U., Cambridge, 1973—, chmn. dept. psychology and social relations, 1978-81; mem. exec. com. NRC Com. on Hearing, Bioacoustics and Biomechanics, 1968-71, chmn., 1970-71; mem. communicative scis. study sect. NIH, 1970-73; mem. sci. adv. group on noise Calif. Environ. Quality Study Council, 1970-72; chmn. CHABA Subcom. 9 Adoption of Composite Noise Scale, 1972—; mem. NRC Assembly of Behavioral and Social Scis., 1973—. Author: (with J.A. Swets) Signal Detection Theory and Psychophysics, 1966, Introduction to Hearing, 1976; Cons. editor: Psychol. Bull, 1965-68, Perception and Psychophysics, 1972—; mem. editorial bd.: Cognitive Psychology, 1970-73. Guggenheim fellow, 1973-74; overseas fellow St. Johns Coll., Cambridge, Eng., 1973-74. Fellow Am. Psychol. Assn., Acoustical Soc. Am. (Biennial award 1966, chmn. com. psychol. and physiol. acoustics 1970-73, exec. council 1972-75, pres. 1981-82), AAAS; mem. Soc. Exptl. Psychologists, Psychonomic Soc., Psychometric Soc., AAUP, Nat. Acad. Scis. Subspecialties: Psychophysics; Sensory processes. Home: 9 Lakeview Terr Winchester MA 01890 Office: 33 Kirkland St Cambridge MA 02138

GREEN, DONALD EUGENE, analytical toxicologist, researcher; b. Napa, Calif., Nov. 25, 1926; s. Joseph and Helen (Rubin) G.; m. Margaret Ann Maurer, July 29, 1951; children: Dennis, Gretchen, Mark, Gary, Sheryl. B.S., U. Calif., Berkeley, 1948; M.S., U. Calif.-San Francisco, 1952, B.S. in Pharmacy, 1955; Ph.D., Wash. State U., 1962. Instr. pharm. chemistry Idaho State U., Pocatello, 1955-57, 58-60, Wash. State U., Pullamn, 1957-58; research biochemist VA Med. Center, Palo Alto, Calif., 1962-64; sr. engr. Varian Assoc., Palo Alto, 1962-70; research asso. Stanford Med. Sch., Palo Alto, 1970-74; sr. research scientist Inst. Chem. Biology, U. San Francisco, 1974-81; research biochemist VA Med. Center, Palo Alto, 1981—; cons. Universal Monitor Corp., Pasadena, Calif., 1971-76. Contbr. articles to profl. jours. Served with USNR, 1944-46, 51-53. Mem. Am. Chem. Soc. (chmn. Santa Clara Valley sect. 1971-72, councillor 1975-81, 83—), Am. Pharm. Assn., Internat. Assn. Forensic Toxicologists, Calif. Assn. Toxicologists, Western Pharmacology Soc., Am. Soc. Pharmacology and Exptl. Therapeutics, U. Calif. Alumni Assn. Club: Santa Clara Country (pres. 1980-82). Subspecialties: Toxicology (medicine); Mass spectrometry. Current work: Development of analysis procedures for drug metabolites, bioinstrumentation, mass spectrometry. Patentee in field. Office: VA Med Center Palo Alto CA 94304

GREEN, DONALD ROSS, research director; b. Holyoke, Mass., Aug. 12, 1924; s. Donald Ross and Constance (McLaughlin) G.; m. Mary Reese, June 16, 1950; children: Alice Angell, Mitchell Reese. B.A., Yale U., 1948; M.A., U. Calif.-Berkeley, 1954, Ph.D., 1958. Instr. math. George Sch., 1948-50; statistician Cancer Research Inst., San Francisco, 1953-56; asst., assoc. in edn. U. Calif.-Berkeley, 1956-57; instr. edn. Emory U., Atlanta, 1957-67, assoc. prof. psychology, 1963-67; dir. research CTB/McGraw-Hill, Monterey, Calif., 1967—. Author: Educational Psychology, 1964, (with R.L. Henderson) Reading for Meaning in the Elementary School, 1967; editor: (with Ford and Flamor) Measusrement and Piaget, 1971, The Aptitude Achievement Distinction, 1974, Achievement Testing of Disadvantaged and Minority Students. Served with U.S. Army, 1943-45. Mem. Am. Psychol. Assn., Am. Edn. Research Assn., Psychometric Soc., Nat. Council Measurement in Edn., Internat. Reading Assn. Subspecialties: Psychometrics; Learning. Current work: Research on applications of item response theory to test construction, ethnic bias in testing, and instructional inputs and test outcomes in schools. Home: PO Box 3284 Monterey CA 93940 Office: CTB/McGraw-Hill 2500 Garden Rd Monterey CA 93940

GREEN, DOROTHY EUNICE, research psychologist; b. Montgomery, Ala., Dec. 18, 1917; d. Cliff and Willie May G. B.S., Auburn U., 1937; M.A., Peabody Coll., 1941; Ph.D., U. Chgo., 1950. Aviation psychologist U.S. Air Force, 1951-54; Research psychologist U.S. Civil Service Commn., Washington, 1954-67; research psychologist USPHS, Washington, 1967-77, Chilton Research Services, Radnor, Pa., 1977-82, Audists and Surveys, Princeton, N.J., 1982—. Contbr. arttticles to profl. jours. Mem. Am. Psychol. Assn., Eastern Psychol. Assn., D.C. Psychol. Assn. Presbyn. Subspecialty: Social psychology. Current work: Research in the dynamics of human behavior, particularly in the health field; prevention of illness through change in lifestyle behavior; survey research. Home: 3509 N Dickerson St Arlington VA 22207

GREEN, HARRY, pharm. co. exec.; b. Phila., Sept. 7, 1917; s. Samuel and Mary (Bogatin) G.; m. Harriett Borten, Oct. 6, 1945; children: Ann Frankel, Jane. A.B. in Chemistry, U. Pa., 1938; M.S., 1939; Ph.D. (Harrison fellow), 1942. Research chemist Lion Oil Refining Co., El Dorado, Ark., 1941-44; sr. research organic chemist Pennsalt Mfg. Co., Phila., 1944-47; research assoc. in physiol. chemistry U. Pa. Med. Sch., Phila., 1947-52, asst. prof., 1954-68; chief biochem. research Wills Eye Hosp., Phila., 1952-58; sr. research biochemist Smith Kline & French Labs., Phila., 1958-61, group leader, 1961-64, head neurobiochemistry, 1964-67, dir. biochemistry, 1967-75, dir. sci. liaison, 1975-80; v.p. sci. liaison and tech. Smith Kline Beckman Corp., Phila., 1980—. Mem. Am. Soc. Biol. Chemists, Am. Soc. Pharmacology and Exptl. Therapeutics, Am. Chem. Soc., AAAS, Assn. Research in Nervous and Mental Diseases, Internat. Soc. Biochem. Pharmacology, N.Y. Acad. Scis., Sigma Xi. Subspecialties: Neurochemistry; Molecular biology. Current work: Intermediary metabolism, ocular biochemistry and physiology, enzymology corticosteroids, neurobiochemistry, drug metabolism, molecular biology, recombinant DNA research and technology, neuropharmacology, monoclonal antibodies. Home: 305 Penbree Terr Bala Cynwyd PA 19004 Office: One Franklin Plaza PO Box 7929 Philadelphia PA 19101

GREEN, HARRY WESTERN, II, geology educator; b. Orange, N.J., Mar. 13, 1940; s. Harry Beutel and Mabel (Hendrickson) G.; m. Maria Manuela Martins, May 15, 1975; children: Mark, Stephen, Carolyn, Alice, Miquel, Maria. B.A., UCLA, 1963, M.S., 1967, Ph.D. with honors, 1968. Postdoctoral assoc. Case Western Res. U., Cleve., 1968-70; asst. prof. geology U. Calif.-Davis, 1970-74, assoc. prof., 1974-80, prof., 1980—, acting chmn. dept., 1983—, chmn. dept., 1984—. NSF grantee, 1970—. Mem. Am. Geophys. Union, Geol. Soc. Am., Mineral. Soc. Am. Subspecialties: Geophysics; High-temperature materials. Current work: The nature and history of the earth's mantle, especially the convective flow responsible for continental drift; utilizing transmission electron microscopy, electron energy-loss, spectroscopy and infrared spectroscopy to examine the deformation of natural mantle rocks and their volatile constituents, plus experimental analogs. Home: 1225 Purdue Dr Davis CA 95616 Office: Dept Geology U Calif Davis CA 95616

GREEN, JACK PETER, med. scientist; b. N.Y.C., Oct. 4, 1925; s. Maurice and Tillie (Herman) G.; m. Arlyne Genevieve Frank, Oct. 25, 1958. B.S., Pa. State U., 1947, M.S., 1949; Ph.D., Yale, 1951, M.D., 1957; postgrad., Poly. Inst., Copenhagen, 1953-55, Inst. de Biologie Physico-Chimique, Paris, 1964-65. Vis. scientist Poly. Inst., Copenhagen, 1953-55, Inst. de Biologie Physico-Chimique, Paris, 1964-65; asst. prof. Yale, 1957-61, asso. prof., 1961-66, Cornell U. Med. Coll., 1966-68; prof., chmn. dept. pharmacology Mt. Sinai Sch. Medicine, 1968—; Mem. research grant rev. com. USPHS; mem. N.Y.C. Health Research Council, Dysautonomia Found., Irma T. Hirsch Trust. Contbr. articles profl. jours.; Mem. editorial bds. profl. jours. Recipient Claude Bernard Vis. Professorship U. Montreal, 1966. Mem. N.Y. Acad. Sci., Am. Chem. Soc., Am. Soc. Biol. Chemists, Soc. Drug Research, N.Y. Acad. Medicine, Harvey Soc., A.A.A.S., Am. Soc. Pharmacology and Exptl. Therapeutics, Internat. Soc. Quantum Biology, Am. Coll. Neuropsychopharmacology, Am. Soc. Neurochemistry, Soc. for Neurosci., Sigma Xi, Alpha Omega Alpha, Phi Lambda Upsilon, Gamma Sigma Delta. Subspecialties: Planetology; Remote sensing (geoscience). Current work: Compilation of encyclopedia of volcanology, economic geology, geothermics; resources, exploration geology of precious metals. Home: 1212 Fifth Ave New York City NY 10029 Office: Mt Sinai Sch Medicine Dept Pharmacology Fifth Ave at 100th St New York City NY 10029

GREEN, LARRY JOY, orthodontist, educator; b. Snyder, N.Y., Jan. 1, 1931; s. Ross and Iola (Borden) G.; m. Rachel E. Chaffin, Dec. 27, 1958; children: Nancy E., Patricia A., Cecilia A. B.S. in Zoology and Chemistry, U. Pitts., 1953, D.D.S., 1956, M.S. in Orthodontics, 1960; Ph.D. in Phys. Growth and Devel, U. Iowa, 1965. Diplomate: Am. Bd. Orthodontics. Clin. practice orthodontics, 1960-62; asst. prof. orthodontics U. Pitts., 1960-62; USPHS postdoctoral fellow U. Iowa, 1962-65; asst. prof. orthodontics SUNY-Buffalo, 1965-66, assoc. prof., 1967-71, prof., 1971—, adj. prof. anthropology, 1978-79. Abstractor: Oral Research Abstracts of Am. Dental Assn, 1977; editorial cons.: Jour. Dental Research, 1979-81; book reviewer: Bull. History of Dentistry, 1977; contr. articles, abstracts and revs. to profl. jours. Bd. dirs. Zool. Soc. Buffalo and Erie County; sec. North Campus adv. council Erie Community Coll; bd. dirs., pres. Buffalo Urban League; sec. bd. dirs. Health Systems Agy. Western N.Y.; pres. Upstate Med. Alliance; mem. Erie County subarea council Boy Scouts Am.; bd. dirs. Amherst Youth Bd., Catchment II, Community Mental Health Agy., Buffalo Equity Found., Alpha Kappa Scholarship Found., Inc. Served to capt. Dental Corps USAF, 1956-58. Grantee USPHS, United Way Buffalo and Erie County, NIH. Fellow Am. Coll. Dentists, Internat. Coll. Dentists; mem. Am. Dental Assn., 8th Dist. Dental Soc. (chmn. council dental health and health planning 1982-83), Internat. Assn. Dental Research (pres. craniofacial biology 1977-78), AAAS, Am. Assn. Orthodontists, Am. Assn. Phys. Anthropologists, Soc. Study of Human Biology, Erie County Dental Soc. (sec., bd. dirs.), Beta Beta Beta, Omicron Kappa Upsilon. Lodge: Rotary. Subspecialties: Orthodontics; Dental growth and development. Home: 380 Berryman Ave Snyder NY 14226 Office: SUNY 235 Farber Hall Buffalo NY 14214

GREEN, LAWRENCE, physicist, nuclear engr.; b. Gelenes, Hungary, May 3, 1937; came to U.S., 1939; s. Frank and Paula (Gottdiener) G.; m. Vera Anne Silverstone, Mar. 17, 1963; children: Miriam Jacqueline, Alisa Roanna. B.S., CCNY, 1959; Ph.D., Pa. State U., 1963. Sr. scientist Westinghouse Corp., Pitts., 1964-72, fellow scientist, 1972-77, 1978—; vis. prof. Ben Gurion U., Beer Sheva, Israel, 1977-78. Contbr. articles on physics to profl. jours. Mem. Pitts. Symphony Soc., 1975—; vol. South Hills Mental Health System, Pitts., 1976-77. State of N.Y. Regents scholar, 1955-59; AEC fellow, 1961-63. Mem. Am. Nuclear Soc., Am. Phys. Soc., AAAS. Subspecialties: Nuclear fission; Nuclear fusion. Current work: Experimental neutronics, radiation transport studies, fusion reactor blanket and shield design. Home: 151 Dutch Ln Pittsburgh PA 15236 Office: Westinghouse Corp Pittsburgh PA

GREEN, MAURICE, microbiology educator, university administrator, biochemist; b. N.Y.C., May 5, 1926; m., 1950; 3 children. B.S., U. Mich., 1949; M.S., U. Wis., 1952, Ph.D. in Biochemistry, 1954. Instr. biochemistry U. Pa., 1955-56; from asst. prof. to assoc. prof. Sch. Medicine, St. Louis U., 1956-63, prof. microbiology, 1963—, dir., 1964—. Recipient Research Career award USPHS, 1962—; Nat. Found. Infantile Paralysis research fellow U. Pa., 1954-55; Lalor fellow, 1955-56; USPHS sr. fellow, 1958-62. Mem. AAAS, Am. Soc. Biol. Chemistry, Am. Chem. Soc., Am. Soc. Microbiology. Subspecialties: Virology (biology); Molecular biology. Office: Inst for Molecular Virology St Louis U Sch Medicine Saint Louis MO 63110

GREEN, MICHAEL ENOCH, chemistry and educator and researcher; b. N.Y.C., Nov. 5, 1938; s. George A. and Esther (Gladstone) G.; m. Nihal Kustimur, Oct. 12, 1974; 1 son, Omar. B.A., Cornell U., 1959; M.S., Yale U., 1961, Ph.D., 1964. Postdoctoral research assoc. Calif. Inst. Tech., 1963-64; vis. lectr. Middle East Tech. U., Ankara, Turkey, 1964-66; asst. prof. chemistry CUNY, 1966-72, assoc. prof., 1972-83, prof., 1984—; vis. lectr. Hacettepe U., Ankara, 1973-74, Lanzhou Inst. Chem. Physics, China, 1980, 82. Author: (with others) Physical Chemistry Laboratory, 1978, Safety in Working with Chemicals, 1978. Mem. health tech. com. N.Y. Com. Occupational Safety and Health, 1979-81. NSF predoctoral fellow, 1960-63. Mem. Am. Chem. Soc., Am. Phys. Soc., N.Y. Acad. Scis., AAAS, Phi Beta Kappa. Subspecialties: Surface chemistry; Biophysics (physics). Current work: Ion transport across membranes, membrane potentials, fluctuation spectroscopy, light scattering. Office: City College of City University of New York Dept of Chemistry 138th St and Convent Ave New York NY 10031

GREEN, PAUL BARNETT, biology educator; b. Phila., Feb. 15, 1931; s. Otis Howard and Mabel (Barnett) G.; m. Margaret E. Barnett, Feb. 2, 1931; children: Robert, Peter, Katherine. B.A., U. Pa., 1952; Ph.D., Princeton U., 1957. Asst. to full prof. U. Pa., Phila., 1957-70; prof. biology Stanford (Calif.) U., 1971—. Contbr. articles to profl. jours.; author: Developmental Order: Its Origin and Regulation, 1982; assoc. editor: Ann. Rev. Plant Physiology, 1972-82. Recipient Darbaker Prize, 1964; Pelton award, 1974. Mem. Soc. for Developmental Biology, Bot. Soc. Am. Democrat. Congregationalist. Subspecialties: Developmental biology; Plant growth. Current work: Morphogenesis - the initiation of new organs in plants; control of cell division planes, control of microtubule orientation. Home: 997 Cottrell Way Stanford CA 94305 Office: Stanford U Dept Biol Sci Stanford CA 94305

GREEN, PAUL ELIOT, JR., electrical engineer; b. Durham, N.C., Jan. 14, 1924; s. Paul Eliot and Elizabeth Atkinson (Lay) G.; m. Dorrit L. Gegan, Oct. 30, 1948; children: Dorrit Green Rodemeyer, Nancy E., Judy J., Paul M., Gordon M. A.B., U. N.C.-Chapel Hill, 1943; M.S., N.C. State U., 1948; Sc.D., MIT, 1953. Group leader MIT Lincoln Lab., Lexington, 1951-69; sr. mgr. IBM Research Div., Yorktown Heights, N.Y., 1969-80, mem. corp. tech. com., Armonk, N.Y., 1980—. Served to lt. comdr. USNR, 1943-60; ret. Named Disting. Engring. Alumnus N.C. State U., 1983. Fellow IEEE (Aerospace Pioneer award 1981); mem. Nat. Acad. Engring. Subspecialty: Distributed systems and networks. Current work: Advanced protocols for architectures for large dynamic computer networks. Home: Roseholm Pl Mount Kisco NY 10549 Office: IBM Old Orchard Rd 3A-57 Armonk NY 10504

GREEN, RICHARD FREDERICK, astronomer; b. Omaha, Feb. 13, 1949; s. Jack Maxwell and Bernice (Bordy) G.; m. Joan Auerbach, June 16, 1974; children: Alexander Simon, Nathaniel Martin. A.B., Harvard Coll., 1971; Ph.D. in astronomy, Calif. Inst. Tech., 1977. Research fellow in astronomy Hale Obs., Calif. Inst. Tech., 1977-79; asst. astronomer Steward Obs., U. Ariz., Tucson, 1979—. Author: The Palomar-Green Catalogue. Mem. Am. Astron. Soc., Astron. Soc. Pacific, Phi Beta Kappa. Jewish. Subspecialties: Optical astronomy; Ultraviolet high energy astrophysics. Current work: Optical and space ultraviolet spectroscopy of quasars and hot stars; studies of faint clusters of galaxies around quasars; wide-field photographic surveying for rare stellar ojects. Office: Steward Observatory University of Arizona Tucson AZ 85721

GREEN, ROBERT EDWARD, JR., educator, physicist; b. Clifton Forge, Va., Jan. 17, 1932; s. Robert Edward and Hazle Hall (Smith) G.; m. Sydney Sue Truitt, Feb. 1, 1962; children: Kirsten Adair, Heather Scott. B.S., William and Mary Coll., 1953; Ph.D., Brown U., 1959; postgrad. (Fulbright grantee), Aachen (Germany) Technische Hochschule, 1959-60. Physicist underwater explosions research div. Norfolk (Va.) Naval Shipyard, 61959; asst. prof. mechanics Johns Hopkins, Balt., 1960-65, asso. prof., 1965-70, prof., 1970—, chmn. mechanics dept., 1970-72, chmn. mechanics and materials sci. dept., 1972-73, chmn. civil engring./materials sci. and engring. dept., 1979—; Ford Found. resident sr. engr. RCA, Lancaster, Pa., 1966-67; cons. U.S. Army Ballistic Research Labs., Aberdeen Proving Ground, Md., 1973-74; physicist Center for Materials Sci., U.S. Nat. Bur. Standards, Washington, 1974—. Author: Ultrasonic Investigation of Mechanical Properties (Treatise on Materials Science and Technology, Vol. 3), 1973; also articles. Mem. Am. Phys. Soc., Acoustical Soc. Am., Am. Inst. Mining, Metall. and Petroleum Engrs., Am. Soc. for Metals, Am. Soc. Nondestructive Testing, A.A.A.S., Sigma Xi, Tau Beta Pi, Alpha Sigma Mu, Sigma Nu. Methodist. Subspecialties: Materials; Materials (engineering). Current work: Nondestructive characterization of materials; nondestructive evaluation of surgical implants, x-ray topographic examination of materials using synchrotron radiation. Research in underwater shock waves, recovery, recrystallization, elasticity, plasticity, crystal growth and orientation, X-ray diffraction, electro-optical systems, linear and non-linear elastic wave propagation, light-sound interactions, high-power ultrasonics, ultrasonic attenuation, dislocation damping, fatigue, acoustic emission, non-destructive testing, polymers, biomaterials. Home: 936 Ellendale Dr Towson MD 21204 Office: Materials Sci and Engring Dept Johns Hopkins U Baltimore MD 21218

GREEN, VICTOR EUGENE, JR., research agronomist, educator; b. De Ridder, La., Sept. 3, 1922; s. Victor Eugene and Laura Mae (Harris) G.; m. Ada Ruth Hellert, June 5, 1945; children: Judy Ellen Green Brewer, Philip Martin. B.S., La. State U., 1947, M.S., 1948; Ph.D., Purdue U., 1951. Cert. Am. Soc. Agronomy. Asst. prof. La. Agrl. Expt. Sta., Baton Rouge, 1948-49; from asst. to assoc. prof. Fla. Agrl. Expt. Sta., Everglades, 1951-65, prof., 1968-70, prof./agronomist, Gainesville, 1970—; prof./adv. AID, San Jose, Costa Rica, 1965-68; cons. World Bank, Jamaica Sch. Agr., others. Recipient diploma of honor Costa Rica Ministry of Agr., 1968, PCCMCA, Guatemala, 1980. Mem. Soil and Crop Sci. Soc. Fla. (pres. 1965), Am. Soc. Agronomy, Internat. Sunflower Assn., Crop Sci. Soc. Am., Soil Sci. Soc. Am. Republican. Lutheran. Subspecialties: Plant physiology (agriculture); Sunflower crop production. Current work: Oil production and fatty acid composition through nuclear magnetic resonance and gas-liquid chromatography techniques. Home: 3915 SW 3d Ave Gainesville FL 32607 Office: U Fla Bldg 857 Gainesville FL 32611

GREEN, WILLIAM ROBERT, immunologist, educator; b. Toledo, Jan. 25, 1950; s. Robert Maynard and Evelyn Marie (Rupp) G.; m. Kathy Ann Hutchinson, Aug. 28, 1971; children: Andrea, Matthew. B.S., U. Mich., 1972; Ph.D., Case Western Res. U., 1977. Asst. mem. Fred Hutchinson Cancer Research Ctr., Seattle, 1980-83; research asst. prof. microbiology U. Wash., Seattle, 1981-83; asst. prof. dept. microbiology Dartmouth Med. Sch., Hanover, N.H., 1983—. Mem. Am. Assn. Immunologists, Phi Beta Kappa. Subspecialties: Immunobiology and immunology; Cancer research (medicine). Current work: Study of T lymphocyte-mediated cellular immune responses to syngeneic murine leukemia-virus-induced tumors. Office: Dept Microbiology Dartmouth Med Sch Hanover NH 03756 Home: 6 Fletcher Circle Hanover NH 03755

GREENAWAY, FREDERICK THOMAS, chemistry educator; b. Rakaia, N.Z., Aug. 18, 1947; came to U.S., 1973; s. Norman and Thelma Beryl (Pluck) G. B.Sc. with honors, U. Canterbury, Christchurch, N.Z., 1969, Ph.D., 1973. Research assoc. Mich. State U., East Lansing, 1973-74; research assoc. Syracuse (N.Y.) U., 1974-78, vis. asst. prof., 1978-80; asst. prof. chemistry Clark U., Worcester, Mass., 1980—. Contbr. articles in field to profl. jours. Mem. Am. Chem. Soc., Internat. Soc. Magnetic Resonance, N.Y. Acad. Scis. Clubs: Rugby (coach 1981-82), Mass. Dance Ensemble (Worcester, Mass.) (dir. 1981-82). Subspecialties: Biophysical chemistry; Biochemistry (biology). Current work: Bioinorganic chemistry with emphasis on metal-drug interactions and metalloenzymes, nuclear magnetic resonance and electron paramagnetic resonance of metal complexes. Office: Dept Chemistry Clark U Worcester MA 01610

GREENBERG, BERNARD *See also* **GREENBERG, NEIL**

GREENBERG, HAROLD PAUL, electronics co. exec.; b. Balt., May 10, 1933; s. Harry and Pauline Henrietta (Levin) G.; m. Lois Ann Lavine, Feb. 2, 1958; 1 dau., Roberta. A.S., Northeastern U., 1961, B.S., 1964, M.B.A., 1970. Registered profl. engr., Mass. Sr. engr. GTE Sylvania, Needham, Mass., 1965-70; sr. engr., mgr. vendor quality assurance Polaroid Corp., Cambridge, Mass., 1971-74; quality assurance sect. head GTE Sylvania, Needham, Mass., 1974-79; dir. reliability and regulatory affairs Analogic Corp., Wakefield, Mass., 1979—; cons. in field. mem. faculty North Shore Community Coll., Beverly, Mass. Served with U.S. Army, 1953-55. Fellow Am. Soc. Quality Control (cert. quality engr., cert. reliability engr.); mem. Engring. Socs. New Eng. (past pres.). Lodge: Masons. Subspecialties: Quality engineering; Electronics. Current work: Developing and implementing quality and reliability systems. Home: 6 Coe Rd Framingham MA 01701 Office: 14 Electronics Ave Danvers Industrial Park Danvers MA 01923

GREENBERG, KENNETH FREEMAN, computer co. exec.; b. Bay Shore, N.Y., Dec. 25, 1948; s. Seymour and June Edith (Freeman) G.; m. Jacqueline Ann O'Dell, Aug. 27, 1977; 1 son, Daniel Patrick. B.A. in Info. Scis, U. Calif., Santa Cruz, 1979. Data base group leader Dynalectron Corp., Norco, Calif., 1974-77; Sci. programmer Itek/Applied Tech., Sunnyvale, Calif., 1978-79; sr. engr. Advanced Micro Computers, Santa Clara, Calif., 1979-81, software engring. mgr., advanced micro devices, 1981—; cons., applications of microprocessors in control systems. Served with USN, 1968-73. Mem. IEEE Computer Soc., Assn. for Computing Machinery, PASCAL Users Group. Subspecialties: Software engineering; Computer architecture. Current work: Microprogramming, software tools, the relationship of high-level languages and computer architecture, software portability, local area networks. Home: 3118 Kermath Dr San Jose CA 95132 Office: 3340 Scott Blvd Santa Clara CA 95051

GREENBERG, LOWELL HERBERT, medical oncologist, hematologist; b. Bklyn., May 11, 1932; s. Max and Sarah (Dreibach) G.; m.; children: Kerin, Bradford, Michelle, Barrie, Brooke, Gregory. B.S. cum laude, Tufts U., 1953; M.D., NYU, 1957. Diplomate: Am. Bd. Internal Medicine. Intern and resident Bellevue Hosp., UCLA Ctr., Los Angeles, 1957-61; fellow New Eng. Med. Ctr., Boston, 1961-62, assoc. clin. prof. medicine UCLA Med. Ctr., 1963—, med. oncologist Hematology-Oncology Assocs., Torrance, Calif., 1963—; cons. physician UCLA, Harbor Gen. Hosp.; dir. dept. oncology Little Co. of Mary Hosp., Torrance. Contbr. articles to med. jours. Fellow ACP; mem Am. Soc. Hematology, Am. Soc. Clin. Oncology. Republican. Clubs: Palos Verdes (Calif.) Tennis; Casa de Vida (Torrance). Subspecialties: Chemotherapy; Hematology. Current work: Research in oncology and hematology with particular interest in the hematologic malignancies. Home: 617 Palos Verdes Dr West Palos Verdes CA 90274 Office: Hematology-Oncology Assocs 3440 W Lomita Blvd Torrance CA 90505

GREENBERG, NEIL (BERNARD GREENBERG), zoologist, educator; b. Newark, Oct. 30, 1941; s. Henry and Norma (Wexelman) G.; m. Alicia Carolyn Berry, June 29, 1969; 1 dau.: Haley Jessica Elise. B.A., Drew U., 1963; M.S., Rutgers U., 1968, Ph.D., 1973. Research ethologist NIMH, Bethesda, Md., 1973-78; research assoc. Mu. Comparative Zoology, Harvard U., 1977-82; asst. prof. zoology U. Tenn., Knoxville, 1978—; cons., lectr. in field. Contbr. articles to profl. jours.; editor: Behavior and Neurology of Lizards, 1978. NIMH Grant Found. fellow, 1973-75; 1st Tenn. Bank scholar, 1981; Danforth assoc., 1981. Mem. Am. Soc. Zoologists, Animal Behavior Soc., Soc. Neurosci., Sigma Xi. Subspecialties: Ethology; Comparative neurobiology. Current work: Reciprocal relationships of social behavior, stress, and repro endocrinology. Home: 1304 Avonmouth Rd Knoxville TN 37914 Office: Dept Zoology U Tenn Knoxville TN 37996

GREENBERG, RICHARD NEIL, medical educator, researcher; b. Washington, Oct. 16, 1947; s. Herman and Betty (Saks (Glassman)) G. B.A., Cornell U., 1968; M.D., Tufts U., 1972. Intern and med. resident Ind. U. Med. Sch., Indpls., 1972-74; epidemic intelligence officer Ctr. Disease Control, Atlanta, 1974-76; med. resident La. State U. Sch. Medicine, New Orleans, 1976-77, fellow in infectious diseases, 1977-79; research fellow U. Va., Charlottesville, 1979-81; asst. prof. medicine and microbiology St. Louis U., 1981—; cons. WHO, 1974-75, Sherwood Industries, St. Louis, 1983—. Contbr. chpts. to med. textbooks, articles to sci. and med. jours. Served as surgeon USPHS, 1974-76. So. Med. Assn. fellow, 1978; recipient Research award Nat. Found. Infectious Disease, 1982; numerous grants. Fellow ACP; mem. Infectious Disease Soc. Am., Am. Soc. Microbiology, Am. Fedn. Clin. Research, Central Soc. Clin. Research. Jewish. Subspecialties: Internal medicine; Infectious diseases. Current work: E. coli enterotoxins, streptococcal pyrogenic exotoxins, antibiotic efficacy both in vivo and in vitro, diarrheal diseases. Patentee in field. Home: 1281 Bentoak St Kirkwood MO 63122 Office: 1515 Lafayette Saint Louis MO 63104

GREENBERG, ROGER PAUL, clinical psychologist and practitioner, educator, researcher; b. N.Y.C., July 2, 1941; s. Matthew Robert and Bertha Sylvia (Judem) G.; m. Vivian Vicki Miller, Aug. 30, 1964; 1 son, Michael David. B.A., Bklyn. Coll., CUNY, 1963; M.S., Syracuse U., 1966, Ph.D., 1968. Cert. clin. psychologist, N.Y. Intern in psychology Syracuse (N.Y.) VA Med. Center, 1966-67; asst. prof. psychiatry SUNY Upstate Med. Center, Syracuse, 1968-71, assoc. prof., 1972-77, prof., 1978—, dir. psychology internship tng., 1971—; pvt. practice clin. psychology, 1969—. Author: (with Seymour Fisher) The Scientific Credibility of Freud's Theories and Therapy (Library Jour. award 1978), 1977 (Psychology Today award 1977); contbr. numerous articles to profl. jours.; editor: (with Seymour Fisher) The Scientific Evaluation of Freud's Theories and Therapy, 1978; manuscript editor various profl. jours., 1973—. USPHS fellow, 1963-64, 67-68; grantee, 1968-70, 73-76; VA grantee, 1975. Mem. Am. Psychol. Assn. (accreditation site visitor 1975—), N.Y. State Psychol. Assn., Central N.Y. Psychol. Assn., Nat. Register Health Service Providers in Psychology. Current work: Research in psychotherapy, personality, psychosomatic issues, psychodiagnosis. Office: Dept Psychiatry SUNY Upstate Med Center 750 E Adams St Syracuse NY 13210

GREENBERG, ROLAND, pharmacologist; b. Winnipeg, Man., Can., June 15, 1935; s. Hymie and Ida (Portnuff) G.; m. Arlene Susan, July 4, 1965; children: Daniel, Deborah. B.Sc., U. Man., 1960, M.Sc., 1964, Ph.D., 1968. Sr. pharmacologist Ayerst Research Labs., Montreal, Que., Can., 1970-71, research asso., 1971-77; sr. research investigator Squibb Inst. Med. Research, Princeton, N.J., 1977-81, research fellow, 1981—, assoc. dir. sci. liaison, 1983—. Contbr. articles to profl. jours. Can. Med. Research postdoctoral fellow, 1968-70. Mem. Pharmacol. Soc. Can., Am. Soc. Pharmacology and Exptl. Therapeutics. Subspecialty: Pharmacology. Current work: Pharmacology of prostaglandins and enkephalins. Home: 67 Randall Rd Princeton NJ 08540 Office: PO Box 4000 Princeton NJ 08540

GREENBERG, STAN SHIMEN, pharmacologist, educator, researcher; b. Bklyn., Sept. 14, 1945; s. Louis Meyer and Anna (Pinckosowitz) G.; m. Patricia Ann Powers, Oct. 13, 1978; 1 son, Jonathan Michael. B.S. magna cum laude, Bklyn. Coll. Pharmacy, L.I. U., 1968; M.S., U. Iowa, Ph.D., 1972. Postdoctoral fellow div. clin. pharmacology U. Iowa, Iowa City, 1972-73; NIH postdoctoral scholar dept. physiology U. Mich., Ann Arbor, 1973-74; instr. pharmacology and myocardial biology Baylor Coll. Medicine, 1974-75; asst. prof. pharmacology Ohio State U. Coll. Medicine, 1975-77; assoc. prof. pharmacology U.S. Ala. Coll. Medicine, Mobile, 1977—. Author: (with T.M. Glenn) Prostanoids in Cardiovascular and Cardiopulmonary Disease, 1981, Physiology of Smooth Muscle, 1982, Procs. Soc. Exptl. Biology and Medicine; mem. editorial bd.: Methods and Findings in Clin. and Exptl. Therapeutics; contbr. articles to profl. jours. NIH grantee, 1977—; Am. Heart Assn. grantee, 1975-78. Mem. Am. Fedn. Clin. research, Internat. Study Group for Research in Cardiac Metabolism, AAAS, Western Pharmacology Soc., Am. Soc. Pharmacology and Exptl. Therapeutics, Am. Physiol. Soc., Soc. Exptl. Biology and Medicine, High Blood Pressure Council Am. Heart Assn. (fellow med. adv. bd.), Mobile Area High Blood Pressure Council (v.p.), Shock Soc., Microcirculatory Soc., Sigma Xi. Jewish. Subspecialties: Pharmacology; Cellular pharmacology. Current work: Vascular smooth muscle function in hypertension, animal research to elucidate the pathogenesis of renal hypertension and coronary artery disease. Home: 47 Lakeview Dr Morris Plains NJ 07950 Office: Dept Physiology U Medicine and Dentistry of NJ Berley Labs Cedar Knolls NJ 07927

GREENBERG, STEPHEN ROBERT, pathology educator; b. Omaha, May 5, 1927; s. Nathan H. and Ruth (Levey) G.; m. Constance Betine Milder, June 4, 1952; children: Andrew Eugene, Nathan Henry. B.S., St. Louis U., 1951, M.S. in Anatomy, 1952, Ph.D. in Pathology, 1954. Assoc. in pathology Clarkson Hosp., Omaha, 1954-55; assoc. in pathology Chgo. Med. Sch., 1955-59, instr. in pathology, 1957-62, asst. prof. pathology, 1962-69, assoc. prof. pathology, 1969—; lectr. in pathology Cook County Grad. Sch. Medicine, Chgo., 1972—. Editor: Toxicologic Pathology; reviewer: Vet. Pathology. Recipient York Cross of Honor Ill. Priory No. 11, 1980. Fellow AAAS; mem. Am. Assn. Pathologists, Internat. Acad. Pathology, Am. Soc. Clin. Pathologists, Inst. of Medicine Chgo. Republican. Jewish. Lodges: Masons; K.T. Subspecialties: Pathology (medicine); Environmental toxicology. Current work: Tissue effects of environmental agents - formalim, fluoride, asbestos, changes induced by them on animal organs and tissues. Office: Dept Pathology Chicago Med School 3333 Green Bay Rd North Chicago IL 60064 Home: 418 Huron St Park Forest IL 60466

GREENBLATT, SAMUEL HAROLD, neurosurgeon; b. Potsdam., N.Y., May 16, 1939; s. Louis and Rose Leah (Clopman) G.; m. Judith Ruth Shapiro, June 23, 1963; children: Rachel, Daniel, Miriam. B.A., Cornell U., 1961, M.D., 1966; M.A., Johns Hopkins U., 1964. Diplomate: Am. Bd. Neurol. Surgery. Intern Boston City Hosp., 1966-67; resident in neurology Boston VA Hosp., 1967-68; resident in neurol. surgery Dartmouth Affiliated Hosp., Hanover, N.H., 1970-74; hon. sr. registrar Nurosurg. unit Guy's, Maudlsey and King's Coll. Hosps., London, 1972; instr. neurol. surgery Albert Einstein Coll. Medicine, Bronx, N.Y., 1974-77; asst. attending neurol. surgeon Bronx Mcpl. Hosp. Ctr., 1974-77; asst. prof. neurol. surgery Med. Coll. Ohio, Toledo, 1977-80, assoc. prof., 1980—, staff neurosurgeon, 1977—; clin. asst. neurol. surgery St. Barnabas Hosp., Bronx, 1975, research asst. neurol. surgery, 1976-77; assoc. staff neurosurgeon Mercy Hosp., Toledo, 1977-80, courtesy staff neurosurgeon, 1980—. Contbr. articles to profl. jours. and books. Served with U.S. Army, 1968-70. USPHS fellow, 1963-64; Tiffany Blake fellow, 1972-73. Fellow A.C.S.; mem. History Sci. Soc., Am. Assn. History Medicine, Internat. Neuropsychol. Soc., Am. Epilepsy Soc., Soc. health and Human Values, Soc. Neurosci., N.Y. Acad. Scis., Acad. Medicine Toledo, Ohio State Med. Assn., Behavioral Neurology Soc., Ohio State Neurosurg. Soc., Am. Assn. Neurol. Surgeons, Congress Neurol. Surgeons. Jewish. Subspecialty: Neuropsychology. Current work: Primary research concerns the anatomical correlates of neurobehavioral abnormalities (especially alexia) and normal brain substratum of cognitive behavior. Office: Medical College of Ohio Department of Neurosciences CS # 10008 Toledo OH 43699

GREENE, ARTHUR EDWARD, physicist; b. Chgo., Dec. 10, 1945; s. Shirley Edward and Ellen Catherine (Tweedy) G.; m. Nancy Ellen Green, Sept. 12, 1970; 1 dau., Ellen Dorothy. Student, Doane Coll., Crete, Nebr., 1963-65; B.S., Ohio State U., 1967, Ph.D. in astronomy, 1971. Staff mem. Los Alamos Nat. Lab., 1975-81, staff mem. thermonuclear applications group, 1981—. Contbr. articles in field to profl. jours. Served with USAF, 1971-75. Mem. Am. Phys. Soc., Phi Beta Kappa. Subspecialties: Atomic and molecular physics; Plasma physics. Current work: Plasma physics and radiation transport, laser physics, stellar atmospheres of evolved stars. Office: MS-B220 Los Alamos National Laboratory Los Alamos NM 87545

GREENE, BRUCE MCGEHEE, research internist, educator; b. Auburn, Ala., Nov. 11, 1945; s. James Etheridge and Mary Linton (McGehee) G.; m. Theo Elmore Newsom, June 13, 1967; children: James Newson, Duncan McGehee, Robert Bruce. B.A. in Chemistry, U. of South, 1967; M.D., Johns Hopkins U., 1971. Diplomate: Am. Bd. Internal Medicine. Intern in medicine Johns Hopkins U., 1971-72, resident in medicine, 1972-73, asst. chief med. service, 1977-78; asst. prof. medicine Johns Hopkins Med. Instns., 1977-79; fellow in infectious diseases Vanderbilt U., 1973-75; asst. prof. medicine Case Western Res. U. and Univ. Hosps., 1979-83, assoc. prof., 1983—. Contbr. articles to profl. jours. Served as lt. comdr. Internal Medicine Service USN, 1975-77. USPHS-NIH grantee, 1979—. Fellow ACP; mem. Am. Assn. Immunologists, Am. Fedn. Clin. Research, Infectious Disease Soc. Am., Am. Soc. Tropical Medicine and Hygiene, Royal Soc. Tropical Medicine and Hygiene. Club: Pithotomy (Balt.). Subspecialties: Infectious diseases; Internal medicine. Current work: Immunology of chronic infections; chemotherapy of onchocerciasis; parasite chemotherapy; infectious diseases; internal medicine. Office: Dept Medicine Univ Hosps Case Western Res U Cleveland OH 44106

GREENE, CHRIS H., physics educator; b. Lincoln, Nebr., Aug. 1, 1954; s. William Henry and Helen (Kiesselbach) G.; m. Christy Ann, Sept. 15, 1977. B.S., U. Nebr., 1976; M.S., U. Chgo., 1977, Ph.D., 1980. Research assoc. Stanford (Calif.) U., 1980-81; asst. prof. physics and astronomy La. State U., Baton Rouge, 1981—. Contbr. articles in field to profl. jours. IBM fellow, 1979-80. Mem. Am. Phys. soc. Subspecialties: Atomic and molecular physics; Theoretical physics. Current work: Research on basic properties of atoms and molecules, electron correlation, photoionization and quantum defect theory. Office: La State U Dept Physics and Astronomy Baton Rouge LA 70803

GREENE, ELIAS LOUIS, medical researcher, consultant; b. N.Y.C.; s. Sidney and Gussie (Triebwasser) G.; m. Lena Greene, Apr. 3, 1953; children: Summer, Francesca. B.A., Bklyn. Coll., 1953; Ph.D., Cornell U., 1964. Registered clin. chemist Nat. Registry Clin. Chemistry; cert. clin. lab. dir. N.Y.C. Research asst. Sloan-Kettering Inst. Cancer Research, N.Y.C., 1955-64; virologist/immunologist Inst. Med. Research, Henry Putnam Meml. Hosp., Bennington, Vt., 1964-65; clin. lab. technologist Mercy Hosp., Rockville Centre, N.Y., 1960-67; lectr. N.Y. Community Coll., Bklyn., 1967; research immunologist L.I. Jewish Hosp., New Hyde Park, 1965-67; instr. dept. pediatrics U. Miami (Fla.) Sch. Medicine, 1967-69, asst. prof., 1969-73; asst. dir. diagnostics evaluation Berhing Diagnostics div. Am. Hoechst Corp., Somerville, N.J., 1973-77; asst. dir. med. ops. Cutter Labs., Inc., Berkeley, Calif., 1977-79, mgr. clin. surveillance and biostats., 1979—; clin. lab. technologist Miami Heart Inst., Miami Beach, 1967-68, cons. immunologist organ transplant program, 1968-70; dir. quality control, assoc. lab. dir. N.Am. Biologicals, Inc., Miami, 1970-73; cons. immunologist United Labs/Biodiagnostics Labs., Pelham Manor, N.Y., 1974-77. Contbr. articles to sci. jours. Served in U.S. Army, 1953-55. Nat. Cystic Fibrosis Research Found. fellow, 1968; USPHS fellow, 1959-64. Mem. AAAS, Am. Soc. Microbiology, Harvey Soc., N.Y. Acad. Sci., Am. Assn. Clin.Chemists, Am. Soc. Epidemiology, Am. Soc. Quality Control, Am. Assn. Bioanalysts, Am. Fedn. Clin. Research, Soc. Epidemioloic Research, Am. Assn. Immunologists, Internat. Assn. Biol. Standardization, Tissue Culture Assn., Assn. Clin. Scientists, Am. Assn. Clin. Histocompatibility Testing, U.S. profl. Tennis Assn. Club: Mira Vista Country (El Cerrito, Calif.). Subspecialties: Immunobiology and immunology; Behaviorism. Home: 2800 Claremont Blvd Berkeley CA 94705 Office: 2200 Powell St Emeryville CA 94608

GREENE, FREDERICK LESLIE, surgeon, educator; b. Norfolk, Va., Dec. 18, 1944; s. William Joseph and Theresa (Davis) G.; m. Donna W. Greene, June 21, 1970; children: Stephanie, Adam. B.A., U. Va., 1966, M.D., 1970. Diplomate: Am. Bd. Surgery. Assoc. prof. surgery U. S.C., Columbia, 1980—; chief surg. service Dorn VA Hosp., Columbia, 1980—; chmn. profl. edn. State S.C. Am. Cancer Soc., 1982—. Contbr. articles to profl. jours. Served with USN, 1976-78. Fellow Southeastern Surg. Congress; mem. Assn. Acad. Surgery, AMA. Subspecialties: Surgery; Oncology. Current work: Clinical and research dealing with breast and colon cancer. Office: Dept Surgery U SC 3321 Medical Park Rd Columbia SC 29203

GREENE, JOHN CLIFFORD, dentist, university dean; b. Ashland, Ky., July 19, 1926; s. G. Norman and Ella R. G.; m. Gwen Rustin, Nov. 17, 1957; children: Alan, Lisa, Laura. A.A., Ashland Jr. Coll., 1947; student, Marshall Coll., 1948; D.M.D., U. Louisville, 1952, Sc.D. (hon.), 1980; M.P.H., U. Calif., Berkeley, 1961; Sc.D. (hon.), U. Ky., 1972, Boston U., 1975. Diplomate: Am. Bd. Dental Public Health (pres.). Intern USPHS Hosp., Chgo., 1952-53, staff, San Francisco, 1953-54; asst. regional dental cons. Region IX, San Francisco, 1954-56, asst. to chief dental officer, Washington, 1958-60; chief epidemiology program Dental Health Center, 1961-66; dep. dir. Div. Dental Health, 1966-70, acting dir., 1970, dir., 1970-73; acting dir. Bur. Health Resources Devel., 1973-74, dir., 1974-75; chief dental officer USPHS, 1974-81, dep. surgeon gen., 1978-81; with Epidemic Intelligence Service, Communicable Disease Center, Altanta and Kansas City, Mo., 1956-57; epidemiology and biometry br. Nat. Inst. Dental Research, NIH, Bethesda, Md., 1957-58; dean. Sch. Dentistry,U. Calif., San Francisco, 1981—; spl. cons. WHO, India, 1957; faculty Calif., U. Mich., U. Pa.; cons. Am. Dental Assn. Council, Nat. Health Professions Placement Network. Contbr. writings to profl. publs. Served with USN, 1945-46. Recipient citation Sch. Grad. Dentistry Boston U., 1971, U. of the Pacific, 1977, Meritorious and Disting. Service awards HEW, 1972, 75, Outstanding Alumnus award U. Louisville, 1980, award of merit FDI, 1978. Fellow Am. Coll. Dentists; mem. ADA, Calif. Dental Assn., San Francisco Dental Soc., Internat. Assn. Dental Research, Am. Assn. Public Health Dentists, Am. Acad. Periodontology, Am. Assn. Dental Schs. (v.p.), Inst. of Medicine of Nat. Acad. Sci., Federation Dentaire Internationale (chmn. commn. on public dental health, mem. WHO panel of experts on dental health), Omicron Kappa Upsilon, Delta Omega. Subspecialties: Preventive dentistry; Epidemiology. Current work: Changes in disease patterns affecting dentistry; dental care delivery systems. Home: 103 Peacock Dr San Rafael CA 94901 Office: U Calif Sch Dentistry: San Francisco CA 94143

GREENE, MURRAY A., cardiologist.; b. Bklyn., May 20, 1927; s. Max and Beatrice (Kolomer) G.; m. Eileen Smolkin, Dec. 19, 1953; children: Barry T., Larry B. A.B., NYU, 1948; M.D., Columbia U., 1952. Diplomate: Am. Bd. Internal Medicine. Intern Maimonides Hosp., Bklyn., asst. resident, 1953-54; resident in cardiology Montefiore Hosp., N.Y.C., 1955-56; research fellow Cardiopulmonary Lab., Maimonides Hosp. and SUNY-Downstate Med. Center, 1954-55, 56-57; chief div. cardiovascular diseases Bronx-Lebanon Hosp., N.Y.C., 1957-71, dir. intensive care unit and Cardiopulmonary Lab., 1957-71, attending physician dept. medicine, 1957—; asst. clin. prof. medicine Albert Einstein Coll. Medicine, N.Y.C., 1972—. Contbr. numerous articles to med. jours. Served with U.S. Army, 1945-47. Fellow ACP, Am. Coll. Chest Physicians, Am. Fedn. Clin. Research; mem. AMA, Am. Physicians Fellowship Assn., N.Y. Acad. Scis., Am. Heart Assn., N.Y. Heart Assn., Am. Soc. Internal Medicine, N.Y. Soc. Internal Medicine, Med. Soc. State N.Y., Phi Beta Kappa. Subspecialties: Cardiology; Physiology (medicine).

GREENE, ROBERT JAY, surgeon, oncologist; b. Phila., Jan. 17, 1930; s. Samuel Robert and Kathryn (Purisch) G.; m. Barbara Lee Bassin, May 21, 1962; 1 son, David; m.; children by previous marriage; Steven, Linda, Karen. A.B., U. Pa., 1950; M.D., Harvard U., 1954. Diplomate: Am. Bd. Surgery. Intern Univ. Hosps. of Clev., 1954-55, resident, 1955-56, Beth Israel Hosp., Boston, 1958-62; practice medicine specializing in gen. surgery and oncology, New Bedford, Mass., 1962—; clin. asst. in surgery Harvard Med. Sch., 1979—; mem. staff N.E. Deaconess Hosp., Boston, 1979—. Contbr. numerous articles to profl. jours. Served to capt. U.S. Army, 1956-58. Fellow ACS; mem. Am. Soc. Clin. Oncology, Mass. Med. Soc., Phi Beta Kappa, Alpha Omega Alpha. Jewish. Clubs: Harvard (Boston); N.B. Yacht (South Dartmouth, Mass.). Subspecialties: Surgery; Chemotherapy. Current work: Clinical research and therapy in cancer, plus active private practice in general surgery. Home: 49 Hawthorn St New Bedford MA 02740 Office: New Bedford Surg Assocs Inc 49 Hawthorn St New Bedford MA 02740

GREENE, ROBERT MORRIS, anatomy educator, college dean; b. Dorpen, Germany, Dec. 15, 1945; came to U.S., 1953, naturalized, 1953; s. Samuel and Henryka (Janssen) G.; m. Charlain Ann Ransom, Aug. 5, 1967; children: Caroline, James. B.A., Utica Coll., Syracuse U., 1967; Ph.D., U. Va., 1974. Postdoctoral fellow NIH, Bethesda, Md., 1974-76, staff fellow, 1976-78; asst. prof. anatomy Thomas Jefferson U., 1978-82, assoc. prof., 1982—, asst. dean, 1982—. Contbr. articles to profl. jours. Research fellowship NIH, 1974, 75; Research Career Devel. awardee, 1981; grantee, 1981, 83. Mem. AAAS, Am. Soc. Cell Biology, Soc. Devel. Biology, Teratology Soc., Internat Assn. Dental Research, Am. Assn. Dental Research, Sigma Xi. Subspecialties: Developmental biology; Teratology. Current work: Analysis of normal and abnormal craniofacial development; role of cyclic nucleotides, prostaglandins and B-adrenergic receptors in regulating development of the palate. Office: Jefferson Med Coll Dept Anatomy 1020 Locust St Philadelphia PA 19107

GREENER, EVAN H., biological materials educator, researcher; b. Bklyn., Sept. 8, 1934; m., 1957; 2 children. B.Met.E., Poly. Inst. Bklyn., 1955; M.S., Northwestern U., 1957, Ph.D. in Materials Sci, 1960. Lab. instr. Poly. Inst. Bklyn., 1955; Aitcheson fellow in metallurgy Northwestern U., 1955-58, research fellow, 1958-59, prin. investigator, 1959-60, assoc. prof. biol. materials, 1964-69, prof., 1969—, chmn. dept. biol. materials, 1964—; from asst. prof. to assoc. prof. materials sci. Marquette U., 1960-64. Fogarty Internat. fellow Turner Dental Sch., U. Manchester, 1977-78. Fellow AAAS; mem. N.Y. Acad. Sci., Am. Soc. Metals, AIME, Internat. Assn. Dental Research. Subspecialty: Biomaterials. Office: Dept Biol Materials Northwestern U 311 E Chicago Ave Chicago IL 60611

GREENFELD, NORTON ROBERT, info. scientist; b. Miami, Fla., June 14, 1945; s. David and Lillian G.; m. Ellen Dewald; 1 dau.: Lee. B.S. in Math, Calif. Inst. Tech., 1967, M.S. in Info. Sci, 1968, Ph.D., 1972. Research fellow, div. humanities and social scis. Calif. Inst. Tech., Pasadena, 1972-73; with Info. Scis. Inst., U. Soc. Calif., Los Angeles, 1973-75, Lab. for Computer Sci., M.I.T., Cambridge, 1975-77; sr. scientist, artificial intelligence dept. Bolt Beranek and Newman Inc., Cambridge, 1977-82; with Applied Expert Systems, Inc., Cambridge, 1982—. Mem. Assn. for Computing Machinery, Am. Assn. for Artificial Intelligence, Sigma Xi. Subspecialty: Artificial intelligence. Current work: Effective knowledge systems. Office: Applied Expert Systems Five Cambridge Ctr Cambridge MA 02142

GREENFIELD, IRWIN GILBERT, college dean; b. Phila., Nov. 30, 1929; s. William and Sara (Baumbor) G.; m. Barbara Shapiro, June 16, 1951; children: Richard, Hermine, Steven. A.B. in Metallurgy, Temple U., Phila., 1951; M.S. in Metall. Engring, U. Pa., 1954, Ph.D., 1962. Registered profl. engr., Del. Metallurgist, Naval Air Exptl. Sta., Phila., 1951-53; sr. scientist Franklin Inst. Labs. Research and Devel., Phila., 1953-63; mem. faculty U. Del., Newark, 1963—, prof. metallurgy and mech./aerospace engring., 1968—, dean Coll. Engring., 1975—, interim dir. Materials Durability Ctr.; vis. lectr. univs. in Japan, 1965; vis. prof. Stanford U., 1969—, Oxford U., 1970—, Tech. U. Eindhoven, Netherlands, 1978; mem. uniform exams. com. Nat. Council Engring. Examiners, 1981—. Author papers in field., Research surfaces, electron microscopy, mech. properties, fatigue, diffused surface layers, erosion, wear, photovoltaic materials. Mem. U.S. Senator Roth's Energy Advisory Com., Del. Energy Resources Commn. conf.; keynote speaker. Chmn. exec. com. Del. Program Minority Engrs., 1974—. Grantee U. Del., 1963-64, NSF, 1964-69, 70-74, 78-81, Air Force Office Sci. Research, 1970-74, NASA, 1969-70; NSF travel grantee, 1965, 73. Mem. Am. Inst. Mining, Metall. and Petroleum Engrs., Electron Microscope Soc. Am., Am. Soc. Metals, Am. Soc. Engring. Edn., Soaring Assn. Am., Del. Assn. Profl. Engrs. (council), Sigma Xi. Subspecialties: Materials (engineering); Materials. Current work: Effect of diffused coatings on dislocation motion; fundamental study of erosion and wear; methods of improving fatigue life of metals. Home: 605 Country Club Rd Newark DE 19711 Office: Coll Engring Univ Del Newark DE 19711

GREENFIELD, STANLEY MARSHALL, consulting company executive, researcher; b. N.Y.C., Apr. 16, 1927; s. Harry William and Millie (Jaller) G.; m. Rhoda Claire Barish, Sept. 1, 1951; children: Diane, David. B.S., NYU, 1950; Ph.D., UCLA, 1967. Head environ. scis. dept. Rand Corp., Santa Monica, Calif., 1950-71; asst. adminstr. for research and devel. EPA, Washington, 1971-74; sr. mem. tech. staff Flow Resources Corp., San Rafael, Calif., pres., 1974-78; sr. v.p. and tech. dir. Teknikron Research Ins., Berkeley, Calif., 1978-81; pres. Systems Applications, Inc., San Rafael, 1981—; Mem. various com. and panels Nat. Acad. Scis., 1960-83; mem. adv. bds. U.S. Air Force Sci. Adv. Bd., 1959-70, Space sci. panels NASA, 1957-70; mem. adv. sci. and tech. panel Calif. State Legislature, Sacramento, 1969-71. Contbr. over 30 sci. articles to profl. publs. Served with USN, 1943-45. Fellow Am. Meteorol. Soc. (Spl. award 1961); mem. Pan Am. Med. Assn., AAAS, Internat. Acad. Environ. Safety. Subspecialties: Meteorologic instrumentation; Air pollution dispersion. Current work: Environmental science with particular emphasis on air pollution; acid deposition; energy impacts; risk assessment; data base management. Home: 133 Knollwood Dr San Rafael CA 94901 Office: 101 Lucas Valley Rd San Rafael CA 94903

GREENFIELD, SYDNEY STANLEY, botany educator; b. Bklyn., Nov. 28, 1915; s. Max and Gussie (Weiss) G. B.A., Bklyn. Coll., 1936; M.A., Columbia U., 1937, Ph.D., 1941. Research assoc. plant physiology Columbia U., N.Y.C., 1941-46; asst. prof. biology Rutgers U., Newark, 1946-49, assoc. prof., 1949-59, chmn. botany dept., 1961-72, prof. botany, 1959—, chmn. Newark campus planning com., chmn. landscaping com. Contbr. articles in botany to profl. jours. Mem. AAAS, Bot. Soc. Am. Subspecialties: Plant physiology (biology); Plant growth. Current work: Teaching courses in biology of seed plants, physiology of plant growth, and economic botany. Office: Dept Botany Rutgers U Newark NJ 07102

GREENGARD, PAUL, biochemist, pharmacology educator; b. N.Y.C., Dec. 11, 1925; m., 1954; 2 children. A.B., Hamilton Coll., 1948; Ph.D. in Biophysics and Biochemistry, Johns Hopkins U., 1953. Dir. dept. biochemistry Geigy Research Labs., N.Y., 1958-67; dir. dept. neuropharmacology Inst. Basic Research in Mental Retardation, N.Y. State Dept. Mental Hygiene, 1967-68; prof. pharmacology Sch. Medicine, Yale U., 1968—; Andrew D. White prof.-at-large Cornell U., 1981—; vis. scientist Nat. Heart Inst., 1958-59; vis. assoc. prof. Albert Einstein Coll. Medicine, 1961-68, vis. prof., 1968—, Vanderbilt U., 1967-68, Lamson Meml. lectr., 1974; Louis B. Flexner lectr. U. Pa., 1976; first Disting. lectr. Soc. Gen. Physiologists, Internat. Congress Physiol. Sci., Paris, 1977; assoc. Neurosci. Research Program; lectr. Harvey Soc., 1980. Assoc. editor: Jour. Cyclic Nucleotide Research, 1974—. Recipient Mayor's Gold medallion City Milan; Dickson prize and medal U. Pitts., 1977; Ciba-Geigy Drew award, 1979; Biol. and Med. Sci. award N.Y. Acad. Sci., 1980; NSF fellow Inst. Psychiatry, U. London, 1953-54; Nat. Found. Infantile Paralysis fellow Molteno Inst., Cambridge U., 1954-55; Paraplegia Found. fellow Nat. Inst. Med. Research Eng., 1956-58. Fellow Am. Coll. Neuropsychopharmacology; mem. Nat. Acad. Sci., Am. Acad. Arts and Sci., Soc. Gen. Physiologists. Subspecialties: Neuropharmacology; Biochemistry (biology). Home: Dept Pharmacology Yale U Sch Medicine New Haven CT 06510

GREENHOUSE, HAROLD MITCHELL, physical chemist; b. Chgo., Nov. 20, 1924; s. Jacob and Frances (Perlman) G.; m. Rosalyne Joseph, June 13, 1948; children: Joel, Vickie, Jack. B.Sc., Ohio State U., 1948, M.Sc., 1951. Research assoc. Ohio State U. Research Found., 1950-53; research chemist Ferroxcube Corp., Saugerties, N.Y., 1953-55; dir. materials research Alladin Electronics, Nashville, 1955-59; sr. staff engr. Bendix Communications, Balt., 1959—. Contbr. articles to profl. jours. Chmn. edn. com. Balt. C. of C., 1968-71. Mem. Internat. Soc. Hybrid Microelectronics, Am. Vacuum Soc., IEEE, Assn. Computing Machinery, Optical Soc. Am., Am. Crystallographic Soc., Sigma Xi. Jewish. Subspecialties: Microelectronics; Electronic materials. Current work: Hybrid microelectronic technology. Patentee in field. Home: 2605 Summerson Rd Baltimore MD 21209 Office: Bendix Communications E Joppa Rd Baltimore MD 21204

GREENKORN, ROBERT ALBERT, chemical engineering educator; b. Oshkosh, Wis., Oct. 12, 1928; s. Frederick John and Sophie (Phillips) G.; m. Rosemary Drexler, Aug. 16, 1952; children: David Michael, Eileen Anne, Susan Marie, Nancy Joanne. Student, Oshkosh State Coll., 1951-52; B.S., U. Wis., 1954, M.S., 1955, Ph.D., 1957. Postdoctoral fellow Norwegian Tech. Inst., 1957-58; research engr. Jersey Prodn. Research Co., Tulsa, 1958-63; lectr. U. Tulsa, 1958-63; asso. prof. theoretical and applied mechanics Marquette U., Milw., 1963-65; asso. prof. Sch. Chem. Engring., Purdue U., Lafayette, Ind., 1965-67; prof., head Sch. Chem. Engring., 1967-72, asst. dean engring., 1972-76; asso. dean engring., dir. Engring. Expt. Sta., 1976-80; v.p., asso. provost, v.p. for programs Purdue Research Found., 1980—. Author: (with D.P. Kessler) Transfer Operations, 1972, (with K.C. Chao) Thermodynamics of Fluids: An Introduction to Equilibrium Theory, 1975, (with D.P. Kessler) Modeling and Data Analysis for Engineers and Scientists, 1980, Flow Phenomena in Porous Media, 1983; Contbr. articles to profl. jours. Served with USN, 1946-51. Decorated D.F.C., Air medal with two oak leaf clusters. Fellow Am. Inst. Chem. Engrs.; Mem. Soc. Petroleum Engrs., Am. Inst. Mining, Am. Soc. Engring. Edn., Metall. and Petroleum Engrs., Am. Chem. Soc., Am. Geophys. Union, Sigma Xi, Phi Eta Sigma, Tau Beta Pi, Phi Gamma Delta. Roman Catholic. Subspecialties: Chemical engineering; Petroleum engineering. Current work: Flow phenomena in porous media; thermo dynamics of fluids; coal liquefaction; environmental modeling. Patentee in field. Home: 151 Knox Dr West Lafayette IN 47906

GREENLEE, LORANCE LISLE, patent lawyer; b. Oskaloosa, Iowa, Apr. 12, 1935; s. Max Russell and Helen Crane (Lisle) G.; m. Carol Lacher, Mar. 1954; children: Russell, Ross; m. Barbara Bowman Greenlee, May 30, 1971; children: Allison, Travis. A.B., U. Colo., 1957; Ph.D., Duke U., 1962; J.D., U. Utah, 1976. Bar: Bar: Utah 1976, D.C 1977, U.S. Supreme Ct 1980. Postdoctoral fellow Duke U., Durham, N.C., 1962-63, Calif. Inst. Tech., Pasadena, 1963-66; asst. prof. biology U. Utah, Salt Lake City, 1966-70, assoc. prof., 1970-73; law clk. Utah Supreme Ct., 1975; assoc. Irons & Sears, P.C., Washington, 1976-79, Keil & Witherspoon, 1979-82; chief patent counsel Agrigenetics Corp., Denver, 1982—. NIH grantee, 1966-71. Mem. ABA, Am. Soc. Microbiology, AAAS, N.Y. Acad. Scis.

Subspecialties: Plant genetics; Molecular biology. Current work: Provide patent protection for research in genetics, molecular biology and other areas of biotechnology. Office: Agrigenetics Corp 14142 Denver West Pkwy Denver CO 80401

GREENLICK, MERWYN RONALD, health services researcher; b. Detroit, Mar. 12, 1935; s. Emanuel and Fay (Ettinger) G.; m. Harriet Cohen, Aug. 19, 1956; children—Phyllis, Michael, Vicki. B.S., Wayne State U., 1957; M.S., U. Mich., 1961, Ph.D., 1967. Pharmacist, Detroit, 1957-60; spl. instr., instr. pharmacy adminstrn. Coll. Pharmacy Wayne State U., 1958-62; dir. of research Kaiser-Permanente Med. Care Program, Oreg. region, Portland, 1964—; v.p. (research) Kaiser Found. Hosps., 1981—; adj. prof. sociology Portland State U., 1965—; asso. clin. prof. preventive medicine and pub. health U. Oreg. Med. Sch., 1971—; mem. study com. on health delivery systems Gov.'s Comprehensive Health Planning Council; cons. Gov.'s Health Manpower Council. Bd. dirs. Washington County Community Action Orgn., 1966-70; pres. Jewish Edn. Assn., Portland, 1976-78; bd. dirs. Jewish Fedn., Portland, 1975-79. USPHS trainee, 1962-63, 63-64. Fellow Am. Pub. Health Assn. (governing council); mem. AAAS, Am. Sociol. Assn., Am. Statis. Assn., Group Health Assn. Am., Nat. Acad. Scis., Inst. Medicine. Jewish. Subspecialty: Health services research. Home: 712 NW Spring Portland OR 97229 Office: 4610 SE Belmont Portland OR 97215

GREENMAN, GREGORY MICHAEL, nuclear engineer; b. Pontiac, Mich., July 22, 1954; s. Harlan Ray and Mary (Barar) G. B.S., Oakland U., Rochester, Mich., 1975; Ph.D., MIT, 1980. Nuclear engr. Argonne (Ill.) Nat. Lab., 1980—. Mem. Am. Nuclear Soc. Subspecialties: Nuclear engineering; Nuclear fission. Current work: Computer modelling of nuclear reactors and nuclear processes. Office: Argonne Nat Lab Bldg 208 Room W213B Argonne IL 60439

GREENSPAN, JOEL DANIEL, research neuroscientist; b. Chgo., Nov. 30, 1952; s. Dan and Helen (Ross) G.; m. Deborah Ann Barringer, Apr. 17, 1982; 1 stepson, Robert Crews. B.S., Rollins Coll., Winter Park, Fla., 1974; M.S., Fla. State U., 1976, Ph.D., 1980. Predoctoral fellow in psychobiology Fla. State U., Tallahasse, 1975-76, grad. research asst. dept. psychology, 1976-80; postdoctoral fellow in neurobiology U. N.C., Chapel Hill, 1981—. Contbr. articles to profl. jours. NIH grantee, 1981—. Mem. Soc. for Neurosci., Internat. Assn. Study of Pain, AAAS. Subspecialties: Neurophysiology; Psychophysics. Current work: Information processing in the somotosensory system, pain, tactile sense, thermal sense, somatosensory system, spinal cord, cerebral cortex. Office: Dept Physiology Univ North Carolina 206-H 54 Med Research Wing Chapel Hill NC 27514

GREENSPAN, KALMAN, physiologist, pharmacologist, cardiac researcher; b. Bariez, Poland, Apr. 27, 1925; s. Samuel and Sarah (Appledorfer) G.; m. Shirley Barbara Baron, June 25, 1950; children: Mark A., Janet H., Carol D. B.S. cum laude, L.I. U., 1948; M.A., Boston U., 1950, Columbia U., 1953; Ph.D., SUNY, Downstate Med. Ctr., Bklyn., 1960. Research asst. E.R. Squibb, 1953-56; instr. physiology SUNY, 1960-62; asst. prof. physiology and medicine Ind. U., Indpls., 1962-65, assoc. prof., 1965-68, prof., 1968—; head sect. physiology Krannert Inst. Cardiology, 1962-74; head div. physiology Terre Haute Center Med. Edn., 1974—; prof. elec. engring. Rose-Hulman Inst., 1976—; Med. com. Hooverwood Hosp., also mem. admissions com. Rev. editor 10 sci. jours.; contbr.: Physiology, 1983; editorial bd.: Jour. Electrocardiology, 1970—; editorial bd. cons.: Am. Jour. Cardiology, 1974-81; contbr. numerous articles profl. jours. Exec. bd. Office of Sec., Congregation Beth-El Zedick. Served with USNR, 1943-46. Fellow AAAS, Am. Coll. Cardiology; mem. Am. Physiology Soc., Am. Soc. Pharmacology and Exptl. Therapeutics, Cardiac Muscle Soc., Biophys. Soc., Soc. Gen. Physiologists, N.Y. Acad. Scis., Am. Heart Assn., Ind. Heart Assn., Am. Fedn. Clin. Research, Cardiac Electrophysiol. Group, Internat. Study Group Research in Cardiac Metabolism, Sigma Xi. Subspecialties: Physiology (medicine); Cellular pharmacology. Current work: Cardiac cellular electrophysiology and electropharmacology, dysrhythmia induction and therapeutic mechanisms, mechanisms of cardiac rhythm disturbances and therapeutic interventions. Office: Ind U Sch Medicine Terre Haute Ctr Med Edn Terre Haute IN 47809

GREENSTEIN, JESSE LEONARD, astronomer, emeritus educator; b. N.Y.C., Oct. 15, 1909; s. Maurice and Leah (Feingold) G.; m. Naomi Kitay, Jan. 7, 1934; children: George Samuel, Peter Daniel. A.B., Harvard U., 1929, A.M., 1930, Ph.D., 1937. Engaged in real estate and investments, 1930-34, Nat. Research fellow, 1937-39; assoc. prof. Yerkes Obs., U. Chgo., 1939-48; research assoc. McDonald Obs., U. Tex., 1939-48; mil. research under OSRD (optical design), Yerkes Obs.), 1942-45; prof. Calif. Inst. Tech., 1948-70, Lee A. DuBridge prof. astrophysics, 1971-81, prof. emeritus, 1981—; also staff mem. Hale Obs., 1949—, Palomar Obs., 1979—, exec. officer for astronomy, 1949-72; chmn. of faculty of inst., 1965-67; mem. obs. com. Hale Observatories; mem. staff Owens Valley Radio Obs.; cons., also com. mem. NASA and NSF on astronomy and radio astronomy; chmn. astronomy survey Nat. Acad. Scis., 1969-72; spl. cons. NASA, 1978—; vis. prof. Princeton, 1955, Inst. for Advanced Studies, 1964, 68-69, U. Hawaii, 1979, Niels Bohr Inst., 1979, NORDITA, Copenhagen, 1972, U. Del., 1981; lectr. in field; cons. Sci. Adv. Bd. USAF; dir. Itek Corp., Hycon Corp.; Chmn. bd. dirs. Associated Univs. Research in Astronomy, 1974-77; bd. overseers Harvard, 1965-71; bd. dirs. Pacific Asia Mus. Author sects. of treatises, 380 tech. papers.; Editor: Stellar Atmospheres, 1960; Contbr. sci. articles; author govt. reports. Named Calif. Scientist of Yr., 1964; recipient Apollo award, Disting. Public Service medal NASA, 1974. Mem. Royal Astron. Soc. (asso., gold medal 1975), Astron. Soc. Pacific (Bruce medalist 1971), Am. Astron. Soc. (councillor 1947-50, v.p. 1955-57, Russell lectr. 1970), Internat. Astron. Union (pres. commn. on spectroscopy 1952-58, chmn. U.S. del. 1969-72, Rennie Taylor award 1982), Nat. Acad. Scis. (councillor, sect. chmn. com. on sci. and pub. policy), Am. Philos. Soc., Am. Acad. Arts and Scis., Phi Beta Kappa. Club: Athenaeum (Pasadena) (bd. govs.). Subspecialties: Optical astronomy; Theoretical astrophysics. Current work: Observation of faint stars at the end of their evolution; composition and cooling of condensed matter. Home: 2057 San Pasqual St Pasadena CA 91107 A long and happy life, in which scientific discovery was like breath. With age one faces the question—was it worth doing? Were the uncomfortable thousand nights at the large telescopes drudgery or drama? Was the blood in the committee-room necessary? Yes, Yes, Yes

GREENSTEIN, TEDDY, chemical engineering educator, researcher; b. Czechoslovakia, Mar. 16, 1937; s. Sam and Serena (Klein) G.; m. Judith Lefkowich, July 6, 1982. B.Ch.E., CCNY, 1960; M.Ch.E., NYU, 1962, Ph.D., 1967. Rating engr. Davis Engring., Wallington, N.J., 1960; research asst. NYU, N.Y.C., 1964-67; asst. prof. chem. engring. N.J. Inst. Tech., Newark, 1967-78, assoc. prof., 1978—. Contbr. articles in field to profl. jours. Grantee NSF, 1964, Inst. Paper Chemistry, 1965-67, Found. Advanced Grad. Study Engring., 1967-69; recipient Founders Day award NYU, 1967. Mem Am. Inst. Chem. Engrs., Am. Soc. Microbiology, Am. Soc. Engring. Edn., Assn. Orthodox Jewish Scientists, N.Y. Acad. Scis., Sigma Xi, Tau Beta Pi, Omega Chi Epsilon. Subspecialties: Chemical engineering; Enzyme technology. Current work: Low Reynolds number hydrodynamics, chemical engineering, biochemical engineering, microbiology. Home: 3000 Ocean Pkwy Brooklyn NY 11235 Office: NJ Inst Tech 323 High St Newark NJ 07102

GREENSTOCK, CLIVE LEWIS, research scientist, consultant; b. High Wycombe, Eng., Aug. 14, 1939; s. George Henry and Clarice Irene (Lewis) G.; m. Gwen Dorothy Johns, July 17, 1965; children: Erica Jane, Andrea Gail. B.Sc. in Physics (scholar), U. Leeds, Eng., 1960; M.Sc. in Radiation Physics, U. London, 1963; Ph.D. in Radiation Biochemistry, U. Toronto, Can., 1968. Hosp. physicist Cardiff (Wales) Radiotherapy Centre, 1960-61; sci. officer Nat. Phys. Lab., Teddington, Eng., 1963-64; postdoctoral fellow Cancer Research Campaign, Mt. Vernon Hosp., Eng., 1969-70; research officer, chmn. long-range research study group Atomic Energy Can., Ltd., Pinawa, Man., Can., 1970—; cons. Radiation Chemistry Data Center of U. Notre Dame, Nat. Cancer Program, HEW, Fed. Strategy for Research of NIH; lectr. radiation protection tng. course; grant reviewer. Assoc. editor: Radiation Research Jour, 1977-80; contbr. articles and revs. to profl. jours., chpts. in books. Nat. coach Cross Country Ski Assn.; chmn. Pinawa Library Bd.; chmn. sch. tchr. Pinawa Christian Fellowship; sci. fair judge. Grantee Sci. Research Council, 1961-63; awardee Can. Cancer Research, 1964-68; Nat. Cancer Inst. fellow, 1969-70. Mem. Am. Assn. Cancer Research, Radiation Research Soc. (chmn. membership com. 1979-80), Biophys. Soc., Antioxidant Soc., Brit. Assn. Cancer Research, Brit. Assn. Radiation Research, Brit. Inst. Radiology, Am. Radio Relay League., Mensa Soc. Anglican. Subspecialties: Cancer research (medicine); Kinetics. Current work: Molecular radiobiology, radiation damage in DNA, lipids and proteins, radioprotection and sensitization, free radical mechanisms in radiation and chemical carcinogenesis and its prevention, redox processes in metabolism, toxicity activated oxygen and its control by superoxide dismutase; pulse radiolysis and chemical kinetics, structure-function relationships in biopolymers. Home: 112 Burrows Rd Pinawa MB Canada ROE 1L0 Office: Medical Biophysics Br Atomic Energy of Canada Ltd Pinawa MB Canada ROE 1L0

GREENWALD, EDWARD S., physician, medical oncologist; b. New Rochelle, N.Y., May 13, 1928; s. Irving and Belle Elizabeth (Jacobson) G.; m. Edith Deborah Greenwald, Dec. 4, 1949; children: David, Daniel, Joel, Joshua. B.A., Amherst Coll., 1948; M.D., N.Y. U., 1952. Diplomate: Am. Bd. Internal Medicine. Practice medicine specializing in med. oncology, New Rochelle, N.Y., 1958—; acting chief dept. oncology Montefiore Hosp., Bronx, N.Y., 1976-82; clin. prof. medicine Albert Einstein Coll. Medicine, Bronx, 1982—. Author: Cancer Chemotherapy, 2d edit, 1973. Served to capt. USAF, 1953-55. Fellow ACP; mem. AMA, Am. Soc. Clin. Oncology, Am. Assn. Cancer Research, Phi Beta Kappa. Democrat. Jewish. Subspecialties: Oncology; Cancer research (medicine). Current work: Cancer chemotherapy, chemotherapy research, cancer epidemiology. Home: 39 Disbrow Circle New Rochelle NY 10804 Office: 838 Pelhamdale Ave New Rochelle NY 10801

GREENWALD, PETER, physician, government medical program administrator; b. Newburgh, N.Y., Nov. 7, 1936; s. Louis and Pearl (Reingold) G.; m. Harriet Reif, Sept. 6, 1968; children: Rebecca, Laura, Daniel. B.A., Colgate U., 1957; M.D., SUNY Coll. Medicine, 1961; M.P.H., Harvard U., 1967, Dr. P.H., 1974. Intern Los Angeles County Hosp., 1961-62; resident in internal medicine Boston City Hosp., 1964-66; asst. in medicine Peter Bent Brigham Hosp., 1967-68; mem. epidemiology and disease control study sect. NIH, 1974-78; mem. N.Y. State Gov.'s Breast Task Force, 1976-78; with N.Y. State Dept. Health, Albany, 1968-81, dir., 1968-76, dir. epidemiology, 1976-81; prof. medicine Albany Med. Coll., 1976-81; attending physician Albany Med. Ctr. Hosp., 1968-81; adj. prof. biomed. engring. Rensselaer Poly. Inst., Troy, N.Y., 1976-81; assoc. scientist Sloan-Kettering Inst. for Cancer Research, N.Y.C., 1977-81; dir. Div. Resources, Ctr. and Community Activities, Nat. Cancer Inst., NIH, Bethesda, Md., 1981—; mem. VA Merit Rev. Bd. Med. Oncology, Washington, 1972-74. Editor in chief: Jour. Nat. Cancer Inst, NIH, 1981—; contbr. articles to profl. jours. Served with USPHS, 1962-64, 81—. Recipient Disting. Service award N.Y. State Dept. Health, 1975; Edway medal and award for med. writing N.Y. State Jour. Medicine, 1977; N.Y. State Gov.'s Citation for pub. health achievement, 1981. Fellow ACP, Am. Coll. Preventive Medicine, Am. Pub. Health Assn. (epidemiology sect. chmn. 1981); mem. Am. Assn. Cancer Research, Am. Coll. Epidemiology (bd. dirs. 1981-82), Am. Cancer Soc., Nat. Com. Cancer Prevention and Detection, Am. Soc. Preventive Oncology, Internat. Cancer Registry Assn., Internat. Epidemiology Soc., Nat. Acad. Scis. (food and nutrition bd.). Subspecialty: Preventive medicine. Office: NIH Bldg 31 Room 4A32 9000 Rockville Pike Bethesda MD 20205

GREENWOOD, ALLAN NUNNS, electrical engineering educator; b. Leeds, Eng.; s. William Nunns and Ethel May (Burrell) G.; m. Grace Ruth Neville, July 24, 1944; children: Janet Penelope, Stephen Richard, Hilary Jane. B.A., Cambridge U., 1943, M.A., 1948; Ph.D., Leeds U., 1952. Devel. engr. Imperial Chem. Industries, Stourport, Eng., 1946-48; lectr. U. Leeds, 1948-54; vis. prof. U. Toronto, Ont., Can., 1954-55; cons. engr. sr. cons. engr. Gen. Electric Co., Phila., 1955-72; Philip Sporn prof., dir. Center Electric Power Engring., Rensselaer Poly. Inst., Troy, N.Y., 1972—; cons., industry, govt. Author: Electrical Transients in Power Systems, 1971, (with Lafferty et al) Vacuum Arcs, 1980, (with Tanaka) Advanced Power Cable Technology, Vols. I and II, 1983; contbr. articles to profl. jours. Served to lt. Royal Navy, 1943-46. Fellow IEEE; mem. Conf. Internat. des Grands Reseaux Electriques, Sigma Xi, Eta Kappa Nu. Unitarian. Subspecialty: Electrical engineering. Patentee in power-switching technology. Office: Rensselaer Poly Inst Troy NY 12181

GREENWOOD, MICHAEL SARGENT, plant physiologist; b. Winthrop, Mass., Nov. 7, 1940; s. Willard Priest and Nancy Hacker (Brown) G.; m. Susan Fowle, June 10, 1961; children: Willard, Davis. B.A. in Botany, Brown U., 1963; M.F., Yale U., 1965, M.S., 1966; Ph.D., 1969. Asst. prof. Middlebury Coll., 1968-74; vis. scientist U. Glasgow, Scotland, 1971-72; tree physiologist Weyerhaeuser Co., Hot Springs, Ark., 1974—; adj. asst. prof. N.C. State U., 1981—; Mem. Union Concerned Scientists. Contbr. articles to profl. jours. AEC fellow, 1965-67. Mem. Am. Bot. Soc., Am. Soc. Plant Physiologists, AAAS. Democrat. Subspecialties: Plant physiology (agriculture); Plant genetics. Current work: Forest tree breeding. Patentee method of inducing flowering in pines, 1979; developed operational procedures for breeding pines more rapidly. Office: PO Box 1060 Hot Springs AR 71901

GREER, JOANNE MARIE G., mathematical psychologist; b. New Orleans, Aug. 24, 1937; d. Carl Matthewson and Sydney (Comeaux) G.; m. Thomas Vernon Greer, Apr. 23, 1966; children: Marc Bernley, Carl Mathieu. B.S. cum laude, St. Mary's Dominican Coll., 1962; M.Ed., La. State U.-Baton Rouge, 1969; Ph.D., U. Md.-College Park, 1974; research affiliate, Washington Psychoanalytic Inst., 1979—. Tchr., adminstr. Sister of St. Joseph of Medaille, New Orleans, 1959-65; mem. faculty La. State U., Baton Rouge, 1965-66, U. Md., College Park, 1969-74; ops. research analyst VA, Washington, 1975; researcher, adminstr. USPHS, Washington, 1975-81; math. statistician advanced techniques staff Office Insp. Gen. Office Sec., HHS, Washington, 1981—; cons. Easter Seal Treatment Ctr., Rockville, Md., 1972-74, Mobile Med. Care, Montgomery County, Md., 1980—. Co-Author: The Sexual Aggressor: Current Perspectives on Treatment, 1982; film Rape: Treatment of the Victim, 1981; contbr. articles to profl. jours. Recipient Outstanding Performance award USPHS, 1978; NDEA fellow, 1969-72; NSF fellow, 1964-66. Fellow Md. Psychol. Assn.; mem. Am. Psychol. Assn., Am. Statis. Assn., Psychometric Soc., Classification Soc. Subspecialties: Psychoanalysis; Mathematical modelling. Current work: Current research interest is application of quantitative models to psychoanalytic theory and empirical tests of psychoanalytic models; current applied work in in developing computer models to identify perpetrators of fraud in government benefit programs. Home: 12420 Kuhl Rd Silver Spring MD 20902 Office: Secretary Dept Health and Human Services 330 Independence AVe SW Suite 5739 Washington DC 20201

GREGERMAN, ROBERT ISAAC, endocrinologist, physician; b. Boston, Apr. 18, 1930; m. Marjorie Libby Bender, June 30, 1957; children: Lisa Claire, Debra Mida. A.B., Harvard U., 1951; M.D., Tufts U., 1955. Diplomate: Am. Bd. Internal Medicine. Intern Tufts-NewEng. Med. Center, Boston, 1955-56; commd. med. officer USPHS, 1956—; resident Washington VA Hosp., 1958-59; resident in medicine, fellow in endocrinology Univ. Hosp., U. Mich., Ann Arbor, 1959-61; investigator Gerontology Research Center, Nat. Inst. on Aging, Balt. City Hosps., 1961—, chief endocrinology sect., 1963—; mem. faculty Sch. Medicine Johns Hopkins U., Balt., 1964—. Contbr. articles to med. jours. Fellow Gerontol. Soc.; mem. Endocrine Soc., Am. Thyroid Assn., Am. Soc. Biol. Chemists. Subspecialties: Endocrinology; Gerontology. Current work: Endocrinology of aging, endocrinology of clinico-pathologic states. Home: 2417 Briarwood Rd Baltimore MD 21209 Office: Gerontology Research Center NIH Nat Inst Aging Balt City Hosps Baltimore MD 21224

GREGG, JOHN BAILEY, health services administrator, medical educator, surgeon; b. Sioux Falls, S.D., June 5, 1922; s. John B. and Anna Elida (Bailey) G.; m. Pauline Benfer Synder, June 1947; children: Michele Lee, John Benfer, Stewart David, Rebecca Jo. B.A., U. Iowa, 1943, M.D., 1946. Diplomate: Nat. Bd. Med. Examiners, Am. Bd. Otolaryngology. Intern U. Md. Hosp., Balt., 1946-47; resident in gen. surgery Univ. Hosps., Iowa City, Iowa, 1949-51, resident in otolaryngology and maxillofacial surgery, 1951-53; instr. otolaryngology U. Iowa, Iowa City, 1953-54, asst. prof., dir. broncho-esophagology clinic, 1959-60; assoc. prof. otolaryngology Sch. Medicine, U. S.D., Sioux Falls, 1955-59, prof., 1962-77, chmn. dept. otolaryngology, 1971-73, prof. surgery, coordinator specialties of surgery, 1977—; assoc. prof. otolaryngology U. Nebr., 1971-80; prof. anthropology (paleopathology and med. anthropology) U. Tenn., Knoxville, 1972; dir. med. services S.D. Dept. Health, Pierre, 1982—; mem. staff McKennen Hosp., Sioux Falls, VA Hosp., St. Mary's Hosp., Pierre; chmn. Deafness Research Found. and Temporal Bone Bank for N.D. and S.D., 1961-69. Served to lt. (j.g.) USNR, 1944-49. Recipient Meritorious Service citation Pres.'s Com. on Employment of Handicapped, 1966. Fellow ACS; mem. Am. Laryngological, Rhinological and Otological Soc., Am. Acad. Ophthalmology and Otolaryngology, AMA, AAAS, Am. Cleft Palate Assn., S.D. State Med. Assn. (pres. 7th dist. 1966-67), Am. Assn. Phys. Anthropologists, Paleopathology Assn., Am. Dermatoglyphic Assn., Alpha Tau Omega, Alpha Kappa Kappa, Sigma Alpha Eta (hon.). Episcopalian. Clubs: Elks, Rotary. Subspecialties: Otorhinolaryngology; Medical anthropology and paleopathology. Current work: Ancient osteopathology, longitudional epidemiology, epidemiology of infectious disease, especially upper respiratory; administration; education; research. Office: Dept Surgery and the Specialties of Surgery Sch of Medicine U SD 2501 W 22d St Sioux Falls SD 57105 Office: SD Dept Health Foss Building Pierre SD 57501

GREGOR, CLUNIE BRYAN, geology educator; b. Edinburgh, Scotland, Mar. 5, 1929; came to U.S., 1968; s. David Clunie and Barbara Mary (Beilby) G.; m. Suzanne Assir, Apr. 19, 1955 (div. 1969); 1 son, Andrew James; m. Anna Bramanti, Apr. 15, 1969; children: Tommaso James, Matteo James. B.A., Cambridge (Eng.) U., 1951, M.A., 1954; D.Sc., Utrecht (Netherlands) U., 1967. Geophys. engr. Ray Exploration Co., Houston, 1957-58; asst. Am. U. of Beirut, 1958-59, instr., 1959-62, asst. prof., 1962-64; assoc. prof., chmn. geology dept., 1966-68; dir. Crystallographic Lab., Delft (Netherlands) Inst. Tech., 1964-66; vis. prof. geology Case-Western Res. U., Cleve., 1968-69; prof. geology West Ga. Coll., Carrollton, 1969-72, Wright State U., Dayton, Ohio, 1972—. Author: Geochemical Behaviour of Sodium, 1967; editor: Geochem. News, 1969—. NSF grantee, 1977-82; Geochem. Soc. grantee, 1982—. Fellow Geol. Soc. London; mem. Am. Geophys. Union, Geochem. Soc. (council 1979—, sec. 1983-86), U.S.A. Work Group on Geochem. Cycles (chmn. 1972—). Subspecialties: Geochemistry; Geology. Current work: Mass-age distributions of rocks; modeling the rock cycle. Home: 136 W North College St Yellow Springs OH 45387 Office: Dept Geol Scis Wright State U Dayton OH 45435

GREGORIADES, ANASTASIA, biologist; b. Macedonia, Greece, May 5, 1940; d. Diamantes and Olympia (Costopoulos) G.; m. Demetrios T. Stavropoulos, July 12, 1970; children: Alexander, Nicholas. B.A., Hunter Coll., 1962; M.A. (Sloan Fund fellow), CUNY, 1964; Ph.D. (Sloan fellow), Cornell U. Med. Coll. Basic Scis., 1968. Postdoctoral fellow Pub. Health Research Inst., N.Y.C., 1969-72, asst. dept. virology, 1972-75, assoc., 1975—. Mem. Am. Soc. Microbiology, Harvey Soc., N.Y. Acad. Scis., Assn. Women in Sci. Democrat. Greek Orthodox. Subspecialty: Virology (biology). Current work: Replication of influenza virus; synthesis and assembly of viral membranes. Home: 21 Stuyvesant Oval Apt 7D New York NY 10009 Office: 455 1st Ave New York NY 10016

GREGORIOU, GREGOR GEORG, aerodynamicist, researcher; b. Athens, Greece, Feb. 5, 1937; emigrated to Germany, 1956; s. Georg and Athina (Koulia) G.; m. Josephine Nievelstein, Aug. 31, 1962; children: Lauretta, Katja. Dipl.-Ing., Tech. U., Aachen, Ger., 1962, Dr.-Ing., 1973. Aerodynamicist Vereinigte Flugtechn Werke, Bremen, Ger., 1963-64; aerodynamicist Messerschmitt-Bolkow-Blohm, Ottobrunn, Ger., 1964-71, mgr. aerodynamics, 1971—; cons. Army Research Ctr., Athens, Greece, 1982—. Contbr. articles to profl. jours. Mem. AIAA, Gesellschaft für Angewandte Mathematik und Mechanik, Deutsche Gesellschaft für Luft und Raumfahrt. Subspecialty: Aeronautical engineering. Current work: Research in missile aerodynamics. Office: Messerschmitt-Bolkow-Blohm Einsteinstrasse Ottobrunn Federal Republic Germany 8012

GREGORY, DONALD CLIFFORD, research physicist; b. Tyler, Tex., Sept. 12, 1949; s. John Clifford and Dorothy (Kingston) G.; m. Jean Wheat, Dec. 28, 1971; children: Eric William, Lauren Elizabeth. B.S., U. Tex., Austin, 1971, M.A., 1973, Ph.D., 1976. Exchange scientist Hungarian Acad. Scis., Budapest, 1974-75; Welch Found. fellow physics and chemistry depts. U. Tex., Austin, 1975-76; research assoc. Joint Inst. for Lab. Astrophysics, Boulder, Colo., 1976-78; asst. physicist Brookhaven Nat. Lab., Upton, N.Y., 1978-79, assoc. physicist, 1979-80; staff scientist Oak Ridge (Tenn.) Nat. Lab., 1980—. Contbr. articles to profl. jours. Mem. Am. Phys. Soc., Amateur Radio Club of Oak Ridge, Sigma Xi, Sigma Pi Sigma. Subspecialty: Atomic and molecular physics. Current work: Experimental studies of electron impact processes on multiply charged ions. Office: Oak Ridge Nat Lab Bldg 6003 PO Box X Oak Ridge TN 37830

GREGORY, FRANCIS J., microbiologist; b. Bklyn., June 21, 1921; s. Frank and Elizabeth (Cavanaugh) G.; m. Elizabeth Jane Bugie, July 2, 1949; children: Patricia Camp, Eleen Ott. B.A., Bklyn. Coll., 1942; Ph.D., Rutgers U., 1954. Research scientist Merck and Co., Rahway, N.J., 1946-51; postdoctoral fellow Rutgers U., 1954-55; unit supr. Wyeth Labs., Inc., Radnor, Pa., 1955-80, mgr. microbiology sect., 1980—. Contbr. articles to profl. jours. Bd. dirs. Friends of Tredyffrin Library, 1962-65, Delaware Valley Lupus Assn., 1979—; pres. Cath. Youth Assn., Berwyn, Pa., 1956- 58, Daylesford Village Civic Assn., 1964-65. Served with U.S. Army, 1942-46. Mem. Am. Soc. Microbiologists, Am. Assn. for Cancer Research, Mycological Soc. Am., Tissue Culture Assn., Am. Soc. Parasitologists. Roman Catholic. Subspecialties: Microbiology; Immunobiology and immunology. Current work: Cancer research, immunology, oncology, chemotherapy, parasitology, cell biology, cell and tissue culture. Home: 11 Cypress Ln Berwyn PA 19312 Office: Wyeth Labs Inc PO Box 8299 Philadelphia PA 19101

GREGORY, GAROLD FAY, plant pathologist; b. Arkansas City, Kans., Aug. 15, 1926; s. John Fay and Birdie Maude (Inman) G.; m. Flossy June Lewman, Dec. 25, 1953; children: Cherylynn Gay, Andrew Fay. B.S., Kans. State U., 1951; M.S., Iowa State U., 1956; Ph.D., Cornell U., 1962. Plant pathologist U.S. Dept. Agr. Forest Service, Delaware, Ohio, 1962—. Contbr. articles to profl. jours. Served with U.S. Army, 1951-53. Allied Chem. and Dye fellow, 1960-61. Mem. Am. Phytopathol. Soc. Mem. Ch. of God. Subspecialties: Plant pathology; Plant physiology (biology). Current work: Urban tree insect and disease problems. Biochemistry of plant disease.

GREGORY, JOHN DELAFIELD, biochemist, educator; b. N.Y.C., May 18, 1923; s. John and Katharine (Crosby) G.; m. Helen Carter Powell, Apr. 12, 1958; children: Paul C., Christopher C. B.S., Yale U., 1944; Ph.D., 1947. Asst. in Chem. pharmacology Rockefeller Inst., N.Y.C., 1947-49; asst. biochemist Mass. Gen. Hosp., Boston, 1949-53, assoc. biochemist, 1953-57; assoc. prof. biochemistry Rockefeller U., 1957—. Editor: Glycoconjugate Research, 1979. Mem. AAAS, Biochem. Soc., Soc. Complex Carbohydrates (pres. 1976), Am. Soc. Biol. Chemists, Sigma Xi. Subspecialty: Biochemistry (biology). Current work: Structure and function of proteoglycans of connective tissues, especially cornea and skin. Office: Rockefeller U 1230 York Ave New York NY 10021

GREGORY, RICHARD LEE, microbiologist, researcher; b. Elmhurst, Ill., Oct. 31, 1954; s. Walter Leonard and Jetta Mae (Jensen) G.; m. Rebecca Jo Brown, Aug. 21, 1981. B.S., Eastern Ill. U., Charleston, 1976; M.T., Decatur Meml. Hosp., 1976; M.S., So. Ill. U., Carbondale, 1979, Ph.D., 1982. Med. technologist Decatur (Ill.) Meml. Hosp., 1975-76; teaching and research asst. So. Ill. U., Carbondale, 1976-82, lab. instr., 1981; research assoc. U. Ala., Birmingham, 1982—; cons. Stolle Found., Birmingham, 1982—. Contbr. articles to profl. jours. Asst. scoutmaster Boy Scouts Am., Birmingham, 1982—. Predoctoral trainee So. Ill. U., 1976-82; postdoctoral trainee NIH, U. Ala.-Birmingham, 1982; NIDR fellow, 1983. Mem. Am. Soc. Med. Technology, Am. Soc. Clin. Pathologists, Am. Soc. Microbiology, Am. Assn. Dental Research, Internat. Assn. Dental Research. Republican. Lutheran. Subspecialties: Microbiology; Immunobiology and immunology. Current work: Mucosal immunobiology of the oral cavity, eyes and mammary glands. Office: U Ala Dept Microbiology Birmingham AL 35294

GREGORY, ROBERT AARON, botanist, researcher; b. Hudson Falls, N.Y., July 27, 1927; s. Edward David and Mary (Bordeau) G.; m. Shirley Joyce Staples, Aug. 10, 1970; 1 dau., Katherine Jane. B.S., Cornell U., 1952; M.F., Yale U., 1954; Ph.D., Oreg. State U., Corvallis, 1968. Researcher Dept. Agr. Forest Service, Juneau and College, Alaska, 1951-67; researcher Forest Physiology Lab., Beltsville, Md., 1968-74; research plant physiologist George D. Aiken Maple Lab., U. Vt., Burlington, 1974—; instr. tree growth and devel. Bullard fellow Harvard U., 1962-63. Contbr. numerous articles on botany to profl. jours. Mem. AAAS, Internat. Assn. Wood Anatomists, Sigma Xi. Subspecialty: Plant growth. Current work: Woody plant meristems, vascularization, circulation, storage, and mobilization of assimilates. Home: PO Box 173 Charlotte VT 05445 Office: PO Box 968 Burlington VT 05401

GREGSON, VICTOR GREGORY, laser scientist; b. East St. Louis, Ill., June 10, 1935; s. Victor Gregory and Dorothy Mae (Glaze) G.; m. Lois Jean Rigden, May 30, 1964; children: Christopher Todd, Jennifer. A.B., Washington U., St. Louis, 1958; M.A., 1959; Ph.D., Stanford U., 1965. Project physicist Poulter Labs., Stanford (Calif.) Research Inst., 1961-64; sr. engr. Ill. Inst. Tech. Research Inst., Chgo., 1964-69; tech. leader Mfg. Staff Gen. Motors Tech. Ctr., 1969-71; product mgr. Western div. GTE Sylvania, Mountain View, Calif., 1977-79; product mgr./mktg. Indsl. div. Coherent, Inc., Palo Alto, Calif., 1979-83; tech. mgr. laser ctr. Marine div. Westinghouse Electric Corp., Sunnyvale, Calif., 1983—. Author: (with others) Guide for Material Processing by Lasers, 1977; contbr. articles to profl. jours. Mem. IEEE, Optical Soc. Am., Laser Inst. Am. (bd. dirs. 1973-82). Republican. Presbyterian. Subspecialties: Industrial laser processing; Materials processing. Current work: Computer modelling of industrial laser processes; industrial laser processing. Home: 10894 Dryden Ave Cupertino CA 95014 Office: Westinghouse Electric Corp Hendy Ave Sunnyvale CA 94088

GREINER, JACK VOLKER, physician, ophthalmology researcher; b. Fountain Hill, Pa., Aug. 25, 1949; s. Harry Sandt and Vera Lilian G.; m. Cynthia Ann Mis, May 17, 1980. A.A., Valley Forge Mil. Acad., 1969; B.A., U. Vt., 1971; M.S. in Anatomy, Purdue U., 1973, Ph.D., U. Toledo, 1975; D.O., Chgo. Coll. Osteo. Medicine, 1982. Research fellow in ophthalmology Howe Lab. of Ophthalmology, Harvard U. Med. Sch. and Mass. Eye and Ear Infirmary, Boston, 1974-76; research fellow in corneal and external diseases of eye Eye Research Inst., Retina Found., 1976-78; research fellow in ophthalmology Harvard U. Med. Sch., Boston, 1976-78; research assoc. in ophthalmology U. Ill. Eye and Ear Infirmary, Chgo., 1979-81, research asst. prof. ophthalmology, 1981-83; med. intern Cook County Hosp., Chgo., 1982-83; resident in ophthalmology Georgetown U. Med. Ctr., 1983—; adj. asst. scientist Eye Research Inst., Retina Found., Boston, 1978; adj. asst. prof. ophthalmic pathology Chgo. Coll. Osteo. Medicine, 1979-82, asst. prof. dept. pathology, 1982-83, assoc. prof., 1983—; co-dir. Eye Research Lab., Chgo. Osteo. Hosp., 1980—. Contbr. chpts. to books, articles to profl. publs. Served to capt. C.E. USAR, 1971-78. Fight For Sight grantee, 1980-82; Nat. Soc. to Prevent Blindness grantee, 1981-82; NIH Nat. Eye Inst. grantee, 1982-85. Mem. Am. Assn. Anatomists, Assn. Research in Vision and Ophthalmology, Soc. Exptl. Biology and Medicine, N.Y. Acad. Scis., AMA, Chgo. Med. Soc., Cook County Med. Soc., Am. Acad. Ophthalmology, Sigma Xi, Phi Kappa Phi, Sigma Sigma Phi. Subspecialties: Ophthalmology; Nuclear magnetic resonance (biotechnology). Current work: Phosphorus-31 nuclear magnetic resonance and histopathology of cornea and crystalline lens cataracts; specific clinical interests include diseases of cornea and external ocular tissues. Office: Dept Ophthalmology Georgetown U 3800 Reservoir Rd NW Washington DC 20007

GREISEN, KENNETH INGVARD, physicist; b. Perth Amboy, N.J., Jan. 24, 1918; s. Ingvard C. and Signa (Nielsen) G.; m. Elizabeth C. Chase, Apr. 12, 1941 (dec.); children: Eric Winslow, Kathryn Elise; m. Helen A. Leeds, Mar. 27, 1976. Student, Wagner Coll., 1934-35; B.S., Franklin and Marshall Coll., 1938; Ph.D., Cornell U., 1942. Instr. Cornell U., 1942-43, asst. prof., 1946-48, asso. prof., 1948-50, prof. physics, 1950—, chmn. dept. astronomy, 1976-79, univ. ombudsman, 1975-77, dean faculty, 1978-83; scientist Manhattan Project, Los Alamos, 1943-46. Fellow Am Phys. Soc.; mem. Am. Astron. Soc., Internat. Astron. Union, Nat. Acad. Sci., AAUP. Subspecialties: Cosmic ray high energy astrophysics; Gamma ray high energy astrophysics. Research cosmic rays. Home: 336 Forest Home Dr Ithaca NY 14850

GRENELL, ROBERT GORDON, psychiatrist, neuroscientist; b. N.Y.C., Apr. 3, 1916; s. Max and Lee (Gordon) G.; m. Dena Schild, May 15, 1979. B.Sc., CCNY, 1935; M.Sc., N.Y. U., 1936; Ph.D., U. Minn., 1943. Research asst., instr. physiology, then instr. neuroanatomy Yale U. Med. Sch., 1943-47; sr. fellow USPHS, U. Pa. and Johns Hopkins U., 1947-49; mem. faculty U. Md. Med. Sch., 1950—, prof. psychiatry, 1959—, dir. div. neurobiology, 1959—; USAID vis. prof. pharmacology, Trivandrum, India, 1961-63, cons. in field. Co-author: Biological Foundations of Psychiatry, 1976; co-editor: Molecular Structure and Functional Activity of the Nervous System, 1956, Neurophys. Pathology, 1962, From Nerve to Mind, 1972; co-editor, contbr.: Psychiatric Foundations of Medicine, 6 vols, 1978; editor-in-chief: Jour. Neurosci. Research, 1975-81; cons. editor, 1981—; contbr. articles to profl. publs. Fellow AAAS; mem. Soc. Biol. Psychiatry (pres. 1978), Am. Physiol. Soc. (travel award 1947), Soc. Neurosci., Assn Research in Nervous and Mental Disease, Soc., Exptl. Biology and Medicine, Eastern Assn. EEG, Am. Assn. EEG, N.Y. Acad. Sci., Physiol. Soc. Phila., Md. Biol. Soc., Am. Acad. Neurology, Biophys. Soc., Internat. Soc. Cell Biology, Md. Psychiat. Soc., Inst. Radio and Electronic Engrs., Sigma Xi. Subspecialties: Neurobiology; Neurophysiology. Current work: Neuron membrane and its role in behavior; brain mechanisms of learning, memory and information processing. Home: 204 E Highfield Rd Baltimore MD 21218 Office: IPHB Univ Hosp Baltimore MD 21201 The goal was to use what ability one had in some creative way, in an attempt to add something to our understanding of our world and ourselves. Ideas are important; power and power driven motivation are destructive.

GRENFELL, RAYMOND FREDERIC, physician, researcher; b. West Bridgewater, Pa., Nov. 23, 1917; s. Elisha Raymond and Pearl (Boll) G.; m. Maude Byrnes Chisholm, Aug. 19, 1944; children: Raymond Frederic, Milton Wilfred, James Byrnes, Robert Chisholm. B.S., U. Pitts., 1939, M.D., 1941. Intern Western Pa. Hosp., Pitts., 1941-42; practice medicine specializing in internal medicine, Jackson, 1946-79, practice medicine specializing in diagnosis and treatment of hypertension, 1979—; mem. staffs Hinds Gen. Hosp., Riverside Hosp., St. Dominic-Jackson Meml. Hosp., Miss. Bapt. Hosp., Doctor's Hosp.; clin. instr. U. Miss. Med. Sch., Jackson, 1955-59, clin. asst. prof. medicine, 1959—, vis. teaching physician, 1977—, head hypertension clinic, 1956-79. Pres. Jackson Symphony Orch. Assn., 1961, Duling PTA, Jackson, 1963; deacon First Baptist Ch., Jackson, 1960—. Served to maj. U.S. Army, 1942-46. Recipient bronze medal Am. Heart Assn., 1963, silver medal, 1965. Fellow Am. Coll. Angiology (gov. 1979—); Internat. Coll. Angiology Am. Coll. Chest Physicians; mem. Am. Soc. Clin. Pharmacology and Therapeutics (dir. 1968, v.p. 1976); Am. Fedn. Clin. Research So. Med. Assn. (councilor 1968-73), Miss. Heart Assn. (pres. 1964-65). Republican. Clubs: Country, Univ. (Jackson) (Dir. 1974—). Subspecialties: Internal medicine. Current work: Investigation of antihypertensive drugs with pioneer work in short and long term double-blind, controlled studies of antihypertensive drugs in premarketing phase. Home: 190 Ridge Dr Jackson MS 39216 Office: 514-H E Woodrow Wilson Jackson MS 39216

GRENIER, EDWARD JOSEPH, engring. specialist; b. Herkimer, N.Y., Apr. 4, 1942; s. Lawrence Edward and Elizabeth Ann (Schaeffer) G.; m. Kathleen Thomas, June 1, 1964; children: Lisa, John, Brandy. B.S. in Pyschology, Physics, U. Rochester, 1968, M.S. in Stats, 1972. Engr., analyst quality control Bausch & Lomb, Rochester, N.Y., 1964-73, sect. head quality engring., 1974-79, quality assurance quality control mgr., 1980-81, tech. engring. specialist, 1982—. Mem. Am. Soc. Quality Control, Am. Statis. Assn. Subspecialties: Quality control; Applied mathematics. Current work: Sampling methods, data analysis, vendor qualification, operator training, quality standards, product liability. Home: 2080 Clover St Rochester NY 14618 Office: 465 Paul Rd Rochester NY 14624

GRESS-GORDON, JEAN ANNE, toxicologist; b. Pa., Jan. 8, 1939; d. Desiderius Edmund and Irene Grace (McClintock) Gress; m. (div.); children: Scott Eugene, April Dawn. B.S. in Microbiology, Calif. State U., Long Beach, 1970, M.S. in Med. Tech, 1978. Med. technologist Pediatric Cardiopulmonary Lab. UCLA, 1972-75, med. technologist, 1975-78, 1978-79, clin. specialist, 1979-80, sr. supervising clin. lab. technologist, 1980—. Mem. So. Calif. Med. Technologists Educators, Calif. Assn. Toxicologists (sec.-treas. 1981-83), Am. Soc. Clin. Pathologists (affiliate), Am. Assn. Clin. Chemistry, Soc. Applied Spectroscopy, Disciples of Christ. Subspecialties: Toxicology (medicine); Clinical chemistry. Current work: Clinical and acute overdose analytical toxicology. Office: UCLA CHS A2-260 10833 Le Conte Ave Los Angeles CA 90024

GREULICH, RICHARD CURTICE, anatomist, gerontologist; b. Denver, Mar. 22, 1928; s. William Walter and Mildred Almena (Libby) G.; m. Betty Brent Mitchell, Dec. 19, 1948 (div. 1955); children: Christopher, Robert; m. Leonora Faye Colleasure, Dec. 27, 1958; children: Jeffrey, Hilary. A.B., Stanford U., 1949; Ph.D. (AEC fellow), McGill U. (Can.), 1953. Instr. Sch. Medicine, UCLA, 1953-55, asst. prof. anatomy, 1955-61, asso. prof. anatomy, 1961-64, prof. anatomy, 1964-66, asso. prof. oral biology Sch. Dentistry, 1961-64, prof. oral biology, 1964-66; sci. dir. Nat. Inst. Dental Research, NIH, Bethesda, Md., 1966-74; acting dir. Nat. Inst. Aging, Bethesda, 1975-76; dir. Gerontology Research Center and sci. dir. Nat. Inst. Aging, Balt., 1976—; staff dir. U.S. Pres.'s Biomed. Research Panel, 1974-75; vis. investigator Karolinska Inst., Stockholm, 1955-57, U. London, 1962-63, McGill U., 1963; vis. prof. anatomy U. Va., 1966-73. Served with F.A., U.S. Army, 1946-48. Recipient award for basic research in oral sci. Internat. Assn. Dental Research, 1963, Superior Service award HEW, 1971; Bank of Am.-Giannini Found. fellow, 1955-57; USPHS spl. fellow, 1962-63. Mem. Am. Assn. Anatomists, Gerontol. Soc., Am. Inst. Biol. Scis., AAAS, Am. Soc. Cell Biology, Sigma Xi. Club: Cosmos (Washington). Subspecialties: Anatomy and embryology; Gerontology. Research, publs. on growth, differentiation and aging at cellular and organismic level. Office: Gerontology Research Center Balt City Hosps Baltimore MD 21224

GREULING, JACQUELIN WREN, psychologist; b. Fort Worth, Sept. 24, 1931; d. Jack and Rose (Clymer) Wren; m. William Nash Greuling, Sr., May 29, 1954; children: William Nash, Jr., Robert Wren. B.A., U. Tex., 1953; M.A., U. Tex. El Paso, 1972; Ph.D., N. Mex. State U., 1979. Paraprofl. William Beaumont Army Med. Ctr., El Paso, 1971-73; psychometrist Child Guidance Ctr., El Paso, 1972-75; cons. psychology Crisis Line, Hobbs, N. Mex., 1972-74; William Beaumont Army Med. Ctr., El Paso, 1981-82, fellow health psychology, 1982—; bd. dirs. El Paso (Tex.) Plating Works, Inc., 1958—; vice-chmn. women and smoking Tex. div. Am. Cancer Soc., 1982. Trainer, coordinator Am. Cancer Soc., El Paso, 1982. Mem. Am Psychol. Assn., Am. Soc. Clin. Hypnosis, El Paso Psychol. Assn. Democrat. Episcopalian. Subspecialty: Health Psychology. Current work: The role of hypnotizability in the surgical procedure and in the post-operative period. Home: 330 Olivia Circle El Paso TX 79912 Office: William Beaumont Army Med Center El Paso TX 79920

GRIBBLE, GORDON WAYNE, chemistry educator; b. San Francisco, July 28, 1941; s. Waldron Boger and Jane Adena (Marl) G.; m. Mary Ellen Braun, June 22, 1963; children: Wayne, Julie, Jon. A.A., San Francisco City Coll., 1961; B.S., U. Calif.-Berkeley, 1963; Ph.D., U. Oreg.-Eugene, 1967; M.A., Dartmouth Coll., 1981. Research assoc. UCLA, 1967-68; asst. prof. Dartmouth Coll., Hanover, N.H., 1968-74, assoc. prof., 1974-80, prof. chemistry, 1980—; cons. Merck, Sharp & Dohme, West Point, Pa., 1971-74. Contbr. articles in field to profl. jours. Recipient Career Devel. award NIH, 1971-76; Sci. Faculty Devel. award NSF, 1977-78. Mem. Internat. Soc. Heterocyclic Chemistry, Am. Chem. Soc., Chem. Soc. London. Subspecialty: Organic chemistry. Current work: Organic synthesis of natural products, anticancer agents, synthetic methodology. Home: Meadowbrook Rd Norwich VT 05055 Office: Dartmouth Coll Steele Chem Bldg Hanover NH 03755

GRIBBONS, WARREN DAVID, psychology educator, researcher; b. Worcester, Mass., June 23, 1921; s. John B. and Mary (Leary) G.; m. Jean C. Cote, Aug. 30, 1947. B.A., Boston U., 1955; Ed.D., Harvard U., 1959. Cert. psychologist, Mass. Asst. prof. psychology Clark U., 1959-62; prof. Regis Coll., 1962—. Author: Emerging Careers, 1968, Careers in Theory and Experience, 1982; contbr. numerous articles to profl. jours. Served to sgt. USAAF, 1942-45. Eng. Office Edn. grantee, 1960-74. Mem. Am. Psychol. Assn. Subspecialty: Counseling psychology. Current work: Career development study—25-year follow-up of original 1958 sample, interviewed every two years. Office: Regis Coll 235 Wellesley St Weston MA 02193

GRICHAR, WILLIAM JAMES, research scientist; b. Houston, Oct. 21, 1949; s. William James and Julia (Janes) G.; m. Dimple Dell Fowler, Aug. 21, 1976; 1 dau., Amy Michelle. Student, Navarro Jr. Coll., 1968-70; B.S. in Botany, Tex. A&M U., 1973, M.Agr., 1975. Lab. asst. Tex. A&M U., College Station, 1973-75; technician Tex. Agr. Expt. Sta., Yoakum, 1975-80, research assoc., 1980-82, research scientist, 1982—. Contbr. articles to profl. jours. Mem. Am. Peanut Research and Edn. Soc., Weed Sci. Soc. Am., Tex. Weed Workers (pres. 1982). Democrat. Roman Catholic. Club: Dewitt-Lavaca County Aggie. Lodge: KC. Subspecialties: Weed control; Plant pathology. Current work: Interested in new types of weed control, chemical or biological; various diseases which affect crops. Home: Box 467 Yoakum TX 77995 Office: Box 755 Yoakum TX 77995

GRIFFIN, CHARLES CAMPBELL, chemistry educator, researcher; b. Phila., July 23, 1938; s. Edward P. and Elizabeth (Campbell) G.; m. Helen McGuire, Sept. 3, 1960; children: Edward M., Lisa M. A.B., Cath. U. Am., 1960; Ph.D., Johns Hopkins U., 1969. Research asst. Armed Forces Inst. of Pathology, Washington, 1960-64; asst. prof. Miami U., Oxford, Ohio, 1968-74, assoc. prof. chemistry, 1974—. Mem. Am. Soc. Biol. Chemists. Lodge: Lions. Subspecialty: Biochemistry (biology). Current work: Carbohydrate transport by yeasts. Office: Hughes Labs Miami U Oxford OH 45056 Home: 710 Melissa Dr Oxford OH 45056

GRIFFIN, DANA GOVE, III, botany educator; b. Ft. Worth, Nov. 9, 1938; s. Dana Gove and Jessie Calwell (Carter) G.; m. Nancy Claire Wilson, Dec. 20, 1964; children: Dana Gove, Sarah Brittain. B.S., Tex. Tech. U., 1961, M.S., 1962; Ph.D., U. Tenn., 1965. Fulbright lectr., Peru, 1965-66; asst. prof. botany U. Tenn., 1966-67; asst. prof. U. Fla., Gainesville, 1967-72, assoc. prof., 1972-79, prof. botany, 1979—; joint curator dept. botany and Fla. State Mus. Contbr. articles to profl. jours. Recipient U. Fla. Grad. Sch. faculty research award for bot. expdn. to Venezuela, 1972. Mem. Am. Bryological and Lichenological Soc., Am. Soc. Plant Taxonomists, Bot. Soc. Am., Brit. Bryological Soc., Nordic Bryological Soc., Internat. Assn. Bryologists, Internat. Assn. Plant Taxonomy, Assn. Tropical Biology, Torrey Bot. Club, Assn. Southeastern Biologists, Sigma Xi. Subspecialties: Taxonomy; Floristics. Current work: Taxonomy and floristics of tropical bryophyte floras. Home: 3859 NW 32d Pl Gainesville FL 32606 Office: Fla State Mus U Fla Gainesville FL 32611

GRIFFIN, DONALD R(EDFIELD), zoology educator; b. Southampton, N.Y., Aug. 3, 1915; s. Henry Farr and Mary Whitney (Redfield) G.; m. Ruth M. Castle, Sept. 6, 1941 (div. Aug. 1965); children: Nancy Griffin Jackson, Janet Griffin Abbott, Margaret, John H.; m. Jocelyn Crane, Dec. 16, 1965. B.S., Harvard U., 1938, M.A., 1940, Ph.D., 1942. Jr. fellow Harvard U., Cambridge, Mass., 1940-41, 46, research assoc., 1942-45, prof., 1953-65; asst. prof. Cornell U., Ithaca, N.Y., 1946-47, assoc. prof., 1947-52, prof., 1952-53, Rockefeller U., N.Y.C., 1965—, trustee, 1973-76; pres. Harry Frank Guggenheim Found., 1979-84. Author: Listening in the Dark, 1958 (Nat. Acad. Scis. Elliot medal 1961), Echoes of Bats and Men, 1959, Animal Structure and Function, 1962, Bird Migration, 1964 (Phi Beta Kappa prize 1966), The Question of Animal Awareness, 1976. Mem. Am. Ornithologists Union, Am. Soc. Zoologists, Am. Physiol. Soc., Ecol. Soc. Am., Am. Acad. Arts and Scis., Nat. Acad. Scis., Am. Philos. Soc., Am. Soc. Naturalists, Phi Beta Kappa, Sigma Xi. Subspecialties: Ethology; Comparative neurobiology. Current work: Animal navigation, sensory physiology; cognitive ethology. Office: Rockefeller U 1230 York Ave New York NY 10021

GRIFFIN, GERALD DUANE, aeronautical engineer; b. Athens, Tex., Dec. 25, 1934; s. Herschel Hayden and Helen Elizabeth (Boswell) G.; m. Sandra Jo Huber, Apr. 19, 1958; children: Kirk Laurence, Gwendolyn Diane. B.S. in Aero. Engring, Tex. A. and M. U., 1956. With Douglas Aircraft Co., Long Beach, Calif., 1956, Lockheed Missiles and Space Co., Sunnyvale, Calif., 1960-62, Gen. Dynamics Co., Ft. Worth, 1962-64; with NASA, 1964-81, 82—; dir. NASA Johnson Space Center, Houston, 1982—; v.p. systems engring. and mgmt. Scott Sci. and Tech., Inc., Lancaster, Calif., 1981-82. Served as navigator USAF, 1956-60. Recipient Exceptional Service medal, Outstanding Leadership medal NASA; Presdl. rank Meritorious Sr. Exec.; Presdl. Medal of Freedom group award; named Old Master Purdue U. Asso. fellow AIAA; mem. Tau Beta Pi. Subspecialty: Aerospace engineering and technology. Office: Office of Dir Johnson Space Ctr Houston TX 77058

GRIFFIN, THOMAS WILLIAM, physician, researcher, educator; b. Glen Ridge, N.J., Apr. 26, 1946; s. Peter C. and Kathleen G.; m. Mary Ellen Rybak, Nov. 30, 1977. B.S. magna cum laude, Boston Coll., 1968; M.D., Cornell U., 1972. Diplomate: Am. Bd. Internal Medicine, also sub-splty. in Oncology. Med. intern Peter Bent Brigham Hosp., Boston, 1972-73, resident in medicine, 1973-75; fellow in med. oncology Sidney Farber Cancer Center, Boston, 1975-77, spl. fellow in med. oncology, 1977-78; clin. fellow in medicine Harvard U. Med. Sch., Boston, 1972-78; asst. prof. medicine U. Mass. Med. Sch., Worcester, 1978—, Am. Cancer Soc. jr. clin. faculty fellow, 1981. Contbr. articles to med. jours. Recipient award Minn. Found., 1980. Mem. ACP (assoc.), AAAS, Mass. Med. Soc., Worcester Dist. Med. Soc., Am. Soc. Clin. Oncology, New Eng. Cancer Soc., Alpha Omega Alpha. Subspecialties: Cancer research (medicine); Oncology. Office: 55 Lake Ave N Worcester MA 01605

GRIFFISS, J(OHN) MCLEOD, bacterial immunochemist; b. Chattanooga, July 9, 1940; s. James Johnston and Mary Virginia (Keating) G.; m. Helle Krarup, Mar. 23, 1966; 1 son, Padrick McLeod. B.A., U. N.C., 1962; M.D., Yale U., 1966. Diplomate: Am. Bd.Internal Medicine. Intern King County Hosp., Seattle, 1966-67; resident in medicine U. Wash. Affiliated Hosps., Seattle, 1967-68; fellow in infectious diseases Walter Reed Army Inst. Research, Washington, 1971-73, med. research officer, 1973-75, acting chief dept. bacterial diseases, 1976, sr. med. research officer, 1977-79; staff physician Infectious Disease Service Walter Reed Army Med. Ctr., 1971-79; asst. prof. medicine Harvard Med. Sch., Boston, 1979—; assoc, in medicine Brigham and Women's Hosp., Boston, 1979, physician, 1982—; cons NIH, Dept. Def.; epidemiologist, cons. in infectious diseases USPHS Hosp., Brighton, Mass., 1980-81, Dana Farber Cancer Inst., Boston. Contbr. articles to profl. jours. Served to maj. USAF, 1968-71; to lt. col. U.S. Army, 1978-79. Recipient Nat. Def. Service award, 1968; research career devel. award, 1979-84. Mem. AMA, Am. Soc. for Microbiology, Am Fedn. for Clin. Research, Am. Assn. Immunologists, Infectious Diseases Soc. Am. Subspecialties: Epidemiology; Infectious diseases. Current work: Epidemiology and immunochemistry of Neisseria meningitidis. Immunological function of serum IGA. Immunology of bacterial polysaccharide vaccines. Immunochemistry of Neisseria gonorrhoeae. Epidemiology and immunology of Gram negative sepsis. Office: Harvard Med Sch Channing Lab 180 Longwood Ave Boston MA 02115

GRIFFITH, BELVER CALLIS, information science educator; b. Hampton, Va., Mar. 28, 1931; m. Carolyn Adams, 1982; children: Caitlin, Wynne, Leigh. B.A., U. Va., 1951; M.A., U. Conn., 1953, Ph.D. in Psychology, 1957. Mem. tech. staff Bell Telephone Labs., Murray Hill, N.J., 1956-57; research assoc. Edward R. Johnstone Tng. & Research Center, Bordentown, N.J., 1957-62; assoc. dir., prin. investigator NSF Project Sci. Info. Exchange in Psychology, Am. Psychol. Assn., Washington, 1961-69; prof. library sci. Grad. Sch. Library Sci., Drexel U., Phila., 1970—; assoc. in communications Annenberg Sch. Communications, U. Pa., Phila., 1966-67; cons. in field. Mem. AAAS, Am. Psychol. Assn., Am. Soc. Info. Sci., Psychonomic Soc., Internat. Union Psychol. Sci. (com. mem. 1968-70). Subspecialty: Information science policy studies. Current work: Communication in science and technology; organization of knowledge and information; technology transfer; information products and services; social structure of science. Office: Coll Info Studies Drexel U Philadelphia PA 19104

GRIFFITH, OWEN WENDELL, biochemistry educator; b. Oakland, Calif., June 19, 1946; s. Charles H. and Gladys C. (Farrar) G. B.A., U. Calif.-Berkeley, 1968; Ph.D., Rockefeller U., 1975. Postdoctoral fellow Cornell U. Med. Coll., N.Y.C., 1974-77, instr., 1977-78, asst. prof., 1978-81, assoc. prof., 1981—. Contbr. articles on biochemistry to profl. jours. NIH grantee, 1979—; March of Dimes grantee, 1979-82; Irma T. Hirschl Found. fellow, 1981-85. Mem. Am. Soc., Biol. Chemists, Am. Chem. Soc., N.Y. Acad. Scis., AAAS. Subspecialties: Biochemistry (medicine); Medicinal chemistry. Current work: Amino acid and fatty acid metabolism, enzyme mechanisms, design and synthesis of enzyme specific substrates and inhibitors. Home: 430 E 63d St Apt 6H New York NY 10021 Office: Cornell U Med Coll 1300 York Ave New York NY 10021

GRIFFITH, WILLIAM SCHULER, mathematical scientist, applied probabilist; b. Bradford, Pa., Oct. 10, 1949; s. William H. and Hazel Marjorie (Schuler) G.; m. Deborah Ann Fragale, Aug. 12, 1972; 1 son, Mark Robert. B.S., Grove City (Pa.) Coll., 1971; M.A., U. Pitts., 1973, Ph.D., 1979. Lectr. Carlow Coll., Pitts., 1976-77; mathematician Westinghouse Research and Devel. Ctr., Pitts., 1977-79; asst. prof. stats. U. Ky.-Lexington, 1979—. Mem. Am. Stats. Assn., Inst. Math. Statis., Biometrics Soc., Ops. Research Soc. Am., Soc. Indsl. and Applied Math. Subspecialties: Probability; Statistics. Current work: Mathematical theory of reliability, applied probability. Home: 3367 Nevius Dr Lexington KY 40513 Office: Dept Stats U Ky Lexington KY 40506 0027

GRIFFITHS, RICHARD EDWIN, astrophysicist, consultant; b. Gt. Britain, Sept. 19, 1947; came to U.S., 1976, naturalized, 1982; s. Edwin Lloyd and Anne Mary (Jones) G.; m. Michele Antoinette Pallatt, Apr. 11, 1978; 1 son, Daniel M. B.Sc. with honors (Univ. entrance scholar), U. London, 1968; Ph.D., U. Leicester, Eng., 1971. Research fellow Centre D'Etudes Nucleaires de Saclay, France, 1971-72; research fellow U. Leicester, 1972-74, research assoc., 1974-76; astrophysicist Harvard-Smithsonian Ctr. for Astrophysics, Cambridge, Mass., 1976-83; assoc. astronomer Space Telescope Sci. Inst., Balt., 1983—; cons. x-ray imaging systems. Contbr. numerous articles to profl. jours. Fellow Royal Astron. Soc.; mem. Am. Astron. Soc., Internat. Astron. Union. Unitarian. Club: Vintage Sports Car. Subspecialties: 1-ray high energy astrophysics; Satellite studies. Current work: Developed x-ray detectors and spectrometers; instrument scientist for charge-coupled device camera on space telescope; optical identification of cosmic x-ray sources. Home: 9 Middleton Ct Baltimore MD 21212 Office: Space Telescope Science Inst Homewood Campus Baltimore MD 21218

GRIFFY, THOMAS ALAN, physicist, educator, researcher; b. Oklahoma City, Dec. 16, 1936; s. Judson H. and Dicie (Johnston) G.; m. Peggy Lynn Walker, June 6, 1958; children: David, Alan, Marjorie. B.A., Rice U., 1959, M.A., 1960, Ph.D., 1961. Asst. prof. physics Duke U., 1961-62; research assoc. High Energy Physics Lab., Stanford U., 1962-65; assoc. prof. physics U. Tex.-Austin, 1965-68, prof., 1968—, chmn. dept. physics, assoc. dean, 1970-73. Contbr. articles to profl. jours. Fellow Am. Phys. Soc. Methodist. Subspecialties: Acoustics; Nuclear physics. Current work: Underwater acoustics, nuclear energy. Office: Dept Physics U Tex Austin TX 78712

GRIM, EUGENE, physiology educator, researcher; b. Stillwater, Okla., July 19, 1922. B.S., Kans. State U., 1945, M.S., 1946; Ph.D. in Physiol. Chemistry, U. Minn., 1950. From instr. to assoc. prof. U. Minn., Mpls., 1952-62, prof. physiology, 1962—, head dept. physiology, 1968—; mem. physiology study sect. NIH, 1967-69, chmn. sect., 1969-71. Sect. editor: Jour. Physiology, 1966-68. Recipient Lederle med. faculty award, 1954; Career Devel. award USPHS, 1963-68; USPHS sr. research fellow U. Minn., 1958-63. Fellow AAAS; mem. Am. Physiol. Soc., Am. Chem. Soc., Am. Gastroent. Soc., Biophys. Soc. Subspecialty: Physiology (biology). Office: Dept Physiology U Minn Minneapolis MN 55455

GRIMES, DALE MILLS, electrical engineering educator; b. Marshall County, Iowa, Sept. 7, 1926; s. LeRoy and Helen (Mills) G.; m. Janet LaVonne Moore, Mar. 22, 1947; children: Prudence Rae, Craig Alan. B.S. in Physics, Math. and Chemistry, Iowa State U., 1950, M.S. in Physics and Math, 1951; Ph.D. in Elec. Engring, U. Mich., Ann Arbor, 1956. From research asso. to prof. elec. engring. U. Mich., 1951-76; chief scientist Conductron Corp., Ann Arbor, 1960-63; prof. elec. engring., chmn. dept. U. Tex., El Paso, 1976-79; prof. elec. engring., head dept. Pa. State U., 1979—; cons. to govt. and industry. Author: Electromagnetism and Quantum Theory, 1969, Automotive Electronics, 1974; also articles on radar, biconical antennas, electromagnetic radiation. Served with USNR, 1943-46. Mem. IEEE, Am. Phys. Soc., AAAS, Am. Soc. Engring. Edn. Patentee ferrite radar absorbing material, magnetic absorbers. Home: 2548 Sleepy Hollow Dr State College PA 16801 Office: Elec Engring Dept Pa State Univ University Park PA 16802

GRIMLEY, LIAM KELLY, educator; b. Dublin, Ireland, Apr. 4, 1936; came to U.S., 1970; s. William and Eileen (Kelly) G.; m. Marie Sadon, Aug. 26, 1973; children: Kevin, Conor. B.A., Nat. U. Ireland, 1960; L.Ph., Faculte Libre, Paris, 1963; H.D.Ed., Clongowes Wood Coll., Ireland, 1964; Th.B., Inst. Philosophy and Theology, Dublin, 1968, S.T.L., 1970; M.Ed., Kent State U., 1971, Ph.D., 1973. Tchr. English Lycee Moderne, LePuy, France, 1961-62; asst. dir. Summer Sch. English, Observatorio del Ebro, Tortosa, Spain, 1961-62; tchr. math and modern langs. Clongowes Wood High Sch., Ireland, 1963-64; tchr. math, classical langs. St. Ignatius Elementary and Secondary Sch., Galway, Ireland, 1964-66; instr. statistics and probability theory Univ. Coll., Galway, 1965-66; prof. theology Conf. Major Religious Superiors, Dublin, 1969-70; counselor Newman Center, Syracuse U., 1970; tchr. social studies Walsh Jesuit High Sch., Cuyahoga Falls, Ohio, 1971; asst. dir. Ohio Soc. Crippled Children and Adults, Tiffin, summer 1971; intern sch. psychologist Kent State U., 1972-73, research and devel. dir. lab. schs., 1972-73; prof. spl. edn. Ind. State U., Terre Haute, 1973—, chmn. dept., 1975-81, dir. Inst. Continuing Edn. in Psychology, 1976-78; cons. Joseph P. Kennedy Found., 1973—; mem. State Adv. Com., Div. Pupil Personnel, 1975-78; Mem. Ind. State Manpower Steering Com., 1977-80; chmn. State Adv. Bd. on Pupil Personnel Services, 1980—; mem. State Council on Edn. of Handicapped, 1979—. Editor: The Sch. Psychology Digest, 1976-79; contbr. articles to profl. jours. Mem. Am. Psychol. Assn., Ind. Psychol. Assn. (pres. div. sch. psychology 1980-81), Nat. Assn. Sch. Psychologists. Roman Catholic. Current work: Developer of Grimley Personality scale for children. Home: 43 Allendale Terre Haute IN 47802 Office: Dept Spl Edn Sch Psychology and Communication Disorders Ind State U Terre Haute IN 47809

GRIMM, ELIZABETH ANN, cellular immunologist; b. Charleston, W.Va., Nov. 24, 1949; d. Harper Granville and Nellie (Simmons) G.; m. Jack Alan Roth, Jan. 29, 1945. A.B. in Chemistry, Randolph-Macon Woman's Coll., 1971; Ph.D. in Microbiology and Immunology, UCLA, 1979. Research assoc. Harvard Med. Sch., 1971, tech. supr. research lab., 1972-73; tech. supr. tumor immunology research labs. Sepulveda (Calif.) VA Hosp., 1973-75; teaching asst. microbiology and immunology UCLA Sch. Medicine, 1975-79; postdoctoral fellow Molecular Biology Inst., UCLA Sch. Arts and Scis., 1979-80; cancer expert surgery br. Nat. Cancer Inst., NIH, Bethesda, Md., 1980—. Contbr. articles to profl. jours. Muscular Dystrophy Assn. fellow, 1979. Mem. Am. Assn. Immunologists, Am. Assn. Cancer Research. Subspecialties: Immunology (medicine); Cancer research (medicine). Current work: Research in basic cellular immunology with direct applications to adoptive immunotherapy of human cancer; specifically immunologic memory and development of cytolic effector cells. Office: Nat Cancer Inst NIH Bldg 10 Rm 10N116 Bethesda MD 20205

GRIMM, LOUIS J., mathematics educator; b. St. Louis, Nov. 30, 1933; s. Louis and Florence (Hammond) G.; m. Barbara Mitko, May 6, 1967; children: Thomas Hammond, Mary Elizabeth. B.S., St. Louis U., 1954; M.S., Ga. Inst. Tech., 1960; Ph.D., U. Minn., 1965. Asst. prof. U. Utah, Salt Lake City, 1965-69; assoc. prof. U. Mo., Rolla, 1969—74; vis. prof. U. Nebr., Lincoln, 1978-79; prof. math. U. Mo., Rolla, 1974—, chmn. math. and stats. dept., 1981—. Contbr. articles in field to profl. jours. NSF research grantee, 1969-73, 76-79; Nat. Acad. Scis. exchange scholar, 1981. Mem. Am. Math. Soc., Soc. Indsl. and Applied Math., Gesellschaft für angewandte Mathematik und Mechanik. Subspecialties: Applied mathematics; Numerical analysis. Current work: Analytic theory of ordinary differential and difference equations. Numerical solution of ordinary and functional differential equations. Office: Dept Math and Stats U Mo Rolla MO 65401

GRIMSLEY, DOUGLAS LEE, psychology educator, consultant; b. San Diego, Calif., Sept. 18, 1939; s. Ralph Waldo and Anna (Gall) G.; m. Mary Margaret Beasley, Apr. 23, 1965; children: Stephen, Cynthia. B.S. with honors, Fla. State U., 1961; Ph.D., Syracuse U., 1964. Asst. prof. Fla. State U., 1964-66; sr. scientist George Washington U., Monterey, Calif., 1966-68; assoc. prof. U. So. Miss., Hattiesburg, 1968-70; prof., chmn, dept. psychology U. N.C., Charlotte, 1970—; speaker sci. meetings. Contbr. articles to profl. jours. USPHS trainee NIMH, Syracuse U., 1961-64; named tchr. of excellence U. N.C.-Charlotte and N.C. Nat. Bank., 1976. Mem. Am. Psychol. Assn., AAAS, Southeastern Psychol. Assn., Biofeedback Soc. Am. Subspecialties: Physiological psychology; Biofeedback. Current work: Biofeedback; research and application; central nervous system, preference behavior and body needs, electrophysiological recordings. Office: Psychology Dept U North Carolina Charlotte NC 28223

GRINDEL, JOSEPH MICHAEL, pharm. co. exec.; b. Kansas City, Mo., Dec. 18, 1946; s. Edward A. and Inez (Weber) G.; m. Cecelia M. Gatson, Aug. 1, 1970; children: Charles, Mary, David. B.Sc., Benedictine Coll., Kans., 1969; Ph.D., U. Kans., 1973. Chief clin. drug metabolism lab. Walter Reed Army Inst. Research, Washington, 1973-76; with McNeil Pharm., Spring House, Pa., 1976—, dir. dept. drug metabolism, 1980-82, exec. dir. research and devel. project planning, 1982—. Contbr. articles to profl. jours. Cubmaster Boy Scouts Am., 1979-82; pres. St. Stanislaus Sch. Bd., 1982—. Served to 1st lt. AUS, 1973-76. Mem. Am. Chem. Soc., Am. Soc. Pharmacology and Exptl. Therapeutics, Drug Metabolism Discussion Group. Roman Catholic. Current work: Drug metabolism and pharmacokinetics in man and animal; analytical methodology in biofluids. Home: 1787 Cindy Ln Hatfield PA 19440 Office: McNeil Pharm McKean Rd Spring House PA 19477

GRISELL, RONALD DAVID, computer scientist; b. Forest Grove, Oreg., May 26, 1944; s. Roll and Opal Evelyn (Spelbrink) G. B.S. in Physics, Harvey Mudd Coll., Claremont, Calif., 1966; M.A. in Math, U. Oreg., Eugene, 1970, Ph.D., 1971. Engr. Tektronix Inc., Beaverton, Oreg., 1965-66; asst. prof. Pahlavi U., Shiraz, Iran, 1971-72; cons. I.M.R. Assocs., Munich, W. Ger., 1973-74, dir. computing div., N.Y.C. and Tex., 1979—; sr. research assoc. U. Tex. Med. Br., Galveston, 1974-78; coordinator computer sci. Adelphi U., Garden City, N.Y., 1978—; reviewer NSF, 1978, 80. Author: Optical and Ultrasonic Data Processing, 1974, Mathematics for Design and Digital Systems, 1982. Asst scout master Boy Scouts Am.; councillor New Life Youth Camps, Eugene, 1966-70. Recipient Bausch & Lamb Sci. award Bausch & Lamb, Inc., 1968. Mem. IEEE, Assn. Computing Machinery, N.Y. Acad. Scis. Subspecialties: Artificial intelligence; Neurobiology. Current work: Pattern recognition, models of vision, neurodynamics of individual neurons and networks, computer aids to data visualization in large databases, cognitive processing. Home: 201 E Starling Dr Austin TX 78753 Office: Adelphi U Garden City NY 11530

GRISSOM, MICHAEL PHILLIP, radiation science specialist, naval officer; b. Warren, Ohio, Oct. 25, 1948; s. Edward Henry and Cala Beatrice (Pickens) G.; m. Linda Marie Gerou, June 25, 1974. B.S. in Biol. Scis, Colo. State U., 1970; M.S.E. in Nuclear Sci. and Engring, Catholic U. Am., 1977. Commd. ensign Med. Service Corps U.S. Navy, 1971, advanced through grades to lt. comdr., 1980; radiation health officer in support of med. div. and Navy Nuclear Power Program, 1972-74, research biophysicist Armed Forces Radiobiology Research Inst., Bethesda, Md., 1974-76, radiation specialist Armed Forces Radiobiology Research Inst., 1977-82, dir. med. records search Navy Nuclear Test Personnel Rev., Arlington, Va., 1982—; cons. health physicist Three Mile Island Nuclear Generating Sta. during recovery ops., 1979, 80. Contbr. articles to profl. jours. Active Mountaingate Homeowners' Assn., Frederick, Md., 1978-80, including acting treas., 1979, bd. dirs., 1980. Decorated Nat. Def. medal, Def. Meritorious Service medal; Recipient Citizenship award Col. SAR, 1966, Order of World Wars award San Diego Naval Hosp., 1971; Nat. Heart, Lung and Blood Inst. grantee, 1974-75. Mem. Health Physics Soc., Soc. Nuclear Medicine, Am. Assn. Physicians in Medicine, AAAS, U.S. Naval Inst. Club: U.S. Navy Officer's (Bethesda, Md.). Subspecialties: Nuclear medicine; Imaging technology. Current work: Development of imaging techniques to research effects of various qualities, quantities of ionizing radiations on animal systems; application of single photon tomographic methods, planar imaging, and nuclear magnetic resonance; use of receptor bound radiopharmaceuticals in nuclear imaging. Home: 1494 Dogwood Dr Frederick MD 21701 Office: CWB Room 756 1300 Wilson Blvd Arlington VA 22209

GRISWOLD, CHARLES EARL, marine equipment designer; b. Davenport, Iowa, June 11, 1935; s. Charles Burton and Dorothea A. (Natzke) G.; m. Joyce Elizabeth Slaughter, Apr. 26, 1968. B.A., U. Wash.-Seattle, 1964. Gen. mgr. Tri-Met Inc., Seattle, 1972-73; inspector Engring. Constrn. Service, Honolulu, 1974-75; cons. Sea-Quest Assocs., Bombay, India, 1976-77; ops. mgr. New Eng. Ocean Service, Boston, 1978-80; dir. Griswold Marine Assocs., Seattle, 1980—; research technician Rocket Research Co., Seattle, 1964-67; submersible instr. N.W. Submarine Inst., Seattle, 1967-68; research vessel operator USPHS, Seattle, 1962-63. Cinematographer for: film Sand Island Outfall, 1974 (Indsl. Film of Yr. 1975). Mem. Marine Tech. Soc., Nat. Assn. Underwater Instrs. Republican. Universalist. Subspecialties: Ocean engineering; Offshore technology. Current work: Directing the development of an acoustically transmitted, present time, underwater television system. Inventor diving helmet, 1972. Office: Griswold Marine Assocs PO Box 2754 Seattle WA 98111

GRISWOLD, DON E., immunopharmacologist; b. Newton, Kans., June 24, 1943; s. Sherwin B. and May (Langerman) G.; m. Karen Diane Krehbiel, Aug. 28, 1965; children: Brian C., Darren C. Ph.D., U. Kansas., 1969. Research fellow Brown U., Providence, 1969-71, instr., 1971-73, asst. prof. medicine, 1973-74; assoc. sr. investigator Smith Kline & French Labs., Phila., 1974-78, sr. investigator, 1978—. Contbr. articles to profl. jours. Mem. Am. Assn. for Cancer Research, AAAS, N.Y. Acad. Sci., Sigma Xi, Beta Beta Beta, Lambda Delta Lambda. Methodist. Subspecialty: Immunopharmacology. Current work: Autoimmunity, immunoregulatory agents, host defense mechanisms, immunoregulatory circuits. Office: 1500 Spring Garden St Philadelphia PA 19101

GRIVETTI, LOUIS EVAN, nutrition educator; b. Billings, Mont, Sept. 13, 1938; s. Rex Michael and Blanche Irene (Carpenter) G.; m. Georgia Mayerakis, May 25, 1967; 1 dau., Joanna. A.B., U. Calif.-Berkeley, 1960, M.A., 1962, Ph.D., 1976. Asst. prof. U. Calif.-Davis, 1976-80, assoc. prof., 1980—; research asst. Vanderbilt U., Nashville, 1964-70; adminstrv. asst. Meharry Med. Coll., Nashville, 1973-75, Ministry of Health, Gaborone, Republic of Botswana, 1973-75; chmn. grad. group in nutrition U. Calif.-Davis, 1983—. Author: The Gift of Osiris, 1977; contbr. articles on nutrition to profl. jours. Served to lt. USPHS, 1962-64. Recipient Nutrition Founds. award, 1977. Mem. Assn. Am. Geographers, Botswana Soc., Soc. Nutrition Edn., West Coast Nutritional Anthropologists (past pres., v.p., dir.). Subspecialties: Nutrition (biology); Nutrition (medicine). Current work: Nutritional anthropology, food habits, food history, cultural determinants of dietary behavior, food fads, diet and nutritional status, ecology of malnutrition. Home: 3308 Monterey Ave Davis CA 95616 Office: Dept Nutrition U Calif Davis CA 95616

GROB, DAVID, physician, educator, researcher; b. N.Y.C., Feb. 23, 1919; s. Hyman and Fannie (Baumwall) G.; m. Elizabeth Nussbaum, Dec. 26, 1948; children: Charles, Susan, Emily, Philip. B.S., CCNY, 1937; postgrad., Columbia U., 1937-38; M.D., Johns Hopkins U., 1942. Med. intern Johns Hopkins Hosp., Balt., 1942-43, asst. resident in medicine, 1945-46, 47-48; fellow in medicine Johns Hopkins U. Sch. Medicine, 1946-47, instr., 1948-51, asst. prof., 1951-55, instr. pharmacology and exptl. therapeutics, 1953-54; assoc. prof. medicine, 955-58; prof. medicine SUNY Coll. Medicine, Bklyn., 1958—, asst. dean, 1962-67, clin. assoc. dean, 1979—; dir. med. services Maimonides Med. Ctr., Bklyn., 1958—, dir. research and edn., 1960-67, dir. med. edn., 1970—. Editor: Myasthenia Gravis, 1976, Myasthenia Gravis: Pathophysiology and Management, 1981; contbr. over 110 articles to sci. jours., chpts. to sci. books. Mem. med. adv. bd. Myasthenia Gravis Found., 1953—, chmn., 1961-63. Served to 1st lt. and capt. M.C., U.S. Army, 1943-45; ETO. Decorated Bronze Star; recipient Townsend D. Harris medal CCNY Alumni Assn., 1964; Honor award Fedn. Jewish Philanthropies N.Y., 1964; Humanitarian award Myasthenia Gravis Found., 1963; Kermit Osserman award, 1982. Fellow ACP; mem. Am. Soc. Clin. Investigation, Am. Physiol. Soc., Am. Soc. Pharmacology and Exptl. Therapeutics, Assn. Am. Physicians, Phi Beta Kappa, Alpha Omega Alpha, Assn. Program Dirs. in Internal Medicine (councilor 1977-81, pres.-elect 1978-79, pres. 1979-80). Subspecialties: Internal medicine; Neurophysiology. Current work: Physiology, pharmacology and immunology of diseases of the neuromuscular system, Key terminology: neuromuscular transmission, muscle function, fatigue, acetylcholine receptor and antibodies, myasthenia gravis, myastenic syndromes, weakness, immunobiology and neuroimmunology. Home: 20 Fern Dr Roslyn NY 11576 Office: 4802 10th Ave Brooklyn NY 11219

GROB, ROBERT LEE, educator; b. Wheeling, W.Va., Feb. 13, 1927; s. William E. and Mary Margaret (Shanley) G.; m. Marjorie D. Sage, Aug. 4, 1928; children: R. Kent, G. Duane, J. Allyson, M. Michele. B.S., Coll. Steubenville, 1957; M.S., U. Va., 1954, Ph. D., 1955. Research analytical chemist Esso Research, Linden, N.J., 1955-57; prof. chemistry Wheeling (W.Va.) Coll., 1957-63; prof. analytical chemistry Villanova (Pa.) U., 1963—. Served with U.S. Army, 1945-47. Pratt fellow U. Va., 1954. Mem. Am. Chem. Soc., Am. Inst. Chemists, Chromatography Discussion Group London, Chromatography Forum Delaware Valley, Sigma Xi. Republican. Roman Catholic. Subspecialties: Analytical chemistry; Ecosystems analysis. Current work: Gas and liquid chromatography, trace metal analysis, environmental chemistry. Office: Chem Dept Villanova U Villanova PA 19085

GRODBERG, MARCUS GORDON, pharmaceutical company executive; b. Worcester, Mass., Jan. 27, 1923; s. Isaac and Rosalie (Hirsch) G.; m. Shirley Florence Merkle, April 15, 1951; children: Joel David, Kim Gordon, Jeremy Daniel. A.B., Clark U., 1944; M.S., U. Ill., 1948. Jr. research chemist Schenley Labs., Inc., Lawrenceburg, Ind., 1944-47; chemist Marine Products Co., Boston, 1948-50, Brewer & Co., Inc., Worcester, 1950-55; tech. dir. Gray Pharm. Co., Inc., Newton, Mass., 1955-58; dir. research and devel. Hoyt Labs., div. Colgate-Palmolive Co., Norwood, Mass., 1958—; cons. case history project Harvard U. Grad. Sch. Bus. Adminstrn., 1957. Asst. leader Cub Scouts Boy Scouts Am., Newton; solicitor United Fund, Newton. Ellis Fund scholar, 1941-42; recipient cert. of merit for disting. service to leadership in pharm. research Dictionary of Internat. Biography, Cambridge, Eng., 1968. Mem. Internat. Assn. for Dental Research,

Am. Soc. Dentistry for Children, Orthopedic research Soc., Am. Pharm. Assn., ADA, AAAS, Acad. Pharm. Scis., N.Y. Acad. Scis., Acad. of Dentistry for the Handicapped, Phi Alpha. Jewish. Subspecialties: Pharmaceutical products research and development management. Current work: Research and development dental drug products, fluoride for dental and bone diseases, antimicrobial agents for periodontal disease, phosphate for calcium metabolic disorders. Patentee in field. Home: 111 Hyde St Newton MA 02161 Office: Hoyt Labs 575 University Ave Norwood MA 02062

GROEBER, EDWARD OTTO, JR., electronics engineer; b. Hempstead, N.Y., June 28, 1946; s. Edward Otto and Evelyn Ruth (Laufle) G.; m. Matilda Brown Eaton, June 23, 1968; children: Jennifer Jane, Jill Ann. B.S. in Nuclear Engring, N.Y. U.-Bronx, 1968; M.E.E., Fairleigh Dickinson U., 1980. Sect. chief U.S. Air Force Weapons Lab., Kirtland AFB, N.Mex., 1970-72; with CS&TA Lab., Ft. Monmouth, N.J., 1972—, br. chief., 1981—; Mem dosimeter task group Am. Nat. Standards Inst., Washington, 1982—. Contbr. numerous articles in field to profl. publs. Served to capt USAF, 1968-72. Recipient Outstanding Performance awards CS&TA Lab., 1977-79. Mem. Am. Nuclear Soc., Eta Kappa Nu. Democrat. Subspecialties: Electronics; Nuclear engineering. Current work: Management of the research, development and initial production of nuclear radiation detection devices and systems for the U.S. Army. Home: 24 Woodland Ln Jackson NJ 08527 Office: CS&TA Lab DELCS-K Fort Monmouth NJ 07703

GROENIER, WILLIAM SAMUEL, chemical engineering executive, researcher; b. Chgo., Feb. 5, 1936; s. Willis Lambert and Bernice E. (Kress) G.; m. Janet Marie Goodenow, June 19, 1960; children; Laurice Ann, Katherine Elizabeth. B.S., Northwestern U., 1958, M.S., 1959. Chem. engr. Tee-Pak, Inc., Chgo., 1954-59; chem. engr. Union Carbide Corp., Oak Ridge, 1959-62, Oak Ridge Nat. Lab., 1962-75, program mgr., 1975-80, sect. head, 1980—. Contbr. numerous articles to govt. reports and profl. jours. Mem Am. Nuclear Soc., Am. Inst. Chem. Engrs., Sigma Xi. Republican. Subspecialties: Chemical engineering; Nuclear fission. Current work: Manage chemistry and chemical engineering activites of research and development nature for reprocessing nuclear breeder reactor fuels. Office: Oak Ridge National Laboratory PO Box X Oak Ridge TN 37830

GROGAN, JAMES BIGBEE, immunologist, educator; b. Edwards, Miss., May 15, 1932; s. Kenneth Forbes and Elfie (Bigby) G.; m. Nita Pauline Young, June 17, 1956; children: Frankie L., Paula D. B.S., Miss. Coll., 1955; M.S. in Bacteriology, U. Wis., 1957; Ph.D., U. Miss., 1963. Teaching asst., research asst. U. Wis.-Madison, 1955-57; supr. surg. research bacteriology lab. U. Miss. Sch. Medicine, 1957-63, fellow dept. microbiology, 1957-63, asst. prof. surgery, 1965-68, asst. prof. microbiology, 1965-72, assoc. prof. surgery, 1968-74, assoc. prof. microbiology, 1972—, prof. surgery, 1974—. Contbr. articles to profl. jours. Served with U.S. Army, 1950-52. NIH grantee. Mem. Am. Soc. Microbiology, Miss. Acad. Scis., Transplantation Soc., Reticuloendothelial Soc., Am. Assn. Immunologists, So. Soc. Clin. Investigation, Sigma Xi. Baptist. Subspecialties: Immunology (agriculture); Microbiology. Home: 5124 N Hill Dr Jackson MS 39211 Office: U Miss Med Center 2500 N State St Jackson MS 39216

GROGAN, WILLIAM MCLEAN, biochemist, educator; b. Knoxville, Tenn., Feb. 4, 1944; s. William McLean and Virginia Isabelle (Petree) G.; m. Virginia Griffith, Aug. 21, 1965; children: Margaret Kathleen, Scott Edward. B.S., Belmont Coll., Nashville, 1967; Ph.D., Purdue U., 1972. Research assoc. Vanderbilt U., Nashville, 1972-75; assoc. prof. Med. Coll. Va., Richmond, 1975—; co-dir. Cansort lab. Med. Coll. Va., Va. U. Cancer Ctr. NIH Prin. Investigator grantee, 1979-82, 82-85. Mem. Am. Soc. Biol. Chemists, Am. Oil Chemists Soc., AAAS, Soc. Analytical Cytology. Baptist. Subspecialties: Biochemistry (medicine); Reproductive biology. Current work: Role of lipids in differentiation and development of the testis; regulation of lipid metabolism; analytical and preparative flow cytometry. Home: 9307 Wishart Rd Richmond VA 13229 Office: Medical Coll Va Dept Biochemistry Richmond VA 23298

GRONHOVD, GORDON HARLAN, coal technology consultant; b. Grand Forks, N.D., Jan. 26, 1927; s. Edwin M. and Helmina (Hendrickson) G.; m. Dorothy H. Hylden, June 19, 1949; children: Paul, Allen, Keith. B.S. in Mech. Engring, U. N.D., 1950. Field engr. J.F. Case Tractor Works, Racine, Wis., 1950-51; project engr. Gen. Motors Corp., Milw., 1951-53; coal tech. researcher U.S. Bur. Mines, ERDA, and Dept. Energy, Grand Forks, 1953-80; dir. Grand Forks Energy Tech. Ctr., 1973-80; cons. coal tech., Grand Forks, 1980—; assoc. mem. grad. faculty U. N.D. Contbr. coal research articles to profl. jours. Recipient Outstanding Service award U.S. Dept. Energy, U. N.D., 1981; Disting. Career Service award Dept. Energy, 1980. Mem. ASME, N.D. Acad. Sci., Sigma Xi. Lutheran. Subspecialties: Coal; Combustion processes. Current work: Low-rank coal research and technology and consultant in field. Home and office: 3004 Belmont Rd Grand Forks ND 58201

GROSCH, CHESTER ENRIGHT, oceanography and computer science educator, consultant; b. Hoboken, N.J., Jan. 13, 1934; s. Chester Enright and Winifred (Gorman) G.; m. Joan Casmere, June 30, 1956; children: Joseph, Liza, Crys, Margaret. M.E., Stevens Inst. Tech., 1956, M.S., 1959, Ph.D., 1967. Research scientist Davidson Lab. Stevens Inst. Tech., 1956-66; research assoc. Hudson Lab. Columbia U., 1966-68; research scientist Teledyne-Isotopes, Westwood, N.J., 1968-69; assoc. prof. physics and computer sci. Pratt Inst., 1969-73; prof. Old Dominion U., 1973—; cons. Rand Corp., Santa Monica, Calif., Inst. Computer Applications Sci. Engring., Hampton, Va. Mem. Am. Phys. Soc., Soc. Indsl. and Applied Math. Subspecialties: Fluid mechanics; Applied mathematics. Current work: Hydrodynamic stability and transition, unsteady viscous flows, geo-physical fluid dynamics, computational fluid dynamics, algorithms for parallel computation. Home: 1130 Manchester Ave Norfolk VA 23508 Office: Old Dominion University Norfolk VA 23508

GROSCH, DANIEL SWARTWOOD, genetics educator; b. Bethlehem, Pa., Oct. 25, 1918; s. E. Samuel and Laura F. (Hoodmaker) G.; m. Edith D. Taft, Mar. 27, 1944; children: Laura D., Barbara T., Douglas T., Robert L., Tustav. B.S., Moravian Coll., 1939; M.S., Lehigh U., 1940; Ph.D., U. Pa., 1944. Instr. zoology U. Pa., 1941-44; instr. meterology Navy Pre-Flight Sch., Phila., 1943-44; asst. prof. zoology N.C. State U., Raleigh, 1946-51, assoc. prof. genetics, 1951-57, prof. genetics, 1957—; co-investigator U.S. Biosatellite Program, 1964-71. Author: Biological Effects of Radiation, 2d edit, 1979; contbr. articles to profl. jours. Served with M.C. U.S. Army, 1944-46. Recipient Comenius Alumni award Moravian Coll., 1964; Lehigh U. scholar, 1939-40. Fellow AAAS; mem. Am. Inst. Biol. Sci., Am. Soc. Naturalists, Entomol. Soc. Am., Genetics Soc. Am., Radiation Research Soc., Soc. Devel. Biology, N.C. Acad. Scis., Sigma Xi, Phi Kappa Phi. Moravian. Subspecialties: Space Biology; Toxicology (agriculture). Current work: Chemical and physical agents which alter fecundity and fertility; comparative sensitivity of the cell types in oogenesis and spermatogenesis to cytotoxic compounds and mutagens; effects of radiation combined with weightlessness, vibration and acceleration profiles of space flight. Office: NC State U Genetics Dept Gardner Hall Raleigh NC 27650

GROSCH, JOSEPHINE CATALDO, microbiologist; b. Elkhart, Ind., July 7, 1938; d. Joseph Samuel and Mary Concetta (Cataldo) Cataldo; m. David Lee Grosch, Sept. 2, 1961; 1 dau., Mary Michelle. A.B., Ind. U., 1962; Ph.D., U. Notre Dame, 1980. Research assoc. Ind. U., Bloomington, 1962-66, Variety Children's Found., Miami, Fla., 1967-69; sr. research scientist, supr. recombinant DNA dept. Miles Labs., Elkhart, 1969—. Contbr. articles to profl. jours. Mem. Am. Soc.Microbiology, AAAS, Sigma Psi. Republican. Roman Catholic. Subspecialties: Genetics and genetic engineering (biology); Molecular biology. Current work: Development of B.subtiles host-vector systems as cloning vehicles for production of gene products by fermentation of genetically engineered micro-organisms. Patentee in field.

GROSH, RICHARD JOSEPH, manufacturing company executive; b. Ft. Wayne, Ind., Oct. 29, 1927; s. Joseph A. and Vera (Vogeding) G.; m. Susan Marie Ankenbruck, June 24, 1950; children: Katherine (Mrs. Craig Johnson), Anton, Richard, John, Jane, Suzanne. B.S., Purdue U., 1950, M.S., 1952, Ph.D., 1953. Registered profl. engr., N.Y., Ind. Research, devel. Capehart Farnsworth Corp., Ft. Wayne, 1950-51; asst. prof. mech. engring. Purdue U., Lafayette, Ind., 1953-56, asso. prof., 1956-58, prof., 1958-71, head, 1961-65, asso. dean engring., 1965-67, dean, 1967-71; pres. Rensselaer Poly. Inst., Troy, N.Y., 1971-76; chmn. bd., chief exec. officer Ranco, Inc., Dublin, Ohio, 1976—, also dir., Dublin; dir. Sterling Drug Inc., Transway Internat. Cons. editor, Charles E. Merrill Book Co.; Contbr. articles to profl. jours. Served with CIC USAAF, 1946-47. Fellow ASME; mem. Nat. Acad. Engring., AIAA, Pi Tau Sigma, Sigma Pi Sigma, Tau Beta Pi. Subspecialty: Mechanical engineering. Home: 818 Bluffview Dr Worthington OH 43085 Office: Ranco Inc 555 Metro Pl N Suite 550 Dublin OH 43017

GROSS, BOB DEAN, oral and maxillofacial surgery educator, researcher; b. Freeman, S.D., Dec. 21, 1942; s. George A. (G.); m. Nancy A. Dufton, May 5, 1979. B.S., Colo. State U., 1964; D.D.S., U. Mo., 1968; M.S., U. Tex.-Houston, 1971. Diplomate: Am. Bd. Oral and Maxillofacial Surgery (examiner). Resident in oral and maxillofacial surgery U. Tex.-Houston, 1971-74; pvt. practice and group practice oral surgery, Vallejo, Calif., 1974-75; asst. prof. U. Conn. Health Center, Farmington, 1975-78; assoc. prof. oral and maxillofacial surgery La. State U. Med. Center, Shreveport, 1978—; part-time practice Drs. Worley, Clark and Gross, Shreveport, 1979—; cons. USAF Barksdale AFB, Bossier City, La., 1980—. Contbr. articles to profl. jours. Served with USAF, 1968-70. Co-investigator NIH grant, 1983. Fellow Am. Soc. Central Anesthesiologists; mem. La. State Assn. Oral and Maxillofacial Surgeons (v.p. 1981-82), ADA, Southeastern Soc. Oral and Maxillofacial Surgeons, Am. Assn. Dental Research, Internat. Assn. Dental Research, Omicron Kappa Upsilon. Republican. Subspecialty: Oral and maxillofacial surgery. Current work: Electrical stimulation of bone healing, bone induction mechanisms, ultrasound analysis of bone healing, biomechanical stress testing of facial bones and fixation devices. Home: 7450 S Lakeshore Dr Shreveport LA 71119 Office: Dept Surgery La State Univ Med Center 1501 Kings Highway Shreveport LA 71130

GROSS, DONALD, operations research educator; b. Pitts., Oct. 20, 1934; s. Frank and Marion (Horovitz) G.; m. Alice Gold, Sept. 20, 1959; children: Stephanie Lynne, Joanne Susan. B.S., Carnegie-Mellon U., 1956; M.S., Cornell U., 1959, Ph.D., 1961. Ops. research analyst Atlantic Refining Co., Phila., 1961-65; from asst. prof. to prof. dept. ops. research George Washington U., Washington, 1965—, chmn. dept., 1976—; cons. industry and fed. agys. Co-author: Fundamentals of Queueing Theory, 1974; contbr. articles to sci. jours. Treas. Williamsburg Civic Assn., Arlington, Va., 1978-79. Served to capt., Signal Corps U.S. Army, 1962-63. Grantee NASA, NSF, Office Naval Research, USAF. Mem. Ops. Research Soc. Am. (council 1982—), Inst. Mgmt. Scis., Inst. Indsl. Engrs., Washington Ops. Research Mgmt. Council (trustee 1969-73, pres. 1974-75), Sigma Xi, Tau Beta Pi. Subspecialties: Operations research (engineering); Probability. Current work: Queuing theory; inventory theory; model development and numerical solution techniques. Home: 3530 N Rockingham St Arlington VA 22213 Office: George Washington U Washington DC 20052

GROSS, ERIC TARAS BENJAMIN, elec. engr.; b. Vienna, Austria, May 24, 1901; came to U.S., 1939, naturalized, 1943; s. Berthold and Sophie (Gerstman) G.; m. Catharine B. Rohrer, Aug. 14, 1942; children—Patrick Walter, Elizabeth Sophia, Margaret Joan. E.E., Tech. U., Vienna, 1923, D.Sc., 1932. Registered profl. engr., Ill., N.Y., Vt.; Chartered engr., U.K. Elec. engr. in industry with emphasis on heavy electric power engring., 1923-42; asst. prof. elec. engring. Cornell U., 1942-45, prof. elec. engring. Ill. Inst. Tech., 1945 62; chmn. electric power engring., 1962-73; Philip Sporn prof. engring. Rensselaer Poly. Inst., Troy, N.Y., 1962—; cons. War Dept., 1942-45; Vis. scholar Va. Polytech. Inst. and State U., 1972. Contbr. numerous articles to profl. jours. Recipient citation Am. Power Conf., 1972; Distinguished Faculty award Rensselaer Poly. Inst., 1972; Western Electric Fund award Am. Soc. for Engring. Edn., 1972; spl. citation Edison Electric Inst., 1976; Austrian Cross of Honor in Sci. and Arts 1st class, 1980. Fellow N.Y. Acad. Scis., IEEE (citation and silver plaque Brazil Council 1974, Northeastern Region award 1976, Power Generation Com. award 1977, Edn. Com. award 1978), AAAS, Inst. Elec. Engrs. (London); mem. Nat. Acad. Engring., Am. Arbitration Assn. (mem. nat. panel), Am. Soc. Engring. Edn., Panamerican Congress on Engring. (v.p. for U.S. 1970-78), Sigma Xi, Tau Beta Pi, Eta Kappa Nu (nat. pres. 1953-54, eminent mem. award 1975). Subspecialties: Solar energy; Wind power. Current work: Electric power engineering. Home: 2525 McGovern Dr Schenectady NY 12309 Office: Rensselaer Poly Inst Troy NY 12181

GROSS, GARRETT JOHN, pharmacology educator; b. Britton, S.D., July 4 1942; s. Maurice John and Frances Marie (Smith) G.; m. Carol Anne King, Mar. 31, 1967. B.S., S.D. State U., 1965, M.S. in Pharmacology, 1967, Ph.D., U. Utah, 1971. Registered pharmacist, S.D. Instr. dept. pharmacology Med. Coll. Wis., Milw., 1973-75, asst. prof., 1975-77, assoc. prof., 1977-80, prof., 1980—; cons. G.D. Searle and Co., Baxter-Travenol Co. Named Researcher of Yr Wis. Heart Assn., 1982. Mem. Am. Soc. Pharmacology and Exptl. Therapeutics, Am. Heart Assn. Methodist. Lodge: Masons. Subspecialty: Pharmacology. Current work: Cardiovascular physiology and pharmacology. Office: Med Coll Wis 8701 W Watertown Plank Rd Milwaukee WI 53226

GROSS, JEROME, medical educator, biologist, physician; b. N.Y.C., Feb. 25, 1917; m., 1947; 3 children. B.S., MIT, 1939; M.D., NYU, 1943. Research fellow in medicine Harvard U. Med. Sch., 1948-50, research assoc., 1950-54, assoc., 1954-57, from asst. prof. to assoc. prof., 1957-69, prof. medicine, 1969—; clin. and research fellow Mass. Gen. Hosp., Boston, 1948-51, assoc. biologist, 1951-66, biologist, 1966—; research assoc. MIT, 1946-55; mem. subcom. skeletal system NRC, 1955-62; mem. sci. adv. com. Helen Hay Whitney Found., 1956—; established investigator Am. Heart Assn., 1956-61; mem. adv. panel molecular biology NSF, 1959-62. Adv. editor: Jour. Exptl. Medicine, 1963; cons. editor: Devel. Biology, 1965. Chmn. Bd. Sci. Counselors Nat. Inst. Dental Research, 1963; bd. dirs. Med. Found, Inc., 1974-80. Recipient Ciba award, 1959; Kappa Delta award Am. Acad. Orthopedic Surgery, 1965. Mem. Nat. Acad. Sci., Am. Acad. Arts and Sci., Am. Soc. Cell Biology, Am. Soc. Biol. Chemistry, Histochem. Soc. (sec. 1956-60). Subspecialty: Developmental biology. Office: Devel Biology Lab Mass Gen Hosp Boston MA 02114

GROSS, JOSEPH FRANCIS, chemical engineering educator, biomedical engineering researcher; b. Plauen, Vogtland, Germany, Aug. 22, 1932; came to U.S., 1932; s. Joseph and Helen (Doelling) G. B.Chem. Engring., Pratt Inst., 1953; Ph.D. in Chem. Engring., Purdue U., 1956. Profl. engr., Calif. Research engr. Rand Corp., Santa Monica, Calif., 1956-72; prof. chem. engring. U. Ariz., Tucson, 1972—. Editor: Mathematics of Micro-circulation, 1980. Fulbright fellow, 1956; Humboldt fellow, 1979. Fellow Am. Inst. Chem. Engrs. (div. chmn. 1973), AIAA (assoc.); mem. Microcirculatory Soc. (pres. 1974), ASME, Am. Inst. Soc. Biorheology (treas. 1982), Internat. Inst. Microcirculation (sec.-treas. 1983). Subspecialties: Biomedical engineering; Pharmacokinetics. Current work: Research in microcirculation, biorheology, pharmaco-kinetics. Home: PO Box 41445 Tucson AZ 85717 Office: Dept Chem Engring U Arizona Tucson AZ 85721

GROSS, KENNETH IRWIN, mathematics educator; b. Malden, Mass., Oct. 14, 1938; s. Harry and June (Maltzman) G.; m. Mary Jou Shannahan, Dec. 18, 1964; children: Laura, Karen. A.B., Brandeis U., 1960, A.M., 1962; Ph.D., Washington U., St. Louis, 1966. Asst. prof. Tulane U., 1966-68, Dartmouth Coll., 1968-73; assoc. prof. U. N.C.-Chapel Hill, 1973-78, prof., 1978-81; prof. math., head dept. math. U. Wyo., 1981—. Research numerous publs. in field; editorial bd.: Soc. Indsl. and Applied Math. Series Frontiers in Applied Mathematics, 1982—. Mem. Soc. Indsl. and Applied Math., Am. Math. Soc., Math. Assn. Am. (Lester Ford prize 1979, Chauvenet prize 1981), AAUP, AAAS. Subspecialty: Applied mathematics. Current work: Lie theory, group representations, harmonic analysis. Home: 1442 Whitman St Laramie WY 82070 Office: Dept Math U Wyo Laramie WY 82071

GROSS, LEO, biophysicist, cons.; b. Bklyn., Feb. 13, 1915; s. Isador and Rose (Lichenstein) G.; m.; children: Alan, Walter, Joan. B.S., Bklyn. Coll., 1934; M.A., Columbia U., 1934; Ph.D., N.Y. U., 1963. Technician inhalation therapy dept. Coll. Physicians and Surgeons, Columbia U., 1935-36; tchr. physics N.Y.C. Bd. Edn., 1936-41; physicist Bur. Ordanance, U.S. Navy, 1941-42; physicist, mem. tech. staff Los Alamos Sci. Lab., U. Calif., 1943-46; mem. tech. staff Bell Telephone Labs., 1946-49; chief systems engr. Polarad Electronics Corp., Lake Success, N.Y., 1949-54; pres. Hub Electronics Corp., White Plains, N.Y., 1954-58; adminstr., dir. biophysics, dir. ednl. programs Waldemar Med. Research Found., Bayside, N.Y., 1958—; nat. Acad. Sci. internat. exchange scholar, 1973. Contbr. articles to profl. jours. Served to 1st lt., C.E. U.S. Army, 1943-46. Mem. Am. Phys. Soc., IEEE, Optical Soc. Am., Soc. Photo-Optical Instrumentation Engrs., N. Y. Acad. Scis., Biophys. Soc., Nat. Sci. Tchrs. Assn. (Sci. Teaching Achievement award 1975), Nat. Assn. Biology Tchrs. Subspecialties: Biophysics (biology); Electronics. Home: 36-11 217th St Bayside NY 11361 Office: Waldemar Med Research Found Bayside NY 11361

GROSS, LUDWIK, physician; b. Cracow, Poland, Sept. 11, 1904; came to U.S., 1940, naturalized, 1943; s. Adolf and Augusta (Alexander) G.; m. Dorothy L. Nelson, Oct. 7, 1943; 1 dau., Augusta H. M.D., Iagellon U., Cracow, 1929; Prix Chevillon, Acad. Medicine, Paris, 1937; Dr.Sci. honoris causa, Mt. Sinai Sch. Medicine, CUNY, 1983. Diplomate: Am. Bd. Internal Medicine. Intern and resident St. Lazar Gen. Hosp., Cracow, 1929-32; part time research exptl. cancer Pasteur Inst., Paris; postgrad. clin. tng. Salpetriere, U. Paris, 1932- 39; cancer research Christ Hosp., Cin., 1941-43; chief cancer research VA Med Center, Bronx, 1946—, Distinguished physician, 1977; research prof. dept. medicine Mt. Sinai Sch. Medicine, N.Y.C., 1971-73, emeritus prof., 1973—; cons. Sloan Kettering Inst., Meml. Center, N.Y.C., 1955-57, assoc. scientist, 1957-60; Distinguished Leukemia lectr. U. So. Ala., 1976; 17th G.H.A. Clowes Meml. lectr., 1977. Author: Oncogenic Viruses, 1961, 3d edit., 1983; author numerous papers on cancer and leukemia in profl. jours. Served from capt. to maj. M.C. AUS, 1943-46. Decorated chevalier Legion of Honor, France; recipient Robert R. De Villiers award for research on leukemia Leukemia Soc. N.Y., 1953, Walker prize Royal Coll. Surgeons Eng., 1962, Pasteur Silver medal Pasteur Inst., 1962, Lucy Wortham James award James Ewing Soc., 1962, WHO UN prize, 1962, The Bertner Found. award, 1963, Albert Einstein Centennial medal, 1965, Albion O. Bernstein award Med. Soc. N.Y. State, 1971, Spl. Virus Cancer Program award Nat. Cancer Inst., 1972, William S. Middleton award VA, 1973, Albert Lasker Basic Med. Research award, 1974, Founders award Cancer Research Inst., 1975; prin. Paul Ehrlich-Ludwig Darmstaedter prize, 1978; Prix Griffuel, Paris, 1978; Exceptional Service award VA, 1979, VA Disting. physician, 1977-81. Fellow A.C.P., AAAS, Internat. Soc. of Hematology, N.Y. Acad. Scis.; mem. Am. Soc. Hematology, AMA, Nat. Acad. Scis., Am. Assn. Cancer Research (dir. 1973-76), Assn. Mil. Surgeons U.S., Soc. of Exptl. Biology and Medicine, Bronx County, N.Y. State med. socs. Subspecialties: Cancer research (medicine); Virology (medicine). Current work: Etiology of cancer and leukemia; oncogenic viruses. Home: 29 Ramona Ct New Rochelle NY 10804

GROSS, PAUL MUNN, medical researcher; b. Chatham, Ont., Can., Aprl 17, 1950; s. Frank George and Lilian Gladys G.; m. Cheryl Ann Collins, Sept. 9. 1978; 1 son, Cameron Stuart. B.S., Ohio State U., 1971; M.S., U. Ill., 1973; Ph.D., U. Glasgow, Scotland, 1981. Research asst. cardiovascular div. U. Iowa, Iowa City, 1976-78; research asst. Wellcome Surg. Inst., U. Glasgow, 1978-81; research fellow Lab. Cerebral Metabolism, NIMH, Bethesda, Md., 1981—. Contbr. articles to profl. jours. NOAA grantee, 1973-74; Sir Halley Stewart Trust scholar, 1978-79; Migraine Trust scholar, 1979-80; Can. Heart Found. grantee, 1980-81; NIH fellow, 1981-83; Can. Heart Found. fellow, 1982-83. Mem. AAAS, Am. Heart Assn., Am. Physiology Soc., Soc. Neurosci., Internat. Soc. Cerebral Blood Flow and Metabolism. Subspecialties: Neurobiology; Physiology (medicine). Current work: Brain metabolism and vascular control; brain imaging; neuroendocrinology; cerebral metabolism; neurotransmitters. Office: NIH Bldg 36 Room 1a-27 9000 Rockville Pike Bethesda MD 20205

GROSS, PAUL RANDOLPH, biologist, laboratory administrator; b. Phila., Nov. 27, 1928; s. Nathan and Kate (Segal) G.; m. Mona Lee Feld, Mar. 27, 1949; 1 dau., Wendy Loren. B.A., U. Pa., 1950, Ph.D. (Harrison and NSF fellow), 1954; M.A., Brown U., 1963; D.Sc., Med. Coll. Ohio, 1979. Asst. prof. biology N.Y. U., 1954-58, asso. prof., 1958-61; asso. prof. biology Brown U., 1962-65; prof. biology M.I.T., 1965-71; prof., chmn. dept. biology U. Rochester, 1972-78, dean grad. studies, 1975-78, chmn. sci. adv. com., 1974-78; pres., dir. Marine Biol. Lab., Woods Hole, Mass., 1978—; prin. investigator research and tng. grants from NSF and NIH, 1955—; chmn. coordinating com. for biology N.Y. State Edn. Dept.; mem. adv. com. cell and developmental biology Am. Cancer Soc.; mem. oversight com. Assn. Am. Colls.; mem. nat. adv. council Nat. Inst. Child Health and Human Devel.; mem. sci. adv. com. Tufts U. Sch. Vet. Medicine; mem. vis. com. dept. biology U. Va. Contbr. sci. articles to profl. jours. Mem. Indsl. Devel. Corp., Town of Falmouth (Mass.); trustee U. Rochester. Lalor fellow, 1954-55; NSF fellow U. Edinburgh, 1961-62. Fellow N.Y. Acad. Scis.; mem. Internat. Soc. for Developmental Biology, Am. Physiol. Soc., Am. Soc. Zoologists (chmn. sect. on developmental biology), AAAS, Am. Soc. Cell Biology. Clubs: Woods Hole Yacht, Cosmos. Subspecialties: Developmental biology;

Biochemistry (biology). Office: Marine Biological Laboratory Woods Hole MA 02543

GROSS, PETER ALAN, physician, researcher; b. Newark, Nov. 18, 1938; s. Meyer P. and Nathalie (Bass) Denburg) G.; m. Regina Teri Gittlin, May 30, 1964; children: Deborah Karen, Michael Philip, Daniel Brian. B.A. cum laude, Amherst Coll., 1960; M.D., Yale U., 1964. Diplomate: Am. Bd. Internal Medicine. NIH fellow virology dept. epidemiology Yale U., New Haven, 1969-71; intern Yale-New Haven Hosp., 1964-65, jr. resident, 1965-66; sr. resident Peter Bent Brigham Hosp., Boston, 1968-69; research and edn. assoc. VA Hosp., West Haven, Conn., 1971-72, acting chief infectious disease sect., 1972-73, chief infectious disease sect., 1973-74, Hackensack (N.J.) Med. Center, 1974—, chmn. dept. medicine, 1980—; prof. medicine N.J. Med. Sch., Newark, 1981—; assoc. clin. prof. medicine Columbia U. Coll. Phys. and Surgs., N.Y.C., 1977-81, asst. clin. prof., 1974-77; asst. prof. medicine Yale U. Sch. Medicine, New Haven, 1971-74; ad hoc reviewer NIH, Nat. Inst. Allergy and Infectious Diseases research grants, 1974—. Author: Gram Strain Recognition, 1975, 2d edit., 1980; editorial bd.: Jour. Clin. Microbiology, 1980—, Infection Control, 1980—. Served to lt. comdr. USPHS, 1966-68. NIH fellow, 1969-71. Fellow Infectious Disease Soc. Am., ACP; mem. Am. Soc. Virology, AAAS, Am. Soc. Microbiology. Republican. Jewish. Subspecialties: Infectious diseases; Epidemiology. Current work: Immune response to influenza vaccine, hospital-acquired infections in the elderly. Home: 807 Morningside Rd Ridgewood NJ 07450 Office: Hackensack Med Ctr Hackensack NJ 07601

GROSS, ROBERT ALFRED, physics educator; b. Phila., Oct. 31, 1927; s. John and Esther (Schwartz) G.; m. Elee B. Kauffmann, Nov. 21, 1952; children: David Andrew, John Henry. B.S., U. Pa., 1949; M.S., Harvard U., 1950, Ph.D., 1952. Chief research engr. Fairchild Engine & Airplane Co., 1954-59; mem. faculty Columbia U., N.Y.C., 1960—, prof. engring. sci., 1960-78, prof. applied physics, 1978—, Vera and Percy Hudson prof., 1980—, dean Sch. Engring. and Applied Sci., 1982—; founder Plasma Physics Lab., 1961, chmn. plasma physics program, 1960-70, chmn. mech. engring. dept., 1970-76, chmn. dept. applied physics and nuclear engring., 1978—; prof. Internat. Sch. Physics, Enrico-Fermi-Varenna, Italy, 1969; Dir. Fairfield Tech., 1962-66, Samson Fund, 1963-70, Fundamatic Investors, Inc., 1970-71; sr. vis. fellow Australian Acad. Scis., U.S.-USSR exchange scientist, 1967; vis. prof. Sydney (Australia) U., 1974, 79; cons. indsl., govt. agencies, 1950—; mem. nat. com. fluid mechanics films NSF, 1964-70; mem. rev. com., mechanics div. Nat. Bur. Standards, 1971-76; mem. fusion power coordinating com. U.S. Dept. Energy, 1980-82, mem. magnetic fusion adv. com., 1982—; mem. Fulbright-Hayes adv. screening com. Council Internat. Exchange Scholars, 1976-78; mem. vis. com. Brookhaven Nat. Lab., 1979—, Lehigh U., 1982—. Asso. editor: Physics of Fluids Jour, 1970-73; corr. editor: Jour. Comments on Plasma Physics and Controlled Fusion, 1978—; Contbr. articles to tech. lit. Served with AUS, 1945-46. Recipient Waverly Gold medal Combustion Inst., London, Eng., 1959; Great Tchr. award Columbia U., 1975; NSF sr. post-doctoral fellow U. Calif. at Berkeley, 1959-60; Guggenheim Found. fellow, Fulbright-Hays fellow, 1966-67; Fulbright-Hayes sr. fellow, 1974. Fellow AIAA (v.p. 1965-66), Am. Inst. Aeros. and Astronautics (editor-in-chief selected reprint series 1968-77, Pendray award 1967), Am. Phys. Soc. (chmn. exec. com. fluid dynamics div. 1970-71, exec. com. plasma physics div. 1974-75); mem. AAAS, Univ. Fusion Assn. (exec. com. 1980-82), Sigma Xi, Tau Beta Pi. Subspecialties: Fusion; Plasma. Current work: Plasma physics; tokamak confinement; fusion engineering; high temperature gas dynamics. Home: 14 Sunnyside Way New Rochelle NY 10804

GROSS, ROBERT H., biology educator, consultant; b. N.Y.C., Mar. 6, 1945; s. Stanley and Annette (Goldenberg) G.; m. Roberta B. Brucks, Sept. 2, 1968; children: Jason, Lisa, Kevin. B.E.S., Rensselaer Poly. Inst., 1967, M.S. in Biomed. Engring, 1968; Ph.D. in Biophysics, Johns Hopkins U., 1974. Asst. prof. biology Dartmouth Coll., Hanover, N.H., 1977—, asst. prof. biochemistry, 1978—; sr. staff Norris Cotton Cancer Center, Hanover, 1980—; research assoc. Arthritis Center./Mary Hitchcock Meml. Hosp., Hanover, 1981—; cons. Verax Corp., Hanover, 1981—. Mem. Upper Valley Support Group for Handicapped Children, 1981—. Grantee Am. Cancer Soc., 1978, NIH, 1980, 83, NSF, 1983. Mem. Am. Soc. Cell Biology, Am. Chem. Soc., N.Y. Acad. Scis., AAUP, AAAS. Subspecialties: Gene actions; Molecular biology. Current work: 1. Using recombinant DNA technology to study changes in gene expression which occur in rheumatoid arthritis; 2. using liposomes and recombinant DNA to study possible role of small nuclear RNAs in splicing of mRNA precursors. Office: Dartmouth College Hanover NH 03755

GROSSBECK, MARTIN LESTER, national laboratory staff member; b. Paterson, N.J., July 5, 1944; s. Lester A. and Mary Emma (Brennan) G.; m. Jane Sharp Powell, Jan. 24, 1981. B.S. in Physics, Rensselaer Poly. Inst., 1966; M.S., Cornell U., 1968; Ph.D. in Metall. Engring, U. Ill.-Urbana, 1975. Instr. reactor physics U.S. Naval Nuclear Power Sch., Mare Island, Calif., 1967-71; research assoc. U. Ill., 1975; research staff mem. Oak Ridge Nat. Lab., 1975—, task leader, 1978—. Bd. dirs. Firends of Library, Oak Ridge, 1980—. Served to lt. USN, 1966-71. Mem. Am. Nuclear Soc., Am. Vacuum Soc., Am. Soc. Metals, Sigma Xi, Sigma Pi Sigma, Tau Beta Pi. Subspecialties: Nuclear fusion; Fusion reactor materials. Current work: Radiation effects research in fusion reactor materials; hydrogen embrittlement in metals. Home: 131 Clemson Dr Oak Ridge TN 37830 Office: Oak Ridge Nat Lab PO Box X Oak Ridge TN 37830

GROSSE, ERIC H., computer scientist; b. Detroit, Feb. 24, 1953; s. Burck E. and Dennise G.; m. Brenda S. Baker, June 27, 1982. B.S., Mich. State U., 1975; Ph.D., Stanford U., 1980. Research mathematician Gen. Motors Research Lab., Warren, Mich., 1976; mem. tech. staff Bell Labs, Murray Hill, N.J., 1980—. Subspecialty: Numerical analysis. Current work: Approximation theory and optimization; tools for scientific computation. Home: 281 Timber Dr Berkeley Heights NJ 07922 Office: Bell Labs Murray Hill NJ 07974

GROSSER, GEORGE SAMUEL, psychology educator, research consultant; b. Boston, Aug. 14, 1929; s. Sidney and Eva Risa (Shapiro) G. A.B. cum laude, Harvard Coll., 1951; M.A., Boston U., 1952, Ph.D., 1957. Exptl. and physiol. psychologist U.S. Army Chem. Corps, Edgewood, Md., 1957-58; asst. prof. Am. Internat. Coll., Springfield, Mass., 1958-64, assoc. prof., 1964—; researcher and pres. Consumer Behavior Research Assn., Springfield, 1980-82. Author: (with W. Zinn) Vitametrics: Human Self-Evaluation Formula, 1980; editor: (with others) General Psychology, 1967. USPHS grantee, 1959-60. Mem. AAAS, Am. Psychol. Assn., N.Y. Acad. Scis., Sigma Xi. Jewish. Club: Springfield Chess. Subspecialties: Physiological psychology; Social psychology. Current work: Vision; memory trace; attitude scaling; learning. Home: 335 Maple Rd Longmeadow MA 01106 Office: Am Internat Coll 1000 State St Springfield MA 01109

GROSSFELD, ROBERT MICHAEL, neurobiologist, educator; b. N.Y.C., Oct. 15, 1943; s. Aaron M. and Anne (Ellman) G.; m. Margaret Anne Sodwith, May 24, 1970. B.S. in Microbiology, U. Wis.-Madison, 1963; Ph.D. in Neurol. Scis, Stanford U., 1968; postgrad., Harvard U. Med. Sch., Boston, 1968-70. Instr. Harvard U. Med. Sch., 1968-70; asst. prof. Cornell U., 1970-74; research scientist VA Hosp., Sepulveda, Calif., 1974-76; vis. asst. prof. U. Tex.-Austin, 1976-79; asst. prof. zoology N.C. State U., 1979—. Phi Kappa Phi nat. scholar, 1963; Steenbock Alpha Zeta scholar, 1962-63; NSF predoctoral fellow, 1963-65; NIH predoctoral fellow, 1965-68; NIH postdoctoral fellow, 1968-70. Mem. Soc. Neurosci., N.C. Soc. Neurosci., Phi Eta Sigma, Phi Kappa Phi. Subspecialties: Neurochemistry; Neurobiology. Current work: Synaptic transmission, degeneration-regeneration, glial-neuronal interactions.

GROSSMAN, DAVID GARY, dental ceramicist; b. Bronx, N.Y., Aug. 20, 1941; s. Abbot A. and Sylvia (Gelman) G.; m. Ann Christine Streeting, June 19, 1971; children: Jeffery Stuart, Sarah Elizabeth. B.S., Rutgers U., 1962; M.S., 1964, Ph.D., U. Sheffield, Eng., 1967. Lectr. Rugby Coll. Eng., 1967-69; research scientist Corning Glass Works, N.Y., 1969-80, research assoc., 1980—. Contbr. articles to profl. jours. Recipient N.J. State scholarship, 1959-63; Foster Research Prize U. Sheffield, 1968. Mem. Am. Ceramic Soc. (govt. liaison com. rep. 1981-84), Soc. Glass Tech., Internat. Assn. Dental Research, ADA (assoc.). Subspecialties: Biomaterials; Ceramics. Current work: class-ceramic systems containing unusually shaped crystals, orientation, low expension phases, strengthening mechanisms; glass-ceramics for dental applications. Patentee various glass ceramics. Office: Sullivan Park Dv-2 Corning Glass Works Corning NY 14831 Home: 200 Wall St Corning NY 14830

GROSSMAN, LAWRENCE, biochemist, educator; b. Bklyn., Jan. 23, 1924; s. Isidor Harry and Anna (Lipkin) G.; m. Barbara Meta Grossman, June 24, 1949; children: Jon David, Carl Henry, Ilene Rebecca. Student, CCNY, 1946-47; B.A., Hofstra U., 1949; Ph.D., U. So. Calif., 1954. Scientist NIH, Bethesda, Md., 1957-62; asst. prof. biochemistry Brandeis U., Waltham, Mass., 1957-62, assoc. prof., 1962-67, prof., 1967-75; E.V. McCollum prof., chmn. dept., biochemistry Johns Hopkins U., Balt., 1975—; sci. adv. com. Am. Cancer Soc.; cons. Author: Methods in Nucleic Acids, 9 vols; contbr. numerous articles to sci. jours. Served to 1t. USNR, 1942-45. Decorated D.F.C. (2), Air medal (3).; Commonwealth Fund fellow, 1963; Guggenheim fellow, 1973; recipient Career Devel. award NIH, 1964-74; research grantee NIH, NSF, Dept. Energy. Mem. Am. Soc. Biol. Chemists. Jewish. Club: Alberg 30 (Annapolis, Md.). Subspecialties: Biochemistry (biology); Genetics and genetic engineering (biology). Current work: Genes and enzymes that repair damaged DNA in bacteria and human cells. Home: 5723 Uffington Rd Baltimore MD 21209 Office: 615 N Wolfe St Baltimore MD 21705

GROSSMAN, LAWRENCE, geochemistry educator, consultant; b. Toronto, Ont., Can., Feb. 2, 1946; s. David Saul and Marian Lillian (Jacobs) G.; m. Karen Lee Fruitman, Aug. 11, 1968; children: Sheryl Gloria, Daniel Martin. B.Sc. with honours in Chemistry and Geology, McMaster U., 1968; M.Phil. in Geochemistry, Yale U., 1970, Ph.D., 1972. Asst. prof. geochemistry U. Chgo., 1972-76, assoc. prof., 1976-81, prof., 1981—; Caswell Silver disting. lectr. U. N.Mex., 1981; Lady Davis vis. prof. Hebrew U. of Jerusalem, 1981-82. Recipient award for advancement basic and applied sci. Yale Sci. and Engring. Assn., 1982; Alfred P. Sloan research fellow, 1976-78. Fellow Am. Geophys. Union (James B. Macelwane award 1980), Mineral. Soc. Am., Meteoritical Soc.; mem. Geochem. Soc. (F.W. Clarke medal 1974), Internat. Astron. Union. Subspecialties: Geochemistry; Planetology. Current work: Chemical processes in the early solar system; origin of meteorites, planets and the solar system; cosmochemistry; mineralogical, chemical and isotopic composition of meteorites. Office: Dept Geophys Scis and Enrico Fermi Inst U Chgo 5734 S Ellis Ave Chicago IL 60637

GROSSMAN, MICHAEL, geneticist; b. N.Y.C., Dec. 21, 1940; s. Benjamin Harry and Alice (Berkowitz) G.; m. Margaret Rosso, June 27, 1970; children: Aaron William, Daniel Benjamin. BS., CCNY, 1962; M.S., Va. Poly. Inst. and State U., 1965; Ph.D., Purdue U., 1969. Asst. prof. U. Ill., Champaign, 1969-74, asso. prof., 1974—; vis. prof. Instituto de Fitotecnia, INTA, Argentina, 1970, Gadjah Mada U., Yogyakarta, Indonesia, 1974. Contbr. articles to sci. jours. Recipient teaching awards U. Ill., 1971, 72; award AMOCO Found. Inc., 1972; Danforth faculty assoc., 1976. Mem. AAAS, Am. Dairy Sci. Assn., Am. Genetic Assn., Am. Soc. Animal Sci., Biometric Soc., Genetics Soc. Am., Gamma Sigma Delta, Phi Sigma, Sigma Xi. Subspecialties: Animal genetics; Statistics. Current work: Theoretical and experimental population and quantitative genetics; estimation of genetic parameters; analysis of dairy goat lactation records. Home: 2206 Valley Brook Dr Champaign IL 61821 Office: 1207 W Gregory Dr Urbana IL 60801

GROSSWEINER, LEONARD IRWIN, physicist, educator; b. Atlantic City, Aug. 16, 1924; s. Jules H. and Rae (Goldberger) G.; m. Bess Tornheim, Sept. 9, 1951; children-Karen Ann, Jane (dec.), James Benjamin, Eric William. B.S., Coll. City N.Y., 1947; M.S., Ill. Inst. Tech., 1950, Ph.D., 1955. Asst. chemist Argonne (Ill.) Nat. Lab., 1947-50, asso. physicist, 1950-57; asso. prof. physics Ill. Inst. Tech., Chgo., 1957-62, prof. physics, 1962—, chmn. dept. physics, 1970-81, Sang Exchange lectr., 1972-73; vis. prof. radiology Stanford U. Sch. Medicine, 1979; cons. Donner Lab. U. Calif., Berkeley, Chgo. Med. Sch., North Chicago, Ill., Hines (Ill.) VA Hosp., Michael Reese Med. Center, Chgo.; Mem. U.S. Nat. Com. Photobiology, 1977-81, chmn., 1980-81. Author: Organic Photoconductors in Electrophotography, 1970; Contbr. articles to profl. jours. Served with AUS, 1944-46. Fellow Am. Phys. Soc. (sec.-treas. div. biol. physics 1973-76, chmn. 1977-78), N.Y. Acad. Scis.; mem. Am. Chem. Soc., AAAS, Radiation Research Soc., Am. Soc. Photobiology (council 1977-80, sec.-treas. 1981—), Biophys. Soc., Inter-Am. Photochem. Soc. (exec. com. 1976-78), Sigma Xi (distinguished faculty lectr. 1970). Subspecialties: Laser photochemistry; Cancer research (medicine). Current work: Laser flash photolysis research in photobiology relevant to clinical phototherapy, especially tumor phototherapy with porphyrins and light, radiant energy damage to the eye, and PUVA phototherapy of human skin diseases. Home: 231 Wentworth Ave Glencoe IL 60022 Office: Ill Inst Tech IIT Center Chicago IL 60616

GROSZMANN, ROBERTO JOSE, physician; b. Buenos Aires, Argentina, Aug. 17, 1939; came to U.S., 1965, naturalized, 1979; s. Jose and Sofia (Hirsch) G.; m. Aida Zugman, May 2, 1965; children: Yvette, Daniel. Bachiller, Buenos Aires U., 1958, M.D., 1964. Intern Mt. Sinai Hosp., Chgo., 1965-66; resident internal medicine VA Hosp., Washington, 1966-68, research fellow hemodynamics program, 1968-71; asst. prof. medicine Buenos Aires, U., 1972-75, Yale U., New Haven, 1975-79, asso. prof., 1979—; dir. hepatic hemodynamic lab. W. Haven VA Hosp., 1980—. Editor: Gastroenterologia y Hepatologia, 2 vols, 1982; author balloon catheter technique for measuring wedged hepatic venous pressure, 1979. NIH Career Devel. award, 1979—; Bonarino Udaondo award Argentina Soc. Gastroenterology, 1975. Mem. Am. Assn. Study Liver Diseases, Internat. Assn. Study Liver Diseases, Am. Fedn. Clin. Research, Am. Soc. Clin. Investigation. Current work: Portal hypertension, hepatic circulation. Home: 33 Pine Ridge Rd Woodbridge CT 06525 Office: VA Med Center West Spring St West Haven CT 06516

GROTA, LEE JAMES, psychologist, neuroendocrinologist, educator; b. Sturgeon Bay, Wis., May 12, 1937; s. Hubert D. and Carrol C. (Bartmann) G.; m. Mary Peterson, June 20, 1959; children: Catherine, Steven, Michael, Carl. B.S., Marquette U., 1959; M.S., Purdue U., Lafayette, Ind., 1961, Ph.D., 1963. Postdoctoral fellow dept. biochemistry U. Utah, 1963-65; asst. prof. psychiatry and psychology U. Rochester, 1965-71, assoc. prof., 1971—; vis. assoc. prof. neuroscis. and psychiatry McMaster U., Hamilton, Ont., Can., 1977—. Contbr. numerous articles, chpts. to profl. publs.; author numerous profl. papers. Mem. Endocrine Soc., Soc. Neurosci., Internat. Soc. Devel. Psychobiology. Subspecialties: Neuroendocrinology; Neuroimmunology. Current work: Neuroendocrinology of indolealkylamines; central nervous system modulates immune reactivity. Office: Dept Psychiatry 1-9045 U Rochester Med Center Rochester NY 14642

GROTENHUIS, MARSHALL, federal comm. administrator; b. Oostburg, Wis., Oct. 17, 1918; s. William and Isabel (Marshall) G.; m. Marilynn Johnson, Dec. 21, 1946; children: Susan Kidd, Alan, Judith Black, Brian. B.S., Milw. State Tchrs. Coll., 1941; M.S., Marquette U., 1948. Instr. Marquette U., Milw., 1947-49; physicist Argonne Nat. Lab., Chgo., 1949-71; div. licensing project mgr. NRC, Washington, 1971—. Author, editor: Compendium of Radiation Shielding, 1967. Pres. P.T.A., Downers Grove, Ill., 1958; chmn. Boy Scouts Am., LaGrange, Ill., 1968. Served with USAAF, 1942-46. Recipient Silver Beaver award Boy Scouts Am., 1971. Mem. Am. Nuclear Soc. (chmn. shielding div. 1961-62, bd. dirs. 1963-66, disting. service award 1975), Sigma Xi. Subspecialties: Nuclear engineering; Nuclear fission. Current work: Regulation of nuclear power plants. Home: 216 Summit Hall Rd Gaithersburg MD 20877 Office: NRC 7920 Norfolk Ave Bethesda MD 20555

GROTH, EDWARD JOHN, III, physics educator; b. St. Louis, May 13, 1946; s. Edward John and Marion Catherine (Winkel) G.; m. Jane Ellen, June 15, 1968; children: Jeffrey, Amy. B.S., Calif. Inst. Tech., 1968; M.A., Princeton U., 1970, Ph.D., 1971. Instr. Princeton (N.J.) U., 1971-72, asst. prof., 1972-78, assoc. prof. physics, 1978—; leader Space Telescope Data and Ops. Team, 1977—; cons. TRW Space Telescope Ground System, Hewlett Packard Personal Computer Software. Alfred P. Sloan fellow, 1973-75. Mem. Am. Phys. Soc., Am. Acad. Sci., AAAS. Subspecialties: Cosmology; Optical astronomy. Current work: Cosmology optical astronomy Space Science, astronomical/image processing software. Home: 9 Alyce Ct Lawrenceville NJ 08648 Office: Physics Dept Judwin Hall Princeton U Princeton NJ 08544

GROTH, RICHARD HENRY, chemistry educator; b. New Britain, Conn., Oct. 14, 1929; s. Henry and Nellie Emma (Otterbein) G.; m. Joyce Lorraine Weaver, June 6, 1959; children: Eileen, Kathleen. B.S., Tchrs. Coll. Conn., 1951; Ph.D., Ohio State U., 1955. Postdoctoral assoc. Duke U., Durham, N.C., 1956-57; asst. prof. U. Hartford, Conn., 1957-64, assoc. prof., 1964-70; assoc. prof. chemistry Central Conn. State U., New Britain, 1970-76, prof., 1976—, chmn. dept., 1975—; cons. Pratt & Whitney Aircraft, 1961—. Author: Chemistry and Environment, 1974, Fundamentals of Chemistry, 1978. Chmn. Cons. Com. for Adult Edn., Meriden, Conn., 1976—; mem. steering com. Econ. Devel. Project, Meriden Community Action Agy., 1982. Recipient Disting. Service award Central Conn. State U., 1982. Fellow AAAS; mem. New Eng. Assn. Chemistry Tchrs., Am. Chem. Soc. (chmn. sect. Conn. Valley 1970, exec. bd. 1961-81), Sigma Xi, Phi Lambda Upsilon, Kappa Delta. Republican. Episcopalian. Lodge: Masons. Subspecialties: Analytical chemistry; Organic chemistry. Current work: Methodology of measurement of engine exhaust emissions. Home: 75 Coe Ave Meriden CT 06450 Office: Central Conn State U New Britain CT 06050

GROTTE, JEFFREY HARLOW, research analyst, consultant; b. Youngstown, Ohio, Jan. 16, 1947; s. Sylvan and Mildred (Backall) G.; m. Margaret Spencer, June 10, 1973; 1 son, Nathaniel Douglas. B.S., MIT, 1969; M.S., Cornell U., 1970, Ph.D., 1974. Mem. research staff. Inst. for Def. Analyses, Alexandria, Va., 1974—; asst. professorial lect. George Washington U., 1976. Served with U.S. Army, 1971-72. Mem. Soc. for Indsl. and Applied Math., Ops. Research Soc. Am. (assoc.), AAAS. Subspecialties: Operations research (mathematics); Applied mathematics. Current work: Defense analysis, applications of game theory, optimization. Home: 30 W Glendale Ave Alexandria VA 22301 Office: Inst for Def Analyses 1801 N Beauregard St Alexandria VA 22311

GROUSE, LAWRENCE DOUGLAS, editor, association executive, biochemical researcher, medical educator; b. Mpls., Nov. 18, 1946; s. Tom and Helene Dorothea (Burnson) G.; m. Jan Ellen Lindtwed, Oct. 27, 1973; children: Eric Roger, Carrie Katherine. B.A., Carleton Coll., 1968; M.D., U. Wash., 1973, Ph.D., 1973. Intern in internal medicine U. Calif.-San Diego, 1973-74; research assoc. NIH, Bethesda, Md., 1974-76, med. officer, 1976-79; sr. editor Jour. AMA, 1971—; dir. dept. sci. activities AMA, Chgo., 1979—; vis. scientist U. Chgo.; vis. asst. prof. preventive medicine and family practice Rush Med. Coll. Contbr. articles to profl. publs. Served to lt. comdr. USPHS, 1976-79. Mem. Am. Fedn. Clin. Research, Soc. Neurosci., AMA, Med. Soc. Va. Current work: Gene expression in nerve cells. Office: AMA 535 N Dearborn Chicago IL 60610

GROVE, ANDREW S., electrical engineer, engineering physicist, corporate executive; b. Budapest, Hungary, Sept. 2, 1936; m., 1958; 2 children. B.S., CCNY, 1960; Ph.D., U. Calif.-Berkeley, 1963. Mem. tech. staff Fairchild Semicondr. Research Lab., 1963-66, head sect. surface and device physics, 1966-67, asst. dir. research and devel., 1967-68; v.p., dir. ops. Intel Corp., Santa Clara, Calif., 1968-75, exec. v.p., 1975-79, pres., 1979—, chief operating officer, 1976—; lectr. dept. elec. engring. and computer sci. U. Calif.-Berkeley, 1966-72. Recipient Medal award Am. Inst. Chemists, 1960; cert. merit Franklin Inst., 1975; Townsend Harris medal CCNY, 1980. Fellow IEEE (Achievement award 1969, J. J. Ebers award 1974); mem. Nat. Acad. Engring. Subspecialties: Electronics; Integrated circuits. Office: Intel Corp 3065 Bowers Ave Santa Clara CA 95051

GROVE, DON J., research physicist; b. Pitts., Oct. 8, 1919; s. Earle M. and Mary (Jones) G.; m. Dane McVay, Sept. 23, 1943; children: Ellen, James, Robert. A.B., Coll. Wooster, 1941; Ph.D., Carnegie Mellon U., Pitts., 1953. Research physicist and cons. Westinghouse Electric Corp., Pitts., 1943-82; prin. research physicist Princeton (N.J.) U., 1982—, mgr. Tokamak Fusion Test Reactor project Plasma Physics Lab., 1982—. Contbr. over 75 sci. articles to profl. publs. Recipient Order of Merit award Westinghouse Electric Corp., 1976; Disting. Assoc. award U.S. Dept. Energy, 1976. Fellow Am. Phys. Soc.; mem. Sigma Xi. Subspecialties: Nuclear fusion; Plasma physics. Patentee in field. Home: 191 Riverside Dr Princeton NJ 08540 Office: Princeton U Plasma Physics Lab PO Box 451 Princeton NJ 08544

GROVE, PATRICIA ANN, biology educator; b. Bronx, N.Y., Oct. 3, 1952; d. Edward J. and Helen E. (Leyko) G.; m. Anthony R. Candela, May 6, 1979. B.S., Coll. Mt. St. Vincent, Bronx, 1974; M.A., CCNY, 1977, M.Phil., 1979, Ph.D., 1981. Adj. lectr. CCNY, 1974-79, Mercy Coll., Dobbs Ferry, N.Y., 1978-79; instr. Coll. Mt. St. Vincent, Bronx, 1979-81, asst. prof. biology, 1981—. Mem. AAAS, Animal Behavior Soc., Assn. Women in Sci., Am. Ornithol. Union, Sigma Xi. Subspecialties: Behavioral ecology; Ethology. Current work: Vertebrate social behavior; behavioral toxicology; avian territoriality; evolution of mating systems. Home: 377 N Broadway Yonkers NY 10701 Office: Coll Mount Saint Vincent Riverdale Ave and W 263d St Bronx NY 10471

GROVER, JOHN HARRIS, fisheries educator; b. Rockville Centre, N.Y., Dec. 21, 1940; s. Roscoe A. and Arlene (Harris) G.; m. Janice E. Jacobson, June 24, 1967; children: Sarah, John, Marion, Ruth. B.S., U. Utah-Salt Lake City, 1964; M.S., Iowa State U.-Ames, 1966, Ph.D., 1969. Lectr. zoology U. Libya, Tripoli, 1969-71; asst. prof. Auburn U., 1969-77, assoc. prof., 1977—; cons. AID, World Bank, U.S. Dept. Agr. Contbr. articles to profl. jours. Mem. Am. Fisheries Soc. Mormon. Subspecialties: Resource conservation. Current work: Aquaculture, international development and academic administration. Home: 233 Willow Creek Rd Auburn AL 36830 Office: Dept Fisheries and Allied Aquacultures Auburn U Auburn AL 36849

GROVER, PUSHPINDER SINGH, research dentist, army officer; b. Amritsar, Punjab, India, Feb. 16, 1946; came to U.S., 1971, naturalized, 1978; s. Mohan Singh and Mohinder (Kaur) G.; m. Surinder K. Moonga, Oct. 2, 1970; children: Amrita K., Ajeet K., Davinder S. B.Sc., Khalsa Coll., Amritsar, 1964; B.D.S., Punjab Govt. Dental Coll., Amritsar, 1969; D.M.D., Tufts U., 1977. Commd. capt. Dental Corps, U.S. Army, 1977, advanced through grades to maj., 1982; gen. dental officer, Ft. Leonard Wood, Mo., 1977-80; research dental officer Inst. Dental Research, Walter Reed Army Med. Center, Washington, 1980-82, Inst. Dental Research at Letterman Army Inst. Research, Presidio of San Francisco, Calif., 1982—; Contbr. numerous articles to profl. jours. Recipient Patient Care award Ft. Leonard Wood, 1979. Mem. ADA, Internat. Assn. Dental Research, Am. Assn. Forensic Dentistry, Am. Assn. Dental Research. Subspecialty: Preventive dentistry. Current work: Predictability of dental emergencies by radiographs; rapid indentification of combat casualties and mass disaster victims using radiographs and computers; co-investigator various research projects (development of a portable field x-ray unit, design and development of a protective facial mask). Home: 419-A Washington Blvd Presidio of San Francisco CA 94129 Office: US Inst Dental Research Letterman Army Inst Research Presidio of San Francisco CA 94129

GROVES, PHILIP MONTGOMERY, psychiatry educator, biological psychologist; b. Washington, June 21, 1944; s. G.P. and Eleanor G.; m. Jennifer Capp, Mar. 16, 1968; children: John Stephen, Andrea Laura. B.A., UCLA, 1966; M.A., San Diego State U., 1967; Ph.D., U. Calif.-Irvine, 1970. Research asst., teaching asst. dept. psychology San Diego State U., 1966-67; research asst. dept. psychobiology U. Calif.-Irvine, 1967-70; asst. prof. psychology U. Colo.-Boulder, 1970-72, assoc. prof., 1972-75, prof. biol. psychology, 1975-80; prof. psychiatry U. Calif.-San Diego, 1980—; mem. neuropsychol. research rev. com. NIMH, 1971-75, research sci. devel. rev. com., 1980-83; mem. drug abuse research rev. com. Nat. Inst. Drug Abuse, 1975-79. Author: (with K. Schlesinger) books, including An Introduction to Biological Psychology, 1979, 2d edit., 1982; brain research numerous publs.; mem. editorial adv. bd.: Neural and Behavioral Biology, 1973—, Biol. Psychiatry, 1977—, Behavioral Neurosci., 1983—, and others. Recipient Research Sci. Devel. award NIMH, 1973, Nat. Inst. Drug Abuse, 1978; NIMH grantee, 1971—; Nat. Inst. Drug Abuse grantee, 1971—; research Scientist award Nat. Inst. Ong. Abuse, 1983—; John D. and Catherine T. MacArthur grantee, 1982—. Fellow Am. Psychol. Assn. (Early Career award 1978); mem. Soc. Biol. Psychiatry, Soc. Neurosci., Internat. Brain Research Orgn., Collegium Itnernationale Neuro-Psychopharmacologicum, Sigma Xi. Subspecialty: Psychobiology. Current work: Brain research. Home: 4761 Lomitas Dr San Diego CA 92116 Office: U Calif-San Diego Dept Psychiatry M-003 Sch Medicine La Jolla Ca 93093

GROW, RICHARD W., electrical engineering educator, researcher; b. Lynndyl, Utah, Oct. 31, 1925; m., 1947; 4 children. B.S., U. Utah, 1948, M.S., 1949; Ph.D. in Elec. Engring, Stanford U., 1955. Electronic scientist radio countermeasures br. Naval Research lab., 1949-50, nucleonics div, 1950-51; research assoc. Stanford Electronics Lab., 1953-58; assoc. research prof., asst. dir. high velocity lab. U. Utah, 1958-59, assoc. research prof., dir. lab., 1959-62, assoc. research prof. microwave device and phys. electronics lab., 1962-64, research prof., 1964-66, dir. microwave device and phys. electronics lab., 1960-80, prof. elec. engring., 1966—, chmn. dept. elec. engring., 1965-80; cons. in field. Mem. IEEE (sr.). Subspecialties: Electronics; Microwave physics. Office: Dept Elec Engring U Utah Salt Lake City UT 84112

GRUBER, DUANE EDWARD, X-ray astronomer; b. Culver City, Calif., Aug. 22, 1940; s. Leo Edward and Irma Caroline (Waller) G.; m. Linda Storch, June 22, 1974; children: Leah, Mark, Brian, Emily, Daniel. A.B., UCLA, 1962; Ph.D., U. Calif., San Diego, 1974. Reactor physicist Atomics Internat., Canoga Park, Calif., 1962-65; X-ray astronomer physics dept. and Ctr. for Astrophysics and Space Sci. U. Calif., San Diego, La Jolla, 1967—. Contbr. papers to tech. jours. and profl. confs. Mem. Am. Astron. Soc. Subspecialties: 1-ray high energy astrophysics; Gamma ray high energy astrophysics. Current work: Astrophysics of close binary stellar systems; cosmology; instrumentation. Home: 440 Nardo Solana Beach CA 92075 Office: Univ Calif San Diego Code C-011 La Jolla CA 92093

GRUBER, HELEN ELIZABETH, medical educator, researcher; b. Wallace, Idaho, Nov. 6, 1946; d. Hugo John and Margaret A. (Dorsey) G. B.S., U. Idaho, 1969; M.S., Oreg. State U., 1974, Ph.D., 1976. NIH postdoctoral fellow U. Iowa, Iowa City, 1976-78; instr. U. Wash., Seattle, 1978-81; research fellow nephrology U. So. Calif., Los Angeles, 1981-82, research asst. prof., 1982—. N.W. Coll. & Univ. Assn. for Sci. fellow, 1972-73. Mem. Am. Soc. Cell Biology, Am. Fedn. Clin. Research, Am. Soc. Bone and Mineral Research, Soc. Exptl. Biology and Medicine, AAAS, Nat. Kidney Found. So. Calif. (mem. sci. adv. council 1981—). Subspecialties: Cell biology (medicine); Endocrinology. Current work: Metabolic bone disease and mineral metabolism; human bone biopsy quantification; bone tumors; quantitative histology. Home: 504 N Louise St Apt 20 Glendale CA 91206 Office: Univ So Calif Sch Medicine Nephrology Div Room 4250 Unit I 2025 Zonal Ave Los Angeles CA 90033

GRUBER, SAMUEL HARVEY, marine biologist, educator; b. Bklyn., May 13, 1938; s. Sidney and Claire (Mednick) G.; m. Mariko Hirata; children: Meegan Minori, Marisa Aya. B.S. in Zoology, U. Miami, 1960; M.S. in Marine Sci., Inst. of Marine and Atmospheric Sci., U. Miami, 1966; Ph.D. in Marine Scis, Inst. of Marine and Atmospheric Sci., U. Miami, 1969. Research scientist Rosenstiel Sch. of Marine and Atmospheric Sci., U. Miami, 1969-73, research asst. prof., 1972-73; asst. prof. U. Miami, 1973-76, assoc. prof., 1976—. Contbr. articles on marine biology to profl. jours. NSF grantee; Delfiner Found. grantee, 1974-75; Binational Sci. Found. (U.S.-Israel) grantee, 1982-85; Office of Naval Research grantee, 1970-83. Mem. Am. Fishery Soc., Am. Inst. Fishery Research Biologists, Assn. for Research in Vision and Opthalmology, Internat. Assn. for Aquatic Animal Medicine, Internat. Assn. for Marine Animal Trainers, Soc. Neurosci., Optical Soc. Am., Sigma Xi. Democrat. Subspecialties: Marine Biology; Behavioral ecology. Current work: Behavioral ecology of apex marine predators; researcher in shark-human interactions. Office: RSMAS U Miami Rickenbacker Causeway Miami FL 33149

GRUBMAN, MARVIN, biochemist; b. Bronx, N.Y., Nov. 4, 1945; s. Abe and Ruth (Weiner) G.; m. Annette, Nov. 21, 1970; children: David, Susan. B.S., CCNY, 1967; Ph.D., U. Pitts., 1972. Research assoc. Albert Einstein Coll. Medicine, 1972-76; research chemist Plum Island Animal Disease Center, 1976—; NIH fellow, 1973. Contbr. articles to profl. jours. Recipient U.S. Dept. Agr. Disting. Service award, 1982. Mem. Am. Soc. Microbiology, Am. Soc. Virology, AAAS (Newcomb-Cleveland award), N.Y. Acad. Sci., Picornavirus Study Group. Subspecialties: Virology (biology); Molecular biology. Current work: Molecular biology of foot and mouth disease virus, bluetongue viruse, study replication of virus and viral specific RNA and proteins. Office: PO Box 848 Greenport NY 11944

GRUNBERG, EMANUEL, biologist; b. Everett, Mass., July 9, 1922; s. Meyer and Sarah (Horowitz) G.; m. Eleanor Muriel Grunberg, Jan. 15, 1946; children: Steven Marc, Neil Everett. Asst., dept. dermatology Barnard Skin & Cancer Hosp., St. Louis, 1944; chief mycologist dept. dermatology New Haven Hosp., 1945-46; sr. bacteriologist dept. chemotherapy Hoffmann-LaRoche, Inc., Nutley, N.J., 1946-57, sr. scientist, 1957-59, asso. dir., 1959-60, dir., 1960-73, dir. diagnostic research dept., 1970-76, asso. dir. biol. research, 1973-76, dir. chemotherapy/diagnostic research, 1976-80, dir. biol. research, 1980-82, asst. v.p. sci. affairs, 1982—; guest lectr. Seton Hall U., Jersey City, 1961-63; cons. Newark Beth Israel Med. Center, Newark, 1962—, St. Michael Hosp., 1968-76, Roche Clin. Labs., Raritan, N.J., 1973—; adj. asso. prof. pharmacology Cornell U. Med. Coll., N.Y.C., 1973-80; mem. Pan Am. Health Orgn. com. on med. mycology, 1973-76; industry liaison Toxicology subcom. FDA Bur. Med. Devices and Diagnostic Products, 1975-77; vis. lectr. Dept. Pharmacology, UCLA, 1978-81; adj. prof. Cornell U., N.Y.C., 1980—, U. Calif. - San Diego, 1980—. Contbr. articles to profl. jours. Fellow N.Y. Acad. Sci., Am. Acad. Microbiology, Internat. Acad. Law and Sci., Am. Coll. Clin. Pharmacology and Chemotherapy; mem. Internat. Soc. Leprosy (hon. mem.), AAAS, Soc. Exptl. Biology and Medicine, Am. Soc. Microbiology, Am. Thoracic Soc., Internat. Soc. Tropical Dermatology, Internat. Soc. Chemotherapy, Am. Soc. Microbiology, Theobald Smith Soc., Mycol. Soc. Am., Pan Am. Health Orgn., N.J. Acad. Sci., Am. Inst. Biol. Sci., Am. Chem. Soc., N.J. Soc. Parasitology, Infectious Diseases Soc. Am., Phi Beta Kappa, Sigma Xi, Phi Eta Sigma. Subspecialties: Microbiology (medicine); Immunopharmacology.

GRUNBERG, NEIL EVERETT, medical psychologist, educator, researcher; b. Newark, Aug. 9, 1953; s. Emanuel and Eleanor (Hoffman) G. B.S., Stanford U., 1975; M.A., Columbia U., 1977, M.Phil., 1979, Ph.D., 1980. Asst. prof. Uniformed Services U. Health Scis., Bethesda, Md., 1979—; research assoc. Nat. Cancer Inst., Bethesda, 1982—. NIH grantee, 1976-79. Mem. Am. Psychol. Assn. Subspecialties: Psychobiology; Social psychology. Current work: Cigarette smoking, appetitive behaviors. Home: 7805 Fairfax Rd Bethesda MD 20814 Office: Uniformed Services U Health Scis Med Psychology Dept 4301 Jones Bridge Rd Bethesda MD 20814

GRUNDER, ALLAN ANGUS, geneticist, educator; b. Kincordine, Ont., Can., Mar. 13, 1935; s. Neil A. and Lillian B. (Norman) G.; m. Linda R. Marshall, July 22, 1972; children: Laura, Vivian. B.S. in Agr, Ont. Agrl. Coll., 1958; M.Sc., U. Alta., 1961; Ph.D., U. Calif.-Davis, 1966. Drainage surveyor Ont. Dept. Agr., 1958; feed rep. Ogilvy & Roses Co., 1958-59; research scientist Animal Research Ctr., 1966—. Mem. Genetics Soc. Am., Genetics Sco. Can., Poultry Sci. Assn., World Poultry Sci. Assn., Sigma Xi. Presbyterian. Subspecialties: Animal genetics; Animal physiology. Current work: Genetic improvement of commercial poultry. Research on egg shell quality, lower fat in poultry, improvement of goose productivity. Office: Animal Research Centre Ottawa ON Canada K1A OC6

GRUNDY, SCOTT MONTGOMERY, medical educator, physician, nutritionist; b. Memphis, Tex., July 10, 1933; s. Allen Clack and Beulah scott (Montgomery) G.; m. Lois Bernice Parker, July 7, 1955; children: Pamela Charlene, Stephanie Scott. B.S., Tex. Tecnol. U.; M.S., M.D., Baylor U.; Ph.D., Rockefeller U. Practice medicine specializing in internal medicine; asst. prof. Rockefeller U., N.Y.C., 1968-71; chief clin. research sect. NIH, Phoenix, 1971-73; prof. medicine U. Calif., San Diego, 1973-81; prof. internal medicine and biochemistry U. Tex. Health Sci. Center, Dallas, 1983—, dir., 1983—. Contbr. articles to profl. jours. Mem. Am. Soc. Clin. Investigation, Am. Assn. Physicians, Am. Physiol. Soc. Subspecialties: Metabolism; Nutrition (medicine). Current work: Cholesterol, triglycerides, lipoproteins, atherosclerosis. Office: U Texas Health Science Center (64-100) Dallas TX 75235

GRUNES, DAVID LEON, research soil scientist, educator, editor; b. Paterson, N.J., June 29, 1921; s. Jacob and Gussie (Griggs) G.; m. Willa Freeman Grunes, June 26, 1949; children: Lee Alan, Mitchell Ray, Rima Louise. B.S., Rutgers U., 1944; Ph.D., U. Calif., 1951. With U.S. Dept. Agr., 1950 , research soil scientist, Ithaca, N.Y., 1964—; assoc. prof. agronomy dept. Cornell U., 1967-76, prof., 1976—; cons. editor soils, agr. McGraw-Hill Ency., Sci. and Tech., 1965—. Contbr. chpts. to books and articles to profl. jours. Served with U.S. Army, 1944-45. Recipient U.S. Dept. Agr. award for research, 1959. Fellow AAAS, Am. Inst. Chemists, Am. Soc. Agronomy, Soil Sci. Soc. Am.; mem. Internat. Soc. Soil Sci., Western Soc. Soil Sci., Council for Agri. Sci. and Tech., Sigma Xi. Subspecialties: Soil chemistry; Plant growth. Current work: Agronomic aspects of crop quality for humans and animals; research in plant nutrition; serve on committees of assns. Office: US Plant Soil and Nutrition Lab Tower Rd Ithaca NY 14853

GRUNEWALD, GARY LAWRENCE, medicical chemistry educator; b. Spokane, Wash., Nov.11, 1937; s. Douglas Laurence and Ruby Anne (Olson) G.; m. Joan O. Dietrich, Dec. 19, 1966. B.S., Wash. State U., Pullman, 1960, B.Pharmacy, 1960; Ph.D., U. Wis.-Madison, 1966. Asst. prof. medicinal chemistry U. Kans.-Lawrence, 1966-72, assoc. prof. medicinal chemistry, 1972-76, prof., 1976—; chmn. cardiovascular research com. Am. Heart Assn., Kans. Affiliate, Inc., Topeka, 1980-82; grant reviewer NIH, Washington, 1982. Editor: Medicinal Research Series of Monographs, 1968—. Mem. Am. Chem. Soc., Acad. Pharm. Scis. (chmn. medicinal chemistry sect. 1983-84), Soc. Neurosci., Royal Soc. Chemistry. Subspecialties: Medicinal chemistry; Neurochemistry. Current work: Design and synthesis of pharmacologically active compounds, particularly those acting in the central and peripheral nervous systems, applications of QuSAr and MO calculations. Home: 1937 Emerald Dr Lawrence KS 66044 Office: U Kans Dept Medicinal Chemistry Malott Hall Lawrence KS 66045

GRUNWALD, ERNEST MAX, chemistry educator; b. Wuppertal, W.Ger., Nov. 2, 1923; m. Esther R. Grunwald, Mar. 17, 1952; 1 dau., Judith. B.S. and B.A., UCLA, 1944, Ph.D. in Chemistry, 1947. Inst. chemistry UCLA, 1947; resident chemist Portland Cement Assn., 1948; Jewel fellow Columbia U., N.Y.C., 1949; from assoc. prof. to prof. chemistry Fla. State U., 1949-61; resident chemist Bell Telephone Labs., Inc., 1961-64; prof. chem. Brandeis U., 1964—. Contbr. articles to profl. jours. Recipient Pure Chemistry award Am. Chem. Soc., 1959; Weizman fellow, 1955; Sloan fellow, 1958-61; Guggenheim fellow, 1975-76. Mem. Nat. Acad. Sci., Am. Acad. Arts and Sci., Am. Chem. Soc., Am. Phys. Soc. Subspecialties: Organic chemistry; Laser-induced chemistry. Current work: Kinetics and equilibrium properties of solutions; infrared laser chemistry. Office: Dept Chemistry Brandeis Univ Waltham MA 02154

GRUNWALD, GERALD BRUCE, developmental neurobiologist; b. Bklyn., Apr. 6, 1954; s. Leo and Ruth (Feldman) G.; m. Barbara Jane Lepak, Aug. 11, 1979. B.A. in Biology cum laude, Cornell U., 1976; M.S. in Zoology-Devel. Biology, U. Wis., 1978, Ph.D., 1981. Nat. Research Service Award fellow Lab. Biochem. Genetics, Nat. Heart, Lung and Blood Inst., NIH, Bethesda, Md., 1981—. Contbr. articles to profl. jours. Mem. Soc. Devel. Biology, Am. Soc. Cell Biology, Soc. for Neurosci., AAAS, N.Y. Acad. Scis., Sigma Xi, Alpha Chi Sigma. Subspecialties: Developmental biology; Neurobiology. Current work: Molecular basis of specific cell-cell interactions during neuronal development. Monoclonal antibodies as probes for ontogenetic and phylogenetic studies of the nervous system. Office: Lab Biochem Genetics NHLBI NIH Bldg 36 Room 1C21 Bethesda MD 20205

GRYC, GEORGE, geologist; b. St. Paul, July 27, 1919; s. Anthony S. and Lillian (Teply) G.; m. Jean L. Funk, Dec. 4, 1942; children: James C., Stefan M., Christina L., Paula Jean, Georgina. B.A., U. Minn., 1940, M.S., 1941; postgrad., Johns Hopkins U., 1947-49. Geologist U.S. Geol. Survey, Washington, 1943-60, staff geologist, 1960-63; chief Alaskan Geol. Br., Menlo Park, Calif., 1963-76, regional geologist, 1976, chief, 1977-82, asst. dir. Western Region, dir.'s rep., 1982—. Recipient Meritorious Service award Dept. Interior, 1973, Disting. Service award, 1978. Fellow Geol. Soc. Am., Arctic Inst. N.Am. (gov. 1966-72, sec. 1970-71); mem. Paleontol. Soc., Am. Assn. Petroleum Geologists (assoc. editor 1976-80), Alaska Geol. Soc., Geol. Soc. Washington. Club: Cosmos. Subspecialties: Sedimentology; Geology. Current work: Administration and supervision of earth science research. Home: 753 Saranac Dr Sunnyvale CA 94087 Office: US Geol Survey 345 Middlefield Rd Menlo Park CA 94025

GUARINO, ANTHONY MICHAEL, toxicologist; b. Framingham, Mass., Dec. 11, 1934; s. Alfred Vincent and Nellie Lucy G.; m. Aida Iris Guarino, Nov. 11, 1957; children. B.S., Boston Coll., 1956; M.S., U. R.I., 1963, Ph.D., 1966. Research assoc. Nat. Heart Inst., NIH, 1966-68; research pharmacologist Lab. Chem. Pharmacology, Nat. Cancer Inst., 1968-73, chief toxicology lab., 1973-80; regulatory pharmacologist/toxicologist Office New Drugs, Nat. Ctr. Drugs and Biologics, FDA, Rockville, Md., 1980—; mem. faculty, chmn. dept. pharmacology and toxicology NIH Grad. Sch., Bethesda, 1970—. Contbr. 90 articles to profl. jours. Served to lt. (j.g.) USN, 1957-60; scientist officer USPHS, 1966—; capt. USPHS, 1975—. Mem. Am. Soc. Pharmacology and Explt. Therapeutics, Soc. Toxicology, Am. Chem. Soc. Subspecialties: Toxicology (medicine); Pharmacology. Current work: Drug (xenobiotic) transport, drug (xenobiotic) Toxicology, biliary transport, drug metabolism, marine pharmacology and toxicology, quantitative toxicology, regulatory toxicology, biochemical toxicology. Home: 5903 Melvern Dr Bethesda MD 20817 Office: 5600 Fishers Ln Rm 17B45 Rockville MD 20857

GUASTAFERRO, ANGELO, federal agency administrator; b. Hoboken, N.J., June 4, 1932; s. Carlo and Rafaela Nancy (Gioffi) G.; m. Eleanor Lago, Sept. 12, 1954; children: Carl, Mark, John Brian. B.S. in Mech. Engring, N.J. Inst. Tech., 1954; M.B.A., Fla. State U., 1963. With NASA, 1963—, dep. project mgr., 1974-76, dir. planetary programs, Washington, 1979-81; dep. dir. Ames Research Center, Moffett Field, Calif., 1981—; v.p., bd. dirs. Langley Fed. Credit Union, 1977-79; cons. in field. Served with USAF, 1955-58. Recipient Langley Spl. Achievement award NASA, 1974, 77, 78, Outstanding Leadership medal, 1977, Superior Performance award, 1980, Exceptional Service medal, 1981, Presdl. Meritorious rank, 1982. Fellow AAAS; Mem. AIAA (Space Systems medal 1982), Planetary Soc., Mars First Landing Soc. (pres. 1978-79). Roman Catholic. Clubs: Toastmasters (Eglin AFB, Fla.) (past pres.); K.C. (grand knight). Subspecialties: Aerospace engineering and technology; Systems engineering. Office: Ames Research Center Code 200-2 Moffett Field CA 94035

GUBAR, GEORGE, clinical psychologist, educator; b. West New York, N.J., Dec. 12, 1921; s. Bernardo and Anna (Starr) G.; m. Beulah Weill, June 28, 1942; children: Norinne Lynn, Stephen Gregory, Bentley. B.A., Rutgers U., 1961, Ed.M., 1963; Ph.D., 1965. Pvt. practice psychology, Englewood, N.J., 1945-63; program dir. Mount Carmel Guild Social Service Center, Paterson, N.J., 1964—; assoc. prof. psychology Seton Hall U., South Orange, N.J., 1965—; psychol. cons. Bergen Pines Hosp., Paramus, N.J., 1970—. Contbr. articles in field to profl. jours. Served to lt. U.S. Army, 1942-45; ETO. Mem. Am. Psychol. Assn., N.J. Psychol. Assn. (bd. dirs. 1970-72), N.Y. Acad. Scis., Sigma Xi, Kappa Delta Pi. Subspecialties: Clinical psychology; Behavioral psychology. Current work: Emotions and facial expressions, addictive personality, drug and alcohol rehabilitation, substance abuse prevention. Home: 6 Cobblewood Rd Livingston NJ 07039 Office: Seton Hall U South Orange Ave South Orange NJ 07079

GUBERMAN, STEVEN LAWRENCE, physicist; b. Bklyn., Dec. 11, 1945; s. Irving and Rosalind (Levine) G. B.A., SUNY, Binghamton, 1967; Ph.D., Calif. Inst. Tech., 1972. Research fellow, lectr. Harvard Coll. Obs., Cambridge, Mass., 1973-78; Nat. Acad. Scis., NRC sr. resident research assoc. Air Force Geophysics Lab., Hanscom AFB, Mass., 1978-79; prin. scientist Phys. Scis., Inc., Woburn, Mass., 1979-80; vis. fgn. scientist fellow Max-Planck Inst. fur Physik und Astrophysik, Munich, W.Ger., 1980; sr. research physicist Boston Coll., Newton, Mass., 1980—. Contbr. articles to profl. jours. NASA grantee, 1981—; Air Force Geophysics Lab. grantee, 1980—. Mem. Am. Phys. Soc., Am. Chem. Soc., Am. Geophys. Union, Sigma Xi. Subspecialty: Theoretical chemistry. Current work: The development of new techniques in quantum chemical physics with application to important processes in the upper atmosphere, interstellar space, lasers. Home: 22 Bonad Rd Winchester MA 01890 Office: Boston Coll 885 Centre St Newton MA 02159

GUCKENHEIMER, JOHN, mathematics educator; b. Baton Rouge, Sept. 26, 1945; s. Ludwig and Gertrude (Goldschmidt) G.; m. Meredith Kusch, June 15, 1974; 1 son, Mathew. B.A., Harvard U., 1966; Ph.D., U. Calif., Berkeley, 1970. Research fellow U. Warwick, Coventry, Eng., 1969-70; mem. Inst. Advanced Study, Princeton, N.J., 1970-72; lectr. MIT, Cambridge, 1972-73; vis. mem. Courant Inst., N.Y.C., 1978; mem. I.H.E.S., Bures-sur-Yvette, France, 1979; asst. prof. to prof. math. U. Calif.-Santa Cruz, 1973—. Author: (with J. Moser, S. Newhouse) Dynamical Systems, 1980, (with P. Holmes) Nonlinear Oscillations, Dynamical Systems and Bifurcation Theory, 1983. Mem. Am. Math. Soc., AAAS. Subspecialties: Applied mathematics; Nonlinear dynamics. Current work: Nonlinear dynamics. Home: 72 Alta Vista Dr Santa Cruz CA 95060 Office: U Calif Dept Math Santa Cruz CA 95064

GUDEHUS, DONALD HENRY, astrophysicist; b. Jersey City, Sept. 13, 1939; s. Herman Andrew and Katherine Pauline (Hirner) G.; m. Linda Hope Gudehus, Sept. 19, 1968. B.S. in Physics, MIT, 1961, A.M., Columbia U., 1963; Ph.D. in Astronomy (NASA predoctoral trainee), UCLA, 1971. Engr., scientist McDonnell Douglas, El Segundo, Calif., 1964-67; postdoctoral scholar astronomy UCLA, 1971-75; asst. prof. physics Los Angeles City Coll., 1974-81; asst. research scientist physics U. Mich., Ann Arbor, 1981—. Contbr. articles to profl. jours. Recipient grant in aid in research Sigma Xi, 1974. Mem. Am. Astron. Soc., Lorquin Entomol. Soc. Subspecialties: Cosmology; Optical astronomy. Current work: Observational cosmology with a CCD camera, clusters of galaxies, redshift, CCD, photometry, spectroscopy, extragalactic astronomy, instrumentation. Office: Randall Lab U Mich Ann Arbor MI 48109

GUDMESTAD, NEIL CARLTON, plant pathologist; b. Balley City, N.D., Dec. 30, 1952; s. George and Carla Marie G.; m. Arne Aarnes Rogalie, Feb. 1, 1977; 1 dau., Aarnes Erin. B.S. in Biology, Valley City State Coll., 1974; M.S., N.D. State U., 1978, Ph.D. in Plant Pathology, 1982. Research asst. dept. plant pathology N.D. State U., Fargo, 1977-78; plant pathologist N.D. State Seed Dept., State University Sta., Fargo, 1978—. Contbr. articles to profl. jours. Mem. Am. Phytopathol. Soc., Potato Assn. Am., Sigma Xi. Subspecialties: Plant cell and tissue culture; Plant virology. Current work: Epidemiology of bacterial, fungal and viral diseases affecting seed potatoes, meristem culture for maintenance of basic seedstocks. Home: 1831 19th St S Fargo ND 58103 Office: ND State Seed Dept State University Station Fargo ND 58105

GUEFT, BORIS, pathologist, educator, consultant; b. Cannes, France, Nov. 10, 1916; came to U.S., 1917, naturalized, 1927; s. Amshel and Nina (Oussoltseff) G.; m. June 25, 1943; children: Nina, Esther, Michael. A.B., Columbia U., 1938; M.D., N.Y.U., 1941. Diplomate: Am. Bd. Anatomic and Clin. Pathology. Intern Sinai Hosp., Balt., 1941-42; resident New Britain (Conn.) Gen. Hosp., 1946-47, Mt. Sinai Hosp., N.Y.C., 1947-50; dir. lab. Fairfield State Hosp., Newtown, Conn., 1950-55, Cin. Va. Hosp., 1955-58; assoc. prof. to prof pathology Albert Einstein Coll. Medicine, N.Y.C., 1958-70; dir. lab. Union Hosp. of the Bronx, N.Y.C., 1970—; cons. electron microscopy Bronx-Lebanon Hosp., 1979—; adj. prof. pathology N.Y. Med. Coll., 1979—. Contbr. articles to profl. jours. Served to capt. USMC, 1943-46. NIH grantee, 1957-67. Mem. Am. Assn. Pathologists, Electron Microscopy Soc. Am., Internat. Acad. Pathology, AMA. Subspecialties: Biophysics (biology); Cell biology. Home: 128 Vernon Dr Scarsdale NY 10583

GUENTER, JOSEPH MARTIN, physicist, educator; b. Little Rock, Mar. 21, 1938; s. Bernard Henry and Mary (Martin) G. B.A., Hendrix Coll., 1960; M.S., U. Ark., 1962; postgrad., U. Mo., Rolla, 1966, La. State U., 1970, Vanderbilt U., 1973. Instr. physics Ark. A&M Coll., Monticello, 1962-71, head dept. physics, 1965-71; mem. faculty U. Ark., Monticello, 1971—, asst. prof. physics, 1965, dir., 1978—, assoc. dir., 1978—; planetarium cons. NSF Summer Insts., 1971, 72. Contbr. articles to sci. jours. Mem. Monticello Community Edn. Adv. Bd., 1977-80; bd. dirs. Drew County Hist. Soc., 1976-82. Named Prof. Grubbs Outstanding Advisor Sigma Tau Gamma, 1979, Outstanding Vol. in Drew County, 1981. Mem. Am. Phys. Soc., Am. Assn. Physics Tchrs., AAUP, AAAS, Ark. Acad. Sci., Internat. Planetarium Soc. Democrat. Presbyterian. Subspecialties: Radiation physics; Astronomy education. Current work: Creative programs in a small planetarium both dealing with astronomy and many other areas. Home: 168 Babin Dr Monticello AR 71655 Office: U Ark Box 3480 Monticello AR 71655

GUERIGUIAN, JOHN LEO, govt. ofcl., cons., researcher; b. Alexandria, Egypt, Sept. 20, 1935; s. Levon Artin and Valentine (Mamigonian) G.; m. Ida Fai-Fong; children: Leo Fong, Vincent John, Florence Marie. B.S., U. Paris, 1958, M.S. in Chemistry, 1964, M.D., 1965. Research fellow Harvard U. Med. Sch., 1965-67; attendant staff Peter Bent Brigham Hosp., Boston, 1965-67; research scientist dept. biochemistry U. Paris Med. Sch., 1967-69; asst. prof. U. N.C., Chapel Hill, 1969-73; assoc. prof. U. Minn. Sch. Medicine, Duluth, 1973-78, adj. asso. prof., 1978—; supervisory med. officer Nat. Center for Drugs and Biologics, FDA, Rockville, Md., 1978—; cons. in field. Contbr. articles to profl. jours., also several editorships. Co-founder, bd. dirs. Mamigonian Found., Rockville. Mem. Endocrine Soc., Am. Soc. Clin. Pharmacology and Therapeutics, Am. Soc. Pharmacology and Exptl. Therapeutics. Mem. Armenian Apostolic Ch. Subspecialties: Pharmacology; Endocrinology. Current work: Reproductive endocrinology; modern methods in communication; heuristics of drug regulation. Home: 14513 Woodcrest Dr Rockville MD 20853 Office: 5600 Fishers Ln Rockville MD 20857

GUERON, HENRI MAXIMILIEN, utility executive; b. Neuilly sur Seine, France, June 3, 1936; came to U.S., 1964; s. Jules and Genevieve (Bernheim) G.; m. Judith Mitchell, Sept. 29, 1963; 6, children: Michele, Nicole. Ingenieur, ENSAE, Paris, 1958; M.A., Brandeis U., 1960; Ph.D., MIT, 1966. Nuclear physicist Comnissariat a'L'Energie Atomique, Paris, 1962-64; utility cons. S.M. Stoller Corp., N.Y.C., 1966-76; dir. nuclear fuel and coal supply ConEdison Co. N.Y., N.Y.C., 1974—. Served to 1st lt. French Air Force, 1960-62. Mem. Am. Nuclear Soc. Subspecialties: Fuels; Coal. Current work: Conversion of oil-burning electrical power plants to coal. Home: 285 Central Park W New York NY 10024 Office: 4 Irving Pl New York NY 10003

GUETTER, HARRY HENDRIK, astronomer; b. Andijk, Netherlands, Feb. 1, 1935; s. John and Neeltje (Cupido) G.; m. Joan Adriana Boodt, July 5, 1963; children: Mark, Adrian, Stephanie. B.Sc. with honors in Physics and Math, Queen's U., Can., 1961; M.A. in Astronomy, U. Toronto, Ont., Can., 1963. Research asso. David Dunlop Obs., Richmond Hill., Ont., 1963-64; astronomer U.S. Naval Obs., Flagstaff (Ariz.) Sta., 1964—. Contbr. articles to profl. jours. Mem. Am. Astron. Soc., Internat. Astron. Union, Astron. Soc. Pacific, Sigma Xi. Democrat. Baptist. Subspecialties: Optical astronomy; Infrared optical astronomy. Current work: Determination of trigonometric stellar parallexes; spectroscopy, polarimetry, near-infrared and optical photometry of young stars in galactic clusters. Home: 526 W Havasupai Rd Flagstaff AZ 86001 Office: PO Box 1149 Flagstaff AZ 86002

GUIDA, PETER MATTHEW, surgeon, educator; b. N.Y.C., July 18, 1927; s. Santo and Anna (Tambury) G.; m. Bernadette Castro, Mar. 10, 1979; children: Patricia, Peter M. B.S., L.I. U., 1949; M.D., Albany Med. Coll., 1954. Diplomate: Am. Bd. Surgery, Am. Bd. Thoracic Surgery. Intern N.Y. Hosp., N.Y.C., 1954-55; resident in surgery, 1955-60; prof. surgery Cornell U. Med. Coll., N.Y.C., 1968—; attending surgeon N.Y. Hosp., 1968—. Served to lt. comdr. USNR, 1943-46; PTO. Recipient Horatio Alger award Horatio Alger Soc., 1981. Fellow ACS, Internat. Coll. Surgeons, Am. Coll. Cardiology. Roman Catholic. Subspecialties: Surgery; Theoretical computer science. Current work: Computers in medicine. Address: New York Hospital Cornell University Med Ctr 525 E 68th St New York NY 10021

GUIDA, WAYNE CHARLES, chemistry educator; b. Tampa, Fla., Mar. 20, 1946; s. Angelo and Violet G.; m. Anne Richards, Dec. 12, 1976; 1 stepdau.: Stephanie Marie D'Angelo. B.A., U. South Fla., 1968, Ph.D., 1976. Tchr. East Bay High Sch., Riverview, Fla., 1968-72; fellow Duke U., Durham, N.C., 1976; vis. asst. prof. Eckerd Coll., St Petersburg, Fla., 1977, asst. prof., 1977-82, assoc. prof. chemistry, 1982—. Contbr. articles in field to profl. jours. Research Corp. grantee, 1979-81; NSF grantee, 1980-82; Petroleum Research Fund grantee, 1982-84. Mem. Am. Chem. Soc., Sigma Xi, Phi Kappa Phi. Subspecialties: Organic chemistry; Synthetic chemistry. Current work: Chemistry of crown ethers. Office: Dept Chemistry Eckerd Coll 4400 54th S Saint Petersburg FL 33733

GUILBAULT, GEORGE GERALD, chemistry educator, researcher; b. New Orleans, Dec. 22, 1936; s. George Robert and Valerie (Kothe) G.; m. Susan Glorgh-Bachman, Mar. 28, 1983; children: George G., Ann Marie, Eve Michelle, Stephen C. B.S., Loyola U., New Orleans, 1958; M.S., Princeton U., 1960, Ph.D., 1961. Sr. research scientist U.S. Army, Edgewood Arsenal, Md., 1961-66; prof. chemistry U. New Orleans, 1966—; vis. prof. U. Denmark, Copenhagen, 1973-75, U. Lund, 1982-83, U. Cl. Bernard, Lyon, France, 1982-83; pres. Universal Sensors, New Orleans, 1981—. Author: Instrumental Analysis, 1969, Modern Quantitative Analysis: Experiments for Non-Majors, 1970, Practical Fluorescence, 1973, Handbook of Enzymatic Methods of Analysis, 1976, Handbook of Immobilized Enzymes, 1983; others; editor: Jour. Analytical Letters, 1967—; contbr. chpts. to books and articles to profl. jours. Served to 1st lt. U.S. Army, 1961-62. Mem. Am. Chem. Soc., Internat. Union Pure and Applied Chemistry (chmn. nomenclature com.), Nat. Clin. Lab. Standards (chmn.), N.Y. Acad. Sci., NRC. Roman Catholic. Subspecialties: Analytical chemistry, Clinical chemistry. Current work: Research in analytical biochemistry/clinical chemistry and environmental analysis; immobilized enzymes, immunochemistry, fluoresence, piezoelectric crystals, immunoanalysis, microbiology, enzyme electrodes. Home: 2300 Edenbron St Apt 376 Metairie LA 70001 Office: Chemistry Dept U New Orleans Lakefront Campus New Orleans LA 70148

GUILD, GREG, biology educator. B.S., N.C. State U., 1972; Ph.D., Rutgers U., 1976. Postdoctoral fellow Stanford U. Sch. Medicine, 1976-79; asst. prof. biology U. Pa., Phila., 1979—. Subspecialties: Genetics and genetic engineering (biology); Molecular biology. Current work: Developmental biology, drosophila, gene regulation, gene expression, recombinant DNA.

GUILD, WALTER RUFUS, biochemistry educator; b. Ann Arbor, Mich., Oct. 25, 1923; s. Stacy Rufus and Florence Ruth (White) G.; m. Ellen Christine Sangster, Feb. 22, 1946; 1 son, Thomas D. Student, Swarthmore Coll., 1941-43, NYU, 1943-44; B.S., U. Tex., 1948, M.A., 1949; Ph.D., Yale U., 1951. Instr. physics Yale U., 1951-54, asst. prof. biophysics, 1954-60, radiation safety officer, 1955-60; assoc. prof. biochemistry Duke U., 1960-65, prof., 1965—; mem. internat. fellowship rev. com. NIH Fogarty Internat. Ctr., 1968-72; mem. microbial genetics study sect. NIH, 1982—. Editorial bd.: Radiation Research, 1964-67, Jour. Bacteriology, 1975—; contbr. articles to sci. jours. Served to 1st lt. USAAF, 1943-46. AEC fellow, 1950-51; NIH sr. postdoctoral fellow, 1970-71. Mem. Am. Soc. Biol. Chemists, Biophys. Soc., Radiation Research Soc., Genetics Soc. Am., AAAS, Am. Soc. Microbiology, Fedn. Am. Scientists, Sigma Xi. Subspecialties: Genetics and genetic engineering (biology); Molecular biology. Current work: Gene transfer mechanisms in Streptococcus pneumoniae, conjugation and the spread of drug resistance among streptococci; transformation and transfection in gram positive bacteria; plasmids, phages, and conjugative transposons. Home: 2625 McDowell St Durham NC 27705

GUILLAUME, GERMAINE GABRIELLE CORNELISSEN, chronobiologist, statistician, researcher; b. Schaerbeek, Belgium, Nov. 22, 1949; came to U.S., 1976; d. Alphonse and Helene A (Minne) Cornelissen; m. Francis M. Guillaume, Nov. 22, 1975. M. Physics, U. Brussels, 1971, M. Ed., 1971, Ph.D. in Physics, 1976. Tchr. sci. Lycee E. Max, Brussels, 1971-73; Irsia fellow U. Brussels, 1974-76; research fellow U. Minn.-Mpls., 1976-82, research assoc., 1982—; chairperson/co-organizer various profl. meetings, 1975—. Contbr. articles to profl. jours.; referee various profl. jours., 1978—. NIH grantee, 1981—. Mem. Groupe D'Etude Des Rythmes Biologiques, Internat. Soc. Chronobiology, Societe Belge de Physique, Am. Phys. Soc., Soc. Indsl. and Applied Math., AAAS. Subspecialties: Chronobiology; Statistics. Current work: Implementation and application of methods of time series analysis to biologic and medical data, with emphasis on cancer treatment and prevention. Home: 2008 Brewster Apt 205 Saint Paul MN 55108 Office: U Minn 420 Washington Ave SE Minneapolis MN 55455

GUILLEMIN, ROGER, physiologist; b. Dijon, France, Jan. 11, 1924; came to U.S., 1953, naturalized, 1963; s. Raymond and Blanche (Rigollot) G.; m. Lucienne Jeanne Billard, Mar. 22, 1951; children—Chantal, Francois, Claire, Helene, Elizabeth, Cecile. B.A., U. Dijon, 1941, B.Sc., 1942; M.D., Faculty of Medicine, Lyons, France, 1949; Ph.D., U. Montreal, 1953; Ph.D. (hon.), U. Rochester, 1976, U. Chgo., 1977, Baylor Coll. Medicine, 1978, U. Ulm, Germany, 1978, U. Dijon, France, 1978, Free U. Brussels, 1979, U. Montreal, 1979. Intern, resident univs. hosps., Dijon, 1949-51; asso. dir., asst. prof. Inst. Exptl. Medicine and Surgery, U. Montreal, 1951-53; asso. dir. dept. exptl. endocrinology Coll. de France, Paris, 1960-63; prof. physiology Baylor Coll. Medicine, 1953—; adj. prof. medicine U. Calif. at San Diego, 1970—; resident fellow Salk Inst., 1970—. Decorated Legion of Honor, France, 1974; recipient Gairdner Internat. award, 1974; U.S. Nat. Medal of Sci., 1977; co-recipient Nobel prize for medicine, 1977; recipient Lasker Found. award, 1975; Dickson prize in medicine, 1976; Passano award med. sci., 1976; Schmitt medal neurosci., 1977; Barren gold medal, 1979; Dale medal Soc. for Endocrinology, U.K., 1980. Fellow AAAS; Mem. Am. Physiol. Soc., Endocrine Soc. (council), Soc. Exptl. Biology and Medicine, A.A.A.S., Internat. Brain Research Orgn., Internat. Soc. Research Biology Reprodn., Soc. Neuro-scis., Nat. Acad. Scis., Am. Acad. Arts and Scis., Club of Rome. Subspecialty: Neuroendocrinology. Office: Salk Inst Box 85800 San Diego CA 92138

GUILLERY, RAINER WALTER, anatomy educator; b. Greifswald, Germany, Aug. 28, 1929; came to U.S., 1964; s. Hermann and Eva (Hackel) G.; m. Margot Cunningham Pepper, Dec. 21, 1954; children: Peter, Edward, Philip, Jane. B.Sc. in Anatomy, U. Coll., London, Eng., 1951; Ph.D., 1954. Asst. lectr. Univ. Coll., London, Eng., 1953-57, lectr., 1957-63, reader, 1963-64; asso. prof. U. Wis. at Madison, 1964-68, prof. anatomy, 1968-77; prof. dept. pharm. and physiol. Scis. U. Chgo., 1977—. Mem. editorial bd.: Jour. Comparative Neurology, 1971—, Jour. Neurocytology, 1972-76, Jour. Neurophysiology, 1975-81, Neurosci, 1979—, Jour. Neurosci, 1980—. Fellow Royal Soc.; Mem. Am. Assn. Anatomists, Soc. for Neurosci., Am. Soc. Cell Biology, Anatom. Soc. Gt. Britain. Subspecialties: Neurobiology; Anatomy and embryology. Current work: Neurobiology; anatomy. Research on central nervous system, synapses, degeneration, devel. visual pathways. Home: 5805 S Blackstone Ave Chicago IL 60637

GUILLET, JAMES EDWIN, chemistry educator, researcher; b. Toronto, Ont., Can., Jan. 14, 1927; m., 1953; 4 children. B.A., U. Toronto, 1948; Ph.D. in Phys. Chemistry, Cambridge (Eng.) U., 1955, Sc.D., 1974. Research chemist Eastman Kodak Co., 1948-50, research chemist, 1950-52, sr. research chemist, 1955-62, research assoc., 1963; assoc. prof. U. Toronto, 1963-69, prof. chemistry, 1969—; cons. in field; vis. prof. Nat. Ctr. Sci. Research, Strasbourg, France, 1970-71, Kyoto (Japan) U., 1974; dir. Ecoplastics Ltd., 1971, pres., 1975—. Fellow Chem. Inst. Can., Royal Soc. Can.; mem. Am. Chem. Soc., The Chem. Soc. Subspecialty: Physical chemistry. Office: Dept. Chemistry U Toronto Toronto ON Canada M4W 2V2

GUILLORY, RICHARD JOHN, biochemist, biophysicist, educator, researcher; b. San Diego, Oct. 3, 1930; s. John Antoine and Margaret Sophia (Zigrapski) G.; m. Jyhlih Stella Jeng, Aug. 24, 1974; children: Cynthia, Olivia, Amber. B.A., Reed Coll., 1952; Ph.D., UCLA, 1962. Am. Heart Assn. postdoctoral fellow U. Amsterdam, Netherlands, 1962-63; asst. prof. chemistry Ariz. State U., 1964-66; asst. prof. biochemistry and molecular biology Cornell U., 1966-68, assoc. prof., 1968-71; prof. biochemistry-biophysics U. Hawaii John A. Burns Sch. Medicine, Honolulu, 1971—, chmn. dept. biochemistry-biophysics, 1973-76; established investigator Am. Heart Assn., 1968-73. Contbr. numerous articles, abstracts to profl. publs. Served with Med. Service Corps. U.S. Army, 1953-55. Am. Heart Assn. grantee, 1964-78; NSF grantee, 1964-80; NIH grantee, 1964-82. Mem. Am. Chem. Soc., Am. Soc. Biochemistry, Brit. Biochem. Soc., Am. Biophys. Soc., AAAS. Republican. Roman Catholic. Subspecialties: Biochemistry (biology); Biophysical chemistry. Current work: Membrane biochemistry; use of photochemistry in evaluating membrane-structure function. Office: U Hawaii John A Burns Sch Medicine Honolulu HI 96822

GUINAN, JOHN JOSEPH, JR., physiologist; b. Phila., Mar. 12, 1941; s. John Joseph and Evelyn ([illegible]) G.; m. Shelley Hathaway Swift, June 6, 1967; children: Ashley, Lindsey; m. Ellen Esther Harrison, Jan. 6, 1980; children: Sarah, Liza. S.B., M.I.T., 1963, S.M., 1964, Ph.D., 1968. Asst. prof. elec. engring. M.I.T., 1968-73; research assoc. dept. elec. engring. and computer sci., 1974—; research assoc. Mass. Eye & Ear Infirmary, Boston, 1968—. Contbr. numerous articles to profl. jours. Mem. Am. Fedn. Scientists, Acoustical Soc. Am., Neurosci. Soc. Subspecialty: Neurophysiology. Current work: Neural encoding of signals in the auditory system. Home: 166 Dickerman Rd Newton MA 02161 Office: Dept Electrical Engring Computer Sci MIT Mass Eye & Ear Infirmary 243 Charles St Boston MA 02114

GUIORA, ALEXANDER ZEEV, psychologist; b. Nyiregyhaza, Hungary, June 13, 1925; came to U.S., 1963; s. Solomon and Theresa (Gottlieb) Goldberg; m. Susie Nira Neuser, Jan 20, 1955; 1 son, Amos. Docteur De l'Universite de Paris, U. Paris-Sorbonne, 1951. Prof. psychiatry, psychology and linguistics U. Mich.-Ann Arbor, 1964—; vis. prof. U. Negev, Israel, 1971, Technion Med. Sch., 1975-76; exec. dir. Lang. Learning, 1979—. Editor: Perspectives in Clinical Psychology, 1968; contbr. articles in field to profl. jours. Mem. Am. Psychol. Assn. Jewish. Subspecialties: Clinical psychology; Psycholinguistics. Current work: The reciprocal relationship between personality development and language behavior. Home: 2115 Londonderry Rd Ann Arbor MI 48104 Office: U Mich 1405 E Ann St Ann Arbor MI 49109

GUJRATI, BITTHAL DAS, mechanical engineer; b. Varanasi, India, June 15, 1942; came to U.S., 1967, naturalized, 1982; s. Baldeo and Krishna (Devi) Das.; m. Meena Bhatia, July 14, 1975; children: Manu, Kusha. B.Sc. with honors, Banaras Hindu U., India, 1960, 1964; M.S., Pa. State U., 1970; Ph.D., U. Mich., 1974. Scientist Indian Inst. Petroleum, Dehradun, 1964-67; research asst. U. Mich., Ann Arbor, 1969-74; research engr. Amoco Chems. Corp., Naperville, Ill., 1974-79, staff research engr., 1979; research scientist Internat. Harvester Co. Sci. and Tech. Lab., Hinsdale, Ill., 1979-81, mgr. tech. planning, 1981-82; project mgr. Wilson Sporting Goods Co., River Groves, Ill., 1982—; mem. tech adv. com. Metal Properties Council, Inc. Contbr. articles to profl. jours. Recipient Prince of Wales gold medal Banaras Hindu U., 1964. Mem. ASME, Soc. for Advancement of Materials and Process Engring., Soc. Plastics Engrs., ASTM (student award 1973). Subspecialties: Composite materials; Materials processing. Current work: Composite material processing facility development; research, and management; material technology assessment; composite and plastic products/processes; friction/wear/lubrication including elastohydrodynamics. Office: 2233 West St River Grove IL 60171

GULARI, ERDOGAN, chem. engr., educator; b. Erzincan, Turkey, Nov. 6, 1946; s. Fahri Hasan and Sahsenem (Buklu) G.; m. Esin Ayse, June 29, 1969; 1 child, Bora. B.S., Roberts Coll., Istanbul, Turkey, 1969; Ph.D. in Chem. Engring, Calif. Inst. Tech., 1973. Postdoctoral assoc. SUNY, Stony Brook, 1973-74; plant mgr. Komili, Inc., Turkey, 1974-76; postdoctoral assoc. SUNY, Stony Brook, 1976-78; assoc. prof. chem. engring. U. Mich., Ann Arbor, 1982—. Contbr. articles to profl. jours. Mem. Am. Chem. Soc., Am. Inst. Chem. Engrs., Catalysis Soc. N.Am. Moslem. Subspecialties: Chemical engineering; Physical chemistry. Current work: Liquid state physics, catalysis, colloidal phenomena; catalysis by clusters, tertiary oil recovery, micellization and microemulsions, infrared, laser and x-ray spectroscopy. Office: U Mich 3168B Dow Bldg Ann Arbor MI 48109

GULARTE, RONALD CARL, ocean engr., cons.; b. Salinas, Calif., July 7, 1937; s. Louis L. and Dorothy M. (Rauch) G.; m. Alice Ann, July 18, 1958. B.S.M.E., U. So. Calif., 1966, M.S.M.E., 1968; O.E., M.I.T., 1972; Ph.D., U. R.I., 1978. Registered profl. engr., Tex., 1982. Mem. tech. staff Mechanics Research Inc., Los Angeles, 1966; staff engr. Truesdal Labs., Inc., Los Angeles, 1968; assoc. prof. ocean engring. Mass. Maritime Acad., Buzzards Bay, 1973-75; asst. prof. ocean engring. U.S. Naval Acad., 1979-82; sr. project engr. VSE Corp., Arlington, Va., 1982—. Contbr. articles on geotechniques, structures, instrumentation to profl. jours. Square D Found. scholar, 1964; Archimedes Circle fellow, 1968; Woods Hole fellow, 1970. Mem. ASME, ASCE, Am. Geophys. Union, Sigma Xi, Tau Beta Phi, Pi Tau Sigma. Current work: Marine geotechniques, primarily erosion of cohesive sediments. Patentee in field.

GULDEN, TERRY D., materials scientist, research and development manager; b. Seattle, May 4, 1938; s. Don D. and Ann (Feagan) G. B.S. in Ceramic Engring, U. Wash.-Seattle, 1960; M.S. in Materials Sci, Stanford (Calif.) U., 1962, Ph.D., 1965. Registered nuclear engr. Calif. Ceramic engr. United Tech., Sunnyvale, Calif., 1960-61; research assoc. Berkeley Nuclear Lab, 1965-67; assoc. scientist General Atomic, San Diego, 1967-68, staff scientist, 1968-71, br. mgr., 1971-73; dept. mgr. GA Technologies Inc., San Diego, 1973—. Co-editor: Nuclear Technology on Coated Particle Fuels, 1977; contbr. articles to tech. jours. Mem. AAAS, Am. Ceramic Soc., Am. Nuclear Soc. Subspecialties: Ceramics; Materials. Current work: Materials science, ceramics, chemistry, coatings, nuclear fuels, research and development management. Office: GA Technologies Inc PO Box 81608 San Diego CA 92138

GULL, DWAIN D., vegetable crops educator; b. Meadow, Utah, Aug. 21, 1923; s. Hyaum Bryant and Lula Mae (Duncan) G.; m. Isabel Orwin, June 4, 1944; children: Valoie, Gayla, LuJean, Mark. B.S., Utah State U.-Logan, 1952-55; M.S., Cornell U., 1957, Ph.D., 1959. Research asst. Cornell U., Ithaca, N.Y., 1955-59; asst. prof. U. Fla., Gainesville, 1959-66, assoc. prof., 1967—; chief-of-party Internat. Programs, El Salvador, San Salvador, 1973-75. Councilman City Council, Meadow, Utah, 1946; mem. Sch. Bd., Gainesville, Fla., 1962-63. Served to 1st ltd. USAF, 1943-45, 50-52. Fla. Tomato Exchange grantee, 1981, 82, 83; grantee Catalylic Generators, Inc., 1982. Mem. Am. Soc. Hort. Sci., Inst. Food Technologists, Fla. State Hort. Soc., Sigma Xi, Gamma Sigma Delta. Democrat. Mem. Ch. of Jesus Christ of Latter-day Saints. Subspecialties: Food science and technology; Plant physiology (agriculture). Current work: Tomato ripening, post harvest quality enhancement and improving nutritional composition of fresh vegetables. Home: 2629 NW 11th Ave Gainesville FL 32605 Office: U Fla Vegetable Crops Dept 1213 HSPP IFAS Gainesville FL 32611

GULL, THEODORE RAYMOND, astrophysicist; b. Hot Springs, S.D., Aug. 17, 1944; s. Albert Henry and Virginia Irene (Sieger) G.; m. Hazel Joy Constantine, July 1, 1967; children: Michael, Matthew. B.S. in Physics, M.I.T., 1966; Ph.D., Cornell U., 1971. Research asst. Yerkes Obs., U. Chgo., Williams Bay, Wis., 1971-72; asst. astronomer Kitt Peak Nat. Obs., Tucson, 1972-75; engr. prin. Lockheed Elec.

Corp., Houston, 1975-77; astrophysicist Goddard Space Flight Ctr., Greenbelt, Md., 1977—; cons. in field; study scientist Space Lab Wide Angle Telescope; mission scientist Astro Shuttle Missions. Contbr. articles to profl. jours. Mem. Am. Astron. Soc., Astron. Soc. Pacific, Internat. Astron. Union, Sigma Xi. Subspecialties: Optical astronomy; Ultraviolet high energy astrophysics. Current work: Interstellar medium, star formation and death, structure of interstellar medium; design/development of space/ground-based astronomical instruments. Office: Goddard Space Flight Center Code 683 Greenbelt MD 20771

GULLINO, PIETRO MICHELE, pathologist; b. Saluzzo, Italy, Mar. 24, 1919; came to U.S., 1957, naturalized, 1961; s. Antonio and Olimpia (Camisassi) G.; m. Marisa I. Bigo, Feb. 1, 1956. M.D. summa cum laude, U. Torino, Italy, 1943, Ph.D., 1952. Assoc. to chair pathologic anatomy U. Torino, 1946-48, vice-chmn. dept. pathologic anatomy, 1948-52; fellow Italian Research Council, Lab. Physiol. Chemistry, U. Naples, Italy, 1952-53, Technische Hochschule, Munich, W. Ger., 1953-54, Italian League Against Cancer, 1954-55; guest worker Nat. Cancer Inst., Bethesda, Md., 1954-55, vis. scientist, 1957-64, med. officer, 1964-68; head tumor physiopath. sect. Lab. Biochemistry, 1968-73; acting chief Lab. Pathophysiology, 1973-74, chief, 1974—; chmn. exptl. biology com. Breast Cancer Task Force, 1971-75, chmn., 1975-79; adj. prof. chem. engring. U. Del., Newark, 1976—, Carnegie-Mellon U., Pitts., 1979—; chmn. White House Conf. on Breast Cancer, 1976. Recipient awards Italian League Against Cancer, 1952, Italian Research Council; Ganassini award U. Torino, 1953; Golden Ambrosino award City of Milan, Italy, 1972; Dirs. award NIH, 1975. Mem. Am. Assn. Cancer Research, Am. Soc. Exptl. Biology, Soc. Exptl. Biology and Medicine, AAAS. Subspecialties: Cancer research (medicine); Pathology (medicine). Current work: Physiological processes controlling mammary tumor formation, growth and regression, with emphasis on angiogenesis. Patentee in field. Home: 8302 Melody Ct Bethesda MD 20817 Office: 9000 Rockville Pike Bldg 10/5B-36 Bethesda MD 20205

GULLION, GORDON WRIGHT, wildlife research biologist; b. Eugene, Oreg., Apr. 16, 1923; s. Omar Ray and Anna Elizabeth (Wright) G.; m. Ardelle Marie Vicary, June 3, 1944; children: Christina Maria, Anne Lorraine, Barbara Elizabeth, Rebecca Susanne. B.S., U. Oreg., 1948; M.A., U. Calif.-Berkeley, 1950. Wildlife technician Nev. Fish and Game Commn., Reno, 1951-56, dist. supr., 1956-58; research fellow U. Minn., St. Paul, 1958-68, research assoc., 1968-76, assoc. prof., 1976-80, prof., 1980—, project leader forest wildlife project, Cloquet, Minn., 1958—. Contbr. articles, to profl. jours. mags., newspapers. Mem. Am. Ornithologists Union (life), Wildlife Soc. (Conservation award Minn. chpt. 1979), Wilson Ornithol. Soc., Soc. Am. Foresters, Ruffed Grouse Soc. (nat. bd. dirs. 1972—), Sigma Xi, Gamma Sigma Delta. Subspecialties: Resource management; Population biology. Current work: Biology and management of ruffed grouse and development of forest management procedures to benefit wildlife inhabiting forested environments. Home: 605 Slate St Cloquet MN 55720 Office: Forest Wildlife Project U Minn 175 University Rd Cloquet MN 55720

GÜLLNER, HANS-GEORG, physician; b. Berlin, Germany, Jan. 9, 1939; came to U.S., 1972; s. Georg and Johanna (Schneider) G. B.S., Georg-Büchner-Schule, W.Ger., 1958; M.D., U.Frankfurt, W.Ger., 1965, Dr.med., 1968. Vis. scientist NIH, Bethesda, Md., 1977-78, guest worker, 1980—; instr. medicine U. Tex.-San Antonio, 1978-80. Author: Norepinephrine, 1982. Recipient Career Devel. award German Reseaarch Assn., 1972-74; German Acad. Exchange Program traveling fellow, 1974. Mem. Am. Physiol. Soc., Endocrine Soc., Am. Soc. Nephrology, Am. Fedn. Clin. Research, Internat. Soc. Chronobiology. Subspecialties: Neuroendocrinology; Endocrinology. Current work: Comparitive endocrinology of neuropeptides, brain-gut-skin tringle, electrolyte disorders. Office: NIH Bldg 10 Rm 8C103 9000 Rockville Pike Bethesda MD 20205

GUMPORT, RICHARD I., biochemistry educator; b. Pocatello, Idaho, June 23, 1937; s. Isaac and Helen Roberta (Burkey) G.; m. Roberta Helene Kugell, Sept. 18, 1960; children: Susan Rachel, William Isaac. B.S., U. Chgo., 1960, Ph.D., 1968; postgrad., Stanford U., 1968-71. Asst. prof. biochemistry U. Ill., Urbana, 1971-77; vis. scholar biochemistry Harvard U., 1979-80; assoc. prof. biochemistry U. Ill., Urbana, 1978—. Contbr. articles to profl. jours. Recipient Career Devel. awards USPHS, 1972-77; Guggenheim fellow, 1979-80; NIH research grantee, 1972—. Mem. Am. Soc. Biol. Chemists, Am. Chem. Soc., Am. Soc. Microbiology, AAAS. Subspecialties: Biochemistry (biology); Genetics and genetic engineering (biology). Current work: Chemical and enzymatic synthesis of defined sequences of DNA, protein-nucleic acid interactions, control of gene expression by attenuation. Home: 2009 S Anderson St Urbana IL 61801 Office: U Ill 506 S Mathews St Urbana IL 61801

GUNAJI, NARENDRA NAGESH, civil engineer, educator; b. Belgaum, India, Jan. 9, 1931; came to U.S., 1954, naturalized, 1962; s. Nagesh V. and Saraswati N. (Manage) G.; m. Georgianna P. Boyle, May 28, 1958; children: Rajini, Monica, Greg, Kim, Shannon. B.C.E., U. Poona, India, 1953; M.S., U. Wis., 1955, Ph.D., 1958. Registered profl. engr., Ohio, N.Mex. Research asst. U. Wis., Madison, 1954-55, Wis. Alumni Research Found. fellow, 1955-58, corr. study instr., 1955-58; asst. prof. civil engring. Ohio No. U., Ada, 1958-60; asst. prof. N.Mex. State U., Las Cruces, 1960-61, assoc. prof., 1961-64, prof., 1964—, dir. Engring. Expt. Sta., 1966-82, dir. Bldg. Materials Reearch and Testing Inst., 1976-82. Mem. Nat. Soc. Profl. Engrs., Internat. Assn. Hydraulic Research, Am. Geophys. Union, ASCE, Water Pollution Control Fedn., Am. Soc. Engring. Edn., AAAS, Am. Meteorol. Soc., Nat. Water Well Assn., Am. Water Works Assn., Sigma Xi, Tau Beta Pi, Chi Epsilon. Club: Rotary. Subspecialties: Civil engineering; Ground water hydrology. Current work: Removal of uranium from drinking water; ground water withdrawals; land subsidence due to large water withdrawals. Office: New Mexico State U Las Cruces NM 88003

GUND, PETER HERMAN, research scientist; b. N.Y.C., Feb. 20, 1940; s. Herman and Nora Lourie (Percival) G.; m. Tamara H. Mladineo, July 15, 1967; 1 dau.: Suzanne. A.B., Columbia Coll., 1961; M.S., Purdue U., 1963; Ph.D., U. Mass., 1967; postdoctoral student, Princeton U., 1970-73. Research chemist Am. Cyanamid Agrl. Research Ctr., Princeton, N.J., 1967-70; NIH research fellow depts. chemistry and biochemistry Princeton U., 1970-73; research scientist Merck Sharp & Dohme Research Labs., Rahway, N.J., 1973—, research fellow, 1973-79, sr. research fellow, 1979—. Contbr. articles to tech. jours., chpts. to books. Recipient Bausch & Lomb Sci. prize, 1957; N.Y. State Regents Am. Chem. Industry Scholarship award, 1957-61; NIH postdoctoral and spl. postdoctoral fellowship awards, 1970-73. Mem. Drug Info. Assn. (chmn. workship on computer assisted chemistry in drug design 1983), Am. Chem. Soc., AAAS, Quantum Chemistry Program Exchange. Subspecialties: Medicinal chemistry; Theoretical chemistry. Current work: Computer assisted chemistry. 3-D molecular modeling for new drug design. Computer graphics systems for chemical research. Home: 14 Cornwall Dr East Windsor NJ 08520 Office: Merck Sharp & Dohme Research Labs PO Box 2000 Rahway NJ 07065

GUNDERSEN, MARTIN ADOLPH, physicist, educator; b. Glenwood, Minn., May 19, 1940; s. Gilbert Theodore and Frances (Iverson) G.; m. Roberta McShirley, Dec. 20, 1963; children: Gilbert, Martin. B.A., U. Calif., Berkeley, 1965; Ph.D. in Physics, U. So. Calif., 1973. Assoc. prof. elec. engring. and physics Tex. Tech. U., 1973-77, assoc. prof., 1977-80; assoc. prof. elec. engring. and physics U. So. Calif., Los Angeles, 1980—. Contbr. articles in field of quantum electronics, pulsed power and laser physics to profl. jours. Subspecialties: Quantum electronics; Optical engineering. Current work: Infrared and ultraviolet lasers, semiconductor physics, spectroscopy, pulsed power physics. Office: Dept Elec Engring and Physics SSC 420 U So Calif Los Angeles CA 90089

GUNDERSON, LESLIE CHARLES, electrical engineer; b. Rahway, N.J., Aug. 1, 1935; s. Anslie Carl and Anna Helen (Jacobsen) G.; m. Edith Van Muiswinkel, Nov. 9, 1957; children: Lynn Suzanne, Diane Joy, David Charles. B.Mech. Engring., Stevens Inst., Hoboken, N.J., 1957, B.Computer Sci., 1960; M.S. in Elec. Engring, Columbia U., 1964, Ph.D., N.C. State U., Raleigh, 1971. Engr. ITT Fed. Labs., Nutley, N.J., 1957-64; sr. engr. Corning Glass Works, Raleigh, 1966-72, research dir., Corning, N.Y., 1972-80, dir. devel., 1980—. Mem. IEEE, Am. Phys. Soc., Optical Soc. Am., Am. Mgmt. Assn. Subspecialties: Fiber optics; Electronics. Current work: Optical components, fiber optics, optical communication. Home: 5 Ridgeway Circle Painted Post NY 14870 Office: Main Plant BH-04 Corning Glass Work Corning NY 14831

GUNN, JAMES EDWARD, astrophysicist; b. Livingstone, Tex., Oct. 21, 1938; s. James Edward and Rhea (Mason) G. B.S., Rice U., Houston, 1961; Ph.D., Calif. Inst. Tech., 1966. Sr. space scientist Jet Propulsion Lab., 1966-69; asst. prof. Princeton (N.J.) U., 1969-70, Eugene Higgins prof. astrophysics, 1980—; asst. prof., then prof. astrophysics Calif. Inst. Tech., 1970-80; dep. prin. investigator space telescope wide field camera NASA, 1977—. Served with C.E. USAR, 1967. Sloan Found. fellow, 1972-76. Mem. Am. Astron. Soc., Astron. Survey Com., Nat. Acad. Scis. Democrat. Subspecialty: Cosmology. Office: Peyton Hall Princeton U Princeton NJ 08544

GUNN, MICHAEL RICHARD, manufacturing and process engineer; b. Buffalo, May 13, 1950; s. Lewis Joseph and Frances Sarah (Thomas) G. B.M.E., SUNY-Buffalo, 1979. With Dresser-Clark, Olean, N.Y., 1979; with Motorola, Inc., Arcade, N.Y., 1979—, supr. assembly of hybrid products, 1982—. Served with USNR, 1969-75. Mem. Nat. Audubon Soc. Roman Catholic. Subspecialties: Mechanical engineering; Materials. Current work: The interrelationships of materials and processes; supervising the assembly processes for thick film. Home: PO Box 59 Sardinia NY 14134 Office: 400 Main St Arcade NY 14009

GUNZBURGER, MAX DONALD, mathematics educator, researcher; b. Buenos Aires, Argentina, Oct. 11, 1945; came to U.S., 1954; s. John and Griseld Christina (Farquharson) G.; m. Barbara Jane McCann, Oct. 21, 1971 (div. 1982); children: Cecilia, Emily, Margaret, Matthew. B.S., N.Y. U., 1966, M.S., 1967, Ph.D., 1969. Research scientist N.Y. U., N.Y.C., 1969-70, asst. prof. math., 1970-71; NRC postdoctoral assoc. Naval Ordnance Lab., Silver Spring, Md., 1971-73; vis. scientist Inst. for Computer Applications of Sci. and Engring., Hampton, Va., 1973-76; prof. math. U. Tenn., Knoxville, 1976-81, Carnegie-Mellon U., Pitts., 1981—; cons. Union Carbide Nuclear Div., Oak Ridge, 1977—, ICASE, Hampton, Va., 1976—. Mem. Indsl. and Applied Math. Subspecialties: Numerical analysis; Applied mathematics. Current work: Numerical solution of partial differential equations by finite element and other methods, with applications in mechanics. Office: Dept Math Carnegie-Mellon U Pittsburgh PA 15213

GUPTA, GIAN CHAND, environmental science educator; b. Delhi, India, Oct. 10, 1939; s. Bhagat Ram and Shanti G.; m. Bindu Hirdesh, Apr. 18, 1972; children: Tarra, Suneal. B.Sc., Panjab U., India, 1959, B.Tech., 1960; M.Sc., Vikaram U., Bhopal, India, 1962; Ph.D., Roorkee (India) U., 1967. Postdoctoral research assoc. U. Miss., 1968-69; assoc. prof. Rust Coll., Hollyspring, Miss., 1969-70; dir. environ. health M.E. Health Ctr., Fayette, Miss., 1970-72, J.H.C. Health Ctr., Jackson, Miss., 1972-77; asst. prof. environ. sci. U. Md. Eastern Shore, Princess Anne, 1977—. NASA grantee, 1979-81; NSF grantee, 1979-82. Mem. Am. Chem. Soc., Am. Soc. Agronomy, Nat. Environ. Health Assn., Soc. Environ. Health and Geochemistry. Subspecialties: Water supply and wastewater treatment; Soil chemistry. Current work: Research in water-air purification, disposal of energy related pollutants in the aquatic sediments, use of aquatic weeds in pollution control. Office: Dept Natural Sci U Md Eastern Shore Princess Anne MD 21853 Home: 411F Woodview Sq Salisbury MD 21801

GUPTA, MOOL CHAND, researcher, scientist; b. Alwar, Rajasthan, India, July 15, 1945; came to U.S., 1969; s. Gordhan Das and Bhagwati (Devi) G.; m. Rita Gupta, Jan. 30, 1977; children: Nita, Varun, Anup. B.Sc., Gujarat U., India, 1966, M.Sc., 1968; Ph.D., Wash. State U., 1973. Lectr. U. Mo., 1975-76; research assoc. Cornell U., 1976-78; sr. research fellow Calif. Inst. Tech., 1978-79; sr. scientist Jet Propulsion Lab., Pasadena, Calif., 1979-82; sr. research scientist Eastman Kodak Co., Rochester, N.Y., 1982—. Contbr. articles to tech. jours. Research Corp. grantee, 1976. Mem. Am. Phys. Soc., N.Y. Acad. Scis., Sigma Xi. Subspecialties: Materials; Laser data storage and reproduction. Current work: Data storage and reproduction using lasers, erasable physical or chemical processes for laser recording, thin films, polymeric materials, laser recording. Patentee in field. Office: Physics Division Eastman Kodak Co Kodak Park Rochester NY 14650 Home: 787 High Tower Way Webster NY 14580

GUPTA, MURARI L., med. physicst; b. Agra, India, Jan. 3, 1937; s. Pati R. and Triveni D. G.; m. Indira Gupta, June 15, 1980; children: Veena, Vandana, Anjana. Ph.D., Howard U., 1968. Sr. med. physicist Radiation Therapy Ctr., Poughkeepsie, N.Y., 1979-82; dir. med. physics div. Greenville (S.C.) Hosp., 1982—; staff radiation therapy dept. St. Francis Hosp., Wichita, Kan., 1977-79; cons. in field. Contbr. articles to profl. jours. Recipient Cert. of Commendation Civil Def., 1981; Princeton U. summer fellow, 1969; Howard U. grad. assistantship, 1962-67; Agra U. bursary fellow, 1957-59; Meml. Sloan-Kettering Cancer Ctr. fellow, 1976-77. Mem. Am. Soc. Therapeutic Radiologists, Am. Assn. Physicians in Medicine. Democrat. Hindu. Current work: Treatment planning of radiation therapy patients, management of radiation accidents, radiation dosimetry, radiation protection. Home: 27 High Acres Dr Poughkeepsie NY Office: Greenville Hosp Grove Rd Greenville SC 09605

GUPTA, PREM KAMAL, cardiologist, educator; b. Jammu, India, Sept. 22, 1941; came to U.S., 1966, naturalized, 1979; s. Bodh R. and Kaushlya (Devi) G.; m. Neelam Mahajan, Jan. 7, 1973; 1 child Sumita. I.Sc., G.G.M. Science, Jammu, 1959; M.B.B.S., Govt. Med. Coll., Srinagar, India, 1964. Diplomate: Am. Bd. Internal Medicine. Intern Bekman Downtown Hosp., N.Y.C., 1966-67; resident VA Hosp., N.Y.C., 1967-68, VA Hosp, Bronx, N.Y., 1968-71, practice medicine specializing in cardiology, Bklyn., 1976-; assoc. dir. cardiology Mt. Sinai-City Hosp., Elmhurst, N.Y., 1971-76; assoc, dir. cardiology Maimonides Hosp., Bklyn., 1976-79, attending cardiology, 1979—; assoc in medicine Mt. Sinai Med. Sch., N.Y.C., 1971-73; asst. prof. clin. medicine, 1973-76; assoc. prof. medicine SUNY-Bklyn., 1976—. Fellow ACP, Am. Coll. Cardiology, Am. Heart Assn., Royal Coll. Physicians and Surgeons, Can.; mem. Am. Fedn. Clin. Research. Subspecialty: Cardiology. Current work: Echo cardiography, including 2-dimentional and Doppler effect. Office: 909 49th St Brooklyn NY 1219

GUPTA, RAJENDRA, physicist; b. Mauranipur, India, Jan. 1, 1943; came to U.S., 1962; s. Mahipal and Ramati (Devi) G.; m. Usha Chand, July 11, 1970; children: Tripti, Sangeet. B.Sc., Agra U., 1959; M.Sc., 1961; Ph.D., Boston U., 1970. Research assoc. Columbia U., N.Y.C., 1970-73, lectr., 1973-74, asst. prof., 1974-78, U. Ark., Fayetteville, 1978-81, assoc. prof., 1981—. Contbr. articles to profl. jours. Research Corp. grantee, 1979-82; USAF grantee, 1980—; NSF grantee, 1980—. Mem. Am. Phys. Soc., Am. Assn. Physics Tchrs., Sigma Xi. Subspecialty: Atomic and molecular physics. Current work: Laser spectroscopy, application of laser spectroscopy to basic and applied problems in physics. Office: Dept Physics Univ Ark Fayetteville AR 72701

GUPTA, SHANTI SWARUP, statistician; b. Saunasi, India, Jan. 25, 1925; s. Sitaram and Ramphali G.; m. Marianne Heinicke, Feb. 20, 1974; 1 dau., Maya Erika. B.A. with honors, St. Stephen's Coll., Delhi, 1946; M.A. in Math, Delhi U., 1949; Ph.D. in Statistics, U. N.C., 1956. Lectr. Delhi Coll., 1949-53; research statistician Bell Telephone Labs., 1956-57, 58-61; asso. prof. math. U. Alta., 1957-58; vis. asso. prof. Stanford U., 1961-62; prof. statistics and math. Purdue U., West Lafayette, Ind., 1962—, head dept., 1968—. Co-author 2 books in field; editor 2 books; contbr. articles to profl. jours. Fulbright fellow, 1953-54. Fellow Am. Statis. Assn., Inst. Math. Statistics, Royal Statis. Soc., Internat. Statis. Inst., AAAS; mem. Bernoulli Soc. for Probability and Statistics, Math. Assn. Am., Sigma Xi. Subspecialty: Statistics. Current work: Multiple decision theory and reliability; order statistics in the broad framework of statistical inference. Home: 104 Tecumseh Park Ct West Lafayette IN 47906 Office: Dept Statistics Purdue Univ West Lafayette IN 47907

GUPTA, SOHAN LAL, biochemist, researcher; b. Aligarh, India, June 20, 1939; came to U.S., 1968; s. Shyam Lal and Dropadi Devi G.; m. Kusum Gupta, July 4, 1964; children: Sanjay, Avneesh. M.S., Aligarh Muslim U., 1960; Ph.D., All-India Inst. Med. Scis., New Delhi, 1967. Sci. officer All-India Inst. Med. Scis., 1963-68; research assoc. Yale U., 1969-74; assoc. Sloan-Kettering Inst. Cancer Research, N.Y.C., 1975-81, asst. mem., 1981—. Contbr. articles on molecular aspects of protein biosynthesis, mechanism of interferon action to profl. jours. NIH grantee, 1976—. Mem. Am. Soc. Microbiology, Am. Soc. Virology, AAAS, N.Y. Acad. Scis. Subspecialties: Biochemistry (biology); Biochemistry (biology). Current work: Mechanism of interferon action, gene expression and regulation. Office: 1275 York Ave New York NY 10021

GUPTA, UDAIPRAKASH INDUPRAKASH, computer science educator; b. Allahabad, India, Feb. 4, 1952; came to U.S., 1974; s. Induprakash and Krishna (Agarwal) G.; m. Shalini (Bhushan) G., July 18, 1980. B.Tech., Indian Inst. Tech., Bombay, 1974; M.S.E., Princeton U., 1975, M.A., 1976, Ph.D., 1978. Asst. prof. computer sci. Northwestern U., Evanston, Ill., 1978—; cons. Bell Labs., Naperville, Ill., summers 1981, 82; fin. chmn. Compsac, Chgo., 1979—. Mem. Assn. for Computing Machinery, IEEE Computer Soc. Hindu. Subspecialties: Database systems; Algorithms. Current work: Research areas of database management, system design, and efficient algorithms for data handling. Home: 4901-D Carol St Skokie IL 60077 Office: Northwestern U Dept Elec Engring and Computer Sci Evanston IL 60201

GUPTA, VIJAY KUMAR, chemistry educator; b. Ambala Cantt, Haryana, India, Apr. 27, 1941; s. Rattan Lal and Sharda Devi (Singal) G.; m. Surjit Mohini Aggarwal, Sept. 5, 1968; children: Sonia, Angela, Ashish. B.Sc., Panjab. U., Chandigarh, India, 1961, M.Sc., 1962, Ph.D., 1969. Lectr. Punjab Engring. Coll., Chandigarh, 1962-64, asst. prof., 1967-68; postdoctoral assoc. Wright State U., Dayton, 1968-69; assoc. prof. chemistry Central State U., Wilberforce, Ohio, 1969—; research chemist Lawrence Livermore (Calif.) Nat. Lab., 1980. Contbr. articles to profl. jours. Mem. constn. com. Hindu Community Orgn., Dayton, 1982. NSF fellow, 1979; U.S. Air Force grantee; NASA grantee, 1976-79. Mem. Electrochem. Soc., Am. Chem. Soc., Ohio Acad. Scis., AAUP. Democrat. Hindu. Club: India (Dayton). Subspecialties: Fuels and sources; Thermodynamics. Current work: Production of synthetic fuels, high energy density battery systems, thermodynamics of solutions, energy conservation and energy conversion technologies. Home: 1447 New Way Dr Xenia OH 45385 Office: Dept Chemistry Central State U Wilberforce OH 45384

GUR, DAVID, radiation physicist; b. Haifa, Israel, Apr. 7, 1947; s. Zeev and Kedma G.; m. Zipora, Aug. 31, 1971; children: Saar, Ilan. B.S. in Physics, Technion Israel Inst. Tech., 1973; M.S. in Radiation Health, Grad. Sch. Pub. Health U. Pitts., 1976, Sc.D., 1977. Sr. pilot, flight instr. Shahaf Aviation Service, Tel-Aviv, Israel, 1971-73; asst. prof. radiation health and radiology U. Pitts., 1977-80, assoc. prof. radiology and radiation health, 1980-83, prof. radiation health, 1983—. Served with Israeli Mil. Forces, 1965-68. Mem. Health Physics Soc., Am. Assn. Physicists in Medicine, Am. Heart Assn., Pa. Acad. Sci. Subspecialties: Imaging technology; Comparative physiology. Current work: Development safe noninvasive techniques for derivation of in vivo functional information with improved anatomic specificity; perform research enhancing understanding of normal and abnormal tissue function resulting from local and regional alterations of tissue function. Office: Univ Pitts Pittsburgh PA 15261

GUR, RUBEN C., neuropsychologist; b. Zagreb, Yugoslavia, Aug. 13, 1947; came to U.S., 1970, naturalized, 1983; s. Moshe Mavro and Ella (Gluck) Cohen; m. Raquel E. Gur, Aug. 19, 1969; 1 dau., Tamar Lea. B.A., Hebrew U. Jerusalem, 1970; M.A., Mich. State U., 1971, Ph.D., 1973. Research assoc. Stanford U., Palo Alto, Calif., 1973-74; asst. prof. psychology U. Pa., Phila., 1974-80, assoc. prof. neuropsychology, 1980—; dir. neuropsychology (Grad. Hosp.), Phila., 1981—, dir. behavioral neurology, 1981—. Contbr. articles to profl. jours. Spencer Found. grantee, 1979, 82; NSF grantee, 1975; NIH grantee, 1976—; NIMH grantee, 1978—; NIA grantee, 1982—. Mem. Am. Psychol. Assn., AAAS, Internat. Neuropsychol. Soc., N.Y. Acad. Scis., Am. Speech-Lang.-Hearing Assn. Subspecialties: Neuropsychology; PET scan. Current work: Regional brain physiology in relation to behavior and psychopathology. Home: 3946 Delancey Pl Philadelphia PA 19104 Office: Grad Hosp U Pa 1 Graduate Plaza Philadelphia PA 19146

GURD, RUTH SIGHTS, physician, biochemist, researcher, educator; b. Chgo., Sept. 17, 1927; d. Warren Preston and Helen (Coleman) Sights; m. Frank Ross Newman Gurd, June 12, 1956; children: Martha Helen, Charles Baillie. B.S., U. Mich., 1949; M.D., Washington U., St. Louis, 1957. Life Ins. Med. Research Fund postdoctoral fellow in physiology Cornell U. Coll. Medicine, 1957; med. cons. Aerospace Research Application Ctr., Ind. U., Bloomington, 1963-66, sr. research assoc. dept chemistry, 1966-71, asst. prof. biochemistry med. sci. program, 1973-77, assoc. prof., 1978—. Contbr. articles to sci. jours. NSF fellow, 1954-55. Mem. Am. Inst. Nutrition, Biophys. Soc.; mem. Am. Soc. Biol. Chemistry; Mem. N.Y. Acad. Scis., Nat. Soc. Arts and Letters, Kappa Kappa Gamma. Subspecialties: Biochemistry (medicine); Receptors. Current work: Structure function relationships of proteins and peptides; peptide structure, function, hormones, glucagon, receptors, diabetes mellitus, physical methods, synthesis, semisynthesis. Home: 2600 Fairoaks Ln Bloomington IN 47401 Office: Dept Chemistry Ind U Bloomington IN 47405

GURLL, NELSON JOSEPH, surgeon, educator, researcher; b. Providence, Jan. 8, 1942; s. Nelson Joseph and Leonora Ann (Capraro) G.; m. Margaret Eleanor Simmons, June 18, 1965. A.B., U. Calif.-Berkeley, 1963, M.D., 1967. Diplomate: Am. Bd. Surgery. Intern Beth Israel Hosp., Boston, 1967-68, resident, 1970-73, Boston Children's Hosp., 1968-69, San Francisco Gen. Hosp., 1969; asst. prof. surgery U. Iowa, Iowa City, 1976-80, assoc. prof., 1980—, mem., 1982—; acting chief surgery VA Med. Ctr., Iowa City, 1978-79, dir., 1977-78. Contbr. articles to profl. jours. Actor Iowa City Community Theater, 1980—; big brother Pals Program, Iowa City, 1978-80. Served with U.S. Army, 1973-76. Recipient Merk Manual award Merck Co., 1967; Mosby scholar book award Mosby Co., 1967. Fellow ACS (liason fellow Commn. on Cancer 1982—); mem. Soc. Univ. Surgeons, The Shock Soc., Am. Physiol. Soc., Am. Gastroent. Assn. Democrat. Club: Hawkeye Area Wrestling (Iowa City). Subspecialties: Surgery; Physiology (medicine). Current work: Endorphins in shock, histamine in gastrointestinal patho-physiology, fibrinolysis, gastrointestinal ion transport.

GURNEY, ELIZABETH TUCKER GUICE, biologist; b. Berkeley, Calif., Apr. 4, 1941; d. Clarence Norman and Elizabeth Lillian (Eichbauer) Guice; m. Theodore Gurney, Jr., June 18, 1966. B.A., U. Chgo., 1962; M.S., U. Calif.-Berkeley, 1970, Ph.D., 1975. Tech. asst. dept. biology M.I.T., 1963-67; postgrad. research biochemist U. Calif., 1970-73; postdoctoral fellow U. Utah, 1975-77, research asst. prof. biology, 1976—. Contbr. articles in field to profl. jours. NIH grantee, 1977—. Mem. AAAS, Am. Soc. Cell Biology, Am. Soc. Microbiology. Subspecialties: Genetics and genetic engineering (biology); Cell and tissue culture. Current work: Control of cellular growth, tumor virus transformation, moncional antibodies. Home: 203 4th Ave Salt Lake City UT 84103 Office: Dept Biology U Utah Salt Lake City UT 84112

GURPIDE, ERLIO, biochemist, educator; b. Buenos Aires, Argentina, Apr. 8, 1927; came to U.S., 1958, naturalized, 1964; s. Miguel and Maria (Eslava) G.; m. Pilar Gloria Carbajal, Sept. 2, 1961; 1 dau., Bettina. Ph.D., U. Buenos Aires, 1955. Asst. prof. Columbia U. dept. ob-gyn, N.Y.C., 1959-69; prof. dept. biochemistry and ob-gyn U. Minn., 1969-72; prof. biochemistry and ob-gyn Mt. Sinai Sch. Medicine, N.Y.C., 1972—. Mem. Am. Soc. Biol. Chemists, Soc. Gynecologic Investigation, Endocrine Soc. Subspecialties: Reproductive endocrinology; Cancer research (medicine). Current work: Biochemistry of hormonal action and metabolism. Office: 1176 Fifth Ave New York NY 10029

GURTOO, HIRA L., research biochemical pharmacologist, geneticist; b. Kashmir, India, Apr. 13, 1938; came to U.S., 1965; s. Rugh Nath and Kamala (Kamala) G.; m. Lalita Durani, Apr. 13, 1937; children: Lalit, Rajeev. B.V.Sc., Madras Vet. Coll., 1959, M.V.Sc., 1962; Ph.D. in Biochemistry, Va. Poly. Inst., 1968. Vet. surgeon, Kashmir, 1959-60; asst. prof. Sch. Agr., Kashmir, 1963-65; postdoctoral tng. Yale U. Med. Sch., New Haven, 1968-70; research pharmacologist Miles Labs., Elkhart, Ind., 1970-78; cancer research scientist III, IV & V Roswell Park Meml. Inst., Buffalo, 1978-80, assoc. chief cancer research sci., 1980—. Contbr. numerous articles and abstracts to sci. jours., also chpts. to books. Grantee NIH, Nat. Cancer Inst., Nat. Inst. Environ. Health Sci., Am. Cancer Soc., U.S. Council for Tobacco Research. Mem. Am. Assn. for Cancer Research, AVMA. Hindu. Subspecialties: Environmental toxicology; Molecular pharmacology. Current work: Chemical carcinogenesis and chemotherapy; cytochrome P-450 cloning, benzo(a)pyrene metabolism, aflatoxin metabolism and carcinogenesis, cyclophosphamide metabolism and chemotherapy.

GURWITZ, ROBERT FREY, computer scientist, consultant; b. Providence, Mar. 17, 1955. Sc.B., Brown U., 1977, Sc.M., 1979. Research asst. Brown U., Providence, 1977-78, research assoc., 1978-80; computer scientist Bolt Beranek and Newman, Cambridge, Mass., 1980—. Mem. IEEE, Assn. Computing Machinery, Sigma Xi. Subspecialties: Distributed systems and networks; Operating systems. Current work: Distributed systems, computer networks, network protocols, operating systems development, distributed operating systems, internetworking, unix, computer graphics, man-machine interaction. Office: Bolt Beranek and Newman 10 Moulton St Cambridge MA 02238

GUSDON, JOHN PAUL, JR., gynecologist/obstetrician, educator; b. Cleve., Feb. 13, 1931; s. John and Pauline (Malencek) G.; m. Marcelle Simone Deiber, June 16, 1956 (div. dec); children: Marguerite, John, Veronique. B.A., U. Va., 1952, M.D., 1959. Diplomate: Am. Bd. Ob-Gyn. Intern, resident Univ. Hosps., Cleve., 1959-64; asst. prof. Western Res. U. Sch. Medicine, 1964-67; asst. prof. ob-gyn Bowman Gray Sch. Medicine, Wake Forest U., Winston-Salem, N.C., 1967-71, asso. prof., 1971-74, prof., 1974—. Contbr. numerous papers to sci. jours. Served with USN, 1952-55. Mem. Am. Coll. Obstetricians and Gynecologists (Pres.'s award 1977, Found. prize 1971), Am. Assn. Immunologists, Soc. Gynecol. Investigation, S. Atlantic Assn. Obstetricians and Gynecologists, So. Med. Assn. Roman Catholic. Subspecialties: Obstetrics and gynecology; Immunology (medicine). Current work: Immunology of reprodn., immunology of cancer and immunotherapy. Office: Bowman Gray Sch Medicine Winston-Salem NC 27103

GUSKEY, THOMAS ROBERT, education educator, consultant; b. Johnstown, Pa., Feb. 15, 1950; s. Robert Charles and Evelyn (Yarnick) G.; m. Jeanette Tanabe, Aug. 26, 1978; children: Jennifer, Michael. B.A., Thiel Coll., 1972; M.Ed., Boston Coll., 1975; Ph.D., U. Chgo., 1979. Math. tchr. St. Andrew Sch., Erie, Pa., 1972-74; research asst. Boston Coll., Chestnut Hill, Mass., 1974-75; instr. U. Chgo., 1975-76; dir. research Chgo. Pub. Schs., 1976-78; prof. edn. U. Ky.-Lexington, 1978—, dir., 1978-80, Teaching Research Ctr., Chgo., 1981-82; ednl. cons. N.Y.C. Pub. Schs., 1977-82. Author: Implementing Mastery Learning, 1984; contbr. articles to profl. jours. Fellow Boston Coll., 1974; scholar U. Chgo., 1976, 77; achievement awards U. Ky., 1980, 81, 82, 83. Mem. Am. Ednl. Research Assn., Am. Psychol. Assn., Nat. Council Measurement in Edn., Nat. Soc. Study of Edn. Club: Gingertree Gym (Lexington) (v.p. 1980—). Subspecialties: Learning; Cognition. Current work: Theories of learning, cognitive and affective; instructional psychology; psychometric applications; theories and models of change. Home: 3505 Adoric Ct Lexington KY 40502 Office: Coll Edn Univ Ky Lexington KY 40506

GUSTAFSON, BO AKE STURE, astrophysicist; b. Karlskrona, Sweden, Mar. 28, 1953; came to U.S., 1977; s. Karl Ake Sture and Elsa Anna Stina (Modigh) G. Ph.D., U. Lund, 1981. Research asst. Space Astron. Lab., SUNY, Albany, 1977-80; asst. research scientist Space Astron. Lab., U. Fla., Gainesville, 1981—; cons. Internat. Astron. Union. Contbr. articles to profl. jours. Mem. Am. Astron. Soc., Am. Assn. Aerosol Research, Swedish Astron. Soc. Club: Miami Yacht. Subspecialties: Planetary science; Satellite studies. Current work: Researcher in light scattering, zodiacal light, comets, cometary dust, interplanetary particles, celestial mechanics. Office: Space Astronom Lab University of Florida 1810 NW 6th St Gainesville FL 32601

GUSTAFSON, DAVID EARL, physicist; b. St. Paul, May 19, 1948; s. George E. and Dorothy S. (Stuart) G.; m. Patricia Ann Snider, June 3, 1978. B.A. in Physics, Hamline U., St. Paul, 1970, Ph.D., U. Va., 1976. Postdoctoral fellow in nuclear physics Fla. State U., 1975-76; postdoctoral fellow in physiology/biophysics Mayo Clinic, Rochester, Minn., 1976-70; asst. prof. radiology U. N.Mex. Sch. Medicine, Albuquerque, 1979-80; prin. research physicist Siemens Gammasonics, Inc., Des Plaines, Ill., 1980—. Contbr. articles to profl. jours. Subspecialties: Nuclear physics; Medical physics. Current work: Research and development of nuclear medicial and radiological imaging services and application software. Office: 2000 Nuclear Dr Des Plaines IL 60018

GUSTAFSON, JOHN PERRY, geneticist; b. Greeley, Colo., Aug. 1, 1944; s. Elmer R. and Barbara N. (Wilson) G.; m. Christine S. McKinstry, Mar. 13, 1977; 1 dau., Kathryn. B.S., Colo. State U., 1967, M.S., 1968; Ph.D., U. Calif., Davis, 1972. Research asst. U. Calif., Davis, 1968-72; research prof. U. Man., Can., 1972-76, research asso., 1976-77, assoc. prof., 1977-82; research geneticist U.S. Dept. Agr., U. Mo., Columbia, 1982—. Contbr. articles to profl. jours. Union Pacific R.R. scholar, 1962; Nat. Sci. and Engring. Research Council Can. grantee, 1973-81. Mem. AAAS, Am. Soc. Agronomy, Am. Genetic Assn., Genetics Soc. Am., Sigma Xi, Gamma Sigma Delta, Phi Kappa Phi. Subspecialties: Plant genetics; Genome organization. Current work: Researcher in plant genetics, plant breeding, cytogenetics and evolution. Home: 3103 Crawford St Columbia MO 65201 Office: U Mo 208 Curtis Hall Columbia MO 65211

GUSTAFSON, LEWIS BRIGHAM, exploration company executive; b. Timmins, Ont., Can., Sept. 4, 1933; s. John Kyle and Elizabeth (Brigham) G.; m. Ursula Lewkert, Jan. 21, 1961; children: Katrin, Kirsten, Irene. B.S.Engring., Princeton U., 1955; M.S., Calif. Inst. Tech., 1959; Ph.D., Harvard U., 1962. Asst. chief geologist primary metals div. Anaconda Co., Tucson, 1971-73, chief geologist research and tech., 1973-75; faculty econ. geology Australian Nat. U., Canberra, 1975-81; research group leader exploration research div. Conoco Inc., Ponca City, Okla., 1981-82; sr. staff geologist Freeport Exploration Co., Reno, 1982—; councillor Australian Mineral Found., Adelaide, 1977-79; lectr. in field, 1969—. Contbr. numerous sci. articles to profl. lit. Fellow Geol. Soc. Am.; mem. Soc. Econ. Geologists (Lindgren award 1962, editorial bd. mem. Jour. 1970-80, Thayer Lindsey vis. lectr. 1973-74, chmn. research com. 1981—), AIME, Geol. Soc. Australia, Am. Assn. Petroleum Geologists. Subspecialties: Geology. Current work: Geology and geochemistry of mineral deposits (base and precious metals, uranium) and application of modern technology to their exploration. Home: 3520 San Mateo Ave Reno NV 89509 Office: Freeport Exploration Co PO Box 1911 Valley Bank Plaza Reno NV 89505

GUSTAFSSON, BORJE KARL, veterinarian, educator; b. Varnamo, Sweden, Feb. 26, 1930; s. Albin Karl and Svea Gertrud (Andersson) G.; m. Gunilla A. Granzelius, July 11, 1958; children: Katarina, Charlotte, Lars. B.Vet. Sci., Royal Vet. Coll., Stockholm, 1953, D.V.M., 1960, Ph.D., 1966. Research assoc., instr., asst. prof. Royal Vet. Coll. Stockholm, 1960-67, tchr., researcher animal reproduction, head clinics dept. Ob-Gyn, 1967-75, acting prof., chmn. dept. Ob-Gyn, 1970-73; vis. prof. U. Minn. Coll. Agr., St. Paul, 1974; prof. theriogenology Coll. Vet. Medicine, 1976-78; dir. grad. edn. in theriogenology U. Minn., 1976-78; prof., head dept. vet. clin. medicine Coll. Vet. Medicine, U. Ill., Urbana-Champaign, 1978—. Contbr. numerous articles in field of animal reproduction to profl. jours. Served with Swedish Vet Corps, 1952-54. Lagerlof's fellow, 1974. Mem. Swedish Vet. Med. Assn., AVMA, Assn. Am. Vet. Med. Colls., Am. Assn. Vet. Clinicians, Soc. Study Reproduction, World Assn. Vet. Physiologists, Pharmacologists and Biochemists. Subspecialties: Theriogenology; Reproductive biology (medicine). Current work: Testicular and epidedymal function; female genital infections; prostaglandins and reproduction; bovine mastitis. Home: 2102 S Race St Urbana IL 61801 Office: 1008 W Hazelwood St Urbana IL 61801

GUTERMAN, SONIA KOSOW, microbiologist, educator; b. Bklyn., June 27, 1944; d. Irvin Lionel and Ruth Carol (Cooper) Kosow; m. Martin Mayr Guterman, June 14, 1964; children: Lila Miriam, Beth Susanna. B.S., Cornell U., 1964, M.S., 1967; Ph.D., MIT, 1971. Instr. Brandeis U., Waltham, Mass., 1971; NIH postdoctoral fellow Tufts U. Med. Sch., Boston, 1972-74, research assoc., 1974-75, instr., 1975-76; asst. prof. Boston U., 1976-82, assoc. prof. biology, 1982—; sr. scientist Biotechnica Internat., Inc., Cambridge, Mass., 1983—; vis. scientist MIT, 1979; vis. scholar Harvard U., 1982. Contbr. articles to profl. jours. NSF grantee, 1975-77; NIH grantee, 1977—; March of Dimes Basil O'Connor grantee, 1976-79. Mem. Am. Soc. Microbiology, AAAS, Genetics Soc. Am. Subspecialties: Microbiology; Biochemistry (biology). Current work: Research in regulation of antibiotic biosynthesis. Office: 2 Cummington St Boston MA 02215

GUTH, ALAN HARVEY, physicist; b. New Brunswick, N.J., Feb. 27, 1947; s. Hyman and Elaine (Cheiten) G.; m. Susan Tisch, Mar. 28, 1971; 1 son, Lawrence David. S.B. and S.M., MIT, 1969, Ph.D. in Physics, 1972. Instr. Princeton (N.J.) U., 1971-74; research assoc. Columbia U., N.Y.C., 1974-77, Cornell U., Ithaca, N.Y., 1977-79, Stanford Linear Accelerator Ctr., Calif., 1979-80; assoc. prof. Physics MIT, Cambridge, 1980—. Alfred P. Sloan fellow, 1981. Mem. Am. Phys. Soc. Subspecialties: Particle physics; Cosmology. Current work: Applications of particle physics (particularly grand unified theories) to the very early universe; consequences of the inflationary universe scenario. Office: Center Theoretical Physics MIT Cambridge MA 02139

GUTH, JOSEPH HENRY, biochemist, research and testing lab. exec.; b. Cleve., May 10, 1942; s. William J. and Sara J. G.; m. Ann Samuel Falls, Oct. 23, 1977; 1 dau., Sabra Denise. B.S., U. Calif., Berkeley, 1970, Ph.D. in Biophysics, 1975. Lab. asst. dept. infectious diseases UCLA, 1960-61, lab. technician dept. biol. chemistry, 1961-63; research chemist U.S. Army Chem. Warfare Ctr., Edgewood Arsenal, Md., 1964-65; chemist, owner Intersci. Research Group, Los Angeles, 1966-69, Palo Alto, Calif., 1970-72; postdoctoral fellow U. Wis.-Madison, 1975-76; asst. prof. chem. sci. Old Dominion U., Norfolk, Va., 1976-79; pres., dir. Intersci. Research, Inc., Norfolk, Va., 1978—; Pres. Found. for the Study in Aging, 1978—. Mem. Tissue Culture Assn., Assn. ofcl. Analytical Chemists, Am. Chem. Soc., Soc. Forensic Toxicologists, Biophysical Soc., Internat. Assn. Arson Investigators, AAAS, Mid-Atlantic Assn. Forensic Scientists, Am. Indsl. Hygiene Assn., ASTM, Soc. Forensic Toxicologists, Am. Assn. Textile Chemists and Colorists, Sigma Xi. Subspecialties: Analytical chemistry; Biochemistry (medicine). Current work: Study of aging, study of cancer, the mechanism of asbestos-induced carcinogenesis; calcium and plasma membrane integrity/odor pattern analysis for forensic studies; hazard-risk assessment methodology for toxic materials. Office: 2614 Wyoming Ave Norfolk VA 23513

GUTH, LLOYD, anatomy educator; b. N.Y.C., Oct. 8, 1929; s. Benjamin G. and Syd (Fischer) G.; m. Josephine Rose Zalewski; children: Michael Walter, Robert William. A.B., NYU, 1949, M.D., 1953. Diplomate: Nat. Bd. Med. Examiners. Staff scientist, lab. of neuroanat. sci. Nat. Inst. Neurol. and Communicative Disorders and Stroke-NIH, Bethesda, Md., 1954-61, head sect. on neural devel. and regeneration, 1961-75, mem. adv. com., 1977—; mem. neurology B study sect. NIH, 1977-81; prof., chmn, dept. anatomy U. Md. Sch. Medicine, Balt., 1975—. Contbr. sects. to books, many sci. articles to profl. jours. Served to med. dir. USPHS, 1954-75. Mem. Am. Assn. Anatomists, Am. Physiol. Soc., AAAS, Assn. Anatomy Chairmen. Club: Cajal. Subspecialties: Regeneration; Anatomy and embryology. Current work: Spinal cord regeneration, trophic nerve function, neurological development and regenration. Home: 7511 Elmore Ln Bethesda MD 20817 Office: Dept Anatomy Sch Medicine U MD 655 W Baltimore St Baltimore MD 21201

GUTHRIE, FRANK EDWIN, entomologist, educator; b. Louisville, Jan. 14, 1922; s. Blaine and Lora May (Waller) G.; m. Bernice Button, Dec. 26, 1947; children—Janet, Caroline. B.S., U. Ky., 1947; M.S., U. Ill., 1949, Ph.D., 1952. Asst. prof. entomology U. Fla., Quincy, 1952-54; asst. prof. entomology N.C. State U., Raleigh, 1954-59; asso. prof., 1959-62, prof., 1962—; asst. dean Grad. Sch., 1962-64, dir. research and tng. programs in pesticide toxicology, 1964—. Co-author: Concepts of Pest Management, 1970, Biochemical Toxicology, 1980, Environmental Toxicology, 1980; contbr. articles to profl. jours. Served with USMCR, 1943-46, 51-52. Mem. Entomol. Soc. Am., Am. Chem. Soc., Soc. Toxicology. Subspecialties: Toxicology (agriculture); Entomology. Current work: Absorption and distribution by blood macromolecules of pesticides. Home: 823 Beaver Dam Rd Raleigh NC 27607 Office: Dept Entomology NC State Univ Raleigh NC 27607

GUTHRIE, HELEN A., nutrition educator, consultant; b. Sarnia, Ont., Can., Sept. 25, 1925; d. David and Helen (Sweet) Andrews; m. George Guthrie, June 4, 1949; children: Barbara, Jane, James. B.A., U. Western Ont., 1946, D.Sc. (hon.), 1982; M.S., Mich. State U., 1948; Ph.D., U. Hawaii, 1968. Cert. registered dietetian. Asst. prof. Pa. State U., University Park, 1949-60, prof., 1972—; dir. Nabisco Brands Inc., Parsipanny, N.J., 1974—; cons. Kraft Inc., Chgo., 1978-81. Author: Introductory Nutrition, 1967, 2d edit., 1983. Chmn. Bd. of Health State Coll. Pa., 1977-82. Recipient Borden award Am. Home Econs. Assn., 1976. Mem. Am. Inst. Nutrition (councillor 1982—), Soc. Nutrition Edn. (pres. 1978-79), Am. Dietetics Assn., Am. Pub. Health Assn., Inst. Food Technology. Subspecialty: Nutrition (biology). Home: 1316 S Garner St State College PA 16801 Office: Pa State U 106 Human Devel University Park PA 16802

GUTIERREZ, FERNANDO JOSE, psychologist; b. Matanzas, Cuba, Mar. 1, 1951; s. Alberto Rodolfo and Mariana Elena (Cartaya) G. B.A., Mich. State U., 1973; M.S. in Edn, Purdue U., 1974; Ed. D., Boston U., 1981. Staff counselor U. Wis.-Stevens Point, 1975-77; counseling psychologist San Francisco State U., 1980-81, U. Santa Clara, 1981—; Western region coordinator profl. workshops Nat. Assn. Minority Students and Educators in Higher Edn., 1982—. Mem. adv. com. to Commr. Mental Health-Children's Com., 1980; pres. Hispanic Orientation, Recreation and the Arts Assn., 1980. Title VII fellow HEW, 1978; recipient Operation Kindness service award Mass. Dept. Edn. and United Community Services, 1968. Mem. Am. Psychol. Assn., Am. Personnel and Guidance Assn., Bicultural Assn. Spanish-speaking Therapists and Advs., Kappa Delta Pi, Pi Lambda Theta. Democrat. Roman Catholic. Subspecialties: Developmental psychology; Social psychology. Current work: Bilcultural personality development; cross-cultural clinical applications, biofeedback training and stress related disorders. Home: 500 King Dr 1008 Daly City CA 94015 Office: University of Santa Clara 208 Benson Center Santa Clara CA 95053

GUTMAN, GEORGE ANDRE, microbiology educator; b. Domme, France, Sept. 15, 1945; s. Peter M. and Frances F. (Reitman) G.; m. Janis Lynn Schonauer, July 26, 1946; 1 son, Pierre Daniel. A.B., Columbia U., 1966; Ph.D., Stanford U., 1973. Postdoctoral fellow Stanford U., 1973-74, Walter and Eliza Hall Inst., Melbourne, Australia, 1974-76; asst. prof. U. Calif.-Irvine, 1976-82, assoc. prof., 1982—. Fulbright scholar, 1966-67; Research Career Devel. awardee NIH, 1978-83. Mem. Am. Assn. Immunologists. Subspecialties: Genetics and genetic engineering (biology); Immunobiology and immunology. Current work: Structure of antibody proteins and genes. Office: Dept Microbiolog U Calif Irvine CA 92717

GUTOWSKY, HERBERT SANDER, chemistry educator; b. Bridgman, Mich., Nov. 8, 1919; s. Otto and Hattie (Meyers) G.; m. Barbara Stuart, June 22, 1949 (div. Sept. 1981); children: Daniel Kurt (dec.), Robb Edward, Christopher Carl.; m. Virginia Warner, Aug. 1982. A.B., Ind. U., 1940, D.Sc. (hon.), 1983; M.S., U. Calif.-Berkeley, 1946; Ph.D., Harvard U., 1949. Mem. faculty U. Ill. at Urbana, 1948—, prof. chemistry, 1956—, head div. phys. chemistry, 1956-63, head dept. chemistry and chem. engring., 1967-70; dir. Sch. Chem. Scis., head dept. chemistry, 1970—; mem. chemistry panel NSF, 1963-66, chmn. panel, 1965-66, mem. adv. com. on planning, 1971-74; mem. Ill. Bd. Natural Resources and Conservation, 1973—; G.N. Lewis Meml. lectr., 1976, G.B. Kistiakowsky lectr., 1980. Mem. adv. bd. Petroleum Research Fund, 1959-61; mem. selection and scheduling com. Gordon Research Conf., 1959-64, 68-72, trustee, 1969-72, chmn. bd. trustees, 1971-72. Served to capt., chem. warfare service AUS, 1941-45. Recipient 1966 $5000 Irving Langmuir award Am. Chem. Soc.; Midwest award St. Louis sect., 1973; 1974 $1000 prize Internat. Soc. Magnetic Research; Peter Debye award in phys. chemistry Am. Chem. Soc., 1975; Nat. medal of Sci., 1977; Guggenheim fellow, 1954-55. Fellow Am. Phys. Soc. (chmn. div. chem. physics 1973-74), AAAS, Am. Acad. Arts and Scis., mem., Nat. Acad. Scis. (mem. com. sci. and pub. policy 1972-75, chmn. panel on atmospheric chemistry 1975-77, mem. com. impacts of stratospheric change 1975-77), Am. Chem. Soc. (chmn. div. phys. chemistry 1966-67, com. on profl. tng. 1969-77, chmn. 1974-77), AAUP, Phi Beta Kappa, Sigma Xi. Subspecialties: Nuclear magnetic resonance (biotechnology); Nuclear magnetic resonance (chemistry). Current work: Magnetic resonance; molecular and solid state structure; relaxation phenomena; dynamic structure of membranes; the role of manganese in photosynthesis. Home: 202 W Delaware Ave Urbana IL 61801 Office: Noyes Lab 505 S Mathews St Urbana IL 61801

GUTTAG, KARL MARION, electrical engineer; b. Washington, May 29, 1954; s. Alvin and Norma (Samons) G. B.S. in Elec. Engring, Bradley U., Peoria, Ill., 1976; M.S. in Elec.Engring, U. Mich., 1977. Design engr. Tex. Instruments, Houston, 1977-79, project engr., 1980-81, graphics strategy mgr., 1982—. Recipient Sr. Mem. Tech. Staff award Tex. Instruments, 1982. Mem. IEEE, Assn. Computing Machinery. Subspecialty: Integrated circuits. Current work: VLSI logic and system design and definition; computer graphics. Patentee in field of video-display processors. Home: 11602 Ensbrook St Houston TX 77099 Office: Tex Instruments PO Box 1443 Houston TX 77001

GUTTMAN, LESTER, chemist, researcher; b. Mpls., Apr. 18, 1919; m., 1955. B.Chem., U. Minn., 1940; Ph.D. in Chemistry, U. Calif., 1943. Asst. in chemistry U. Calif., 1940-42; assoc. scientist Manhattan Engring. Dist., U.S. War Dept., N.Mex., 1943-46; research assoc. Inst. Study Metals, U. Chgo., 1946-47, from instr. to asst. prof., 1947-55; Guggenheim fellow U.K. Atomic Energy Authority, 1955-56; phys. chemist research lab. Gen. Electric Co., 1956-60; sr. chem. scientist Argonne (Ill.) Nat. Lab., 1960—. Assoc. editor: Jour. Applied Physics, 1965-73; editor, 1974—; assoc. editor: Applied Physics Letter, 1965-73. Fellow Am. Phys. Soc. Subspecialty: Solid state physics. Office: Solid State Sci Div Argonne Nat Lab 9700 S Cass Ave Argonne IL 60439

GUTTORP, PETER MALTE, statistician, educator; b. Lund, Sweden, Mar. 10, 1949; came to U.S., 1975; s. Nils Malte and Inga Anna Erika (Nilsson) G.; m. June Gloria Morita, Aug. 9, 1982. B.Journalism, Stockholm Sch. Journalism, 1969; B.A., U. Lund, 1974; M.A., U. Calif.-Berkeley, 1976, Ph.D., 1980. Asst. prof. dept. stats. U. Wash., Seattle, 1980—. Recipient Evelyn Fix award dept. statistics, U. Calif.-Berkeley, 1980. Mem. Inst. Math. Statistics, Soc. Indsl. and Applied Math. Subspecialties: Statistics; Probability. Current work: Time series, point processes, statistical methods for stochastic processes, stochastic modelling and statistical computing. Office: Dept Stats GN-22 Univ Wash Seattle WA 98195

GUY, JAMES DAVID, JR., clinical psychologist, educator; b. Milw., Dec. 18, 1952; s. James David and Dorothy Ann (Adams) G.; m. Margaret Paul, June 15, 1975. B.A. in Psychology, Wheaton Coll., Ill., 1975; M.A. in Theology, Fuller Theol. Sem., Pasadena, Calif., 1978; Ph.D. in Clin. Psychology, Fuller Sch. Psychology, Pasadena, 1981. Lic. clin. psychologist, Calif. Prin. investigator NIMH Ctr. for Study of Schizophrenia, Camarillo (Calif.) State Hosp., 1979-80; psychology clk., intern Harbor UCLA Med. Ctr., Torrance, 1979-80; psychology intern Milwaukee County Mental Health Complex, Milw., 1980-81; postdoctoral resident Northwestern Inst. Psychiatry, Chgo., 1981-82; staff psychologist Rosemead (Calif.) Counseling Service, 1982—; asst. prof. Rosemead Sch. Psychology, La Mirada, Calif., 1982—; pvt. practice clin. psychology, Encino, Calif., 1982—; cons. community service dept. City of Pasadena, 1979; cons. Milw. Pub. Sch. System, 1979-80, Trinity Bapt. Ch., Westminster, Calif., 1978-79; asst. dir. treatment Sycamore Residential Treatment Center, Altadena, Calif., 1976-79. Contbg. author: Ency. of Psychology, 1983; Contbr. articles to profl. jours. Mem. Am. Psychol. Assn., Calif. State Psychol. Assn. Presbyterian. Subspecialties: Neuropsychology; Psychotherapy. Current work: Research focusing on the role of genetic and teratogenic factors in the development of schizophrenia, role of neuropsychological impairment in the predisposition towards the development of schizophrenia. Office: Rosemead Sch Psychology 13800 Biola Ave La Mirada CA 90639

GUYER, PAUL QUENTIN, animal scientist; b. Linn County, Mo., Mar. 31, 1923; s. Quinn E. and M. Eleanor (McDonald) G.; m. Reatha J. Jones, June 18, 1948; children: Laure, Scott, Gregory. B.S. in Agr, U. Mo., 1948, M.A., 1949, Ph.D., 1954. Extension beef specialist U. Nebr., 1954—. Served with AUS, 1944-46. Mem. Am. Soc. Animal Sci. (extension award 1973), Gamma Sigma Delta, Sigma Xi. Republican. Mem. Ch. of Christ. Subspecialty: Animal nutrition. Current work: Beef feeding and mgmt. Home: 230 Parkvale Dr Lincoln NE 68510 Office: 204 Marvel Baker Hall Lincoln NE 68523

GUZE, SAMUEL BARRY, psychiatrist, educator, univ. ofcl.; b. N.Y.C., Oct. 18, 1923; s. Jacob and Jenny (Berry) G.; m. Joy Lawrence Campbell, June 7, 1946; children—Jonathan, Ann. Student, Coll. City N.Y., 1939-41; M.D., Washington U., 1945. Diplomate: Am. Bd. Internal Medicine, Am. Bd. Psychiatry and Neurology. Faculty Washington U. Sch. Medicine, St. Louis, 1951—, prof. psychiatry, asso. prof. medicine, 1964—, asst. to dean, 1965-71, vice chancellor for med. affairs, 1971—, co-head dept. psychiatry, 1974-75, head dept., 1975—, Spencer T. Olin prof., 1974—; pres. Washington U. Med. Center, 1971—; staff Barnes Hosp., St. Louis, 1951—, psychiatrist-in-chief, 1975—; staff Renard Hosp., 1953—, psychiatrist-in-chief, 1975—; asst. dir. Psychiatry Clinic, Washington U. Sch. Medicine, 1951-55, dir., 1955-75. Contbr. articles to profl. jours. Fellow A.C.P., Am. Psychiat. Assn., Royal Coll. Psychiatry, Am. Coll. Psychiatry; mem. Am. Fedn. for Clin. Research, Psychiat. Research Soc., AMA, Am. Psychosomatic Soc., Assn. for Research in Nervous and Mental Diseases, Am. Psychopathol. Soc., Soc. Biol. Psychiatry, Soc. Neurosci., Inst. of Medicine of Nat. Acad. Scis., Sigma Xi, Alpha Omega Alpha. Subspecialty: Psychiatry. Current work: Phychiatric diagnosis and classification; alcoholism; criminality; psychiatric genetics; predicting course and outcome of illness. Home: 17 Ridgemoor Dr St Louis MO 63105 Office: 4940 Audubon Ave St Louis MO 63110

GWAZDAUSKAS, FRANCIS CHARLES, endocrinologist, educator; b. Waterbury, Conn., July 25, 1943; s. Francis Julius and Agnes Eva (Lizauskas) G.; m. Judy Keller, Mar. 20, 1971; children: Jennifer, James, John, Peter. B.S. in Animal Husbandry, U. Conn., 1966; M.S. in Dairy Sci, U. Fla., 1972; Ph.D. in Animal Sci, U. Fla., 1974. Asso. prof. dept. dairy sci. Va. Poly. Inst. & State U., Blacksburg, 1980—, asst. prof., 1974-80. Contbr. to profl. jours. Served with U.S. Army, 1967-68. Mem. Am. Dairy Sci. Assn., Am. Soc. Animal Sci., Soc. Study Reprodn., Soc. Exptl. Medicine and Biology. Roman Catholic. Subspecialties: Animal physiology; Animal breeding and embryo transplants. Current work: Uterine protein contribution to early embryo development; timing of artifical insemination; environmental effects on fertility. Office: Va Inst and State U 2070 Animal Sci Bldg Blacksburg VA 24061

GYFTOPOULOS, ELIAS PANAYIOTIS, mechanical and nuclear engineering educator, consultant, researcher; b. Athens, Greece, July 4, 1927; came to U.S., 1953, naturalized, 1961; s. Panayiotis Elias and Despina (Louvaris) G.; m. Artemis E. Scalleri, Sept. 3, 1962; children: Vasso, Maro, Rena. Diploma in Mech. and Elec. Engring, Tech. U., Athens, 1953; Sc.D. in Elec. Engring, M.I.T., 1958. Registered profl. engr., Mass. Research asst. in elec. engring. M.I.T., 1953-55, instr., 1955-58, asst. prof. elec. engring, 1958-60, asst. prof. elec. and nuclear engring, 1960-61, assoc. prof. nuclear engring., 1961-64, prof., 1964-70, Ford prof. engrin., 1970—; cons. in field; dir. Thermo Electron Corp. Author: Thermionic Engery Conversion, vol. 1, 1973, vol. 2, 1979, Potential Fuel Effectiveness in Industry, 1974; editor: Manuals on Energy Conservation, vols. 1-17, 1982. Served with Greek Navy, 1948-51. Fellow Am. Acad. Arts and Scis., Am. Nuclear Soc., Acad. Athens, Nat. Acad. Engring.; mem. Am. Nuclear Soc., ASME, Am. Phys. Soc., AAAS. Subspecialties: Thermodynamics; Energy conservation. Current work: Foundations of quantum mechanics and thermodynamics; energy conservation; nuclear reactor safety. Office: MIT Room 24-109 Cambridge MA 02139

GYR, JOHN WALTER, research scientist; b. Switzerland, July 20, 1923; emigrated to U.S., 1947; s. Walter Otto and Adrienne (Van Bennekom) G.; m. Marian E. Strickland, Aug. 28, 1948; children: Walter R., Kim, John D., Duff, Kaj, Drew W. Student, Tech. U., Delft (Netherlands), 1942-43; Demie lic., U. Gen., 1947; M.A., Miami U., 1948; Ph.D., U. Mich., 1953. Research assoc. U. Colo., Boulder, 1954-57; asst. research scientist to research scientist U. Mich., Ann Arbor, 1957-82; prof. Psychiatrische Universitaetsklinik, Bern, Switzerland, 1982—, assoc. dir., 1982—. Contbr. chpts. to books, numerous articles to profl. jours. Active Democratic Party. Fellow Am. Phychol. Assn. Subspecialties: Cognition; Self-organizing systems. Current work: At the moment interested in schizophrenic communication through an extension of mathematical linquistics that includes multivalued logic and infinite algebraic self-referential systems. Home: Giacometti Strasse 16 Bern Switzerland 3006 Office: Psychiatrische Universitaetsklinik Bolligen Strasse 111 Bern Switzerland 3072

HAACKE, EWART MARK, research scientist, consultant; b. Toronto, Ont., Can., Jan. 24, 1951; came to U.S., 1978, naturalized, 1980; s. Ewart Mortimer and Helena Doris (Davies) H.; m. Linda Theresa Clarke, July 19, 1975; 1 son, Bryon Clarke. B.Sc., U. Toronto, 1973, M.Sc., 1975, Ph.D., 1978. Postdoctoral fellow Toronto (Ont., Can.) U., 1978; research assoc. Case Western Res. U., Cleve., 1978-80, instr., 1980-81, sr. research assoc., 1980-83; research scientist Gulf Research & Devel. Co., Pitts., 1981-83; sr. research scientist Picker Internat., Cleve., 1983—; cons. dept. medicine Case Western Res. U., Cleve., 1981-83. Mem. geneal. com. Western Res. Hist. Soc., Cleve., 1981—. Burton fellow U. Toronto, 1977; Ont. grad. fellow, 1975-76; Stevens fellow, 1974; Victoria Coll. fellow, 1970; recipient Ont. scholar award Ont. Govt., 1969. Mem. Am. Phys. Soc., Soc. Exploration Geophysicists, Soc. Indsl. and Applied Math., Ont. Geneal. Soc. Club: Case Western Res. U. Table Tennis (Cleve.) (sec., faculty supr.). Subspecialties: Imaging technology; Particle physics. Current work: Tomographic and wave equation inversion techniques, NMR and CT medical imaging, gauge field theories and parton properties, pulmonary mechanics and statistical analyses. Home: 2312 Glendon Rd University Heights OH 44118 Office: Picker Internat 595 Miner Rd Highland Heights OH 44143

HAAK, RICHARD ARLEN, biophysicist, researcher, educator; b. Fairmont, Minn., Sept. 14, 1944; s. Rudolph A. and Lenora (Becker) H.; m. Mary Rebecca Kerlin, Dec. 27, 1979; children: Tim, Kathy, Kristin. B.A., MacMurray Coll., 1966; M.A., So. Ill. U.-Carbondale, 1968, Ph.D., 1972. Prof. dept. microbiology Ind. U. Sch. Medicine, Indpls., 1972—. Contbr. articles to publs. in field. NIH research grantee, 1978—. Mem. Biophys. Soc., Am. Soc. Microbiology. Lutheran. Subspecialties: Biophysics (biology); Microbiology (medicine). Current work: Membrane biophysics of pathogenic microorganisms and host cells. Home: 7926 Ridgegate W Dr Indianapolis IN 46268 Office: Dept Microbiology Ind U Sch Medicine 635 Barnhill Dr Indianapolis IN 46223

HAAKE, EUGENE VINCENT, nuclear engineering administrator; b. Cleve., Sept. 25, 1921; s. Eugene Louis and Vincenta Mildred (Hettinger) H.; m. Elsie Warren Burton, Jan. 20, 1951; children: Barbara, Janet, Ronald. B.S. in Physics, Western Res. U., 1943, M.A., UCLA, 1948. Registered profl. engr., nuclear engring., Calif. Assoc. physicist Fairchild Engine/Airplane Corp., Oak Ridge, Tenn., 1948-50; physicist Oak Ridge Nat. Lab., 1950-52; sr. nuclear engr. Convair/Gen. Dynamics, Ft. Worth, 1952-55, group supr., 1955-62; sr. staff mem. Gen. Atomic/Gen. Dynamics, San Diego, 1962-73; project engr. Gen. Atomic Co., San Diego, 1973-76; br. mgr. GA Techs. Inc., San Diego, 1976—. Contbr. articles to profl. jours. Served as 1st lt. U.S. Army, 1943-46. Mem. Am. Phys. Soc., Am. Nuclear Soc. (San Diego sect. treas. 1971-72, dir. 1975-78), Nat. Mgmt. Assn., Phi Beta Kappa. Republican. Lutheran. Subspecialty: Nuclear engineering. Current work: Nuclear power plant control and dynamics, nuclear systems analysis, technical management. Patentee Nuclear Reactor Improvements, 1967. Home: 3703 Brandywine St San Diego CA 92117 Office: GA Techs Inc PO Box 81608 San Diego CA 92138

HAAN, CHARLES THOMAS, agricultural engineering educator; b. Randolph County, Ind., July 10, 1941; s. Charles Leo and Dorothy Mae (Smith) H.; m. Janice Kay Johnson, June 3, 1967; children: Patricia Kay, Christopher Thomas, Pamela Lynn. B.S. in Agrl. Engring, Purdue U., 1963, M.S., 1965; Ph.D. in Agrl. Engring, Iowa State U., 1967. Grad. asst. Purdue U., W. Lafayette, Ind., 1963-64; research asso. Iowa State U., Ames, 1964-67; asst. prof., asso. prof., prof. U. Ky., Lexington, 1967-78; prof., head agrl. engring. dept. Okla. State U., Stillwater, 1978—; cons. in area of hydrology various firms and govtl. orgns. Author: Statistical Methods in Hydrology, 1977, Hydrology and Sedimentology of Surface Mined Lands, 1978; editor: Hydrologic Modeling of Small Watersheds, 1981; contbr. tech. papers and reports to publs. and confs. Recipient and or adminstr. various research grants. Mem. Am. Soc. Agrl. Engrs. (Young Researcher of 1975, research paper award 1969), Am. Geophys. Union, Nat. Soc. Profl. Engrs., Okla. Soc. Profl. Engrs., Am. Soc. for Engring. Edn., Am. Inst. Hydrologists, Sigma Xi, Tau Beta Pi, Alpha Epsilon, Gamma Sigma Delta, Phi Kappa Phi. Roman Catholic. Subspecialties: Agricultural engineering; Hydrology. Current work: Hydrologic modeling. Home: 720 Lakeshore Dr Stillwater OK 74075 Office: Oklahoma State Univ Stillwater OK 74078

HAAN, DAVID CHARLES, mechanical engineer, consultant; b. Hammond, Ind., July 1, 1948; s. Jacob and Grace Martha (Specking) H. B.S. in Mech. Engring, Purdue U., 1970, M.S. in Engring, 1975. Registered profl. engr., Ill. Mech. engr. Sargent & Lundy, Chgo., 1970-73, mech. project engr., 1973-77, project mgr., 1977-81, assoc. project mgr., 1981—. Mem. ASME, Am. Nuclear Soc., Western Soc. Engrs. (Charles Ellet Award 1981). Subspecialties: Mechanical engineering; Nuclear engineering. Home: 4250 Saratoga Downers Grove IL 60515 Office: Sargent & Lundy 55 E Monroe Chicago IL 60603

HAAS, PAUL ARNOLD, chemical engineer; b. Rolla, Mo., Aug. 11, 1929; s. Arnold G. and Mary M. (Sachs) H.; m. Betty Jane Turner, May 31, 1958; children: Barry, Janet, Robert, Alan. B.S., U. Mo., Rolla, 1950; M.S., Mont. State U., 1951; Ph.D., U. Tenn., 1965. Registered profl. engr., Tenn. Devel. engr. Oak Ridge Nat. Lab., 1952-57, group leader, 1957—, sr. staff scientist, 1975—. Contbr. articles to profl. jours. Mem. Am. Inst. Chem. Engrs., Am. Nuclear Soc., Am. Chem. Soc. Republican. Subspecialties: Chemical engineering; Nuclear fission. Current work: Nuclear fuel preparation and fabrication. Nuclear fuel conversion processes; gel-sphere processes for nuclear fuel fabrication, nuclear waste disposal. Patentee in field. Home: 8000 Bennington Dr Knoxville TN 37919 Office: Oak Ridge Nat Lab PO Box X Oak Ridge TN 37830

HAAS, VIOLET BUSHWICK, electrical engineer, educator, researcher; b. N.Y.C., Nov. 23, 1926; d. Morris and Cynthia (Cherkoff) Bushwick; m. Felix Haas, Apr. 16, 1948; children: Richard A., Elizabeth A., David R. A.B., Bklyn. Coll., 1947; S.M., MIT, 1949, Ph.D., 1951. Lectr. math. Immaculata (Pa.) Coll., 1952-55; instr. math. U. Conn., Storrs, 1955-56, Wayne State U., Detroit, 1956-57; asst. to assoc. prof. math. U. Detroit, 1957-62; asst. to assoc. prof. elec. engring. Purdue U., West Lafayette, Ind., 1962-78, prof. elec. engring., 1978—. Author: Analog Computer Handbook, 1982. Bd. dirs. Lafayette Symphony, 1982—; active YWCA, 1980—. AAUW Vassie James Hill fellow, 1951-52; NSF fellow, 1960-61; grantee, 1976-78, 80-82; recipient D.D. Ewing award Sch. Elec. Engring., Purdue U., 1977, H.B. Schleman award Assn. Women Students, 1978. Mem. IEEE (sr.), Soc. Women Engrs. (sr.), Soc. for Indsl. and Applied Math., Am. Soc. Engring. Edn., Assn. for Women in Math., Assn. for Women in Sci., LWV, NOW. Subspecialties: Electrical engineering; Applied mathematics. Current work: Optimal control, optimal estimation. Home: 132 Arrowhead Dr West Lafayette IN 47906 Office: Sch Elec Engring Purdue U West Lafayette IN 47907

HABASHI, WAGDI GEORGE, mechanical engineering educator, consultant; b. Port-Said, Egypt, June 29, 1946; emigrated to Can., 1964; s. George and Iris (Bassili) H.; m. Yvette Hanna, June 25, 1967; children: Jenny, Stephanie-Anne, Andrew Glenn. B.S., McGill U., 1967, M.E., 1969; Ph.D., Cornell U., 1975. Asst. prof. Stevens Inst. Tech., Hoboken, N.J., 1974-75; asst. prof. Concordia U., Montreal, Que., Can., 1975-79, assoc. prof. mech. engring., 1979—; aerodynamics cons. Pratt & Whitney Aircraft Can., Longueuil, Que., 1977—. Editorial bd.: Internat. Jour. Numerical Methods in Fluids, 1982; editor: books Advances in Computational Transonics, 1984, Viscous Flow Computational Methods, 1984. Recipient Brit. Assn. medal for gt. distinction in mech. engring., 1967. Mem. ASME, AIAA, Sigma Xi. Subspecialties: Aeronautical engineering; Fluid mechanics. Current work: Transonic flows over wing and bodies and in turbomachinery; finite element and other applied numerical methods. Office: Mech Engring Dept Concordia U 1455 de Maisonneuve Blvd W Montreal PQ Canada H3G 1M8

HABER, BERNARD, neurobiology educator; b. Lodz, Poland, July 20, 1934; s. Michael and Maria (Gewurz) H.; m. Christine Tancred, Apr. 24, 1982; children: Leslie Sue, Harold Allan, Debra Beth. B.Sc., McGill U., 1956, M.Sc., 1957, Ph.D., 1962. Research assoc. Galesburg (Ill.) Research Hosp., 1962-65; assoc. research scientist City of Hope Med. Center, Duarte, Calif., 1965-70; assoc. prof. biochemistry, neurology U. Tex. Med. Br., Galveston, 1971—; chief neurochemistry sect. Marine Biomed. Inst., 1971—; mem. behavioral and neurol. scis. study sect. NIH. Editor: (with M. Aprison) Neuropharmacology and Behavior, 1978, (with G. Cohen, A. Goldstein) Neurochemical and Immunological Components in Schziophrenia, 1978; contbr. articles to profl. jours. Welch grantee, 1971—; Nat. Found. March of Dimes grantee, 1981-82; NIH grantee, 1971—. Mem. Am. Soc. Neurochemistry, Internat. Soc. Neurochemistry, Soc. Neuroscis., AAAS. Jewish. Subspecialties: Neurochemistry; Regeneration. Current work: Cellular neurobiology, neuroimmunology, regeneration neurons, cell-cell interactions, trophic factors regeneration. Office: 200 University Blvd #519 Galveston TX 77550

HABER, SEYMOUR, mathematician, consultant; b. Bklyn., Oct. 7, 1929; s. Morris and Rose (Soller) H.; m. Blossom Delman, Aug. 1954; children: Melanie, Miriam, Jesse, Eli. B.S., Yeshiva Coll., 1950; M.A., Syracuse U., 1951; Ph.D., MIT, 1954. Research assoc. Courant Inst., N.Y.U., 1955-56; lectr. math. Bar-Ilan U., Ramat Gan, Israel, 1955-56; mathematician Weizmann Inst., Rehovoth, Israel, 1956-57; asst. prof. math. Poly. Inst. N.Y., Bklyn., 1957-60; mathematician Nat. Bur. Standards, Washington, 1960—. Mem. Am. Math. Soc., Soc. Indsl. and Applied Math., Spl. Interest Group Numerical Analysis. Subspecialties: Applied mathematics; Numerical analysis. Current work: Numerican evaluation of integrals, mathematical graphics. Home: 1106 N Belgrade Rd Silver Spring MD 20902 Office: Nat Bur Standards Washington DC 20234

HABERFIELD, PAUL, chemist, educator; b. Carnuntum, Czechoslovakia, May 29, 1933; s. Andrew and Terry (Steiner) H.; m. Mamie Birnbaum, May 8, 1966; children: Shulamith, Rebecca, Licia, Saul. B.S., M.I.T., 1955; Ph.D., UCLA, 1960. Postdoctoral research fellow Purdue U., West Lafayette, Ind., 1960-61; instr. to assoc. prof. chemistry Bklyn. Coll., CUNY, 1961-72, prof., 1972—. Contbr. articles to profl. jours. Mem. Am. Chem. Soc., Sigma Xi. Subspecialties: Organic chemistry; Photochemistry. Current work: The effect of proximate charges on reaction centers; solute-solvent interactions in ground and electronic excited states. Home: 1666 52nd St Brooklyn NY 11204 Office: Dept Chemistry Bklyn Coll Brooklyn NY 11210

HABERMAN, CHARLES MORRIS, engineering educator; b. Bakersfield, Calif., Dec. 10, 1927; s. Carl Morris and Rose Marie (Braun) H. B.S., UCLA, 1951; M.S. in Mech. Engring., U. So. Calif., 1954, Engr., 1957, M.S. in Aero. Engring., 1961. Lead, sr., group engr. Northrop Aircraft Corp., Hawthorne, Calif., 1951-59; mem. faculty Calif. State U.-Los Angeles, 1959—, prof. mech. engring., 1967. Author: Engineering Systems Analysis, 1965, Use of Computers for Engineering Applications, 1966, Vibration Analysis, 1968, Basic Aerodynamics, 1971. Served with AUS, 1946-47. Mem. Am. Acad. Mechanics, AAUP, AIAA, Am. Soc. Engring. Edn. Democrat. Roman Catholic. Subspecialties: Mechanical engineering; Aeronautical engineering. Current work: Heat transfer, fluid mechanics, aerodynamics, vibrations and systems analysis. 90032

HABERMAN, RICHARD, mathematics educator; b. Bklyn., June 27, 1945; s. Henry and Jane (Lainer) H.; m. Elizabeth Hart, Feb. 1, 1969; children: Ken, Vicki. B.S., M.I.T., 1967, Ph.D., 1971. Asst. researchgeophysicist U Calif.-San Diego, La Jolla, 1971-72; asst. prof. math. Rutgers U., New Brunswick, N.J., 1972-77, Ohio State U., Columbus, 1977-78; assoc. prof. math. So. Meth. U., Dallas, 1978—. Author: Mathematical Models: Mechanical Virbrations, Population Dynamics, and Traffic Flow, 1977, Elementary Applied Partial Differential Equations with Fourier Series and Boundary Value Problems, 1983. Jewish. Subspecialty: Applied mathematics. Current work: Nonlinear wave motion, asymptotic and perturbation methods for ordinary and partial differential equations. Office: Dept Math So Meth U Dallas TX 75275

HABERMAN, WILLIAM LAWRENCE, physicist, consultant; b. Vienna, Austria, May 5, 1922; s. Isaac and Fannie Anne (Rathaus) H.; m. Florence H. Frank, Sept. 19, 1957. B.M.E., Cooper Union, 1949; M.S., U. Md., 1961, Ph.D., 1956. Registered profl. engr., N.J. Engr. Bur. Ships, USN, 1949-50; physicist Naval Ship Research & Devel. Ctr., Carderock, Md., 1950-63; mgr. Manned Space Flight, NASA, Washington, 1963-71; prof. U. Md., College Park, 1956-71; prof., chmn. mech. engring. Newark Coll. Engring., 1971-73; lectr. Montgomery Coll., 1973-78; engring. cons. ERC Co., Rockville, Md., 1973—. Author: Fluid Mechanics, 1971, 2nd edit. 1980, Engineering Thermodynamics, 1980, Heat Transfer, 1982; contbr. articles to profl. jours. Commr. Rockville Energy Comm., 1974—. Served with C.E. U.S. Army, 1942-45. Recipient NASA Apollo award, 1970. Mem. Am. Phys. Soc., Am. Geophys. Union, ASME, Assn. Energy Engrs., AIAA, Am. Soc. Engring. Edn., Soc. Naval Architects and Marine Engrs., Refrigeration Service Engrs. Subspecialties: Thermodynamics; Solar energy. Current work: Research on absorption thermodynamics, applicable to absorption refrigeration utilizing solar energy; research on thermodynamics of solar energy for heating. Office: PO Box 1723 Rockville MD 20850

HABERMANN, ARIE NICOLAAS, computer science educator; b. Groningen, Netherlands, June 26, 1932; m., 1956; 4 children. B.S., Free U. Amsterdam, 1953, M.S., 1958; Ph.D. in Computer Sci, Eindhoven Tech. U., 1967. Tchr. high sch., Netherlands, 1954-62; from lectr. to asst. prof. math. programming Eindhoven Tech. U., 1962-68; vis. research scientist programming systems Carnegie-Mellon U., Pitts., 1968-69, assoc. prof. computer sci., 1969-73, prof., 1973—, acting head dept., 1979-80, head dept., 1980—; software cons. to govt. agys.; vis. prof. computer sci. U. Newcastle, Eng., 1973, Tech. U. Berlin, 1976. Editor: Acta Info, ACM Trans. on Program Lang. and Systems. Mem. Assn. for Computing Machinery, N.Y. Acad. Scis. Subspecialty: Programming languages. Office: Dept Computer Sci Carnegie-Mellon U Pittsburgh PA 15213

HACH, EDWIN ELLISON, JR., physical chemist; b. Shippenville, Pa., Jan. 5, 1934; s. Edwin Ellison and Sara R (Reed) H.; m. Betty Jane Orcutt, Aug. 4, 1955; children: Edwin Ellison III, Sarah A. B.S. in Sci. Edn, Clarion State Coll., 1959; M.S.T., U. N.H., 1967, Ph.D. in Chemistry, 1967. Tchr. chemistry Bradford (Pa.) Area High Sch., 1959-62; assoc. prof. chemistry St. Bonaventure (N.Y.) U., 1974—. Contbr. articles to profl. jours. Bd. dirs. Otto-Eldred Schs., 1981—; coach Little League., Otto Twp., Pa., 1979-82. Served with USN, 1955-57. Pratt grantee, 1963; St. Bonaventure U. grantee, 1970, 77. Mem. ; Am. Chem. Soc. Ruffed Grouse Soc., Sigma Xi. Lodge: Lions. Subspecialties: Physical chemistry; Numerical analysis. Current work: CAI program development, simulations, educational software generation. Home: Box 236 Duke Center PA 16729 Office: Dept Chemistry St Bonaventure U St Bonaventure NY 14778

HACKEL, LLOYD ANTHONY, physicist; b. Little Chute, Wis., Oct. 14, 1949; s. Lawrence Andrew and Margaret Mary (Williams) H.; m. Linda Kay, Sept. 1, 1971; children: Catherine, Laura. B.S., U. Wis., 1971; M.S., MIT, Sc.D., 1974. Mem. staff MIT Research Lab. of Electronics, 1974-75; project leader integrated expt. group laser isotope separation program Lawrence Livermore (Calif.) Nat. Lab., 1976—. Contbr. articles to profl. jours. Mem. Am. Phys. Soc., Phi Beta Kappa. Roman Catholic. Subspecialties: Spectroscopy; Plasma physics. Current work: Laser isotope separation, electron beam vapor generation.

HACKEN, GEORGE, computer-based automation company executive, consultant; b. Alma-Ata, USSR, Mar. 4, 1942; s. Emanual and Vera (Altman) H.; m. Tana Cohn, June 18, 1978; children: Diane Lee Carlson, Erica Lynn Carlson. A.B., Columbia U., 1964, Ph.D., 1971. Research assoc., instr. nuclear physics Columbia U., N.Y.C., 1971-76; sr. tech. cons. ADP Network Services, 1976-77; sr. softward engr., math. cons. Decision Systems, Inc., Mahwah, N.J., 1977-78; control systems tech. project mgr., supr. Am. Can Co., Fairlawn, N.J., 1978-81; v.p. Intelligent Indsl. Systems, Inc., Secaucus, N.J., 1982—. Contbr. articles in field to profl. jours. Mem. Thomas Jefferson Inst., San Diego, 1982-83. Recipient 100% Club award ADP Network Services, 1977; Spl. Recognition award Am. Can Co., 1981. Mem. Am. Math. Soc., Am. Phys. Soc., Am. Nuclear Soc., Assn. Computing Machinery, IEEE, Robotics Internat., Soc. Indsl. and Applied Math., N.Y. Acad. Scis., Sigma Xi. Subspecialties: Real-time, concurrent systems; Software engineering. Current work: Mathematical software for optimal distributed control systems, efficient schemes for concurrent multiprocessing. Home: 12 Shepard Dr Wanaque NJ 07465 Office: 1 Harmon Plaza 3d Floor Secaucus NJ 07094

HACKER, MILES PAUL, pharmacologist; b. Melrose Park, Ill., May 6, 1947; s. Frank Paul and Lucille Florence (Wendt) H.; m. Patricia Ann Hacker, May 29, 1971. B.S., Murray State U., 1970; Ph.D., U. Tenn., 1975. Postdoctoral assoc. Yale U., 1975-77; sr. toxicologist Midwest Research Inst., Kansas City, 1977-80; asst. prof. dept. pharmacology U. Vt., 1980—; Chmn. Human Experimentation com. U. Vt., 1981—. Contbr. articles in field to profl. jours. Nat. Cancer Inst. grantee, 1982—; NIH grantee, 1981—. Mem. Am. Assn. Cancer Research, Am. Soc. Pharmacology and Exptl. Therapeutics, N.Y. Acad. Sci., Soc. Toxicology. Subspecialties: Pharmacology; Cancer research (medicine). Current work: Cancer chemotherapy, immunology, toxicology; research in the area of organ specific toxicity of anticancer drugs such as bleomycin, adriamycin, cis-platinum. Office: Department Pharmacology Given Bldg University Vermont Burlington VT 05405

HACKETT, CHARLES JOSEPH, research scientist, educator; b. Detroit, Nov. 25, 1947; s. Victor Garl and Josephine (Kaminski) H.; m. Risa Grossman, Apr. 22, 1977. B.A., Wayne State U., 1970, Ph.D., 1977. Summer fellow in physiology Marine Biol. Lab., Woods Hole, Mass., 1973; course instr. biology dept. Wayne State U., Detroit, summer 1977; research assoc., 1977-78; fellow in cancer research Nat. Inst. Med. Research, London, 1978-80, mem. sci. staff, 1980-81; research assoc. Wistar Inst. Anatomy and Biology, Phila., 1981—. Contbr. articles to profl. publs. Recipient Wilhelmine L. Haley award Wayne State U., 1973; Damon Runyon-Walter Winchell Cancer Fund fellow, 1978-80; Wellcome Trust fellow, 1980. Mem. Genetics Soc. Am., Brit. Soc. Immunology. Subspecialties: Immunobiology and immunology; Cell biology. Current work: The cellular basis of immune response to virus infection. Cloned lines of T-lymphocytes are being used in vitro to study recognition and stimulation by specific viral and accessory cell antigenic molecules, and collaboration with B-lymphocytes in antibody production to influenza virus. Office: Wistar Inst 36th St at Spruce St Philadelphia PA 19104

HACKETT, JOHN TAYLOR, physiologist, educator; b. Chgo., July 24, 1941; s. Glenn Leonard and Valrie (White) H.; m. Jane Ann Svetlik, Aug. 22, 1972; children: Jeffrey Colin, Keith Gordon. Ph.D., U. Ill.-Chgo., 1970. Asst. prof. dept. physiology Med. Sch., U. Va., Charlottesville, 1973-77, assoc. prof., 1977—. Author: Structure and Function of Muscle, 1973; contbr. articles to profl. publs. Postdoctoral fellow Muscular Dystrophy Can., 1969-72, Nat. Inst. Neurol. Disorders and Stroke, 1972-73; Nat. Research Service award, 1981; recipient Research Sci. Devel award Nat. Inst. Drug Abuse, 1975-80. Mem. Soc. for Neurosci., Am. Physiol. Soc. Subspecialties: Neurophysiology; Neuropharmacology. Current work: Central nervous system integration and synaptic transmission, physiology and pharmacology of in vitro brain slices, metamorphosis of amphibian brainstem. Office: Dept Physiology Sch Medicine U Va Charlottesville VA 22908

HACKETT, JOSEPH LEO, microbiologist; b. Springfield, Ohio, Jan. 11, 1937; s. John Roger and Alice Pearl (Parker) H.; m. Phyllis Ann Boice, Apr. 27, 1963; children: Amy, Ron, Beth, Susan. B.A., Ohio State U., 1959, M.S., 1963, Ph.D. in Clin. Pathology, 1968. Registered technologist, Calif. Microbiology sect. head Reference Lab., Abbott Labs., North Hollywood, Calif., 1969-72; supr. microbiology quality control Pfizer Diagnostics, Maywood, N.J., 1972-74; supervisory microbiologyst standards div. FDA, Silver Springs, Md., 1974-77, investigational device exemption coordinator, 1977-80, chief microbiology/immunology standards br., 1980—. Contbr. articles to profl. jours. Mem. Am. Soc. Microbiology. Republican. Roman Catholic. Subspecialties: Microbiology (medicine); Infectious diseases. Current work: Standardization of clinical tests; clinical microbiology, clinical immunology. Office: FDA 8757 Georgia Ave Silver Spring MD 20910

HACKETT, PETER ANDREW, research chemist; b. Havering, Eng., July 16, 1948; s. Cyril Charles and Alice Rosina (Peck) H.; m. Nuala Ann Farrell, July 25, 1971; children: Clare, Joanna, Alison. B.Sc. in Chemistry, Southampton U., Eng., Ph.D. Postgrad. fellow NRC, Ottawa, Ont., Can., 1972-75, mem. staff, 1975—, head laser chemistry group, 1980—. Contbr. articles to prof. jours. Mem. Chem. Inst. Can., Inter-Americas Photochemistry Soc. Subspecialty: Laser-induced chemistry. Current work: Isotope separation, selective photochemistry, reaction dynamics. Patentee in field. Office: Div Chemistry Nat Research Council 100 Sussex Dr Ottawa ON Canada K1A 0R6

HACKMAN, JOHN CLEMENT, neurophysiologist, educator; b. Dayton, Ohio, May 16, 1947; s. Clem Frank and Martha Virginia (Schneble) H.; m. Susan Joan Pollard, June 3, 1968; children: Dawn, Jeff, Mark. Ph.D. in Biology, U. Miami, 1979. Adj. prof. neurology U. Miami (Fla.) Sch. Medicine, 1979-80, research asst. prof. neurology, 1980-82, asst. prof., 1982—, asst. prof. pharmacology, 1983—; research physiologist VA Med. Ctr., Miami, 1979—. Contbr. articles to sci. jours. Mem. attendance boundary com. Dade County Sch. Bd., 1981-82; v.p. Devonaire Elem. Sch. PTA, 1981-82. United Way grantee, 1980—; Nat. Parkinson Found. grantee, 1981—. Mem. AAAS, Soc. for Neursci., Am. Physiol. Soc. Republican. Roman Catholic. Subspecialties: Neuropharmacology; Neurophysiology. Current work: Currently studying the modulation of sensory input into the signal cord through the use of neurophysiological, neuropharmacological technique. Home: 12244 SW 105 Ln Miami FL 33186 Office: Dept Neurology U Miami Sch Medicine PO Box 16189 Miami FL 33101

HACYAN, SHAHEN, physicist; b. Istambul, Turkey, Oct. 24, 1947; s. Migirdic and Adrine (Saleryan) H.; m. Deborah Dultzin; children: Arturo, Esther. B.Sc., U. Mex., 1968; Ph.D., U. Sussex (Eng.), 1972. Assoc. researcher Universidad Nacional Autonoma de Mexico, 1973-76, researcher, 1976—, physics tchr., 1973—. Contbr. articles to sci. jours. Mem. Internat. Astron. Union, Am. Astron. Soc., Internat. Soc. Gen. Relativity and Gravitation, Academia de la Investigacion Cientifica. Subspecialty: General relativity. Current work: Relativistic astrophysics and general relativity. Home: Monserrat 157-6 Mexico DF Mexico 04330 Office: Instituto de Astronomia A P 70-264 Mexico DFMexico 04510

HADDAD, GEORGE ILYAS, research scientist, educator; b. Aindara, Lebanon, Apr. 7, 1935; came to U.S., 1952, naturalized, 1961; s. Elias Ferris and Fahima (Haddad) H.; m. Mary Louella Nixon, June 28, 1958; children—Theodore N., Susan Anne. B.S. in Elec. Engring, U. Mich., 1956, M.S., 1958, Ph.D., 1963. Mem. faculty U. Mich., Ann Arbor, 1963—, asso. prof., 1965-69, prof. elec. engring., 1968—, dir. electron physics lab., 1968-75, chmn. dept. elec. and computer engring., 1975—; cons. to industry. Contbr. articles to profl. jours. Recipient Curtis W. McGraw research award Am. Soc. Engring. Edn., 1970. Fellow IEEE (editor proc. and trans.); mem. Am. Soc. Engring. Edn., Am. Phys. Soc., Sigma Xi, Phi Kappa Phi, Eta Kappa Nu, Tau Beta Pi. Subspecialties: 3emiconductors; Microelectronics. Current work: Microwave solild-state devices and circuits; microwave and millimeter-wave monolithic integrated circuits. Office: Dept Elec and Computer Engring Univ Mich Ann Arbor MI 48109

HADDEN, EDWARD LEAL, JR., biology educator; b. Summit, N.J., Aug. 2, 1946; s. Edward Leal and Marjorie (Harvey) H.; m. Evelyn Blackman, Apr. 1, 1972; children: Christi, Katie. B.S. in Biology, Muhlenberg Coll., 1968, M.A., Wake Forest U., 1970, Ph.D., 1974. Assoc. prof. biology Wingate (N.C.) Coll., 1975—, chmn. div. sci. and math., 1976-80. Mem. Electron Microscopy Soc. Am., Southeastern Electron Microscopy Soc., Assn. Southeastern Biologists, Sigma Xi. Republican. Subspecialties: Microbiology; Developmental biology. Current work: Electron microscopy. Home: PO Box 689 111 Smith St Wingate NC 28174 Office: Wingate College Wingate NC 28174

HADDEN, JOHN WINTHROP, research physician, consultant; b. Berkeley, Calif., Oct. 23, 1939; s. David Rodney and Joanna Russel (Jennings) H.; m. Elba Luz Mas, July 31, 1964; children: John Winthrop, II, Paul Jennings. B.A., Yale U., 1961; M.D., Columbia U., 1965; cert. internal medicine, Roosevelt Hosp., N.Y.C., 1968. Intern Roosevelt Hosp., N.Y.C., 1965-66, resident in internal medicine, 1966-69; spl. fellow immunology U. Minn. Hosp., Mpls., 1969-72, asst. prof. pathology, 1972-73; assoc. prof., assoc. attending Cornell Grad. Sch. Med. Sci., Meml. Hosp., N.Y.C., 1973-82; assoc. mem., dir. lab. immunopharmacology Sloan Kettering Inst. Cancer Research, N.Y.C., 1973-82; dir. immunopharmacology program, prof. medicine, med. microbiology and immunology U. South Fla. Med. Coll., Tampa, 1982—; cons. in field. Editor: Immunopharmacology, 1977, Advances in Immunopharmacology, 1981, Lymphokines, 1982; assoc. editor: Internat. Jour. Immunopharmacology, 1978—; contbr. numerous chpts. to books and articles to profl. jours. V.p. 150 East Tenants Corp., N.Y.C., 1976-78, pres., 1978-80. Grantee in field; recipient Angier Research prize Yale U., 1961, Kellog Research prize Roosevelt Hosp., 1967, 68, 69; Established investigator Am. Heart Assn., 1972-77. Mem. Am. Assn. Immunlogists, Am. Soc. Exptl. Pathology, Internat. Soc. Immunopharmacology (v.p. 1982—), N.Y. Acad. Sci. Subspecialties: Immunology (medicine); Immunopharmacology. Patentee in field. Office: U South Fla Med Coll 12901 N 30th St Tampa FL 33612

HADDOCK, FREDERICK THEODORE, JR., radio astronomer, observatory administrator; b. Independence, Mo., May 31, 1919; m. (div.); 2 children. B.S., MIT, 1941; M.S., U. Md., 1950; D.Sc. (hon.), Southwestern at Memphis, 1965, Ripon Coll., 1966. Physicist, electronic scientist U.S. Naval Research Lab., 1941-56; assoc. prof. astronomy and elec. engring. U. Mich., Ann Arbor, 1956-59, prof. elec. engring., 1959-67, prof. astronomy, 1959—; dir. Radio Astron. Obs., 1961—; trustee Assoc. Univs. Inc., 1964-68; mem. vis. com. Nat. Radio Astron. Obs., 1956-58, 63-64; mem. adv. panel on astronomy NSF, 1957-60, 63-66; mem. adv. panel Internat. Yr. Quiet Sun, 1963-66; mem. facilities panel Office Instrn. Programs, 1963-64, 67-70; mem. U.S. Navy Eclipse Expdns., Aleutian Islands, 1950, Khartoum, 1952; mem. space sci. bd. ad hoc com. on astronomy Nat. Acad. Sci., 1958-62, researcher on radio frequency requirement sci., 1961-68, mem. Whitford panel on astron. facilities, 1963-64; mem. ad hoc panel on U.S. Navy 600 foot radio telescope President's Sci. Adv. Com., 1962; mem. evaluation panel on Arecibo (P.R.) Obs. Air Force Office Sci. Research, 1962-69; mem. adv. com. on planetary and interplanetary sci. Office Space Sci., NASA, 1961-62, mem. ad hoc working group on Apollo sci. expts. and tng., 1962-63, mem. hdqrs. astron. subcom., 1967-69, cons., 1970—, mem. astron. missions bd. radio astron. panel, 1968-71, mem. ad hoc com. on, 1970, mem. outer planets grand tour mission, 1971—; vis. research assoc. Calif. Inst. Tech., 1966; mem. adv. com. for Owens Valley Obs., 1969—; mem. commn. on radio astronomy Internat. Astron. Union, 1955—, mem. exec. and organizing coms., 1964-67, mem. commn. on space astronomy, 1964—; chmn. commn. 5 Internat. Sci. Radio Union, 1954-57, del. to assemblies, 1958-61, mem. nat. com., 1958-61. Fellow IEEE, Royal Astron. Soc.; mem. Am. Astron. Soc. (v.p. 1961-63), AIAA. Subspecialty: Radio and microwave astronomy. Office: Radio Astronomy Observatory U Mich Ann Arbor MI 48109

HADIDI, AHMED FAHMY, microbiologist; b. Damanhur, Egypt, Mar. 1, 1937; came to U.S., 1959, naturalized, 1970; s. Mohammed Abdel-Salam and Madeha (Abdel-Hammeed) H.; m. Agnes Catherine Evans, Aug. 8, 1964; children: Suzanne, Frederick, Jon. B.S. cum laude, Cairo U., 1958; M.S., U. Minn., 1962; Ph.D., Kans. State U., 1967. Research fellow U. Ky., Lexington, 1967-68, Purdue U., West Lafayette, Ind., 1968-69; research asso. Baylor Coll. Medicine, Houston, 1970, U. Calif.-Berkeley, 1970-73; virologist Litton Bionetics, Inc., Bethesda, Md., 1973-75; microbiologist U.S. Dept. Agr., Beltsville, Md., 1975—. Contbr. articles in field to profl. jours. Mem. AAAS, Am. Phytopath. Soc., Am. Soc. Virology, N.Y. Acad. Sci., Sigma Xi. Subspecialties: Plant virology; Molecular biology. Current work: Researcher in transcription and replication of viruses and viroids. Office: US Dept Agriculture Plant Virology Lab Beltsville MD 20705

HADLER, HERBERT ISAAC, biochemist, educator; b. Toronto, Ont., Aug. 22, 1920; s. Moses and Annie (Rosenberg) H.; m. Miriam Celia Perenson, Feb. 23, 1947; children: Laurie Coppe, Mitchell Reuben, Kenneth. B.A.Sc., U. Toronto, 1942; Ph.D., U. Wis., 1952. Postdoctoral fellow McArdle Lab., U. Wis.-Madison, 1953-60; asst. prof. Enzyme Inst., 1960-66; assoc. prof. div. oncology Chgo. Med. Sch.; prof. So. Ill. U., Carbondale, 1966—. Contbr. articles to profl. jours. Mem. Am. Soc. Biol. Chemists, Am. Assn. Cancer Research., Am. Coll. Toxicology, Sigma Xi. Subspecialties: Biochemistry (biology); Cosmic ray high energy astrophysics. Current work: Oxidative phosphorylation, mitochondrial genes. Office: Dept Chemistry and Biochemistry So Ill U Carbondale IL 62901 Home: 302 Wedgewood Ln Carbondale IL 62901

HADLER, NORTIN M., physician; b. N.Y.C., Nov. 13, 1942; s. Morris H. and Lucille (Hochberg) H.; m. Carol S. Spiegel, June 20, 1965; children: Jeffrey, Elana. A.B., Yale U., 1964; M.D., Harvard U., 1969. Diplomate: Am. Bd. Internal Medicine with subspltys. in rheumatology, allergy and immunology. Med. resident Mass. Gen. Hosp., Boston, 1968-70; clin. assoc. ARB-NIAMDD, NIH, Bethesda, Md., 1970-72; resident, rheumatology fellow Mass. Gen. Hosp.-Harvard U., 1972-73; guest scientist Clin. Research Ctr., Harrow, Eng., 1973-74, 79-80; asst. prof. medicine and bacteriology U. N.C., 1974-78, assoc. prof., 1978—. Contbr. articles to profl. jours. Served with USPHS, 1970-72. Investigator Am. Heart Assn., 1976-81. Fellow ACP; mem. Am. Rheumatism Assn., Am. Assn. Immunologists, Am. Soc. Clin. Investigation. Subspecialties: Internal medicine; Rheumatology. Current work: Biology of joints; industrial rheumatology. Office: U NC 932 Floor 231H Dept Medicine Chapel Hill NC 27514

HADLEY, FRED JUDSON, chemistry educator; b. Kansas City, Mo., July 18, 1946; s. Hugh Gordon and Freda May (Brooks) H.; m. Kathleen Ann Hamilton, Mar. 23, 1968 (div.); m. Maxine Seip Parks, Dec. 29, 1982. B.A., U. Kans., 1968; postgrad., Tex. A&I U., 1970-73; Ph.D., Rice U., 1977. Vol. Peace Corps, 1968-70; mem. faculty U. Ill.-Champaign, 1976-78; asst. prof. chemistry Wabash Coll., 1978—. Author: (with N. Grallick) General Chemistry Experiments, 1978. Mem. Community Chorus. Mem. Am. Chem. Soc., Sigma Xi. Democrat. Subspecialties: Physical chemistry; Kinetics. Current work: Kinetics of permanganate oxidations in non-aqueous media. Home: RR 1 Traction Rd Crawfordsville IN 47933 Office: Dept Chemistry Wabash Coll Crawfordsville IN 47933

HADLOCK, DANIEL C., oncologist, medical administrator, educator; b. Wilmington, Del., June 5, 1941; s. Canfield and Josephine (Cook) H.; children: Timothy, Amy, Joel, Karen. B.A. in English, Dartmouth Coll., 1962, B.M.S., 1964; M.D., U. Pa., 1966; M.S., U. Minn., 1975. Diplomate: Am. Bd. Internal Medicine (medical oncology). Intern Johns Hopkins Hosp., Balt., 1966-67, resident in internal medicine, 1967-70; practice medicine specializing in oncology, Orlando, Fla., 1975-80; med. dir. Hospice Orlando, Fla., 1976-80, Riverside Hospice, Boonton, N.J., 1980-81; med. dir./prof. edn. Hospice Inc., Miami/Ft. Lauderdale, Fla., 1981, med. dir., 1981-83; asst. clin. prof. oncology U. Miami, Fla., 1982—. Contr. chpt. to book, articles to profl. jours. Served as maj. U.S. Army, 1970-72. Am. Cancer Soc. fellow, 1974-75; recipient cert. merit Madigan Gen. Hosp., 1972; Hospice Pioneer plaque Fla. State Hospice Orgn., 1980. Fellow ACP; mem. Am. Soc. Clin. Oncology, AMA, Am. Geriatric Soc., N.Y. Acad. Sci., Nat. Hospice Orgn. (v.p. 1979-80, pres. 1980-81). Republican. Subspecialties: Internal medicine; Oncology. Current work: Palliative care. Office: Hospice Inc 111 NW 10th Ave Miami FL 33128

HAEBERLE, FREDERICK ROLAND, oil company scientist; b. Phila., Oct. 6, 1919; s. Frederick Edward and Faye Vivian (Davis) H.; m. Cynthia Lee Davis, Feb. 22, 1946; children: Cynthia Faye, Frederick Edward. B.S., Yale U., 1947, M.S, 1948; M.B.A., Columbia U., 1962. Geologist Standard Oil Calif., Houston, 1948-52; chief geologist J. J. Lynn Oil Div., Abilene, Tex., 1952-53; div. mgr. Mayfair Minerals, Abilene, 1953-54; cons. geologist, Abilene, 1954-57; chief subsurface geologist Atlantic Refining Co., Caracas, Venezuela, 1957-60; geol. specialist Mobil Oil Corp., Dallas, 1962-83; cons. geologist, Dallas, 1983—; asst. prof. U. Houston, 1948-50; prof. McMurray Coll., Abilene, Tex., 1954-57. Contbr. articles to profl. jours. Served to 1st lt U.S. Army, 1941-46; PTO. Fellow Geol. Soc. Am.; mem. Am. Assn. Petroleum Geologists, Soc. Profl. Well Log Analysts, Assn. Profl. Geol. Scientists, Tex. Acad. Sci. Republican. Presbyterian. Club: Brook Haven Country (Dallas). Subspecialties: Geology; Information systems (information science). Current work: Statistical studies of exploration drilling activity and reserves recovered, applications of computers to geological work. Office: Mobil Oil Corp Box 900 Dallas TX 75221

HAENSEL, VLADIMIR, energy co. exec.; b. Freiburg, Germany, Sept. 1, 1914; came to U.S., 1930; s. Paul and Nina (Tugenhold) H.; m. Mary Magraw, Aug. 28, 1939; children—Mary Ann (Mrs. Michael J. Ahlen), Katherine (Mrs. C.K. Webster). B.S., Northwestern U., 1935, Ph.D., 1941, D.Sc., 1957; M.S., Mass. Inst. Tech., 1939; D.Sc. (hon.), U. Wis., Milw., 1979. Research chemist Universal Oil Products Co. (name changed to UOP Inc.), Des Plaines, Ill., 1937-64, v.p., dir. research, 1964-72, v.p. sci. and tech., 1972-79, cons., 1979—; prof. U. Mass., 1979—. Contbg. author several sci. books.; Contbr. numerous articles to profl. jours. Recipient award Chgo. Jr. C. of C., 1944, Precision Sci. Co., 1952, Profl. Progress award Am. Inst. Chem. Engrs., 1957, Indsl. and Engring. Chemistry award Esso Research & Engring. Co., 1965, Modern Pioneers in Creative Industry award N.A.M., 1965, Perkin medal, 1967, Nat. medal Sci., 1973. Mem. Nat. Acad. Scis., Nat. Acad. Engring., Am. Chem. Soc., Indsl. Research Inst., Catalysis Soc., Sigma Xi, Phi Lambda Upsilon, Tau Beta Pi. Club: Hinsdale (Ill.) Golf. Subspecialties: Catalysis chemistry; Chemical engineering. Current work: Energy conversion processes and heterogeneous catalysis. Patentee in field. Home: 924 Oakwood Terr Hinsdale IL 60521 Office: 10 UOP Plaza Des Plaines IL 60016

HAFER, MARILYN DURHAM, psychology educator; b. Guthrie, Okla., Feb. 10, 1924; d. Walker and Elizabeth (Gooch) Durham. B.A., Tex. Woman's U., 1966; Ph.D., Tex. Tech. U., 1971. Registered psychologist, Ill. Asst. prof. dept. psychology Ill. Inst. Tech., Chgo., 1971-77, Adminstrn. Ctr., DePaul U., 1977-78; psychologist U.S. Civil Service Commn., Chgo., 1977-79; assoc. prof. Rehab. Inst., So. Ill. U., Carbondale, Ill., 1979—; cons. Commn. on Rehab. Counselor Certification, 1981—; chmn. legis. com. Nat. Rehab. Adminstrn. Assn., 1979-80. Contbr. articles to profl. jours.; cons. editor: Research & The Retarded, 1978-79, Jour. Rehab. Adminstrn, 1978, 81—; editorial adv. bd.: Vocat. Evaluation and Work Adjustment Bull, 1980—. NSF fellow, 1966-70. Mem. Am. Psychol. Assn., Nat. Rehab. Assn., Nat. Rehab. Adminstrn. Assn. Democrat. Subspecialties: Rehabilitation; Psychometrics, test construction, personnel selection. Current work: Research design and statistics, tech. writing, psychometrics, burnout in rehab. personnel; personnel selection. Home: 2021 A Woodriver Dr Carbondale IL 62901 Office: Rehab Inst So Ill U Carbondale IL 62901

HAGAN, RAYMOND DONALD, exercise physiologist, researcher; b. Los Angeles, Mar. 14, 1943; s. Raymond Thomas and Muriel Ann (Guth) H.; m. Margaret Riley, May 23, 1981. Student, Brigham Young U., 1961-62; B.A., Calif. State U.-Northridge, 1966; M.A., U. Calif.-Santa Barbara, 1968; Ph.D., U. Oreg., 1975. Tchr. biology James Madison Jr. High Sch., North Hollywood, Calif., 1969-71; NIH postdoctoral fellow Inst. Environ. Stess, Santa Barbara, Calif., 1975-77; asst. dir. exercise physiology Inst. Aerobics Research, Dallas, 1977-81; adj. prof. exercise physiology Tex. Woman's U., Denton, 1981—; dir. exercise physiology Inst. Aerobics Research, Dallas, 1981—; sci. adv. Hydra-Fitness Industries, Belton, Tex., 1981—. Contbr. articles in field to profl. jours. Fellow Am. Coll. Sports Medicine; mem. Am. Physiol. Soc., AAAS. Democrat. Roman Catholic. Lodge: Rotary. Subspecialties: Physiology (medicine); Preventive medicine. Current work: Human studies of physiological response and adaptation to aerobic, anaerobic, and muscular strength exercises and the relation of these exercises to preventive medicine. Office: Inst Aerobics Research 12200 Preston Rd Dallas TX 75230

HAGEDORN, DONALD JAMES, phytopathologist, educator, agrl. cons.; b. Moscow, Idaho, May 18, 1919; s. Frederick William and Elizabeth Viola (Scheyer) H.; m. Eloise Tierney, July 18, 1943; 1 son, James William. B.S., U. Idaho, 1941, D.Sc. (hon.), 1979; M.S., U. Wis., 1943, Ph.D., 1948. Prof. agronomy and plant pathology U. Wis.-Madison, 1948-64, prof. plant pathology, 1964—; courtesy prof. plant pathology Oreg. State U., Corvallis, 1972-73; vis. scientist DSIR Lincoln Research Center, Christchurch, N.Z., 1980-81. Contbr. chpts. to books, articles to profl. jours. Served with USAAF, 1943-46. Recipient Campbell award AAAS, 1961; CIBA-Geigy award, 1974; Meritorious Service award Nat. Pea Improvement Assn., 1979, Bean Improvement Coop., 1979; Forty-Niners award, 1983; NSF sr. postdoctoral fellow, 1957. Fellow Am. Phytopath. Soc.; mem. Sigma Xi, Gamma Sigma Delta, Alpha Zeta. Methodist. Lodge: Kiwanis. Subspecialties: Plant pathology; Plant genetics. Current work: Diseases of peas and beans; breeding peas and beans for multiple disease resistance. Home: 927 University Bay Dr Madison WI 53705 Office: U Wis 1630 Linden Dr 583 Russell Labs Madison WI 53706

HAGEL, ANDREW RICHARD, computer scientist; b. N.Y.C., Sept. 14, 1949; s. Morris and Florence Lyla (Brodsky) H. M.S. in Computer Sci, SUNY-Albany, 1979. Systems analyst Sperry Univac, Albany, 1979-81; telecommunications analyst NBC, N.Y.C., 1981; sr. software engr. Wang Labs., Inc., Lowell, Mass., 1981—. Served with U.S. Army, 1972-75. Mem. Assn. for Computing Machinery, Soc. Indsl. and Applied Math., IEEE. Subspecialties: Information systems, storage, and retrieval (computer science); Software engineering. Current work: Office automation. Home: 27 Boylston Ln Lowell MA 01852 Office: 1 Industrial M/S 1393 Lowell MA 01861

HAGEMAN, GILBERT ROBERT, physiologist, educator; b. Covington, Ky., May 21, 1947; s. Gilbert Charles and Dorothy (Schmidt) M.; m. Mary Eileen Stewart, June 7, 1968; children: Gilbert Leo, Anthony Edward, Christopher R., Matthew. A.B., Thomas More Coll., 1968; Ph.D., Loyola U., Chgo., 1975. Research assoc. Cardiovascular Research and Tng. Ctr., U. Ala.-Birmingham, 1974—, instr. in medicine, 1977—, asst. prof. physiology, 1977-82, assoc. prof. physiology, 1982—. Contbr. articles on neural regulation of circulation to profl. jours. Served with U.S. Army, 1969-71; Thailand. Established investigator Am. Heart Assn., 1980-85. Fellow Circulation Soc. of Am. Heart Assn., Cardiovascular Sect. of Am. Physiol Soc.; mem. Ala. Acad. Sci., Sigma Xi (pres. U. Ala.-Birmingham chpt. 1983-84). Democrat. Roman Catholic. Subspecialties: Physiology (biology); Cardiology. Office: U Ala Z311 Birmingham AL 35294

HAGEMAN, LOUIS ALFRED, mathematician; b. Danville, Ill., Oct. 8, 1932; s. Louis W. and Ann E. (Larson) H.; m. Jane C. Pohelia, Sept. 21, 1959 (div. 1970); 1 dau., Heidi; m. Marilyn C. Cupps, June 26, 1971. B.A., DePauw U., 1955; B.S., Rose-Hulman Inst. Tech., 1955; Ph.D., U. Pitts., 1962. With Westinghouse Electric, West Mifflin, Pa., 1955—, jr. engr., 1955-57, assoc. mathematician, 1957-60, mathematician, 1960-62, sr. mathematician, 1962-67, fellow mathematician, 1967-72, adv. mathematician, 1972—; sr. lectr. Carnegie-Mellon U., Pitts., 1973-76. Author: Applied Iterative Methods, 1981; contbr. articles to jours. Mem. Soc. Indsl. and Applied Math., Tau Beta Pi. Subspecialties: Numerical analysis; Mathematical software. Current work: Numerical solution of partial differential equations; iterative methods for solving large sparse matrix problems. Home: 1628 Citation Dr Library PA 15129 Office: Bettis Atomic Power Lab PO Box 79 West Mifflin PA 15122

HAGEMAN, RICHARD HARRY, crop physiology educator, researcher; b. Powell, Wyo., Apr. 14, 1917; s. Frank Roy and Creda Delema (Wright) H.; m. Margaret Elizabeth Catlett, Aug. 14, 1941; children: James Howard, Janet Ann, Peggy Diane. B.S., Kans. State U., 1938; M.S., Okla. State U., 1940; Ph.D., U. Calif.-Berkeley, 1954. Asst. chemist U. Ky., Lexington, 1940-41, 46-47; chemist U.S. Dept. Agr., Mayaguez, P.R., 1947-50; asst. prof. U. Ill.-Urbana, 1954-57, assoc. prof., 1957-61, prof. crop physiology, 1961—; vis. prof. Mich. State U., 1967-68. Contbr. chpts. to books, articles to profl. publs. Served to capt. U.S. Army, 1941-46. AEC predoctoral fellow, 1950-53; Rockefeller Found. fellow, Long Ashton, Eng., 1960-61; recipient Funk Found. research award U. Ill., 1974; sr. Fulbright research fellow, Melbourne, Australia, 1975-76; one of six plant scientists listed among 1000 scientists most cited by peers during period 1965-78 Current Contents, 1981. Fellow Am. Soc. Agronomy (Crop Sci. Achievement award 1967, assoc. editor. 1965-68, 68-70); mem. Am. Soc. Plant Physiology (editorial bd. 1969-70, 74—, life mem.), Plant Biochem. Soc. India (editorial bd. 1973—), Am. Soc. for Exptl. Biology, Mil. Order of World Wars, Phi Kappa Phi. Republican. Baptist. Subspecialties: Plant physiology (agriculture); Plant genetics. Current work: Use of biochemical traits as selection criteria for enhancing productivity and improvement of nitrogen use efficiency by corn. Interaction of carbon and nitrogen metabolism as related to productivity of crop plants. Home: 1302 E McHenry Urbana IL 61801 Office: Agronomy Dep U Ill 1102 S Goodwin Urban IL 61801

HAGEMARK, KJELL INGVAR, laboratory administrator; b. Mysen, Norway, Oct. 2., 1934; came to U.S., 1965; s. Karl Ingvar and Marie Elisabeth (Tangvik) H.; m. Unni Hoff, Aug. 26, 1961; children: Bent, Petter, Erik. B.S., Tech. U. Norway, 1960, Ph.D., 1965. Research assoc. Tech. U. Norway, 1961-65; postdoctoral fellow N. Am. Aviation Sci. Ctr., Thousand Oaks, Calif., 1965-67; sr. physicist 3M Central Research Lab., St. Paul, 1967-73, supr., mgr., 1973-79, lab. mgr. optical recording project, 1979—. Chmn. Parent Adv. com. High Sch., WhiteBear Lake, Minn., 1977; mem. council Norwegian Lutheran Meml. Ch., Mpls., 1980—. Mem. Am. Phys. Soc., Norwegian Phys. Soc., Norwegian Chem. Soc., Norwegian Am. Tech. Soc. (v.p. 1981-82). Club: Torskeklubben (Mpls.). Subspecialties: Information systems, storage, and retrieval (computer science); Condensed matter physics. Current work: Lases recording, optical disc, video disc, audio disc, replication, solid state physics, thermodynamics. Patentee efficient luminescent ZnO. Home: 2568 Oak Dr White Bear Lake MN 55110 Office: 3M Opitcal Recording Project 3M Ctr 260-4A-8 Saint Paul MN 55144

HAGEN, ARNULF PEDER, chemist, educator; b. Tacoma, Wash., June 6, 1942; s. Carl A. and Marjorie W. (Black) H. B.S., U. Wash., 1964; Ph.D., U. Pa., 1968. Assoc. prof. U. Okla., 1967-83, prof., 1983—. Contbr. articles to profl. jours. Mem. Am. Chem. Soc., Assn. Asphalt Paving Technologists. Subspecialties: Inorganic chemistry; Coal. Current work: Chemical interactions which will lead to an in-situ mining of coal, reactions at high pressures, asphalt technology. Office: Dept Chemistry U Okla Norman OK 73019

HAGEN, ARTHUR AINSWORTH, pharmcologist; b. Hot Springs, S.D., Oct. 9, 1933; s. Arthur and Gussie (Ainsworth) H.; m. Laurin J. Kirley, June 1, 1957; children: Kristen, Karol, Sandr, Sharon. Student, Cornell Coll., Iowa, 1951-52; B.A., U. S.D., 1955, M.A., 1957; Ph.D., U. Tenn., 1961. Postdoctoral trainee U. Utah, Salt Lake Cty, 1961-63; Swedish Med. Research Council fellow Karolinska Hosp., Stockholm, 1963-64; mem. faculty U. Tenn. Center Health Scis., Memphis, 1965—, prof. pharmacology, 1980—. Mem. Am. Soc. Exptl. Pharmacology and Therapeutics, Endocrine Soc., Soc. Study of Reprodn. Methodist. Subspecialties: Pharmacology; Endocrinology. Current work: Prostaglandins and reprodn.; prostaglandins and cerebrovasospasm.

HAGEN, CHARLES ALFRED, microbiologist; b. East Rutherford, N.J., Feb. 1, 1925; s. Charles Alfred and Lina Dorothea (Scharch) H.; m. Alice Diana Wiltse, Dec. 14, 1951; children: Erich Christoph, Kristine Ann, Susan Lynn. A.B., U. Chgo., 1952, M.S., 1956. Microbiologist U. Chgo., 1954-55; sr. med. technologist Inst. Tb Research, Chgo., 1955-56; research microbiologist Kraftco, Glenview, Ill., 1956-62; microbiologist Ill. Inst. Tech., Chgo., 1962-69; lab. mgr. Avco Jet Propulsion Lab., Pasadena, Calif., 1967-72, Bionetics Corp. Jet Propulsion Lab., Pasadena, 1972-76; chief bacteriologist Becton Dickinson Lab., Oxnard, Calif., 1976-79, regulatory affairs officer, 1979—. Contbr. articles to profl. jours. Served with USN, 1943-46. Mem. AAAS, Am. Assn. Lab. Animal Sci., Am. Soc. Microbiology, Soc. Indsl. Microbiology, Am. Soc. Quality Control. Republican. Lutheran. Subspecialty: Microbiology. Current work: Assure compliance with state, federal and corporate medical device manufacturing regulations; sterilization procedures-ethylene oxide and radiation; clean room techniques and contamination control procedures. Home: 2085 Lyndhurst Ave Camarillo CA 93010 Office: 1950 Williams Dr Oxnard CA 93010

HAGEN, DANIEL RUSSELL, reproductive physiologist; b. Springfield, Ill., Sept. 29, 1952; s. Robert W. and Russella (Lane) H.; m. Rosemary Simonetta, Mar. 25, 1978; children: Matthew, Mark, Lane. B.S., U. Ill., 1974, Ph.D., 1978. Research assoc. Cornell U., 1978; asst. prof. Pa. State U., 1978—. Mem. Am. Soc. Animal Soc., Soc. Study Reprodn., Soc. Study Fertility, AAAS, Sigma Xi, Phi Kappa Phi, Alpha Zeta. Subspecialties: Reproductive biology; Animal physiology. Current work: Fetal/maternal relationships, fertility testing. Office: 304 WL Henning Bldg Pa State U University Park PA 16802

HAGER, EUGENE RANDOLPH, engineering executive; b. Omaha, Aug. 3, 1930; s. Eugene Hayes and Annabella F. (Kise) (Bowlin)) H.; m. Pauline Papacalos; children: Christopher Randolph, Barry Eugene. B.S. in Mech. Engring, San Diego State U., 1968, M.S., 1971. Registered profl. engr., S.C., Calif. Asst. engr., design supr. Gen. Atomic Co. (name now GA Techs. Inc.), San Diego, 1954-68, staff assoc., 1968-70, mgr. engring. support fusion, 1975-79, mgr. mech. engring. fusion, 1979—; nuclear mech. engr., tech. mgr. Allied Gen. Nuclear Service, Barnwell, S.C., 1970-75. Contbr. articles to profl. jours. Chrm. program com. La Jolla Profl. Men's Soc., 1983; mem. Eclectics, 1979—. Mem. Am. Nuclear Soc. (mem. fusion subcom. 1979, chmn. remote systems tech. div. tech. session 1978-79), ASME. Republican. Methodist. Subspecialties: Nuclear engineering; Remote handling. Current work: Computer programming. Home: 5776 Desert View Dr La Jolla CA 92037 Office: GA Techs Inc PO Box 81608 San Diego CA 92138

HAGER, LOWELL PAUL, educator, biochemist; b. Girard, Kans., Aug. 30, 1926; s. Paul William and Christine (Selle) H.; m. Frances Erea, Jan. 22, 1949; children—Paul, Steven, JoAnn. A.B., Valparaiso U., 1947; M.A., U. Kans., 1950; Ph.D., U. Ill., 1953. Postdoctoral fellow Mass. Gen. Hosp., 1953-55; asst. prof. biochemistry Harvard, 1955-60; Guggenheim fellow Oxford (Eng.) U., also Max Planck Inst. Zellchemie, 1959-60; mem. faculty U. Ill. at Urbana, 1960—, prof. biochemistry, 1965—, head biochem. div., 1967—; Chmn. physiol. chemistry study sect. NIH, 1965—; vis. scientist Imperial Cancer Research Fund, 1974; cons. NSF, 1976. Editor: life scis. Archives Biochemistry and Biophysics, 1966—; asso. editor: Biochemistry, 1973—; mem. editorial bd.: Jour. Biol. Chemistry, 1974—. Served with USAAF, 1945. Mem. Am. Chem. Soc., Am. Soc. Biol. Chemists, Am. Soc. Microbiology (chmn. physiology div. 1967). Subspecialty: Biochemistry (biology). Research enzyme mechanisms, intermediary metabolism, tumor virus. Home: 801 W Delaware St Urbana IL 61801

HAGER-RICH, JEAN CAROL, cancer research scientist; b. Avalon, Calif., Aug. 11, 1943; d. Herbert Frank and Marion Arlene (Hammer) Hager; m. Marvin A. Rich, Jan. 1, 1981. B.S., Bates Coll., Lewiston, Maine, 1965; M.S., U. Ill., Urbana, 1969, Ph.D., 1974. Research assoc. U. Ill., 1974-75; instr. Brown U., Providence, 1975-78, asst. prof., 1978-79; research assoc. Roger Williams Gen. Hosp., Providence, 1975-79; scientist Mich. Cancer Found., Detroit, 1979-80, asst. mem., 1981; asst. prof. pathology Wayne State U., 1979-81; scientist AMC Cancer Research Center, Lakewood, Colo., 1981—. Contbr. chpts. to books and articles in field to profl. jours. Mem. Internat. Assn. Breast Cancer Research (asst. sec. gen. 1980—), AAUW, Am. Assn. Cancer Research, Tissue Culture Assn., Electron Microscopy Soc. Am. Subspecialties: Cancer research (medicine); Immunology (medicine). Current work: Cancer cell biology, metastatic process in breast cancer, role of immunological mechanisms in breast cancer and leukemia. Home: 1733 S Sand Lily Dr Golden Co 80401 Office: 6401 W Colfax Ave Lakewood CO 80214

HAGGAG, FAHMY MAHMOUD, nuclear engineer, researcher; b. Alexandria, Egypt, Jan. 1, 1948; came to U.S., 1979; s. Mahmoud Mohamed Haggag and Farida (Abd El-Aziz) El-Basyouni; m. Rabea Derdar, Dec. 26, 1977 (div. May 1979); 1 son, Mohammed-Nazim; m. Hassnaa Abd El-Kader, Aug. 4, 1981; 1 dau., Mona Fahmy. B.S. in Nuclear Engring, U. Alexandria, 1970, M.S., 1976, M.S., U. Calif.-Santa Barbara, 1980. Mil. engr. Egyptian Air Force, 1970-72; research engr. AEC, Cairo, 1972-76; instr. U. Algiers, Algeria, 1976-79; research asst. U. Calif.-Santa Barbara, 1979-80; sr. engr. EG&G Idaho, Inc., Idaho Falls, 1980—; cons. Centre Des Sciences Et La Technologie Nucleaires, Algiers, 1977-79. Mem. Egyptian Gymnastics Olympic Team, Egyptian Gymnastics Fedn., 1965-72; coach YMCA, Idaho Falls, 1981-82. Mem. Am. Nuclear Soc. (Best Contributed Paper in Materials Sci. for 1981), Am. Soc. Metals. Subspecialties: Materials (engineering); Nuclear engineering. Current work: Nuclear fuel behavior and manufacturing, materials engineering, and nuclear safety analysis, including severe reactor core accidents. Inventor semi-pilot prodn. of uranium metal, 1972-76. Home: 724 Saturn Ave Apt 8 Idaho Falls ID 83042 Office: EG&G Idaho Inc PO Box 1625 Idaho Falls ID 83415

HAGGERTY, ROBERT JOHNS, physician, educator; b. Saranac Lake, N.Y., Oct. 20, 1925; s. Gordon Abbott and Nina (Johns) H.; m. Muriel Ethel Protzmann, Oct. 29, 1949; children: Robert, Janet, Richard, John. A.B., Cornell U., 1946, M.D., 1949; A.M. (hon.), Harvard U., 1975. Intern Strong Meml. Hosp., Rochester, N.Y., 1949-51; from resident to chief resident pediatrics Children's Hosp. Med. Center, Boston, 1953-55; med. dir. family health care program Harvard Med. Sch., also asst. prof. pediatrics, 1953-64; prof. pediatrics, chmn. dept. U. Rochester Sch. Medicine, 1964-75; Roger I. Lee prof. health services, chmn. dept. health services Harvard Sch. Pub. Health, 1975-78; prof. pediatrics Harvard Med. Sch., Boston, 1975-78, clin. prof., 1978-80; pres. Wm. T. Grant Fedn., N.Y.C., 1980—; clin. prof. pediatrics Cornell U. Med. Sch., N.Y.C., 1980—; dir. gen. pediatrics acad. devel. program Robert Wood Johnson Found., 1978—; mem. health services research sect. USPHS, 1964-70, 82—, chmn., 1968-70, 82—; mem. N.Y. State Health Planning Adv. Council, Carnegie Council on Children, 1972-77; chmn. panel health scis. research, com. on nat. needs for biomed. and behavioral research personnel NRC, 1975-78; Mem. bd. U.S. Com. on UNICEF, 1981—. Editor: (with M. Green) Ambulatory Pediatrics, 1968, 2d edit., 1977, 3d edit., 1983; Co-editor: (with J. Lucey) Pediatrics, 1973-80; editor-in-chief: Pediatrics in Rev, 1978—; contbr. articles to med. jours. Mem. U.S. com. UNICEF. Served to capt. USAF, 1951-53. Recipient Martha M. Eliot award Am. Public Health Assn., 1976; Markle scholar acad. medicine, 1962-67; fellow Center for Advanced Study Behavioral Scis., Stanford, Calif., 1974-75. Mem. Assn. Med. Sch. Pediatric Dept. Chairmen (pres. 1969-70), Am. Assn. Poison Control Centers (pres. 1962-64), Am. Acad. Pediatrics (Grulee award 1981, v.p., pres.-elect 1983-85), Am. Pediatric Soc., Ambulatory Pediatric Assn. (chmn. 1963-64), Assn. Am. Med. Colls., Internat. Epidemiological Assn., Soc. Pediatric Research (v.p. 1970-71), Inst. Medicine (council 1974-77, chmn. steering com. nat. study quality assurance programs 1975-76), Phi Beta Kappa, Alpha Omega Alpha. Subspecialty: Pediatrics. Current work: Behavioral pediatrics; stress and coping.

HAGIWARA, SUSUMU, physiologist; b. Hokkaido, Japan, Nov. 6, 1922; came to U.S., 1960, naturalized, 1972; s. Ichiro and Isao (Sato) H.; m. Satoko Ohara, May 30, 1953; 1 son, Kazunari. M.D., Tokyo U., 1946, Ph.D. in Physiology, 1951. Intern Tokyo Univ. Hosp., 1946-47; asso. prof. physiology Tokyo Med. and Dental U., 1950-59, prof., chmn. dept. physiology, 1959-62; research zoologist UCLA, 1961-65, prof., 1969—, E. I. Leslie prof. neurosci., 1978—; prof. U. Calif., San Diego, 1965-69. Recipient Kenneth Cole award in biophysics Am. Biophys. Soc., 1976. Fellow AAAS; mem. Nat. Acad. Sci., Am. Acad. Art and Sci., Am. Physiol. Soc., Am. Soc. Zoologists, Internat. Brain Research Orgn., Soc. Neuroscis., Biophys. Soc., Soc. Cell Biology. Subspecialty: Neurophysiology.

HAGLUND, WILLIAM ARTHUR, plant pathology educator, agricultural consultant; b. Mpls., May 3, 1930; s. Knute Martin and Corinne A. (Carlson) H.; m. Jean Christine Rallis, June 21, 1952; children: Michael Martin, Lynne Ann. B.S., U. Minn., 1953, M.S., 1958, Ph.D., 1960. With Minn. Mining and Mfg. Co., Mpls., 1955; mem. faculty Wash. State U., Mt. Vernon, 1960—, prof. plant pathology, 1971—; pres./cons. Cascade Agrl. Service Co., 1963—. Contbr. articles to sci. jours. Bd. dirs. Mt. Vernon Sch.; mem. comml. panel Am. Arbitration Assn. Served in U.S. Army, 1953-55. Mem. Am. Phytopath. Soc., Am. Soc. Nematologists, Pea Improvement Assn., Bean Improvement Assn., Am. Arbitration Assn. Lutheran. Club: Skagit Golf. Lodges: Elks; Lions; Masons. Subspecialties: Plant pathology; Plant genetics. Current work: Control of soil-borne diseases of agricultural crops, chemical and genetic; varietal development of peas and beans. Patentee plant variety devel. Home: 110 Claremont Pl Mount Vernon WA 98273 Office: 1468 Memorial Hwy Mount Vernon WA 98273

HAGMANN, DEAN BERRY, engineering company executive; b. Cleve., Sept., 24, 1934; s. Vern and Elizabeth (Berry) H.; m.; children: DeAnn, Lynda, Teresa. B.S.M.E., U. Wyo., 1959; postgrad., U. Idaho, 1959-70. Engr. Phillips Petroleum Co., Idaho Falls, Idaho, 1959-63; supr. Idaho Nuclear Corp., Idaho Falls, 1963-69; project engr. Argonne Nat. Lab., Idaho Falls, 1969-75, assoc. dir., 1976-82; v.p. Remotec, Oak Ridge, Tenn., 1982-83; staff engr. GA Technologies, San Diego, 1983—. Mem. Am. Nuclear Soc. (bd. dirs. 1978—). Club: Ski (Knoxville, Tenn.). Subspecialties: Mechanical engineering; Nuclear fission. Current work: Application of remote technology-radioactive materials handling and remote maintenance applied to both fission and fusion plants. Home: 2714 Caminito Verdugo San Diego CA 92014 Office: GA Technologies Inc PO Box 85608 San Diego CA 92138

HAGNI, RICHARD DAVIS, geology educator; b. Howell, Mich., Apr. 29, 1931; s. Walter Davis and Merle (Whitaker) H.; m. Rachael Anne Stutzman, June 16, 1953; children: John, Sandra, Ann, David. Student, Hope Coll., 1949-51; B.S., Mich. State U., 1953, M.S., 1954; Ph.D., U. Mo.-Columbia, 1962. Asst. prof. geology U. Mo.-Rolla, 1960-66, assoc. prof., 1966-77, prof., 1977—. Editor: Process Mienralogy II, 1983. Served with U.S. Army, 1954-56. U.S. Bur. Mines grantee, 1974-81; U.S. Geol. Survey grantee, 1975-79. Fellow Geol. Soc. Am.; mem. Soc. Econ. Geology (life), Am. Inst. Profl. Geologists (pres. Mo. sect. 1981-82), AIME (chmn. procs. mineralogy com. 1981-82), Internat. Assn. on Genesis of Ore Deposits (Best Paper award, Tbilisi, USSR 1982), Internat. Geol. Congress. Methodist. Subspecialties: Mineralogy; Petrology. Current work: Reflected light, microscopy of ores and beneficiation products. Home: 27 Johnson Rolla MO 65401 Office: Dept Geology and Geophysics U Mo Rolla MO 65401

HAGSTRUM, HOMER DUPRE, physicist; b. St. Paul, Mar. 11, 1915; s. Andrew and Sadie Gertrude (Fryckberg) H.; m. Bonnie Doone Cairns, Aug. 29, 1948; children—Melissa Billings, Jonathan Tryon. B.E.E., U. Minn., 1935, B.S., 1936, M.S., 1939, Ph.D., 1940. Teaching and research asst. U. Minn., 1935-40; research physicist Bell Telephone Labs., Inc., Murray Hill, N.J., 1940—, head surface physics research dept., 1954-78; gen. chmn. Phys. Electronics Conf. Com., 1976-80. Fellow Am. Phys. Soc. (chmn. div. electron and atomic physics 1957, Davisson-Germer prize 1975); mem. Am. Vacuum Soc. (dir. 1976-78, Welch award 1974), Nat. Acad. Scis., Sigma Xi, Tau Beta Pi, Eta Kappa Nu. Subspecialties: Condensed matter physics; Surface chemistry. Current work: Surface physics and chemistry; interaction with surfaces of atoms that carry potential energy(ions; metastable atoms); surface electronic structure- surface sensitive electron spectroscopies. Author research articles, rev. chpts. mass spectrometry, microwave magnetrons, surface physics, electron spectroscopy, interaction of ions and metastable atoms with surfaces. Home: 30 Sweetbriar Rd Summit NJ 07901 Office: 600 Mountain Ave Murray Hill NJ 07974

HAHN, BEVRA H(ANNAHS), internist, rheumatologist, educator; b. Wheeling, W.Va., Dec. 9, 1939; m. Theodore J. Hahn; 2 children. B.S. summa cum laude, Ohio State U., 1960; M.D., Johns Hopkins U., 1964. Intern in medicine Washington U., Barnes Hosp., St. Louis, 1964-65, asst. resident in medicine, 1965-66; fellow in medicine div. connective tissue Johns Hopkins U., 1966-69; research assoc. VA, St. Louis, 1969-70; instr. in medicine and preventive medicine Washington U., 1969-71, asst. prof. medicine and preventive medicine, 1971-78, assoc. prof. medicine, 1978—; asst. physician Barnes Hosp., 1971-78, assoc. physician, 1978; cons. rheumatology Montebello State Hosp., Balt., 1967-69; clin. investigator St. Louis VA Hosp., 1970-73; bd. dirs. Eastern Mo. chpt. Arthritis Found., 1974, mem. med. and sci. com., chmn. med. and sci. adv. com., 1977; mem. arthritis ctr. rev. com. Nat. Arthritis Found., 1979; mem. adv. bd. Cooperating Clinics for Systematic Studies in Rheumatic Diseases, 1979; mem. rheumatology exam. sect. Am. Bd. Internal Medicine, 1980; mem. immunolo. scis. study sect. NIH, 1981; mem. Nat. Arthritis Adv. Bd., 1981, Mo. Arthritis Adv. Bd., 1982. Contbr. chpts., numerous articles to profl. publs. Elected Tchr. of Yr. med. housestaff Barnes Hosp., 1979. Mem. Am. Rheumatism Assn. (Nat. Lupus Council 1976—, chmn. Nat. Lupus Council 1981-82, pres. Central region 1982), St. Louis Rheumatism Soc. (pres. 1974), St. Louis Internal Medicine Soc., Am. Fedn. Clin. Research, Central Soc. Clin. Research (chmn. Rheumatology sect. 1982), Phi Beta Kappa, Alpha Omega Alpha. Subspecialties: Immunopharmacology; Rheumatology. Office: Washington U Med Sch Dept Rheumatology Saint Louis MO 63110

HAHN, ERIC WALTER, scientist, radiation biologist, cons., educator; b. N.Y.C., June 1, 1932; s. Eric G. and Sarah F. (Steele) H.; m. Janet M. Krobak, Mar. 3, 1956; children: Eric C., Corina K. Hahn Mitchell, Charles G. B.S., U. Ga., 1954, M.S., 1957; Ph.D., U. Ill., 1960. Instr. U. Rochester, N.Y., 1960-64, asst. prof., 1964-68, U. Minn., Mpls., 1968-69; assoc. mem. Sloan Kettering Inst., N.Y.C., 1969-80; attending radiation biologist Meml. Sloan Kettering Cancer Ctr., 1979—; prof. radiotherapy Mt. Sinai Med. Sch., N.Y.C., 1980—; cons. USPHS Hosp., S.I., N.Y., 1977-82, Nassau County (N.Y.) Med. Ctr., 1979—. Chmn. Boy Scouts Am. Served with U.S. Army, 1952-54. Mem. Am. Physiol. Soc., Endocrine Soc., Radiation Research Soc., Am. Soc. Therapeutic radiology, N.Y. Acad. Scis. Republican. Lutheran. Subspecialties: Radiation biology; Radiology. Current work: Radiation biology and hyperthermia for cancer treatment. Basic and applied radiation biology research combined with hyperthermia for use in the treatment of cancer. Home: 826 Eastfield Rd Westbury NY 11598 Office: Mt Sinai Med Center New York City NY 10029

HAHN, ERWIN LOUIS, physicist, educator; b. Sharon, Pa., June 9, 1921; s. Israel and Mary (Weiss) H.; m. Marian Ethel Failing, Apr. 8, 1944 (dec. Sept. 1978); children—David L., Deborah A., Katherine L.; m. Natalie Woodford Hodgson, Apr. 12, 1980. B.S., Juniata Coll., 1943, D.Sc., 1966; M.S., U. Ill., 1947, Ph.D., 1949; D.Sc., Purdue U., 1975. Asst. Purdue U., 1943-44; research asso. U. Ill., 1950; NRC fellow Stanford, 1950-51, instr., 1951-52; research physicist Watson IBM Lab., N.Y.C., 1952-55; asso. Columbia, 1952-55; faculty U. Calif. at Berkeley, 1955—, prof. physics, 1961—; asso. prof. Miller Inst. for Basic Research, Berkeley, 1958-59, prof., 1966-67; vis. fellow Brasenose Coll., Oxford (Eng.) U., 1981-82; cons. Office Naval Research, Stanford, 1950-52, AEC, 1955—; spl. cons. USN, 1959; adv. panel mem. Nat. Bur. Standards, Radio Standards div., 1961-64; mem. Nat. Acad. Sci./NRC com. on basic research; adv. to U.S. Army Research Office, 1967-69. Author: (with T.P. Das) Nuclear Quadrupole Resonance Spectroscopy, 1958. Served with USNR, 1944-46. Recipient Oliver E. Buckley prize Am. Phys. Soc., 1971; prize Internat. Soc. Magnetic Resonance, 1971; award Humboldt Found., Germany, 1976-77; Guggenheim fellow, 1961-62, 69-70; NSF fellow, 1961-62; vis. fellow Brasenose Coll., Oxford, 1969-70. Fellow Am. Phys. Soc. (past mem. exec. com. div. solid state physics); mem. Am. Acad. Arts and Scis., Nat. Acad. Scis., Slovenian Acad. Scis. and Arts (fgn.). Home: 69 Stevenson Ave Berkeley CA 94708 Office: Dept Physics U Calif Berkeley CA 94720

HAHN, FRED ERNST, chemist, scientific editor; b. Allenstein, Germany, Sept. 22, 1916; s. Georg Hans-Joachim and Ruth (Prusse) H.; m. Rosemarie Elisabeth Weyland, July 12, 1945; children: Patricia Monica, Michael Gregory, Christina Andrea. Diplomated chemist, U. Kiel, Ger., 1943, Ph.D., 1948. Head biochem. lab. Inst. Virus Research, U. Heidelberg, Germany, 1946-49; research biochemist dept. virus diseases Walter Reed Army Inst. Research, Washington, 1949-56, asst. chief dept. rickettsial diseases, 1956-59, chief dept. molecular biology, 1959-76, asst. to dir. chem. warfare def. research, 1976—. Editor sci. monograph collections.; Contbr. 200 articles to nat. and internat. profl. jours. Served with German Air Force, 1939-43. Recipient Meritorious Service award Dept of Army, 1960; Bronze Medallion Army Sci. Conf., 1972. Mem. Am. Soc. Microbiology, Am. Soc. Biol. Chemists, Am. Soc. Pharmacology and Exptl. Therapeutics. Club: Kenwood Golf and Country (Bethesda, Md.). Subspecialties: Biochemistry (biology); Molecular biology. Current work: Mechanisms of action of chemical warfare agents and antidotes, implmentation and conduct of multifaceted research program in medical defense against chemical warfare. Office: Walter Reed Army Inst of Research Washington DC 20012

HAHN, GEORGE LEROY, agrl. engr., biometeorologist; b. Muncie, Kans., Nov. 12, 1934; s. Vernon Leslie and Marguerite Alberta (Breeden) H.; m. Clovice Elaine Christensen, Dec. 3, 1955; children—Valerie, Cecile, Steven, Melanie. B.S., U. Mo., Columbia, 1957, Ph.D., 1971; M.S., U. Calif., Davis, 1961. Agrl. engr., project leader and tech. advisor Agrl. Research Service, U.S. Dept. Agr., Columbia, Mo., 1957, Davis, Calif., 1958-61, Columbia, 1961-78, Clay Center, Nebr., 1978—. Contbr. articles to tech. jours. and books on impact of climatic and other environ. factors on livestock prodn. and evaluation of methods of reducing impact. Recipient award Am. Soc. Agrl. Engrs.-Metal Bldgs. Mfrs. Assn., 1976. Mem. Am. Meteorol. Soc. (award for outstanding achievement in bioclimatology 1976), Am. Soc. Agrl. Engrs., Internat. Soc. Biometeorology, AAAS. Subspecialties: Agricultural engineering. Current work: Evalution of livestock responses(production; reproduction; efficiency; behavior, health)to environmental factors, primarily thermal factors; and methods of reducing their impact. Office: US Meat Animal Research Center PO Box 166 Clay Center NE 68933

HAHON, NICHOLAS, microbiologist, educator; b. N.Y.C., Mar. 24, 1924; s. Samuel Alexander and Catherine (Wolsak) H.; m. Katheryn E., Jan. 31, 1948; 1 dau., Nicolette K. Hahon-Granack. B.S., Davis and Elkins Coll., 1948; Sc.M. in Hygiene, Johns Hopkins U., 1950, Kellogg scholar, 1951. Supervisory virologist U.S. Army Labs., Frederick, Md., 1951-65, chief aerobiology br., 1965-71; research microbiologist Appalachian Labs. for Occupational Safety and Health, USPHS, Morgantown, W.Va., 1971—; Mem. U.S. Civil Service Exam. Bd. Microbiology, 1961-71; asst. prof. microbiology W.Va. U. Med. Sch., 1971-74. Contbr. numerous articles to profl. publs.; editor: Selected Papers On Virology, 1964, Selected Papers on Pathogenic Rickettsiae, 1968. Served with U.S. Army, 1943-46. Decorated Bronze Star; recipient cert. achievement U.S. Army, 1971, Alumni Disting. Service award David and Elkins Coll., 1980. Fellow Am. Acad. Microbiology, AAAS; mem. Am. Soc. Microbiology, Am. Soc. Virology, Am. Soc. Rickettsiology, N.Y. Acad. Sci., Tissue Culture Assn., Sigma Xi. Roman Catholic. Subspecialties: Microbiology (medicine); Virology (medicine). Current work: Aerobiology, immunology, and assay of viruses; role of interferon in occupational-related diseases; coal workers' penumoconiesis; asbestosis; metals. Office: 944 Chestnut Ridge Rd Morgantown WV 26505

HAIER, RICHARD JAY, research psychologist, hospital administrator, educator; b. Buffalo, Arp. 30, 1949; s. I. Dan and Dorothy (Nyman) H. B.A., SUNY-Buffalo, 1971; M.A., Johns Hopkins U., 1973, Ph.D., 1975. Staff fellow NIMH, Bethesda, Md., 1975-79; asst. prof. psychiatry and human behavior Brown U., 1979—; dir. psychology Butler Hosp., 1979—. Book rev. editor: Psychiatry Research, 1982. Mem. Am. Psychol. Assn., AAAS, Soc. Biol. Psychiatry, Am. Psychopath. Assn., Phi Beta Kappa. Subspecialties: Neurophysiology; Personality. Current work: Evoked potentials and personality; biological markers in psychopathology. Home: 70 Manning Providence RI 02906 Office: Butler Hos/Brown U 345 Blackstone Providence RI 02906

HAIGLER, HENRY JAMES, neuropharmacologist; b. Columbia, S.C., July 23, 1941; s. Harry Delk and Evelyn (Talton) H.; m. Jean Smith, July 25, 1964; children: Henry James, Elizabeth Ashley. B.S., Wake Forest U., 1963; Ph.D., Bowman Gray Sch. Medicine, 1969. Postdoctoral fellow Mental Health Research Inst., U. Mich., 1969-71; research asso. dept. psychiatry Yale U. Sch. Medicine, New Haven, 1971-74; asst., then assoc. prof. dept. pharmacology Emory U., Atlanta, 1974-82, dir. grad. studies dept. pharmacology, 1980-82; sect. head CNS pharmacology Searle Research & Devel., Skokie, Ill., 1982—; Magistral lectr. Neurol. Congress, 1980. Contbr. articles to profl. jours., chpts. in books. Deacon Clairmont Presbyn. Ch., 1981-82. Nat. Inst. Drug Abuse grantee, 1975-82. Mem. Soc. for Neurosci. (pres. Atlanta chpt. 1981-82), Am. Soc. Pharmacology and Exptl. Therapeutics, AAAS. Subspecialties: Cellular pharmacology; Neurobiology. Current work: Neuropharmacological studies of the mechanism of action of neurotropic drugs such as narcotic analgesics and psychotomimetics. Home: 4901 Searle Pkwy Skokie IL 60077

HAILMAN, JACK PARKER, zoology eductor; b. St. Louis, May 6, 1936; s. David E. and Katharine Lillard (Butts) H.; m. Elizabeth Bailey Davis, Aug. 26, 1958; children: Karl Andrew, Peter Eric. A.B., Harvard U., 1958; Ph.D., Duke U., 1964. NIH postdoctoral fellow U. Tubingen, Germany, 1964, Rutgers U., 1964-66; asst. prof. zoology U. Md., 1966-69; assoc. prof. U. Wis.-Madison, 1969-72, prof., 1972—; hon. research assoc. Smithsonian Instn., Washington, 1966-69. Author: Ontogeny of an Instinct, 1967, Optical Signals, 1977; co-author: Introduction to Animal Behavior, 1967; co-editor: Fascinating World of Animals, 1971. Bd. Dirs. County chpt. ACLU, 1968. Served with USNR, 1958-61. James P. Duke fellow, 1961; NIH fellow, 1962, 64; NIH research grantee, 1966-69; NSF research grantee, 1970-78. Fellow AAAS, Am. Ornithologists Union; mem. Animal Behavior Soc. (pres. 1981-82). Subspecialties: Ethology; Evolutionary biology. Current work: Vision and visually guided behavior; animal communication; behavioral development; speciation and evolution. Home: 3205 Tally Ho Ln Madison WI 53705 Office: Dept Zoology Birge Hall U Wis Madison WI 53706

HAINES, CHARLES WILLS, mechanical engineering educator, dean; b. Phila., Apr. 4, 1939; s. J. Edward and Ella (Peck) H.; m. Carolyn Hanna Anderson, June 17, 1961; children: Marie, Karen. A.B., Earlham Coll., 1961; M.S., Rensselaer Poly. Inst., 1963, Ph.D., 1965. Asst. prof. math. Clarkson Coll., Potsdam, N.Y., 1966-71; mem. faculty Rochester (N.Y.) Inst. Tech., 1971—, assoc. prof. math. and mech. engring., 1971-83, asst. provost, 1973-81, assoc. dean engring., 1982—; cons. Xerox Corp., Webster, N.Y., 1974, 78, 80. Author: Analysis for Engineering, 1974; contbr. articles in field to profl. publs. NASA fellow Rensselaer Poly. Inst., 1963-65; NASA-Am. Soc. Engring. Edn. fellow, Langley, Va., 1966-67. Mem. Soc. Indsl. and Applied Math., ASME, Am. Soc. Engring. Edn., Sigma Xi (pres. Rochester chpt. 1980-81). Subspecialties: Applied mathematics; Theoretical and applied mechanics. Current work: Dynamics of physical systems, aircraft control, and integration of computers into education. Office: Rochester Inst Tech Coll Engring Rochester NY 14623

HAINLINE, LOUISE, psychology educator; b. New London, Conn., Apr. 22, 1947; d. Wilmer Roger and Ina Lucien (Kart) H. A.B., Brown U., 1969; A.M., Harvard U., 1971, Ph.D., 1973. Asst. prof. psychology Bklyn. Coll. CUNY, 1972-80, assoc. prof., 1980—; asst. prof. Grad. Ctr. CUNY, 1978-80, assoc. prof., 1980—; lectr. pediatrics SUNY-Downstate Med. Ctr., Bklyn., 1979—. NIH grantee, 1973—; NSF grantee, 1973-74; CUNY grantee, 1977—; Nat. Found. grantee, 1979-81. Mem. Am. Psychol. Assn., Soc. Research Child Devel., AAAS, Sigma Xi, Phi Beta Kappa. Subspecialties: Developmental psychology; Sensory processes. Current work: Development of the human visual system, infant development, perceptual development. Home: 764 E 22d St Brooklyn NY 11210 Office: Dept Psychology Brooklyn Coll CUNY Brooklyn NY 11210

HAIRE, MARVIN JONATHAN, nuclear engineer, research scientist; b. Jackson, Ala., Sept. 18, 1943; s. Marvin Reynolds and Abalene (Creson) H.; m. Janet Gewn Newsom, June 20, 1965; children: Sarah Elizabeth, Rebecca Anne. B.S., N.C. State U., 1965, M.S., 1967, Ph.D., 1970. Registered profl. engr., Calif. Staff engr., section leader Gen. Atomic Co., La Jolla, Calif., 1970-76; asst. prof. nuclear engring. Ga. Inst. Tech., Atlanta, 1976-78; staff mem. Oak Ridge Lab., Tenn., 1978—; cons. Contbr. articles to profl. jours. Mem. Am. Nuclear Soc., Soc. for Risk Analysis, Personal Computers Club, Sigma Xi, Tau Beta Pi, Sigma Pi Sigma. Subspecialties: Nuclear fission; Nuclear engineering. Current work: Risk analysis, systems analysis economic evaluations, radioactivity source terms, heat transfer, ventillation systems. Office: Oak Ridge Nat Lab PO Box X Oak Ridge TN 37830

HAISCH, BERNHARD MICHAEL, astrophysicist; b. Stuttgart-Bad Canstatt, W.Ger., Aug. 23, 1949; came to U.S., 1952, naturalized, 1961; s. Friedrich Wilhelm and Gertrud Paula (Dammbacher) H.; m. Pamela S. Eakins, July 29, 1977; children: Katherine Stuart, Christopher Taylor. Student, St. Meinrad Coll., 1967-68; B.S. in Astrophysics magna cum laude, Ind. U., Bloomington, 1971; M.S. (Wis. Alumni Research Found. fellow), U. Wis., Madison, 1973; Ph.D. in Astrophysics, U. Wis., Madison, 1975. Research assoc. Joint Inst. for Lab. Astrophysics, U. Colo., Boulder, 1975-77, 78-79; vis. scientist Rijksuniversiteit Utrecht (Netherlands), Astron. Inst., 1977-78; research scientist Space Scis. Lab., Lockheed Palo Alto Research Lab., Palo Alto, 1979-83, staff scientist, 1983—. Research, publs. in field. Fellow Royal Aston. Soc.; Mem. Internat. Astron. Union, Am. Astron. Soc., Am. Astronautical Soc., Astron. Soc. Pacific, Phi Beta Kappa, Sigma Xi, Phi Kappa Phi. Club: Lockheed Tae Kwon Do (Karate) (Sunnyvale, Calif.). Subspecialties: Theoretical astrophysics; 1-ray high energy astrophysics. Current work: Development of concepts and technologies for future space research; studies of solar and stellar atmospheres, coronae and flares. Home: 1441 Nilda Ave Mountain View CA 94040 Office: Div 52-12 Bldg 25 Lockheed Palo Alto Research Lab 3251 Hanover St Palo Alto CA 94304

HAKALA, REINO WILLIAM, mathematics and chemistry educator; b. Albany, N.Y., Aug. 25, 1923; s. Toivo Wiljami and Emma Liisa (Kujanpaa) H.; m. Eunice Irma Kazanowski, June 17, 1950; children: Jonathan, Lisamaria, Christina. A.B., Columbia Coll., 1946; M.A., Columbia U., 1947; Ph.D., Syracuse U., 1965. Assoc. prof. chemistry and math. Mich. Tech. U., Houghton, 1964-67; chmn. math. and physics Oklahoma City U., 1967-70, prof. math., 1967-72; dean Sch. Sci. and Tech., Lake Superior State Coll., Sault Ste Marie, Mich., 1973-77, prof. chemistry, math, and physics, 1973-80, asst. to v.p. acad. affairs, 1977; dean Coll. Arts and Scis. Govs. State U., Park Forest South, Ill., 1980-81, spl. asst. to provost, 1982, prof. math., phys. scis. and environ. scis., 1982—; cons. Nat. Bur. Standards, Washington, 1962-63. Contbr. articles in field to profl. jours. Co-chmn. Adv. Com. to Dir. of Bd. Health on PBB Problem, Chippewa County, Mich., 1975; chmn. Air Quality Control Variance Bd. City of Oklahoma City, 1970-72. Served with U.S. Army, 1943-44. Danforth assoc. Danforth Found., St. Louis, 1979-85; Washington Acad. Scis. fellow, 1960; NSF faculty fellow, 1963-64. Mem. Soc. Indsl. and Applied Math., Assn. Computing Machinery, Sigma Xi, Phi Lambda Upsilon, Sigma Pi Sigma. Mem. Soc. of Friends. Club: Columbia U. Alumni (Chgo.) (bd. dirs. 1981—). Subspecialties: Numerical analysis; Condensed matter physics. Current work: Mathematical modeling, iteration acceleration techniques, equations of state, critical region of fluids, teaching computer programming, numerical analysis, statistics, physical chemistry, calculus, advanced inorganic chemistry. Home: 2945 Chayes Park Dr Homewood IL 60430 Office: Govs State U Park Forest S IL 60466

HAKEL, MILTON DANIEL, psychology educator, consultant; b. Hutchinson, Minn., Aug. 1, 1941; s. Milton Daniel and Emily Ann (Kovar) H.; m. Lee Ellen Pervier, Sept. 1, 1962; children: Lane, Jennifer. B.A., U. Minn., 1963, Ph.D., 1966. Lic. Psychologist, Ohio. Research asst., research fellow U. Minn., 1963-66, asst. prof., 1966-68; prof. psychology Ohio State U., 1968—; pres. Organizational Research and Devel., Inc., Columbus, Ohio, 1977—. Co-author: Making It Happen: Doing Research with Implementation in Mind, 1982; Contbr. articles to profl., tech. jours.; Editor: Personnel Psychology Jour, 1973—. Coordinator AFS Internat.-Intercultural Exchange Program, Upper Arlington, Ohio, 1982-83; mem. Alverno Coll. Adv. Com. on Validation, 1978—; bd. dirs. Found. for Study in Devel. in Indsl. Psychology in Europe, Amsterdam, Netherlands, 1974-80. Fulbright-Hays Sr. Scholar, 1978; NSF grantee, 1966-73. Fellow Am. Psychol. Assn.; mem. Acad. Mgmt., Soc. Organizational Behavior, Internat. Assn Applied Psychology. Presbyterian. Subspecialties: Behavioral psychology. Current work: Research interests include the employment interview; interpersonal perception; sterotyping; decision making; appraisal of performance and development of the integrated personnel system. Home: 2069 Fairfax Rd Columbus OH 43221 Office: Ohio State U 404-C W 17th St Columbus OH 43210

HALARIS, ANGELOS, psychiatrist, educator; b. Athens, Greece, Nov. 30, 1942; came to U. S., 1971, naturalized, 1979; s. Eleftherios and Elli (Georgiadou) H.; m. Ann L. Lyons, June 15, 1980. M.D., Ph.D., U. Munich, 1967. Diplomate: Am. Bd. Psychiatry and Neurology. Resident, fellow U. Chgo., 1971-77, asst. prof. psychiatry, 1975-77, assoc. prof., 1978-80, UCLA Sch. Medicine, 1980-82, prof., 1982—; chief Biol. Psychiatry Lab.; cons. Cedars-Sinai Med. Ctr., Los Angeles. Contbr. numerouus articles on psychiatry to profl. jours. Served with Greek Army, 1970-71. Fulbright Scholar, 1955-60; Founds. Fund for Research in Psychiatry fellow, 1971-73. Mem. Am. Hellenic Inst., N.Y. Acad. Scis., AMA, Am. Psychiat. Assn., Am. Coll. Neuropsychopharmacology, Internat. Coll. Neuropsychopharmacology. Subspecialties: Psychiatry; Psychopharmacology. Current work: Research in biological psychiatry and psychopharmacology. Office: 11301 Wilshire Blvd Los Angeles CA 90073

HALBERSTAM, HEINI, mathematician; b. Most, Czechoslovakia, Sept. 11, 1926; came to Eng., 1939, naturalized, 1947; s. Michael and Judith (Honig) H.; m. Heather M. Peacock, Mar. 11, 1950 (dec. 1971); children: Naomi Deborah, Judith Marion, Lucy Rebecca, Michael Welsford; m. Doreen Bramley, Sept. 28, 1972. B.S. with honours, Univ. Coll., London U., 1946, M.S., 1948, Ph.D., 1952. Lectr. math. U. Exeter, 1949-57; reader Royal Holloway Coll., London U., 1957-62; Erasmus Smith prof. Trinity Coll., Dublin, Ireland, 1962-64; prof. Nottingham U., England, 1964-80; prof. math., head dept. U. Ill., Urbana-Champaign, 1980—; vis. lectr. Brown U., 1955-56; vis. prof. U. Mich., 1966, U. Tel Aviv, 1973, U. Paris-South, 1972; mem. Nat. Com. Math., 1972—. Co-author: Sequences, 1966, Sieve Methods, 1975; co-editor math. papers of, W.R. Hamilton, H. Davenport; contbr. articles to profl. jours. Mem. London Math. Soc. (v.p. 1962-63, 74-77), Math. Assn. Am., Edinburgh Math. Soc., Am. Math Soc. Subspecialty: Number Theory.

HALBOUTY, MICHEL THOMAS, geologist, petroleum engineer, petroleum operator; b. Beaumont, Tex., June 21, 1909; s. Tom Christian and Sodia (Monnelly) H.; m.; 1 dau., Linda Fay. B.S., Tex. A. and M. U., 1930, M.S., 1931, Profl. Degree in Geol. Engring, 1956; D.Eng. (hon.), Mont. Coll. Mineral Sci. and Tech., 1966. Geologist, petroleum engr. Yount-Lee Oil Co., Beaumont, 1931-33, chief geologist, petroleum engr., 1933-35; v.p., gen. mgr., chief geologist and petroleum engr. Glenn H. McCarthy, Inc., Houston, 1935-37; owner firm cons. geologists and petroleum engrs., Houston, 1937—, discoverer numerous oil and gas fields, La. and Tex.; adj. prof. Tex. Tech U.; vis. prof. Tex. A. and M. U. Author several books.; Contbr. numerous papers on geology and petroleum engring. to profl. jours. Served as lt. col. AUS, 1942-45. Recipient Tex. Mid-Continent Oil and Gas Assn. distinguished service award for an ind., 1965; named engr. of year Tex. Soc. Profl. Engrs. and Engrs. Council, 1968; Distinguished Alumni award Tex. A. and M. U., 1968; Michel T. Halbouty Geoscis. Bldg. named for him, 1977; DeGolyer Distinguished Service medal Soc. Petroleum Engrs. of Am. Inst. Mining, Metall. and Petroleum Engrs., 1971; hon. mem. Spindletop sect., 1972; hon. mem. inst., 1973; Anthony F. Lucas Gold medal, 1975; Pecora award NASA, 1977; Horatio Alger award Am. Schs. and Colls. Assn., 1978. Mem. Am. Assn. Petroleum Geologists (hon., pres. 1966-67, Human Needs award 1975, Sidney Powers Meml. medal 1977), Am. Soc. Oceanography, Internat. Assn. Sedimentology, Inst. Petroleum, London, Am. Petroleum Inst., Am. Inst. Mining and Metall. Engrs., Soc. Paleontologists and Mineralogists, Soc. Econ. Geologists, Mineral. Soc. Am., Geol. Soc. Am., Soc. Exploration Geophysicists, Nat. Acad. Engring., Houston Geol. Soc. (hon.), N.Y., Tex. acads. scis., A.A.A.S., Am. Inst. Profl. Geologists, Am. Geol. Inst., Tex., Nat. socs. profl. engrs. Episcopalian. Clubs: Ramada, Houston, Petroleum, River Oaks Country (Houston); Dallas Petroleum; Eldorado Country (Palm Desert, Calif.); New Orleans Petroleum; Cosmos (Washington); Broadmoor, Kissing Camels (Colorado Springs, Colo.). Subspecialties: Geology; Petroleum engineering. Home: 49 Briar Hollow Houston TX 77027 Office: Halbouty Center 5100 Westheimer Rd Houston TX 77056

HALBREICH, URIEL, psychiatrist; b. Jerusalem, Israel, Nov. 23, 1943; came to U.S., 1978, naturalized, 1982; s. Mordechai and Zipora (Tennenbaum) H.; m. Tatiana Or, Feb. 1, 1966; 1 child, Jasmin. M.D., Hebrew U., 1969. Diplomate: Tel Aviv U. Psychiatry and Psychotherapy. Vice chief med. officer Israel Navy, 1970-72; second, then first asst. Hadassah U. Hosp., Jerusalem, 1972-78, temp. chief physician, 1978; chief psychiatrist Israel Navy, 1977-78; asst. prof., research psychiatrist Columbia U., N.Y.C., 1978-80; asso. prof. dir. Div. Biol. Psychiatry, Albert Einstein Coll. Medicine, N.Y.C., 1982—. Contbr. articles to profl. jours.; editor: Transient Psychosis, 1983. Served to comdr. Israeli Navy, 1970-78. Recipient Ben Gurion award Gen. Fedn. Labor, 1976; Yair Gon award Hebrew U. Hadassah Med. Sch., 1978; Nat. Research Service award NIH, 1978; NIMH grantee, 1982. Mem. Am. Coll. Neuropsychopharmacology, Am. Psychopathology Assn., Soc. Biol. Psychiatry, Endocrine Soc., others. Jewish. Subspecialties: Psychopharmacology; Neuroendocrinology. Current work: Research is in biol. psychiatry especially behavioral endocrinology and psychopharmacology of affective disorders and hormone related mood and behavior in women. Home: 2166 Broadway Apt 178 New York NY 10024 Office: Albert Einstein Coll of Medicine 1300 Morris Park Ave Bronx NY 10461

HALE, MAYNARD GEORGE, plant physiologist, educator, researcher; b. Mentor, Ohio, Apr. 5, 1920; s. William Jonathan and Elizabeth Ann (Raynor) H.; m. Pollyanna Pozniko, Oct. 15, 1943; children: Patricia Jane, Arthur Lynn. B.S., Ohio State U.-Columbus, 1947, M.S., 1949, Ph.D., 1951. Asst. botany Ohio State U., Columbus, 1946-51; assoc. prof. plant physiology Va. Poly. State U., Blacksburg, 1951—. Agrl. editorial bd., AVI Books, Westport, Conn., 1980—; contbr. articles in field to profl. jours., chpts to books. Leader Boy Scouts Am., Blacksburg and Roanoke, Va., 1956—. Served with AUS, 1941-44; ETO. Grantee U.S. Dept. Agr., 1967-74, 80-84, U. Santiago de Compostelo, Spain, 1981-83, NRC, Sweden, 1981. Mem. Am. Soc. Plant Physiologists, Plant Growth Regulation Soc., Assn. Gnotobiologists, Va. Acad. Sci., Sigma Xi. Republican. Presbyterian. Subspecialties: Plant physiology (agriculture); Plant pathology. Current work: Allelopathy, gnotobiology, plant stress physiology.

Office: Va Poly Inst and State U Dept Plant Pathology and Physiology Blacksburg VA 24061

HALE, WILLIAM KENT, operations research analyst; b. Charleston, W. Va., Aug. 24, 1941; s. Joseph William and Eloise (Keely) H.; m. Callie Jones, Aug. 22, 1964; children: Clifford Wade, Christine Calore. B.S., W. Va. U., 1966, M.S., 1967. Prof. math. Radford (Va.) U., 1968-73; mathematician Def. Communications Agy., Reston, Va., 1973-1979; ops. research analyst, 1982—, electronics engr. Inst. for Telecommunications Scis., Boulder, Colo., 1979-81; ops. research analyst Dept. Army, Washington, 1981-82. Author: Bibliography-Frequency Assignment Methodology, 1981; contbr. research articles to publs. Organizer Nat. Soc. Profs., Radford, Va., 1970-72; faculty sponsor ACLU, Radford, 1970-72; coach Little League Baseball, Summit Point, W.Va., 1975-79. Recipient Meritorious Service Award Def. Communications Agy., Reston, 1979; research grantee NSF, Chgo., 1971; research fellow, 1968. Mem. IEEE, Ops. Research Soc. Am., Math. Assn. Am. Baptist. Subspecialties: Algorithms; Distributed systems and networks. Current work: Application of recent developments in discrete mathematics and computational complexity to problems in communication systems engineering. Home: Box 208 Summit Point WV 25446 Office: Def Communications Agy 1860 Wiehle Ave Reston VA 22090

HALER, LAWRENCE EUGENE, nuclear reactor company analyst; b. Iowa City, Jan. 24, 1951; s. Eugene Hilbert and Mary Elizabeth (Hans) H.; m. Jenifer Lea Leitz, June 1, 1974. B.A. in Liberal Arts, Pacific Lutheran U., 1974. Pacific Luth. U. intern Wash. State Ho. of Reps., Olympia, 1972; in public adminstrn. City of Tacoma, 1974; with UNC-United Nuclear Corp., Richland, Wash., 1974—, mgr. tng. devel. and adminstrn., 1982, investigative analyst, 1982—. Chmn. Benton County (Wash.) Republican Party, 1976-78. Mem. Am. Nuclear Soc. Lutheran. Subspecialty: Nuclear fission. Current work: Analyze unusual events and occurrences at nuclear reactor; also analyze operational trends as they occur at nuclear reactor. Home: 4265 Ironton Dr West Richland WA 99352 Office: UNC-United Nuclear Corp PO Box 490 Richland WA 99352

HALES, ALFRED WASHINGTON, mathematics educator, consultant; b. Pasadena, Calif., Nov. 30, 1938; s. Raleigh Stanton and Gwendolen (Washington) H.; m. Virginia Dart Greene, July 7, 1962; children: Andrew Stanton, Lisa Ruth, Katherine Washington. B.S., Calif. Inst. Tech., 1960, Ph.D., 1962. NSF postdoctoral fellow Cambridge (Eng.) U., 1962-63; Benjamin Peirce instr. Harvard U., 1963-66; faculty mem. UCLA, 1966—, prof. math., 1973—; cons. Jet Propulsion Lab., La Canada, Calif., 1966-70, Inst. for Def. Analyses, Princeton, N.J., summers 1964, 65, 76, 79-82; vis. lectr. U. Wash., Seattle, 1970-71; vis. mem. U. Warwick Math. Inst., Coventry, Eng., 1977-78. Co-author: Shift Register Sequences, 1967, 82; contbr. articles in field to profl. jours. Mem. Am. Math. Soc., Math. Assn. Am., Soc. for Indsl. and Applied Math. (Polya prize in combinatorics 1972), Sigma Xi. Clubs: Pasadena Badminton, Manhattan Beach (Calif.) Badminton. Current work: Research in algebra - structure of groups, modules and lattices, group rings; research in combinatonics - Ramsey theory and related problems. Office: Dept Math U Calif Los Angeles 405 Hilgard Ave Los Angeles CA 90024

HALEVY, SIMON, physician, educator; b. Bucharest, Romania, June 5, 1929; came to U.S., 1963, naturalized, 1970; s. Meyer Abraham H. and Rebecca (Landau) H.; m. Hilda M. Valdes, 1968; 1 son, Daniel Abraham. M.D., U. Bucharest, 1953. Diplomate: Am. Bd. Anesthesiology. Intern Univ. Hosp., Coltzea, Romania, 1952-53, resident, 1953-54; practice medicine specializing in anesthesiology, 1955—; instr. anesthesia Postgrad. Inst. Medicine, Bucharest, 1955-57, chief lab. in anesthesia, 1957-60; preparator, instr. anatomy U. Bucharest Med. Sch., 1950; attending anesthesiologist Univ. Hosp., Fundeni, Bucharest, 1960-63; intern Community Hosp., Glen Cove, N.Y., 1964-65; resident Mt. Sinai Hosp., N.Y.C., 1965-67; asst. prof. anesthesiology Mt. Sinai Sch. Medicine, 1967-68; asst. prof. Albert Einstein Coll. Medicine, 1969-74; assoc. prof. Coll. Physicians and Surgeons, Columbia U., 1974-75; prof. SUNY, 1976—; asst. attending anesthesiologist Mt. Sinai Hosp. Services and Bronx Mcpl. Hosp. Center, 1967-71, attending anesthesiologist, 1973-74; attending anesthesiologist, dir. obstet. anesthesiology Nassau County Med. Center, 1976—; Chmn. com. on sci. exhibits Postgrad. Assembly in Anesthesiology, N.Y.C., 1971-80. Mem. editorial bd.: Microcirculation; mem. editorial bd.: Convergences Médicales; Contbr. articles to sci. jours. Fellow Am. Coll. Anesthesiologists; mem. AMA, Am. Soc. Anesthesiologists, Assn. des Anesthesiologistes Français, Deutche Gesellschaft für Anaesthesiologie und Intensivmedizin; Fellow Société Française d'Anesthésve et de Réanimation; mem. Association Internationale des Anesthésioloz gistes d'Expression Française (v.p., mem. adminstrv. council), N.Y. Acad. Scis., AAAS, Am. Soc. Pharmacology and Exptl. Therapeutics, N.Y. Acad. Scis. Current work: Pathophysiology of shock syndrome, pharmacology of thyroid hormones, histamine and antihistamines, perinatology. Office: Nassau County Med Center 2201 Hempstead Turnpike East Meadow NY 11554

HALL, BARRY GORDON, molecular biologist, researcher, educator; b. N.Y.C., July 17, 1942; s. S. Henry and Helen (Norton) H.; m. Susan Marie Hall, May 2, 1964; children: Steven John, Scott Owen, Rebecca Anne. B.S. in Genetics, U. Wis., Madison, 1968; Ph.D., U. Wash., 1971. Research assoc. Inst. Molecular Biology, U. Oreg., Euegene, 1971-72, postdoctoral fellow, 1972-73, U. Minn., 1973-74; asst. prof. molecular biology Faculty of Medicine, Meml. U. Nfld. (Can.), St. John's, 1974-77; asst. prof. biology U. Conn., Storrs, 1977-80, assoc. prof., 1980—. Contbr. articles to sci. jours. NIH fellow, 1968-73; Career Devel. awardee, 1980—; Med. Research Council Can. grantee, 1974-77; others. Mem. Genetics Soc. Am., Am. Soc. Naturalists, Am. Soc. Biol. Chemists, Am. Soc. Microbiology, Sigma Xi, Gamma Sigma Delta. Subspecialties: Evolutionary biology; Genetics and genetic engineering (biology). Current work: Evolution of new enzymes in the laboratory; experimental evolution. Office: Biol Scis U Conn Storrs CT 06268

HALL, BENJAMIN DOWNS, genetics and biochemistry educator; b. Berkeley, Calif., Dec. 9, 1932; m., 1954; 2 children. A.B., U. Kans., 1954; Ph.D. in Chemistry, Harvard U., 1959. From instr. to assoc. prof. chemistry U. Ill., 1958-63; from assoc. prof. to prof. chemistry U. Wash., Seattle, 1963-70, prof. genetics and biochemistry, 1970—; mem. microbial chemistry study sect. NIH, 1971-75. Guggenheim fellow, 1962. Mem. Genetics Soc., Am. Soc. Biol. Chemists, Biophys. Soc. Subspecialty: Genetics and genetic engineering (biology). Office: Dept Genetics U Wash Seattle WA 98105

HALL, BRIAN KEITH, biology educator; b. Port Kembla, New South Wales, Australia, Oct. 28, 1941; s. Harry James and Doris Grace (Garrad) H.; m. June Denise Priestley, May 21, 1966; children: Derek Andrew, Imogen Elizabeth. B.Sc., U. New Eng., Armidale, New South Wales, 1963, 1965, Ph.D., 1968, D.Sc., 1977. Teaching fellow U. New Eng., Armidale, New South Wales, Australia, 1966-68; asst. prof. Dalhousie U., Halifax, N.S., Can., 1968-72, assoc. prof., 1972-75, prof. biology, 1975—, chmn. dept., 1978—. Author: Developmental and Cellular Skeletal Biology, 1978; editor: Cartilage Vols. 1-3, 1983; contbr. numerous articles to profl. jours. Named Young Scientist of Yr. Atlantic Provinces Interuniv. Com. on the Scis., 1974; Nuffield Found. fellow, U.K., 1982; Med. Research Council Can. vis. prof., 1980. Mem. AAAS (mem. nominating com. 1973-76), Canadian Com. Univ. Biology Chmn. (mem. nat. exec. com. 1978-80). Club: Dalhousie Faculty (Halifax (N.S.)). Subspecialties: Developmental biology; Oral biology. Current work: Development and evolution of the skeleton, craniofacial development, biology of the neural crest. Home: 2384 Armcrescent E Halifax NS Can B3L 3C7 Office: Biology Dept Dalhousie U Halifax NS Canada B3H 3J1

HALL, CHARLES A., medical educator and researcher; b. Castine, Maine, Mar. 7, 1920; s. William D. and Letitia A. (Hatch) H.; m. Jane E. Dempsey, Sept. 14, 1944 (div. 1973); children: Nancy A., Christine M., Peter L.; m. Mary E. Rappazzo, Nov. 23, 1974; children: C. Andrew, Dominic W. B.A., U. Maine, 1941; M.D., Yale U., 1944. Diplomate: Am. Bd. Internal Medicine. Instr. Yale U. Sch. Medicine, New Haven, 1947-48; mem. faculty Albany (N.Y.) Med. Coll., 1951—, prof. medicine, 1966—; med. investigator VA, Albany, 1971—. Author: Blood in Disease, 1968; editor: The Cobalamins, 1983; contbr. chpts. to books. Served to capt. U.S. Army, 1945-47. Fulbright fellow, Turku, Finland, 1960-61. Mem. Am. Soc. Hematology, Am. Fedn. Clin. Research., Am. Soc. Clin. Nutrition, Soc. Exptl. Biology and Medicine, Am. Inst. Nutrition. Subspecialties: Hematology; Nutrition (medicine). Current work: Hematology resaerch-cobalamin transport; nutrition-clinical and research. Office: VA Med Center Nutrition Lab Clin Assessment and Research Room 151E Albany NY 12208

HALL, CHARLES ALLAN, numerical analyst; b. Pitts., Mar. 19, 1941; s. George Orbin and Minnie (Carter) H.; m. Mary Katherine Harris, Aug. 11, 1962; children—Charles, Eric, Katherine. B.S., U. Pitts., 1962, M.S., 1963, Ph.D., 1964. Sr. mathematician Bettis Atomic Power Lab., West Mifflin, Pa., 1966-70; assoc. prof. math stats. U. Pitts., 1970-78, prof., 1978—; exec. dir. Inst. Computational Maths. and Applications, 1978; cons. Gen. Motors Research, 1971—, Westinghouse Electric Corp., Pitts., 1974-81, Pitts. Corning Co., 1980—. Contbr. articles to profl. jours. Served to 1st lt. AUS, 1964-66. Mem. Math. Asn. Am., Soc. Indsl. and Applied Math. Subspecialties: Applied mathematics; Mathematical software. Current work: Large scale scientific computing; fine element methodology and computational fluid dynamics. Home: Box 83 Murrysville PA 15668 Office: ICMA Univ Pitts Pittsburgh PA 15621

HALL, CHARLES FREDERICK, space scientist, govt. adminstr.; b. San Francisco, Apr. 7, 1920; s. Charles Rogers and Edna Mary (Gibson) H.; m. Constance Vivienne Andrews, Sept. 18, 1942; children—Steven R., Charles Frederick, Frank A. B.S., U. Calif., Berkeley, 1942. Aero. research scientist NACA (later NASA), Moffett Field, Calif., 1942-60, mem. staff space projects, 1960-63, mgr., 1963-80. Recipient Disting. Service medal NASA, 1974, Achievement award Am. Astronautical Soc., 1974, Spl. Achievement award Nat. Civil Service League, 1976, Astronautics Engr. award Nat. Space Club, 1979. Subspecialties: Aerospace engineering and technology; Aeronautical engineering. Research, reports on performance of wings and inlets at transonic and supersonic speeds, on conical-cambered wings at transonic and supersonic speeds, 1942-60; Pioneer Project launched 4 solar orbiting, 2 Jupiter and 2 Venus spacecraft. Home: 817 Berry St Los Altos CA 94022

HALL, EDWARD DALLAS, pharmacologist, biol. scientist, med. educator; b. Bedford, Ohio, June 16, 1950; s. Edward Ellis and Martha Elaine (Johnston) H.; m. Marilynn Frances Gay, Sept. 12, 1970; children: Edward William, Christian David. B.S., Mt. Union Coll., Alliance, Ohio, 1972; Ph.D. (Nat. Inst. Gen. Med. Sci. fellow), Cornell U., 1976. Postdoctoral fellow dept. pharmacology Cornell U. Med. Coll., N.Y.C., 1976-77; asst. prof. pharmacology Northeastern Ohio Univs. Coll. Medicine, Rootstown, 1978-82, asso. prof., 1982; asst. prof. Kent State U., 1978-82; research scientist The Upjohn Co., Kalamazoo, 1982—; cons. on drug-related issues. Contbr. articles and abstracts to sci. and med. jours., chpt. to book. Elder Randolph (Ohio) Christian Ch., 1981-82; v.p. Portage County Combined Gen. Health Dist. Bd. NIMH grantee, 1978-79, 80-82; Amyotrophic Lateral Sclerosis Soc. Am. grantee, 1978-81, 81-82. Mem. Am. Soc. for Pharmacology and Exptl. Therapeutics, Soc. for Neurosci., Soc. for Exptl. Biology and Medicine, AAAS, N.Y. Acad. Scis., Sigma Xi, Phi Sigma. Lodge: Randolph Lions. Subspecialties: Pharmacology; Neuropharmacology. Current work: Pathophysiology and pharmacol. treatment of acute brain and spinal cord injury and stroke; pharmacol. treatment of neuromuscular diseases; hormonal involvement in psychiat. disease. Home: 1432 Woodland Ave Portage MI 49002 Office: CNS Research Unit The Upjohn Co Kalamazoo MI 49001

HALL, FRANCIS RAMEY, hydrology educator, consultant; b. Salmon, Idaho, Feb. 25, 1925; s. Francis Ramey and Dorothy (Quarles) H.; m. Carman Frye, July 23, 1960. B.S. in Geology, Stanford U., 1949, M.A., UCLA, 1953, Ph.D., Stanford, U., 1961. Ground-water geologist U.S. Geol. Survey, Newport, Ky., 1951-56; grad. tech. asst. Stanford U., Calif., 1956-58; ground-water geologist Stanford Research Inst., Menlo Park, Calif., 1958-61; Assoc. hydrologist N. Mex. Inst. Mining and Tech., Socorro, N.Mex., 1961-64; assoc. prof. U. N.H., Durham, 1964-71, prof., 1971—; pres. Hydrosci. Assocs., Inc., Durham, 1980—; bd. dirs. Environ. Hazards Mgmt. Inst., Portsmouth, N.H., 1981—. Active N.H. Govs. Hazardous Waste Task Force, Concord, 1980-81. Served with U.S. Army, 1943-45. Recipient ground-water pollution study Allied Found., 1981-82. Fellow Geol. Soc. Am.; mem. Am. Geophys. Union, Am. Water Resource Assn., Am. Water Works Assn., Nat. Water Well Assn. Subspecialty: Hydrogeology. Current work: Ground-water - surface water interactions, ground-water recharge, chemical transport as related to waste disposal. Home: PO Drawer A Durham NH 03824 Office: INER -James Hall U NH Durham NH 03824

HALL, HERBERT JOSEPH, physicist, consultant; b. Springfield, Mass., Sept. 30, 1916; s. Herbert C. and Jospheine A. (Whalan) H.; m. Georgine Fleming, Sept. 21, 1949; children: Molly J., John L., Stephen W.; m. Jean Cummings, Nov. 4, 1962. B.S., Trinity Coll., Conn., 1939; M.S., U. Mich., 1940, doctoral candidate, 1939-41. Mem. staff radiation lab. MIT, Cambridge, 1941-45; mem. staff for Bikinitests Los Alamos Lab., 1946; physicist, sr. physicist, then asst. dir. research Research Cottrell, Inc., Bound Brook, N.J., 1946-58; sci. cons., N.J., 1959-61; dir. research and devel. Research Cottrell, Inc., 1962-68; v.p. Recon Systems, Inc., Princeton, N.J., 1969-72; pres. H.J. Hall Assocs., Inc. (Sci. Cons.), Princeton, 1973—; dir. corps. Contbr. articles to sci. jours. Served with OSRD; U.K. and Europe; liaison Brit. Air Ministry and USAF, 1943-45. H.E. Russell fellow, 1939. Mem. Am. Phys. Soc., AAAS, NY. Acad. Scis., Air Pollution Control Assn. Republican. Subspecialties: Gas cleaning systems; Condensed matter physics. Current work: Electrostatic precipitation; air pollution control systems and equipment; frontier technology; aerosol and gaseous discharge physics; litigation problems. Patentee in field. Office: 1250 State Rd Princeton NJ 08540

HALL, HOWARD RALPH, psychology educator, researcher, clinician; b. Yokohama, Japan, Sept. 14, 1950; s. Howard Ralph and Dorothy Lillian (Johns) H.; m. Clara Jean Mosley, Aug. 12, 1978; Ilea Elizabeth Mosley. B.S., Del. State Coll., 1973; M.A., Princeton U., 1975, Ph.D., 1978; Psy.D., Rutgers U., Piscataway, N.J., 1982. Postdoctoral fellow Center Alcohol Studies, Rutgers U.-Piscataway, 1978-80; intern in psychology Coll. Medicine and Dentistry N.J./Rutgers U. Med Sch., Piscataway, 1979-80; asst. prof. psychology Pa. State U., 1980—; mem. adv. bd. Claremont Sch. Theology Inst. for Religion and Wholeness, 1982. Vice pres. Pa. State U. Forum on Black Affairs, 1981-82. Nat. Inst. Alcohol Abuse and Alcoholism grantee, 1979. Mem. Am. Psychol. Assn., Soc. Clin. and Exptl. Hypnosis, Kappa Alpha Psi. Subspecialty: Behavioral psychology. Current work: Effects of hypnosis and imagery on lymphocyte functioning. Home: 510 Galen Dr State College PA 16801 Office: Pa State U 311 More Bldg University Park PA 16802

HALL, LEROY BROOKS, JR., veterinary pathologist, educator; b. Pensacola, Fla., Oct. 7, 1950; m. Earnestine Cotton, May 19, 1972. B.S., Tuskegee Inst., 1975, D.V.M., 1976; M.S., Iowa State U., 1980, Ph.D., 1982. Intern Tuskegee Inst., 1977; instr. vet. pathology Iowa State U., 1977—. Mem. AVMA, Fla. Vet. Med. Assn. Subspecialty: Pathology (veterinary medicine). Current work: Pathogenesis of pseudorabies (Aujeszky's disease), virus infection of the reproductive tract of boars and gilts. Office: Dept Vet Pathology Iowa State U Ames IA 50011

HALL, LINDA M., neurogeneticist, researcher, educator, consultant; b. Wilkes-Barre, Pa., Aug. 6, 1943; d. Elbert H. and Annette Kay (Owens) McIntyre. B.S. in Biology and Chemistry, Bucknell U., 1965; Ph.D. in Biochemistry, U. Wis.-Madison, 1970. Postdoctoral fellow dept. zoology U. B.C. (Can.), Vancouver, 1970-73; asst., then assoc. prof. biology MIT, 1973—; assoc. prof. neurosci., assoc. prof. genetics Albert Einstein Coll. Medicine; mem. neurol. scis. study sect. NIH, 1977-81; instr. in neurobiology Woods Hole Marine Biol. Lab., summers 1980—. Contbr. articles to profl. publs.; editor: (with Sattelle and Hildebrand) Receptors for Neurotransmitters, Hormones and Pheromones in Insects, 1980. Recipient McKnight Scholars award in neurosci. Mc Knight Found., 1977-80, Ira T. Hirschl-Monique Weill-Caudier Career Scientist award, 1983-87; Med. Research Council Can. postdoctoral fellow, 1970-73; NIH; NSF; Council Tobacco Research grantee. Mem. Genetics Soc., Soc. Neurosci., AAAS, N.Y. Acad. Sci., LWV, Appalachian Mountain Club, Audubon Soc. Club: N.Y. Rd. Runners (N.Y.C.). Subspecialties: Neurobiology; Genetics and genetic engineering (biology). Current work: Genetic Analysis of ion channels and neurotransmitter receptors in Drosophila melanogaster. Home: 515 Minnieford Ave City Island NY 10464 Office: Dept Genetics Albert Einstein Coll Medicine Bronx NY 10461

HALL, MADELINE MOLNAR HALL, researcher, educator; b. Cleve., June 6, 1936; d. Louis P. and Irene Molnar; m. Robert E. Hall, Aug. 24, 1957; children: Ann Christine, Roy Louis; m. Stephen Cverna, Dec. 26, 1977. B.S., Ohio State U., 1964, Ph.D., 1970. NIMH postdoctoral fellow U. Chgo., 1970-72; NIH postdoctoral fellow Cleve. Clinic Found., 1972-74; asst., then asso. prof. Cleve. State U., 1974—; adj. prof. Ohio Coll. Podiatric Medicine, 1978-80. Contbr. articles, chpts. to profl. lit. Recipient lower prize Cleve. Clinic Found., 1973. Mem. Am. Assn. Pharmacology and Exptl. Therapeutics, N.Y. Acad. Scis., AAUP, AAAS, Am. Chem. Soc. (sect. medicinal chemistry). Democrat. Roman Catholic. Subspecialties: Pharmacology; Biochemistry (biology). Current work: Prostaglandin in smooth muscle. Teaching, research, consulting. Home: 3793 Bushnell Rd University Heights OH 44118 Office: 1983 E 24th St Cleveland OH 44115

HALL, NANCY ROSE, software engineer; b. Englewood, N.J., Jan. 16, 1942; d. Stephen R. and Lucy A. (Pastor) Gardiner; m. John L. Hall, Apr. 23, 1966 (div. 1971); 1 dau., Ilea Elizabeth Mosley. B.A., N.Y. U., 1963, M.S., 1967; Ph.D., Poly. Inst. N.Y., 1983. Programmer IBM, N.Y.C., 1966-70, project programmer, mgr., Morristown, N.J., 1972-76, mgr., devel. programmer, Wayland, Mass., 1976-80; adv. programmer fed. systems div., Bethesda, Md., 1980—; teaching cons. Poly. Inst. N.Y., N.Y.C., 1979-80; mem. adj. faculty Fairleigh Dickinson U., Madison, N.J., 1975-76. N.Y. U. Alumni scholar, 1959-63. Mem. Assn. Computing Machinery, Soc. Indsl. and Applied Math. Subspecialties: Software engineering; Distributed systems and networks. Current work: Software engineering education for distributed systems and networks. software complexity measurement for networks and architecture. Office: IBM 6600 Rockledge Dr Bethesda MD 20817

HALL, R. VANCE, psychology researcher, educator, administrator, consultant, business executive; b. Gt. Falls, Mont., Dec. 4, 1928; s. Robert Nelson and Marie (Woodruff) H.; m. Alice Gertrude Judd, Dec. 23, 1950 (dec. Jan. 1976); children: Alison, Darien, Laurel; m. Marilyn Lee Hawkins, Feb. 5, 1977; stepchildren: Douglas, Debra Julie. B.A., Central Wash. State U., 1951; M.Ed., U. Wash., 1962, Ph.D., 1966. Cert. profl. psychologist, Kans. Tchr. Highline Sch. Dist., Seattle, 1951-55, prin., 1955-62, Exptl. Edn. Unit, U. Wash., 1962-65; research assoc. Bur. Child Research, U. Kans.-Kansas City, 1965-69, assoc. prof. human devel. and spl. edn., 1969-72, sr. scientist, prof., 1972—, dir. juniper gardens children's project, 1967—; pres., chmn. bd. H & H Enterprises (pub. firm), Lawrence, Kans., 1970—, Responsive Mgmt., Inc. (psychol. and ednl. clinic), Overland Park, Kans., 1980—; pres. Performance Techs. (bus. cons. firm), Overland Park, 1982—. Author: book series Management Behavior series, 1970, rev., 1978, How To Change Behavior series, 1980, 4 books, 1980; co-editor: 16 books How To Change Behavior series, 1980-82; research numerous publs. in field. Mem. adv. bd. Princeton (N.J.) Child Devel. Inst., 1974—, May Inst. for Autistic Children, Chatham, Mass., 1978—; v.p. Uriel F. Owens Sickle Cell Anemia Found., Kansas City, Kans., 1981—. Recipient Citation Classics award Current Contents mag., 1979; grantee Nat. Inst. Child Health and Human Devel., 1967—, NIMH, 1972-70, Bur. Edn. of Handicapped, 1972—, Office Edn., 1978-82. Fellow Am. Psychol. Assn.; mem. Am. Ednl. Research Assn., Assn. Advancement Behavior Therapy, Assn. Applied Behavior Analysis. Subspecialties: Behavioral psychology; Developmental psychology. Current work: Use of systematic measurement and observation procedures and behavior modification; behavior analysis and behavior therapy procedures in education, community and business settings. Home: 470 Hillcrest E Lake Quiviva KS 66106 Office: Bur Child Research U Kans Lawrence KS 66045

HALL, ROBERT DEAN, geology educator, consultant; b. Warren, Ohio, Mar. 21, 1941; s. Thomas Augustus and Helen (Logan) H.; m. Shannon Leslie Smith, July 22, 1976 (div. Dec. 1981). B.S., Purdue U., 1963; M.S., U. Colo., 1966; Ph.D., Ind. U., 1973. Geologist Texaco Petroleum Co., Quito, Ecuador, 1966-69; asst. prof. geology Ind.-Purdue U.-Indpls., 1974—; pvt. practice cons. geologist, Indpls., 1973—. NSF grantee, 1982; Amax Coal Co. grantee, 1982. Mem. Geol. Soc. Am., AAAS, Am. Quaternary Assn. Subspecialties: Geology; Hydrology. Current work: Glacial chronology and Quaternary history of middle Rocky Mountains; weathering and soil development as function of petrology. Office: Dept Geology Ind-Purdue U 425 Agnes St Indianapolis IN 46202

HALL, ROBERT DICKINSON, entomology educator; b. Washington, Mar. 6, 1947; s. David Goodsell, Sr. and Pauline (Overholt) H.; m. Sarah J. Dinsmore, Nov. 29, 1968; children: Emily D., Pamela O. B.A., U. Md., 1973; M.S., Va. Poly. Inst. and State U., 1975, Ph.D., 1977. Cert. Am. Registry of Profl. Entomologists. Asst. prof. entomology U. Mo.-Columbia, 1977—. Served to staff sgt. USAF, 1968-72; capt. USAR, 1978—. Mem. Am. Registry Profl.

Entomologists, Entomol. Soc. Am., Sigma Xi, Phi Kappa Phi, Phi Sigma. Club: Cosmos (Washington). Current work: Biology and control of arthropods affecting livestock and humans. Home: 512 Poplar St Boonville MO 65233 Office: Dept Entomology 1 87 Agriculture U Columbia MO 65211

HALL, ROBERT DILWYN, psychologist, educator; b. Phila., Jan. 19, 1929; s. Joseph Howard and Fern Elizabeth (Trempe) H.; m. Sandra Lee Terry, July 15, 1962; children: Jennifer Ann, Peter Joseph, Matthew Robert. A.B., Dartmouth, Coll., 1956; M.A. in Psychology, Brown U., 1958; Ph.D., 1959; instr. psychology Brown U., Providence, 1959-60; research assoc., staff scientist M.I.T. Research Lab. of Electronics, Cambridge, 1960-74; staff scientist Neurosci. Research Program, 1974-76, Worcester Found. for Exptl. Biology, Shrewsbury, Mass., 1976-83; vis. prof. Bowdoin Coll., Brunswick, Maine, 1983—. Contbr. articles on psychology to profl. jours. Served with U.S. Army, 1951-53. NSF fellow, 1956-59; NIH fellow, 1960-63. Mem. Am. Psychol. Assn., Psychonomic Soc., Internat. Soc. for Developmental Psychobiology, Soc. for Neurosci., AAAS, Phi Beta Kappa. Subspecialties: Physiological psychology; Learning. Current work: Research on the effects of early malnutrition on the nervous system and behavior, research on neural mechanisms of learning. Home: 30 Possum Ln Sudbury MA 01776 Office: Dept Psychology Bowdoin Coll Brunswick ME 04001

HALL, ROBERT NOEL, physicist; b. New Haven, Dec. 25, 1919; m., 1941; 2 children. B.S., Calif. Inst. Tech., 1942, Ph.D. in Physics, 1948. Lab. asst. Calif. Inst. Tech., 1942-46, research assoc., 1948-52; physicist research lab. Corp. Research and Devel. Ctr., Gen. Electric Co., Schenectady, 1952—. Fellow Am. Phys. Soc., IEEE (David Sarnoff award 1963, Jack A. Morton award 1976); mem. Nat. Acad. Engring., Nat. Acad. Sci., Electrochem. Soc. (Solid State Sci. and Tech. award 1977). Subspecialty: Solid state physics. Office: Gen Electric Corp Research and Devel Center PO Box 8 Schenectady NY 12301

HALL, STANTON HARRIS, orthodontic educator, orthodontist; b. Boise, Idaho, Apr. 8, 1940; s. Perce and Orpha (Harris) H.; m. Sharon V. Price, June 30, 1962; children: Jennifer A., Camille E., Matthew R. D.D.S., Northwest U., 1967, M.S., 1967; Ph.D., U. Wash., 1974; cert. orthodontics, U. Wash-Seattle, 1979. Research assoc. Lab. Molecular Genetics, Nat. Inst. Child Health and Human Devel., NIH, Bethesda, Md., 1974-77; acting asst. prof. U. Wash., Seattle, 1979-80, asst. prof. orthodontics, 1980—. Served to comdr. USPHS, 1974-77. NIH grantee, 1981; NIH predoctoral research fellow Northwestern U., 1964. Mem. ADA, Internat. Assn. Dental Research, Craniofacial Biology Study Group, Am. Assn. Orthodontists. Democrat. Mem. Ch. Jesus Christ of Latter-day Saints. Subspecialties: Cell and tissue culture; Dental growth and development. Current work: Differentiation of osteoprogenitor cells, sutural morphogenesis. Home: 4549 Thackeray Pl NE Seattle WA 98105 Office: Dept Orthodontic U Wash Seattle WA 98195

HALL, STEPHEN WILLIAM, physician, medical educator, cancer researcher; b. Virginia, Minn., Oct. 20, 1946; s. William Frank and Nelma Loraine (Taipale) H.; m. Annabelle Whiting Hall, July 1, 1971; children: Ryan K., Reid S. B.S., Ariz. State U., 1967; M.D., U. Ariz., 1971. Intern U. Ariz., Tucson, 1971-72; resident, 1972-74; fellow in med. oncology Cancer Ctr., U. Tex., 1974-76; asst. prof. clin. pharmacology M.D. Anderson Hosp. and Tumor Inst., 1976-79; asst. prof. medicine U. Tex., Houston, 1977-83; assoc. prof. medicine U. Nev., Reno, 1979-83, prof., 1983—, dir. oncology hematology, 1979—. Contbr. articles to profl. jours. Am. Cancer Soc. fellow, 1974-75, 78-79. Mem. Am. Soc. Clin. Oncology, Am. Assn. Cancer Research, Am. Fedn. Clin. Research. Subspecialties: Cancer research (medicine); Chemotherapy. Current work: Cancer research, clinical pharmacology of antineoplastic agents, clinical trials cancer research. Home: 15425 Cherrywood Dr Reno NV 89511 Office: 1000 Locust St Reno NV 89520

HALL, THOMAS CHRISTOPHER, pharmacologist, educator, medical program administrator; b. N.Y.C., Nov. 26, 1921; s. John Clarence and Theresa (McDonald) H.; m. Lorina A. Friesen, July 30, 1978; children: Christopher, Thomas, Seth, Amity, Bronwen, Nathan, Jinny, Nicholas. M.D., Harvard U., 1949. Diplomate: Am. Bd. Internal Medicine. Intern Peter Bent Brigham Hosp., Boston, 1949-50, assoc. in medicine, 1964; research fellow Harvard Med. Sch., 1950-53, teaching fellow, 1954, asst. in medicine, 1955; clin. and research fellow Mass. Gen. Hosp., Boston, 1951-53, resident, 1954, asst. in medicine, 1955; sr. research assoc. in biochemistry Brandeis U., Waltham, Mass., 1959, adj. assoc. prof., 1961; asst. physician Children's Hosp. Med. Ctr., Boston, 1961, research assoc. in pathology, 1961, sr. assoc. in medicine, 1963; sr. assoc. Children's Cancer Research Found., 1961, chief clin. and biochem., pharmacology, 1964; dir. div. oncology U. Rochester (N.Y.), 1968-72, prof. medicine and pharmacology, 1968-72; physician Strong Meml. Hosp., 1968-72; dir. medicine and biochemistry Los Angeles County/U. So. Calif. Cancer Hosp. and Research Center, 1972-75; dir. Cancer Control Agy. of B.C., 1975-76; prof. medicine U. B.C. Faculty of Medicine, 1975-77; scientist Pasadena (Calif.) Found for Med. Research, 1977; clin. prof. U. Calif., Irvine, 1978—; dir. cancer control Cancer Ctr. of Hawaii, U. Hawaii, 1978—; cons. in internal medicine and oncology to various hosps. Contbr. articles to profl. jours. Recipient grants in field, various lectureships. Mem. AAAS, Am. Assn. for Cancer Edn., Am. Assn. for Cancer Research (dir. 1971-74), Am. Soc. Clin. Pharmacology and Chemotherapy, A.C.P., Internat. Soc. for Biochem. Pharmacology, Am. Fedn. for Clin. Research, Am. Soc. Clin. Oncology, Am. Soc. Hematology, Am. Soc. for Pharmacology and Exptl. Therapeutics, AAUP, Endocrine Soc., James Ewing Soc., Histochem. Soc., Internat. Soc. Chemotherapy, Radiation Research Soc., Internat. Lung Cancer Working Party, Soc. for Cryobiology, Soc. for Devel. Biology, Western Assn. Physicians, Soc. for Exptl. Biology and Medicine, Western Soc. for Clin. Research, Can. Oncologic Soc., Sigma Xi, Alpha Omega Alpha. Subspecialties: Cancer research (medicine); Chemotherapy. Current work: Pharmacology and cancer control. Home: 263 Kakahiaka St Kailua HI 96734 Office: Cancer Control Edn and Outreach Program Cancer Center Hawaii U Hawaii 1236 Lauhala St Honolulu HI 96813

HALL, WILLIAM BARTLETT, geology educator; b. Cin., July 12, 1925; s. Rufus Bartlett and Frances Glenn (Ebersole) H.; m. Elizabeth Carson, June 15, 1951; children: Molly V., Patricia D., David E. A.B., Princeton U., 1950; M.S., U. Cin., 1951; Ph.D., U. Wyo., 1961. Cert. profl. geologist, Am. Inst. Profl. Geologists; registered profl. geologist, Idaho. Asst. geologist Stanolind Oil & Gas, Wichita Falls, Tex., 1950; geologist Pure Oil Co., Billings, Mont., 1951-54; grad. teaching asst. dept. geology U. Wyo., Laramie, 1954-57, instr., 1957-58; asst. prof. geol. engring. Mont. Sch. Mines, Butte, 1958-65; prof. geology U. Idaho, Moscow, 1965—. Served to sgt. U.S. Army, 1943-46. Sinclair fellow, 1955. Fellow Geol. Soc. Am. (outstanding paper award 1969); mem. Am. Assn. Petroleum Geologists, Am. Quaternary Assn., Am. Soc. Photogrammetry, Am. Inst. Profl. Geologists. Subspecialties: Remote sensing (geoscience); Geology. Current work: Geomorphology and structural geology from combined remote sensing and field mapping; landsat interactive digital image processing. Developed stereo projector remote controlled stand, 1966. Home: 2343 Wallen Rd Moscow ID 83843 Office: Dept Geology U Idaho Moscow ID 83843

HALL, WILLIAM FRANKLIN, JR., consulting company executive; b. Bellefonte, Pa., Jan. 8, 1931; s. William Franklin and Margaret Spotts H.; m. Dorothy Ann Young, June 15, 1957 (div. 1974); children: William W., Jonathon A., Stephanie Y. B.S., Pa. State U., 1952; M.S., Iowa State U., 1953; Ph.D., SUNY-Buffalo, 1972. Lic. nuclear reactor sr. operator. Contract engr. Foster-Wheeler Corp., N.Y.C., 1956-58; ops. mgr. Curtiss-Wright Research Reactor, Quehanna, Pa., 1958-59; ops. mgr., pres. Western N.Y. Nuclear Research Ctr., Buffalo, 1959-70, 70-72; program mgr. N.Y. State Atomic and Space Devel. Authority, N.Y.C., 1972-73; v.p. Ecology and Environment Internat., Buffalo, N.Y., 1973—. Served with U.S. Army, 1954-55. Mem. Ops. Research Soc. Am., Am. Nuclear Soc., Sigma Xi, Phi Eta Sigma, Sigma Tau Alpha. Subspecialties: Operations research (engineering); Environmental consulting. Current work: Heat transfer processes, nuclear reactor dynamics, stochastic processes in the court system. Home: 104 Fruehauf Ave Snyder NY 14226 Office: Ecology and Environment International PO Box D Buffalo NY 14225

HALL, WILLIAM JACKSON, statistician, educator; b. Beltsville, Md., Nov. 13, 1929; s. Reginald Foster and Lily (Hambleton) H.; m. Helen Bloxom Cox, Mar. 27, 1954 (div. 1981); children: Jacqueline Arden, Rebecca Clayton, Bryan Hambleton, Kay Randall.; m. Nancy T. Hufsmith, Jun. 1, 1982. A.B., Johns Hopkins U., 1950; M.A., U. Mich., 1951; Ph.D., U. N.C., 1955; postgrad., Manchester (Eng.) U., 1953, Cambridge (Eng.) U., 1954. Statistician Bell Telephone Labs., N.Y.C., 1954-55; asst. chief Polio Surveillance Unit, Communicable Disease Center, USPHS, Atlanta, 1955-57; lectr. U. Calif. at Berkeley, 1957, vis. prof., 1969; asst. prof. U. N.C., 1957-61, assoc. prof., 1961-66, prof. statistics, 1966-69; vis. prof. Stanford, 1967-69; prof. dept. stats. and div. biostats. U. Rochester, N.Y., 1969—, chmn. dept. stats., 1969-81; vis. prof. stats. and biostats. U. Washington, 1982. Asso. editor: Annals of Mathematical Statistics, 1968-73, Jour. Am. Statis. Assn, 1976-78. Fellow AAAS, Am. Statis. Assn., Inst. Math. Stats. (council 1973-76); mem. Royal Statis. Soc. Subspecialty: Statistics. Current work: Statistical theory; biostatistics. Home: 75 Chelmsford Rd Rochester NY 14618

HALL, WILLIAM JOEL, educator, consulting civil engineer; b. Berkeley, Calif., Apr. 13, 1926; s. Eugene Raymond and Mary (Harkey) H.; m. Elaine Frances Thalman, Dec. 18, 1948; children—Martha Jane, James Frederick, Carolyn Marie. Student, U. Calif. at Berkeley, 1943-44, U.S. Mcht. Marine Acad., 1944-45; B.S. in Civil Engring, U. Kans., 1948; M.S., U. Ill., Urbana, 1951, Ph.D., 1954. Teaching asst. U. Kans., 1947-48; engr. Sohio Pipe Line Co., 1948-49; mem. faculty U. Ill., Urbana, 1949—, prof. civil engring., 1959—; cons. structural dynamics seismic, materials to govt. orgns. and industry. Author books, articles, chpts. in books, revs. Recipient A. Epstein Meml. award U. Ill., 1958; Halliburton Engring. Edn. Leadership award, 1980. Fellow ASCE (pres. Central Ill. sect. 1967-68, chmn. structural div. exec. com. 1973—, chmn. tech. council on lifeline earthquake engring. exec. com. 1982—, Kan. sect. award 1948, Walter L. Huber award 1963), AAAS; mem. Nat. Acad. Engring., Am. Concrete Inst., ASME, Am. Welding Soc. (Adams Meml. membership award 1967), Internat. Assn. Bridge and Structural Engrs., Earthquake Engring. Research Inst., Seismol. Soc. Am., ASTM, Soc. Exptl. Stress Analysis, Am. Soc. Engring Edn., Ill., Nat. socs. profl. engrs., Sigma Xi, Tau Beta Pi, Sigma Tau, Chi Epsilon, Phi Kappa Phi. Subspecialties: Civil engineering; Theoretical and applied mechanics. Current work: Earthquake engineering, structural dynamics, and research/design activity on military systems hardness development. Home: 3105 Valley Brook Dr Champaign IL 61820 Office: 1245 Newmark Civil Engring Lab 208 N Romine St Urbana IL 61801

HALL, WILLIAM STERLING, psychology educator; b. Lonoke County, Ark., July 6, 1934; s. Joseph William and Mattie (Brock) H. A.B., Roosevelt U., 1957; Ph.D., U. Chgo., 1968. Instr., asst. prof. ednl. psychology N.Y. U., 1966-68; assoc. research psychologist Ednl. Testing Service, Princeton, N.J., 1968-70; asst. prof. psychology Princeton (N.J.) U., 1970-73; assoc. prof. psychology Vassar Coll., Poughkeepsie, N.Y., 1973-74, Rockefeller U., N.Y.C., 1974-78; prof. psychology U. Ill., Urbana-Champaign, 1978-81, U. Md., College Park, 1981—; mem. study sect. NIMH, 1977-81; mem. grad. evaluation panel NRC. Bd. dirs. Lazarus Awards Com., N.Y.C., 1975—; bd. dirs. Nat. Coll. Adv. Service, N.Y.C., 1982—. Carnegie Corp. grantee, 1975, 77; Ford Found. grantee, 1975. Fellow N.Y. Acad. Scis.; mem. AAAS (sci. fellows selection com.), Am. Psychol. Assn., Soc. Research Child Devel., Sigma Xi, Alpha Phi Alpha. Republican. Subspecialty: Developmental psychology. Current work: Cultural context of language development; psycholinguistics; developmental psycholinguistics. Home: 1140 23d St NW Washington DC 20037 Office: Dept Psychology U Md College Park MD 20742

HALLER, GARY LEE, chemical engineer; b. Loup City, Nebr., July 10, 1941; m., 1962; 2 children. B.A., Kearney State Coll., 1962; Ph.D. in Chemistry, Northwestern U., 1966. NATO fellow in phys. chemistry Oxford U., 1966-67; asst. prof. engring. and applied sci., 1967-72, assoc. prof., 1972-80; prof. chem. engring. Yale U., New Haven, 1980—. Mem. Am. Chem. Soc., Am. Inst. Chem. Engrs., Catalysis Soc. Subspecialty: Chemical engineering. Office: Dept Chem Engring Yale U New Haven CT 06520

HALLETT, JOHN, physicist; b. Bristol, Eng.; came to U.S., 1966; m. Joan Terry Collar, July, 1960; children: Jennifer, Joyce, Elaine, Rosemary. B.Sc., U. Bristol, Eng., 1953; Ph.D., Imperial Coll., U. London, 1958. Asst. prof. meteorology UCLA, 1960-62; lectr. physics U. London, 1962-66; Marston prof. atmospheric physics Desert Research Inst., U. Nev., Reno, 1978—. Mem. Am. Meteorol. Soc., Royal Meteorol. Soc., Internat. Glaciology Soc. Subspecialties: Condensed matter physics; Meteorology. Current work: Cloud physics; atmospheric electricity, crystal growth. Office: Desert Research Inst U Nev PO Box 60220 Reno NV 89506

HALLETT, MARK, physician, researcher, educator; b. Phila., Oct. 22, 1943; s. Joseph W. and Estelle (Barg) H.; m. Judith Peller, June 26, 1966; children: Nicholas, Victoria. A.B. magna cum laude, Harvard U., 1965, M.D. cum laude, 1969. Diplomate: Am. Bd. Psychiatry and Neurology, Am. Bd. EEG. Medicine intern Peter Bent Brigham Hosp., Boston, 1969-70; research fellow NIH, 1970-72, Inst. Psychiatry, London, 1975-76; resident in neurology Mass. Gen. Hosp., Boston, 1972-75; dir. clin. neurophysiology Brigham and Women's Hosp., Boston, 1976—; asst. prof. neurology Harvard U. Med. Sch., Boston, 1977-83, assoc. prof., 1983—. Contbr. numerous articles to sci. jours. Served with USPHS, 1970-72. Mem. Am. Acad. Neurology, Am. Assn. Electromyography and Electrodiagnosis, Am. EEG Soc., Soc. for Neurosci., Phi Beta Kappa, Alpha Omega Alpha. Subspecialties: Neurology; Neurophysiology. Current work: Neurophysiology of control of movement in man, neurology, neurophysiology, electromyography, electroencephalography. Home: 5147 Westbard Ave Bethesda MD 20816 Office: Brigham and Women's Hosp Boston MA 02115

HALLICK, RICHARD BRUCE, biochemist, educator, consultant; b. Glendale, Calif., Jan. 15, 1946; s. John and Rosa Elizabeth (Coates) H.; m. Lesley M. Moore, Aug. 4, 1968; m. Elizabeth Ann Crocker, June 3, 1978; children: Deborah Marie, Matthew Richard, Christopher Andrew. B.A., Pomona Coll., 1967; Ph.D. in Biochemistry, U. Wis.-Madison, 1971. NIH undergrad. trainee Pomona Coll., summer 1965, 65-67, Max Planck Inst., Munich, W.Ger., summer 1966; NIH grad. trainee U. Wis., 1967-71; Am. Cancer Soc. postdoctoral fellow U. Calif.-San Francisco, 1971-73; asst. prof. chemistry U. Colo., Boulder, 1973-78, assoc. prof., 1978-83, prof., 1983—; mem. molecular cytology study sect. USPHS/NIH, 1981-85. Contbr. articles to profl. jours.; editorial bd.: Plant Sci. Letters, 1980—, Jour. Cell Biology, 1981-84, Jour. Biol. Chemistry, 1984—. Recipient Research Career Devel. award NIH, 1977-82; NSF postdoctoral fellow, 1971; NIH grantee, 1973-75, 76-82, 82-87, 80-83. Mem. Plant Molecular Biology Assn., Am. Chem. Soc., Am. Soc. Cell Biology, Am. Soc. Biol. Chemists. Episcopalian. Subspecialties: Genetics and genetic engineering (agriculture); Genome organization. Current work: Chloroplast molecular biology; investigation of chloroplast gene orgn. and expression. Office: U Colo Dept Chemistry Campus Box 215 Boulder CO 80309

HALLIGAN, JAMES EDMUND, university administrator, chemical engineer; b. Moorland, Iowa, June 23, 1936; s. Raymond Anthony and Margaret Ann (Crawford) H.; m. Ann Elizabeth Sorenson, June 29, 1957; children: Michael, Patrick, Christopher. M.S. in Chem. Engring, Iowa State U., 1962, 1965, Ph.D., 1968. Process engr. Humble Oil Co., 1962-64; mem. faculty Tex. Tech U., 1968-77; dean engring. U. Mo., Rolla, 1977-79, U. Ark., Fayetteville, 1979-82, vice chancellor for acad. affairs, 1982-83, interim chancellor, 1983—; v.p. engring. Kandahar Cons. Ltd.; mem. Gov. Tex. Energy Adv. Council, 1972-74. Served with USAF, 1954-58. Recipient Disting. Teaching award Tex. Tech U., 1972, Disting. Research award, 1975, 76; Disting. Teaching award U. Mo., Rolla, 1978. Mem. Am. Inst. Chem. Engrs., Am. Soc. Engring. Edn., Tau Beta Pi, Phi Kappa Phi, Pi Mu Epsilon. Roman Catholic. Club: Rotary. Subspecialty: Chemical engineering. Office: Adminstrn Bldg 422 U Ark Fayetteville AR 72701

HALLOIN, JOHN MCDONELL, plant pathologist; b. Green Bay, Wis., Aug. 14, 1938; s. Joseph Emil and Dorothy Belle (McDonell) H.; m. Jeanne Muyskens, Mar. 17, 1967; children: David Muyskens, Joseph Muyskens. B.S., U. Wis., 1960; M.S., U. Minn., 1964; Ph.D., Mich. State U., 1968. Research assoc. dept. plant pathology U. Wis., Madison, 1968-71; asst. prof. botany and plant pathology Mich. State U., East Lansing, 1971; research asso. biochemistry Iowa State U., Ames, 1971-72; research plant physiologist Nat. Cotton Pathology Research Lab., U.S. Dept. Agr., College Station, Tex., 1972—. Contbr. articles in field to profl. jours. Mem. Am. Phytopath. Soc., Am. Soc. Plant Physiologists, Am. Soc. Agronomy, Crop Sci. Soc. Am., Cotton Disease Council, Sigma Xi. Subspecialties: Plant pathology; Plant physiology (agriculture). Current work: Researcher on mechanisms of deterioration and resistance to deterioration of seeds, with primary emphasis on improving quality of cotton planting seed. Home: RR5 Box 1257 College Station TX 77840 Office: National Cotton Pathology Research Lab PO Drawer JF College Station TX 77841

HALME, JOUKO KALERVO, endocrinologist; b. Helsinki, Finland, Oct. 23, 1942; came to U.S., 1977; s. Kalervo A.A. and Helmi E.A. (Tuominen) H.; m. Anja-Pirkko Larkio, Oct. 29, 1976; children: Anna, Orvokki; m.; 1 dau. by previous marriage: Kati. M.B., U. Helsinki, 1964, M.D., 1968, Ph.D., 1970, docent, 1977. Lic. physician, Finland, Mo., N.C. Postdoctoral fellow in biochemistry U. Miami, Fla., 1970-71; resident in obstetrics/gynecology U. Helsinki, 1972-75, acting asso. prof., 1976; instr. medicine Washington U., St. Louis, 1977-80; fellow reproductive endocrinology U. N.C., Dept. Obstetrics and Gynecology, Chapel Hill, 1980-82, asst. prof., 1982—. Contbr. articles to profl. jours. NIH grantee, 1983—. Fellow Am. Coll. Obstetricians and Gynecologists; mem. Am. Fedn. Clin. Research, N.C. Med. Soc., Finnish Med. Assn., Finnish Gynecol. Assn. (sec. 1973-74). Subspecialties: Reproductive biology (medicine); Immunology (medicine). Current work: Endometriosis and cell-mediated immunology; mechanism of infertility in endometriosis. Home: 149 Dixie Dr Chapel Hill NC 27514 Office: Univ NC 214 Macnider Bldg 202H Chapel Hill NC 27514

HALONEN, MARILYN JEAN, immunologist, educator; b. Duluth, Minn.; d. George and Helmi E. (Aalto) Wainio; m. Michael A. Cusanovich, 1 son, Darren Anthony. B.S., U. Minn., 1963; M.S., Iowa State U., 1968; Ph.D., U. Ariz., 1974. Research asso. dept. microbiology U. Ariz. Coll. Medicine, Tucson, 1974-77, adj. asst. prof. medicine, div. respiratory scis., 1977-83, research assoc. prof. medicine, 1983—, mem. faculty grad. program molecular biology, 1977—; guest worker Nat. Inst. Allergy and Infectious Disease, NIH, Bethesda, Md., 1981-82. Contbr. articles to sci. jours. NIH grantee, 1975, 77—. Mem. Am. Assn. Immunologists. Subspecialties: Immunobiology and immunology; Immunopharmacology. Current work: Research in animal models of acute allergic reactions. Office: Div Respiratory Scis Coll Medicine U Ariz Tucson AZ 85724

HALPERIN, BERTRAND ISRAEL, physics educator; b. Bklyn., Dec. 6, 1941; s. Morris and Eva (Teplitsky) H.; m. Helena Stacy French, Sept. 23, 1962; children: Jeffery Arnold, Julia Stacy. A.B., Harvard U., 1961; A.M., U. Calif., 1963, Ph.D., 1965; vis. grad. student, Princeton U., 1964-65. NSF postdoctoral fellow U. Paris, 1965-66; mem. tech. staff Bell Labs., Murray Hill, N.J., 1966-76; lectr. Harvard U., 1969-70, prof. physics, 1976—; cons. Bell Labs. Assoc. editor: Revs. Modern Physics, 1973-80. Fellow Am. Phys. Soc. (Oliver Buckley prize 1982), Am. Acad. Arts and Scis.; mem. Nat. Acad. Scis. Subspecialty: Condensed matter physics. Research in solid state theory, statis. physics. Office: Dept Physics Harvard U Cambridge MA 02138

HALPERIN, JOHN JACOB, neurologist, educator; b. Montreal, Que., Can., Jan. 25, 1950; s. David and Maizie (Pottel) H.; m. Toula Jaravinos, June 15, 1975; 1 son, Daniel Mark. B.S. in Physics, MIT, 1971; M.D., Harvard U., 1975. Diplomate: Am. Bd. Internal Medicine, Am. Bd. Psychology and Neurology. Intern and resident in medicine U. Chgo. Hosp. and Clinics, 1975-77; resident in neurology Mass. Gen. Hosp., Boston, 1977-80, fellow in neurology, 1980-83; instr. neurology Harvard U. Med. Sch., 1981-83; asst. prof. neurology SUNY-Stony Brook, 1983—; cons. in neurology McLean Hosp., Belmont, Mass., 1980—. Contbr. articles to profl. jours. NIH fellow, 1980-82. Mem. ACP, Am. Acad. Neurology, Soc. Neurosci., AAAS, N.Y. Acad. Scis. Subspecialties: Neurology; Neurobiology. Current work: Ultrastructure of axons and synapses, development of synapses neurology of neuromuscular diseases, clinical electrophysiology; ultrastructure of synapses in development and regeneration. Office: Neuro Ambulatory Uni Mass Gen Hosp Boston MA 02114

HALPERN, ARTHUR M(ERRILL), chemist, educator; b. Bayonne, N.J., Aug. 4, 1943; s. Maurice and Edna (Green) H.; m. Janis M. Kaye; children: Sharon, Alison, David, Maura. B.A., Rutgers U., 1964; Ph.D., Northeastern U., 1968. Research assoc. U. Minn., 1968-70; mem. tech. staff Bell Labs., 1970-71; asst. prof. chemistry N.Y.U., 1970-73, Northeastern U., Boston, 1973-77, assoc. prof., 1977-81, prof., 1981—; indsl. cons. Author books, articles and revs. Alfred P. Sloan fellow, 1974-76; NATO sr. scientist fellow, 1981; grantee NSF, NIH, Research Corp., Petroleum Research Fund Am Chemistry, Air Force Office Sci. Research. Mem. Am. Chem. Soc., Am. Phys. Soc., AAAS, European Photochem. Assn. Subspecialties: Physical chemistry; Photochemistry. Current work: Photochemistry and photophysics of organic molecules; photoassociation, photokinetics; thermodynamics and kinetics of excited dimers and complexes. Office: 360 Huntington Ave Boston MA 02115

HALPERN, JACK, chemistry educator; b. Poland, Jan. 19, 1925; came to U.S., 1962; s. Philip and Anna (Sass) H.; m. Helen Peritz, June 30, 1949; children: Janice Deborah, Nina Phyllis. B.Sc., McGill U., 1946, Ph.D., 1949. Postdoctorate overseas fellow NRC, U. Manchester (Eng.), 1949-50; instr. chemistry U. B.C., 1950, prof., 1961-62; Nuffield Found. traveling fellow Cambridge (Eng.) U., 1959-60; prof. chemistry U. Chgo., 1962-71, Louis Block prof. chemistry, 1971—; x; vis. prof. U. Minn., 1962, Harvard, 1966-67, Calif. Inst. Tech., 1968-69, Princeton U., 1970-71, Max. Planck Institut, Mulheim, W. Ger, 1970-71, U. Copenhagen, 1978; Sherman Fairchild Disting. scholar Calif. Inst. Tech., 1979; guest scholar Kyoto U., 1981; Firth vis. prof. U. Sheffield, 1982; numerous guest lectureships; cons. editor Macmillan Co., 1963-65, Oxford U. Press; cons. Am. Oil Co., Monsanto Co., Argonne Nat. Lab., IBM, Air Products Co.; mem. adv. bd. Am. Chem. Soc. Petroleum Research Fund, 1972-74; mem. medicinal chemistry sect. NIH, 1975-78, chmn., 1976-78; mem. chemistry adv. council Princeton U., 1982—; Mem. Art Inst. Chgo., 1964—. Asso. editor: Inorganica Chimica Acta, Jour. Am. Chem. Soc; co-editor: Collected Accounts of Transition Metal Chemistry, vol. 1, 1973, vol. 2, 1977; editorial bd.: Jour. Organometallic Chemistry; Contbr.: articles to research jours. Ency. Brit; Ency. Britannica; contbr.: Accounts of Chem. Research; Catalysius Revs.; Jour. of Catalysis, Jour. Molecular Catalysis, Jour. Coordination Chemistry, Gazzetta Chimica Italiana. Trustee Gordon Research Confs., 1968-70. Recipient Young Author's prize Electrochem. Soc., 1953; award in inorganic chemistry Am. Chem. Soc., 1968; award in catalysis Noble Metals Chem. Soc., London, 1976; Humboldt award, 1977; Richard Kokes award Johns Hopkins U., 1978; Alfred P. Sloan research fellow, 1959-63. Fellow Royal Soc. (London), AAAS, Am. Acad. Arts and Scis., Chem. Inst. Can., Royal Soc. Chemistry (London), N.Y. Acad. Scis.; mem. Am. Chem. Soc. (editorial bd. Advances in Chemistry series 1963-65, 78-81, chmn. inorganic chemistry div. 1971), Sigma Xi. Subspecialties: Inorganic chemistry; Catalysis chemistry. Current work: Inorganic, organometallic and bioinorganic chemistry; kinetics and mechanisms of inorganic and organometallic reactions; catalysis. Home: 5630 Dorchester Ave Chicago IL 60637 Office: U Chgo Dept Chemistry Chicago IL 60637

HALPERN, MARTIN, company administrator; b. Montreal, Que., Can., Dec. 24, 1937; came to U.S., 1959; s. Harry and Rachel (Cohen) H.; m.; children: Andrea, Adam. Sc.B., McGill U., 1959; M.S., U. Wis., 1961, Ph.D., 1963. Research scientist S.W. Ctr. Advanced Studies, Dallas, 1964-70; prof. U. Tex.-Dallas, 1970-81, assoc. dean, 1976-77; sr. staff geologist Enserch Exploration, Inc., Dallas, 1981-82, exploration mgr., Houston, 1982—; vis. research fellow U. Leeds, Eng., 1971-72; cons. Mex. Petroleum Inst., Mexico City, 1979-80. Recipient Antarctic Service medal NSF, 1964-80. Fellow Geol. Soc. Am.; mem. Am. Geophys. Union, Am. Assn. Petroleum Geologists. Subspecialties: Geology; Geochemistry. Current work: Petroleum exploration. Home: 722 B Country Place Dr Houston TX 77079 Office: Exserch Exploration Inc 10375 Richmond Ave Suite 1102 Houston TX 77042

HALPERN, MIMI N(AOMI), anatomy and cell biology educator, university dean, psychologist; b. Antwerp, Belgium, June 19, 1938; came to U.S., 1941, naturalized, 1948; m. (married); 2 children. A.B. in Psychology; Oberlin Coll., 1960; Ph.D. in Psychology, Adelphi U., 1964. Teaching asst. dept. psychology Oberlin Coll., 1959-60; teaching and research asst. dept. psychology Adelphi U., 1960-61, research asst. NSF grant, 1961-63, instr., 1963-64; assoc. research scientist SUNY Downstate Med. Ctr., 1964-67, instr., 1967-69, asst. prof. anatomy and cell biology and grad. program in biol. psychology, 1969-74, assoc. prof. anatomy and cell biology and grad. program in biol. psychology, 1974-79, prof. anatomy and cell biology, 1979—, asst. dean, 1975-82, assoc. dean, 1982—, dir. grad. program in biol. psychology, 1976—; vis. prof. Rockefeller U., 1980—; mem. organizing com. Internat. Soc. Neuroethology; ad hoc mem. biol. and neurosci. subcom., mental health research rev. com. NIMH. Contbr. chpts., numerous articles to profl. publs.; author profl. papers; reviewer for profl. jours.; book rev. editor, editorial bd.: Brain, Behavior and Evolution, 1972-76. NIH grantee, 1969-72, 74-80, 75-77, 78-82, 80-83; SUNY Research Found. grantee-in-aid, 1974-75. Mem. Am. Psychol. Assn., Eastern Psychol. Assn., Am. Assn. Anatomists, Soc. Neuroscis., AAAS, Am. Soc. Zoologists, Assn. Women in Sci., Cajal Club, Sigma Xi. Club: Oberlin of N.Y. (dir.). Current work: Biological basis of behavior. Home: 262 Central Park W New York NY 10024

HALPIN, DANIEL WILLIAM, civil engineering educator, consultant; b. Covington, Ky., Sept. 29, 1938; s. Jordan W. and Gladys E. (Moore) H.; m. Maria Kirchner, Feb. 8, 1963; 1 son, Rainer. B.S., U.S. Mil. Acad., 1961; M.S.C.E., U. Ill., 1969, Ph.D., 1973. Research analyst Constrn. Engring. Research Lab., Champaign, Ill., 1970-72; faculty U. Ill.-Urbana, 1972-73; mem. faculty Ga. Inst. Tech., Atlanta, 1973—, prof., 1981—; cons. constrn. mgmt.; vis. assoc. prof. U. Sydney, Australia, 1981. Author: Design of Construction and Process Operations, 1976, Construction Management, 1980, Planung und Kontrolle von Bauproduktionsprozessen, 1979, Constructo - A Heuristic Game for Construction Management, 1973. Served with C.E. U.S. Army, 1961-67. Decorated Bronze Star; recipient Walter L.Huber prize ASCE, 1979; grantee NSF, Dept. Energy. Mem. ASCE (past sect. pres. 1981-82), Am. Soc. Engring. Edn., Sigma Xi. Methodist. Subspecialties: Civil engineering; Systems engineering. Current work: Simulation of construction operations using computers; applications of microcomputers in construction management.

HALSEY, NORMAN DOUGLAS, aerodynamicist; b. St. Petersburg, Fla., May 17, 1947; s. Norman Cockrem and Virginia Ann (Knighton) H. B.S., U. Fla., 1969; M.S., Calif. State U.-Long Beach, 1978. Sr. engr./scientist Douglas Aircraft Co., Long Beach, Calif., 1970—; lectr. Calif State U., Long Beach, 1980-81. Recipient Recognition cert. NASA, 1980. Mem. AIAA, Soc. Indsl. and Applied Math. Democrat. Clubs: Windsurfer Fleet (Long Beach); Astronomy (Whittier). Subspecialties: Aeronautical engineering; Numerical analysis. Current work: Development of efficient and accurate numerical methods for computing flow over complex configuartions, particularly aircraft high-lift systems. Home: 916 Stevely Ave Long Beach CA 90815 Office: Douglas Aircraft Co 3855 Lakewood Blvd Long Beach CA 90846

HALSEY, WILLIAM GUY, nuclear engineer, researcher; b. Battle Creek, Mich., Sept. 23, 1953; s. Leroy W. and Margaret E. (Wood) H. Student, Mich. Technol. U., Houghton, 1971-73; B.S. in Nuclear Engring, U. Mich., 1975, M.S., 1976, 1978, Ph.D. in Nuclear Engring, 1980. Physicist Lawrence Livermore (Calif.) Lab., 1980-81, acting group leader, 1981—. Mem. Am. Nuclear Soc., Am. Soc. for Metals. Current work: Material science in support of fusion energy research. Home: 99 Mozden Ln Pleasant Hill CA 94523 Office: Lawrence Livermore Lab L-482 PO Box 5508 Livermore CA 94550

HALSTEAD, BRUCE WALTER, biotoxicologist; b. San Francisco, Mar. 28, 1920; s. Walter and Ethel Muriel (Shanks) H.; m. Joy Arloa Mallory, Aug. 3, 1941; children: Linda, Sandra, David, Larry, Claudia, Shari. A.A., San Francisco City Coll., 1941; B.A., U. Calif.-Berkeley, 1943; M.D., Loma Linda U., 1948. Research asst. in ichthyology Calif. Acad. Scis., 1935-43; instr. Pacific Union Coll., 1943-44; mem. faculty Loma Linda U., 1948- 58; research asso. Lab. Neurol. Research, Sch. Medicine, 1964—; dir. World Life Research Inst., Colton, Calif., 1959—, Internat. Biotoxicol. Center; research asso. in ichthyology Los Angeles County Mus., 1964—; instr. Walla Walla Coll., summers 1964—; Cons. to govt. agys., pvt. corps; mem. editorial staff Exerpta Medica, 1959—, Toxicon, 1962—; mem. joint group experts on sci. aspects marine pollution UN; Dir. Nat. Assn. Underwater Instrs., Internat. Underwater Enterprises, Internat. Bots., Inc. Author: Poisonous and Venomous Marine Animals of the World, 4 vols., 1966; others.; contbr. numerous articles to profl. jours. Fellow AAAS, Internat. Soc. Toxicology (a founder), N.Y. Acad. Scis., Royal Soc. Tropical Medicine and Hygiene; mem. Am. Inst. Biol. Scis., Am. Micros. Soc., Am. Soc. Ichthyologists and Herpetologists, Am. Soc. Limnology and Oceanography, numerous others. Subspecialties: Preventive medicine; Biotoxicology. Current work: Natural products research, immunology and degenerative diseases. Address: 23000 Grand Terrace Rd Colton CA 92324

HALVERSON, THOMAS GEORGE, nuclear facility safety exec.; b. Madison, Wis., Apr. 14, 1948; s. Arthur John and Mary Jane (Hoffman) H.; m. Linda Sue Vandine, Feb. 17, 1977; children: Aaron, Brian, Wendy, Margot. B.S. in Nuclear Engring, U. Wis., 1971. Tech. engr. Commonwealth Edison Co., Zion, Ill., 1971-74; design engr. Westinghouse Hanford Co., Richland, Wash., 1974-76, engring. sect. mgr., 1976-80; mgr. safety Fast Flux Test Facility, 1980-81, mgr. nuclear facility safety, 1981—. Pres. Lower Columbia Basin Search and Rescue, Kennewick, Wash., 1982; reservist Benton County Sheriff's Office, Kennewick, 1982. Mem. Am. Nuclear Soc. (progrma chmn. 1980-82). Lutheran. Club: Atomic Ducks Dive (Kennewick) (pres. 1981-82). Subspecialties: Nuclear engineering; Nuclear fission. Current work: Liquid metal fast reactor nuclear safety. Application of nuclear safety principles to operation of the Fast Flux Test Facility and advanced fuel research laboratories. Office: Westinghouse Hanford Co PO Box 1970 W/C 75 Richland WA 99352

HAMACHER, HORST WILHELM, Mathematical programming educator; b. Buir, Germany, Apr. 21, 1951; came to U.S., 1981; s. Aloys G. and Helene G. (Hanussek) H.; m. Renate K. Kremer, Aug. 22, 1972; children; Elke, Jens. Pre-diploma, U. Cologne, Germany, 1973, diploma, 1977, Ph.D., 1980. Lectr. Maths. U. Cologne, 1973-77, asst. prof., 1977-81; instr. Coll. Engring., Ger., 1980-81; asst. prof. U. Fla., Gainesville, 1981—; Speaker Sch. Parliament, Bergheim, Ger., 1967-69; treas. Rudolf Steiner Edn. Assn., Gainesville, 1982. Author: Flows in Regular Matroids, 1981. Mem. Math. Programming Soc., Soc. Indsl. and Applied Math., Am. Math. Assn., Ops. Research Soc. Roman Catholic. Subspecialties: Operations research (mathematics); Algorithms. Current work: Research and teaching in mathematical programming, combinatorial optimization and applied mathematics. Home: 5229 NW 26th Pl Gainesville FL 32606 Office: U of Fla 303 Weil Hall Gainesville FL 32611

HAMAD, CHARLES DEAN, psychologist, educator; b. Danbury, Conn., Oct. 17, 1949; s. and Doris (Hamed) H. B.A., Quinnipiac Coll., 1972; M.A., C.W. Post Coll., L.I.U., 1974; Ph.D., U. Kans.-Lawrence, 1977. Chief psychologist Walter E. Fernald State Sch., Belmont, Mass., 1977—, dir. psychology, 1979—; assoc. psychologist Eunice Kennedy Shriver Ctr., Belmont, 1978—; vis. asst. prof. Northeastern U., 1978—. Contbr. articles to profl. jours., papers to profl. cons. Mem. Am. Psychol. Assn., Assn. Behavior Analysis. Democart. Subspecialty: Behavioral psychology. Current work: Mental retardation research; behavior analysis. Office: Walter E Fernald State Sch Dept Psychology 200 Trapelo Rd Belmont MA 02178 Home: 11 Centre St Cambridge MA 02139

HAMASAKI, SEISHI, plasma physicist; b. Yamaguchi-ken, Japan, Mar. 10, 1937; came to U.S., 1961; s. Tamotsu and Shizue (Kawano) H.; m. Miyako T. Kanetani, Mar. 30, 1970; 1 dau., Sonya. B.S., Waseda U., Tokyo, Japan, 1961; M.A., Wesleyan U., 1963; Ph.D., U. Wis.-Madison, 1967. Research assoc. U. Md.-College Park, 1969-72, Cornell U., Ithaca, N.Y., 1972-73, U. Md.-College Park, 1973-74; sr. scientist SAI, La Jolla, Calif., 1974-79, Jaycor, San Diego, 1979—; cons. Los Alamos (N.Mex.) Nat. Lab., 1980—. Mem. Am. Phys. Soc., Phys. Soc. of Japan. Subspecialties: Nuclear fusion; Plasma. Current work: Microinstability and its related transport phenomena of plasma, R.F. heating of plasma. Home: 2203 14th St Olivenhain CA 92024 Office: Jaycor 11011 Torreyana Rd San Diego CA 92138

HAMBLEN, JOHN WESLEY, computer scientist; b. Story, Ind., Sept. 25, 1924; s. James William and Mary Etta (Morrison) H.; m. Brenda F. Harrod, Mar. 1, 1947 (div. 1979); 1 son, James. A.B., Ind. U., 1947; M.S., Purdue U., 1952; Ph.D., 1955. Tchr. math and sci. Kingsbury (Ind.) High Sch., 1946-48, Bluffton (Ind.) High Sch., 1948-51; asst. prof. math. Okla. State U., Stillwater, 1955-57; cons. in statis. methods for research staff Agrl. Expt. Sta., 1955-56, asso. prof. math., 1957-58; dir. Computing Center, 1957-58; asso. prof. stats., dir. Computing Center, U. Ky., Lexington, 1958-61; prof. math and technology Southern Ill. U., Carbondale, 1961-65; dir. Data Processing and Computing Center, 1961-65; project dir. computer scis. So. Regional Edn. Bd., Atlanta, 1965-72; prof. U. Mo., Rolla, 1972—, chmn. dept. computer sci., 1972-81; mem. tech. adv. com. Creative Application of Tech. to Edn., Tex. A. and M. U., 1966-68; mem. tech. adv. panel Western Interstate Commn. for Higher Edn., 1969-70; vis. scientist Ctr. for Applied Math. Nat. Bur. Standards, 1981-83; program chmn. World Conf. Computers in Edn., 1985; cons. FTC, 1978—, NSF, 1975-76. Editor: Ednl. Data Processing Newsletter, 1964-65; asso. editor: Jour. Ednl. Data Processing, 1965-67; editor: Jour. Assn. Ednl. Data Systems, 1967-68; asso. editor, 1968—; contbr. articles to profl. jours. Purdue Research Found. fellow, 1954-55; NSF grantee, 1966-81. Fellow AAAS; mem. Assn. Computing Machinery (sec. 1972-76, chmn. curriculum com. computer sci. 1976-80, gen. chmn. 1981 Computer Scis. Conf. 1979-81, chmn. Disting. Ser. Award com. 1980-81), IEEE Computer Soc., Inst. Math Stats., Data Processing Mgmt. Assn., Assn. Ednl. Data Systems (chmn. conv. adv. com. 1977-80, pres. 1968-69, sec. 1976-77, dir. 1965-70, 76-79), Am. Fedn. Info. Processing Socs. (dir. 1981—, chmn. edn. com. 1971-72, 79—), Soc. Indsl. and Applied Math, Am. Statis. Assn., Math. Assn. Am., Sigma Xi, Pi Mu Epsilon, Theta Chi, Upsilon Pi Epsilon, Alpha Chi Sigma. Club: Rotary. Home: Route 1 Box 256A Saint James MO 65557 Office: Dept Computer Science Univ Mo Rolla MO 65401 It is difficult to improve upon the popular version of the "Golden Rule" for a succinct guide in life. A clear conscience and a good insurance program contribute greatly to a good night's sleep. With moderation in food and drink plus a good night's rest we should be able to handle most anything that comes our way.

HAMBRECHT, FREDERICK TERRY, med. research dir., researcher; b. Galesburg, Ill., Aug. 18, 1939; s. Frederic Emerson and Mary Ellen (Walpole) H.; m. Gloria Jean, Dec. 7, 1965; children: Christopher Emerson, Kevin Matthew. B.S., Purdue U., 1961; M.S., M.I.T., 1963; M.D., Johns Hopkins U., 1968. Lic. physician, Md. Research asso. dept. elec. engring. M.I.T., 1961-63; intern in surgery Duke U., Durham, N.C., 1968-69; commd. officer USPHS, 1969; research asso. lab. neural control Nat. Inst. Neurol. and Communicative Disorders and Stroke, Bethesda, Md., 1969—, asst. project officer sensory prosthesis program., 1970-72, head neural prosthesis program, 1972—. Co-editor: Functional Electrical Stimulation: Applications in Neural Prostheses, 1977. Bd. dirs. Centers for Handicapped Devel. Corp., 1981—. Recipient Commendation medal USPHS, 1975, Meritorious Service medal, 1980. Mem. Soc. Neurosci., Biomed. Engring. Soc. (chmn. publ. bd. 1982-83). Subspecialties: Biomedical engineering; Artificial organs. Current work: Neural Prostheses, application of biomedical engineering to diagnosis and therapy of human disorders, physiology of neural control. Home: 14015 Manorvale Rd Rockville MD 20853 Office: 7550 Wisconsin Ave Room 916 Bethesda MD 20205

HAMBURG, DAVID A., psychiatrist; b. Evansville, Ind., 1925. M.D., Ind. U., 1947, D.Sc. (hon.), 1976, Rush U., 1977. Diplomate: in Psychiatry, Am. Bd. Psychiatry and Neurology. Intern Michael Reese Hosp., Chgo., 1947-48, resident in psychiatry, 1949-50; asst. resident in psychiatry Yale U.-New Haven Hosp., 1948-49; practice medicine, specializing in psychiatry, 1950—; staff psychiatrist Brooke Army Hosp., 1950-52; research psychiatrist Army Med. Service Grad. Sch., 1952-53; asso. dir. Psychosomatic and Psychiat. Inst., Michael Reese Hosp., Chgo., 1953-56; fellow Center for Advanced Study in Behavioral Scis., Palo Alto, Calif., 1957-58, 67-68; chief Adult Psychiat. br. NIMH, Bethesda, Md., 1958-61; asst. in pathology Ind. U., 1946-47; asst. in psychiatry Yale U., 1948-49; prof., exec. head dept. psychiatry Stanford U. Med. Sch., 1961-72, Reed-Hodgson prof. human biology, 1972-76; Sherman Fairchild Distinguished scholar Calif. Inst. Tech., 1974-75; pres. Inst. Medicine Nat. Acad. Scis., Washington, 1975-80; dir. div. health policy research and edn. Harvard U., Cambridge, Mass., 1980—. Served as capt. M.C. AUS, 1950-53. Recipient numerous awards including; Pres.'s medal Michael Reese Med. Center, 1974; A.C.P. award, 1977; Mass. Inst. Tech. Bicentennial medal, 1977. Mem. Am. Psychiat. Assn. (Vestermark award 1977), AAAS, Am. Psychosomatic Soc., Assn. Research Nervous and Mental Disease (pres. 1967-68), Internat. Soc. Research on Aggression (pres. 1976-78), Internat. Soc. Research in Psychoneuroendocrinology, Psychiat. Research Soc. (chmn. 1965-66, 67-68), Am. Acad. Arts and Scis., Phi Beta Kappa, Alpha Omega Alpha. Subspecialties: Health services research; Psychiatry. Address: Center Health Policy Kennedy Sch Govt Harvard U 79 Boylston St Cambridge MA 02138

HAMBURGER, MAX I., physician. educator; b. Long Branch., N.J., June 20, 1947; s. Aaron and Dorothy (Friedl) H.; m. Frances Marsha, Nov. 21, 1971; children: Jordan, Nicole. B.A., Rutgers U., 1969; M.D., Albert Einstein Coll. Medicine, 1973. Intern Bellevue Hosp. Ctr., N.Y.C., 1973-74, resident in medicine, 1974-76; clin. assoc. clin. immunology sect. Lab. Clin. Investigation, Nat. Inst. Allergy and Infectious Diseases, NIH, Bethesda, Md., 1976-79; asst. prof. dept. medicine div. allergy and rheumatology, dir. therapeutic pheresis SUNY, Stony Brook, 1979—. Sr. editor: Plasma Therapy and Transfusion Tech, 1982—; editor: Jour. Clin. Apheresis, 1982—; Contbr. articles to profl. jours. Chmn. profl. adv. com. L.I. Arthritis Found.; mem. exec. council N.Y. Arthritis Found. Served to lt. comdr. USPHS, 1976-79. Mem. Am. Fedn. Clin. Research, Am. Assn. Immunologists, Am. Rheumatism Assn., Suffolk County Med. Soc., N.Y. Rheumatism Assn., Am. Soc. Apheresis (dir.), Phi Beta Kappa, Delta Phi Alpha. Subspecialties: Immunology (medicine); Infectious diseases. Current work: Immune complex diseases, complement proteins. Office: 222 E Main St Suite 115 Smithtown NY 11787

HAMBY, DRANNAN C., electrochemist, educator; b. Duncan, Okla., Nov. 16, 1933; s. Wellington V. and Dessie A. H.; m. Beverly Reinhart, Mar. 1, 1952; children: Mark, Marcy. B.A., Linfield Coll., 1954; M.A., Oreg. State U., 1962, Ph.D., 1971. Chemist Linfield Research Inst., McMinnville, Oreg., 1956—; prof. chemistry Linfield Coll., McMinnville, 1962—. Contbr. articles to profl. jours. Mem. McMinnville City Council, 1976-80. Fulbright fellow, 1954-55. Mem. Am. Chem. Soc., Electrochem. Soc., Sigma Xi. Democrat. Baptist. Subspecialties: Electrochemistry; Physical chemistry. Current work: Batteries, high temperature electrolysis. Patentee in field.

HAMERTON, JOHN LAURENCE, pediatrics educator, researcher; b. Eng., Sept. 23, 1929; s. Bernard John C. and Nora (Casey) H.; m. Irene Tuck; children: Katharine, Sarah. B.Sc. in Zoology with honors, London U., 1951, D.Sc. in Human Genetics, 1968. Sci. staff Med. Research Council/Radiobiol. Research Unit Council, Harwell, Eng., 1951-56; sci. staff Brit. Mus., 1956-59, Brit. Empire Cancer Campaign, Kings Coll., London U., 1959-60; sr. lectr., head cytogenetics sect., pediatric research unit Guy's Hosp. Med. Sch., U. London, 1960-69; head sect. genetics dept. pediatrics U. Man., Can., 1969-79, prof. pediatrics, 1972—, dir. dept. genetics, Winnipeg, 1972-79, assoc. dean, 1977-81; dir. dept. genetics Children's Hosp. Winnipeg, Man., Can.; vis. prof. dept. genetics Hebrew U. Jerusalem, 1975; organizer profl. confs. Author 3 books, numerous sci. articles; mem. editorial bds. Recipient Robert Roessler de Villiers award Leukemia Soc. Am., 1956; Huxley Meml. medal Imperial Coll. Sci. and Tech., 1958; Med. Research Council Can. research prof., 1981-82; grantee, 1970—. Mem. Am. Soc. Human Genetics, N.Y. Acad. Scis., Genetics Soc. Can., Royal Soc. Medicine, Linnean Soc. Clubs: Alpine of Can., Man. Naturalists. Subspecialties: Genetics and genetic engineering (biology); Molecular biology. Current work: Human cytogenetics, gene mapping and recombinant DNA technology. Home: Box 111 Rural Route 2 Dugald MB Canada R0E 0K0 Office: 250-770 Bannatyne Winnipeg MB Canada R3E 0W3

HAMILL, ROBERT WALLACE, neurologist, research neurobiologist; b. Hartford, July 30, 1942; s. Robert Francis and Sarah (Wallace) H.; m. Donna Gail Kole, June 26, 1966; children: Kara, Heidi, Meghan. B.S. in Biology, Springfield (Mass.) Coll., 1964; M.D., Bowman Gray Sch. Medicine, 1968. Diplomate: Am. Bd. Neurology and Psychiatry. Intern U. Rochester Med. Center-Strong Meml. Hosp., 1968-69, resident, 1969-70; resident in neurology N.Y. Hosp.-Cornell Med. Ctr., 1973-76; asst. prof. neurology U. Rochester, N.Y.; dir. neurology unit Monroe Community Hosp., Rochester, 1980—. Contbr. articles to profl. jours. Served to lt. comdr. U.S. Navy, 1970-73. Nat. Inst. Neurol., Communicative Disorders and Stroke fellow, 1978-80; Sloan Found. fellow, 1975-76; Jordan fellow, 1977. Mem. Am. Acad. Neurology, Soc. Neurosci., AAAS. Subspecialties: Neurology; Neurobiology. Current work: Autonomic nervous system; degenerative neurological diseases; paraplegia; hormonal regulation of sympathetic neurons; development of sympathetic neurons. Office: Dept Neurology U Rochester Sch. Medicine and Dentistry 601 Elmwood Ave Rochester NY 14642

HAMILTON, ANGUS CAMERON, survey engineer, educator; b. Listowel, Ont., Can., Apr. 18, 1922; s. Angus and Annie (McClure) H.; m. Margaret Claire Fisher, June 25, 1949; children: Anne, Elizabeth, Stuart, Nancy, James. B.Sc.E., U. Toronto, 1949, M.Sc.E., 1951; P.Eng., N.B., 1971. Commd. land Surveyor N.B. Surveyor, Shoran sect. Geodetic Survey of Can., Dept. of Energy, Mines and Resources, Ottawa, Ont., 1951-58; sr. sci. officer gravity div. Dominion Obs., Ottawa, 1958-67, coordinator research and tng. surveys and mapping br., 1967-70; acting chief Geodetic Survey of Can., Ottawa, 1970-71; prof., chmn. dept. surveying engring. U. N.B. (Can.), Fredericton, 1971—; cons. and lectr. in field. Contbr. articles to profl. jours. and lectr. to profl. confs. Served as sgt. R.C.A.F., 1941-45. Mem. Assn. Profl. Engrs. N.B., Canadian Inst. Surveying, Am. Congress on Surveying and Mapping (Earl Fennell award 1983). Subspecialties: Information systems (information science); Surveying engineering. Current work: Searching for understanding of land information; land information systems; survey systems. Home: Rural Route 4 Fredericton NB Canada E3B 4X5 Office: Dept Surveying Engring U NB Fredericton NB Canada E3B 5A3

HAMILTON, BRUCE KING, biochemical engineer; b. Easton, Pa., May 26, 1947. B.S., MIT, 1974, Ph.D. in Biochem. Engring, 1974. Research assoc. MIT, 1974-75; group leader Frederick (Md.) Cancer Research Ctr., 1975-77, sect. head fermentation tech., 1978-80; dir. biotech. Genex Corp., Gaithersburg, M. Md., 1980-81, v.p. biotech., 1981—. Mem. AAAS, Am. Inst. Chem. Engrs., Am. Chem. Soc., Am. Soc. Microbiology, Soc. Indsl. Microbiology. Subspecialties: Enzyme technology; Biochemical engineering. Office: Genex Corp 16020 Industrial Dr Gaithersburg MD 20877

HAMILTON, BYRON BRUCE, pharmacology researcher; b. Brighton, Pa., July 6, 1934; m., 1958; 2 children. A.B., Syracuse U., 1956; M.D., SUNY, 1959, Ph.D., 1971. Intern in medicine Boston City Hosp., 1959-60; asst. prof. pharmacology and rehab. SUNY-Upstate Med. Ctr., 1967-70; asst. prof. Med. Sch. Northwestern U., Chgo., 1970-75, assoc. prof. clin. rehab., 1975—; dir. research Nat. Inst. Handicapped Research, Rehab. Research and Tng. Ctr., Northwestern U.-Rehab. Inst., Chgo., 1970—. USPHS fellow, 1960-61, 63-67; recipient Licht award. Mem. AAAS, Am. Rheumatism Assn., Am. Congress Rehab. Medicine. Subspecialty: Physical medicine and rehabilitation. Office: Rehab Inst 345 E Superior St Chicago IL 60611

HAMILTON, CARLOS ROBERT, JR., endocrinologist, consultant; b. Houston, June 12, 1939; s. Carlos Robert and Berta (Denman) H.; m. Carolyn Frances Burton, Aug. 12, 1961; children: Carlos Robert, Patricia Frances. B.A., U. Tex., 1961; M.Sc., Baylor U. Coll. Medicine, 1966, M.D., 1966. Cert. Am. Bd. Internal Medicine; cert. endocrinology, metabolism. Intern Johns Hopkins Hosp., Balt., 1966-67, resident in endocrinology, 1966-69, fellow, 1966-69; postdoctoral research fellow Mass. Gen. Hosp., Boston, 1969-70; chief resident, instr. Johns Hopkins Hosp., 1970-71; asst. prof. Johns Hopkins U., 1971-72; dir. endocrine research Wilford Hall, USAF Med. Ctr., San Antonio, 1972-74; clin. asst. prof., cons. Baylor Coll. and Med. Clinic, Houston, 1974—. Served as lt. col. USAF, 1972-74. Fellow ACP; mem. Endocrine Soc., Am. Thyroid Assn., AMA, Am. Soc. Internal Medicine, Explorers Club. Baptist. Subspecialties: Endocrinology; Internal medicine. Current work: Metabolic bone disease. Home: 3713 Chevy Chase Houston TX 77019 Office: Med Clinic Houston 1707 Sunset Blvd Houston TX 77005

HAMILTON, CHARLES LEROY, geology educator; b. Nyack, N.Y., Feb. 19, 1932; s. Walter Leroy and Jeanne Almira (Conover) H.; m. Mary Lou Jones, June 5, 1954; children: Brian, Donald, Deborah. B.A., Lehigh U., 1953; M.A., Dartmouth Coll., 1954; Ph.D., Va. Poly. Inst., 1964. Exploration geologist N.J. Zinc Co., Mineral, Va., 1956-58; teaching asst. Va. Poly. Inst., 1958-60; intr. Rutgers U.-Newark, 1960-64, asst. prof., 1964-69; assoc. prof. geology Montclair State coll., 1969-80, porf., 1980-. Served to 1st lt. U.S. Army, 1954-56. Mem. Nat. Assn. Geology Tchrs., Mineral. Soc. Am., Sigma Xi. Subspecialties: Mineralogy; Petrology. Current work: Mineralogy and petrology of New Jersey igneous and metamorphic rocks. Home: 3 Lafayette lCt Wayne NJ 07470 Office: Montclair State Coll Normal Ave Upper Montclair NJ 07043

HAMILTON, EUGENE PHILLIP, research mathematician; b. Wilmington, Del., Dec. 22, 1947; s. Eugene Cook and Phyllis (Brinkman) H. B.S., U. Del., 1968; M.S., Cornell U., 1970, Ph.D., 1973. Asst. prof. math. Vanderbilt U., 1973-77; ops. analyst Ctr. Naval Analyses, Arlington, Va., 1977-78; asst. prof. math. Washington Coll., Chestertown, Md., 1978—. Mem. Math. Assn. Am., Soc. Indsl and Applied Math., Am. Math. Soc., N.Y. Acad. Scis., Sigma Xi. Democrat. Subspecialty: Applied mathematics. Current work: Construction of variational principles for differential equations, quantum field theory. Home: 121A Washington Ave Apt 2 Chestertown MD 21620 Office: Washington College Chestertown MD 21620

HAMILTON, JOSEPH HANTS, JR., educator, physicist; b. Ferriday, La., Aug. 14, 1932; s. Joseph Hants and Letha (Gibson) H.; m. Jannelle Jauree Landrum, Aug. 5, 1960; children: Melissa Claire, Christopher Landrum. B.S., Miss. Coll., 1954, D.Sc. (hon.), 1982; M.S., Ind. U., 1956, Ph.D., 1958. Mem. faculty Vanderbilt U., 1958—, prof. physics, 1966—, Landon C. Garland prof. physics, 1981—, chmn. dept., 1979—; NSF postdoctoral fellow U. Uppsala, Sweden, 1958-59; research fellow Inst. Nuclear Studies, Amsterdam, 1962; vis. prof. U. Frankfort, 1979-80; mem. adv. panel Nat. Heavy Ion Labs., 1971-73; mem. nat. policy bd. Holifield Heavy Ion Facility, 1974—; organizer, chmn. exec. com., prin. investigator Univ. Isotope Separator, Oak Ridge, 1970—; cons. Oak Ridge Nat. Lab, 1972—; cons.; mem. council Oak Ridge Asso. Univs., 1974-80; organizer, dir. Joint Inst. for Heavy Ion Research, Oak Ridge, 1980—; chmn. Internat. Conf. Internal Conversion Processes, 1965, Internat. Conf. Radioactivity in Nuclear Spectroscopy, 1969; Internat. Conf. Future Directions in Studies Nuclei far from Stability, 1979. Co-author: Science: Faith and Learning, 1972, ORAU from the Beginning, 1980; co-author, editor: Internal Conversion Processes, 1966, Radioactivity in Nuclear Spectroscopy, 1972, Reactions Between Complex Nuclei, 1974, Future Directions in Studies of Nuclei Far from Stability, 1980; contbr. articles to profl. jours., chpts. to books. Mem. Mayor Nashville Citizens Adv. Com. Housing, 1970-74; bd. dirs. Vineyard Conf. Center, Louisville, 1972-77, Danforth asso., 1965—, So. Bapt. Conv. Hist. Commn., 1983—. Harvie Branscomb Disting. Prof. award Vanderbilt U., 1983-84; NSF grantee, 1959-76; ERDA-Dept. Energy grantee, 1975—; Humbolt prize W. Ger., 1979. Fellow Am. Phys. Soc. (vice chmn. Southeastern sect. 1972-73, chmn. Southeastern sect. 1973-74 1975, Jesse Beams gold medal for research 1975); mem. Sigma Xi (chpt. pres. 1970). Subspecialties: Nuclear physics; Water supply and wastewater treatment. Current work: Experimental studies of nuclear structures via heavy ion Coulomb excitation, in-beam gamma-ray spectroscopy, on-line mass separator studies of nuclei far from stability ; nuclear reaction mechanisms; standardization of radioactivity measurements; measurements of absolute alpha and beta radioactivites ofdrinking water as required by Tennessee law. Address: 305 Hildreth Ct Nashville TN 37215

HAMILTON, LEONARD DERWENT, physician, molecular biologist; b. Manchester, Eng., May 7, 1921; came to U.S., 1949, naturalized, 1964; s. Jacob and Sara (Sandelson) H.; m. Ann Twynam Blake, July 20, 1945; children: Jane Derwent, Stephen David, Robin Michael. B.A., Balliol Coll., Oxford (Eng.) U., 1943, B.M., 1945, M.A., 1946, D.M., 1951; M.A., Trinity Coll., Cambridge (Eng.) U., 1948, Ph.D., 1952. Diplomate: Am. Bd. Pathology. USPHS research fellow U. Utah, 1949-50; mem. staff Sloan-Kettering Inst., N.Y.C., 1950—, head isotope studies sect., 1957-64, asso. scientist, 1965—; mem. staff Meml. Hosp., N.Y.C., 1950-65, asst. attending physician dept. medicine, 1958-65; mem. faculty Sloan-Kettering div. Grad. Sch. Med. Scis., Cornell U. Med. Coll., 1956-64; sr. scientist, head div. microbiology Med. Research Center, Brookhaven Nat. Lab., Upton, N.Y., 1964-76; head Office Environ. Policy Analysis, 1976—; also attending physician Hosp. Med. Research Center, 1964—; prof. medicine Health Sci. Center, SUNY-Stony Brook, 1968—; cons. HEW Center for Disease Control, Nat. Inst. Occupational Safety and Health Epidemiologic Study of Portsmouth Naval Shipyard, 1978—; vis. fellow St. Catherine's Coll., Oxford U., 1972-73; mem. internat. panel of experts on fossil fuel UN Environment Programme, 1978, panel on nuclear energy, 1978-79, panel on renewable sources, 1980, panel on comparative assessment of different sources, 1980; mem. various coms. Nat. Acad. Sci.-NRC, Washington, 1975-80; mem. N.Y.C. Mayor's Tech. Adv. Com. on Radiation, 1963-77, N.Y.C. Commr. of Health Tech. Adv. Com. on Radiation, 1978—. Editor: Gerrard Winstanley, Selections From His Works, 1944, Physical Factors and Modification of Radiation Injury, 1964, The Health and Environmental Effects of Electricity Generation—a Preliminary Report, 1974. Am. Cancer Soc. scholar, 1953-58; Commonwealth Fund grantee, 1955-62. Mem. Am. Assn. Cancer Research, Am. Soc. Clin. Investigation, Am. Assn. Pathologists, Brit. Med. Assn., Harvey Soc. Subspecialties: Molecular biology; Environmental effects of energy technologies. Current work: Research on life-span of lymphocytes and their function in immunity; collaborator on proof of three-dimensional structure of DNA; effects of various chemicals and ionizing radiation on cells and man; the health and environmental effects of different energy technologies. Home: Childs Ln Old Field Setauket NY 11733 Office: Brookhaven Nat Lab Upton NY 11973

HAMILTON, LEONARD W., psychology educator; b. Hedrick, Iowa, May 9, 1943; s. Everett B. and Julia L. (DeRuiter) H.; m. Carol Robin Timmons, Apr. 1, 1980; children: Erika, Emily. B.S. in Psychology with honors, U. Iowa, 1965; Ph.D. in Biopsychology, U. Chgo., 1968. Mem. faculty Rutgers U., New Brunswick, N.J., 1968—, chmn. psychology dept., 1976-79, prof. psychology, 1979—. USPHS grantee, 1968-72, 75-76, 79-80; Rutgers U. Research Council grantee, 1969-70, 72-73; Biol. Scis. Support grantee, 1968-70, 71-73; Busch Fund grantee, 1975-76, 80-81; Nutrition Found. grantee, 1977-79. Mem. Eastern Psychol. Assn., AAAS, AAUP, Soc. Neuroscis., Psychonomic Soc., Sigma Xi. Subspecialties: Physiological psychology; Learning. Current work: Brain and behavior; behavioral inhibition, learning development, limbic system. Office: Psychology Dept Rutgers U 22 Tillett Hall New Brunswick NJ 08903

HAMILTON, ROBERT BRUCE, neuroscientist; b. Ft. Belvoir, Va., Feb. 17, 1950; s. Howard Edmund and Dona Sue (Finnen) H.; m. Sandra Jean Halibrand, June 22, 1980. A.A., Long Beach (Calif.) City Coll., 1970; B.A., Calif. State U.-Long Beach, 1972, M.A., 1975; Ph.D., U. Miami, 1980. Teaching asst. U. Miami, Coral Gables, Fla., 1976-79, predoctoral trainee, 1979-80; postdoctoral research assoc. Rockefeller U., N.Y.C., 1980-82, postdoctoral fellow, 1982—. Contbr. articles to profl. jours. NIH grantee, 1979-80. Mem. Soc. for Neurosci., Phi Kappa Phi, Psi Chi, Sigma Xi. Democrat. Roman Catholic. Subspecialties: Neurophysiology; Neuroanatomy. Current work: Organization of the visceral afferent nervous system with special emphasis on the gustatory and cardiovascular subsystems. Home: 500 E 63d St Apt 4A New York NY 10021 Office: Rockefeller U 1230 York Ave New York NY 10021

HAMILTON, ROBERT WILLIAM, physiologist, consultant; b. Stanton, Tex., June 5, 1930; s. Robert William and Lois (Rogers) H.; m. Beverly Luth Cooper, Jan. 23, 1954 (dec. 1970); children: Kitty Hamilton Amat, Lucy Hamilton Kantor, Sally; m. Kathryn Ann Faulkner, Apr. 22, 1972. B.A., U. Tex.-Austin, 1951; M.S., Tex. A&M U., 1958; Ph.D., U. Minn., 1964. Lab. dir. Union Carbide Corp., Tarrytown, N.Y., 1964-75; v.p. research and devel. Tarrytown Labs., Ltd., 1975-76; prin. cons. Hamilton Research Ltd., Tarrytown, 1976—; cons. Shell, London, 1982, Norsk Hydro, Oslo, Bergen, Norway, 1978-80, Swedish Navy, Stockholm, 1979-82, Norwegian Underwater Inst., Bergen, 1977-82. Editor: symposium Decompression from Deep Dives, 1976; co-author: program and manual Decompression Computation and Analysis Program, 1980; author: sect. Ency. Britannica; editor: Hyperberic Oxygen in Emergency Medical Care, 1983. Vestryman, warden Christ Ch., Tarrytown, 1970-73, 77-80, 81-83. Served to maj. USAF, 1951-55, 68-69. Recipient award Aerospace Indsl. Life Scis. Assn., 1972. Mem. Undersea Med. Soc. (sec. 1981-82, Stover-Link Award 1977), Am. Physiol. Soc., Aerospace Med. Assn., Human Factors Soc., N.Y. Acad. Scis. (sect. chmn. 1973-74), Phi Kappa Psi. Democrat. Episcopalian. Club: Mensa. Subspecialties: Physiology (medicine); Biomedical engineering. Current work: Studies and problem solving in physiology of deep sea diving and aerospace; decompression, gases, life support, performance, human factors. Patentee neon as a diving gas. Home and Office: 80 Grove St Tarrytown NY 10591

HAMILTON, VELDA MARIE, microbiologist; b. Henryetta, Okla., June 14, 1938; d. James Ira and Pauline E. (Endres) Youngblood; m. Louis E. Hamilton, Nov. 28, 1957. B.A., U. Calif.-Berkeley, 1966. With Cutter Labs., Inc., Berkeley, 1966—, now regulatory affairs specialist. Mem. Am. Soc. Quality Control, Am. Soc. Microbiology. Subspecialties: Microbiology; Tissue culture. Office: Cutter Labs 4th and Parker Sts Berkeley CA 94710

HAMILTON, WILLIAM OLIVER, physicist, educator; b. Lawrence, Kans., Sept. 5, 1933; s. Francis Corbin and Bernraine (Winegar) H.; m. Mary Helen Kelson, June 23, 1956; children: Eric William, Christopher David, Ann Elizabeth. B.S., Stanford U., 1955, Ph.D., 1963. NSF fellow Stanford (Calif.) U., 1963-65, asst. prof. physics, 1965-70; assoc. prof. physics La. State U., Baton Rouge, 1970-76, prof. physics, 1976—; vis. prof. physics U. Rochester (N.Y.), 1977-78; cons. naval studies bd. Nat. Acad. Sci., Washington, 1979-82, Lawrence Livermore (Calif.) Nat. Lab., 1982—. Contbr. numerous sci. articles to profl. publs. Served to lt. USN, 1955-58. Stanford Univ. fellow, 1961; NSF grantee, 1970-84; Air Force Office of Sci. Research grantee, 1970-76. Mem. Am. Phys. Soc., Am. Assn. Physics Tchrs., AAAS, Phi Beta Kappa. Subspecialties: Relativity and gravitation; Low temperature physics. Current work: Experimental general relativity: search for gravitational waves; fundamental limits of measurement. Co-inventor detectors for gravitational radiation and infrared radiation. Home: 644 Castle Kirk Dr Baton Rouge LA 70808 Office: Physics and Astronomy Dept La State U Baton Rouge LA 70803

HAMMER, CARL, computer scientist, former computer co. exec.; b. Chgo., May 10, 1914; s. Karl Heinrich and Kaethe (Patzig) H.; m. T. Jeannette George, Sept. 23, 1944. Dipl. Math. Statistics, U. Munich, 1936, Ph.D. magna cum laude, 1938. Mathematician, statistician Tex. Co. Research Labs., Beacon, N.Y., 1938-43; statistician Pillsbury Mills Inc., N.Y.C., 1944-47; chmn. div. tech. edn. Walter Hervey Jr. Coll., 1947-51; sr. staff engr. Franklin Inst., Phila., 1951-55; dir. UNIVAC European Computing Center Sperry Rand Corp., Frankfurt/Main, Germany, 1955-57; staff cons., acting mgr. programming and analysis dept. Sylvania Electronic Products, Inc., Needham, Mass., 1957-59; sr. engring. scientist surface communications div. RCA, N.Y.C., 1959-61, mgr. sci. computer applications, Washington, 1961-63; dir. computer sci. UNIVAC, Washington, 1963-81; instr. German for staff officers U.S. Mil. Acad., summer 1942; instr. math. Pratt Inst., 1945-46; instr. math. and statistics Sch. Gen. Studies, Hunter Coll., 1947-52; adj. prof. Am. U., 1962-80; vis. prof. Indsl. Coll. Armed Forces, Washington, 1967—. Author: Viscosity Index Tables, 1941, Rank Correlation of Cities, 1951, Univac Programming with Compilers, 1956, Computers and Simulation, Vol. IV, Number 4, 1961, High-Speed Digital Communication Networks, 1963, Statistical Validation of Mathematical Computer Routines, 1967, Signature Simulation and Certain Cryptographic Codes, 1971, Space Communications Procs., Panel Sci. and Tech, 1972, Computers in Research, Procs. Internat. Symposium, 1974; contbr. articles in field of computer tech. to profl. jours. Mem. Nat. Def. Exec. Res., 1970—. Recipient Computer Sci. Man of Year award Data Processing Mgmt. Assn., 1973. Fellow AAAS, N.Y. Acad. Scis., Assn. Computer Programmers and Analysts; mem. IEEE (sr.), AAUP, Am. Math. Soc., Am. Soc. for Cybernetics (sec. 1967, v.p. 1968, pres. 1969-72), Am. Statis. Assn., Assn. for Computing Machinery (chpt. chmn. Washington 1966-68, rep. Capital region 1968—, chmn. accreditation com. 1968-70, nat. lectr. 1969-70, chmn. nominating com. 1971-73, Disting. Service award 1979), N.Y. Acad. Scis., Assn. Systems Mgmt. (dir. Washington chpt. 1969-71), Inst. Math. Statistics, Math. Assn. Am., Soc. Indsl. and Applied Math. (treas. 1953-55), Research Soc. Am. Subspecialties: Computer architecture; Cryptography and data security. Home: 3263 O St NW Washington DC 20007

HAMMER, JACOB MEYER, physicist; b. N.Y.C., Sept. 14, 1927; s. Joseph Israel and Miriam (Silverman) H.; m.; children: Daniel, Jonathan, Miriam. B.S., N.Y. U., 1950, Ph.D., 1956; M.S., U. Ill., 1951. Mem. tech. staff Bell Telephone Labs., Murray Hill, N.J., 1956-59, RCA Labs., Princeton, N.J., 1959-68, 69—; sr. visitor Cavendish Lab., Cambridge, Eng., 1968-69; adj. prof. elec. engring. Poly. Inst. N.Y., 1981. Contbr. sect. to book, articles to profl. jours. in field. Served with AUS, 1946-47. Recipient Founders Day award N.Y. U., 1956; outstanding achievement award RCA Labs., 1962, 64, 73. Mem. Am. Phys. Soc., IEEE (sr.), AAAS. Subspecialties: Fiber optics. Current work: Application of optical waveguides to communication and technology. Patentee. Office: RCA Labs Princeton NJ 086540

HAMMES, GORDON G., chemistry educator; b. Fond du Lac, Wis., Aug. 10, 1934; s. Jacob and Betty (Sadoff) H.; m. Judith Ellen Frank, June 14, 1959; children: Laura Anne, Stephen R., Sharon Lyn. A.B., Princeton, 1956; Ph.D., U. Wis., 1959. NSF postdoctoral fellow Max Planck Inst. fur physikalische Chemie, Göttingen, Germany, 1959-60; from instr. to asso. prof. Mass. Inst. Tech., Cambridge, 1960-65; prof. Cornell U., Ithaca, N.Y., 1965—, chmn. dept. chemistry, 1970-75, Horace White prof. chemistry and biochemistry, 1975—; Mem. physiol. chemistry study sect., tng. grant com. NIH; bd. counselors Nat. Cancer Inst., 1976-80; mem. adv. council chemistry dept. Princeton, 1970-75, Poly. Inst. N.Y., 1977-78, Boston U., 1977—. Author: Principles of Chemical Kinetics, (with I. Amdur) Enzyme Catalysis and Regulation, Chemical Kinetics: Principles and Selected Topics; (with I. Amdur) also articles. NSF sr. postdoctoral fellow, 1968-69; NIH Fogarty scholar, 1975-76. Mem. Am. Chem. Soc. (award biol. chemistry 1967, editorial bd. jours., exec. com. div. phys. chemistry 1976-79, exec. com. div. biol. chemistry 1977—), Am. Soc. Biol. Chemists (editorial bd. jour.), Nat. Acad. Scis., Am. Acad. Arts and Scis., Phi Beta Kappa, Sigma Xi, Phi Lambda Upsilon. Subspecialties: Biophysical chemistry; Biochemistry (biology). Current work: Enzyme catalysis and regulation; membrane bound enzymes; multienzyme complexes; ion transport across membranes. Home: 107 Warwick Pl Ithaca NY 14850

HAMMILL, TERRENCE MICHAEL, biologist, educator; b. Potsdam, N.Y., Dec. 28, 1940; s. Jeremiah James and Margaret Mae (Blanchard) H.; m. Martha Lois Trembley, Aug. 24, 1963; children: Michael Sean, Jeffery Terrence. B.S., SUNY, Potsdam, 1963; M.Ed., U. Ga., 1968; Ph.D., SUNY Coll. Forestry, Syracuse, 1971, Syracuse U., 1971. Tchr. sci. Indian River Central High Sch., Phila., 1963-67; asst. prof. biology SUNY, Oswego, 1971-74, asso. prof., 1974—. Contbr. articles to research jours. Active Oswego Little League Baseball. NSF grantee, 1974-76, 80-82; Research Corp. grantee, 1974; SUNY Awards Council grantee, 1974. Mem. AAAS, Am. Inst. Biol. Sci., Bot. Soc. Am., Brit. Mycol. Soc., Electron Microscopy Soc. Am., Mycol. Soc. Am. Democrat. Subspecialties: Microbiology; Developmental biology. Current work: Developmental biology of fungi; research on devel. biology and ultrastructural cytology. Home: 69 W 5th St Oswego NY 13126 Office: SUNY B-18 Piez Hall Oswego NY 13126

HAMMITT, FREDERICK GNICHTEL, nuclear engineer; b. Trenton, N.J., Sept. 25, 1923; s. Andrew Baker and Julia (Stevenson Gnichtel) H.; m. Barbara Ann Hill, June 11, 1949; children: Frederick, Harry, Jane. B.S. in Mech. Engring., Princeton U., 1944; M.S., U. Pa., 1949, Stevens Inst., 1956; Ph.D. in Nuclear Engring, U. Mich., 1958. Registered profl. engr., N.J., Mich. Engr. John A. Roebling Sons Co., Trenton, 1946-48, Power Generators Ltd., 1948-50; project engr. Reaction Motors Inc., Rockaway, N.J., 1950-53, Worthington Corp., Harrison, N.J., 1953-55; research asso. U. Mich., Ann Arbor, 1955-57, asso. research engr., 1957-59, asso. prof., 1959-61, prof. nuclear engring., 1961—, mech. engring., 1965—, also prof. in charge Cavitation and Multiphase Flow Lab., 1967—; cons. govt. and industry; vis. scholar Electricité de France, Paris, 1967, Société Grenobloise Hydrauliques, Grenoble, France, 1971; Fulbright sr. lectr. French Nuclear Lab., Grenoble, 1974; Polish Acad. Sci. lectr. Inst. Fluid Mechanics, Gdansk, 1976. Author: (with R.T. Knapp, J.W. Daily) Cavitation, 1970, Cavitation and Multiphase Flow Phenomena, 1980; contbr. 400 articles to profl. jours., 5 chpts. to books. Served with USN, 1943-46. Fellow Inst. Mech. Engrs. (U.K.), ASME (past chmn. cavitation com. fluids div.), ASTM (past chmn. cavitation and liquid impingement); mem. Am. Nuclear Soc. (past chmn. S.E. Mich. sect.), Internat. Assn. Hydraulic Research (chmn. cavitation scale effects com.), Phi Beta Kappa, Sigma Xi, Tau Beta Pi. Republican. Presbyterian (elder). Patentee in field (5). Home: 1306 Olivia St Ann Arbor MI 48104

HAMMOND, BENJAMIN FRANKLIN, microbiologist, educator; b. Austin, Tex., Feb. 28, 1934; s. Virgil Thomas and Helen Marguerite (Smith) H. B.A., U. Kans., 1954; D.D.S., Meharry Med. Coll., 1958; Ph.D., U. Pa., 1962. Mem. faculty U. Pa. Sch. Dental Medicine, Phila., 1958—, prof. microbiology, 1970—, chmn. dept., 1972—; Pres.'s lectr. U. Pa., 1981; Mem. oral medicine study sect. NIH, 1972-75; mem. Nat. Adv. Dental Research Council, 1975—; cons. in field. Recipient USPHS Research Career Devel. award, 1965, Lindback award U. Pa., 1969; Médaille d'Argent, City of Paris, 1978; NIH grantee, 1981—. Mem. Am. Soc. Microbiology, Internat. Assn. Dental Research (E.H. Hatton award 1959), Am. Assn. Dental Research (pres. 1978-79). Subspecialties: Microbiology. Current work: Oral microbial ecology; bacterial physiology, periodontal microbiology. Home: 560 N 23d St Philadelphia PA 19130

HAMMOND, DONALD L., physical science research administrator; b. Kansas City, Mo., Aug. 7, 1927; s. Clark E. and Laila G. (Morris) H.; m. Phyllis E. Whitmore, Aug. 21, 1949; children: Deborah Ruth, Katherine Ilene, Carol Linda, Nancy Linda, Paul David. B.S. in Physics, Colo. State U., 1950, M.S., 1952, D.Sc. (hon.), 1974. Chief crystal research U.S. Army Electronics Command, Fort Monmouth, N.J., 1952-56; dir. research Scientific Electronic Products, Fort Collins, Colo., 1956-59; mgr. precision quartz crystal devel. Hewlett-Packard, Palo Alto, Calif., dir. phys. electronics lab., now dir. phys. research ctr.; bd. dirs. Lexel Corp., Palo Alto, Calif.; Lectr. CB Sawyer Frequency Control Symposium, 1970. Mem. bd. edn. Palo Alto Unified Sch. Dist., 1971-81, also pres. Served as ensign USN, 1945. Fellow IEEE; mem. Am. Inst. Physics. Subspecialties: Electronics; Physics research administrator. Current work: Physical acoustics; electron optics; chemical instrumentation; medical electronics; submicron lithography; research and development management.

Home: 12660 Corte Madera Lane Los Altos Hills CA 94022 Office: 1501 Page Mill Rd Bldg 3U Palo Alto CA 94304

HAMMOND, GEORGE SIMMS, chemist; b. Auburn, Maine, May 22, 1921; s. Oswald Kenric and Marjorie (Thomas) H.; m. Marian Reese, June 8, 1945 (div. 1977); children: Kenric, Janet, Steven, Barbara, Jeremy; m. Eva L. Menger, May 22, 1977; stepchildren—Kirsten Menger-Anderson, Lenore Menger-Anderson. B.S., Bates Coll., 1963; M.S., Ph.D., Harvard, 1947; D.Sc., Wittenberg U., 1972, Bates Coll., 1973; Dr. honoris causa, U. Ghent, 1973. Postdoctoral fellow U. Calif. at Los Angeles, 1947-48; mem. faculty Iowa State Coll., 1948-58, prof. chemistry, 1956-58; vis. asso. prof. U. Ill., summer 1953; prof. organic chemistry Calif. Inst. Tech., Pasadena, 1958-72, div. chemistry and chem. engring., 1968-72; Arthur Amos Noyes prof. chemistry; vice chancellor natural scis. U. Calif. at Santa Cruz, 1972-74, prof. chemistry, 1972-78; dir. Integrated Chem. Systems Lab. Allied Chem. Co., Morristown, N.J., 1978—; mem. chem. adv. panel NSF, 1962-65; fgn. sec. Nat. Acad. Scis., 1974-78. Author: (with J. S. Fritz) Quantitative Organic Analysis, 1956, (with D.J. Cram) Organic Chemistry, 1958, (with J. Osteryoung, T. Crawford and H. Gray) Models in Chemical Science, 1971; Editor: Advances in Photochemistry, 1961; Editorial bd.: Jour. Am. Chem. Soc, 1967—. Guggenheim fellow; NSF sr. fellow Oxford (Eng.) U. and U. Basel, Switzerland; Calif. Inst. Tech., 1956-57; Mem. Maine N.G., 1938-40; Recipient James Flack Norris award in phys. organic chemistry, 1968. Mem. Nat. Acad. Scis., Am. Chem. Soc. (award in petroleum chemistry 1960, Priestly medal 1976), Chemistry Soc. (London), Am. Acad. Arts and Scis., Phi Beta Kappa, Sigma Xi. Subspecialties: Organic chemistry; Physical chemistry. Current work: Chemical processes, new materials chemistry, chemical systems. Home: 43 Noe Ave Madison NJ 07940

HAMMOND, THOMAS JOSEPH, optical engineer; b. Warren, Ohio, Dec. 21, 1940; s. Francis Raymond and Anne (Birskovich) H.; m. Nancy Ann Kines, Apr. 14, 1962; children: Scott Alan, Brian Marshall, Jonathan Kendall. A.E.E., DeVry Tech. Inst., 1964; B.S. in Math. and Physics, U. Rochester, 1969; postgrad. in Mech. Engring., U. Rochester, 1969-70. Research aide, physics lab. Xerox Corp., Rochester, N.Y., 1965-67, sr. physicist, 1969-80, tech. specialist, project mgr. optical tech., 1980—. Contbr. articles to profl. jours. Mem. Penfield (N.Y.) Zoning Com., 1971-72. Served with USN, 1958-62. Mem. Optical Soc. Am., Am. Wine Soc. (pres. chpt. 1980). Republican. Lutheran. Subspecialties: Plasma physics; Optical engineering. Current work: Develop and carry from concept to product issue, light sources for use in electrophotography. Patentee in field. Home: 108 Henderson Dr Penfield NY 14526 Office: Xerox Sq 147 Rochester NY 14644

HAMNING, RICHARD RUDOLPH, clinical psychologist, consultant researcher; b. Harvey, Ill., July 2, 1953; s. Rudolph and E. and Virginia (Phelps) H. B.S., Iowa State U., 1975; M.S., George Peabody Coll., 1978; Ph.D., Peabody Coll., Vanderbilt U., 1981. Lic. psychologist, Tenn. Psychol. examiner Middle Tenn. Mental Health Inst., Nashville, 1978-80; clin. psychologist children and youth programs, 1981-82; dir., clin. psychologist Brentwood (Tenn.) Counseling Ctr., 1982—; adj. prof. psychology George Peabody Coll., Vanderbilt U., 1982—; cons. Giles County Schs., Pulaski, Tenn., 1982—, Nashville, 1982—; dir. spl. needs unit Spencer Youth Ctr., 1983—. Mem. Am. Psychol. Assn. Methodist. Current work: Social adaptation in psychopathology; role attribution in deviant family systems; neuroanatomical repair and psychological functioning; epistemology. Office: Brentwood Counseling Ctr 783 Old Hickory Blvd Brentwood TN 37027

HAMON, DANNY JOE, entomologist, botanical consultant; b. Bakersfield, Calif., Mar. 9, 1947; s. Daniel Boone and Josemae (Glover) H.; m. Veronica Lynn Hamon, Dec. 6, 1969; children: Jennifer Lynn, Deborah Jo. A.A., Porterville (Calif.) City Coll., 1973; B.A., Calif. State U., Fresno, 1977, M.A. candidate. Cert. U.S. Dept. Agr. Profl. Devel. Sch., 1981. Botanist, Sierra Nat. Forest, Calif., 1978-80; plant protection biologist U.S. Dept. Agr., Stockton, Calif., 1980—; speaker. Contbr. articles to profl. jours. Served with U.S. Army, 1966-69; Vietnam. Recipient Lillian Wells award in Botany Calif. State U., 1979. Mem. Calif. Bot. Soc., No. Calif. Entomology Soc., Nature Conservancy, Calif. Native Plant Soc. Democrat. Subspecialties: Resource management; Species interaction. Current work: Rare plant populations and plant community interactions; plant quarantine and pest control.

HAMPAR, BERGE, research facility exec.; b. Rockaway, N.Y., Aug. 20, 1932; s. Yervant and Dikranouhi H.; m. Nancy C. Landes, June 7, 1976; children: Adrienne, Natalie. B.A., Columbia U., 1954, D.D.S., 1960; postgrad., Balt. U. Sch. Law, 1980-82. Postdoctoral fellow dept. microbiology Columbia U., 1960-62; sr. scientist Nat. Inst. Dental Research, 1962-67; asst. chief. head microbiology sect. Lab. Molecular Oncology NIH, 1967-81; gen. mgr. Frederick (Md.) Cancer Research Facility, Nat. Cancer Inst., 1981—. Contbr. articles to profl. jours. Served with USN, 1954-56; Served with USPHS, 1962—. Mem. Am. Assn. Immunologists, Am. Assn. Cancer Research, Am. Soc. Virology, Omicron Kappa Upsilon. Subspecialties: Virology (medicine); Cancer research (medicine). Current work: Studies on human herpes viruses and their association with cancer. Office: Nat Cancer Inst NCI-FCRF Bldg 427 Frederick MD 21701

HAMPEL, NEHEMIA, physician, urologist, educator; b. Radomsko, Poland, Mar. 27, 1941; came to U.S., 1975; s. Symcha and Nechama (Justman) H.; m. Nitza Pollack, Aug. 6, 1963; children: Ori Z., Amit M., Anat. M.D., Haddasah Med. Sch. Hebrew U., Jerusalem, 1969. Intern Rambam Med Ctr., Haifa, Israel, 1968; resident in urology/surgery, 1969-75; chief sect. urology VA Med. Ctr., Cleve., 1977—. Mem. Am. Urol. Assn. Subspecialty: Urology. Current work: Microsurgery in urology; urodynamics. Office: 20620 North Park Blvd #210 Shaker Heights OH 44118

HAMPSON, BRADFORD ELLSWORTH, computer scientist; b. Taunton, Mass., Sept. 9, 1953; s. Frank R. and Paula H. (Hathaway) H.; m. Odette M. Hebert, Sept. 30, 1979; children: Kenneth B., Ashley L. B.S., M.S. in Computer Sci, MIT, 1977. Engring. intern Hewlett Packard Med. Products Group, Waltham, Mass., 1973-75; sr. tech. cons. operating systems devel. Prime Computer, Inc., Farmingham, Mass., 1977—; Phillips Exeter Acad. regional chmn. MIT Ednl. Council, 1979—. Mem. Assn. Computing Machinery, Tau Beta Pi, Eta Kappa Nu. Subspecialties: Operating systems; Distributed systems and networks. Current work: Distributed operating systems, security in distributed systems, data abstraction-based file systems, programming language support for distributed systems, computer architectures for distributed systems and data abstraction languages. Home: 55 Apple D'Or Rd Framingham MA 01701 Office: 500 Old Connecticut Path Framingham MA 01701

HAMPTON, JAMES WILBURN, physician; b. Durant, Okla., Sept. 15, 1931; s. Hollis Eugene and Ouida (Mackey) H.; m. Carol McDonald, Feb. 22, 1958; children: Jaime, Clay, Diana, Neal. B.A., U. Okla., 1952, M.D., 1956. Intern U. Okla. Hosps., 1956-57; also resident; instr. to prof. U. Okla., Oklahoma City, 1959-77, head hematology/oncology, 1972-77; head hematology research Okla. Med. Research Found., Oklahoma City, 1972-77; dir. cancer program and med. oncology Baptist Med. Center, 1977—; chmn. med. adv. com. Hospice of Central Okla., 1981. Contbr. over 100 articles to profl. jours. Bd. dirs. Heritage Hills, Oklahoma City, 1972, Am. Cancer Soc., 1982; co-chmn. Save St. Paul's Episcopal Cathedral com., 1983, others. NIH Career Devel. Award., 1966-67. Mem. Am. Fedn. Clin. Research (pres. 1970-71), Central Soc. Clin. Research (asso. editor jour. 1975-76), Okla. County Med. Soc. (editor bull.), ACP (fellow), Internat. Soc. Thrombosis and Hematosis. Clubs: Oklahoma City Golf and Country, Blue Cord, Chaine des Rotisseurs. Subspecialties: Cancer research (medicine); Hematology. Current work: Hyperviscosity syndrome; preleukemia syndrome, hospice and palliative care. Home: 1414 N Hudson St Oklahoma City OK 73112 Office: Bapt Med Center 3300 NW Expressway Oklahoma City OK 73112

HAMPTON, RICHARD OWEN, research plant pathologist; b. Dalhart, Tex., Feb. 17, 1930; s. C. C. and M. M. (Wise) H.; m. Willa Mae Johnson, June 12, 1954; children: Kevin Ray, Audrey Camille. B.S., U. Ark, 1951; M.S., Iowa State U., 1954, Ph.D., 1957. Plant pathologist Wash. State U.-Prosser, 1957-61; research plant pathologist, Prosser, 1961-65, Agrl. Research Service, U.S. Dept. Agr. at Oreg. State U., 1965—; invited speaker for nat., internat. symposia, seminars, workshops. Contbr. numerous articles to profl. jours. Mem. Internat. Working Group on Legume Viruses (exec. com. 1975—, exec. sec. 1978-81), Am. Phytopathol. Soc., Internat. Soc. Plant Pathology, Nat. Pea Improvement Assn., Pisum Genetics Assn., Am. Sci. Affiliation. Republican. Baptist. Subspecialties: Plant pathology; Plant virology. Current work: identification and characterization of viral pathogens of edible legumes (Pisum, Phaseolus, Lens); epidemiology and ecology of plant viruses; seed-transmission of viruses; crop germplasm health. Home: 1370 Greeley Ave NW Corvallis OR 97330 Office: Dept Botany and Plant Pathology Oreg State U Corvallis OR 97331

HAMRICK, JAMES LEWIS, III, biologist, educator; b. Hopewell, Va., Feb. 26, 1942; s. James Lewis and Frances Louise (Gray) H.; m. Patricia Marie Rhodes, Dec. 21, 1968 (div.); 1 dau., Jennifer Rose; m. Karen Jane Pomeroy, Sept. 14, 1974. B.S., N.C. State U., 1964; M.S. (NSF fellow), U. Calif.-Berkeley, 1966, Ph.D., 1970. NIH fellow U. Calif.-Davis, 1970-71; mem. faculty U. Kans.-Lawrence, 1971—, assoc. prof. botany, systematics and ecology, 1974-79, prof., 1979—; panel mem. NSF, 1979-82. Mng. editor: Evolution, 1976-79; contbr. articles to sci. jours. Vice chmn. Kans. Sierra Club, 1976-77. Mem. Genetics Soc. Am., Soc. for Study Evolution, Am. Naturalist Soc., Ecol. Soc. Am., Brit. Ecol. Soc. Democrat. Subspecialties: Evolutionary biology; Population biology. Current work: The evolution and ecology of natural plant populations. Home: 1005 W 20th St Lawrence KS 66044 Office: Dept Botany U Kans Lawrence KS 66045

HAMRICK, JOSEPH THOMAS, mechanical engineer, corporate executive; b. Carrollton, Ga., Mar. 20, 1921; s. James Mayfield and Mattie Almon (Gaston) H.; m. Dorothy Elizabeth Jones, June 19, 1948; children: Jane Elizabeth Hamrick Kneisley, Nancy Ann Hamrick Owen, Thomas Mayfield. B.M.E., Ga. Inst. Tech., 1946, M.S.M.E., 1948. With NACA, Cleve., 1948-55, Thompson Ramo Wooldridge, Euclid, Ohio, 1955-61; pres. Aerospace Research Corp., Roanoke, Va., 1961—; dir. Biomass Energy Systems, Inc., Ft. Worth; ltd. partner World Energy Systems, Ft. Worth. Contbr. articles to profl. jours. Pres. North Franklin County Pub. Park, Inc. Served to 1st lt. USAAF, 1943-46; PTO. Dept. of Energy grantee, 1978-80; NSF grantee, 1980. Mem. ASME. Republican. Unitarian. Subspecialties: Biomass (energy science and technology); Combustion processes. Current work: Research on fueling gas turbines with wood, operation of 500-hp gas turbine with wood fuel. Patentee in field. Home: 6364 JAE Valley Rd SE Roanoke VA 24014 Office: 5454 JAE Valley Rd SE Roanoke VA 24014

HAN, CHARLES CHIH-CHAO, research chemist; b. Szuchuan, China, Jan. 18, 1944; came to U.S., 1967; s. Teh-Hei and Pao-Shu (Chu) H.; m. Sally L. Lau, Sept. 6, 1944; children: Ivan, Ada, Ina. B.S., Nat. Taiwan U., 1966; M.S. in Physical Chemistry, U. Houston, 1969; Ph.D., U. Wis.-Madison, 1973. Teaching asst. U. Houston, 1967-68, U. Wis.-Madison, 1969-73; research chemist Nat. Bur. Standards, Washington, 1974—. Author: (with others) Application of Photon Correlation Spectroscopy, 1983; contbr. articles to profl. jours. Recipient Bronze medal U.S. Dept. Commerce, 1980, Silver medal, 1982. Fellow Am. Phys. Soc.; mem. Am. Chem. Soc. Subspecialties: Polymer chemistry; Physical chemistry. Current work: Polymer characterization, static and dynamic light and neutron scattering; experimental statistical mechanics of polymers. Home: 19512 Burlingame Way Gaithersburg MD 20879 Office: National Bureau of Standards Washington DC 20234

HAN, MOO-YOUNG, physicist; b. Seoul, Korea, Nov. 30, 1934; came to U.S., 1954; s. Sunghoon and Kiejer (Kim) H.; m. Changki Hong, Aug. 29, 1959; children: Grace, Chris, Tony. B.S., Carroll Coll., Waukesha, Wis., 1957; Ph.D., U. Rochester, 1964. Research assoc. Syracuse U., 1964-65; asst. prof. U. Pitts., 1965-67; asst. prof. physics Duke U., Durham, N.C., 1967-71, assoc. prof., 1971-77, prof., 1977—; vis. prof. Kyoto U., 1974. Recipient Outstanding Prof. award Duke U., 1971, Disting. Teaching award, 1972; Disting. Fgn. Scholar award Kyoto U., 1974. Mem. Am. Phys. Soc. Subspecialties: Particle physics; Theoretical physics. Current work: nature and symmetry of fundamental building blocks of matter, leptons and quarks. Home: 615 Duluth St Durham NC 27705 Office: Dept Physics Duke U Durham NC 27706

HANAFUSA, HIDESABURO, virologist; b. Nishinomiya, Japan, Dec. 1, 1929; came to U.S., 1961; s. Kamehachi and Tomi H.; m. Teruko Inoue, May 11, 1958; 1 dau., Kei. B.S., Osaka (Japan) U., 1953, Ph.D., 1960. Research asso. Research Inst. for Microbial Diseases, Osaka U., 1958-61; postdoctoral fellow virus lab. U. Calif., Berkeley, 1961-64; vis. scientist College de France, Paris, 1964-66; asso. mem., chief dept. viral oncology Public Health Research Inst. of City N.Y. Inc., 1966-68, mem., 1968-73; prof. Rockefeller U., 1973—. Mem. editorial bd.: Internat. Jour. Cancer, 1974—, Jour. Virology, 1975—, BBA Rev. Cancer, 1973—, Intervirology, 1972—, Jour. Exptl. Medicine, 1976—, Cell, 1979—; contbr. articles to profl. jours. Recipient Howard Taylor Ricketts award, 1981, Albert Lasker Basic Med. Research award, 1982; Nat. Cancer Inst. grantee, 1966—; Am. Cancer Soc. grantee, 1976—. Mem. Am. Soc. Microbiology, AAAS, N.Y. Acad. Sci. Subspecialties: Molecular biology; Cancer research (medicine). Research on RNA tumor viruses. Home: 500 E 63d St New York NY 10021 Office: Rockefeller U 1230 York Ave New York NY 10021

HANCE, ANTHONY JAMES, pharmacologist, educator, researcher; b. Bournemouth, Eng., Aug. 19, 1932; s. Walter Edwin Stanley and Jessie Irene (Finch) H.; m. Ruth Anne Martin, July 17, 1954; children: David, Peter, John. B.S., Birmingham (Eng.) U., 1953, Ph.D., 1956. Research fellow in electrophysiology Birmingham (Eng.) U., 1957-58; research pharmacologist UCLA, 1959-62; research assoc. pharmacology Stanford (Calif.) U., 1962-65, asst. prof. pharmacology, 1965-68; assoc. prof. pharmacology U. Calif.-Davis, 1968—; cons. Riker Labs. Inc., 1959-60, Stanford Research Inst., 1963-64, Ampex Corp. Spl. Products and Videofile, 1964-68, Time Data Corp., 1966-68. Contbr. writings in field to sci. jours. Rockefeller research scholar, 1953-56. Mem. Am. Soc. Pharmacology and Exptl. Therapeutics, Assn. for Computing Machinery, Biomed. Engring. Soc., AAAS. Subspecialties: Pharmacology; Neuropharmacology. Current work: Central and autonomic nervous system pharmacology; electrophysiology of central nervous system; computer analysis of neurophysiological electrical signals. Office: Dept Pharmacology Med Sch U Calif Davis CA 95616

HANCE, ROBERT LEE, chemistry educator; b. El Paso, Tex., Mar. 1, 1943; s. William Henry and Mary Francis (Jordan) H.; m. Nedra Faye Drake, Aug. 29, 1963; children: Bryon, Kirk, Clint, Holly. A.A., York Coll., 1963; B.S., Abilene Christian U., 1966; Ph.D., M.I.T., 1970. Asst. prof. Abilene (Tex.) Christian U., 1970-75, assoc. prof., 1975-78, prof. chemistry, 1979—; vis. assoc. prof. U. Tex., Austin, 1978-79, research assoc., 1981. Contbr. articles to profl. jours. Active Taylor County Foster Parents Assn., 1970-82, Nat. Foster Parents Assn., 1974-82. Robert A. Welch Found. grantee, 1972-82. Mem. Am. Chem. Soc., Am. Phys. Soc., Am. Vacuum Soc., Tex. Acad. Sci. Mem. Ch. of Christ. Subspecialties: Surface chemistry; Atomic and molecular physics. Current work: Transition metal catalysis, electron spectroscopy, molecular beam scattering and intermolecular forces. Office: Sta ACU PO Box 8127 Abilene TX 79699

HANCOCK, EVERETT BRADY, dental educator, researcher; b. Centralia, Ill., Feb. 27, 1941; s. Everett Oral and Constance May (Brady) H.; m. Caryl Rae Ramstadt, June 18, 1966; children: Heidi Lynne, Janna Rene. B.S.D., U.-Ill.-Chgo., 1964, D.D.S., 1967; M.S.D., Ind. U.-Indpls., 1974. Commd. lt. U.S. Navy, 1967, advanced through grades to capt., 1980; dental officer U.S.S. Iwo Jima, San Diego, 1969-71; asst. dental officer, fellow in periodontics Naval Dental Clinic, Long Beach, Calif., 1971-72; dental officer U.S.S. Sperry, San Diego, 1974-76; research officer Navy Dental Research Inst., Gt. Lakes, Ill., 1976-80; mem. teaching staff Nat. Naval Dental Ctr., Bethesda, Md., 1980-83; oral and dental health program mgr. Naval Med. Research and Devel. Command, 1983—; asst. prof. U. Ill.-Chgo., 1976-80; assoc. prof. periodontics U. Md., 1980—; cons. in field. Contbr. articles to dental jours. Decorated Navy Achievement medal. Mem. ADA, Am. Assn. Dental Research, Am. Acad. Periodontology, Western Soc. Periodontology, Greater Washington Soc. Periodontology. Subspecialties: Periodontics; Oral biology. Current work: Wound healing-the role of the diseased root surface and its treatment in the surgical repair of periodontal defects. Home: 12044 Cheviot Dr Herndon VA 22070 Office: Oral and Dental Health Program Mgr Naval Med Research and Devel Bethesda MD 20814

HANCOCK, JOHN CHARLES, pharmacologist; b. Lockwood, Mo., Aug. 20, 1938; s. Daniel L. and Cordelia (Oats) H. B.S., U. Mo., 1962; M.S., U. Tex., 1965, Ph.D., 1967. Postdoctoral fellow U. Conn., Hartford, 1967-68, instr., 1968-69, asst. prof., 1969-71, La. State U., New Orleans, 1971-73, assoc. prof., 1973-77; prof. pharmacology, dir. neurosci. tng. program East Tenn. State U., Johnson City, 1977—. Contbr. articles nat. and internat. jours. Grantee U. Conn. Research Found., 1971-72, La. State U. Research Found., 1972-73, AMA, 1971-73, NIMH, 1974-75, Nat. Inst. Neurol. Diseases and Stroke, 1974-77, E. Tenn. Research Found., 1979-80; Tenn affiliate Am. Heart Assn., 1979-81, Biomed. Research Devel., 1979-80, 81-82. Mem. Am. Soc. Pharmacology and Exptl. Therapeutics, Soc. Neurosci. (pres. Appalachian chpt.), AAAS, Sigma Xi. Subspecialties: Pharmacology; Neuropharmacology. Current work: Central nervous system regulation of the heart and blood pressure, neuronal activity, electrophysiology, neuroendocrinology baroreceptor function, cardiovascular regulation. Home: 1306 Althea St Johnson City TN 37601 Office: Dept of Pharmacology Quillen Dishner College of Medicine East Tennessee State University Johnson City TN 37614

HANCOCK, JOHN COULTER, electrical engineer, educator, university dean; b. Martinsville, Ind., Oct. 21, 1929; s. Floyd A. and Katherine (Coulter) H.; m. Betty Jane Holden, Feb. 6, 1949; children: Debbie, Dwight, Marilyn, Virginia. B.S. in Elec. Engring., Purdue U., 1951, M.S., 1955, Ph.D., 1957. Research engr. U.S. Naval Avionics Facility, Indpls., 1951-57; asst. prof. elec. engring. Purdue U., West Lafayette, Ind., 1957-60, asso. prof., 1960-63, prof., 1963—, head Sch. Elec. Engring., 1965-72, dean Schs. Engring., 1972—; dir. CTS Corp., Elkhart, Ind., Pub. Service Co. Ind., McClure Research Park, Lafayette, Ind., Schwab Safe Co., Lafayette, Ransburg Corp., Indpls., Hillenbrand Industries, Batesville, Ind. Author: An Introduction to the Principles of Communication Theory, 1961, Signal Detection Theory, 1966, An Introduction to Electrical Design, 1972. Bd. dirs. United Community Services, Lafayette, 1969-73; trustee Christian Theol. Sem., 1977—; chmn. commn. on new ch. devel. Christian Ch. Ind., 1975-78. Fellow IEEE (chmn. field awards com. 1979—); mem. Nat. Acad. Engring., Nat. Engring. Consortium (formerly Nat. Electronics Conf., dir. 1966-67, 74-77); fellow Am. Soc. Engring. Edn. (sec. elec. engring. div. 1969-70, vice chmn. 1971-72, chmn. 1972-73, exec. com. council profl. and tech. edn. 1973-74, dir. 1977-79, chmn. longrange planning com. 1977-79, pres.-elect 1982-83, pres. 1983-84); mem. Midwestern Program Minorities in Engring. (exec. com.), Eta Kappa Nu (nat. pres. 1969-70); Mem. Christain Ch. (elder, trustee 1974—). Subspecialties: Computer-aided design. Home: 3829 Windward Pl West Lafayette IN 47906 Office: Engring Adminstrn Bldg Purdue U West Lafayette IN 47906

HANCOCK, JOSEPH GRISCOM, JR., plant pathology educator; b. Bridgeton, N.J., Apr. 8, 1938; m., 1960; 2 children. B.S., Rutgers U., 1960; M.S., Cornell U., 1963, Ph.D. in Plant Pathology, 1964. Asst. prof., asst. plant pathologist U. Calif.-Berkeley, 1964-70, assoc. prof. and assoc. plant pathologist, 1970-76, prof. plant pathologist, 1976—, chmn dept. conservation and resource studies., 1974-76, 83-84; vis. prof. Imperial Coll. (U. London), 1970. Mem. Am. Phytopath. Soc., Mycol. Soc. Am., Am. Soc. Plant Physiologists. Subspecialty: Plant pathology. Office: Dept Plant Pathology U Calif 147 Hilgard Hall Berkeley CA 94720

HANCOX, WILLIAM THOMAS, nuclear engineer; b. New Westminster, C., Can., Mar. 19, 1940; s. Joseph T. and Ruby (Cameron) H.; m. Ida Kathleen Patenaude, July 20, 1963; children: Kathleen, Kirsten. B.Eng., Carleton U., 1966, M.Eng., 1967; Ph.D., U. Waterloo, 1971. Research engr. Westinghouse Can., Inc., Hamilton, Ont., 1967-73; sect. head Atomic Energy Can., Pinawa, Man., 1973-76, br. head, 1976-78, dir. applied sci., 1978—; adj. prof. U. Waterloo, Ont., 1971-73. Mem. Can. Nuclear Soc., Am. Nuclear Soc. Subspecialties: Nuclear fission; Fluid mechanics. Current work: Fluid mechanics and heat transfer processes associated with fault conditions in water-cooled thermal nuclear reactors; reprocessing of advanced fuels for thermal nuclear reactors and fuel cycle strategies. Home: 23 Lansdowne Pinawa MB Canada ROE 1L0 Office: Atomic Energy Can Ltd Pinawa MB Canada

HAND, ARTHUR RALPH, biomedical research scientist, government research administrator; b. Los Angeles, May 15, 1943; s. Arthur Vaughn and Hazel Marie (Bashaw) H.; m. Kathi Rintoul, June 24, 1961; children: Gregg Arthur, Kristen. Student, UCLA, 1961-64, D.D.S. summa cum laude, 1968. Lic. dentist, Calif. Commd. USPHS, 1968; research investigator lab. biol. structure Nat. Inst. Dental Research, NIH, Bethesda, Md., 1968-76, acting chief lab., 1977-78, chief lab. biol. structure, 1978-82, chief lab. oral biology and physiology, 1982—, acting chief Mineralized Tissue Research Br.,

1982—; vis. prof. anatomy McGill U., Montreal, Que., Can., 1976-77; tchr., lectr. Georgetown U., Howard U., U. Md. Contbr. numerous articles, rev. articles, chpts. to profl. publs. Decorated Commendation medal; UCLA scholar, 1964-68. Mem. AAAS, Am. Soc. Cell Biology, Histochem. Soc., Internat. Assn. Dental Research (Basic Research in Oral Sci. award 1978), Chesapeake Soc. Electron Microscopy, Sigma Xi. Subspecialties: Cell biology; Cytology and histology. Current work: Structure and function of bells and organelles; mechanisms of exocrine secretion. Office: Nat Inst Dental Research NIH Bldg 30 Room 211 Bethesda MD 20205

HAND, BRYCE MOYER, geology educator; b. Jersey City, Mar. 22, 1936; s. Horace B. and Edna Mae (Moyer) H.; m. Judith Kuder, Sept. 14, 1963; children: Briana L., William B. A.B., Antioch Coll., 1958; M.S., U. So. Calif., 1961; Ph.D., Pa. State U., 1964. Asst. prof. geology Amherst (Mass.) Coll., 1964-69; asst. prof. geology Syracuse (N.Y.) U., 1969—. Fellow Geol. Soc. Am.; mem. Am. Assn. Petroleum Geologists, Soc. Econ. Paleontologists and Mineralogists. Subspecialty: Sedimentology. Current work: Dynamics of bedforms and sediment transport. Home: 132 Lynn Circle Syracuse NY 13205 Office: Dept Geology Syracuse U Syracuse NY 13210

HAND, PETER JAMES, neurobiologist, educator; b. Oak Park, Ill., Jan. 5, 1937; s. James Harold and Edna Mae (Watson) H.; m. Mary Minnis, Sept. 16, 1958; children: Katherine Patricia, Carol Jane, Margaret Anne, Robin Lynn, Stephen Douglas, Peter James; m. Carol Louise Corson, Oct. 23, 1976; stepchildren: Scott Curtis Carlson, Glenn Arthur Carlson. V.M.D., U. Pa., 1961, Ph.D., 1964. Mem. faculty U. Pa., Phila., 1964—, prof. anatomy, 1979—, head, 1980—. Contbr. numerous neurobiol. articles to profl. publs. Pres. USO Council, Cape May, N.J., 1972-73, nat. del.; trustee Mid-Atlantic Ctr. for the Arts, Cape May, 1973-74; bd. dirs. Cape May Taxpayers Assn., 1972-74, University City Hist. Soc., Phila., 1978-80. NIH grantee, 1970-82. Mem. Am. Assn. Anatomists, Am. Assn. Vet. Anatomists, Soc. Neurosci., Internat. Brain Research Orgn., World Assn. Vet. Anatomists, Sigma Xi, Alpha Psi (trustee 1965—). Democrat. Subspecialties: Neurobiology; Neurophysiology. Current work: Central processing of somatosensory information, including pain; analgesia mechanisms, including acupuncture; functional and anatomical plasticity of the central nervous system. Office: U Pa Sch Vet Medicine Philadelphia PA 19104

HAND, ROGER, physician, educator; b. Bklyn., Sept. 25, 1938; s. Morton and Angela (Belevedere) H.; m. Abby Lippman, Dec. 24, 1961; children: Christopher, Jessica. B.S., NYU, 1959, M.D., 1962. Intern, then resident in internal medicine NYU Med. Center, 1962-68; postdoctoral fellow, asst. prof. Rockefeller U., N.Y.C., 1968-73; clin. asst. prof. medicine Cornell U. Med. Coll., N.Y.C., 1970-73; asst. prof., then assoc. prof. medicine McGill U., Montreal, Que., Can., 1973-80; prof. medicine, dir. McGill Cancer Center, 1980—; sr. physician Royal Victoria Hosp., Montreal, 1980—. Contbr. articles to profl. jours. Served with AUS, 1964-65. Decorated Air medal. Med. research grantee, fellow. Fellow ACP, Royal Coll. Physicians and Surgeons (Can.); mem. Am. Soc. Clin. Investigation, Am. Soc. Biol. Chemists, Am. Assn. Cancer Research, Am. Soc. Clin. Oncology, Infectious Disease Soc. Am., Can. Soc. Clin. Investigation, others. Subspecialties: Molecular biology; Cancer research (medicine). Current work: Regulation of DNA regulation, general mechanism of action of virus tumor proteins. Office: McGill Cancer Center McGill U 3655 Drummond St Montreal PQ H3G 1Y6 Canada

HANDE, KENNETH ROBERT, physician; b. Mpls., Jan. 20, 1946; s. Edwin Kenneth and Evelyn Dorothy (Ogrosky) H.; m. Mary Saunders, Aug. 22, 1970; 1 child. A.B., Princeton U., 1968; M.D., Johns Hopkins U., 1972. Intern, resident Barnes Hosp., St. Louis, 1972-74; clin. assoc., career expert Nat. Cancer Inst., Bethesda, Md., 1974-78; asst. prof. medicine and pharmacology Vanderbilt U., Nashville, 1978-82, assoc. prof., 1982—. Subspecialties: Chemotherapy; Pharmacology. Current work: Antineoplastic drug pharmacology. Home: 502 W Hillwood Nashville TN 37205 Office: Dept Medicine Div Oncology Vanderbilt U Sch Medicine Nashville TN 37232

HANDLER, ROBERT ALPHONSE, mechanical engineer, researcher; b. Newark, Aug. 17, 1951; s. Herbert and Bernice (Zoppi) H. B.E., Stevens Inst. Tech., 1973; M.S.E., U. Mich., 1974; S.M., M.I.T., 1976; Ph.D., U. Minn., 1980. Research asst. M.I.T., Cambridge, 1974-76; teaching assoc. U. Minn., Mpls., 1976-79; mech. engr. Naval Ship Research and Devel. Ctr., Bethesda, Md., 1980-81, Naval Research Lab., Washington, 1981—. Mem. ASME, AIAA, Acoustical Soc. Am. Subspecialties: Fluid mechanics; Theoretical and applied mechanics. Current work: Experimental and theoretical work in hydroacoustics; use of optical methods (fiber optics, laser doppler velocimetry) in fluid mechanics. Home: 11978 Home Guard Dr Woodbridge VA 22192 Office: Naval Research Lab 4555 Overlook Ave SW Washington DC 22303

HANDORF, CHARLES RUSSELL, pathologist, medicinal chemist; b. Memphis, Jan. 9, 1951; s. Everett Charles and Lucille (Preston) H.; m. Miriam Howard Fulmer, Dec. 28, 1976; children: Charles Russell II, Jennifer Anne. B.A., Rice U., 1973; M.D., U. Tenn., 1977, Ph.D., 1981. Diplomate: Am. Bd. Pathology. Teaching asst. Rice U., Houston, 1972-73; research asst. VA Hosp., Memphis, 1977-78; resident in pathology U. Tenn.-Memphis, 1978-80, Meth. Hosp., Memphis, 1980-82; assoc. pathologist Duckworth Pathology Group, Memphis, 1982—; adj. instr. U. Tenn., 1978-80; jr. med. staff Meth. Hosp., 1982—, Eastwood Hosp., Memphis, 1982—. Marion Labs. fellow, 1973-74; recipient Lange award U. Tenn., 1977. Mem. So. Med. Assn., Coll. Am. Pathologists, Am. Soc. Clin. Pathologists, AMA, Sigma Xi, Phi Chi, Rho Chi. Methodist. Subspecialties: Pathology (medicine); Medicinal chemistry. Current work: Development and implementation of novel approaches to problems in clinical toxicology and therapeutic drug monitoring. Home: 470 Greenfield Rd Memphis TN 38117 Office: Duckworth Pathology Group 1331 Union St Suite 1005 Memphis TN 38104

HANDY, LYMAN LEE, chemist, educator; b. Payette, Idaho, Aug. 4, 1919; s. Clarence Lee and Lillie (Hall) H.; m. Lenore E. Ross, Aug. 28, 1948; children—Mark Ross, Gail Eileen. Student, Western Wash. Coll., 1938-40; B.S., U. Wash., 1942, Ph.D., 1951. With Chevron Oil Field Research Co., 1951-66; mem. faculty U. So. Calif., 1966—, prof. chem. and petroleum engring., chmn. petroleum engring., 1966—, chmn. chem. engring., 1969-76, Omar B. Milligan prof. petroleum engring., 1976—; cons. in field. Mem. editorial bd., Trans. Am. Inst. Mining Engrs., 1960, 68, 69; Contbr. articles to profl. jours. Served to lt. USNR, 1942-46. Mem. Am. Chem. Soc. (chmn. Orange County sect. 1969), Soc. Petroleum Engrs. (dir. Los Angeles basin sect. 1971-75, chmn. 1974, nat. dir.-at-large 1978—), Am. Inst. Chem. Engrs., A.A.A.S., Phi Beta Kappa, Sigma Xi, Phi Lambda Upsilon, Tau Beta Pi. Subspecialties: Petroleum engineering; Surface chemistry. Current work: Fluid flow through porous materials and new methods for enhanced recovery of oil. Home: 1401 Dana Pl Fullerton CA 92631 Office: University Park Los Angeles CA 90007

HANEBRINK, EARL LEE, biology educator; b. Cape Girardeau, Mo., Mar. 24, 1924; s. Harry Harrison and Augusta (Fornkohl) H.; m. Sue Robinson, Mar. 1, 1972; children: from previous marriage: Lisa Ann, Kay Lynn. B.S. in Edn, S.E. Mo. State U., 1948; M.S., U. Miss., 1955; Ed.D., Okla. State U., 1965. Tchr. sci. high schs., Parma and Kennett, Mo., 1948-58; instr. biology Ark. State U., Jonesboro, 1958-60, asst. prof., 1960-62, assoc. prof., 1962-66, prof., 1966—; environ. cons. U.S. C.E., 1970—. Author books in field. Served with AUS, 1943-46. Mem. Am. Pigeon Fanciers Council (pres. 1979-80, award 1981), Ark. Audubon Soc. (past v.p., pres.), N.E. Ark. Audubon Soc. (pres. 1974-83), Soc. Southwestern Naturalists, Tenn. Ornithol. Soc., Sigma Xi (past pres. chpt.). Methodist. Lodge: Elks. Subspecialties: Ecology; Zooplankton limnology. Current work: Heronries, pigeon behavior, natural history of birds. Home: 4112 Oakhill Ln Jonesboro AR 72401 Office: Ark State U Box 67 State University AR 72467

HANIFIN, LEO EUGENE, manufacturing and research administrator, mechanical engineer; b. Binghamton, N.Y., Aug. 2, 1946; s. Leo Francis and Mary Hanna (McDonald) H.; m. Angela Papa, Aug. 2, 1980; children: Jacqueline, Sonia, Leo Daniel. Student, St. Bonaventure U., 1964-66; B.M.E., U. Detroit, 1969, M. in M.E., 1972, Dr. Engring. and Mfg. Systems, 1976; postgrad., UCLA, 1969-71. Hughes fellow Hughes Aircraft Co., 1969-70, mem. staff, 1971; engr. Aerojet Gen. Corp., Culver City, Calif., 1970-71; systems coordinator gen. mfg. div. Chrysler Corp., Detroit, 1971-78, project mgr., Syracuse, N.Y., 1972-80; dir. Center for Mfg. Productivity and Tech. Transfer, Rensselaer Poly. Inst., Troy, N.Y., 1980—; adj. assoc. prof. U. Mich., 1978; cons. for mfg. productivity improvements. Mem. Soc. Mfg. Engrs. (vice chmn. mfg. mgmt. council, outstanding young mfg. engr. of yr. 1981, lectr.), Am. Inst. Indsl. Engrs. Democrat. Roman Catholic. Subspecialties: Manufacturing productivity; Robotics. Current work: Improving productivity and application of advanced technologies, education for manufacturing engineers, robotics-inspection, manufacturing systems design and analysis, control systems, quality systems.

HANIN, ISRAEL, pharmacologist, educator, researcher; b. Shanghai, China, Mar. 29, 1937; s. Leo and Rebecca (Lubarsky) H.; m. Leda Toni, June 12, 1960; children: Adam, Dahlia. B.S., UCLA, 1962, M.S., 1965, Ph.D. in Pharmacology, 1968. Vis. scientist dept. toxicology Karolinska Inst., Stockholm, 1968; pharmacologist Lab. Preclin. Pharmacology NIMH, Bethesda, Md., 1969-73; from asst. prof. to assoc. prof. psychiatry and pharmacology U. Pitts. Sch. Medicine, 1973-81, prof., 1981—; also dir. psychopharmacology program Western Psychiat. Inst. and Clinic; mem. research grant rev. com. NIMH, 1979-82. Editor 7 books; contbr. over 150 articles to sci. jours. Served to 2d lt. Armored Corps Israeli Army, 1955-58. NIMH grantee, 1965—. Mem. Pitts. Neurosci. Soc. (pres. 1982-83), Am. Chem. Soc., Am. Soc. Pharmacology and Exptl. Therapeutics, Am. Soc. Neurochemistry, Am. Coll. Neuropsychopharmacology. Subspecialties: Neuropharmacology; Neurochemistry. Current work: Elucidation of factors controlling neurotransmitter synthesis with particular emphasis on cholinergic system; clin. correlates; animal models. Office: 3811 O'Hara St Pittsburgh PA 15213

HANKES, LAWRENCE VALENTINE, clinical biochemical researcher; b. Chgo., Nov. 24, 1919; s. Michael John and Matilda Ann (Bachman) H.; m. Mary Catherine Hamm, Sept. 16, 1951; children: Lawrence Michael, Catherine Ann, Matthew William. A.B., DePaul U., 1942; M.S., Mich. State U., 1943; Ph.D., U. Wis., 1949, 1950. Registered clin. lab. dir., N.Y. Instr. biochemistry Mich. State U., East Lansing, 1942-43; instr. biochemistry Northwestern U. Dental Sch., Chgo., 1943-44; indsl. research fellow U. Wis., Madison, 1947-49; head allergy group VA Hosp., Aspinwall, Pa., 1950; head clin. chemistry med. ctr. Brookhaven Nat. Lab., Upton, N.Y., 1951—, sr. scientist, 1968—; cons. and researcher in field. Contbr. articles in field to profl. jours. Fellow Brit. Chem. Soc., Nat. Acad. Clin. Biochemists; mem. Am. Chem. Soc., Am. Soc. Biol. Chemists, Soc. Exptl. Biology & Medicine, Am. Assn. Clin. Chemists, Alpha Chi Sigma, Sigma Xi, Phi Sigma. Club: Explorer's. Subspecialties: Clinical chemistry; Nutrition (medicine). Current work: Research on tryptophan metabolism in cancer, scleroderma, scurvy and pellagra synthesis of labeled metabolites; clinical chemistry methods. Home: 11 Maple Rd Setauket NY 11733 Office: Med Research Ctr Brookhaven Nat Lab Upton NY 11973

HANKINS, TIMOTHY HAMILTON, educator, radioastronomer; b. Miami, Fla., Mar. 13, 1941; s. Frank Hamilton, Jr. and Anne Chapin (Hudson) H.; m. Mary E. Nutt, Oct. 1, 1977; 1 son, Samuel Clark. B.A., Dartmouth Coll., 1962, M.S., 1967; Ph.D., U. Calif., San Diego, 1971. Research physicist, instr. U. Calif., San Diego, 1971-74; research assoc. Arecibo (P.R.) Obs., 1974-81; Alexander von Humboldt fellow Max Planck Inst. for Radioastronomy, Bonn, W.Ger., 1978-79; assoc. prof. Thayer Sch. Engring., Dartmouth Coll., Hanover, N.H., 1981—. Contbr. tech. papers to sci. publs. Served to lt. USN, 1962-64. Various research grants NSF. Mem. Am. Astron. Soc., Internat. Astron. Union, Internat. Sci. Radio Union. Subspecialties: Radio and microwave astronomy; Computer engineering. Current work: Radio pulsar studies, signal processing, microprocessors. Office: Dartmouth Coll HB 8000 Hanover NH 03755

HANLEY, KEVIN JOSEPH, orthodontist, researcher, dental educator; b. Utica, N.Y., Oct. 25, 1952; s. Richard Joseph and Mary Teresa (Cain) H.; m. Carmella Marie Rosetti, June 24, 1978. B.A., SUNY, Buffalo, 1974, D.D.S., 1978. Cert. orthodontics U Conn., 1980. Clin. asst. prof. in residence orthodontics U. Conn. Sch. Dental Medicine, Farmington, 1980—. Author: (C.J. Burstone) Syllabus-Modern Edgewise Mechanics Segmented Arch Technique, 1982. Mem. Am. Assn. Orthodontists, ADA, Northeastern Soc. Orthodontists, Conn. State Dental Assn., Hartford Dental Soc. Republican. Roman Catholic. Clubs: Energy Independent (Simsbury, Conn.); Am. Mus. Natural History (N.Y.C.). Subspecialties: Orthodontics; Cell and tissue culture. Current work: Effects of pulsating electromagnetic fields on osteoblasts in culture, clinical application of biomechanics in orthodontics. Home: 7 Cornfield Rd Simsbury CT 06070 Office: 345 N Main St West Hartford CT 06117

HANLIN, RICHARD THOMAS, mycologist; b. Hammond, Ind., May 10, 1931; s. Arthur M. and Mary E. (Hedges) H.; m. Elba I. Bueno, Aug. 15, 1955 (div.); children: Maria, Janath, Richard. B.S., U. Mich., 1953, M.S., 1955, Ph.D., 1960. Asst. plant pathologist Ga. Expt. Sta., Experiment, Ga., 1960-66; assoc. prof. U. Ga., Athens, 1967-74, prof., 1974—. Mem. Bot. Soc. Am., Mycological Soc. Am., Am. Phytopathol. Soc., Sociedad Mexicana de Micologia, Mycological Soc. Japan, Sociedad Venezolana de Fitopathologia, Brit. Mycological Soc. Subspecialties: Taxonomy; Plant pathology. Current work: Studies on the developmental morphology and taxonomy of the perithecial ascomycetes and their conidial states. Home: Route 1 Box 459E Hwy 106 Hull GA 30646 Office: Dept Plant Pathology U Ga Athens GA 30602

HANNAH, DAVID, JR., space launch service company executive; b. Houston, Apr. 9, 1922; s. David and Ethel May (Bloomfield) H.; m. Catherine Coburn, June 26, 1943; children: David III, Douglas, Glen Hannah Cole. B.A., Rice U., 1944. Pres. Ayrshire Corp., Houston, 1946-75, Castlewood Corp., Denver, 1975-80; pres. Space Services of Am., Houston, 1981—, chmn. bd., 1982—. Trustee Hermann Hosp. Estate, 1980-83; chmn. Houston Com. Fgn. Relations, 1982-83. Presbyterian. Club: Houston Country. Lodge: Kiwanis. Subspecialty: Space Launch Service. Current work: Chairman of the board of company engaged in developing private commercialization of space launch services. Office: Space Services of Am 2028 Buffalo Terr Houston TX 77019

HANNAN, GARY LOUIS, biologist; b. San Mateo, Calif., Nov. 18, 1951; s. Robert Stanley and Frances Taylor (Lucas) H.; m. Laurianne Lochhead, Aug. 17, 1974. B.A., U. Calif., Santa Barbara, 1973, Ph.D. in Botany, 1979. Lectr. U. Calif., Berkeley, 1980, Calif. State U., Fresno, 1980; lectr. U. Calif., Santa Barbara, 1980-82, research assoc., 1980-82; asst. prof. biology Eastern Mich. U., 1982—. Mem. AAAS, Bot. Soc. Am., Soc. Study Evolution, Am. Soc. Plant Taxonomists, Calif. Bot. Soc. Subspecialties: Reproductive biology; Systematics. Current work: Pollination and reproductive biology of plants; plant systematics; systematics of papaveraceae and woody hydrophyllaceae; reproductive biology of species polymorphic for flower color; research on platystomen californicus (papaveraceae). Office: Dept Biology Eastern Mich U Ypsilanti MI 48197

HANNAY, N(ORMAN) BRUCE, chemist; b. Mt. Vernon, Wash., Feb. 9, 1921; s. Norman Bond and Winnie (Evans) H.; m. Joan Anderson, May 27, 1943; children—Robin, Brooke. B.A., Swarthmore Coll., 1942, D.Sc. (hon.), 1979; M.S., Princeton U., 1943, Ph.D., 1944; Ph.D. (hon.), Tel Aviv U., 1978, D.Sc., Poly. Inst. N.Y., 1981. With Bell Telephone Labs., Murray Hill, N.J., 1944-82, exec. dir. materials research div., 1967-73, v.p. research and patents, 1973-82, ret., 1982; dir. Plenum Pub. Co., Gen. Signal Corp., Rohm and Haas Co.; Chmn. sci. adv. council Atlantic Richfield Corp.; Regents' prof. UCLA, 1976, U. Calif., San Diego, 1979; cons. Alexander von Humboldt Found. Author: Solid State Chemistry, 1967, also articles.; Mem. numerous editorial bds.; editor: Semiconductors, 1959, Treatise on Solid State Chemistry, 1974. Recipient Acheson medal, 1976, Perkin medal, 1983. Fellow Am. Phys. Soc.; mem. Nat. Acad. Engring. (fgn. sec.), Nat. Acad. Scis., Am. Acad. Arts and Scis., Mexican Nat. Acad. Engring., Am. Chem. Soc., Electrochem. Soc. (past pres.), Indsl. Research Inst. (past pres., medal 1982), Dirs. of Indsl. Research (past chmn.). Subspecialties: Materials; Electronics. Current work: Business and research consultant in high technology fields: electronics, materials, communications, specialty chemicals, and energy resources. Research on dipole moments and molecular structure; thermionic emission, mass spectroscopy; analysis of solids, solid state chemistry, semiconductors, superconductors. Home: Mitchell Point Friday Harbor WA 98250 Office: Nat Acad Engring 2101 Constitution Ave Washington DC 20418

HANNUM, WILLIAM HAMILTON, federal government departmental administrator; b. Menominee, Mich., Apr. 22, 1933; s. Robert Henry and Amy (Taylor) Guthrie; m. Irene Kunkle, Sept. 3, 1960; children: William Hamilton IV, Audrey Lynn, Susan Faye. A.B., Princeton U., 1954; M.S., Yale U., 1956, Ph.D., 1958. Sr. scientist Westinghouse Co. Bettis Atomic Power Lab., West Miflin, Pa., 1958-63; sr. scientist, asst. group leader Los Alamos Sci. Lab., 1963-68; asst. dir. nuclear safety U.S. Dept. Energy, Washington, 1968-76, dep. mgr., Idaho Falls, 1976-77, chief reactor physics, 1982—, dir. West Valley Project, 1982—; dep. dir. gen. OECD-Nuclear Energy Agy., Paris, 1977-82. Chmn. Princeton Alumni Schs. Com. for France, Paris, 1978-82; pres. Am. sect. Lycee Internat., St. Germaine-en-Laye, France, 1979-82. Mem. Am. Nuclear Soc. (bd. dirs. La Grange, Ill. 1974-77). Subspecialties: Nuclear fission; Nuclear engineering. Home: 9 Bruce Dr Orchard Park NY 14127 Office: West Valley Project Office US Dept Energy Box 191 West Valley NY 14171

HANOVER, PAUL NORDEN, electronic engineer; b. Hartford, Conn., Sept. 13, 1927; s. Adrian Norden and Ruth H.; m. Margaret Manning, Sept. 11, 1954; children: Nancy, Diane. B.E.E., Rensselaer Poly. Inst., 1952; postgrad., U. Ariz., 1969. Circuit designer Douglas Aircraft Co., Santa Monica, Calif., 1952-54; electron devices engr. Hughes Aircraft Co., Culver City, Calif., 1954-63, sr. project engr., Tucson, Ariz., 1964—, sr. project engr. missiles, 1973—. Author: (with others) Star Spangled Speakers, 1982; author, narrator: cassette Space Shuttle, 1982. Bd. dirs. Space Series Tucson (Ariz.) Pub. Library, 1981. Served with USN, 1946-48. Fellow AIAA (assoc., chmn. 1982-83); mem. IEEE (chmn. 1972-73), Fedn. Aerospace Socs. Tucson (chmn. 1981-83), Nat. Speakers Assn., Nat. Space Inst. Republican. Christian Scientist. Club: Hughes Tucson Mgmt. (pres. 1982-83). Lodge: Masons. Subspecialties: Aerospace engineering and technology; Electronics. Current work: Keynote addresses at conferences and banquets on the space programs of America. Home: 3551 Winslow Dr Tucson AZ 85715 Office: Hughes Aircraft Co Tucson AZ 85734

HANRATTY, THOMAS JOSEPH, chemical engineer, educator; b. Phila., Nov. 9, 1926; s. John Joseph and Elizabeth Marie (O'Connor) H.; m. Joan L. Hertel, Aug. 25, 1956; children: John, Vincent, Maria, Michael, Peter. B.S. Chem. Engring., Villanova U., 1947, 1979; M.S., Ohio State U., 1950; Ph.D., Princeton U., 1953. Engr. Fischer & Porter, 1947-48; research engr. Battelle Meml. Inst., 1948-50; engr. Rohm & Haas, Phila., summer 1951; research engr. Shell Devel. Co., Emeryville, Calif., 1954; faculty U. Ill., Urbana, 1953—, asso. prof., 1958-63, prof. chem. engring., 1963—; cons. in field; vis. asso. prof. Brown U., 1962-63. Contbr. articles to profl. jours. Mem. U.S. Nat. Com. on Theoretical and Applied Mechanics. NSF sr. postdoctoral fellow, 1962; recipient Curtis W. McGraw award Am. Soc. Engring. Edn., 1963, Sr. Research award, 1979. Fellow Am. Phys. Soc., Am. Acad. Scis.; mem. Nat. Acad. Engrs., Am. Inst. Chem. Engrs. (Colburn award 1957, Walker award 1964, Profl. Progress award 1967), Am. Chem. Soc. Roman Catholic. Club: Serra Internat. Subspecialties: Chemical engineering; Fluid mechanics. Home: 1019 W Charles St Champaign IL 61820 Office: 205 Roger Adams Lab U Ill Urbana IL 61801

HANSBOROUGH, LASH DEVOUS, mechanical engineer; b. Tyler, Tex., Mar. 31, 1943; s. John W. and Eloise O. (Garrard) H. B.S.M.E., U. Tex.-Austin, 1966, M.S.M.E., 1973. Registered profl. nuclear engr., Calif. Staff mem. Los Alamos Nat. Lab., 1974-78, mech. sect. leader, 1978-80, staff mem., 1980—; vis. scientist Chalk River Nuclear Labs., Ont., Can., 1982. Served to lt. USN, 1966-70. Mem. Am. Nuclear Soc., ASME, Am. Vacuum Soc., Naval Res. Assn. Republican. Presbyterian. Club: Los Alamos Ski. Subspecialty: Accelerator Engineering. Current work: Particle accelerator technology; linear accelerator design; radio-frequency cavity development. Developer inventor radio-frequency quadruple linear accelerator. Office: Los Alamos Nat Lab AT-1, MS H817 Los Alamos NM 87545

HANSCH, CORWIN HERMAN, educator; b. Kenmare, N.D., Oct. 6, 1918; s. Herman William and Rachel (Corwine) H.; m. Gloria J. Tomasulo, Jan. 8, 1944; children—Clifford, Carol. B.S., U. Ill., 1940; Ph.D., N.Y.U., 1944. Research chemist Manhattan project E.I. du Pont de Nemours & Co., Inc., 1944-45, research chemist, 1945-46; prof. chemistry Pomona Coll., 1946—; spl. research relationship chem. structure and drug action. Guggenheim fellow Fed. Inst. Tech., Zurich, Switzerland, 1952-53, Pomona Coll., 1966-67; Petroleum Research Fund fellow U. Munich, Germany, 1959-60; Recipient medal Italian Soc. Pharm. Sci., 1967; Coll. Chemistry Teaching award Mfg. Chemists Assn., 1969; Research Achievement award Am. Pharm. Assn., 1969; E.A. Smissman award Medicinal Chemistry Am. Chem. Soc., 1975; Tolman award Los Angeles sect., 1976. Subspecialty: Organic chemistry. Home: 4070 Olive Knoll Pl Claremont CA 91711

HANSCHE, BRUCE DAVID, electrical engineer; b. Albuquerque, Dec. 8, 1947; s. George Everett and Helen Barbara (Johnson) H.; m. Christina Cripps Husted, Dec. 21, 1968; children: Heather, Jena. B.S.E.E., Mich. State U., 1969; M.S.E.E., Stanford U., 1970; Ph.D. in Elec. Engring., U. Mich., 1976. Mem. staff nondestructive testing tech. Sandia Labs., Albuquerque, 1969—. Contbr. articles to profl. jours. Mem. Optical Soc. Am., Eta Kappa Nu, Tau Beta Pi. Subspecialties: Optical engineering; Graphics, image processing, and pattern recognition. Current work: Advanced non-destructive testing technology, image processing for radiography, holographic interferometry. Home: Box 962 Tijeras NM 87059 Office: Sandia Labs Div 7551 Albuquerque NM 87185

HANSEN, CARL JOHN, physics educator, researcher; b. Bklyn., Dec. 21, 1933; s. Carl H. and Ada J. (Michaelis) H.; m.; 1 son, Ethan R. B.S., Queens Coll., 1956; M.S., Yale U., 1961, Ph.D., 1966. Reactor analyst Combustion Engring., Windsor, Conn., 1956-60; asst. prof. U. Colo., Boulder, 1968-70, assoc. prof., 1970-74, prof. physics, 1974—. NSF grantee, 1972-83. Mem. Internat. Astron. Union, Am. Astron. Soc. Democrat. Subspecialty: Theoretical astrophysics. Current work: Stellar structure, evolution and stability. Office: U Colo JILA Box 440 Boulder CO 80309

HANSEN, CARL TAMS, geneticist; b. Greeley, Colo., July 22, 1929; s. Jens and Marie (Andersen) H.; m. Janet Coyle, June 8, 1962; children: Alan, Carol. B.S. in Agr, Colo. State U., 1951; M.S., S.D. State U., 1959; Ph.D. in Genetics, U. Wis., 1967. Geneticist Vet. Resources br. Div. Research Services, NIH, Bethesda, Md., 1964—. Contbr. articles on genetic models for biomedresearch to profl. publs. Served to capt. USMC, 1951-53. Recipient Superior Service award HEW, 1975. Mem. Genetics Soc. Am., Am. Genetics Assn., AAAS. Subspecialties: Genetics and genetic engineering (medicine); Animal breeding and embryo transplants. Current work: The use of genetics, genetic engineering and animal breeding techniques to develop mammalian model systems of human diseases. Office: NIH Bldg 14A Room 102 Bethesda MD 20205

HANSEN, GRANT LEWIS, aerospace and info. systems executive; b. Bancroft, Idaho, Nov. 5, 1921; s. Paul Ezra and Leona Sarah (Lewis) H.; m. Iris Rose Heyden, Apr. 21, 1945; children: Alan Lee, Brian Craig, Carol Margaret, David James, Ellen Diane. B.S. in Elec. Engring., Ill. Inst. Tech., 1948; postgrad. engring. and mgmt., UCLA, Calif. Inst. Tech.; D.Sc., Nat. U., 1978. With Douglas Aircraft Co., 1948-60; v.p., program dir. for Centaur (Convair div.), 1960-65; v.p. launch vehicle programs Convair div. Gen. Dynamics Corp., 1965-69; asst. sec. air force for research and devel., 1969-73; v.p. Gen. Dynamics Corp., San Diego, 1974-78, v.p., gen. mgr., 1973-78; exec. v.p. System Devel. Corp., Santa Monica, Calif., 1978—; also pres. SDC Systems Group, 1978—; U.S. del. NATO (Adv. Group for Aerospace Research and Devel.), 1969-73; U.S. mem. sci. com. for nat. reps. SHAPE Tech. Center, The Hague, Netherlands, 1969-73; mem. research and tech. adv. council NASA, 1971-73; mem. sci. adv. bd. Dept. Air Force, 1976—. Served with USNR; World War II. Decorated Purple Heart; recipient Pub. Service award NASA, 1966, Disting. Pub. Service award, 1975; Alumni Recognition award Ill. Inst. Tech., 1967; USAF Exceptional Civilian Service medal, 1973, 83. Fellow AIAA (nat. pres. 1975), Am. Astronautical Soc., AAAS, Internat. Acad. Astronautics; mem. IEEE (sr.), German Soc. Air and Space Travel (corr.), Nat. Alliance Businessmen (nat. bd. dirs., dir. region IX), Nat. Acad. Engring. (aeros. and space engring. bd.), NRC, Eta Kappa Nu, Tau Beta Pi. Subspecialties: Aerospace engineering and technology; Information systems (information science). Current work: Aeronautical and space vehicles and technology; computer based information, command and control and intelligence systems. Home: 10737 Fuerte Dr LaMesa CA 92041 Office: System Devel Corp 2500 Colorado Ave Santa Monica CA 90406 I've given my whole self to each challenge I've accepted, believing that what's best for my future is an honest day's effort today. I have great faith in my God and my country.

HANSEN, JAMES VERNON, computer science and information systems educator, researcher; b. Idaho Falls, May 31, 1936; s. Heber Lorenzo and Myrtle Jane (Simmons) H.; m. Diane Lynne Bradbury, Sept. 18, 1963; children: Tamsin, Jeffrey, Dale, Peter. B.S., Brigham Young U., 1963, M.S., 1966; Ph.D., U. Wash., 1973. Systems analyst TRW, Redondo Beach, Calif., 1966-69; sr. research scientist Battelle Meml. Inst., Richland, Wash., 1972-74; asst. prof. Ind. U., Bloomington, 1974-77, assoc. prof., 1978-81; prof. Brigham Young U., Provo, Utah, 1982—; prin. Phil Johnson & Assocs., Phoenix, 1980—; cons. Battelle Meml. Inst. Author: Controls in Microcomputer Systems, 1984; also monographs. Advanced Systems Analysis. Mem. dist. scout council Boy Scouts Am., Ind., 1978. Served in U.S. Army, 1959-62. Grantee Peat, Marwick, Mitchell Found., 1982, 83, Inst. Internal. Auditors, 1983, Kellogg Found., 1972. Mem. Assn. Computing Machinery, Inst. Mgmt. Sci., Ops. Research Soc. Am., Am. Assn. Artificial Intelligence, Sierra Club. Republican. Mem. Ch. Jesus Christ of Latter-day Saints. Subspecialties: Information systems (information science); Artificial intelligence. Current work: Expert systems in computer auditing; use of inductive methods for knowledge engineering design to analysis and design; operations research methods in information systems design. Office: Brigham Young U Provo UT 84602

HANSEN, KENT FORREST, nuclear engineering educator; b. Chgo., Aug. 10, 1931; s. Kay Frost and Mary (Cummins) H.; m. Katherine Elizabeth Kavanagh, June 13, 1959 (dec. Dec. 1975); children—Thomas Kay, Katherine Mary; m. Deborah Lea Hill, June 26, 1977. S.B., Mass. Inst. Tech., 1953, Sc.D., 1959. Mem. faculty Mass. Inst. Tech., 1960—, prof. nuclear engring., 1969—, exec. officer nuclear engring. dept., 1972-76, acting head dept., 1975—, assoc. dean engring., 1979-81; dir. EG&G, Inc.; cons. to industry. Co-author: Numerical Methods of Reactor Analysis, 1964, Advances in Nuclear Science and Technology, Vol. 8, 1975. Ford postdoctoral fellow, 1960-61. Fellow Am. Nuclear Soc. (dir., Arthur Holly Compton award 1978); mem. Soc. Indsl. and Applied Math., Assn. Computing Machines, Am. Soc. Engring. Edn., Nat. Acad. Engring., Sigma Xi, Sigma Chi. Subspecialty: Nuclear engineering. Current work: Nuclear reactor analysis and safety; nuclear fuel management; energy systems. Home: Baker Bridge Rd Lincoln MA 01773 Office: Mass Inst Tech Massachusetts Ave Cambridge MA 02139

HANSEN, MORRIS HOWARD, statistician, former govt. ofcl.; b. Thermopolis, Wyo., Dec. 15, 1910; s. Hans C. and Maud Ellen (Omstead) H.; m. Mildred R. Latham, Aug. 31, 1930; children—Evelyn Maxine, Morris Howard, James Hans, Kristine Ellen. B.S., U. Wyo., 1934; M.A., Am. U., 1940; LL.D., U. Wyo., 1959. Statistician Wyo. Relief Adminstrn., 1934; statistician U.S. Bur. of Census, Washington, 1935-43, statis. asst. dir., 1944-49, asst. dir. statis. standards, 1949-61; asso. dir. research and devel., 1961-68; sr. v.p. Westat, Inc., 1968—; instr. statistics grad. sch. Dept. Agr., 1945-50; formerly statis. cons. Nat. Analysts, Inc. Co-author: Sample Survey Methods and Theory, 2 vols, 1953; Contbr. articles to statis. jours. Recipient Rockefeller Pub. Service award, 1962. Fellow Am. Statis. Assn. (pres. 1960), Royal Statis. Soc. (hon.), A.A.A.S., Inst. Math. Statistics (pres. 1953); mem. Internat. Statis. Inst. (hon. mem.), Inter-Am. Statis. Inst., Population Assn. Am., Nat. Acad. Sci. (com. nat. statistics 1972-76), Internat. Assn. Survey Statisticians (pres. 1973-77), Sigma Xi, Alpha Tau Omega, Phi Kappa Phi. Subspecialties: Statistics; Operations research (mathematics). Current work: Design of system for obtaining and utilizing information. Home: 5212 Goddard Rd Bethesda MD 20014

HANSEN, RICHARD OLAF, staff scientist; b. Ottawa, Ont., Can., Oct. 4, 1946; came to U.S., 1968; s. Hyllard Olaf and Muriel (Helson) H.; m. Kathleen Jean Thoms, June 15, 1968. B.S., Carleton U., 1968; M.S., U. Chgo., 1969, Ph.D., 1973. Research assoc. U. Pitts., 1973-75; research asst. U. Oxford, Eng., 1975-76; lectr. U. Calif.-Berkeley, 1976-78; numerical analyst EG&G Geometrics, Sunnyvale, Calif., 1979-81, staff scientist, 1981—. Mem. Am. Phys. Soc., Am. Math. Soc., Soc. Indsl. and Applied Math., Soc. Exploration Geophysicists (assoc.), Assn. for Computing Machinery (assoc.). Subspecialties: Geophysics; Numerical analysis. Current work: Analysis of geophysical potential field data. Office: EG&G Geometrics 395 Java Dr Sunnyvale CA 94086

HANSEN, ROBERT C(LINTON), electrical engineering consultant; b. St. Louis, Aug. 3, 1926; m., 1952; 2 children. B.S., U. Mo., 149; M.S., U. Ill., 1950, Ph.D., 1955. Research assoc. antenna lab. U. Ill., 1950-55; sr. staff engr. microwave lab. Hughes Aircraft Co., 1955-59; telecommunications lab. Space Technol. Labs., 1959-0; dir. test mission analysis office Aerospace Corp., Calif., 1960-67; head electronics div. KMS Technol. Ctr., 1967-71; pres., cons., R.C. Hansen, Inc., Tarzana, Calif., 1971—; Mem. commn. VI Internat. Sci. Radio Union, chmn. profl. group antennas and propagation, 1964, chmn. comm. VI, 1968-70; mem. U.S. nat. com., 1970-72. Editor: Microwave Scanning Antennas, 1964-65. Fellow IEEE, mem., Am. Phys. Soc. Subspecialty: Electrical engineering. Office: PO Box 215 Tarzana CA 91356

HANSEN, STANLEY SEVERIN, II, astronomer, computer scientist; b. St. Joseph, Mo., Sept. 16, 1945; s. Stanley Severin and Gertrude (Campbell) H. B.S. in Physics, U. Mo., 1967; M.S. in Astronomy, U. Mass., 1972, Ph.D., 1980. Researcher, Oak Ridge Associated Univs., 1966; sci. staff IBM, Poughkeepsie, N.Y., 1967-70; faculty Mount Holyoke Coll., South Hadley, Mass., 1971-73, U. Mass., Amherst, 1974; astronomer Onsala Space Obs., Rao, Sweden, 1974, Nat. Radio Astronomy Obs., Charlottesville, Va., 1974-81; mem. tech. staff Aerospace Corp., El Segundo, Calif., 1982—. Contbr. articles to profl. jours. Mem. Am. Astron. Soc., Tau Beta Pi. Subspecialties: Radio and microwave astronomy; Operating systems. Current work: Astrophysics, star formation, very long baseline interferometry, molecular masers, scientific computing, system programming, computer simulation, signal precossing, image formation; aerospace technology, embedded computerr systems. Home: 420 S Catalina Ave Apt 208 Redondo Beach CA 90277 Office: The Aerospace Corp 2350 E El Segundo Blvd El Segundo CA 90245

HANSEN, THOR ARTHUR, geology educator; b. Washington, Feb. 22, 1951; s. Odd Edwin and Mary Sue (Williams) H.; m. Penny Shawn Chambers, Mar. 13, 1982; 1 son, Shawn. B.S., George Washington U., 1974; Ph.D., Yale U., 1978. Asst. prof. geology U. Tex.-Austin, 1978—. Mem. Paleontological Soc., Paleontological Research Inst., Internat. Paleontological Assn. Subspecialties: Paleobiology; Paleoecology. Current work: Evolutionary rates and patterns of Tertiary Molluscs, larval dispersal of fossil molluscs, Cretaceous/Tertiary extinction events. Home: 4506 Bridlewood Dr Austin TX 78759 Office: Dept Geol Sci U Tex Austin TX 78712

HANSFORD, RICHARD GEOFFREY, research biochemist; b. Market Bosworth, Leicestershire, Eng., Apr. 29, 1944; came to U.S., 1973; s. Charles John and Elizabeth (Brookes) H.; m. Linda Christine Strange, Dec. 8, 1967; children: Thomas, Katherine, Benjamin. B.Sc. with honors, U. Bristol, U.K., 1965, Ph.D., 1968. Postdoctoral fellow Johns Hopkins U., Balt., 1969-70; lectr. U. Coll., Cardiff, Wales, U.K., 1970-73; vis. assoc./scientist Nat. Inst. Aging, Balt., 1973-79, sr. investigator, 1979—. Editorial bd.: Jour. Bioenergetics Biomembranes, 1983—; contrb. articles in field to profl. jours. Mem. Am. Soc. Biol. Chemists. Democrat. Subspecialties: Biochemistry (biology); Gerontology. Current work: Research on control of mitochondrial metabolism, especially by Ca^2 , impact of aging on bioenergetics. Home: 106 Gorsuch Rd Timonium MD 21093 Office: Nat Inst on Aging Gerontology Research Ctr Balt City Hosps Baltimore MD 21224

HANSL, NIKOLAUS RUDOLF, pharmacologist, executive; b. Wiener Neustadt, Austria, Oct. 24, 1923; came to U.S., 1948; s. Rudolf and Luise (Fulop) H.; m. Adele Y. Bertagnolli, Jan. 23, 1971; 1 dau., Liesl Marie. Ph.D., Vienna U., 1946. Asst. dir. research Sahyun Labs., Santa Barbara, Calif., 1957-64; pres. Pacific Research Labs., Santa Barbara, and Omaha, 1964—; prof. Creighton U., Omaha, 1967—; cons. Hoffmann-LaRoche, Nutley, N.J., 1971-75. Mem. Am. Chem. Soc., AAAS, Santa Barbara Research Council, N.Y. Acad. Scis., Nebr. Acad. Scis. (co-chmn. biology and medicine sect.). Republican. Roman Catholic. Subspecialties: Neuropharmacology; Cognition. Current work: Inception and development of novel compounds facilitating cognition in man, the normal adult and the geriatric with impaired cognitive function, biochemistry of cognition. Patentee in field. Home: 7815 Pine St Omaha NE 68124 Office: Creighton U 2500 California St Omaha NE 68178

HANSLER, RICHARD LOWELL, research physicist; b. Mpls., Aug. 21, 1924; s. George C. and Dorothy (Schultheiss) H.; m. Wanda Hansler, Aug. 20, 1949; children: James, Mark, Stephen, Susan. Student, Capital U., 1945-47; B.S., U. Chgo., 1948; Ph. D., Ohio State U., 1952. With Lighting Research Lab., Gen. Electric Co., Cleve., 1952—, sr. research adviser physics and optics, 1981—. Pres. Lutheran Housing Corp., 1974-81. Served with USAAF, 1942-45. Mem. Optical Soc. Am. Democrat. Lutheran. Subspecialties: Laser processing; Plasma physics. Current work: Devel. of new light sources and methods of mfr. Home: 28120 Belcourt Rd Cleveland OH 44124 Office: Nela Park Department 8431 Cleveland OH 44112

HANSON, DOUGLAS M., biochemist, consultant; b. Clinton, Mass., July 1, 1942; s. Joseph Perry and Noel Neal (Picket) H.; m. Lorraine Nancy Haigh, July 20, 1968; children: Michael, Jeff, Brian, Laura. B.A., Nasson Coll., Springvale, Maine, 1964; Ph.D., Mich. State U., 1968. Nat. Inst. Aging postdoctoral research fellow Boston U. Sch. Medicine, 1968-70; research scientist VA Hosp., Bedford, Mass., 1970-77, acting asso. chief of staff, 1975-76; exec. v.p. Bioassay Systems Corp., Wobrun, Mass., 1977—; mem. faculty Boston U. Sch. Medicine; tchr. continuing edn. in biochemistry, immunology, microbiology VA Hosp., Bedford. Contbr. articles to profl. jours; author numerous govt. contract project reports. Mem. Am. Soc. Microbiology, AAAS, Genetic Toxicology Assn., Alpha Chi. Democrat. Subspecialties: Toxicology (medicine); Biochemistry (medicine). Current work: Developement short-term vitro toxicology assay methods; corporate developement and project management in biotechnology, toxicology and in-vitro research and developement. Office: 225 Wildwood Ave Woburn MA 01801

HANSON, FLOYD BLISS, applied mathematician, mathematical biologist; b. Bklyn., Mar. 9, 1939; s. Charles Keld and Violet Ellen (Bliss) H.; m. Ethel Louisa Hutchins, July 27, 1962; 1 dau., Lisa Kirsten. B.S., Antioch Coll., 1962; M.S., Brown U., 1964, Ph.D., 1968. Space technician Convair Astronautics, San Diego, 1961; applied mathematician Arthur D. Little, Inc., Cambridge, Mass., 1961; physicist Wright-Patterson AFB, Dayton, Ohio, 1962; assoc. research scientist Courant Inst., N.Y.C., 1967-68; asst. prof. U. Ill.-Chgo., 1969-75, assoc. prof., 1975-83, prof., 1983—. Contbr. articles in field to profl. jours. NSF research grantee, 1970-83; NSF equipment grantee, 1973. Mem. Soc. Indsl. and Applied Math., Am. Math. Soc., AAAS. Subspecialties: Applied mathematics; Numerical analysis. Current work: Asymptotic analysis, singular perturbations, numerical analysis, mathematical biology, functional differential equations, harvesting of renewable resources. Home: 5435 East View Park Chicago IL 60615 Office: U Ill Dept Math Stats and Computer Sci PO Box 4348 Chicago IL 60680

HANSON, GEORGE PETER, botanist; b. Conde, S.D., July 20, 1933; s. George Henry and Rosa Wilhelmina (Peterson) H.; m. Gloria Ann Gauntt, June 1, 1969; children: Heather, Peter; m. Barbara Jean Graves, Aug. 20, 1958; children: David, Carol. B.S., S.D. State U., 1956; M.S., 1958; Ph.D., Ind. U., 1962. Asst. prof. biology Thiel Coll., Greenville, Pa., 1962-65, Butler U., Indpl., 1965-68; biologist Los Angeles Arboretum, Arcadia, Calif., 1968-82; botanist J.W.D. Agritech Co., Los Angeles, 1982—. Contbr. articles to profl. jours. EPA grantee, 1969-73; NSF grantee, 1976-80; USDA grantee, 1980—. Mem. Genetics Soc. Am., Bot. Soc. Am., Am. Inst. Biol. Scis., AAAS. Republican. Methodist. Subspecialty: Plant genetics. Current work: Research in Guaynle breeding to domesticate this rubber crop to U.S. semiarid climate. Home: 1345 W Haven Rd San Marino CA 91108 Office: 301 N Baldwin Ave Arcadia CA 91006

HANSON, GILBERT NIKOLAI, geochemistry educator; b. Mpls., Apr. 30, 1936; s. John Jacob Hanson and Evelyn Bernice (Huikko) Mattson; m. Janet Rae Hillman, Sept. 15, 1963; children: Lynn, Kevin, Darlene. B.A., U. Minn.-Mpls., 1958, M.S., 1962, Ph.D., 1964. Research assoc. Minn. geol. Survey, Mpls., 1964-65; Swiss Fed. Inst. Tech., Zurich, 1965-66; asst. prof. geochemistry SUNY-Stony Brook, 1966-70, assoc. prof., 1970-75, prof., 1975—. Contbr. articles on geochemistry and geochronology to profl. jours. Fellow Geol. Soc. India; mem. Geol. Soc. Am., Am. Geophys. Union, AAAS. Democrat. Unitarian. Subspecialties: Geochemistry; Petrology. Current work: Use of isotopes, trace elements, and major element in the petrogenesis of igneous and carbonate rocks. Home: 7 Pilgrim Dr Port Jefferson NY 11777 Office: Dept Earth and Space Scis SUNY Stony Brook NY 11794

HANSON, JOE ALLAN, researcher; b. Los Angeles, July 8, 1928; s. Albert John and Reva Miriam (Silver) H.; m. Sarah Lee Daley, Mar. 15, 1956 (div. Nov. 1968); children: Leslie Marie, Kurt Allan; m. Carol Rosalea Cooper, Feb. 14, 1969; children: Christopher James, Jeffrey Charles, Mark Steven. B.S., UCLA, 1956. Mem. tech. staff RAND/SDC, Santa Monica, Calif., 1957-58, TRW, Inc., Redondo Beach, Calif., 1958-63; sr. assoc. Planning Research Corp., Honolulu, 1963-70, dir., 1968-70; sr. systems scientist The Oceanic Inst., Waimanalo, Hawaii, 1970-77; mem. tech. staff Jet Propulsion Lab., Pasadena, Calif., 1977—; adj. prof. Hawaii Loa Coll., Kailua, 1976-77; affiliate faculty U. Hawaii, Honolulu, 1971-72; cons. Walter E. Dizney Enterprises, Burbank, Calif., 1979-82; instr. U. So. Calif., Los Angeles, 1975. Editor, author: Open Sea Mariculture, 1974; Editor: Shrimp and Prawn Farming in the Western Hemisphere, 1977. Dir. Hawaii Child Centers, 1972-77; exec. sec. Nat. Council on Seaward Advancement, Washington, 1974-76. Mem. AAAS, World Mariculture Soc., N.Y. Acad. Sci. Clubs: Athenaeum Caltech (Pasadena, Calif.); Kaneohe Yacht (Kaneohe, Hawaii) (port capt. 1975-76). Subspecialties: Ecology; Ecosystems analysis. Current work: One of three originators of materially closed, laboratory scale ecosystems capable of long term persistance, now initiating research; manage research and development related to advanced life support systems for military and space applications. Home: 1100 Madre Vista Rd Altadena CA 91001 Office: Jet Propulsion Lab 4800 Oak Grove Dr Pasadena CA 91109

HANSON, JOHN EDWARD, nuclear power engineer, consultant; b. Thomaston, Maine, Nov. 5, 1929; s. John Olaf and Anna Lydia (Johnson) H.; m. Darlene June White, July 26, 1950; children: Laura Jean, Rick Edward. B.S. in Mech. Engring, U. Idaho, 1956, M.S., 1960. Registered profl. mech. engr., Wash., Calif. Profl. nuclear engr. Calif. Engr. Gen. Electric Co., Richland, Wash., 1956-61, profl. nuclear engr., San Jose, Calif., 1962-66; mgr. Westinghouse Hanford Co., Richland, 1966-79; mgr./prin. engr. EG&G Idaho, Inc., Idaho Falls, 1979-82; program mgr. Los Alamos Nat. Lab., 1983—. Served with USAF, 1947-52. Mem. Am. Nuclear Soc. Republican. Lodges: Elks; Masons. Subspecialties: Nuclear engineering; High-temperature materials. Current work: Program manager space and military reactor program. Home: 15 Loma Vista Dr Los Alamos NM 87544 Office: Los Alamos Nat Lab Box 1633 Los Alamos NM 87545

HANSON, JOHN ELBERT, educator; b. Toledo, Ohio, Mar. 5, 1935; s. John E. and Ruth (Sylvia) (Fike) H.; m. Esther Ruth Johnson, June 13, 1959; children: Heidi, Heather. B.A., Olivet Nazarene Coll., 1957; postgrad., Washington U., 1957-58; Ph.D., Purdue U., 1964. Mem. faculty Olivet Nazarene Coll., Kankakee, Ill., 1961—, prof. chemistry, 1970—; research assoc. U. Chgo., 1974. Precinct committeeman, 1972-74; mem. Village Task Force for Impact of Hazardous Waste Incinerator, 1981. Named Tchr. of Year Olivet Nazarene Coll., 1966, Outstanding Alumnus, 1970. Mem. Am. Chem. Soc., Ill. Acad. Sci. Mem. Ch. of the Nazarene. Subspecialty: Inorganic chemistry. Current work: Organosilicon chemistry, cobalt complexes, synthetic inorganic and organometaloid chemistry, environmental science. Office: Chemistry Dep Olivet Nazarene Coll Kankakee IL 60901

HANSON, KENNETH WARREN, horticulturist; b. Graceville, Minn., July 15, 1922; s. Edwin William and Esther Dorothy (Lundquist) H.; m. Margaret Lucile Stokes, Apr. 28, 1946; children: Kenneth Warren, Linda L., Robert G., Edwin William II. B.S., U. Minn., 1948, M.S., 1951, Ph.D., 1952. With USDA, various locations, 1942-43, 46-51; assoc. prof. U. Ga., 1952-54; asst. prof. Cornell U., 1954-60; asso. prof. Am. U. Beirut, 1960-63; dir. sta., horticulturist Mo. Fruit Expt. Sta., S.W. Mo. State U., Mountain Grove, 1963—. Served to lt. USNR, 1943-46. Fellow AAAS; mem. Am. Soc. Hort. Sci., Internat. Soc. Hort. Sci., Am. Pomol. Soc., Mo. Hort. Soc., Ark. Hort. Soc., Am. Inst. Biol. Scis., Minn. Hort. Soc., Mountain Grove C. of C. (pres. 1967), Sigma Xi, Gamma Alpha. Republican. Methodist. Clubs: Rotary (pres. 1967-68, dir. 1965-67. Subspecialties: Plant genetics; Plant physiology (agriculture). Home: Rt 3 Box 63 Mountain Grove MO 65711 Office: Mo Fruit Expt Sta Rt 3 Box 63 Mountain Grove MO 65711

HANSON, MARVIN HAROLD, mathematics and physical science educator; b. Centerville, S.D., Oct. 21, 1927; s. Gunnar and Beata (Risanger) H.; m. Betty L. Carrico, June 23, 1951; children: Merry L., Mikal G. B.A., Augustana Coll., 1949; M.A., U. S.D., 1954. Cert. ednl. administr., S.D. Instr. pub. schs., Hills, Minn., 1949-53; assoc. prof. dept. math. Huron (S.D.) Coll., 1953—. Dir. United Fund, Huron, 1973-78. Mem. S.D. Acad. Sci. (pres. 1964-65), Nat. Sci. Tchrs. Assn., S.D. Council Tchrs. of Math., Augustana Coll. Alumni. Republican. Lutheran. Lodges: Lions.; Sons of Norway. Subspecialty: Applied mathematics. Current work: Teacher preparation. Home: 670 Oregon Ave Huron SD 57350 Office: Huron Coll Dept Math 9th Ohio Huron SD 57350

HANSON, ROBERT D(UANE), civil engineering educator; b. Albert Lea, Minn., July 27, 1935; m., 1959; 2 children. B.S.E., U. Minn., 1957, M.S.C.E., 1958; Ph.D., Calif. Inst. Tech., 1965. Asst. prof. civil engring. U. N.D., 1959-61; asst. prof. U. Calif.-Davis, 1965-66; from asst. prof. to prof. civil engring. U. Mich., Ann Arbor, 1966—, chmn. dept., 1976—; UNESCO expert Internat. Inst. Seismology and Earthquake Engring., Tokyo, 1970-71. Mem. Earthquake Engring. Research Inst., ASCE (Reese award 1980), Am. Concrete Inst. Subspecialty: Civil engineering. Office: Dept Civil Engring U Mich Ann Arbor MI 48109

HANSON, ROY EUGENE, science administrator; b. Grand Forks, N.D., Dec. 26, 1922; s. Henry Elmer and Dagna Helen (Thone) H.; m. Marcella Ann Krajcovic, Mar. 17, 1946; children: Gail, Wayne, Gregg. B.S., U. N.D., 1947; Ph.D., St. Louis U., 1952. Program dir. geophysics NSF, 1957-80; sr. staff officer Nat. Acad. Scis., Washington, 1980—; scientist Office Naval Research, London, 1969-71. Served with USAF, 1943-46, 50-57; maj. Res.; ret. Mem. Am. Geophys. Union, Seismol. Soc. Am., Soc. Exploration Geophysicists. Subspecialties: Geoscience Program Administration; Geophysics. Home: 208 Salisbury St Seabreeze Rehoboth Beach DE 19971 Office: Nat Acad Sciences 2101 Constitution Ave NW Washington DC 20418

HANSON, TREVOR RUSSELL, computer scientist; b. Cambridge, Eng., Sept. 24, 1955; s. Norwood Russell and Frances Fay (Kenney) H. Student, U. Chgo., 1973-75. Systems programmer Sears, Roebuck & Co., Chgo., 1975-77; systems programmer Nat. CSS, Inc., Wilton, Conn., 1977-79; founder, prin. Hanson-Smith, Ltd., Shelton, Conn., 1979—. Mem. Assn. Computing Machinery, IEEE. Subspecialties: Database systems; Operating systems. Current work: Non-programmer application development systems, dataflow, actor systems, compiler design, fifth generation systems, extensible systems, distributed databases, delivering technology as products. Office: 58 Martinka Dr Shelton CT 06484

HANSON, WILLIAM BERT, physics educator, researcher in space physics; b. Warroad, Minn., Dec. 30, 1923; s. Bert and Viola Mae (Carlquist) H.; m. Wenonah Ann Dalquist, 1946 (dec.); children: Bryan, Craig, David, Karyn. B.S., U. Minn., 1944, M.S., 1949; Ph.D. in Physics, George Washington U., 1954. Research physicist Nat. Bur. Standards, Washington, 1949-54, Boulder, Colo., 1954-56; research scientist Lockheed Missiles and Space Co., Palo Alto, Calif., 1956-57, head atmospheric physics, 1957-59, head Ionospheric physics, 1959-62; prof. Southwest Ctr. for Advanced Studies, Dallas, 1962-69; prof., head div. atmospheric and space scis. Inst. for Phys. Scis., U. Tex., Dallas, 1969-75, prof. physics, 1975—, dir., 1975—; mem. adv. bd. and vis. com. Nat. Astronomy and Ionosphere Ctr., Arecibo, P.R.; mem. atmospheric scis. research-evaluation panel Air Force Office of Sci. Research; mem. mgmt. and ops. working group Planetary Atmospheres NASA. Served to lt. (j.g.) USN, 1944-46. Recipient Alexander Von Humboldt Sr. Scientist award, W. Ger., 1981. Fellow Am. Geophys. Union. Club: Canyon Creek Country (Richardson, Tex.). Subspecialties: Aeronomy; Planetary atmospheres. Office: U Tex PO Box 688 Mail Stop FO22 Richardson TX 75080

HAPKE, BRUCE WILLIAM, astronomer, educator; b. Racine, Wis., Feb. 17, 1931; s. William E. and Blanche V. H.; m. Joyce Zellinger, June 18, 1954; children: Kevin, Jeffrey, Cheryl. B.S., U. Wis., 1953; Ph.D., Cornell U., 1962. Research assoc., then sr. research assoc. Ctr. for Radiophysics and Space Research, Cornell U., Ithaca, N.Y., 1960-67; assoc. prof. dept. geology and planetary sci. U. Pitts., 1967-79, prof. planetary scis., 1979—; research assoc. Carnegie Mus. Natural History, 1980—. Contbr. articles to profl. jours. Served to lt. USNR, 1953-55. NASA grantee. Mem. AAAS, Am. Astron. Soc., Am. Geophys. Union, Planetary Soc. Subspecialty: Planetary science. Current work: Scattering of electromagnetic radiation from surfaces; nature and genesis of planetary surfaces and interiors. Office: U Pitts 321 Old Engring Hall Pittsburgh PA 15260

HAPP, STAFFORD COLEMAN, research geologist; b. Sparrowbush, N.Y., Sept. 16, 1905; s. Conrad and Hattie Adella (Coleman) H.; m. Inez Ellen Hale, Dec. 26, 1935; 1 dau., Ellen Coleman Happ Hill. Student, Wesleyan U., 1926-27; A.B. cum laude, Marietta Coll., 1931; Ph.D. in Geology, Columbia U., 1939. Cert. geologist and engring. geologist, Calif. Head stream and valley sediment research U.S. Soil Conservation Service, Washington, 1935-43; dist. geologist U.S. Army Engrs., Ocala, Fla. and Kansas City, Mo., 1943-55; chief prodn. services br. U.S. AEC, Grand Junction, Colo., 1955-64; research geologist U.S. Geol. Survey, Denver, 1964-65, U.S. Dept. Agr. Sediment Lab., Oxford, Miss., 1965—. Author in field. Recipient Disting. Service award Trempealeau County (Wis.) Land Conservation Dept., 1982; Outstanding Service award Wis. Assn. Conservation Dists., 1982. Fellow Geol. Soc. Am. (chmn. engring. div. 1959-60); mem. ASCE, AIMME, AAAS. Republican. Methodist. Subspecialties: Sedimentology; Environmental engineering. Current work: Fluvial sedimentation related to agricultural soil erosion, including recovery of sampling cross sections established 1936-41, in some 50 valleys representing severe soil erosion effects, but abandoned during World War II and records largely lost. Measurements and interpretations evaluate changing erosion rates, and sediment effects. Home: 503 N 14th St Oxford MS 38655 Office: IS Dept Agr Sediment Lab PO Box 1157 Oxford MS 38655

HAQ, BILAL U., research scientist; b. Gorakhpur, India, Oct. 8, 1943; came to U.S., 1969; s. Fazl-I and Sorraya (Rabbani) H. Sc.B., U. Panjab, Pakistan, 1961, M.Sc., 1963; Ph.D., U. Stockholm, 1967, D. Sc., 1972. Research fellow Geol. Survey Austria, Vienna, 1964-65; research assoc. U. Stockholm, 1965-69; vis. investigator Woods Hole (Mass.) Oceanographic Inst., 1969-71, asst. scientist, 1971-75, assoc. scientist, 1975-81; sr. research specialist Exxon Prodn. Research Co., Houston, 1982—. Author: Introduction to Marine Micropaleontology, 1978, Bio-Stratigraphy, 1983; contbr. numerous articles to profl. jours.; chief editor: Marine Micropaleontology. Fellow Geol. Soc. Am.; mem. Am. Geophys. Union, AAAS, Swedish Geol. Soc., Am. Assn. Petroleum Geologists. Subspecialties: Geology; Oceanography. Current work: Reconstruction of past climates and oceanographic conditions and basin analysis. Office: Exxon Prodn Research Co PO Box 2189 Houston TX 77001

HAQUE, PROMOD, electrical engineer; b. Simla, India, Apr. 20, 1948; came to U.S., 1972, naturalized, 1984; s. Alexander and Phulwanti (Gangaram) H.; m. Dorcas A. Daniels, July 15, 1978. B.S.E.E., U. Delhi, 1969; M.S.E.E., Northwestern U., 1974, Ph.D., 1976, M.B.A., 1983. Sales engr. Siemens, New Delhi, 1969-72; research asst. Bio-Med. Center, Northwestern U., Evanston, Ill., 1972-76; lab. dir. EMI Med. Inc., Northbrook, Ill., 1976-81; v.p., chief operating officer Emergent Corp., Anaheim, Calif., 1981-82; v.p. R&D engring and mktg. Omnimed. Corp., Anaheim, 1982-83; pres., chief exec. officer Dimensional Medicine Inc., Minnetonka, Minn., 1983—. Contbr. articles to profl. jours. Mem. IEEE, Am. Assn. Physicists in Medicine. Republican. Evangelical Christian. Subspecialties: Electronics; Biomedical engineering. Current work: CAT scanners, digital fluoroscopy, radiography, ultrasound, nuclear magnetic resonance, computer based graphics, array processors and signal processing, analog and digital instrumentation. Home: 10311 Cedar Lake Rd #309 Minnetonka MN 55343 Office: 10999 E Bren Rd Minnetonka MN 55343

HARAF, FRANK JOSEPH, oncologist; b. Chgo., May 1, 1945; s. Walter Stanley and Sophie (Malinski) H.; m. Janice R. Sala, June 13, 1970; children: Frank Jr., Paul A. B.S., No. Ill. U., 1967; M.D., U. Ill., 1970. Diplomate: Am. Bd. Internal Medicine (medical oncology, hematology). Intern Michael Reese Hosp., Chgo., 1971, resident in internal medicine, 1972; resident U. N.C.-Chapel Hill, 1973, fellow, 1974. Served with USN, 1974-76. Mem. Am. Soc. Clin. Oncology, Alpha Omega Alpha. Subspecialties: Chemotherapy; Hematology. Current work: Private practice. Office: Knoxville Hematology Oncology Assocs 1114 Weisgarber Rd Knoxville TN 37919

HARAKAL, CONCETTA, pharmacologist; b. Chieti, Italy, Nov. 25, 1923; d. Francesco and Maria (Gagliardi) De Leo, m. Michael P. Harakal, Oct. 16, 1948. Ph.D. in Pharmacology, Temple U., 1962. Instr. Temple U. Sch. Medicine, Phila., 1962-64, asst. prof., 1964-68, asso. prof., 1968-76, prof. pharmacology, 1976—. Contbr. articles to profl. jours. Recipient Golden Apple Teaching award Temple U. Sch. Medicine, 1975, 77, 80; Sowell Meml. award for excellence in basic sci. teaching, 1979; Chapel of Four Chaplains Legion of Honor award, 1977; Lindback Found. award, 1980. Mem. Am. Soc. Pharmacology and Exptl. Therapeutics, AAAS, N.Y. Acad. Scis., Sigma Xi. Roman Catholic. Club: 25-Year Faculty Temple U. Subspecialty: Pharmacology. Current work: Cardiovascular pharmacology.

HARBAUGH, JOHN WARVELLE, geological educator, researcher; b. Madison, Wis., Aug. 6, 1926; s. Marion Daight and Marjorie (Warvelle) H.; m. Josephine Taylor, Nov. 24, 1951; children: Robert, Dwight, Richard. B.S., U. Kans.-Lawrence, 1948, M.S., 1949; Ph.D., U. Wis.-Madison, 1955. Asst. prof. to prof geology Stanford U., 1955—; chmn. bd. Terrascis., Inc., San Francisco, 1977—. Co-author: Computer Simulation in Geology, 1970, Probability Methods in Oil Exploration, 1977. Recipient Haworth Disting. Alumni award U. Kans., 1968. Fellow Geol. Soc. Am.; mem. Am. Assn. Petroleum Geologists (A. I. Levorsen award 1970). Republican. Club: Cosmos (Washington). Subspecialties: Mathematical geology; Mathematical software. Current work: Development of statistical petroleum resource forecasting procedures; computer simulation of geological processes. Home: 683 Salvatierra Stanford CA 94305 Office: Stanford U Geology Bldg Stanford CA 94305

HARBAY, EDWARD WILLIAM, nuclear engineer; b. Johnstown, Pa., May 16, 1937; s. Edward F. and Helen V. H. B.S., U. Pitts., 1961, postgrad., 1963-64. Rocket devel. engr. Hercules Allegany Ballistics Lab., Cumberland, Md., 1961-66; reactor core engr. Westinghouse Electric Corp. Bettis Atomic Power Lab., West Mifflin, Pa., 1966-74; nuclear engr. Detroit Edison Co., 1974—. Mem. Nat. Soc. Profl. Engrs., Am. Nuclear Soc. Democrat. Roman Catholic. Subspecialties: Nuclear engineering; Fuels. Current work: Boiling Water Reactor nuclear fuel cycle engineering. Office: Detroit Edison Co 2000 2d Ave Detroit MI 48226 Home: 22271 Derby Rd Woodhaven MI 48183

HARBISON, RAYMOND DALE, toxicology educator, adminstr.; b. Peru, Ill., Jan. 1, 1943; s. Raymond C. and Marie (Haupt) H.; m. Holly Hosutt, Sept. 9, 1962; children: John, Matthew, Andrew, Stephen. B.S., Drake U., 1965; M.S., U. Iowa, 1967, Ph.D., 1969. Cert. toxicologist Acad. Toxicol. Scis. Asst. prof. Tulane Med. Sch., New Orleans, 1969-72; assoc. prof. Vanderbilt U., Nashville, 1972-81; prof. and dir. toxicology U. Ark. for Med. Scis., Little Rock, 1982—; cons. NIH, 1970—, EPA, Washington, 1975—, Nat. Inst. Occupational Safety and Health, 1980—. Contbr. numerous articles on toxicology to prof. jours. Mem. Soc. Toxicology (achievement award 1978), Am. Soc. Pharmacology and Exptl. Therapeutics, Teratology So., AAAS. Subspecialties: Toxicology (medicine); Teratology. Current work: Study of effects of environmental and occupational pollutants on reproduction and organ function. Office: U Ark for Med Scis 4301 W Markham Little Rock AR 72207

HARD, JAMES ELLSWORTH, government engineering agency official; b. Duluth, Minn., Dec. 22, 1929; s. Victor Ellsworth and Beryl (Winter) H.; m. Patricia Deon Anderson, Nov. 4, 1953 (div. 1972); children: Steven James, Amy Diane. B.S.Ch.E., U. Wis.-Madison, 1952. Registered profl. engr., Ill., Md. Reactor engr. Gen. Electric Co., Richland, Wash., 1952-64; reactor insp. U.S. AEC, Chgo., 1964-67; sr. staff asst. U.S. AEC Adv. Com. on Reactor Safeguards, Washington, 1967-73; tech. adv. U.S. AEC and U.S. NRC, Washington, 1973-79; sr. officer Internat. Atomic Energy Agy., Vienna, 1979-80; tech. adv. Atomic Safety and Licensing Bd. Panel, U.S. NRC, Washington, 1980-83, sr. resident insp., Red Wing, Minn., 1983—. Recipient Spl. Achievement award U.S. NRC, 1977. Mem. AAAS, Am. Nuclear Soc., Nat. Soc. Profl. Engrs., Md. Soc. Profl. Engrs. Republican. Lutheran. Clubs: Nat. Capital Soaring Assn. (Washington) (chief flight instr. 1970-72, v.p. 1983. Subspecialties: Nuclear fission; Nuclear engineering. Current work: Regulation of nuclear power plants and other nuclear facilities. Home: 5940 Pioneer Rd S Saint Paul Park MN 55071 Office: US Nuclear Regulatory Commn Washingto DC 20555

HARDIN, HILLIARD FRANCES, microbiologist, educator; b. Columbia, S.C., Dec. 12, 1917; d. Lawrence Legare and Addria Eugenia (Chrestzberg) H. B.A., Duke U., 1939, M.A., 1949, Ph.D., 1953. Research assoc. microbiology Duke U. Med Ctr., Durham, 1958-63; chief mycology tng. unit CDC, Atlanta, 1963-68; supr. chief sect. microbiology VA Hosp., Little Rock, 1968—; assoc. prof. U. Ark. Med. Ctr., Little Rock, 1968—; guest lectr. E. Tenn. State Med. Sch., 1980—; mycology cons.; assoc. prof. U. Ark. Med. Sch., 1968—. Served with WAVES, 1942-45. Mem. N.Y. Acad. Sci., Med. Mycology Soc. Am., Am. Bus. Women's Assn., Sigma Xi. Republican. Methodist. Subspecialties: Infectious diseases; Microbiology (medicine). Current work: Immuno-serology, human mycoses. Office: Microbiology Dept Lab Service 300 E Roosevelt Rd Little Rock AR 72206

HARDING, JOHN DELANO, biology educator, researcher; b. Detroit, Feb. 20, 1946; s. James Harold and Martha Marie (Mulju) H.; m. Sarah Fulton, Aug. 1, 1971; children: David, Philip. B.A., Stanford U., 1968; Ph.D., Columbia U., 1974. Research assoc. U. Calif.-San Francisco, 1974-78; asst. prof. biology Columbia U., N.Y.C., 1978—. Subspecialties: Genetics and genetic engineering (biology); Genome organization. Current work: Organization and regulation of eucaryotic genes. Office: Columbia U 922 Fairchild Ctr New York NY 10027

HARDORP, JOHANNES C., astronomer; b. Bremen, Germany, 1929; came to U.S., 1965; s. Gerhard D. and Clara (Karstensen) H.; m. Ingeborg M. S. Arndt, 1954; children: Detlef, Agnes. Ph.D. in Astronomy, U. Hamburg, W.Ger., 1960. Tchr., researcher Hamburg U., 1958-68; research physicist U. Calif., San Diego, 1965-66; research assoc. U. Mich., Ann Arbor, 1966-67; mem. staff Inst. Theoretical Astronomy, Cambridge (Eng.) U., 1968-69; assoc. prof. astronomy SUNY-Stony Brook, 1969—; professeur associéu College de France, Paris, 1976-. Contbr. numerous articles to sci. jours. Mem. Deutsche Astronomische Gesellschaft, Internat. Astron. Union. Subspecialty: Optical astronomy. Current work: Stellar atmospheres, stellar rotation, peculiar stars, solar twins. Office: Dept Earth and Space Sci SUNY Stony Brook NY 11794

HARDY, JUDSON, JR., physicist; b. New Orleans, Nov. 8, 1931; s. Charles Judson and Helena (Chalaron) H.; m. Dorris Montgomery, June 12, 1954; children: Christopher, Janet, Nancy. B.S., U. N.C.-Chapel Hill, 1953; Ph.D., Princeton U., 1958. Sr. scientist Westinghouse Bettis Lab, West Mifflin, Pa., 1958-68, adv. scientist, 1968—; mem. cross sect. evaluation working group Brookhaven Nat. Lab., 1971—. Contbr. articles to profl. jours. Mem. Am. Phys. Soc., Am. Nuclear Soc. Republican. Episcopalian. Subspecialties: Nuclear fission; Nuclear physics. Current work: Nuclear reactor physics, nuclear data for reactor design, data testing. Home: 3226 Kennebec Rd Pittsburgh PA 15231 Office: Westinghouse Bettis Lab PO Box 79 West Mifflin PA 15122

HARDY, MARK ADAM, surgeon, immunologist; b. Lwow, Poland, Jan. 5, 1938; s. Paul and Rose (Pomeranz) H.; m. Ruth C. Komisarow, Jan. 14, 1967; children: Peter, Arthur, Karen. A.B., Columbia U., 1959; M.D., Albert Einstein Coll. Medicine, 1963. Diplomate: Am. Bd. Surgery. Asst. instr. surgery U. Rochester, 1962-64; asst. prof. Albert Einstein Coll. Medicine, 1971-75; assoc. prof. Columbia Coll. Physicians and Surgeons, 1975-80, prof., 1980—, dir. transplantation. Served to lt. comdr. USNR, 1964-70. NIH scholar. Mem. Soc. Univ. Surgeons, Transplantation Soc., ACS, Am. Assn. Immunology. Subspecialties: Transplant surgery; Transplantation. Current work: Immunosuppression and Immunostimulation radiobiology.

HARDY, MILES WILLIS, psychology educator, cons.; b. Ft. Fairfield, Maine, Dec. 28, 1926; s. Miles Willis H.; m. Patricia Kelly, Aug. 12, 1973. B.S., Ball State U., 1951; M.S., Fla. State U., 1954, Ph.D., 1960. Grad. instr., research asst. Fla. State U., Tallahassee, 1954-55; clin. psychol. intern VA trainee program, Gulfport and Biloxi, Miss., 1955-58; staff psychologist MacDonald Tng. Center, Tampa, Fla., 1958-60; prof. U. South Fla., Tampa, 1960—; founder, dir. Assocs. Psychol. Service, Tampa, 1962—. Mem. Am. Psychol. Assn., Am. Soc. Clin. Hypnosis, Fla. Psychol. Assn. (pres.), Nat. Register Health Providers, Bay Region Soc. Clin. Psychologists. (pres.). Subspecialties: Behavioral psychology; Learning. Current work: Diagnosis, research and treatment of emotional disorders, director Ph.D. clinical training clinic, private outpatient clinic. Office: 10320 N 56th St Suite C Temple Terrace FL 33617

HARDY, RALPH W.F., biochemist; b. Lindsay, Ont., Can., July 27, 1934; s. Wilbur and Elsie H.; m. Jacqueline M. Thayer, Dec. 26, 1954; children: Steven, Chris, Barbara, Ralph, Jon. B.S.A., U. Toronto, 1956; M.S., U. Wis.-Madison, 1958, Ph.D., 1959. Asst. prof. U. Guelph, Ont., Can., 1960-63; research biochemist DuPont deNemours & Co., Wilmington, Del., 1963-67, research supr., 1967-74, assoc. dir., 1974-79, dir. life scis., 1979—; mem. exec. com. bd. agr. Nat. Acad. Sci.; editorial bd. sci. jours. Author: Nitrogen Fixation, 1975, A Treatise on Dinitrogen Fixation, 3 vols, 1977-79; contbr. over 100 articles to sci. jours. Recipient Gov. Gen.'s Silver medal, 1956; WARF fellow, 1956-58; DuPont fellow, 1958-59. Mem. Am. Chem. Soc. (exec. com. biol. chemistry div., Del. award 1969), Am. Soc. Biol. Chemists, Am. Soc. Plant Physiology (exec. com.), Am. Soc. Agronomy, Am. Soc. Microbiology. Episcopalian. Subspecialties: Plant physiology (agriculture); Nitrogen fixation. Current work: Biochemistry of key plant processes -N_2 fixation, photosynthesis. Patentee (2). Home: Box 364 Unionville PA 19375 Office: EI DuPont Co Wilmington DE 19898

HARE, JAMES FREDERIC, biochemistry educator; b. Bryn Mawr, Pa., June 4, 1945; s. James Frederic Jr. and Maybella (Badgely) H.; m. Phoebe Rich, Aug. 4, 1945; 1 child, Quinn Rich. A.B., Lafayette Coll., Easton, Pa., 1967; M.S., U. N.H.-Durham, 1969; Ph.D., Purdue U., 1973. Research assoc. Harvard U. Med Sch., Boston, 1973-75, Calif. Inst. Tech., Pasadena, 1975-77; asst. prof. biochemistry Oreg. Health Scis. U., Portland, 1977-82, assoc. prof., 1982—. NIH fellow, 1972, 82; Damon Runyon fellow, 1975. Mem. Am. Chem. Soc., Am. Soc. Biol. Chemists. Subspecialties: Biochemistry (biology); Cell and tissue culture. Current work: Biosynthesis and turnover of cellular membranes. Home: 536 SW Westwood Dr Portland OR 97201 Office: Oreg Health Scis U Portland OR 97201

HARFORD, AGNES GAYLER, biologist; b. St. Louis, June 27, 1941; d. Carl Gayler and Mary Broadhead (Cowan) H. A.B., Harvard U., 1963; Ph.D., Johns Hopkins U., 1971. Postdoctoral fellow Yale U., 1971-73; asst. prof. biology SUNY-Buffalo, 1973-80, assoc. prof., 1980—; vis. assoc. prof. U. Colo.-Boulder, 1981-82. Contbr. articles on molecular biology to profl. jours. NIH grantee, 1974-83. Mem. Am. Soc. Cell Biology, Genetics Soc. Am. Subspecialties: Genome organization; Molecular biology. Current work: DNA sequence organization, Drosophila, ciliated protozoa. Office: Dept Biol Sci SUNY Buffalo NY 14260

HARFORD, CARL GAYLER, microbiologist, educator; b. St. Louis, June 27, 1906; s. Edwin Marvin and Agnes (Gayler) H.; m. Mary Cowan (dec. Aug. 5, 1933) (dec. ; 1980); children: John, Gayler, Carolyn; m. Viola Graves, May 23, 1982. A.B., Amherst Coll., 1928; M.D., Washington U., St. Louis, 1933; postgrad., Rockefeller Inst. for Med. Research, 1936-38. Diplomate: Am. Bd. Internal Medicine, Nat. Bd. Med. Examiners. Instr. bacteriology and clin. medicine Washington U., 1938-43, asst. prof. medicine and preventive medicine, 1943-49, assoc. prof. medicine, 1949-63, acting head dept. microbiology, 1952,59, prof. medicine, 1964-74, prof. emeritus, 1974—; assoc. mem. Commn. on Acute Respiratory Diseases, Armed Forces, 1950-71. Mem. editorial bd.: Procs. Soc. Exptl. Biology and Medicine, 1974—; author articles. Recipient Founders Day award Washington U., 1973, Alumni/Faculty award, 1982. Mem. Am. Soc. Microbiologists, Central Soc. for Clin. Research (sec.-treas. 1944-45), Am. Soc. for Clin. Investigation, Assn. Am. Physicians, St. Louis Med. Soc., St. Louis Soc. Internal Medicine, Common Cause, Sigma Xi, Alpha Omega Alpha. Subspecialties: Infectious diseases; Virology (medicine). Current work: Aim to make cultured cells susceptible to viruses.

HARGIS, BETTY JEAN, immunologist; b. Madison, Ind., Aug. 14, 1925; d. Carleton Edwin and Margaret Emma (Geile) H. B.S., Purdue U., 1947; A.M., Boston U., 1958, Ph.D., 1967. Research Asst. Peter Bent Brigham Hosp., Boston, 1955-62, Children's Cancer Research Found., 1963-78; research asst. in pathology Children's Hosp. Med. Ctr., Boston, 1965-71, research assoc., 1971—; assoc. in pathology (immunology) Harvard Med. Sch., Boston, 1974—, research assoc., 1978-81; assoc. in pathology Sidney Farber Cancer Inst., Boston, 1981—. Contbr. articles to sci. jours. Mem. AAAS, Am. Assn. Immunologists, Reticuloendothelial Soc., N.Y. Acad. Sci., Boston Cancer Research Assn., Sigma Xi. Club: Appalachian Mountain (Boston). Lodge: Order Eastern Star. Subspecialties: Immunobiology and immunology; Immunology (medicine). Current work: Alteration of IgG and IgE responses of mice to L-asparaginase in its native and chemically modified form. Hybridoma studies. Home: 115 Park St Apt 5 Brookline MA 02146 Office: Sidney Farber Cancer Inst 44 Binney St Boston MA 02115

HARGRAVES, ROBERT BERO, geology educator, researcher; b. Durban, South Africa, Aug. 11, 1928(parents Am. citizens); (parents Am. citizens); ; s. John and Maud Alice (Gager) H.; m. Isabel V. E. Sinclair, Aug. 30, 1955; children: Monica Jane, Allison Maud, Colleen Sinclair. B.Sc., U. Natal, 1948, M.Sc., 1952; Ph.D., Princeton U., 1959. Geologist Uruwira Minerals Ltd., Mpanda, Tanganyika, 1949-50, Union Corp. Ltd., Johannesburg, South Africa, 1950-52, Newmont Mining Corp., N.Y., 1953-54; research fellow U. Witwatersrand, Johannesburg, 1959-61; asst. prof. geology Princeton U., 1961-64, assoc. prof., 1964-71, prof., 1971—; vice chmn. Working Group 4

Internat. Lithosphere Project, 1981—. Author: Physical Geology, 1976; editor: Physics of Magmatic Processes, 1980. Served to cpl. U.S. Army, 1954-56. Recipient Exceptional Sci. Achievement award NASA, 1977; NATO postdoctoral fellow, Munich, W.Ger., 1967-68; Indo-U.S. Subcommn. fellow, Hyderabad, India, 1980-81. Fellow Geol. Soc. Am.; mem. Geol. Soc. South Africa, Am. Geophys. Union. Club: Springdale (Princeton, N.J.). Subspecialties: Geology; Petrology. Current work: Precambrian geology, paleomagnetism and tectonics, rock magnetism and petrology. Home: 747 The Great Rd Princeton NJ 08540 Office: Dept Geol and Geophys Scis Princeton U Princeton NJ 08544

HARGROVE, LOGAN EZRAL, physicist; b. Spiro, Okla., Mar. 3, 1935; s. Logan E. and Ila Mae (Borum) H; m. (div.); children: Paul Hamilton, David Hamilton. B.S. in Physics, Okla. State U., 1956, M.S., 1957; Ph.D. in Physics, Mich. State U., 1961. Research assoc. Mich. State U., 1961-62; mem. tech. staff Bell Labs., Murray Hill and Holmdel, N.J., 1962-76; with physics div. Office of Naval Research, Arlington, Va., 1976—, sci. officer phys. acoustics task area, 1976—. Contbr. articles to profl. jours. Fellow Acoustical Soc. Am. (Biennial award 1970), Am. Phys. Soc.; mem. Optical Soc. Am., IEEE, AAAS, Sigma Xi, Sigma Pi Sigma. Methodist. Subspecialties: Acoustics; Optical physics. Current work: Management and scientific direction of sponsored research in physical acoustics. Patentee in field. Office: Office of Naval Research 800 N Quincy St Arlington VA 22217

HARKINS, STEPHEN WAYNE, gerontology, psychiatry and psychology educator, researcher; b. Pitts., Aug. 20, 1942; s. Ralph Watson and Elizabeth (Barrington) H.; 1 dau., Emily. Ph.D., U. N.C.-Chapel Hill, 1974. Research asst. Ctr. Study of Aging & Human Devel. Duke U., Durham, 1969-74; fellow dept. psychiatry and behavioral scis. U. Wash., Seattle, 1974-76, asst. prof. anesthesiology and psychiatry, 1976-79; assoc. prof. gerontology, psychiatry, psychology Med. Coll. Va., Va. Commonwealth U., Richmond, 1979—. Editor: (with L. Poon) Aging in the 1980's: Psychological Issues, 1980; mem. editorial bd.: Jour. Gerontology, 1981; contbr. chpts. to books, articles in field to profl. jours. Founder Alzheimer's Disease Support Group, Richmond, 1981. NIA fellow, 1974-76. Mem. Am. Psychol. Assn., Internat. Assn. Study Pain, Soc. Psychophysiol. Research, Gerontol. Soc. Am. Subspecialties: Psychobiology; Cognition. Current work: Gerontology, psychophysiology. Home: 3455 Northview Pl Richmond VA 23225 Office: Medical College Virginia Box 228 MCV Station Richmond VA 23298

HARKNESS, SAMUEL DACKE, III, research manager; b. Richmond, Va., Oct. 28, 1940; s. Samuel Dacke and Jane Margaret (Morin) H.; m. Christine Lee, Dec. 28, 1963; children: Samuel, Laura, Matthew. B.S. in Metall. Engring, Cornell U., 1963; Ph.D. in Material Sci, U. Fla., 1967. Research engr. Atomics Internat., Canoga Park, Calif., 1963-64; leader radiation effects group Argonne Nat. Lab., Argonne, Ill., 1967-73; mgr. materials tech. sect. Combustion Engring. Co., Windsor, Conn., 1973-76; assoc. dir. fusion power Argonne Nat. Lab., 1976-79; mgr. fuel devel. Bettis Atomic Power Lab., West Mifflin, Pa., 1979—. Editor: Radiation Damage in Metals, 1976; Contbr. articles in field to profl. jours. Recipient Robert Lansing Hardy gold medal award AIME, 1969. Mem. Am. Soc. for Metals (sustaining), Am. Nuclear Soc. Subspecialties: Materials; Nuclear fission. Current work: Development of fuel, cladding and poison materials for advanced nuclear reactors; effects of radiation on the corrosion of metals; relationship of processing to properties of reactor materials. Patentee in field of radiation damage resistant metal and advanced fusion reactor design. Office: Bettis Atomic Power Lab West Mifflin PA 15122

HARKNESS, WILLIAM LEONARD, statistics educator; b. Lansing, Mich., June 25, 1934; m., 1956 and 1980; 3 children. B.S., Mich. State U., 1955, M.A., 1956, Ph.D., 1959. Research mathematician Inst. Air Weapons Research, 1957; research asst. in stats. Mich. State U., East Lansing, 1957-59, instr., 1959; from asst. prof. to assoc. prof. math. Pa. State U., University Park, 1956-69, prof., 1969—, head dept., 1970—; vis. assoc. prof. stats. Calif. State Coll.-Hayward, 1967; cons. NIH, 1977—; vis. prof. stats. Stanford U., 1979-80. Fellow Am. Statis. Assn., Inst. Math. Stats. (program sec. 1974-80); mem. Biometric Soc., Royal Statis. Soc., Math. Assn. Am. Subspecialty: Statistics. Office: Pa State U 218 Pond Lab University Park PA 16802

HARLAND, BARBARA FERGUSON, research biologist; b. Chgo., Apr. 16, 1925; d. Frank Clevel and Dorothy Sargeant (Brown) Ferguson; m. James Wallace Harland, Sept. 6, 1947; children: Jospeh, Jane, Janet. B.S., Iowa State U., Ames, 1946; M.S., U. Washington-Seattle, 1949; Ph.D., U. Md., 1971. Registered dietitian. Chief dietitian Lakeview Meml. Hosp., Stillwater, Minn., 1946-47; grad. asst. U. Washington, 1947-49; nutrition instr. U. Ind.-Jeffersonville, 1964-65, U. Md., 1967-70; research biologist FDA, Washington, 1971—; bd. dirs. and sec. Am. Digestive Disease Soc., Washington, 1979—; grad. nutrition instr. USDA Grad. Sch., 1971—; vis. nutrition instr. 3 univs., Bogota, Colombia, 1981; adj. assoc. prof. nutrition Howard U., 1982—. Pres. PTA George Rogers Clark Elem. Sch., Louisville, Ky., 1960-64. Mem. Omicron Nu, Phi Sigma, Sigma Xi., Am. Inst. Nutrition (scholarship, auditing, awards coms.), Am. Dietetic Assn., Soc. Exptl. Biology & Medicine, Soc. Nutrition Edn., Am. Chem. Soc., Assn. Ofcl. Analytical Chemists. Republican. Methodist. Club: Congressional Country. Subspecialties: Physiology (biology); Biochemistry (medicine). Current work: Developed method for analysis of phytate by ion-exchange chromatography; coinstigator of collaborative study for total dietary fiber, human study to determine effects of high fiber diet; nutrient assessment; bioavailability and interaction of nutrients. Performer/developer audio and video tapes "Trace Elements in Foods," "Food Safety," "Food Fibre," 1975-80; coordinator FDA's Selected Minerals in Foods program, 1975-82; developer column chromatography analysis of phytate in foods, 1977; coinstigator gravimetric procedure analysis of fiber in foods, 1982-83. Home: 7929 Robison Rd Bethesda MD 20817 Office: Div Nutrition FDA 200 "C" St SW Washington DC 20204

HARLEMAN, DONALD ROBERT FERGUSSON, civil engineering educator; b. Palmerton, Pa., Dec. 5, 1922; s. Robert Roy and Nora (Curry) H.; m. Martha Havens, Oct. 21, 1950; children: Kathleen T., Robert I.H., Anne C. B.S. in Civil Engring., Pa. State U., 1943; M.S., MIT, 1947, D.Sc., 1950. Design engr. Curtiss-Wright Corp., Columbus, O., 1944-45; research asst., research asso. Hydrodynamics Lab., Mass. Inst. Tech., 1945-50, asst. prof. hydraulics, 1950-56, asso. prof. hydraulics, 1956-62, prof. civil engring., 1963-75, Ford prof. engring., 1975—, head water resources and hydrodynamics div., 1971—, dir., 1973—; vis. prof. Cal. Inst. Tech., 1962-63; del. U.S.-Japan Joint Sci. Seminar on Coastal Engring., 1964, 1974; sr. visitor applied math. and theoretical physics Cambridge (Eng.) U., 1968-69; vis. scientist Internat. Inst. Applied Systems Analysis, Vienna, 1977-78; Mem. Water Pollution Control Fedn.; mem. U.S. Nat. Com. Internat. Assn. Water Pollution Research. Bd. editors: Jour. Hydraulic Research. Recipient Desmond Fitzgerald medal Boston Soc. Civil Engrs., 1967; named Outstanding Alumnus Coll. Engring., Pa. State U., 1979; Guggenheim fellow, 1968-69. Mem. ASCE (research prize 1960, Karl Hilgard Hydraulic prize 1971, 73, J.C. Stevens award 1973), Am. Geophys. Union, Internat. Assn. for Hydraulic Research, Am. Soc. Limnology and Oceanography. Subspecialties: Environmental engineering; Surface water hydrology. Current work: Water quality of lakes, reservoirs, estuaries. Research in fluid transport process and water quality control, waste heat disposal and power plant siting. Home: 100 Memorial Dr Cambridge MA 02142 Office: 48-335 Parsons Lab for Water Resources and Hydrodynamics Mass Inst Tech Cambridge MA 02139

HARMAN, DENHAM, physician, educator; b. San Francisco, Feb. 14, 1916; s. Leslie and Ruth F. (Wright) H.; m. Helen C. Harman, Oct. 3, 1943; children: Douglas, David, Mark, Robin. B.S. in Chemistry with honors, U. Calif.-Berkeley, 1940, Ph.D., 1940; M.D., Stanford U., 1954. Intern San Francisco City and County Hosp., 1953-54; resident Stanford U., 1956-57, VA Hosp., San Francisco, 1957-58; research chemist Shell Devel. Co., Emeryville, Calif., 1943-49; research assoc. Donner Lab. Med. Physics U. Calif., 1954-56; with Coll. Medicine U. Nebr., Omaha, 1958—, Millard prof. medicine, 1972—, prof. biochemistry, 1968—. Contbr. in field. Mem. Omaha Mayor's Commn. on Aging, 1970-74. Fellow ACP, Radiation Research Soc., Gerontol. Soc., Am. Geriatric Soc., Am. Coll. Nutrition; mem. Am. Aging Assn., Am. Chem. Soc., Am. Fedn. Clin. Research, Am. Heart Assn., AMA, Sigma Xi. Subspecialties: Gerontology; Free radical chemistry. Current work: Role of free radical reactions in aging and disease. Office: U Nebr Med Ctr 42d St and Dewey Ave Omaha NE 68105

HARMAN, GARY ELVAN, seed microbiologist, educator; b. La Junta, Colo., Nov. 13, 1944; s. Ivan D. and Ruth A. (Bloyd) H.; m. B. Jean Wilkinson, Sept. 22, 1965; children: Douglas, Jeffrey, Trieu. B.S., Colo. State U., 1966; Ph.D., Oreg. State U., 1970. Postdoctoral assoc. N.C. State U., 1969-70; asst. prof. seed microbiology N.Y. State Agr. Expt. Sta., Cornell U., Geneva, 1970-76, assoc. prof., 1976-83, prof., 1983—. Contbr. articles to sci. jours. Grantee Snap Bean Research Assn., 1978-79, Rockefeller Found., 1978-81, Am. Seed Research Found., 1979-82, N.Y. Seed Assn., 1980-81, U.S.-Israel Binat. Agr. Research Devel. Orgn., 1981-84, U.S. Dept. Agr., 1982-84. Mem. Am. Phytopathol. Soc., AAAS, Sigma Xi. Presbyterian. Subspecialties: Plant pathology; Microbiology. Current work: Biological control of seed and root-rotting fungi; detection and control of seed-borne plant pathogens, including fungi, bacteria, and viruses; mechanisms of aging in seeds. Office: Dept Horticultural Sciences NY State Agr Expt Station Cornell U Geneva NY 14456

HARMON, DAVID E., JR., oil company executive, consultant; b. Kittanning, Pa., Aug. 12, 1932; s. David E. and Martha E. (Cochran) H.; m. Paula Ann Younker, Oct. 8, 1960; children: David Michael, Sabrina Marie, John Matthew Wills, Patrick Kevin, Paul Charles, Robert Jude. A.B. in Geology, Marietta Coll., 1954. Cert. profl. engr., Ky., Ind.; profl. land surveyor, Ind.; profl. geologist, Ga. Geologist B.H. Putnam, Marietta, Ohio, 1954-58, F.E. Moran Oil Co., Owensboro, Ky., 1958-64; chief engr. Dept. Pub. Works Jefferson County, Louisville, 1967-69; chief geologist Guernsey Petroleum Corp. Atlanta, 1969-72, cons., 1972-73; v.p. Johnston Petroleum Corp., Cambridge, Ohio, 1973-75; v.p., dir. O'Neal Petroleum, Inc., New Concord, Ohio, 1975-83, Belleville, Ill., 1976-83; pres. Concord Energy, Inc., New Concord, Ohio, 1983—; cons. Ind. Glass Sand Corp., Elizabeth, Ind., 1964-66, Duchsherer and Assoc., Louisville, 1966-67, Ballard and Cordell, Atlanta, 1972-73. Pres. and dir. East Muskingum Swimming Pool Assn., 1976; dir. S.E. Ohio Symphony Orch., 1978; nat. research assoc. Smithsonian Instn., 1980. Served with U.S. Army, 1955-57. Fellow Geol. Soc. Am.; mem. Am. Assn. Petroleum Geologists, Soc. Petroleum Engrs., Am. Inst. Profl. Geologists. Roman Catholic. Lodge: KC (4th degree). Subspecialties: Geology; Fuels. Home: Morgan House Route #2 New Concord OH 43762 Office: Concord Energy Inc 22 W Main St New Concord OH 43762

HARMON, GARY R., engineer; b. Ravenna, Ohio, Jan. 7, 1950; s. Raymond A. and Betty J. (Hassler) H.; m. Jacquelyn Mitchell, Sept. 6, 1981. B.Sc. in Physics, Ohio State U., 1972, M.Sc., 1974. Commd. 2d lt. USAF, 1974, advanced through grades to capt., 1977; space program mgr. Hdqrs. Aerospace Def. Command, Colorado Springs, Colo., 1977-79; policy and issue analyst Office ofSec. of Air Force, Washington, 1979-81; devel. engr. Sec. of Air Force Office of Spl. Projects, Los Angeles, 1981—. Decorated Air Force Commendation medal with oak leaf cluster, Meritorious Service medal, Def. Meritorious Service medal. Mem. IEEE, AIAA, Air Force Assn. Republican. Baptist. Club: Colo. Mountain (Colorado Springs). Subspecialties: Plasma physics; Laser-induced chemistry. Current work: Aerospace technology. Home: 1711 E Pine Ave #4 El Segundo CA 90245

HARMON, JOHN WATSON, surgeon; b. White Plains, N.Y., Aug. 22, 1943; s. Frederick W. and Annie (Page) H.; m. Gail McGreevy, June 11, 1969; children: James, Eve. B.A., Harvard U., 1965; M.D., Columbia U., 1969. Intern Harvard U. Service, Boston City Hosp., 1969-70; surg. resident Harvard U. Med. Sch., 1969-75; commd. 1st lt. U.S. Army, 1969, advanced through grades to lt. col., 1983; investigator Walter Reed Army Inst. Research, Washington, 1975—; assoc. prof. surgery Uniformed Services U., Bethesda, Md., 1979—; U.S. Army rep. surg. study group NIH, Bethesda, 1982. Editor: Mechanisms of Muocal Injury, 1982. Recipient William Beaumont award Fed. Gastroenterologists, 1982. Fellow ACS; mem. Am. Physiol. Soc., Soc. Univ. Surgeons, Am. A Gastroent. Soc., Assn. Acad. Surgery, Alpha Omega Alpha. Episcopalian. Club: St. Marys River Yacht (St. Marys City, Md.). Subspecialties: Surgery; Comparative physiology. Current work: Surgical gastroenterology and endocrinology. Office: Dir Div Surgery Walter Reed Army Research Inst Walter Reed Army Med Center Washington DC 20307

HARMONY, MARLIN DALE, chemistry educator, researcher; b. Lincoln, Nebr., Mar. 2, 1936; s. Philip and Helen Irene (Michal) H. A.A., Kans. City Mo. Jr. Coll., 1956; B.S. in Chem. Engring, U. Kans., 1958; Ph.D., U. Calif.-Berkeley, 1961. Asst. prof. U. Kans., Lawrence, 1962-67, assoc. prof., 1967-71, prof., 1971—, chmn., 1980—. Contbr. articles on chemistry to profl. jours. NSF fellow, 1961-62. Mem. Am. Chem. Soc., Am. Phys. Soc., AAUP, AAAS, Sigma Xi, Alpha Chi Sigma, Phi Lambda Upsilon. Democrat. Subspecialty: Physical chemistry. Current work: Microwave spectroscopy and molecular structure. Home: 1033 Avalon Lawrence KS 66044 Office: Chemistry Dept Kansas U Lawrence KS 66045

HARMS, ARCHIE ARKADIUS, engineering physics educator, researcher; b. Nova-Dwor, Poland, Apr. 18, 1934; emigrated to Can., 1948; s. Paul and Wilhelmina (Kliewer) H.; m. Ursula Margareta Claassen, Aug. 17, 1957; children: Trudy, Dolores, Theodore. B.Sc., U. B.C., Can.), Vancouver, 1963; M.Sc.Eng., U. Wash., 1965, Ph.D., 1969. Engr. Internat. Power Engring. Cons., Vancouver, 1963-65; asst. prof. engring. physics McMaster U., Hamilton, Ont., 1969-72, assoc. prof., 1972-79, prof., 1979—; cons. in field; vis. scholar Internat. Inst. Applied Systems Analysts, Laxenburg, Austria, 1980. Author: Nuclear Energy Synergetics, 1982. Mem. Am. Phys. Soc., Am. Nuclear Soc., Can. Assn. Physicists. Subspecialties: Nuclear fission; Nuclear fusion. Home: 245 Taylor Rd Ancaster ON Canada L9G 1P6 Office: McMaster U Hamilton ON Canada L8S 4M1

HARN, STANTON DOUGLAS, anatomist, educator; b. Pomona, Calif., Mar. 14, 1945; s. Clifford M. and Minnie D. (Dickey) H.; m. Pamela S. Clarke, Oct. 1, 1965 (div. Oct. 1981); children: Deborah Sue, Megan Julia. B.A., LaVerne Coll., 1967; Ph.D., U. Utah, 1972. Asst. prof. U. Nebr., Lincoln, 1972-78, assoc. prof., 1978—; curator Coll. Dentistry Mus., Lincoln, 1978—. Contbr. sci. articles to profl. jours. Recipient Outstanding Tchr. awards Coll. Dentistry Students, 1972-78, U. Nebr. Alumni Assn., 1976. Mem. Am. Assn. Anatomists, Am. Acad. History of Dentistry. Subspecialties: Anatomy and embryology; Oral biology. Current work: Anatomy of deep face, anatomy of local anesthesia, history of dentistry. Office: U Nebr Coll Dentistry 40th and Holdrege Sts Lincoln NE 68503 Home: 1631 Brent Blvd Lincoln NE 68506

HARNED, JOSEPH WILLIAM, consultant; b. Allentown, Pa., Jan. 26, 1940; s. William Biechele and Mary Martha (Baily) H.; m. Antoinette Falquier, July 20, 1963; children: Sheilah, Margret, Jennifer. B.A., Yale U., 1961. Project dir. Atlantic Inst., Paris, 1963-65, dep. dir. studies, 1965-68; asst. dir. Atlantic Council U.S., Washington, 1968-75, dep. dir. gen., 1975—; U.S. rep. Atlantic Inst. Internat. Affairs, Paris, 1968—; sr. assoc. cons. Internat. Energy Assocs. Ltd., Washington, 1975—; pres. Falquier & Harned Ltd., Alexandria, Va., 1976—. Author, editor: Nuclear Fuels Policy, 1976, Nuclear Power and Nuclear Weapons Proliferation, 1978, The Common Security Interests of Japan, United States and NATO, 1981, U.S. Energy Policy and U.S. Foreign Policy in the 1980's, 1982, others; contbr. articles in field to profl. jours. Served to lt. (j.g.) USNR, 1961-63. Mem. Am. Nuclear Soc., Acad. Ind. Scholars. Lutheran. Subspecialty: Policy formulation. Current work: Director of policy formulation programs in the fields of energy policy, strategic security, trade, technology transfer, military manpower. Home: 2504 Culpeper Rd Alexandria VA 22308 Office: Atlantic Council US 1616 H St NW Washington DC 20006

HARNEY, ROBERT CHARLES, laser technologist, educator; b. Pasadena, Calif., Sept. 28, 1949; s. Ervin Charles and Ethel Josephine (Erickson) H.; m. Jane Withers, June 23, 1972; children: Elizabeth, Catherine. B.S. in Chemistry, Harvey Mudd Coll., 1971, 1971; M.S. in Applied Sci, U. Calif.-Davis, 1972, Ph.D., 1976. Participating guest physicist Lawrence Livermore Lab., Livermore, Calif., 1971-76; research engr. U. Calif.-Davis, 1976; staff scientist M.I.T. Lincoln Lab., Lexington, Mass., 1976-82; mem. profl. staff Martin Marietta Aerospace, Orlando, Fla., 1982—; cons., lectr. in field. Editor: Physics and Technology of Coherent Infrared Radar, 1982; contbr. articles to profl. jours. Recipient 1st prize Laser Inst. Am. Laser Quiz, 1977; Dept. Def. exec. intern, 1969; Fannie and John Hertz Found. fellow, 1972-76. Mem. Am. Chem. Soc., Am. Phys. Soc., Optical Soc. Am., Soc. Photo-Optical Instrumentation Engrs., IEEE, Astron. Soc. Pacific, Am. Assn. Physics Tchrs., Am. Def. Preparedness Assn., Sigma Xi. Republican. Mem. Ch. of Christ. Subspecialties: Optical radar; Spectroscopy. Current work: Application of lasers and optics to fundamental and applied research problems in physics, chemistry, biology, and engineering. Development of coherent infrared radar systems for solution of civilian and mil. imaging, detection, and remote sensing problems. Patentee in field. Home: 6852 Parson Brown Dr Orlando FL 32811 Office: PO Box 5837 Orlando FL 32855

HARPAVAT, GANESH LAL, scientist, engineering consultant; b. Nai (Udaipur), Rajasthan, India, May 13, 1944; came to U.S., 1965; s. Dadam Ch and Najar Devi (Darari) H.; m. Kiran Kumari Singhvi, Dec. 6, 1970; children: Manisha, Sanjiv. B.Engring. with honors, U. Jodhpur, India, 1965; M.S., U. Rochester, 1967, Ph.D., 1968; M.B.A., U. Dallas, 1978. Research assoc. U. Rochester, N.Y., 1965-68; assoc. scientist Xerox Corp., Rochester, 1968-70, scientist, 1970-75, mgr. bus. strategy, Dallas, 1982, sr. scientist, 1975-82, mgr. bus. strategy, 1982-83, regional support mgr., 1983—. Contbr. research articles to publs. Pres. Goodwill Assn. Am., Rochester, 1972—. Recipient Gold medal U. Jodhpur, 1965. Mem. ASME, AIAA, IEEE (hon. mention 1981), Sigma Xi, Sigma Iota Epsilon. Subspecialties: Fluid mechanics; Numerical analysis. Current work: Fluid mechanics, computer modeling and simulations, structural analysis, heat transfer, xerography, inkjet technology, piezoelectricity, electrostatics, applied mechanics, business planning and forecasting, microcomputers. Patentee. Home: 921 Angela Dr Lewisville TX 75067 Office: Curtis 1000 Info Systems 4560 Beltline Rd Suite 210 Dallas TX 75067

HARPER, ALFRED EDWIN, educator; b. Lethbridge, Alta., Can., Aug. 14, 1922; came to U.S., 1952, naturalized, 1957; s. Alexander and Frances (Bradley) H.; m. Naila Evelyn Jwaideh, Apr. 17, 1948; children—Shareen Frances, Gwendolyn Ann. B.Sc., U. Alta., 1945, M.Sc., 1947; Ph.D., U. Wis.-Madison, 1953; NRC research fellow, U. Cambridge, Eng., 1955-56. Lectr., then asst. prof. biochemistry U. Alta., 1948-54; asst., then asso. prof. biochemistry U. Wis.-Madison, 1954-61; prof. nutrition Mass. Inst. Tech., 1961-65, prof. biochemistry U. Wis.-Madison, 1965—, chmn. dept. nutritional scis., 1968—, E.V. McCollum prof. nutritional sci., 1974—; Mem. Food and Nutrition Bd., NRC-Nat. Acad. Scis., 1961-71, chmn., 1978—, mem. exec. com., 1968-71; mem. nutrition tng. com. Nat. Inst. Gen. Med. Scis., 1967-71, chmn., 1968-71; bd. dirs. Nat. Nutrition Consortium, 1973-76; mem. sci. adv. com. Nutrition Found., 1965-72, Am. Inst. Baking, 1961-78; NIH Spl. fellow, 1972-73; vis. prof. INCAP, Guatemala, 1972; mem. expert com. energy and protein needs FAO-WHO, 1971-72. Asso. editor: Canadian Jour. Biochemistry, 1968-71; editorial bd.: Am. Jour. Physiology, 1969-74, Physiol. Revs, 1973-79, Jour. Nutrition, 1959-63, The Profl. Nutritionist, 1976-79; contbr. numerous articles to profl. jours. Fellow A.A.A.S.; mem. Am. Inst. Nutrition (council 1966-70, pres. 1970-71, Borden award 1965), Am. Physiol. Soc., Am. Soc. Biol. Chemists, Sigma Xi. Democrat. Unitarian. Subspecialty: Nutrition (biology). Home: 3447 Edgehill Pkwy Madison WI 53705

HARPER, DOYAL ALEXANDER, JR., astronomer, educator; b. Atlanta, Oct. 9, 1944; s. Doyal Alexander and Emily (Brown) H.; m. Carolyn James, Mar. 11, 1967; children: Scott Alexander, Nathan Todd, Amy Claire, Evan James. B.A. in Elec. Engring, Rice U., Houston, 1966, Ph.D. in Space Sci, 1971. Asst. prof. astronomy and astrophysics U. Chgo., 1971, assoc. prof., 1976-80, prof., 1980—; dir. Yerkes Obs., 1982—. Mem. Am. Astron. Soc. (Newton Lacy Pierce prize o61979), Astron. Soc. Pacific. Subspecialty: Infrared optical astronomy. Current work: Infrared observations of galaxies, stars, planets, star formation regions, far infrared detectors, cryogenics, optical systems. Home: Yerkes Observatory Williams Bay WI 53191

HARPER, KIMBALL TAYLOR, botany educator; b. Oakley, Ida., Feb. 15, 1931; s. John Mayo and Mary Ella (Overson) H.; m. Caroline Frances Stepp, June 7, 1958; children: Ruth Lynn (Mrs. Steven V. Long), James K., Gay A., Denise C., Karla D., Steven S. B.S., Brigham Young U., 1958, M.S., 1960; Ph.D., U. Wis.-Madison, 1963. Range technician U.S. Forest Service, Manti-LaSal Nat. Forest, 1957, range scientist, Ogden, Utah, 1958-59; research asst. U. Wis.-Madison, 1959-63; asst. prof. botany U. Utah-Salt Lake City, 1963-70, assoc. prof., 1970-73; prof., chmn. botany and range sci. dept. Brigham Young U., 1973-76, prof. botany and range sci., 1976—; Chmn., mem. tech. adv. bd. Native Plants, Inc., Salt Lake City, 1975. Contbr. numerous articles to sci. jours. Served with AUS, 1953-54. Recipient Karl G. Maeser Disting. Research award Brigham Young U., 1979. Fellow AAAS; mem. Bot. Soc. Am., Ecol. Soc. Am., Am. Inst. Biol. Scis., Soc. Range Mgmt., Brit. Ecol. Soc., Soc. Study of Evolution, Am. Soc. Naturalists. Republican. Mormon. Subspecialties: Ecology; Resource management. Current work: Ecology of plant reproduction; floral biology; pollination ecology; sex expression in plants; environmental impacts on plant reproduction. Home: 410 S 300 E Spanish Fork UT 84660

HARPER, MICHAEL JOHN KENNEDY, medical educator, cons.; b. London, Feb. 25, 1935; U.S., 1964; s. John Kennedy and Helen Malvina (Koeller) H.; m. Marian Kennedy Wedd, July 23, 1960 (div. Feb. 1982); children: Charlotte G.K., Tristram J. K., Felicity W. K. B.A. in Agr, U. Cambridge, Eng., 1957, Ph.D., 1962, Sc.D., 1979; diploma agr., U. Reading, Eng., 1958. Tech. officer Imperial Chem. Industries, Alderley Edge, Cheshire, Eng., 1961-64, 1965-66; vis. scientist Worcester Found. for Exptl. Biology, Shrewsbury, Mass., 1964-65, staff scientist, 1966-68, sr. scientist, 1968-72, scientist WHO, Geneva, 1972-75; assoc. prof. ob-gyn U. Tex. Health Sci. Ctr., San Antonio, 1975-81, prof., 1981—; mem. task force com. WHO, 1972—, adv. group human reproduction, 1982—; cons. Am. Pub. Health Assn., 1982—, NSF, 1983—; mem. ad hoc adv. com. contraceptive devel. br. Nat. Inst. Child Health and Human Devel., 1975—. Author: Birth Control Technologies, 1983; contbr. numerous articles on reprodn.to profl. jours. Recipient Woodman prize U. Cambridge, 1957; Agrl. Food Products prize U. Reading, 1958; NIH grantee, 1969-83; others. Fellow Inst. Biology (Eng.); mem. Endocrine Soc. U.S.A., Soc. Endocrinology (U.K.), Soc. Study of Reproduction, Soc. Study of Fertility (U.K.), Am. Fertility Soc., Am. Assn. Anatomists, Soc. Gynecol. Investigation. Republican. Subspecialties: Reproductive biology; Reproductive biology (medicine). Current work: Pre-implantation stages of pregnancy, prostaglandins, contraceptive research and development. Inventor tamoxifen, 1963; patentee in field. Home: 3410 Turtle Village San Antonio TX 78230 Office: U Tex Health Sci Center Dept Ob-Gyn 7703 Floyd Curl Dr San Antonio TX 78284

HARPIN, RAOUL EDWARD, behavioral science educator, researcher; b. Lynn, Mass., Feb. 23, 1947; s. Raoul Edward and Mary Louise (Pyne) H.; m. June Lam, Aug. 22, 1970; 1 dau., Bettina Lam. B.A. cum laude, Boston Coll., 1969; M.A., Stanford U., 1972; Ph.D., SUNY-Stony Brook, 1978. Lic. clin. psychologist, Calif. Fellow Mass. Gen. Hosp.-Harvard Med. Sch., 1977-79; asst. prof. U. Md., Balt., 1979-82; dir. behavioral sci. edn. Family Practice/Kern Med. Ctr., Bakersfield, Calif., 1982—. Author: Behavior Therapy for Depression, 1982; contbr. articles to profl. jours., chpts. to books. Mem. Am. Psychol. Assn., Assn. Advancement Behavior Therapy, Soc. Tchrs. Family Medicine, Cross and Crown Honor Soc. Subspecialties: Behavioral psychology; Behavior therapy. Current work: Research in behavior therapy for depression, behavioral science for physicians in training, behavior therapy for clinical psychosocial problems. Home: 3913 Fairmount St Bakersfield CA 93306 Office: Dept Family Practice Kern Med Ctr 1830 Flower St Bakersfield CA 93305

HARPSTER, JOSEPH W. C., business executive, scientist, consultant; b. Sewickley, Pa., June 29, 1932; s. Benjamin and Ruth (Ramer) H.; m.; children: Jodean Harpster Foringer, Brad C., Brian K., Timothy J.; m. Marilyn Yueh-Chin Fu, Jan. 4, 1975. B.S., Geneva Coll., 1959; M.S. in Physics, Case Western Res. U., 1964; Ph.D. in Nuclear Engring, Ohio State U., 1971. Plant engr./research physicist Valvoline Oil Co., Freedom, Pa., 1954-60; research assoc. Harshaw Chem. Co., Cleve., 1960-63; research supr. dept. elec. engring. Ohio State U., Columbus, 1964-73; v.p. research and devel. Ohio Semitronics, Inc., Columbus, 1965-76; pres., chmn. bd. Intek, Inc., Columbus, 1976—; cons. mfg. process control instrumentation to industry, govt., and military. Contbr. over 60 articles to sci. jours. Mem. com. SBA, 1981—; mem. Brookview Civic Assn., 1976—. Served with USN, 1950-54; ETO. Recipient C.E. McArtney Sci. award Geneva Coll., 1959; NASA Group Achievement award, 1975; Apollo Soyuz medallion NASA, 1975. Mem. NAM, Instrument Soc. Am. Republican. Roman Catholic. Club: Wickertree Tennis (Columbus). Subspecialties: Electronic materials; Atomic and molecular physics. Current work: Physical sensors, instrumentation, semiconductor materials, atomic and material structure. Patentee in field (6). Home: 11450 Overbrook Ln Galena OH 43021 Office: 515 Schrock Rd Columbus OH 43229

HARRELL, RUTH FLINN, mental-nutritional research educator; b. Americus, Ga., Apr. 19, 1900; d. Daniel and Neva (Poley) Flinn; m. William Lee Harrell, Nov. 24, 1928 (dec. Apr. 1972); children: Ruth Harrell Capp. B.S., Wesleyan Coll., Macon, Ga., 1920; M.A., Columbia U., 1924, Ph.D., 1942. Diplomate: Am. Psychol. Assn. Bd. Pub. sch. psychologist, Norfolk, Va., 1926-39; rehab. psychologist neuro-surgery Johns Hopkins Hosp., Balt., 1936-47; faculty mem. Old Dominion U., Norfolk, 1955-70, prof. psychology, to 1970, research prof., 1976—. Author: Effect of Added Thiamin on Learning, 1943, Further Effects of Added Thiamin on Learning, 1947, Effect of Mothers' Diets on Intelligence of Offspring, 1955, Can Nutritional Supplements Help Mentally Retarded Children?, 1981. Recipient Ann. award for advancing research on mental retardation-nutrition Atlanta Med. Research Found., 1982. Mem. Am. Psychol. Assn., N.Y. Acad. Scis., Delta Kappa Gamma. Presbyterian. Subspecialties: Nutrition (biology); Learning. Current work: Study of Epilepsy as related to nutrition - i.e. seizures may be a manifestation of nutritional deficiencies. Home: 6411 Powhatan Ave Norfolk VA 23508 Office: Research Found Old Dominion U Norfolk VA 23508

HARRIMAN, PHILIP DARLING, geneticist, government official; b. San Rafael, Calif., Nov. 24, 1937; s. Theodore Darling and Luciel Harriet (Muller) H.; m. Jenny Elizabeth Flack, June 12, 1959; 1 son, Marc Stuart. B.S. in Physics, Calif. Inst. Tech., 1959; Ph.D. in Biophysics, U. Calif., Berkeley, 1964. Postdoctoral fellow U. Cologne, W.Ger., 1964-65, Pasteur Inst., Paris, 1965-66, Cold Spring Harbor Lab., N.Y., 1966-68; asst. prof. biochemistry Duke U. Med. Center, Durham, N.C., 1968-75; assoc. prof. biology U. Mo., Kansas City, 1975-77; program dir. for genetic biology NSF, Washington, 1977—, sr. scientist office of Asst. Dir. Biology, Behavioral and Social Scis., 1981-82. Contbr. articles to profl. jours. Legis. asst. Congressman Dave McCurdy of Okla., 1980-81. Served to 1st lt. USAFR, 1962—. Congressional fellow U.S. Office Personnel Mgmt., 1980-81. Mem. AAAS, Genetics Soc. Am. Unitarian. Subspecialties: Genetics and genetic engineering (biology); Molecular biology. Current work: I evaluate and fund research proposals in area of genetics and genetic engineering. Home: 2606 Arcola Ave Wheaton MD 20902 Office: Nat Sci Found Room 326 Washington DC 20550

HARRINGTON, J(AMES) PATRICK, astronomer; b. Salem, Ohio, Dec. 21, 1939; s. Joseph Edwin and Margaret E. (Brobander) H.; m. Marianne Jordan Harrington, June 18, 1966; children: Catherine, Jordan, Nora. B.S., U. Chgo., 1961; M.S., Ohio State U., 1964, Ph.D., 1967. Asst. prof. astronomy U. Md., 1967-73, assoc. prof., 1973—. Mem. Am. Astron. Soc., Royal Astron. Soc., Internat. Astron. Union. Subspecialties: Ultraviolet high energy astrophysics; Theoretical astrophysics. Current work: Planetary nebulae, radiative transfer. Office: Astronomy Program University Maryland College Park MD 20742

HARRINGTON, ROBERT SUTTON, astronomer; b. Newport News, Va., Oct. 21, 1942; s. Jean Carl and Virginia Hall (Sutton) H.; m. Betty Jean, July 23, 1976; children: Amy Lucille, Ann Charon. B.A. in Physics, Swarthmore Coll., 1964; Ph.D. in Astronomy, U. Tex., 1968. Astronomer U.S. Naval Obs., Washington, 1967—. Contbr. tech. articles to profl. publs. Mem. Internat. Astron. Union, Am. Astron. Soc., AAAS, Sigma Xi. Subspecialties: Astrometry; Celestial mechanics. Current work: Stellar distances and motions, solar system dynamics. Office: Naval Obs Washington DC 20390

HARRINGTON, RODNEY ELBERT, biochemist; b. Mayville, N.D., Jan. 9, 1932; s. Elbert Wellington and Marjorie H.; m.; children: Tiffany Anne, Jennifer Ellen; m. Ilga Butelis, Jan. 26, 1979. B.A., U. S.D., 1953; Ph.D., U. Wash., 1960. Research chemist Ames Lab., Iowa State U., 1953; research asst. U. Wash., 1957; research assoc. U. Calif.-San Diego, 1960; asst. prof. U. Ariz., 1962; assoc. prof. U. Calif.-Davis, 1964; prof. chemistry U. Nev., Reno, 1972-82, chmn. dept. chemistry, 1970-76, prof. biochemistry, 1982—. Bd. dirs. Nev. Opera Co., 1973-78; mem. Community Relations Bd., U. Nev., 1977-79. NIH and NSF research grantee, 1961—. Mem. Am. Chem. Soc., Am. Phys. Soc., Biophys. Soc., AAAS, Sigma Xi. Lutheran. Subspecialties: Biochemistry (biology); Molecular biology. Current work: Chromatin structure and gene regulation, fundamental research. Office: Dept Biochemistry U Nev Reno NV 89557

HARRINGTON, ROY VICTOR, chemist; b. Bklyn., Sept. 28, 1928; s. Victor Earl and Karen (Hanson) H.; m. Catherine Elisabeth Wiese, June 14, 1952; children—Bruce Allan, Karen Jane, Thomas Andrew. B.S. in Chemistry, Poly. Inst. Bklyn., 1952; Ph.D., U. Colo., 1955. Chemist Gen. Foods Corp., Hoboken, N.J., 1949-52, Corning Glass Works, N.Y., 1955-68; with Ferro Corp., Independence, Ohio, 1968—, asso. dir., then dir. research and devel., now v.p. research and devel.; pres. Cleve. Assn. Research Dirs., 1975; mem. panel radioactive waste disposal Nat. Acad. Scis., 1979-80. Author. Founder, pres. Corning Sci. Seminars Gifted High Sch. Students, 1960-67; Mem. steering com. Case Assos., Case Western Res. U. Mem. Am. Chem. Soc. (sect. chmn. 1978), Am. Ceramic Soc. Club: Lakeside Yacht (Cleve.). Subspecialties: Inorganic chemistry; Materials. Current work: Broad materials interests, glass, ceramics, composites, electronic materials, plastics, high temperature chemistry radiochemistry. Patentee in field. Office: 7500 E Pleasant Valley Rd Independence OH 44131

HARRINGTON, STEVEN JAY, research scientist; b. Portland, Oreg., Nov. 28, 1947; s. Everett J. and Virginia (Stovall) H.; m. E. Christine Stout, Mar. 29, 1969. B.S., Oreg. State U., 1968; M.S., U. Wash., 1976, Ph.D., 1976. Assoc. research instr. U. Utah, 1976-78; asst.prof. SUNY-Brockport, 1978-81; mem. research staff Xerox Corp., Rochester, N.Y., 1981—. Author: Computer Graphics: A Programming Approach, 1982; Contbr. numerous articles to profl. jours. Served with AUS, 1969-71. Mem. Assn. Computing Machinery, IEEE, Soc. Indsl. and Applied Math., Am. Phys. Soc. Subspecialties: Graphics, image processing, and pattern recognition; Mathematical software. Current work: Design of high resolution electronic printing software systems; image representations and data compression. Home: 76 S Main St Holley NY 14470 Office: Xerox Corp Xerox Sq 128 Rochester NY 14644

HARRINGTON, WILLIAM FIELDS, biochemist, educator; b. Seattle, Sept. 25, 1920; s. Ira Francis and Jessie Blanche (Fields) H.; m. Ingeborg Leuschner, Feb. 24, 1947; children: Susan, Eric, Peter, Robert, David. B.S., U. Calif. at Berkeley, 1948, Ph.D., 1952. Research chemist virus lab. U. Calif. at Berkeley, 1952-53; Nat. Found. Infantile Paralysis postdoctoral fellow Cambridge (Eng.) U., 1953-54; Nat. Cancer Inst. postdoctoral fellow Carlsberg Lab., Copenhagen, Denmark, 1954-55; asst. prof. chemistry Iowa State U., 1955-56; biochemist Nat. Heart Inst., 1956-60; prof. biology Johns Hopkins, Balt., 1960—, chmn. dept. biology, 1973-83, Henry Walters prof. biology, 1975—; dir. McCollum Pratt Inst., 1973-83; vis. scientist Wiezmann Inst., Rehovot, Israel, 1959, vis. prof., 1970, Oxford U., 1970; mem. adv. panel physiol. chemistry NIH, 1962-66, adv. panel biophys. chemistry study sect., 1968-72; bd. sci. councillors Nat. Inst. Arthritis and Metabolic Diseases, 1968-72; mem. vis. com. for biology Brookhaven Nat. Lab., 1969-73; adv. bd. Fedn. Advanced Edn. in the Scis., 1975—. Co-editor: Monographs on Physical Biochemistry, 1970—; Bd. editors: Jour. Biol. Chemistry, 1963-68, Mechanochemistry and Motility, 1970-76, Biochemistry, 1971-77, Jour. Phys. Biochemistry, 1973—. Fellow Am. Acad. Arts and Sci.; mem. Biophysics Soc., Nat. Acad. Scis., Soc. Biol. Chemists, Phi Beta Kappa, Sigma Xi. Subspecialty: Molecular biology. Home: 2210 W Rogers Ave Baltimore MD 21209

HARRIS, ALAN WILLIAM, scientist, researcher; b. Portland, Oreg., Aug. 3, 1944; s. James Stewart and Jane Ann (Gordon) H.; m. Rose Marie Spitt, Aug. 22, 1970; children: W. Donald, David S., Catherine R. B.S., Calif. Inst. Tech., 1966; M.S., UCLA, 1967, Ph.D., 1974. Geophysicist space div. Rockwell Internat., Downey, Calif., 1967-70; physics tchr. Immaculate Heart High Sch., Hollywood, Calif., 1970-74; instr. physics Santa Monica (Calif.) Coll., 1970-71; research scientist, earth and space scis. div. JPL, Pasadena, Calif., 1974—; prin. investigator, lunar and planetary program NASA, 1976—. Contbr. articles to profl. jours. Mem. Internat. Astron. Union, Am. Astron. Soc., Am. Geophys. Union. Subspecialties: Planetary science; Optical astronomy. Current work: Dynamical theory of plant formation, observation and theoretical research on asteroid rotation and collisional processes. Home: 4603 Orange Knoll La Canada CA 91011 Office: JPL 4800 Oak Knoll Dr 183-501 Pasadena CA 91109

HARRIS, CHARLES LEON, biology educator; b. Christiansburg, Va., Jan. 3, 1943; s. Garnett Johnson and Dora Alice (Shelor) H.; m. Mary Jane Zewe, Aug. 5, 1971. B.S. in Physics, Va. Poly. Inst., 1966; M.S. in Biophysics, Pa. State U., 1967, Ph.D., 1969. From asst. prof. to prof. biology SUNY-Plattsburgh, 1970—. Author: Evolution—Genesis and Revelations, 1981; editor: Procs. of ABLE, 1983. Mem. N.Y. Acad. Sci., Sigma Xi. Subspecialties: Neurobiology; History and philosophy of biology. Current work: Functioning of the nervous system of cockroach. Design of teaching experiments. Office: Dept Biol Scis SUNY Plattsburgh NY 12901

HARRIS, CYRIL MANTON, electrical engineering and architecture educator, consulting acoustical engineer; b. Detroit, June 20, 1917; s. Bernard O. and Ida (Moss) H.; m. Ann Schakne, July 12, 1949; children: Nicholas Bennett, Katherine Anne. B.A., UCLA, 1938, M.A., 1940; Ph.D., MIT, 1945; Sc.D. (hon.), N.J. Inst. Tech., 1981. Teaching asst. UCLA, 1939-40; research fellow MIT, 1940; war research OSRD, 1941-44, teaching fellow, 1943-45; war research Carnegie Instn. Washington, 1941; mem. staff Bell Telephone Labs., 1945-51; cons. Office Naval Research, London, Eng., 1951; Fulbright lectr. Tech. U., Delft, Holland, 1951-52; now Charles Batchelor prof. elec. engring., prof. architecture and chmn. div. archtl. tech. Columbia U.; vis. Fulbright prof. U. Tokyo, 1960; acoustical cons. Met. Opera House, N.Y.C., John F. Kennedy Center for Performing Arts, Washington, Krannert Center for Performing Arts, U. Ill., Powell Symphony Hall, St. Louis, Nat. Acad. Scis. Auditorium, Washington, Minn. Orch. Hall, Mpls., Nat. Centre for Performing Arts, Bombay, India, new Avery Fisher Hall, State Theater reconstruction Lincoln Center, N.Y.C., Symphony Hall, Salt Lake City; past dir. U.S. Inst. Theatre Tech.; mem. noise control group, com. undersea warfare NRC, 1955-57; mem. council hearing and bio-acoustics Armed Forces-NRC, 1953-55; mem. NRC adv. panel 213 to Nat. Bur. Standards, 1966-69, chmn., 1969-71; mem. bldg. research adv. bd. NRC, 1977-79. Author: (with V.O. Knudsen) Acoustical Designing in Architecture, 1950, rev., 1980, Handbook of Noise Control, 1957, 2d edit., 1979, (with C.E. Crede) Shock and Vibration Handbook, 1961, 2d edit., 1976, Dictionary of Architecture and Construction, 1975, Historic Architecture Sourcebook, 1977; Contbr.: articles to profl. jours. Historic Architecture Sourcebook; Editorial adv. bd.: Physics Today, 1955-66. Bd. dirs. Armstrong Meml. Research Found., 1976—; hon. v.p. St. Louis Symphony Soc., 1977—; mem. nat. adv. bd. Utah Symphony Orch., 1976—. Recipient Franklin medal, 1977, Emile Berliner award, 1977, Hon. award U.S. ITT, 1977, Wallace Clement Sabine medal, 1979, AIA medal, 1980. Fellow Acoustical Soc. Am. (pres. 1964-65, asso. editor jour. 1959-70), I.E.E.E. (chmn. profl. group ultrasonic engring. 1957-58, profl. group audio 1961-62), Audio Engring. Soc. (hon. mem.); mem. Am. Inst. Physics (governing bd. 1965-66), Nat. Acad. Scis., Nat. Acad. Engring., Sigma Xi, Tau Beta Pi. Subspecialties: Acoustics; Acoustical engineering. Current work: Architectural acoustics, noise control. Office: Mudd Bldg Columbia U New York NY 10027

HARRIS, DENNY OLAN, microbiology educator; b. Louisville, May 2, 1937; s. George E. and Lena (Wooldridge) H.; m. Patricia Ann Appling, Dec. 29, 1962; children: Lisa, Bradley, Gregory. B.S., U. Louisville, 1961, M.S., 1963; Ph.D., Ind. U., Bloomington, 1967. Prof. microbiology U. Ky., Lexington, 1967—. Contbr. articles to profl. jours. Mem. Phycol. Soc. Am., Internat. Phycol. Soc. Subspecialties: Microbiology; Morphology. Current work: Toxin production in fresh water algae; especially those which can be used to control the indiscriminate growth of nuisance aquatic organisns. Home: 612 Edgewater Dr Lexington KY 40502 Office: Sch Biol Sci U Ky Lexington KY 40506

HARRIS, DEVERLE PORTER, mineral economics educator; b. Lovell, Wyo., Jan. 21, 1931; s. David Lel and Winiferd (Porter) H.; m. Jean Marie Partridge, Nov. 23, 1949 (div. Oct. 1966); children: Randall Scott, David Vernon, Richard Partridge; m. Sandra Ellen Hall, Jan. 21, 1967; children: Kirstan Ellen, Brett DeVerle, Todd William. B.S., Brigham Young U., 1956, M.S., 1958; Ph.D., Pa. State U.-University Park, 1965. Grad. asst. Pa. State U., University Park, 1960-62; geologist Geophoto Services, Denver, Calgary, Alta., 1957-60; research asst. Pa. State U., University Park, 1962-65; geostatistician Union Oil Co., Brea, Calif., 1965-66; prof. mineral econs. Pa. State U., University Park, 1966-74; prof. mineral econs., prof. geol. engring. U. Ariz., Tucson, 1974—; cons. Inter-Am. Devel. Bank, Washington, 1982—, U.S. Dept. Energy, Grand Junction, Colo., 1972-81. Author: Mineral Resources Appraisal, 1983. Mem. AIME (chmn. sessions of nat. meeting 1977-83), Am. Econ. Assn., Soc. Econ. Geologists (program chmn. nat. meeting 1980). Republican. Mem. Ch. of Jesus Christ of Latter-day Saints. Subspecialty: Mineral and energy resources. Current work: Formalization of geoscience for the probabilistic estimation of mineral and energy endowments, and the design and use of engineering-economic systems for the estimation of potential supply of minerals and energy. Home: 3330 N Jackson St Tucson AZ 85719 Office: U Ariz Dept Mining and Geol Engring Coll Mines Tucson AZ 85721

HARRIS, DON NAVARRO, biochemist, biochem. pharmacologist, cons.; b. N.Y.C., June 17, 1929; s. John Henry and Margaret Vivian (Berkeley) H.; m. Regina G. Brooks, July 29, 1954; children: Donna Michele, John Craig, Scott Anthony. A.B., Lincoln U., 1951; M.S., Rutgers U., 1959, Ph.D., 1963. Sr. research investigator dept. pharmacology Squibb Inst. for Med. Research, Princeton, N.J., 1965—; coadj. assoc. prof. biochemistry Univ. Coll. of Rutger U., New Brunswick, N.J., 1975-77; mem. U.S. Army Sci. Bd. Contbr. articles on biochemistry and pharmacology to profl. jours. Mem. adv. coms. Frank Twp.Sch. Bd. and Library Bd. Served with U.S. Army, 1951-53. Mem. AAAS, Am. Chem. Soc., N.Y. Acad. Scis., Physiol. Soc. of Phila., Am. Heart Assn., Am. Soc. Pharmacology and Exptl. Therapeutics, Sigma Xi, Alpha Phi Alpha. Democrat. Baptist. Subspecialties: Biochemistry (biology); Molecular pharmacology. Current work: Biochemistry of Arachidonic acid metabolites, blood platelet function, cyclic nucleotide function. Patentee in field. Home: 26 Summerall Rd Somerset NJ 08873 Office: Dept Pharmacolog Squibb Inst Med Research Princeton NJ 08540

HARRIS, GALE ION, radiology and physics educator; b. Arlington, Calif., Aug. 7, 1935; s. Albert I. and Carmine A. (Waters) H.; m. Bonnie J. Hazlett, Mar. 31, 1956; children: Gayla Jean, Nathan Ward. B.S., U. Kans., 1957, M.S., 1959, Ph.D., 1962; S.M., M.I.T., 1973. Project leader nuclear physics Aerospace Research Lab. GS-15, Wright-Paterson AFB, Ohio, 1965-72, dep. dir., sr. scientist, 1973-74; asst. prof. adminstrn. dept. radiology, dir. Office Mgmt. Research, Johns Hopkins Med. Sch., Balt., 1974-75; assoc. prof. radiology and physics Mich. State U., East Lansing, 1975—; pres. Mich. Research Ctr., Inc., East Lansing, 1975—. Contbr. articles to profl. jours. Served from 1st lt. to capt. USAF, 1962-65. Named Outstanding Profl. Employee Dayton Met. Area Soc. Personnel Adminstrn., Dayton C. of C., 1970; Sloan fellow Mass. Inst. Tech., 1972-73. Mem. Am. Phys. Soc., Soc. Nuclear Medicine, AAAS, Am. Assn. Physicists in Medicine. Subspecialties: Nuclear physics; Nuclear medicine. Current work: Diagnostic imaging systems, nuclear spectroscopy. Home: 1312 Basswood Circle East Lansing MI 48864 Office: Mich State U B-220 Clin Center East Lansing MI 48824-1315

HARRIS, HAROLD HART, chemist, educator; b. Council Bluffs, Iowa, Mar. 12, 1940; s. Arthur A. and Opal E. (Hart) H.; m. Mary E. Cline, June 25, 1966; children: Matthew M., Jill E. B.S. in Chemistry, Harvey Mudd Coll., 1962; Ph.D. in Phys. Chemistry, Mich. State U., 1967. Postdoctoral fellow U. Calif., Irvine, 1966-67, instr. in chemistry, 1967-70; asst. prof. U. Mo., St. Louis, 1970-75, assoc. prof. chemistry, 1975—. Contbr. articles to profl. jours. Mem. Am. Phys. Soc., Am. Chem. Soc., AAAS, Am. Soc. Mass Spectrometry, Fedn. Am. Scientists, Sigma Xi. Subspecialties: Spectroscopy; Physical chemistry. Current work: Teaching chemistry and spectroscopy; polarization spectroscopy and quasielastic laser light scattering.

HARRIS, HARRY, genetics educator; b. Manchester, Eng., Sept. 30, 1919; m., 1948; 1 child. B.A., Cambridge U. (Eng.), 1941, M.B., B.Chir., 1943, M.D., 1949. Prof. and chmn. biochemistry Kings Coll. U. London, 1960-65, Galton prof. human genetics Univ. Coll., 1965-76; Harnwell prof. human genetics U. Pa. Med. Sch., Phila., 1976—; Dir. human biochem. genetics unit Med. Research Council, 1967-76. Mem. Nat. Acad. Sci. (fgn.), Am. Soc. Human Genetics, Genetics Soc.; fellow Royal Soc. Subspecialty: Genetics and genetic engineering (medicine). Office: Dept Human Genetic Sch Medicine U P Philadelphia PA 19104

HARRIS, HUGH COURTNEY, astronomer; b. New Rochelle, N.Y., Dec. 21, 1947; s. Hugh C. and Louise (Mallory) H. B.S., Cornell U., 1970; Ph.D., U. Wash., Seattle, 1980. Research fellow Dominion Astrophys. Obs., Hertzberg Inst. Astrophysics, Victoria, B.C., Can., 1980-82; research fellow dept. physics McMaster U., 1982—. Served with USCGR, 1970-74. Mem. Am. Astron. Soc., Astron. Soc. Pacific. Current work: Stellar motions and compositions; galactic structure and evolution; star clusters. Office: Dept Physics McMaster U Hamilton ON Canada L8S4M1

HARRIS, JANE ELLEN, toxicologist; b. N.Y.C., Feb. 26, 1945; d. Daniel and Fay (Gherstein) H.; m. Joseph Longino, Sept. 5, 1976; children: Moira Harris, Tristan Harris. B.S. in Biochemistry, Cornell U., 1965; Ph.D. in Pharmacology, Yale U., 1971. Research fellow in psychiatry Mass. Gen.Hosp., dept. psychiatry Harvard U. Med. Sch., Boston, 1971-73; instr., asst. prof. pharmacology Emory U. Med. Sch., Atlanta, 1973-78; toxicologist Bur. Foods, FDA, Washington, 1978—. Contbr. writings in field to publs. NIH grantee, 1975-78; NIMH grantee, 1974-75; Scottish Rite research grantee, 1974-75; recipient merit award FDA, 1982. Mem. Soc. for Neurosci., Am. Soc. Pharmacology and Exptl. Therapeutics, Sigma Xi. Subspecialties: Toxicology (medicine); Neurochemistry. Current work: Regulatory toxicology, specialization in carcinogenesis, teratology and reproduction and neurotoxicity. Home: 410 Third St SE Washington DC 20003 Office: FDA HFF-156 200 C St SW Washington DC 20204

HARRIS, LEE ERROL, engineering consultant and educator; b. Ft. Pierce, Fla., Oct. 29, 1953; s. Kenneth Albert and Betty Patricia (Patterson) H.; m. Jo Ann King, June 16, 1975; 1 son, Jeffrey Lee. Student, Fla. Inst. Tech., Melbourne, 1971-72, postgrad., 1981—; B.S. in Ocean Engring, Fla. Atlantic U., Boca Raton, 1972-74; M.E. in Coastal Engring, U. Fla., Gainesville, 1975. Registered profl. engr., Fla. Research asst. U. Fla. Coastal and Oceanographic Engring. Lab., Gainesville, 1974-75; coastal engr. U.S. Army C.E., Jacksonville, Fla., 1975-77; cons. engr. Coastal Data and Engring., Inc., Jensen Beach, Fla., 1977—; asst. prof., head dept. engring. scis. and oceanographic tech. Fla. Inst. Tech., Jensen Beach, 1980—; cons. engr. Harris Engring. and Surveying, Jensen Beach, 1977—. Contbr. articles and reports to profl. jours. Mem. Fla. Oceanographic Soc. (dir. 1979-83, pres. 1980—), Internat. Oceanographic Found., Marine Tech. Soc., Fla. Shore and Beach Preservation Assn., Oceanic Soc. Subspecialties: Ocean engineering; Civil engineering. Current work: Hydrodynamics of coastal tidal inlets; beach erosion and coastal processes; education in fields of engineering studies, especially oceanographic technology. Office: Fla Inst Tech 1707 NE Indian River Dr Jensen Beach FL 33457

HARRIS, LOUIS SELIG, pharmacologist, cons.; b. Boston, Mar. 27, 1927; s. Max and Pearl (Oppochinsky) H.; m. Ruth I. Schaufus, Aug. 25, 1952; 1 son, Charles Allan. B.A. in Chemistry, Harvard U., 1954, M.S. in Med. Scis, 1956, Ph.D. in Pharmacology, 1958. Research asst. in anesthesiology Mass. Gen. Hosp., Boston, 1951-52; research biologist Sterling-Winthrop Research Inst., 1960-61, assoc. mem., 1961-62, sect. head, sr. research biologist, 1962-66; assoc. prof. U. N.C., Chapel Hill, 1966-70, prof., 1970-73; cons. in field; Harvey Haag prof., chmn. dept. pharmacology Med. Coll. Va./Va. Commonwealth U., Richmond, 1972—; dir. SISA, Inc., SISA Inst. Research, Com. of One Thousand, Quintox, Inc. Contbr. numerous articles to profl. jours. Bd. dirs. Human Services, Inc. Fellow Am. Coll. Neuropsychopharmacology; mem. Am. Pharm. Found., Acad. Pharmacol. Scis., Am. Soc. Pharmacology and Exptl. Therapeutics, Am. Chem. Soc., Am. Assn. Med. Sch. Pharmacology, AAAS, Assn. Harvard Chemists, Internat. Anesthesia Research Soc., Internat. Narcotic Enforcement Officers Assn., AAUP, Assn. Neurosci., Internat. Soc. Biochem. Pharmacology, Internat. Soc. Study Pain, Va. Acad. Sci., Am. Pain Soc. Club: Harvard. Subspecialties: Pharmacology; Neuropharmacology. Current work: Pharmacology of substances which effect the central nervous system and drug development. Home: 7830 Rockfalls Dr Richmond VA 23225 Work: Dept Pharmacology Va Commonwealth U Richmond VA 23298

HARRIS, MARY BIERMAN, educator; b. St. Louis, Feb. 9, 1943; d. Norman and Margaret (Loeb) Bierman; m. Richard J. Harris, June 14, 1965; children: Jennifer, Christopher. B.A., Radcliffe Coll., 1964; M.A., Stanford U., 1965, Ph.D., 1968. Instr. Talladega (Ala.) Coll., 1965-66; asst. prof. to prof. ednl. founds. U. N.Mex., Albuquerque, 1968—; vis. prof. U. New South Wales, Sydney, Australia, 1981-82; vis. assoc. prof. Ohio State U., Columbus, 1974-75. Contbr. articles to profl. jours.; editor: Classroom Uses of Behavior Modificaion, 1972. Mem. Am. Psychol. Assn., Am. Ednl. Research Assn., Phi Beta Kappa, Sigma Xi. Subspecialties: Behavioral psychology; Developmental psychology. Current work: Weight control, obesity, sex roles and stereotypes, modeling, social learning, altruism, aggression, social, developmental and educational psychology. Office: U N Mex Ednl Founds Dept Albuquerque NM 87131 Home: 1719 Rita Dr NE Albuquerque NM 87106

HARRIS, MILTON, chemist; b. Los Angeles, Mar. 21, 1906; s. Louis and Naomi (Granish) H.; m. Carolyn Wolf, Mar. 30, 1934; children—Barney Dreyfuss (adopted), John. B.Sc., Oreg. State Coll., 1926; Ph.D., Yale, 1929; Dr. Textile Sci., Phila. Textile Inst., 1955. Research asso. Am. Assn. Textile Chemists and Colorists, Nat. Bur. Standards, 1931-39; dir. research Textile Found., 1939-45; pres. and founder Harris Research Labs., 1945-61; dir. research Gillette Co. (its subsidiaries), 1956-66, v.p. corp., 1957-66; dir., chmn. exec. com. Sealectro Corp.; dir. Warner Lambert Co.; adv. bd. Jour. Polymer Sci.; asso. editor Textile Research Jour.; cons. Exec. Office of Pres., Office Sci. and Technology, 1962-65; Adv. bd., cons. O.Q.M.G., World War II; chmn. com. on textiles and cordage, tropical deterioration project Nat. Def. Research Com.; sec. com. on clothing NRC, World War II; chmn. Wool Conservation Bd., World War II; mem. panel on clothing Research and Devel. Bd., World War II; mem. Yale Council, 1964-69, Yale Devel. Bd., 1964-67; exec. bd. Yale Grad. Sch. Assn., 1965—; mem. adv. com. Nat. Bur. Standards, 1971—, chmn. vis. com., 1973—; mem. sub-com. Food and Agrl. Orgn. UN; mem. Utilization research and devel. adv. com. U.S. Dept. Agr., 1966—; mem. adv. com. planning NSF, 1968—; mem. Pres.'s Sci. Adv. Com. Panel on Environment, 1968—; observer-cons. Task Group Nat. Systems Sci. and Tech. Information, Fed. Council Sci. and Tech., 1968—. Contbr. articles to tech. jours.; Editor: Chemistry in the U.S. Economy. Trustee Phila. Textile Inst., 1956-60; dir. Dermatology Found., Textile Research Inst.; bd. dirs. Sci. Service, Acorn Fund, Chgo. Recipient award Wash. Acad. Sci., 1943, Olney medal for textile chemistry research, 1945; honor award Am. Inst. Chemists, 1957; Harold DeWitt Smith Meml. medal, 1966; Distinguished Service award Oreg. State U., 1967; Perkin Medal award Soc. Chem. Industry, 1970; Wilbur Lucius Cross medal Yale, 1974. Fellow Textile Inst., N.Y. Acad. Sci.; mem. Am. Assn. Textile Tech., Yale Chemists Assn. (past pres.), Am. Assn. Biol. Chemists, Nat. Acad. Engring., Am. Inst. Chemists (pres. 1960-61), N.A.M., Textile Research Inst., Am. Assn. Textile Colorists and Chemists, Am. Oil Chemists Soc., Soc. Cosmetic Chemists, Fiber Soc. (hon.), Am. Chem. Soc. (chmn. bd. dirs. 1966-70, dir.-at-large, treas. 1973—, priestly medal 1980), A.A.A.S. (editorial bd. publ. Sci. 1968-70), Soc. Chem. Industry, Wash. Acad. Sci., Sigma Xi, Tau Beta Pi, Phi Lambda Upsilon, Phi Kappa Phi, Gamma Alpha. Clubs: Cosmos (Washington); Chemists (N.Y.C.). Subspecialties: Polymer chemistry; Natural and synthetic fibers.. Home: 4101 Linnean Ave Washington DC 20008 Office: 3300 Whitehaven St NW Washington DC 20007

HARRIS, NORMAN OLIVER, dental educator, researcher; b. Shinglehouse, Pa., May 27, 1917; s. Theodore Clifton and Amelia (Rappenecker) H.; m. Grace Haynes, June 23, 1954; 1 son, Gary. D.D.S., Temple U., Phila., 1939; M.S.D., Ohio State U., 1952. Prof. (hon.) U. Pernambuco, Recife, Brazil, 1974; Dental intern USPHS, 1939-40; commd. 1st lt. U.S. Army, 1941; advanced through grades to lt. col. U.S. Air Force, 1963; dental officer U.S. Army, various locations, 1941-45, UNRRA, China, 1946-47, U.S. Air Force, 1947-63; ret., 1963; prof. U. P.R. Dental Sch., San Juan, 1963-75, U. Tex. Dental Sch., San Antonio, 1975—. Author, editor: Enviromental Protection in the Dental Operatory, 1978, Primary Preventive Dentistry, 1982. Decorated Meritorious Service medal U.S. Air Force, 1963; decorated numerous others. Fellow Am. Coll. Dentists; mem. Am. Dental assn., Internat. Assn. for Dental Research, Am. Chem. Soc., Am. Assn. Pub. Health Dentists. Subspecialties: Preventive dentistry; Cariology. Current work: Dental health delivery systems; primary preventive dentistry. Research in early detection of caries and remineralization. Home: Box 1622 Route 1 Boerne TX 78006 Office: U Tex Dental Sch 7703 Floyd Curl Dr San Antonio TX 78284

HARRIS, RALPH R., animal nutritionist, educator; b. Winfield, Ala., Mar. 9, 1929; s. David Ralph and Mary Louise (Kelley) H.; m. Eleanor A. Armbrester, July 6, 1951; children: Mary E., Ralph R., Joel C. B.S. in Agr, Auburn U., 1951, M.S., 1952; Ph.D. in Dairy Nutrition, Tex. A&M U., 1959. Asst. prof. Tex. A&M U., College Station, 1959-60; asso. prof. animal and dairy scis. Auburn (Ala.) U., 1960-68, prof., 1968—. Served to lt. col. USAF, 1951-54. Mem. Am. Soc. Animal Sci., Gamma Sigma Delta, Alpha Zeta. Methodist. Subspecialty: Animal nutrition. Current work: Forage utilization; systems of production; nutrition research. Home: 631 Delwood Dr Auburn AL 36830 Office: Dept Animal Sci Auburn U Auburn AL 36830

HARRIS, ROBERT ALLISON, biochemistry educator; b. Boone, Iowa, Nov. 10, 1939; s. Arnold E. and Marie A. (Wilcox) H.; m. Karen K. Dutton, Dec. 27, 1960; children: Kelly, Chris, Heidi, Shawn. B.S., Iowa State U., 1962; M.S., Purdue U., 1964, Ph.D., 1965. Postdoctoral fellow U. Wis.-Madison, 1966-69, asst. research prof., 1969-70; assoc. prof. biochemistry Ind. U.-Indpls., 1970-74, prof., 1974—, assoc. dir, 1982—. Contbr. chpts., articles to prfl. publs.; editor: (with N.W. Cornell) Isolation, Characterization and Use of Hepatocytes, 1983. Recipient Carrie Wolf award Ind. Heart Assn., 1976, Research award Ind. Diabetes Assn., 1976, Teaching award AMOCO, 1981; established investigator Am. Heart Assn., 1969-74. Mem. Biochem. Soc. Am., Soc. Biol. Chemists, Am. Oil Chemists, Am. Heart Assn., Am. Diabetes Assn. Subspecialty: Biochemistry (medicine). Current work: Hormonal regulation of metabolic processes; control of enzymes by covalent modification. Office: Dept Biochemistry Ind U Sch Medicine 635 Barnhill Dr Indianapolis IN 46223

HARRIS, ROY M(ARTIN), animal geneticist and physiologist, educator, researcher, cons.; b. Ogden, Utah, July 7, 1927; s. LeRoy and Cornelia (Sanders) H.; m. Mary E.; children: Mark (adopted), Chad. B.S., Utah State Agrl. Coll., 1952; M.S., Utah State U., 1954, Ph.D., 1970. Researcher Utah State U., 1952-54; asst. prof. animal genetics and reproductive physiology Calif. Poly. State U., 1954-62, asso. prof., 1962-68, prof., 1968—; cons. research div Eli Lilly, 1979-81, Govt. Mexico, 1980—. Author: The Occurrence of Estrus in the Domestic Ewe, 1974; Contbr. articles to profl. jours. Served with U.S. Army, 1944-46. Eli Lilly grantee, 1979-81; Syntex grantee, 1978-79; CARE grantee, 1977-78. Mem. Am. Soc. Animal Sci., Am. Registry Cert. Animal Scientists. Subspecialties: Genetics and genetic engineering (agriculture); Animal physiology. Current work: Chemotherapy of bio-rhythms in cattle, sheep and horses; genetic studies of horse motion as it relates to speed, development coring mechanism for muscle tissue. Home: 934 Longhorn Ln Arroyo Grande CA 93420 Office: Dept Animal/Vet Sci Calif Poly State U San Luis Obispo CA 93407

HARRIS, SIGMUND PAUL, physics educator, space sciences consultant; b. Buffalo, Oct. 12, 1921; s. Nathan N. and Ida H.; m. Florence Katcoff, Sept. 19, 1948; 1 dau., Roslyn. B.A., SUNY-Buffalo, 1941, M.A., 1943; Ph.D., Ill. Inst. Tech., 1954. Jr. scientist Los Alamos Sci. Lab., 1944-46; assoc. physicist Argonne (Ill.) Nat. Lab., 1946-53; sr. physicist Tracer Lab., Inc., Boston, 1954-56; sr. research engr. Atomics Internat., Canoga Park, Calif., 1956-64; head physics sect. Maremont Corp. Research Lab., Pasadena, Calif., 1964-66; faculty Los Angeles Pierce Coll., Woodland Hills, Calif., 1966—, prof. physics, 1979—; cons. Space Scis., Inc., Monrovia, Calif., 1978—. Author: Introduction to Air Pollution, 1973 2d edit., 1983. Mem. Am. Phys. Soc, Am. Nuclear Soc., Am. Assn. Physics Tchrs., Phi Beta Kappa, Sigma Xi. Subspecialties: Nuclear physics; Environmental engineering. Current work: nuclear science, air pollution, mass spectrometer measurements. Patentee method for measuring power level of a nuclear reactor. Office: Dept Physics Los Angeles Pierce Coll 6201 Winnetka Ave Woodland Hills CA 91371

HARRIS, STEPHEN ERNEST, elec. engr., physicist; b. N.Y.C., Nov. 29, 1936; s. Henry and Anne H.; m. Frances Joan Greene, June 7, 1959; children—Hilary, Craig. B.S.E.E., Rensselaer Poly. Inst., 1959; M.S.E.E., Stanford U., 1961, Ph.D. in Elec. Engring, 1963. Mem. tech. staff Bell Telephone Labs., Murray Hill, N.J., 1959-60; asst. prof. elec. engring. Stanford U., 1963-67, asso. prof. elec. engring., 1967-71, prof. elec. engring. and applied physics, 1971—; founder Chromatix, Inc., 1968. Recipient Alfred Noble prize, 1965; Curtis W. McGraw Research award Am. Soc. Engring. Edn., 1973; Guggenheim fellow, 1976. Fellow Am. Phys. Soc., IEEE (David Sarnoff award 1978), Optical Soc. Am.; mem. Nat. Acad. Engring., Nat. Acad. Scis. Subspecialty: Electrical engineering. Home: 880 Richardson Ct Palo Alto CA 94303 Office: Edward L Ginzton Lab Stanford U Stanford CA 94305

HARRIS, SUZANNE STRAIGHT, nutritional biochemist, educator; b. Miami, Nov. 18, 1944; d. William Marcellus and Sara Louise (Buford) S.; m. William David Harris, Apr. 6, 1974; children: William David, Jonathan Lindley. B.A., Vanderbilt U., 1966; Ph.D., U. Ala.-Birmingham, 1976. Assoc. biochemist So. Research Inst., Birmingham, 1966-74; postdoctoral fellow U. Ala., Birmingham, 1977; investigator Inst. Dental Research, 1978—, instr. dept. nutrition scis., 1978-80, asst. prof. dept. nutrition scis. and comparative medicine, 1980—. Mem. Birmingham Beautification Bd., 1972-74; chmn. Jefferson County Republican Party, Birmingham, 1975—; mem. Ala. Rep. Exec. Com., 1974—. Mem. Internat. Assn. Dental Research, Sigma Xi. Republican. Episcopalian. Subspecialties: Nutrition (medicine); Dental growth and development. Current work: Vitamin A and mineralized tissues, Vitamin A and salivary gland cancer. Office: Inst Dental Research Sch Dentistry U Ala Birmingham AL 35294

HARRIS, WESLEY LEROY, aeronautics educator, consultant; b. Richmond, Va., Oct. 29, 1941; s. William M. and Rosa P. (Minor) H.; m. Myrtle A. Satterwhite, June 14, 1960; children: Wesley, Zelda, Kamau, Kalomo. B.Aero.Engring., U. Va., 1964; M.A., Princeton U., 1965, Ph.D., 1968. Asst. prof. U. Va., Charlottesville, 1968-70; assoc. prof. So. U., Baton Rouge, 1970-71, U. Va., 1971-72; prof. aeronautics and astronautics MIT, Cambridge, Mass., 1972—; mem. U.S. Army Sci. Bd., Washington, 1979—; bd. dirs. Nat. Tech. Assocs., Inc., Washington, 1978-82. Editor: Jour. Nat. Tech. Assocs, 1978—. Recipient M.L. King Jr. achievement award, 1978; Black Achiever award YMCA, 1979; Sizer award MIT, 1979. Mem. Am. Helicopter Soc., AIAA, Nat. Tech. Assn., N.Y. Acad. Scis., Sigma Xi, Tau Beta Pi, Phi Beta Sigma. Subspecialties: Aeronautical engineering; Acoustical engineering. Current work: Theoretical transonics: analytical and exptl. aeroacoustics. Home: 19 Rangeley Rd Newton MA 02165 Office: MIT Room 37-435 77 Massachusetts Ave Cambridge MA 02139

HARRIS, WILLIAM FRANKLIN, III, science administrator; b. Jacksonville, Fla., Sept. 27, 1942; s. William Franklin and Elizabeth Putman (Guy) H.; m. Marilyn Wholleber, June 11, 1966; children: James Guy, Steven Franklin. B.A., Wabash Coll., 1964; M.S., U. Tenn., 1966, Ph.D., 1970. Research ecologist Oak Ridge Nat. Lab., 1970-73, research group leader, 1973-76, sect. head terrestrial ecology div. environ. scis., 1976-80; program dir. ecosystems studies NSF, Washington, 1980-81, Dep, dir. div. biotic systems and resources, 1981—; adj. asst. prof. U. Tenn., 1977-80. Contbr. articles to profl. jours. Oak Ridge Associated Univs. fellow, 1968-70; NDEA fellow, 1965-68; recipient Norman Treves award Wabash Coll., 1963. Mem. AAAS, Am. Inst. Biol. Scis., Ecol. Soc. Am., Soil Sci. Soc. Am., Sigma Xi. Subspecialties: Ecosystems analysis; Ecology. Current work: Grant administration, science policy; research interests include forest ecology, especially primary productivity and nutrient element cycling. Office: 1800 G St NW Room 1140 Washington DC 20550

HARRIS, WILLIAM JAMES, JR., research administrator; b. South Bend, Ind., June 17, 1918; s. William James and Elizabeth M. (Scott) H.; m. Ruth Laubinger, Aug. 26, 1944 (dec. 1977); children: June Elizabeth Sherren, William James III, Debbie Shafer Hayden, Britta Shafer Kreuger, Barkley Shafer.; m. Elizabeth Dotten Shafer, June 24, 1978. B.S. in Chem. Engring, M.S. in Engring, Purdue U., 1940; D.Engring. (hon.), Purdue U., 1978; Sc.D., M.I.T., 1948. Head ferrous alloys br. metallurgy div. Naval Research Lab., 1947-51; exec. sec. materials adv. bd. Nat. Acad. Sci.-NRC, 1951-54, exec. dir., 1957-60, asst. sec., planning div. engring., 1960-62; asst. to dir. Battelle Meml. Inst., 1954-57, asst. to v.p., 1962-67; asst. dir. tech. Columbus Labs., 1967-69; v.p. research and test dept. Assn. Am. Railroads, 1970—; Pres., chmn. bd. Piscataway Co., Accokeek, Md., 1958-63; mem. Nat. Exec. Res. Dept. Transp., 1983—. Editor: (with others) Perspectives in Materials Research, 1963; Contbr. articles to tech. publs. Mem. nat. materials adv. bd. Nat. Acad. Sci., 1967—, chmn., 1969-70; sec. Pres.'s Com. on Hwy. Safety, 1969; mem. high speed ground transp. adv. com. U.S. Dept. Transp., 1972-74, Md. Gov.'s Sci. Adv. Com., 1972-76, Md. Gov.'s Energy Council, 1974-76; Pres. Moyoane Assn., 1951-53, 58; pres., chmn. bd. Alice Ferguson Found., 1966-68. Served to lt. comdr. USNR, 1941-45. Decorated Naval Letter of Commendation.; Recipient Distinguished Alumnus award Purdue U., 1965; Disting. Service award Transp. Research Bd.-NRC, 1977; named Railroad Man of Year, 1976. Fellow Am. Soc. Metals, Metall. Soc. (pres. 1970); mem. Am. Inst. Mining, Metall. and Petroleum Engrs. (dir. 1964-69, v.p. 1964—, chmn. inst. metals div. 1960, Mathewson medal 1950), Nat. Acad. Engring., Engrs. Joint Council (bd. dirs. 1965—, pres. 1968-70), Engring. Found. (chmn. research conf. com. 1964-67, bd. dir. 1968-70), Am. Ordnance Assn. (chmn. materials div. 1966-68), ASME (mem. transp. dir. 1973—), Nat. Security Indsl. Assn. (chmn. exec. planning com. 1965-67, chmn. research and devel. adv. com. 1967-69), Sigma Xi, Alpha Sigma Mu, Tau Beta Pi, Phi Lambda Upsilon, Sigma Delta Chi. Subspecialties: Materials; Metallurgy. Current work: Track-train dynamics;improvements in design; specification and use of the elements composing tracks;bridges; rolling stock, and shipping containers, application of new development in technology to railway uses. Home: 1200 N Nash St Apt 1140 Arlington VA 22209 Office: 1920 L St NW Washington DC 20036

HARRIS, ZELLIG SABBETTAI, mathematical linguist, researcher; b. Balta, Russia, Oct. 23, 1909; m. Hyman and Rachel H. B.A., U. Pa., 1930, Ph.D., 1934. Instr. U. Pa., Phila., 1931-37, asst. prof., 1937-43, assoc. prof., 1943-47, prof., 1947-80, Benjamin Franklin U. prof., 1966—, prof. emeritus, 1980—; sr. research scientist Center for Social Scis., Columbia U., N.Y.C., 1980—. Author: Structural Linguistics, 1951, String Analysis of Sentence Structure, 1962, Discourse Analysis Reprints, 1963, Mathematical Structures of Language, 1968, Papers in Structural and Transformational Linguistics, 1970, Notes du cours de syntaxe, 1978, Papers on Syntax, 1981, Grammar of English on Mathematical Principles, 1982. Mem. Am. Philos. Soc., Nat. Acad. Scis. (applied phys. and math. scis. sect.), Am. Acad. Arts and Scis. Subspecialty: Information representation. Current work: Mathematical theory of language, formulas of science information derived directly from science reports. Office: Center Social Scis Columbia U New York NY 10027

HARRISON, DONALD CAREY, cardiologist; b. Blount County, Ala., Feb. 24, 1934; s. Walter Carey and Sovola (Thompson) H.; m. Laura Jane McAnnally, July 24, 1955; children—Douglas, Elizabeth, Donna Marie. B.S. in Chemistry, Birmingham So. Coll., 1954; M.D., U. Ala., 1958. Diplomate: Am. Bd. Internal Medicine (cardiovascular disease). Intern, asst. resident Peter Bent Brigham Hosp., 1958-60; fellow in cardiology Harvard U., 1961, NIH, 1961-63; mem. faculty Stanford U. Med. Sch., 1963—, chief div. cardiology, 1967—, prof. medicine, 1971—; chief cardiology Stanford U. Hosp., 1967—, William G. Irwin prof. cardiology, 1972—; cons. to local hosps., industry, govt. Editorial bd.: Am. Jour. Cardiology, 1970-78, Chest, Heart and Lung Jour, 1973-78, Drugs; Practical Cardiology, Clin. Cardiology; Contbr. articles to med. jours., chpts. to books. Served with USPHS, 1961-63. Fellow Interam. Soc. Cardiology (v.p. 1980—), Am. Coll. Cardiology (membership chmn., v.p. 1972-73, sec. 1969-70, trustee 1972-78), Am. Heart Assn. (fellow council circulation, clin. cardiology and basic sci., chmn. program com. 1972-76, chmn. publs. com. 1976-80, pres.-elect 1980-81, pres. 1981—); mem. Am. Soc. Clin. Investigation, Am. Fedn. Clin. Research, Am. Soc. Pharmacology and Exptl. Therapeutics, Am. Assn. Physicians, Am. Physiol. Soc., Calif. Acad. Medicine, A.C.P., Assn. U. Cardiologists. Subspecialty: Cardiology. Current work: Design of computer based arrhyghmia program; pharmacology of antiarrhythmic drugs. Home: 151 Mountain View Ave Los Altos CA 94022 Office: Room C-248 Stanford Univ Med Sch Stanford CA 94305

HARRISON, GUNYON M., pediatrics and rehabilitation medicine educator; b. Fredericksburg, Va., Mar. 6, 1921; m. (div.); 2 children. B.S., Va. Mil. Inst., 1943; M.D., U. Va., 1946. Diplomate: Am. Bd. Pediatrics, 1956. Intern St. Joseph's Hosp., Balt., 1946-47; pediatric intern Duke Univ. Hosp., Durham, N.C., 1951-52, resident, 1952-53, fellow, 1953; resident Jefferson Davis Hosp., Houston, 1953-54, Baylor U. Coll. Medicine and Tex. Children's Hosp., 1954-55; instr. pediatrics Baylor U. Coll. Medicine, 1955-57, asst. prof. pediatrics and rehab., 1957-64, assoc. prof., 1964-77, prof. pediatrics and rehab., 1977—; fellow Polio Respiratory Ctr. Nat. Found. Infantile Paralysis, Houston, 1955-56; dir. cystic fibrosis div. Tex. Children's Hosp., 1955-61, chief respiratory therapy, 1968—; also mem. active staff; dir. Cystic Fibrosis and Related Pulmonary Disease Ctr. Baylor U. Coll. Medicine (Inst. Rehab. and Research and Tex. Children's Hosp.), 1961—; mem. staff Ben Taub Gen. Hosp., Methodist Hosp., St. Luke's Hosp. Mem. Am. Acad. Pediatrics, Am. Thoracic Soc., Am. Assn. Respiratory Therapy, Assn. Advancement of Med. Instrumentation. Subspecialties: Pediatrics; Pulmonary medicine. Office: Inst. Rehab and Research 1333 Moursund Houston TX 77030

HARRISON, JOHN CHRISTOPHER, geophysics educator; b. County Durham, Eng., May 20, 1929; came to U.S., 1953, naturalized, 1966; s. Charles Frederick Reed and Anne Lucy (Middleton) H.; m. Elaine Cassels Paul, Aug. 12, 1960; children: Kirsteen Elizabeth, Fiona Anne, Keith Thomas. B.A., Cambridge U., 1950, M.A., 1953, Ph.D., 1953. Asst. research geophysicist Inst. Geophysics, UCLA, 1953-55, assoc. research geophysicist, 1957-61; mem. tech. staff Hughes Research Labs., Malibu, Calif., 1961-65; assoc. prof. dept. geol. scis. U. Colo., Boulder, 1965-70, prof., 1970—; assoc. dir. Coop. Inst. for Research in Environ. Scis., U. Colo./NOAA, 1968—; mem. com. on geodesy Nat. Acad. Scis., 1978—. Contbr. articles to profl. jours. Served to lt. Royal Naval Vol. Res., 1955-57. Recipient U.S. Sr.

Scientist award Alexander von Humboldt Found., 1975; Japanese Soc. for Promotion of Sci. fellow, 1982. Fellow Royal Astron. Soc.; mem. Am. Geophys. Union, Soc. Exploration Geophysicists. Subspecialty: Geophysics. Current work: Earth tilt, earth tides, gravity instrumentation, modern geodesy, geodynamics. Office: Cooperative Institute for Research in Environmental Sciences Box 449 University of Colorado Boulder CO 80309

HARRISON, JOHN HENRY, IV, chemistry and biology educator, research scientist; b. Pitts., Aug. 8, 1936; s. John Henry, III and Jeanne (Leach) H.; m. Judith Henrietta Cline, Dec. 27, 1962; children: John Henry, Robert Nathaniel. B.S., U. Tex.-Austin, 1958, Ph.D., 1964. Postdoctoral research fellow Harvard Med. Sch., Boston, 1964-67; asst. prof. chemistry U. N.C., Chapel Hill, 1967-72, assoc. prof. chemistry, 1972-76, prof. chemistry and biology, 1976—; vis. scientist dept. biochemistry Oxford (Eng.) U., 1978-79; vis. prof. chemistry U. P.R., Mayaguez, 1982. Contbr. writings in field to sci. publs. Served to capt. USMC, 1958-60. Recipient Career Devel. Award U.S. NIH, 1974-79; Jane Coffin Childs fellow Jane Coffin Childs Meml. Fund for Med. Research, 1965-67. Mem. Am. Soc. Biol. Chemists, N.Y. Acad. Scis., Am. Chem. Soc., Sigma Xi, Alpha Chi Sigma. Republican. Subspecialties: Biochemistry (medicine); Biophysical chemistry. Current work: Physicochemical investigation of mechanism of action of pyridine nucleotide dependent dehydrogenase enzymes. Home: 806 Kenmore Rd Chapel Hill NC 27514 Office: Univ NC Dept Chemistry Venable Hall Chapel Hill NC 27514

HARRISON, LESLIE SHERMAN, energy engineer, general contractor, consultant; b. Tilghman, Md., Jan. 23, 1925; s. Leslie Dobson and Jenny Ruth (Butler) H.; m. Julia Barbara Harrison, Aug. 29, 1947; children: Linda Leslie, Larry Kirk, Edward Allen. Student, U. Wis., 1942-43, Capitol Radio Engring. Inst., 1950, Indsl. Coll. Armed Forces, 1968-69. Lic. amateur radio operator FCC, 1956; comml. flight instr. FAA, 1982. Gen. contractor, 1950-52; communications elec. engr., also factory mgr., warehouse ops. mgr., program mgr. Dept. of Defense, 1952-78; pres. Energy Masters of Md., Inc., Beltsville, 1977—. Active Prince George's County C. of C., Bd. Trade; past treas. Civic Assn. Served with USNR, 1942-50; PTO. Recipient Civilian Meritorious Service award Dept. Def., 1978; Spl. Service award C. of C., 1980. Mem. Aircraft Owners and Pilots Assn., Assn. Energy Engrs., Air Safety Found., Silver Wings, Nat. Assn. Flight Instr. Methodist. Clubs: VFW, Am. Legion, Potomac Antique Aero Squadron, Antique Airplane Assn. Current work: Investigative, testing and marketing. Home: 6617 Bowie Dr Springfield VA 22150 Office: 12021 Old Gunpowder Rd Beltsville MD 20705

HARRISON, MARTIN BERNARD, plant pathology educator; b. N.Y.C., Dec. 8, 1924; s. Harold and Belle (Drucker) H.; m.; children: Amy R., Juliet R. B.S., Cornell U., 1950; Ph.D., 1955; M.S., Kans. State U., 1951. Mem. faculty Cornell U., 1955—, assoc. prof. plant pathology, 1960—. Contbr. articles to profl. jours. Served with AUS, 1943-46. Mem. Am. Phytopathol. Soc., Soc. Nematologists, European Soc. Nematologists. Subspecialty: Plant pathology. Current work: Plant parasitc nematode biology and control. Office: Plant Pathology Cornell U Ithaca NY 14852 Home: 516 Warren Rd Ithaca NY 14850

HARRISON, MICHAEL ALEXANDER, computer scientist, educator; b. Phila., Apr. 11, 1936; s. Milton and Mamie May (Gross) H.; m. Evalee Solomon, Aug. 23, 1959; m. Susan Graham, Nov. 16, 1971; 1 son, Craig. B.S., Case-Western Res. U., 1958, M.S., 1959; Ph.D., U. Mich., 1963. Asst. Prof. U. Calif.-Berkeley, 1963-66, assoc. prof., 1966-71, prof., 1971—; vis. prof. MIT, Cambridge, 1969-70, Hebrew U., 1970, Stanford U., 1981; cons. in field; mem. Computer Sci. and Tech. bd. Nat. Acad. Sci. Editor, contbr. articles to profl. jours. Guggenheim fellow, 1969-70; NSF grantee, 1965—; recipient U. Helsinki Univ. medal, 1980. Mem. Assn. for Computing Machinery (v.p.), IEEE Computer Soc., AAAS, European Assn. Theoretical Computer Sci. Subspecialties: Software engineering; Theoretical computer science. Current work: Security of computer systems; models of protection in operating systems; office automation theoretical computer science. Office: University of California Computer Science Division Berkeley CA 94720

HARRISON, ROBERT WALKER, III, medical educator; b. Natchez, Miss., Oct. 13, 1941; s. Robert Walker and Charlotte (Mackel) H.; , June 1, 1963; children: Robert Walker, William Seth. B.S., Tougaloo So. Christian Coll., 1961; M.D., Northwestern U., 1966. Diplomate: Am. Bd. Internal Medicine. Fellow in endocrinology Vanderbilt U. Sch. Medicine, Nashville, 1972-74, asst. prof. medicine, 1974-81, assoc. prof., 1981—; investigator Howard Hughes Med. Inst., Nashville, 1975-82. Served to lt. comdr. USN, 1970-72. Josiah Macy Found. fellow, 1973. Mem. ACP, Endocrine Soc., Am. Soc. Cell Biologists, Am. Soc. Biol. Chemists. Subspecialties: Endocrinology; Receptors. Current work: Biochemical mechanisms of steroid hormone action. Home: 2817 White Oak Dr Nashville TN 37215 Office: Vanderbilt U Sch Medicine 21st and Garland Aves Nashville TN 37232

HARRISON, STEADMAN DARNELL, JR., toxicologist, cons.; b. New Albany, Miss., Apr. 13, 1947; s. Steadman Darnell and Peggy Anne (Caldwell) H.; m. Anita Page, June 6, 1969; 1 son, Steadman Darnell III. B.S., Miss. State U., 1969; M.S., Ind. U., 1972, Ph.D. (NIH predoctoral fellow), 1973. Diplomate: Am. Bd. Toxicology. Research pharmacologist to head pharmacology sect. So. Research Inst., Birmingham, Ala., 1973-80; adj. prof. optometry Ind. U., 1978—; assoc. prof. toxicology U. Ky., 1980—; researcher, cons. Contbr. articles and abstracts to profl. jours. Mem. Soc. Toxicology, Am. Assn. Cancer Research, Royal Soc. Chemistry, Nat. Acad. Clin. Biochemistry, Am. Assn. Clin. Chemistry, Soc. Risk Analysis, Sigma Xi. Baptist. Subspecialties: Toxicology (medicine); Pharmacology. Current work: Mechanisms of toxicity, vitamin A analogs, arachidonate metabolism, metals and alkylating agts., calcium. Patentee in field. Office: U Ky 202 Medical Center Annex 5 Lexington KY 40536 0078

HARRISON, WILLARD WAYNE, analytical chemistry educator; b. McLeansboro, Ill., July 28, 1937; m., 1959; 4 children. B.A., So. Ill. U., 1958, M.A., 1960; Ph.D. in Atomic Absorption, U. Ill., 1964. Asst. prof. chemistry U. Va., Charlottesville, 1964-69, assoc. prof., 1969-74, prof. analytical chemistry, 1974—, chmn. dept. chemistry, 1978—. Mem. Am. Chem. Soc. Subspecialty: Analytical chemistry. Office: Dept Chemistry U Va Charlottesville VA 22901

HARRISON, WILLIAM DOUGLAS, physicist; b. Saint John, N.B., Can., Oct. 23, 1936; s. Arthur T.G. and Thelma (Spence) H.; m. Anne K. Sloss; 2 children. Ph.D., Calif. Inst. Tech. Mem. faculty U. Alaska, Fairbanks, 1972-83, prof. physics, 1982—; research in glacier and permafrost. Contbr. articles in field to profl. jours. Mem. Am. Geophys. Union, Am. Phys. Soc., Internat. Glaciological Soc. Subspecialties: Geophysics; Hydrology. Current work: Glacier dynamics and hydrology; heat and mass transport processes in subsea permafrost. Office: Geophysical Institut University of Alaska Fairbanks AK 99701

HARRISON, YVONNE ELOIS, pharmacologist; b. Norfolk, Va., Apr. 29, 1939; d. Herman H. and Georgia M. (Hall) H.; m. Melvin C. Johnson, Sept. 27, 1975. B.S., Howard U., 1959, M.S., 1970, Ph.D., 1972. Research asst. Burroughs Wellcome & Co., 1964-69; research assoc. dept. pharmacology Howard U., 1970-72; biol. research coordinator Hoffmann-La Roche, Inc., 1972-73, asst. dir. dept. exptl. therapeutics, 1974-79, asst. dir. pharm. research and devel., 1980-82, dir. pharm. research coordination, 1983—; Bd. dirs. Consumer Health Info. and Resource Center. Named Black Achiever in Industry YMCA, 1974; recipient Twin Tribute to Women in Industry YMCA, 1975; recipient Disting. Corporate Alumni award Nat. Assn. Equal Opportunity in Higher Edn., 1982. Mem. AAAS, Am. Pharm. Assn. Acad. Pharm. Scis., Am. Physiol. Soc., Am. Soc. Pharmacology and Exptl. Therapeutics, Assn. Women in Sci., Fedn. Am. Scientists, Fedn. Am. Socs. Exptl. Biology, Internat. Soc. Ecotoxicity and Environ. Safety, N.Y. Acad. Scis., Sigma Xi. Subspecialties: Pharmacology; Biochemistry (medicine). Current work: Evaluation of research programs and priorities for a major division of 600 people in both preclinical and clinical areas. Office: 340 Kingsland St Nutley NJ 07110

HARRIS-WARRICK, RONALD MORGAN, neurobiologist, educator; b. Berkeley, Calif., July 28, 1949; s. Morgan and Marjorie Ruth (Mason) Harris-W.; m. Rebecca Lamar, Apr. 5, 1975; children: Sheridan, Thomas. B.A., Stanford U., 1970, Ph.D. in Genetics, 1976. NIH fellow dept. neurobiology Stanford U., 1976-78; Muscular Dystrophy Assn. fellow dept. neurobiology Harvard U., 1978-80; asst. prof. sect. neurobiology and behavior Cornell U., 1980—; mem. faculty Marine Biol. Lab., Woods Hole, Mass., 1982. Contbr. articles to profl. jours. NIH grantee, 1981—; Muscular Dystrophy Assn. grantee, 1981—. Mem. Soc. Neurosci., AAAS, Sierra Club, Audubon Soc., Phi Beta Kappa. Subspecialties: Neurophysiology; Neurochemistry. Current work: Mechanism of action of neuromodulators; modulation of motor activity by biogenic amines. Office: Section of Neurobiology and Behavior Cornell University Ithaca NY 14850

HARROLD, ROBERT LEE, animal nutrition educator; b. New Castle, Ind., Oct. 5, 1940; s. James Richard and Virginia Joyce (Draper) H.; m. Marilyn Jean Garrigus, Sept. 3, 1965; children: Patrick Lee, Ericka Kathleen, Bonnie Colleen. B.S., Purdue U., 1962, M.S., 1964, Ph.D., 1967. Postdoctoral fellow Ohio Agrl. Research and Devel. Ctr., Wooster, 1967-68; asst. prof. dept. animal sci. N.D. State U., Fargo, 1968-74, assoc. prof., 1974-80, prof., 1980—. Contbr. articles to profl. jours. N.D. Pork Producer's Council grantee, 1980. Mem. Am. Soc. Animal Sci., Council Agrl. Sci. and Tech., Animal Nutrition Research Council, Am. Assn. Lab. Animal Sci., Sigma Xi. Subspecialties: Animal nutrition; Food science and technology. Current work: Bioavailable energy content of foodstuffs and feedstuffs; protein quality including methodology comparison and phosphorus bioavailability. Home: 801 Elm St West Fargo ND 58078 Office: Dept Animal Sci ND State U Fargo ND 58105

HARSHA, PHILIP THOMAS, research company executive, combustion researcher; b. N.Y.C., Feb. 22, 1942; s. Palmer and Catherine (Redinger) H.; m. Jean Ann Quinn, Oct. 23, 1965; children: Peter Charles, Evan Michael. B.S. in Engring. Sci, SUNY-Stony Brook, 1962, M.S., 1964; Ph.D. in Aero. Engring, U. Tenn., 1970. Combustion research engr. Gen. Electric Co., Cin., 1964-67; research engr. ARO, Inc., Arnold AFB, Tenn., 1969-74; researcher R & D Assocs., Marina del Rey, Calif., 1974-76; staff scientist Sci. Applications Inc., Chatsworth, Calif., 1976-81, div. mgr., 1981—. Mem. AIAA (recipient H.H. Arnold award Tenn. sect. 1972), ASME, N.Y. Acad Sci., Sigma Xi. Republican. Methodist. Subspecialties: Combustion processes; Fluid mechanics. Current work: Turbulent combustion analysis and experimental research; gas-phase and two-phase combustion processes; turbulent flow modeling. Home: 7235 Cirrus Way Canoga Park CA 91307 Office: Science Applications Inc 9760 Owensmouth Ave Chatsworth CA 91311

HARSHBARGER, JOHN CARL, JR., pathobiologist; b. Weyers Cave, Va., May 9, 1936; s. John Carl and Myn Alma (Baker) H. B.A., Bridgewater Coll., 1957; M.S., Va. Poly. Inst., 1959; Ph.D., Rutgers U., 1962. NSF postdoctoral research assoc. Insect Pathology Lab., U.S. Dept. Agr., Beltsville, Md., 1962-64; asst. research pathobiologist U. Calif., Irvine, 1964-67; dir. registry of tumors in lower animals Nat. Mus. Natural History, Smithsonian Instn., Washington, 1967—. Editor: (with others) Neoplasms and Related Disorders of Invertebrate and Lower Vertebrate Animals, 1969, Aquatic Pollutants and Biologic Effects with Emphasis on Neoplasia, 1977, Phyletic Approaches to Cancer, 1981; contbr.: articles to profl. jours. Phyletic Approaches to Cancer. Mem. AAAS, Am. Assn. for Cancer Research, Southeastern Cancer Research Assn., Interagy. Collaborative Group on Environ. Carcinogenesis, Internat. Assn. for Comparative Research on Leukemia and Related Diseases (world com. 1979—), N.Y. Acad. Scis., Soc. for Invertebrate Pathology (sec. 1974-76), Sigma Xi (sec. D.C. chpt. 1976-78, v.p. 1978-80, pres. 1980-82). Club: Cosmos (Washington). Subspecialties: Pathobiology; Oncology. Current work: Pathology of neoplasms in ectothermic vertebrate and invertebrate animals. Home: 2038 Columbia Pike Apt 3 Arlington VA 22204 Office: Nat Mus Natural History Smithsonian Inst Room W216A Washington DC 20560

HART, DAVID ANDERSON, mechanical design engineer; b. Bklyn., Dec. 20, 1918; s. David Anderson and Ethel Adeline (Stilson) H. B.S. in Mech. Engring, Ga. Inst. Tech., 1940, M.S., 1943. Registered profl. engr., Pa. Mech. design engr. Central Sta. Turbine, Westinghouse Electric Corp., Lester, Pa., 1942-54, marine turbine and gear mech. design engr., 1954-65, rotor creep design engr., 1965-80; solar energy design cons. R.B. Luckenbach Co., Glen Mills, Pa., 1980—. Mem. ASME, Pi Tau Sigma. Subspecialties: Mechanical engineering; Fluid mechanics. Current work: Super efficient condensing heat transfer equipment design. Patentee in sealing arrangement for a rotor shaft.

HART, ELWOOD ROY, entomology educator; b. Sioux City, Iowa, Mar. 6, 1938; s. Roy Charles and Ida Caroline (Cox) H.; m. Nancy Louise Fues, June 2, 1979; 1 son by previous marriage: Curtis Brian. B.A., Cornell U., 1959; M.Ed., Tex. A&M U., 1965, Ph.D., 1972. Postdoctoral fellow Tex. A&M U., College Station, 1972-74; asst. prof. dept. entomology Iowa State U., Ames, 1974-78, assoc. prof., 1978—. Group facilitator Youth & Shelter Services, Substance Abuse Program, Ames, 1980—. Mem. Entomol. Soc. Am., Sigma Xi, Xi Sigma Pi, Gamma Sigma Delta. Democrat. Mem. Christian Ch. Lodge: Masons. Subspecialties: Integrated pest management; Ecosystems analysis. Current work: Urban forest entomology, host-pest interactions; system analysis, pheromone systems. Office: Iowa State U Dept Entomology 403 Science II Ames IA 50011 Home: 1122 Johnson St Ames IA 50010

HART, FRED CLINTON, environmental consultant; b. Sharon, Pa., July 5, 1940; s. Fred Clinton and Elizabeth (Innis) H.; m. Elizabeth H. Semple, Mar. 13, 1982; 1 dau., Meredith. B.C.E., Cornell U., 1963; M.S., Stanford U., 1964; M.B.A., U. Conn., 1970. Registered profl. engr., N.Y. With Dorr-Oliver, Inc., Stamford, Conn., 1966-68, Internat. Paper Co., N.Y.C., 1969-70; commr. Dept. Air Resources, City of N.Y., 1970-74; pres. Fred C. Hart Assocs., Inc., N.Y.C., 1975—, dir., 1975—; dir. Innova, Inc., Clearwater, Fla., Innova Tech. Inc., Clearwater; lectr. toxicology Sch. Journalism, NYU. Contbr. articles to profl. jours. Served with U.S. Army, 1964-66. Decorated Bronze Star medal; named Outstanding Grad. U. Conn. Bus. Sch., 1972. Mem. Air Pollution Control Assn., ASCE, ASME. Subspecialty: Environmental engineering. Office: Fred C Hart Associates Inc 530 Fifth Ave New York NY 10036

HART, JOHN BIRDSALL, physics educator; b. Hamilton, Ohio, Aug. 24, 1924; s. John Wilson and Elizabeth (Birdsall) H.; m. Agnes Marie Roegner, June 17, 1948 (dec. Oct. 1960); 1 dau., Mary Agnes. B.S., Xavier U., Cin., 1948, M.S., 1950. Instr. physics Xavier U., 1950-56, asst. prof., 1956-62, assoc. prof., 1962-68, prof., 1968—, chmn. physics, 1958-71, 82-83; vis. prof., cons. Fla. State U., 1967-68; cons. prof. Ohio U., 1970-71, Miami U., summer 1974. Author: (with Richard Gadske) The National Curriculum in Navigation for U.S. Naval Reserve, 1957, Lectures in Atomic Physics; contbr. articles to profl. jours. Chmn. exhibits com. Cin. Ctr. Sci. and Industry, 1966-67. Served with USNR, 1943-46. Mem. Am. Assn. Physics Tchrs., Ohio Acad. Sci., Am. Phys. Soc., Space Studies Inst., Sigma Pi Sigma. Subspecialties: Operational general physics; Psychophysics. Current work: Foundations of operational general physics. Patentee. Home: 3836 Ledgewood Dr Cincinnati OH 45207 Office: Xavier University 3800 Victory Pkwy Cincinnati OH 45207

HART, JOHN HENDERSON, forestry pathology educator, cons.; b. Kansas City, Mo., June 18, 1936; s. Creighton Carlton and Virginia (Page) H.; m. Dora Lucy Barnes, Feb. 28, 1959; children: Mark, Kit. B.A., Dartmouth Coll., 1958; Ph.D., Iowa State U., 1963. Prof. forest pathology Mich. State U., East Lansing. Mem. Am. Phytopath. Soc., Soc. Am. Foresteres, Wildlife Soc., Walnut Council, Audubon Soc., Nat. Rifle Assn., Ducks Unltd., Mich. United Conservation Clubs, Mich. Wild Turkey Fedn. (v.p.). Subspecialties: Plant pathology; Resource management. Current work: Forest pathology, vertebrates as vectors of plant pathogens-damage to woody plants by vertebrates. Office: Dept Botany and Plant Pathology Mich State U East Lansing MI 48824

HART, RAYMOND KENNETH, materials consultant; b. Newcastle, N.S.W., Australia, Feb. 15, 1928; s. William Kenneth and Olive (Palmer) H.; m. Betty Joyce Bingemann, Sept. 5, 1952; children: Timothy Kenneth, Rowena Jane. A.S.T.C., Sydney Tech. Coll., 1949; D.I.C., Imperial Coll., London, 1952; Ph.D., U. Cambridge, Eng., 1955. Research officer Aeronautical Research Labs., Melbourne, Australia, 1955-58; research scientist Argonne Nat. Lab., Ill., 1958-66, sr. research scientist, 1967-69; prin. research scientist Ga. Inst. Tech., 1970-75; pres., dir. research Pasat Research Assocs., Inc., Atlanta, 1976—; materials cons., expert witness. Contbr. articles to profl. jours. Com. chmn. Boy Scouts Am., 1965-67, com. mem., 1971-72. Recipient NASA cert. of recognition, 1976. Fellow Inst. Physics (London), Royal Australian Chem. Inst.; mem. Am. Acad. Forensic Scis., Electron Microscopy Soc. Am., Am. Phys. Soc., Sigma Xi. Republican. Episcopalian. Subspecialties: Metallurgy; Microscopy. Current work: High resolution analytical electron microscopy, material defect and failure analysis, environ. metal toxins, forensic sci. Home and Office: 585 Royervista Dr Atlanta GA 30342

HART, RICHARD ALLAN, biology educator; b. Nora Springs, Iowa, Dec. 6, 1930; s. Oliver Rosewll and Minnie Pearl (Schilling) H.; m. Margaret Jane Lee, Aug. 31, 1958; children: Jeffrey, James, Mary-Louise. B.S. in Chemistry, U. Mo., 1958, M.S. in Entomology, 1960, Ph.D., 1964; student in acctg., Hamilton Sch. Commerce, 1949. Clk. Internat. Harvester, Mason City, Iowa, 1949-50; teller First State Bank, Nora Springs, Iowa, 1950-51; entomology asst. U. Mo. - Columbia, 1962-64; prof. biology N.W. Mo. State U., Maryville, 1964—; research entomologist U.S. Dept. Agr., Honolulu, 1966-68; cons. U.S. EPA, Denver, 1978-81. Contbr. articles to profl. jours.; author: manual Mosquito Control Projects, 1981. Campus rep. Am. Lung Assn., Maryville, 1970—. Served with USAF, 1951-55. Recipient award for outstanding performance EPA, 1980. Mem. AAAS, Entomol. Soc. Am., Am. Mosquito Control Assn., Sigma Xi. Baptist. Subspecialties: Integrated pest management. Current work: Mosquitoes of Missouri; computer evaluation of test question performance, instructional performance. Home: 114 Memory Ln Maryville MO 64468 Office: NW Mo State U Maryville MO 64468

HART, RICHARD C., physicist, sci. acad. adminstr.; b. Bklyn., Nov. 7, 1945; s. Michael A. and Muriel (Cullen) H.; m. Anna Maria Biagioni, Aug. 7, 1980; children: Jennifer Meesun, David Kyusun. B.S. in Physics, Wagner Coll., 1967; M.A., Boston U., 1969, Ph.D., 1972. Instr. Boston U., 1972-74, Bentley Coll., 1971-74; staff space sci. bd. Nat. Acad. Scis., Washington, 1974—. Mem. AAAS, Am. Astron. Soc., Astron. Soc. Pacific. Current work: Science management and consulting. Office: Space Sci B Nat Acad Sci 2101 Constitution Ave Washington DC 20418

HART, RONALD WILSON, radiobiologist, toxicologist, government research executive; b. Syracuse, N.Y., Mar. 23, 1942; s. Wilson and Annabell H.; m. Teresa Leigh Hoskins, Aug. 31, 1974. B.S., Syracuse U., 1967; M.S., U. Ill., 1969, Ph.D., 1971; postgrad. (Nat. Cancer Inst. trainee), Oak Ridge Nat. Lab., 1973. USPHS trainee, 1970-71; Asst. prof. Ohio State U., Columbus, 1971-75, dir. radiation biology research div., 1971-82, asso. prof. depts. biology, biophysics, 1975-78, asso. prof. depts. biology, biophysics dept. preventive medicine, 1976-78, asso. prof. pharmacology, medicinal chemistry dept. preventive medicine, 1977-78, dir. chem., biomed. environ. research group dept. preventive medicine, 1977-82, prof. depts. radiology, preventive medicine, pharmacology, medicinal chemistry, vet. pathobiology, 1978-82; dir. Nat. Center for Toxicological Research, Jefferson, Ark., 1980—; prof. U. Poona, India, 1978—, Cairo U., 1979—; cons. Oak Ridge Nat. Lab., 1971-75, Brookhaven Nat. Lab., 1975-78, Argonne Nat. Lab., 1975-78, EPA, 1976, 78, Am. Indsl. Health Council, 1978, PPG Industries, 1978, Informatics, 1978-80, FDA, 1980; mem. Nat. Scis./NRC Bd. on Toxicology and Environ. Health Hazards, 1976-82; mem. interagy. staff group Office Sci. and Tech. Policy Exec Office of Pres., 1982—, chmn., 1983—; bd. dirs. Ark. Sci. and Tech. Authority, 1983—. Contbr. chpts. to books, numerous articles to profl. jours. Recipient Hopkins award for grad. research, 1971, Japanese Med. Assn. award, 1978, Karl-August-Forester award, Germany, 1980, W. Ger., 1980, award of merit FDA, 1982, Sr. Exec. Service award, 1982, Superior Service award USPHS, 1983; named Syracuse U. Outstanding Alumnus, 1976. Mem. AAAS, Radiation Research Soc., Biophys. Soc., Photochem. and Photobiol. Soc., Gerontol. Soc., Am. Coll. Toxicology (pres. 1981), Sigma Xi. Subspecialties: Toxicology (medicine); Radiation biology. Office: Nat Center for Toxicological Research US Dept Health and Human Services Jefferson AR 72079

HARTER, DONALD HARRY, medical educator; b. Breslau, Germany, May 16, 1933; came to U.S., 1940, naturalized, 1945; s. Harry Morton and Leonor Evelyn (Goldmann) H.; m. Lee Grossman, Dec. 18, 1960 (div. 1976); children: Kathryne, Jennifer, Amy, David. A.B., U. Pa., 1953; M.D., Columbia U., 1957. Diplomate: Am. Bd. Psychiatry and Neurology. Intern in medicine Yale-New Haven Med. Center, 1957-58; asst. resident, then resident neurology N.Y. Neurol. Inst., 1958-61; guest investigator Rockefeller U., 1963-66; mem. faculty Columbia Coll. Physicians and Surgeons, 1960-75, prof. neurology and microbiology, 1973-75; vis. fellow Clare Hall, Cambridge, Eng., 1973-74; attending neurologist N.Y. Neurol. Inst., Presbyn. Hosp., 1973-75; Charles L. Mix prof., chmn. dept. neurology Northwestern U., 1975—; chmn. dept. neurology Northwestern Meml. Hosp., Chgo., 1975—; mem. adv. com. on fellowships Nat. Multiple Sclerosis Soc., 1976-79, chmn., 1977-79; mem. Nat. Commn. on

Venereal Disease, HEW, 1970-72; mem. med. adv. bd. Am. Parkinsons' Disease Assn., 1976—, Myasthenia Gravis Found., 1980—; mem. sci. adv. council Nat. ALS Found., 1978—; mem. neurol. disorders program project rev. A com. Nat. Inst. Neurol. and Communicative Disorders and Stroke, NIH, HHS, 1981—. Editorial bd.: Neurology, 1976-82, Anns. of Neurology, 1983—; adv. bd.: Archives of Virology, 1975-81. USPHS spl. fellow, 1963-66; Am. Cancer Soc. scholar, 1973-74; Guggenheim fellow, 1973; recipient Joseph Mather Smith prize Columbia U., 1970, Lucy G. Moses award, 1970, 72. Mem. Am. Soc. Clin. Investigation, Am. Neurol. Assn. (membership adv. com. 1980-82), Am. Assn. Neuropathologists, Soc. Exptl. Biology and Medicine, Assn. Univ. Profs. Neurology, Infectious Disease Soc. Am., Soc. Neurosci., Am. Acad. Neurology (alt. rep. to Council of Med. Splty. Socs. 1979-82), Am. Assn. Immunologists, Am. Soc. Microbiology, Soc. Gen. Microbiology, Am. Epilepsy Soc., Am. Assn. for Study Headache, Am. Assn. for History Medicine, Pan Am Med. Assn., Phi Beta Kappa, Sigma Xi. Subspecialties: Neurology; Virology (medicine). Current work: Virus-nerve cell interactions; neurobiology of viral infections; slow viral diseases of the nervous system. Home: 900 Lake Shore Dr Chicago IL 60611 Office: Dept Neurology Northwestern U Med Sch 303 E Chicago Ave Chicago IL 60611

HARTER, H(ARMAN) LEON, statistics educator and researcher; b. Keokuk, Iowa, Aug. 15, 1919; s. Harman Theodore and Mary Josie (Hough) H.; m. Alice Lauretta Madden, Oct. 23, 1943. A.B., Carthage Coll., 1940; A.M., U. Ill.-Urbana, 1941; Ph.D., Purdue U., 1949. Grad. asst. math. U. Ill.-Urbana, 1941-43; prof. physics Mo. Valley Coll., Marshall, 1943-44; instr. math. Purdue U., West Lafayette, 1946-48, research fellow, 1948-49; asst. prof. math. Mich. State U., East Lansing, 1949-52; math. statistician Aerospace Research Labs., Wright-Patterson AFB, Ohio, 1952-75, Air Force Flight Dynamics Lab., 1975-76, mathematician, 1976-78; research prof. math. and stats. Wright State U., Dayton, Ohio, 1979—; disting. vis. prof. math. stats. Air Force Inst. Tech., Wright-Patterson AFB, Ohio, 1982—; vis. lectr. stats. Inst. Math. Stats., 1970-73; cons. U. Dayton Research Inst., Aerospace Med. Research Lab., Wright-Patterson AFB, Ohio, 1979-81. Author: New Tables of the Incomplete Gamma-Function Ratio, 1964, Order Statistics and Their Use in Testing and Estimation, Vols. 1-2, 1970, A Chronological Annotated Bibliography on Order Statistics, Vols. 1-2, 1978, 83; co-editor: (with D.B. Owen) Selected Tables in Mathematical Statistics, Vols. 1-3, 1970, 74, 75. Served with USN, 1944-46. Carthage Coll. scholar U. Ill., 1940; Office Naval Research research fellow Purdue U., 1948-49. Fellow Am. Statis. Assn. (council 1963-64, 72-73), Inst. Math. Stats. (co-editor 1967-75); mem. Internat. Statis. Inst., Soc. Indsl. and Applied Math. (chpt. pres. 1959-60), Math. Assn. Am., Ops. Research Soc. Am. Democrat. Mem. Christian Ch. Lodge: Masons. Subspecialties: Statistics; Probability. Current work: Order statistics and their applications, bibliography on order statistics, least squares and alternatives, modifications of goodness-of-fit tests. Home: 32 S Wright Ave Dayton OH 45403 Office: Air Force Inst Tech Bldg 640 Wright-Patterson AFB OH 45433

HARTER, ROBERT DUANE, chemist, educator; b. Muskegon, Mich., July 6, 1936; s. Maurice D. and Rachel V. (Baker) H.; m. Nancy B. Bradshaw, Feb. 22, 1969; children: Carl W., Eric D. B.S. in Agr, Ohio State U., 1961, M.S., 1962; Ph.D., Purdue U., 1966. Research fellow Ohio State U., 1961-62; grad. teaching asst. Purdue U., West Lafayette, Ind., 1962-66; asst. soil scientist Conn. Agrl. Research Sta., New Haven, 1966-68; assoc. research scientist N.Y. U., N.Y.C., 1968-69; asst. prof. soil chemistry U. N.H., Durham, 1969-75, assoc. prof., 1975—; cons., speaker environ. problems. Contbr. articles to profl. jours. Mem. Soil Sci. Soc. Am., Am. Soc. Agronomy, Sigma Xi. Baptist. Subspecialty: Soil chemistry. Current work: Solid-solution interfacial chemistry; forest nutrient cycling; heavy metal adsorption by soils, forest litter decomposition, soil reaction kinetics, role of forest clear cutting on nutrient cycling. Office: Inst Nat Environ Resources U NH Durham NH 03824

HARTER, WILLIAM GEORGE, physicist, educator; b. Lancaster, Pa., July 18, 1943; s. John W. and Elisabet (Richards) H.; m. Margot J. Williams, June 22, 1974; 1 son, Alexander. A.B., Hiram Coll., 1964; Ph.D., U. Calif.-, Irvine, 1967. Asst. prof. U. So. Calif., 1968-73; adj. prof. Universidade de Campinas, Sao Paulo, Brazil, 1974-76; vis. fellow U. Colo., Boulder, 1976-78; asst. prof. physics Ga. Inst. Tech., Atlanta, 1979-81, assoc. prof. physics, 1982—, also dir.; cons. Los Alamos Nat. Lab. Author monograph, also articles. Grantee Research Corp., 1980, NSF, 1980-82, 82-84. Mem. Am. Phys. Soc. Subspecialties: Theoretical physics; Atomic and molecular physics. Current work: Theory of atomic and molecular laser spectroscopy. Home: 1186 Standard St Atlanta GA 30317 Office: Sch Physics Ga Inst Tech Atlanta GA 30332

HARTFIEL, DARALD JOE, mathematics educator, researcher; b. Ray Point, Tex., Oct. 10, 1939; s. Richard Gustav and Ella Laura (Drosche) H.; m. Verna Faye Melton, Aug. 20, 1962; children: Andra Christine, Simone Charmaine. B.S., S.W. Tex. State U., 1962; M.S., U. Houston, 1966, Ph.D., 1969. Asst. prof. Tex. A&M U., College Station, 1969-74, assoc. prof., 1974-81, prof. math., 1981—. Contbr. articles to profl. jours. Mem. Am. Math. Soc., Soc. Indsl. and Applied Math. Lutheran. Subspecialty: Applied mathematics. Current work: Research in matrix theory and combinatorics. Home: 1220 Haines College Station TX 77840 Office: Math Dept Texas A&M U College Station TX 77843

HARTH, ERICH, physicist; b. Vienna, Austria, Nov. 16, 1919; s. Martin Nassau and Sofie H.; m. Dorothy E. Feldmann, Feb. 4, 1951; children: Peter, Rick. Ph.D., Syracuse U., 1951. Scientist U.S. Naval Research Lab., Washington, 1951-54; research assoc. dept. physics Duke U., Durham, N.C., 1954-57; asst. prof. dept. physics Syracuse (N.Y.) U., 1957-60, assoc. prof., 1960-65, prof., 1965—. Author: Windows on the Mind: Reflections on the Physical Basis of Consciousness, 1982; contbr. articles to profl. jours. Served with U.S. Army, 1944-46. Mem. Am. Phys. Soc., Soc. Neuroscience, AAAS, AAUP, N.Y. Acad. Scis. Subspecialties: Biophysics (physics); Sensory processes. Current work: Visual Information Processing and Sensory Motor Interactions; higher information processing; dynamic properties of neural systems. Home: 4451 Lafayette Rd Jamesville NY 13078 Office: Dept Physics Syracuse U Syracuse NY 13210

HARTIG, ELMER OTTO, aerospace co. exec.; b. Evansville, Ind., Jan. 23, 1923; s. Otto E. and Frieda K. (Sunderman) H.; m. Evelyn Ann Cameron, Aug. 21, 1949; children—Pamela Ann, Jeffery C., Gregory W., Bradley A. B.S.E.E., U. N.H., 1946, M.S. in Physics, 1947; Ph.D., Harvard U., 1950. With Goodyear Aerospace Corp., Akron, Ohio, 1950—, dir. research and engring., Litchfield Park, Ariz., 1976, v.p. research and engring., Akron, 1976-81, v.p. ops., def. and energy, 1981—; Mem. U.S. Army Sci. Bd., 1979—. Fellow IEEE; mem. AIAA. Subspecialties: Aerospace engineering and technology; Electronics. Office: 1210 Massillon Rd Akron OH 44315

HARTL, DANIEL LEE, geneticist, educator; b. Marshfield, Wis., Jan. 1, 1943; s. James W. and Catherine E. (Stieber) H.; m. Carolyn Teske, Sept. 5, 1964; children: Dana Margaret, Theodore James; m. Christine Blazynski, July 23, 1980. B.S. in Zoology, U. Wis., 1965, Ph.D. in Genetics, 1968; postdoctoral student, U. Calif., Berkeley, 1968-69. From asst. prof. to assoc. prof. U. Minn., St. Paul, 1969-74; from assoc. prof. to prof. Purdue U., West Lafayette, Ind., 1974-81; prof. genetics Washington U. Sch. Medicine, St. Louis, 1982—; mem. genetics study sect. NIH, 1976-80, chmn., 1978-80; vis. prof. U. Zurich, 1978. Author: Priciples of Population Genetics, 1980, Human Genetics, 1983, others, also articles. Recipient Samuel Weiner award U. Wis., 1963; NASA trainee, 1965-68; NIH fellow, 1968-69; NIH career devel. awardee, 1974-79. Mem. Genetics Soc. Am., Genetics Soc. Brazil, AAAS, Am. Inst. Biol. Scis., Phi Beta Kappa. Subspecialties: Gene actions; Population biology. Current work: Functional effects of naturally occurring genetic variants. Use of genetic, molecular and continuous-culture technology

HARTLINE, HALDAN KEFFER, educator, physiologist; b. Bloomsburg, Pa., Dec. 22, 1903; s. Daniel Schollenberger and Harriet Franklin (Keffer) H.; m. Mary Elizabeth Kraus, Apr. 11, 1936; children—Daniel Keffer, Peter Haldan, Frederick Flanders. B.S., Lafayette Coll., Easton, Pa., 1923, D.Sc. (hon.), 1959; M.D., Johns Hopkins, 1927, LL.D. (hon.), 1969, Sc.D., U. Pa., 1971, M.D., U. Freiburg i/B, 1971, Sc.D., Rockefeller U., 1976, U. Md., 1978, Syracuse U., 1979; Eldridge Johnson traveling research scholar, U. Leipzig and Munich, 1929-31. Nat. Research fellow med. scis. Johns Hopkins, 1927-29; fellow in med. physics Eldridge Johnson Research Found., U. Pa., 1931-36, asst. prof. biophysics, 1936-40, 41-42, asso. prof. biophysics, 1943-48, prof., 1948-49; asso. prof. physiology Cornell Univ. Med. Coll., N.Y.C., 1940-41; prof. biophysics, chmn. dept. Johns Hopkins, Balt., 1949-53; prof. Rockefeller U., N.Y.C., 1953-74, emeritus, 1974—. Recipient William H. Howell award physiology, 1927; Howard Crosby Warren medal exptl. psychology, 1948; A.A. Michelson award Case Inst., 1964; Nobel prize in physiology or medicine, 1967; Lighthouse award, N.Y.C., 1969. Mem. Nat. Acad. Scis., Am. Physiol. Soc., Am. Philos. Soc., Am. Acad. Arts and Scis., Royal Soc. (fgn. mem; London), Biophys. Soc., Optical Soc. Am. (hon.), Physiol. Soc. (U.K.) (hon.), Phi Beta Kappa, Sigma Xi. Subspecialty: Physiology (medicine). Address: Rockefeller U 66th St and York Ave New York NY 10021 also Patterson Rd Hydes MD 21082

HARTMAN, PATRICK JAMES, mechanical engineer, researcher; b. Ann Arbor, Mich., Dec. 5, 1944; s. Norman James and Mary Jane (Cottrill) H.; m. Lee Ann Walraff, Oct. 5, 1968; children: Elizabeth Marie, Suzanne Caroline. B.M.E., Marquette U., 1968; M.S., U. R.I., 1974, Ph.D., 1976. Researcher U. R.I., Kingston, 1972-76; research engr. E. I. duPont de Nemours Co., Wilmington, Del., 1976-79; sr. ocean engr. Gould, Inc., Glen Burnie, Md., 1979-80; sr. mech. engr. USN, Washington, 1980—. Organizer, Community Assn. Tasks, Columbia, Md., 1982. Served to lt. (j.g.) USCG, 1969-72. Recipient Sci. award Bausch and Lomb, 1963, Vigil Honor award Boy Scouts Am., 1963; M. Kollinski Found. scholar, 1967; U. R.I. fellow, 1972-74. Mem. ASME (chmn. Ocean engring. div. 1982-83), Soc. Naval Architects and Marine Engrs., Nat. Soc. Profl. Engrs., Tau Beta Pi, Pi Tau Sigma. Subspecialties: Mechanical engineering; Ocean engineering. Current work: Mechanical, ocean and reliability engineering using analyses on vector supercomputers to improve shipboard equipment design. Home: 5070 Durham Rd W Columbia MD 21044 Office: Naval Sea Systems Command Code 05MR Washington DC 20362

HARTMAN, PAUL ARTHUR, microbiology educator; b. Balt., Nov. 23, 1926; s. Carl G. and Eva M. (Rettenmeyer) H.; m. Marjorie Ann Stewart, Aug. 19, 1950; children: Philip, Helen, Mark. B.S., U. Ill., 1949; M.S., U. Ala., 1951; Ph.D., Purdue U., 1954. Asst. prof. microbiology Iowa State U., Ames, 1954-58, assoc. prof., 1958-62, prof., 1962-72, disting. prof., 1972—, chmn. dept., 1974-81; cons. Hach. Co., Loveland, Colo., 1980—; collaborator U.S. Dept. Agr., 1975—. Author: Miniaturized Microbial Methods, 1968. Pres. Ames Community Presch. Ctr., 1967—; rep. Ames Child Care Fedn. Bd., 1981—. Fellow Am. Acad. Microbiology; mem. AAAS, Inst. Food Technologists, Internat. Assn. Milk, Food Environ. Sanitation, Am. Soc. Microbiology (pres. North Central bf. 1971), Soc. Indsl. Microbiology (chmn. membership 1978-79), Soc. Gen. Microbiology, Soc. Applied Bacteriology, Alpha Delta Phi, Gamma Sigma Delta. Subspecialties: Microbiology; Food science and technology. Current work: Agricultural and food microbiology; emphasis on rapid methods and automation; detection of coliforms, salmonellae, enterococci, antibiotics; microbial amylases, rumen microbiology. Home: 2300 Timberland Rd Ames IA 50010 Office: Dept Microbiology Iowa State U Ames IA 50011

HARTMAN, PHILIP EMIL, geneticist, microbiologist; b. Balt., Nov. 23, 1926; s. Carl G. and Eva M. (Rettenmeyer) H.; m. Zlata E. Demerec, Aug. 4, 1955; children: Paul Stuart, Sherry Lynn, Frederick Davor. B.S., U. Ill., Urbana, 1949; Ph.D., U. Pa., 1953. Postdoctoral fellow dept. genetics Carnegie Inst. Washington, Cold Spring Harbor, Maine, 1954-55; dept. bacteriology and immunology Harvard U., 1955-56; dept. biology U. Brussels, 1956-57; asst. prof. dept. biology Johns Hopkins U., 1957-61, assoc. prof., 1951-65, prof., 1965—, William D.Gill prof., 1975—. Author: (with S.R. Suskind) Gene Action, 1965, 2d edit., 1969; contbr. articles to profl. jours. Mem. exec. bd. Com. to preserve Assateague, Md.-Va., 1970—. Served to lt. U.S. Navy, 1944-46. Mem. Am. Soc. Microbiology, Genetics Soc. Am., Can. Soc. Genetics, Environ. Mutagen Soc., Am. Soc. Biol. Chemists, Am. Assn. Cancer Research, AAAS. Democrat. Subspecialties: Genetics and genetic engineering (biology); Cancer research (medicine). Current work: Microbiology, mutagenesis, tumor initiation, metaplasias, nitrous acid, nitrites, nitrates, bacteria, stomach cancer, genetics, nitrosation, diet. Home: 1604 Ralworth Rd Baltimore MD 21218 Office: Dept Biology Johns Hopkins U Charles St and 34th St Baltimore MD 21218

HARTMAN, ROBERT DALE, plant pathologist; b. Miami, Mar. 31, 1948; s. John Robert and Audrey Deloris (McGrady) H.; m. Linda Diane Rockwell, Oct. 25, 1969; children: Sandee, Dawn, Kristy, Robert. B.S., U. Fla., 1970, Ph.D., 1974. Research dir. Pan Am. Plant Co., West Chicago, Ill., 1974-77; partner, mgr. Hartman's, Palmdale, Fla., 1977—. Contbr. articles to profl. jours. Served with Army NG, 1970-76. NDEA fellow, 1971-72; U. Fla., 1971-72. Mem. Tissue Culture Assn., Am. Phytopathol. Soc., Am. Soc. Hort. Sci., Internat. Assn. Plant Tissue Culture. Subspecialties: Plant cell and tissue culture; Plant virology. Current work: In vitro propagation of certified plants; in vitro maintenance of genetic lines for breeding; disease elimination. Home: PO Box 154 Palmdale FL 33944 Office: PO Box 90 County Rd 733 Palmdale FL 33944

HARTMANIS, JURIS, computer scientist, educator; b. Riga, Latvia, July 5, 1928; came to U.S., 1950, naturalized, 1956; s. Martins and Irma (Liepins) H.; m. Ellymaria Rehwald, May 16, 1959; children—Reneta, Martin, Audrey. Student, U. Marburg, 1947-49; M.A., U. Kansas City, 1951; Ph.D., Calif. Inst. Tech., 1955. Instr. Cornell U., Ithaca, N.Y., 1955-57, prof., 1965—, Walter R. Read prof. engring., 1980—, chmn. dept. computer sci., 1965-71, 77—; asst. prof. Ohio State U., 1957-58; research mathematician Gen. Electric Research & Devel. Center, Schenectady, 1957-65. Author: (with R.E. Stearns) Algebraic Structure Theory For Sequential Machines, 1966, Feasible Computations and Provable Complexity Properties, 1978; Editor: SIAM Jour. Computing; asso. editor: Jour. Computer and Systems Scis, 1966—, Jour. Math. Systems Theory, 1966—; co-editor: Springer-Verlag Lecture Notes in Computer Sci, 1973—. Mem. Am. Math. Soc., Math. Assn. Am., Assn. Computing Machinery, Sigma Xi. Subspecialties: Theoretical computer science; Foundations of computer science. Current work: Theory of computation, computatonal complexity. Home: 324 Brookfield Rd Ithaca NY 14850

HARTMANN, BRUCE, research physicist; b. St. Louis, June 30, 1938; s. Maurice Milton and Helen Louise (Diebels) H.; m. Judith Ann Bryan, June 30, 1962; children: Eric, Lisa, Kevin. A.B., Cath. U. Am., 1960; M.S., U. Md., 1966; Ph.D., Am. U., 1971. Physicist Naval Ordnance Lab., Silver Spring, Md., 1960-75; head polymer physics Naval Surface Weapons Ctr., Silver Spring, 1975—. Contbr. articles profl. jours. Recipient Performance award Weapons Ctr., 1981. Mem. Am. Phys. Soc., ASTM, Phi Beta Kappa. Clubs: Wheaton Boys (commr. 1977-80, bd. dirs. 1978-80. Subspecialties: Polymer physics; Polymers. Current work: Relating molecular structure and intermolecular potential to macroscopic physical properties for polymers. Home: 10614 Dunkirk Dr Silver Spring MD 20902 Office: Naval Surface Weapons Ctr White Oak Silver Spring MD 20910

HARTMANN, ERNEST LOUIS, psychiatrist, educator; b. Vienna, Austria, Feb. 25, 1934; s. Heinz and Dora (Karplus) H.; m. Barbara Hengst, Dec. 26, 1961; m. Eva Neumann, Aug. 20, 1975; children: Jonathan, Katherine. A.B., U. Chgo., 1954; M.D., Yale U., 1958. Diplomate: Am. Bd. Psychiatry and Neurology. Clin. assoc. NIMH, Bethesda, Md., 1962-64; mem. faculty dept. psychiatry Tufts U. Sch. Medicine, Boston, 1964—, prof., 1975—; sr. psychiatrist, dir. Sleep Lab. Boston State Hosp., 1964-80; sr. psychiatrist, dir. Sleep Lab. and Sleep Disorders Center West-Ros Park Mental Health Center, Lemuel Shattuck Hosp., 1980—; practice medicine specializing in psychiatry and sleep disorders, Boston, 1964—. Author: The Biology of Dreaming, 1967, Adolescents in a Mental Hospital, 1968, The Functions of Sleep, 1973, The Sleeping Pill, 1978, The Nightmare, 1983-84; contbr. numerous articles to profl. jours. Served to lt. comdr. USPHS, 1962-64. Recipient Holt prize Yale U., 1956; Psychopharmacology prize Am. Psychol. Assn., 1970. Mem. Am. Psychiat. Assn., Boston Psychoanalytic Soc. and Inst., AAAS, Am. Coll. Neuropsychopharmacology, Assn. Psycholphysiol. Study of Sleep. Subspecialties: Psychopharmacology; Neuropharmacology. Current work: Sleep, schizophrenia. Office: 170 Morton St Jamaica Plain (Boston) MA 02130

HARTNETT, GEORGE JOSEPH, JR., chemical company executive; b. Bayonne, N.J., Dec. 2, 1915; s. George Joseph and Alice Helene (Butler) H.; m. Gertrude Cleary, Jan. 5, 1946; children: George, Marie, Carol, James, Christopher. B.S.M.E., Purdue U., 1936; grad., Advanced Mgmt. Program, Harvard U., 1965. Field constrn. mgr. Babcock & Wilcox, 1936-42, supt. planning, Barberton, Ohio, 1946-51, gen. mgr., 1951-58; gen. mgr. mfg. Air Products & Chems., Allentown, Pa, 1958-61, v.p. mfg., 1968—; dep. mng. dir. Air Products Ltd., London, 1961-67. Contbr. articles to profl. jours. Active Wiley House home for disturbed children, Allentown, Lehigh County (Pa.) Indsl. Devel. Authority, Allentown C. of C. Served with U.S. Army, 1942-45. Mem. ASME, Am. Def. Preparedness Assn. Roman Catholic. Subspecialties: Cryogenics; Mechanical engineering. Current work: Mechanical engineering and manufacturing development. Patentee mech. engring. Home: 3316 Congress St Allentown PA 18104 Office: PO Box 538 Allentown PA 18105

HARTSHORNE, ROBERT (ROBIN) COPE, mathematics educator; b. Boston, Mar. 15, 1938; m., 1969; 2 children. A.B., Harvard U., 1959; Ph.D. in Math, Princeton U., 1963. Jr. fellow Harvard U., 1963-66, from asst. prof. to assoc. prof. math., 1966-72; assoc. prof. math. U. Calif.-Berkeley, 1972-74, prof., 1974—; vis. prof. Tata Inst. Fundamental Research, Bombay, 1969-70, Kyoto (Japan) U., 1975-76. A.P. Sloan Found. fellow, 1970-72. Mem. Am. Math. Soc. (Steele prize 1979). Office: Dept Math U Calif Berkeley CA 94707

HARTUNG, JACK BURDAIR, planetary scientist; b. Des Moines, Mar. 10, 1937; s. Robert Burdair and Edythe Marie (Nervig) H.; m. Ann Lupton Daniel, June 30, 1962; children: Ann Elizabeth, Jack Burdair. B.S., Iowa State U., 1959; postgrad., Coll. William and Mary, 1960-61; Ph.D., Rice U., 1968. Aerospace engr. ops. div. NASA Space Task Group, Langley Field, Va., 1959-62; aerospace engr. Apollo Spacecraft Project Office, NASA Manned Spacecraft Ctr., Houston, 1962-64; planetary scientist planetary and earth sci. div., 1967-72; guest scientist Max-Planck-Institut fur Kernphysik, Heidelberg, W.Ger., 1972-74; adj. assoc. prof. SUNY, Stony Brook, 1974-79; vis. assoc. prof. Hunter Coll., CUNY, N.Y.C., 1979-80; NRC sr. research assoc. NASA Johnson Space Ctr., Houston, 1980-81; NRC research mgmt. assoc. NASA Hdqrs., Washington, 1981-82; adj. prof. George Mason U., Fairfax, Va., 1982—; pres. Solar System Assocs. Ltd., Gt Falls, Va., 1980—. Vice-pres. bd. dirs. Clear Creek Basin Authority, 1970-72. Mem. Meteoritical Soc., Am. Geophys. Union, Am. Astron. Soc., Internat. Assn. Planetology, Explorers Club. Unitarian-Universalist. Subspecialties: Planetary science; Geology. Current work: Effect of space environment on exposed surfaces, terrestrial impact craters and basins, asymmetric distbn. lunar maria, radiometric dating of rocks, space and planetary sci. research mgmt. Office: 9108 Potomac Ridge Rd Great Falls VA 22066

HARTWIG, NATHAN LEROY, weed science educator; b. Monroe, Wis., Aug. 10, 1937; s. Lawrence Edward and Viola (Schroeder) H.; m. Elfriede Marie Dietz, Aug. 3, 1963; 1 dau., Susan Ellen. B.S., U. Wis., 1959; Ph.D., 1970; M.S., U. Ariz., 1965. Asst. prof. weed sci. Pa. State U., State College, 1969-76, assoc. prof., 1976—. Served to 1st lt., inf. U.S. Army, 1960-63. Republican. Methodist. Subspecialty: Integrated pest management. Current work: Weed control methods in agronomic crops. No-tillage crop production with living mulches. Office: Pa State U 119 Tyson Bldg University Park PA 16802

HARTZ, KENNETH E., environmental engineer, educator; b. Phila., Dec. 21, 1935; s. Kenneth H. and Eugenie M. (Von Zieber) H.; m. Linda T. Thomas, Sept. 1, 1962; children: David T., Leslie S. B.S., U. Maine, 1962; M.S., Pa. State U., 1965; Ph.D., U. Wis., 1979. Project leader Ingersol-Rand Corp., Nashua, N.H., 1967-70; Mgr. Jeffrey Mfg. Co., Columbus, Ohio, 1970-72; gen mgr. Allis Chalmers Corp., Appleton, Wis., 1972-75; pres. Environ. Tech Inc., Menasha, Wis., 1975-77; head environ. engring. sect. Wash. State U., Pullman, 1979—; cons. Coffman Engrs., Olympic Assoc. Co.; arbitrator Wash. Superior Ct., Spokane, 1980-81; expert witness Wash. Dept. Transp., Pasco, 1981. Contbr. articles to profl. jours. Commr. pub. works, Nashua, N.H., 1969. Served with USAF, 1954-58. Mem. Water Pollution Control Fedn., Am. Water Works Assn., Am. Soc. Engring. Edn., Sierra Club, Sigma Xi, Tau Beta Pi. Lodge: Elks. Subspecialties: Environmental engineering; Fuels and sources. Current work: Biofuels and treatment of toxic waste from organic chemical processes. Patentee in field. Office: Dept Civil and Environ Engring Wash State U Pullman WA 99164

HARTZELL, CHARLES ROSS, III, research scientist, administrator; b. Butler, Pa., Aug. 12, 1941; s. Charles Ross Jr. and Ada Grace (Giles) H.; m. Marguerite K. Getty, Aug. 16, 1963; children: Scott David, Amy Lynette. B.S., Geneva Coll., 1963; Ph.D., Ind. U., 1967. Research assoc. Ind. U., Bloomington, 1966-67; research scientist in protein chemistry Commonwealth Sci. and Indsl. Research Orgn., Melbourne, Australia, 1967-68; research assoc., asst. research prof. U. Wis.-Madison, 1968-71; asst. prof. to assoc. prof. Pa. State U., 1971-78; sr.

research scientist A.I. duPont Inst., Wilmington, Del., 1978—, research dir., 1981—. Contbr. numerous articles to profl. publs. USPHS/Ind. U. pre—doctoral fellow, 1966; USPHS/U. Wis. postdoctoral fellow, 1968-70; established investigator Am. Heart Assn., 1970-75. Mem. Am. Soc. Biol. Chemists, Biophysics Soc., Am. Chem. Soc. Republican. Subspecialties: Biochemistry (medicine); Cell biology (medicine). Current work: Muscle growth and development from neonatal period to maturity; cell culture techniques, nerve/muscle coculture, microanalytical approaches. Office: Alfred I duPont Inst PO Box 269 Wilmington DE 19899

HARVEY, CHARLES ARTHUR, mathematical researcher; b. Gering, Nebr., Aug. 14, 1929; s. Walter Carlton and Marcia (Hilliker) H.; m. Margaret Ruth Stone, Aug. 1, 1952; children: Jenny Beth, Charles Arthur, Peter John. A.B., Nebr. Wesleyan U., 1951; M.A., U. Nebr., 1953; Ph.D., U. Minn., 1960. Design engr. Honeywell Aero Div., Mpls., 1955-57; research scientist Honeywell Systems and Research Ctr., Mpls., 1960—. Contbr. articles to profl. jours. Served with AUS, 1953-55. Recipient Alumni Achievement award Nebr. Wesleyan U., 1977. Mem. IEEE (chmn. applications com. 1979-81), Am. Math. Soc., Soc. Indsl. and Applied Math. Methodist. Subspecialty: Applied mathematics. Current work: Development and application of automatic control theory. Co-patentee in field. Home: 3843 Zenith Ave S Minneapolis MN 55410 Office: Honeywell 2600 Ridgway Pky Minneapolis MN 55413

HARVEY, JOHN ADRIANCE, educator; b. N.Y.C., Oct. 14, 1930; s. John A. and Paula (Truhar) H.; m. Rhoda S. Sadigur, Dec. 20, 1958; children: David A., Andrew M., Michael A. A.B., U. Chgo., 1954, Ph.D., 1959. Research assoc. U. Chgo., 1959-61, asst. prof., 1961-67, assoc. prof., 1967-68; prof. psychology and pharmacology U. Iowa, Iowa City, 1968—; guest research worker Maudsley Hosp., London, 1966-67; cons. NIMH, NIH. Contbr. articles to profl. jours.; author: Behavioral Analysis of Drug Action, 1971; editor various jours. Served with U.S. Army, 1952-54. NIMH grantee, 1963-68; Research Scientist award, 1969-74. Fellow Am. Psychol. Assn.; mem. Am. Soc. Pharmacology (editorial bd.), Soc. Neuroscis. (pubs. com.), Am. Coll. Neuropsychopharmacology (communications com.). Subspecialties: Neuropharmacology; Neuropsychology. Current work: Effects of drugs on learning incl. anatomical systems, synaptic transmitters, neurochem. changes. Home: 227 Magowan Ave Iowa City IA 52240 Office: U Iowa Dept Psychology SSH E117 Iowa City IA 52242

HARVEY, LEONARD A., chem. co. exec.; b. St. Catharines, Ont., Can., Aug. 20, 1925; came to U.S., 1952, naturalized, 1960; m. Shirley Williams, Oct. 7, 1950; children—Brian, Bruce, Christopher. B.Sc. with honors, Queens U., 1950. With Borg Warner Chems. Inc., 1952—, pres., Parkersburg, W.Va., 1976—; dir. McGean Chem. Co., Parkersburg Nat. Bank. Served with RCAF, World War II. Mem. Soc. Plastics Industry, Chem. Mfrs. Assn., Parkersburg C. of C. (pres. 1981-82). Subspecialty: Polymer chemistry. Address: Box 1868 Parkersburg WV 26101

HARVEY, PAUL M., astronomer; b. Phila., Feb. 6, 1947; s. Maurice E. and Eugenie (Kidawa) H.; m. Susan R. Terapane, June 13, 1970; children: Amber J., Karen L. B.A., Conn. Wesleyan U., 1968; Ph.D., Calif. Inst. Tech., 1973. Research asso. U. Ariz., Steward Observatory, Tucson, 1973-79, asst. astronomer, 1979-80; asst. prof. astronomy dept. U. Tex., Austin, 1980—. Contbr. numerous articles to sci. jours. Mem. Am. Astron. Soc., Internat. Astron. Union. Subspecialties: Infrared optical astronomy; Radio and microwave astronomy. Current work: Infrared astrophysics, particularly as applied to studies of star formation. Office: Astronomy Dept U Tex Austin TX 78712

HARVEY, ROBERT DARNELL, engineering educator, consultant; b. Mpls., May 28, 1919; s. Percy James and Lulu Bessie (Darnell) H.; m. JoAnne Potts, Sept. 12, 1945; children: Wayne Norman, Roy Leland. B.M.E., U. Minn., 1939, M.E., 1954; M.S.M.E., MIT, 1947; C.A.S., No. Ill. U., 1975. Registered profl. engr., Mich., Ill. With Gen. Motors Corp., Detroit, 1947-56, Ford Motor Co., 1956-61, Borg-Warner Corp., 1962-68; prof. engring. Coll. of DuPage, 1968—; cons. mech. engr. Served to capt. AUS, 1941-45. Mem. Am. Soc. Engring. Edn., ASME, Nat. Soc. Profl. Engrs. Subspecialty: Mechanical engineering. Current work: Mechanical Design, materials, performance, quality, failure analysis. Address: 3932 Forest St Western Springs IL 60558

HARVEY, RONALD GILBERT, cancer researcher; b. Ottawa, Ont., Can., Sept. 9, 1927; came to U.S., 1948, naturalized, 1952; s. Gilbert and Adeline (Leclaire) H.; m. Helene Szpara, May 18, 1952; 1 son, Ronald E. B.S. in Biology, UCLA, 1952; M.S. in Chemistry, U. Chgo., 1955, Ph.D., 1960. Project leader Sinclair Research Labs., Harvey, Ill., 1956-58; from instr. to prof. Ben May Lab. Cancer Research, U. Chgo., 1960-75; prof. chemistry U. Chgo., 1975—; cons. in field. Author. Sr. postdoctoral fellow U. London, 1963; grantee Nat. Cancer Inst., Am. Cancer Soc. Mem. Am. Assn. Cancer Research, Am. Chem. Soc., Royal Chem. Soc., AAAS, Am. Inst. Chemists, AAUP, Sigma Xi. Subspecialties: Cancer research (medicine); Organic chemistry. Current work: Cancer research, mechanism of chem. carcinogenesis, chemistry of polycyclic aromatic hydrocarbons. Patentee in field. Home: 7350 Choctaw Rd Palos Heights IL 60463 Office: 950 E 59th St Box 424 Chicago IL 60637

HARVEY, WILLIAM DONALD, aerospace engr., researcher; b. Roanoke, Va., Oct. 12, 1934; s. William Howard and Hattie Virginia (Harris) H.; m. Billie Jo Sensing, Dec. 22, 1952; children: Alice Anne, Lisa Lynne, William Donald Jr. B.S. in Physics, Murray State U., Ky., 1961; M.S. in Mech. Engring, Old Dominion U., Norfolk, Va., 1975. Reactor test engr. Newport News (Va.) Shipping and Dry Dock Co., 1961-62; aerospace engr. NASA Langley Research Ctr., Hampton, Va., 1962-77; mgr. laminar flow control Airfoil Expt. Office, 1962-79; asst. head airfoil aerodynamics br. Transonic Aerodynamics Div., 1979-80, head, 1980—. Contbr. 40 tech. reports and articles to profl. jours. Served with U.S. Army, 1954-52; Japan. Recipient Outstanding Performance award NASA Langley Research Ctr., 1978, 79, Group Achievement award, 82. Fellow AIAA. Lutheran. Lodge: Moose. Subspecialties: Aerospace engineering and technology; Aeronautical engineering. Current work: Responsible for administrative and technical direction of research programs and scheduling and operations of facilities. Home: 606 Windemere Rd Newport News VA 23602 Office: NASA Langley Research Ctr Hampton VA 23665

HARWOOD, JULIUS J., automotive co. exec.; b. N.Y.C., Dec. 3, 1918; s. Louis and Rebecca (Schwartz) Horowitz; m. Dorothy Ginsberg, Sept. 14, 1941; children—Dane L., Gail A., Caren L. B.S., CCNY, 1939; M.S., U. Md., 1953. Materials engr. U.S. Naval Gun Factory, Washington, 1941-46; head metallurgy br. Office Naval Research, Washington, 1946-60; with Ford Motor Co., Dearborn, Mich., 1960—, dir. phys. scis., 1974-75, dir. materials scis. lab., 1975—; adj. prof. Wayne State U.; mem. bd. control Mich. Tech. U.; chmn. vis. com. Faculty Engring. Rensselaer Poly. Inst.; chmn. nat. materials adv. bd. Nat. Acad. Sci.-Nat. Acad. Engring., 1977—; cons. Office Tech. Assessment, U.S. Congress. Contbr. numerous articles to profl. jours.; editor: Molybdenum, 1958, Effects of Radiation on Materials, 1958, Refractory Metals and Alloys, 1961, Perspectives in Materials Research, 1961. Served with USN, 1945-46. Named Ford Citizen of Year, 1968. Fellow Am. Soc. Metals (John H. Shoemaker award of distinction 1977), AAAS, Metall. Soc., Engring. Soc. Detroit; mem. AIME (hon.; pres. 1976-77), Engring. Soc. Detroit, Mich. Acad. Scis., Nat. Acad. Engring., Am. Assn. Engring. Socs., Sigma Xi. Jewish. Subspecialty: Materials science research and development administration.. Home: 2258 Shorehill Dr West Bloomfield MI 48033 Office: Ford Motor Co PO Box 2053 Dearborn MI 48121

HASEGAWA, RYUSUKE, physicist; b. Nagoya, Japan, Feb. 7, 1940; s. Tokusaburo and Teru (Hanai) H.; m. Alice Pamela Quayle, Dec. 17, 1967; children: Sergei Pol, Linnea Marie. B.Eng., Nagoya U., 1962, M.Eng., 1964; M.S., Calif. Inst. Tech., 1968, Ph.D., 1969. Research fellow Calif. Inst. Tech., Pasadena, 1969-72; postdoctoral fellow IBM Thomas J. Watson Research Ctr., Yorktown Heights, N.Y., 1973-75; sr. staff physicist Allied Corp., Morristown, N.J., 1976-78, group leader, 1978-80, research assoc. group leader, 1980—; mem. mgmt. com., program com., publs. com. Ann. Magnetism and Magnetic Materials Conf., 1981—. Editor: (with others) Amorphous Magnetism II, 1977; editor: Glassy Metals: Magnetic, Chemical and Structural Properties, 1983; Contbr. articles to profl. jours. RCA David Sarnoff scholar for Japan, 1961; RCA David Sarnoff fellow, 1964-65; Fulbright scholar, 1964-69. Mem. Am. Phys. Soc., IEEE (chmn. magnetics chpt. Princeton sect. 1980—, chmn. amorphous magnetic tech. com. magnetic soc.), AAAS, Materials Research Soc., Sigma Xi. Subspecialties: Magnetic physics; Amorphous metals. Current work: Conduct research in area of local structure-property relationship in amorphous metallic materials and lead a group of scientists engaged in research and development of amorphous materials. Patentee in amorphous materials. Office: PO Box 1021R Morristown NJ 07960

HASEGAWA, TONY SEISUKE, computer graphics consultant; b. Tokyo, Dec. 21, 1941; U.S., 1973, naturalized, 1982; s. Sukesaburo and Chiyo (Sano) H. B.S., U. Electro-Communications, Tokyo, 1965; M.S., U. Santa Clara, 1979. Systems analyst, project mgr. Control Data Far East, Inc., Tokyo, 1967-74; computer graphics cons. Control Data Corp., 1974-82; advance project cons. NASA Ames Research Center, Moffett Field, Calif., 1975-82; pres. Fine Tech. Corp., 1983—. Recipient Spl. Scholarship Japanese Govt., 1961-65. Mem. Assn. Computing Machinery., AIAA. Republican. Roman Catholic. Club: Ski (Moffett Field, Calif.). Subspecialties: Graphics, image processing, and pattern recognition; Aerospace engineering and technology. Current work: Bible to Sartre, math. to Chopin, computer graphics to Renoir, super-computers to computational fluid dynamics. Office: Fine Tech Corp 2083 Landings Dr Mountain View CA 94043

HASELHORST, DONALD DUANE, manufacturing company executive; b. Northville, S.D., May 21, 1930; m. Nancy G. Wilz, Aug. 29, 1953; children: Stephen, Linda, Lynn. B.S. in Elec. Engring, S.D. State U., Brookings, 1956. Engr. Univac div. Sperry Rand Corp., St. Paul, 1956-59; v.p. Fabri-Tek Inc., Amery, Wis., 1959-67; chmn., chief exec. officer Nicolet Instrument Corp., Madison, Wis., 1967—; dir. M & I Bank of Madison, Demco Inc., Viking Ins. Co., Realist Inc., Sigma-Aldrich Chem. Co., Union State Bank, Central Life Assurance Co. Pres. Madison Service Clubs Council, 1976-77; bd. dirs. Oakwood Luth. Homes, Madison, 1976—, Madison Gen. Hosp. Served with USAF, 1948-52. Recipient Disting. Engr. award S.D. State U., 1981, Disting. Alumni award S.D. State U., 1982. Mem. IEEE, Am. Mgmt. Assn. (president's council), Sales and Mktg. Assn. (Exec. of Yr. award 1981). Lutheran. Clubs: West Kiwanis (past pres.), Nakoma Golf, Madison. Subspecialties: Bioinstrumentation; Nuclear magnetic resonance (chemistry). Office: 5225 Verona Rd Madison WI 53711

HASELTINE, WILLIAM ALAN, biologist; b. St. Louis, Oct. 17, 1944; s. William Reed and Jean (Ellsberg) H.; m. Patricia Eileen Gercik, June 12, 1966; children: Mara, Alexander M. B.A., U. Calif., Berkeley, 1966; Ph.D., Harvard U., 1973. Vis. prof. U. Copenhagen Inst. Microbiology, 1973; postdoctoral fellow M.I.T., 1973-76; asst. prof. microbiology, pathology Harvard Med. Sch., Boston, 1975-78, 1978-79, assoc. prof. pathology, 1979—, assoc. prof. microbiology, 1979—; chief lab. molcular studies of cancer cause and treatment Sidney Farber Cancer Inst., 1980-81, chief lab. biochem. pharmacology, 1981—. Contbr. articles to profl. jours. Nat. exec. bd. Leukemia Soc. Am. Helen Hay Whitney fellow, 1973-75; recipient award Ruth Estrin Goldberg Meml. Soc., 1975-76, Faculty research award Am. Cancer Soc., 1978—. Mem. Am. Soc. Microbiology, Am. Soc. Biol. Chemists, Internat. Leukemia Soc., Radiation Research Soc., Mutagenesis Assn. New Eng., Am. Soc. Virology, Phi Beta Kappa. Subspecialties: Molecular biology; Cell study oncology. Office: Dana-Farber Cancer Inst 44 Binney St Boston MA 02115

HASHIM, GEORGE A., immunologist, educator; b. Lebanon, May 28, 1931; m. Audrey E. Mailniot; children: Laura E., Charles E., Sami G. B.A., Northeastern U., 1962; M.S., Columbia U., 1963, Ph.D., 1967. Trainee Salk Inst., La Jolla, Calif., 1967-69; dir. biol. research Continental Research Inst., N.Y.C., 1969-71; dir. Jane Forbes Clark Surg. Research Lab., St. Luke's-Roosevelt Hosp. Ctr., N.Y.C., 1971—; assoc. prof. microbiology and surgery Columbia U., N.Y.C., 1976—. Assoc. editor: Neurochem. Research, 1980—, Jour. Neurosci. Research, 1979—; editor: books including Clinical and Biological Aspects of Peripheral Nerve Diseases, 1983; contbr. articles to profl. jours. NIH research grantee, 1977; Nat. Multiple Sclerosis Soc. research grantee, 1973; Biddle Found. research grantee, 1969. Mem. Am. Assn. Immunologists, Am. Soc. Biol. Chemists, Am. Soc. Neurochemistry, Internat. Soc. Neurochemistry, Transplantation Soc., Am. Chem. Soc. Subspecialties: Immunobiology and immunology; Neuroimmunology. Current work: Chemical structure and immunological functions of antigens known to induce diseases; brain antigens, myelin, tumor antigens and transplant antigens. Office: St Luke's-Roosevelt Hospital Center 421 W 114th St New York NY 10025

HASHIMOTO, SHUICHI, molecular biologist; b. Tokyo, Jan. 8, 1940; U.S., 1973; s. Tsurukichi and Kin H.; m. Akiko, Nov. 28, 1940. D.Sc., Nagoya (Japan) U., 1969. Research assoc. Cancer Inst., Tokyo, 1970-71; instr. Tokushima (Japan) U., 1971-73; research assoc. St. Louis U., 1974-76, asst. research prof. molecular virology, 1977-81, assoc. research prof., 1981—. Mem. Am. Assn. Microbiology. Subspecialties: Molecular biology; Virology (medicine). Current work: Gene expression of eukaryotic cells; expression and function of viral oncogene. Office: Saint Louis U Med Center 3681 Park Ave Saint Louis MO 63110

HASKINS, FRANCIS ARTHUR, geneticist; b. Omaha, Aug. 20, 1922; s. William Forrest and Lona Abbie (Davis) H.; m. Dorothy Genevieve Masters, Dec. 3, 1951; children: John, Ann Olney, Katherine, William. B.Sc., U. Nebr., 1943; M.Sc., 1948; Ph.D., Calif. Inst. Tech., 1951. Research fellow Calif. Inst. Tech., 1951-52; research scientist U. Tex., 1952-53; asst. prof. U. Nebr., Lincoln, 1953-55, assoc. prof., 1955-58, prof., 1958—, Regents prof., 1967—. Contbr. articles to profl. jours. Served with AUS, 1943-45. NSF grantee; AEC grantee; U.S. Dept. Agr. grantee. Fellow AAAS, Am. Soc. Agronomy; mem. Genetics Soc. Am., Am. Soc. Plant Physiologists, Am. Inst. Biol. Scis., Phytochem. Soc. N.Am. Republican. Presbyterian. Subspecialty: Plant genetics. Current work: Researcher in genetics, breeding and biochemistry of Melilotus and Sorghum. Home: 820 Robert Rd Lincoln NE 68510 Office: Dept Agronomy U Nebr Lincoln NE 68583

HASKINS, JOHN THOMAS, neurophysiologist, neuropharmacologist, researcher; b. Vinita, Okla., Apr. 27, 1947; s. Billy Dean and Janie Imogene (Harrison) H.; m. Vera Sue Tucker, Feb. 5, 1950; children: Kerri Deanne, Sarah Marie. B.S. in Zoology, U. Tulsa, 1969; Ph.D. in Physiology, Okla. State U., 1976. Postdoctoral fellow Marine Biomed. Inst., U. Tex. Med. Br., Galveston, 1976-79; postdoctoral fellow dept. physiology U. Tex. Health Sci. Center, Dallas, 1979-81; supr. neurophysiology/neuropharmacology Wyeth Labs., Inc., Phila., 1982—. Contbr. articles to sci. jours. Served with U.S. Army, 1969-71. NSF summer research award, 1974; Nat. Inst. Neurol. and Communicative Disease and Stroke research fellow, 1977-79. Mem. AAAS, Soc. for Neurosci., N.Y. Acad. Scis. Subspecialties: Neuropharmacology; Neurophysiology. Current work: Neurophysiol. techniques (EEG, single unit recording, iontophoresis) are employed to assess the activity of new compounds with the CNS. Home: 1258 Tanager Ln West Chester PA 19380 Office: Wyeth Labs Inc PO Box 8299 Philadelphia PA 19101

HASLER, ARTHUR DAVIS, educator; b. Lehi, Utah, Jan. 5, 1908; s. Walter Thalmann and Ada (Broomhead) H.; m. Hanna Prusse, Sept. 6, 1932 (dec.); children: Sylvia (Mrs. Gilbert Thatcher), A. Frederick, Mark, Bruce, Galen, Karl; m. Hatheway Minton, July 24, 1971. B.A., Brigham Young U., 1932; Ph.D., U. Wis., 1937; D.Sc., U. Nfld., 1967. Aquatic biologist U.S. Fish and Wildlife Service, 1935-37; instr., prof. U. Wis., Madison, 1937-78, prof. emeritus, 1979—, chmn. dept. zoology, 1953, 55-57, dir. Lab. Limnology, 1963-79; chmn. com. freshwater productivity internat. biol. program, 1964-74, Nat. Acad. Sci.-NRC, 1963-70; chmn. nat. com. Internat. Union Biol. Scis., 1965-69, chmn. com. ecology, 1966-69; pres. Internat. Congress Limnology, 1962, Internat. Assn. Ecology, 1962-74; dir. The Inst. Ecology, 1971-74; Disting. prof. U. Va., 1981, Tex. A & M U., 1979. Author: Underwater Guideposts, 1966; Contbr. articles to profl. jours. French horn player, mem. exec. com. Madison Civic Music Assn., 1937-65; chmn. Lake Mendota Problems Com., 1965-72. Fulbright research scholar, Germany, 1955, Finland, 1963; recipient Disting. Service award Am. Inst. Biol. Scis., 1980, Soil Sci. Soc. Am., 1980, Nat. Sea Grant Assn., 1980. Fellow Societas Zoologica Botanica Fennica Finland, Phila. Acad. Sci., Am. Acad. Arts and Sci., Royal Netherlands Acad. Sci.; mem. Am. Behavioral Soc., A.A.A.S. (past v.p. div. F), Am. Soc. Limnology and Oceanography (past pres.), Ecol. Soc. Am. (past pres.), Am. Fisheries Soc. (award of excellence 1977), Am. Soc. Naturalists (past v.p.), Am. Soc. Zoologists (pres. 1971), Nat. Acad. Scis. U.S., Internat. Assn. Limnology, Phi Kappa Phi (hon.). Mem. Ch. of Jesus Christ of Latter-day Saints. Subspecialty: Ecology. Home: 1233 Sweet Briar Rd Madison WI 53705

HASLER, ARTHUR FREDERICK, research meteorologist; b. Madison, Wis., Aug. 21, 1940; s. Arthur Davis and Hanna (Prusse) H.; m. Mary Louise Huebner, Oct. 8, 1965; children: Anneliese, Marta, Katherine, Matthew. B.S., U. Wis.-Madison, 1963, M.S., 1965, Ph.D., 1971. Research asst. meteorology dept. U. Wis.-Madison, 1963-71; sr. scientist Nat. Center for Atmospheric Research, Boulder, Colo., 1971-74; guest scientist Laboratoire de Meteordogie Dynamique, Paris, 1975; research meteorologist NASA, Greenbelt, Md., 1974—; guest scientist German Aerospace Agy., 1982. Scoutmaster Balt. area council Boy Scouts Am., 1981-82; Eagle Scout. Mem. Am. Meteorol. Soc., AIAA. Democrat. Mormon. Club: South River Ski (v.p. 1978, 80, 82). Subspecialties: Meteorology; Remote sensing (atmospheric science). Current work: Meteorology, research, satellites, remote sensing, stereoscopy, interactive computer display systems, aircraft measurements. Home: 638 Hillmeade Rd Edgewater MD 21037 Office: NASA Code 914 Greenbelt MD 20771

HASSAN, MARK DAVID, psychologist; b. Los Angeles, Jan. 16, 1949; s. Abraham Herman and Joyce Bernice (Ginsberg) H.; m. Elizabeth Worsley Toole, Sept. 19, 1976. A.B., U. Calif.-Berkeley, 1971; M.A., Calif. Sch. Profl. Psychology, 1974, Ph.D., 1976. Psychologist, assoc. dir. narcotics rehab. program Los Angeles Suicide Prevention Ctr. and Inst. Studies of Destructive Behavior, Los Angeles, 1971-79; pvt. practice clin. psychology, Pasadena, 1978—; Psychol. cons. LaVina Hosp. Respiratory Diseases, Altadena, Calif., 1978, Los Angeles Police Dept., 1980; asst. prof. Fuller Grad. Sch. Psychology, Pasadena, 1978—. Contbr. articles to profl. jours. Recipient Outstanding Contbn. award Los Angeles Suicide Prevention Center, 1974. Mem. Am. Psychol. Assn. Subspecialty: Psychoanalytic psychotherapy. Current work: Psychoanalytic psychotherapy; psychology in pulmonary medicine. Office: Arroyo Psychol Group 696 E Colorado Blvd Suite 207 Pasadena CA 91101

HASSETT, CAROL ALICE, psychologist; b. Bklyn., Apr. 19, 1947; d. Joseph and Anna (Portanova) Lusardi; m. John J. Hassett, June 29, 1968; 1 son, John J. B.S., St. John's U., 1968; M.A. in Edn, Hofstra U., 1974, Hofstra U., 1978, Ph.D., 1981. Tchr. N.Y.C. Bd. Edn., Bklyn, 1968-69; grad. teaching asst. Hofstra U., Hempstead, N.Y., 1978-79, adj. asst. prof., 1980-81; asst. psychologist South Nassau Community Hosp., Oceanside, N.Y., 1982, Nassau County Dept. Drug and Alcohol, East Meadow, N.Y., 1981—. Contbr. articles to profl. jours. Bd. dirs. Malverne Vol. Ambulance Corps, N.Y., 1972—. Mem. Am. Psychol. Assn., Nassau County Psychol. Assn., N.Y. Psychol. Assn., Adj. Assn. of Hofstra U. Republican. Roman Catholic. Home: 105 Franklin Ave Malverne NY 11565 Office: Nassau County Dept Drug and Alcohol Addiction Nassau County Med Ctr Hempstead Turnpike East Meadow NY 11554

HASSOUN, HUSSEIN ALI, state energy agency administrator; b. Beirut, Lebanon, Jan. 1, 1946; came to U.S., 1963, naturalized, 1975; s. Ali and Mashaal (Houdruj) H.; m. Judith Carol Bedea, Dec. 21, 1968; children: Ali, Tarek. B.S. in Engring, Oreg. State U., 1969, M.B.A., 1970, M.S. in Mgmt. Sci. and Ops. Research, 1972. Asst. Tech. dir. Evans Products Co., Corvallis, Oreg., 1972-74; dep. project mgr. CAT Inc., Al Khobar, Saudi Arabia, 1974-75; cons. Office of Pres., Oreg. State U., Corvallis, 1975-76; staff engr. Oreg. Dept. Energy, Salem, 1976-79, adminstr. planning div., 1979—. Contbr. Articles to profl. jours. Subspecialties: Energy planning; Operations research (mathematics). Current work: Construct models that simulate energy consuming systems and energy producing systems. These models are then applied to develop long term forecasts of energy demand for the state. Office: Oreg Dept Energy 102 L&I Bldg Salem OR 97306

HASTIE, JOHN WILLIAM, research chemist; b. Eng., Mar. 29, 1941; came to U.S., 1966, naturalized, 1983; s. John Victor and Dorothy (Walker) H.; m. Hilary Joan Asten, May 14, 1966; children: Duncan, Brendan. B.S., U. Tasmania, Australia, 1963, B.S.with 1st class honors, 1964, Ph.D., 1967, D.Sc. (hon.), 1973. Postdoctoral fellow Rice U., Houston, 1966-69; research chemist Nat. Bur. Standards, Washington, 1966-76, group leader, high temperature processes, 1976—. Author: High Temperature Vapors: Science and Technology, 1975; contbr. numerous articles to profl. publs.; editor: Characterization of High Temperature Vapors and Gases, vols. I and II, 1979. Recipient Silver medal Dept. Commerce, 1974; Disting. Young Scientist award Md. Acad. Sci., 1975. Mem. Am. Chem. Soc., Am. Ceramic Soc., Electrochem. Soc., Combustion Inst., Am. Assn. Mass Spectrometry, Internat. Union Pure and Applied Chemistry, NRC. Subspecialties: High temperature chemistry; High-temperature materials. Current work: Characterization of high temperature species present in vapor phase of high temperature systems involving materials processing and use. Co-inventor transpiration mass spectrometry (Indsl. Research Mag. IR-100 award 1980). Office: Nat Bur Standards Washington DC 20234

HASTINGS, ROBERT CLYDE, physician, educator; b. Tipton County, Tenn., Apr. 23, 1938; s. Robert Simpson and Margaret Marie (Peterson) H.; m. Virginia Ruth Thomas, Jan. 3, 1981; children: Cynthia Margaret, Robert Clyde, Jeffrey Scott. Ph.D., Tulane U., 1971; M.D., U. Tenn., 1962. Lic. physician, La., Tenn. Intern City of Memphis hosps., 1963-64; staff physician USPHS Hosp., Carville, La., 1964-68; chief pharmacology research dept. Nat. Hansen's Disease Center, 1971—; adj. clin. prof. Tulane U., 1974-83, assoc. staff, 1977—. Editor: Internat. Jour. Leprosy, 1979—; Contbr. articles to profl. jours. Served with USPHS, 1964—. Mem. Southeastern Pharmacology Soc., Am. Soc. Clin. Pharmacology and Therapeutics, N.Y. Acad. Scis., Soc. Exptl. Biology and Medicine, Am. Soc. Pharmacol. and Exptl. Therapeutics, Am. Fedn. Clin. Research, Am. Chem. Soc., Am. Coll. Clin. Pharmacology, Reticuloendothelial Soc., Am. Soc. Tropical Medicine and Hygiene, Internat. Leprosy Assn., USPHS Commd. Officers Assn., Council Biology Editors, Sigma Xi, Alpha Omega Alpha. Democrat. Methodist. Subspecialties: Pharmacology; Immunopharmacology. Current work: Immunology and pharmacology of leprosy. Office: Nat Hansen's Disease Center Carville LA 70721

HATCH, FREDERICK TASKER, biomedical research administrator, scientist; b. Boston, Aug. 27, 1924; s. Frederick Southard and Beatrice (Tasker) H.; m. Virginia Weeks, Mar. 3, 1946; children: Daniel F., Daphne A., Deborah J., Douglas E. B.A., Dartmouth Coll., 1944; M.D., Harvard U., 1948; Ph.D., M.I.T., 1960. Diplomate: Nat. Bd. Med. Examiners. Intern medicine Roosevelt Hosp., N.Y.C., 1948-49; research fellow Columbia U., N.Y.C., 1949-52; chief arteriosclerosis Mass. Gen. Hosp., Boston, 1960-65; sr. biomed. scientist Lawrence Livermore Lab., Calif., 1965—, asst. assoc. dir., 1980—; mem. research com. VA Med. Center, Livermore, 1970-80; mem. adv. com. Nat. Heart Lung Inst., Bethesda, Md., 1973-76. Assoc. editor: Lipids, 1969-73. Mem. steering com. Gen. Environ. Toxicology Assn. No. Calif., 1983-84. Served to capt. U.S. Army, 1952-55. Fellow Arteriosclerosis Council of Am. Heart Assn. (exec. com. 1971-73), Am. Inst. Chemists; mem. Am. Soc. Biol. Chemists, Environ. Mutagen Soc., Am. Chem. Soc., Am. Soc. Cell Biology. Subspecialties: Biochemistry (biology); Environmental toxicology. Current work: Genetic toxicology, DNA damage, mutation and repair, mutagens from cooking foods, toxicology of synfuels, structure of DNA types, chromatin and chromosomes. Office: Lawrence Livermore Nat Lab PO Box 5507 Livermore CA 94550 Home: 1130 Crellin Rd Pleasanton CA 94566

HATCH, NORMAN LOWRIE, JR., research geologist; b. Boston, May 27, 1932; s. Norman Lowrie and Marion Stearns (Jenney) H.; m. Sara Ballantyne, Sept. 7, 1957; children: Kirstin Wylie, Andrew James Ballantyne. B.A., Harvard U., 1956, M.A., 1958, Ph.D., 1961. Research geologist U.S. Geol. Survey, Beltsville, Md., 1961-70, supervisory geologist, Washington, 1970-73, Reston, Va., 1973-78, research geologist, 1978—. Contbr. articles to profl. jours. Served with U.S. Army, 1952-54. Fellow Geol. Soc. Am. (chmn. N.E. sect. 1976-77); mem. Mineral. Soc. Am., Geol. Soc. Washington. Subspecialty: Tectonics. Current work: Stratigraphy, structure, tectonics, and plate-tectonic history of the New England crystalline Appalachians. Home: 8514 Irvington Ave Bethesda MD 20817 Office: Mail Stop 926 US Geol Survey Reston VA 22092

HATCH, ROGER CONANT, pharmacologist, toxicologist, veterinarian; b. St. Joseph, Mich., Jan. 23, 1935; s. Conant Hopkins and Helen Ann (First) H.; m. Judith Earleen Hatch, Nov. 23, 1956; children: Roger Stephen, Timothy Paul. B.S., Mich. State U., 1957, D.V.M., 1959; M.S., Purdue U., 1964, Ph.D., 1966. Gen practice vet. medicine, Berwyn, Ill., 1959-60; instr. pharmacology and toxicology Purdue U., West Lafayette, Ind., 1962-66; assoc. prof. U. Guelph, Ont., Can., 1966-73; prof. U. Ga., Athens, 1973—. Contbr. articles to profl. jours. Served to capt. U.S. Army, 1960-62. Fellow Am. Acad. Vet. Pharmacology and Therapeutics, Soc. Toxicology; mem. N.Y. Acad. Sci., Phi Zeta. Republican. Subspecialties: Toxicology (medicine); Pharmacology. Current work: Beneficial and harmful interactions between drugs and chemicals; research in pharmacology and toxicology; teach these 2 subjects to veterinary students; diagnostic toxicology service for veterinarians; consultations. Home: 310 Hickory Hill Dr Watkinsville GA 30677 Office: College Vet Medicine Athens GA 30602

HATCH, STEPHAN LAVOR, plant taxonomist, educator, researcher; b. Logan, Utah, July 22, 1945; s. LaVor Joseph and Elva (McBride) H.; m. Nora Lee Cooper, July 7, 1967; children: Stephanie, Jocelynn, Sherrie, Shayne, Jennifer, Sean. B.S., Utah State U., 1970, M.S., 1972; Ph.D., Tex. A&M U., 1975. Vis. asst. prof. Tex. A&M U., 1974-75, postdoctoral fellow, 1975 76, asst. prof., curator Tracy Merbarium, 1979-83, assoc. prof., curator Tracy Merbarium, 1983—; asst. prof. N.Mex. State U., 1976-79. Contbr. to profl. jours. Tom Slick fellow, 1973-74. Mem. Am. Soc. Plant Taxonomists, Southwestern Assn. Naturalists, Soc. Range Mgmt., Sigma Xi, Phi Sigma. Mormon. Subspecialties: Systematics; Taxonomy. Current work: Taxonomy, systematics, morphology, cytology, anatomy, population studies of the Poaceae. Office: Dept Range Sci Texas A&M U College Station TX 77843

HATCHER, JOHN CHRISTOPHER, psychology educator; b. Atlanta, Sept. 18, 1946; s. John William and Kay (Carney) H. B.A., U. Ga., 1968, M.S., 1970, Ph.D., 1972. Psychologist Clayton Mental Health Ctr., Atlanta, 1971-72; dir. intern tng. psychology service Beaumony Med. Ctr., El Paso, 1972-74; assoc. clin. prof. psychology U. Calif.-San Francisco, 1974—; faculty U.S. Fire Acad., U.S. Fire Adminstrn., 1981—; cons. Fed. Emergency Mgmt. Agy., others; spl. asst. to mayor San Francisco, 1975—. Author: (with Brooks) Handbook of Gestalt Therapy, 1976, (with Himmelstein) Innovation in Psychology, 1977; editor: Am. Jour. Family Therapy, 1980. Chmn. Mayor's Commn. on Family Violence, San Francisco, 1978-80; adv. Arson Task Force, San Francisco Fire Dept., 1977—, Calif. Gov.'s Task Force on Earthquake Preparedness, 1981-82. U.S. Army Med. Research and Devel. grantee, 1971; Western Army Inst. Grantee, 1972; Maxicare Health Found. grantee, 1977; Nat. Inst. Corrections grantee, 1983; others. Mem. Am. Psychol. Assn., Western Psychol. Assn., Calif. Psychol. Assn., Phi Kappa Phi. Subspecialty: Social psychology. Current work: Psychology of crisis and emergency, assistance of state, federal and foreign government in management of disasters both natural and man-made. Office: U Calif Psychiatry Dept Langley Porter Inst PO Box 33C 401 Parnassus Ave San Francisco CA 94143

HATCHER, ROBERT DEAN, JR., geology educator, researcher, editor, consultant; b. Madison, Tenn., Oct. 22, 1940; s. Robert Dean and Elva Louise (Harris) H.; m. Velma Diana Simpson, June 7, 1965; children: Laura, Melinda. B.A., Vanderbilt U., 1961, M.S., 1962; Ph.D., U. Tenn.-Knoxville, 1965. Geologist Exxon USA, New Orleans, 1965-66; asst. prof. Clemson U., 1966-70, assoc. prof., 1970-76, prof., 1976-78; prof. geology Fla. State U., 1978-80, U.S.C., 1980—; sec. Internat. Geol. Correlation Program-Caledonide Project U.S. Working Group, Washington, 1977-84. Author: Physical Geology Principles-Processes and Problems, 1976; editor: Tectonics and Geophysics of Mountain Chains, 1983. NSF grantee, 1968-84; Duke Power Co. and Westinghouse grantee, 1974; NRC grantee, 1978. Fellow Geol. Soc. Am. (council 1980-83, editor bull. 1981-84), Geol. Assn. Can.; mem. Am. Geophys. Union, AAAS, Sigma Xi. (pres. chpt. 1975-76). Methodist. Subspecialties: Tectonics; Geophysics. Current work: Structural geology, tectonics, regional geophysics. Office: Dept Geology U SC Columbia SC 29208

HATCHER, S(TANLEY) RONALD, nuclear engineer; b. Salisbury, Eng., Aug. 20, 1932; m., 1955; 4 children. B.Sc., U. Birmingham, 1953, M.Sc., 1955; Ph.D. in Chem. Engring, U. Toronto, 1958. Chem. engr. Chalk River Nuclear Labs. Atomic Energy Can. Ltd., Ont., Can., 1958-63, sr. chem. engr., 1963-65, head chem. tech. br., 1965-74, dir. applied sci. div., 1974-78; v.p. and mgr. Whiteshell Nuclear Research Establishment, Pinawa, Man., 1978—. Mem. Can. Soc. Chem. Engrs., Chem. Inst. Can. Subspecialties: Nuclear engineering; Nuclear energy research administration. Office: Whiteshell Nuclear Research Establishment Box 411 Bldg 400 Sta 1 Pinawa MB Canada R0E 1L0

HATCHER, VICTOR BERNARD, biochemistry educator; b. Ormstown, Que., Can., Apr. 1, 1943; m. Barbara Joan Palaisy, Aug. 27, 1966; children: Jennifer Anne, Julie Kristin. B.S., Bishop's U., Lennoxville, Que., Can., 1964; M.S., McGill U., Montreal, Que., Can., 1966, Ph.D., 1969. Postdoctoral fellow U. Helsinki, Finland, 1968-70; research fellow Harvard U. Med. Sch., Boston, 1970-73; asst. prof. biochemistry and medicine Albert Einstein Coll. Medicine, N.Y.C., 1973-79, assoc. prof. biochemistry and medicine, 1979—. N.Y. Heart Assn. Sr. Investigator, 1977-81; established fellow, 1982-87; Cystic Fibrosis Found. Ann Weinberg scholar, 1981-83. Mem. Am. Soc. Cell Biology, Am. Assn. Immunologists, Am. Fedn. Clin. Research. Subspecialties: Biochemistry (biology); Cell biology. Current work: studies on mechanism of cell adhesion; cell-cell interaction and cell aging. Office: Albert Einstein Coll Medicine Montefiore Med Center 111 E 210th St Bronx NY 10467

HATFIELD, G. WESLEY, molecular biologist, inst. ofcl., educator; b. Avant, Okla., Aug. 2, 1940; s. Guy Wesley and Erna May (Estill) H.; m.; children: Lianna Karin, Jessica Chiara. B.S. in Analytical Biology and Biochemistry, U. Calif., Santa Barbara, 1964; Ph.D. in Molecular Biology and Biochemistry (NIH predoctoral fellow), Purdue U., 1968; NIH postdoctoral fellow, Sch. Medicine, Duke U., 1968-70. Asst. prof. med. microbiology Coll. Medicine, U. Calif., Irvine, 1970-74, assoc. prof., 1974-78, prof. microbiology, 1978—, dir. Contbr. articles to publs. USPHS career devel. awardee, 1971-77; Eli Lilly research awardee, 1975; grantee Am. Cancer Soc., 1970-76, NIH and NSF, 1970—. Mem. Am. Soc. Exptl. Biol. Chemists, Am. Soc. Microbiology, AAAS. Subspecialties: Genetics and genetic engineering (biology); Cell and tissue culture. Current work: Molecular genetics of gene expression in procaryotes and eucaryotes; molecular genetic basis of human genetic diseases.

HATFIELD, JERRY LEE, biometeorologist; b. Wamego, Kans., May 1, 1949; s. Virgil Hiram and Elsie Louise (Fischer) H.; m. Patricia JoAnn Reigle, Sept. 1, 1968; children: Mark Edward, Andrew James. B.S., Kans. State U., 1971; M.S., U. Ky., 1972; Ph.D., Iowa State U., 1975. Research asst. U. Ky., Lexington, 1971-72; teaching asst. Iowa State U., Ames, 1973-75; biometeorologist U. Calif., Davis, 1975—. Editor: (with I.J. Thomason) Biometeorology and Integrated Pest Management, 1982; assoc. editor: Agronomy Jour, 1981-84. Research Grantee U.S. Dept. Agr., NASA, Dept. Energy, USGS, State of Calif., 1975—. Mem. Am. Soc. Agronomy (div. chmn. 1980-82), Am. Meteorol. Soc. (chmn. com. on agr. and forest meteorology 1982-84). Republican. Mem. Covenant Ch. Subspecialties: Micrometeorology; Remote sensing (atmospheric science). Current work: Evaluation of energy exchanges in soil-plant-atmosphere systems with emphasis on plant adaptation and water use through measurements with remote sensing and ground based technologies. Home: 1706 Balsam Pl Davis CA 95616 Office: U Calif 119 Veihmeyer Davis CA 95616

HATHAWAY, DAVID HENRY, astronomer; b. Bangor, Maine, Aug. 29, 1951; s. Henry L. and Ruth M. (Roberts) H.; m. Janet T. Baril, June 2, 1973; 1 son, Adam David. B.S. in Astronomy, U. Mass., 1973; M.S., U. Colo., Boulder, 1975, Ph.D. in Astrophysics, 1979. Research asst. Lab. Atmospheric and Space Physics, Boulder, 1975-76; dept. astrogeophysics U. Colo., 1976-79; postdoctoral fellow advanced study program Nat. Center Atmospheric Research, Boulder, 1979-81; asst. astronomer Sacramento Peak Nat. Obs., Sunspot, N.Mex., 1981—. Contbr. articles to profl. jours. Mem. AAAS, Am. Astron. Soc., Am. Meteorol. Soc., Royal Astron. Soc. Subspecialties: Theoretical astrophysics; Solar physics. Current work: Astrophys. fluid dynamics; interactions between convective motions, rotation, shear flows, and magnetic fields in astrophys. objects; dynamo theory for generation magnetic fields. Home and Office: Sacramento Peak Obs Sunspot NM 88349

HATHAWAY, WILLIAM HOWARD, programmer, analyst; b. Saratoga Springs, N.Y., June 14, 1948; s. Harry M. and Mildred I. (Wright) H.; m. Amy Marie DeBrower, May 8, 1981. B.S. in Physics, Rensselaer Poly. Inst., 1970, M.S. in Astronomy, 1974. Mem. tech. staff Computer Sci. Corp., 1978—, Goddard Space Flight Ctr., NASA, Greenbelt, Md., 1978—; telescope operator Internat. Ultraviolet Explorer, 1978-82, programmer/analyst, 1982—. Mem. Am. Astron. Soc., Astron. Soc. Pacific, Planetary Soc., NOW, Nat. Abortion Rights Action League. Subspecialties: Ultraviolet high energy astrophysics; Space observatory operations and calibration. Current work: Photometric calibration; data management; ultraviolet stellar spectra; ultraviolet stellar spectroscopy; spectral data management and analysis. Office: Goddard Space Flight Ct NASA Code 6859 Obs Greenbelt MD 20769

HATHEWAY, ALLEN WAYNE, geological engineering educator, consultant; b. Los Angeles, Sept. 30,1937; s. Clarence Wilman and Marie Elizabeth (Sisto) H.; m. Anne Sellars, Apr. 4, 1959; children: Shannon, Brian, Steven. A.B., UCLA, 1961; M.S., U. Ariz., 1966, Ph.D., 1971, Profl. Degree in Geol. Engring, 1982; diploma, U.S. Army War Coll., 1980. Registered profl. engr., Ariz., Calif., Mass.; registered geologist, Calif., Maine; cert. engring. geologist, Calif. Project engr. U.S. Forest Service, Arcadia, Calif., 1971-72; sr. engr. Woodward-Clyde Cons., Los Angeles, 1971-74; project geologist Shannon & Wilson, Inc., San Francisco, 1974-76; v.p., chief geologist Haley & Aldrich Inc., Cambridge, Mass., 1976-81; prof. geol. engring. U. Mo.-Rolla, 1981—; cons. N.Y. Environ. Facilities Corp., Albany, 1978-81, Camp, Dresser & McKee, Inc., Boston, 1976—, Fenix & Scisson, Inc., Tulsa, 1980—; Mo. River Div. U.S. Army C.E., 1981—. Author: Guidelines for Professional Practice, 1981. Served with U.S. Army, 1961-63; Served with USAR, 1963—. Fellow Geol. Soc. Am. (chmn. engring. geology div. 1980); mem. Assn. Engring. Geologists (treas. 1981-83, v.p. 1984), ASCE, Am. Geophys. Union, Soc. Am. Mil. Engrs. Republican. Subspecialties: Environmental engineering; Civil engineering. Current work: Cleanup of uncontrolled hazardous waste sites; engineering seismology and earthquake risk; machine tunneling in rock; siting of hazardous waste management facilities; disposal of utility generation wastes; environmental permitting. Home: RFD 4 Box 66 Rolla MO 65401 Office: Dept Geol Engring 125 Mining Bldg U Mo-Rolla Rolla MO 65401

HATOFF, ALEXANDER, pediatrics educator, cons.; b. Bklyn., Feb. 14,1916; s. Isadore Charles and Gertrude (Feigelman) H.; m. Esther Lydia Gross, May 23, 1942; children: Ann Elizabeth, David Edward, Brian William. B.S., Cornell U., 1935; M.P.H., U. Mich., 1936; M.D., U. Rochester, 1940. Diplomate: Am. Bd. Pediatrics. Intern U. Mich. Hosp., Ann Arbor, 1940-41, resident in pediatrics, 1941-43, instr., 1942-43; chief pediatrics dept. Permanente Hosp., Oakland, Calif., 1943-46, Highland Gen. Hosp., Oakland, 1969-74; pvt. practice medicine specializing in pediatrics, Oakland, 1946-69; chmn. dept. chpts. Am. Acad. Pediatrics, Evanston, Ill., 1974-78; assoc. clin. prof. pediatrics Northwestern U., Chgo., 1974-78; cons. med. services Calif. Dept. Health Services, San Francisco, 1978—; instr. to asst. clin. prof. pediatrics U. Calif.-San Francisco, 1969—. Contbr. sci. articles to profl. publs. Recipient Ho Nun De Kah award Cornell U., 1935. Fellow Am. Acad. Pediatrics (chmn. chpt. 1969-70, Outstanding Chpt. award 1970); mem. Am. Fedn. Clin. Research (sr.). Democrat. Jewish. Clubs: Rossmoor Jewelry, Rossmoor Lapidary (Walnut Creek). Subspecialties: Health services research; Pediatrics. Current work: Medical consultant. Home: 1221 Avenida Sevilla Apt 3C Walnut Creek CA 94595 Office: Calif Dept Health Services Surveillance & Utilization Review PO Box 5500 San Francisco CA 94101

HATSOPOULOS, GEORGE NICHOLAS, mechanical engineer, thermodynamicist, educator; b. Athens, Greece, Jan. 7, 1927; came to U.S., 1948, naturalized, 1954; s. Nicholas and Maria (Platsis) Hatzopoulos; m. Daphne Phylactopoulos, June 14, 1959; children: Nicholas, Marina. Student, Nat. Tech. U., Athens, 1945-47; B.S., M.S., M.I.T., 1950, M.E., 1954, Sc.D., 1956; Sc.D. (hon.), N.J. Inst. Tech., 1982. Instr. M.I.T., 1954-56, asst. prof. mech. engring., 1956-58, assoc. prof., 1959-62, sr. lectr. in mech. engring., 1962—; founder, pres., chief exec. officer Thermo Electron Corp., developer, mfr. and marketer of products based on thermodynamic technologies of heat transfer and energy conversion, Waltham, Mass., 1956—; chmn. bd. Coll. Yr. in Athens; mem. adv. bd. Energy Productivity Center, Carnegie-Mellon Inst., 1978-81; tech. witness numerous Senate and Congl. hearings; mem. engring. edn. com. M.I.T. Center for Policy Alternatives, 1973; mem. policy com. for innovation program NSF at M.I.T., 1973-74; mem. ad hoc com. on air quality and power plant emissions NRC, 1974-75; mem. environ. ad. com. FEA, Washington, 1974-75; dir. Fed. Res. Bank of Boston. Author: Principles of General Thermodynamics, 1965, Thermionic Energy Conversion, vol. 1, 1973, vol. 2, 1979; contbr. numerous articles to profl. jours. Recipient Corp. Leadership award M.I.T., 1980. Fellow IEEE, Am. Acad. Arts and Scis., Nat. Acad. Engring., ASME (chmn. exec. com. div. energetics 1968-69); mem. Am. Acad. Achievement (Golden Plate award 1961), AIAA, Sigma Xi, Pi Tau Sigma (Gold Medal award 1960). Greek Orthodox. Subspecialties: Thermodynamics; Mechanical engineering. Current work: Quantum thermodynamics. Home: Tower Rd Lincoln MA 01773 Office: PO Box 459 101 First Ave Waltham MA 02254

HATTMAN, STANLEY MARTIN, molecular biology educator; b. Bklyn., July 19, 1938; s. Herman and Sarah (Goldberg) H.; m. Rosemarie Winter, Aug. 6, 1963; children: Heidi, Ursula, Rebecca. B.S., CCNY, 1960; Ph.D., M.I.T., 1965. Helen Hay Whitney Found. fellow Max-Planck Inst. Biochemistry, Munich, Germ., 1965-67, Albert Einstein Coll. Medicine, Bronx, N.Y., 1967-68; mem. faculty U. Rochester, N.Y., 1968—, prof. biology, 1981—. Recipient Research Career Devel. award NIH, 1972. Mem. Am. Soc. Microbiology, Am. Soc. Biol. Chemists. Subspecialties: Gene actions; Molecular biology. Current work: Study of DNA modification systems in pro- and eucaryotes with emphasis on nature, specificity and biological function. Office: Univ Rochester Biology Dept Rochester NY 14627 Home: 242 Dorchester Rd Rochester NY 14610

HATTON, GLENN IRWIN, neurobiology educator, academic administrator; b. Chgo., Dec. 12, 1934; s. Irwin Alfred and Anita Claussen (Richter) H.; m. Patricia J. Dougherty, Oct. 16, 1954; children: James D., William G., Christopher J., Jennifer K., Trent D., Tracey E. B.A., North Central Coll., Naperville, Ill., 1960; M.A., U. Ill., 1962, Ph.D., 1964. NIMH postdoctoral fellow U. Ill., Urbana, 1964; asst. prof. neurobiology Mich. State U., East Lansing, 1965-68, assoc. prof., 1968-73, prof., 1973—, dir. neurosci. program, 1979—; vis. prof. U. Calif.-Irvine, 1976; sr. research scholar Corpus Christi Coll., U. Cambridge, Eng., 1982-83; vis. scientist Agrl. Research Council Inst. Animal Physiology, Babraham, Cambridge, 1982-83. Contbr. chpts. to books, articles to profl. publs. Recipient Career Devel. award NIH, 1970-75; NIH Fogarty Sr. Internat. fellow, 1982-83; NIH grantee, 1967—. Mem. Soc. Neurosci., Am. Assn. Anatomists, Am. Soc. Zoologists, N.Y. Acad. Scis., AAAS. Subspecialties: Neurophysiology; Neuroendocrinology. Current work: Neural control of hypothalamic neuroendocrine cells; electrophysiology; electron microscopy; immunocytochemistry; brain slices. Office: Neurosci Program Mich State U East Lansing MI 48824

HATTOX, SUSAN ELLEN, biochemist; b. Wellesley, Mass., Aug. 29, 1939; d. James Grady and Doris May (Sanborn) H. B.A. (scholar), Wellesley Coll., 1961; Ph.D. (NIH fellow), Baylor Coll. Medicine, 1973. Asst. scientist Polaroid Corp., Cambridge, Mass., 1961-64; postdoctoral fellow U. Colo. Health Sci. Ctr., Denver, 1973-75, vis. asst. prof., 1981; research assoc. Yale U. Sch. Medicine, New Haven, 1975-81; sr. prin. biochemist Boehringer Ingelheim Ltd., Ridgefield, Conn., 1981—; cons. peer rev. process NIH. Contbr. articles to sci. jours. Mem. AAAS, Am. Soc. Mass Spectroscopy, Soc. Neurosci., LWV. Subspecialties: Biochemistry (medicine); Mass spectrometry. Current work: Application of mass spectrometry to biological systems; metabolism of endogenous and exogenous compounds. Office: 175 Briar Ridge Rd Ridgefield CT 06877

HATZIOS, KRITON KLEANTHIS, plant physiology educator, weed scientist; b. Florina, Greece, Aug. 6, 1949; came to U.S., 1976; s. Kleanthis Matthews and Adamantia Vasil (Tsougos) H.; m. Maria Kriton Grammatikakis, Sept. 8, 1979; children: Adamantia Kriton, Artemis Kriton. B.S. in Agr, Aristotelian U., Thessaloniki, Greece, 1972; M.S. in Crop Sci, Mich. State U., 1977; Ph.D. in Plant Physiology, Mich. State U., 1979. Research asst. Mich. State U., East Lansing, 1976-79; asst. prof. plant physiology Va. Poly. Inst. and State U., Blacksburg, 1979—; cons. Breton Pubs., 1981, 82. Author: Metabolism of Herbicides, in Higher Plants, 1982; mem. editorial bd.: Weed Sci, 1982—. Advisor Hellenic Assn. Va. Poly. Inst. and State U., 1980-82. Served to 1st lt. Greek Air Force, 1973-75. Named Outstanding Scientist dept. plant physiology Va. Poly. Inst. and State U., 1982. Mem. Am. Soc. Plant Physiologists, AAAS, Am. Soc. Photobiology, Weed Sci. Soc. Am., Am. Chem. Soc. (pesticide chemistry div.). Greek Orthodox. Subspecialties: Plant physiology (agriculture); Photosynthesis. Office: Dept Plant Pathology and Physiolog VA Poly Inst and State U Blacksbur VA 24061

HAUBRICH, ROBERT RICE, biology educator; b. Claremont, N.H., May 4, 1923; s. Frederick William and Marian Norma (Rice) H. B.S. in Forestry, Mich. State U., 1949, M.S. in Zoology, 1952; Ph.D. in Biology, U. Fla., 1957. Asst. prof. biology E. Carolina U., Greenville, N.C., 1947-61; asst. prof. biology Oberlin (Ohio) Coll., 1961-62; asst. to assoc. prof. biology Denison U., Granville, Ohio, 1962-68, prof., 1968—; assoc. dir. Earlham Biology Sta., Richmond, Ind., 1967-72; Sci. Coordinating rep. Great Lakes Coll. Assn., Ann Arbor, Mich., 1966-67; Faculty rep. Denison U. Jour. Biol. Sci., 1963—. Served as sgt. USAF, 1943-46. Fellow Ohio Acad. Sci. (v.p. zoology sect. 1972), AAAS; mem. Animal Behavior Soc., Nat. Wildlife Fedn., Sigma Xi. Subspecialties: Sociobiology; Evolutionary biology. Current work: Behavior and ecology of the starhead topminnow; conceptual structure of biology. Office: Dept Biology Denison U Granville OH 43023

HAUCK, JAMES PIERRE, research engineer; b. St. Cloud, Minn., Jan. 23, 1946; s. Harold and Marie Teresa (Pollard) H.; m. Gail Elaine Norfolk, June 16, 1968; m. Linda Lehman, Jan. 28, 1973; children: Thomas, Tiffany, Barara, Beverly. B.S. in Physics, Calif. State Poly. U., 1968; M.A. in Physics (NDEA Title IV fellow), U. Calif., Irvine, 1970; Ph.D. in Physics (NDEA TITLE IV fellow), U. Calif., Irvine, 1976. NOAA fellow U. Colo., 1971; research asst. U. Calif., Irvine, 1971-76, instr. physics and engring., 1971—; mem. tech. staff Rockwell Internat., Anaheim, Calif., 1976-79, staff scientist, 1980-82; research engr. Northrop Corp., Hawthorne, Calif., 1983—; dir. engring. Illumination Industries Inc., Sunnyvale, Calif., 1979-80; adj. prof. physics Calif. State U., Fullerton, 1982—; instr. physics and astronomy, Long Beach, 1977—. Contbr. articles to profl. jours. Active YMCA Indian Guides, YMCA Indian Princesses; treas. River Nation, 1982. Mem. Am. Phys. Soc. (div. plasma physics), IEEE (div. nuclear and plasma sci.), Optical Soc. Am. Subspecialties: Plasma physics; Laser gyroscopes and radars. Current work: Plasmas and lasers, especially laser gyroscopes and laser radars, laster physics, electrophoresis, cataphoresis, laster resonators, spectroscopy, laser systems. Patentee apparatus for generating temporally shaped laster pulses. Home: 1391 Longmont Pl Santa Ana CA 92705 Office: 2301 W 120th Street Hawthorne CA 90250

HAUEISEN, WILLIAM DAVID, business educator, consultant, author; b. Columbus, Ohio, July 10, 1943; s. William Frank and Martha M. (Meyerholtz) H.; m. Janice I. Ferne, June 1967; 1 dau., Lisa R. B.A., Capital U., 1966; M.Div., Lutheran Theol. Sem., 1970; M.A., Ohio State U., 1977, M.B.A., 1978, Ph.D., 1978. Research asst. Mgmt. Horizons, Columbus, 1973-76, v.p., 1977-80; Asst. prof. Sch. Bus., Pacific Lutheran U., 1976-77; asst. prof. Coll. Adminstrv. Sci., U South Fla., 1980—; prin. Haueisen Assocs. (cons. group.). Author: Business Systems for Microcomputers, 1981; contbr. articles to profl. jours.; editor: Cosmetic Jour, 1971-72. Mem. Acad. Mgmt., Am. Psychol. Assn., Inst. Decision Scis., Am. Mktg. Assn. Republican. Lutheran. Club: Carrollwood Country (Tampa, Fla.). Subspecialty: Information systems, storage, and retrieval (computer science). Current work: Systems design for micro-computers; software adaptation for small business computers; software ealuation of systems packages. Home: 5015 Chattam Ln Tampa FL 33624 Office: U South Fla 4202 Fowler Ave TAmpa FL 33620

HAUER, JEROME MAURICE, physiologist; b. N.Y.C., Oct. 31, 1951; s. Milton and Rose (Muscatine) H.; m. Glenda Reed, Sept. 27, 1980. B.A., NYU, 1976; M.H.S., Johns Hopkins U., 1978; doctoral candidate, Tufts U. Sch. Medicine. Assoc. adminstr. ARC Blood Services, Boston, 1970-80; research assoc. Johns Hopkins Sch. Medicine, Balt., 1976-78; asst. dir. transfusion services U. Md. Sch. Medicine, Balt., 1978-79; research assoc. Beth Israel Hosp. Harvard Med. Sch., Boston, 1980—. Contbr. articles to profl. jours. Served to lt. USAR, 1981—. Mem. Am. Heart Assn., Assn. Advancement Med. Instrumentation, Am. Pub. Health Assn., Am. Fedn. Clin. Research, Am. Assn. Blood Banks, Internat. Soc. Blood Transfusion, N.Y. Acad. Scis. Democrat. Jewish. Subspecialties: Hematology; Physiology (medicine). Current work: Investigations in the nature of coagulation defects in massive transfusion and trauma and alternates to homolgous blood transfusions. Home: PO Box 1267 Hightstown NJ 08520 Office: Beth Israel Hosp Harvard Med Sch 330 Brookline Ave Boston MA 02166

HAUGH, CLARENCE GENE, agrl. engr., engring. cons., educator; b. Spring Mills, Pa., Oct. 11, 1936; s. Clarence Glenn and Estella Jane (Baney) H.; m. Patricia Ann Breon, Apr. 30, 1942; children: Amy Elizabeth, Jennifer Lea, Mitchell Breon. B.S. in Agrl. Engring, Pa. State U., 1958, M.S., U. Ill., 1959; Ph.D., Purdue U., 1964. Registered profl. engr., Fla. Instr. agrl. engring. Purdue U., 1961-64; asst. prof. agrl. engring. U. Fla., Gainesville, 1964-65, Purdue U., 1965-68, assoc. prof., 1968-72, prof., 1972-79; head agrl. engring. Va. Poly. Inst. and State U., Blacksburg, 1979—. Contbr. articles profl. jours. Bd. trustees Chippokes Plantation State Park. Served to 1st lt. USAFR, 1958-64. Mem. Am. Soc. Agrl. Engrs. (Young Researcher award 1976), Inst. Food Technolgoists, Am. Soc. Engring. Education. Republican. Methodist. Lodges: Rotary; Masons; Shriners. Subspecialties: Agricultural engineering; Food science and technology. Current work: Prof. biol. materials, food engring. Patentee in field. Home: 406 Murphy St Blacksburg VA 24060 Office: Agrl Engring Dept Va Poly Inst and State U Blacksburg VA 24061

HAUKE, RICHARD LOUIS, botanist, educator; b. Detroit, Apr. 28, 1930; s. Henry George and Elanor Anne (Duquette) H.; m. Kathleen Armstron, Sept. 20, 1958; children: Katherine, Nellie, Andrew, Henry. B.S., U. Mich., 1952, Ph.D., 1960; M.A., U. Calif.-Berkeley, 1954. Instr. U. R.I., Kingston, 1959, asst. prof., 1960-64, assoc. prof., 1965-69, prof., 1970—; Fulbright lectr. U. Jordan, 1973-74. Served with U.S. Army, 1954-56. Mem. Am. Fern Soc., Bot. Soc. Am., Am. Inst. Biol. Scis. Roman Catholic. Subspecialty: Developmental biology. Current work: Developmental anatomy of Equisetum, transition from vegetative to reproductive phase of the shoot apex. Home: 3 Diane Dr Kingston RI 02881 Office: Dept Botany U R I Kingston RI 02881

HAUN, ROBERT DEE, JR., electric products manufacturing company research exec.; b. Lexington, Ky., Apr. 3, 1930; s. Robert Dee and Edna May (Minor) H.; m. Shirley Anne Porter, June 20, 1954 (div.); m. Karen Blash, June 4, 1977; children: Barbara L., Lynn Ellen Haun Betts, Janet D. B.S., U. Ky., 1952, Ph.D., MIT, 1957. With Westinghouse Electric, 1957—, dir. applied physics and math., Pitts., 1969-74, mgr. applied scis. research and devel., 1974-81, dir. industry products research and devel., Pitts., 1981—. Fellow AAAS; mem. Am. Phys. Soc., Optical Soc. Am., IEEE, Phi Beta Kappa, Sigma Xi. Subspecialties: Atomic and molecular physics; Laser research. Current work: Optical information processing; research managment. Designer cesium atomic beam frequency standard thesis apparatus on exhibit in Smithsonian Museum Sci. and Industry, Washington. Office: Westinghouse Research and Devel Center 1310 Beulah Rd Pittsburgh PA 15235

HAUPT, H. JAMES, mechanical design engineer; b. Palmerton, Pa., Jan. 3, 1940; s. Harry C. and Mary L. (Patrick) H.; m. Betty S. Niemi, Sept. 5, 1970; children: Nadine R., Heather J. A.A. Sci., Broome Tech. Coll., 1964; B.S., Ill. Inst. Tech., 1971. Registered profl. engr., Ill. Mech. engr. Argonne Nat. Lab., Ill., 1964—. Mem. Am. Nuclear Soc., Phi Theta Kappa, Tau Beta Pi. Subspecialties: Mechanical engineering; Nuclear engineering. Current work: Nuclear - liquid metal fast breeder; reactor technology - design, planning and assembly activities of large test vehicles for national reactor safety programs. Office: Argonne Nat Lab 9700 S Cass Ave Argonne IL 60439 Home: 3215 Saddle Dr Joliet IL 60435

HAURI, PETER J., sleep researcher, educator; b. Sirnach, Switzerland, June 25, 1933; came to U.S., 1960, naturalized, 1970; s. Rudolph and Verena (Wirz) H.; m. Deborah Rea, Aug. 29, 1961; children: Heidi, David, Katie. B.A., North Central Coll., Naperville, Ill., 1961; M.A., U. St. Gallen, Switzerland, 1956; Ph.D. in Clin. Psychology, U. Chgo., 1966. Staff psychologist Dewitt State Hosp., Auburn, Calif., 1966-68; asst. prof. psychiatry clin. psychology U. Va., Charlottesville, 1968-71; assoc. prof. psychiatry/clin. psychology Dartmouth Med. Sch., Hanvoer, N.H., 1971-78, prof. psychiatry/clin. psychology, 1978—, assoc. prof. psychology, 1978—. USPHS fellow, 1962-66; Founds. Fund for Psychiatry fellow, 1978-79. Mem. Sleep Research Soc. (exec. sec.-treas. 1981—), Am. Psychol. Assn. Congregationalist. Lodge: Rotary. Current work: Research on insomnia and sleep disorders. Office: Dartmouth-Hitchcock Sleep Disorders Center 703 Remsen Bldg Hanover NH 03756

HAUROWITZ, FELIX, biochemist, educator; b. Prague, Czechoslovakia, Mar. 1, 1896; came to U.S., 1948, naturalized, 1952; s. Rudolf and Emilie (Russ) H.; m. Gina Perutz, June 23, 1925. M.D., German U., Prague, 1922, Sc.D., 1923; M.D. (hon.), U. Istanbul, Turkey, 1973, Ph.D., Ind. U., 1974. Asst. prof. physiol. chemistry Med. Sch. German U. Prague 1922-30, asso. prof., 1930-39; head dept. biol chemistry, also prof. Med. Sch. U. Istanbul, Turkey, 1939-48; prof. chemistry Ind. U., 1948—, Distinguished prof., 1958—. Author: Biochemistry, 1955, Progress in Biochemistry, since 1949, 1959, Chemistry and Function of Proteins, 1963; Immunochemistry and the Biosynthesis of Antibodies, 1968. Recipient Paul Ehrlich prize and gold plaquette Paul Ehrlich Fund, Frankfurt, Germany, 1960. Fellow Am. Acad. Arts and Sci.; mem. Am. Chem. Soc. (chmn. div. biol. chemistry 1962-63), Leopoldina Acad. Scis., Am. Soc. Biol. Chemists, Am. Assn. Immunologists, Nat. Acad. Scis., Am. Soc. Microbiology (hon.), Societe de Chimie Biologique (hon.), Societe Immunologique (hon.). Subspecialty: Biophysical chemistry. Current work: Mechanism of antibody production. Spl. research protein chemistry and immunochemistry. Home: 910 Juniper Pl Bloomington IN 47401

HAURWITZ, BERNHARD, educator; b. Glogau, Germany, Aug. 14, 1905; came to the U.S., 1941, naturalized, 1946; s. Paul and Betty (Cohn) H.; m. Eva Schick, May 11, 1934 (div. Nov. 1946); 1 son, Frank David; m. Marion B. Wood, Jan. 16, 1961. Ph.D., U. of Leipzig, 1927. Privatdozent U. of Leipzig; 1931-32; research asso. Harvard, 1932-35; lectr. U. Toronto, 1935-37; meteorologist, Dominion, Can., 1937-41; asso. prof. meteorology Mass. Inst. Tech., 1941-47; asso. Woods Hole Oceanographic Inst., 1947- 59; prof., chmn. dept. meteorology and oceanography N.Y.U., 1947-59; prof. astrogeophysics U. Colo., 1959-64, prof. geophysics, 1960; with Nat. Center Atmospheric Research, Boulder, Colo., 1964—, dir. advanced study program, 1968-69; prof. atmospheric scis. U. Tex., 1966-68, also Colo. State U.; prof. U. Alaska, 1970—. Contbr. tech. articles to numerous publs. Decorated Cross of Merit 1st class Fed. Republic of Germany; recipient Recipient Rossby award Am. Meteorol. Soc., 1962. Mem. Nat. Acad. Sci., Deutsche Akademie der Naturforscher Leopoldina, Royal Meteorol. Soc., Am. Meteorol. Soc. (hon. mem.), Am. Geophys. Union (Bowie award 1970), ACLU, Sigma Xi. Subspecialty: Dynamic meteorology. Home: 2523 Constitution Ave Fort Collins CO 80526 Office: Dept Atmospheric Sci Colo State Univ Fort Collins CO 80523

HAUS, HERMANN ANTON, electrical engineering educator; b. Ljubljana, Yugoslavia, Aug. 8, 1925; came to U.S., 1948, naturalized, 1956; s. Otto Maxmilian and Helene (Hynek) H.; m. Eleanor Laggis, Jan. 24, 1953; children: William Peter, Stephen Christopher, Cristina Ann, Mary Ellen. Student, Technische Hochschule, Graz, 1946-48, 1948; B.S., Union Coll., 1949; M.S., Rensselaer Poly. Inst., 1951; Sc.D., Mass. Inst. Tech., 1954. Asst. prof. Mass. Inst. Tech., Cambridge, 1954-58, asso. prof., 1958-62, prof. elec. engring., 1962-73, Elihu Thomson prof. elec. engring., 1973—; vis. prof. Technische Hochschule, Vienna, 1959-60, Tokyo Inst. Tech., 1980; vis. MacKay prof. U. Calif. at, Berkeley, summer 1968; cons. Raytheon Co., 1956—, Lincoln Labs., 1963—; mem. Nat. Acad. Scis. adv. panel, Radio Propagation Lab. Nat. Bur. Standards, 1965-67. Author: (with R.B. Adler) Circuit Theory of Linear Noisy Networks, 1959, (with L.D. Smullin) Noise in Electron Devices, 1959, (with P. Penfield, Jr.) Electrodynamics of Moving Media, 1967; Mem. editorial bd.: Jour. Applied Physics, 1960-63, (with P. Penfield, Jr.) Electronics Letters, 1965-73, Internat. Jour. Electronics, 1975—. Guggenheim fellow, 1959-60. Fellow IEEE, Am. Acad. Arts and Scis.; mem. Nat. Acad. Engring., Am. Phys. Soc., Sigma Xi, Eta Kappa Nu., Tau Beta Pi, Phi Delta Theta. Subspecialties: Electronics; Optical signal processing. Current work: Waveguide optics, picosecond optics; electrodynamics. Home: 3 Jeffrey Terr Lexington MA 02173 Office: 77 Massachusetts Ave Cambridge MA 02139

HAUSER, RAY LOUIS, materials engr.; b. Litchfield, Ill., Apr. 16, 1927; s. A. Vernon and Grace Baker (Gregg) H.; m. Consuelo W. Minnich, Sept. 2, 1951; children: Beth, Cynthia, Dewi, Chris. B.S. in Chem. Engring, U. Ill., 1950; M. Engring., Yale U., 1952; Ph.D., U. Colo., 1957. Registered profl. engr., Colo., Calif. Project engr. Conn. Hard Rubber Co., New Haven, 1950-52; mem. research staff U. Colo., Boulder, 1954-57; head materials engring. Martin Co., Denver, 1957-61; dir. research Hauser Labs., Boulder, 1961—; mem. faculty U. Colo., 1954-70. Contbr. articles to profl. jours. Pres. Boulder Civic Opera, 1969-71. Served with U.S. Navy, 1945-46; Served with U.S. Army, 1952-54. Mem. AAAS, Am. Inst. Chem. Engrs., Soc. Plastics Engrs., ASTM, Soc. Advancement Materials and Process Engrs. Presbyterian. Subspecialties: Polymers; Chemical engineering. Current work: Adhesives and bonding processes; expert testimony. Patentee in field. Office: PO Box G Boulder CO 80306

HAUSMAN, HERSHEL JUDAH, physics educator; b. Pitts., Aug. 19, 1923; s. David and Sophie H.; m. Korene Brenner, May 18, 1944; children: Herbert A., Sally Z., William B. B.S., Carnegie-Mellon U., 1948, M.S., 1949; Ph.D., U. Pitts., 1952. Asst. prof. physics Ohio State U., Columbus, 1952-57, assoc. prof., 1957-63, prof., 1963—, supr. cyclotron lab., 1952-63; supr. Van de Graaff Accelerator Lab., 1963—. Contbr. articles to profl. jours. Served to 1st lt. USAAF, 1943-45. Decorated Air medal with 3 oak leaf clusters. Fellow Am. Phys. Soc.; mem. AAAS, AAUP. Subspecialties: Nuclear physics; Nuclear astrophysics. Current work: Research in medium-energy particle-capture reactions; nuclear astrophysics. Office: Ohio State University Van de Graaff Accelerator Laboratory 1302 Kinnear Rd Columbus OH 43212

HAUSMANN, WERNER KARL, pharmaceutical company executive; b. Edigheim, Germany, Mar. 9, 1921; came to U.S., 1948, naturalized, 1954; s. Carl and Johanna (Sprenger) H.; m. Helen Margaret Vas, Sept. 29, 1949; 1 son, Gregory. M.S. in Chem. Engring, Swiss Fed. Inst. Tech., 1945; D.Sc., 1947. Cert. quality engr. Research fellow U. London, 1947-48; research asso. Rockefeller Inst. for Med. Research, N.Y.C., 1949-57; research group leader Lederle Labs., Pearl River, N.Y., 1957-66; ass. dir. quality control Ayerst Labs., Rouses Point, N.Y., 1966-71; dir. quality control Stuart Pharms., Pasadena, Calif., 1971-74; dir. quality assurance, analytical research and devel. Adria Labs. Inc., Columbus, Ohio, 1974—. Contbr. articles to profl. jours. Pres. Ednl. TV Assn., 1970-71; radiation officer CD, 1962-66. Served to 1st lt. Swiss Army, 1939-46. Fellow N.Y. Acad. Scis., AAAS, Am. Soc. Quality Control (chmn. Columbus sect.); mem. Acad. Pharm. Scis., Am. Soc. Biol. Chemists, Am. Chem. Soc., Am. Soc. Microbiology, Parenteral Drug Assn., Federation Internationale Pharmaceutique. Presbyterian. Club: Sawmill Athletic (Columbus). Subspecialties: Analytical chemistry; Organic chemistry. Current work: research and development of analytical methods of drug substances and dosage forms as such and in biological fluids, with emphasis on high performance liquid chromatography and gas chromatography. Patentee antibiotics. Office: 5000 Post Rd Dublin OH 43017

HAUSTEIN, PETER EUGENE, research chemist; b. Detroit, Jn. 17, 1944. B.S., U. Calif., 1966; M.S., and Ph.D., Iowa State U., 1970. Fellow in nuclear chemistry Brookhaven Nat. lab., Upton, N.Y., 1970-72; asst. prof. chemistry Yale U., New Haven, 1972-75; assoc. chemist Brookhaven Nat. Lab., 1975-78, chemist, 1978—; vis. scientist Lawrence Berkeley Lab. (Calif.), 1979. Mem. Am. Chem. Soc., Am. Phys. Soc., Sigma Xi. Subspecialty: Nuclear chemistry. Office: Dept Chemistry Brookhaven Nat Lab Upton NY 11973

HAUTALUOMA, JACOB EDWARD, organizational behavior educator, consultant; b. Chatham, Mich., June 28, 1933; s. Toivo Jack and Iria Aurora (Nikkinen) H.; m. Betty Lou Johnson, Mar. 24, 1956; children: Jodi, Grey. B.A., U. Minn.-Duluth, 1955; M.S., U. Colo., 1963, Ph.D., 1967. Indsl. engr. U.S. Steel Corp., Duluth, 1956-60; asst. prof. indsl.-organizational psychology Colo. State U., 1965-71, assoc. prof., 1971-81, prof., 1981—, chmn. sect. indsl.-organizational psychology, 1972-76, 82—; project dir., trainee devel. officer, field assessment officer Peace Corps, Malaysia, Morocco, Micronesia and Afghanistan, summers 1967, 68, 71, 75, spring 1968; Disting. vis. prof. and vis. lectr. Oreg. State U., U. Hawaii, U. Minn.-Duluth, summers 1972, 73, 77; expert social scientist AID, Washington, summers 1974, 75; Fullbright lectr. U. Iceland, Reykjavik, 1977, Istanbul (Turkey) U. and Aegean U., Izmir, Turkey, 1980, Helsinki (Finland) Sch. Econs., spring 1980; lectr. and cons. U. Iceland and Icelandic Mgmt. Assn., Reykjavik, 1981; postdoctoral research fellow Yale U., 1971-72; cons. organizational behavior to numerous orgns. in, U.S., Afghanistan, Finland and Iceland, 1967—. Author tech. reports; contbr. articles to profl. publs. Democratic precinct capt., Ft. Collins, Colo., 1967-68, mem. county exec. com., fin. chmn., 1969-71; mem. ch. council, stewardship chmn. Trinity Lutheran Ch., Ft. Collins, 1979-82; planner and leader workshop to help produce collaboration between County, Fed. Exec. Assn. and C. of C., Ft. Collins, 1982. Recipient Grad. Advising award Dept. Psychology Colo. State U., 1981; USPHS fellow, 1960-61, 63-64. Mem. Am. Psychol. Assn., Acad. Mgmt., Colo. Wyo. Assn. Indsl./Organizational Psychologists (a founder; chmn. 1973—), Internat. Council Psychologists (Colo. and Idaho rep.), Rocky Mountain Psychol. Assn., Soc. Advancement Social Pschology, Internat. Assn. Applied Psychologists, Fulbright Alumni Assn., Orgn. Behavior Teaching Soc. Club: Gt. Books (Ft. Collins) (a founder). Subspecialties: Organizational behavior; Social psychology. Current work: Research on conflict resolution, overeducated workers, organizational change, group entry, etc.; consulting in organizational development, interdisciplinary and intercultural teamwork. Home: 701 Dartmouth Tr Fort Collins CO 80525 Office: Dept Psychology Colo State U Fort Collins CO 80523

HAVA, MILOS, pharmacologist, pharmaceutical company executive; b. Prague, Czechoslovakia, Oct. 15, 1927; came to U.S., 1968, naturalized, 1983; s. Emanuel and Eta (Dorman) H.; m. Maria M. Kovac, Sept. 5, 1951; 1 dau., Nadia. M.D., Charles U., Prague, 1952, Ph.D., 1955. Asst. prof. pharmacology Med. Sch., Charles U., Prague, 1952-55; sr. research worker Czechoslovakian Acad. Scis., Prague, 1955-58; dir. dept. pharmacology Research Inst. Natural Products, Prague, 1958-68; assoc. prof. pharmacology Sch. Medicine, U. Kans., Kansas City, 1968-73, Sch. Medicine, U. Ill., Peoria, 1973-75; asst. med. dir. Marion Labs., Inc., Kansas City, 1975-79; adj. prof. dept. pharmacology U. Kans. Med. Sch., Kansas City, 1977-79; dir. clin. research Carter-Wallace, Inc., Cranbury, N.J., 1979-80, exec. dir. clin. drug devel., 1980-82; regional dir. clin. research Wyeth Internat. Ltd., Phila., 1982—; adj. prof. diagnostic imaging Temple U. Sch. Medicine, Phila., 1983—. Contbr. numerous articles on pharmacology and clin. pharmacology to sci. jours. Recipient prize Czechoslovakian Acad. Sci., 1956; prize for best publ. of yr. in Czechoslovakian pharmacology, 1964, 67. Mem. Am. Soc. Pharmacology and Exptl. Therapeutics, AAAS, AMA, Am. Assn. Clin. Pharmacology. Subspecialties: Pharmacology; Cellular pharmacology. Current work: Clinical research and drug development; pharmacology of the gastrointestinal smooth muscle; pharmacology of steroids. Office: PO Box 8616 Philadelphia PA 19101

HAVAS, HELGA FRANCIS, immunologist, educator, researcher; b. Vienna, Austria, Nov. 26, 1915; came to U.S., 1941, naturalized, 1948; d. Franz and Bertha (Tarnay) Hollering; m. Peter Havas; children: Eva Catherine, Stephen, Walter. M.A., Columbia U., 1944; Ph.D., Lehigh U., 1950. Research asso. Inst. Cancer Research, Phila., 1950-63; asso. prof. immunology Tempe U. Sch. Medicine, 1963-72, prof., 1972—, USPHS spl. fellow, 1964-65; lectr. in field. Recipient Career Devel. award NIH USPHS, 1967-71. Mem. Am. Assn. Immunologists, Am. Cancer Soc. Subspecialties: Immunology (medicine); Immunobiology and immunology. Current work: Cancer research; immunology. Office: Temple U Sch Medicine Braod and Tioga Philadelphia PA 19140

HAVENER, WILLIAM HENRY, medical educator, ophthalmologist; b. Portsmouth, Ohio, June 2, 1924; s. Gilbert and Laura M. (Braunlin) H.; m. Phyllis Ann, Jan. 26, 1946; children: Ann, Michael, Mark, Gail, John, Amy, Neal. B.A., Wooster Coll., 1944, D.Sc. (hon.), 1982; M.D., Western Res. U., 1948; M.S. in Ophthalmology, U. Mich., 1953. Diplomate: Am. Bd. Ophthalmology. Intern in medicine Univ. Hosps., Cleve., 1948-50; resident in ophthalmology U. Mich., Ann Arbor, 1951-53; asst. prof. dept. ophthalmology Ohio State U., Columbus, 1953-56, assoc. prof., acting chmn. dept., 1956-59, prof., 1959—, chmn. dept., 1959-61, 72—; hon. bd. dirs. Ohio Soc. for Prevention of Blindness. Author: books including Atlas of Diagnostic Techniques & Treatment of Retinal Detachment, 1967, Atlas of Cataract Surgery, 1972. Bd. dirs. Columbus chpt. Mothers Against Drunk Drivers. Recipient teaching awards Ohio State U., 1978, 79, 80. Mem. Am. Acad. Ophthalmology, Retina Soc., Ohio Med. Soc., Phi Beta Kappa, Alpha Omega Alpha. Subspecialty: Ophthalmology. Office: Ohio State University 456 Clinic Dr Columbus OH 43210

HAVENS, A. VAUGHN, meteorology educator; b. Fairfield, Conn., Sept. 10, 1922; s. J. Gilbert and Anna (Anderson) H.; m. Charlotte Stone, July 16, 1943; children: Eileen Marie, Patricia Mary. B.Sc., N.Y. U., 1944; M.S., Rutgers U., 1948. Instr. meteorology Rutgers U., New Brunswick, N.J., 1948-50, asst. prof., 1950-56, assoc. prof., 1956-60, prof., 1960—; cons. U.S. Weather Service, 1956-61, N.J. Dept. Environ. Protection, 1965-75, Environplan Inc., West Orange, N.J., 1979—. Contbr. chpt. to book; editorial bd.: Internat. Jour. Agrl. Meteorology, 1962—. Served to lt. col. USAF, 1943-46, 51-53. Fellow N.Y. Acad. Sci (sect. chmn. 1973-74); mem. Am. Meteorol. Soc., Am. Geophys. Union, Air Pollution Control Assn., Sigma Xi. Subspecialties: Climatology; Micrometeorology. Current work: Agricultural meteorology and climatology; air pollution meteorology and climatology. Home: 2 Haywood Ave Piscataway NJ 08854 Office: Cook Coll Rutgers U PO Box 231 New Brunswick NJ 08903

HAVILAND, JAMES WEST, physician; b. Glens Falls, N.Y., July 18, 1911; s. Morrison LeRoy and Mabel Eva (West) H.; m. Marion Cranston Bertram, Oct. 23, 1943; children—James Marshall, Elizabeth Bullard, Donald Sherman, Martha Adams. A.B., Union Coll., Schenectady, 1932; M.D., Johns Hopkins, 1936. Intern medicine Johns Hopkins Hosp., 1936-37, intern, asst. resident, chief outpatient dept. pediatrics, 1937-38, asst. resident medicine, 1939-40, New Haven Hosp., 1938-39; instr. medicine Yale Med. Sch., 1938-39, Johns Hopkins Sch. Medicine, 1939-40; chief services crippled children Wash. Dept. Social Security, also Dept. Health, 1940- 42; lectr.

medicine U. Wash. Sch. Nursing, 1946-60; practice medicine, Seattle, 1946—; clin. asst. prof., to clin. prof. U. Wash. Sch. Medicine, 1947—, asst. dean, 1949-53, 1954-59, acting dean, 1953-54, asso. dean, 1972-76. Trustee Seattle Artificial Kidney Center, Seattle Symphony Orch. Served as lt. comdr., M.C. USNR, 1942-46. Fellow Am. Geog. Soc. N.Y., Am. Heart Assn.; mem. Wash. State Med. Assn. (sec.-treas. 1948-51), Seattle Acad. Internal Medicine (pres. 1952-53), King County Med. Soc. (pres. 1962), AMA (council med. edn. 1966-76, chmn. 1974-76), Pacific Interurban Clin. Club, AAAS, Am. Fed. Med. Research, Western Soc. Clin. Research, North Pacific Soc. Internal Medicine, A.C.P. (pres. 1970), Am. Clin. and Climatol. Assn. (pres. 1981), Am. Assn. History Medicine, Nat. Acad. Scis. (Inst. Medicine), Phi Beta Kappa, Sigma Xi, Alpha Omega Alpha, Kappa Alpha. Subspecialty: Internal medicine. Home: 8208 SE 30th St Mercer Island WA 98040 Office: 721 Minor Ave Seattle WA 98104

HAVLEN, ROBERT JAMES, astronomer; b. Utica, N.Y., Sept. 16, 1943; s. Frank James and Marian (Briggs) H.; m. Carolyn Wolf, Sept. 2, 1967. B.S. in Astrophysics, U. Rochester, 1965; Ph.D. in Astronomy, U. Ariz., 1969. Staff astronomer European So. Obs., Santiago, Chile, 1970-77; vis. lectr. U. Va., 1977-79; asst. to dir./assoc. scientist Nat. Radio Astronomy Obs., Charlottesville, Va., 1979—. Contbr. articles on astronomy, astrophysics to profl. jours., 1972-78. Mem. Am. Astron. Soc., Internat. Astron. Union. Methodist. Subspecialties: Optical astronomy; Radio and microwave astronomy. Current work: Stellar assns., galactic structure, clusters of galaxies. Office: Nat Radio Astronomy Obs Edgemont Rd Charlottesville VA 22901

HAWK, HAROLD WILLIAM, physiologist, researcher, government research leader; b. Meadville, Pa., Dec. 29, 1927; s. Stewart Dean and Anna Belle (Smith) H.; m. Donna Gail Haney, July 31, 1953; children: Sharon, Susan, Kevin. B.S., Pa. State U., 1951; M.S., U. Wis., 1953, Ph.D., 1956. Research physiologist Dept. Agr., 1956-64; leader physiology investigations unit Beltsville (Md.) Agrl. Research Ctr., 1964-72; chief Animal Reprodn. Lab., 1972—; Upjohn lectr. Am. Fertility Soc., 1969. Recipient Superior Service award Dept. Agr., 1966, Physiology and Endocrinology award Am. Soc. Animal Sci., 1972. Subspecialties: Animal physiology; Reproductive biology. Current work: Reproductive physiology of domestic animals. Office: Agrl Research Center Beltsville MD 20705

HAWKE, RONALD SAMUEL, electronic engr.; b. Oakland, Calif., Mar. 4, 1940; s. Samuel Henry and Marian Anrionette (Howland) H.; m. Nancy Lee Wirth, May 23, 1975; children: Laura, Augusta, Veronica, Theresa, Monisa. B.S. in E.E, U. Calif., Berkeley, 1962. Elec. engr. Lawrence Livermore Nat. Lab., Livermore, Calif., 1961—; guest scientist Max Planck Inst., Stuttgart, Germany, 1973-74. Contbr. articles to profl. jours. Mem. Am. Phys. Soc., AAAS. Subspecialties: Applied magnetics; Magnetic physics. Current work: Electromagnetic railgun development and application to high pressure research, fusion and space. Patentee in field. Office: L-355 LLNL Box 808 Livermore CA 94550

HAWKES, GRAHAM SIDNEY, ocean engineer; b. London, Dec. 23, 1947; U.S., 1981; s. Sidney Charles and Winifred Florence (Brooks) H. H.N.D. Mech. Engring. with honors, Borough Poly., London, 1969. Engr. Plessy Underwater Weapons Unit, U.K., 1971-75; engr. D. H. B. Constrn., Ltd., U.K., 1975-76; co-founder, mng. dir. Offshore Submersible, Ltd., Great Yarmouth, 1977, Osel Mantis, Ltd., 1978, Osel Group, 1979-81; co-founder, pres., chmn. Deep Ocean Tech., Inc., Oakland, Calif., 1981—, Deep Ocean Engring., Inc., Oakland, 1982—; adv. Oceanic Soc., Stamford, Conn., 1982; mem. adv. bd. Ocean Trust Found., San Francisco, 1980—; indsl. adv. Sci. Research Council U.K., 1979—. Recipient Canadian Forces cert. mil. achievement, 1978, Charles A. Lindbergh award Charles A. Lindbergh Found., 1981. Mem. ASME, Soc. Underwater Tech., Marine Tech. Soc., Oceanic Soc. Clubs: Yacht, Single-handed Sailing Soc. (San Francisco). Subspecialties: Robotics; Systems engineering. Current work: Management of research and development and offshore operations companies, design and development of subsea technology including sensory manipulator systems, robotic and manned submersible systems for ocean exploration and industrial uses. Patentee in field. Office: Deep Ocean Tech Inc 12812 Skyline Blvd Oakland CA 94619

HAWKINS, NEIL MIDDLETON, educator, civil engineer; b. Sydney, Australia, Jan. 31, 1935; s. Cecil Alfred and Sybil Mabel (Ralph) H.; m. Saundra Ann Youmans, Sept. 15, 1961; children: Susan Elizabeth, David Clark. B.Sc., U. Sydney, 1955, B.E., 1957; M.S., U. Ill., 1959, Ph.D., 1961. Cons. engr., Sydney, 1958; lectr. U. Sydney, 1962-65, sr. lectr., 1966-68; devel. engr. Portland Cement Assn., Chgo., 1965-66; asso. prof. U. Wash., Seattle, 1968-72, prof., 1972—, chmn. dept. civil engring., 1978—; prin. investigator NSF projects on seismic resistance of structures, 1973—. Contbr. articles to profl. jours. Served to 2d lt. Australian Citizen Mil. Forces, 1953-62. Fellow Am. Concrete Inst. (dir. 1982—, Wason medal 1970, Raymond C. Reese award 1978); mem. Australian Instn. Engrs. (Edward Noyes prize 1967), ASCE (State of the Art award 1974, Raymond C. Reese award 1976), Earthquake Engring. Research Inst., Post-Tensioning Inst. Subspecialty: Civil engineering. Current work: Reinforced and prestressed concrete structures,mixed steel and concret structures subject to dynamic and repeated loads. Home: 18204 NE 28th St Redmond WA 98502 Office: U Wash Seattle WA 98195

HAWKINS, RICHARD ALBERT, physiologist; b. Greenwich, Conn., Mar. 27, 1940; s. Albert Rice and Florence Marie Elizabeth (Hansen) H.; m. Enriqueta Maria Antonia Elias-Mondeja, May 9, 1964; children: Richard Alfred, Paul Andres. B.Sc., San Diego State U., 1963; M.A., Harvard U., 1969, Ph.D. in Physiology, 1969. Research fellow Oxford (Eng.) U., Metabolic Research Lab., Radcliffe Infirmary, 1969-71; sr. staff fellow NIMH, St. Elizabeth's Hosp., Washington, 1971-74; chief phys. sci. br. Bur. Med. Devices, HEW, Rockville, Md., 1974-76; asso. prof. exptl. neurosurgery and physiology N.Y. U. Med. Center, N.Y.C., 1975-77; prof. physiology, chief div. anesthesia and metabolic research Milton S. Hershey Med. Center, Pa. State U. Coll. Medicine, 1977—; mem. neurology B. study sect. NIH, Bethesda, 1981—. Contbr. articles to profl. jours. USPHS fellow, 1966, 69-71; Nat. Inst. Neurol. and Communicative Disorders grantee, 1976—; Am. Diabetes Assn. grantee, 1979-80. Fellow Am. Heart Assn.; mem. Biochem. Soc., Soc. Neurosci., Am. Physiol. Soc., Harvey Soc., Am. Soc. Neurochemistry, Internat. Soc. Cerebral Blood Flow and Metabolism (founding mem.). Subspecialties: Physiology (medicine); Neurochemistry. Current work: Cerebral physiology and nutrition; cerebral metabolism, cerebral blood flow, blood-brain barrier. Office: 500 University Dr Hershey PA 17033

HAWKINS, ROBERT C(LEO), mech. engr.; b. Bedford,County, Va., Mar. 20, 1927; m. Dorothy Maxine, Dec. 12, 1925; children. B.S.M.E. magna cum laude, Va. Poly. Inst., 1950; postgrad. Am. studies for exec. programs, Williams Coll., 1980; grad., Advanced Mgmt. Program, Harvard U., 1982. With Gen. Electric Co., 1957—, mgr. Evendale prodn. engring. ops., Cin., 1980-82, gen. mgr. advanced tech. ops., 1982—. Served with USN, 1944-46. Mem. AIAA, Cin. Engring. Soc., Herman Schneider Found., Tau Beta Pi, Pi Tau Sigma. Republican. Club: Maketewah Country (Cin.). Subspecialties: Mechanical engineering; Aerospace engineering and technology. Current work: Advanced aircraft gas turbine design, devel. and prodn. techs.; mgmt. advanced tech. research and devel.

HAWKINS, ROBERT DRAKE, neuroscience researcher, educator; b. Washington, May 2, 1946; s. Edward Russell and Hermione Helene (Hunt) H. B.A., Stanford U., 1968; Ph.D., U. Calif.-San Diego, 1973. Research assoc. U. Calif.-San Diego, 1974; postdoctoral fellow Centre National de la Recherche Scientifique, Gif-sur-Yvette, France, 1974-75, Columbia U. Coll. Physicians and Surgeons, 1975-79, staff assoc., 1979-82, asst. prof. neurobiology and behavior, 1982—. Contbr. articles, chpts. to profl. publs. Mem. AAAS, Soc. Neurosci., Union Concerned Scientists, Phi Beta Kappa. Subspecialties: Neurobiology; Learning. Current work: The cellular mechanisms of learning and memory; cellular and behavioral studies of the neural mechanism of classical conditioning in Aplysia californica. Home: 104 Riverside Dr Apt 4A New York NY 10024 Office: 722 W 168th St New York NY 10032

HAWKINS, WALTER LINCOLN, engineer; b. Washington, Mar. 21, 1911; s. William Langston and Catherine Elizabeth (Johnson) H.; m. Lilyan Varina Bobo, Aug. 19, 1939; children: W. Gordon, Philip L. Chem.E., Rensselaer Poly. Inst., Troy, N.Y., 1932; M.S., Howard U., Washington, 1934; Ph.D., McGill U., Montreal, Que., 1938; LL.D., Montclair State Coll., 1974, Kean State Coll., 1983; D.Eng., Stevens Inst. Tech., 1979. Sessional lectr. McGill U., 1938-41; NRC fellow Columbia, 1941-42; with Bell Telephone Labs., Inc., Murray Hill, N.J., 1942—. Editor: Polymer Stabilization; Contbr. articles to profl. jours., chpts. to books. Trustee Montclair State Coll. Recipient Honor scroll Am. Inst. Chemists, 1970. Mem. Nat. Acad. Engring. Subspecialties: Polymer chemistry; Polymers. Current work: Stabilization of synthetic polymers against environmental degradation; recycling of plastic scrap. Patentee in field. Home: 26 High St Montclair NJ 07042 Office: Bell Telephone Labs Murray Hill NJ 07971 There are many measures of success, but none to be more cherished than the role one may have played in encouraging others to follow in your footsteps.

HAWKINS, WILLIS MOORE, JR., aircraft company executive; b. Kansas City, Mo., Dec. 1, 1913; s. Willis Moore and Elizabeth (Daniels) H.; m. Anita E. Stanfil, June 22, 1940 (dec. Nov. 5, 192); children—Nancy Gay, Willis Moore, James Walter. Student, Ill. Coll., 1932-34, D.Sc., 1966; B.S. in Aero. Engring, U. Mich., 1937, D.Eng., 1964. Engr. trainee Grumman Aircraft Co., 1936-37; with Lockheed Aircraft Co., 1937-54, dir. advanced design, 1942-54; with Lockheed Missiles and Space Div., Sunnyvale, Calif., 1954-63, asst. gen. mgr., 1957-61, v.p., gen. mgr. space systems, 1961-62, corporate v.p. engring., 1962-63; lectr. aerospace scis. and mgmt. U. Calif. at, Los Angeles, 1954-55; asst. sec. army for research and devel., 1963-66; v.p. sci. and engring. Lockheed Aircraft Corp., Burbank, Calif., 1966-70, sr. v.p., 1970-74, sr. adviser, 1974-76, dir, 1972-80; pres. Lockheed Calif. Co., 1976-79, sr. v.p. (aircraft), 1979-80, sr. advisor, 1980—; dir. Wackenhut Corp., Billings Corp., AVEMCO; mem. Army Sci. Adv. Panel, 1957-74; adviser NACA, 1952-54; mem. adv. council NASA, 1978-83, chmn. safety adv. panel, 1981—. Recipient Disting. Pub. Service medal for contbns. to Polaris fleet ballistic missile system Navy Dept., 1961; Disting. Civilian Service medal Dept. Army, 1965; with laurel, 1966; Disting. Civilian Service medal NASA, 1975; Wright Bros. Meml. Trophy, 1982. Fellow AIAA, Royal Aero. Soc.; mem. Nat. Acad. Engring., Tau Beta Pi. Subspecialty: Aeronautical engineering. Home: 4249 Empress Ave Encino CA 91436 Office: Lockheed Corp Burbank CA 91520

HAWROT, EDWARD, biochemist, educator; b. Hamburg, W.Ger., Aug. 18, 1948; came to U.S., 1949, naturalized, 1955; s. Frank and Eugenia (Mokwinski) H.; m. Donna Pierson, Nov. 26, 1970; children: Aimee Caroline, Jacquelyn Renee. A.B. with honors, U. Detroit, 1970; Ph.D., Harvard U., 1976. Postdoctoral fellow dept. neurobiology Harvard U. Med. Sch., Boston, 1976-80; asst. prof. dept. pharmacology Yale U. Sch. Medicine, New Haven, 1980—. Contbr. articles to sci. jours. Helen Hay Whitney fellow, 1976-79. Mem. AAAS, Soc. for Neurosci. Subspecialties: Neurobiology; Cell and tissue culture. Current work: Studying biochemistry of cell surface receptors. Office: PO Box 3333 New Haven CT 06510

HAWRYLKO, EUGENIA ANNA, biomedical researcher, physician; b. N.Y.C., Feb. 7, 1942; d. Nicholas and Ludmila (Pavlov) H.; m. Raymond J. Aab, Aug. 17, 1974; children: Allison, Elizabeth. A.B., Vassar Coll., 1962; M.D., N.Y. U., 1966. Diplomate: Am. Bd. Allergy and Immunology, 1979. Intern, resident in medicine Bellevue Hosp., N.Y.C., 1966-68; research fellow in medicine Harvard Med. Sch., 1968-69; research fellow Trudeau Inst., Saranac Lake, N.Y., 1969-73, asst. mem., 1973-74; research assoc. Sloan-Kettering Inst., N.Y.C., 1974-75, assoc., 1975-81; clin. fellow in allergy and immunology N.Y. Hosp.-Cornell Med. Center, N.Y.C., 1978-80; chief immunology labs. dept. allergy-immunology L.I. Coll. Hosp., Bklyn., 1982—; mem. adv. com. clin. investigations immunology and immunotherapy Am. Cancer Soc., 1978-82. Contbr. articles to profl. jours. USPHS trainee, 1968-73; Am. Cancer Soc. Faculty Research awardee, 1976-81. Mem. Am. Assn. Immunologists, Am. Assn. Cancer Research, Am. Acad. Allergy, N.Y. Allergy Soc. Subspecialties: Immunology (medicine); Cancer research (medicine). Current work: Regulation of cellular interactions in antitumor responses. Home: 116 25 Union Turnpike Forest Hills NY 11375 Office: Long Island Coll Hosp 340 Henry St Brooklyn NY 11201

HAWTHORNE, MARION FREDERICK, chemistry educator; b. Ft. Scott, Kans., Aug. 24, 1928; s. Fred Elmer and Colleen (Webb) H.; m. Beverly Dawn Rempe, Oct. 30, 1951 (div. 1976); children: Cynthia Lee, Candace Lee; m. Diana Baker Razzaia, Aug. 14, 1977. B.A., Pomona Coll., 1949; Ph.D. (AEC fellow), U. Calif. at Los Angeles, 1953; D.Sc. (hon.), Pomona Coll., 1974. Research asso. Iowa State Coll., 1953-54; research chemist Rohm & Haas Co., Huntsville, Ala., 1954-56, group leader, 1956-60, lab. head, Phila., 1961; vis. lectr. Harvard, 1960, Queen Mary Coll., U. London, 1963; vis. prof. Harvard U., 1968; prof. chemistry U. Calif. at Riverside, 1962-68, U. Calif. at Los Angeles, 1968—; vis. prof. U. Tex., Austin, 1974; Mem. sci. adv. bd., USAF, 1980—. Editor: Inorganic Chemistry, 1969—; Editorial bd.: Progress in Solid State Chemistry, 1971—, Inorganic Syntheses, 1966—, Organometellics in Chemical Synthesis, 1969—, Synthesis in Inorganic and Metalorganic Chemistry, 1970—. Recipient Chancelors Research award, 1968; Herbert Newby McCoy award, 1972; Am. Chem. Soc. award in inorganic chemistry, 1973; Nebr. sect. award, 1979; Sloan Found. fellow, 1963-65. Fellow AAAS, Am. Acad. Arts and Scis.; mem. Aircraft Owners and Pilots Assn., Nat. Acad. Scis., Sigma Xi, Alpha Chi Sigma, Sigma Nu. Club: Cosmos. Subspecialties: Inorganic chemistry; Immunocytochemistry. Current work: Homogeneous catlysis of organic reactions; labeling of tumor-seeking antibodies for cancer therapy and diagnosis. Home: 3415 Green Vista Dr Encino CA 91316

HAY, DONALD IAN, dental research scientist, biochemistry educator; b. Peterborough, Huntingdonshire, Eng., Nov. 5, 1933; came to U.S., 1965; s. George Henry and Margaret Irene H.; m. Valerie Anne Butterworth, Mar. 7, 1969; 1 son, Ian Michael. B.Sc., U. London, 1959, Ph.D., 1972. Research scientist Unilever Ltd., Bedford, Eng., 1959-65; research scientist Forsyth Dental Center, Boston, 1965—; asst. prof. biochemistry Harvard U. Sch. Dental Medicine, Boston, 1975-83, assoc. prof., 1983—. Fellow Royal Soc. Chemistry (London); mem. Internat. Assn. Dental Research, Am. Chem. Soc. Subspecialties: Biochemistry (biology); Oral biology. Current work: Structure-function relationships of salivary proteins; biochemical bases of dental caries. Home: 12 Snake Brook Rd Wayland MA 01778 Office: Forsyth Dental Center 140 Fenway Boston MA 02115

HAY, WILLIAM WINN, educator; b. Dallas, Oct. 12, 1934; s. Stephen John and Avella (Winn) H. Student, Universitaet Muenchen, Germany, 1953-54; B.S., So. Methodist U., 1955; postgrad., Universitaet Zuerich, Switzerland, 1955-56; M.S., U. Ill., 1958; Ph.D., Leland Stanford Jr. U., 1960; NSF postdoctoral fellow, Universitaet Basel, Switzerland, 1959-60. Asst. prof. U. Ill., Urbana, 1960-63, asso. prof., 1963-68, prof., 1968-73; adj. prof. Inst. Marine Scis., U. Miami, Fla., 1966-68; prof. Rosenstiel Sch. Marine and Atmospheric Sci., 1968—, chmn. div. marine geology and geophysics, 1974-76, interim dean, 1976-77, dean, 1977-80; pres. Joint Oceanographic Instns. Inc., Washington, 1980—; mem. Joides Planning Com., 1969-76, chmn., 1972-74; mem. Joides Exec. Com., 1976-80; chmn. SEPM Research Symposium, 1970; hon. research fellow U. Coll., U. London, Eng.; mem. ocean sci. bd. NRC, 1977-80; mem. Am. Commn. Strategic Nomenclature, 1975-78; mem. code com. Am. Stratigraphic Commn., 1978—. Contbr. articles to profl. jours. Trustee Internat. Oceanographic Found. Fellow AAAS, Geol. Soc. Am., Geol. Soc. (London); mem. Am. Assn. Petroleum Geologists (ad hoc com. on revision and updating stratigraphic corr. charts N.Am. 1974, Internat. Stratigraphic Commn. working group for establishing biostratigraphic zonation of Cretaceous-Tertiary deep-sea beds 1975—), Am. Geophys. Union, Am. Micros. Soc., European Geophys. Soc., Deutsche Geol. Gesellschaft (Leopold von Buch medal 1976), Nat. Assn. Geology Tchrs., Geol. Vereinigung, Paleontol. Research Instn., Internat. Assn. Sedimentologists, Internat. Assn. Math. Geologists, Soc. Econ. Paleontologists and Mineralogists (pres. Gulf Coast sect. 1971-72), Marine Council, Paleontol. Soc., Paleontol. Assn., Schweiz. Geol. Ges., Schweiz. Paleontol. Ges., Soc. Geol. France, Phi Beta Kappa, Sigma Xi, Phi Eta Sigma, Delta Phi Alpha, Omicron Delta Kappa (gold key). Clubs: Whitehall (Chgo.); Cosmos (Washington); Dial (Urbana); Ocean Reef (Key Largo, Fla.). Subspecialties: Sedimentology; Paleontology. Current work: Interaction of tectonics, paleoclimatology, sedimentology and distbn. of organisms in time and space. Office: Joint Oceanographic Instns Inc 2600 Virginia Ave NW Suite 512 Washington DC 20037

HAYDEN, FREDERICK GLENN, medical educator; b. Madison, Wis., Jan. 11, 1948; s. Glen Emil and Mary Elizabeth (Reichardt) H.; m. Mary Lou Williams, June 19, 1970; children: Melissa, Geoffrey. Student, U. Wis.-Madison, 1966-69; B.A., Stanford U., 1972, M.D., 1973. Intern Strong Meml. Hosp., Rochester, N.Y., 1973-74, asst. resident in medicine, 1974-75, assoc. resident in medicine, 1975-76; trainee, fellow in infectious diseases U. Rochester, 1976-78; asst. prof. internal medicine and pathology U. Va. Sch. Medicine, Charlottesville, 1978—, assoc. dir. clin. microbiology lab., dir. clin. virology lab., 1978—. Mem. ACP, Am. Soc. Microbiology, Infectious Diseases Soc. Am., Phi Beta Kappa, Alpha Omega Alpha. Subspecialties: Virology (medicine); Internal medicine. Current work: Antiviral chemotherapy; diagnostic virology. Home: 745 Lockridge Ln Earlsville VA 22936 Office: Dept Internal Medicine U Va Med Center Box 473 Charlottesville VA 22908

HAYDEN, HOWARD CORWIN, JR., physicist, educator; b. Pueblo, Colo., June 20, 1940; s. Howard Corwin and Virginia Dayle (Burr) H.; m. (div.); children: Alexis, Vanessa. B.S. in Physics, U. Denver, 1962, M.S., 1964, Ph.D., 1967. Research asst. U. Conn., 1967; vis. asst. prof. U. Tenn., 1974, 75; asst. prof. physis U. Conn., 1968-75, asso. prof., 1975—; cons. ion implantation. Mem. Am. Phys. Soc., Am. Assn. Physicists Tech., Sigma Xi. Subspecialties: Atomic and molecular physics; Ion implantation. Current work: Atomic and ionic collisions, ion implantation, energy, atomic collision spectroscopy, surface modification through ion implantation, textbook writing. Office: Dept Physics U Conn Storrs CT 06868

HAYEK, THEODORE CRAIG, human resources development manager, organizational and industrial psychologist; b. N.Y.C., July 15, 1953; s. Henry and Ann (Spadaro) H. B.A. in Psychology, Bard Coll., 1975, M.A., New Sch. Social Research, 1977, 1982—. Supr. rehab. services Community Mental Health Program, Trenton, N.J., 1977; orgnl. devel. specialist Ins. Services Office, N.Y.C., 1978-79; mgmt. devel. specialist SCM Corp., N.Y.C., 1979-81, mgr. human resources devel., 1981—; cons., N.Y.C., 1979—. N.Y. State Regents scholar, 1971-75; Bard Coll. scholar, 1971-75; New Sch. scholar, 1975-77; N.Y. State scholar, 1971 75; N.Y. State Tuition scholar, 1975 77. Mem. Am. Psychol. Assn., Soc. Orgnl. and Indsl. Psychology, Am. Human Resources Planning Soc., N.Y. Human Resource Planners Assn. Subspecialties: Human resources management and development; Organizational and industrial psychology. Current work: Succession and manpower planning; organizational planning and development; climate and attitude survey research; training and development; internal staffing; performance appraisal systems; selection and assessment. Home: 14 W 88th St Apt 2A New York NY 10024 Office: Human Resources Devel SCM Corp 299 Park Ave New York NY 10171

HAYES, A. WALLACE, toxicologist, educator; b. Corning, Ark., Aug. 21, 1939; s. Andrew W. and Helen (Latimer) H.; m. Sandra June Smith, June 11, 1942; children—Andrew, Cathleen, Benjamin. A.B., Emory U., 1961; M.S., Auburn U., 1964, Ph.D., 1967. Diplomate: Am. Bd. Toxicology, Am. Acad. Toxicol. Scis. Prof. toxicology U. Ala., 1968-75, U. Miss. Med. Sch., Jackson, 1975-80, Temple U., Phila., 1981—; dir. toxicology research Rohm and Hass Co., Spring House, Pa., 1980—. Author: Mycotoxin Teratogenicity and Muta-genicity, 1981, Principles and Methods in Toxicology, 1982; contbr. over 150 articles to profl. jours. Mem. Soc. Toxicology, Am. Soc. Pharmacology and Exptl. Therapeutics, Am. Chem. Soc., Am. Inst. Nutrition. Mem. Ch. of Christ. Subspecialties: Toxicology (medicine); Teratology. Office: Rohm and Hass Co 727 Norristown Rd Spring House PA 19477

HAYES, DORA KRUSE, research chemist, chronobiologist; b. Kindred, N.D., June 16, 1931; d. Martin George and Dorothy (Strehlow) Kruse; m. John Clifford Hayes, Nov. 22, 1953 (dec. 1979); children: Robert Martin, John Wallace. B.S., Hamline U., 1952; M.S., U. Wis., 1953; Ph.D., U. Minn., 1961. Chemist Gen. Mills, Mpls., 1953-54; jr. scientist U. Minn., Mpls., 1955-57, teaching asst., 1957-61; research chemist U.S. Dept. Army, Dugway Proving Ground, Utah, 1961-65; lab. chief U.S. Dept. Agr. (Agrl. Research Service, Agrl. Environ. Quality Inst., Livestock Insects Lab.), Beltsville, Md., 1965—; cons. UN Indsl. Devel. Orgn., N.Y.C., 1981—; Chmn. Gordon Research Conf., Andover, N.H., 1981. Contbr. over 80 sci. articles to profl. publs. Wis. Alumni Research Assn. fellow, 1952; USPHS fellow, 1959, 60. Mem. Internat. Soc. Chronobiology (bd. dirs. 1982—, life. mem.), AAAS, Am. Chem. Soc., Am. Soc. Photobiology, Entomol. Soc. Am., Am. Soc. Biol. Chemistry, Federally Employed Women. Lutheran. Club: Bus. and Profl. Women's (Falls Church). Subspecialties: Chronobiology; Biochemistry (biology). Current work: Insect hibernation-diapause; chronobiology; aircraft disinsection; insect hormones and neurotransmitters. Home: 9105 Shasta Ct Fairfax VA 22031 Office: US Dept Agr Agrl Research Service Agrl Environ

Quality Inst LivestockInsects Lab Room 120 Bldg 307 BARC-East Beltsville MD 20705

HAYES, EDWARD J(AMES), mechanical engineer, business executive; b. Bklyn., Apr. 8, 1924; m., 1950; 3 children. B.S., MIT, 1950; M.S., U. Md., 1955. Engr. design and devel. E.I. duPont de Nemours & Co., N.Y.C., 1950-51, dir. electronics lab., 1951-52; project supr. applied mechanics and hydraulics Johns Hopkins U., Balt., 1952-58; dir. R&D Kelsey-Hayes Co., Romulus, Mich., 1958-64, v.p. research, engring. and quality control, 1964-65, v.p. corp. R&D, 1966—67, pres. and gen. mgr. wheel, drum and brake div., 1968-76, corp. v.p. R&D and engring., 1976—, Fruehauf Corp., 1980—; Hon. mem. sr. staff Johns Hopkins U.; v.p. Nat. Friction Prducts, 1965—. Mem. AAAS, ASME, AIAA, Am. Ordnance Assn., N.Y. Acad. Sci. Subspecialty: Mechanical engineering. Office: Kelsey-Hayes Co 38481 Huron River Dr Romulus MI 48174

HAYES, JOHN BERNARD, oil company geologist, researcher; b. Omaha, Nov. 30, 1934; s. Bernard Johnson and Dorothy Kathleen (Kinkade) H.; m. Patricia Ann Dengel, Aug. 19, 1961; children: Michael John, Steven Patrick. Student, U. Omaha, 1952-53; B.S., Iowa State U., 1956, M.S., 1957; Ph.D., U. Wis.-Madison, 1961. Assoc. prof. geology U. Iowa, 1960-68; sr. research geologist Marathon Oil Co., Littleton, Colo., 1968—; vis. lectr. U. Colo.-Boulder, 1973; adj. prof. Colo. Sch. Mines, 1976. Contbr. numerous articles to profl. jours. NSF fellow, 1956-60. Fellow Geol. Soc. Am.; mem. Clay Minerals Soc. (pres. 1977-78), Am. Assn. Petroleum Geologists (lectr. 1978-83), Soc. Econ. Paleontologists and Mineralogists (Best Paper award 1981), Rocky Mountain Assn. Geologists. Subspecialties: Petrology; Sedimentology. Current work: Analysis of sedimentary basins worldwide. Patentee in field; a decade of research on sandstone diagenesis led to original breakthrough. Office: Marathon Oil Co PO Box 269 Littleton CO 80160

HAYES, WALLACE D(EAN), aeronautical engineering educator; b. Peking, China, Sept. 4, 1918; m., 1948; 3 children. B.S., Calif. Inst. Tech., 1941; Ae.E., 1943; Ph.D. in Physics, 1947. Jr. stress analyst Consol. Vultee Aircraft Corp., Calif., 1939; jr. stress engr. Lockhead Aircraft Corp., Calif., 1940, aerodynamicist, 1943-45; Jewett fellow Princeton U., 1947-48; asst. prof. applied math. Brown U., 1948-49, assoc. prof., 1949-51; sci. liaison officer Office Naval Research, London, 1952-54; assoc. prof. aero. engring. Princeton (N.J.) U., 1954-57, prof., 1957—; cons. N.Am. Aviation, Inc., 1946-48, Space Technol. Labs., Thompson-Ramo-Wooldridge, Inc., 1954-58, Aero. Research Assocs., Princeton, Inc., 1958—; Fulbright vis. lectr. Delft U., 1951-52; NSF fellow and assoc. prof. U. Paris, 1964-65. Mem. Nat. Acad. Engring., Am. Phys. Soc., AIAA. Subspecialty: Aeronautical engineering. Office: Dept Aerospace Engring Princeton U Princeton NJ 08540

HAYES, WAYLAND JACKSON, JR., toxicologist, educator; b. Charlottesville, Va., Apr. 29, 1917; s. Wayland Jackson and Mary Lula (Turner) H.; m. Barnita Donkle, Feb. 1, 1942; children: Marie Hayes Sarneski, Maryetta Hayes Hacskaylo, Lula Hayes McCoy, Wayland, Roche del Hayes Moser. B.S., U. Va., 1938, M.D., 1946; M.A., U. Wis., 1940, Ph.D., 1942. Chief vector-transmission investigations USPHS, Savannah, Ga., 1947-48, chief toxicology sect., 1949-60, Atlanta, 1960-67, chief toxicologist, 1967-68; prof. biochemistry Vanderbilt U. Sch. Medicine, Nashville, 1968—; Vol. asso. prof. pharmacology Emory U., Atlanta, 1962-68; cons. WHO, 1950—, Nat. Acad. Scis.-NRC, 1964—. Author: Clinical Handbook on Economic Poisons, 1963, Toxicology of Pesticides, 1975, Pesticides Studied in Man, 1982; Mem. editorial bds.: Jour. Pharmacology and Exptl. Therapeutics, 1962-64, Archives Environmental Health, 1965-72, 76—, Food and Cosmetics Toxicology, 1967-78, Essays in Toxicology, 1972-76; Contbr. sci. papers to profl. lit. Served with AUS, 1943-46. Recipient Meritorious Service medal USPHS, 1964. Mem. Soc. Toxicology (charter, pres. 1971-72), Am. Soc. Pharmacology and Exptl. Therapeutics, Am. Soc. Tropical Medicine and Hygiene, Am. Conf. Govtl. Indsl. Hygienists. Current work: Toxicology of pesticides. Home: 2317 Golf Club Ln Nashville TN 37215

HAYFLICK, LEONARD, educator, administrator, cell biology researcher; b. Phila., May 20, 1928; s. Nathan Albert and Edna (Silbert) H.; m. Ruth Louise Hayflick, Apr. 11, 1926; children: Joel, Deborah, Susan, Rachel, Anne. B.A., U. Pa., 1951, M.S., 1953, Ph.D., 1956. Research asst. Merck, Sharp and Dohme, Inc., 1951-52; assoc. mem. Wistar Inst., Phila., 1958-68; asst. prof. research medicine U. Pa., 1966-68; prof. med. microbiology Stanford U., 1968-76; research cell biologist Children's Hosp. Med. Ctr., Oakland, Calif., 1976-81; prof. zoology, microbiology and immunology U. Fla., Gainesville, 1981—, dir. Ctr. Gerontol. Studies, 1981—. Contbr. articles to profl. jours. Served with U.S. Army, 1946-48. James W. McLaughlin fellow, 1956-58; recipient Career Devel. award Nat. Cancer Inst., 1962-70, Biomed. Sci. and Aging award U. So. Calif., 1974, Kesten award, 1974. Fellow Gerontol. Soc. Am. (pres. 1982—, Robert W. Kleemeier award 1972, Brookdale award 1980); mem. Am. Soc. Microbiology, AAAS, Tissue Culture Assn., Soc. Exptl. Biology and Medicine, Am. Cancer Soc., Western Gerontol. Soc. (dir. 1981—), Am. Fedn. Aging Research (dir.), Federated Socs. Exptl. Biology, Am. Assn. Cancer Research, Am. Assn. Cell Biology, Am. Assn. Pathologists, Am. Longevity Assn. Subspecialties: Cell and tissue culture; Gerontology. Current work: Cell biology of aging, transformation of normal human cells, cell biology, transformation, aging, gerontology, cell fusion, mycoplasmology, virus vaccines, cancer biology. Office: U Fla Center for Gerontological Studies 3357 GPA Gainesville FL 32611

HAYMOND, HERMAN RALPH, medical physicist; b. Salt Lake City, Aug. 29, 1924; s. Shelby Stanton and Orissa (Minchey) H.; m. Patricia Hindley, Apr. 23, 1949; children: Charles, Janice,Robert, Patricia, Philip, Elizabeth. A.B., U. Calif.-Berkeley, 1944, Ph.D., 1955. Diplomate: Am. Bd. Radiology. Physicist U. Calif. Radiation Lab., Berkeley, 1947-55; asst. prof. dept. radiology U. So. Calif., Los Angeles, 1955-57, assoc. prof., 1957-68, prof., 1968—. Served with USN, 1943-46. Mem. Am. Assn. Physicists in Medicine, Am. Chem. Soc., Biophys. Soc., IEEE, Radiation Research Soc., Radiol. Soc. N.Am. Subspecialties: Imaging technology; Cancer research (medicine). Current work: Nuclear magnetic resonance imaging in medicine; effects of radiation therapy on immune response in cancer patients. Office: 1200 N State St Box 304 Los Angeles CA 90033

HAYNER, DENIS ROBERT, mech. engr.; b. Onaway, Mich., Mar. 14, 1938; s. William Henry and Vera Adelia (Roberts) H.; m. Ruth Ann Black, Sept. 4, 1965; children: Stephanie M., Thompson W., Michael R. B.S.M.E., Mich. Technol. U., 1961; M.S.M.E., Calif. Inst. Tech., 1965. Registered profl. engr., Mass., 1972. Designer Pratt & Whitney Aircraft, East Hartford, Conn., 1961-64; design engr. Continental Aviation and Engring., Detroit, 1964-66; mech. design engr. Gen. Electric Co., Lynn, Mass., 1966-78, sr. designengr., 1978—. Mem. ASME, Vols. in Tech. Assistance, Phi Eta Sigma, Phi Kappa Phi, Tau Beta Pi. Roman Catholic. Subspecialty: Mechanical engineering. Current work: Mech. design of advanced aircraft gas turbine engine components, including compressors and turbines of subsonic and supersonic engines. Home: 40 Winter St Newburyport MA 01950 Office: 1000 Western Ave Lynn MA 01910

HAYNES, MACK W., JR., computer scientist; b. Lafayette, La., Mar. 29, 1948; s. Mack William and Violet Elizabeth (Henry) H.; m. Patsy Jean Fontenot (div.); m. Gretchen Hoffer, Nov. 20, 1977. B.S., Northwestern State U., Natchitoches, La., 1973, postgrad., 1973-75. Sr. research technician Exxon Research, Houston, 1975-81; sr. programmer analyst Tenneco Oil, Houston, 1981-82, computer system mgr., 1982—. Mem. Soc. Exploration Geophysics, Assn. Computing Machinery, Digital Equipment Users Group, Integraph Users Group. Subspecialties: Graphics, image processing, and pattern recognition; Operating systems. Home: 11523 Ensbrook St Houston TX 77099 Office: Tenneco Oil 1100 Louisiana Suite 2956 Houston TX 77001

HAYNES, MARTHA PATRICIA, astronomer, site dir.; b. Boston, Apr. 24, 1951; d. William Veech and Louise Mary (Healy) H. B.A. in Astronomy and Physics, Wellesley Coll., 1973, M.A., Ind. U., 1975, Ph.D., 1978. Assoc. instr. Ind. U., Bloomington, 1974-76; instr. Piedmont Va. Community Coll., Charlottesville, 1978; jr. research assoc. Nat. Radio Astronomy Obs., Charlottesville, 1976-78, asst. dir. for, 1981—; research assoc. Nat. Astronomy & Ionosphere Center, Arecibo, P.R., 1978-80, staff research assoc., 1980-81. Mem. Am. Astron. Soc., AAAS, Sigma Xi. Subspecialty: Radio and microwave astronomy. Current work: Extragalactic radio astronomy. Home and Office: PO Box 2 Green Bank WV 24944

HAYNES, ROBERT HALL, biophysicist; b. London, Ont., Can., Aug. 27, 1931; s. James Wilson and Lillian May (Hall) H.; m. Nancy Joanne May, Sept 23, 1954; children—Mark Douglas, Geoffrey Alexander, Paul Robert; m. Charlotte Jane Banfield, June 2, 1966. B.Sc., U. Western Ont., 1953, Ph.D., 1957. British Empire Cancer Campaign fellow dept physics St. Bartholomew's Hosp., Med. Coll., U. London, 1957-58; asst. prof. biophysics U. Chgo., 1958-64; asso. prof. biophysics U. Calif., Berkeley, 1964-68; prof. biology York U., Toronto, Ont., 1968—; mem. Nat. Research Council Can., 1975-82; chmn. ministerial com. on mutagenesis Can. Dept. Nat. Health & Welfare, 1978—; mem. tech. adv. com. on nuclear fuel waste mgmt. program Atomic Energy of Can. Ltd., 1979—. Contbr. articles to profl. jours.; editor: The Molecular Basis of Life, 1968, The Chemical Basis of Life, 1973, Man and the Biological Revolution, 1976. 0ecorated Queen Elizabeth II Silver Jubilee medal; USSR Acad. Scis. exchange visitor, 1972, 78; Brit. Council exchange visitor, 1973; Japan Soc. for Promotion of Sci. exchange visitor, 1979; Academia Sinica exchange visitor, 1980. Fellow Royal Soc. Can., Royal Soc. Arts (U.K.); mem. Genetics Soc. Can. (pres. 1983-85), Environ. Mutagen Soc. (councillor), Sigma Xi, Beta Theta Pi. Club: Univ. Toronto. Subspecialties: Genetics and genetic engineering (biology); Biophysics (biology). Current work: DNA repair, mutation and recombination in yeast; environ. mutagenesis; radiation microbiology. Home: 15 Queen Mary's Dr Toronto ON Canada M8X 1S1 Office: 4700 Keele St 306 Farquharson Bldg Toronto ON Canada M3J 1P3

HAYNES, ROBERT RALPH, botany educator; b. Minden, La., Feb. 24, 1945; s. Thomas A. and Ruby L. (McDonald) H.; m. Elizabeth Edna Zappa, Aug. 1, 1946; children: Robert Charles, Roxann. B.S., La. Poly. Inst., 1967; M.S., U. Southwestern La., 1969; Ph.D., Ohio State U., 1973. Lectr. botany Ohio State U., Columbus, 1973-74; asst. prof. La. State U., Shreveport, 1974-76, U. Ala., University, 1976-80, assoc. prof. biology, 1980—, dir. univ. herbarium, 1976—. Contbr. articles to sci. jours. Grantee NSF, 1978—. Mem. Bot. Soc. Am., Internat. Assn. Plant Taxonomy, Am. Soc. Plant Taxonomists, Assn. Southeastern Biologists, So. Appalachian Bot. Club. Democrat. Baptist. Subspecialties: Systematics; Taxonomy. Current work: Systematics of aquatic monocots. Office: U Ala University AL 35486

HAYNES, SUZANNE G., epidemiology educator/researcher; b. Huntington Park, Calif.; married, 1970. B.A., U. Tenn., 1969; M.A., U. Tex.-Austin, 1970; M.P.H., U. Tex-Houston, 1972; Ph.D. in Epidemiology, U. N.C.-Chapel Hill, 1975. Research economist Houston Mayor's Manpower Planning Comm., 1970-71; research assoc. epidemiologist U. N.C.-Chapel Hill, 1974-76; epidemiologist Nat. Heart, Lung and Blood Inst., NIH, Bethesda, Md., 1975-80; research asst. prof. epidemiology U. N.C., 1980—; Mem. epidemiology contract rev. com. Nat. Cancer Inst., 1979-83; mem. exec. com. Am. Heart Assn. Council on Epidemiology, 1982-85; mem. adv. com. Western Ctr. Behavioral and Preventive Medicine, 1980—; cons. epidemiology br. Nat. Heart, Lung & Blood Inst., 1980—; project coordinator Community Surveillance Cardiovascular Disease, Southeastern U.S., 1981—. Editor: Jour. Gerontology, 1978-81. Fellow Acad. Behavioral Medicine, Am. Coll. Epidemiology. Subspecialty: Epidemiology. Office: Sch Pub Health U NC Chapel Hill NC 27514

HAYS, ESTHER FINCHER, physician; b. Lexington, Ky., Apr. 18, 1927; m. Daniel M. Hays, Sept. 15, 1951; children: Sarah Margaret, Jonathan Fincher, Margaret LeVan, Elizabeth Colby. B.A., Cornell U., 1948, M.D., 1951. Diplomate: Am. Bd. Internal Medicine. Asst. dept. medicine Cornell U., N.Y.C., 1952-54; intern N.Y. Hosp., N.Y.C., 1951-52, resident, 1952-54; instr. in residence UCLA dept. medicine, 1955-57, asst. prof., 1957-62, assoc. prof., 1962-72, prof. medicine, 1972—; chief med. div. Lab. Nuclear Medicine/Radiation Biology, 1958-63. Contbr. articles to profl. jours. Recipient Woman of Sci. award UCLA Med. Center Aux., 1951. Fellow A.C.P.; mem. Western Soc. Clin. Research, Western Assn. Physicians, Internat. Soc. Exptl. Hematology, Am. Soc. Hematology, Am. Assn. Cancer Research., Alpha Omega Alpha. Subspecialties: Hematology; Internal medicine. Current work: Etiology and pathogenesis of leukemia. Home: 181 S Las Palmas Ave Los Angeles CA 90004 Office: UCLA 900 Veteran Ave Los Angeles CA 90024

HAYS, JOHN BRUCE, chemistry educator, researcher; b. Springfield, Ill., June 21, 1937; s. Loren Eastman and Mary Elizabeth (Russell) H.; m. Judith Gail Gumm, Sept. 1, 1961; children: Elinor, Stephen, Laura. Student, Deep Springs Coll., 1954-56; B.S., U. N.Mex., 1960; Ph.D., U. Calif.-San Diego, 1968; postgrad., Johns Hopkins U., 1969-72. Asst. prof. dept chemistry U. Md. Baltimore County, Catonsville, 1972-77, assoc. prof., 1977-82, prof. dept. chemistry, 1982—. Contbr. articles to pubis. Vestryman St. Bartholomew's Episcopal Ch., Balt., 1974-77. Served to lt. (j.g.) USN, 1960-63. NIH predoctoral fellow, 1965-68; Am. Cancer Soc. postdoctoral fellow, 1969-71; faculty research awardee, 1981—; various research grants NIH, Am. Cancer Soc. Mem. Am. Soc. Biol. Chemists. Subspecialty: Molecular biology. Current work: Research in biochemistry and genetics of genetic recombination and DNA repair. Home: 6309 Mount Ridge Rd Catonsville MD 21228 Office: Dept Chemistry Univ Md Balt County 5401 Wilkens Ave Catonsville MD 21228

HAYWARD, JAMES LLOYD, biology educator; b. Melrose, Mass., Sept. 30, 1948; s. James Lloyd and Jane Beverly (Watson) H.; m. Cheryl Lynn Kirkpatrick, May 26, 1974; 1 dau., Shanna Marie. B.S., Walla Walla Coll., 1972; M.A., Andrews U., 1975; Ph.D., Wash. State U., 1982. Instr. biology Andrews U., Berrien Springs, Mich., 1975-76, Southwestern Union Coll., Keene, Tex., 1976-77; vis. instr. biology WWC Marine Sta., Anacortes, Wash., 1977, 79, 83; instr. biology Walla Walla Coll., 1980-81; assoc. prof. biology Union Coll., Lincoln, Nebr., 1981—; research assoc. Andrews U., 1975-76. Contbr. articles to profl. jours. Named Outstanding Grad. Biology Walla Walla Coll., 1972; recipient Chapman Fund award Am. Mus. Natural History, 1979-81; Sigma Xi grantee, 1979, 80. Mem. Am. Ornithologists Union, Animal Behavior Soc., Nebr. Acad. Sci., Sigma Xi, Phi Kappa Phi. Subspecialties: Ethology; Paleobiology. Current work: Ethology, specifically behavioral ecology of nesting gulls; paleobiology, specifically taphonomic studies of tephra-buried eggs. Home: 3909 S 52d St Lincoln NE 68506 Office: Div Sci and Math Union Coll Lincoln NE 68506

HAYWARD, WILLIAM S., molecular biologist, educator; b. Riverside, Calif., May 29, 1941; s. Herman E. and Jean H. (Port) H.; m. Linda C., June 19, 1965; 1 dau., Danya Lys. B.A., U. Calif.—Riverside; Ph.D., U. Calif.—San Diego, 1969. Postdoctoral fellow Albert Einstein Coll. Medicine, 1969-71; asst. Pub. Health Research Inst., City N.Y., 1971-73; asst. prof. Rockefeller U., 1973-78, assoc. prof., 1978-82; prof. molecular biology Sloan-Kettering div. Cornell U. Grad. Sch. Medicine, N.Y.C.; and mem. Meml. Sloan-Kettering Cancer Ctr., N.Y.C., 1982—; mem. sci. adv. com. on personnel for research Am. Cancer Soc. Contbr. numerous articles on molecular biology, virology to profl. jours.; editorial bd.: Virology, 1983—, Jour. Virology, 1979—. Recipient Flora E. Griffin Fund award, 1981; Am. Cancer Soc. grantee, 1969; NIH grantee, 1976. Mem. Am. Soc. Microbiology, Am. Soc. Virology (charter), N.Y. Acad. Scis., AAAS. Subspecialties: Molecular biology; Cell study oncology. Current work: Molecular biology of RNA tumor virus; mechanisms of viral and non-viral oncogenesis; control of gene expression; viral genetics. Office: 1275 York Ave Suite 406K New York NY 10021

HAYWOOD, ANNE MOWBRAY, biochemist, virologist, pediatrician, educator; b. Balt., Feb. 5, 1935; d. Richard Mansfield and Margaret (Mowbray) H. B.A. in Chemistry, Bryn Mawr Coll., 1955; M.D., Harvard U., 1959. Diplomate: Am. Bd. Pediatrics. Intern in pediatrics U. Calif. Med. Center, San Francisco, 1959-60; postdoctoral fellow div. biology Calif. Inst. Tech., 1960-61, 62-64; postdoctoral fellow in biochemistry Columbia U., 1961-62; asst. prof. microbiology Northwestern U. Med. Sch., Chgo., 1964-66, Yale U. Med. Sch., 1966-73; vis. scientist biophys. unit Agrl. Research Council, Cambridge, Eng., 1972-74; resident in pediatrics U. Wash., 1974-75, fellow in pediatric infectious diseases, 1975-76, Vanderbilt U., 1976-77; asso. prof. pediatrics and microbiology U. Rochester Med. Center, 1977—; vis. asst. prof. Rockefeller U., 1971-72; cons. in field. Contbr. articles to prfl. publs. Am. Cancer Soc. postdoctoral fellow, 1960-62; NIH spl. fellow, 1971-73; European Molecular Biology Orgn. fellow, 1973-74. Mem. Am. Soc. Biol. Chemists, Am. Soc. Microbiology. Democrat. Quaker. Subspecialties: Virology (biology); Membrane biology. Current work: Virus replication with emphasis on membrane-related aspects; persistent viral infections; interactions between membranes. Office: Dept Pediatric U Rochester Rochester NY 14642

HAYWOOD, H(ERBERT) CARL(TON), psychologist; b. Taylor County, Ga., July 2, 1931; s. Howard Chapman and Rosebud (Smith) H.; m. Nancy Patricia Roberts, Oct. 5, 1951 (div. Mar. 1971); children: Carlton, Terence, Elizabeth, Kristin. A.B., San Diego State Coll., 1956, M.A., 1957; Ph.D., U. Ill., 1961. Mem. faculty George Peabody Coll. (merged with Vanderbilt U. 1979), Nashville, 1962—, prof. psychology, 1969—, prof. spl. edn., 1975-79, dir. mental retardation research tng. program, 1968-70; dir. Inst. Mental Retardation and Intellectual Devel., 1970-73, Office Research Adminstrn., 1974-76, John F. Kennedy Center Research Edn. and Human Devel., 1971-83; prof. neurology Vanderbilt U. Sch. Medicine, 1971—; vis. prof. U. Toronto, 1965-66; sr. fellow Vanderbilt Inst. Pub. Policy Studies, 1983—; chmn. Nat Mental Retardation Research Center Dirs., 1979-82; adv. bd. Ill. Inst. Developmental Disabilities, Chgo., 1970-78, Eunice Kennedy Shriver Center Mental Retardation, Waltham, Mass., 1973—, Tenn. Dept. Mental Health, 1964—; cons. President's Com. on Mental Retardation, 1968-73; mem. sci. rev. com., health research facilities br., div. edn. and research facilities NIH, 1967-71. Editor: Brain Damage in School Age Children, 1968, Social Cultural Aspects of Mental Retardation, 1970, (with Begab and Garber) Prevention of Retarded Development in Psychosocially Disadvantaged Children, 1981, (with J.R. Newbrough) Living Environments for Developmentally Retarded Persons, 1981; editor: Am. Jour. Mental Deficiency, 1969-79; editorial bd.: Jour. Abnormal Child Psychology, 1973—, Contemporary Psychology, 1982—; Contbr. articles on child devel., motivation and mental retardation to profl. jours. Served with USN, 1950-54. Fellow Am. Assn. Mental Deficiency (v.p. psychology 1975-77, 1st v.p. 1978-79, pres. 1980-81), Am. Psychol. Assn. (pres. Div. 33 1978-79, mem. Council of Reps. 1980-82); mem. Soc. Research Child Devel., Inst. Medicine, Psychonomic Soc. Democrat. Episcopalian. Club: Cosmos (Washington). Subspecialties: Mental retardation; Neurology. Office: Inst for Public Policy Studies Peabody Coll Vanderbilt U Nashville TN 37203 Dominant values include enthusiasm for scholarship, equal parts of dedication to science for its own sake and concern for social progress, and the conviction that self-concern and self-seeking constitute the most dangerous threat to the collective goals of humanity. The future lies in education designed to stretch minds and develop processes of critical thought rather than to impart job-oriented skills.

HAYWOOD, L. JULIAN, educator, physician; b. Reidsville, N.C., Apr. 13, 1927; s. Thomas Woodly and Louise Viola (Hayley) H.; m. Virginia Elizabeth Paige, Dec. 3, 1953; 1 son, Julian Anthony. B.S., Hampton Inst., 1948; M.D., Howard U., 1952. Intern St. Mary's Hosp., Rochester, N.Y., 1952-53; resident Los Angeles County Hosp., 1956-58; fellow cardiology White Meml. Hosp., 1959-61; traveling fellow U. Oxford, Eng., 1963; instr. medicine Loma Linda (Calif.) U., 1960-61, asst. prof., 1961-72, asso. clin. prof., 1973—; asst. prof. medicine U. So. Calif., 1963-68, asso. prof., 1968-76, prof., 1976—; dir. comprehensive sickle cell ctr. Los Angeles County-U. So. Calif. Med. Center, dir. coronary care unit; past dir. physicians tng. program (Regional Med. Programs), 1970-75; cons. Los Angeles County Coroner, Indsl. Accident Bd. Calif., Health Care Tech. Div., USPHS, Nat. Heart and Lung Inst.; past mem. cardiology adv. com. div. heart and vascular diseases. Bd. dirs., pres. Sickle Cell Disease Research Found. Contbr. articles profl. jours.; Mem. editorial bds.: Jour. Nat. Med. Assn. Served with M.C. USNR, 1954-56. Recipient award of merit Los Angeles County Heart Assn., 1968, 69, 73, 75. Fellow Los Angeles Acad. Medicine, A.C.P., Am. Coll. Cardiology, Am. Heart Assn. (fellow council on clin. cardiology; mem. council on other osclerosis; mem. exec. com. council on epidemiology; mem. long-range planning com., dir., past sec., v.p. Greater Los Angeles affiliate, now pres.); mem. Am. Fedn. Clin. Research, AAAS, Soc. Clin. Investigation, Western Soc. Clin. Research, Assn. Advancement Med. Instrumentation, AMA, Nat. Med. Assn. (Charles Drew Med. Soc.), N.Y. Acad. Scis., Hampton Inst. Alumni Assn. (past pres. Los Angeles chpt.), Med. Faculty Assn. U. So. Calif. Sch. Medicine (past pres.), Los Angeles Soc. Internal Medicine (past pres.), Western Assn. Physicians, AAUP, Fedn. Am. Scientists, Alpha Omega Alpha. Subspecialty: Cardiology. Home: 3551 Lowry Rd Los Angeles CA 90027 Office: 1200 N State St Los Angeles CA 90033

HAZDRA, JAMES J., health care educator; b. Chgo., Aug. 30, 1933; s. James J. and Beatrice A. H.; m. Mary A. Rojas, Aug. 27, 1955 (div. Jan. 1982); children: James, Michael, Margaret, Richard, Steven. B.S., Ill. Benedictine Coll., 1955; Ph.D., Purdue U., 1959. Accredited prof. chemist. Research chemist Olin Chem., Joliet, Ill., 1959-60; dir. product devel. Continental Can Co., Plainfield, Ill., 1960-63; prof. chemistry Ill. Benedictine Coll., Lisle, 1961—, head dept. chemistry, 1964-77, dir. health care edn., 1977—; cons. Argonne (Ill.) Nat. Lab., 1967-75, Packer Engring., Naperville, Ill., 1963-80. Bd. dirs.

Community Convalescent Ctr., Naperville, 1979—, Health Systems Agy., Oak Park, Ill., 1981—; pres. St. Mary's Sch. Bd., 1972-74; mem. health com. Dupage Sr. Citizens, Wheaton, Ill., 1981—. Namde Alumnus of Yr. Ill. Benedictine Coll., 1970, Educator of Yr., 1976. Fellow Am. Coll. Nutrition (affiliate); mem. AAAS, Am. Assn. Allied Health Professions. Roman Catholic. Lodges: KC; Cath. Order Foresters. Subspecialties: Nutrition (medicine); Health services research. Current work: Study of trace element in blood cells. Patentee in field. Office: Ill Benedictine Coll 5700 College Rd Lisle EL 60532

HAZELBAUER, GERALD LEE, molecular biology educator; b. Chgo., Sept. 27, 1944; s. Carl F. and Margaret J. (Ort) H.; m. Linda L. Randall, Aug. 29, 1970. B.A., William Coll., 1966; M.S., Case Western Res. U., 1968; Ph.D., U. Wis.-Madison, 1971. Postdoctoral fellow U. Wis.-Madison, 1971, Institut Pasteur, Paris, 1971-73, U. Uppsala, Sweden, 1973-75, asst. prof., 1975-81; assoc. scientist Wash. State U., Pullman, 1981-82, assoc. prof. biochemistry, 1982—. Mem. Am. Soc. Microbiology. Subspecialties: Molecular biology; Membrane biology. Current work: Molecular biology of bacterial chemotaxis, membrane proteins, receptors and sensory transducers. Home: 920 D St Pullman WA 99163 Office: Biochemistry/Biophysics Wash State U Pullman WA 99164-4630

HAZELWOOD, ROBERT L(EONARD), physiology educator; b. Oakland, Calif., July 11, 1927; s. Percy John and Gertrude M. (Hanley) H.; m. Barbara W. Schultz, Aug. 20, 1955; 1 dau., Anna Marie. B.A., U. Calif.-Berkeley, 1950, M.S., 1952; Ph.D., U. Calif.-Davis, 1958. Instr., asst. prof. Sch. Medicine, Boston U., 1958-61; asst. prof. Sch. Medicine, U. Calif.-San Francisco, 1961-63; assoc. prof. U. Houston, 1963-71, prof. dept. biology, 1971—. Contbr. chpts. to books, articles to profl. publs. Served with USNR, 1945-47. NSF research grantee, 1961-82. Mem. Endocrine Soc., Am. Physiol. Soc., Soc. Exptl. Biology and Medicine. Roman Catholic. Subspecialties: Physiology (biology); Comparative physiology. Current work: Regulation of hormone release from the vertebrate endocrine pancreas; physiology of pancreatic polypeptide. Office: Dept Biology Univ Houston 4700 Calhoun Houston TX 77004

HAZEN, DAVID COMSTOCK, aerospace and mechanical engineering educator; b. Greenburg, N.Y., July 3, 1927; m. Mary Ann Shipherd, Nov. 27, 1948; children: George Shipherd, Thomas Coe, Anne Scott. Grad. cum laude, Choate Sch., 1944; B.S. magna cum laude in Engring, Princeton U., 1948, M.S., 1949. Instr. Princeton U., 1949-51, research asso., 1951-53, asst. prof. aerospace and mech. engring., 1953-56, asso. prof., 1956-63, prof., 1963-83, prof. emeritus, 1983—, asso. dean faculty, 1966-69, asso. chmn. dept. aerospace and mech. Scis., 1969-74; mem. various engring. sch. coms. and univ. coms. coms.; exec. dir. Assembly of Engring. NRC, 1980-82—, exec. dir. Commn. on Engring. and Tech. Systems, 1982—; cons. in field; mem. Research Adv. Com. Lab. Adv. Bd. for Naval Ships, 1968-78; mem. (Naval Research Adv. Com. Marine Corps Panel), 1969-78, chmn., 1970-78, mem., 1971-74, 1971-78, vice chmn., 1972-75, chmn., 1975-77; mem. naval studies bd. NRC, 1978—. Trustee Sterling Sch., Craftsbury Common, Vt. (chmn. 1965-70, vice chmn. 1970-72), Robert Coll. Istanbul, Turkey, 1969-81, Univ. Petroleum and Minerals, Saudi Arabia, 1973—; mem. Princeton-in-Asia Found., 1964-80, v.p., 1968-80. Recipient Dept. Navy Distinguished Service award, 1977, certificate of commendation USMC, 1978. Fellow AIAA (chmn. edn. com. 1965-68, v.p. edn. 1971-73); mem. Aero. Soc. India (governing bd. Kanpur br. 1964-65), Wingfoot Lake Lighter Than Air Soc. (hon.), Am. Ordnance Assn., Engrs. Council for Profl. Devel. (mem. bd. 1971-77), Am. Soc. Engring. Edn. (exec. com. aerospace div. 1971), Sigma Xi, Phi Beta Kappa. Subspecialty: Aeronautical engineering. Adress: 1657 B S Hayes St Arlington VA 22202

HEACOCK, E(ARL) LARRY, electronics engineer; b. Tuscola, Ill., Jan. 27, 1935; s. Earl Rice and Helen Irene (Kaga) H.; m. Nancy Louise Voelkel, Sept. 2, 1956; children: Douglas (dec.), Gregory, Kent, Christopher. B.S.E.E., U. Ill., 1957, M.S.E.E., 1966. Engr. Ill. Bell Telephone Co., Springfield, 1957-62; chief electronics br. Nat. Weather Satellite Ctr., Washington, 1962-69; staff engr. European Space Research Orgn., Noordwijk, Holland, 1970-72; project mgr. European Space Agy., Toulouse, France, 1972-76; dir. office devel. and planning Nat. Environ. Satellite, Data and Info. Service, NOAA, Washington, 1976—; del UN. Com. on Peaceful Uses of Outer Space, N.Y.C., 1978, World Adminstrv. Radio Conf., Geneva, 1971. Adminstrv. bd. St. Mathews Methodist Ch., Bowie, Md., 1969; treas. Boy Scouts Am., Hague, Holland, 1971. Served to capt. USAF, 1958-60. Fellow Brit. Interplanetary Soc.; mem. AIAA, Am. Astron. Soc. (dir. 1979—). Subspecialties: Aerospace engineering and technology; Infrared spectroscopy. Current work: Responsible for design, development, manufacture, and launch of all U.S. meteorological operational satellites and launch of remote sensing satellites. Home: 363 Kingsberry Dr Annapolis MD 21401 Office: Nat Environmental Satellite Data and Information Service Washington DC 20233

HEAD, JAMES W., educator; b. Richmond, Va., Aug. 4, 1941. B.S., Washington and Lee U., 1964; Ph.D., Brown U., 1969. With Bellcomm, Inc., Washington, 1968-72; interim dir. Lunar Sci. Inst., Houston, 1973-74; ednl. vis. scientist Space Shuttle Astronaut Tng. Program, Lunar and Planetary Inst., Houston, 1978; asst. prof. research Brown U., 1973-74, asso. prof., research, 1974-75, asso. prof., 1975-80, prof., 1980—; Mem. Office of President-Elect Transition Team, 1980; mem. space and terrestrial applications adv. com. subcom. on geodynamics and geology NASA, NRC; mem. NASA Solar System Exploration Com.-Sci. Working Group on Inner Plant Missions, 1981—. Contbr. articles to sci. jours. Mem. Geol. Soc. Am. (spl. commendation for participation in Apollo program 1973). Subspecialties: Planetology. Current work: Study of the processes operating to form and modify planetary surfaces and lithospheres - volcanism, tectonism and impact cratering. Office: Dept Geol Scis Brown U Providence RI 02912

HEADLEY, R. PAUL, aerospace engineer; b. Oconto, Nebr., May 30, 1937; s. Carl Leslie and Lois Irene (Blakeslee) H.; m. Mary Kristin Cave, Nov. 24, 1965 (div. July 1982); children: Heather Eileen, Kirstin Irene. B.S. in Aero. Engring, U. Colo., 1959, 1959; M.S. in Engring Sci, Rensselaer Poly. Inst., 1961. Registered profl. engr., Colo. Mech. engr. Hamilton Standard, Windsor Locks, Conn., 1959-61; sr. engr. Martin-Marietta Corp., Denver, 1961-63,1965-66; research mathematician U. Colo., Boulder, 1963-65, 66-72; asst. prof. Met. State Coll., Denver, 1976-77, adj. prof., 1980—; staff engr. Bendix Corp., Denver, 1972-77; sr. group engr. Martin-Marietta Corp., Denver, 1977—. Mem. AIAA, Planetary Soc. Republican. Roman Catholic. Club: Pinehurst Country (Denver). Subspecialties: Aerospace engineering and technology; Electronics. Current work: Real time simulation of spacecraft dynamics, attitude determination and attitude control algorithms. Use of flight hardware or flight equivalent hardware in simulations (digital multiprocessor implementation). Home: 4725 W Quincy Ave Apt 1109 Denver CO 80236 Office: Martin Marietta Corp PO Box 179 Denver CO 80201

HEALY, ALICE FENVESSY, research psychologist, educator; b. Chgo., June 26, 1946; d. Stanley J. and Doris (Goodman) F.; m. James Bruce Healy, May 9, 1970. A.B. summa cum laude, Vassar Coll., 1968; Ph.D., Rockefeller U., 1973. Asst. prof. Yale U., New Haven, 1973-78, assoc. prof., 1978-81, U. Colo., Boulder, 1981—; research assoc. Haskins Labs., New Haven, 1976-81; mem. basic behavioral processes research rev. com. NIMH, 1979-81. Contbr. articles to psychology to profl. jours.; assoc. editor: Jour. Exptl. Psychology, 1981—; editorial bd.: Memory and Cognition, 1976—. NIMH grantee, 1975-77; Spencer Found. grantee, 1978-80; NSF grantee, 1977-84. Mem. Psychonomic Soc., Am. Psychol. Assn., Soc. Math. Psychology, Cognitive Sci. Soc., Rocky Mountain Psychol. Assn., Eastern Psychol. Assn., Phi Beta Kappa, Sigma Xi. Club: University (Boulder). Subspecialty: Cognition. Current work: Human information processing, including reading, learning, memory, speech, language, decision making. Home: 840 Cypress Dr Boulder CO 80303 Office: U Colo Dept Psychology Campus Box 345 Boulder CO 80309

HEARD, HARRY GORDON, computer scientist, executive; b. Reines, Tenn., Sept. 23, 1922; s. Pascal Harrison and Cliffie Muse (Page) H.; m. Allison Louise Norcross, Aug. 27, 1947; children: Pamela Suzanne, Todd Addison Crandall. B.S., U. Calif.-Berkeley, 1949, M.S., 1951. Research engr. Lawrence Berkeley Labs., 1951-59; chief engr. Levinthal Electronic Products, 1959-60; v.p. Radiation, Inc., 1959-60, Energy Systems, Inc., 1960-64, HNU Systems, div. Ohio Steel, 1964-68; pres. Resalab, Inc., Menlo Park, Calif., 1968-70; v.p. Info. Systems, MBA, Menlo Park, 1970-74; scientist Inst. Advanced Computation, TDC, Mountain View, Calif., 1975—; dir. Intronex, Inc., Radiation at Stanford, Inc., Resalab Sci., Menlo Park, 1975—. Author: Laser Parameter Measurements Handbook, 1966; contbr. articles in field to profl. jours. Vice-pres., chmn. Berkeley Flower and Hobby Show. Served with USAF, 1943-46. Mem. Phys. Soc., Research Soc. Am. (chpt. pres. 1961), Optical Soc. Am., IEEE, Assn. Computing Machinery, Jaycees, Sigma Xi, Tau Beta Pi, Eta Kappa Nu. Subspecialties: Computer architecture; Information systems, storage, and retrieval (computer science). Current work: Computer architecture, data storage and retrieval, software engineering. Home: 50 Skywood Way Woodside CA 94062

HEARTH, DONALD PAYNE, aero. engr., NASA ofcl.; b. Fall River, Mass., Aug. 13, 1928; s. Alvin George and Hildreth (Fogwell) H.; m. Joan Hall Smith, Dec. 30, 1950; children—Susan Hall, Douglas Payne, Anne Hall, Janet Hall. B.S. in Mech. Engring, Northeastern U., Boston, 1951; postgrad., U. Calif., Los Angeles, U. So. Calif.; grad., Fed. Exec. Inst., 1973. With NASA, 1951-57, mgr. advanced programs, 1962-67, dir. planetary programs, Washington, 1967-70; dep. dir. Goddard Space Flight Center, Greenbelt, Md., 1970-75; dir. Langley Research Center, Hampton, Va., 1975—; project mgr., dept. mgr. Marquardt Corp., Van Nuys, Calif., 1957-62. Author articles. Recipient Exceptional Service medal NASA, 1969; Exec. Performance award, 1975; Distinguished Service award, 1975. Fellow Am. Inst. Aeros. and Astronautics, Am. Astronautical Soc.; mem. Am. Soc. Pub. Adminstrn. Subspecialties: Aerospace engineering and technology; Astronautics. Office: Langley Research Center NASA Hampton VA 23665

HEATH, COLIN ARTHUR, company administrator, nuclear engineer; b. London, Eng., July 25, 1939; s. Eric Arthur Douglas Heath and Margaret Elizabeth (Jones) Young; m. Margaret Gordon, Jan. 20, 1962; children: Michael, Richard. B.S. in Chem. Engring., Case Inst. Tech., 1960; M.S. in Nuclear Engring., U. Wash., 1962, Ph.D., 1964. Staff engr. Gen. Atomic Co., San Diego, 1967-71, regional mgr., Washington, 1971-73, mgr. reprocessing dept., San Diego, 1973-76; dir. waste isolation U.S. Dept. Energy, Washington, 1977-81; North ops. mgr. cons. div. NUS Corp., Gaithersburg, Md., 1981—. Served as 1st lt. USAF, 1964-67. Mem. Am. Nuclear Soc. (chmn. San Diego chpt. 1976). Subspecialties: Nuclear engineering; Chemical engineering. Current work: Radioactive waste management and disposal, nuclear fuel reprocessing, power reactor operations, nuclear fuel cycle. Home: 9605 Sunset Dr Rockville MD 20850 Office: NUS Corp 910 Clopper Rd Gaithersburg MD 20878

HEATH, HUNTER, III, biomedical research scientist, clinical endocrinologist, educator; b. Dallas, June 8, 1942; s. Hunter and Velma M. (Brandon) H.; m. Glenna A. Witt, July 25, 1965; 1 son, Ethan Ford. B.A. in Chemistry, Tex. Tech. U., 1964; M.D., Washington U., 1968. Intern, resident in internal medicine U. Wis. Hosps., Madison, 1968-70; fellow in endocrinology and metabolism Walter Reed Army Med. Ctr., Washington, 1970-72; chief endocrinology sect. Letterman Army Med. Ctr., San Francisco, 1972-74; instr. internal medicine U. Calif.-San Francisco, 1973-74; fellow in mineral metabolism Mayo Grad. Sch. Medicine, Rochester, Minn., 1974-76, cons. in endocrine research, 1976—, asst. prof. medicine, 1976-80, assoc. prof., 1980—. Mem. editorial bd.: Jour. Clin. Endocrinology and Metabolism, 1979-83, Am. Jour. Physiology, 1982—; author numerous sci. publs. Served to maj. USAR, 1970-74. Fellow ACP; mem. Am. Soc. Clin. Investigation, Endocrine Soc. (publ. com. 1983—), Am. Soc. Bone and Mineral Research (chmn. edn. com. 1981-84). Republican. Unitarian-Universalist. Subspecialties: Endocrinology; Physiology (biology). Current work: Endocrinology, physiology, and pathophysiology of calcium and skeletal homeostasis; parathyroid hormone; calcitonin; radioimmunoassay; hormone action. Office: Mayo Clinic 200 SW 1st St Rochester MN 55905

HEATH, JAMES EDWARD, physiologist; b. Evansville, Ind., May 3, 1935; s. Max Levy and Mae B. (McNutt) H.; m. Maxine Shoemaker, Apr. 2, 1955; children—Cynthia Maxine, Pamela Diane, Jessica Scott. Student, Los Angeles City Coll., 1953-55; B.A. (Calif. State scholar), UCLA, 1957, M.A., 1958, Ph.D., 1962. Fellow mental health tng. program Brain Research Inst., UCLA, 1958-60; postdoctoral fellow UCLA, 1962-64; asst. prof. physiology and biophysics U. Ill., Urbana, 1964-67, asso. prof., 1967-73, prof., 1973—, head dept. physiology and biophysics, 1975—; prof. zoology, chmn. dept. zoology U. Fla., 1974-75. Contbr. numerous articles to profl. jours.; mem. council, editorial bd.: Ecol. Monographs, 1972-75; editorial bd.: Am. Midland Naturalist, 1971-77, Ann. Revs. Physiology, 1979—; editor: Jour. Thermal Biology, 1975—; co-editor: Physiol. Zoology, 1975—. Fellow AAAS; mem. Am. Physiol. Soc., Ecol. Soc. Am., Soc. Study Evolution, Soc. Neurosci., Assn. Chairmen Depts. Physiology, Sigma Xi. Subspecialties: Comparative neurobiology; Comparative physiology. Current work: Temperature reglation; locomotive enregetic; temperature adaptation; comparative nervous systems. Office: Dept Physiology and Biophysics 524 Burrill Hall U Ill Urbana IL 61801

HEATH, MICHELE CHRISTINE, botany educator; b. Bournemouth, Eng., Sept. 22, 1945; emigrated to Can., 1971, naturalized, 1977; d. Percy and Winifred Iris Lily (Downes) Roy; m. Ian Brent Heath, Sept. 23, 1967; 1 dau., Lorraine. B.Sc., Westfield Coll., U. London, 1966; Ph.D., D.I.C., Imperial Coll., U. London, 1969. Postdoctoral fellow U. Ga., 1969-71; lectr. dept. botany U. Toronto, Ont., 1972-73, asst. prof., 1973-76, assoc. prof., 1976-81, prof., 1981—. Contbr. articles to sci. jours. Recipient Huxley Meml. medal, 1979; E.W.R. Steacie Meml. fellow, 1982. Fellow Am. Phytopathol. Soc.; mem. Can. Phytopathol. Soc., Mycol. Soc. Am., Am. Soc. Plant Physiologists. Subspecialty: Plant pathology. Current work: Biochemical basis of the host specificity of plant pathogenic fungi. Office: Department of Botany University of Toronto Toronto ON Canada M5S 1A1

HEAVNER, JAMES EDWARD, pharmacologist, veterinarian; b. Cumberland, Md., Apr. 25, 1944; s. Douglas B. and Grace E. (Frantz) H.; m. Betsey C. Clark, Sept. 11, 1967; children—Matthew, Kori, Benjamin. Student, U. Md., 1962-64; D.V.M., U. Ga., 1968; Ph.D., U. Wash., 1971. Diplomate: Am. Coll. Vet. Anesthesiology.; lic. veterinarian, Md., Wash. Mem. faculty U. Wash., 1968-80, asst. prof., 1972-75, assoc. prof., 1975-80; chief pharmacology-toxicology br. FDA, Beltsville, Md., 1980-82; vis. scientist U. Edinbrugh, Scotland, 1978. Contbr. articles to books, profl. jours. Mem. sch. com. St. Martin's Luth. Ch., Annapolis, Md., 1981-82. NIH fellow, 1971-74; grantee, 1970-80. Mem. AMVA, AAAS, Am. Soc. Pharmacology and Exptl. Therapeutics, Phi Zeta, Alpha Zeta, Gamma Sigma Delta. Subspecialties: Neuropharmacology; Veterinary Neurology. Current work: Meachanism of action and toxicity of anesthetics and organophosph. Home: 4503 7th St Lubbock TX 79416 Office: Anesthesia Research Labs Tex Tech U Lubbock TX 79430

HECHT, ELIZBAETH ANNE, research neuropsychologist; b. Spokane, Wash., June 23, 1939; d. Adolph Hyman and Eleanor (Pearlstein) Fink; m. Mervyn Leonard Hecht, Jan. 31, 1960; children: Matthew, Spencer, Rachel. B.A., Boston U., 1961; Ed.M., Harvard U., 1962; postgrad., Law Sch., 1962-63; M.S., UCLA, 1974, Ph.D., 1980. Mem. research staff dept. psychiatry UCLA, 1974-80, research psychologist, 1980—. Contbr. chpts. to books, articles to profl. jours. Mem. AAAS, Am. Psychol. Assn., N.Y. Acad. Scis., Internat. Neurol. Soc., ACLU. Subspecialties: Neuropharmacology; Neuropsychology. Current work: Cognitive effects of marijuana smoking in moderate users. The neurochemical effects of tobacco smoking that make smoking pleasurable. Home: 311 Amalfi Dr Santa Monica Canyon CA 90402 Office: Dept Psychiatry UCLA 760 Westwood Plaza Los Angeles CA 90024

HECHT, FREDERICK, research institute executive, geneticist; b. Balt., July 11, 1930; m., 1977; 6 children. B.A., Dartmouth Coll., 1952; M.D., U. Rochester, 1960. Pediatric intern and resident Strong Meml. Hosp., Rochester, N.Y., 1960-62; from asst. prof. to prof. pediatrics, med. genetics and perinatal medicine Med. Sch., U. Oreg., Eugene, 1965-78; pres. S.W. Biomed. Research Inst. and dir. Genetics Ctr., Tempe, Ariz., 1978—; research fellow med. genetics U. Wash., 1962-64, Nat. Inst. Child Health, Genetics and Pediatrics spl. fellow, 1964-65; Nat. Inst. Child Health and Human Devel. spl. research fellow, genetics unit Mass. Gen. Hosp.; vis. assoc. prof. pediatrics Harvard Med. Sch.; adj. prof. zoology Ariz. State U., 1978—; adj. prof. pediatrics U. Ariz. Health Sci. Ctr., 1980-81; mem. pediatrics staff Ariz. Children's Hosp., St. Joseph's Hosp., Good Samaritan Hosp., Desert Samaritan Hosp.; mem. med. staff Maricopa County Hosp.; Royal Soc. Medicine traveling fellow, U.K., 1971-72. Mem. Am. Soc. Human Genetics, Am. Pediatric Soc., Am. Fedn. Clin. Research, Soc. Pediatric Research. Subspecialties: Genetics and genetic engineering (medicine); Biomedical research administration. Address: 123 E University Dr Tempe AZ 85281

HECHT, LEE MARTIN, artificial intelligence company executive; b. Phila., May 11, 1942; s. Hymen Nathan and Anne Rosalee (Brodsky) H.; m. Kenney Jean Schowalter, June 17, 1967; 1 dau., Kimberley Kenney. M.S. in Physics, U. Chgo., 1965, M.B.A. (NDEA fellow), 1969. Teaching asst. physics, research asst. U. Chgo., lectr. physics, 1966-67, applied maths., 1967-69, policy studies, 1973-80; pres., chief exec. officer Phoenix-Hecht Inc. (computer services co.), Chgo., 1968-75, dir., 1968-76; pres., chief exec. officer Phoenix-Hecht Cash Mgmt. Services Inc., Chgo., 1973-75, dir., 1973-76; pres., chief exec. officer Kenwood-Pacific Corp., San Francisco, 1973—; dir. Holloway Health Mgmt. Group, Ltd., Chgo., 1973-82, chmn., 1976-82; pres., chief exec. officer Electron Storage Ring Corp., San Francisco, 1977—; chmn., chief exec. officer Teknowledge, Inc., Palo Alto, Calif., 1981—; pres. Middlefield Group Inc., Middlefield Capital Corp., 1981—; dir. Digital Pathways Inc., Palo Alto, Calif.; chmn. Kenwood Group Inc., 1978-82; lectr. bus. adminstrn. U. Calif.-Berkeley, 1975-77; vis. lectr. mgmt. Stanford U., 1976; v.p. Nat. Vidiograph Inc. (motion picture prodn.), Berkeley, 1975-77. Trustee Anne R. Hecht Trust. Mem. Am. Phys. Soc. Clubs: Economic, Tavern (Chgo.). Subspecialty: Artificial intelligence. Address: 37 Irving Ave Atherton CA 94025

HECHT, NORMAN BERNARD, reproductive biologist, molecular biologist, educator, researcher; b. Newark, Dec. 14, 1940; s. Samuel and Lea H.; m. Mary Regnier, June 18, 1968; children: David, Rachelle. B.S., Rensselaer Poly. Inst., 1962; Ph.D., U. Ill.-Champaign, 1967. Lectr. biology U. Calif.-San Diego, La Jolla, 1967-70; asst. prof. Tufts U., 1970-76, assoc. prof., 1976—, prof., 1983—, dir. lab., 1970—. Contbr. articles to profl. jours. NSF grantee, 1970—; NIH grantee, 1970—. Mem. AAAS, Soc. Study Reproduction, Am. Soc. Cell Biology, Am. Soc. Andrology. Subspecialties: Genetics and genetic engineering (biology); Molecular biology. Current work: Research in regulation of gene expression during mammalian spermatogenesis using recombinant DNA technology, biogenesis and differentation of mitochondria during spermatogenesis. Office: Tufts U Dept Biology Medford MA 02155

HECK, DANIEL CURTIS, JR., mechanical engineer; b. Chgo., Jan. 8, 1951. B.S.M.E., U. Ill., 1973. Hybrid microelectronics engr. Sperry Flight Systems, Phoenix, 1973-78, prin. mech. actuators for autopilots engr., 1978—. Bd. elders Orangewood Presbyterian Ch., 1982—. Republican. Subspecialties: Mechanical engineering; Microelectronics. Current work: Product development integrating the new super plastics into traditionally aluminum/stainless servomotor devices. Office: 5353 W Bell Rd Glendale AZ 85308

HECK, HENRY D'ARCY, biochemical toxicologist; b. Bryn Mawr, Pa., Apr. 18, 1939; s. Harold Joseph and Lydia Suzanne (Holt) H.; m. Mary Montgomery, June 21, 1963; children: Katherine, Julia. A.B., Princeton U., 1962; Ph.D., Northwestern U., 1966. Postdoctoral fellow Max-Planck Inst. Physikalische Chemie, Göttingen, W.Ger., 1966-68; asst. prof. chemistry U. Calif.-Berkeley, 1968-72; analytical biochemist Stanford Research Inst., Menlo Park, Calif., 1972-77; scientist Chem. Ind. Inst. Toxicology, Research Triangle Park, N.C., 1977—. NSF fellow, 1963-66, 66-67; European Molecular Biology Orgn. fellow, 1967. Mem. AAAS, Soc. Toxicology (Frank R. Blood award 1983), Am. Soc. Biol. Chemists, Am. Soc. Mass Spectrometry, Am. Chem. Soc. Subspecialties: Toxicology (medicine); Biochemistry (biology). Current work: Biochemical and inhalation toxicology, analytical chemistry. Home: 311 Elliott Rd Chapel Hill NC 27514 Office: Chem Ind Inst Toxicology PO Box 12137 Research Triangle Park NC 27709

HECK, WALTER WEBB, research plant physiologist; b. Columbus, Ohio, May 28, 1926; s. Arch Oliver and Frances Margaret (Agnew) H.; m. Corinne Ruth Schiller, Dec. 23, 1959; children: Carolyn R., Ninon, Frederick R., Lee A., Frances E. B.S. in Edn, Ohio State U., 1947; M.S., U. Tenn., 1950; Ph.D., U. Ill., 1954. Postdoctoral trainee in radiocarbon techniques U. Ill., 1954-55; asst. prof. biology Ferris State Coll., Big Rapids, Mich., 1955-58, assoc. prof., 1959; assoc. prof. plant physiology Tex A&M U., College Station, 1959-63; research plant physiologist Dept. Agr., Cin., 1963-69, supervisory plant physiologist, 1969—; prof. dept. botany N.C. State U., Raleigh, 1969—, research leader air quality effects on crop, prodn. Contbr. articles on plant physiology to profl. jours. EPA grantee, 1973-82; NASA grantee, 1976-79. Mem. Air Pollution Control Assn. (Frank A. Chambers award 1981), Am. Soc. Plant Physiologists, Bot. Soc. Am., AAAS, Am. Inst. Biol. Scis., Soil Conservation Soc. Am. Subspecialties: Plant physiology (agriculture); Environmental toxicology. Current work: Determine response of plants and plant systems to atmospheric constituents-primarily air pollutants, plant growth and development. Home: 3612 Browning Pl Raleigh NC 27609 Office: Botany Dept NC State U Raleigh NC 27650

HECKMAN, RICHARD AINSWORTH, nuclear engrineer; b. Phoenix, July 15, 1929; s. Harris and Ann (Sells) H.; m. Olive Anne Biddle, Dec. 17, 1950; children: Mark, Bruce. B.S. in Chem. Engring, U. Calif.-Berkeley, 1950. Chem. engr U. Calif. Radiation Lab., 1950-51; Calif. Research and Devel. Co., Livermore, 1951-53; program leader Lawrence Livermore Nat. Lab., 1953-81, project leader, 1981—. Co-author: Nuclear Waste Management, 1982. Bd. dirs. Calif. Industries for the Blind, Inc., Los Angeles, 1977, sec. corp., 1978; bd. dirs. Hands of Tri-Valley, Inc., Livermore, 1980. Mem. AAAS, Am. Nuclear Soc., N.Y. Acad. Sci. Democrat. Clubs: Island Yacht (Alameda, Calif.) (commodore 1971); Midget Ocean Racing (San Francisco) (commodore 1972). Subspecialties: Nuclear fission; Chemical engineering. Current work: Research and development, next generation nuclear reactor fuel cycle, utilizing laser isotope separation in front andback of the fuel cycle. Patentee nuclear engring. (6). Home: 5683 Greenridge Rd Castro Valley CA 94546 Office: Lawrence Livermore Nat Lab PO Box 808 Livermore CA 94550

HECKMAN, THOMAS PAUL, mechanical engineer; b. Kokomo, Ind., July 28, 1917; s. Thomas S. and Mary (Grosswege) H.; m. Jane C.Stewart, Mar. 27, 1955; children: Eric, Jan, Mark. B.S.M.E., Purdue U., 1940. Registered profl. engr., Wash. Project engr., classification officer, tech. info. officer AEC, 1956-70; pres. T.P. Heckman Assocs., Lombard, Ill., 1970—; cons. Argonne Nat. Lab. Served with USN, 1943-46. Mem. ASME, Am. Nuclear Soc. Unitarian. Subspecialty: Mechanical engineering. Current work: Research and development of permanent magnet connectors and levitation; home and yard devices and procedures for handicapped persons. Patentee in field. Home and Office: 20 W 533 Edgewood Rd Lombard IL 60148

HECKMAN, TIMOTHY MARTIN, astronomer, educator; b. Toledo, Ohio, Oct. 11, 1951; s. Dale H. and Joan E. (Martin) H.; m. Joanne E. Orsini, Oct. 19, 1978. B.A., Harvard U., 1973; Ph.D., U. Wash., 1978. Research asst. Harvard Coll. Obs., 1972-73, 74; dept. astronomy U. Wash., 1973-78; postdoctoral fellow Leiden Obs., 1978-80; Bok fellow Steward Obs., U. Ariz., 1980-81; asst. prof. astronomy U. Md., College Park, 1982-. Contbr. articles to profl. publs. Nat. Merit scholar, 1969-73; NATO research grantee, 1979-82; NASA grantee, 1980-81; U. Md. grantee, 1982. Mem. Am. Astron. Soc., ACLU, NOW, Sierra Club. Subspecialties: Optical astronomy; Radio and microwave astronomy. Current work: Observations (particularly in optical and radio regimes) of extragalactic objects. Special interest in active galaxies and quasars. Home: 2906 56th Pl Cheverly MD 20785 Office: Astronomy Program U Md College Park MD 20742

HECTOR, MINA L., chemistry educator; b. Oak Park, Ill., Oct. 15, 1947; d. Charles Ramon and Mary Jane (Downing) Fisher; m. Gary Bruce Hector, June 27, 1975; 2 sons, Charles James, Jamie Lee. B.A., Lake Forest (Ill.) Coll., 1969; Ph.D., U. Colo., 1975. Assoc. prof. chemistry Calif. State U.-Chico, 1975—. Contbr. articles to profl. jours. Bd. dirs. North Valley Sci. Fair, Chico, 1977-80. Mem. Am. Chem. Soc., Am. Soc. Microbiology, Internat. Union Biochemistry. Democrat. Subspecialties: Biochemistry (biology); Genetics and genetic engineering (biology). Current work: Investigation of the isoprenoid degradative pathway in pseudomonads by genetical analysis to genetically engineer bacteria capable of oil degradation and single cell protein synthesis. Office: Dept Chemistry Calif State U Chico CA 95929

HEDBERG, KENNETH WAYNE, chemistry educator; b. Portland, Oreg., Feb. 2, 1920; s. Gustave N. and Ruth H. (Haagsma) H.; m. Lise Smedrik, Aug. 11, 1954; children: Erik, Katrina. Ph.D., Calif. Inst. Tech., 1948. Research fellow Calif. Inst. Tech., 1948-56; mem. faculty dept. chemistry Oreg. State U., Corvallis, 1956—, prof., 1964—. Contbr. articles to profl. jours. Guggenheim-Fulbright fellow, 1952-53. Mem. Am. Chem. Soc., Am. Phys. Soc., Norwegian Acad. Sci. (fgn.), Sigma Xi. Subspecialties: Physical chemistry; Molecular structure. Current work: Gas-phase molecular structure. Office: Dept Chemistry Oreg State U Corvallis OR 97331

HEDGCOTH, CHARLES, biochemist, educator; b. Graham, Tex., Jan. 29, 1936; s. Charlie and Edna Mae Pearl (Pirkle) H.; m. Barbara Anne Graham, June 20, 1956; children: Kelli, Kimberly, Charles Michael. B.S. in Chemistry, U. Tex., Austin, 1961, Ph.D. in Chemistry (Rosalie B. Hite fellow, NIH fellow), 1965. Asst. prof. biochemistry Kans. State U., Manhattan, 1965-68, assoc. prof., 1968-76, prof., 1976—, ancillary prof. div. biology, 1979—; vis. assoc. prof. U. B.C. (Can.), Vancouver, 1975. Served with USCG, 1954-58. NATO/NSF sr. fellow, 1975; NSF, NIH, Nat. Cancer Inst. grantee. Mem. Am. Soc. Biol. Chemists, Am. Chem. Soc., AAAS, Sigma Xi. Subspecialties: Cancer research (medicine); Genetics and genetic engineering (agriculture). Current work: Characterization of wheat storage protein genes; wheat mitochondrial DNA: perturbations of lysine transfer ribonucleic acids in transformed cells. Home: 1305 Waters St Manhattan KS 66502 Office: Dept Biochemistry Kans State U Manhattan KS 66506

HEDGES, HARRY GEORGE, educator; b. Lansing, Mich., Oct. 7, 1923; s. Charles William and Elsie (Frost) H.; m. Mary J. Corbishley, June 14, 1944 (dec.); children—Susan, Martha. B.S., Mich. State U., 1949, Ph.D., 1960; M.S., U. Mich., 1954. Electronics engr. USAF Wright Air Devel. Center, Dayton, Ohio, 1949-51; research asso. U. Mich., 1951-54; instr. Mich. State U., East Lansing, 1954-60, asst. prof., 1960-63, asso. prof., 1963-69, prof., chmn. dept. computer sci., 1969—; Dir. Nat. Electronics Conf., Inc., 1968—. Tech. editor: Analysis of Discrete Physical Systems, 1967; mem.: Computer Sci. Bd, 1973—; chmn., 1974-75. Chmn. Selective Service Bd. 264, Lansing, 1970-76. Served with AUS, 1943-46; PTO. NSF sci. faculty fellow, 1960. Mem. Am. Soc. Engring. Edn. (chmn. N.Central sect. 1968-69), IEEE (dir. 1967-69, treas. 1969, vice chmn. 1973, chmn. 1974, Southeastern Mich. sect.). Subspecialty: Computer engineering. Home: 1623 Woodside Dr East Lansing MI 48823

HEDSTROM, JOSEPH CHARLES, operations research analyst; b. Dothan, Ala., Sept. 21, 1952; s. Robert Allan and Elizabeth (Menchion) H. B.S. in Indsl. Engring, U. Ala-Tuscaloosa, 1975, M.S., Ga. Tech. U., 1976, now postgrad. Indsl. engr. Ford Motor Co., Sterling Heights, Mich., 1977-78; ops. research analyst Lockheed-Ga. Co., Marietta, Ga., 1978-80, sr. ops. research analyst, 1980—. Recipient Indsl. Engring. award U. Ala., 1975. Mem. Ops. Research Soc. Am., Am. Inst. Indsl. Engrs. (chpt. v.p. 1974-75), AIAA, Tau Beta Pi. Roman Catholic. Subspecialties: Operations research (engineering); Industrial engineering. Current work: Operations research. Home: 1805 Roswell Rd Apt 33-A Marietta GA 30062 Office: Lockheed-Georgia Company Marietta GA 30063

HEER, CLIFFORD V., physicist, educator; b. Fulton County, Ohio, May 31, 1920; s. Nelson Veer and Minnie May (Leu) H.; m. Esther Jean Leonard, Dec. 17, 1949; children: Barbara Jean, Deborah Ann, Daniel Nelson. B.Sc. in physics, Ohio State U., 1942, Ph.D., 1949. Faculty dept. physics Ohio State U., 1949—, prof. physics, 1961—; cons. Ramo-Wooldridge Corp., 1956-58, Space Tech. Lab., 1958-65; Honeywell, Inc., 1964-65, TRW, 1965-70, Litton Corp., 1981-82. Author, 1972, also articles.; Inventor laser gyroscope. Served to lt. Signal Corps U.S. Army, 1942-46. Fellow Am. Phys. Soc. Republican. Methodist. Subspecialties: Atomic and molecular physics; Statistical physics. Current work: Current research is related to laser physics.

HEERWAGEN, DEAN REESE, consulting building engineer, educator; b. Summit, N.J., Dec. 27, 1942; s. Arthur Robert and Jane Roberts (Thoman) H.; m. Judith Hannula, Aug. 21, 1971; 1 dau., Margaret Jane Reese. B. Metall. Engring., Cornell U., 1965; M.S., MIT, 1967, B.Arch., 1971. Asst. prof. architecture Cornell U., 1971-73; pvt. practice bldg. engring., Boston, 1974; mem. faculty U. Wash., Seattle, 1975—, now assoc. prof. architecture; energy cons. City of Seattle and; Wash. State govt. Contbr. bldg. energy articles to profl. jours. Wash. State Govt. grantee, 1976-79; NSF grantee, 1980 others. Mem. Am. Solar Energy Soc., ASHRAE, Council on Tall Bldgs. and Urban Habitat. Subspecialties: Solar energy; Heat transfer. Current work: Teaching and research on developing design guidelines and tools for achieving energy-efficient and cost-effective building operation and insuring occupant thermal comfort. Home: 2716 NE 91st St Seattle WA 98115 Office: Dept Architecture U Wash 208 Gould Hall JO-20 Seattle WA 98105

HEESCHEN, DAVID SUTPHIN, astronomer, educator; b. Davenport, Iowa, Mar. 12, 1926; s. Richard George and Emily (Sutphin) H.; m. Eloise St. Clair, June 11, 1950; children: Lisa Clair, David William, Richard Mark. B.S., U. Ill., 1949, M.S., 1951; Ph.D., Harvard U., 1954; Sc.D. (hon.), W.Va. Inst. Tech., 1974. Instr. Wesleyan U., Middletown, Conn., 1954-55; lectr., research assoc. Harvard U., 1955-56; scientist Nat. Radio Astronomy Obs., 1956-77, sr. scientist, 1977—, dir., 1962-78; research prof. astronomy U. Va., 1980—; Cons. NASA, 1960-61, 68-72. Contbr. sci. jours. G.R. Agassiz fellow Harvard Obs., 1953-54; Recipient Disting. Public Service award NSF, 1980. Fellow AAAS; mem. Am. Astron. Soc. (v.p. 1969-71, pres. 1980-82), Internat. Astron. Union (v.p. 1976-82), Internat Sci. Radio Union, Nat. Acad. Sci., Am. Acad. Arts and Sci., Am. Philos. Soc. Subspecialty: Planetary science. Current work: Variability of extra galactic sources; star formation in galaxies; nature of quasars. 22901

HEFFERLIN, RAY, physicist, educator; b. Paris, May 2, 1929; s. Milo and Ruth (Streiff) H.; m. Inelda Phillips, Sept. 5, 1954; children—Lorelei, Heidi, Melissa, Jennifer. Ph.D., Calif. Inst. Tech., 1955. Prof. physics So. Coll., Collegedale, Tenn., 1955—. Mem. Am. Phys. Soc., Am. Astron. Soc., Internat. Astron. Union, Tenn. Acad. Sci. Subspecialty: Periodic systems of small molecules. Current work: Periodic systems of diatomic and other small molecules. Home and Office: PO Box H Collegedale TN 37315

HEFFERREN, JOHN JAMES, dental research administrator, educator, consultant; b. Chgo., Aug. 12, 1928; s. John and Josephine (Walsh) H.; m. Sandra A. Mancewicz, Feb. 6, 1965; children: Anne W., Aileen C., Neal R., Clare J. B.S., Loyola U., Chgo., 1950; M.S., U. Wis.-Madison, 1952, Ph.D., 1954. Chemist AMA, Chgo., 1953-59; dir. div. chemistry ADA, Chgo., 1959-64; dir. div. biochemistry ADA Health Found., 1964-75, dir. research inst., 1975—; mem. faculty Northwestern U. Dental Sch., Chgo., 1977—; cons. VA Lakeside Hosp., Chgo., 1970—; lectr. U. Mich. Sch. Pharmacy, Ann Arbor, 1972—. Contbr. over 100 articles to profl. jours.; editor books in fild. Pres. Montessori of Holy Family, Chgo., 1970—. Am. Found. Pharm. Edn. fellow, 1950-53. Mem. Internat. Assn. Dental Research, Am. Chem. Soc., Am. Pharm. Assn., N.Y. Acad. Scis., European Orgn. Caries Research, Rho Chi, Phi Sigma. Roman Catholic. Club: KC. Subspecialties: Oral biology; Biochemistry (biology). Current work: Role of food, nutrition and consumption patterns in dental health, trace element metabolism, preclinical and clinical test methodology. Home: 3001 Normandy Pl Evanston IL 60201 Office: 211 E Chicago Ave Chicago IL 60201

HEFFNER, HENRY EDWARD, psychologist, researcher; b. Harvey, Ill., Sept. 22, 1944; s. Henry Edward and Lillian (Ferguson) H.; m. Rickye JoAnn Sarno, Nov. 23, 1967; children: Henry E., Peter E. B.A., Trinity Coll., 1966; postgrad., Vanderbilt U., 1966-67; M.S., Fla. State U., 1969, Ph.D., 1973. Research fellow U. Birmingham, Eng., 1971-72; research assoc. Johns Hopkins U. Sch. Medicine, 1972-73, U. Kans. Bur. Child Research Parsons State Hosp and Tng. Ctr., 1973—. Contbr. articles to profl. jours. Mem. Am. Psychol. Assn., Acoustical Soc. Am., Neurosci. Soc. Subspecialties: Physiological psychology; Sensory processes. Current work: Behavioral study of auditory nervous system; comparative study of mammalian hearing. Office: Bur Child Research U Kans Parsons State Hosp and Tng Center Parsons KS 67357 Home: 1715 Morgan Ave Parsons KS 67357

HEFFNER, REID RUSSELL, JR., pathologist, neuropathologist, educator; b. Phila., Apr. 16, 1938; s. Reid Russell and Katherine (Dewey) H.; m. Elenora Markunas, July 24, 1965; children: Honora, Reid. B.A., Yale U, 1960, M.D., 1965. Diplomate: Am. Bd. Pathology. Chief neuromuscular disease div. Armed Forces Inst. Pathology, Washington, 1972-76; prof. pathology and neurology SUNY, Buffalo, 1976—, chief div. neuropathology, 1976—; dir. dept. pathology Erie County Med. Center, Buffalo, 1979—; med. adv. Buffalo Assn. Neurologic Disease; mem. med. adv. bd. Western N.Y. Alzheimer's Disease Assn.; cons. Buffalo Gen. Hosp., VA Med. Center, Millard Fillmore Hosp., all Buffalo. Contbr. articles to profl. jours. Served to maj. M.C. U.S. Army, 1970-72. Mem. Am. Assn. Neuropathologists, Am. Acad. Neurology, Am. Soc. Clin. Pathologists, Internat. Acad. Pathology, N.Y. Acad. Scis. Republican. Episcopalian. Club: Wanakah (N.Y.) Country. Subspecialty: Pathology (medicine). Current work: Neuromuscular disease; inflammatory myopathy. Office: 462 Grider St Buffalo NY 14215

HEFFNER, THOMAS GARY, psychopharmacologist, researcher; b. Salem, Ohio, Sept. 20, 1949; s. Fred Curtis and Irma Virginia (Buchanan) H. B.S., U. Pitts., 1971, Ph.D., 1976. Teaching asst. U. Pitts., 1971-74, teaching fellow, 1974-76; postdoctoral fellow U. Chgo., 1976-79, research asst. prof. pharmacology and physiol. scis., 1979-83; research assoc. (group leader) antipsychotic drug research Warner Lambert/Parke-Davis Pharm. Research, Ann Arbor, Mich., 1983—. Contbr. articles to profl. jours. Brain Research Found. grantee, 1982. Mem. Soc. Neurosci., Am. Psychol. Assn., AAAS, Sigma Xi. Subspecialties: Pharmacology; Neuropharmacology. Current work: Psychopharmacology, mechanism of action of drugs which alter behavior, use of animal models to understand psychoactive drug action in humans with mental disorders. Home: 636 Duane Ct Ann Arbor MI 48103 Office: Dept Pharmacology Warner Lambert/Parke-Davis 2800 Plymouth Rd Ann Arbor MI 48105

HEFZY, MOHAMED SAMIR W. M., engineering educator; b. Cairo, Feb. 26, 1951; U.S., 1976; s. Wahba Mohamed and Samya Mohamed (El-Mahdi) H.; m. Nabila A. H. Gomaa, July 28, 1976; 1 dau., Hebah. B.Sc. in Civil Engring, Cairo U., 1972, Ain Shams U., 1974; M.S. in Applied Mechanics, U. Cin., 1977, Ph.D., 1981. Instr. dept. math. and phys. engring. Cairo U., 1972-76; teaching and research asst. aero. engring. dept. U. Cin., 1976-81, asst. prof. evening coll., 1981-83, research assoc. biomechanics lab., dept. orthopedic surgery, 1981-83; asst. prof. engring. dept. physics and engring. Grand Valley State Coll., Allendale, Mich., 1983—. Recipient NASA award, 1981. Mem. Am. Soc. Engring. Edn., ASME, Am Soc. Engring. Edn., Am. Acad. Mechanics, Soc. Engring. Sci. Subspecialties: Theoretical and applied mechanics; Biomedical engineering. Current work: Continuum modeling of the mechanical and thermal behavior of discrete repetitive large structures; geometric modeling and analysis of large flat segmented latticed surfaces; modeling the knee joint using finite elements; kinematics of the knee joint. Office: Dept Physics and Engring Grand Valley State Coll Allendale MI 49401

HEGEMAN, GEORGE DOWNING, microbiologist, educator, consultant; b. Glen Cove, N.Y., Aug. 31, 1938; s. George Downing and Bonnie (Blair) H.; m. Sally Lofgren, Aug. 26, 1961; children: Susan Elizabeth, Adrian Daniel. A.B., Harvard Coll., 1960; Ph.D., U. Calif.-Berkeley, 1965. Instr. U. Calif.-Berkeley, 1965, asst. prof. bacteriology and immunology, 1966-72; USPHS postdoctoral fellow Institut d'Enzymologie, Cente National pour la Recherche Scientifique, Gif-sur-Yvette, France, 1965-66; assoc. prof. microbiology Ind. U.-Bloomington, 1972-78, prof., 1978—. Contbr. numerous articles, abstracts to profl. jours.; editorial bd.: Jour. Bacteriology, 1970-79, Applied and Environ. Microbiology, 1979—. NIH Dept. health and Human Services grantee, 1966-78; NSF grantee. Mem. AAAS, Am. Soc. Microbiology, Am. Soc. Biochemistry, Sigma Xi. Subspecialties: Microbiology; Biochemistry (biology). Current work: Microbial metabolism and physiology; metabolic control; enzymology; aerobic catabolism by bacteria. Patentee in field. Office: Biology Dept Ind U 138 Jordan Hall Bloomington IN 47405

HEGGESTAD, HOWARD EDWIN, research plant pathologist; b. Stoughton, Wis., July 24, 1915; s. Erick A. and Gunda (Veium) H.; m. Dolores Kathryn Andersen, June 10, 1939; children: Arnold, Margot, David. Ph.B., U. Wis.-Madison, 1940, Ph.D., 1944. Instr. U. Wis., 1944-46; agronomist U.S. Dept. Agr., Greeneville, Tenn., 1946-55, Beltsville, Md., 1955-63, acting leader tobacco investigations, 1963-64, leader tobacco breeding and disease investigation, 1964-66, chief air pollution lab., 1966-75, research plant pathologist plant stress lab., 1975—; cons. Electric Power Research Inst., 1979-81. Contbr. chpts. to books, articles to profl. jours. Recipient Disting. Sci. Research award Cigar Industry Am., 1961; Environ. Quality Research award Am. Soc. Hort. Sci., 1973. Mem. Am. Phytopathological Soc., AAAS, Am. Inst. Biol. Sci., Air Pollution Control Assn. Lutheran. Subspecialties: Environmental toxicology; Plant pathology. Current work: Determine effects of ozone and sulfur dioxide on crop productivity; determine the combined effects of ozone and soil moisture stress on soybean yields under field condition; identify and develop cultivars with tolerance to pollutants. Home: 3112 Castleleigh Rd Silver Spring MD 20904 Offic: Plant Stress La US Dept Agr Beltsville MD 20705

HEGSTED, DAVID MARK, biochemistry educator, research administrator; b. Rexburg, Idaho, Mar. 25, 1914; s. John and Edna Margaret (Porter) H.; m. Maxine Scow, May 26, 1941; children: Christina, Eric John. B.S., U. Idaho, 1936; M.S., U. Wis., 1938, Ph.D., 1940; A.M. (hon.), Harvard U., 1962. Research asst. U. Wis., 1936-41; research chemist Abbott Labs., 1941-42; assoc. nutrition Harvard Schs. of Medicine and Pub. Health, 1942-43, asst. prof., 1943-48, assoc. prof., 1948-62, prof., 1962-80, prof. nutritron emeritus, 1980—; assoc. dir. research New Eng. Regional Primate Research Ctr. Harvard Sch. Medicine, 1982—; adminstr. Human Nutrition Center, U.S. Dept. Agr., Washington, 1978-82; Cons. nutrition to Colombian Govt., 1946; nutritionist Inst. Inter-Am. Affairs, Peru, 1950-51; cons. UN FAO, Chile, 1956, Rome, 1961, 69, WHO, 1962, 70, NIH, 1958—; mem. food and nutrition bd. NRC, 1955-72, chmn. food and nutrition bd., 1968-72. Editor: Nutrition Revs, 1968-78; contbr. articles, chpts. profl. jours. and books. Named to U. Idaho Hall of Fame, 1976. Mem. Am. Inst. Nutrition (Osborne Mendel award 1965, Conrad A. Elvehjem award 1979, pres. 1972-73), Am. Chem. Soc., Am. Pub. Health Assn., N.Y. Acad. Scis., Peruvian Pub. Health Soc., A.M.A. (council foods and nutrition 1960-68), Nat. Acad. Scis., Am. Dietetic Assn. (hon.), Sigma Xi, Alpha Chi Sigma, Sigma Alpha Epsilon. Clubs: Harvard, Cosmos. Subspecialties: Nutrition (medicine); Nutrition (biology). Home: 58 Boulder Rd Wellesley Hills MA 02115

HEGYI, DENNIS JEROME, physicist, educator; b. Reading, Pa., Dec. 23, 1942; s. Frank and Adeline A. (Mazzolla) H.; m. Michelle A. Aminoff, Jan. 10, 1982. B.S., M.I.T., 1963; Ph.D., Princeton U., 1968. Postdoctoral assoc. (Inst. Space Studies), N.Y.C., 1968-70; asst. prof. Boston U., 1970-73, Bartol Research Found., Swarthmore, Pa., 1973-75, U. Mich., 1975-78, assoc. prof. physics, 1978—. Contbr. articles to profl. jours. on primordial element abundances, relativistic equations of state, optical pulsars, cosmic background radiation, change-coupled devices, massive halos of spiral galaxies. Mem. Am. Phys. Soc., Am. Astron. Soc. Subspecialties: Cosmology; High energy astrophysics. Current work: Studying the nature of cosmological "missing mass"; researching and teaching. Home: 1512 Morton Ave Ann Arbor MI 48104 Office: Randall La U Mich Ann Arbor MI 48109

HEIDENBERG, WILLIAM JAY, cardiologist; b. N.Y.C., Oct. 27, 1936; s. Benjamin and Fannie Mollie (Reiff) H.; m. Iris Ellen Lilienfeld, June 18, 1967; children: Lawrence Lee, Michael Stephen. B.S., Bklyn. Coll., CUNY, 1956; M.D., U. Pitts., 1960. Diplomate: Am. Bd. Internal Medicine (Cardiovascular Disease). Intern and resident Motefiore Hosp., Bronx, 1960-62; resident VA Hosp., Bronx, 1962-63; resident in cardiology Montefiore Hosp., 1963-64; fellow in cardiology Yale U., New Haven, 1966-67; practice medicine specializing in cardiology, White Plains, N.Y., 1967—; coordinator nuclear cardiology White Plains Hosp., 1977—; clin. asst. prof. medicine (cardiology) N.Y. Med Coll., Valhalla, 1970—. Served as lt. comdr. USPHS, 1964-66. Fellow ACP, Am. Coll. Cardiology, Am. Heart Assn. Council Clin. Cardiology; mem. Am. Fedn. Clin. Research (sr.), Soc. Nuclear Medicine. Subspecialties: Cardiology; Internal medicine. Current work: Clinical cardiology and internal medicine, nuclear cardiology. Office: 85 Old Mamaroneck Rd White Plains NY 10605

HEIDERSBACH, ROBERT HENRY, JR., chemical engineering educator; b. El Paso, Tex., Dec. 30, 1940; s. Robert Henry and Dorothy Mae (Dunteman) H.; m. Dianne Katherine Shrum, Oct. 5, 1963; children: R. Scott, Krista. Met.E., Colo. Sch. Mines, 1963; M.E., U. Fla., 1968, Ph.D., 1971. Metallurgist Constrn. Engring. Research Lab., Champaign, Ill., 1971-74; prof. U. R.I., Kingston, 1974-81; prof. chem. engring. Okla. State U., Stillwater, 1981—; cons. Dow Chem. Co., Midland, Mich., 1980—, Ocean Tech., Inc., Newport, R.I., 1977—, USN, Boston, 1974-81, Gen. Motors Co., Dayton, Ohio, 1980. Served to capt. U.S. Army, 1963. INCO fellow U. Fla., 1970; NACE fellow, 1968-70. Mem. Nat. Assn. Corrosion Engrs. (com. chmn. 1973—), Am. Inst. Chem. Engrs. (OTC com. 1981—), Am. Soc. Metals, Electrochem. Soc., Microbeam Analysis Soc. Republican. Methodist. Subspecialties: Corrosion; Metallurgical engineering. Current work: Corrosion, raman spectroscopy, electrochemistry, failure analysis, corrosion of metals in concrete, cathodic protection. Office: Okla State Chem Engring Dept Stillwater OK 74078

HEIDRICK, LEE EDWARD, plant pathologist, agrl. researcher; b. Little Valley, N.Y., June 23, 1921; s. Otto Paul and Martha (Frenz) H.; m. Esther Marie Lisdell, Sept. 6, 1952. B.S., Cornell U., 1943, M.S., 1950; Ph.D. W.Va. U, 1955. Research specialist Rockefeller Found., Mexico City, 1950-53, Bogota$Z, Colombia, 1953-63; agrl. research specialist Chevron Chem. Co., Mt. Laurel, N.J., 1963—. Research publs. in phytopathology. Served to lt. USN, 1943-46. Mem. AAAS, Am. Phytophatol. Soc., Sigma Xi. Congregationalist. Lodge: Elks.

Subspecialties: Plant pathology; Plant physiology (agriculture). Current work: Pesticide research on fungicides, insecticides, herbicides, plant growth regulators, activity, efficacy, toxicology, mode of action. Patentee pesticide synergists. Home: 119 Leeds Rd Mount Laurel NJ 08054 Office: Box 118 Moorestown NJ 08057

HEILMEIER, GEORGE HARRY, research electrical engineer; b. Phila., May 22, 1936; s. George C. and Anna I. (Heineman) H.; m. Janet S. Faunce, June 24, 1961; 1 dau., Elizabeth. B.S. in Elec. Engring., U. Pa., 1958; M.S. in Engring., Princeton U., 1960, A.M., 1961, Ph.D., 1962. With RCA Labs., Princeton, N.J., 1958-70, dir. solid state device research, 1965-68, dir. device concepts, 1968-70; White House fellow, spl. asst. to sec. def., Washington, 1970-71; asst. dir. def. research and engring. Office Sec. Def., 1971-75; dir. Def. Advanced Projects Agy., 1975-77; v.p. research, devel. and engring. Tex. Instruments Inc., 1978-83, sr. v.p., chief tech. officer, 1983—; mem. vis. com. Moore Sch., U. Pa., Stanford U.; also elec. engring. dept. Princeton U.; mem. Def. Sci. Bd., Air Force Sci. Adv. Bd., Army Sci. Adv. Bd.; mem. adv. group on electron devices Dept. Def. Author. Recipient IEEE David Sarnoff award RCA, 1969; IR-100 New Product award Indsl. Research Assn., 1968, 69; Sec. Def. Disting. Civilian Service award, 1975, 77; Arthur Flemming award U.S. Jaycees, 1974. Fellow IEEE (Sarnoff Field award 1976); mem. U. Pa., Princeton U. Grad. alumni assns., Nat. Acad. Engring., Sigma Xi, Tau Beta Pi, Eta Kappa Nu (Outstanding Young Engr. in U.S. award 1969). Methodist. Subspecialties: Artificial intelligence; Semiconductors. Patentee in field. Office: Tex Instruments 13500 North Central Expy Box 225474 MS 400 Dallas TX 75265

HEIM, LYLE RAYMOND, clinical immunologist, editor and publisher; b. Lignite, N.D., Apr. 30, 1933; s. John Ludvig and Anna Josephine (Kleppe) H.; m. Julia Roorda, June 28, 1957; children: Bradley Lyle, Denise Lynal, Brian Jonathan. B.A., U. Minn., 1963, M.S., 1966, Ph.D., 1969. Postdoctoral fellow Baylor Coll. Medicine, Houston, 1969-72; asst. prof. Columbia Hosp. and Med. Coll. Wis., Milw., 1972-78; assoc. prof. Tex. Tech. U., 1978-82; editor-in-chief Exptl. Hematology, El Paso, Tex., 1971—, pub., 1980—; cons. Internat. Soc. Exptl. Hematology.; Chmn. Immunology Research Found.; pres. Council Biology Editors; dir. Pet Facilitation Therapy; dir., pres. pro-tem Habitat for Humanity of El Paso. Contbr. articles and abstracts to profl. jours., chpts. in books. Chmn. Troop 2 com. Boy Scouts Am. Served with USAF, 1952-56. Mem. AAAS, Am. Med. Writers Assn., Am. Soc. Microbiology, Council Biology Editors, Internat. Soc. Exptl. Hematology, Minn. Alumni Assn., N.Y. Acad. Scis., Soc. Scholarly Publs., So. Soc. Pediatric Research, S.W. Sci. Forum, Planetary Soc., Smithsonian Assocs., Sigma Xi. Subspecialties: Immunology (medicine); Marrow transplant. Current work: Immunologic response augmentation; bone-marrow transplantation, immunology and experimental hematology. Home and Office: 6365 Los Robles Dr El Paso TX 79912

HEIMER, RALPH, biochemistry educator; b. Vienna, Austria, Nov. 11, 1921; came to U.S., 1938, naturalized, 1944; s. Oskar and Irene (Ziegler) H.; m. Caryl P. Biren, June 13, 1943 (dec. Feb. 1973); children: Robert, Paul; m. Phyllis M. Sampson, Aug. 23, 1974. B.S., CCNY, 1948; M.A., Columbia U., 1951, Ph.D., 1957. Research assoc. Hosp. Spl. Surgery, N.Y.C., 1956-62; asst. prof. Cornell Med. Coll., N.Y.C., 1958-62; assoc. prof. N.J. Coll. Medicine, Jersey City, 1963-66; prof. biochemistry Thomas Jefferson U., Phila., 1967—. Served with U.S. Army, 1942-46. Recipient awards NIH, 1960-76, 80—, Am. Cancer Soc., 1970-72. Mem. Am. Assn. Immunologists, Am. Soc. Biol. Chemists, Am. Rheumatism Assn. Subspecialties: Biochemistry (biology); Immunology (medicine). Current work: Detection of antigens in immune complexes, autoantibodies in rheumatic diseases. Office: Thomas Jefferson U 1020 Locust St Philadelphia PA 19107 Home: 235 S 3d St Philadelphia PA 19106

HEIN, JAMES RODNEY, geologist; b. Santa Barbara, Calif., Mar. 15, 1947; s. Warren C. and Beatrice E. (Gale) H.; m.; children: Lanee T., Tasha R. A.A., Santa Barbara City Coll., 1967; B.Sc., Oreg. State U., 1969; Ph.D., U. Calif.-Santa Cruz, 1973. Lectr. earth sci. U. Calif-Santa Cruz, 1972-79, lectr. earth scis., 1980; geologist U.S. Geol. Survey, Menlo Park, Calif., 1974—; internat. group leader UNESCO-IUGS, 1975—; convenor Internat. Confs., 1978, 81; chmn. research group Soc. Econ. Paleontologists and Mineralogists, 1975-79; cons. in field. Contbr. articles to profl. jours.; editor: Silicous Deposits of the Pacific Region, 1982. U. Calif. teaching fellow, 1971-72; NRC postdoctoral assoc., 1972-73. Fellow Geol. Soc. Am.; mem. Am. Geophys. Union, AAAS, Internat. Assn. Sedimentologists, Soc. Econ. Paleontologists and Mineralogists. Subspecialties: Geology; Sedimentology. Current work: Sedimentary ore deposits; geology of volcanic arcs; deep sea sedimentology/minerology; low temperature geochemistry. Office: 345 Middlefield Rd Menlo Park CA 94025 Home: 511 Grant St Santa Cruz CA 95060

HEIN, JOHN WILLIAM, dentist, educator; b. Chester, Mass., Sept. 29, 1920; s. Rudolf Jacob and Mercedes Viola H.; m. Jeanette Marie BeVier, Dec. 16, 1944. B.S., Am. Internat. Coll., 1941; D.M.D., Tufts U., 1944; Ph.D., U. Rochester, 1952; A.M. (hon.), Harvard, 1962, D.Sc., Am. Internat. Coll., 1979. Student instr. oral pathology Tufts Coll. Dental Sch., 1943-44; head div. dental research U. Rochester, 1948-52, sr. fellow dental research, 1949-52, instr. pharmacology, 1951-53, asst. prof. dental research, 1952-55, asst. prof. pharmacology, 1954-55, chmn. dept. dentistry and dental research, 1952-55; instr. anatomy and physiology Eastman Sch. Dental Hygiene, 1950-55, lectr. dental research, 1953-55; research specialist Bur. Biol. Research, Rutgers U., 1955-59; dental dir. Colgate Palmolive Co., 1955-59; prof. preventive dentistry, dean Sch. Dental Medicine, Tufts U., 1959-62; dir. Forsyth Dental Center, 1962—; prof. dentistry Harvard Dental Sch., 1962-67. Trustee Am. Internat. Coll., 1960-76. Served to capt. AUS, 1942-47. Fellow AAAS, Internat. Coll. Dentists (regent 1967-72, pres. U.S. 1975-76, internat. pres. 1983-84); mem. ADA, Mass. Dental Soc. (pres. 1964-65), Internat. Assn. Dental Research (treas. 1978-82), Am. Acad. Dental Sci., New Eng. Dental Soc. (hon. pres. 1978), Am. Soc. Dentistry for Children, Assn. Ind. Research Insts. (1st v.p. 1980, pres. 1981-83), Sigma Xi, Omicron Kappa Upsilon, Delta Sigma Delta. Club: Harvard (Boston). Subspecialty: Cariology. Home: Bridge St Medfield MA 02052 Office: 140 The Fenway Boston MA 02115

HEINBERG, PAUL JULIUS, communication educator, consultant; b. Birmingham, Ala., Aug. 25, 1924; s. Benjamin Fries and Juliette Helen (Isaacs) H.; m. Joyce Suwal, July 8, 1945; 1 dau., Juliette Caye Anson. B.S., Columbia U., 1949, M.A. Tchrs. Coll., 1950; Ph.D., U. Iowa, 1956. Fashion prediction analyst Amos Parrish, Inc., N.Y.C., 1948-50; instr. speech and drama Tex. Women's U., Denton, 1950-52; asst. prof. speech Okla. State U., Stillwater, 1952-57; asst. prof. speech and drama U. Iowa, Iowa City, 1957-65; assoc. prof. speech and communication U. Hawaii, Honolulu, 1965-69, prof. communication, 1969—, dir. Speech-Communication Ctr., 1965-70; cons. AT&T, N.Y.C., 1957-66, HawTel, Honolulu, 1970-76, Inst. Ednl. Research, Washington, 1958-62. Author: (blank verse) Quadrennial Olympiad of the Arts, 1964 (first prize). Disting. pres. affiliate of Optimist Internat., Iowa City, 1965. Served to 1st lt. USAF, 1941-47. Grantee Okla. Research Found., 1959; U. Iowa Research Found., 1962; AT&T, 1963; FAA, 1980. Mem. Soc. Gen. Systems Research, Am. Psychol. Assn., Internat. Communication Assn., Speech Communication Assn. Democrat. Jewish. Club: Optimists (Iowa City) (pres. 1964-65). Subspecialties: Learning; Social psychology. Current work: Development of 2-person and 3-person interactive learning systems to produce effective and efficient changes in language skills, creativity, consulting, and communication. Patentee audio-visual teaching machine, automated instructional device, lang. teaching method. Home: 1530 Ahuawa Loop Honolulu HI 96816 Office: Dept Communication Univ Hawaii 2560 Campus Rd Room 313 Honolulu HI 96822

HEINE, URSULA INGRID, microbiologist; b. Berlin, Feb. 19, 1926; d. Georg Frederick and Alice Frieda (Gunhold) H. Diplom-biologist, Berlin, 1950; Dr. rer. nat., Berlin, 1953. Research assoc. German Acad. Scis., Berlin, 1950-59; assoc. Sch. Medicine Duke U., 1959-68; microbiologist Nat. Cancer Inst. NIH, Bethesda, Md., 1968-72, head virus studies sect., 1972-76, head ultrastructural studies sect., Frederick, Md., 1976—; cons. in field. Contbr. articles to profl. jours. Mem. Am. Assn. Cancer Research, AAAS, Am. Soc. Cell Biology, Electron Microscopy Soc. Am. Subspecialties: Cancer research (medicine); Cell study oncology. Current work: Cell transformation, viral and chemical carcinogenesis. Home: Box 97 Olney MD 20832 Office: NCI-FCRF Bldg 538 Room 205 E Frederick MD 21701 Home: Box 97 Olney MD 20832

HEINEMANN, EDWARD H., aircraft design engineer, consultant; b. Saginaw, Mich., Mar. 14, 1908; s. Gustave Christian and Margaret (Schust) H.; m. Zell Shewey, 1959; 1 dau. by previous marriage, Joan Heinemann Coffee. Ed. pub. schs., Mich., Calif.; D.Sc. (hon.), Northrop U., 1976. Designer, project engr., also chief engr. Moreland Aircraft Corp., Internat. Aircraft Corp., Northrop Aircraft Corp., 1931-36; chief engr. El Segundo (Calif.) div. Douglas Aircraft Co., Inc., 1936-58, corp. v.p. in charge combat aircraft engring., 1958-60; v.p. European sales Douglas Aircraft Co., 1960; exec. v.p. Summers Gyroscope Co., 1960-62; v.p. engring. and program devel. Gen. Dynamics Corp., N.Y.C., from 1962, ret., 1973; now with Heinemann Assocs., Rancho Santa Fe, Calif.; cons. to numerous govt. coms. Recipient Sylvanus Albert Reed award, 1952; Collier trophy, 1953; So. Calif. Aviation Man of Yr. award, 1954; Paul Tissandier diploma Fedn. Aeronautique Internationale, 1955; Nat. medal Sci., 1983. Fellow AIAA, Soc. Aero. Weight Engrs. (hon.), Inst. Aero. Scis. (hon.), Royal Aero. Soc., Am. Astronautic Soc.; mem. Soc. Naval Architects and Marine Engrs., Am. Soc. Naval Engrs., Nat. Acad. Engring., Soc. Automotive Engrs., Tau Beta Pi (hon.). Subspecialty: Aeronautical engineering. Current work: Aircraft design. In charge design, devel. Navy BT, SBD Dauntless dive bombers, Navy R3D transport, A-20 Air Force Havoc attack bomber, DB-7 Boston attack bomber for Brit. and French, A-26 Air Force Invader attack bomber, Navy AD Skyraider attack bomber series, A2D Skyshark attack bomber, F3D land 2 Skyknight night fighters, F3D-3 night fighters, F4D Skyray interceptor, D-558 Skystreak, D-458-2 Skyrocket, D-558-3 research airplanes, A3D attack bomber, A4D Skyhawk attack bomber. Office: Heinemann Assocs PO Box 1795 Rancho Santa Fe CA 92067

HEINEMANN, HEINZ, chemical engineer; b. Berlin, Aug. 21, 1913; m., 1948; 2 children. B.S., Technische Hochschule, Berlin, 1935; Ph.D. in Chemistry, U. Basel, Switzerland, 1937. Chief research chemist Rodessa Oil & Refining Corp., 1938-39; research chemist Danciger Oil & Refining, 1939-41; research fellow Carnegie Inst. Tech., 1941; lab. supr. Attapulgus Clay Co., 1941-48; sect. chief process research Houdry Process Corp., 1948-57; asst. to v.p. R&D, assoc. dir. research M.W. Kellogg Co., 1957-61, mgr., 1961-67, dir. chem. and engring. research, 1967-9; sr. research assoc. central research lab. Mobil R&D Corp., 199-70, 1969-70, mgr. catalysis research, 1970-76, mgr. research contracts, 1976-78; staff sr. scientist Lawrence Berkeley Lab., U. Calif.-Berkeley, 1978—, lectr. chem. engring., 1979—; Pres. Internat. Congress Catalysis, 1956-60; mem. Council Sci. Research, Spain, 1964—. Fellow Am. Inst. Chemists, Royal Soc.; mem. Nat. Acad. Engring., Am. Chem. Soc. (V Murphree medal 1971), Catalysis Soc. N.Am. (E.J. Houdry award 1975). Subspecialty: Physical chemistry. Address: 1588 Campus Dr Berkeley CA 94708

HEINER, DOUGLAS CRAGUN, pediatrics and immunology educator, physician; b. Salt Lake City, July 27, 1925; s. Spencer and Eva Lillian (Cragun) H.; m. Joy Luana Wiest, Jan. 8, 1946; children: Susan, Craig, Joseph, Marianne, James, David, Andrew, Carolee, Pauli. B.S., Idaho State U., 1947; M.D., U. Pa., 1950; Ph.D., McGill U., 1969. Diplomate: Am. Bd. Pediatrics, Am. Bd. Allergy and Immunlogy. Teaching asst. dept. pediatrics Harvard Med. Sch., 1954-55; instr. dept. pediatrics U. Ark., 1956-57, asst, prof., 1957-60; asst. prof. dept. pediatrics U. Utah, 1960-63, assoc. prof., 1963-66; USPHS spl. research fellow McGill U., 1966-69; prof. pediatrics, head immunology-allergy Harbor UCLA Med. Ctr., 1969—; cons. pediatricians, allergists, immunologists, sci. jours. Author: Allergies to Milk, 1980; Contbr. numerous articles to sci. jours.; chpts. to books. Boy Scout and Explorer Scout Leader Los Angeles council Boy Scouts Am., 1973-80. Served with AUS, 1951-53. Decorated Bronze Star.; Recipient Research prize New Eng. Pediatric Soc., 1955; Research award Western Soc. Pediatric Soc., 1961. Mem. Calif. Med. Assn., Am. Acad. Pediatrics, Soc. Pediatric Research, Am. Pediatric Soc., Los Angeles Allergy Soc., Am. Acad. Allergy, Am. Assn. Immunologists. Mormon. Subspecialties: Immunology (medicine); Allergy. Current work: IgD, IgE, IgG4, food allergy, development of allergy, immunology of parasites. Office: 1000 W Carson St Torrance CA 90509

HEININGER, CLARENCE GEORGE, JR., chemistry educator; b. Rochester, N.Y., July 30, 1928; s. Clarence George and Elvira Monica (Nier) H.; m. Katherine Agnes Kress, Oct. 11, 1952; children: Joseph, Laurence, Ann, Ellen, Peter. B.S., Villanova U., 1950; Ph.D. in Phys. Chemistry, U. Rochester, 1954. Research asst. Princeton, (N.J.) U., 1954-55; asst. prof. chemistry Villanova (Pa.) U., 1955-58; mem. faculty St. John Fisher Coll., Rochester, N.Y., 1958—, prof. chemistry, 1964—, dean Faculty and instrn., 1969-72; vis. prof. U. Lyon, France, 1967-68, U. Del., Newark, 1982-83. Mem. Penfield (N.Y.) Planning Bd., 1961-67, chmn., 1966-67; mem. sch. bd. Penfield Central Sch. Dist., 1974-77, pres., 1976-77. Mem. Am.Chem. Soc., AAAS, AAUP, Sigma Xi. Subspecialties: Analytical chemistry; Physical chemistry. Current work: Effective teaching of analytical and physical chemistry to undergraduates; analytical procedures and techniques. Home: 2048 Five Mile Line Rd Penfield NY 14526 Office: 3690 East Ave Rochester NY 14618

HEINRICH, MILTON ROLLIN, space life scientist, research biochemist; b. Linton, N.D., Nov. 25, 1919; s. Fred and Emma (Becker) H.; m. Ramona G. Cavanagh, May 31, 1966. A.B., U. S.D., 1941; M.S., U. Iowa, 1942, Ph.D., 1944. Postdoctoral fellow in biochemistry U. Pa., Phila., 1947-49; research assoc. Amherst (Mass.) Coll., 1949-58; spl. postdoctoral fellow U. Calif.-Berkeley, 1958-60; asst. prof. biochemistry U. So. Calif., Los Angeles, 1960-63; research scientist NASA Ames Research Ctr., Moffett Field, Calif., 1963—, space sta. study scientist, 1982—; lectr. space life scis. Editor: Extreme Environments, 1976; Contbr. articles to sci. jours. Served to lt. (j.g.) USN, 1944-47; PTO. USPHS fellow, 1947, 58; grantee, 1960-63; recipient Cosmos Group award NASA, 1981. Mem. Am. Soc. Biol. Chemists, Am. Chem. Soc., AAAS, Explorers Club, Phi Beta Kappa, Phi Eta Sigma, Sigma Xi, Phi Lambda Upsilon. Subspecialties: Gravitational biology; Biochemistry (medicine). Current work: Study scientist for planning life sciences on future space station, biochemical research on enzymes, enzyme technology, immobilized enzymes. Home: 27200 Deer Springs Way Los Altos Hills CA 94022 Office: NASA Ames Research Center 236-5 Moffett Field CA 94035

HEINRICHS, DONALD FREDERICK, geophysicist; b. Shafter, Calif., Nov. 8, 1938; m., 1962; 2 children. B.S., Stanford U., 1960, Ph.D., 1966. Research asst. geophysicist Stanford U., 1961-66; instr. physics Menlo Coll., 1964-66; asst. prof. geophys. oceanography Oreg. State U., 1966-75; program mgr. submarine geology and geophysics NSF, Washington, 1975—; mem. staff Office Naval Research, 1972-75. Mem. Am. Geophys. Union, Soc. Exploration Geophysicists., Geol. Soc. Am. Subspecialty: Marine geophysics. Office: NSF Washington DC 20550

HEINRICHS, WALTER EMIL, JR., mining company and exploration company executive, cons.; b. Superior, Ariz., Jan, 16, 1919; s. Walter Emil and Mary Gertrude (Smith) H.; m. Marcella Jean Heath, Aug. 1, 1941; children: Heath Douglas, Frederick Walter. Registered profl. engr., Colo., Ariz.; registered geophysicist, Calif. Geol. Engr. Colo. Sch. Mines, 1940; Asst. chief geophysicist U.S. Bur. Reclamation, Denver, 1946-47, Newmont Mining Corp., N.Y.C., 1947-49; sr. ptnr. United Geophys. Corp. Mining div., Tucson, 1949-54; mgr. Pima Mining Co., Tucson, 1954-55, Minerals Exploration Co., Los Angeles, 1955-58; pres., dir. Adit Resources Corp., Tucson, 1980—, Heinrichs GeoExploration Co., 1958—; mem. and chmn. bd. govs. Ariz. Dept. Mineral Resources. Contbr. geophys. articles to profl. jours. Adviser to trustee Colo. Sch. Mines, 1962-68; precinct committeeman and alt. del. Republican Nat. Conv. Served to lt. (j.g.) USNR, 1944-46; PTO. DFI (Sweden) grantee Van Diest award Colo. Sch. Mines, 1955. Mem. AIME (Peele award 1955), Am. Inst. Profl. Geologists (cert.), Soc. Econ. Geologists. Presbyterian (deacon). Subspecialties: Geology; Geophysics. Home: 2943 E Chula Vista Dr Tucson AZ 85716 Office: Heinrichs Geo Exploration Co Box 5964 Tucson AZ 85703

HEINZ, DON J, agricultural association executive; b. s. Rexburg, Idaho, Oct. 29, 1931; s. William Fred and Bernice A. (Steiner) H.; m. Marsha B. Hegsted, Apr. 16, 1956; children: Jacqueline, Grant, Stephanie, Karen, Ramona, Amy. Student, Ricks Coll., 1955, 56; B.S., Utah State U., 1958, M.S., 1959; Ph.D., Mich. State U., 1961; postgrad., Hartford Exec. Course, 1982. With Hawaiian Sugar Planters Assn. Experiment Sta., Aiza, 1961—, v.p., dir., 1979—; affiliate grad. faculty U. Hawaii; vice chmn. Hawaii, chmn. germplasm com. Internat. Soc. Sugar Cane Technologists; cons., Geypt, Philippines. Contbr. articles to profl. jours. Served with USAF, 1951-54. Mem. Internat. Soc. Sugar Cane Technologists, Pacific Sci. Assn., Hawaiian Sugar Technologists, Agronomy Soc., AAAS, Sigma Xi. Mormon. Subspecialty: Cell and tissue culture. Current work: Private agricultural research devoted to improving yields of sugarcane in Hawaii. Office: Hawaiian Sugar Planters Assn 99-193 Aiea Hgights Dr PO Box 1057 Aiea HI 96701

HEINZ, ERICH, physical biochemist, researcher; b. Essen, Rheinland, West Germany, Jan. 10, 1912; came to U.S., 1953, naturalized, 1958; s. Peter and Margarete (Gubener) H.; m. Ursula M. Staak, Sept. 2, 1941 (div. 1977); children: Bettina, Agnes, Peter; m. Ursula Loewe, Dec. 1, 1977. M.D., U. Kiel, Germany, 1941, dozent, 1949. Intern, dozent U. Kiel, Ger., 1949-55; assoc. prof. Tufts U., Boston, 1955-58; research prof. George Washington U., Washington, 1958-59; prof., dept. head J. W. Goethe U., Frankfurt, Ger., 1959-78; prof. dept. physiology Cornell U. Med. Coll., N.Y.C., 1978—; mem. sponsoring com. Internat. Conf. Biol. Membranes, 1970—; mem. adv. com. NSF, Washington, 1982—. Author: Mechanics and Energetics of Biological Transport, 1978, Electrical Potentials in Biological Transport, 1980; editor: Na-linker Transport, 1972; co-editor: Gastric Secretion (W.G. Sachs and K.J. Ullrich), 1972; editorial bd.: Biochem. Biophys. Acta, 1971—, Am. Jour. Physiology, 1980—. Brit. Council fellow, 1949-50; USPHS sr. research fellow, 1958; NSF, USPHS research grantee, 1954—. Mem. Am. Soc. Biol. Chemists, Biophys. Soc., Gesellsch Biol. Chemie. Club: Health (N.Y.C.). Subspecialties: Biochemistry (medicine); Physiology (medicine). Current work: Biological membrane transport and permeability, biochemical energetics. Office: Cornell U Med Coll 1300 York Ave New York NY 10021 Home: 24 W 70th St New York NY 10023

HEIRTZLER, JAMES RANSOM, geophysics researcher; b. Baton Rouge, Sept. 16, 1925; s. William Ransom and Jimmie Lemon (Clark) H.; m. Phyllis Virginia Trossen, Feb. 7, 1951; children: Fenton Ransom, Jason Dean. B.S. in Physics, La. State U., 1947, M.S., 1948, Ph.D., N.Y. U., 1953. Asst. prof. physics Am. U. Beirut, 1953-60; sr. physicist Gen. Dynamics Corp., 1956-60; sr. research assoc. Lamont Geol. Obs., Palisades, N.Y., 1960-67; dir. Hudson Labs., Columbia U., Dobbs Ferry, N.Y., 1967-69; chmn. dept. geology and geophysics Woods Hole (Mass.) Oceanographic Inst., 1969-76, sr. scientist, 1969—; chmn. planning com. Joint Oceanographic Instns. for Deep Earth Sampling, Woods Hole, 1978-79, also chmn. various coms., 1969-79; dir. sci. research Joint Oceanographic Inst., Washington, 1979-80; com. reporter U.S. Geodynamics Com., 1971—, U.S. chief scientist for French-Am. Mid-Ocean Undersea Study project, 1971-75; compiler for marine area N.Am. Magnetic Anomaly Map Com., 1980—. Editor: books, most recent being Initial Reports of DSDP, vol. 46, 1978; editor: Revs. of Geophysics and Space Physics, 1984-88. Served with USNR, 1944-46; PTO. NSF grantee, 1960—. Fellow Am. Geophys. Union (pres. GP sect. 1982-84), Geol. Soc. Am., AAAS; mem. Soc. Exploration Geophysicist, Am. Phys. Soc. Subspecialties: Geophysics; Sea floor spreading. Current work: Marine geophysics, marine geomagnetism, deep-sea drilling, submersible science. Home: 24 Cumloden Dr Falmouth MA 02540 Office: Woods Hole Oceanographic Instn Woods Hole MA 02543

HEISER, ARNOLD MELVIN, astronomer; b. Bklyn., Feb. 9, 1933; s. Hyman Samuel and Sadie (Kretchmer) H.; m. Vivian Carol Jacobs, June 6, 1964; children—Naomi Elizabeth, David Alan. A.B., Ind. U., 1954, M.A., 1956; Ph.D., U. Chgo., 1961. Research asst. Ind. U., 1954-56; research fellow U. Chgo., 1956-61; asst. prof. physics and astronomy Vanderbilt U., Nashville, 1961-66, asso. prof., 1966—; dir. A.J. Dyer Obs., 1972—; H. Shapley vis. prof. Am. Astron. Soc., 1969—. Contbr. articles to astron. jours. Mem. Am. Astron. Soc., Internat. Astron. Union, Royal Astron. Soc., Tenn. Acad. Sci., AAAS, Sigma Xi. Subspecialty: Optical astronomy. Current work: Photoelectric photometry of variable stars; observational studies of galactic clusters. Home: 6132 Gardendale Dr Nashville TN 37215 Office: A J Dyer Observatory Vanderbilt University Nashville TN 37235

HEJTMANCIK, MILTON RUDOLPH, medical educator, physician; b. Caldwell, Tex., Sept. 27, 1919; s. Rudolph Joseph and Millie (Jurcak) H.; m. Myrtle Lou Erwin, Aug. 21, 1943; children—Kelly Erwin, Milton Rudolph, Peggy Lou; m. Myrtle Frances McCormick, Nov. 27, 1976. B.A., U. Tex., 1939, M.D., 1943. Diplomate: in cardiovascular diseases Am. Bd. Internal Medicine. Intern Phila. Gen. Hosp., 1943; resident internal medicine U. Tex., 1946-49, instr. internal medicine, 1949-51, asst. prof., 1951-54, assoc. prof., 1954-65, prof. internal medicine, 1965—, dir. heart clinic, 1949—, dir. heart sta., 1965—; chief of staff John Sealy Hosp., 1957-58; chief staff U. Tex. Hosps., 1977-79. Contbr. articles profl. jours. Served from 1st lt. to capt., M.C. AUS, 1944-46; ETO. Fellow ACP, Am. Coll. Chest

Physicians, Am. Coll. Cardiology; mem. Am. Heart Assn. (fellow council clin. cardiology), Tex. Heart Assn. (pres. 1979-80), Galveston Dist. Heart Assn. (pres. 1956), A.M.A. (Billing's Gold medal 1973), Am. Fedn. Clin. Research, AAAS, Tex. Acad. Internal Medicine (gov. 1971-73, v.p. 1973-74, pres. 1976-77), N.Y. Acad. Scis., Tex. Club Cardiology (pres. 1972), Galveston County Med. Assn. (pres. 1971), Tex. Med. Assn. (del. 1972-80), Am. Heart Assn. (pres. Tex. affiliate 1979-80), Phi Beta Kappa, Sigma Xi, Alpha Omega Alpha, Phi Eta Sigma, Mu Delta, Phi Rho Sigma. Subspecialty: Cardiology. Current work: Electrophysiology and ultrasound. Home: 6198 Alton Ln Beaumont TX 77706 Office: VA Med Ctr 3385 Fannin Beaumont TX 77707

HEKMATPANAH, JAVAD, neurosurgery educator; b. Isfahan, Iran, Mar. 25, 1934; came to U.S., 1957, naturalized, 1973; s. Namatola and Tuba H.; m. Lyra Van Wien, Aug. 15, 1959; children: Daria, Kevin, Cameron. M.D., U. Tehran, 1956. Diplomate: Am. Bd. Neurosurgery, Am. Bd. Psychiatry and Neurology. Resident in neurology U. Wis. Gen. Hosp., Madison, 1958-61; resident in neurosurgery U. Chgo., 1961-64, asst. prof. neurosurgery, 1964-70, assoc. prof., 1970-75, prof., 1975—. Editor: Gliomas: Current Concepts in Biology, Diagnosis and Theory, 1975. Fellow A.C.S., Am. Acad. Neurology; mem. Am. Assn. Neurol. Surgeons, Chgo. Neurol. Soc., Chgo. Surg. Soc. Subspecialties: Neurosurgery; Neurology. Current work: Cerebral circulation, cerebral compression (cardiac arrest) and brain tumors (clinical and laboratory). Office: U Chgo Hosp 950 E 59th St Box 405 Chicago IL 60637

HELD, RICHARD MARX, educator, psychologist; b. N.Y.C., Oct. 10, 1922; s. Lawrence W. and Tessie (Klein) H.; m. Doris F. Bernays, June 29, 1951; children: Lucas D.B., Julia B., Andrew L.B. B.A., Columbia U., 1943, B.S., 1944; M.A., Swarthmore Coll., 1948; Ph.D., Harvard U., 1952. Research asst. Swarthmore Coll., 1946-48; research fellow Jackson Hole Wildlife Park, summer 1948; research asst. psycho-acoustic lab. Harvard U., 1949-52, teaching fellow, 1950-51; NIH postdoctoral fellow, 1952-53; instr., then asst. prof. psychology Brandeis U., 1953-58, asso. prof., then prof., chmn. dept. psychology, 1958-62; mem. Inst. Advanced Study, Princeton, 1955-56; sr. research fellow NSF, 1962-63; vis. prof. Mass. Inst. Tech., Cambridge, 1962-63, prof. psychology, 1963—, chmn. dept., 1977—; dir. Sinauer Assocs., Transkinetics Inc.; Mem. com. vision Armed Forces-NRC; exec. com. mem.; assoc. Neuroscis. Research Program.; mem. vision research program com. Nat. Eye Inst. Editorial bd.: Psychol. Research, Perception. Bd. dirs. Founds.' Fund for Research in Psychiatry. Recipient Glenn A. Fry award, 1978. Fellow Am. Acad. Arts and Scis., Am. Acad. Optometry, AAAS, Am. Psychol. Assn.; mem. Nat. Acad. Sci., Eastern Psychol. Assn., Assn. for Research in Vision and Ophthalmology, Internat. Brain Research Orgn., Soc. Neuroscis., Soc. Exptl. Psychologists, Psychonomic Soc., Old Cambridge Shakespeare Assn., Sigma Xi. Subspecialties: Sensory processes; Developmental psychology. Current work: Study of the development of vision in human infants. Home: 102 Appleton St Cambridge MA 02138

HELD, RONALD DENNIS, systems analyst; b. Bklyn., Oct. 29, 1951; s. Arthur G. and Rosalind (Ribak) H. B.S., Rensselaer Poly. Inst., 1973, M.S., 1976, Ph.D., 1978. Systems analyst Space Systems div. Gen. Electric Co., King of Prussia, Pa., 1980—. Mem. Am. Astron. Assn., Sigma Pi Sigma. Subspecialties: Mathematical software; Theoretical astrophysics. Current work: Math modelling, algorithm devel., stellar structure, gen. relativity and gravitation, cosmology. Home: 959 Penn Circle C210 King of Prussia PA 19406

HELDERMAN, J. HAROLD, physician, medical educator and researcher; b. Newark, Feb. 5, 1945; s. Jacob Leo and Shirley Ethel (Applebaum) H.; m. Phyllis E. Koppel, Jan. 29, 1967; children: Alexander S., Ira Philip, Rosalind Sarah. A.B. with highest honors, U. Rochester, 1967; M.D. summa cum laude, SUNY Downstate Med. Ctr., Bklyn., 1971. Intern Johns Hopkins Hosp., Balt., 1971-72, resident, 1972-73; clin. assoc. NIH, Balt., 1973-75; postdoctoral fellow Harvard Med. Sch., Boston, 1975-77; asst. prof. U. Tex. Health Sci. Ctr., Dallas, 1977-81, assoc. prof. dept. internal medicine, 1981—, dir. renal immunology lab., 1977—. Contbr. articles to profl. publs. Served to lt. comdr. USPHS, 1973-75. Research grantee NIH, 1978—, Am. Diabetes Assn., 1978-80, Upjohn Co., 1977-79. Fellow ACP; mem. Am. Fedn. Clin. Research, So. Soc. Clin. Investigation, Internat. Soc. Transplantation, Am. Assn. Immunology, Am. Soc. Nephrology, Internat. Soc. Nephrology, AAAS, Am. Heart Assn., Am. Diabetes Assn. Subspecialties: Transplantation; Nephrology. Current work: Hormonal immunoregulation, transplantation immunity, mechanisms of glomerolonephritis, hormone receptor chemistry. Office: Univ Tex Health Sci Center 5323 Harry Hines Blvd Dallas TX 75235

HELFAND, DAVID JOHN, astrophysicist; b. New Bedford, Mass., Dec. 7, 1950; s. Barney and Mary (Griffin) H.; m. Sarah Florence Spencer, June 4, 1972; m. Jada Rowland, May 22, 1982. B.A., Amherst Coll., 1973; M.S., U. Mass., 1977, Ph.D., 1977. Research asst. Five Coll. Radio Astronomy Obs., U. Mass., Amherst, 1971-77; research assoc. Astrophysics Lab., Columbia U., N.Y.C., 1977-78, asst. prof. astronomy, 1978-82, assoc. prof. physics, 1982—; v.p. N.Y. Astronomy Corp., N.Y.C., 1981—. Contbr. articles to profl. jours. Recipient Porter prize in astronomy Amherst Coll., 1971; NASA research grantee, 1979—; Air Force Office Sci. Research grantee, 1981—; Alfred P. Sloan research fellow, 1983—. Mem. Internat. Astron. Union, Am. Astron. Soc., N.Y. Acad. Scis., Astron. Soc. N.Y., Sigma Xi. Subspecialties: Radio and microwave astronomy; 1-ray high energy astrophysics. Current work: Observational astrophysics of supernova remnants, pulsars, active galaxies, stars, quasars, etc. Home: 853 7th Ave PH-B New York NY 10019 Office: Columbia Astrophysics Lab 538 W 120th St New York NY 10027

HELFAND, EUGENE, research chemist; b. Bklyn., Jan. 8, 1934; s. and Sondra Yoskowitz; m., Nov. 17, 1957; children: Robin H., Dawn A., Russ D. B.S., Poly. Inst. Bklyn., 1955; M.S., Yale U., 1957, Ph.D., 1958. Mem. tech. staff Bell Labs., Murray Hill, N.J., 1958-60, supr., 1960—; adj. prof. Yeshiva U., N.Y.C., 1960-62, Poly. Inst. Bklyn., 1963-64; vis. prof. N.Y. U., N.Y.C., 1963; mem. panel on polymer sci. and engring. NRC, Washington, 1979-81. Recipient (Disting. Tech. Staff award 1983); Contbr. sci. articles to tech. publs. Recipient Disting. Tech. Staff award, 1983; Guggenheim fellow, 1969-70. Fellow Am. Phys. Soc.; mem. Am. Chem. Soc., Sigma Xi, Phi Lambda Upsilon. Subspecialties: Polymer physics; Statistical mechanics. Current work: Statistical mechanics of polymers, particularly motion in polymers, rheology, surfaces and interfaces, block copolymers. Office: Bell Lab 1A 361 600 Mountain Ave Murray Hill NJ 07974

HELFFENSTEIN, DENNIS ALAN, clinical neuropsychologist; b. Cin., Oct. 3, 1952; s. Frederick John and Opal Jane (Silvers) H.; m. Karyn Brown, May 5, 1981. B.A., U. Va., 1975, M.Ed., 1976, Ph.D., 1981. Lic. psychologist, Tex. Predoctoral intern Woodrow Wilson Rehab. Ctr., Fishersville, Va., 1978-80; asst. psychologist Western Res. Psychiat. Hab. Center, Northfield, Ohio, 1980-81; clin. neuropsychologist Med. Ctr., DelOro Hosp., Houston, 1981—; pvt. practice psychology Mary Ellen Hayden & Assos., Houston, 1981—. Contbr. articles to profl. jours. Mem. Am. Psychol. Assn., Internat. Neuropsychol. Soc. Subspecialty: Neuropsychology. Current work: Head trauma rehabilitation; cognitive retraining following cerebral insult; interpersonal and communication skills following head injury. Home: 10134 Pear Limb Dr Houston TX 77099 Office: Med Center DelOro Hosp 8081 Greenbriar Dr Houston TX 77096

HELGESON, HAROLD CHARLES, geochemistry educator, consultant; b. Mpls., Nov. 13, 1931; s. Harold Rollin and Phoebe Dorothy (Kildahl) H.; m. Velda Fay Fennell, Mar. 19, 1956 (div. Nov. 1977); children: Christopher, Kimberley; m. Suzanne Tuxen, Aug. 31, 1978 (div. June 1982). B.Sc., Mich. State U., 1953; Ph.D., Harvard U., 1962. Exploration/mining geologist Tech. Mine Cons. Uranium City, Sask., Can., 1953-54; mining/exploration geologist Anglo-American Corp., South Africa, S.W. Africa, Zambia, 1956-59; research chemist Shell Devel. Co. Houston and Imperial Valley, Calif., 1962-65; asst. prof. Northwestern U., 1965-68, assoc. prof., 1968-70; prof. geochemistry U. Calif.-Berkeley, 1970—, Miller research prof., 1974-75. Served to 1st lt. USAF, 1954-56; Germany. Guggenheim fellow, 1977-78. Fellow Geol. Soc. Am., Mineral Soc. Am.; mem. Geochem. Soc. (councilor 1973-76), Am. Chem. Soc., Internat. Assn. Geochemistry and Cosmochemistry. Democrat. Subspecialties: Geochemistry; Thermodynamics. Current work: High temperature/pressure solution chemistry; thermodynamics of hydrothermal/geothermal systems; chemical interaction of minerals and aqueous solutions in geochemical processes. Office: Dept Geology and Geophysics U Calif Berkeley CA 94720

HELKE, CINDA JANE, neuroscientist, educator; b. Waterloo, Iowa, Feb. 27, 1951; d. Gerald A. and Lorna R. (Smith) Pieres; m. Joel E. Helke, Aug. 10, 1974. B.S., Creighton U., 1974; Ph.D., Georgetown U., 1978. Registered pharmacist, Nebr. Research assoc. NIH, Bethesda, Md., 1978-80; asst. prof. Uniformed Services U. Health Sci., Bethesda, 1980—. Contbr. articles to sci. publs. Recipient Dean Calvert Meml. award Med. Coll. Wis., 1978. Mem. Soc. Neurosci., Am. Soc. Pharms. and Exptl. Therapeutics, AAAS, Assn. Women in Sci., NOW, Phi Beta Kappa, Rho Chi, Alpha Sigma Nu. Subspecialties: Neurobiology; Neuropharmacology. Current work: Research interests: neurobiology of central nervous system pathways controlling cardiovascular function. Office: 4301 Jones Bridge Rd Bethesda MD 20814

HELLER, ALLEN HARVEY, neurologist, pharmacologist, educator; b. Providence, Nov. 30, 1947; s. Gerald S. and Betty D. (Steinberg) H.; m. Elizabeth R. Blackman, Oct. 23, 1976; 2 sons, David J, Richard A. A.B., Brown U., 1969; M.D., Johns Hopkins U., 1973. Diplomate: Am. Bd. Internal Medicine. Resident in internal medicine Brown U., 1973-75; resident in neurology Boston U., 1975-76; research fellow in pharmacology Harvard U. Med. Sch., 1976-78, resident and chief resident in neurology, 1977-79, instr. neurology, 1979-82, asst. prof. neurology, 1982—; research assoc. in neurosci. Children's Hosp. Med. Center, Boston, 1982—; asst. neurologist Beth Israel Hosp., Brigham and Women's Hosp., Boston, 1979—. Contbr. articles to profl. jours. Grantee NIH, 1981, Epilepsy Found. Am., 1981-82, William F. Milton Fund,., 1982. Mem. Soc. Neurosci., Am. Epilepsy Soc., AAAS. Subspecialties: Neurobiology; Neuropharmacology. Current work: Genetic control of neurotransmitter function. Research neuroscientist using biochemical and neuropharmacological techniques to investigate genetic determinants of neurotransmitter receptor function.

HELLER, BARBARA RUTH, statistician, educator; b. Milw., May 15, 1931; d. Mitchell and Sophie Lerner (Goldberg) Steigman; m. Alfred Heller, July 22, 1956; 1 son, Daniel. Ph.B., U. Chgo., 1950; B.S., Roosevelt U., 1953; M.S., U. Chgo., 1965, Ph.D., 1979; B.S., Roosevelt U., Chgo., 1953. Asst. prof. math. Ill. Inst. Tech., Chgo., 1980—. Contbr. papers to profl. jours. Mem. Inst. Math. Stats., Am. Statis. Assn., Am. Math. Soc., Soc. Indsl. and Applied Math. Subspecialties: Statistics; Probability. Current work: Characterization of probability distributions, application of statistics in stereological problems; size distributions. Office: Dept Mat Ill Inst Tech Chicago IL 60616

HELLER, GERALD S., solid state physics educator; b. Detroit, Sept. 5, 1920; m., 1943; 2 children. Sc.B., Wayne U., 1942; Sc.M., Brown U., 1946, Ph.D. in Physics, 948. Mem. staff radiation lab. MIT, 1942-45; asst. prof. physics Brown U., Providence, 1948-52; group leader resonance physics Lincoln Lab., MIT, 1954-62, vis. prof. elec. engring., 1962-63; prof. engring. Brown U., 1963—, dir. materials research lab., 1968—; Mem. staff radiophysics lab. Sydney U., 1944-45. Fellow IEEE; mem. Am. Phys. Soc. Subspecialties: Magnetic physics; Solid state physics. Office: Box M Brown U Providence RI 02912

HELLER, JACK, computer science educator; b. Bklyn., Sept. 11, 1922; s. Michael and Lillian (Shohart) H.; m. Myra Minna Levine, Aug. 8, 1946; children: Glen, Cathy, Douglas, Adam. B.AeE., M.A., Ph.D. in Physics, Poly. Inst. Bklyn. Instr. Newark Coll. Engring., 1949-51; research scientist Princeton (N.J.) U., 1952-54, Courant Inst.-NYU, N.Y.C., 1954-63; assoc. prof. NYU, 1963-66, prof., 1966-70; prof. computer sci. SUNY-Stony Brook, 1970—, chmn. dept. computer sci., 1978—; cons. Mus. Computer Network, 1960—, UN Indexing Project, 1968-76. Mem. Assn. Computing Machinery. Subspecialties: Database systems; Software engineering. Current work: Education and research in database systems, programming environments and software engineering. Home: 5 Tallmadge Gate Setauket NY 11733 Office: Dept Computer Sci SUNY Stony Brook NY 11794

HELLIGE, JOSEPH BERNARD, psychology educator; b. Ft. Madison, Iowa, Sept. 3, 1948; s. Bernard Frances and Mary Margaret (Moffitt) H.; m. Colleen Staudt, Dec. 18, 1971; children: Erin, Katherine. B.A., St. Mary's Coll., 1970; M.A., U. Wis., 1972, Ph.D., 1974. Asst. prof. U. So. Calif., 1974-80, assoc. prof., 1980—. Editor: Hemispheric Asymmetry, 1983; contbr. articles to profl. jours. NIMH research grantee, 1975; NSF Research grantee, 1976, 79, 83. Mem. Psychonomic Soc., Am. Psychol. Assn., Internat. Neuropsychol. Soc., AAAS, Sigma Xi. Subspecialties: Neuropsychology; Cognition. Current work: Hemispheric specialization and consequences for normal human cognition; visual and auditory information processing; attention. Home: 1612 Crest Dr Los Angeles CA 90035 Office: Dept Psychology MC 1061 U So Calif University Park Los Angeles CA 90089

HELLING, ROBERT B., molecular biologist, educator; b. Madelia, Minn., Mar. 21, 1936; s. Siver Hage and Lila Evelyn (Wiborg) H.; m. (div.); children: Eric Robert, Julie Anne. B.S., St. Olaf Coll., 1958; M.S., U. Pitts., 1960, Ph.D., 1963. Postdoctoral scholar Karolinska Inst., Stockholm, Sweden, 1963-65; mem. faculty Div. Biol. Scis., U. Mich.-Ann Arbor, 1965—, now molecular biologist, prof.; vis. prof. dept. microbiology U. Calif.-San Francisco, 1972-73; vis. prof. Inst. Molecular and Cellular Biology, Strasbourg, France, 1982. Subspecialty: Genetics and genetic engineering (agriculture). Current work: Molecular genetics. Office: Div Biol Sci U Mich Ann Arbor MI 48109

HELLIWELL, ROBERT ARTHUR, electrical engineering educator; b. Red Wing, Minn., Sept. 2, 1920; s. Harold Harlowe and Grace (Robson) H.; m. Jean Perham, Apr. 5, 1942; children: Bradley Athearn, David Robson, Richard Perham, Donna Marie. B.E.E., Stanford U., 1942, M.A., 1943, E.E., 1944, Ph.D. in Elec. Engring, 1948. Mem. faculty Stanford U., 1946—, prof. elec. engring., 1958—, dir. Ctr. for Space Sci. and Astrophysics, 1983—. Author: Whistlers and Related Ionospheric Phenomena, 1965, also articles. Recipient Antarctica Service medal Royal Soc., London, Appleton prize, 1972. Fellow IEEE, Internat. Sci. Radio Union; mem. Am. Geophys. Union, AAAS, AAUP, Nat. Acad. Scis., Phi Beta Kappa, Sigma Xi, Tau Beta Pi. Subspecialties: Radioscience; Electrical engineering. Home: 2240 Page Mill Rd Palo Alto CA 94304 Office: Radio Sci Lab Stanford Univ Stanford CA 94305

HELLMAN, MARTIN EDWARD, electrical engineering educator; b. N.Y.C., Oct. 2, 1945; m., 1967; 2 children. B.E., NYU, 1966; M.S., Stanford U., 1967, Ph.D., 1969. Mem. staff Thomas J. Watson Research Ctr., IBM, 1968-69; asst. prof. elec. engring. MIT, 1969-71; prof. elec. engring. Stanford (Calif.) U., 1971—; Vinton Hayes fellow, 1969-71. Mem. IEEE, Inst. Math. Stats. Subspecialties: Computer engineering; Cryptography and data security. Office: Stanford U Durand 135 Stanford CA 94305

HELLMAN, RHONA PHYLLIS, hearing science researcher, educator; b. N.Y.C., May 22, 1935; d. David and Florence (Schlesinger) Rosenberg; m. William S. Hellman, Nov. 4, 1954; children: Ronald Bruce, Adrian David. B.A. cum laude, Bklyn. Coll., 1955; M.S., Syracuse U., 1960. Research audiologist VA, 1963-64; research asst. Auditory Perception Lab., Northeastern U., Boston, 1965-66; sr. research asst. Lab. Psychophysics, Harvard U., 1967-76; adj. asst. prof. communication disorders Boston U., 1975-81, adj. assoc. prof., 1981—, establisher Communcation Scis. Lab, 1982. Reviewer: Jour. Acoustical Soc. Am; Contbr. articles to sci. jours., chpts. to books. Sponsor Arlington (Mass.) Philharm. Soc.; mem. Arlington Civil Rights Com. NASA grantee, 1979—. Mem. Acoustical Soc. Am. (psychol. and physiol. acoustics com. 1979-82, chmn. com. on auditory magnitudes), Soc. for Neurosci., AAAS, Mass. Speech and Hearing Assn. Subspecialties: Acoustical engineering; Psychobiology. Current work: Psychoacoustics: psychophysiology of normal and impaired hearing; sensory processes. Office: 40 Cummington St Boston MA 02215

HELLMANN, ROBERT A., biological sciences educator, environmental consultant; b. July 7, 1927. Student, Harvard U., 1947-49; B.S., Cornell U., 1954, M.S., 1957; ED.D., Columbia U., 1966. Instr., lectr. Am. Mus. Natural History, N.Y.C., 1956-61; instr. Tchrs. for East Africa project of Columbia U., Makerere U. Coll., Kampala, Uganda, 1961-62; assoc. prof. biol. scis. SUNY, Brockport, 1962—; dir. Fancher Forest Conservation Ctr. Served with U.S. Army, 1945-46. Mem. AAAS, Am. Inst. Biol. Scis., Am. Assn. Bot. Gardens and Arboreta, Ecol. Soc. Am. Clubs: Harvard, Cornell (Rochester, N.Y.). Subspecialties: Resource conservation; Resource management. Current work: Conceived, established, and developed Fancher Forest Conservation Center, consisting of an arboretum of native New York State trees and shurbs and a demonstration managed woodlot. Home: PO Box 231 Brockport NY 14420 Office: Dept Biol Scis SUNY Brockport NY 14420

HELLQUIST, CARL BARRE, biology educator; b. Summit, N.J., Dec. 8, 1940; s. Francis Ludvick and Marion Lithgow (Hall) G.; m. Marion Temple, Oct. 10, 1970; children: Carl Eric, Paul Temple. A.A.S., Paul Smith's (N.Y.) Coll., 1962; B.S., U. N.H., 1965, M.S., 1966, Ph.D. in Botany, 1975. Mem. faculty dept. biology Boston State Coll., 1967-82; prof. biology North Adams (Mass.) State Coll., 1982—; cons. U.S. Fish and Wildlife Service. Contbr. articles to profl. jours. Mem. Hopkinton (Mass.) Conservation Comm., 1974-82. Mem. Am. Soc. Plant Taxonomists, Am. Fern Soc., New Eng. Bot. Club, Internat. Assn. Aquatic Vascular Plant Biologists. Subspecialties: Taxonomy; Aquatic freshwater ecology. Current work: Taxonomy and floristics of vascular aquatic plants. Home: 305 W Main St Hopkinton MA 01748 Office: Dept Biology North Adams State Coll North Adams MA 02147

HELLUMS, JESSE DAVID, univ. dean; b. Stamford, Tex., Aug. 19, 1929; s. John V. and Fannie May (Beauchamp) H.; m. Marilyn Biel, July 13, 1957; children—Mark William, Jay David, Robert James. B.S., U. Tex., 1950, M.S., 1957; Ph.D., U. Mich., 1960. Registered profl. engr., Tex. Process engr. Mobil Oil Co., Beaumont, Tex., 1950-54; mem. faculty Rice U., Houston, 1960—, prof. chem. engring., 1968—, dir. biomed. engring. lab., 1968-80, chmn. dept., 1969-75, dean engring., 1980—; adj. prof. Baylor U. Coll. Medicine, 1960—, U. Tex. Med. Sch., 1977—; NSF sci. faculty fellow Cambridge (Eng.) U., 1967-68; vis. prof. Imperial Coll., London, 1973-74. Author papers in field. Served to 1st lt. USAF, 1954-56. Mem. Am. Inst. Chem. Engrs., Am. Chem. Soc., AAAS, AAUP, Am. Soc. Artificial Internal Organs, Microcirculation Soc., Soc. Rheology, Bioelectromagnetics Soc. Subspecialties: Biomedical engineering; Chemical engineering. Home: 2202 Albans Rd Houston TX 77005 Office: PO Box 1892 Houston TX 77001

HELLWARTH, ROBERT WILLIS, physicist, educator; b. Ann Arbor, Mich., Dec. 10, 1930; s. Arlen Roosevelt and Sarah Matilda (Townsend) H.; m. Abigail Gurfein, Sept. 20, 1957 (div. 1979); children: Benjamin John, Margaret Eve, Thomas Abraham. B.S., Princeton U., 1952; D.Phil. (Rhodes scholar), St. John's Coll., Oxford (Eng.) U., 1955. Sr. scientist, mgr. Hughes Research Labs., Malibu, Calif., 1956-70; vis. asso. prof. elec. engring. and physics U. Ill., Urbana, 1964-65; research asso., sr. research fellow Calif. Inst. Tech., Pasadena, 1966-70; NSF sr. postdoctoral fellow Clarendon Lab.-St. Peter's Coll., Oxford (Eng.) U., 1970-71; George Pfleger prof. physics and elec. engring. U. So. Calif., 1970—. Author monograph, articles in field; asso. editor: IEEE Jour. Quantum Electronics, 1964-76. Grantee NSF, Dept. Energy, Air Force Office Sci. Research, U.S. Army Research Office. Fellow IEEE, Am. Phys. Soc., AAAS; mem. Nat. Acad. Engring., AAUP, Phi Beta Kappa, Sigma Xi, Eta Kappa Nu. Subspecialty: Optical physics. Patentee Q-switched laser, nonlinear optical microscope, phase conjugate mirror. Home: 921 12th St Santa Monica CA 90403 Office: SSC 303 Physics Dept U So Calif Los Angeles CA 90007

HELMER, RICHARD GUY, nuclear physics researcher; b. Homer, Mich., Feb. 19, 1934; s. Hurshul Guy and Edith Maude (Putnam) H; m. Mary Joan Scrivens, June 10, 1956; children: Gary Allen, Carl William. Research assoc. Argonne Nat. Lab. (Ill.), 1958-61; physicist Phillips Petroleum Co., Idaho Nat. Engring. Lab., Idaho Falls, 1961-65, Idaho Nuclear Corp., Idaho Nat. Engring. Lab., 1965-70, Aerojet Nuclear Co., Idaho Nat. Engring. Lab., 1970-76; sr. scientist EG&G Idaho, Inc., Idaho Nat. Engring. Lab., Idaho Falls, 1976—; asst. prof., collaborator Utah State U., Logan, 1965—; mem. task group on gamma ray energies Internat. Union Pure and Applied Physics, 1972—. Contbr. articles to profl. jours. Bd. dirs. Child Devel. Ctr., Idaho Falls, 1968-74; mem. Regional Council Christian Ministry, Idaho Falls, 1970—; trustee Sch. Dist. 91, Idaho Falls, 1979—. Recipient Outstanding Contbn. to Edn. award Idaho Falls Edn. Assn., 1971; Disting. Service award Kiwanis Club, East Idaho Falls, 1972; cert. recognition Gov. C. Andrus, 1972. Fellow Am. Phys. Soc.; mem. Am. Nuclear Soc., N.Y. Acad. Sci., AAAS. Subspecialty: Nuclear physics. Current work: Nuclear structure research; precise gamma-ray spectrometry; evaluation of nuclear structure data. Home: 792 Sonja Ave Idaho Falls ID 83402 Office: EG&G Idaho Inc PO Box 1625 Idaho Falls ID 83415

HELOU, GEORGE, astronomer; b. Chayah, Lebanon, Feb. 16, 1954; came to U.S., 1975; s. Tanios and Georgette (Aziz) H.; m. Andree Harfouche, July 27, 1978; 1 dau., Ariane Nada. B.S. in Physics, Am. U.

Beirut, 1975; M.S. in Astrophysics, Cornell U., 1978, Ph.D., 1980. Internat. fellow Arcetri Astrophys. Obs., Florence, Italy, 1980-81; research assoc. Center for Radiophysics and Space Research, Cornell U., 1981-82, instr. astronomy dept., 1981—. Contbr. articles to tech. jours. Recipient Am. U. Beirut Philip K. Hitti prize for academic excellence, 1975; Dudley Obs. research grantee, 1982. Mem. Am. Astron. Soc., Sigma Xi. Subspecialties: Cosmology; Galaxies. Current work: Observations: Extra-galactic astronomy, radio astronomy, properties of galaxies, clusters of galaxies, binary galaxies, interpretation of data and some theoretical investigations. Office: Cornell U 222 Space Scis Bldg Ithaca NY 14853

HELSLEY, CHARLES EVERETT, geophysics educator, research institute executive; b. Oceanside, Calif., June 24, 1934; m., 1955; 4 children. B.S., Calif. Inst. Tech., 1956, M.S., 1957; Ph.D. in Geology, Princeton U., 1960. Asst. prof. geology Calif. Inst. Tech., 1960-62; asst. prof. Case Western Res. U., 1962-63; from asst. prof. to assoc. prof. S.W. Ctr. Advanced Studies, U. Tex.-Dallas, 1963-69, prof. geosci., 1969-76, assoc. head geosci. div., 1971-72, head geosci. program and dir. Inst. Geosci., 1972-75; prof. geology and geophysics U. Hawaii-Manoa and dir. Hawaii Inst. Geophysics, Honolulu, 1976—; adj. prof. So. Meth. U., 1963-76, U. Tex. Marine Sci. Inst., Galveston, 1973-76. Mem. Geol. Soc. Am., Am. Geophys. Union. Subspecialty: Marine geophysics. Office: 2525 Correa Rd Honolulu HI 96822

HELSLEY, GROVER CLEVELAND, pharm. co. exec.; b. Strasburg, Va., Sept. 26, 1926; s. Grover Clevel and Vallie Mae (Putnam) H.; m. Betty Jean Midkiff, Oct. 30, 1949; children—Grover Cleveland, Linda Suzanne, Robert Christopher. B.S. with honors, Shepherd Coll., 1954; M.S., U. Va., 1956, Ph.D. (Philip Francis duPont fellow), 1958. Research chemist E.I. duPont de Nemours & Co., Inc., Richmond, Va., 1958-62; research chemist A.H. Robins Co., Richmond, 1962-64, group leader, 1964-68, asso. dir. chem. research, 1968-70; dir. research Hoechst-Roussel Pharms. Inc., Somerville, N.J., 1970-72, v.p. pharm. research, 1972—. Contbr. sci. articles to profl. jours. Served with USAAF, 1945-47. Mem. Am. Chem. Soc., Pharm. Mfrs. Assn. (editorial adv. bd. drug devel. research 1980), Indsl. Research Inst. Mem. Disciples of Christ Ch. Subspecialties: Organic chemistry; Medicinal chemistry. Current work: Chemistry of heterocyclic compounds; fluoride displacement on aromatic rings. Patentee in field. Home: PO Box 117 Pottersville NJ 07979 Office: Hoechst-Roussel Pharms Inc Route 202-206 N Somerville NJ 08876

HELSON, LAWRENCE, oncologist; b. N.Y.C., Mar. 21, 1931; s. Julius and Minerva H.; m. Christiane Spahr Ducommon, Feb. 6, 1963; children: Jean Luc, Janique, Guy Alain. B.S., CCNY, 1953; M. Med., U. Geneva, 1959, M.D., 1962; M.S., NYU, 1958. Intern L.I. Coll. Hosp., Bklyn.; med. pediatric oncologist Meml. Sloan Kettering Cancer Ctr., N.Y.C., 1967—, head lab. pediatric cancer research, 1973—. Contbr. numerous articles to med. jours.; editorial bd.: Internat. Jour. Cancer Research. Served with Chem. Corps. U.S. Army, 1953-55. Subspecialties: Chemotherapy; Cell and tissue culture. Current work: Heterotransplantation of human neural tumors, specific studies with chemotherapy, vitamin therapy of neuroblastoma, retinoblastoma and brain tumors, autologous bone marrow transplantation, monoclonal antibodies. Home: 27 E 95th St New York NY 10128 Office: Meml Sloan Kettering Cancer Center 1275 York Ave New York NY 10021

HELSTAD, DONALD DEAN, med. educator, physician; b. Chgo., Apr. 2, 1940; m. Sandra Jansen, Jan. 25, 1964; children: Dean, Wendy. M.D., U. Chgo., 1963. Diplomate: Am. Bd. Internal Medicine, subspecialty bd. in cardiovascular disease. Intern U. Chgo., 1963-64, resident, 1964-66; cardiovascular trainee U. Iowa, 1966-67; research internist U.S. Army Research Inst. of Environ. Medicine, Natick, Mass., 1967-70; asst. prof. medicine U. Iowa Coll. Medicine, Iowa City, 1970-73, assoc. prof., 1973-76, prof., 1976—. Contbr. numerous articles on internal medicine to profl. jours.; assoc. editor: Circulation Research; mem. editorial bd.: Jour. Cerebral Blood Flow and Metabolism. Royal Soc. Med. Found. fellow, 1973; recipient Cecile Lehman Mayer Research award Am. Coll. Chest Physicians, 1973; Research Career Devel. award, 1975; Irving S. Wright award Am. Heart Assn., 1976; Harry Goldblatt award Council for High Blood Pressure Research AMA, 1980; NIH grantee, 1973-86; VA grantee, 1971—84. Mem. Am. Fedn. Clin. Research, Am. Physiol. Assn., Central Soc. Clin. Research, Soc. Pharmacology and Exptl. Therapeutics, Am. Soc. Clin. Investigation, Assn. Am. Physicians, Am. Heart Assn., Council Basic Sci., Council Circulation, Council on Clin. Cardiology, Stroke Council, Council on High Blood Pressure Research, ACP. Subspecialties: Cardiology; Physiology (medicine). Current work: Cardiovascular physiology and pharmacology. Home: 435 Lexington Ave Iowa City IA Office: Dept Internal Medicin Cardiovascular Disease Div Univ Iow Iowa City IA 52242

HELTON, AUDUS WINZLE, plant pathologist; b. Bethel, Okla., Oct. 5, 1922; s. Leonard Huston and Berniece Gladys (Wright) H.; m. Christina Raza, July 29, 1978; children—Rebecca, John, Kathy, Thomas, Carolyn. B.A., Ohio Wesleyan U., 1947, M.S., 1948; Ph.D., Oreg. State U., 1951. Plant pathologist U. Idaho, Moscow, 1951—, prof. plant pathology, 1956—. Served with USN, World War II. Chem. Co. research grantee, 1960—. Mem. Am. Phytopath. Soc., Am. Inst. Biol. Scis., AAUP, AAAS, Idaho Acad. Scis., Wildlife Fedn., Sigma Xi. Lodges: Lions; Masons. Subspecialties: Plant pathology; Plant physiology (agriculture). Current work: Fungal diseases of tree fruits, effects of pesticides on nitrogen fixation in legumes. Office: Dept Plant Soil and Entomol Scis U Idaho Moscow ID 83843

HEMENWAY, MARY KAY, astronomer, cons.; b. Akron, Ohio, Nov. 20, 1943; d. Ralph Elwood and Mary Esther (Keegan) Meacham; m. Paul Derek Hemenway, June 1, 1968; children: Anne, Sara. B.S., Notre Dame Coll., Cleve., 1965; M.A. (NDEA fellow), U. Va., 1967, Ph.D., 1971. Asst. prof. Mary Baldwin Coll., Staunton, Va., 1970-71; asst. prof., instr. U. Tex., Austin, 1975-80, dir., 1980—, ednl. cons. dept. sci. edn. and extension, corr. studies. Author: (with R. Robert Robbins) Modern Astronomy-An Activities Approach, 1982. Active LWV. U. Va. fellow, Sterrewacht, Leiden, Netherlands, 1968-69. Mem. Am. Astron. Soc., Sigma Xi. Roman Catholic. Subspecialties: Optical astronomy. Current work: Galactic structure. Home: 3205 Skylark Dr Austin TX 78757 Office: Univ Tex RLM 15.220 Austin TX 78712

HEMINGWAY, PETER, psychologist, researcher; b. N.Y.C., June 29, 1940; s. Leicester Clarence and Patricia (Shedd) H.; m. Rosalie Louise Weiss, Apr. 11, 1965; children: Patrick Edward, Daniel Eric, Ian Walter, Paul Christopher. B.A., George Washington U., 1962, M.A., 1964; Ph.D., U. Sask., 1970. Cert. sch. psychologist, N.Y.; registered psychologist, Sask. Research asst. U.S. VA, Washington, 1962-64; clin. psychologist Syracuse (N.Y.) U., 1964-66; research psychologist Govt. Sask., Regina, Can., 1967-68; instr., asst. prof., assoc. prof. U. Regina, 1968-82, prof. psychology, 1982—. Mem. Sask. Psychol. Assn. (pres. 1980-82), Can. Psychol. Assn., Am., Psychol. Assn. Subspecialties: Cognition; Social psychology. Current work: Research in criminal justice system, classification and diagnosis, psychological tests and measurement, methodology. Office: Univ Regina Regina SK Canada S4S 0A2 Home: 2728 Assiniboine Ave Regine SK Canada S4S 1C6

HEMMENDINGER, LISA M., neurobiologist; b. New Haven, Mar. 9, 1954; d. Laurence and Doris (Nelson) H.; m. Lawrence C. Uhteg, Oct. 18, 1980. A.B., Smith Coll., 1976; Ph.D., U. Chgo., 1980. Postdoctoral fellow assoc. dept. neurology SUNY, Stony Brook, from 1980; now research fellow dept. neurology Mass. Gen. Hosp., Boston, Eunice Kennedy Shriver Ctr., Waltham, Mass. Contbr. articles to profl. jours. Mem. Soc. for Neurosci., AAAS, Sigma Xi. Subspecialties: Neurobiology; Developmental biology. Current work: Organization of the adult and developing mammalian central nervous system. Office: Southard Unit Eunice Kennedy Shriver Ctr 200 Trapelo Rd Waltham MA 02254

HEMMINGS, ROBERT LESLIE, chemical engineer; b. Edmonton, Alta., Can., May 2, 1940; came to U.S., 1981. B.Sc. in Chem. Engring, U. Alta., 1962, Ph.D., U. London, 1965; diploma, Imperial Coll. Sci. and Tech., 1965. Research chem. engr. Atomic Energy Can., Ltd., Pinawa, Man., 1965-69, commissioning engr., Gentilly, Que., 1969-71, design engr., Mississanga, Ont., 1971-75, engring. mgr., Montreal, Que., 1975-78; engring. specialist, Canatom, Montreal, 1978-81; mgr. tech. services London Nuclear, Niagara Falls, N.Y., 1981-83, gen. mgr., 1983—. Mem. Order Engrs. Que., Nat. Assn. Corrosion Engrs., Am. Nuclear Soc., Can. Nuclear Soc. Subspecialties: Nuclear engineering; Materials (engineering). Current work: Decontomination; radiation exposure reduction at nuclear facilities. Office: London Nuclear 2 Buffalo Ave Niagara Falls NY 14303

HEMMINGSEN, BARBARA BRUFF, microbiologist, educator; b. Whittier, Calif., Mar. 25, 1941; d. Stephen Cartl and Susanna Jane (Alexander) Bruff; m. Edvard Alfred Hemmingsen, Aug. 5, 1967; 1 dau., Grete Anne. B.A., U. Calif.-Berkeley, 1962, M.A., 1964; Ph.D., U. Calif.-San Diego, 1971. Microbiologist Ames Research Ctr., NASA, Moffitt Field, Calif., 1964-65, summer 1966; vis. asst. prof. Ecology Lab., Zoology Inst., Aarhus (Denmark) U., 1971-72; lectr. in microbiology San Diego State U., 1973-77, asst. prof. microbiology, 1977-81, assoc. prof., 1981—. Author: (with T. Fenchel) Manual of Microbial Ecology, 1974, Japanese transl., 1975; contbr. chpts., articles to profl. publs. Mem. Planned Parenthood Assn. San Diego, 1973—; mem. La Jolla (Calif.) Elem. Sch. PTA, 1980—. Mem. AAAS, Am. Soc. Microbiology, Soc. Gen. Microbiology, Soc. Protozoologists, Fedn. Am. Scientists. Democrat. Subspecialty: Microbiology. Current work: Effects of extreme gas supersaturations on microorganisms and animal cells; conditions that promote gas bubble nucleation intracellularly; taxonomy and physiology of marine bacteria. Office: Dept Microbiology San Diego State San Diego CA 92182-0067

HEMP, GENE W(ILLARD), mech. engr., univ. exec., educator; b. Mpls., Dec. 6, 1938; s. Willard H. and Ann (Thompson) H.; m. Evelyn H. Ploetz, Mar. 19, 1960; children: Barbara Jean, Suzanne Marie. B.S. in Aero. Engring, U. Minn., 1961, B.S.B. in Bus. Adminstrn, 1962, M.S. in Mechanics and Materials, 1963, Ph.D., 1967. Asst. prof. engring. scis. U. Fla., 1967-70, assoc. prof., 1970-76, prof., 1976—, asst. dean engring., 1972-74, asst. v.p. acad. affairs, 1974-76, assoc. v.p., 1976—; cons. Harry Diamond Labs., 1970-72. Mem. ASME, AIAA, Am. Soc. Engring. Edn., Soc. Engring. Scis. Subspecialties: Theoretical and applied mechanics; Applied mathematics. Current work: Nonlinear oscillation, dynamic material properties, vibrations of discrete and continuous media, rigid body mechanics. Home: 9909 NW 59th Pl Gainesville FL 32606 Office: U Fla 233 Tigert Hall Gainesville FL 32611

HEMSTOCK, GLEN ALTON, corporation research and development executive; b. Owen Sound, Ont., Can., Oct. 5, 1925; s. Herman Edwin and Ada May (Tyler) H.; m. Evelyn Christina Beaton, Oct. 6, 1951; children: Carol Marie, Bruce Edward. B.S., U. Toronto, 1948; Ph.D., Purdue U., 1951. Asst. prof. agronomy Purdue U., 1951-52; research assoc. Inst. Paper Chemistry, 1952-56; research supr. minerals and chemicals Philipp Corp., 1956-62; mgr. fundamental research minerals and chemicals, 1967, from asst. dir. research to dir. research, 1967-74; v.p. research and devel. Engelhard Corp., Edison, N.J., 1974—; group leader Union-Bag Camp Corp., 1962-67. Sect. editor: Phys. Chemistry of Pigments and Paper Coatings, 1974-77. Mem. TAPPI, Indsl. Research Inst., Assn. Research Dirs., Am. Chem. Soc., Clay Minerals Soc., Research and Devel. Council N.J. Republican. Presbyterian. Subspecialties: Surface chemistry; Catalysis chemistry. Current work: Chemistry and surface properties of clay minerals. Patentee in field. Home: 137 Balcort Dr Princeton NJ 08540 Office: Minerals and Chemicals Div Engelhard Corp Menlo Park CN 28 Edison NJ 08818

HENDERER, WILLARD E(VERETT), III, mech. engr.; b. Wilmington, Del., June 20,1948; s. Willard Everett and Nacny B. H.; m. Maureen D.; children: Daniel, Jeffrey. B.S., U. Vt., 1972, M.S., 1974, Ph.D., 1976. Mgr. metall. lab./research Vt. Tap & Die Co. div. Vt. Am. Corp., Lyndonville, 1976—. Contbr. articles on materials processing, tool steel metallurgy to profl. jours. Mem. ASME, Am. Soc. Metals, ASTM, Tau Beta Pi. Subspecialties: Metallurgical engineering; Materials processing. Current work: Research and development of tool steel alloy, deposition of refractory coatings for cutting tool application.

HENDERSON, ANN SHIRLEY, geneticist, researcher; b. Honea Path, S.C., Aug. 29, 1938; s. William Campbell and Annie Sue (Anderson) H. B.A., Winthrop Coll., 1960; M.A., U. N.C., 1963, Ph.D., 1967. Staff researcher Internat. Lab. Genetics and Biophysics, Naples, Italy, 1968-70; research assoc. Columbia U., N.Y.C., 1971-74, asst. prof., 1975-82; assoc. prof. biol. scis. Hunter Coll., CUNY, N.Y.C., 1983—; cons. Clin. Testing-Research, Hohokus, N.J., 1981—. Contbr. articles to profl. jours. Leukemia scholar Leukemia Soc. Am., 1981—. Subspecialties: Genome organization; Molecular biology. Current work: Studies of molecular organization of the human chromosome. Office: Hunter Coll CUNY 695 Park Ave New York NY 10021

HENDERSON, DALE B(ARLOW), nuclear explosives physics researcher; b. Tulsa, June 6, 1941; s. Eugene M. and Nellie (Barlow) H.; m. Denise R. Newell, Aug. 29, 1964; children: Leslie, Gregory, Claire, Stephen. B.Engring. Physics, Cornell U., 1963, Ph.D. in Applied Physics, 1967. With Los Alamos Nat. Lab., 1966—, mgr. nuclear explosive physics research, 1980—. Mem. Am. Phys. Soc., AIAA, AAAS. Subspecialties: Plasma physics; Fusion. Current work: Inertial fusion, laser development, nuclear explosive physics numerical computation of physical processes. Home: 723 Kris Ct Los Alamos NM 87544 Office: Los Alamos Nat Lab X-DO Mail Stop B 218 Los Alamos NM 87545

HENDERSON, DONALD AINSLIE, univ. dean; b. Cleve., Sept. 7, 1928; s. David Alexander and Grace Eleanor (McMillan) H.; m. Nana Irene Bragg, Sept. 1, 1951; children—Leigh Ainslie, David Alexander, Douglas Bruce. B.A., Oberlin (Ohio) Coll., 1950, D.Sc. (hon.), 1979; M.D., U. Rochester, N.Y., 1954, D.Sc. (hon.), 1977; M.P.H., Johns Hopkins U., 1970; LL.D. (hon.), Marietta (Ohio) Coll., 1978, D.Sc., U. Ill., 1979, U. Md., 1980, M.D., U. Geneva, 1980, L.H.D., SUNY, 1981. Diplomate: Am. Bd. Preventive Medicine. Intern, then resident Mary Imogene Bassett Hosp., Cooperstown, N.Y., 1954-55, 57-59; chief epidemic intelligence service Center Disease Control, USPHS, Atlanta, 1955-57, chief surveillance sect., 1960-66; chief med. officer smallpox edn. WHO, Geneva, 1966-77; dean Johns Hopkins U. Sch. Hygiene and Pub. Health, 1977—. Contbr. articles to med. jours. Recipient Commendation medal USPHS, 1962, Distinguished Service medal, 1976; Ernst Jung prize, 1976; award Govt. India-Indian Soc. Malaria and Other Communicable Diseases, 1975; Rosenthal internat. award for excellence, 1975; George MacDonald medal London Sch. Hygiene and Tropical Medicine, Royal Soc. Tropical Medicine and Hygiene, 1976; Health medal Govt. Afghanistan, 1976; Spl. Albert Lasker Pub. Health Service award for WHO, 1976; Public Welfare medal Nat. Acad. Scis., 1978; Joseph C. Wilson award in internat. affairs, 1978; James D. Bruce Meml. award, 1978; 50th Anniversary Disting. Service award Blue Cross-Blue Shield, 1979; medal for contbns. to health Govt. of Ethiopia, 1979; Outstanding Alumnus award Delta Omega, 1980. Hon. fellow Am. Acad. Pediatrics, Royal Coll. Physicians (U.K.); mem. Inst. Medicine (Nat. Acad. Scis.), Am. Public Health Assn., Internat. Epidemiol. Assn., Royal Soc. Tropical Medicine and Hygiene, Indian Soc. Malaria and Other Communicable Diseases. Subspecialty: Public health education administration. Current work: Public health education; international health and development. Home: 3802 Greenway Baltimore MD Office: 615 N Wolfe St Baltimore MD 21205

HENDERSON, EDWARD GEORGE, biophys. pharmacologist, educator, researcher; b. Bridgeport, Conn., Apr. 6, 1935; s. George and Elizabeth (Danczak) H.; m. Alice T. Troy, Aug. 16, 1958; children: Dana Lynn, Laurie Ann. Ph.D. in Biophysics (NIH fellow), Sch. Medicine, U. Md., 1966. Group leader dept. radiochemistry Wyeth Labs., 1965-68; asst. prof. pharmacology U. Conn. Health Center, Farmington, 1968-73, asso. prof., 1973—. Contbr. articles to profl. jours. Served with USN, 1953-57. Mem. Biophys. Soc., Am. Soc. Pharmacology and Exptl. Therapeutics, Soc. for Neurosci., Sigma Pi Sigma. Subspecialties: Neurophysiology; Neuropharmacology. Current work: Synaptic transmission, electrophysiology, neurochemistry, receptor pharmacology, single channel currents, voltage clamp. Office: U Conn Health Center Farmington Ave L6002 Farmington CT 06032

HENDERSON, ISAAC CRAIG, medical oncologist, researcher; b. Paullina, Iowa, Aug. 10, 1941; s. Isaac C. and Ora E. (Tjossem) H.; m. Mary Turner Henderson, June 11, 1966; children: Isaac Craig, Amy Hudson. A.B., Grinnell (Iowa) Coll., 1963; M.D., Columbia U., 1970. Cert. internal medicine, 1977, med. oncology, 1979. Intern Presbyterian Hosp., N.Y.C., 1970-71, resident, 1971-72; research assoc. NIH, 1972-74; instr. medicine Harvard U. Med. Sch., Boston, 1975-76, asst. prof., 1976—; dir. Breast Evaluation Center, Dana Farber Cancer Inst., 1980—. Contbr. articles to profl. jours. Trustees, Cambridge Friends Sch. Served with USPHS, 1972-74. Fulbright Research scholar, 1964-65; Merck, Sharpe and Dohme Internat. fellow, 1966; recipient Columbia Presbyterian Med. Soc. Research prize, 1970. Mem. Am. Soc. Clin. Oncology, Am. Assn. Cancer Research. Mem. Soc. of Friends. Subspecialties: Oncology; Cancer research (medicine). Current work: Clinical protocols evaluating new treatment modalities for the treatment of breast cancer; short term tissue culture assay systems for human breast cancer; clinical studies on adriamycin cardiotoxicity. Home: 8 Clengarry Rd Winchester MA 01890 Office: 44 Binney St Dana 1720 Boston MA 02115

HENDERSON, JAMES HENRY MERIWETHER, plant physiologist, cons., adminstr.; b. Falls Church, Va., Aug. 10, 1917; s. Edwin Bancroft and Mary Ellen (Meriwether) H.; m. Betty Alice Francis, Mar. 28, 1948; children: Edith Ellen, Dena R., James F., Edwin B. B.S., Howard U., 1939; M.Ph., U. Wis., 1941, Ph.D., 1943. Research fellow Calif. Inst. Tech., Pasadena, 1938-50; research assoc. Tuskegee Inst., Ala., 1950-68, from asst. prof. to prof., 1957-68; dir. Carver Research Found., 1968-75, sr. research prof. biology, chmn. div. natural sci., 1975—, dir. minority biomed. research support program. Author sci. articles. Vice-chmn. Macon County (Ala.) Bd. Edn., 1968-74; chmn. Carver dist. Boy Scouts Am., 1956-61, 75-82. Recipient Alumni Achievement award Howard U., 1964, 75; Faculty Achievement award Tuskegee Inst., 1976, 80; Silver Beaver award Boy Scouts Am., 1961. Fellow AAAS; mem. Am. Soc. Plant Physiologists, Tissue Culture Assn., Sigma Xi, Phi Beta Kappa, Beta Kappa Chi. Presbyterian (elder). Club: Optimists. Subspecialties: Plant physiology (biology); Tissue culture. Current work: In vitro culture of sweet potato roots to produce cell suspensions in order to produce regeneration of plantlets with several desirable characteristics. Home: PO Box 247 Tuskegee Institute AL 36088 Office: Carver Research Labs Tuskegee Institute AL 36088

HENDERSON, JAMES STUART, pathologist; b. Dundee, Scotland, May 26, 1928; emigrated to U.S., 1951; s. James Duncan and Caroline (Fowlie) H.; m. Ursula Offenbacher, Mar. 15, 1956; children: Benjamin Duncan, Adam Stuart. M.B.Ch.B., St. Andrews U., Scotland, 1951. House staff Muhlenberg Hosp., Plainfield, N.J., 1951-52; mem. Atomic Bomb Casualty Commn., Hiroshima, 1954; resident Duke U., Durham, N.C., 1955-57; research assoc. Rockefeller Inst., 1957-60; asst. prof. Rockefeller U., N.Y.C., 1960-70; prof. U. Man. (Can.), Winnipeg, 1970—; dir. research Deer Lodge Vets. Hosp., 1970—, acting dir. labs., 1979—; cons. pathologist Winnipeg Health Scis. Center, 1981—. Contbr. articles to profl. jours. Served to capt., M.C. Royal Army, 1952-54. Mem. Am. Assn. Pathologists, Path. Soc. Gt. Britain, Am. Assn. Cancer Research, Reticuloendothelial Soc., Am. Soc. Photobiology. Subspecialties: Pathology (medicine); Cancer research (medicine). Current work: Very early growth of murine solid tumors, pathology, neoplasia, edn. Office: 770 Bannatyne Ave Winnipeg MB Canada R3E 0W3

HENDERSON, MAUREEN MCGRATH, medical educator; b. Tynemouth, Eng., May 11, 1926; came to U.S., 1960; d. Leo E. and Helen (McGrath) H. M.B. B.S, U. Durham, Eng., 1949, D.P.H., 1956. Prof. preventive medicine U. Md. Med. Sch., 1968-75, chmn. dept. social and preventive medicine, 1971-75; asso. epidemiology Johns Hopkins U. Sch. Hygiene and Pub. Health, 1970-75; asso. v.p. health scis., prof. medicine, prof. health services U. Wash. Med. Sch., 1975-81, prof. epidemiology and medicine, 1981—; chmn. epidemiology and disease control study sect. NIH, 1969-72; chmn. clin. trials rev. com. Nat. Heart, Lung and Blood Inst., 1975-79; mem. Nat. Cancer Adv. Bd., 1979-84. Contbr. med. publs. Luke-Armstrong scholar epidemiology, 1956-57; John and Mary Markle scholar acad. medicine, 1963-68. Mem. Inst. Medicine (council 1981-85), Assn. Tchrs. Preventive Medicine (pres. 1972-73), Soc. Epidemiol. Research (chmn. 1969-70), Internat. Assn. Epidemiol. Assn. (exec. officer 1971-76), Am. Epidemiol. Assn., Royal Soc. Medicine. Subspecialty: Epidemiology. Home: 5309 NE 85th St Seattle WA 98115 Office: School of Public Health and Community MED SC-30 Univ Wash Seattle WA 98195

HENDERSON, ROBERT EDWARD, state research agency administrator; b. Kokomo, Ind., Feb. 28, 1925; s. Chester Ellsworth and Nellie B. (Ackerson) H.; m. Shirley S. Shroyer, Dec. 30, 1967; children: Ann, Elizabeth , Carol, Katherine, Craig, James. B. A., Carleton Coll., 1949; M.A., U. Mo., 1951, Ph.D. in Physics, 1953. With Allison div. Gen. Motors Corp., 1953-73, dir. research., 1968-73; pres., chief exec. officer Indpls. Ctr. Advanced Research, 1973-83; dir. S.C. Research Authority, Columbia, 1983—. Served with U.S. Army, 1943-45. Decorated Purple Heart. Mem. IEEE, ASHRAE, Am. Phys. Soc., AIAA (assoc. fellow), Am. Solar Energy Soc., Soc. Mfg. Engrs. Presbyterian. Subspecialties: Solar energy; Photosynthesis. Office: SC Research Authority PO Box 12025 Columbia SC 29211

HENDLER, RICHARD WALLACE, biochemist; b. Phila., Mar. 3, 1927; s. Myer and Reba (Gordon) H.; m. Nancy Gloria Cohen, Nov.

20, 1948; children: Peter Laurence, Michael Dana. B.S., Pa. State U., 1948; M.S., U. Pa., 1949; Ph.D., U. Calif.-Berkeley, 1952. Chief sec. membrane enzymology, lab. cell biology Nat. Heart, Lung and Blood Inst., NIH, Bethesda, Md., 1952—. Author: Protein Biosynthesis and Membrane Biochemistry, 1968; contbr. articles to sci. jours. NIH Polio Found. fellow, 1952; USPHS spl. fellow, Belgium, 1955. Mem. Phi Beta Kappa, Sigma Xi, Pi Mu Epsilon. Subspecialties: Biochemistry (biology); Membrane biology. Current work: Bioenergetics; membrane chemistry. Office: NHLBI Bldg 3 Room B106 NIH Bethesda MD 10105

HENDLEY, EDITH D., physiology educator; b. N.Y.C., Sept. 5, 1927; d. Michael and Rose (Tarillo) diPasquale; m. Daniel Dees Hendley, Apr. 21, 1952; children: Jane Alice, Joyce Louise, Paul Daniel. A.B., Hunter Coll., CUNY, 1948; M.Sc., Ohio State U., 1950; Ph.D., U. Ill., Chgo., 1954. Asst. lectr. Sheffield (Eng.) U., 1956-57; instr. Johns Hopkins U., Balt., 1963-66, research assoc., 1966-72; sr. investigator Friends Med. Sci. Research Ctr., Balt., 1972-73; assoc. prof. physiology U. Vt., Burlington, 1973-83, prof., 1983—, clin. assoc. prof. dept. psychiatry, 1975—. Contbr. chpts. to books, articles to profl. jours. NIH grantee, 1975-81; NIMH grantee, 1982—; Vt. Heart Assn. grantee, 1977, 82. Mem. Soc. Neurosci., Am. Physiol. Soc., Am. Soc. Pharmacology and Exptl. Therapeutics, Assn. Women in Sci. (treas. 1972-74), Phi Beta Kappa, Sigma Xi. Subspecialties: Neurochemistry; Neuropharmacology. Current work: Biogenic amines in the brain. Neurochemical correlates of behavior. Home: 10 Highland Terr South Burlington VT 05401 Office: Dept Physiology and Biophysics U Vt Burlington VT 05405

HENDLEY, JOSEPH OWEN, pediatrician; b. Chattanooga, Aug. 18, 1937; s. Flavius Josephus and Cornelia Adelaide (Smartt) H.; m. Sarah P., May 26, 1978; children. B.A., Vanderbilt U., 1959; M.D., U. Pa., 1963. Diplomate: Am. Bd. Pediatrics. Intern Duke U., 1963-64, resident, 1964-67; research fellow dept. pediatrics U. Va., 1965-67, asst. prof. pediatrics, 1970-76, assoc. prof., 1976-82, prof., 1982—; research fellow Harvard U. Sch. Public Health, Boston, 1968-70. Mem. Am. Acad. Pediatrics, Am. Epidemiol. Soc., Am. Pediatrics Soc., Am. Soc. Microbiology, Infectious Disease Soc. Am., Soc. Pediatrics Research, So. Soc. Pediatrics Research, Va. State Med. Soc., Alpha Omega Alpha. Subspecialties: Pediatrics; Infectious diseases. Current work: Characterization of the capsule on N. gonorrhoeae; transmission of rhinovirus colds; study of mechanisms of human infection by and immunity to B. pertussis. Office: U Va PO Box 386 Charlottesville VA 22908

HENDRICK, HAL WILMANS, human factors educator, management consultant; b. Dallas, Mar. 11, 1933; s. Harold E. and Audrey S. (Wilmans) H.; m. Jytte Lauridson, Sept. 9, 1972; children: Hal, David, John, Jennifer; m.; stepchildren: Sharon, Debra, Jacquiline. B.A., Ohio Wesleyan U., 1955; M.S., Purdue U., 1961, Ph.D., 1966. Cert. psychologist, Tex. Mgmt. trainee Timkin Roller Bearing Co., Canton, Ohio, 1955-56; commd. 2d lt. U.S. Air Force, 1956, advanced through grades to lt. col., 1973; radar weapons controller USAF Air Def. Command, N.Y. State, 1956-60; human performance engr. USAF Systems Command, Wright-Patterson AFB, Ohio, 1961-66; assoc. prof. psychology USAF Acad., 1966-72; chmn. behavior sci. div. Def. Race Relations Inst., Partick AFB, Fla., 1972-76; ret., 1976; assoc. prof. human factors U. So. Calif., 1976-82, chmn. dept. human factors, 1983—; mgmt. cons., Honolulu, 1976-82, Los Angeles and Honolulu, 1983—. Author: Behavioral Research and Analysis, 1981; research publs. in field, 1978-82; editor: Symposium Procs. Psychology in the Air Force, 1970, 71, 72, Individual and Group Behavior, 1975. Decorated Meritorious Service medal with oak leaf cluster, Air Force Commendation medal; named U. So. Calif. Outstanding Tchr. U. So. Calif. Assocs., 1969. Mem. Human Factors Soc. (chmn. Tech. Group 1981—), Internat. Ergonomics Assn. (exec. council 1980—), Am. Psychol. Assn., Organizational Devel. Network, Acad. Mgmt., Western Psychol. Assn., Sigma Xi. Democrat. Unitarian. Lodge: Rotary. Current work: Leadership, personality, management decision-making, organizational design, human factors. Home: 849 Hahaione St Honolulu HI 96825 Office: Dept Human Factors Inst Safety and Systems Mgmt U So Calif Los Angeles CA 90089

HENDRICK, IRA GRANT, aerospace company executive; b. Kansas City, Mo., Feb. 10, 1913; m., 1934; 2 children. B.S., U. Ark., 1936; grad. degree, Princeton U., 1937. Designer Waddell & Hardesty, Cons. Engrs., 1937-42; chief structural engr. Eritrean Project, Johnson, Drake & Piper, 1942-43; structural engr. U.S. Army Dept. Engrs., 1943; project stress analyst Grumman Aircraft Engring. Copr., 1943-46, chief of structure, 1946-57, chief tech. engr., 1957-63, v.p. engring., 1963-70, sr. v.p. and dir. tech. ops., 1970-73, sr. vp. and dir. advanced systems tech., 1973-75; sr. v.p. and presdl. asst. for corp. tech. Grumman Aerospace Corp., Bethpage, N.Y., 1975—; Mem. vis. com. dept. physics Lehigh U., 1966-74, dept. aeros. MIT, 1967-73; mem. research and tech. adv. council, chmn. adv. com. materials and structure NASA, 1971-77; mem. Aero. and Space Engring. Bd., NRC, 1977—, USAF Sci. Adv. Bd., 1977—. Recipient Spirit of St. Louis award ASME, 1967. Fellow AIAA (Sylvanus Albert Reed award 1971); mem. Nat. Acad. Engring., Soc. Exptl. Stress Analysis. Subspecialty: Aerospace engineering and technology. Address: 250 Mount Joy Ave Freeport NY 11520

HENDRICK, JAMES G., III, agricultural engineer; b. Birmingham, Ala., June 15, 1931; s. James G. and Leslie (Polk) H.; m. Joy West, June 5, 1972. B.S., Ala. Poly. Inst., 1958; M.S., Auburn U., 1960; Ph.D., Mich. State U., 1962. Registered profl. engr., Ala. Assoc. prof. dept. agrl. engring. Auburn (Ala.) U., 1962-68; agrl. engr. Nat. Tillage Machine Lab., Agrl. Research Service, U.S. Dept. Agr., Auburn, 1968—. Contbr. articles to profl. jours. Aux. police officer Auburn Police Dept., 1974—. Served to 1st lt. USAF, 1952-57. Mem. Am. Soc. Agrl. Engrs., Nat. Soc. Profl. Engrs., Ala. Soc. Profl. Engrs., Sigma Xi, Gamma Sigma Delta. Methodist. Subspecialty: Agricultural engineering. Current work: Soil dynamics, soil-metal friction, powered tillage tools, soil-metal wear, electrical discharge in soil. Office: Nat Tillage Machinery Lab PO Box 792 Auburn AL 36830

HENDRICKS, JOHN STANLEY, scientist; b. Hollywood, Calif., June 9, 1949; s. Richard and Grayce Ann (Houseman) H.; m. Dianne Marie Coane, Aug. 18, 1972; children: Dianne, Suzanne, James, David. A.A., Los Angeles Valley Coll., 1969; B.S. in Engring, UCLA; M.S. in Nuclear Engring, UCLA, 1972, Ph.D., M.I.T., 1975. Research asst. Los Alamos Nat. Lab., 1971-73, staff scientist, 1975—. NSF fellow, 1972-75. Mem. Am. Nuclear Soc. (chmn. trinity sect. 1979-80, meritorious sect. 1980). Club: Toastmasters (area gov. 1978). Subspecialties: Nuclear engineering; Mathematical software. Current work: Monte Carlo method software development and applications. Office: Los Alamos Nat Lab X-6 MS B226 Los Alamos NM 87545

HENDRICKSON, WALDEMAR FORRSEL, nuclear engineer; b. Tobias, Nebr., June 7, 1934; s. Walter Bernodotte and Stella Fay (Maxson) H.; m. Annette Marilyn Syverson, Feb. 14, 1957; children: Jene Diane, Olaf Knute. B.S., U. Idaho-Moscow, 1956, M.S. Ch.E., 1958; M.S. in Nuclear Engring, Wash. State U., Pullman, 1964; Ph.D. in Engring. Sci, Wash. State U., Pullman, 1971. Nuclear engr. Wash. State U., Pullman, 1958-72; research physicist USN Ordnance Lab., Silver Spring, Md., 1972-74; chem. engr. AEC, Idaho Falls, 1974-79; nuclear engr. fast flux test facility U.S. Dept. Energy, Richland, Wash., 1979—. Mem. Am. Nuclear Soc., Health Physics Soc. Subspecialties: Chemical engineering; Nuclear engineering. Current work: Program management in core physics, ultrasonic testing for inservice inspection, fast reactor fuel fabrication, reprocessing and waste solidification. Office: Dept Energy Box 550 Richland WA 99352

HENDRIE, JOSEPH MALLAM, physicist, nuclear engineer, government official; b. Janesville, Wis., Mar. 18, 1925; s. Joseph Munier and Margaret Prudence (Hocking) H.; m. Elaine Kostell, July 9, 1949; children: Susan Debra, Barbara Ellen. B.S., Case Inst. Tech., 1950; Ph.D., Columbia U., 1957. Registered profl. engr., N.Y., Calif. Asst. physicist Brookhaven Nat. Lab., Upton, N.Y., 1955-57, asso. physicist, 1957-60, physicist, 1960-71, sr. physicist, 1971—, chmn. steering com., project chief engr. high flux beam reactor design and constrn., 1958-65, acting head exptl. reactor physics div., 1965-66, project mgr. pulsed fast reactor project, 1967-70, asso. head engring. div., dept. applied sci., 1967-71, head, 1971-72, chmn. dept. applied sci., 1975-77, spl. asst. to dir., 1981—; dep. dir. licensing for tech. rev. U.S. AEC, 1972-74; chmn. U.S. Nuclear Regulatory Commn., Washington, 1977-79, 81, commr., 1980; lectr. nuclear power plant safety MIT, Ga. Inst. Tech., Northwestern U., summers 1970-77; cons. radiation safety com. Columbia U., 1964-72; mem. adv. com. reactor safeguards AEC, 1966-72, chmn., 1970; U.S. mem. sr. adv. group on reactor safety standards IAEA, 1974-78. Mem. editorial adv. bd.: Nuclear Tech, 1967-77. Served with AUS, 1943-45. Recipient E.O. Lawrence award, 1970; decorated comdr. Order of Leopold II (Belgium), 1982. Fellow Am. Nuclear Soc. (dir. 1976-77, v.p. 1983-84, pres. 1984-85), ASME; mem. Nat. Acad. Engring., Am. Phys. Soc., Am. Concrete Inst., IEEE, Nat. Soc. Profl. Engrs., Sigma Xi, Tau Beta Pi. Subspecialties: Nuclear physics; Nuclear engineering. Research, publs. on physics nuclear reactors, nuclear power plant safety, engring. design reactors, elec. power transmission, chem. physics nitrogen dissociation process, structure oxygen molecule. Co-inventor high flux beam reactor. Office: Brookhaven Nat Lab Upton NY 11973

HENDRIX, JOHN EDWIN, botany educator; b. Van Nuys, Calif., Aug. 30, 1930; s. John Edwin and Leana (Paul) H.; m. Joan Beverly Haas, Apr. 10, 1954; children: Janet L., James A. A.S., Pierce Jr. Coll., Los Angeles, 1951; B.S., Fresno State Coll., 1956; M.S., Ohio State U., 1963, Ph.D., 1967. Instr. Ohio State U., Columbus, 1964-67; asst. prof. botany and plant pathology Colo. State U., Ft. Collins, 1967-73, assoc. prof., 1973—. Served to 1st lt. U.S. Army, 1951-53. NSF grantee, 1977, 81. Mem. Am. Soc. Plant Physiologists (editorial staff 1980-83), Bot. Soc. Am., AAAS, Am. Inst. Biol. Scis. Subspecialty: Plant physiology (biology). Current work: Translocation and utilization of carbohydrates in plants. Office: Dept Botany Colo State U Fort Collins CO 80523

HENDRON, JOHN ALDEN, quality assurance engineer, cons., nondestructive testing engr.; b. Granite Falls, Wash., Mar. 23, 1920; s. John Alden and Grace Athena (Muirhead) H.; m. Mary Ann Wick, Oct. 6, 1953; children—Susan, Gail, Heather, John. Student, U. Wash., 1940-41, 46-52. Registered profl. engr., Calif. Research and devel. engr., motion picture x-ray systems for physiology research U. Wash., 1948-52; design engr., early field x-ray equipment Indsl. X-ray Engrs., Seattle, 1952-55; supr. inspection team Esso Research & Devel. Co., Fawley, Eng., 1956-59; head nondestructive test engring. group for devel. tests and equipment Polaris missile motors Aerojet Strategic Propulsion Co., Sacramento, 1959-70, cons. nondestructive testing engring., 1970-73, 77—; quality assurance mgr. Alaskan Copper Co., Seattle, 1973-77. Contbr. articles on motion pictures x-ray research in physiology and indsl. inspection techniques to profl. jours.; nondestructive test devels. Occasional speaker in support of nuclear power to Republican groups. Served to sgt. U.S. Army, 1941-45; ETO. Recipient Tech. Utilization award AEC/NASA Nuclear Rocket Program, 1967, 1969, 1971. Mem. ASME, Am. Welding Soc., Am. Soc. Nondestructive Testing. Club: U. Wash. Alumni. Lodges: Shriners (Seattle); Masons (Eng.). Subspecialties: Aerospace engineering and technology; Nuclear engineering. Current work: Nondestructive test development, quality assurance programs management. Office: Dept 6430 Aerojet Strategic Propulsion Co PO Box 15699C Sacramento CA 95813

HENDRY, ARCHIBALD WAGSTAFF, physics educator; b. Darvel Ayrshire, Scotland, Nov. 18, 1936; s. William and Maggie (Noble) H.; m. Jeanette M. Brown, June 20, 1964; children: Diana M., Andrew W., Gordon A. B.Sc., Glasgow U., 1958, Ph.D., 1962. Research assoc. U. Calif.-San Diego, 1962-64; research scientist Rutherford Lab., Oxford (Eng.) U., 1964-67; vis. guest prof. U. Heidelberg, W.Ger., 1967; research asst. prof. U. Ill., Urbana, 1967-69; asst. prof. physics Ind. U., Bloomington, 1969-72, assoc. prof., 1972-76, prof., 1976—. Mem. Am. Phys. Soc., AAUP. Democrat. Subspecialties: Particle physics; Theoretical physics. Current work: Theoretical high energy particle physics, theories of quarks and leptons. Home: 4311 Wembley Ct Bloomington IN 47401 Office: Dept Physics Ind U Bloomington IN 47405

HENKEL, CRAIG KENNETH, neuroscientist, educator; b. Canton, Ohio, Mar. 8, 1949; s. George E. and Dorothy Mae (Barto) H.; m. Janice Ellen Heffer, July 12, 1975; 1 son, Brian Robert. B.S., Wheaton (Ill.) Coll., 1971; Ph.D., Ohio State U., 1975. Post-docotoral research fellow dept. anatomy U. Va. Sch. Medicine, 1975-78; asst. prof. anatomy Wake Forest U. Bowman Gray Sch. Medicine., Winston-Salem, N.C., 1978—. NSF grantee, 1980-82; Nat. Inst. Neurol. Communicable Disease and Stroke, 1982-84. Mem. Soc. Neurosci., Am. Assn. Anatomists, So. Soc. Anatomists, N. C. Soc. Neurosci., Cajal Club. Mem. Christian and Missionary Alliance Ch. Subspecialty: Neuroanatomy. Current work: Orgnization of brain pathways related to acoustico-and-visual-motor behaviors.

HENKEL, JAMES G., medicinal chemistry educator; b. Santa Monica, Calif., Nov. 22, 1945; s. Otto Theodore and Dorothy Luella (Harris) H.; m. Diana Pepin, May 11, 1970. B.S., UCLA, 1967; Ph.D., Brown U., 1973. Postdoctoral fellow U. Minn., Mpls., 1972-74, asst. prof., 1974-77; asst. prof. medicinal chemistry U. Conn., Storrs, 1977-80, assoc. prof., 1982—. Author: Essentials of Drug Product Quality, 1978; patentee in field. Recipient Morse-Amoco award for undergrad. teaching U. Minn., 1977; NIH fellow, 1973. Mem. Am. Chem. Soc., Acad. Pharm. Scis., Am. Assn. Colls. Pharmacy, Sigma Xi, Alpha Chi Sigma. Subspecialty: Medicinal chemistry. Current work: Biological alkylations and cancer chemotherapy; synthesis and study of CNS-active compounds; chemistry of bridged polycyclic systems; computer applications in drug design. Patentee in field. Home: 45 Hamilton Dr Manchester CT 06040 Office: U Conn Sch Pharmacy U-92 Storrs CT 06268

HENLEY, ERNEST MARK, physics educator; b. Frankfurt, Germany, June 10, 1924; came to U.S., 1939, naturalized, 1944; s. Fred S. and Josy (Dreyfuss) H.; m. Elaine Dimitman, Aug. 21, 1948; children: M. Bradford, Karen M. B.E.E., Coll. City N.Y., 1944; Ph.D., U. Calif. at Berkeley, 1952. Physicist Lawrence Radiation Lab., U. Calif. at Berkeley, 1950-51; research asso. physics dept. Stanford, 1951-52; lectr. physics Columbia, 1952-54; mem. faculty U. Wash., Seattle, 1954—, prof. physics, 1961—, chmn. dept., 1973-76, dean, 1979—. Author: (with W. Thirring) Elementary Quantum Field Theory, 1962, (with H. Frauenfelder) Subatomic Physics, 1974, Nuclear and Particle Physics, 1975. F.B. Jewett fellow, 1952-53; NSF sr. fellow, 1958-59; Guggenheim fellow, 1967-68; NATO sr. fellow, 1976-77. Mem. Am. Phys. Soc. (chmn. div. nuclear physics 1979-80), Nat. Acad. Scis., Sigma Xi. Subspecialties: Theoretical physics; Nuclear physics. Research and numerous publs. on symmetries, nuclear reactions and high energy particle interactions. Office: Physics Dept FM 15 U Wash Seattle WA 98195

HENLEY, MELVIN BRENT, educator; b. Hickory Valley, Tenn., Aug. 25, 1935; s. Jesse C. and Effie B. (Mosely) H.; m. Eva Mohler, Feb. 28, 1954; children: Sonny, Stanley, Stuart, Steven, Shane. Student, U. Nev., Reno, 1954-57, Valley Coll., San Bernardino, Calif., 1957; B.S., Murray State U., 1961; Ph.D., U. Miss., 1964. Assoc. prof. Murray (Ky.) State U., 1967—, asst. prof. chemistry, 1964-67; mayor City of Murray, 1978-82; chief exec. officer Henley Devel. Corp., Murray, 1968—; owner, chief exec. officer Mobile Home Park Devel., Murray, 1968—. Mem. Ky. Mcpl. Statutes Revision Commn., 1978-81; bd. dirs. Ky. Mcpl. League, 1978-82; mem. econ. issues com. Nat. League Cities, 1979-82; bd. dirs. Purchase Area Devel. Dist., 1978—; mem. Murray Airport Bd., Murray Hosp., 1978-82, Murray Headstart Bd., 1979—. Served with USAF, 1954-58. Named Citizen of Yr. Murray, Calloway County C. of C., 1982; NDEA fellow U. Miss., 1961-64. Mem. Am. Chem. Soc. (chmn. Ky. lake sect. 1975). Club: Civitan (Murray, Ky.). Subspecialty: Water supply and wastewater treatment. Current work: Trihalomethanes in public water supplies, methodology of trihalomethane removal and prevention of formation. Home: Fox Meadows Murray KY 42071 Office: Dept Chemistry Murray State U Murray KY 42071

HENLEY, WALTER L., physician, educator; b. Germany; s. Fred S. and Josy (Dreyfuss) H.; m. Edith Hertz, Sept.7, 1952; children: Madeline, Betsy. B.S., UCLA, 1947; M.D., U. Pa., 1951. Intern Hosp. U. Pa., 1951-52; resident in pediatrics, researcher in virology Mt. Sinai Hosp., N.Y.C., 1953-57; guest investigator immunology Rockefeller U., 1966-68; acting dir. pediatrics Beth Israel Med. Ctr., N.Y.C., 1981-82; chief pediatric medicine Orthopedic Inst., Hosp. Joint Disease, N.Y.C., 1982—; cons. in field. Contbr. articles to profl. jours. Mem. Riverdale Community Council, Community Planning Bd. Health Com., 1970-75. Served with AUS, 1943-46. Decorated Bronze Star, Purple Heart. Mem. Am. Acad. Sci., Am. Assn. Immunology, Am. Acad. Pediatrics, Am. Pediatric Soc., N.Y. Pediatric Soc., Assn. Research Vision and Ophthalmology. Subspecialties: Pediatrics; Immunology (medicine). Current work: Humoral and cellular immunity in diseases of cornea.

HENNEBERGER, WALTER CARL, physicist, educator; b. Bradley, Ill., Jan. 17, 1930; s. Peter and Gertrude (Link) H.; m. Gerlinde Emma Henneberger, Aug.17, 1967; children: Bernard, Petra. B.S., Purdue U., 1952, M.S., 1956; Dr. rer. nat., U. Gottingen, Ger., 1959. Research assoc. U. Notre Dame, 1959-60; asst. prof. Fordham U., 1960-62; postdoctoral scholar Dublin Inst. Advanced Studies, 1962-63; asst. prof. So. Ill. U., Carbondale, 1963-67, assoc. prof., 1967-72, prof. physics, 1972—. Contbr. articles to profl. jours. Served with U.S. Army, 1954-55. Mem. Am. Phys. Soc., Am. Assn. Physics Tchrs. Subspecialties: Atomic and molecular physics; Theoretical physics. Current work: Theoretical investigation of Aharonov-Bohm effect and its relation to quantum electrodynamics. Home: 1516 Taylor Dr Carbondale IL 62901 Office: Dept Physics So Illinois U Carbondale IL 6—901

HENNEKENS, CHARLES H., physician, epidemiologist, researcher, educator; b. N.Y.C., June 12, 1942; s. Charles H. and Pauline H.; m. Deborah Cole, July 29, 1975; children: Charles, Jennifer, Alissa. B.S., Queens Coll., 1963; M.D., Cornell U., 1967; M.P.H., Harvard U., 1972, M.S., 1973, D.P.H., 1975. Diplomate: and fellow Am. Bd. Preventive Medicine. Intern in medicine Cornell U., 1967-68, resident in medicine, 1968-69; EIS med. epidemiologist Center for Disease Control, Atlanta, 1969-71; fellow in epidemiology Harvard U., 1971-73; asst. prof. U. Miami Sch. Medicine, 1973-75; asst. prof. medicine Harvard U., 1975-81, assoc. prof., 1981—; assoc. in medicine Peter Bent Brigham Hosp., Boston, 1975—; asst. vis. physician Boston City Hosp.; vis. epidemiologist dept. of Regius Prof. of Medicine Radcliffe Infirmary, Oxford (Eng.) U., 1978-79; vis. prof. epidemiology and public health U. Miami, 1980—; vis. assoc. prof. div. biostats. and epidemiology Boston U., Sch. Public Health, 1981—; cons. in field. Contbr. numerous articles, rev. articles to profl. publs.; editorial bd.: Stats. in Medicine, 1981. Served to lt. comdr. USPHS, 1969-71. Recipient Research Career Devel. award NIH, 1977-82; Butcher scholar, 1959-60; N.Y. State Regents scholar, 1959-60; Teagle scholar, 1963-67. Fellow Am. Heart Assn. Council Epidemiology, Am. Coll. Epidemiology; mem. AAAS, Am. Epidemiol. Soc., Am. Fedn. Clin. Research, Am. Public Health Assn., Am. Soc. Preventive Oncology, Assn. Tchrs. Preventive Medicine, Internat. Epidemiol. Assn., Mass. Heart Assn., Mass. Public Health Assn., N.Y. acad. Sci., Internat. Soc. and Fedn. for Cardiology (Sci. Council on Epidemiology and Prevention), Soc. Clin. Trials, Soc. Epidemiologic Research (Postdoctoral Prize for paper 1975), Mass. Med. Soc., Norfolk dist. Med. Soc. Subspecialties: Epidemiology; Preventive medicine. Current work: Epidemiology of chronic disease: cardiovascular disease and cancer; clinical epidemiology: research and training. Office: 55 Pond Ave Brookline MA 02146

HENNESSY, EDWARD LAWRENCE, JR., chemical, and technical and energy company executive; b. Boston, Mar. 22, 1928; s. Edward Lawrence and Celina Mary (Doucette) HE.; m. Ruth Frances Schilling, Aug. 18, 1951; children: Michael E., Elizabeth R. B.S., Fairleigh Dickinson U., 1955; postgrad., NYU Law Sch. With Heublein, Inc., Hartford, Conn., 1965-72, v.p. fin., 1965-68, sr. v.p. adminstrn., fin., 1969-72; sr. v.p. fin. and adminstrn. United Techs. Corp., Hartford, 1972-77, chief fin. officer, group v.p., 1977, exec. v.p., 1978-79; chmn., pres., chief exec. officer Allied Corp., Morris Township, N.J., 1979—; dir. Fed. Res. Bank N.Y.; Trustee St. Joseph Coll., Fairleigh Dickinson U., USCG Acad. Found. Served with USNR, 1949-55. Mem. Fin. Execs. Inst., Econ. Club N.Y. Roman Catholic. Clubs: Cat Cay (Bahamas); N.Y. Yacht; Ocean Reef (Key Largo, Fla.); Baltusrol.

HENNEY, CHRISTOPHER SCOT, immunologist, educator; b. Sutton Coldfield, Warwickshire, Eng., Feb. 4, 1941; came to U.S., 1966; s. William Scot and Rhoda Agnes (Bateman) H.; m. Janet Barnsley, June 20, 1964; children: James Scot, Samantha Jane. B.Sc. with honors, U. Birmingham, Eng., 1962, Ph.D., 1965, D.Sc., 1973. Immunologist WHO, Lausanne, Switzerland, 1968-70; asst. prof. medicine and microbiology Johns Hopkins U. Sch. Medicine, Balt., 1970-73, assoc. prof., 1973-78; prof. microbiology and immunology U. Wash. Med. Sch., Seattle, 1978-82; head program in basic immunology Fred Hutchinson Cancer Research Ctr., Seattle, 1978-82; sci. dir., exec. v.p. Immunex Corp., Seattle, 1982—; mem. pathology B study sect. NIH, 1978-82; chmn. immunology rev. com. Am. Cancer Soc., N.Y.C., 1982—; mem. sci. adv. bd. Damon Runyon Cancer Found., N.Y.C., 1974-78, Pacific Sci. Ctr., Seattle, 1982—. Assoc. editor: Jour. Immunology, 1972-79; sect. editor, 1979-82, Jour. Reticuloendothelial Soc, 1978-79. Med. Research Council Eng. scholar, 1962-65; Nat. Inst. Allergy and Infectious Disease grantee, 1972; Sci. in Medicine lectr. Case Western Res. U. Med. Sch., 1980; U. Wash. Med. Sch., 1980. Mem. AAAS, Reticuloendothelial Soc., Am. Acad. Allergy, Am. Assn. Immunologists, Am. Assn. Exptl. Pathology, Brit. Soc. Immunology. Subspecialties: Immunobiology and immunology; Cell and tissue culture. Home: 1115 22nd Ave East Seattle WA 98112 Office: Immunex Corp 51 University Bldg Seattle WA 98101

HENNING, JAMES SCOTT, clinical psychologist, administrator; b. Duluth, Minn., Dec. 6, 1942; s. Roland John and Margaret Louise (Lynch) Breuer. A.B., Gonzaga U., 1966; M.Div., Jesuit Sch. Theology, Berkeley, Calif., 1973; M.S., U. Wis.-Milw., 1970, Ph.D., 1974. Clin. psychologist William Greenleaf Eliott Research Ctr., Washington U., 1976-77; liaison psychologist Harbor/UCLA Med. Sch., 1977-79; psychologist, adminstr Kaiser Hosp., Los Angeles, 1979—, coordinator psychol. services, 1983—; teaching asst. U. Wis.,-Milw., 1968-70, 73-74; clin. fellow med. psychology Harvard U. Med. Sch., 1974-75; instr. med. psychology Washington U., 1975-76; asst. clin. prof. med. psychology UCLA, 1976-82; child abuse task force So. Calif. Permanente Med. Group., 1980—; psychol. cons. Labor Health Inst., St. Louis U., 1975-76. Contbr. chpts. in books, articles profl. jours.; editorial bd.: Jour. Clin. Psychology, 1977—, Merrill-Palmer Quar, 1979—. Sec., trustee Rossi Fund, Los Angeles, 1981—. Fellow Am. Orthopsychiat. Assn.; mem. Am. Psychol. Assn. Subspecialties: Developmental psychology; Clinical psychology. Current work: Mental Health and public health administration, chronic ulcers and eating disorders. Home: 1315 W Adams Blvd North University Park CA 90-07 Office: Southern California Permanente Medical Group 9449 E Imperial Hwy Downey CA 90242

HENNINGFIELD, JACK EDWARD, research pharmacologist, behavioral biology educator; b. St. Paul, Aug. 23, 1952; s. Donald B. and Vivian P. (Nankivell) H. B.A. summa cum laude, U. Minn., 1974, Ph.D., 1977. Research fellow Nat. Council Alcoholism, Mpls., 1977-78; research assoc. Balt. City Hosp., 1978-80; instr. Sch. Medicine Johns Hopkins U., Balt., 1978-80, asst. prof., 1980—; staff fellow scientist Nat. Inst. Drug Abuse/Addiction Research Ctr., Balt., 1980-82, pharmacologist, 1982—; grant rev. com. Nat. Inst. Drug Abuse, Rockville, Md., 1982—; rev. com. Nat. Cancer Inst., Bethesda, 1982; sci. cons. pub. info. program Surgeon Gen.'s Office on Smoking and Health, Rockville, 1982—. Contbr. chpts. to books, articles to publs. USPHS fellow, 1974; Nat. Council Alcoholism fellow, 1977. Mem. Am. Psychol. Assn., Behavioral Pharmacology Soc., Internat. Study Group Investigating Drugs as Reinforcers, Hastings Center Inst. Soc., Ethics and Life Scis. Subspecialties: Behavioral psychology; Pharmacology. Current work: Behavioral pharmacologic studies in human and nonhuman subjects to determine commonalities that underly dependence to tobacco, opioids, stimulants and sedatives. Designer, inventor; primate drinking device, human cigarette puff monitor. Office: NIDA/Addiction Research Center PO Box 5180 Baltimore MD 21224 Home: 3115 Tyndale Ave Baltimore MD 21214

HENRICKSON, JAMES SOLBERG, botany educator; b. Eau Claire, Wis., Oct. 15, 1940; s. Henry Marvel and Grace Isabel (Solberg) H.; m.; children: Jonathan James. M.S., Claremont Grad. Sch., 1964, Ph.D., 1968. Asst. prof. dept. biology Calif. State U.-Los Angeles, 1966-71, assoc. prof., 1971-77, prof., 1979—; dir. Ind. Environ. Cons., Los Angeles, 1970—. Mem. AAAS, Am. Assn. Plant Taxonomists, Bot. Soc. Am., Calif. Bot. Soc., Internat. Assn. Plant Taxonomists. Subspecialty: Systematics. Current work: Flora of Chihuahuan Desert (Mexico), flora of Mojave Desert. Home: 1409 Oneonta Knoll South Pasadena CA 91030 Office: Dept Biology Calif State U Los Angeles CA 90032

HENRIKSEN, RICHARD NORMAN, physics educator, astrophysicist; b. St. John, N.B., Can., Dec. 14, 1940; s. Norman Victor and Gertrude May (Snow) H.; m. Sandra Maureen Metcalf, Aug. 25, 1962 (div.); children: Stephen Myles, Selaine May, Erica Emma. B.Sc. in Math and Physics with honors, McGill U., 1962; Ph.D. in Astronomy, U. Manchester, Eng., 1965. Asst. prof. physics Queen's U., Kingston, Ont., 1966-70, assoc. prof., 1970-78, prof., 1978—; vis. prof. Tubingen, Ger., 1972; sr. research fellow U. Sussex, Eng., 1973, Stanford U., 1980. Contbr. articles to profl. jours. Alexander von Humbldt fellow, 1972; Commonwealth scholar, 1962-65. Mem. Can. Astron. Soc., Am. Astron. Soc., Internat. Astron. Union. Subspecialties: Theoretical astrophysics; High energy astrophysics. Current work: Currently pursuing the astrophys. of jets from radio galaxies; theory of quasar energy sources, pulsar magnetospheres, cosmology and the early universe, solar physics. Office: Astronomy Group Dept Physics Queens U Kingston ON Canada K7L 3N6

HENRY, J. PATRICK, astronomy educator; b. Mojave, Calif., Apr. 16, 1947; s. Bernard F. and Mary A. (Gonder) H. B.A., La Salle Coll., Phila., 1969; Ph.D., U. Calif., Berkeley, 1974. Physicist, Smithsonian Astrophys. Obs., Cambridge, Mass., 1974 81; asst. prof. astronomy U. Hawaii, Honolulu, 1981-83, assoc. prof., 1983—. Contbr. articles to profl. jours. Judge Hawaii State Sci. Fair. Woodrow Wilson fellow, 1969-74. Mem. Am. Astron. Soc.; mem. Astron. Soc. Pacific; Mem. Soc. Photo-Optical Instrumentation Engrs. Subspecialties: 1-ray high energy astrophysics; Optical astronomy. Office: 2680 Woodlawn Dr Honolulu HI 96822

HENRY, PATRICK M., chemistry educator; b. Joliet, Ill., Sept. 29, 1928; m., 1956; 3 children. B.S., DePaul U., 1951, M.S., 1953; Ph.D. in Chemistry, Northwestern U., 1956. Research chemist Hercules Inc., 1956-71; from assoc. prof. to prof. chemistry U. Guelph, Ont., Can., 1971-81; prof., chmn., dept. chemistry Loyola U., Chgo., 1981—. Mem. Am. Chem. Soc. Subspecialties: Catalysis chemistry; Biomass (energy science and technology). Office: Loyola U 6525 N Sheridan Rd Chicago IL 60626

HENSHAW, EDGAR CUMMINGS, biochemical researcher, educator; b. Cin., Dec. 14, 1929; s. Lewis Johnson and Dorothy (Cummings) H.; m. Betty Ann Barnes, Jan. 11, 1936; 1 son: Daniel. A.B., Harvard U., 1952, M.D., 1956. Lic. physician, N.Y. Intern Harvard Med. Service-Boston City Hosp., 1956-57, resident, 1959-60, fellow dept. bacteriology, 1960-62; instr. Harvard U., 1962-64, assoc., 1964-69, asst. prof., 1969-71, assoc. prof., 1971-76; prof. oncology in medicine and biochemistry U. Rochester, N.Y., 1976—, asst. dir. basic sci., 1978—; chmn. research adv. com. United Cancer Council, Rochester. Contbr. articles to profl. jours. Served with M.C. U.S. Navy, 1957-59. Nat. Cancer Inst. grantee, 1971—. Mem. Am. Assn. Cancer Research, Am. Soc. Biol. Chemists, Am. Assn. Cancer Edn., Blochem. Soc., N.Y. Acad. Sci, Rochester Acad. Medicine, Monroe County Med. Soc. Episcopalian. Subspecialties: Biochemistry (medicine); Cancer research (medicine). Current work: Biochemistry of cancer; regulation of cell growth; regulation of protein synthesis; effects upon cell metabolism of alteratiopns in cellular nutrition. Home: 542 Allens Creek Rd Rochester NY 14618 Office: U Rochester Cancer Center PO Box 704 Rochester NY 14642

HENSLEY, JOHN HIGGINS, psychology educator; b. Asheville, N.C., Nov. 1, 1947; s. Roy Wilson and Martha (Higgins) H.; m. Valerie Lynn Burson, June 4, 1969 (div. 1975); m. Candace Ann Cusenbarry, Jan. 7, 1976; children: Nathan Edward, Rachel Elizabeth. Student, N.C. State U., 1965-67; A.B., U. N.C., 1969; M.S., Tulane U., 1972, Ph.D., 1973. Cert. psychologist, Tex. Asst. prof. Midwestern State U., Wichita Falls, Tex., 1973-79, assoc. prof. psychology, 1979—. Contbr. articles to profl. jours. NSF grad. fellow, 1972-73. Mem. Am. Psychol. Assn., Soc. Research on Child Devel., Southwestern Psychol. Assn., AAAS. Subspecialties: Behavioral psychology; Developmental psychology. Current work: Research on age-related changes in perception and learning. Office: Div Social and Behavioral Sci Midwestern State Uni 3400 Taft Wichita Falls TX 76308 Home: 2313 Tulanar Ln Wichita Falls TX 76301

HENSON, BOB LONDES, physics educator; b. Pierce City, Mo., July 28, 1935; s. Arthur Londes and Perna Frances (Burnett) H.; m. Rayma Louise Hammer, Sept. 4, 1960. B.S. in Elec. Engring, U. Mo.-Columbia, 1957, M.S. in Physics, 1960; Ph.D., Washington U., 1964. Teaching asst. U. Mo.-Columbia, 1958-60; research asst. in physics Washington U., St. Louis, 1960-64, research assoc., 1964; asst. prof. physics U. N.D., Grand Forks, 1964-66, U. Mo., St. Louis, 1966-70, assoc. prof., 1970—. Contbr. articles to profl. jours. Served to 2d lt. U.S. Army, 1957. Mem. Am. Phys. Soc., AAAS, Soc. Indsl. and Applied Math., Internat. Assn. Math. Modeling, Sigma Xi. Subspecialties: Plasma physics; Mathematical modeling. Current work: Mathematical modeling, electrical discharges, charge transport; non linear differential equations, glows, coronas, mobility, ions, electrons, fluids, rivers. Home: 2721 Surrey Hill Dr Saint Charles MO 63301 Office: Dept Physics U Mo Saint Louis MO 63121

HEPPEL, LEON ALMA, biochemist; b. Granger, Utah, Oct. 20, 1912; s. Leon George and Rosa (Zimmer) H.; m. Adelaide Keller, June 6, 1944; children: David E., Alan B. B.S. in Chemistry, U. Calif. at Berkeley, 1933, Ph.D. in Biochemistry, 1937; M.D., U. Rochester, 1941. Intern Strong Meml. Hosp., Rochester, N.Y., 1941-42; officer USPHS, 1942—, med. dir., 1956; research indsl. toxicology, 1942-47, enzymology, 1948—; now specializing in membrane transport and energy transduction; specialist in studies membrane structure and function NIH, 1958-67; prof. biochemistry Cornell U., Ithaca, N.Y., 1967—. Recipient 3M award in life scis. FASEB, 1977; Guggenheim fellow, Cambridge, Eng., 1953, Imperial Cancer Research Fund, London, 1975; Fogarty scholar NIH, 1982-83. Mem. Am. Soc. Biol. Chemists, Am. Chem. Soc. (Hillebrand award Washington sect. 1959), Nat. Acad. Scis. Subspecialties: Cell and tissue culture; Membrane biology. Current work: Membrance properities of tumor cells; F-ATPase of Escherichia Coli. Office: Dept Biochemistry Cornell U Ithaca NY 14850

HEPPNER, GLORIA HILL, foundation executive; b. Great Falls, Mont., May 30, 1940; s. Eugene M. and Georgia M. (Swanson) H.; m.; 1 son, Michael. B.A., U. Calif.-Berkeley, 1962, M.A., 1964, Ph.D., 1967. Research fellow U. Wash., Seattle, 1967-69; asst. prof., assoc. prof. Brown U./Roger Williams Gen. Hosp., Providence, 1969-79; chmn., sci. dir. Mich. Cancer Found., Detroit, 1979—; chmn. oncology bd. VA, 1979-83. Contbr. articles to profl. jours.; editorial bd.: Cancer Research, 1980—, Invasion and Metastasis, 1982, Cancer Metastasis Rev, 1982. NIH research grantee, 1974. Mem. Am. Assn. Cancer Research, Am. Assn. Immunology. Democrat. Club: Wayne State U. Faculty (Detroit). Subspecialties: Immunobiology and immunology; Cancer research (medicine). Current work: Breast cancer research, immunology, tumor heterogeneity. Office: Mich Cancer Found 110 E Warren Ave Detroit MI 48201

HEPWORTH, HARRY KENT, engineering educator, mechanical engineering consultant; b. Phoenix, Aug. 14, 1942; s. Clare and Lilla (Webb) H.; m. Paula Ann Molczyk, June 2, 1974; 1 son, Donald Thomas. B.S. in Mech. Engring, Okla. State U., 1964; M.S.E., Ariz. State U., 1966, Ph.D., 1969. Registered profl. engr., Ariz. Research assoc. semiconductor products and devices Tex. Instruments Inc., Dallas, 1965-70; with No. Ariz. U., 1970—, prof. engring., 1979—; cons. in field. Contbr. articles to profl. jours. Mem. ASME, AIAA, Am. Soc. Engring. Edn., Tau Beta Pi, Phi Kappa Phi, Sigma Xi. Republican. Lutheran. Subspecialties: Fluid mechanics; Nuclear fission. Current work: Internal flow in a gas turbine engine with particular interest in tip leakage; gas dynamic flow in compressor and turbine engine sections; nuclear fuel enrichment. Home: 3260 S Gillenwater Dr Flagstaff AZ 86001 Office: Coll of Engineering No Ariz U PO Box 15600 Flagstaff AZ 86011

HERB, RAYMOND GEORGE, physicist, business executive; b. Navarino, Wis., Jan. 22, 1908; s. Joseph and Annie (Stadler); ; s. Joseph and Annie (Herb); m. Anne Williamson, Dec. 26, 1945; children: Stephen, Rebecca, Sara, Emily, William. Ph.D., U. Wis., 1935; hon. degrees, U. Basel, 1960, U. Sao Paulo, 1959. Research assoc. physics U. Wis., 1935-39, research assoc., asst. prof., Madison, 1939-40, assoc. prof., 1941-45, prof., 1945-51, Charles Mendenhall prof. physics, 1961-72; founder, pres., chmn. bd. Nat. Electrostatics Corp., Middleton, Wis., 1972—. Contbr. articles to profl. jours.; patentee high voltage electrostatic and ultra high vacuum equipment. Recipient Tom W. Bonner award, 1968, Disting. Service citation Coll. Engrs. U. Wis., 1976. Fellow Am. Phys. Soc.; mem. Nat. Acad. Scis. Subspecialty: Nuclear physics. Current work: Developer electrostatic accelerator insulated by high pressure gas. Office: 7240 Graber Rd Middleton WI 53562

HERBER, ROLFE H., chemistry educator; b. Dortmund, Westfalia, Ger., Mar. 10, 1927; came to U.S., 1938; s. Paul and Thea H.; m. Rita Joan Goldstein, June 27, 1954; children: Sharon Anne, Karen Sue, Sandra Beth. B.S., UCLA, 1949; Ph.D., Oreg. State U., 1952. Research assoc. dept. chemistry MIT, Cambridge, 1952-55; asst. prof. chemistry U. Ill., Urbana, 1955-59; assoc. prof. chemistry Rutgers U., New Brunswick, N.J., 1959-64, prof., 1964-75, disting. prof., 1975—; vis. scientist French AEC, 1974; K.T. Compton prof. physics Technion, Israel, 1981-82. Author: Principles of Chemistry, 1959; author, editor: Inorganic Isotopic Systheses, 1962, Chemical Applications of Mossbauer Spectroscopy, 1968, Mossbauer Spectroscopy, 1970. NSF fellow, 1965-66. Fellow Am. Phys. Soc.; mem. Am. Chem. Soc. (cons. 1976-80), Internat. Com. Mossbauer Spectroscopy (exec. com.), Com. Concerned Scientists (exec. com.). Subspecialties: Solid state chemistry; Condensed matter physics. Current work: Mossbauer spectroscopy; bonding and structure of inorganic, organometallic and intercalation compounds; chemical consequences of nuclear transformations. Office: Dept Chemistry Rutgers U PO Box 939 Piscataway NJ 08903

HERBERMAN, RONALD BRUCE, medical researcher; b. Bklyn., Dec. 26, 1940; s. Louis Leon and Sunny Fay (Miller) H.; m. Harriett Linda Muster, Apr. 7, 1963; children: Holly, Steven. B.A., NYU, 1960, M.D., 1964. Diplomate: Am. Bd. Internal Medicine. Intern, resident in internal medicine Mass. Gen. Hosp., 1964-66; clin. assoc. Nat. Cancer Inst., Bethesda, Md., 1966-68, sr. investigator immunology br., 1968-69, head cellular and tumor immunology sect., 1969-73, chief lab. immunodiagnosis, 1973-81, chief biol. therapeutics br. Lab Cell Biology, 1981—, acting dir. biol. response modifiers program, 1983—. Author: Immunodiagnosis of Cancer, 1977, Natural Cell-Mediated Immunity Against Tumors, 1980, NK Cells and Other Natural Effector Cells, 1982; mem. editorial bds. numerous sci. publs. Served to capt. USPHS, 1966-69, 74—. Recipient USPHS Commendation medal, 1976. Mem. Am. Soc. Clin. Investigation, Am. Assn. Immunologists, Am. Assn. Cancer Research, Transplantation Soc. Democrat. Jewish. Subspecialty: Immunology (medicine). Current work: Natural killer cells, cancer immunology, biological response modifiers. Home: 8528 Atwell Rd Potomac MD 20854 Office: National Cancer Institute Frederick MD 21701

HERBERT, EDWARD, chemistry educator; b. Hartford, Conn., Jan. 28, 1926; m., 1948; 1 child. B.S., U. Conn., 1949; Ph.D. in Cell Physiology, U. Pa., 1953. USPH-S fellow biochemistry U. Wis., 1953-55; from instr. to assoc. prof. biology MIT, 1955-63; assoc. prof. chemistry U. Oreg., Eugene, 1963-67, prof., 1967—; Reviewer research proposals NSF, 1961—. Grantee NSF, NIH. Mem. Am. Soc. Biol. Chemists. Subspecialties: Cell biology; Molecular biology. Office: Dept Chemistry U Oreg Eugene OR 97403

HERBERT, FLOYD LEIGH, physicist; b. Orange, Calif., Apr. 16, 1942; s. Ralph and Irene (Leigh) H. B.S., Calif. Inst. Tech., 1964; Ph.D. in Physics, U. Ariz., 1975. Research asst. Kitt Peak Nat. Obs., 1965-74; research asst. Planetary Sci. Inst., Tucson, 1974; research assoc. lunar and planetary lab. U. Ariz., 1975-79, sr. research assoc., 1979—. Mem. Am. Astron. Soc., Am. Geophys. Union. Subspecialty: Planetary science. Current work: Origin and evolution of solar system. Home: 3319 E Bermuda Ave Tucson AZ 85716 Office: Lunar and Planetary Lab U Ariz Tucso AZ 85721

HERBERT, GEORGE RICHARD, research executive; b. Grand Rapids, Mich., Oct. 3, 1922; s. George Richard and Violet (Wilton) H.; m. Lois Anne Watkins, Aug. 11, 1945; children—Gordon, Patricia, Alison, Douglas, Margaret. Student, Mich. State U., 1940-42; B.S., U.S. Naval Acad., 1945; D.Sc. (hon.), N.C. State U., 1967, LL.D., Duke U., 1978. Line officer USN, 1945-47; instr. elec. engring. Mich. State U., 1947-48; asst. to dir. Stanford Research Inst., 1948-50, mgr. bus. ops., 1950-55, exec. asso. dir., 1955-56, asst. sec., 1950-56; treas. Am. & Fgn. Power Co., Inc., N.Y.C., 1956-59; pres. Research Triangle Inst., 1959—; chmn., dir. Microelectronics Center N.C.; dir. Central Carolina Bank & Trust Co., Duke Power Co.; Mem. N.C. Sci. and Tech., 1963-79; mem. tech. adv. bd. U.S. Dept. Commerce, 1964-69, N.C. Atomic Energy Adv. Com., 1964-71; mem. Korea-U.S. joint com. for sci. cooperation Nat. Acad. Scis., 1973-78; mem. bd. sci. and tech. for internat. devel. Nat. Acad. Sci., 1978—. Bd. dirs. Oak Ridge Assoc. Univs., 1971-74, 78-82. Mem. Sigma Alpha Epsilon. Clubs: Cosmos (Washington); Hope Valley Country. Home: 46 Beverly Dr Durham NC 27707 Office: Box 12194 Research Triangle Park NC 27709

HERBICH, JOHN BRONISLAW, engineering educator; b. Warsaw, Poland, Sept. 1, 1922; came to U.S., 1953, naturalized, 1962; s. Henry Pawel and Jadwiga Eleonora (Lopienski) H.; m. Margaret Pauline Boylan, Jan. 27, 1951; children: Ann (dec.), Barbara K., Gregory J., Patricia J. B.Sc., U. Edinburgh, Scotland, 1949; M.S. in C.E, U. Minn., 1957; Ph.D., Pa. State U., 1973; postgrad., U. Calif., Berkeley, 1964, Utah State U., 1966. Registered profl. engr., Tex. Field engr. John Laing & Son, London, Eng., 1948; research engr. U. Delft, Netherlands, 1949-50; research fellow, intermediate engr. Aluminum Co. Can., Ltd., 1950-53; research fellow U. Minn., 1953-57; asst. prof. Lehigh U., 1957-60, asso. prof., 1960-65, prof., 1965-67; prof. civil engring., head ocean and hydraulic engring. group, head ocean engring. program, dir. Center for Dredging Studies, Tex. A. and M. U. College Station, 1967—; on leave as UN project mgr. Central Water and Power Research Sta., Govt. of India, Khadakwasla, Poona, 1972-73; lectr. in, Venezuela, India, China, other countries; dir. Ocean Pollution Control, Inc., Dallas; v.p. Cons. and Research Services, Inc., Bryan, Tex. Author: Coastal and Deep Ocean Dredging, 1975, Offshore Pipelines: Design Elements, 1981, Scour Around Offshore Structures, 1983; also numerous articles and reports; contbr.: chpts. to Studies in Marine Environmental Pollution, 1980. Pres. PTA Hamilton Sch., Bethlehem, Pa., 1965-66. Served with Brit. Army, 1940-45. Recipient Karl Emil Hilgard Hydraulic Prize Am. Soc. C.E., 1965-66; NSF Faculty-Sci. fellow, 1963-64. Mem. Internat. Assn. Hydraulic Research, World Dredging Assn., Am. Soc. Engring. Edn., ASCE, Marine Tech. Soc., Permanent Internat. Assn. Nav. Congresses, Sigma Xi, Phi Kappa Phi, Chi Epsilon. Subspecialties: Coastal Engineering; Ocean Engineering. Current work: Coastal engineering and dredging technology, ship channel design, interaction between ships and channels, wave forces on structures. Patentee in field. Home: 764 S Rosemary Dr Bryan TX 77802 Office: Ocean Engring Program Tex A and M U College Station TX 77843

HERBST, ERIC, physicist, educator; b. Bklyn., Jan. 15, 1946; s. Stuart Karl and Dorothy (Polakoff) H.; m. Judith Strassman, Oct. 15, 1972; children: Elisabeth, Andrea. M.A., Harvard U., 1969, Ph.D., 1972. Asst. prof. chemistry Coll. William and Mary, Williamsburg, Va., 1974-79; assoc. prof., 1979-80, Duke U., 1980—; cons. NASA. Contbr. articles to profl. jours. NSF fellow, 1966-71; Woodrow Wilson fellow, 1966; Alfred P. Sloan fellow, 1976-80; grantee in field. Mem. Am. Phys. Soc., Am. Astron. Soc., Am. Chem. Soc., Internat. Astron. Union. Subspecialties: Radio and microwave astronomy; Atomic and molecular physics. Current work: Radio and microwave astronomy; chem. and phys. processes in interstellar clouds. Home: 5022 Raintree Rd Durham NC 27712 Office: Dept Physics Duke U Durham NC 27706

HERGET, WILLIAM FREDERICK, infrared spectroscopist, consultant; b. Wheeling, W.Va., Sept. 29, 1931; s. Frederick Welton and Katherine Ernestine (Scharf) H; children: Frederick J., Catherine E., Allen N., Nancy L. B.S., U. Richmond, 1952; M.S., Vanderbilt U., 1955; Ph.D. in Physics, U. Tenn., 1962. Research specialist Rocketdyne div. Rockwell Internat., Canoga Park, Calif., 1962-70; chief spl. techniques group EPA, Research Triangle Park, N.C., 1970-81; mgr. spl. projects group Nicolet Instrument Corp., Madison, Wis., 1981—; cons. on infrared spectroscopy. Contbr. articles to profl. jours. Recipient Bronze medal for meritorious service EPA, 1978. Mem. Optical Soc. Am., Air Pollution Control Assn., Soc. Automotive Engrs., AAAS. Subspecialties: Infrared spectroscopy; Remote sensing (atmospheric science). Current work: Analysis of gaseous chemical systems using infrared spectroscopy; research on the applicatiopn of Fourier transform infrared spectroscopy to the species quantification of a variety of chemical systems. Home: 5146 Anton Dr Apt 208 Madison WI 53719 Office: 5225-1 Verona Rd Madison WI 53711

HERLYN, MEENHARD FOLKEUS, immunologist, researcher; b. Upleward, Lower Saxony, Germany, Aug. 8, 1944; came to U.S., 1976; s. Meenhard and Gretchen (Ockinga) H.; m. Dorothee Schmidt-Ruppin, Apr. 25, 1970; 1 dau., Anjye. D.V.M., U. Munich, 1972. Research asst. Inst. Med. Microbiology, Munich, Germany, 1971-76; assoc. scientist Wistar Inst., Phila., 1976-79, research assoc., 1979-81, asst. prof., 1981—; cons. M.D. Anderson Cancer Inst., Houston, 1982—. Contbr. articles in field to profl. jours. Mem. Am. Assn. Cancer Research, Am. Assn. Pathologists, Am. Assn. Microbiology, Tissue Culture Assn., Soc. Immunology. Subspecialties: Cancer research (medicine); Cell biology. Current work: Cancer research, monoclonal antibodies, cancer diagnosis, tumor markers, tumor progression. Home: 1223 Knox Rd Wynnewood PA 19096 Office: The Wistar Inst 36th at Spruce St Philadelphia PA 19104

HERMAN, MARC STEVEN, child psychologist; b. N.Y.C., Aug. 28, 1951; s. Steven and Elaine Jean (Meltzer) H. B.A., SUNY-Buffalo, 1974; M.A., Wayne State U., 1978, Ph.D., 1981. Tchr. West Seneca (N.Y.) State Sch., 1971-74; supervisory psychologist Geriatric Screening, Detroit, 1977-80; child psychologist New Center Community Mental Health Center, Detroit, 1980-81; staff psychologist dept. child psychiatry Mount Carmel/Mercy Hosp., Detroit, 1982—; pvt. practice psychology Westside Mental Health, Dearborn Heights, Mich., 1982—. Author, presenter: Social Skills of

Retarded Children, 1979, Impulsivity of Preschoolers, 1982. Wayne State U. fellow, 1975-76. Mem. Am. Psychol. Assn., Mich. Psychol. Assn. Clubs: Civic Chorus (Livonia, Mich.); Audubon Soc. (Oakland, Mich.). Subspecialty: Developmental psychology. Current work: Parent-child interactions, effects on social and linguistic development, exceptional children, developmental disabilities in young children. Office: Dept Child Psychiatry Mount Carmel Hosp 6071 W Outer Dr Detroit MI 48235

HERMAN, ROBERT, physicist; b. N.Y.C., Aug. 29, 1914; m., 1939; 3 children. B.S., CCNY, 1935; M.S. and Ph.D. in Physic, Princeton U., 1940. Fellow physics CCNY, 1935-36; research assoc. differential analyzer Moore Sch. Elec. Engring., U. Pa., 1940-41, instr. physics CCNY, 1941-42; supr. chem. physics group, physicist, asst. to dir. applied physics lab. Johns Hopkins U., Balt., 1942-55; vis. prof. U. Md., 1955-56; head theoretical physics dept., traffic sci. dept. research labs. Gen. Motors Corp., Detroit, 1956—; Assoc. editor Rev. Modern Physics, 1953-55, Ops. Research Soc. Am., 1960-75; editor Transp. Sci., 1966-73. Recipient Lanchester prize Johns Hopkins U., 1959; Magellanic Premium Am. Philos. Soc., 1975; prix Georges Vanderlinden Belgian Acad., 1975; Regents lectr. U. Calif.-Santa Barbara, 1975; George E. Kimball medal Ops. Research Soc. Am., 1976. Mem. Nat. Acad. Engring., NRC, Sigma Xi; fellow Am. Phys. Soc. Subspecialties: Theoretical physics; Operations research (mathematics).

HERMAN, WILLIAM SPARKES, zoology educator; b. Seattle, Oct. 12, 1931; s. William Sparkes and Jane Ione (Ardery) Weidel; m. Charlotte Katherine Meyer, Oct. 5, 1962; children: Alexandria, Max, Carter. B.S., Portland State Coll., 1958; M.S., Northwestern U., 1960, Ph.D., 1964. NIH postdoctoral fellow U. Calif., Berkeley, 1964-66; asst. prof. U. Minn., Mpls., 1966-70, asso. prof., 1970-75, prof. zoology, 1976—, head genetics and cell biology dept., 1981—; cons. in field. Contbr. articles to profl. jours. Served with USAF, 1954-58. Mem. AAAS, Am. Soc. Zoologists. Subspecialties: Neuroendocrinology; Cell biology. Current work: Anthropod neuroendocrinology. Home: 79 Clarence Ave SE Minneapolis MN 55414 Office: Genetics and Cell Biology Dept Univ of Minn Saint Paul MN 55455

HERNANDEZ, LINDA LOUISE, neuroscience researcher, psychology educator; b. Wakefield, Mass., May 21, 1952; d. Robert Hathaway and Helen Barbara (Barker) Clarke; m. Miguel Hernandez, Jan. 31, 1975. Student, N.Y. U., 1970-72; B.S., U. Fla., 1973; M.S., U. S.C., 1978, Ph.D., 1981. Chemist State of Fla., Gainesville, 1974, Campbell Soup Co., Sumter, S.C., 1974; instr. U. S.C. and Midlands Tech. Coll., 1975-81; biol. lab. technician Williams Jennings Bryan Dorn Vets. Hosp., Columbia, S.C., 1977-81, research psychologist, 1981—; adj. asst. prof. psychology U. S.C., 1982—; research assoc. behavioral pharmacology lab. pharmacology lab, 1981-82. Reviewer: Psychopharmacology Jour., 1980—, Developmental Pharmacology and Therapeutics Jour., 1982; contbr. articles to profl. jours. Neidich fellow, 1980-81; Distilled Spirits Council grantee, 1982-83; Sigma Xi grantee, 1982-83; VA grantee, 1983—. Mem. N.Y. Acad. Scis., Nature Conservancy, Assn. Women in Sci., AAAS, Soc. Neurosci., Am. Psychol. Assn., Am. Chem. Soc., Sigma Xi. Subspecialties: Physiological psychology; Neuropharmacology. Current work: Brain neurotransmitters; neuromodulators and behavior; mechanisms of action of abused drugs; pain and stress; autonomic and somatomotor learning mechanisms. Home: 2816 Wesley Dr Columbia SC 29204 Office: Neuroscience Research 151 William Jennings Bryan Dorn Veterans Hosp Columbia SC 29201

HERNANDEZ, SAMUEL P., chemistry educator; b. Aguadilla, P.R., July 10, 1951; s. Samuel A. and Maria J. (Rivera) H.; m. Maria del Carmen Reinat, Aug. 31, 1974; children: Samuel E., Ricardo J. B.S., U. P.R., 1981, M.A., 1982; Ph.D., Johns Hopkins U., 1983. Asst. prof. chemistry U. P.R. Contbr. articles to profl. jours. Ford Found. fellow, 1977-81; Nat. Hispanic scholar, 1981-82. Mem. Am. Chem. Soc., Am. Inst. Chemists, Am. Phys. Soc., AAAS, N.Y. Acad. Sci. Roman Catholic. Subspecialties: Physical chemistry; Satellite studies. Current work: Electron impact and laser spectroscopy of supersonic beams. Office: Dept Chemistry U PR Mayaguez PR 00708

HERNDON, JAMES HENRY, orthopedic educator; b. Los Angeles, Oct. 31, 1938; s. James Greene and Kathleen Theresa (Murphy) H.; m. Geraldine Grace Armiger, Feb. 26, 1971; children: Jennifer, Jonathan. B.S., Loyola U., Los Angeles, 1961; M.D., UCLA, 1965; M.A. ad eundum, Brown U., 1979. Diplomate: Am. Bd. Orthopedic surgery. From asst. clin. prof. to assoc. clin. prof. surgery Mich. State U., Grand Rapids, 1974-78; prof., chmn. dept. orthopedics Brown U., Providence, 1979—; orthopedic surgeon in chief R.I. Hosp., Providence, 1979—; cons. Crippled Children's Service of R.I., 1980—. Co-author: Scoliosis and other Deformities of the Axial Skeleton, 1975; editor: New Developments in Orthopedic Surgical Clinics of North America, 1983. Bd. govs. Arthritis Found. R.I., Providence, 1980; mem. joint adv. com. Palestine Temple and R.I. Hosp., Providence, 1978. Served to maj. U.S. Army, 1971-73. Recipient Edith and Carl Lasky Meml. award UCLA Med. Sch., 1965; Bronze award Am. Congress Rehab. Medicine, 1972; 1st award for clin. research N.Y. Med. Soc., 1974; OAS traveling fellow, 1978. Fellow Am. Orthopedic Assn., Am. Acad. Orthopedic Surgeons, A.C.S., Scoliosis Research Soc.; mem. Orthopedic Research Soc. Republican. Roman Catholic. Clubs: Agawam Hunt, Hope (Providence). Subspecialty: Orthopedics. Current work: Fat Embolism syndrome; artificial joints; implants in the hand; child amputee; orthopedic education. Home: 86 Taber Ave Providence RI 02906 Office: Dept Orthopedic Surgery RI Hosp 233 Eddy St Providence RI 02902

HERR, EARL BINKLEY, JR., pharm. research co. exec.; b. Lancaster, Pa., Apr. 14, 1928; s. Earl B. and Irene (Zeamer) H.; m. Elizabeth Sydney Hook, June 17, 1950; children—Audrey, Linda. B.S., Franklin and Marshall Coll., 1948; M.S. in Chemistry, U. Del., 1950; Ph.D. in Biochemistry, U. Del., 1953; postgrad., Cornell U., 1953-55, Brookhaven Nat. Labs., 1955-57. With Lilly Research Labs., Indpls., 1957—, mgr. antibiotic purification devel., 1963-64, head pharm. research, 1964-65, asst. dir. prodn. devel., 1965, dir. antibiotic ops., 1965-68, exec. dir. biochem. and biol. ops., 1968-69, v.p. biochem. ops., 1969-70, v.p. indsl. relations, 1970, v.p. research, devel. and control, 1970-73, pres., 1973—; dir. Eli Lilly and Co. Bd. dirs. Ind. Sci. Edn. Fund. Mem. AAAS, Am. Chem. Soc., Sigma Xi. Subspecialty: Biochemistry (biology). Home: 12011 Eden Glen Dr Carmel IN 46032 Office: 307 E McCarty St Indianapolis IN 46206

HERR, JOHN MERVIN, JR., biology educator, cons., researcher; b. Charlottesville, Va., July 26, 1930; s. John Mervin and Belva (Byrd) H.; m. Sue Highfield, Aug. 30, 1952; children: Susan Rebecca Herr Reich, Rachel Lynn; m. Lucrecia Linder, Dec. 30, 1974; 1 stepson, F. Brent Wahl. B.A., U. Va., 1951; M.A., 1952; Ph.D. (Coker Fellow), U. N.C., 1957. Instr. Washington and Lee U., Lexington, Va., 1952-54; Fulbright postdoctoral fellow U. Delhi, India, 1957-1958; asst. prof. Pfeiffer Coll., Misenheimer, N.C., 1958-59, U. S.C., Columbia, 1959-63, assoc., prof., 1963-69, prof., 1969—. Contbr. articles to profl. jours. Mem. Bot. Soc. Am., Internat. Soc. Plant Morphologists, Assn. Southeastern Biologists, Sigma Xi. Baptist. Subspecialties: Reproductive biology; Systematics. Current work: Development of the ovule and female gametophyte in flowering plants for use in solution of taxonomic problems, development of clearing techniques to replace paraffin section techniques primarily for the above mentioned studies.

HERR, LEONARD JAY, plant pathologist; b. Orrville, Ohio, Dec. 21, 1928; s. Roy Albert and Orpha (Shoup) H.; m. Lucille Alice Adelsberger, Sept. 18, 1954; children: Lynn Allen, Karen Marie, Melissa Ann. Student, Antioch Coll., 1946-49; B.S., Ohio State U., 1952, M.S., 1953, Ph.D., 1956. Instr. Ohio Agrl. Expt. Sta., Ohio State U.-Ohio Agrl. Research and Devel. Center, Wooster, 1956-57, asst. prof., 1957-63, asso. prof., 1963-77, prof. plant pathology, 1977—. Contbr. articles to profl. jours. Mem. Am. Phytopath. Soc., AAAS, Am. Inst. Biol. Sci., Ohio Acad. Sci., N.Y. Acad. Sci., AAUP, Assn. Applied Biologists, Sigma Xi. Roman Catholic. Subspecialty: Plant pathology. Current work: Soil-borne pathogens; sunflower diseases; ecology of soil-borne pathogens. Office: Dept Plant Pathology Ohio Agrl Research and Devel Center Wooster OH 44691

HERR, RICHARD BAESSLER, astronomer, educator; b. Phila., Mar. 3, 1936; s. Daniel Irwin and Edna Elizabeth (Baessler) H.; m. Mary Dilling, Sept. 6, 1958; 1 son, Daniel Dilling. B.S. in Physics, Franklin and Marshall Coll., 1957; M.S., U. Del., 1960; Ph.D. in Astronomy, Case Inst. Tech., 1965. Asst. prof. physics and astronomy U. Del., Newark, 1964-70, assoc. prof., 1970—; research astronomer Mt. Cuba Astron. Obs.; cons. in computer assisted instrn. Mem. Am. Astron. Soc., Internat. Astron. Union, Astron. Soc. Pacific, Royal Astron. Soc. Can. Subspecialty: Optical astronomy. Current work: Photoelectric photometry, computer assisted education. Home: 913 Pickett Ln Newark DE 19711 Office: Dept Physics U Del Newark DE 19711

HERRANS, LAURA LETICIA, psychology educator; b. Vega Baga, P.R., June 16, 1935; d. Juan B. and Maria (Perez) H. B.A., U. P.R., 1955; M.A., Cath. U. Am., 1957, Ph.D., 1969. Psychologist Mental Health Clinic, Rio Piedras, 1957-60; asst. prof. psychology U. P.R., Rio Piedras, 1960-65, assoc. prof., 1966-77, prof., 1977—, chmn. dept. psychology, 1981—; cons. P.R. Dept. Edn., 1977—; prin. investigator Mental Health Secretariat P.R., 1981-82. Author: Psicologia y Medicion: El, 1980. Mem. Am. Psychol. Assn. (chmn. equal opportunities in psychology com. 1977-79), P.R. Psychologists Assn. (pres. 1971). Subspecialty: Educational psychology. Current work: Research on test standardization.

HERRING, H(UGH) JAMES, research scientist; b. Boston, Aug. 3, 1939; s. Pendleton and Katherine Sedgwick (Channing) H.; m. Carol Rose Parter, Aug. 28, 1960; children: James Pendleton, Matthew Eugene, Katherine Ruth. B.A., Harvard U., 1961, B.S., 1962; M.S., Princeton U., 1966, Ph.D., 1967. Mem. research staff dept. aerospace and mech. scis. Princeton U., 1967-76; pres. Dynalysis of Princeton, 1970—. Assoc. editor Jour. Fluids Engring, 1978-81. Mem. ASME (sec. fluid machinery com. 1976-79), AIAA (treas. 1977-79), Central Jersey Engring. Council (treas. 1977-79), AAAS, Sigma Xi. Subspecialties: Fluid mechanics; Oceanography. Home: 350 Riverside Dr Princeton NJ 08540 Office: Dynalysis of Princeton 20 Nassau St Princeton NJ 08540

HERRING, WILLIAM CONYERS, physicist; b. Scotia, N.Y., Nov. 15, 1914; s. William Conyers and Mary (Joy) H.; m. Louise C. Preusch, Nov. 30, 1946; children—Lois Mary, Alan John, Brian Charles, Gordon Robert. A.B., U. Kans., 1933; Ph.D., Princeton, 1937. NRC fellow Mass. Inst. Tech., 1937-39; instr. Princeton, 1939-40, U. Mo., 1940-41; mem. sci. staff Div. War Research, Columbia, 1941-45; prof. applied math. U. Tex., 1946; research physicist Bell Telephone Labs., Murray Hill, N.J., 1946-78; prof. applied physics Stanford (Calif.) U., 1978—; mem. Inst. Advanced Study, 1952-53. Recipient Army-Navy Cert. of Appreciation, 1947; Distinguished Service citation U. Kans., 1973; J. Murray Luck award for excellence in sci. reviewing Nat. Acad. Scis., 1980; von Hippel award Materials Research Soc., 1980. Fellow Am. Phys. Soc. (Oliver E. Buckley solid state physics prize 1959), Am. Acad. Arts and Scis.; mem. AAAS, Nat. Acad. Scis. Subspecialties: Condensed matter physics; Theoretical physics. Current work: Theory of electronic and atomic structures of solids, and their magnetic and transport properties. Home: 3945 Nelson Dr Palo Alto CA 94306 Office: Dept Applied Physics Stanford U Stanford CA 94305

HERRMANN, DOUGLAS J., psychology, educator; b. Wilmington, Del., Aug. 1, 1941; s. Carl Victor and Ruth Naomi (Ice) H.; m. Donna Lynn Shellenberger, Mar. 21, 1969; 1 dau., Amanda. B.S., U.S. Naval Acad., 1964; M.S., U. Del., 1972, Ph.D., 1974. Research assoc. Stanford U., 1972-73; assoc. prof. Hamilton Coll., Clinton, N.Y., 1973—; research scholar applied psychology unit Med. Research Council, Eng., 1972-73; cons. in field. Author: Inventory of Memory Experience, 1978; Contbr. articles to profl. jours. Head Refugee Sponsorship Group, Utica, N.Y., 1975—. Served to capt. USMC, 1964-67. Recipient Excellence in teaching award U. Del., 1972; Social Sci. Research Council fellow, 1972. Mem. Psychonomic Soc., Sigma Xi. Unitarian. Subspecialty: Cognition. Current work: Research problems in psychology of language, memory and thought. Office: Dept Psychology College Hill Rd Clinton NY 13323 Home: RD 1 Harding Rd Apt 519 Clinton NY 13323

HERRMANN, GEORGE, educator; b. USSR, Apr. 19, 1921. Dipl. C.E., Swiss Fed. Inst. Tech., 1945, Ph.D. in Mechanics, 1949. Asst., then asso. prof. civil engring. Columbia, 1950-62; prof. civil engring. Northwestern U., 1962-69; prof. applied mechanics Stanford, 1969—; cons. SRI Internat., 1970-80. Contbr. 200 articles to profl. jours; editorial bd. numerous jours. Fellow ASME (Centennial medal 1980); mem. ASCE (Th. v. Karman medal 1981), Nat. Acad. Engring., AIAA. Subspecialty: Theoretical and applied mechanics. Address: Div Applied Mechanics Durand Bldg Stanford U Stanford CA 94305

HERRMANN, RAYMOND, geologist; b. Chgo., July 16, 1941; s. Raymond B. and Josephine (Rickman) H.; m. Emilie L. Juestrich, July 17, 1965; children: Stefanie, Michelle. B.S., Columbia U., 1968; M.S., U. Wyo., 1972, Ph.D., 1972. Registered profl. geologist, Ga. Research asst. U. Wyo., Laramie, 1968-72; groundwater geologist State of Oreg., Salem, 1972; regional hydrologist S.E. region NPS, Atlanta, 1973-74, regional chief scientist, 1974-79, chief air water resources, Washington, 1979-81; dir. Water Resources Field Support Lab., Ft. Collins, Colo., 1981—; chmn. task group Interagy. Task Force on Acid Precipitation, 1981—; adj. assoc. prof. U. Tenn., Knoxville, 1980-82, faculty assoc., 1982—; assoc. faculty Colo. State U., 1981—. Contbr. articles to profl. jours. Coach, referee Youth Sports, Atlanta and Springfield, Va., 1977, 79-81, Ft. Collins, 1982—. Served with USMC, 1958-62. Fellow Geol. Soc. Am.; mem. Am. Water Resources Assn. (dir.), AAAS, Sigma Xi. Subspecialties: Geology; Hydrogeology. Current work: Remote natural area monitoring program development; directs interdisciplinary environmental and water research, acid precipitation, others. Office: Nat Park Service Water Resources Field Support Lab Colo State Univ 107C Natural Resources Fort Collins CO 80523

HERRMANN, ROBERT ARTHUR, mathematics educator; b. Balt., Apr. 29, 1934; s. Ernest Carl and Catherine (Brostrum) H.; m. Sandi A. Baldi, Feb. 1, 1969; children: Kimberley, Laura, Diana. B.A., Johns Hopkins U., 1963; M.A., Am. U., 1968, Ph.D., 1973. Prof. math. U.S. Naval Acad., Annapolis, Md., 1968—; dir. Inst. for Math. Philosophy, Annapolis, 1980—; adviser U.S. Congress, Washington, 1980—. Author: Nonstandard Analysis, 1977, the G-Model, 1980, The Miraculous Model, 1982; contbr. articles to profl. jours. Served with U.S. Army, 1955-57. Mem. Am. Math. Soc., Math. Assn. Am., Am. Sci. Affiliation, Phi Beta Kappa, Sigma Xi, Phi Kappa Phi. Subspecialties: Applied mathematics. Current work: Mathematical philosophy, nonstandard logic and modelling of natural systems, convergence space theory. Office: Dept Math US Naval Acad Annapolis MD 21402

HERRMANN, ROBERT LAWRENCE, science educator and adminstr.; b. N.Y.C., July 17, 1928; s. Philip Charles and Florence Gertrude (Benn) H.; m. Elizabeth Ann Cook, Aug. 12, 1950; children: Stephen, Karen, Holly, Anders. B.S. in Chemistry, Purdue U., 1951; Ph.D. in Biochemistry, Mich. State U., 1956. Postdoctoral fellow MIT, 1956-59; from asst. prof. to assoc. prof. biochemistry Boston U. Sch. Medicine, 1959-76; prof., chmn. dept. biochemistry Oral Roberts U. Sch. Medicine and Dentistry, Tulsa, 1976-81, assoc. dean biomed. sci., 1978-79; lectr. chemistry Goldon Coll., Wenham, Mass, 1981, adj. prof., 1982; exec. dir. Am. Sci. Affiliation, Ipswich, Mass., 1981—. Contbr. chpts. to books, articles to profl. jours. Trustee Christian Med. Soc., 1976-79, Barrington Coll., 1975-78; mem. Bd. Health, Bedford, Mass., 1975-76. Served with USN, 1946-48, 51-52. Fellow AAAS, Gerontol. Soc.; mem. Am. Soc. Biol. Chemists, Soc. for Health and Human Values, Victoria Inst., Inst. for Soc. Ethics and Life Scis. Evang. Christian. Subspecialties: Biochemistry (biology). Current work: Bioethics. Home: 12 Spillers Ln Ipswich MA 01938 Office: Box J Ipswich MA 01938 Dept Chemistry Gordon Coll Wenham MA 01984

HERRNKIND, WILLIAM FRANK, biology educator; b. Bayshore, N.Y., Oct. 15, 1940; s. Henry W. and Doris (Lambie) H.; m. Lynne G., Aug. 11, 1962; 1 dau., LeAnn. B.S., SUNY-Albany, 1961; M.S., U. Miami, Fla., 1965, Ph.D., 1968. Instr. to assoc. prof. dept. biol. scis. Fla. State U., Tallahassee, to 1967, now prof. biol. scis., dir. marine lab. Contbr. chpts. to books, articles to profl. jours. NSF grantee, 1971-81; NOAA grantee, 1970-71, 74-75, 83—. Mem. Am. Soc. Zoologists, Animal Behavior Soc., Am. Inst. Fishery Research Biologists, Crustacean Soc. Subspecialties: Ethology; Behavioral ecology. Current work: Study of behavioral and ecological adaptations of marine animals; teaching; research; graduate training program development; administration of marine research facilities. Office: Dept Biol Sci Fla State Univ Tallahassee FL 32306

HERSCHBACH, DUDLEY ROBERT, educator, chemist; b. San Jose, Calif., June 18, 1932; s. Robert Dudley and Dorothy Edith (Beer) H.; m. Georgene Botyos, Dec. 26, 1964; children—Lisa Marie and Brenda Michele. B.S. in Math, Stanford U., 1954, M.S. in Chemistry, 1955; A.M. in Physics, Harvard U., 1956; Ph.D. in Chem. Physics, Harvard U., 1958; D.Sc. (h.c.), U. Toronto, 1977. Jr. fellow Harvard U., 1957-59; faculty U. Calif.-, Berkeley, 1959-63; prof. chemistry Harvard U., 1963-76, Frank B. Baird prof. sci., 1976—, chmn. chemistry dept., 1977-80, mem. faculty council, 1980-83, master Currier House, 1981—; cons. editor W.H. Freeman; lectr. Haverford Coll., 1962; Falk-Plaut lectr. Columbia, 1963; vis. prof. Göttingen (Germany) U., summer 1963, U. Calif. at Santa Cruz, 1972; Harvard lectr. Yale, 1964; Debye lectr. Cornell U., 1966; Rollefson lectr. U. Calif. at Berkeley, 1969; Guggenheim fellow U. Freiburg, Germany, 1968; vis. fellow Joint Inst. for Lab. Astrophysics U. Colo., 1969; Reilly lectr. U. Notre Dame, 1969; Phillips lectr. U. Pitts., 1971; Distinqguished vis. prof. U. Ariz., 1971, U. Tex., 1977, U. Utah, 1978; Gordon lectr. U. Toronto, 1971; Clark lectr. San Jose State U., 1979; Fairchild Distinguished scholar Calif. Inst. Tech., 1976; Sloan fellow., 1959-63, Exxon Faculty fellow, 1980—. Asso. editor: Jour. Phys. Chemistry. Recipient pure chemistry award Am. Chem. Soc., 1965, Centenary medal, 1977, Pauling medal, 1978; Spiers medal Faraday Soc., 1976; Polanyi medal, 1981; Langmuir prize, 1983. Fellow Am. Phys. Soc. (chmn. chem. physics div. 1971-72), Am. Acad. Arts and Scis; mem. Am. Chem. Soc., AAAS, Nat. Acad. Scis., Phi Beta Kappa, Sigma Xi. Subspecialty: Physical chemistry. Office: 12 Oxford St Cambridge MA 02138

HERSCOWITZ, HERBERT BERNARD, immunologist, educator; b. N.Y.C., June 19, 1939; s. Michael and Sarah (Sussman) H.; m. Ellen Carol Levine, Aug. 26, 1961; children—Robert, Stefanie, Andrew. B.S., Bklyn, Coll., 1961; M.S. cum laude, L.I. U., 1963; Ph.D., Hahnemann Med. Coll., 1968. Asst. instr. U. Pa. Sch. Dental Medicine, Phila., 1963-65; predoctoral fellow Hahnemann Med. Coll., Phila., 1965-68; postdoctoral fellow Case-Western Res. U., Cleve., 1968-70; asst. prof. microbiology Georgetown U. Schs. Medicine and Dentistry, Washington, 1970-76, asso. prof., 1976-81, prof., 1981—. Contbr. articles to profl. jours. NIH fellow, 1967-68, 68-70; named Alumnus of Year. Hahnemann Med. Coll., 1979. Mem. Am. Assn. Immunologists, Am. Soc. Microbiology, AAAS, Reticuloendothelial Soc., Sigma Xi. Subspecialties: Immunology (medicine); Immunobiology and immunology. Current work: Cellular interactions, alveolar macrophage function. Office: 3900 Reservoir Rd Washington DC 20007

HERSHBERGER, CHARLES LEE, research scientist; b. Louisville, Ill., May 1, 1942; s. Merrill George and Betty Hormel (Walsh) H.; m. Maria K., June 30, 1962; children: Mardi, C. Douglas, Jamie. B.S. in Chemistry, Eureka Coll., 1964; Ph.D. in Biochemistry, U. Ill. Coll. Medicine, 1968. Research assoc. Molecular Biology Inst. U. Wis., Madison, 1967-70; faculty dept. microbiology U. Ill., Urbana, 1970-76; sr. scientist Eli Lilly and Co., Indpls., 1976-80, research scientist, 1981—. Contbr. articles to profl. publs. Active Crossroads of Am. council Boy Scouts Am., 1967-70, 76—, scoutmaster, 1979—; mem. Social Ministry Commn. Ind.-Ky. Synod Lutheran Ch. Am., 1981—. Mem. Am. Soc. Biol. Chemists, Am. Soc. Microbiologists., AAAS, Sigma Xi. Subspecialties: Genetics and genetic engineering (biology); Biochemistry (biology). Current work: Expression of mammalian proteins in bacteria, cloning genes for biosynthesis of antibiotics. Office: Eli Lilly Co 307 E McCarty St Indianapolis IN 46285

HERTEL, RICHARD JAMES, software designer consultant; b. Stillwater, Minn., Nov. 28, 1942; s. Roland Kermit and Dorothy Margaret (Mueller) H.; m. Mary A. Brinksneader, Nov. 18, 1976. B.A., Lewis and Clark Coll., 1965; M.S., Cornell U., 1969. Engr. electro-optical products div. ITT, Ft. Wayne, Ind., 1968-73, sr. project engr. aerospace/optical div., 1973-82; v.p. engring., treas. Software Cons. Specialists, Ft. Wayne, 1982—, dir., 1981—. Contbr. papers to profl. publs. Mem. AIAA, IEEE, Soc. Photo-Optical Instrumentation Engrs. Methodist. Subspecialties: Software engineering; Graphics, image processing, and pattern recognition. Current work: Robotic vision systems; creation of a means to measure size and shape of objects in 3 space, in real time for purposes of robot navigation or control. Inventor, patentee electro-optics. Office: Software Consulting Specialists Inc PO Box 15367 Fort Wayne IN 46885

HERTING, DAVID CLAIR, nutritional biochemist, researcher; b. Pottstown, Pa., Sept. 29, 1928; s. George Claire and Ruth Mowry (Davidheiser) H.; m. Martha Elizabeth Schaeffer, Aug. 30, 1952; children: David Allen, Kenneth Edward, Philip Martin, Carl Andrew. B.S., Pa. State U., 1950; M.S., U. Wis.-Madison, 1952, Ph.D., 1954. Cert. prof. chemist Am. Inst. Chemists. Research biochemist Distillation Products Industries, Rochester, N.Y., 1954-58, sr. research biochemist, 1959-64; research assoc. Eastman Kodak Co., Rochester, 1965-75, sr. research assoc., 1976—. Contbr. writings to ency. and profl. jours. Adult leader Otetiana council Boy Scouts Am., Rochester,

1956—. Basic Research awardee Glycerine Producers Assn., 1960. Fellow Am. Inst. Chemists, AAAS; mem. Am. Inst. Nutrition, Am. Chem. Soc., N.Y. Acad. Scis., Sigma Xi, Alpha Chi Sigma. Republican. Lutheran. Subspecialties: Animal nutrition; Biochemistry (biology). Current work: Nutrition, metabolism and analysis of fats, fat-soluble vitamins, sterols, carotenoids; physiological disposition of coating agents, antifungal agents, nutrients useful in animal science and agriculture. Patentee. Home: 2778 Nichols St Spencerport NY 14559 Office: Eastman Kodak Co PO Box 1911 Rochester NY 14603

HERTZ, RICHARD CORNELL, rabbi; b. St. Paul, Oct. 7, 1916; s. Abram J. and Nadine (Rosenberg) H.; m. Mary Louise Mann, Nov. 25, 1943 (div. July 1971); children: Nadine (Mrs. Michael Wertheimer), Ruth Mann (Mrs. Alain Joyaux); m. Renda Gottfürcht Ebner, Dec. 3, 1972. A.B., U. Cin., 1938; M.H.L., Hebrew Union Coll., 1942, D.D. (hon.), 1967; Ph.D., Northwestern U., 1948. Ordained rabbi, 1942; asst. rabbi N. Shore Congregation Israel, Glencoe, Ill., 1942-47; asso. rabbi Chgo. Sinai Congregation, 1947-53; sr. rabbi Temple Beth El, Detroit, 1953-82, rabbi emeritus, 1982—; adj. prof. Jewish thought U. Detroit, 1970—, disting. prof. Jewish studies, 1980—; spl. cons. to pres. Cranbrook Ednl. Community, 1983—; del. to internat. conf. World Union for Progressive Judaism, London, 1959, 61, Amsterdam, 1978, bd. dirs. union, 1973—; Lectr. Jewish Chautauqua Soc.; former mem. plan bd. Synagogue Council Am.; mem. chaplaincy commn., former bd. dirs. Nat. Jewish Welfare Bd.; former mem. exec. com., vice chmn. Citizen's Com. for Equal Opportunity; mem. Mich. Gov.'s Com. on Ethics and Morals, 1963-69; mem. Mich. adv. council U.S. Commn. on Civil Rights, 1979—; mem. nat. bd. dirs. Religious Edn. Assn.; adv. bd. Joint Distbn. Com.; former mem. nat. rabbinical council United Jewish Appeal; mem. rabbinic cabinet Israel Bonds, 1972—; pres. Hyde Park and Kenwood Council Chs. and Synagogues, Chgo., 1952. Author: Rabbi Yesterday and Today, 1943, This I Believe, 1952, Education of the Jewish Child, 1953, Our Religion Above All, 1953, Inner Peace for You, 1954, Positive Judaism, 1955, Wings of the Morning, 1956, Impressions of Israel, 1956, Prescription for Heartache, 1958, Faith in Jewish Survival, 1961, The American Jew in Search of Himself, 1962, What Counts Most in Life, 1963, What Can A Man Believe, 1967, Reflections for the Modern Jew, 1974, Israel and the Palestinians, 1974, Roots of My Faith, 1980, also articles in sci., popular publs. Dir. Am. Jewish Com., mem. nat. exec. bd., former hon. vice-chmn. Detroit chpt.; past dir. Mich. Soc. Mental Health, Jewish Family and Children's Services, United Community Services, Jewish Welfare Fedn. Detroit, Jewish Community Council Detroit; dir. United Found., Boys Clubs, Mich. region Anti-Defamation League; chmn. bd. overseers Hebrew Union Coll.-Jewish Inst. Religion, 1968-72; bd. govs. Detroit Inst. Tech., 1955-70. Served as chaplain AUS, 1943-46. Fellow Am. Sociol. Soc.; mem. Detroit Hist. Soc., Central Conf. Am. Rabbis (former nat. chmn. com. on Jews in Soviet orbit), Am. Jewish Hist. Soc., Am. Legion (dept. chaplain 1956-57), Jewish War Vets. (dept. chaplain 1958-59, 72—), Alumni Assn. Hebrew Union Coll.-Jewish Inst. Religion (past dir.). Clubs: Rotary, Economic (Detroit) (dir.); Wranglers (past pres.), Great Lakes, Standard, Franklin Hills, Knollwood, Tam O'Shanter. Subspecialties: Aerospace engineering and technology; Laser fusion. Current work: Aerodynamics, heat transfer, energy conversion, propulsion, fluid mechanical aspects of lasers, high powered lasers, fusion research, advanced energy conversion techiques, laser applications, space laser concepts, advanced energy conversion concepts. Went on spl. mission for White House to investigate status Jews and Judaism in USSR 1959, mission for chief chaplains Def. Dept. to conduct retreats for Jewish chaplains and laymen, Berchtesgaden, Germany, 1973; mem. mission to Arab countries and Israel, Nat. Council Chs.-Am. Jewish Com., 1974; 1st Am. rabbi received in pvt. audience at Papal Palace by Pope Paul VI, 1963. Home: 4324 Knightsbridge Ln West Bloomfield MI 48033 Office: Temple Beth El 7400 Telegraph Rd at 14 Mile Birmingham MI 48010

HERTZ, ROY, pharmacology educator, physician; b. Cleve., June 19, 1909; m., 1934; 2 children. A.B., U. Wis., 1930, Ph.D., 1933, M.D., 1939; M.P.H., Johns Hopkins U., 1941. Asst. zoology U. Wis., 1930-34; instr. pharmacology Med. Sch., Howard U., Washington, 1934-35; intern Wis. Gen. Hosp., 1939-40; res. officer USPHS, 1941-47; endocrinologist, chmn. sect. and mem. study sect. endocrinology and metabolism Nat. Cancer Inst., NIH, 1947-65; sci. dir. Nat. Inst. Child Health and Devel., NIH, 1965-66; prof. ob-gyn Sch. Medicine, George Washington U., 1966-67; dir. reprodn. br. Nat. Inst. Child Health and Devel., 1967-69; sr. physician, assoc. dir. biomed. div. Population Council Rockefeller U., N.Y.C., 1969-72; prof. ob-gyn and medicine N.Y. Med. Coll., 1972-73; prof. pharmacology and ob-gyn Med. Sch. George Washington U., 1974—; Mem. adv. council Am. Cancer Soc. Fellow ACP, Am. Assn. Obstetricians and Gynecologists (hon.); mem. Nat. Acad. Sci., Am. Physiol. Soc., Soc. Exptl. Biology and Medicine. Subspecialties: Physiology (medicine); Endocrinology. Office: 2300 Eye St NW Washington DC 20037

HERWALD, SEYMOUR W(ILLIS), electronics company executive; b. Cleve., Jan. 17, 1917; m., 1941; 3 children. B.S., Case Inst. Tech., 1938; M.S., U. Pitts., 1940, Ph.D. in Math, 1944. Engr. Westinghouse Electric Corp., Pitts., 1939-46, spl. products engr., 1946-47, mgr. dvel. sect., 1947-51, mgr. engring, dept., air armaments div., 1951-56, mgr. div., 1956-59, v.p. research, 1959-62, v.p. electronic components and splty. products group., 1962-68, v.p. engring., 1968-70, v.p. engring, and devel., 1970-77, v.p. services, 1977—; Mem. Air Force Sci. Adv. Bd., 1956-71; cons. NASA, 1960-64, NSF, 1974-76. Mem. Nat. Acad. Engring., ASME, IEEE (pres. 1968), AIAA, Sigma Xi. Subspecialty: Electronics. Address: 2282 Elmhill Rd Pittsburgh PA 15221

HERZ, FRITZ, biochemist, researcher, educator; b. Heilbronn, Württemberg, Germany, July 16, 1930; came to U.S., 1959; s. Ludwig and Ida Jella (Oppenheimer) H.; m. Vona Kern, May 8, 1966; children: Eric Nathan, Lisa Sonia. B.A., Colegio Vicente Rocaferte, Guayaquil, Ecuador, 1950; Chemist, Guayaquil U., 1954, Ph.D., 1955. Assoc. dept. head Inst. de Higiene, Guayaquil, 1955-57; Humboldt fellow Free U. Berlin, W.Ger., 1957-59; USPHS fellow Sinai Hosp., Balt., 1959-62, research assoc., 1962-67, assoc. pediatric research dir., 1967-73; head tissue culture div., dept. pathology Montefiore Med. Ctr., Bronx, N.Y., 1973—; assoc. prof. pathology Albert Einstein Coll. Medicine, Bronx, 1973—. Contbr. articles to sci. jours. and philatelic publs. Mem. Am. Soc. Biol. Chemists, Am. Soc. Cell Biology, Am. Chem. Soc., Tissue Culture Assn., Soc. Exptl. Biology and Medicine. Democrat. Jewish. Club: Harrison (N.Y.) Stamp. Subspecialties: Biochemistry (biology); Cancer research (medicine). Current work: Biochemistry of cultured cells, regulation of enzyme activity, oncodevelopmental gene expression, cell structure and function. Office: Montefiore Med Center 111 E210th St Bronx NY 10467 Home: 45 Dora Ln New Rochelle NY 10804

HERZ, NORMAN, geology educator; b. N.Y.C., Apr. 12, 1923; s. Julius Edel and Vivian Molly (Becker) H.; m. Rhoda Judith Salzmann, Dec. 24, 1950; children: David S., Jonathan A., Sara J. B.S., CCNY, 1943; postgrad., U. Chgo., 1946-47; Ph.D., Johns Hopkins U., 1950. Instr. geology dept. Wesleyan U., Middletown, Conn., 1950-51; research scholar Fulbright program, Athens, Greeece, 1951-52; geologist U.S. Geol. Survey, Boston, Washington, Brazil, 1952-70; prof., head geology dept. U. Ga., Athens, 1970-77, prof. geology, 1977—; vis. prof. U. Sao Paulo, Brazil, 1962-64, George Washington U., Washington, 1969. Served to 2d lt. USAF, 1943-46. Nat. Acad. Sci. exchange scientist Romania, Bulgaria, 1971, 73, 76; recipient Creative Research medal U. Ga. Research Found., 1981; grantee Am. Philos. Soc., 1977, Nat. Geog. Soc., 1981—. Fellow Geol. Soc. Am. (chmn. archaeol. geology div. 1981-82), Geol. Soc. Brazil, Am. Geophys. Union, Mineral. Soc. Am., Geochem. Soc.; mem. Ga. Geol. Soc. (councilor, pres. 1972-73, 1979-80), Soc. Econ. Geologists (chmn. ann. meeting 1980). Subspecialties: Geology; Petrology. Current work: Archaeometry, applications of petrology and stable isotope geochemistry to problems of archaeology, mineral deposits, titanium, petrology, anorthosites and related rocks. Home: 250 Terrell Dr Athens GA 30606 Office: Dept Geology U Ga Athens GA 30602

HERZBERG, GERHARD, physicist; b. Hamburg, Ger., Dec. 25, 1904; emigrated to Can., 1935, naturalized, 1945; s. Albin and Ella (Biber) H.; m. Luise H. Oettinger, Dec. 29, 1929 (dec.); children: Paul Albin, Agnes Margaret; m. Monika Tenthoff, Mar. 21, 1972. Dr. Ing., Darmstadt Inst. Tech., 1928; postgrad., U. Goettingen, U. Bristol, 1928-30; D.Sc. hon causa, Oxford U., 1960, U. Chgo., 1967, Drexel U., 1972, U. Montreal, 1972, U. Sherbrooke, 1972, McGill U., 1972, Cambridge U., 1972, U. Man., 1973, Andhra U., 1975, Osmania U., 1976, U. Delhi, 1976, U. Bristol, 1975, U. Western Ont., 1976; Fil. Hed. Dr., U. Stockholm, 1966; Ph.D., Weizmann Inst. Sci., 1976; LL.D., St. Francis Xavier U., 1972, Simon Fraser U., 1972, others. Lectr., chief asst. physics Darmstadt Inst. Tech., 1930-35; research prof. physics U. Sask., Saskatoon, 1935-45; prof. spectroscopy Yerkes Obs., U. Chgo., 1945-48; prin. research officer NRC Can., Ottawa, 1948, dir. div. pure physics, 1949-69, disting. research scientist, 1969—; Bakerian lectr. Royal Soc. London, 1960; holder Francqui chair U. Liege, 1960. Author books including: Spectra of Diatomic Molecules, 1950; Electronic Spectra and Electronic Structure of Polyatomic Molecules, 1966, The Spectra and Structures of Simple Free Radicals, 1971, (with K.P. Huber) Constants of Diatomic Molecules, 1979. Recipient Faraday medal Chem. Soc. London, 1970, Nobel prize in Chemistry, 1971; named companion Order of Can., 1968, academician Pontifical Acad. Scis., 1964. Fellow Royal Soc. London (Royal medal 1971), Royal Soc. Can. (pres. 1966, Henry Marshall Tory medal 1953), Hungarian Acad. Sci. (hon.), Indian Acad. Scis. (hon.), Am. Phys. Soc., Chem. Inst. Can.; mem. Internat. Union Pure and Applied Physics (past v.p.), Am. Acad. Arts and Scis. (hon. fgn. mem.), Am. Chem. Soc. (Willard Gibbs medal 1969, Centennial fgn. fellow 1976), Nat. Acad. Sci. India, Indian Phys. Soc. (hon.), Japan Acad. (hon.), Chem. Soc. Japan (hon.), Royal Swedish Acad. Scis. (fgn., physics sect.), Nat. Acad. Sci. (fgn. asso.), Faraday Soc., Am. Astron. Soc., Can. Assn. Physicists (past pres., Achievement award 1957), Optical Soc. Am. (hon., Frederic Ives medal 1964). Subspecialties: Atomic and molecular physics; Infrared spectroscopy. Current work: Molecular spectroscopy: study of molecular structure, especially of simple free radicals. Recent studies: traiatomic hydrogen (H3)and ammonium(NH4). Home: 190 Lakeway Dr Rockcliffe Park Ottawa ON Canada Office: Nat Research Council Ottawa ON K1A 0R6 Canada

HERZENBERG, LEONARD ARTHUR, genetics educator; b. Bklyn., Nov. 5, 1931; m., 1953; 4 children. B.A., Bklyn. Coll., 1952; Ph.D. in Biochemistry, Calif. Inst. Tech., 1956. Asst. prof. Stanford (Calif.) U., 1959-64, assoc. prof., 1964-69, prof. genetics, 1969—; Am. Cancer Soc. fellow Pasteur Inst., Paris, 1955-57; mem. genetics study sect. NIH. Mem. AAAS, Genetics Soc. Am., Am. Assn. Immunologists, Biomed. Engrs. Soc., Soc. Developmental Biology. Subspecialties: Genetics and genetic engineering (biology); Immunobiology and immunology. Office: Dept Genetics Stanford U Med Ctr Stanford CA 94305

HERZIG, DAVID JACOB, pharm. co. exec. researcher; b. Cleve., Dec. 13, 1936; s. Marvin Lawrence and Lillian Gertrude (Blaine) H.; m. Phyllis Glicksberg, Sept. 2, 1962; children—Michael, Pamela, Roberta, Karen. B.A., Oberlin Coll., 1958; Ph.D. in Chemistry, U. Cin., 1963. Vis. scientist, Damon Runyon fellow lab. chem. biology NIH, 1963-65; staff fellow lab. gen. and comparative biology NIMH, 1965-67; sr. research assoc. N.Y. U. Med. Sch., 1967-68; sr. scientist Warner-Lambert Co., Ann Arbor, Mich., 1968-74, sr. research assoc., 1974-76, dir. immunoinflammatory/pulmonary pharmacology, 1977-81, dir. sci. devel. pharm. research div., 1981—; vis. scholar U. Mich., 1980—. Contbr. numerous articles to profl. jours. Mem. Am. Soc. Pharmacology and Exptl. Therapeutics, Am. Acad. Allergy, AAAS. Club: Fencers (N.Y.C.). Subspecialties: Immunopharmacology; Pharmacology. Current work: Pharmacology of allergy and immunological diseases; find, evaluate and license acquisition opportunities for Parke-Davis division of Warner-Lambert. Home: 3540 Windemere Ann Arbor MI 48105 Office: 2800 Plymouth Rd Ann Arbor MI 48105

HERZIG, GEOFFREY PETER, physician, educator; b. Cleve., Dec. 6, 1941; s. David Jack and Sylvia Cecelia (Ehrlich) H.; m. Ricki Jane Trepner, Aug. 28, 1964; 1 son, Andrew. B.S., U. Cin., 1963; M.D., Case Western Res. U., 1967. Intern Bronx Mcpl. Hosps., 1967-68, resident, 1968-69; clin. assoc. Nat. Cancer Inst., Bethesda, 1969-72; fellow in hematology Washington U., St. Louis, 1972-73; sr. investigator Nat. Cancer Inst., 1973-75; asst. prof. medicine Washington U., 1975-80, assoc. prof., 1980—, dir., 1975—; chmn. marrow transplant com. Southeastern Cancer Study Group, Birmingham, Ala., 1979—. Nat. Cancer Inst. grantee, 1979. Mem. Am. Fedn. Clin.Research, Am. Assn. Cancer Research, Am. Soc. Clin. Oncology, Am. Soc. Hematology, Phi Beta Kappa, Alpha Omega Alpha. Subspecialties: Marrow transplant; Chemotherapy. Current work: Clinical studies of bone marrow transplantation for cancer therapy; marrow cryopreservation and autologous transplantation; chemotherapy of acute leukemia; granulocyte transfusion therapy. Office: Washington University Box 8125 660 S Euclid Ave Saint Louis MO 63110

HESS, CHARLES E., horticultural educator, university dean; b. Paterson, N.J., Dec. 20, 1931; m., 1953; 4 children. B.S., Rutgers U., 1953; M.S., Cornell U., 1954, Ph.D., 1957. From asst. prof. to prof. horticulture Purdue U., 1958-66; research prof., chmn. dept. horticulture and forestry Rutgers U., New Brunswick, N.J., 1966-70, assoc. dean Coll. Agr. and Environ. Sci., 1970, acting dean, 1971-72; dean Cook Coll., 1972-75, (Coll. Agr. and Environ. Sci.), U. Calif.-Davis, 1975—; dir. N.J. Agr. Expt. Sta., 1970; cons. AID, 1965, Office Technol. Assessment, 1977—. Editor: Internat. Plant Propagators Soc, 1962-72. Recipient Norman Jay Coleman award, 1967; Jackson Dawson Meml. medal, 1971. Fellow Am. Soc. Hort. Sci. (pres. 1972), AAAS; mem. Am. Soc. Plant physiologists, Internat. Plant Propagators Soc. (award 1963, pres. 1970), Sigma Xi. Subspecialties: Plant physiology (agriculture); Horticulture. Office: Coll Agr U Calif Davis CA 95616

HESS, EARL HOLLINGER, lab. exec.; b. Lancaster County, Pa., June 16, 1928; s. Abram Myer and Ruth (Stoner) H.; m. Anita F. Swords, Sept. 2, 1951; children: Kenneth Earl, Bonita Sue, Carol Denise. B.S. cum laude, Franklin and Marshall Coll., 1952; Ph.D. in Organic Biochemistry, U. Ill., 1955. Teaching asst. U. Ill., Urbana, 1952-54; Socony-Mobil research fellow, 1954-55; asst. prof. chemistry Franklin and Marshall Coll., Lancaster, Pa., 1955-57; group leader chem. research Gen. Cigar Co., Lancaster, 1957-61; pres., chief exec. officer Lancaster Labs., Inc., 1961—; Pres.-elect, dir. govt. relations Am. Council Ind. Labs., 1978—; mem. Pa. Tech. Assistance Program, 1979-82. Contbr. chpt. to book. Mem., moderator Conestoga Ch. of the Brethren; bd. dirs. Bethany Theol. Sem., 1981—. Receipient Spl. award Am. Council Ind. Labs., 1979. Mem. Am. Assn. Small Research Cos. (dir.), Am. Assn. for Lab Accreditation (dir.), ASTM, Am. Chem. Soc., Am. Pub. Health Assn., AAAS, N.Y. Acad. Scis., Nat. Fedn. Ind. Businesses, Am. Water Works Assn., Nat. Profl. Services Firm, U.S. C. of C. (small bus. council 1980—), Pa. C. of C., Lancaster C. of C. (vice Chmn. pub. affairs 1981—), Phi Beta Kappa, Sigma Xi. Current work: Agricultural products utilization research; new food processes and techniques; basic biochemistry studies in agriculture, food products, and environmental issues. Patentee tobacco processing. Home: 4 Forest Hill Rd Leola PA 17540 Office: 2425 New Holland Pike Lancaster PA 17601

HESS, MARILYN E., pharmacology educator; b. Erie, Pa., Dec. 31, 1924; d. James Adair and Florence Margaret (Blass) H. B.S., Villa Maria Coll., 1946; M.S., U. Pa., 1949, Ph.D., 1957. Mem. faculty U. Pa., Phila., 1957—, assoc. prof. pharmacology, 1968-76, prof., 1976—; cons. U.S. Govt. Contbr. articles to profl. jours. Am. Heart Assn. fellow, 1960-62; established investigator, 1962-67; USPHS Research Career Devel. awardee, 1967-72. Mem. Am. Soc. Pharmacology and Exptl. Therapeutics, N.Y. Acad. Scis., Council Basic Scis. Am. Heart Assn. Subspecialties: Pharmacology; Molecular pharmacology. Current work: Effect of hormones and drugs on cardiac function, myocardial neurotransmitters and heart metabolism. Office: Med Sch U Pa Philadelphia PA 19104

HESS, ORVAN WALTER, obstetrician, gynecologist, educator; b. Bayoba, Pa., June 18, 1906; s. Philip O. and Effie F. (Shoemaker) H.; m. Carol Woodruff Maurer, Aug. 31, 1928; children: Katherine Hess Halloran, Carolyn Hess Westerfield. B.S., Lafayette Coll., 1927; M.D., SUNY-Buffalo, 1931. Diplomate: Am. Bd. Ob-Gyn. Intern Children's Hosp., Buffalo, 1931-32; intern New Haven Hosp., 1932-33, asst. resident, 1934-36, resident in ob-gyn, 1936-37; research fellow in surgery and gynecology Yale U. Sch. Medicine, New Haven, 1933-35, instr., 1936-37, clin. prof., 1975—, dir. regional perinatal monitoring program, div. perinatology, dept. ob-gyn, 1971—; med. dir. Conn. Welfare Dept., 1967-70, dir. bur. health services, 1970-71; pres., Conn. Med. Inst; chmn. Conn. Com. on Maternal Mortality and Morbidity; pres. Conn. Health Assn., Inc.; mem. exec. com., mem. adv. com. Conn. Regional Med. Program; dir. Corometrics Electronics, Inc.; mem. staff Yale-New Haven Med. Ctr., Hosp. of St. Raphael, Middlesex Hosp., Middletown, Conn. Contbr. articles to med. jours. Served with U.S. Army, 1942-45; col. Res. (ret). Decorated Bronze star (5); recipient Kidd award Lafayette Coll., 1980; fellow Morse Coll. Yale. Fellow , ACS, Am. Coll. ob-Gyn.; mem. AMA (sci. achievement award 1979, del.), Assn. Mil. Surgeons U.S., IEEE, Conn. Med. Soc. (pres. 1966-67), Sigma Xi, Nu Sigma Nu. Presbyterian. Clubs: Lawn, Yale, Graduate (New Haven). Subspecialties: Neonatology; Maternal and fetal medicine. Current work: Fetal monitoring, telemetry, ultrasound. Office: Yale University School of Medicine 333 Cedar St New Haven CT 06510

HESS, WILMOT NORTON, sci. adminstr.; b. Oberlin, Ohio, Oct. 16, 1926; s. Walter Norton and Rachel Victoria (Metcalf) H.; m. Winifred Esther Lowdermilk, June 16, 1950; children—Walter Craig, Alison Lee, Carl Ernest. B.S. in Elec. Engring, Columbia, 1946; M.A. in Physics, Oberlin Coll, 1949; D.Sc., 1970; Ph.D., U. Calif., Berkeley, 1954. Staff Lawrence Radiation Lab., U. Calif., Berkeley and Livermore, 1954-59, head plowshare div., Livermore, 1959-61; dir. theoretical div. Goddard Spaceflight Center (NASA), Greenbelt, Md., 1961-67; dir. sci. and applications Manned Spacecraft Center, Houston, 1967-69; dir. NOAA Research Labs. (Commerce Dept.), Boulder, Colo., 1969-80, Nat. Center for Atmospheric Research, 1980—; adj. prof. U. Colo., 1970-78. Contbr. articles to profl. jours.; editor: Introduction to Space Science, 1965; author: Radiation Belt and Magnetosphere, 1968, (with others) Weather and Climate Modification, 1974; asso. editor: Jour. Geophys. Research, 1961-67, Jour. Atmospheric Sci, 1961-67, Jour. Am. Inst. Aeros. and Astronautics, 1967-69. Served with USN, 1944-46. Fellow Am. Geophys. Union, Am. Phys. Soc.; mem. Nat. Acad. Engring. Club: Cosmos (Washington). Subspecialty: Meteorology. Current work: Oil spills, acid rain. Home: 4927 Idylwild Trail Boulder CO 80301 Office: Nat Center Atmospheric Research Boulder CO 80307

HESSELTINE, CLIFFORD WILLIAM, microbiologist; b. Brighton, Iowa, Apr. 4, 1917; s. Merlin Jerome and Charlotte Jane (Owen) H.; m. Harriet Elsie Herm, Aug. 8, 1941; children: Christopher, Nancy, Anna, Rise. B.S., U. Iowa, 1940; Ph.D., U. Wis., 1950. Research mycologist Am. Cyanamid Co., Pearl River, N.Y., 1947-53; head ARS Culture Collection, Fermentation Lab., No. Regional Research Center, Peoria, Ill., 1953-68, chief, 1968—. Contbr. articles to profl. jours. Served with M.C. U.S. Army, 1942-46. Recipient Superior Service award U.S. Dept. Agr., 1959, Disting. Service award, 1981. Mem. AAAS, Am. Chem. Soc., Am. Phytopath. Soc., Bot. Soc. Am., Mycol. Soc. Am. (Disting. Mycologist award), Soc. Am. Microbiology (Pasteur award Ill. br. 1978), Soc. Ill. Bacteriologists, Soc. for Indsl. Microbiology (Charles Thom award 1980), Torrey Bot. Club, Am. Acad. Microbiology. Subspecialties: Microbiology; Plant pathology. Current work: Mycology, microbiology, mycotoxins, fermented foods, mucorales. Patentee in field. Home: 5407 Isabell St Peoria IL 61614 Office: 1815 N University St Peoria IL 61604

HESSER, JAMES E(DWARD), astronomer; b. Wichita, Kans., June 23, 1941; s. J(ames) Edward and Ina (Lowe) H.; m. Betty Louise Hinsdale, Aug. 24, 1963; children: Nadja Lynn, Rebecca Ximena, Diana Gillian. B.A. in Astronomy, U. Kans., 1963; Ph.D. in Astrophys, Scis, Princeton U., 1966. Research assoc. Princeton U. Obs., 1966-68; jr. asst., assoc. astronomer Observatorio Interamericano de Cerro Tololo, La Serena, Chile, 1968-77, asst. dir., 1972-74; assoc. research officer Dominian Astrophys. Observatory, Victoria, B.C., Canada, 1977-81, sr. research officer, 1981—; vis. prof. U. Chile, 1973-74; chairperson Can. Starlab Working Group, 1979—. Contbr. articles to profl. jours. Mem. Internat. Astron. Union, Am. Astron. Soc., Astron. Soc. Pacific (dir. 1981—, bd. editors publs. 1982—), Can. Astron. Soc. Subspecialties: Optical astronomy; Space astronomy. Current work: Chemical evolution of galaxies, star clusters, interstellar lines, variable stars; research. Home: 1874 Ventura Way Victoria BC Canada V8N 1R3 Office: 5077 W. Saanich Rd Victoria BC Canada V8X 4M6

HESTER, JARRETT CHARLES, mechanical engineer, educator, research and development official; b. Mt. Vernon, Tex., Dec. 14, 1938; s. Jarrett B. and Edith L. (Hutson) H.; m. Marjorie M. Hester, Sept. 10, 1980; children: James, Michael, Laura. B.S.M.E., Arlington State Coll., 1962; M.S.M.E., Okla. State U., 1964, Ph.D.M.E., 1966. Registered profl. engr., Tex. Engr. LTV Aerospace Corp. Research Center, 1962-66, sr. engring. specialist, 1967-70; asst. prof. mech. engring. U. Tex., Austin, 1966-67; assoc. prof. mech. engring. Clemson U., 1970-71, dept. head dept. mech. engring., 1971-74, assoc. dean, 1974-77, prof. mech. engring., 1977—; dir. S.C. Energy Research and Devel. Center, 1982—; cons. J.E. Sirrine Co.; mgr. advanced tech. projects, 1979-80. Served with USAFR, 1956-64. Mem. ASME, ASHRAE, Am. Soc. Engring. Educators, Sigma Xi, Tau Beta Pi, Pi Tau Sigma. Presbyterian. Subspecialties: Mechanical engineering; Energy systems. Current work: Energy systems. Patentee in field.

Home: 206 Mountain View Ln Clemson SC 29631 Office: Mechanical Engineering Clemson Clemson U SC 29631

HESTER, REID KEVIN, clinical psychologist; b. Seattle, Sept. 20, 1951; s. Lawrence Pershing and Eileen (Henson) H. B.S., U. Wash., 1973; M.S., Wash. State U., 1976, Ph.D., 1979. Cert. psychologist, N.Mex. Dir. psychol. services Raleigh Hills Hosp., Glendale, Calif., 1979-80; rehab. psychologist St. Joseph Hosp., Albuquerque, 1981—; profl. advisor N.Mex. Head Injured Assn., Albuquerque, 1981—. Contbr. articles to profl. jours. USPHS trainee, 1975-76. Mem. Am. Psychol. Assn., Soc. Behavioral Medicine, Western Psychol. Assn., Assn. Advancement of Behavior Therapy. Democrat. Subspecialty: Behavioral psychology. Current work: Treatment outcome for alcohol abuse; behavioral medicine; rehabilitation psychology. Home: 5223 Zurich Pl NE Albuquerque NM 87111 Office: La Mesa Med Ctr 700 Cutler NE Albuquerque NM 87111

HESTER, RICHARD KELLY, pharmacologist, med. educator; b. Austin, Tex., July 30, 1947; s. Glenn Richard and Doris Pernell (Clanahan) H.; m. Joan Christine Rydman, Mar. 25, 1973; 1 son: Kasey Clanahan. B.A. in Biology (Ford Found. grantee), Austin Coll., 1969; Ph.D. in Pharmacology (NIH fellow), U. Tex. Health Sci. Ctr., San Antonio, 1975. NIH postdoctoral fellow dept. pharmacology U. Miami (Fla.) Sch. Medicine, 1975-76; research fellow depts. pharmacology and surgery U. Tex. Health Sci. Ctr., Dallas, 1976-77, instr., 1977-78, asst. prof., 1978-79; asst. prof. med. pharmacology Coll. Medicine, Tex. A&M U., College Station, 1979—; cons. Contbr. numerous articles to profl. jours. Am. Heart Assn. grantee, 1977-80; Nat. Heart, Lung and Blood Inst. grantee, 1980-83. Fellow Am. Heart Assn.; mem. AAAS, Western Pharmacol. Soc., Microcirculatory Soc., Am. Soc. Pharmacology and Exptl. Therapeutics, Sigma Xi. Episcopalian. Subspecialties: Cellular pharmacology; Membrane biology. Current work: Ca^{++} and excitation/contraction (relaxation) in vascular smooth muscle, vascular smooth muscle in vitro and microcirculation in vivo. Home: 3106 Hummingbird Circle Bryan TX 77801 Office: Dept Med Pharmacology Coll Medicine Tex A&M U College Station TX 778434

HESTON, LEONARD LANCASTER, psychiatrist, educator; b. Burns, Oreg., Dec. 16, 1930; s. Alexander W. and Florence (Woodhouse) H.; m. Renate Heston, Mar. 1, 1936; children; William, Steven, Diane, Gwendolen, Ardis, Barbara. B.S., U. Oreg., 1955, M.D., 1961. Intern Bernalillo County-Indian Hosp., Albuquerque, 1961-62; resident in psychiatry U. Oreg. Med. Sch., 1962-65; guest worker Inst. Psychiatry, London, 1965-66; asst. prof. U. Iowa, 1966-70; assoc. prof. psychiatry U. Minn., Mpls., 1970-74, prof., 1974—, dir. adult psychiatry, 1980—. Author: The Medical Casebook of Adolf Hitler, 1980; contbr. over 70 med. articles to profl. publs. Served with U.S. Army, 1949-51. USPHS spl. fellow, 1965-66; NIH grantee, 1965—. Fellow Am. Psychiat. Assn.; mem. AMA, Behavior Genetic Soc., Am. Psychopath. Assn. Subspecialty: Psychiatry. Current work: Genetics of brain diseases; medical history; clinical practice of psychiatry; teaching of clinical psychiatry; research in genetics. Office: Univ Minn Univ Hosp Box 392 Minneapolis MN 55455

HETTCHE, LEROY R., engineering educator, researcher; b. Balt., Mar. 24, 1938; m., 1965; 3 children. B.S., Bucknell U., 1961; M.S., Carnegie-Mellon U., 1963, Ph.D. in Civil Engring, 1965. Asst. prof. civil engring. Rutgers U., 1964-66; NRC research assoc., Nat. Bur. Standards, 1966-68; structural engr. Naval Research Lab., 1968-71, phys. scientist, 1971-74, supt. materials sci. and tech. div., 1974-81; dir. applied research lab., prof. engring. research Pa. State U., State College, 1981—. Mem. ASCE, ASTM, ASME. Subspecialty: Solid mechanics. Office: Pa State U PO Box 32 State College PA 16801

HETTINGER, WILLIAM PETER, JR., petroleum company executive; b. Aurora, Ill., Sept. 13, 1922; s. William Peter and Gertrude Kathryn (Schomer) H.; m. Alice May Mietz, Apr. 20, 1944; children: Diana Lee, William Peter, Scott Edward, Sally Ann (dec.). B.S. in Chemistry, Purdue U., 1947; Ph.D., Northwestern U., 1951. Dir. corp. devel. Ga. Kaolin Co., 1972-74; v.p. research and devel., N.L. Industries, 1974-76; mem. spl. staff Arthur D. Little, Inc., 1976-77; dir. research and devel. Ashland Petroleum Co., Ky., 1977-79, v.p. research and devel. automotive and product application labs., 1979—. Contbr. articles to profl. jours. Rep. Greater Severna Park council, 1966-71, v.p., 1969-70; mem. St. Paul's Lutheran Day Sch., Glen Burnie, Md., 1967-72, chmn., 1969-71; bd. mgrs. Luth. Home for Aged, Jersey City. Served with USAAF, 1943-45; ETO. Decorated Air Medal with oak leaf cluster; advanced tng. fellow NIH, 1968-71. Fellow Am. Inst. Chemists; mem. N.Y. Acad. Scis. (hon. life), Am. Chem. Soc., AAAS, Indsl. Research Inst., Catalyst Club Chgo., Catalyst Club Phila., Catalyst Club N.Y., Catalyst Club Tri-State (pres., founder 1978-80), Am. Inst. Chem. Engrs., Gerontol. Soc., Sci. Research Soc. Am., Research Dirs. Assn. N.Y. (pres. 1976-77), TAPPI, Assn. Advancement Invention and Innovation, Sigma Xi, Phi Lambda Upsilon. Lutheran. Clubs: Chartwell Country of Severna Park (dir. 1970-72), Echo Lake Country (Westfield, N.J.); Nashawtuc Country (Concord, Mass.); Bellefonte (Ky.) Country. Subspecialties: Catalysis chemistry; Petroleum engineering. Patentee in field. Home: 203 Meadowlark Rd Russell KY 41169 Office: PO Box 391 Ashland KY 41114

HEUER, MICHAEL ALEXANDER, dentist, educator; b. Grand Rapids, Mich., Apr. 27, 1932; s. Harold Maynard and Gwendolyn Ruth (Kremer) H.; m. Barbara Margaret Naines, Nov. 23, 1955; children—Kristan M., Karin E., Katrina A. D.D.S., Northwestern U., 1956; M.S., U. Mich., 1959. Practice dentistry specializing in endodontics, Chgo., 1959—; asst. prof. Northwestern U., 1960-66; asso. prof. Loyola U., Chgo., 1968-73; prof., chmn. dept. endodontics Northwestern U., 1974-83, assoc. dean acad affairs, 1983—; dir. Am. Bd. Endodontics, 1971-77, sec.-treas., 1973-76, pres., 1976-77; chmn. subcom. Am. Nat. Standards Inst.; mem. com. on advanced edn. Commn. on Accreditation of Dental Edn., 1974-77. Contbr. articles in field to profl. jours. Served with USNR, 1956-58. Fellow Am. Coll. Dentistry, Internat. Coll. Dentistry, Am. Assn. Endodontists (exec. council 1967-71, sec. 1979—); mem. ADA (mem. council dental materials and devices 1972-78, chmn. 1977-78), AAAS, Internat. Assn. Dental Research, Am. Assn. Dental Schs., Chgo. Odontographic Soc. (pres. 1982-84), Edgar D. Coolidge Endodontic Soc. (trustee), Phi Eta Sigma, Omicron Kappa Upsilon, Chi Psi, Delta Sigma Delta. Subspecialties: Endodontics; Biomaterials. Current work: Endodontic instruments, instrumentation and materials. Home: 156 Timber Ridge Lake Barrington Shores Barrington IL 60010 Office: Dental Sch Northwestern U Chicago IL 60611

HEUFT, RICHARD WILLIAM, computer engr., educator; b. Toronto, Ont., Can., Aug. 15, 1951; s. William Ernest and Louise Elenor (Wenhardt) H. B.A.Sc. in Elec. Engrining, U. Waterloo, 1975, M.A.Sc., 1977, Ph.D., 1980. Registered profl. engr. Programmer Ministry of Transport, Ottawa, Ont., 1973; engring. asst. Microsystems Ltd., Ottawa, 1974; research asst. U. Waterloo, Ont., 1975-80; asst. prof. computing sci. U. Alta. (Can.), Edmonton, 1980-82, assoc. prof., 1982—. Contbr. articles to profl. jours. Mem. IEEE, Assn. for Computing Machinery. Subspecialties: Computer engineering; Computer architecture. Current work: High-speed, special-purpose computers for signal processing research. Teaching: microcomputers, VLSI design, logic design, computer engineering, multiprocessors. Office: Computer Engring 238 Elec Engring Bldg Edmonton AB Canada T6G 2G7

HEUSCH, CLEMENS AUGUST, physicist; b. Aachen., Germany, Apr. 19, 1932; s. Hermann and Elisabeth (Pauli) H.; m. Karin von Gilgenheimb, July 6, 1968; children: Marina, Bettina. Student, Bowdoin Coll., 1951-52; diploma in Physics, Tech. U., Aachen, 1955; student, U. Paris, 1955-56; Dr. rer. nat., Tech. U. Munich, 1959. Research asst. Tech. U. Munich, 1956-59; project leader Allgemeine Elektrizitäts-Gesellschaft, Frankfurt, W.Ger., 1960-61; staff scientist Deutsches Elektronen-Synchrotron, Hamburg, W.Ger., 1961-63; research fellow to assoc. prof. Calif. Inst. Tech., Pasadena, 1963-69; prof., prin. investigator particle physics U. Calif-Santa Cruz, 1969—; vis. prof. Calif. Inst. Tech., 1969-71, U. Munich, 1974-75, Tech. U. Aachen, 1980; sci. assoc. Max-Planck Inst. Physics, Munich, 1974, European Ctr. for Nuclear Research, CERN, Geneva, Switzerland, 1974-75, 83-84; cons. in field. Contbr. chpts. to books, articles to profl. jours. Roman Catholic. Subspecialty: Particle physics. Current work: Photon-hadron interactions, lepton-hadron scattering, electron-positron annihilation, instrumentation. Office: Inst for Particle Physic Univ Calif Santa Cruz CA 9506

HEVEZI, JAMES MICHAEL, radiological physicist; b. Gary, Ind., June 21, 1940; s. James Emery and Margaret (Olah) H.; m. Suzanne Landig, Feb. 5, 1978; children: Julie, Lloyd, Matthew, Lisa, Martin, Jane. B.S., St. Procopius Coll., 1962; Ph.D., U. Notre Dame, 1969. Cert. radiol. physicist, Am. Bd. Radiology, 1975. Asst. prof. radiology U. Wis., Madison, 1969-71, U. Tex., Houston, 1971-79; staff physicist U. Ariz., Tucson, 1979—; v.p. Tex. Radiol. Equip. Corp., Houston, 1973-77; pres. Landig-Hevezi Assoc., Rapid City, S.D., 1978-79; cons. Radiol. Phys. Service, Inc., Phoenix, 1982—. Nat. Cancer Inst. grantee, 1974, 76. Mem. Am. Assn. Physicist in Medicine (asso. editor 1976-82), Am. Coll. Radiology, Am. Soc. Therapeutic Radiologists, Radiol. Soc. N. Am. Current work: Radiol, physics, diagnostic and therapeutic radiology, hyperthermia, implant radiation oncology. Inventor patentee "The Inverse Pinhole Camera, 1974, Light Beam Readout, 1976. Home: 2223 E Camino Rio Tucson AZ 85718 Office: Radiation Oncology Health Sci Center Univ Ariz Tucson AZ 85724

HEVNER, ALAN RAYMOND, computer science educator, consultant; b. Marion, Ind., Dec. 9, 1950; s. Raymond L. and Pauline H. (Roach) H. B.S., Purdue U., 1973, M.S., 1976, Ph.D., 1979. Grad. instr. Purdue U., West Lafayette, Ind., 1975-79; prof. U. Minn., Mpls., 1979-81, U. Md., College Park, 1981—; research cons. Honeywell, Bloomington, Minn., 1979—, Software Systems Tech., Silver Spring, Md., 1982—; bd. dirs. Center Automation Research, U. Md., 1982—. Editor: Database Engring. jour, 1981—; contbr. chpts. to books, articles to profl. jours. Served as lt. U.S. Army, 1973-75. Research grantee, 1981-82. Mem. Assn. Computing Machinery, IEEE, IEEE Computer Soc. Subspecialties: Database systems; Distributed systems and networks. Current work: Research on distributed database systems, database systems design, database system performance evaluation, database machine architecture, and office information systems. Home: 9005 Walden Rd Silver Spring MD 20901 Office: U Md Coll Business College Park MD 20742

HEWARD, WILLIAM LEE, special education researcher; b. Michigan City, Ind., Nov. 22, 1949; s. Joe William and Helen Mae (Jensen) H.; m. Jill Carolyn Dardig, Aug. 21, 1976; 1 son, Lee Dardig. B.A., Western Mich. U., 1971; Ed.D., U. Mass., 1974. Dir. Project Change, Greenfield, Mass., 1973; research asst. N.E. Regional Media Ctr. for Deaf, U. Mass., Amherst, 1972-74; asst. prof. Ohio State U. Coll. Edn., Columbus, 1975-79, assoc. prof., 1979—; dir. Visual Response System Project, Ohio State U.-U.S. Office Edn., 1977-82; co-dir. Project Interaction, 1979-82. Author: Sign Here, 1976, 2d edit., 1981, Working with Parents of Handicapped Children, 1979, Exceptional Children, 1980, Voices: Interviews with Handicapped People, 1981. Bd. dirs. St. Joseph's Group Home, Westerville, Ohio, 1980—, Franklin County Residential Services, Franklin County, Ohio, 1982—; chmn. awards com. Ohio Spl. Olympics, Columbus, 1977—. Recipient Waldo-Sangren award Western Mich. U., 1970-71. Mem. Am. Psychol. Assn., Assn. Behavior Analysis (research award 1980), Assn. Spl. Edn. Tech., Council Exceptional Children. Subspecialties: Behavioral psychology; Special education. Current work: Applied behavior analysis, instructinal technology for handicapped learners, analysis and design of systems for group instruction. Office: Ohio State Coll Edn 1945 N High St Columbus OH 43210

HEWITT, ANTHONY VICTOR, astronomer; b. Witney, Eng., Feb. 2, 1943; s. Victor Alfred and Edna Jane (Price) H.; m. Audrey Eileen Harris, June 24, 1965. B.A., Oxford U., 1964, D.Phil., 1967. Astronomer U.S. Naval Obs., Flagstaff, Ariz., 1967—. Recipient Johnson Meml. prize Oxford U., 1963. Mem. Internat. Astron. Union, Am. Astron. Soc., Sigma Xi. Subspecialties: Optical astronomy; Graphics, image processing, and pattern recognition. Current work: Photometry using panoramic detectors; globular clusters, astrometry. Home: PO Box 1752 Flagstaff AZ 86002 Office: PO Box 1149 Flagstaff AZ 86002

HEYDE, JOHN BRADLEY, dental researcher, dentist; b. Marion, Ill., Dec. 27, 1926. B.S. in Chemistry, U. Mich., 1950, M.S. in Biol. Chemistry, 1952, D.D.S., 1957. Cert. Dental Bds., Pa., Mich., Del. Dir. profl. research L. D. Caulk Co. div. Dentsply Internat., Inc., Milford, Del.; lectr. on dental materials and restorative dentistry. Contbr. articles on dentistry to profl. jours. Fellow Royal Soc. Health (Eng.), Acad. Dentistry Internat.; mem. Fedn. Dentaire Internationale, Internat. Assn. for Dental Research (dental materials and pulp groups), Am. Prosthodontic Soc., Pierre Fauchard Acad., ADA, Del. State Dental Soc., Kent-Sussex Dental Soc. (hon. life mem., past pres. and sec.), Acad. Operative Dentistry, Acad. Gen. Dentistry, Am. Acad. Plastics Research (past pres., meritorious service award 1972-73), Am. Soc. Dentistry for Children, Fedn. Prosthodontic Orgns. (del. 1970, 71, 72), Sigma Alpha Epsilon, Psi Omega. Subspecialty: Dental materials. Current work: New systems and techniques for use in orthodontics, periodontics, endodontics and operative dentistry. Home: 508 Kings Hwy Milford DE 19963 Office: L D Caulk Co Div Dentsply Internat Inc Lakeview and Clarke Aves Milford DE 19963

HEYDEGGER, H(ELMUT) ROLAND, phys. chemist, educator, researcher, cons.; b. Phila., Dec. 3, 1935; s. Helmut and Allyse (Paulich) H. B.S., Queens Coll., CUNY, 1956; M.S., U. Ark., Fayetteville, 1958; Ph.D. (Gen. Electric Found. fellow), U. Chgo., 1968. Phys. chemist U.S. Bur. Mines, Bartlesville, Okla., 1958; instr. Prairie State Coll., 1961-62; asst. prof. chemistry Purdue U. Calumet, Hammond, Ind., 1970-75, assoc. prof., 1975-81, prof., 1981—, head dept. chemistry and physics, 1979—; research assoc. Enrico Fermi Inst., U. Chgo., 1968-78, sr. research asso., 1978—; cons. Argonne Nat. Lab., 1973-74; vis. fellow Australian Nat. U., 1976-77; vis. staff mem. Los Alamos Nat. Lab., 1978—. Contbr. articles to profl. jours. Mem. Am. Chem. Soc., Am. Phys. Soc., Am. Geophys. Uion, Geochem. Soc., Internat. Assn. Geochemistry and Cosmochemistry, Meteoritical Soc. Subspecialties: Analytical chemistry; Geochemistry. Current work: Application of nuclear sci. to geo- and cosmochem. problems. Office: Dept Chemistry and Physics Purdue U Calumet Hammond IN 46323

HEYDEMANN, PETER LUDWIG MARTIN, physicist; b. Gottingen, Ger., Nov. 10, 1928; m., 1958; 2 children. Ph.D. in Physics, Chemistry and Physiology, U. Göttingen, 1958. Asst. prof. physics U. Göttingen, 1957-61, asst. to dir., 1961-64; physicist Nat. Bur. Standards, Washington, 1964-70, chief pressure and vacuum sect., 1970-78, program analyst, 1978-80, dir. ctr. chem. physics, 1980-81, assoc. dir. bur., 1980—; mem. Indo-U.S. subcom. Sci. and Tech., 1975—. Recipient IR-100 award. Mem. Am. Phys. Soc. Subspecialty: Chemical physics research administration. Office: Nat Bur Standards Washington DC 20234

HEYER, ERIC JOHN, neurologist, educator; b. N.Y.C., Feb. 10, 1946; s. John W. and Dora (Kaplan) H.; m. Diana A. Steele, June 8, 1980. B.S., U. Chgo., 1968; postgrad., Rockefeller U., 1968-69; Ph.D., Albert Einstein Coll. Medicine, 1974, M.D., 1975. Instr. neurology U. Mich., 1979-80; asst. prof. neurology Mt. Sinai Med. Sch., N.Y.C., 1981—. Contbr. articles to profl. jours. R.S. Morison fellow Grass Found., 1979-80. Mem. Am. Acad. Neurology, Am. Epilepsy Soc., AAAS, N.Y. Acad. Scis., Soc. Neurosci. Subspecialties: Neurophysiology; Neurology. Current work: Mechanism of action of dopamine on cultured basal ganglia neurons. Home: 275 W 96th St Apt 22E New York NY 10025 Office: Dept Neurology Mt Sinai Sch Medicine One Gustave L Levy Pl New York NY 10029

HEYER, MIRIAM HARRIET, training coordinator, behavioral consultant; b. Long Beach, Calif., May 9, 1942; d. George Herbert and Harriet Laura (Rollwagen) Muedeking; m. William Ronald Heyer, June 16, 1964; children: Laura Miriam, Elena Diane. B.A., Pacific Lutheran U., 1964, M.A., 1973; Ph.D., Cath. U. Am, 1982. Elem. tchr. Los Angeles County Schs., 1964-66; field research asst. Smithsonian Instn., Washington, 1974-77; grad. teaching asst. Cath. U. Am., Washington, 1979-81; program coordinator for tng. Community Teaching Homes, Alexandria, Va., 1982—; cons. trainee Insts. for Behavioral Resources, Washington, 1982—. Univ. scholar Cath. U., 1978-79. Mem. Am. Psychol. Assn., Am. Assn. Marriage and Family Therapists. Subspecialties: Social psychology; Behavioral psychology. Current work: Family counseling parent training, training private and public agency personnel in behavioral techniques for specialized foster care. Office: Community Teaching Homes 623 S Pickett St Alexandria VA 22304

HEYING, ROBERT HILARIUS, psychologist; b. Ossian, Iowa, Aug. 4, 1944; s. Joseph and Rose Marie (Bohr) H. B.S., Loras Coll., Dubuque, Iowa, 1966; M.S., Marquette U., Milw., 1968; Ph.D., St. Louis U., 1973. Lic. psychologist, Calif. Psychometrist DePaul Rehab. Center, Milw., 1967; vocat. rehab. counselor Curative Workshop, Milw., 1969; psychology trainee John Cochrane VA Hosp., St. Louis, 1969-72; pvt. practice clin. psychology, St. Louis, 1973-74; staff psychologist Napa (Calif.) State Hosp., 1974—; pvt. practice, 1976—; clin. psychologist med. staff Vallejo (Calif.) Gen. Hosp., 1982—; teaching, tng. in Gestalt therapy U. Calif.-Santa Cruz, 1979—. Contbr. articles to profl. jours., newspapers. NIMH fellow, 1968-69. Mem. Am. Psychol. Assn., Western Psychol. Assn., Napa State Hosp. Psychology Assn. (pres. 1980), Calif. Orgn. Stazte Hosp. Psychologists (acting pres. 1979-80, v.p. No. sect. 1980—), Napa Valley Psychol. Assn. (v.p., treas. 1978). Democrat. Subspecialties: Behavioral psychology; Social psychology. Current work: Current focus is in the application of Gestalt Therapy to sports behavior: Gestalt Therapy with schizophrenics, neurotics, and normals; teach Gestalt to members of the helping profession including doctoral candidates developing as therapists. Office: 1220 Laurel St Napa CA 94558

HEYMANN, FRANK JOSEPH, mechanical engineer; b. Frankfurt am Main, Germany, Aug. 11, 1927; s. Frederick G. and Edith E. (Auerbach) H.; m. Irene R., Apr. 12, 1969; children: Leslie J. Sharkey, Christopher B. Sharkey, Robert E. Ordinary Nat. Cert. in Aero. Engring, Northampton Poly., London, 1946; B.M.E. magna cum laude, CCNY, 1951; S.M. in Mech. Engring, M.I.T., 1953. Registered profl. engr., Del. Machinist, draftsman, 1943-47; mem. research staff M.I.T., 1953-54; engr. in devel. engring. Steam-Turbine-Generator div. Westinghouse Elec. Corp., Lester, Pa., 1954—, Orlanda, Fla., 1954—, sr. engr., 1963—; cons. in field. Contbr. articles, mainly on erosion by liquid impact, valve noise, steam turbine noise, vibration, and terminology to profl. jours., confs. Former mem. playground com., budget com. Village of Arden, Wilmington, Del. Fellow ASTM (Dudley medal 1968, award of Merit 1976); mem. ASME. Club: Buck Ridge Ski (Phila.). Subspecialties: Mechanical engineering; Acoustical engineering. Current work: Steam turbines, valves, noise control, acoustical engrning., tribology, erosion, wear. Patentee in field. Home: 4823 Staghorn Ct Winter Springs FL 32708 Office: Westinghouse STGD Mail Code 101 The Qadrangle Orlando FL 32817

HEYS, RONALD JAY, dentist, educator, researcher; b. Grand Rapids, Mich., Aug. 27, 1946; s. John Louis and Kathryn (Breen) H.; m. Susan Holdaway, June 17, 1978; children: David Benjamin, Stephanie Lynne. B.S., Calvin Coll., 1964-68; D.D.S., U. Mich.-Ann Arbor, 1972, M.S., 1975. Instr. dentistry U. Mich, Ann Arbor, 1972-73, asst. prof., 1973-79, assoc. prof., 1979—. Contbr. articles to profl. jour., chpts. to books. Mem. Ann Arbor C. of C., Internat. Assn. Dental Research (Pulp Biology Group). Republican. Subspecialty: Oral biology. Current work: Restorative dentistry, pulp biology, pulpal response, mechanisms of pulp, histopathology, clinical research. Home: 2605 Powell St Ann Arbor MI 48104 Office: U Mich Dental Sch N University St Ann Arbor MI 48109

HEYSSEL, ROBERT MORRIS, physician, hospital administrator; b. Jamestown, Mo., June 19, 1928; s. Clarence D. and Meta and (Reusser) H.; m. Maria McDaniel, Aug. 7, 1955; children: James Olin, Maria Lisa, Robert Morris, Kurt Frederick, Helen Perrier. B.S., U. Mo., 1951; M.D., St. Louis U., 1953. Postgrad. tng. St. Louis U. Hosp., 1953-56, Barnes Hosp., St. Louis, 1953-56; hematologist, acting dir. dept. medicine Atomic Bomb Casualty Commn., Nagasaki and Hiroshima, Japan, 1956-58; mem. faculty Sch. Medicine, Vanderbilt U., Nashville, 1959-68, dir. div. nuclear medicine, 1962-68, asso. prof. medicine, 1964-68; asso. dean Sch. Medicine, Johns Hopkins U., Balt., 1968-72, dir. health care programs and outpatient services, 1968-72, prof. medicine, 1971—, prof. health care orgn., 1972—; exec. v.p., dir. Johns Hopkins Hosp., 1972-83, pres., 1983—; dir. Union Trust BanCorp, Union Trust Bank, Balt.; chmn. Commonwealth Fund on Acad. Health Ctrs., 1983. Contbr. articles to profl. jours. Pres. Columbia (Md.) Hosp. and Clinics, 1969-74; chmn. health services com. Assn. Am. Med. Colls., 1971-74, mem. gen. assembly, 1974—, mem. exec. council; chmn. com. on emergency med. services Nat. Acad. Scis., 1973—; mem. Joint Commn. on Prescription Drugs, 1976; chmn. council teaching hosps. Assn. Am. Med. Colls., 1978, numerous other local, state and nat. coms. on health, medicine and med. edn.; Bd. dirs. East Balt. Community Corp. Recipient USPHS Career Devel. award, 1962; Distinguished Alumnus award U. Mo., 1972. Fellow ACP, Internat. Soc. Hermatology; mem. Inst. Med. Nat. Acad. Scis., Assn. Am. Med. Colls. (chmn. elect 1983), Assn. Am. Physicians, Soc. Med. Adminstrs., numerous other sci. assns. Club: Elk Ridge (Balt.). Subspecialties: Internal medicine; Hematology. Current work: Hospital administration. Home: 200 Ridgewood Rd Baltimore MD 21210 Office: 601 N Broadway Baltimore MD 21205

HIATT, HOWARD H., educator, physician; b. Patchogue, N.Y., July 22, 1925; s. Alexander and Dorothy (Askinas) H.; m. Doris Bieringer, Nov. 29, 1947; children—Jonathan, Deborah, Frederick. M.D., Harvard, 1948. Intern, then resident medicine Beth Israel Hosp.,

Boston, 1948-50; research fellow Cornell Med. Coll., 1950-53; clin. investigator USPHS, 1953-55; mem. faculty Harvard Med. Sch., 1955—, H.L. Blumgart prof. medicine, 1963-72, prof. medicine, 1972—; physician-in-chief Beth Israel Hosp., 1963-72; dean Harvard Sch. Pub. Health, 1972—. Mem. Am. Soc. Clin. Investigation, Assn. Am. Physicians, Am. Acad. Arts and Scis., Inst. Medicine, Alpha Omega Alpha. Subspecialties: Health services research; Environmental health scinence. Home: 22 Hyslop Rd Brookline MA 02146 Office: 677 Huntington Ave Boston MA 02115

HIATT, THOMAS ANDREW, genetic engineering company executive; b. Indpls., Feb. 24, 1948; s. Herbert Dale and Elmire (Freshley) H.; m. Nora McKinney, Aug. 15, 1979. B.A., Wabash Coll., 1970; M.Sc., MIT, 1972. Asst. to rep. Ford Found., Islamabad, Pakistan, 1972-75; mgr. bus. planning Elanco Products Co. (div. Eli Lilly Co.), Indpls., 1975-80; pres. Sungene Techs. Corp., Palo Alto, Calif., 1981—. Author: The Young Internationalists, 1972, Genetic Engineering and Implications for Grain Production, 1982; editor: Sloan Bus. Rev, 1971—72. Mem. Phi Beta Kappa. Subspecialties: Plant genetics; Plant cell and tissue culture. Current work: Management, agriculture, genetic engineering marketing, planning administration. Office: Sungene Technologies Corp 3330 Hillview Ave Palo Alto CA 94304

HIBBARD, EMERSON, biology educator; b. Syracuse, N.Y., Jan. 21, 1929; s. Elmer Arthur and Esther Twilton (Emerson) H. B.S., Cornell U., 1950; M.S., U. Mich., 1957; Ph.D., 1959. Fulbright fellow Inst. Animal Genetics U. Edinburgh, 1959-60; biologist Lab. Perinatal Physiology, NIH, Bethesda, Md., 1960-63; research fellow biology Calif. Inst. Tech., Pasadena, 1963-68; assoc. prof. biology Pa. State U., 1968-74, prof., 1974—. Served with USAF, 1950-54. Mem. Soc. Neurosci., Am. Assn. Anatomists, Soc. Devel. Biology, Am. Soc. Zoologists. Subspecialties: Developmental biology; Regeneration. Current work: Optic Nerve regeneration, neuronal specificity in lower vertebrates and birds. Office: Pennsylvania State 325 Mueller Lab University Park PA 16802

HIBBEN, CRAIG RITTENHOUSE, plant pathology researcher; b. Montclair, N.J., May 25, 1930; m., 1958; 2 children. B.S., Pa. State U., 1953; M.S., Cornell U., 1959, Ph.D., 1962. Researcher plant pathology Kitchawan Research Lab. Bklyn. Bot. Garden, 1962—. Mem. Internat. Shade Tree Conf., Am. Phytopath. Soc. Subspecialty: Plant pathology. Office: 712 Kitchawan Rd Ossining NY 10562

HIBBS, JOHN WILLIAM, dairy sci. educator; b. Cleve., June 2, 1917; s. Edwin Gerome and Erma Stella (Thompson) H.; m. Marie Maxwell, June 18, 1914; children: David William, Samuel Edwin. B.S. in Agr, Ohio State U., 1940, M.S., 1941, Ph.D., 1947. With Ohio Agri. Research and Devel. Center, Wooster, Ohio, 1940—, prof. dept. dairy sci., 1953—, assoc. chmn. dept., 1971—. Contbr. numerous articles to profl. jours. Mem. Am. Dairy Sci. Assn. (Borden award 1952, Am. Feed Mfg. award 1962), Am. Soc. Animal Sci., Am. Inst. Nutrition. Methodist. Lodge: Kiwanis. Subspecialties: Animal nutrition; Animal physiology. Current work: Animal nutrition and physiology; research and dept. adminstrn. Home: 185 Cherry Ln Wooster OH 44691 Office: Ohio Agrl Research and Devel Center Dept Dairy Sci Wooster OH 44691

HIBSHMAN, HENRY JACOB, chem. engr., bus. exec., cons.; b. Ambler, Pa., June 28, 1914; s. Frank Peter and Anna May (McConnel) H.; m. Anna Burnett, Apr. 24, 1940; children: Jane Norine, John Gerhardt, David Burnett, Martha Collins, Peter Sandusky. B.S. in Chem. Engring, Pa. State U., 1936, M.S. in Physics, 1937; Ph.D., Purdue U., 1950. Registered profl. engr., N.Y. Sate. With Exxon Research & Engring. Co., Linden, N.J., 1940-70; prof. U. Libya, Tripoli, 1970-73; process engr. Jacobs Engring. Co., Mountainside, N.J., 1973-75; cons. engr. contracting with U.S. Dept. Energy, Plainfield, N.J., 1975-80; prof. Algerian Inst. Petroleum, Boumerdes, 1980-81; pres. Desert Reclamation Industries, Plainfield, 1981—; cons. Contbr. articles to profl. jours. Mem. Friends of Plainfield Library. Subspecialties: Energy consulting; Environmental control systems. Current work: Supplying technology: for storing heat and cold in aquifers for processes and space conditioning; and for conserving rain water and soil in semi-arid areas. Patentee in petroleum processing (over 30). Address: 6 Crabapple Ln Plainfield NJ 07060

HICKEY, DONAL ALOYSIUS, biologist, educator; b. County Kerry, Ireland, July 13, 1948; s. William and Margaret Mary (Daly) H. B.Sc., Nat. U. Ireland, 1970; Ph.D., Harvard U., 1977. Asst. prof. biol. scis. Brock U., St. Catharines, Ont., Can., 1978-81; assoc. prof. biology U. Ottawa, Ont., 1981—. Mem. Genetics Soc. Am., AAAS, Soc. Study Evolution, Am. Soc. Naturalists. Subspecialties: Gene actions; Genome organization. Current work: Eucaryotic genetics; population genetics and evolutionary theory; evolutionary dynamics of transposable genetic elements; control of amylase gene expression in drosophila melanogaster. Home: 36 Burnham Rd Ottawa ON Canada K1S 0J8 Office: Dept Biology U Ottawa Ottawa ON Canada K1N 6N5

HICKEY, LEO JOSEPH, paleontologist; b. Phila., Apr. 26, 1940; s. James Joseph and Helen Marie (Schwarz) H.; m. Judith McKendry, June 29, 1968; children: Geoffrey Alan, Damian Michael, Jason Alexander. B.S., Villanova U., 1962; M.A., Princeton U., 1964, Ph.D., 1967. Postdoctoral research assoc. NRC-Smithsonian Instn., Washington, 1966-69; assoc. curator Mus. Natural History, 1969-80, curator, 1980-82; chmn. Mus. Natural History Exhibits Com., 1973-75; dir. Peabody Mus. Natural History, Yale U., New Haven, 1982—; adj. prof. U. Md., 1979—, U. Pa., 1982—; prof. biology Yale U., 1982—, geology and geophysics, 1982—. Contbr. articles on botany and paleontology to profl. jours. Smithsonian Research Found. grantee, 1972-76; Nat. Geog. Soc. grantee, 1979; recipient Best Paper award Geol. Soc. Washington, 1981; Henry Alan Gleason award N.Y. Bot. Garden, 1977. Mem. Geol. Soc. Am., Bot. Soc. Am., Torrey Bot. Club, AAAS, Paleontol. Soc., Yellowstone Bighorn Research Assn. (pres. 1981-83). Subspecialties: Paleobiology; Morphology. Current work: Leaf architecture of the flowering plants, dinosaurian extinction, Cretaceous and early Cenozoic paleoecology of Arctic and Rocky Mountains, early angiosperm evolution.

HICKIS, CHARLES FRANCIS, psychologist; b. Bronx, Aug. 13, 1947; s. Charles Joseph and Marion (Streng) H.; m. Judy C. Johnson, Sept. 8, 1979; children: Gregory, Rebecca, Matthew. B.A., Hofstra U., 1970, M.A., 1974; Ph.D., U. Colo., 1978. Teaching asst. psychology Hofstra U., Hempstead, N.Y., 1970-72; research asst. U. Colo., Boulder, 1972-76; asst. prof. psychology Weber State Coll., Ogden, Utah, 1977-79; postdoctoral fellow UCLA, 1979-80; clin. dir. Chem. Dependency Center, Cody, Wyo., 1980—; pvt. practice psychology, Cody, 1981—. Contbr. articles to profl. jours. NIMH fellow, 1975; NSF grantee, 1978; NIMH research service award UCLA, 1979; State of Wyo. grantee, 1981. Mem. Am. Psychol. Assn., AAAS, Rocky Mt. Psychol. Assn. Democrat. Reorganized Ch. of Jesus Christ of Latter Day Saints. Clubs: Absaroka Flycasters, N.Am. Hunting. Subspecialties: Behavioral psychology; Psychobiology. Current work: Application of behavioral and physiol. principles to treatment of substance abuse problems. Home: 2001 29th St Cody WY 82414 Office: West Park Hosp Chemical Dependency Center Cody WY 82414

HICKMAN, JACK WILLIAM, engineering executive; b. Ponca City, Okla., Oct. 4, 1936; s. Clyde A. and Hazel L. Hutson (Gilchrist) H.; children: Kirt Clyde, Lin Anne. B.S. in Elec. Engring., Okla. State U., Stillwater, 1962, M.S., U. N.Mex., 1964. Mem. staff Sandia Nat. Labs., Albuquerque, 1962-74, supr., 1974—; lectr. George Washington U., 1972-82. Mem. Am. Nuclear Soc. Subspecialty: Nuclear fission. Current work: Probabilistic risk assessment, nuclear power, nuclear safety. Home: 9508 Avenida de la Luna NE Albuquerque NM 87111 Office: Sandia Nat Labs Albuquerque NM 87185

HICKS, DARRELL LEE, mathematician, computer scientist, consultant; b. Clovis, N.Mex., July 3, 1937; s. Jason and Jesse Winona (Pierce) H.; m. Kathryn Jean Chaney, Mar. 3, 1979; children: April Lee, Rachel Elizabeth, Jason Chaney. B.S., U. N.Mex., 1961, Ph.D., 1969. Grad. asst. U. N.Mex., Albuquerque, 1961-63; research mathematician Air Force Weapons Labs., Albuquerque, 1962-69; mem. tech. staff Sandia Nat. Labs., Albuquerque, 1969-81, Idaho Nat. Engring. Lab., Idaho Falls, 1982-83; prof. math. U. Colo., Denver, 1981-83, dir. math./computer sci. clinic, 1982, chmn. applied math./computer sci. com., 1981-83; prof. dept. math. and computer sci. Mich. Technol. U., Houghton, 1983—; cons. KMS Fusion Inc., Ann Arbor, Mich., 1983—. Author: Numerical Analysis with Software Considerations, 1983; contbr. articles to profl. jours. Judge N.Mex. Sci. Fair, 1974-81. Recipient cert. of Proficiency Nat. Council Math. Tchrs., 1954; N.Mex. Legislature scholar, 1955-56; Outstanding Performance award Air Force Weapons Lab., 1967. Mem. Am. Math. Soc., Math. Assn. Am., Am. Phys. Soc., Assn. Computing Machinery. Subspecialties: Numerical analysis; Applied mathematics. Current work: Applied mathematics/computer science/numerical analysis, partial differential equations, algorithms for parallel computers, two-phase flow. Home: 4 Woodland Rd Houghton MI 49931 Office: Dept Math and Computer Scis Mich Technol U Houghton MI 49931

HICKS, ROBERT ALVIN, psychology educator; b. San Francisco, July 25, 1932; s. James B. and Vera L. (Br) H.; m. Maralee Jefferies, June 15, 1957; 1 son, Gregory J. B.A., U. Calif.-Santa Barbara, 1955; M.A., San Jose State U., 1960; Ph.D., U. Denver, 1964. Psychometrist San Jose (Calif.) State U., 1957-61; teaching fellow U. Denver, 1961-63, asst. prof., 1963-66, San Jose State U., 1966-68, assoc. prof., 1968-70, prof. psychology, 1970—; cons. Atari Corp., Sunnyvale, Calif., 1982. Contbr. articles to profl. jours. Bd. dirs. San Jose State U. Found., 1981-84. Grantee NIH, NIMH. Mem. Am. Psychol. Assn., Psychonomic Soc., AAAS (v.p. div. J, Pacific div. 1982-84), Western Psychol. Assn. (exec. bd. 1980—), Sigma Xi. Current work: Research on functions of REM sleep and the behavioral consequences of sleep selected sleep habits. Office: Dept. Psychology San Jose State U San Jose CA 95192 Home: 1118 Littleoak Circle San Jose CA 95129

HICKSON, EDWARD LILLIOTT, software engineer, consultant; b. Bayonne, N.Y., Apr. 6, 1948; s. Edward Lilliott and Julia Bird (Paschal) H.; m. Kathleen Patricia Hehir, Dec. 22, 1973; 1 dau., Sara Kathleen. Student, De Pauw U., Greencastle, Ind., 1966-69; B.S. in Telecommunications, Kent State U., 1971; postgrad., No. Va. Community Coll., Alexandria, 1978-79, U. S.C., 1980. Program dir. Warner Cable, Canton, Ohio, 1972-73; dir. audiovisual services Canton Art Inst., 1974-77; dir. computer systems S.C. Ednl. TV, Columbia, 1979-82; software engr. Raytheon Data Systems, Norwood, Mass., 1982—; cons. Bisso and Assocs., Columbia, 1980-81. Photographer: book Architecture in Canton, 1976. Pres. Clear Spring Neighborhood Assn., Inc., Columbia, 1980-81; student del. campus council No. Va. Community Coll., 1978-79; mem. Downtown Entertainment Com., Canton, 1976-77; pres., youth advisor Community Devel. Council Greencastle, 1968-69. Mem. Assn. Computing Machinery, S.C. Assn. Data Processsing Dirs. (tech. tng. com. 1981-82, telecommunications com. 1982). Subspecialties: Distributed systems and networks; Information systems, storage, and retrieval (computer science). Current work: Systems support engineering for distributed processing, network software-an enlargement of past efforts in design and development of complex application systems. Office: Raytheon Data Systems 1415 Providence-Boston Turnpike Norwood MA 02062

HICKSON, PAUL, astronomer, astrophysicist. B.Sc. with honors in Physics, U. Alta., 1971; Ph.D. in Astronomy, Calif. Inst. Tech., 1976. Fellow U. B.C., Vancouver, 1978-81, Natural Scis. and Engring. Research Council Can. fellow, 1981—. Contbr. articles to profl. jours. Recipient Sam Fefferman Meml. Gold medal U. Alta., 1971. Mem. Am. Astron. Soc., Internat. Astron. Union, Astron. Soc. Pacific. Subspecialties: Cosmology; Optical astronomy. Current work: Cosmology, galaxy clusters, study of galaxies and groups, development of low light electronic imaging and spectrophotometric systems. Office: Dept Geophysics and Astronomy U BC 2219 Main Mall Vancouver BC Canada V6T I W5

HIERHOLZER, JOHN CHARLES, microbiologist, cons., researcher; b. Gravenhurst, Ont., Can., July 1, 1938; s. Leo Newman and Cathrine Ann (Picker) H.; m. Connie Louise McArthur, Oct. 21, 1967; children: Jack, Karl, Mike. B.S., Spring Hill Coll., 1960; M.S., U. Fla., 1962; Ph.D., U. Md., 1966. Lab asst. dept. bacteriology U. Fla., 1961-62; research microbiologist U.S. Dept. Agr., Beltsville, Md., 1962-67; supervisory research microbiologist Ctr. Disease Control, Dept. Health and Human Resources, Atlanta, 1967—; cons. Fernbank Sci. Ctr., Atlanta, 1974—. Contbr. numerous articles on med. virology and epidemiology, viral biochemistry to profl. jours. Served to lt. comdr. USN, 1967-69. Mem. Am. Soc. Microbiology, Am. Chem. Soc., N.Y. Acad. Scis., AAAS, Sigma Xi. Subspecialties: Virology (medicine); Biochemistry (biology). Current work: Viral biochemistry and immunology; human respiratory virus infections. Office: 1600 Clifton Rd NE Atlanta GA 30333

HIETBRINK, BERNARD EDWARD, pharmacology educator; b. Strasburg, N.D., Nov. 23, 1930; s. Bernard Dirk and Esther E. (VanDenTak) H.; m. A. Elaine Jonker, Mar. 16, 1951; children: Deanne Rae, Beth, Laura Jean, Bernard Dale. B.S., S.D. State U., 1958; Ph.D., U. Chgo., 1961. Research assoc., asst. prof. U. Chgo., 1961-64; asst. prof. pharmacology S.D. State U., Brookings, 1964-66, assoc. prof., head pharmacology, 1966-71, prof., head pharmacology, 1971-83, head pharm. sci., 1983—. Served with USAF, 1951-55. Mem. Soc. Toxicology, Soc. Pharmacology and Exptl. Therapeutics, S.D. Pharm. Assn., Rho Chi, Phi Lambda Upsilon. Republican. Mem. Reformed Ch. Subspecialties: Pharmacology; Toxicology (medicine). Current work: Drug metabolism, influence of chemical agents on microsomal enzyme activity, mechanism of action of toxic substances. Office: Dept Pharmacoloty SD State U Brookings SD 57007 Home: 1613 Buffalo Trail Brookings SD 57006

HIGASHI, GENE ISAO, immunoparasitologist; b. Gardena, Calif., Nov. 6, 1938; s. Kay Kasutaro and Takeko (Ogo) H.; m. Elizabeth Lee, Aug. 20, 1966; 1 dau., Misao Elizabeth. B.A., Swarthmore Coll., 1960; M.D., Yale U., 1964; Sc.D., Johns Hopkins U., 1973. Intern Grace-New Haven Community Hosp., Yale U. Sch. Medicine, 1964-65; asst. resident anatomic pathology Yale-New Haven Hosp. and Med. Ctr., Yale U. Sch. Medicine, 1972-73, research fellow in parasite immunology dept. pathology, 1972-73, asst. prof. dept. pathology, 1973-75, clin. asst. prof., 1975-78; head immuno-parasitology div. U.S. Naval Med. Research Unit, Cairo, 1969-72, head parasite immunology div., 1974-76, head immunology dept., 1977-79; assoc. prof. dept. epidemiology U. Mich. Sch. Pub. Health, Ann Arbor, 1979—. Contbr. articles to profl. jours. Served to lt. comdr. USNR, 1969-72. Mem. AAAS, Am. Soc. Parasitologists, Am. Soc. Tropical Medicine and Hygiene, Royal Soc. Tropical Medicine and Hygiene, Am. Assn. Immunologists, Alpha Omega Alpha. Subspecialties: Parasitology; Infectious diseases. Office: 109 Observatory St Ann Arbor MI 48109

HIGGINS, CHARLES GRAHAM, geology educator; b. Oak Park, Ill, Nov. 18, 1925; s. Charles Graham and Frances Anne (Henderson) H.; m. Rosalie Darleen Trew, Feb. 9, 1974; children by previous marriage: Kimberley Frances Higgins Tolley, Lesley Vivian. S.B., U. Chgo., 1946, S.M., 1947; Ph.D., U. Calif.-Berkeley, 1950. Field instr. geology U. Mich., summer, 1949, instr. geology, 1950-51; asst. prof. geology U. Calif., Berkeley and Davis, 1951-53, U. Calif.-Davis, 1953-58, assoc. prof., 1958-66, prof., 1966—. Fellow Geol. Soc. Am.; mem. Nat. Assn. Geology Tchrs. (vice chmn. sect. 1953-55, treas. 1954-57), Brit. Geomorphology Research Group, AAAS, Am. Quaternary Assn., AAUP, Phi Beta Kappa, Sigma Xi. Subspecialties: Geology; Hydrogeology. Current work: Role of subsurface water in shaping landforms and landscapes of the Earth's surface; role of terracette microrelief formed by grazing animals in modifying hillslopes and hillslope-forming processes; provenance of ancient marble. Office: Dept Geolog U Calif Davis CA 95616

HIGGINS, EDWIN STANLEY, biochemistry educator; b. N.Y.C., Mar. 12, 1925; s. John Thomas and Nettie Viola (Stein) H.; m. Barbara Jean Wilson, June 20, 1958; children: Elizabeth Jean, James Edwin, Glenn David. B.A. cum laude, Alfred (N.Y.) U., 1952; Ph.D., SUNY-Syracuse, 1956. Teaching, research asst. dept. biology Alfred (N.Y.) U., 1951-52; teaching, research fellow SUNY Med. Ctr., Syracuse, 1952-56; asst. prof. biochemistry Med. Coll. Va., Va. Commonwealth U., Richmond, 1956-62, assoc. prof. biochemistry, 1962-68, acting chmn, dept. biochemistry, 1976-77, prof. biochemistry, 1968—; research dir. State Health Dept., Alcohol Studies, Richmond, Va., 1958-73; cons. and lectr. in field. Contbr. articles to profl. jours.; reviewer and editor. Served to cpl. U.S. Army, 1943-46. Named Outstanding Tchr. Med. Coll. Va., 1970; A.D. Williams Fund grantee, 1956-57; NIH grantee, 1956-69; Va. State Health Dept. grantee, 1959-73. Mem. Soc. Exptl. Biology and Medicine (region cons. 1961-68), Am. Chem. Soc. (exec. com. 1961-62), Am. Inst. Nutrition, AAAS, Sigma Xi (chpt. pres. 1967-68). Presbyterian. Subspecialties: Biochemistry (biology); Nutrition (biology). Current work: Research in bioenergetics and function of mitochondria; teaching, medicine, dentistry, pharmacy, institutional research. Home: 5928 Old Orchard Rd Richmond VA 23227 Office: Dept Biochemistry Med Coll V Va Commonwealth U Richmond VA 23298

HIGGINS, IRWIN RAYMOND, biochemist, chemical company executive; b. Mapleton, Maine, feb. 15, 1919; s. Raymond and Hazel (Foss) H.; m. Pauline Lewis, May 4, 1943; children: Peggy, Judith, Mellissa. B.S. in Biochemistry, U. Maine, 1942. Process engr. E.I. DuPont deNemours, Ill., Tenn., Ala., Wash., N.J., 1942-46; group leader ion exchange Oak Ridge Nat. Lab., Tenn., 1946-58; v.p., tech. dir. Chem. Separations Corp., Oak Ridge, 1958-82, C/S Assocs., 1982—. Fellow Am. Inst. Chem. Engrs.; mem. AAAS, Am. Nuclear Soc., AIME. Republican. Baptist. Subspecialties: Water supply and wastewater treatment; Metallurgical engineering. Current work: Continuous ion exchange process development. Office: C/S Assocs 101 Midway Ln Oak Ridge TN 37830

HIGGINS, PAUL JOSEPH, molecular biologist, cancer researcher; b. Bklyn., July 30,1946; s. Vincent John and Lucille Theresa (Gendus) H.; m. Denise Laura Cote, Feb. 24, 1973; children: Jennifer Ann, Stephen Paul, Craig Evan, Erik James, Sean Patrick. B.S., Iona Coll., New Rochelle, N.Y., 1969; M.S., L.I. U., 1973; Ph.D., NYU, 1976. Research asso. Meml. Sloan-Kettering Cancer Center, 1976-79, assoc., head hepatic carcinogensis program, 1979—; asst. prof. genetics and molecular biology Cornell U., 1980—. Author numerous sci. papers. Recipient Young Investigators award Nat. Cancer Inst., 1980-83. Mem. Harvey Soc., Am. Assn. Cancer Research, Am. Soc. Cell Biology, N.Y. Acad. Scis., Am. Soc. Microbiology. Subspecialties: Cancer research (medicine); Cell biology. Current work: Molecular biology of cell differentiation and transformation; cancer research; gene cloning. Home: 1858 Byrd Dr East Meadow NY 11554 Office: 1275 York Ave New York NY 10021

HIGGINSON, JOHN, pathologist, consultant; b. Belfast, No. Ireland, Oct. 16, 1922; came to U.S., 1958, naturalized, 1963; s. William and Ellen Margaret (Rogers) H.; m. Nan McKee, Nov. 29, 1949; children: Jaqueline Higginson Huntly, Wendy Higginson Dunne. B.A., U. Dublin, 1945, M.B., B.Ch., 1946, M.D., 1961. Intern Sir Patrick Duns Hosp., Dublin, Ireland, 1946; resident dept. pathology Western Infirmary, U. Glasgow, Scotland; head geog. pathology unit South African Inst. Med. Research, Johannesburg, 1950-58; Am. Cancer Soc. prof. geog. pathology U. Kans., Kansas City, 1958-66; dir. Internat. Agy. for Research on Cancer (WHO), Lyon, France, 1966-81; sr. scientist Univs. Assoc. for Research and Edn. in Pathology, Bethesda, Md., 1982—; bd. dirs. Chem. Industries Inst. of Toxicology; cons. to univs., govt., industry. Contbr. articles on health, disease and enviro. to profl. jours. Recipient 1st ann. award Am. Council on Sci. and Health, 1981. Fellow Royal Coll. Physicians (London), Royal Acad. Medicine Ireland (hon.); mem. AAAS, Am. Assn. Cancer Research, Am. Assn. Pathologists, Soc. Toxicology, Sigma Xi. Presbyterian. Club: Cosmos (Washington). Subspecialties: Cancer research (medicine); Environmental toxicology. Current work: Environment, health and cancer; environment and social facts; lifestyle and disease; epidemiology. Office: Univs Assoc for Research and Edn in Pathology 9650 Rockville Pike Bethesda MD 20814

HIGGS, LLOYD ALBERT, astronomer; b. Moncton, N.B., Can., June 21, 1937; s. Maxwell Lemert and Reta Mae (Jollymore) H.; m. Kathleen Mary Fletcher, Jan. 15, 1966; children: Kevin, Scott, Michelle. B.Sc., U. N.B., Fredericton, 1958; Ph.D., U. Oxford, Eng., 1961. Research scientist NRC Can., Herzberg Inst. Astrophyics, Dominion Radio Astrophys. Obs., Ont., 1961-81, dir., Penticton, B.C., 1981—; research scientist Leiden U., Holland, 1964-65. Editor: Jour. Royal Astron. Soc. Can, 1976-80; contbr. numerous articles to profl. jours. Rhodes scholar, 1958. Mem. Can. Astron. Soc., Royal Astron. Soc., Am. Astron. Soc., Royal Astron. Soc. Can. Subspecialty: Radio and microwave astronomy. Current work: Physics of interstellar medium, supernova remnants, applications of computers in astronomy. Home: 2515 Dartmouth Rd RR 2 Penticton BC Canada V2A 6J7 Office: PO Box 248 Penticton BC Canada V2A 6K3

HIGH, LEE RAWDON, JR., petroleum geologist; b. Pinebluff, Ark., Feb. 6, 1941; s. Lee Rawdon and Audelia Vaugine (Watson) H.; m. Marilyn Gay Clarkson, Dec. 7, 1968; children: Terry Loren, Matthew Lee, Jason Scott. A.B., Princeton U., 1963; Ph.D., Rice U., 1967. Instr. U. Nebr.-Lincoln, 1966-67; assoc. prof. Oberlin Coll., 1967-78, research status, 1973-74; geologist Mobil Oil Co., Dallas, 1978—. Author: Sedimentary Structures of Ephemeral Streams, 1973; contbr. articles to profl. jours. Active Circle Ten council Boy Scouts Am., 1981—. NDEA fellow, 1963; Am. Chem. Soc. grantee, 1967-69; Research Corp. grantee, 1969-71. Fellow Geol. Soc. Am.; mem. Am. Assn. Petroleum Geologists, Soc. Econ. Paleontologists and Mineralogists (assoc. editor 1977—). Subspecialties: Geology; Petroleum geology. Current work: Petroleum exploration in frontier areas; sedimentology of lacustrine rocks. Home: 2109 Lymington Carrollton TX 75007 Office: Mobile Oil Co PO Box 900 Dallas TX 75221

HIGHBERGER, PAUL FEIGHTNER, chemical engineer; b. Mt. Vernon, N.Y., Aug. 26, 1933; s. William Webber and Dorothy (Feightner) H.; m. Sydney Young, Feb. 10, 1957; children: William, Steven, Kathleen. B.S. in Chem. Engring, Lafayette Coll., 1958. With Allied Chem. Corp., Morristown, N.J., 1958—, mgr. comml. devel., 1962-71; regional mktg. mgr. Allied Gen. Nuclear Services, Barnwell, S.C., 1971-77, spent fuel program mgr., 1977-81, dir. advanced programs, 1981—. Mem. Am. Nuclear Soc. (chmn. topical meeting 1982). Republican. Presbyterian. Subspecialty: Nuclear fission. Current work: Spent fuel storage, transportation and disposal. Home: 109 Hartwell Dr Aiken SC 29801 Office: Allied Gen Nuclear Services Osborn Rd Barnwell SC 29812

HIGHLEY, TERRY LEONARD, forest products pathologist; b. Anamosa, Iowa, July 3, 1940; s. Charles Leonard and Muriel Frieda (Hanssen) H.; m. Barbara Lee McKnight, June 17, 1967; children: Christopher, Stephanie. B.S., Iowa State U., 1962; M.S., Oreg. State U., 1964, Ph.D., 1967. Researcher Oreg. State U., Corvallis, 1962-67; research forest products pathologist Forest Products Lab., Madison, Wis., 1967—. Contbr. articles to profl. jours. Recipient fellowships in bark and wood chemistry and plant pathology. Mem. Am. Phytopathol. Soc. Democrat. Methodist. Subspecialties: Plant pathology; Microbiology. Current work: Mechanisms of wood decay and development of non-toxic control methods. Home: 910 Hathaway Dr Madison WI 53711 Office: Gifford Pinchot Dr Madison WI 53705

HIGHTON, RICHARD TAYLOR, zoologist; b. Chgo., Dec. 24, 1927; s. Albert Henry and Helen Irene (Taylor) H.; m. Kathryn Ann Adams, June 23, 1950; children: Barbara, Kim, Scott, Caitlin Ann. A.B., N.Y. U., 1950; M.S., U.Fla., 1953, Ph.D., 1956. Asst. prof. zoology U. Md., College Park, 1956-62, assoc. prof., 1962-73, prof., 1973—. Contbr. articles to profl. jours. Served with AUS, 1946-48. Fellow AAAS; mem. Am. Soc. Ichthyologists and Herpetologists (sec. 1967-73, pres. 1976), Am. Soc. Naturalists, Genetics Soc. Am., Soc. Systematic Zoology, Ecol. Soc. Am., Animal Behavior Soc., Soc. Study of Evolution, Herpetologists League, Soc. Study Amphibians and Reptiles, Sigma Xi. Subspecialties: Evolutionary biology; Systematics. Current work: research in evolutionary biology, population genetics, ecology and systematics of amphibians and reptiles, especially salamanders of the genus Plethodon. Home: 3613 Van Ness St NW Washington DC 20008 Office: Dept Zoology U Md College Park MD 20742

HIGHTOWER, JOE WALTER, chemical engineering educator, consultant; b. Morrilton, Ark., Sept. 14, 1936; s. Walter Eugene and Verda Mae (Poindexter) H.; m. Sallie Turner, Sept. 5, 1959; 1 dau., Amy Margaret; m. Ann Grekel, May 11, 1980. B.S., Harding Coll., 1959; M.S., Johns Hopkins U., 1961, Ph.D., 1963. NSF postdoctoral fellow, Belfast, No. Ireland, 1963-64; research fellow Mellon Inst., Pitts., 1964-67; assoc. prof. dept. chem. engring. Rice U., Houston, after 1967, now prof.; cons. to several cos.; mem. sci. adv. bd. Haldor Topsoe, Copenhagen. Contbr. articles to sci. tech. jours.; editor. Co-founder, pres. Human Resources Devel. Found., 1968—. Mem. Am. Chem. Soc. (nat. award in petroleum chemistry 1973, Southeastern Tex. sect. award 1976), Am. Inst. Chem. Engrs., Southwest Catalysis Soc., Catalysis Soc., ASTM. Republican. Mem. Ch. of Christ. Subspecialties: Catalysis chemistry; Chemical engineering. Current work: Heterogeneous catalysis, catalytic reaction mechanisms, adsorption, heterogeneous kinetics, surface chemistry. Office: Department of Chemical Engineering Rice University Houston TX 77001

HIGUCHI, TAKERU, chemistry educator; b. Los Altos, Calif., Jan. 1, 1918; s. Iekichi and Chiye (Shiki) H.; m. Aya Toki, Jan. 1, 1944; children: Kenji W., Junji H., Chie S., Peter T. A.B., U. Calif. at Berkeley, 1939; Ph.D., U. Wis., 1943; D.Sc. (hon.), U. Mich., 1967, Eidgenössische Technische Hochschule, Zurich, 1978, U. Ill., 1980, Phila. Coll. Pharmacy and Sci., 1982. Research asso. U. Wis., 1943-44; research chemist Office Rubber Research, U. Akron, 1944-47; mem. faculty U. Wis., 1947-67, prof. pharm. chemistry, 1954-64, Edward Kremers prof., 1964-67; Regents distinguished prof. chemistry and pharmacy U. Kans., Lawrence, 1967—; Pres. INTERx Research Corp., 1972—; v.p. Merck Sharp & Dohme Research Labs.; revision com. U.S. Pharmacopoeia, 1960-70; David E. Guttman Meml. lectr. U. Ky., 1978; Rachelle lectr. Calif. State U., Long Beach, 1979; Allen I. White lectr. Wash. State U., 1982. Author numerous papers in field. Co-recipient Ebert prize Am. Pharm. Assn., 1951, 52, winner, 1954; recipient Sturmer Lectr. award PCPS chpt. Rho Chi, 1956; Research Achievement award phys. pharmacy Am. Pharm. Assn. Found., 1962; Justin Power award pharm. analysis, 1964; research achievement award in stimulation of research, 1967; hon. citation U. Wis., 1969; award for advancement indsl. pharmacy, 1974; Internat. Surfactant Chemistry prize Italian Oil Chemists' Soc., 1974; Scheele lectr. award Pharm. Soc. Sweden, 1970; Rho Chi lectr., 1971; Recipient Kolthoff Gold medal award Am. Pharm. Assn. Acad. Pharm. Scis., 1977; Volwiler award Am. Assn. Colls. Pharmacy, 1978; Citation for Disting. Service U. Kans., 1982, Roland T. Lakey award Wayne State U., 1982. Fellow Acad. Pharm. Scis.; mem. Am. Chem. Soc. (Midwest award 1975), Am. Pharm. Assn. (life mem.; past chmn. sci. sect., Remington Honor medal 1983), Am. Oil Chemists Soc., Internat. Assn. Dental Research, Chem. Soc. (London, Eng.), Acad. Pharm. Scis. (pres. 1965-67), Japanese Pharm. Soc. (hon.), Victorian Pharm. Soc. Australia (hon.), Mexican Assn. Students Pharmacy (hon.), Phi Beta Kappa (hon.), Sigma Xi, Rho Chi. Subspecialties: Physical chemistry; Thermodynamics. Current work: Physical chemistry of dosage forms and drug administration; analytical chemistry of pharmaceutical and organic species. Home: 2811 Schwarz Rd Lawrence KS 66044

HILBERT, MORTON SHELLY, environmental health educator; b. Pasadena, Calif., Jan. 3, 1917; s. George Lewis and Mary (Shelly) H.; m. Stephanie Mayer, July 3, 1972; children: Kathy, Barbara, Stephen. B.S. in Civil Engring, U. Calif.-Berkeley, 1940; M.P.H., U. Mich., 1946. Registered profl. engr., Mich. San. engr. Barry County (Mich.) Health Dept., 1940-41, Chippewa County (Mich.) Health Dept., 1941-42; dir. environ. health County of Wayne (Mich.), 1944-61; prof. environ. health Sch. Pub. Health U. Mich., Ann Arbor, 1961—, chmn. dept. environ. and indsl. health, 1968—, also dir.; mem. Comprehensive Health Planning Council for Southeastern Mich., 1982—. Recipient spl. recognition award Mich. Environ. Health Assn., 1976; Outstanding Service award State of Mich., 1976. Fellow Am. Acad. Environ. Engrs.; mem. Am. Pub. Health Assn. (pres. 1977), Engring. Soc. Detroit (pres. 1971), Mich. Pub. Health Assn. (pres. 1955, Disting. Service award 1976). Subspecialties: Environmental engineering; Ecology. Current work: Impact of environment on human health; on site waste disposal; environmental health in developing countries. Office: School of Public Health University of Michigan Ann Arbor MI 48109

HILBERTZ, WOLF HARTMUT, marine resources company executive; b. Gutersloh, Westfalia, Germany, Apr. 16, 1938; came to U.S., 1965; s. Rudolf Robert and Erna Charlotte (Uslat) H.; m. Frances Louise Carvey, Aug. 20, 1982; 1 dau., Erna Navassa; m.; children by previous marriage: Kai Hannes, Derrick Max August, Halona Cordula. Werkarchitekt HBK Berlin, Staatliche Hochschule Fuer Bildende Kuenste, Berlin, 1965; M.Arch., U. Mich.-Ann Arbor, 1966. Designer Max Urbain Architects, N.Y.C., 1965-66; sr. designer Smith, Hinchman, Grylls, Detroit, 1966-67; asst. prof. dept. arch. So. U., Baton Rouge, 1967-68, assoc. prof., 1968-70, Sch. Arch., U. Tex.-Austin, 1970-81, research scientist marine sci., 1978-80; pres. Marine Resources Co., Galveston, Tex., 1979—, Sea-Crete Resources La., Lake Charles, 1982—; cons. Bekaert, Zvewegem, Belgium, 1982, Dresser Industries, Houston, 1972, Maxwell A.G., Zurich, Switzerland, 1981—, Dome Petroleum, Calgary, Alta., Can., 1981; dir. Sea-Crete Resources of Japan, Oklahoma City, 1982—, The Symbiotic Processes Lab., Austin, Tex., 1970-81. Contbr. articles in field to profl. jours. Served with German Air Force, 1959-60. So. Consumers Coop. grantee, 1969; Sea Grant Office grantee, 1976; U. Tex.-Port Aransas grantee, 1965-79; Nat. Endowment for Arts mineral accretion workshop grantee, 1979. Mem. AAAS, Am. Inst. Ocean Architecture (pres. 1981—), Assn. Study of Man-Environment Relations, Oceanic Soc., Marine Tech. Soc., Coastal Soc. Subspecialties: Marine Architecture. Current work: Evolutionary morphological systems, ocean architecture, OTEC cold water pipes, aquaculture facilities, energy harnessing systems, shore reclamation and protection; protection of wooden structures in marine environments. Patentee in field. Home: 819 Ball Ave Galveston TX 77550 Office: Marine Resources Co 819 Ball Ave Galveston TX 77550

HILBORN, ROBERT CLARENCE, physicist, educator; b. Norristown, Pa., June 24, 1943; s. Clarence L. and Dorothy (Ditzler) H.; m. Shirley A. Antosiewicz, June 27, 1970; children: Stephen, Kurt. B.A., Lehigh U., 1966; M.A., Harvard U., 1967; Ph.D., 1971. Research assoc., lectr. SUNY-Stony Brook, 1971-73; vis. researcher U. Calif.-Santa Barbara, 1979-80; assoc. prof. Oberlin (Ohio) Coll., 1973—; cons. on optics and spectroscopy. Contbr. articles to profl jours. Research Corp. grantee, 1974-82; NSF grantee, 1981—. Mem. Am. Phys. Soc., Am. Assn. Physics Tchrs., Optical Soc. Am., Phi Beta Kappa, Sigma Xi. Subspecialties: Atomic and molecular physics; Spectroscopy. Current work: Atomic and molecular laser spectroscopy. Home: 56 Spring St Oberlin OH 44074 Office: Dept Physics Oberlin Coll Oberlin OH 44074

HILDEBRAND, JOHN G(RANT), III, educator, researcher, musician; b. Boston, Mar. 26, 1942; s. John G. and Helen Mathilda (Swedberg) H.; m. Gail Deerin Burd, July 24, 1982. A.B., Harvard U., 1964; Ph.D., Rockefeller U., 1969. Fellow in neurobiology Harvard U. Med. Sch., 1969-72, asst. prof., 1972-77, assoc. prof., 1977-80, tutor in biochem. scis., 1970-80, vis. prof., 1980-81, assoc. in behavioral biology, 1980—; prof. biol. scis. Columbia U., 1980—; adj. prof. Rockefeller U., 1981—; trustee, mem. exec. com., co-dir. neurobiol. course Marine Biol. Lab., Woods Hole, Mass. Editor: Chemistry of Synaptic Transmission, 1974, Receptors for Neurotransmitters, Hormones and Pheromones in Insects, 1980; mem. editorial bd.: Jour. Neuroscience, 1981—, Jour. Neurochemistry, 1981—, Trends in Neurosciences, 1981—, Neuroscience Commentaries, 1981—, Insect Biochemistry, 1977—; contbr. articles to profl. jours. Helen Hay Whitney Found. fellow, 1969-72; Alfred P. Sloan Found. fellow, 1973-77. Mem. Am. Soc. Neurochemistry, Assn. Chemoreception Scis., Am. Soc. Biol. Chemists, Entomol. Soc. Am., Soc. Exptl. Biology, Soc. Neuroscience, Phi Beta Kappa, Sigma Xi. Club: Cambridge (Mass.) Entomol. Subspecialties: Neurobiology; Neurochemistry. Current work: Developmental neurobiology, synaptic neurochemistry, biology, behavior, and neurobiology of arthropods (especially insects), chem. senses, insects as vectors of parasitic diseases and as agricultural pests. Office: Dept Biological Sciences Columbia U Fairchild 913 New York NY 10027

HILDEBRANDT, THOMAS OWEN, health physicist; b. Detroit, Apr. 27, 1954; s. Nestor Owen and Elaine Theresa (Najewski) H.; m. Donna Lynn Bennet, Oct. 21, 1972; 1 dau., Kelly Suzanne. A.S., SUNY-Albany, 1979. Dosimetry specialist Miss. Power & Light Co., Port Gibson, 1980-82, supr. health physics, 1982—. Served with USN, 1972-80. Mem. Am. Nuclear Soc., Health Physics Soc. Republican. Roman Catholic. Subspecialty: Nuclear fission. Current work: Health physics, radiation protection, dosimetry, radioactive waste. Office: Mississippi Power & Light Co PO Box 756 Port Gibson MS 39150

HILDNER, ERNEST GOTTHOLD, III, research scientist; b. Jacksonville, Ill., Jan. 23, 1940; s. Ernest Gotthold Jr. and Jean Johnston (Duffield) H.; m. Sandra Whitney Shellworth, June 22, 1968; children: Cynthia Whitney, Andrew Duffield. B.A., Wesleyan U., 1961; M.A., Colo. U., 1964, Ph.D., 1971. Scientist NOAA, Boulder, Colo., 1961-67, Nat. Ctr. for Atmospheric Research, Boulder, 1967-80; br. chief Solar Scis. br. NASA/Marshall Space Flight Ctr., Huntsville, Ala., 1980—. Contbr. articles to profl. jours. Recipient Tech. Advancement award Nat. Ctr. Atmospheric Research, 1974; Group Achievement award NASA, 1975. Mem. Am. Astron. Soc. (solar physics div.), Internat. Astron. Union, Am. Geophys. Union, AAAS, Sigma Xi. Subspecialties: Solar physics; Optical astronomy. Current work: Coronal transients, MHD; managing solar physics research group with both observational and theoretical work. Office: ES 52 NASA/Marshall Space Flight Ctr Huntsville AL 35812

HILDRETH, EUGENE A., physician, educator; b. St. Paul, Mar. 11, 1924; s. Eugene A. IV and Lila K. (Clator) H.; m. Dorothy Anne Myers, Mar. 23, 1946; children: Jeffrey Reed, William Myers, Anne Sarver, Katherine Clator. B.S., Washington Jefferson Coll., 1943; M.D., U. Va., 1947. Diplomate: Am. Bd. Internal Medicine (mem. 1969-72, 75—, cons., com. mem. 1972-75), Am Bd. Allergy and Immunology (founding com. 1970, mem. 1970-72, 1st co-chmn.). internal medicine 1970-71, cons. 1972—). Intern Johns Hopkins, 1947-48; resident in medicine Hosp. U. Pa., 1948-49, USPHS Postdoctoral Research fellow in cardio-vascular disease, 1949-51, chief resident in medicine, 1953-54, fellow in allergy and immunology, 1954-58, faculty, 1954-69, 71—; instr. medicine U. Pa., Phila., 1953-54, asso. medicine, 1954-55, asst. prof. medicine, 1955-60, asso. prof., 1960-69, asso. dean, 1964-69, prof. clin. medicine, 1971—, acting chmn. dept. research medicine, 1960-64; chmn. dept. medicine Reading (Pa.) Hosp. and Med. Center.; Cons. project site visits USPHS, 1965-70, rev. devel. new methods research in chronic pulmonary disease, 1967-69; cons. VA Hosp. Phila., 1955—; nat. adv. com. Medic Alert Found. Internat., 1964-83; cons. Citizens' Com. to Study Grad. Med. Edn., 1966; Am. Bd. Med. Spltys. rep. of subsplty. Bd. Allergy and Immunology of Am. Bd. Internal Medicine, 1969-72; chmn. certifying exam. com. Am. Bd. Internal Medicine, 1978-81, mem. exec. com., 1978-82; chmn., 1981-82; mem. rep. Am. Bd. Med. Spltys., 1976—, chmn. nominating com., 1979-80, mem. evaluation procedures study com., 1979—; mem. med. adv. bd. Lupus Found. Del. Valley, 1979—; chmn. Federated Council Internal Medicine; appeals bd. liaison Council of Grad. Med. Edn., 1980—. Co-author: Low Fat Diet, 1953, also research articles, chpts. in textbooks.; Editorial bd.: Annals Internal Medicine, 1960-68, Postgrad. Medicine, 1969-75, Jour. Berks County Med. Soc, 1969-73, Internal Medicine Digest, 1971-75. Served with USNR, 1943-45, 51-53. John and Mary R. Markle scholar in acad. medicine, 1958-63; USPHS Research grantee. Fellow Am. Clin and Climatologic Assn., A.C.P.; mem. Peripatetic Soc., AAAS, Fedn. Am. Socs. for Exptl. Biology, N.Y. Acad. Scis., Am. Heart Assn., Inst. Medicine of Nat. Acad. Scis. (nominating com. 1982—), Pa. Thoracic Soc., Phila. Art Mus., AMA, Am. Acad. Allergy, Nat. Kidney Found. Subspecialties: Internal medicine; Immunology (medicine). Current work: Research in evaluation of medical education. Home: RD 3785 Mohnton PA 19540 Office: Reading Hosp and Medical Center Reading PA 19603

HILER, EDWARD ALLAN, agrl. engr.; b. Hamilton, Ohio, May 14, 1939; s. Earl and Thelma H.; m. Patricia Ann Burke, Jan. 30, 1960; children—Karen, Richard, Scott. B.S., Ohio State U., 1963, M.S., 1966, Ph.D. (USPHS fellow 1964-65), 1966. Registered profl. engr., Tex. Instr. Ohio State U., 1962-64, Ohio Agrl. Research and Devel. Center, 1965-66; Mem. faculty Tex. A. and M. U., 1966—, prof. agrl. engring., 1973—, chmn. dept., 1974—. Contbr. articles profl. jours. Recipient Disting. Service award Tex. A&M U., 1974, Faculty Disting. Achievement award, 1973; named Disting. Alumnus Ohio State U. Coll. Engring., 1978. Fellow AAAS; mem. Am. Soc. Agrl. Engrs. (Paper award 1972, 74, Young Researcher award 1977, Disting. Young Agrl. Engr. award 1975), Am. Soc. Engring. Edn., Am. Geophys. Union, Council Sgrl. Sci. and Tech., Tex. Soc. Profl. Engrs., Sigma Xi (chpt. Disting. Mem. award 1975). Presbyterian. Subspecialties: Agricultural engineering; Biomass (energy science and technology). Current work: Irrigation and drainage crop requirements, irrigation and drainage system design, biomass energy. Office: Dept Agrl Engring Tex A&M U College Station TX 77843

HILFER, SAUL ROBERT, biology educator; b. Quakertown, Pa., June 12, 1931; s. Samuel and Hannah (Frankel) H.; m. Eva Martha Kuhn, Dec. 11, 1959; children: Eric Stefan, Ada Martha, Susanna Cindy. B.S., Queens Coll., 1954; M.A., Amherst Coll., 1955; Ph.D., Yale U., 1960. Postdoctoral fellow Harvard Med. Sch., Boston, 1959-61; asst. prof. dept. biology Temple U., Phila., 1961-66, assoc. prof., 1966-74, prof., 1974—; co-organizer Symposia on Ocular-Visual Devel., Pa. Coll. Optometry and Temple U., Phila., 1974—. Editor (with J.B. Sheffield); book series Cell and Developmental Biology of the Eye, 1981. Mem. Am. Soc. Zoologists, Am. Soc. Cell Biology, Internat. Soc. Devel. Biology, Soc. Development Biology. Subspecialties: Developmental biology; Cell biology. Current work: Role of contractile proteins during early organ development; extracellular matrix changes during development and cell interaction; functional stability of cells in culture. Office: Temple Univ Dept Biology Philadelphia PA 19122

HILGARD, ERNEST ROPIEQUET, psychologist; b. Belleville, Ill., July 25, 1904; s. George Engelmann and Laura (Ropiequet) H.; m. Josephine Rohrs, Sept. 19, 1931; children—Henry Rohrs, Elizabeth Ann. B.S., U. Ill., 1924; Ph.D., Yale, 1930; D.Sc., Kenyon Coll., 1964; LL.D., Centre Coll., 1974. Asst. instr. in psychology Yale U., 1928-29, instr., 1929-33; successively asst. prof., asso. prof., prof. psychology Stanford, 1933-69, emeritus prof., 1969—, exec. head dept., 1942-50, dean grad. div., 1951-55; Bd. dirs., pres. Ann. Reviews, Inc., 1948-73; With USDA, Washington, 1942, OWI, 1942-43, Office Civilian Requirements, WPB, 1943-44; Collaborator, div. child devel. and tchr. personnel Am. Council Edn., 1940-41; nat. adv. mental health council USPHS, 1952-56; fellow (Center Advanced Study Behavioral Scis.), 1956-57; Mem. U.S. Edn. Mission to Japan, 1946. Author: several books, latest Theories of Learning, 1948, rev. edit., 1981, Introduction to Psychology, 1953, revised edit., 1983, Hypnotic Susceptibility, 1965, Hypnosis in the Relief of Pain, 1975, Divided Consciousness, 1977, American Psychology in Historical Perspective, 1978. Bd. curators Stephens Coll., Mo., 1953-68. Recipient Warren medal in exptl. psychology, 1940; Wilbur Cross medal Yale U., 1971; Gold medal Am. Psychol. Found., 1978. Hon. fellow Brit. Psychol. Assn.; mem. Am. Psychol. Assn. (pres. 1948-49), Am. Acad. Arts and Scis., Nat. Acad. Edn., Soc. Psychol. Study Social Issues (chmn. 1944-45), AAAS, Nat. Acad. Scis., Am. Philos. Soc., Internat. Soc. Hypnosis (pres. 1973-76, Benjamin Franklin gold medal 1979), Sigma Xi. Subspecialties: Cognition. Current work: History of American psychology. Home: 850 Webster Palo Alto CA 94301

HILGEMAN, THEODORE WILLIAM, Laboratory administrator in optical physics; b. Amityville, N.Y., Dec. 21, 1942; s. Roy John and Erma Louise (Fabian) H.; m. Susan Elenor Goldsmith, Sept. 6, 1963; children: Leslie Ann, Erica Lynn. B.S. in Physics, M.I.T., 1964, Ph.D., Calif. Inst. Tech., Pasadena, 1970. Research scientist Grumman Aerospace Corp., Bethpage, N.Y., 1970-76, head optical physics br., 1976-79, head optical physics lab., 1979—; ednl. counselor M.I.T., 1972-80. Editor: Episcopal Expression of Marriage Encounter, 1974; author: (with Walter Egan) Optical Properties of Inhomogeneous Materials, 1979; contbr. articles in field to profl. jours. Recipient M.I.T. Presidential citation, 1972. Mem. Am Astron. Soc., Optical Soc. Am., Sigma Xi. Subspecialties: Remote sensing (atmospheric science); Optical engineering. Current work: Airborne and spaceborne electro-optical and photonic systems remote sensing, optical properties, electro-optical systems, photonics, infrared, atmospheric properties, background radiation, detectors. Office: Grumman Aerospace Corporation M/S A01-26 Bethpage NY 11714

HILL, A. LEWIS, psychologist, researcher; b. Glen Cove, N.Y., Nov. 5, 1940; s. James C. and Ruth M. (Lewis) H.; m. Patricia Munday, June 24, 1964 (div. 1976); children: Melissa, James; m. Florence Duguid, Jan. 10, 1980; children: Gareth Bramley, Lymond Bramley. B.A., Rollins Coll., 1963; Ph.D., Yeshiva U., 1968; profl. diploma in clin. psychology, Adelphi U., 1982. Lic. psychologist, N.Y. Sr. research scientist N.Y. Inst. Basic Research, S.I., 1967-79, research scientist IV, 1980; cons. editor profl. jours.; research cons. U. Md., 1972-75; asst. prof. CUNY, 1969-72. Contbr. articles to sci. jours. Chmn. neighborhood adv. bd. Page Ave. Group Home for Retarded, S.I., 1980—. Mem. Am. Psychol. Assn., Psychonomic Soc., Eastern Psychol. Assn. Subspecialties: Neuropsychology; Psychology. Current work: Idiot savants, developmental perception, aging, psychotherapy, biofeedback, psychometrics, mental retardation. Home: 824 Page Ave Staten Island NY 10309 Office: 1050 Forest Hill Rd Staten Island NY 10314

HILL, ALAN EUGENE, physicist; b. Durango, Colo., Sept. 4, 1939; s. Glenn Worland and Minnie Willard (Hermsmeier) H.; m. Carol Ann Havens, Mar. 26, 1960; children: Larry Glenn, Roy Leon. B.S.E.E., U. Mich., 1964, M.S., 1965. Leader laser research br. Lear Siegler Corp., Ann Arbor, Mich., 1965-67; sr. scientist, tech. dir. elec. laser div. USAF Weapons Lab., Kirtland AFB, Albuquerque, 1967-77; pres., chief scientist Plasmatronics, Albuquerque, 1977—; adj. prof. U. Ariz., Tucson, 1975—. Contbr. articles to prof. jours. Active Cave Research Found., 1963—, Nat. Speleological Soc., 1963—. Recipient sci. achievement award Air Force Systems Command, 1971; citation of honor Air Force Assn., 1972; outstanding peformance awards U.S. Air Force, 1968-72. Subspecialty: Laser Research and Development. Current work: Generation and control of large volume, high pressure plasmas; application of pulsed power technology and gas dynamic principles to high energy lasers and accelerators; control and application of plasma-acoustic interactions. Patentee, inventor in field. Office: 2460 Alamo SE Suite 101 Albuquerque NM 87106

HILL, SISTER ANN GERTRUDE, chemistry educator, computer center director; b. Cleve., Mar. 15, 1922; d. William J. and Gertrude Margaret (Kuehner) H. B.S., Ursuline Coll, 1944; M.S., U. Notre Dame, 1952, Ph.D., 1957. Joined Ursuline Order, Roman Catholic Ch., 1943; prof. chemistry Ursuline Coll., Cleve., 1946—, prof., dir. computer ctr., 1980—. Vol. Sunny Acres Skilled Nursing Facility. AEC fellow, 1955; NSF postdoctoral fellow, summer 1964; NSF grantee; Cleve. Commn. Higher Edn. grantee, 1970. Mem. Am. Chem. Soc., Sci. Edn. Council of Ohio, N.Y. Acad. Scis., Iota Sigma Pi. Subspecialties: Inorganic chemistry; Foundations of computer science. Current work: Teaching of inorganic and analytical chemistry and introductory computer programming, directing of small computer center. Home: 2600 Lander Rd Cleveland OH 44124 Office: 2550 Lander Rd Cleveland OH 44124

HILL, DONALD LYNCH, biochemist; b. Decherd, Tenn., June 24, 1937; s. James Bransford and Bertha Olivia (Mooney) H.; m. Alma Elaine Martin, May 29, 1960; children: John, Gerald, Stephen. B.S. in Sci, Middle Tenn. State Coll., 1960, postgrad., 1960-61; Ph.D. in Biochemistry, Vanderbilt U., 1964, U. Calif., Berkeley, 1964-65. Research biochemist So. Research Inst., Birmingham, Ala., 1965-67, sr. biochemist, 1967-71, head membrane biochemistry sect., 1971-76, head biochem. pharmacology div., 1976—; assoc. adj. prof. pharmacology U. Ala., Birmingham, 1977—. Author: The Biochemistry and Physiology of Tetrahymena, 1972, A Review of Cyclophosphamide, 1975. Served with USAF, 1954-58. Mem. Am. Soc. Pharmacology and Exptl. Therapeutics, Am. Assn. for Cancer Research, AAAS, Am. Soc. Biol. Chemists, Southeastern Cancer Research Assn. Mem. Ch. of Christ. Subspecialties: Cancer research (medicine); Pharmacology. Current work: Metabolism and site of action of antitumor agents and carcinogens. Home: 1349 Wilshire Dr Birmingham AL 35213 Office: Box 55305 Birmingham AL 35255

HILL, E(LGIN) ALEXANDER, chemistry educator; b. Pitts., May 16, 1935; s. E. Alexander and Barbara (Ziegler) H.; m. Barbara Bell, June 21, 1958; children: David, Andrea, Roger. B.S., Allegheny Coll., 1957; Ph.D., Calif. Inst. Tech., 1961. Postdoctoral fellow Pa. State U., State College, 1960-61; asst. prof. chemistry U. Minn., Mpls., 1960-66, U. Wis.-Milw., 1966-69, assoc. prof., 1969-80, prof., 1980—, chmn. dept., 1970-72. Contbr. articles to profl. jours. Petroleum Research Found. grantee, 1962-68, 80-83; NSF grantee, 1965-68. Mem. Am. Chem. Soc. (councilor 1978-80). Subspecialties: Organic chemistry; Kinetics. Current work: Rearrangements in organometallic chemistry; radical and carbanionic rearrangements; secondary deuterium isotope effects, structure-reactivity relationships. Office: Dept Chemistry U Wis Milwaukee WI 53201

HILL, ERNEST ELWOOD, engineering company executive; b. Oakland, Calif., May 15, 1922; s. George Leslie and Ollie (Morel) H.; m. Bettejean Schaegelen, Mar. 27, 1942; children: Eric Evan, Steven Richard, Lawrence Martin. B.S., M.E., U. Calif.-Berkeley, 1943, M.S. in Nuclear Engring, 1959. Cert. profl. engr., mech. engr., nuclear engr., Calif. Reactor supr. Lawrence Livermore Lab., Calif., 1955-64, div. leader, 1967-82; branch chief AEC, Berkeley, 1964-67; pres. Hill Assocs., Danville, Calif., 1982—; adminstrv. judge Nuclear Regulatory Commn., Washington, 1972—. Served to capt. USAF, 1943-46. Mem. Am. Nuclear Soc. (sect. chmn. 1966-67). Subspecialties: Nuclear engineering; Human factors engineering. Current work: Engineering consulting in nuclear engineering, mechanical engineering; human factor engineering and probabalistic risk assessment. Home: 210 Montego Dr Danville CA 94526 Office: Hill Assocs 210 Montego Dr Danville CA 94526

HILL, HENRY ALLEN, educator, physicist; b. Port Arthur, Tex., Nov. 25, 1933; s. Douglas and Florence (Kilgore) H.; m. Ethel Louise Eplin, Aug. 23, 1954; children—Henry Allen, Pamela Lynne, Kimberly Renee. B.S., U. Houston, 1953; M.S., U. Minn., 1956, Ph.D., 1957; M.A. (hon.), Wesleyan U., 1966. Research asst. U. Houston, 1952-53; teaching asst. U. Minn., 1953-54, research asst., 1954-57; research asso. Princeton, 1957-58, instr., then asst. prof., 1958-64; asso. prof. Wesleyan U., Middletown, Conn., 1964-66, prof. physics, 1966-74, chmn. dept., 1969-71; prof. physics U. Ariz., 1966—. Contbr. articles to profl. jours. Sloan fellow, 1966-68. Mem. Am. Phys. Soc., Am. Astron. Soc., Royal Astron. Soc., Optical Soc. Am. Subspecialties: General relativity; Solar physics. Current work: Solar oscillations; their utility in studies of internal structure of sun; tests of gravitation, and determination of solar; terrestrial climate relationship. Research on nuclear physics, relativity and astrophysics. Home: 340 S Avenida de las Palmas Tucson AZ 85716

HILL, IDEN NAYLOR, retired dental surgery educator; b. Emporia, Kans., July 29, 1911; s. Charles Willard and Myrtle Elizabeth (Naylor) H.; m. Dorothy Jane Cheney, Sept. 16, 1936. D.D.S., Northwestern U., 1937. Dentist Dept. Health Chgo., 1937-38; pvt. practice dentistry, Evergreen Park, Ill., 1938-42; with U. Chgo., 1945-75; ret., 1975; cons. Research Inst., ADA, 1977-82. Served to maj., AC U.S. Army, 1942-45. Mem. ADA, Internat. Assn. Dental Research, Ill. Dental Soc., Chgo. Dental Soc. Subspecialty: Preventive dentistry. Current work: Fluorine and dental caries. Home: 231 Newton Ave Glen Ellyn IL

HILL, JANE FOSTER, botanical researcher; b. Portland, Oreg., Apr. 15, 1946; d. Ellery Alonzo and Marion C. (Gold) Foster; m. William Albert Hill, Sept. 4, 1971. B.A., Carleton Coll., 1968; M.S., George Washington U., 1976, Ph.D., 1980. Botanist water resources div. U.S. Geol. Survey, Reston, Va., 1978-80; vis. scientist forest physiology lab. U.S. Forest Service, U.S. Dept. Agr., Beltsville, Md., 1980—. Mem. Bot. Soc. Am., Sigma Xi. Subspecialties: Plant growth; Ecology. Current work: Relationships among leaf morphology (entire and serrated margins, pinnately compound leaves, leaf shape), certain wood-anatonomical characteristics, and latitute of distribution in the diciduous forest of eastern North America. Office: USDA Forest Service Forest Physiology Lab Beltsville MD 20705

HILL, JIM TOM, scientific association administrator, biochemist; b. Cushing, Okla., Apr. 27, 1939; s. Wilburn C. and Susie (Ruckman) H.; m. Linda Archer, Aug. 30, 1963; children: Sheri, David, Susan. B.S. in Chemistry, Abilene Christian U., 1961; M.S. in Biochemistry, U. Tenn., 1964, Ph.D., 1968. Sr. research scientist E.R. Squibb & Sons, New Brunswick, N.J., 1968-69; sr. research scientist Lakeside Labs., Milw., 1969-75; sr. research specialist Monsanto Corp., St. Louis, 1975-78; dir. chemistry Hazelton Labs. Am., Vienna, Va., 1978-80; mgr. toxicology Phelps Dodge, Washington, 1980-81; dir. sci. affairs Chem. Specialties Mfrs. Assn., Washington, 1981—. Contbr. articles to profl. jours. Mem. Am. Coll. Toxicology, Am. Soc. Pharmacology and Exptl. Therapeutics, Environ. Mutagen Soc., Am. Chem. Soc., AAAS, Greater Washington Area Soc. Toxicology, Sigma Xi. Mem. Ch. of Christ. Subspecialties: Toxicology (medicine); Pharmacokinetics. Current work: Toxicology, Biochemical pharmacology, mutagenesis, carcinogenesis, environmental toxicology, biochemistry. Home: 2477 Freetown Dr Reston VA 22091 Office: 1001 Connecticut Ave NW Suite 1120 Washington DC 20036

HILL, JOHN HEMMINGSON, plant pathologist, plant virologist, educator; b. Evanston, Ill., Feb. 19, 1941; s. Robert Kermit and Adelaide (Nyden) H.; m. Laani Fong, Aug. 5, 1967; children: Brent, Bryce, Bjork. B.A., Carleton Coll., 1963; M.S., U. Minn.-St. Paul, 1966; Ph.D., U. Calif.-Davis, 1971. Postgrad. research plant pathologist U. Calif., Davis, 1971; postdoctoral plant pathologist Iowa State U., Ames, 1971-72, asst. prof. plant pathology, 1972-78, assoc. prof., 1978-82, prof., 1982—. Contbr. articles in field to profl. jours. Deacon, session mem. Collegiate Presbyn. Ch., Ames.; Boy Scout leader, Ames, Iowa. NSF grad. trainee, 1967-68. Mem. Am. Phytopath. Soc., Am. Soc. Virology, Sigma Xi, Gamma Sigma Delta. Subspecialties: Plant virology; Immunology (agriculture). Current work: Serological methods for detection of plant viruses, monocolonal antibodies, biochemical characterization of plant viruses—protein and nucleic acid, epidemiology of plant viruses, aphid transmission of plant viruses. Office: Dept Plant Pathology Seed and Weed Sci Iowa State U Ame IA 50011

HILL, ORVILLE FARROW, scientist; b. Decatur, Ill., Jan. 6, 1919; s. Edgar D. and Lucy May (Jones) H.; m. Alta Lee Meeker, July 1, 1944; children: Diane Louise, James Michael, Barbara Jean. B.S., Millikin U., 1940, D.Sc. (hon.), 1963; M.S., U. Ill,-Urbana, 1941, Ph.D., 1948. Prin. chem. engr. Gen. Electric Co., Richland, Wash., 1948-65, Isochem., Inc., Richland, 1965-67, Atlantic Richfield Hanford Co., 1967-77; staff scientist Battelle-N.W., Richland, 1977—. Contbr. numerous articles to profl. publs. Fellow AAAS; mem. Am. Chem. Soc. (Richland Sect. award 1982, chmn. Richland sect. 1946-48, councilor 1955-69, mem.-at-large exec. com. div. nuclear chemistry and tech. 1966, councilor 1972-81), Am. Inst. Chem. Engrs., Am. Nuclear Soc. Republican. Episcopalian. Clubs: Wash. State Fedn. Square and Round Dance Clubs (pres. Blue Mountain council 1982-83), Melody Mixers Round Dance (pres. 1972), Prairie Shufflers Round and Square Dance (del. 1980-82). Subspecialties: Nuclear fission; Inorganic chemistry. Current work: Technology and environmental impacts of nuclear fuel cycle activities. Home: 2315 Camas Ave Richland WA 99352 Office: Battelle NW PO Box 999 Richland WA 99352

HILL, PERCY HOLMES, mechanical engineering educator, consultant; b. Norfolk, Va., Feb. 19, 1923; s. Percy Holmes and Theresa Rose (Freitas) H.; m. Charlotte Hamilton Hall, Sept. 7, 1946; children: Mary Ann, Elizabeth H. B.M.E., Rensselaer Polytechnic Inst., 1944; S.M., Harvard U., 1951. Instr. engring. and math. Va. Poly Inst., Norfolk, 1946-48; lectr. applied mechanics Northeastern U., Boston, 1951-62; mem. engring. design faculty Tufts U., Medford, Mass., 1948—, now prof.; cons. to industry; pres. Applied Ergonomics Corp., Winchester, Mass., 1976—, Stratford Labs., Inc., Englewood Cliffs, N.J., 1982—. Author: Science of Engineering Design, 1970; co-author: Analysis & Design of Mechanisms, 1960, Making Decisions, 1979. Founder, commodore Silver Lake Sailing Club; vice-chmn. Silver Lake Environ. Assn., 1982. Served to lt. USNR, 1944-6; PTO. Recipient Excellence in Design award Indsl. Design, 1977. Mem. Human Factors Soc. (awards com. 1969-72), ASME, Am. Soc. Engring. Edn. (chmn. 1971-72, Frank Oppenheimer award 1968, Disting. Service award 1977, Merryfield Design award 1982), Sigma Xi, Tau Beta Pi. Republican. Unitarian. Clubs: Silver Lake (N.H.) Sailing, Tennis (North Conway, N.H.). Subspecialties: Mechanical engineering; Human factors engineering. Current work: research and innovative design and reduction to practice of consumer products in the health and recreation areas that have a strong human factors basis. Co-inventor REACH toothbrush, 1976. Home: East Shore Dr Silver Lake NH 03875 Office: Tufts Univ Medford MA 02155

HILL, RAY ALLEN, botanist, cell biologist, educator; b. Houston, Sept. 16, 1942; s. Cal and Ann Mae (Stewart) H. B.S., Howard U., 1964, M.S., 1965; Ph.D., U. Calif., Berkeley, 1977. Cert. coll. chief adminstrv. officer, Calif. Instr. So. U., Baton Rouge, 1965-66, Howard U., Washington, 1966-73; asst. prof. Fisk U., Nashville, 1977-80; assoc. prof. biology Community Coll. Balt., 1980-82, Morgan State U., Balt., 1982—; staff scientist NASA, summer 1979, Lawrence Berkeley Lab., summer 1980. NSF fellow, 1974; Ford Found. awardee, 1975-77. Member. AAAS, Bot. Soc. Am., Am. Soc. Cell Biology, N.Y. Acad. Sci., Tenn. Acad. Sci., Sigma Xi, Beta Kappa Chi. Roman Catholic. Subspecialties: Plant physiology (biology); Cell and tissue culture. Current work: Plant development/morphogenesis; ultrastructure. Home: PO Box 462 Baltimore MD 21203

HILL, REBA MICHELS, physician neonatologist; b. Houston, Oct. 8, 1930; d. John Robert and Bessie Mae (Puddy) Michels; m. L. Leighton Hill, July 5, 1958; 3 children. B.S., Baylor U., 1952, M.D., 1955. Instr. Baylor Coll. Medicine, 1960; dir. neurol. study clinic Jefferson Davis Hosp., Houston, 1960-62; chief newborn research St. Lukes Episcopal Hosp., Houston, 1962—; asst. prof. pediatrics Baylor Coll. Medicine, 1965-70, asso. prof., 1970-79, prof., 1979—, chief developmental toxxcology dept. pediatrics, 1968—. Author: Breast Feeding a Passage in Life, 1978; contbr. articles to profl. jours. Pres. bd. dirs. West Briar Sch., 1980-82. Recipient Alumnus award Alpha Omega Alpha, 1969. Mem. Soc. Pediatric Research, Am. Pediatric Soc., Am. Soc. Pharmacology and Exptl. Therapeutics. Episcopalian. Subspecialties: Neonatology; Pediatrics. Current work: Research in the effects of maternal drugs on the fetus, transfer of drugs into breast milk, effect of intrauterine malnutrition on the fetus. Home: 4906 Tilbury Houston TX 77056 Office: 6720 Bertner Houston TX 77030

HILL, STEVEN RICHARD, botanist, educator, range land consultant; b. Hartford, Conn., Dec. 30, 1950; s. Richard James and Margaret Elizabeth (Stevens) H. B.S. in Biology, Bates Coll., 1972, M.A., H. H. Lehman Coll. of CUNY, 1975; Ph.D. in Botany, Tex. A&M U., 1979. Herbarium fellow N.Y. Boty. Garden, Bronx, N.Y., 1972-74; bot. collector N.Y. Bot. Garden, 1972-74; adj. lectur. dept. biology H. H. Lehman Coll., CUNY, 1974; herbarium asst. Fairchild Tropical Garden, Miami, Fla., 1975, bot. collector, 1975; teaching asst. Tex. A&M U., College Station, 1975-79; curator herbarium instr. dept. botany U. Md., College Park, 1979—. Contbr. articles on botany to profl. jours. Alfred P. Sloan Found. grantee, 1977. Mem. Am. Fern Soc., Am. Soc. Plant Taxonomists, Bahamas Nat. Trust, Bot. Soc. Am., Conn. Bot. Soc., Internat. Assn. Plant Taxonomy, Washington Bot. Soc., New Eng. Bot. Club, So. Appalachian Bot. Club, Sigma Xi, Phi Kappa Phi. Subspecialties: Systematics; Taxonomy. Current work: Systematics, taxonomy, floristics. Office: Dept Botany U Md College Park MD 20742

HILL, TERRELL LESLIE, chemist, biophysicist; b. Oakland, Calif., Dec. 19, 1917; s. George Leslie and Ollie (Moreland) H.; m. Laura Etta Gano, Sept. 23, 1942; children: Julie Lisbeth Eden, Carolyn Jo (Mrs. Gary Lineburg), Ernest Evan. A.B., U. Calif. at Berkeley, 1939, Ph.D., 1942; postgrad., Harvard U., 1940. Instr. chemistry Western Res. U., 1942-44; research asso. radiation lab. U. Calif. at Berkeley, 1944-45; research asso. chemistry, then asst. prof. chemistry U. Rochester, 1945-49; chemist U.S. Naval Med. Research Inst., 1949-57; prof. chemistry U. Oreg., 1957-67, U. Calif. at Santa Cruz, 1967-71, adj. prof., 1977—, vice chancellor for scis., div. natural scis., 1968-69; research chemist NIH, Bethesda, Md., 1971—; Mem. biophysics study sect. USPHS, 1954-57; chemistry panel NSF, 1961-64. Author: Statistical Mechanics, 1956, Statistical Thermodynamics, 1960, Thermodynamics of Small Systems, Vol. I, 1963, Vol. II, 1964, Matter and Equilibrium, 1965, Thermodynamics for Chemists and Biologists, 1968, Free Energy Transduction in Biology, 1977, also research papers. Guggenheim fellow Yale, 1952-53; recipient Arthur S. Flemming award U.S. Govt., 1954; Distinguished Civilian Service award U.S. Navy, 1955; award Washington Acad. Scis., 1956; Disting. Service award USPHS, 1981, U. Oreg., 1983; Sloan Found. fellow, 1958-62. Mem Nat. Acad. Scis., Am. Chem. Soc. (Kendall award 1969), Biophys. Soc., NAACP, ACLU, Phi Beta Kappa. Subspecialties: Biophysics (biology); Biophysical chemistry. Current work: Application of theoretical chemistry and physics to various problems in bio energetics and cell biology. Home: 9626 Kensington Pkwy Kensington MD 20895

HILL, THOMAS JOHNATHAN, nuclear engineer; b. Seattle, Jan. 1, 1944; s. Jack Evenman and Janet Madeline (Williamson) H.; m. Claire-Ann Middel, Aug. 26, 1967; children: Michelle Leigh, Nicole Marie. A.A., Skagit Valley Coll., 1964; B.S.M.E., Wash. State U., 1967. Registered profl. engr., Idaho. Engr. Phillips Petroleum Co., Idaho Falls, 1967-69; sr. engr. Idaho Nuclear Co., Idaho Falls, 1969-71; engring. supr. Aerojet Nuclear Co., Idaho Falls, 1971-76; mgr. fast breeder reactor EG&G Idaho, Idaho Falls, 1976—. Mem. ASME (com. chmn. 1968—), Nat. Soc. Profl. engrs., Am. Nuclear Soc. Subspecialties: Nuclear fission; Nuclear engineering. Home: 929 12th St Idaho Falls ID 83401 Office: EG&G Idaho PO Box 1625 Idaho Falls ID 83415

HILL, WALTER ENSIGN, phys. biochemist; b. Bottineau, N.D., July 25, 1937; s. Armin J. and Virginia A. (Nelson) H.; m. Annette Smith, June 8, 1961; children: Heber, David, Virginia, Kathryn, Kenneth, Elizabeth, Bryce. Student, Pomona Coll., 1955-57; B.S., Brigham Young U., 1961; M.S., U. Wis., 1964, Ph.D., 1967. Postdoctoral assoc. Oreg. State U., 1967-69; asst. prof. chemistry U. Mont., 1969-73, assoc. prof., 1973-77, prof., 1977—; Mem. governing bd. Missoula Gen. Hosp. Contbr. articles in field to profl. jours. NIH fellow, 1968-69; NIH grantee, 1970—; recipient NIH Career Devel. award, 1971-76. Mem. AAAS, Biophysical Soc., Am. Soc. Biol. Chemists. Republican. Mormon. Subspecialties: Biophysical chemistry; Biophysics (physics). Current work: Phys. studies of ribosomes and biological macromolecules, studies of ribosomes and membranes, immunogenicity of ribosomes. Office: Chemistry Department University of Montana Missoula MT 59812

HILL, WALTER ERNEST, microbiologist; b. San Francisco, July 24, 1945; s. Ernest Joseph and Norma May (Letroadec) H.; m. Linda Marie Bigger, Aug. 24, 1968; children: Daniel, Alice. A.B., U. Calif.-Berkeley, 1967; Ph.D., U. Wash., 1972. Postdoctoral fellow U. Wis.-Madison, 1972-74; asst. prof. biology U. Wis.-Whitewater, 1974; postdoctoral fellow U. Chgo., 1974-75; asst. prof. microbiology U. Southwestern La., Lafayette, 1975-78; microbiologist FDA, Washington, 1978—; assoc. referee Assn. Ofcl. Analytical Chemists. Contbr. articles to profl. jours. Mem. Genetics Soc. Am., Am. Soc. Microbiology, AAAS, Sigma Xi. Subspecialties: Genetics and genetic engineering (biology); Molecular biology. Current work: Genetic methods to detect food-borne microbial pathogenics, role of plasmids in microbial virulence. Office: 200 C St SW HFF-234 Washington DC 20204

HILLEL, DANIEL, soil physics and hydrology educator, researcher, author, consultant; b. Los Angeles, Sept. 13, 1930; s. Morris and Sarah (Fromberg) Bugeslov. B.S., U. Ga.-Athens, 1950; M.S., Rutgers U., 1951; Ph.D., Hebrew U., 1958. Research fellow U. Calif.-Davis and Berkeley, 1959-61; head soil tech. div. Agrl. Research Orgn., Rehovot-Bet Dagan, Israel, 1952-66; prof., head soil and water dept. Hebrew U., Jerusalem-Rehovot, 1966-74; vis. prof. soil physics Tex. A&M U., College Station, 1974-75; vis. prof. environ. scis. U. Va., Charlottesville, 1975-77; research fellow Internat. Food Policy Research Inst., Washington, 1977; prof. soil physics and hydrology U. Mass., Amherst, 1977—; Chmn. Israel Nat. Com. on Soil Pollution Research, Israel, 1971; mem. Group on Water Mgmt. Research-Internat. Devel. Research Ctr., Can., 1978, Adv. Panel on Water Conservation, Dept. Water Resources Calif., 1979, Joint U.S. Dept. Agr.-Land Grant Univs. New Eng. Regional Research Priorities, Mass., 1981—; irrigation cons. World Bank, 1982-83; mem. U.S. Soil Sci. del. to People's Republic of China, 1983. Author: Soil and Water: Physical Principles and Processes, 1971, Computer Simulation of Soil Water Dynamics, 1977, Fundamentals of Soil Physics, 1980, Applications of Soil Physics, 1980, Negev: Land, Water and Civilization, 1982, Introduction to Soil Physics, 1982; editor: Optimizing the Soil Physical Environment, 1972, Advances in Irrigation, vol. 1, 1982, vol. 2, 1983. NSF research grantee, 1978, 82. Fellow Am. Soc. Agronomy, Soil Sci. Soc. Am.; mem. Internat. Soil Sci. Soc. (vice-chmn. soil physics 1964-68, 82-86), AAAS, Am. Geophys. Union. Subspecialties: Integrated systems modelling and engineering; Ground water hydrology. Current work: Soil-water-solute-energy dynamics in agriculture and in the environment. Patentee in field. Office: U Mass Stockbridge Hall Amherst MA 01003

HILLER, FREDERICK CHARLES, medical educator; b. Kansas City, Kans., Jan. 30, 1942; s. Frederick Charles and Beulah (Hackler) H.; m. Michelle Steele, Mar. 20, 1965; children: David, Christopher, Anne, John. B.A., U. Kans.-Lawrence, 1964; M.D., U. Kans.- Kansas City, 1968. Diplomate: Am. Bd. Internal Medicine (Pulmonary Disease). Intern U. Kans. Med. Ctr., Kansas City, 1968-69, resident, 1970-73; asst. prof. medicine U. Ark. Med. Sci., Little Rock, 1975-80, assoc. prof., 1980—; cons. Ark. Health Dept., Little Rock, 1975—, Little Rock VA Hosp., 1975—. Contbr. articles to med. jours. Mem. com., patron Gibbs Intermediate Sch., Little Rock. Served to capt. USAR, 1969-71. NIH grantee, 1976-82; EPA grantee, 1981-83; Nat. Cystic Fibrosis Found. grantee, 1979-81. Fellow Am. Coll. Chest Physicians; mem. Am. Thoracic Soc. Subspecialties: Pulmonary medicine; Internal medicine. Current work: Physical properties of inhaled aerosols and deposition in the human respiratory tract, oxygen delivery to tissue, respiratory failure, shot. Office: Pulmonary Div Univ Ark for Med Scis 4301 Markham Slot 555 Little Rock AR 72205

HILLER, JACOB MOSES, biochemical pharmacologist; b. N.Y.C., Dec. 12, 1939; s. Nathan and Ilse (Katzman) H.; m. Deborah Tamar Posen, May 2, 1965; children: Simona, Sarah, Rachel. B.S., CCNY, 1961; M.S., NYU, 1967, Ph.D., 1970. Research asst. N.Y. Zool. Soc. expdn. to Antarctica, 1964-65; asst. research scientist dept. biology NYU, N.Y.C., 1966-69, research scientist, 1969-70; assoc. research scientist NYU Med. Ctr., 1970-73, 1973-74, research asst. prof. dept. medicine, 1974-80, research assoc. prof., 1980—. Contbr. articles to sci. jours. N.Y. State scholar, 1959-61; recipient U.S. Congressional medal for Antarctic Service, 1965; Founders Day award NYU, 1970. Mem. AAAS, Am. Soc. Microbiology, Internat. Narcotics Research Conf., Sigma Xi. Democrat. Jewish. Subspecialties: Neuropharmacology; Molecular pharmacology. Current work: Elucidation of molecular mechanisms of acute and chronic effects of opiate narcotic analgesics via the study of the endogenous opioid peptide system, with particular emphasis on the characterization, solubilization and purification of opiate receptors. Home: 17 Carlton Rd Monsey NY 10952 Office: NYU Med Ctr 550 First Ave New York NY 10016

HILLER, JUDITH IRENE, chemical engineering educator, researcher; b. San Francisco, May 19, 1937; d. Martin Keller and Constance Marie (Dunne) Blaine; m Douglas Hiller, Aug. 2, 1961; children: Abigail, Diana, Gregory. B.S., U. Ill., 1958; Ph.D., Princeton U., 1962. Asst. prof. chem. engring. Purdue U., West Lafayette, Ind., 1962-66, assoc. prof., West Layfayette, Ind., 1966-69, U. Akron, 1969-71, prof. chem. engring., 1971—. Mem. AAAS, Am. Inst. Chem. Engring., Am. Chem. Soc., Am. Assn. Aerosol Research, Am. Assn. Engring. Edn., Sigma Xi. Subspecialty: Chemical engineering. Home: Werik Place 50 S Maple Akron OH 44303

HILLERS, JOE KARL, animal science educator; b. Centerville, Iowa, July 20, 1938; s. Christe and Rita (King) H.; m. Virginia A. Nerlin, June 1, 1963; children: Julie, Ken. B.A., Northwest Mo. State Coll., 1960; M.S., Iowa State U., 1963, Ph.D., 1965. Asst. prof. Wash. State U., Pullman, 1965-71, assoc. prof., 1971-. Mem. Am. Dairy Sci. Assn.,

Am. Soc. Animal Sci., Biometrics Soc. Methodist. Subspecialties: Animal breeding and embryo transplants; Animal genetics. Current work: Selection programs and computer systems analysis of dairy cattle and production traits. Home: NW 515 Irving Pullman WA 99163 Office: Wash State U 141 Clark Hall Pullman WA 99164

HILLEY, JAMES ROGER, JR., nuclear engineer; b. Augusta, Ga., Nov. 3, 1957; s. James R. and Lynn T. (Thurman) H.; m. Sonia Calderon, June 27, 1981. B.N.E., Ga. Inst. Tech., 1979; M.S.N.E., U. Ill., 1981. Engr. assoc. Duke Power Co., Charlotte, N.C., 1981—; summer tech. assignments E.I. duPont Savannah River Plant & Lab., 1978-80. Mem. Am. Nuclear Soc. Republican. Methodist. Subspecialty: Nuclear engineering. Current work: Radiation shielding analysis of nuclear power plants, corrosion product characterization of LWRs. Home: 3310 Thaxton Pl Matthews NC 28105 Office: Duke Power Co PO Box 33189 Charlotte NC 28242

HILLIARD, ASA GRANT, III, psychology educator, researcher; b. Galveston, Tex., Aug. 22, 1933; s. Asa Grant and (Lois) Lowe; ; s. Asa Grant and (Lois) Williams; m. Patsy Jo Morrison, Nov. 16, 1957; children: Asa Grant IV, Robi Nsenga, Nefertari Patricia, M. Hakim. B.A., U. Denver, 1955, M.A., 1961, Ed.D., 1963. Teaching fellow U. Denver, 1961-63; asst. prof. San Francisco State U., 1963-67, assoc. prof., 1967-70, prof., 1970-83, dean, 1972-80; Fuller E. Calloway prof. Ga. State U., 1980—; dir. Nguzo Saba Film Co., San Francisco; chief desegregation cons. Portland Pub. Schs., 1979—; nat. adv. Panel Head Start Measures Project, Westport, Conn., 1980—; research in field. Bd. dirs. Mental Health Assn. Met. Atlanta, 1982. Served to 1st lt. U.S. Army, 1955-57. Knighted Govt. Liberia, 1974; recipient Alice Miel award Columbia U., 1978; cert. honor San Francisco Mayor, 1980. Mem. Am. Psychol. Assn. (bd. ethnic affairs 1980—), Nat. Assn. Black Psychologists, Nat. Assn. Edn. Young Children, Omega Psi Phi, Phi Delta Kappa. Democrat. African Methodist Episcopalian. Subspecialties: Cognition; Cultural bias and testing. Current work: Cultural bias and testing; research and teaching in culture, testing, cognition and learning. Home: 3350 Sir Henry St East Point GA 30344 Office: Georgia State University University Plaza PO Box 243 Atlanta GA 30303

HILLIS, WILLIAM DANIEL, SR., biology educator, physician; b. Paris, Ark., June 12, 1933; s. Charles Raymond and Carra Elizabeth Daniel (Coffee) H.; m. Argye Idell Briggs, Dec. 23, 1952; children: William Daniel Jr., David Mark, Argye Elizabeth Hillis Trupe. B.S., Baylor U., 1953; M.D., Johns Hopkins U., Balt., 1957. Intern Johns Hopkins Hosp., 1957-58; research fellow Johns Hopkins U., 1958-60, asst. prof. to assoc. prof. pathobiology, 1965-82, asst. to assoc. prof. medicine, 1972-83; prof., chmn. dept. biology Baylor U., Waco, Tex., 1981—; chmn. med. adv. bd. Endstage Renal Disease, Network 31, Balt., 1978-82; mem. Md. Kidney Disease Commn., 1977-82; dir. med. research tng. program Johns Hopkins Sch. Medicine, 1978-82. Pres. Bapt. Conv. Md., 1976-78; Md. rep. exec. com. So. Baptist Conv., Nashville, 1977-82. Served to col. USAFR, 1960—. Recipient Seaman prize Assn. Mil. Surgeons U.S., 1979; Christian Citizen of Year award United Christian Citizens Md., 1978. Mem. Am. Soc. Microbiology, Am. Assn. Immunologists, N.Y. Acad. Sci., Soc. for Exptl. Biology and Medicine, Sigma Xi, Phi Beta Kappa, Alpha Omega Alpha. Club: Johns Hopkins (Balt.). Subspecialties: Immunobiology and immunology; Virology (biology). Current work: Virology, immunology, and epidemiology of human viral hepatitis; immunopathology of mouse hepatitis virus; viral etiology of human renal disease; primate viruses and their interrelationship; respiratory viruses of children; nutritional effects in infection. Office: Dept Biology Baylor U Waco TX 76798

HILLMAN, ELIZABETH ANN, biologist, researcher; b. Ft. Edward, N.Y., Oct. 23, 1938; d. Albert R. and Martha A. (Hoag) H.; m. Edwin J. Matthews, Jan. 29, 1972; 1 dau., Susan Elizabeth. B.A. with honors, Russel Sage Coll., 1960; Ph.D in Microbiology/Pathology, Duke U., 1972. Research asso. dept. pathology Med. Coll. Va., 1963-67; research asst. dept. surgery Duke U. Med. Center, 1967-71; dept. pathology U. Va., 1972; scientist II Viral Resources Lab., Frederick Cancer Research Center, Frederick, Md., 1972-76; asso. prof. pathology U. Md., 1976—, head electron microscopy, dept. pathology, 1976-82; cons. in electron microscopy. Contbr. articles, abstracts to profl. jours. Mem. Am. Soc. Cell Biology, Am. Assn. Cancer Research, Electron Microscopy Soc. Am., Chesapeake Soc. Electron Microscopy, Assn. Women in Sci. Subspecialties: Cancer research (medicine); Microscopy. Current work: Research in breast cancer—in vitro systems, pathology, esophageal carcinogenesis—two stage model, x-ray microanalysis. Office: U Md Sch Medicine 660 W Redwood St Baltimore MD 21201

HILLMAN, GILBERT ROTHSCHILD, pharmacologist, educator, researcher; b. New Haven, May 1, 1943; s. Jacob and Clara (Rothschild) H.; m. Rachel Read, Aug. 27, 1965; 1 dau., Laura. B.A., Harvard U., 1965; Ph.D., Yale U., 1969. Asst. prof. Brown U., Providence, 1970-76; assoc. prof. pharmacology U. Tex. Med. Br., Galveston, 1976—. Contbr. articles to profl. jours. NIH grantee, 1976—. Mem. Am. Soc. Pharmacology and Exptl. Therapeutics, Am. Soc. for Tropical Medicine and Hygiene, AAAS, N.Y. Acad. Scis. Subspecialties: Neuropharmacology; Graphics, image processing, and pattern recognition. Current work: Neurochemistry; drug receptors; computer interpretation of histochemical images. Home: 172 San Marino Dr Galveston TX 77550 Office: Dept Pharmacology U Tex Med Br Galveston TX 77550

HILLS, FREDERICK JACKSON, agronomist, biostatistician; b. Oakland, Calif., Feb. 27, 1919; s. Frederick Bertrum and Alamah (Hollenback) H.; m. Juanita Alberta Wood, June 23, 1943; children: D. Jackson, Christine E., Cynthia A., Diane L. B.S., U. Calif., Berkeley, 1941, M.S., 1951, Ph.D., 1961. Agronomist Spreckels Sugar Co., Woodland, Calif., 1946-51; extension agronomist U. Calif., Davis, 1951—. Author: Agricultural Experimentation: Design and Analysis, 1978, also articles. Served to 1st lt. USMC, 1942-46. Mem. Am. Soc. Agronomy, Am. Phytopathol. Soc., Am. Soc. Sugarbeet Technologists, Phi Kappa Phi. Democrat. Subspecialties: Agronomy; Statistics. Current work: Sugarbeet agronomy and pathology. Design and analysis of experiments. Biomass. Home: 848 Oeste Dr Davis CA 95616 Office: Agronomy Extension U Calif Davis CA 95616

HILLS, JACK GILBERT, theoretical astrophysicist; b. Keflavik, Iceland, May 15, 1943; came to U.S., 1949, naturalized, 1965; s. Cleon Ralph and Kristin Laura H. A.B., U. Kans., 1966, M.A., 1966; Ph.D., U. Mich., 1969. Instr. astronomy dept. U. Mich., Ann Arbor, 1967-70, asst. prof. dept. astronomy, 1970-76, U. Ill., Urbana, 1976-77; assoc. dept. physics Mich. State U., East Lansing, 1977-82, prof. dept. physics, 1982—; mem. staff theoretical astrophysics group theoretical div. Los Alamos (N.Mex.) Nat. Lab., 1981—. Contbr. articles to profl. jours. Nat. winner Westinghouse Sci. Talent Search, 1962. Mem. Am. Phys. Soc., Am. Astron. Soc., Royal Astron. Soc., Internat. Astron. Union, AAAS. Republican. Subspecialty: Theoretical astrophysics. Current work: Dynamics of stellar systems; quasars; supernovae. Home: 11 Loma Vista Dr Los Alamos NM 87544 Office: Theoretical Div Los Alamos Nat Lab T-6 MS B288 Los Alamos NM 87545

HILLS, WILLARD ANDREW, engineer, consultant; b. Balt., Jan. 28, 1934; s. Henry Willard and Marjorie Doris (Colton) H.; m. Lillian Morrow, Nov. 9, 1957; children: Cynthia Elaine, Susan Doris. B.S., Bates Coll., 1955; M.B.A., U. Hawaii, 1974. Program engr. Gen. Electric Co., Cin., 1955-56; nuclear submarine officer U.S. Navy, Charleston, S.C., Groton, Conn., 1956-77; sr. engr. Carolina Power and Light, Raleigh, N.C., 1977; station supt. Hawaiian Electric Co., Honolulu, 1977-81; ops. service engr. Stone and Webster Engring. Corp., Boston, 1981—. Mem. Am. Nuclear Soc., ASME. Methodist. Club: Bates College (Lewiston, Maine). Subspecialties: Mechanical engineering; Nuclear engineering. Current work: Application of advanced techniques to provide energy efficient large scale electric production. Office: Stone and Webster Engring Corp 245 Summer St Boston MA 02107 Home: 1720 Chateau Ct Baton Rouge LA 70815

HILLYER, GEORGE VAN ZANDT, biology educator, researcher, cons.; b. San Juan, P.R., Dec. 8, 1943; s. William V. and Ruth L. H.; m. Josefina Gomez, June 15, 1968; children: George V., Julian F. B.S. in Biology, U. P.R., Rio Piedras Campus, 1968; Ph.D. in Microbiology, U. Chgo., 1972. Asst. prof. Lab. Parasite Immunology dept. biology U. P.R. (Rio Piedras Campus), 1972-75, asso. prof., 1975-80, prof., 1980, chmn., 1982—; adj. prof. pathology U. P.R. Med. Scis. Campus; adj. prof. tropical medicine Tulane U. Sch. Public Health and Tropical Medicine; cons. NIH Tropical Medicine and Parasitology Study Sect. Contbr. numerous articles to profl jours. Mem. Am. Assn. Immunologists, Am. Soc. Tropical Medicine and Hygiene (councilor 1983—), Am. Soc. Parasitologists (Henry Baldwin Ward medal 1982), Royal Soc. Tropical Medicine and Hygiene, Am. Soc. Microbiology, Soc. Exptl. Biology and Medicine, Sociedad de Microbiologos de Puerto Rico (pres. 1979-80), Sigma Xi (pres. U. P.R. San Juan Club 1979-81, 81-83). Subspecialties: Immunology (medicine); Magnetic physics. Current work: Immunology of parasitic infections. Home: 254 Himalaya Urbanization Monterrey Rio Piedras PR 00926 Office: Dept Biology U PR Rio Piedras PR 00931

HILTNER, WILLIAM ALBERT, astronomy educator; b. Continental, Ohio, Aug. 27, 1914; s. John Nicholas and Ida Lavina (Schafer) H.; m. Ruth Moyer Kreider, Aug. 12, 1939; children—Phyllis Anne, Kathryn Jo, William Albert, Stephen Kreider. B.S., U. Toledo, 1937; M.S., U. Mich., 1938, Ph.D., 1942. Mem. faculty U. Chgo., 1943-70, prof. astronomy, 1955-70; dir. Yerkes Obs., 1963-66; acting dir. Cerro Tololo Inter-Am. Obs., 1966-67; prof. U. Mich. at Ann Arbor, 1970—, chmn. dept. astronomy, 1970-82; Bd. dirs. Assn. Univs. for Research Astronomy, 1959-71, 74—, pres. bd., 1968-71. Co-author: Photometric Atlas of Stellar Spectra, 1946; Editor: Astronomical Techniques, 1962. NRC fellow, 1942-43. Mem. Astron. Soc. Pacific, Am. Astron. Soc. (councilor 1962-65), A.A.A.S. Subspecialty: Optical astronomy. Current work: Optical counterpart of x-ray sources; development of a 2.4 meter telescope and other astronomical research instruments. Home: 801 Berkshire Ann Arbor MI 48104

HILTON, THOMAS FREDERICK, research psychologist, naval officer; b. Cin., Apr. 12, 1947; s. Frederick and Julia (Burns) H. B.A., Elmhurst Coll., 1970; M.A., Fla. Atlantic U., 1976; Ph.D., Tex. Christian U., 1980. Commd. ensign U.S. Navy, 1968, advanced through grades to lt. MSC, 1980; grad. research fellow Inst. Behavioral Research, Ft. Worth, 1977-80; research coordinator psychiatry Tarrant County Hosp. Dist., Ft. Worth, 1980-82; instr. psychiatry and psychology Southwestern Med. Sch., Dallas, 1980-82; head health psychology projects Naval Health Research Ctr., San Diego, Calif., 1982—; research cons. Ctr. Orgn. Research and Evaluation Studies, 1981-82, Inst. Behavioral Research, 1980-82; dept. phys. medicine St. Joseph's Hosp., Ft. Worth, 1982; Research fellow Inst. Behavioral Research, 1977; research asst. Fla. Atlantic U., 1976. Mem. Am. Psychol. Assn., Internat. Assn. Applied Psychology, Soc. Advancement Social Psychology (steering com. feature editor newsletter), AAAS, Psi Chi, Sigma Xi. Subspecialties: Health psychology; Social psychology. Current work: Occupational health, healthful lifestyles, health service delivery, client-practitioner fit, health information systems, organizational climate and performance and health. Office: Naval Health Research Center PO Box 85122 San Diego CA 92138

HILTY, JAMES WILLARD, plant pathology educator, researcher; b. Wadsworth, Ohio, Oct. 11, 1936; s. Willard Daniel and Margaret N. (Dowd) H.; m. Harriette H. Henninger, Dec. 19, 1959; children: Ellen E., James E. B.S. in Agr, Ohio State U., 1958, M.Sc. in Botany and Plant Pathology, 1960, Ph.D., 1960. Asst. prof. plant pathology U. Tenn., 1965-69, assoc. prof., 1969-76, prof., 1976—. Active Boy Scouts Am. Mem. Am. Phytopathology Soc., Am. Inst. Biol. Scis., Sigma Xi, Phi Kappa Phi, Gamma Sigma Delta. Subspecialty: Plant pathology. Current work: Diseases of field crops, soybeans. Home: U Tenn 212 Plant Sci Bldg Knoxville TN 37901

HILU, KHIDIR WANNI, botany educator; b. Baghdad, Iraq, Jan. 1, 1946; came to U.S., 1972, naturalized, 1983; s. Wanni and Kafhin (Kasim) H.; m. Kieran Maggard Ford, Feb. 5, 1977; children Yasamine Julia, Enmar Mark. B.S., U. Baghdad, 1966, M.S., 1971; Ph.D., U. Ill.-Urbana, 1976. Asst. prof. Oklahoma City U., 1977-78; research biosystematist U. Calif.-Riverside, 1978-79; research assoc. U. Ill., 1979-81; asst. prof. botany Va. Poly. Inst. and State U., 1981—; cons. FAO. Contbr. articles to profl. jours. NSF Tavel award, 1980; Va. Poly. Inst. and State U. grantee, 1981, 82. Mem. Am. Soc. Naturalist, AAAS, Bot. Soc. Am., Internat. Assn. Plant Taxonomists, Am. Soc. Plant Taxonomists, Soc. Econ. Botany, Assn. Southeastern Biologists. Subspecialties: Taxonomy; Evolutionary biology. Current work: Biosystematics and evolution of higher plants, particularly grasses; chloroplast DNA in grass systematics; genetic basis of phenetic characters in plants; origin, evolution and systematics of domesticated plants. Office: Biology Dept Va Poly Inst and State U Blacksburg VA 24061

HIMELICK, EUGENE BRYSON, plant pathologist, educator; b. Summitville, Ind., Feb. 11, 1926; s. Virgil B. and Madalene M. (Bryson) H.; m. Elizabeth Ann Oyler, June 17, 1951; children: David E., Kirk J., Douglas N. B.S., Ball State U., 1949; M.S., Purdue U., 1952; Ph.D., U. Ill., 1959. Asst. plant pathologist Ill. Natural History Survey, Urbana, 1952-58, assoc. plant pathologist, 1959-64, plant pathologist, 1965—; prof. plant pathology U. Ill., Urbana, 1973—. Contbr. articlers to profl. jours., chpts. to books. Scoutmaster Boy Scouts Am., 1960-75, dist. chmn. Arrowhead council, 1975-76; chmn. Urbana Tree Commn., 1972—. Served with USN, 1945-46. Recipient Ken Frederick's award Boy Scouts Am., 1971, also; Scouters Key; Silver Beaver award. Mem. Internat. Soc. Arbiriculture (hon. life; exec. dir. 1969-79, authors citation 1979, award for research 1983), Am. Phytopath. Soc., Met. Tree Improvement Alliance, Arbiriculture Research and Edn. Acad., Ill. Arborists Assn. (pres.-elect 1983), Sigma Xi, Gamma Sigma Delta. Methodist. Lodge: Masons. Subspecialty: Plant pathology. Current work: Research on the causes and control of forest and urban tree diseases. Home: 601 Burkwood Ct E Urbana IL 61801 Office: Ill Natural History Survey 384 Natural Resources Bldg Champaign Il 61820

HIMES, FRANK LAWRENCE, soil chemistry educator; b. Crawfordsville, Ind., July 30, 1927; s. Ralph and Mary (Herr) H.; m. Dorothy L. Hostetter, Dec. 23, 1951; children: Laura, Caroline, Glenn. A.B., Wabash Coll., 1949; M.S., Purdue U., 1951, Ph.D., 1956. Tchr. Remington (Ind.) High Sch., 1951-53; asst. prof. Middle Tenn. State U., Murfreesboro, 1956-57; mem. faculty Ohio State U., Columbus, 1957—, now prof. soil chemistry; vis. scientist Rothamsted Exptl. Sta., Harpenden, Eng., 1966, U. Birmingham, Eng., 1981. Author: A-T Notes for Soils, 4th sedit, 1979. Named Outstanding Tchr. Gamma Sigma Delta, 1978. Fellow Am. Soc. Agronomy, Soil Sci. Soc. Am.; mem. Internat. Soil Sci. Soc., Am. Chem. Soc. Lodge: Masons. Subspecialty: Soil chemistry. Current work: Reactions of humic acids with heavy metals, clay surfaces, and other soil organic materials. Office: Agronomy Dept Ohio State U 2021 Coffey Rd 410C Columbus OH 43210

HINATA, SATOSHI, astrophysicist, plasma physicist; b. Tokyo, Japan., Aug. 6, 1944; s. Takao and Kikuyo (Nagano) H.; m. Yoshiko, Oct. 30, 1968; children: Kaoru, Kaede, Taroh. B.E. in Nuclear Engring, U. Tokyo, 1967; Ph.D. in Physics, U. Ill., 1973. Research assoc. in physics U. Ill., Urbana, 1973-74; postdoctoral fellow Harvard Coll. Obs., Cambridge, Mass., 1974-76; instr. Yale Obs., New Haven, 1976-78; scientist Sacramento (N.Mex.) Peak Obs., 1978-80; assoc. prof. physics Auburn (Ala.) U., 1980—. Mem. Am. Astron. Soc. Subspecialties: Theoretical astrophysics; Solar physics. Current work: Magnetic field generation, plasma, radiation, coronal heating, pulsar; theoretical plasma physics.

HINCKLEY, CONRAD CUTLER, chemistry educator, researcher; b. Ft. Worth, May 8, 1934; s. Herbert McDonald and Sadie Martha (Burton) H.; m. Nelda Jean Williams, Nov. 17, 1956; children: Russell C., Kristin A. Hinckley Bradfield. B.S., North Tex. State U., 1959, M.S., 1960; Ph.D., U. Tex., 1964. Postdoctoral fellow dept. chemistry U. Tex., Austin, 1964-66; asst. prof. chemistry So. Ill. U. Carbondale, 1966-71, assoc. rof., 1971-76, prof., 1976—; vis. assoc. prof. U. Tex., summer, 1976. Served with U.S. Army, 1954-56. Recipient Kaplan Research award So. Ill. chpt. Sigma Xi., 1972; Sloan fellow, 1973-75. Mem. Am. Chem. Soc., Chem. Soc. (London), AAAS. Subspecialties: Inorganic chemistry; Bioinorganic chemistry. Patentee in field. Office: Dept. Chemistr So Ill U Carbondale IL 62901

HINDS, THOMAS EDWARD, research plant pathologist; b. Albuquerque, Nov. 30, 1922; s. Frank Leslie and Ruth Amelia (Weiser) H.; m. Carmelita June Benander, June 13, 1957; children: Carl, Kevin, Keith. B.S., Colo. State U., 1958, M.S. in Plant Pathology, 1968. Biol. aide U.S. Bur. Plant Industry, Albuquerque, 1950-54; forestry aide U.S. Forest Service, Ft. Collins, Colo., 1954-58; plant pathologist, 1958-70, research plant pathologist, 1970—. Served with USN, 1943-46. Mem. Am. Phytopath. Soc., AAAS, Mycological Soc. Am., Sigma Xi. Subspecialty: Plant pathology. Current work: Forest tree diseases, tree cankers, tree decay fungi. Office: 240 W Prospect St Fort Collins CO 80526

HINES, ROBIN HINTON, electrical engineer; b. Nashville, July 18, 1937; s. C. Vernon and Laurabel (Carter) H.; m. L. Gail Vester, June 4, 1960; children: Phyllis, Miriam, Bobby, Mary Beth. B.Engring., Vanderbilt U., 1959; M.S., U. Tenn., 1963. Registered profl. engr., Tenn. Supr., engr. Aro, Inc. (Arnold Engring. Devel. Ctr.), 1963-68; v.p. Laser Systems & Electronics, 1968-71, Precision Internat., Inc., Tullahoma, Tenn., 1971-82; pres. Quantime, Inc., Tullahoma, 1982—, also dir.; dir. 1st Fed. Savs. & Loan Assn. Tullahoma. Scoutmaster, com. chmn. Boy Scouts Am.; active Chattanooga Track Club. Served to lt. (j.g.) USNR, 1959-62. Mem. Soc. Photog. and Instrumentation Engrs., Optical Soc. Am. Methodist. Club: Fellowship Christian Athletes. Lodge: Rotary. Subspecialties: Electronics; Optical engineering. Current work: Electronic distance measuring equipment.

HINGTGEN, JOSEPH NICHOLAS, psychologist, educator; b. Dubuque, Iowa, Nov. 18, 1936; s. Joseph Theodore and Clara Adelaide (Jungers) H.; m. Eleanor Anita Quinn, June 26, 1965; children: Cynthia, Christina, Charles. B.S., Loras Coll., 1958; Ph.D., Loyola U., 1963. Cert. psychologist, Ind. Instr. Ind. U. Sch. Medicine-Indpls., 1963-66, asst. prof. psychiatry, 1966-70, assoc. prof. psychology, 1970-77, prof. psychology/neurobiology, 1977—, research assoc. Inst. Psychiat. Research, 1962-74, chief sect. research service Inst. Psychiat. Research, 1974-79, mem. exec. bd. inst., 1974—; lectr. in psychology Marian Coll., 1963—. Author: (with C.Q. Bryson) Early Childhood Psychoses, 1971; contbr. chpts., numerous articles to prof. publs.; editor: Biol. Psychiatry, 1975—. Mem. Soc. Biol. Psychiatry (editorial bd. 1975—), Am. Psychol. Assn., Soc. Neurosci. Roman Catholic. Subspecialties: Psychopharmacology; Neurochemistry. Current work: Neurochemical correlates of behavior; animal models of depression; biological psychiatry; infantile autism; behavioral pharmacology; pre-and postdoctoral teaching in medical neurobiology. Office: Inst Psychiat Research Ind U Med Center Indianapolis IN 46223

HINKLEY, EVERETT DAVID, JR., physicist; b. Augusta, Maine, Nov. 19,1936; s. Everett David and Julina Margaret (Nolan) H.; m. Christine Marie Caso, June 18, 1960; children: Anne, Mark, Kristin, David. B.S. in Engring. Physics, Washington U., St. Louis, 1958; M.S. in Physics, Northwestern U., 1961, Ph.D., 1963. Staff mem. Lincoln Lab., MIT, Cambridge, Mass., 1963-76; v.p. Laser Analytics, Inc., Lexington, Mass., 1976; sr. research scientist Jet Propulsion Lab., Calif. Inst. Tech., Pasadena, 1976—, mgr. atmospheric scis. sect., 1980-83, program mgr. sensor tech., 1980—. Editor: Laser Monitoring of the Atmosphere, 1976; contbr. numerous chpts. to books, articles in field to profl. publs. Mem. clean air com. Pasadena Lung Assn., 1978—. Recipient Cert. of Recognition NASA, 1981. Fellow Optical Soc. Am.; mem. IEEE, AAAS. Subspecialties: Remote sensing (atmospheric science); Spectroscopy. Current work: Remote sensing of the earth's atmosphere and surface from space; lasers; molecules; gases; meteorology; winds; clear-air turbulence. Patentee semiconductor heterojunction diode, driving arrangement for magnetic devices. Office: 4800 Oak Grove Dr MS 180-701 Pasadena CA 91109

HINMAN, EDWARD JOHN, health care administrator; b. New Orleans, Nov. 10, 1931; s. E. Harold and Katharine (Fradenburgh) H.; m. Emma Jean Richmond, June 15, 1954; children: Cynthia, Alan, David. B.A., U. Okla., 1951; M.D., Tulane U., 1955; M.P.H., Johns Hopkins U., 1971. Diplomate: Am. Bd. Preventive Medicine. Intern USPHS Hosp., New Orleans, 1955-56, resident, Balt., 1958-61; dir. spl. research and devel. project Nat. Center Health Service Research, Rockville, Md., 1973-74; asst. surgeon gen., dir. div. hosps. and clinics USPHS, 1974-78; exec. dir. Group Health Assn., Washington, 1978-83, bd. mem., 1972—; bd. mem. Am. Assn. Med. Systems and Informatics, Bethesda, Md., 1980—, Coop. League U.S.A., Washington, 1982—. Contbr. articles in field to profl. jours. Served with USPHS, 1955-78. Recipient Bronze Letzeiser medal U. Okla., 1951; Meritorious Service award USPHS, 1976; Outstanding Achievement award Md. chpt. Federally Employed Women, 1976. Fellow ACP, Am. Coll. Preventive Medicine (v.p. 1976-77), Am. Pub. Health Assn., Soc. Advanced Med. Systems (dir. 1968-82, pres. 1976-77), Am. Fedn. Clin. Research, Phi Beta Kappa. Subspecialty: Preventive medicine. Home: Box 48D Cape Leonard Dr St Leonard MD 20685 Office: Group Health Assn Inc 2021 L St NW Washington DC 20036

HINNOV, EINAR, physicist; b. Estonia, Mar. 17, 1930; m., 1956; 1 child. B.A., St. Olaf Coll., 1952; Ph.D. in Physics, Duke U., 1956. Research assoc. physics U. Md., 1956-58, asst. research prof., 1958-59; research physicist (Project Matterhorn), 1959-63, Plasma Physics Lab.,

Princeton U., 1963-74, sr. research physicist, 1974-80, prin. research physicist, 1980—, lectr. dept. astros. sci., 1965-74; mem. NRC Com. on Atomic and Molecular Scis. Fellow Am. Phys. Soc. Subspecialty: Plasma physics. Office: Plasma Physics Lab Princeton U Princeton NJ 08544

HINRICHS, JAMES EDWARD, periodontal educator, clinical researcher; b. Omaha, Sept. 17, 1949; s. Edward John and Helen Clarice (Lapcheska) H.; m. Linda Kathleen Mason, June 15, 1972; children: Scott Jeffery, David James. B.S., U. Nebr., 1967-71, D.D.S., 1975; M.S., Coll. Dentistry, 1979. Gen. practice residential dentistry Wood VA Hosp., Omaha, 1976-77; clin. instr. Creighton Sch. Dentistry, Omaha, 1976-77; periodontal resident, dentistry Lincoln (Nebr.) VA Hosp., 1977-78; clin. instr. U. Nebr. Coll. Dentistry, 1978-79; asst. prof. U. Minn., Mpls., 1980—; clin. dir. periodontology U. Minn. Hosp. Dental Clinic, Mpls., 1980—; pvt. practice gen. dentistry, Omaha, 1976-77, pvt. practice periodontology, Bloomington, Minn., 1980—. Active Minn.-Am. Heart Assn., Mpls., 1980—. Grantee U. Nebr. Med. Center, Omaha, 1979; recipient Scholarly Achievement award Internat. Coll. Dentists, 1975. Mem. Am. Acad. Periodontology (Orban Meml. award 1979), Mpls. Dist. Dental Soc., Bloomington Dental Study Club (pres. 1981—), Am. Dental Soc. Anesthesiology (state pres. 1983), Internat. Assn. Dental Research. Roman Catholic. Subspecialty: Periodontics. Current work: Relationship between periodontal diagnostic parameters, oral microbiology and treatment modalities. Office: Health Sci Unit A 7-368 Univ Minn 515 Delaware St SE Minneapolis MN 55455

HINRICHS, JAMES VICTOR, psychology educator; b. Harlan, Iowa, May 19, 1941; s. James Victor and Leona Gladys (Van Der Stoep) H.; m. Charlene Ruth Becker, Dec. 29, 1964; children: Susan Elaine, Kristine Ann. B.A., U. Iowa, 1963; A.M., Stanford U., 1964, Ph.D., 1967. Asst. prof. psychology U. Iowa, 1967-70, assoc. prof., 1970-78, prof., 1978—. NIMH grantee, 1969-72, 79—. Mem. Am. Psychol. Assn., Psychonomic Soc. Subspecialties: Cognition; Neuropsychology. Current work: Benzodiazapines and human cognition; mathematical cognition. Home: 5 Ridgewood Ln Iowa City IA 52240 Office: Dept Psychology U Iowa Iowa City IA 52242

HINTHORNE, JAMES ROSCOE, geology educator; b. Los Angeles, Dec. 23, 1942; s. Gilbert Theodore and Esther Marian (Williams) H.; m. Jean Marie Mendenhall, Aug. 22, 1964; children: Kathleen, Michael. A.A., Palomar Coll., 1962; B.A., U. Calif.-Santa Barbara, 1965, Ph.D., 1974; M.S., U. Mass., 1967. Scientist Applied Research Labs., Goleta, Calif., 1969-74, sr. scientist, 1974-77, mgr. research and devel., Sunland, Calif., 1977-80; prof. geology Central Wash. U., Ellensburg, 1980—. Fellow Geol. Soc. Am.; mem. Mineral. Soc. Am., Microprobe Soc. Am. Subspecialties: Mineralogy; Analytical chemistry. Current work: Theoretical treatment of ion microprobe analytical data for the trace element analysis of minerals; automation development for analytical chemistry hardware, especially microprobes. Office: Dept Geology Central Wash U Ellensburg WA 98926

HINTZE, LEHI FERDINAND, geology educator; b. Denver, Apr. 14, 1921; s. Ferdinand F. and Henrietta (Jones) H.; m. Ione Nelson, Nov. 20, 1942; children: Sharon, David, Paul, Wayne. A.B., U. Utah, 1941; M.A., Columbia U., 1949, Ph.D., 1951. Asst. prof. Oreg. State U., Corvallis, 1949-55; assoc. prof. Brigham Young U., Provo, 1955-60, chmn. dept. geology, 1962-69; cons. oil cos., 1950-70; geologist U.S. Geol. Survey, Menlo Park, Calif., part-time 1970—; Chmn. Utah Earthquake Awareness Com., Provo, 1975-78; mem. adv. bd. Utah Geol. Survey, 1963-72, Utah County Master Plan, Provo, 1973-75; regional coordinator COSUNA-Great Basin, Tulsa, 1978-82. Author: Lower Ordovician Trilobites, 1951, Geologic History of Utah, 1973; maps Geology of Utah, 1963, 80. Served to capt. U.S. Army, 1942-48; Japan. Grantee NSF, 1963-65, 1962. Fellow Geol. Soc. Am.; mem. Sigma Xi. Subspecialties: Tectonics; Remote sensing (geoscience). Current work: Structural geologic mapping in Western Utah. Office: Dept Geology Brigham Young Univ Provo UT 84602 Home: 1835 N 1450 E St Provo UT 84604

HINZE, WILLIE LEE, chemist, educator, cons; b. Burton, Tex., Jan. 17,1949; s. Willie L. H. and Alma (Tresseler) H.; m. Wen-wen Chu, Dec. 14, 1980. A.A., Blinn Coll., Brenham, Tex., 1969; B.S., Sam Houston State U., 1970, M.A., 1972; Ph.D., Tex. A&M U., 1974. Lectr. in chemistry, NIH postdoctoral fellow Tex. A&M U., 1974-75; instr. in chemistry Blinn Coll., 1974-75; asst. prof. chemistry Wake Forest U., 1975-80, assoc. prof., 1980—; cons analytical chemistry. Contbr. articles to profl. jours. NRC travel award grantee 28th Internat. Union Pure and Applied Chemistry Congress, Vancouver, B.C., Can., 1981; Wake Forest Research and Publ. Fund grantee, 1976-82; Am. Chem. Soc. Petroleum Research Fund grantee, 1976-80, 80-82, 82; Research Corp. Cottrell Coll. Sci. grantee, 1977-78; Sigma Xi grantee, 1980; Research Corp. Susan Greenwall Found. grantee, 1980-81; NSF grantee, 1980-83. Mem. Am. Chem. Soc., Assn. Ofcl. Analytical Chemists, Am. Oil Chemists Soc., Am. Inst. Chemists (cert. profl. chemist), Sigma Xi. Subspecialties: Analytical chemistry; Clinical chemistry. Current work: Use of micellar and cyclodextrin systems in chemical analysis (in ultraviolet-vis, fluorescence, phosphorescence and chemiluminescence determinations, and chromatographic separations). Home: 2200 Faculty Dr Apt 5-D Winston-Salem NC 27106 Office: Dept Chemistry Wake Forest U PO Box 7486 Winston-Salem NC 27109

HIRLEMAN, EDWIN DANIEL, JR., mech. engr., educator, researcher, inventor; b. Wichita, Kans., Dec. 1, 1951; s. Edwin Daniel and Marcille Parker (Wohlgemuth) H.; m. Laura Kay Kennedy, Sept. 13, 1975; children: Daniel Garth, Emily Diane, Mark David. B.S.M.E., Purdue U., 1972, M.S.M.E. (NSF fellow, Hughes Aircraft fellow), 1974, Ph.D. in Mech. Engring, 1977. Mem. engring. staff Hughes Aircraft Co., 1974-77; vis. researcher Tech. U. Denmark, Copenhagen, 1974-75; asst. prof. mech. engring. Ariz. State U., Tempe, 1977-81, assoc. prof., 1981—; cons. Argonne Nat. Lab., Phelps Dodge Corp., Danker and Wohlk, Garrett Turbine Engine Co., EPA, NSF, Spectron Devel. Labs. Contbr. articles to profl. jours. Recipient award for significant accomplishment in research Engring. Coll., Ariz. State U., 1980, Prof. of Yr. award mech. engring. dept., 1982. Mem. Optical Soc. Am., Combustion Inst., ASME, AIAA, IEEE, Tau Beta Pi, Pi Tau Sigma. Subspecialties: Laser instrumentation; Optical instrumentation. Current work: Laser/optical instrumentation for combustion and fluid mechanics; microprocessor-based instrumentation. Patentee in field. Office: Mech Engring Dept Ariz State U Tempe AZ 85287

HIROSE, TERUO TERRY, surgeon; b. Tokyo, Jan. 20, 1926; s. Yohei and Seiko (Ogushi) H.; m. Tomiko Kodama, June 1, 1976; 1 son, George Philamore. B.S., Tokyo Coll., 1944; M.D., Chiba U., Japan, 1948, Ph.D., 1958. Diplomate: Am. Bd. Surgery, Am. Bd. Thoracic Surgery. Intern Chiba U. Hosp., 1948-49, resident in surgery, 1949-52, Am. Hosp., Chgo., 1954; resident in thoracic surgery Hahnemann Med. Coll., Phila., 1955-56, N.Y. Med. Coll., N.Y.C., 1961-62; practice medicine specializing in surgery, Chiba, Japan, 1952-53; chief of surgery Tsushimi Hosp., Hagi, Japan, 1958-59; asst. prof. surgery Chiba U., 1959; research fellow advanced cardiovascular surgery Hahnemann Hosp., Phila., 1959; teaching fellow surgery N.Y. Med. Coll., 1959-60, instr., 1961-62; practice medicine specializing in surgery, N.Y.C., 1965—, N.J., 1975—; dir. cardiovascular lab. St. Barnabas Hosp., N.Y.C., 1965—, sr. attending surgeon, 1965—; chief vascular surgery Union Hosp., Bronx, N.Y., 1966-67; attending surgeon Flower and Fifth Ave Hosp., N.Y.C., 1973—, Jewish Hosp. Med. Center, Bklyn., 1976—, St. Vincent Hosp., N.Y.C., 1976—, Mamonides Hosp., Bklyn., 1976-78, Passaic Gen. Hosp., 1977—, Westchester (N.Y.) County Hosp., 1977-78, Yonkers (N.Y.) Profl. Hosp., 1978-79, Westchester Sq. Hosp., 1978—, Yonkers Gen. Hosp., 1980—, St. Joseph Hosp., Yonkers, 1980—; clin. prof. surgery N.Y. Med. Coll., 1974—. Contbr. articles in field of cardiovascular surgery to Am. and Japanese med. jours. Recipient Hektoen Bronze medal AMA, 1965, Gold medal, 1971. Fellow Am. Coll. Angiology, Am. Coll. Chest Physicians, A.C.S., Am. Coll. Cardiology, Internat. Coll. Surgeons, N.Y. Acad. Medicine; mem. Am. Assn. Thoracic Surgery, N.Y. Soc. Thoracic Surgery, Pan-Pacific Surg. Assn., Internat. Cardiovascular Soc., Am. Geriatric Soc., Am. Fedn. Clin. Research. Subspecialties: Cardiac surgery; Surgery. Inventor single pass low prime oxygenator; pioneer aortocoronary direct bypass surgery, open heart surgery without blood transfusion. Office: 1625 St Peters Ave Bronx NY 10461 One should respect another's religion or creed and offer assistance regardless of whether or not one is in agreement with the other's belief, provided that belief harms no other.

HIRSCH, ANN MARY, biologist, educator; b. Milw., June 2, 1947; d. Clifford and Irene (Janes) H.; m. Stefan J. Kirchanski, Sept. 26, 1970. B.S. (scholar), Marquette U., 1969; Ph.D., U. Calif.-Berkeley, 1974. Asst. prof. U. Minn., St. Paul, 1974-76; Cabot fellow Harvard U., Cambridge, Mass., 1976-78; asst. prof. Wellesley (Mass.) Coll., 1978-83, assoc. prof., 1983—; vis. scholar Harvard U., 1981-82. Contbr. articles to sci. jours. Brachman-Hoffman fellow, 1982-84; NSF trainee, 1972-73; other grants. Mem. Am. Soc. Plant Physiologists, Plant Molecular Biology Assn., Assn. for Women in Sci., Bot. Soc. Am., Am. Soc. Cell Biology, Sigma Xi. Subspecialties: Plant physiology (biology); Nitrogen fixation. Current work: Analysis of symbiotically defective nodules of alfalfa infected with transposon-generated mutants of Rhizobium; localization of Rhizobium host range and nodulation genes and determination of gene products. Office: Wellesley College Wellesley MA 02181

HIRSCH, HELMUT VILLARD BUNTENBROICH, devel. neurobiologist, educator; b. Chgo., Sept. 22, 1943; s. Helmut and Eva (Buntenbroich) H.; m. Natalie Gans, June 13, 1966 (div.); 1 son, Paul Devin. A.B. in Math, U. Chgo., 1965; Ph.D. in Psychology, Stanford U., 1971. Postdoctoral fellow Johns Hopkins U., Balt., 1970-72; asst. prof. biology SUNY, Albany, 1972-77, assoc. prof., 1977—; Alfred P. Sloan Found. fellow in neurobiology, 1975-79. Contbr. articles to sci. jours. NIH grantee; NSF grantee. Mem. AAAS, Soc. for Neurosci., Assn. for Research in Vision and Ophthalmology, Sigma Xi. Subspecialties: Neurobiology; Neurophysiology. Current work: Development of nervous system; sensory physiology; electrophysiology; neuronal plasticity; visual system electrophysiology; image processing by visual systems. Office: Dept Biology SUNY Albany Albany NY 12222

HIRSCH, HENRY RICHARD, physiology and biophysics educator; b. N.Y.C., Mar. 27, 1933; s. Samuel Hirsch and Edna (Gurner) Hirsch S.; m. Genevieve Tenen, July 17, 1954; children: Steven A., Samuel. S.B., M.I.T., 1954, Ph.D., 1960. Elec. engr. Bell Telephone Labs., Murray Hill, N.J., 1954-57; physicist NIH, Bethesda, Md., 1961-63; mem. faculty U. Ky. Coll. Medicine, Lexington, 1963—, prof. physiology and biophysics, 1976—. Reviewer: Zentralblatt fur Mathematik, Berlin, 1967—; contbr. numerous articles to profl. publs. Fellow Gerontol. Soc. Am.; mem. IEEE (sr.), Am. Physiol. Soc. (pub. affairs com. 1982—), Fedn. Am. Socs. Exptl. Biology (pub. affairs com. 1979—), Am. Phys. Soc., Biophys. Soc., AAAS. Subspecialties: Physiology (medicine); Gerontology. Current work: Mathematical biology with special application to problems in physiology and gerontology. Office: Univ Ky Physiology and Biophysics Dept Lexington KY 40536-0084

HIRSCH, JERRY, educator, biologist, psychologist; b. N.Y.C., Sept. 20, 1922; s. Samuel M. and Mollie (Barnett) H.; m. Marjorie J. Barrie, July 29, 1950; 1 son, Wesley M. Student, Johns Hopkins, 1938-40, U. Paris, Sorbonne, France, 1949-50; B.A., U. Cal. at Berkeley, 1952, Ph.D., 1955. With Cavendish Trading Corp., 1940, Cohn, Hall, Marx Co., 1940-41; pres. Jostex Corp., 1941-49; NSF fellow U. Calif. at Berkeley, 1955-57; asst. prof. psychology Columbia, 1956-60; NIH fellow Center For Advanced Study in Behavioral Scis., Stanford, Calif., 1960-61; asso. prof. U. Ill., Urbana, 1960-63, prof. psychology, 1963—, prof. zoology, 1966—, prof. ecology, ethology and evolution, 1976—; zoology research asso. U. Edinburgh, Scotland, 1968; NSF-A.A.A.S. Chautauqua course lectr., 1973-74, 74-75; mem. com. genetics and behavior Social Sci. Research Council, 1962-65; mem. behavioral scis. tng. com. Nat. Inst. Gen. Med. Scis., NIH, 1966-70; mem. U.S. nat. com. Internat. Union Biol. Scis., Assembly Life Scis. NRC-NAS, 1975—; mem. U.S. com. XIV and XV Internat. Ethological Conf., 1975, 77, 79, 81; dir. tng. program for research on instnl. racism NIMH, 1977—. Author, editor: Behavior-Genetic Analysis, 1967; Am. editor: Animal Behaviour, 1968-72; Contbr. articles to profl. jours. Served with USAAF, 1942-43. Recipient Aux. Research award Social Sci. Research Council, 1962. Mem. A.A.A.S., A.A.U.P., Animal Behavior Soc. (exec. com. 1967—, pres. 1975), Am. Eugenics Soc., Am. Genetic Assn., Am. Psychol. Assn., Ecol. Soc., Genetics Soc. Am., Behavior Genetics Assn. (charter), Soc. for Neurosci., Am. Soc. Human Genetics, Sigma Xi. Club: Cosmos. Subspecialty: Behavioral psychology. Home: 2012 Zuppke Circle Urbana IL 61801

HIRSCH, PHILIP FRANCIS, pharmacologist; b. Stockton, Calif., June 24, 1925; s. Harold and Elsa (Frohman) H.; m. Eugenia Isaeff, Sept. 21, 1956; children—Steven, Lisa, Kenny, Nancy. B.S. in Chemistry, U. Calif., Berkeley, 1950, Ph.D. in Physiology, 1954. Lectr. physiology U. Calif., Berkeley, 1954-55; instr. pharmacology Sch. Dental Medicine, Harvard U., Cambridge, Mass., 1955-57, asso. in pharmacology, 1957-63, asst. prof. pharmacology, 1964; physiologist Lawrence Livermore Lab., 1964-66; asso. prof. pharmacology Sch. Medicine, U. N.C., Chapel Hill, 1966-70, prof., 1970—, dir., 1975—; mem. gen. medicine B study sect. NIH, 1974-78. Contbr. articles to profl. jours. Bd. dirs. YMCA, Chapel Hill, 1981—. Served with AUS, 1943-46. Mem. AAAS, Endocrine Soc., Am. Soc. Pharmacology and Exptl. Therapeutics, Internat. Assn. Dental Research, Sigma Xi. Subspecialties: Endocrinology; Pharmacology. Current work: Hormones and other substances affecting calcium metabolism. Home: 2008 S Lake Shore Dr Chapel Hill NC 27514 Office: Dental Research Center 210-H U of NC Chapel Hill NC 27514

HIRSCHBERG, ALBERT IRWIN, chemistry educator; b. N.Y.C., May 9, 1934; s. Aaron and Julia (Schlosberg) H.; m. Carol T. Levine, June 11, 1960; children: Joyce, Julie. B.S., Bklyn. Coll., 1954; M.S., Poly. Inst. N.Y., 1956, Ph.D., 1960. Asst. prof. chemistry L.I. U., 1962-65, assoc. prof., 1966-71, prof., 1971—, chmn. dept. chemistry, 1979—. Author audio visual programs. NIH fellow, 1960. Fellow Royal Soc. Chemistry; mem. Am. Chem. Soc. Republican. Jewish. Subspecialty: Organic chemistry. Current work: Chemistry education; heterocyclic chemistry, medicinal chemistry. Office: Dept Chemistry LI U University Plaza Brooklyn NY 11201

HIRSCHBERG, CARLOS BENJAMIN, biochemistry educator; b. Santiago, Chile, Feb. 1, 1943; came to U.S., 1965, naturalized, 1977; s. Erich and Gertrud (Landsberger) H.; m. Lois Barnes, Sept. 5, 1973. M.S., Rutgers U., 1966; Ph.D., U. Ill., 1970. Biochemist U. Chile, Santiago, 1965; Grad. research asst. U. Ill.-Urbana, 1966-70; research fellow Harvard U., 1970-72; research assoc. M.I.T., 1972-74; asst. prof. biochemistry St. Louis U., 1974-78, assoc. prof., 1978-82, prof., 1982—; mem. cell biology study sect. NIH, Bethesda, Md., 1978-82. Jane Coffin Childs Fund fellow, 1970; recipient Faculty Research award Am. Cancer Soc., 1979. Mem. Am. Soc. Biol. Chemists, Am. Soc. Cell Biology, AAAS. Subspecialties: Biochemistry (biology); Cell biology. Current work: Membrane biogenesis; biosynthesis of glycoproteins in normal and transformed cells. Office: Saint Louis U Sch Medicine 1402 S Grand Blvd Saint Louis MO 63104

HIRSCHFELDER, JOSEPH OAKLAND, chemistry educator; b. Balt., May 27, 1911; m., 1953. B.S., Yale U., 1931; Ph.D., Princeton U., 1936. Fellow Inst. Advanced Study, 1936; asst. Princeton U., 1936-37; research fellow U. Wis.-Madison, 1937-39, instr. physics and chemistry, 1940, asst. prof. chemistry, 1941-42; head interior ballistics groups, geophys. lab. and cons. rockets Nat. Def. Research Com., 1942-45; chief phenomenologist Bikini Bomb Test, 1946; prof. chemistry U. Wis-Madison, 1946-62, Homer Adkins prof., 1962—, dir. theoretical chemistry lab., 1959-62, dir., 1962-77; Group leader Los Alamos Sci. Lab., 1943-46; head theoretical physics, Naval Ordnance Test Sta, China Lake, Calif., 1945-46; chmn. div. phys. chemistry NRC, 1958-61; mem. computer panel NSF, 1962-65; mem. policy adv. bd. Argonne Nat. Lab., 1962-66. Mem. Nat. Acad. Sci, Am. Chem. Soc. (Debye award 1966), Am. Phys. Soc., Am. Acad. Arts and Sci., Combustion Inst. (Egerton Gold medal 1966). Subspecialty: Theoretical chemistry. Office: Dept Chemistry U Wis Madison WI 53706

HIRSCHHORN, IRA DANIEL, neuropharmacologist, educator; b. Bklyn., Jan. 2, 1946; s. Harold Norman and Norma Irma (Krause) H.; m. Susan B. Wood, 1976; m. Suzanne Jo Feldberg, June 5, 1981. B.S. in Biology, U. Pitts., 1967; Ph.D. in Pharmacology, SUNY-Buffalo, 1971. Fellow dept. pharmacology Med. Coll. Va., Richmond, 1972-74; asst. prof. dept. pharmacology N.Y. Med. Coll., Valhalla, 1974-76; asst. prof. N.Y. Coll. Podiatric Medicine, N.Y.C., 1976-77; assoc. dept. biochemistry Albert Einstein Coll. Medicine, N.Y.C., 1978—; cons. asst. dir. Bayside Labs. Contbr. articles to profl. jours. NIH fellow, 1967-71; trainee, 1978-81; grantee, 1982—; AMA fellow, 1972-74; Pharm. Mfrs. Assn. grantee, 1975-77. Mem. AAAS, Behavioral Pharmacology Soc., N.Y. Acad. Sci., Soc. Neuroscience, Soc. Stimulus Properties of Drugs, Sigma Xi. Democrat. Jewish. Subspecialties: Neuropharmacology; Neurochemistry. Current work: Biochemical and behavioral actions of CNS drugs, neurotrasmitters receptors.

HIRSCHHORN, JOEL S(TEPHEN), metallurgist, public policy analyst; b. N.Y.C., Sept. 8, 1939; m., 1961; 2 children. B.Met.E., Poly. Inst. Bklyn., 1961, M.S., 1962; Ph.D., Rensselaer Poly. Inst, 1965. Research metallurgist adv. materials R&D lab. Pratt & Whitney Aircraft, 1962-63; from asst. prof. to prof. metall. engring. U. Wis-Madison, 1965-78; project dir. Office Technol. Assessment, U.S. Congress, Washington, 1978-83, sr. assoc., 1983—; Cons. Friction Products Co., 1967-68, dir. research, 1968-72; Cons. Stellite div. Cabot Corp., 1970-71, Advanced Products Corp., 1971-73, Que. Metal Powders, 1975-76. Mem. AAAS, AIME, Am. Soc. Engring. Edn., Am. Powder Metallurgy Inst. Subspecialty: Science and technology policy. Current work: steel industry; strategic planning; environmental regulations; technology assessment; science and technology policy; hazardous waste management. Office: Office Technol Assessment US Congress Washington DC 20510

HIRSCHMAN, LYNETTE, computer scientist; b. Huntington, W.Va., Nov. 22, 1945; d. Isidore I. and Miriam (Diamant) H. B.A. in Chemistry, Oberlin (Ohio) Coll., 1966; M.A. in German Lit, U. Calif.-Santa Barbara, 1966-68; Ph.D. in Computational Linguistics, U. Pa., Phila., 1972. Assoc. research scientist, linguistic string project N.Y. U., 1975-78, research scientist, 1978-81; project engr. research and devel. fed. and spl. systems Burroughs Corp., Paoli, Pa., 1981—, mgr. artificial intelligence group, 1982; mgr. artificial intelligence group research and devel. System Devel. Corp., Paoli, 1983—. Mem. Assn. Computing Machinery, Am. Assn. Artificial Intelligence, Assn. Computational Linguistics, IEEE, Phi Beta Kappa. Subspecialties: Artificial intelligence; Automated language processing. Current work: Natural language processing of texts; representation of time in narrative; construction of rule based assistants; logic and programming and natural language processing. Home: 264 W Harvey St Philadelphia PA 19144 Office: Research and Devel Activity System Devel Corp Box 517 Paoli PA 19301

HIRSH, IRA JEAN, scientist, educator; b. N.Y.C., Feb. 22, 1922; s. Ellis Victor and Ida (Bernstein) H.; m. Shirley Helene Kyle, Mar. 21, 1943; children—Eloise, Richard, Elizabeth, Donald. A.B., N.Y. Coll. for Tchrs., 1942; A.M., Northwestern U., 1943; M.A., Harvard, 1947, Ph.D., 1948. Research asst. psycho-acoustic lab. Harvard, Cambridge, Mass., 1946-47, research fellow, 1947-51; with Central Inst. for Deaf, St. Louis, 1951—, asst dir. research, 1958-65, dir., 1965-83; dir. emeritus Central Inst. Deaf, 1983—; mem. faculty or adminstrn. Washington U., St. Louis, 1951—, prof. psychology, 1961—, dean faculty arts and scis., 1969-73; vis. prof. U. Paris, France, 1962-63; U.S. del Internat. Standards Orgn., 1962-76; mem. Internat. Acoustics Commn., 1969-75; chmn. behavioral scis. and edn. NRC, 1982—. Author: The Measurement of Hearing, 1952; Contbr. articles to profl. jours. Served with USAAF, 1943-45; Served with AUS, 1945-46. Recipient Biennial award Acoustical Soc. Am., 1956, Assn. Honors Am. Speech and Hearing Assn., 1968. Fellow Acoustical Soc. Am. (pres. 1967-68), Am. Psychol. Assn., Am. Speech and Hearing Assn. (exec. council 1958-61, 65-68); mem. Nat. Acad. Sci. Subspecialties: Sensory processes; Psychophysics. Current work: Hearing; auditory perception; speech. Home: 6629 Waterman Ave Saint Louis MO 63130

HIRSH, KENNETH ROY, pharmacologist, pharmacist; b. Bronx, N.Y., Nov. 7, 1944; s. Lester and Sara H.; m. Phyllis Pomerantz, Nov. 24,1965; children: Jeffrey, Jennifer. Ph.D. in Pharmacology, Columbia U., 1972. Prin. scientist Gen. Foods Corp., Cranbury, N.J., 1972—. Mem. Am. Soc. Pharmacology and Exptl. Therapeutics. Jewish. Subspecialties: Neuropharmacology; Pharmacology. Current work: Central nervous system stimulants-mechanisms. Home: 33 Scott Dr Morganville NJ 07751 Office: Prospect Plains Rd 025/C Cranbury NJ 08512

HIRST, WILLIAM CHARLES, psychology educator, researcher; b. Chester, Pa., Oct. 4, 1950; s. William Henry and Anna Jean (Lamey) H. B.A., Carnegie-Mellon U., 1972; Ph.D., Cornell U., 1976. Postdoctoral fellow Rockefeller U., 1976-78, asst. prof., 1978-80, adj. asst. prof., 1980-82; asst. prof. psychology Princeton U., 1980—; vis. lectr. New Sch. Social Research Grad. faculty, 1976-78; cons. in field. NIH postdoctoral fellow, 1976-78; grantee, 1979-81; Nat. Inst. Edn. grantee, 1978-80; Nat. Inst. Neurol. and Communicative Disorders and Stroke grantee, 1982—. Mem. Am. Psychol. Assn. Subspecialties: Cognition; Neuropsychology. Current work: Interest in cognitive science has led to study of memory and attention, not only from a traditional framework, but also through examination of brain damaged patients. Office: Dept Psychology Princeton U Princeton NJ 10014

HIRTH, JOHN PRICE, metallurgical engineering educator; b. Cin., Dec. 16, 1930; s. John Willard and Betty Ann (Price) H.; m. Martha Joan Davis, Nov. 28, 1953; children: John Marcus, Laura Ellen, James Gregory, Christina Louise. B. Metall. Engring., Ohio State U., 1953; M.S., Carnegie-Mellon U., 1953, Ph.D., 1957. Asst. prof. metall. engring. Carnegie-Mellon U., Pitts., 1958-61; Mershon prof. Ohio State U., 1961-67; vis. prof. Stanford, 1967-68; prof. Ohio State U., Columbus, 1967—, Alcoa vis. prof. Nat. U. Mex., Mexico City, 1976; cons. in field; vis. adv. com. Nat. Bur. Standards, 1969-72, Argonne Nat. Lab., 1970-73, Cornell U., 1971-74, Carnegie-Mellon U., 1967-76; bd. overseers Acad. for Contemporary Problems, 1971-76. Author: Condensation and Evaporation, 1964, Theory of Dislocations, 1968, 82; editor: Scripta Metallurgica, 1974—. Served with USAF, 1953-55. Fulbright fellow Bristol U., Eng., 1957-58. Fellow Am. Soc. Metals (Stoughton award 1964, Campbell lectr. 1972), AIME (Hardy medal 1960, Mehl medal 1980, Mathewson medal 1982), Am. Soc. Engring. Edn. (McGraw award 1967); mem. Nat. Acad. Engring., Sigma Xi. Club: Ohio State Rugby Football. Subspecialties: Metallurgy; Materials. Current work: Dislocation theory; physical metallurgy. Home: 4062 Fairfax Dr Columbus OH 43220 Office: Dept Metall Engring Ohio State U Columbus OH 43210

HISCOCK, ROBERT RUSSELL, radiol. health physicist, hosp. adminstr.; b. Balt., July 22, 1938; s. Raymond and Mildred Ruth (Shaffer) H.; m. Linda M. Wheat, July 25, 1959; children: Lori, Richard, Beth. A.A., Balt. Jr. Coll., 1958; M.S., Johns Hopkins U., 1960; postgrad., Loyola Coll., Balt., 1963-64. With Sinai Hosp., Balt., 1960—, radiation control officer, 1963—, adminstrv. dir., 1968—; cons. radiol. health, Balt., 1965—; instr. radiobiology Essex Community Coll., Balt., 1975—; cons. in field. Mem. editorial bd.: Radiologic Tech, 1978-82. Mem. Am. Assn. Physicists in Medicine, Am. Soc. Radiol. Technologists (chmn. radiation safety com.), Health Physics Soc., Internat. Radiation Protection Assn., Soc. Nuclear Medicine. Subspecialties: Nuclear medicine; Biophysics (physics). Current work: Adminstrative and technical director of radiation therapy services, medical health physics, teaching and consulting. Office: 2401 W Belvedere Ave Baltimore MD 21215

HISER, HOMER WENDELL, environmental engineer; b. Ava, Ill., Nov. 21, 1924; m., 1953. B.S., U. Ill., 1951, M.S., 1954; D.Sc. in Environ. Engring, Washington U., St. Louis, 1972. Research assoc., research radar meteorologist Ill. State Water Survey (U. Ill.), 1950-55; prof., dir. radar meteorol. lab. Rosenstiel Sch. Marine and Atmospheric Sci. (U. Miami), Fla., 1955-73, prof., dir. remote sensing lab., 1974—; Cons. Battelle Columbus Labs., 1974; mem. adv. panel NASA, 1975. Recipient cert. of appreciation Nat. Weather Service, 1971. Mem. Am. Meteorol. Soc., Am. Geophys. Union, IEEE (sr.), Air Pollution Control Assn. Subspecialty: Remote sensing (atmospheric science). Office: U Miami PO Box 248003 Coral Gables FL 33124

HITCHCOCK, MARGARET, medical researcher, educator; b. Harrow, Middlesex, Eng., July 17, 1940; came to U.S., 1965, naturalized, 1975; d. Frank and Olga Mary (Warburton) H.; m. John Joseph Burns, May 11, 1974. B.Sc., U. London, 1962, Ph.D., 1965. Postdoctoral fellow Harvard U. Sch. Public Health, 1965-67; cons., research specialist Calif. Dept. Public Health, 1967-69; asst. prof. dept. epidemiology and public health Yale U. Sch. Medicine, New Haven, 1969-75, asso. prof., 1975-78, sr. research asso., lectr., 1978-83, lectr. dept. pharmacology, 1978—; mem. study sects. NIH, 1972; mem. bd. sci. councillors Nat. Toxicology Program, 1979—; mem. environ. studies bd. Nat. Acad. Sci.-NRC, 1982—; cons., 1983—. Contbr. articles on pharmacology, toxicology, environ. health to profl. publs. NIH grantee, 1969—. Mem. Am. Soc. Pharmacology and Exptl. Therapeutics, Soc. Toxicology (council 1979-81), Soc. Occupational and Environ. Health, N.Y. Acad. Sci. Club: Fairfield County Hunt (Westport, Conn.). Subspecialties: Pharmacology; Toxicology (medicine). Current work: Biochemical and immunological mechanisms underlying response of the lung to environmental and therapeutic agents; risk assessment of health effects of environmental agents. Office: 480 Catamount Rd Fairfield CT 06430

HITCHCOCK-DEGREGORI, SARAH ELLEN, biologist, educator; b. Washington, Nov. 19, 1943; d. Ethan W. and Mary (Kibbe) Hitchcock; m. Alessandro DeGregori, June 7, 1980. A.B., Smith Coll., 1965; M.A., Wesleyan U., 1967; Ph.D., Case Western Res. U., 1970. Research assoc. Brandeis U., Waltham, Mass., 1970-73; vis. scientist Med. Research Council Lab. Molecular Biology, Cambridge, Eng., 1973-76; asst. prof. biol. scis. Carnegie-Mellon U., Pitts., 1976-82, assoc. prof., 1982—. Author research papers. Muscular Dystrophy research fellow, 1971-73; Brit.-Am. heart fellow, 1973-75; NIH spl. fellow, 1975-76; research career devel. award, 1982—. Mem. Am. Soc. Cell Biology, Biophys. Soc. Subspecialties: Biochemistry (medicine); Cell biology. Current work: Regulation of contraction in muscle and blood platelets. Study of contractile proteins and their mechanism of regulation. Office: Mellon Inst 4400 5th Ave Pittsburgh PA 15213

HITCHINGS, GEORGE HERBERT, pharm. co. exec., biochemist; b. Hoquiam, Wash., Apr. 18, 1905; s. George Herbert and Lillian Bell (Matthews) H.; m. Beverly Reimer, June 24, 1933; children: Laramie Hitchings Brown, Thomas E. B.S., U. Wash., 1927, M.S., 1928; Ph.D., Harvard U., (hon. degrees), 1933; D.Sc., U. Mich., 1971, U. Strathclyde, 1977. Teaching fellow U. Wash., 1926-28, Harvard U., 1928-34, 34-36; sr. instr. Western Res. U., 1939-42; biochemist Burroughs Wellcome Co., Tuckahoe, N.Y., 1942-46, chief biochemist, 1946-55, asso. research dir., 1955-63, research dir., 1963-67, v.p. in charge research, 1967-75; also dir.; dir. Burroughs Wellcome Fund, 1968-80, pres., 1971—; adj. prof. Duke U., 1970—, U. N.C., Chapel Hill, 1972—; vis. prof. Chuang-Ang U., Seoul, Korea, 1974-77; vis. lectr., Pakistan and Iran, 1976, Japan and India, 1980; staff dept. medicine Roger Williams Gen. Hosp., Brown U., 1980; mem. vis. com. drug research bd. Nat. Acad. Scis.-NRC, 1974-75; mem. research and eval. adv. com. N.C. State Dept. Corrections, 1974-76; vis. lecturer, Republic of South Africa, 1981. Contbr. articles to profl. jours. Mem. drug devel. com. Cancer Research, Internat. Transplantation Soc., NRC, Westchester Chem. Soc. (chmn. 1959-60), Phi Beta Kappa, Sigma Xi, Phi Lambda Upsilon. Subspecialty: Biochemistry (biology). Home: 4022 Bristol Rd Durham NC 27707 Office: 303 Cornwallis Rd Research Triangle Park NC 27709

HITTINGER, WILLIAM CHARLES, electronics co. exec.; b. Bethlehem, Pa., Nov. 10, 1922; s. John Tilghman and Pearl (Heimbach) H.; m. Elizabeth Herman, July 9, 1944; children—Patricia, William, David, Nancy. B.S. with honors in Metall. Engring, Lehigh U., 1944, D.Engring. (hon.), 1973. Engr. Western Electric Co., 1946-52; prodn. mgr. Semiconductor div. Nat. Union Radio Corp., 1952-54; exec. dir. Bell Telephone Labs., 1954-66; pres. Bellcomm Inc., Washington, 1966-68, Gen. Instrument Corp., N.Y.C., 1968-70; v.p., gen. mgr. RCA Corp., Somerville, NJ., 1970-72, exec. v.p., N.Y.C., 1972—, also dir.; dir. Am. Fletcher Nat. Bank, Am. Fletcher Corp., Indpls. Bd. dirs. Bethlehem (Pa.) Fgn. Policy Assn., 1960-62, Nat. Action Council for Minorities in Engring., Inc.; trustee Lehigh U. Served to capt. AUS, 1943-46. Named hon. citizen, Bethlehem, 1966. Fellow I.E.E.E.; mem. Nat. Acad. Engring., Omicron Delta Kappa, Phi Gamma Delta. Home: 149 Bellevue Ave Summit NJ 07901 Office: David Sarnoff Research Center Princeton NJ 08540

HIVELY, RAY MICHAEL, physics and astronomy educator; b. Muskogee, Okla., June 2, 1944; s. Sherman E. and Rose Marie (Senft) H.; m. Susan Quest Wallin, June 26, 1976; children: Laura Evans, Andrea Quest, Ericka Wallin. B.S., U. Okla., 1966; Ph.D., Harvard U., 1972. With Johnson Spaceflight Center, Houston, summers, 1966-69; prof. physics and astronomy Earlham Coll., 1969—, Clifford Crump prof., 1972—. NSF fellow, 1966-71; Woodrow Wilson fellow, 1966. Mem. Am. Phys. Soc., Am. Assn. Physics Tchrs., Ind. Acad. Scis., Phi Beta Kappa. Subspecialties: Cosmology; Archaeoastronomy. Current work: Cosmology; galaxy formation and origin of microwave background; archaeoastronomy; interpretation of Hopewellian Earthworks. Home: 901 College Ave Richmond IN 47374 Office: Physics Dept Earlham Coll Richmond IN 47374

HJELLMING, ROBERT MICHAEL, astrophysicist, educator; b. Gary, Ind., Dec. 21, 1938; s. Lester Alpheus and Valera Amelia (Gurauskaus) H.; m. Carol Ann Johnson, June 13, 1959; children: Michael, Marya, Thomas, Peter, Teresa. B.Sc., U. Chgo., 1960, M.S., 1961, Ph.D., 1965. Asst. prof. astronomy Case Western Res. U., Cleve., 1965-68; assoc. scientist Nat. Radio Astronomy Obs., Charlottesville, Va., 1968-71, scientist, 1971-76, Socorro, N.Mex, 1976—; adj. prof. astrophysics N.Mex. Inst. Mining and Tech., 1976—. Contbr. articles to profl. jours. Served with USMCR, 1956-62. Mem. Am. Astron. Soc., Internat. Astron. Union, Astron. Soc. Pacific, N.Mex. Acad. Sci. Roman Catholic. Subspecialties: Radio and microwave astronomy; Theoretical astrophysics. Current work: Radio emission from stars, x-ray sources, interstellar gas, radio behavior of stars and other active objects.

HJELMFELT, ALLEN TALBERT, JR., agricultural research administrator, researcher; b. Holdrege, Nebr., Oct. 21, 1937; s. Allen Talbert and Doris (Hauber) H.; m. Marian Park; children: Allen, Eric, Allison, Joel. B.S.C.E., Kans. State U., 1959; M.S.E.M., U. Kans.-Lawrence, 1961; Ph.D., Northwestern U., 1965. Registered profl. engr., Mo. Devel. engr. Union Carbide, Oak Ridge, 1962-63; research engr. Northwestern U., 1963-64; prof. civil engring. U. Mo.-Columbia, 1965-78; research engr. Agrl. Research Service, U.S. Dept. Agr., Columbia, 1979—. Author: Hydrology for Engineers and Planners, 1975. Recipient chpt. honor Chi Epsilon, Kans. State U., 1981. Mem. ASCE (J.C. Stevens award 1983), Am. Geophys. Union, Internat. Assn. Hydraulic Research. Subspecialties: Surface water hydrology; Fluid mechanics. Current work: Fluid mechanics of surface water hydrology. Home: 1004 Maplewood Columbia MO 65201 Office: Agrl Research Service 207 Buis Loop 70E Columbi MO 65201

HJERTAGER, BJORN HELGE, research scientist; b. Bergen, Norway, Feb. 26, 1947; s. Harald A. and Annie E. (Hovdenes) H.; m. Inger-Lill Storhaug, Dec. 31, 1969; children: Hilde B., Lene K., Nina. Ing., Bergen Engring. Coll., 1968; M.Sc., U. Trondheim, Norway, 1972, Ph.D., 1979. Sci. asst. U. Trondheim, 1973, asst. prof., 1976-78; head dept. Norwegian Underwater Inst., Bergen, 1979; sr. scientist Christian Michelsen Inst., Bergen, 1980—. Author: Flow Heat Transfer and Combustion in Three-Dimensional Enclosures, 1979; contbr. articles to profl. jours. Served with Norwegian Navy, 1968-69. Mem. Combustion inst., AIAA, Internat. Assn. Math. and Computers in Simulation, Internat. Centre Heat and Mass Transfer, Polyteknisk Forening. Subspecialties: Combustion processes; Fluid mechanics. Current work: Gas explosion, turbulent flow, computer simulation. Home: Vakleiva 101 Bergen Norway N-5062 Office: Chr Michelsen Inst Fantoftvegen 38 Bergen Norway N-5036

HLASS, I. JERRY, govt. ofcl.; m. Helen Mae Diller, Dec. 21, 1963; 1 son, George O. B. Mech. Engring., N.C. State U., 1949; M. Engring. Adminstrn., George Washington U., 1971. Dir. space shuttle facilities NASA, Washington, 1971-76, mgr. nat. space tech. labs., Bay St. Louis, Miss., 1976—. Mem. Nat., Miss. socs. profl. engrs. Methodist. Subspecialty: Space technology management. Office: Nat Space Tech Labs NSTL Station MS 39529

HLAVACEK, VLADIMIR, chemical engineering educator, researcher; b. Prague, Czechoslovakia, Feb. 28, 1939; came to U.S., 1981; s. Josef and Marie (Zalud) H.; m. Jirina Lehka, Aug. 22, 1970. M.S. in Chemistry, Inst. Chem. Tech., Prague, 1961, Ph.D. in Chem. Engring, 1965. Asst. prof. Inst. Chem. Tech., Prague, 1961-70, sr. scientist, research fellow, 1969-70, head, 1970-80; prof. Catholic U., Leuven, Belgium, 1980-81, SUNY, Buffalo, 1981—; cons. in field. Author: Numerical Solution of Nonlinear Boundary Value Problems with Applications, 1982; editor: Sci. Jour. Chem. Engring. Sci, 1975-81. Recipient Nat. Czechoslovak award for outstanding results in sci. Czechoslovak Govt., 1974. Mem. Am. Inst. Chem. Engrs., Soc. Indsl. and Applied Math. Subspecialties: Numerical analysis; Chemical engineering. Office: Dept Chemical Engineering SUNY Amherst Campus Buffalo NY 14260

HNATIUK, BOHDAN TARAS, engineering educator, researcher, consultant; b. Zaliszczyki, Tarnopil, Ukraine, July 25, 1915; came to U.S., 1949; s. Wasyl T. and Anastazja R. (Schuch) H.; m. Irene M. Tomkiw, Jan. 30, 1944; children: Bohdanna W., Irene R., Oleh W. B.S., State Seminary, Zalszczyki, Ukraine, 1935; Diploma, Tech. U., Danzig Free City, 1943, D. Ing., 1945. Sci., research asst. Tech. U., Danzig and Vienna, 1942-45; research scientist Dornier Werke, Friedrichshafen, Germany, after 1945; asst., then assoc. prof. U. Notre Dame, South Bend, Ind., 1951-57; prof. W. Va. U., Morgantown, W. Va., 1957-60, Drexel U., Phila., 1960—; cons. analysis Bendix Aviation Corp., Mishawaka, Ind., 1955-57; cons. propulsion Allegany Ballistics Lab., Pinto, W. Va., 1959-72, Pneumo Dynamics Corp., Bethesda, Md., 1961-63; cons. aero. NASA (MSFC), Huntsville, Ala., 1967-69. Mem. Bala Cynwyd (Pa.) PTA, Merion Park (Pa.) Civic Assn. Recipient Alexander von Humboldt award Edn. Ministerium, Berlin, 1939-43; Faculty fellow awards, 1967,68. Assoc. fellow AIAA (Outstanding Faculty Advisor award 1972, past mem. council; past chmn. edn., student affairs); mem. Am. Soc. Engring. Edn., Am. Assn. Univ. Profs., AAAS, Sigma Gamma Tau, Pi Tau Sigma, Tau Beta Phi. Republican. Ukrainian Catholic. Club: Engring. (Phila.). Subspecialties: Aerospace engineering and technology; Astronautics. Current work: analytical and numerical study on reducing missile drag; wind energy potential for electrical power generation; propulsion noise reduction. Office: Drexel U 32d & Chestnut Sts Philadelphia PA 19104

HO, CHIEN, biological sciences educator, academic administrator; b. Shanghai, China, Oct. 23, 1934; s. Ping Yin and Chin Hwa (Chiu) H.; m. Nancy Tseng, Dec. 21, 1963; children: Jeanette H., Carolyn Y. B.A., Williams Coll., 1957; Ph.D., Yale U., 1961. Research chemist Linde Co. subs. Union Carbide Corp., Tonawanda, N.Y., 1960-61; research assoc. MIT, Cambridge, Mass., 1961-64; from asst. prof. to prof. biol. scis. U. Pitts., 1965-79; prof. biol. scis. Carnegie-Mellon U., Pitts., 1979—; cons. in field. Contbr. articles to profl. jours. John Simon Guggenheim fellow, 1970-71. Mem. Am. Chem. Soc., Am. Soc. Biol. Chemists, Biophys. Soc., AAAS, Soc. Magnetic Resonance in Medicine. Subspecialties: Biochemistry (biology); Molecular biology. Current work: Structure-function relationships in proteins, enzymes, and biological membranes; applications of nuclear magnetic resonance spectroscopy to biochemical and biomedical problems; human abnormal hemoglobins, hemoglobinopathies, sickle cell anemia; membrane transports. Office: Dept Biol Scis Carnegie-Mellon U 4400 5th Ave Pittsburgh PA 15213

HO, CHIH-MING, aeronautical engineering educator, consultant; b. Chunking, China, Aug. 16, 1945; came to U.S., 1968, naturalized, 1980; s. Shao-Nan and I-Chu H.; m. Shirley Chang, Mar. 4, 1972; 1 son, Dean. B.S.M.E., Nat. Taiwan U., 1967; Ph.D., Johns Hopkins U., 1974. Assoc. research scientist Johns Hopkins U., Balt., 1974-75; asst. prof. U. So. Calif., Los Angeles, 1976-81, assoc. prof. dept. aero. engring., 1981—; cons. Rockwell Internat., 1980, Dynamics Tech., 1978—, Flow Industries, 1982—. Mem. Am. Physiol. Soc., AIAA, Sigma Xi, Phi Beta Kappa, Tau Beta Pi. Subspecialties: Fluid mechanics; Theoretical and applied mechanics. Current work: Turbulence, unsteady fluid mechanics, instrumentation. Inventor fiber optics pressure transducer, 1977; patentee in field. Address: Univ So Calif Dept Aero Engring Hoover St Los Angeles CA 90089

HO, HON HING, mycologist, plant pathologist, educator; b. Hong Kong, May 3, 1939; s. Wa-ching and Yin-hing (Kong) H.; m. Lucinda Choy-ho Chui, June 28, 1968; children: Cynthia, Nancy. Demonstrator dept. botany U. Western Ont., London, Can., 1964-66, postdoctoral research fellow, 1966-67; asst. prof. dept. botany and bacteriology Ohio Wesleyan U., Delaware, 1967-68; asst. prof. dept. biology SUNY-New Paltz, 1968-71, assoc. prof., 1971-79, prof., 1979—; vis. research assoc. dept. plant pathology U. Calif.-Riverside, 1975-76; vis. prof. Inst. Mcrobiology, Acadameia Sinica, Beijing, China, 1982-83, Dept. plant protection Nanjing Agrl. Coll. (China), 1982-83. Contbr. articles to profl. jours. NSF grantee, 1970-74. Mem. Am. Phytopath. Soc., Mycol. Soc. Am. Subspecialties: Plant pathology; Microbiology. Current work: Biology and taxonomy of Phytophthora. Home: 11 Bonticouview Dr New Paltz NY 12561 Office: Dept Biology SUNY New Paltz NY 12561

HO, ING K., pharmacologist-toxicologist, researcher; b. Taiwan, May 7, 1939; came to U.S., 1964, naturalized, 1975; s. Wu Chao and Quei Kin (Lin) H.; m. Patricia Leu, Oct. 4, 1965; children: Hubert, Deborah, Christine. B.S., Nat. Taiwan U., Taipei, 1962; Ph.D., U. Calif.-San Francisco, 1968. Research assoc. dept. anesthesiology Baylor Coll. Medicine, Houston, 1968-69, instr., 1969-70; research pharmacologist and asst. prof. depts. psychiatry and pharmacology U. Calif.-San Francisco, 1970-75; assoc. prof. pharmacology and toxicology U. Miss. Med. Ctr., Jackson, 1975-78, prof., 1978—, chmn. dept., 1982—; cons. EPA-sponsored Miss. Epidemiologic Studies Program. Contbr. chpts., numerous articles, abstracts to profl. publs.; editorial bd.: Neurotoxicology, 1980—; reviewer for profl. jours. Recipient Faculty Devel. award in basic pharmacology Pharm. Mfrs. Assn. Found., 1974-76. Mem. Am. Soc. Pharmacology and Exptl. Therapeutics, Soc. Neurosci., Soc. Toxicology, Internat. Soc. Biochem. Pharmacology, AAAS, Western Pharmacology Soc., Chinese Biochem. Soc., Internat. Union Pharmacology, Sigma Xi. Subspecialties: Neuropharmacology; Toxicology (medicine). Current work: Biochemical mechanisms of tolerance to and dependence on narcotics and sedative-hypnotics; mechanisms of organophosphate-induced neurotoxicity. Home: 5050 Meadow Oaks Park Dr Jackson MS 39211 Office: Dept Pharmacology and Toxicology U Miss Med Center Jackson MS 39216

HO, JOHN TING-SUM, physics educator; b. Hong Kong, July 5, 1942; came to U.S., 1965, naturalized, 1977; s. Sui-Ming and Ping-Chun (Yang) H.; m. Martha C. Leung, Aug. 8, 1970. B.Sc., U. Hong Kong, 1964, 1965; Ph.D., M.I.T., 1969. Asst. prof. physics U. Pa., 1969-74; assoc. prof. physics U. Houston, 1974-75, SUNY-Buffalo, 1975—. Contbr. numerous articles to sci. jours. NSF grantee, 1978—; NIH grantee, 1978—. Mem. Am. Phys. Soc., Biophys. Soc., N.Y. Acad. Sci., Sigma Xi. Subspecialties: Condensed matter physics; Biophysics (physics). Current work: Phase transitions, liquid crystals, light scattering, biomembranes. Office: Dept Physics SUNY Buffalo NY 14260

HO, JU-SHEY, biology educator; b. Taipei, Taiwan, Dec. 20, 1935; came to U.S., 1962; s. Kai-sheng and Keh (Cheng) H.; m. Pao-Hsi Yeh, Aug. 28, 1965; children: Min-min, Phi-lip. B.S., Nat. Taiwan U., 1958; M.A., Boston U., 1965, Ph.D., 1969. Teaching fellow Boston U., 1962-65, research assoc., 1968-70; asst. prof. biology Calif. State U.-Long Beach, 1970-74, assoc. prof., 1974-79, prof., 1979—. Author: Laboratory Manual for Invertebrate Zoology, 1978. Mem. Soc. Systematic Zoology, AAAS, Biol. Soc. Washington, Marine Biol. Assn. U.K., Sigma Xi. Subspecialties: Biogeography; Systematics. Current work: Copepod parasites of marine animals, their implication on the phylogeny and biogeography of host. Office: Dept Biology Calif State U 1250 Bellflower Blvd Long Beach CA 90840

HO, KANG-JEY, pathologist, researcher; b. Tainan, Taiwan, China, Aug. 2, 1937; came to U.S., 1964; s. Ruey-Lin and Su-Fung (Tsui) H.; m. Le-Hong Cheng, Dec. 27, 1969; children: Yen-ching, Yen-Dong. M.D., Nat. Taiwan U., Taipei, 1963; Ph.D., Northwestern U., 1968. Diplomate: Am. Bd. Pathology. Intern Nat. Taiwan U. Hosp., Taipei, 1962-63; intern Evanston (Ill.) Hosp., 1964-65, resident, 1965-68; instr. pathology Northwestern U., Evanston, 1968-69, asst. prof., 1969-70; asst. prof. pathology U. Ala., Birmingham, 1970-71, assoc. prof., 1972-75, prof., 1976—; dir. pathology research Evanston Hosp., 1969-70; chief pathology research VA Med Center, Birmingham, Ala., 1977—. Served with Chinese Marine Corps, 1963-64. Grantee NIH, 1970-76, NSF, 1976-77, VA, 1977—; Schweppe Found. fellow, 1973. Fellow Am. Heart Assn.; mem. Am. Assn. Pathologists, Am. Soc. Exptl. Pathology, Am. Soc. Clin. Nutrition, Am. Inst. Nutrition, Soc. Exptl. Biology and Medicine, Internat. Soc. Chronobiology. Subspecialty: Pathology (medicine). Current work: Lipid metabolism and its relationship to atherosclerosis and cholelithiasis. Home: 1583 Mountain Gap Dr Birmingham AL 35226 Office: U Ala 619 S 19th St Birmingham AL 35294

HO, RAYMOND HOW-CHEE, anatomist, educator; b. Hong Kong, Sept. 21,1947; s. Cho Um Ho and Wan Sou Chan. B.S. cum laude, St. John's U., Minn., 1971; Ph.D., U. Minn., 1978. Asst. prof. anatomy Ohio State U., Columbus, 1977-83, assoc. prof., 1983—. Contbr. articles to profl. jours. NIH grantee, 1978—; Ohio State U. grantee, 1977-78; Upjohn Co. grantee, 1978-79. Mem. Soc. Neuroscience, Midwest Assn. Anatomists, Am. Assn. Anatomists, Ohio Acad. Sci. Roman Catholic. Subspecialties: Neurobiology; Immunocytochemistry. Current work: Chemically identified neuronal elements in the spinal cord; immunocytochemistry.

HO, THOMAS INN MIN, computer scientist, educator; b. Honolulu, Oct. 17, 1948; s. Herbert Low Seu and Rose (Lee) H.; m. Jean Joan Kwan, Aug. 26, 1971; 1 son, Brian Koon Leong. B.S., Purdue U., 1970, M.S., 1971, Ph.D., 1974. Asst. prof. computer sci. and mgmt. Purdue U., West Lafayette, Ind., 1975-78, asso. prof., 1978—, head computer tech., 1978—; cons. in field. Contbr. articles to profl. jours. NSF fellow, 1970-72. Mem. Assn. Computing Machinery, Data Processing Mgmt. Assn., Soc. for Info. Mgmt., Phi Kappa Phi. Office: Computer Tech Dept Purdue U West Lafayette IN 47907

HOAG, ARTHUR ALLEN, astronomer; b. Ann Arbor, Mich., Jan. 28, 1921; s. Lynne Arthur and Wilma S. (Wood) H.; m. Marjorie Paulison Beers, Dec. 17, 1949; children—Stefanie (Mrs. William J. Hoel),

Thomas. A.B., Brown U., 1942; Ph.D., Harvard, 1952. Physicist U.S. Naval Ordnance Lab., Washington, 1942-45; astronomer U.S. Naval Obs., Washington, 1950-55, dir., Flagstaff, Ariz., 1955-65, Stellar Program, Kitt Peak Nat. Obs., Tucson, 1965-75, astronomer, 1975-77; dir. Lowell Obs., Flagstaff, Ariz., 1977—. Served with USNR, 1943-45. Fellow AAAS (councillor sect. D 1970-71), Ariz. Acad. Sci. (Councillor 1961-64, pres. 1965-66); mem. Am. Astron. Soc. (council 1966-69, v.p. 1974-76), Royal Astron. Soc., Astron. Soc. Pacific, Internat. Astron. Union. Subspecialty: Optical astronomy. Home and Office: Lowell Observatory PO Box 1269 Flagstaff AZ 86002

HOAG, DAVID GARRATT, elec. engr.; b. Boston, Oct. 11, 1925; s. Alden Bomer and Helen Lucy (Garratt) H.; m. Grace Edward Griffith, May 10, 1952; children—Rebecca Wilder, Peter Griffith, Jeffrey Taber, Nicholas Alden, Lucy Seymour. B.S., MIT, 1946, M.S., 1950. Staff engr. Mass. Inst. Tech. Instrumentation Lab., Cambridge, 1946-57; tech. dir. Polaris Missile Guidance, 1957-61; tech. dir., program mgr. Apollo Spacecraft Guidance, 1961-74; advanced system dept. head C.S. Draper Lab., Inc., 1974—; Reviewer children's sci. book com. Harvard U., 1965—. Incorporator, bd. dirs. Medway Community Nursery Sch. Served with USN, 1943-46. Recipient Pub. Service award NASA, 1969, Spl. award Royal Inst. Navigation, Britain, 1970. Fellow Am. Inst. Aeros. and Astronautics (Louis W. Hill Space Transp. award 1972, chmn. New Eng. sect. 1979-80); mem. Nat. Acad. Engring., Inst. Navigation (Thurlow award 1969, pres. 1978-79), Internat. Acad. Astronautics (asso. editor ACTA Astronautica 1973—). Subspecialties: Systems engineering; Aerospace engineering and technology. Current work: Guidance, navigation, control and pointing in aeronautics and astronautics. Home: 116 Winthrop St Medway MA 02053 Office: CS Draper Lab Inc 555 Technology Sq Cambridge MA 02139

HOAG, ROLAND BOYDEN, JR., exploration geochemist; b. Boston, Sept. 3, 1945; m., 1968; 2 children. B.S., U. N.H., 1967; Ph.D., McGill U., 1975. Geologist stream sediment geochemistry N.Am. Exploration, 1967; geologist element study of pegmatite veins U. Pa., 1967-68; cons. geologist Scandia Mining & Exploration Co. and Labrador Mining & Exploration Co., 1971; research assoc. McGill U., 1976; cons. geologist, 1976-77; v.p. BCI Geonetics, Inc., 1977—. Mem. Geol. Soc. Am., Soc. Exploration Geochemists, Soc. Econ. Geologists, Newcomen Soc., Sigma Xi. Subspecialty: Hydrogeology.

HOAGLAND, GORDON WOOD, mathematics educator; b. Nampa, Idaho, Oct. 22, 1936; s. Clyde Mackay and Clara (Wood) H.; m. Byrnina Louise Burningham, Aug. 1, 1962; children: David, Daniel, Deborah. B.S., Brigham Young U., 1966, M.S., 1968; postgrad., Oreg. State U., 1969. Prof. math. Ricks Coll., Rexburg, Idaho, 1969—, chmn. dept., 1979—. Bishop Ch. Jesus Christ of Latter-day Saints. Mem. Am. Math. Soc., Soc. Indsl. and Applied Math. Republican. Club: Upper Valley Square Dance. Subspecialties: Applied mathematics; Numerical analysis. Current work: Sampling techniques in forestry; gamma ray spectroscopy. Home: 206 E 2d St S Rexburg ID 83440 Office: Dept Math Ricks Coll Rexburg ID 83440

HOAGLAND, MAHLON BUSH, educator, biochemist; b. Boston, Oct. 5, 1921; s. Hudson and Anna (Plummer) H.; m. Olley Virginia Jones, Jan. 10, 1961; children by previous marriage—Judith, Mahlon Bush, Robin. Student, Williams Coll., 1940-41, Harvard U., 1941-43, M.D., 1948; Sc.D. (hon.), Worcester Poly. Inst., 1973. From research fellow to asst. prof. medicine Harvard Med. Sch. at Mass. Gen. Hosp., 1948-60; asso. prof. bacteriology and immunology (Med. Sch.), 1960-67; prof. biochemistry, chmn. dept. Dartmouth Med. Sch., 1967-70; research prof. U. Mass. Med. Sch., 1970—; dir. Worcester Found. for Exptl. Biology, Shrewsbury, Mass., 1970—; research asso. Carlsberg Labs., Copenhagen, Denmark, 1951-52, Cavendish Labs., Cambridge, Eng., 1957-58; Exec. sec. com. research Mass. Gen. Hosp., 1954-57; cons. NIH, 1961-64, Am. Cancer Soc., 1965-68; bd. sci. counsellors Nat. Heart and Lung Inst., 1972-74; Bd. dirs. Mass. div. Am. Cancer Soc.; Scholar cancer research Am. Cancer Soc., 1953-58. Contbr. articles to profl. jours. Fellow Am. Acad. Arts and Scis.; mem. Am. Soc. Biol. Chemists (Franklin medal 1976). Subspecialties: Biochemistry (biology); Cell and tissue culture. Current work: Research institute administration; science writing for public; advisory liaison with goverment to strengthen medical science. Research on mechanism of carcinogenic action of beryllium, mechanism of synthesis of coenzyme A; discovery mechanism of amino acid activation and role of transfer ribonucleic acid in protein synthesis. Home: 234 Gulf St Shrewsbury MA 01545 Office: Worcester Found Shrewsbury MA 01545

HOBART, ROBERT H.B.W.S., JR., physicist, mathematician; b. N.Y.C., Feb. 2, 1932; s. Robert H.B.W.S. and Edith Child (Sorensen) H. S.B., M.I.T., 1954; M.S., Stanford U., 1955; Ph.D., U. Ill., 1961. Asst. prof. physics Dalhousie U., Halifax, N.S., Can., 1961-63; sr. physicist/ prin. investigator Battelle Meml. Inst., Columbus, Ohio, 1963-70; prof., head physics dept. Fort Hare U., Alice, South Africa, 1973-75, Kenyatta U. Coll., Nairobi, Kenya, 1976-78; prof. physics and math. Middle Ga. Coll., Cochran, 1980-82; physicist Fgn. Sci. and Tech Ctr., Charlottesville, Va., 1983—. Served to capt., Signal Corps USAR, 1954-66. NSF fellow, 1954-55; Gulf Oil fellow, 1956-57. Mem. Am. Phys. Soc. Anglican. Subspecialties: Theoretical physics; Applied mathematics. Current work: Theory of solitons and other nonlinear phenomena. Home: 611 Rugby Rd Apt 311A Charlottesville VA 22903

HOBBIE, JOHN EYRES, marine scientist; b. Buffalo, June 5, 1935; m., 1959; 3 children. B.A., Dartmouth Coll., 1957; M.A., U. Calif.-Berkeley, 1959; Ph.D. in Zoology, Ind. U., 1962. Research assoc. zoologist U. Calif.-Davis, 1962-63; NIH fellow Inst. Limnology, Uppsala, Sweden, 1963-65; from asst. prof. to assoc. prof. zoology N.C. State U., 1965-71, prof., 1971-75; sr. scientist Marine Biol. Lab., Woods Hole, Mass., 1976—; NSF sr. fellow Norwegian Inst. Water Research, Oslo, 1971-72; dir. tundra biometric aquatic program U.S. Inst. Biol. Programs, 1971-74. Mem. Am. Soc. Limnology and Oceanography, Ecol. Soc. Am., Internat. Assn. Theoretical and Applied Limnology. Subspecialty: Ecology. Office: Marine Biol Lab Woods Hole MA 02543

HOBBIE, RUSSELL KLYVER, physicist; b. Albany, N.Y., Nov. 3, 1934; s. John Remington and Eulin Pomeroy (Klyver) H.; m. Cynthia Ann Borcherding, Dec. 28, 1957; children—Lynn Katherine, Erik Klyver, Sarah Elizabeth, Ann Stacey. B.S. in Physics, Mass. Inst. Tech., 1956; A.M., Harvard U., Ph.D., 1960. Research asso. U. Minn., 1960-62, mem. faculty, 1962—, prof. physics, 1972—, dir., 1979—. Author: Intermediate Physics for Medicine and Biology, 1978. Mem. Am. Assn. Physics Tchrs. (exec. bd. 1980-83), Am. Phys. Soc., Am. Assn. Physicists in Medicine, AAAS., Biophys. Soc. Subspecialties: Biophysics (biology); Biophysics (physics). Current work: Research in medical computing and biophysics. Home: 2151 Folwell St St Paul MN 55108 Office: 103 Shepherd Labs U Minn 100 Union St SE Minneapolis MN 55455

HOBBS, CARL HEYWOOD, III, geologist, educator; b. New Haven, May 3, 1946; s. Carl Heywood and Lydia Browning (Hewitt) H.; m. Elizabeth Ann Reaney Moncure, Mar. 8, 1974 (div. Sept. 1980); 1 dau., Catherine Elizabeth. B.S., Union Coll., Schenectady, 1968; M.S., U. Mass., 1972. Cert. profl. geologist, Va. Assoc. marine scientist, asst. prof. Va. Inst. Marine Sci., Coll. William and Mary, Gloucester Point, 1972—. Contbr. articles to profl. jours. Mem. Gloucester County Transp. Safety Commn., 1981-82, Abingdon Vol. Res. Squad, 1982—. Mem. Soc. Econ. Paleontologists and Mineralogists, Geol. Soc. Am., Am. Assn. Petroleum Geologists. Subspecialties: Sedimentology; Geology. Current work: Marine gology, sedimentology and stratigraphy, design and mgmt. of research projects, environmental management, graduate education. Office: Va Inst Marine Science Gloucester Point VA 23062

HOBBS, LEWIS MANKIN, astronomer; b. Upper Darby, Pa., May 16, 1937; s. Lewis Samuel and Evangeline Elizabeth (Goss) H.; m. Jo Ann Faith Hagele, June 16, 1962; children: John, Michael, Dara. B.Engring. Physics, Cornell U., 1960; M.S., U. Wis., 1962, Ph.D. in Physics, 1966. Jr. astronomer Lick Obs., U. Calif., Santa Cruz, 1965-66; mem. faculty U. Chgo., 1966—, prof. astronomy and astrophysics, 1976—; also dir. Yerkes Obs., Williams Bay, Wis., 1974-82; bd. dirs. Assn. Univs. Research in Astronomy, Tucson, 1974—; mem. astronomy com. of bd. trustees Univs. Research Assn., Inc., Washington, 1979—, chmn., 1979-81. Contbr. to profl. jours. Bd. dirs. Mil. Symphony Assn. of Walworth County, 1972—. Alfred P. Sloan scholar, 1955-60. Mem. Am. Astron. Soc., Am. Phys. Soc., Internat. Astron. Union, Wis. Acad. Scis. Arts and Letters. Subspecialties: Optical astronomy; Ultraviolet high energy astrophysics. Current work: Interstellar matter; galactic structure; atomic and molecular physics; interferometric spectroscopy. Home: 18 Highland St Williams Bay WI 53191 Office: Yerkes Obs U Chgo Williams Bay WI 53191

HOBBS, PETER VICTOR, atmospheric sciences educator, consultant; b. London, May 3, 1936; U.S., 1963, naturalized, 1969; s. Victor George and Daisy Francis (Kincaid) H.; m. Sylvia Helen Wood, Jan. 18, 1963; children: Stephen, Julian, Rowland. B.Sc. with honors, U. London, 1960, Ph.D., diploma, 1963. Asst. prof. atmospheric scis. U. Wash., Seattle, 1963-65, assoc. prof., 1965-70, prof., 1970—, dir. cloud and aerosol research group, 1963—. Author: Ice Physics, 1974, (with J. M. Wallace) Atmospheric Sciences: An Introductory Survey, 1977; contbr. articles to profl. jours. Served with RAF, 1955-57. Recipient Alexander von Humboldt sr. award, 1983-84; Fulbright grantee, 1983-84. Fellow Am. Meteorol. Soc. (editor's award 1970, Charney award 1984), AAAS, Royal Meteorol. Soc.; mem. Am. Geophys. Union. Subspecialties: Meteorology; Atmospheric chemistry. Current work: Cloud and precipitation processes, mesoscale meteorology, atmospheric aerosol, cloud chemistry, trace gases, air pollution and acid rain.

HOBSON, J. ALLAN, psychiatrist, educator; b. Hartford, Conn., June 3, 1933. A.B., Wesleyan U., 1955; M.D., Harvard U., 1959. Intern in medicine Bellevue Hosp., N.Y.C., 1959-60; resident in psychiatry Mass. Mental Health Center, Boston, 1960-61, 64-66; spl. fellow NIMH dept. physiology U. Lyon, France, 1963-64; research assoc. dept. physiology Harvard Med. Sch., Boston, 1964-67, asst. in psychiatry, 1965-66, instr., 1966-67, asso. in psychiatry, 1967-69, asst. prof. psychiatry, 1969-74, assoc. prof., 1974-78, prof., 1978—; dir. lab. neurophysiology Mass. Mental Health Center, Boston, 1967, prin. psychiatrist, 1967—, dir. group psychotherapy tng. program, 1972; vis. scientist, lectr. U. Bordeaux, France, 1973; Sandoz lectr. dept. psychiatry U. Edinburgh, Scotland, 1975; vis. prof. Japan Soc. for Promotion of Sci., 1980; lectr. Italian NIH, 1980; vis. prof. Instituto di Psicologia U. Dogli Studi, 1983; bd. sci. counselors NIH, 1979; dir. Dream stage, and Exptl. Portrait of the Sleeping Brain, 1977-81; cons. in electroencephalography Peter Bent Brigham Hosp., Boston, 1974-80, Sleep Disorders Clinic, Beth Israel Hosp., 1980; in electroencephalography Brigham and Women's Hosp., Boston, 1981; chmn. sci. adv. com. Boston Mus. Sci. Brain Exhibit. Author, editor books in field; contbr. chpts. to books, articles to profl. jours.; mem. editorial bd.: Jour. Cellular and Molecular Neurobiology, 1980; contbg. editor: Sleep Revs, 1970-72; assoc. editor, 1972-73; editor-in-chief, 1973-74; book rev. editor, 1975-76; cons. editor: Dreamworks, 1980—; creator: Dreamscreen, 1982. Recipient Benjamin Rush Gold medal for best sci. exhibit Am. Psychiat. Assn., 1978. Mem. Assn. Psychophysiol. Study of Sleep (program chmn. 1968-69, program co-chmn. 2d internat. sleep congress and 14 an. meeting 1975), Soc. Neurosci. (program com. 1974-76), Internat. Brain Research Orgn. (co-organizer and sci. chmn. Joint IBRO/Soc. Neurosci. internat. symposium on reticular formation revisited 1979), Sigma Xi. Subspecialties: Psychiatry; Neurophysiology. Current work: Neurophysiol. basis of mind and behavior; sleep and dreaming. Office: Dept Psychiatry Harvard Med School 74 Fenwood Rd Boston MA 02115

HOCH, GEORGE EDWARD, biologist, educator; b. Brookings, S.D., Mar. 11, 1931; s. Alfred A. and Rose Mary (Sullivan) H.; m. Kathleen Ann McCullough, Apr. 9, 1953; children—Gregory, Marie, Lambert, Ellen, James, Ann, Kathryn, George. B.S. in Chemistry, S.D. State U., 1952; Ph.D. in Biochemistry, U. Wis., 1958. Scientist Research Inst. for Advanced Studies, Balt., 1958-64; vis. asso. prof. biophysics Johnson Research Found., U. Pa., 1964-65; asso. prof. biology U. Rochester, N.Y., 1965-70, prof., 1970—, chmn. dept., 1978—; vis. research biologist U. Calif., San Diego, 1979; sr. fellow Carnegie Inst., Washington, 1980. Served with U.S. Army, 1954-55. Subspecialties: Photosynthesis; Biochemistry (biology). Current work: Energy migration among photosynthetic pigments and the mechanism of oxygen evolution. Office: Dept Biology U Rochester Rochester NY 14627

HOCH, JAMES ALFRED, microbiologist, consultant; b. Brookins, S.D., Jan. 22, 1939; s. Alfred A. and Rosemary (Sullivan) H.; m. Sallie Robinson O'Neil, May 17, 1969; children; Fred, Patrick. B.Sc., S.D. State U., 1961; Ph.D., U. Ill.-Urbana, 1965. USPHS fellow Center Molecular Genetics, Gif-sur-Yvette, France, 1965-67; fellow Scripps Clinic and Research Found., La Jolla, Calif., 1967-68, assoc., 1968-72, assoc. mem., 1972-80, mem., 1980—, head div. cellular biology, 1982—; mem. sci. adv. bd. Cetus Corp., Syntro Corp., P-L Biochems. Recipient Faculty Research Assoc. award AM Cancer Soc., 1969-74. Mem. AAAS, Am. Soc. Microbiology, Am. Soc. Biol. Chemists. Subspecialties: Genetics and genetic engineering (biology); Microbiology (medicine). Current work: Genetics and regulation of development. Office: Scripps Clinic and Research Found 10666 N Torrey Pines Rd La Jolla CA 92037

HOCHBERG, MARC CRAIG, rheumatologist, educator; b. N.Y.C., June 29, 1949; s. Irving Ira and Elenore (Hyman) H.; m. Susan Rose Newhouse, June 10, 1973; children: Francine, Jennifer. A.B., Franklin and Marshall Coll., 1969; M.D., Johns Hopkins U., 1973, M.P.H., 1979. Diplomate: Am. Bd. Internal Medicine. Instr. medicine Johns Hopkins U., Balt., 1977-79, joint appt. in epidemiology, 1979—, asst. prof., 1979—, dir. statistical core unit, 1980—. Bd. dirs., mem. med. and sci. com. Arthritis Found., Balt., 1979—; bd. dirs. Jewish Big Brother and Big Sister League, Balt., 1981—. Arthritis Found. fellow, 1981. Fellow ACP; mem. Am. Rheumatism Assn., Am. Fedn. Clin. Research, Soc. for Epidemiologic Research, Md. Soc. Rheumatic Disease (pres. 1980-81). Democrat. Jewish. Lodge: B'nai B'rith. Subspecialties: Internal medicine; Epidemiology. Current work: Epidemiology of rheumatic diseases. Factors associated with the occurrence as well as the prognosis of chronic rheumatic diseases. Home: 6305 Green Meadow Pkwy Baltimore MD 21209 Office: Johns Hopkins U Sch Medicine 720 Rutland Ave Baltimore MD 21205

HOCHHAUSER, MARK, management consultant; b. Pitts., June 22, 1946; s. Leonard and Anna (Malackany) H.; m. Barbara Lynn Segal, July 10, 1968; 1 dau., Marcy Laura. B.S., U. Pitts., 1968, M.S., 1970, Ph.D., 1973. Acting head dept. psychology Morris Harvey Coll., Charleston, W.Va., 1972-73; asst. prof. psychology W.Va. State Coll., 1974-77; research assoc. U. Minn., Mpls., 1977-81; dir. research Human Devel. Systems, Mpls., 1983—; cons. Nat. Inst. Drug Abuse, Rockville, Md., 1978—. Manuscript reviewer: Jour. Nervous and Mental Disease, Towson, Md., 1982; Author book chpts.; contbr. articles to profl. jours. Mem. Am. Psychol. Assn., Soc. Psychologists in Addictive Behaviors (sec.-treas. 1983—), N.Y. Acad. Scis., AAAS. Subspecialties: Learning; Developmental psychology. Current work: Management consulting, employee selection procedures, organizational research, research in addictive behaviors, ethical issues in research. Home: 3344 Scott Ave N Golden Valley MN 55422 Offic: 3400 W 66th S Suite 280 Minneapoli MN 55435

HOCHMAN, BENJAMIN, genetics educator, geneticist; b. N.Y.C., Apr. 8, 1925. A.B., U. Calif.-Berkeley, 1949, M.A., 1952, Ph.D., 1956. Research assoc. City Hope Med. Ctr., Duarte, Calif., 1956-57; asst. prof. genetics U. Utah, Salt Lake City, 1957-63; research participant Oak Ridge Nat. Lab., 1960-61; prof. U. Tenn., Knoxville, 1963—. Contbr. chpts. to books and articles to profl. jours. Served with U.S. Army, 1944-46; ETO. NIH grantee, 1962-65; NSF grantee, 1965-75. Mem. Genetics Soc. Am., Soc. Study Evolution, N.Y. Acad. Scis., AAUP, Sigma Xi. Subspecialties: Genome organization; Gene actions. Current work: Analysis of the genetic content and developmental significance of the fourth chromosome in drosophila melanogaster; induction of mutations and cytogenetic studies. Office: Dept Zoology U Tenn Knoxville TN 37996

HOCHMAN, ROBERT F(RANCIS), metallurgical engineering educator; b. Chgo., May 1, 1928; m., 1960; 1 child. B.S., U. Notre Dame, 1950, M.S., 1954, Ph.D. in Metallurgy, 1959. Foundry metallurgist Dodge Mfg. Corp., Indpls., 1950-51; metallurgist Bendix Aviation Corp., 1954-55; spl. instr. evening div. Mich. State U., 1958-59; from asst. prof. to assoc. prof. metallurgy Ga. Inst. Tech., Atlanta, 1959-68, prof., 1968-80, assoc. dir. metallurgy, 1971-82; cons. Zimmer Mfg. Co., 1957—, Lockheed Ga., 1964—, Exxon Power Industries, others. Mem. Am. Soc. Metals (nat. dir. 1973-76), Nat. Assn. Corrosion Engrs., Am. Soc. Nondestructive Testing, ASTM. Subspecialty: Metallurgy. Office: Metallurg Program Dept Chem Engring Ga Inst Tech Atlanta GA 30332

HOCHMANN, PETR TOMÁS, chemistry educator, researcher; b. Pardubice, Czechoslovakia, Oct. 13, 1934; came to U.S., 1968, naturalized, 1977; s. Pavel and Marie (Neumann) H.; m. Jana Vit, May 15, 1966; children: David, Barbara. Student, Leningrad State U., 1954-58; M.S., Charles U., Prague, 1961; Ph.D., Czechoslovak Acad. Sci., Prague, 1967. Resercher Czechoslovak Acad. Sci., 1967-68; research assoc. Mich. State U., 1968-70; asst. prof. chemistry La. State U., 1970-75; assoc. prof. U. Tex., San Antonio, 1975—; adj. prof. U. Houston, 1980—. Subspecialties: Theoretical chemistry; Solid state chemistry. Current work: High energy molecular excitations, interactions of molecular beams with solid surfaces. Office: Dept Earth Sciences U Texas San Antonio TX 78285

HOCHWALT, CARROLL ALONZO, JR., chemist; b. Dayton, Sept. 29, 1923; s. Carroll Alonzo and Pauline (Burkhardt) H.; m. Rita Ann Dineen, Nov. 13, 1948; children: Deberah Ann, Pauline Dineen, Carolyn Rose, Carroll Alonzo, Anne Elizabeth. B.S., Princeton U., 1944; Ph.D. in Organic Chemistry, Ohio State U., 1949. Staff chem. engr. Manhattan Dist. Project, Los Alamos, 1944-46; sales mgr. DuPont, Wilmington, Del., 1949-58; research lab. asst. dir. Inco Ltd., Sterling Forest, N.Y., 1958-67; gen. mgr. CORD Group, Inc., CPC Internat., Englewood Cliffs, N.J., 1967-73; v.p. Martin-Marietta Chems., N.Y.C., 1973-74; dir. corp. planning-bus. devel. Allied Chem. Corp., Morristown, N.J., 1974-77; gen. mgr. COGAS Devel. Co., Princeton, N.J., 1977—. Contbr. articles to profl. jours. Mem. Am. Chem. Soc., Comml. Devel. Assn., Chem. Market Research Assn., Am. Inst. Chemists, Soc. Chem. Industry. Clubs: Nassau, Dial (Princeton). Subspecialties: Combustion processes; Coal. Current work: Energy and the need for synthetic fuels. Patentee in field. Home: 136 Inwood Ave Upper Montclair NJ 07043 Office: PO Box 8 Princeton NJ 08540

HOCKING, WILLIAM GRAY, medical educator, physician; b. San Francisco, Apr. 3, 1947; s. Harold Benn and Carol Afton (Finn) H.; m. Karen Seitz; 1 dau., Erin. B.S., Tulane U., 1969, M.D., 1973. Intern in internal medicine UCLA, 1973-74, resident in internal medicine, 1974-76, fellow in hematology/oncology, 1976-79, chief resident in medicine, 1978-79, adj. asst. prof. medicine, 1978-79, asst. prof. in residence, dept. medicine, 1979-80, assoc. chief div. hematology/oncology, 1982-83; physician Marshfield Clinic (Wis.), 1983—. Contbr. chpts., articles to profl. publs.; editor, contbg. author: Practical Hematology, 1983. Named Outstanding Resident in Medicine UCLA, 1976, Tchr. of Yr. Dept. Medicine, UCLA, Full-time Faculty, 1981; Am. Cancer Soc. clin. fellow, 1977-78. Fellow ACP; mem. AAAS, Am. Soc. Hematology, Internat. Soc. Exptl. Hematology, Am. Soc. Clin. Oncology, Am. Fedn. Clin. Research, Am. Assn. Cancer Research, Physicians for Social Responsibility, Sigma Xi, Alpha Omega Alpha. Subspecialties: Hematology; Oncology. Current work: Normal human hematopoiesis and bone marrow microenvironment; bone marrow microenvironment in myeloproliferative and leukemic disorders. Office: Marshfield Clinics 1000 N Oak Ave Marshfield WI 54449

HODEL, RICHARD EARL, mathematics educator; b. Winston-Salem, N.C., Sept. 24, 1937; m., 1970. B.S., Davidson Coll., 1959; Ph.D. in Math, Duke U., 1962. Asst. prof. math. Duke U., Durham, N.C., 1965-70, assoc. prof., 1970—. Mem. Am. Math. Soc., Math. Assn. Am. Office: Dept Math Duke U Durham NC 27706

HODES, LOUIS, research mathematician; b. N.Y.C., June 19, 1934; s. Morris B. and Anna (Magid) H.; m. Susan B. Levine, Nov. 23, 1967. B.E.E., Bklyn. Poly. Inst., 1956, M.S., 1958; Ph.D., MIT, 1962. Mem. staff IBM Research Inst., Yorktown Heights, N.Y., 1961-65; vis. scientist NYU Courant Inst., N.Y.C., 1965-66; research mathematician NIH, Bethesda, Md., 1966—; vis. scientist Hebrew U., Jerusalem, Israel, 1964. Mem. Soc. Indsl. and Applied Math., N.Y. Acad. Scis., Am. Chem. Soc. (affiliate). Subspecialties: Applied mathematics; Chemotherapy. Current work: Computer methods to select chemicals for anti-cancer screening. Inventor radiation-treatment planning. Home: 10201 Grosvenor Pl Rockville MD 20852 Office: Nat Cancer Inst 830D Colesville Rd Silver Spring MD 20910

HODGE, DAVID CHARLES, psychologist; b. Ft. Worth, July 5, 1931; s. Charles S. and Mildred Mae (Wise) H.; m. Sara Margaret Sappenfield, June 6, 1957. A.A., Mars Hill Coll., 1951; B.A., Hardin-Simmons U., 1953; M.A., Tex. Tech. Coll., 1959; Ph.D., U. Rochester, 1963. Research psychologist U.S. Army Human Engring. Lab., Aberdeen Proving Ground, Md., 1980—, research and team leader, 1962-80; cons. Gen. Dynamics/Electron, Rochester, N.Y., 1961; research asst. U. Rochester, 1959-62; HEL rep. S-12 Acoustical Soc. Am., N.Y.C., 1981—; CHABA com. Nat. Acad. Sci., Washington, 1963—; U.S. prin. rep. NATO RSG-4, Brussels, 1972-78. Contbr. chpts. to books. Fin. sec./deacon/trustee Calvary Bapt. Ch., Bel Air, Md., 1962—; trustee, pub. relations dir. Bapt. Home of Md., Owings

Mills, 1978—; photographer Md. Hist. Trust, Balt., 1975-80; v.p. Hist. Soc. Harford County, 1979-81. Served with U.S. Army, 1953-56. Recipient U.S. Army Research and Devel. award, 1970; commendation Sec. Def., 1979; Spl. Act and Service award Dir. USAHEL, 1982. Fellow AAAS; mem. Acoustical Soc. Am., Sigma Xi, Steppingstone Mus. Assn. (dir.), Harford Artists Assn. Subspecialties: Human factors engineering; Acoustics. Current work: Effects of noise on hearing and behavior; effects of hearing losses on soldiers performance; criteria for limiting noise exposure. Address: US Army Human Engring Lab Bldg 459 Aberdeen Proving Ground MD 21005

HODGE, GORDON KARL, psychology educator; b. Medford, Oreg., July 20, 1949; s. Karl Richard and Wilma Lucille (Rarick) H. B.A., U. Denver, 1971; M.A., UCLA, 1972, Ph.D., 1977. Teaching assoc. UCLA, 1975-76; asst. prof. psychology U. N.Mex., Albuquerque, 1976-82, assoc. prof., 1982—; cons. S.W. Resource Ctr. for Sci. and Engring., Albuquerque, 1982—. Editor: U. New Mexico Psychology Reader, 1979, Psychology: Contemporary Concepts, 1981; contbr. articles to profl. jours. NIH grantee, 1979; NIMH grantee, 1982-83. Mem. Am. Psychol. Assn., Soc. for Neurosci., N.Y. Acad. Sci., AAAS (pres. div. 1981-82), Biofeedback Soc. Am. Subspecialties: Neuropharmacology; Physiological psychology. Current work: Investingating the physiological and biochemical substrates of childhood hyperactivity by using an animal model of the disorder. Home: 6205 Buenos Aires Pl NW Albuquerque NM 87120 Office: Dept Psychology U NMex Albuquerque NM 87131

HODGE, IAN MOIR, industrial chemist; b. Auckland, New Zealand, Jan. 28, 1946; came to U.S., 1969; s. Gordon James and Agnes Mary (Edlington) H. B.Sc., U. Auckland, 1966, M.Sc., 1967; Ph.D., Purdue U., 1974. Postdoctoral fellow U. Aberdeen, Scotland, 1974-75, McGill U., Montreal, Que., Can., 1975-76, Purdue U., West Lafayette, Ind., 1977-78; research and devel. chemist B. F. Goodrich Co., Brecksville, Ohio, 1978—. Contbr. articles to profl. jours. Mem. Am. Chem. Soc., Am. Phys. Soc.; mem. N.Y. Acad. Sci.; Mem. N.Am. Thermal Analysis Soc. Subspecialties: Physical chemistry; Polymer physics. Current work: Relaxation effects in glassy polymers; the glass transition; conducting polymers. Office: BF Goodrich Research and Devel Center 9921 Brecksville St Brecksville OH 44141

HODGE, MILTON HOLMES, JR., psychology educator, researcher; b. Detroit, Feb. 10, 1929; s. Milton Holmes and Vera Jane (McCormick) H.; m. Elizabeth Jane Garrett, June 7, 1952; children: Bruce Allen, Steven Jeffrey. B.A., U. Va., 1953, M.A., 1955, Ph.D., 1957. Asst. prof. Mary Washington Coll., 1957-59; research psychologist USAF, Washington, summer 1959; asst. prof. psychology U. Ga., 1959-63, assoc. prof., 1963-68, prof., 1968—, head dept. psychology, 1974-76. Contbr. articles to psychology jours. Served to cpl. USAF, 1946-49. USPHS grantee, 1963-68; research contract USAF, 1960-61, Bell Telephone Labs., 1968-71, U.S. Army Research Inst., 1977-78. Fellow AAAS; mem. Am. Psychol. Assn., Psychonomic Soc., Soc. Math. Psychology, So. Soc. Philosophy and Psychology (council 1967-70), Raven Soc. Subspecialty: Cognition. Current work: Experimental and Theoretical studies of human memory and attentional processes. Home: 195 Clyde Rd Athens GA 30605 Office: U Ga Dept Psychology Athens GA 30602

HODGE, PHILIP GIBSON, JR., mechanical engineering educator; b. New Haven, Nov. 9, 1920; s. Philip Gibson and Muriel (Miller) H.; m. Thea Drell, Jan. 3, 1943; children: Susan E., Philip T., Elizabeth M. A.B., Antioch Coll., 1943; Ph.D., Brown U., 1949. Research asst. Brown U., 1947-49, asso., 1949; asst. prof. math. UCLA, 1949-53; asso. prof. applied mechs. Poly. Inst. Bkly., 1953-56, prof., 1956-57; prof. mechanics Ill. Inst. Tech., 1953-71, U. Minn., Mpls., 1971—; Russell Severance Springer vis. prof. U. Calif., 1976; sec. U.S. Nat. Com./Theoretical and Applied Mechanics, 1982—. Author: 5 books, the most recent being Limit Analysis of Rotationally Symmetric Plates and Shells, 1963, Continuum Mechanics, 1971; research numerous publs. in field, 1949—; tech. editor: Jour. Applied Mechanics, 1971-76. NSF sr. postdoctoral fellow, 1963. Mem. ASME (hon.), Nat. Soc. Profl. Engrs., Nat. Acad. Engring. Mem. Democratic Farm Labor Party. Subspecialties: Solid mechanics; Theoretical and applied mechanics. Current work: Theory of plasticity; numerical methods in plastic analysis. Home: 2962 W River Pkwy Minneapolis MN 55406 Office: 107 Akerman Hall U Minn Minneapolis MN 55455

HODGEN, GARY DEAN, biomedical scientist, endocrinologist, government agency executive; b. Frankfort, Ind., May 10, 1943; s. Noble Fee and Mary Eleanor (Stern) H.; m. Linda Kay Hufford, Oct. 12, 1942; children: Michele, Amy. B.S., Purdue U., 1965, M.S., 1966; Ph.D., Ohio State U., 1969. NIH staff fellow Nat. Inst. Child Health and Human Devel., NIH, Bethesda, Md., 1969-73, sr. investigator, 1973-74, head. endocrinology sect., 1974-77, chief, 1977—. Contbr. over 200 med. articles to profl. publs. Ayerst lectr. Am. Fertility Soc., 1982. Recipient Line Mgr.'s award, 1980, EEO award Nat. Inst. Child Health and Human Devel., 1981, Research award Soc. for Study of Reprodn., 1981. Mem. Laurentian Hormone Conf. (Pincus lectr. 1982), Endocrine Soc., Am. Fertility Soc., Soc. Gynecologic Investigation. Subspecialties: Reproductive endocrinology; Fetal surgery. Current work: Reproductive biology; fertility; fetal development; menstrual cycle; ovulation; in vitro fertilization, neuroendocrinology. Home: 7517 Coddle Harbor Ln Potomac MD 20854 Office: Nat Inst Child Health and Human Devel NIH Reprodn Research Br Bethesda MD 20205

HODGES, CARL NORRIS, dir. research lab.; b. New Braunfels, Tex., Mar. 19, 1937; s. Roy Ranal and Martha Marie (Meyer) H.; m. Marilyn Ottinger, July 16, 1960 (div. Jan. 1982); children: Roy Malone, Mary Martha. B.S. in Math, U. Ariz., 1959. Research assoc. U. Ariz., Tucson, 1961-63, supr., 1963-67, dir., 1967—; dir. Superior Farming Co., Bakersfield, Calif.; chmn. Ariz. Solar Energy Commn.; mem. adv. com. tech. innovation Nat. Acad. Scis.; mem. tech. assessment adv. council Office Tech. Assessment, U.S. Congress; mem. (Ariz.-Mexico Commn.). Contbr. articles to profl. jours. Bd. dirs. Tucson Airport Authority. Recipient Man of Year award Elec. League Ariz., 1977; Creative Tech. award U. Ariz. Found., 1977; Govt. Abu Dhabi grantee, 1969-76; Rockefeller Found. grantee, 1967-70, 70-77, 78-80, 82-83; Oil Services Co. Iran grantee, 1972-78; Coca-Cola Co. grantee, 1975-81; F.H. Prince & Co. grantee, 1975—; W.R. Grace & Co. grantee, 1981—. Fellow AAAS, Ariz.-Nev. Acad. Sci.; mem. Ariz. Solar Energy Assn., Internat. Solar Energy Soc. (past treas.), World Mariculture Soc., Tucson Tomorrow. Republican. Subspecialties: Food science and technology; Solar energy. Current work: Controlled environment agr.; controlled environment aquaculture of crustaceans; research in seawater-irrigated, salt tolerant plants; solar heating and cooling. Home: 2965 N Santa Rosa Pl Tucson AZ 85712 Office: Environ Research Lab Tucson Internat Airport Tucson AZ 85706

HODGES, ROBERT EDGAR, physician, educator; b. Marshalltown, Iowa, July 30, 1922; s. Wayne Harold and Blanche Emma (McDowell) H.; m. Norma Lee Stempel, June 8, 1946; children: Jeannette Louise, Robert William, Karl Wayne, James Wolter. B.A., State U. Iowa, 1944, M.D., 1947, M.S. in Physiology, 1949. Diplomate: Nat. Bd. Med. Examiners, Am. Bd. Internal Medicine. Intern Meml. Hosp., Johnstown, Pa., 1947-48; fellow physiology, also obstetrics and gynecology, then resident in internal medicine State U. Iowa Hosp., 1948-52, dir. metabolic ward, 1952-71; mem. faculty State U. Iowa Med. Sch., 1952-71, prof. internal medicine, 1964-71, chmn. com. nutritional edn., adminstrn. grad. ednl. program nutrition, 1968-71; mem. liaison com. Maximum Security Hosp., Iowa City, 1966-71; prof. internal medicine, chief sect. nutrition U. Calif. Med. Sch., Davis, 1971-80, U. Nebr. Coll. Medicine, Omaha, 1980-82; prof. and dir. nutrition program, dept. family medicine, prof dept. internal medicine U. Calif. Irvine Sch. Medicine, 1982—; mem. nutrition study sect. NIH, 1964-68; chmn. subcom. ascorbic acid and pantothenic acid ARC, 1966-68; mem. com. nutrition overview and adjustment of food on demand Nat. Acad. Scis.-NRC, 1976; cons. to hosps., other govt. agencies. Author: Nutrition in Medical Practice, 1980, also articles.; Editor: Human Nutrition, A Comprehensive Treatise, 1980; Mem. editorial bds. med. jours. Served to capt. M.C. AUS, 1943-46, 54-56. Fellow ACP; mem. AMA, Am. Heart Assn. (fellow councils atherosclerosis, epidemiology; chmn. com. nutrition 1966-68), Am. Bd. Nutrition (pres. 1973-74), Internat. Soc. Parenteral Nutrition, Am. Soc. Parenteral and Enteral Nutrition, Soc. Exptl. Biology and Medicine, Am. Fedn. Clin. Research, Am. Inst. Nutrition, Am. Soc. Clin. Nutrition (pres. 1966-67), Nutrition Soc. (London). Subspecialty: Nutrition (medicine). Home: 200 4A S Circle View Dr Irvine CA 92715 Office: Dept Internal Medicine U Calif Irvine Sch Medicine 101 City Dr South Orange CA 92668

HODGES, THOMAS KENT, plant physiologist; b. Bedford, Ind., Oct. 18, 1936; s. Ollie Russell and Frances M. (Foster) H.; m. Sharon Ann Fultz, June 9, 1957; children: Christine Ann, Cynthia Lynne, Scott Russell. B.S., Purdue U., 1958; M.S., U. Calif., Davis, 1960, Ph.D., 1962. Postdoctoral fellow U. Ill., Urbana, 1962-63, asst. prof., then asso. prof. plant physiology, 1963-71; vis. prof. botany U. Calif., Davis, 1968-69; mem. faculty Purdue U., 1971—, prof. plant physiology, 1973—, head dept. botany and plant pathology, 1977-82; program mgr. plant biology grants Dept. Agr., 1981. Author papers in field. Recipient Daryl Snyder award Farm House Frat., 1976. Mem. Am. Soc. Plant Physiologists (Charles Albert Shull award 1975), Sigma Xi, Pi Alpha Xi, Phi Sigma, Alpha Zeta. Subspecialties: Plant physiology (agriculture); Plant cell and tissue culture. Current work: Regeneration of agronomic crops (i.e.corn)from proto plasts;developmental physiology of plants. Office: Lilly Hall Purdue U West Lafayette IN 47907

HODGES, WILLIAM FITZGERALD, psychologist, educator; b. Pensacola, Fla., Dec. 26, 1940; s. Paul and Lilian Banks (Fitzgerald) H.; m. Linda Joan Etheredge, Sept. 1, 1962; children: Heather Lynn, Wendy Lee. B.A., Vanderbilt U., 1963; Ph.D., 1967. Lic. psychologist, Colo. Intern Duke U. Med. Ctr., 1966-67; asst. prof. U. Colo., 1967-72, assoc. prof., 1972—; cons. Boulder (Colo.) Valley Pub. Schs., 1968-73, 80—; mem. Colo. Bd. Psychologist Examiners, 1979-82. Editor: (with others) The Field of Mental Health Consultation, 1983; assoc. editor: Am. Jour. Community Psychology, 1973—; contbr. articles to profl. jours. NIMH grantee, 1978-81. Mem. Am. Psychol. Assn., Rocky Mountain Psychol. Assn., Colo. Psychol. Assn. Democrat. Methodist. Subspecialties: Developmental psychology; Clinical psychology. Current work: Research on identifying children at risk for maladjustment as a function of the divorce of parents; primary focus is on preschool children and the family and demographic variables; developing intervention technqiues for this population. Home: 4975 Durham St Boulder CO 80301 Office: U Colo Dept Psychology Box 345 Boulder CO 80309

HODGMAN, JOAN ELIZABETH, neonatologist; b. Portland, Oreg., Sept. 7, 1923; d. Kenneth E. and Ann (Vannet) H.; m. Amos N. Schwartz, Jan. 30, 1949; children—Ann Vannet, Susan Lynn. B.A., Stanford U., 1943; M.D., U. Calif., San Francisco, 1946. Intern in pediatrics U. Calif. Hosp., San Francisco, 1946-47; resident in pediatrics Harbor Gen. Hosp., Torrance, Calif., 1947-48, Los Angeles County-U. So. Calif. Med. Center, 1948-50; practice medicine specializing in pediatrics, S. Pasadena, Calif., 1950-52; mem. faculty U. So. Calif. Med. Sch., 1952—, prof. pediatrics, 1969—; dir. newborn div. Los Angeles County-U. So. Calif. Med. Center, 1955—; chmn. med. adv. com. Nat. Found.-March of Dimes, 1972-75; adv. com. Western sect. UNICEF, 1975; med. adv. com. Calif. Legislature, 1970; cons. Calif. Health Dept. Author articles in field, chpts. in books. Recipient cert. appreciation Am. Cancer Soc., 1964, Cameo of Committent award B'nai B'rith, 1969, Meritorious award Nat. Found.-March of Dimes, 1969; named Woman of Year Calif. Museum Sci. and Industry, 1974, Los Angeles Times, 1976. Mem. Am. Pediatric Soc., Am. Acad. Pediatrics, Am. Thoracic Soc., Western Soc. Pediatric Research, Southwestern Pediatric Soc., Calif. Perinatal Assn., Calif. Med. Assn., Los Angeles County Med. Assn., Los Angeles Pediatric Soc. Subspecialty: Neonatology. Home: 494 Stanford Dr Arcadia CA 91006 Office: 1240 Mission Rd Los Angeles CA 90033

HODGSON, GORDON WESLEY, engineering educator, space scientist, environmentalist, geochemist; b. Islay, Alta., Can., May 25, 1924; s. Wesley White and Olive Beatrice (Trevithick) H.; m. Jeannette Fairbairn Doull, May 25, 1953; children: Patricia Coats, Kathryn Falck, Robin, Lauren, Shannon. B.Sc., U. Alta., 1946, M.Sc., 1947; Ph.D., McGill U., 1949. Head petroleum research Research Council Alta., Edmonton, 1958-67; NAS sr. research assoc. NASA Ames Research Center, Calif., 1967-68; research assoc. genetics dept. Stanford U., Calif., 1968-69; dir. Kananaskis Centre for Environ. Research, Calgary, Alta., 1973-82; prof. faculty engring. U. Calgary, 1982—; hon. research assoc. Arctic Inst. N.Am., 1979—. Contbr. articles to sci. jours.; patentee in field. Recipient grants Nat. Acad. Scis., 1967, Petroleum Research Fund, 1970, NRC Can., 1975-82. Mem. Am. Chem. Soc., Am. Astron. Soc., Geochem. Soc., Chem. Inst. Can., Sigma Xi. Presbyterian. Subspecialties: Resource management; Organic geochemistry. Current work: Geochemical modelling of energy resources, environmental processes of conversion and transport, theory and practice of technology transfer. Office: Arctic Inst NAm U Calgary 2500 University Dr NW Calgary AB Canada T2N 1N4 Home: 18 Varbay Pl NW Calgary AB Canada T3A OC8

HODGSON, RICHARD HOLMES, plant physiology researcher; b. Orange, N.J., Aug. 30, 1929; m., 1951; 1 child. B.S., Duke U., 1951, M.A., 1954; Ph.D in Botany, Ohio State U., 1959. Plant physiologist Denver Fed. Ctr., 1959-64; research plant physiologist, metabolism and radiation research lab. U.S. Dept. Agr., Fargo, N.D., 1964-78, research leader weed sci. research lab., sci. and edn. adminstrn., Frederick, Md., 1978—. Mem. O2AAAS Weed Sci. Soc. Am., Am. Soc. Plant Physiology, Scandinavian Soc. Plant Physiology, Am. Soc. Hort. Sci. Subspecialty: Plant physiology (agriculture). Office: US Dept Agr Bldg 1301 Ft Detrick Frederick MD 210701

HODOSH, MILTON, dentist, educator, researcher; b. Providence, Jan. 22, 1926; s. Samuel and Eva (Schoenfield) H.; m. Jacqueline Ann Cohen; children: Steven Harris, Gail Susan, Alex Jay. B.A., Brown U., 1950; D.M.D., Tufts U., 1954. Resident dept. oral surgery R.I. Hosp., Providence, 1954-55, dir. dental research and edn., 1971—; instr. dept. periodontology Sch. Dental Medicine Tufts U., Medford, Mass., 1954-60, research mem. dept. oral pathology, 1963-69, asst. research prof. oral pathology, 1969; pvt. practice dentistry Milton Hodosh, D.M.D., Inc., Providence, 1955—; research mem. Inst. Health Sci. Brown U., Providence, 1960-74, adj. clin. assoc. prof. dental surgery, 1975—; instr. dept. periodontology Harvard Sch. Dental Medicine, 1971, asst. clin. prof. oral pathology and oral medicine, 1971, assoc. clin. prof. oral pathology, 1974—; mem. staff R.I., Hane Brown, St. Joseph, Lady of Fatima, Miriam hosps., 1955—; mem. adv. com. Joseph Samuels Dental Ctr. for Children, 1971—; sec. R.I. Dental Examiners, 1963-69; mem. Nat. Bd. Oral Pathology and Radiography Test Constrn. Com., 1976-81; Prin. investigator various grants Nat. Inst. Dental Research. Contbr. articles to publs. in field. Served with USN, 1944-46. Fellow Acad. Dentistry Internat., Pierre Fauchard Acad., AAAS; mem. Am. Acad. Dental Medicine, Am. Cancer Soc. (bd. dirs. R.I. div.), Am. Acad. Dental Sci., Internat. Assn. Dental Sci., Internat. Assn. Dental Research, Northeastern Soc. Periodontists, Internat. Platform Soc., ADA, R.I. Dental Soc., Tufts Dental Alumni Assn., Am. Assn. Dental Examiners, Am. Acad. Implant Dentistry, New Eng. Dental Soc., Fedn. Prosthodontic Orgns., Royal Soc. Health, Internat. Coll. Oral Implantologists, Soc. Am. Inventors, Soc. Biomaterials, Am. Acad. Implant Dentistry Research Found., Omicron Kappa Upsilon. Jewish. Subspecialty: Implantology. Current work: Dentinal hypersensitivity, ulcerative oral lesions, implantology (dentistry), developer of injection devices. Inventor toothpaste for sensitive teeth; originator polymer implant concept; discovered potassium nitrate for treatment of dentinal hypersensitivity, apthous stomatitis and pulpitis. Home: 72 Overhill Rd Providence RI 02906 Office: 145 Whitmarsh St Providence RI 02907

HOEFERT, LYNN LUCRETIA, botanist, researcher; b. Billings, Mont., July 7, 1935; d. Arthur Carl and Lucretia Brown (McMullen) H. B.S., Mont. State Coll., 1958, M.S., 1961; Ph.D., U. Calif., Davis, 1965. Postgrad. research botanist U. Calif., Santa Barbara, 1965-66; botanist Agrl. Research Service, U.S. Dept. Agr., Logan, Utah, 1966-69, Salinas, Calif., 1969—, dir., 1971—. Contbr. articles to profl. jours. Dept. Agr. research grantee, 1980-83. Mem. Bot. Soc. Am., Am. Phytopathol. Soc., Electron Microscopy Soc. Am., Am. Soc. Sugar Beet Technologists, Am. Inst. Biol. Scis., Federally Employed Women, Sigma Xi. Subspecialties: Plant virology; Plant pathology. Current work: Electron microscopy of plant virus diseases; devel. ultrastructure of plants. Home: 405 Lena Gilroy CA 95020 Office: PO Box 5098 1636 E Alisal Salinas CA 93915

HOEG, DONALD FRANCIS, chemist; b. Bklyn., Aug. 2, 1931; s. Harry Herman and Charlotte (Bourke) H.; m. Patricia Catherine Fogarty, Aug. 30, 1952; children—Thomas Edward, Robert Francis, Donald John, Marybeth, Susan Catherine. B.S. in Chemistry summa cum laude, St. John's U., N.Y., 1953, Ph.D., Ill. Inst. Tech., 1957. Fellow in chemistry and chem. engring. Armour Research Found., 1953-54; grad. research asst. Ill. Inst. Tech., 1954-56; research chemist W.R. Grace & Co., 1956-58, sr. research chemist, 1958-61; group leader addition polymer chemistry Roy C. Ingersoll Research Center, Borg-Warner Corp., Des Plaines, Ill., 1961-64, mgr. polymer chemistry, 1964-66, asso. dir., head chem. research dept., 1966-75, dir., 1975—; mem. solid state scis. adv. bd. Nat. Acad. Scis.; Bd. overseers Lewis Coll. Scis. and Letters of Ill. Inst. Tech., 1980—; bd. dirs. Ill. Inst. Tech. Alumni, 1979-82, Mt. Prospect Combined Appeal, 1963-65. Bd. editors: Research Mgmt. Mag, 1979—; contbr. numerous articles tech. publs., chpts. in books. TaPing Lin scholar, 1955-56; AEC asst., 1954; Armour Research Found. fellow, 1953-54; Ill. Inst. Tech. Achievement award, 1983. Mem. Am. Chem. Soc., AAAS, N.Y. Acad. Scis., Dirs. Indsl. Research, Am. Mgmt. Assn., Research Dirs. Assn. Chgo. (pres. 1977-78), Sigma Xi. Subspecialties: Polymer chemistry; Organometallic Chemistry. Current work: Advanced materials, power transmission, fluid mechanics, manufacturing technology. Patentee in field. Office: Roy C Ingersoll Research Center Wolf and Algonquin Rds Des Plaines IL 60018 I've counseled myself that all ideas and concepts, no matter how seemingly difficult, are products of man's mind, and, therefore fundamentally understandable.

HOEGERMAN, STANTON FRED, cytogeneticist, educator; b. Bklyn., May 13, 1944; s. Fred and Edith (Rost) H.; m. Georgeanne Stengele, Mar. 5, 1966; children: Elizabeth, David. B.S., Cornell U., 1965; M.S., N.C. State U., 1968, Ph.D., 1972. Instr. dept. biology Lincoln U., 1970-72; asst. biologist RER div. Argonne Nat. Lab., ILL., 1972-76; assoc. prof. biology Coll. William and Mary, Williamsburg, Va., 1976—. Contbr. numerous articles on human cytogenetics and radiation to cytogenetics to profl. publs. Mem. Am. Soc. Human Genetics, Health Physics Soc., Genetics Soc. Am., Bot. Soc. Am., AAAS. Unitarian. Subspecialties: Cell and tissue culture; Genetics and genetic engineering (biology). Current work: Cytogenetics of fragile sites and autism. Human cytogenetics, fragile sites, autism. Home: 367 Hiden Blvd Newport News VA 23606 Office: Biology Dept Coll William and Mary Williamsburg VA 23185

HOEL, DAVID GERHARD, statistician; b. Los Angeles, Nov. 18, 1939; m., 1961; 3 children. A.B., U. Calif.-Berkeley, 1961; Ph.D. in Stats, U. N.C.-Chapel Hill, 1966. Fellow Stanford U., 1966-67; sr. mathematician Westinghouse Research Labs., Pitts., 1967-68; research statistician Oak Ridge Nat. Labs., 1968-70; chief environ. biometric br. Nat. Inst. Environ. Health Sci., NIH, Research Triangle Park, N.C., 1973-81, math. statistician, 1970—, dir. biometric and risk assessment program, 1981—. Mem. Internat. Statis. Inst., Royal Statis. Soc., Biometric Soc.; fellow Am. Statis. Assn. Office: PO Box 12233 Research Triangle Park NC 27709

HOEL, LESTER A., civil engineering educator; b. Bklyn., Feb. 26, 1935; s. Johannes and Julia (Michelsen) H.; m. Unni Sonja Blegen, Jan. 24, 1959; children: Julie Britt, Sonja Leslie, Lisa Maureen. B.C.E., City Coll., N.Y., 1957; M.S. in Civil Engring, Bklyn. Poly. Inst., 1960; D.Eng., U. Calif. at Berkeley, 1963. Registered profl. engrs. Calif., Pa., Va. Asst. prof. engring. San Diego State Coll., 1962-64; Fulbright research scholar Inst. Transport Economy, Oslo, Norway, 1964-65; prin. engr. Wilbur Smith & Assos., San Francisco, 1965-66; faculty Carnegie-Mellon U., Pitts., 1966-74, prof. civil engring., 1970-74; asso. dir. Transp. Research Inst., 1966-74; Hamilton prof., chmn. dept. civil engring. U. Va., 1974—; staff cons. Gen. Analytics, Inc., 1971-78; Cons. P.R. Planning Bd., 1969-70. Editor: Public Transportation: Planning, Operations and Management, 1979; Mem. editorial bd.: Transp. Research; Author tech. papers, books and articles. Chmn. bd. mgmt. YMCA, 1968-69; mem. Churchill Boro Planning Commn., 1972-74. Recipient Alumni award in Civil Engring. Coll. City N.Y., 1957; Pyke Johnson award Transp. Research Bd., 1977; Fulbright travel grantee, 1964-65. Mem. ASCE (Huber research prize 1976), Am. Scandinavian Found., Inst. Transp. Engrs., Transp. Research Bd. (exec. com.), Am. Soc. Engring. Edn., Sigma Xi, Chi Epsilon, Tau Beta Pi. Subspecialty: Civil engineering. Home: 1703 Old Forge Rd Charlottesville VA 22901

HOELLER, LOUISE, nurse educator, nun, consultant; b. Fond du Lac, Wis., Dec. 5, 1928; d. Francis R. and Grace A. (Gilkey) H. R.N., St. Joseph Sch. Nursing, Chgo., 1953; B.S., DePaul U., 1954; M.Nursing, La. State U., 1977. Mem. Daus. of Charity of St. Vincent dePaul; supr. operating room-recovery room DePaul Hosp., St. Louis, 1956-62; Supr. operating room-recovery room Charity Hosp., New Orleans, 1962-65; clin. instr. U. Wis., St. Mary's Hosp., Milw., 1967-69; program dir. surg. tech. Forest Park Coll., St. Louis, 1970-75, Charity Hosp., New Orleans, 1975-80; asst. prof. perioperative nursing La. State U. Med. Center, Shreveport, 1981—; cons. Mo. Profl. Liability Ins. Assn., Columbia, 1980; site visitor, com. mem. Assn. Surg. Technologists, Denver, 1981-82. Author: Operating Room Technician, 1965, 2d edit., 1968, Surgical Technology: Basis for Clinical Practice, 1974. Mem. Am. Nurses Assn. (sect. chmn. 1962-65), Nat. League Nursing (nat. com. 1961-65), Assn. Operating Room

Nurses, La. State U. Nursing Alumni, Sigma Theta Tau. Subspecialties: Health services research; Human factors engineering. Current work: Competency-based learning-modular programmed instruction in surgical instruments. Home: PO Box 21976 Shreveport LA 71120 Office: La State U Med Center 1541 Kings Hwy Shreveport LA 71130

HOFELDT, FRED DAN, physician, researcher; b. Rock Springs, Wyo., Sept. 11, 1936; s. Fred D. and Cecilia (Roniker) H.; m. Ardyce Martin, July 14, 1957 (div. 1976); children: Fred Dan, Scott; m. Laurie Hendricks, Dec. 13, 1977; children: Leann, Stacy. B.S., Coll. Idaho, Caldwell, 1959; M.D., U. Wash., 1963. Commd. 2d lt. U.S. Army, 1961; advanced through grades to col., 1978; intern Tripler Army Med. Ctr., Honolulu, 1963-64, resident in internal medicine, 1964-67; fellow in endocrinology U. Calif.-San Francisco, 1969-71; chief dept. hosp. clinics Tripler Army Med. Ctr., 1967-68; asst. chief med. cons. Office Surgeon Gen., Washington, 1968-69; chief endocrine metabolic service Fitzsimons Army Med. Ctr., Aurora, Colo., 1970—, dir. endocrine fellowship program, 1978—; clin. prof. medicine U. Colo. Health Scis. Ctr., Denver, 1978—. Contbr. chpts. to textbooks and numerous articles to profl. jours. Chmn. Colo. Diabetes Inst., Denver, 1981—. Decorated Meritorious Service award, Army Commendation medal. Fellow ACP; mem. Assn. Mil. Surgeons, Am. Fedn. Clin. Research, Am. Diabetes Assn., Endocrine Soc., AAUP. Republican. Roman Catholic. Subspecialty: Endocrinology. Current work: Hypoglycemia, diabetes, thyroid. Home: 5513 S Kenton Ct Englewood CO 80111 Office: Endocrine Service Dept Medicine Fitzsimons Army Med Ctr Aurora CO 80045

HOFER, KURT GABRIEL, biologist, scientist, educator; b. Feldkirchen, Carinthia, Austria, March 2, 1939; s. Stefan and Katharina (Mark) H.; m. Maria G. Geyer, Nov. 30, 1965; 1 dau., Andrea Marie. B.Edn., Tchr. Tng. Coll., Klagenfurt, Austria, 1959; Ph.D. in Biology, U. Vienna, Austria, 1965. Postdoctral fellow Tufts U. Med. Sch., Boston, 1966-70; asst. prof. Ohio State U., Columbus, 1970-71; mem. faculty Fla. State U., Tallahassee, 1971—, prof. biol. sci., 1979—, dir. program of molecular biophysics, 1976-79. Contbr. numerous sci. articles to profl. pbls. Recipient Pres. Teaching award Fla. State U., 1980; NIH grantee, 1971-82; Am. Cancer Soc. grantee, 1978-80. Mem. Radiation Research Soc., Am. Assn. Cancer Research, Cell Kinetics Soc., Am. Cancer Soc. (bd. dirs. 1972-81), Sigma Xi. Subspecialties: Cell biology; Cancer research (medicine). Current work: The experimental radiotherapy of animal cancers, toxicity of radionuclides, cell kinetics, tissure culture studies, and chronobiology. Office: Fla State Univ Inst Molecular Biophysics Tallahassee Fl 32306

HOFER, MYRON ARMS, psychiatrist, educator; b. N.Y.C.; s. Philip and Frances Louise (Heckscher) H.; m. Adeline Van Nostrand Paul, June 12, 1954; children: Timothy Philip, Adeline Van Nostrand, Andrew Paul. A.B., Harvard U., 1954, M.D., 1958. Diplomate: Am. Bd. Neurology and Psychiatry. Inter and resident in medicine Mass. Gen. Hosp., 1958-60; research assoc. N.Y. Hosp., Clin. Center NIMH; resident in psychiatry Columbia Presbyn. Hosp., N.Y.C., 1960-66; asst. prof., prof. dept. psychiatry and neurosci. Albert Einstein Coll. Medicine, Bronx, N.Y., 1967—; cons. NIMH. Contbr. numerous articles on psychobiology to profl. jours.; author: Roots of Human Behavior, 1981. Served with USPHS, 1962-64. Recipient Boylston Prize, 1958; NIMH Research Scientist awardee, 1968, 73, 78, 82. Mem. Internat. Soc. for Devel. Psychobiology (pres. 1980-81), Am. Psychosomatic Soc. (pres. 1982), Psychiat. Research Soc. Democrat. Club: Century Assn. (N.Y.C.). Subspecialties: Psychobiology; Psychophysiology. Current work: Developmental psychobiology, the contribution of the Parent-Infant Interaction to the development of young. Office: Montefiore Med Ctr 111 E 210th St Bronx NY 10467

HOFF, JULIAN THEODORE, neurosurgeon, educator; b. Boise, Idaho, Sept. 22, 1936; s. Harvey O. and Helen (Boraas) H.; m. Diane Shanks, June 3, 1962; children: Paul, Allison, Julia. B.A., Stanford U., 1958; M.D., Cornell U., 1962. Intern N.Y. Hosp., 1962-63, resident, 1963-64, 66-70; asst. prof. surgery U. Calif., San Francisco, 1970-74, assoc. prof., 1974-78, prof., 1978-81; prof. surgery U. Mich., Ann Arbor, 1981—, head sect. neurosurgery, 1981—. Editor: ann. revision Practice of Surgery, 1977—; editor: Current Surgical Management of Neurological Disease, 1980, Foundations of Neurosurgery, 1983. Served to capt. U.S. Army, 1964-66. Recipient Tchr.-Investigator Devel. award NIH, 1972-77; Josiah Macy scholar, 1979. Fellow A.C.S.; mem. Am. Assn. Neurol. Surgeons, Soc. Neurol. Surgeons, Am. Acad. Neurol. Surgeons, AMA, Alpha Omega Alpha. Republican. Presbyterian. Subspecialty: Neurosurgery. Current work: Cerebral Ischemia. Office: Department of Neurosurgery University Hospital Ann Arbor MI 48109 Home: 2120 Wallingford St Ann Arbor MI 48104

HOFF, MARCIAN EDWARD, JR., electronics engineer; b. Rochester, N.Y., Oct. 28, 1937; s. Marcian Edward and Mary Elizabeth (Fitzpatrick) H.; m. Judith Schless Rytand, May 19, 1977; children: Carolyn, Lisa, Jill. B.E.E., Rensselaer Poly. Inst., Troy, N.Y., 1958; M.S., Stanford U., 1959, Ph.D., 1962. Research asso. Stanford U., 1962-68; mgr. applications research Intel Corp., Santa Clara, Calif., 1968-83; v.p. research and devel. Atari Inc., Sunnyvale, Calif., 1983—. Author articles on adaptive systems, microcomputers. NSF fellow, 1958—; recipient Stuart Ballantine medal Franklin Inst., 1979. Mem. IEEE (Clido Brunetti award 1980), Sigma Xi, Eta Kappa Nu, Tau Beta Pi. Subspecialty: Electronics industry research and development management.. Patentee track circuits, electrochem. memory, digital filters, integrated circuits. Home: 1075 Astoria Dr Sunnyvale CA 94087 Office: 1196 Borregas Ave Sunnyvale CA 94086

HOFF, NICHOLAS JOHN, mechanical and aerospace engineer; b. Magyarovar, Hungary, Jan. 3, 1906; came to U.S., 1939, naturalized, 1944; s. Miklos and Lenke H.; m. Vivian Church, July 20, 1940 (dec. Apr. 1969); m. Ruth Kleczewski, Nov. 17, 1972; 1 dau., Karen Brandt. M.E., Fed. Poly. Inst. Zurich, Switzerland, 1928, Dipl.-Ing., 1928; Ph.D. in Engring. Mechs, Stanford U., 1942. Airplane designer, Hungary, 1928-38; research asst. Stanford U., 1939-40, head dept. aeros., 1957-71, prof. aeros., 1957-71, prof. emeritus, 1971—; instr. aeros. Poly. Inst. Bklyn., 1940-41, asst. prof., 1941-43, asso. prof., 1943-46, prof., 1946-57, head dept. aeros., 1950-57; vis. prof. Monash (Australia) U., 1971, Ga. Inst. Tech., 1973, Cranfield (Eng.) Inst. Tech., 1974-75, Poly. Inst. Zurich, 1975; Clark/Crossan prof. engring. Rensselaer Poly. Inst., 1976-79, vis. Disting. prof., 1979-81; cons. to govt. and industry. Author: The Analysis of Structures, 1956; editor: books, including High Temperature Effects in Aircraft Structures, 1958, Creep in Structures, 1962; contbr. numerous articles on structural and stress analysis, aeros. to profl. jours. Fellow AIAA (hon., Pendray award 1971, Structures, Structural Dynamics, and Materials award 1971), ASME (hon., Worcester Reed Warner medal 1967, ASME medal 1974), Royal Aero. Soc. (Gt. Britain); mem. ASCE (life, von Karman medal 1972), Aero. Soc. India (hon.), U.S. Nat. Acad. Engring., Internat. Acad. Astronautics (corr.). Subspecialties: Aeronautical engineering; Aerospace engineering and technology. Current work: Research on composite structures and history of aeronautics. Office: Dept Aeros and Astronautics Stanford U Stanford CA 94305

HOFFERT, MARVIN JAY, neurophysiologist, educator; b. Phila., Jan. 28, 1949; s. Paul W. and Rosolyn (Sheiman) H.; m. Leslie Sloan, June 9, 1968; children: Alexander S., Jason M.; m. Cara Orben, July 27, 1980. Pre-med. student, Yale U., M.D., 1974. Diplomate: Am. Bd. Med. Examiners, Am. Bd. Psychiatry and Neurology. Resident, then chief resident in neurology Cleve. Met. Gen. Hosp., 1975-78, fellow in neuropathology, 1978-79; sr. staff fellow in neurobiology and anesthesiology NIH, Bethesda, Md., 1979-83; asst. prof. depts. neurology and neurophysiology U. Wis. Med. Sch., Madison, 1983—; adj. prof. psychology Am. U.; clin. asst. prof. neurology Uniformed Services U. Health Scis. Contbr. articles to profl. jours. Mem. Internat. Assn. Study Pain, Soc. Neurosci., Am. Acad. Neurology, Sigma Xi. Jewish. Subspecialties: Neurophysiology; Neurology. Current work: Pain mechanisms; experimental and clinical pain; intracellularly HRP filled neurons. Home: 2505 Balder St Madison WI 53713 Office: U Wis Med Sch Neurology Dept Rm H6/552 600 Highland Ave Madison WI 53792

HOFFLEIT, ELLEN DORRIT, astronomer; b. Florence, Ala., Mar. 12, 1907; d. Fred and Kate (Sanio) H. A.B., Radcliffe Coll., 1928, M.A., 1932, Ph.D., 1938. From research asst. to astronomer Harvard Coll. Obs., 1929-56; mathematician Ballistic Research Labs., Aberdeen Proving Ground, Md., 1943-48; tech. expert, 1948-62; lectr. Wellesley Coll., 1955-56; mem. faculty Yale U., 1956—, sr. research astronomer, 1974—; dir. Maria Mitchell Obs., Nantucket, Mass., 1957-78; mem. Hayden Planetarium Com., N.Y.C., 1975—; editor Meteoritical Soc., 1958-68. Author: Some Firsts in Astronomical Photography, 1950, Yale Bright Star Catalogue, 4th edit, 1982; also research papers. Recipient Caroline Wilby prize Radcliffe Coll., 1938, Grad. Soc. medal, 1964; certificate appreciation War Dept., 1946; alumnae recognition award Radcliffe Coll., 1983. Mem. Internat. Astron. Union, Am. Astron. Soc., AAAS, Am. Geophys. Union, Meteoritical Soc., Am. Assn. Variable Stars Observers, Am. Def. Preparedness Assn., Nantucket Maria Mitchell Assn., Nantucket Hist. Soc. (hon.), Phi Beta Kappa, Sigma Xi. Subspecialty: Optical astronomy. Current work: Variable stars, astrometry, star catalogues. Home: 255 Whitney Ave New Haven CT 06511 Office: Yale U Obs Box 6666 New Haven CT 06511 The guiding motto of my life has been, "Work for the work's sake and it will become a part of you./Work for the sake of worldly gain and you sell your soul to the Devil." Love for research and boundless perseverance have enabled me to achieve, not all that I might have wished, but far more than I would ever have dared to expect on the basis of mediocre high school grades.

HOFFMAN, ALAN JEROME, mathematician, educator; b. Bklyn., May 30, 1924; s. Jesse and Muriel (Schrager) H.; m. Esther Walker, May 30, 1947. A.B., Columbia U., 1947, Ph.D., 1950. Mem. Inst. Advanced Study, Princeton, N.J., 1950-51; mathematician Nat. Bur. Standards, Washington, 1951-56; sci. liason officer Office Naval Research, London, 1956-57; cons. Gen. Electric Co., N.Y.C., 1957-61; fellow IBM Research Ctr., Yorktown Heights, N.Y., 1961—; vis. prof. CUNY, Yale U., Stanford U. Served with AUS, 1943-46. Mem. Nat. Acad. Sci., N.Y. Acad. Sci., Am. Math. Soc. Current work: Interplay among combinatorics, linear algebra and linear programming. Office: IBM Research Center PO Box 218 Yorktown Heights NY 10598

HOFFMAN, DARLEANE CHRISTIAN, nuclear chemist; b. Terril, Iowa, Nov. 8, 1926; d. Carl Benjamin and Elverna E. (Kuhlman) Christian; m. Marvin Morrison Hoffman, Dec. 26, 1951; children: Maureane R., Daryl K. B.S. in Chemistry, Iowa State U., 1948, Ph.D. in Nuclear Chemistry, 1951. NSF sr. postdoctoral fellow Institut for Atomenergi, Kjeller, Norway, 1964; staff mem. radiochemistry group Los Alamos Sci. Lab., 1953-71, assoc. group leader, 1971-79, div. leader chemistry and nuclear chemistry, 1979-82, div. leader isotope and nuclear chemistry, 1982—; Guggenheim fellow Lawrence Berkeley Lab., 1978-79. Contbr. articles to profl. jours. Recipient Citation of Merit Coll. Scis. and Humanities, Iowa State U., 1978. Fellow Am. Inst. Chemists; mem. Am. Chem. Soc. (award for nuclear chemistry 1983, John Dustin Clark award Central N.Mex. chpt. 1976), Am. Phys. Soc., AAAS. Subspecialty: Nuclear chemistry. Current work: Mechanisms of nuclear fusion; transuranium elements; science administration; nuclear chemistry; isotope chemistry. Office: Los Alamos Nat Lab INC DO MS J515 Los Alamos NM 87545

HOFFMAN, DENNIS MARK, polymer scientist; b. Huntingdon, Pa., July 22, 1947; s. Dennis M. and Ida Gene (Lane) H.; m. Pamela Joyce Jackson, Feb. 13, 1982. B.S., Juniata Coll., 1969; Ph.D., U. Mass., 1979. Polymer scientist Lawrence Livermore Labs., U. Calif., Livermore, 1979—. Served with AUS, 1970-72. Mem. Golden Gate Soc. Plastic Engrs. (bd. dirs. plastic analysis div. 1981—), Am. Chem. Soc., AAAS, Soc. Plastics Engrs., Am. Phys. Soc. Subspecialties: Polymer chemistry; Polymer engineering. Current work: Polymer science and engineering; adhesives, blends, coatings, foams, polymer physics; crystallinity, structure-property relationships. Office: Lawrence Livermore Nat Labs U Calif L-338 7000 East Ave Livermore CA 94550

HOFFMAN, DONALD RICHARD, immunologist, educator, consultant; b. Boston, Aug. 25, 1943; s. William Maurice and Laura (Rodman) H.; m. Valeria Anne Mossey, Sept. 21, 1971; 1 son, Avram Joseph. A.B., Harvard U., 1965; Ph.D., Calif. Inst. Tech., 1970. Cancer research scientist Roswell Park Meml. Inst., Buffalo, 1970-71; asst. prof. pediatrics U. So. Calif. Sch. Medicine, Los Angeles, 1971-75; assoc. prof. pathology Creighton U. Sch. Medicine, Omaha, 1975-77; assoc. prof pathology and lab. medicine E. Carolina U. Sch. Medicine, Greenville, N.C., 1977-82, prof., 1982—; cons. Pharmacia Diagnostics, T and M Immunodiagnostics. Contbr. articles to profl. jours. Mem. Am. Assn. Immunologists, AAAS, N.Y. acad. Scis., AAUP; fellow Am. Acad. Allergy and Immunology. Subspecialties: Allergy; Immunology (medicine). Current work: In vitro diagnosis of allergy; allergens; radioimmunoassay. Home: 213 N Jarvis St Greenville NC 27834 Office: E Carolina U Sch Medcine Greenville NC 27834

HOFFMAN, EDWARD JACK, energy consultant; b. Marion, Kans., June 10, 1925; s. Chiles Edward and Mary (Lynch) H.; m. Suzanne Pierson, Mar. 7, 1953; children: Kathy Sue, Cal Edward, Paul Brian. B.S. in Chem. Engring, Okla. State U., 1944; M.S., U. Mich., 1950; postgrad., U. Colo., 1954-57. Refinery chemist Continental Oil Co., Ponca City, Okla., 1946-47; research chemist Carter Oil Co., Tulsa, 1947-48; engr. Black, Sivalls & Bryson, Oklahoma City, 1951-52; instr. Okla. State U., Stillwater, 1952-54; asst. prof. U. Colo., 1954-57; assoc. prof. U. Tulsa, 1957-62; research engr. Heat Transfer Research Inc., Alhambra, Calif., 1963-65; assoc. prof. U. Wyo., Laramie, 1965-72; energy cons., Laramie, 1972—. Author: Azeotropic and Extractive Distillation, 1964, 77, The Concept of Energy: An Inquiry into Origins and Applications, 1977, Coal Conversion, 1978, Heat Transfer Rate Analysis, 1980, Coal Gasifiers, 1981, Phase and Flow Behavior in Petroleum Production, 1981, Synfuels: The Problems and the Promise, 1982. Served to lt. USN, 1944-46. Mem. Am. Inst. Chem. Engrs., Am. Chem. Soc., ASME, Inst. Briquetting and Agglomeration, Sigma Xi. Democrat. Methodist. Subspecialties: Fuels; Chemical engineering. Current work: Development and commercialization of the Hoffman process for the direct catalytic conversion of coal or other carbonaceous materials with steam to coproduce synthetic fuels and carbon dioxide for enhanced oil recovery. Patentee in field. Address: PO Box 1352 Laramie WY 82070

HOFFMAN, ERIC ALFRED, cardiopulmonary physiologist; b. Rochester, Minn., Sept. 5, 1951; s. Murray Stanley Hoffman and Doris (Creamer) Kal; m. Georgia Ann Milan, July 10, 1982. B.A., Antioch Coll, Yellow Springs, Ohio, 1974; Ph.D., U. Minn.-Mayo Grad. Sch. Medicine, 1981. Research cons. Royal Berkshire Hosp., Reading, Eng., 1972; research asst. Fels Research Inst., Yellow Springs, Ohio, 1973-74, U. Colo. Med. Ctr., Denver, 1974-75; pre-doctoral fellow Mayo Clinic Found., 1975-81, postdoctoral fellow, 1981-83, instr. physiology, 1982-83, assoc. cons., 1983—, asst. prof., 1983—; mem. research com. Colo. Heart Assn., 1974-75; new investigator Nat. Heart, Lung and Blood Inst., Washington, 1983—. Mem. Am. Thoracic Soc., Am. Heart Assn. Council Cardiopulmonary Disease, Am. Physiol. Soc. (long range planning com.), AAAS, Union Concerned Scientists. Subspecialties: Physiology (biology); Imaging technology. Current work: Developing and validating new imaging techniques associated with synchronous volumetric x-ray scanning computed tomography. Office: Biodynamics Research Unit Mayo Clinic 200 1st St SW Rochester MN 55905

HOFFMAN, FREDERICK, mathematics educator; b. Cleve., Nov. 10, 1937; s. Earl Aaron and Barbara (Epstein) H.; m. Elisabeth Moffat Clark, Dec. 15, 1968; children: Sara Millen, Amanda Ann, David Clark, Jessica Thorn. B.S., Georgetown U., 1958; postgrad., Columbia U., 1958-59; Ph.D., U. Va., 1963. Instr. to asst. prof. math. U. Ill., Urbana, 1962-66; mathematician Inst. for Def. Analyses, Princeton, N.J., 1966-67; asst. prof. math. Drexel Inst. Tech., Phila., 1967-68; asst. to assoc. prof. math. Fla Atlantic U., Boca Raton, 1968-77, prof. math., 1977—; dir. Southwestern Conf. on Combinetorics/Graph Theory/Computing, Boca Raton, 1972—; vis. assoc. prof. U. Waterloo, Ont., Can., 1975; faculty assoc. IBM, Boca Raton, 1982-83. Vice pres. Palm Beach County PTA Council, West Palm Beach, Fla., 1979—; pres. South Palm Beach County chpt. Fla. Assn. for Gifted, 1978—. NSF fellow, 1958-61; Woodrow Wison Found. hon. fellow, 1958. Mem. Am. Math. Soc., Math. Assn. Am. (chmn. Fla. chpt. 1979-80), Soc. Indsl and Applied Math., AAAS, IEEE, N.Y. Acad. Scis., Assn. Computing Machinery, Am. Assn. Artificial Intelligence. Democrat. Jewish. Subspecialties: Applied mathematics; Artificial intelligence. Current work: Theoretical research in applications of algebra and combinatorics. Research in applications of artificial intelligence to VLSI design and to configuraton problems. Home: 4307 NW 5th Ave Boca Raton FL 33431 Office: Dept Math Fla Atlantic U Boca Raton FL 33431

HOFFMAN, GEORGE ROBERT, geneticist, educator; b. Jersey City, June 6, 1946; s. George and Rose (Barchi) H.; m. Linda Wagner, June 8, 1969. B.A., Rutgers U., Newark, 1967; M.S., U. Tenn., Knoxville, 1969, Ph.D., 1972. Instr. biology Roane State Community Coll., Harriman, Tenn., 1972-73; postdoctoral fellow Nat. Inst. Environ. Health Scis., Research Triangle Park, N.C., 1973-74, geneticist, 1977-78; asst. prof. biology Meredith Coll., Raleigh, N.C., 1974-77; sr. staff officer Nat. Acad. Scis., Washington, 1979-81; asst. prof. biology Holy Cross Coll., Worcester, Mass., 1981—; cons. com. on chem. environ. mutagens Nat. Acad. Scis.; cons. L'Oreal Research Labs., Aulnay-sous-Bois, France, EPA. Mem. editorial bd.: Mutation Research; book rev. editor: Environ. Mutagenesis; contbr. articles to sci. jours. Mem. Genetics Soc. Am., Environ. Mutagen Soc., Am. Soc. Microbiology, Bot. Soc. Am., Nat. Audubon Soc., Sigma Xi. Subspecialties: Genetics and genetic engineering (biology); Toxicology (medicine). Current work: Bacterial and fungal genetics; mutagenesis in microorganisms; genetic toxicology testing. Home: 115 Elmwood St Auburn MA 01501 Office: Biology Dept Holy Cross College Worcester MA 01610

HOFFMAN, JOHN ROBERT, JR., engring. cons. co. exec.; b. Tuscaloosa, Ala., Nov. 3, 1945; s. John R. and Dorothy M. H.; m. Patricia J. Meyer, Aug. 5, 1967; children: Sheryl Ann, Michael Andrew. B.S.M.E., Cooper Union for Advancement Sci. and Art, 1967; M.S.N.E., U. Lowell, 1977. Registered profl. engr., Mass., N.H., Vt. Primary coolant system engr. Westinghouse Electric Corp., West Mifflin, Pa., 1967-69; test engr. Pratt & Whitney Aircraft, East Hartford, Conn., 1969-71; asst. to v.p. Yankee Atomic Electric Co., Framingham, Mass., 1971—. Mem. ASME. Subspecialties: Mechanical engineering; Nuclear engineering. Home: 33 Woodlawn Dr Sturbridge MA 01566 Office: 1671 Worcester Rd Framingham MA 01701

HOFFMAN, KENNETH CHARLES, research and consulting company executive; b. N.Y.C., Nov. 29, 1933; s. Arthur A. and Helen M. (McElearney) H.; m. Ann Theresa Hynes, Aug. 20, 1955; children: Kenneth M., Theresa A., Charles M. B.M.E., NYU, 1954; M.S., Adelphi U., 1968; Ph.D. in Systems Engring, Poly. Inst. Bklyn., 1972. Registered profl. engr., N.Y. Mem. sci. staff Brookhaven Nat. Lab., 1956-73, chmn. dept. energy and environ., 1973-78, dir., 1975, sr. v.p. Math tech Inc., Arlington, Va., 1979—. Served with USAF, 1954-56. Fellow AAAS; mem. ASME, N.Am. Soc. Corp. Planners, Internat. Assn. Energy Economists. Roman Catholic. Subspecialties: Systems engineering; Resource policy. Current work: Resource policy. Office: Mathtech Inc 1401 Wilson Blvd Suite 930 Arlington VA 22209

HOFFMAN, L. RICHARD, psychology educator; b. Taunton, Mass., June 6, 1930; s. Harry and Leah (Stampel) H.; m.; children: Cynthia A., Karen E., Elizabeth D., Valerie J. B.S., Queens Coll., 1952; M.A., U. Mich., 1953, Ph.D., 1957. Lic. psychologist, Ill. Asst. to study dir. Survey Reseach Ctr., U. Mich., Ann Arbor, 1954-57, lectr. to assoc. prof. dept. psychology, 1957-65; prof. Grad. Sch. Bus., U. Chgo., 1965-81, Grad. Sch. Mgmt., Rutgers U., Newark, 1981—; cons. in field. Author: (with F.C. Mann) Automation and the Worker, 1960, (with others) Superior-Subordinate Communication in Management, 1961, Group Problem Solving, 1979; contbr.: articles to profl. jours. Group Problem Solving. Fellow Am. Psychol. Assn.; mem. N.Y. Acad. Scis. Subspecialties: Organizational psychology; Social psychology. Current work: Research on leadership, group problem-solving process. Home: 555 Mount Prospect Ave Newark NJ 07104 Office: Grad Sch Mgmt Rutgers U 92 New St Newark NJ 07102

HOFFMAN, MARK PETER, animal scientist, livestock producer, educator; b. West Reading, Pa., Feb. 4, 1941; s. Mark Webber and Pearl Matilda (Troutman) H.; m. Lorraine Johnson, Aug. 24, 1969; children: Kourtney Katherine, Royelle Marka. B.S., Delaware Valley Coll., 1963; M.S., Iowa State U., 1967, Ph.D., 1969. Asst. prof. animal sci. Iowa State U., 1969-75, assoc. prof., 1976-80, prof., 1981—. Author: manual Introduction to Animal Science, 1976, Basic Principles of Animal Nutrition, 1972, Reproductive Physiology of Livestock, 1978; contbr. numerous articles on environ. factors affecting beef cattle performance to profl. jours. Mem. Am. Soc. Animal Sci. (cert. animal scientist), Am. Dairy Sci. Assn., Am. Inst. Biol. Scis., AAAS. Republican. Lutheran. Subspecialties: Animal nutrition; Animal physiology. Current work: Research and teaching beef cattle production. Home: Rural Route 2 Ames IA 50010 Office: Iowa State U 119 Kildee Ames IA 50011

HOFFMAN, MYRON ARNOLD, mechanical engineering educator; b. Chgo., Nov. 15, 1930; m. s. Sharna F. Rosenthal, Oct. 8, 1963; children: Steven, Douglas. S.B., MIT, 1952, S.M., 1952, Sc.D., 1955. Asst. prof. MIT, Cambridge, 1955-56, assoc. prof., 1959-68; prof. U. Calif.-Davis, 1968—. Served to 1st lt. USAF, 1956-59. Mem. Am. Nuclear Soc. Subspecialties: Fluid mechanics; Nuclear fusion. Home: 627 Cordova Pl Davis CA 95616 Office: U Calif Davis CA 95616

HOFFMAN, NEIL ROBERT, physician; b. Mpls., Apr. 19, 1938; s. Joseph and Ussie (Bernstein) H.; m. Diane Lava, Aug. 20, 1961; children: Leslie, Lisa, Michael. B.A., U. Minn.-Mpls., 1960, B.S., 1962, M.D., 1964. Intern Hennepin County Med. Ctr., U. Minn., Mpls., 1964-65, resident, 1965-68, U.S. Naval Med. Ctr., Oakland, Calif., 1968-70; practice medicine specializing in internal medicine and med. oncology Mpls. Med. and Diagnostic Ctr., 1970—; cons. med. oncology Hennepin County Med. Ctr., Mpls.; dir. profl. edn. Am. Cancer Soc., 1974-79. Contbr. articles to profl. jours. Served to lt. comdr. USN, 1968-70. Fellow ACP; mem. Am. Soc. Clin. Oncology, Minn. State Med. Assn., Calif. Med. Assn., Hennepin County Med. Soc. (exec. bd. 1979-82), T.C. Oncology Club. Jewish. Subspecialties: Internal medicine; Oncology. Office: Mpls Med and Diagnostic Ctr 2219 Chicago Ave Minneapolis MN 55404

HOFFMAN, NORMAN EDWIN, chemistry educator; b. Chgo., Oct. 26, 1928; s. Edwin Valentine and Ruth H.; m. Margaret Mary Cunningham, May 16, 1953. B.S., Loyola U., Chgo., 1950; Ph.D., Northwestern U., 1954. Research chemist Standard Oil Ind., Whiting, 1953-56; instr. chemistry Marquette U., Milw., 1956-58, asst. prof., 1958-63, assoc. prof., 1963-67, prof., 1967—, chmn., 1972-81; cons. VA Hosp., Wood, Wis., 1965—, ADA, Chgo., 1974—, Baxter Travenol, Deerfield, Ill., 1981—. Contbr. chpts. to books, articles to profl. jours. Mem. Am. Chem. Soc. (councilor 1981-83, mem., Milw. sect. award 1983); Mem. Soc. for Applied Spectroscopy, Am. Assn. Clin. Chemistry, AAAS, AAUP. Subspecialties: Analytical chemistry; Clinical chemistry. Current work: High performance liquid chromatography, gas chromatography, analytical methodology applied to biological systems. Home: PO Box 32 Wales WI 53183 Office: Marquette U Todd Wehr Chemistry Bldg Milwaukee WI 53233

HOFFMAN, PAUL NED, physician, educator, researcher; b. Cornwall, N.Y., Sept. 16, 1947; s. Harold Louis and Selma Evelyn (Ushman) H.; m. Carolyn M. Van Schoik, July 10, 1983. B.S., Cornell U., 1969; Ph.D., Case Western Res. U., 1974, M.D., 1976. Diplomate: Am. Bd. Ophthalmology. USPHS fellow dept. anatomy Case Western Res. U., 1969-74; Grass Found. fellow, Woods Hole, Mass., 1974; postdoctoral fellow ophthalmology Johns Hopkins U., Balt., 1976-77, resident, 1977-80, fellow neuropathology, 1980-81, asst. prof., 1981—. Contbr. articles to profl. jour. Sloan Found. fellow, 1981—; Hart Found. fellow, 1981—. Mem. Soc. Neruosci., Am. Soc., Neurochemistry. Subspecialties: Ophthalmology; Neurobiology. Current work: Neurobiology, axonal transport, nerve regeneration, neurofilaments, neuro-ophthalmology. Home: 630 W Timonium Rd Timonium MD 21093 Office: Wilmer Ophthal Inst Johns Hopkins U 601 N Wolfe St Baltimore MD 21205

HOFFMAN, WILLIAM CHARLES, applied mathematics educator, researcher; b. Portland, Oreg., Aug. 11, 1919; s. William C. and Myra (Mayo) H.; m. Ruth Ann Ketler, June 10, 1950 (div. Jan. 1975); children: Nancy E. Hoffman McKellar, Robert T.; m. Dorothea M. Hanlon, May 15, 1975; 1 son, Brian F. B.A., U. Calif.-Berkeley, 1943; M.A., UCLA, 1947, Ph.D., 1953. Head analysis staff U.S. Navy Electronics Lab., San Diego, 1947-48, 52-55; mathematician RAND Corp., Santa Monica, Calif., 1955-60; mem. Math. Research Lab., Boeing Sci. Research Labs., Seattle, 1961-66; prof. math. Oreg. State U., 1966-69; prof. math. sci. Oakland U., Rochester, Mich., 1969—; research fellow U. Melbourne, Australia, 1975; research assoc. Centre Nationale de Recherche Scientifique, Institut de Neurophysiologie, Marseilles, France, 1976. Author: Statistical Methods in Radio Wave Propagation, 1960; contbr. numerous articles on geometric neuropsychology and Lie transformation group theory of neuropsychology to profl. jours. Mem. Am. Math. Soc., Soc. Indsl. and Applied Math., London Math. Soc., Soc. Neurosci. Republican. Subspecialties: Applied mathematics; Neuropsychology. Current work: Mathematical (geometric) neuropsychology; dynamical systems and differential equations. Home: 585 McGill Dr Rochester MI 48063 Office: Dept Math Scis Oakland U Rochester MI 48063

HOFFMAN, WILLIAM FLOYD, animal scientist, educator; b. Alvo, Nebr., Jan. 16, 1968; s. William George and Julia Ann (Paul) H.; m. Mailyn Cecile Elliott, June 10, 1968; 1 dau., Linda Ann. B.S., Iowa State U., 1960; M.S., U. Mo., 1962, Ph.D. in Dairy Sci, 1965. Farmer, Akron, Iowa, 1946-52; processed foods insp. U.S. Dept. Agr., San Jose, Calif., 1960-61; research asst. U. Mo., Columbia, 1961-64; chmn. dept. agrl. sci. U. Wis., Platteville, 1964—. Served with USAF, 1952-56. Gund scholar, 1959. Mem. Am. Soc. Animal Sci., Am. Dairy Sci. Assn., Am. Inst. Biol. Scis. Methodist. Lodges: Masons; Lions. Subspecialties: Animal breeding and embryo transplants; Animal physiology. Current work: Applied in reproductive physiology; nutrition in calves and stress on calves. Home: 1100 Hollman St Platteville WI 53818 Office: 221 Ullrich Hall U Wis Platteville WI 53818

HOFFMANN, JAMES ALLEN, research plant pathologist; b. Breese, Ill., Dec. 18, 1928; s. Harrison Walter and Byrdie Anne (Cross) H.; m. Sharon Lynne Smock, Nov. 14, 1974; children: Allen, Jennifer, Hillary. Student, Washington U., 1948-51; B.S. in Forest Mgmt, Utah State U., 1954; Ph.D. in Plant Pathology, Wash. State U., 1961. Research asst. Wash. State U., Pullman, 1954-58; research plant pathologist Cereal Disease Research Lab., Agrl.-Reseach Service, U.S. Dept. Agr., Wash. State U., Pullman, 1958-72; research plant pathologist, research leader Crops Research Lab., Agrl. Research Service, U.S. Dept. Agr., Utah State U., Logan, 1972—. Contbr. articles to profl. jours. Served with U.S. Army, 1946-48. Mem. Am. Phytopath. Soc., AAAS, Sigma Xi. Subspecialties: Plant pathology; Integrated pest management. Current work: Research on the biology and control of the smut fungi, particularly the bunt diseases of wheat; diseases of cereal crops. Home: 586 N Maple Dr Smithfield UT 84335 Office: Crops Research Lab Utah State U Logan UT 84322

HOFFMANN, LOUIS GERHARD, microbiology educator, sex therapist; b. Bloemendaal, Netherlands, July 12, 1932; came to U.S., 1950, naturalized, 1956; s. Gerhard Hendrik and Louise Gertrude (Tobi) H.; m. Georgianne Grace Stracke, Nov. 4, 1955; children: Juliann Tobi, Eugenie Claire. B.A., Wesleyan U., 1953; Sc.M. in Hygiene, Johns Hopkins U., 1958, Sc.D., 1960. Postdoctoral fellow U. Calif.-Berkeley, 1960-62; instr. microbiology Johns Hopkins U., 1962-63, asst. prof., 1963-64; asst. prof. microbiology U. Iowa, 1964-67, assoc. prof., 1967-73, prof., 1973—; pvt. practice sex therapy. NIH grantee, 1964-67, 80-83; NSF grantee, 1968-72; Iowa Heart Assn. grantee, 1969-72, 77-79. Mem. Am. Assn. Immunologists, AAAS, N.Y. Acad. Scis., Soc. Exptl. Biology and Medicine, Sigma Xi. Subspecialties: Immunocytochemistry; Immunology (medicine). Current work: Research on relationship between antibody structure and biological function. Office: Dept Microbiology U Iowa Iowa City IA 52242

HOFFMANN, ROALD, chemist, educator; b. Zloczow, Poland, July 18, 1937; came to U.S., 1949, naturalized, 1955; s. Hillel and Clara (Rosen) Safron; m. Eva Börjesson, Apr. 30, 1960; children: Hillel Jan, Ingrid Helena. B.A., Columbia U., 1958; M.A., Harvard U., 1960, Ph.D., 1962; D.Tech. (hon.), Royal Inst. Tech., Stockholm, 1977, D.Sc., Yale U., 1980, Columbia U., 1982, Hartford U., 1982, CUNY, 1983, U. P.R., 1983. Jr. fellow Soc. Fellows Harvard, 1962-65; asso. prof. Cornell U., Ithaca, N.Y., 1965-68, prof., 1968-74, John A. Newman prof. phys. sci., 1974—. Author: (with R.B. Woodward) Conservation of Orbital Symmetry, 1970. Recipient award in pure chemistry Am. Chem. Soc., 1969, Arthur C. Cope award, 1973; Fresenius award Phi Lambda Upsilon, 1969; Harrison Howe award Rochester sect. Am. Chem. Soc., 1970; ann. award Internat. Acad. Quantum Molecular Scis., 1970; Pauling award, 1974; Nobel prize in chemistry, 1981; inorganic chemistry award; Am. Chem. Soc., 1982. Mem. Nat. Acad. Scis., Am. Acad. Arts and Scis., Internat. Acad. Quantum Molecular Scis. Subspecialty: Theoretical chemistry. Current work: Electronic structure of organic; inorganic and organometallic molecules; and of extended solid state structures. Home: 4 Sugarbush Ln Ithaca NY 14850

HOFMANN, JULIUS JOSEPH, psychologist; b. Providence, Sept. 7, 1925; s. Julius Joseph and Lillian Margaret (Whitaker) H.; m. Mary Holst, July 28, 1974; 1 dau., Carolyn Marie. A.B., Woodstock Coll., 1950, S.T.B., 1958; M.Ed., Iona Coll., 1969; profl. diploma, St. John's U., 1972. Lic. marriage counselor, N.J., Calif. Instr. St. Peter's Prep Sch., Jersey City, 1950-53; chaplain Kings County Hosp. Ctr., Bklyn., 1958-73; counselor, employee adv. service N.J. Dept. Civil Service, Newark, 1974-78; adminstr., employee assistance program Jersey Central Power & Light Co., Morristown, 1978—; EAP cons. Edison Electric Inst., 1980—. Contbr. articles to profl. jours. Charter mem., subcom. chmn. N.J. Occupational Adv. Com., Trenton, 1978—; charter mem., past chmn. Morris County C. of C. EAP Com., 1979—. Mem. Assn. Labor-Mgmt. Adminstrs. and Cons. on Alcoholism, Am. Assn. Marriage and Family Therapists, Am. Psychol. Assn., Am Personnel and Guidance Assn., Phi Delta Kappa. Democrat. Roman Catholic. Subspecialty: Social psychology. Current work: Codify principles and practice of new profession of industrial employee assistance counseling. Home: 26 Van Winkle St Bloomfield NJ 07003 Office: Jersey Central Power and Light Co Madison Ave and Punch Bowl Rd Morristown NJ 07960

HOFMANN, KLAUS HEINRICH, experimental medicine educator; b. Karlsruhe, Ger., Feb. 21, 1911; m., 1936 and 1964; 1 child. Ph.D. in Organic Chemistry, Swiss Fed. Inst. Tech., 1936. Fellow Rockefeller Found., 1938-40; asst. in biochemistry Med. Coll., Cornell U., 1940-42; guest scientist Ciba Pharm. Products, Inc., N.J., 1942-44; from asst. research prof. to assoc. research prof. organic biochemistry Sch. Medicine, U. Pitts., 1944-47, research prof. chemistry, 1947-52, chmn. dept. biochemistry, 1953-64, prof. biochemistry, 1952—, Salk Commonwealth prof. exptl. medicine and dir. protein research lab., 1964—, Univ. prof. exptl. medicine and biochemistry, 1972—. Mem. Nat. Acad. Sci., AAAS, Am. Chem. Soc., Am. Soc. Biol. Chemists, Endocrine Soc. Subspecialty: Biochemistry (medicine). Office: U Pitts Sch Med Pittsburgh PA 15261

HOFMANN, LORENZ MARTIN, clinical pharmacologist, pharmacist; b. Chgo., Jan. 10, 1937; s. John C. and Margaret E. (Moews) H.; m. Victoria T. Knefel, May 3, 1968; children: Lorenz Martin II, Paul. B.S. in Pharmacy, U. Ill., Chgo., 1959, M.S. in Pharmacology, 1961, Ph.D., 1964. Registered pharmacist, Ill., Ohio. Research scientist Searle Labs., 1964-76, mgr. regulatory affairs, 1976-77; asso. dir. clin. pharmaology research Ross Labs., 1977-81; asso. dir. med. info. Adria Labs., Columbus, Ohio, 1982—; adj. asst. prof. depts. pharmacology Sch. Medicine and Sch. Pharmacy, Ohio State U., Columbus; lectr. in field. Contbr. articles to sci. jours. Mem. Am. Soc. Clin. Pharmacology and Therapeutics, Am. Soc. Pharmacology and Exptl. Therapeutics. Subspecialties: Pharmacology; Pharmacokinetics. Current work: Research on safety and effectiveness of antihypertensives, diuretics, analgesics and dermatologics. Home: 6987 Wethersfield Worthington OH 43085 Office: PO Box 16529 Columbus OH 43216

HOFMANN, PETER LUDWIG, nuclear engineer, systems analyst; b. Vienna, Jan. 25, 1925; U.S., 1939; s. Arthur Oliver and Luise (Kamhuber) H.; m. Garda S. Steiner, May 27, 1950; children: Mark Eric, Monica Louise. B.E.E., Cooper Union, 1950; M.S., Union Coll., 1954; D. Eng. Sci., R.P.I., 1960. Mgr. nuclear design Gen. Electric Co., Schenectady, 1950-61, mgr. engring. physics, Richland, Wash., 1961-65; mgr. nuclear analysis Battelle Meml. Inst., Richland, 1965-70; mgr. systems analysis Westinghouse Co., Richland, 1970-74; assoc. dir. planning and analysis Battelle Meml. Inst., Columbus, Ohio, 1974-79, mgr. tech., engring. econs., 1979—. Contbr.: The Physics of Intermediate Spectrum Reactors, 1955, Advances in Nuclear Science and Technology, 1969, Solar Energy Technology Handbook, 1980; editor: The Technology of High-Level Nuclear Waste Management, 1981; Contbr. numerous articles to profl. jours. Served with U.S. Army, 1943-46. Mem. Am. Nuclear Soc., Am. Phys. Soc., Sigma Xi, Tau Beta Pi. Subspecialties: Nuclear fission; Nuclear engineering. Current work: Nuclear fuel cycle, reactor physics, disposal of radioactive waste, cost and systems analysis. Home: 5080 Dublin Rd Columbus OH 43220 Office: Battelle Meml Inst 505 King Ave Columbus OH 43201

HOFSTADTER, DOUGLAS RICHARD, computer scientist, educator; b. N.Y.C., Feb. 15, 1945; s. Robert and Nancy (Givan) H. B.S. in Math. with distinction, Stanford U., 1965; M.S., U. Oreg., 1972, Ph.D. in Physics, 1975. Asst. prof. computer sci. Ind. U., Bloomington, 1977-80, asso. prof., 1980—. Author: Gödel, Escher, Bach: an Eternal Golden Braid, 1979; editor: (with Daniel C. Dennett) The Mind's I, 1981; columnist: Metamagical Themas in Sci. Am., 1981-83. Recipient Pultizer prize for gen. nonfiction, 1980; Am. Book award, 1980; Guggenheim fellow, 1980-81. Mem. Assn. for Computing Machinery, Assn. Computational Linguistics. Subspecialty: Artificial intelligence. Office: Computer Sci Dept Ind Univ Bloomington IN 47405

HOFSTETTER, KENNETH JOHN, radiochemist; b. Moline, Ill., Aug. 18, 1940; s. Glenn Arthur and Maxine Lorraine (Glancey) H.; m. Marilyn Jean Palmer, June 5, 1965; children: Eric Glenn, Kelli Rae. A.B., Augustana Coll., Rock Island, Ill., 1962; Ph.D., Purdue U., 1967. Postdoctoral Tex. A&M U., 1967-69; asst. prof. chemistry U. Ky., Lexington, 1969-74; chemistry supr. Allied Gen. Nuclear Services, Barnwell, S.C., 1974-80; radiochemical engring. supr. GPU Nuclear Corp., Middletown, Pa., 1980—. Am. Chem. Soc. grantee, 1970; NSF grantee, 1973. Mem. Am. Chem. Soc., Am. Phys. Soc. Democrat. Lutheran. Subspecialties: Nuclear fission; Analytical chemistry. Current work: Research and development on radioactive decontamination; development of radiochemical analysis methods. Office: GPU Nuclear Corp PO Box 480 Middletown PA 17057

HOGAN, CLARENCE LESTER, electrical engineer; b. Great Falls, Mont., Feb. 8, 1920; s. Clarence Lester and Bessie (Young) H.; m. Audrey Biery Peters, Oct. 13, 1946; 1 dau., Cheryl Lea. B.S. in Chem. Engring., Mont. State Coll., 1942; M.S. in Physics, Lehigh U., 1947, Ph.D., 1950; A.M. (hon), Harvard U., 1954; D.Eng. (hon.), Mont. State U., 1967, Lehigh U., 1971, D.Sc., Worcester Poly. Inst., 1969. Research chem. engr. Anaconda Copper Mining Co., 1942-43; instr. physics Lehigh U., 1946-50; tech. staff Bell Telephone Labs., 1950-52, sub. dept. head, 1952-53; asso. prof. Harvard U., 1953-57, Gordon McKay prof. applied physics, 1957-58; gen. mgr. semi-conductor products div. Motorola, Inc., Phoenix, 1958-60, v.p., 1960-66, exec. v.p., 1966-68; pres., chief exec. officer Fairchild Camera & Instrument Corp., Mountain View, Calif., 1968-74, vice chmn. bd., 1974-79, tech. adviser to pres., 1979—; dir. First Interstate Bank, Rolm Corp., Tab Products Co., Timeplex, Inc., Varian Assocs., Osborne Computer Corp., Semicondr. Specialists.; Gen. chmn. Internat. Conf. on Magnetism and Magnetic Materials, 1959, 60; materials adv. bd. Dept. Def., 1955-57; adv. council dept. elec. engring. Princeton U., 1962-68; adv. council NASA-Electronic Research Council, 1967-70; adv. bd. Coll. Engring., U. Calif.-Berkeley, 1974—; mem. nat. adv. bd. Desert Research Inst., 1976-80; vis. com. Lehigh U., 1966, trustee, 1971-80; vis. com. dept. elec. engring. and computer sci. MIT, 1975—; adv. council div. engring. Stanford U., 1976—; mem. sci. and ednl. adv. com. Lawrence Berkeley Lab., 1978-82; mem. Pres.'s Export Council, 1976-80; mem. adv. panel to tech. adv. bd. U.S. Congress, 1976-80; Trustee Western Electronic Edn. Fund; governing bd. Maricopa County Jr. Coll., 1966-68; bd. regents U. Santa Clara. Served from ensign to lt. (j.g.) USNR, 1943-46. Recipient Community Service award NCCJ, 1978, medal of achievement Am. Electronics Assn., 1978; named Bay Area Outstanding Businessman of Year San Jose State Coll., 1977, One of 10 Greatest Innovators in Past 50 Yrs. Electronics Mag., 1980; Berkeley citation U. Calif., Berkeley, 1981. Fellow IEEE (exec. v.p. 1978, Frederik Philips medal 1976), AAAS, Instn. Elec. Engrs. (London, hon.); mem. Am. Phys. Soc., Nat. Acad. Engring., Sigma Xi, Tau Beta Pi, Phi Kappa Phi, Kappa Sigma. Subspecialty: Electronics. Home: 36 Barry Ln Atherton CA 94025 Office: 464 Ellis St Mountain View CA 94042

HOGAN, JOHN DANIEL, psychologist, educator; b. Tarrytown, N.Y., Feb. 24, 1939; s. Francis J. and Victoria (Marcinek) H. B.S., St. John's U., 1960; M.S., Iowa State U., 1962; Ph.D., Ohio State U., 1970. Asst. prof. psychology St. John's U., Jamaica, N.Y., 1970—; counselor N.Y. Foundling Hosp., N.Y.C., 1974-76. Contbr. articles to profl. jours. Mem. Am. Psychol. Assn., Soc. for Research in Child Devel., Internat. Council of Psychologists. Subspecialty: Developmental psychology. Current work: Research in death anxiety in physicians, treatment of attention deficit disorder with hyperactivity. Home: 222 Martling Ave Tarrytown NY 10591 Office: Dept Psychology Saint Johns U Jamaica NY 11439

HOGAN, YVONNE HOLLAND, immunoparasitologist, biology educator; b. Bay City, Tex., Oct. 24, 1937; d. Andrew and Daisy (Revis) (McCoy)) Holl; m. John E. Perry II, May 3, 1962 (div. Sept. 1967); 1 son, John E. III; m. Booker T. Hogan, Jr., July 5, 1968 (div. 1974). B.S., Howard U., 1958, M.S., 1960, Ph.D., 1981. Grad. asst. Howard U., Washington, 1958-60; research biologist NIH, Bethesda, Md., 1961-62; bacteriologist Howard U. Med. Sch., Washington, 1962; biology tchr. Houston Ind. Sch. Dist., 1964-65; biology instr. Tex. So. U., Houston, 1965-69, asst. prof. biology, 1970-83, assoc. prof., 1983—, coordinator, 1981-83. MARC predoctoral fellow Nat. Inst. Gen. Med. Scis., NIH, 1977-81; Smith-Noirs scholar Howard U., 1959-60. Mem. Am. Inst. Biol. Sci., Tex. Acad. Scis., Sigma Xi, Beta Beta Beta, Beta Kappa Chi. Democrat. Roman Catholic. Subspecialties: Parasitology; Immunobiology and immunology. Current work: Effect of trace metal contaminants on immune responses in rats infected with trypanosomes. Home: 5407 Burkett St Houston TX 77004 Office: Tex So U Dept Biology 3201 Wheeler Ave Houston TX 77004

HOGE, ARTHUR FRANKLIN, JR., oncologist, educator; b. Ft. Smith, Ark., Jan. 25, 1923; s. Arthur F. and Lilly Arabella (Boyd) H.; m. Barbara Standingbear, July 27, 1977; children by previous marriage: Rickert F., Libe, Arthur, Maurie, Elizabeth. B.S., Tulane U., 1945, M.D., 1949. Research fellow U. Okla. and Okla. Med. Research Found., 1971-73; intern, resident U. Md. and Bon Secours Host., Balt., 1949-51, 54-56; pvt. practice med. oncology, Ft. Smith, 1956-71; clin. and pathol. research U. Ark., Ft. Smith, 1961-71; chmn. dept. Ob-Gyn St. Edwards Hosp., Ft. Smith, 1970-71; prin. investigator, dir. Ark.-Okla. Cancer Program, 1969-71, 1974-78; mem. cancer control intervention programs review com. NCI, 1977-80; prin. investigator Cancer Coop. Group N.W. Ark., 1979-81, Washington Regional Med. Center, 1980-81; staff St. Edwards Hosp., Ft. Smith, 1956-71, Sparks Regional Med. Center, 1956-71; cons. staff Crawford County Hosp., Van Buren, Ark., 1956-71; staff VA Hosp., Oklahoma City, 1971-78, Univ. Hosp., 1971-78; cons., lectr. Bapt. Med. Center, Oklahoma City, 1972-77; attending physician, assoc. mem. Okla. Med. Research Found., 1973-77; staff Children's Meml. Hosp., 1973-77, Presbyn. Hosp., Oklahoma City, 1974-77, Wash. Regional Med. Center., Fayetteville, Ark., 1977-81, Springdale Meml. Hosp., 1977-81, Rogers (Ark) Meml. Hosp., 1977-81, City of Faith Med. and Research Center, Tulsa, 1981—; assoc. prof. medicine Oral Roberts U., Tulsa, 1981—. Contbr. articles to profl. jours. Pres. Community Concers of Ark., 1965-69; pres. Noon Civics Club, 1965-66; pres. Ark. div. Am. Cancer Soc., 1969-70; mem. Arts Council Ft. Smith, 1965-71, others. Served to capt. U.S. Army, 1952-54. Fellow S.W. Surg. Congress, Am. Coll. Ob-Gyn.; mem. Am. Assn. Cancer Research, S.W. Oncology Group, Am. Soc. Clin Oncology, AMA, Okla. County Med. Soc., Okla. State Med. Soc., Stewart Wolf Soc., Okla. Gynecol. Soc., Gynecol. Soc. for Study of Breast Disease, others, Sigma Xi. Roman Catholic. Current work: Clinical oncology and cancer immunology/endocrinology research. Office: 8181 S Lewis St Tulsa OK 74136

HOGENKAMP, HENRICUS PETRUS CORNELIS, biochemist, educator; b. Doesburg, Netherlands, Dec. 20, 1925; came to U.S., 1963, naturalized, 1974; s. Johannes Hermanus and Maria Margaretha Johanna (Abeln) H.; m. Lieke Ter Haar, Apr. 25, 1953; children—Harry Peter, Derk John, Margaret Angelina. B.S.A., U. B.C., 1957, M.S., 1958; Ph.D., U. Calif. at Berkeley, 1961. Research biochemist U. Calif. at Berkeley, 1961-62; asso. scientist Fisheries Research Bd. of Can., Vancouver, B.C., 1962-63; asst. prof. dept. biochemistry U. Iowa, Iowa City, 1963-67, assoc. prof., 1967-71, prof., 1971-76; prof., chmn. dept. biochemistry U. Minn., Mpls., 1976—; vis. prof. Australian Nat. U., Canberra, Australia, 1966-67; guest scientist U. Calif., Los Alamos Sci. Lab., Los Alamos, N.Mex., 1974-75. Served with Royal Netherlands Army, 1946-50. U. Iowa Research Council awardee, 1966-67; Coll. Medicine Compensation Plan fellow, 1966-67; Guggenheim fellow, 1974-75. Mem. Am. Chem. Soc., Am. Soc. Biol. Chemists. Home: 2211 Marion Rd Roseville MN 55113 Office: U Minn Med Sch Dept Biochemistry Minneapolis MN 55455

HOGG, DAVID CLARENCE, physicist; b. Vanguard, Sask., Can., Sept. 5, 1921; came to U.S., 1953, naturalized, 1964; s. Francis Sandison and Frances Katherine (Gadsby) H.; m. Jean E. MacMillan, Feb. 15, 1947; children—David Randal, Rebecca Jean. B.Sc., U. Western Ont. (Can.), London, 1949; M.Sc., McGill U., Montreal, Que., Can., 1951, Ph.D., 1953. With Bell Telephone Labs., 1953-77, head atmospheric physics research, 1966-72, head antenna and propagation research, Holmdel, N.J., 1972-77; chief environ. radiometry wave propagation lab. Environ. Research Lab., NOAA, Boulder, Colo., 1977-83, chief radio metrology, 1983—. Research, numerous publs. on microwaves, optics, satellite communications and remote sensing; patentee microwave antennas. Served with Can. Army, 1940-45. Fellow IEEE; mem. AAAS, Nat. Acad. Engring., Union Radio Scientifique Internationale. Episcopalian. Subspecialty: Radiophysics. Office: NOAA WAve Propagation Lab Environ Research Lab Boulder CO 80302

HOGG, DAVID EDWARD, research scientist; b. Newmarket, Ont., Can., Jan. 18, 1936; s. Frank S. and Helen B. (Sawyer) H.; m. Carol J. McCabe, May 30, 1959; children: Brian E., Douglas S. B.A., Queen's U., Kingston, Ont., 1957, M.Sc., 1959; Ph.D., U. Toronto, Ont., Can., 1962. Asst., then assoc. scientist Nat. Radio Astronomy Obs., Charlottesville, Va., 1961-69, scientist, 1969—, asst. dir., 1970-74, assoc. dir., 1974-79; mem. astronomy adv. com. NSF, 1980-83, chmn.,

1982-83. Mem. Am. Astron. Soc. (councilor 1980-82), Internat. Astron. Union, Internat. Union Radio Sci., Can. Astron. Soc. Subspecialty: Radio and microwave astronomy. Current work: Stellar evolution, mass loss and winds; supernova remnants; extragalactic radio sources. Office: Nat Radio Astronomy Obs Edgemont Rd Charlottesville VA 22901

HOGG, ROBERT VINCENT, JR., educator, mathematical statistician; b. Hannibal, Mo., Nov. 8, 1924; s. Robert Vincent and Isabelle Frances (Storrs) H.; m. Carolyn Joan Ladd, June 23, 1956; children—Mary Carolyn, Barbara Jean, Allen Ladd, Robert Mason. B.A., U. Ill., 1947; M.S., U. Iowa, 1948, Ph.D., 1950. Asst. prof. math. U. Iowa, Iowa City, 1950-56, asso. prof., 1956-62, prof. math., 1962-65, chmn. dept. statistics, prof. statistics, 1965—. Co-author: Introduction to Mathematical Statistics, 1959, 4th edit., 1978, Finite Mathematics and Calculus, 1974, Probability and Statistical Inference, 1977; Asso. editor: Am. Statistics, 1971-74, Jour. Am. Statis. Assn, 1978-80; Contbr. articles to profl. jours. Served with USNR, 1943-46. NIH research grantee, 1966-68, 75-78; NSF research grantee, 1969-74. Fellow Inst. Math. Statistics (program sec., exec. bd. 1968-74), Am. Statis. Assn. (pres. Iowa sect. 1962-63, council 1965-66, 73-74, vis. lectr. 1965-68, 77—, chmn. tng. sect. 1973); mem. Math. Assn. Am. (pres. Iowa sect. 1964-65, gov. Iowa dist. 1971-74, vis. lectr. 1976-81), Internat. Statis. Inst., Sigma Xi (pres. Iowa chpt. 1970-71), Pi Kappa Alpha. Episcopalian (vestryman 1958-60, 66-68). Club: Rotarian. Home: Rural Route 6 Box 219A Iowa City IA 52240

HOGGAN, M. DAVID, microbiologist, former army officer, researcher; b. Salt Lake City, Feb. 28, 1930; s. Malcolm David and Beatrice (Riley) H.; m. Joyce Oliphant, May 20, 1950; children: David Arden, Melissa Ann. B.S., U. Utah, 1952, M.S., 1953; Sc.D., Johns Hopkins U., 1959. Commd. officer U.S. Army, 1953, advanced through grades to capt., 1963; asst. chief bacteriology sect. 6th Army Med. Lab., Calif., 1953-56; virologist histobacteriology lab. Armed Forces in Pathology, Washington, 1959-63, ret., 1963; virologist lab. infectious diseases NIH, Bethesda, Md., 1963-66, virologist lab. virus sect., 1966—, staff scientist; vis. prof. Brigham Young U. Served with USPHS, 1963—. Mem. Soc. Am. Microbiologists, Electronmicroscopic Soc. Am., Am. Assn. Immunologists, Am. Soc. Virologists. Mormon. Subspecialties: Genetics and genetic engineering (biology); Virology (biology). Current work: Molecular biology and genetics of endogenous retroviral sequences and their associated cellular oncogenes. Office: LVD NIAID NIH Bldg 7 Bethesda MD 20205

HOGNESS, DAVID SWENSON, biochemistry educator; b. Oakland, Calif., Nov. 17, 1925; s. Thorfin Rusten and Phoebe (Swenson) H.; m. Judith Gore, Sept. 18, 1948; children: Peter Swenson, Christopher Gore. B.S., Calif. Inst. Tech., 1949, Ph.D., 1952. Prof. dept. biochemistry Stanford (Calif.) U. Sch. Medicine, 1966—. Served in USNR, 1944-46. Mem. Nat. Acad. Scis., Am. Acad. Arts and Scis. Subspecialties: Developmental biology; Genetics and genetic engineering (biology). Office: Stanford U Sch Medicine Stanford CA 94305

HOGNESS, JOHN RUSTEN, assn. exec.; b. Oakland, Calif., June 27, 1922; s. Thorfin R. and Phoebe (Swenson) H. Student, Haverford Coll., 1939-42, D.Sc. (hon.), 1973; B.S., U. Chgo., 1943, M.D., 1946; D.Sc. (hon.), Med. Coll. Ohio at Toledo, 1972; LL.D., George Washington U., 1973; D.Litt., Thomas Jefferson U., 1980. Diplomate: Am. Bd. Internal Medicine. Intern medicine Presbyn. Hosp., N.Y.C., 1946-47, asst. resident, 1949-50; chief resident King County Hosp., Seattle, 1950-51; asst. U. Wash. Sch. Medicine, 1950-52, Am. Heart Assn. research fellow, 1951-52, mem. faculty, 1954-71, prof. medicine, 1964-71, med. dir. univ. hosp., 1958-63, dean, chmn. bd. health scis., 1964-69, exec. v.p. univ., 1969-70; dir. Health Scis. Center, 1970-71; pres. Inst. Medicine, Nat. Acad. Scis., 1971-74, mem., 1971—; prof. medicine George Washington U., 1972-74; pres. U. Wash., Seattle, 1974-79, prof. medicine, 1974-79; pres., Assn. Acad. Health Centers, 1979—; mem. commr.'s adv. com. on exempt orgns. IRS, 1969-71; mem. adv. com. for environ. scis. NSF, 1970-71; adv. com. to dir. NIH, 1970-71; mem. Nat. Cancer Adv. Bd., 1972-76, Nat. Sci. Bd., 1976—; trustee China Med. Bd.; mem. selection com. for Rockefeller pub. service awards Princeton U., 1976—; chmn. med. injury compensation study steering com. Inst. Medicine, Nat. Acad. Scis.; mem. council for biol. scis. Pritzker Sch. Medicine, U. Chgo., 1977—; chmn. adv. panel on cost-effectiveness of med. techs. Office Tech. Assessment, U.S. Congress, 1978-80; mem. Council Health Care Tech., HEW; adv. panel for study fin. grad. med. edn. Dept. Health and Human Services, 1980—. Contbr. articles to profl. jours. Served with AUS, 1947-49. Recipient Disting. Service award Med. Alumni Assn. U. Chgo., 1966; Convocation medal Am. Coll. Cardiology, 1973; Cartwright medal Columbia U. Coll. Physicians and Surgeons, 1978. Fellow A.C.P., Am. Acad. Arts and Scis.; mem. Assn. Am. Med. Colls. (exec. council, chmn.-elect council of deans 1968-69); Assn. Am. Physicians Alpha Omega Alpha. Subspecialties: Endocrinology; Internal medicine. Current work: Full-time administration. Office: 11 DuPont Circle Washington DC 20036

HOGNESTAD, EIVIND, civil engineer; b. Time, Norway, July 17, 1921; came to U.S., 1947, naturalized, 1954; s. Hans E. and Dorthea (Norheim) H.; m. Andree S. Hognestad, Apr. 4, 1964; children: Hans E., Marta Marie, Kirsten Andree. M.Sc., Norwegian Tech. U., 1947, D.Sc., 1952; M.Sc., U. Ill., 1949. Research asst. to asso. prof. U. Ill., 1947-53; mgr. structural devel. sect. Portland Cement Assn., Skokie, Ill., 1953-66, dir. engring. devel. dept., 1966-74, dir. tech. and sci. devel., 1974—; cons. offshore devel. petroleum fields various oil cos. and contractors., condr. field and lab. investigations of concrete structures. Contbr. to: Ency. Brit, 1966, also over 100 articles on structural engring. and concrete tech. to tech. jours. Served with Royal Norwegian Navy, 1944-46. Fellow Am. Concrete Inst. (chmn. com. 357 offshore concrete structures, Wason medal 1956, Henry L. Kennedy award 1971, hon. mem. 1976, Alfred E. Lindau award 1977, Delmar L. Bloem award 1980), ASCE (past chmn. adminstrv. com. on masonry and reinforced concrete, Research prize 1956, Chgo. Civil Engr. of Yr. award 1977, Arthur J. Boase award 1981); mem. Nat. Acad. Engring., Prestressed Concrete Inst. o3(past chmn. tech. activities com.), European Concrete Com., Internat. Prestressing Fedn., Structural Engring. Soc. P.R. (hon.), Norwegian Acad. Engring., Royal Norwegian Acad. Sci. Subspecialty: Civil engineering. Home: 2222 Prairie St Glenview IL 60025 Office: 5420 Old Orchard Rd Skokie IL 60077

HOGUE, DOUGLAS EMERSON, animal science educator; b. Holdredge, Nebr., Aug. 8, 1931; m., 1955; 2 children. B.S., U. Calif., 1953; M.S., Cornell U., 1955, Ph.D., 1957. From asst. prof. to assoc. prof. Cornell U., Ithaca, NY., 1957-73, prof. animal husbandry and nutrition, 1973—. Mem. Am. Soc. Animal Sci., Am. Dairy Sci. Assn., Am. Inst. Nutrition. Subspecialty: Animal nutrition. Office: Cornell U Morrison Hall Ithaca NY 14853

HOHAM, RONALD WILLIAM, botany educator, phycologist, researcher; b. Omaha, July 10, 1942; s. Robert Ellsworth and Anna Frances (Rolan) H.; m. Ruth Kennedy Sibole, Mar. 21, 1970; children: Erik Michael, Ross Alan. B.A., Mcpl. U. Omaha, 1964; M.S. in Botany, Mich. State U., 1966; Ph.D., U. Wash., 1971. Asst. prof. biology Colgate U., Hamilton, N.Y., 1971-78, assoc. prof., 1978—83, prof., 1983—; vis. research assoc. No. Ariz. U., Flagstaff, spring 1978; vis. prof. phycology U. Mont. Biol. Sta., Flathead Lake, summers 1980, 82. Assoc. editor: Phycologia, 1978-83; editorial bd.: Jour. Phycology, 1982—; contbr. chpts. to books, articles to sci. jours. Recipient Dimond award Bot. Soc. Am., 1981; NSF grantee, 1981. Mem. Bot. Soc. Am. (sec. phycol. sect. 1981-83), Phycol. Soc. Am. (Bold Award com., Prescott Award com. 1975-83), Internat. Phycol. Soc., Inst. Arctic and Alpine Research, B.C. Provincial Mus., Brit. Lichen Soc., New Eng. Bot. Club, Sigma Xi. Current work: Research on physiological ecology, life histories, nutrition and systematics of cryophilic algae that live in snow. Office: Colgate U 210 Olin Bldg Hamilton NY 13346

HOKAMA, YOSHITSUGI, pathologist, immunologist; b. Niulii, Kohala, Hawaii, Oct. 25, 1926; s. Royei and Kamada (Matsudo) H.; m. Haruko Yoshimoto, Feb. 3, 1951; children: Jon Keith Yoshimoto, Julie Lynn R. Yoshimoto. A.B., UCLA, 1951, M.A. in Microbiology, 1953, Ph.D., 1957. Jr. research, asst. research microbiologist UCLA, 1952-66; assoc. prof. Calif. State U., 1964-66; assoc. prof. pathology U. Hawaii, 1966-68, prof., 1968—; cons. Courtland Labs., Los Angeles, 1964-66; cons. immunologist, assoc. dir. SKCL-Accupath Labs., Honolulu, 1974—. Author: Immunology, Immunopathology, Basic Concepts, 1982; Contbr. articles to profl. jours. Served with AUS, 1945-48. Nat. Cancer Inst. Grantee, 1968-75—; FDA grantee, 1979-82; NOAA grantee, 1979-81. Mem. Am. Soc. Microbiologists, Am. Assn. Immunologists, Am. Soc. Pathologists, Reticuloendothelial Soc., N.Y. Acad. Scis., Am. Soc. Zoologists, Hawaii Soc. Pathologists, Sigma Xi. Episcopalian. Subspecialties: Immunobiology and immunology; Cancer research (medicine). Current work: Acute phase protein (C-Reactive Protein).

HOLADAY, JOHN WALDRON, neuropharmacologist; b. N.Y.C., June 9, 1945; s. Beverley Eli and Alta (Waldron) H.; m. Camilla Canty, June 1, 1968. B.S., U. Ala-Tuscaloosa, 1966, M.S., 1969; Ph.D., U. Calif.-San Francisco, 1977. Research chemist Walter Reed Army Inst. Research, Washington, 1969-74; research pharmacologist, 1976-81, chief neuropharmacology br., 1981—; assoc. prof. Uniformed Services U. Health Scis., Bethesda, Md., 1979—. Contbr. articles to profl. jours. Served to capt. U.S. Army, 1969-72. Recipient U.S. Army Sci. Conf. award Dept. Def., 1980, U.S. Army Research and Devel. Achievement award, 1980. Mem. Am. Soc. Pharmacology and Exptl. Therapeutics (Dean N. Calvert award 1977), Shock Soc., N.Y. Acad. Scis., Internat. Narcotics Research Conf., Sigma Xi. Subspecialties: Neuropharmacology; Neuroendocrinology. Current work: Role of endogenous opiates and other neuro-modulators in the etiology of circulatory shock, spinal trauma, stress reactions, seizure disorders, depression, temperature dysregulation. Patentee in narcotic antagonists in therapy of shock and spinal injury. Home: 4 Vallingby Circle Rockville MD 20850 Office: Neuropharmacology Br Dept Med Neuroscience Walter Reed Army Inst Research Washington DC 20012

HOLBERG, JAY BRIAN, research scientist; b. Spokane, Wash., Nov. 29, 1945; s. Walter Roy and Alice Hilma (Westersund) H.; m. Catharine Jean, Nov. 18, 1979. B.S. in Physics, Wash. State U., 1969, Ph.D., U. Calif., Berkeley, 1974. Lectr. in physics U. Nairobi, Kenya, 1974-77; expt. rep. Jet Propulsion Lab., Pasadena, Calif., 1978-79; research scientist Space Scis. Inst., Tucson (Ariz.) Labs., U. So. Calif., 1979—. Recipient NASA cert. appreciation, 1981, NASA group achievement awards. Mem. Am. Astron. Soc., Am. Geophys. Union, AAAS, Phi Beta Kappa. Subspecialties: Ultraviolet high energy astrophysics; Planetary science. Current work: Analysis planetary and astronomical data returned by Ultraviolet Spectrometer on board Voyager Spacecraft.

HOLCOMB, GORDON ERNEST, plant pathologist; b. Monroe, Wis., July 6, 1932; s. Ernest and Florence (Henneman) H.; m. Alice Harriet Duff, Jan. 25, 1964; children: Janette Lynn, Amy Florence. B.S., U. Wis.-Platteville, 1959; Ph.D., U. Wis.-Madison, 1965. Research asst. U. Wis., 1959-65; asst. prof. plant pathology La. State U., 1965-70, asso. prof., 1970-78, prof., 1978—. Contbr. articles to profl. jours. Served with USAF, 1951-55. Recipient Dave Feathers award Am. Camellia Soc., 1982. Mem. AAAS, Am. Phytopath. Soc., Am. Inst. Biol. Scis., La. Acad. Scis., Sigma Xi. Subspecialties: Plant pathology; Plant cell and tissue culture. Current work: Diseases of ornamental plants and turf, mycology, plant cell culture and biocontrol of weeds with plant pathogens. Home: 779 Rodney Dr Baton Rouge LA 70808 Office: Dept Plant Pathology and Crop Physiology La State U Baton Rouge LA 70803

HOLDEN, FREDERICK THOMPSON, petroleum geologist; b. N.Y.C., June 23, 1915; s. Frederick Barlow and Maude L. (Thompson) H. A.B., Denison U., 1937, Sc.D. (hon.), 1980; postgrad., Kans. U.-Lawrence, 1938; Ph.D., U. Chgo., 1941. Geologist Carter Oil Co., Shreveport, La., 1941-46, dist. geologist, Jackson, Miss., 1946-52, staff geologist, Tulsa, 1952-54; sr. geologist and sr. profl. geologist Carter Oil Co. and Humble Oil and Refining Co., Shreveport and Oklahoma City, 1954-71; exploration geologist, geol. scientist Exxon Co., U.S.A., Midland, Tex., 1971-78, sr. geol. scientist, 1978—. Contbr. articles to profl. jours. Fellow Geol. Soc. Am.; mem. Am. Assn. Petroleum Geologists, Am. Inst. Profl. Geologists (v.p. Okla. sect. 1951-52), Miss. Geol. Soc., Shreveport Geol. Soc., Oklahoma City Geol. Soc., West Tex. Geol. Soc. Republican. Presbyterian. Subspecialties: Geology; Tectonics. Current work: Applicaton of plate tectonics to southern and southwestern United States structural geology. Home: 1601 Midkiff Rd #214 Midland TX 79701 Office: Exxon Co USA PO Box 1600 Midland TX 79702

HOLDEN, JOSEPH THADDEUS, research biochemist, research institute administrator; b. Bklyn., Jan. 18, 1925; s. Joseph and Stella (Golas) Kubikowski; m. Nancy Lyn Collier, Jan. 10, 1948; children: Jon Christohper, Wendy Ellen, Lauri Kassana, Todd Joseph Miles. B.S. in Chemistry, Bklyn. Poly. Inst., 1944; M.S. in Biochemistry, U. Wis.-Madison, 1948, Ph.D., 1951. Postdoctoral fellow Calif. Inst. Tech., 1951-53; research scientist E. I. duPont de Nemours & Co., Wilmington, Del., 1953-54, Beckman Research Inst. of City of Hope, Duarte, Calif., 1954—, assoc. chmn. neurosci., 1957—, asst. dir. research, 1981—. Contbr. numerous articles to profl. jours.; editor, contbg. author: Amino Acid Pools, 1962; editorial bd.: Jour. Bacteriology, 1970-75. Served to cpl. U.S. Army, 1944-46. USPHS postdoctoral fellow, U.S., 1951, France, 1974; USPHS grantee, 1957-81; NSF grantee, 1970-73; Office Naval Research grantee, 1958-70. Mem. Am. Soc. Biol. Chemists (Travel award Japan 1967), Am. Soc. Microbiology (Travel award Israel 1973), Am. Soc. Cell Biology, Am. Soc. Neurochemistry, Internat. Soc. Neurochemistry. Subspecialties: Membrane biology; Biochemistry (biology). Current work: mechanism of amino acid transport; membrane structure-function relations; nutritional, genetic modification of membrane function. Home: 440 La Loma Rd Pasadena CA 91105 Office: Beckman Research Inst. 1450 E Duarte Rd Duarte CA 91010

HOLDEN, PALMER JOSEPH, swine specialist, consultant; b. Breckenridge, Minn., Apr. 10, 1943; s. Palmer and Rozella (Schultz) H.; m. V. Sheryl Holden, Oct. 1, 1943; 1 son, Daniel. B.S., N.D. State U., 1965, M.S., Iowa State U., 1967, Ph.D., 1970. Extension swine specialist Iowa State U., 1972—, prof., 1982—; cons. in field. Contbr. articles to profl. jours. Served to capt. AUS; Vietnam. Decorated Bronze Star. Mem. Am. Soc. Animal Sci., Alpha Zeta, Gamma Sigma Delta, Epsilon Sigma Phi. Subspecialties: Animal nutrition; Animal physiology. Current work: Production-oriented swine nutrition and management. Office: Iowa State 109 Kildee Hall Ames IA 50011

HOLDER, NADINE DUGUID, chemical engineer, researcher; b. Lusk, Wyo., Oct. 11, 1932; d. James Otto and Marian Hope (Elkenberry) Duguid; m. Fred Wayne Holder, 1952 (div. 1960); m. William Vade Fountain, 1962 (div. 1967); children: Holly Lynn Holder Sharp, Kenneth William. B.A. in Anthropology, San Diego State U., 1972, M.P.A., 1982. Registered mfg. engr., Calif. Bus. analyst GA Technologies (formerly Gen. Atomic), San Diego, 1968-72, adminstrv. coordinator, 1972-74, sr. engr., 1974-80, staff engr., 1980-83, tech. coordinator spent fuel treatment, 1981-83; U.S. rep. to W. Ger. for high temperature gas-cooled reactor spent fuel treatment, 1980, 82. Author, co-author tech. publs. for, Dept. Energy, 1974-82. Mem. Am. Inst. Chem. Engrs. (profl. devel. cert. 1982), Am. Nuclear Soc., Nat. Mgmt. Assn., Mensa. Republican. Subspecialties: Chemical engineering; Nuclear fission. Current work: Nuclear fuel processing equipment design and development.

HOLDER, WALTER DALTON, JR., general surgeon, oncology researcher; b. Glendive, Mont., May 3, 1945; s. Walter Dalton and Valeria Marie (Meyer) H.; m. Frances Lou Smith, June 7, 1969; children: Jane Elizabeth, Emily Blair. B.A. in Chemistry and Zoology, U. N.C., Chapel Hill, 1967, M.D., 1971. Lic. physician, N.C., Calif., diplomate: Am. Bd. Surgery. Intern in surgery Barnes Hosp., Washington U., St. Louis, 1971-72; research assoc. Viral Biology Br., Nat. Cancer Inst., Bethesda, Md., 1972-74; NIH postdoctoral fellow in surgery Duke U. Med. Center, Durham, N.C., 1974-76, resident in surgery, 1976-81, Am. Cancer Soc. clin. fellow, 1976-78; asst. prof. surgery Stanford U., 1981—; cons.; researcher surg. oncology. Contbr. articles to profl. jours. Served with USPHS, 1972-74. Recipient Deborah C. Leary Research award, 1971. Mem. Am. Soc. Microbiology, Am. Assn. Cancer Research, Electromicroscopy Soc. Am., Santa Clara County Med. Soc., Sigma Xi, Alpha Epsilon Delta. Subspecialties: Surgery; Oncology. Current work: Oncology, cell study, cancer research, immunology. Home: 863 Highland Circle Los Altos CA 94022 Office: Stanford University Medical Center S067 Stanford CA 94305

HOLDREN, JOHN PAUL, energy and environment educator, researcher, consultant; b. Sewickley, Pa., Mar. 1, 1944; s. Raymond Andrew and Virginia (Fuqua) H.; m. Cheryl Edgar, Feb. 5, 1966; children: John Craig, Jill Virginia. S.B., MIT, 1965, S.M., 1966; Ph.D., Stanford U., 1970; Sc.D. (hon.), U. Puget Sound, 1975. Aerodynamics engr. Lockheed Missiles and Space Co., Palo Alto, Calif., 1966-67; theoretical physicist Lawrence Livermore Lab., Livermore, Calif., 1970-71, cons. fusion energy, 1974—; sr. research fellow Calif. Inst. Tech., Pasadena, 1972-73; asst. prof. U. Calif.-Berkeley, 1973-75, assoc., 1975-78, prof. energy, 1978—, lectr. elec. engring., 1970-72; vis. fellow East-West Ctr., Honolulu, 1979, 80; sr. investigator Rocky Mountain Biol. Lab., Crested Butte, Colo., 1974—. Author: Energy, 1971, Human Ecology, 1973, Ecoscience, 1977; editor: Man and the Ecosphere, 1971. Recipient MacArthur prize MacArthur Found., 1981; Gustavsen lectr. U. Chgo., 1978. Mem. Fedn. Am. Scientists (council 1974-78, treas. 1979-80, pub. service award 1979, vice chmn. 1981—), Pugwash Confs. on Scis. and World Affairs (council 1982—), Am. Phys. Soc. Democrat. Subspecialties: Energy technology assessment; Nuclear fusion. Current work: Comparative environmental assessment of energy technologies, energy's effects on ecosystems, fusion energy technology, nuclear forces and arms control. Office: Energy and Resources Grou U Calif Berkeley 100 T-4 Berkeley CA 94720

HOLE, FRANCIS DOAN, soil science educator; b. Muncie, Ind., Aug. 25, 1913; married, 1941; 2 children. A.B., Earlham Coll., 1933; M.A., Haverford Coll., 1934; Ph.D., U. Wis.-Madison, 1943. Instr. in German, Friends Century Country Day Sch., 1934-35; instr. French, Westtown Friends Boarding Sch., 1935-38; instr. geology Earlham Coll., 1940-41, 43-44; asst. prof. soil sci. U. Wis.-Madison, 1946-51, assoc. pof prof., 1952-57, prof. soil sci. and geography, 1958-83, prof. emertius, 1983—; vis. prof. U. Ariz., 1981. Editor: Soil Survey Horizons, 1960-63; author, co-author: 4 books in field. Recipient Excellence in Teaching award U. Wis. Extension, 1981. Fellow Soil Sci. Soc. Am., AAAS, Soil Conservation Soc., Geol. Soc. Am., Wis. Acad. Sci., Arts and Letters; Mem. Assn. Am. Geographers, Sigma Xi. Subspecialty: Soil science. Address: 2201 Center Ave Madison WI 53704

HOLEMAN, GEORGE ROBERT, health physicist; b. Danville, Ky., Dec. 23, 1937; s. Ernest Ray and Rosa Mae (Tuggle) H.; m. Pamela Reed, Sept. 5, 1959; children: David Reed, Heather Leigh. A.B., Centre Coll. Ky., 1960; A.M., Harvard U., 1961. Diplomate: Am. Bd. Health Physics. AEC health physics fellow Harvard U., Cambridge, Mass., 1960-61, Brookhaven Nat. Lab., Upton, L.I., N.Y., 1961; health physicist Gen. Electric Co., Knolls Atomic Power Lab., Schenectady, 1961-63; health physicist dept. univ. health Yale U., New Haven, 1963-71, lectr. pub. health dept. epidemiology and pub. health, 1963—, co-dir. grad. radiol. health tng. project, 1964-70, dir. health physics div., 1971—, program dir. radioisotope facility, 1974-77; attending health physicist VA Med. Ctr., West Haven, Conn., 1976—; cons. State N.Mex. Environ. Evaluation Group, Santa Fe, 1979; mem. ind. risk assessment team State Conn. Gov's. Office, Hartford, 1980—; mem. sci. com. dosimetry of of neutrons from med. accelerators Nat. Council Radiation Protection and Measurement, 1979—; mem. Task Force on Low-Level Waste, Sci. Com., Radioactive Waste Disposal, 1982—. Contbr. articles in field to profl. jours. Vice-pres. Guilford Keeping Soc. Served with Ky. N.G., 1956-64. Mem. Health Physics Soc. (sec. 1980—, mem. exec. com. 1980—), Am. Pub. Health Assn. (chmn. radiol. health sect. nominating com. 1969-70), Am. Nuclear Soc. (exec. council Conn. chpt. 1975-77), Am. Assn. Physicists in Medicine (chmn. radiation protection com., mem. sci. council 1980—). Subspecialty: Health physics. Current work: Applied health physics, dosimetry and medical radiation protection.

HOLLAENDER, ALEXANDER, biophysicist; b. Samter, Germany, Dec. 19, 1898; came to U.S., 1921, naturalized, 1927; s. Heymann and Doris (Rotholz) H.; m. Henrietta Wahlert, Oct. 10, 1925. A.B., U. Wis., 1929, M.A., 1930, Ph.D., 1931, D.Sc. honoris causa, 1969, U. Vt., 1959, U. Leeds, Eng., 1962, Marquette U., 1967; M.D. honoris cause, U. Chile Med. Sch., 1970; Prof. honoris causa, Fed. U. Rio de Janeiro. Asst. phys. chemistry U. Wis., 1929-31; NRC fellow in biol. scis., 1931-33; investigator Rockefeller Found., 1934; investigator charge radiation work NRC project, Wis., 1934-37; asso. biophysicist Washington Biophysics Inst., NIH, USPHS, 1937-38, biophysicist, 1938-41, sr. biophysicist, 1941-45, prin. biophysicist, 1945-46, head biophysicist, 1946-50; dir. biol. div. Oak Ridge Nat. Lab., 1946-66, sr. research adviser, 1967-73; prof. radiation biology U. Tenn., 1957-66; prof. biomed. scis. U. Tenn.-Oak Ridge Grad. Sch. Biomed. Scis., 1966—; dir. Archives Radiation Biology, U. Tenn., 1966—; Messenger lectr. Cornell U., 1962; cons. Oak Ridge Nat. Lab., Brookhaven Nat. Lab., Nat. Inst. for Environ. Health Scis., Nat. Cancer Inst., Plenum Pub. Co., EPA; organizer fgn. and domestic workshops and tng. courses in environ. mutagenesis and carcinogenesis. Civilian with AEC, OSRD, Office Surgeon Gen., USN; mem. com. radiation biology, mem. com. photobiology, div. biol. and agr., chmn. and mem. subcom. radiobiology, div. phs. scis. NRC. Editor: Radiation Biology (3 vols.), Vol. III, 1956, Radiation Protection and Recovery, 1960,

Chemical Mutagens: Principles and Methods for their Detection, 7 vols, 1971-81, Genetic Engineering for Nitrogen Fixation, 1977, Limitations and Potentials for Biological Nitrogen Fixation in the Tropics, 1978, The Biosaline Concept: An Approach to the Utilization of Underexploited Resources, 1979, (with J.K. Setlow) Genetic Engineering: Principles and Methods, Vol. I, 1979, Vol. II, 1980, Vol. III, 1981, Vol. IV, 1982, (with Rains and Valentine) Genetic Engineering of Osmoregulation, 1980, (with others) Trends in the Biology of Fermentations for Fuels and Chemicals, 1981, Engineering of Microorganisms for Chemicals, 1981, (with R.A. Fleck) Genetic Toxicology: An Agricultural Perspective, 1982, (with others) Biological Basis of New Developments in Biotechnology, 1983, (with Kosuge and Meredith) Genetic Engineering of Plants: An Agricultural Perspective, 1983. Recipient AEC citation for outstanding service to atomic energy program, 1966, Finsen medal 5th Internat. Congress on Photobiology, 1968; E.M.S. award, 1975; decorated Order Merit Republic Italy, 1961. Fellow AAAS, Am. Acad. Arts and Sci., Indian Nat. Sci. Acad. (fgn.), Brazilian Acad. Sci. (fgn.); mem. Solar Energy Soc., Am. Physiol. Soc., Radiation Research Soc. (pres. 1954-55), Am. Soc. Cell Biology, Internat. Assn. Radiation Research (pres. 1962-66), Nat. Assn. de Photobiologie (pres. 1954-60, hon. pres. 1964, exec. com. 1960-66), Genetics Soc. Am. (citation 1979), Am. Soc. Naturalists (v.p. 1952-53), Soc. Gen. Physiologists, U.S. Nat. Acad. Scis. (award for environ. quality 1979), Am. Soc. Microbiology, Am. Physiol. Soc., Environ. Mutagen Soc. (pres. 1969-71), Internat. Environmental Mutagens Soc. (pres. 1973-77), Knoxville Acad. Medicine (hon.). Subspecialties: Biophysics (physics); Genetics and genetic engineering (agriculture). Home: 2540 Massachusetts Ave NW Washington DC 20008

HOLLAND, HEINRICH DIETER, geochemist, educator; b. Mannheim, Germany, May 27, 1927; came to U.S., 1940, naturalized, 1948; s. Otto and Jeanette (Liebrecht) H.; m. Alice Tilghman Pusey, June 20, 1953; children: Henry Lawrence, Anne Liebrecht, John Pusey, Matthew Tilghman. B.A., Princeton, 1946; M.S., Columbia, 1948, Ph.D., 1952; M.A., Harvard, 1972. Mem. faculty Princeton, 1950-72, prof. geology, 1966-72; prof. geochemistry Harvard, 1972—; dir. Center for Earth and Planetary Scis., 1978-80; NSF postdoctoral fellow Oxford (Eng.) U., 1956-57; Fulbright lectr. Durham (Eng.) U., 1963-64; vis. prof. U. Hawaii, 1968-69, 81. Author: (with R.A. Rich, U. Petersen) Hydrothermal Uranium Deposits, 1977, The Chemistry of the Atmosphere and Oceans, 1978, The Chemical Evolution of the Atmosphere and Oceans, 1984; also articles on geochemistry, ore forming fluids, ocean atmosphere evolution. Pres. sch. bd., Rocky Hill, N.J., 1961-63, 67-68; mem. Winchester (Mass.) Sch. Com., 1977-80, vice chmn., 1978-80; mem. Winchester Town Meeting, 1977-80; chmn. No. N.J. chpt. Scientists and Engrs. for Johnson, 1964. Served with AUS, 1946-47. Recipient Humboldt prize, 1980; Guggenheim fellow, 1975-76. Fellow Geol. Soc. Am., Mineral. Soc. Am., Geochem. Soc. (v.p. 1969-70, pres. 1970-71), Am. Geophys. Union; mem. Nat. Acad. Scis., Am. Acad. Arts and Scis. Subspecialties: Geochemistry. Current work: Controls on the composition of the atmosphere and oceans; chemical history of the atmosphere and oceans; nature of hydreothermal solutions. Home: 14 Rangeley Rd Winchester MA 01890 Office: Dept Geol Scis Harvard Univ Cambridge MA 02138

HOLLAND, JOHN JOSEPH, biology educator; b. Pitts., Nov.16, 1929; m., 1960. B.S., Loyola U., New Orleans, 1953; Ph.D. in Microbiology, UCLA, 1957. From instr. to asst. prof. bacteriology U. Minn., 1957-60; from asst. prof. to assoc. prof. microbiology U. Wash., 1960-64; prof. U. Calif.-Irvine, 1964-68; prof. biology U. Calif.-San Diego, 1968—. Recipient Eli Lilly award, 1963. Mem. AAAS, Soc. Exptl. Biology and Medicine, Am. Soc. Microbiology. Subspecialty: Cell biology. Office: Dept Biology U Calif La Jolla CA 92037

HOLLAND, RICHARD DARLAN, biology educator; b. Bayou La Batre, Ala., Feb. 24, 1944; s. Jordan Richard and Rae Elizabeth (Ates) H.; m. Rebecca Ann Guy, Aug. 24, 1968. B.S., Livingston U., 1965, M.S., 1967; Ph.D., U. Tenn., 1974. Instr. natural scis. Livingston (Ala.) U., 1967-69, asst. prof. biology, 1969-70; asst. prof. botany U. Tenn., Knoxville, 1975; assoc. prof. Livingston U., 1976—. Contbr. articles on biology and phycology to profl. jours. Mem. Beautification Bd. Livingston, 1975—; mem. hort. com. Sumter County Extension Council, Livingston, 1981—. NIH grantee, 1972-74; NDEA fellow, 1973-74. Mem. Ala. Acad. Sci., AAAS, Am. Inst. Biol. Scis., Assn. Southeastern Biologists, Bot. Soc. Am., Brit. Phycological Soc., Internat. Assn. Plant Taxonomy, Nat. Assn. Biology Tchrs., Phycological Soc. Am., AAUP. Methodist. Subspecialties: Taxonomy; Ecology. Current work: Teaching, research in algal taxonomy, plant ecology and taxonomy. Home: 16 A Shady Heights Circle PO Box 297 Livingston AL 35470 Office: Livingston U Station 7 Livingston AL 35470

HOLLAND, ROBERT CAMPBELL, anatomist, educator; b. Bushnell, Ill., Aug. 16, 1923; s. Harvey Howard and Lois Sarah (Campbell) H.; m. Nancy J. Hallenbeck, Dec. 10, 1982; children from previous marriage: Jonathan Robert, Heather Ann. B.S., U. Wis., 1945, M.S., 1946, Ph.D. (Wis. Alumni Research Found. fellow), 1955. Instr. Northwestern U. Dental Sch., Chgo., 1949-51; asst. prof. U. N.D., 1955-60; assoc. prof. anatomy U. Ark., Little Rock, 1960-66; mem. staff Rockefeller Found., 1966-77; prof., chmn. dept. anatomy Mahidol Sch. Medicine, Bangkok, Thailand, 1966-76; vis. prof. UCLA, 1977; prof., chmn. dept. anatomy Morehouse Sch. Medicine, Atlanta, 1977—. Contbr. articles to profl. jours. Served with AUS, 1943-46. Nat. Found. Infantile Paralysis fellow, 1957-58; NIH grantee. Mem. Am. Acad. Neurology, Soc. Neurosci., Am. Assn. Anatomists, Soc. Exptl. Biology and Medicine, Sigma Xi. Subspecialties: Anatomy and embryology; Neurophysiology. Current work: Anatomy and physiologyof limbic system and hypothalamus; maternal behavior and reproduction. Home: 1501 Kingfisher Dr Marietta GA 30062 Office: Morehouse Med Sch 720 Westview Dr Atlanta GA 30310

HOLLAND, ROBERT LOUIS, aerospace engineer; b. Athens, Ala., June 6, 1931; s. Aubrey Eli and Ruth Marie (Daniel) H.; m. Nancy Hill Hagood, Aug. 30, 1955; children: Neysa Sue, James Brandon. B.S., Athens Coll., 1957; M.A., U. Ala., 1967. Asst. prof. math. Athens (Ala.) State Coll., 1953-55; indsl. engr. Wolverine Tube, Decatur, Ala., 1956-61; aerospace engr. NASA, Huntsville, Ala., 1961—, mathematician, 1961-63, br. chief, 1965-75, space scientist, 1975—. Contbr. articles to profl. jours.; editor: Athenian newspaper, 1951-52. Chmn. Tenn. Valley Old Time Fiddlers, 1967. Served with U.S. Army, 1951-53. Mem. Jr. C. of C. (Athens chmn. 1964), Soc. Indsl. and Applied Math. Democrat. Subspecialties: Applied mathematics; Astronautics. Current work: Numerical analysis for weather prediction, solar activity prediction and its affect on weather. Developer gravity gradiometer, 1969. Home: 609 Sudie Athens AL 35611 Office: Marshall Space Flight Ctr Huntsville AL 35812

HOLLANDER, MILTON BERNARD, corp. exec.; b. Bayonne, N.J., Nov. 29, 1928; s. Harry and Lena (Hutner) H.; m. Betty Ruth Grodberg, June 6, 1952; children—Eva Lynn, J. Steven, Aaron Phillip, Joel Daniel. B.S., Purdue U., 1951; M.S., Mass. Inst. Tech., 1953; Ph.D., Columbia, 1959. Dir. engring. center Am. Machine & Foundry Co., Springdale, Conn., 1956-67; v.p. tech. Am.-Standard, Inc., N.Y.C., 1967-72; chmn. bd. Gulf & Western Invention Devel. Corp., N.Y.C., 1975—; v.p. tech. Gulf & Western Industries, Inc., N.Y.C., 1972—; dir. PolyGulf Corp., Bklyn., 1973—; cons. electronics lab. Columbia U., 1955-57. Author tech. papers temperature measurement, metal cutting, instrumentation. Com. chmn. local Boy Scouts Am. Served with C.E. AUS, 1946-48; Korea. Research fellow Mass. Inst. Tech., 1952-53; duPont research fellow Columbia U., 1955-57; research fellow Am. Soc. Tool and Mfg. Engrs., 1954-55; recipient Outstanding Alumnus award Purdue U., 1972. Mem. ASME, Am. Welding Soc., Indsl. Research Inst., Sigma Xi. Current work: Research management, automated equipment; energy storage: energy conversion. Patentee in field. Office: 1 Gulf & Western Plaza New York NY 10023

HOLLANDER, PHILIP BEN, pharmacologist, cardiologist, educator; b. Chgo., May 4, 1924; s. Max and Rose (Sadowska) H.; m.; children: Marc Joel, Carole Ruth, Daniel Charles. B.S., UCLA, 1948; M.Sc., U. So. Calif., 1955, Ph.D., 1960. Life ins. fellow Los Angeles County Heart Assn., 1953-59; mem. staff RCA, 1961-64; adj. asst. prof. Woman's Med. Coll., Phila., 1961-64; assoc. prof. depts. pharmacology, biomed. engring. and coll. biol. scis. Ohio State U., Columbus, 1964-69, prof., 1970—, dep. chmn. dept. pharmacology, 1979—; biophysics cons. in membrane activities; dir. pharmacology tng. for cert. R.N. anesthetist, 1969—. Contbr. chpts. to 4 books, numerous articles to profl. jours. Served with U.S. Army, 1941-45. NIH predoctoral fellow, 1959-61; NIH career devel. fellow, 1964-74. Mem. AAAS, Am. Soc. Pharmacology and Exptl. Therapeutics, Biophys. Soc., Biomed. Engring. Soc., Am. Coll. Clin. Pharmacology, N.Y. Acad. Scis. Jewish. Subspecialties: Cellular pharmacology; Biophysics (biology). Current work: Direct intracellular determinations of physiological and pharmacological ions, pH, pO2, pCO2, substrates and enzymes within the living viable intact system. Office: Dept Pharmacology Ohio State U Coll Medicine 333 W 10th Ave Columbus OH 43210

HOLLENBECK, ALBERT RUSSELL, psychologist, educator; b. San Francisco, Apr. 6, 1948; s. Albert Russell and Mary Margaret (Allread) H.; m. Barbara Ellen Belmont, Oct. 10, 1981. B.S., U. Calif.-Davis, 1970; Ph.D., U. Wash.-Seattle, 1976. Predoctoral assoc. U. Wash.-Seattle, 1970-76; predoctoral research asst. Battelle-Human Affairs Research Centers, Seattle, 1974-75; staff fellow NIMH, Bethesda, Md., 1976-79; asst. prof. George Mason U., Fairfax, Va., 1978—; cons. Child's Hosp. Nat. Med. Center, Washington, 1980; prof. Gov.'s Sch. for Gifted, 1981. Contbr. articles in field to profl. jours.; guest editor: Animal Behavior, 1980, Behavioral Assessment, 1980, Devel. Psychology, 1977-80, Jour. Applied Devel. Psychology, 1982; guest reviewer, NSF, Washington, 1982. Calif. state scholar, 1969; Soc. Psychol. Study Social Issues grantee-in-aid, 1979; NSF grantee, 1980. Fellow Am. Orthopsychiat. Assn.; mem. Internat. Soc. Study Behavioral Devel., Am. Psychol. Assn., Soc. Research in Child Devel., AAAS, Calif. Scholarship Fedn. (life). Lodge: Internat. Order Foresters. Subspecialty: Developmental psychology. Current work: Developmental methodology, applied developmental, infancy, socialization. Home: 5432 Cabot Ridge Ct Fairfax VA 22032 Office: George Mason U 4400 University Dr Fairfax VA 22030

HOLLENBERG, PAUL FREDERICK, biochemist, educator; b. Phila., Sept. 18, 1942; s. Frederick Henry and Catherine (Dentzer) H.; m. Emily Elizabeth Vanootighem, May 6, 1967; children: Kathryn Mary, David Paul. B.S. in Chemistry, Wittenberg U., Springfield, Ohio, 1964; M.S., U. Mich., 1966, Ph.D. in Biol. Chemistry, 1969. Teaching asst. in biochemistry U. Mich., Ann Arbor, 1964-69, postdoctoral fellow in biochemistry, 1969, U. Ill., Urbana, 1969-72; asst. prof. biochemistry Northwestern U. Med. Sch., Chgo., 1972-80, assoc. prof. pathology and pharmacology, 1980-81, assoc. prof. pathology and molecular biology, 1981—. Contbr. articles to profl. jours. Schweppe Found. fellow, 1974-77; USPHS grantee, 1974—. Mem. AAAS, Am. Assn. Cancer Research, Am. Chem. Soc., Am. Soc. Biol. Chemists, Am. Soc. Pharmacology and Exptl. Therapeutics, Biophys. Soc., N.Y. Acad. Scis., Sigma Xi, Phi Kappa Phi, Phi Lambda Upsilon. Subspecialties: Biochemistry (medicine); Toxicology (medicine). Current work: Mechanisms of enzyme action, chemical carcinogenesis and xenobiotic toxicity. Home: 3811 Louise St Skokie IL 60076 Office: Dept Pathology Northwestern U Med Sch 303 E Chicago Ave Chicago IL 60611

HOLLEY, CHARLES DEWAYEN, medical educator, researcher, psychologist; b. Borger, Tex., Dec. 26, 1945; s. James Wesley HOlley and Thelma Pearl (Crenshaw) Sparkman; m. b. Penny Lynn Barnes, Apr. 20, 1966; children: Shane Lee, Shawn DeWayne. B.S., Am. Tech. U., 1975; M.Ed., West Tex. State U., 1976; Ph.D., Tex. Christian U., 1979. Cert. psychologist, Tex. Research asst. West Tex. State U., 1975-76; research assoc. Tex. Christian U., 1976-79; sr. analyst Data Design Labs., Arlington, Va., 1979-80; asst. prof. med. edn., dir. evaluation services, acting dir. acad. info. systems Tex. Coll. Osteo. Medicine, 1980—; adj. asst. prof. basic health scis. North Tex. sTate U.; adj. asst. prof. math. Tex. Christian U.; cons. in field. Author: Performance Appraisal in Medical Schools, 1983; editor: Spatial Learning Strategies, 1983. Coach basketball team, Ft. Worth, 1981-83. Served to maj. U.S. Army, 1966-74; Served to maj. USAR, 1974—. Recipient 1st place award for article Soc. Motivational Exptl. Psychology, 1978; Nat. Inst. Edn. grantee, 1979; Dallas Ind. Sch. Dist. grantee, 1981; Am. Osteo. Assn. grantee, 1982. Mem. Am. Psychol. Assn., Southwestern Soc. Multivariate Psychology (v.p. 1979-80), Am. Ednl. Research Assn., Res. Officers Assn., Southwestern Ednl. Research Assn. Subspecialties: Cognition; Statistics. Home: 4832 Lubbock Fort Worth TX 76115 Office: Tex Coll Osteo Medicine Camp Bowie at Montgomery Fort worth TX 76107

HOLLEY, CHARLES H., electronics engineer; b. Apr. 15, 1919. B.E.E., Duke U., 1940. Mem. staff Gen. Electric Co., Schenectady, 1941-74, gen. mgr. elec. utility systems engring. dept., 1974-80, turbine tech. assessment ops., turbine bus. group, 1980—. Fellow IEEE; mem. Nat. Acad. Engring. Subspecialty: Electrical engineering. Office: Gen Electric Co Bldg 2 Rm 600 Schenectady NY 12345

HOLLEY, ROBERT WILLIAM, educator, scientist; b. Urbana, Ill., Jan. 28, 1922; s. Charles E. and Viola (Wolfe) H.; m. Ann Dworkin, Mar. 3, 1945; 1 son, Frederick. A.B., U. Ill., 1942; Ph.D., Cornell U., 1947. Am. Chem. Soc. fellow State Coll. Wash., 1947-48; asst. prof., then asso. prof. organic chemistry N.Y. State Agr. Expt. Sta., Cornell U., 1948-57; research chemist plant, soil and nutrition lab. U.S. Dept. Agr., Cornell U., 1957-64; prof. biochemistry Cornell U., 1964-69, chmn. dept. biochemistry, 1965-66; resident fellow Salk Inst. Biol. Studies, La Jolla, Calif., 1968—; mem. biochemistry study sect. NIH, 1962-66; vis. fellow Salk Inst. Biol. Studies; vis. prof. Scripps Clinic and Research Found., La Jolla, 1966-67. Recipient Distinguished Service award U.S. Dept. Agr., 1965, Albert Lasker award basic med. research, 1965; U.S. Steel Found. award in molecular biology Nat. Acad. Scis., 1967; Nobel prize for medicine and physiology, 1968; Guggenheim fellow Calif. Inst. Tech., 1955-56. Fellow AAAS; mem. Am. Acad. Arts and Scis., Am. Soc. Biol. Chemists, Am. Chem. Soc., Nat. Acad. Scis., Phi Beta Kappa, Sigma Xi. Subspecialty: Cell and tissue culture. Current work: Control of growth of normal and malignant cells. Home: 7381 Rue Michael La Jolla CA 92037 Office: PO Box 85800 San Diego CA 92138

HOLLIDAY, GEORGE HAYES, environ. specialist; b. Toledo, Oct. 26, 1922; s. Carl and Alice S. (Bauer) H.; m. Doris S. Holliday, Nov. 24, 1944; children: William, Kathy, Karen. B.S.M.E., U. Calif., Berkeley, 1948; M.S.C.E., U. So. Calif., 1962, E.M.E., 1965; Ph.D., U. Houston, 1970. Registered profl. engr., Tex., La. Drilling engr. Shell Oil Co., Ventura and Bakersfield, Calif., 1950-53, prodn. engr., Farmington, N.Mex. and Salt Lake City, 1953-57, asst. chief mech. engr., Los Angeles, 1957-60, steam flood devel. engr., 1960-65, sr. staff research engr., Houston, 1965-70, sr. staff environ specialist, 1970—; Disting. lectr. Soc. Petroleum Engrs., 1980-81. Contbr. articles profl. jours. Mem. Tex. Acid Rain Research Com., 1981—. Served in AUS, 1942-46. Fellow ASME. Republican. Presbyterian. Subspecialty: Environmental engineering. Home: 11303 Dement St Tomball TX 77375 Office: Box 2463 Houston TX 77001

HOLLIMAN, RHODES BURNS, biology educator; b. Birmingham, Ala., Feb. 28, 1928; s. Cecil Rhodes and Ruby Eugenia (Burns) H.; m.; children: Cecil James, Daniel Rhodes. B.S., Samford U., 1951; M.S., U. Miami, Fla., 1953; Ph.D., Fla. State U., 1960. Fellow in tropical medicine La. State U. Med. Sch., New Orleans, 1953; asst, prof. Jacksonville (Fla.) State U., 1960-61, Fla. State U., Tallahassee, 1961-62; asst. prof., tropical medicine Va. Poly Inst. and State U., 1962-64, assoc. prof., 1964-71, prof., 1971—; archaeologist Can. Govt. (Louisbourg Fortress), N.S., 1975, U.S., Nat. Park Service (Canyon de Chelly Nat. Monument), 1972, 1976, State of Va., 1965-67; med. practicioner Hosp. Behrhorst, Chimaltenango, Guatemala, 1975-78. Contbr. articles to profl. jours.; author: Marine Parasitology, 1960 (Phi Sigma Research award 1960). Counselor Boy Scouts Am., Blacksburg, 1968. Recipient Wine Award for Teaching Faculty of Va. Tech., 1971, Sporn Award, 1968; Alpha Phi Omega award, 1972. Mem. Am. Soc. Parsitologists, Am. Soc. Tropical Medicine and Hygiene, Soc. Paleopathologists, Archaeol. Soc. Va. Subspecialties: Parasitology; Epidemiology. Current work: Transplacental transmission of trichinella spiralis; trichinella spiralis in wild hosts; the etiology of diarrheal disease in the tropical Americas; diagnosis of onchocerciasis. Home: 1306 Hillcrest Dr SE Blacksburg VA 24060 Office: Dept Biology Va Poly Inst and State U Blacksburg VA 24061

HOLLINGER, MANNFRED ALAN, pharmacologist-toxicologist, educator; b. Chgo., June 28, 1939; s. Walter Henry and Theda Mae (Miner) H.; m. Georgia Lee Hastings, Sept. 2, 1961; children: Randy, Christopher. B.S., North Park Coll., 1961; M.S., Loyola U., Chgo., 1965, Ph.D., 1967. Asst. pharmacologist Baxter Labs., Morton Grove, Ill., 1961-63; instr. Stanford U., 1967-69; asst. prof. pharmacology U. Calif.-Davis, 1969-75, assoc. prof., 1975—. Asst. editor: Jour. Pharmacology and Exptl. Therapeutics, 1977—; pulmonary field editor, 1977—. Mem. Soc. Toxicology, Am. Soc. Pharmacology and Exptl. Therapeutics. Subspecialties: Cellular pharmacology; Toxicology (medicine). Current work: Pulmonary pharmacology-toxicology. Home: 405 Iris Pl Davis CA 95616 Office: Dept Pharmacology U Calif Med Sch Davis CA 95616

HOLLINGSWORTH, CHARLES ALVIN, chemist, educator; b. Earl, Colo., June 22, 1917; s. Lewis Edmund and Mary Gertrude (Kremser) H.; m. Dorothy L. Barton, Dec. 23, 1944; children: Diane, June, Mark. B.A., Western State Coll. Colo., 1941; Ph.D., State U. Iowa, 1946. Instr. State U. Iowa, 1943-46; chemist duPont Co., Waynesboro, Va., 1946-48; instr. U. Pitts., 1948-51, asst. prof. chemistry, 1951-56, assoc. prof., 1956-67, prof., 1967—. Author: Vectors Matrices and Group Theory for Scientist and Engineers, 1967; contbr. numerous articles to profl. jours. Mem. Am. Chem. Soc., Sigma Xi. Subspecialties: Physical chemistry; Theoretical chemistry. Current work: Applications of multilinear analysis, tensor analysis, and perturbation theory to chem. physics. Home: 2901 Broadway St Pittsburgh PA 15234 Office: Dept Chemistry U Pitts Pittsburgh PA 15260

HOLLIS, JAN MICHAEL, astronomer, systems software analyst; b. Martinsburg, W.Va., June 5, 1941; s. Delbert I. and Betty Ruth (Collier) H.; m. Carol Getz, Feb. 1, 1964; children: David Collier, Mary Morgan. A.B. in Math, Duke U., 1963; M.A. in Astronomy, U. Va., 1972, Ph.D., 1976. Asst. prof. physics Va. Mil. Inst., Lexington, 1973; astronomer, software analyst Nat. Radio Astronomy Obs., Tucson, 1973-79; astronomer NASA Goddard Space Flight Center lab. for Astronomy and Solar Physics, 1979—, supervisory astrophysicist, head observatories and data analysis br., 1982—. Contbr. articles to profl. jours. Served to lt. U.S. Navy, 1963-69. Mem. Am. Astron. Soc., Internat. Astron. Union. Subspecialties: Radio and microwave astronomy; Operating systems. Current work: Research in spectral line radio astronomy and how such observations relate to the chemical and physical conditions existing within interstellar molecular clouds, comets, stellar and extragalactic objects. Home: 11318 Johns Hopkins Rd Clarksville MD 21029 Office: NASA Goddard Space Flight Center Code 685 Greenbelt MD 20771

HOLLIS, JOHN PERCY, JR., microbiologist, plant pathologist, pest control co. exec.; b. Jennings, Okla., Nov. 1, 1981; s. John Percy and Eva Francis (Ham) H.; m. Mary Catharine Lee Billings, Sept. 14, 1943; children: Teresa Sue, Jan Claire, Barbara Ann. Megan Elizabeth, John Wesley. Ph.D., U. Nebr., 1949. Asst. plant pathologist Conn. Agr. Experiment Sta., New Haven, 1948-49; asst. prof. U. Mo., Columbia, 1949-53; plant pathologist U. Fruit Co., La Lima, Honduras, 1953-54; prof. plant pathology La. State U., Baton Rouge, 1954—; pres. Gulfpestco., Biotron; dir. research Biotron Tropical Agrl. Experiment Sta., Belize. Contbr. articles to profl. jours. Served to 2d lt. USAAF, 1943-45. Decorated Air medal; Sr. Fulbright research scholar, Kenya, 1961-62; NATO sr. fellow, 1973. Mem. Am. Phytopath. Soc., Helminthological Soc., Sigma Xi. Subspecialty: Plant pathology. Current work: Nematode swarming rice nematode damage, nematicide, herbicide, fungicide research by contract; crawfish research, crawfish farming.

HOLLISTER, ALAN SCUDDER, clinical pharmacologist; b. Balt., Feb. 28, 1947; s. William Gray and Frances Flora (Scudder) H.; m. Susan Blair, Aug. 29, 1970; children: Rebecca L., Nathan A. B.A. in Chemistry, Swarthmore Coll., 1970; Ph.D. in Pharmacology, U. N.C., 1976, M.D., 1977. Diplomate: Am. Bd. Internal Medicine. Intern U. Fla. Hosp., Gainesville, 1977-78, resident in internal medicine., 1978-80; sr. research fellow div. clin. pharmacology depts pharmacology and medicine Vanderbilt U., Nashville, 1980—, also asst. prof., assoc. physician Clin. Research Center. Contbr. articles to profl. jours. Chmn. social concerns com., bd. dirs. 1st Unitarian-Universalist Ch., Nashville; active Physicians for Social Responsibility. Recipient Michiko Kuno Research award U. N.C. Sch. Medicine, 1976; Pharm. Mfrs. Assn. Found. research fellow, 1974; Burroughs Wellcome fellow, 1982. Mem. Soc. for Neurosci., ACP, Am. Soc. for Pharmacology and Exptl. Therapeutics, Am. Fedn. Clin. Research., Am. Heart Assn., AAAS. Subspecialties: Internal medicine; Pharmacology. Current work: Clinical pharmacology; hypertension research; adrenergic receptor pharmacology. Office: Dept Pharmacology Vanderbilt Univ Nashville TN 37232

HOLLISTER, LEO EDWARD, physician; b. Cin., Dec. 3, 1920; s. William B. and Ruth V. (Appling) H.; m. Louise Agnes Palmieri, Feb. 1, 1950; children—Stephen, David, Cynthia, Matthew. B.S., U. Cin., 1941, M.D., 1943. Diplomate: Am. Bd. Internal Medicine. Intern Boston City Hosp., 1944, VA Hosp., San Francisco, 1947-49; chief med. service Palo Alta (Calif.) VA Hosp., 1952-60, asso. chief of staff, 1960-70, med. investigator, 1970-76, dir. psychopharmacology research, 1976-82, sr. med. investigator, 1982—; prof. medicine, psychiatry, and pharmacology Stanford U. Sch. Medicine, 1970—.

Contbr. over 300 articles to sci. jours., chpts. to books; author 3 books in field. Served to comdr. USNR, 1945-46, 50-51. Recipient Meritorious Service award VA, 1960; William S. Middleton award, 1966; Taylor Manor award, 1974. Mem. Am. Soc. Clin. Pharmacology and Therapeutics (pres. 1972), Am. Coll. Neuropsychopharmacology (pres. 1974), Collegium Internationale Neuropsychopharmacologicum (pres. 1978). Presbyterian. Subspecialty: Neuropharmacology. Current work: Drug treatment of mental disorders; biol. aspects of mental disorders. Home: 1800 University Ave Palo Alto CA 94301 Office: 3801 Miranda Ave Palo Alto CA 94304

HOLLOMON, JOHN HERBERT, educator; b. Norfolk, Va., Mar. 12, 1919; s. John Herbert and Pearl (Twiford); m. Margaret Knox Wheeler, Aug. 12, 1941 (dec.); children—Jonathan Bradford, James Martin, Duncan Twiford, Elizabeth Wheeler Vrugtman, Peter Heinz Richter; m. Nancy Elizabeth Gade, Dec. 27, 1970. Grad., Augusta Mil. Acad., Ft. Defiance, Va., 1936; B.S., M.I.T., 1940, Sc.D., 1946; hon. doctorates, Worcester Poly. Inst., 1964, Mich. Tech. U., 1965, Rensselaer Poly. Inst., 1966, Carnegie-Mellon U., 1967, Northwestern U., U. Akron, 1967. Instr. Harvard U. Grad. Sch. Engring., 1941-42; research asso. Gen. Electric Research Lab., Gen. Electric Co., 1946-49, asst. mgr. metallurgy research dept., 1949-52, mgr. metallurgy and ceramics research dept., 1952-60, gen. mgr., 1960-62; asst. sec. for sci. and tech. Dept. Commerce, 1962-67; also acting under sec.; pres. U. Okla., 1967-70; cons. to pres. and to provost M.I.T., 1970-72; dir. Center Policy Alternatives, 1972—, 1st Japan Steel Industry prof., 1975—; adj. prof. Rensselaer Poly. Inst., 1950-62; dir. Bell & Howell Corp.; cons. Pres.'s Sci. Adv. Com.; also chmn. Atmospheric Scis. Com., 1963-67; Mem. Commerce Tech. Adv. Bd., 1962-67. Author: (with Leonard Jaffe) Ferrous Metallurgical Design, 1947; Contbr. articles to profl. jours. Served as maj. AUS, 1942-46; chief phys. metallurgy sect.; Watertown (Mass.) Arsenal. Decorated Legion of Merit; recipient Rossiter W. Raymond award AIME, 1946, Alfred Nobel prize Combined Engring. Socs., 1947, Rosenhain medal Brit. Inst. Metals, 1958. Fellow Am. Phys. Soc., Am. Inst. Chemists, AAAS; mem. Soc. for History of Tech., Am. Acad. Arts and Scis., Am. Soc. Metals (trustee 1957), Royal Swedish Acad. Engring. Sci. (fgn.), Mid-Am. State Univs. (pres. 1969), Acta Metallurgica (sec.-treas.), Nat. Acad. Engring. (a founder), Nat. Planning Assn. (bus. adv. council), Sigma Xi, Kappa Sigma. Clubs: Harvard (Boston); Cosmos (Washington). Subspecialty: Metallurgy. Current work: Science, technology, and public policy. Address: 121 Carlton St Brookline MA 02146

HOLLOSZY, JOHN OTTO, preventive medicine educator, scientist, physician; b. Vienna, Austria, Jan. 2, 1933; came to U.S., 1950, naturalized, 1960; s. Alfred A. and Gabrielle E. (Kovarick) H.; m. Constance E. Penfield, June, 1957. M.D., Washington U., 1957. Intern, resident Washington U. Med. Ctr., 1957-61; surgeon USPHS, Urbana, Ill., 1961-63; research fellow biochemistry Washington U., St. Louis, 1963-65, asst. prof. preventive medicine, 1965-70, assoc. prof., 1970-73, prof., 1973-83, prof. medicine, 1983—; mem. Aging Rev. Com., Nat. Inst. on Aging NIH, Bethesda, Md. Contbr. numerous articles to profl. jours. Served with USPHS, 1961-63. NIH grantee, 1965-83; Career Devel. award NIH, 1968-73; Am. Heart Assn. grantee, 1977-80; Muscular Dystrophy grantee, 1973-83. Mem. Am. Physiol. Soc., Soc. Clin. Investigation, Am. Inst. Nutrition, Am. Diabetes Assn., Sigma Xi. Episcopalian. Subspecialties: Physiology (medicine); Preventive medicine. Current work: My research deals with the acute and long term metabolic and physiological adaptations to exercise, and the role of exercise in prevention of disease and maintaining function during aging. Home: 1434 Reauville St Warson Woods MO 63122 Office: Washington U Sch Medicine 4566 Scott St Saint Louis MO 63110

HOLLOWAY, CAROLINE TOBIA, biochemistry educator; b. N.Y.C., July 29, 1937; d. Martin Bartholomew and Margaret (Giolito) Tobia; m. Peter William Holloway, Oct. 25, 1963 (div. 1981); children: Philippa Elizabeth, Kenneth William. B.S., CCNY, 1959; Ph.D., Duke U., 1964. Research assoc. Shell Agrl. Chems., Sittingbourne, Kent, Eng., 1964-67, Duke U., Durham, N.C., 1967-69, U. Va., Charlottesville, 1975-76, research asst. prof., 1976—; ad hoc reviewer Lipids, 1980—. Pres., founder. Jackson Via Community Edn. Program, Charlottesville, Va., 1977; mem. Charlottesville (Va.) Sch. Bd., 1978-81, Spl. Edn. Adv. Com., Charlottesville, 1979-81. First Milstead research fellow, 1964-67. Mem. Am. Soc. Biol. Chemists, Phi Beta Kappa. Roman Catholic. Subspecialties: Biochemistry (medicine); Nutrition (medicine). Current work: Regulation of membrane properties in animals: in particular the way lipid components of membranes can influence membrane function, nutritional aspects of dietary lipid and their role in cardiovascular disease. Office: U Va Sch Medicine Dept Biochemistry Charlottesville VA 22908

HOLLOWAY, DENNIS ROBERT, architect, urban designer, solar architecture researcher; b. Owosso, Mich., Mar. 26, 1943; s. Robert Edwin and Antonia Louise H.; m. Bess Ann, Aug. 26, 1966; children: Adam David, Daniel Robert, Lauren Denise. B. Arch., U. Mich., 1966; M. Arch. in Urban Design, Harvard U., 1967; postgrad. (Fulbright scholar), U. Liverpool, Eng., 1968. Registered architect, Mich., Colo., Nat. Council Archtl. Registration Bds. Designer Robert Metcalf (Architect), Ann Arbor, Mich., 1962-66, Tufts-New Eng. Med. Center Planning Office, Boston, 1967; architect Conklin & Rossant (Architects), N.Y.C., 1968-70; assoc. prof. architecture U. Minn., Mpls., 1970-77; dir. Energy Self-sufficient House, 1973-77; assoc. prof. environ. design U. Colo., Boulder, 1977-80; prin. Dennis Holloway Architects and Solar Designers, Boulder, 1980—; Am. Collegiate Schs. Architecture exchange grantee, London, summer 1965; designer Farmer & Dark (Chartered Architects), London, 1965; cons., lectr. in field. Bd. dirs. Colo. Coalition for Full Employment, Denver, 1980-82. Recipient Environ. Quality award Region V EPA, 1976. Mem. Nat. Center Appropriate Tech. (founding dir. 1976-78), Colo. Solar Energy Assn. (dir. Denver chpt. 1980-81), Phi Kappa Phi, Tau Sigma Delta. Subspecialty: Solar energy. Current work: Appropriate tech. design; passive solar architecture; passive solar energy application to all sacles of architecture and community design. Home and Office: 2336 Pearl St Boulder CO 80302

HOLLOWAY, FRANK ALBERT, biological psychology educator, researcher; b. Jan. 1, 1940; m. Joan A. Buchanan; children: Karen, Benjamin, Jason. B.S., U. Houston, 1961, M.A., 1964, Ph.D., 1966. Instr. med. psychology U. Okla. Health Sci. Ctr., Oklahoma City, 1966-67, asst. prof., 1967-72, assoc. prof., 1972-77, prof., 1977—, dir. grad. studies, 1972-77, dir. biol. psychology Ph.D. program, 1977—. Contbr. articles to profl. jours. Pres. 1st Unitarian Ch., Oklahoma City, 1981-82. Nat. Inst. Alcoholism and Alcohol Abuse grantee, 1978—; Nat. Inst. Drug Abuse grantee, 1980—. Fellow Am. Psychol. Assn.; mem. Okla. Psychol. Assn. (dir. 1980-81), Soc. Neurosci., AAAS, N.Y. Acad. Scis. Democrat. Subspecialties: Psychobiology; Neuropsychology. Current work: Alcohol and substance abuse (psychopharmacology); neural basis of behavior (learning and memory); biological rhythms. Office: U Okla Health Scis Ctr 800 NE 13th St PO Box 26901 Oklahoma City OK 73190

HOLLOWAY, HARRY LEE, JR., biological, polar and water diversion parasitologist; b. York County, Va., May 22, 1926; s. Harry L. and Beatrice (Insley) H.; m. N. Juanita Yow, July 31, 1948; children: Harry Lee III, M. Ellen Holloway Betting, Ralph J., D. Bryan. B.S., Randolph Macon Coll., Ashland, Va., 1948; M.A., U. Richmond, 1951; Ph.D., U. Va., Charlottesville, 1956. Grad. teaching asst. U. Va., 1951-53; prof., chmn. dept. Roanoke Coll., Salem, Va., 1953-69; dean faculty Western Md. Coll., Westminster, 1969-71; chmn. dept. biology U. N.D., Grand Forks, 1971-74, prof. biology, 1971—; biology com. Internat. Garrison Diversion Study Bd., 1975-77; Wagner Coll. evaluation team assoc. Middle States Commn. Higher Edn., 1970. Contbr. articles profl. jours. Served with USNR, 1944-46. Recipient Antarctic medal U.S. Congress, 1968; Mt. Holloway named in his honor, 1967; research grantee NSF, Bur. Reclamation, Nat. Marine Fisheries Service. Fellow AAAS; mem. Am. Microscopical Soc., Washington Helminthological Soc., Am. Inst. Biol. Scis., Am. Soc. Parasitologists, Sigma Xi, Kappa Alpha Order, Beta Beta Beta. Methodist. Club: Kiwanis (pres. 1980-81). Subspecialties: Parasitology; Ecology. Current work: Ecology of fish parasites in North America and Antarctica, animal parasites. Home: 1305 Chestnut St Grand Forks ND 58201 Office: 111 Starcher Hall U of ND Grand Forks ND 58202

HOLLOWAY, JOHN THOMAS, physicist; b. Cape Girardeau, Mo., June 19, 1922; s. Herbert Henry and Addie Mae (Cahill) H.; m. Kay Vickers, Nov. 11, 1965; children—Linda, Kim. A.B., Millikin U., Decatur, Ill., 1943; Ph.D., Iowa State U., 1957. With nuclear physics br. Office Naval Research, Washington, 1946-53, head br., 1951-52; research asst. Ames (Iowa) Lab., AEC, 1954-57; with Office Dir. Def. Research and Engring., Washington, 1958-61, dep. dir., 1959-61; with NASA, 1961-68, dep. dir. grants and research contracts, 1961-67, chief advanced programs and tech., space applications div., 1967-68; dir. Nat. Hwy. Safety Research Center, Dept. Transp., 1968-69; v.p. research Ins. Inst. Hwy. Safety, 1969-72; asso. dir. ops. Interdisciplinary Communications Program, Smithsonian Instn., 1972-77, program mgr. internat. program population analysis, 1972-77, research and devel. cons. in hwy. safety, biomed. electronics, energy conservation, 1977-78; sr. staff officer bd. on radioactive waste mgmt. Nat. Acad. Scis.-NRC, 1978—; dir. Interdisciplinary Communications Assos., Inc.; mem. conf. com. Nat. Conf. Advancement Research, 1971-75. Author papers in field; adviser documentary films. Served with USNR, 1944-46. Mem. Am. Phys. Soc., Philos. Soc. Washington, Sigma Xi. Clubs: Cosmos (Washington); Army-Navy Country (Arlington, Va.). Subspecialties: Radioactive Waste Management; Nuclear physics. Current work: Management (including disposal) of high-level radioactive waste. Home: 2220 Cathedral Ave NW Washington DC 20008

HOLLOWAY, THOMAS THORNTON, chemist; b. Dallas, Mar. 6, 1944; s. Thomas Thornton and Sadie Miller (Lawton) H.; m. Vivian Elaine Matula, June 8, 1968. B.A., Rice U., 1966, Ph.D., 1969. Postdoctoral research assoc. Johns Hopkins U., 1969-71; Welch research fellow Tex. Tech U., 1971-73; asst. prof. chemistry William Jewell Coll., 1973-79, assoc. prof., 1979-80; chemist EPA, Kansas City, Kans., 1980—. Contbr. articles to profl. jours., numerous presentations in field. Recipient Spl. Achievement award EPA, 1982, 83; Oak Ridge Assoc. Univs. grantee Summer Faculty Inst. on Energy Conservation, 1978. Mem. Am. Chem. Soc., Air Pollution Control Assn., Sigma Xi. Baptist. Subspecialties: Atmospheric chemistry; Graphics, image processing, and pattern recognition. Current work: Air pollution monitoring and abatement; computer mapping of enviromental quality data; implementation of air pollution regulations; evaluation and improvement of state environmental quality monitoring programs. Home: 820 NE 73d Pl Gladstone MO 64118 Office: EPA Region VII Lab 25 Funston Rd Kansas City KS 66115

HOLLWEG, JOSEPH VINCENT, research physicist, educator, space science center executive; b. N.Y.C., Mar. 20, 1944; s. Joseph Julius and Rose Marie (Novak) H.; m. Leslie Francine, Dec. 23, 1976. Ph.D. in Plasma Physics and Space Sci., MIT, 1968. Research fellow Max Planck Inst., 1968-70, Calif. Inst. Tech., Pasadena, 1970-72; scientist Nat. Ctr. for Atmospheric Research, Boulder, Colo., 1972-80; research prof. U. N.H., Durham, 1980—; dir. Space Sci. Ctr., 1982—. Contbr. chpts. to books, articles to publs. in field. Recipient publ. prize Nat. Ctr. for Atmospheric Research, 1974; Henry Webb Salisbury award MIT, 1964; Wayne B. Nottingham prize Am. Physics Soc., 1967. Mem. Am. Geophys. Union, Am. Astron. Soc. Subspecialties: Solar physics; Plasma physics. Current work: Theoretical studies of energy and momentum balance of solar atmosphere. Office: Demeritt Hall Univ NH Durham NH 03824

HOLLY, FRANK JOSEPH, medical educator, consultant; b. Budapest, Hungary, Dec. 3, 1934; came to U.S., 1956, naturalized, 1962; s. Sandor and Maria (Acsay) H.; m. Katalin Maria Szalay, May 1, 1969; children: Thomas Ferenc, Kathleen Maria, Gloria Anna, Paul Gyorgy. Student, Tech. U., Budapest, 1953-56; Ph.D., Cornell U., 1962. Research phys. chemist Procter & Gamble Co., Cin., 1962-65; prof. chemistry U. El Salvador, San Salvador, 1965-66; research scientist Thermo Electron Corp., Waltham, Mass., 1966-68; assoc., sr. scientist Eye Research Inst., Boston, 1968-78; assoc. prof. Tex. Tech. U. Health Sci. Center, Lubbock, 1978-80, prof. ophthalmology and biochemistry, 1980—; v.p. Holles Labs., Inc., Cohasset, Mass., 1978—; cons. Burton, Parsons Co., Washington, 1972-76, Allergan Pharms., Irvine, Calif., 1978-79, Dow Corning Co., Midland, Mich., Ciba Vision Care, Atlanta, 1977-80, Bausch and Lomb, 1983—. Editor, Internat. Ophthalmol. Clinic, 1973, 80-83. Scoutmaster Hungarian Scout Assn. in Exile, Garfield, N.J., 1958-62. Nat. Eye Inst. spl. research fellow, 1972-74; Nat Eye Inst. grantee, 1975-77, 78-81, 81-84. Mem. Internat. Soc. Contact Lens Research (council mem. 1978—), Internat. Soc. Colloids and Interface Scientists, Am. Chem. Soc., Assn. Research in Vision and Ophthalmology, Brit. Soc. Cell Biology, Sigma Xi. Roman Catholic. Club: Mensa. Subspecialties: Physical chemistry; Surface chemistry. Current work: Application of interface science and physical chemistry to medicine, ocular physiology, lacrimal physiology, basic dental research, cancer research, biomaterials such as hydrogels, contact lenses, and their biocompatibility, collyria, life sciences in general. Patentee in field. Office: Dept Ophthalmology Tex Tech U Health Scis Center Lubbock TX 79430

HOLMAN, B. LEONARD, nuclear medicine physician, educator; b. Sheboygan, Wis., June 26, 1941; s. Max and Sophia (Penn) H.; m. Dale Elyse Barkin, Jan. 22, 1971; children: Amy Lynn, Allison Stacy. B.S., U. Wis., 1963; M.D., Washington U., St. Louis, 1966. Diplomate: Am. Bd. Nuclear Medicine, Am. Bd. Radiology. Intern Mt. Zion Hosp., San Francisco, 1966-67; resident Mallinckrodt Inst. Radiology, St. Louis, 1967-70; nuclear radiologist Peter Bent Brigham Hosp., Boston, 1970-75; from instr. to assoc. prof. Harvard U. Med. Sch., Boston, 1970-82, prof. radiology, 1982—; nuclear radiologist Children's Hosp., Boston, 1970—, Dana-Farber Cancer Ctr., 1976—; attending nuclear radiologist West Roxbury (Mass.) VA Hosp., 1973—; cons. radiology New Eng. Deaconess Hosp., Boston, 1981—, Beth Israel Hosp., 1982—; dir. clin. nuclear medicine services Brigham and Women's Hosp., Boston, 1975—; Mem. med. adv. com. U.S. NRC, Bethesda, Md., 1980—; bd. dirs. Am. Bd. Nuclear Medicine, Los Angeles, 1980—. Author: (with J.A. Parker) Computer Assisted Cardiac Nuclear Medicine, 1981; editor: (with P.J. Ell) Emission Computed Tomography, 1982; others; contbr. articles to profl. jours., chpts. to books. Nat. Inst. Gen. Med. Sci. fellow, 1968-70. Fellow Am. Coll. Chest Physicians, Am. Heart Assn. (council cardiovascular radiology, council circulation), Am. Coll. Cardiology (trustee 1980-83); mem. Soc. Nuclear Medicine (trustee 1976, 80-83, sec. 1983—), Am. Coll. Radiology, Sigma Xi. Subspecialties: Nuclear medicine; Imaging technology. Current work: Cardiac nuclear medicine, single photon emission computed tomography, regional cerebral blood flow. Home: 25 Nancy Rd Chestnut Hill MA 02167 Office: Brigham and Women's Hosp 75 Francis St Boston MA 02115

HOLMAN, GORDON DEAN, astrophysicist, educator; b. Fort Lauderdale, Fla., July 15, 1949; s. Raymond Carson and Doris Ruth (Herman) H.; m. Mary Virginia Spanburgh, Nov. 24, 1979; 1 dau., Trisha Michelle. B.S., Fla. State U., 1971; M.S., U. N.C., 1973, Ph.D., 1977. Fellow Center for Theoretical Physics, U. Md., College Park, 1977-79, research scientist, research assoc., lectr., 1979—. Contbr. articles to profl. jours. Mem. Am. Astron. Soc., Am. Geophys. Union. Subspecialties: Theoretical astrophysics; Solar physics. Current work: Researcher in radio and x-ray sources in clusters of galaxies, plasma astrophysics, cosmic ray propagation and acceleration; solar flare research. Office: U Md Astronomy Program College Park MD 20742

HOLMAN, RALPH THEODORE, educator, biochemist; b. Mpls., Mar. 4, 1918; s. Alfred Theodore and May (Nilson) H.; m. Karla Calais, Mar. 26, 1943; 1 son, Nils Teodor Calais. A.A., Bethel Jr. Coll., 1937; B.S., U. Minn., 1939, Ph.D., 1944; M.S., Rutgers U., 1941. Instr. U. Minn., 1944-46, asso. prof. physiol. chemistry, 1951-56; prof. Hormel Inst., 1956—, dir. inst., 1975—; prof. biochemistry Mayo Med. Sch., 1977—; asso. prof. Tex. A. and M. U., 1948-51; Mem. adv. bd. Deuel Conf. Lipids, 1958—, chmn., 1972; mem. nutrition study sect. NIH, 1960-63; mem. com. fats Food and Nutrition Bd., Nat. Acad. Scis.-NRC, 1956-62; pres. Golden Jubilee Congress on Essential Fatty Acids and Prostoglandins, 1980. Editor: (with W.O. Lundberg and T. Malkin) Progress in the Chemistry of Fats and Other Lipids, vols. 1-6, 1951-63; sole editor, vols. 7-16, 1963—; assoc. editor: Lipids, 1966-74; editor, 1974—; editorial bd.: Jour. Lipid Research, 1959-61, Jour. Nutrition, 1962-66, Jour. Parenteral and Enteral Nutrition, 1977-82, Jour. Lab. and Clin. Medicine, 1979—. Pres. Mower County Council Chs., 1954-58. Recipient Fachini medal Italian Oil Chemistry Soc., 1974; NRC fellow Med. Nobel Inst., Stockholm, Sweden, 1946-47; Am. Scandinavian Found. fellow U. Uppsala, Sweden, 1947; spl. fellow NIH, U. Gothenburg, Sweden, 1962. Mem. Am. Soc. Biol. Chemists, Am. Inst. Nutrition (Borden award 1966), Am. Oil Chemists Soc. (gov. bd. 1968-70, sec. 1972, v.p. 1973, pres. 1974, Bailey award 1972, award in lipid chemistry 1978), Soc. Exptl. Biology and Medicine, Nat. Acad. Scis., Hormel Found., Am. Orchid Soc., Sigma Xi. Democrat. Conglist. Subspecialties: Biochemistry (medicine); Nutrition (medicine). Current work: Lipids, oxidative deterioration of fats, lipoxidase, essential fatty acid metabolism, polyunsaturated fatty acids, methods of lipids analysis, mass spectrometry, in virto metabolism of fatty acids, quantitative chemical taxonomy. Research, over 300 publs. on spectrophotometric studies fat oxidation, isolation and characterization lipoxidase, displacement chromatography lipids, biochem. characterization essential fatty acid deficiency; established nutritional requirements essential fatty acids; research on metabolism polyunsaturated fatty acids, relationship of essential fatty acid abnormalities to diseases in humans, near-infrared spectra lipids, mass spectrometry lipids; analysis of odors; fragrance and taxonomy; developed methods for lipid analysis, quantitative chem. taxonomy magnolia and orchids based on floral odor, effect of double bond structure upon metabolism of unsaturated acids, effect of partially hydrogenated fats upon nutrition and metabolism of essential fatty acids. Home: 1403 2d Ave SW Austin MN 55912 Office: Hormel Inst U Minn Austin MN 55912

HOLMES, DALE ARTHUR, optical scientist; b. Biwabik, Minn., Dec. 31, 1937; s. Arthur Emil and Saimie Amanda (Luoma) H.; m. Joan Christine Holmes, May 4, 1962; children: Kevin, Camille. B.S. in E.E, Purdue U., 1960, M.S., Carnegie Inst. Tech., 1961, Ph.D., 1965; M.S. in optics, U. Rochester, 1969. Asst. prof. Carnegie Inst. Tech., 1965-66; dir. research and tech. Rocketdyne/Rockwell Internat., Canoga Park, Calif., 1974-76, sr. staff scientist optics, 1977—. Contbr. articles to profl. jours. Served to capt. USAF, 1966-74. Recipient USAF Research and Devel. award, 1970. Mem. Optical Soc. Am., Nat. Rifle Assn. Am. Republican. Clubs: Rocketdyne Rifle & Pistol, Internat. Handgun Metallic Silhouette Assn. Subspecialties: Laser optics; high energy laser systems. Current work: Design analysis test of optical subsystems for high energy laser systems. Home: 2307 E Knollhaven St Simi Valley CA 93065 Office: 6633 Canoga Ave FA44 Canoga Park CA 91304

HOLMES, DYER BRAINERD, corporation executive; b. N.Y.C., May 24, 1921; s. Marcellus B. and Theodora (Pomeroy) H.; m. Roberta M. Donohue. B.S. in Elec. Engring. (McMullen scholar), Cornell U., 1943; postgrad., Bowdoin Coll., M.I.T., 1943-44; hon. degrees, U. N.Mex., 1963, Worcester Poly. Inst., 1978. Registered profl. engr., N.J. Design engr. Western Electric Co.; also mem. tech. staff Bell Labs., 1945-53; gen. mgr. maj. def. systems div., 1961; project mgr. Navy Talos land based missile system devel., 1954-57, Air Force Atlas launch control and checkout equipment devel., 1957, USAF ballistic missile early warning system, 1958-61; dep. asso. adminstr. manned space flight NASA, 1961-63; sr. v.p., dir. Raytheon Co., Lexington, Mass., 1963-69, exec. v.p., 1969-75, pres., 1975—, dir., 1969—; dir. Wyman-Gordon Co., Worcester, Mass., Bank of Boston Corp. (and subsidiary), Kaman Corp., Bloomfield, Conn.; chmn. bd. Beech Aircraft Corp. (subs. Raytheon Co.). Author articles, papers in field. Mem. corp. Northeastern U. Served with USNR, 1942-45. Recipient Outstanding Leadership medal NASA; Paul T. Johns award Arnold Air Soc. Fellow IEEE, AIAA; mem. Nat. Acad. Engring., Aerospace Industries Assn. U.S. (exec. com.), Am. Def. Preparedness Assn. (dir.), Nat. Security Indsl. Assn. (trustee), Navy League, Tau Beta Pi, Eta Kappa Nu. Clubs: Nat. Space, Metropolitan (Washington); Algonquin (Boston). Subspecialties: Electrical engineering; Aerospace engineering and technology. Initiated, developed first precision rec. transmission measuring set, other test equipment; participated devel. long distance coaxial telephone and TV systems, RCA, 1953-61. Office: Raytheon Co 141 Spring St Lexington MA 02173

HOLMES, JOHN THOMAS, solar energy engineer; b. Oak Park, Ill., Aug. 10, 1936; s. Glenn T. and Olive C. H.; m. Edith M. Kramer, July 13, 1963; children: Ann E., Mark T. B.S. in Chem. Engring, U. Wis., 1958, M.S., U. Calif.-Berkeley, 1960. With Argonne Nat. Lab., Argonne, Ill., and Idaho Falls, Idaho, 1960-76; solar energy engr. Sandia Nat. Labs., Albuquerque, 1976—. Contbr. articles to profl. jours. Recipient IR 100 award, 1974. Democrat. Unitarian. Subspecialties: Solar energy; Chemical engineering. Current work: Solar Power Tower System development, component testing, safety, materials behavior and solar chemistry in high intensity sunlight. Co-patentee devices for sodium chemistry and nuclear fuel chem. processes. Office: Div 6222 Sandia National Laboratories Albuquerque NM 87185

HOLMES, KING KENNARD, medical educator, physician; b. St. Paul, Sept. 1, 1937; m.; children: Kimberly, Heather, King. A.B. cum laude, Harvard U., 1959; M.D., Cornell U., 1963; Ph.D. in Microbiology, U. Hawaii, 1967. Diplomate: Nat. Bd. Med. Examiners, Am. Bd. Internal Medicine. Intern in medicine Vanderbilt U., Nashville, 1963-64; resident in medicine U. Wash., Seattle, 1967-68, chief resident in medicine, 1968-69, instr., 1969-70, asst. prof., 1970-74, assoc. prof., 1974-78, prof., 1978—, adj. prof. microbiology and immunology, 1978—, adj. prof. epidemiology, 1980—, affiliate, 1976—, 1978—; asst. to dir. sexually transmitted diseases control div. ctr. for Prevention Services Ctrs. for Disease Control, Atlanta, 1981—; asst. chief dept. medicine, head div. infectious diseases Seattle Pub. Health Hosp., 1981—; mem. staff Harborview Med. Ctr.; mem. expert adv. panel WHO. Author: (with others) books including Chlamydia Infections, 1982, Sexually Transmitted Diseases, 1982; contbr.: articles to med. jours. Sexually Transmitted Diseases; mem. editorial bd.: Sexually Transmitted Diseases, 1974-80, Drugs, 1977—, Revs. Infectious Diseases, 1978—, Jour. AMA, 1980—. Fellow ACP; mem. Assn. Am. Physicians, Am. Fedn. for Clin. Research, Am. Pub. Health Assn., Am. Soc. for Clin. Investigation, Am. Soc. Microbiology, Am. Venereal Disease Assn. (pres. 1975), Infectious Disease Soc. Am. (Squibb award 1978), Med. Soc. Study of Venereal Diseases (London), USPHS Commd. Officer Assn., Western Assn. Physicians, Western Soc. for Clin. Research, Am. Social Health Assn. (dir. 1980—). Subspecialties: Infectious diseases; Internal medicine. Office: Department of Medicine University of Washington School of Medicine Seattle WA 98195

HOLMES, NEAL J., chemistry educator; b. Mercer, Mo., Aug. 2, 1931; s. Allie O. and Marie (Goddard) H.; m. Mary L. Liggett, Dec. 20, 1953; children: Tom, Tim, Ted, Todd. B.S.Ed., N.E. Mo. State U., 1957; M.A., Washington U., St. Louis, 1962; Ed.D., Okla. State U.-Stillwater, 1967. Secondary tchr. Washington (Mo.) Pub. Schs., 1957-61; sci. cons. Parkway Schs., Chesterfield, Mo., 1962-67; mem. faculty Central Mo. State U., Warrensburg, 1967—, now prof. chemistry; cons. McGraw Hill Book Co., Manchester, Mo., 1965-67. Author: Science, People Concepts Prove, 1974, Gateways to Science, 1979, 82, 83. Served with USAF, 1952-56. NSF fellow, 1960-61, 65. Subspecialties: Soil chemistry; Microchip technology (engineering). Current work: Electro chemical soil analysis. Home: Route 4 Box 370 Warrensburg MO 64093 Office: Central Mo State U Warrensbury MO 64093

HOLMQUIST, GERALD PETER, geneticist, researcher; b. Chgo., Feb. 22, 1942; s. Gunnar A. and Elaine (Roehr) H.; m. Dorothy Jane, Dec. 29, 1969; 1 son, Peter Crittenden. B.S. in Physics, U. Chgo., 1964, M.S., 1967; Ph.D., U. Ill., 1971. Cert. clin. cytogenetics Am. Bd. Med. Genetics. Postdoctoral fellow Karolinska Inst., Stockholm, 1972, Harvard Med. Sch., 1973-74, City of Hope, Duarte, Calif., 1974, 76; asst. prof. medicine Baylor U. Coll. Medicine, Houston, 1976—, dir. cytogenetics lab., 1980—. Active Republican politics. NIH Career devel. awardee, 1979—. Mem. Am. Soc. Cell Biology, Genetics Soc. Am., Sigma Xi. Club: Briar (Houston). Subspecialties: Genome organization; Genetics and genetic engineering (medicine). Current work: Basic research on chemical mechanisms responsible for mammalian chromosome banding. Home: 2056 Branard Houston TX 77098 Office: 1200 Moursund Houston TX 77030

HOLMSTROM, VALERIE LOUISE, clinical psychologist, educator; b. Seattle, June 13, 1948; d. Frank Gottfried and Laura (Lofthus) H. A.B., Boston U., 1968; Ph.D., U. Wash., 1972. Lic. clin. psychologist, S.C. Staff psychologist VA Med. Ctr., Brockton, Mass., 1972-75, asst. chief psychology service, Charleston, S.C., 1975-81, chief psychology service, 1981—; asst. prof. dept. psychiatry and behavioral scis. Med. U. S.C., Charleston, 1975—; cons. Mass. Rehab. Commn., Boston, 1974-75, S.C. Div. Vocat. Rehab., Charleston, 1979—; lectr. Boston U., 1974, Grad. Sch. Edn., The Citadel, Charleston, 1979-80. USPHS fellow, 1968-69; recipient Outstanding Performance award VA Med. Ctr., Brockton, 1975, VA Med. Ctr., Charleston, 1982. Mem. Am. Psychol. Assn., Internat. Assn. Group Psychotherapy, VA Chiefs Assn. Democrat. Subspecialties: Behavioral psychology; Neuropsychology. Current work: Neuropsychological and personality characteristics of epilepsy patients; coping patterns of psychiatric patients with terminal illness. Home: 1115 Sea Oats Ct Mount Pleasant SC 29464 Office: VA Med Ctr 109 Bee St Charleston SC 29464

HOLONYAK, NICK, JR., electrical engineering educator; b. Ziegler, Ill., Nov. 3, 1928; s. Nick and Anna (Rosoha) H.; m. Katherine R.A. Jerger, Oct. 8, 1955. B.S., U. Ill., 1950, M.S., 1951, Ph.D. (Tex. Instruments fellow), 1954. Mem. tech. staff Bell Telephone Labs., Murray Hill, N.J., 1954-55; physicist, unit mgr., mgr. advanced semiconductor lab. Gen. Electric Co., Syracuse, N.Y., 1957-63; prof. elec. engring. and materials research lab. U. Ill., Urbana, 1963—; mem. Center Advanced Study, 1977—; series editor Prentice-Hall, Inc., 1962—; cons. Monsanto Co., 1964—, Nat. Electronics Co., 1963-70, Skil Corp., 1967, GTE Labs. Tech. Adv. Council, 1973. Author: (with others) Semiconductor Controlled Rectifiers, 1964. Served with U.S. Army, 1955-57. Recipient Cordiner award Gen. Electric Co., 1962, John Scott medal City of Phila., 1975, GaAs Conf. award, 1976. Fellow Am. Phys. Soc., IEEE (Morris Liebmann prize 1973, Jack A. Morton award 1981); mem. Electrochem. Soc. (Solid State Sci. and Tech. award 1983), Math. Assn. Am., AAAS, Nat. Acad. Engring. Subspecialties: 3emiconductors; Microelectronics. Current work: Semiconductor materials and devices (thyristors, transistors.LEDs, lasers, ICs). Home: 2212 Fletcher St Urbana IL 61801 Office: Dept Elec Engring U Ill Urbana IL 61801

HOLOWKA, DAVID ALLAN, biochemist; b. Rochester, N.Y., Aug. 21, 1948; s. Michael James and Frances G. (Purcell) H.; m. Barbara Baird, Sept. 1, 1979. B.S. summa cum laude, St. John Fisher Coll., Rochester, N.Y., 1970; Ph.D., Tufts U., 1975. Postdoctoral fellow Cornell U., Ithaca, N.Y., 1975-77, research assoc. dept. chemistry, 1980—; postdoctoral fellow NIH, Bethesda, Md., 1977-80. Contbr. articles to profl. jours. Arthritis Found. fellow, 1977; NIH awardee, 1975, 82. Mem. Am. Assn. Immunologists. Subspecialties: Biophysical chemistry; Immunobiology and immunology. Current work: Structure and function of immunoreceptors, fluorescence and biochemical studies of the cell surface receptor for immunoglobulin E. Office: Dept Chemistry Baker Lab Cornell U Ithaca NY 14853

HOLST, ROBERT WEIGEL, plant physiologist; b. Mineola, N.Y., Feb. 13, 1944; s. Gustave E. and Rose M. (Weigel) H.; m. Rebecca Ann Roesener, June 13, 1970; children: Daniel, Philip. A.B in Biology, U. Rochester, 1966; M.S. in Botany, So. Ill. U., 1973, Ph.D., 1977. Postdoctoral fellow Boyce Thompson Inst., Yonkers, N.Y., 1976-77; plant physiologist Office Pesticide Programs, EPA, Washington, 1977—. Contbr. articles on plant nitrogen metabolism, mineral uptake, plant growth, pesticide spray drift to profl. jours., presentations with abstracts to profl. meetings. Served with USN, 1966-70. Mem. Am. Soc. Plant Physiologists, Am. Soc. Agronomy, Weed Sci. Soc. Am., Naval Res. Assn., Sigma Xi. Lutheran. Subspecialties: Plant physiology (agriculture); Integrated systems modelling and engineering. Current work: Evaluation of pesticide plant protection; design and evaluation of models for determination quantities of pesticides in environment; estimation exposures for humans and other organisms to pesticides. Office: EPA (TS-769) Washington DC 20460

HOLSTEIN, THEODORE DAVID, physics educator; b. N.Y.C., Sept. 18, 1915; s. Samuel and Ethel (Stein) H.; m. Beverlee Ruth Roth, Aug. 31, 1945; children: Lonna Beth Holstein Smith, Stuart Alexander. B.S. cum laude, N.Y. U., 1935; M.S., Columbia U., 1936; Ph.D., NYU, 1940. Instr. CCNY, 1940; research physicist Westinghouse Research Lab., Pitts., 1941-60; prof. U. Pitts., 1960-65, UCLA, 1965—. Author articles on solid-state theory. Fellow Am. Acad. Arts and Scis.; mem. Nat. Acad. Sci., Am. Phys. Soc. Jewish. Subspecialties: Condensed matter physics; Theoretical physics. Current work: Theory of transport phenomena in solids; dynamical properties of many-particle systems. Office: UCLA Dept Physics Los Angeles CA 90024

HOLT, ALAN CRAIG, government space agency administrator; b. Camp Lejeune, N.C., Mar. 16, 1945; s. Floyd Marshall and Bernice Ann (Schmidt) H.; m. Susan Carol Darnall, Aug. 8, 1970; 1 son, Christopher Scott. B.S., Iowa State U.-Ames, 1967; M.S., U. Houston, Clear Lake, 1979. Expt. procedures specialist NASA, Manned Spacecraft Ctr., Houston, 1967-70, Skylab crew proceedings and tng. specialist, 1970-74; Spacelab crew ops. specialist Johnson Space Ctr., 1974-78, Spacelab systems tng. supr., 1978-80, payload group leader, 1980—; pres. Holt Research & Devel. Co., 1983—; bd. dirs. Vehicle Internal Systems Investigative Team, Inc., Friendswood, Tex., 1978—. Recipient Sustained Superior Performance award NASA, Johnson Space Center, 1973, 82. Mem. Am. Astron. Soc. (assoc.), AIAA, Soc. Photo-Optical Instrumentation Engrs., N.Y. Acad. Scis. Lutheran. Club: AMORC (regional monitor 1979-81). Subspecialties: Theoretical astrophysics; Magnetic physics. Current work: Hyperspatial theoretical models and advanced propulsion systems, crystal physics, anomalous aerial phenomena, parapsychology. Inventor field resonance propulsion system. Office: NASA Johnson Space Center Code CG 6 Houston TX 77058 Office: Holt Research and Devel Co PO Box 5892 Houston TX 77258

HOLT, DONALD ALEXANDER, university administrator, agronomist, consultant, researcher; b. Joliet, Ill., Jan. 29, 1932; s. Cecil B. and Helen E. (Eichoff) H.; m. Marilyn Louise, Sept. 6, 1953; children: Kathryn Holt Stichnoth, Steven, Jeffrey, William. Student, Joliet Jr. Coll., 1950-53; B.S. in Agrl. Sci, U. Ill., Urbana, 1954; M.S. in Agronomy, U. Ill., Urbana, 1956, Ph.D., Purdue U., 1967. Teaching asst. U. Ill., 1954-56, head dept. agronomy, 1982—; farmer, Minooka, Ill., 1956-63; instr. Purdue U., 1963-67, asst. prof. agronomy, 1967-73, assoc. prof., 1973-77, prof., 1977-82; cons. Contbr. numerous articles to profl. jours. Fellow Am. Soc. Agronomy; mem. Crop Sci. Soc. Am., Am. Forage and Grassland Council. Methodist. Subspecialty: Plant physiology (agriculture). Current work: Computer simulation of crop growth; agrl. adminstrn. in land-grant coll. Home: 706 Evergreen Circle Urbana IL 61801 Office: 1301 W Gregory Dr 211 Mumford Hall Urbana IL 61801

HOLT, RICHARD A., physicist, educator; b. N.Y.C., Sept. 4, 1942; s. Benjamin and Mary (Weisthal) H.; m. Renee Silberman, June 18, 1978; children: Benjamin J., Daniel M. A.B. cum laude, Harvard U., 1964, M.A., 1966, Ph.D., 1973. Postdoctoral fellow Brown U., 1973-74; research scientist U. Western Ont., 1974-76, asst. prof. physics, 1976-82, assoc. prof. physics, 1982—. Recipient Harvard U. Robbins prize in physics, 1964. Mem. Am. Phys. Soc., Can. Assn. Physicists, Sigma Xi. Subspecialties: Atomic and molecular physics; Spectroscopy. Current work: Fast beam laser spectroscopy; foundations of physics. Office: Physics Dept U Western Ont London ON Canada N6A 3K7

HOLT, STEPHEN S., astrophysicist; b. N.Y.C., May 17, 1940; s. Aaron J. and Faye E. (Schwartz) H.; m. Carol Ann Weissman, June 3, 1961; children: Peter David, Eric Lawrence, Laura Kimberly. B.S., N.Y. U., 1961, Ph.D. in Physics, 1966. Instr. physics N.Y. U., 1964-66; astrophysicist Goddard Space Flight Center, Greenbelt, Md., 1966—; chief high energy astrophysics NASA Hdqrs., 1980-81; dir. Lab. for High Energy Astrophysics Goddard Space Flight Ctr., Greenbelt, Md., 1983—; lectr. physics U. Md., 1967—. Contbr. articles to profl. jours. Recipient medal for exceptional sci. achievement NASA, 1977, 80. Mem. Am. Phys. Soc., Am. Astron. Soc., Sigma Xi, Tau Beta Pi, Sigma Pi Sigma. Subspecialties: High energy astrophysics; 1-ray high energy astrophysics. Current work: Investigator; scientist space-borne scientific investigations including Einstein Observaroty(the first X-ray astronomy telescope). Home: 1207 Mimosa Ln Silver Spring MD 20904 Office: Code 660 Goddard Space Flight Center Greenbelt MD 20771 The most important intrinsic requisites for success in experimental science are probably imagination and diligence. Very few individuals possess these in sufficient quantities to dominate the extrinsic variables which shape their careers in research, however. I consider myself fortunate to have been able to capitalize on whatever talent I possess by having my research interests aligned with funding priorities, and by being blessed with the cooperation of unselfish and stimulating colleagues.

HOLT, WILLIAM R., math. statistician, entomologist; b. Phila., Feb. 17, 1930; s. William and Mabel (Price) H.; m. Betty Boal, Aug. 18, 1926. B.S. in Forestry, Pa. State U., 1952, M.F., 1956; B.A. in Math, Ohio Wesleyan U., 1981. Forester U.S. Forest Service, Gainesville, Ga., 1952-56, research entomologist, Delaware, Ohio, 1960-69; statistician Merck Sharpe & Dohme Research Lab., Rahway, N.J., 1969-72, Boehringer Ingelheim Ltd., Elmsford, N.Y., 1972-77; math. statistician U.S. Army, Ft. Rucker, Ala., 1978—. Mem. Am. Statis. Assn., Math. Assn. Am., Math. Soc. Am., Biometric Soc. Subspecialties: Statistics; Applied mathematics. Current work: General linear regression model theory and practice. Consultation and data analysis. Office: D-8 care Commander PO Box 577 USAARL Fort Rucker AL 36362

HOLTER, MARVIN ROSENKRANTZ, physicist; b. Fairport, N.Y., July 4, 1922; m., 1956; 2 children. B.S., U. Mich., 1949, M.S., 1951, 58. Research engr. Aero. Research Ctr. U. Mich., Ann Arbor, 1947-53, asst. supr. Project Wizard, missile def. systems, 1953, supr. Project Wizard, missile def. systems, 1954; research engr. Willow Run Labs., 1954-56; assoc. prof. engring. mechanics U. Toledo, 1956-57; mem. tech. ops. and long range planning staffs, lab. ops. and planning Willow Run Labs., U. Mich., 1957-58, head sensory devices group infrared lab., 1958-64; head infrared and optical sensor lab. Inst. Sci. and Tech., Willow Run Ctr., 1964-70; prof. Sch. Natural Resources, U. Mich., 1968-70; chief earth observations div. NASA Manned Spacecraft Ctr., 1970-72; dep. dir. Willow Run Labs., U. Mich., 1972-73; exec. v.p. Environ. Research Inst. Mich., Ann Arbor, 1973—; Mem. com. remote sensing for agrl. purposes Nat. Acad. Sci.-NRC, 1961-70; mem. div. adv. group aero. systems div. USAF, 1964; mem. USAF Sci. Adv. Bd., 1975-77, Joint U.S.-USSR Working Group for Sensing of Environ., 1971-77; mem. space applications com. NASA, 1971-77; mem. adv. com. Def. Intelligence Agy., 1978-81. Recipient Exceptional Sci. Achievement award NASA, 1973, Exceptional Service award Air Force Sci. Adv. Bd., 1979. Mem. Explorers Club. Subspecialty: Remote sensing (geoscience). Office: PO Box 8618 Ann Arbor MI 48107

HOLTKAMP, DORSEY E(MIL), medical research scientist; b. New Knoxville, Ohio, May 28, 1919; s. Emil and Caroline (Meckstroth) H.; m. Marianne Church Johnson, Mar. 20, 1942 (dec. May 1956); m. Marie P. Bahm Roberts, Dec. 20, 1957 (dec. Apr. 1982); 1 son, Kurt Lee; stepchildren: Charles Timothy Roberts, Michael John Roberts. Student, Ohio State U., 1937-39; A.B., U. Colo., 1945, M.S., 1949, Ph.D., 1951, Sch. Medicine, U. Colo., 2-1/2 yrs. Sr. research scientist biochemistry Smith, Kline & French Labs., Phila., 1951-57, endocrine-metabolic group leader, 1957-58; head endocrinology dept. Merrell Nat. Labs. div. Richardson-Merrell, Inc., Cin., 1958-70, group dir. endocrine clin. research, 1970-81; group dir. med. research dept. Merrell Dow Pharms. (subs. Dow Chem. Co.), Cin., 1981—; cons. Contbr. articles to profl. jours. Fellow AAAS, Am. Inst. Chemists; mem. Am. Chem. Soc., Am. Soc. Pharmacology and Exptl. Therapeutics, Soc. Exptl. Biology and Medicine, Endocrine Soc., Reticuloendothelial Soc., Pacific Coast Fertility Soc., Acad. Medicine Cin. (asso.), Am. Inst. Biol. Sci., Am. Soc. Zoologists, N.Y. Acad. Sci., Ohio Acad. Scis., Am. Assn. Lab. Animal Sci., AMA (affiliate), Soc. Study of Reprodn., Am. Soc. Clin. Pharmacology and Therapeutics, Am. Fertility Soc., Internat. Family Planning Research Assn., Sigma Xi. Republican. Presbyterian. Subspecialties: Endocrinology; Pharmacology. Current work: Research and development of prescription drugs for humans in field of endocrinology. Patentee in field. Home: 9464 Bluewing Terr Cincinnati OH 45241 Office: 2110 E Galbraith Rd Cincinnati OH 45215

HOLTON, JAMES R., atmospheric science educator; b. Spokane, Wash., Apr. 16, 1938; m., 1962; 2 children. B.A., Harvard U., 1960; Ph.D. in Meteorology, MIT, 1964. NSF fellow U. Stockholm, 1964-65; from asst. prof. to assoc. prof. U. Wash., Seattle, 1965-73, prof. atmospheric sci., 1973—. Co-chief editor: Jour. Atmospheric Sci, 1978-83. Mem. Am. Meteorol. Soc. (Meisinger award 1973); mem. Am. Meteorol Soc. (2d half-century award 1982); Mem. Am. Geophys. Union. Subspecialty: Meteorology. Office: Dept Atmospheric Sci U Wash Seattle WA 98195

HOLTON, WILLIAM COFFEEN, physicist; b. Washington, July 24, 1930; s. William Bultman and Esther M. (Coffeen) H.; m. Mary Schaeffer, Aug. 5, 1953; children: Elizabeth Ashe, William Andrew, Sarah Anne. B.S. in Physics, U. N.C., 1952; M.S., U. Ill., 1958, Ph.D., 1960. Mem. tech. staff Central Research Labs., Tex. Instruments, Dallas, 1960-66, mgr. quantum electronics br. Central Research Labs., 1966-71, dir. advanced devel. lab., 1971-78, research and devel. mgr. Seemicondr. Group, 1978—. Contbr. articles to profl. jours. Served to lt. (j.g.) USN, 1952-54. Recipient Josephus Daniels award U. N.C., 1952. Fellow Am. Phys. Soc., IEEE; mem. AAAS, Phi Beta Kappa, Phi Eta Sigma. Club: Rush Creek Yacht (Dallas). Subspecialties: 3emiconductors; Condensed matter physics. Current work: Solid state physics/semiconductors. Patentee in field. Home: 12106 Shiremont Dallas TX 75230 Office: PO Box 225012 MS 72 Dallas TX 75265

HOLTZER, MARILYN EMERSON, chemist; b. Belleville, Ill., July 22, 1938; d. Robert August and Ethel Ruth (Hodges) Emerson; m. Alfred Melvin Holtzer, June 24, 1969; children: Rachel, Dan. A.B., Washington U., St. Louis, 1960, A.M., 1963, Ph.D., 1966. Research assoc. Washington U., St. Louis, 1966-67; asst. prof. Webster Coll., St. Louis, 1967-69; instr. John Burroughs Sch., St. Louis, 1969-70; vis. asst. prof. U. Mo., St. Louis, 1971-72; instr. Washington U., St. Louis, 1972-80, research assoc., 1980—. Recipient DuPont Research award, 1963, Shell fellow, 1964. Mem. Handweavers Guild Am., St. Louis Weavers Guild, St. Louis Artists Guild (Arachne prize 1977, 79, Crawford prize 1980). Subspecialties: Physical chemistry; Biophysical chemistry. Current work: Investigation of the stability of hetero-helical, two-chain coiled coils, focusing particularly on strength and specificity of interactions between the two helical chains in completely helical proteins, tropomyosin and paramyosin. Home: 6636 Pershing Ave Saint Louis MO 63130 Office: Dept Chemistry Washington U Saint Louis MO 63130

HOLTZMAN, JORDAN LOYAL, clinical pharmacologist; b. Chgo., July 12, 1933; s. Samuel and Fannie (Greenspon) H.; m. Joyce Elaine Modey, Aug. 31, 1958; children: Jeremy, Julie, Jon. B.A., U. Chgo., 1952, M.S., 1955, M.D., 1959, Ph.D., 1964. Intern U. Ill. Research and Edn. Hosps., Chgo., 1959-60; Staff Student Health Service, U. Chgo. Clinics, 1962-63; pharmacologist sect. on chemotherapy Lab. Parasite Chemotherapy, Nat. Inst. Allergy and Infectious Disease, NIH, Bethesda, Md., 1964-65; sec. drug enzyme interaction Lab. Chem. Pharmacology, 1965-67; sr. scientist sect. biochemistry, lab. pharmacology Nat. Cancer Inst., Balt. Cancer Research Ctr., 1967-71; resident internal medicine U. Minn., Mpls., 1972-73, assoc. prof. dept. medicine, 1974—, assoc. prof. dept. pharmacology, 1971—; cons. clin. pharmacology VA Med. Ctr., Mpls., 1971—. Served with USPHS. Mem. Am. Soc. Pharmacology and Exptl. Therapeutics, Am. Soc. Biol. Chemistry, Central Soc., Sigma Xi. Subspecialties: Molecular pharmacology; Internal medicine. Current work: The biochemical pharmacology correlation of effects, toxicity and metabolism of drugs to their concentrations in the body and to their biochemical actions. Home: 4710 Girard Ave Minneapolis MN 55409 Office: VA Med Center 4801 E 54th St Minneapolis MN 55417

HOLTZMAN, NEIL ANTON, pediatrician; b. Bklyn., Apr. 8, 1934; m., 1955; 4 children. B.A., Saw Swarthmore Coll., 1955; M.D., NYU, 1959. Diplomate: Am. Bd. Pediatrics, 1965. From intern to sr. asst. resident Harriet Lane Home, Johns Hopkins Hosp., Balt., 1959-62; USPHS fellow biophysics Sch. Medicine, Johns Hopkins U., Balt., 1962-64, asst. prof., 1964-70, assoc. prof. pediatrics, 1970—; pediatrician Johns Hopkins Hosp., 1964—; Resident fellow Nat. Acad. Sci., 1974-75; cons. staff John F. Kennedy Inst., Balt., 1969—; mem. com. study inborn errors of metabolism Nat. Acad. Sci.-NRC, 1972-76; mem. Md. State Commn. Hereditary Disease, 1974—; coordinator hereditary disorders service Md. Dept. Health and Mental Hygiene; dir. Robert Wood Johnson Gen. Pediatric Acad. Devel. Program, Johns Hopkins U. Joseph P. Kennedy Jr. Meml. Found. research scholar, 1972-74; USPHS career devel. awardee, 1969-73; Md. Dept. Health grantee, 1968—, USPHS, 1968-70, 71-73. Fellow Am. Acad. Pediatrics (com. genetics 1978—); mem. AAAS, Am. Soc. Human Genetics, Am. Soc. Biol. Chemists, Am. Pediatric Soc., Soc. Pediatric Research. Subspecialties: Genetics and genetic engineering (medicine); Pediatrics. Office: Johns Hopkins Hosp Dept Pediatrics Baltimore MD 21205

HOLTZMAN, SAMUEL, human-machine decision systems researcher, consultant; b. Mexico City, Feb. 9, 1955; U.S., 1973; s. Aaron Holtzman and Flora Dantus. S.B., M.I.T., 1977, S.M., 1980, E.E., 1980; postgrad, Stanford U., 1980—. Research and teaching asst. MIT, 1977-80; research and teaching asst. Stanford U., 1980-83, instr. engring.-econ. systems, 1983-84; assoc. Strategic Decisions Group, Menlo Park, Calif., 1981-82; cons. artificial intelligence and decision analysis; pres. Engring.-Econ. Systems, Student Assn., Stanford U., 1981. Inventor non-uniform time-scale modification system for speech and music, 1980. Mem. IEEE, Assn. Computing Machinery, AAAS, Mensa, Sigma Xi, Eta Kappa Nu. Subspecialties: Artificial intelligence; Operations research (engineering). Current work: Human-machine decision systems; application of decision analysis and artificial intelligence to design and implementation of intelligent decision aids. Home: PO Box 5405 Stanford CA 94305 Office: Stanford U 301 Terman Engring Center Stanford CA 94305

HOLTZMAN, WAYNE HAROLD, psychologist, educator; b. Chgo., Jan. 16, 1923; s. Harold Hoover and Lillian (Manny) H.; m. Joan King, Aug. 23, 1947; children: Wayne Harold, James K., Scott E., Karl H. B.S., Northwestern U., 1944, M.S., 1947; Ph.D., Stanford, 1950; L.H.D. (hon.), Southwestern U., 1980. Asst. prof. psychology U. Tex., Austin, 1949-53, asso. prof., 1953-59, prof., 1959—; dean Coll. Edn., 1964-70, prof. psychology and edn., 1965—; also dir. Hogg Found. Mental Health, 1955-64, pres., 1970—; Dir. Social Sci. Research Council, 1957-63, Centro de Investigationes Sociales, Mex., 1960—; cons. USAF, also mem. sci. adv. bd., 1969-71; mem. com. basic research com. NRC, 1968-71; mem. behavioral sci. study sect. USPHS, 1957-59, mental health study sect., 1960, chmn. personality and cognition research rev. com., 1968-72; research adv. panel Social Security Adminstrn., 1961-62. Author: (with B. M. Moore) Tomorrow's Parents, 1964, Computer Assisted Instruction Testing and Guidance, 1971, (with R. Diaz-Guerrero and J. Swartz) Personality Development in Two Cultures, 1975, Introduction to Psychology, 1978; Editor: Jour. Ednl. Psychology, 1966-72. Trustee Ednl. Testing Service, Princeton, 1972-74; dir. Sci. Research Assos., 1975—; bd. dirs. Southwest Ednl. Devel. Lab., pres., 1974-75; mem. adv. com. computing activities NSF, 1970-73; mem. computer sci. and engring. bd. Nat. Acad. Scis., 1971-73, chmn. panel on selection and placement of mentally retarded students, 1979—; chmn. interdisciplinary cluster on social and behavioral devel. Pres.'s Biomed. Research Panel, 1975-76; bd. dirs. Found.'s Fund for Research in Psychiatry, 1973-77, chmn., 1976-77; dir. Conf. of S.W. Found., 1976—, pres., 1978-79; trustee Ednl. Testing Service, 1977-80, J.W. and Cornelia Scarborough Found., 1977-82, Center for Applied Linguistics, 1978-80, Salado Inst. Humanities, 1980—, Population Inst., 1979—, Menninger Found., 1982—, Population Resource Center, 1980—; mem. nat. adv. mental health council Alcohol, Drug Abuse, and Mental Health Adminstrn., 1978-81; mem. acad. info. systems adv. council IBM, 1982—. Served from ensign to lt. (j.g.) USNR, 1944-46. Faculty Research fellow Social Sci. Research Council, 1953-54, Center Advanced Study Behavioral Scis., 1962-63. Fellow Am. Psychol. Assn.; mem. Tex. Psychol. Assn. (pres. 1957), S.W. Psychol. Assn. (pres. 1958), Am. Statis. Assn., AAAS, Interam. Soc. Psychology (pres. 1966-67), Am. Ednl. Research Assn., Internat. Union Psychol. Scis. (sec.-gen. 1972—), Philos. Soc. Tex. (pres. 1982-83), Sigma Xi. Methodist. Subspecialties: Social psychology; Developmental psychology. Current work: Cross-cultural studies of personality, cognitive and perceptual development in children; techniques of personality assessment; mental health and educational technology. Home: 3300 Foothill Dr Austin TX 78731

HOLTZMANN, OLIVER VINCENT, plant pathologist; b. Highmore, S.D., June 26, 1922; s. Alphonse J. and Mary Gertrude Lona (St. Pierre) H.; m. Cecelia Catherine Lucas, June 8, 1957; children: Fredrick, Kathryn, Nicholas, Joseph, Eleanor. Research asst. prof. N.C. State U., Raleigh, 1955-56; asst. prof. U. Hawaii, Honolulu, 1956-66, assoc. prof., 1966-71, prof., 1971—, chmn. dept., 1970-82. Contbr. articles to profl. jours. Served with AUS, 1942-45. Mem. Am. Phytopath. Soc., Soc. Nematologists. Roman Catholic. Lodge: KC. Subspecialty: Plant pathology. Current work: Biology and control of plant parasitic nematodes, diseases of tropical crops; turf diseases. Office: Dept Plant Pathology U Hawaii 3190 Maile Way Honolulu HI 96822

HOLUB, RICHARD ANTHONY, neurophysiologist; b. Bklyn., Jan. 30, 1948; s. Andrew Alexander and Mary Evalyn (Nolan) H.; m. Rebecca Ann Keebler, Aug. 29, 1981. M.Sc. in Physiol. Psychology, Brown U., 1973; Ph.D. in Neurophysiology, U. Wis., Madison, 1977. Postdoctoral fellow Nat. Inst. Neurol. Disease and Stroke, 1977-78; vis. postdoctoral fellow in ophthalmology Baylor Coll. Medicine, 1978; research asso. Boston U. Sch. Medicine, 1978-81, asst. research prof. physiology, 1981—, adj. asst. prof. systems, computer, and elec. engring., 1981—. Contbr. articles to profl. jours. NIH grantee. Mem. Assn. Research Vision and Ophthalmology, Soc. Neurosci., Union Concerned Scientists. Subspecialties: Neurophysiology; Computer architecture. Current work: Research on vertebrate visual system; computer-microprocessor applications to vision and visual research. Home: 265 Mount Auburn St Cambridge MA 02138 Office: 80 E Concord Boston MA 02118

HOLZBACH, R. THOMAS, gastroenterologist, researcher; b. Salem, Ohio, Aug. 19, 1929; s. Raymond T. and Nelle Agnes (Conroy) H.; m. Lorraine Cozza, May 26, 1956; children: Ellen, Mark, James. B.Sc., Georgetown U., 1951; M.D., Case Western Res. U., 1955. Diplomate: Am. Bd. Internal Medicine. Intern U. Ill., Chgo., 1955-56; asst. resident VA Hosp., Chgo., 1956-57; sr. asst. resident in medicine Cleve. Met. Gen. Hosp., 1959-60; fellow in gasteroenterology Case Western Res. U., 1960-62; asst. chief gastroenterology VA Med. Ctr., Cleve., 1961-63; physician GI clinic Univ. Hosps. Cleve., 1961-63; instr. medicine Case Western Res. U. Sch. Medicine, Cleve., 1961-64, clin. instr. medicine, 1964-71; dir. med. edn. St. John's Hosp., Cleve., 1966-67; head gastrointestinal research unit, assoc. physician div. medicine St. Luke's Hosp., Cleve., 1967-73, dir. div. gastroenterology, 1970-73; head gastrointestinal research unit, dept. medicine Cleve. Clinc Found., 1973—, mem. com. clin. research projects, 1973—; vis. prof. Mayo Med. Sch., 1974, Washington U., St. Louis, 1976, U. Calif.-San Diego, 1977, Med. Coll. Ohio, 1977, U. Ind. Med. Ctr., 1977, Wayne State U., 1978, U. London, 1978, U. Zurich, 1973, Heidelberg U., 1979, Bowman Gray Sch. Medicine, 1979, U. Ark. Med. Ctr., 1980, U. Munich, Ger., 1982; lectr. KROC Found., Santa Ynez, Calif., 1972, 78, also throughout world. Reviewer numerous profl. jours.; contbr. chpts. to books, numerous articles to sci. jours. Alexander von Humboldt fellow, 1978, 82. Fellow ACP; mem. Am. Gastroent. Assn., Central Soc. Clin. Research, Am. Assn. Study of Liver Diseases, AAAS, Am. Soc. Biol. Chemists, Am. Physiol. Assn., Biophys. Soc., Internat. Assn. Study of the Liver, Am. Fedn. Clin. Research, Midwest Gut Club, Am. Soc. Clin. Nutrition, Cleve. Acad. Medicine (profl. edn. com. 1972—), Ohio State Med. Assn., AMA, Sigma Xi, Alpha Omega Alpha. Subspecialties: Internal medicine; Gastroenterology. Current work: Bile, cholesterol gallstone pathogenesis. Home: 4152 Giles Rd Chagrin Falls OH 44022 Office: 9500 Euclid Ave Cleveland OH 44106

HOLZER, JOSEPH MANO, utility company manager, engineer; b. Somers Point, N.J., Jan. 18, 1952; s. Juda and Esther (Ziegler) H. B.S., Rennsselaer Poly. Inst., 1974, M.Eng., 1975. Physicist Combustion Engring., Inc., Windsor, Conn., 1975-78; engr. Yankee Atomic Electric Co., Framingham, Mass., 1978-82; dir. nuclear fuel Cin. Gas & Electric Co., 1982—. Contbr. articles to profl. jours. Mem. Am. Nuclear Soc. Jewish. Subspecialties: Nuclear engineering; Nuclear fission. Current work: Excore fuel management, incore fuel management; operational support. Home: 9455 Hunters Creek Rd Blue Ash OH 45242 Office: Cin Gas & Electric PO Box 960 Cincinnati OH 45201

HOLZER, THOMAS LEQUEAR, geologist; b. Lafayette, Ind., June 26, 1944; s. Oswald Alois and Ruth Alice (Lequear) H.; m. Mary Elizabeth Burbach, June 13, 1968; children: Holly Christine, Elizabeth Alice. B.S.E., Princeton U., 1965; M.S., Stanford U., 1966, Ph.D., 1970. Asst. prof. U. Conn., Storrs, 1970-75; geologist U.S. Geol. Survey, Menlo Park, Calif., 1975-82, dep. asst. dir. research, Reston, Va., 1982—; adj. environmentalist Griswold & Fuss, Manchester, Conn., 1973-75. Contbr. articles in field to profl. jours. Recipient geology honors award Stanford U., 1970; Superior Service award U.S. Geol. Survey, 1981. Fellow Geol. Soc. Am.; mem. Am. Geophys. Union, AAAS, Nat. Water Well Assn., Sigma Xi. Republican. Presbyterian. Subspecialties: Geology; Hydrogeology. Current work: Land subsidence, conducts research on ground failure associated with ground-water withdrawal from unconsolidated sediments. Home: 2129 Cabots Point Ln Reston VA 22091 Office: US Geol Survey 104 Nat Center Reston VA 22092

HOLZWARTH, GEORGE MICHAEL, biophysicist; b. Dusseldorf, Germany, May 7, 1937; s. Hans Theodor and Matilde (Hellendall) H.; m. Natatie Ann Weiss, June 20, 1970; children: Edward, Clara. B.A., Wesleyan U., 1959; M.S., Harvard U., 1961, Ph.D., 1964. Postdoctoral fellow Fla. State U., Tallahassee, 1966-67; asst. prof. U. Chgo., 1968-74; research scientist Exxon Research and Engring. Co., Linden, N.J., 1974—. Mem. Biophys. Soc., Am. Chem. Soc. Subspecialties: Biophysical chemistry; Polymer chemistry. Current work: Physical chemistry of macromolecules, especially polysaccharides and nucleic acids. Structure and role of cellular membranes. Opitcal activity (electronic and vibrational). Office: Exxon Research and Engring Co Box 45 Linden NJ 07036

HOMANN, PETER HINRICH FRITZ, plant physiology educator; b. Wittenberge, Germany, Apr. 3, 1933; s. Wilhelm Hinrich and Lotte Paula (Luethke) H.; m. Ursel Hofmann, Feb. 7, 1964; children: Philip Peter, Oliver Robin. Diploma in Chemistry, Tech. U. Karlsruhe, W.Ger., 1959, Dr. rer. nat., 1962. Asst. prof. dept. biol. sci. Fla. State U., Tallahassee, 1966-70, assoc. prof., 1970-78—, prof., 1978—. Contbr. articles to prof. jours. Mem. AAAS, Am. Soc. Plant Physiologists, Am. Soc. Photobiology, Biophys. Soc., Deutsche Ornithologen Gesellschaft, Sigma Xi. Subspecialties: Plant physiology (agriculture); Photosynthesis. Current work: Structure-function relationships of chloroplast thylakoids; variations of properties of chloroplast as related to functional adaptations. Home: 117 Ridgeland Rd Tallahassee FL 32312 Office: 202 Inst Molecular Biophysics Fla State U Tallahassee FL 32306

HOMBURGER, FREDDY, physician, scientist, artist; b. St. Gall, Switzerland, Feb. 8, 1916; came to U.S., 1941, naturalized, 1952; s. Ludwig and Cécile (Gaille) H.; m. Regina Thûrlimann, Nov. 8, 1939. Student, U. Vienna, Austria, 1936-37; M.D., U. Geneva, Switzerland, 1941. Diplomate: Nat. Bd. Med. Examiners., Am. Bd. Toxicology. Research fellow, intern pathology Yale Med. Sch. and New Haven Hosps., 1941-43; intern, research fellow in medicine Harvard Med. Sch., Thorndike Meml. Lab., Boston City Hosp., 1943-45; fellow in medicine Meml. Hosp., N.Y.C., 1946-48; chief clin. investigation Sloan-Kettering Inst. Cancer Research, N.Y.C., 1945-48; instr. medicine Cornell U. Med. Coll., 1946-48, research prof. medicine, 1948-57; dir. cancer research and control unit Tufts U. Sch. Medicine, Boston, 1948-57; mem. courtesy staff Mt. Desert Island Hosp., Bar Harbor, Maine, 1955-73, Eastern Meml. Hosp., Ellsworth, Maine, 1957-60; sci. asso. Jackson Lab., Bar Harbor, 1951-60; research prof. oncology, div. basic scis. Sch. Grad. Dentistry, Boston U., 1973—; research prof. pathology Sch. Medicine, 1974—; mem. sci. staff Mallory Inst. Pathology, Boston City Hosp., 1979—; mem. Grad. Sch. Faculty Boston U., 1981—; Mem. corp. Gesell Inst. Child Devel., 1960-78; chmn. adv. com. Am. Students U. Geneva; pres., dir. Bio-Research Inst., Inc., 1957—, Bio-Research Cons., Inc., 1957—; pres. Trenton Exptl. Lab. Animal Co., Bar Harbor, 1969-81; treas., dir. Cambridge Coordinating Com. Drugs, 1972-74; hon. consul of Switzerland in, Boston, 1964—; neutral mem. mixed med. commn. War Dept., 1944-46. Author: The Medical Care of the Aged and Chronically Ill, 3d edit, 1973, The Biological Basis of Cancer Management, 1957; also numerous sci. papers.; Editor: The Physiopathology of Cancer, 3d edit, 1974-76, Progress in Experimental Tumor Research, vols. I-XXVII, 1960—; sr. editor: Symposia on Research Advances Applied to Medical Practice, Current Concepts in Toxicology; Exhibited paintings one-man shows, N.Y.C., Paris, Zurich, Geneva, Boston. Mem. overseers com. to visit Harvard, 1965-71, 76—; bd. dirs. Cambridge Soc. Early Music, 1970; trustee Opera Co., Boston, 1967—; chmn. Friends Busch-Reisinger Mus., 1974—; visitor paintings Boston Mus. Fine Arts, 1974—; mem. adv. bd. Lachaise Found. Fellow AAAS, N.Y. Acad. Scis. (ednl. adv. com. 1967); mem. Nat. Hypertension Assn. (nat. adv. council 1978—), AMA, Endocrine Soc., Am. Assn. Cancer Research, Am. Fedn. Clin. Research, N.Y. Acad. Medicine, Soc. Exptl. Biology and Medicine, Am. Writers Assn., Am. Assn. Pathologists, Soc. Toxicology, Am. Soc. Pharmacology and Exptl. Therapeutics, Royal Soc. Health, Brit. Soc. Toxicology, Soc. Pharmacological and Environ. Pathologists, Soc. Study Reproduction, Endocrine Soc., Am. Assn. Lab. Animal Scis., New Eng. Soc. Pathologists, Acad. Toxicological Scis., Cambridge C. of C. (dir. 1969-73), Sigma Xi. Clubs: Harvard (Boston); Yale (N.Y.C.); Cosmos (Washington). Subspecialties: Otorhinolaryngology; Pathology (medicine). Current work: Method improvement for in vico chronic toxicity and carcinogenesis bioassays; development of animal models of human disease through inbreeding of Syrian hemsters. Home: 759 High St Dedham MA 02026 also Trenton ME 04605 Office: 9 Commercial Ave Cambridge MA 02141

HOMER, LOUIS DAVID, physiologist; b. Washington, Mar. 21, 1935; s. David and Louise Berthe (Perreau) H.; m. Margie Winifred Parker, June 17, 1961; children: Margie Lou, David Louis, Mary Jane. A.B., Columbia U., 1955; Ph.D., Med. Coll. Va., 1962, M.D., 1963. Asst. prof. Emory U., Atlanta, 1963-65, assoc. prof., 1965-67, Brown U., Providence, 1969-71; med. officer Naval Med. Research Inst., Bethesda, Md., 1971—. Stats. cons.: Jour. Applied Physiology, 1982—. Served to lt. comdr. USN, 1967-69. Mem. Biometrics Soc., Am. Physiol. Soc., Sigma Xi. Subspecialty: Physiology (medicine). Current work: Mathematical physiology, theoretical biology. Office: Naval Med Research Inst Nat Naval Med Ctr Bethesda MD 20814

HOMER, PERCY ALBERT, mechanical engineer; b. Birmingham, Ala., Jan. 26, 1925; s. Percy A. and Janetta D. (Woodard) H.; m. Dorothy Mae Taylor, Mar. 8, 1947; m. Betty May Smith, July 25, 1970; adopted children: Randy, Mary, Lesa. B.S.M.E., Tri State Coll., Angola, Ind., 1950. Registered profl. engr., Ill. Project engr. Barber Colman, Rockford, Ill., 1950-64, sr. devel. engr., 1966-82, prin. engr. in research and devel., 1982—; staff engr. Landis Machine Co., Waynesboro, Pa., 1965. Served to cpl. USAAF, 1943-46; PTO. Mem. ASME, Abrasive Engring. Soc. Methodist. Lodges: Am. Legion; Moose; Masons. Subspecialties: Mechanical engineering; Applied mathematics. Current work: Development of computer numerically controlled gear and sharer cutter grinding machinery and inspection machine. Research and devel. on precision machine tools; design projects include lathes, hobbing machines, form tool grinder, gear rolling machine, computer controlled gear grinder, metrication machine. Home: 1643 Lawney Ct South Beloit IL 61080 Office: 1351 Windsor Rd Loves Park IL 6111

HOMMES, FRITZ AUKUSTINUS, biochemistry educator, researcher; b. Bellingwolde, Groningen, Netherlands, May 28, 1934; came to U.S., 1979; s. Aukustinus and Anje (Wester) H.; m. Grietje Renes, June 14, 1958; children: Peter, Anneliek. M.S., U. Groningen, 1958; Ph.D., U. Nijmegen, 1961. Research asst. dept. biochemistry U. Nijmegen, Netherlands, 1959-61; Fulbright postdoctoral fellow dept. biochemistry U. Pa., Phila., 1961-63; head lab. dept. pediatrics U. Groningen, Netherlands, 1963-72, assoc. prof. dept. pediatrics, 1972-79; prof. dept. cell molecular biology Med. Coll. Ga., Augusta, 1979—, dir. biochem. genetics lab., 1980—; cons. genetic diseases Dutch Health Council, 1974-79, chmn. bioenergetics study group, 1975-77; co-chmn. comprehensive genetics systems State of Ga., 1982—. Editor: Inborn Errors of Metabolism, 1973, Normal and Pathological Development of Energy Metabolism, 1975, Models for the Study of Inborn Errors of Metabolism, 1979; contbr. articles to research publs. Bd. dirs. Parents Assn., Huize Maartenswouden, Drachten, Netherlands, 1975-79, Fountainhead Condominium Assn., Augusta, 1982-83. Genetics lectr. Japanese Govt., 1981. Mem. European Soc. Pediatric Research, Soc. Study Inherited Metabolic Diseases, N.Y. Acad. Scis., Am. Soc. Human Genetics, Soc. Study Inborn Errors of Metabolism, AAAS, Am. Soc. Biol. Chemists (lectr. 1965). Roman Catholic. Clubs: Round Table (Groningen) (chmn. 1970-71, chmn. No. Dist. 1973-75); Round Table (Groningen) (nat. bd. 1974-75). Lodge: Rotary (Augusta). Subspecialties: Biochemistry (medicine); Genetics and genetic engineering (medicine). Current work: Biochemistry of inborn errors in metabolism. Patentee detection of lactaciduria. Home: 793 Brookfield Pkwy Augusta GA 30907 Office: Dept Cell Molecular Biology Med Coll Georgia Augusta GA 30912

HOMSY, CHARLES ALBERT, biomaterials researcher, educator; b. Boston, June 21, 1932; m., 1956; 3 children. S.B., MIT, 1953, Sc.D. in Chem. Engring., 1960. Asst. chem. engring. MIT, 1953-54, asst. dir. chem. engring. practice sch., 1954-55, asst., 1955-59; research engr. E.I. DuPont de Nemours & Co., 1959-61, tech. rep., 1961-64, mktg. mgr., 1964-66; coordinator prosthetic materials and devel., dir. orthopedic prosthesis lab. Methodist Hosp., Houston, 1966-67, dir. lab., 1967—; research assoc. prof. Baylor U. Coll. Medicine, Houston, 1977—; pres. Vitek, Inc. Mem. Orthopedic Research Soc., Soc. Biomaterials., Am. Inst. Chem. Engrs., Am. Chem. Soc. Subspecialties: Prosthetics; Biomedical engineering. Office: Methodist Hosp 6560 Fannin Suite 2080 Scurlock Tower Houston TX 77030

HOMYK, THEODORE, JR., geneticist; b. Miami, Fla., June 6, 1944; s. Theodore and Georgia (Hardeman) H.; m. Mona Zaklama, June 8, 1974; children: David Emmanuel, Andrew Peter. B.S., U. Ga., 1966; Ph.D., Vanderbilt U., 1974. Postdoctoral fellow U. B.C. (Can.), Vancouver, 1974-80; postdoctoral fellow dept. biol. scis. Purdue U., West Lafayette, Ind., 1980—. Contbr. articles to profl. jours. Mem. Genetics Soc. Am. Democrat. Methodist. Subspecialties: Developmental biology; Genetics and genetic engineering (biology). Current work: Developmental biology and the role of genes in the function and development of neural and muscular tissue in Drosophila. Home: 400 N River Rd Apt 1316 West Lafayette IN 47906 Office: Dept Biol Sci Purdue U 3-301-B West Lafayette IN 47907

HONG, SU-DON, material scientist, research director; b. Chang-Hua, Taiwan, Republic of China, June 23, 1941; came to U.S., 1970; s. Ten-Sheng and Lee-Tze (Lee) H.; m. Grace C. Wang, Jan. 30, 1971; children: Emily W., Jennifer J. B.S., Nat. Taiwan U., 1966; M.S., U. Waterloo, Ont., Can., 1970; Ph.D., U. Mass., 1975. Teaching asst. Nat. Taiwan U., Taipei, Repulic of China, 1967-68; research assoc. U. Calif.-Berkeley, 1976-77; sr. scientist Jet Propulsion Lab., Pasadena, Calif., 1977-78, tech. group leader, 1978—; cons. Three Bond Corp., Gardena, Calif., 1980. Mem. Am. Chem. Soc., Am. Phys. Soc., Soc. Rheology, Soc. Plastic Engrs., Sigma Xi. Subspecialties: Polymer physics; Composite materials. Current work: Relationship between molecular relaxation mechanism and mechanical properties of polymers, physical aging and long-term performance of polymers, advanced composite materials of improved toughness. Patentee double-beam optical method and apparatus for measuring thermal diffusibility. Home: 10551 E Danbury St Temple City GA 91780 Office: Calif Inst Tech Jet Propulsion Lab 4800 Oak Grove Dr Pasaden CA 91109

HONIG, ARNOLD, physics educator; b. N.Y.C., Feb. 28, 1928; m., 1947 and 1979; 3 children. B.A., Cornell U., 1948; M.A., Columbia U., 1950, Ph.D. in Physics, 1953. Asst. microwave spectroscopy Columbia U., 1951-53; research physicist solid state physics U. Calif., 1953-54; research fellow molecular physics Ecole Normale Superieure, Paris, 1954-56; from asst. prof. to assoc. prof. physics Syracuse (N.Y.) U., 1956-62, prof., 1962—. Mem. Am. Phys. Soc., Fedn. Am. Scientists. Subspecialty: Low temperature physics. Office: Dept Physics Syracuse U Syracuse NY 13210

HONIG, CARL R., physiologist, educator; b. N.Y.C., July 15, 1925; s. Samuel Earl and Edith Mildred (Sussman) H.; m. Betty Rita London, July 4, 1947; children: David Earl, Amy Susan. B.S., U. Rochester, 1947; M.D., L.I. Coll. Medicine, 1949. Diplomate: Am. Bd. Internal Medicine. Staff physiology sect. Wright-Patterson Air Devel. Ctr., 1952-54; mem. faculty U. Rochester (N.Y.) Sch. Medicine, 1957—, assoc. prof., 1963-67, prof. physiology, 1967—; adj. prof. physiology U. Hawaii Sch. Medicine, 1979—. Author: Modern Cardiovascular Physiology, 1980; Contbr. articles to med. jours. Served with USNR, 1943-45; Served with USAFR, 1952-54. USPHS Research Career Devel. awardee, 1961-71. Mem. Am. Physiol. Soc., Microcirculatory Soc., Internat. Soc. for O2 Transport to Tissue, Cardiac Muscle Soc. Subspecialties: Physiology (biology); Microcirculation. Current work: Microcirculation, O_2 transport, neural control of circulation; intracellular O_2 gradients are measured by use of cryomicrospectrophotometry of myglobin. Home: 70 Stoneham Dr Rochester NY 14625 Office: 601 Elmwood Ave Box 642 Rochester NY 14642

HONIGBERG, BRONISLAW MARK, zoology educator; b. Warsaw, Poland, May 14, 1920; came to U.S., 1941, naturalized, 1948; s. Zachary Z. and Mary (Laks) H.; m. Rhoda Springer, Feb. 7, 1948; children: Paul Mark, Martin Philip. A.B., U. Calif., Berkeley, 1943, M.A., 1946, Ph.D. (A. Rosenberg research fellow 1949-50), 1950. Instr. to prof. zoology U. Mass., Amherst, 1950—, chancellor's medalist, lectr., 1975, faculty fellow, 1981-82; asst. prof. Columbia U., summer 1954; research asso. in pathobiology Sch. Hygiene, Johns Hopkins U., 1958-59; asst. prof. Harvard U., summer 1959; guest investigator lab. parasitic diseases Nat. Inst. Allergy and Infectious Diseases, NIH, 1965-66; guest investigator, hon. fellow Centre for Tropical Veterinary Medicine, U. Edinburgh, 1973-74; dir. tng. grants NIH, 1973—; mem. Internat. Commn. Protozoology, 1965—, tropical medicine parasitology study sect. NIH, 1973-77; v.p., chmn. sci. program V, Internat. Congress Protozoology, 1977; v.p. Internat. Symposiumon Trichomoniasis, Bialystok, Poland, 1981. Editor: Jour. Protozoology, 1971-80; asso. editor: Trans. Am. Micros. Soc., 1966-71; editor N.Am.,: Zeitschrift für Parasitenkunde, 1974—; bd. reviewers: Acta Tropica, 1977-82; contbr. articles to profl. jours. Trustee Am. Type Culture Collection, 1966-72, mem. exec. com., 1971-72. Recipient Gold medal for human trichomoniasis studies Med. Faculty Comenius U., Bratislava, Czechoslovakia, 1977; Alexander von Humboldt Found. sr. scientist award, 1982; NIH research grantee, 1955—; USPHS spl. research fellow, 1965-66, 73-74. Fellow AAAS, N.Y. Acad. Sci., Royal Soc. Tropical Medicine and Hygiene; mem. Soc. belge de Médicine Tropicale (corr.), Deutsche Gesellschaft für Parasitologie (corr.), Am. Soc. Zoologists, Am. Soc. Parasitologists, Soc. Protozoologists (pres. 1965-66, hon.), Am. Micros. Soc. (pres. 1964-65), Am. Soc. Tropical Medicine and Hygiene, Biol. Stain Commn., Soc. Systematic Zoology, Phi Beta Kappa, Sigma Xi, Phi Kappa Phi. Subspecialties: Parasitology. Current work: Immunology and pathogenicity mechanisms of parasitic protozoa, especially flagellates (trichomonads and trypanosomes.). Home: 95 Red Gate Ln Amherst MA 01002 Office: Zoology Dept Morrill Sci Center Univ Mass Amherst MA 01003

HONMA, SHIGEMI, horticulture educator; b. Haina, Hawaii, Feb. 14, 1920; s. Hichitaro and Tatsumi (Kawashima) H.; m. Isao Okajima, Oct. 5, 1945; children: Valerie Edith, Alan Kern. B.A., Cornell U., 1949; Ph.D., U. Minn., 1953. Research assoc. U. Minn., Mpls., 1950-52; asst. horticulturist U. Nebr., Lincoln, 1953-55; prof. horticulture Mich. State U., East Lansing, 1955—. Served with U.S. Army, 1943-45; ETO. Recipient Meritorious Service award Bean Improvement Coop., 1975; Disting. Alumni Lectr. Dept. Horticulture, U. Minn., 1981. Fellow Am. Soc. Hort. Sci.; mem. Am. Genetic Assn.

Subspecialties: Plant genetics; Hydroponics. Current work: Vegetable geneticist and breeder, greenhouse crop production. Office: Dept Horticulture Mich State U East Lansing MI 48824

HOOD, LEROY EDWARD, biologist; b. Missoula, Mont., Oct. 10, 1938; s. Thomas Edward and Myrtle Evylan (Wadsworth) H.; m. Valerie Anne Logan, Dec. 14, 1963; children—Eran William, Marqui Leigh Jennifer. B.S., Calif. Inst. Tech., 1960, Ph.D. in Biochemistry, 1968; M.D., Johns Hopkins U., 1964. Med. officer USPHS, 1967-70; sr. investigator Nat. Cancer Inst., 1967-70; asst. prof. biology Calif. Inst. Tech., Pasadena, 1970-73, asso. prof., 1973-75, prof., 1975—, Bowles prof. biology, 1977—, chmn. div. biology, 1980—. Author: (with others) Biochemistry, a Problems Approach, 1974, Molecular Biology of Eukaryotic Cells, 1975, Immunology, 1978, Essential Concepts of Immunology, 1978. Mem. Am. Assn. Immunologists, Am. Assn. Sci., Sigma Xi. Subspecialties: Immunobiology and immunology; Genetics and genetic engineering (biology). Current work: Molecular biology of antibody genes and genes of the major histocompatability complex; microchemical instrumentation. Home: 1453 E California Blvd Pasadena CA 91106 Office: Div Biology Calif Inst Tech Pasadena CA 91125

HOOD, RONALD DAVID, biology educator; b. El Paso, July 5, 1941; s. Estil Dunlap and Cecile Lenore (Russie) H.; m. Paula Nell Shaw, June 2, 1963 (div. 1970); m. Barbara Kay Owen, Jan. 8, 1972; 1 dau., Rebecca Ann. B.S., Tex. Tech. U., 1963, M.S., 1965; Ph.D., Purdue U., 1969. Asst. prof. dept. biology U. Ala., Tuscaloosa, 1968-73, assoc. prof., 1973-78, prof., 1978—; prin. assoc. Ronald D. Hood & Assocs. (Cons. Toxicologists), Northport, Ala., 1978—. Contbr. articles to profl. jours. Tex. Tech. U. Pantex Research fellow, 1964; NASA research traineeship, 1966; NIOSH grantee, 1979; March of Dimes Birth Defects Found. grantee, 1982. Mem. Teratology Soc., Behavioral Teratology Soc. (charter), Soc. for Study of Reprodn. (charter), Am. Assn. Lab. Animal Sci., AAAS. Unitarian. Club: Bama Bassmasters. Subspecialties: Teratology; Toxicology (medicine). Current work: Teratogenic effects of arsenicals; arsenic metabolism; development of new screening systems for teratogens and reproductive toxins. Home: 12 N Northwood Lake Northport AL 35476 Office: Dept Biology U Ala Tuscaloosa AL 35486

HOOK, JERRY BRUCE, toxicologist, educator; b. Elk City, Okla., Sept. 7, 1937; m.; children: Bruce, Marilyn. B.S. with honors in Pharmacy (teaching and research asst.), Wash. State U., 1960; M.S., U. Iowa, 1964, Ph.D. in Pharmacology (USPHS fellow), 1966. Diplomate: Am. Bd. Toxicology. Asst. prof. pharmacology Mich. State U., East Lansing, 1966-71, asso. prof., 1971-75, prof., 1975-78, prof. pharmacology and toxicology, 1978—, coordinator basic sci., office interdeptl. curriculum, 1970-72; adj. prof. U. Mich., 1977; vis. scientist Imperial Chem. Industries, Alderley Park, Eng., 1979-80; dir. Center Environ. Toxicology, Mich. State U., 1980—. Editorial bd.: Fundamental and Applied Toxicology, 1981—, Jour. Pharmacology and Exptl. Therapeutics, 1972—, Jour. Pharmacol. Methods, 1977—, Jour. Toxicology and Applied Pharmacology, 1978—, Life Sci, 1978—, Pediatric Pharmacology, 1979—, Procs. Soc. Exptl. Biology and Medicine, 1979—; contbr. numerous articles to sci. jours. Nat. Inst. Environ. Health Scis. grantee, 1980-81. Mem. Am. Physiol. Soc., Am. Soc. Nephrology, AAAS, Am. Soc. Pharmacology and Exptl. Therapeutics, Brit. Toxicology Soc., Internat. Soc. Nephrology, Mich. Heart Assn., Mich. Kidney Found., Nat. Kidney Found., Soc. Environ. Toxicology and Chemistry, Soc. Exptl. Biology and Medicine, Soc. Toxicology. Subspecialties: Environmental toxicology; Toxicology (medicine). Current work: Target organ toxicology; renal toxicology.

HOOK, JOHN WILLIAM, exploration geologist, consultant; b. Capon Bridge, W.Va., June 11, 1922; s. John William and Beulah Francis (Orndorff) Oates) H.; m. Charlotte Ruth Winans, Aug. 27, 1949; children: John William III, Thomas Leslie, Richard Charles, James Garvin. Student, Elon Coll., 1941-43; A.B., U. Tenn., 1947. Geologist Am. Zinc Co., Tenn., Okla., Mo., 1947-54, Reynolds Metals Co., Ky., Oreg., Wash., 1954-75, John W. Hook & Assocs., Salem, Oreg., 1974—. Served with USAAF, 1943-45. Fellow Geol. Soc. Am.; mem. Soc. Econ. Geologists, Geothermal Resources Council. Democrat. Methodist. Subspecialty: Geology. Current work: Publications on zinc deposits in Tenn., flourite deposits in Ky., bauxite deposits in Oreg. and Wash., and geothermal energy in the Pacific Northwest; discoverd the Young Mine in Tenn. Home: 7315 Battle Creek Rd SE Salem OR 97301 Office: John W Hook & Assocs Inc PO Box 3133 Salem OR 97302

HOOKER, ARTHUR LEE, geneticist, plant pathologist; b. Lodi, Wis., Oct. 12, 1924; s. Robert Lee and Dora Magdalena (Leuth) H.; m. Ellen Margaret Zimmerman, July 5, 1950; children: David Lee, Margaret Ann. B.S., U. Wis.-Madison, 1948, M.S., 1949, Ph.D., 1952. Project assoc. U. Wis., Madison, 1951-52, asst. prof. plant pathology, 1954-58; asst. prof. botany and plant pathology Iowa State U., Ames, 1952-54; plant pathologist U.S. Dept. Agr., 1954-58; mem. faculty U. Ill., Urbana, 1958-80, prof. plant pathology and genetics, 1963-80; biosci. dir. Pfizer Genetics, Inc., St. Louis, 1980-82; biosci. dir DeKalb-Pfizer Genetics, DeKalb, Ill., 1982—; cons. in field. Contbr. articles to profl. jours. Served with U.S. Army, 1944-46. Recipient Funk award U. Ill., 1974; Guggenheim Found. fellow, 1964. Fellow AAAS, Am. Phytopath. Soc., Am. Soc. Agronomy; mem. Crop Sci. Soc. Am. Republican. Subspecialties: Plant genetics; Plant pathology. Current work: Genetics and physiology of plant parasite interactions, plant breeding, plant genetic resources and utilization, world food and agriculture. Home: 39W749 Deerhaven Trail Saint Charles IL 60174 Office: 1300 Sycamore Rd DeKalb IL 60115

HOOPER, GARY RAY, plant pathologist; b. Belvedere, Calif., Aug. 28, 1937; s. Claude Gerald and Rachael Ann (Palmer) H.; m. Karen Anne Nicol, Aug. 23, 1960; children: Laurie Ann, Michael C. and Boyd C. (twins). B.S., Brigham Young U., 1963; Ph.D., U. Calif.-Riverside, 1968. Plant pathologist, asst. prof. Mich. State U., East Lansing, 1968-71; dir. (Central Electron Microscope Lab.), 1972-80; asst. prof. Calif. Poly. State U., San Luis Obispo, 1971-72; head dept. plant pathology and physiology Va. Poly. Inst. and State U., Blacksburg, 1980—. Contbr. articles to profl. jours. Mem. sports bd. YMCA, East Lansing, 1972-80; bishop Ch. Jesus Christ of Latter-day Saints, East Lansing, 1976-80. Served with USAFR, 1956-64. Mem. Am. Phytopath. Soc., Weed Sci. Soc. Am., Electron Microscopy Soc. Am., Sigma Xi. Subspecialties: Plant pathology; Plant virology. Current work: Ultrastructure of diseased plants, fungal electron microscopy, plant pathology. Office: Dept Plant Pathology and Physiology Va Poly Inst and State U Blacksburg VA 24061

HOOPER, ROBERT GEORGE, physician, medical administrator; b. New Brunswick, N.J., Dec. 15, 1943; s. John H. and Katheryn A. (Behrens) H.; m. Dana Dean Glasgow, July 24, 1965. B.S., U. Okla., 1965, M.D., 1968. Diplomate: Am. Bd. Internal Medicine, Am. Bd. Pulmonary Disease. Intern Maricopa County Hosp., Phoenix, 1968-69; resident U. Okla. Med. Center, 1969-71, Tripler Army Med. Center, 1971-72, Walter Reed Army Med. Center, 1972-74; practice medicine, specializing in pulmonary disease, Phoenix, 1979—; dir. pulmonary function lab. Walter Reed Army Med. Ctr., Washington, 1974-79, asst. chief pulmonary disease service, 1975-79; dir. dept. respiratory care services St. Luke's Hosp. Med. Ctr., Phoenix, 1979—. Contbr. articles to sci. jours. Served to lt. col. U.S. Army, 1971-79. Fellow Am. Coll. Chest Physicians, ACP; mem. Am. Thoracic Soc., Am. Heart Assn., Am. Fedn. Clin. Research, Ariz. Thoracic Soc. Subspecialty: Pulmonary medicine. Current work: Clinical evaluation of bronchogenic carcinoma; clinical physiology; respiratory care. Office: 525 N 18th St Phoenix AZ 85006

HOORY, SHLOMO, Med. radiation physicist; b. Baghdad, Iraq, June 1, 1935; came to U.S., 1973, naturalized, 1980; s. Ezra and Salha (Dabby) H.; m. Hadassah Gelbgarb, Feb. 8, 1968 (div.); children: Itai, Eyal. Ph.D., The Hebrew U. Jerusalem, Israel, 1970. Chief physicist Hadassah U. Med. Center, 1962-63; chief physicist, researcher Tel-Hashomer Tel-Aviv U. Hosp., 1964; in-charge bldg. sci. and med. equipment Div. X-Rays and Highly Ionized Spectra, Hebrew U., 1964-70; chief solar observatory Tel-Aviv U., Israel, 1972-73; chief physicist div. nuclear medicine L.I. Jewish Hillside Med. Center, 1974—; asst. prof. radiation physics and radiobiology L.I. Jewish Hillside Med. Center, Stony Brook U. Served with Israeli Def. Forces. Mem. Soc. Nuclear Medicine, Am. Assn. Physicists in Medicine, Health Physics Soc., AAAS., Am. Phys. Soc. Jewish. Subspecialty: Nuclear medicine. Current work: Reconstruction of images from projections; solar physics; astrophysics, biophysics, computerization of radionuclide inventory and radioactive waste disposal; spectra of highly ionized atoms and soft x-rays, nuclear medicine dosimetry. Home: 3E Mill Dr 1F Great Neck NY 11021 Office: Dept Nuclear Medicine L I Jewish Hosp New Hyde Park NY 11042

HOOTMAN, HARRY EDWARD, nuclear engineer, consultant; b. Oak Park, Ill., June 5, 1933; s. Merle Albert and Rachel Edith (Atkinson) H.; m. Linda Pearl Smith, Nov. 23, 1963; children: David, Holly, John. B.S. in Chemistry, Mich. Technol. U., 1957, M.S. in Nuclear Engring, 1962. Registered profl. engr., S.C. Research assoc. Argonne (Ill.) Nat. Lab., 1959-62; process engr. Savannah River Plant, Aiken, S.C., 1962-65; research staff engr. Savannah River Lab., Aiken, 1965—; cons. transuranic waste disposal and incineration. Bd. dirs. Central Savannah River Area Sci. and Engring. Fair, Inc., Augusta, Ga., 1972—. Served to sgt. USAF, 1953-57. Mem. Am. Acad. Environ. Engrs., Nat. Soc. Profl. Engrs. (local chmn. 1978-79), Am. Nuclear Soc. (local chmn. 1979-80), Am. Phys. Soc., Sigma Xi. Baptist. Subspecialties: Nuclear engineering; Laser-induced chemistry. Current work: Process and production plant design for production, separation, shielding and disposal of radioisotopic sources and products. Inventor alpha waste incinerator. Home: 820 Brandy Rd Aiken SC 29801 Office: Savannah River Lab Aiken SC 29808

HOOVER, DONALD BARRY, pharmacology educator; b. Sunbury, Pa., July 20, 1950; s. Robert T. and Helen V. (Bartholomew) H.; m. Joyce Ann Hoover, June 19, 1971; 1 son, Bryan P. B.S., Grove City Coll., 1972; Ph.D., W.Va., U., 1976. Research assoc. NIH, Bethesda, Md., 1976-78; asst. prof. pharmacology East Tenn. State U., 1978—. Mem. Am. Soc. Pharmacology and Exptl. Therapeutics, Soc. Neurosci. Subspecialties: Pharmacology; Neuropharmacology. Current work: Localization of cholinergic neurons; role of peptide neurotransmitters in central and peripheral cardiovascular regulation. Office: Dept Pharmacology East Tennessee State University Johnson City TN 37614

HOOVER, L. JOHN, engineering executive; b. Altoona, Pa., July 16, 1942; s. Leo John and Fannie (Waxler) H.; m. Margaret Lytle, June 5, 1965; children: Todd Duncan, Beth Ann. B.S. in Engring. Mechanics, Pa. State U., 1964, Ph.D. in Nuclear Engring, 1968. Dir. integrated assessments Argonne (Ill.) Nat. Lab., 1968-80; sr. v.p. Energy Impact Assocs., Pitts., 1980-82; mgr. energy and environ. systems Advanced Tech., McLean, Va., 1982—; advisor Great Lakes Basin Commn., Mich., 1979-80. Mem. Darien (Ill.) Planning Commn., 1970-71. Mem. Am. Nuclear Soc., AAAS, Sigma Xi. Subspecialties: Fuels; Systems engineering. Current work: Technology assessment of cost, performance and reliability of engineering systems. Office: Advanced Technology 7923 Jones Branch McLean VA 22102

HOOVER, WILLIAM LEICHLITER, forestry educator, researcher; b. Brownsville, Pa., July 29, 1944; s. Aaron Jones and Edith (Leichliter) H.; m. Peggy Jo Spangler, Aug. 30, 1976; children: Jennifer Mary, Monica Susan, Samuel Spangler. B.S., Pa. State U., 1966, M.S., 1971; Ph.D., Iowa State U., 1977. Research asst. Pa. State U., Iowa State U., 1970-74; asst. prof. Purdue U., West Lafayette, Ind., 1974-79, assoc. prof. dept. forestry and natural resources, 1980—; sec./treas., dir. econ. and fin. analysis Tim Tech., Inc., West Lafayette, 1978—. Asst. editor: Timber Tax Jour, 1979—; author: A Guide to Federal Income Tax for Timber Owners, 1982, Timber Tax Management, 1978, Timber Taxation and Investment Management, 1983. Served to 1st lt., C.E. U.S. Army, 1967-69. Decorated Bronze Star medal. Mem. Forest Products Research Soc., Am. Econ. Assn., Nat. Assn. Pub. Accts., Soc. Am. Foresters. Republican. Presbyterian. Subspecialties: Composite materials; Materials processing. Current work: Design and development of new wood based composite materials for structural applications, including marketing and strategic considerations. Home: 206 Connolly St PO Box 2257 West Lafayette IN 47906 Office: Dept Forestry Purdue U West Lafayette IN 47907

HOOVIS, MARVIN LORIN, physician; b. N.Y.C., Apr. 22, 1929; s. Philip E. and Pansy Sarah (Kripser) H.; m. Beverly Ann Vigneault, Jan 6, 1959; children: Michelle, Keith, Dana, Hilary, Blaine. B.S., Bklyn. Coll., 1950; postgrad., U. Fribourg, Switzerland, 1950-52; M.D., U. Lausanne, Switzerland, 1956. Intern Coney Island Hosp., Bklyn., 1956-57, resident in internal medicine, 1957-59; chief resident in internal medicine St. Mary's Hosp., Waterbury, Conn., 1959-60, mem. staff, 1961-68, Roger Williams Gen. Hosp., Providence, 1968-74; dir. oncology Wesson Meml. Hosp., Springfield, Mass., 1974-80; mem. attending staff Union Hosp., Terre Haute, Ind., 1980—, Terre Haute Regional Hosp., 1980—; cons. med. oncology Lakeview Med. Center, Danville, Ill., 1980—, Clay County Hosp., Brazil, Ind., 1981—, Vermillion County Hosp., Clinton, Ind., 1981—, St. Elizabeth Hosp., Danville, 1981—; asst. prof. medicine Brown U., Providence, 1970—; asst. clin. prof. medicine Yale U. Sch. Medicine, 1975—; asst. prof. medicine U. Mass., Worcester, 1976—. Contbr. articles to profl jours. Mem. Am. Cancer Soc. (v.p. Vigo County unit 1981-82), Am. Soc. Clin. Oncology, Am. Assn. Cancer Edn., Am. Assn. Cancer Research, Mass. Med. Soc., Am. Soc. Preventive Oncology, Vermilion County Med. Soc., Ill. State Med. Assn., AMA. Subspecialties: Oncology; Hematology. Current work: Research on cell culture, biochemical pharmacology, and drug mechanisms of action. Clinical research and cancer education for community physicians and nurses. Office: 806 N Logan Ave Danville IL 61832

HOPCROFT, JOHN EDWARD, educator; b. Seattle, Oct. 7, 1939; s. Horace George and Catherine Matilda (Nist) H.; m. Judith Ann Bridgewater, June 20, 1964; children: AnnMarie, Michael, Thomas. B.S., Seattle U., 1961; M.S., Stanford U., 1962, Ph.D., 1964. Asst. prof. Princeton (N.J.) U., 1964-67; assoc. prof. Cornell U., Ithaca, N.Y., 1967-72, prof., 1972—; vis. assoc. prof. Stanford (Calif.) U., 1970-71; mem. NSF adv. panel on computer sci. and engring., 1971-74; sci. rev. panel U.S. Israel Binat. Sci. Found., 1974-78; advisor Ont. Council on Grad. Studies, 1974; mem. NRC Computer & Tech. Bd., 1980-82; cons. Army Electronics Command, 1966-67, Bell Labs., 1966-68, Systems Devel. Found., 1966-69, IBM, 1981—. Editor: SIAM Jour. on Computing, 1972-74; mng. editor (1974-77); editor (1980); assoc. editor: Jour. Computer & System Scis, 1980—; editorial adv. bd.: Info. & Control, 1982—; author: (with Ullman) Formal Languages and Their Relation to Automata, 1969, (with Aho and Ullman) The Design and Analysis of Computer Algorithms, 1974, (with Ullman) Introduction to Automata Theory, Language and Computation, 1979, (with Aho and Ullman) Data Structures and Algorithms, 1983. Mem. Assn. for Computing Machinery, IEEE, Soc. Indsl. and Applied Math. Subspecialties: Algorithms; Robotics. Home: 123 Christopher Circle Ithaca NY 14850 Office: Dept Computer Sci Cornell U 308 Upson Hall Ithaca NY 14853

HOPE, GEORGE MARION, research scientist; b. Waycross, Ga., Jan. 24, 1938; s. George Marion and Jessie (Norman) H.; m. Dorothy Marie Hendrix, Aug. 4, 1956; 1 son, Stephen Richard. Student, Armstrong Coll., 1957-58, Ga. Inst. Tech., 1958-60; A.B., Mercer U., 1965; M.A., U. Fla., 1967, Ph.D., 1971. Research asst. dept. ophthalmology Washington U., St. Louis, 1970-71; instr. dept. ophthalmology U. Louisville, 1972-73, asst. prof., 1974-79; assoc. research scientist dept. oththalmology U. Fla., 1980—, dir. low vision service dept. ophthalmology, 1980—, dir., 1981—; cons. def. div. Brunswick Corp., Deland, Fla., 1980; dir. Low Vision Clinic, dept. ophthalmology U. Louisville Sch. Medicine, 1973-79. Contbr. articles in field to profl. jours. Active Old Louisville Neighborhood Council, 1400 Block 4th St. Orgn., Chakala Neighborhood Orgn. Recipient Med. Faculty Research award U. Louisville, 1973-74; USPHS postdoctoral fellow, 1972; grantee So. Med. Assn., 1973-74; grantee-in-Aid Fight for Sight, Inc., 1973-74, Eye Inst., NIH, 1975-78, Brunswick Corp., 1980-81, Multi Optics Corp., 1982-83. Mem. Assn. Research in Vision and Ophthalmology (dir. placement service 1972—), Soc. Neurosci., AAAS, So. Soc. Philosophy and Psychology, N.Y. Acad. Scis., Sigma Xi. Democrat. Methodist. Subspecialties: Visual neuroscience; Sensory processes. Current work: Visual system structure and function, low vision, vision research low vision. Home: 1930 SW 19th Way Gainesville FL 32608 Office: Dept Ophthalmology Box J-284 JHMHC U Fla Coll Medicine Gainesville FL 32610

HOPEN, HERBERT J., horticulture educator; b. Madison, Wis., Jan. 7, 1934; s. Alfred O. and Amelia R. (Sveum) H.; m. Joanne C. Emmel, Sept. 12, 1959; children: Timothy, Rachel. B.S., U. Wis., 1956, M.S., 1959; Ph.D., Mich. State U., 1962. Asst. prof. horticulture U. Minn., Duluth, 1962-64; from asst. prof. to prof. vegetable crops in horticulture U. Ill.-Urbana, 1965—. Served to sgt. U.S. Army, 1958. Recipient Campbell Research award Am. Phytopath. Soc., 1980. Mem. Weed Sci. Soc. Am., Am. Soc. for Hort. Sci., North Central Weed Control Conf. Lutheran. Subspecialty: Integrated pest management. Current work: Cultural and biological weed control and mode of action of herbicides. Home: 409 Evergreen Ct W Urbana IL 61801 Office: 206A Vegetable Crops Bldg U Ill 1103 W Dorner Dr Urbana IL 61801

HOPFER, ULRICH, anatomy educator; b. Burg, Germany, Apr. 7, 1939; came to U.S., 1974, naturalized, 1980. M.D., U. Goettingen, W.Ger., 1966; Ph.D., Johns Hopkins U., 1970. Research assoc. Mass. Gen. Hosp., Boston, 1971-72; oberassistent, privatdozent Swiss Fed. Inst., Zurich, 1972-74; asst. prof. dept. anatomy Case Western Res. U., Cleve., 1974-77, assoc. prof., 1977-83, prof., 1983—; vis. scientist Max-Planck Inst. for Biophysics, Frankfurt, W.Ger., 1980-81; cons. Am. Heart Assn., 1974-77, Cystic Fibrosis Found., 1979-82, NIH, 1982—. Mem. Am. Biophys. Soc., AAAS, Am. Soc. Biol. Chemists, Am. Physiol. Soc., Brit. Biochem. Soc. Subspecialties: Physiology (medicine); Biochemistry (medicine). Current work: Mechanisms of epithelial transport; patholphysiology of exocrine secretion in cystic fibrosis. Home: 2915 Coleridge Rd Cleveland Heights OH 44118 Office: Dept Developmental Genetics and Anatomy 2119 Abington Rd Cleveland OH 44106

HOPFIELD, JOHN JOSEPH, biophysicist, educator; b. Chgo., July 15, 1933; s. John Joseph and Helen (Staff) H.; m. Cornelia Fuller, June 30, 1954; children—Alison, Jessica, Natalie. A.B., Swarthmore Coll., 1954; Ph.D., Cornell U., 1958. Mem. tech. staff Bell Telephone Labs., 1958-60, 73—; vis. research physicist Ecole Normale Superieure, Paris, France, 1960-61; asst. prof., then asso. prof. physics U. Calif. at Berkeley, 1961-64; prof. physics Princeton U., 1964-80, Eugene Higgins prof. physics, 1978-80; Dickinson prof. chemistry and biology Calif. Inst. Tech., Pasadena, 1980—. Guggenheim fellow, 1969. Fellow Am. Phys. Soc. (Oliver E. Buckley prize 1968); mem. Nat. Acad. Scis., Am. Acad. Arts and Scis., Neuroscis. Research Program, Phi Beta Kappa, Sigma Xi. Subspecialties: Theoretical chemistry; Neural Modeling. Current work: The relation between structure and function in biology; electron and transfer processes and solar energy; collective properties of neural networks. Home: 1728 San Pasqual St Pasadena CA 91106

HOPKIN, JOHN ALFRED, agricultural economist, educator; b. Fairview, Wyo., Dec. 23, 1918; s. Alfred and Maeline (Tolman) H.; m. Bonita Gardner, June 11, 1943; children: Tana H., Arden, Mark, Kerry, Jeffrey. B.S., U. Wyo., 1942, M.S., 1949; Ph.D., Iowa State U., 1954. Instr., asst. prof. U. Wyo., 1946-54; agrl. economist, chief agr. and commodity research, v.p. agribus. Bank of Am., San Francisco, 1955-67; prof. agrl. fin. U. Ill., Urbana, 1967-70; now Stiles prof. agrl. fin. Tex. A&M U., College Station. Co-author: Financial Management in Agriculture; Contbr. articles to profl. jours. Vice chmn. Family Farm Council Tex.; mem. Food and Agr. Com. U.S. C. of C.; mem. Nat. Brucellosis Tech. Commn. Served to capt. U.S. Army, 1942-46. Recipient Disting. Alumni award Grad. Sch. Credit and Fin. Mgmt. Dartmouth Coll. Mem. Am. Agrl. Econs. Assn., So. Agrl. Econs. Assn., Western Agrl. Econs. Assn., Internat. Agrl. Econs. Assn., Sigma Xi, Phi Kappa Phi, Alpha Zeta. Mormon. Subspecialty: Agricultural economics. Current work: Agricultural finance, production economics, livestock economics international development. Home: 1012 Holt College Station TX 77840 Office: Tex A&M U College Station TX 77843

HOPKINS, DONALD LEE, plant pathology educator; b. Sacramento, Ky., Mar. 13, 1943; s. Henry Lee and Virginia Evelyn (Bowman) H.; m. Paula Dean Lawrence, Dec. 29, 1967; children: Todd Anthony, Mauna Ree, Lee Anna. B.S., Western Ky. U., Bowling Green, 1965; Ph.D., U. Ky., 1968. Dept. Agr. postdoctoral research assoc. U. Wis.-Madison, 1968-69; asst. prof. plant pathology U. Fla., 1969-74, assoc. prof., 1974-79, prof., 1979—. Mem. Am. Phytopath. Soc., AAAS. Democrat. Methodist. Subspecialties: Microbiology; Plant physiology (biology). Current work: Gram-negative, xylem-limited bacteria in plants; etiology and control of watermelon and grape diseases; physiology of plant disease. Office: PO Box 388 Leesburg FL 32748

HOPKINS, HARVEY CHILDS, JR., nuclear engineer, chemical engineer; b. Quincy, Ill., Oct. 15, 1928; s. Harvey C. and Ruth M. (Michelmann) H.; m. Lucy T. Small, July 24, 1954; children: Stephen M., Harvey S. A.B. in Chemistry, Amherst Coll., 1951; B.S. in Chem. Engring, MIT, 1951, M.S., 1952. Cert. profl. nuclear engr., Calif.; profl. chem. engr., Calif. Asst. project engr. Pratt & Whitney Aircraft, East Hartford, Conn., 1952-57; sr. staff engr. Gen. Atomic Co., San Diego, 1957-82, cons. engr., 1983—. Mem. Am. Nuclear Soc. (standard com. 1958—), Sigma Xi, Kappa Kappa Sigma. Subspecialties: Nuclear engineering; Chemical engineering. Current work: Nuclear safety, equipment qualification, probabilistic risk assessment, synthetic fuel plant evaluation, computer design analysis, and data base construction. Office: PO Box 833 Rancho Santa Fe CA 92067

HOPKINS, STEPHEN WILLIAM, mechanical engineer; b. Brockton, Mass., July 23, 1943; s. Royal Carlton and Bertha Irene (Crowell) H.; m. Sandra Grippen, June 23, 1963; children: Stephen R., Susan M., Scott C., Spencer M. A.S. in Mech. Engring, Wentworth Inst., 1963, B.S., U. New Haven, 1971; M.S. in Applied Mechanics, Rensselear Poly. Inst., 1974. Registered profl. engr., Calif. Exptl. research asst. Pratt & Whitney Aircraft Co., North Haven, Conn., 1963-69, sr. materials engr., Middletown, Conn., 1969-76; mng. engr. Failure Analysis Assocs., Palo Alto, Calif., 1976-82; dir. Exptl. and Materials Lab., 1982—; invited lectr. on material behavior and structural life predictions. Contbr. articles to profl. jours. Mem. Am. Acad. Mechanics, ASME, Soc. Exptl. Stress Analysis (chmn. No. Calif. sect. 1978-79), ASTM, Alpha Sigma Lambda. Subspecialties: Mechanical engineering; Fracture mechanics. Current work: Experimental stress analysis, instrumentation, data acquisition and mechanical lifetime predictions analysis as they related to structures and component behavior to real service loads. Office: 2225 E Bayshore Rd Palo Alto CA 94303

HOPKINS, WILLIAM CHRISTOPHER, computer scientist; b. Winchester, Mass., Mar. 7, 1946; s. William McNeil and Lysbeth Hope H.; m. Susan Schopf, Feb. 15, 1975; 1 dau., Katherine Ruth. B.S., Yale U., 1969; postgrad., SUNY-Buffalo, 1969—. Instr. SUNY-Buffalo, part-time, 1972-75; vis. asst. prof. U. Pa., 1976-77; staff programmer Burroughs Corp., Downingtown, Pa., 1977—; chmn. 16th Ann. Workshop on microprogramming; vice chmn. 15th Ann. Workshop. Mem. Assn. Computing Machinery (newsletter editor), IEEE, Sigma Xi. Subspecialties: Software engineering; Computer architecture. Current work: Microprogramming, firmware engineering, software engineering, programming languages, instruction set design and analysis.

HOPLIN, HERMAN PETER, information science, educator, technical management consultant; b. Brandon, Minn., June 23, 1920; s. Peter Nelson and Edna Viola (Larson) H.; m. Eleanor Irene Johnson, Dec. 20, 1943; 1 dau., Barbara Lee. B.S., St. Cloud (Minn.) State U., 1942; M.B.A., Syracuse U., 1955; M.A., George Washington U., 1963, D.B.A., 1975. Commd. 2d lt. U.S. Army, 1942, advanced through grades to col., 1965; ret., 1972; mem. staff Joint Chiefs of Staff, Washington, 1967-72; research assoc. George Washington U., 1972-75; sr. mgmt. cons. Howard Finley Corp., Houston, 1973-75; professional lectr. Am. U., Washington, 1976-78, asst. prof. computers and mgmt. info. systems, 1978-79; assoc. prof. mgmt. info. systems Syracuse U., 1979—; lectr. Soc. for Mgmt. Info. Systems, 1975—; dir. Univ. Scis. Forum, Falls Church, Va., 1977-81; mem. exec. bd. Internat. Cons. Found., Washington, 1982—; mem. info. systems program and curriculum com. Data Processing Mgmt. Assn. Edn. Found., Park Ridge, Ill., 1982—. Author, contbr. papers and articles to profl. jours., symposia and confs.; referee: Systems Research Jour. Mem. adv. bd. University Coll., Syracuse, 1982—. Decorated Legion of Merit, D.S.M.; recipient Eagle scout award Boy Scouts Am., 1938; named Indian Chief Confederated Tribes of Umatilla Reservation, Oreg., 1967. Mem. Syracuse Systems Execs. Group (acad. mem.), Assn. for Computing Machinery (chmn. Syracuse chpt. 1982—), Soc. for Gen. Systems Research, Inst. Mgmt. Scis., IEEE Computer Soc. Lodge: Rotary. Subspecialties: Information systems (information science); Database systems. Current work: User-oriented and designed information systems, organization and management of information centers, high technology performance evaluation, and data base research in eight significant areas included distributed data base. Home: 8188 Pembroke Dr Manlius NY 13104 Office: Syracuse Univ 116 College Pl Syracuse NY 13210

HOPPER, DARREL GENE, chem. physicist; b. Stillwater, Okla., June 10, 1944; s. Doyle Houston and Dorthea Eleen (Randolph) H.; m. Chahira Metwally, Aus. 31, 1969; children: Dalia Lee Elizabeth, Paul Darrel Alexander. B.S. with honors, Okla. State U., 1966, Ph.D., 1971. Fellow NRC, Washington, 1972-74; faculty level appointee Argonne Nat. Lab., 1974-77; sr. scientist Sci. Applications, Inc., La Jolla, Calif., 1977-79; research prof. Wright State U., Dayton, Ohio, 1977—; theoretical chemist JAYCOR, La Jolla, 1979-80; dir. Mattergy Research Inst., Dayton, 1982—; Mem. Zoning Bd. Appeals, City of Beavercreek, Ohio, 1982—; physicist USAF, Wright - Patterson AFB, Ohio, 1982—. Contbr. articles to profl. jours. Mem. Am. Phys. Soc., Am. Chem. Soc., Am. Soc. Mass Spectroscopy, Laser Inst. Am., Assn. Old Crows. Republican. Presbyterian. Club: Officers (Dayton). Subspecialties: Atomic and molecular physics; Theoretical chemistry. Current work: Molecular structure, atmospheric chemistry, light-matter interactions. Quantum mechanical theory, ab initio methods, reaction mechanisms, molecular dynamics, photochemistry, ozone layer. Office: Mattergy Research Inst Overlook Br Box 31385 Dayton OH 45431

HOPPER, GRACE BREWSTER MURRAY, mathematician; b. N.Y.C., Dec. 9, 1906; d. Walter Fletcher; ; d. Mary Campbell and (Van Horne) Murray; m. Vincent Foster Hopper, June 15, 1930 (div. 1945). B.A., Vassar Coll., 1928; M.A. (Vassar fellow, Sterling scholar), Yale, 1930, Ph.D., 1934; postgrad. (Vassar Faculty fellow), N.Y. U., 1941-42; D.Eng. (hon.), Newark Coll. Engring., 1972, D.Sc., C.W. Post Coll. L.I. U., 1973, Pratt Inst., 1976, Linkoping U., Sweden, 1980, Bucknell U., 1980, Acadia U., Can., 1980, So. Ill. U., 1981, Loyola U. Chgo., 1981; LL.D., U. Pa., 1974; D.Public Service, George Washington U., 1981. From instr. to asso. prof. math. Vassar Coll., 1931-44; asst. prof. math. Barnard Coll., summer 1943; research fellow engring. scis., applied physics Computation Lab., Harvard, 1946-49; sr. mathematician Eckert-Mauchly Computer Corp., Phila., 1949-50; sr. programmer Eckert-Mauchly div. Remington Rand, 1950-59; systems engr., dir. automatic programming devel. UNIVAC div. Sperry Rand Corp., Phila., 1959-64, staff scientist systems programming, 1964-71; ret., 1971; vis. lectr. Moore Sch. Elec. Engring., U. Pa., 1959-63, vis. asso. prof. elec. engring., 1963-74, adj. prof., 1974; professorial lectr. George Washington U., 1971—. Contbr. numerous articles to profl. jours. Served to comdr. WAVES, 1944-46, 67—; capt. USNR, 1973; presently serving active duty NAVDAC. Decorated Legion of Merit, Meritorious Service award; recipient Naval Ordnance Devel. award, 1946, Connelly Meml. award, 1968, Wilbur L. Cross medal Yale, 1972, Sci. Achievement award Am. Mother's Com., 1970, others. Distinguished fellow Brit. Computer Soc.; fellow Assn. Computer Programmers and Analysts, IEEE (McDowell award 1979), AAAS; mem. Nat. Acad. Engring., Assn. Computing Machinery, Data Processing Mgmt. Assn. (Man of Yr. award 1969), Am. Fedn. Information Processing Socs. (Harry Goode Meml. award 1970), Soc. Women Engrs. (Achievement award 1964), Franklin Inst., U.S. Naval Inst., Internat. Oceanographic Found., DAR, Dames Loyal Legion, Hist. Soc. Pa., Geneal. Soc. Pa., N.H. Hist. Soc., New Eng. Hist.-Geneal. Soc., Valley Forge Hist. Assn., Ret. Officers Assn., Huguenot Soc. Pa., Nat., N.Y. geneal. socs., Pechin Soc., Phi Beta Kappa, Sigma Xi. Home: 1400 S Joyce St Arlington VA 22202 Office: Dept Navy NAVDAC Washington DC 20374 A ship in port is safe; but that is not what ships are built for.

HORAKOVA, ZDENKA ZAHUTOVA, toxicologist, researcher; b. Jindrichuv Hradec, Czechoslovakia, Apr. 6, 1925; d. Josef and Aloisie (Sohajova) Zahut; m. Vaclav Horak, Sept. 26, 1949; 1 son: David. Magister of Pharmacy, Charles U., Prague, 1949, Dr. of Natural Scis., 1952; Ph.D. in Pharmacology, Czechoslovakian Acad. Sci., 1961. Teaching asst. dept. pharmacology Med. Faculty, Charles U., Prague, 1949-50; research pharmacologist Research Inst. for Pharmacy and Biochemistry, Prague, 1950-58, head pharmacol. dept., 1958-68; vis. guest Zambon Pharm. Research Inst., Bresso-Milano, Italy, 1968; research pharmacologist exptl. therapeutics br. Nat. Heart and Lung Inst., NIH, Bethesda, Md., 1969-74, Lab. of Cellular Metabolism and Pulmonary Br., 1974-78; toxicologist, researcher Food Safety and Inspection Service, Residue Evaluation, Sci. div. Dept. Agr., Washington, 1978—; mem. Nat. Toxicology Program Working Group. Contbr. numerous articles on toxicology and pharmacology to profl. jours. Recipient diploma for Sci. Discovery on Inhibin Ministry of Health Prague, 1960. Mem. Am. Soc. of Pharmacology and Exptl. Therapeutics, Soc. Toxicology, Internat. Union of Pharmacology (organizing com. 2d Internat. Congress on Pharmacology 1963), Internat. Soc. for Study of Xenobiotics, Soc. for Exptl. Biology and Medicine (D.C. sect.), Cell and Molecular Biology in Space (NASA), Inflammation Research Assn., Internat. Soc. for Biochem Pharmacology, Internat. Inflammation Club, European Biol. Research Assn., Sigma Delta Epsilon. Roman Catholic. Subspecialties: Pharmacology; Toxicology (medicine). Current work: Pharmacological testing, methodology, inflammation, toxicology-residue evaluation. Home: 5508 Oakmont Ave Bethesda MD 20817 Office: 300 12th St SW Washington DC 20250

HORECKER, BERNARD LEONARD, biochemist; b. Chgo., Oct. 31, 1914; s. Paul and Bessie (Bornstein) H.; m. Frances Goldstein, July 12, 1936; children: Doris Colgate, Marilyn Diamond, Linda Lally. B.S., U. Chgo., 1936, Ph.D., 1939; Laureate honoris causa in Biol. Scis., U. Urbino (Italy), 1982. Research asso. chemistry U. Chgo., 1939-40; examiner U.S. Civil Service Commn., 1940-41; biochemist USPHS, NIH, Bethesda, Md., 1941-59; chief lab. of biochemistry and metabolism Nat. Inst. Arthritis and Metabolic Disease, 1956-59; professorial lectr. enzyme chemistry George Washington U., 1950-57; guest research-worker Pasteur Inst., Paris, France, 1957-58; prof. microbiology, chmn. dept. N.Y. U. Coll. Medicine, 1959-63; prof. molecular biology, chmn. dept. Albert Einstein Coll. Medicine, 1963-72, asso. dean for sci. affairs, 1971-72; mem. Roche Inst. Molecular Biology, Nutley, N.J., 1972—, head Lab. Molecular Enzymology, 1977—; vis. prof. Albert Einstein Coll. Medicine, 1972—; adj. prof. Cornell U. Med. Coll., 1972—; vis. prof. biochemistry U. Calif., 1954, U. Parana, Brazil, 1960, 63; vis. lectr. U. Ill., 1956; Ciba lectr. Rutgers U., 1962; Phillips lectr. Haverford Coll., 1965; vis. prof. Kyoto (Japan) U., 1967; vis. prof. biochemistry and molecular biology Cornell U., 1965—; vis. prof. U. Ferrara, Italy; Reilly lectr. Notre Dame U., 1969; vis. lectr. U. Rotterdam, 1970; prof. honoris causa Fed. U. Parana, Curitiba, Brazil, 1981—; mem. sci. adv. bd. Roche Inst. Molecular Biology, Nutley, N.J., 1967-72, chmn., 1971-72; dir. Academic Press, Inc., 1968-73; mem. Research Career Award com. Nat. Inst. Gen. Med. Scis., 1966-70; mem. personnel com. Am. Cancer Soc., 1968-72, mem. sci. adv. com. for biochemistry and chem. carcinogenesis, 1974-78; mem. biology div. adv. com. Oak Ridge Nat. Lab., 1976-80; mem. Med. Scientist Tng. Program Sect. NIH, 1970—. Editor: Biochem. and Biophys. Research Communications, 1959—, Current Topics in Cellular Regulation, 1969—, Archives Biochemistry and Biophysics, 1960-68; chmn. editorial bd.: Archives of Biochemistry and Biophysics, 1968—; contbr. articles to sci. publs. Recipient Paul Lewis Labs. award in enzyme chemistry, 1952; Superior Accomplishment award Fed. Security Agy., 1952; Rockefeller Pub. Service award, 1957; Hillebrand prize Am. Chem. Soc., 1954; award in biol. scis. Washington Acad. Sci., 1954; Fulbright Travel award, 1963; Commonwealth Fund fellow, 1967. Fellow AAAS, Am. Acad. Arts and Scis.; mem. Am. Chem. Soc. (vice chmn. div. biol. chemistry 1975-76, chmn. 1976—), Biochem. Soc. (Eng.), Swiss Biochem. Soc. (hon. mem.), Spanish Biochem. Soc., hon. mem.), Japanese Biochem. Soc. (hon. mem.), Hellenic Biochem. and Biophys. Soc. (hon. mem.), Am. Soc. Biol. Chemists (pres. 1967-68, chmn. editorial com. 1962-63, Merck award 1981, Neuberg medal 1981), Nat. Acad. Scis., Harvey Soc. (v.p. 1969-70, pres. 1970-71), Brazilian Acad. Sci. (hon.), PanAm. Assn. Biochem. Socs. (vice chmn. 1971, chmn. 1972, mem. exec. com. 1971—), Instituto de Investigaciones Citologicas (corr.), Phi Beta Kappa, Sigma Xi. Subspecialties: Biochemistry (biology); Molecular biology. Current work: Intracellular proteinosis: their role in regulation of metabolism. Peptides and peptide hormones: structure, function and biosynthesis. Office: Roche Inst Molecular Biology Nutley NJ 07110

HORENSTIEN, DAVID, clinical psychologist, hospital program administrator; b. N.Y.C., Apr. 1, 1948; s. Edward and Ann (Oshman) H.; m. Linda Allen, May 29, 1969; children: Devin, Geri. B.A., L.I.U., 1969; M.A., U. Kans., 1971, Ph.D., 1973. Lic. psychologist, N.Y. Research asst. U. Kans., 1969-70, 70-71, psychologist in tng., 1970-72; clin. psychology intern Wyandot Mental Health Ctr., Kansas City, Kans., 1972, 73; asst. prof. counseling psychology and student devel. SUNY-Albany, 1973-79; pvt. practice psychology, Clifton Park, N.Y., 1975—; cons. clin. psychology Albany County Community Mental Health Clinic, 1973-75, Albany Med. Ctr., 1974; consulting psychologist Pittman Hall-St. Francis Home for Girls, Londonville, N.Y., 1977, United Cerebral Palsy of Schenectady, 1977—; med. lectr. St. Clare's Hosp., Schenectady, 1978-79, mem. faculty family practice residencey program, 1979—; consulting clin. psychology Lexington Ctr., Johnstown, N.Y., 1981—. Editorial cons.: Human Relations Jour. Tavistock Inst. of Human Relations, London, Psychol. Bull. Juor. Am. Psychol. Assn; contbr. articles to profl. jours. Bd. dirs. The Clin. Sch., Albany, Albany City Hostel, Inc. Mem. Am. Psychol. Assn., N.Y. State Psychol. Assn., Nat. Rehab. Assn., Council Rehab. Educators, Nat. Rehab. Counseling Assn. Subspecialties: Clinical psychology; Family practice. Home: 70 Blue Spruce Ln Ballston Lake NY 12019 Office: Drs Horenstein & Wishnoff 873 Route 146 Clifton Park NY 12019

HORIUCHI, KENSUKE, molecular biologist; b. Tokyo, Sept. 21, 1933; s. Makoto and Yoshiko (Ishizaka) H.; m. Atsuko Yuyama, Apr. 15, 1962; children: Kentaro, Junjiro, Yozo. B.S., Tokyo U., 1957, M.S., 1959, Ph.D., 1962. Postdoctoral fellow dept. microbiology Yale U., New Haven, 1963-64; research assoc. Rockefeller Inst., 1964-66; asst. Tokyo U., 1966-69; research assoc. Rockefeller U., N.Y.C., 1969-73, asst. prof. molecular biology, 1973-78, assoc. prof., 1978—. Contbr. articles to profl. jours. Matsunaga Sci. Found. grantee, 1967; Fulbright grantee, 1963. Mem. Am. Assn. Microbiology, Am. Soc. for Virology. Subspecialties: Genetics and genetic engineering (biology); Molecular biology. Current work: Molecular genetics of bacteriophage, DNA replication, gene expression, restriction and modification enzymes. Home: 500 E 63d St New York NY 10021 Office: Dept Molecular Biology Rockefeller U 1230 York Ave New York NY 10021

HORN, HENRY STAINKEN, biology educator; b. Phila., Nov. 12, 1941; s. Henry Eyster and Catherine Hedwig (Stainken) H.; m. Elizabeth Ruth Gates, Sept. 7, 1963; children: Jennifer Downing, Eric Bailey. A.B., Harvard U., 1962; Ph.D., U. Wash.-Seattle, 1966. Teaching asst. Harvard U., 1962, U. Wash.-Seattle, 1962-66; asst. prof. Princeton (N.J.) U., 1966-71, assoc. prof., 1971-78, prof. biology, 1978—. Author: Adoptive Geometry of Trees, 1971. Bullard fellow Harvard U., 1975; NSF fellow U. Wash., 1962. Fellow AAAS; mem. Ecol. Soc. Am., Am . Soc. Naturalists, Lepidopterists Soc. Subspecialties: Population biology; Ethology. Current work: Ecology, ethology and population biology of birds, trees and butterflies. Office: Dept Biology Princeton U Princeton NJ 08544

HORN, JOHN LEONARD, psychology educator; b. St. Joseph, Mo., Sept. 7, 1928; s. John Leonard and Nellie Rae (Weldon) H.; m. Bonnie Colleen Hoskins, July 30, 1955; children: John Leonard, James Bryan, Julia Lynn, Jennifer Lee. B.A., U. Denver, 1956; postgrad., U. Melbourne, Australia, 1956-58; M.A., U. Ill., 1961, Ph.D., 1965. Asst. prof. psychology U. Denver, 1961-65, assoc. prof. psychology, 1965-69, prof. psychology dept., 1969—; vis. lectr. psychology U. Calif.-Berkeley, 1967; research assoc. Inst. Psychiatry, U. London, 1972. Fulbright fellow, 1956; NIH career devel. awardee, 1968-73; Fulbright-Hays speaker awardee, Yugoslavia, 1973; univ. lectr. awardee U. Denver, 1980. Fellow Soc. Multivariate Exptl. Psychology (pres. 1976, ann. disting. publs. award 1972), AAAS, Am. Psychol. Assn. Subspecialties: Developmental psychology; Cognition. Current work: Theory and technology human abilities, development, alcoholism, motivation and personality structure and dynamics. Home: 196 S Corona St Denver CO 80209 Office: Dept Psychology U Denver Denver CO 80208

HORN, WADE FREDRICK, psychology educator; b. Coral Gables, Fla., Dec. 3, 1954; s. John David and Daisy (Anderson) H.; m. Claudia Watson, Jan. 7, 1977; children: Christiana Watson. B.A., Am. U., 1975; M.A., So. Ill. U., 1978, Ph.D., 1981. Research asst. NIH, Bethesda, Md., 1973-75; psychol. technician Chestnut Lodge, Rockville, Md., 1975-76; behavioral cons. Wabash & Ohio Valley Spl. Edn. Dist., Norris City, Ill., 1978-79; pre-doctoral intern Children's Hosp. Nat. Med. Ctr., Washington, 1980-81; postdoctoral research fellow Children's Hosp., 1981-82; asst. prof. Mich. State U., East Lansing, 1982—; tech. cons. Ultivisions Corp., Bound Brook, N.J., 1982—. Contbr. articles to profl. jours. AURI grantee, 1982; So. Ill. U. fellow, 1977-78; Dissertation Research awardee, 1979-80. Mem. Am. Psychol. Assn., Midwestern Psychol. Assn., Phi Kappa Phi. Subspecialties: Behavioral psychology; Developmental psychology. Current work: Conductor clinical research with hyperactive and learning disabled populations from bio-behavioral perspective. Home: 4741 Woodcraft Rd Okemos MI 48864 Office: Mich State Dept Psycholog East Lansing MI 48824

HORNBACK, JOSEPH MICHAEL, chemist, educator, researcher; b. Middletown, Ohio, Sept. 16, 1943; s. Cletus Edward and Margaret (Long) H.; m. Margaret Ann Ruffing, July 17, 1965 (div.); children: Joseph Michael, Patrick W. B.S. magna cum laude, U. Notre Dame, 1965; Ph.D., Ohio State U., 1968. Research assoc. U. Wis., Madison, 1968-70; asst. prof. chemistry U. Denver, 1970-75, assoc. prof., 1975—. Contbr. numerous articles to profl. jours. NSF grantee, 1978-80. Mem. Am. Chem. Soc., Sigma Xi. Subspecialties: Photochemistry; Organic chemistry. Current work: Organic photochemistry; organic synthesis and mechanisms; synthetic use of organic photochem. reactions,especially photochemically generated transient intermediates. Home: 2325 S Linden Ct Apt 112N Denver CO 80222 Office: Dept Chemistry U Denver Denver CO 80208

HORNBEIN, THOMAS F., physician; b. St. Louis, 1930. M.D., Washington U., St. Louis, 1956. Diplomate: Am. Bd. Anesthesiology. Intern King County Hosp., Seattle, 1956-57; resident in anesthesiology Washington U. Hosp., 1957-59, research fellow in respiratory physiology, 1959-61; asst. in anesthesiology Barnes Hosp., St. Louis, 1960-61; asst. prof. anesthesiology U. Wash., 1963-67, asso. prof., 1967-70, prof. anesthesiology, 1970—, prof. physiology and biophysics, 1970—, vice chmn. dept. anesthesiology, 1972-74, chmn. dept., 1978—. Served to lt. comdr. USNR, 1961-63. Subspecialty: Anesthesiology. Current work: Respiratory physiology, control of breathing, brain acid-base regulation, brain hypoxia-limits. Office: U Wash Sch of Medicine RN-10 Seattle WA 98195

HORNBROOK, KENT ROGER, pharmacology educator; b. New Martinsville, W.Va., Oct. 23, 1936; s. Kent Maidlow and Marjorie Honore (Wakefield) H.; m. Lois Jane Schupbach, Sept. 6, 1958; 1 dau., Jane Catherine. B.S. in Pharmacy, W.Va. U., 1958; Ph.D., U. Mich., 1963. Postdoctoral fellow dept. pharmacology Washington U., St. Louis, 1963-65; asst. prof. dept. pharmacology Emory U., Atlanta, 1965-72; assoc. prof. dept. pharmacology U. Okla. Health Sci. Ctr., Oklahoma City, 1972-77, prof., 1977—. Mem. Am. Soc. Pharmacological Exptl. Therapeutics, Endocrine Soc., Soc. for Exptl. Biology and Medicine. Subspecialties: Cellular pharmacology; Receptors. Current work: Biochemical pharmacology, carbohydrate metabolism adrenergic mechanisms. Office: Dept Pharmacology PO Box 26901 Oklahoma City OK 73190

HORNER, HARRY THEODORE, botany educator; b. Chgo., Jan. 28, 1936; s. Harry Theodore and Eloise Gertrude (Blum) H.; m. Cecilia Astrid Midthun, Mar. 27, 1937; children: Kevin Scott, Amy Lynn, Allison Lee. B.A., Northwestern U., 1959, M.S., 1961, Ph.D., 1964. Pub. health postdoctoral fellow Iowa State U., Ames, 1964-66, asst. prof., 1966-69, assoc. prof., 1969-73, prof. dept. botany, 1973—; guest prof. U. Konstanz, W.Ger., 1979-80; dir. Bessey Microscopy Facility, 1970—. Contbr. articles in field to profl. jours. Recipient Centennial award Iowa Acad. Sci., 1975, Adviser award Iowa State U., 1982. Mem. Bot. Soc. Am., Am. Soc. Cell Biology, Am. Inst. Biol. Sci., Am. Microscopical Soc., Am Soc. Electron Microscopy, Iowa Acad. Sci., Sigma Xi. Republican. Presbyterian. Subspecialties: Cell and tissue culture; Developmental biology. Current work: Plant calcification/mineralization, male and female sterility. Office: Dept Botany Iowa State U Ames IA 50011

HORNER, JOHN ROBERT, paleontologist; b. Shelb, Mont., June 15, 1946; s. John Henry and Mirium Whitted (Stith) H.; m. Virginia Lee Horner, Mar. 30, 1970 (div.); 1 son, Jason James. Student, U. Mont., 1964-66, 68-72. Research asst. U. Mont., Missoula, 1969-73; research asst. dept. geol. and geophys. sciss. Princeton (N.J.) U., 1976-79; asst. curator Mus. Natural History, 1979-82; curator paleontology Mus. of Rockies, Mont. State U., Bozeman, 1982—. Served with USMC, 1966-68. NSF grantee. Mem. Soc. Vertebrate Paleontology. Club: Explorers (N.Y.C.). Subspecialties: Paleobiology; Paleoecology. Current work: Paleosociobiologic and evoluationary aspects of extinct vertebrate animals. Office: Museum of the Rockies Montana State University Bozeman MT 59717

HORNG, WAYNE J(ING-WEI), immunologist/biochemist, researcher; b. Taiwan, China, June 22, 1942; came to U.S., 1969, naturalized, 1978; s. Hsi-Chun and Chu (Liu) H.; m. Show-Jen, Nov. 30, 1968; children: Felix, Eric, Andrew. Ph.D. in Biochemistry, U. Tex. Med. Br., Galveston, 1973. Asst prof. microbiology and immunology U. Ill. Med. Ctr., Chgo., 1976-82; research pathologist U. Tex. Health Sci. Ctr., Houston, 1982—. Contbr. articles to profl. jours. USPHS grantee, 1978-81. Mem. Am. Assn. Immunologists. Subspecialties: Immunogenetics; Immunobiology and immunology. Current work: Immunology, immunochemistry, molecular genetics, idiotype regulation. Home: 1835 Indian Wells Dr Missouri City TX 77459 Office: 6431 Fannin St Houston TX 77025

HORNIG, DONALD FREDERICK, scientist; b. Milw., Mar. 17, 1920; s. Chester Arthur and Emma (Knuth) H.; m. Lilli Schwenk, July 17, 1943; children: Joanna, Ellen, Christopher, Leslie. B.S., Harvard U., 1940, Ph.D., 1943; LL.D., Temple U., 1964, Boston Coll., 1966,

Dartmouth Coll., 1974; D.H.L., Yeshiva U., 1965; D.Sc., U. Notre Dame, 1965, U. Md., 1965, Rensselaer Poly. Inst., 1965, Ripon Coll., 1966, Widener Coll., 1967, U. Wis., 1967, U. Puget Sound, 1968, Syracuse U., 1968, Princeton U., 1969, Seoul Nat. U., Korea, 1973, U. Pa., 1975, Lycoming Coll., 1980; D.Eng., Worcester Poly. Inst., 1967. Research asso. Woods Hole (Mass.) Oceanographic Instn., 1943-44; scientist, group leader Los Alamos Lab., 1944-46; asst. prof. chemistry Brown U., 1946-49, asso. prof., 1949-51, prof., 1951-57; dir. Metcalf Research Lab., 1949- 57, asso. dean grad. sch., 1952-53, acting dean, 1953-54; vis. prof. Princeton U., 1957, prof. chemistry, 1957-64, chmn. dept., 1958-64, Donner prof. sci., 1959-66; spl. asst. sci. and tech. to Pres. U.S., 1964-69; dir. Office Sci. and Tech., 1964-69; chmn. Fed. Council Sci. and Tech., 1964-69; v.p., dir. Eastman Kodak Co., 1969-70; prof. chemistry U. Rochester, 1969-70; pres. Brown U., Providence, 1970-76; hon. research asso. in applied physics Harvard U., 1976-77, prof. chemistry in pub. health, dir. Interdisciplinary Programs in Health, 1977—, Alfred North Whitehead prof. chemistry (public health), 1981—; dir. Upjohn Co., Westinghouse Electric Corp.; Mem. Pres.'s Sci. Adv. Com., 1960-69, chmn., 1964-69, Project Metcalf, Office of Naval Research, 1951-52. Author articles sci. jours. Bd. overseers Harvard U., 1964-70; bd. dirs. Overseas Devel. Council, 1969-75; trustee George Eastman House, 1969-71, Manpower Inst., 1969-76. Decorated Disting. Civilian Service medal, Korea; Guggenheim fellow, 1954-55; Fulbright fellow, 1954-55; recipient Charles Lathrop Parsons award Am. Chem. Soc., 1967, Engring. Centennial award, 1967, Mellon Inst. award, 1968. Fellow Am. Phys. Soc., Am. Acad. Arts and Scis.; mem. Nat. Acad. Scis., Am. Chem. Soc., AAAS, Am. Philos. Soc., Romanian Acad. (fgn.), Sigma Xi. Subspecialties: Environmental toxicology; Physical chemistry. Current work: Molecular spectroscopy; engrgy transfer; environmental health and public policy. Home: 16 Longfellow Park Cambridge MA 02138

HORNING, EVAN CHARLES, chemistry educator; b. Phila., June 6, 1916. B.A., U. Pa., 1937; Ph.D. in Organic Chemistry, U. Ill., 1940; D.h.c., Karolinska Inst., Sweden, 1973, U. Ghent, Belgium, 1978. Instr. chemistry Bryn Mawr Coll., 1940-41, U. Mich., 1941-43; from asst. prof. to assoc. prof. U. Pa., 1945-50; chief lab. chem. natural products Nat. Heart Inst., NIH, 1950-61; dir. Lipid Research Ctr., Baylor U. Coll. Medicine, Houston, 1961-66, Inst. Lipid Research, 1966—, prof. chemistry, 1961—, chmn. dept. biochemistry, 1962-66; Research assoc. com. med. research Nat. Def. Research Com., U. Mich., 1943-45. Guggenheim fellow, 1958; recipient Bergman award Swedish Chem. Soc., 1973; Tswett medal in Chromatography, 1975; Warner-Lambert award Am. Assn. Clin. Chemists, 1976; Scheele award Pharm. Soc. Sweden, 1976; S.S. Dal Nogare award, 1980. Mem. Am. Chem. Soc. (award 1978), Brit. Biochem. Soc. Subspecialties: Organic chemistry; Biochemistry (medicine). Office: Inst Lipid Research Baylor Coll Medicine Houston TX 77030

HORNOF, WILLIAM J., veterinarian, educator; b. San Francisco, Aug. 22, 1946; s. William L. and Irene J. H.; m. Kristine T. Hornof, Oct. 10, 1968; 1 son, Kiby J. B.S., U. Calif.-Davis, D.V.M., M.S., 1981. Diplomate: Am. Coll. Vet. Radiology. Pvt. practice vet. medicine, 1975-76; resident in vet. radiology U. Calif.-Davis, 1976-79; asst. prof. (Sch. Vet. Medicine), 1979—. Contbr. articles to profl. jours. Pres. PTA, Fairfield Sch., Davis, 1983—. Served with U.S. Army, 1965-69. Mem. AVMA, Am. Coll. Vet. Radiology, Internat. Vet. Radiology Assn. Republican. Lodge: Masons. Subspecialties: Veterinary Medical Research; Nuclear medicine. Current work: Application of digital nuclear image processing techniques to the diagnosis of veterinary diseases. Clinical research in these diseases as well as studying the pathophysiology of human diseases in appropriate animal models. Home: 1919 Barry Rd Davis CA 95616 Office: Radiological Scis Veterinary Medicine U Calif Davis CA 95616

HORNUNG, DAVID EUGENE, physiologist; b. Pitts., Apr. 30, 1945; s. George Gates and Esther Irene (Hooks) H.; m. Susan E. Burkhart, Nov. 22, 1972 (div. 1981). B.S., Geneva Coll., 1967; M.S., Kent State U., 1969; Ph.D., SUNY, Syracuse, 1975. Instr. St. Lawrence U., Canton, N.Y., 1969-73, asst. prof., 1975-78, assoc. prof., 1978-83, prof., 1983—; research asst. Upstate Med. Center, Syracuse, N.Y., 1973-75, research asst. prof. physiology, 1976-79, research assoc. prof., 1979—. Author articles. Ruling elder Canton Presbyn. Ch., 1977-83; bd. dirs. Am. Lung Assn. of Central N.Y., 1983—. NIH grantee, 1976-81, 81—. Mem. Am. Physiol. Soc., European Chemoreception Orgn., Assn. Chemoreception Scis. Subspecialties: Neurobiology; Physiology (medicine). Current work: Mechanisms and clinical implications of taste and smell. Home: PO Box 211 Canton NY 13617 Office: Dept Biology St Lawrence U Canton NY 13617

HOROVITZ, ZOLA PHILIP, pharm. co. exec.; b. Pitts., Oct. 12, 1934; s. Reuben and Jean (Liff) H.; m. Marlene Davis, Aug. 24, 1958; children: Bonna Lynn, Reid Alan. B.S. in Pharmacy, U. Pitts., 1955, M.S. in Pharmacology, 1958, Ph.D., 1960. Registered pharmacist, Pa. Vis. investigator VA Research Lab. Neuropsychiatry, Pitts., 1958-59; sr. research scientist Squibb Inst. Med. Research, E.R. Squibb & Sons Inc., Princeton, N.J., 1959-65, clin. assoc., 1965-66, dir. dept. pharmacology, 1967-72, assoc. dir. research, 1972-79, v.p. drug devel., 1979—; vis. prof. psychiatry Rutgers Sch. Medicine, 1974-76; vis. prof. pharmacology Rutgers Coll. Pharmacy, 1979—, trustee, mem. adv. bd., 1980—; mem. sci. adv. council Princeton U., 1978—. Mem. East Brunswick (N.J.) Bd. Edn., 1967-70; pres. Jewish Center at Princeton, 1980-82. Served with U.S. N.G., 1953-62. Recipient A.E. Bennett prize in neuropsychopharmacology, 1965; Am. Found. Pharm. Edn. fellow, 1958-60. Fellow Acad. Pharm. Scis.; mem. Am. Pharmacology Soc., Brit. Pharmacology Soc. Subspecialty: Pharmacology. Current work: Administrator of pharmaceutical research and development. Patentee in field. Home: 30 Philip Dr Princeton NJ 08540 Office: Box 4000 Princeton NJ 08540

HOROWITZ, BRUCE RICHARD, microbiologist; b. N.Y.C., Jan. 31, 1950; s. Mac and Adele (Winter) H.; m. Helen Lebin, June 2, 1979. B.S., Bklyn. Coll., 1971. Microbiologist E. Fougera & Co., 1973-76; sr. microbiologist Becton-Dickinson, Inc., 1976-78; sr. quality assurance engr. Searle, Inc., 1978-79; quality assurance mgr. MG Med. Systems, 1979-80; mgr. Intermedics, Freeport, Tex., 1980—. Mem. Am. Soc. Quality Control (chmn. local br.), Am. Soc. for Microbiology, Regulatory Affairs Profl. Soc. Jewish. Subspecialties: Microbiology; Toxicology (medicine). Current work: Microbiology, biocompatibility. Office: PO Box 617 Freeport TX 77541

HOROWITZ, CARL, chemist, chem. co. exec.; b. Lvov, Poland, Aug. 10, 1923; s. Nathan and Amalie (Roth) H.; m. Irene Mandel, Dec. 9, 1946; children: Alice, Terry. B.S., Columbia U., 1950; M.S. in Chemistry, Bklyn. Poly. Inst., 1961. Vice-pres. Yardney Electric Co., N.Y.C., 1951-63; pres. Polymer Research Corp., Bklyn., 1963—. Mem. Am. Chem. Soc., Electrochem. Soc., Am. Inst. Chem. Engrs., Am. Assn. Textile Chemists and Colorists. Democrat. Jewish. Subspecialty: Polymer chemistry. Current work: Chemical grafting of polymers; silver-zinc highpower batteries. Patentee in field. Office: 2186 Mill Ave Brooklyn NY 11234

HOROWITZ, ELLIS, computer science educator; b. N.Y.C., Feb. 11, 1944; m., 1968; 2 children. B.S., Bklyn. Coll., 1964; M.S., U. Wis., 1967, Ph.D., 1970. Assoc. prof. computer sci. U. So. Calif., Los Angeles, 1973—. Editor: Trans. Math. Software, 1976—. Mem. ACM, IEEE, Soc. Indsl. and Applied Math., AAAS, Sigma Xi. Office: Dept Computer Sci U So Cli Calif Los Angeles CA 90007

HOROWITZ, JACK, biochemistry educator, researcher; b. Vienna, Austria, Nov. 25, 1931; came to U.S., 1938; s. Joseph and Florence (Gutterman) H.; m. Carole Ann Sager, June 11, 1961; children: Michael Joseph, Jeffrey Frederick. B.S., CCNY, 1952; Ph.D., Ind. U., 1957. Asst. prof. Iowa State U., Ames, 1961-64, assoc. prof., 1964-71, prof., 1971—, chmn. dept. biochemistry and biophysics, 1971-74, coordinator molecular, cellular, devel. biology program, 1977-80; vis. scientist Rockefeller U., N.Y.C., 1968; vis. prof. Yale U., New Haven, 1974-75. Precinct leader Democratic Party, Ames, 1980—, mem. county central com. NSF predoctoral fellow, 1952-54; postdoctoral fellow, 1957-59. Mem. Am. Soc. Biol. Chemists, AAAS, Phi Beta Kappa, Phi Kappa Phi. Jewish. Subspecialties: Biochemistry (biology); Molecular biology. Current work: Structure-function relationships of transfer RNA; protein biosynthesis; ribosome structure and function. Home: 2014 Country Club Blvd Ames IA 50010 Office: Dept Biochemistry and Biophysics Iowa State U Ames IA 50011

HORROCKS, DONALD LEONARD, research chemist; b. Dearborn, Mich., July 14, 1929; s. Wilfrid and Deena Laurene (Hendrick) H.; m. Margaret A. Powell, July 30, 1955; children: Andrea Jean, Cynthia Kay. B.A., Reed Coll., 1951; PH.D, Iowa State U., 1955. Research chemist Nat. Carbon Research Lab., Parma, Ohio, 1955-56; assoc. chemist ARgonne (Ill.) Nat. Lab., 1956-71; prin. staff scientist Beckman Instrument, Inc., Irvine, Calif., 1971—; speaker at internat. confs., 1967-82. Author: Applications of Liquid Scintillation Counting, 1974; editor: Organic Scintillators, 1967, Liquid Scintillation Counting, 1980; contbr. numerous articles to profl. jours. Mem. Am. Chem. Soc., Am. Nuclear Soc. Democrat. Methodist. Subspecialties: Nuclear physics; Bioinstrumentation. Current work: Liquid scintillation systems; use of radionuclides. Patentee in field. Home: 324 Bagnall Ave Placentia CA 92670

HORROCKS, LLOYD ALLEN, neurochemical researcher, biochemistry educator; b. Cin., July 13, 1932; s. Robert A. and Martha (Keeler) H.; m. Marjorie Werstler, June 30, 1956; children: Richard A., Rebecca A. Haxe. B.A. with honors in Chemistry, Ohio Wesleyan U., 1953; M.Sc., Ohio State U., 1955, Ph.D., 1960. Project engr. Wright Air Devel. Ctr., U.S. Air Force, Dayton, Ohio, 1955-58; research assoc. Lab. Neurochemistry, Cleve. Psychiat. Inst., 1960-68; asst. prof. dept. physiol. chemistry Ohio State U., 1968-70, assoc. prof., 1970-73, prof. dept. physiol. chemistry coll. medicine, 1973—, dir. postdoctoral tng. program in neurochem. pathology, 1980—; neurol. disorders program project rev. com. NIH, 1981—. Editor-in-chief: Neurochem. Pathology, 1983—; mem. editorial bd.: Jour. Lipid Research, 1976-83, Jour. Neurochemistry, 1978—; contbr. articles to profl. jours., chpts. to books. Served with USAF, 1953-58. Nat. Inst. Neurol. Diseases and Blindness spl. fellow, 1964-65; Faculty scholar Josiah Macy Jr. Found., Universite Louis Pasteur, Strasbourg, France, 1974-75. Mem. AAUP, Am. Soc. Biol. Chemists, Am. Soc. Neurochemistry, Biochem. Soc., European Soc. Neurochemistry, Internat. Soc. Neurochemistry, Soc. Neurosci. Methodist. Subspecialties: Neurochemistry; Biochemistry (medicine). Current work: Phosphoglyceride metabolism of central nervous system membranes; prostaglandins; current studies include metabolism and analysis of brain lipids, mechanisms of membrane injury including phospholipases during dymelinating disease and spinal cord injury; effects of dietary choline on neurochemistry behavior. Home: 328 W 6th Ave Columbus OH 43201 Office: Physiol Chemistry 1645 Neil Ave Columbus OH 43210

HORST, RALPH KENNETH, educator; b. Massillon, Ohio, June 22, 1935; s. Ralph Emerson and Florence Ellen (Huff) H.; m. Nancy J. Vernon, June 11, 1960 (div.); children: Jeffrey Todd, Bradley Craig, Timothy; m. Hope T. Thorn, July 11, 1969; 1 stepdau, Anne Elizabeth. B.S., Ohio U., 1957; M.S., Ohio State U., 1959, Ph.D., 1961. Dir. plant pathology lab. Yoder Bros., Inc., Barberton, Ohio, 1962-68; asst. prof. plant pathology Cornell U., Ithaca, N.Y., 1968-74, asso. prof., 1974-80, prof., 1980—. Mem. Internat. Soc. Plant Pathology, Internat. Soc. Hort. Sci., Am. Phytopath. Soc., AAAS, Soc. Am. Florists, Tissue Culture Assn., Sigma Xi, Gamma Sigma Delta. Methodist. Club: Rotary. Subspecialties: Plant virology; Plant cell and tissue culture. Current work: Teacher and researcher in detection and therapeutic procedures for viroid diseases and diseases caused by mycoplasmalike organisms. Home: 107 Birchwood Dr Ithaca NY 14850 Office: Cornell University Dept Plant Pathology 323 Plant Science Bldg Ithaca NY 14853

HORSTMANN, DOROTHY MILLICENT, physician, educator; b. Spokane, Wash., July 2, 1911; d. Henry J. and Anna (Hunold) H. A.B., U. Calif., 1936, M.D., 1940; D.Sc. (hon.), Smith Coll., 1961, M.A., Yale, 1961, Dr. Med. Scis., Women's Med. Coll. of Pa., 1963. Intern San Francisco City and County Hosp., 1939-40, asst. resident medicine, 1940-41, Vanderbilt U. Hosp., 1941-42; Commonwealth Fund fellow, sect. preventive medicine Sch. Medicine, Yale U., New Haven, 1942-43, instr. preventive medicine, 1943-44, 45-47, asst. prof., 1948- 52, asso. prof., 1952-56, asso. prof. preventive medicine and pediatrics, 1956-61, prof. epidemiology and pediatrics, 1961—, John Rodman Paul prof. epidemiology, prof. emeritus, 1982—; instr. medicine U. Calif., 1944-45. Recipient Albert Coll. award, 1953; Gt. Heart award Variety Club Phila., 1968; Modern Medicine award, 1974; James D. Bruce award ACP, 1975; Thorvald Madsen award State Secum Inst. (Denmark), 1977; Maxwell Finland award Infectious Disease Soc.-Am., 1978; Disting. Alumni award U. Calif. Med. Sch., 1979; NIH fellow Nat. Inst. Med. Research, London, 1947-48. Master A.C.P.; hon. asso. fellow Am. Acad. Pediatrics; mem. Am. Soc. Clin. Investigation, Am. Epidemiol. Soc., Am. Pediatric Soc., Am. Soc. Virology (council 1983-84), Assn. Am. Physicians, Infectious Diseases Soc. Am. (pres. 1974-75), Am. Assn. Immunologists, Soc. Epidemiologic Research, Pan Am. Med. Assn., Internat. Epidemiol. Assn., Royal Soc. Medicine (hon. mem. sect. epidemiology and preventive medicine), Nat. Acad. Scis., Conn. Acad. Sci. and Engring., European Assn. Against Virus Diseases., South African Soc. Pathologists. Subspecialties: Epidemiology; Microbiology (medicine). Current work: Clinical virology, rubella, Epstein-Barr virus infections. Home: 11 Autumn St New Haven CT 06511

HORTON, CHARLES E., plastic surgeon; b. Purdy, Mo., June 27, 1925; s. Ray B. and Beatrice (McCraw) H.; m. Geraldine; children: Mary Anne, Elizabeth, Charles E., Nancy, Kathrine. A.B., U.Ark.; B.S., U. Mo.; M.D., U. Va. Diplomate: Am. Bd. Plastic Surgery (past chmn.). Practice medicine specializing in plastic surgery, Norfolk, Va., 1955—; dir. Eastern Va. Grad. Sch. Medicine, Norfolk, 1976—; assoc. dean clin. relations Eastern Va. Med. Sch., Norfolk, 1982—. Served with U.S. Army, 1944-45; Served with USN, 1950-52. Named Med. Alumni of Year U. Va., 1972; Stein meml. lectr. Aâhaus, Denmark, 1978; James Barrett Brown lectr., 1978; Zovikian lectr., 1980; C. J. Devine lectr., 1980; Gerber lectr., 1980; Thuss lectr., 1981; Keihn lectr., 1982. Fellow Royal Coll. Surgeons; mem. AMA (physicians recognition award 1978/81), Am. Soc. Plastic and Reconstructive Surgery, Am. Assn. Plastic Surgery, Plastic Surgery Research Council, Am. Soc. Aesthetic Plastic Surgery, Soc. Pediatric Urology, South African Soc. Plastic Surgery (hon.), Canadian Soc. Plastic Surgery (hon.). Episcopalian. Subspecialty: Surgery. Current work: Genital reconstruction and aesthetic surgery. Office: 400 W Brambleton Ave Norfolk VA 23510

HORTON, FRANK O., III, physician, educator; b. Cin., Feb. 18, 1947; s. Frank O. and Sylvia (Smith) H.; m. Mary A. Golden, Nov. 16, 1974; 1 son, Frank O. IV. B.S., Morehead State U., 1969; M.D., Med. Coll. Ohio, 1973. Diplomate: Am. Bd. Internal Medicine, subcert. in pulmonary disease. Chief pulmonary sect., dir. pulmonary function lab., dir. sleep disorder lab., assoc. dir. respiratory therapy Toledo Hosp., 1981—; clin. asst. prof. Med. Coll. Ohio, Toledo, 1981—. NIH grantee, 1976-80. Fellow Am. Coll. Chest Physicians, ACP; mem. Am. Thoracic Soc., AMA, Northwestern Ohio Lung Assn. (dir. 1983—). Subspecialty: Pulmonary medicine. Current work: Adult sleep apnea. Pulmonary stress testing; Enhanced alveolar macrophage killing in presence of alveolar lining material. Glucose metabolism in human alveolar macrophages. Office: Toledo Hosp 2142 N Cove Blvd Toledo OH 43606

HORTON, JOHN, oncologist, educator; b. Sheffield, Eng., June 25, 1934; came to U.S., 1960, naturalized, 1966; s. Leonard and Myra H.; m.; children: Russell Stephen, Nicholas Jon, Danielle Alexa. M.B., Ch.B., U. Sheffield, Eng., 1957. Diplomate: Am. Bd. Internal Medicine; lic. physician, N.Y., Okla. Rotating intern Albany (N.Y.) Med. Ctr. Hosp., 1957-58; sr. house officer Royal Hosp., Sheffield, Eng., 1958-59; practice medicine, Sheffield, 1959-60; asst. resident in medicine Albany Med. Ctr. Hosp., 1960-62; clin. instr. medicine Albany Med. Coll., 1960-63, research fellow in oncology, 1962-63, instr. medicine, 1963-66, asst. prof., 1966-69, assoc. prof., 1969-74, prof., 1974-75, 76—, head div. oncology, 1976—; prof. medicine Tulsa Med. Coll., 1975-76; attending physician, attending oncologist Albany Med. Ctr. Hosp., 1969-75, 76—; attending physician Albany VA Hosp., 1964-75, 76—; dir. Natalie Warren Bryant Cancer Ctr., Tulsa, 1975-76; cons. Fox Meml. Hosp., Oneonta, N.Y., 1967—, Del. Valley Hosp., Walton, N.Y., 1972—, Leonard Hosp., Troy, N.Y., 1974—, St. Mary's Hosp., Troy, 1978—, Samaritan Hosp., 1978-80, Cohoes (N.Y.) Hosp., 1980—, St. Peter's Hosp., Albany, 1981—; prin. investigator Eastern Coop. Oncology Group, 1972—, Gastrointestinal Tumor Study Group, 1975—; dir. Regional Breast Cancer Program for Upstate N.Y. and Western Mass., 1975-80. Author: Malignant Effusions, 1982, Adjuvant Chemotherapy in the Treatment of Primary Breast Cancer, 1982; editor: (with G.J. Hill) Clinical Oncology, 1977; med. editor: Current Concepts in Oncology, 1979—, Clin. Cancer Briefs, 1979—; contbr. numerous articles and abstracts to profl. jours. Pres. Albany County unit Am. Cancer Soc., 1969-71, bd. dirs., 1963—; bd. dirs. N.Y. State div., 1968-75, 76—, pres., 1980. Fellow ACP, Am. Coll. Gastroenterology; mem. Albany County Med. Soc., N.Y. State Med. Soc., N.Y. State Cancer Programs Assn. (pres. 1979), Am. Soc. Clin. Oncology, Am. Assn. Cancer Research, N.Y. Acad. Scis., Soc. Surg. Oncology, Internat. Assn. Study of Lung Cancer, Am. Assn. Cancer Edn. (pres. 1982), Alpha Omega Alpha. Subspecialties: Cancer research (medicine); Oncology

HORTON, JOHN EDWARD, periodontist educator, researcher; b. Brockton, Mass., Dec. 30, 1930; s. Harold Ellsworth and Anita Helen (Samuelson) H.; m.; children: John Edward, Janet E., James E., Jeffrey E., Joseph E. B.S., Providence Coll., 1952; D.M.D., Tufts U., 1957; M.S.D., Baylor U., 1965; M.A., George Washington U., 1978. Commd. 1st lt. U.S. Army, 1957, advanced through grades to col., 1972; chief depts. microbiology and immunology (U.S. Army Inst. Dental Research, Walter Reed Army Med. Ctr.), Washington, 1973-77, ret., 1977; guest scientist Nat. Inst. Dental Research, 1970-73; lectr. Johns Hopkins U. Sch. Pub. Health, 1975-79; professorial lectr. George Washington U., 1976-77; assoc. prof., chmn. dept. periodontics Harvard U. Sch. Dental Medicine, Boston, 1977-81; prof. Ohio State U. Coll. Dentistry, Columbus, 1981—, chmn. dept. periodontics, 1981—; cons. NIH, 1973—, NASA, 1979, VA, Columbus, 1981—, USAF Med. Ctr., Wright-Patterson, Ohio, 1982—. Editor: Mechanisms of Localized Bone Loss, 1978; contbr. articles to profl. jours., chpt. to book. Decorated Legion of Merit, Meritorious Service medal, Commendation medal. Fellow AAAS, Am. Pub. Health Assn., Internat. Coll. Dentists, Royal Soc. Health; mem. Am. Assn. Dental Research (pres. Columbus sect. 1982-83), Omicron Kappa Upsilon, Phi Delta Kappa. Subspecialties: Periodontics; Immunology (medicine). Current work: Investigations in cell-mediated immunologic mechanisms as related to the host response to inflammation and the regulation of bone resorptive processes. Office: Ohio State U Coll Dentistry 305 W 12th Ave Columbus OH 43210

HORTON, WENDELL CLAUDE, JR., physics educator; b. Houston, Feb. 3, 1942; s. Claude Wendell and Louise (Walthall) H.; m. Elisabeth Alice Becker, June 1, 1963; children: John W., Mike A. B.S., U. Tex., 1963; Ph.D., U. Calif.-San Diego, 1967. Mem. faculty dept. physics U. Tex.-Austin, 1969—, prof., 1969—. Editor: Long Time Prediction in Dynamics, 1983. Mem. Am. Phys. Soc., Am. Geophys. Soc. Subspecialties: Nuclear fusion; Plasma physics. Current work: Plasma turbulence and plasma instabilities. Home: 6501 Mesa Dr Austin TX 78731 Office: Inst Fusion Studies U Tex Austin TX 78712

HORVÁTH, CSABA GYULA, chemical engineering educator; b. Szolnok, Hungary, Jan. 25, 1930; m., 1963; 2 children. Dipl. chem. engring., Budapest Tech. U., 1952; D.Phil., U. Frankfurt, Ger., 1963. Asst. prof. chem. tech. Budapest (Hungary) Tech. U., 1952-56; research scientist Farbwerke Hoechst AG, Ger., 1956-60; research fellow Mass. Gen. Hosp. and Harvard U., Boston, 1963-64; research assoc. Sch. Medicine, Yale U., New Haven, 1964-70, lectr., 1967-72, assoc. prof., 1970-75, from assoc. prof. to prof. engring. and applied sci., 1972-81, prof. chem. engring., 1981—. Recipient Humboldt Sr. U.S. Scientist award, 1982. Mem. Am. Chem. Soc. (award o61983), Inst. Food Technologists, Am. Inst. Chem. Engrs., Deutsche Gesellschaft fuer Chemisches Apparatewesen. Subspecialties: Chemical engineering; Biomedical engineering. Office: Dept Chem Engring Yale U PO Box 2159 Yale Sta New Haven CT 06520

HORVATH, WILLIAM JOHN, health systems educator, research scientist; b. N.Y.C., Sept. 13, 1917; s. John and Anna (Horvath) H.; m. Rebecca Sue Badger, Feb. 23, 1963; children: Susan, John. B.S., CCNY, 1936; M.S., NYU, 1938, Ph.D., 1940. Physicist U.S. Navy-Bur. Ordnance, Washington, 1940-43; ops. analyst Chief of Naval Ops., Washington, 1943-49; sci. advisor Weapons Systems Evaluation Group, Sec. of Def., Washington, 1949-52; staff cons. Sylvania Electric Co., Bayside, N.Y., 1952-55; sect. head. Airborne Instrument Lab., Mineola, N.Y., 1955-58; prof. health systems, research scientist U. Mich., Ann Arbor, 1958—; cons. Nat. Acad. Sci., 1952-55, NIH, 1962-70, NIMH, 1967-72. Author numerous chpts. and articles in nuclear physics, ops. research, math. sociology, med. physics, health care delivery, health behavior. Bd. dirs. Washtenaw Community Services, Ann Arbor, 1978-80, Community Systems Found., Ann Arbor, 1978-81. Recipient Naval Ordnance Devel. award U.S. Dept. Navy, 1945; Presdl. Cert. of Merit, 1947. Fellow AAAS, Am. Pub. Health Assn.; mem. Am. Phys. Soc., Ops. Research Soc. Am. (Kimball medal 1977), Royal Soc. Health. Club: Cosmos (Washington). Subspecialties: Operations research (engineering); Health services research. Current work: Currently studying population health behavior, using survey methods to develop preventive measures for the promotion of better health. Home: 2451 Trenton Ct Ann Arbor MI Office: Mental Health Research Inst U Mich Ann Arbor MI 48109

HORVITZ, LEO, geochemist; b. Central Falls, R.I., Aug. 12, 1909; s. Samuel and Pearl H.; m. Sarah Deluty, June 2, 1936; children: E. Philip, Sigmund Alan, Ira Samuel. Sc.B., Brown U., 1931, Sc.M., 1932;

Ph.D., U. Chgo., 1935. Chief chemist Subterrex, Houston, 1936-41; group leader Nat. Def. Research Com., U. Chgo., 1942-43; ptnr., owner Horvitz Research Labs., Houston, 1943-69, pres., 1970; cons. geochemist, 1941—. Contbr. articles to profl. jours. Fellow AAAS, Tex. Acad. Sci.; mem. Am. Chem. Soc., Am. Assn. Petroleum Geologists, Soc. Exploration Geophysicsts, Geochem. Soc. Subspecialties: Organic geochemistry; Geochemistry. Current work: Geochemical prospecting techniques. Patentee in geochem. exploration for petroleum. Home: 5207 S Braeswood Blvd Houston TX 77096 Office: Horvitz Research Labs 8116 Westglen Dr Houston TX 77063

HORWITT, MAX KENNETH, biochemist, educator; b. N.Y.C., Mar. 21, 1908; s. Harry and Bessie (Kenitz) H.; m. Frances Levine, 1933 (dec.); children: Ruth Ann Horwitt Singer, Mary Louise Horwitt Goldman; m. Mildred Gad Weitzman, Jan. 1, 1974. B.A., Dartmouth Coll., 1930; Ph.D., Yale U., 1935. Diplomate: Am. Bd. Nutrition, Am. Bd. Clin. Chemistry. Research fellow physiol. chemistry Yale U., 1935-37, lab. asst., 1932-34, asst., 1934-35; dir. biochem. research lab. Elgin (Ill.) State Hosp., 1937-59, L.B. Mendel Research Lab., 1960-68, dir. research, 1966-68; asso. dept. biol. chemistry U. Ill. Coll. Medicine, Chgo., 1940-43, asst. prof., 1943-51, asso. prof., 1951-62, prof., from 1962; prof. dept. biochemistry St. Louis U. Sch. Medicine, 1968-76, prof. emeritus, 1976—, cons. in nutrition div. endocrinology dept. internal medicine, 1976—, chmn. univ. instl. rev. bd., 1981-82; acting dir. div. research services Ill. Dept. Mental Health, Chgo., 1967-68; cons. human nutrition Rush Med. Sch., Chgo., 1967-83, vis. prof. dept. internal medicine, 1979-82; mem. expert group on Vitamin E WHO, 1981-82; field dir. Anemia and Malnutrition Research Center, Chiang Mai Med. Sch., Thailand, 1968-69; cons., 1976—. Contbr. numerous articles on clin. nutrition, biochemistry and psychopharmacology to profl. publs.; editorial bd.: Jour. Nutrition, 1967-71; co-editor: Am. Jour. Clin. Nutrition, 1974. Pres. Kneseth Israel Congregation, Elgin, 1965. Recipient Osborne and Mendal award Am. Inst. Nutrition, 1961. Fellow AAAS, N.Y. Acad. Scis., Am. Inst. Chemists, Gerontol. Soc.; mem. Am. Soc. Biol. Chemists, Am. Soc. Clin. Nutrition, NRC (food and nutrition bd. 1980—, com. dietary allowances), Soc. Exptl. Biology and Medicine, Soc. Biol. Psychiatry, Assn. Vitamin Chemists, Am. Chem. Soc. Subspecialties: Plant physiology (biology); Biochemistry (biology). Current work: Recommended dietary allowances of National Research Council, with specific assignments on Vitamin E, Thiamine, Riboflavin and Niacin-tryptophan. Office: St Louis Univ Sch Medicine 1402 S Grand Blvd Saint Louis MO 63104

HORWITZ, BARBARA ANN, physiologist, educator, cons.; b. Chgo., Sept. 26, 1940; d. Martin and Lillian (Knell) H.; m. John M. Horowitz, Aug. 17, 1970. B.S., U. Fla., 1961, M.S., 1962; Ph.D. (USPHS predoctoral fellow), Emory U., 1966. Asst. research physiologist U. Calif., Davis, 1968-72, asst. prof. physiology, 1972-75, assoc. prof., 1975-78, prof., 1978—; cons. Contbr. articles profl. jours. USPHS postdoctoral fellow, 1966-68; recipient Disting. Teaching award U. Calif., Davis, 1981-82. Mem. Am. Physiol. Soc., Am. Soc. Zoologists, N.Y. Acad. Scis., Phi Beta Kappa, Sigma Xi, Phi Kappa Phi, Phi Sigma. Subspecialties: Animal physiology; Physiology (biology). Current work: Cellular basis of adaptation to environment; cell basis of obesity, cellular metabolism and thermogenesis, cell physiology. Office: Dept of Animal Physiology U Calif Davis CA 95616

HORWITZ, DAVID ALLEN, physician; b. Chgo., Mar. 23, 1937; s. Joseph and Ethel (Levy) H.; m. Essie Lois Waxler, June 12, 1962; children: Mitchell, Bruce, Julie. B.A., U. Mich, 1958; M.D., U. Chgo., 1962. Diplomate: Am. Bd. Internal Medicine. Intern Michael Reese Hosp., Chgo., 1962-63, resident, 1963-66; fellow U. Tex. Southwestern Med. Sch., Dallas, 1966-68, instr. internal medicine, 1968-69; asst. prof. internal medicine U. Va. Sch. Medicine, Charlottesville, 1969-72, asso. prof., 1973-79, prof., 1979-80; vis. prof. clin. research Centre Harrow, Eng., 1976-77; prof. internal medicine U. So. Calif Sch. Medicine, 1980—; chmn. clin. immunology and rheumatic disease sect. Los Angeles County-U. So. Calif. Med. Center, 1980—. Contbr. articles for profl. jours. Served with U.S. Army, 1963-69. Arthritis Found. fellow, 1971-74; scholar, 1975-80; NIH grantee, 1977—. Mem. Am. Rheumatism Assn., Am Fedn. Clin. Research, AAAS, Am Soc. for Clin. Investigation, AAUP, Reticuloendothelial Soc. Subspecialty: Immunology (medicine). Current work: The identification of human blood monocyte precursors and clinical implications. The development of a method to evaluate blood monocyte chemotaxis. The description of an immunoregulatory IgG factor. The demonstration, clinical significance and mechanism of impaired cell mediated immunity in systemic lupus erythematosus. The utilization of immunologic methods to titrate dosage of cyclophosphamide in patients with connective tissue diseases. The characterization of a third human blood mononuclear population. Office: 2025 Zonal Ave HMR 711 Los Angeles CA 90033

HORWITZ, DAVID LARRY, health care medical executive, biomedical researcher; b. Chgo., July 13, 1942; s. F. Milton and Dorothy (Glass) H.; m. Gloria Jean Madian. B.A., Harvard U., 1963; M.D., U. Chgo., 1967, Ph.D., 1968. Diplomate: Am. Bd. Internal Medicine. Med. resident U. Chgo., 1971-74, asst. prof., 1974-79; assoc. prof. medicine U. Ill., Chgo., 1979—; med. dir. Travenol Labs., Deerfield, Ill., 1982—; dir. Am. Diabetes Assn. (no. Ill. affiliate), Chgo., 1977—. Contbr. articles in field to profl. jours. Served to lt. comdr. USNR, 1969-71. Recipient Research and Devel. award Am. Diabetes Assn., 1973-75; Research Career Devel. award NIH, 1975-81; named Outstanding Young Citizen Ill. Jr. C. of C., 1976. Fellow ACP; mem. Am. Diabetes Assn. (vice-chmn. research com. 1981-83), Endocrine Soc., Central Soc. Clin. Research. Subspecialties: Endocrinology; Nutrition (medicine). Current work: Insulin secretion, insulin infusion systems, nutritional aspects of metabolic diseases, exercise physiology. Office: Travenol Labs Inc 1425 Lake Cook Rd Deerfield IL 60015

HOSICK, HOWARD LAWRENCE, biol. researcher; b. Champaign, Ill., Nov. 1, 1943; s. Arthur Howard and Eunice Irma (Miller) H.; m. Cynthia Ann Jacobson, June 15, 1968; children: Steven C., Anna E., Rachel V. B.A., U. Colo., 1965; Ph.D., U. Calif., Berkeley, 1970. Postdoctoral research fellow Karolinska Inst., Stockholm, Sweden, 1970-72; asst. research biochemist Cancer Research Lab., U. Calif., Berkeley, 1972-73; asst. prof. Wash. State U., 1973-77, assoc. prof., 1978—, dir. cancer research program, 1979—; mem. steering com. Program on Aging, 1981—; vis. scientist U. Reading, Eng., 1979. Contbr. articles to jours. Recipient R.G. Gustavson award, 1965; Harry S. Boyce award, 1981; Damon Runyon Fund fellow, 1970-72; Anna Fuller Fund fellow, 1973-74; Am. Cancer Soc. Eleanor Roosevelt fellow, 1979; NIH grantee; others. Mem. Am. Soc. Cell Biology, Tissue Culture Assn., Am. Assn. Cancer Research, Internat. Assn. Breast Cancer Research. Democrat. Club: Palouse Runner. Subspecialties: Cell and tissue culture; Cancer research (medicine). Current work: Cell growth and differentiation, cancer and aging. Home: NE 1185 Lake St Pullman WA 99163

HOSKINS, JOHNNY DURR, veterinarian, educator; b. Dallas, Sept. 19, 1943; s. Durr Gary and Kathryn Alice (Barnett) H.; m. Ramonia Kay Hertel, Dec. 27, 1964; children: Holly Kay, Paul Matthew, Jason Ryan. B.A., Okla. State U., 1967, D.V.M., 1968; Ph.D., Iowa State U., 1977. Diplomate: Am. Coll. Veterinary Internal Medicine, 1979. Intern dept. small animal medicine and surgery Iowa State U., Ames, 1968-69, instr., 1970-72, asst. prof., extension veterinarian, 1972-76; veterinarian Mt. Prospect (Ill.) Animal Clinic, 1969-70; assoc. prof. vet. clin. scis. La. State U., Baton Rouge, 1976-80, prof. vet. clin. medicine, 1980—; radio and TV speaker on care and mgmt. of dogs and cats. Contbr. chpts. to textbooks, articles to profl. jours. Deacon Broadmoor Bapt. Ch., Baton Rouge, 1979—. Mem. AVMA, Am. Animal Hosp. Assn., La. Vet. Med. Assn., Am. Assn. Vet. Clinicians, Assn. Am. Vet. Med. Colls., Am. Coll. Vet. Internal Medicine, Phi Kappa Phi, Phi Zeta, Gamma Sigma Delta. Democrat. Baptist. Subspecialties: Internal medicine (veterinary medicine); Gastroenterology. Current work: Electron microscopic studies in gastroenterologic patients; veterinary drug experience reporting program in cooperation with Food and Drug Administration. Home: 11843 Parkmeadow Ave Baton Rouge LA 70816 Office: Small Animal Clinic La State U Baton Rouge LA 70803

HOSKO, MICHAEL J., JR., pharmacology educator; b. Scranton, Pa., Aug. 28, 1929; s. Michael J. and Mary (Evanso) H.; m. W. Alice Krayer, Nov. 26, 1953; children: Michael F., Mark E. B.S., U. Scranton, 1951; A.M., U. Mo., 1953; Ph.D., Jefferson Med. Coll., 1964. Head neuropharm. group Wyeth Labs., Phila., 1957-64; asst. prof. pharmacology Marquette U. Sch. Medicine, Milw., 1964-72; mem. faculty Med. Coll. Wis., Milw., 1972—, prof. pharm., 1978—, vice chmn. pharm. and toxicology dept., 1979—. Contbr. over 80 articles in field to profl. publs. Nat. Inst. Drug Abuse grantee, 1970—. Mem. AAAS, Soc. Neurosci., Am. Soc. Pharmacology and Exptl. Therapeutics, Biomed. Engring. Soc., Sigma Xi, Gamma Alpha. Subspecialties: Pharmacology; Neurophysiology. Current work: Effect of drugs on the central nervous system. Home: 2930 Burlawn Pkwy Brookfield WI 53005 Office: PO Box 26509 Milwaukee WI 53226

HOSLER, CHARLES LUTHER, JR., meteorologist, educator, university dean; b. Honey Brook, Pa., June 3, 1924; s. Charles Luther and Miriam Deichley (Stauffer) H.; m. Gladys Cheesbrough, 1947 (div.); children:Sharon Elizabeth, David Charles, Lynn Rebecca, Peter William; m. Anna R. Stahel, 1971. Student, Bucknell U., 1943-44, Mass. Inst. Tech., 1944-45; B.S., Pa. State U., 1947, M.S., 1948, Ph.D., 1951. Faculty Pa. State U., University Park, 1948—, prof. meteorology, 1960—, head dept., 1961-65, dean, 1965—; Hydrographer Pa. Dept. Forests and Waters, 1949-59; meteorol. cons., 1950—, vis. prof. colls., lectr. civic and profl. groups; condr. daily TV weather program, 1957-67; spl. research microphysics of clouds; adv. com. on meteorology EPA; chmn. storm fury adv. panel Nat. Acad. Scis.; mem. nat. adv. com. on oceans and atmosphere.; mem., chmn. bd. trustees Univ. Corp. for Atmospheric Research, Boulder, Colo., 1981—. Contbr. articles to profl. jours. Served to lt. (j.g.) USNR, 1943-46; lt. comdr. Res. Fellow Am. Meteorl. Soc. (councilor, pres. 1976); mem. Nat. Acad. Engring., Am. Geophys. Union (award outstanding paper hydrology 1955), ASCE (weather modification com.), Am. Chem. Soc. (regional lectr. 1971-72), AAUP (v.p. Pa. State U. 1961), AAAS, Sigma Xi (pres. Pa. State U. 1958, nat. lectr. 1972), Tau Beta Pi, Sigma Gamma Epsilon. Subspecialty: Meteorology. Home: 1229 Smithfield Circle State College PA 16801 Office: Earth and Mineral Scis Pa State U University Park PA 16802

HOSS, WAYNE P., biochemist, educator; b. Paso Robles, Calif., Dec. 11, 1943; s. Donald D. and Rosalie M. (Gauthier) H.; m. Dorothy M. Hart, Sept. 3, 1967. B.S. in Chemistry, U. Idaho, 1966, Ph.D., U. Nebr., 1971. NSF postdoctoral fellow dept. chemistry U. Rochester, 1970-72; NIH postdoctoral fellow Center for Brain Research, 1970-72, asst. prof., 1975-80, assoc. prof., 1980—, dir. undergrad. neurosci. program, 1980—; vis. assoc. prof. biochemistry Nagoya City (Japan) U., 1977. Contbr. articles to profl. jours. NIMH research career devel. awardee, 1976-81. Mem. Am. Chem. Soc., Am. Soc. Neurochemistry, Biophys. Soc., N.Y. Acad. Scis., Soc. Neurosci. Subspecialties: Neurochemistry; Neuropharmacology. Current work: Interaction of drugs, neurotransmitters and neuroregulators with cellular membranes and their components including receptors, enzymes, lipids and ion channels. Home: 1748 Blossom Rd Rochester NY 14610 Office: Center for Brain Research Rochester Med Center Rochester NY 14642

HOSSNER, KIM LEE, physiologist; b. Idaho Falls, Idaho, Jan. 3, 1949; s. Ralph F. and Ruth (Merrick) H.; m. Linda Timmons, Aug. 21, 1971; children: Nathan Robert, Matthew Colin. B.S., U. Idaho, 1971; Ph.D., U. Tenn., 1976. Teaching asst. U. Tenn.-Knoxville, 1971-76; postdoctoral fellow Case Western Res. U., 1976-80; instr., research assoc. U. Idaho-Moscow, 1980—. Contbr. articles profl. jours. Tenn. Heart Assn. predoctoral fellow, 1973-74; U. Idaho Biomed. research grantee, 1980-81. Mem. AAAS, Sigma Xi, Phi Sigma, Mu Epsilon Delta. Subspecialties: Physiology (biology); Endocrinology. Home: 605 Ekes Rd Moscow ID 83843 Office: Dept Animal Scis Idaho Moscow ID 83843

HOSTETLER, KARL YODER, physician, educator, biochemist; b. Goshen, Ind., Nov. 17, 1939; s. Carl Milton and LaVerne (Yoder) H.; m. Margaretha Steur, Dec. 17, 1971; children: Saskia Emma, Kirsten Cornelia, Carl Martijn. B.A., DePauw U., 1961; M.D., Western Res. U., 1965. Diplomate: Am. Bd. Internal Medicine, Sub-Bd. Endocrinology and Metabolism. Intern Univ. Hosp., Cleve., 1965-66, resident, 1968-69; fellow in endocrinology metabolism Cleve. Clinc Found., 1969-70; postdoctoral fellow in biochemistry and medicine Case Western Res. U., Cleve., 1966-68, U. Utrecht, Netherlands, 1970-72; dir. Metabolism Clinic, VA Med. Ctr., San Diego, 1973—; asst. prof. medicine U. Calif.-San Diego, La Jolla, 1973-79, assoc. prof., 1979—; exchange scientist USSR Acad. Sci., Moscow, 1974. Contbr. articles to profl. jours. Guggenheim Found. fellow, 1980. Mem. Am. Fedn. Clin. Research; mem. Western Soc. Clin. Investigation; Mem. Am. Soc. Biol. Chemists, AAAS, Am. Soc. for Clin. Investigation, Am. Diabetes Assn. (dir. So. Calif. affiliate 1976-83, pres. San Diego chpt. 1982-83, service award 1976). Subspecialties: Endocrinology; Biochemistry (medicine). Current work: Physician, scientist, diseases of endocrinology and metabolism, regulation of intracellular lipid metabolism. Office: VA Med Ctr (111G) 3350 La Jolla Village Dr San Diego CA 92161

HOTES, LAWRENCE STEVEN, internist, endocrinologist; b. N.Y.C., Nov. 14, 1951; s. Edward and Jean Hotez; m. Linda Fine, Nov. 12, 1978. B.S., Trinity Coll., Hartford, Conn., 1973; M.D., George Washington U., 1977. Diplomate: Am. Bd. Internal Medicine. Intern Univ. Hosp., Boston, 1977-78, resident, 1978-80; internist Associated Internists of Randolph, Inc., Mass., 1980—; researcher Harvard Med. Sch., Boston, 1972, George Washington U., 1973-74. Fellow ACP; mem. Mass. Med. Soc., Am. Fedn. Clin. Research, Alpha Omega Alpha. Subspecialties: Internal medicine; Endocrinology. Current work: Endocrinology. Office: 999 N Main St Randolph MA 02368

HOTTA, YASUO, research biologist; b. Nagoya, Aichi-ken, Japan, Jan. 12, 1932; came to U.S., 1960; s. Seiichiro and Sumiko (Moriyama) H.; m. Sachiko Tsurumi, Oct. 10, 1962; children: Eugene, Naomi. B.S., Biol. Inst., Nagoya U., 1954, M.S., 1956, Dr.Sci., 1959. Lectr. Kinjo Woman's U., Nagoya, Japan, 1959; postdoctoral fellow Can. Dept. Agr., Ottawa, Ont., 1959-60; research assoc. in biology U. Ill.-Urbana, 1960-65; assoc. research biologist U. Calif.-San Diego, La Jolla, 1965-73, research biologist, 1973—; cons. Hirokawa Pub. Co., Tokyo, 1970—, M.B.L., Nagoya, Japan, 1980—, Japan Pharm. Devel. Co., Osaka, 1979—. Author: Cell Cycle, 1983, Cell Biology, 1967, Biology Today, 1975. Pres. P.T.A., San Diego Japanese Sch., 1973—; pres. New Liberal Democratic Club U.S.A., San Diego, 1982—. Recipient Research Travel award Bourghs Welcome, 1979; Research Grant Nat. Inst. Child and Human Devel., 1979; Lectr. Travel award Naito Found., 1981; Research Travel award Acad. Senate U Calif San Diego, 1980. Mem. Am. Soc. Biol. Chemists, Am. Soc. Cell Biology, Japanese Soc. Cell Biology, Japanese Soc. Plant Physiologists, Sigma Xi. Subspecialties: Cell biology; Reproductive biology. Current work: Meiosis, genetic recombination during reproduction, genetic transformation under the specific activation of gene. Home: 6635 Fisk Ave San Diego CA 92122 Office: U Calif San Diego Dept Biology La Jolla CA 92093

HOUCK, LAURIE GERALD, plant pathologist; b. Tucson, Aug. 13, 1928; s. Gerald Wesley and Laura Lee (Baker) H.; m. Marlene Moore, Apr. 17, 1970 (dec.); children: Lorna Jeanne, Marlys Lee; m. Margaret Victoria Evers, July 28, 1978. B.S., U. Ariz, Tucson, 1952, M.S., 1954; Ph.D., Oreg. State U., Corvallis, 1962. Plant physiologist U.S. Dept. Agr., Phoenix, 1954-55; asst. horticulturist U. Ariz., Mesa, 1955-57; research asst. plant pathology Oreg. State U., Corvallis, 1958-62; research plant pathologist U.S. Dept. Agr., Pomona, Calif., 1962-73, Riverside, Calif., 1973-77, Fresno, Calif, 1977—; instr. Calif. State Poly. U., 1963, lectr., 1964, 66. Served with AUS, 1946-48. Johnson Wax fellow, 1953-54. Mem. Am. Soc. Plant Physiologists, AAAS, Am. Soc. Horticulturists, Mycol. Soc. Am., Am. Phytopath. Soc., Food Distbn. Research Soc. Unitarian. Club: Toastmasters. Subspecialties: Plant pathology; Plant physiology (agriculture). Current work: Post harvest horticulture, storage and transportation. Home: 1758 S Waldby Ave Fresno CA 93727 Office: U S Dept Agr PO Box 8143 Fresno CA 93747

HOUGAS, ROBERT WAYNE, genetics educator; b. Blythedale, Mo., June 17, 1918; m., 1943. B.S., U. Wis., 1942, Ph.D. in Genetics, 1949. From asst. prof. to assoc. prof. U. Wis.-Madison, 1949-62, prof. genetics, 1962—, asst. dir. exptl. sta., 1962-65, asst. dean and dir. Coll. Agr., 1965-66, leader inter-regional potato introduction project, 1950—, assoc. dean and dir. Coll. Agr., 1966—. Mem. Genetics Soc. Am., Am. Phytopath. Soc., Potato Assn. Am. (sec. 1952-55, v.p. 1956, pres. 1957), Am. Genetic Assn. Subspecialty: Plant genetics. Office: U Wis 140 Agr Hall Madison WI 53706

HOUGH, DAVID GRANVILLE, computer programmer; b. Ft. Sill, Okla., Jan. 22, 1949. B.A., Carleton Coll., Northfield, Minn., 1968; Ph.D., U. Calif.-Berkeley, 1977. Astronomer U.S. Naval Obs., Washington, 1968-71; computer programmer Tektronix, Wilsonville, Oreg., 1976-80, Apple Computer, Cupertino, Calif., 1980—. Mem. Assn. for Computing Machinery, Soc. for Indsl. and Applied Math., IEEE, Sierra Club. Subspecialties: Mathematical software; Numerical analysis. Current work: Computer arithmetic, mathematical software, numerical analysis. Home: PO Box 561 Cupertino CA 95015 Office: Apple Computer Cupertino CA 95014

HOUGH, JAMES EMERSON, consulting engineer, geologist; b. Paducah, Ky., July 25, 1930; s. Winfred Cyril and Goldie B. (Stewart) H.; m. Valeska Marie Runge, Aug. 11, 1956; children: Jerome Kevin, Christopher Kendal. A.S., Paducah Jr. COll., 1951; B.S., U. Ky., 1953, M.S., 1958. Registered profl. engr., Ohio, Ky., Ill., Ind., Va., Pa. Asst. geologist Ky. Geol. Survey, Lexington, Ky., 1957-58; geologist TVA, Lexington, 1958-59; soil engr. Ky. Dept. Hwys., Lexington, 1959-60, Ill. Dept. Hwys., Paris, 1960-63; chief engr. Earth Sci. Labs., Cin., 1963; prin. James E. Hough & Assocs., Cin., 1963—. Author books in field of engring. geology. Elder United Presbyterian Ch., 1966-70; co-chmn. Earthwork Regulations Task Force, City of Cin., 1973-74; adv. Earth Movement Task Force, Hamilton County, 1982; mem. Earthwork Regulations Com., Hamilton County, 1982—, Mapping Com., Hamilton County, 1982—. Served with U.S. Army, 1953-55. Fellow Geol. Soc. Am.; mem. Assn. Soil and Found. Engrs. (publs. com. 1981—), Am. Cons. Engrs. Council (geotech. engring. com. 1982), ASCE, Assn. Engring. Geologists, Am. Soc. Photogrammetry, Am. Arbitration Assn., Internat. Soc. Soil Mechancis and Found. Engrs., Internat. Assn. Engring. Geologists, Cons. Engrs. Ohio (sec., trustee 1974-77), Ohio Assn. Cons. Engrs. (bd. dirs. 1977-78), DAV, Sigma Gamma Epsilon. Lodge: Masons. Subspecialties: Civil engineering; Geology. Current work: Consultant in the applied earth sciences, lecturer in engineering geology at University of Cincinnati. Home: 10936 Gosling Rd Cincinnati OH Office: James E Hough and Assocs 3398 W Galbraith Rd Cincinnati OH 45239

HOUGH-EVANS, BARBARA RAYMOND, research biologist; b. Hackensack, N.J., Aug. 28, 1924; d. Landon Thomas and Dorothy (Curtis) Raymond; m. Paul VanCampen Hough, Oct. 21, 1945 (div. Sept. 1972); children: David S., Judith A.; m. John William Evans, Dec. 18, 1975. B.A., Swarthmore Coll., 1945; M.A., Cornell U., 1948; Ph.D., SUNY-Stony Brook, 1968. Sr. research fellow Calif. Inst. Tech., Pasadena, 1971-81, sr. research assoc., 1981—. Democrat. Subspecialties: Developmental biology; Molecular biology. Current work: Molecular biology of development, using sea urchin embryos; oogenesis and nuclear RNA and maternal RNA research. Home: 397 S Santa Anita Ave Pasadena CA 91107 Office: 120 1 E California Blvd Pasadena CA 91125

HOUGHTON, ALAN HOURSE, JR., medical oncologist, researcher; b. Boston, Apr. 12, 1947; s. Alan N. and Elizabeth (Jones) H.; m. Meryl Kauffman, June 29, 1975. B.A., Stanford U., 1970; M.D., U. Conn.-Farmington, 1974. Diplomate: Am. Bd. Internal Medicine. Intern Thomas Jefferson U., 1974-75; resident U. Conn.-Farmington, 1975-77; med. oncology fellow Meml. Sloan Kettering Inst., N.Y.C., 1977-80, asst. attending physician, 1980—. Contbr. articles to profl. jours. Am. Cancer Soc. fellow, 1980-83. Mem. Am. Assn. Cancer Research, Am. Soc. Clin. Oncologists, AAAS, N.Y. Acad. Scis. Subspecialties: Cancer research (medicine); Oncology. Current work: Research in cell biology, immunology. Office: Po Box 465 1275 York Ave New York NY 10021

HOUGHTON, DAVID DREW, meteorologist; b. Phila., Apr. 26, 1938; s. Willard Fairchild and Sara Nancy (Holmes) H.; m. Barbara Flora Coan, June 22, 1963; children: Eric Brian, Karen Jeanette, Steven Andrew. B.S., Pa. State U., 1959; M.S., U. Wash., 1961, Ph.D., 1963. Research scientist Nat. Center Atmospheric Research, Boulder, Colo., 1963-68; exchange scientist USSR Acad. Scis., Moscow, 1966; vis. scientist Courant Inst. Math. Scis., N.Y.C., 1966; asst. prof. dept. meteorology U. Wis., Madison, 1968-69, asso. prof., 1969-72, prof., 1972—, chmn. dept., 1976-79; scientist Internat. Sci. and Mgmt. Group for Global Atmospheric Research Program, Bracknell, Eng., 1972-73; lectr. Nanjing U., People's Republic of China, 1980. Contbr. articles to profl. jours.; editor-in-chief: Handbook of Applied Meteorology. Vice chmn. Planning Commn., Town of Dunn, Wis., 1977-81. NSF fellow, 1960-63. Fellow Am. Meteorol. Soc.; mem. AAAS, Phi Beta Kappa, Sigma Xi, Phi Kappa Phi. Quaker. Subspecialty: Meteorology. Office: Dept Meteorology U Wis Madison WI 53706

HOUGHTON, JANET ANNE, cancer research scientist; b. Grantham, Lincolnshire, Eng., May 21, 1952; came to U.S., 1977; d. George Edward and Viola Mary (Grant) Geeson; m. Peter James Houghton, July 21, 1973. B.Pharmacy, U. Bradford, Eng., 1973; Ph.D., Inst. Cancer Research, 1977. Postdoctoral fellow St. Jude Children's

Research Hosp., Memphis, 1977-80, research assoc., 1980-82, asst. mem., 1982—; cons. VA. Contbr. articles to profl. jours. Am. Cancer Soc. grantee, 1980—; Nat. Cancer Inst. grantee, 1982—. Mem. Pharm. Soc. Gt. Britain, Am. Assn. Cancer Research. Subspecialties: Cancer research (medicine); Chemotherapy. Current work: Study of characteristics of human solid tumors maintained in immune-deprived mice, relating to cell biology, biochemical pharmacology, cancer chemotherapy. Home: PO Box 381 Route 1 Somerville TN 38068 Office: 332 N Lauderdale St Memphis TN 38101

HOUGHTON, PETER JAMES, cancer research scientist; b. London, Feb. 1, 1949; U.S., 1977; s. George Douglas Stanley and Ruby (Power) H.; m. Janet Anne Houghton, July 21, 1973. B.Pharmacy with first class honors, U. Bradford, Eng., 1972; Ph.D., Inst. Cancer Research, 1976. Postdoctoral fellow Inst. Cancer Research, Surrey, Eng., 1976-77; postdoctoral fellow St. Jude Children's Hosp., Memphis, 1977-79, research assoc., 1979-81, asst. mem., 1981—. Contbr articles to profl. jours. Am. Cancer Soc. grantee, 1980—; Nat. Cancer Inst. grantee, 1982—. Mem. Am. Assn. Cancer Research, Pharm. Soc. Gt. Britain. Subspecialties: Cancer research (medicine); Chemotherapy. Current work: Study of the characteristics of human solid tumors maintained in immune-deprived mice, relating to cell biology, biochemical pharmacology, cancer chemotherapy. Home: Route 1 PO Box 381 Somerville TN 38068 Office: 332 N Lauderdale St Memphis TN 38101

HOUGLAND, ARTHUR ELDON, microbiologist, educator; b. Omaha, Jan. 5, 1935; s. Arthur J. and Marie B. (Perkins) H.; m. Margaret Webb, June 4, 1961; children: Amy, Jon, Kriss. B.A., State U. Iowa, 1958; M.S., Brigham Young U., 1961; Ph.D., U. S.D., 1975. Research microbiologist Dept. Army, Ft. Detrick, Md., 1963-69; assoc. prof. biol. scis. East Tenn. State U., 1973—; asst. prof. microbiology Quillen Dishner Sch. Medicine, 1978—. Served with Med. Service Corps U.S. Army, 1961-63. Recipient S.D. br. Am. Cancer Soc. Research award, 1969-71, Oak Ridge Associated Univs. Research Contract and Faculty Participation award, 1977. Mem. Am. Soc. Microbiology, Tissue Culture Assn., Mil. Surgeons, AAAS, Res. Officers Assn. Subspecialties: Virology (medicine); Biochemistry (medicine). Current work: Lipids of cells and plasma membranes; enzymes. Home: Route 19 Box 267 Johnson City TN 37601 Office: East Tenn State U PO Box 22690A Johnson City TN 37614

HOUK, JAMES CHARLES, physiologist, educator; b. Northville, Mich., June 3, 1939; s. James Charles and Elowene Elsie (Tower) H.; m. Antoinette, Dec. 28, 1963; children—Philip Adler, Nadia Rosella, Peter Charles. B.S. in Elec. Engring, Mich. Technol. U., 1961; M.S., M.I.T., 1963; Ph.D. in Physiology, Harvard U., 1966. Research asso. Sch. Medicine, San Juan, P.R., summers 1962, 64; postdoctoral fellow Faculte de Medecine, Toulose, France, 1966-67, Harvard U. Med. Sch., 1967-68, instr., 1967-69, asst. prof. physiology, 1969-73; lectr. dept. elec. engring. M.I.T., 1971-73; assoc. prof. physiology Johns Hopkins U. Sch. Medicine, 1973-78; adj. assoc. prof. physiology U. N.C., 1975; Nathan Smith Davis prof., chmn. dept. physiology Northwestern U. Sch. Medicine and Dentistry, Chgo., 1978, prof. engring. sci. Technol. Inst., dir. behavioral neurosci. tng. program, 1979—, dir. tng. program in neural control, 1983—; mem. neurology B study sect. NIH, 1978-82. Contbr. to: Med. Physiology, 14th edit, 1980, Handbook of Physiology, Motor Control, 1981. Mem. IEEE, AAAS, Am. Physiol. Soc., Soc. for Neuroscience, Sigma Xi, Tau Beta Pi, Eta Kappa Nu, Phi Kappa Phi. Subspecialty: Physiology (medicine). Home: 2900 Lincoln St Evanston IL 60201 Office: Northwestern U Med Center Dept Physiology Ward 5-319 303 E Chicago Ave Chicago IL 60611

HOULIHAN, JOHN FRANK, physics educator, researcher; b. Springfield, Ill., Sept. 9, 1942; s. William John and Alice Laura (Vetter); m. Elena Sue Hiatt, Sept. 5, 1968. Ph.D. in Solid State Physics, Pa. State U., 1973. Engr. solid state device group Avco Electronics, Cin., 1967; prof. physics Pa. State U., Sharon, 1968—. Contbr. articles to profl. jours. Area chmn. Republican Party, Farrell, Pa., 1974-80. Served in USMCR, 1960-66. Recipient award U.S. Citizens Congress, 1976; grantee NSF, Gulf Oil Corp., Pa. State U. Mem. Am. Phys. Soc., Am. Assn. Physics Tchrs. Methodist. Subspecialties: Condensed matter physics; Solar energy. Current work: Photoelectrochemical cells - materials and device fabrication. Home: 1330 Roemer Blvd Farrell PA 16121 Office: 147 Shenango Ave Sharon PA 16146

HOUSE, EDWIN WESLEY, biologist, physiologist, researcher; b. Corning, N.Y., Oct. 20, 1939; s. Alvin J. and Alice E. (Huntsman) H.; m. Janet G. Gryth, Sept. 4, 1964; children: Naomi A., Matthew P. B.S., Western Mont. Coll, 1960; M.A., U. Mont., 1962; Ph.D., U. N.D., 1965. Research fellow U. N.D., 1962-65; research asso. Loyola U., Chgo., 1964; asst. prof. U. Mont., 1965-66; asst. prof. to prof. biology Idaho State U., Pocatello, 1966—, dir. animal care facilities, 1972—. Contbr. articles to profl. jours; author lab. manuals. Mem. Inter-Varsity Christian Fellowship. Grantee Am. Heart Assn., 1974-83; named Disting. Tchr. of Yr. Idaho State U., 1979, Disting. Alumnus We. Mont. Coll., 1980. Mem. AAAS, Am. Phys. Soc., N.W. Sci., Sigma Xi, Phi Sigma. Presbyterian. Subspecialties: Physiology (biology); Physiology (medicine). Current work: Cardiovascular physiology, pathology, arteriosclerosis, cardiovascular control, hormonal mechanisms affecting arteriosclerosis, steelhead trout model for study of coronary artery disease. Home: 119 Princeton St Pocatello ID 83201 Office: Dept of Biology Idaho State University Pocatello ID 83209

HOUSE, VERL LEE, geneticist; b. Wellsville, Mo., Apr. 10, 1919; s. Ralph and Annie Lee (Tipton) H.; m. Marylin Grace, Sept. 11, 1948; children: Jeffrey, Garth, Tipton. A.B., U. Calif.-Berkeley, 1941, M.A., 1948, Ph.D., 1950. Asst. prof. dept. biology Johns Hopkins U., Balt., 1950-54; prof., chmn. dept. biology Radford (Va.) Coll., 1954-58; assoc. prof. zoology and entomology Ohio State U., Columbus, 1958-66; prof. genetics, 1967-81, prof. emeritus, 1981—; vis. prof. zoology U. Calif.-Berkeley, 1960, 64, 67; assoc. dir. fellowships Nat. Acad. Sci.-NRC, Washington, 1966-67. Contbr. articles to profl. jours. Served with Signal Corps U.S. Army, 1943-46. Sigma Xi grantee, 1951-52; NSF grantee, 1954-58; Ohio State U. grantee, 1960-70. Fellow Ohio Acad. Scis.; mem. Am. Genetics Soc., Sigma Xi, Phi Beta Kappa, Alpha Zeta, Xi Sigma Pi. Subspecialties: Gene actions; Developmental biology. Current work: Physiological genetics with particular emphasis on developmental aspects of gene expression. Home: 378 Walhalla Dr Columbus OH 43202 Office: Dept Genetics Ohio State U 1735 Neil Ave Columbus OH 43210

HOUSEPIAN, EDGAR MINAS, neurosurgeon; b. N.Y.C., Mar. 18, 1928; s. Moses Minas and Makroulie (Ashiian) H.; m. Marion Lyon, Sept. 8, 1954; children: David Minas, Stephen Lyon, Jean Carleton. A.B., Columbia Coll., 1949; M.D., Columbia U., 1953. Diplomate: Am. Bd. Neurosurgery, 1961. Intern, gen surg. resident Lakeside Hosp., Cleve., 1953-55; resident Columbia Presbyn. Med. Center, N.Y.C., 1955-59; faculty Columbia U., N.Y.C., 1961—, prof. neurol. surgery, 1978—. Served to lt. comdr. M.C. USNR. Fellow A.C.S.; mem. AMA, AAAS, N.Y. Med. Soc., Am. Assn. Neurol. Surgeons, Am. Neurol. Surgery Congress U.S., Neurol. Surgery Research Soc., N.Y. Neurosurg. Soc., Phi Beta Kappa. Subspecialties: Surgery; Neurology. Current work: Researcher in tumors of the nervous system and vascular surgery. Office: Columbia Dept Neurosurgery 710 W 168th St New York NY 10032

HOUSER, RONALD EDWARD, veterinarian, researcher; b. Fairbury, Nebr., Aug. 11, 1949; s. Edward Erle and Lois Charlotte (Dux) H.; m. Linda Marie Webber, June 13, 1971; children: Angela Marie, Brian Edward, Darren James. D.V.M., U. Mo., Columbia, 1974; M.S., Ohio State U., Columbus, 1979. Diplomate: Am. Coll. Vet. Preventive Medicine. Staff veterinarian Florence Animal Hosp. and Harless Animal Clinic, Omaha, 1976-77; grad. teaching assoc. Ohio State U., 1977-79; asst. instr. U. Nebr., Lincoln, 1979—; mem. Nebr. State Bd. Health. Del. plains regional conf. Young Ams. for Freedom; mem. Henry Doorly Zoo. Served to capt. USAF, 1974-76. Prodn. Credit Assn. scholar, 1967-68; Union Pacific R.R. scholar, 1967-68; U. Nebr. Regents fellow, 1981-82. Mem. AVMA, Am. Coll . Vet. Preventive Medicine, Nat. Assn. State Pub. Health Veterinarians, Nebr. Vet. Med. Assn. (dist. pres.), Sigma Xi. Republican. Methodist. Club: Paddock Rd. Community (Omaha). Subspecialties: Preventive medicine (veterinary medicine); Virology (veterinary medicine). Current work: Pathogenesis of bovine respiratory syncytial virus in gnotobiotic calves, immune response of gnotobiotic calves to bovine respiratory syncytial virus. Home: 1611 North Gate Rd Lincoln NE 68521 Office: 230 VBS U Nebr Lincoln NE 68583

HOUSER, STEVEN ROBERT, physiologist; b. Camden, N.J., Mar. 29, 1951; s. Robert H. and Ruth E. (Hagy) H.; m. Janet Phinney, Sept. 29, 1972. B.S., Eastern Coll., St. Davids, Pa., 1973; Ph.D., Temple U., 1977. Research fellow dept. cardiology Temple U., Phila., 1977-79, asst. prof. dept. physiology, 1979—, exercise cons., 1979—. Contbr. articles to profl. jours. NIH grantee, 1979-81. Mem. Am. Heart Assn., Am. Physiol. Soc., Biophys. Soc., AAAS, Sigma Xi. Baptist. Subspecialties: Physiology (medicine); Cell biology. Current work: Control of ion metabolism in cardiac muscle. The role of abnormalities in ionic homeostatic mechanisms in cardiac hypertrophy. Home: 2512 Hillcrest Rd Drexel Hill PA 19026 Office: Temple U Sch Medicine 3400 N Broad St Philadelphia PA 19140

HOUSEWORTH, LLOYD DOUGLAS, research plant pathologist; b. Carrollton, Mo., Apr. 11, 1944; s. Paul E. and Fern (Warner) H.; m. Linda Sue Helm, June 13, 1965; children: Cheryl Lynn, Laura Kaye. Student, Central Mo. State Coll., 1962-64; B.S., U. Mo., 1966, M.S., 1968, Ph.D., 1973. Research specialist U. Mo., Columbia, 1968-73; field research rep. Rohm and Haas, Searcy, Ark., 1973-76; responsible for residue programs Ciba-Geigy, Greensboro, N.C., 1976-79, tech. planning specialist, 1979-80, biol. research staff, Greensboro, N.C., 1980—. Contbr. writings in field to profl. jours. Mem. Am. Phytopath. Soc., So. Soybean Disease Workers, Soil Borne Disease Workers. Subspecialty: Plant pathology. Current work: Development of new fungicides; research and development of experimental fungicides for commercial use. Office: Ciba- Geigy Corp PO Box 11422 Greensboro NC 27409

HOUSTON, DAVID ROYCE, research plant pathologist; b. Worcester, Mass., Sept. 18, 1932; s. Bradley Royce and Margaret Pelton (Brown) H.; m. Janet Haburay, Dec. 19, 1954; children: Leslie Carol, Linda Kay, Bradley Keitz. B.S., U. Mass., 1954; M.F., Yale U., 1955; Ph.D., U. Wis., 1961. Instr. dept. plant pathology U. Wis., 1958-61, asst. prof., 1961; prin. plant pathologist, project leader Ctr. for Biol. Control of Northeastern Forest Insects and Diseases, U.S. Dept. Agr. Forest Service, Hamden, Conn., 1961—. Author: Understanding the Game of the Environment, 1979; contbr. articles to profl. jours. Mem. Soc. Am. Foresters, Am. Phytopath. Soc. Subspecialties: Plant pathology; Ecology. Current work: Ecological relationships of forest tree diseases with special emphasis on biological control; stress-triggered tree diseases, dieback/decline diseases. Home: 404 Opening Hill Rd Madison CT 06443 Office: Ctr for Biol Control of Northeastern Forest Insects and Diseases 51 Mill Pond Rd Hamden CT 06514

HOUSTON, L.L., biochemist, educator; b. Wichita, Kans., June 3, 1940; s. Floyd E. and Lois (DeHaven) H.; m. Teri L. DeHon, June 5, 1962; children: Leslie Ann, Scott Eric. B.S., Kans. State U., 1962; Ph.D., U. Wash., 1967. Research assoc. U. Calif.-Berkeley, 1967-69; asst. prof. biochemistry U. Kans.-Lawrence, 1969-73, assoc. prof., 1973-78, prof., 1978—; vis. lectr. U. Ill.-Urbana, 1973; guest prof. U. Basel, Biozentrum, Basel, Switzerland, 1975-76; vis. scientist Fred Hutchinson Cancer Research Ctr., Seattle, 1979-80. Recipient Research Career Devel. award Nat. Cancer Inst., 1975-80. Mem. Am. Soc. Biol. Chemists. Subspecialties: Biochemistry (biology); Chemotherapy. Current work: Immunotoxins; mechanism of action of plant toxins. Office: Dept Biochemistry U Kans Lawrence KS 66045

HOVER, GERALD ROBERT, clinical psychologist; b. San Francisco, June 14, 1947; s. Walter William and Doris Helen (Lynch) H. B.A., Gonzaga U., 1969, M.A., 1970; Ph.D. in Psychology, U. B.C., Vancouver, 1980. Counselor, instr. Central Oreg. Community Coll., Bend, 1970-71; dir. recreation and student activities Central Wash. U., Ellensburg, 1971-73; psychotherapist Vancouver (Can.) Gen. Hosp., 1973-78, Valley Gen. Hosp., Renton, Wash., 1973-78; psychologist U. Wash., Harborview Community Mental Health Ctr., Seattle, 1978—. Author: psychotherapy invention The Goal Attainment Scaling Process, 1980. Mem. Am. Psychol. Assn. (assoc.), Western Psychol. Assn., Wash. Psychol. Assn. Roman Catholic. Subspecialty: Behavioral psychology. Current work: Psycho-education delivery to the chronically mentally ill. Home: 32422 2d Ave SW Federal Way WA 98003 Office: U Wash Harborview Community Mental Health Ctr Seattle WA 98104

HOVINGH, JACK, nuclear engineer; b. Grand Rapids, Mich., May 5, 1935; s. Peter and Hermina (Kraker) H.; m. Patsy VanderKam, June 22, 1956 (div. 1978); children: Mary, Mark. B.S. in Mech. Engring, U. Mich., 1958, 1958; M.S. in Engring. Sci, U. Calif.-Berkeley, 1973. Nuclear engr. Lawrence Livermore (Calif.) Nat. Lab., 1958—; Mem. rev. panels Electric Power Research Inst., Palo Alto, Calif., 1980—. Contbr. sci. articles to profl. publs. Mem. gen. plan rev. com. City of Pleasanton, Calif., 1983. Mem. Am. Nuclear Soc. Subspecialties: Nuclear fusion; Mechanical engineering. Current work: Pioneering work on the design and analyses of both magnetically and inertially confined fusion reactors for energy applications. Home: 4250 Muirwood Dr Pleasanton CA 94566 Office: Lawrence Livermore Nat Lab PO Box 5508 L-480 Livermore CA 94550

HOWALD, REED ANDERSON, chemist, educator; b. Pitts., Nov. 23, 1930; s. Arthur Mark and Katharine Whitesell (Anderson) H.; m. Elaine Marie Sheperd, Dec. 29, 1961; children: Glenn, Shere, Craig. B.A., Oberlin Coll., 1952; Ph.D., U. Wis., Madison, 1955. Instr. UCLA, 1955-56, Harvard U., 1956-59; asst. prof. Oberlin (Ohio) Coll., 1959-60, St. John's U., 1960-63, Mont. State U., 1963-65, assoc. prof., 1965-69, prof. chemistry, 1969—. Mem. Am. Chem. Soc., Phi Beta Kappa, Sigma Xi. Methodist. Subspecialties: Thermodynamics; High temperature chemistry. Current work: Computer calculation of phase diagrams; silicates; slag; geochemistry; multicomponent phase equilibria; computers; kinetics. Home: 265 Story Hill Rd Bozeman MT 59715 Office: Dept Chemistry Mont State U Bozeman MT 59717

HOWARD, DARLENE VAGLIA, psychology educator, researcher; b. Pitts., Mar. 7, 1947; d. Lawrence Joseph and Mary Jane (Nessenthaler) Vaglia; m. James H. Howard, Jr., Sept. 11, 1971; 1 son, Jeffrey Matthew. B.S., Juniata Coll., 1969; M.A., Brown U., 1972, Ph.D., 1974. Asst. prof. Georgetown U., 1973-79, assoc. prof. psychology, 1979—. Author: Cognitive Psychology: Memory, Language and Thought, 1983; contbr. numerous articles to profl. jours.; editorial bd.: Jour. Gerontology, 1981-82; ad-hoc reviewer: Exptl. Aging Research. Nat. Inst. Aging research grantee, 1978—. Mem. Am. Psychol. Assn., Eastern Psychol. Assn., Gerontol. Soc. Am., Psychonomic Soc., Soc. Research in Child Devel., Sigma Xi. Subspecialties: Cognition; Developmental psychology. Current work: Research on adult age differences in memory and language processing. Office: Georgetown U Psychology Dept 37th and O Sts Washington DC 20057

HOWARD, IAN PORTEOUS, psychology educator; b. Rochdale, Lancashire, Eng., July 20, 1927; emigrated to Can., 1966, naturalized, 1975; s. Thomas and Annie (Porteous) H.; m. Antonie Eber, Feb. 28, 1928; children: Ruth, Neil, Martin. B.Sc., Manchester (Eng.) U., 1952; Ph.D., Durham (Eng.) U., 1966. Lectr. Durham U., 1952-65; assoc. prof. NYU, N.Y.C., 1965-66; assoc. prof. psychology York U., Toronto, Ont., Can., 1966-68, prof., 1968—. Author: Human Spatial Orientation, 1966, Human Visual Orientation, 1982; contbr. chpts. to books, articles to jours. Fellow Can. Psychol. Assn.; mem. Am. Psychol. Assn., Exptl. Psychology Soc. Subspecialties: Psychophysics; Sensory processes. Current work: Psychophysical investigation of mechanisms of human space perception, including depth perception, and perception of direction and orientation of objects. Home: 49 Dove Ln Thornhill ON Canada L3T 1W1 Office: Dept Psychology York U Downsview ON Canada M3J 1P3

HOWARD, JAMES LAWRENCE, pharmacologist; b. Glen Ellyn, Ill., Nov. 30, 1941; s. Ralph Orson and June Virginia (Underwood) H.; m. Judith Bennett Howard, Aug. 25, 1963; children: David Lawrence, Erin Kendra. B.A., U. N.C., 1963; M.S., Tulane U., 1966, Ph.D., 1968. Asst. prof. dept. psychiatry U. N.C., Chapel Hill, 1968-74; dir. behavioral pharmacology Burroughs Wellcome Co., Research Triangle Park, N.C., 1974—; adj. prof. U. N.C.-Chapel Hill, and Raleigh, 1974—; cons. N.C. Alcohol. Research Authority, NSF. Contbr. articles to profl. jours. NDEA fellow, 1964-68. Fellow Am. Psyschol. Assn.; mem. Soc. Neurosci., AAAS, Behavioral Pharmacology Soc., Soc. Stimulus Properties of Drugs. Republican. Presbyterian. Subspecialties: Neuropsychology; Physiological psychology. Current work: Behavioral pharmacology, aging and senility. Home: Route 1 Old Morrow Mill Rd Chapel Hill NC 27514 Office: Burroughs Wellcome Co Research Triangle Park NC 27709

HOWARD, JAY LLOYD, appliance manufacturing company executive; b. Olivia, Minn., June 24, 1952; s. Joe Earl and Jacqueline May (Gilmore) H., Jr.; m. Pamela Sue Dougan, Sept. 15, 1979; children: Melissa Ann, Joe Earl III, Heather Lea. B.A. in Psychology, U. Mo., Kansas City, 1972. Cert. mfg. technologist Soc. Mfg. Engrs. Chief pre-grind insp. Coors Porcelain Co., Golden, Colo., 1973-74; quality assurance dir. Dazey Products Co., Kansas City, Kans., 1974—, mgr. info. systems, 1974—, mfg./acctg. product info. control systems v.p., 1980—; instr. data based systems. Served with U.S. Army, 1970-71; Vietnam. Decorated Air medal, Purple Heart, Army Commendation medal. Mem. Am. Soc. for Quality Control, Am. Soc. for Prodn. and Inventory Control, Nat. Rifle Assn. Subspecialties: Database systems; Information systems (information science). Current work: Establish and modify manufacturing computer database and manage its use. Office: Industrial Airport 1 Dazey Circle KS 66031

HOWARD, JOHN HAYES, computer scientist; b. Lexington, Mass., Jan. 23, 1942; s. John Hayes and Nancy (Stark) H.; m. Nancy Wright, Aug. 14, 1965; children: Gregory Charles, Jackson Alexander. B.S., M.I.T., 1965; Ph.D. in Computer Sci, U. Tex.-Austin, 1970. Systems analyst M.I.T., Cambridge, 1965-67; computer programmer U. Tex., Austin, part-time, 1968-70, asst. prof. computer sci., 1970-77; mem. research staff IBM Research, San Jose, Calif., 1977-83, IBM/CMU Info. Tech. Center, 1983—. Contbr. articles to profl. jours. Bd. dirs. Rovers Soccer Club, San Jose, 1981-83; registrar South San Jose Youth Soccer League, 1982-83. Nat. Merit scholar, 1960-65; NSF trainee, 1967-69. Mem. Assn. for Computing Machinery. Subspecialties: Operating systems; Information systems, storage, and retrieval (computer science). Current work: Storage systems, file servers, operating systems, performance evaluation, management. Office: IBM/CMU Info Tech Center Schenley Park Pittsburgh PA 15213

HOWARD, KEITH ARTHUR, geologist; b. Price, Utah, Sept. 12, 1939; s. Reginald George and Betty (Richardson) H.; m. Jean Elliott, Sept. 9, 1961 (div. 1981); 1 dau., Crystal. B.S., U. Calif.-Berkeley, 1961, M.S., 1962; Ph.D., Yale U., 1966. Registered geologist, Calif. Geologist U.S. Geol. Survey, Menlo Park, Calif., 1966—. Fellow Geol. Soc. Am.; mem. Am. Geophys. Union. Subspecialties: Geology; Tectonics. Current work: Regional geology and tectonics of Southwest U.S., volcanology, planetology. Office: 345 Middlefield Rd Menlo Park CA 94002

HOWARD, KENNETH IRWIN, psychologist, educator; b. Chgo., Oct. 19, 1932; s. Simon and Florence (Bergman) H.; m. April Rose Zweig, Dec. 15, 1979; children: Deborah, Peter, Lisa, David, Rebecca. A.B., U. Calif.-Berkeley, 1954; Ph.D., U. Chgo., 1959. Research assoc. U. Chgo., 1960-63; research scientist Inst. for Juvenile Research, Chgo., 1963-75; prof. Northwestern U., Evanston, Ill., 1967—. Author: Varieties of Psychotherapeutic Experience, 1976, The Adolescent: A Psychological Self-Portrait, 1981; Contbr. articles to profl. jours. Served to 1st lt. U.S. Army, 1954-56. Fellow Am. Psychol. Assn.; mem. Soc. for Psychotherapy Research (pres. 1969-70), Soc. for Multivariate Exptl. Psychology (sec.-treas. 1971-73). Subspecialty: Psychotherapy Research. Current work: Research on the process and outcome of psychotherapeutic interventions. Home: 2626 Sheridan Rd Evanston IL 60201 Office: Dept Psychology Northwestern U Evanston IL 60201

HOWARD, LAUREN DAVIS, biology and botany educator, phytosociologist and plant taxonomist; b. Nashua, N.H., Mar. 12, 1950; s. Lauren Alvin and Eleanor (Davis) H.; m. Judith Saunders, June 24, 1972; children: Lauren Fredrick, Scott Davis. B.A. magna cum laude in Biology, Hartwick Coll., 1971; Ph.D. with distinction in Botany, U. Vt., 1979. Lectr. biology St. Michael's Coll., Winooski, Vt., 1976; asst. prof. biology Norwich U., Northfield, Vt., 1976-83, assoc. prof., 1983—; curator Norwich U. Herbarium; lectr. Vt. Council Humanities and Pub. Issues. Contbr. articles on botany to profl. jours.; author texts. Recipient Homer L. Dodge award Norwich U., 1979; Outstanding Young Alumnus award Hartwick Coll, 1981. Mem. Bot. Soc. Am., Ecol. Soc. Am., Am. Forestry Assn., New Eng. Bot. Club, Nature Conservancy, Nat. Wildlife Fedn. Republican. Subspecialties: Sociobiology; Taxonomy. Current work: Plant community structure and succession in New England, documentation of floras and development of floristic keys for the Northeast. Office: Cabot Sci Bldg Norwich U Northfield VT 05663

HOWARD, ROBERT ERNEST, chemist, educator; b. Tulsa, Oct. 29, 1947; s. Robert Carey and Dorothy Curtis (Hendrickson) H.; m. Marilyn Ann Thomas, May 29, 1969; children: Matthew Thomas, Zachary Richard. B.A., Cornell Coll., 1969; Ph.D. in Chem. Physics, Ind. U., 1975. Postdoctoral research assoc. IBM Corp., Watson Research Ctr., Yorktown Heights, N.Y., 1975-76; NSF energy-related postdoctoral fellow U. Calif., Berkeley, 1976-77; asst. prof. U. Tex. of the Permian Basin, Odessa, 1977-81; assoc. prof. chemistry U. Tulsa, 1982—. Contbr. papers in field to profl. jours. R.A. Welch Found. sci. research grantee, 1978-82. Mem. Am. Chem. Soc., Am. Phys. Soc., Assn. for Computing Machinery, AAAS. Subspecialty: Theoretical chemistry. Current work: Quasi-classical trajectory calculations for gas phase reactive scattering. Home: 3319 S Darlington Tulsa OK 74135 Office: 600 S College Tulsa OK 74104

HOWARD, SETHANNE, astronom. researcher; b. Coronado, Calif., Feb. 2, 1944; d. Seth Thomas and Margaret Anne (McCarthy) H.; m. Donald Scott Hayes, June 11, 1966. B.S., U. Calif.-Davis, 1965; M.S., Rensselaer Poly. Inst., 1972. Astron. researcher Lick Obs. U. Calif.-Santa Cruz, 1965-67, Rensselaer Poly. Inst., 1967-72, Smithsonian Astrophys. Obs., 1972, Kitt Peak Nat. Obs., Tucson, 1972-81; computer programmer analyst, tng. specialist in ADP, geophysics Fleet Numerical Oceanography Ctr., Monterey, Calif., 1981-82, research assoc., 1982—. Contbr. articles to research jours. Mem. Am. Astron. Assn. Subspecialties: Numerical analysis; Optical astronomy. Current work: T Tauri stars, young clusters of stars. Office: Fleet Numerical Oceanography Ctr Monterey CA 93940

HOWARD, WILLIAM EAGER, III, astronomer, science administrator; b. Washington, Aug. 25, 1932; s. William Eager, Jr. and Frances (Bacon) H.; m. Miriam Rachael Sitler, June 22, 1957; children: William Eager IV, Jennifer Miriam. B.S. in Physics (scholar), Rensselaer Poly. Inst., 1954; A.M. in Astronomy (G.R. Agassiz fellow 1955-56), Harvard U., 1956; Ph.D. in Astronomy (J.E. Thayer scholar 1957), Harvard U., 1958. Successively research assoc., instr., asst. prof., assoc. prof. U. Mich., Ann Arbor, 1959-64; asst. to dir., assoc. scientist, then scientist, asst. dir. Nat. Radio Astronomy Obs., Charlottesville, Va., and Green Bank, W. Va., 1964-77; div. dir. astron. scis. NSF, Washington, 1977-82; assigned Office Tech. Assessment, U.S. Congress, 1982, Fed. govt., 1982—. Contbr. articles in field to publs. Served to 1st lt. Signal Corps USAR, 1954-59. Mem. Internat. Astron. Union, Am. Astron. Soc. (treas. 1975-77), Internat. Sci. Radio Union, AAAS, Astron. Soc. Pacific, Sigma Xi. Lutheran. Subspecialties: Radio and microwave astronomy; Aerospace engineering and technology. Current work: Radio astronomy; galactic structure; scientific administration, management analysis. Home: 4061 N Stuart St Arlington VA 22207

HOWARD-PEEBLES, PATRICIA NELL, clin. cytogeneticist, educator; b. Lawton, Okla., Nov. 24, 1941; d. John Marion and Reba Leona (Prestidge) Howard; m. Thomas Marvin Peebles, Aug. 16, 1975. B.S.Ed., Central Okla. State U., 1963; Ph.D. in Genetics, U. Tex., Austin, 1969. Diplomate: Am. Bd. Med. Genetics, 1982. Tchr. Piedmont (Okla.) Pub. Schs., 1963-64; biochem. technician Oak Ridge Nat. Lab., 1964-66; instr. depts. pediatrics and cytotech. Okla. U. Med. Ctr., Oklahoma City, 1971-72; asst. prof. dept. microbiology Inst. Genetics, U. So. Miss., Hattiesburg, 1973-77, assoc prof., 1977-80; assoc. prof. and clin. cytogeneticist dept. pub. health and Lab. Med. Genetics U. Ala., Birmingham, 1980-81; assoc. prof. dept. pathology, dir. cytogenetics lab. U. Tex. Health Scis. Ctr., Dallas, 1981—; genetic cons. Ellisville (Miss.) State Sch., 1973-80; attending staff dept. pathology Parkland Meml. Hosp., Dallas, 1981—; cytogenetic cons. El Paso Rehalb. Center. Contbr. articles and abstracts to profl. lit. Fellow Am. Cancer Soc., NIH, NSF, NIMH, Nat. Inst. Child Health and Human Devel.; Mem. Am Soc. Human Genetics, Genetics Soc. Am., AAAS, So. Soc. Pediatric Research, Am. Assn. Mental Deficiency, Tex. Genetics Soc., Sigma Xi, Alpha Omicron Pi. Baptist. Subspecialties: Genetics and genetic engineering (medicine); Cell biology (medicine). Current work: Cytogenetics of fragile sites, fragile X chromosomes and causes of mental retardation. Cytogenetics research, director of clinical cytogenetics laboratory. Office: Dept Pathology U Tex Health Scis Center 5323 Harry Hines Blvd Dallas TX 75235

HOWARDS, STUART S., urologist, physiologist, educator; b. Milw., Mar. 29, 1937; s. Harvey H. and Anne (Levin) Horwitz; m. Mary Carter Randolph Nelson, Aug. 20, 1966; children: Penny, Hugh. B.A., Yale U., 1959; M.D., Columbia U., 1963. Intern Peter Bent Brigham Hosp., Boston, 1963-64, resident, 1968-71, Boston Children's Hosp., 1964-65, NIH, Bethesda, Md., 1965-68; asst. prof. urology and physiology U. Va., 1971-73, assoc. prof., 1973-76, prof., 1976—. Editor: Infertility in the Male, 1983; editor: Yearbook of Urology, 1978—; assoc. editor: Investigative Urology, 1977—. Served to lt. comdr. USPHS, 1965-71. Recipient Career Devel. award NIH, 1976-81, Golden Cytoscope award AM. Urologic Assn., 1980. Mem. ACS, Am. Urology Assn., Am. Fertility Soc., Am. Soc. Genito-Urinary Surgeons, Am. Soc. Andrology, Phi Beta Kappa, Alpha Oemga Alpha. Subspecialties: Urology; Reproductive biology. Current work: Urology and male reproductive physiology. Home: 1150 W Leigh Dr Charlottesville VA 22901 Office: U Va Hosp Box 422 Charlottesville VA 22908

HOWARTH, ALAN JACK, research scientist; b. Kellogg, Idaho, Aug. 1, 1952; s. Jack Sperry and Evelyn (Killpack) H.; m. Valerie Adams, Jan. 4, 1980; children: Catherine Lee, Jodie Lynn. Student, Lassen Coll., Susanville, Calif., 1970-71; B.S., U. Calif.-Davis, 1974, M.S., 1975, Ph.D., 1981. Research asst. U. Calif.-Davis, 1974-75, 78-81; postdoctoral research assoc. U. Ill., Urbana, 1982—. Contbr. articles to profl. jours. Jastro Shields research scholar, 1978-79, 79-80. Mem. Am. Soc. for Virology. Republican. Mormon. Subspecialties: Plant virology; Genetics and genetic engineering (agriculture). Current work: Viral genome structure and expression, molecular basis of viral functions. Molecular cloning, nucleotide sequencing, restriction endonuclease mapping, viral genome organization, viral gene expression, genetic engineering. Office: 1102 S Goodwin Ave Suite N519 Turner Hall Urbana IL 61801

HOWD, FRANK HAWVER, geologist, educator; b. Delmar, N.Y., Dec. 2, 1924; s. Albert I. and Helen (Koch) H.; m. Barbara M. Campbell, June 23, 1951; children: Peter A., Thomas C., Christopher A. A.B., U. Rochester, 1951, M.S., 1953; Ph.D., Wash. State U., 1956. Geologist Bear Creek Mining Co., Eureka, Utah, 1956-59; cons. geologist Kennecott Copper Corp., Salt Lake City, 1960-70, U.S. Geol. Survey, Denver, 1979-82; asst. prof. geology U. Maine Orono, 1959-65, assoc. prof., 1965—. Mem. Orono Sch. Bd., 1976-82. Served with USN, 1943-45; PTO. Fellow Geol. Soc. Am.; mem. Soc. Econ. Geologists, Maine Mineral Resources Assn. (dir. 1980-82). Subspecialties: Geology; Geochemistry. Current work: Relationship between hydrothermal alteration, trace element dispersion and ore deposits; interpretation of stream sediment geochemistry; experimental hydrothermal replacement by sulfide minerals. Home: 14 Spencer St Orono ME 04473 Office: Dept Geol Sci U Maine Orono ME 04469

HOWDEN, DAVID GORDON, welding engineering educator; b. Scarborough, Eng., Aug. 22, 1937; m., 1962; 1 child. B.Sc., U. Eng., 1959, Ph.D. in Metallurgy, 1962. Asst. prof. and researcher Central Tech. Aeronautic, Sao Jose dos Campos, Brazil, 1963-65; sr. scientist welding metallurgy Can. Dept. Energy Mines and Resources, 1965-67; assoc. mgr. welding and fabric Battelle Columbus Ls Labs., Ohio, 1967-77; assoc. prof. welding engring. Ohio State U., Columbus, 1977—; U.S. del. Internat. Inst. Welding, 1977—. Recipient Oxygen of Brazil prize, 1975. Mem. Am. Welding Soc., Am. Soc. Metals, Welding Research Concil. Subspecialty: Welding technology. Office: 190 W 19th Ave Columbus OH 43210

HOWE, CHIN C., molecular biologist, educator; b. Taipei, Taiwan; d. (Twent W.) and Shien (Lin) Chen; m. King L. Howe, June 18, 1960. Ph.D., U. Pa., 1972. Postdoctoral trainee Wistar Inst., Phila., 1973-75, research assoc., 1976-78, asst. prof., 1979—. Contbr. articles to profl. publs. Mem. Soc. Devel. Biology. Subspecialties: Molecular biology; Developmental biology. Current work: Control of gene expression, gene cloning, recombinant DNA technology protein chemistry and structure, teratocarcinoma cells (Marine), embryogenesis, basement membrane, laminin. Office: Wistar Inst 26th and Spruce Sts Philadelphia PA 19104

HOWE, HENRY BRANCH, JR., educator, microbiologist; b. Atlanta, Aug. 5, 1924; s. Henry Branch and Grace Lee (Mann) H.; m. Margaret Anne Haden, Sept. 1, 1951; children: Stephen Jeffrey, Barbara Lynn, Alan Haden. A.B., Emory U., 1948, M.A., 1950; Ph.D., U. Wis., 1955. Asst. prof. biology Union Coll., 1954-57; asst. prof. biology Wake Forest U., 1957-59; asst. prof. microbiology U. Ga., 1959-64, assoc. prof., 1964-70, prof., 1970—, assoc. dean Grad. Sch., 1981—; cons. McGraw-Hill Info. Systems, Prentice-Hall, Inc., NSF. Contbr. numerous articles to profl. jours. Served with U.S. Army, 1944-46. Decorated Purple Heart (2); recipient M.G. Michael Research award U. Ga., 1965. Mem. Genetics Soc. Am., Am. Soc. Microbiology, Mycological Soc. Am., Am. Ornithologists Union, Assn. Southeastern Biologists (research prizes 1967, 71), Ga. Acad. Sci. (editor 1963-74), Sigma Xi. (excellence in research award 1965, 77), Phi Beta Kappa, Phi Delta Theta. Methodist. Subspecialties: Microbiology; Gene actions. Current work: Genetics and development of the genus Neurospora. Home: 130 Bishop Dr Athens GA 30606 Office: Dept Microbiology U Ga Athens GA 30602

HOWE, HENRY FRANKLIN, ecologist, zoology educator; b. Gardner, Mass., Dec. 24, 1946; s. Volney Webster and Aileen (O'Brien) H.; m. Ann Wheeler, Jan. 7, 1971 (div. 1976). A.B., Earlham Coll., 1968; A.M., U. Mich., 1969, Ph.D., 1977. Biology instr. Phillips Acad., Andover, Mass., 1971-72; Smithsonian fellow Smithsonian Tropical Research Inst., Balboa, Panama, 1977-78; asst. prof. dept. zoology U. Iowa, Iowa City, 1978-82, assoc. prof., 1982—. Contbr. articles to profl. publs. in field. Active Resource Conservation Commn., Iowa City, 1983—. Served with U.S. Army, 1969-71. Grantee World Wildlife Fund, Costa Rica, 1975, Am. Mus. Natural History, 1975, NSF, 1979, 82. Mem. Am. Ornithologists Union, Am. Soc. Naturalists, Ecol. Soc. Am., AAAS, Soc. for Study of Evolution. Subspecialties: Population biology; Ecology. Current work: Ecology of mutualisms, seed dispersal by animals, vertebrate reproduction. Ecology and management of tropical forests. Office: Dept Zoology U Iowa Iowa City IA 52242

HOWE, JOHN EDWARD, computer scientist; b. Ventura, Calif., Mar. 21, 1953; s. Harold Byron and Catherine (Phillips) H.; m. Nancy Bonacich, June 26, 1976. B.A. in Math, UCLA, 1975; M.A., U. So. Calif., 1976, M.S. in Computer Sci, 1977; M.B.A., U. Santa Clara, 1983. Sr. software devel. engr. Intel Corp., San Jose, Calif., 1977-80, mgr. software evaluation, 1980-81, application software mgr., 1981—. Mem. IEEE, ACM, Soc. Indsl. and Applied Math. Lutheran. Subspecialties: Computer architecture; Software engineering. Current work: Computer architecture, software engineering, office automation software; strategy development for application business software for new multi-user microprocessor-based commercial computers. Office: Intel Corp 2360 Bering Dr San Jose CA 95131

HOWE, TIMOTHY MAX, nuclear engineering manager, researcher; b. Grand Rapids, Mich., Dec. 14, 1947; s. Clifford Max and Billie Lue (Barnhart) H.; m. Beverly Lue Magee, Jan. 25, 1969; 1 dau., Amy Lauren. B.S. in Engring. Physics, Tex. Tech. U., 1970; M.S. in Nuclear Engring, Purdue U., 1972. Engr. Gen. Electric, San Jose, Calif., 1972-74; engring. supr. Aerojet Nuclear, Idaho Falls, Idaho, 1974-78; sr. engr. TVA, Chattanooga, 1978-79; br. mgr. EG&G Idaho, Inc., Idaho Falls, 1979—. ARMCO scholar, 1966. Mem. Am. Nuclear Soc., ASME, Phi Kappa Phi, Tau Beta Phi. Republican. Mem. Church of Christ. Club: Idaho Falls Ski. Subspecialties: Mechanical engineering; Numerical analysis. Current work: Light water reactor safety; development of computer models that simulate light water reactor system, including thermal-hydraulics and fuel performance. Office: EG&G Idaho Inc PO Box 1625 Idaho Falls ID 83415

HOWELL, JOHN REID, mechanical engineering educator, researcher; b. Columbus, Ohio, June 13, 1936; s. Frederick Edward and Hilma Lavilla (Kief) H.; m. Arlene Elizabeth Pollitt, June 20, 1959 (div. 1974); children: John Reid, Keli Diane, David Lee; m. Susan G. Conway, May 20, 1979. B.S., Case Inst. Tech., 1958, M.S., 1960, Ph.D., 1962. Registered profl. engr. Researcher NASA Lewis Research Ctr., Cleve., 1961-68; assoc. prof. U.Houston, 1968-71, prof. mech. engring., 1971-78; prof. mech. engring U. Tex.-Austin, 1978-82, E.C.H. Bantel prof., 1982—; cons. various indsl. firms. Author: Catalog of Radiation Configuration Factors, 1982; co-author: Thermal Radiation Heat Transfer, 1980, Solar Thermal Energy Systems, 1982. Commr. Renewable Energy Resources Commn., City of Austin, 1979-80. Served to 1st. lt. USAF, 1962-65. Recipient spl. service award NASA Lewis Research Ctr., 1964. Fellow AIAA (assoc.), ASME; mem. Internat. Solar Energy Soc. Subspecialties: Mechanical engineering; Solar energy. Current work: Radiative heat transfer in absorbing-emitting-scattering media, with application to energy systems. Home: 3200 Kerbey Ln Austin TX 78703 Office: Dept Mech Engring U Tex Austin TX 78712

HOWELL, ROBERT WAYNE, agronomy educator; b. Houlka, Miss., Nov. 26, 1916; s. Raleigh Wayne and Frances Ethel (Stacy) H.; m. Elizabeth Virginia Blair, Sept. 25, 1940; children: Jacqueline (Mrs. Charles Frank Choate), Richard James, Wayne Davis. Student, George Washington U., 1934-37; B.S., Miss. Coll., 1949; M.S., U. Wis., 1951, Ph.D., 1952. Clk., adminstrv. asst. U.S. Dept. Agr., Washington, Cheyenne, Wyo., Ithaca, N.Y., 1934-43; bus. mgr. Pineapple Research Inst., Hawaii, 1947; plant physiologist U.S. Regional Lab., Urbana, Ill., 1952-65; leader soybean investigations U.S. Dept. Agr., Beltsville, Md., 1965, chief oilseed and indsl. crops research br., 1966-71; prof. agronomy U. Ill., Urbana, 1971—, chmn. dept., 1971-82. Editor: Crop Science, 1969-71. Served to capt. AUS, 1943-46. Recipient award of merit Am. Soybean Assn., 1972. Fellow Am. Soc. Agronomy; mem. Crop Sci. Soc. Am., Am. Soc. Plant Physiologists, A.A.A.S., Am. Soybean Assn. (hon. life), Sigma Chi. Subspecialty: Agronomy. Home: 2012 S Cottage Grove Urbana IL 61801

HOWELL, STEPHEN BARNARD, medical educator, administr., researcher; b. Shirley, Mass, Sept. 29, 1944; s. Wallace E. and Christine G. H.; m. Julianne R. Howell, 1968; children: Justin, Brett. A.B. in Biology with honors, U. Chgo., 1966; M.D. in Immunology magna cum laude, Harvard U., 1970. Diplomate: Am. Bd. Internal Medicine. Intern Mass. Gen. Hosp., Boston, 1970-71, resident, 1971-72; research assoc. Lab. Cell Biology, Nat. Cancer Inst., Bethesda, Md., 1972-74; resident U. Calif. Hosps., San Francisco, 1974-75; fellow in oncology Sidney Farber Cancer Inst., Boston, 1975-77; asst. prof. medicine U. Calif.-San Diego, 1977-80, assoc. prof., 1981—; dir. Lab. Pharmacology and Cytokinetics, 1978—; mem. biochem. modulation adv. group div. cancer treatment Nat. Cancer Inst., 1979—. Contbr. numerous articles to profl. jours. Mem. Am Assn. Cancer Research, Am. Soc. Clin. Oncology, Am. Fedn. Clin. Research, Cell Kinetics Soc., N.Y. Acad. Sci., Phi Beta Kappa, Alpha Omega Alpha. Subspecialties: Chemotherapy; Cancer research (medicine). Current work: Clinical pharmacology and molecular pharmacology of antineoplastic agents. Office: Department of Medicine University of California San Diego La Jolla CA 92093

HOWERTON, ROBERT JAMES, research physicist, engineer; b. Hammond, Ind., Sept. 27, 1923; s. William Columbus and Kathryn Inez (Coutchure) H.; m. Beatrice Elizabeth Harkins, Mar. 2, 1946. B.S., Northwestern U., 1946, M.S., 1947. Asst. prof. Regis Coll., Denver, 1948-54; physicist Materials Testing Reactor, Idaho Falls, Idaho, 1956-57; asst. div. leader div. theoretical physics Lawrence Livermore Nat. Lab., Livermore, Calif., 1957—; cons. CIA, Washington, 1959-63, Am. Inst. Physics, N.Y.C., 1961-67; mem. adv. com. Def. Nat. Agy., Washington, 1966-75; chmn. methods and formats com. U.S. Cross Sect. Evaluation Working Group, Upton, N.Y., 1967—. Contbr. articles to profl. jours. Mem. Diocesan Bd. Edn., Oakland, Calif., 1977—; cons. Diocesan Vicar Gen., Oakland, 1978—; mem. adv. bd. Family Aid to Catholic Edn., Oakland, 1979—. Served with USN, 1942-45. Fellow Am. Nuclear Soc.; mem. Scientists for Secure Energy, Sigma Xi, Tau Beta Pi, Pi Mu Epsilon. Subspecialties: Nuclear physics; Nuclear engineering. Current work: Phenomenology of nuclear reactions induced by neutrons and light charged particles. Home: 11594 Ladera Ct Dublin CA 94568 Office: Lawrence Livermore Nat Lab Box 808 Livermore CA 94550

HOWLAND, HOWARD CHASE, neurobiologist, educator; b. Lafayette, Ind., May 26, 1933; s. Warren E. and Elizabeth G. (Giddings) H.; m. Monica, May 23, 1965; children: Frank, Jacob, David. B.A., U. Chgo., 1952; M.S., Tufts U., 1958; Ph.D., Cornell U., 1968. Instr. in biol. sci. SUNY-Oyster Bay, 1960-66, asst. prof. biol. sci., 1966-67; asst. prof. neurobiology and behavior Cornell U., 1968-74, assoc. prof., 1974—. Contbr. numerous articles, chpts., abstracts to profl. publs. Served with U.S. Army, 1953-55. Mem. Am. Physiol. Soc., Soc. Neurosci., Optical Soc. Am., Assn. Research in Vision and Opthalmology, AAAS, Am. Soc. Zoologists, Am. Soc. Ichthyologists and Herpetologists. Subspecialties: Physiology (biology); Comparative neurobiology. Office: Cornell U W201 Mudd Hall Ithaca NY 14850

HOWLAND, ROBERT ALDEN, JR., engineer, educator; b. Bridgeport, Conn., May 7, 1943; s. Robert Alden and Winifred Leona (Goodell) H. B.A., Yale U., 1965, M.S., 1967; Ph.D., N.C. State U., 1974. Vis. instr. N.C. State U., Raleigh, 1974-78; asst. prof. Rose-Hulman Inst. Tech., Terre Haute, Ind., 1978-81, U. Notre Dame, South Bend, Ind., 1981—. Percussionist, South Bend Symphony Orch.; Contbr. articles to profl. publs. Grantee NSF, 1980-82. Mem. Am. Acad. Mechanics, Soc. Engring. Sci., Am. Astron. Soc. (div. dynamical astronomy), Sigma Xi. Republican. Presbyterian. Subspecialties: Theoretical and applied mechanics; Applied mathematics. Current work: accelerated asymptotic analysis of nonlinear systems; celestial mechanics. Home: 1523 Turtle Creek Dr South Bend IN 46637 Office: Aerospace/Mech Engring U Notre Dame Notre Dame IN 46556

HOWLEY, THOMAS PATRICK, dentistry educator, researcher; b. Deep River, Ont., Cam., Aug. 27, 1951; s. James Thomas and Anne (Halabuza) H. B.S., McGill U., Montreal, Que., Can., 1972; M.A., U. Toronto, Ont., Can., 1974, Ph. D., 1979. Statistician Ont. Ministry of Natural Resources, Toronto, 1979—; asst. prof. dentistry U. Toronto, 1979—, asst. prof. nursing, 1981 (part-time); statis. cons. Recipient Career Scientist award Ont. Ministry Health, Toronto, 1982-87; Can. Council doctoral fellow, 1975-77. Mem. Internat. Assn. Dental Research. Roman Catholic. Subspecialties: Statistics; Experimental design in dentistry. Current work: Dental manpower, oral biology, dental public health, non-normal data, mathematical models of caries incidence. Office: Biometrics Sect U Toronto Faculty Dentistry 124 Edward St Toronto ON Canada M5G 1G6

HOWSE, HAROLD DARROW, instn. exec.; b. Poplarville, Miss., Nov. 8, 1928; s. William Jefferson and Artie Mittie (Smith) H.; m. Mittie Hazel Gibson, Dec. 18, 1960; children: Trijetta Lynn Howse Cropp, Claude Demitris Gibson. A.A., Pearl River Jr. Coll., Poplarville, 1947; B.S., U. So. Miss., 1959, M.S., 1960; Ph.D., Tulane U., 1967. Instr. zoology Miss. Coll., Jackson, 1960; instr. biology U. So. Miss., Hattiesburg, 1960-63; NIH trainee Tulane U., New Orleans, 1963-67; head sect. microscopy Gulf Coast Research Lab., Ocean Springs, Miss., 1967-79, asst. dir., 1971, acting dir., 1971-72, dir., 1972—; prof. biology U. Miss., 1972—; prof. zoology Miss. State U.; prof. biology U. So. Miss., 1972—. Contbr. articles to profl. jours. Chmn. Miss. Coastal Zone Mgmt. Adv. Com., 1981—, Miss. Coastal Energy Impact Com., 1981 . Served with U.S. Navy, 1949-54. NIH trainee, 1963-67. Mem. Am. Assn. Anatomists, AAAS, Am. Microscopical Soc., Am. Soc. for Cell Biology, Am. Soc. Zoologists, Assn. Southeastern Biologists, Electron Microscopy Soc. Am., Miss. Acad. Sci., N.Y. Acad. Scis., others. Baptist. Subspecialties: Cell biology; Cytology and histology. Current work: Comparative histology, histochemistry and ultrastructure of the several organ systems of marine invertebrates and vertebrates. Home: PO Box AG Ocean Springs MS 39564 Office: Gulf Coast Research Lab Ocean Springs MS 39564

HOY, MARJORIE ANN, entomology educator; b. Kansas City, Kans., May 19, 1941; d. Dayton and Marjorie Jean (Acker) Wolf; m. James B. Hoy, Dec. 22, 1961; 1 son, Benjamin Lee. A.B., U. Kans., 1963; M.S., U. Calif.-Berkeley, 1966, Ph.D., 1972. Research geneticist dept. genetics U. Calif.-Berkeley, 1964-66; research entomologist Conn. Agr. Expt. Sta., New Haven, 1973-75, U.S. Forest Service, Hamden, Conn., 1975-76; asst. prof. entomology U. Calif.-Berkeley, 1976-80, assoc. prof., 1980-82, prof., 1982—. Editor: Recent Advances in Knowledge of the Phytoseiidae, 1982; co-editor: Genetics in Relation to Insect Management, 1979. Mem. Entomol. Soc., Am., Acarology Soc. Am., Internat. Orgn. Biol. Control, Soc. Study of Evolution, Phi Beta Kappa, Sigma Xi. Subspecialties: Animal genetics; Integrated pest management. Current work: Genetic improvement of biological control agents to enhance pest control in agricultural crops; selection for pesticide resistance in a predator of spider mites. Home: 1004 Grizzly Peak Blvd Berkeley CA 94708 Office: Dept Entomol Scis U Calif 201 Wellman Hall Berkeley CA 94720

HOYE, ROBERT EARL, systems science educator; b. Warwick, R.I., Jan. 12, 1931; s. S. Earl and Alice M. (Landry) H.; m. Patricia Buswell, Aug. 20, 1955; children: Robert Earl, Jr., Joanne D., Peter M., Kathleen B. A.B., Providence Coll., 1953; M.S., St. John's U., 1955; Ph.D., U. Wis.-Madison, 1973. Dean Champlain Coll., Burlington, Vt., 1957-58; supt. Frontier Regional Sch., Deerfield, Mass., 1958-60; northeast dist. dir. IBM Sci. Research, Chgo., 1960-66; dir. learning systems Xerox Corp., N.Y.C., 1965; dir. instructional media U. Wis., 1966-73; prof. systems sci. U. Louisville, 1974—, mem. commn. on excellence, 1977; mem. adv. panel U.S. Congress OTA, Wshington, 1981-82. Author: Index To Computer Based Learning, 1973; editor: Education Jour, 1969-73, College Student Jour, 1969-73; contbr. numerous articles to profl. jours. Mem. Econ. Opportunity Program, Taunton, Mass., 1963-65; mem. Dighton (Mass.) Sch. Bd., 1965. Mem. Am. Psychol. Assn., Hosp. Mgmt. Systems Soc. Roman Catholic. Subspecialties: Information systems (information science); Information systems, storage, and retrieval (computer science). Current work: Systems science, health systems, information systems, instructional technology. Home: 2238 Wynnewood Circle Louisville KY 40222 Office: University of Louisville Louisville KY 40292

HOYER, LEON WILLIAM, medical educator, physician; b. Mpls., Mar. 6, 1936; s. Ludolf J. and Inez (Fuglesteen) H.; m. Diane Desmond Lawrence, Dec. 30, 1960; children: Helen Kristin, Sharon Anne, Erik William. A.B., Harvard U., 1958; M.D., U. Minn., 1962. Diplomate: Am. Bd. Internal Medicine. Intern Presbyn. Hosp., N.Y.C., 1962-63, resident, 1963-64, Strong Meml. Hosp., Rochester, N.Y., 1966-67; asst. prof. medicine U. Rochester (N.Y.) Sch. Medicine and Dentistry, 1968-70, assoc. prof., 1970-74; prof. medicine and lab. medicine U. Conn. Sch. Medicine, Farmington, 1974—, co-head div. hematology/oncology; mem. hematology study sect. NIH, 1976-80; chmn. med. and sci. adv. council Nat. Hemophilia Found.; co-chmn. working party on factor VIII antigens Internat. Com. on Thrombosis and Hemostasis; exec. com. council on thrombosis Am. Heart Assn. Contbr. articles on hematology to profl. jours. Bd. dirs., exec. com. ARC, Hartford (Conn.) chpt., 1976-78. Served with USPHS, 1964-66. Josiah Macy Jr. Found. scholar, 1978-79; recipient Alumni Assn. award U. Conn., 1979; Thelin Research award Nat. Hemophilia Found., 1981. Mem. Am. Assn. Immunologists, Am. Soc. Hematology, Am. Soc. for Clin. Investigation, AAAS, N.Y. Acad. Scis. Subspecialties: Hematology; Immunology (medicine). Current work: Biochemical and Immunologic studies of the factor VIII complex, investigation of the moleculr basis of hemophilia and von Willebrand's disease. Home: 23 White Oak Rd Farmington CT 06032 Office: Dept Medicine U Conn Health Ctr Farmington CT 06032

HOYT, DOUGLAS VINCENT, astronomer, climatologist; b. Farmington, Maine, Oct. 7, 1946; s. Winston S. and Esther (Manning) H. B.S., Rennselaer Poly. Inst., 1968; M.S., U. Colo., 1971. Physicist NOAA, Boulder, Colo., 1972-79; vis. scientist Coop. Inst. for Research on Environ. Scis., Boulder, 1979-80, World Radiation Ctr., Davos, Switzerland, 1980; astronomer High Altitude Obs., Nat. Ctr. Atmosphere Research, Boulder, 1980—. Contbr. articles to profl. jours. Recipient Bausch & Lomb Sci. award, 1964-. Mem. Am. Astron. Soc, Am. Meteorol. Soc. Club: Nat. Steroscopic Assn. Subspecialties: Solar physics; Climatology. Current work: Researcher in variations in solar constant, climatic change, history of solar activity, solar physics. Home: 994 55th St Boulder CO 80303 Office: NCAR PO Box 3000 Boulder CO 80307

HOYT, WILLIAM HENRY, oceanography and geology educator; b. Evanston, Ill., July 5, 1952; s. Nelson Landon and Barbara (McAdams) H.; m. Denise Selders, Aug. 21, 1976. B.S., Middlebury Coll., 1974; M.S., SUNY-Albany, 1976; Ph.D., U. Del., 1982. Grad. research asst. SUNY-Albany, 1974-76, U. Del., 1977-80; asst. prof. oceanography and geology U. No. Colo., Greeley, 1981—; cons. Nat. Park Service, Wellfleet, Mass., 1979-80. Mem. Geol. Soc. Am., Soc. Econ. Paleontologists and Mineralogists, Colo. Sci. Soc., Am. Geophys. Union, Sigma Xi (sec.-treas. chpt.). Congregationalist. Subspecialties: Sedimentology; Oceanography. Current work: Study of sedimentary deposits, both ancient and modern; development of vibracoring systems for high quality yet inexpensive core samples. Home: 1508 11th Ave Greeley CO 80631 Office: Dept Earth Sci U No Colo Greeley CO 80639

HOYUMPA, ANASTACIO MANINGO, physician, medical educator; b. Baybay, Leyte, Philippines, July 4, 1937; came to U.S., 1962; s. Anastacio Bandalan and Lamberta (Maningo) H.; m. Joan Maureen Howland, June 22, 1963; children: Rebecca, Danilo, Amelia, Benjamin. M.D., U. Santo Tomas, Manila, 1961. Rotating intern USAF Hosp., Clark AFB, Philippines, 1960-61, med. resident, 1961-62; med. intern Sinai Hosp., Balt., 1962-63, med. resident, 1963-65; gastroenterology fellow U. Cin., 1965-67; prof. medicine U. Tex. Health Sci. Center, San Antonio, 1982—; asst. prof. medicine U. Cin., 1968-72, Vanderbilt U., Nashville, 1972-76, assoc. prof., 1976-82; chief gastroenterology VA Hosp., San Antonio, 1982—. Contbr. articles in field to profl. jours. Served to lt. col. M.C. USAR. Grantee VA, 1973-75, 76-78, 79-81. Mem. Internat. Assn. Study Liver Diseases, Internat. Soc. Biomed. Research on Alcoholism, Am. Assn. Study Liver Disease, Am. Gastroenterol. Assn., Am. Soc. Gastrointestinal Endoscopy, Am. Fedn. Clin. Research, Central Soc. Clin. Research. Roman Catholic. Subspecialties: Internal medicine; Gastroenterology. Current work: Thiamin, intestinal transport, drug metabolism, fetal/alcohol syndrome. Home: 203 Fawn Dr San Antonio TX 78231 Office: U Tex Health Sci Center 7703 Floyd Curl Dr San Antonio TX 78284

HRAZDINA, GEZA, biochemistry educator; b. Letenye, Zala, Hungary, Mar. 16, 1939; came to U.S., 1966; s. Geza and Maria (Volgyi) H.; m. Helga M. Stritzke, Apr. 29, 1964; 1 son, Geza Karoly. Dipl.ing.agr., Swiss Fed. Inst. Tech., Zurich, 1963, Dr.sc.tech., 1966. Asst. prof. biochemistry Cornell U., Geneva, N.Y., 1968-73, assoc. prof., 1973-80, prof., 1981—; vis. prof. U. Freiburg, Germany, 1974-75, Tech. U. Budapest, Hungary, 1979, U. Cologne, Germany, 1981. Editor: (with others) Cellular and Sub Cellular Localizations in Plant Metabolism, 1982. A. von Humboldt fellow, 1974; 81; Nat. Acad. Scis. fellow, 1979. Mem. Phytochem. Soc. N.Am. (pres. 1982-83), Am. Soc. Plant Physiologists, Am. Chem. Soc., Phytochem. Soc. Europe. Subspecialties: Plant physiology (agriculture); Food science and technology. Current work: Enzymology of secondary plant metabolites; flavonoids, anthocyanins; subcellular localizations in plant metabolism. Home: 992 E Lake Rd Romulus NY 14541 Office: Cornell University Agricultural Experiment Station Geneva NY 14456

HRDY, SARAH BLAFFER, anthropology educator; b. Dallas, July 11, 1946; (married), 1972; 2 children. B.A., Radcliffe Coll., 1969; Ph.D. in Behavioral Biology, Harvard U., 1975. Instr. anthropology U. Mass., 1973; lectr. biol. anthropology Harvard U., 1975-76, fellow in biology, 1977-78, assoc., 1979-82; vis. assoc. prof. anthropology Rice U., Houston, 1981-84; prof. anthropology U. Calif.-Davis, 1984—. Author: The Black-Man of Zinacantan, 1972, The Langurs of Abu, 1977, The Woman that Never Evolved, 1981; cons. editor: Am. Jour. Primatology. Subspecialty: Sociobiology. Office: Dept Anthropology U Calif Davis CA 95616

HREJSA, ALLEN FRANCIS, physicist; b. Chgo., Aug. 12, 1942; s. Francis and Rose (Tonyan) H.; m. Lenore Jane Kerzenski, June 25, 1966; children: David, Julie, Jeffrey. B.S., U. Ill., 1964; M.S., So. Ill. U., 1966; Ph.D., U. Notre Dame, 1972. Diplomate: Am. Bd. Radiology. NIGAMS research fellow U. Tex.-M.D. Anderson Hosp., Houston, 1972-73; med. physicist Univ. Radiologists, Springfield, Ill., 1973-78; with Fermi Lab., Batavia, Ill., 1978-79; dir. med. physics Luth. Gen. Hosp., Park Ridge, Ill., 1979—; faculty Nat. Coll. Edn., 1981—, U. Health Scis., Chgo. Med. Sch., 1981—. Contbr. articles to profl. jours. Mem. Am. Phys. Soc., Am. Assn. Physicists in Medicine, Health Physics Soc., Am. Coll. Radiology, Ill. Radiol. Soc., Sigma Xi. Roman Catholic. Office: Radiotherapy 1775 W Dempster St Park Ridge IL 60068

HRONES, JOHN ANTHONY, engineering educator; b. Boston, Sept 28, 1912; s. Emil and Olga Victoria (Cech) H.; m. Margaret Baylis, June 17, 1938; children: Janet H. Roach, Stephen Baylis, Mary H. Parsons, John Anthony. S.B., MIT, 1934, S.M., 1936, Sc.D., 1942. Asst. factory mgr. Coldwell Lawnmower Co., Newburgh, N.Y., 1937-39; asst. mech. engring. dept. MIT, 1934-36, instr., 1936-37, 39-41, asst. prof., 1941-45, asso. prof., 1945-48, prof. mech. engring., 1948, head machine design div., 1946, dir. Dynamic Analysis and Control Lab., 1950; v.p. acad. affairs Case Inst. Tech., Cleve., 1957-67, provost, 1964-67; provost sci. and tech. Case-Western Res. U., 1967-76, provost emeritus, prof. engring., 1976—; cons. automatic control and machine design, 1939—; chmn. Univ. Circle Research Center Corp., 1967-73; pres. ChiCorp., 1967-68, chmn., 1967-77; research adv. com. AID, 1978-82. Author: (with Nelson) Analysis of the Four Bar Linkage, 1951; contbr. articles to engring. publs. Bd. dirs. Cleve. Mus. Nat. History; trustee Asian Inst. Tech.; pres. A.I.T. Found. Mem. Newcomen Soc., ASME, Am. Soc. Engring. Edn., Am. Acad. Arts and Scis., Inst. for Def. Analyses (trustee), Nat. Acad. Engring., Sigma Xi, Tau Beta Pi, Pi Tau Sigma. Club: Cleveland Skating (trustee 1970-73). Subspecialty: Mechanical engineering. Home: 9397 Midnight Pass Rd Apt 306 Sarasota FL 33581 Office: Case Western Res U Cleveland OH 44106

HRUBAN, ZDENEK, physician, pathologist; b. Prerov, Czechoslovakia, June 15, 1921; s. Jaroslav and Aloisie (Rieger) H.; m. Jarmila Stanek, Aug. 27, 1955; children: Paul Y., Ralph H., Diana J. M.D., U. Chgo., 1956, Ph.D., 1963. Diplomate: Am. Bd. Pathology, 1963. Intern Presbyn. Hosp., Chgo., 1956-57; resident in pathology U. Chgo. Hosps., 1957-62; asst. prof. U. Chgo., 1963-67, assoc. prof., 1967-73, prof. pathology, 1973—. Author: Microbodies and Related Particles, 1969; contbr. articles to profl. publs. Mem. Faculty of Discussants, C.L. Davis Found., 1976—; v.p. Council on Higher Edn., 1967; bd. dirs. Czechoslovak Soc. of Arts and Scis. in Exile, 1962. Mem. Am. Assn. Pathologists and Bacteriologists, Am. Soc. Cell Biology, Electron Microscopy Soc. Am., Am. Assn. for Study of Liver Diseases, Sigma Xi. Unitarian. Subspecialty: Pathology (medicine). Current work: Cellular reaction to injury.

HRUSHESKY, WILLIAM JOHN MICHAEL, scientist, physician; b. Poughkeepsie, N.Y., Nov, 9, 1947; s. William Michael and Mary Margaret (Burns) H.; m. m Donna Marie Turchetti, Apr. 7, 1947. A.B., Syracuse U., 1969; M.D., SUNY-Buffalo, 1973. Diplomate: Am. Bd. Internal Medicine, in med. oncology, 1979. Cancer researcher Roswell Park Meml. Inst., Buffalo, 1970; intern Balt. City Hosps. and Johns Hopkins, 1973-74; clin. assoc. NIH, 1974-76; resident, fellow in ocnology U. Minn., 1976-79, asst. prof. medicine and lab. medicine, 1979—. Contbr. articles to profl. jours. Served to lt. comdr. USPHS, 1974-76. Grantee in field. Mem. AAAS, Am. Fed. Clin. Research, N.Y. Acad. Scis., Am. Soc. Clin. Oncology, Am. Assn. Cancer Research, Internat. Soc. Chronobiology. Club: Provocateurs (pres.). Subspecialties: Cancer research (medicine); Chronobiology. Current work: Application of principals of biologic time structure to study of etiology and treatment of malignant disease. Home: 3538 Fremont Ave S Apt 1 Minneapolis MN 55408 Office: U Minn Hosp Box 414 May Meml Bldg Minneapolis MN 55455

HRYCAK, PETER, mechanical engineer, educator; b. Przemysl, Poland, July 8, 1923; came to U.S., 1949, naturalized, 1956; s. Eugene and Ludmyla (Dobrzanska) H.; m. Rea Meta Limberg, June 13, 1949; children—Michael Paul, Orest W.T., Alexandra Martha. Student, U. Tubingen, Germany, 1946-48; B.S. with high distinction, U. Minn., 1954, M.S., 1955, Ph.D., 1960. Registered profl. engr., N.J. Instr. U. Minn., Mpls., 1955-60; mem. tech. staff Bell Telephone Labs., Murray Hill, N.J., 1960-65; sr. project engr. Curtiss-Wright Corp., Woodridge, N.J., 1965; asso. prof. mech. engring. N.J. Inst. Tech., 1965-68, prof., 1968—. Contbr. articles to profl. jours. NASA grantee, 1967-68. Sr. mem. Inst. Environ. Scis.; mem. ASME, AIAA, Am. Soc. Engring. Edn., Ukrainian Engrs. Soc. Am. (pres. 1966-67), N.Y. Acad. Scis., Am. Geophys. Union, AAUP, Shevchenko Sci. Soc., Ukrainian Acad. Arts and Scis. in U.S.A., Pi Tau Sigma, Tau Beta Pi, Sigma Xi. Subspecialties: Mechanical engineering; Fluid mechanics. Current work: Experimental investigation and analytical studies of fluid flow and heat transfer characteristics of impinging jets; world wide regulation mechanism of carbon dioxide; man-made climate changes. Home: 19 Roselle Ave Cranford NJ 07016 Office: 323 High St Newark NJ 07102 Looking back over my professional career, it seems to me that there is no substitute for interest in, and curiosity for, new developments, and hard work to generate new ideas and to update oneself in the rapidly shifting environment of today. First comes, however, one's responsibility to maintain one's body in good mental and physical health. This may be achieved through a life filled with physical activities and hobbies, but also through an ability to "take it easy" at times, to recover from life's strain and to contemplate. All that should be filled with feelings for social justice and awareness of one's social responsibility which, in itself, may temper and blunt the inevitable desires, conflicts, and frustrations of the highly competitive modern life. Last but not least is perhaps the ability to laugh at oneself and be able to see both sides of the story.

HSIA, HENRY TAO-SZE, consulting firm executive, consultant; b. Peking, Hepei, China, June 16, 1923; came to U.S., 1947, naturalized, 1955; s. Ching and Wen Ling (Chen) H.; m. Alice C. Chung, Dec. 21, 1947; children: Victor Kai, Jean Mei, Alexander Hao. B.S., Chiao Tung U., Shanghai, China, 1944; M.S., Harvard U., 1948; Engr., Stanford U., 1964, Ph.D., 1966. Research specialist Lockheed Missile & Space Co., Sunnyvale, Calif., 1957-62; sr. staff scientist United Technologies, Sunnyvale, 1962-73; sr. engr. EDS Nuclear, San Francisco, 1973-74; sr. staff engr., cons. MB Assocs., San Ramon, Calif., 1974-76; program mgr. Gen. Electric Co., San Jose, Calif., 1976-82; pres. Tecon Services, Palo Alto, Calif., 1982—; cons., lectr. Chungshan Inst. Sci. and Tech., Taiwan, China, 1968—; lectr. Profl. Engrs. Tech. Inst., Menlo Park, Calif., 1982—; chmn. energy group Chinese Inst. Engrs., San Francisco, 1982-83. Author: Fundamentals of Rocket Propulsion, 1968; contbr. articles to profl. jours. Bd. dirs. Chinese Culture Assn., Palo Alto, Calif., 1966-76, Soc. Chinese Performing Arts, San Francisco, 1976—, Chinese Am. Assn. Sci. and Culture, Palo Alto, 1982—. Served to maj. Chinese Army, 1944-45. Recipient Quality Performance award Gen. Electric Co., 1979. Mem. ASME, AIAA, Am. Nuclear Soc. Republican. Subspecialties: Nuclear engineering; Aeronautical engineering. Current work: Nuclear power plant safety related analysis and design, alternate energy sources, rocket propulsion, technology transfer to developing nations. Patentee in field. Home: 865 Robb Rd Palo Alto CA 94306 Office: Tecon Services 865 Robb Rd Palo Alto CA 94306

HSIA, MONG-TSENG S., toxicology educator; b. Shanghai, China, Sept. 5, 1946; came to U.S., 1969; s. Mou-Siu M. and Li-Yen (Chang) H.; m. Hui-Erh V. Cheng, July 28, 1971; children: Rhoda J., Shelly H. B.S. in Chemistry, Nat. Cheng Kung U., Taiwan, 1968; M.S., U. Calif.-San Diego, 1971, Ph.D. in Organic Chemistry, 1974. Instr. Air Force Communications Electronics Sch., Taiwan, 1968-69; research fellow U. Calif.-San Diego, 1969-74; postdoctoral research assoc. U. Wis.-Madison, 1974-76, asst. scientist, 1974-78, asst. prof. toxicology, 1978-82, assoc. prof., 1982—; cons. Nat. Sci. Council Taiwan, 1980—. Pres. Madison Chinese Christian Fellowship, Madison, 1976, 77, 81, trustee, 1977—. Recipient Young Environ. Health Scientist award Nat. Inst. Environ. Health Scis., 1978. Mem. Am. Soc. Toxicology, Am. Coll. Toxicology, Am. Chem. Soc., Internat. Soc. Study Xenobiotics, Soc. Environ. Toxicology and Chemistry (charter). Subspecialties: Toxicology (agriculture); Toxicology (medicine). Current work: Pesticide toxicology, insect growth regulators, chemical carcinogenesis, halogenated aromatic hydrocarbons, short-term carcinogenicity-mutagenicity testings. Home: 2510 Homestead Rd Madison WI 53711 Office: Dept Entomology U Wis 1630 Linden Dr Madison WI 53706

HSIA, YU-PING, chemist, educator; b. China, May 16, 1936; came to U.S., 1961, naturalized, 1972; s. Hsiang-ming and Shu-chen (Wu) H.; m. Ting-mei Chen, Dec. 23, 1961; children: Irene, Richard. B.S., Tunghai U., 1959; M.S., U. Santa Barbara, 1963; Ph.D., Ill. Inst. Tech., 1967. Asst. prof. U. Bridgeport, Conn., 1967-68; asst., then assoc. prof. Calif. State Poly. U., Pomona, 1968-76, prof. chemistry, 1976—, also dir. grad. program. Contbr. articles to profl. jours. Petroleum research fellow, 1963-67; NSF grantee, 1971. Mem. Am. Chem. Soc., Chinese Am. Profl. Soc. in So. Calif. Subspecialties: Physical chemistry; Coal. Current work: Energy science; research in energy field; teaching in graduate and undergraduate level courses. Office: Calif Poly U 3801 W Temple Ave Pomona CA 91768

HSIAO, TING HUAN, insect physiologist; b. Hangchow, Chekiang, China, Feb. 6, 1936; came to U.S., 1958, naturalized, 1972; s. Tze Yuan and Mou C. (Yang) H.; m. Catherine Tang, Mar. 21, 1961. M.S. in Entomology, U. Minn., St. Paul, 1961; Ph.D. in Insect Physiology, U. Ill., Urbana, 1966. Research assoc. dept. entomology U. Ill., 1966-67; asst. prof. zoology Utah State U., 1967-72, assoc. prof. biology, 1972-79, prof. biology, 1979—; vis. prof. entomology Agrl. U., Wageningen, Netherlands, 1975, 77, 78, 81. Contbr. numerous articles to profl. publs. Mem. Entomol. Soc. Am., Am. Soc. Zoologists, Am. Inst. Biol. Sci., AAAS, Sigma Xi. Subspecialties: Animal physiology; Evolutionary biology. Current work: Insect/host plant relationships as related to feeding behavior, chemical interactions, nutrition and host plant resistance; biotypes of insect pests; physiology; feeding behavior; nutrition; ecological genetics. Office: Utah State U Dept Biology UMC 53 Logan UT 84322

HSIEH, JEN-SHU, radiation physicist; b. Taipei, Taiwan, Apr. 6, 1936; s. Yung-Ho and Shin-Fu (Li) H. B.S., Nat. Taiwan U., 1960; M.S., N.Mex. Highlands U., 1964; Ph.D. in Physics, Ohio U., 1968. Postdoctoral research assoc. Ohio U., 1968-70; NIH fellow, med. physicist UCLA, 1971-76; research physicist, cons. Internat. Sensor Tech., Santa Ana, Calif., 1973-76; health physicist radiation health div. Radiol. Control Office, Norfolk Naval Shipyard, 1976-78; physicist Armed Forced Radiobiology Research Inst., Bethesda, Md., 1978—. Contbr. articles to profl. jours. Mem. Am. Phys. Soc., AAAS, Am. Assn. Physicists in Medicine, Health Physics Soc., Radiation Research Soc., Sigma Xi, Sigma Pi Sigma, Phi Kappa Phi. Subspecialty: Medical/health physics. Current work: Effects of ionizing radiation; radiation transport codes for man and animal models; radiobiological dosimetry; thermoluminescense response from irradiated human teeth; Monte Carlo simulation of neutron spectra and dose distribution in phantom. Home: 10620 Weymouth St Apt 201 Bethesda MD 20814 Office: Armed Forces Radiobiology Research Inst Bethesda MD 20814

HSIEH, TSUYING CARL, aerospace engineer; b. Fukien, China, May 20, 1936; came to U.S., 1960; s. Kong-Young and Sui-Ching H.; m. Lydia C., Sept. 19, 1964; children: Mae, Stephen. B.S., Cheng-Kung U., 1956; M.S., U. Iowa, 1962; Ph.D., U. Md., 1970. Research scientist Hydronautics, Inc., Laurel, Md., 1962-68; research asst. prof. U. Md., College Park, 1968-73; aerospace engr. ARO, Inc., Arnold AFS, Tenn., 1973-79, Naval Surface Weapons Ctr., Silver Spring, Md., 1979—; dir. The Polygon, Inc., Knoxville, Tenn., 1978—. Contbr. articles to profl. jours. Served to 2d lt. Chinese Air Force, 1956-58. Mem. AIAA, Sigma Xi. Democrat. Subspecialties: Aerospace engineering and technology; Aeronautical engineering. Current work: Fluid dynamics with applications to aerospace engineering; special areas of research include computational fluid dynamics for transonic, supersonic and hypersonic flows over vehicles, bodies of revolution at high angle of attack, three-dimensional flow separation, and internal flow for ramjet inlets and diffusers. Home: 13 Stonegate Dr Silver Spring MD 20904 Office: Naval Surface Weapons Ctr Silver Spring MD 20910

HSIUNG, EDITH *See also* **HSIUNG, GUEH-DJEN**

HSIUNG, GUEH-DJEN (EDITH HSIUNG), microbiologist, educator; b. Hupeh, China, Sept. 16, 1918; d. Chu-yun and Bao-yu (Wu) H. B.S., Ginling Coll., Nanking, China, 1942; M.S., Mich. State U., 1948, Ph.D., 1951. Instr. Yale U. Sch. Medicine, 1954-57, research assoc., 1957-62, asst. prof., 1962-65; assoc. prof. N.Y. U. Sch. Medicine, 1965-67; chief virology lab. VA Med. Ctr., West Haven, Conn., 1967—; assoc. prof. dept. lab. medicine Yale U. Sch. Medicine, 1969-74, prof., 1974—. Author: Recent Advances in Clinical Virology, 1981, Diagnostic Virology, 1982; contbr. articles to profl. jours. Recipient Woman of Yr. award Fed. Exec. Assn. Greater New Haven, 1978; Career Scientist award VA, 1978. Mem. Am. Assn. Immunologists, Am. Soc. Microbiology, Am. Acad. Microbiology (fellow), Infectious Diseases Soc. Am. (fellow). Methodist. Subspecialties: Virology (medicine); Virology (biology). Current work: Medical virology; diagnostic virology; virus recognition and characterization; pathogenesis and epidemiology of virus infection; animal models for genital herpes and cytomegalovirus; endogenous virus infection and viral latency. Home: 30 W Haycock Point Branford CT 06504 Office: Virology La VA Med Center West Haven CT 06516

HSU, CLEMENT C.S., research physician, clin. immunologist; b. Taiwan, Oct. 9, 1937; came to U.S., 1965, naturalized, 1976; s. Ma-Wong and Yui-Ju (Chen) H.; m. Yui-Li Wu, Nov. 20, 1965; children: Felix S.W., Ben S.L. M.D., Nat. Taiwan U., 1963. Diplomate: Am. Bd. Internal Medicine. Intern Jersey City Med. Center, 1965-66; resident Montefiore Hosp., N.Y.C., 1966-67, Boston City Hosp., 1967-68; fellow in liver disease N.J. Coll. Medicine and Dentistry, East Orange, 1968-69; tng. in immunology research Inst. Cancer Research, Columbia Presbyn. Med. Center and Mt. Sinai Hosp., N.Y.C., 1969-72; mem. faculty Northwestern U. Med. Sch., Chgo., 1972—, asso. prof. medicine, 1978—; chief infectious disease sect. Columbus-Cuneo-Cabrini Med. Center, Chgo., 1977—. Contbr. articles to profl. jours. Named Outstanding New Citizen Citizenship Council Met. Chgo., 1976; Leukemia Research Found. grantee, 1974, 75, 77; NIH grantee, 1974-77; Nat. Cancer Inst. grantee, 1977-80. Mem. Am. Assn. Immunologists, Am. Assn. for Cancer Research, Infectious Disease Soc. Am., Central Soc. for Clin. Research, AAAS, N.Am. Taiwanses Profs. Assn. Subspecialties: Immunobiology and immunology; Infectious diseases. Current work: Membrane immunoglobulin isotypes on normal and neoplastic human lymphocytes; circulating inhibitor of lymphocyte responses in vitro; rabbit antibody inhibitor in human serum. Office: 2520 N Lakeview Ave Chicago IL 60614

HSU, HEI-TI, plant virologist, educator; b. Taipei, Taiwan, Sept. 3, 1939; came to U.S., 1968, naturalized, 1980; s. Chung-kuang and Cheng-Mei (Kuo) H.; m. Hsing Wu, Dec. 23, 1966; children: Elvin, Marvin. Ph.D., U. Ill., Champaign-Urbana, 1971. Research asso. U. Ill., Urbana, 1971-75; plant virologist Am. Type Culture Collection, Rockville, Md., 1975—; adj. prof. U. Md., 1981—; mem. plant virus subcom. Internat. Com. on Taxonomy of Viruses and chmn. study group Internat. Collection Plant Virus Type Cultures, 1978-84. Contbr. numerous articles to profl. publs.; asso. editor: Plant Diseases of Am. Phytopathol. Soc, 1978-82. Am. Soc. Microbiology travel awardee 3d Internat. Congress for Virology, 1975. Mem. Am. Phytopathol. Soc., Soc. Gen. Microbiology. Subspecialties: Plant virology; Immunobiology and immunology. Current work: Plant virus characterization; monoclonal antibodies; virus systematics.

HSU, IH-CHANG, biochemist; b. Taiwan, China, Aug. 3, 1938; s. Sae and Tsu-Ying (Wong) H.; m. Chang-Mei Tung, Aug. 16, 1968; children Alexander, Daniel. B.Pharmacy, Kaohsiung Med. Coll., 1964; Ph.D. (Peterson fellow), U. Wis.-Madison, 1972. Research assoc. dept. pathology U. Wis.-Madison, 1972-75; research fellow in cancer program Oak Ridge Nat. Lab., 1975-77; expert investigator Cancer Cause and Prevention div. Nat. Cancer Inst., Bethesda, Md., 1977-82; assoc. prof. toxicology U. Md., Balt., 1982—. Contbr. sci. articles to profl. publs. Mem. Am. Assn. Cancer Research, Environ. Mutagen Soc., Sigma Xi. Presbyterian. Subspecialties: Cancer research (medicine); Biochemistry (medicine). Current work: To establish systems for identification of environmental carcinogens. Research in the molecular basis of carcinogenesis and mutagenesis. Evaluation of gentoxic hazards of environmental carcinogens. Patentee in field; co-inventor ultra-sensitive enzymatic radioimmunoassay. Home: 3863 Spencer Ct Ellicott City MD 21403 Office: 10 S Pine St Baltimore MD 21201

HSU, MING-TEH, nuclear engineer; b. Kaoshung, Taiwan, China, June 10, 1946; came to U.S., 1969, naturalized, 1981; s. Kuai-teng and Chin-ju (Tong) H.; m. Chun-mei Lin, Jan. 27, 1979; 1 son, Johnny Meng. B.S. in Nuclear Engring, Nat. Tsing-Hua U., Taiwan, 1968; M.S., U. Md., 1974, Ph.D., 1974. Engr. Singer Co., Silver Spring, Md, 1975; sr. application engr. Control Data Corp., Rockville, Md., 1975-78; sr. engr. Idaho Nat. Engring. Lab., Idaho Falls, 1978-81, Bechtel, Norwalk, Calif., 1981—. Mem. Am. Nuclear Soc., Sigma Xi, Phi Kappa Phi. Subspecialties: Nuclear engineering; Fluid mechanics. Current work: The safety analysis of nuclear power plant, such as loss of coolant analysis or pipe break analysis. Home: 513 Lyons Way Placentia CA 92670 Office: Bechtel 12400 E Imperial Hwy Norwalk CA 90650

HSU, SHAW LING, physics educator; b. Shanghai, China, July 14, 1948; came to U.S., 1961; s. Pa Ho and Quan Fong (Wu) H.; m. Lilian Ming-Te, Aug. 15, 1970; children: David, Jennifer. B.A., Rutgers U., 1970; Ph.D., U. Mich., 1975. Research chemist Allied Chem. Co., Morristown, N.J., 1976-78; asst. prof. U. Mass., Amherst, 1978-82, assoc. prof., 1982—. Mem. Am. Phys. Soc., Am. Chem. Soc., Phi Beta Kappa. Subspecialties: Polymer physics; Polymers. Home: 6 Campbell Ct Amherst MA 01002 Office: U Mass Amherst MA 01003

HSU, SHU YING, immunologist; b. Peking, China, Aug. 16,1920; s. Chiao Ping and Hsin Fu (Chen) Li; m. Hsi Fan Hsu, Apr. 18, 1954. B.S. in Biology, Nat. Peking Normal U., 1940; Ph.D. in Preventive Medicine, U. Iowa, 1957. Asst. prof., head parasitology Nat. Shenyang (China) Med. Coll., 1947-48; assoc. prof. zoology Nat. Taiwan U., Taipei, 1949-52, prof., 1952; instr. bacteriology Mich. State U., East Lansing, 1952-53; research assoc. tropical pub. health Harvard Sch. Pub. Health, Boston, 1953-54; asst. prof. preventive medicine U. Iowa, Iowa City, 1957-63, research assoc. prof., 1963-73, prof., 1973—. Contbr. articles to profl. jours. Mem. Am. Assn. Immunologists, Soc. Exptl. Biology and Medicine, Am. Soc. Tropical Medicine and Hygiene, Am. Soc. Parasitologists, Royal Soc. Tropical Medicine and Hygiene. Subspecialties: Immunobiology and immunology; Parasitology. Current work: Immunity and vaccination against schistosomiasis. Office: Coll Medicine U Iowa Iowa City IA 52242

HSU, WALTER HAW, pharmacologist, educator; b. Fu-Chien, China, July 10, 1946; came to U.S., 1971, naturalized, 1983; s. Han-Po and Hua-Eng (Yuan) H.; m. Rou-Jean Shiaw, Aug. 21, 1971; children: Karen S-Hon, Susan S-Shien. D.V.M., Nat. Taiwan U., 1969; Ph.D., U. N.C., 1975. Research assoc. Purdue U., West Lafayette, Ind., 1975-77; asst. prof. pharmacology Iowa State U., Ames, 1977-82, assoc. prof., 1982—. Mem. AVMA; mem. Soc. Exptl. Biology and Medicine; Mem. Am. Soc. Vet. Pharmacology and Exptl. Therapeutics. Subspecialty: Veterinary Pharmacology. Current work: Endocrine pharmacology, neuropharmacology, veterinary pharmacology. Office: Iowa State U 2066 Vet Medicine Ames IA 50011

HU, CHUNG-HONG, dermatology educator; b. Taipei, Taiwan, Jan. 1, 1942; came to U.S., 1968; s. Sway-Wang and Yoh-Nee (Lin) H.; m. Mimi Wang; children: Michael, Mario. M.D., Taipei Med. Coll., 1966; postgrad., Mayo Grad. Sch. Medicine, 1970-74. Diplomate: Am. Bd. Dermatology, Am. Bd. Dermatopathology. Fellow Mayo Clinic-Mayo Found., Rochester, Minn., 1970-74; instr. U. Minn. Mayo Med. Sch., Rochester, 1974-75; asst. prof. Case Western Res. U., Cleve., 1975-79, Stanford (Calif.) U., 1979-82, assoc. prof., 1982—; asst. dermatologist Univ. Hosp., Cleve., 1975-79, dir. dermatopathology lab., 1977-79; chief dermatology Palo Alto (Calif.) VA Hosp., 1979—; sr. research assoc. Internat. Psoriasis Research Found., 1981—. Author: Diagnostic Electron Microscopy, 1980, Vasculitis, 1980. Chmn. bd. Chinese Acad. Cleve., 1979-79; v.p. Friends of Chinese Acad. Cleve., 1976-78; mem. med. council Ohio Lupus Found., 1978-79. Served as ensign Taiwan Navy, 1967-68. NIH grantee, 1976-79. Fellow ACP, Am. Acad. Dermatology, Am. Soc. Dermatopathology; mem. Soc. Investigative Dermatology, Internat. Soc. Tropical Dermatology. Subspecialties: Dermatology; Pathology (medicine). Current work: Diagnosis and treatment of blistering diseases of the skin and skin tumors; pathogenesis and new therapeutic approaches for psoriasis. Office: 300 Pasteur Dr Stanford CA 94305

HU, PATRICK HUNG-SUN, physicist; b. Shangshai, China, Apr. 5, 1943; came to U.S., 1966, naturalized, 1980; s. S.Y. and L.F. (Chiao) H.; m. Fanny Tai Hu, June 27, 1970; children: Anita, Kevin. B.S., Nat. Taiwan U., 1965; M.S., Columbia U., N.Y.C., 1967, Ph.D., 1972. With Bell Lab., Murray Hill, N.J., 1972—, mem. tech. staff, 1972—. Contbr. articles to profl. jours. Mem. Am. Phys. Soc. Subspecialties: Condensed matter physics; Atomic and molecular physics. Current work: Observation of first raman echoes, study of spectral diffusion high energy phonan physics, pico-second spectroscopy. Office: Bell Laboratory Murray Hill NJ 07974

HUA, HSICHUN MIKE, aeronautical industry executive; b. China, Dec. 6, 1925; m. Margaret Chow, Jan. 1, 1954. M.S., Purdue U., 1965, Ph.D., 1968; postdoctoral, Harvard U., 1979. Enlisted Republic of China Air Force, 1944, advanced through ranks to lt. gen., 1983; fighter pilot, 1949-64; aerodynamicist Cessna Aircraft Co., Wichita, Kans., 1968-69; aerodynamics engr. Lockheed Aircraft Co., Burbank, Calif., 1969-70; chief aircraft design Aero Industry Devel. Ctr., Taichung, Taiwan, Republic of China, 1970-74, dep. dir. engring. and research, 1974-82, dir., 1982—; assoc. prof. Cheng-Kung U., Taiwan, 1970-72; prof. Tunghai U., Taiwan, 1972-74; v.p. Internat. Turbine Engine Corp., Phoenix, 1982—. Contbr. numerous articles to profl. jours. Decorated D.F.C. Assoc. fellow AIAA; mem. Aero. and Astronautical Soc. (dir. 1972—), Soc. Theory and Applied Mechanics (dir. 1978—), Sigma Xi. Clubs: Am. Univ., Taipei, Taiwan, Harvard U. Subspecialty: Aeronautical engineering. Current work: Supervising development and fabrication of aircrafts and related products.

HUANG, CHENG-CHUN, biochemistry educator; b. Taipei, Taiwan, Feb. 2, 1938; came to U.S., 1964, naturalized, 1975; s. Wan-Chih and Hao (Chuang) H.; m. Cynthia Yu-Chih Yu, June 16, 1967; children: Tina, Eric. B.S., Taiwan U.-Taipei, 1963; M.S., Vanderbilt U., 1966; Ph.D., U. Iowa-Iowa City, 1970. Research assoc. U. Iowa, Iowa City, 1970-71, research scientist, 1973-77; research assoc. Yale U., New Haven, 1971-73; asst. prof. biochemistry Columbia U., N.Y.C., 1977—, dir., 1977—. Contbr. articles to profl. jours. Recipient 1st prize award Am. Acad. Opthalmology and Otolarynology, 1978; NIH grantee, 1978—. Mem. Am. Chem. Soc., Assn. Research Otolaryngology, N.Y. Acad. Sci., Am. Soc. Biol. Chemists, Sigma Xi. Subspecialties: Biochemistry (medicine); Otorhinolaryngology. Current work: Pathogenesis and bone resorption in chronic middle ear disease; proteases and invasion of head and neck tumors. Office: Columbia U Dept Otolaryngology 630 W 168th St New York NY 10032

HUANG, CHIA MING, physician; b. Taiwan, July 2, 1941; came to U.S., 1968, naturalized, 1979; s. Zong Ho and Kin (Lin) H.; m. Duen Mei Wung, Mar. 21, 1970; 4 children. M.D., Taipei Med. Coll., 1966. Diplomate: Am. Bd. Internal Medicine. Intern Grant Hosp., Chgo.; resident Northwestern U. Med. Sch.; practice medicine, specializing in internal medicine; assoc. prof. medicine Northwestern U., Chgo., 1975—. Mem. Am. Fedn. Clin. Research, Am. Soc. Nephrology, Internat. Soc. Nephrology, Am. Soc. Clin. Pharmacology and Therapeutics, ACP. Subspecialty: Nephrology. Office: VA Lakeside Med Center 333 E Huron St Chicago IL 60646

HUANG, CHIN PAO, environmental engineering educator, researcher; b. Changhua, Taiwan, Oct. 4, 1941; came to U.S., 1966, naturalized, 1975; s. Wang-Chi and You-Liang H.; m. Yu-Chu Chang, Apr. 1, 1971; children: Catherine Kailyn, Calvin Kaiming. B.S., Nat. Taiwan U., Taipei, 1965; M.S., Harvard U., 1967, Ph.D., 1971. Registered profl. engr., Del. Asst. prof. Wayne State U., 1971-74; asst. prof. civil engring. U. Del., 1974-77, assoc. prof., 1977-81, prof., 1981—; cons. in field. Editor: Industrial Waste, 1981. Sec. Orgn. Chinese Ams., Newark, Del., 1978. NSF grantee, 1974-76, 75-77, 78-83; EPA grantee, 1977-82. Mem. Am. Chem. Soc., ASCE, Am. Soc. Limnology and Oceanography, Isnternat. Assn. Colloid Scientists, Water Pollution Control Fedn. Subspecialties: Water supply and wastewater treatment; Industrial engineering. Current work: Conduct research in the areas of colloid and surface sciences, physical-chemical treatment of water and wastewater, and the fate of pollutants in the environment. Home: 621 Normna's Ln Newark DE 19711 Office: Dept Civil Engring U Del Newark DE 19711

HUANG, CHING-HSIEN, biochemistry educator, researcher; b. Tien-tsin, China, Oct. 24, 1935; came to U.S., 1961, naturalized, 1973; s. P.T. and S.C. (King) H.; m. Laura C. Shen, July 29, 1972; children: Tien-tsin, Tien-wei. B.S., Tunghai (Taiwan) U., 1959; Ph.D., Johns Hopkins U., 1965. Asst. prof. biochemistry U. Va., Charlottesville, 1967-73, assoc. prof., 1973-77, prof., 1977—, vice chmn. biochemistry, 1982—; research chemist NIH, Bethesda, Md., 1981-82. Served with Taiwan Air Force, 1959-61. Recipient spl. award United Bd. for Christian Higher Edn. in Asia, 1961; Helen Hay Whitney fellow, 1966-69. Subspecialties: Biochemistry (biology); Biophysical chemistry. Current work: Membrane biophysics and biochemistry, biophysical chemistry of phospholipids, molecular structure and mechanism of protein kinase. Office: Dept Biochemistry U Va Sch Medicine 1300 Jefferson Park Ave Charlottesville VA 22908

HUANG, CLARK *See also* **HUANG, ENG-SHANG**

HUANG, ENG-SHANG (CLARK HUANG), microbiology educator; b. Chia-Yi, Taiwan, Mar. 17, 1940; came to U.S., 1968; s. Jong-Sun and King-Fa (Ong) H.; m. Shu-Mei Huong, Dec. 26, 1965; children: David, Benjamin. B.S., Nat. Taiwan U., 1962, M.S. in Pub. Health, 1964; Ph.D. in Bacteriology and Immunology, U. N.C., 1971. Vis. asst. prof. U. N.C.-Chapel Hill, 1973-74, asst. prof., 1974-78, assoc. prof., 1978—; mem. virology study sect. research and grant div. NIH, Bethesda, Md., 1979-83. Contbr. articles on microbiology and virology to profl. jours. NIH Research Career Devel. awardee, 1978-83; USPHS fellow, 1972-73. Mem. Am. Soc. Microbiology, Am. Soc. Virology, N.Y. Acad. Scis., Am. Acad. Microbiology. Democrat. Subspecialties: Virology (medicine); Genetics and genetic engineering (medicine). Current work: Research and teaching in the molecular biology of human tumor viruses, research in molecular epidemiology of virus infection, viral oncology, antiviral and genetic engineering. Office: Cancer Research Ct U NC Sch Medicine Chapel Hill NC 27514

HUANG, HAI CHOW, nuclear engineer; b. Hankow, China, Dec. 10, 1927; came to U.S., 1963; s. Yung Ting and Chih Chuen (Shieh) H.; m. Leejen Hsueh, Dec. 21, 1961; children: Lucie, James. B.S., Nat. Taiwan U., 1955; M.S., U. Colo., 1964; Ph.D. in Nuclear Sci. and Engring., Carnegie Inst. Tech., Pitts., 1973. Registered profl. engr., Pa., Colo. Sr. engr. Rust Engring. Co., Pitts., 1964-66; chief engr. Salvucci Engrs., Inc., Pitts., 1966-69; sr. engr. Westinghouse Nuclear Energy Systems, Pitts., 1970-73; prin. engr. Advanced Reactors div. Westinghouse Electric Corp., Pitts., 1974-75, mgr. licensing standards, 1975-76; mgr. Clinch River breeder reactor plant licensing, 1976-80, mgr. Clinch River breeder reactor plant licensing and in-vessel safety analysis, 1980-81, mgr. licensing advanced reactors div., Madison, Pa., 1981—, mem. seismic criteria com., 1976—. Author tech. papers. Mem. Nat. Soc. Profl. Engrs., Am. Nuclear Soc., ASME, N.Y. Acad. Sci., AAAS, Sigma Xi. Subspecialties: Nuclear engineering; Nuclear fission. Current work: Development of nuclear safety positions and safety and design evaluations of various concepts of liquid-metal fast breeder reactor nuclear power plants. Reactor system and in-core accident analyses. Radiological analyses for protection of public safety. Home: 594 Trotwood Ridge Dr Pittsburgh PA 15241 Office: Westinghouse Electric Corp PO Box 158 Madison PA 15663

HUANG, HUEY-WEN, physics educator, consultant; b. Tokyo, Feb. 22, 1940; U.S., 1963; s. Chin-Chiang and Shin (Lin) H.; m.; 1 dau., Emily Pei-Hsin. B.S., Nat. Taiwan U., 1962; Ph.D., Cornell U., 1967. Research assoc. Columbia U., N.Y.C., 1967-69; lectr. Yale U., New Haven, 1969-71; asst. prof. So. Ill. U., Carbondale, 1971-73; successively asst. prof., assoc. prof., prof. dept. physics Rice U., Houston, 1973—; cons. Exxon Research Corp., Linden, N.J., 1982. Grantee NIH, 1974-77, Office Naval Research, 1974—, Robert A Welch Found., 1977—. Mem. Am. Phys. Soc., AAAS, Biophysics Soc. Subspecialties: Biophysics (physics); Statistical physics. Current work: Structural determination of protein dynamics by time-resolved x-ray absorption spectroscopy; statistical problems of biological systems. Home: 3710 Aberdeen Way Houston TX 77025 Office: Physics Dept Rice U Houston TX 77251

HUANG, JENG-SHENG, plant pathologist, educator; b. Taiwan, May 11, 1940; s. Ku-I and Pi-yuan (Shih) H.; m. Pi-yu Lin, Sept. 6, 1967; children: Jennifer C., Robert H. B.S., Chung-Hsing U., Taiwan, 1962; M.S., U. Mo., 1969, Ph.D., 1972. Teaching asst. Chung-Hsing U., 1963-67; research asst. U. Mo., Columbia, 1967-72, research microbiologist, 1972-75; asst. prof. plant pathology N.C. State U., Raleigh, 1975-80, assoc. prof., 1980—. Contbr. articles to profl. jours. U.S. Dept. Agr. grantee, 1981—; Union Carbide grantee, 1982—. Mem. Am. Phytopath. Soc., AAAS. Subspecialties: Plant pathology; Plant cell and tissue culture. Current work: Host-parasite interactions in plant tissue culture systems; effects of parasite on plant physiology. Home: 3325 Boulder Ct Raleigh NC 27607 Office: Dept Plant Pathology NC State U Raleigh NC 27650

HUANG, JU-CHANG, civil engineering educator; b. Kaohsiung, Taiwan, Jan. 3, 1941; m., 1965; 2 children. B.S., Nat. Taiwan U., 1963; M.S., U. Tex.-Austin, 1966, Ph.D. in Environ. Health Engring, 1967. From asst. prof. to assoc. prof. U. Mo.-Rolla, 1967-75, prof. civil engring., 1975—, dir., 1975—; Cons. to govt. and industry. Grantee U.S. Dept. Interior, EPA, NSF. Mem. ASCE (Walter L. Huber prize 1979), Water Pollution Control Fedn., Am. Soc. Engring. Edn., Am. Acad. Environ. Engrs., Assn. Environ. Engring. Profs. Subspecialties: Environmental engineering; Water supply and wastewater treatment. Office: Dept Civil Engring U Mo Rolla MO 65401

HUANG, KEH-NING, physicist; b. Nanking, China, Dec. 6, 1947; came to U.S., 1969; s. Han-Liang Huang and Chu-Chiu (Hu); m. Ying Kao Huang, June 18, 1972; children: Wei-Hwa, Wei-Chung. B.S., Nat. Cheng-Kung U. Tainan, Taiwan, 1968; Ph.D., Yale U., 1974. Postdoctoral fellow U. Oreg., Eugene, 1974-76; research assoc. U. Nebr., Lincoln, 1976-78; vis. asst. prof. U. Notre Dame, 1978-81; physicist Argonne (Ill.) Nat. Lab., 1981—; cons. in field. Author: Infinity and Set Theory, 1968; contbr. articles in field to profl. jours. NSF grantee, 1979-81; Cottrell grantee, 1980-82. Mem. Am. Phys. Soc., Sigma Xi. Subspecialties: Theoretical physics; Atomic and molecular physics. Current work: Relativistic many-body theory and its applications; relativistic collision theory; plasma fusion related atomic processes. Office: Argonne National Laboratory Argonne IL 60439

HUANG, LEAF, biochemistry educator, consultant; b. Chang-sa, Hunan, China, Sept. 23, 1946; came to U.S., 1969; s. Pi-cheng and Kwan (Wu) H.; m. Shilling Feng, 1972; children: Benjamin, Jennifer. B.S., Nat. Taiwan U., 1968; Ph.D., Mich. State U., 1974. Postdoctoral fellow Carnegie Inst. of Washington, Balt., 1974-76; asst. prof. U. Tenn., Knoxville, 1976-80, assoc. prof. dept. biochemistry, 1980—; cons. NIH, 1982—. Contbr. chpts. to books, articles to jours. in field. NIH research grantee, 1974—; research career devel. awardee, 1981—; Muscular Dystrophy Assn. Am., grantee, 1980-81. Mem. Am. Soc. Biol. Chemists, Biophys. Soc., Am. Soc. Cell Biologists. Subspecialties: Biochemistry (medicine); Drug carriers. Current work: Development of target-specific biocompatable drug carriers, with special emphasis on liposomes (or lipid vesicles). Office: Dept Biochemistry U Tenn Knoxville TN 37996-0840

HUANG, RICHARD SHIH-CHIU, mechanical engineer; b. Peking, Feb. 28, 1932; U.S., 1946, naturalized, 1955; s. Fang-Kang and Viola Johnson (Misner) H.; m. Adele Marrie Farren, June 4, 1960; children: William Farren, Michael Edward. B.S.M.E., Duke U., 1955; postgrad., George Washington U., 1974. Propulsion engr. Vought Corp., Dallas, 1955-60, aeroballistics engr., 1960-66, engring. mgr., 1966—. Fellow AIAA (assoc.). Republican. Clubs: Acad. Model Aeronautics (Washington) (dir. 1964); Soc. Antique Modelers (San Jose, Calif.) (rules com. 1982). Subspecialties: Systems engineering; Aerospace engineering and technology. Current work: Orbital flight mechanics, cryogenics, electro-optics, and systems engineering. Home: 4032 Deep Valley Dr Dallas TX 75234 Office: Vought Corp PO Box 225907 Dallas TX 75265

HUANG, SUN-YI, polymer chemist, researcher; b. Su-Ao, Ilan, Taiwan, Sept. 14, 1940; came to U.S., 1966; s. Su Sen and Shaw Lin H.; m. Misa Lin, June 9, 1968; 1 son, Herman Lin. B.S., Nat. Cheng Keng U.; M.S., N.Mex. Highlands U., 1968; Ph.D., U. Mo.-Kansas City, 1973. Research assoc. U. Mo.-Kansas City, 1973-74; postdoctoral U. Akron, 1974-76; research chemist Am. Cyanamid Co., Stamford, Conn., 1976-78, sr. research chemist, 1978—. Contbr. articles to profl. jours. Served to 2d lt. Chinese Army, 1963-64. Recipient Sci. Achievement award Am. Cyanamid Co., 1981; research grantee. Mem. Am. Chem. Soc., Am. Phys. Soc., Chinese-Am. Chem. Soc., Chinese-Am. Polymer Soc. Subspecialties: Polymer chemistry; Synthetic chemistry. Current work: Elastomers, engineering plastics, block and Graft copolymers, water soluble polymers, novel polyelectrolytes for enhanced oil recovery, paper additives and water treating. Patentee in field U.S. and fgn. countries; co-inventor long-last electrochromic device watch. Home: 17 Loughran Ave Stamford CT 06902 Office: American Cyanamid Co 1937 W Main St Stamford CT 06904

HUARD, THOMAS KING, immunologist; b. Wauseon, Ohio, Jan. 20, 1947; s. C. Paul and Donna S. H.; m. Rebecca S. Spangler, Aug. 28, 1971; 1 son, Thomas C. Student, DePauw U., 1965-69; B.S., U. Ill.-Chgo., 1970; M.S., 1976; Ph.D., 1979. Research asst. Am. Dental Assn., Chgo., 1970-72; clin. lab. technician Rush-Presbyn.-St. Luke's Med. Ctr., Chgo., 1974-78; instr. Ill. Coll. Podiatric Medicine, Chgo., 1974-78; postdoctoral scholar U. Mich., Ann Arbor, 1978-80, asst. research scientist, 1980—; cons. Internat. Biotech. Found., Denton, Tex., 1982—. Recipient Young Research award Internat. Assn. for Dental Research, 1974; NIH grantee, 1980. Mem. Am. Fedn. Clin. Research, Reticuloendothelial Soc., Sigma Xi. Repbulican. Methodist. Subspecialties: Immunopharmacology; Immunobiology and immunology. Current work: Cell-mediated immunity and cancer, monocyte macrophage-immunobiology, biological response modifiers (interferon) effects on immune function. Home: 319 S Revena Ann Arbor MI 48103 Office: U Mich Simpson Meml Research Inst 102 Observatory Ann Arbor MI 48109

HUBA, GEORGE JOHN, psychologist; b. Bronx, Feb. 19, 1951; s. George J. and Barbara (Wentworth) H. B.S., Fordham U., 1972; M.S., Yale U., 1975, Ph.D., 1977. Lic. psychologist, Calif. Asst. prof. U. Minn., Mpls., 1976-77; asst. to assoc. research psychologist UCLA, 1977—. Author: Drugs, Daydreaming, Personality, 1980, Antecedents and Consequences of Adolescent Drug Use, 1983; editor: Drug Abuse Theory, Research and Practice, 1983, Assessing Marijuana Consequences, 1981. Mem. Am. Psychol. Assn., Am. Statis. Assn., Western Psychol. Assn., Soc. Multivariate Exptl. Psychology. Subspecialty: Psychology. Current work: Causal modeling, multivariate applications in psychology, adolescent devel. and drug use, computer applications, psychol. assessement. Home: 6325 Green Valley Circle Culver City CA 90230 Office: Dept Psychology UCLA 405 Hilgard Ave Los Angeles CA 90024

HUBBARD, HAROLD MEAD, research company executive; b. Beloit, Kans., Apr. 16, 1924; s. Clarence Richard and Elizabeth (Mead) H.; m. Doreen J. Wallace, Aug. 13, 1948 (div. 1975); children—Stuart W., David D.; m. Barbara Bell Czarnecki, May 9, 1976. B.S., U. Kans., 1948, Ph.D., 1951. Instr. chemistry U. Kans., Lawrence, 1949-51; research chemist, research mgr., lab. mgr. E. I. dUpont de Nemours & Co., Inc., Wilmington, Del., 1951-69; dir. phys. sci. Midwest Research Inst., Kansas City, Mo., 1970-75, v.p. research, 1976-78, sr. v.p. ops., 1979-82; exec. v.p., dir. Solar Energy Research Inst., 1982—; dir. Guaranty State Bank, Percy Kent Inc. Mem. adv. com. U. Kans. Sch. Engring.; trustee U. Kansas City. Served with U.S. Army, 1942-45. Mem. Mo. Acad. Sci. (councillor at large 1977—), Tech. Transfer Soc. (v.p. 1978—), Am. Chem. Soc., AAAS, World Future Soc., Sigma Xi, Delta Upsilon. Unitarian. Club: Rockhill Tennis. Subspecialties: Analytical chemistry; Solar energy. Current work: Exploratory research and advanced development in all principal solar-related technologies; biomass, combustion, wind power, materials, photo voltaics, ocean thermal energy, photosynthesis. Home: 2605 Vivian St Lakewood CO 80215 Office: 1617 Cole Blvd Golden CO 80401

HUBBARD, LINCOLN BEALS, radiological physicist; b. Hawkesbury, Ont., Can., Sept. 8, 1940; s. Carroll Chauncey and Mary

Lunn (Beals) H.; m. Nancy Ann Krieger, Apr. 3, 1961; children: Jill, Katrina. B.S., U. N.H., 1961; Ph.D., M.I.T., 1967; postgrad. (fellow), Argonne Nat. Lab., 1966-68. Diplomate: Am. Bd. Health Physics, Am. Bd. Radiology. Asst. prof. math., physics Knoxville (Tenn.) Coll., 1968-70; asst. prof. physics Furman U., Greenville, S.C., 1970-74; chief physicist Mt. Sinai Hosp., Chgo., 1974-75, 81—; partner Fields, Griffith, Hubbard & Assocs., Inc., Glencoe, Ill., 1975—; chief physicist Cook County Hosp., Chgo., 1979—; cons. in field; assoc. prof. med. physics Rush U., 1983—. Author: Mathematics for Technologists in Radiology, 1979. NSF fellow, 1962-65; Research Corp. grantee, 1972-74. Mem. Am. Assn. Physicists in Medicine, Am. Phys. Soc., Am. Coll. Radiology, Health Physics Soc. Subspecialty: Radiology. Current work: Application of radiation physics to all aspects of radiology imaging, nuclear medicine and radiation therapy including related safety and instructional aspects. Home: 4113 W End Rd Downers Grove IL 60515 Office: PO Box 367 Hines IL 60141

HUBBARD, WALTER CLYDE, pharmacologist, pharmacist; b. Ben Lomond, Ark., June 5, 1942; s. Clyde and Maggie (Harrison) H.; m. Bertha E. Veazey, Nov. 19, 1965; m. Laura E. Thompson, April 11, 1970; children: Amy E., David A. B.S. in Pharmacy, U. Ark., 1966; Ph.D. in Pharmacolgy, Vanderbilt U., 1974. Registered pharmacist, Ark., Tenn. Research asso., then asso. prof. Vanderbilt U., Nashville, 1974—; pharmacy intern Kavanaugh Pharmacy, Little Rock, 1966-67; research asso. Med. Coll. Ga., Augusta, 1972-73. Contbr. articles to profl. jours. Recipient faculty key award U. Ark., 1966; NDEA/NIH trainee, 1967, 72; NIH grantee, 1981—. Mem. Am. Soc. Mass Spectrometry, Am. Soc. Pharmacology and Exptl. Therapeutics. Baptist. Lodge: Elks (Nashville). Subspecialties: Pharmacology; Biochemistry (medicine). Current work: Physiology, pharmacology and biochemistry of polyunsaturated fatty acids; arachidonic acid, role of cyclooxygenase, lipoxygenase and peroxidase products in physiol. and pathophsiol. conditions. Office: Vanderbilt U Nashville TN 37232

HUBBARD, WILLIAM BOGEL, JR., astronomer, educator, cons.; b. Liberty, Tex., Nov. 14, 1940; s. William Bogel and Marie (Young) H.; m. Jean North Gilliland, June 8, 1963; children: Lynne, Laurie. B.A. in Physics, Rice U., 1962; Ph.D. in Astronomy, U. Calif., Berkeley, 1967. Postdoctoral fellow Calif. Inst. Tech., Pasadena, 1967-68; asst. prof. astronomy U. Tex., Austin, 1968-72; assoc. prof. planetary scis. U. Ariz., 1972-75, prof., 1975—; dir. Lunar and Planetary Lab., 1977-81; cons. in field. Contbr. numerous articles to profl. jours. Mem. Am. Geophys. Union, Am. Astron. Soc., Internat. Astron. Union. Subspecialties: Planetary science; Condensed matter physics. Current work: Planetary interiors, high-pressure physics, structure Jovian planets, gravity fields. Home: 2618 E Devon St Tucson AZ 85716 Office: Lunar and Planetary Lab U Ariz 417 Kuiper Bldg Tucson AZ 85721

HUBBELL, FLOYD ALLAN, med. educator, researcher; b. Waco, Tex., Nov. 13, 1948; s. F. E. and Margaret (Fraser) H.; m. Nancy Cooper, May 23, 1975. B.A., Baylor U., 1971, M.D., 1974; M.S. in Pub. Health, UCLA, 1983. Diplomate: Am. Bd. Internal Medicine. Intern U. Calif.-Irvine-Long Beach Med. Program, 1975-76, resident, 1976-78; staff physician VA Med. Ctr., Long Beach, Calif., 1978—; asst. prof. medicine U. Calif.-Irvine, 1979—, assoc. residency program dir., 1979—. Contbr. articles to med. jours., HHS, 1980-83. Fellow ACP; mem. Am. Fedn. Clin. Research, Assn. Program Dirs. in Internal Medicine, Soc. for Research and Edn. in Primary Care Internal Medicine. Democrat. Subspecialties: Internal medicine; Health services research. Current work: Health services research. Office: VA Med Ctr 5901 E 7th St Long Beach CA 90822

HUBBELL, JOHN HOWARD, radiation physicist; b. Ann Arbor, Mich., Apr. 9, 1925; s. Howard Adams and Mildred Jeanetta (Lipe) H.; m. Jean Garber Norford, June 11, 1955; children: Anne Virginia Cooper, Shelton Eric, Wendy Jean. B.S.E. in Engring. Physics, U. Mich., 1949, M.S. in Physics, 1950. Physicist Nat. Bur. Standards, Washington, 1950—, dir. x-ray, ionizing radiation data ctr., 1965—; mem. Shielding subcom. cross sect. evaluation U.S. Dept. Energy, 1965—; cons. Lawrence Berkeley Lab., 1981—. Author: Photon Cross Sections, Attenuation Coefficients and Energy Absorption Coefficients from 10 kev to 100 Gev, 1969 (citation classic Inst. Sci. Info. 1982); contbr. chpts. to books, articles in field. Served with AUS, 1943-45. Recipient faculty medal Tech. U., Prague, 1982. Fellow Am. Nuclear Soc. (chmn. ANS-6 SI units com. 1974—); mem. Health Physics Soc. (chmn. liaison com. 1980-81), Radiation Research Soc., Am. Phys. Soc., Soc. Nuclear Medicine, Philos. Soc. Washington, D.C., Am. Forestry Assn., Internat. Union Crystallography (sec. task group 1978—), Internat. Radiation Physics Soc. (sec. 1982—). Methodist. Clubs: Potomac Appalachian Trail, Astronomy (Washington); Appalachian Trail Conf. (Harpers Ferry). Subspecialties: Radiation physics; Applied mathematics. Current work: Photon attenuation coefficients, cross sections, transport, atomic photoeffect, coherent and incoherent scattering, pair and triplet production, form and buildup factors, distributed sources, x-ray crystallography, radiation gauging. Home: 11830 Rocking Horse Rd Rockville MD 20852 Office: Nat Bur Standards Ctr Radiation Research Washington DC 20234

HUBBERT, MARION KING, geologist, geophysicist; b. San Saba, Tex., Oct. 5, 1903; s. William Bee and Cora Virginia (Lee) H.; m. Miriam Graddy Berry, Nov. 11, 1938. Student, Weatherford Coll., 1921-23; B.S., U. Chgo., 1926, M.S., 1928, Ph.D., 1937; D.Sc. (hon.), Syracuse U., 1972, Ind. State U., 1980. Asst. geologist Amerada Petroleum Corp., Tulsa, summer 1926, 27-28; teaching asst. geology U. Chgo., 1928-30; instr. geophysics Columbia, 1930-40; geophysicist Ill. Geol. Survey, summers 1931-32, 35-37; assoc. geologist U.S. Geol. Survey, summer 1934; pvt. research, writing, 1940-41; sr. analyst Bd. Econ. Warfare, Washington, 1942-43; research geophysicist Shell Oil Co., Houston, 1943-45, assoc. dir. research, 1945-51, chief cons. gen. geology, 1951-55; cons. gen. geology Shell Devel. Co., 1956-64; vis. prof. geology and geophysics Stanford U., 1962-63, prof., 1963-68, prof. emeritus, 1968—; vis. prof. geography Johns Hopkins U., spring 1968; regents prof. U. Calif. at Berkeley, spring 1973, mem. adv. bd., 1974-77; research geophysicist U.S. Geol. Survey, 1964-76, cons., 1976—; mem. U.S. delegation UN Sci. Conf. Conservation and Utilization Resources, Lake Success, N.Y., 1949; mem. com. geophysics Nat. Research Council; adviser Office Naval Research, 1949-51; mem. com. Disposal Radioactive Waste Products, 1955-63; mem. Adv. Selection Com. for Allowing Grants under Fulbright Act, 1950-51; mem. vis. com. earth scis. Mass. Inst. Tech., 1958-60; mem. earth scis. adv. panel NSF, 1953-57, chmn., 1954-57; vis. lectr. M.I.T., 1959; regents lectr. UCLA, 1960; mem. com. natural resources Nat. Acad. Scis., 1961-62; chmn. div. earth scis. Nat. Acad. Scis.-NRC, 1963-65; nat. adv. bd. U. Nev. Desert Research Inst., 1967-73; mem. com. resources and man NRC, 1966-70. Author: The Theory of Groundwater Motion and Related Papers, U.S. Energy Resources, A Review as of 1972; co-author: Resources and Man, Structural Geology; Editor: Geophysics, 1947-49; assoc. editor: Jour. Geology, 1958-82, Bull. Am. Assn. Petroleum Geologists, 1955-74; Contbr. articles to profl. jours. Trustee, sec. Population Reference Bur., 1966-72; lectr. exec. seminars U.S. Civil Service, Office of Personnel Mgmt., 1971—; USIA lectr., Europe, 1975, 77. Recipient Lucas medal Am. Inst. Mining, Metall. and Petroleum Engrs., 1971; Rockefeller Pub. Service award, 1977; William Smith medal Geol. Soc. London, 1978; Elliott Cresson medal for outstanding work in field of geology Franklin Inst., Phila., 1981; Vetlesen gold medal and cash award Columbia U., 1981. Fellow Am. Acad. Arts and Scis., AAAS, Geol. Soc. Am. (Day medal 1954, Penrose medal 1973, council 1947-49, pres. 1962), Internat. Union Geol. Scis. (U.S. nat. com. 1961-64, com. on geosci. and man 1972-76); mem. Am. Assn. Petroleum Geologists (hon., Distinguished lectr. U.S. and Can. 1945, 52, 73-74), Am. Geophys. Union, Soc. Petroleum Engrs. (hon., Distinguished lectr. 1963-64), Soc. Exploration Geophysicists (hon., life), Canadian Soc. Petroleum Geologists (hon.), Nat. Acad. Scis., Sigma Xi, Gamma Alpha. Club: Cosmos (Washington). Subspecialties: Geology; Geophysics. Current work: World energy and mineral resources and their implications for human society. Home: 5208 Westwood Dr Bethesda MD 20816

HUBBS, CLARK, biology educator and editor; b. Ann Arbor, Mich., Mar. 15, 1921; s. Carl L. and Laura C. (Clark) H.; m. Catherine V. Symons, Sept. 10, 1949; children: Laura Ellen, John Clark, Ann Frances. B.A., U. Mich., 1942; Ph.D., Stanford U., 1951. Instr. U. Tex.-Austin, 1949-52, asst. prof., 1952-57, assoc. prof., 1957-63, prof. zoology, 1963—, chmn., 1974-76, 78-8-; vis. prof. U. Okla., Kingston, 1970-83; mng. editor Am. Soc. Icthyologists and Herpetologists, 1971—; sci. adv. com. Tex. Utilities Generating Co., 1971—; leader Rio Grande Fishers Recovery Team U.S. Dept. Interior, 1978—, adv. com., 1975-77. Natural resources preservation com. Capitol Area Planning Council, Travis County, Tex., 1977-78; mem. fish adv. com. Union Internationale pour la Conservation de la Nature, London, 1976—; chmn. Inland Fish Force, Power Plant Siting Com., Gov.'s Office, Austin, 1971-72. Served with U.S. Army, 1942-46. Mem. Southwestern Assn. Naturalists (pres. 1966-67), Tex. Acad. Sci. (pres. 1972-73), Tex. Orgn. Endangered Species (pres. 1978-79), Soc. Systematic Zoology (councillor 1962-64), Soc. Study of Evolution (councillor 1968-70). Subspecialties: Evolutionary biology; Systematics. Current work: Why are animals (fishes) able to live in their environments? How did they become adapted to succeed there. Home: 5719 Marilyn Dr Austin TX 78731 Office: Dept Zoology Univ Texas Austin TX 78712

HUBE, DOUGLAS PETER, astronomer, physicist, educator; b. St. Catharines, Ont., Can., May 19, 1941; s. Clarence C. and Dorothy H. (Jago) H.; m. Joan O. Rieck, Oct. 16, 1965; children: Sharon, Susanne. B.Sc., U. Toronto, 1964, M.A., 1965, Ph.D., 1968. Vis. astronomer Radcliffe Obs., S. Africa, 1966-67; lectr. U. Toronto, 1967-68; Nat. Research Council Can. Postdoctoral fellow Kitt Peak Nat. Obs., Tucson, 1968-69; asst. prof., then assoc. prof. U. Alta., Edmonton, 1969-82, prof. physics, 1982—; founding mem. Edmonton Space Scis. Found. Contbr. articles to profl. jours., mags., newsletters and newspapers. Mem. Internat. Astron. Union, Can. Astron. Soc., Am. Astron. Soc., Royal Astron. Soc. Can. (service award 1982), Brit. Interplanetary Soc., Soc. Sci. Exploration (founding). Subspecialty: Optical astronomy. Current work: Spectroscopic and photometric observations and analysis of close binary stars, and peculiar A-type stars. Office: Dept Physics U Alt Edmonton AB Canada T6G 2J1

HUBEL, DAVID HUNTER, educator, physiologist; b. Windsor, Ont., Can., Feb. 27, 1926; s. Jesse Hervey and Elsie (Hunter) H.; m. Shirley Ruth Izzard, June 20, 1953; children—Carl Andrew, Eric David, Paul Matthew. B.Sc., McGill U., 1947, M.D., 1951, D.Sc. (hon.), 1978, A.M., Harvard, 1962. Intern Montreal Gen. Hosp., 1951-52; asst. resident neurology Montreal Neurol. Inst., 1952-53, fellow clin. neurophysiology, 1953-54; asst. resident neurology Johns Hopkins Hosp., 1954-55; sr. fellow neurol. scis. group Johns Hopkins, 1958-59; faculty Harvard Med. Sch., 1959—, George Packer Berry prof. physiology, chmn. dept., 1967-68, George Packer Berry prof. neurobiology, 1968—; George H. Bishop lectr. exptl. neurology Washington U., St. Louis, 1964; Jessup lectr. biol. scis. Columbia, 1970; James Arthur lectr. Am. Mus. Natural History, 1972; Ferrier lectr. Royal Soc. London, 1972; Harvey lectr. Rockefeller U., 1976; Weizmann meml. lectr. Weizmann Inst. Sci., Rehovot, Israel, 1979. Served with AUS, 1955-58. Recipient Trustees Research to Prevent Blindness award, 1971; Lewis S. Rosenstiel award for disting. work in basic med. research, 1972; Karl Spencer Lashley prize Am. Philos. Soc., 1977; Louisa Gross Horwitz prize Columbia U., 1978; Dickson prize in Medicine U. Pitts., 1979; Ledlie prize Harvard U., 1980; Nobel prize, 1981; Sr. fellow Harvard Soc. Fellows, 1971—. Fellow Am. Acad. Arts and Scis.; mem. Nat. Acad. Sci., Am. Physiol. Soc. (Bowditch lectr. 1966), Deutsche Akademie der Naturforscher Leopoldina, Soc. for Neurosci. (Grass lecture 1976), Assn. for Research in Vision and Ophthalmology (Friedenwald award 1975), Johns Hopkins U. Soc. Scholars. Subspecialty: Neurobiology. Research brain mechanisms in vision; bd. syndics Harvard U. Press, 1979. Home: 98 Collins Rd Waban MA 02168 Office: 25 Shattuck St Boston MA 02115

HUBER, DON MORGAN, plant pathologist; b. Mesa, Ariz., Mar. 19, 1935; s. Albert Elmo and Emma Lapreel (Davis) H.; m. Paula Elese Towery, Feb. 19, 1959; children: Brenda, Joyce, Aaron, Louise, Lynette, Sharon, Sarah, Elese, Natalie, Kevin, Derek. B.S., U. Ida., 1957, M.S., 1959; Ph.D., Mich. State U., 1963. Asst. prof. U. Ida., Moscow, 1963-68, asso. prof., 1968-71; asso. prof. plant pathology Purdue U., West Lafayette, 1971-81, prof., 1981—; research cons., 1965—; dir. Decah Mfg. Co., 1979—. Contbr. numerous articles to profl. jours. Served with AUS, 1959. Recipient Dow Chem. Research award, 1980. Mem. Am. Phytopathol. Soc., Western Soil Sci. Soc., Internat. Plant Pathology Soc., Sigma Xi, Alpha Zeta. Mormon. Subspecialties: Plant pathology; Microbiology. Current work: Basic and applied research on ecology of soil organisms, biological and cultural disease control, biology of the nitrogen cycle and physiology of pathogenesis. Patentee in field.

HUBER, DONALD JOHN, postharvest physiologist; b. Hamilton, Ohio, Sept. 11, 1951; s. Richard Robert and Vivian Maria (Gase) H.; m. Joyce Ann Abney, June 23, 1973 (div. Apr. 1982). B.S., Miami U., Oxford, Ohio, 1973, M.S., 1976; Ph.D., Iowa State U., 1980. Postdoctoral assoc. Iowa State U., Ames, 1980-81; asst. prof. vegetable crops dept. U. Fla., Gainesville, 1981—. Contbr. articles to profl. jours. Recipient Acad. Excellence award Iowa State U., 1976, 77. Mem. Am. Soc. Plant Physiologists, Am. Soc. Hort. Sci. Subspecialties: Plant physiology (agriculture); Biochemistry (biology). Current work: Enzymology of fruit softening, physiology of ripening and senescence. Office: Vegetable Crops Dept U Fla Gainesville FL 32611

HUBER, H. STEPHEN, physicist; b. Phila., May 20, 1950; s. Harvey H. and Kathryn S. H. B.A., Earlham Coll., 1972; M.S., Drexel U., 1975, Ph.D., 1978. Asst. prof. physics Beaver Coll., Glenside, Pa., 1977—; adj. asst. prof. physics and math. Drexel U., 1975—. Contbr. articles to profl. jours. Mem. Am. Phys. Soc., AAAS, AAUP, Sigma Xi. Subspecialties: Theoretical physics; Biophysics (physics). Current work: Biomagnetism, theoretical nuclear and molecular physics. Office: Dept Chemistry and Physics Beaver Coll Glenside PA 19038

HUBER, R. JOHN, psychologist, educator; b. Cleve., Oct. 26, 1940; s. Rudolph Joseph and Jennette (Kemper) H.; m. Pauline Rita Poirer, Aug. 27, 1966; children: Jennifer Ketching, Beth Ely, Emily Atkinson. B.A., Kent State U., 1962; M.A., U. Vt., 1965; Ph.D., U. N.H., 1970. Resident counselor U. Vt., Burlington, 1962-64; instr. SUNY-Plattsburgh, 1965-67; asst. prof. psychology Skidmore Coll., Saratoga Springs, N.Y., 1970-74; chmn., assoc. prof. dept. psychology Meredith Coll., Raleigh, N.C., 1974—, prof., 1980—; resident III Sch. Pastoral Care, Baptist Hosp., Winston-Salem, N.C., 1982-83. Contbr. chpts. to books, articles to profl. jours.; cons. editor: Individual Psychology. Vice chmn. parish council Sacred Heart Cathedral, Raleigh, 1977; chmn. Life Enrichment Ctr., Raleigh, 1982; bd. dirs. Plattsburgh Little Theatre, 1965-67. NIMH grantee, 1970; Shell grantee, 1976. Mem. Am. Psychol. Assn., Eastern Psychol. Assn., Associated Heads Depts. Psychology, Southeastern Psychol. Assn., N.Am. Soc. Adlerian Psychology. Republican. Roman Catholic. Subspecialty: Personality. Current work: Research on the relation of Adlerian psychology to perceptual processing, animal behavior and literature. Home: 5413 Maple Ridge Rd Raleigh NC 27609 Office: Meredith Coll Dept Psychology Raleigh NC 27607

HUBER, WILLIAM GEORGE, veterinarian, educator; b. Hinsdale, Ill., Jan. 10, 1924; s. Otto and Alice Ruth (Linde) H.; m. Joyce Flack, Mar. Aug. 14, 1948; children: Constance, Carrie, Jane. B.S., U. Ill., 1950, D.V.M., 1958, Ph.D., 1960. Pvt. practice vet. medicine, Ill., 1952-58; in charge large animal research Chas. Pfizer Co., Terre Haute, Ind., 1958-60; from asst. prof. to prof., assoc. dean U. Ill., Urbana, 1960-73; dir. research agrichems. and animal health Hoffmann-LaRoche, Nutley, N.J., 1973-77; prof., assoc. dean Wash. State U., Pullman, 1977-82; prof., dir. research Miss. State U., 1982—; cons. in field. Contbr. articles to profl. jours.; author textbook. Active Am. Heart Assn. Served with USN, 1942-46; PTO. Grantee NIH, FDA, U.S. Dept. Agr. Mem. Am. Soc. Pharmacology and Exptl. Therapeutics, AVMA, Am. Acad. Vet. Pharmacology and Therapeutics. Lodges: Masons; Shriners; Rotary. Subspecialties: Pharmacology; Toxicology (medicine). Current work: Antimicrobial drugs, antibacterial drug residues. Home: 218 Edgewood Dr Starkville MS 39759 Office: Miss State U Coll Vet Medicine Drawer V MS 39762

HUBERMAN, ELIEZER, microbiologist, researcher, educator; b. Lukow, Poland, Feb. 8, 1939; came to U.S., 1976, naturalized, 1982; s. Samuel and Mina (Slushni) H.; m. Lily Ginsburg, May 11, 1967; children: Ilan, Ron. M.Sc., Tel Aviv U., 1964; Ph.D., Weizmann Inst. Sci., 1969. Postdoctoral fellow McArdle Lab., U. Wis, 1969-71; scientist dept. genetics Weizmann Inst. Sci., 1971-73, sr. scientist, assoc. prof., 1973-77; sr. scientist, group leader biology div. Oak Ridge (Tenn.) Nat. Lab., 1976-81; dir. biology and med. research div. Argonne (Ill.) Nat. Lab., 1981—; prof. dept. microbiology U. Chgo., 1982—. Mem. Am. Assn. for Cancer Research. Subspecialties: Cancer research (medicine); Cell and tissue culture. Current work: Cell differentiation and mutagenesis. Home: 424 Sunset Ave LaGrange IL 60525 Office: 9700 S Cass Ave Argonne IL 60439

HUBERT, JAY MARVIN, physicist; b. Denver, Apr. 11, 1944; s. Marvin A. and Elizabeth Ann (Cowan) H.; m. Mary Weed, Aug. 31, 1968; children: George, Mary Clare, Julia. B.A. in Physics, Reed Coll., 1966, M.S., Tex. A&M U., 1968, Ph.D., 1970. Research physicist Chevron Research Co., Richmond, Calif., 1970-75, sr. research physicist, 1975-81, sr. research assoc., 1981-82; mgr. Tech. Info. Center, 1982—. Contbr. articles on low temperature physics, thermal analysis to profl. jours. Mem. Am. Phys. Soc., AAAS. Subspecialties: Condensed matter physics; Surface chemistry. Current work: Surface analysis; electron microscopy; technical information services management energy storage technology. Office: 576 Standard Ave Richmond CA 94802

HUBSCHMANN, OTAKAR RUDOLF, neurosurgeon; b. Prague, Czechoslovakia, Jan. 20, 1944; s. Otakar and Marketa (Hausser) H.; m. Catherine Sullivan, Aug. 28, 1973; children: O. Gregory, Andrew Raymond. B.A., Acad. Gymnasium, Prague, 1961; M.D., Charles U. Prague, 1967. Diplomate: Am. Bd. Neurol. Surgery. Intern Albert Einstein Coll. Medicine, 1968-69, resident in surgery, 1969-70, resident in neurosurgery, 1970-76; staff neurosurgeon VA Med. Ctr., East Orange, N.J., 1976—; attending neurosurgeon Coll. Hosp., Newark, 1979—, Martland Hosp., 1976-78; attending staff St. Michael's Med. Ctr., Newark, 1977—, Presbyn. Hosp., 1978—; asst. in pediatrics Children's Hosp. of Newark, 1978—; attending neurosurgeon Newark Beth Israel Med. Ctr., 1979—; assoc. prof. neurosurgery Univ. Medicine and Dentistry-N.J. Med. Sch., 1981—, asst. prof., 1976-81; cons. head trauma HEW, Washington, 1981. Contbr. articles and abstracts to profl. jours. Served to maj. USAR, 1969—. VA grantee; Am. Heart Assn. grantee. Fellow A.C.S.; mem. Am. Assn. Neurol. Surgeons, Congress Neurol. Surgeons, AMA, Soc. Neurosci., Soc. Neurol. Intensive Care and Neuroanesthesia, Southeastern Surg. Congress, Essex County Med. Soc., AAUP, Brit. Brain Research Assn. (hon.), European Brain and Behavior Soc. (hon.). Club: Maplewood (N.J.) Country. Subspecialties: Microsurgery; Neurosurgery. Office: Suite 217 22 Old Short Hills Rd Livingston NJ 07039

HUCHRA, JOHN PETER, astronomer; b. Jersey City, Dec. 23, 1948; s. Mieczyslaw Piotr and Helen Ann (Lowicka) H. B.S., M.I.T., 1970; Ph.D., Calif. Inst. Tech., 1976. Center fellow Harvard-Smithsonian Center for Astrophysics, 1976-78; astronomer Smithsonian Astrophys. Obs., 1978—; lectr. dept astronomy Harvard U., 1979—. Contbr. articles to profl. jours. Smithsonian scholar, 1980—; NASA grantee, 1979-83. Mem. Am. Astron. Soc., Am. Phys. Soc., Internat. Astron. Union, Sierra Club, Sigma Xi, Gamma Nu. Subspecialties: Infrared optical astronomy; Cosmology. Office: 60 Garden St Cambridge MA 02138

HUDAK, WILLIAM JOHN, pharmacologist; b. Duquesne, Pa., Jan. 3, 1929; s. Stephen and Anna (Dvorznak) H.; m. Cecelia A. Byers, Sept. 11, 1954; children: Theresa, William, Cindy, Kathy, Steve, Ralph, Agnes, Paul, Brian. B.S., U. Pitts., 1954, M.S., 1956, Ph.D. in Pharmacology (George A. Kelly fellow), 1959. Registered pharmacist, Pa. Cardiovascular sect. head Merrell Nat. Labs., Cin., 1959-70, asst. group dir., 1970-71, asso. group dir. cardiovascular clin. research, 1971-80, asst. to v.p. research ops., 1980-82; mgr. research info. Merrell Dow Pharms., Inc., Cin., 1982—. Contbr. articles to sci. jours. Sec. Sharonville (Ohio) Recreation Commn., 1972-75, chmn., 1972-75, chmn., 1975-77; pres. Merrell Employees Fed. Credit Union, 1973—. Nat. Heart Inst. grantee, 1958-59. Fellow AAAS; mem. Am. Soc. Clin. Pharmacology and Therapeutics, N.Y. Acad. Scis., Am. Heart Assn., Am. Soc. Pharmacology and Exptl. Therapeutics, Am. Chem. Soc., Sigma Xi, Phi Sigma, Phi Delta Chi, Rho Chi. Roman Catholic. Subspecialty: Pharmacology. Current work: Devel. of drugs for treatment of cardiovascular disease. Patentee in field (2). Home: 10476 Wintergreen Ct Sharonville OH 45241 Office: 2110 E Galbraith Rd Cincinnati OH 45215

HUDDLESTON, PHILIP LEE, research physicist; b. St. Louis, Jan. 12, 1947; s. Joseph Berl and Myrtle (Craig) H.; m. Angela Jeaninne, Aug. 10, 1973. B.S. in Physics, Wash. U., 1967, M.A., Boston U., 1969, Ph.D., 1974. Instr. math. Tex. Tech. U., Lubbock, 1974-75; asst. prof. math and physics Edward Waters Coll., Jacksonville, Fla., 1975-76; asst. prof. physics Parks Coll. St. Louis U., Cahokia, Ill., 1976-79; sci. programmer McDonnell Douglas Corp., St. Louis, 1979-81, research scientist, 1981—. Mem. AAAS, Am. Math. Soc., Am. Phys. Soc., Soc. Indsl. and Applied Math., Sigma Xi. Subspecialties: Theoretical physics; Applied mathematics. Current work: Current research interests include computational and applied electromagnetics and inverse scattering problems. Home: PO Box 222 Barnhart MO 63012 Office: McDonnell Douglas Research Labs PO Box 516 Saint Louis MO 63166

HUDECKI, MICHAEL STEPHEN, biologist, researcher; b. Ft. Bragg, N.C., Nov. 7, 1943; s. Stephen Edward and Veronica Aileen (Kwolek) H.; m. Rajmohini Sebastian, June 9, 1973. B.S., Niagara U., 1965, M.S., 1967, D.Sc. (hon.), 1981; M.A., SUNY, Buffalo, 1970, Ph.D., 1973. Lectr., research asso. SUNY, Buffalo, 1979-80, research asst. prof., 1980—; cons. AAAS project on handicapped in sci., U.S. Ho. of Reps. com. on opportunities in sci. program, NSF handicapped in sci. project; adv. Office Services to Handicapped, SUNY, Buffalo. Contbr. articles in field to profl. jours. Muscular Dystrophy Assn. fellow, 1972-76; grantee, 1977—; NIH grantee, 1980-83; recipient USPHS Research Career Devel. award NIH, 1980—. Mem. AAAS, Soc. Neuroscience, N.Y. Acad. Sci., Am. Soc. Cell Biology, Fedn. Sci. and the Handicapped, Delta Epsilon Sigma. Roman Catholic. Subspecialties: Muscular dystrophy; Muscle pathology. Current work: Drug therapy trials using chickens with inherited muscular dystrophy; muscle protein turnover in dystrophic chickens. Office: State University of New York at Buffalo 670 Cooke Department of Biological Sciences Amherst NY 14260

HUDGIN, DONALD EDWARD, chemical research and development executive; b. Greenville, S.C., Aug. 11, 1917; s. Thomas and Virginia H.; m. Charlotte Hass, Jan. 14, 1943; children: Richard Henry, Frederick William, Charlotte Dott. B.S., Clemson U., 1938; M.S., Purdue U., 1940, Ph.D., 1947. Research chemist Procter & Gamble, Cin., 1947-52; research project leader Mallinckrodt Chem., St. Louis, 1952-55; research sect. head Celanese Corp., Summit, N.J., 1955-60; dir. research and devel. Gary Chems., East Brunswick, N.J., 1960-61; research dir. Diamond Alkali Co., Cleve., 1961-66; tech. advisor to v.p. research Exxon Research & Engring. Co., Linden, N.J., 1966-67; dir. research Princeton Chem. Research, Princeton, N.J., 1967-70; v.p. Princeton Polymer Labs., Plainsboro, N.J., 1970-80, pres., 1980—; lectr. chemistry U. Mass., 1965, Princeton U., 1972; speaker in field. Editor: Polymer Engineering Book Series, 1980; assoc. editor: Internat. Jour. Polymer Process Engring, 1980; contbr. articles to profl. jours. Exec. dir. Northeast Ohio council Boy Scouts Am., 1962-66; bd. dirs. Civil Def. Council, Summit, N.J., 1957-61. Served to maj. AUS, 1942-46; ETO; Served to maj. USAR, 1946-66. Fellow Am. Inst. Chemists; mem. Am. Chem. Soc., AAAS, N.Y. Acad. Sci., Chemists Club, Soc. Plastics Engrs., Am. Council Ind. Labs., Assn. Small Research Companies, Assn. Research Dirs. (pres. 1974-75), Plastics Edn. Found. (bd. trustee 1974-75), Sigma Xi. Republican. Unitarian. Clubs: West Windsor Tennis Assn., Ret. Officers Assn. Subspecialties: Polymer chemistry; Organic chemistry. Current work: Polymer science, technical market studies. Patentee in field. Office: 501 Plainsboro Rd Plainsboro NJ 08536

HUDGINS, PATRICIA MONTAGUE, physiology educator, researcher; b. Buckhannon, W.V, Jan. 31, 1938; d. Richard Wells and Clella Mae (Barger) Montague; m. Aubrey Clude Hudgins, Dec. 12, 1964; children: Audrey Dale, Monica Sue; m. Guy Hugh Bond, June 30, 1975. B.S. with high honors, W.Va. U., 1959, M.S., 1960, Ph.D. in Pharmacology, 1966. Instr. Med. Coll. of Va., Va. Commonwealth U., Richmond, 1966-68, asst. prof., 1968-72, assoc. prof, 1972-75; assoc. prof. Kirksville (Mo.) Coll. Osteo. Medicine, 1975-80, prof. dept. physiology, 1980—; dir. Adair County Nursing Home; mem. Mo. Community Health Corp. Contbr. articles on physiology to profl. jours. Mem. Am. Soc. for Pharmacology and Exptl. Therapeutics, Am. Physiol. Soc., Am. Heart Assn. (research com. Mo. Affiliate, Inc.). Subspecialties: Physiology (medicine); Pharmacology. Current work: Cation transport enzymes and the mechanism of ATP hydrolysis, mechanism of the sodium pump, trace element inhibition of transport. Home: 301 Hillcrest Dr Kirksville MO 63501 Office: Dept Physiology Kirksville Coll Osteo Medicine Kirksville MO 63501

HUDRLIK, PAUL FREDERICK, chemistry educator, researcher; b. Portland, May 10, 1941; s. Otto L. and Claudia (Bergman) H.; m. Anne Marie Bachmann, Sept. 9, 1967; children: Janet, Carol. B.S., Oreg. State U.-Corvallis, 1963; M.A., Columbia U., 1964, Ph.D., 1968. Asst. prof. chemistry Rutgers U., 1969-76; assoc. prof. Howard U., 1977-81, prof., 1981—. Contbr. articles to profl. jours. NIH fellow, 1968-69; recipient Louis P. Hammett award Columbia U., 1967. Mem. Am. Chem. Soc., Royal Soc. Chemistry, AAAS. Subspecialties: Organic chemistry; Synthetic chemistry. Current work: Organosilicon chemistry and synthetic organic chemistry; development of new synthetic methods for organic chemistry. Office: Dept Chemistry Howard U Washington DC 20059

HUDSON, DONALD ELLIS, civil engineering educator; b. Alma, Mich., Feb. 25, 1916; s. Albert W. and Ruth (Ellis) H. B.S., Calif. Inst. Tech., 1938, M.S., 1939, Ph.D., 1942. Prof. mech. engring. and applied mechanics Calif. Inst. Tech., Pasadena, 1941-81, prof. emeritus, 1981—; prof. and chmn. dept. civil engring. U. So. Calif., Los Angeles, 1981—, Fred Champion prof. engring., 1982—. Author: (with G.W. Housner) Applied Mechanics, Statics and Dynamics, 1949-50. Fellow ASME; mem. ASCE, Am. Soc. for Engring. Edn., Soc. Exptl. Stress Analysis, Am. Geophys. Union, Seismol. Soc. Am. (past pres., editorial com.), Internat. Assn. Earthquake Engring. (pres. 1980—), Earthquake Engring. Research Inst., Nat. Acad. Engring. Subspecialties: Earthquake Engineering; Solid mechanics. Current work: Structural dynamics; tests of full-scale structures. Measurements and instrument design in earthquake engineering. Office: Dept Civil Engring U So Calif University Park Los Angeles CA 90089

HUDSON, JOHN LESTER, chemical engineering educator; b. Chgo., June 19, 1937; s. John J. and Linda M. (Panozzo) H.; m. Janette Caton, June 29, 1963; children: Ann, Barbara, Sarah. B.S., U. Ill., 1959; M.S.E., Princeton U., 1960; Ph.D., Northwestern U., 1962. Registered profl. engr., Ill. Assoc. prof. dept. chem. engring. U. Ill., 1963-74; mgr. air pollution control div. Ill. EPA, 1974-75; prof., chmn. dept. chem. engring. U. Va., Charlottesville, 1975—. Contbr. articles to profl. jours. Fulbright fellow, 1962-63, 82-83. Mem. Am. Inst. Chem. Engrs., Am. Chem. Soc., Air Pollution Control Assn. Subspecialty: Chemical engineering. Current work: Research on the dynamics of chemical reactors and on air pollution control. Office: University of Virginia Department of Chemical Engring Thornton Hall Charlottesville VA 22901

HUDSON, ROBERT FRANK, geologist; b. Bridgeport, Conn., Aug. 14, 1931; s. Harold Charles and Mary Agnes (Lefon) H.; m. Hazel Elizabeth Chilson, Sept. 10, 1955; children: Michael, Ann, Neil, Peter, Roberta, Amy, David, Paul. B.A., Colby Coll., 1954; M.S., U. Iowa, 1957, Ph.D., 1965. Geologist Texaco Inc., Corpus Christi, Tex., 1959-65, Houston, 1966-78, La. Land & Exploration Co., 1978-81, MCOR Oil Devel. Inc., 1981-82, Donald C. Slawson (oil producer), 1982-83, Sohio Petroleum Co., 1983—. Contbr. articles to profl. jours. Active Holy Ghost Ch., Houston, 1966—, Boy Scouts Am., Houston, 1968-72; coach and ofcl. Sharpstown Little League Baseball, Houston, 1968-74, Southwest Girls Softball Assn., Houston, 1972—. Fellow Geol. Soc. Am. (Penrose Fund grantee 1957); mem. Am. Assn. Petroleum Geologists, Houston Geol. Soc., Sigma Xi. Roman Catholic. Subspecialty: Petroleum geology. Current work: Petroleum exploration. Home: 6833 Hazen Houston TX 77074 Office: Donald C Slawson Oil Producer One Allen Ctr Suite 580 Houston TX 77002

HUDSON, WILLIAM DONALD, JR., natural sciences educator; b. St. Johnsbury, Vt., June 28, 1950; s. William Donald and Marguerite (McConnell) H.; m. Josephine Wilcox Ewing, June 16, 1979. A.B. in French, Dartmouth Coll., 1972; M.S. in Botany, U. Vt., 1979; Ph.D. in Biology, Ind. U., 1983. Instr. Chewonki Found., Wiscasset, Maine, 1972-76, biologist, coordinator natural scis. edn., 1982—. Ind. Acad. Scis. grantee, 1980-81; Sigma Xi grantee, 1980-81. Mem. Bot. Soc. Am., Am. Soc. Plant Taxonomists, Soc. for Study of Evolution, Soc. Econ. Botany, Ecol. Soc. Am., Brit. Ecol. Soc., Inst. Biology, Am. Ornithol. Union, AAAS, New Eng. Bot. Soc., Northeastern Bird-Banding Assn., Sigma Xi. Subspecialties: Evolutionary biology; Taxonomy. Current work: Investigations of the origin and evolution of domesticated plants, notably members of the potato family. Solanaceae Physalis. Office: Chewonki Found Wiscasset ME 04578

HUESER, JAMES NICHOLAS, physician, medical oncologist; b. Clinton, Mo., Dec. 6, 1938; s. Edward John and Geneva Catherine (Putthoff) H.; m. Lorraine A. Buchanan, Oct. 23, 1965; children: Michael, Michelle, Christopher, Mark. B.S., Rockhurst Coll., 1959; M.D., U. Mo.-Columbia, 1963. Diplomate: Am. Bd. Internal Medicine and Med. Oncology. Intern San Francisco Gen. Hosp., 1963-64; resident U. Mo., Columbia, 1964-67; internist Ellis Fischel State Cancer Hosp., Columbia, 1967-69, cons. med. oncology, 1969—; attending physician Bonne Hosp. Center, Columbia, 1969—; internist, med. oncologist Columbia Clinic, 1972—; attending physician Columbia Regional Hosp., 1974—; dir. Cancer Detection Clinic Cancer Research Center, Columbia, 1971-75; cons. med. oncology Bothwell Meml. Hosp., Sedalia, Mo, 1978—, Audrain Med. Center, Mexico, Mo., 1980—, Fitzgibbon Hosp., Marshall, Mo., 1981—. Contbr. articles to profl. jours. Bd. dirs. Mo. Div. Am. Cancer Soc., Jefferson City, 1975—, pres., 1979—; bd. Ronald McDonald House, Columbia, 1982—; del. mem. Assn. Community Cancer Centers, 1982—. Investigator Western Cancer Study Group, Los Angeles, 1969-75, Cancer and Acute Leukemia Group B, N.Y.C., 1981—, Nat. Surg. Adjuvant Breast Project, Pitts., 1981—. Fellow Am. Soc. Clin. Oncology, Am. Assn. Cancer Research, ACP; mem. N.Y. Acad. Scis., AAAS. Republican. Roman Catholic. Subspecialty: Cancer research (medicine). Current work: Clinical research in cancer chemotherapy. Home: 604 W Broadway Columbia MO 65201 Office: Columbia Clinic 401 Keene St Columbia MO 65201

HUFF, KENNETH O., consulting geologist; b. Daleville, Ind., Dec. 17, 1926; s. George Byron and Mary Ethel (Smith) H.; m. Donna M. Huff, Mar. 25, 1957; children: John, Robert, Doald, Patricia. B.S., Ind. U., 1956; postgrad., Ball State U., 1947-48, Purdue U., 1944-45. Cert. Am. Inst. Profl. Geologists. Well logging engr., lab. mgr. Core Labs., Inc., Williston, N.D. and Farmington, N.Mex., 1956-64; lab. mgr., sales engr., Casper, Wyo. and Farmington, 1964-67, Rocky Mountain dist. supr., 1967-69, cons. geologist, 1969—; pres. Adventures, Inc., Casper, 1972—. Mem. dist. export council U.S. Dept. Commerce, 1977—. Served with U.S. Army, 1944-46, 50-51; Korea. Mem. Soc. Petroleum Engrs., Am. Assn. Petroleum Geologists, Wyo. Geol. Assn., Rocky Mountain Assn. Petroleum Geologists, Casper Petroleum Club. Subspecialties: Geology; Petroleum engineering. Current work: Corporate management; sample recovery equipment for drill rigs. Patentee automated sample recovery equipment for rotary drill rigs (3). Office: 535 N Lenox St Casper WY 82601

HUFFMAN, JOHN WILLIAM, chemistry educator; b. Evanston, Ill., July 21, 1932; s. John W. and Florence (Kearns) H.; m. Dana A. Holderby, Dec. 5, 1975; children from previous marriage: Paul W., James R., George R., John E. B.S., Northwestern U., 1954; A.M., Harvard U., 1956, Ph.D., 1957. Asst. prof. chemistry Ga. Inst. Tech., Atlanta, 1957-60; asst. prof. chemistry Clemson (S.C.) U., 1960-62, assoc. prof., 1962-67, prof., 1967—. NIH grantee, 1965-70. Mem. Am. Chem. Soc., Sigma Xi. Democrat. Roman Catholic. Subspecialties: Organic chemistry; Synthetic chemistry. Current work: Synthesis of natural products, synthetic methodology, stereochemistry of carbocyclic molecules. Home: PO Box 614 Clemson SC 29633 Office: Dept Chemistry Clemson U Clemson SC 29631

HUFFMAN, RONALD DEAN, research pharmacologist; b. Vandergrift, Pa., Dec. 13, 1937; s. Hudson Monroe and Helen L. (Dinger) H. B.S., Pa. State U., 1959; M.S., Purdue U., 1961, Ph.D. in Pharmacology, 1967. Research scientist S.W. Found. for Research and Edn., San Antonio, 1967-68; lectr. dept. physiology U. B.C., Vancouver, 1967-68; asst. prof. dept. anatomy and pharmacology U. Tex. Med. Sch., San Antonio, 1968-72, asso. prof., 1972-73; asso. prof. dept. pharmacology U. Tex. Health Sci. Center, San Antonio, 1972—. NIH grantee, 1970-71, 74-77, 77-80, 78-81; NSF grantee, 1978-80; Morrison Trust Found. San Antonio grantee. Mem. Soc. Neurosci., Am. Soc. Pharmacology and Exptl. Therapeutics, Am. Assn. Anatomists, AAAS, San Antonio Audubon Soc. Subspecialties: Neuropharmacology; Neurophysiology. Current work: Pharmacology of synaptic transmission, pharmacology of opiate receptors, synaptic connections of basal ganglia, role of the endogenous opioid peptides in the globus pallidus, biogenic amine supersensitivity in chronic spinal cord injury. Home: 9840 Tower View Helotes TX 78023 Office: U Tex Health Sci Center Dept Pharmacology 7703 Floyd Curl Dr San Antonio TX 78284

HUFT, MICHAEL JOHN, botanist; b. Highland Park, Mich., Jan. 15, 1949; s. John Francis and Louise Myrle (Day) H.; m. Eva Birgitta Svensson, Jan. 15, 1972 (div. May 1983); children: John Eric, Mary Margaret. B.S. in Physics, U. Notre Dame, 1971; Ph.D. in Botany, U. Mich., 1979. Reseach assoc., acting curator herbarium U. Okla., Norman, 1980; postdoctoral curatorial trainee Mo. Bot. Garden, St. Louis, 1980-81, research botanist, 1981—; vis. asst. curator dept. botany Field Mus. Natural History, Chgo., 1981—. Mem. Am. Soc. Plant Taxonomists, Bot. Soc. Am., Sigma Xi. Subspecialty: Taxonomy. Current work: Systematics of the Euphorbiaceae, particularly of the New World tropics; flora of Central America and Mexico. Office: Field Mus Natural History Roosevelt Rd at Lake Shore Dr Chicago IL 60605

HUGGINS, CHARLES, surgical educator; b. Halifax, N.S., Can., Sept. 22, 1901; s. Charles Edward and Bessie (Spencer) H.; m. Margaret Wellman, July 29, 1927; children: Charles Edward, Emily Wellman Fine. B.A., Acadia U., 1920, D.Sc., 1946; M.D., Harvard U., 1924; M.Sc., Yale, 1947; D.Sc., Washington U., St. Louis, 1950, Leeds U., 1953, Turin U., 1957, Trinity Coll., 1965, U. Wales, 1967, U. Mich., 1968, Med. Coll. Ohio, 1973, Gustavus Adolphus Coll., 1975, Wilmington (Ohio) Coll., U. Louisville; LL.D., U. Aberdeen, 1966, York U., Toronto, U. Calif. at Berkeley, 1968; D.P.S., George Washington U., 1967, Bologna U., 1964. Intern in surgery U. Mich., 1924-26, instr. surgery, 1926-27; with U. Chgo., 1927—, instr. surgery, 1927-29, asst. prof., 1929-33, asso. prof., 1933-36, prof. surgery, 1936—, dir. Ben May Lab. for Cancer Research, 1951-69, William B. Ogden Distinguished Service prof., 1962—; chancellor Acadia U., Wolfville, N.S., 1972-79; Macewen lectr. U. Glasgow, 1958, Ravdin lectr., 1974, Powell lectr., Lucy Wortham James lectr., 1975, Robert V. Day lectr., 1975, Cartwright lectr., 1975. Trustee Worcester Found. Exptl. Biology; bd. govs. Weizmann Inst. Sci., Rehovot, Israel, 1973—. Decorated Order Pour le Mérite, Germany; Order of The Sun, Peru).; Recipient Am. Urol. Assn. award for research on male genital tract, 1948; Francis Amory award for cancer research, 1948; AMA gold medals for research, 1936, 40, Société Internationale d'Urologie, 1948; Am. Cancer Soc. award, 1953; Bertner award M.D. Anderson Hosp., 1953; award Am. Pharm. Mfrs. Assn., 1953; gold medal Am. Assn. Genito Urinary Surgeons, 1955; Borden award Assn. Am. Med. Colls., 1955; FRCS (hon.), Edinburg, 1958, London, 1959; Comfort Crookshank award Middlesex Hosp., London, 1957; Charles Mickle fellow Toronto U., 1958; Cameron prize Edinburg U., 1958; Valentine prize N.Y. Acad. Medicine, 1962; Hunter award Am. Therapeutic Soc., 1962; Lasker award for med. research, 1963; gold medal for research Rudolf Virchow Soc., 1964; Laurea and award Am. Urol. Assn., 1966; gold medal in therapeutics Worshipful Soc. Apothecaries of London, 1966; Gairdner award, Toronto, 1966; Nobel prize for medicine, 1966; Chgo. Med. Soc. award, 1967; Centennial medal Acadia U., 1967; Hamilton award Ill. Med. Soc., 1967; Bigelow medal Boston Surg. Soc., 1967; Distinguished Service award Am. Soc. Abdominal Surgeons, 1972; Sheen award A.M.A., 1970. Fellow A.C.S. (hon.), Royal Coll. Surgeons Can. (hon.), Royal Soc. Edinburgh (hon.); mem. Am. Philos. Soc., Nat. Acad. Scis. (Charles L. Meyer award for cancer research 1943), Am. Assn. Cancer Research, Canadian Med. Assn. (hon.), Alpha Omega Alpha. Subspecialty: Cancer research (medicine). Home: 5807 Dorchester Ave Chicago IL 60637 Office: 950 E 59th St Chicago IL 60637

HUGGINS, WILLIAM HERBERT, educator; b. Rupert, Idaho, Jan. 11, 1919; s. William John and Alafretta Evelyn (Roraback) H. B.S., Oreg. State Coll., 1941, M.S., 1942; Sc.D., Mass. Inst. Tech., 1953. Instr. elec. engring. Oreg. State Coll., 1942-44; spl. research asso. radio research lab. Harvard, 1944-46; supervising scientist Air Force Research Center, Cambridge, Mass., 1946-54; research asso. Mass. Inst. Tech., 1949-54; prof. elec. engring. Johns Hopkins, 1954—, chmn. dept., 1970-74; Cons. editor Addison-Wesley Pub. Co., 1957-60, Blaisdell Pub. Co., 1961-65; cons. Rand Corp., 1955-73. Recipient decoration for exceptional civilian service USAF, 1954; Browder J. Thomson Meml. prize Am. I.R.E., 1948; Lindback Found. award for distinguished teaching, 1961; Western Electric Fund award Am. Soc. Engring. Edn., 1965. Fellow I.E.E.E. (Edn. medal 1966), Acoustical Soc. Am., A.A.A.S.; mem. Nat. Acad. Engring., Soc. Indsl. and Applied Math., Phi Beta Kappa, Sigma Xi. Subspecialty: Electrical engineering. Home: One E University Pkwy 1005 Baltimore MD 21218

HUGHES, CARROLL GARVIN, III, physicist; b. Pelzer, S.C., Apr. 4, 1936; s. Carroll Garvin and Mary Louise (Lawson) H.; m. Connie K. Bannister, July 29, 1962; children: Elizabeth, Lesley. B.S. in Physics, Clemson U., 1958. Registered profl. engr., Calif. Physicist Philco Corp., Vandenberg AFB, Calif., 1962-64; mem. tech. staff Hughes Aircraft Co., Culver City, Calif., 1964-67; fellow engr. Westinghouse DEC Standards Lab., Balt., 1967—. Author sci. articles. Served with U.S. Army, 1958-62. Mem. Am. Inst. Physics, Am. Phys. Soc., Optical Soc. Am., AAAS, ASTM. Democrat. Methodist. Subspecialties: Optical radiation measurements; Statistics. Current work: Development of optical radiation measurement standards; measurement techniques; statistical methods; computer-assisted measurements. Home: 1196 Tanager Dr Millersville MD 21108 Office: Box 746 MS550 Baltimore MD 21203

HUGHES, JOHN HENRY, medical virologist, educator; b. Cleve., Jan. 7, 1942; s. James W. and Mary E. (Kostelia) H.; m. Laura Jo Hughes, July 10, 1965; children: Paula Jo, Jennifer, Darrell, Eric. M.A., Bowling Green State U., 1967; Ph.D., Ohio State U., 1972—. Clin. microbiologist Children's Hosp., Columbus, Ohio, 1971; virologist dept. med. microbiology Ohio State U., 1972—, assoc. dir. diagnostic virology, 1981—. Contbr. articles to profl. jours. Mem. AAAS, Am. Soc. Microbiology. Subspecialties: Animal virology; Microbiology (medicine). Current work: Diagnostic virology, antivirals, viral genetics and molecular biology of RNA viruses. Home: 4326 Oak View Dr Columbus OH 43204 Office: 700 Children's Dr Ross Hall Columbus OH 43205

HUGHES, KAREN WOODBURY, geneticist, educator; b. Madison, Wis., Aug. 15, 1940; d. Lowell Angus and Dorothy (Naylor) Woodbury; m. Philip D. Hughes, Dec. 26, 1969. B.S. with honors, U. Utah, 1962, M.S., 1964, Ph.D. in Genetics, 1972. Postdoctoral fellow U. Utah, 1972-73; asst. prof. U. Tenn., Knoxville, 1973-79, assoc. prof. dept. botany, 1979—; assoc. program dir. genetic biology NSF, 1980-81; U.S. Dept. Agr. grantee, 1979-83; mem. adv. bd. Selby Bot. Gardens. Contbr. articles to sci. jours. NIH fellow, 1964. Mem. Tissue Culture Assn. (sec. plant div. 1978-80, v.p. 1980-82, pres. 1982—, exec. bd. 1980-82), Genetics Soc. Am., Am. Genetic Assn., AAAS, Internat. Plant Tissue Culture Assn., Am. Soc. Plant Physiology, Sigma Xi, Phi Kappa Phi, Phi Sigma. Subspecialties: Tissue culture; Genetics and genetic engineering (biology). Current work: Isolation of higher plant mutants which are resistant to high levels of free radicals. Metabolic and genetic characterization of these mutants. Home: 1705 Tonalea Rd Knoxville TN 37919 Office: U Tenn Knoxville TN 37996

HUGHES, RAYMOND HARGETT, physics educator; b. Walla Walla, Wash., June 1, 1927; s. Clifford P. and Frances E. (Hargett) H.; m. Olive J. Wipson, Feb. 8, 1952; children: Diane F. Hughes Huston, Marshall R., Clayton W., Randall C. A.B., Whitman Coll., 1949; M.S., U. Wis., 1951, Ph.D., 1954. Asst. prof. physics U. Ark., Fayetteville, 1954-59, assoc. prof., 1959-65, prof., 1965—; research assoc. Air Force Weapons Lab., Kirtland AFB, N.Mex., 1983—. Contbr. articles to profl. jours. Grantee Air Force, NASA, NSF. Fellow Am. Phys. Soc., Sigma Xi, Phi Beta Kappa. Lodge: Elks. Subspecialty: Atomic and molecular physics. Current work: Laser producediions; propagation of relativistic electrons. Office: Dept Physics U Ark Fayetteville AR 72701 Home: 2510 Wedington Dr Fayetteville AR 72701

HUGHES, THOMAS JOSEPH, mechanical engineering educator, consultant; b. Bklyn., N.Y., Aug. 3, 1943; s. Joseph Anthony and Mae (Bl) H.; m. Susan Elizabeth Weh, July 1, 1972; children: Emily, Ian, Elizabeth. B.S., Pratt Inst., 1965, M.S., 1967; M.S., U. Calif.-Berkeley, 1974, Ph.D., 1974. Mech. design engr. Grumman Aerospace, Bethpage, N.Y., 1965-66; research engr. Gen. Dynamics, Groton, Conn., 1967-69; research engr., lectr. U. Calif., Berkekely, 1974-76; assoc. prof. Calif. Inst. Tech., 1976-80; prof. Stanford U., 1980—. Author: Mathematical Foundation of Elasticity, 1983, A Short Course in Fluid Mechanics, 1976; editor: Computational Methods in Transient Analysis, 1983, Computer Methods in Applied Mechanics and Engineering, 1980. Recipient Bernard Friedman prize U. Calif., Berkeley, 1975. Fellow Am. Acad. Mechanics; mem. ASME (Melville medal 1979), Am. Soc. Civil Engrs. (Huber Research prize 1978), Soc. Engring. Sci., AIAA. Subspecialties: Mechanical engineering; Civil engineering. Current work: Computational methods in solid, fluid, structural and soil mechanics. Office: Division of Applied Mechanics Durand Bldg Stanford University Stanford CA 94305

HUGHES, VICTOR AUGUSTINE, astronomer, educator; b. Manchester, Eng., Mar. 17, 1925; s. Augustine A. and Bessie I. (Robinson) H.; m. Patricia M. Blakely, Mar. 22, 1958 (dec.); m. Joan M. Baird, Sept. 11, 1970; children: Kenneth, Richard, Joanne, Kathryn. B.Sc., U. Manchester, 1944, M.Sc., 1950, D.Sc., 1977. Registered profl. engr., Ont. Sci. officer Army Operational Research Group U.K., 1949-52; Sr./prin. sci. officer Royal Radar Establishment, Malvern, Eng., 1952-60; prin. sci. officer Radio and Space Research Sta., Slough, Eng., 1960-63; prof. physics (astronomy) Queen's U., Kingston, Ont., Can., 1963—; dir. Westar. Natural Scis. and Engring. Contbr. articles to prof. jours. Research Council Can. grantee. Fellow Royal Astron. Soc. (London); mem. Internat. Astron. Union, Am. Astron. Soc., Can. Astron. Soc., Can. Assn. Physicists.

Subspecialties: Radio and microwave astronomy; Star formation. Current work: Solar oscillations, galactic structure star formation. Home: 165 Ontario St Apt 906 Kingston ON Canada K7L 2Y6 Office: Queens U Dept Physics Kingston ON Canada K7L 3N6

HUGHES, WILLIAM PERRY, packaging engineer; b. N.Y.C., Aug. 16, 1946; s. Thomas and Edyth (McKewen) H.; m. Josephine Maria Sutera, Sept. 7, 1968; children: Michelle, Charlene, Thomas. B.S.M.E., Fairleigh Dickinson U., 1968; M.S.M.E., Columbia U., 1973. Registered profl. engr., N.J. Package engr., sr. engr., then project engr. Becton-Dickinson & Co., Rutherford, N.J., 1968-76; mgr. packaging CBS Records div. CBS, Inc., Milford, Conn., 1976-78; mgr. package ops. Wyeth Internat. Ltd., Phila., 1978—. Mem. Packaging Inst., ASME. Lodge: KC. Subspecialties: Mechanical engineering; Systems engineering. Current work: Package systems; machinery technology materials; design, development and installation of materials and machinery to package pharmaceuticals. Home: RD 2 12 Carriage Dr Downingtown PA 19335 Office: PO Box 8616 Philadelphia PA 19101

HUGHETT, PAUL WILLIAM, computer image generation consultant; b. San Rafael, Calif., May 19, 1950; s. Earl Howard and Shirley Helen (Hoitela) H. Student, MIT, 1966-73. Mem. tech. staff Hewlett-Packard, Palo Alto, Calif., 1973-77; system engr. Singer-Link, Sunnyvale, Calif., 1977-81; cons. Dragoncraft, Palo Alto, 1981—. Mem. Spl. Interest Group on Computer Graphics, IEEE, Profl. & Tech. Cons. Assn. Subspecialties: Graphics, image processing, and pattern recognition; Aeronautical engineering. Current work: Real-time computer image generation for flight simulation; pilot visual cues and training value; visual data base design. Home: PO Box 60 Palo Alto CA 94302

HUGUENIN, ROBERT LOUIS, astronomy educator, university administrator; b. East Stroudsburg, Pa., Mar. 26, 1946; s. George Louis and Evelyn (Riday) H.; m. Nancy Hoffman, June 21, 1969. B.S. in Earth and Planetary Scis, U. Pitts., 1969; Sc.D. in Planetary Sci, M.I.T., 1972. NASA trainee M.I.T., 1969-70, grad. research asst., 1970-71, Sloan fellow, 1971-72, research asso., 1971-73, sr. scientist dept. earth and planetary scis., 1973-77; assoc. prof. physics and astronomy U. Mass., 1977—, dir. Remote Sensing Center, 1979—; vis. asso. prof. geol. scis. Brown U., 1977-78; panelist, speaker in field; prin. investigator NOAA/Nat. Marine Fisheries Service contracts, 1981, 82 (2). Contbr. numerous articles to profl. publs.; asso. editor: Proc. Eighth Ann. Lunar and Planetary Sci. Conf., Geochemica et Cosmochema Acta, 1977, Jour. Geophys. Research, 1979-82. NASA grantee, 1976-77, 77-78 (2), 78-82, 78— (2), 80-81; NSF grantee, 1982-84; Am. Sci. and Tech. Corp. grantee, 1982. Mem. AAAS, Am. Astron. Soc., Am. Chem. Soc., Am. Geophs. Union, Am. Phys. Soc., Am. Soc. Photogrammetry, Geochm. Soc., Optical Soc. Am., Soc. Photooptical Instrumentation Engrs., N.E. Area Remote Sensing Systems Assos. (charter asso.), Planetary Remote Sensing Research Consortium, Planetary Spectroscopy and Computing Network, Sigma Xi. Democrat. Methodist. Club: M.I.T. of Boston. Subspecialties: Planetology; Remote sensing (geoscience). Current work: Planetary sci.: research and devel. in chem. weathering and corrosion, heterogeneous catalysis, optical and photo-electronic properties of transition metal solids, remote sensing by reflectance spectroscopy and multispectral imaging, remote sensing instrumentation and commercialization. Home: PO Box 347 Groton MA 01450 Office: Remote Sensing Center U Mass Amherst MA 01003

HUI, CHIU SHUEN, educator; b. Hong Kong, June 7, 1942; U.S., 1966, naturalized, 1981; s. Kwong Chiu and Yee Wan (Choy) H.; m. Siu Lui Hui, Mar. 27, 1975. B.S., U. Hong Kong, 1966; Ph.D., MIT, 1973. Research assoc. Center for Theoretical Studies, U. Miami, Fla., 1973-74; postdoctoral asso. Yale U. Med. Sch., New Haven, 1974-79; asst. prof. physiology Purdue U., W. Lafayette, Ind., 1979—. NIH fellow, 1977-79; Research Career Devel. award, 1983—. Mem. Biophys. Soc., Am. Heart Assn. Subspecialties: Physiology (medicine); Biophysics (biology). Current work: Physiology and biophysics of excitable cells, excitation-contraction coupling in muscle, gating properties of ionic channels. Office: Purdue Univ Dept Biological Scis West Lafayette IN 47907

HUIZENGA, JOHN ROBERT, nuclear chemist, educator; b. Fulton, Ill., Apr. 21, 1921; s. Harry M. and Josie B. (Brands) H.; m. Dorothy J. Koeze, Feb. 1, 1946; children—Linda J., Jann H., Robert J., Joel T. A.B., Calvin Coll., 1944; Ph.D., U. Ill., 1949. Lab. supr. Manhattan Wartime Project, Oak Ridge, 1944-46; instr. Calvin Coll., Grand Rapids, Mich., 1946-47; asso. scientist Argonne Nat. Lab., Chgo., 1949-57, sr. scientist, 1958-67; professorial lectr. chemistry U. Chgo., 1963-67; prof. chemistry and physics U. Rochester, 1967-78, Tracy H. Harris prof. chemistry and physics, 1978—, chmn. dept. chemistry, 1983—; vis. prof. Joliot-Curie Lab., U. Paris, 1964-65, Japan Soc. for Promotion of Sci., 1968; chmn. Nat. Acad. Sci.-NRC Com. on Nuclear Sci., 1974-77. Author: (with R. Vandenbosch) Nuclear Fission, 1973; Contbr. articles to profl. jours. Fulbright fellow, Netherlands, 1954-55; Guggenheim fellow, Paris, 1964-65, Berkeley, Calif., 1973, Munich, W.Ger., 1974, Copenhagen, 1974; recipient E.O. Lawrence award AEC, 1966; named Disting. Alumnus Calvin Coll., 1975. Fellow Am. Phys. Soc., AAAS; mem. Nat. Acad. Sci., Am. Chem. Soc. (award for nuclear applications in chemistry 1975), Phi Beta Kappa, Phi Kappa Phi, Sigma Xi. Current work: Nuclear chemistry; nuclear reactions and structure, including nuclear fission and heavy-ion reaction mechanisms. Home: 51 Huntington Meadow Rochester NY 14625 Office: Dept Chemistry U Rochester Rochester NY 14627

HULBERT, LLOYD CLAIR, plant ecology educator; b. Lapeer, Mich., June 27, 1918; s. L. Claire and Mary (Hungerford) H.; m. Jean Elizabeth Smaltz, June 28, 1952; children: Steven L., Mark J., Thomas A., John R. B.S., Mich. State Coll., 1940, student, 1940-41; student, Cornell U., 1941-42; Ph.D., Wash. State U., 1953. Instr. math U. Minn., 1946-47; instr. botany Mont. State U., 1947-49, U. Minn., 1951-55; asst. prof. div. biology Kans. State U.-Manhattan, 1955-64, assoc. prof., 1964-72, prof., 1972—, dir. Konza Prairie Research Natural Area, 1972—. Contbr. articles to sci. jours.; Co-author: (with F. W. Oehme) Plants Poisonous to Livestock, 3d edit, 1968. NSF Research grantee, 1980—; Biol. Field Facility grantee, 1981; recipient The Nature Conservancy's Oak Leaf award, 1976, Pres.'s Stewardship award, 1978. Subspecialty: Ecology. Current work: Productivity and composition of prairie under various treatments and effects of fire on vegetation. Home: 2323 Bailey Dr Manhattan KS 66502 Office: Div Biology Kans State U Manhattan KS 66506

HULCHER, FRANK HOPE, biochemistry educator, consultant, machinist; b. Hampton, Va., Mar. 12, 1926; s. Frank A. and Elizabeth J. (Keithley) H.; m. Louise Long, June 10, 1953; children: Mark Charles, Alex David, Frank Alan. B.S., Va. Poly. Inst. and State U., 1950, M.S. in Microbiology, 1953, Ph.D. in Biochemistry, 1957; postdoctoral fellow, Yale U., 1957-59. Research asst. Va. Agrl. Expt. Sta., 1952-53; instr. Va. Poly. Inst., Blacksburg, 1953-55; research asst. Yale U., New Haven, 1957-58, research assoc., 1958; instr. neurochemistry Bowman Gray Sch. Medicine, Winston-Salem, N.C., 1958-60, asst. prof., 1960, asst. prof. biochemistry, 1960-68, assoc. prof., 1968—; vis. scientist, research assoc. dept. biology Brookhaven Nat. Lab., Upton, N.Y., 1960; cons. hazardous chems. Contbr. articles to profl. jours. Mem. PTA, 1965—, Reynolda Manor Civic Club, 1974—, Union of Concerned Scientists, Common Cause; mem. Codocil Club, Wake Forest U. Served with USAAF, 1944-46. Grantee NSF, NIH, Nat. Multiple Sclerosis Soc., Kettering Found., United Way. Mem. Am. Chem. Soc., Am. Soc. Microbiology, AAAS, Am. Inst. Chemists, Am. Heart Assn., Am. Soc. Biol. Chemists, Sigma Xi, Phi Sigma, Phi Lambda Upsilon. Episcopalian. Subspecialties: Biochemistry (medicine); Enzyme technology. Current work: Enzymology, kinetics and mechanism, enzymes of cholesterol metabolism, regulation and control. Home: 4149 Wycliff Dr Winston-Salem NC 27106 Office: Bowman Gray Sch Medicine Winston-Salem NC 27103

HULKA, BARBARA SORENSON, medical educator, epidemiologist; b. Mpls., Mar. 1, 1931; d. Herbert Fritchof and Mabel Adelia (Alquist) Sorenson; m. Jaroslav F. Hulka, Nov. 13, 1954; children: Carol Ann, Gregory Fabian, Bryan Herbert. B.A., Radcliffe Coll., 1952; M.S., Juilliard Sch. Music, N.Y.C., 1954; M.D., Columbia U., 1959; M.P.H., 1961. Diplomate: Am. Bd. Preventive Medicine, Nat. Bd. Med. Examiners. Intern USPHS Hosp., S.I., N.Y., 1959-60; asst. pub. health physician Pa. Health Dept., Pitts., 1961-62; research instr. dept. ob-gyn U. Pitts., 1962-65, research asst. prof., 1966-67; research assoc. dept. preventive medicine (Sch. Medicine, U.N.C.), Chapel Hill, 1967-68; asst. prof. dept. epidemiology Sch. Pub. Health, 1967-71, assoc. prof., 1972-76, prof., 1977—, acting chmn. dept., 1982, asst. prof. dept. family medicine, 1968-76, clin. assoc. prof., 1977—; clin. assoc. prof. community and family medicine Duke U. Med. Center, Durham, N.C., 1976—. Contbr. articles to med. jours. USPHS grantee, 1962-67; Nat. Cancer Inst. grantee, 1970-72, 80—; Nat. Inst. Occupational Safety and Health grantee, 1980-82. Mem. Soc. for Epidemiol. Research (exec. com. 1973-77, pres. 1975-76), Am. Pub. Health Assn. (chmn. epidemiology sect. council 1976-77, mem. governing council 1976-78), Am. Epidemiologic Soc., Am. Soc. Preventive Oncology, Am. Coll. Preventive Medicine, Assn. Tchrs. Preventive Medicine, N.C. Pub. Health Assn., Delta Omega (nat. pres. 1979—). Subspecialties: Epidemiology; Health services research. Current work: Epidemiology of cancer, evaluation of health care delivery.

HULL, BRUCE LANSING, vet. surgeon, educator; b. Albany, N.Y., Feb. 15, 1941; s. Lansing and Esther (White) H.; m. Karen Bruner, July 3, 1965; children: Kristina, Kimberly. D.V.M., Cornell U., 1965; M.S., Iowa State U., 1971. Diplomate: Am. Coll. Vet. Surgeons. Pvt. practice vet. medicine, Delmar, N.Y., 1965-66; with Iowa State U., Ames, 1968-76; prof. surgery Ohio State U., Columbus, 1976—. Contbr. to books. Served to capt. AUS, 1966-68. Named Outstanding Tchr. Ohio State U., 1977, Norden Outstanding Tchr. Norden Labs., 1978. Mem. AVMA, Am. Assn. Bovine Practitioners, Assn. Vet. Clinicians. Presbyterian. Lodge: Lions. Subspecialties: Surgery (veterinary medicine); Internal medicine (veterinary medicine). Current work: Experimental surgery in food animals. Office: 1935 Coffey Rd Columbus OH 43210

HULL, CLARK RAMSEY, environmental coordinator, marine biologist, petroleum engineer; b. Harlingen, Tex., Sept. 16, 1951; s. Lawrence Carl and Martha Dean (Ramsey) H. B.S., Tex. Tech. U., 1973. Tchr. Lyford (Tex.) Ind. Sch. Dist., 1973-76; research assoc. Tex. A. and M. U., Galveston, 1976-80; environ. coordinator frontier region Cities Service Co., Houston, 1980—. Mem. Marine Tech. Soc., Soc. Petroleum Industry Biologists. Methodist. Subspecialties: Environmental engineering; Ecology. Current work: Alaska frontier marine mammal population dynamics, oil and gas exploration environmental impact, oil and gas exploration and production severe frontier areas. Office: Cities Service Co 5100 SW Freeway Box 27570 Houston TX 77227

HULL, DIANA, psychologist; b. Lawrence, N.Y., Dec. 27, 1969; d. Louis Albert and Roslyn (Diamont) Jaffree; m. David Pershing Hull, Dec. 27, 1969; children: Marcy Burton, Allison Langdon Boomer. B.A., CUNY, 1946; M.S.W., U. Mich., 1954; Ph.D., U. Tex.-Houston, 1975. Asst. prof. psychology dept. psychiatry Baylor Coll. Medicine, Houston, 1976-80, clin. instr., 1966-76, cons. group psychotherapy program Child Psychiatry Clinic, 1975-80; pvt. practice psychology, Santa Barbara, Calif.; group therapist milieu treatment program and substance abuse program VA Hosp., Houston, 1962-67; cons. Child Guidance Clinic of Houston, 1967-69; editorial cons. Issue on Migration and Health Internat. Migration Rev., 1981. Contbr. articles to profl. jours. Bd. dirs. Phoenix House of Santa Barbara. Recipient sustained superior performance award VA, 1963. Fellow Am. Group Psychotherapy Assn. (Master Instr. award 1976); mem. Am. Psychol. Assn., AAAS, N.Y. Acad. Scis., Tex. Psychol. Assn., Calif. Psychol. Assn., Santa Barbara Area Psychol. Assn., Assn. Media Psychology (sec.), Southwestern Group Psychotherapy Soc., Houston Group Psychotherapy Soc. (mem. faculty and adv. bd. group psychotherapy and family therapy tng. program 1976-80, pres. 1967-69), So. Calif. Demographic Forum, Sierra Club (nat. population com.). Clubs: Birnam Wood Golf, Coral Casino. Subspecialties: Social psychology; Population and environment, media psychology. Current work: Migration, population and environment; health psychology, media psychology, group process. Address: 815 Cima Linda Ln Santa Barbara CA 93108

HULL, HARVARD LESLIE, corp. ofcl.; b. Holstein, Neb., Oct. 23, 1906; s. Joel Leslie and Caroline Evangeline (Larsen) H.; m. Alta Zera Jones, June 9, 1928; children—Gwen Alta Hull Quackenbush, Janet Barbara Hull Clark. A.B. with distinction, Nebr. Wesleyan U., 1927; Ph.D. in Physics, Columbia, 1933. Project engr. Sperry Gyroscope Co., Bklyn, 1933-35, research engr., 1935-40, dir. remote control devel., 1940-43; introduced new equipment Sperry Gyroscope Co., Ltd., Eng., 1934, 35-36; dir. process improvement electromagnetic process of separation Uranium 235 Tenn. Eastman Corp., Oak Ridge, 1943-46; asso. dir. Argonne Nat. Lab., Chgo., 1946-49, dir. remote control engring. devel., 1949-53; v.p. research and devel. div. Capehart-Farnsworth Co., Ft. Wayne, 1953-54; pres. Farnsworth Electronics Co., Co. (div. ITT), Ft. Wayne, 1954-56; v.p. Litton Industries, Beverly Hills, Calif., 1956-57; pres. Hull Assos., Chgo., 1957—; dir., pres. Chgo. Aerial Industries, Inc., Barrington, Ill., 1962-64, Internat. Tech. Corp., Western Springs, Ill.; dir. Central Research Labs., Inc.; Dir. research Aero-Space Inst., Chgo., chmn. bd., 1973—. Fellow AIAA (asso.); mem. IEEE (life), Am. Nuclear Soc. (cert. of appreciation 1978), AAAS, Am. Phys. Soc., Nat. Telemetering Conf. (chmn. 1956), Sigma Xi, Phi Kappa Phi, Theta Chi. Conglist. Club: Executives. Subspecialties: Electronics; Robotics. Current work: Application of electronics, hydraulics and stereo TV to control of automated machinery and other systems. Address: 5223 Caroline Ave Western Springs IL 60558

HULL, RICHARD BAXTER, computer scientist, computer consultant; b. Abington, Pa., Sept. 18, 1953; s. John Laurence and Margaret Elizabeth (Baxter) H. B.A. in Philosophy, U. Calif.-Santa Barbara, 1975, 1975; M.A., U. Calif.-Berkeley, 1977, Ph.D., 1979. Research assoc. dept. computer sci. U. So. Calif., Los Angeles, 1979-80, vis. asst. prof., 1980-81; vis. asst. prof. dept. computer sci. UCLA, 1981-82; asst. prof. dept. computer sci. U. So. Calif., Los Angeles, 1982—; cons. Jet Propulsion Lab./NASA, Pasadena, Calif., 1980-82. Mem. Am. Math. Soc., Assn. Computing Machinery, European Assn. Theoretical Computer Sci., Phi Beta Kappa. Subspecialties: Foundations of computer science; Database systems. Current work: Application of theoretical techniques to study various database issues, including semantic data models, data access and restructuring, and data integrity; data structures; computational complexity; automata theory. Office: Dept Computer Sci U So Calif Los Angeles CA 90089-0782

HULLINGER, RONALD LORAL, veterinary anatomy educator; b. Des Moines, Feb. 4, 1941; m., 1959; 3 children. B.S., Iowa State U., 1964, D.V.M., 1965, M.S., 1966, Ph.D. in Vet. Anatomy, 1968. Asst. prof. vet. anatomy Iowa State U., 1968-69; asst. prof. Purdue U., West Lafayette, Ind., 1969-72, assoc. prof. vet. anatomy, 1972—; vis. assoc. prof. Cornell U., 1974-75, U. Utrecht, Netherlands, 1976-77, U. Ill., 1978-79. Mem. Am. Assn. Vet. Anatomy, AVMA, World Assn. Vet. Anatomy. Subspecialty: Veterinary anatomy. Office: Dept Vet Anatomy Purdue U West Lafayette IN 47907

HULM, JOHN KENNETH, physicist; b. Southport, Eng., July 4, 1923; came to U.S., 1949, naturalized, 1957; s. James and Frances Elizabeth (Goodall) H.; m. Joan Audrey Beatrice Askham, Sept. 25, 1948; children—Clair Frances, Carol Anne, Cherie, Megan, John Allen. B.A. in Natural Scis, Gonville and Caius Coll., Cambridge (Eng.) U., 1943; M.S. in Physics, Cambridge (Eng.) U., 1948; Ph.D. (Dept. Sci. and Indsl. Research research fellow), Cambridge (Eng.) U., 1949. Union Carbide research fellow U. Chgo., 1949-51, asst. prof. physics, 1951-54; mem. staff Westinghouse Research Labs., 1954-74, 76—, dir. systems research, 1972-74, dir. chem. scis. div., 1976—, dir. corp. research, 1980—; sci. attache Am. Embassy, London, 1974-76. Served as officer RAF, 1943-46. Recipient John Price Wetherill medal Franklin Inst. Fellow Am. Phys. Soc. (Internat. prize for new materials 1979); mem. Nat. Acad. Engring., Pitts. Phys. Soc. (pres. 1962-63). Subspecialties: Condensed matter physics; Low temperature physics. Current work: Superconductivity, materials, superconducting machinery, integration of research and development with business strategy. Home: 5636 Woodmont St Pittsburgh PA 15217 Office: Westinghouse Research Labs Pittsburgh PA 15235

HULSE, RUSSELL ALAN, physicist; b. N.Y.C., Nov. 28, 1950; s. Alan Earle and Betty Joan (Wedemeyer) H. B.S., Cooper Union, 1970; M.S., U. Mass., 1972, Ph.D., 1975. Research assoc. Nat. Radio Astronomy Obs., Charlottesville, Va., 1975-77; mem. tech. staff Plasma Physics Lab., Princeton U., 1977-80, staff research physicist II, 1980—. Contbr. articles to profl. jours. Mem. Am. Astron. Soc., Am. Phys. Soc. Subspecialties: Plasma physics; Atomic and molecular physics. Current work: Atomic processes of impurity ions in high-temperature, controlled thermonuclear fusion plasmas, computer modeling of impurity transport, radiation and associated effects in fusion plasmas. Office: Princeton U Plasma Physics Lab Princeton NJ 08544

HULST, DAVID CLARK, agricultural chemical company executive; b. Malone, N.Y., Nov. 13, 1950; s. Edward M. and Dorothy C. (Clarke) H.; m.; children: Bethanie Chantal, Brandon Clark. A.S., Modesto Jr. Coll., 1972; B.S. with honors, Calif. Poly. Inst., 1974; M.S. in Plant Protection, U. Ariz., 1980. Lic. pest control adviser, Calif. Product developer asst. Occidental Chem. Co., summers 1975-77; research asst. U. Ariz., 1976-78; tech. asst. Upjohn Co., then tech. extension assoc., 1978—; ranch mgr. almond and walnuts orchard, Hughson, Calif., 1976—. Mem. Entomology Soc. Am., Am. Phytopathology Soc. Republican. Mem. Evang. Free Ch. Subspecialties: Plant pathology; Integrated pest management. Current work: Fungicide devel. and IPM program management. Address: Route 2 4449 Tully Rd Hughson CA 95326

HUMAYDAN, HASIB SHAHEEN, plant pathologist; b. Ain-Anoub, Lebanon, Mar. 6, 1945; came to U.S., 1971, naturalized, 1982; s. Shaheen Hamad and Amria (Magid) H.; m. Anna Maria DiLiegro, Mar. 8, 1980; 1 son: Michael James. Ph.D. in Plant Pathology and Plant Genetics, U. Wis., 1974. Dir. plant pathology and tissue culture Joseph Harris Co., Inc., Rochester, N.Y., 1974—. Recipient Edgecombe award Am. U. Beirut, 1969. Mem. Am. Phytopathol. Soc., Internat. Soc. Plant Pathologists, N.Y. Acad. Scis., AAAS. Subspecialties: Plant pathology; Plant genetics. Current work: Long term disease control through breeding for multiple disease resistances in the major vegetable crops by utilizing traditional techniques as well as tissue culture, eletrophoresis and induced mutations. Home: 34 Regina Dr Rochester NY 14606 Office: 3670 Buffalo Rd Rochester NY 14624

HUMBERTSON, ALBERT O., JR., anatomist; b. Cumberland, Md., Sept. 30, 1933; s. Albert O. and Eloise Virginia (Daniels) H.; m. Esther I. Cain, June 4, 1956; children: Debra Susan, Allison Ann, Melinda Elise. B.S., Marietta Coll., 1956; postgrad., U. Kans., 1956-58; Ph.D., Ohio State U., 1962. Teaching asst. U. Kans., 1956-58; asst. instr. anatomy Ohio State U., 1958-62, instr., 1962-63, asst. prof., 1963-70, assoc., prof., 1970—. Mem. Am. Assn. Anatomists, Soc. for Neurosci., Soc. Developmental Neursci., AAAS, Sigma Xi. Subspecialties: Neurobiology; Anatomy and embryology. Current work: neuroanatomy, neurohistochemistry. Office: 1645 Neil Ave Columbus OH 43210

HUMES, PAUL EDWIN, animal science educator, researcher, college dean; b. Colville, Wash., Nov. 19, 1942; s. Thornton James and Dorothy Louise (Graeber) H.; m. Tamra Fae Lickfold, June 19, 1965; children: Steven Roy, Krista Leigh. B.S., Wash. State U., 1964; Ph.D., Oreg. State U., 1968. NDEA Title IV fellow Genetics Inst., Oreg. State U., 1965-68; mem. faculty dept. animal sci. La. State U., 1968—, asst. prof. animal sci., 1968-72, assoc. prof., 1972-77, prof., 1977—, assoc. dean, 1983—. Troop treas. Istrouma Area council Boy Scouts Am., 1981-83, troop instl. rep., 1981-83. Mem. Am. Soc. Animal Sci., Nat. Assn. Colls. and Tchrs. AGr., Sigma Xi, Gamma Sigma Delta (La. State U. chpt. Teaching award 1976), Omicron Delta Kappa, Phi Kappa Phi. Republican. Lutheran. Club: La. State U. Sci. Subspecialties: Animal breeding and embryo transplants; Animal genetics. Current work: Germ plasm evaluation of recently introduced cattle breeds; genotype-environment interactions in cattle; embryo transplants and embryo splitting in cattle and sheep. Home: 7516 Cardiff Baton Rouge LA 70808 Office: Dept Animal Sci La State U Baton Rouge LA 70803

HUMI, MAYER, mathematics educator, consultant; b. Baghdad, Iraq, Sept. 29, 1944; came to U.S., 1971, naturalized, 1978; s. Joshua and Salima (Mizrahi) H. M.Sc., Hebrew U., Jerusalem, 1964; Ph.D., Weizmann Inst., Rehovoth, Israel, 1969; also various advanced courses and seminars in computers and math. Asst. prof. U. Toronto, Ont., Can., 1969-71; assoc. prof. math. Worcester (Mass.) Poly. Inst., 1971—; vis. asst. prof. Clark U., Worcester, 1975-76; vis. assoc. prof. Ga. Inst. Tech., 1978-79; cons. in field; program organizer and presenter Scientific and Technological Views program (ednl. TV), Worcester, 1975-76. Author: Methods of Applied Mathematics, 1983; contbr. articles to profl. jours. Mem. Am. Math. Soc., Soc. Indsl. and Applied Math., Am. Phys. Soc., Internat. Assn. Math. Physics. Subspecialties: Applied mathematics; Information systems, storage, and retrieval (computer science). Current work: Research projects: simulation for energy balance of Earth; interactive numerical algorithms and numerical algorithms for differential equations; exact solutions of Einstein Equations; symmetry principles and differential equations. Office: Worcester Poly Inst Institute Rd Worcester MA 01609

HUMMEL, MYRON FLOYD, solar systems designer; b. Sioux City, Iowa, Dec. 26, 1949; s. Floyd Walter Dee and Ruth Harriet Elizabeth (Johnson) H.; m. Joette Louis Hirsch, Oct. 11, 1970; children: Donald, Cassandra, Candace. Student, S.D. Sch. Mines and Tech., 1968-70; A.S., Scott Community Coll., 1981; A.A. with honors, Marycrest Coll., 1982. Cert. engring. technician Nat. Inst. Cert. Engring. Technicians, 1977. Solar systems designer Solaron Corp., Englewood, Colo., 1975; solar designer and estimator Natural Energy Corp., Lakeville, Minn., 1976; tool design checker Sundstrand Hydro-Transmission, Ames, Iowa, 1977-79; solar instr., program chmn. Scott Community Coll., 1979-81; solar designer Solar-Pak Design Div. Town & Country Lumber, Bettendorf, Iowa, 1981-82; mech. technician Shive-Hattery Engrs., Davenport, IA, 1983—; cons. to solar firms. Served with Army ROTC, 1968-70. Recipient Passive Solar award Iowa Energy Policy Council, 1982. Mem. ASHRAE (assoc.), Am. Solar Energy Soc., MECI Soc. Mfg. Engrs. (cert. mfg. technologist), Computer and Automated Systems Assn. Mem. Ch. of God. Subspecialties: Solar energy; Software engineering. Current work: CAD/CAM systems; HVAC systems; solar energy tech.; devel. low cost active passive solar systems and related applied computer analysis and monitoring methods. Home: 526 Wisconsin St Le Claire IA 52753

HUMPHREY, ARTHUR EARL, university administrator; b. Moscow, Idaho, Nov. 9, 1927; s. Samuel Earl and Iris May (Rowe) H.; m. Sheila Claire Darwin, June 13, 1951; children: Andrea Lynn, Allyson Dawn. B.S. in Chem. Engring, U. Idaho, 1948, M.S., 1950, D.Sc. (hon.), 1974; Ph.D., Columbia U., 1953; M.S. in Food Tech, Mass. Inst. Tech., 1959. Mem. faculty U. Pa., Phila., 1953-80, prof. chem. engring., 1961-80, dir., 1962-72, dean, 1972-80; provost, v.p., prof. biochem. engring. Lehigh U., Bethlehem, Pa., 1980—; also co-dir. Center for Biotechnology Research; NSF sci. tchr. fellow Mass. Inst. Tech., 1957-58; Fulbright lectr. U. Tokyo, Japan, 1963, U. New South Wales, Australia, 1970; guest lectr. Inst. Biology, Czechoslovakian Acad. Sci., 1964, Tech. Inst., Budapest, 1966; I.I.T. Delhi, New Delhi, India, 1970, Tungai U., Taichung, Taiwan, 1968; cons. Merck Sharp & Dohme, 1957-63, Merck Chem. Co., 1963-64, 80—, Sun Oil Co., 1961-68, Bioferm, 1964-67, Cryotherm, 1966-67, Fermentation Design, 1967-74, E.R. Squibb, 1967-73, Air Products, 1971—. Author: ann. Fermentation Rev., 1960-64; author textbooks on biochem. engring.; Contbr. articles to profl. jours. Pres. Phila. Trail Club, 1960-61; councilor Appalachian Trail Conf., 1961-67; chmn. space sci. panel Nat. Acad. Sci.; co-chmn. 3d Internat. Fermentation Symposium; mem. engring. adv. bd. NSF; mem. single cell protein working group, protein adv. group WHO-FAO-UN; chmn. group on prodn. substances by microbial means U.S.-USSR Cooperation in Sci. and Tech. Recipient Outstanding Tchr. award U. Pa., 1959. Mem. Nat. Acad. Engring., Internat. Assn. Microbiol. Socs. (sec.-gen. econ. and applied microbiology), Nat. Acad. Engring., Am. Chem. Soc. (chmn. div. microbial. chem. and tech. 1967, Div. Disting. Service award 1979), Am. Inst. Chem. Engrs. (chmn. food and bioengring. div. 1972, Profl. Progress award 1972, Food and Bio-Engring award 1973, Inst. lectr. 1975), Franklin Inst., Japanese Soc. Fermentation Tech., Am. Soc. Microbiology, Sigma Xi, Sigma Tau. Clubs: Trail, Horse-Shoe Trail (Phila.); Appalachian Mountain (Boston). Subspecialties: Enzyme technology; Chemical engineering. Office: Alumni Hall Lehigh U Bethlehem PA 18015 I have found that in many things I do in today's world it is much better to be approximately correct than exactly wrong.

HUMPHREY, FLOYD BERNARD, electrical engineering educator; b. Greeley, Colo., May 20, 1925; m., 1955; 4 children. B.S., Calif. Inst. Tech., 1950, Ph.D. in Chemistry, 1956. Mem. tech. staff Bell Telephone Labs., 1955-60; research group supr. Jet Propulsion Lab., Calif. Inst. Tech., 1960-64, assoc. prof. elec. engring., 1964-71, prof. elec. engring. and applied physics, 1971-80; prof. and head dept. elec. engring. Carnegie-Mellon U., Pitts., 1980—. Fellow IEEE; mem. Am. Phys. Soc., Am. Vacuum Soc. Subspecialty: Applied magnetics. Office: Dept Elec Engring Carnegie-Mellon U Pittsburgh PA 15213

HUMPHREY, JOHN P., chemical engineer; b. Lewistown, Mont.; s. Claude and Mable E. (McGuire) H.; m. Eleanor P. Paulson, June 20, 1954; children: Michael, David, Valerie. B.S. in Chem. Engring, Mont. State U., 1959, Ph.D. Chem. Engring., 1962. Research engr. Dow Chem. Co., Midland, Mich., 1962-65, group leader, 1965-68, research mgr., 1968-78; exec. v.p. Highland Energy, Olive Hill, Ky., 1979—; owner Humphrey Energy Enterprises, Lewistown. Author numerous articles on synthetic fuels. Mem. Am. Inst. Chem. Engrs., Am. Small Research Cos., Lewistown C. of C. (dir.), Sigma Xi, Tau Beta Pi. Presbyterian. Club: Kiwanis (dir.). Subspecialties: Chemical engineering; Mathematical software. Current work: Synthetic fuels; computer science, chemical process technology. Home: 216 7th Ave S Lewistown MT 59457

HUMPHREY, RONALD DE VERE, microbiology educator; b. Denver, Mar. 31, 1938; s. Walter D. and Edna (Carroll) H.; m. Julia Frances Hamlett, Feb. 6, 1960; children: Sean, Lara. B.S. in Biol. Sci, Colo. State U., Ft. Collins, 1960, M.S. in Bacteriology, 1963; Ph.D. in Microbiology, U. Tex.-Austin, 1970. Teaching assoc. U. Tex.-Austin, 1970; assoc. prof. biology Prairie View (Tex.) A&M U., 1970—; summer research appointee Argonne (Ill.) Nat. Lab., 1978. Referee Washington County Youth Soccer Assn. Served to lt. USNR, 1963-66. Grantee Sci. and Edn. Adminstrn. of U.S. Dept. Agr., 1978-83. Mem. Am. Soc. Microbiology, AAAS, N.Y. Acad. Scis., Nat. Inst. Sci., Sigma Xi. Lodge: Brenhem (Tex.) Evening Lions. Subspecialties: Microbiology; Nitrogen fixation. Current work: Physiology of nitrogen fixing legume root nodule bacteria, nitrogen fixation, legume, root nodules, pesticide. Office: Dept of Biology Prairie View A&M University Prairie View TX 77445

HUMPHREYS, ROBERTA MARIE, astronomer, educator; b. Indpls., May 20, 1944; d. Robert R. and Mary C. (Furnas) H.; m. Kris D. Davidson, June 10, 1976; 1 son: Rowan M.H. A.B., Ind. U., 1965; M.S., U. Mich., 1967, Ph.D., 1969. Research assoc. Dyer Obs., Vanderbilt U., Nashville, 1969-70, Steward Obs., U. Ariz., Tucson, 1970-72; asst. prof. U. Minn., Mpls., 1972-76, assoc. prof. astronomy, 1976—; mem. NSF astronomy adv. com., 1981—. Contbr. articles to profl. jours. Alfred P. Sloan Found. fellow, 1976-80. Fellow AAAS; Mem. Am. Astron. Soc., Internat. Astron. Union, Astron. Soc. of Pacific, Assn. Univs. for Research in Astronomy (bd. dirs. 1981—), Phi Beta Kappa. Subspecialties: Optical astronomy; Infrared optical astronomy. Current work: Observational stellar astrophysics; stellar spectroscopy; stellar evolution; massive stars; galactic structure; infrared astronomy. Office: 116 Church St SE Minneapolis MN 55455

HUMPHRIES, JOAN ROPES, psychology educator; b. Bklyn., Oct. 17, 1928; d. Lawrence Gardner and Adele Lydia (Zimmermann) Ropes; m. Charles C. Humphries, Apr. 6, 1957; children: Peggy Ann, Charlene Adele. B.A., U. Miami, 1950; M.S., Fla. State U., 1955; Ph.D., La. State U., 1963. Psychol. cons. East La. State Hosp., Jackson, 1961-62; part-time instr. U. Miami, Coral Gables, 1964; assoc. prof. Miami-Dade Community Coll., 1965—. Prin. author: The Application of Scientific Behaviorism to Humanistic Phenomena, 1975, rev. edit., 1979; contbr. articles to profl. jours. Recipient award of appreciation for continuous, dedicated service to women in community AAUW, 1977. Mem. Internat. Platform Assn. (bd. govs.), Am. Psychol. Assn., AAUP (v.p. 1980, sec. 1981, pres. chpt.), Dade County Psychol. Assn., Phi Lambda. Subspecialties: Learning; Behavioral psychology. Current work: Use of biofeedback techniques to help students learn. Home: 1311 Alhambra Circle Coral Gables FL 33134 Office: 11380 NW 27th Ave Miami FL 33167

HUMPHRIES, THOMAS JOEL, gastroenterologist, pharmaceutical company executive, researcher; b. Worcester, Mass., Mar. 1, 1945; s. Joseph Edward Dozier and Elizabeth Agnes (Grinkis) H.; m. Paula Frances Callahan, June 26, 1971; children: Juliette Hyun Joo, Joshua Chul, Alicia Kelley. A.B., Harvard U., 1966; M.D., Tufts U., 1970. Diplomate: Am. Bd. Internal Medicine, Am. Bd. Gastroenterology. Intern Naval Regional Med. Center, Phila., 1970-71, resident in internal medicine, 1971-73, resident in gastroenterology, 1973-75; commd. ensign U.S. Navy, 1966, advanced through grades to comdr., 1976; dir. student tng. Naval Reg. Med. Center, Phila., 1976-78, head clin. investigation, 1976-78, dir. gastroenterology div., 1976-78; ret., 1978, practice medicine specializing in gastroenterology, Falmouth, Mass., 1978-81; dir. GI Group med. affairs Smith Kline & French Labs., Phila., 1981—; asst. prof. medicine Thomas Jefferson U., Phila., 1976-78; vis. clin. assoc. prof. Hahnemann U., Phila., 1981—. Contbr. articles to med. jours. Served to lt. col. Air NG, 1981—. Fellow ACP, Am. Coll. Gastroenterology; mem. Am. Fedn. Clin. Research, Am. Gastroent. Assn., Am. Soc. for Gastrointestinal Endoscopy. Roman Catholic. Club: Harvard (Boston). Subspecialties: Internal medicine; Gastroenterology. Current work: H2-receptor antagonists and their clinical application to gastrointestinal disease, excellence in endoscopy and its photographic documentation. Home: 6600 Wissahickon Ave Philadelphia PA 19119 Office: Smith Kline & French Labs 1500 Spring Garden St Philadelphia PA 19101

HUNDLEY, LOUIS REAMS, physiology educator, anatomist; b. Albemarle County, Va., May 22, 1926; s. Elijah Dupuy and Louise Agnes (Reams) H.; m. Katheryne Leigh Tindall, July 18, 1953; 1 dau., Mary Louise. B.S., Va. Mil. Inst., Lexington, 1950; M.S., Va. Poly. Inst. now Va. Poly. Inst. and State U., Blacksburg, 1953, Ph.D., 1956. Asst. prof. biology Va. Mil. Inst., 1955-60, assoc. prof., 1960-65, prof., 1965—, acting head dept., 1965-67; vis. prof. Washington and Lee U., 1982, Va. Poly. Inst. and State U., 1980, 81, 82. Crusade chmn. Am. Cancer Soc., Lexington, 1982. Mem. Va. Acad. Sci., Assn. Southeastern Biologists, AAAS, Am. Physiol. Soc., Am. Inst. Biol. Sci., Va. Assn. Biol. Edn. Subspecialty: Genetics and genetic engineering (biology). Current work: Gross body composition and fat free body weight. Home: 4 Ringneck Rd Lexington VA 24450 Office: Dept Biology Va Mil Inst Lexington VA 24450

HUNEYCUTT, JAMES ERNEST, JR., weapon system analyst; b. Gastonia, N.C., Dec. 14, 1942; s. James Ernest and Ida (Cobb) H.; m. Mary Jane Shelton, Dec. 23, 1964. B.S., U. N.C., 1963, M.A., 1965, Ph.D., 1968. Asst. prof. math N.C. State U., Raleigh, 1968-74, assoc. prof., 1974-78; sr. software engr. Hadron, Inc., Vienna, Va., 1978-82; weapon system analyst Applied Physics Lab., Scaggsville, Md., 1982—. Author: Introduction to Probability, 1972. Mem. Am. Math. Soc., Math. Assn. Am., Soc. Indsl. and Applied Math, Sigma Xi. Subspecialties: Applied mathematics; Probability. Current work: Kalman filtering, linear estimation, software engineering, accuracy analysis for trident missiles. Home: 15205 Watergate Rd Silver Spring MD 20904 Office: Applied Physics Lab Scaggsville MD

HUNG, KUEN-SHAN, anatomy educator; b. Chia-Yi, Taiwan, Jan. 13, 1938; came to U.S., 1964, naturalized, 1976; s. Tien-Won and Kuam (Tsai) H.; m. Shirley H. Hwang, Nov. 30, 1968; children: Irene, Melissa. B.S., Nat. Taiwan U., 1960; Ph.D., U. Kans., 1969. Asst. prof. U. S.D., Vermillion, 1969-71, U. So. Calif., 1971-74; asst. prof. U. Kans., Kansas City, 1974-76, assoc. prof. dept. anatomy, 1976-83, prof., 1983—. Pres. Taiwanese Club Greater Kansas City (Mo.), 1976. Served to 2d lt. Taiwanese Army, 1960-61. Am. Lung Assn. research grantee, 1974-76; NIH research grantee, 1976-81; Am. Heart Assn. research grantee, 1981—. Mem. Am. Assn. Anatomists, Am. Soc. Cell Biology, Electron Microscopy Soc. Am., Central States Electron Microscopy Soc. (sec. 1979). Subspecialties: Anatomy and embryology; Cell biology (medicine). Current work: Innervation and development of respiratory system, animal models of pulmonary hypertension, electron microscopy, histochemistry, fluorescent microscopy. Office: U Kans Med Center Dept Anatomy 39th and Rainbow Blvd Kansas City KS 66103

HUNG, PAUL P(ERWEN), pharmaceutical research director, educator; b. Taipei, Taiwan, Republic of China, Sept. 30, 1933; came to U.S., 1955, naturalized, 1963; s. Y. H. and S. C. (Wu) H.; m. Nancy C. Clark, May 4, 1956; children: Pauline E., Eileen K., Clark D. B.S., Millikin U., 1956; M.S., Purdue U., 1958, Ph.D., 1960. Head molecular virology and biology lab. Abbott Labs., North Chicago, Ill., 1960-81, gen. mgr. genetics div. Bethesda Research Labs., Gaithersburg, Md., 1981-82; dir. microbiology div. Wyeth Labs., Radnor, Pa., 1982—; adj. prof. Northwestern U. Med. Sch., Chgo., 1975—; Spl. leave to Stanford U. Med. Sch., Abbott Labs., 1969. Contbr. chpt. to books, acticles to profl. jours. Pres. Chinese Community Orgn., Lake County, Ill., 1976. Mem. Am. Biol. Chemists Soc., Am. Assn. Cancer Research, Am. Soc. Microbiology, Am. Assn. Adv. Sci., Am. Chem. Soc. Subspecialties: Gene actions; Molecular biology. Current work: Genetic engineering of human genes, oncogenic viruses, molecular biology of bacterial viruses, antibiotic biosynthesis. Patentee chemotherapy of cancer. Home: 506 Ramblewood Dr Bryn Mawr PA 19010 Office: Wyeth Labs PO Box 8299 Philadelphia PA 19101

HUNG, TIN-KAN, civil and biomedical engineer, educator; b. Nanking, Republic of China, June 12, 1936; came to U.S., 1961, naturalized, 1976; s. Mao-Hsiang and Yu-Hwa (Cheng) H.; m. Shu-Nan Cho, Feb. 14, 1971; children: Chee-Hahn, Chee-Ming, Chee-Yuen. B.S., Taiwan Cheng-Kung U., 1959; M.S. in Civil Engring, U. Ill., 1962; Ph.D. in Mechanics and Hydraulics, U. Iowa, 1966. Research engr. Inst. Hydraulic Research, U. Iowa, Iowa City, 1966-67; asst. prof. civil and bioengring. Carnegie-Mellon U., Pitts., 1967-71, assoc. prof., 1971-75; research prof. civil and biomed. engring., research prof. neurol. surgery U. Pitts., 1975—. Contbr. articles to profl. jours. Recipient award Nat. Research Council Republic of China, 1979; NSF grantee; NIH grantee. Mem. ASCE (Walter L. Huber prize 1978), Internat. Assn. Hydraulic Research, Urodynamics Soc., Soc. Neurosci., ASME, Sigma Xi, Chi Epsilon. Christian. Subspecialties: Biomedical engineering; Fluid mechanics. Current work: Unsteady viscous flow, hemodynamics, computational fluid mechanics, hydraulics, biomechanics of spinal cord injury. Home: 3900 Glencoe Ct Murrysville PA 15668 Office: U Pitts 836 Benedum Engring Hall Pittsburgh PA 15261

HUNGERFORD, GORDON DOUGLAS, neuroradiology educator, physician; b. Sydney, New South Wales, Australia, June 29, 1940; came to U.S., 1975; m. Sara Cornelia McMillan, Apr. 12, 1980. M.B., B.S., Sydney U., 1963. Intern Royal North Shore Hosp., Sydney, 1964-65, resident, 1966-67; resident in diagbostic radiology Prince Henry Hosp., Sydney, 1968-72; research scholar U. New South Wales, Sydney, 1968-69; Elec. and Musical Industries research fellow Nat. Hosp. for Nervous Diseases, London, 1975, sr. registrar, 1973-74, asst. cons., 1974-75; assoc. prof. radiology dept., chief neuroradiology div. Med. U.S.C., Charleston, 1975—. Contbr. articles in field to profl. jours. Mem. Royal Australasian Coll. Radiologists, Am. Soc. Neuroradiology (sr.), Assn. Univ. Radiologists, Thoracic Soc. Australia, Southeastern Neuroradiol. Soc. (founding sec. 1977-78). Methodist. Club: Snee Farm (Mt. Pleasant, S.C.). Subspecialty: Radiology. Current work: Computed tomography of clinical disorders, embolization techniques in neuroradiology, side-effects of metrizamide dye in the central nervous system. Home: 27 Rutledge Blvd Charleston SC 29401 Office: Neuroradiology Div Med Univ SC 171 Ashley Ave Charleston SC 29425

HUNNINGHAKE, DONALD BERNARD, medical researcher, educator; b. Corning, Kans., 1934. M.D., U. Kans., 1959. Diplomate: Am. Bd. Internal Medicine. Intern U. Kans. Med. Ctr., 1959-60, resident in medicine, 1962-65, fell in clin. pharmacology, 1964-67; mem. staff U. Hosps. of U. Minn., Mpls.; instr. dept. medicine U. Kans., 1965-67; asst. prof. dept. medicine and pharmacology U. Minn., Mpls., 1967-70, assoc. prof., 1970—; also dir. Lipid Research Ctr. Served to capt., M.C. U.S. Army, 1960-62. Fellow ACP; mem. Am. Soc. Pharmacology and Exptl. Therapeutics. Subspecialty: Internal medicine. Office: Lipid Research Center U Minn Minneapolis MN 55455

HUNT, BOBBY RAY, electrical engineer, educator; b. McAlester, Okla., Aug. 24, 1941; s. George Clifford and Shirley Mason (Core) H.; m. Susan Elizabeth Caldwell, Aug. 21, 1965; children: Vicki, Lori. B.Sc., Wichita State U., 1964; M.S., Okla. State U., 1964; Ph.D., U. Ariz., 1967. Systems engr. Sandia Lab., 1967-68; mem. tech. staff Los Alamos Sci. Lab., 1968-75; mem. faculty U. Ariz., Tucson, 1975—, prof. elec. engring. and optical scis., 1978—; chief scientist Sci. Applications, Internat., 1981—; dir. Internat. Imaging Systems. Author book in field; contbr. chpts. to books, articles to profl. jours.; mem. editorial bd. 4 profl. jours. Fellow IEEE; Mem. Optical Soc. Am. Subspecialties: Graphics, image processing, and pattern recognition; Artificial intelligence. Current work: Research, consultation in image processing and pattern recognition. Office: Univ Arizona Dept Elec Engring Tucson AZ 85721

HUNT, ELIZABETH HOPE, psychologist, author, researcher, educator; b. Hattiesburg, Miss., Oct. 14, 1943; d. Emory Spear and Ida Elizabeth (Burkette) H.; m. John Volney Allcott III, Sept. 9, 1978; 1 son, Hunt. A.B., Sweet Briar Coll., 1965; M.S.W., U. Pa., 1971; Ph.D., U. Oreg., 1980. Vol. Peace Corps, Santiago, Chile, 1967-69; civil rights specialist HEW, Phila., 1971-74; research asst. U. Oreg., Eugene, 1978-79; lectr., 1980, pvt. practice psychology, Eugene, 1980—; workshop presented Oreg. Mental Health Assn., Eugene, 1980. Contbr. articles to profl. jours. Bd. dirs. Lane County Relief Nursery for Abused and Neglected Children, Eugene, 1982—; co-chmn. Speakers Bur., Physicians for Social Responsibility, Eugene, 1982. Fellow U. Oreg., 1974-77; Nat. Inst. Handicapped Research grantee, 1977-79. Mem. Am. Psychol. Assn., Oreg. Psychol. Assn., Lane County Psychologists Assn. (pub. affairs com. 1981—), Profl. Women's Network. Current work: Past writing chiefly behavioral; current research and writing on theory of human development, life cycle problems, diversity possible within normal range of development. Home: 2650 Cresta De Ruta Eugene OR 97403 Office: Oreg Family Center 3225 Willamette St Suite 3 Eugene OR 97405

HUNT, HAROLD RUSSELL, chemistry educator; b. Evansville, Ind., Mar. 22, 1932; s. Harold Russell and Verna Louise (Hasseries) H.; m. Doris June Stephens, July 14, 1956; children: Gregory Charles, Rebecca Louise. A.B., Harvard U., 1953; Ph.D., U. Chgo., 1957. Asst. prof. Ga. Inst. Tech., Atlanta, 1957-64, assoc. prof., 1964—, coordinator undergrad. studies in chemistry; cons. in field. Author: Introductory Chemistry Laboratory Manual, 1974. Mem. Am. Chem. Soc., Sigma Xi. Lutheran. Subspecialties: Inorganic chemistry; Kinetics. Current work: Kinetics and mechanisms of reactions of coordination compounds, chemical education, curriculum planning, administration. Office: Sch Chemistry Ga Inst Tech Atlanta GA 30332

HUNTER, CHRISTOPHER, mathematician, educator, researcher; b. Manchester, Eng., May 28, 1934; came to U.S., 1960, naturalized, 1971; s. William and Helen Mary (Potts) H.; m. Hilda May Salmon, Dec. 23, 1961; children: James, Alison, Rosemary, Andrew. B.A., Cambridge (Eng.) U., 1957, Ph.D., 1960. Research assoc. M.I.T., Cambridge, 1960-61, lectr., 1961-62, asst. prof., 1964-65, assoc. prof., 1965-70; research fellow Trinity Coll., Cambridge U., 1962-64; prof., dir. applied math. Fla. State U., Tallahassee, 1970—; vis. fellow Joint Inst. Lab. Astrophysics, Boulder, Colo., 1976-77. Served to 2d lt. Brit. Army, 1952-54. Recipient Mayhew prize Cambridge U., 1957, Smith's prize, 1959; NSF grantee, 1970—. Fellow Royal Astron. Soc.; mem. Soc. Indsl. and Applied Math. (pres. regional sect. 1980), Math. Assn. Am., Am. Astron. Soc. Episcopalian. Subspecialty: Applied mathematics. Current work: Dynamics of fluids and stellar systems, asymptotic methods and applied complex analysis, computer methods. Home: 1540 Marion Ave Tallahassee FL 32303 Office: Dept Math and Computer Sci Fla State Univ Tallahassee FL 32306

HUNTER, ERIC, microbiologist, educator; b. Guisborough, Yorkshire, Eng., Aug. 18, 1948; came to U.S., 1972; s. Alan Mason and Gladys Emily (Lamb) H.; m. Deborah Williams, Apr. 6, 1979. B.Sc. with honors, U. Birmingham, Eng., 1969; Ph.D., Brunel U., Eng., 1972. Postdoctoral fellow U. So. Calif., Los Angeles, 1972-76; asst. prof. microbiology U. Ala., Birmingham, 1976-80, assoc. prof., 1980—, assoc. scientist, 1976—; ad hoc grant reviewer NIH. Contbr. chpts. to books, articles to profl. jours. NIH grantee. Mem. Am. Soc. for Virology, Am. Soc. Biol. Chemists. Subspecialties: Molecular biology; Virology (biology). Current work: Molecular genetics and biochemistry of tumor viruses. Office: 520 11th St S Birmingham AL 35294

HUNTER, HARRY LAYMOND, physician, pharmaceutical company executive; b. Girard, Kans., Mar. 7, 1923; s. Adolphus Osborne and Mary Elizabeth (White) H.; m. Louise R. Leone, Aug. 19, 1949 (dec. July 25, 1982); children: John Patrick, Mary Anne. A.B., U. Ill., Urbana, 1944; B.S., U. Ill.-Chgo., 1944, M.D., 1946. Diplomate: Am. Bd. Internal Medicine. Intern Gorgas Hosp., C.Z., 1946-47, resident in internal medicine, 1947-48; resident Ill. Central Hosp., Chgo., 1949-50, U. Mich., Ann Arbor, 1950-51; assoc. chief medicine Blanchard Valley Hosp., Findlay, Ohio, 1951-52; dir. exec. health Ill. Central Hosp., Chgo., 1953-57, assoc. chief medicine, 1957-64, chief med. officer, 1968-74; clin. assoc. prof. medicine U. Ill. Coll. Medicine, Chgo., 1953-76; assoc. dir. clin. pharmacology Abbott Labs., North Chicago, Ill., 1965-67, med. dir., 1975-76; dir. clin. studies Mead Johnson & Co., Evansville, Ind., 1976—. Contbr. numerous articles to profl. jours. Bd. dirs. Ill. Council on Alcoholism, Chgo., 1970-74, Am. Cancer Soc., Evansville, 1978-80. Served to capt. U.S. Army, 1943-49; Panama. Fellow ACP, Am. Soc. Clin. Oncology, Chgo. Soc. Internal Medicine, Chgo. Inst. Medicine; mem. AMA. Subspecialties: Internal medicine; Oncology. Current work: involves clinical development of new drugs and new therapeutic indications. Home: 4141 Orchard Rd Evansville IN 47721 Office: Mead Johnson & Co 2404 Pennsylvania Evansville IN 47721

HUNTER, JAMES EDWARD, nutritionist; b. Phila., May 4, 1945; s. James Bruce and Ruth Moyer (Lenker) H.; m. Marilyn Kay Jones, Aug. 24, 1968; children: Melanie Kay, Timothy Edward. B.S. in Chemistry, Lehigh U., 1967; M.S. in Biochemistry, U. Wis.-Madison, 1969, Ph.D., 1974. Lab. technician Campbell Soup Co., Camden, N.J., summers, 1966-67; staff nutritionist Procter & Gamble Co., Cin., 1974—; mem. biol. subcom., tech. com. Inst. Shortening and Edible

Oils, Washington, 1981—; assisted in organizing symposium on diet and exercise AMA, Orlando, Fla., 1981. Contbr. articles to profl. jours.; editor: booklet Food Fats and Oils, 1982. Mem. phys. edn. com. Powel Crosley Jr. YMCA, Cin., 1978—, mem. com. mgmt., 1980—, mem. exec. com. and rec. sec. com. mgmt., 1982—. Served with U.S. Army, 1969-71. NIH trainee, 1967-69, 71-74. Mem. Am. Chem. Soc. (chmn. profl. relations com. cin. chpt. 1981—), Am. Inst. Nutrition, Am. Heart Assn., Greater Cin. Nutrition Council, Phi Beta Kappa, Sigma Xi, Tau Beta Pi. Clubs: Masters Swim Team (pres. 1978—), Pacemakers Running, Lehigh U. Alumni (v.p. 1981—), Cincinnati). Subspecialties: Nutrition (biology); Food science and technology. Current work: Lipid nutrition, fats and oils processing, dietary fat and health, iron availability, diet/exercise. Home: 423 Flemridge Ct Cincinnati OH 45231 Office: Procter & Gamble Co 6071 Center Hill Rd Cincinnati OH 45224

HUNTER, JOHN STUART, statistician, consultant; b. Holyoke, Mass., June 3, 1923; s. John and Irene (Robinson) H.; m. Edna Taylor Martz, Sept. 19, 1952; 1 dau., Jean Bartlett; m. T.J. Hirasuna, Aug. 13, 1977; children: William Mark, Anne Robinson. Ph.D. in Exptl. Stats., N.C. State U., 1954, M.S. in Engring. Math, 1949, B.S. in Elec. Engring, 1947. Staff statistician Am. Cyanamid Co., 1954-59; with Statis. Techniques Research Group, 1957-59, Math. Research Center, U. Wis., 1959-61; assoc. prof. Princeton, 1962-67; prof. engring. Princeton U., 1968-82, prof. emeritus, 1982—; statistician in residence U. Wis., 1967-68; lectr. Korean Standards Research Inst., 1979, Nat. Center: Indsl. Sci. and Tech. Mgmt. Devel., Dalian, People's Republic of China, 1981, 82; Mem. staff com. nat. statistics Nat. Acad. Scis., 1975-76, mem. com., 1976—, chmn. com. pres.'s of statis. socs., 1976-79; chmn. panel Nat. Bur. Standards, 1977-80. Author, cons., lectr. in field.; Founding editor: Technometrics, 1959-63. Served with AUS, 1942-46. Fellow Am. Statis. Assn., Am. Soc. Quality Control (Shewhart medal, 1971, Youden award 1977, Ott award 1979), AAAS (council mem. 1974-77, chmn. com. on fellows 1977); mem. Biometrics Soc., Inst. Math. Statistics, Am. Inst. Chem. Engrs., ASTM, Am. Soc. Indsl. Engrs., ASCE, Royal Statis. Soc., Internat. Statistics Inst. Episcopalian. Club: Cosmos. Subspecialties: Statistics; Operations research (mathematics). Current work: Statistical design and analysis of experiments industrial process and quality enhancement. Home: 100 Bayard Ln Princeton NJ 08540

HUNTER, RICHARD EDMUND, research plant pathologist; b. Jersey City, Jan. 26, 1923; s. Frederick William and Margaret (Dahlgren) H.; m. Earline Clark; children: Catherine Hunter Hays, Margaret Hunter Roberts, Richard Clark. B.S., Rutgers U., 1949; M.S., Okla. State U., 1951, Ph.D., 1968. Asst. in biology N.Mex. State U., 1951-55; instr. Okla State U., Stillwater, 1958-68, asst. prof., 1968-71, assoc. prof., 1971-72; research plant pathologist U. S. Dept. Agr., Stillwater, 1968-72, Nat. Cotton Pathology Lab., U.S. Dept. Agr., College Station, Tex., 1972-75; research plant pathologist, nut prodn. unit Southeastern Fruit and Nut Tree Lab., U.S. Dept Agr., Byron, Ga., 1975-79, research leader, 1976-79; research leader, location leader W.R. Poage Pecan Field Sta., U.S. Dept. Agr., Brownwood, Tex., 1979—. Contbr. articles to profl jours. Active Presbyn. Ch., Stillwater, 1958-72; ch. sch. tchr., mem. adminstrv. bd. First United Meth. Ch., Brownwood, 1979—. Served to capt. USAAF, 1943-46. Mem. Am. Phytopathol. Soc., Am. Soc. Hort. Scientists, Sigma Xi, Alpha Zeta, Phi Sigma. Subspecialties: Plant pathology; Plant genetics. Current work: Host resistance and genetic vulnerability to pecan diseases; in charge Department of Agriculture pecan genetics and improvement program. Home: 3903 Glenwood Dr Brownwood TX 76801 Office: WR Poage Pecan Field Sta 701 Woodson Rd Brownwood TX 76801

HUNTER, ROBERT L., pathology educator; b. Chgo., Jan. 27, 1939; m., 1965; 2 children. A.B., Harvard U., 1961; M.S. and M.D., U. Chgo., 1965, Ph.D. in Pathology, 1969. Staff assoc. immunology Nat. Inst. Allergy and Infectious Diseases, NIH, 1967-69; from instr. to asst. prof. pathology U. Chgo., 1970-76, assoc. prof., 1976—, Seymour Coman fellow, 1969-71. Am. Cancer Soc. fellow, 1970-72; Schweppe Found. fellow, 1971-74; recipient 3d award SAMA-Mead Johnson Sci. Forum, 1965; Sheard Sanford award Am. Assn. Clin. Pathologists, 1965; Joseph A. Capps prize Inst. Medicine Chgo., 1969; Hektoen award Chgo. Path. Soc., 1970. Mem. AAAS, Reticuloendothelial Soc. Subspecialty: Immunology (medicine). Office: 4837 S Kenwood Ave Chicago IL 60615

HUNTER, ROBERT OLIN, JR., laser scientist; b. Riverside, Calif., Nov. 1, 1946; m., 1974; 2 children. B.S., Stanford U., 1967; Ph.D. in Physics, U. Calif.-Irvine, 1981. Nuclear research officer USAF Aero. Propulsion Lab., 1968-70, Weapons Lab., 1970-72; sr. staff scientist Maxwell Labs., Inc., 1973-78; pres. Western Research Corp., San Diego, 1978—. Subspecialty: Laser physics. Address: 8616 Commerce Ave San Diego CA 92121

HURD, EDWARD NELSON, III, management consultant, nuclear engineer; b. Sewickley, Pa., Apr. 14, 1926; s. Austin Avery and Hannah (Shaffer) H.; m. Marjorie L. Harrison, Sept. 5, 1949; children: Marion H. Leonard, Edward N. B.S., U. Pitts., 1950. Testing mgr. Naval Reactor Facility, Idaho Falls, Idaho, 1961-72; U.S. rep. to Japan's Liquid Metal Fast Breeder Reactor program Westinghouse Co., Dept. of Energy, Mito, Japan, 1978-79; prin. engr. Fast Flux Test Facility, Richland, Wash., 1972-77, 79-81, Nuclear Quality Assurance Program Office, Westinghouse Hanford Co., Richland, 1981—; cons. Dept. Energy, Nuclear Quality Assurance Program Office, Richland, 1981—. Chmn. Local Homeowners Assn., Richland, 1980—. Served with U.S. Army, 1944-46; ETO. Mem. Am. Nuclear Soc. Republican. Episcopalian. Club: Meadow Springs Country (Richland). Subspecialties: Nuclear fission; Nuclear engineering. Current work: Nuclear cons. Dept of Energy Hdqrs. and field office programs. Home: 517 Greenbrook Pl Richland WA 99352 Office: Westinghouse Hanford Co PO Box 1970 Richland WA 99352

HURD, RALPH EUGENE, chemist; b. Columbus, Ohio, May 19, 1950; m., 1971; 2 children. B.A., Calif. State Coll.-San Bernardino, 1973; Ph.D. in Biochemistry, U. Calif.-Riverside, 1978. Research asst. biochemist U. Calif.-Riverside, 1979-80; applied chemist Nicolet Magnetics Corp., 1980—. Mem. Am. Chem. Soc. Subspecialty: Nuclear magnetic resonance (chemistry). Office: 145 E Dana St Mountain View CA 194041

HURD, WALTER LEROY, JR., aerospace engineer, consultant; b. Columbus, Mont., July 8, 1919; s. Walter Leroy and Mary Daisy (Gibbon) H.; m. Ann Vivian Cornell, Sept. 26, 1941; children: David, Caroline, Drew, Bruce, Kevin. B.A., Morningside Coll., 1940; M.A., San Jose State U., 1977. Registered profl. engr., Calif. Vice-pres., gen. ops. mgr., pilot Philippine Airlines, Manila, 1946-54; quality mgr. Nat. Motor Bearing, Inc., Redwood City, Calif., 1955-58; reliability and quality engring. mgr. Lockheed Missile & Space Corp., Sunnyvale, Calif., 1959-65, electronics product assurance dir., 1966-67; dir. space systems product assurance Lockheed Corp., Burbank, Calif., 1967-77, corp. dir. product assurance, 1977—; dir. Assurance Scis. Found., San Jose, Calif., 1976—, Laura T. Barbour Flight Safety Award, Washington, 1982—; ex-officio dir. Flight Safety Found., Washington, 1981—; chmn. quality assurance com. Aerospace Industries Assn., Washington, 1983. Contbr.: chpt. to Reliability Handbook, 1966. Served to lt. col. USAAF, 1941-46; to brig. gen. Res.; ret. Decorated Air medal (4), D.F.C. (3). Fellow Am. Soc. Quality Control (B.L. Lubelsky award 1971, E.L. Grant award 1974, nat. pres. 1977-78), Brit. Interplanetary Soc.; academician Internat. Acad. Quality; assoc. fellow AIAA; assoc. mem. European Orgn. Quality Control; mem. Space Studies Inst., Nat. Space Inst., Nat. Soc. Profl. Engrs.; hon. mem. Philippine, Australian, N.Z. quality control socs. Democrat. Unitarian. Lodge: Elks. Subspecialties: Aerospace engineering and technology; Space colonization. Current work: quality, reliability and maintainability assurance for Lockheed products and services including spacecraft, missiles, aircraft, ships and electronic systems; space industrialization. Office: Lockheed Corp PO Box 551 Burbank CA 91520 Home: 18120-82 Oxnard St Tarzana CA 91536

HURFORD, GORDON JAMES, astrophysicist; b. Montreal, Que., Can., Dec. 24, 1942; s. James G. and Alice M. (Murray) H.; m. Sheryl May Scherer, July 17, 1970; 1 son: Patrick L. B.Sc., McGill U., 1963; M.A., U. Toronto, 1964; postgrad., M.I.T., 1964-66; Ph.D., Calif. Inst. Tech., 1974. Lectr. Xavier Coll., Sydney, N.S., 1966-68; research fellow Calif. Inst. Tech., Passdena, 1974-77, assoc. scientist, 1977-79, scientist, 1979-80, mem. profl. staff, 1980—. Contbr. articles to profl. jours. Mem. Am. Astron. Soc. (solar physics div.), Can. Assn. Physicists. Subspecialties: Solar physics; Radio and microwave astronomy. Current work: Development and use of observational techniques, particularly microwave interferometry, for study of solar activity and the structure of solar atmosphere. Office: MC 264-33 Caltech Pasadena CA 91125

HURKMAN, WILLIAM JAMES, II, plant physiologist, cell biologist; b. Oakland, Calif., Sept. 8, 1947; s. William James and Marian Theresa (Meyer) H.; m. Marion Frances McGuckin, June 6, 1970; 1 son, Robert William. B.S. in Biology, U. Wis., Whitewater, 1970, M.S. in Botany, 1974; Ph.D. in Plant Physiology, Purdue U., 1979. Grad. research asst. and teaching asst. dept. botany U. Wis., Milw., 1973-74; dept. botany and plant pathology Purdue U., W. Lafayette, Ind., 1974-79; postdoctoral research assoc. dept. botany and microbiology U. Okla., Norman, 1979-82; plant physiologist Dept. Agr. Western Regional Research Ctr., Berkeley, Calif., 1982—. Contbr. articles to profl. jours. Served with U.S. Army, 1970-72. Decorated Bronze Star. Mem. Am. Soc. Plant Physiologists. Roman Catholic. Subspecialties: Plant physiology (agriculture); Molecular biology. Current work: Characterization of cellular membranes and the identification of mechanisms of plant tolerance to environmental stresses, particularly that of salinity. Office: USDA Western Regional Research Center 800 Buchanan St Berkeley CA 94710

HURLEY, THOMAS JOHN, JR., chemical manufacturing corporation physicist; b. Denver, Sept. 27, 1928; s. Thomas John and Emma (Rose) H.; m. Louanne Kathleen Murphy, Nov. 7, 1953; children: Thomas John III, Regina M., Elizabeth T., Roberta L., David D., Patrick M., Anne B. B.S., St. Marys (Calif.) Coll., 1950; M.S., U. Notre Dame, 1953. Engr. Sperry Gyroscope Co., Great Neck, N.Y., 1953-54; physicist E.I. DuPont de Nemours & Co., Inc., Aiken, S.C., 1956—. Umpire Boys Baseball of Aiden, 1967—; troop com. chmn. Boy Scouts Am., Aiken, 1969-75. Served with Signal Corps AUS, 1954-56. Mem. Am. Nuclear Soc. (treas. Aiken chpt. 1977-76, exec. com. 1980-81). Subspecialties: Nuclear fission; Nuclear reactor process control. Current work: Computer applications for safety and control of all phases of nuclear reactor operations. Home: 1140 Williams Dr Aiken SC 29801 Office: E I DuPont de Nemours & Co Inc Savannah River Plant Aiken SC 29801

HURT, VALINA KAY, science educator, researcher; b. Bowling Green, Ky., Jan. 3, 1953; d. Ottis C. and Geraldine (Andrew) H. B.S. (Regents' scholar), Western Ky. U., 1975, M.S., 1979; postgrad., U. Okla., 1979—. Dental asst., Bowling Green, 1974, 75; grad. teaching asst. Western Ky. U., Bowling Green, 1978, 79, U. Okla., Norman, 1979-83; instr. sci. Hazard Community Coll., (Ky.), 1983—. Contbr. articles on botany to profl. jours. Okla. Mining and Mineral Resources Research Inst. scholar, 1980-83. Mem. AAUP, Ecology Soc. Am., Ky. Acad. Sci., Am. Genetic Assn., AAAS, Assn. for Women in Sci., Bot. Soc. Am., Southeastern Assn. Biologists, Okla. Acad. Sci., Southwestern Assn. Naturalists, Sigma Xi, Alpha Epsilon Delta, Beta Beta Beta, Beta Sigma Phi. Republican. Subspecialties: Ecology; Resource management. Current work: Strip mine revegetation and reclamation, allelopathy, old field succession, ecotypes, family size and sex preference, heterogeneity of variances. Home: PO Box 250 Combs KY 41729 Office: Hazard Community Coll Hazard KY 73069

HURWITZ, JERARD, microbiology educator, researcher; b. N.Y.C., Nov. 20, 1928; m., 1950; 2 children. B.A., Ind. U., 1949; Ph.D. in Biochemistry, Case Western Res. U., 1953. Research asst. biochemist Western Res. U., 1949-50; Am. Cancer Soc. fellow Nat. Inst. Med. Research, London, 1953-56; instr. Sch. Medicine, Washington U., St. Louis, 1956-58; asst. prof. Sch. Medicine, NYU, 1958-59, assoc. prof. microbiology, 1960-63; prof. molecular biology Albert Einstein Coll. Medicine, Bronx, N.Y., 1963-67, prof. developmental biology and cancer, chmn. dept., 1967—. USPHS sr. fellow, 1956—; Charles Mickle fellow, Can., 1967; Guggenheim fellow Inst. Pasteur, Paris, 1968; recipient Eli Lilly award, 1962; Am. Cancer Soc. Prof. award, 1963; Brown-Hazen lectr., 1972. Fellow N.Y. Acad. Sci.; mem. Nat. Acad. Sci., Am. Soc. Biol. Chemists, Brit. Biochem. Soc., Sigma Xi. Subspecialties: Molecular biology; Microbiology. Office: Albert Einstein Coll Medicine Bronx NY 10461

HUSAIN, SYED, pharmacologist; b. Hyderabad, India, Dec. 9, 1939; came to U.S., 1962, naturalized, 1978; s. Syed Fazal and Ruquia (Begum) H.; m. Dilnaz Moinuddin, Dec. 25, 1968; children: Adil, Asif, Dilshad. B.Sc., Osmania U., Hyderabad, India., 1959; B.S., U. Wis.-Madison, 1965; M.S., U. Mo., Columbia, 1968, Ph.D., 1971. Postdoctoral research assoc. Ind. U. Med. Sch., Indpls., 1971-74; biochem. pharmacologist Stanford Research Inst., Menlo Park, Calif., 1974-76; asst. prof. pharmacology U. N.D., Grand Forks, 1976-81, assoc. prof., 1981—, mem. spl. review com., 1982—; cons. Valley Family Planning Center, Grand Forks, 1981; mem. admissions com. Sigma Xi, 1982—. Recipient Scholarship U. Wis., Madison, 1963; named Hon. Citizen City of Indpls., 1972; Nat. Inst. Drug Abuse grantee, 1974, 80; NIH grantee, 1974, 82. Mem. Am. Soc. Pharmacology and Exptl. Therapeutics, Am. Council for Drug. Edn., Soc. Exptl. Biology and Medicine, Internat. Soc. Biochem. Pharmacology, Sigma Xi. Subspecialties: Pharmacology; Toxicology (medicine). Current work: Possible antidote for phencyclidine (PCP) toxicity, phencyclidine, methaqualone and marijuana interaction, biochemical bases for the effects of marijuana on male reproductive functions. Office: Dept Pharmacology U ND Grand Forks ND 58201

HUSAIN, ZAKIUD DIN, petroleum research engineer; b. Dhaka, Bangladesh, Jan. 23, 1947; came to U.S., 1976, naturalized, 1982; s. Muhammad Jasimud Din and Husn Ara Jasmin; m. Shamsun Nahar Husain, Dec. 2, 1970; 1 son, Zeeshan Shahriar. B.S. in Mech. Engring, Bangladesh U. Engring. and Tech., Dhaka, 1968, M.S., 1974; Ph.D. in Mech Engring, U. Houston, 1982. Lectr. mech. engring. Bangladesh U. Engring. and Tech., 1969-73, asst. prof., 1973-75; research fellow dept. mech. engring. Aerodymaics and Turbulence Lab., U. Houston, 1976-82, lectr. central campus, 1982—; sr. research engr., corp. research and devel. Daniel Industries, Inc., Houston, 1982—; cons.; organizing sec. Tchrs. Assn., Bangladesh U. Engring. and Tech., 1973. Author in field. Vice pres. Bangladesh Assn., Houston, 1976. Nat. Merit scholar, Bangladesh, 1957-62, 62-64; with distinction, 1964-68. Assoc. mem. ASME, AIAA, Inst. Engrs. Bangladesh; mem. Tau Beta Pi. Moslem. Subspecialties: Fluid mechanics; Mechanical engineering. Current work: Fluid mechanics; flow products development; instrumentation; computer-aided data acquisition; analysis; experimental fluid mechanics research. Office: Corp Research and Devel Daniel Industries Inc 9720 Old Katy Rd Houston TX 77055

HUSKEY, ROBERT JOHN, geneticist; b. San Antonio, Dec. 29, 1938; s. William Forrest and Irene (Staubus) H.; m. Marilyn Bee Reynolds, June 3, 1960; children: Margot Susan, Samuel Jonathan. B.S., U. Okla., 1960, M.S., 1962; Ph.D., Calif. Inst. Tech., 1968. Asst. prof. microbial genetics Syracuse (N.Y.) U., 1967-69; asst. prof. biology U. Va., Charlottesville, 1969-74, assoc. prof., 1974—, assoc. dean, 1982—. Mem. AAAS, Am. Soc. Cell Biology, Genetics Soc. Am., Soc. Developmental Biology. Subspecialties: Genetics and genetic engineering (biology); Developmental biology. Current work: Researcher in genetic regulation of cellular determination and differentiation. Office: U Va Dept Biology Charlottesville VA 22901

HUSSAIN, A.K.M. FAZLE, mechanical engineering educator, researcher; b. Dhaka, Bangladesh, Jan. 20, 1943; came to U.S., 1965; s. Mohammed Tarik Ullah and Begum Farhatunesa; m. Rehana, Nov. 10, 1968; 1 dau., Shama. B.S. in Mech. Engring, Bangladesh U. Engring and Tech., Dhaka, 1963, M.S., Stanford U., 1966, Ph.D., 1969. Design engr. Bangla Dock Ltd. Dhaka, 1959-63; lectr. Bangladesh U., 1963-65; vis. asst. prof. John Hopkins U., Balt., 1969-71; asst. prof. U. Houston, 1971-73, assoc. prof., 1973-76, prof., 1976—. Assoc. editor: Physics of Fluids, Am. Inst. Physics, 1980—; asst. editor: Turbulence in Liquids, Biennial Symposium, 1975—. Recipient Eckert prize Stanford U., 1971; Outstanding Researcher award Cullen Coll. Engring, U. Houston, 1979; Fulbright scholar, 1965; August Berner Honor fellow, 1966; Freeman scholar ASME, 1984. Assoc. fellow AIAA; mem. Am. Phys. Soc., ASME. Democrat. Moslem. Subspecialties: Mechanical engineering; Aerospace engineering and technology. Current work: Fluid mechanics, turbulence, aerodynamics, aeroacoustics, hydrodynamic stability, biofluid dynamics. Office: Dept Mech Engring U Houston TX 77004

HUSSAIN, MOAYYED A., mathematician; b. Poona, India, Feb. 25, 1937; came to U.S., 1960; s. Abdulhussain and Bujabai (Amiralt) H.; m. Jan. 9, 1966; children: Naveed, Amena. B.E., U. Poona, 1959; M.S., Rensselaer Poly. Inst., 1962, Ph.D., 1965. Research mech. engr. U.S. Army, Watervliet, N.Y., 1962-79; research mathematician Gen. Electric Research, Schenectady, 1979—. Recipient Achievement award U.A. Army, 1972-73. Mem. ASME, Am. Math. Soc., Soc. Applied Mat., Sigma Xi. Subspecialties: Applied mathematics; Fracture mechanics. Current work: Applied mathematics, numerical analysis, symbolic computation. Home: 4 Dennin Dr Menands NY 12204 Office: Gen Electric Co Research and Devel PO Box 8 Schenectady NY 12301

HUSSEY, RICHARD SOMMERS, plant pathologist, educator; b. Wheeling, W.Va., Dec. 18, 1942; s. Walter John and Eleanor (Sommers) H.; m. Griscelda Gray, Jan. 20, 1944; children: Matthew, Kristen. A.B., Miami U., 1965; M.S., U. Md., 1968, Ph.D., 1970. Chief nematologist N.C. Dept. Agr., Raleigh, 1973-74; research assoc. dept. plant pathology N.C. State U., Raleigh, 1970-73; prof. nematology, dept. plant pathology U. Ga., Athens, 1974—. U.S. Dept. Agr. research grantee, 1979-83; U.S. Dept. Agr./AID internat. tng. grantee, 1978-79. Mem. Soc. Nematologists, Sigma Xi, Phi Sigma, Gamma Sigma Delta. Subspecialty: Plant pathology. Current work: Control of plant parasitic nematodes - chem. and plant resistance, biology of nematodes, host-parasite relations. Office: Dept Plant Pathology U Ga Athens GA 30602

HUSTED, RUSSELL FOREST, researcher, educator; b. Lafayette, Ind., Apr. 4, 1950; s. Robert Forest and Miriam Ruth (Jackson) H.; m. Nancy Lee Driscoll, Oct. 25, 1969; children: Jacqueline Marie, Randall Forest. B.S., Colo. State U., 1972; Ph.D., U. Utah, 1976; postgrad., U. Iowa, 1976-79. Asst. research scientist U. Iowa, 1979-81, asst. research scientist, 1982—; asst. prof. in residence U. Conn., Hartford, 1981-82. Mem. Soc. Gen. Physiology, Am. Physiol. Soc., Am. Soc. Nephrology, N.Y. Acad. Sci., Sigma Xi. Subspecialties: Physiology (medicine); Nephrology. Current work: Mechanisms of ion transport and acidification across urinary epithelia. Home: Rural Route 1 Box 8M North Liberty IA 52317 Office: U Iowa 317 Med Labs Iowa City IA 52242

HUSTON, DAVID PAUL, physician, researcher; b. Tallahassee, Fla., June 11, 1947; s. Ralph B. and Elsie (Beall) H.; m. Marilyn McConnell, Dec. 20, 1978; children: Kimberly B., Virginia B. B.S., Wofford Coll., 1969; M.D., Bowman Gray Sch. Medicine, 1973. Intern, resident in medicine Baylor Affiliated Hosps., Baylor Coll. Medicine, Houston, 1973-76; chief med. resident Meth. Hosp., Baylor Coll. Medicine, Houston, 1976-77; research fellow in immunology Baylor Affiliated Hosps., 1977; clin. assoc. arthritis and rheumatism br. Nat. Inst. Arthritis, Metabolism and Digestive Diseases, NIH, Bethesda, Md., 1977-80, co-dir. plasma-cytophoresis unit, 1979-80; asst. prof. medicine clin. immunology sect. Baylor Coll. Medicine, 1980—; dir. immunotherapy Meth. Hosp., Houston, 1980—, asst. prof. microbiology and immunology, immunology sect., 1983—. Contbr. articles to profl. jours. Duke Found. scholar, 1971. Mem. Am. Rheumatology Assn., Am. Assn. Immunologists, AAAS, ACP, Harris County Med. Soc., Tex. Med. Assn., Am. Fedn. Clin. Research. Methodist. Subspecialties: Immunology (medicine); Allergy. Current work: Autoimmune diseases, regulation cytotoxic lymphocyte responses. Home: 2715 Talbot Houston TX 77005 Office: 6565 Fannin Houston TX 77030

HUSTON, NORMAN EARL, nuclear engineering educator; b. Jefferson, Iowa, Jan. 24, 1919; s. Sherburn Sherwood and Helen Isadore (Briggs) H.; m. Mary Belle Felton, June 27, 1943; children—Norman Earl, Anne Marie (Mrs. Thomas Daniel Sigerstad), Susan Deane (Mrs. Alan Braddock). Student, Los Angeles City Coll., 1938-40; A.B., U. Calif. at Berkeley, 1943; Ph.D., U. So. Calif., 1952. Registered profl. engr., Wis., Calif. Physicist U. Calif. Radiation Lab., Berkeley, 1943-44; research engr., dir. dept. Atomics Internat. div. N.Am. Aviation Co., Los Angeles, 1950-65; sr. sci. adviser, asst. to v.p. Autonetics div., Anaheim, Calif., 1966; prof. nuclear engring., dir. Instrumentation Systems Center, U. Wis., Madison, 1966—; dir. Adv. Center for Med. Tech. and Systems, 1972-78; also dir. Ocean Engring. Labs., 1967-70, U. Wis.-NASA Biomed. Application Team, 1974-77; mem. subcom. of com. on interaction of engring. in biology and medicine Nat. Acad. Engring.; reviewer, cons. NSF, AID, project leader mission to Cairo, Mgmt. Lab. Instrn., 1976; project leader Egypt Sci. and Tech. Project, Nat. Research Center for AID and NSF, Cairo, 1977; UN expert on mission Singapore Inst. Standards and Indsl. Research, 1972; mem. NSF task force on instrumentation requirements to Nat. Research Centre, Egypt, 1974, NSF team mission to Cairo, Rome, workshops sci. intruments, 1975. Contbg. author: Summary of Reactor Design, 1955, Progress in Nuclear Energy, Series VI, Vol. 3, 1959; editor: Management Systems for Lab. Instrument Services, 1980; mem. editorial bd.: Jour. Internat. Measurement Confedn., 1982—; contbr. articles to profl. jours. Served to lt. (j.g.) USNR, 1944-47. Levi Strauss scholar, 1941-43. Fellow AAAS (council mem.), Instrument Soc. Am. (v.p. edn. and research 1971-72, v.p. publs. 1974-76, sec. 1977-78, pres. 1978-79), Inst. Measurement and

Control; mem. Am. Nuclear Soc., Am. Soc. Engring. Edn., Am. Phys. Soc., Nat. Mgmt. Assn. (v.p. NAA Valley chpt., treas. 1958-59), Western Boys Baseball Assn. (pres. Woodland Hills 1959), Sigma Xi, Alpha Delta Sigma, Alpha Tau Omega. Club: Rotary (Madison). Subspecialties: Instrumentation; Nuclear engineering. Current work: Scientific instruments-management systems and research and development. Patentee in field. Home: 4556 Winnequah Rd Monona WI 53716 Office: 1500 Johnson Dr Madison WI 53706

HUSTON, RONALD LEE, engineering educator; b. Somerset County, Pa., Aug. 5, 1937; s. Charles Virgil and Pauline (Brubaker) H.; m. Barbara Ann Howe, July 30, 1956; children: Thomas Richard, Dryver Roy, Suzanne Huston Phillips. B.S., U. Pa., 1959, M.S., 1961, Ph.D., 1962. Registered profl. engr., Ohio. Asst. prof. U. Cin., 1962-66, assoc. prof., 1966-70, prof., 1970—, dir., 1970—, acting sr. v.p., provost, 1982—; div. dir. NSF, Washington, 1978-79; dir. Inst. Space Sci., U. Cin., 1970-78, Inst. Applied Interdisciplinary Research, 1978—; reviewer Applied Mechanics Reviews, N.Y.C., 1970—. Editor: Aircraft Crashworthiness, 1975, Recreational Vehicle Dynamics, 1980, Wheeled Recreational Vehicles, 1981. Recipient Profl. Accomplishment award Tech. Soc. Cin., 1980; NSF grantee, 1975-83; NASA grantee, 1977-84; Office Naval Research grantee, 1972-83. Mem. ASME, Soc. Automotive Engrs. (Teetor award 1980), Soc. Indsl. and Applied Math., AIAA, Am. Soc. Engring. Edn. Democrat. Subspecialties: Theoretical and applied mechanics; Mechanical engineering. Current work: Dynamics, biomechanics, robotics, multibody systems, crash victim simulation. Home: 6242 Savannah Ave Cincinnati OH 45224 Office: U Cin Location 72 Cincinnati OH 45221

HUT, PIET, astrophysicist; b. Utrecht, Netherlands, Sept. 26, 1952; came to U.S., 1981; s. Jan Lambertus and Jenneke Johanna (Broekroelofs) H.; m. Helen Haskell Northrop, Dec. 26, 1981. M.S., U. Utrecht, 1977; Ph.D., U. Amsterdam, 1981. Researcher Inst. for Theoretical Physics, U. Utrecht, 1977-78, Astron. Inst., U. Amsterdam, 1978-81; mem. Inst. for Advanced Study, Princeton, N.J., 1981—. Contbr. articles to profl. jours. Mem. Am. Astron. Soc., Netherlands Astronomers Club. Mem. Zen Community N.Y. Subspecialties: Theoretical astrophysics; Cosmology. Current work: Stellar dynamics, especially gravitational scattering in the three-body problem. Home: 109H S Olden Ln Princeton NJ 08540 Office: Inst for Advanced Study Princeton NJ 08540

HUTCHINGS, JOHN BARRIE, astronomer; b. Johannesburg, South Africa, July 18, 1941. B.Sc., Witwatersrand U., 1962, 1963, M.Sc., 1964; Ph.D., Cambridge U., 1967. NRC fellow Dominion Astrophys. Obs., Victoria, B.C., Can., 1967-69, research scientist, 1969—; sr. research officer NCR Can., 1979—. Contbr. articles to profl. jours. Mem. Am. Astron. Soc., Internat. Astron. Union, Royal Acad. Scis. Subspecialties: Optical astronomy; Ultraviolet high energy astrophysics. Current work: Optical, UV, X-Ray, radio investigation of stellar X-ray sources, massive stars, quasars and active galaxies.

HUTCHINGS, WILLIAM FRANK, mech. engr.; b. Rochester, N.Y.; s. Frank N. and Margaret (Remmely) H.; m. Marcia Jean Hutchings; children: Lori Ann, Michael, Susan. B.S. in Mech. Engring, Case Western Res. U., 1960; M.S. in Mech. and Aero. Scis, U. Rochester, 1969. Design engr. Calspan Corp., Buffalo, 1961-66; supervising engr. Consol. Vacuum Corp., Rochester, N.Y., 1966-70; dir. product devel. engring. Mixing Equipment Co., Rochester, 1970—. Mem. ASME, ASTM. Democrat. Subspecialties: Mechanical engineering; Solid mechanics. Current work: Rotating machinery design. Patentee fluid and solids mixing devices. Office: 135 Mt Read Blvd Rochester NY 14611

HUTCHINSON, DONALD PATRICK, plasma physicist; b. Laurel, Miss., Jan. 15, 1947; s. Ira Julius and Mamie Esther (Bush) H.; m. Beverly Jean Hooker, Sept. 2, 1967; children: Donald Christopher, Julie Rebecca. B.S.E.E., U. Miss., 1968; M.S. in Nuclear Engring, MIT, 1970; Sc.D. in Applied Plasma Physics, MIT, 1974. Physicist Oak Ridge Nat. Lab., 1974—, group leader advanced plasma dianostics, 1979—. Contbr. articles to profl. jours. Mem. Am. Phys. Soc., Sigma Xi. Baptist. Subspecialties: Plasma physics; Laser design. Current work: Development of far-infrared lasers for plasma diagnostics. Office: Oak Ridge Nat Lab PO Box X Bldg 6003 Oak Ridge TN 37830

HUTCHINSON, FRANK DAVID, III, engineering company executive; b. Bronxville, N.Y., June 10, 1929; s. Frank David and Frances Viola (Garfield) H.; m. Diane Mary O'Keefe, Oct. 28, 1950; children: Diane Mary, Peter John, Donna Marie. B.C.E., Cooper Union U., 1953; diploma, Internat. Inst. Nuclear Sci. and Engring., 1961. Registered profl. engr., 23 states. With Gibbs & Hill, Inc., N.Y.C., 1950—, v.p. engring., 1975-79, sr. v.p., 1979—; chmn. adv. com., nuclear engring. dept. Rensselaer Poly. Inst., Troy, N.Y., 1978—. Recipient Volme Freis Lecture award Renssalear Poly. Inst., 1979. Mem. ASME (chmn. exec. com. 1975-76, Leadership award 1977), Am. Nuclear Soc., AAAS, ASCE, ASTM. Republican. Presbyterian. Subspecialties: Nuclear engineering; Graphics, image processing, and pattern recognition. Current work: Management of operations, sales and marketing. Home: 536 Pine Acres Blvd Brightwaters NY 11718 Office: Gibbs & Hill Inc 393 7th Ave New York NY 10001

HUTCHINSON, WILLIAM BURKE, surgeon, research foundation executive; b. Seattle, Sept. 6, 1909; s. Joseph Lambert and Nona Bernice (Burke) H.; m. Charlotte Martha Rigdon, Mar. 25, 1939; children: Charlotte Hutchinson Reed, William Burke, John L., Stuart, Mary Hutchinson Wiser. B.S., U. Wash., 1931; M.D., McGill U., 1936; H.H.D. (hon.), Seattle U., 1982. Diplomate: Am. Bd. Surgery. Active staff surgery Swedish Hosp., Seattle, from 1941; pres., founding dir. Pacific N.W. Research Found., 1956—, Fred Hutchinson Cancer Research Ctr., Seattle, 1972-81, pres., chmn. bd., 1981—; clin. prof. surgery U. Wash. Med. Sch., 1974-83; now prof. emeritus; mem. Yarborough Com., 1970; pres. 13th Internat. Cancer Congress, 1978-82. Contbr. articles to med. jours. Recipient First Citizen of Seattle award, 1976. Mem. ACS, Pacific Coast Surg. Assn., King County Med. Soc., Western Surg. Assn., N. Pacific Surg. Assn., AMA, Seattle Surg. Soc. Clubs: Men's Univ., Seattle Golf. Subspecialty: Surgery. Current work: Cancer research. Home: 7126 55th Ave S Seattle WA 98118 Office: 1124 Columbia St Seattle WA 98104 also: 1309 Summitt Seattle WA 98104

HUTCHISON, GEORGE BARKLEY, epidemiologist, educator; b. Lexington, Ky., Oct. 18, 1922; s. George Barkley and Aliena Hale (Hunter) Hutchinson. A.B., Harvard U., 1943, M.D., 1951, M.P.H., 1960. Intern Mass. Meml. Hosps., Boston, 1951-52; resident in medicine Lahey Clinic, Boston, 1952-55; asst. med. dir. Equitable Life Assurance Soc. U.S., N.Y.C., 1955-56, Health Ins. Plan Greater N.Y., 1956-58; dir. research N.Y.C. Dept. Health, 1958-59; epidemiologist Michael Reese Hosp., Chgo., 1966-71; assoc. prof. radiology U. Chgo., 1970-71; asst. to assoc. prof. epidemiology Harvard U. Sch. Pub. Health, Boston, 1960-66, prof., 1972—; cons. NIH, 1982-83, Nat. Acad. Scis., 1982-83. Contbr. over 40 sci. articles to profl. publs. Served to capt. U.S. Army, 1943-46. NIH grantee, 1961-74, 67-85, 78-82; WHO grantee, 1961-74. Mem. Am. Pub. Health Assn., N.Y. Acad. Sci., N.Y. Acad. Medicine, Am. Statis. Assn. Democrat. Subspecialties: Epidemiology; Preventive medicine. Current work: Epidemiology of radiation effects including malignancy; epidemiology; radiation effects; oncology; preventive medicine. Office: 677 Huntington Ave Room 928 Boston MA 02115

HUTCHISON, JAMES ROBERT, chemist, educator; b. Wooster, Ohio, July 23, 1940; s. Virgil O. and Lola H. H.; m. Margaret Llewellyn, Aug. 6, 1940; children: Claire, Robert, David, Matthew. B.S., Wittenberg U., 1962; Ph.D., Princeton U., 1967. Instr. Swarthmore Coll., 1966-68, asst. prof. chemistry, 1968-73; research assoc. M.I.T., 1969-70; assoc. prof. Alma Coll., 1973-79, prof., 1979—, chmn. dept. chemistry, 1978—; adj. prof. Mich. State U., 1979-81. Contbr. articles to profl. jours. Mem. Am. Chem. Soc., Sigma Xi. Presbyterian. Subspecialties: Inorganic chemistry; Physical chemistry. Current work: Nuclear magnetic resonance of metal nuclides; rapid intramolecular rearrangement reactions. Office: Alma Coll Alma MI 48801

HUTTLESTON, DONALD GRUNERT, plant taxonomist; b. Homer, N.Y., June 28, 1920; s. Carey Frank and Helen Augusta (Grunert) H. B.S., Cornell U., 1942, M.S., 1948, Ph.D., 1953. Curator of herabarium Bklyn. Bot. Garden, 1950-55; taxonomist Longwood Gardens, Inc., Kennett Square, Pa., 1955—; adj. prof. U. Del., 1968—. Served with U.S. Army, 1942-46. Mem. Internat. Soc. Plant Taxonomists, Am. Soc. Plant Taxonomists, Am. Fern Soc., Bot. Soc., Am. Inst. Biol. Scis. Subspecialty: Taxonomy. Current work: Taxonomy of ornamental plants. Home: 409 Greenwood Rd Kennett Square PA 19348 Office: Longwood Gardens Kennett Square PA 19348

HUYSER, EARL STANLEY, chemist, educator, cons.; b. Holland, Mich., May 27, 1929; s. Stanley Q. and Gertrude (Westra) H.; m. Barbara Lou VanKolken, June 26, 1952; children: Nancy, Thomas, David, Gretchen. A.B., Hope Coll., 1951; Ph.D., U. Chgo., 1954. Postdoctoral fellow Columbia U., 1956-57; research chemist Dow Chem. Co., 1957-59, cons., 1959—; asst. prof. U. Kans., Lawrence, 1959-63, asso. prof., 1963-67, 1967—. Author: research, numerous publs. in field General College Chemistry; editor: Methods of Free Radical Chemistry. 5 vols, 1969-74. Served with U.S. Army, 1955-57. Recipient Teaching award Amoco, 1975; NSF sr. postdoctoral fellow, 1964-65. Mem. Am. Chem. Soc. Republican. Presbyterian. Club: Cosmopolitan (Lawrence). Lodge: Elks (Lawrence). Subspecialty: Organic chemistry. Current work: Reactions of free radicals. Patentee in field. Home: 1821 W 21st St Lawrence KS 66044 Office: Dept Chemistry U Kans Lawrence KS 66045

HWANG, DAH-MIN DAVID, physicist, educator; b. Taiwan, Nov. 3, 1946; s. Ying-shu and Shih-Jau (Yang) H.; m. Huey-Luen Ma, June 26, 1971; children: Guang-Yue, Guang-Iong. B.S., Nat. Taiwan U., 1968; Ph.D., U. Chgo., 1973. Assoc. prof. physics Nat. Tsing Hua U., Taiwan, 1973-78; vis. scholar U. Chgo., 1978-80; asst. prof. U. Ill., Chgo., 1980—. Contbr. articles to profl.jours. Mem. Am. Phys. Soc. Subspecialty: Condensed matter physics. Current work: Experimental condensed matter physics; graphite, intercalated graphite, acoustics, optics, electron microscopy, Raman spectroscopy. Office: Dept Physics U Ill Box 4348 Chicago IL 60680

HWANG, WILLIAM G., communications systems engineer, operations researcher; b. Shanghai, China, June 17, 1949; came to U.S., 1956. B.S., Calif. Inst. Tech., 1971; M.S., Cornell U., 1974, Ph.D., 1978. Assoc. mgr. THe BDM Corp., McLean, Va., 1978-82; sr. engr. M/A-COM LINKABIT, Inc., McLean, 1982—. Mem. IEEE, Soc. Indsl. and Applied Math., Am. Math. Soc., Armed Forces Communications-Electronics Assn. Subspecialties: Communications systems engineering; Operations research (engineering). Current work: Satellite communications, communications systems, command, control and communications, systems control, system identification. Office: M/A Com Linkabit Inc 1517 Westbranch Dr 4th Floor McLean VA 22102

HYDE, BEAL BAKER, botany educator; b. Dallas, June 26, 1923; s. Mark Powell and Alice Beal (Baker) H.; m. Margaret Lynn Powe, Aug. 20, 1947; children: Margaret Lynn, Thomas Beal, Alice Robbins. Student, Amherst Coll., 1941-43; A.B., Harvard U., 1948, Ph.D., 1952. Research assoc. Ind. U., Bloomington, 1951-54; asst. prof. botany U. Okla., 1954-60; research fellow Calif. Inst. Tech., Pasadena, 1960-63; vis. assoc. prof. U. Tex., 1963-64; prof., chmn. dept. botany U. Vt., Burlington, 1965—. Contbr. articles to sci. jours. Trustee Unitarian Ch., Children's Ctr.; active United Way. Served to 2d lt. USAAF, 1943-46. NSF fellow, 1961-62. Mem. AAAS, Am. Soc. Cell Biology, Genetics Soc. Am., Sigma Xi. Democrat. Unitarian. Subspecialties: Genome organization; Plant genetics. Current work: Chloroplast DNA and inheritance ultrastructure of pollen. Home: 334 S Union St Burlington VT 05401 Office: Life Science U Vt 230 Marsh Burlington VT 05405

HYDE, JANET SHIBLEY, psychology educator; b. Akron, Ohio, Aug. 17, 1948; d. Grant Ohio and Dorothy Mae (Reavy) Shibley; m. Clark Hyde, June 2, 1969; children: Margaret, Luke. B.A., Oberlin Coll., 1969; Ph.D., U. Calif.-Berkeley, 1972. Asst. prof. psychology Bowling Green State U., 1972-76, assoc. prof., 1976-78; assoc. prof. psychology Denison U., 1979-83, prof., 1983—. Author: Half the Human Experience: The Psychology of Women, 1980, Understanding Human Sexuality, 1982. NIMH grantee, 1975-78; recipient Academic Excellence award Bowling Green State U., 1976. Mem. Am. Psychol. Assn., Soc. Research in Child Devel., Am. Assn. Sex Educators, Counselors and Therapists, Behavior Genetics Assn., Sigma Xi. Democrat. Episcopalian. Subspecialties: Developmental psychology; Psychology of women. Current work: Psychology of women; role of language in gender-role development. Home: 98 W Central Ave Delaware OH 43015 Office: Denison U Granville OH 43023

HYDE, KENNETH EDWIN, chemist, educator, computer engineer; b. McKees Rocks, Pa., July 26, 1941; s. Kenneth Edwin and Helen (Czajkowski) H.; m. Elaine Wenderholm, Aug. 11, 1973; 1 dau., Valerie. B.S. in Chemistry, Carnegie-Mellon U., 1963, Ph.D., U. Md., 1968. Instr. Univ. Coll., U. Md., 1967-68; asst. prof. chemistry SUNY Coll., Oswego, 1968-71, asso. prof., 1971-79, prof., 1979—; applications programming, computer applications Gen. Electric Co., Syracuse, N.Y.; vis. asso. prof. SUNY Univ. Center, Buffalo, 1974-76; vis. scientist Inst. Phys. Chemistry, U. Frankfurt, W.Ger., summer 1976. Contbr. articles primarily in inorganic chemistry to jours. SUNY Research Found. grantee, 1969-71, 73-74, 77-80; NSF grantee, 1971-73; Am. Chem. Soc. Petroleum Research Fund grantee, 1973-76; Research Found. Cottrell grantee, 1976-78; recipient Chancellor's award for Excellence in Teaching, 1981. Mem. Am. Chem. Soc., N.Y. Acad. Sci., Nat. Sci. Tchrs. Assn., Sigma Xi. Subspecialties: Inorganic chemistry; Software engineering. Current work: Applications of micro-computers in the laboratory environment; inorganic reaction mechanism; magnetic properties of transition metal complexes. Home: 214 Riverdale Rd Liverpool NY 13088 Office: Dept Chemistry or Dept Computer Sci SUNY Coll Oswego NY 13126

HYDE, RICHARD MOOREHEAD, immunologist, educator; b. Pierre, S.D., Feb. 11, 1933; s. Charles Lee and Florence (Moorehead) H.; m. Ruth Marie Curry, Aug. 21, 1953; children: Roderick M., Scott R., Dawn Marie Hyde Smith. B.A., U. S.D., 1955, M.A., 1956; Ph.D., U. Minn., 1962. Asst. prof. San Francisco State Coll., 1960-62, U. Mo., Columbia, 1962-65; prof. microbiology, immunology Okla. U., Oklahoma City, 1965—; cons. Inst. Microecology, Herborn, W.Ger., 1978—. Author: Immunology, 1978; contbr. articles to profl. jours. USPHS fellow, 1959-60; NSF fellow, 1955-56; USPHS spl. fellow, 1968-69. Mem. AAAS, Am. Soc. Microbiology, Am. Assn. Immunologists, Soc. Exptl. Biology and Medicine, Reticuloendothelial Soc., Sigma Xi. Republican. Methodist. Lodge: Shriners. Subspecialties: Immunology (medicine); Infectious diseases. Current work: Innate immunity; autoimmune diseases.

HYMAN, ALBERT LEWIS, physician; b. New Orleans, Nov. 10, 1923; s. David and Mary (Newstadt) H.; m. Neil Steiner, Mar. 27, 1964; 1 son, Albert Arthur. B.S., La. State U., 1943; M.D., 1945; postgrad., U. Cin., U. Paris, U. London, Eng. Diplomate: Am. Bd. Internal Medicine. Intern Charity Hosp., 1945-46, resident, 1947-49, sr. vis. physician, 1959-63; resident Cin. Gen. Hosp., 1946-47; instr. medicine La. State U., 1950-56, asst. prof. medicine, 1956-57; asst. prof., Tulane U, 1957-59, asso. prof., 1959-63, asso. prof. surgery, 1963-70, prof. research surgery in cardiology, 1970—, adj. prof. pharmacology, 1974—; dir. Cardiac Catheterization Lab., 1957—; sr. vis. physician Touro Hosp., Touro Infirmary, Hotel Dieu; chief cardiology Sara Mayo Hosp.; cons. in cardiology USPHS, New Orleans Crippled Children's Hosp., St. Tammany Parish Hosp., Covington La. area VA, Hotel Dieu Hosp., Mercy Hosp., East Jefferson Gen. Hosp., St. Charles Gen. Hosp.; electrocardiographer Metairie Hosp., 1959-64, Sara Mayo Hosp., Touro Infirmary, St. Tammany Hosp.; cons. cardiovascular disease New Orleans VA Hosp.; cons. cardiology Baton Rouge Gen. Hosp.; Barlow lectr. in medicine U. So. Calif., 1977. Contbr. articles to profl. jours. Recipient award for research of the Hadassah, 1980. Fellow ACP, Am. Coll. Chest Physicians, Am. Coll. Cardiology, Am. Fedn. Clin. Research; mem. Am. Heart Assn. (fellow council on circulation, mem. council on cardiopulmonary medicine, regional rep. council clin. cardiology, chmn. sci. com. of cardiopulmonary council 1981, chmn. cardiopulmonary council, bd. dirs.), La. Heart Assn. (v.p. 1974, Albert L. Hyman Ann. Research award), Am. Soc. Pharmacology and Exptl. Therapeutics, So. Soc. Clin. Investigation, So. Med. Soc., Am. Physiol. Soc., N.Am. Soc. Pacing and Electrophysiology, N.Y. Acad. Scis., AAUP. Subspecialties: Cardiology; Pharmacology. Current work: Director of surgical cardiopulmonary research laboratory at Tulane Medical School engaged in research in pulmonary circulation; interest is diverted toward pharmacology; physiology, and biochemistry of pulmonary blood flow. Research in cardiopulmonary circulation. Home: 5550 Jacquelyn Ct New Orleans LA 70124 Office: 3629 Prytania St New Orleans LA 70115

HYMAN, CAROL BRACH, pediatric hematologist-oncologist; b. South Orange, N.J., Jan. 21, 1923; d. Leon S. and Madeline E. (Rosenthal) Brach; m. Maurice M. Hyman, Apr. 14, 1951; children: Gayle, J. Madeline, Celia. B.A., Cornell U., 1943, M.D., 1947. Diplomate: Am. Bd. Pediatrics with subsplty in pediatric hematology-oncology. Intern Newark City Hosp., 1947-49; resident Cornell-N.Y. Hosp., 1949-50, Mt. Sinai Hosp., N.Y.C., 1950-51, Children's Hosp. of Los Angeles, 1952-54, staff, 1954-80; faculty U. So. Calif., 1954—, asso. prof. clin. pediatrics, 1980—; pediatric hematologist/oncologist Cedars-Sinai Med. Center, Los Angeles, 1980—; also dir. Thalassemia-Chronic Ironload Overload Program. Author articles, book chpts. in pediatric hematology-oncology. Recipient 25 Yr. Service award Children's Hosp. of Los Angeles, 1980. Mem. AMA, Am. Assn. Cancer Research, Am. Soc. Hematology, Calif. Med. Assn., Internat. Soc. Hematology, Los Angeles County Med. Soc., Los Angeles Pediatric Soc., Western Soc. Pediatric Research, Women in Bus., Sierra Club, Nat. Audubon Soc., Cornell U. Alumni Assn., Cornell U. Med. Sch. Alumni Assn. Jweish. Subspecialties: Hematology; Oncology. Current work: Pediatric hematology-oncology. Present research in Thalassemia (Cooley's Anemia) and chronic iron overload. Office: 8635 W 3d St Los Angeles CA 90048

HYMAN, EDWARD JAY, information scientist, psychologist, consultant, researcher; b. Roslyn, N.Y., Oct. 25, 1947; s. Herbert H. and Edith (Tannenbaum) H. A.B., Columbia Coll., 1969; postgrad., Harvard U., 1970; C.Phil., U. Calif.-Berkeley, 1974, Ph.D., 1975. Lectr. U. Calif.-Berkeley, 1976-77; chmn. bd. Assn. for Advanced Tng. in Behavioral Scis., Berkeley and Los Angeles, 1977-79; prin. investigator Center for Social Research, Berkeley, 1977-79; asst. prof. U. San Francisco, 1979-81; sci. dir. Center for Social Research, Berkeley, 1981—; clin. dir. Occpational Stress Ctr., Inst. for Labor and Mental Health, 1983—; cons. Union of Am. Hebrew Congregations, 1975, Standard Oil of Calif., San Francisco, 1976, Exxon, U.S.A., Houston, 1976-77, Edison Electric Inst., Washington, 1977-80, Random House, Pubs., 1982—. Author: Random House Information Support System, 1984; contbr. on info. system, occupational stress, attitudes toward tech.and sci. and life stress chpts to books, articles to popular media and profl. jours. Co-chmn. voter registration Democratic Nat. Com., Washington, 1972; mem. Berkeley Citizens Action, 1979—; chmn. Berkeley Global Modeling Group, U. Calif.-Berkeley, 1976-80. Recipient Disting. Research award U. San Francisco, 1981; Regents fellow U. Calif.-Berkeley, 1972-75; Sr. fellow Center for Social Research, 1975; Edison Electric Inst. grantee, 1977. Mem. Assn. Info. and Decision Sci., AAAS, N.Y. Acad. Sci., Am. Acad. Polit. and Social Scis., Internat. Soc. Applied Psychology, Internat. Soc. Polit. Psychol., Internat. Congress Psychology, Am. Psychol. Assn. Subspecialties: Information systems, storage, and retrieval (computer science); Social psychology. Current work: Information storage and retrieval systems application to behavioral correlates of social perceptions in innovative technologies, scientific changes particularly environmental and energy issues, prevention, treatment and diagnosis of occupational stress. Office: Center for Social Research 439 Boynton Ave Berkeley CA 94707

HYMAN, EDWARD SIDNEY, internist, consultant, researcher; b. New Orleans, Jan. 22, 1925; s. David and Mary (Newstadt) H.; m. Jean Simons, Sept. 29, 1956; children: Judith, Sydney, Edward David, Anne. B.S., La. State U., 1944; M.D., Johns Hopkins U., 1946. Diplomate: Am. Bd. Internal Medicine. Intern Barnes Hosp., Washington U., St. Louis, 1946-47; fellow in medicine Stanford U., San Francisco, 1949-51, asst. resident in medicine, 1950-51, Peter Bent Brigham Hosp., Boston, 1951-53; teaching fellow in medicine Harvard U., Boston, 1952-53; practice medicine specializing in internal medicine, New Orleans, 1953—; dir. kidney unit Charity Hosp., New Orleans, 1953-55; investigator Touro Research Inst., New Orleans, 1959—; mem. staff Sara Mayo Hosp., 1954-79, chief of staff, 1968-70, trustee, 1970-78; mem. staff Touro Infirmary, New Orleans, St. Charles Hosp.; panelist Pres.'s Commn on Health Needs of Nation, 1952; cons. water quality New Orleans Sewerage and Water Bd., 1978; mem. research adv. com. Cancer Assn. New Orleans, 1976—, La. Bd. Regents, 1983. Contbr. articles to profl. jours. NIH grantee, 1960-81; Am. Heart Assn. grantee, 1962-65. Fellow ACP; mem. Am. Fedn. Clin. Research, Am. Soc. Artificial Internal Organs, Am. Physiol. Soc. Biophys. Soc. (chmn. local arrangements 1971, 77, 81), Pvt. Doctors Am. (co-founder 1968, v.p. 1968—, Dist. Service award 1981), Orleans Parish Med. Soc. (gov. 1972-80), La. State Med. Soc. (ho. of dels. 1970-81). Jewish. Subspecialties: Internal medicine; Biophysics (physics). Current work: Clinical internal medicine, biochemistry, biophysics, nephrology, artificial organs, water quality, government in medicine, cause of death in renal failure, significance of bacteria in urine. Isolated aldosterone, 1949; patentee sheet plastic oxygenator (artificial heart), oil detection device; inventor telephone tranmission of electrocardiogram, early data transmission; inventor hydrogen-platinum detection of heart shunts. Office: 3525 Prytania St Suite 200 New Orleans LA 70115

HYMAN, HOWARD ALLAN, physicist; b. Montreal, Que., Can., Dec. 5, 1943; s. Harold and Marcia (Rohr) H.; m. Judith Lynn Seltzer, Oct. 19, 1969; 1 son: David. B.S.E.E., U. So. Calif., 1965; Ph.D., Yale U., 1970. Research fellow Queen's U., Belfast, No. Ireland, 1970-72; prin. research scientist Avco Everett Research Lab., Inc., Mass., 1972—. Contbr. articles to profl. jours. Mem. Am. Phys. Soc. Subspecialties: Atomic and molecular physics; Laser physics. Current work; Theory of atomic collisions; spectroscopy; physics of visible uv/x-ray lasers. Patentee in field. Office: Avco Everett Research Lab Inc Everett MA 02149

HYMAN, IRWIN A., psychology educator; b. Neptune, N.J., Mar. 22, 1935; s. Henry Meltzer and Harriet (Grenetz) H.; m. Nada Pospishil, Feb. 4, 1982; children by previous marriage: Nadine Meltzer, Deborah Nan. B.A., U. Maine, 1957; M.Ed., Rutgers U., 1961, Ed.D., 1964. Diplomate: in sch. psychology Am. Bd. Profl. Psychology. Tchr. Millstone Twp. (N.J.) Sch., 1957-61; sch. psychologist Lawrence Twp. (N.J.) Schs., 1962-66; chief clin. services Tng. Sch., Vineland, N.J., 1966-67; prof. spl. edn. Kean Coll., Union, N.J., 1967-68; dir. Nat. Ctr. Study of Corporal Punishment and Alternatives in the Schs., Temple U., Phila., 1977—, prof. sch. psychology, 1968—; cons. Nat. Inst. Edn., Washington, 1977—, ACLU, 1977—, Trenton (N.J.) Pub. Schs., 1980—, Ednl. Testing Service, Princeton, N.J., 1978-79. Co-editor: Corporal Punishment in American Education, 1979, School Consultation, 1977; editor: Sch. Psychologist, 1971-74. Recipient D.H. Hughes award N.Y. U., 1982. Fellow Am. Psychol. Assn. (pres. dist. 1977-78, bd. ethical and social research 1977-79, council of reps. 1974-77); mem. N.J. Assn. Sch. Psychologists. Subspecialties: School psychology; Discipline, behavioral management of children. Current work: Research and advocacy in work related to use of corporal punishment and discipline in schools. Office: National Center for Study of Corporal Punishment and Alternatives in Schools Temple U Philadelphia PA 19122 Home: 207 Carter Rd Princeton NJ 08540

HYMAN, RICHARD WALTER, virologist, educator; b. San Francisco, Oct. 22, 1941; s. Maurice J. and Fay S. (Samuel) H. B.S., U. Calif., Berkeley, 1962; M.S., Cornell U., 1964; Ph.D., Calif. Inst. Tech., 1970. Research assoc. Yale U. Med. Sch., 1970-73; asst. prof. microbiology Pa. State U. Coll. Medicine, Hershey, 1973-78, assoc. prof., 1978-82, prof., 1982—; mem. adv. com. on microbiology and virology Am. Cancer Soc., 1978-82. Recipient Faculty Research award Am. Cancer Soc., 1977-82. Mem. Am. Soc. Biol. Chemists, Am. Soc. Virology, Am. Soc. Microbiology, Biophys. Soc. Subspecialties: Virology (medicine); Genome organization. Current work: Herpes virus molecular biology. Office: Dept Microbiology Pa State U Coll Medicine Hershey PA 17033

HYNDMAN, ARNOLD GENE, neurobiologist; b. Los Angeles, Oct. 16, 1952; s. Joseph E. and Inez E. (Streets) H.; m. Juliet Masculino, Aug. 12, 1978; 1 son, Aaron-Eugene. A.B., Princeton U., 1974; Ph.D., UCLA, 1978. Cert. secondary tchr., N.J. Lectr. UCLA, 1978; postdoctoral fellow Ohio State U., Columbus, 1978-79; postdoctoral researcher U. Calif., San Diego, 1979-81; asst. prof. biol. scis. Rutgers U., New Brunswick, N.J., 1981—. Contbr. articles to profl. publs. Recipient cert. for sci. achievement Com. Advanced Sci. Tng., 1969; Mental Health Tng. grantee UCLA, 1975-78. Mem. Assn. for Research in Vision and Ophthalmology (grantee 1981), Soc. Neurosci., Sigma Xi. Subspecialties: Neurobiology; Cell and tissue culture. Current work: Purified cultures of neural retina are analyzed for neuronal survival, neurite growth, cell proliferation, and cellular responses to various cytotoxins. Office: Biol Sci Dept Campus Kilmer Rutgers U Piscataway NJ 08854

HYNES, MARTIN DENNIS, III, neuropharmacologist; b. Albany, N.Y., Dec. 23, 1949; s. Martin Dennis and Mary Guilfoil (Lynch) H., Jr.; m. Lynn Miller, Apr. 17, 1982. B.A. in Psychology, Providence Coll., 1972; M.S. in Pharmacology and Toxicology, U. R.I., 1972-75, Ph.D., 1978. Grad. teaching asst. dept. pharmacology and toxicology U. R.I., Kingston, 1972-76, USPHS predoctoral fellow, 1976-77, instr. dept. psychology Extension Div., Providence, 1976-77; postdoctoral fellow dept. physiol. chemistry and pharmacology Roche Inst. Molecular Biology, Nutley, N.J., 1977-79; sr. pharmacologist Lilly Research Labs., Indpls., 1979—. Mem. editorial bd.: Drug Devel. Research, 1979—; contbr. articles and abstracts to sci. jours. Explorer post adv. Boy Scouts Am., 1980—. Mem. AAAS, Soc. Neurosci., Soc. Exptl. Biology and Medicine, Internat. Narcotic Research Conf.; mem. Am. Soc. Pharmacology and Exptl. Therapeutics; Mem. N.Y. Acad. Scis., Sigma Xi. Subspecialties: Pharmacology; Neuropharmacology. Current work: The major emphasis of my scientific research activities is the discovery and development of unique analgesic and antipsychotic agents. Home: 8601 Amy Ln Indianapolis IN 46256 Office: 307 E McCarty St Indianapolis IN 46285

HYSLOP, NEWTON EVERETT, JR., physician, researcher; b. Newton, Mass., Oct. 14, 1935; s. Newton Everett and Mary Edna (Ross) H.; m. Deborah Boyer, Sept. 7, 1957; children: Marcia Elizabeth, Ross Harrison. A.B., Harvard U., 1957, M.D., 1961. Diplomate: Am. Bd. Internal Medicine, (Infectious Diseases), Am. Bd. Allergy and Clin. Immunology. Intern, asst. resident and research fellow in medicine Mass. Gen. Hosp., Boston, 1961-63, 66-68, asst. in medicine, 1968-71, asst. physician, 1971-76, assoc. physician, 1977—, dir. lab. infectious disease immunology, 1969—; instr. medicine Harvard Med. Sch., Boston, 1967-70, asst. prof., 1971—; sr. resident in medicine Peter Bent Brigham Hosp., Boston, 1965-66; research assoc. immunology Nat. Inst. Allergy and Infectious Diseases, NIH, Bethesda, Md., 1963-65; vis. scientist immunochemistry, dept. biochemistry Oxford U., 1968-69; investigator Howard Hughes Med. Inst., 1972-76. Contbr. articles to profl. jours. Served to lt. comdr. USPHS, 1963-65. Moseley Travelling fellow, 1968-69. Fellow ACP, Infectious Disease Soc. Am.; mem. Am. Assn. Immunologists, Am. Soc. Microbiology. Subspecialties: Cell biology (medicine); Infectious diseases. Current work: Mechanisms of immunogen formation by B-lactam antibiotics and the nature of the human immune response to antibiotics; also factors governing chemotactic responses of human polymorphonuclear leukocytes. Office: Infectious Disease Unit Mass Gen Hosp Boston MA 02114

HYSON, MICHAEL TERRY, biologist, consultant; b. Rockford, Ill., July 5, 1948; s. Harvey Eugene and Norma Audry (Allison) H.; m. Emillia Ann Smith, June 15, 1970 (div. Sept. 1971). B.S. in Biology, U. Miami, Fla., 1970, M.S., 1973, Ph.D., 1976. Research assoc. Calif. Inst. Tech., 1977-80, vis. assoc., 1982—; NASA fellow NASA Robotics Study, Santa Clara, Calif., 1980; cons. G. C. H. Astronautics, Sunnyvale, Calif., 1980—; pres. Hy Tech Glendale, Calif., 1980—. Editor, pub.: The Caltech Space Settlement Conference, 1981; software programmer. Mem. citizens adv. com. nat. space policy L-5 Soc., Los Angeles, 1981-82. Served with USNR, 1970. Recipient Nat. Res. Service award NIH, 1977-80; NSF trainee, 1971-72; Maytag Grad. fellow, 1973-75. Mem. L-5 Soc. (speaker 1st nat. conf. 1982), Lighter-than-air Soc. (life), Assn. Computing Machinery, Am. Soc. Photogrammetry, Soc. Photo-Optical and Illumination Engrs. Libertarian. Subspecialties: Neurobiology; Space colonization. Current work: Eye movements in binocular vision; neural models of vision; robot vision; application of robotics and teleoperation and biology to colonization of space; consulting on space-related topics; would like to see space colonized and moon and asteroids mined; interested in dolphins and interspecies communication. Performed experiments in weightlessness aboard NASA KC-135. Home and Office: Hy Tech 1155 N Verdugo Rd Apt C Glendale CA 91206

IACHELLO, FRANCESCO, physics educator; b. Francofonte, Italy, Jan. 11, 1942; s. Giovanni and Gaetana Laura (Marziano) I.; m. Irena Hana Holemarova, Jan. 16, 1971; 1 son, Giovanni. Dott. Ing., Politecnico di Torino, Italy, 1964; Ph.D., MIT, 1969. Asst. prof. Politecnio di Torino, 1971-74; sr. scientist Kernfysisch Versneller Inst., Groningen, Netherlands, 1974-76, prof. physics U. Groningen, 1976-82, Yale U., New Haven, 1978—. Contbr. articles to profl. jours. Recipient Akzo prize Dutch Acad. Scis., 1981. Mem. Am. Phys. Soc., European Phys. Soc., Italian Phys. Soc., Dutch Phys. Soc. Subspecialty: Nuclear physics. Current work: Nuclear structure. Office: 272 Whitney Ave New Haven CT 06511

IANNA, PHILIP A., astronomer, educator; b. Phila., May 27, 1938; s. Michael and Jeanette E. (Russell) I.; m. Susan Osborne, Dec. 28, 1968. B.A., Swarthmore Coll., 1960, M.A., 1962; Ph.D., Ohio State U., 1968. Research assoc. Center for Advanced Studies, U. Va., Charlottesville, 1968-70; research assoc. Leander McCormick Obs., U. Va., 1970-73; asst. prof. dept. astronomy U. Va., 1973-79, assoc. prof., 1979—. Author: (with Roger B. Culver) The Gemini Syndrome, 1979; also articles. Recipient Group Achievement award NASA, 1981. Mem. Am. Assn. Variable Star Observers, Am. Astron. Soc., Internat. Astron. Union, Sigma Xi. Subspecialties: Optical astronomy; Astrometry. Current work: Establishment of fundamental properties of nearby stars, studies of stellar open clusters. Astrometry, trigonometric parallaxes, proper motions, stellar luminosities, stellar masses. Home: PO Box 3818 Charlottesville VA 22903 Office: Dept Astronomy Cabell Dr Charlottesville VA 22903

IANNACCONE, PHILIP MONROE, biologist, pathologist, educator; b. Syracuse, N.Y., May 3, 1948. B.Sc. in Biochemistry, Syracuse U., 1968, SUNY, Syracuse, 1968; M.D., SUNY Upstate Med. Center, 1972; Ph.D., U. Oxford (Eng.), 1977. Diplomate: Am. Bd. Pathology. Intern SUNY Upstate Med. Center, 1972-73, resident in pathology, 1973-74; asst. prof. U. Calif.-San Diego, 1978-79; asst. prof. pathology Northwestern U. Med. Sch. and Cancer Center, Chgo., 1979—. Contbr. sci. articles to profl. publs. Mem. AAAS, Internat. Acad. Pathologists, Am. Assn. Cancer Research, Am. Assn. Pathologists. Subspecialties: Cell and tissue culture; Cancer research (medicine). Current work: Biology of cancer, reproductive toxicology. Office: 303 E Chicago Ave Chicago IL 60611

IATRIDIS, PANAYOTIS GEORGE, physician, physiology and medicine educator, university administrator; b. Alexandira, Egypt, Dec. 10, 1926; came to U.S., 1962, naturalized, 1975; s. George E. and Ioanna (Nicholaides) I.; m. Catherine P. Mouzouris, Apr. 27, 1957; children: Ioanna P., Mary P. Grad., Greek Gymnasium Ambetios Sch., Cairo, 1944; M.D., U. Athens, 1951, D.Sc. with honors in Physiology, 1968. Med. lic. Greece, Egypt, N.C., Ind., Ill. Resident Univ. Med. Clinic, Athens, 1951-53; resident Greek Hosp., Alexandria, 1953-55, asst. dir. dept. medicine, 1959-62; research assoc. dept. physiology U. N.C., Chapel Hill, 1963-66, asst. prof. physiology, 1969-72; assoc. prof. physiology Ind. U., Gary, 1972-76, prof., 1976—, prof. medicine, 1981—, dir., 1975—, asst. dean, 1979—; vis. research scientist Protein Found. and Harvard Sch. Pub. Health, 1966; research scientist dept. physiology U. Athens, 1967-69; cons. St. Mary Med. Ctr., Meth. Hosp., St. Catherine Hosp., Porter Meml. Hosp., St. Anthony Hosp. Contbr. numerous articles in field to profl. jours. Bd. dirs. Porter County (Ind.) Mental Health Assn., 1973-77; bd. dirs. Am. Heart Assn., Ind. affailiate, 1975—, Am. Lung Assn., Northwest Ind. chpt., 1977—, Am. Cancer Soc., Northwest Ind. chpt., 1978—; mem. council SS Constantine and Helen Greek Orthodox Cathedral, Merrillville, Ind., 1979-81; vice chmn. Community Health Assn., Lake County, Ind., 1979—, chmn. med. adv. com., 1979—; founder, pres. Greek Orthodox Ch. of Porter County, 1980-81. Grantee in field. Mem. Internat. Soc. Hematology, Internat. Soc. Thrombosis and Hemostasis, N.C. Acad. Sci., N.Y. Acad. Sci., Am. Physiol. Soc., AAAS, Med. Soc. Athens, Greek Hematological Soc., World Fedn. Hemophilia, Am. Thoracic Soc., Ind. Thoracic Soc., Assn. Am. Med. Colls., Assn. Ind. Dirs. Med. Edn., Sigma Xi. Subspecialties: Internal medicine; Hematology. Current work: Effects of 2,3-DPG on platelet aggegation and prostaglandin synthesis, blood coagulation, fibrinolysis, thrombosis, and related topics. Home: 603 Hastings Terr Valpariso IN 46383 Office: Indiana U Sch Medicine 3400 Broadway St Gary IN 46408

IBEN, ICKO, JR., astrophysicist, educator; b. Champaign, Ill., June 27, 1931; s. Icko and Kathryn (Tomlin) I.; m. Miriam Genevieve Fett, Jan. 28, 1956; children: Christine, Timothy, Benjamin, Thomas. B.A., Harvard, 1953; M.S., U. Ill., 1954, Ph.D., 1958. Asst. prof. physics Williams Coll., 1958-61; sr. research fellow in physics Calif. Inst. Tech., 1961-64; asso. prof. physics Mass. Inst. Tech., 1964-68, prof., 1968-72; vis. prof. astronomy Harvard, 1966, 68, 70; vis. fellow Joint Inst. for Lab. Astrophysics, U. Colo., 1971-72; vis. prof. astronomy and astrophysics U. Calif. at Santa Cruz, 1972; prof. astronomy and physics, head dept. astronomy U. Ill. at Urbana-Champaign, 1972—; vis. prof. physics and astronomy Inst. for Astronomy, U. Hawaii, 1977; mem. adv. panel, astronomy sect. NSF, 1972-75; mem. vis. com. Aura Observatories, 1979-82. Contbr. articles to profl. jours. Mem. Am. Astron. Soc. (councilor 1974-77). Subspecialty: Theoretical astrophysics. Current work: Evolution of asymptotic giant branch stars (dredge up and nucleosynthesis); evolution of central stars of planetary nebulae and cooling white dwarfs. Home: 3910 Clubhouse Dr Champaign IL 61820 Office: Astronomy Bldg 1011 W Springfield Urbana IL 61801

IBERS, JAMES ARTHUR, chemist; b. Los Angeles, June 9, 1930; s. Max Charles and Esther (Imerman) I.; m. Joyce Audrey Henderson, June 10, 1951; children—Jill Tina, Arthur Alan. B.S., Calif. Inst. Tech., 1951, Ph.D., 1954. NSF post-doctoral fellow, Melbourne, Australia, 1954-55; chemist Shell Devel. Co., 1955-61, Brookhaven Nat. Lab., 1961-64; mem. faculty Northwestern U., 1964—, prof. chemistry, 1964. Mem. Am. Chem. Soc. (inorganic chemistry award 1979), Am. Crystallographic Assn. Subspecialties: Inorganic chemistry; Solid state chemistry. Current work: Synthesis, characterization and properties of new inorganic, organometallic and solid state materials; correlation of properties with structure. Home: 2657 Orrington Ave Evanston IL 60201 Office: Dept Chemistry Northwestern Univ Evanston IL 60201

IBRAHIM, SHAWKI AMIN, radiology educator; b. Cairo, Egypt, Mar. 27, 1942; came to U.S., 1970, naturalized, 1975; s. Amin and Venice (Michael) I.; m. Janette Yousef, Mar. 21, 1967; children: Maria, Daniel. B.S., Alexandria (Egypt) U., 1962; M.S., L.I. U., 1972; Ph.D., N.Y. U., 1980. Lab. instr. Alexandria (Egypt) U., 1962-70; chief technologist Coney Island Hosp., N.Y.C., 1972-75; research scientist N.Y. U., 1975-80; asst. prof. Colo. State U., 1980—. Contbr. articles to profl. jours. Mem. Health Physics Soc. (student award 1978). Subspecialties: Nuclear fission; Ecology. Current work: Radiochemistry and environmental aspects of nuclear waste management. Home: 3349 Pepperwood Ln Fort Collins CO 80525 Office: Colo State U Dept Radiation Biology Fort Collins CO 80523

ICERMAN, LARRY, univ. research adminstr.; b. Muncie, Ind., Sept. 22, 1945; s. Charles and Janelyn (Mock) I. B.S., M.I.T., 1967; M.S., U. Calif., San Diego, 1968, Ph.D., 1976; M.B.A., San Diego State U., 1976. Research engr. U. Calif., San Diego, 1976; asst. prof. tech. and human affairs Washington U., St. Louis, 1976-79, asso. prof., 1979-80; dir. N.Mex. Energy Inst., Las Cruces, 1980-81, N.Mex. State U. Energy Inst., 1982—. Author: (with S. S. Penner) Energy: Demands, Resources, Impact, Techonolgy, and Policy, 1974, 81, Energy: Non-Nuclear Energy Technologies, 1976, (with R. P. Morgan) Renewable Resources Utilization for Development, 1981. Subspecialties: Geothermal power; Solar energy. Current work: Renewable energy tech. and policy, including research, engring., and commercialization. Home: 4212 Colt Rd Las Cruces NM 88001 Office: N Mex State U Box 3EI Las Cruces NM 88003

ICHIKI, ALBERT T., research scientist, educator; b. Lahaina, Hawaii, Sept. 21, 1936; s. Shigeru and Toyoko (Fujioka) I. B.S., Purdue U., 1958, M.S., 1961; Ph.D., UCLA, 1969. USPHS and NIH fellow John Curtin Sch. Med. Research, Australian Nat. U., Canberra, 1969-71, research assoc., 1971-72, research asst. prof., 1972-80; asst. prof. U. Tenn. Coll. Medicine, Knoxville, 1978-80, assoc. prof. med. biology, 1980—. Served to capt. USAF, 1960-65. Mem. AAAS, Am. Assn. Cancer Research, Am. Assn. Immunologists, Am. Soc. Hematology, Internat. Soc. Hematology. Subspecialties: Immunobiology and immunology; Hematology. Current work: Lymphocyte differentiation. Home: 3113 Ginn Dr Apt 4 Knoxville TN 37920 Office: Dept Med Biology UTMRCH 1924 Alcoa Hwy Knoxville TN 37920

ICHIYE, TAKASHI, oceanographic educator, physical oceanography researcher; b. Kobe, Japan, Oct. 1, 1921; came to U.S., 1957, naturalized, 1972; s. Mankichi and Toyo (Yumoto) I.; m. Chiyoko Nagao, Oct. 6, 1952; children: Toshiko, Keiko. Cert., First Coll., Tokyo, 1942; B.S., U. Tokyo, 1944, D.Sc., 1952. Research officer Kobe Marine Obs., 1944-54; assoc. chief oceanography sect. Japan Meteorol. Agy., Tokyo, 1954-57; vis. scientist Woods Hole (Mass.) Oceanographic Instn., 1957-58; research assoc., asst. prof. Fla. State U., Tallahassee, 1958-63; sr. research scientist Lamont Geol. Obs., Columbia U.-Palisades, N.Y., 1963-68; prof. oceanography Tex. A&M U., College Station, 1968—; cons. EG&G Co., Waltham, Mass., 1973-76; vis. researcher U. So. Calif., Los Angeles, 1978; cons. Exxon Research & Devel. Co., Houston, 1980-82; coordinator (Japan and East China Sea Study Program), 1981—. Author: Illustrative Oceanography (in Japanese), 1953; editor: Proc. of Diffusion Symposium, 1964, Proc. of JECSS-I Workshop, 1982. Recipient award for disting. research on Tsunami Japan Meteorol. Agy., 1950. Mem. Am. Geophys. Union, Am. Meteorol. Soc., Am. Soc. Limnology and Oceanography, Oceanog. Soc. Japan, Sigma Xi. Subspecialty: Laser medicine. Current work: Circulation in marginal seas, upper ocean layer dynamics, diffusion, turbulence, themohaline fine structures, ocean pollution. Home: Route 5 Box 1357 College Station TX 77840 Office: Dept Oceanography Tex A&M U College Station TX 77843

IDLER, DAVID RICHARD, biochemist, marine scientist, educator; b. Winnipeg, Man., Can., Mar. 13, 1923; s. Ernest and Alice (Lydon) I.; m. Myrtle Mary Betteridge, Dec. 12, 1956; children: Louise, Mark. B.A., U. B.C., Can.), Vancouver, 1949, M.A., 1950; Ph.D., U. Wis., 1954. With Fisheries Research Bd. of Can., 1953-71; dir., investigator in charge of steroid biochemistry Halifax (N.S.) Lab., 1961-69, Atlantic regional dir. research, Halifax, 1969-71; dir. Marine Sci. Research Lab.; prof. biochemistry Meml. U. Nfld., Can., St. John's, 1971—. Editor: Steroids in Nonmammalian Vertebrates, 1972; editorial bd.: Steroids, 1963—, Gen. and Comparative Endocrinology, 1966-82, Endocrine Research Communications, 1974—, Can. Jour. Zoology, 1979—; mem. bd. corr. editors: Jour. Steroid Biochemistry. Served with RCAF, 1942-45. Decorated D.F.C. Fellow Royal Soc. Can.; mem. European Soc. Comparative Endocrinologists (founding), Can. Biochem. Soc., Am. Chem. Soc., AAAS, Am. Zool. Soc., Endocrine Soc., N.Y. Acad. Scis. Subspecialty: Biochemistry (biology). Home: 44 Slattery Rd St John's NF A1A 1Z8 Canada Office: Marine Scis Research Lab Meml U Nfld St John's NF A1C 5S7 Canada

IFJU, GEZA, wood technologist, educator; b. Azeged, Hungary, Jan. 26, 1931; s. Jeno John and Margaret (Bedy) I.; m.; children: Katherine, Peter, Paul. B.S., U. B.C., 1959; M.F., Yale U., 1960; Ph.D., U. B.C., 1963. Asst. specialist U. Calif.-Berkeley, 1963-64, asst. prof., 1964-67; assoc. prof. dept. forest products Va. Poly. Inst. and State U., Blacksburg, 1967-71, prof., 1971—, head dept., 1979—; Nat. Acad. Sci. vis. scientist, Hungary, 1972. Contbr. articles to profl. jours. Recipient Wood award Forest Products Research Soc., 1963; Diamond award Mo. Bot. Gardens, 1975. Mem. Forest Products Research Soc., Soc. Wood Sci. and Tech., Internat. Soc. Stereology. Roman Catholic. Subspecialties: Biomaterials; Composite materials. Current work: Quantitative characterization of wood and wood composites; structure property relationships for wood and wood composites. Home: 803 Broce Dr Blacksburg VA 24060 Office: Va Tech Inst 210 Cheatham Hall Blacksburg VA 24061

IGLEWSKI, WALLACE JOSEPH, microbiologist, researcher; b. Cleve., Aug. 17, 1938; s. Wallace Frank and Marie Ann (Sech) I.; m. Barbara Hotham Iglewski, Apr. 23, 1965; children: Eric, William. B.S., Western Res.U., 1961; M.S., Pa. State U., 1963, Ph.D., 1965. Postdoctoral fellow U. Colo. Med. Center, 1965-66, Pub. Health Research Inst., N.Y.C., 1966-68; asst. prof. microbiology U. Oreg., 1968-72; assoc. prof. Oreg. Health Sci. U., 1972—. Contbr. articles to profl. jours. NIH grantee, 1968-79; NSF grantee, 1979-82. Mem. Am. Soc. Microbiology, Am. Soc. Virology. Subspecialties: Virology (medicine); Molecular biology. Current work: Biochemistry of replilation of RNA viruses; structure and function of elongation factor-2; ADP-ribosylation of proteins, molecular biology of herpes viruses. Office: Department Microbiology and Immunology 3181 SW Sam Jackson Park Rd Portland OR 97201

IGNOFFO, ROBERT JOHN, pharmacy educator, oncologist; b. San Francisco, Nov. 17, 1944; s. Anthony Joseph and Inez (Garioto) I.; m. Ruth Ann Blanz, June 4, 1967; children: Michele, Andrea, Toni. B.A., Calif. State U.-San Francisco, 1967; Pharm. D., U. Calif.-San Francisco, 1971. Assoc. prof. pharmacy U. Calif.-San Francisco, 1972—. Author, editor: Manual on Oncology Therapeutics, 1981. Mem. Am. Soc. Hosp. Pharmacists (oncology fellowship preceptor 1980, 83), Am. Soc. Clin. Oncology. Subspecialties: Chemotherapy; Oncology. Current work: The use of implantable pump technology in the treatment of human cancer; the management of acute and chronic toxicity from cancer chemotherapy. Office: University of California San Francisco CA 94143

IHLER, GARRET MARTIN, medical biochemist, educator; b. Milw., Nov. 4, 1939; s. Gerald J. and Marjorie M. I.; m. Karin A., May 3, 1970; 2 children. B.S., Calif. Inst. Tech., 1961; Ph.D., Harvard U., 1967; M.D., U. Pitts., 1976. Asst. prof. U. Pitts., 1969-72, assoc. prof., 1972-76; prof., head dept. med. biochemistry Coll. Medicine, Tex. A&M U., College Station, 1976—. Mem. Am. Soc. Biol. Chemists, Am. Soc. Microbiology, Am. Soc. Cell Biologists, Am. Chem. Soc., Acad. Clin. Lab. Physicians and Scientists, Tex. Genetics Soc. Subspecialties: Biochemistry (medicine); Genetics and genetic engineering (biology). Current work: DNA, cloning, electron microscopy, drug delivery systems.

ILG, RONALD JON, marine biology researcher, educator; b. Cleve., Dec. 10, 1945; s. John and Helen May (Eschuk) I.; m. Susanne Audrey Warren/Bye, Apr. 12, 1970 (div. Dec. 1978); m. Patricia Ellen McCoy, May 19, 1982. B.Sc., Ohio State U., 1966; Ph.D., U. Tex., Austin, 1980. Shipboard biologist Lamont Geol. Obs., Columbia U., Palisades, N.Y., 1966-67; communications cons. Ohio Bell Telephone Co., Cleve., 1967-74; research asst. U. Tex., Port Aransas, 1977-80; asst. prof. biology, prin. investigator McNeese State U., Lake Charles, La., 1980—; cons. Port of Corpus Christi, Tex., 1980, Weyerhaeuser Co., Tacoma, Wash., 1981, PPG Industries, Lake Charles, La., 1981-82, S.W. La. Shrimping and Fishing Assn., Vinton, La., 1982—. Mem. Calcasieu Commn. on Arts and Humanities, Lake Charles, La., 1982. Ohio State U. stadium scholar, 1963-64. Mem. AAAS, Union Concerned Scientists, Gulf Estuarine Research Fedn., Estuarine Research Fedn., Am. Fisheries Soc., Am. Soc. Zoologists. Subspecialties: Ecology; Resource management. Current work: Physiological ecology of fish particularly regarding salinity and pollution, coastal zone management, innovative fisheries management and economical repercussions. Home: 625 W LaGrange St Lake Charles LA 70605 Office: McNeese State U PO Box 1468 Lake Charles LA 70609

IMBEMBO, ANTHONY LOUIS, surgeon, educator, educational administrator; b. N.Y.C., Nov. 8, 1942; s. Emil Anthony and Theresa (Rippert) I. A.B., Columbia U., 1963, M.D., 1967. Diplomate: Am. Bd. Surgery Am. Bd. Thoracic Surgery, Nat. Bd. Med. Examiners. Intern Mass. Gen. Hosp., 1967-68, resident, 1968-73; asst. prof. surgery Johns Hopkins U., 1973-77, assoc. prof., 1977-83, surgeon, 1973-83; prof. surgery, vice chmn. dept. Case Western Res. U., 1983—; dir. dept. surgery Cleve. Met. Gen. Hosp., 1983—; cons. Walter Reed Army Med. Ctr., 1982—. Johns Hopkins Hosp. grantee, 1976; recipient George J. Stuart award Johns Hopkins U. Sch. Medicine, 1977-82. Fellow A.C.S.; mem. Soc. Univ. Surgeons, Assn. Surg. Edn. (pres. 1982-83), Soc. Surgery of the Alimentary Tract, Internat. Cardiovascular Soc. Subspecialties: Surgery; Artificial organs. Current work: Surgeon (general, vascular and thoracic surgery); educator; president of Association for Surgical Education and numerous educational teaching awards; researcher implantable programmable infusion pump for delivery of insulin (human trials beginning); research role of amino acid analogues in nutrition. Home: 20901 Claythorne Rd Shaker Heights OH 44122 Office: Cleve Met Gen Hosp 3395 Scranton Rd Cleveland OH 44109

IMBERSKI, RICHARD BERNARD, zoology educator; b. Amsterdam, N.Y., Nov. 5, 1935; s. Bernard Anthony and Josephine Amelia (Nasadoski) I.; m. Marijke van Oordt, Aug. 31, 1967; children: Monique, Corinne. B.S., U. Rochester, 1959, Ph.D., 1966. Research/teaching asst. U. Rochester, 1960-63, research trainee, 1963-65; research assoc. Johns Hopkins U., 1965-67; asst. prof. dept. zoology U. Md.-College Park, 1967-73, assoc. prof., 1973—. Editor newsletter: Jour. Ephestia Newsletter, 1970—; contbr. articles to profl. jours.; author: (with J. Potter and A. Barnett) Experiments in Genetics, 1973, 2nd edit. 1976, Experiments in Genetics and Development, 1980, 2d edit., 1981. U. Md. Faculty Research awardee, 1968, 71, 73, 76; research grantee, 1968-70, 74-77; NSF grantee, 1969-70. Mem. AAAS, Genetics Soc. Am., Soc. Developmental Biology, Internat. Soc. Developmental Biologists, Am. Soc. Zoologists, Sigma Xi. Subspecialties: Gene actions; Developmental biology. Current work: Genetic analysis of hormone prodn.; developmental genetics of enzymes and specific non-enzymatic proteins. Office: Dept Zoology Univ Md College Park MD 20742

IMBRIE, JOHN, geologist; b. Penn Yan, N.Y., July 4, 1925; s. Charles Kisselman and Margaret (Fleming) I.; m. Barbara A. Zeller, Oct. 11, 1947; children—Katherine Palmer, John Zeller. Student, Coe Coll., 1942-43; B.A., Princeton U., 1948; M.S., Yale U., 1950, Ph.D., 1951. Asst. prof. geology U. Kans., 1951-52; asst. prof. geology Columbia U., 1952-54, asso. prof., 1955-60, prof., 1961-66; prof., dept. geology Brown U., 1967-75, H. L. Doherty prof. oceanography, 1976—; faculty fellow NSF, 1959; exec. com. CLIMAP project, 1971—; mem. U.S. Climate Bd., 1977—; earth scis. bd. NRC, 1976—; mem. U.S.-USSR Bilateral Agreement on the Environment, 1974—, U.S. Com. INQUA, 1975-77; mem. earth sci. bd. Nat. Research Council, 1976—; sr. vis. research assoc. Lamont-Doherty Geol. Obs., 1967—; adj. prof. oceanography U. R.I., 1976—. Co-author: Oceanography, 1978, Ice Ages, 1979; co-editor: Approaches to Paleoecology, 1964. MacArthur fellow, 1981. Fellow Geol. Soc. Am.; mem. Soc. Econ. Paleontologists and Mineralogists (sec.-treas. 1959-60), Am. Geophys. Union, Paleontol. Soc., Nat. Acad. Scis., Nat. Acad. Arts and Scis., Am. Meteorol. Soc., Am. Quaternary Assn., Phi Beta Kappa, Sigma Xi. Subspecialties: Paleoecology; Oceanography. Home: 55 Pamden Ln Seekonk MA 02771 Office: Dept Geol Sci Brown U Providence RI 02912

IMIG, CHARLES JOSEPH, physiologist, researcher; b. Waterloo, Iowa, Oct. 14, 1922; s. Louis Henry and Margaret Ann (Miller) I.; m. Donna Mae Collins, Sept. 19, 1944; children: John, Marcia, Douglas, Louise, Michael, Kathleen. A.B., Coe Coll., 1944; R.P.T., U. Iowa, 1947, M.S., 1948, Ph.D., 1951. Instr. physiology U. Iowa, 1951-52, research assoc., 1952-54, asst. prof., 1954-58, assoc. prof., 1958—. Contbr. articles in field to profl. jours. Served with USN, 1942-46; PTO. Arthritis and Rheumatism Found. fellow, 1952-55; recipient Med. Faculty award Lederle Corp., 1955-57; Gold medals Am. Congress Phys. Medicine, 1948, 53. Fellow Iowa Acad. Sci.; mem. Am. Physiol. Soc., Soc. Exptl. Biology and Medicine, Sigma Xi, Am. Legion. Roman Catholic. Lodges: Elks; Eagles. Subspecialties: Physiology (biology); Physiology (medicine). Current work: Peripheral blood flow, vascular reactivity, vascular muscle energetics. Home: 536 S Dodge St Iowa City IA 52240 Office: U Iowa BSB Iowa City IA 52242

IMONDI, ANTHONY ROCCO, pharmacologist; b. Providence, Aug. 21, 1940; s. Anthony and Phyllis (Cacchiotti) I.; m. Lucy Ann Corsini, July 6, 1963; children: Lisa, Michael, Gina, Gail. B.S., U. R.I., 1962; M.S., U. Maine, 1964, Ph.D., 1966. USPHS postdoctoral fellow Cornell Med. Coll., N.Y.C., 1966-69; postdoctoral fellow Sloan-Kettering Inst. for Cancer Research, N.Y.C., 1967-69; sr. pharmacologist Warren-Teed Pharms., Columbus, Ohio, 1969-74; project leader Rohm & Haas Co., Phila., 1974-77; mgr. pharmacology Adria Labs., Columbus, 1977—. Contbr. articles to profl. jours. Pres. Adria Labs. Credit Union, 1981—; v.p. Rolling Ridge Co., residents assn. Mem. Am. Soc. Pharmacology and Exptl. Therapeutics, Soc. Exptl. Biology and Medicine, AAAS. Subspecialties: Pharmacology; Pharmacokinetics. Current work: Development of new drugs for human use, research in pharmacology including drug metabolism, cancer chemotherapy, cardiovascular, central nervous system drugs and gastrointestinal pharmacology. Office: PO Box 16529 Columbus OH 43216

IMSANDE, JOHN D., genetics educator; b. Grass Range, Mont., June 14, 1931; s. Louis H. and Freda M. (Dengel) I.; m. Marcia F. Rohrbach, Aug. 13, 1976; children: Carol Lynn, Louis Daniel. B.A., U. Mont., 1953; M.S., Mont. State U., 1956; Ph.D., Duke U., 1960. Lectr., research scientist Princeton U., 1961-62; asst. prof. biology Case Western Res. U., Cleve., 1962-64, assoc. prof., 1964-68; vis. scientist U. Edinburgh, Scotland, 1968-69; prof. genetics and biochemistry Iowa State U., Ames, 1969-75, prof. genetics, 1975—; lectr. numerous univs. Contbr. chpts. to books, articles to profl. jours. Served with U.S. Army, 1953-55. NIH grantee. Mem. Am. Soc. Plant Physiologists. Subspecialties: Nitrogen fixation; Plant genetics. Current work: Identification of plant (soybean) genotype that promotes a high level of nitrogen fixation; enhanced dinitrogen fixation. Home: 1121 N Hyland Ave Ames IA 50010 Office: Dept Genetics Iowa State U Ames IA 50011

INAGAMI, TADASHI, biochemistry, educator; b. Kobe, Japan, Feb. 20, 1931; came to U.S., 1954; s. Yoshio and Yoshi (Hoshi) I.; m. Masako Araki, Nov. 12, 1961; children: Sanae, Mary. B.S., Kyoto (Japan) U., 1953; M.S., Yale U., 1955, Ph.D., 1958. Research assoc. Kyoto U., 1959-62; instr. biochemistry Nagoya City U., 1962; research assoc. Yale U., 1962-66; asst. prof. biochemistry Vanderbilt U., Nashville, 1966-69, asso. prof., 1969-75, prof., 1975—, dir., 1979—; mem. cardiovascular and renal disease study sect. NIH, 1976-80. Contbr. over 120 articles and revs. to sci. jours., chpts. to books. Mem. Am. Soc. Biol. Chemists, Endocrine Soc., Am. Heart Assn., Am. Chem. Soc., Soc. for Neurosci., Internat. Soc. Hypertension, Am. Soc. Cell Biology. Subspecialties: Biochemistry (medicine); Cell biology (medicine). Current work: Biochemistry of high blood pressure; renin, angiotensin, cell biology. Home: 2029 Kingsbury Dr Nashville TN 37215 Office: Dept Biochemistry Vanderbilt U Nashville TN 37232

INCE, LAURENCE PETER, psychologist; b. N.Y.C., June 29, 1937; s. Eugene and Ernestine (Goldstein) I.; m. Mariene Sandra Rosenberg, Aug. 3, 1963; children: Valerie, Elizabeth. B.A., Hobart Coll., 1959; M.S., L.I.U., 1961; Ph.D., Fla. State U., 1965. Cert. psychologist, N.Y. State. Sr. psychologist N.Y.U. Med. Center, Goldwater Meml. Hosp., 1966-79, now supr. psychology, dir. psychophysiology and biofeedback lab.; dir. Center for Learning Disabilities, N.Y.C.; adj. prof. psychology Queens Coll., CUNY, N.Y.C. Contbr. articles to profl. publs. Mem. Am. Psychol. Assn., Am. Congress Rehab. Medicine, AAAS. Subspecialties: Behavioral psychology; Physiological psychology. Current work: Biofeedback, spinal cord functioning. Home: 295 Winthrop Rd Teaneck NJ 07666 Office: Psychology Service Goldwater Meml Hosp Roosevelt Island NY 10044

INDERBITZEN, ANTON LOUIS, JR., scientific programs manager, consultant; b. Sacramento, Dec. 9, 1935; s. Anton Louis and Gertrude (Ramus) I.; m. Patricia Ann Noland, June 20, 1959; children: Daniel Robert, Heidi Marie, Rebecca Sue, Jennifer Ann. B.S., Stanford U., 1957; Ph.D., 1970; M.A., U. So. Calif., 1960. Registered geologist, Calif., Del.; registered engring. geologist, Calif. Engring. geologist Maurseth & Howe Engrs., Los Angeles, 1958-61; sr. scientist-oceanography Lockheed Aircraft Corp., San Diego, 1961-72; dir. marine ops. U. Del.-Lewes, 1972-77; sr. staff scientist MAR, Inc., Rockville, Md., 1977-78; program mgr. NSF, Washington, 1978—; NSF rep. Dept. Interior Outer Continental Shelf Mining Policy Task Force, Washington, 1979-81; mem. marine mining panel U.S.-Japan Coop. Program in Natural Resources, Washington, 1982—. Contbr. articles on marine and engring. geology to profl. jours.; editor: Deep Sea Sediments: Physical and Mechanical Properties, 1974. Fellow Geol. Soc. Am.; mem. Marine Tech. Soc. (v.p. ocean engring. systems 1973-76), AAAS, Am. Geophys. Union, Am. Oceanic Orgn. Mem. Christian Missionary Alliance Ch. (Disciples of Christ). Lodge: Rotary. Subspecialties: Oceanography. Current work: Marine mineral deposits and their origin; reactions of the sea floor sediments to induced stresses. Patentee sediment corer, marine instruments. Office: NSF 1800 G St NW Washington DC 20550 Home: 17805 Shady Mill Rd Derwood MD 20855

INFANGER, ANN, geneticist, biologist, educator; b. Newark, Dec. 20, 1933; d. Adolph Omega and Louise Elizabeth Catherine (Stuerm) I. B.A., Seton Hill Coll., 1955; Ph.D., Cornell U., 1963. Joined Sisters of Charity of Seton Hill, Roman Catholic Ch., 1956; mem. faculty Seton Hill Coll., Greensburg, Pa., 1958—, prof. biology, 1972—. NIH grantee, 1963-69; NSF grantee, 1969-73. Mem. Genetics Soc. Am., AAAS, Nat. Assn. Biology Tchrs., Sigma Xi. Subspecialty: Genetics and genetic engineering (biology). Current work: Mitochondrial genetics of Neurospora; analysis of DNA of mitochondria of mutant Neurospora. Home and Office: Seton Hill Coll Greensburg PA 15601

INGARD, KARL UNO, educator; b. Gothenburg, Sweden, Feb. 24, 1921; came to U.S., 1952, naturalized, 1959. E.E., Chalmers Inst. Tech., 1944, tech. licentiate, 1948; Ph.D. (Sweden-Am. Found. fellow, Acoustical Materials Assn. fellow), MIT, 1950; hon. degree, Chalmers Inst. Tech., 1979. Research asst. Research Lab. Electronics, Chalmers Inst. Tech., Gothenburg, 1943-45, dir., 1946-48, lectr., 1947-51, docent, dir. acoustics lab., 1951-52; research engr. Nat. Lab., Stockholm, 1945-46; faculty MIT, Cambridge, 1952—, prof. dept. physics, 1966—, prof. dept. aeros. and astronautics, 1971—; vis. prof. Royal Inst. Tech., Stockholm, 1960, Tech. U. Berlin, 1966; mem. com. on hearing, bioacoustics and biomechanics Nat. Acad. Scis.-NRC. Armstrong Cork fellow, 1950; Guggenheim fellow, 1960; recipient Rayleigh medal Inst. Acoustics, U.K., 1981. Fellow Acoustical Soc. Am. (Biennial award 1954, mem. exec. council 1970-73), Am. Phys. Soc.; mem. Chalmers Engring. Assn. (Gustaf Dalen medal 1970), Am. Soc. Swedish Engrs. (John Ericsson medal 1972), Inst. Noise Control Engring. (pres. 1973, dir. 1974), Nat. Acad. Engring., Sigma Xi. Subspecialty: Acoustics. Office: MIT 77 Massachusetts Ave Cambridge MA 02139

INGBER, LESTER, physicist; b. Bklyn., Mar. 26, 1941; s. Phillip and Helen (Felder) I.; m. Louise Frazer, Feb. 14, 1981. B.S., Calif. Inst. Tech., 1962; postgrad., Niels Bohr Inst., Copenhagen, 1964; Ph.D., U. Calif.-San Diego, La Jolla, 1966. NSF postdoctoral fellow UCLA, U. Calif.-Berkeley, 1967-69; asst. prof. SUNY-Stony Brook, 1969-70; research physicist U. Calif.-San Diego, La Jolla, 1970-72, hon. research assoc., 1972-74, 81—; pres. Phys. Studies Inst., Solana Beach, Calif., 1970—; cons. RAND Corp., Santa Monica, Calif., 1965-66; instr. karate. Author: Karate: Kinematics and Dynamics, 1981. Bd. dirs. Conservatory Ballet Arts Co., Solana Beach, 1981—. NSF fellow, 1967-69; Kelman scholar, 1958-62. Mem. Am. Phys. Soc., Biophys. Soc., Soc. Indsl. and Applied Math., Soc. Math. Biology, Soc. Neurosci. Subspecialties: Biophysics (physics); Statistical physics. Current work: Statistical mechanics of neocortical interactions; global and local human information processing, Riemannian contributions to nuclear forces. Home: Drawer W Solana Beach CA 92075 Office: Phys Studies Inst Drawer W Solana Beach CA 92075

INGLE, L. MORRIS, horticulture educator, researcher; b. Covina, Calif., July 28, 1929; s. Luther and Kaye O. (Traywick) I.; m. Marta Eugenie Salas, Nov. 3, 1960; children: Paul, Kaye, Mark. A.B., U. Calif.-Santa Barbara, 1951; A.M., U. Calif.-Davis, 1956; Ph.D., Purdue U., 1960. Plant physiologist United Fruit Co., La Lima, Honduras, 1959-63; now prof. horticulture W.Va. U., Morgantown. Subspecialty: Plant physiology (agriculture). Current work: Fruit maturation; temperature management; physiological disorders; calcium management. Home: 728 Spring Branch Morgantown WV 26505 Office: WVa U Morgantown WV 26506

INGLIS, JAMES, psychologist, educator; b. Edinburgh, Scotland, Aug. 12, 1927; emigrated to Can., 1959, naturalized, 1965; s. Neil McNeil and Jean (Rourke) I.; m. Lily Brandl, Sept. 5, 1953; children: Jane, Katrin. M.A. with 1st class honors, U. Edinburgh, 1952; Dip.Psychology, U. London, 1953, Ph.D., 1958, D.Sc., 1971. Registered psychologist, Ont. Asst. lectr. to lectr. Inst. Psychiatry, U. London, 1953-59; asst. to assoc. prof. dept. psychology Queens U., Kingston, Ont., 1959-65, prof. psychology, 1968—; assoc. prof. to prof. Dept. Behavioral Sci., Temple U. Med. Sch., Phila., 1965-68; med. research scientist Eastern Pa. Psychiat. Inst., 1965-68; sci. officer Med. Research Council Can., Ottawa, 1974-83. Author: Scientific Study of Abnormal Behavior, 1966; editor: Canadian Jour. Behavioural Sci, 1970-74. Served with RAF, 1945-48. Recipient Annual award Ont. Psychol. Assn., 1982. Fellow AAAS, Am. Psychol. Assn., Brit. Psychol. Soc., Canadian Psychol. Assn. (pres. 1978-79), Gerontol. Soc. Am. Subspecialties: Neuropsychology; Gerontology. Current work: Studies of rehabilitation and cognitive functioning in stroke patients; sex differences in cognitive effects of brain damage. Office: Dept Psychology Queens Univ Kingston ON Canada K7L 3N6 Home: 23 Sydenham St Kingston ON Canada K7L 3G8

INGRAHAM, JOHN LYMAN, bacteriology educator; b. Berkeley, Calif., Sept. 22, 1924; s. Dean Clement and Velma Etta (Lewis) I.; m. Marjorie Frances Mitchell, June 30, 1950; children: Catherine Ann, Thomas Mitchell. B.S., U. Calif.-Berkeley, 1947, Ph.D., 1951. With DuPont Co., Newark, Del., 1951-56, Western Regional Research Lab., U.S. Dept. Agr., Albany, Calif., 1956-58; prof. bacteriology U. Calif.-Davis, 1958—; cons. in field. Author: Microbial World, 1976, Introduction to the Microbial World, 1979; author: Growth of the Bacterial Cell, 1983; contbr. articles to profl. jours. Served with USNR, 1944-46. Guggenheim fellow, 1965-66; NIH grantee, 1962-81; NSF grantee, 1977-83; U.S. Dept. Agr. grantee, 1981-83. Mem Soc. Gen. Microbiology, Am. Soc. Microbiology, Genetics Soc. Am., AAAS, Spanish Microbiol. Soc. (hon.). Democrat. Subspecialties: Microbiology; Genetics and genetic engineering (biology). Current work: Genetic study of bacteria. Office: Dept Bacteriology Univ Calif Davis CA 95616

INGRAM, ALVIN JOHN, surgeon; b. Jackson, Tenn., Mar. 31, 1914; s. Alvin Hill and Margaret (Gallagher) I.; m. Catherine Davis, Feb. 7, 1943; children: Mildred Ingram Dyer, Catherine Ingram Doyle, Peggy Ingram Tagg. B.S., U. Tenn., 1939, M.D., 1939, M.S. in Orthopaedic Surgery, 1947. Diplomate: Am. Bd. Orthopaedic Surgery (dir. 1972-78, v.p. 1976, pres. 1976-78; mem. residency rev. commn. orthopedic surgery 1972-76, chmn. 1975-76). Intern Univ. Hosp., Ann Arbor, Mich., 1939-40, asst. resident surgery, 1940-41; fellow orthopaedic surgery Campbell Clinic, Memphis, 1941-42, 46-47, mem. staff, 1947—, dep. chief of staff, 1967-69, chief of staff, 1970-78, chief of staff emeritus, 1979—; pvt. practice orthopaedic surgery, Memphis, 1947—; med. dir. Crippled Children's Hosp., 1948-61, chief staff, 1961-70; med. dir. Les Passes Cerebral Palsy Treatment Center, 1953-56; med. adv. com. Memphis and W. Tenn. chpt. Nat. Found. Infantile Paralysis, 1947-57, chmn., 1947-55; med. adv. com. Shrine Sch. Crippled Children, 1947-56; med. adv. bd. Variety Club Convalescent Hosp., 1952-56; asso. prof. orthopaedic surgery U. Tenn. Coll. Medicine, 1960-71, prof., chmn. dept., 1971-79, prof. emeritus, 1979—; mem. staff Bapt. Meml. Hosp., exec. com. med. staff, 1969-70, chmn. orthopaedic dept., 1970-74, pres. med. staff, 1973; mem. staff St. Joseph Hosp.; cons. orthopedics Richards Med. Co., 1983—; mem. staff LeBonheur Children's Hosp. (trustee 1968-71); cons. staff Meth. Hosp. Program; chmn. 2d Tenn. Conf. Handicapped Children, 1958; chmn. med. div. United Fund Shelby County, 1961, mem. budget com., 1963-65; dir. at large Nat. Assn. Blue Shield Plans, 1965-70; mem. Gov. Tenn. Adv. Bd. Crippled Children's Service, 1961-77, chmn., 1967-77; mem. exec. com. Am. Bd. Med. Specialties, 1980-83; mem. Tenn. Bd. Med. Examiners, 1981—. Contbr. to books. Bd. dirs. Front St. Theatre, Memphis, 1963-64. Served to maj. M.C., AUS, 1942-46. Mem. Am. Acad. Orthopaedic Surgeons (chmn. program com. 1954, 71, mem. manpower com. 1974-81), Am. Orthopaedic Assn. (chmn. program com., pres. 1973), Central Orthopaedic Club (charter), Tenn. Orthopaedic Soc. (pres. 1963-64), Willis C. Campbell Club (pres. 1967), Internat. Soc. Orthopaedics and Traumatology, Am. Acad. Cerebral Palsy (chmn. program com. 1955, publs. com. 1957, exec. com. 1958, pres. 1958-59), ACS (mem. grad. edn. com. 1974-76), AMA (ho. of dels. 1961-64, trustee 1964-70, sec. treas. 1968-70, sec. bd. trustees 1968-70), So. Med. Assn., Tenn. Med. Assn., Memphis and Shelby County Med. Soc. (pres. 1962, bd. censors 1963-65, ho. of dels. 1965), Nat. Acad. Sci. Inst. Medicine (council 1972-75), Memphis Ind. Practice Assn. (med. dir. 1983—), U.S. C. of C. Methodist (ofcl. bd. 1952—, vice chmn. ofcl. bd. 1965, 66, 69, 70, chmn. 1971-72, gen. chmn. every mem. canvass 1955-57, 63, pres. men's club 1958, sec. stewardship 1964-65). Subspecialties: Orthopedics; Health Care Delivery. Current work: Orthopedic surgery; health care delivery—medical administration. Home: 190 Belle Meade Ln Memphis TN 38117 Office: Campbell Clinic 869 Madison Ave Memphis TN 38173

INGRAM, GLENN R., applied mathematician; b. Terry, Mont., Apr. 25, 1928; m., 1981; 2 children. B.S., Mont. State Coll., 1952, M.S., 1954; Ph.D. in Math, Wash. State U., 1962. From instr. to assoc. prof. math. Mont. State Coll., 1952-65; dir. computer ctr., 1962-65; engr. Mont. Hwy. Dept., 1954-55; asst. computer analyst Wash. State U., 1957-62; assoc. program dir. computer sci. NSF, Washington, 1965-66, program dir. computer facilities, 1966-69, acting head office computer activities, 1969-70; assoc. prof. computer sci. and dir. computer ctr. Wash. State U., 1970-73; chief tech. applications div. Nat. Inst. Edn., 1973-75; chief computer service and tech. ERDA, 1975-78; assoc. dir. computing Nat. Bur. Standards, 1978—; Cons. NSF; mem. steering com. Computers in Undergrad. Curricula. Mem. ACM, IEEE, Soc. Indsl. and Applied Math. Subspecialty: Numerical analysis. Address: 18417 Kingshill Rd Germantown MD 20874

INGRAM, LONNIE O'NEAL, microbiologist, educator; b. S.C., Dec. 30, 1947; s. Thomas Belk and Jean (Weeks) I.; m. Vickie Webb, Oct. 18, 1968; children: Thomas O'Neal, Kenneth Paul. B.S. in Biology, U. S.C., Columbia, 1969; Ph.D. in Biol. Scis, U. Tex., Austin, 1971. NSF summer fellow U. Tex., 1969; postdoctoral researcher Oak Ridge Nat. Lab., 1971-72; asst. prof. microbiology and cell sci., immunology and med. microbiology U. Fla., 1972-76, assoc. prof., 1976-82, prof., 1982—; speaker in field; cons. to industry. Contbr. numerous articles to profl. publs. Recipient undergrad. research award NSF, 1969; Career Devel. award Nat. Inst. Alcohol Abuse and Alcoholism, 1979-83; grantee U. Fla., 1972, 76, Am. Cancer Soc., 1973-75, NSF, 1975-77, 82-84, Distilled Spirits Council Am., 1975-76, NIH, 1975, 78-82. Mem. AAAS, Am. Soc. Microbiology (O. B. Williams award Tex. br. 1970), Sigma Xi. Subspecialties: Biomass (energy science and technology); Genetics and genetic engineering (biology). Current work: Biochemical determinents of alcohol tolerance in bacteria, yeast and mammalian cells; cloning of alcohol resistance genes.

INHABER, HERBERT, risk analyst; b. Montreal, Que., Can., Jan. 25, 1941; came to U.S., 1962; s. Samuel and Mollye (Blumenfeld) I.; m. Elizabeth Rose Bowen, Dec. 21, 1964 (div. 1981). B.Sc., McGill U., Montreal, 1962; M.S., U. Ill.-Urbana, 1964; Ph.D., U. Okla., 1971. Physicist U.S. Steel Research, Monroeville, Pa., 1964-65; sci. adv. Sci. Council Can., Ottawa, 1971-72; policy analyst Can. Dept. Environ., Ottawa, 1972-77; vis. lectr. Yale U., New Haven, 1975; sci. adv. Atomic Energy Control Bd., Ottawa, 1977-80; sr. research scientist Oak Ridge Nat. Lab., 1980—; lectr. Carleton U., Ottawa, 1974-80; cons. Computer Horizons Inc., Cherry Hill, N.J., 1975, OECD, Paris, 1979. Author: Environmental Indices, 1976, Physics of the

Environment, 1978, Energy Risk Assessment, 1982; columnist: Oak Ridger, 1980—; editorial bd.: Scientometrics, 1978—, Risk Analysis, 1980—; contbr. numerous articles to profl. publs. Mem. Soc. Risk Analysis (ad hoc pres. Oak Ridge chpt. 1982-83), Am. Nuclear Soc., Can. Assn. Physicists, Am. Phys. Soc., Friends of Library (v.p. 1982—); mem. Mensa; Mem. Sigma Xi. Club: Geneva (Oak Ridge). Subspecialties: Energy Science and Technology; Environmental Science. Current work: Risk to human health and environment from a variety of sources, ranging from energy to that incurred in transferring from larger to smaller cars; integrating data and assumptions from a wide group of sciences and engineering fields. Home: 28 Montclair Rd Oak Ridge TN 37830 Office: Oak Ridge Nat Lab PO Box X Oak Ridge TN 37830

INHORN, STANLEY LEE, medical educator; b. Phila., Aug. 1, 1928; s. Charles and Nan (Ostrow) Einhorn; m. Shirley Gertrude Sherburne, Aug. 22, 1954; children—Lowell Frank, Marcia Claire, Roger Charles. B.S., Western Res. U., 1949; M.D., Columbia, 1953. Diplomate: Am. Bd. Pathology, Nat. Bd. Med. Examiners. Intern U. Wis. Hosp., Madison, 1953-54, resident, 1956-60; mem. faculty Med. Sch. U. Wis., Madison, 1959—, prof. pathology and preventive medicine, 1969—, chmn. dept. pathology, 1978-81; asst. dir. Wis. Lab. Hygiene, Madison, 1960-66, dir., 1966-79, med. dir., 1979—; cons. medicare div. HEW, 1968-69, 73-74, Center Disease Control, 1968-79. Violinist, Madison Symphony Orch., 1967-74. Bd. dirs. Wis. div. Am. Cancer Soc., Wis. Youth Symphony Orch., 1974—. Served with M.C. USNR, 1954-56. Mem. Am. Soc. Clin. Pathologists, Am. Pub. Health Assn., Am. Soc. Cytology. Subspecialties: Pathology (medicine); Cytology and histology. Current work: Laboratory method in clinical cytology and clinical and environmental cytogenetics. Research in cytogenetics of congenital anomalies, diagnostic lab. practice. Home: 210 Ozark Trail Madison WI 53705

INMAN, BOBBY RAY, electronics executive; b. Rhonesboro, Tex., Apr. 4, 1931; s. Herman H. and Mertie F. (Hinson) I.; m. Nancy Carolyn Russo, June 14, 1958; children: Thomas, William. B.A., U. Tex., 1950; grad., Nat. War Coll., 1972. Commd. ensign U.S. Navy, 1952, advanced through grades to adm., 1981; asst. naval attache, Stockholm, 1965-67, exec. asst., sr. aide to vice chief naval ops., Washington, 1972-73, asst. chief staff intelligence on staff comdr. in chief U.S. Pacific Fleet, 1973-74; dir. Naval intelligence Dept. Navy, Washington, 1974-76; vice dir. Def. Intelligence Agy., 1976-77; dir. Nat. Security Agy., Ft. Meade, Md., 1977-81; dep. dir. CIA, 1981-82; chmn., pres., chief exec. officer Microelectronics and Computer Tech. Corp., Arlington, Va., 1982—. Decorated Def. D.S.M., Navy D.S.M., Legion of Merit, Def. Superior Service medal, Meritorious Service medal, Nat. Security medal, Joint Services Commendation medal. Subspecialty: Computer technology research and development management. Office: 1501 Wilson Blvd 12th Floor Arlington VA 22209

INMAN, JOHN K., research biochemist; b. St. Louis, May 21, 1928; s. Chelsea O. and Dorothy C. (Keith) I.; m. Nancy Jeanne Jaques, Apr. 24, 1954; children: Nancy Jeanne, Louise Anne, Keith Griscom. B.S., Calif. Inst. Tech., 1950; Ph.D. in Biochemistry, Harvard U., 1956. Research group leader Mich. Dept. Health Labs., Lansing, 1956-60; dir. div. biochemistry Ortho Pharm. Corp., Raritan, N.J., 1960-63; USPHS postdoctoral fellow dept. biophysics Johns Hopkins U. Sch. Medicine, Balt., 1963-65; sr. investigator Lab. Immunology, Nat. Inst. Allergy and Infectious Diseases, NIH, Bethesda, Md., 1965—. Contbr. articles to profl. jours. Recipient Dir.'s award NIH, 1978. Mem. Am. Chem. Soc., Am. Assn. Immunologists, AAAS, Sigma Xi. Quaker. Subspecialties: Biochemistry (biology); Immunobiology and immunology. Current work: Principles underlying immunological specificity; multispecificity of antibody combining regions in relation to antibody diversity; development of methods for high-sensitivity, primary structure analysis of antibodies and cell membrane proteins; chemical synthesis and modification of antigens for use in studies on the mechanisms of immune responses. Office: NIH Bldg 10 Bethesda MD 20205

INOKUTI, MITIO, research physicist; b. Tokyo, Japan, July 6, 1933; s. Haruhisa and Takako (Kure) I.; m. Makiko, Mar. 12, 1960; 1 child: Mika. B.A., U. Tokyo, 1956, M.S., 1958, Ph.D., 1962. Instr. U. Tokyo, Japan, 1960-65; research assoc. Northwestern U., Evanston, Ill., 1962-63; research asso. Argonne (Ill.) Nat. Lab., 1963-65, assoc. physicist, 1965-73, sr. physicist, 1973—, head fundamental molecular physics and chemistry sect., 1971—; vis. fellow Joint Inst. for Lab. Astrophysics, U. Colo., Boulder, 1969-70; vis. prof. Japan Soc. for Promotion of Sci., 1978-79, Nordisk Institut fur Teoretisk Atomfysik, 1980. Contbr.: papers to profl. jours. including Jour. Physics B: Atomic and Molecular Physics. Fellow Am. Phys. Soc., Inst. of Physics (London); mem. Phys. Soc. Japan, Radiation Research Soc. Club: Internat. House of Japan (Tokyo). Subspecialties: Atomic and molecular physics; Radiation physics. Current work: Interactions of ionizing radiation with matter; electronic and atomic collisions; theoretical studies. Home: 6481 Blackhawk Trail La Grange IL 60525 Office: Argonne Nat Lab 9700 S Cass Ave Argonne IL 60439

INOUYE, DAVID WILLIAM, ecologist, researcher, educator; b. Phila., Jan. 7, 1950; s. William Yoshio and Eleanor (Ward) I.; m. Bonnie Ann Gregory, May 31, 1969; children: Brian David, Kevin Scott. B.A., Swarthmore Coll., 1971; Ph.D., U. N.C., 1976. Asst. prof. dept. zoology U. Md., College Park, 1976-81, assoc. prof., 1981—; NATO postdoctoral fellow U. Vienna, Austria, 1978; trustee Rocky Mountain Biol. Lab., Crested Butte, Colo., 1978-87. NSF grantee, 1974-84. Mem. Soc. Am. Naturalists, Ecol. Soc. Am. (bd. editors 1979-82), Bot. Soc. Am., Fedn. Am. Scientists (corr.), Sigma Xi. Quaker. Subspecialty: Ecology. Current work: Population biology of plants, pollination biology, ant-plant mutualisms. Home: 3304 Gumwood Dr Hyattsville MD 20783 Office: Dept Zoology Univ Md College Park MD 20742

INSEL, PAUL ANTHONY, internist; b. N.Y.C., Nov. 22, 1945; s. Herman Herbert and Ruth Leona (Friedman) I.; m. Louise Rausa, Dec. 29, 1977; m. Lola Sarah Steinbaum, June 9, 1968; children: Rachel Lauren, Sarah Rebecca Jo. Student, George Washington U., 1962-64; M.D., U. Mich., 1968. Diplomate: Am. Bd. Internal Medicine. Intern, resident in medicine Harvard Med. unit Boston City Hosp., 1968-70; clin. fellow in medicine Harvard Med. Sch., 1969-70; clin. assoc., med. officer Gerontology Research Ctr., NIH, Balt., 1970-74; research fellow U. Calif.-San Francisco, 1974-77, asst. prof. in residence, 1977-78; asst. prof. medicine U. Calif.-San Diego, 1978-81; assoc. prof., 1981—; cons. NIH, NSF, Am. Cancer Soc., VA; mem. pharmacology study sect. NIH, 1982—; Investigator Am. Heart Assn., 1977-82. Contbr. articles to profl. jours. Served with USPHS, 1970-74. U.S. France Coop. Program in Cancer Research awardee, 1981; NSF grantee, 1977—; NIH grantee, 1980—; Am. Heart Assn. grantee, 1977-82; Am. Cancer Soc. grantee, 1982—. Mem. AAAS, Am. Fedn. Clin. Research, Western Soc. Clin. Research, Am. Soc. Pharmacology and Exptl. Therapeutics, Am. Soc. Cell Biology, Endocrine Soc., Am. Soc. Clin. Investigation, Am. Soc. Biol. Chemists, Am. Heart Assn. (basic sci. and hypertension councils). Subspecialties: Molecular pharmacology; Cell biology (medicine). Current work: Mechanisms of catecholamine action and regulation of catecholamine receptors, alpha adrenergic receptors, beta-adrenergic receptors, cyclic AMP, Adenylate cyclase, cultured cells, human platelets. Office: Dept Medicine M-013-H U Calif San Diego LaJolla CA 92093

INSKEEP, EMMETT KEITH, endocrinologist, educator; b. Petersburg, W.Va., Jan. 11, 1938; s. Emmett Vanmeter and June Marie (Clower) I.; m. Ansusan Presby, Aug. 27, 1960; children: Todd Keith, Thomas Clower. B.S., W.Va. U., Morgantown, 1959; M.S., U. Wis.-Madison, 1960, Ph.D., 1964. Asst. prof. animal physiology W.Va. U., 1964-69, assoc. prof., 1970-73, prof., 1974—, chmn. faculty reproductive physiology, 1965—; mem. reproductive biology study sect NIH, 1978-82. Sect. editor physiology: Jour. Animal Sci, 1972-74; contbr. articles profl. jours., chpts. in books. Mem. Morgantown Planning Commn., 1974-76. Recipient research award Nat. Assn. Animal Breeders, 1981, Faculty Cert. of Merit Gamma Sigma Delta, 1973. Mem. Endocrine Soc., Soc. Study of Fertility, Soc. Study of Reprodn. (sec. 1981-83), Am. Soc. Animal Sci. (Outstanding Young Scientist award N.E. region 1977). Republican. Presbyterian. Club: Hilltop Investors (Morgantown). Subspecialties: Animal physiology; Reproductive biology. Current work: Roles of pituitary hormones and ovarian steroids in the lifespan and function of the corpus luteum, anestrus, puberty. Office: G044 Agricultural Science Bldg W Va U Morgantown WV 26506

INSKO, CHESTER ARTHUR, JR., psychology educator; b. Augusta, Ky., June 20, 1935; s. C. Arthur and Elizabeth (Mathews) I.; m. Verla Clemens, Nov. 18, 1961; children: Erik Kenton, Kurt Brian. B.A., U. Calif.-Berkeley, 1957, Ph.D., 1963; M.A., Boston U., 1958. Asst. prof. U. Hawaii, Honolulu, 1963-65; asst. prof. U. N.C.-Chapel Hill, 1965-68, assoc. prof., 1968-72, prof. psychology, 1972—. Author: Theories at Attitude Change, 1967, Experimental Social Psychology, 1972, Introductory Statistics for Psychology, 1977; assoc. editor: Jour. Exptl. Social Psychology, 1968-72; editorial cons.: Jour. Personality and Social Psychology, 1981—. NSF grantee, 1980—. Fellow Am. Psychol. Assn.; mem. Soc. Exptl. Social Psychology, Phi Beta Kappa. Democrat. Subspecialty: Social psychology. Current work: Attitude change, outgroup rejection, interpersonal attraction, balance theory. Home: 610 Surry Rd Chapel Hill NC 27514 Office: Univ NC Davie Hall Chapel Hill NC 27514

IONA, MARIO, physics educator; b. Berlin, June 17, 1917; m., 1949; 2 children. Ph.D. in Physics, U. Vienna, (1939.), Austria. Internat. Student Service fellow U. Uppsala, Sweden, 1939-41; fellow U. Chgo., 1941-42, from research asst. physics to instr., 1941-46; from asst. prof. to assoc. prof. U. Denver, 1946-61, prof. physics, 1961—, coordinator Inter-Univ. High Altitude Labs., 1948-62, coordinator labs., 1962-82; Cons. Denver Pub. Schs., 1962-65, 76—, Denver Research Inst., 1955-77, Jefferson County Pub. Schs., 1973, Jour. Sci. and Children., Jour. ScienceScope. Fellow AAAS; mem. Am. Phys. Soc., Am. Assn. Physics Tchrs. (Disting. Service citation 1971, assoc. editor jour. 1963-66, sect. editor 1970—), Am. Geophys. Union, Nat. Sci. Tchrs. Assn. Subspecialty: Cosmic ray physics. Office: Dept Physics U Denver Denver CO 80208

IORIO, LOUIS CARMEN, research pharmacologist, educator; b. Bronx, N.Y., July 15, 1930; s. Frank and Marie Assunta (Sullo) I.; m. Mary A. Condercuri, Nov. 28, 1953; m. Kathryn A. Whiteman, Mar. 6, 1979; children: Louis F., Teresa, Philip, Michelle, Michael, Stephanie. B.A., Am. Internat. Coll., Springfield, Mass., 1960; Ph.D., SUNY-Buffalo, 1965. Scientist Charles Pfizer, Inc., Groton, Conn., 1965-67; group leader Sandox Pharms., East Hanover, N.J., 1967-73; sect. leader Schering-Plough, Inc., Bloomfield, N.J., 1973—; adj. prof. Fairleigh-Dickinson Coll. Dental Sch. Contbr. numerous articles to profl. jours. Served to sgt. USAF, 1951-55. U.S. Govt. tng. grantee, 1960-65. Mem. Am. Soc. Pharmacology and Exptl. Therapeutics, Am. Pain Soc., N.Y. Acad. Scis., AAAS. Subspecialties: Neuropharmacology; Medicinal chemistry. Current work: Major activity in program for development of pain suppression drugs. Home: RD 1 Box 165 Lebanon NJ 08837 Office: 60 Orange St Bloomfield NJ 07003

IPPEN-IHLER, KARIN ANN, microbiologist, educator; b. Fountain Hill, Pa., Mar. 13, 1942; d. Arthur Thomas and Elisabeth Anne (Wagenplatz) Ippen; m. Garret Martin Ihler, May 2, 1970; children: Elisabeth Emma, Alexander Thomas. B.A., Wellesley Coll., 1963; Ph.D., U. Calif.-Berkeley, 1967. Postdoctoral fellow in microbiology and immunology Harvard Med. Sch., Boston, 1967-69; MRC molecular genetics unit U. Edinburgh, Scotland, 1969-70; asst. prof. biophysics and microbiology U. Pitts., 1970-76; assoc. prof. med. microbiology and immunology Tex. A&M U., College Station, 1977—. Contbr. articles to profl. jours. Nat. Inst. Allergy and Infectious Disease grantee. Mem. AAAS, Am. Soc. Microbiology. Democrat. Subspecialties: Genetics and genetic engineering (biology); Microbiology. Current work: Genetic and biochemical basis of transfer by conjugal plasmids, F factor, conjugation, F-pili, transfer operon, pili assembly. Home: 1115 Langford St College Station TX 77840 Office: Dept Med Microbiology and Immunolog Tex A&M U College Station TX 77843

IQBAL, ZAFAR, biochemist, researcher; b. Lucknow, India, July 12, 1946; came to U.S., 1972, naturalized, 1979; s. Shujaat Ali and Saleha (Begum) Siddiqi; m. Bernida Lucile Iqbal, Nov. 27, 1974; children: Jameel, Shareen. Ph.D., All-India Inst. Med. Scis., New Delhi, 1971. Asst. research officer All India Inst. Med. Scis., New Delhi, 1968-71; research assoc. physiology Ind. U., Indpls., 1972-77, asst. prof. biochemistry, med. biophysics, 1977-82; asst. prof. neurology Northwestern U. Sch. Medicine, Chgo., 1982—. Contbr. numerous articles to profl. jours. Council Sci. and Indsl. Research fellow, 1963-66; Juvenile Diabetes Found. fellow, 1981; Am. Cancer Soc. grantee, 1979; NSF grantee, 1980; Ind. Acad. Sci. research grantee, 1979; Ind. U. research grantee, 1978; Am. Diabetes Assn. grantee, 1980. Mem. Am. Physiol. Soc., Soc. Neurosci., Internat. Brain Research Orgn., Internat. Soc. Neurochemistry, Am. Soc. Neurochemistry, Biophys. Soc., Soc. Exptl. Biology and Medicine, N.Y. Acad. Sci., Ind. Acad. Sci. (chmn. cell biology 1982-83), Soc. Biol. Chemists, Indian Acad. Neurosci., Sigma Xi. Subspecialties: Biochemistry (medicine); Neurochemistry. Current work: Neurochemistry, analysis of axoplasmic transport, role of calcium-binding protein in neuronal function, role of calcium and polyamines in signal transduction. Office: Medical Science Bldg Indianapolis IN 46223

IQBAL, ZAFAR MOHD, biochemist, pharmacologist, cancer researcher, consultant; b. Hyderabad, India, Dec. 12, 1938; came to U.S., 1965, naturalized, 1973; s. M.A. and Haleemunissa (Begum) Rahim. B.Sc., Osmania U., 1958; M.Sc., Osmonia U., 1962; Ph.D., U. Md., 1970. Fogarty Internat. fellow Nat. Cancer Inst., Bethesda, Md., 1970-71, staff fellow, 1971-74; asst. prof. pharmacology Case Western Res. U., Cleve., 1974-76; assoc. dir. ERC programs in occupational toxicology U. Ill. Med. Center, Chgo., 1980-81, assoc. prof. microbiology, 1977-80, assoc. prof. occupational and environ. health, 1976—, assoc. prof. preventive medicine, 1982—; cons.; grant reviewer; lectr. continuing edn. Contbr. articles profl. jours. Council Sci. and Indsl. Research of India research fellow, 1963-65; Fogarty Internat. fellow NIH, 1970-71; NSF exchange scientist, 1981. Mem. Am. Assn. Cancer Research, Am. Pancreatic Assn., N.Y. Acad. Scis., Am. Chem. Soc., Ill. Cancer Council, AAUP, AAAS, Soc. Toxicology, Sigma Xi. Subspecialties: Biochemistry (medicine); Cancer research (medicine). Current work: Cancer research, chemical and environmental/occupational carcinogenesis, drug metabolism, DNA damage/repair in cell culture and organs, toxicology (environmental and occupational), molecular and biochemical pharmacology, science reviewer federal, state and other public interest agencies, science consultant private and governmental agencies.

IRELAND, JOHN RICHARD, mechanical and nuclear engineer; b. Hereford, Tex., June. 25, 1951; s. Richard and Patsy Ann (Slagle) I.; m. Judith Ann Shipman, Dec. 23, 1969; children: Travis Ryle, Shannon Marie. B.S. in Mech. Engring, N.Mex. State U., 1974; M.S. in Engring, U. Calif.-Berkeley, 1977. Program engr. Gen. Electric Co., San Jose, Calif., 1974-77; sect. leader Los Alamos Nat. Lab., 1977-80, project leader, 1980-82, assoc. group leader, 1982-83, project mgr. for nuclear saftey programs, 1983—; cons. Presdl. Commn. Accident Three Mile Island. Mem. Am. Nuclear Soc., ASME. Democrat. Methodist. Subspecialties: Nuclear engineering; Nuclear fission. Current work: Briefly, I manage a large number of programs associated with nuclear reactor safety. Home: 329 Rover Blvd Los Alamos NM 87544 Office: Los Alamos Nat Lab PO Box 1663 MSK5566 Los Alamos NM 87545

IRGOLIC, KURT JOHANN, chemist, educator; b. Hartberg, Austria, Sept. 28, 1938; s. Johann and Aloisia (Staude) I.; m. Gerlinde Zillich, Feb. 1, 1964; 1 dau., Birgit Petra. Cert. elem. edn., Tchrs. Tng. Coll., Graz, Austria, 1957; postgrad., Karl Franzens U., Graz, 1957-63, Ph.D. in Inorganic and Analytical Chemistry, 1964. Postdoctoral fellow Tex. A&M U., College Station, 1964-66, asst. prof., 1966-72, assoc. prof., 1972-77, prof. chemistry, 1977—, research coordinator, 1972-75, assoc. dir., 1975—; cons. to industry. Author: (with F. L. Kolar) Chemistry for Liberal Arts Majors, 1973, The Organic Chemistry of Tellurium, 1974, (with R. O. O'Connor) Fundamentals of Chemistry in the Laboratory, 1974, 2d edit., 1981, (with F. L. Kolar) Chemistry 106 Laboratory, 1975; contbr. chpts. to books, articles to profl. jours. R.A. Welch Found. grantee, 1969, 71, 72, 73, 75, 76, 77, 78, 79, 80, 81, 82; NIH grantee, 1974, 75, 76, 77, 78; EPA grantee, 1978; NATO grantee, 1979, 80, 82; numerous others. Mem. Am. Chem. Soc., Internat. Assn. Hydrogen Energy, Sigma Xi, Phi Lambda Upsilon, Phi Kappa Phi, Brazos Valley Gem and Mineral Soc. (pres. 1971). Roman Catholic. Subspecialties: Inorganic chemistry; Analytical chemistry. Current work: Synthesis of organic compounds of arsenic, selenium, tellurium; biological transformation of arsenic, selenium compounds; trace element analysis; inductively coupled plasma emission spectrometry. Home: 1819 Hondo College Station TX 77840 Office: Tex A&M U College Station TX 77843

IRGON, JOSEPH, researcher, energy cons., research and devel. exec.; b. Polonnoe, Ukraine, Russia, Dec. 30, 1919; came to U.S., 1922, naturalized, 1942; s. Joseph and Ida (Galperin) I.; m. Thelma Pugach, Apr. 11, 1948; children: Deborah L., Judith M., Adam E. B.S. in Chem. Engring, Northeastern U., Boston, 1943; Ph.D. in Phys. Chemistry, M.I.T., 1948. Industry project dir. or dept. head Gen. Foods Corp., 1948-52, Reaction Motors, Inc., 1952-56; founder, tech. head Fulton-Irgon Corp., 1956-60; Fulton-Irgon div. and Hydro-Space div. Lithium Corp. Am., 1960-63, Proteus, Inc., 1963-69, Ocean Recovery Systems, Inc., 1969-73; dir. Joseph Irgon Assocs., Fairfield, N.J., 1979—; cons. dir. research Energy Tech., Inc. (div. Gen. Machine & Instrument Corp.), 1979—; energy and materials researcher; cons. Author: The Most Far-reaching: Supersonic Aircraft Escape Principle, 1954, First Handbook of Rocket Propellants, 1956, First Handbook of Ocean Materials, 1964, An Emergency Buoyancy System for Deep-diving Submarines, 1966. Mem. Am. Chem. Soc., ASME. Subspecialties: Solar energy; Biomass (energy science and technology). Current work: New light-focusing principle and large-scale conversion of biomass, invention and preliminary research followed by cooperative programs with industry or govts. Patentee in field. Home and Office: 144 Emmans Rd Flanders NJ 07836

IRONS, EDGAR T(OWAR), computer science educator; b. Detroit, Oct. 1, 1936. B.S.E., Princeton U., 1958; M.S., Calif. Inst. Tech., 1959. Research asst. computer programming Princeton U., 1959-60; mem. tech. staff comuter research Inst. Def. Analyses, 1961-69; assoc. prof. computer sci. Yale U., New Haven, 1969-72, prof., 1972—; vis. lectr. Princeton U., 1965-66. Mem. ACM. Subspecialties: Programming languages; Operating systems. Office: Dunham Lab Yale U 10 Hillhouse Ave New Haven CT 06520

IRVINE, CYNTHIA EMBERSON, astronomer; b. Washington, Aug. 14, 1948; d. Richard Maury and Virginia Burke Nicol Emberson; m. Nelson James Irvine, June 5, 1971; 2 daus., Alice Kathleen, Laura Elizabeth. B.A., Rice U., 1970; Ph.D. in Astronomy, Case Western Res. U., 1975. Research assoc. U.S. Naval Postgrad. Sch., Monterey, Calif., 1975-81; founding mem. Monterey Inst. Research in Astronomy, 1972, exec. dir., 1981-82, pres., 1982—; instr. Lyceum of Monterey. Contbr. articles to profl. jours. Mem. Astron. Soc. Pacific, Am. Astron. Soc., Pacific Grove Mus. Natural History, Optical Soc. Am. Subspecialty: Optical astronomy. Office: 900 Major Sherman Ln Monterey CA 93940

IRVINE, EILEEN M., research scientist, consultant; b. Albuquerque, Nov. 22, 1948; d. James and Mildred Patricia (Jones) I.; m. Gene Winkler, Oct. 10, 1982. B.A. with honors, U. Calif.-Santa Barbara, 1970, M.A. in Environ. Biology, 1972. Jr. math. analyst Dynalectron Corp., Point Mugu, Calif., 1972; assoc. engr. Raytheon Co., Santa Barbara, 1972-74; research scientist ERT, Inc., Westlake Village, Calif., 1977-81, cons., 1981—. Contbr. articles to profl. jours. Mem. Ecol. Soc. Am., Bot. Soc. Am., Mortar Board. Subspecialties: Ecosystems analysis; Integrated systems modelling and engineering. Current work: Computer simulations—environmental impact research with emphasis on mathematical modeling. Studies on toxics in food chain model, accidental spill of hazardous chemical models, air quality dispersion models. Home and office: 3040 Gibralter Rd Santa Barbara CA 93105

IRVING, DONALD J., university dean; b. Arlington, Mass., May 3, 1933; m. Jewel P. Irving; children: Kevin William, Todd Lawrence. B.A., Mass. Coll. Art, 1955; M.A., Columbia U. Tchrs. Coll., 1956, Ed.D., 1963. Tchr. art White Plains (N.Y.) High Sch., 1958-60; instr. art SUNY-Oneonta, 1960-62; prof. art, dean Moore Coll. Art, Phila., 1963-67; chmn. art dept., dir. Peabody Mus. Art, George Peabody Coll. Tchrs., Nashville, 1967-69; dir. Sch. Art Inst., Chgo., 1969-82; dean Faculty Fine Arts U. Ariz., Tucson, 1982—; mem. U.S. del. Conf. Nat. Soc. Edn. Through Art, Prague, Czechoslovakia, 1966; cons. ednl. TV series Art Now, WRCV-TV, Phila. Author: Sculpture Material and Process, 1970; contbr. articles in field to profl. jours. Mem. Nat. Assn. Schs. Art (treas., dir. 1975-77), Union Ind. Colls. Art (chmn., dir.), Nat. Council Art Adminstrs. (dir.), Fedn. Ind. Ill. Colls. and Univs. (dir.), Nat. Art Edn. Assn. (officer Eastern region 1966-68), Eastern Arts Assn. (council 1964-66, mgr. conv. 1959-64), Nat. Council Arts in Edn., Internat. Soc. Edn. Through Art, Coll. Art Assn., Phi Delta Kappa. Subspecialties: Planetary science; Radio and microwave astronomy. Home: 5810 E Paseo San Valentine Tucson AZ 85715 Office: Univ of Arizona Coll of Fine Arts Tucson AZ 85721

IRVING, EDWARD, geophysicist; b. Colne, Eng., May 27, 1927; s. George Edward and Nellie (Petty) I.; m. Sheila Ann Irwin, Sept. 23, 1957; children: Kathryn Jean, Susan Patricia, Martin Edward, George Andrew. B.A., Cambridge (Eng.) U., 1950, M.A., 1957, Sc.D., 1965;

D.Sc. (hon.), Carleton U., 1979. Research fellow Australian Nat. U., Canberra, 1954-57, fellow, 1957-59, sr. fellow, 1959-64; sci. officer Can. Dept. Mines and Tech. Surveys, Ottawa, Ont., 1964-66; prof. geophysics U. Leeds, Eng., 1966-67; research scientist Can. Dept. Energy Mines and Resources, Ottawa, 1967-81, Sidney, B.C., 1981—; adj. prof. geology Carleton U., Ottawa. Author: Paleomagnetism, 1964; asso. editor: Tectonphysics, Physics of Earth and Planetary Interiors; contbr. articles on paleomagnetism and related geol. topics to profl. publs. Served with Brit. Army, 1945-48. Recipient Chestien Mica Gondwana medal Mining, Geol. and Metall. Inst. India, 1965. Fellow Royal Soc. Can., Royal Soc. (London), Am. Geophys. Union (Walter H. Bucher medal 1979), Royal Astron. Soc. (U.K.), Geol. Soc. Am., Geol. Assn. Can. (Logan medal 1975). Mem. United Ch. Can.. Subspecialties: Geophysics; Tectonics. Current work: Paleomagnetic studies in Canadian Cordillera, the Canadian shield and in Arctic Islands to determine their tectonic history. Office: Pacific Geosci Ctr PO Box 6000 Sidney BC V8L 4B2 Canada

IRWIN, GEORGE RANKIN, physicist, mechanical engineering educator; b. El Paso, Tex., Feb. 26, 1907; s. William Rankin and Mary (Ross) I.; m. Georgia Shearer, June 10, 1933; children: Joseph Ross, Mary Susan Irwin Gillett, Sarah Belle Irwin Lofgren, John Shearer. A.B., Knox Coll., 1930; M.S., U. Ill., 1933, Ph.D., 1937; D.Eng. (hon.), Lehigh U., 1977. Asso. prof. physics Knox Coll., 1935-36; fellow physics U. Ill., 1936-37; physicist U.S. Naval Research Lab., 1937-67; prof. mechanics Lehigh U., 1967-72; prof. mech. engring. U. Md., 1972—; vis. prof. U. Ill., 1961, 70; hon. lectr. Internat. Congress on Fracture, 1981. Contbr. articles to profl. jours. Recipient Navy Disting. Civilian Service award, 1947; Knox Coll. Alumni Assn. Achievement award, 1949; Navy Conrad award, 1969; Grand medal French Metall. Soc., 1976; Clamer award Franklin Inst., 1978; Md. Gov.'s citation, 1982. Fellow ASTM (Dudley medal 1960, hon. mem. 1974), Washington Acad. Sci.; mem. Washington Philos. Soc., Soc. Exptl. Stress Analysis (Murray lectr. 1973, Lazan award 1977), ASME (Thurston lectr. 1966, Nadai award 1977), Am. Soc. Metals (Sauveur award 1974), Nat. Acad. Engring. Subspecialties: Fracture mechanics; Materials (engineering). Current work: Fracture mechanics applications to structural safety, materials research and engineering education. Pioneer devel. fracture mechanics. Home: 7306 Edmonston Ave College Park MD 20740

IRWIN, LOUIS NEAL, biologist; b. Big Spring, Tex., Jan. 8, 1943; s. Mac Neal and Anna Beth (Friedrichs) I.; m. Kay Lanette Bullock, Jan. 26, 1963; m. Carol Lee Irwin, Nov. 24, 1967; children: Sean Stewart, Anthony Irwin, Brian Irwin. B.A., Tex. Tech. U., 1965; Ph.D., U. Kans., 1969. Research fellow Kans. Bur. Child Research, Parsons, 1969-70; asst. prof. biology Coll., Pharm. Scis., Columbia U., N.Y.C., 1970-73; asst. prof. physiology Wayne State U. Med. Sch., Detroit, 1973-76; assoc. biochemist Eunice Kennedy Shriver Center, Waltham, Mass., 1977—; assoc. prof. biology Simmons Coll., Boston, 1980—. Contbr. articles to profl. jours. Recipient 1st prize writing competition Perspectives in Biology and Medicine, 1978. Mem. Internat. Soc. Neurochemistry, Am. Soc. Neurochemistry, Soc. Neuroscience, Herpetologists' League, AAAS, Sigma Xi. Subspecialties: Neurobiology; Neurochemistry. Current work: Developmental and evolutionary neurobiology; developmental biology; brain evolution; molecular evolution; neural plasticity. Home: 420 Lowell Ave Newton Ma 02160 Office: 300 The Fenway Boston MA 02115

IRWIN, MARY JANE, computer scientist, educator; b. Cairo, Ill., July 14, 1949; d. Darrell D. and Mary Elizabeth (Bastin) Simmons; m. Vernon Charles Irwin, July 1, 1966; 1 son, John Darrell. B.S. magna cum laude, Memphis State U., 1971; M.S., U. Ill., Urbana-Champaign, 1975, Ph.D., 1977. Grad. teaching asst. dept. computer sci. U. Ill., Urbana-Champaign, 1972-74, grad. research asst., 1974-77; asst. prof. dept. computer sci. Pa. State U., University Park, 1977-83, assoc. prof., 1983. Contbr. articles to profl. jours. Research grantee Pa. State U., 1978-79; NSF grantee, 1978-80, 80-82, 83-84; Office Naval Research grantee, ARO grantee, 1983-85, 1983-84. Mem. IEEE Computer Soc., Tech. Com. Computer Architecture (sec. 1980—), Assn. Computing Machinery, Spl. Interest Group in Computer Architecture. Subspecialties: Computer architecture; Distributed systems and networks. Current work: Computer arithmetic, supercomputers, VLSI systems design, multiprocessor orgns. Office: 305 Whitmore Lab Pa State U University Park PA 16802

ISAACSON, ROBERT LEE, psychologist, research scientist, center administrator, educator; b. Mich., Sept. 26, 1928; s. Emil Alfred and Evelyn Edna (Johnson) I.; m. Susan Doherty, Aug. 1, 1956; m. Ann W. Braden, Dec. 31, 1974; children: Gunnar, Lars, Mary Ingrid, Mary Christina. A.B., U. Mich., 1950, M.S., 1954, Ph.D., 1958. Asst. prof. psychology U. Mich., 1958-63, assoc. prof., 1963-67, prof., 1967-68, U. Fla., 1968-76, grad. research prof., 1976-78; Disting. prof. psychology SUNY, Binghamton, 1978—, dir., 1978—. Author: books, the most recent being The Limbic System, 1974, 2d edit., 1982, (with N. E. Spear) The Expression of Knowledge, 1982; contbr. numerous articles to profl. publs. Pres. Alachua County (Fla.) Assn. Retarded Children, 1970-72; mem. Fla. Gov.'s Blue Ribbon Com. for Mental Retardation, 1974. Served to lt. USN, 1950-53. NSF and NIH awardee. Fellow Am. Psychol. Assn., AAAS, Am. Physiol. Soc., Soc. Neurosci., Internat. Brain Research Orgn., Sigma Xi. Subspecialties: Neuroscience; Psychobiology. Current work: Limbic system of brain; recovery from brain damage; neuropeptides and behavior. Office: Dept Psychology SUNY Binghamton NY 13901

ISAKOFF, SHELDON ERWIN, chem. engr.; b. Bklyn., May 25, 1925; s. Harry and Rebecca I.; m. Anita Ginsburg, Aug. 18, 1946; 1 son, Peter D. B.S., Columbia U., 1945, M.S., 1947, Ph.D., 1952. Guest fellow Brookhaven Nat. Lab., Upton, N.Y., 1949-50; with E.I. duPont de Nemours & Co., Inc., Wilmington, Del., 1951—, dir. engring. research and devel., 1975—; mem. Nat. Materials Adv. Bd., 1980—. Served with USNR, 1943-46. Fellow Am. Inst. Chem. Engrs. (past dir., Founders award 1980), AAAS; mem. Am. Chem. Soc., Nat. Acad. Engring., Sigma Xi, Tau Beta Pi, Phi Lambda Upsilon. Subspecialty: Chemical engineering. Address: RD 1 Box 361 Chadds Ford PA 19317

ISENBURGER, HERBERT R., radiation laboratory executive. Pres. St. John X-Ray Lab., radiation cons., engring., tng., and ct. work, Califon, N.J.; mem. N.J. Gov.'s Adv. Com. on Radiation Protection. Author: Bibliography on X-Ray Stress Analysis, 1966, Bibliography on Industrial Radiology, 1942-70, (with Ancel St. John) Industrial Radiology, 2nd edit, 1943, Bibliography on Filmbadge Monitoring, 1961; contbr. to: Ency. Chem. Tech; contbr. numerous articles to profl. jours., confs. Mem. ASME (com. on equipment for non-med. radiation applications), Am. Soc. Metals, ASTM, Soc. Nondestructive Testing. Subspecialty: Radiology. Current work: Radiation monitoring. Office: Box 192 Route 2 Califon NJ 07830

ISGANITIS, research chemist; b. Bklyn., Nov. 5, 1954; s. Peter P. and Anita (Sirianni) I.; m. Jamie K. Costello, May 22, 1976. Student, Northwestern U., 1972-73; B.S. in Chemistry, Fordham U., 1975; M.S., U. Rochester, 1977; Ph.D. in Chemistry, U. Rochester, 1981. Mem. tech. staff Xerox Corp., Webster, N.Y., 1981—. Sherman Clarke fellow, 1976-79. Mem. Am. Chem. Soc., Am. Phys. Soc., Optical Soc. Am. Subspecialties: Physical chemistry; Spectroscopy. Current work: Spectroscpic characterization of materials, bulk and thin films, UV-visible, IR, Raman, colorimetry on polymeric materials and additives. Home: 112 Brunswick St Rochester NY 14607 Office: Xerox Corp 800 Philips Rd W 130 Webster NY 14580

ISHII, DOUGLAS NOBUO, pharmacologist, neurobiologist; b. Santa Ana, Calif., July 30, 1942; s. William T. and Fusaye (Hatanaka) I.; m. (div.); children: Jordon T., Gregory S., Aaron T. B.A., U. Calif., Berkeley, 1967; Ph.D. in Pharmacology, Stanford U., 1974; postdoctoral study in neurobiology, Stanford U., 1974-76. Instr. pharmacology U. Pacific, San Francisco, 1968-70; asst. prof. dept. pharmacology Columbia U. Coll. Physicians and Surgeons, N.Y.C., 1977-83, assoc. prof. dept. pharmacology, 1983—; also researcher Cancer Research Center. Contbr. articles to profl. publs. Recipient career devel. award NIH, 1978-83; Damon Runyon-Walter Winchell Cancer Research postdoctoral fellow, 1974-75. Mem. Am. Soc. Pharmacology and Exptl. Therapeutics, Soc. Neurosci., Am. Assn. Cancer Research, Amnesty Internat. Subspecialties: Neurobiology; Cancer research (medicine). Current work: Biology of human neuroblastoma and pheochromocytoma; mechanism of nerve growth factor action; mechanism of tumor promoter action. Office: Pharmacology Dept Columbia U 630 W 168th St New York NY 10032

ISHIMARU, AKIRA, electrical engineering educator, consultant; b. Rukuoka, Japan, Mar. 16, 1928; came to U.S., 1952, naturalized, 1963; s. Shigezo and Yumi (Yamada) I.; m. Yuko Kaneda, Nov. 21, 1956; children: John, Jane, James, Joyce. B.S. in Elec. Engring, U. Tokyo, 1951, Ph.D., U. Wash, 1958. Engr. Electro-Tech. Lab., Tokyo, 1951-52; mem. tech. staff Bell Telephone Lab., Holmdel, N.J., summer 1956; asst. prof. elec. engring. U. Wash., 1958-61, assoc. prof., 1961-65, prof., 1965—; vis. assoc. prof. U. Calif.-Berkeley, 1963-64; cons. Jet Propulsion Lab., Pasadena, Calif., 1964—. Author: Wave Propagation and Scattering in Random Media, 1978; editor: Radio Sci. jour, 1978—; assoc. editor: Jour. Optical Soc. Am, 1973—; editorial bd. Procs. IEEE, 1973—. Fellow IEEE (Region VI Achievement award 1968), Optical Soc. Am.; mem. Internat. Union Radio Sci. (vice chmn. Commn. B.). Subspecialties: Electrical engineering; Optical scattering. Current work: Research on wave propagation and scattering, random media, remote sensing, imaging, optical and acoustic scattering. Home: 2913 165th Pl NE Bellevue WA 98008 Offic: U Was Dept Elec Engring FT-10 Seattle WA 98195

ISLER, RALPH CHARLES, physicist; b. Pitts., Apr. 23, 1933; s. Ralph C. and Sadie (White) I.; m. Nancy J., Sept. 13, 1958; children: Susan, Janet, Robert. B.S., U. Pitts., 1955; Ph.D., John Hopkins U., 1963. Research assoc. Columbia U., N.Y.C., 1964-66; faculty U. Fla., Gainesville, 1966-78, prof. physics, 1976-78; sr. research staff mem. Oak Ridge Nat. Lab., 1978—. Served to lt. U.S. Army, 1955-57. Mem. Am. Phys. Soc. Subspecialties: Plasma physics; Atomic and molecular physics. Current work: Spectroscopic research on impurities in thermonuclear plasmas. Home: 110 Canterbury Rd Oak Ridge TN 37830 Office: Oak Ridge National Lab Fusion Energy Div Oak Ridge TN 37830

ISMAIL, MOURAD E. H., mathematician, educator; b. Cairo, Egypt, Apr. 27, 1944; came to Can., 1968, naturalized, 1979; s. El-Houssieny M. and Aisha (El-Shourbagy) I.; m. Thanaa M.T. Rashed, June 6, 1969. B.Sc., Cairo U., 1964; M.A., U. Alta., Edmonton, 1969, Ph.D., 1974. Asst. scientist U. Wis.-Madison, 1974-75; vis. lectr., postdoctoral fellow U. Toronto, 1975-76; asst. prof. applied math. McMaster U., Hamilton, Ont., 1976-81; asst. prof. math. Ariz. State U., Tempe, 1979-80, assoc. prof. math., 1980-82, prof. math., 1982—. Contbr. research papers to math. jours. Grantee, 1976—. Mem. Am. Math. Soc., Math. Assn. Am., Soc. for Indsl. and Applied Math., Can. Soc. for History and Philosophy of Math. Muslim. Subspecialties: Applied mathematics; Special functions and approximation theory. Current work: Orthogonal polynomials, spectra of Jacobi matrices, Bessel functions, basic hypergeometric functions, hypergeometric functions, asymptotics, infinite divisibity. Office: Dept Math Ariz State Univ Tempe AZ 85287

ISRAEL, HERBERT WILLIAM, botanist, educator; s. Elmer William and Mildred Elsa (Koepp) I.; m. Ruth Mary Goetz., June 20, 1953; children: Lynn M. Israel Walter, Carla A., Willaim J. B.S. in Elem. Edn, Concordia Tchrs. Coll., 1953; postgrad., Northwestern U., 1951-53, U. Chgo., 1954-57; M.S. in Botany-Zoology, U. Wis., 1960, Ph.D., U. Fla., 1962. Registered referee Am. Jour. Botany, Nature, Protoplasma, Sci., Virology. Chmn. dept. sci. Luther High Sch. South Chicago, 1953-60; NSF fellow U. Wis.-Madison, 1959-60; NDEA fellow, Turtox scholar, grad. teaching asst. dept. botany U. Fla., Gainesville, 1960-62, research assoc., 1963; NIH vis. postdoctoral fellow dept. botany Cornell U., Ithaca, N.Y., 1963-64, research assoc. lab. cell physiology, growth and devel., 1964-69, sr. research assoc., 1969-72, sr. research assoc. dept. plant pathology, 1972—; mem. SEM policy com., 1977—; lectr. in field. Contbr. numerous articles on botany to profl. jours.; assoc. editor: Phytopathology; editorial adv. bd.: Protoplasma, 1973-81. Baseball coach Kiwanis, 1975-80. Mem. AAAS, Am. Inst. Biol. Scis., Am. Phytopath. Soc. (disease and pathogen physiology com. 1974-77, editorial bd. 1981—). Republican. Lutheran. Club: Finger Lakes Cycling (Ithaca) (treas.). Subspecialties: Plant cell and tissue culture; Plant pathology. Current work: Plant cell biology, plant pathology. Home: 954 Snyder Hill Rd Ithaca NY 14850 Office: 303 Plant Sci Bldg Cornell U Ithaca NY 14853

ISRAEL, MARTIN HENRY, physicist, educator; b. Chgo., Jan. 12, 1941; s. Herman and Anna Catherine I.; m. Margaret Ellen Mitouer, June 20, 1965; children: Elisa, Samuel. S.B., U. Chgo., 1962; Ph.D., Calif. Inst. Tech., 1969. Asst. prof. physics Washington U., St. Louis, 1968-72, assoc. prof., 1972-75, prof., 1975—, assoc. dir. McDonnell Ctr. for Space Scis., 1982—. Contbr. articles to profl. jours. Recipient Exceptional Sci. Achievement award NASA, 1980; Alfred P. Sloan fellow, 1970-72. Mem. Am. Phys. Soc., Am. Astron. Soc., AAAS, AAUP. Subspecialty: Cosmic ray high energy astrophysics. Current work: Composition and origin of cosmic rays. Office: Dept Physics Washington U Saint Louis MO 63130

ISSELBACHER, KURT JULIUS, educator, physician; b. Wirges, Germany, Sept. 12, 1925; came to U.S., 1936, naturalized, 1945; s. Albert and Flori (Strauss) I.; m. Rhoda Solin, June 22, 1955; children: Lisa, Karen, Jody, Eric. A.B., Harvard U., 1946, M.D. cum laude, 1950. Intern, then resident Mass. Gen. Hosp., Boston, 1950-53; investigator NIH, 1953-56; chief gastrointestinal unit Mass. Gen. Hosp., 1957, chmn. com. research, 1967; prof. medicine Harvard Med. Sch., 1966—, chmn. exec. com. depts. medicine, 1968—, Mallinckrodt prof. medicine, 1972—, chmn. univ. cancer com., 1972—. Editor-in-chief: (Harrison) Principles of Internal Medicine, 1976. Fellow Am. Acad. Arts and Scis., ACP; mem. Nat. Acad. Scis., Assn. Am. Physicians (pres. 1977-78). Subspecialties: Gastroenterology; Cell biology (medicine). Current work: Studies of membrane transport of sugars and amino acids in normal and malignant cells; intestinal cell structure and function, biosynthesis of membrane glycoproteins; viral hepatitis diagnosis and treatment. Monoclonal antibodies in the diagnosis and treatment of cancer. Research in structure and function of intestinal cells, membrane changes in malignant cells and serologic tests for malignancy. Discovered cause of galactosemia as 1st definitely proven disease due to hereditary enzyme defect; elucidated mechanism of intestinal fat absorption and causes of fatty liver; described genetic disturbance of amino acid and lipid metabolism (isovaleric acidemia); developed new serologic tests for diagnosis of cancer and viral hepatitis. Home: 20 Nobscot Rd Newton Center MA 02159 Office: Mass Gen Hosp Boston MA 02114

ITANO, HARVEY AKIO, biochemist; b. Sacramento, Nov. 3, 1920; s. Masao and Sumako (Nakahara) I.; m. Rose Nakako Sakemi, Nov. 5, 1949; children—Wayne Masao, Glenn Harvey, David George. B.S., U. Calif., Berkeley, 1942; M.D., St. Louis U., 1945; Ph.D., Calif. Inst. Tech., 1950. Research fellow Calif. Inst. Tech., 1950-52, sr. research fellow, 1952-54; biochem. researcher NIH, Bethesda, Md., 1954-70, chief sec. chem. genetics, 1962-70; prof. pathology U. Calif. Sch. Medicine, San Diego, 1970—; vis. prof. U. Chgo., 1965, U. Calif., San Francisco, 1967, Osaka (Japan) U., 1961-62; Minot lectr. AMA, 1955. Mem. Nat. Acad. Scis., Am. Chem. Soc. (Eli Lilly award 1954), Am. Soc. Biol. Chemists, Am. Soc. Hematology, Internat. Soc. Hematology, AAAS. Subspecialty: Hematology. Research on sickle cell anemia and other hematological conditions, 1946. Office: U Calif San Diego La Jolla CA 92093

ITO, MICHIO, cancer scientist, researcher, virologist; b. Niigata, Japan, Sept. 24, 1930; came to U.S., 1970; s. Takeo and Hide (Hattori) I.; m. Fumiko, Feb. 1, 1933; children: Yuri, Mana. M.D., Niigata Med. Coll., 1954; Ph.D., Inst. Med. Sci., U. Tokyo, 1960. Scientist dept. enteroviruses NIH of Japan, Tokyo, 1960-67, chief smallpox sect., 1967-70; research fellow dept. virology and epidemiology Baylor Coll. Medicine, 1965-67, Govt. Japan fellow, 1965-66; research asst.prof. microbiology SUNY-Buffalo, 1970-74, asst. research prof., 1976—; asst. prof. microbiology and immunology U. Ark. Med. Scis., 1974-75; sr. cancer research scientist Roswell Park Meml. Inst., Buffalo, 1976—. Recipient Henry C. and Bertha H. Buswell Fellowship award dept. microbiology Sch. Medicine, SUNY-Buffalo, 1970-73. Mem. Am. Soc. Microbiology, AAAS. Subspecialties: Cell biology (medicine); Infectious diseases. Current work: Biological effects of interferons; anticellular and antitumor effects; effects of ammonia on cell cultures and virus replication; virus-induced cell membrane antigens; tumor-specific antigens. Home: 38 Cottonwood Dr Williamsville NY 14221 Office: Roswell Park Inst 666 Elm St Buffalo NY 14263

ITO, YOICHIRO, government research scientist, physician; b. Osaka, Japan, Dec. 22, 1928; came to U.S., 1968, naturalized, 1978; s. Taichi and Ai (Kubota) I.; m. Ryoko Tanioka, Dec. 23, 1963; children: Koichi, Shin. M.D., Osaka City U., 1958. Rotating intern U.S. Yokosuka (Japan) Naval Hosp., 1958-59; resident in pathology Cleve. Met. Gen. Hosp., 1959-61, Michael Reese Hosp., Chgo., 1961-63; instr. physiology Osaka City U. Med. Sch., 1963-68; vis. scientist Nat. Heart, Lung and Blood Inst., NIH, Bethesda, Md., 1968-78, med. officer, 1978—. Mem. Japanese Am. Citizens League, Kenshinkai. Recipient 1st place award ann. sci. research presentation at Cleve. Met. Gen. Hosp., 1960, Tech. Excellence award for devel. blood cell separator, 1979; Fulbright exchange scholar, 1959-63; WHO Research Tavel Fund grantee Nat. Inst. Med. Research, London, 1968. Subspecialties: Analytical chemistry; Chemical engineering. Current work: Innovation in separation science, including continuous development of countercurrent chromatography, cell separation methods. Initiated and developed countercurrent chromatography; patentee coil planet centrifuge, rotating-seal-free flow-through centrifuge. Office: 9000 Rockville Pike NIH Bldg 10 Room 5D-12 Bethesda MD 20205

ITOGA, STEPHEN YUKIO, scientist educator; b. Honolulu, June 14, 1943. B.S., Cornell U., 1965, M.S., 1966; Ph.D., UCLA, 1973. Mem. tech. staff TRW, Redondo Beach, Calif., 1966-65; sr. staff engr. Fairchild Co., San José, Calif., 1981-82; assoc. prof. U. Hawaii, 1975—. Mem. Assn. for Computing Machinery, IEEE, Soc. for Indsl. and Applied Math. Subspecialties: Algorithms; Database systems. Current work: Theory of Algorithms, Data base management systems, formal languages. Office: Univ Hawaii 2565 The Mall Honolulu HI 96822

ITURRIAN, WILLIAM BEN, pharmacologist, educator, cattleman; b. Hudson, Wyo., May 17, 1939; s. Benito and Dorothy June (Eley) I.; m. Rosanne Connolly, Mar. 22, 1965; children—Kit, Andre, Novajean, Peter. B.S. in Pharmacy, U. Wyo., 1962; Ph.D., Oreg. State U., 1968. Registered pharmacist, Wyo., Oreg. Instr. Oreg. State U., 1964-66, research asso., 1967; asst. prof. pharmacology U. Ga., Athens, 1968-72, assoc. prof., 1972—; research investigator in epilepsy Med. Coll. Ga., 1979—; cattleman, 1972—. Contbr. articles to sci. jours. Vice pres. Ga. chpt. A.G. Bell Parents of Deaf Children, 1970-76; pres. Ga. Assn. Parents of the Deaf, 1981. Named Jaycees Outstanding Young. Ga. Educator, 1973; AMA teratology fellow, 1966. Mem. Soc. for Neurosci., Internat. Soc. Devel. Psychobiology, Ga. Acoustical Soc., Ga. Acad. Sci. (chmn. biomedicine sect.), Ga. Soc. for Neurosci. (pres. 1981), Oglethorpe Cattlemen (treas. 1978—), Sigma Xi. Subspecialties: Neuropharmacology; Toxicology (medicine). Current work: Effects of noise on the developing brain; research and teaching. Home: 125 Appleby Dr Athens GA 30605 Office: Pharmacology Dept U Ga Athens GA 30602

IVERSON, GILBERT MICHAEL, immunobiologist, educator; b. San Diego, May 1, 1938; s. Gilbert and Julia Ann (Persons) I.; m. Nora Antonette Keolker, Apr. 20, 1968; children—Peter, Robert, Sven. B.A., San Jose State U., 1964; Ph.D., Nat. Inst. Med. Research, London, 1972. Fellow tumor immunology unit Imperial Cancer Research Fund, Univ. Coll., London, 1971-73; staff fellow cellular immunology sect. lab. microbiology and immunology Nat. Inst. Dental Research, Bethesda, Md., 1974-74; mem. acad. staff-research dept. genetics Stanford (Calif.) U. Sch. Medicine, 1974-78; research assoc. in pathology, lectr. biology Yale U., New Haven, 1978-80; sr. assoc. lab. cellular immunology Howard Hughes Med. Inst., Yale U. Sch. Medicine, 1980—. Contbr. articles to profl. jours. Bd. dirs. Madison (Conn.) ABC Program; bd. dirs. Youth Soccer, Madison. Served with USMC, 1956-58. Recipient Disting. Alumni award San Jose State U., 1982; NIH grantee. Mem. Am. Assn. Immunologists, Brit. Soc. Immunology, Brit. Transplantation Soc. Subspecialties: Immunobiology and immunology; Immunocytochemistry. Current work: Understanding the molecular basis of immunoregulation. Home: 501 Opening Hill Rd Madison CT 06443 Office: 310 Cedar St New Haven CT 06510

IVERSON, LOUIS ROBERT, ecology research scientist; b. Jamestown, N.D., June 25, 1954; s. Norris V. and Virginia L. I.; m. Margaret Grace Saethre, May 6, 1978; 1 dau., Heather Renee. B.S., U. N.D., 1976; Ph.D. (Chester Fritz scholar), 1981; postgrad. (Fulbright-Hays scholar), U. York, Eng., 1979-80. Postdoctoral fellow Mining and Minerals Resources Research Inst., U. N.D., Grand Forks, 1981-82, grad. research asst., 1976-77, NSF trainee, dept. biology, 1977-79, grad. teaching asst., 1981; asst. profl. scientist Ill. Natural History Survey, Champaign, 1982—. Contbr. articles on ecology to profl. jours. Mem. Ecol. Soc. Am., Brit. Ecol. Soc., N.D. Acad. Sci., Soil Sci Soc. Am., Agronomy Soc. Am., Phi Beta Kappa, Sigma Xi. Lutheran. Subspecialties: Ecology. Current work: Reclamation of mined land ecosystems, plant ecology, plant-plant interactions, plant-soil relationships, autecology of Chenopodiaceae, nitrogen fixation, allelochemics, succession, physiological plant ecology. Office: Ill Natural History Survey 607 E Peabody Dr Champaign IL 61820

IVES, PHILIP TRUMAN, genetics researcher; b. Amherst, Mass., Aug. 15, 1909; s. Frank Truman and Nellie Persis (Gage) I.; m.

Dorothy Ann Mays, June 30, 1940; children—Richard T., Elinor Ives Goff, Donald F., Carolyn D. Ives Dingman. B.A., Amherst Coll., 1932, M.A., 1934; Ph.D., Calif. Inst. Tech., 1938. Faculty Amherst (Mass.) Coll., 1938—, research asso., 1938-75, emeritus, 1975—. Contbr. articles in field to profl. jours. Fellow AAAS; mem. Am. Soc. Zoologists, Genetics Soc. Am., Soc. for Study of Evolution. Republican. Congregationalist. Subspecialties: Genetics and genetic engineering (agriculture); Climatology. Current work: Research genetic structure of natural Drosophila populations. Office: Amherst College Box 1773 Amherst MA 01002

IZAWA, SEIKICHI, biology educator, plant physiologist; b. Yokohama, Japan, Sept. 28, 1926; came to U.S., 1963; s. Hiroshi and Shika (Oshige) I.; m. Toyoko Tsukada, Nov. 27, 1968; 1 dau., Eri. B.S., U. Tokyo, Japan, 1950, D.Sc., 1961. Asst. prof. Tokyo Inst. Tech., 1961-63; research assoc. Mich. State U., 1963-66; assoc. prof. Queen's U., Kingston, Ont., Can., 1966-68; assoc. prof. biology Mich. State U., 1970-74; prof. biology Wayne State U., Detroit, 1974—. Contbr. numerous articles in profl. jours.; mem. editorial bd.: Plant Physiology, 1981—, Plant and Cell Physiology, 1983—. C. F. Kettering Found. Internat. fellow, 1963-64; NSF Research grantee, 1968—; Japan Soc. Promotion of Disting. Visitor fellow, 1981. Mem. Am. Soc. Biol. Chemists, Am. Soc. Plant Physiologists, Am. Soc. Photobiologists. Subspecialty: Photosynthesis. Current work: Research activity centered on the biochemistry and physiology of plant chloroplasts related to the mechanisms of energy transduction and oxygen evolution. Home: 16974 Lauderdale Dr Birmingham MI 48009 Office: Wayne State U Detroit MI 48202

IZZO, JOSEPH L., JR., physician; b. Rochester, N.Y., July 25, 1946; s. Joseph L. and Helen E. (Pedersen) I.; m. Margaret J. Sussmann, May 20, 1972. A.B., Princeton U., 1968; M.D., Johns Hopkins U., 1972. Diplomate: Am. Bd. Internal Medicine. Resident in medicine Washington U., St. Louis, 1972-74; staff assoc. Nat. Heart, Lung and Blood Inst., NIH, Bethesda, Md., 1975-79; asst. prof. medicine U. Rochester, 1979—, dir. hypertension service dept. medicine, 1980—, nephrologist, 1980—. Served to lt. comdr. USPHS, 1975. Buswell fellow, 1979; Sinsheimer Found. fellow, 1982. Mem. Am. Fedn. Clin. Research, Am. Soc. Nephrology, Internat. Soc. Nephrology, InterAm. Soc. for Hypertension, AAAS. Subspecialties: Neuroendocrinology; Pharmacology. Current work: Blood pressure control mechanisms and pharmacotherapy. Sympathetic nervous pathophysiology; catecholamine metabolism. Home: 85 Palmerston Rd Rochester NY 14618 Office: U Rochester Box MED 601 Elmwood Ave Rochester NY 14642

JACKEL, JANET LEHR, physicist; b. N.Y.C., May 30, 1947; d. Paul and Adele (Fred) Lehr; m. Lawrence David Jackel, 1969; children: David Aaron, Robert Abraham. B.A. in Physics magna cum laude, Brandeis U., 1969; M.S., Cornell U., 1973, Ph.D., 1976. Mem. tech. staff guided wave research lab. Bell Labs., Holmdel, N.J., 1976—. Contbr. articles to profl. jours. Mem. Optical Soc. Am. Subspecialty: Guided-wave optics. Office: Bell Labs Holmdel NJ 07733

JACKLET, JON W., biologist, educator; b. L.I., N.Y., Apr. 16, 1935; s. Clyde W. and Hilda I. (Hague) J.; m. A.C. Carleton, Dec. 21, 1962; children: Alan, Ben, Jessica. Ph.D. in Biology, U. Oreg., 1967. Postdoctoral fellow Calif. Inst. Tech., Pasadena, 1967-68; asst., then assoc. prof. biology SUNY, Albany, 1968-81, prof., 1981—. Contbr. articles to sci. publs. Served with USMC, 1954-57. Research grantee NIH, 1969-78, NSF, 1978—. Mem. Soc. Neurosci. Subspecialties: Neurobiology; Chronobiology. Current work: Neural correlates of behavior. Neuronal mechanisms of rhythmic activity, especially circadian. Office: Dept Biology SUNY Albany NY 12222

JACKSON, BERNARD VERNON, astrophysicist, astronomer; b. Peoria, Ill., Nov. 7, 1942; s. Ervin W. and Martha L. (Kappler) J.; m.; children: David, Mark. B.S., U. Ill., 1964; M.A., Ind. U., 1967, Ph.D., 1970. Postdoctoral fellow dept. geophysics and space sci. UCLA, 1970-73, sr. scientist, South Pole Sta., Antarctica, 1970-71; sci. programmer Arecibo (P.R.) Obs., 1973-75; postdoctoral fellow Skylab Solar Workshops, High Altitude Obs., Boulder, Colo., 1975-78; research asst. Colo. U., 1978-80; research asst. dept. elec. engring. and computer sci. U. Calif. at San Diego, LaJolla, 1980—. Contbr. articles to profl. jours. Jackson Glacier named in his honor; NSF grantee, 1975. Mem. Am. Astron. Soc., Am. Geophys. Union. Republican. Methodist. Clubs: U. Calif. at San Diego Sailing (pres.), Hanohano (San Diego) (treas.). Subspecialties: Solar physics; Radio and microwave astronomy. Current work: Solar physics, interplanetary scintillation, radio astronomy, optical solar astronomy. Office: U Calif San Diego Dept EE/CS Code C-014 La Jolla CA 92903

JACKSON, CHARLES EUGENE, physician, educator; b. Bluffton, Ind., Aug. 17, 1925; s. Charles Spurgeon and Lois Marie (Kyle) J.; m. Norma Lea Wampler; children: Penny Rupley, Janice Byrd, Jeffrey A. A.B. in Chemistry, Ind. U., 1944; M.D., Ind. U. Sch. Medicine, Indpls., 1946. Diplomate: Am. Bd. Internal Medicine, Am. Bd. Med. Genetics. Intern Scott and White Hosp., Temple, Tex., 1946-47; resident Tulane U. Hosp., Charity Hosp., New Orleans, 1949-51; dir. clin. research Caylor Nickel Clinic, Bluffton, Ind., 1956-70; staff internal medicine Henry Ford Hosp., Detroit, 1970—, chief clin. genetics div. dept. medicine, 1975—; clin. prof. medicine U. Mich. Sch. Medicine, Ann Arbor, 1978—. Contbr. articles on clin. genetics and endocrinology to profl. publs. Served to maj. M.C. U.S. Army, 1947-49. Mem. Am. Soc. Human Genetics, Endocrine Soc., Central Soc. Clin. Research. Methodist. Subspecialties: Internal medicine; Genetics and genetic engineering (medicine). Current work: Clinical genetics, particularly endocrine neoplasia syndromes. Home: 17111 E Jefferson St Grosse Pointe MI 48230 Office: Henry Ford Hosp Detroit MI 48202

JACKSON, DAVID ARCHER, molecular biologist, business executive; b. N.Y.C., Apr. 29, 1942; m., 1966. A.B., Harvard U., 1964; Ph.D., Stanford U., 1969. USPHS fellow Med. Sch., Stanford U., 1969-70, Nat. Cystic Fibrosis Research Found. fellow, 1970-72; asst. prof. microbiology U. Mich., Ann Arbor, 1972-76, assoc. prof., 1977-81; chmn. sci. adv. bd. Genex Corp., Rockville, Md., 1977—, v.p., sci. dir., 1980—; Cons. Pres.'s Commn. for Study of Ethical Problems in Medicine, Bio-medicine and Behavioral Research, 1981—; mem. ad hoc com. nat. issues in genetic engring. NSF, 1981—; mem. adv. panel Office Technol. Assessment, 1981—; adj. prof. applied molecular biology U. Md.-Balt. County, 1981—. Mem. Am. Soc. Microbiology, Genetic Soc. Am., Assn. Research Dirs. Subspecialties: Genetics and genetic engineering (biology); Molecular biology. Office: Genex Corp Labs Gaithersburg MD 20877

JACKSON, DAVID PHILLIP, research physicist, educator; b. Toronto, Ont., Can., Oct. 2, 1940; s. Peter Morris and Bertha Belle (Morris) J.; m. Susan Jane Bell, May 30, 1975; children: Scott, Timothy. B.Sc., U. Toronto, 1962, M.A., 1964, M.A. Sci., 1966, Ph.D., 1968. Sci. programmer IBM, 1962-63; research asst. Inst. Aerospace Studies, 1965-68; sr. research officer Atomic Energy Can., Chalk River Nuclear Labs., 1968—; vis. scientist Max Planck Inst. Plasma Physics, Munich, W.Ger., 1964-65, U. Calgary, 1969, U. Leeds, 1972; prof. engring. physics McMaster U., 1979—; mem. Inst. Materials Research, 1979—, Inst. Energy Studies, 1980—; cons. vis. prof. Bell Labs., Murray Hill, N.J., 1982—; Can. mem. Internat. Energy Agy., 1982—. Editor: Atomic Collisions in Solids, 1980; contbr. articles to profl. jours. Served with Royal Can. Naval Res., 1958-75. Awardee research grants. Mem. Am. Phys. Soc., Chem. Inst. Can., Can. Assn. Physicists. Subspecialties: Condensed matter physics; Fusion. Current work: ion solid interactions, first wall problem in fusion systems. Home: 3 Hammond Ct PO Box 1338 Deep River ON Canada KOJ 1PO Office: Atomic Energy Can Chalk River ON Canada KOJ 1JO

JACKSON, DON VERNON, JR., oncologist; b. Portsmouth, Va., Sept. 11, 1946; m. Davie Ann Haga, June 29, 1968; children: Robert William, Carrie Ann, Don Vernon III. B.A., Emory and Henry Coll., 1968; M.D., U. Va., 1973. Diplomate. Am. Bd. Internal Medicine. Intern Roanoke (Va.) Meml. Hosp., 1972-73; resident N.C. Bapt. Hosp., 1973-75; fellow in hematology, oncology Bowman Gray Sch. Medicine, Winston-Salem, 1975-78, asst. prof. medicine, 1977-82, asoc. prof., 1982—. Contbr. articles to profl. jours. Am. Cancer Soc. fellow, 1978-80. Mem. Am. Assn. Cancer Research, Am. Fedn. Clin. Research, Am. Soc. Clin. Oncology, N.C. Med. Soc., AMA, Forsyth County Med. Soc. Methodist. Subspecialties: Cancer research (medicine); Oncology. Current work: Vinca alkaloids, clin. cancer research. Home: 3071 Magazine Dr Winston-Salem NC 27106 Office: 300 S Hawthorne Rd Winston-Salem NC 27103

JACKSON, ETHEL NOLAND, geneticist; b. Geneva, N.Y., Apr. 27, 1944; d. Lloyd H. and Ethel (Beare) Nol; m. David Archer, Jackson, June 17, 1966. A.B. cum laude, Radcliffe Coll., 1966; Ph.D., Stanford U., 1973. Postdoctoral fellow dept. human genetics U. Mich., Ann Arbor, 1972-74, asst. prof. microbiology, 1974-81; research dir. vectors and expression systems dept. Genex Corp., Gaithersburg, Md., 1981—. Contbr. articles to profl. jours. Am. Cancer Soc. fellow, 1974; NIH grantee, 1974-81; NSF grantee, 1978-81; Robert Steling Clarke Found. grantee, 1974-77; H.H. Rackham Sch. Grad. Studies grantee, 1978. Mem. Am. Soc. Microbiology, Genetics Soc. Am., AAAS, Research Club U. Mich., Am. Women in Sci. Club (Washington). Subspecialties: Genetics and genetic engineering (biology); Gene actions. Current work: Researcher in regulation of gene expression, chromosome structure, bacteriophage morphogenesis. Home: 11612 Danville Dr Rockville MD 20852 Office: 16020 Industrial Dr Gaithersburg MD 20877

JACKSON, GEORGE JOHN, microbiologist; b. Vienna, Austria, Dec. 10, 1931; s. Arthur and Louise (Wagner) J. A.B., U. Chgo., 1951, M.S., 1954, Ph.D., 1958. Instr. U. Chgo., 1958-59; guest investigator Rockefeller Inst./U., N.Y.C., 1959-62, faculty, 1962-72; head parasitology lab., food and cosmetics microbiology br. Div. Microbiology, Bur. Foods, FDA, Washington, 1972-80, acting br. chief, 1980-82, chief, 1982—; guest investigator Amazon Research Inst., Manaus, Brazil, 1963; U.S. del. to Food Hygiene Com. WHO/FAO Codex, Alimentarius Commn., 1980—. Editorial staff: Chgo. Rev, 1954-58; editor: Experimental Parasitology; mem. editorial bd.: Ballet rev; contbr. poetry, lit., dance criticism to publs., U.S., Europe. USPHS fellow, career devel. awardee, research grantee, 1959-72; La. State U. travelling fellow in tropical medicine to Central Am., 1958. Mem. Am. Soc. Parasitologists, Am. Soc. Tropical Medicine and Hygiene, Soc. Protozoologists, Helminthol. Soc. Washington, N.Y. Soc. Tropical Medicine, Assn. Ofcl. Analytical Chemists, Dance Critics Assn. Subspecialty: Microbiology (medicine). Office: HFF-234 FDA Washington DC 20204

JACKSON, (JOHN) DAVID, physics educator; b. London, Ont., Can., Jan. 19, 1925; came to U.S., 1957; s. Walter David and Lillian Margaret (Ferguson) J.; m. Barbara Cook, June 26, 1949; children: Ian, Nan, Maureen, Mark. B.Sc., U. Western Ont., 1946; Ph.D., M.I.T., 1949. From asst. prof. to assoc. prof. math. McGill U., Montreal, Que., Can., 1950-56; from assoc. prof. to prof. physics U. Ill., Urbana, 1957-67; prof. physics U. Calif., Berkeley, 1967—, chmn. dept., 1978-81, assoc. dir., head physics div., 1982-83; mem. high energy physics adv. panel Dept. Energy, Washington, 1982—; chmn. vis. com. for Fermilab, Batavia, Ill., 1979-83; mem. sci. policy com. Stanford Linear Accelerator Ctr., 1980—. Author: Physics of Elementary Particles, 1957, Classical Electrodynamics, 2d edit, 1975; editor: Ann. Rev. Nuclear and Particle Sci, 1977—; contbr. articles to profl. jours. Guggenheim fellow, 1956-57; vis. fellow Clare Hall, Cambridge, Eng., 1970. Fellow Am. Phys. Soc.; mem. AAUP, Berkeley Faculty Assn., ACLU. Subspecialties: Particle physics; Theoretical physics. Current work: Theoretical research in particle physics. Office: Dept Physics U Calif Berkeley CA 94720

JACKSON, JOHN EDWIN, electro-optics design engr., cons.; b. Delaware, Ohio, Apr. 26, 1941; s. Edwin Milton and Wilma Francis (Krichbaum) J.; m. Sharon Kay Holladay, Aug. 11, 1942; children: Jason J., Jeremiah B. B.S. in Physics, Ohio State U., 1964, M.S., 1965. Engr. McDonnell Douglas Co., St. Louis, 1965-72, sr. tech. specialist, 1972—, mgr. submarine laser communications receiver contract, 1981—; cons. Ninetronics, St. Louis. Contbr. articles on optics and laser communications to profl. jours. Sponsoring treas. Boy Scouts Am., Hazelwood, Mo. Mem. Optical So. Am., Phi Eta Sigma, Sigma Pi Sigma. Subspecialties: Optics research; Laser research. Current work: Manager of advanced development submarine laser communications receiver. Home: 8357 Latty Ave Hazelwood MO 63042 Office: Box 516 Saint Louis MO 63166

JACKSON, KENNETH LEE, radiological health educator; b. Berkeley, Calif., Jan. 6, 1926; m., 1948; 4 children. A.B., U. Calif.-Berkeley, 1949, Ph.D. in Physiology, 1954. Research asst. Donner Lab., U. Calif.-Berkeley, 1949-51; research physiologist Office Naval Research Unit One, 1951-53; sr. investigator biochem. br. U.S. Naval Radiol. Def. Lab., 1954-60; head radiobiology group Boeing Co., 1960-63; from asst. prof. to assoc. prof. U. Wash., Seattle, 1963-76, prof. environ. health, head div., 1979—, chmn. radiol. sci. group., 1966—. Mem. AAAS, Am. Physiol. Soc., Radiation Research Soc., Health Physics Soc., Am. Pub. Health Assn. Subspecialty: Radiation biology. Office: Radiol Sci SB-75 U Wash Seattle WA 98195

JACKSON, PETER STERLING, nuclear engineer; b. Lower Merion, Pa., Jan. 14, 1956; s. Sterling Warren and Roxanna Page (Hood) Tillotson J.; m. Kathleen Ann Esper, Aug. 18, 1979; 1 son, Daniel Peter. B.S., Rensselaer Poly. Inst., 1978, M.Engr., 1979, Ph.D., 1981. Registered profl. engr., N.Y. Sr. engr. Comubstion Engring., Windsor, Conn., 1981-83; prin. engr. Combstion Engring., Windsor, 1983—. Mem. Am. Nuclear Soc., IEEE, Am. Statis. Assn., Soc. Risk Analysis, Sigma Xi (assoc.). Republican. Subspecialties: Nuclear engineering; Probability. Current work: Development of analytical models and methods for demonstrating the reliability and safety of nuclear power plant systems and components. Office: Combustion Engring Inc 1000 Prospect Hill Rd Windsor CT 06095

JACKSON, RAYMOND CARL, cytogeneticist, biosystematist, educator, cons; b. Medora, Ind., May 7, 1928; s. Thornton Comadore and Flossie Oliva (Booker) J.; m. T. June Jackson, Oct. 24, 1947; children: Jeffrey, Rebecca. A.B., Ind. U., 1952, A.M., 1953; Ph.D., Purdue U., 1955. Instr. biology U. N.Mex., Albuquerque, 1955-57, asst. prof., 1957-58; asst. prof. botany U. Kans., 1958-60, assoc. prof., 1961-64, prof., 1964-71, chmn. dept. botany, 1969-71; prof. biol. scis. Tex. Tech U., Lubbock, 1971—, chmn. biol. scis., 1971-78; vis. assoc. prof. U. Iowa, summer 1962, Ind. U., summer 1963. Contbr. articles to sci. jours. Served in USAF, 1946-49. Mem. Am. Soc. Naturalists, Soc. Study of Evolution, Bot. Soc. Am., AAAS, Internat. Assn. Plant Taxonomy, Am. Soc. Plant Taxonomists, Internat. Orgn. Plant Biosystematists. Subspecialties: Genome organization; Systematics. Current work: Genome relationships; mathematical models to predict meiotic configurations and their frequencies in diploids and polyploids. Home: 3726 64th Dr Lubbock TX 79413 Office: Tex Tech U Lubbock TX 79409

JACKSON, THOMAS LARRY, psychologist, educator; b. Antioch, Calif., Aug. 9, 1951; s. Thomas Larry and Mary Kathryn (Donlon) J.; m. Patricia Ann Petretic, July 28, 1978. B.A., U. Pacific, 1973; M.A., Bowling Green State U., 1975, Ph.D., 1978. Lic. psychologist, S.D. Asst. prof. psychology U. S.D., Vermillion, 1978-82, coordinator psychol. services, 1981—; cons. Neurol. Inst. and Pain Center, Sioux City, Iowa, 1978—. Contbr. articles to profl. jours. Mem. Am. Psychol. Assn., S.D. Psychol. Assn. (pres. 1982-83), Sigma Xi. Subspecialties: Behavioral psychology; Neuropsychology. Current work: Attribution of incest blame; psychotherapy; neuropsychology; expert witness testimony; professional ethics. Home: 14 Willow St Vermillion SD 57069 Office: Dept Psychology U SD Vermillion SD 57069

JACKSON, WILLIAM DAVID, research executive; b. Edinburgh, Scotland, May 20, 1927; came to U.S., 1955, naturalized, 1968; s. Joseph and Margaret (Johnston) J.; m.; children—Margaret Eleanor, David Foster. B.Sc., U. Glasgow, Scotland, 1947, Ph.D., 1960; postgrad., U. Strathclyde, Glasgow, 1948. Apprentice English Electric Co., Stafford, 1945-47; research asst. elec. engring. dept. U. Strathclyde, Glasgow, 1948-51; lectr. elec. engring. U. Manchester, Eng., 1951-55, 57-58; vis. lectr. dept. elec. engring. Mass. Inst. Tech., 1955-57, asst. prof., 1958-62, asso. prof., 1962-66, lectr. elec. engring., 1968-73; vis. prof. Tech. U., Berlin, Germany, 1966; prof. elec. engring., dept. energy engring. U. Ill., Chgo., 1966-67; prin. research scientist, dir. tech. edn. Avco-Everett Research Lab., Everett, Mass., 1967-72; prof. elec. engring. U. Tenn. Space Inst., Tullahoma, 1972-73; mgr. Electric Power Research Inst., Palo Alto, Calif., 1973-74; mgr. office coal research Interior Dept., Washington, 1974-75; dir. magnetohydrodynamic div. ERDA, Washington, 1975-77; dir. tech. analysis div. Office Energy Research, Dept. Energy, Washington, 1977-79; pres. Energy Cons., Inc., 1979—, HMJ Corp., 1982—; professorial lectr. George Washington U., 1979—; cons. numerous indsl. firms and govt. agys., 1948—; chmn. internat. magnetohydrodynamic liaison group Internat. Atomic Energy Agy./UNESCO, 1966—, chmn., 1969-74; coordinator coop. program magnetohydrodynamic power generation, U.S.-USSR, 1974-79, mem. numerous govt. and internat. coms. and panels. Editor: Electricity From MHD, 1968; editorial bd.: Internat. Jour. Elec. Engring. Edn, 1962-70. U.K. Fulbright scholar, 1955-57. Fellow Instn. Elec. Engrs. (U.K.) (past com. sec., chmn.), IEEE (sr.; sec.-treas. profl. group biomed. electronics Boston sect. 1962-63, energy devel. subcom. 1973—, congl. fellow 1984); mem. AIAA, AAUP, Am. Phys. Soc., AAAS, ASME (past chmn. energetics div.), Am. Soc. Engring. Edn., Sigma Xi. Subspecialty: Electrical engineering. Home and office: 3509 McKinley St NW Washington DC 20015 Office: 11508 Reguid Dr Silver Spring MD 20902

JACKSON, WILLIAM MORGAN, chemist, physicist, educator, researcher, cons.; b. Birmingham, Ala., Sept. 24, 1936; s. William Morgan and Claudia C. (Haygood) J.; m. Ethel Barnes, June 19, 1959; children: Eric Morgan, Cheryl Lynn; m. Lydia Catherine Bourne, May 5, 1982. B.S., Morehouse Coll., 1956; PhD., Cath. U. Am., 1961. Chemist Nat. Bur. Standards, Washington, 1959-61, NRC postdoctoral assoc., 1963-64; research scientist Martin-Marietta Co., 1961-63, Goddard Space Flight Ctr., 1964-69, 70-74; vis. prof. physics U. Pitts., 1969-70; prof. chemistry Howard U., Washington, 1974—. Contbr. articles to sci. jours. Recipient Percy L. Julian award Howard U., 1979, Outstanding Research award Coll. Liberal Arts, 1982. Mem. Am. Chem. Soc., AAAS, Optical Soc. Am., Am. Phys. Soc., Nat. Orgn. Black Chemists and Chem. Engrs. (Disting. Service award 1978), Phi Beta Kappa, Sigma Xi. Subspecialties: Laser photochemistry; Laser-induced chemistry. Current work: Photodissociation dynamics using laser induced fluorescence detection of free radicals, astrochemistry, chemical kinetics. Office: Chemistry Dept Howard U Washington DC 20059

JACO, E. GARTLY, sociomedical science educator, medical sociologist, researcher; b. Memphis, Oct. 5, 1923; s. Oscar Hubert and Delzell (Simpson) J.; m. Adele Marie Bolles, May 28, 1947; children: Linda Dell, Jerry Monroe, Andrew Richard, John Douglas. B.A., U. Tex.-Austin, 1949, M.A., 1950; postgrad., U. Chgo., summer 1951; Ph.D., Northwestern U., 1954. Instr. U. Tex.-Austin, 1952-55; cons. Hogg Gound. Mental Health, 1978—; assoc. prof. neuropsychiatry and preventive medicine and community health U. Tex. Med. Br., Galveston, 1955-59, prof., 1978—, dir. div. health behavior, 1978-80; dir. Lab. Socialenviron. Studies, Cleve. Psychiat. Inst., 1959-61; assoc. prof. Western Res. U., 1959-62; prof. hosp. adminstrn. Sch. Public Health, U. Minn. Med. Center, Mpls., 1962-66; and research dir. and dir. Ph.D. program Sch. Public Health U. Minn.-Mpls., 1962-66; prof. sociology U. Calif.-Riverside, 1966-78, chmn. dept. sociology, 1966-70. Author: books, including Patients, Physicians and Illness, 1958, 3d edit., 1979, Social Epidemiology of Mental Disorders, 1960; contbr. articles to profl. publs.; co-founder, editor-in-chief: Jour. Health and Human Behavior, 1960-67. Served to sgt. USAAF, 1943-46; CBI. NIH grantee, 1959, 60, 62; Kellogg Found. grantee, 1960-62, 82-83; Russell Sage Found. award, 1959-61; Hill Family Found. award, 1964-67. Fellow Am. Public Health Assn., Am. Sociol. Assn.; mem. AAAS, Am. Psychol. Assn., Internat. Sociol. Assn., Sigma Xi. Presbyterian. Subspecialties: Sociobiology; Behavioral ecology. Current work: Research in psychosocial factors in health and illness, analysis of health care delivery system, ethnic/racial factors in illness; social epidemiology of illness; relationships of human society to health status. Office: U Tex Med Br Dept Preventive Medicine and Community Health Galveston TX 77550

JACOB, ROBERT JOSEPH KASSEL, computer scientist, educator; b. Bklyn., Nov. 11, 1950; s. Ezekiel Joseph and Ethel Charlotte (Behr) J.; m. Kathryn Allamong, June 9, 1973; 1 dau., Charlotte Allamong. B.A., Johns Hopkins U., 1972, M.S.E., 1974, Ph.D., 1976. Research asst. Johns Hopkins U., Balt., 1972-76; vis. asst. prof. Towson State U., Balt., 1976; computer scientist Naval Research Lab., Washington, 1977—; assoc. professorial lectr. George Washington U., Washington, 1978—. Contbr. chpts. to books, writings to profl. publs. Hopkins fellow, 1973-75; recipient Kiwanis Layman's award Bklyn., 1968. Mem. IEEE, Assn. Computing Machinery, Am. Statis. Assn., Human Factors Soc., Sigma Xi. Jewish. Subspecialties: Software engineering; Human factors engineering. Current work: Human factors in computer systems. Investigating formal specification techniques for describing user-computer interaction and rapid prototyping, study of faces to represent multidimensional data. Office: Naval Research Lab Code 7590 Washington DC 20375 Home: 911 Portner Pl Alexandria VA 22314

JACOBS, DAVID, mech. engr.; b. Bklyn., Sept. 28, 1951; s. Bernard George and Selma Ann (Monoson) J. B.S.M.E., Northeastern U., 1974. Project engr. Barry Controls, Watertown, Mass., 1974-77, sr. project engr., 1977-81, mgr. new products devel., 1981—. Mem. ASME, Boston Rubber Group, Pi Tau Sigma. Clubs: Old Gold Rugby (Boston); New Bedford (Mass.) Yacht. Subspecialties: Mechanical engineering; Elastomerics. Current work: Development of new products in field of shock/vibration isolation systems with research in highly damped, high temperature resistant elastomeric materials. Patentee in field of elastomer design. Office: 700 Pleasant St Watertown MA 02172

JACOBS, DIANE MARGARET, microbiologist; b. Port of Spain, Trinidad and Tobago, Mar. 24, 1940; d. Saul and Eleanor (Rosenberger) J. A.B., Radcliffe Coll., 1961; Ph.D., Harvard U., 1967. Instr., lectr. Hadassah Med. Sch., Jerusalem, 1967-71; research fellow U. Calif.-San Diego, 1971-74; research assoc. Salk Inst. Biol. Studies, LaJolla, Calif., 1974-76; assoc. prof. SUNY-Buffalo, 1976-80, prof. microbiology, 1980—. Contbr. articles to profl. jours. NIH grantee, 1974—; Am. Cancer Soc. grantee, 1977-80. Mem. Am. Assn. Immunologists, Am. Soc. Microbiologists, AAAS, Assn. Women in Sci. Subspecialty: Immunobiology and immunology. Current work: Immunomodulatory agents of bacterial origin particularly lipopoly saccharide; lymphocyte membrane receptors, lymphocyte triggering. Office: 321 Sherman Hall Dept Microbiology SUNY Buffalo NY 14214

JACOBS, DURAND FRANK, psychologist; b. Chgo., Sept. 28, 1922; s. Joseph and Jeannette (Friedman) J.; m. Lenore M. Sobel, Sept. 4, 1949; children: Beth Anne, William Walder, Timothy. B.S., U. Mich., 1947, M.S., 1949; Ph.D., Mich. State U., 1953. Diplomate: in clin. psychology Am. Bd. Profl. Psychology. Chief counseling psychol. service VA Hosp., Marion, Ind., 1953-57, chief clin. psychol. service, Tomah, Wis., 1958-60, research psychologist, exec. officer, Washington, 1960-61, chief psychology service, Brecksville, Ohio, 1961-73, Cleve., 1973-77, Loma Linda, Calif., 1977—; prof. psychiatry Loma Linda U. Med. Sch., 1978—; clin. prof. Fuller Sch. Profl. Psychology, Pasadena, Calif., 1980—; research prof. U. Calif., Riverside, 1982—; cons. State Dept. Mental Health, 1981—. Contbr. chpts. to books, articles to encys. Served to sgt. U.S. Army, 1942-46. Recipient various service awards. Fellow Am. Paychol. Assn. (pres. rehab. psychol. div. 1976-77, pres. psychologists in pub. service 1982-83, council of reps. 1977-80), Nat. Rehab. Assn. Democrat. Jewish. Subspecialties: Behavioral psychology; Health services research. Current work: Research on etiology, treatment and prevention of addictive behaviors; life style management approaches to chronic' illness. Home: 432 E Crescent Ave Redlands CA 92373 Office: Jerry L Pettis Meml Vets Hosp 11201 Benton St Loma Linda CA 92357

JACOBS, JEROME BARRY, experimental pathologist, electron microscopist, researcher; b. Worcester, Mass., Dec. 15, 1942; s. Maxwell Sol and Ruth (Jacobson) J.; m. Lois Ann Halfen, Aug. 30, 1964; children: Harry, Douglas, Rachel. B.S., U. Vt., 1965, M.S. (research fellow), 1967; teaching fellow, Clark U., 1967-68, Ph.D. (research fellow), 1971. Research fellow U. Mass., 1968-69, instr. pathology, 1973—; affiliate prof. Worcester Poly. Inst., 1975—; dir. electron microscopy St. Vincent Hosp., Worcester, 1971—. Contbr. articles to profl. jours. Gov.'s appointee Mass. Rehab. Commn., 1978—, Mass. Devel. Disability Council, 1978—. Mem. Am. Assn. Pathologists, Am. Assn. Cancer Research, AAAS, Am. Soc. Cell Biology, European Assn. Cancer Research, Sigma Xi. Subspecialties: Cancer research (medicine); Cell biology (medicine). Current work: Diagnostic electron microscopy, cancer research, pathology services.

JACOBS, JOHN EDWARD, educator; b. Kansas City, June 15, 1920; s. Charles Hawley and Lucille Hartman (Boetjer) J.; m. Elizabeth Anne Brazell, Feb. 23, 1945; children—Patricia, Robert, William, Thomas, Marie, Stephen. B.S. in Elec. Engring, Northwestern Tech. Inst., 1947, M.S., 1948, Ph.D., 1950; Sc.D. (hon.), U. Strathclyde, 1972. Mgr. advance devel. X-ray dept. Gen. Electric Co., Milw., 1950-58, engring. scientist research labs., Schenectady, 1958-60; prof. elec. engring. Northwestern U., Evanston, Ill., 1960—, Walter P. Murphy prof. elec. engring. and engring. scis., 1969—; Exec. dir. Bio-med. Engring. Center, 1962-69, dir., 1969—; pres. Biomed. Engrs. Resource Corp., 1968—; cons. to industry and govt., 1961—; Spl. adviser Nat. Inst. Gen. Med. Scis., HEW, 1966-72; health manpower cons. Pres.'s adv. council mgmt. improvement, 1971-72. Mem. Com. Study Nat. Needs for Biomed. and Behavioral Research Personnel, NRC, 1975—; Mem. research com. Chgo. Heart Assn., 1962-65. Served to lt. (j.g.) USNR, 1942-46. Mem. IRE (chmn. Milw. sect. 1957-58, asso. editor bio-med. trans. 1960-65), Bio-Med. Engring. Soc. (founder 1968, treas. 1968-72), Nat. Acad. Engring. Subspecialties: Biomedical engineering; Electrical engineering. Patentee in field. Home: 631 Milburn St Evanston IL 60201

JACOBS, JOSEPH DONOVAN, engineering firm executive; b. Motley, Minn., Dec. 24, 1908; s. Sherman William and Edith Mary (Donovan) J.; m. Virginia Mary O'Meara, Feb. 8, 1937; 1 son, John Michael. B.S. in Civil Engring, U. Minn., 1934. Civil engr., constrn. supr. Walsh Constrn. Co., N.Y.C. and San Francisco, 1934-54; chief engr. Kaiser-Walsh-Perini-Raymond, Australia, 1954-55; founder, sr. officer Jacobs Assos., San Francisco, 1955—; Chmn., U.S. nat. com. on tunnelling tech. Nat. Acad. Scis., 1977. Recipient Golden Beaver award for engring., 1980; Non-Mem. award Moles, 1981. Fellow ASCE, Instn. Engrs. Australia; mem. Nat. Acad. Engring., Am. Inst. Mining and Metall. Engrs., Nat. Soc. Profl. Engrs., World Trade Club San Francisco, Engrs. Club San Francisco, Delta Chi. Club: Corinthian Yacht (San Francisco). Subspecialty: Construction Engineering. Inventor in field of mining and tunnel excavation. Home: 84 Almenar Dr Greenbrae CA 94904 Office: 500 Sansome St San Francisco CA 94111

JACOBS, KEITH WILLIAM, psychologist, educator; b. Ames, Iowa, Feb. 24, 1944; s. Cyril William and Sylvia (Woodrum) J. B.A., U. No. Iowa, 1968; M.A., Eastern Ill. U., 1972; Ph.D., U. So. Miss., 1975. Lic. psychologist, La. Adj. instr. U. So. Miss., Natchez, 1974-75; asst. prof. psychology Loyola U. New Orleans, 1975-79, assoc. prof. psychology, 1979—; lectr. Holy Cross Coll., New Orleans, 1976-78; aux. faculty William Carey Sch. Nursing, New Orleans, 1979-80; cons. in field. Contbr. articles to profl. jours. Bd. dir. Oak Harbor Homeowners Assn., Pearlington, Miss., 1979-80. Recipient Teaching award Am. Psychol. Assn., 1980. Mem. Am. Psychol. Assn., Am. Assn. Sex Educators, Counselors and Therapists, La. Acad. Sci., Southeastern Psychol. Assn., Nat. Rifle Assn., ACLU, Childbirth Edn. Assn., Childbirth Edn. Assn. New Orleans. Subspecialties: Behavioral psychology; Information systems, storage, and retrieval (computer science). Current work: Applications of computers to edn., psychol. effects of color, psychophysiology, personality variables in prediction of sexual behavior. Home: P O Box 102 Pearlington MS 39572 Office: 6363 Saint Charles Ave New Orleans LA 70118

JACOBS, LAWRENCE DAVID, neurologist; b. Buffalo, July 20, 1938; s. Louis Melvin and Genevieve (Bibby) J.; m. Pamela Regis; children: Christopher, Luke, Lawrence, Jessica, Elizabeth. B.S., Niagara U., 1961; M.D., St. Louis U., 1965. Diplomate: Am. Bd. Psychiatry and Neurology. Intern Downstate Med. Ctr., Bklyn., 1966, resident in neurology, 1966-69; NIH spl. fellow Mt. Sinai Med. Ctr., N.Y.C., 1969-73; assoc. prof. neurology and physiology SUNY-Buffalo, 1973—; attending neurologist Dent Neurol. Inst., Buffalo. Author: Computerized Tomography of the Orbit and Sella Turcica, 1980; contbr. articles to profl. jours. Served with USNR, 1966—. CUNY fellow, 1970; Western N.Y. grantee. Mem. Am. Acad. Neurology (sec. treas.), Am. Soc. Neuroimaging (sec. treas.), Nat. Multiple Sclerosis Soc., Am. Neurol. Assn. (trustee). Subspecialty: Neurology. Current work: Neuroophthalmology, CT scanning, multiple sclerosis, vision, control of eye movements, neuroimaging, multiple sclerosis. Office: Millard Fillmore Hosp 3 Gates Circle Buffalo NY 14209

JACOBS, MARK, plant physiologist, educator; b. Princeton, N.J., May 19, 1950; s. William Paul and Jane (Shaw) J.; m. Candace Margaret Clarke, Dec. 29, 1973; children: Jeffrey, Robinson, Patrick. B.A. magna cum laude, Harvard U., 1971; Ph.D., Stanford U., 1975. Asst. prof. dept. biology Swarthmore (Pa.) Coll., 1975-81, assoc. prof., 1981—. Contbr. articles to profl. jours. State of Calif. fellow, 1972-75; NATO fellow, 1976-77; NSF grantee, 1975—. Mem. Am. Soc. Plant Physiologists, Sigma Xi. Subspecialties: Plant physiology (biology); Plant growth. Current work: The role and biochemical mode of action of plant hormones in plant development. Home: 401 Dickinson Ave Swarthmore PA 19081 Office: Department of Biology Swarthmore College Swarthmore PA 19081

JACOBS, MERLE E(MMOR), scientific researcher, educator; b. Hollsopple, Pa., Nov. 30, 1918; s. Paul A. and Trella Emma (Risch) J.; m. Elizabeth B. Beyeler, Dec. 19, 1937. B.A., Goshen Coll., 1948; Ph.D., Ind. U., 1953. Instr. Goshen (Ind.) Coll., 1953-54, research prof. zoology, 1964—; research assoc., instr. Duke U., Durham, N.C., 1954-57; assoc. prof. Bethany (W.Va.) Coll., 1957-61; prof. Eastern Mennonite Coll., Harrisonburg, Va., 1961-64. Contbr. article to profl. jour. Eigenmann fellow, 1952; Lalor Found. awardee, 1955; recipient Sigma Xi award AAAS, 1962; NIH grantee, 1964—. Mennonite. Subspecialties: Gene actions; Behaviorism. Current work: Research in behavioral and biochemical genetics. Patentee composition and process for producing pigmentation in hair. Home: 2214 S Main Goshen IN 46526 Office: Goshen College Goshen IN 46526

JACOBS, RONALD MICHAEL, computer scientist; b. Cleve., July 6, 1956; s. Leonard Paul and Gertrude Delores (Michael) J. B.S. in Computer Sci, Northwestern U., 1978, M.S. in Computer Sci., 1980. Sr. applications programmer Fin. Computer Services, Evaston, Ill., 1978-80; sr. systems programmer Northwestern U., Evanston, 1980; regional systems analyst Modular Computer Systems, Elmhurst, Ill., 1980-82; sr. systems analyst Quadrex Corp., Campbell, Calif., 1982—. Mem. Assn. for Computing Machinery, IEEE Computer Soc. Subspecialties: Computer architecture; Distributed systems and networks. Current work: Data- and demand-driven computer architectures and languages. Office: 1700 Dell Ave Campbell CA 95008

JACOBS, STEPHEN FRANK, optical scientist, educator; b. N.Y.C., Oct. 1, 1928; s. Henry Lawrence and Josette (Frank) J.; m. Kathleen Mitchell, Feb. 9, 1963; children: Henry, Tom, Jane. B.S. in Physics, Antioch Coll., 1951, Ph.D., Johns Hopkins U., 1956. Physicist Perkin-Elmer Corp., Norwalk, Conn., 1956-60; sr. physicist TRG, Inc., Farmingdale, N.Y., 1960-65; mem. faculty dept. optical scis. U. Ariz., Tucson, 1965—, prof., 1965—. Fellow Optical Soc. Am.; mem. Am. Phys. Soc. Subspecialties: Laser metrology; Laser optics. Current work: Developed optically pumped CS laser, 1962; using lasers to make ultraprecise measurements of dimensional stability of materials with time and temperature. Office: Dept Optical Scis U Ariz Tucson AZ 85721

JACOBS, STEVEN JAY, biochemist, physician, researcher; b. N.Y.C., Apr. 28, 1946; s. Benjamin and Sally (Blond) J.; m. Pamela Goode, Mar. 25, 1974; 1 son, James Bigby. B.A., Boston U., 1968, M.D., 1968. Diplomate: Am. Bd. Internal Medicine. Intern Mt. Sinai Hosp., N.Y.C., 1968-69, resident, 1969-72; fellow dept. pharmacology Johns Hopkins U. Sch. Medicine, Balt., 1974-75; research scientist Wellcome Research Labs., Research Triangle Park, N.C., 1975-81, group leader, 1981—; assoc. prof. dept. medicine U. N.C. Sch. Medicine, Chapel Hill, 1975—. Contbr. articles to profl. jours. Served to maj., M.C. U.S. Army, 1972-74. Mem. Endocrine Soc., Am. Soc. Biol. Chemists, Am. Soc. for Pharmacology and Exptl. Therapeutics. Subspecialty: Receptors. Current work: Insulin receptors—structure, function and regulation. Home: 611 Kensington Dr Chapel Hill NC 27514 Office: 3030 Cornwalis Rd Research Triangle Park NC 27709

JACOBSOHN, GERT MAX, biochemistry educator, researcher; b. Berlin, Aug. 1, 1929; U.S., 1948, naturalized, 1963; s. Martin and Margaret (Bach) J.; m. Myra Kramer, June 21, 1959; children: Hannah G., Jamie A., Diane R., Alice P. A.B., Ill. Coll., Jacksonville, 1952; M.S., Purdue U., 1955, Ph.D., 1957. Post doctorate Columbia U., 1957-60; asst. mem. Einstein Med. Ctr., Phila., 1960-62; asst. prof. biochemistry Hahnemann U., 1962-67, assoc. prof., 1967-77, prof., 1977—. Fellow AAAS; mem. Am. Soc. Biol. Chemists, Endocrine Soc., Am. Chem. Soc., Phytochemistry Soc. N.Am. Subspecialties: Biochemistry (medicine); Nutrition (medicine). Current work: Steroid transformation, enzymology, metabolism and function of plant sterols, human nutrition. Home: 1108 Clark Rd Wyndmoor PA 19118 Office: Dept Biol Chemistry Hahnemann U 230 N Broad St Philadelphia PA 19102

JACOBSON, ALLAN STANLEY, astrophysicist, computer scientist; b. Chattanooga, June 18, 1932; s. Mack and Anne (Shavin) J.; m. Edith Lieberman, June 24, 1956. A.B., UCLA, 1962; M.Sc., U. Calif., San Diego, 1964, Ph.D., 1968. Asst. research physicist U. Calif., San Diego, 1968-69; sr. scientist, mem. tech. staff Jet Propulsion Lab., Calif. Inst. Tech., Pasadena, 1969-73; prin. investigator HEAO-C1, 1970, sr. research scientist, 1981, tech. group supr., 1973—, instr. space physics, 1980-82. Contbr. articles to profl. jours. Served with USAF, 1951-55. Recipient Exceptional Sci. Achievement medal NASA, 1980. Mem. Am. Astron. Soc., Phi Beta Kappa. Subspecialties: Gamma ray high energy astrophysics; 1-ray high energy astrophysics. Current work: High resolution gamma-ray spectroscopy of celestial sources, conducted from stratospheric balloons and spacecraft.

JACOBSON, BARRY MARTIN, chemistry educator; b. N.Y.C., Apr. 23, 1945; s. Samuel and Edna (Fried) J. S.B., U. Chgo., 1965; Ph.D., Harvard U., 1971. Postdoctoral fellow Rice U., Houston, 1970-71, instr., 1971-72; research assoc. Brookhaven Nat. Lab., Upton, N.Y., 1972-73; vis. asst. prof. SUNY-Stony Brook, 1973-74; asst. prof. Barnard Coll., N.Y.C., 1974-80, assoc. prof., chmn. chemistry dept., 1980—. NSF grad. fellow, 1965-70; Cottrell Coll. sci. grantee Research Corp., 1975-77, 79-81. Mem. Am. Chem. Soc., AAAS, AAUP, Sigma Xi. Subspecialty: Organic chemistry. Current work: Physical-organic chemistry, reaction mechanisms, the Diels-Alder and Alderene reactions, mechanisms of ketene formation. Office: Chemistry Dept Barnard Coll 606 W 120th St New York NY 10027

JACOBSON, E. PAUL, agricultural engineer; b. Harcourt, Iowa, Oct. 23, 1909; s. Frank A. and Pauline (Jacobson) J.; m. Marion E. Jensen, June 3, 1934; children: Ann Jacobson Arden, Mary Jacobson Kimsey, Jean Jacobson Marx, Paul A. B.S. in Agrl. Engring, Iowa State U., 1932. With Soil Conservation Service, U.S. Dept. Agr., 1934-64, state conservation engr., Iowa, 1954-64; cons. drainage, flood control and conservation Harza Engring. Co., Chgo., 1964—. Author: (with C.B. Richey and Carl W. Hall) Agricultural Engineering Handbook, 1961. Active Common Cause, SANE, Iowa Freeze Network, Nature Conservancy. Mem. Am. Soc. Agrl. Engring. (Hancor Soil and Water Engring. award 1973, Doerfer Engring. Concept of Yr. award 1981), Soil Conservation Soc. Am., Iowa Engring. Soc. Subspecialties: Flood Control; Agricultural engineering. Current work: Developing terrace systems adapted to mechanized farming. Developer system of parallel terraces with tile outlets. Adre: Rural Route 1 Dow City IA 51528

JACOBSON, GEORGE ROBERT, psychologist; b. Chgo., Apr. 1, 1941; s. Daniel and Dinah (Farell) J.; m. Mary L. Ruskin, Aug. 26, 1962; children: David R., Daniel S. B.A., U. Ariz., 1963; M.A., Coll. William and Mary, 1965; Ph.D., Ill. Inst. Tech., 1971. Lic. psychologist, Wis., Ill. Psychologist City of Chgo., 1968-72; clin. assoc. prof. psychology U. Wis., Milw., 1977—; asst. prof. psychiatry Med. Coll. Wis., Milw., 1978—; dir. research and tng. DePaul Rehab. Hosp., Milw., 1972—; faculty Wis. Sch. Profl. Psychology, Milw.; adj. faculty St. Mary's Coll., 1979—; cons. State of Wis., 1977-79. Author: The Alcoholisms, 1976, (with others) Handbook on Alcohol and Drug Abuse, 1983; contbr. articles to profl. jours. Mem. Am. Psychol. Assn., Wis. Psychol. Assn. (exec. council), Research Soc. on Alcoholism, AAAS, Wis. Biofeedback Soc. Subspecialty: Addictive disorders. Current work: Alcoholism, drug abuse, altered states of awareness, hypnosis, innovative and unorthodox psychotherapies, biofeedback, perception and cognition. Home: 5062 N Diversey Blvd Whitefish Bay WI 53217 Office: 4143 S 13th St Milwaukee WI 53221

JACOBSON, GUNNARD KENNETH, microbiologist; b. Phoenix, Feb. 19, 1947; s. Kenneth Brown and Mary Jane (Bloomquist) J.; m. Claudia Lenore Hale, Aug. 19, 1948. B.S., U. Ariz., 1969; Ph.D., Oreg. State U., 1972. Postdoctoral fellow Nat. Inst. Gen. Med. Scis., U. Chgo., 1972-75; research assoc. Argonne Nat. Lab., Ill., 1975-77; sr. research microbiologist Universal Foods Corp., Milw., 1977-83, sr. scientist, 1983—. Contbr. articles to profl. jours. Mem. Am. Soc. Microbiology, AAAS, Genetics Soc. Am. Subspecialties: Microbiology; Genetics and genetic engineering (biology). Current work: Genetics and physiology of indsl. yeasts. Office: 6143 N 60th St Milwaukee WI 53218

JACOBSON, HAROLD GORDON, physician, educator; b. Cin., Oct. 12, 1912; s. Samuel and Regina (Dittman) J.; m. Ruth Enenstein, Aug. 10, 1941; children: Richard, Arthur. B.S., U. Cin., 1934, M.B., 1936, M.D., 1937. Diplomate: Am. Bd. Radiology (trustee 1971-82, chmn. written exams. com. in diagnostic radiology 1973-81, co-chmn. 1981—, treas. 1976-78, v.p. 1978-80, pres. 1980-82, mem. residency rev. com. 1976-82, vice-chmn. 1979-80, chmn. 1980—, exec. com. 1976—). Intern Los Angeles County Gen. Hosp., 1936-38; fellow in pathology Longview Hosp., Cin., 1938; resident Mt. Sinai Hosp., N.Y.C., 1939-41, Associated Hosps. U. Tex., 1941-42; asst. in radiology U. Tex., 1941-42; assoc. radiologist New Haven (Conn.) Hosp.; also instr. Yale U., 1952; asst. chief, asso. radiologist VA Hosp., Bronx, N.Y., 1946-50, chief radiology service, 1950-53, cons., 1958—; asst. clin. prof. N.Y. U., 1952-53, clin. prof., 1953-59, prof. clin. radiology, 1959-64; prof. radiology Albert Einstein Coll. Medicine, 1964-71; prof., chmn. Albert Einstein Coll. Medicine of Montefiore Hosp. and Med. Center, N.Y.C., 1972—; dir. dept. roentgenology Hosp. for Spl. Surgery, N.Y.C., 1953-55; radiologist-in-chief Montefiore Hosp. and Med. Center, N.Y.C., 1955—; sr. cons. in radiology Nat. Bd. Med. Examiners, 1975—, mem. bd., 1979-83; vis. prof. radiology Inst. Orthopaedics, U. London, 1975—; vis. prof., lectr., U.S.A., Israel, Brazil, Finland.; named lectures include Felson Lecture, Carman Lecture, Baylin Lecture, Beeler Lecture, Freedman Lecture, Pfahler Lecture, Chamberlain Lecture, Evans Lecture, Sampson Lecture, Wolf Meml. Lecture, Caffey Lecture, Grubbe Lecture, Myron Melamed Lecture. Author: (with Clarence Schein, William Z. Stern) The Common Bile Duct, 1967, Neuroradiology Workshop, Vol. III, 1968, (with Ronald O. Murray) Radiology of Skeletal Disorders: Exercises in Diagnosis, 1971, 2d edit., 1977; co-author: Bone Disease Syllabus, 1972, 2d series, 1976, 3d series, 1980, Index for Roentgen Diagnosis, 3d edit, 1975; co-editor in chief: Jour. Internat. Skeletal Soc, 1976—; editorial bd.: Excerpta Medica, 1974—, Jour. AMA, 1979—; coordinator: topics in radiology Jour. AMA, 1977-79; editor, 1979—; mem. editorial bd. for radiology, 1979—; contbr. articles to profl. jours. Served as maj. M.C AUS, 1942-46. Recipient Gold medal Assn. Univ. Radiologists, 1982, Phi Lambda Kappa, 1983. Fellow Am. Coll. Radiology (councilor 1960—, bd. chancellors, chmn. com. on radiol. coding 1967—, mem. commn. on credentials 1968—, chmn. commn. on affairs Am. Inst. Radiology 1971—, co-chmn. com. on diagnostic coding index and thesaurus 1973—, Gold medal 1978), Royal Coll. Radiologists (London) (hon.); mem. N.Y. Roentgen Soc. (pres. 1959-60, historian 1967—), AMA, N.Y. State, N.Y. med. socs., Soc. of Chairmen Acad. Radiology Depts. (mem. exec. council 1972—, pres. 1973-74), Radiol. Soc. N.Am. (pres. 1966-67, mem. bd. censors 1968—, gold medal 1972), Am. Roentgen Ray Soc. (Cert. of Appreciation 1983), Royal Soc. Medicine (hon.), Internat. Skeletal Soc. (co-founder, pres. 1974-75, chmn., mem. exec. com. 1976—), Alpha Omega Alpha (Rigler lectr. 1964, 70, Crookshank lectr. London 1974, Holmes lectr. Boston 1974). Subspecialties: Radiology; Diagnostic radiology. Current work: Radiology of Skeletal Disorders. Home: 3240 Henry Hudson Pkwy New York NY 10463

JACOBSON, KARL BRUCE, biochemist, educator; b. Manning, Iowa, Mar. 5, 1928; s. David Joel and Ruth (Wood) J.; m. Phyllis Jean Greenler, June 23, 1951; children: Deborah, Paul, Steven, Daniel. B.S. in Chemistry, St. Bonaventure Coll., 1948; Ph.D. in Biochemistry, Johns Hopkins U., 1956. Postdoctoral fellow Calif. Inst. Tech., 1956-68; staff scientist Oak Ridge Nat. Lab., 1958—; sr. staff mem. I U. Tenn.-Knoxville, 1968; prof. 1969; Mem. Human Resources Adv. Bd., Oak Ridge, 1981-83; bd. dirs. Planned Parenthood, 1978-80. Served in U.S. Army, 1950-52. Am. Cancer Soc. fellow, 1956-58. Mem. Am. Soc. Biol. Chemists, Am Chem. Soc., AAAS, N.Y. Acad. Sci. Methodist. Subspecialties: Biochemistry (biology); Molecular biology. Current work: Nucleic acid structure-function relationships; mechanism of genetic suppression; pterdine biosynthesis; toxic metals effects in biological systems. Home: 940 W Outer Dr Oak Ridge TN 37830 Office: Biology Div Oak Ridge Nat Lab Box Y Oak Ridge TN 37830

JACOBSON, LEON ORRIS, physician; b. Sims, N.D., Dec. 16, 1911; s. John and R. Patrine (Johnson) J.; m. Elizabeth Benton, Mar. 18, 1938; children: Eric Paul, Judith Ann. B.S., N.D. State Coll., 1935, D.Sc. (hon.), 1966; M.D., U. Chgo., 1939; D.Sc., Acadia U., N.S., 1972. Intern U. Chgo., 1939-40, asst. resident medicine, 1940-41, asst. in medicine, 1941-42, instr., 1942-45, asst. prof., 1945-48, asso. dean, div. biol. scis., 1945-51, asso. prof., 1948-51, prof. medicine, 1951—, Joseph Regenstein prof. biol. and med. scis., 1965—, chmn. dept. medicine, 1961-65, dean div. biol. scis., 1966-75; head hematology sect. U. Chgo. Clinics, 1951-61; mem. Inst. Radiobiology and Biophysics, 1949-54; dir. Franklin McLean Meml. Research Inst., 1974-77; asso. dir. health Plutonium project Manhattan Dist., 1943-45, dir. health, 1945-46; dir. Argonne Cancer Research Hosp., U. Chgo., 1951-67; U.S. rep. 1st and 2d UN Conf. on Peaceful Uses Atomic Energy, Geneva, 1955, 58, WHO conf. Research Radiation Injury, 1959; cons. biology div. Argonne Nat. Lab.; mem. adv. com. on isotope distbn. AEC, 1952-56; mem. nat. adv. com. radiation USPHS, 1961, mem. com. radiation studies, cons. hematology study sect.; mem. com. cancer diagnosis and therapy NRC, 1949-55; mem. bd. sci. counselor Nat. Cancer Inst., 1963-67; mem. nat. adv. cancer council, nat. cancer adv. bd. NIH, 1968-72; chmn. sci. adv. bd. Council for Tobacco Research; lectr. Internat. Soc. Hematology and Internat. Congress Radiology, Eng., France, Norway, Sweden, 1950, 5th Internat. Cancer Congress, Paris, 1950, Internat. Soc. Hematology, Argentina, 1952, Paris, 1954, others. Author book on erythropoletin; contbr. chpts. on specialized items to various med. books, articles to med. jours.; Book editor: Perspectives in Biology and Medicine, 1979—. Recipient Janeway medal, 1953; Robert Roesler de Villiers award Leukemia Soc.; Borden award med. scis. Assn. Am. Med. Colls., 1962; Modern Med. and Am. Nuclear Soc. awards, 1963; John Phillips Meml. award, 1975; Theodore Roosevelt Rough Riders award State of N.D., 1977; Lincoln Laureate State of Ill., 1979; Kennecott lectr., 1963. Mem. A.C.P. (master), Am. Soc. Clin. Investigation, Assn. Am. Physicians, Soc. Exptl. Biology and Medicine, Central Soc. Clin. Research, Am. Assn. Cancer Research, Internat. Soc. Hematology, AMA, Nat. Acad. Sci., Central Clin. Research Club, AAAS, Radiation Research Soc., Am. Soc. Exptl. Pathology, Sigma Xi, Theta Chi, Nu Sigma Nu, Phi Key, Alpha Omega Alpha. Subspecialties: Hematology; Transplantation. Current work: Mechanisms of red cell production and steady state maintenance; transplantation of blood-forming tissue to radiation induced or aplasia from other known mechanisms. Home: 5801 Dorchester Ave Chicago IL 60637 Office: 950 E 59th St Chicago IL 60637

JACOBSON, NATHAN, mathematics educator; b. Warsaw, Poland, Sept. 8, 1910; m., 1942; 2 children. A.B., U. Ala., 1930; Ph.D. in Math, Princeton U., 1934; D.Sc. (hon.), U. Chgo., 1972. Asst. math. Inst. Advanced Study, Princeton, N.J., 1933-34, Procter fellow, 1934-35; lectr. Bryn Mawr (Pa.) Coll., 1935-36; Nat. Research Found. fellow U. Chgo., 1936-37; from instr. to asst. prof. U. N.C., 1937-40; vis. assoc. prof. Johns Hopkins U., Balt., 1940-41; assoc. prof. U. N.C., 1941-42, Johns Hopkins U., 1943-47; from assoc. prof. to prof. math. Yale U., New Haven, 1947-67, Henry Ford II prof. math., 1967—; vis. prof. U. Chgo., 1964, U. Tokyo, 1965, Tata Inst. Fundamental Research, India, 1970, U. Rome, 1971, Hebrew U., Jerusalem, 1971, Australian Nat. U., 1978. Editor bull.: Am. Math. Soc, 1948-53. Guggenheim fellow, 1951-52; Fulbright grantee U. Paris, 1951-52. Mem. Nat. Acad. Sci., AM. Math. Soc. (v.p. 1957-58, pres. 1971-73), French Math. Soc., Math. Soc. Japan, Am. Acad. Arts and Sci. Office: Dept Math Yale U Box 2155 Yale Sta New Haven CT 06520

JACOBSON, RICHARD OTTO, biology educator, consultant; b. Monroe, Wis., Apr. 5, 1946; s. Otto and Violet (Rickey) J.; m. Laura Ofelia Duran, June 5, 1971; children: Tanya, Natasha, Jessica. B.S., U. Wis.-Platteville, 1969, M.A.T., 1974; M.S., N.Mex. State U., 1977; grad. student, U. Mo.-Columbia, 1982—. Instr. N.Mex. State U., 1976; instr. Dakota Wesleyan U., 1977-78; asst. prof. biology Central Meth. Coll., 1978—; cons. Addison Biol. Lab., Fayette, Mo., 1981—. Served with USAF, 1969-73. Recipient NSF Instructional Sci. Equipment Program grant, 1979. Mem. Am. Soc. Microbiology, Mo. Acad. Sci. Subspecialties: Microbiology; Molecular biology. Current work: Pili of Bordetella bronchiseptica; isolation and identification of halobacterium sp. Home: 321 Corprew Ave Fayette MO 65248 Office: 411 Central Methodist Sq Fayette MO 65248

JACOBY, GEORGE ALONZO, JR., physician, microbiologist; b. Bronxville, N.Y., Sept. 20, 1932; s. George Alonzo and Ruth Bookwalter (Burtner) J.; m. Ann Freeman Austin, 1957 (div. 1968); children: Gregory Austin, Douglas Burtner, Alison Freeman; m. Lee Breckenridge, Apr. 30, 1969; children: Robert Breckenridge, Sarah Hieatt. B.A., Yale U., 1954; M.D., Harvard U., 1958. Diplomate: Am. Bd. Internal Medicine. Intern Mass. Gen. Hosp., Boston, 1958-59, resident, 1959-60, 64; research assoc. Nat. Inst. Arthritis and Metabolic Diseases, Bethesda, Md., 1960-62; USPHS postdoctoral fellow Nat. Inst. for Med. Research, Mill Hill, London, 1962-63; research fellow in bacteriology and immunology Harvard Med. Sch. Boston, 1965-67, research assoc., 1967, instr. medicine, 1968, assoc. in medicine, 1968-70, asst. prof., 1970-76, assoc. prof., 1976—; assoc. physician Mass. Gen. Hosp., 1974—, also infectious disease cons. Editor: Antimicrobial Agents and Chemotherapy; assoc. editor: Plasmid; author articles. Served with USPHS, 1960-62. Mem. Am. Soc. Microbiology, Mass. Med. Soc., Infectious Diseases Soc. Am. Episcopalian. Subspecialties: Infectious diseases; Microbiology (medicine). Current work: Plasmids and antibiotic resistance in P. aeruginosa. Plasmids, B-lactamases, transposons, aminoglycoside modifying enzymes, evolution of antibiotic resistance, restriction-modification systems. Office: Mass Gen Hosp Boston MA 02114

JACOBY, HENRY DONNAN, economist, educator; b. Dallas, June 25, 1935; s. Henry Harris and Margaret Cameron (Miller) J.; m. Martha Hughes Jacoby, Apr. 4, 1959; children—Daniel Donnan, Caroline Hughes. B.S. in Mech. Engring, U. Tex., Austin, 1957; Ph.D. in Econ, Harvard U., 1967. Systems analyst Tudor Engring. Co., San Francisco, 1959-61; economist Harvard Devel. Adv. Service, Argentina Project, 1963-65; asst. prof. dept. econs. Harvard U., Cambridge, Mass., 1965-69; assoc. prof. polit. economy John F. Kennedy Sch. Govt., 1969-73; prof. mgmt. MIT, Cambridge, 1973—; dir. Center for Energy Policy Research, 1978-83; vis. scholar London Bus. Sch., 1983-84; chmn. Mass. Gov.'s Emergency Energy Tech. Adv. Com., 1973-74; mem. Nat. Petroleum Council, 1975-83. Author: (with F.S. Brooman) Macroeconomics, 1970, (with R. Dorfman and H.A. Thomas, Jr.) Models for Managing Regional Water Quality, 1973, (with J.D. Steinbruner) Clearing The Air, 1973, Analysis of Investment in Electric Power, 1979, (with R. deLucia) Energy Planning for Developing Countries, 1982. Served with USN, 1957-59. Mem. Am. Econ. Assn., Tau Beta Pi. Democrat. Episcopalian. Subspecialties: Resource management. Office: MIT Sloan Sch of Mgmt 50 Memorial Dr Cambridge MA 02139

JACOBY, HENRY I., gastrointestinal pharmacologist; b. Scranton, Pa., Aug. 26, 1936; s. Milton and Pearl (Kluger) J.; m. Gloria Joyce Yablonsky, Apr. 2, 1967; children: Melissa Beth, Seth Adam. B.S., Phila. Coll. Pharmacy and Sci., 1958; Ph.D., U. Mich., 1963. Research fellow Merck, Sharpe & Dohme Research Labs., West Point, Pa., 1963-72; group leader McNeil Pharms., Ft. Washington, Pa., 1972-79, research fellow, 1979—. Contbr. articles to profl. jours. Bd. dirs. friends of Abington Free Library, 1982—; chmn. sch. parents com. Abington Friends Sch., 1980-81, chmn. community scholarship com., 1981—. Recipient Philip B. Hofman Research Scientist award Johnson & Johnson, 1979. Mem. Am. Gastroenterological Assn., Am. Soc. Pharmacology and Exptl. Therapeutics, Phila. Gastrointestinal Research Forum, Gastrointestinal Pharmacology Group. Jewish. Subspecialty: Pharmacology. Current work: Gastric anti-secretory drugs. Home: 910 Meetinghouse Rd Rydal PA 19046 Office: McNeil Pharm Spring House PA

JACOBY, JACOB HERMAN, psychiatrist; b. Bklyn., Mar. 3, 1943; s. Philip and Shirley (O'Berkowitz) J.; m. Judita Stransky, June 7, 1970; children: Jonathan Alexander, Sanford, Deborah. B.S., Bklyn. Coll., 1964; M.S., Rutgers U., 1970; Ph.D., Albert Einstein Coll. Medicine, 1972; M.D., SUNY-Buffalo, 1980. Research asso. M.I.T., Cambridge, 1972-75; asst. prof. pharmacology Coll. Medicine and Dentistry of N.J., Newark, 1975-78; asst. instr. psychiatry Western Psychiatric Inst. & Clinic, Pitts., 1978—. Mem. Am. Soc. Exptl. Pharmacology and Therapeutics, Internat. Soc. Neuroendocrinology, Am. Coll. Clin. Pharmacology, Endocrine Soc., Soc. for Neurosci. Jewish. Subspecialties: Psychopharmacology; Neuroendocrinology. Current work: Researcher in relationships between biogenic amine metabolism and neuroendocrine functions. Office: Western Psychiatric Institute & Clinic 3811 O'Hara St Pittsburgh PA 15213

JACOBY, ROBERT OTTINGER, comparative pathology educator; b. N.Y.C., June 20, 1939. D.V.M., Cornell U., 1963; M.Sc., Ohio State U., 1968, Ph.D. in Pathology, 1969. Asst prof. pathology Ohio State U., 1969; NIH fellow U. Chgo., 1969-71; asst. prof. Yale U. Sch. Medicine, New Haven, 1971-77, assoc. prof. comparative medicine, 1977—, chmn. sect. comparative medicine and dir. animal care, 1978—. Mem. AAAS, Am. Coll. Vet. Pathologists, Internat. Acad. Pathology, AVMA, Sigma Xi. Subspecialties: Immunology (medicine); laboratory animal medicine. Office: Yale U 375 Congress Ave New Haven CT 06510

JACOBY, WILLIAM RICHARD, ceramic engineer; b. Columbia, N.J., Sept. 8, 1926; s. Albert Dewey and Evangeline (Vesseller) J.; m. Jeanne Lois Rasmussen, Aug. 20, 1949; children: Barry Alan, Blaine Evan, Blake Dawson. B.S., Dickinson Coll., 1950; M.S., Rutgers U., 1953, Ph.D., 1956. Asst. instr. Rutgers U., New Brunswick, N.J., 1954-56; research assoc. KAPL, Gen. Electric Co., Schenectady, 1957-59; lab. supr. 3M Co., St. Paul, 1959-64; mgr. fuel labs. Westinghouse Electric Corp., Piits., 1965-81; mgr. D&D engring. West Valley Nuclear Service Co., Inc., N.Y., 1981—. Contbr.: Neutron Absorber Materials for Reactor Central, 1962. Trustee Parkview Congregational Ch., 1960-63; v.p. PTA, 1963-64; active Republican party. Served with U.S. Army, 1944-46. Fellow Am. Ceramic Soc. (program com. 1967-70, chmn. 1974-76, trustee 1976-79, v.p. 1979-80). Subspecialties: Nuclear fission; Materials (engineering). Current work: Decontamination and decommissioning of nuclear fuel facilities. Home: 5517 George Dr Hamburg NY 14075 Office: West Valley Nuclear Services Co Inc PO Box 191 West Valley NY 14171

JACOVITCH, JOHN, physicist, nuclear engineer; b. Hemphill, W.Va., Feb. 8, 1930; s. Nicolai and Domnica J.; children: John David, Michael Alan, Daniel Nicolas. B.S., Roanoke Coll., 1958; postgrad., Vanderbilt U., 1958-59; M.S., Lynchburg Coll., 1968. AEC fellow Vanderbilt U., Oak Ridge Nat. Labs., 1958-59; scientist Edgerton Germeshausen & Grier, Inc., Las Vegas, Nev., Goleta, Calif., 1959-62; nuclear research physicist Naval Civil Engring. Lab., Port Hueneme, Calif., 1962-63; radiol. engr. Atomic Internat., Canoga Park, Calif., 1963-65; health physicist, project engr. Babcock & Wilcox Co., Lynchburg, Va., 1965-70; health physicist United Nuclear Corp., Wood River Junction, R.I., 1970-71; mgr. health physics U. Mo. Research Reactor Facility, Columbia, Mo., 1971-76; cons., Columbia, 1976-78; asst. prof. med. radiology U. Ill. Med. Center, Chgo., 1978-80; project engr. Wis. Electric Power Co., Milw., 1980—. Contbr. articles to profl. jours. Served with USN, 1947-51. Mem. Internat. Radiation Protection Assn., Health Physics Soc., Am. Assn. Physicists in Medicine, Sigma Pi Sigma. Subspecialties: Nuclear physics; Nuclear engineering. Current work: Radiological design. Home: 1009 N Jackson St Apt 2407-B Milwaukee WI 53202 Office: 231 W Michigan St Suite 381 Milwaukee WI 53201

JACQUET, YASUKO FILBY, research scientist; b. San Francisco; d. Atow and Fude (Okumura) Matsuoka; m. Royston, H. Filby, Feb. 28, 1958 (div. 1962); 1 dau., Mariko; m. Herve M. Jacquet, May 28, 1969. Diploma, Tsuda Coll., Tokyo, 1951; B.A., U. Iowa, 1953; Ph.D., Ind. U., 1962. Research fellow Yale U., 1963-64; staff fellow NIMH, Washington, 1965; research scientist N.Y. State Dept. Mental Health, N.Y.C., 1966—; vis. scientist NIMH, Washington, 1976, Nat. Inst. Med. Research, Mill Hill, London, 1977, Uppsala (Sweden) U., 1979, Mass. Gen. Hosp., Boston, 1982. Mem. Soc. Neurosci., Am. Soc. Pharmacology and Exptl. Therapeutics, AAAS, N.Y. Acad. Sci., Sigma Xi. Subspecialty: Neuropharmacology. Current work: Neuropeptides. Home: Heritage Hills 270B Somers NY 10589

JACQUEZ, JOHN ALFRED, physiologist, biomathematician; b. Pfastatt, France, June 26, 1922; came to U.S., 1929; s. Francois Albert and Victorine (Oestermann) J.; m. Marianne Rose Reibel, Mar. 20, 1948; children: Albert R., Nicholas P., Geoffrey M., Phillip F. M.D., Cornell U., N.Y.C., 1947. Research fellow to asst. mem. Sloan Kettering Inst., N.Y.C., 1947-53, asst. mem. to assoc. mem., 1956-62; intern Kings County Hosp., Bklyn., 1955-56; assoc. prof. physiology and biostatistics U. Mich., Ann Arbor, 1962-69, prof., 1969—; cons. RAND Corp., Santa Monica, Calif., 1959-64; chmn. computer research study sect. NIH, 1966-70. Author: A First Course in Computing and Numerical Methods, 1970, Compartmental Analysis in Biology and Medicine, 1972, Respiratory Physiology, 1979; Editor: The Diagnostic Process, 1964, Computer Diagnosis and Diagnostic Methods, 1972; Assoc. editor: Math. Bioscis, 1967-75; editor, 1975—. Served to capt. U.S. Army, 1953-55. Research fellow Nat. Cancer Inst., 1948-50; scholar in residence Fogarty Internat. Center, 1983. Mem. AAAS, Am. Physiol. Soc., Biophys. Soc., Soc. Indsl. and Applied Math., Soc. Math. Biology. Subspecialties: Membrane biology; Mathematical biology. Current work: Transport of glucose across cell membranes; mathematical modeling of physiological systems. Home: 490 Huntington Dr Ann Arbor MI 48104 Office: Dept Physiology Univ Michigan 7712 Med Sci II Ann Arbor MI 48109

JAEGER, ROBERT GORDON, ecologist, biology educator; b. Balt., Dec. 16, 1937; s. Charles and Grace (Manchey) J. B.S., U. Md., 1960, Ph.D., 1969; M.A., U. Calif.-Berkeley, 1963. Postdoctoral assoc. U. Wis.-Madison, 1971-74; asst. prof. SUNY-Albany, 1974-80, adj. asst. prof., 1980-81; asst. prof. U. Southwestern La., Lafayette, 1981-82, assoc. prof., 1982—. Editor: Jour. Herpetologia, 1982—; contbr. articles to profl. jours. Mem. Herpetologists League (trustee 1982—), Am. Soc. Ichthyologists and Herpetologists (gov. 1976-81). Subspecialties: Behavioral ecology; Species interaction. Current work: Amphibian behavioral ecology. Home: 318 Monroe St Lafayette LA 70501 Office: Univ Southwestern La Lafayette LA 70504

JAEGER, THEODORE B., psychologist, educator; b. Cape Girardeau, Mo., Jan. 4, 1951; s. Theodore Andrew and Margaret Bruce (McDonald) J.; m. Susan Mayo Yost, Aug. 26, 1978. B.A., Washington and Lee U., 1973; M.A., Appalachian State U., 1974; Ph.D., U. Ga., 1977. Postdoctoral researcher U. Ga., Athens, 1977-78; instr. psychology Meth. Coll., Fayetteville, N.C., 1978-79, asst. prof., 1979-82, assoc. prof., 1982—. Contbr. articles to profl. jours. Recipient various grants. Mem. Am. Psychol. Assn., Eastern Psychol. Assn., So. Soc. Philosophy and Psychology, Sigma Xi., Psi Chi. Subspecialties: Cognition; Sensory processes. Current work: Research on visual perception with special interest in the interaction of sensory and cognitive processes underlying human size perception. Home: 5201 Southport Rd Fayetteville NC 28301 Office: Methodist Coll Raleigh Rd Fayetteville NC 28301

JAFFE, ARTHUR MICHAEL, physicist, mathematician, educator; b. N.Y.C., Dec. 22, 1937; s. Henry and Clarisse J.; m. Nora Frances Crow, July 24, 1971. A.B., Princeton U., 1959, Ph.D., 1965; B.A., Cambridge U., 1961. Acting asst. prof. math. Stanford U., 1966-67; asst. prof. physics Harvard U., Cambridge, Mass., 1967-69, asso. prof., 1969-70, prof. physics, 1970-77, prof. math. physics, 1977—; research fellow Princeton U., 1965-66, Stanford Linear Accelerator Center, 1966-67; mem. Inst. for Advanced Study, 1967; vis. prof. Eidgenössische Technische Hochschule, Zurich, 1968; vis. prof. math. physics Princeton U., 1971; vis. prof. Rockefeller U., 1977. Author: Vortices and Monopoles, 1980, Quantum Physics, 1981; Asso. editor: Jour. Math. Physics, 1970-72; editorial council: Annals of Physics, 1975-77; asst. editor, 1977—; editor: Communications Math. Physics, 1976—; chief editor, 1979—; mem. adv. bd.: Letters in Math. Physics, 1975—; editor: Progress in Physics, 1979—, Selecta Mathematica Sovetica, 1980—; contbr. articles to profl. jours. Alfred P. Sloan Found. fellow, 1968-70; Guggenheim Found. fellow, 1977-78; award Math. and Phys. Scis., N.Y. Acad. Sci., 1979; Dannie Heineman prize for Math. Physics, 1980. Fellow Am. Phys. Soc., Am. Acad. Arts and Scis.; mem. Am. Math. Soc., AAAS, Internat. Assn. Math. Physics. Subspecialty: Mathematical physics. Home: 27 Lancaster St Cambridge MA 02140

JAFFE, BERNARD MORDECAI, physicist, educator; b. N.Y.C., Mar. 7, 1917; s. Allen and Ida (Slavin) J.; m. Emily Landes, Apr. 20, 1939; B.S., CCNY, 1939; Ph.D., NYU, 1950. Fellow dept. physics CCNY, 1938-39, tutor, 1949-51, tutor, lectr., 1952-56; engr. Amperex, N.Y.C., 1940-41; physicist Signal Corps Labs., 1941-44; engr. Lloyd Rogers Co., N.Y.C., 1945-46; AEC fellow, 1949-51; instr. dept. physics Stevens Inst. Tech., Hoboken, N.J., 1956-59; vis. asst. prof. to prof. dept. physics Columbia U., summers 1959-68, NYU, summers 1961-63; asst. prof. dept. physics Adelphi U., Garden City, N.Y., 1959-62, assoc. prof., 1962-65, prof., 1965—. Translator: (M. Francon) Optical Image Formation and Processing. Mem. Am. Phys. Soc., Optical Soc. Am., Am. Assn. Physics Tchrs. Subspecialties: Holography; Optical image processing. Current work: Writing in optics, particularly lasers and holography. Patentee flash analyzer, flash azimuth locator having image scanning means. Office: Dept Physics Adelphi Univ Garden City NY 11530

JAFFE, ERNST RICHARD, internist, medical school executive; b. Chgo., Jan. 4, 1925; s. Richard Hermann and Berta (Kohn) J.; m. Anne Jane Sylvestre, Aug. 5, 1950; children: Stephanie Anne Jaffe Green, Richard Sheridan. B.S., U. Chgo., 1945, M.D., 1948, M.S. in Pathology, 1948. Diplomate: Am. Bd. Internal Medicine (hematology). Intern Presbyn. Hosp., N.Y.C., 1948-49; resident in medicine, 1949-50, 53, 54-55; clin. asst. vis. physician Bronx (N.Y.) Mcpl. Hosp. Center, 1955-56, asst. vis. physician, 1956-62, assoc. vis. physician, 1962-63, attending physician, 1963—; mem. faculty Albert Einstein Coll. Medicine, Bronx, 1956—, prof. medicine, 1969—, head div. hematology, 1970-82, acting dean, 1972-74, sr. assoc. dean, 1983—, assoc. dean for faculty, 1983—, dir. Belfer Inst. Advanced Biomed. Studies, 1983—, acting dean, 1983—. Editor: Jour. Blood, 1975-77; co-editor: Jour. Seminars in Hematology, 1968—; contbr. numerous sci. articles to profl. publs. Served to capt. USAF, 1951-53. Health Research Council City N.Y. grantee, 1961-71; NIH grantee, 1965—; Nat. Found. N.Y. grantee, 1955-57. Mem. Am. Fedn. Clin. Research, Am. Soc. Clin. Investigation, Assn. Am. Physicians, Am. Physiol. Soc., Soc. Exptl. Biology and Medicine, Internat. Soc. Hematology (counselor 1980-86), Am. Soc. Hematology (pres. 1983), U. Chgo. Alumni Assn. (Disting. Service award 1981), Sigma Xi, Phi Beta Kappa, Alpha Omega Alpha. Subspecialties: Hematology; Internal medicine. Current work: Investigations of human erythrocyte metabolic disorders with enzyme deficiencies associated with hemolytic disease and methemoglobinemia. Office: Albert Einstein Coll Medicine 1300 Morris Park Ave Bronx NY 10461

JAFFE, NORMAN, pediatrician; b. Johannesburg, South Africa, Sept. 25, 1933; s. Solly and Helen (Soreson) J.; m. Louise Carr, Jan. 7, 1961; children: Saul, Mark, Debra, Deena. M.D., U. Witwatersrand, 1956, diploma in pediatrics, 1962. Fellow Sidney Faber Cancer Inst.; Boston adminstrv. chief out-patient tumor therapy service; now prof. pediatrics, div. chief solid tumor service M.D. Anderson Hosp. and Tumor Inst. U. Tex. System Cancer Center, Houston. Recipient Nicholas Andry award, 1976. Mem. Am. Assn. Cancer Research, Am. Soc. Clin. Oncology. Jewish. Subspecialty: Pediatrics. Current work: Pediatric oncology. Office: M D Anderson Hospital Texas Medical Center Houston TX 77030

JAFFE, NORMAN J., systems programmer; b. Vancouver, C., Can., Jan. 1, 1954; s. Jack and Marsha (Lewis) J. B.Sc. in Computing Sci. and Math, Simon Fraser U., Burnaby, B.C., 1977. Computer systems programmer Canadian Forest Products, 1973-77, systems programmer, 1977-79, Macmillan Bloedel Ltd., 1979-81; ops. mgr. Canadian Car (Pacific), Vancouver, 1981—. Recipient B.C. Profl. Engrs. award, 1971; TAPPI Found. scholar, 1971. Mem. Assn. for Computing Machinery, Pascal Users Group, HP 1000 Internat. Users Group. Subspecialties: Information systems, storage, and retrieval (computer science); Programming languages. Current work: Multilevel mesh-structured database systems, Actor systems, recursive systems, extensible languages, message based information systems, non-hierarchic databases, threaded languages. Home: 1848 Venables St Vancouver BC Canada V5L 2H7 Office: PO Box 4200 Vancouver BC Canada V6B 4K6

JAFFEE, ROBERT ISAAC, research metallurgist; b. Chgo., July 11, 1917; s. Louis Robert and Sadie (Braidman) J.; m. Edna Elspeth Winram, June 2, 1945; children: William Louis, Michael David. B.S., Ill. Inst. Tech., 1939; S.M., Harvard U., 1940; Ph.D., U. Md., 1943. Lectr. U. Md., 1942; metallurgist Leeds & Northrup, Phila., 1943, U. Calif., 1944; with Battelle Meml. Inst., Columbus, Ohio, 1944-75, asso. mgr., 1960-64, sr. fellow, 1964-73, chief materials scientist, 1973-75; sr. tech. advisor Electric Power Research Inst., Palo Alto, Calif., 1975—; cons. prof. Stanford, 1975—; Mem. Nat. Material Adv. Bd., 1970-74; mem. Acta Metall. Bd. Govs., 1969-74, chmn., 1974—; cons. PSAC, 1966; chmn. NASA Adv. Com. Materials, 1966-71; mem. NATO-AGARD Structure and Materials Panel, 1961-63, 69—; Gillett lectr. ASTM, 1976. Author: The Science, Technology and Application of Titanium, 1970, Refractory Metals and Alloys III, Applied Aspects, 1966, Refractory Metals and Alloys IV, Research and Development, 1967, Phase Stability in Metals and Alloys, 1967, Dislocation Dynamics, 1968, Inelastic Behavior of Solids, 1969, Molecular Processes on Solid Surfaces, 1970, Critical Phenomena in Alloys, Magnets and Superconductors, 1971, Interatomic Potentials and Simulation of Lattice Defects, 1972, Defects and Transport in Oxides, 1973, Titanium Science and Technology, 1973, Physical Basis of Heterogeneous Catalysis, 1974, Fundamental Aspects of Structural Alloy Design, 1975, also articles. Fellow Inst. Metallurgists (London), Metall. Soc., Am. Inst. Metall. Engrs. (hon. mem.; pres. 1978), Am. Soc. Metals (Campbell lectr. 1977, James Douglas gold medal 1983); mem. Nat. Acad. Engring., Am. Phys. Soc., AAAS, Harvard Soc. Sci. and Engring., Sigma Xi, Tau Beta Pi, Phi Lambda Upsilon. Club: Stanford Golf. Subspecialties: Alloys; Metallurgy. Current work: Materials for steam power plants; design of steels for rotors; application of titanium in turbine blading. Research non-ferrous phys. metallurgy, particularly titanium and refractory metals. Patentee in field. Home: 3851 May Ct Palo Alto CA 94303 Office: 3412 Hillview Ave Palo Alto CA 94304

JAGENDORF, ANDRE TRIDON, plant physiologist; b. N.Y.C., Oct. 21, 1926; s. Moritz Adolph and Sophie Sheba (Sokolsky) J.; m. Jean Elizabeth Whitenack, June 12, 1952; children—Suzanne E., Judith C., Daniel Z.S. B.A., Cornell U., 1948; Ph.D., Yale U., 1951. Merck postdoctoral fellow UCLA, 1951-53; from asst. prof. to prof. Johns Hopkins U., 1953-66; prof. plant physiology Cornell U., 1966—, Liberty H. Bailey prof. plant physiology, 1981—. Author papers, revs. in field. Recipient Outstanding Young Scientist award Md. Acad. Sci., 1961, Kettering Research award, 1963; Weizmann fellow, 1962. Fellow Am. Acad. Arts and Scis., AAAS; mem. Nat. Acad. Sci., Am. Soc. Plant Physiologists (pres. 1967, C.F. Kettering award in photosynthesis 1978), Am. Soc. Biol. Chemists, Am. Soc. Photobiology (councilor 1980), Soc. Gen. Physiologists, Am. Soc. Cell Biology, Japanese Soc. Plant Physiologists. Jewish. Subspecialties: Plant physiology (biology); Biochemistry (biology). Current work: Mechanism of photosynthetic energy transduction; from electron flow to vectorial proton translocation to ATP synthesis. Biogenesis of chloroplasts including organellar transcription and translation. Office: Plant Biology Sect Plant Sci Bldg Cornell Univ Ithaca NY 14853

JAHAN-PARWAR, BEHRUS, neurobiologist; b. Ghotchan, Iran, May 26, 1938; s. Hossein Agha and Turan (Milanian Faruji) J-P.; m. Shayda Jafarzadeh, July 23, 1966; children: Babak, Sasan. M.D., U. Gottingen, W.Ger., 1964, D.Sc., 1965. Research assoc. physiology dept. U. Gottingen, 1964-66; asst. research neurophysiologist Mental Health Research Inst., U. Mich., 1966-68; asst. prof. physiology Clark U., Worcester, Mass., from 1968, assoc. prof., to 1973; sr. scientist neurobiology Worcester Found. Exptl. Biology, 1973—; adj. prof. U. R.I., 1981—. Contbr. articles to profl. jours. NIH research career devel. award, 1970; prin. investigator NIH grants, 1969—; Grass Found. grantee, 1975-78; NSF grantee, 1978—. Mem. Soc. Neurosci., Am. Physiol. Soc., Am. Soc. Zoologists, European Chemoreception Research Orgn., Assn. Chemoreception Scis. Subspecialties: Neurophysiology; Neuropsychology. Current work: Principles of neuronal organization underlying processing of sensory information generation and modification of behavior with particular emphasis on the mechanisms of chemoreception, generation and modification of rhythmic behaviors such as locomotion and feeding. Office: 222 Maple Ave Shrewsbury MA 01545

JAHN, ROBERT GEORGE, educator; b. Kearny, N.J., Apr. 1, 1930; s. George E. and Minnie (Holroyd) J.; m. Catherine Seibert, June 20, 1953; children—Eric George, Jill Ellen, Nina Marie, Dawn Anne. B.Sc. in Mech. Engring. with highest honors, Princeton U., 1951, M.A. in Physics, 1953, Ph.D., 1955. Teaching asst. Princeton U., 1953-55; instr. Lehigh U., Bethlehem, Pa., 1955-56, asst. prof., 1956-58; asst. prof. jet propulsion Calif. Inst. Tech., Pasadena, 1958-62; asst. prof. aero. engring. Princeton U., 1962-64, asso. prof., 1964-67, prof. aerospace scis., 1967—, dir. grad. studies aerospace and mech. scis. dept., 1968-71, dean. sch. engring. and applied sci., 1971—, exec. com. council univ. community, 1969-71, research bd., 1971—; Cons. editor Am. Scientist, 1966-70; mem. research adv. com. on fluid mechs. NASA, 1965-68, mem. research and tech. adv. subcom. on electrophysics, 1968-71, mem. research and tech. adv. com. on space propulsion and power, 1971-72; mem. com. on space propulsion and power NASA Research and Tech. Adv. Council, 1976-77, mem. space systems and tech. adv. com. nat. adv. council, 1978—; mem. ad hoc adv. com. minority engring. edn. Alfred P. Sloan Found., 1974-79; mem. com. edn. and employment women in sci. and engring. of commn. on human resources NRC, 1975-79. Author: Physics of Electric Propulsion, 1968; also contbr. articles in field. Trustee Asso. Univs., Inc., 1971—, chmn. bd., 1977-79; chmn. council on energy and environ. studies Princeton U., 1973—. Recipient Shuichi Kusaka Meml. prize in physics, 1951, Curtis W. McGraw Research award Am. Soc. for Engring. Edn., 1969. Fellow Am. Phys. Soc., Am. Inst. Aeros. and Astronautics (lectr. electric propulsion ednl. programs 1971—, electric propulsion tech. com. 1963-67, 71-74); mem. AAUP (asso.), Am. Soc. Engring. Edn., Phi Beta Kappa, Sigma Xi. Subspecialties: Aerospace engineering and technology; Plasma physics. Home: 60 Monroe Ln Princeton NJ 08540

JAHNS, HANS O(TTO), research scientist; b. Kamen, Ger., Sept. 4, 1931; m.; 4 children. Diploma, Clausthal Tech. U., 1955, 56, Dr.Ing., 1961. Research asst. petroleum engr. Inst. Drilling and Petroleum Prodn., Clausthal Tech. U., 1956-59; reservoir engr. Wintershall AG, Ger., 1959-62; research engr. prodn. research Jersey Prodn. Research Co., Standard Oil Co., N.J., Okla., 1962-65, research engr., 1965-68, research assoc., 1968-73, research adv., 1973-77, sr. research adv., 1977-80; research scientist Exxon Prodn. Research Co., Houston, 1980—; Mm. permafrost com. NRC, 1975-80, mem. polar research bd., 1976—; mem. adv. bd. Geophys. Inst., U. Alaska, 1978—. Mm. Soc. Petroleum Engrs., AAAS, Am. Petroleum Inst. Subspecialty: Petroleum engineering. Office: Exxon Prodn Research Co PO Box 2189 Houston TX 77001

JAIN, MAHENDRA KUMAR, chemist, educator, researcher; b. Ujjain, India, Oct. 12, 1938; came to U.S., 1967, naturalized, 1972; s. Hira Lal and Chintamani J.; m.; 1 dau., Dipti. Ph.D., Weizman Inst., Rehovoth, Israel, 1967. Research assoc. Ind. U., 1967-73; asst. prof. chemistry U. Del., 1973-75, assoc. prof., 1975-81, prof., 1981—. Author: books, including The Bimolecular Lipid Membrane, 1972, Introduction to Biological Membranes, 1980, Handbook of Enzyme Inhibitors, 1982; contbr. numerous articles, revs. to profl. jours. Mem. Fedn. Am. Scientists for Exptl. Biology, Biophys. Soc. Subspecialty: Biophysical chemistry. Current work: biological membranes; phospholipases. Home: 759 Arbour Dr Newark DE 19711 Office: U Del Newark DE 19711

JAIN, NARESH KUMAR, microbiologist; b. Sunam, India, Feb. 28, 1948; came to U.S., 1981; s. Sohan L. and Swarn J.; m. Santosh Goyal, Sept. 29, 1976; children: Anshika, Vishal. B.Sc. with honors, Panjab U., Chandigarh, India, 1971, M.Sc., 1973; Ph.D., S.N. Med. Coll., Agra, India, 1981. Asst. prof. Sch. Pharmacy, Delhi U., 1973-75, Med. Sch., Agra (India) U., 1975-80; assoc. prof. Med. Sch., Meerut (India) U., 1980—; research assoc. SUNY-Buffalo, 1981—. Mem. Indian Assn. Med. Microbiology. Subspecialties: Immunology (medicine); Infectious diseases. Current work: Purification and characterization of antigen structure of infectious mononucleosis. Home: 450-A Allenhurst Rd Amherst NY 14226 Office: Oral Biology SUNY-Buffalo Main St Campus Buffalo NY 14214

JAIN, PARVEEN KUMAR, nuclear engineer; b. New Delhi, June 18, 1953; U.S., 1975; s. Rikhi Ram and Shanti Devi (Jain) J.; m. Neeraj Jain, Jan. 16, 1979; 1 son, Puneet Rikhi. B.Sc. in Physics, Delhi (India) U., 1972; B.E. in Elec. Engring, Indian INst. Sci., Bangalore, 1975; M.S. in Nuclear Engring, U. Cin., 1976, Ph.D., U. Ill.-Urbana, 1981. Engr. Systems Control, Inc., Palo Alto, Calif., 1981-83, sr.,engr., 1983—. Contbr. articles to profl. publs. Mem. Am. Nuclear Soc. (Best Paper award Student Conf. 1977), Sigma Xi (assoc.). Subspecialties: Nuclear engineering; Nuclear fission. Current work: Power plant dynamic analysis; mathematical modeling; analytical and experimental research in heat transfer, two-phase flow and thermal-hydraulics. Home: 4882 Scarletwood Terr San JOse CA 95129 Office: Systems Control Inc 1801 Page Mill Rd Palo Alto CA 95303

JAIN, SUBODH KUMAR, biology educator, researcher; b. Nanauta, India, Dec. 11, 1934; s. Jambu Prasad and Jagwati Devi J.; m. Saroj Kumari, July 6, 1957; children: Sudhanshu, Vinoo, Sarita. B.S. with honors, U. Delhi, India, 1954; M.S., Indian Agrl. Research Inst., Delhi, 1956; Ph.D., U. Calif., 1960. Sr. fellow Govt. India, New Delhi, 1961-63; asst. geneticist U. Calif.-Davis, 1963-67, assoc. prof., 1967-72, prof. agronomy, 1972—; cons. Internat. Crops Research Inst. Semi-Arid Tropics, Hyderabad, India, Instituto Argentino de Investigación de las Zonas Áridas, Mendoza, Argentina; mem. genetic resources coms. Internat. Bd. Plant Genetic Resources, Crop Soc. Am.; mem. Amaranth panel U.S. Nat. Acad. Scis. Mem. editorial com.: Ann. Rev. Ecology and Systematics; assoc. editor: Evolution, 1981-83; contbr. chpts. to books, tech. papers and rev. to profl. publs.; co-editor: Topics in Plant Population Biology, 1979, Indian Genetic Resources, 1983; editor book on vernal pools, their ecology and conservation, 1976. Active Nature Conservancy and other conservation groups; campaigner for germ plasm and vernal pool conservation and new crop devel. Recipient Govt. India Prize for Agr., 1954; grad. fellow Tata Found., 1956, U. Calif., 1958-60; Guggenheim fellow, 1971-72; sr. fellow Commonwealth Sci. and Indsl. Research Orgn., Australia, 1971-72; Fulbright Indo-Am. fellow, 1978. Mem. Soc. Econ. Botany, Soc. Study of Evolution, Am. Genetic Assn., Bot. Soc. Am., Am. Soc. Naturalists. Democrat. Jainist. Subspecialties: Evolutionary biology; Population biology. Current work: Evolution in plant populations; genetics and ecology of annual plants in vernal pools; plant breeding theory; genetic resources and new crop development.

JAIN, SULEKH CHAND, mfg. technologist; b. Raksehra, India, Mar. 5, 1937; came to U.S., 1969, naturalized, 1974; s. Chater Sain and Parmeshwari (Devi) J.; m. (married); children: Anudeep, Vandana. B.S. in M.E, Punjab U., 1960; M.S., Indian Inst. Tech., Khargpur, 1962; Ph.D., Birmingham (Eng.) U., 1969; M.B.A., Clark U., 1977. Cert. mfg. engr., Calif. Various tech. positions, India, 1960-66; sr. research engr. Wyman Gordon Co., North Grafton, Mass., 1969-74, research group leader, 1974-80; tech. dir. Beaumont Well Works, Houston, 1980—; lectr. mech. engring. M.I.T., 1976-79; adj. prof. mech. engring. Worcester Poly. Inst., 1980-81. Contbr. articles to profl. jours. Pres. Jain Soc. Houston, 1982-83; mem. exec. com. Jain Center Greater Boston, 1973-79. Mem. ASME, Am. Soc. Metals, Soc. Mfg. Engrs., India Assn. Worcester (sec. 1971-72). Jain. Lodge: Lions. Subspecialties: Materials processing; Mechanical engineering. Current work: Plastic deformation of metals; powder metallurgy; metalworking lubrication, friction, tool design. Patentee in field. Home: 3419 Palm Desert Ln Missouri City TX 77459 Office: 4710 Bellaire Houston TX 77001

JAIN, SUSHIL KUMAR, biochemistry educator, researcher; b. Nabha, Punjab, India, Mar. 31, 1950; came to U.S., 1977; s. Gian Chand and Parkash (Devi) J.; m. Shubh Laxmi, Apr. 2, 1980. B.Sc. with honors, Punjab U., 1970; M.Sc., Postgrad., Inst. Med. Edn. and Research, Chandigarh, India, 1972, PH.D., 1976. Cert. clin. chemist. Tutor in biochemistry Postgrad. Inst., Chandigarh, 1976-77; postdoctoral fellow U. So. Calif., 1977-79, U. Calif.-San Francisco, 1979-81; instr. biochemistry La. State U. Med. Ctr., Shreveport, 1981-82, asst. prof., 1983—. Contbr. articles profl. jours. Vice pres. Indian Assn., Shreveport, 1983—. Recipient Beecham award So. Blood Group, Houston, 1982, Ross award So. Soc. Pediatric Research, 1982, Biomed. research award Nat. Inst. Health, 1982, Stiles award La. State U., 1983; recipient grants from pvt. drug cos. Mem. Am. Chem. Soc., Am. Soc. Hematology, N.Y. Acad. Scis., Am. Fedn. Clin. Research, AAAS. Subspecialties: Hematology; Nutrition (medicine). Current work: Red blood cell structure and functions, lipid metabolism, lecithin-cholesterol acyl transferase and hyperlipidemias, iron-deficiency anemia, lipid peroxidation role in cellular injury, membrane lipid peroxidation, sickle cell pathology. Home: 1509 Westbury Dr Shreveport LA 71105 Office: Louisiana State University Medical Center 1501 Kings Hwy Shreveport LA 71130

JAINCHILL, JEROME, biochemist, researcher, educator; b. N.Y.C., Jan. 27, 1932; s. Nathan and Frieda (Fried) J.; m. Roberta Cohen, Apr. 21, 1940; children: Charles, Melissa, Susan. Ph.D., NYU, 1963. Research assoc. Sloan Kettering Inst., N.Y.C., 1963-65; assoc. research biochemist NYU Med. Ctr., N.Y.C., 1965-67; sr. research biochemist Endo Labs. Dupont, Garden City, N.Y., 1967-77; dir. lab. ops. Cornell U., 1977-80; research dir. North Star Research, N.Y.C., 1980—; adj. prof. N.Y. Inst. Tech., 1977-78. Contbr. articles to profl. publs. Served with U.S. Army, 1954-56. Mem. N.Y. Acad. Sci., AAAS, Am. Chem. Soc., Am. Soc. Pharmacology and Exptl. Therapeutics. Subspecialties: Pharmacokinetics; Biochemistry (medicine). Current work: Genetics, pharmacology, drug detection. Home: 2362 Garfield St North Bellmore NY 11710

JALURIA, RAJIV, communications engineering manager, planner; b. Lucknow, India, Nov. 23, 1952; s. Jagdishwar and Maya (Verma) J.; m. Anjali Bajaj, Jan. 1, 1979; 1 son, Samir. M.Sc., U. Delhi, India, 1974; Ph.D. candidate, Purdue U., 1976; M.B.A., Pepperdine U., 1982. Engr. ITT, Galion, Ohio, 1976-78; mem. tech. staff Bell Labs., Naperville, Ill., 1978-79; mgr. advanced network planning SP Communications, Burlingame, Calif., 1979—. Contbr. articles to profl. jours. Mem. IEEE, Assn. Computing Machinery. Subspecialties: Computer engineering; Systems engineering. Current work: Voice switching, networking and optimization; data communications; digital networks and enhancements. Home: 983 Lurline Dr Foster City CA 94404 Office: SP Communications 1 Adrian Ct Burlingame CA 94010

JAMDAR, SUBHASH CHANDRASHEKHAR, biochemistry educator; b. Nagpur, Maharastra, India, Apr. 11, 1937; came to U.S., 1968; s. Chandrashekhar M. and Anandibai J.; m. Sujata Katdare, Jan. 7, 1967; children: Sachin, Niteen. B.Sc., Nagpur U., 1958, M.Sc., 1960; Ph.D., 1966. Research assoc. U. N.C., Chapel Hill, 1971-74; asst. prof. biochemistry Med. Coll. Va., Richmond, 1974-79; head div. lipid biochemistry Med. Research Inst., Fla. Inst. Tech., Melbourne, 1979—. NIH research grantee, 1978-81; Weight Watchers Found. research grantee, 1982-84. Mem. Am. Assn. Biol. Chemists, Soc. Exptl. Biology and Medicine, N.Y. Acad. Sci., Sigma Xi. Subspecialties: Biochemistry (medicine); Cell biology (medicine). Current work: Research on regulation of adipose lipid metabolism. Home: 771 Atlantic Dr Satellite Beach FL 32901 Office: Med Research Inst Fla Inst Tech 3325 W New Haven Melbourne FL 32901

JAMES, MARGARET O., medicinal chemistry educator, researcher; b. Wales, U.K.; d. L. Keith and Olive M. (Bound) J. B.S. in Chemistry, Univ. Coll. London, U.K., 1969; Ph.D. in Biochemistry, St. Mary's Med. Sch., U. London, 1972. Postdoctoral fellow Nat. Inst. Environ. Health Scis., Research Triangle Park, N.C., 1972-75, sr. staff fellow, St. Augustine, Fla., 1975-80; asst. prof. dept. medicinal chemistry U. Fla., Gainesville, 1980—; cons. in field. Contbr. articles to sci. jours. Mem. Am. Soc. Pharmacology and Exptl. Therapeutics, Am. Chem. Soc., Biochem. Soc. (U.K.), AAAS. Democrat. Episcopalian. Subspecialties: Biochemistry (medicine); Molecular pharmacology. Current work: Factors affecting the metabolism and toxicity of drugs in different animal species; mechanisms of drug metabolism and toxicity. Office: Hillis Miller Health Center Univ Fla Box J4 Gainesville FL 32610

JAMES, PHILIP BENJAMIN, physicist, astronomer, educator; b. Kansas City, Mo., Mar. 18, 1940; s. Benjamin and Catharine (Bagley) J.; m. Sharon Lynn James, Aug. 28, 1965; children: Eric, Kevin, Kirsten. B.S., Carnegie-Mellon U., 1961; Ph.D., U. Wis., 1966. Research assoc. physics U. Ill., Urbana, 1966-68; asst. prof. physics and astronomy U. Mo., St. Louis, 1968-72, assoc. prof., 1972-78, prof., 1978—; NRC sr. assoc. Jet Propulsion Lab., 1977-78. Contbr. articles to sci. jours.; editor: Am. Jour. Physics. NASA grantee, 1978—. Mem. Am. Phys. Soc., Am. Geophys. Union, Am. Astron. Soc., AAAS. Lutheran. Subspecialties: Planetary science; Climatology. Current work: Climate variability on Mars and its implications for terrestrial climate, modeling, data analysis. Office: 530 Benton Hall U MO Saint Louis MO 63121

JAMES, SHERMAN ATHONIA, epidemiology educator; b. Hartsville, S.C., Oct. 25, 1943; s. jerome and Helen (Bachus) J.; m. Jacqueline Anita Blaylock, Apr. 16, 1965; children: Sherman Alexander, Scott Anthony. A.B., Talladega Coll., 1964; Ph.D., Washington U., St. Louis, 1973. Instr. U. N.C.-Chapel HIll, 1973-74, asst. prof. epidemiology, 1974-80, assoc. prof., 1980—; mem. epidemiology study sect. NIMH, 1979-83. Editorial bd.: Med. Care, 1980-83. Served to capt. USAF, 1964-69. Recipient Research Career Devel. award NIH Heart, Lung and Blood Inst., 1982-87. Mem. Am. Pyschol. Assn., Am. Public Health Assn., Delta Omega. Democrat. Subspecialties: Epidemiology; Social psychology. Current work: Role of behavioral stressors in cardiovascular disease with special applications to black Americans. Office: Sch Public Health U NC Chapel Hill NC 27514

JAMES, STEPHANIE LYNN, immunoparasitologist; b. Little Rock, Feb. 20, 1950; d. Vernal and Gene J. B.A., Hendrix Coll., Conway, Ark., 1972; Ph.D., Vanderbilt U., 1976. Research fellow in medicine Harvard Med. Sch., Boston, 1977-80; research assoc. Lab. Parasitic Diseases, NIH, Bethesda, Md., 1980—. Contbr. articles to profl. jours. Nat. Research Service fellow, 1979-81. Mem. Am. Assn. Immunologists, Am. Soc. Tropical Medicine and Hygiene. Subspecialties: Infectious diseases; Parasitology. Current work: Immunology of parasitic diseases, particularly schistosomiasis. Office: Room 740C Ross Hall George Washington U Med Ctr 2300 Eye St NW Washington DC 20037

JAMES, THOMAS L., biochemistry educator; b. North Platte, Nebr., Sept. 8, 1944; s. James J. and Guinevere (Richards) J.; m. Joyce E. Voge, July 8, 1964 (div. 1975); m. Jacqueline L. Volz, Jan. 28, 1978; children: Marc, Tristan. B.S., U. N.Mex., 1965; Ph.D., U. Wis.-Madison, 1969. Research chemist Celanese Chem. Co., Corpus Christi, Tex., 1969-71; research assoc. U. Pa., Phila., 1971-73; prof. U. Calif.-San Francisco, 1973—, dir., 1975—. Author: Nuclear Magnetic Resonance in Biochemistry, 1975; contbr. articles to profl. jours. NIH fellow, 1966, 72, 79. Mem. Internat. Soc. Magnetic Resonance (nat. adv. com. 1983), Soc. Magnetic Resonance in Medicine, Am. Biophys. Soc., Am. Chem. Soc. Mem. Reorganized Ch. Jesus Christ of Latter-day Saints. Subspecialties: Biophysics (biology); Nuclear magnetic resonance (chemistry). Current work: Nuclear magnetic resonance applications to biochemical and biological systems, especially nucleic acid structure and dynamics, nucleic acid interaction with proteins and drugs, in vivo NMR of animals and humans, glycoproteins. Office: U Calif Dept Pharm Chemistry San Francisco CA 94143 Home: 234 San Carlos Way Novato CA 94947

JAMES, THOMAS NAUM, cardiologist, educator; b. Amory, Miss., Oct. 24, 1925; s. Naum and Kata J.; m. Gleaves Elizabeth Tynes, June 22, 1948; children—Thomas Mark, Terrence Fenner, Peter Naum. B.S., Tulane U., 1946, M.D., 1949. Diplomate: Am. Bd. Internal Medicine (mem. subsplty. bd. cardiovascular diseases 1972-78, bd. govs. 1982—). Intern Henry Ford Hosp., Detroit, 1949-50, resident in internal medicine and cardiology, 1950-53; practice medicine specializing in cardiology, Birmingham, Ala., 1968—; mem. staff Henry Ford Hosp., 1959-68, U. Ala. Hosps., 1968—; instr. medicine Tulane U., New Orleans, 1955-58, asst. prof., 1959; prof. medicine U. Ala. Med. Center, Birmingham, 1968—, prof. pathology, 1968-73, assoc. prof. physiology and biophysics, 1969-73, dir. Cardiovascular Research and Tng. Center, 1970-77, chmn. dept. medicine, 1973—, Mary Gertrude Waters prof. cardiology, 1976—; Disting. prof. of univ. U. Ala., 1981—; mem. adv. council Nat. Heart Lung and Blood Inst., 1975-79; mem. cardiology del. invited by Chinese Med. Assn. to, People's Republic of China, 1978. Author: Anatomy of the Coronary Arteries, 1961, The Etiology of Myocardial Infarction, 1963; Mem. editorial bd.: Circulation, 1966-83; mem. editorial bd.: Am. Jour. Cardiology, 1968-76; assoc. editor, 1976—; mem. editorial bd.: Am. Heart Jour, 1976-79; Contbr. articles on cardiovascular diseases to med. jours. Served as capt. M.C. U.S. Army, 1953-55. Mem. ACP (gov. Ala. 1975-79, master 1983), AMA, Am. Clin. and Climatological Assn., Assn. Am. Physicians, Am. Soc. Clin. Investigation, Assn. Univ. Cardiologists (pres. 1978-79), Am. Heart Assn. (pres. 1979-80), Am. Coll. Cardiology (v.p. 1970-71, trustee 1970-71, 76-81, First Disting. Scientist award 1982), Am. Soc. Pharmacology and Exptl. Therapeutics, Soc. Exptl. Biology of Medicine, Am. Coll. Chest Physicians, Central Soc. Clin. Research, So. Soc. Clin. Investigation, Am. Fedn. Clin. Research, Phi Beta Kappa, Sigma Xi, Omicron Delta Kappa, Ala. Acad. Honor, Alpha Omega Alpha, Alpha Tau Omega, Phi Chi. Presbyterian. Clubs: Cosmos, Mountain Brook. Subspecialties: Internal medicine; Cardiology. Current work: Research in internal medicine and cardiology. Office: Dept Medicine U Ala Med Center Birmingham AL 35294

JAMESON, DOROTHEA, sensory psychologist; b. Newton, Mass., Nov. 16, 1920; d. Robert and Josephine (Murray) J.; m. Leo M. Hurvich, Oct. 23, 1948. B.A., Wellesley Coll., 1942; M.A. (hon.), U. Pa., 1973. Research asst. Harvard, 1941-47; research psychologist Eastman Kodak Co., Rochester, N.Y., 1947-57; research scientist, N.Y. U., 1957-62; vis. scientist Venezuelan Inst. Sci. Research, 1965; research asso. to prof. Psychol. and Inst. Neurol. Scis., U. Pa., 1962-74; Univ. prof. U. Pa., 1975—; vis. prof. Center Visual Sci., U. Rochester, 1974, Columbia U., 1974-76; cons. in field. Mem. Nat. Acad. Sci.-NRC Commn. on Human Resources, 1977-80, chmn. com. on vision, 1980-81. Co-author: The Perception of Brightness and Darkness, 1966; co-author: (E. Hering) introduction and English translation Outlines of a Theory of the Light Sense, 1964; Co-editor, author: Visual Psychophysics: Handbook of Sensory Physiology, vol. VII/4, 1972; Contbr. articles to profl. jours. Recipient I.H. Godlove award Inter-Soc. Color Council, 1973; Alumnae Achievement award Wellesley Coll., 1974; fellow Center for Advanced Study in the Behavioral Scis., 1981-82. Mem. Soc. Exptl. Psychologists (Howard Crosby Warren medal 1971), Internat. Brain Research Orgn., Am. Psychol. Assn. (Distinguished Sci. Contbn. award 1972), Nat. Acad. Scis., Am. Acad. Arts and Scis., AAAS, Assn. Research in Vision and Ophthalmology, Biophys. Soc., Internat. Research Group Color Vision Deficiencies, Optical Soc. Am. (Tillyer medal 1982), Psychonomic Soc., Soc. Neurosci., Sigma Xi. Subspecialties: Sensory processes; Psychophysics. Current work: Vision and perception; color and spatial vision; sensory mechanisms; brain mechanisms and art. Home: 286 St James Pl Philadelphia PA 19106 Office: 3815 Walnut St Philadelphia PA 19104

JAMESON, WILLIAM JAMES, JR., telecommunications cons.; b. Billings, Mont., June 8, 1930; s. William James and Mildred T. (Lore) J.; m. Margaret Sue Randall, Sept. 9, 1953; children: Margaret Elizabeth, Katharine Rebecca. B.A. in Math, U. Mont., 1952; M.A. in Physics, U. Tex., Austin, 1954; Ph.D., Iowa State U., 1962. Physicist Lockeed Missile Co., Palo Alto, Calif., 1958-59; engr./mathematician Collins Radio Co., Cedar Rapids, Iowa, 1962-72; cons. Spectra Assocs., Inc., Cedar Rapids, 1972—; adj. assoc. prof. Iowa State U., 1962—; adj. staff CMC Colls., Cedar Rapids, 1982—. Chmn. Linn County Metro Pub. Safety Telecom. Bd., Cedar Rapids, 1974—; bd. dirs. Hawkeye Area council Boy Scouts Am., 1967-80. Served to 1st lt. USAF, 1952-58; PTO. Mem. IEEE, Soc. Indsl. and Applied Math. (sec. 1964-69, v.p. programs 1969-74), Assn. Computing Machinery. Lodge: Kiwanis. Subspecialties: Applied mathematics; Systems engineering. Current work: Numerical computation, telecommunications systems. Office: 330 1st St SE Cedar Rapids IA 52401 Home: 328 Forest Dr SE Cedar Rapids IA 52403

JAMISON, HARRISON CLYDE, oil company executive; b. St. Louis, Jan. 15, 1925; s. William Clyde and Katherine Maurice

(Fitzgerald) J.; m. Beverly Joy Johnson, June 26, 1946; children: Susan, David, Leslie, Daniel, Dale, Nancy, Sara. B.A. cum laude, UCLA, 1949, postgrad., 1949-50. Geologist Richfield Oil Corp., Bakersfield, Calif., 1950-52, Olympia, Wash., 1952-55, Los Angeles, 1955-60, regional exploration supr., 1961-65; Alaska dist. mgr. Atlantic Richfield Co., Anchorage, 1966-69, Alaska coordinator, Dallas, 1969-70; Alyeska pipeline mgr. ARCO Pipeline, Bellevue, Wash., 1970-72; v.p., chief geologist Atlantic Richfield Co., Dallas, 1972-80, western dist. mgr., Denver, 1980-81; pres. ARCO Exploration Co. (div. Atlantic Richfield Co.), Dallas, 1981—; sr. v.p. Atlantic Richfield Co., Los Angeles, 1981—; vice-chmn. ARCO Alaska, Inc., Anchorage, 1982—; chmn. bd. Resolution Seismic Services, Inc., Wilmington, Del., 1981—. Nat. chmn. Amigos de Ser, Jobs for Progress, Inc., Dallas, 1983-84; bd. dirs. Tex. Research League, Austin, 1981-84. Served with USN, 1943-46. Fellow Geol. Soc. Am.; mem. Am. Assn. Petroleum Geologists, Am. Inst. Petroleum Geologists, Am. Petroleum Inst., Geol. Soc. Am. Found. (trustee 1982). Republican. Clubs: Petroleum (Dallas); (Anchorage) (pres. 1969). Subspecialty: Geology. Current work: Multi-billion dollar petroleum exploration program in North America over the next five years, utilizing a staff of 1800 people, of whom about 60% are professional explorationists. Home: 6415 Forest Creek Dr Dallas TX 75230 Office: Arco Exploration Co PO Box 2819 Dallas TX 75221

JAMPLIS, ROBERT WARREN, surgeon, medical foundation executive; b. Chgo., Apr. 1, 1920; s. Mark and Janet (McKenna) J.; m. Roberta Cecelia Prior, Sept. 5, 1947; children: Mark Prior, Elizabeth Ann Jamplis Halliday. B.S., U. Chgo., 1941, M.D., 1944; M.S., U. Minn., 1951. Diplomate: Am. Bd. Surgery, 1952, Am. Bd. Thoracic Surgery, 1953; Lic. physician, Calif., Minn., Ill. Asst. resident in surgery U. Chgo., 1946-47; fellow in thoracic surgery Mayo Clinic, Rochester, Minn., 1950-52; chief thoracic surgery Palo Alto (Calif.) Med. Clinic, 1958—, exec. dir., 1965-81; clin. prof. surgery Stanford U. Sch. Medicine, 1958—; dir. Coopers Labs., Inc.; Mem. council SRI Internat.; vice-chmn. bd. TakeCare Corp.; charter mem., bd. regents Am. Coll. Physician Execs.; mem. staff Stanford Univ. Hosp., Santa Clara Valley Med. Center, San Jose, VA Hosp., Palo Alto, Sequoia Hosp., Redwood City, Calif., El Camino Hosp., Mountain View, Calif., Harold D. Chope Community Hosp., San Mateo, Calif.; pres., chief exec. officer Palo Alto Med. Found.; vice-chmn. Fedn. Western Clinics; mem. physician adv. com. Blue Cross Calif.; varsity football team physician Stanford U. Contbr. numerous articles to profl. jours.; author: (with G.A. Lillington) A Diagnostic Approach to Chest Diseases, 1965, 3d edit., 1984. Trustee Santa Barbara Med. Found. Clinic; pres. Calif. div. Am. Cancer Soc.; chmn. bd. Group Practice Polit. Area Com.; mem. athletic bd. Stanford U.; mem. cabinet U. Chgo.; bd. dirs. Herbert Hoover Boys' Club; past trustee No. Calif. Cancer Program; past bd. dirs. Core Communications in Health, Community Blood Res., others. Served to lt. USNR, 1944-46, 52-54. Recipient Alumni citation U. Chgo., 1968; Nat. Div. award Am. Cancer Soc., 1979; Med. Exec. award Am. Coll. Med. Group Adminstrs., 1981; Russel V. Lee award lectr. Am. Group Pratice Assn., 1982. Mem. Inst. Medicine of Nat. Acad. Scis., ACS, Am. Assn. Thoracic Surgery, Samson Thoracic Surg. Soc. (pres.), Western Surg. Assn., Pacific Coast Surg. Assn., San Francisco Surg. Soc. (past pres.), Portland Surg. Soc. (hon.), Doctors Mayo Soc., Am. Coll. Chest Physicians, Calif. Acad. Medicine, Am. Fedn. Clin. Research, Am. Group Practice Assn. (pres.), AMA, Calif. Med. Assn., Santa Clara County Med. Assn., Sigma Xi. Republican. Roman Catholic. Clubs: Bohemian, Commonwealth of California (San Francisco); Menlo Country (Woodside, (Calif.)); Menlo Circus (Atherton, Calif.); Stanford (Calif.) Golf; Rancheros Visitadores (Santa Barbara, Calif.). Subspecialty: Surgery. Current work: Innovative and progressive administration of health care delivery. Office: 400 Channing St Palo Alto CA 94301

JANDA, LOUIS HUGO, clinical psychologist; b. Durango, Colo., Nov. 24, 1946; s. Stanley and Lorraine (Stock) J.; m. Meredith Luras, Sept. 20, 1968; children: Christopher, Michael. B.S., Colo. State U., Ft. Collins, 1969; M.S., Ariz. State U., Tempe, 1971, Ph.D., 1972. Lic. clin. psychologist, Va. Assoc. prof. Old Dominion U., Norfolk, Va., 1973—; clin. psychologist Briddel & Assoc., Virginia Beach, Va., 1981—. Author: Human Sexuality, 1980, Exploring Human Sexuality, 1981, Personal Adjustment, 2d edit, 1981, Psychology: Its Study and Uses, 1982. Served with U.S. Army, 1966-68. Mem. Am. Psychol. Assn. Current work: Clinical applications of restricted environmental stimulation technique. Home: 1080 Willowbrook Ct Virginia Beach VA 23464 Office: Old Dominion Univ Norfolk VA 23508

JANDHYALA, BHAGAVAN SRIKRISHNA, pharmacology educator, researcher; b. India, Aug. 3, 1938; came to U.S., 1961, naturalized, 1976; s. Dakshina Murty and Kanaka Durgamba (Chavali) J.; m. Marie Louis Steenberg, Nov. 16, 1967; 1 son, Murty Dakshina. B. Pharmacy, Andhra U., India, 1961; M.S., U. Pitts., 1963, Ph.D., 1966. Sr. research grup leader U.S.V. Pharm. Corp., N.Y.C., 1966-67; asst. prof. U. Pitts., 1967-72, assoc., prof., 1972-73, U. Houston, 1973—; cons. in field. Contbr. numerous articles on pharmacology to profl. jours. Recipient facultl devel. award U. Houston, 1979; NIH grantee, 1974-77. Mem. Am. Soc. Pharmacology and Exptl. Therapeutics, Soc. Exptl. Biology, Internat. Soc. Hypertension, Indian Assn. of Physiologists and Pharmacologists, Sigma Chi, Rho Chi. Subspecialties: Pharmacology; Psychiatry. Current work: Cardiovascular research, pathogenesis of hypertension, salt and hypertension, central mechanisms, anti hypertensive drug research and mechanisms. Home: 14011 Britoak Houston TX 77079 Office: 460 SR2 4800 Calhoun Houston TX 77004

JANE, JOHN ANTHONY, neurosurgeon; b. Chgo., Sept. 21, 1931; s. Kamil and and Serrita (Jane) Schulhof; m. Noella Fortier; children: Serrita, Jennifer, Catherine, John. B.A. cum laude, U. Chgo., 1951, M.D., 1956, Ph.D., 1967. Diplomate: Am. Bd. Neurol. Surgeons. Instr. dept. psychology U. Chgo., 1951-52, research asst. dept. psychology, 1953-56; intern Royal Victoria Hsop., McGill U., Montreal, Que., Can., 1956-57; sr. fellow neurophysiology, 1959-60; resident in neurosurgery U. Chgo. Clinics, 1957, Royal Vistoria Hosp., 1957-58; fellow neurophysiology Montreal Neurol. Inst., NIH, 1958-59; research asst. neurosurgery St. George's Hosp. and Nat. Hosp. Queen Square, London, 1961; research assoc. Duke U., 1962; sr. resident neurosurgery U. Ill. Research and Edn. Hosps., Ill. Neuropsychiat. Inst., 1963-64; sr. instr. neurosurgery Case Western Res. U., 1965-66, asst. prof., 1967-68, assoc. prof., 1968-69; asst. neurosurgeon Univ. Hosps., Cleve., 1965-69; chief neurosurg. div. Cleve. VA Hosp., 1965-69; alumni prof., chmn. dept. neurosurgery U Va., Charlottesville, 1969—. Author: (with D. Yashon) Cytology of Tumors Affecting the Nervous System, 1969; contbr. chpts. to books, numerous articles to profl. jours. Fellow ACS, Royal Coll. Surgeons Can.; mem. Am. Assn. Anatomists, Am. Psychol. Soc., Am. Assn. Neurol. Surgeons, AMA, Albemarle Med. Assn., Can. Neurosurg. Soc., Med. Soc. Va., Neurosurg. Soc. Vas., Neurosurg. Soc. Am., Pavlovian Soc., Research Soc. Neurol. Surgeons, Royal Coll. Physicians and Surgeons, Soc. Brit. Neurol. Surgeons, Soc. Neurol. Surgeons, Soc. Neurosci., So. Neurosurg. Soc. Club: Cajal. Subspecialty: Neurosurgery. Current work: Craniofacial anomalies, head trauma. Home: 1902 Blue Ridge Rd Charlottesville VA 22902 Office: U Va Dept Neurosurgery Box 212 Charlottesville VA 22908

JANECKE, JOACHIM WILHELM, physics educator; b. Heidelberg, Germany, Feb. 5, 1929; came to U.S., 1965; m. Christa M. Hawner; 2 children. Dipl.Phys., U. Heidelberg, W.Ger., 1952, Dr. rer.nat., 1955. Research physicist Max-Planck Inst., Heidelberg, 1955-60; research assoc., lectr. U. Mich., Ann Arbor, 1960-62, assoc. prof. physics, 1965-69, prof., 1969—; research physicist Nuclear Research Ctr., Karlsruhe, W.Ger., 1962-65; vis. prof. Max-Planck Inst., 1972, Kernysisch Versneller Inst., Groningen, Netherlands, 1979-80. Fellow Am. Phys. Soc.; mem. Sigma Xi. Subspecialty: Nuclear physics. Current work: Experimental nuclear physics; nuclear reactions with light and heavy ions; nuclear structure; nuclear astrophysics. Office: Dept Physics U Mich Ann Arbor MI 48109

JANICK, JULES, horticulture educator, editor, researcher; b. N.Y.C., Mar. 16, 1931; s. Carl and Frieda (Tullman) J.; m. Shirley Reisner, June 15, 1952; children: Peter, Robin. B.S., Cornell U., 1951; M.S., Purdue U., 1952, Ph.D., 1954. Instr. Purdue U., West Lafayette, Ind., 1954-56, asst. prof., 1956-59, assoc. prof., 1959-63, prof., 1963—; hon. research assoc. Univ. Coll., London, 1962; horticulturist Rural Univ. Minas Gerais, Brazil, 1963-65. Author: Horticultural Science, 1979, Plant Science: An Introduction to World Crops, 1981, Advances in Fruit Breeding, 1975; Methods in Fruit Breeding, 1983; contbr. numerous articles to profl. jours.; editor: HortScience, 1970-83, Jour. Am. Soc. Hort. Sci, 1976-83, Hort. Revs, 1979—, Plant Breeding Revs, 1982—; contbg. editor: Horticulture, 1982—. Fellow Am. Soc. Hort. Sci. (Marion Meadows award 1971, Stark award 1978, 82, Wilson Popenoe award 1980, Kenneth Post award 1981, N.F. Childers award 1982), Portuguese Hort. Assn.; mem. Am. Pomological Soc. (Paul Howe Shepard award 1960, 70), Internat. Soc. Hort. Sci., AAAS, Sigma Xi. Jewish. Subspecialties: Plant genetics; Plant cell and tissue culture. Current work: Horticulture, genetics, plant breeding, tissue culture. Patentee in field. Home: 420 Forest Hill Dr West Lafayette IN 47906 Office: Dept Horticulture Purdue U West Lafayette IN 47907

JANIS, RONALD ALLEN, pharmacologist, physiologist, educator; b. Mossbank, Sask., Can., Oct. 11, 1943; came to U.S., 1968; m. Adriana Beukers, Jan. 23, 1968; children: Mary Alice, Joseph Walter. B.S.P., U. B.C., 1966, M.S.P. (univ. grad. fellow), 1968; Ph.D., SUNY, Buffalo, 1972. Asst. prof. physiology Northwestern U., Chgo., 1974-80, asso. prof., 1980; head Biochem. Pharmacology Lab., Miles Inst. for Preclin. Pharmacology, New Haven, 1980—; asso. prof. in residence dept. medicine U. Conn. Health Center, Farmington, 1981—; cons. in field. Contbr. articles to pharmacology, biochemistry and physiology jours., chpts. to books. Med. Research Council of Can. fellow, 1973; Can. Heart Found. fellow, 1973, 74; NIH grantee, 1978—. Mem. Am. Physiol. Soc., Am. Soc. for Pharmacology and Exptl. Therapeutics, Biophys. Soc., Brit. Pharm. Soc., Pharmacology Soc. Can. Subspecialties: Molecular pharmacology; Membrane biology. Current work: Pharmacology of calcium antagonists; new drug development; regulation of smooth muscle contraction; role of blood vessels in hypertensive diseases; calcium transport; ligand binding; protein phosphorylation; cyclic nucleotides. Home: 656 High Ridge Rd Orange CT 06477 Office: Box 1956 New Haven CT 06509

JANKELSON, BERNARD, dentist, research director; b. Bloemfontein, Orange Free State, Republic of South Africa, Sept. 8, 1902; came to U.S., 1920; s. Maurice and Sophia (Asher) J.; m. Agnes Jane Neighbors, Sept. 24, 1938; children: Robert Reed, Roland Clark. D.N.D., U. Oreg., 1924. Diplomate: Am. Bd. Prosthodontics. Pvt. practice dentistry, Seattle, 1924—; clin./research assoc. U. Wash., Seattle, 1950-60; cons. UPSHS, 1955-60, U.S. Dept. Army, 1956-58, U.S. VA, 1955-75; dir. research and devel. Myo-tronics Research, Inc., Seattle, 1967-83. Author: numerous publs. including Biophysics and Physiology of the Mandible, 1983. Recipient Diploma de Honor Colegio Chirujanos Dentistas, P.R., 1953; Cercle Paradontie, Paris, 1971. Fellow Internat. Coll. Dentists, Am. Coll. Prosthodontics; mem. Internat. Coll. Craniomandibular Orthopedics, Am. Acad. Gen. Dentistry (hon.), Accademia Italiana di Paradontologia (hon.), Am. Assn. Dental Research, Am. Dental Assn., Omicron Kappa Upsilon. Subspecialties: Biophysics (biology); Physiology (biology). Current work: Basic science research in biophysics and physiology of musculoskeletal system of head and neck; applied clinical research for diagnosis and treatment of musculoskeletal dysfunction of head and neck. Patentee in field of biomed. instrumentation and chemical formulae. Home: 1100 University St Apt 16E Seattle WA 98101 Office: Myotronics Research Inc 720 Olive Way Suite 800 Seattle WA 98101

JANNASCH, HOLGER WINDEKILDE, marine scientist; b. Holzminden, Ger., May 23, 1927; m., 1956; 1 child. Ph.D. in Microbiology, U. Göttingen, Ger., 1955. Research asst. microbiologist Max Planck Soc., 1957-60; privat docent U. Göttingen, 1963; sr. scientist Woods Hole Oceanographic Instn. (Mass.), 1963—; fellow Scripps Inst. Oceanography, U. Calif.-San Diego, 1957-58; research assoc. U. Wis., 1958-59; mem. panel water criteria Nat. Acad. Sci.-NSF, 1975-78. Recipient Henry Bryant Bigelow medal, 1980, Fischer Sci. Award, 1982. Mem. Am. Soc. Microbiology, Am. Soc. Limnology and Oceanography, Internat. Assn. Theoretical and Applied Limnology. Subspecialties: Deep-sea biology; Microbiology. Current work: Physiology and ecology of marine bacteria; steady state culture; bacterial processes at deep sea hydrothermamal vents. Office: Woods Hole Oceanographic Instn Woods Hole MA 02543

JANSEN, GEORGE JAMES, consulting mineralogist; b. Canton, Ohio, Apr. 22, 1925; s. George Bernard and Caroline Agnes (Wilkinson) J.; m. Patricia Jean Wood, Feb. 14, 1953; children: George James, Kenneth V.; m. Marjorie Ann Molloy, June 12, 1971. B.S. in Geology, U. Notre Dame, 1951, M.A., Bryn Mawr Coll., 1952. Geologist U.S. Geol. Survey, 1952-57; prin. geologist Battelle Meml. Inst., 1957; supr. Republic Steel Co., 1958-69; sr. scientist Amax, 1969-75; coal petrographer CT & E., 1976-77; v.p., prin. investigator Rocky Mountain Coal Petrography, Inc., Golden, Colo., 1978—. Contbr. articles to profl. jours. Mem. Colo. Right to Life, Internat. Host Family Program of Colo. Sch Mines. Served with AUS, 1943-46. Fellow Geol. Soc. Am.; mem. Mineral. Soc. Am., Mineral Assn. Can., Denver Coal Club, Colo. Mining Assn., Soc. Ind. Profl. Earth Scientists. Republican. Roman Catholic. Subspecialties: Mineralogy; Coal. Current work: Microscopic and x-ray characterization of raw materials. Office: Box 88 Golden CO 80402

JANSEN, GUSTAV RICHARD, food science and nutrition educator, nutritionist; b. N.Y.C., May 19, 1930; s. Gustav Enoch and Ruth Miriam (Olson) J.; m. Coerene Miller, July 5, 1953; children: Norman, Barbara, Kathryn, Ellen. B.A., Cornell U., 1950, Ph.D., 1958. Jr. and assoc. chemist Am. Cyanamid, Stamford, Conn., 1953-54; research biochemist E. I. duPont de Nemours & Co., Wilmington, Del., 1958-62; research fellow Merck Inst., Rahway, N.J., 1962-69; prof., head dept. food sci. and nutrition Colo. State U., Fort Collins, 1969—; teaching asst. dept. biochemistry Cornell U., 1954-58, lab. technician dept. animal husbandry, 1950; program mgr. U.S. Dept. Agr. Competitive Grants Research Program Human Nutrition, 1982; project dir. Expt. Sta. Project 87, Energy and Protein Utilization, 1970—. Contbr. articles in field to profl. jours.; editorial bd.: Jour. Nutrition, 1977-81, Nutrition Reports Internat, 1979—, Plant Foods for Human Nutrition, 1979—. Served with USAF, 1950-53. Recipient Disting. Service award Colo. State U., 1980. Mem. Am. Inst. Nutrition, Am. Soc. Biol. Chemists, Inst. Food Technologists (chmn. nutrition div.), Am. Dietetic Assn., AAAS, Soc. Nutrition Edn., Latin Am. Nutrition Soc., Sigma Xi, Phi Kappa Phi, Gamma Sigma Delta. Republican. Methodist. Subspecialties: Nutrition (biology); Nutrition (medicine). Current work: Protein-energy relationships during lactation with emphasis on protein synthesis, turnover and hormonal relationships. Patentee in field. Home: 1804 Seminole Dr Fort Collins CO 80525 Office: Dept Food Sci and Nutrition Colo State U Fort Collins CO 80523

JANSKI, ALVIN MICHAEL, biochemist; b. Braham, Minn., May 27, 1949; s. Norbert LeRoya and Frances (Salitros) J.; m. Rebecca Gudrun Kraft, Nov. 27, 1971; children: Frederick Linn. B.A., St. Cloud State U., 1971; Ph.D., N.D. State U., 1975. Research assoc. Iowa State U., Ames, 1975-78, NIH fellow, 1978; staff fellow Nat. Insts. on Alcohol Abuse and Alcoholism, Rockville, Md., 1978-79, sr. staff fellow, 1979-81; research scientist Internat. Minerals and Chem. Corp., Terre Haute, Ind., 1981-82, mgr. biochem. research and devel., 1982—. Contbg. author: Symposium Contribution Internat. Symposium on Isolation Characterization and Use of Hepatocytes, 1983; contbr. articles on biochemistry to profl. jours. NIH fellow, 1978. Mem. Am. Soc. Biol. Chemists, Am. Diabetes Assn. (research com.), N.Y. Acad. Scis., Am. Chem. Soc., AAAS, Sigma Xi. Roman Catholic. Subspecialties: Biochemistry (biology); Animal products. Current work: Research in animal endocrinology and metabolism. Home: R R 51 PO Box 734 Terre Haute IN 47805 Office: Internat Minerals and Chem Corp 1331 S 1st St PO Box 207 Terre Haute IN 47808

JANSSON, PETER ALLAN, engring. physicist; b. Hackensack, N.J., May 20, 1942; s. John Harry and Gudrun Friedeborg (Lindquist) J.; m. Muriel Jansson; children: Karen, Jonathan. B.S., Stevens Inst. Tech., Hoboken, N.J., 1964; Ph.D., Fla. State U., 1968. With E.I. DuPont de Nemours and Co., Inc., Wilmington, Del., 1968—, sr. research assoc., 1980—. Contbr. articles to profl. jours. Mem. Am. Phys. Soc., Optical Soc. Am., Soc. Photo-Optical Instrumentation Engrs. Subspecialties: Optics research; Graphics, image processing, and pattern recognition. Current work: Research and development of instrumentation in optics, digital image processing and spectroscopy. Home: 19 Arthur Dr RD 1 Hockessin DE 19707 Office: DuPont Exptl Station Bldg 357 Wilmington DE 19898

JANZOW, EDWARD FRANK, nuclear engineer; b. St. Louis, Mar. 19, 1941; s. Carl Paul and Helen (Lea) J.; m. Treva Lorraine Barbre, Sept. 2, 1967; 1 son, Lee Alan. B.S.M.E., Washington U., St. Louis, 1963; M.S. in Nuclear Engring, U. Mo., Columbia, 1964; M.B.A., U. Dayton, Ohio, 1981; Ph.D. in Nuclear Engring, U. Ill., 1970. Engr. Lawrence Radiation Lab., Livermore, Calif., summers 1963, 64; desalination researcher U. Ill., 1964-70; sr. engr. Montsanto Research Corp., Dayton, 1971-72, group leader nuclear engring. design and devel., 1972-76, mgr. engring. design and devel., 1976-81; ops. mgr., 1981—. Served to capt. USAR, 1963-71. Mem. Am. Nat. Standards Inst. (mem. subcom. on sealed radioactive sources), Am. Nuclear Soc., ASTM. Republican. Subspecialties: Nuclear engineering; Mechanical engineering. Current work: Development and design of radiation and heat sources using plutonium or trans-plutonium isotopes, especially neutron sources. Development and design of equipment, facilities, and techniques for fabricating such sources. Inventor, patentee desalination process by controlled freezing, polyester resin composition with improved neutron shielding properties, radiation sources and process. Home: 2671 Crone Rd Xenia OH 45385 Office: Monsanto Research Corp 1515 Nicholas Rd Dayton OH 45418

JAQUESS, JAMES FLETCHER, nuclear power generation, quality engineering and management consultant; b. Evansville, Ind., Mar. 25, 1948; s. John Roberts Jaquess and Sybil L. (Rye) Kaho; m. Karen Louise Byrd, July 29, 1972; 1 dau.; Karen Renee. Student, Ind. U.-Bloomington, 1966-67; B.S., Ind. State U.-Evansville, 1971; M.B.A., Ind. State U.-Terre Haute, 1975. Cert. quality engr. Navy nuclear quality assurance engr. Babcock & Wilcox, Mt. Vernon, Ind., 1971-73, resident quality engr., 1973-76, lead quality assurance engr., Lynchburg, Va., 1976-79, internat. project mgr., 1979-82; lead sr. engr. EDS Nuclear Inc., Atlanta, 1982—. Mem. Am. Soc. for Quality Control, Am. Nuclear Soc., Soc. Mfg. Engrs., Ind. State U. Alumni Assn. Subspecialties: Nuclear fission; Nuclear Power generation-quality engineering. Current work: International quality assurance consultant to electric utilities. Home: 3647 Frederica Rd Duluth GA 30136 Office: EDS Nuclear Inc 333 Technology Park Norcross GA 30092

JARBOE, CHARLES HARRY, pharmacologist, educator, conultant; b. Louisville, Oct. 3, 1928; s. Charles Harry and Mary Elizabeth (O'Daniel) J.; m. Carla A., June 29, 1982; children: Jamisene, Charles H., Richard J., Herman H., Nancy H., Elizabeth. B.S., U. Louisville, 1951, Ph.D., 1956. Research asst. prof. chemistry U. Louisville, 1957-58; chief scientist Brown and Williamson Tobacco Corp., 1958-61; assoc. prof. pharmacology and toxicology U. Louisville, 1962-72, prof. pharmacology and toxicology, 1972—, dir.Therapeutics and Toxicology Lab., 1972—; prof. pharmacology King Faisal U., Dammam, Saudi Arabia, 1982—; also cons. Contbr. numerous articles to profl. publs. Mem. formulary com. Ky. Med. Assistance Program; mem. Ky. Environ. Quality Commn., Jefferson County Air Pollution Control Bd. Served with USMC, 1945-58; Served with USAR, 1951-53. Decorated Purple Heart, Silver Star.; Research grantee public agys., also pvt. industry. Mem. N.Y. Acad. Sci., Am. Soc. Pharmacology and Exptl. Therapeutics, Am. Acad. Clin. Toxicology, Sigma Xi. Democrat. Episcopalian. Subspecialties: Molecular pharmacology; Toxicology (medicine). Current work: Human toxicology and human pharmacokinetics. I am a teaching medical school professor, director of my university Institutional Review Board and investigator in human toxico- and pharmacokinetics. Patentee in field. Home: PO Box 4053 Louisville KY 40204 Office: Dept Pharmacology and Toxicology U Louisville Louisville KY 40292

JARON, DOV, biomedical engrineer, educator; b. Tel Aviv, Oct. 29, 1935; U.S., 1958, naturalized, 1972; s. Meir and Sara (Levit) Yarovsky; m. Brooke E. Boberg, Sept. 16, 1978; children: Shulamit, Tamara. B.S. magna cum laude, U. Denver, 1961; Ph.D., U. Pa., 1967. Sr. research asso. Maimonides Med. Center, Bklyn., 1967-70; dir. surg. research Sinai Hosp. of Detroit, 1970-73; asso. prof. elec. engring. U. R.I., Kingsston, 1973-77, prof., 1977-79, coordinator biomed. engring., 1973-79; prof., dir. Biomed. Engring. and Sci. Inst., Drexel U., Phila., 1979—; vis. prof. elec. engring. Rutgers U., New Brunswick, N.J., 1968-73; adj. prof. biomed. engring. Wayne State U., 1971-73; adj. prof. physiology Temple U. Sch. Medicine, 1980—; adj. prof. radiology Jefferson Med. Coll., 1983—. Contbr. articles to sci. jours. NSF, NIH, Office Naval Research, pvt. founds. research grantee. Mem. Biomed. Engring. Soc., Am. Soc. Engring. Edn., Assn. Advancement Med. Instrumentation, Internat. Soc. Artificial Organs, Am. Soc. Artificial Internal Organs, Biophys. Soc., N.Y. Acad. Scis., IEEE, AAAS, AAUP, Sigma Xi, Tau Beta Pi, Eta Kappa Nu. Subspecialties: Biomedical engineering; Artificial organs. Current work: Development, control and optimization of cardiac assist devices; computer technologies applied to cardiovascular dynamics and diagnosis; bio-medical instrumentation. Researcher in cardiac assist devices, cardiovascular modeling, biomed. instrumentation. Home: 122 Bethlehem Pike Philadelphia PA 19118 Office: Drexel U Philadelphia PA 19104

JARRETT, MARK PAUL, physician; b. Bklyn., Dec. 29, 1949; s. Irving J. and Claire (Rockower) J.; m. Michele Jonas, Aug. 15, 1974; children—Matthew, Nicole. B.S., Muhlenberg Coll., 1971; M.D., N.Y.U., 1975. Diplomate: Am. Bd. Internal Medicine, Am. Bd. Rheumatology. Resident in medicine Montefiore Hosp., Bronx, N.Y., 1975-78, fellow in rheumatology, 1978-80; assoc. in medicine Northwestern U. Med. Sch., Chgo., 1980-81, asst. prof. medicine, 1981-82; clin. asst. prof. medicine Downstate Med. Sch., Bklyn., 1982—. Contbr. articles to profl. jours. Fellow A.C.P.; Mem. Am. Fedn. for Clin. Research, Am. Rheumatism Assn., AMA, AAAS, Phi Beta Kappa. Subspecialty: Immunology (medicine). Current work: Immunology, cellular and humoral and its effect on collagen vascular diseases. Office: 1460 Victory Blvd Staten Island NY 10301

JARRETT, NOEL, chemical engineer; b. Long Eaton, Eng., Nov. 17, 1921; came to U.S., 1926, naturalized, 1946; s. John Richard and Lena Eliza (Hexter) J.; m. Violet E. Dipner, Sept. 24, 1949; children: Robert, Kenneth, James, Thomas. B.S. in Chem. Engring, U. Pitts., 1949, M.S., U. Mich., 1951. Lubrication sales engr. Freedom-Valvoline Co., Freedom, Pa., 1949-50; with Alcoa Labs., Aluminum Co. Am., 1951—, chief div. process metallurgy, 1969-81, asst. dir. metal prodn. labs., 1981-82, tech. dir. chem. engring., 1982—. Served with U.S. Army, 1942-45. Fellow Am. Soc. Metals; mem. Nat. Acad. Engrs., Am. Inst. Chem. Engrs., AIME, Electrochem. Soc., Sigma Xi. Episcopalian. Clubs: Masons, Elks. Subspecialties: Electrochemical Engineering; High-temperature materials. Current work: Development of new or composite construction materials for anodes, cathodes, containment of materials in fused salt electrolysis to permit reduction of capital intensity and improved power efficiency through bipolar cell design. Patentee smelting and melting of aluminum. Home: 149 Jefferson Ave Lower Burrell PA 15068 Office: Alcoa Labs Alcoa Center PA 15069 I have found that the one who performs the tasks immediately at hand so well that his work cannot be ignored will reap society's rewards without asking.

JARVIK, ROBERT K., artificial organs researcher, physician, educator; b. Midland, Mich., May 11, 1946. B.A. in Zoology, Syracuse U., 1968, D.Sc. (hon.), 1983; postgrad., U. Bologna Sch. Medicine; M.A. in Occupational Biomechanics, NYU, 1971; M.D., U. Utah, 1976. Research asst. div. artificial organs U. Utah., 1971-77, acting dir. exptl. labs., div. artificial organs, 1977-78, asst. dir. exptl. labs., 1978—, also asst. research prof. surgery. Contbr. articles, abstracts, chpts. to profl. publs.; editorial reviewer for profl. jours. USPHS grantee. Mem. Am. Soc. Artificial Internal Organs, Internat. Soc. Artificial Organs. Subspecialties: Biomedical engineering; Artificial organs. Patentee in field; designer Jarvik 3 and Jarvik artificial hearts, 1971-77, Jarvik 7 heart, 1977-78. Office: Dept Surgery U Utah Coll Medicine 50 N Medical Dr Salt Lake City UT 84132

JARZEMBSKI, WILLIAM BERNARD, biomedical engineer, educator, consultant; b. Little Rock, June 25, 1923; s. Thaddeus and Elsie (Sachs) J.; m.; children: Nancy, Susan, Donna. B.S. in E.E, Northwestern U., (1947); Ph.D., Marquette U., (1970.). Registered profl. engr., Tex., Ill., Ky.; cert. clin. engr. Chief elec. engr. Advanced Research Sunbeam, Chgo., 1961-64; dir. product engring. Appleton Electric, Chgo., 1965-66; pvt. practice cons., Chgo., 1966-68; research fellow Marquette U., 1968-70; assoc. prof. biomed. engring. Med. Coll. Wis., Milw., 1970-74; prof. Tex. Tech. U., Lubbock, 1974—. Editor: Proceedings Computer Workshop, 1978. Served with U.S. Army, 1942-45; ETO. NIH fellow, 1968. Mem. IEEE (pace coordinator 1982-83), Engrs. in Medicine and Biology, Assn. Advancement Med. Inst., Am. Soc. Hosp. Engring., Sigma Xi. Subspecialties: Biomedical engineering; Bioinstrumentation. Current work: Electrical properties of neural tissue, electrical stimulation for therapeutic purposes. Patentee electronic thermometer, electromagnetic transducer, electronic oven, electrostatic copier. Home: 4208 49th St Lubbock TX 79413 Office: Texas Technical University School of Medicine Lubbock TX 79430

JASINSKI, DONALD ROBERT, government scientist and scientist adminstrator, physician; b. Chgo., Aug. 27, 1938; s. Mitchell M. and Leona B. (Danowski) J.; m. Diane Irene Navigato, Sept. 26,1964; children: Donald, Denise, Douglas, Daniel. Student, Loyola U., Chgo., 1956-59; M.D., U. Ill., 1963. Intern U. Ill. Resarch and Ednl. Hosps., Chgo., 1963-64; staff physician NIMH Addiction Research Ctr., Lexington, Ky., 1965-67, chief opiate unit, 1967-68, acting dir. ctr., 1977-78; chief clin. pharmacology sect. Nat. Inst. Drug Abuse Addiction Research Ctr., Balt., 1969—, dir. ctr., 1978-81, sci. dir., 1981—; adj prof pharmacology and exptl therapeutics U. Md. Sch. Medicine; assoc. prof. Johns Hopkins U. Sch. Medicine, 1980—; expert cons. WHO, 1974—. Assoc. editor: Jour. Pharmacology and Exptl. Therapeutics, 1971—; mem. editorial adv. bd., 1971—; author articles. Mem. Am. Soc. Pharmacology and Exptl. Therapeutics, Am. Soc. Clin. Pharmacology and and Therapetics, Soc. Neurosci., Internat. Brain Research Orgn., AAAS, Am. Coll. Neuropsychopharmacology, Sigma Xi. Roman Catholic. Subspecialty: Neuropharmacology. Current work: Neuropsychopharmacologic research into causes and treatment and prevention of substance abuse.

JASIULEK, JOACHIM NORBERT, mathematics educator; b. Kassel, West Germany, Apr. 7, 1952; came to U.S., 1973; s. Heinz and Marga (Engels) J. M.A., U. Calif.-Davis, 1974, M.S.E.E., 1975, Ph.D., 1980. Lectr. research engring. U. Calif.-Davis, 1980-81; vis. asst. prof. Simon Fraser U., B.C., Can., 1981-82; vis. asst. prof. math. Case Western Res. U., Cleve., 1982—; cons. Prentice Hall, Englewood, N.J., 1982. Contbr. articles in field to profl. jours. Fulbright scholar, 1973-75. Mem. Soc. Indsl. and Applied Math., IEEE. Subspecialties: Algorithms; Numerical analysis. Current work: Development of fast algorithms for applications in signal and image processing, implementation as software packages. Home: 1700 E 13th St Apt 16W Cleveland OH 44114 Office: Case Western Res U Math Dept Cleveland OH 44106

JASTROW, ROBERT, physicist; b. N.Y.C., Sept. 7, 1925; s. Abraham and Marie (Greenfield) J. A.B., Columbia, 1944, M.A., 1945, Ph.D., 1948; post-doctoral fellow, Leiden U., 1948-49, Princeton Inst. Advanced Study, 1949-50, 53, U. Calif. at Berkeley, 1950-53; D.Sc. (hon.), Manhattan Coll., 1980. Asst. prof. Yale, 1953-54; cons. nuclear physics U.S. Naval Research Lab., Washington, 1958-62; head theoretical div. Goddard Space Flight Center NASA, 1958-61, chmn. lunar exploration com., 1959-60, mem. com., 1960-62; dir. Goddard Inst. Space Studies, N.Y.C., 1961—; adj. prof. geology Columbia, 1961-81, dir., 1962-70, adj. prof. astronomy, 1977-82; adj. prof. earth sci. Dartmouth, 1973-82. Author: The Evolution of Stars, Planets and Life, 1967, Astronomy: Fundamentals and Frontiers, 1972, Until the Sun Dies, 1977, God and the Astronomers, 1978, Red Giants-White Dwarfs, 1979, The Enchanted Loom, 1981; Editor: Exploration of Space, 1960; co-editor: Jour. Atmospheric Scis, 1962-74, The Origin of the Solar System, 1963, The Venus Atmosphere, 1969. Recipient Medal of Excellence Columbia, 1962, Grad. Faculties Alumni award, 1967; Arthur S. Flemming award, 1965; medal for exceptional sci. achievement NASA, 1968. Fellow Am. Geophys. Union, A.A.A.S., Am. Phys. Soc.; mem. Internat. Acad. Astronautics, Council Fgn. Relations, Leakey Found. Clubs: Cosmos, Explorers, Century. Subspecialties: Astronautics; Planetary atmospheres. Home: 22 Riverside Dr New York NY 10023

JASZCZAK, RONALD JACK, physicist; b. Chicago Heights, Ill., Aug. 23, 1942; s. Jacob and Julia (Gudowicz) J.; m. Nancy Jane Bober, Apr. 15, 1967; children: John, Monica. B.S. with highest honors, U. Fla., Gainesville, 1964, Ph.D., 1968. AEC postdoctoral fellow Oak Ridge (Tenn.) Nat. Lab., 1968-69, staff physicist, 1969-71; prin. research scientist Searle Diagnostics, Inc., Des Plaines, Ill., 1971-73, sr. prin. research scientist, 1973, research group leader, 1973-77, chief scientist, 1977-79; asso. prof. radiology Duke U. Med. Center, Durham, N.C., 1979—; pres., chmn. bd. dirs. Data Spectrum Corp., Chapel Hill, N.C.; cons. Technicare Corp., 1980—, Siemens Gammasonics, Inc., 1981—. Contbr. articles in field to profl. jours. NIH sr. research fellow, 1980-82; NASA predoctoral fellow, 1964-67. Mem. Am. Phy. Soc., AAAS, Soc. Nuclear Medicine, IEEE, Am. Assn. Physicists in Medicine, Soc. Photo-Optical Instrumentation Engring., Phi Beta Kappa, Phi Kappa Phi, Sigma Tau Sigma, Sigma Pi Sigma. Subspecialties: Bioinstrumentation; Imaging technology. Current work: Medical imaging, research in design and application Single Photon Emission Computed Tomography. Research interests include physical aspects of imaging and display technologies; positron emission tomography, nuclear magnetic resonance and ultrasound. Patentee in field. Home: 2307 Honeysuckle Rd Chapel Hill NC 27514 Office: Duke U Med Center PO Box 3949 Durham NC 27710

JAUHAR, PREM PRAKASH, cytogeneticist, researcher, consultant; b. W. Punjab, India, Sept. 15, 1939; came to the U.S., 1976; s. Ram Lal and Maya Devi (Bhatla) J.; m. Raj Trehan, May 15, 1965; children: Rajiv, Sandeep, Suneeta. M.Sc., Agra U., India, 1959; Ph.D., New Delhi U., 1965. Assoc. prof. Indian Agr. Research Inst., New Delhi, 1963-72, postgrad. faculty, 1965-72; sr. sci. officer Welsh Plant Breeding Sta., U. Wales, Aberystwyth, 1972-75; research assoc. U. Ky., Lexington, 1976-78; research cytogeneticist U. Calif.-Riverside, 1978-81; cytogeneticist City of Hope Nat. Med. Ctr., Duarte, Calif., 1981; research dir. U.S. Agri. Labs. Research and Devel. Corp., Riverside, 1982—; Cons. genetics, cytogenetics and plant breeding; lectr. European labs. Area capt Am. Heart Assn., Riverside. Author: Cytogenetics and Breeding of Pearl Millet and Related Species, 1981; contbr. articles to profl. jours., chpts. to books. Mem. Genetics Soc. Am., Crop Sci. Soc. Am., Am. Genetic Assn., Am. Soc. Agronomy; fellow Linnean Soc. (London), Indian Soc. Genetics and Plant Breeding, Tissue culture Assn. Am. Subspecialties: Plant genetics; Plant cell and tissue culture. Current work: Cytogenetics and plant breeding to genetically reprogram plants; plant cells and tissue culture for improving and cloning genetically superior plants; genetic control of chromosome pairing. Discovered regulatory mechanism controlling chromosome pairing in polyploid species of Festuca. Home: 230 W Campus View Drive Riverside CA 92507 Office: US Agri Labs Research and Development Corp 12155 Magnolia Ave Bldg 4C Riverside CA 92503

JAVAID, JAVAID IQBAL, biochemical pharmacologist, researcher; b. Lahore, Pakistan, Oct. 1, 1942; s. Mohammad and Sarwar (Begum) Latif; m. Marie M. Dunwody; 1 son, Naser Latif. Ph.D. in Biochemistry, SUNY-Buffalo, 1972; B.S. with honors in Chemistry, U. Panjab, Lahore, Pakistan, 1964; M.S. in Biochemistry, U. Panjab, Pakistan, 1965. Lab. asst. Pakistan Council Sci. and Indsl. Research, Lahore, 1961-62, research asst., 1966-67; research fellow Ill. State Psychiat. Inst., Chgo., 1972-74, research scientist III, 1974-78, research scientist IV, assoc. dir. biol. research, 1978—; asst. prof. dept. psychiatry U. Chgo. Sch. Medicine, 1980—. Contbr. numerous articles to sci. jours. Mem. AAAS, Am. Soc. Neurochemistry, Am. Soc. Pharmacology and Exptl. Therapeutics, Duncliff Chem. Soc., Soc. Neurosci. Subspecialties: Psychopharmacology; Neuropharmacology. Current work: Psychoactive drugs in relation to their pharmacokinetics and pharmacodynamics; effects on biogenic amines. Office: 1601 W Taylor St Chicago IL 60612

JAVAN, ALI, educator, physicist; b. Tehran, Iran, Dec. 27, 1926; came to U.S., 1948, naturalized, 1963; s. Moosa and Jamileh (Azarbaghi) J.; m. Marjorie Browning, July 12, 1962; children—Maia Azar, Lila Hamideh. Student, Tehran U., 1947-48; Ph.D., Columbia U., 1954. Mem. research staff Bell Telephone Labs., Murray Hill, N.J., 1958-61; mem. faculty Mass. Inst. Tech., 1960—, prof. physics, 1964—, Francis Wright Davis prof., 1978—; cons. to industry and govt., 1960—. Recipient John and Fanny Hertz Found. award, 1964, Ballantine medal Franklin Inst., 1962, Sepas medal, Iran, 1971; Frederic Ives medal Optical Soc. Am., 1975; Outstanding Patent award N.J. Research and Devel. Council, 1977; U.S. Sr. Scientist award Humboldt Found., 1980; Guggenheim fellow, 1967. Fellow Am. Phys. Soc., Optical Soc. Am., Nat. Acad. Sci.; mem. Am. Acad. Arts and Scis., Royal Acad. Sci. In Iran (hon.), Sigma Xi. Subspecialty: Quantum electronics. Spl. research fundamental quantum electronics. Inventor first gas laser, 1960. Home: 12 Hawthorne St Cambridge MA 02138 Office: Mass Inst Tech Cambridge MA 02139

JAVID, MANUCHER J., neurosurgeon; b. Tehran, Iran, Jan. 11, 1922; came to U.S., 1944, naturalized, 1957; s. Asdolah and Touba (Ahdiyeh) J.; m. Lida Emma Fabbri, Oct. 19, 1951; children—Roxane, Daria, Jeffrey, Claudia. M.D., U. Ill., 1946. Diplomate: Am. Bd. Neurosurgery. Intern Augustana Hosp., Chgo., 1946-47, resident gen. surgery, 1947-48, resident neurosurgery, 1948-49; asst. in neuropathology Ill. Neuropsychiat. Inst., Chgo., 1948-49; fellow in neurosurgery Lahey Clinic, Boston, 1949; resident neurosurgery New Eng. Med. Center, Boston, 1950; clin. research fellow neurosurgery Mass. Gen. Hosp., Boston, 1950, asst. resident, 1951, sr. resident neurosurgery, 1952; teaching fellow in surgery Harvard, 1952; instr. Med. Sch. U. Wis., Madison, 1953-54, asst. prof., 1954-57, asso. prof., 1957-62, prof. neurosurgery, 1962, chmn. div. neurosurgery, 1963—; Cons. neurosurg VA Hosp., Madison, Mercy Hosp., Janesville, Wis., 1956—. Contbr. articles profl. jours. Mem. AMA, ACS, Soc. Neurol. Surgeons, Am. Assn. Neurol. Surgeons, AAAS, Am. Assn. Med. Colls., AAUP, Am. Trauma Soc., Pan Am. Med. Assn., Soc. for Neurosci., Central Neurosurg. Soc. (pres. 1964), N.Y. Acad. Scis., Xeiron, Sigma Xi, Phi Beta Pi. Mem. Baha'i Faith. Club: Rotarian. Subspecialty: Neurosurgery. Introduced clin. use of urea for reduction intracranial and intraocular pressure. Home: 4750 Lafayette Dr Madison WI 53705 Office: Univ Wis Hosp and Clinics 600 Highland Ave Madison WI 53792 Since I was a small child, I wanted to be a doctor and help the sick. As I grew older, the Baha'i Faith, served as a guideline to achieve this goal. Its teachings have helped me to appreciate the oneness of God, the oneness of religion, the oneness of humanity, and the sanctity of life.

JAVID, NIKZAD SABET, prosthodontics educator, dentist; b. Kashan, Iran, May 24, 1934; came to U.S., 1969, naturalized, 1983; s. Salam and Pika (Farhang) Javid-S.; m. Mahnaz Zolfaghari, Oct. 22, 1942; children: Nikrooz, Behrooz, Farnaz. D.M.D., U. Tehran, Iran, 1958; cert., U. Chgo., 1970; M.Sc., Ohio State U., 1971; M.Ed., U. Fla., 1981. Asst. prof. U. Tehran, 1959-69, prof., dean, 1975-79; asst. prof. Ohio State U., 1971-73, assoc. prof., 1973-74; assoc. prof. removable prosthodontics U. Fla., 1974-75, prof., 1980—; pvt. practice dentistry specializing in prosthodontics, Gainesville, Fla., 1980—; cons. in field. Author books, including: Stress Breaker in Partial Denture, 1966, Cleft Palate Prosthetics, 1968, Complete Denture Construction, 1974, (with Sara Nawab) Essentials of Complete Denture Prosthodontics, 1979; contbr. numerous articles to profl. jours. Named Outstanding Clin. Instr. of Yr. Student Dental Council, Columbus, Ohio, 1973. Fellow Internat. Coll. Dentists, Royal Soc. Health (Eng.); mem. Iranian Dental Assn. (dir. 1975-78), ADA, Internat. Assn. Dental Research (sec.-treas. Iran div. 1978). Lodge: Lions. Subspecialty: Prosthodontics. Current work: Teaching and research in clinical dentistry. Home: 3865 NW 38th Pl Gainesville Fl 32605 Office: U Fla JHMHC Box J-435 Gainesville FL 32610

JAVITT, NORMAN B., medical educator; b. N.Y.C., Mar. 9, 1928; m., 1955; 4 children. A.B., Syracuse U., 1947; Ph.D. in Physiology, U.N.C., 1951; M.D., Duke U., 1954. Diplomate: Am. Bd. Internal Medicine, 1962. Intern medicine Mt. Sinai Hosp., N.Y.C., 1954-55, asst. resident, 1957-58; Am. Heart Assn. fellow Coll. Physicians and Surgeons, Columbia U., N.Y.C., 1958-59; chief resident in medicine Mt. Sinai Hosp., 1960, research assoc., 1961-62; from instr. to asst. prof. Sch. Medicine, NYU, N.Y.C., 1962-68; assoc. prof. medicine Med. Coll., Cornell U., N.Y.C., 1968-72, prof., 1972—, head div. gastroenterology, 1970—; USPHS spl. fellow Mt. Sinai Hosp., 1961-62. Fellow ACP; mem. Am. Fedn. Clin. Research, Am. Soc. Clin. Investigation, Am. Gastroent. Assn., Am. Assn. Study of Liver Disease. Subspecialty: Gastroenterology. Office: Cornell U Med Coll NY Hosp New York NY 10021

JAWAD, MAAN H(AMID), mechanical engineer, educator; b. Baghdad, Iraq, Dec. 3, 1943; came to U.S., 1964, naturalized, 1972; s. Hamid M. and Makia R. (Chalabi) J.; m. Dixie L., Aug. 24, 1968; children: Jennifer, Mark. M.S.C.E., U. Kans., Lawrence, 1965; Ph.D. in Structural Engring., Iowa State U., 1968. Registered profl. engr., Ind., Iowa, Mo. Bridge engr. Iowa State Hwy. Commn., Ames, 1967-68; design engr. Nooter Corp., St. Louis, 1968-77, mgr. engring. design, 1977—; adj. prof. Grad. Engring. Ctr., U. Mo.-Rolla. Co-author: Structural Analysis and Design of Process Equipment, 1983. Mem. ASME (chmn. subgroup materials sect. VIII), ASCE. Republican. Unitarian. Clubs: Rotary (Collinsville, Ill.); Mo. Athletic. Subspecialties: Mechanical engineering; Civil engineering. Current work: Theoretical and exptl. analysis of pressure vessels. Patentee pressure vessel head. Home: 590 Watch Hill Collinsville IL 62234 Office: PO Box 451 Saint Louis MO 63166

JAWAD, SARIM NAJI, engineering mechanics educator, researcher; b. Baghdad, Iraq, July 1, 1953; s. Naji Mohammad and Radiea Ali (al-Alawiee) J.; m. Ann Marie Roddy, Dec. 2, 1982. B.S., Baghdad U., 1976; M.S., Liverpool Poly. Inst., Eng., 1977; Ph.D., Hatfield Poly. Inst., Eng., 1982. Registered profl. engr. Asst. engr. Nat. Co. Vegetable Oil Products, Iraq, 1975; research assoc. Hatfield Poly. Inst., 1978-82, vis. research fellow, 1982—; cons. devl., cons. engr. Emirates Cons. Co., Abu Dhabi, United Arab Emirates, 1982-83; lectr. engring. mechanics and thermodynamics United Arab Emirates U., Al Ain, 1983—; cons. engr. Dominan Ltd., Madrid, Spain, 1981-83. United Arab Emirates Ministry Edn. grantee, 1980-82. Mem. ASME, AIAA, Am. Nuclear Soc., Royal Aero. Soc. Moslim. Subspecialties: Fluid mechanics; Robotics. Current work: Research in the aero-thermodynamic and structural design of high pressure ratio, radial flow, centrifugal compressors. Home: PO Box 4349 Abu Dhabi United Arab EmiratesOffice: Engring Coll United Arab Emirates University PO Box 15551 Al Ain United Arab Emirates

JAWETZ, PINCAS, energy policy consultant; b. Czernowitz, Rumania, Dec. 20, 1935; s. Herman and Pepi (Frankel) J.; m. Irith Brenner, May 30, 1972; children: Gil-Shalom, Tom-Tsvi. M.Sc. in Chemistry, Hebrew U., Jerusalem, 1961; M.I.M., Am. Grad. Sch. Internat. Mgmt., 1975; postgrad., Rutgers U., 1968-74. Conducted studies on energy policy for Hudson Inst., GAO; Conducted studies on energy policy for countries of Colombia, Costa Rica, New Zealand , Israel, others, ind. cons. on energy policy, N.Y.C., congl. witness on energy policy. Contbr. articles profl. jours. Mem. Am. Chem. Soc., Am. Inst. Chem. Engrs., AAAS, N.Y. Acad. Scis., Internat. Assn. Energy Economists, Am. Econs. Assn. Subspecialties: Fuels; Agricultural economics. Current work: Biofuels, integration of farm policy with energy policy, octane levels and petroleum refineries, alcohol fuels in U.S. and overseas. Home: 425 E 72d St New York NY 10021 Office: 235 E 54th St New York NY 10022

JAWORSKI, ERNEST GEORGE, biochemist; b. Mpls., Jan. 10, 1926; s. Leon and Miecieslawa (Tchorzawska) J.; m. Pauline B. Robinson, July 8, 1950; children: Diane, David, Christopher. B.S. in Chemistry, U. Minn., 1948; M.S. in Biochemistry, Oreg. State U., 1950, Ph.D., 1952. Research biochemist Monsanto Co., St. Louis, 1952-60, dir. molecular biology program, corp. research labs., 1980; Disting. Sci. fellow, dir. molecular biology program, corp. research labs., 1980-83; sci. fellow Monsanto Agrl. Products Co., St. Louis, 1960-62, sr. sci. fellow, 1962-70, disting. sci. fellow, 1970—; chmn. bd. trustees Gordon Research Confs., Inc. Contbr. chpts. to books; mem. editorial bd., Ann. Revs., Inc., 1976-81, Jour. Am. Soc. Plant Physiologists, 1973-78; editor: Trends in Biotechnology, 1983—; contbr. articles to profl. jours. Mem. Florissant (Mo.) Sch. Bd., 1957-61. Served with USN, 1944-46. Mem. Am. Chem. Soc., AAAS, N.Y. Acad. Scis., Am. Soc. Plant Physiologists, Internat. Assn. Plant Tissue Culture, Tissue Culture Assn. Subspecialties: Genetics and genetic engineering (agriculture); Molecular biology. Current work: Plant and microbial genetic transformation; direction of research, plant molecular biology and cellular transformation, animal growth hormone, cellular mediators. Patentee.

JAY, RICHARD MARTIN, pediatric orthopedist, administrator, educator; b. N.Y.C., Dec. 23, 1946; s. Don F. and Blanche J.; m. Roslyn Kricheff, Mar. 15, 1980; 1 dau., Kate Jenny. B.S., Bethany Coll., 1969; postgrad. in Biochemistry, California (Pa.) State U., 1969-70; D.Podiatric Medicine, Pa. Coll. Podiatric Medicine, 1976. Diplomate: Am. Bd. Podiatric Surgery. Intern J.F. Kennedy Meml. Hosp. Med. Center, Phila., 1976-77, resident, 1977-78; dir. pediatric orthopedics Pa. Coll. Podiatric Medicine, Phila., 1978—, asst. prof. pediatric foot orthopedics, 1978—, dir. pediatric orthopedics, 1978—; dir. foot surgery residency J.F. Kennedy Hosp., Phila., 1978—; surg. cons. Inter County Hosp. Plan, Foxcroft, Pa., 1982—. Pediatrics editor: Jour. Current Podiatry, 1982; editor: Jour. Am. Podiatry Assn, 1983; contbr. chpt. to book, article to jour. in field. Recipient Annual Foot Surgery award Am. Coll. Foot Surgeons, 1976, award Am. Coll. Foot Radiologists, 1976, Radiologists in Podiatry award Pa. Coll. Podiatric Medicine, 1976. Fellow Am. Coll. Foot Orthopedics, Am. Coll. Podopediatrics; mem. Am. Coll. Podiatric Sports Medicine, Am. Podiatry Assn. Subspecialties: Pediatrics; Orthopedics. Current work: Research in pediatric foot orthopedics for internal tibial torsion - implant designs and production for flatfeet via sinus tarsi plugs. Office: 7915 Frankford Ave Philadelphia PA 19136 Pa Coll Podiatric Medicine 8th and Race St Philadelphia PA 19106

JAYAWEERA, KOLF, geophysics educator; b. Kalutara, Sri Lanka, Dec. 2, 1938; came to U.S., 1970, naturalized, 1979; s. Sirisena and Kalyanawath (Kulesekark) Fernando-Jayaweera; m. Irma A. Kalliomaki, Aug. 25, 1965; children: Anita, Eric, Tina. B.S., U. Sri Lanka, 1960; Ph.D., U. London, 1965. Lectr. U. Sri Lanka, Colombo, 1960-62, sr. lectr., 1965-67; research scientist C.S.I.R.O., Sydney, Australia, 1967-70; asst. prof. dept. geophysics U. Alaska, Fairbanks, 1970-74, assoc. prof., 1974-81, prof., 1981—; assoc. program dir. NSF, 1978-79. Contbr. articles to profl. jours. Commonwealth scholar, 1962-65. Fellow Royal Meteorol. Soc.; mem. Am. Meteorol. Soc., Am. Geophys. Soc. Subspecialties: Meteorology; Remote sensing (atmospheric science). Current work: Cloud physics, sea ice, radar.

Home: SR 10828 Fairbanks AK 99701 Office: Geophys Inst U Alaska Fairbanks AK 99701

JAYNE, BENJAMIN ANDERSON, educator; b. Enid, Okla., Oct. 10, 1928; s. Albert and Bertha Elizabeth (Anderson) J.; m. Betty Lu Bailey, Aug. 10, 1950; children—David N., Kristie A., Summer L. A.A., Boise Jr. Coll., 1949; B.S., U. Ida., 1952; M.F., Yale, 1953, Ph.D., 1955. Asst. prof. Yale, 1955-58; asso. prof. Wash. State U., 1959-62; sr. postdoctoral fellow U. Cal. at San Diego, 1963; prof. forestry N.C. State U., 1963-66, U. Wash., Seattle, 1966—, asso. dean, 1966-71, dir., 1971-76; dean Sch. Forestry and Environmental Studies Duke U., 1976—. Editor: Wood and Fiber, Jour. Soc. Wood Sci. and Tech, 1969-72. Mem. A.A.A.S., Soc. Wood Sci. and Tech. (pres. 1968), Soc. Am. Forestry, Sigma Xi, Xi Sigma Pi. Subspecialties: Resource management; Composite materials. Current work: Systems analysis applied to natural resource management. Home: 2610 Sevier St Durham NC 27705 Office: Sch Forestry and Environmental Studies Duke U Durham NC 27706

JEDLINSKI, HENRYK, research plant pathologist; b. Bialystok, Poland, Feb. 15, 1924; s. Tomasz and Jadwiga (Weglewska) J.; m. Helena Malinowska, Dec. 24, 1958; children: Michael T., Janine V. B.S., U. Nebr., 1950, M.A., 1954, Ph.D., 1959. Research asst. U. Nebr., Lincoln, 1951; research plant pathologist Agrl. Research Service U.S. Dept. Agr., Urbana, Ill., 1959—; assoc. prof. dept. plant pathology U. Ill., Urbana, 1979—. Mem. Am. Phytopathological Soc., Internat. Soc. Plant Pathology, Am. Soc. Virology, Sigma Xi, Gamma Sigma Delta. Republican. Roman Catholic. Subspecialties: Plant pathology; Plant virology. Current work: Diseases of cereals-oats-wheat; resistance, insect transmission of viruses; virus interactions with other pathogens. Office: 1102 S Goodwin Ave Urbana IL 61801

JEDRUCH, JACEK, nuclear engineer, computer scientist, writer; b. Warsaw, Poland, Feb. 22, 1927; came to U.S., 1951, naturalized, 1957; s. Alexander and Ann (Borsuk) J.; m. Eva Christina Hoffman, Apr. 15, 1972. B.S. in Mech. Engring, Northeastern U., 1956; M.S. in Nuclear Engring, MIT, 1958, Ph.D., Pa. State U., 1966. Metall. Analyst Acme Type Metals Co., Everett, Mass., 1952-53; engring. trainee H.B. Smith Co., Inc., Westfield, Mass., 1953-56; ind. cons. Columbia Nat. Co., Cambridge, Mass., 1957; scientist Westinghouse Electric Co., Pitts, 1957-65, fellow scientist, 1966—. Author: Constitutions, Elections and Legislatures of Poland, 1493-1977, 1982. Devel. chmn. Am. Youth Hostels, 1958-64. Served with Polish Army, 1944-46; Served with Brit. Army, 1946-48. Mem. Am. Soc. Mech. Engrs., Am. Nuclear Soc., AAAS. Republican. Roman Catholic. Club: MIT of Western Pa. (treas. 1978-82). Subspecialties: Nuclear engineering; Numerical analysis. Current work: Conceptual designs and computational methods for neutronics, photonics, economics and material performance of fission and fusion power reactors. Home: 377 Maize Dr Pittsburgh PA 15236 Office: Westinghouse Electric Co Nuclear Fuel Div PO Box 3912 Pittsburgh PA 15230

JEE, WEBSTER SHEW SHUN, anatomy educator, researcher; b. Oakland, Calif., June 25, 1925; s. Chueck Kwan and Shee (Jong) J.; m. Alice Shew, Nov. 7, 1951; 1 son, Kenneth Wesley. B.A., U. Calif., Berkeley, 1949, M.A., 1951; Ph.D., U. Utah, 1959. Faculty U. Utah, Salt Lake City, 1959—, instr. anatomy, 1959-60, asst. research prof. anatomy, 1960-61, asst. prof. anatomy, 1961-63, assoc. prof. anatomy, 1963-67, prof. anatomy, 1967—, bone group leader, radiobiology, 1959—; cons. Procter & Gamble Distbg., Cin., 1976-79, Colgate Palmolive, Piscataway, N.J., 1978-80, Upjohn Co., Kalamazoo, 1978—. Editor, co-editor: Some Aspects of Internal Irradiation, 1962, Delayed Effects of Bone Seeking Radionuclides, 1969, Health Effects of Plutonium and Radium, 1976, Bone Histomorphometry, 1981, Osteoporosis: Recent Advances in Pathogenesis and Treatment, 1981; assoc. editor: jours. Anatomical Record, 1969—, Calcified Tissue Research/Internat, 1977-81, Jour. Metabolic Bone Disease and Related Diseases, 1978—. Served with USAAF, 1943-46. Grantee U.S. AEC, Dept. Energy, 1956-74, 74—, NIH, 1964-74, NASA, 1981—. Mem. Radiation Research Soc., Am. Assn. Anatomists, Internat. Assn. Dental Research (pres. F.S. McKay Sect. 1968-70), Orthopeadic Research Soc., Am. Soc. Bone and Mineral Research, Gerontology Soc., Sigma Xi. Subspecialties: Morphology; Toxicology (medicine). Current work: Bone Histomorphology; skeletal physiology and diseases; pathogenesis and treatment; toxicity of internal irradiation and bone seeking substances on hard tissues. Home: 1948 E 5150 S Salt Lake City UT 84117 Office: Univ Utah Div Radiobiology Dept Pharmacology Bldg 351 Salt Lake City UT 84112

JEFFCOAT, MARJORIE, periodontology educator, clinic director; b. Boston, June 14, 1951; d. Jack and Miriam (Taylor) Kaplan; m. Robert L. Jeffcoat, Aug. 12, 1973. S.B., MIT, 1972; D.M.D., Harvard Sch. Dental Medicine, 1976. Cert. periodontologist. Instr. in periodontology Harvard Sch. Dental Medicine, Boston, 1978-79, asst. prof. periodontology, 1979—, dir. faculty practice, 1981—; cons. Children's Hosp. Med. Ctr., Boston, 1981—. Mem. Am. Acad. Periodontology, ADA, Harvard Odontol. Soc. (exec. com. 1982). Subspecialty: Periodontics. Current work: Bone resorption due to periodontal disease; nuclear medicine and oral disease. Office: Harvard Sch Dental Medicine 188 Longwood Ave Boston MA 02115

JEFFERIES, JOHN TREVOR, astrophysicist; b. Kellerberrin, Western Australia, Apr. 2, 1925; s. John and Vera (Healy) J.; m. Charmian Candy, Sept. 10, 1949; children—Stephen R., Helen C., Trevor R. B.Sc., U. Western Australia, 1947, D.Sc., 1961; M.A., Cambridge U., 1949. Research officer Commonwealth Sci. and Indsl. Research Orgn., Sydney, Australia, 1949-60; cons. to dir. Nat. Bur. Standards, Boulder, Colo., 1960-62; fellow Joint Inst. Lab. Astrophysics, Boulder, 1962-64; prof. physics and astronomy, dir. Inst. Astronomy, U. Hawaii, 1964—; prof. Coll. de France, 1970, 77; bd. dirs. Associated Univs. for Research in Astronomy, 1976—. Author: Spectral Line Formation, 1967; Contbr. articles profl. jours. Guggenheim fellow, 1970. Fellow Royal Astron. Soc., AAAS; mem. Internat. Astron. Union (pres. commn. X 1970-73), Am. Astron. Soc. (chmn. solar phys. div. 1971-72). Subspecialty: Solar physics. Home: 2800 Pacific Heights Rd Honolulu HI 96813

JEFFERS, JOHN BRYANT, engineering consulting firm official, nuclear consultant; b. Waukesha, Wis., Feb. 28, 1937; s. Howard B. and Verginia A. (Bass) J.; m. Joyce A. Oestreich, Sept. 16, 1961; children: Jennifer E., Julie A., Jill C. B.S. in Mech. Engring., U. Wis., 1961; M.S. in Mgmt., Rensselaer Poly. Inst., 1979. Cert. profl. mgr. Engr. Mercury program NASA, Cape Kennedy, Fla., 1961-65, supr., 1965-74; mgr. nuclear tng. Gen. Dynamics, Groton, Conn., 1974-78; mgr. engring. cons. Quadrex Corp., Pitts., 1978-83, United Energy Services Corp., Atlanta, 1983—. Named outstanding engr. Kennedy Space Ctr., 1970; recipient Apollo achievement award NASA, 1969. Mem. Am. Mgmt. Assn. (cert. profl. mgr.), Am. Nuclear Soc., ASME. Clubs: Montour Heights Country., Indian Hills Country. Subspecialties: Nuclear engineering; Nuclear fission. Home: 227 Rockwood Ct Marietta GA 30067 Office: United Energy Services Corp 1640 Powers Ferry Rd Atlanta GA 30067

JEFFERY, WILLIAM RICHARD, zoology educator; b. Chgo., June 9, 1944; m. Mary Ann Rakkin, Dec. 27, 1975. B.S., U. Ill., 1967; Ph.D., U. Iowa, 1971. Research assoc. U. Wis., Madison, 1971-72; research assoc. Tufts U. Sch. Medicine, Boston, 1972-74; asst. prof. U. Houston, 1974-77; asst. prof. dept. zoology U. Tex.-Austin, 1977-80, assoc. prof., 1980—; dir. embryology course Marine Biol. Lab., Woods Hole, Mass., 1983—. Editor: Time, Space and Pattern in Embryonic Development, 1983. Am. Cancer Soc. grantee, 1975-78; NIH grantee, 1975—; NSF grantee, 1976-79; Muscular Dystrophy Assn. grantee, 1982—. Mem. Soc. Devel. Biology, Am. Soc. Cell Biology, AAAS. Subspecialties: Developmental biology; Cell biology (medicine). Current work: Molecular basis of cell organization and determinative events of early embryonic development. Office: Dept Zoology U Tex Austin TX 78712

JEFFORDS, RUSSELL MACGREGOR, biostratigrapher; b. Shinglehouse, Pa., May 11, 1918; s. Harrison Morton and Irene Mary (MacGregor) J.; m. Ann Lucille Hill, May 15, 1943; 1 son, Russell Morton. A.B., Syracuse U., 1939; M.A., U. Kans., 1941, Ph.D., 1946; postgrad, U. W.Va., 1942-43. Research asst. Kans. Geol. Survey, Lawrence, 1939-42; mil geologist U.S. Geol. Survey, Washington, 1942, geologist in charge, Morgantown, W.Va., 1943-45; instr. Brown U., Providence, 1946-47; asst. prof. U. Tex., Austin, 1947-48; geologist U.S. Geol. Survey, Iowa City, 1948-54; research geologist to sr. research adviser Exxon Prodn. Research Co., Houston, 1954-79; ind. geol. cons., Houston, 1979—. Contbr. articles to profl. jours.; assoc. editor: Kans. Paleontol. Inst, 1968—. U. Kans. fellow, 1945-46; recipient Haworth Disting. Alumni citation U. Kans., 1964, Winchell Disting. Alumni award Syracuse U., 1979; hon. curator Houston Natural History Mus., 1976—. Fellow Geol. Soc. Am.; mem. Am. Geophys. Union, Soc. Econ. Paleontologists (chmn. research com. 1965), Peleontol. Soc., Soc. Tech. Writers and Editors, Am. Quaternary Assn., Paleobotanical Soc. Subspecialties: Paleontology; Ground water hydrology. Current work: Application of little-used fossil groups (e.g., macrofossils, crinoid columnals, chitinozoans) to age determinations and environmental interpretations. Home: 8002 Beverly Hill Houston TX 77063

JEFFREY, ALAN MILES, biochemist, educator, researcher; b. Fording Bridge, U.K., Nov. 4, 1944. B.Sc. in Biochemistry, U. Hull, Eng., 1966; Ph.D., U. Bangor, N. Wales, 1970. NATO fellow Nat. Inst. Arthritis and Metabolis Diseases, NIH, Bethesda, MD., 1970-72, research fellow, 1972-74; research assoc. Inst. Cancer Research, Columbia U., N.Y.C., 1974-76, asst. prof., 1976—. Contbr. articles to profl. jours. Mem. Am. Assn. Cancer Research, Biochem. Soc. (U.K.), Am. Coll. Toxicology. Subspecialties: Cancer research (medicine); Environmental toxicology. Current work: Metabolic activation of and reaction with DNA of polycyclic aromatic hydrocarbons. Office: Inst Cancer Research 701 W 168th St New York NY 10032

JEFFREY, GEORGE ALAN, scientist, educator; b. Cardiff, Eng., July 28, 1915; came to U.S., 1953, naturalized, 1962; s. George F. and Beatrice (H) J.; m. Maureen Ward, Sept. 5, 1942; children—Susan M., Paul D. B.Sc., U. Birmingham, Eng., 1936, Ph.D., 1939, D.Sc., 1953. Crystallographer Brit. Rubber Producers Assn., 1939-45; lectr. U. Leeds, Eng., 1945-53; prof. chemistry and physics, dir. crystallography lab. U. Pitts., 1953-69, prof. crystallograpy, 1965—, chmn. dept. crystallography, 1969-74, 76—; sr. scientist Brookhaven Nat. Lab., Upton, N.Y., 1974-76; Sec. U.S. Nat. Com. Crystallography, 1956-58, chmn., 1965-66. Mem. Am. Crystallographic Soc. (pres. 1963), Am. Inst. Physics (bd. govs. 1969-74). Subspecialties: Crystallography; Organic chemistry. Current work: Carbohydrate chemistry, x-ray and neutron crystal structure determination, hydrogen bonding, liquid crystal studies. Home: 1220 King Ave Pittsburgh PA 15206 Office: U Pitts Pittsburgh PA 15260

JEFFREYS, JAMES VICTOR, aero-mech. engr., air force officer; b. Nashville, June 28, 1938; s. James Terry and Jean Young (Stewart) J.; m. Carolyn Virginia Beam, Sept. 7, 1960; children: Mark, Kathryn, Clara. B.M.E., Vanderbilt U., 1961; M.S. in Aero-Mech. Engring, Air Force Inst. Tech., 1967. Registered profl. engr., Calif. Commd. 2d lt. U.S. Air Force, 1961, advanced through grades to lt. col., 1981; mech. engr. (USAF Security Service), Washington, 1961-65, aero. engr., Robins AFB, Ga., 1967-69, dep. comdr., Saigon, Vietnam, 1969-70, chief, Robins AFB, 1970-74, asst. chief div. tech. services, Columbus, Ohio, 1974-78, group leader, directorate of intertial engring., Newark Air Force Sta., Ohio, 1978-82, chief div. engring. tech., Wright-Patterson AFB, Ohio, 1982—. Active Central Ohio council Boy Scouts Am. Decorated Bronze Star, Meritorious Service medal with oak leaf cluster, Joint Services Commendation medal, Air Force Commendation medal, U.S.; Cross of Gallantry with palm, Vietnam). Mem. Nat. Assn. Corrosion Engrs. (accredited), ASME, Soc. Am. Mil. Engrs. (pres. Columbus post 1975), Nat. Rifle Assn. (life), Nat. Eagle Scout Assn., Air Force Assn., SCV, Mil. Order Stars and Bars, Sigma Nu. Methodist. Lodges: Masons (32 degree); KT; Shriners. Subspecialties: Aerospace engineering and technology; Corrosion. Home: 306 Jennie Dr Gahanna OH 43230 Office: Hdqrs Air Force Logistics Command Wright-Patterson AFB OH 45433

JEFFRIES, HARRY PERRY, oceanography educator; b. Newark, Apr. 15, 1929; s. Harry Perry and Miranda (Thomas) J.; m. Margaret Fuller Wood, July 24, 1951; children: Christie, Ann, Elizabeth, Janet, Matthew. B.S., U. R.I., 1951, M.S., 1955, Ph.D., 1959. Pharmacologist Ciba Pharm. Products, Summit, N.J., 1955-56; mem. faculty dept. oceanography U. R.I., Kingston, 1959—, prof., 1974—; vis. investigator Skidaway Inst. Oceanography, Savannah, 1981; advisor U.S. Dept. Energy, 1980-82; mem. R.I. Statewide Planning Com., 1978-80. Author: Estuarine Zooplankton, 1967-82; assoc. editor: Estuaries, 1977-79; editor: Coastal Oceanography Newsletter, 1978-83. Bd. dirs. Save the Bay, Providence, 1983—; v.p. South County Art Assn., Kingston, 1981—; trustee Kingston Free Pub. Library, 1972-74; mem. South Kingstown (R.I.) Conservation Commn., 1974-80. EPA Grantee, 1972; Sea Grant Program grantee, 1972; NOAA grantee, 1982; U.S. Dept. Energy grantee, 1982—. Fellow AAAS; mem. Estuarine Research Fedn. (pres. 1974-76), New Eng. Estuarine Research Assn. (pres. 1975-75), Am. Assn. Limnology and Oceanography. Episcopalian (former sr. warden). Subspecialties: Theoretical ecology; Zooplankton limnology. Current work: Biological oceanography of coastal waters, with emphasis on plankton production and long-term changes in abundance of fishes; responses of communities to environmental change. Office: Grad Sch Oceanography U Rhode Island Kingston RI 02881 Home: 83 Biscuit City Rd Kingston RI 02881

JEFFS, GEORGE W., aerospace exec.; b. Stockton, Calif., Mar. 9, 1925; s. George Andrew and Evelyn Kathleen (MacFarlane) J.; m. Christina Kelly, Apr. 15, 1950; children—Georgia, James, William. B.S., U. Wash., 1945, M.S. in Aero. Engring, 1947. Aerophysics and design engr. N. Am. Aviation, 1947-55, chief engr. missile div., 1955-59, mgr. tech. planning, 1959-63, v.p. paraglider program, 1963-65; chief engr. Apollo CSD, 1965-69; v.p. N. Am. Rockwell, 1969-73; pres. space div. Rockwell Internat., 1974-76, corp. v.p., 1976—; pres. N. Am. Space Ops., El Segundo, Calif., 1976—. Served with U.S. Navy, 1942-49. Recipient Disting. Service medal NASA, Presdl. medal of Freedom, Golden Knight Mgmt. award Nat. Mgmt. Assn. Fellow AIAA, Am. Astronautical Soc.; mem. Nat. Acad. Engring., Aircraft Owners and Pilots Assn., Alpha Delta Phi. Clubs: Bel Air Country, Masons. Subspecialty: Aerospace engineering and technology. Office: 2230 E Imperial Hwy El Segundo CA 90245

JEN, PHILIP HUNG SUN, biology educator; b. Hung, Hunan, China, Jan. 11, 1944; s. Shou-shon and Yun In (Kuo) J.; m. Betty Yu Lee, Feb. 20, 1971. B.Sc., Tunghai U., 1967; M.A., Washington U., 1971, Ph.D., 1974. Instr. biology Chinese Air Preparatory Sch., Tung Kung, Taiwan, 1967-68; tchr. biology St. Dominics High Sch., Kaohsiung, 1968-69; teaching asst. Washington U., St. Louis, 1969-74, research assoc., 1974-75; asst. prof. U. Mo., Columbia, 1975-80, assoc. prof. dept. biol. sci., 1981—. Contbr. articles to profl. jours. Served to 2nd lt. Chinese Army, 1967-68. Tunghai U. scholar, 1963-67; NSF research grantee, 1978, 80; NIH Research Career Devel. awardee, 1980—. Mem. Am. Soc. Zoologists, AAAS, N.Y. Acad. Sci., Soc. for Neurosci., Acoustic Soc. Am. Subspecialties: Neurophysiology; Neurobiology. Current work: Behavior, neurophysiology and neuroanatomy of bat's biosonar system. Office: Div Biol Sci U Mo Columbia MO 65211

JENCKS, WILLIAM PLATT, educator, biochemist; b. Bar Harbor, Maine, Aug. 15, 1927; s. Gardner and Elinor (Melcher) J.; m. Miriam Ehrlich, June 3, 1950; children—Helen Esther, David Alan. Grad., St. Paul's Sch., Balt., 1944; student, Harvard, 1944-47, M.D., 1951. Intern Peter Bent Brigham Hosp., Boston, 1951-52; postdoctoral fellow Mass. Gen. Hosp., Boston, 1952-53, 55-56; postdoctoral fellow chemistry Harvard, 1956-57; mem. faculty Brandeis U., 1957—, prof. biochemistry, 1963—. Served as 1st lt., M.C. AUS, 1953-55. Mem. Am. Chem. Soc. (award biol. chemistry 1962), Am. Soc. Biol. Chemists, Am. Acad. Arts and Scis., AAAS, Nat. Acad. Scis., Alpha Omega Alpha. Subspecialties: Biochemistry (medicine); Organic chemistry. Current work: Reaction mechanics and catalysis in chemistry and enzymology. Home: 11 Revere St Lexington MA 02173 Office: Grad Dept Biochemistry Brandeis Univ Waltham MA 02254

JENKINS, ARNOLD MILTON, management information systems educator, consultant, researcher; b. Milford, Mass., June 12, 1938; s. Arnold Milton and Evelyn Estalla (Williams) J.; m. Jane Monica Biedugnis, July 20, 1964; children: Arnold Milton, Michelle Evelyn. A.Mech. Engring., Worcester Coll., 1961; B.S. in Engring, U. Albuquerque, 1969; M.B.A., U. N.Mex., 1971; Ph.D. in Mgmt. Info. Sci, U. Minn., 1977. Design engr. Whitin Machine Works, Whitinsville, Mass., 1957-62; engr., mgr. Sandia Labs., Albuquerque, 1962-72; assoc. mem. grad. faculty U. Minn., Mpls., 1972-76; prof. mgmt. info. systems Ind. U., Bloomington, 1976—; tchr./cons. AID, U.S. Dept. State, Washington, 1978—; cons. Owens Corning Fiberglass, Toledo, 1982—, Procter & Gamble Co., Cin., 1980—, Gen. Mills, Inc., Mpls., 1980—. Author: Program of Research for Investigating MIS, 1982; editor for theory and research: Mgmt. Info. Systems Quar, 1977—. Mem. Soc. Mgmt. Info. Systems, Am. Inst. Decision Scis. (thesis competition award 1977), Assn. Systems Mgmt., Assn. Computing Machinery. Subspecialties: Information systems (information science); Graphics, image processing, and pattern recognition. Current work: Methodologies and procedures for design, development and implementation of computerized information systems. Development of theory for user-system interface/ergonomics. Techniques for increasing productivity of knowledge workers through effective information utilization. Office: Operations and Systems Mgmt Dep Grad Sch Busines Ind U 650 Business Bldg Bloomington IN 47405 Home: 414 Meadowbrook Ave Bloomington IN 47401

JENKINS, EDWARD BEYNON, astronomer; b. San Francisco, Mar. 20, 1939; s. Francis Arthur and Henrietta Beynon (Smith) J.; m. Myrna Dean Stewart, June 29, 1963; children: Brian Francis, Eric Dean. A.B., U. Calif., Davis, 1962; Ph.D., Cornell U., 1966. Research assoc. Princeton U., 1966-67; research staff mem. Princeton U. Obs., 1967-73, research astronomer, 1973-79, sr. research astronomer, 1979—. Contbr. articles to profl. jours. Mem. Am. Astron. Soc., Internat. Astron. Union, Sigma Xi. Democrat. Unitarian. Subspecialties: Optical astronomy; Space astronomy (ultraviolet). Current work: Composition and physical state of the interstellar medium. Home: 1851 Stuart Rd W Princeton NJ 08540 Office: Princeton U Observatory Princeton NJ 08544

JENKINS, W(INBORNE) TERRY, chemist, educator; b. Waupun, Wis., Mar. 23, 1932; s. Winborne Terry and Priscilla F. (Fletcher) J.; m. Alexia V. Hanitsch, July 19, 1958; children: Christopher, Mary, Mark. B.A., Cambridge (Eng.) U., 1953, M.A., 1957; Ph.D., M.I.T., 1957. Instr. M.I.T., Cambridge, 1957-58; instr., then asst. prof. U. Calif., Berkeley, 1958-66; successively assoc. prof., then prof. chemistry Ind. U., Bloomington, 1966—. Contbr. articles to profl. jours. Recipient NIH Career Devel. award, 1967-72. Mem. Am. Soc. Biol. Chemists. Episcopalian. Subspecialties: Biochemistry (biology); Biomaterials. Current work: Purification and characterization of enzymes; mechanisms of enzyme catalysis especially those enzymes with Vitamin B-6. Office: Ind U Room 221 Chemistry Bldg Bloomington IN 47405

JENNINGS, AARAON AUSTIN, civil engineering educator; b. Coeur d' Alene, Idaho, Jan. 25, 1949; s. Author Austin and Shirley (Shelp) J.; m. Viluna B. Jennings, Jan. 11, 1969. A.A.S., Adirondack Community Coll., 1970-72; B.E.T., Rochester (N.Y.) Inst. Tech., 1975; M.S.CE., U. Mass., 1977, Ph.D., 1980. Cert. profl. engr., Ind. Designer Hershey, Malone & Assocs., Rochester, 1972-75; research assoc. dept. civil engring. U. Mass., Amherst, 1975-79, univ. fellow, 1979-80; asst. prof. dept. civil engring. U. Notre Dame, Ind., 1980—; cons. Mitre Corp., Bedford, Mass., 1978-80; mem. chem. and physics rev. panel EPA, 1980—. Contbr. articles to profl. jours. Served to sgt. USMC, 1967-70. Named Disting. Alumnus Rochester Inst. Tech., 1980; grantee EPA, NSF, Dept. Energy. Mem. Am. Geophys. Union, ASCE, Assn. Environ. Engring. Profs., Am. Water Resources Assn., Phi Kappa Phi. Subspecialties: Civil engineering; Environmental engineering. Current work: Problems of groundwater quality and hazardous waste management. Office: Dept Civil Engrin U Notre Dame Notre Dam IN 4655 Home: 18677 Welworth Ave South Bend IN 46637

JENNINGS, BURGESS HILL, educator, mech. engr.; b. Balt., Sept. 12, 1903; s. Henry Hill and Martha (Burgess) J.; m. Etta M. Crout, Nov. 7, 1925; 1 son, Robert Burgess. B.E., Johns Hopkins, 1925; M.S., Lehigh U., 1928, M.A., 1935. Test engr. Consol. Gas & Electric Co., Balt., 1925; mem. faculty Lehigh U., 1926-40; prof. dept. mech. engring. Northwestern U., 1940—, chmn. dept., 1943-57, asso. dean, 1962-70; research investigator U.S. OSRD, 1942-45; dir. research labs. Am. Soc. Heating, Refrigerating and Air Conditioning Engrs., Cleve., 1957-60; cons. and gen. research and writing relating to refrigeration, air conditioning and energy usage, 1930—. Author: books and articles on engring., heating, air conditioning Heating and Air Conditioning, 1956, Environmental Engineering, 1970, The Thermal Environment, 1978; co-author: Gas Turbine Analysis and Practice, 1953, Air Conditioning and Refrigeration, 1958. Recipient Richards Meml. award in Mech. Engring., 1950; Merit award Chgo. Tech. Socs. Council, 1963; Worcester Reed Warner medal, 1972. Fellow ASHRAE (pres. 1948-49, F. Paul Anderson medal 1981); mem. ASME (hon.), Nat. Acad. Engring., Am. Soc. Lubricating Engrs. (v.p. 1947-50), Am. Soc. Engring. Edn., Internat. Inst. Refrigeration (v.p. 1958-67), Sigma Xi, Pi Tau Sigma (pres. 1948-50), Tau Beta Pi. Club: Michigan Shores (Wilmette, Ill.). Subspecialties: Environmental engineering; Solar energy. Current work: Studies in technology of energy control and conservation and more effective usage of solar and wind energy. Home: 1500 Sheridan Rd Wilmette IL 60091

JENNINGS, LARRY EUGENE, computer scientist; b. Oklahoma City, Aug. 3, 1947; s. Eugene Edgar and Helen Mary (Maune) J.; m. Linda Sue Dries, Dec. 22, 1970; children: Amy, Libby, Meggin. A.S., Okla. State U., 1968; B.S. in Math, U. Albuquerque, 1978; M.S., U. N.M., 1982. Materials technician Standard Testing & Engring., Oklahoma City, 1966-68; draftsman Sandia Nat. Lab., Albuquerque, 1968-72, electro-mech. designer, 1972-76, software engr., 1976-81, database designer, 1981-82, computer scientist CAD/CAM, 1982—. Mem. Inst. Interconnecting and Packaging Electronic Circuits, Assn. Computing Machinery. Subspecialties: Database systems. Current work: Research concerning the communication of computer based product definition and engring. data among the various agencies within the nuclear weapons complex. Home: 6104 Case de Vida NE Albuquerque NM 87111 Office: Sandia Nat Labs Div 9345 Albuquerque NM 87185

JENNINGS, STEVEN FLETCHER, sofware engineer; b. Leesburg, Fla., Dec. 18, 1954; s. Warren Abner and Gladys Irene (Crane) J.; m. Mary Carol Stone, Dec. 27, 1980; 1 son, Jeremiah Steven. B.S., Graceland Coll., Lamoni, Iowa, 1975; M.S., Iowa State U., 1978, Ph.D., 1981. Asst. prof. computer sci. Colo. State U., Ft. Collins, 1980-83; sr. software engr. Burroughs Corp., Boulder, Colo., 1983—. Contbr. articles to profl. jours. Mem. Assn. Computing Machinery, IEEE, IEEE Computer Soc. Democrat. Mormon. Subspecialties: Distributed systems and networks; Operating systems. Current work: Operating systems, parallel processing, program modeling, distributed systems, data flow architectures. Home: 8363 West Fork Rd Boulder CO 80302 Office: Burroughs Corp Advanced Systems Group 6655 Lookout Rd Boulder CO 80301

JENSEN, ARTHUR ROBERT, educator; b. San Diego, Aug. 24, 1923; s. Arthur Alfred and Linda (Schachtmayer) J.; m. Barbara Jane DeLarme, May 6, 1960; 1 dau., Roberta Ann. B.A., U. Calif. at Berkeley, 1945; Ph.D., Columbia, 1956. Asst. med. psychology U. Md., 1955-56; research fellow Inst. Psychiatry, U. London, 1956-58; prof. ednl. psychology U. Calif. at Berkeley, 1958—. Author: Genetics and Education, 1972, Educability and Group Differences, 1973, Educational Differences, 1973, Bias in Mental Testing, 1979, Straight Talk about Mental Tests, 1981; Contbr. to profl. jours., books. Guggenheim fellow, 1964-65; fellow Center Advanced Study Behavioral Scis., 1966-67. Fellow Am. Psychol. Assn., Eugenics Soc., AAAS; mem. Am. Ednl. Research Assn. (v.p. 1968-70), Psychonomic Soc., Am. Soc. Human Genetics, Soc. for Social Biology, Behavior Genetics Assn., Psychometric Soc., Sigma Xi. Subspecialties: Cognition; Learning. Current work: Research on measurement and causes of individual and group differences in human mental abilities, particulary intelligence, using psychometric, experimental, and behavior-genetic approaches. Home: 30 Canyon View Dr Orinda CA 94563

JENSEN, BARBARA LYNNE, physicist; b. Salt Lake City, July 12, 1939; d. Howard D. and Lucile (Miner) J.; m. William D. Jensen, Sept. 18, 1965. B.S. in Physics, U. Utah, 1964, postgrad., 1964-66; M.A. in Physics, Columbia U., 1972, Ph.D., 1973. Research asst. IBM Thomas J. Watson Research Center, Yorktown Heights, N.Y., 1970-73; instr. in physics U. Lowell, Mass., 1974-77; asst. prof. physics Boston U., 1978—. Contbr. articles to profl. jours. Mem. Am. Phys. Soc., IEEE, Am. Vacuum Soc., AAAS. Subspecialties: Condensed matter physics; Materials. Current work: Semiconductors, solid state physics, optical communications, submillimeter waves, microelectronics. Office: Dept Physics Boston U 111 Cummington St Boston MA 02215

JENSEN, BETTY KLAINMINC, utility company research physicist; b. Lodz, Poland, June 20, 1949; came to U.S., 1962; d. Sam and Gussie (Alter) Klainminc; m. Richard Alan Jensen, Dec. 19, 1971; children: David Jonathan, Andrew Michael, Penelope Judith. B.S., Bklyn. Coll., 1970; M.A., Columbia U., 1972, M.Ph., 1973, Ph.D., 1976; M.B.A., St. John's U., 1981. Teaching fellow Columbia U., 1970-76; adj. lectr. CUNY, 1973-76; sr. physicist-research Pub. Service Electric & Gas Co., Newark, 1976-79, prin. staff physicist-research, 1979-81, prin. physicist-research, 1981—; adv. Electric Power Research Inst., Palo Alto, Calif., 1976—, Helium Breeder Assocs., LaJolla, Calif., 1977-81, Gas Cooled Reactor Assocs., LaJolla, 1978—, MIT Energy Lab., Princeton (N.J.) Plasma Physic Lab., 1976—. Author: Physics of Fusion Plasma Boundary Layers, 1976; contbr. articles in field to profl. jours. Mem. energy com. Am. Jewish Com., N.Y.C., 1978. Mem. IEEE/Engring. Mgmt. Soc. (chmn. 1981-82), Am. Nuclear Soc., Am. Phys. Soc., AAAS, Phi Beta Kappa, Sigma Xi, Omicron Delta Epsilon, Beta Gamma Sigma. Jewish. Subspecialties: Nuclear fission; Nuclear fusion. Current work: Energy, nuclear energy, thermonuclear fusion, materials research for energy applications. Office: Nuclear Research Div Public Service Electric & Gas Co 80 Park Plaza Newark NJ 07101

JENSEN, ELWOOD VERNON, biochemist; b. Fargo, N.D., Jan. 13, 1920; s. Eli A. and Vera (Morris) J.; m. Mary Welmoth Collette, June 17, 1941 (dec. Nov. 1982); children: Karen Collette, Thomas Eli. A.B., Wittenberg U., 1940, D.Sc. (hon.), 1963; Ph.D., U. Chgo., 1944; D.Sc. (hon.), Acadia U., 1976. Faculty U. Chgo., 1947—; assoc. prof. biochemistry Ben May Lab. Cancer Research, 1954-60, prof., 1960-63, Am. Cancer Soc. research prof. physiology, 1963-69; dir. Ben May Lab., 1969-82; med. dir. Ludwig Inst. Cancer Research, 1983—; prof. physiology Ben May Lab., 1969-73, 77—, prof. biophysics, 1973—, prof. biochemistry, 1980—; med. dir. Ludwig Inst. for Cancer Research, 1983—; dir. Biomed. Center for Population Research, 1972-75; Vis. prof. Max-Planck-Inst. für Biochemie, Munich, Germany, 1958; chmn. endocrinology panel Cancer Chemotherapy Nat. Service Center, 1960-62; mem. chemotherapy rev. bd. Nat. Cancer Inst., 1960-62, bd. sci. counselors, 1969-72; mem. Nat. Adv. Council Child Health and Human Devel., 1976-80; mem. adv. com. biochemistry and chem. carcinogenesis Am. Cancer Soc., 1968-72, council for research and clin. investigation, 1974-77; mem. assembly life scis. NRC, 1975-78; mem. com. on sci., engring. and public policy Nat. Acad. Scis., 1981-82. Editorial bd.: Perspectives in Biology and Medicine, 1966—, Archives of Biochemistry and Biophysics, 1979—; editorial adv. bd.: Biochemistry, 1969-72, Life Scis, 1973-78, Breast Cancer Research and Treatment, 1980—; asso. editor.: Jour. Steroid Biochemistry, 1974—; Contbr. articles to profl. jours. Guggenheim fellow, 1946-47; recipient D.R. Edwards medal, 1970; La Madonnina prize, 1973; G.H.A. Clowes award, 1975; Papanicolaou award, 1975; prix Roussel, 1976; Nat. award Am. Cancer Soc., 1976; Amory prize, 1977; Gregory Pincus Meml. award, 1978; Gairdner Found. award, 1979; Lucy Wortham James award, 1980; Charles F. Kettering prize, 1980; Nat. Acad. Clin. Biochemistry award, 1981; Pharmacia award, 1982; Hubert H. Humphrey award, 1983; Rolf Luft medal, 1983. Mem. Am. Acad. Arts and Scis., Am. Soc. Biol. Chemists, Am. Chem. Soc., Am. Assn. Cancer Research, Endocrine Soc. (pres. 1980-81), AAAS, Soc. Study Reprodn. Clubs: Chicago Literary, Cosmos. Subspecialties: Cancer research (medicine); Receptors. Current work: Responsibility for supervision and development of cancer research programs at the nine research branches of the Ludwig Institute in six countries. Home: Kanalweg 5a CH-8714 Feldbach Switzerland Office: Stadelhoferstrasse 22 CH-8001 Zurich Switzerland

JENSEN, EMRON ALFRED, zoology educator; b. Richfield, Utah, Jan. 5, 1925; s. Alfred Oscar and Julia Reina (Ogilvie) J.; m. Melva Bringhurst; children: Kenneth B., Brian F., Eldon C., Paul R., Diana, Elyse, Clark E., Charlotte. B.S., Utah State U., 1950, M.S., 1961, Ph.D., 1963. Tchr. Blackfoot (Idaho) High Sch., 1950-52; chemistry lab. technician Am. Cyanamid Co., Arco, Idaho, 1952-53; tchr. Sevier Sch. Dist., Richfield, 1952-59; asst. prof. zoology Weber State Coll., Ogden, Utah, 1963-67, assoc. prof. zoology, 1967-70, prof., chmn. dept. zoology, 1970—. Author: Introductory Physiology, 1973. Mem. steering com. Utah Conf. Higher Edn., Salt Lake City, 1968-71; County del. Rep. party, 1979-80. Served with U.S. Army, 1944-46. NDEA fellow, 1959-62; NIH fellow, 1962-63. Mem. AAAS, Sigma Xi (local pres. 1971-72). Mormon. Subspecialties: Parasitology; Physiology (medicine). Current work: Parasitology, protozoa. Office: Dept Zoology Weber State Coll 3750 Harrison Blvd Ogden UT 84408

JENSEN, KEITH EDWIN, scientific researcher, consultant; b. Council Grove, Kans., Sept. 6, 1924; s. Adolph George and Irma Mae (Alexander) J.; m. Betty Mae Gardner, Dec. 2, 1943; children: Dennis, Diana Jensen Marsh, Karen Jensen Manix, Michael. A.B., U. Kans., 1948, M.A., 1949; Ph.D., Jefferson Med. Coll. Phila., 1951. Diplomate: Am. Bd. Med. Microbiology. Asst. prof. Sch. Pub. Health, U. Mich., 1951-56; mgr. respiratory disease unit Ctr. Disease Control, USPHS; also dir. WHO Influenza Ctr., 1956-58; dir. biols. research and devel. Pfizer, Inc., Terre Haute, Ind., 1958-65, exec. dir. virology and cancer research, 1965-80, sr. sci. adviser, 1980—. Contbr. numerous articles to profl. jours. Chmn. Bd. Edn., Old Lyme, Conn., 1967-70; chmn. civic action program Pfizer/Groton Employees, 1981—. Served with AUS, 1943-45. Fellow Am. Acad. Microbiology; mem. Am. Assn. Immunologists, Am. Soc. Microbiology, N.Y. Acad. Sci., Sigma Xi. Republican. Methodist. Subspecialties: Immunopharmacology; Cancer research (medicine). Current work: Human medicinals, animal health products. Home: Trumbull Rd Waterford CT 06385 Office: Pfizer Central Research Groton CT 06340

JENSEN, LEO STANLEY, animal nutritionist, educator; b. Bellingham, Wash., Feb. 28, 1925; s. Peter Hansen and Magdalena Petrea (Christensen) J.; m. Sylvia Brown, Oct. 30, 1930; children: Peter, Eric, Kristin, Carol. B.S., Wash. State U., 1949; Ph.D., Cornell U., 1954. Asst. prof. animal nutrition Wash. State U., Pullman, 1954-60, assoc. prof., 1960-66, prof., 1966-70, chmn. grad. program in nutrition, 1970-73; prof. animal nutrition U. Ga., Athens, 1973—. Served with U.S. Army, 1943-45. Decorated Purple Heart; Nat. Turkey Fedn. research awardee, 1963; Am. Feed Mfrs. Assn. awardee, 1966. Mem. Poultry Sci. Assn. (pres. 1982, Merck award 1979), Am. Inst. Nutrition, Soc. for Exptl. Biology and Medicine. Presbyterian. Subspecialty: Animal nutrition. Current work: Research on unidentified nutritional factors affecting hormone levels and lipid metabolism in avian species; mineral nutrition; disease-nutrition interaction. Home: 195 Horseshoe Circle Athens GA 30605 Office: Dept Poultry Sci U Ga Athens GA 30602

JENSEN, MEAD LEROY, geology educator, consultant; b. Salt Lake City, June 11, 1925; s. Joseph Mead and Gertrude (Grayson) J.; m. Lou Deen Davis, Oct. 10, 1947; children: Joseph, Pamela, Patricia, Janice, Robert. B.S., U. Utah, 1948; Ph.D., MIT, 1951. Instr. Yale U., New Haven, 1951-52, asst. prof., 1952-58, assoc. prof., 1958-65; prof. dept. geology and geophysics U. Utah, Salt Lake City, 1965—; pres. Geoventures, Inc., Salt Lake City, 1962—. Author: Economic Mineral Deposits, 3d edit, 1980. Served with USN, 1944-46; PTO. Australian Acad. Scis. vis. scholar, 1963; Sigma Xi disting. lectr., 1967. Fellow Geol. Soc. Am., Soc. Econ. Geologists (program chmn. 1954-55), Soc. Exploration Geophysicists, Mineral Soc. Am.; mem. Mining and Metall. Soc. Am. Democrat. Mormon. Club: Bonneville Knife and Fork (Salt Lake City). Subspecialties: Geology; Geochemistry. Current work: Geology of mineral deposits, development and use of exploration tools in the field and laboratory. Patentee in field. Office: Dept Geology and Geophysic U Utah Salt Lake City 84112 Home: 1354 S Ambassador Way Salt Lake City UT 84108

JENSEN, RICHARD GRANT, biochemist, educator; b. Los Angeles, Apr. 16, 1936; s. Frank Richard and Ruth (Grant) J.; m. Annette Anderson, Sept. 1, 1961; children: Karl Glenn, Jennifer Ruth, Bruce Duane, Byron Davis. B.A., Brigham Young U., 1961, Ph.D. in Biochemistry, 1965. USPHS postdoctoral fellow chem. biodynamics lab. U. Calif.-Berkeley, 1965-67; asst. prof. chemistry U. Ariz., 1967-72, assoc. prof. biochemistry, 1972-79, prof. biochemistry and plant scics., 1979—; vis. prof. Chem. Inst., Tech. U. Munich, W.Ger., 1974-75, Bot. Inst., U. Berne, Switzerland, 1975. Contbr. articles to profl. jours. Nat. Cancer Inst. fellow, 1965-67; Fulbright-Hays scholar, 1974-75. Fellow AAAS; mem. Am. Soc. Biol. Chemists, Am. Soc. Plant Physiology, Am. Chem. Soc. Mormon. Subspecialties: Photosynthesis; Plant physiology (agriculture). Current work: Regulation of photosynthesis and carbon metabolism in plants; biochemistry of the chloroplast, properties of the ribulose bisphosphate carboxylase and its role in CO2 fixation in plants. Office: Dept Biochemistry U Ariz Tucson AZ 85721

JENSEN, RICHARD JORG, educator; b. Erie County, Ohio, Jan. 17, 1947; s. Aksel Carl and Margaret Odenbaugh (Wolfe) J.; m. Faye Ellen Robertson, May 30, 1970. B.S., Austin Peay State U., 1970, M.S., 1972; Ph.D., Miami U., Oxford, Ohio, 1975. Teaching fellow dept. botany Miami U., Oxford, Ohio, 1972-74, dissertation fellow, 1974-75; asst. prof. dept. biol. scis. Wright State U., Dayton, Ohio, 1975-79; guest assoc. prof. dept. biology U. Notre Dame, Ind., 1980—; assoc. prof. dept. biology St. Mary's Coll., Notre Dame, Ind., 1979—. Contbr. articles in field to profl. jours. Recipient award for outstanding teaching Wright State U., 1978; NSF predoctoral grantee, 1973-75; Willard Sherman Turrell Herbarium grantee Miami U., 1974-75; NSF research grantee, 1978-82. Mem. Am. Soc. Plant Taxonomists, Bot. Soc. Am., Internat. Assn. Plant Taxonomy, Numerical Taxonomy Assn., Torrey Bot. Club, Ind. Acad. Sci., Sigma Xi (research grantee 1974-75), Phi Kappa Phi. Subspecialties: Systematics; Taxonomy. Current work: Numerical analysis of systematic and taxonomic relationships, especially in plants, numerical taxonomy, plant systematics, plant taxonomy, phenetics, phylogenetics. Home: 2044 Carrbridge Ct South Bend IN 46614 Office: Dept Biology St Mary's Coll Notre Dame IN 46556

JENSEN, ROY A., biology educator; b. Racine, Wis., Apr. 8, 1936. B.A., Ripon Coll., 1958; Ph.D. in Biochemistry and Genetics, U. Tex., 1963; postdoctoral, U. Wash. Sch. Medicine, 1964-66. Asst. prof. dept. biology SUNY-Buffalo, 1966-78; prof. dept. biology SUNY-Binghamton, 1976—; mem. grad. faculty Tex. A&M U., 1971-77; assoc. prof. dept. microbiology Baylor U. Coll. Medicine, Houston, 1986-73; assoc. prof. dept. biology M.D. Anderson Hosp. and Tumor Inst., U. Tex., 1974-76; mem. panel genetic biology NSF, 1978-80; mem. biocatalysis unit Monsanto Chem. Co., 1980—; mem. site visit rev. teams U.S. Dept. Energy, 1978—; geneticist biomed. programs AEC, 1973-75; molecular biologist for biomed. programs ERDA, 1975-76; dir. Ctr. Somatic-cell Genetics and Biochemistry, SUNY, 1976—; participant profl. symposia. Contbr. articles to sci. jours. John Wax Found. fellow, 1954-58; Snelling scholar, 1954-55; Ripon Coll. scholar, 1954-58; USPHS fellow, 1964-66. Subspecialties: Plant cell and tissue culture; Plant physiology (agriculture). Home: 2067 Cheshire Rd Binghamton NY 13903 Office: SUNY Dept Biol Sci Binghamton NY 13901

JENSEN, WILLIAM AUGUST, botanist, educator; b. Chgo., Aug. 22, 1927; s. William McKinley and Gertrude (Hild) J.; m. Joan Nancy Sell, June 20, 1948; children—Scott William, Christina Cathrine. Ph.B., U. Chgo., 1948, M.S., 1950, Ph.D., 1953. NIH fellow Carlsberg Lab., Copenhagen, Denmark, 1952-53, Calif. Inst. Tech., Pasadena, 1953-55; NSF fellow U. Brussels, Belgium, 1955-56; asst. prof. dept. biology U. Va., Charlottesville, 1956-57; faculty dept. botany U. Calif. at Berkeley, 1957—, prof., 1963—, chmn. dept. botany, 1971-73, chmn. dept. biology, 1974—; asso. dean Coll. Letters and Sci., 1963-66; program dir. developmental biology NSF, Washington, 1973-74; pres. Biology Media, Berkeley, 1974—. Author: Botanical Histochemistry, 1962, The Plant Cell, 1964, 71, (with R. Park) Cell Utrastructure, 1967, (with F. Salisbury) Botany, 1972, (with B. Heinrick, D. Wake, H. Wake) Biology, 1979; Contbr. articles to profl. jours.; Author, producer teaching modules and multi-image lectures. Recipient Disting. Teaching award U. Calif. at Berkeley, 1960, N.Y. Bot. Garden award for bot. research, 1964, Ohaus award Nat. Sci. Tchr. Assn., 1976. Mem. Soc. for Study Devel. and Growth (past sec.), Bot. Soc. Am. (program dir. 1963-67, v.p. 1976, pres. 1978), AAAS, Am. Inst. Biol. Scis., Soc. Developmental Biology, Am. Soc. Cell Biology, Histochem. Soc. Subspecialty: Plant physiology (biology). Devel. bot. histochem. procedures and application of these procedures to problems of early cell devel. in plants especially root tips and embryos; research on fertilization and early embryo devel. in flowering plants. Home: 280 Los Altos Dr Kensington CA 94708

JENSH, RONALD PAUL, anatomy educator; b. N.Y.C., June 14, 1938; s. Werner Gunther and Dorothy (Hensle) J.; m. Ruth-Eleanor Dobson, Aug. 18, 1962; children: Victoria Lynn, Elizabeth Whitney. B.A., Bucknell U., 1960, M.A., 1962; Ph.D., Jefferson Med. Coll., Phila., 1966. Instr. dept. anatomy Jefferson Med. Coll., Phila., 1966-68, asst. prof., 1968-74, assoc. prof., 1974-82, prof., 1982—; cons. in field. Contbr. articles in field to profl. jours. Mem. Teratology Soc., Behavioral Tertology Soc., Am. Coll. Toxicology, Am. Assn. Anatomists, Sigma Xi. Republican. Methodist. Subspecialties: Teratology; Developmental biology. Current work: Experimental embryology; teratology; radiation developmental biology, experimental psychology. Home: 230 E Park Ave Haddonfield NJ 08033 Office: Dept Anatomy Jefferson Med Coll Thomas Jefferson U 1020 Locust St Phila Pa 19107

JEON, KWANG WU, educator; b. Yong-ju, Korea, Nov. 10, 1934; came to U.S., 1965, naturalized, 1970; s. Y.R. and C.K. (Woo) J.; m. Myong Sheila Kim, May 7, 1958; children: Iang, Grace. B.S., Seoul (Korea) Nat. U., 1957, M.S., 1959; Ph.D., U. London, 1964. Research assoc. Nat. Chem. Labs., Seoul, 1957-59; lab. instr. King's Coll., London, 1961-63; research fellow Middlesex Hosp. Med. Sch., London, 1964-65; research asst. prof. SUNY-Buffalo, 1966-70; asso. prof. U. Tenn., Knoxville, 1970-76, prof., 1976—. Asst. editor: Internat. Rev. Cytology, 1967—; editor: The Biology of Amoeba, 1973, Interacellular Symbiosis, 1983. NSF grantee, 1974—; NIH grantee, 1971-73; Am. Heart Assn. grantee, 1971-73; U. Tenn. Research Sch. award, 1981. Mem. Am. Soc. Cell Biology, AAAS, N.Y. Acad. Sci., Soc. Protozoologists, Soc. Developmental Biologists. Subspecialties: Cell biology; Cell and tissue culture. Current work: Cell growth and div., organelle structure and function; interorganelle relations; endosymbiosis leading to acquisition of new cell components. Home: 912 Sky Blue Dr Knoxville TN 37923 Office: Univ Tenn Dept Zoology Knoxville TN 37996

JEREB, MARJAN JOSIP, physician; b. Ljubljana, Slovenia, Oct. 5, 1930; came to U.S., 1973; s. Peter and Olga (Vodisek) J.; m. Berta Verzun, Sept. 21, 1961. M.D., U. Ljubljana, 1954. Radiologist Karolinska Hosp., Stockholm, 1963-73; asst. prof. radiology SUNY Downstate Med. Ctr., Bklyn., 1973-75, assoc. prof., 1978—; radiologist Clin. Ctr., Ljubljana, Yugoslavia, 1975-77; head chest radiology Kings County Hosp., Bklyn., 1977—. Mem. Am. Assn. Univ. Radiologists, Swedish Radiol. Soc. Subspecialty: Diagnostic radiology. Current work: Application of new technologies in the diagnosis of chest, especially mediastinal, disease. Home: 504 E 81st St Apt 3 J New York NY 10028 Office: Downstate Med Ctr 450 Clarkson Ave Brooklyn NY 11203

JERINA, DONALD MICHAEL, chemist, pharmacologist; b. Chgo., Jan. 17, 1940; s. Anthony and Mary J.; m. Colleen B. Burgess, Aug. 8, 1964; children: Julianne, Derek. B.A., Knox Coll., 1962; Ph.D., Northwestern U., 1966. With NIH, 1966—, chief sect. oxidation mechanisms, 1975—. Contbr. numerous articles to profl. jours. Recipient USPHS award, 1975; Hildebrant prize, 1980; B. B. Brodie award, 1982. Mem. Am. Chem. Soc., Am. Soc. Biol. Chemists, Federated Am. Soc. Exptl. Medicine and Biology, Am. Soc. Pharmacology and Exptl. Therapeutics. Subspecialties: Synthetic chemistry; Cancer research (medicine). Current work: Oxidative drug metabolism, chemical carcinogenesis, bay-region theory. Home: 9717 Brixton Ln Bethesda MD 20817 Office: NIH Bldg 4 Room 216 Bethesda MD 20205

JERISON, HARRY JACOB, neurobiologist educator, researcher; b. Bialystok, Poland, Oct. 13, 1925; came to U.S., 1929; s. Eli and Esther (Rasky) Jerusalimsky; m. Irene Landkof, Dec. 17, 1950; children: Jonathan, Andrew, Elizabeth. B.S., U. Chgo., 1947; Ph.D., 1954. Research psychologist Aero. Med. Lab., Dayton, Ohio, 1949-57; lectr. Ind. U., South Bend, 1951-52; from assoc. prof. to prof. Antioch Coll., Yellow Springs, Ohio, 1957-69; prof. UCLA, 1969—; dir. Behavior Research Lab., Antioch Coll., 1957-68; cons. Nat. Library Medicine, Washington, 1977-78; vis. scientist Applied Psychology unit Med. Research Council, Cambridge, Eng., 1978-79. Author: Evolution of Brain and Intelligence, 1973, (with M. Wittrock and others) The Human Brain, 1977; editor: (with Irene Jerison) Paleoneurology, 1978; contbr. chpts. to books, articles to profl. jours. Hon. research assoc. Los Angeles Mus. Natural History, 1969—. Served to cpl. U.S. Army, 1944-46. Fellow Ctr. Advanced Study in the Behavioral Scis., 1967-68. Fellow AAAS, Am. Psychol. Assn.; mem. Am. Soc. Naturalists, Psychonomic Soc., Soc. Neurosci., Soc. Vertebrate Paleontology. Democrat. Jewish. Subspecialties: Comparative neurobiology; Psychobiology. Current work: Evolution of brain and behavior with emphasis on cognitive processes, language, and attention; studies of brain evolution emphasize allometric brain/body relations in fossil and living species. Home: 503 W Rustic Rd Aanta Monica CA 90402 Office: Dept Psychiatry and Biobehavioral Sci UCLA Med Sch 760 Westwood Plaza Los Angeles CA 90024

JERNIGAN, ROBERT LEE, research scientist; b. Portales, N.Mex., May 4, 1941; s. Frank H. and Jimmie M. (Loyd) J.; m. Marydee A. Becker, June 15, 1968; 1 son, Alexander L. B.S., Calif. Inst. Tech., 1963; Ph.D., Stanford U., 1967. Postdoctoral fellow Stanford (Calif.) U., 1967-68; NIH postdoctoral fellow U. Calif., San Diego, 1968-70; sr. staff fellow NIH, Bethesda, Md., 1970-75, theoretical phys. chemist, 1975—. Contbr. articles to profl. jours. Recipient Spl. Achievement award NIH, 1982. Mem. Biophys. Soc., Am. Chem. Soc., Am. Phys. Soc., AAAS. Subspecialties: Molecular biology; Polymer chemistry. Current work: Elucidation of molecular details of biochemical and biophysical processes by computer studies of macromolecular conformations. Home: 3700 Dunlop St Chevy Chase MD 20815 Office: NIH Nat Cancer Inst Bldg 10 Rm 4B-56 Bethesda MD 20205

JERRY, LAURENCE MARTIN, med. oncologist, cancer ctr. adminstr.; b. Toronto, Ont., Can., Jan. 2, 1937; s. Carman Alfred and Merle Amanda (Martin) J.; m. Marian Bernice Hyatt, Sept. 3, 1960; children: Paul Alexander, Marc Philip. M.D., U. Toronto, 1961;

Ph.D., Rockefeller U., 1971. Asst. prof. medicine McGill U., Montreal, 1971-73, assoc. prof., 1973-77; assoc. physician Royal Victoria Hosp., Montreal, 1971-77; sr. mem. McGill Cancer Research Group, 1973-77; dir. Tom Baker Cancer Ctr. and So. Alta. Cancer Program, Calgary, 1977—; prof. medicine, chmn. oncology research group U. Calgary, 1977—. Contbr. articles to profl. jours. Med. Research Council Can. research scholar, 1972-77. Fellow Royal Coll. Physicians Can., ACP; mem. Can. Immunol. Soc., Am. Immunol. Soc., Can. Soc. Clin. Investigation, Am. Soc. Clin. Investigation, Am. Assn. Cancer Research. Lodge: Rotary. Subspecialties: Cancer research (medicine); Oncology. Current work: Adjuvant therapy in breast and colorectal cancer and melanoma-clinical trials-studies of immune regulation and tumor antigens in melanoma; studies of membranes, antigens and differentiation markers in leukemia. Office: 1331-29 St NW Calgary AB Canada T2N 4N2

JERVIS, HERBERT HUNTER, geneticist, educator; b. Wilmington, Del., June 25, 1942; s. Herbert Willard and Dorothy Grotz J.; m. Mary Gregory, June 9, 1974. B.S., Springfield Coll., 1964, M.Ed., 1966; M.S., Fla. State U., 1971, Ph.D., 1974. Postdoctoral fellow dept. biochemistry Va. Inst. Tech., 1974-75; asst. prof. genetics Adelphi U., 1975-80, asso. prof., 1980—; cons. biotech. patents Scully, Scott, Murphy & Presser, 1982—; cons. investment group Arthur Merrill Assos., 1982—. Contbr. articles in field to profl. jours. Served with U.S. Army, 1966-69. Mem. AAAS, Genetics Soc. Am., Am. Soc. Microbiology, N.Y. Acad. Sci., Springfield Coll. Alumni Assn., Sigma Xi. Subspecialties: Genetics and genetic engineering (biology); Biotechnology patents. Current work: Fungal, genetics, developmental genetics, biotech. patents. Home: 124 Brixton Rd S West Hempstead NY 11552 Office: Dept Biology Adelphi U Garden City NY 11530 200 Garden City Plaza Garden City NY 11530

JESEN, JAMES BURT, pasasitology educator, researcher; b. Los Angeles, Mar. 8, 1943; s. Marvin James and Ottilla (McKnight) J.; m. Jane Gunderson, Feb. 26, 1965; children: Christian, Stephanie, Davidson, Levi, Leah, Joshua. B.Sc., Brigham Young U., 1970, M.Sc., 1972; postgrad., Utah State U., 1972-74; Ph.D., Auburn U., 1976; postdoctoral student, Rockefeller U., 1976-77. Asst. Prof. Rockefeller U., 1977-79; asst. prof. parasitology Mich. State U., East Lansing, 1979-82, assoc. prof., 1982—; cons. on malaria WHO, 1978—; mem. internat. bd. Kuvin Ctr. for Study Infectious and Tropical Diseases, Hadassah Med. Sch, Hebrew U., Jerusalem, 1983. Editor: Cultivation of Protozoan Parasites, 1983. Research grantee AID, WHO, NIH. Mem. Am. Soc. Tropical Medicine and Hygiene, Soc. Protozoologists, Am. Soc. Parasitologists, N.Y. Soc. Tropical Medicine. Republican. Mormon. Subspecialties: Parasitology; Infectious diseases. Current work: Immunologic reactions to malarial parasites in human infections of Plasmodium spp; biochemical-pharmacology of huan malarial parasites; epidemiology of human malaria in Africa. Home: 636 Charles St East Lansing MI 48823 Office: Dept of Microbiology and Public Healt Giltner Hall Michigan State University East Lansing MI 48824

JESKA, EDWARD L(AWRENCE), immunologist, educator; b. Erie, Pa., Aug. 6, 1927; s. Francis A. and Martha I. (Nowak) J.; m. Elizabeth E. Ahlgren, July 24, 1931. B.A., Gannon U., 1951; M.S., Marquette U., 1954; Ph.D., U. Pa., 1966. Asst. prof. vet. pathobiology U. Pa., Phila., 1966; asst. prof. vet. pathology Iowa State U., 1967-70, assoc. prof., 1970-75, prof., 1975—; mem. expert panels NSF, Dept. Agr. Contbr. numerous articles to tech. jours., chpts. to books. Mem. Am. Assn. Immunology, Am. Soc. Parasitology, Conf. Research Workers in Animal Diseases, Reticuloendothelial Soc. Subspecialties: Infectious diseases; Pathology (medicine). Current work: Immunobiology of mucosalvimmunity and functions of macrophages during infection. Office: Coll Vet Medicine Iowa State U Ames IA 50017

JESPERSEN, NEIL DAVID, chemist, educator, cons.; b. Bklyn., Mar. 5, 1946; s. Jack William Eric and Evelyn Lillian (Nelson) J.; m. Marilyn Josephine Zak, Apr. 25, 1970; children: Lisa Marie, Kristen Marie. B.S., Washington and Lee U., 1967; Ph.D., Pa. State U., 1971. Asst. prof. U. Tex., 1971-77; assoc. prof. St. John's U., Jamaica, N.Y., 1977—. Contbr. articles on chemistry to profl. jours. Mem. Citizens Adv. Com., 1980-82. Mem. Am. Chem. Soc., AAAS, Sigma Xi. Subspecialties: Analytical chemistry; Ecosystems analysis. Current work: Analytical chemistry of biological systems. Office: St Johns U Grand Central and Utopia Jamaica NY 11439

JESSER, WILLIAM AUGUSTUS, materials scientist, educator; b. Waynesboro, Va., Dec. 20, 1939; s. Richard Alexander and Margaret Leonora (Fry) J.; m. Barbara Lee Schwab, Aug. 18, 1962; children: William Augustus, Nicole E. B.A., U. Va., 1962, M.A., 1964, Ph.D., 1966. Lectr. dept. physics U. Witwatersrand, Johannesburg, South Africa, 1966-67; prof. dept. materials sci. U. Va., Charlottesville, 1968—; dir. High Voltage Electron Microscopy Facility, 1976—; vis. prof. dept. metallurgy Nagoya U., Japan, 1978; dept. physics U. Pretoria, South Africa, 1982. Contbr. articles to profl. jours. Chmn. bd. dirs. Hearthstone Children's House, 1980-82; comdr. Monticello Squadron, CAP, 1980—. Recipient Alan T. Gwathmey Prize for research U. Va., 1966. Mem. Am. Soc. Metals, Electronic Microscope Soc. Am., Va. Acad. Sci., Sigma Xi. Subspecialties: Alloys; Fusion. Current work: Electron microscopy, radiation damage, phase transformations. Home: Montvue Charlottesville VA 22901 Office: Thornton Hall U Va Charlottesville VA 22901

JESSUP, JOHN MILBURN, surgeon, educator; b. New Haven, Aug. 4, 1946; s. John Baker and Dorothy Milburn J.; m. Kathleen Foxen, May 7, 1977; children: Katherine Elizabeth, John Milburn. B.A., Yale U., 1968; M.D., N.Y. Med. Coll., 1972. Diplomate: Am. Bd. Surgery. Resident N.Y. Hosp., 1972-74; clin. assoc. surg. br. Nat. Cancer Inst., 1974-76; vis. scientist Basic Research Program, Frederick Cancer Research Ctr., 1976-77; resident in surgery U. Tex. Med. Sch., Houston, 1977-80; faculty assoc. dept. gen. surgery M.D. Anderson Hosp., Houston, 1980-81, asst. prof., asst. surgeon depts. gen. surgery and clin. immunology and biol. therapy, 1981—. Contbr. articles to profl. jours. Served to lt. comdr. USPHS, 1974-76. Mem. Aerospace Med. Assn., AAAS, Am. Assn. Cancer Research, Assn. Acad. Surgery, Alpha Omega Alpha. Subspecialties: Immunology (medicine); Cancer research (medicine). Current work: Image analysis, cloning t-cells, oncologic surgery. Office: 6723 Bertner Ave Houston TX 77030

JETTE, ARCHELLE NORMAN, physicist, researcher; b. Portland, Oreg., May 15, 1934; s. Archelle Norman, Sr. and Muriel Edith (Fisher) J.; m. Jamie Drago, June 16, 1972; 1 dau., Andrea Nichelle. Student, San Bernardino Valley Coll., 1956-59; A.B., U. Calif., Riverside, 1961, M.A., 1963, Ph.D., 1965. Research technician dept. chemistry U. Calif., Riverside, 1959-61, research asst. dept. physics, 1961-64; physics instr. San Bernardino Valley Coll., 1963-64; research assoc. Columbia Radiation Lab., Columbia U., N.Y.C., 1965; physicist Applied Physics Lab., Johns Hopkins U., Laurel, MD., 1965—; vis. prof. physics Cath. U., Rio de Janeiro, 1972; vis. scientist Ctr. for Interdisciplinary Research, U. Bielefeld, W.Ger., 1980. Contbr.: articles to profl. publs. including Physics Rev. Letters. Mem. Am. Phys. Soc., Am. Vacuum Soc. Subspecialties: Materials; Atomic and molecular physics. Current work: Surface science, electron diffraction, gas surface interactions, theoretical solid state physics, theoretical atomic and molecular physics. Home: 4021 Arjay Circle Ellicott City MD 21043 Office: Applied Physics Lab Johns Hopkins Univ Johns Hopkins Rd Laurel MD 20707

JEWELEWICZ, RAPHAEL, educator, physician; b. Nowogrodek, Poland, Dec. 26, 1932; came to U.S., 1963; s. Chaim and Chaia (Tawticky) J.; m. Ronnie Oved, July 23, 1955; children: Rachel, Dov, Daniel, Dory. M.D., Hebrew U., Israel, 1961. Intern Hadassah Hebrew U. Hosp., Jerusalem, Israel, 1960-61; resident Bellevue Hosp.-NYU Med. Ctr., N.Y.C., 1964-68; practice medicine specializing in ob.-gyn.; asst. prof. Columbia U., N.Y.C., 1969-75, assoc. prof., 1975—; dir. div. reproductive endocrinology Columbia Presbyn. Med. Center, 1975—. Contbr. articles to profl. jours. Fellow Am. Coll. Ob.-Gyn., A.C.S.; mem. Soc. for Gynecol. Investigation, Endocrine Soc., Am. Fertility Soc. Subspecialties: Obstetrics and gynecology; Endocrinology. Current work: Control of menstrual cycle, induction of ovulation, infertility, in vitro fertilization and embryo transfer. Home: Church St Alpine NJ 07620 Office: Columbia U Dept Obstetrics and Gynecology 630 W 168 St New York NY 10032

JEWELL, WILLIAM SYLVESTER, engineering educator; b. Detroit, July 2, 1932; s. Loyd Vernon and Marion (Sylvester) J.; m. Elizabeth Gordon Wilson, July 7, 1956; children—Sarah, Thomas, Miriam, William Timothy. B.Engring. Physics, Cornell U., 1954; M.S. in Elec. Engring, MIT, 1955, Sc.D., 1958. Assoc. dir. mgmt. scis. div. Broadview Research Corp., Burlingame, Calif., 1958-60; asst. prof. dept. indsl. engring. and operations research U. Calif.-Berkeley, 1960-63, asso. prof., 1963-67, prof., 1967—, chmn. dept., 1967-69, 76-80; dir. Teknekron Industries, Inc., Berkeley, 1968—; cons. operations research problems, 1960—; guest prof. Eidgenössiches Technische Hochschule, Zurich, 1980-81. Contbr. articles to profl. jours. Fulbright research scholar, France, 1965; research scholar Internat. Inst. Applied Systems Analysis, Austria, 1974-75. Mem. Ops. Research Soc. Am., Inst. Mgmt. Scis., Am. Risk and Ins. Assn., Actuarial Assn. Netherlands, Assn. Swiss Actuaries, Internat. Actuarial Assn., Mensa, Triangle., Sigma Xi, Triangle. Subspecialties: Operations research (engineering); Statistics. Current work: Hardware and software reliability; Bayesian statistical models; risk analysis; actuarial science. Home: 67 Loma Vista Orinda CA 94563 Office: U Cal Dept Indsl Engring and Operations Research Berkeley CA 94720

JINKS, GORDON MCMILLAN, pediatric dentist, educator; b. Vancouver, C., Can., Jan. 28, 1921; s. Frederick Thomas and Sarah Anna (McMillan) J.; m. Norma Jean Wilson, June 6, 1946; children: Cynthia Margaret, Barry Wilson. D.D.S., U. Toronto, 1946. Spl. lectr. U. Wash., Seattle, 1960-66; lectr. U. Oreg., Portland, 1956-58; staff dentist Children's Hosp., Vancouver, 1948-60; clin. instr. U. B.C., Vancouver, 1978—; chmn. 3d Can. Congress on Dentistry for Children, Vancouver, 1977-79; lectr. dental research socs. Contbr. articles to profl. jours. Active United Way Campaign, Vancouver, 1982. Recipient Cert. Achievement Wash. State Dental Assn., 1961; Outstanding service to Dentistry award Wash. Soc. Dentistry for Children, 1971; Teaching and Guidance award Portland Pediatric Research Soc., 1958. Fellow Royal Coll. Dentists Can., Am. Acad. Pedodontics; mem. Can. Soc. Dentistry for Children (pres 1969—), Can. Acad. Pedodontics (past pres.), Can. Dental Assn., B.C. Dental Assn. Club: Western Gate. Current work: clinical research into dental caries prevention. Office: 301-1701 W Broadway Vancouver BC Canada V6J 1Y3 Home: P.H 3A 2450 Cornwall St Vancouver BC Canada V6K 1B8

JOBS, STEVEN PAUL, computer corporation executive; b. 1955; adopted s. Paul and Clara J. Student, Reed Coll. With Hewlett-Packard, Palo Alto, Calif., summers while high-sch. student; designer video games Atari Inc., 1974; now chmn. bd. Apple Computer, Inc., Cupertino, Calif. Subspecialties: Electronics; Computer company management. Co-designer (with Stephen Wozniak) Apple I microcomputer, introduced 1976. Office: Apple Computer Inc 10260 Bandley Dr Cupertino CA 95014

JOBST, JOEL EDWARD, nuclear physicist, researcher; b. South Milwaukee, Wis., May 13, 1936; s. Edward John and Cecilia Teresa (Kugel) J.; m. Corinne Mary Lawler, June 20, 1959; children: Brian, Kevin, Erin. B.S., Marquette U., 1959; M.S., U. Wis.-Madison, 1961, Ph.D., 1966. Sci. specialist EG&G Energy Measurements, Inc., Las Vegas, 1966—; pvt. practice solar cons., Las Vegas, 1974—. Author numerous research reports. Mem. Am. Nuclear Soc. (gen. chmn. ann. meeting 1980), Am. Phys. Soc., Am. Solar Energy Soc., Nev. Solar Energy Advocates (pres. 1981,82). Subspecialties: Nuclear physics; Remote sensing (atmospheric science). Current work: Gamma radiation measurements to determine natural terrestrial background and determine strength and distribution of manmade radiation sources. Home: 3013 Bryant Ave Las Vegas NV 89102 Office: EG&G Energy Measurements Inc PO Box 1912 Las Vegas NV 89125

JOEL, AMOS EDWARD, JR., electrical engineer; b. Phila., Mar, 12, 1918; m.; 3 children. B.S., MIT, 1940, M.S., 1942. Switching systems devel. engr. Bell Telephone Labs., Holmdel, N.J., 1954-60, head electronic switching planning dept., 1960-61, dir. switching systems devel. lab., 1961-62, local switching lab., 1962-67, switching cons., 1967—. Recipient Outstanding Patent award N.J. Council R&D, 1972, Stuart Ballantine medal Franklin Inst., 1981, 2d Centenary prize Internat. Telecommunication Union, 1983. Life fellow IEEE. (Alexander Graham Bell medal 1976); fellow Nat. Acad. Engring.; Mem. ACM, AAAS. Subspecialty: Electrical engineering. Office: Bell Telephone Labs Holmdel NJ 07733

JOESTEN, RAYMOND, metamorphic petrologist; b. San Francisco, Sept. 12, 1944; s. Leonard L. and Joan (Mabie) J.; m. Jane Elizabeth Joesten, June 24, 1967; children: Peter, Robert. B.S. with gt. distinction and honors in Geology, San Jose State Coll., 1966; Ph.D., Calif. Inst. Tech., 1974. Instr. dept. geology and geophysics U. Conn., 1971-73, asst. prof., 1973-77, assoc. prof., 1977—, head dept., 1983—. Commr. planning and zoning Town of Coventry, Conn., 1981—. NSF grantee; recipient Leon M. Miller award, 1966. Mem. Geol. Soc. Am. Democrat. Congregationalist. Subspecialty: Petrology. Current work: Mineral equilibria and mass transfer in metamorphic rocks, applications of non-equilibrium thermadynamics to modelling of multicomponent diffusion, studies of annealing and grain growth. Office: Dept of Geology and Geophysics U Conn Storrs CT 06268

JOFFE, ANATOLE, mathematics educator/researcher; b. Belgium, Sept. 1, 1932; m.; 2 children. Lic.Sc. and advanced teaching degree, U. Brussels, 1954, 1955; Ph.D., Cornell U., 1959. Asst. prof. math. McGill U., 1960-61; from asst. prof. to assoc. prof. math. U. Montreal, Que., Can., 1961-73, prof., 1973—; dir. Math. Research Ctr., 1973—. Mem. com. NRC, 1974-77, 74—. Mem. Am. Math. Soc., Inst. Math. Stats., Math. Soc. Can. Subspecialty: Statistics. Office: Math Research Ctr U Montreal Montreal PQ Canada H3C 3J4

JOFFE, JUSTIN MANFRED, psychology educator, researcher; b. Johannesburg, Transvaal, South Africa, Oct. 29, 1938; came to U.S., 1967, naturalized, 1978; s. Harry and Rita (Naar) J.; m. Daryll Elizabeth Eagle, May 22, 1963; children: Emily, Natasha, Chantal, Jasper. B.A., U. Witwatersrand, Johannesburg, 1959, 1960, M.A., 1962; Ph.D., U. London, 1965. Jr. lectr. U. Witwatersrand, 1962-63; research worker U. London, 1963-65; lectr. Birmingham U., Eng., 1965-67; asst. prof. U. Vt., 1969-70, assoc. prof., 1970-74, prof. psychology, 1974—; v.p. Vt. Conf. on Primary Prevention of Psychopathology, Burlington, 1975—; trustee Psychol. Research Found. Vt., Burlington, 1971—. Author: Prenatal Determinants of Behavior, 1969; editor: The issues: An Overview of Primary Prevention, 1975, Prevention Through Political Action and Social Change, 1981, Facilitating infant and Early Childhood Development, 1982; spl. features editor: Jour. Primary Prevention, 1979—. NIH postdoctoral fellow NIH, Stanford U., 1968-69; research scientist Nat. Acad. Scis., Yugoslavia, 1980-81; Nat. Inst. Child Health and Human Devel., Nat. Inst. Drug Abuse research grantee, 1972-75, 75-78. Mem. Psychonomic Soc., Internat. Soc. Devel. Psychobiology. Subspecialties: Teratology; Developmental psychology. Current work: Effects of paternal exposure to drugs and chemicals on reproductive outcome and progeny, effects of prenatal and perinatal hormonal factors on offspring development and behavior. Office: Dept Psychology U Vt Dewey Hall Burlington VT 05405

JOHANNSEN, ULMER JAMES, researcher, medical technologist; b. Clinton, Iowa, Apr. 22, 1945; s. Ulmer William and Evelyn Mary (Wight) J. B.A., Coe Coll., Cedar Rapids, Iowa, 1976. Cert. med. technologist. Med. technologist Victory Hosp., Waukegan, Ill., 1976; mem. research staff U. Iowa, Iowa City, 1976—, sr. research asst., 1981—. Contbr. chpts. to books. Mem. N.Y. Acad. Scis., Am. Physiol. Soc. (assoc. mem.), Am. Soc. Clin. Pathologists (assoc.). Subspecialties: Cardiology; Physiology (medicine). Current work: Neural regulation of the coronary circulation. Office: Univ Iowa 200A Med Lab Iowa City IA 52240

JOHANSON, JERRY RAY, cons. engring. co. exec., mech. engr., cons., researcher; b. Murray, Utah, Aug. 29, 1937; s. Albert F. and Elizabeth (Cox) J.; m. Harlean Marie Shepherd, July 12, 1957; children: Kerry, Bryan, Michael, Cynthia, Elizabeth. Ph.D. in Mech. Engring, U. Utah, 1962. Registered profl. engr., Mass. Sr. technologist U.S. Steel Applied Research Lab., Monroeville, Pa., 1962-66; pres. Jenike & Johanson, Inc., North Billerica, Mass., 1966—; cons. in field; seminar tchr. Contbr. numerous articles on storage and flow of solids to profl. jours. Mem. ASME (Henry Hess award 1966, chmn. div. materials handling 1972-73). Mormon. Subspecialties: Theoretical and applied mechanics; Fuels. Current work: Consulting and research in flow of bulk solids, materials handling, process design, two-phase solid-fluid flow analysis. Patentee blending apparatus for bulk solids, air permeation, material blending apparatus. Office: 3485 Empressa Dr San Luis Obispo CA 93401

JOHNS, DEARING WARD, internist, researcher; b. Springfield, Mo., Aug. 8, 1941; d. Peter Otey and Annie (Boyd) Ward; m. Thomas Richards Johns, July 1, 1978; 1 dau., Sarah Dearing. A.B. in Physics, Sweet Briar Coll., 1963; M.D., U. Va., 1977. Asst. prof. internal medicine U. Va., 1983—. Mem. ACP, Am. Fedn. Clin. Research, Am. Coll. Cardiology. Subspecialties: Cardiology; Internal medicine. Current work: Electrophysiology and pharmacology of vascular smooth muscle with special emphasis on its relationship to hypertension. Home: 19 Old Farm Rd Charlottesville VA 22901 Office: University of Virginia Medical Center Box 257 Charlottesville VA 22908

JOHNS, RICHARD JAMES, physician; b. Pendleton, Oreg., Aug. 19, 1925; s. James Shanard and Pearl (McKenna) J.; m. Carol Greacen Johnson, June 27, 1953; children: James Ashmore, Richard Clark, Robert Shanard. B.S., U. Oreg., 1947; M.D., Johns Hopkins U., 1948. Diplomate: Am. Bd. Internal Medicine. Intern Johns Hopkins Hosp., 1948-49, asst. resident, 1951-53, resident, 1955-56, physician, 1956—; asst. in medicine Johns Hopkins U., 1951-53, fellow in medicine, 1953-55, instr., 1955-57, asst. prof., 1957-61, assoc. prof., 1961-66, asst. dean admissions, 1962-66, prof. medicine, 1966—, dir. subdept. biomed. engring., 1966-70, prof., dir. dept. biomed. engring., 1970—; mem. adv. bd., prin. profl. staff Applied Physics Lab., 1967—. Sec., vice chmn., chmn. med. bd. Myasthenia Gravis Found.; trustee Am. Bd. Clin. Engring., pres., 1976—. Served from 1st lt. to capt. M.C. AUS, 1949-51. Fellow ACP, AAAS; mem. Am. Clin. and Climatol. Assn. (v.p. 1977-78), Am. Soc. Clin./Investigation, Assn. Am. Physicians, Biomed. Engring. Soc. (dir. 1972-75, pres. 1978-79), IEEE (pres. group on engring. in medicine and biology 1970-72), Johns Hopkins Med. Soc. (pres. 1968-69), Interurban Clin. Club, Peripatetic Club, Caduceus Club, Sigma Xi, Alpha Omega Alpha, Phi Kappa Psi, Nu Sigma Nu. Clubs: Annapolis Yacht, Hamilton St., Johns Hopkins (v.p. 1969-70). Subspecialty: Biomedical engineering. Office: Dept Biomed Engring Johns Hopkins U 34th and Charles St Baltimore MD 21218

JOHNSEN, EUGENE CARLYLE, mathematics educator, researcher; b. Mpls., Jan. 27, 1932; s. Bernhardt Thorwald and Esther Elvira (Eklund) J.; m. Marjorie Marie Wacklin, Aug. 31, 1957. B.Chem., U. Minn., 1954; Ph.D., Ohio State U., 1961. Research assoc. Nat. Bur. Standards, Washington, 1962-63; lectr. in math. U. Calif.-Santa Barbara, 1963-64, asst. prof. math., 1964-68, assoc. prof. math., 1968-74, prof. math., 1974—; vis. lectr. in math. U. Mich.-Ann Arbor, 1968-69; mathematician Sperry Rand, St. Paul, summers 1956, 57; instr. chemistry and math. U. Minn., 1956-57; instr. math. Ohio State U., Columbus, 1962. Recipient numerous research awards and grants. Mem. Am. Math. Soc., Math. Assn. Am., Soc. for Indsl. and Applied Math., AAAS, Internat. Network for Social Network Analysis, Phi Beta Kappa, Sigma Xi, Phi Lambda Upsilon, Pi Mu Epsilon. Clubs: Faculty, Sons of Norway. Subspecialty: Applied mathematics. Current work: Combinatorial designs, matrices, and algebraic structures and their applications in the biological and social sciences. Office: Dept Math U Calif Santa Barbara CA 93106

JOHNSGARD, PAUL AUSTIN, biology educator, writer; b. Fargo, N.D., June 28, 1931; s. Alfred Bernard and Yvonne Margarette (Morgan) J.; m. Lois Miriam Lampe, June 25, 1955; children: Jay Erik, Scott Kenneth, Anne Yvonne, Karin Luisa. B.S., N.D. State U., 1953; M.S., Wash. State U.-Pullman, 1955; Ph.D., Cornell U., 1959. Postdoctoral fellow Bristol (Eng.) U., 1959-61; instr. ornithology U. Nebr.-Lincoln, 1961-62, asst. prof., 1962-65, assoc. prof., 1965-68, prof., 1968-80, Found. prof., 1980—. Author: Grouse & Quails of North America, 1973, Waterfowl of North America, 1975, Ducks, Geese & Swans of the World, 1979, Plovers, Sandpipers & Snipes of World, 1981. Guggenheim fellow, 1970; recipient Disting. teaching award U. Nebr., 1968, Fulbright Hays award Council Internat. Exchange, 1980. Fellow Am. Ornithologists Union; mem. Cooper Ornithol. Soc., Wilson Ornithol. Soc. Unitarian. Subspecialties: Ethology; Taxonomy. Current work: Current research involves evolutionary and comparative behavioral studies of the pheasants of the world, the grouse of the world and the cranes of the world. Home: 7341 Holdrege St Lincoln NE 68505 Office: School of Life Sciences U Nebr Lincoln NE 68588

JOHNSON, ARMEAD HOWARD, immunologist; b. Waco, Tex., Dec. 16, 1942; s. Edwin and Armead (Howard) J. B.S., U. Tex., 1964; Ph.D., Baylor U., 1971. Postdoctoral fellow dept. microbiology and immunology Baylor Coll. Medicine, 1971-72; Postdoctoral fellow dept. microbiology, div. immunology Duke U. Med. Ctr., 1972-74, research assoc. dept. microbiology and immunology, 1974-75, asst. research prof. dept. microbiology and immunology, 1975-80; asst. prof. dept. pediatrics, div. immunologic oncology Georgetown U. Sch. Medicine, Washington, 1980—. Nat. Cancer Inst. grantee, 1975-80; Office Naval Research grantee, 1980-83; NIH grantee, 1982—. Mem. Am. Assn.

Immunologists, Genetics Soc., Am. Assn. Clin. Histocompatibility, Sigma Xi. Episcopalian. Subspecialty: Histocompatibility. Current work: Research in immunogenetics of the major histocompatibility comples of man. Home: 5411 Harwood Rd Bethesda MD 20814 Office: Georgetown U Sch Medicine Lombardi Cancer Center Div Immunologic Oncology Reservoir Rd Washington DC 20007

JOHNSON, BENJAMIN MARTINEAU, chemical engineer, educator; b. Chiralla, India, Oct. 28, 1930; s. Benjamin Martineau and Inez Jenny (Jones) J.; m. Mary Margaret Anderson, July 31, 1954; children: Daniel Paul, Judith Kay. B.S. in Chem. Engring, Cornell U., 1952; M.S., U. Wis., 1953, Ph.D., 1956. Sr. engr. Gen. Electric Co., Richland, Wash., 1956-61, research assoc., 1961-65; sect. mgr. Battelle-N.W., Richland, 1965-72, assoc. dept. mgr., 1973-77, program mgr. and sr. staff engr., 1977—; affiliate prof., chem. engring. program coordinator Joint Ctr. Grad. Study, Richland, 1959—. NSF fellow, 1953, 54. Fellow Am. Soc. Chemists; mem. ASME, Am. Nuclear Soc., Am. Inst. Chem. Engrs. (local sect. chmn. 1964), Sigma Xi, Tau Beta Pi. Subspecialties: Chemical engineering; Mechanical engineering. Current work: Thermal and fluid sciences; heat, mass and momentum transfer; advanced air cooling and evaporate cooling for power plants; solar thermal technology, solar ponds, isotope separation, nuclear reactor fuel design, safety evaluation, fuel reprocessing, radioactive waste disposal. Patentee in field. Home: 6928 Willamette Kennewick WA 99336 Office: Battelle-NW PO Box 999 Richland WA 99352

JOHNSON, BRANT MONTGOMERY, research physicist; b. Houston, Aug. 25, 1949; s. William Delmo and Alice Constance (Brain) J.; m. Marcia O'Shee Prosser, Dec. 28, 1973; children: Anna Montgomery, Austin Prosser. B.S. in Physics, U. Tex.-Austin, 1971, M.A., 1974, Ph.D., 1975. Research sci. assoc. II, Robert A. Welch Found. research fellow U. Tex.-Austin, 1971-75; research assoc. physics dept. Brookhaven Nat. Lab., Upton, N.Y., 1975-77, asst. physicist, 1977-79, assoc. physicist, 1979-80, physicist, 1981—; vis. scientist Lawrence Berkeley (Calif.) Lab., 1977, Oak Ridge Nat. Lab., 1977, Triumf Lab., Vancouver, B.C., Can., 1980; co-prin. investigator div. basic energy scis. U.S. Dept. Energy, 1979—. Contbr. articles to sci. jours. Mem. Am. Phys. Soc. Subspecialties: Atomic and molecular physics; Plasma physics. Current work: Atomic physics experimental research on ion-atom, ion-electron, ion-photon (synchrotron radiation) collisions. Emphasis on atomic processes relevant to high temperature plasmas. Home: RR 2 28 Zophar Mills Rd Wading River NY 11792 Office: Brookhaven Nat Lab Tandem Bldg 901A Upton NY 11973

JOHNSON, BRIAN WEATHERRED, computer science educator; b. Dallas, Jan. 21, 1944; s. Enoch Mather and Grace Eicke (Edwards) J.; m. Margaret Susan Jeffreys, Sept. 8, 1967. B.S., Carnegie-Mellon U., 1967, M.S., 1968, Ph.D., 1971. Mem. Faculty U. Tex., Dallas, 1970—, assoc. prof. computer sci., 1977—. Contbr. articles to profl. jours. Mem. Assn. Computing Machinery, IEEE. Republican. Subspecialties: Distributed systems and networks; Electronics. Current work: Asynchronous interaction in hardware and software, design of asynchronous systems and circuits. Home: 1525 Westlake Dr Plano TX 75075 Office: PO Box 688 M/S F023 Richardson TX 75080

JOHNSON, CARL MAURICE, forestry educator, extension forester; b. Cedar City, Utah, Oct. 22, 1918; s. John Henry and Florence (Smith) J.; m. Doris Naomi Hughes, Oct. 23, 1944; children: Alan, Kathryn, Linda, Carl, Sheila, David. B.S. in Forestry, Utah State Agrl. Coll., 1940, M.S., Utah State U., 1963; Ed.D. in Conservation Edn, U. No. Colo., 1980. Fireguard U.S. Forest Service, Hebo, Oreg., 1940; commd. 2d lt. U.S. Marine Corps, 1941, advanced through grades to lt. col., 1961; served in, U.S., South Pacific, Japan and Korea, ret., 1961; assoc. prof., extension forester Utah State U., 1963—. Author: Common Native Trees of Utah, 1970, Growing Trees and Shrubs in Utah, 1966, How to Know Trees of Region 4, 1980. Recipient Dist. award of merit Boy Scouts Am., 1979. Mem. Soc. Am. Foresters, Xi Sigma Pi. Republican. Mormon. Lodge: Logan Lions. Subspecialties: Resource conservation; Taxonomy. Current work: Natural resource multiple-use conservation, conservation and environmental education. Office: Utah State University Logan UT 54322 Home: 630 N 3d E Logan UT 54321

JOHNSON, CAROL WILLIAM, physicist, design engr.; b. Chgo., Apr. 8, 1946; s. Carl Alvin and Catherine Ann (Opyt) J.; m. Judi Lorraine Prather, Sept. 3, 1972. B.A. in Physics, U. Chgo., 1967; postgrad., Purdue U., 1967-69. Tchr. Thornridge High Sch., Dolton, Ill., 1969-72; cons. mech. and elec. engring., Knox, Ind., 1972-73; founder JUCA, Inc., North Judson, Ind., 1974—, pres., chief exec. officer, 1975—; cons. in field. Contbr. energy related articles to profl. publs. Subspecialties: Theoretical and applied mechanics; Numerical analysis. Current work: Applying computer tech. to alt. energy applications: in solid fuel devices, solar heating devices, and in wind and water power. Patentee in field. Home: Route 3 Box 93 North Judson IN 46366 Office: 1400 Lake St Laporte IN 46350

JOHNSON, (CHARLES) BRUCE, photoelectronics research and devel. dir.; b. Sioux City, Iowa, Aug. 5, 1935; s. David Henry and Alice Katherine Nickolina (Anderson) J.; m. June Margaret Graham, Sept. 16, 1956; children: Kimberly Jill, Kirsten June. B.S., Iowa State U., 1957; M.S.E.E., U. Minn., 1963, Ph.D., 1967. Engr. RCA Electronic Components, Lancaster, Pa., 1968-70; sr. staff engr. Bendix Research Labs., Southfield, Mich., 1970-74; tech. dir. (ITT Electro-Optical Products div.), Ft. Wayne, Ind., 1974—; teaching asst., research fellow U. Minn., 1961-67; instr. Lawrence Inst. Tech., 1971-74. Contbr. articles to profl. jours. Served with USAF, 1953-58. Mem. IEEE, AAAS, Am. Phys. Soc., Optical Soc. Am., Soc. Photo-optical Instrumentation Engring. United Methodist. Club: Ft. Wayne Scandinavian. Subspecialties: Photoelectronics; Robotics. Current work: Assists the director of the ITT tube and sensor laboratories in technical planning and operations. Patentee image tube; device for converting an acoustic pattern into a visual image; coplanar dynode electron multiplier; scan-converted tube system; corona shield for an image-tubing photocathode; an oblique streak-tube; magnetically focused streak-tube. Home: 6521 Centerton Dr Fort Wayne IN 46815 Office: PO Box 3700 Fort Wayne IN 46801

JOHNSON, DALE ROBERT, mechanical engineer, consultant; b. Murdock, Nebr., Feb. 23, 1939; s. Percival Theodore and Lucy Jane (Miller) J.; m. Barbara Jean Palmer, Sept. 18, 1955; children: Lindy Jo, Lori Jean, Mark Dale. B.S. In Mech. Engring, U. Colo., 1956. Registered profl. engr., Colo., Ariz. Mech. engr. Los Alamos Nat. Lab., 1956-59; project engr. Martin Marietta Corp., Denver, 1959-73, Boulder, Colo., 1973-83; now owner, pres., chmn. bd., gen. mgr. Nelson and Johnson Engring., Boulder. Served to 1st lt. C.E. U.S. Army. Mem. Nat. Soc. Profl. Engrs. Republican. Presbyterian. Subspecialty: Aerospace engineering and technology. Patentee control system for motor vehicles.

JOHNSON, DAVID ALFRED, chemistry educator; b. Muskegon, Mich., Mar. 13, 1938; s. Alfred and Theresa (Johnston) J.; m. Janice Arlene Derscheid, Aug. 13, 1960; children: Jacquelynn, Elizabeth, Craig. A.B., Greenville (Ill.) Coll., 1960; postgrad., U. Kans., 1960-62; Ph.D., La. State U., 1966. Chemist Pet Milk Co., Greenville, Ill., 1960; grad. asst. U. Kans., Lawrence, 1960-62; asst. prof. chemistry Greenville Coll., 1962-64; instr. La. State U., Baton Rouge, 1964-66; vis. asst. prof., 1970-71; mem. faculty dept. chemistry Spring Arbor (Mich.) Coll., 1966—, prof., 1973—; cons. Council Ind. Colls., 1978—, Sparton Electronics, 1981—. Contbr. articles to profl. jours. Pres. bd. trustees Pineview Childrens Home, Evart, Mich., 1966—; bd. dirs. Free Methodist Ch., Spring Arbor, 1966—. NSF grantee, 1963, 70-71; NASA grantee, 1980-83. Mem. Am. Chem. Soc., Electrochem. Soc. Subspecialties: Physical chemistry; Inorganic chemistry. Current work: Electrochemistry of chromium solutions, thermodynamics of mixed solvent systems, five and six coordinate complexes, computer application in chemistry. Home: 128 Melody Ln Spring Arbor MI 49283 Office: Dept Chemistry Spring Arbor Coll Spring Arbor MI 49283

JOHNSON, DAVID ANDREW, biochemistry educator, researcher; b. Memphis, May 27, 1944; s. Carl E. and Evelyn (Petty) J.; m. Judith Alice Moth, Dec. 16, 1967; children: Colin Andrew Moth, Nicholas Edward Moth, Susannah Evelyn. B.S., Memphis State U., 1967, Ph.D., 1973. Research assoc. U. Ga., Athens, 1973-76, asst. biochemist, 1976-78; asst. prof. East Tenn. State U. Coll. Medicine, Johnson City, 1978—. Contbr. articles to sci. jours. NIH fellow, 1974. Mem. Am. Soc. Biol. Chemists, Am. Thoracic Soc. Subspecialty: Biochemistry (medicine). Office: Dept Biochestmiry Coll Medicine E Tenn State U Johnson City TN 37614 Home: Rt 11 Hickory Estates Jonesboro TN 37658

JOHNSON, DAVID EDWIN, physicist, high energy physics; b. Newark, Sept. 3, 1944; s. Lester Leonard and Alice Maude (Batchelder) J. A.B. in Physics, U. Calif.-Berkeley, 1966; Ph.D. in High Energy Physics, Iowa State U., 1972. Research and teaching asst. Ames Lab. and Iowa State U., Ames, 1966-72; postdoctoral fellow Ames Lab., 1972-73; physicist Tevatron I Project, Fermilab, Batavia, Ill., 1973—. Contbr. writings in field to profl. publs. Mem. Am. Phys. Soc., AAAS, Sigma Xi. Subspecialties: Particle physics; Accelerator Physics. Current work: Particle physics at very high energies. Design and construction of Tevatron I - intense proton-antiproton collisions at 1000 GeV x 1000GeV. Designer electron cooling ring and neutron cancer therapy beam, Fermilab. Office: Fermilab Tevatron I PO Box 500 Batavia IL 60510

JOHNSON, DAVID NORSEEN, neuropharmacologist; b. Bronx, N.Y., Sept. 28, 1938; s. Einar Victor and Eunice Marie (Norseen) J.; m. Carolyn, Oct. 8, 1960; children: Lauren Dale, Brian David. B.S. in Biology, North Park Coll., Chgo., 1960; M.S. in Psychopharmacology, U. Louisville, 1967; Ph.D. in Neuropharmacology, Med. Coll. Va., 1975. Research pharmacologist A.H. Robins Co. Inc., Richmond, Va., 1966-69, sr. research pharmacologist, 1970-75, mgr. neuropharmacology, 1976—. Contbr. articles on treatment of mental disease, sleep research, animal models for mental disease to sci. jours., also chpts. in books. Mem. Am. Soc. Pharmacology and Exptl. Therapeutics, Soc. for Neurosci., Am. Chem. Soc. (div. medicinal chemistry). Presbyterian. Subspecialties: Neuropharmacology; Psychopharmacology. Current work: Development of new agents for treatment of mental disease; development of animal models of mental illness; effects of drugs on sleep/waking patterns in laboratory animals. Home: 11406 Blendon Ln Richmond VA 23233 Office: 1211 Sherwood Ave Richmond VA 23233

JOHNSON, DONALD REX, sci. lab. mgr.; b. Tacoma, July 19, 1938; s. Richard Carl and Freida Maria (Dahlstrom) J.; m. Karen Yvonne Neswoog, July 24, 1959; children: Eric R., Brad A. B.S. in Physics, U. Puget Sound, 1960, M.S., U. Idaho, 1962; Ph.D. in Exptl. Physics, U. Okla., 1967. Physicist Nat. Bur. Standards, Washington, 1967—, sr. mgmt. trainee, 1976-78, dep. dir. for resources and ops., 1978—. Contbr. articles to tech. publs. Bd. dirs. West Riding Citizens Assn., 1971-74, pres., 1972-73; bd. dirs. Gaithersburg Area Civic Coalition, 1972-74; mem. Shady Grove Sector Plan Adv. Com., 1974-75; vice chmn. Gaithersburg Bicentennial Coordinating Com., 1975-76; sec. Gaithersburg City Planning Commn., 1977, chmn., 1978—. NSF grad. trainee in physics, 1965-67; recipient Silver medal U.S. Dept. Commerce, 1973, Gold medal, 1977; Arthur S. Flemming Award, 1976; Sr. Exec. Service meritorious exec., 1981. Fellow Am. Phys. Soc.; mem. ASTM, Am. Astron. Soc., AAAS, Internat. Astronomers Union, Sigma Xi. Subspecialties: Physical chemistry; Radio and microwave astronomy. Student stellar atmospheres; co-discoverer various new interstellar molecules.

JOHNSON, DOUGLAS WILLIAM, astronomer, physicist; b. Marion, Ind., Sept. 15, 1953; s. William Russell and Mary Ann (Scott) J.; m. Maryfran Peabody, June 28, 1975; 1 dau., Erin Alexandra. B.S. in Physics, Rensselaer Poly. Inst., 1975; Ph.D. in Radioastronomy, U. Fla., 1980. Instr. Undergrad. Physics Lab., U. Fla., Gainesville, 1975-77, instr. physics recitations, 1977-80; postdoctoral research fellow Battelle Meml Inst., Richland, Wash., 1980-82; research scientist Battelle N.W. Labs., 1982—. Author articles. Mem. Am. Astron. Soc., AAAS, Astron. Soc. Pacific, Am. Geophys. Union, N.Y. Acad. Sci., Sigma Xi. Subspecialties: Radio and microwave astronomy; Remote sensing (atmospheric science). Current work: Structure and evolution of nearby galaxies from radio observations. Upper atmospheric remote sensing at microwave and infrared wavelengths. Office: Battelle Obs PO Box 999 Richland WA 99352

JOHNSON, EDWARD MICHAEL, molecular biologist; b. Kenosha, Wis., Apr. 9, 1945; s. Edward and Mary Margaret (Pratch) J.; m. Elizabeth Buckingham Childs, June 14, 1969; 1 son, Nathaniel Livingston. B.A., Pomona Coll., 1967; Ph.D., Yale U., 1971. Postdoctoral fellow Rockefeller U., N.Y.C., 1971-73, asst. prof., 1975-81, assoc. prof., 1981—; research assoc. Meml. Sloan-Kettering Cancer Ctr., 1973-75, assoc. scientist, 1975-81; mem. genetics field Cornell U. Grad. Sch. Med. Scis., 1979—. Contbr. numerous articles to profl. jours. Jane Coffin Childs fellow, 1971-73; Leukemia Soc. Am. spl. fellow, 1974-76; Faculty Research award Am. Cancer Soc., 1982—. Mem. Am. Soc. Cell Biology; mem. Am. Soc. Biol. Chemists; Mem. N.Y. Acad. Scis., Am. Soc. Pharmacology and Exptl. Therapeutics. Subspecialties: Gene actions; Genome organization. Current work: Regulation of gene activity by chromosomal proteins; structure activity and chromosomal packaging of ribosomal genes, chromosome construction. Home: 500 E 63d St Apt 7A New York NY 10021 Office: Rockefeller U New York NY 10021

JOHNSON, ELMER MARSHALL, reproductive toxicologist and teratologist; b. Midlothian, Ill., June 16, 1930; s. Burt and Gertrude Esther (Miller) J.; m. Sharon Ann Coyle, May 9, 1976; children—Mark Dee, Kim Lea, Erik Marshall, Lora Marlys. Student, U. Mex., 1948; diploma, Thornton Jr. Coll., 1950; B.S. (teaching asst.), Tex. A. and M. U., 1954, M.S., 1955; Ph.D. (teaching asst.), U. Calif., Berkeley, 1959. Research asst. U.S. Army Surgeon Gen./Tex. A. and M. U. Research Found., College Station, 1955; instr. anatomy and physiology Contra Costa Coll., San Pablo, Calif., 1958-59; instr. U. Fla. Coll. Medicine, Gainesville, 1960-61, asst. prof., 1961-65, asso. prof., 1965-68, prof., 1968-70, acting chmn. dept. anatomy, 1969-70; prof., chmn. dept. anatomy, prof. dept. developmental and cellular biology U. Calif., Irvine, 1970-72; prof., chmn. dept. anatomy, dir. Daniel Baugh Inst., Jefferson Med. Coll., Thomas Jefferson U., Phila., 1972—; chmn. bd. Argus Research Labs., Inc., Perkasie, Pa.; cons. Allied Chem. Co., Dow Chem. Co., Johnson & Johnson, Argus Research Labs., Public Utilities, EPA, U.S. Naval Hosp., Phila., Kirkland & Ellis, Esq., Hoffman-LaRoche, Merck, Inc., McGraw-Hill, Inc., Sterling-Winthrop Research Inst., Interagy. Regulatory Liaison Group, Nat. Acad. Scis., NRC, Columbia Nitrogen Co. Asso. editor: Teratology, 1974-81, Jour. Environ. Pathology and Toxicology, 1979—, Fundamental and Applied Toxicology, 1981—. Served to 2d lt. U.S. Army, 1959. USPHS predoctoral fellow, 1953-55; March of Dimes Nat. Found. research grantee, 1962-63; NIH research grantee, 1963—; Growth Soc. research grantee, 1972-74; NIH teratology predoctoral tng. grantee, 1955-59. Mem. AAAS, Teratology Soc., Am. Assn. Anatomists, Assn. Anatomy Chairmen, Genetic Toxicology Assn., Soc. Toxicology, Am. Coll. Toxicology, So. Soc. Anatomists, Mid Atlantic Reproduction and Teratology Assn., Sigma Xi. Republican. Unitarian. Subspecialties: Teratology; Toxicology (medicine). Office: 1020 Locust St Philadelphia PA 19107

JOHNSON, ELWIN LEROY, ceramic engineer; b. Hillsboro, Ill., July 20, 1935; m., 1967; 3 children. B.S., U. Ill., 1956, M.S., 1957, Ph.D., 1960. From research asst. to research assoc. ceramic engring. U. Ill., 1956-59; mem. tech. staff materials sci. Central Research Lab., Tex. Instruments, Inc., Dallas, 1960-65, engring. sect. mgr. advanced circuits, semicondr. div., 1965-68, FCC mgr. circuits and packaging, 1968-75, program mgr. new products, central research labs., 1975—. Owens-Corning Fiberglass fellow, 1957-60. Mem. Am. Ceramic Soc., Nat. Inst. Ceramic Engrs., Electronic Industry Assn., ASTM. Subspecialties: Ceramic engineering; Solar energy. Office: Tex Instruments PO Box 225303 MS 158 Dallas TX 75265

JOHNSON, ERIC FOSTER, research scientist; b. Cedar Rapids, Iowa, May 3, 1946; s. Robert Charles and Gladys (Barnes) J.; m. Brenda Roland, Oct. 5, 1962; children: Erik Allen. B.S. in Chemistry with honors, U. Tex., Austin, 1967, Ph.D., U. Ill., Urbana, 1972. Research assoc. Scripps Clinic and Research Found., La Jolla, Calif., 1973-77, asst. mem. I, 1977-79, asst. mem. II, 1979-83, assoc. mem., 1983—. Mem.: editorial bd. Chemico-Biol. Interactions, Molecular Pharmacology; contbr. articles to profl. jours. NIH fellow, 1972; Calif Cancer Soc. fellow, 1975-77; NIH grantee, 1978—. Mem. Am. Soc. Biol. Chemists, Am. Chem. Soc., AAAS, Am. Soc. Pharmacology and Exptl. Therapeutics. Subspecialty: Biochemistry (biology). Current work: Structure, function and regulation of the cytochrome P-450 monooxygenases. Office: 10666 N Torrey Pines Rd La Jolla CA 92037

JOHNSON, ERIC VAN, biology educator; b. Medford, Mass., Mar. 11, 1943; s. Van Loran and Marjorie Jean (Carr) J.; m. Laurette Denise Fournier, Aug. 25, 1962 (div.); children: Kristina Marie, Eric Laurent. A.B., Brown U., 1964; Ph.D., Cornell U., 1969. Lectr. biol. scis. dept. Calif. Poly. State U., San Luis Obisop, 1969-70, asst. prof., 1970-74, assoc. prof., 1974-79, prof., 1979—; biol. cons., 1974—. Mem. Am. Ornithologists Union, Cooper Ornithol. Soc., Wilson Ornithol. Soc., Western Field Ornithologists, Western Bird-Banding Assn. Subspecialties: Population biology; Endangered species conservation and management. Current work: Research on California condor; co-developer photographic technique for censusing and individually identifying California condors by molt pattern and feather damage. Office: Dept Biol Sci Calif Poly State U San Luis Obispo CA 93407

JOHNSON, F. BRENT, microbiologist; b. Monroe, Utah, Mar. 31, 1942; s. Horace Jay and Ida (Christiansen) J.; m. Paula Dawn Forbush, June 18, 1965; children: Brian Kay, Matthew Glen, Christopher Jay, Wesley Terin, Stephanie Dawn. Student, Coll. So. Utah, 1960-61; B.S., Brigham Young U., 1966, M.S., 1967, Ph.D., 1970. NIH predoctoral fellow Brigham Young U., 1966-70, asst. prof. microbiology, 1972-75, assoc. prof., 1975-70, prof., 1980—; NIH postdoctoral fellow, Bethesda, Md., 1970-72, research assoc., 1977-78. Contbr. numerous articles on parvoviruses and herpesviruses to sci. jours. NIH grantee, 1973-76; USAF summer faculty fellow, 1977; grantee, 1978-79, 79-82. Mem. Am. Soc. Microbiology, N.Y. Acad. Scis., AAAS, Sigma Xi, Phi Kappa Phi. Mormon. Subspecialties: Microbiology; Virology (biology). Current work: The structure and biology of parvoviruses and herpes virus oncogenicity. Office: Brigham Young U 887 WIDB Provo UT 84602

JOHNSON, FRANCIS SEVERIN, physicist; b. Omak, Wash., July 20, 1918; s. Ralston Severin and Elizabeth (Gruenes) J.; m. Maurine Marie Green, Sept. 12, 1943; 1 dau., Sharan Kaye. B.Sc. with honors in Physics, U. Alta., Can., 1940; M.A. in Physics and Meteorology, UCLA, 1942; Ph.D. in Meteorology, UCLA, 1958. Head, high atmosphere research sect. U.S. Naval Research Lab., Washington, 1946-55; mgr. space physics research Lockheed Missiles & Space Co., 1955-62; head, atmospheric and space scis. div. S.W. Center Advanced Studies, Dallas, 1962-64, dir. earth and planetary scis. lab., 1964-69; acting pres. U. Tex. at Dallas, 1969-71; dir. Center for Advanced Studies, 1971-74, Cecil H. and Ida M. Green honors prof. natural sci., 1974—, exec. dean grad. studies and research, 1976-79; asst. dir. astron., atmosphere, earth and ocean scis. NSF, Washington, 1979-83; cons. ionospheric physics subcom., space scis. steering com. NASA, 1960-62, mem. planetary atmospheres subcom., space scis. steering com., 1962-67, chmn. lunar atmospheric measurements team. Apollo sci. planning teams, 1964-67; mem. adv. bd. Mars space missions, 1964-67, mem. lunar and planetary missions bd., 1967-71; mem. adv. panel atmospheric scis. NSF, 1962-67; mem. working group IV COSPAR, 1965-80, v.p., 1975-80; mem. Nat. Acad. Scis. panel adv. to central radio propagation lab. Nat. Bur. Standards, 1962-65; mem. panel weather and climate modification Nat. Acad. Scis., 1964-70; mem. adv. com. research to coordinating bd. Tex. Coll. and Univ. System, 1966-67; mem. space sci. bd. Nat. Acad. Scis., 1969-81, mem. geophysics research bd., 1971-77; mem. Nat. Acad. Scis com. advisory to NOAA, 1966-71; mem. sci. advisory bd. USAF, 1968-79; mem. nat. adv. com. Oceans and Atmosphere, 1971-73; mem. Climate Research Bd., Nat. Acad. Scis., 1977-79; pres. Spl. Com. on Solar Terrestrial Physics, 1974-77. Author: Satellite Environment Handbook, 1965; also numerous articles. Served with USAAF, 1942-46. Decorated Bronze Star medal; recipient Henryk Arctowski award Nat. Acad. Scis., 1972; Exceptional Sci. Achievement medal NASA, 1973; Meritorious Civilian Service award USAF, 1979. Fellow Am. Geophys. Union (vice chmn. sect. geomagnetism and aeronomy 1964-68, pres. sect. solar planetary relationships 1970-72, John Adam Fleming award 1977), AAAS (council mem. 1968-72), Am. Meteorol. Soc. (councilor 1976-78), IEEE; asso. fellow AIAA (chmn. tech. com. space and atmospheric physics 1961-64, Space Sci. award 1966); mem. Am. Phys. Soc., Am. Astron. Soc., Internat. Assn. Geomagnetism and Aeronomy (exec. com. 1967-71), Internat. Union Radio Sci. (chmn. U.S. Commn. IV 1964-67, sec. U.S. nat. com. 1967-70, vice chmn. 1970-73, chmn. 1973-76), Internat. Union Geodesy and Geophysics (U.S. nat. com. 1973-76), Sigma Xi. Subspecialties: Aeronomy; Satellite studies. Current work: Upper atmosphere structure, magneto-spheric physics. Office: U Tex at Dallas PO Box 688 Richardson TX 75080

JOHNSON, GEORGE ROBERT, educator; b. Caledonia, N.Y., Aug. 2, 1917; s. Arthur E. and Mary J. (Sinclair) J.; m. Beatrice E. Caton, Nov. 7, 1942; children—Diane K., Jane A. Eiden, Rosemary E. Johnson Kurek, Martha L. Brinkman. B.S., Cornell U., 1939; M.S., Mich. State U., 1947, Ph.D., 1954. Tchr. Corfu-East Pembroke Central Sch., Corfu, N.Y., 1939-42; asst. agrl. country agt., St. Lawrence (N.Y.) County, 1942-43; instr. animal husbandry Cornell U., 1943-47, asst. prof., 1947-48, asso. prof., 1948-55, Ohio State U., 1955-58, prof., chmn. dept. animal sci., 1958—. Mem. Am. Soc. Animal Sci., Sigma Xi, Alpha Zeta, Gamma Sigma Delta. Subspecialty: Animal Science Administration. Home: 251 Fairlawn Dr Columbus OH 43214

JOHNSON, GLEN ERIC, mech. engr., educator, cons., researcher; b. Rochester, N.Y., May 29, 1951; s. Ray Clifford and Helen Francis (Lindgren) J.; m. Kathryne Ann DeLoach, May 3, 1975; children: Edward Lindgren, Eric Anders. B.S.M.E., Worcester Poly. Inst., 1973; M.S.M.E. (Pres.'s fellow), Ga. Inst. Tech., 1974; Ph.D. (Harold Stirling Vanderbilt scholar), Vanderbilt U., 1978. Registered profl. engr., Va., Tenn. Mech. engr. Tenn. Eastman Co., Kingsport, 1974-76; asst. prof. mech. engring. Vanderbilt U., 1978-79, assoc. prof., 1981—; asst. prof. U. Va., 1979-81; cons. mech. design, noise control, optimization. Contbr. articles on optimization, mech. design, noise control to profl. jours. Mem. ASME (assoc. editor Jour. Mech. Design 1981—), Math. Programming Soc., Acoustical Soc. Am. Baptist. Subspecialties: Mechanical engineering; Acoustical engineering. Current work: Development and application of optimization techniques; mech. design synthesis, computer aided design, optimization, noise control. Home: 742 Templeton Dr Nashville TN 37205 Office: Vanderbilt U Box 8-B Nashville TN 37235

JOHNSON, HAROLD ARTHUR, manufacturing companies executive; b. Warren, Pa., May 17, 1924; s. Oscar William and Alvina Victoria (Nelson) J.; m. Alice Meredith Jones, June 15, 1955; children: Mark, Thomas. B.S. in Indsl. Engring, Pa. State U., 1950. With Pa. Furnace & Iron Co., Warren, 1941-72, sales and engring. mgr., 1961-63, sec.-treas., 1963-68, v.p., 1968-72; also dir.; exec. v.p. Allegheny Valve Co., Allegheny Coupling Co., Warren, 1972-82, pres., treas., 1982—; also dir.; exec. v.p. Rand Machine Products, Inc., Jamestown, N.Y., 1972—; also dir.; pres., treas., dir. DeFrees Family Found.; instr. short courses Ill. Inst. Tech., 1955, Mich. State U., 1956; past cons. in field; mem. U.S. del. Conferences Conseil International Pour l'Organization Scientifique, Munich, W.Ger., 1972, Caracas, Venezuela, 1975. Contbr. articles to profl. jours. Active Warren County (Pa.) Sch. Bldg. Authorities, 1961—, chmn., 1980—; chmn. Warren County Hosp. Authority, 1971—; mem. exec. bd., chief cornplanter Boy Scouts Am., 1972—, v.p., 1983—; sr. warden Trinity Meml. Ch., Warren, 1963-64, vestryman, 1962-64, treas., 1972—; trustee Erie Diocese, Episcopal Ch., 1975-81. Served with U.S. Army, 1943-46. Decorated Bronze Star. Mem. ASME, Truck Trailer Mfrs. Assn. (chmn. tank conf. engring. com. 1961-62), Tau Beta Pi, Phi Kappa Phi, Phi Eta Sigma, Phi Sigma Kappa. Republican. Clubs: Conewango Valley Country, Conewango, Grotto, Am. Legion (past officer). Lodges: Rotary; Masons; Shriners. Subspecialties: Industrial engineering; Mechanical engineering. Current work: Valves and equipment for liquid transportation tanks. Patentee in field; designed fueling trailers for Project Van Guard, 1st U.S. Space Satellite; early work in transp. of liquified gases at low temperatures. Home: 103 Memorial Pl Rd Warren PA 16365 Office: 419 3d Ave Warren PA 16365

JOHNSON, HERBERT HARRISON, materials science educator, researcher; b. Cleve., July 16, 1931; m., 1955; 4 children. B.S., Case Inst. Tech., 1952, M.S., 1954; Ph.D., 1957. Asst. prof. metallurgy Lehigh U., 1957-60; assoc. prof. materials sci. and engring. Cornell U., Ithaca, N.Y., 1950-67, prof., 1967—, dir. dept., 1970-74; dir. Materials Sci. Ctr., 1974—. Mem. AIME, Am. Soc. Metals, Am. Phys. Soc., AAAS. Subspecialty: Materials science research administration. Office: Bard Hall Cornell U Ithaca NY 14850

JOHNSON, HOLLIS RALPH, astronomer; b. Tremonton, Utah, Dec. 2, 1928; s. Ellwood Lewis and Ida Martha (Hansen) J.; m. Grete Margit Leed, June 3, 1954; children: Carol Ann Johnson Watson, Wayne L., Lyle David, CharlotteJohnson Willian, Lise Marie, Richard L. B.A. in Physics, Brigham Young U., 1955, M.A., 1957; Ph.D. in Astrophysics, U. Colo., 1960. NSF postdoctoral fellow Paris Obs., 1960-61; research asso. Yale U., 1961-63; asso. prof. astronomy Ind. U., 1963-67, prof., 1967—, chmn. dept. astronomy, 1978-82; Nat. Acad. Scis./NRC sr. fellow NASA Ames Research Ctr., 1982-83. Contbr. articles to profl. jours. Served with U.S. Army, 1951-53. Recipient Vis. Scientist award High Altitude Obs., Boulder, Colo., 1971-72. Mem. Internat. Astron. Union, Am. Astron. Soc., AAAS, AAUP, Sigma Xi. Mormon. Subspecialties: Ultraviolet high energy astrophysics; Theoretical astrophysics. Current work: Ultraviolet spectra of red giant stars; calculation of model atmospheres for cool stars; calculation of line opacities. Office: Astronomy Dept Swain W 319 Ind U Bloomington IN 47405

JOHNSON, HOWARD A(RTHUR), Sr., research executive, operations research consultant; b. Shelby County, Ind., Dec. 16, 1923; s. Arthur and Inez Elizabeth (Smiley) J.; m. Joy Ann Nelson, July 19, 1947; children: Howard Arthur, Kraig N. Student engring., Rose Poly. Inst., 1942; A.B. in Math, Franklin Coll. of Ind., 1949; M.A. in Physics, Wesleyan U., 1950. Chief ops. analysis 3d Air Force, Eng., 1958; dep. chief ops. analysis U.S. Air Force Europe, Germany, 1958-61; dir. operation model evaluation Project Omega, U.S. Air Force, Washington, 1961-63; mgr. comparative effect research dept. Spindletop Research, Inc., Lexington, Ky., 1963-68; sci. asst. Armament Devel. and Test Ctr., Electronic Test, Eglin AFB, Fla., 1968-73; ops. research consl. Tactical Air Warfare Ctr., U.S. Air Force, Eglin AFB, Fla., 1973—; cons. Gulf South Research Inst., Baton Rouge, 1968—, U. Ky. Med. Ctr., 1966-67, Ministry of Def., W.Ger., 1960-61, Weapon Systems Phasing Group, Supreme Allied Powers Europe Hdqrs., 1959-61. Advisor Okaloosa County Planning and Zoning, 1979-80; mem. Okaloosa-Walton Jr. Coll. Sports Assn., 1967-77. Served to capt. USAAF, 1943-45; PTO. Alt. scholar Rose Poly. Inst., 1942; fellow Wesleyan U., 1948-50; recipient numerous govt. incentive awards, 1954-73. Mem. Ops. Research Soc. Am., Mil. Ops. Research Soc., Internat. Test and Evaluation Assn., Washington Ops. Research Council, Armed Forces Communications and Electronics Assn. Presbyterian. Lodges: Masons; Shriners. Subspecialties: Operations research (mathematics); Resource management. Current work: C^3I (command control communications intelligence) training effectiveness and C^3I effectiveness evaluations to include weapon systems effectiveness trade-offs with C^3I. Home: 309 Yacht Club Dr Fort Walton Beach FL 32548 Office: USAFTAWC Eglin Air Force Base FL 32542

JOHNSON, HOWARD ERNEST, toxicologist, biology educator; b. Livingston, Mont., Sept. 21, 1935; m., 1959. B.S., Mont. State U., 1959, M.S., 1961; Ph.D. in Fisheries, U. Wash., 1967. Asst. prof. to assoc. prof. Mich. State U., East Lansing, 1967-75, prof. fisheries, 1975—, coordinator environ. contamination, 1978—; mem. panel com. water quality criteria Nat. Acad. Sci., 1971-72; coordinator toxic substances Mich. Service and Edn. Adminstrn. Grant Program, 1978—. Mem. Am. Fisheries Soc., Am. Inst. Fisheries Research Biologists. Subspecialties: Environmental toxicology; Ecology. Office: Pesticide Research Center Mich State Univ East Lansing MI 48823

JOHNSON, JACK DONALD, university program administrator, college dean; b. Huntington, Oreg., Aug. 23, 1931; s. Emanuel and Josephine (Coats) J.; m. Shirley Annette Offord, June 27, 1958; children: Paula, Vincent, Erik, Gregory. B.A., San Diego State U., 1959; M.S., U. Minn., 1967, Ph.D., 1970. Engr., draftsman Humphrey Inc., 1956-58, project engr., 1958-60; engr. Martin Marietta Corp., 1960-61, sect. chief, 1961-63; engr. Jet Propulsion Lab., Pasadena, Calif., 1963-66; research fellow U. Minn.-Mpls. Sch. Pub. Health, 1966-70; dir. Office Arid Lands Studies, U. Ariz., 1971—, assoc. prof., 1971-78, prof., 1979—, dir., 1974-79, asst. coordinator interdisciplinary programs, 1975-78, assoc. dean for interdisciplinary programs, 1981—; lectr. UN Grad. Sch., Geneva, 1976; spl. asst. to coordinator UN Conf. on Desertification, Dept. State, 1978; mem. energy com. Gov.'s Commn. on Ariz. Environment, 1972—, chmn., 1979-82; mem. UNESCO Man and Biosphere, 1975—, U.S.-Mex. Joint Com. on Desertification, 1976-79; U.S. del. UN Conf. on Desertification, 1977. Contbr. articles to profl. jours. Mem. Am. Water Resources Assn., AAAS, Ariz. Acad. Sci., Assn. Energy Engrs., Council Agrl. Sci. and Tech., Am. Soc. Agrl. Engrs., Sigma Xi. Subspecialties: Arid lands agriculture; Biomass (energy science and technology). Home: 4536 N Rocky Creek Circle Tucson AZ 85715 Office: 845 N Park Ave Tucson AZ 85719

JOHNSON, JAMES ALLEN, archtl. and engring. co. exec.; b. Stoughton, Wis., June 2, 1924; s. Martin Helberg and Sophia Amunda (Christensen) J.; m. Kathleen Clare Smith, Nov. 26, 1949; children—Pamela Marie, James Allen, Mark Thomas, Stephen Victor. Student, U. Wis., 1942-43; B.S., U.S. Mil. Acad., 1944-47; M.S., Stanford U., 1957. Registered profl. engr., Vt. Commd. 2d lt. U.S. Army, 1947, advanced through grades to maj. gen., 1973; comdg. gen. U.S. Army Engr. Center, Ft. Belvoir, Va., 1974-77; div. engr. N. Atlantic (U.S. Army C.E.), N.Y.C., 1977-79, dep. chief engrs., Washington, 1979-80, ret., 1980; sr. v.p. Washington ops. Kuljian Corp., 1980—; dir. Eastern Indemnity Corp., Rockville, Md.; Chmn. Engring. Bd. for Rivers and Harbors, Washington, 1979-80; chmn. research devel. and rev. bd. C.E., 1979-80. Bd. dirs. Fairfax (Va.) chpt. Salvation Army, 1977. Decorated D.S.M. (2), Silver Star, Legion of Merit (2), Bronze Star (2), Purple Heart (2). Fellow Soc. Am. Mil. Engrs. (pres. Phila. post 1968-70, pres. N.Y.C. post 1977-78); mem. Am. Assn. Cost Engrs., Washington Soc. Profl. Engrs. Republican. Methodist. Clubs: Univ. (Washington); Army-Navy Country (Arlington, Va.); Sons of Norway. Subspecialties: Civil engineering; Water supply and wastewater treatment. Current work: Water supply to include rehabilitation of present systems; port development; flood control and ocean science. Home: 11000 Henderson Rd Fairfax Station VA 22039 Office: Kuljian Corp 1435 G St NW Suite 1010 Washington DC 20005

JOHNSON, JAMES HARMON, psychology educator, researcher; b. Martin, Tenn., Mar. 30, 1943; s. Harold and Sarah (Ossmer) Remus; m. Theda Fay Cruse, Dec. 22, 1967; children: Jamie Lee, Trey Allen. B.S., Murray State U., 1966, M.S., 1968; Ph.D., No. Ill. U., 1976. Lic. psychologist, Fla. Psychology intern U. Tex. Med. Ctr., 1971-72, instr., 1972; acting asst. prof. psychology U. Wash., Seattle, 1975-76, asst. prof., 1976-79; assoc. prof. psychology U. Fla., Gainesville, 1979—. Co-author: Psychopathology of Childhood, 1981; assoc. editor: Jou. Clin. Child Psychology, 1982—; mem. editorial bd., 1977-82; editorial cons. profl. jours.; contbr. articles to profl. jours. U. Wash. grantee, 1977-78; Children's Village U.S.A. grantee, 1982-83. Mem. Am. Psychol. Assn., Southeastern Psychol. Assn., Western Psychol. Assn. Presbyterian. Subspecialties: Clinical psychology; Developmental psychology. Current work: Studies of the effects of stress on health and adjustment; temperament and child psychopathology; multivariate approaches to classification. Home: 8522 NW 4th Pl Gainesville FL 32610 Office: U Fla Dept Clin Psychology Box J-165 Gainesville FL 32610

JOHNSON, JAMES ROBERT, ceramic engineer, educator; b. Cin., Jan. 2, 1923; s. Charles William and Della Ramona (Schubert) J.; m. Virginia M. Bowen, Apr. 3, 1945; children—Cathy (Mrs. John Whitman), Barbara (Mrs. Charles Kallusky), Randy, John, Jamie (Mrs. J.R. Myers), Brian. B.S., Ohio State U., 1947, M.S., 1948, Ph.D., 1950. Asst. prof. U. Tex., 1950-51; tech. adviser ceramics Oak Ridge Nat. Lab., 1951-56; lab. mgr., dir., exec. scientist Minn. Mining & Mfg. Co., St. Paul, 1956-79, cons., 1979—; adj. prof. U. Wis.-Stout; U. Minn., 1979—. Contbr. articles to profl. jours. Served with C.E. AUS, 1943-46. Recipient Distinguished Alumnus award Ohio State U., 1970, 3M Carlton award, 1970. Fellow Am. Ceramic Soc. (pres. 1973-74, disting. life mem.); mem. Nat. Acad. Engring., Nat. Inst. Ceramics Engrs. (Pace award 1959), Research Engring. Soc. Am. Subspecialties: Ceramic engineering; Materials (engineering). Current work: Consultant in materials; physical sciences. Patentee in field. Home: Route 1 Box 231B River Falls WI 54022

JOHNSON, JAMES RUSSELL, research physicist; b. Granite Falls, Minn., May 17, 1943; s. Carl Russell and Ida Caroline (Trygstad) J.; m. Kathryn Ann Krider, Dec. 19, 1965; children: Erik Sven, Elsa Linnea. B.A., St. Olaf Coll., Northfield, Minn., 1965; M.Sc., Stanford U., 1967, Ph.D., 1970. Research assoc. Stanford (Calif.) Linear Accelerator Ctr., 1970-71; research assoc. Fermi Nat. Accelerator Lab., Batavia, Ill., 1971-74; asst. prof. Northeastern U., Boston, 1974-76; assoc. scientist high energy physics U. Wis., Madison, 1976—. Contbr. articles to profl. jours. Subspecialty: Particle physics. Current work: Experimental particle physics using accelerators and storage rings. Office: Bin 94 SLAC PO Box 4349 Stanford CA 94305

JOHNSON, JOHN ANTHONY, psychology educator; b. Ann Arbor, Mich., Nov. 11, 1953; s. John Charles and Lorraine Irene (Ura) J.; m. Carolyn Jean Hickey, July 9, 1977; 1 son, Martin Dean. B.S. in Psychology, Pa. State U., 1976; M.A., Johns Hopkins U., 1979, Ph.D., 1981; postgrad., SUNY-Brockport, 1976-77. Research analyst Md. State Dept. Edn., Balt., 1980-81; instr. Johns Hopkins Evening Coll., Balt., 1981, Towson State U., Md., 1980; asst. prof. psychology Pa. State U., DuBois, Pa., 1981—; research cons. Personnel Assessment & Selection Service, Balt., 1979-81, Johns Hopkins Evening Coll., 1979-81, Center for Met. Planning and Research, Balt., 1980. Fund for Acad. Excellence grantee DuBois Ednl. Found., 1982—. Mem. Am. Psychol. Assn., Eastern Psychol. Assn., Soc. for Personality and Social Psychology, Pa. State U. Alumni Assn., Phi Kappa Phi. Current work: Sociobiological approach to personality; vocational choice and job performance; factors affecting validity of personality self-reports. Home: 528 1st St DuBois PA 15801 Office: Pa State U College Pl DuBois PA 15801

JOHNSON, JOHN LOWELL, plasma physicist; b. Butte, Mont., Mar. 18, 1926; Lowell Wallace and and Esther (Thornwall) J.; m. Barbara Marion Hynds, June 30, 1951; children: Lowell John, Lesley Jean, Jennifer Ruth. B.S., Mont. State U., 1949; M.S., Yale U., 1950, Ph.D., 1954. Sr. scientist Westinghouse Comml. Atomic Power, Pitts., 1965-64; vis. research staff Princeton U. Plasma Physics Lab., 1955-68, vis. prin. research physicist, 1968—; fellow engr. Westinghouse Research and Devel. Ctr., Pitts., 1964-68, adv. scientist, 1968-78, consulting scientist, 1978—. Scoutmaster George Washington council Boy Scouts Am., Princeton, 1971—. Served with USN, 1944-46. Fellow Am. Phys. Soc. (sec. plasma physics div. 1974-78); mem. Am. Nuclear Soc., AAAS, Sigma Xi (treas. Princeton chpt. 1981-83, sec. 1983—). Methodist. Subspecialties: Plasma physics; Fusion. Current work: Theory of plasma confinement in toroidal magnetic systems with emphasis on analytic and computational programs to determine magnetohydrodynamic properties concerning equilibrium, stability and transport in stellarators, tokamks and RFP's. Office: Princeton University Plasma Physics Lab PO Box 451 Princeton NJ 08544 Home: 540 Ewing St Princeton NJ 08540

JOHNSON, JOHN MORRIS, biologist, educator; b. Boise, Idaho, Mar. 16, 1937; s. Carl T. and Fannie Margaret (King) J.; m. Margaret May Horton, June 13, 1959; children: Mori Kay, Stephen Wade. B.S., Coll. Idaho, 1959; M.S., Oreg. State U., 1961, Ph.D., 1964. Postdoctoral fellow U. Chgo., 1965-66; asst. prof. biology Central Coll., Pella, Iowa, 1964-65, assoc. prof., 1966-69; prof. biology Oreg. Coll. Edn. (named changed to Western Oreg. State Coll.), Monmouth, 1969—. Author: Handbook of Uncommon Plants on the Salem BLM District, 1980; contbr. articles to profl. jours. Recipient Faculty Honors award Western Oreg. State Coll., 1979; Silver Beaver award Boy Scouts Am., 1981. Mem. AAAS, Am. Soc. Cell Biology, Bot. Soc. Am., Sigma Xi, Phi Kappa Phi. Democrat. Methodist. Subspecialties: Plant cell and tissue culture; Cell and tissue culture. Current work: Plant tissue culture, somatic cell genetics, nucleolus and nucleolar vacuoles.

JOHNSON, JOSEPH ALAN, research physiologist, educator; b. West Palm Beach, Fla., Feb. 1, 1933; s. Eli Allen Johnson and Emily Kate (Percy) Wilcher; m. Janice L. Van de Water, July 21, 1956; children: Robert Alan, Gary Francis. B.A., Butler U., 1963; Ph.D., Ind. U. Med. Ctr., 1968. Postdoctoral fellow in physiology U. Mo., Columbia, 1968-71, asst. prof. physiology, 1971-78, assoc. prof. physiology, 1978—; research physiologist H.S. Truman Meml. VA Hosp., Columbia, 1974—. Editor: Comparative Endocrinology of Prolactin, 1977, The Renin-Angiotensin System, 1980; contbr. articles to sci. jours. NIH predoctoral fellow, 1965; postdoctoral fellow, 1969; Research Career Scientist VA, 1979. Fellow Council for High Blood Pressure Research, Am. Heart Assn.; mem. Am. Physiol. Soc., Endocrine Soc., Am. Soc. Nephrology. Subspecialties: Physiology (medicine); Endocrinology. Current work: Research in mechanisms of hypertension by use of animal models. Home: 3100 Timberhill Trail Columbia MO 65201 Office: Research Service H S Truman Meml VA Hosp 800 Stadium Rd Columbia MO 65201

JOHNSON, KARL OTTO, JR., chemical engineer, consultant; b. Bklyn., Jan. 8, 1920; s. Karl O. and Olga (Jensen) J.; m. Opal Blair, Dec. 20, 1944; children: Richard Karl, Mark Edward. B.Ch.E., U. Fla., 1941, M.S. in Engring, 1943. Registered profl. engr., Tenn. Chemist Pan Am. Airways, Miami, Fla., 1942-43; chem. engr. Tenn. Eastman Corp., Oak Ridge, 1943-47; engr./writer/info. engr. Oak Ridge Nat. Lab., 1947-73; sr. engr. Gaseous Diffusion Plant, Oak Ridge, 1973-80; info. engr. Oak Ridge Nat. Lab., 1980-82; info. cons., Oak Ridge, 1982—. Mem. Am. Inst. Chem. Engrs., Am. Chem. Soc., Am. Nuclear Soc., Sigma Xi, Tau Beta Pi. Subspecialties: Information systems (information science); Chemical engineering. Current work: New methods for storage, retrieval and communication of engineering information. Patentee on preparation of VF-3-, 1951. Address: 110 Pomona Rd Oak Ridge TN 37830

JOHNSON, KENNETH ALLEN, biochemistry educator; b. Davenport, Iowa, Mar. 10, 1949; s. Wayne Joseph and Donna (Lee) J.; m. Linda Joan Illian, Aug. 15, 1970; children: Elizabeth, Amanda. B.S., U. Iowa, 1971; Ph.D., U. Wis., 1975. Postdoctoral fellow U. Chg., 1975-79; asst. prof. Pa. State U., University Park, 1979—. NIH grantee, 1979, 83; Am. Heart Assn. investigator, 1983. Mem. Biophys. Soc., Am. Soc. for Cell Biology, AAAS. Subspecialties: Biochemistry (biology); Cell biology. Current work: Mechanisms of microtubule-dependent motility and related force transducing systems; structural and kinetic analysis of the ciliary dynein ATPase. Office: Pa State U 301 Althouse Lab University Park PA 16802

JOHNSON, KENNETH HARVEY, veterinary pathologist; b. Hallock, Minn., Feb. 17, 1936; s. Clifford H. and Alma (Anderson) J.; m., Sept. 17, 1960; children—Jeffrey, Gregory, Sandra. B.S., U. Minn., 1958, D.V.M., 1960, Ph.D., 1965. Jr. asst. health officer NIH, Bethesda, Md., 1958; practice vet. medicine, Edina, Minn., 1960; USPHS-NIH non-service fellow U. Minn., St. Paul, 1960-65, asst. prof. dept. vet. pathology and parasitology, 1965-69, asso. prof., 1969-73, prof. dept. vet. biology, 1973-76, head, sect. pathology, dept. vet. biology, 1974-76; chmn. dept. vet. pathobiology Coll. Vet. Medicine, 1976-83; cons. Minn. Mining & Mfg. Co., Medtronic Inc.; Co-investigator several NIH grants, 1965—. Contbr.: chpts. to Veterinary Clinics of North America, 1971, Spontaneous Animal Models of Human Disease, 1979; articles to sci. jours. Councilman Nativity Lutheran Ch., St. Anthony Village, Minn., 1972-75. Recipient Tchr.-of-Year award, 1968-69, Norden award for disting. tchr. in vet. medicine, 1970. Mem. AAUP, Electron Microscopy Soc. Am., Am. Assn. Feline Practitioners, Minn. Vet. Med. Assn., Minn. Electron Microscopy Soc., Sigma Xi, Phi Zeta, Gamma Sigma Delta. Subspecialty: Animal pathology. Current work: Amyloidosis; foreign body tumorigenesis. Home: 3510 Skycroft Dr Minneapolis MN 55418 Office: Dept Vet Pathobiology Coll Vet Medicine U Minn Saint Paul MN 55108

JOHNSON, LEONARD ROY, physiology educator; b. Chgo., Jan. 31, 1942; s. Leonard W. and Pearl (Anderson) J.; m. Sally Seiders; children: Melinda, Ashley, Matthew. A.B., Wabash Coll., 1963; Ph.D., U. Mich., 1967. Instr. UCLA Med. Sch., 1967-69; asst. prof. U. Okla. Med. Sch., Oklahoma City, 1969-71, assoc. prof., 1971-72; prof. dept. physiology U. Tex. Med. Sch., Houston, 1972—; mem. Nat. Bd. Med. Examiners, 1983—. Editor: textbooks Gastrointestinal Physiology, 1st edit, 1977, 2d edit. 1981, Physiology of Gastrointestinal Tract, 1981; editor: Am. Jour. Physiology: Gastrointestinal and Liver Physiology, 1979—. NIH research career devel. awardee, 1972-77; research grantee, 1970—, 75—. Mem. Am. Physiol. Soc. (Hoffman-LaRoche Prize 1979), Am. Gastroent. Assn., Endocrine Soc. Subspecialties: Physiology (medicine); Receptors. Current work: Regulation of growth of gastrointestinal mucosa especially trophic effects of gastrointestinal hormones and their receptor interactions. Home: 5631 Hummingbird Houston TX 77096 Office: Dept Physiolog U Tex Med Sch Houston 6431 Fannin Houston TX 77030

JOHNSON, LEWIS BENJAMIN, JR., research scientist, educator; b. Charlottesville, Va., May 23, 1921; s. Lewis Benjamin and Sarah (Grayson) J.; m. Alice Lewis Tucker, Dec. 27, 1943; children: Lewis Benjamin III, Walter Tucker, Mary Amie. B.S., U. Va., Charlottesville, 1943, Ph.D., 1951; M.S., Calif. Inst. Tech., Pasadena, 1944. Research scientist U. Va., 1951-56, sr. scientist, 1956-66, research prof., 1966—; cons. Nat. Inst. Dental Research, Bethesda, Md., 1970—. Served with USN, 1943-46. Mem. Internat. Assn. for Dental Research, Am. Assn. for Dental Research, Va. Acad. Sci. Democrat. Methodist. Subspecialties: Dental materials; Corrosion. Current work: Improvements in dental amalgam alloys; high current, long wear electrical brushes. Patentee gold-containing amalgam, plated amalgams. Home: Mooreland Route 1 Charlottesville VA 22901 Office: U Virginia Thornton Hall Charlottesville VA 22901

JOHNSON, MELVIN LAWRENCE, utility company nuclear plant executive; b. Roff, Okla., Feb. 18, 1931; s. William Clay and Ruby Ellen (Wilder) J.; m. Susan Crawford Whitt, Oct. 14, 1955; children: Laura, Robert, Elizabeth, Jennifer, Amy, Sarah. B.S. in Mech. Engring, Okla. State U., 1954; M.S., U. Idaho, 1962. Registered profl. engr., Calif. With Union Electric Co., St. Louis, 1954; engr./mgr. Gen. Electric Co., San Jose, Calif., 1956-73, Idaho Falls, Idaho, 1956-73, Fayetteville, Ark., 1956-73; mgr. nuclear plant engring. Kans. Gas & Electric Co., Wichita, Kans., 1973—. Contbr. reports, papers in field to profl. sci. lit. Clk. of session Presbyn. Ch., San Jose, Calif., 1965-67; ministry, counsel Friends Ch., Wichita, Kans., 1974-80. Served to 1st lt. USAF, 1954-56. Recipient several awards Gen. Electric Co., 1967-72. Mem. Nat. Soc. Profl. Engrs., Kans. Engring. Soc., Wichita Profl. Engring. Soc., Am. Nuclear Soc., ASTM. Republican. Quaker. Clubs: Lake Waltanna Flyers (treas. 1978-82), Lake Waltanna Stables (pres. 1977-79), Lake Waltanna Stables (treas. 1982—). Subspecialty:

Nuclear engineering. Current work: Design, procurement and construction of pressurized water reactor for electric power production. Patentee fast reactor core hardware. Home: 14 Lakeridge Dr Goddard KS 67052 Office: Nuclear Dept Kans Gas & Electric Co 201 N Market St Wichita KS 67201

JOHNSON, OLIN GLYNN, computer science educator; b. Waxahachie, Tex., Dec. 21, 1935; s. Olin and Ruth (Sutton) J.; m. Ferol Marie Gibson, Jan. 20, 1961; 1 son, Olin Wesley. B.S., So. Met. U., 1957, M.S., 1960; M.A., U. Calif.-Berkeley, 1966, Ph.D., 1968. System engr. IBM, Dallas, 1960-64, grad. fellow, Berkeley, 1964-68, research scientist, Houston, 1968-70; exec. v.p. Internat. Math. Statis. Libraries, Houston, 1970-73; assoc. prof. computer sci. U. Houston, 1973-79, prof., 1979—; cons. in field; Disting. Faculty visitor IBM Research, 1979; participating guest faculty Lawrence Livermore Labs., 1981—. Author: Computer Organization and Programming-the VAX—11, 1983; research publs. in field. Mem. Assn. Computing Machinery (council 1972-75, gen. chmn. ann. conf. 1976). Democrat. Methodist. Subspecialties: Vector processing/processors; Numerical analysis. Current work: Vector algorithms, processors, languages. Home: 10914 Willowisp Houston TX 77035 Office: U Houston Dept Computer Sci Houston TX 77004

JOHNSON, PETER CHARLES, physician, neuropathologist, educator; b. Glendale, Calif.; s. Alfred le Roy and Eileen J.; m. MaryAnn MacKenzie, Dec. 20, 1966. B.S., Loyola U., 1963; M.A., U. Calif.-San Francisco, 1967; M.D., 1967. Diplomate: Am. Bd. Pathology. Intern San Francisco Gen. Hosp., 1967-68; resident in pathology/neuropathology U. Calif., San Francisco, 1968-71, instr., 1973-74; fellow in neuropathology Mass. Gen. Hosp., Boston, 1971-72; staff pathologist Letterman Army Med. Ctr., San Francisco, 1972-74; neuropathologist U. Ariz., Tucson, 1974—, assoc. prof., 1974—. Author: Pathology of Peripheral Nerve, 1978. Served to maj. AUS, 1972-74. NIH prin. investigator, 1979. Mem. Am. Assn. Pathologists, Am. Assn. Neuropathologists. Subspecialty: Pathology (medicine). Current work: morphology of peripheral nerve diseases; diabetic neuropathy; Alzheimer's disease. Office: Dept. Pathology U Ariz Tucson AZ 85724

JOHNSON, PETER DEXTER, phys. chemist, patent agt.; b. Norwich, Conn., July 1, 1921; s. Philip Adams and Edith Todd (Dexter) J.; m. Jessie Lois Jones, Oct. 3, 1943 (div. 1978); children: Peter Dexter, Carol, William Todd; m. Mary Grace Wood, Jan. 2, 1982. S.B., Harvard U., 1942; M.A., U. N.C., 1948, Ph.D., 1949. Registered patent agt. U.S. Patent and Trademark Office, 1981. Supr. ballistic testing Hercules Powder Co., Radford, Va., 1942-43; phys. chemist Gen. Electric Research and Devel. Center, Schenectady, 1949—; vis. assoc. prof. physics Cornell U., 1958-59. Contbr. articles to profl. jours. Mem. Niskayuna (N.Y.) Planning and Zoning Commn., 1958-67; mem. Niskayuna Bd. Assessment Rev., 1970-75, chmn., 1972-75. Served with USAAF, 1943-46. Fellow Am. Phys. Soc.; mem. Am. Chem. Soc., Optical Soc. Am., Am. Inst. Chemists, AAAS, Eastern N.Y. Patent Law Assn., Sigma Xi. Subspecialties: Atomic and molecular physics; Optical engineering. Current work: Optical properties of solids, gases, plasmas; innovations in light and ultraviolet radiation sources. Home: 1100 Merlin Dr Schenectady NY 12309 Office: Gen Electric Research and Devel Center PO Box 8 Schenectady NY 12301

JOHNSON, PHILIP LEWIS, research and ednl. assn. exec.; b. Oneonta, N.Y., May 26, 1931; s. Robert A. and Hazel S. (Shaffer) J.; m. Judy Rodgers, Nov. 17, 1973. B.S. in Agr, Purdue U., 1953, M.S. in Natural Resources, 1955; Ph.D. in Ecology, Duke U., 1961. Agrl. economist fruit and vegetable div., sect. program analysis Dept. Agr., 1955; instr. botany U. Wyo., Laramie, 1959-61; botanist U.S. Forest Service, Laramie, 1961-62; ecologist U.S. Cold Regions Research and Engring. Lab., Hanover, N.H., 1962-67; asst. prof. biology Dartmouth Coll., 1963-67; asso. prof. botany and forestry U. Ga., 1967-70; exec. dir. Environ. Center, 1970; program dir. ecosystem analysis program NSF, 1968-69; dep. head Office Interdisciplinary Research, 1970-71, dir. div. environ. systems and resources, 1971-74; exec. dir. Oak Ridge Asso. Univs., 1974-81, John E. Gray Inst., Lamar U., Beaumont, Tex., 1981—; also pres. John E. Gray Found.; research collaborator Brookhaven Nat. Lab., 1963-65; mem. N.H. Pesticide Control Bd., 1965-67; mem. primary productivity com. Internat. Biol. Program, 1967-68, adv. com. tundra biome, 1968-70, deciduous forest biome coordinating com., 1968, 70; mem. environ. panel fgn. currency program Smithsonian Instn., 1969-70; adv. council Public Broadcast Environ. Center, Washington, 1970; vice chmn. interagy. com. ecol. research Fed. Council Sci. and Tech./Council Environ. Quality, 1972; mem. U.S. com. Man and Biosphere Program, 1973-74; exec. com. E. Tenn. Cancer Research Center, Knoxville, 1975-77; regional com. Southeastern Plant Environ. Lab., 1975-80; fellowship adv. panel environ. affairs Rockefeller Found., 1974-78; chmn. com. on environ. research and devel. Nat. Acad. Scis., 1978-79, mem. polar research bd., 1981—. Editorial bd.: Ecol. Monographs, 1968-70, Jour. Remote Sensing of Environ, 1971-75. Trustee Inst. of Ecology, 1976-79, chmn. bd., 1980-81; bd. dirs. Center for Natural Areas, 1979-81; mem. U.S. Commn. for UNESCO, 1978-80. Served with AUS, 1955-57. Recipient Commendation award Cold Regions Research and Engring. Lab., 1964, 66, Meritorious Sch. Achievement award, 1966; Meritorious Service award NSF, 1973; James B. Duke fellow 1957-59. Fellow Arctic Inst. N. Am.; mem. Ecol. Soc. Am., Brit. Ecol. Soc., N.Y. Acad. Scis., AAAS, Sigma Xi, Phi Eta Sigma, Alpha Zeta, Kappa Delta Pi, Xi Sigma Pi. Club: Cosmos (Washington). Subspecialty: Ecology. Home: 5815 Honeysuckle St Beaumont TX 77706 Office: John E Gray Inst Lamar U Beaumont TX 77710

JOHNSON, PHYLLIS ELAINE, research chemist; b. Grafton, N.D., Feb. 19, 1949; d. Donald Gordon and Evelyn Lorraine (Svaren) Lanes; m. Robert S.T. Johnson, Sept. 12, 1969; children: Erik, Sara. B.S. in Chemistry, U.N.D., 1971, Ph.D., 1976. Lab. instr. Mary Coll., Bismarck, N.D., 1971-72; postdoctoral fellow U.N.D., Grand Forks, 1975-76, chemist, 1976-79; research chemist U.S. Dept. Agr. Human Nutrition Research Ctr., Grand Forks, 1979—. Contbr. articles to sci. publs. Bd. dirs. Dist. IV Sons of Norway. Mem. Am. Chem. Soc., Am. Inst. Nutrition, Soc. Exptl. Biology and Medicine, Assn. Women in Sci. (chmn., founder Grand Forks chpt.), Am. Soc. Mass Spectrometry. Lutheran. Subspecialties: Analytical chemistry; Nutrition (medicine). Current work: trace metal absorption and bioavailability in humans; lactation; mass spectrometry. Home: 4809 4th Ave N Grand Forks ND 58201 Office: USDA Human Nutrition Research Ctr Box 7166 University Station Grand Forks ND 58202

JOHNSON, RANDALL K., pharmacologist, cons., researcher; b. Mpls., Aug. 19, 1946; s. Dennis Garfield and Dorothy Eleanor (Pearson) J.; m. Marcia Ann Hewitt, Sept. 15, 1967; children: Hugh Witt, Maranda. B.Sc. cum laude (Internat. Milling Co. scholar), U. Minn., 1968; Ph.D. (NSF fellow), George Washington U., 1972. Staff fellow Lab. Exptl. Chemotherapy Nat. Cancer Inst., NIH, Bethesda, Md., 1972-73, sect. head, 1973-75, NIH postgrad. fellow, 1972-74; sect. head Lab. Chem. Pharmacology, 1975-77; sr. scientist Arthur D. Little Inc., Cambridge, Mass., 1977-81, mgr. exptl. therapeutics sect., 1981-82; assoc. dir. antineoplastic program Smith Kline Beckman, Phila., 1982—. Contbr. articles to sci. jours. Mem. Am. Assn. Cancer Research, Internat. Soc. for Study Xenobiotics. Subspecialties: Cancer research (medicine); Chemotherapy. Current work: Development of new anticancer agents; tumor biology, biochemical pharmacology of cytotoxic agents, DNA damage and repair, mechanism of drug resistance. Home: 71 Llanfair Circle Ardmore PA 19003 Office: Smith Kline Beckman 1500 Spring Garden St Philadelphia PA 19101

JOHNSON, RICHARD DAMERAN, government research and development manager, chemist; b. Zanesville, Ohio, Oct. 28, 1934; s. Earl G. and Merlie (Damerau) J.; m. Sandra Wishart (div. 1968); children: Laurana W., Karen D.; m. Catherine Collins, Dec. 29, 1969; children: Eric C., Gregory N. A.B., Oberlin Coll., 1957; M.S., Carnegie Mellon U., 1961, Ph.D., 1962; S.M., MIT, 1982. Postdoctoral fellow UCLA, 1961-62; sr. scientist Jet Propulsion Lab., Pasadena, Calif., 1963; research scientist NASA-Ames Research Ctr., Moffett Field, Calif., 1963-76, research and devel. mgr., 1976—; prin. investigator NAS-Apollo 11 Lunar Sampel Analysis, 1969. Editor: Space Settlements, 1977; contbr. articles to profl. jours. Coach Am. Youth Soccer, Los Altos, Calif., 1977-82. Recipient Exceptional Service medal NASA, 1977; Sloan fellow MIT, 1981-82. Mem. AAAS, Am Chem. Soc., AIAA. Subspecialties: Space medicine; Space colonization. Current work: Research and development management in life sciences related space activities; extraterrestrial life detection and chemical analysis; space biomedical and biological research; space colonization studies; space station; space commercialization. Home: 11564 Arroyo Oaks Los Altos Hills CA 94022 Office: Biosystems Div NASA-Ames 236-5 Moffett Field CA 94035

JOHNSON, RICHARD FREDERICK, research psychologist, consultant; b. Boston, July 11, 1943; s. Frederick and Alice (Kullen) J.; m. Sharyn Lois Doyle, Sept. 11, 1965; children: Wendy, Adam. B.A. with honors, Northeastern U., 1966; M.A., Brandeis U., 1968, Ph.D., 1970. Diplomate: Lic. psychologist, Mass. Commd. 2d lt. U.S. Army Res., 1966, advanced through grades to capt., 1970; active duty, 1970-72, resigned, 1974; sr. research psychologist Medfield (Mass.) Found., 1972-76; research psychologist (U.S. Army Natick Research and Devel. Labs.), Mass., 1976—; hypnosis cons. Blaisdell Psychol. Services, Medway, Mass., 1978-81; psychophysiology cons. Medfield Found., 1976-78; lectr. Northeastern U., Boston, 1972-76. Editorial cons.: Jour. Cons. and Clin. Psychology, 1973—; corr. assoc. commentator: The Behavioral and Brain Scis, 1978—; contbr. articles to profl. jours. Chmn. human rights com. Mass. Dept. Mental Health Region V, 1974-76. Woodrow Wilson fellow, 1969-70; NASA trainee, 1966-69; NIMH grantee, 1972-76. Mem. Am. Psychol. Assn., Human Factors Soc., Am. Soc. Clin. Hypnosis, AAAS, Sigma Xi, Eastern Psychol. Assn., Soc. Personality and Social Psychology, Southeastern Psychol. Assn., Sigma Xi. Subspecialties: Behavioral psychology; Human Factors Psychology. Current work: Study interaction of natural environment (cold, heat, humidity) and protective systems (e.g., cold weather clothing) on man's ability to function (mental abilities, manual dexterity, sensory performance, etc.); study bias in experimentation with humans. Home: 15 Sahlin Circle Franklin MA 02038 Office: US Army Natick R&D Labs Natick MA 01760

JOHNSON, RICHARD TIDBALL, neurology educator, neurologist; b. Grosse Point Farms, Mich., July 16, 1931; s. Horton A. and Katharine (Tidball) J.; m. Frances W., Sept. 18, 1954; children: Carlton, Erica, Matthew, Nathan. A.B. cum laude, U. Colo.-Boulder, 1953; M.D., U. Colo.-Denver, 1956. Diplomate: Am. Bd. Psychiatry and Neurology. Intern in medicine Stanford U. Hosps., San Francisco, 1956-57; asst. resident in neurology Mass. Gen. Hosp., 1959-60, clin. fellow in neuropathology, 1960-61, sr. resident in neurology, 1961-62; teaching fellow in neurology Harvard U. Med. Sch., 1959-62; teaching fellow in neurology on exchange as first asst. in neurology Med. Sch. of Kings Coll., U. Durham (Eng.) and Royal Victoria Infirmary, Newcastle-Upon-Tyne, Eng., 1962; hon. fellow (USPHS spl. fellow) dept. microbiology John Curtin Sch. Med. Research Australian Nat. U., Canberra, 1962-64; asst. prof. neurology Case Western Res. U., 1964-68, assoc. prof., 1968-69; assoc. neurologist Cleve. Met. Gen. Hosp., 1964-69; assoc. prof. microbiology Johns Hopkins U. Sch. Medicine, 1969-74, prof. microbiology, 1974—, Eisenhower prof. neurology, 1969—, Dean's lectr., 1980; neurologist Johns Hopkins Hosp., 1969—; cons. neurologist Balt. City Hosps., 1974—; vis. prof. Universidad Peruana Cayetano Heredia, Lima, Peru, 1971, Imperial Coll. Health Scis., Pahlavi Med. Center, Teheran, Iran, 1974, Institut fur Virologie und Immunobiologie, U. Wurzburg, Germany, 1976; hon. prof. Universidad Peruana Cayetano Heredia, 1980; numerous spl. lectureships, including Andrew Mark Lippard Meml. lectr. Columbia U., 1976; Zimmerman lectr. Stanford U., 1978; Brain Centenary lectr. Royal Coll. Physicians, London, 1978, Dr. Ena Thomas Meml. lectr., Kingston, Jamaica, 1978; Joseph J. Gitt vis. professorship and lectr. Washington U., St. Louis, 1979; Litchfield lectr. U. Oxford, Eng., 1983; cons. on neurol. diseases Bur. Biologics FDA, 1979—; mem. adv. com. to neurologic disorders program Nat. Inst. Neurol. and Communicative Disorders and Stroke, 1979-82, mem. sci. adv. com., 1982—; mem. external adv. com. Multiple Sclerosis Center, U. Pa., 1974—; mem. ind. assessors panel Nat. Health and Med. Research Council of Australia, 1976—; mem. adv. council James. A. Baker Inst. for Animal Health, Cornell U., 1977—; mem. Nat. Com. for Research in Neurol. and Communicative Disorders, 1982—; chmn. med. adv. bd. Md. chpt. Nat. Multiple Sclerosis Soc., 1970—; mem. med. adv. bd. Nat. Multiple Sclerosis Soc., 1971—, mem. exec. com. to med. adv. bd., 1981—, mem. research programs com., 1979—, chmn., 1981—; mem. med. adv. bd. Internat. Fedn. Multiple Sclerosis Socs., 1980—. Author: Viral Infections of the Nervous System, 1982; contbr. numerous articles, abstracts to profl. jours.; editor: (with J. M. Andrews and M. A. B. Brazier) Amyotrophic Lateral Sclerosis: Recent Research Trends, 1976; editorial bd. jours., currently: Jour. Neuropathology and Applied Neurobiology, 1974—, Multiple Sclerosis Indicative Abstracts, 1975—, Advances in Neurology, 1978—, Revs. Infectious Disease, 1979—, Sci. 80—, 1980—. Served to capt. M.C. U.S. Army, 1957-59. Decorated comendador Order Hipolito Unanue, Peru; recipient Sydney Farber Research award United Cerebral Palsy Research and Ednl, Found., 1974, 76, Humboldt prize Alexander von Humboldt Found., 1976, Weinstein-Goldenson award United Cerebral Palsy Assns., 1979, Gordon Wilson medal Am. Clin. and Climatological Assn., 1980, Frank R. Ford award Neurology Residents Johns Hopkins Hosps., 1981. Fellow Am. Acad. Neurology (2d v.p. 1975-77); mem. Am. Soc. Clin. Investigation, Am. Neurol. Assn., Balt. Neurol. Soc. (pres. 1973-74), Johns Hopkins Med. Soc. (pres. 1970-71), Am. Assn. Neuropathologists (Arthur Weil award 1967), Phi Beta Kappa, Alpha Omega Alpha, and others; hon. mem. Australian Assn. Neurologists, Instituto de Medicina Tropical Alexander von Humboldt. Subspecialties: Neurology; Virology (medicine). Current work: Virology, Immunology, Neurology. Home: 107 Ridgewood Rd Baltimore MD 21218 Office: Johns Hopkins Sch Medicine 600 N Wolfe St Meyer 6-109 Baltimore MD 21205

JOHNSON, ROBERT CHANDLER, research physicist; b. Detroit, Oct. 19, 1930; s. Robert Stephen and Veda Vivian (Manning) J.; m. Mary Jane Wood, Aug. 6, 1955; children: Andrew, Douglas, Sarah, Michael. B.S. in Physics, U. Mich.-Ann Arbor, 1952, M.A., U. Iowa-Iowa City, 1957, Ph.D., Stanford (Calif.) U., 1962. With E. I. DuPont Co., Wilmington, Del., 1962—, research physicist, 1975-78, supr., 1978—. Served with USN, 1952-55. Mem. N.Am. Thermal Analysis Soc. (sec. 1979-81, chmn. conf. 1982-83), Am. Phys. Soc., Am. Chem. Soc. Mem. United Ch. of Christ. Subspecialties: Physical chemistry; Analytical chemistry. Current work: Thermal measurements; electron and acoutsic microscopy; analytical instrmentation. Patentee method and apparatus for calorimetric differential thermal analysis, 1982. Home: 2526 Blackwood Rd Wilmington DE 19810 Office: E I DuPont Central Research and Devel Dept Bldg 228 Exptl Sta Wilmington DE 19898

JOHNSON, ROBERT DENNIS, psychology educator; b. Jersey City, N.J., Oct. 28, 1944; s. John Oscar and Elizabeth (Person) J.; m. Janet Lois (Green), Aug. 26, 1966; children: Travis Robert, Michael Jon. B.S., Widener Coll., 1966; M.S., U. Ga., 1972, Ph.D., 1975. Lic. psychologist, Ark. Asst. prof. psychology Ark. State U., State University, 1975-81, assoc. prof., 1981—. Contbr. articles to profl. jours. Bd. dirs. PACES, Jonesboro, Ark., 1979—. Served to 1st lt. U.S. Army, 1967-69. N.J. Bd. scholar, 1962-66. Mem. Am. Psychol. Assn., AAAS, Southeastern Psychol. Assn., Southwestern Psychol. Assn., Sigma Xi, Psi Chi. Subspecialty: Social psychology. Current work: Sex-roles; deindividuation. Home: 3300 Candlewood Dr Jonesboro AR 72401 Office: Ark State U PO Box 2779 State University AR 72467

JOHNSON, ROBERT E., physicist, educator; b. Chgo., July 3, 1939; s. Theodore J. and Elsie M. (Johnson) J.; m. Barbara F. MacCallum; children: Amanda F. L. MacCallum, Sarah M. B.A. in Math, Colo. Coll., 1961; M.A. in Physics, Wesleyan U., 1963; Ph.D., U. Wis., 1968. With Sandia Corp., 1965; fellow Queen's U., Belfast, No. Ireland, 1968-69; asst. prof. physics So. Ill. U., Carbondale, 1969-71; asst. prof. engring. physics U. Va., Charlottesville, 1971-76, assoc. prof., 1976—, asst. dean, 1982—; vis. researcher Center for Planetary Physics, Harvard, 1977-78, Enrico Fermi Inst., U. Chgo., 1981-82; cons. Bell Telephone Labs., Murray Hill, N.J., 1979-81. Author: An Introduction to Atomic and Molecular Collisions, 1982; contbr. chpt. to book, articles to profl. jours. NATO fellow U. Copenhagen, 1976; summer fellow Argonne Nat. Lab., 1982. Mem. Am. Inst. Physics, Am. Astron. Soc., Am. Geophys. Union. Subspecialties: Atomic and molecular physics; Electronic materials. Current work: Interaction of charged particlesradiations with matter: atomic and molecular collisions; interaction of magnetospheric plasmas with satellites of Jupiter and Saturn. Home: 135 Bollingwood Rd Charlottesville VA 22903 Office: Thornton Hall U Va Charlottesville VA 22901

JOHNSON, ROBERT LOUIS, astronautics company executive; b. Winslow, Ariz., May 16, 1920; s. Ernest Conrad and Carrie Arora (Saunders) J.; m. Betty Tuft, Oct. 24, 1942; children: Jeanne Johnson Dillon, Robert C., Louise Buck, Bruce T., Kirk T. B.S. in Mech. Engring., U. Calif., Berkeley, 1941, M.S., 1942. Vice pres. research and engring., then v.p. manned orbit lab. Douglas Aircraft Co., 1946-69; asst. sec. army for research and devel., 1969-73; v.p. engring. and research McDonnell Douglas Corp., 1973-75; pres. McDonnell Douglas Astronautics Co., Huntington Beach, Calif., 1975—; corp. v.p.-group exec. McDonnele Douglas Corp., 1980—; mem. engring. adv. council U. Calif., Berkeley. Served to lt. USNR, 1942-46. Recipient James H. Wyld Meml. award Am. Rocket Soc., 1960. Fellow Am. Inst. Aeros. and Astronautics; mem. Nat. Acad. Engring., Nat. Space Club, Assn. U.S. Army (dir.). Episcopalian. Club: El Niguel Country. Subspecialty: Aerospace engineering and technology. Address: McDonnell Douglas Corp 3855 Lakewood Blvd Long Beach CA 90846

JOHNSON, ROBERT SHEPARD, mathematician; b. Wilkinsburg, Pa., Nov. 24, 1928; s. George Henry and Hellen Beatrice (Shepard) J.; m. Patricia Elizabeth Pierstorff, Apr. 11, 1959; children: Warren, Barbara, Neil. B.S., Northwestern U., 1950, M.S., 1951; Ph.D., U. Pa., 1959. Research assoc. Inst. Coop. Research, Phila., 1953-59; prin. mem. engring. staff RCA, Moorestown, N.J., 1959—. Served with U.S. Army, 1951-53. Mem. Am. Math. Soc., Soc. Indsl. and Applied Math., Phi Beta Kappa, Pi Mu Epsilon. Republican. Subspecialties: Applied mathematics; Mathematical software. Current work: Stiff differential equations, numerical math. Home: 2102 Brandeis Ave Cinnaminson NJ 08077 Office: RCA Barton Landing Rd Moorestown NJ 08077

JOHNSON, RONALD ROY, agricultural research executive, educator; b. De Smet, S.D., Dec. 8, 1928; s. Roy L. and Edna F. (Harrison) J.; m. Sally Jeanne Shamel, Oct. 9, 1955; children: Denetia, Jennifer, Melissa, J. Scott. B.S., S.D. State Coll., 1950, M.S., 1952; Ph.D., Ohio State U., 1954. Mem. faculty Ohio State U., Columbus, 1955-69, prof. animal sci., 1955-69, Ohio Agrl. Exptl. Sta., Wooster, 1955-69; prof., head dept. animal sci. U. Tenn., Knoxville, 1974-81; prof. nutritional biochemistry Okla. State U., Stillwater, 1969-74; assoc. dir. Okla. State U. Okla. Agrl. Exptl. sta., 1981—. Contbr. articles to profl. jours. USPHS fellow, 1965-66. Mem. Am. Inst. Nutrition, Am. Soc. Animal Sci., Am. Dairy Sci. Assn. Subspecialties: Animal nutrition; Biochemistry (biology). Current work: Lipid metabolism; non-protein nitrogen metabolism; research administration. Home: 1523 Skyline Dr Stillwater OK 74074 Office: Okla Agrl Exptl Sta Okla State Univ 139 Agr Hall Stillwater OK 74078

JOHNSON, ROY MELVIN, bacteriology educator; b. Chgo., Sept. 8, 1926; s. Carl Henning and Ebba Dorothy (Helstrom) J.; m. Betty Lou Schutt, Sept. 6, 1952; children: Renee M., Roberta M., Rhonda M., Regan M. A.B., U. Chgo., 1949; M.S., 1951; Ph.D., U. N.Mex., 1955. Asst. prof. bacteriology Ariz. State U., Tempe, 1955-58, assoc. prof., 1958-64, prof., 1964—; cons. Good Samaritan Hosp., Phoenix, 1965-75, Motorola Corp., Mesa, Ariz., 1971-73. Served with U.S. Army, 1944-46. Fellow AAAS, Am. Acad. Microbiology, Ariz.-Nev. Acad. Sci. Republican. Lutheran. Subspecialties: Microbiology; Bacterial ecology. Current work: Bacterial systematics and ecology. Home: 26405 Lakeview Dr Sun Lakes AZ 85224 Office: Dept Microbiology Ariz State U Tempe AZ 85287

JOHNSON, STEPHEN THOMAS, tool engineer; b. Washington, May 31, 1954; s. Glenn Elmer and Marie Veronica (Rando) J.; m. Joand Marie Wagner, Apr. 16, 1983. B.M.E.T. with honors, Northeastern U., 1978. Draftsman Hollingsworth & Vose, East Walpole, Mass., 1973-74; tech. aide U.S. Army Natick Research and Devel. Command, Natick, Mass., 1975-78; tool designer Boeing Aircraft Co., Renton, Wash., 1978-81, propulsion engr., 1981; sr. tool designer Sikorsky Aircraft, Stratford, Conn., 1981—. Mem. Am. Soc. Metals, ASME. Subspecialty: Mechanical engineering. Patentee direct ohmnic heating device. Home: 1004 Stratford RD Stratford CT 06497

JOHNSON, TERRY CHARLES, biology educator; b. St. Paul, Minn., Aug. 8, 1936; s. Roy A. and Catherine (McKigen) J.; m. Mary Ann Wilhelmy, Nov. 23, 1957; children: James, Gary, Jean. B.S., Hamline U., 1958; M.S., U. Minn., Mpls., 1961, Ph.D., 1963; Postdoctoral fellow, U. Calif.-Irvine, 1963-65. Asst. prof. Northwestern U. Med. Sch., Chgo., 1966-69, assoc. prof., 1969-73, prof., 1973-77; prof., dir. div. biology Kans. State U., Manhattan, 1977—. Contbr. numerous articles to profl. jours. Mem. AAAS, Am. Soc. Cell Biology, Am. Soc. Microbiology, Am. Soc. Neurochemistry. Subspecialties: Cancer research (medicine); Cell biology (medicine). Current work: Cell growth regulation, tumorigenesis, virology, infectious diseases, genetic engineering, neurochemistry, cell surfaces. Home: 205 S Drake Dr Manhattan KS 66502 Office: Div Biology Kans State U Manhattan KS 66506

JOHNSON, TERRY R., research manager, chemical engineer; b. Chgo., Nov. 16, 1932; s. Earl William and Margaret (Fisher) J.; m. Janet Roberts, June 16, 1956; children: Kenneth, Martin, Karen, Jennifer. B.S. in Chem. Engring, Rice U., 1955; M.S., U. Mich., 1956; Ph.D., 1959. Chem. engr. Argonne (Ill.) Nat. Lab., 1958-73, 75—; chief process engr. Aglomet, Inc., Chgo., 1974-75; vis. prof. Iowa State U., Ames, 1969-70; pres. Symposium on Engring. Aspects of MHD, 1983. Contbr. articles to profl. jours. Bd. dirs. Jaycees, Glen Ellyn, Ill., 1962-69; mem. Village Recreation Commn., Glen Ellyn, 1964-69. Mem. Am. Inst. Chem. Engring., Am. Nuclear Soc., Sigma Xi, Tau Beta Pi. Presbyterian. Subspecialties: Chemical engineering; Combustion processes. Home: 1424 S Main St Wheaton IL 60187 Office: Argonne Nat Lab Bldg 205 9700 S Cass Argonne IL 60439

JOHNSON, THOMAS EUGENE, geneticist, educator; b. Denver, June 19, 1948; s. Albert L. and Barbara J. (Bickle) J.; m. Victoria J. Simpson, Apr. 24, 1982. S.B. in Life Scis, M.I.T., 1970; Ph.D. in Genetics, U. Wash., 1975. Research assoc. in genetics and devel. Cornell U., Ithaca, N.Y., 1975-77; research assoc. in molecular biology U. Colo., 1977-82; faculty fellow Inst. Behavoral Genetics, 1981-82; asst. prof. molecular biology and biochemistry U. Calif, Irvine, 1982—; also. cons. Contbr. articles and chpts. to profl. lit. Mem. Genetics Soc. Am., Soc. Developmental Biology, AAAS, Am. Gerontol. Soc. Sufist. Subspecialties: Gene actions; Gerontology. Current work: Genetics and molecular biology of senescence in C. elegans. Genetics, quantitative genetics, aging, senescence, Co. elegans; behavior, development and aging; biochemistry and molecular biology of aging. Office: Molecular Biology and Biochemistry U Calif Irvine CA 92717

JOHNSON, THOMAS FOLSOM, psychologist, educator; b. Milw., July 29, 1918; s. Elmer George and Margaret A. (Eversz) J.; m. Eunice W. Steger, Dec. 24, 1943 (div. 1974); children: Kevin, Melissause, Christopher; m. Jean Bramble Leech; June 1, 1976. B.A., U. Wis., 1942; M.S., Purdue U., 1951, Ph.D., 1953. Clin. psychologist VA, Phila., 1946-50; chief psychologist Osawatomie State Hosp., Kans., 1953-56; dir. psychol. tng. Eastern Pa. Psychiat. Inst., Phila., 1956-57, dir. adult psychology, 1957-72; dir. family intervention services Del. County Juvenile Ct., Media, Pa., 1972—; assoc. prof. Jefferson Med. Coll., Phila., 1978—; adj. prof. Union Grad. Sch., Ohio, 1979—; sr. clin. instr. Hahnemann Med. Sch., Phila., 1964-69. Contbr.: books Therapeutic Intervention, 1982, Family Pathology and its Therapy, 1983. Fellow Pa. Psychol. Assn.; mem. Am. Assn. Family Therapists, Am. Psychol. Assn. Subspecialty: Family therapy. Current work: Developing techniques for working with juvenile delinquents; Developing theory on nature and origins of delinquency. Home: 3616 Haywood St Philadelphia PA 19129 Office: 2nd and North Ave Media PA 19063

JOHNSON, VIRGIL ALLEN, agronomist; b. Newman Grove, Nebr., June 28, 1921; s. Oscar Johannas and Fairy Bell (Johnson) J.; m. Betty Ann Tisthammer, July 27, 1943; children—Karen (Mrs. Ronald Fakes), Leslie (dec.), Reed, Scott. B.S. with distinction, U. Nebr., 1948, Ph.D. (Regents Grad. fellow, Ak-Sar-Ben grad. fellow, Sears, Roebuck Grad. fellow), 1952. Agt. Agrl. Research Service, U.S. Dept. Agr., Lincoln, Nebr., 1951-52, research agronomist, 1954-75, supervisory research agronomist, 1975-78; leader wheat research, asst. agronomist U. Nebr., Lincoln, 1952-54, asso. prof., 1954-63, prof. agronomy dept., 1963—; cons. Gt. Plains Wheat, Inc.; mem. Nat. Wheat Improvement Com.; mem. tech. com. Wheat Quality Council. Contbr. articles to profl. publs. Served in inf. U.S. Army, 1940-43, AC; Served in inf. U.S. Army, 1943-45. Decorated D.F.C., Air medal with 3 oak leaf clusters; recipient Agrl. Achievement award Ak-Sar-Ben, 1970, Disting. Service award Dept. Agr., 1981; AID grantee, 1966-79. Fellow Am. Soc. Agronomy, AAAS; mem. Crop Sci. Soc. Am. (Crop Sci. award 1975, pres. 1978), Am. Genetics Assn., Sigma Xi, Gamma Sigma Delta (Internat. award 1969), Alpha Zeta. Lutheran. Subspecialties: Plant genetics; Plant physiology (agriculture). Current work: Genetic and physiological bases of grain protein variation in Triticum aestivum L. Co-developer 22 varieties of hard red winter wheat. Home: 3849 Dudley St Lincoln NE 68503 Office: 324 Keim Hall East Campus U Nebr Lincoln NE 68583

JOHNSON, WALTER CURTIS, JR., biophysics educator; b. Princeton, N.J., Feb. 11, 1939; s. Walter Curtis and Caroline (Shirk) J.; m. Susan Scheller Johnson, Aug. 28, 1960; children: Walter C., Heather L. B.A. in Chemistry, Yale U., 1961; Ph.D. in Phys. Chemistry, U. Wash., 1966. Asst. prof. biochemistry and biophysics Oreg. State U., 1968-72, assoc. prof., 1972-78, prof., 1978—. Contbr. articles to profl. jours. Ford Found. fellow, 1961-62; NSF fellow, 1963, 66-68; NSF grantee, 1968—; NIH grantee, 1970—. Mem. Biophys. Soc., Western Spectroscopy Assn. Subspecialties: Biophysics (biology); Biophysical chemistry. Current work: Electronic properties of biopolymers, principally their circular dichroism, to study their conformation and resulting biological function. Office: Dept Biochemistry and Biophysics Oregon State U Corvallis OR 97331

JOHNSON, WILLIAM K., health physicist, state ofcl.; b. St. Joseph, Mo., Feb. 6, 1934; s. Maxwell McCole and Kittie Ellen (Jackson) J.; m. Barbara Ann Terrell, June 23, 1980. Student, S.W. Mo. State U., 1961-62, Mich. Tech. U., 1975-77, N.E. Mo. State U., 1982—. Cert. Nat. Registry Radiation Protection Technologists. Supr. licensing Mo. Div. Ins., Jefferson City, 1969-73; supr. radiation systems Disaster Planning and Ops., Jefferson City, 1973-81; state radiol. def. officer State Emergency Mgmt. Agy., Jefferson City, 1981—; Radiation safety officer Mo. N.G., Jefferson City, 1977—; state coordinator Mo. Nuclear Emergency Team, Jefferson City, 1978—; cons. dir. Radiation Safety Consultants, Jefferson City, 1982—. Author emergency procedure publs. Active United Way, Boy Scouts Am. Recipient Silver Beaver award Boy Scouts Am., 1972. Mem. Health Physics Soc., Am. Assn. Physicists in Medicine, Am. Nuclear Soc., Mo. Acad. Scis., Radiol. Def. Officers Assn., Am. Legion. Lodges: Eagles; Masons; K.T. Subspecialties: Radiology and Nuclear Medicine Safety Procedures. Current work: Radiation safety and radiological hazard mitigation from peacetime industrial accidents and/or nuclear attack. Home: 1215 W Miller St Jefferson City MO 65101 Office: State Emergency Management Agy PO Box 116 Jefferson City MO 65102

JOHNSON, WILLIAM LAWRENCE, animal science educator, researcher; b. Keene, N.H., Aug. 28, 1936; s. Stephen Guy and Elsie May (Prentice) J.; m. Nancy Lona Crane, June 7, 1958 (div.); children: Warren Eliot, Susan Louise, Steven Lawrence. B.S. in Agr, U. N.H., 1958; M.S., Cornell U., 1964, Ph.D., 1966. Animal husbandry vol. Internat. Voluntary Services, Laos, 1958-61; research asst. U. Philippines, Los Banos, 1964-66; animal nutritionist N.C. State U. Agrl. Mission to Peru, 1966-72; asst. prof. animal sci. N.C. State U., 1966-74, assoc. prof., 1974—; vis. lectr. Center for Research and Teaching in Tropical Agr., Turrialba, Costa Rica, 1977, 80, 81; prin. investigator U. Calif.-Davis subgrant for sheep and goat nutrition research in N.E. Brazil, West Java, and Morocco, as part of AID-funded Small Ruminants Collaborative Research Support Program, 1978—. Contbr. articles, numerous abstracts on ruminant nutrition and feed utilization to profl. jours. Mem. Am. Soc. Animal Sci., Internat. Goat Assn., Latin Am. Soc. Animal Prodn., Brazilian Zootechnic Soc., Philippine Soc. Animal Sci. Subspecialty: Animal nutrition. Current work: Research on practical feeding systems for ruminants in humid and semiarid tropics; forage, byproduct and crop residue utilization by goats, sheep and cattle. Office: Dept Animal Sci NC State U PO Box 5127 Raleigh NC 27650

JOHNSON, WILLIAM LEWIS, physics educator; b. Bowling Green, Ohio, July 28, 1948; s. Melvin Carl and Martha Maxine (Roller) J. B.A., Hamilton Coll., 1970; Ph.D., Calif. Inst. Tech., 1974. Mem. research staff T.J. Watson Research Center, IBM, Yorktown Heights, N.Y., 1975-77; asst. prof. Calid. Inst. Tech., Pasadena, 1977-80, assoc. prof. material sci., 1980—; cons. to research labs. and pvt. industry. Contbr. articles to profl. jours. U.S. Steele fellow, 1970-72. Mem. Am. Phys. Soc., Metals Soc. AIME, Phi Beta Kappa, Sigma Xi. Democrat. Lutheran. Subspecialties: Materials; Amorphous metals. Current work: Electronic properties, superconductivity, thermodynamics; research on the synthesis and properties of non-equilibrium materials. Patentee. Home: 3546 Mountainview Ave Pasadena CA 91107 Office: California Institute of Technology Pasadena CA 91125

JOHNSONBAUGH, ROGER EARL, pediatric endocrinologist, researcher; b. Chgo., May 1, 1934; s. Sanford Leonard and Dorothy Lucille (Thursby) J.; m. Elizabeth Ann Borgman, Sept. 8, 1956 (div. 1982); children: Kim E., David S., Nancy L., Roger E. III. B.A., Denison U., 1956; M.S., U. Cin., 1958, M.D., 1962; Ph.D., U. N.C., 1973. Diplomate: Am. Bd. Pediatrics. Intern Wayne County Gen. Hosp., Eloise, Mich., 1962-63; resident in pediatrics Cin. Children's Hosp., 1963-65; fellow in pediatric endocrinology and metabolism, 1966-67; postdoctoral fellow in pediatrics Coll. Medicine and Univ. Hosps., Ohio State U., Columbus, 1966-67; USPHS postdoctoral research fellow Nat. Inst. Child Health and Human Devel., NIH, Bethesda, Md., 1967; fellow in pediatric endocrinology U. N.C. Sch. Medicine, 1969-72, instr. pediatrics, 1970-72; asst. prof. child health and devel. George Washington U., 1973-77, assoc. prof., 1977—; assoc. prof. dept. pediatrics Uniformed Services U. of the Health Scis., Bethesda, 1976-79, dir. div. pediatric endocrinology dept. pediatrics, 1977—, prof., 1979—; dir. research Naval Hosp., Naval Med. Command, Bethesda, 1976—, chief pediatric endocrinology, 1972—; med. care cons. pediatric endocrinology Nat. Inst. Child Health and Human Devel., NIH, 1976—; cons. pediatric endocrinology Walter Reed Army Med. Ctr., Washington, 1973—; vis. scientist Karolinska Inst., Stockholm, Sweden, 1981-82. Fellow Am. Acad. Pediatrics; mem. Lawson Wilkins Pediatric Endocrine Soc., Endocrine Soc., Soc. Pediatric Research, Am. Fedn. Clin. Research, AAAS, Assn. Mil. Surgeons U.S. Subspecialties: Endocrinology; Pediatrics. Current work: General pediatric endocrinology with a major interest in growth problems in children. Office: Naval Hosp Bethesda MD 20814

JOHNSTON, BRUCE GILBERT, civil engr.; b. Detroit, Oct. 13, 1905; s. Sterling and Ida (Peake) J.; m. Ruth Elizabeth Barker, Aug. 5, 1939; children—Sterling, Carol Anne. Snow, David. B.S. in Civil Engring, U. Ill., 1930; M.S., Lehigh U., Bethlehem, Pa., 1934; Ph.D. in Sci, Columbia U., 1938. Engaged in engring. constrn. Coolidge Dam, Ariz., 1927-29; with design office Roberts & Schaefer Co., Chgo., 1930; instr. civil engring. Columbia U., 1934-38; charge structural research Fritz Engring. Lab., Lehigh U., 1938-50, asst. dir. lab., 1938-47, dir., 1947-50, mem. univ. faculty, 1938-50, prof. civil engring., 1945-50; prof. structural engring. U. Mich., 1950-68, emeritus, 1968—; prof. civil engring. U. Ariz., Tucson, 1968-70; engr. Johns Hopkins Applied Physics Lab., Silver Spring, Md., 1942-45; chmn. Column Research Council, 1956-62. Author: Basic Steel Design, 2d edit, 1980, also tech. papers.; Editor: Column Research Council Design Guide, 3 edits, 1960-76. Recipient Alumni Honor award for disting. service in engring. U. Ill., 1981. Hon. mem. ASCE (chmn. structural div. 1965-66, chmn. engring. mechanics div. 1961-62, J.J.R. Croes medal 1937, 54, Ernest E. Howard medal 1974); mem. Nat. Acad. Engring., Sigma Xi, Phi Kappa Phi, Tau Beta Pi, Chi Epsilon. Methodist. Subspecialties: Civil engineering; Structural Engineering. Current work: Stability of cell columns and steel structures. Torsion of structural members. History of development of column buckling theory. Address: 5025 E Calle Barril Tucson AZ 85718

JOHNSTON, DANIEL, neurophysiologist; b. Passaic, N.J., Dec. 9, 1947; s. Vivian Daniel and Elizabeth Dorothy (Booth) J.; m. Jean Boxley, Dec. 21, 1973; children: Lisa Danielle, Lauren Blaire. B.S.E.E., U. Va., 1970; Ph.D., Duke U., 1974. Postdoctoral fellow U. Minn., Mpls., 1974-75, instr., 1975-76, asst. prof., 1976-77, Baylor Coll. Medicine, Houston, 1977-81, assoc. prof. dept. neurology, 1981—; cons. Dagan Corp., Mpls., 1971—. Editor: Cellular and Molecular Neurobiology, 1981—. Served with U.S. Army, 1969-71. NIH grantee, 1979—; Epilepsy Found. Am. grantee, 1980-82; McKnight Found. fellow, 1982-84. Mem. Am. Physiol. Soc., Biophys. Soc., Soc. Neurosci., AAAS, N.Y. Acad. Sci. Subspecialties: Neurophysiology; Biomedical engineering. Current work: Cellular basis of epilepsy, design of antiepileptic drugs, cellular and molecular aspects of memory and memory disfunctions. Home: 4943 Valkeith St Houston TX 77096 Office: Baylor Coll Medicine Neurology Dept 1200 Moursund Ave Houston TX 77030

JOHNSTON, DOROTHY MAE, psychologist; b. Mutual, Okla., Feb. 22, 1924; d. Harry Homer and Lena (Barnes) J. B.S. in B.A, U. Ark., 1949; M.A., U. Denver, 1952; Ph.D., N.C. State U., 1971. Engring. psychologist Douglas Aircraft Co., Tulsa, 1956-57, N.Am. Rockwell, Columbus, Ohio, 1957-60, 66-67, Boeing Co., Wichita, Kans. and; Seattle, 1960-61, 62-66, Martin Marietta, Orlando, Fla., 1961-62, Melpar, Inc., Falls Church, Va., 1968; scientist doing ind. research, Prairie Grove and Lincoln, Ark., 1971—. Contbr. articles to profl. jours.; author: Beyond the Limelight, 1983. Mem. Am. Psychol. Assn., AAAS, Soc. Women Engrs., IEEE, Assn. for Women in Sci., Optical Soc. Am., Human Factors Soc. Democrat. Methodist. Subspecialties: Human factors engineering; Sensory processes. Current work: Research in visual perception. Address: Route 2 P O Box 79A Lincoln AR 72744

JOHNSTON, EUGENE BENEDICT, microbiologist; b. Bklyn., June 25, 1951; s. Eugene B. Johnson and Alice Theresa J.; m. Jeanne-Marie Toomey, Oct. 14, 1978; 1 dau., Meghan Anne. B.S., U. Dayton, 1974. Supr. quality control biology Center Labs., Port Washington, N.Y., 1974-75; microbiologist Revlon, Inc., Bronx, N.Y., 1975-76, Becton Dickinson, Rutherford, N.J., 1976-78; sterilization quality assurance engr. C.R. Bard, Inc., Murray Hill, N.J. and Covington, Ga., 1978-81; mgr. dept. microbiology Bard Urol. Div., Covington, 1981-83; chief microbiologist Becton Dickinson Respiratory Systems, Ocala, Fla., 1983—. Mem. Am. Soc. Microbiology, Soc. Indsl. Microbiology, Am. Soc. Quality Control, Parenteral Drug Assn. Inst. for Environ. Scis. Subspecialty: Microbiology. Current work: Industrial sterilization processes, contamination control, microbiological quality control. Office: 1909 NE 25th Ave Ocala FL 32670 Home: 505 SE 46th Ct Ocala FL 32671

JOHNSTON, HAROLD SLEDGE, chemistry educator; b. Woodstock, Ga., Oct. 11, 1920; s. Smith L. and Florine (Dial) J.; m. Mary Ella Stay, Dec. 29, 1948; children—Shirley Louise, Linda Marie, David Finley, Barbara Dial. A.B., Emory U., 1941, D.Sc., 1965; Ph.D., Calif. Inst. Tech., 1948. From instr. to asso. prof. chemistry Stanford, 1947-56; asso. prof. Calif. Inst. Tech., 1956-57; prof. chemistry U. Calif. at, Berkeley, 1957—, dean, 1966-70. Author: Gas Phase Reaction Rate Theory, 1966, Gas Phase Kinetics of Neutral Oxygen Species, 1968, Reduction of Stratospheric Ozone by Nitrogen Oxide Catalysts from Supersonic Transport Exhaust, 1971; Contbr. articles to profl. publs. Recipient Tyler prize for environ. achievement, 1983. Mem. Am. Chem. Soc. (Gold medal Calif. sect. 1956, Pollution Control award 1974), Am. Phys. Soc., Nat. Acad. Scis., Am. Acad. Arts and Scis., Am. Geophys. Union. Subspecialties: Atmospheric chemistry; Laser photochemistry. Current work: Theoretical atmospheric chemistry and laboratory photochemical studies related to problems in stratosphere. Home: 132 Highland Blvd Berkeley CA 94708

JOHNSTON, JAMES BENNETT, biochemist, researcher; b. San Diego, Dec. 31, 1943; s. Thomas Frazier and Mary Hamilton (Meads) J.; m. Margaret Jean Rosenberry, June 7, 1969; children: Mary Elizabeth, Amy Rose. B.S., U. Md., 1966; Ph.D., U. Wis., 1970. NSF-NATO postdoctoral fellow Pasteur Institut, Paris, 1970-72; research fellow U. Kent (Eng.), Canterbury, 1972-74; vis. asst. prof. U. Ill., Champaign/Urbana, 1974-76, asst. prof. environ. biochemistry, 1976-82; v.p. for research AgroBiotics Corp., Champaign, 1981—; cons. in field. Contbr. articles and chpts. to profl. publs. Sec. Champaign Environ. Adv. Commn., 1976-78. Mem. Am. Chem. Soc., Am. Soc. Microbiology, Environ. Mutagen Soc., AAAS. Democrat. Subspecialties: Biochemistry (biology); Genetics and genetic engineering (biology). Current work: Environmental mutagens; genetic aspects of pollution control; development of new pollution controls through applied genetics; development of novel crop plants through genetic manipulations. Home: 2310 Glenoak Champaign IL 61820 Office: U Ill 201 Environ Research Lab 1005 W Western Ave Urbana IL 61801

JOHNSTON, KENNETH J., radio astronomer; b. N.Y.C., Oct. 9, 1941; s. John James and Marion (Nugent) J.; m. Therese Clasen, June 25, 1966. B.E.E., Manhattan Coll., 1964; Ph.D. in Astronomy, Georgetown U., 1969. Nat. Acad. Sci./NRC postdoctoral fellow Naval Research Lab., Washington, 1969-71, radio astronomer, 1971-80, supervisory physicist, 1980—. Mem. Internat. Astron. Union, Am. Astron. Soc., Royal Astron. Soc., URSI. Subspecialty: Infrared optical astronomy. Current work: Astrometry; astrophysics; supervisor of Naval Research Lab. infrared and radio astronomy program.

JOHNSTON, MARILYN F. M., pathologist, educator; b. Buffalo, Mar. 30, 1937. B.S. summa cum laude, Dameon Coll., Buffalo, 1966; Ph.D. in Biochemistry, St. Louis U., 1970, M.D., 1975. Diplomate: Nat. Bd. Med. Examiners, Am. Bd. Pathology. Fellow in immunology Washington U., St. Louis, 1970-72; instr. biochemistry St. Louis U., 1972-75; resident in pathology Washington U. Hosps., 1975-77, St. John's Mercy Med. Ctr., St. Louis, 1977-79; research fellow in hematology St. Louis U. Sch. Medicine, 1979-80, asst. prof. pathology and dir. blood bank, 1980—; insp. Am. Assn. Blood Banks. Goldberger fellow, 1975. Mem. Coll. Am. Pathologists, Am. Assn. Immunologists, Am. Assn. Blood Banks, Am. Chem. Soc., Sigma Xi. Subspecialties: Pathology (medicine); Immunology (medicine). Current work: Immunohematology, structure and genetics of red cell surface antigens. Office: 1325 S Grand Blvd Saint Louis MO 63104

JOHNSTON, MARY HELEN, metallurgical engineer; b. West Palm Beach, Fla., Sept. 17, 1945. B.S., Fla. State U., 1966, M.S., 1969; Ph.D. in Metall. Engring, U. Fla., 1973. Materials engr. George C. Marshall Space Flight Center, NASA, Huntsville, Ala., 1969—; mem. staff metallurgy U. Ala., Huntsville, 1980—; prin. investigator Marshall Space Flight Ctr., NASA, 1976—; pres. Metall. Engring. Tech. Ala., Inc., 1982—. Mem. Am. Soc. Metals, Nat. Soc. Profl. Engrs. Subspecialty: Metallurgy. Office: Mat Sci EH 22 Marshall Space Flight Center Huntsville AL 35812

JOHNSTON, NORMAN PAUL, animal scientist, nutrition consultant; b. Salt Lake City, Apr. 5, 1941; s. Norman James and Olivia Harriet (Wilson) J.; m. Irene Hiller, Sept. 8, 1966; children: Julie, Cherie, Richard, Clark, David, Jed. B.A., Brigham Young U., 1966; M.S., Oreg. State U., 1967, Ph.D., 1971; M.B.A., U. Utah, 1969. Nutrition and mgmt. cons. Brookfield Products, 1968—; prof. animal sci. Brigham Young U., Provo, Utah, 1971—; mem. Utah Egg Mktg. Bd. Contbr. articles to profl. jours. Chester Wilcox Meml. scholar, 1966, 69-70. Mem. Am. Poultry Sci. Assn., Utah Feed Mfrs. (bd. dirs.), World Poultry Sci. Assn., Am. Agronomy Sci. Assn., Am. Rabbit Sci. Assn. (pres.), Sigma Xi, Phi Kappa Phi, Am. Rabbit Breeders Assn. Subspecialties: Animal nutrition; Animal breeding and embryo transplants. Current work: Poultry, rabbits and swine nutrition, poultry reproduction, small scale agriculture, turkey artificial insemination, goat nutrition. Home: 1795 S 340 E Orem UT 84057 Office: 343 WIDB Brigham Young U Provo UT 84602

JOHNSTON, ROBERT HOWARD, plant pathologist; b. Providence, May 29, 1946; s. Lawrence Adrian and Mildred Elizabeth (Drescher) J.; m. Susan Cook, June 8, 1970; 1 son, Jeffrey. B.S., Mont. State U., 1969, M.S., 1972. Research assoc. Mont. State U., 1972—. Contbr. articles on soil-borne diseases of small grains to profl. jours. Subspecialty: Plant pathology. Current work: Soil-borne plant diseases, chemical control of plant diseases. Home: 732 S Tracy St Bozeman MT 59715 Office: Dept of Plant Pathology Montana State University Bozeman MT 59717

JOHNSTON, STEPHEN ALBERT, scientist; b. Ft. Dodge, Iowa, Mar. 1, 1950; s. Richard Edward and Carol Ann (Coyle) J. B.S., U. Wis., 1975, M.S., 1977, Ph.D. in Genetics and Plant Breeding/Plant Genetics, 1980. Postdoctoral fellow M.S. Hershey Med. Center, Hershey, Pa., after 1981; now with dept. botany Duke U., Durham, N.C.; founder Sci. for Progress, 1979; writer, radio commentator on sci. issues. Contbr. articles to profl. jours. Rockefeller postdoctoral fellow, 1981; NIH fellow, 1982—. Mem. AAAS, Genetics Soc. Am., N.Y. Acad. Scis., Sigma Xi. Subspecialties: Molecular biology; Evolutionary biology. Current work: Investigation of genetic regulation in eukaryotes and its evolution. Office: Dept Botany Duke U Durham NC

JOISHY, SURESH K., oncologist and hematologist; b. Udipi, India, Mar. 12, 1944; U.S., 1970, naturalized, 1978; s. Keshav K. and Sushila J.; m. Muktha S. Pau, June 28, 1972; children: Mahima, Mahanth. M.B., B.S., Jawaharlal Inst. Post Grad. Med. Edn. and Research, India, 1970. Diplomate: Am. Bd. Internal Medicine. Clin. instr. U. Rochester, N.Y., 1976-78; cons. staff VA Med. Ctr., Batavia, N.Y., 1976-78; research asst. U. Calif., San Francisco, 1978-80; clin. asst. prof. Ind. U., Bloomington, 1981—; cons. Bloomington Hosp., 1980—, Dunn Meml. Hosp., Bedford, Ind., 1981—, Bedford Med. Ctr., 1981—. Author: Tissue Healing and Regeneration Folia tramatologica Geicy, 1979. Active Am. Cancer Soc. U. Calif. grantee, 1978-80; Asian Pacific Congress Gastroenterology travel grantee, 1980. Mem. Am. Fedn. for Clin. Research, Am. Soc. Hematology, Am. Soc. Clin. Oncology, AMA, Malaysian Med. Assn. Subspecialties: Hematology; Oncology. Current work: Cancer chemotherapy, cancer in twins, hemoglobinopathies, cancer prevention studies, cancer care through patient education, medical arts-graphic and painting. Office: Oncology and Hematology 822 W 1st St Bloomington IN 47401

JOISON, JULIO, surgeon; b. Cordoba City, Cordoba, Argentina, Oct. 2, 1932; came to U.S., 1960, naturalized, 1971; s. Moises and Sofia (Moses) J. B.S. summa cum laude, Dean Funes Coll., Cordoba, 1949; M.D., Nat. U. Cordoba, 1959; Ph.D., U. Buenos Aires, 1970. Diplomate: Am. Bd. Surgery. Intern and research asst. in surgery Sinai

Hosp., Balt., 1960-61; resident in surgery Met. Hosp., N.Y.C., 1961-62; assoc. fellow in surgery Lahey Clinic, Boston, 1962; resident and chief resident in surgery Boston City Hosp., 1962-65; sr. teaching fellow in surgery Boston U., 1964-65; research fellow in surgery Harvard U. Med. Sch., Boston, 1965-66, 68-70; practice medicine specializing in surgery, Boston, 1970—; assoc. staff St. Elizabeth's Hosp., Boston, 1976—; mem. staffs Brookline (Mass.) Hosp., 1971—, Hosp. at Parker Hill, Boston, 1970—, Hahnemann Hosp., 1976—. Mem. Mass. Med. Soc., Am. Coll. Gastroenterologist, N.Y. Acad. Scis., Am. Coll. Angiology, Am. Soc. Contemporary Medicine and Surgery. Subspecialties: Surgery; Transplantation. Current work: Surgery and transplantation. Discoverer metabolic factor produced by the pancreas; developer surg. technique for transplantation of pancreas. Home: 216 St Paul St Brookline MA 02146 Office: 1180 Beacon St Suite 3A Brookline MA 02146

JOIST, JOHANN HEINRICH, hematologist, educator; b. Bergisch Gladbach, W. Germany, Jan. 9, 1935; came to U.S., 1972; s. Heinrich and Katharina (Hasbach) J.; m. Nancy Lee Mexeiner, July 25, 1966; children: Bettina Lynn, Catherine Anne, Heidi Elaine. M.D., U. Cologne, W. Germany, 1962; Ph.D., McMaster U., Hamilton, Ont. Can., 1977. Lic. physician and surgeon Mo., 1973. Sr. Research fellow McMaster U., Hamilton, Ont., Can., 1970-72; asst. prof. medicine Washington U., St. Louis, 1972-78, assoc. prof. medicine/pathology, 1978-82, prof. medicine/pathology, 1982—; dir. Hemostasis Lab., Barnes Hosp., St. Louis, 1972-78, Div. Hematology St. Louis U. Med. Ctr., 1978-83, 1983—. Editor: Venous and Arterial Thrombosis, 1979. Mem. adv. com. Mo. Div. Health; chmn. med. adv. com. Mo./Ill. region ARC; assembly del. Am. Heart Assn. Council Thrombosis, Dallas, 1982. NIH research fellow, 1964-65; Ont. Heart Found. research fellow, 1970-72; recipient NIH individual research grant, 1982. Fellow ACP; mem. Am. Heart Assn., Am. Soc. Hematology, Central Soc. Clin. Research, Am. Assn. Pathologists, St. Louis Soc. Internal Medicine. Subspecialties: Hematology; Pathology (medicine). Current work: normal and abnormal hemostasis; thrombosis; mechanisms of platelet activation in flowing blood; interaction of platelets with red cells and endothelial cells in flowing blood; normal and abnormal blood coagulation. Home: 716 S Central Ave Clayton MO 63105 Office: St Louis U Med Center Div Hematology-Oncology 1402 S Grand Blvd St Louis MO 63104

JOKIPII, JACK RANDOLPH, astrophysicist, educator; b. Ironwood, Mich., Sept. 10, 1939; s. Toivo Sulo and Aili Maria (Juoni) J.; m. Bonnie Jeanne, Sept. 5, 1964; children: Galen, Eron, Kevin. B.Sc., U. Mich., 1961; Ph.D., Calif. Inst. Tech., 1965. Research asso. U. Chgo., 1965-67, asst. prof. physics, 1967-69; asso. prof. theoretical physics Calif. Inst. Tech., 1969-73; prof. astronomy and planetary scis. U. Ariz., 1973—; cons. RAND Corp., 1962-70. Research numerous publs. in field. Alfred P. Sloan Found. fellow, 1969-73; NASA grantee, 1974—; NSF grantee, 1976—. Mem. Am. Phys. Soc., Am. Astron. Soc., Am. Geophys. Union, Internat. Astron. Union. Subspecialties: Cosmic ray high energy astrophysics; Theoretical astrophysics. Current work: Theoretical astrophysics, with emphasis on cosmic rays, astrophys, plasma physics and turbulence, solar-terrestrial relations.

JOKLIK, WOLFGANG KARL, biochemist, virologist, educator; b. Vienna, Austria, Nov. 16, 1926; s. Karl F. and Helene (Giessl) J.; m. Judith Vivien Nicholas, Apr. 9, 1955 (dec. Apr. 1975); children: Richard G., Vivien H.; m. Patricia Hunter Downey, Apr. 23, 1977. B.Sc. with 1st class honors, U. Sydney, Australia, 1948, M.Sc., 1949; D.Phil. (Australian Nat. U. scholar), U. Oxford, Eng., 1952. Australian Nat. U. research fellow, Copenhagen, Denmark, 1953, Canberra, Australia, 1954-56, fellow, 1957-62; assoc. prof. cell biology Albert Einstein Coll. Medicine, Bronx, N.Y., 1962-65, prof. cell biology, 1965-68, Siegfried Ullmann prof. biochem. virology, 1966-68; prof., chmn. dept. microbiology and immunology Duke U. Med. Center, Durham, N.C., 1968—, James B. Duke Distinguished prof. microbiology and immunology, 1972—. Sr. author: Zinsser Microbiology, 15th, 16th, 17th and 18th edits; Contbr. articles profl. jours. Mem. Am. Soc. Cell Biology, Am. Soc. Microbiology, Am. Soc. Biol. Chemists, Am. Soc. Immunology, Nat. Acad. Scis., Inst. Medicine of Nat. Acad. Scis. Subspecialties: Virology (veterinary medicine); Microbiology. Current work: Properties of viruses; virus multiplication in molecular terms; arrangement of genetic material; expression of genetic information and how it is controlled. Address: Dept Microbiology and Immunology Duke U Med Center Durham NC 27710

JOKSIMOVICH, VOJIN, nuclear company executive, consultant; b. Belgrade, Serbia, Yugoslavia, Apr. 17, 1936; came to U.S., 1970, naturalized, 1976; s. Bozidar Joksimovic and Sylvia Turner; m. Nada Ajh, Nov. 26, 1961 (div. Apr. 1969); m. remarried, Oct. 1969; 1 dau., Natasha. Dipl.Ing., Electrotech. Faculty, Belgrade, 1961; Ph.D, Imperial Coll. Sci. and Tech., London, 1970. Registered profl. engr. in nuclear engring., Calif. Cons. engr. Energoprojekt, Belgrade, 1961-65; group leader Atomic Power Constrns., London, 1965-70; lead engr. Westinghouse Electric Corp., Pitts., 1970-72; br. mgr. Gen. Atomic, San Diego, 1973-80, dept. mgr., 1980-81; mgr. San Diego office NUS Corp., 1981—; chmn. various profl. conf. sessions. Contbr. articles to profl. jours. Chmn. Serbians for Reagon, San Diego, 1980; mem. Presdl. Task Force, Washington, 1982. Served with Yugoslavian Army, 1962-63. Mem. Am. Nuclear Soc. Republican. Serbian Orthodox. Club: Tennis Escondido (Calif.). Subspecialties: Nuclear engineering; Nuclear fission. Current work: Risk assessment, nuclear safety, nuclear waste management, nuclear and non-nuclear risks. Home: 406 Hidden Hills Ln Escondido CA 92025

JOLLES, MITCHELL IRA, mechanical engineer, educator, researcher; b. N.Y.C., Feb. 10, 1953; m.; 1 son, Matthew Ian. B.S. in Aerospace Engring, Poly. Inst. Bklyn., 1973, M.S., 1973; Ph.D. in Engring. Mechanics, Va. Poly. Inst., 1976. Instr. Va. Poly. Inst., 1973-76; asst. prof. U. Notre Dame, Ind., 1976-79; assoc. prof. U. Mo., Columbia, 1979-82; head fracture mech. sect. U.S. Naval Research Lab., Washington, 1982—. Mem. ASTM, Soc. Exptl. Stress Analysis (sec. com. fracture 1978-80), Soc. Engring. Scis., Am. Soc. Engring. Edn. (exec. com. edn. research and methods div. 1979-81, new engring. educator com. 1978-81). Subspecialties: Fracture mechanics; Solid mechanics. Current work: Basic and applied research on a broad spectrum of fracture mechanics; primary interest in stress analysis, fatigue propagation and elasticplastic fracture of 3-dimensional crack geometries of structural interest. Home: 6622 10th St Alexandria VA 22307 Office: Naval Research Lab Code 6382 Washington DC 20375

JONA, FRANCO PAUL, physicist, educator; b. Pistoia, Italy, Oct. 10, 1922; s. Frederico S. and Gabriella (Fenoglio) J.; m.; children: Frederico A., Franco S. Diplom, E.T.H., Zurich, 1946, Ph.D., 1949. Asst. prof. Pa. State U., 1952-57; research scientist Westinghouse Research Lab., Pitts., 1957-59, IBM Research Center, Yorktown Heights, N.Y., 1959-69; prof. materials sci. SUNY-Stony Brook, 1969—. Fellow Am. Phys. Soc.; mem. Italian Phys. Soc., Swiss Phys. Soc., Japanese Phys. Soc. Subspecialty: Condensed matter physics. Current work: Surface crystallography, surface physics and chemistry, computer applications. Office: Dept Materials Sci and Engring SUNY Stony Brook NY 11794

JONAS, JIRI, chemistry educator, researcher; b. Prague, Czechoslovakia, Apr. 1, 1932; m., 1968. B.S., Tech. U. Prague, 1956; Ph.D. in Chemistry, Czechslovak Acad. Sci., 1960. Research assoc. in chemistry Czech Acad. Sci., 1960-63; vis. scientist U. Ill.-Urbana, 1963-65, asst. prof. to assoc. prof., 1966-72, prof. chemistry, 1972—, sr. staff mem. materials research, 1970—; Alfred P. Sloan found. fellow, 1967-69, Guggenheim fellow, 1972-73; assoc. mem. Ctr. Advanced Study, U. Ill., 1976-77. Mem. Am. Chem. Soc., Am. Phys. Soc., AAAS. Subspecialties: Nuclear magnetic resonance (chemistry); Physical chemistry. Office: Dept Chemistry Univ Ill Urbana IL 61801

JONES, ALAN ANTHONY, chemist, educator; b. Jamestown, N.Y., Nov. 15, 1944; s. Elliott L. and Lois Katherine (Patterson) J.; m. Eunice Li, Mar. 1, 1972. B.A., Colgate U., 1966; Ph.D., U. Wis., 1972. Research instr. Dartmouth Coll., 1972-74; asst. prof. chemistry Clark U., 1974-78, asso. prof., 1978—, chmn. dept., 1980—. Contbr. articles to profl. jours. Mem. Worcester Conservation Commn., 1980—. NSF grantee, 1976-82; Petroleum Research Fund grantee, 1974; Research Corp. grantee, 1976; Gen. Electric Found. grantee, 1981; Dreyfus Found. grantee, 1982. Mem. Am. Chem. Soc. Subspecialties: Polymer chemistry; Nuclear magnetic resonance (chemistry). Current work: Polymer physical chemistry, especially chain dynamics. Home: 91 Wildwood Ave Worcester MA 01603 Office: Dept Chemistry Clark U Worcester MA 01610

JONES, ALAN LEE, plant pathologist; b. Albion, N.Y., June 23, 1939; m. Cara Collier, Aug. 26, 1967; children: M. Beatrix, Floyd A. B.S., Cornell U., 1961; M.S., 1963; Ph.D., N.C. State U., 1968. Asst. prof. botany and plant pathology Mich. State U., East Lansing, 1968-73, assoc. prof., 1973-77, prof., 1977—. Contbr. articles to profl. jours. Recipient Ciba-Geigy nat. award Am. Phytopath. Soc., 1978. Mem. Am. Phytopath. Soc., Can. Phytopath. Soc. Subspecialties: Plant pathology; Integrated pest management. Current work: Epidemiology and control of tree fruit diseases; fungicide evaluation, mycoplasma diseases. Office: Dept Botany and Plant Pathology Mich State U East Lansing MI 48824

JONES, ALFRED WELWOOD, mathematician, computer scientist; b. N.Y.C., July 6, 1915; s. Adam Leroy and Lily (Murray) J.; m. Pierette Jeanine Petas, Feb. 22, 1962; children: K. Darcy Jones Fuguet, Laurie Jones Bergamini, Alison Murray, Leroy Welwood, Bruce McKinley. B.A., Columbia U., 1937, M.A., 1939, Ph.D., 1944. Instr. U. Maine, Orono, 1939-42; instr. Yale U., New Haven, 1942-44; asst. prof. Mich. State U., East Lansing, 1944-47; assoc. prof. Rensselaer Poly. Inst., Troy, N.Y., 1947-57; systems engr. Bell Telephone Labs., Holmdel, N.J., 1957-64; researcher Inst. Def. Analysis, Arlington, Va., 1964-69; prof. engring. Wayne State U., Detroit, 1969-83; prof. computer systems Fla. Atlantic U., Boca Raton, 1983—; vis. prof. Inst. Advanced Study, Princeton, N.J., 1950-51; cons. Inst. Def. Analysis, 1969-72, Urban Inst., Washington, 1976—, U.S. Army Corps Engrs., 1977-79. Arthor: (with C. Eringen) Continuum Physics, 1971; author: Research Simulation, 1981, 83. Flutist various symphony orchs., 1945-69; Scoutmaster Boy Scouts Am., 1973-81. Recipient Disting. Service citation USAAF, 1945. Fellow AAAS; mem. Engring. Soc. Detroit (chmn. affiliate council 1982-83), Ops. Research Soc. Am. (chmn. edn. 1980-81), Soc. Computer Simulation, Human Factors Soc., IEEE (chmn. 1973-74), Phi Beta Kappa, Sigma Xi, Psi Upsilon, Phi Mu Alpha. Episcopalian. Subspecialties: Operations research (mathematics). Current work: Design and analysis of stochastic systems using systems models which are simluatedon a high speed computer. Dept Computer and Info Systems Fla Atlantic U Boca Raton FL 33431 Home: 2616 NW 37th St Boca Raton FL 33434

JONES, ANITA KATHERINE, computer science educator; b. Ft. Worth. A.B., Rice U., 1964; M.A., U. Tex., 1966; Ph.D. in Computer Sci, Carnegie-Mellon U., 1973. Programmer IBM, 1966-68; asst. prof. Carnegie-Mellon U., Pitts., 1973-78, assoc. prof. computer sci., 1978—; cons. NSF, 1973, Rand Corp., 1975—. NRC travel grantee, 1974. Mem. ACM, IEEE. Subspecialty: Operating systems. Office: Dept Computer Sci Carnegie-Mellon Univ Pittsburgh PA 15213

JONES, BEN MORGAN, research psychologist, adminstr.; b. Lawton, Okla., Aug. 1, 1943; s. Ben Greenleaf and Cynthia (Morgan) J.; m. Marilyn Kaye Miller, May 16, 1975. B.S., Okla. State U., 1965; Ph.D., U. Okla., 1972. Asst. prof. psychiatry and behavioral scis. U. Okla Health Scis. Ctr., Oklahoma City, 1972-77, assoc. prof., 1977-78; research psychologist Okla. Ctr. Alcohol and Drug-Related Studies, Oklahoma City, 1972-78, assoc. dir., 1973-78; mem. contbg. faculty Sch. Applied and Profl. Psychology Rutgers State U. of N.J., New Brunswick, 1978-80; research psychologist, cons. in clin. psychology Carrier Clinic Found., Belle Mead, N.J., 1978-80; research assoc. prof. psychiatry SUNY-Buffalo, 1981—, adj. prof. psychology, 1983—; cons. Alcohol Rev. Bd., 1981—; assoc. dir. Research Inst. on Alcoholism, Buffalo, 1980—; cons. clin. psychology VA, Buffalo, 1980—; bd. dirs. Research Found. for Mental Hygiene, Albany, N.Y., 1980—. Editor: Jour. Biol. Psychology Bull, 1971-72, Alcohol Tech. Reports, 1973-78. Mem. Erie County Com. of Alcoholism Profls., Buffalo, 1980—; mem. alcoholism/substance abuse subcom. of Erie County Mental Hygiene Community Services Bd., Buffalo, 1981—. NSF grantee, 1969-71. Mem. Am. Psychol. Assn., Internat. Neuropsychology Soc., N.Y. Acad. Scis., Research Soc. on Alcoholism, Sigma Xi (sec. chpt. 1981-82). Subspecialties: Neuropsychology; Cognition. Current work: Neuropsychological functioning of male and female alcoholics as a function of age, sex differences in social drinkers, acute effects of alcohol on cognitive abilities as a function of sex hormone status and resultant ethanol and acetaldehyde levels. Home: 6862 Old Lakeshore Rd Derby NY 14047 Office: Research Inst on Alcoholism 1021 Main St Buffalo NY 14203

JONES, CHARLES, mech. engr.; b. N.Y.C., Feb. 27, 1926; s. Leo and Mathilda (Stein) J.; m. Gisele Alice Guerin, Sept. 25, 1946; children: Corinne Gail, Leslie Muriel. B.S.M.E., Columbia U., 1950, M.S.M.E., 1953. With Curtiss-Wright Corp., Wood Ridge, N.J., 1950—, beginning as engine test engr., successively design engr., head stress and applied mechanics sect., chief design engr. rotating combustion engines, chief engr. rotating combustion engines, dir. engring. rotating combustion engines, 1950-81, dir. research rotating combustion engines, 1981—; cons. Wankel-type rotating combustion engines to Curtiss-Wright licensees and govt. agys. Contbr. numerous articles to profl. jours. Served with USAAF, 1944-46. Named No. 1 Rotary Wankel Engr. Wards Wankel Report, 1972. Mem. Soc. Automotive Engrs., Engring. Soc. Detroit, Tau Beta Pi. Subspecialties: Mechanical engineering; Combustion processes. Current work: Advanced Rotary Wankel-type engine research and tech. Holder 70 patents on Wankel-type engines. Home: 208 Forest Dr Hillsdale NJ 07642 Office: 1 Passaic St Wood Ridge NJ 07075

JONES, CLARIS EUGENE, JR., botany educator, writer; b. Columbus, Ohio, Dec. 15, 1942; s. Claris Eugene and Clara Elizabeth (Elliott) J.; m. Teresa Diane Wagner, June 26, 1966; children: Douglas Eugene, Philip Charles, Elizabeth Lynne. B.S., Ohio U., 1964; Ph.D., Ind. U., 1969. Asst. prof. botany Calif. State U., Fullerton, 1969-73, assoc. prof., 1973-77, prof., 1977—, dir. Fullerton Arboretum, 1970-80, dir. Arboretum Project, 1970-76; dir. Faye McFadden Herbarium, 1969—; vis. asst. prof. U. Mich., summer, 1972; part-time faculty Orgn. for Tropical Studies, 1970, 72; pres. Arboretum Bd. dirs. Arboretum Soc., Inc., Calif. State U., Fullerton, 1972-78. Co-author: A Dictionary of Botany, 1980; sr. editor: Handbook of Experimental Pollination Biology, 1983; Contbr. articles on botany to profl. jours. NSF grantee, 1973. Mem. Am. Inst. Biol. Sci., AAAS, Bot. Soc. Am., Am. Soc. Plant Taxonomists, Internat. Assn. for Plant Taxonomy, Soc. for the Study of Evolution, Am. Soc. Naturalists, So. Calif. Botanists, Soc. Calif. Acad. Sci., Systematics Assn., Soc. for Econ. Botany, Ecol. Soc. Am., Calif. Bot. Soc., Sigma Xi (grantee 1970). Democrat. Methodist. Subspecialties: Evolutionary biology; Systematics. Current work: Pollination ecology and plant biosystematics. Office: Dept Biol Sci Calif State U Fullerton CA 92634

JONES, DANIEL TODD, human factors project leader, human factors/ systems researcher; b. San Antonio, May 18, 1950; s. Daniel Burr and Bettie Marsh (Garrison) J.; m. Susanne Elaine Miller, July 19, 1975; 1 dau., Sarah Susanne. B.S. in Engring., U. Central Fla., 1972. Registered profl. engr., Md. Indsl. engr. Naval Air Rework Facility, Jacksonville, Md., 1973, Naval Ordnance Sta., Indian Head, Md., 1973-77; human factors engr. U.S. Coast Guard, Washington, 1977—; cons. in field., 1981—. Served with USN, 1968-69. Mem. Human Factors Soc., Inst. Indsl. Engrs., AIAA, AAAS, Nat. Soc. Profl. Engrs., Phi Kappa Alpha. Republican. Methodist. Club: Methodist Men (Friendly, Md.) (sec. 1982-83). Subspecialties: Human factors engineering; Systems engineering. Current work: Human factors and systems engineering research in commercial vessel and coast guard systems; problems include noise, vibration, human performance, systems design and operation, simulators. Home: 120 Indian Ct Waldorf MD 20601 Office: USCG 2100 2d St SW Washington DC 20593

JONES, DAVID JOSEPH, pharmacology/anesthesiology educator, consultant; b. Greasby, Cheshire, Eng., Mar. 12, 1945; came to U.S., 1952; s. William Leslie and Amy Catherine (Halsall) J.; m. Jo Lynn Braun, June 13, 1970; 1 dau., Leslie Anne. B.S., U. Tex.-Austin, 1968, M.S., 1970; Ph.D., U. Tex.-San Antonio, 1974. Teaching asst. U. Tex.-Austin, 1968-70, research scientist, 1970-71; teaching asst. Health Sci. Ctr., U. Tex.-San Antonio, 1971-74, asst. prof. anesthesiology, 1974-80, assoc. prof., 1980—; cons. Bexar County Hosp. Dist., San Antonio, 1974—, Audie Murphy VA Hosp., 1974—. Pres. Whispering Oaks Homeowners Assn., San Antonio, 1982; bd. dirs. Colonies North PTA, San Antonio, 1982. Mem. Am. Soc. Pharmacology and Exptl. Therapeutics, Am. Soc. Neurochemistry, Soc. Neurosci. (pres. San Antonio chpt. 1981-82); assoc. mem. Am. Soc. Anesthesiologists, Internat. Soc. Neurochemistry, Internat. Soc. Anesthesia Research. Republican. Episcopalian. Lodge: Kiwanis. Subspecialties: Neuropharmacology; Anesthesiology. Current work: Spinal cord receptor pharmacology, spinal cord adrenergic systems, development of spinal cord, spinal cord injury, receptor pharmacology. Office: Dept Anesthesiology U Tex Health Sci Ctr 7703 Floyd Curl Dr San Antonio TX 78284 Home: 2818 Whisper Fawn San Antonio TX 78230

JONES, DAYTON LOREN, astronomer, instrument designer; b. Phila., Sept. 10, 1951; s. Loren F. and Mary (Larzelere) J.; m. Debra Kay Grubb, June 19, 1981. B.A. in Physics, Carleton Coll., 1974; M.S. in Sci. Instrumentation, U. Calif.-Santa Barbara, 1976, Cornell U., 1979, Ph.D., 1981; Summer student, Nat. Radio Astronomy Obs., Green Bank, W.Va., 1974-75. Research fellow in radio astronomy Calif. Inst. Tech., 1981—. Contbr. articles to profl. jours. Active Amnesty Internat., liberal polit. orgns., various environ./conservation groups. Mem. Am. Astron. Soc., Sigma Xi. Democrat. Subspecialties: Radio and microwave astronomy; Optical astronomy. Current work: Very-Long-Baseline Interferometry observations of extragalactic radio sources; optical emission from radio jets; image processing; constructing cryogenically-cooled receivers for Calif. Inst. Tech.'s 40-meter telescope at Owens Valley Radio Obs. Office: Calif Inst Tech 105-24 Pasadena CA 91125

JONES, DONALD GEORGE, chemistry educator; b. Bridgeton, N.J., June 10, 1937; s. Donald Booth and Edna Ware (Parsons) J.; m. Joycelyn Goncz, June 16, 1959; children: Donald Laurence, Julie Joycelyn. B.A., Columbia Union Coll., 1957; Ph.D., U. Md., 1961. Faculty Columbia Union Coll., Takoma Park, Md., 1961—, now chmn. dept. biology and chemistry. Mem. Am. Chem. Soc., Sigma Xi. Republican. Seventh-day Adventist. Subspecialties: Organic chemistry; Physical chemistry. Current work: Organic reaction kinetics, curriculum devel. Home: 8317 Flower Ave Takoma Park MD 20912 Office: Columbia Union Coll Takoma Park MD 20912

JONES, DOUGLAS LINWOOD, mech. engr., educator, cons.; b. Limeton, Va., Dec. 26, 1937; s. Charlie Baxley and Irma Kathelle (Murphy) J.; m. Mary Kay O'Brien, Jan. 2, 1975. B.M.E., George Washington U., 1963, M.S. in Engring. (NASA fellow 1963-66), 1965, D.Sc., 1970. Registered profl. engr., Va. Aerospace technologist Goddard Space Flight Center, NASA, Greenbelt, Md., 1963; teaching fellow George Washington U., Washington, 1966-67, instr. engring. and applied sci., 1967-68, asst. prof., 1968-71, asst. research prof. engring., 1971-74, assoc. research prof., 1974-77, assoc. prof. engring., 1977-82, prof., 1982—; cons. on fracture mechanics, failure analysis, structural analysis and design Comsat Labs., ENSCO Corp., Battelle Meml. Inst., Systems Tech. Labs., duPont Corp. Contbr. articles to profl. jours. Bd. dirs. Wesley Found., George Washington U., 1966-67, vice-chmn., 1969-73, chmn., 1973-76; vice chmn. Commn. on Higher Edn. and Campus Ministry, Balt. Conf. of United Meth. Ch., 1977. Dept. Def. grantee, 1976—; NASA grantee, 1977-80. Mem. Am. Acad. Mechanics, ASME, Am. Soc. for Engring. Edn., Soc. for Exptl. Stress Analysis, ASTM, George Washington U. Engr. Alumni Assn. (dir. 1963—, treas. 1964-66,/69-70, pres. 1970-72, Engr. Alumni Service award 1976), Gen. Alumni Assn. George Washington U. (governing bd. 1970-72, 73-82, Alumni Service award 1974), Sigma Xi, Tau Beta Pi, Pi Tau Sigma. Subspecialties: Fracture mechanics; Composite materials. Current work: Research in fracture, fatigue and mechanical behavior of engineering materials; practical development of the edge-sliding mode of fracture; studies into the fatigue behavior of composite materials. Home: 1818 N Cleveland St Arlington VA 22201 Office: George Washington U Washington DC 20052

JONES, ERIC MANNING, astronomer; b. Goldsboro, N.C., Mar. 25, 1944; s. Thomas Morgan and Jean (Manning) J.; m. Sandie Turner, Sept. 8, 1968; m. Helen Shane, May 16, 1980. B.S. in Astronomy, Calif. Inst. Tech., 1966, Ph.D., U.Wis.-Madison, 1969. Mem. staff Los Alamos Nat. Lab., 1969-75, group leader, 1976-80, mem. staff, 1980—; mem. Dept. Energy Containment Eval. Panel, 1976-81. Contbr. articles to profl. jours. Mem. Los Alamos Little Theater. Mem. Am. Astron. Soc., Internat. Astron. Union. Subspecialties: Theoretical astrophysics; Theoretical physics. Current work: Numerical hydrodynamics, supernova remnants, atmospheric nuclear weapons effects, human migrations, space colonization. Home: 4266 Ridgeway Dr Los Alamos NM 87544 Office: Los Alamos Nat Lab Mail Stop F665 Los Alamos NM 87545

JONES, FRANCIS THOMAS, chemist, educator; b. Pottsville, Pa., Oct. 19, 1933; s. Francis Thomas and Marion A. (Kagel) J.; m. Nuran Kumbaraci, Jan. 3, 1981. B.S., Pa. State U., 1955; Ph.D., Poly. Inst. Bklyn., 1960; M. Engring. (hon.), Stevens Inst. Tech., 1975. Postdoctoral fellow U. Leeds, Eng., 1960-62; research chemist Union Carbide Nuclear Co., Tuxedo, N.Y., 1962-64; faculty participant Oak Ridge Nat. Lab., summer 1965; successively asst. prof., assoc. prof., prof. chemistry Stevens Inst. Tech., Hoboken, N.J., 1964—, chmn. dept. chemistry and chem. engring., 1979—. Author research publs.; designer buildings at, Stevens Inst. and in, Algeria. Mem. Am. Chem.

Soc., Am. Inst. Chem. Engrs. Subspecialties: Physical chemistry; Chemical engineering. Current work: Research in mass spectrometry. Patentee hydrazine formation by ionizing radiation. Home: 692 Steward St Ridgefield NJ 07657 Office: Dept Chemistry Stevens Inst Tech Castle Point Sta Hoboken NJ 07030

JONES, HOWARD WILBUR, JR., gynecologist; b. Balt., Dec. 30, 1910; s. Howard Wilbur and Ethel Ruth (Marling) J.; m. Georgeanna Emory Seegar, June 22, 1940; children—Howard Wilbur III, Georgeanna S., Lawrence M. A.B., Amherst Coll., 1931; M.D., Johns Hopkins, 1935; Dr. Honoris Causa, Cordoba, 1968. Intern, asst. resident, resident gynecology Johns Hopkins Hosp., 1935-37, 46-48; asst. resident, resident surgery Ch. Home and Hosp., Balt., 1937-40; practice medicine, specializing in obstetrics and gynecology, Balt., 1948—; instr., asst. prof., asso. prof., prof. gynecology and obstetrics Sch. Medicine Johns Hopkins, 1948-79, prof. emeritus, 1979—; prof. obstetrics and gynecology Eastern Va. U. Med. Sch., 1978—; nat. cons. USAF, 1968-78; Dir. William & Wilkins Co. Author: (with W.W. Scott) Genital Anomalies and Related Endocrine Disorders, 1958, rev. edit., 1971, (with G.S. Jones) Textbook of Gynecology, 1965, 10th edit., 1981, (with R. Heller) Pediatric and Adolescent Gynecology, 1968, (with J.A. Rock) Reparative and Constructive Surgery of the Female Generative Tract, 1983; Editor in chief: (with G.E.S. Jones) Obstetrical and Gynecological Survey, 1957—; Contbr. articles to profl. jours. Served to maj. M.C. AUS, 1943-46. Decorated Bronze Star medal. Mem. AMA, Am. Assn. Cancer Research, Am. Cancer Soc. (dir. Md. div.), Am. Coll. Obstetrics and Gynecology, Soc. Pelvic Surgeons, Sociedad de Obstetricia Y Gynecologia die Buenos Aires, Sociedad Peruana de Obstetricia Y Ginecologia. Subspecialty: Obstetrics and gynecology. Current work: Reproductive biology. Home: 7506 Shirland Norfolk VA 23505 Office: 603 Med Tower Norfolk VA 23507

JONES, JAMES THOMAS, JR., air force officer; b. LaGrange, Ga., Feb. 5, 1943; s. James Thomas and Louise (Priddy) J.; m. Jeanne Louise Jones, June 20, 1981; children: Andrea P., James Thomas. B.S., U. Ga., 1965; M.S., So. Meth. U., 1970. Commd. U.S. Air Force, 1965, advanced through grades to lt. col., 1981—; ops. research analyst (Tac. Fighter Weapons Group), Nellis AFB, Nev., 1970-72, Edwards AFB, Calif., 1972-75, Eglin AFB, Fla., 1975-78, Washington, 1978-82; asso. prof. ops. research Def. Intelligence Sch., Washington, 1982, vice dean, 1982—; mem. Washington Ops. Research & Mgmt. Sci. Council, 1983—. Contbr. articles to profl. jours. Leader Boy Scouts Am., Niceville, Fla., 1977-78; coach Boy's Little League, Edwards, Calif., 1972-76. Mem. Ops. Research Soc. Am., Mil. Ops. Research Soc. Subspecialties: Operations research (engineering); Probability. Current work: Applications for quantitative methods in intelligence bus. Home: 6149 N Morgan St Alexandria VA 22312 Office: Def Intelligence Coll Washington DC 20743

JONES, JAMES W., agrl. engring. educator; b. Ripley, Miss., Feb. 28, 1944; s. Noverta L. and Ollie D. (Muse) J.; m. Jean W. Williams, Nov. 24, 1966; children: Janene, Joanna, Jennifer. B.S., Tex. Tech U., 1966; M.S., Miss. State U., 1970; Ph.D., N.C. State U., 1975. Agrl. engr. Agrl. Research Service, U.S. Dept. Agr., Starkville, Miss., 1967-72, Raleigh, N.C., 1972-73; asst. prof. Miss. State U., Starkville, 1974-77; assoc. prof. engring. in agr. U. Fla., Gainesville, 1977-80, prof., 1980—. Co-editor: Predicting Photosynthesis for Ecosystem Models, 2 vols, 1980. Mem. Am. Soc. Agrl. Engrs., Am. Soc. Agronomy, Crop Soc. Am., AAAS, Sigma Xi, Tau Beta Pi. Republican. Baptist. Subspecialties: Agricultural engineering; Systems engineering. Current work: Modeling agricultural production systems, optimization of farm production, research on transpiration, photosynthesis, and water stress in crops, crop modeling, micro-computer implementation of models. Office: Dept Agrl Engring U Fla Gainesville FL 32611

JONES, JOHN PAUL, educator, plant pathologist; b. Warren, Ohio, Dec. 10, 1924; s. Robert Paul and Sueanna Florence (Atchison) J.; m. Joyce Shoemaker, May 7, 1950; children: Karen, Stephen, Lynn. B.S., Ohio U., 1950, M.S., 1953; Ph.D., U. Nebr., 1956. Research technician Ohio Agrl. Expt. Sta., 1949-50; research asst. U. Nebr., 1950-55; research plant pathologist U.S. Dept. Agr., Stoneville, Miss., 1955-60; prof. plant pathology U. Ark., 1960—; project leader plant protection Rice Research and Tng. Project, FAO, Dokki, Cairo, 1981—. Contbr. numerous articles to profl. jours. Served with AUS, 1943-46. NSF research grantee, 1963-65; teaching grantee, 1966, 69; U. Ark. Endowment Fund grantee, 1976. Mem. Am. Phytopathol. Soc., Assn. So. Agrl. Workers, Ark. Acad. Sci., Internat. Soc. Plant Pathology, Gamma Sigma Delta. Presbyterian. Lodge: Masons. Subspecialties: Plant pathology; Integrated pest management. Current work: Etiology and control of cereal diseases, primarily wheat, oats and rice. Home: 4 E Sycamore St Fayetteville AR 72701

JONES, JOHN PAUL, plant pathology educator; b. Stockdale, Ohio, Feb. 24, 1932; s. Paul Herdman and Corda Ellen (Cameron) J.; m. Peggy Jean Baker, Jan. 21, 1959; children: Kevin Edward, Andrew Wade, Carisa Lynn. B.S., Ohio State U., 1953, M.S., 1955, Ph.D., 1958. Research asst. Ohio State U., Columbus, 1954-58; asst. prof. U. Fla., Agrl. Research and Edn. Ctr., Bradenton, 1958-65, assoc. prof., 1965-68, prof. plant pathology, 1968—. Recipient ann. research award Fla. Fruit and Vegetable Soc., 1971. Mem. Am. Phytopath. Soc., Fla. State Hort. Soc. (vegetable paper award 1973), Sigma Xi, Phi Epsilon Phi. Subspecialties: Plant pathology; Integrated pest management. Current work: Vegetable disease control by integration of genetic, cultural and chemical factors. Home: 332 Greenwood Ave Sarasota FL 33580 Office: 5007 60th St E Bradenton FL 33508

JONES, LORELLA MARGARET, physics educator; b. Toronto, Ont., Can., Feb. 22, 1943; d. Donald Cecil and F. Shirley (Patterson) J. B.A., Harvard U., 1964; M.Sc., Calif. Inst. Tech., 1966, Ph.D. in Physics, 1968. Postdoctoral fellow Calif. Inst. Tech., 1967-68; asst. prof. physics U. Ill., Urbana, 1968-70, assoc. prof., 1970-78, prof., 1978—. Author: Introduction to Mathematical Methods of Physics, 1974; contbr. articles to profl. jours. Fellow Am. Phys. Soc. Subspecialty: Particle physics. Current work: Phenomenological calculations in elementary particle theory. Office: Dept Physics U Ill 315 Loomis Lab 1110 W Green St Urbana IL 61801

JONES, LYLE VINCENT, psychology educator; b. Grandview, Wash., Mar. 11, 1924; s. Vincent F. and Matilda M. (Abraham) J.; m. Patricia Edison Powers, Dec. 17, 1949 (div. 1979); children: Christopher V., Susan E., Tad W. Student, Reed Coll., 1942-43; B.S., U. Wash., 1947, M.S., 1948; Ph.D., Stanford, 1950. Nat. Research fellow, 1950-51; asst. prof. psychology U. Chgo., 1951-57; vis. asso. prof. U. Tex., 1956-57; asso. prof. U. N.C., 1957-60, prof., 1960-69, Alumni disting. prof., 1969—, dir., 1957-74, 79—, vice chancellor, dean, 1969-79; pres. Assn. Grad. Schs., 1976-77; cons. in field. Author: (with others) Studies in Aphasia: An Approach to Testing, 1961, The Measurement and Prediction of Judgment and Choice, 1968, (with others) An Assessment of Research-Doctorate Programs in the United States, 5 vols.; Mng. editor: Psychometrika, 1956-61; Editorial com. for psychology, Mc-Graw-Hill, 1965-77; Contbr. articles to profl. jours. Served with USAF, 1943-46. Recipient Thomas Jefferson award U. N.C., 1979; Fellow Center Advanced Study in Behavioral Scis., 1964-65, 81-82; grantee NIH, 1957-63, NSF, 1960-63, 71-74, NIMH, 1963-74, 79—. Fellow AAAS, Am. Psychol. Assn. (pres. div. 1963-64); mem. Psychometric Soc. (pres. 1962-63), Am. Statis. Assn., Am. Ednl. Research Assn. Subspecialty: Psychometrics. Current work: Assessing scholastic achievement of nation's youth. Home: Rt 1 Pittsboro NC 27312 Office: Davie Hall U NC Chapel Hill NC 27514

JONES, M(ARVIN) THOMAS, chemistry educator, university administrator; b. St. Louis, Apr. 20, 1936; s. Thomas S. and Margaret V. (Evans) J.; m. Patricia Adel Lenz, June 13, 1958; children: Jonathan Thomas, Jennifer Tracy. A.B., Washington U., St. Louis, 1958, Ph.D., 1961. Mem. research staff central research dept. E.I. duPont de Nemours & Co., Wilmington, Del., 1961-66; assoc. prof. chemistry St. Louis U., 1966-69, U. Mo.-St. Louis, 1969-71, prof., 1971—, acting dean, 1978-79, assoc. dean, 1976—; research assoc. U. Groningen, Netherlands, 1975. Contbr. chpts. to books, articles to profl. jours. Fulbright-Hays alt. fellow, 1975-76. Mem. Am. Chem. Soc. (nat. councilor 1981-85, chmn. St. Louis sect. 1978, award St. Louis sect. 1983), Am. Phys. Soc., AAAS, AAUP, Phi Kappa Phi. Subspecialties: Physical chemistry; Solid state chemistry. Current work: Long term interest in the use of electron spin resonance spectroscopy to study problems of chemical and physical interest; studies of low dimensional synthetic electrical conductors. Home: 286 Elm Ave Glendale MO 63122 Office: Dept Chemistry U Mo Saint Louis 8001 Natural Bridge Rd Saint Louis MO 63121

JONES, MAURICE (MO), JR., energy service co. research and development executive, consultant; b. Pasadena, Tex., Nov. 29, 1951; s. Maurice and Yvonne (Ferguson) J. B.S. in Biology, U. Houston, 1974; M.S. in Environ. Sci, U. Tex.-Houston, 1977. Biologist, technician S.W. Research Inst., Houston, 1974-76; supr. Environ. Lab., Dames & Moore, Houston, 1978-79; supr. environ. affairs IMCO Services, Houston, 1979-82, mgr. environ. services, 1982-83, mgr. research and devel., 1983—; cons. Marine Bd., NRC, Washington, 1980, 83; chmn. drilling fluids task force Petroleum Equipment Suppliers Assn., Houston, 1980-81; chmn. bioassay protocol Am. Petroleum Inst., Dallas, 1982—; co-chmn. environ. sect. Soc. Petroleum Engr./Internat. Assn. Drilling Contractors Symposium, New Orleans, 1983. Contbr. numerous articles to profl. jours.; editor: Perspective on Drilling Fluids and their Environmental Impact, 1984. USPHS trainee, 1974; Marine Biol. Lab. scholar, 1978. Mem. Ecol. Soc. Am. (cert.), Nat. Assn. Environ. Profls., Soc. Ecotoxicology and Environ. Safety, Soc. Petroleum Industry Biologists, Marine Tech. Soc. Democrat. Subspecialties: Environmental toxicology; Offshore technology. Current work: Drilling fluids science and technology; estuarine biology; environmental law and regulations; bioassay procedures; biometry. Home: 1717 Hazard Houston TX 77019 Office: IMCO Services 2400 W Loop S Houston TX 77227

JONES, MOLLY MODRALL, clin. psychologist; b. Albuquerque, Apr. 20, 1940; d. James Ritchie and Constance (Connor) M.; m. Lawrence L. Burckmyer, Feb. 22, 1963 (div. 1968); children: Elizabeth Loring, Mary Constance. B.A., U. Colo., 1961; M.A., Harvard U., 1964, C.A.S., 1965; Ph.D., Union Grad. Sch., Cin., 1978. Licensed psychologist, N.M., D.C., Md. Sch. diagnostician Santa Fe (N.Mex.) pub. schs., 1969-72; clin. psychologist USPHS, Santa Fe, 1974-79; asso. prof. Coll. of Santa Fe, 1972-79; health sci. adminstrs. NIH, Bethesda, Md., 1979-81; clin. psychologist in pvt. practice, Chevy Chase, Md., 1981—; cons. Nat. Presbyn. Counseling Center, 1980—, N.Mex. Dept. Health & Social Services, 1974-79, Indian Health Service, Santa Fe, 1974-79. Contbr. articles to profl. jours. Cons. Rape Crisis Center, Santa Fe, 1972-79; trustee Santa Fe Prep. Sch., 1975-79; precinct capt. Democratic Party, Santa Fe, 1972-74, del., 1972-74; del. others. Am. Council on Edn. fellow, 1979-81; U. N.M. fellow, 1978-79; Indian Health Service predoctoral fellow, 1976-78; research fellow, 1973-78; Shell Oil Co. fellow, 1972. Mem. Am. Psychol. Assn., AAUP, Md. Soc. Clin. Hypnosis (treas.), Washington Psychologists for Psychoanalysis (program chmn.), Md. Psychol. Assn., D.C. Psychol. Assn. Democrat. Presbyterian. Clubs: Kenwood Golf and Tennis, Sangre de Cristo Racquet, Harvard. Subspecialties: Developmental psychology; Physiological psychology. Current work: Pain control, psychosomatic-psychophysiologic conditions psychodynamic psychotherapy, clin. hypnosis, pain control, relaxation tng. Home: 4964 Allan Rd Bethesda MD 20816 Office: 5454 Wisconsin Ave Suite 600 Chevy Chase MD 20815

JONES, ORVAL ELMER, research and development company executive, researcher; b. Ft. Morgan, Colo., Apr. 9, 1934; s. Lawrence Roswell and Hazel Mildred (Stemen) J.; m. Pauline Anna Lunka, May 29, 1934; children: Carol Leslie, Sharon Lynn, Lawrence Anthony. B.S. in Mech. Engring. (Boettcher Found. scholar), Colo. State U., 1956; M.S. in Mech. Engring. (Hughes Aircraft fellow), Calif. Inst. Tech., 1957; Ph.D. in Mech. Engring. (NSF fellow), Calif. Inst. Tech., 1961. Mem. staff Hughes Aircraft Co., Culver City, Calif., 1956-57; research engr. Hydrodynamics Lab., Calif. Inst. Tech., 1960-61; with Sandia Labs. (name changed to Sandia Nat. Labs. 1980), Albuquerque, 1961—, dir. nuclear waste and environ. programs, 1977-78, dir. engring. scis., 1978-82; v.p. responsible for engring., tech. support and testing (name changed to Sandia Nat. Labs. 1980), 1982-83; v.p. responsible Sandia Labs., def. programs name changed to Sandia Nat. Labs., 1983—; adj. prof. U. N. Mex., 1963-64, 67-68. Contbr. articles on theoretical and exptl. investigations of elastic and inelastic wave propagation in solids, phys. protection safeguards for nuclear materials to profl. jours. Com. chmn. cub scout pack 395, 1972-73; troop 395 council Boy Scouts Am., 1974-78, Webelos leader, 1973-74; mem. ednl. futures long-range planning com. Albuquerque Pub. Schs., 1981-83. Mem. Am. Phys. Soc., ASME, AAAS, Sigma Xi. Episcopalian. Subspecialties: Solid mechanics; Systems engineering. Office: Sandia Nat Labs Orgn 7000 Albuquerque NM 87185

JONES, RICHARD VICTOR, physics educator; b. Oakland, Calif., June 8, 1929; married. A.B., U. Calif., 1951, Ph.D. in Physics, 1956; M.A. (hon.), Harvard U., 1961. Sr. engr. Shockley Semiconductor Lab., Beckman Instruments, Inc., 1955-57; asst. prof. to assoc. prof. applied physics Harvard U., Cambridge, Mass., 1957-71, assoc. dean div. engring. and applied physics, 1969-71, dean, 1971-72, prof. applied physics, 1971—; vis. MacKay prof. U. Calif.-Berkeley, 1967-68. Guggenheim fellow, 1960-61. Mem. Am. Phys. Soc., IEEE. Subspecialty: Solid state physics. Office: Gordon McKay Lab Harvard Univ Cambridge MA 02138

JONES, ROBERT EDWIN, JR., physicist, solar energy scientist, educator, researcher; b. Kansas City, Mo., Jan. 4, 1944; s. Robert Edwin and Carol Nadine J.; m. Karen Kay, Mar. 1, 1970; children: Erik Michael, Stephanie Lynn. B.A. in Physics, Kans. State Coll., 1965; Ph.D. in Solid State Physics, Iowa State U., 1971. Tchr. Bismark Schule, Hamburg, W.Ger., 1971-72; postdoctoral fellow Lakehead U., Thunder Bay, Ont., Can., 1972-75, asst. prof., 1975-78; asst. prof. physics and energy sci. U. Colo., Colorado Springs, 1978-80, assoc. prof., 1980—; cons. solar energy, semicondr. materials. Contbr. articles to profl. jours. NRC Can. grantee, 1976-79, 78; Colo. Energy Research Inst. grantee, 1979; NSF grantee, 1980-81, 80-82. Mem. Am. Phys. Soc., Internat. Solar Energy Soc., Am. Solar Energy Soc. Methodist. Subspecialties: Solar energy; Condensed matter physics. Current work: Solar energy optics, including shading effects and concentrators; semiconductor materials for solar cells and microelectronics. Home: 2722 Inspiration Dr Colorado Springs CO 80917 Office: U Colo Austin Bluffs Pkwy Colorado Springs CO 80907

JONES, ROBERT THOMAS, aero. scientist; b. Macon, Mo., May 28, 1910; s. Edward Seward and Harriet Ellen (Johnson) J.; m. Barbara Jeanne Spagnoli, Nov. 23, 1964; children—Edward, Patricia, Harriet, David, Gregory, John. Student, U. Mo., 1928; Sc.D. (hon.), U. Colo., 1971. Aero. research scientist NACA, Langley Field, Va., 1934-46; research scientist Ames Research Center NACA-NASA, Moffet Field, Calif., 1946-62; sr. staff scientist Ames Research Center, NASA, 1970-81, research asso., 1981—; scientist Avco-Everett Research Lab., Everett, Mass., 1962-70; cons. prof. Stanford U., 1981. Author: (with Doris Cohen) High Speed Wing Theory, 1960, Collected Works of Robert T. Jones, 1976; contbr.: articles to profl. jours. Collected Works of Robert T. Jones. Recipient Reed award Inst. Aero. Scis., 1946; Inventions and Contbns. award NASA, 1975; Prandtl Ring award Deutsche Gesellschaft für Luft and Raumfahrt, 1978; Pres.'s medal for disting. fed. service, 1980; Langley medal Smithsonian Instn., 1981. Fellow Am. Inst. Aeros. and Astronautics (hon.); mem. Am. Acad. Arts and Scis., Nat. Acad. Scis., Nat. Acad. Engring. Subspecialties: Aerospace engineering and technology; Aeronautical engineering. Home: 25005 La Loma Dr Los Altos Hills CA 94022 Office: NASA Ames Research Center Moffett Field CA 94035

JONES, RUSSELL ALLEN, oncologist; b. Waycross, Ga., Nov. 20, 1940; s. Russell Allen and Margaret (Blount) J.; m. Kate Mattison, July 3, 1967; children: Jennifer, Jason. M.D., Emory U., 1965. Diplomate: Am. Bd. Internal Medicine. Intern Grady Meml. Hosp., Atlanta, 1965-66, resident, 1966-67, U. Wash., Seattle, 1969-70; fellow in cancer and blood disease Emory U., 1970-71; practice medicine specializing in med. oncology, Clifton Forge, Va., 1971-77, Chattanooga, 1978—. Served to lt. comdr. U.S. Navy, 1967-69. Fellow A.C.P.; mem. Am. Soc. Clin. Oncology, AMA, Alpha Omega Alpha, Alpha Epsilon Upsilon. Presbyterian. Club: Signal Mountain (Tenn.) Golf and Country. Subspecialties: Chemotherapy; Cancer research (medicine). Current work: Cancer chemotherapy. Office: Medical Oncology Associates 975 E 3d St Box 144 Chattanooga TN 37403

JONES, STEPHEN WALLACE, neurobiologist; b. Steubenville, Ohio, Nov. 7, 1953; s. Frederick and Muriel (Squier) J. B.S., Mich. State U., 1974; Ph.D., Cornell U., 1980. Postdoctoral fellow Cornell U., Ithaca, N.Y., 1980-82; research assoc. SUNY, Stony Brook, 1982—. Contbr. articles to profl. jours. NSF fellow, 1971-74; Muscular Dystrophy Assn. postdoctoral fellow, 1980-82. Mem. Soc. for Neurosci., AAAS. Subspecialties: Neurophysiology; Comparative neurobiology. Current work: Effects of neurotransmitters on vertebrate neurons. Home: 180 Norwood Ave Port Jefferson Station NY 11776 Office: Neurobiology and Behavio SUN Stony Brook NY 1179

JONES, WALTER ALLAN, ecologist; b. Gomer, Ohio, July 3, 1941; s. Albert Howell and Carrie Virginia (Shaffon) J.; m. Myrna M. Stewart, June 6, 1964; children: Tamera Lynn, Lloyd Allan, Craig Llewellyn. A.B./B.S., Bluffton Coll., 1966; postgrad., Rutgers U., 1975—. Surg. orderly Lima Meml. Hosp., 1960-61; outdoor edn. coordinator Big Bros. Assn., Columbus, Ohio, 1961-66; dist. fish mgmt. supt. Ohio Div. Wildlife, 1966-67; dir. Seven Ponds Nature Center, Mich. Audubon Soc., 1967-69; chief naturalist Somerset County Park Commn., Basking Ridge, N.J., 1969-79, adminstr. div. natural resources, 1979—; assoc. conservation Barnard Coll., 1972-74; cons. Columbia U., 1972-74. Contbr. articles to profl. jours.; assoc. editor: Jour. Interpretation, 1978—. Mem. Bernards Twp. Environ Com., 1976—. Recipient Outstanding Young Alumnus award Bluffton Coll., 1977; Nat. Conservation award DAR, 1977. Fellow Assn. Interpretive Naturalists (pres., assoc. wildlife biologist); mem. Nat. Outdoor Edn. Assn., Nat. Wildlife Fedn., Nat. Soc. for Youth Found., AAAS, Nat. Audubon Soc., Wildlife Soc., Ecol. Soc. Am., N.J. Outdoor Edn. Assn., N.J. Parks and Recreation Assn., Mich. Audubon Soc., Mich. Assn. Conservation Ecologists. Methodist. Club: Tower. Subspecialties: Ecology; Solar energy. Current work: Alternate energy education; ecology of proscribed burn areas for land management; natural resource inventories. Office: 190 Lord Stirling Rd Basking Ridge NJ 07920

JONES, WILBUR DOUGLAS, JR., research microbiologist; b. July 3, 1927; s. Wilbur Douglas and Alberta B. (Thompson) J.; m. Buell Keith, Dec. 20, 1952; children: Wilbur Keith, Robert Wayne. B.S., Emory U., 1949; M.S., W.Va. U., 1951; Ph.D., Med. Coll. Ga., 1968. Instr. biology and chemistry Truett McConnell Jr. Coll., Celveland, Ga., 1951-52; microbiologist Ga. State Health Dept., Atlanta, 1952-59, sect. chief, 1959-62; instr. biology Ga. State U., Atlanta, 1955-62; research microbiologist mycobacteriology br. Center for Disease Control, Atlanta, 1962—. Served with U.S. Army, 1946-47. Mem. Am. Soc. Microbiologists, Am. Acad. Microbiology, Internat. Working Group on Phageto Typing Mycobacteria, Internat. Com. on Taxonomy of Bacterial Virus (adviser for mycobacteriophages). Baptist. Subspecialties: Microbiology (medicine); Mycobacteriophages. Current work: Phage typing mycobacteria and restriction and modification of mycobacteriophages. Mycobacteria; mycobacteriophage; phage typing; mycobacterium tuberculosis. Office: CDC 1600 Clifton R Bldg 7 Chamblee Atlanta GA 30333

JONES, WILLIAM BARCLAY, physicist; b. San Francisco, Aug. 18, 1919; s. William Francis and Bernice (Rosenquist) J.; m. Patricia Walsh, Dec. 18, 1939; children: Kathleen Dale, William Barclay Jr., Michael David. A.B., U. Calif.-Berkeley, 1947, Ph.D., 1964. Accelerator supr. Crocker Lab., U. Calif.-Berkeley, 1955-62; nuclear physicist Tech. Measurement Corp., San Mateo, Calif., 1962-67; research assoc. physicist Yale U., 1968-75; physicist Brookhaven Nat. Lab., Upton, N.Y., 1975—. Served with USN, 1944-45. Mem. Am. Phys. Soc., Am. Nuclear Soc., AAAS. Subspecialty: Nuclear physics. Current work: Accelerator research and development; nuclear reactions and decay schemes. Office: Brookhaven Nat Lab Bldg 901 Upton NY 11973 Home: Box 174 Creek Rd Wading River NY 11792

JONG, SHUNG CHANG, microbiologist, cons.; b. Taipei, Taiwan, Nov. 12, 1936; came to U.S., 1965, naturalized, 1975; s. Mu-shuey and Fuen (Cheng) J.; m. Chiu-hwa Kou, July 12, 1946; children: Maria, Cynthia, Victoria. Student, Nat. Taiwan U., 1956-60; M.S., Western Ill. U., 1966; postgrad., Wash. State U., 1966-69. Plant pathologist Taiwan Agrl. Research Inst., 1961-63; instr. forest pathology Nat. Taiwan U., 1963-65; teaching asst. Western Ill. U., 1965-66; research asst. Wash. State U.-Pullman, 1966-69; sr. mycologist Am. Type Culture Collection, Rockville, Md., 1969-71, curator fungi, 1971-73, acting head applied microbiology lab., 1976-80, head mycology dept., 1973—; research cons. microbiology dept. George Washington U., Washington, 1975-80, dissertation dir. dept. microbiology, 1975-80; cons. Biotech Research Labs., Inc., Rockville, Md., 1981—. Brown-Hazen grantee, 1974-76; NSF grantee, 1975-79, 80-84. Mem. Internat. Soc. Human and Animal Mycology, Am. Phytopath. Soc., Am. Soc. Microbiology, Plant Molecular Biology Assn., Mycol. Soc. Am., U.S. Fed. Culture Collections, World Fed. Culture Collections, Mycol. Soc. Japan, Washington Acad. Scis., Sigma Xi. Subspecialties: Microbiology; Culture collections. Current work: Preservation and industrial applications of microbial cultures. Office: 12301 Parklawn Dr Rockville MD 20852

JOOS, RICHARD WILLIAM, dentistry researcher; b. Cologne, Minn., Sept. 22, 1934; s. Richard A. and Irene K. (Hoen) J.; m. Joan M. Buesgens, June 18, 1960; children: Teresa, Judith, Richard,

Timothy. B.S., Coll. St. Thomas, 1958; Ph.D., U. Minn., 1964. Teaching asst. U. Minn., Mpls., 1958-1962; researcher VA, Mpls., 1960-64, 3M, St. Paul, 1964-83; cons. Geri, Inc., North St. Paul, Minn., 1982-83. Mem. Internat. Assn. Dental Research (rep. 1980-82). Subspecialties: Dental materials; Preventive dentistry. Current work: Technical service on dental products. Home: 790 Diffley Rd Eagan MN 55123 Office: 3 M Co St Paul MN 55144

JOPPA, LEONARD ROBERT, research plant geneticist; b. Billings, Mont., Sept. 29, 1930; s. Carl and Grace (Shoop) J.; m. Catherine Ann Osborn, June 7, 1959; children: Teresa, William, Barbara, Margaret. B.S. in Agronomy, Mont. State U., 1957; M.S., Oreg. State U., 1962; Ph.D. in Genetics, Mont. State U., 1967. Asst. in agronomy Mont. State U., 1957-64, fellow in agronomy, 1964-67; research geneticist Agrl. Research Service, Dept. Agr., Fargo, N.D., 1967—. Contbr. numerous articles to profl. jours. Served with USAF, 1950-54. Mem. Genetics Soc. Am., Genetics Soc. Can., Crop Sci. Soc. Am., AAAS (fellow), Am. Soc. Agronomy, Sigma Xi, Phi Sigma. Subspecialties: Plant genetics; Plant cell and tissue culture. Current work: Cytogenetics, tissue culture, pest resistance. Home: 90 24th Ave N Fargo ND 58102 Office: Agronomy Dept ND State U Fargo ND 58105

JORDAN, ANGEL GONI, engineering educator; b. Pamplona, Spain, Sept. 19, 1930; came to U.S., 1956, naturalized, 1966; s. Hilario and Perpetua (Goni) J.; m. Nieves Alfonso Cuartero, July 8, 1956; children: Xavier, Edward, Arthur. M.S., U. Zaragoza, Spain, 1952, Carnegie Inst. Tech., 1959, Ph.D., 1959. With NavalOrdnance Lab., Madrid, 1952-56; instr. elec. engring. Carnegie-Mellon U., 1956-58, asst. prof. elec. engring., 1959-62, asso. prof., 1962-65, prof., 1965—, U.A. and Helen Whitaker prof., 1972-80, head dept., 1969-79, dean engineering, 1979-82, provost, 1983—; research fellow Mellon Inst. Indsl. Research, 1958-59; cons. to industry.; chmn. bd. MPC Corp., 1983—. Contbr. articles to profl. jours. Chmn. Pitts. High Tech. Council, 1983—; bd. dirs. Pa. Sci. and Engring. Found, 1981-83, Keithley Inst., Inc., 1983—. NATO sr. scientist fellow, 1976. Fellow IEEE (past chmn. profl. tech. group electron devices Pitts. chpts.); mem. Am. Phys. Soc., Am. Inst. Physics, Am. Soc. Engring. Edn., Sigma Xi, Eta Kappa Nu, Phi Kappa Phi, Tau Beta Pi. Subspecialties: 3emiconductors; Engineering education administration. Home: 5874 Aylesboro Ave Pittsburgh PA 15217 Office: Office of Provost Carnegie Inst Tech Carnegie-Mellon U Pittsburgh PA 15213

JORDAN, ARTHUR KENT, electronics engineer; b. Phila., Dec. 28, 1932; s. Arthur H. and Mary (Schoff) J.; m. Mary Frances Baily, July 10, 1965; children: Thomas B., Edward M., Elizabeth A. B.Sc. in Physics, Pa. State. U., 1957; M.Sc., U. Pa., 1971, Ph.D. in Electronics Engring, 1972. Research engr. Philco Corp., Blue Bell, Pa., 1957-62; electronics engr. Radio Corp. Am., Princeton, N.J., 1962-64; aerospace physicist Gen. Electric Co., Valley Forge, Pa., 1964-69; research assoc., fellow U. Pa., Phila., 1969-73; electronics engr. Naval Research Lab., Washington, 1973—; cons. Contbr. articles to sci. jours.; contbr. to books; asso. guest editor: IEEE Spl. Issue on Inverse Methods for Electromagnetics, 1981. Pres. Washington Mill Sch. PTA, 1979-80. Mem. IEEE, Antennas and Propagation Soc. (chmn. Washington chpt. 1978-79). Episcopalian (vestryman 1981—). Subspecialties: Electronics; Inversion methods. Current work: Inversion methods for electromagnetic wave propagation and scattering, remote sensing, digital communications. Office: Naval Research Lab Washington DC 20375

JORDAN, BYRON DALE, physicist; b. Akron, Ohio, Jan. 24, 1947; s. Augustus L. and Helene (Bendell) J.; m. Katherine E. Ronsheim, June 14, 1969; children: Crispin, Alayne. B.A., Hiram Coll., 1969; Ph.D., McMaster U., 1975. Research physicist Welwyn Research, London, Ont., Can., 1975-77; head optics group Pulp and Paper Research Inst., Pointe Claire, Que., Can., 1977—. Mem. Sci. for Peace, Toronto, Ploughshares, Toronto; del. Commn. Internationale de L'Eclairage. Mem. Am. Phys. Soc., Optical Soc. Am., Soc. Photo-optical Instrumentation Engrs., TAPPI. Presbyterian. Subspecialties: Graphics, image processing, and pattern recognition; Condensed matter physics. Current work: Use of optical techniques to study the behavior paper and its constituents. Home: 72 Sedgefield St Pointe Claire PQ Canada H9R 1N5 Office: Pulp and Paper Inst Can 570 St John's Blvd Pointe Claire PQ Canada H9R 3J9

JORDAN, CARL FREDERICK, ecologist; b. New Brunswick, N.J., Dec. 10, 1935; s. Emil Leopold and Ethel Anabel (Augustine) J.; m. Carmen S. Vega Rivera, Jan. 14, 1967; children—Anabel, Christopher. B.S., U. Mich., 1958; M.S., Rutgers U., 1964, Ph.D., 1966. Asst. scientist, asso. scientist P.R. Nuclear Center, San Juan, 1966-69; ecologist Argonne (Ill.) Nat. Lab., 1969-74; adj. prof. Inst. Ecology U. Ga., Athens, 1974—, sr. ecologist, 1979—; vis. scientist Centro de Ecologia Insituto Venezelano de Investigaciones Cientificas, Caracas, 1974—; coordinator Internat. Study of Ecosystems., Amazon Basin, 1974—. Contbr. articles to profl. jours. Served with USNR, 1958-62. NSF grantee, 1974—; Man and Biosphere grantee, 1979—. Mem. AAAS, Ecol. Soc. Am., Sigma Xi. Subspecialties: Ecosystems analysis; Biomass (agriculture). Current work: Structure and function of tropical forests, tropical forestry and land mgmt. nutrient cycling and productivity. Address: Inst of Ecology U Ga Athens GA 30602

JORDAN, CHARLES RALPH, shipbuilder, civil engineer; b. Kinston, N.C., Dec. 21, 1927; s. George Lyman and Sallie Ballou (Herndon) J.; m. Shirley Jean Wynne, Jan. 14, 1951; children: Cynthia Karen, Charles Ralph, Pamela Jean. B.C.E., N.C. State U., Raleigh, 1951. Designer Newport News Shipbldg. & Dry Dock C., Va., 1951-56, sr. designer, 1956-64, design supr., 1964-66, sr. design supr., 1966-69, assoc. chief, 1969-72, tech. mgr., 1972-74, engring. project mgr., 1974-79, mgr. engring. research dept., 1979—. Contbr. articles profl. jours. Adminstrv. bd. 1st Methodist Ch., 1958-62, 63-74, chmn. commn. on edn., 1964-65; deacon 1st Presbyterian Ch., 1979, chmn. fin. com., 1982—. Mem. Robotics Internat., Soc. Mfg. Engrs., Soc. Naval Architects and Marine Engrs. Subspecialties: Naval engineering; Materials (engineering). Current work: Advancements in state of the art in marine techonolgy, new product development, design of structural welds, Artic marine transportation. Home: 34 N Greenfield Ave Hampton VA 23666 Office: 4101 Washington Ave Bldg 600 Newport News VA 23607

JORDAN, DARRYL ALLYN, farmer, agronomist, agricultural consultant; b. Peru, Ind., Jan. 5, 1946; s. Clifford Darryl and Betty Helen (Robinson) J.; m. Leslie Meredith Tyler, May 13, 1971; children: Irma Yolanda, Jake Isaak, Helen Grace. B.A., Northwestern U., 1968. Farming crew The Farm, Summertown, Tenn., 1971—; pioneer Plenty Agrl. Project, Guatemala, 1976-80; cons. agr., agrl. research. Dictionary editor, Consol. Pub., Chgo., 1969; writer, editor, Explorations Inst., Berkeley, Calif., 1970. Bd. dirs. Plenty Internat.; Mem. Farm Town Planning Com., 1971—, mem. sewage treatment biomass utilization com., 1980—. Recipient Northwestern U. faculty award for fiction, 1968. Mem. Soc. Econ. Botany, N.Am. Fruit Explorers, No. Nut Growers, Seed Savers Exchange, Tenn. Viticultural and Oenol. Soc., Tenn. Native Plant Soc., The Nature Conservancy. Subspecialties: Integrated systems modelling and engineering; Appropriate agriculture. Current work: High altitude tropical soybean and amaranth production and use; no-till rotations without herbicides; sewage treatment; forest biomass use; germplasm collection and preservation; non-photosensitive soybean breeding. Address: The Farm 156 Drakes Ln Summertown TN 38483

JORDAN, RICHARD CHARLES, engineering educator; b. Mpls., Apr. 16, 1909; s. C. and Estelle R. (Martin) J.; m. Freda M. Laudon, Aug. 10, 1935; children: Mary Ann, Carol Lynn, Linda Lee. B. Aero. Engring., U. Minn., 1931, M.S., 1933, Ph.D., 1940. In charge air conditioning div. Mpls. br. Am. Radiator & Standard San. Corp., 1933-36; instr. petroleum engring. U. Tulsa, 1936-37; instr. engring. expt. sta. U. Minn., 1937-41, asst. dir., 1941-44, asso. prof., 1944-45, prof., asst. head mech. engring. dept., 1946-49, prof., head dept. mech. engring., 1950-77, prof., head, 1966-77, acting asso. dean, 1977-78, asso. dean, 1978—; dir. Onan Corp. of McGraw-Edison; cons. various refrigeration and air conditioning cos., 1937—; cons. NSF, U.S. Dept. State, Control Data Corp., others.; Mem. engring. sci. adv. panel NSF, 1954-57, chmn., 1957; mem. div. engring. and indsl. research NRC, mem. exec. com., 1957-69, chmn., 1962-65; del. OAS Conf. on Strategy for Tech. Devel. Latin Am., Chile, 1969; chmn. U.S.-Brazil Sci. Coop. Program Com. on Indsl. Research, Rio de Janeiro, 1967, Washington, 1967, Belo Horizonte, 1968, Houston, 1968; del. World Power Conf., Melbourne, 1962; v.p. sci. council Internat. Institut de du Froid, 1967-71; cons. to World Bank on alternative energy for Northeastern Brazil, 1976. Author: (with Priester) Refrigeration and Air Conditioning, 1948, rev. edit., 1956, also numerous publs. on mech. engring., environ. control, solar energy, energy resources, engring. edn., tech. transfer.; Contbr. Mech. Engring. Recipient F. Paul Anderson medal ASHRAE, 1966, E.K. Campbell award, 1966, Outstanding Publs. Golden Key award, 1949; Outstanding Achievement award U. Minn., 1979; elected to Solar Energy Hall of Fame, 1980. Fellow ASME, AAAS, ASHRAE (presdl. mem.); mem. Nat. Acad. Engring., Assn. Applied Solar Energy (adv. council 1958-61), Am. Soc. Refrigerating Engrs. (1st v.p. 1952, pres. 1953, dir., council mem. 1946-53), Am. Soc. Engring. Edn., AAAS, Nat., Minn. (Engr. of Yr. award 1972), socs. profl. engrs., Internat. Inst. Refrigeration (hon. mem., del. NRC to exec. com. 1957-76, v.p. exec. com. 1959-63, v.p. sci. council 1963-71), Engr. Council Profl. Devel. (chmn. regional edn. and accreditation com.), Sigma Xi, Tau Beta Pi, Pi Tau Sigma, Sigma Chi. Club: Campus. Subspecialties: Mechanical engineering; Solar energy. Current work: Alternative energy research and development and energy conservation research as related to the built environment. Office: Dept Mech Engring U Minn Minneapolis MN 55455

JORDAN, SCOTT WILSON, pathologist, educator; b. Iola, Kans., Aug. 22, 1934; s. Olin Lynn and Ferne (Scott) J.; m. Reita R. West, Aug. 29, 1955; children: Scott Wilson, Carrie, Carla. B.A., U. Kans.-Lawrence, 1956; M.D., U. Kans.-Kansas City, 1959. Intern U. Kans. Med. Ctr., Kansas City, 1959-60, resident in pathology, 1960-63; pathologist Atomic Bomb Casualty Commn., Hiroshima, Japan, 1963-65; asst. prof. pathology U. New Mexico, 1965-69, assoc. prof., 1969-82, prof., 1982—; cons. pathology Midwest Research Inst., Kansas City, Mo., 1962-63; mem. cancer control rev. com. Nat. Cancer Inst., Bethesda, Md., 1975-77. Served to lt. comdr. USPHS, 1963-65; Japan. Mem. Internat. Acad. Pathology, Am. Assn. Pathologists, Radiation Research Soc., Am. Soc. Cytology (exec. com. 1982-85). Subspecialties: Pathology (medicine); Graphics, image processing, and pattern recognition. Current work: Use of digital image processing in quantitating cellular and histologic changes, especially as related to radiation damage. Office: Dept Pahtology U N Mex Albuquerque NM 87131

JORDAN, STUART DAVIS, astrophysicist; b. St. Louis, July 25, 1936; s. Davis Irwin and Lillian (Maenner) J.; m. Elizabeth Susan Roemer, June 24, 1961; children: John Stuart, James William. B.S., Washington State U., 1958; Ph.D., U. Colo., 1968. Research asst. JILA (NBS), Boulder, Colo., 1963-68; research astrophysicist Goddard Space Flight Center, Greenbelt, Md., 1968—, chief of solar physics, 1973-74, project scientist, 1980—, assoc. chief, 1981—. Contbr. articles to profl. jours.; Author: The Sun as a Star, 1981. Served with USAF, 1960-63. Rhodes Scholar, 1958-59. Mem. Internat. Astron. Union, Am. Phys. Soc., Am. Astronon. Soc., AAAS, Fedn. Am. Scientists, Sigma Xi, Tau Beta Pi. Clubs: Washington Area Sierra (speaker), Woods.). Subspecialties: Solar physics; Solar, stellar astrophysics. Current work: Structure and dynamics of solar and stellar atmospheres. Home: 17 Lakeside Dr Greenbelt MD 20770 Office: Goddard Space Flight Center Greenbelt MD 20771

JORDAN, THERESA JOAN, psychologist, educator; b. Irvington, N.J., Sept. 17, 1949; d. Ernest Anthony and Helen Joan (Debski) Balazs; m. Edward Todd Jordan, Sept. 12, 1970. B.A., Washington Square Coll., 1971; M.A., NYU, 1972, Ph.D., 1979. Diplomate: Lic. psychologist, N.Y. Adj. asst. prof. ednl. psychology NYU, 1975—, assoc. dir. research Inst. Devel. Studies, 1979-82; asst. prof. medicine N.J. Med. Sch., Newark, 1982—; cons. Bd. Edn., N.Y.C., 1979; mem. adv. bd. evaluation sect. N.Y. State Dept. Social Services, Albany, 1979-81. N.J. State scholar, 1967-71; NYU scholar, 1969-71; Nat. Inst. Occupational Safety and Health fellow, 1971-72, 74; Inst. Rational-Emotive Therapy fellow, 1983—. Mem. Am. Psychol. Assn., Eastern Ednl. Research Assn. (chmn. self concept spl. interest group 1979—, dir. spl. interest groups 1983—), N.Y. Acad. Scis. (adv. bd. psychology sect. 1981—), Phi Delta Kappa. Subspecialties: Developmental psychology; Cognition. Current work: Effects of early childhood intervention programs on life-span development; role of instrumental enrichment (including computer-assisted instruction) on the growth of competence; cognitive model of self-concept development and change; losses to self-esteem incurred in medical illness. Home: 2 Washington Sq Village Apt 16-J New York NY 10012 Office: Office Primary Health Care Edn NJ Med Sch Univ Medicine and Dentistry NJ 100 Bergen St Newark NJ 07103

JORDAN, V. CRAIG, endocrine pharmacologist, educator; b. New Braunfels, Tex., July 25, 1947; s. Geoffrey Webster and Sybil Cynthia (Mottram) J.; m. Marion Yvonne Williams, July 29, 1969; children: Helen Melissa Yvonne, Alexandra Katherine Louise. B.Sc. with honors, U. Leeds, Eng., 1969, Ph.D. in Pharmacology, 1972. Research assoc. Worcester Found. for Exptl. Biology, Shrewsbury, Mass., 1972-73, vis. scientist, 1973-74; lectr. pharmacology U. Leeds, 1973-79; head endocrinology unit Ludwig Inst. for Cancer Research, U. Berne, Switzerland, 1979-80; asst. prof. human oncology and pharmacology U. Wis., Madison, 1978-81, assoc. prof., 1981—, also leader pharmacology group dept. human oncology. Contbr. numerous articles to profl. jours. Served to capt. Intelligence Corps Brit. Army, 1971-76; Served to capt. Spl. Air Service, 1976-78. Med. Research Council scholar, 1969-72; co-recipient Boston Obstet. Soc. prize, 1974; UICC Internat. Cancer Research Tech. Transfer grantee, 1981. Fellow Am. Inst. Chemists; mem. Am. Assn. for Cancer Research, Am. Soc. for Pharmacology and Exptl. Therapeutics, Endocrine Soc., Biochem. Soc., Brit. Pharm. Soc., Soc. for Endocrinology, Royal Soc. for Chemistry. Conservative. Mem. Ch. of England. Subspecialties: Cancer research (medicine); Endocrinology. Current work: Mechanism of action of antiestrogens as anticancer agents. Antiestrogen structure-activity relationships, molecular pharmacology of antiestrogens, metabolism of antiestrogens in animals and man. Office: 600 Highland Ave Madison WI 53792

JORGENSEN, JAMES HARTLEY, clinical microbiologist, educator, researcher; b. Dallas, July 11, 1946; s. Keith H. and Doris E. (Wolters) J.; m. Jane E. Drummond, Feb. 18, 1978. B.A., North Tex. State U., 1969, M.S., 1970; Ph.D., U. Tex. Galveston, 1973. Specialist in pub. health and med. lab. microbiology. Research assoc. Shriners Burns Inst., Galveston, 1970-73; instr. microbiology U. Tex. Health Sci. Ctr., San Antonio, 1973-75, asst. prof., 1975-78, assoc. prof., 1978—; dir. microbial pathology Bexar County Hosp., San Antonio, 1975—; cons. Audie Murphy VA Hosp., San Antonio, 1975—. Mem. editorial bd.: Antimicrobial Agts. Chemotherapy, 1982—, Diagnostic Microbiology and Infectious Diseases, 1982-83. Mem. Am. Soc. Microbiology (O.B. Williams award Tex. br. 1972), Tex. Infectious Disease Soc., South Tex. Assn. Microbiology Profls. Roman Catholic. Subspecialties: Microbiology (medicine); Pathology (medicine). Current work: Development of rapid methods and automation in clinical microbiology, methods for testing antimicrobial agents. Home: Route 3 Box 3854 Boerne TX 78006 Office: Department Pathology University of Texas Health Science Center San Antonio TX 78284

JORY, VIRGINIA VICKERY, mathematician, research scientist; b. Union City, Ga., Jan. 4, 1934; d. Earl Lee and Mildred (Nolan) Cimerro Vickery; m. Philip Douglas Jory, July 11, 1953; children: Victoria Jory Dennard, Philip Douglas. B.S., Ga. Inst. Tech., 1971, M.S., 1974, Ph.D., 1979. Engring. technician Northwestern U. Aerial Measurements Lab., Patuxent River, Md., 1954-55; programmer Lockheed, Marietta, Ga., 1956-58; grad. teaching asst., instr. Ga. Inst. Tech., Atlanta, 1971-80, research scientist, 1980—; dir. Jory Concrete, Atlanta, 1970—. Named Outstanding Woman Grad. Am. Soc. Women Engrs., 1971; recipient Outstanding Math. Grad. Book award Pi Mu Epsilon, 1971. Mem. IEEE, Am. Math. Soc., Soc. Indsl. and Applied Math., Math. Assn. Am., Sigma Xi. Subspecialties: Applied mathematics; Remote sensing (atmospheric science). Current work: Antenna design and analysis, inverse problems in electromagnetic theory, regularization of ill-posed problems. Office: Ga Inst Tech Engring Expt Sta Atlanta GA 30332

JOSELYN, JO ANN, space scientist, physicist; b. St. Francis, Kans., Oct. 5, 1943; d. James Jacob and Josephine Felzien (Firkins) Cram; m. Robert Joselyn, June 5, 1965. B.S. in Applied Math, U. Colo., 1965, M.S. in astrogeophysics, 1967, Ph.D. in Astrogeophysics, 1978. Research asst. NASA/manned Space Center, Houston, 1966; physicist NOAA/Space Environment Lab., Space Environment Services Center, Boulder, Colo., 1967-78, space scientist, 1978—; mem. U.S. study group 6 of Internat. Radio Consultative Com., 1980—. Contbr. articles on physics to profl. jours. Recipient Unit Citation NOAA, 1971, 80. Mem. Am. Geophys. Union, Union Radio Sci. Internat., Internat. Union Geodesy and Geophysics, Assn. Geomagnetism and Aeronomy, AAAS, Am. Women in Sci., AAUW, Assn. Fed. Prof. and Adminstrv. Women, Sigma Xi, Tau Beta Pi, Sigma Tau. Republican. Methodist. Club: PEO. Subspecialties: Solar physics; Solar-Terrestrial relationships. Current work: Forecasting solar-terrestrial disturbances, understanding solar-terrestrial relationships, solar wind studies, impact of geomagnetic disturbances on technological systems, magnetospheric and ionospheric physics. Office: NOAA R/E/SE2 325 Broadway Boulder CO 80303

JOSEPH, EARL CLARK, futurist; b. St. Paul, Nov. 1, 1926; m., 1955; 4 children. B.A., U. Minn., 1951. Mathematician, programmer and analyst Univac Sperry Rand Corp., 1951-55, systems mgr., 1958-63; staff scientist Sperry Univac, from 1963; pres. Anticipatory Scis., Inc.; vis. lectr. U. Minn., 1971—. Mem. AAAS, IEEE (sr. mem.), Assn. Computing Machinery. Subspecialties: Information systems (information science); Futures. Office: Anticipatory Scis Inc 365 Summit Ave Saint Paul MN 55102

JOSEPH, JOHN LOUIS, computer graphics specialist, programmer; b. St. Louis, June 1, 1955; s. Samuel Richard and Lois Catherine (Schmidt) J. B.S. in Computer Sci, No. Ariz. U., 1977. Cons. bus. coll. No. Ariz. U., Flagstaff, 1973-75, asst. system mgr., 1975-77; program developer Honeywell Info Systems, Los Angeles, 1977-79, customer support rep., 1979-81, project leader computer graphics, 1981—; observor X3H3. Mem. Assn. for Computing Machinery. Subspecialties: Graphics, image processing, and pattern recognition; Operating systems. Current work: Large system implementation of integrated device - independent graphics subsystem. Office: Honeywell Info Systems 5250 W Century Blvd Los Angeles CA 90045

JOSEPH, PETER MARON, physicist, educator; b. Ridley Park, Pa., Mar. 26, 1939; s. Joseph Maron and Doris Helen (Hagens) J.; m. Susan L. Rittenhouse, June 28, 1980. B.S., Lafayette Coll., 1959; M.A. (NSF predoctoral fellow, 1959-61), Harvard U., Ph.D., 1967. Cert. physics of radiology Am. Bd. Radiology. Instr. physics Cornell U., Ithaca, N.Y., 1967-70; asst. prof. physics Carnegie Mellon U., Pitts., 1970-72, med. physics fellow Meml. Hosp., N.Y.C., 1972-73; instr., asst. prof. radiology Columbia U., N.Y.C., 1973-80; assoc. prof. diagnostic imaging physics U. Md., Balt., 1980—. Contbr. articles to publs. Active Balt. Choral Arts Soc. Mem. Am. Phys. Soc., Am. Assn. Physicians in Medicine, Phi Beta Kappa, Sigma Xi. Democrat. Unitarian. Subspecialty: Diagnostic radiology. Current work: Physics of radiology, CT scanning; image quality of radiological images, physical factors, X-ray physics, reconstruction, mathematics. Patentee. Office: Radiology Dept Univ Md Hosp Baltimore MD 21201

JOSHI, ARAVIND KRISHNA, computer and information science educator; b. Poona, India, Aug. 5, 1929; m., 1963. B.Engring., U. Poona, 1951; diploma, Indian Inst. Sci., Bangalore, 1952; M.A., U. Pa., 1958, Ph.D. in E.E, 1960. Research asst. electronics Indian Inst. Sci., Bangalore, 1952-53, and; Tata Inst. Fundamental Research, Bombay, 1953; profl. engr. RCA, N.J., 1954-58; assoc. linguistics analyst U. Pa., Phila., 1958-61, asst. prof. to assoc. prof. elec. engring. and linguistics, 1961-72, prof. computer and info. sci. and dept. chmn., 1972—; assoc. transformations and discourse analysis project Western Res., 1964—; mem. Inst. Advanced Study, 1971-72. Guggenheim fellow, 1971-72. Fellow IEEE, Am. Math. Soc.; mem. ACM. Subspecialties: Artificial intelligence; Natural language analysis and processing. Office: Moore Sch Elec Engring Univ Pa Philadelphia PA 19104

JOSHI, BHAIRAV DATT, chemistry educator; b. Dungrakot, India, Mar. 5, 1939; came to U.S., 1961, naturalized, 1972; s. Liladhar and Jaianti J.; m. Barbara J. Ravenell. M.S., U. Chgo., 1963, Ph.D., 1964. Research assoc. U. Chgo., 1964-66; lectr. chemistry IIT/Kampur, India, 1966-67; reader chemistry Delhi U., 1967-69; teaching fellow SUNY-Stony Brook, 1969-70; asst. prof. chemistry Geneseo, 1970-78, assoc. prof., 1978—. Contbr. articles in field to profl. jours. Recipient L. Banarasi Das Trust prize U. Delhi, 1961; K. Rajeswari Razdan Meml. prize, 1961; grantee in field. Mem. Am. Chem. Soc., Am. Phys. Soc., AAAS, Assn. Computers in Math. and Sci., Sigma Xi. Subspecialties: Theoretical chemistry; Programming languages. Current work: Quantum mechanical studies of atoms and molecules, use of computers in chemical education. Home: 3 Mohawk Ave Geneseo NY 14454 Office: Chemistry Dept SUNY Geneseo NY 14454

JOSHI, JAI HIND, oncologist, hematologist, educator; b. Tavoy, Burma, June 8, 1947; s. Bhavani S. and Ratna J.; m. Paramjit K. Toor, Mar. 31, 1974; children: Amit, Arif. M.D., Christian Med. Coll., Ludhiana, India, 1971. Postdoctoral fellow in hematology and oncology U. Colo. Med. Ctr., Denver, 1976-78; postdoctoral fellow in oncology Nat. Cancer Inst., NIH, Balt., 1980-81; asst. prof. medicine and oncology sect. infectious diseases and microbiology U. Md. Sch.

Medicine and Cancer Ctr., Balt., 1982—. Author numerous book chpts., articles and research papers. Mem. Am. Soc. Clin. Oncology, Am. Soc. Microbiology, Am. Fedn. Clin. Research, Assn. Gnotobiotics. Hindu. Subspecialties: Oncology; Infectious diseases. Current work: Research in various areas of epedimiology diagnosis, prevention and treatment of infections in cancer patients. Home: 2123 Fountain Hill Dr Timonium MD 21093 Office: Sect Infectious Diseases and Microbiology U Md Cancer Center 22 S Greene St Baltimore MD 21203

JOSHI, MADAN MOHAN, microbiologist; b. Almora, India, July 13, 1947; came to U.S., 1969, naturalized, 1978; s. Krishna Kant and Durga Devi (Pandey) J.; m. Lynne Diane, Aug. 9, 1974; children: Kamini Lynne. B.Sc., Agra U., 1964; M.Sc., Banaras Hindu U., 1966; M.S., Auburn U., 1972; Ph.D., La. State U., 1974. NSF research asso. La. State U., Baton Rouge, 1974-75; microbiologist, plant pathologist Kalo Labs., Inc., Quincy, Ill., 1975-76, mgr. product devel. and biol. research, 1976-79, tech. dir., 1979-80; sr. research biologist DuPont Exptl. Sta., Wilmington, Del., 1980-82, group leader plant disease control, 1982—. Contbr. articles in field to profl. jours. Mem. Am. Phytopath. Soc., Sigma Xi, Gamma Sigma Delta. Subspecialties: Plant pathology; Microbiology. Current work: Researcher in development of fungicides to control plant diseases; pesticide biodegradation. Home: 2036 Floral Dr Wilmington DE 19810 Office: DuPont Experimental Station Bldg 268 Wilmington DE 19898

JOSLYN, DENNIS JOSEPH, geneticist, educator; b. Chgo., Apr. 29, 1947; s. Harold Francis and Marguerite (Hart) J. B.S., Ill. Benedictine Coll., 1969; M.S., U. Ill., Urbana, 1973, Ph.D., 1978. Asst. research scientist dept. entomology and nematology U. Fla., Gainesville, 1976-79; research assoc. U.S. Dept. Agr., Gainesville, 1976-79; asst. prof. zoology Rutgers U., Camden, N.J., 1979—. Contbr. articles to sci. jours. Served in U.S. Army, 1970-72. N.J. State Mosquito Control Commn. grantee, 1981-82, 82-83; others. Mem. Genetics Soc. Am., Am. Genetics Assn., AAAS, Am. Mosquito Control Assn., N.J. Mosquito Control Assn., N.J. Acad. Sci., Soc. Invertebrate Pathology, N.Y. Entomol. Soc. (trustee). Subspecialties: Genetics and genetic engineering (biology); Animal genetics. Current work: Evolutionary cytogenetics of Diptera of medical and veterinary significance; genetic control of mosquitoes. Office: Rutgers U Camden NJ 08102

JOY, EDWIN DOUGLAS, JR., dentist, oral surgery educator; b. Bridgeport, Conn., June 15, 1933; s. Edwin Douglas and Bernadette (Fagan) J.; m. Beverly Anne Edwards, Aug. 30, 1953; children: Edwin Douglas III, David Michael. B.A., Yale U., 1954; D.D.S., U. Pa., 1958. Diplomate: Am. Bd. Oral and Maxillofacial Surgery. Practice dentistry specializing in oral surgery, Norfolk, Va., 1965-71; mem. faculty Med. Coll. Va., Richmond, 1971-74; prof., chmn. dept. oral surgery Med. Coll. Ga., Augusta, 1975—; cons. VA, Augusta, 1971—, U.S. Navy, Portsmouth, Va., 1973, U.S. Army, Ft. Jackson, S.C., 1975—. Producer: film Reconstruction after Cancer Surgery, 1977; contbr. articles to profl. jours. Chmn. CPR Am. Heart Assn., Augusta, 1982. Served to comdr. USN, 1958-62. Recipient Disting. Service award Am. Cancer Soc., Richmond, 1974, Am. Heart Assn., Augusta, 1978, 82; Tchr. of Yr. award Med. Coll. Va., 1975. Fellow Am. Assn. Oral Sugeons (committeeman of yr. award 1982); mem. Ga. Soc. Oral Surgeons (pres. 1983—), ADA. Republican. Roman Catholic. Subspecialty: Oral and maxillofacial surgery. Current work: Research in pain control and analgestic drugs. Home: 9 Woodbridge Circle Evans GA 30809 Office: Med Coll Ga Laney-Walker Blvd Augusta GA 30912

JOY, ROBERT MCKERNON, educator, research scientist; b. Troy, N.Y., May 9, 1941; s. Edward M. and Rita Hannah (Sedgwick) J.; m. Mary F., Apr. 5, 1969. Ph.D., Stanford U., 1970. Diplomate: Am. Bd. Toxicology. Research assoc. Stanford Research Inst., Menlo Park, Calif., 1963, 64; pharmacologist U. Calif., Davis, 1969-70, asst. prof., 1970-77, assoc. prof., 1977—, co-dir. health scis. neurotoxicology unit, 1981—; vis. assoc. prof. Harvard U. Med. Sch., 1977-78; vis. research assoc. Children's Hosp., Boston, 1977-78. Contbr. numerous articles to profl. jours. Mem. Am. Soc. Pharmacology and Exptl. Therapeutics, Soc. Toxicology, Soc. Neuroscis., AAAS, Western Pharmacology Soc. Subspecialties: Neuropharmacology; Toxicology (medicine). Current work: Neurotoxicology, neuropharmacology, kindled seizures, epilepsy. Home: 3104 N El Macero Dr El Macero CA 95618 Office: Sch Vet Medicine U Calif Davis CA 95616

JUDD, BRIAN RAYMOND, educator, physicist; b. Chelmsford, Eng., Feb. 13, 1931; s. Harry and Edith (Saltmarsh) J. B.A., Brasenose Coll., Oxford U., 1952, M.A., 1955, D.Phil., 1955. Fellow Magdalen Coll., Oxford U., 1955-62; instr. U. Chgo., 1957-58; asso. prof. U. Paris, 1962-64; staff mem. Lawrence Radiation Lab., Berkeley, Cal., 1964-66; prof. physics Johns Hopkins, Balt., 1966—, chmn. dept., 1979—; Vis. Erskine fellow U. Canterbury, Christchurch, New Zealand, 1968; vis. fellow Australian Nat. U., Canberra, 1975. Author: Operator Techniques in Atomic Spectroscopy, 1963, Second Quantization and Atomic Spectroscopy, 1967, (with J.P. Elliott) Topics in Atomic and Nuclear Theory, 1970, Angular Momentum Theory For Diatomic Molecules, 1975. Subspecialties: Atomic and molecular physics; Theoretical physics. Office: Physics Dept Johns Hopkins Baltimore MD 21218

JUDSON, SHELDON, geology educator; b. Utica, N.Y., Oct. 18, 1918; s. Salmon Sheldon and Dorothy (Eurich) J.; m. Anne Perrin Galpin, Feb. 13, 1943; children: Stephanie Dean, Anne Perrin, Lucy Sheldon. A.B., Princeton U., 1940; M.A., Harvard U., 1946, Ph.D., 1948. Faculty U. Wis., 1948-55, asso. prof. geology, 1955-64; Knox Taylor prof. geology, Princeton, 1964—, chmn. dept., 1970-82; dir. Princeton Coop. Sch. Program, 1964-66; pres. Princeton Jr. Mus., 1964-67; trustee Daily Princetonian; chmn. Univ. Research Bd., Princeton U., 1972-77. Author articles in field; asso. editor: Am. Scientist, 1956-69. Served to lt. USNR, 1942-46. Faculty fellow Fund Advancement Edn., 1954-55; Guggenheim fellow, 1960-61, 66-67; Fulbright fellow, 1960-61. Fellow AAAS, Geol. Soc. Am.; mem. Arctic Inst., Sigma Xi. Clubs: Nassau (Princeton); Princeton of N.Y., Century Assn. N.Y. Subspecialties: Geology; Geomorphology. Current work: Geology of Pleistocene and Holocene with special reference to human activity. Home: 18 Aiken Ave Princeton NJ 08540

JULIANO, RUDOLPH L., pharmacologist, educator; b. N.Y.C., July 18, 1941; s. Rudolph L. and Vera (Aquino) J.; m. Eve-lynn May, Sept. 15, 1963; children: Gregory, Jonathan. B.S. in Physics (Regents scholar), Cornell U., 1963; Ph.D. in Biophysics (NIH fellow), U. Rochester, 1971. Engr. RCA, Princeton, N.J., 1963-64; tchr., community devel. organizer Peace Corps, Philippines, 1964-66; asst. prof. biology SUNY, Buffalo, part-time 1971-72; investigator Hosp. for Sick Children, Toronto, Ont., Can., 1972-78; asst. prof. dept. med. biophysics U. Toronto, 1973-78; assoc. prof. dept. pharmacology U. Tex. Med. Sch., Houston, 1978-82, prof., 1982—; former cons. various pharm. cos.; mem. bd. sci. dirs. Liposome Co., Princeton, 1981—; mem. cell biology adv. panel NSF, 1980—; mem. ad hoc study sects. NIH, 1978, 81; participant numerous profl. meetings, confs., symposia in, U.S. and fgn. countries. Editor: Characteristics and Biomedical Applications, 1980, (with A. Rothstein) Cell Surface Glycoproteins: Structure, Biosynthesis and Biological Functions, 1979; assoc. editor: Cancer Research, 1982—; contbr. articles to profl. jours., also revs., chpts. to books. Med. Research Council Can. and Nat. Cancer Insts. Can. grantee, 1973-78; NSF grantee, 1979-82; NIH grantee, 1980-82; Internat. Union against Cancer travel fellow, 1978; also others, 1980—. Mem. Biophys. Soc., Am. Soc. Cell Biology, Am. Soc. Pharmacology and Exptl. Therapeutics. Subspecialties: Cell biology; Pharmacology. Current work: Cell membranes, cell surface glycoproteins, liposomes, drug delivery systems. Office: Dept Pharmacology U Tex Med Sch Health Scis Center PO Box 20708 Houston TX 77025

JULIENNE, PAUL S., research chemist; b. Spartanburg, S.C., May 8, 1944; s. Adolphe L. and Lelia E. (Bisanar) J.; m. Marietta Lenear, Aug. 31, 1968; children: Marianne E., Alicia K. B.S. in Chemistry, Wofford Coll., 1965, Ph.D., U. N.C., 1969. NRC postdoctoral research assoc. Nat. Bur. Standards, Washington, 1969-73, research chemist, 1974—; research physicist Naval Research Lab., Washington, 1973-74. Contbr. articles to profl. jours. Mem. Am. Phys. Soc., Am. Geophys. Union. Episcopalian. Subspecialties: Atomic and molecular physics; Theoretical chemistry. Current work: Research using quantum mechanical theoretical techniques to study spectroscopy and dynamics of small molecules. Special interest in laser-induced collisional phenomena. Home: 5444 Crows Nest Ct Fairfax VA 22032 Office: Nat Bur Standards B268 Physics Washington DC 20234

JUMP, J. ROBERT, electrical engineering educator; b. Kansas City, Mo., Feb. 15, 1937; m., 1962. B.S., U. Cin., 1960, M.S., 1962; M.S., U. Mich., 1965, Ph.D. in Computer Sci, 1968. Elec. engr. Avco Corp., 1960-61, IBM Corp., 1962-64; asst. prof. Will Rice Coll., Rice U., Houston, 1968-76, assoc. prof., 1976-79, prof. elec. engring., 1979—, chmn. computer sci. program. Mem. IEEE (assoc. editor Trans. Computs 1979-81), ACM. Subspecialty: Computer engineering. Office: Dept Elec Engring Rice Univ Houston TX 77001

JUMPER, ERIC JOHN, air force officer, aerospace engineering educator; b. Washington, Aug. 18, 1946; s. George Yount and Anita (Speranza) J.; m. Marjorie Lee Stanko; children: Eric John Jr., Christine Ann. B.S. in Mech. Engring., U. N.Mex., 1968; M.S., U. Wyo., 1969; Ph.D. in Mech. Engring. and Laser Physics, Air Force Inst. Tech., 1975. Registered profl. engr., Colo. Commd. 2d lt. U.S. Air Force, 1968, advanced through grades to maj., 1980; research mech. engr. (6570th Aeromed. Research Lab., Wright-Patterson AFB), Dayton, Ohio, 1969-73, research aerodynamicist, Albuquerque, 1974-76, assoc. prof. dept. aeros., Colo., 1976-81, assoc. prof. dept. aeros. and astronautics, Wright-Patterson AFB, 1981—; co-dir. Shroud of Turin Research Project, Amston, Conn., 1977—; cons. Broadcast div. Harris Corp., Quincy, Ill., 1979-81, Copland Co., Sidney, Ohio, 1982. Editor, assoc. editor: Aeronautics Digest, 1978—; contbr. articles to profl. jours. Decorated Air Force Commendation medal with oak leaf cluster, Meritorious Service medal. Mem. AIAA (mem. ednl. activites com. 1980—), Sigma Xi, Tau Beta Pi. Subspecialties: Catalysis chemistry; Aeronautical engineering. Current work: Dynamic stall, effect of turbulence on laser propogation, laser interactions with a plasma, supersonic drag prediciton, heat transfer from electrical components. Home: 1700 Radcliffe Rd Dayton OH 45406 Office: Dept Aeronautics and Astronautics Air Force Inst Tech Wright-Patterson AFB OH 45433

JUNCOSA, ADRIAN MARTIN, tropical botanist; b. Balt., Nov. 10, 1951; s. Mario Leon and Vera (Pabo) J. A.B., Harvard U., 1972; Ph.D., Duke U., 1982. Postdoctoral fellow Mo. Bot. Garden, St. Louis, 1982—; collector, Choco Region, Colombia, 1982—. Contbr. articles to profl. jours. NSF/OTS RIAS grantee, 1979; NSF grantee, 1980-82. Mem. AAAS, Bot. Soc. Am. Subspecialties: Morphology; Systematics. Current work: Systematic and functional anatomy/morphology of plants; biology of mangroves; angiosperm biogeography. Office: Mo Bot Garden PO Box 299 St Louis MO 63166

JUNEJA, HARINDER SINGH, medical educator; b. Ambala City, Punjab, India, Nov. 25, 1947; came to U.S., 1975; s. Sudershan Singh and Harbans Kaur J.; m. Ranjeet K. Sondh, May 4, 1975; 1 child, Kavita. Student, Delhi (India) U.; M.B., B.S., India Inst. Med. Scis., 1971, M.D., 1979. Diplomate: Am. Bd. Internal Medicine. Resident St. Mary's Hosp., Rochester, N.Y.; instr. internal medicine U. Tex. Med. Br., Galveston, 1979-80, asst. prof. internal medicine, 1980—. Contbr. articles to med. jours. Mem. Internat. Soc. Exptl. Hematology, Am. Soc. Hematology, Am. Fedn. Clin..Research. Subspecialty: Hematology. Current work: Microenvironment of bone marrow, bone marrow fibroblast, effect of hormones, in both animals and humans. Office: John Sealy Hosp U Tex Med Br MW 404 R-E65 Galveston TX 77551

JUNG, CHAN YONG, biophysicist; b. Choongju, Choong-Book, Korea, Aug.12, 1928; came to U.S., 1960; s. Tae W. and Kwi E. (Lee) J.; m. Soo J. Yoon, Aug. 5, 1982; children: Edward, Soya, Jaiwon. Ph.D., U. Rochester, 1964; M.D.E., R.O.K. P.H.S., Seoul, Korea, 1959. Biophysicist VA Med. Ctr., Buffalo, 1967-75, chief biophysics lab., 1975—; prof. biophysics SUNY-Buffalo, 1982—. NIH grantee, 1968—. Mem. Am. Biophys. Soc., Am. Soc. Biol. Chemists, N.Y. Acad. Sci. Subspecialties: Biophysics (biology); Cell biology (medicine). Current work: Membrane transport, hormone-effects, receptors, radiation target size measurement; reconsititution of membrane-associated functions, membrane biochemistry, endocrinology, growth control. Home: 35 Aster Pl East Amherst NY 14051 Office: VA Med Ctr 3495 Bailey Ave Buffalo NY 14215

JURJEVICH, RATIBOR-RAY M., clinical psychologist; b. Ristovac, Jugoslavia, Aug. 24, 1915; came to U.S., 1950; s. Momchilo and Ljubica (Milovanovic) Djurdjevic; m. Vera Kerecki Petnicki, Oct. 31, 1947. B.Sc. in Forestry, U. Edinburgh, 1938; M.S. in Social Group Work, George Williams Coll., 1953; Ph.D. in Clin. Psychology, U. Denver, 1958. Diplomate: Am. Bd. Profl. Psychology. Dir. Northside Community Ctr., Denver, 1952-58; psychologist State Tng. Sch. for Girls, Morrison, Colo., 1958-60; chief psychologist psychiat. clinic Lowry AFB, Denver, 1961-70; psychologist court services psychiatry dept. Denver Gen. Hosp., 1971-72; chief psychologist Denver County Jail, 1972-73; pvt. practice clin. psychology, Denver and Glenwood Springs, Colo., 1973—; cons. Personal Arts Ctr., Golden, Colo., 1975-76. Author: The Hoax of Freudism, 1974, No Water in My Cup, 1968; editor: Direct Psychotherapy: 28 American Originals, 1973. Mem. Am. Psychol. Assn., Christian Assn. Psychol. Studies. Republican. Eastern Orthodox. Subspecialty: Psychotherapy. Current work: Nihilistic-satanic phenomena in Western culture. Home and Office: 916 Red Mountain Dr Glenwood Springs CO 81601

JURMAIN, JACOB HARRY, medical laboratory executive; b. Boston, Sept. 23, 1923; s. Abraham Jair and Bess (Kaplan) J.; m. Cecile Colin, Jan. 31, 1943; children: Peter Colin, Elizabeth Ann, Pauline, Richard Nesbit. B.S., Harvard U., 1943; M.S., Tufts U., 1947; Ph.D., Boston U., 1955. Project engr. Baird Atomic Corp., Cambridge, Mass., 1947-55; mgr. systems test dept. Space Tech. Labs., TRW, Los Angeles, 1956-60; program mgr. EG&G Inc., Las Vegas, Nev., 1960-66, group gen. mgr., Bedford, Mass., 1966-73; v.p., gen. mgr. Litton Bionetics, Inc., Kensington, Md., 1973—. Served with AUS, 1943-46. Fellow AIAA; mem. IEEE (sect. pres. 1964-65), Am. Meteorol. Soc., Am. Clin. Lab. Assn. Subspecialties: Clinical chemistry; Bioinstrumentation. Current work: Management and systems techniques for biomedical testing businesses. Home: 11401 Grundy Ct Potomac MD 20854 Office: Litton Bionetics Inc 5516 Nicholson Ln Kensington MD 20895

JUSKO, WILLIAM JOSEPH, pharmacologist, educator, cons. pharmacokinetics; b. Salamanca, N.Y., Oct. 26, 1942; s. Joseph C. and Pauline H. (Wrona) J.; m. Laura J. Gillett, May 30, 1964; children: Suzanne, Marjorie, Katherine. B.S. in Pharmacy, SUNY-Buffalo, 1965, Ph.D., 1970. Lic. pharmacist, N.Y. State. Research pharmacologist Boston VA Hosp., 1969-72; asst. prof. pharmacology Boston U. Med. Sch., 1970-72; asst. prof. pharmaceutics Sch. Pharmacy SUNY-Buffalo, 1972-74, assoc. prof., 1974-77, prof., 1977—; con. pharm. industry NIH. Author: Applied Pharmacokinetics, 1980, Frontiers in Clinical Pharmacokinetics, 1980; contbr. articles on pharmacology to profl. jours. Fulbright grantee, 1978. Mem. Am. Pharm. Assn., Acad. Pharm. Sci., Am. Soc. Clin. Pharmacology, Am. Soc. Pharm. Exptl. Therapeutics, Am. Coll. Clin. Pharmacy, AAAS, Am. Assn. Coll. Pharmacy. Subspecialty: Pharmacokinetics. Current work: Clinical pharmacokinetics, drug metabolism, corticosteroids, theophylline, antidepressants, protein binding, computers. Office: Sch Pharmacy SUNY Cooke Hall 319 Buffalo NY 14260

JUSTESEN, DON ROBERT, biomedical investigator, experimental psychology educator; b. Salt Lake City, Mar. 8, 1930; s. Richard Carvel and Elizabeth Agnes (Gustafson) J.; m. Patricia Ann Larson, Feb. 14, 1958; children: Lyle Richard, Jonille Jacelin, Tracy Ann, Anthony Raymond. B.A. in Psychology and Philosophy, U. Utah, 1955, M.A., 1957, Ph.D. in Psychology and Philosophy with distinction, 1960. Field service engr. Rocky Mountain Region Zenith Radio Corp., 1955-56; asst. prof. psychology, chmn. dept. psychology Westminster Coll., 1959-62; dir. Behavioral Radiology Labs., VA Med. Center, Kansas City, Mo., 1962—; asst. prof. psychiatry U. Kans.-Kansas City, 1963-66, assoc. prof., 1966-71, prof., 1971—; lectr. U. Mo.-Kansas City, 1963-66, assoc. prof., 1966-68, prof. psychology, 1968-75; vis. prof. psychology U. Colo., 1965; mem. commn. on metrology Internat. Union Radio Sci., Washington, 1976—; mem. subcom. Am. Nat. Standards Inst., 1976; mem. sci. com. 53 Nat. Council on Radiation Protection and Measurements, Washington, 1977—; keynote speaker for U.S. and Nat. Acad. Scis.'del. to XIXth Gen. Assembly of Internat. Union Radio Sci., Helsinki, Finland, 1978; career research scientist VA, 1980. Contbr. numerous articles, revs., abstracts to profl. publs.; co-editor or editor (biology): Special Supplements on Biological Effects of Electromagnetic Waves, Radio Science, 1977—; assoc. editor: Jour. Microwave Power, 1975—; editorial bd.: Bioelectromagnetics, 1979-83. Served to petty officer 1st class USN, 1948-52; Atlantic. Named Prof. of Yr. Students and Faculty of Westminster Coll., 1962; USPHS predoctoral research fellow, 1958; FDA grantee, 1973-80; Nat. Inst. Environ. Health Sci. grantee, 1980-85. Fellow Am. Psychol. Assn. (First Cash prize 1968), AAAS; mem. IEEE (sr.; chmn. com. on man and radiation 1979-80, mem. exec. com. 1980-83), Bioelectromagnetics Soc. (pres. 1984-85), Sigma Xi, Psi Chi. Subspecialties: Physiological psychology; Biophysics (biology). Current work: Behavioral, developmental and physiological response to radio-frequency electro-magnetic fields, principally microwaves and low-frequency magnetic fields. Introduced dosimetry to biol. study microwaves, 1969; research on dose-determinate conditioning of animals in microwave field, 1970; demonstration of body temperature as an evoked response to discrete stimulation, 1974; inventor behavioral experimentation involving microwaves, 1970. Home: 12416 Ewing Circle Grandview MO 64030 Office: Behavioral Radiology Labs VA Med Center 4801 Linwood Blvd Kansas City MO 64128

JUVET, RICHARD SPALDING, JR., scientist, chemistry educator; b. Los Angeles, Aug. 8, 1930; s. Richard Spalding and Marion Elizabeth (Dalton) J.; m. Martha Joy Myers, Jan. 29, 1955 (div. Nov. 1978); children: Victoria, David, Stephen, Richard P. B.S., UCLA, 1952, Ph.D., 1955. Research chemist Dupont, 1955; instr. U. Ill., 1955-57, asst. prof., 1957-61, asso. prof., 1961-70; prof. analytical chemistry Ariz. State U., Tempe, 1970—; vis. prof. U. Calif. at Los Angeles, 1960, U. Cambridge, Eng., 1964-65, Nat. Taiwan U., 1968, École Polytechnique, France, 1976-77; Mem. air pollution chemistry and physics adv. com. EPA, HEW, 1969-72; cons. R.J. Reynolds Industries, 1966-72; mem. adv. panel on advanced chem. alarm tech., devel. and engring. directorate Def. Systems div. Edgewood Arsenal, 1975. Author: Gas-Liquid Chromatography, Theory and Practice, 1962; Editorial advisor to: Jour. Chromatographic Sci, 1969—, Jour. Gas Chromatography, 1963-68, Analytica Chimica Acta, 1972-74, Analytical Chemistry, 1974-77; biennial reviewer in, 1962-76. NSF sr. postdoctoral fellow, 1964-65; Sci. Exchange Agreement awardee, Czechoslovakia, Hungary, Romania and Yugoslavia, 1977. Fellow Am. Inst. Chemists; mem. Am. Chem. Soc. (nat. chmn. div. analytical chemistry 1972-73, nat. sec.-treas. div. analytical chemistry 1969-71, councilor 1978—, chmn. U. Ill. sect. 1968-69, sec. 1962-63), AAAS, Internat. Platform Assn., Am. Radio Relay League, Sigma Xi, Phi Lambda Upsilon, Alpha Chi Sigma. Presbyn. (deacon 1960—, ruling elder 1972—, commr. Grand Canyon Presbytery 1974-76). Subspecialties: Analytical chemistry; Photochemistry. Current work: Development of ultra-sensitive detectors and high-accuracy, flow moitors for liquid chromatography and and improved liquid chromatography/mass spectrometry interface; computer interfacing to chemical instrumentation; chromatographic peak deconvolution. Research on gas and liquid chromatography, instrumental analysis, computer interfacing. Patentee in field. Home: 4821 E Calle Tuberia Phoenix AZ 85018 Office: Dept Chemistry Arizona State Univ Tempe AZ 85287

JUVINALL, ROBERT CHARLES, mechanical engineer, consultant, educator; b. Danville, Ill., Apr. 11, 1917; s. James Robert and Edna Margaret (Jester) J.; m. Doris Irene Denman; m. Arleene Juvinall; children: Margaret Lee Juvinall Robertson, Nancy Jo. B.S.M.E., Case Inst. Tech., 1939; M.A.E., Chrysler Inst. Engring., 1941; M.S.M.E., U. Ill., 1950. Research and devel. engr. Chrysler Corp., Detroit, 1939-48; asst. prof. mech. engring. U. Ill., 1948-50; assoc. prof. Ill. Inst. Tech., 1950-51; asst. dir. engring. Ransburg Electro-Coating Corp., Indpls., 1951-57; assoc. prof. mech. engring. U. Mich., 1957-63, prof., 1963—; cons. in field. Author: (with C. Lipson) Handbook of Stress and Strength, 1963, Engineering Considerations of Stress, Strain and Strain, 1967, Fundamentals of Machine Component Design, 1983. Mem. ASME, Am. Soc. Engring. Edn., Sigma Xi, Tau Beta Pi, Eta Kappa Nu, Pi Tau Sigma. Subspecialty: Mechanical engineering. Current work: Rehabilitation engineering; teaching, consulting and writing in field of mechanical engineering design. Patentee in field. Home: 3545 Daleview Dr Ann Arbor MI 48103 Office: U Mich Dept Mech Engring GG Brown Ann Arbor MI 48109

KABACK, DAVID B., molecular geneticist, research scientist, educator; b. N.Y.C., May 4, 1950; s. Irving and Rose (Silverman) K. B.S., SUNY-Stony Brook, 1971; Ph.D., Brandeis U., 1976. Postdoctoral fellow Calif. Inst. Tech., 1976-79; asst. prof. microbiology UMDNJ-New Jersey Med. Sch., 1979—. Contbr. articles to profl. jours. Walter Winchell-Damon Runyon Research fellow, 1976; USPHS Nat. Research Service awardee, 1977-78; NIH grantee, 1980—. Subspecialties: Genome organization; Genetics and genetic engineering (biology). Current work: Molecular genetics of yeast and eucaryotic chromosome structure; recombinant DNA; yeast genetics; electron microscopy of nucleic acids. Office: Dept Microbiology UMDNJ-New Jersey Med Sch 100 Bergen St Newark NJ 07103

KABACK, H. RONALD, researcher, physician; b. Phila., June 5, 1936; s. Joseph J. and Evelyn (Bronstein) K.; m. Mollie Schreibman, June 9, 1957; children: Elizabeth, George, Joshua. B.A. in Biology, Haverford Coll., 1958; M.D., Albert Einstein Coll. Medicine, N.Y.C., 1962. Intern in pediatrics Bronx (N.Y.) Mcpl. Hosp., 1962-63; postdoctoral fellow Albert Einstein Coll. Medicine, N.Y.C., 1963-64; staff assoc. Lab. of Biochemistry, NIH, Bethesda, Md., 1964-66, sr. research investigator, 1966-69; asso. mem. Roche Inst. Molecular Biology, Nutley, N.J., 1970-72, mem., 1972—, head lab. of membrane biochemistry, 1977-83, chmn. dept. biochemistry, 1983—; adj. prof. Columbia U., Grad. Sch. and Univ. Ctr. CUNY, N.Y.C.; Lady Davis vis. prof. Hebrew U., Jerusalem, Israel, 1980; Albert Alberman vis. prof. Technicon-Israel Inst. Tech., Haifa, 1981. Contbr. articles to profl. jours. Served as commd. officer USPHS, 1964-66. Recipient third ann. Lewis Rosenstiel award Brandeis U., 1974. Mem. Am. Soc. Biol. Chemists, Harvey Soc., N.Y. Acad. Scis., Am. Soc. Microbiology (Selman A. Waksman hon. lectureship award 1973), Biophys. Soc., Soc. Gen. Physiologists. Subspecialties: Biochemistry (biology); Membrane biology. Office: Dept Biochemistr Roche Inst Molecular Biology Nutley NJ 07110

KABADI, UDAYA MANOHAR, physician, endocrinologist; b. Bombay, Maharashtra, India, Jan. 6, 1942; came to U.S., 1971; s. Manohar Bapu and Suniti Manohar (Sumati) K.; m. Mary Udaya Cheruvillil, Aug. 9, 1970; children: Sajit, Rajit. M.B.B.S., Seth G.S. Med. Coll., U. Bombay, India, 1965, M.D., 1970. Diplomate: Am. Bd. Internal Medicine. Intern Jewish Meml. Hosp., N.Y.C., 1971, resident, 1971-72, Beth Israel Med. Center, N.Y.C., 1972-73; asst. physician dept. medicine, 1975-76; fellow VA Med. Center, Bronx, N.Y., 1973-74; attending physician dept. medicine Gouverneur Hosp., N.Y.C., 1975-78, chief walk-in clinic, 1976-78; staff physician VA Med. Center, Bklyn., 1978-80, chief endocrinology sect., Des Moines, 1980—, research and devel. coordinator, 1981—; cons. endocrinology Broadlawns Med. Center, Des Moines, 1982—; mem. adv. bd. Central Iowa Diabetes Edn. Center, Des Moines, 1981—. Recipient Research Grants VA Research Service, 1981—, Am. Legion, 1981-82. Fellow Royal Coll. Physicians, ACP; mem. Endocrine Soc., Am. Fedn. Clin. Research, Am. Diabetes Assn., European Assn. Study Diabetes, Am. Assn. Lab. Animal Sci. Subspecialties: Endocrinology; Internal medicine. Current work: Thyroid physiology and disease, carbohydrate metabolism of Des Moines, physiology of glucose metabolism, physiology of insulin and glucagon secretion. Office: VA Med Center 30th and Euclid Sts Des Moines IA 50310

KABALKA, GEORGE WALTER, chemistry educator, consultant; b. Wyandotte, Mich., Feb. 1, 1943; s. Walter George and Rose Marie (Witkowski) K.; m. Beth Ann Swaim, Aug. 31, 1968; children: Stephen, Kathy. B.S., U. Mich.-Ann Arbor, 1965; Ph.D., Purdue U., 1970. Asst. prof. U. Tenn., Knoxville, 1970-76, assoc. prof., 1976-81, prof., 1981—; cons. Oak Ridge Nat. Lab., Tenn., 1976—, Oak Ridge Assoc. U., 1977—, Mt. Sinai Hosp., Miami Beach, Fla., 1981—, U. Mich. Hosp., Ann Arbor, 1980—. Contbr. articles in field to profl. jours. Founder Appalachian Zool. Soc., Knoxville, Tenn., 1972; coach Knoxville Am. Youth Soccer, 1975; treas. troop 20 Boy Scouts Am., Knoxville, 1980. Mem. Am. Chem. Soc. (local chmn.), Soc. Nuclear Medicine, Soc. Magnetic Resonance Imaging, Sigma Xi, Phi Lambda Upsilon. Democrat. Roman Catholic. Subspecialties: Synthetic chemistry; Nuclear medicine. Current work: Synthetic chemistry, radiopharmaceutical chemistry, NMR. Office: Chemistry Dept U Tenn Circle Dr Knoxville TN 37996-1600

KABAT, ELVIN ABRAHAM, immunochemist, biochemist, educator; b. N.Y.C., Sept. 1, 1914; s. Harris and Doreen (Otis) K.; m. Sally Lennick, Nov. 28, 1942; children: Jonathan, Geoffrey, David. B.S., Coll. City N.Y., 1932; M.A., Columbia U., 1934, Ph.D., 1937; LL.D. (hon.), U. Glasgow, 1976, Doctoral degree, U. Orleans (France), Ph.D., Weizmann Inst. Sci., Rehovot, Israel. Lab. asst. immunochemistry Presbyn. Hosp., 1933-37; Rockefeller Found. fellow Inst. Phys. Chem., Upsala, Sweden, 1937-38; instr. pathology Cornell U., 1938-41; mem. faculty Columbia U., N.Y.C., 1941—, asst. prof. bacteriology, 1946-48, asso. prof., 1948-52, prof. microbiology, 1952—, prof. human genetics and devel., 1969—; mem. adv. panel on immunology WHO, 1965—; expert cons. Nat. Cancer Inst., 1975-82, Nat. Inst. Allergy and Infectious Disease, 1983—; Alexander S. Wiener lectr. N.Y. Blood Center, 1979. Author: (with M.M. Mayer) Experimental Immunochemistry, 1948, 2d edition, 1961, Blood Group Substances, Their Chemistry and Immunochemistry, 1956, Structural Concepts in Immunology and Immunochemistry, 1968, 2d edit., 1976, (with T.T. Wu and H. Bilofsky) Variable Regions of Immunoglobulin Chains, 1976, Sequences of Immunoglobulin Chains, (with others) Sequences of Proteins of Immunological Interest, 1983; Mem. editorial bd.: Jour. Immunology, 1961-76, Transplantation Bull, 1957-60. Recipient numerous awards including; Ann. Research award City of Hope, 1974; award Center for Immunology, State U. N.Y., Buffalo, 1976; Louisa Gross Horwitz award Columbia U., 1977; R.E. Dyer lectr. award NIH, 1979; Townsend Harris medal CCNY, 1980, Philip Levine award Am. Soc. Clin. Pathology, 1982, award for excellence Grad. Faculties Alumni Columbia U., 1982; Fogarty scholar NIH, 1974-75. Fellow AAAS, Am. Acad. Allergy (hon.); mem. Nat. Acad. Scis., Am. Acad. Arts and Scis., Am. Assn. Immunologists (past pres.), Am. Soc. Biol. Chemists, Am. Chem. Soc., Harvey Soc. (pres. 1976-77), Am. Soc. Microbiology, Internat. Assn. Allergists, Societe Franciaise d'Allergie (hon.), Biochem. Soc. (Eng.), Assn. for Research in Nervous and Mental Diseases, AAUP, Assn. de Microbiologists de Langue Francaise, Sociêtê de Biologie, Sociêtê de Immunologie (hon.), Phi Beta Kappa, Sigma Xi. Subspecialty: Immunology (medicine). Current work: Size and structure of antibody and lectin combining sites, blood group substances, dextrans, monoclonal antibodies, myeloma and hybridoma antibodies. Home: 70 Haven Ave New York NY 10032 Office: Dept Microbiology Coll Physicians and Surgeons Columbia U 701 W 168th St New York NY 10032

KAC, MARK, educator, mathematician; b. Krzemieniec, Poland, Aug. 3, 1914; came to U.S., 1938, naturalized, 1943; s. Bencion and Chana (Rojchel) K.; m. Katherine Elizabeth Mayberry, Apr. 4, 1942; children—Michael Benedict, Deborah Katherine. Magister of Philosophy, U. Lwow, Poland, 1935, Ph.D., 1937; D.Sc. (hon.), Case Inst. Tech., 1966. Teaching asst. U. Lwow, 1935-37; jr. actuary Phoenix Co., Lwow, 1937-38; fellow Parnas Found., Johns Hopkins, 1938-39; instr. Cornell U., 1939-43, asst. prof., 1943-46, prof. math., 1947-61, Andrew D. White prof.-at-large, 1965-71; mem. Inst. Advanced Study, Princeton, 1951-52; prof. Rockefeller U., N.Y.C., 1961-81, U. So. Calif., Los Angeles, 1981—; H.A. Lorentz vis prof. U. Lelden, Netherlands, 1963; vis. fellow Brasenose Coll.; sr. vis. fellow Oxford (Eng.) U., spring, 1969; Solvay lectr. U. Brussels, Belgium, 1971. Contbr. articles to profl. jours. Guggenheim fellow, 1946-47; recipient Chauvenet prize for paper Random Walk and the Theory of Brownian Motion Math. Assn. Am., 1950, Chauvenet prize for paper Can One Hear the Shape of A Drum, 1968; Alfred Jurzykowski Found. award in sci., 1976; Birkhoff prize Am. Math. Soc.-Soc. Indsl. and Applied Math., 1978. Mem. Am. Acad. Arts and Scis., Am. Philos. Soc., Am. Math. Soc., Math. Assn. Am., Nat. Acad. Scis., Inst. Math. Stats., Royal Netherlands Acad. Arts and Sci. (fgn.), Royal Norwegian Acad. (fgn.), Sigma Xi. Subspecialties: Probability; Statistical physics. Current work: Statistical physics and non-linear problems. Home: 3980 Astaire Ave Culver City CA 90230 Office: Dept Math U So Calif Los Angeles CA 90007

KACHMAR, JOHN FREDERICK, clinical biochemist, educator; b. Akron, Jan. 10, 1916; s. Michael and Katherine (Waytowich) K.; m. Jessie Kallman, Dec. 23, 1946; 1 dau., Carlajean Ginnis. B.S., U. Akron, 1936; M.S., U. Minn., 1947, Ph.D., 1951. Registered clin. chemist. Jr. chemist Victor Gasket Co., Chgo., 1936-37, USPHS, 1937-39; asst. chemist, 1939-42, 46; research asst. dept agrl. biochemistry U. Minn., 1947-51, research asst. dept. ob-gyn, 1951-53; clin. chemist Einstein Med. Center So. div., Phila., 1953-58; clin. biochemist Rush-Presbyn.-St. Luke's Med. Center of Chgo., 1958—; assoc. prof. biochemistry Rush-Presbyn.-St. Luke's Med. Center Chgo., 1972—, also sr. attending biochemist; cons. in field. Contbg. author, assoc. editor: Fundamentals of Clinical Chemistry, 1970, 76. Served with U.S. Army, 1942-46. Mem. Nat. Acad. Clin. Biochemistry, Am. Chem. Soc., Am. Assn. Clin. Chemists (recipient Natelson award 1980). Subspecialties: Biochemistry (medicine); Clinical chemistry. Current work: Clinical enzymology, bilirubin chemisty.

KACHUCK, BEATRICE, educator; b. N.Y.C., Jan. 3, 1926; d. Joseph and Lydia (Greenberg) K.; m.; children: Paul Alan, Dan David. B.A., Bklyn. Coll., 1948; M.A., NYU, 1955, Ph.D., 1972. Tchr., dir. Day Care Nursery Schs., N.Y.C., 1945-50, 53-55; tchr. Baldwin, Plainview Pub. Schs., 1953-55, 59-64; reading cons. Lawrence Pub. Schs., Cedarhurst, N.Y., 1964-68; prof. edn. and co-coordinator women's studies program Bklyn. Coll., 1968—; cons. in field. Contbr. articles to profl. jours. Commn. mem. Sex Equity Adv. Commn. for Vocat. Edn., N.Y.C., 1981, 82; mem. Chancellor's Adv. Commn. for Equal Opportunity, N.Y.C., 1980-82. Mem. NOW (chmn. edn. N.Y. State 1980-82), N.Y. Acad. Sci. (vice chmn.), Am. Psychol. Assn., Internat. Reading Assn., Nat. Conf. on Research in English. Subspecialties: Cognition; Learning. Current work: Comprehension of written text-research and writing; psychology and education of women. Office: Women's Studies Program Brooklyn Coll Brooklyn NY 11210

KACZKOWSKI, HENRY RALPH, educator; b. Milw., Jan. 29, 1924; s. Bronislaw and Ludwika (Miesobenska) K.; m. Maryalyce Hornby, June 20, 1953; children: Ann, Thomas, Jane, Mary. B.S., U. Wis., 1948, M.S., 1951, Ph.D., 1954. Tchr., pub. schs., Wis., 1948-50; prof. DePaul U., Chgo., 1954-57, U. Wis., Milw., 1957-62; prof. Coll. Edn., U. Ill.-Urbana, 1962—; psychometrist to industry, Ill., 1962—; cons. Ill. Office Edn., 1962—. Author: Guidance Practices, 1960, Elementary School Guidance, 1975; contbr. articles to profl. jours. Served to lt. USAAF, 1943-45. Mem. Am. Psychol. Assn., Am. Personnel and Guidance Assn. Subspecialties: Behavioral psychology; Learning. Current work: Research in group therapy, instructional design. Home: 713 W Healey St Champaign IL 61820 Office: U Ill Coll of Edn Urbana IL 61801

KACZMAREK, LEONARD KONRAD, neurobiologist; b. Edinburgh, Scotland, Aug. 2, 1947; s. Mieczyslaw and Irena (Garlinska) K.; m. Sheila Frances Hayman, Dec. 11, 1971; children: Zoe Tara, Konrad Eric. B.Sc., U. London, 1968; M.Sc., Imperial Coll., U. London, 1969; Ph.D., Charing Cross Hosp. Med. Sch., U. London, 1971. Research asst. dept. neurochemistry Inst. Neurology, U. London, 1971-72; asst. research anatomist Brain Research Inst., UCLA, 1972-74; European Sci. Exchange Program fellow U. Brussels, 1974-76; sr. research fellow div. biology Calif. Inst. Tech., Pasadena, 1976-81; asst. prof. pharmacology and physiology Yale U., New Haven, 1981—. Contbr. articles to profl. jours. Spencer Found. fellow, 1976-79. Mem. Soc. Neurosci., Soc. Math. Biology, AAAS, N.Y. Acad. Scis. Subspecialties: Neurobiology; Neuropharmacology. Current work: Research into biochemical basis of long lasting changes that occur in the electrical properties of neurons; research using theoretical models in neurobiology. Home: 139 Laurel Brook Dr Guilford CT 06437 Office: Department Pharmacology Yale University School Medicine 333 Cedar St New Haven Ct 06510

KADAMBI, NARASIMHA PRASAD, nuclear engineer; b. Secunderabad, India, Apr. 20, 1944; came to U.S., 1966; s. Narasimha Murthy and Rukmini Devi K.; m. Sheela Setlur, Oct. 20, 1974; children: Nandini, Vijaya Simha. B.E., Osmania U., Hyderabad, India, 1965; M.S., Pa. State U., 1968, Ph.D., 1971. Registered profl. engr., Pa. Sci. officer Bhabha Atomic Research Ctr., Bombay, India, 1972-73; research assoc. Pa. State U., University Park, 1973-74; sr. engr. Westinghouse Elec. Corp., Madison, Pa., 1974-82; project mgr. U.S. NRC, Washington, 1982—. Advisor Jr. Achievement, Greensburg, Pa., 1979. The Nizam of Hyderabad Charitable Trust scholar, 1965. Mem. Am. Nuclear Soc., AAAS. Subspecialty: Nuclear engineering. Current work: Implement regulatory policy toward expeditious licensing of a nuclear power plant. Home: 14530 Woodcrest Dr Rockville MD 20853 Office: US NRC Mail Stop 128 Washington DC 20555

KADANE, JOSEPH BORN, statistics educator; b. Washington, Jan. 10, 1941. B.A., Harvard U., 1962; Ph.D. in Stats, Stanford U., 1966. Asst. prof. stats. Yale U., 1966-69, acting dir. grad. studies, 1967-68; assoc. prof. social sci. and stats. Carnegie-Mellon U., Pitts., 1969-72, prof. stats., 1972—, head grad. dept.; research staff mem. Cowles Found. Research Econs., Yale U., 1966-69; mem. staff Ctr. Naval Analysis, 1968-71; cons. L.I. Lighting Co., 1968, Nat. Devel. Corp., Govt. Tanzania, 1968, Ctr. Naval Analysis, 1971—, Bur. Labor Stats., U.S. Dept. Labor, 1972-73. Grantee NSF, Office Naval Research. Fellow Am. Statis. Assn., Inst. Math. Stats. Royal Statis. Soc., AAAS; mem. Ops. Research Soc. Subspecialty: Statistics. Office: Dept Statistics Carnegie-Mellon Univ Pittsburgh PA 15213

KADANOFF, LEO PHILIP, physicist; b. N.Y.C., Jan. 14, 1937; s. Abraham and Celia (Kibrick) K.; m.; children: Marcia, Felice, Betsy. A.B., Harvard U., 1957, M.A., 1958, Ph.D., 1960. Fellow Neils Bohr Inst., Copenhagen, 1960-61; from asst. prof. to prof. physics U. Ill., Urbana, 1961-69; prof. physics and engring., univ. prof. Brown U., Providence, 1969-78; prof. physics U. Chgo., 1978-82, John D. MacArthur Disting. Service prof., 1982—; Mem. tech. com. R.I. Planning Program, 1972-78, mem. human services rev. com., 1977-78; pres. Urban Obs. R.I., 1972-78. Author: Electricity Magnetism and Heat, 1967; co-author: Quantum Statistical Mechanics, 1963; Adv. bd.: Sci. Year, 1975-79; editorial bd.: Statis. Physics, 1972-79, Nuclear Physics, 1980—, Annals of Physics, 1982—; contbr. articles to profl. jours. NSF fellow, 1957-61; Sloan Found. fellow, 1963-67; recipient Wolf Found. prize, 1980. Fellow Am. Phys. Soc. (Buckley prize 1977), Am. Acad. Arts and Scis.; mem. Nat. Acad. Scis. Subspecialties: Theoretical physics; Condensed matter physics. Current work: Transitions to chaos, phase transitions, many body theory. Home: 5424 S Eastview Park Chicago IL 60615 Office: James Franck Inst U Chgo Chicago IL 60637

KADI, KAMAL SIF-EL, petroleum engineering consultant; b. Batna, Aures, Algeria, June 23, 1941; came to U.S., 1963; s. Said and Fatma (Agoune) K. B.S., Stanford U., 1966; M.S., U. Tulsa, 1974, Ph.D., 1977. Field supt. Sinclair Oil Co., Algiers, Algeria, 1966-69; asst. dir. project engring. Alcore Co., Dallas, 1969-72; simulation engr. Chevron Oil Co., Houston, 1977-79; sr staff specialist Schlumberger, Houston, 1979-81; sr. staff cons. Intercomp, resource devel. and engring., Houston, 1981-82; cons. KSK Enterprises, Houston, 1982—. Mem. Soc. Petroleum Engrs., Soc. Soc. Profl. Well Log Analysts, Soc. Indsl. and Applied Math. Subspecialties: Numerical analysis; Petroleum engineering. Current work: Petroleum reservoir behavior prediction through use of reservoir simulation, including enhanced hydrocarbon recovery.

KADIS, BARNEY MORRIS, medical educator, biochemist; b. Omaha, Dec. 26, 1927; s. Max and Nellie (Lederman) K. B.A., U. Nebr.-Omaha, 1952; Ph.D., Iowa State U., 1957. Asst. prof. U. Dubuque, Iowa, 1957-58; research chemist U. Nebr. Coll. Medicine, Omaha, 1958-60; asst. prof., 1961-66; assoc. prof. SUNY, Albany, 1960-61; prof. So. Ill. U. Sch. Dental Medicine, Alton, 1969-81; prof. div. basic med. scis. Sch. Medicine, Mercer U., Macon, Ga., 1982—; postgrad. fellow Syntex, Palo Alto, Calif., 1966-67, Stanford U. Sch. Medicine, Palo Alto, 1967-68. Served with U.S. Army, 1946-48. Mem. Endocrine Soc., Am. Biol. Chemists, Am. Soc. Bone and Mineral Research, AAAS, Am. Chem. Soc., Am. Soc. Cell Biology. Subspecialties: Biochemistry (medicine); Endocrinology. Current work: Bone metabolism in cell culture in the normal and diabetic state. Office: Sch Medicine Mercer U 1400 Coleman Ave Macon GA 31207

KADIS, SOLOMON, microbiology educator; b. Balt., May 17, 1923; s. Samuel and Minnie (Greenstein) K.; m. Mindel Gardenhire, Jan. 28, 1958 (div. 1972); children: Michael, Alan, Deborah. B.A., St. John's Coll., 1951; M.A., U. Va., 1953; Ph.D., Vanderbilt U., 1957. Research assoc. U.S. Vitamin & Pharm., Yonkers, N.Y., 1957-60, Gerontol. Research Inst., Phila., 1961; assoc. mem. Albert Einstein Med. Ctr., Phila., 1961-72; prof. microbiology U. Ga., Athens, 1972—. Editor: Microbiology Toxins, 8 vols, 1970-72. Served with U.S. Army, 1944-46. Fellow Am. Acad. Microbiology; mem. Am. Soc. Microbiology, Am. Soc. Cell Biology, AAAS, Conf. of Research Workers in Animal Disease, Ga. Acad. Sci., Sigma Xi. Subspecialty: Microbiology. Current work: Pathogenesis of enterotoxic colibacillosis, role of iron on susceptibility of mammals to bacterial disease, mechanisms by which iron influences susceptibility to disease. Home: 265 Brookstone Dr Athens GA 30605 Office: U Ga Dept Med Microbiology Coll Vet Medicine Athens GA 30602

KADLEC, ROBERT HENRY, chemical engineering educator, consultant; b. Racine, Wis., June 11, 1938; s. Henry and Alice Blanche (Chernohorsky) K.; m. D. Kay Ferris, Sept. 28, 1979; children: Debra, Christopher, Jonathan. B.S., U. Wis., 1958; M.S., U. Mich., 1959, Ph.D., 1962. Registered profl. engr., Mich. Asst. prof. chem. engring. dept. U. Mich., Ann Arbor, 1961-65, assoc. prof., 1965-70, prof., 1970—; dir. (Wetland Ecosystem Research Group), 1975—. Mem. Am. Inst. Chem. Engrs., AAUP, Nat. Soc. Profl. Engrs., Am. Soc. for Engring. Edn., Internat. Peat Soc., Mich. Soc. Profl. Engrs. Subspecialties: Chemical engineering; Water supply and wastewater treatment. Current work: Environmental chemistry and wetlands, modeling and computer simulation.

KADUSHIN, PHINEAS, psychotherapist, researcher; b. N.Y.C., Oct. 18, 1925; s. Max and Evelyn (Garfiel) K. B.A., Columbia U., 1945; rabbi, M.H.L., Jewish Theol. Sem., 1950; M.S. in Psychology, Yeshiva U., 1961. Psychology intern N.J. Dept. Instns. and Agencies, Bordentown, 1960-61; sr. psychologist Mental Health Ctr., Perth Amboy, N.J., 1961-66; research assoc. Family Treatment and Study Unit dept. psychiatry N.Y. Med. Coll., 1966-70; psychotherapist in pvt. practice, N.Y.C., 1971—. Mem. Am. Psychol. Assn. (assoc.), Am. Group Psychotherapy Assn. (assoc.), Eastern Group Psychotherapy Assn. (assoc.), Rabbinical Assembly Am. Current work: Dyadic group therapy, a method of improving the man-woman relationship by therapy in a group of two people previously unknown to each other.

KADZIELAWA, KRZYSZTOF, physician, neuropharmacologist, researcher; b. Luck, Poland, May 13, 1936; came to U.S., 1976; s. Stanislaw and Maria K.; m. Renata Gorecka, Aug. 15, 1961; 1 son, Krzysztof. M.D., Acad. Medicine, Cracow, Poland, 1960, Ph.D., 1965. Intern Clinics Acad. Medicine, Cracow, 1960-61; fellow Polish Acad. Scis., Cracow and Warsaw, 1961-65; asst. prof. Acad. of Medicine, Warsaw, Poland, 1965-70, assoc. prof., docent, 1970-73; fellow U. Aarhus, Denmark, 1973-74; scientist Addiction Research Found., Toronto, Ont., Can., 1975-76; assoc. prof. Coll. Medicine U Fla., Gainesville, 1976-82. Contbr. research articles in neuropharmacology to publs. Research grantee NIH, 1979-82, Merck, Sharp & Dohme, 1979, 81. Mem. Am. Soc. Pharmacology and Exptl. Therapeutics, Soc. Neurosci., AAAS. Roman Catholic. Subspecialties: Neuropharmacology; Neurophysiology. Current work: Neurotransmitters; catecholaminergic mechanisms in brain and spinal cord. Home: 3611 B SW 28Terr Gainesville FL 32608

KAESBERG, PAUL JOSEPH, educator; b. Engers, Germany, Sept. 26, 1923; came to U.S., 1926, naturalized, 1933; s. Peter Ernst and Gertrude (Mueller) K.; m. Marian Lavon Hanneman, June 13, 1953; children—Paul Richard, James Kevin, Peter Roy. B.S. in Engring., U. Wis., Madison, 1945, Ph.D. in Physics, 1949; D. Natural Scis. (hon.), U. Leiden, The Netherlands, 1975. Instr. biometry and physics U. Wis., 1949-51, asst. prof. biochemistry, 1956-58, asso. prof., 1958-60, prof., 1960-63, prof. biophysics and biochemistry, 1963—; chmn. Biophysics Lab., 1970—, Wis. Alumni Research Found. prof., 1981—; cons. in field. Contbr. chapts. to books and articles to profl. jours. Subspecialty: Biophysics (physics). Home: 5002 Bayfield Terr Madison WI 53705 Office: 1525 Linden Dr Madison WI 53706

KAFATOS, FOTIS C., biology educator; b. Iraclion, Greece, Apr. 16, 1940; m., 1967. B.A., Cornell U., 1961; M.A., Harvard U., 1962, Ph.D. in Biology, 1965. Tutor Harvard U., Cambridge, Mass., 1962-63, instr. to asst. prof., 1965-69, prof. biology, 1969—. Mem. Nat. Acad. Sci., AAAS, Am. Soc. Zoology, Am. Soc. Cell Biology, Soc. Devel. Biology. Subspecialty: Developmental biology. Office: Biol Labs Harvard Univ Cambridge MA 02138

KAFKA, MARIAN ADELE STERN, research physiologist; b. Richmond, Va., Mar. 30, 1927; d. Henry S. and Adele (Lewit) Stern; m. John S. Kafka, Oct. 3, 1952; children: David Egon, Paul Henry, Alexander Charles. A.B., Conn. Coll., 1948; Ph.D., U. Chgo., 1952. Research asst. dept. physiol. chemistry Emory U., Atlanta, 1952-53; research ass. Ill. Neuropsychiat. Inst., U. Ill. Sch. Medicine, 1953-54; research asst. dept. internal medicine Yale U., 1954-57; USPHS fellow Nat. Heart, Lung and Blood Inst., 1965-68, physiologist, 1968-74; physiologist adult psychiatry br. NIMH, Bethesda, Md., 1974-82, physiologist clin. neurosci. br., 1982—. Contbr. chpts. to books and articles in field to profl. jours. Marie J. Mergier fellow U. Chgo., 1950. Mem. Am. Physiol. Soc., Endocrine Soc., Biophys. Soc., Soc. for Neurosci., Phi Beta Kappa, Sigma Xi. Subspecialties: Chronobiology; Neurobiology. Current work: Neurotransmitter receptors; molecular interaction between neurotransmitters, receptors and cell membranes; central nervous system control of circadian rhythms.

KAGAN, IRVING GEORGE, parasitologist, researcher, administrator; b. N.Y.C., June 1, 1919; s. Harry and Miriam (Slatoff) K.; m. Mildred Reese, June 17, 1940; children: Mila Rosalyn, Jule Greer. A.B., Bklyn. Coll., 1940; M.A., U. Mich., 1948, Ph.D., 1950. Asst. prof. zoology U. Pa., 1952-57; chief helminthology unit Ctrs. for Disease Control, Atlanta, 1957-66, dir. parasitology div., 1967-81, asst. dir. lab. sci. div. parasitic diseases, 1981—; cons. parasitic diseases WHO, 1959—. Contbr. numerous articles to sci. jours. Served to 1st lt. USAAF, 1942-45. Decorated Purple Heart, Air medal, D.F.C.; recipient Superior Service award USPHS, 1965, Behring-Bilharz prize Hoechst Aktiengesellshaft, Cairo, 1982. Mem. Am. Acad. Microbiology, AAAS, Royal Soc. Tropical Medicine and Hygiene, Am. Soc. Tropical Medicine and Hygiene, Am. Soc. Parasitology

(Henry Baldwin Ward medal 1965), Am. Assn. Immunologists, Am. Micros. Soc., Helminthological Soc. Washington, Sci. Research Soc. Am. Jewish. Subspecialties: Parasitology; Immunogenetics. Home: 1074 Oakdale Rd NE Atlanta GA 30307 Office: 1600 Clifton Rd NE Atlanta GA 30333

KAHLER, ALEX LEROY, research geneticist/plant breeder, educator; b. Scottsbluff, Nebr.; s. Alex and Frieda K.; m. Judith Ann Kahler, June 15, 1963; children: Alexander Lucas, Jonathan LeRoy. B.S., U. Calif., Davis, 1965, M.S., 1967, Ph.D., 1973. Research assoc. dept. genetics U. Calif., Davis, 1965-80; geneticist U.S. Dept. Agr., Brookings, S.D., 1980—; assoc. prof. dept. plant sci. S.D. State U., Brookings, 1980—. Contbr. articles to sci. jours. Mem. Am. Soc. Naturalists, Evolution Soc., Genetics Soc. Am., AAAS, Crop Sci. Soc. Am., Am. Soc. Agronomy, Am. Genetics Assn., Sigma Xi, Alpha Zeta. Subspecialties: Plant genetics; Population biology. Current work: Genetics, plant breeding, plant population genetics, plant evolution. Office: US Dept Agr RR 3 Brookings SD 57006

KAHN, ARNOLD S., assn. exec.; b. Sioux City, Iowa, June 14, 1942; s. Lester and Gladys (Eagles) K.; m. Ronnetta Bisman, Aug. 29, 1964; children: Gregory, Jessica. B.A., U. Mo., 1964; M.A., So. Ill. U., 1967, Ph.D., 1969. Asst. prof. Iowa State U., Ames, 1968-72, assoc. prof., 1972-76; vis. prof. Universtat Mannheim, W. Ger., 1975; prof. Iowa State U., 1976—; adminstrv. officer Am. Psychol. Assn., Washington, 1982—. Contbr. articles to profl. jours.; assoc. editor guar.: Psychology of Women, 1981—; cons. editor: Social Psychology Quar, 1981—. Nat. Inst. Edn. research grantee, 1972. Fellow Am. Psychol. Assn. (div. 35 sec.-treas. 1977-80, div. 9 program chmn. 1983). Democrat. Jewish. Subspecialties: Social psychology. Current work: Socially and ethically responsible use of psychology and behavior of psychologists; the role of sex and gender in social interaction; power. Office: 1200 17th St NW Washington DC 20036 Home: 5506 Cornish Rd Bethesda MD 20814

KAHN, BERND, radiochemist, educator; b. Pforzheim, Ger.; came to U.S., 1938; s. Eric Herman and Alice Dora (Meyer) K.; m. Gail Pressman, Aug. 6, 1961; children: Jennifer, Elizabeth. B.S. in Chem. Engring, Newark Coll. Engring., 1950; M.S. in physics, Vanderbilt U., 1951; Ph.D. in Chemistry, MIT, 1960. Health physicist, radiochemist Oak Ridge Nat. Lab., 1951-54; radiochemist, san. engr. USPHS, Oak Ridge, Montgomery (Ala.), and Cin., 1954-74, assigned to EPA, 1971-74; prof. Ga. Inst. Tech., Atlanta, 1974—, dir., 1974—. Contbr. numerous articles to profl. jours. Served to capt. USPHS, 1954-74. Mem. Am. Chem. Soc., Am. Phys. Soc., Health Physics Soc., Nat. Council Radiation Protection and Measurements. Jewish. Current work: Environmental radioactivity measurements and transport; radioactive waste treatment and disposal. Office: Environmental Resources Ctr Georgia Inst Technology Atlanta GA 30332

KAHN, C. RONALD, research physician, endocrinologist; b. Louisville, Jan. 14, 1944. B.A., U. Louisville, 1964, M.D., 1968. Diplomate: Am. Bd. Internal Medicine. Chief sect. on cellular physiology NIH, Bethesda, Md., 1979-81; head div. diabetes Brigham and Women's Hosp., Boston, 1981—; dir. research Joslin Diabetes Ctr., Boston, 1981—; mem. Nat. Diabetes Adv. Bd., Bethesda, 1981—. Contbr. articles to profl. jours. Served to col. USPHS, 1970-81. Recipient Rumbough award Juvenile Diabetes Found., 1977; Lilly award Am. Diabetes Assn., 1981; Ciba-Geigy-Drew award, 1981. Mem. Am. Diabetes Assn., Am. Fedn. for Clin. Research, Am. Soc. for Clin. Investigation. Subspecialties: Internal medicine; Endocrinology. Current work: Diabetes, insulin action. Home: 64 Valentine Park West Newton MA 02165

KAHN, ROY MAX, psychology educator; b. N.Y.C., June 14, 1926; s. Morris Hirsch and Muriel (Frumes) K.; m. Arlene Judy Kahn, Oct. 25, 1968; 1 dau., Jennifer M. A.B., Cornell U., 1949; M.A., Boston U., 1951, Ph.D., 1964. Lic. psychologist, Mass., Calif. Mental health coordinator Dept. Mental Health, Mass., Boston, 1957-63, dir. psychol. research, 1963-68; supr. therapy Harvard U. Phillips Brooks Mental Health Program, Cambridge, 1965-68; sr. psychologist Counseling Center U. Calif.-Berkeley, 1968-70; prof. lectr. Calif. State U., San Francisco, 1969-75; prof. psychology Pacific Grad. Sch. Psychology, Palo Alto, Calif., 1980—; cons. Stanford U., 1966; lectr. in field; dept. head psychology Gaebler Children's Unit, Waltham, Mass., 1965-68; cons. in field. Contbr. articles to profl. jours. Served with U.S. Army, 1944-46; ETO. Fellow Am. Orthopsychiat. Assn., Mass. Psychol. Assn.; mem. Am. Psychol. Assn., Western Psychol. Assn., Mass. Psychol. Assn. (founder, v.p. 1956). Current work: Psychotic child and cybernetic brain models. Home: 95 Sonia St Oakland CA 94618

KAHN, THOMAS, nephrologist, medical center administrator, educator; b. Offenburg, Germany, June 23, 1938; came to U.S., 1947, naturalized, 1952; s. Ludwig and Helen (Kaufman) K.; m. Si Mi Park, Nov. 16, 1968; children: Diana, David, Philip. B.A., NYU, 1958, M.D., 1962. Mem. faculty Mt. Sinai Sch. Medicine, N.Y.C., 1969—, assoc. prof. medicine, 1977—; chief renal sect. Bronx VA Med. Ctr., N.Y.C., 1979—. Contbr. articles to profl. jours. Served to maj. U.S. Army, 1967-69; Korea. N.Y. State Dept. Health grantee, 1975-77; VA grantee, 1981—. Fellow ACP; mem. N.Y. Soc. Nephrology (grantee 1972-74, sec.-treas. 1982-83, pres.-elect 1983-84), Am. Physiol. Soc., Am. Soc. Nephrology, Am. Soc. Artificial Internal Organs. Jewish. Subspecialties: Nephrology; Internal medicine. Current work: Electrolyte and acid-base physiology and pathophysiology. Office: Bronx VA Med Center 130 W Kingsbridge Rd Bronx NY 10021 Home: 511 E 80th St New York NY 10021

KAHNE, STEPHEN JAMES, systems engineer, educator; b. N.Y.C., Apr. 5, 1937; s. Arnold W. and Janet (Weatherlow) K.; m. Irena Nowacka, Dec. 11, 1970; children: Christopher, Katarzyna. B.E.E., Cornell U., 1960; M.S., U. Ill., 1961, Ph.D., 1963. Asst. prof. elec. engring. U. Minn., Mpls., 1966-69, asso. prof., 1969-76; dir. Hybrid Computer Lab., 1968-76; founder, dir., cons. InterDesign Inc., Mpls., 1968-76; prof. dept. systems engring. Case Western Res. U., Cleve., 1976-83, chmn. dept., 1976-80; dir. div. elec., computer and systems engring. NSF, Washington, 1980-81; cons. in field; exchange scientist Nat. Acad. Scis., 1968, 75. Editor: IEEE Transactions on Automatic Control, 1975-79; hon. editor: Internat. Fedn. of Automatic Control, 1975-81; editorial bd.: IEEE Spectrum, 1979-82; dep. chmn. editorial bd.: Automatica, 1976-82; dep. chmn. mng. bd.: Internat. Fedn. Automatic Control Publs, 1976—; contbr. articles to sci. jours. Active Mpls. Citizens League, 1968-75. Served with USAF, 1963-66. Recipient Amicus Poloniae award POLAND Mag., 1975; John A. Curtis award Am. Soc. Engring. Edn.; Case Centennial scholar, 1980. Fellow IEEE (pres. Control Systems Soc. 1981, bd. dirs. 1982-83), AAAS. Subspecialty: Systems engineering. Office: Office of Dean of Engring Poly Inst NY 333 Jay St Brooklyn NY 11201

KAI, MICHAEL S., nuclear engineer; b. N.Y.C., Oct. 22, 1949; s. Robert Yorito and Malo Fusako (Nakahara) K.; m. Joyce Katherine Moore, Aug. 4, 1973; children: Eleanor Marie, Rachel Emily. B.S., NYU, 1970; M.S., U. Ill., 1974; postgrad., Calif. State U., Los Angeles, 1976-79. Nuclear engr. Atomics Internat., Canoga Park, Calif., 1970; research asst. U. Ill., Champaign, 1972-74; nuclear engr. Bechtel Power, Norwalk, Calif., 1974-79, Babcock & Wilcox, Lynchburg, VA., 1979—. Served with U.S. Army, 1971-72. Mem. Am. Nuclear Soc., Tau Beta Pi. Democrat. Subspecialties: Nuclear fission; Nuclear engineering. Current work: Supervisor of fluid and transient analysis of nuclear power plants. Home: 402 Pine Dr Lynchburg VA 24502 Office: Babcock & Wilcox Old Forest Rd Lynchburg VA 24505

KAIGHN, MORRIS EDWARD, cancer researcher; b. Camden, N.J., Aug. 6, 1922. B.S., Bklyn. Coll., 1956; Ph.D., MIT, 1962. Asst. investigator Carnegie Instn., Balt., 1963-67; assoc. investigator N.Y. Blood Ctr., 1967-72; sr. scientist Cell Center, Lake Placid, N.Y., 1972-75; dir. Prostatic Cancer Lab., Pasadena, Calif., 1975-82; acting chief tissue culture sect. Nat. Cancer Inst., Frederick, Md., 1982—. Mem. Tissue Culture Assn. Am., Assn. for Cancer Research, Am. Assn. Cell Biology. Subspecialties: Cell and tissue culture; Cancer research (medicine). Current work: Culture of mammalian epithelial cells, defined media, oncogenic transformation of epithelial cells. Office: Nat Cancer Inst Frederick MD 21701

KAIN, RICHARD Y(ERKES), electrical engineering educator; b. Chgo., Jan. 20, 1936; s. Richard Morgan and Louise Kinsey (Yerkes) K.; m. Helen Buchanan (div.); children: Helen, Karen, Susan; m. Katherine S. Frank. B.S., MIT, 1957, M.S., 1959, Sc.D., 1962. Asst. prof. elec. engring. MIT Cambridge, 1962-66; assoc. prof. elec. engring. U. Minn., Mpls., 1966-77, prof., 1977—; Cons. Sperry Univac, Roseville, Minn., 1969-74, Honeywell, Mpls., 1974—. Author: Automata Theory, 1972. Ford postdoctoral fellow, 1962-64. Mem. IEEE (sr.), Assn. for Computing Machinery, Soc. for Indsl. and Applied Math., AAAS, Sigma Xi, Eta Kappa Nu. Subspecialties: Computer architecture; Distributed systems and networks. Current work: Computer system architecture, specializing in distributed systems and processor design. Office: U Minn 123 Church St SE Minneapolis MN 55455

KAISER, ARMIN DALE, biochemist; b. Piqua, Ohio, Nov. 10, 1927; s. Armin Jacob and Elsa Catherine (Brunner) K.; m. Mary Eleanor Durrell, Aug. 9, 1953; children—Jennifer Lee, Christopher Alan. B.S., Purdue U., 1950; Ph.D., Calif. Inst. Tech., 1955. Postdoctoral research fellow Inst. Pasteur, Paris, 1954-56; asst. prof. microbiology Washington U., St. Louis, 1956-59; mem. faculty Stanford U., 1959—, prof. biochemistry. Served with AUS, 1945-47. Recipient molecular biology award U.S. Steel Corp., 1971; Lasker award in basic med. sci., 1980. Mem. Nat. Acad. Scis., Am. Acad. Arts and Scis., Am. Soc. Biochemists, Genetic Soc. Am. Subspecialties: Genetics and genetic engineering (biology); Biochemistry (biology). Research on virus multiplication, microbial devel. Home: 832 Santa Fe Ave Stanford CA 94305 Office: Biochemistry Dept Stanford Univ Stanford CA 94305

KAISER, C. WILLIAM, surgeon; b. Troy, N.Y., Dec. 7, 1939; s. Carl William and Marjorie Smith (Patterson) K.; m. Mary Elisabeth Padlon, July 31, 1966; children: Holly Anne, Courtney Lee, Christopher William. A.B., Colgate U., 1961; M.D., Tufts Coll., 1965. Cert. Am. Bd. Surgery. Intern Charity Hosp., New Orleans, 1965-66; resident in surgery Boston City Hosp., 1968-72; assoc. vis. surgeon, 1972-76; assoc. chief surgery VA Med. Ctr., Northport, N.Y., 1976-78; chief surgery Pondville Hosp., Walpole, Mass., 1978-81, VA Med. Ctr., Manchester, N.H., 1981—; asst. prof. surgery Harvard U. Contbr. articles to profl. jours. Bd. dirs. Am. Cancer Soc., Boston, 1981—. Served as capt. M.C. U.S. Army, 1966-68. Fellow A.C.S., Royal Coll. Physicians and Surgeons Can. Subspecialty: Oncology. Home: 8 Middle St Concord MA 01742 Office: VA Med Ctr 718 Smyth Rd Manchester NH 03104

KAISER, CHARLES FREDERICK, psychologist; b. N.Y.C., Dec. 30, 1942; s. Alexander and Etta K.; m. Judy Kaiser, Aug. 21, 1966; children: Edward, Michael. B.S., CUNY, 1964, M.A., 1967; Ph.D., U. Houston, 1973. Teaching fellow, NIMH research fellow dept. psychology U. Houston, 1966-72; assoc. prof. psychology Coll. Charleston, S.C., 1972—; adj. prof. phys. medicine Med. U. S.C., Charleston, 1981—. Contbr. articles to profl. jours. Mem. S.C. Psychol. Assn. (trea. 1979-81), Charleston Area Psychol. Assn. (Pres. 1979-80), Biofeedback Soc. S.C. (pres. 1981—). Subspecialties: Behavioral psychology; Psychobiology. Current work: Research in behavioral medicine and biofeedback for pain management, research on depression, research on psychological traits of gifted adolescents. Home: 1416 Birthright St Charleston SC 29407 Office: Dept Psycholog Coll of Charleston 66 George St Charleston SC 29424

KAISER, HANS ELMAR, comparative pathologist, scientific writer; b. Prague, Czechoslavakia, Feb. 16, 1928; came to U.S., 1961; s. Rudolf J. and Charlotte B. (Thiel) K.; m. Charlotte Mohring, Oct. 12, 1961. D.Sc., Eberhard Karls U., Tubingen (W.Ger.), 1958. Exptl. worker dept. pathology Vet. Acad., Hanover, W. Ger., 1958-60; vis. investigator Meml. Sloan-Kettering Cancer Center, N.Y.C., 1960-61; spl. lectr. dept. anatomy U. Saskatchewan, Saskatoon, Can., 1962-63; cons. Biol. Scis. Communications Project, George Washington U., 1964-66, asst. prof. dep. anatomy, 1965-69; cons. exptl. work Forest Physiology Lab., U.S. Dept. Agr., Beltsville, Md., 1972-75; dir. research div. Gt. Lakes Area Paleontol. Mus., Traverse City, Mich., 1975—; research assoc. dept. pathology U. Md. Sch. Medicine, Balt., 1974-77, asst. prof., 1977-79, assoc. prof., 1979—; cons. FDA, 1983; pub. sci. books. Contbr. numerous articles on comparative pathology to profl. jours. Mem. Am. Assn. of Anatomists, Am. Assn. Cancer Research, AAAS, Am. Cancer Soc., Am. Chem. Soc., Am. Micros. Soc., Am. Soc. Exptl. Pathology, Am. Soc. Zoologists, Asociacion Panamerican de Anatomia, Biol. Soc. of Washington, Fedn. Am. Socs. for Exptl. Biology, N.Y. Acad. Scis., 2d Turkish Cancer Soc. (hon. mem., Ankara), Soc. Invertebrate Pathology, Washington Acad. Scis., Vet. Cancer Soc. Lutheran. Club: Starwood Home Owners Assn. (Aspen, Colo.). Subspecialties: Comparative pathology; Oncology. Current work: Comparative pathology, comparative oncology, comparative histology, scientific writing at present about progressive stages of neoplastic growth, investigation of plant promotors. Home: 433 S West Dr Silver Spring MD 20901 Office: 10 S Pine St Baltimore MD 21201

KAISTHA, KRISHAN, KUMAR, toxicologist, pharmacist, clinical chemist; b. Sulah (Kandgra), Himachal Pradesh, India, Apr. 6, 1926; came to U.S., 1959, naturalized, 1974; s. Mangat Ram and Tara Devi (Mahajan) K.; m. Swarn Lata, Feb. 22, 1948; children: Anita, Vivek, Vinek. B.S. in Chemistry, Punjab (India) U., 1947, 1951, M.S., 1955; Ph.D., U. Fla., 1962. Diplomate: Am. Bd. Forensic Toxicology; cert. clin. chemist Nat. Registry Clin. Chemistry; registered pharmacist, Ill. Analytical chemist Punjab govt. Med. Directorate, 1952-57, chief pharmacist, 1957-59; research fellow SUNY, Buffalo, 1962-63; head pharm. services and phytochem. research lab. Punjab Govt. Postgrad. Med. Research Inst., 1964-66; research scientist food and drug directorate Dept. Nat. Health and Welfare, Ottawa, Ont., Can., 1966-69; research assoc. dept. psychiatry U. Chgo., 1969-75; dir. toxicology labs. State of Ill. Dept. Mental Health Drug Abuse Programs, Chgo., 1969-74, chief toxicologist, 1974—, acting adminstr., 1981—. Contbr. numerous articles on toxicology and analytical methodology to profl. jours. Recipient 1st Prize Lunsford-Richardson award, 1962; Gov.'s Economy Incentive award State of Ill., 1973. Fellow N.Y. Acad. Scis., Am. Acad. Forensic Scis., Nat. Acad. Clin. Biochemistry; mem. Am. Acad. Clin. Toxicology, Am. Assn. Clin. Chemists, Am. Soc. Pharmacology and Exptl. Therapeutics, Rho Chi, Rho Pi Phi, Phi Kappa Phi, Gamma Sigma Epsilon.; Mem. Vivekanada Vedanta Soc. Subspecialties: Toxicology (medicine); Biochemistry (medicine). Current work: Analytical methodology in toxicology, clinical chemistry and phar. Home: 542 N Ashbury Ave Bolingbrook IL 60439 Office: I I T Research 10 W 35th St Chicago IL 60616

KAKUNAGA, TAKEO, microbiologist; b. Keijo, Japan, Nov. 8, 1937; came to U.S., 1973; s. Tchiro and Kimiko (Nakamura) K.; m. Mariko Aburaki, Dec. 1, 1963; children: Shino, Sbigeki. B.S. in Pharmacology, Kanazawa U., 1960; M.S., Osaka U., 1962, Ph.D., 1966. Asst. prof. microbiology dept. tumor viruses Research Inst., Microbial Diseases, Osaka U., 1966-73, asso. prof., 1973-78; chief sect. onvology, cell genetics sect. Lab. Molucular Carcinogensis Nat. Cancer Inst., NIH, 1977—; cons. in field. Author numerous papers in field. Recipient Princess Takanatsu prize, 1971. Mem. Am. Assn. Cancer Research, Am. Soc. Cell Biology, Japanese Cancer Assn. Subspecialties: Cancer research (medicine); Cell and tissue culture. Current work: Chem. carcinogenesis, cell biology (cytoskeleton differentiation), cell genetics, tumor promoting agts., (membrane biology). Office: Nat Cancer Inst Bldg 37 Room 3E08 Bethesda MD 20205

KALATA, KENNETH, physicist, inventor; b. Detroit, Aug. 22, 1947; s. Faust Walter and Christine (Ankofski) K. Student, M.I.T., 1964-67, 67-71. Research asst. M.I.T., Cambridge, 1966-67, postdoctoral research assoc., 1971-73, research physicist, 1973-74; physicist Smithsonian Astrophys. Obs., Cambridge, 1974-75; cons., 1978-79; research assoc. Steward Obs., U. Ariz., Tucson, 1976-78; research contractor Brandeis U., Waltham, Mass., 1978-80, sr. research assoc., 1980—; research assoc. Harvard U., 1980—; pres Mimbres Research Inst., San Lorenzo, N.Mex., 1981—, Quanticon Corp., San Lorenzo, 1981—. Contbr. articles to profl. jours. NSF grantee, 1980, 82. Mem. Am. Phys. Soc., Am. Astron. Soc., Soc. Photo-Optical Instrumentation Engrs. Democrat. Zen Buddhist. Club: Mimbres Hot Springs Ranch (San Lorenzo). Subspecialties: 1-ray high energy astrophysics; Biophysics (physics). Current work: Detector systems, X-ray astrophysics, structural biology. Design and development of area X-ray, light and particle detector systems for structural biology, X-ray astrophysics, and astronomy. Home: 53 Sacramento St Cambridge MA 02138 Office: Rosenstiel Center Brandeis U Waltham MA 02254

KALDOR, ANDREW PETER, research chemist; b. Budapest, Hungary, Oct. 11, 1944; s. Erno and Lenke K.; m. Sandra Jean Lewis, June 18, 1967; children: Eric Clay, Tamara. B.S., U. Calif.-Berkeley, 1966; Ph.D., Cornell U., 1970. Research chemist Nat. Bur. Standards, Washington, 1970-74; sr. research chemist Exxon Research and Engring. Co., Linden, N.J., 1974-76, head chem. physics group, 1976-81, head surface sci. group, 1978-81, dir. chem. physics scis. lab., corp. research lab., 1981—; lectr. Case Western Res. U., 1978, U. Calif.-Irvine, 1981. Contbr. articles to profl. jours. Subspecialties: Laser-induced chemistry; Surface chemistry. Current work: Study role of molecular energetics in chemical reactions and mechanisms of surface chemical reactions and at liquid and gas solid interfaces. Patentee in field. Office: PO Box 45 Linden NJ 07036

KALER, JAMES B., astronomer; b. Albany, N.Y, Dec. 29, 1938; s. Earl B. and Hazel A. (Holmgren) K.; m. Maxine Ellen Grossman, June 15, 1960; children: Lauren Lynn, Bruce Jeffrey, Lisa Suzanne, Jill Lenore. A.B., U. Mich., 1961; postgrad., Kiel (W.Ger.) U., 1961-62; Ph.D., UCLA, 1964. Asst. prof. astronomy U. Ill., Urbana, 1964-68, assoc. prof., 1968-76, prof., 1976—. Contbr. articles to profl. jours. Fulbright fellow, 1961-62; Guggenheim fellow, 1972-73. Mem. Am. Astron. Soc., Internat. Astron. Union, AAAS, Astron. Soc. Pacific. Club: Ill. Track (Champaign). Subspecialty: Optical astronomy. Current work: Planetary nebulae, nebular spectrophotometry. Home: 907 Sunnycrest St Urbana IL 61801 Office: 341 Astronomy Bldg 1011 W Springfield St Urbana IL 61801

KALET, IRA JOSEPH, med. physicist, educator; b. Stamford, Conn., Apr. 27, 1944; s. Bernard and Miriam (Pivnick) K.; m. Teresa Lynn Steele, Apr. 7, 1973; children: Nathan, Alan, Brian. A.B. in Physics, Cornell U., 1965; M.A., Princeton U., 1967, Ph.D. in Theoretical Physics, 1968. Research assoc. physics U. Wash., Seattle, 1968-69; lectr. math. edn. U. Pa., Phila., 1974-75; asst. prof. Sonoma State Coll., Rohnert Park, Calif., 1969-70; sr. fellow med. radiation physics U. Wash., Seattle, 1978-80, research assoc., 1980-82, asst. prof. radiation oncology, 1982—. Contbr. articles to profl. jours. NSF fellow, 1965-68. Mem. Am. Assn. Physics Tchrs., Am. Assn. Physicists in Medicine, Radiation Research Soc., Assn. Computing Machinery, Physicians for Social Responsibility. Subspecialties: Medical physics; Artificial intelligence. Current work: Computer graphics simulation of radiation therapy treatments; expert computer systems for radiotherapy treatment planning. Office: Med Radiation Physics RC-08 U Wash Seattle WA 98195

KALIA, RAVINDRA NATH, mathematics educator, researcher; b. Lucknow, India, Sept. 8, 1940; came to U.S., 1980; s. Diwan Ch and Kamla (Sharma) K.; m. Sneh Jarrel, Mar. 13, 1974; 1 dau., Sukanya. B.Sc., Lucknow U., 1959, M.Sc., 1962, 63, Ph.D., 1972. Lectr. Lucknow Christian Coll., 1963-80; postdoctoral fellow U. Hamburg, W. Ger., 1972-75; vis. prof. Adrian (Mich.) Coll., 1980-81; asst. prof. math. Southwest State U., Marshall, Minn., 1981, assoc. prof., 1982—. Sec. World Assn. World Federalists, Lucknow, 1971, sec.-gen., 1978; sec. All-India body, 1979. Mem. Am. Math. Soc., Soc. for Indsl. and Applied Math., Ops. Research Soc. Subspecialties: Applied mathematics. Current work: Applications of mathematics to probability distribution and to bioeconomics, transportation science, modelling. Algorithms in computer science. Office: Dept Math and Computer Sci Southwest State U Marshall MN 56258

KALICK, SHELDON MICHAEL, psychology educator; b. Bklyn., Aug. 27, 1948; s. Stanley Albert and Ruth (Powitz) K. A.B., Princeton U., 1969; M.A., Harvard U., 1971, Ph.D., 1977. Asst. prof. psychology U. Mass., 1977—; instr. psychology Harvard Med. Sch., Boston, 1978—. Contbr. articles to profl. jours. Recipient Bowdoin Grad. prize Harvard U., 1977; Thesis award Soc. for Psychol. Study of Social Issues, 1977. Mem. Am. Psychol. Assn., Soc. for Psychol. Study of Social Issues, Phi Beta Kappa. Subspecialty: Social psychology. Current work: Social perception; body image and psychol. effects of plastic surgery; social values and culture. Home: 10 Dana St (NUM)310 West Cambridge MA 02138 Office: Dept Psychology Univ Mass Harbor Campus Boston MA 02125

KALIMI, MOHAMMED YAHYA, physiologist, educator, researcher; b. Surat, India, Nov. 3, 1939; came to U.S., 1971; s. Yahya A. and Safiya H. (Saifuddin) K. B.S., Bombay U., 1961, M.S., 1964, Ph.D., 1970. Research trainee Columbia, U., 1972-74; research assoc Baylor Coll., 1974-75; assoc. Albert Einstein Coll. Medicine, 1975-79; asst. prof. physiology Med, Coll. Va., 1979-81, assoc. prof., 1981—. Author: Receptor Methodology, 1983; contbr. articles to profl. jours. NIH Research Career Devel. awardee, 1980—. Mem. Am. Physiol. Soc., Endocrine Soc. Islam. Subspecialties: Physiology (biology); Molecular biology. Current work: Glucocrticoids, mechanism of action, receptor, gene expression. Office: Med Col Va Box 551 Richmond VA 23298

KALINAUSKAS, GEDIMINAS LEONARDAS, nuclear engr., cons.; b. Toronto, Ont., Can., Mar. 6, 1954; came to U.S., 1976; s. Leonardas and Maryte (Baranauskas) K.; m. Susan McGann, Dec. 30, 1978; 1 son, Paulius. B.A., U. Toronto, 1976; M.S., U. Wash., 1978. Nuclear engr. Bechtel Power, San Francisco, 1978-80; chem. engr. London

Nuclear, Niagara Falls, 1980-82; sr. engr. E.D.S. Nuclear Inc., Atlanta, 1983—. Mem. Am. Nuclear Soc. (assoc. mem.). Roman Catholic. Subspecialties: Nuclear engineering; Nuclear fission. Current work: Design of radioactive waste processing systems, decontamination of nuclear reactor systems and subsystems using dilute regenerative chemicals. Home: 9123 Branch Valley Way Roswell GA 30076 Office: E D S Nuclear Inc 333 Technology Park Norcross GA 30092

KALLMAN, ROBERT FRIEND, radiobiologist, educator; b. N.Y.C., May 21, 1922; s. Morris and Eva (Cohn) K.; m. Frances Lou Green, June 4, 1948; children: Timothy Raymond, Robin Lou, Lars Peter; m. Ingrid Moelhede Christensen, Apr. 19, 1969. A.B. in Biology, Hofstra U., 1943; M.S. in Biology (teaching fellow), N.Y.U., 1949; Ph.D. in Physiology, N.Y.U., 1952. Instr. biology Bklyn. Coll., 1947-48; asst. research physiologist U. Calif.-San Francisco, 1952-56; research assoc. Stanford U., 1956-59, asst. prof., 1959-64, dir., 1959—, assoc. prof., 1964-72, prof., 1972—, dir., 1978—; cons. U. Calif. Lawrence Berkeley Lab., U. Rochester Cancer Center; former mem. rev. com. Cancer Research Center. Editor: Research in Radiotherapy—Approaches to Chemical Sensitization, 1961, United-States-Italy Cooperative Seminar on Radiation Sensitivity: Facts and Models, 1979; sr. editor biology: Internat. Jour. Radiol. Oncology, Biology, and Physics, 1981—; assoc. editor: Lab. Animal Sci, 1963-78, Cancer Research, 1972-79; bd. editors: Radiation Research, 1971-74, Yearbook of Cancer, 1964-71; contbr. articles to sci. jours. Served with AUS, 1943-46. Recipient Research Career Devel. award USPHS, 1962-72, research grantee, 1956—, tng. grantee, 1978—. Mem. Radiation Research Soc. (pres. 1976-77), AAAS, Am. Assn. Cancer Research, Am. Assn. Lab. Animal Sci., Assn. Gnotobiotics, N.Y. Acad. Scis., Radiation Research Soc., Am. Soc. Therapeutic Radiologists, Sigma Xi. Subspecialties: Radiation biology; Cancer research (medicine). Current work: Mammalian radiobiology, general radiobiology, cell biology, experimental oncology. Office: Stanford U Med Center Radiobiology Research A 038 Stanford CA 94305

KALLOK, MICHAEL JOHN, research scientist, consultant; b. Gary, Ind., Apr. 9, 1948. B.S., U. Colo., 1970; M.S., Purdue U., 1974; Ph.D., U. Minn., 1978. Registered profl. engr., Minn. Research asst. U. Minn., Mpls., 1974-77; part-time engring. cons. Indsl. Health Engring. Assocs., Hopking, Minn., 1976-77; research fellow Mayo Clinic, Rochester, Minn., 1977-79; physiology instr. Mayo Med. Sch., Rochester, 1978-79; sr. staff engr. Medtronic, Inc., Mpls., 1979-81, sr. staff scientist, 1981—. Guest reviewer: Jour. Applied Physiology, 1981—; contbr. articles to profl. jours. Recipient Young Investigator award NIH, 1979; named Young Mech. Engr. of Yr., Minn. sect. ASME, 1981, Young Engr. of Yr. Minn. Soc. Profl. Engrs., 1981. Mem. Am. Heart Assn. (cardiopulmonary council), Am. Physiol. Soc., Am. Thoracic Soc., Biomed. Engring. Soc. (sr.), N.Y. Acad. Scis., Sigma Gamma Tau. Subspecialties: Physiology (medicine); Biomedical engineering. Current work: Cardiovascular physiology, pulmonary physiology, cardiac arrhythmias. Patnetee transvenous defibrillating lead. Office: Medtronic Inc 3055 Old Highway 8 Minneapolis MN 55440

KALMAN, GABOR JENO, physicist, educator, researcher; b. Budapest, Hungary, Dec. 12, 1929; came to U.S., 1965, naturalized, 1976; s. Geza G. and Ilona (Adler) K.; m. Susan Zeizler; children: Katalin, Ron. Diploma in Elec. Engring, Tech. U., Budapest, 1953; D.Sc. in Physics, Technion, Israel Inst. Tech., 1961. Research scientist Central Research Inst. for Physics, Budapest, 1952-56; lectr. Technion, Israel Inst. Tech., 1957-61; prof. associe U. Paris, 1961-65, exchange prof., 1976; dir. research Centre National de la Recherche Scientifique, France, 1965-67; vis. prof. physics Brandeis U., 1966-70; research prof. physics Boston Coll., 1970—; dir. Advanced Study Inst., NATO, 1966, 77; expert Air Force Cambridge Research Labs., 1966-67; asso. Center for Astrophysics, Harvard U., 1973-77, vis. scientist, 1974; vis. fellow Joint Inst. Lab. Astrophysics, U. Colo., 1965-66; research dir. Internat. Center for Theoretical Physics, Trieste, 1981; vis. scientist Obs. of Paris, Meudon, France, 1973-74, Groupe de Recherches Ionospheriques, Orleans, France, 1974, Oxford U., 1975. Editor: (with M. Feix) Nonlinear Effects in Plasmas, 1967, (with P. Carini) Strongly Coupled Plasmas, 1978; contbr. numerous articles to sci. jours. Various research and travel grants including NSF, Dept. Energy, Dept. Def. Fellow N.Y. Acad. Scis.; mem. Am. Phys. Soc., European Phys. Soc., Am. Astron. Soc., AAAS, Societe Francaise de Physique. Subspecialties: Plasma physics; Statistical physics. Current work: Strongly coupled plasmas; ultrastrong magnetic fields; plasma spectroscopy. Home: 357 Clinton Rd Brookline MA 02146 Office: Dept Physics Boston College Chestnut Hill MA 02167

KALMAN, RUDOLF EMIL, research mathematician, system scientist; b. Budapest, Hungary, May 19, 1930; s. Otto and Ursula (Grundmann) K.; m. Constantina Stavrou, Sept. 12, 1959; children: Andrew E.F.C., Elisabeth K. S.B., Mass. Inst. Tech., 1953, S.M., 1954; D.Sc., Columbia U., 1957. Staff engr. IBM Research Lab., Poughkeepsie, N.Y., 1957-58; research mathematician Research Inst. Advanced Studies, Balt., 1958-64; prof. engring. mech. and elec. engring. Stanford U., 1964-67, prof. math. system theory, 1967-71; grad. research prof., dir. Center for Math. System Theory, U. Fla., 1971—; prof. math. system theory Swiss Fed. Inst. Tech., Zurich, 1973—; Sci. adviser Ecole Nationale Superieure des Mines de Paris, 1968—; mem. sci. adv. bd. Laboratorio di Cibernetica, Naples, 1970-73. Author: Topics in Mathematical System Theory, 1969; over 100 sci. and tech. papers.; Mem. editorial bd.: Jour. Math. Modelling, Math. Systems Theory, Jour. Computer and Systems Scis.; mem. editorial bd.: Jour. Nonlinear Analysis, Advances in Applied Math. Named outstanding young scientist Md. Acad. Sci., 1962; recipient IEEE medal of honor, 1974; Guggenheim fellow IHES Bures-sur-Yvette, 1971; Rufus Oldenburger medal ASME, 1976. Fgn. hon. mem. Hungarian Acad. Scis. Subspecialties: Applied mathematics; Robotics. Office: Dept Math U Fla Gainesville FL 32611 It is good to do everything as it was done yesterday, but it is better to examine all accepted assumptions. This is the key to scientific progress as well as to happier interpersonal relations.

KALOS, MALVIN HOWARD, physics educator, researcher; b. N.Y.C., Aug. 5, 1928; s. Sol Kalos and Mollie (Timan) Greenberg; m. Elaine Eberstark, Aug. 20, 1949; children: Stephen, Lauren. B.S., Queens Coll., 1948; M.S., U. Ill., 1949, Ph.D., 1952. Research assoc. U. Ill.-Urbana, 1952-53, Cornell U., Ithaca, N.Y., 1953-55; sci. adv. United Nuclear Corp., White Plains, N.Y., 1955-64; research prof. Courant Inst. Math. Sci., N.Y.C., 1964—; tech. dir. MAGI, Elmsford, N.Y., 1966-76; cons. Los Alamos Sci. Lab., 1973—, Lawrence Livermore Lab., Livermore, Calif., 1978—; mem. U.S. Nuclear Data Commn., 1966-74. Fellow Am. Nuclear Soc., N.Y. Acad. Scis.; mem. Am. Phys. Soc., AAAS. Subspecialties: Statistical physics; Computer architecture. Current work: Quantum many-body problem, non-equilibrium statistical physics, Monte Carlo methods, computer architecture. Office: Courant Inst Math Sci 251 Mercer St New York NY 10012

KALRA, SATYA PAUL, obstetrics and gynecology educator, researcher; b. Mari Indus, India, Jan. 1, 1939; came to U.S., 1968; s. Ishar Dass and Sushila (Malik) K.; m. Pushpa Seth, Sept. 14, 1969; 1 son, Anjay N. B.S., U. Delhi, India, 1960, M.S., 1962, Ph.D., 1966. Research assoc. U. Delhi, 1966-68; postdoctoral fellow UCLA, 1968-69, U. Tex.-Dallas, 1969-71; asst. prof. U. Fla., Gainesville, 1971-76, assoc. prof., 1976-82, prof. ob-gyn, 1982—. Grantee Population Council, 1971-74, NIH, 1975—. Mem. Endocrine Soc., Internat. Neuroendocrine Soc., Am. Physiol. Soc., Soc. for Study Reprodn., Soc. for Gynecol. Investigation, Am. Soc. Andrology. Subspecialties: Neuroendocrinology; Reproductive biology. Current work: Neuroendocrine control of gonadotropin secretion gonadal feedback; neurotransmitter — neuropeptide interaction, hypothalamus. Office: Dept Ob-Gyn U Fla Box J-294 J Hillis Miller Health Ctr Coll Medicine Gainesville FL 32610

KALTENBACH, JOHN PAUL, biochemist, researcher, educator; b. Rockford, Ill., Feb. 28, 1920; s. John and Anne (Kruger) K.; m. Merle H. Kaltenbach, Oct. 15, 1923; 1 son, John C. B.S., Beloit Coll., 1944; M.S., U. Iowa, 1947, Ph.D., 1950. Chief cell metabolism VA Research Hosp., Chgo., 1954-56, sr. biochemist, 1956-57; Prof. pathology Northwestern U., 1957-83, acting dir. admissions, 1978-79, affiliated profl. staff, 1973-82. Contbr. articles to profl. jours. Brittingham fellow, 1950-52; USPHS fellow Karolinska Inst., Stockholm, 1953-54. Mem. AAAS, Am. Assn. Cancer Research, Am. Soc. Exptl. Pathology. Subspecialties: Biochemistry (medicine); Cell biology (medicine). Current work: Induction and protection of D-serine kidney tubular necrosis, pathogenesis of irreversible myocardial injury.

KALU, DIKE NDUKWE, physiology educator and researcher; b. Abiriba, Imo, Nigeria, Jan. 1, 1938; came to U.S., 1971; s. Ndukwe and Oyediya K.; m. Carolyn A. Rotibi, Aug. 24, 1967; children: Nneji, Ngozi, Ndukwe. B.Sc., U. London, 1967; Ph.D., 1971. Fellow Johns Hopkins U., Balt., 1972-75; sci. officer U. London, 1967-71; asst. prof. U. Tex. Health Sci. Ctr., San Antonio, 1975-79; assoc. prof. physiology, 1979—. NIH grantee, 1979, 80. Mem. AAAS, Endocrine Soc., Am. Physiol Soc., Fedn. Am. Soc. Exptl. Biology, N.Y. Acad. Scis. Subspecialties: Physiology (medicine); Endocrinology. Current work: Influence of the endocrines and aging on the regulation of calcium and skeletal metabolism. Home: 6822 Brookvale St San Antonio TX 78238 Office: U Tex Health Sci Ctr Dept Physiology 7703 Floyd Curl Dr San Antonio TX 78284

KALVINSKAS, JOHN JOSEPH, chemical engineer, consultant; b. Phila., Jan. 14, 1927; s. Anthony and Anna (Slezute) K.; m. Louanne Marie Adams, Sept. 3, 1955; 1 son, Adrian John. B.S., M.I.T., 1951, M.S. in Chem. Engring, 1952; Ph.D. (Stauffer Found. fellow), Calif. Inst. Tech., Pasadena, 1959. Chem. engr. duPont, Gibbstown, N.J., 1952-55, 59-60; research specialist Rockwell Internat., El Segundo, Calif., 1960-61, supr. advanced projects, 1961-64, mgr. propellant engring., 1964-67, mgr. environ. mgmt. systems, 1967-68, dir. environ. mgmt. systems, 1968-70; corp. research dir. Resource Dynamics Corp., Los Angeles, 1970-74, Monogram Industries, 1972; project mgr. Holmes & Narver, Inc., Anaheim, Calif., 1974, Jet Propulsion Lab, 1974—, also group supr; cons. Rockwell Internat., 1972-74, Kinetics Tech. Internat., 1980-82. Mem. Town Hall Calif., 1967—. Served with USN, 1944-46; PTO. Recipient Recognition certs. NASA, 1976-82. Mem. Am. Inst. Chem. Engrs., Am. Chem. Soc., Am. Water Resources Assn., N.Y. Acad. Scis., Sigma Xi, Kappa Kappa Sigma. Republican. Roman Catholic. Club: M.I.T. (Los Angeles). Subspecialties: Chemical engineering; Biomass (energy science and technology). Current work: Project management energy programs; consultant on energy programs. Patents, publs. in chem. engring. field. Home: 316 Pasadena Ave South Pasadena CA 91030 Office: Jet Propulsion Lab 4800 Oak Grove Dr Pasadena CA 91109

KALYONCU, RUSTU SUMER, materials scientist, researcher, researcher adminstr., cons.; b. Izmir, Turkey, Apr. 1, 1943; came to U.S., 1962; s. Hilmi and Sabriye (Abuz) K.; m. Aydan A., Aug. 21, 1970; 1 dau., Sevim. B.S. in Ceramic Engring, Alfred U., 1966, M.S., 1967, Ph.D. in Materials Sci, 1970; M.B.A., U. Dayton, 1978. Research scientist Martin Marietta Labs., Balt., 1971-76; prin. research scientist Battelle Columbus Labs., Ohio, 1977-80; programs mgr. U.S. Bur. Mines, University, Ala., 1980—; cons. in field. Contbr. articles in ceramics, cements, and minerals scis. to profl. jours. Mem. Am. Ceramic Soc., ASME, AIME. Subspecialty: Ceramics. Home: Route 3 Box 613 Cottondale AL 35453

KAMAL, ABDUL NAIM, physicist; b. Dacca, Bangladesh, Oct. 28, 1935, s. Abdul and Sarah (Khatoon) K., m. Anita Balodis, Sept. 8, 1962; children: David, Sarah, Katherine. B.Sc. with honors, Dacca U., 1955, M.Sc., 1956; Ph.D. in Theoretical Physics, Liverpool U., 1962. Asst. prof. physics U. Alta., Edmonton, 1964-68, assoc. prof. physics, 1969-73, prof. physics, 1973—, chmn. dept. physics, 1980—; sr. scientist Rutherford Lab., U.K., 1968-69. Author: Problems in Particle Physics, 1966; Contbr. articles to profl. jours. Natural Scis. and Research Council Can. grantee, 1966—. Mem. Am. Phys. Soc., Can. Assn. Physicists. Subspecialties: Particle physics; Theoretical physics. Current work: Theoretical high energy physics. Home: 5503 115th St Edmonton AB Canada T6H 3P4 Office: Dept Physics Univ Alberta Edmonton AB Canada T6G 2J1

KAMAN, CHARLES HENRY, computer software developer. A.B., Harvard U., 1964; M.S., Poly. Inst. N.Y., Bklyn., 1967, Ph.D., 1974. Architect, system cons. Digital Equipment Corp., Maynard, Mass., 1969-81; self-employed software developer, Newtown Highlands, Mass., 1981—. Contbr. articles in field to profl. jours. Mem. Assn. for Computing Machinery, IEEE, Am. Math. Soc., Math. Assn. Am., Soc. Indsl. and Applied Math. Subspecialties: Computer architecture; Applications software. Current work: Computer design and implementation, computer architecture and the tradeoffs between them. Patentee computer implementation. Home: 274 Dedham St Newton Highlands MA 02161

KAMATH, SAVITRI KRISHNA, nutrition educator; b. Kanhangad, Kerala State, India, Sept. 22, 1930; d. Ramarao and Kamala (Ramarao) Kotecheri; m. Krishna I. Kamath, Jan. 2, 1969. I.Sc., Madras U., 1948; B.Sc. with honors, Bombay U., 1950; M.S., U. Baroda, India, 1963; Ph.D., Iowa State U., 1967. Cert. Am. Bd. Nutrition. Reader dept. food and nutrition U. Baroda, 1967-69; research biochemist Hektoen Med. Research Inst., Chgo., 1970-72; from asst. prof. to prof. nutrition and med. dietetics U. Ill. Health Scis. Center, Chgo., 1972—, head dept. nutrition and med. dietetics, 1979—. Contbr. articles to profl. jours. Ford Found. scholar, 1963-67; recipient Amoco award for teaching excellence, 1979. Mem. Am. Dietetic Assn. (registered), Chgo. Dietetic Assn., Chgo. Nutrition Assn., Soc. Nutrition Edn., N.Y. Acad. Scis., Sigma Xi, Iota Sigma Pi, Omicron Nu. Hindu. Subspecialties: Nutrition (biology); Biochemistry (biology). Current work: Nutrition and cancer, diabetes and taste acuity, nutrition/dietetic curriculum development and evaluation. Home: 1519 W Polk St Chicago IL 60607 Office: Dept of Nutrition and Medical Dietetics Room 881 808 S Wood St Chicago IL 60612

KAMB, WALTER BARCLAY, educator, geologist; b. San Jose, Calif., Dec. 17, 1931; s. Karl Walter and Eleanor (Williams) K.; m. Linda Helen Pauling, Sept. 8, 1957; children: Barclay James, Carl Alexander, Anthony Pauling, Linus Peter. B.S. in Physics, Calif. Inst. Tech., 1952, Ph.D. in Geology, 1956. Mem. faculty Calif. Inst. Tech., 1956—, prof. geology and geophysics, 1961—; chmn. div. geol. and planetary scis., 1972-83. Guggenheim fellow, 1960; Sloan fellow, 1964. Fellow Geol. Soc. Am., Mineral. Soc. Am. (award 1968); mem. Am. Geophys. Union, Am. Assn. Petroleum Geologists. Subspecialties: Petrology; Geophysics. Home: 3500 Fairpoint St Pasadena CA 91107

KAMBACK, MARVIN CARL, psychologist; b. Yankton, S.D., July 15, 1939; s. Carl Melvin and Pauline Elizabeth (Albrecht) K.; m. Genevieve Lowthian, Sept. 2, 1962 (div. 1971); children—Elizabeth Farrell, Christopher John. B.A., U. S.D., 1961, M.A., 1962; Ph.D., Vanderbilt U., 1965. Lic. psychologist, Md., Wyo., Calif. Postdoctoral fellow Stanford U. Med. Sch., 1965-66; lectr. U. Calif., Santa Barbara, 1966-67; instr. psychology U. S.D., Vermillion, 1962, asst. prof., 1967-71; asst. prof. psychology Johns Hopkins U. Med. Sch.; sr. clin. psychologist Balt. City Hosps., 1971-74; assoc. prof. U. Md. Med. Sch., Balt., 1974-78; dir. Washakie County (Wyo.) Mental Health Services, 1978-79; dir. psychol. services Alcohol Care Center, Buena Park (Calif.) Hosp., 1979—, Raleigh Hills Hosps., Newport Beach, Calif., 1979—; clin. psychologist Behavior Therapy and Research Inst., 1979-82; dir. psychol. services alcoholism program Advanced Health Systems, Newport, Beach, 1979-82; pvt. practice clin. psychology, San Clemente and Buena Park, 1979—. Contbr. chpts. to books, articles to profl. jours. Mem. Am. Psychol. Assn., AAAS, Md. Psychol. Assn., Soc. for Gen. Systems Research, Sigma Xi. Republican. Lodge: Rotary. Subspecialties: Behavioral psychology; Neuropsychology. Current work: Addictions research. General systems theory as it applies to health care. Brain function research. Office: Buena Park Community Hosp 6850 Lincoln Ave Buena Park CA 90620 Home: 372 Calle Guymas San Clemente CA 92672

KAMBOUR, ROGER PEABODY, research chemist; b. Wilmington, Mass., Apr. 1, 1932; s. George Contantine and Ada (Mattraw) K.; m.; children: Annaliese S., Christian R. B.A., Amherst Coll., 1954; Ph.D., U. N.H., 1960. Research chemist Gen. Electric Research and Devel. Ctr., Schenectady, N.Y., 1960—. Contbr. articles to profl. jours. Ward supr. Schenectady County Bd. Suprs., 1964-65; mem. Schenectady County Charter Commn., 1964-65; bd. dirs. Freedom Forum, 1968-76; mem. Schenectdy Hist. Dist. Commn., 1976-81. Recipient Union Carbide Chems. award Am. Chem. Soc., 1968; Coolidge fellow, 1979. Fellow Am. Phys. Soc.; mem. Am. Chem. Soc. Democrat. Unitarian. Subspecialties: Polymer chemistry; Polymer physics. Current work: Mechanisms of crazing and fracture in thermoplastics, thermodynamics and mechanics of polymer blends. Patentee in field. Home: 1197 Hillside Ave B29 Schenectady NY 12309 Office: General Electric Research and Devel Ctr Schenectady NY 12309

KAMEN, JOSEPH M., marketing educator, consultant; b. Chgo., Mar. 26, 1929; s. Abraham and Sarah (Baime) K.; m. Beatrice Gershowitz, Dec. 5, 1965; children: Paula, Rebecca, Michael. B.S., Roosevelt U., 1950; M.A., U. Ill.-Urbana, 1951, Ph.D., 1955. Registered psychologist, Ill. Research asst. and assoc. U. Ill.-Urbana, 1951-55; supervisory research assoc. Amoco Oil Co., Chgo., 1963-68; prof. mktg. Ind. U.-Gary, 1968—; self-employed mktg. cons., Flossmoor, Ill., 1968—. Contbr. articles to profl. jours. Served with AUS, 1955-57. Fellow Am. Psychol. Assn. (pres. consumer psychology 1971-72, mem. exec. com. 1971-80, mem. council reps. 1968-69); mem. Am. Mktg. Assn. Jewish. Subspecialties: Sensory processes; Consumer behavior. Current work: Psychobiological aspects of esthetics; impact of nonverbal stimuli on behavior and attitudes; physical and chemical correlated of monotony and satiation. Home: 3509 Oak St Flossmoor IL 60422 Office: Ind U Northwest 3400 Broadway Gary IN 46408

KAMEN, PAUL RAPHAEL, dental educator, researcher; b. N.Y.C., Oct. 18, 1945; s. Saul and Helen Claire (Whitehorn) K.; m. Nancy Carol Rifkin, June 28, 1970; children: Ashley, Maxine. B.A., SUNY-Stony Brook, 1968; D.D.S., Columbia U., 1975; Cert. in periodontics and oral medicine, Harvard U., 1979. Resident L.I. Jewish Hosps., New Hyde Park, N.Y., 1975-76; asst. prof. dentistry and periodontology Columbia U., 1979—. Mem. Tissue Culture Assn., N.Y. Acad. Scis., Am. Acad. Periodontology, Sigma Xi, Omicron Kappa Upsilon. Subspecialty: Periodontics. Current work: Cell response to parasites and cell-cell interactions in periodontal disease and chronic inflammation. Office: Columbus University 630 W 168th St New York NY 10032

KAMIN, HENRY, biochemistry educator; b. Warsaw, Poland, Oct. 24, 1920; came to U.S., 1926, naturalized, 1932; s. Benjamin and Paula (Mirkowicz) K.; m. Dorothy Lee Lingle, Oct. 30, 1943. B.S., CCNY, 1940; Ph.D., Duke U., 1948. Prin. scientist VA Hosp., Durham, N.C., 1953-68; instr., assoc. biochemistry Duke U. Med. Center, Durham, 1950-55, asst. prof., 1955-59, assoc. prof., 1959-65, prof., 1965—; cons. in biochemistry, mem. numerous profl. coms. Editor: Flavins and Flavoproteins III, 1971; contbr. articles to profl. jours.; editorial bd.: Jour. Biol. Chemistry, 1974-79, 82—. Served to lt. San. Corps. U.S. Army, 1943-46. USPHS postdoctoral fellow, 1948-50; NIH grantee, 1955—; NSF grantee, 1966-69, 80—. Mem. Am. Soc. Biol. Chemists, Am. Inst. Nutrition, AAAS, Am. Chem. Soc. Jewish. Clubs: Duke U. Golf, Killarney Golf and Fishing. Subspecialties: Biochemistry (biology); Membrane biology. Current work: Flavoproteins, hemoproteins, mechanism of oxidative enzymes, enzymes of nitrogen and sulfur assimilation, biological hydroxylation. Home: 2417 Perkins Rd Durham NC 27706 Office: Dept Biochemistr Duke U Med Center Durham NC 27710

KAMINSKAS, EDVARDAS, physician, educator; b. Kaunas, Lithuania, Sept. 12, 1935; m. Yuriko Otsuka, July 1, 1965. A.B., Seton Hall U., 1955; M.D., Yale U., 1959. Intern Michael Reese Hosp., Chgo., 1959-60, resident in medicine and hematology, 1960-61, Northwestern U., 1961-63; fellow in molecular biology M.I.T, Cambridge, 1965-68; asst. physician Beth Israel Hosp., Boston, 1968-74; asst. prof. medicine Harvard U. Med. Sch., Boston, 1968-74, assoc. prof. medicine, 1981—; head hematology and oncology Mt. Sinai Med. Center, Milw., 1974-81; prof. medicine U. Wis. Med. Sch., Milw., 1974-81; physician-in-chief Hebrew Rehab. Center for Aged, Boston, 1981—. Contbr. sci. articles to profl. publs. Served as capt. USAF, 1963-65. Am. Cancer Soc. Research fellow, 1965-68; Med. Found. fellow, 1968-70; Am. Cancer Soc faculty awardee, 1970-73. Mem. AAAS, Am. Soc. Hematology, Am. Assn. Cancer Research, Am. Soc. Biol. Chemists, Am. Fedn. Clin. Research, Geriatrics Soc. Am. Roman Catholic. Subspecialties: Internal medicine; Cancer research (medicine). Current work: Medical administration; also geriatric medicine and hematology-oncology; biochemical research in cancer and aging. Office: 1200 Centre St Boston MA 02131

KAMIYAMA, MIKIO, immunochemist; b. Kyoto, Japan, Mar. 25, 1936; came to U.S., 1967; s. Seiryo and Tome (Watanabe) K.; m. Minako Toyoguchi, Sept. 30, 1971; children: Eugene, Kay, June. B.S., Kyoto Prefectural U., 1962; D.M.Sc., Ph.D., U. Tokyo, 1967. Postdoctoral fellow Princeton U., 1967-68; research assoc. SUNY-Buffalo, 1969-70; sr. researcher Institut de Puthologie Moléculaire, U. Paris, 1971-72, 73-74; vis. lectr. U. Marburg., W. Ger., 1972-73; assoc.

staff St. Luke's-Roosevelt Hosp. Center, N.Y.C., 1974—. Contbr. articles to profl. jours. Mem. Am. Assn. Immunologists, Am. Fedn. for Clin. Research (sr.), N.Y. Acad. Sci., Harvey Soc., Japanese Biochem. Soc. Subspecialties: Cancer research (medicine); Immunology (medicine). Current work: Immunological and biochemical characterization of tumor-associated antigens, cell membrance markers and receptors. Home: 560 Riverside Dr New York NY 10027 Office: St Luke's-Roosevelt Hosp Center/Columbia U 421 W 113th St New York NY 10025

KAMMERER, RICHARD CRAIG, pharmacology educator, researcher, consultant; b. Richmond, Va., Aug. 23, 1943; so Richard Harrison and Doreen (Kehoe) K.; m. Helen Louise Johnson, Sept. 23, 1967; children: Michael Robert, Julia Erin. B.A., Drew U., 1966; Ph.D., UCLA, 1972, postdoctoral fellow in biol. chemistry and psychiatry, 1974-75, postdoctoral fellow in pharmacology, 1976-79; postdoctoral fellow in radiology, U. Calif.-San Diego, 1975-76. Acting asst. prof. UCLA Med. Sch., 1979-80, asst. prof., 1980—, asso. dir. 1984 Olympics CLin. Pharmacology Lab., 1982—; cons. VA Hosps.; Research, Los Angeles, 1980—; invited speaker joint French-U.S. seminar, France, 1982. Contbr. chpt. to books, articles to profl. publs. Grantee Jonsson Comprehensive Cancer Ctr., UCLA, 1982. Mem. Am. Chem. Soc., Am. Soc. Pharmacology and Exptl. Therapeutics. Roman Catholic. Subspecialties: Pharmacology; Toxicology (medicine). Current work: Drug metabolism; metabolic activation; carcinogenicity; mechanism of action of drugs of abuse. Home: PO Box 7110 Van Nuys CA 91409 Office: UCLA Pharmacology Dept Sch Medicine Los Angeles CA 90024

KAMMLER, DAVID WILLIAM, mathematics educator, consultant; b. Belleville, Ill., Oct. 29, 1940; s. Rugen W.J. and Edith B. (Montgomery) K.; m. Ruth Ann Kuhnert, Aug. 28, 1965; children: Timothy, Daniel. B.A. in Chemistry, So. Ill. U.-Carbondale, 1962; M.S. in Physics, So. Meth. U., 1969; Ph.D. in Math, U. Mich., 1971. Mem. tech. staff Tex. Instruments Inc., Dallas, 1962-68; NSF fellow U. Mich., Ann Arbor, 1968-71; asst. prof. So. Ill. U., Carbondale, 1971-74, assoc. prof., 1974-78, prof. math., 1978—; postdoctoral fellow Rome Air Devel. Ctr., Griffiss AFB, N.Y., 1980-81. Contbr. articles to profl. publs. Mem. bd. edn. Dist. 130, Carbondale, 1980—. Mem. Soc. Indsl. and Applied Math., Am. Math. Soc., Math. Assn. Am., Sigma Xi (pres. 1981-82), Phi Kappa Phi. Subspecialties: Applied mathematics; Numerical analysis. Current work: Numerical analysis, nonlinear approximation theory, transient analysis, approximation with sums of exponentials. Home: Route 3 Carbondale IL 62901 Office: Southern Ill Univ Carbondale IL 62901

KAMP, WILLIAM PAUL, geophysicist; b. St. Paul, Aug. 8, 1949; s. Irving Maywood and Flora E. K.; m. Jan Mary Miesbauer, Mar. 20, 1971; children: Jesse John, Heidi Joy. B.Math., U. Minn.-Mpls., 1971, Ph.D., 1975. Sr. research geophysicist Phillips Petroleum Co., Bartlesville, Okla., 1975—. Elder Presbyn. Ch., Bartlesville, 1978-81. Named Outstanding Young Man of Am. Jr. C. of C., 1981. Mem. Soc. Indsl. and Applied Math. (adv. council Tex.-Okla. sect. 1981—), Soc. Exploration Geophysicists, Am. Math. Assn., Nat. Computer Graphics Soc., Geophys. Soc. Tulsa (treas.). Presbyterian. Subspecialties: Applied mathematics; Geophysics. Current work: Numerical solution of differential equations, image enhancement, data encryption, vector computers. Home: 1125 SE Lariat Dr Bartlesville OK 74003 Office: Phillips Petroleum Co 270 GB Bartlesville OK 74004

KAMPSCHMIDT, RALPH FRED, biomed. scientist, researcher; b. Franklin County, Mo., May 6, 1923; s. Louis and Alma K.; m. Frances Irene Jackson, Oct. 13, 1954; children: Kimberly, Kit, Coby, Kerry. B.S., U. Mo., 1947, M.S., 1948, Ph.D., 1951. Asst. instr. U. Mo., 1948-51; research chemist Research div. Armour & Co., 1951-54; head tumor-host sect. Biomed. div. Samuel Roberts Noble Found., Inc., Ardmore, Okal., 1954—; speaker, lectr. Contbr. articles to profl. jours. Bd. dirs. Carter County chpt. Am. Cancer Soc., 1956-60, pres., 1958, bd. dirs. Okla. div., 1958-62; pres. YMCA, 1974. Served with USAAF, 1942-45. Decorated Air medal with 3 oak leaf clusters, D.F.C. with oak leaf cluster. Mem. AAAS, Am. Assn. Cancer Research, Am. Chem. Soc., Soc. Exptl. Biology and Medicine, Reticuloendothelial Soc., N.Y. Acad. Sci., Okla. Acad. Sci., Sigma Xi, Alpha Zeta, Gamma Alpha. Republican. Methodist. Lodge: Kiwanis. Subspecialties: Biochemistry (medicine); Immunopharmacology. Home: 919 15th St NW Ardmore OK 73401 Office: Route 1 Ardmore OK 73401

KAN, ALLA, research chemist; b. Kiev, Ukraine, USSR, Mar. 15, 1952; came to U.S., 1975, naturalized, 1982; d. Yuli and Natalya (Tamarowa) Glazman; m. Sergei Kan, Dec. 19, 1976. M.S. in Chemistry, Kiev State U., 1974. Research asst. Paper Industry Reseach Inst., Kiev, 1974-75; research asst. Forsyth Dental Ctr., Boston, 1976-77; research chemist COE Labs., Inc., Chgo., 1977-79; research assoc. Albany Internat., Dedham, Mass., 1980—. Mem. Am. Chem. Soc., Internat. Assn. Dental Research. Jewish. Subspecialties: Analytical chemistry; Physical chemistry. Current work: Evaluation and various applications of reverse osmosis membranes; gas separation. Co-inventor treatment for improving reverse osmosis membrane rejections, 1981. Home: 30 High Apt 10A Dedham MA 02026 Office: Albany Internat Route 128 at US 1 Dedham MA 02026

KAN, JOSEPH R., geophysics educator; b. Shanghai, China, Feb. 10, 1938; s. John H. S. and Mary (Chen) K.; m. Rosalind J. Chen, May 18, 1961; children: Christina, Deborah, Steven. B.S., Nat. Cheng-Kung U., Taiwan, 1961; M.S., Wash. State U., 1966; Ph.D., U. Calif.-San Diego, 1969. Postdoctoral fellow radiophysics lab. Dartmouth Coll., Hanover, N.H., 1970-72; asst. prof. Geophys. Inst., U. Alaska, Fairbanks, 1972-76, assoc. prof., 1976-81, prof., 1981—; cons. space physics lab. Aerospace Corp., Los Angeles, 1980-81; vis. assoc. geophysicist UCLA, 1980-81. Editor: Physics of Auroral Arc Formation, 1981. Mem. Am. Geophys. Union, Am. Phys. Soc. Subspecialties: Solar physics; Plasma physics. Current work: Space plasma physics, magnetaspheric physics and auroral physics. Home: 2568 Talkeetna Ave Fairbanks AK 99701 Office: Geophys Inst U Alaska Fairbanks AK 99701

KAN, YUET WAI, biochemistry and biophysics educator; b. Hong Kong, June 11, 1936; m., 1964. M.B., B.S., U. Hong Kong, 1958, D.Sc., 1980; M.S. (hon.), U. Cagliari, Italy, 1981, D.Sc., Chinese U. Hong Kong, 1981. Asst. prof. pediatrics Harvard Med. Sch., 1970-72; chief hematology sect. San Francisco Gen. Hosp., 1972-79; prof. medicine U. Calif.-San Francisco, 1977—, prof. biochemistry and biophysics, 1979—, assoc. prof. medicine, dept. medicine and lab. medicine, 1972-77; Harvey Soc. lectr., 1980-81; investigator Howard Hughes Med. Inst. Lab., 1976—. Recipient Damashek award Am. Soc. Hematology, 1979; Stratton lectr. Internat. Soc. Hematology, 1980. Fellow Royal Soc. London; mem. Am. Soc. Hematology, Am. Fedn. Clin. Research, Am. Soc. Clin. Investigation, Assn. Am. Physicians. Subspecialties: Genetics and genetic engineering (medicine); Hematology. Office: Univ Calif Dept Med HSE 1504 San Francisco CA 94143

KANA, TIMOTHY WILLIAM, consulting scientist, coastal oceanography educator; b. Glen Cove, N.Y., Mar. 28, 1949; s. Milan and Nancy (Austin) K.; m. Patricia May, Dec. 30, 1972 (div. 1976); 1 son, Christopher Townsend; m. Julia Lucas Lumpkin, May 9, 1981. B.A., Johns Hopkins U., 1971; M.S., U. S.C., 1976, Ph.D., 1979. Open water scuba instr. Research assoc. Chesapeake Bay Inst., Balt., 1969-73; research asst. Geology Dept, U. S.C., Columbia, 1974-78; founding ptnr. Research Planning Inst., Inc., Columbia, 1977—, dir. coastal dynamics div., 1979—; adj. prof. U. S.C., Columbia, 1981—; scuba instr., Fla., S.C., 1972—. Co-editor: Classic Depositional Environments, 1976; designer oceanographic sampler, 1976; contbr. writings in field to profl. publs. Served with USNR, 1966-72. Research scholar Amoco Oil Co., 1977; research grantee Army Research Office, U.S. Army, C.E., NOAA, USCG, UN. Mem. ASCE (affiliate), Am. Shore and Beach Preservation Soc., Profl. Assn. (Scuba) Diving Instrs., Coastal Soc., Sigma Xi. Subspecialties: Oceanography. Current work: Innovative soft engineering solutions to beach erosion, shore protection planning and design, surf zone dynamics and sedimentation, prediction of shoreline change. Home: 1386 Kathwood Dr Columbia SC 29206 Office: Research Planning Inst Inc 925 Gervais St Columbia SC 29201

KANAL, LAVEEN NANIK, computer science educator; b. Bombay, India, Sept. 29, 1931; s. Nanik H. and Ganga (Gandhi) K.; m. Agnes Raclare Cordis, Aug. 6, 1960; children: Shobhana, Jayanti, Gyan. B.S.E.E., U. Wash., 1951, M.S.E.E., 1953; Ph.D., U. Pa., 1960. Registered profl. engr., Ont. Devel. engr. Can. Gen. Electric, 1953-55; research engr., instr. Moore Sch. Elec. Engring., U. Pa., 1955-60; mgr. machine intelligence lab. Gen. Dynamics/Electronics, Rochester, N.Y., 1960-62; mgr. info. sci. dept. Philco-Ford Research Lab., Blue Bell, Pa., 1962-65; mgr. advanced engring. and research communications div. Philco-Ford Corp., Blue Bell, Willow Grove, Pa., 1965-69; mng. dir., treas. L.N.K. Corp., Silver Spring, Md., 1969—; adj. prof. Wharton Grad. Sch. Bus., U. Pa., 1963-70, vis. prof., 1970-74; adj. prof. elec. engring Lehigh U., 1965-70; prof. computer sci. U. Md., College Park, 1970—. Editor: Pattern Recognition, 1968; co-editor: Pattern Recognition in Progress, 1981, Progress in Pattern Recognition, 1982, Handbook of Statistics, Vol. 2: Classification, Pattern Recognition and Reduction of Dimensionality, 1982. Mem. College Park Bd. Trade, 1982—. Fellow IEEE (adminstrv. com., bd. govts. info. theory group), AAAS; mem. Am. Soc. for Photogrammetry, Am. Statis. Assn., Assn. Computing Machinery, Pattern Recognition Soc., Nat. Acad. Sci. Army Robotics and Artificial Intelligence Com., Forum for Interdisciplinary Math. (v.p.), Sigma Xi. Subspecialties: Artificial intelligence; Graphics, image processing, and pattern recognition. Current work: Pioneered in automated pattern recognition systems and image analysis and their applications in automation and remote sensing. Patantee in field. Office: Computer Sci Dept U Md College Park MD 20742 also LNK Corp PO Box 136 College Park MD 320740

KANAMORI, HIROO, educator; b. Tokyo, Japan, Oct. 17, 1936; came to U.S., 1972; s. Tokujiro and Saki (Sakurai) K.; m. Keiko Ihara, Apr. 21, 1964; children—Atsushi, Tadashi. B.S., Tokyo U., 1959, M.S., 1961, Ph.D., 1964. Research fellow Calif. Inst. Tech., Pasadena, 1965-66, prof. geophysics, 1972—; asso. prof. geophysics Tokyo U., 1966 69, prof., 1970-72; vis. asso. prof. Mass. Inst. Tech., 1969-70. Author: (with Hitoshi Takeuchi, Seiya Uyeda) Debate About the Earth, 1967. Fellow Am. Geophys. Union; mem. Sigma Xi. Subspecialties: Geophysics. Current work: Strong-motion seismology. Home: 375 S Bonnie Ave Pasadena CA 91106

KANASEWICH, ERNEST RAYMOND, physics educator; b. Eatonis, Sask., Can., Mar. 4, 1931; m., 1969. B.S., U. Alta., 1952, M.S., 1960; Ph.D. in Geophysics, U. B.C., 1962. Seismologist Tex. Instruments-Geophys. Service, Inc., 1953-58; fellow U. B.C., 1962-63; asst. prof. to assoc. prof. U. Alta., Edmonton, Can., 1963-71, asst. chmn. to acting chmn. dept., 1969-74, prof. physics, 1971—, dir. geophys. obs.; mem. subcom. glaciology NRC, 1963—, subcom. seismology, 1967—; mem. adv. com. exploration tech. No. Alta. Inst. Tech., 1964—; mem. subcom. phys. methods applied to geol. problems Nat. Adv. Com. Research Geol. Sci., 1965—; vis. assoc. prof. dept. earth and planetary sci. Calif. Inst. Tech., 1970-71. Mem. Am. Geophys. Union, Soc. Exploration Geophysics, Seismol. Soc. Am., Can. Assn. Physicists. Subspecialty: Geophysics. Office: Dept Physics Univ Alta Edmonton AB Canada T6G 2E1

KANDEL, ERIC RICHARD, neurobiologist; b. Vienna, Austria, Nov. 7, 1929; came to U.S., 1939, naturalized, 1945; s. Harris Z. and Charlotte (Zimels) K.; m. Denise Bystryn, June 10, 1956; children: Paul Iser, Michelle Deborah. B.A., Harvard U., 1952; M.D., N.Y. U., 1956. Intern Montefiore Hosp., N.Y.C., 1956-57; resident psychiatry Mass. Mental Health Center, 1960-62, 63-64; instr. psychiatry Harvard U. Med. Sch., 1964-65; asso. prof., then prof. physiology and psychiatry N.Y. U. Med. Sch., 1965-74; chief dept. neurobiology and behavior Pub. Health Research Inst. City N.Y., 1969-74; prof. physiology and psychiatry, dir. Ctr. Neurobiology and Behavior Columbia Coll. Physicians and Surgeons, 1974—. Author: Cellular Basis of Behavior, 1976, A Cell Biological Approach to Learning, 1978, Behavioral Biology of Aplysia, 1979, Principles of Neural Science, 1981; also articles. Served to sr. asst. surgeon USPHS, 1957-60. Recipient Career Devel. and Scientist awards NIMH, 1967—, Hofheimer award, 1977, Lucy G. Moses award, 1977, Karl Spencer Lashley award, 1981, Dixon prize in biology and medicine, 1982, N.Y. Acad. Scis. prize in biology and medicine, 1982, Lasker award in basic sci., 1983. Mem. Nat. Acad. Scis., Am. Acad. Arts and Scis., Soc. Neurosci. (pres. 1980-81), Neuroscis. Research Program. Subspecialties: Psychiatry; Neurobiology. Home: 9 Sigma Pl Riverdale NY 10471 Office: Coll Physicians and Surgeons Columbia U New York NY 10032

KANDELMAN, DANIEL, dental educator; b. Marseille, France., Apr. 26, 1946; s. Marcel and Nelly K.; m. Isabelle Zebrowski, June 17, 1972; children: Stanislas, Severine. Docteur en Chirurgie Dentaire, U. Marseille, 1968; D.M.D., U. Geneva, 1973; M.P.H., Harvard, U., 1975. Chief of clinic Faculty of Dentistry, U. Geneva, 1975; asst. prof. dentistry U. Montreal, Que., Can., 1976-80, assoc, prof., 1980—; coordinator dental program Montreal Gen. Hosp., 1979-82. Contbg. author: Workshop on Prevention, 1979. Mem. Internat. Assn. Dental Research, Fedn. Dentaire Internationale, Can. Dental Assn. Subspecialties: Preventive dentistry; Epidemiology. Current work: Preventive dentistry and epidemiology. Office: Faculty of Dentistry U Montreal CP 6209 SUCC A Montreal PQ Canada H3C 3T9

KANE, HARRISON, civil engineering educator; b. Bklyn., Jan. 2, 1925; m., 1952. B.C.E., CCNY, 1947; M.S., Columbia U., 1948; Ph.D. in Soil Mechanics, U. Ill., 1961. Design engr. Parsons, Brinckerhoff Hall & Macdonald, 1948-51, 53-56; planning engr. U.S. Dept. Army, Ger., 1951-53; lectr. structures CCNY, 1956; asst. prof. civil engring. Pa. State U.; 956-61 U. Ill., 1961-64; assoc. prof. U. Iowa, Iowa City, 1964-77, prof. civil and materials engring., 1977—, chmn. dept., 1971—. NSF grantee, 1967-69. Mem. ASCE, Am. Soc. Engring. Edn. Subspecialty: Civil engineering. Office: Dept Civil Engring Univ Iowa Iowa City IA 52242

KANESHIRO, KENNETH YOSHIMITSU, biology researcher; b. Honolulu, Dec. 15, 1943; s. Kenichi and Yoshiko (Takushi) K.; m. Betty Fuyuko Muraoka, July 22, 1967; Kevin T., Jennifer H. B.A., U. Hawaii, 1965, M.S., 1968, Ph.D., 1974. Grad. research asst. U. Hawaii, Honolulu, 1965-68, grad. teaching asst., 1970-71, asst. researcher, 1972—; grad. research asst. U. Tex.-Austin, 1968-69; vis. asst. prof. U. São Paulo, Brazil, 1979; coordinator (Hawaiian Drosophila Project), Honolulu, 1972—. Contbr. articles to profl. jours. Commr. Hawaii Natural Areas Res. System Commn., Honolulu, 1980, 81, 82. NSF grantee U. Hawaii, 1982. Mem. Entomol. Soc. Am., Soc. Am. Naturalists, Soc. Study Evolution, Hawaiian Entomol. Soc., Sigma Xi, Gamma Sigma Delta. Subspecialties: Evolutionary biology; Population biology. Current work: Mate recognition system in relation to factors controlling population dynamics, e.g. behavior and effective reproductive community. Office: U Hawaii Dept Entomology 3050 Maile Way Honolulu HI 96822

KANG, MIN HO, optical scientist, educator; b. Kyungnam Province, Korea, July 20, 1946; s. Ji Jung and Ok Hee (Lee) K.; m. Ae Soon Choi, July 10, 1971; children: Jeannie, Soo Young. B.S.E.E., Seoul Nat. U., 1969; M.S.E.E., U. Mo.-Rolla, 1973; Ph.D., U. Tex.-Austin, 1977. Research asst. U. Tex., 1973-77; mem. tech. staff Bell Telephone Labs., Holdel, N.J., 1977-78; head lab. Korea Electrotech. and Telecommunication Research Inst., Seoul, 1978-82, head opto-electronics research sect., 1982—; lectr. Grad. Sch., Seoul Nat. U., 1979—. Contbr. articles to profl. jours. Recipient Nat. Medal of Honor Govt. of Korea, 1982. Mem. IEEE, Optical Soc. Am., Korean Inst. Electronic Engrs., Korean Inst. Elec. Engrs., Phi Kappa Phi, Eta Kappa Nu. Subspecialties: Fiber optics; Electronics. Current work: High speed optical fiber communication systems; research including optoelectronics, telecommunication systems. Office: KPO Box 125 Gwanghwamoon Seoul Korea Home: 6 Dong-512 Mizu Apt Chongryang Ri Seoul Korea

KANNEL, WILLIAM BERNARD, cardiovascular epidemiologist; b. Bklyn., Dec. 13, 1923; s. Joseph M. and Sarah M. (Golden) K.; m. Rita R. Lefkowitz, May 29, 1943; children: Linda J. Kannel Isaacson, Steven Michael, Patricia M. Kannel Hoffman, Forrest S. M.D., Ga. Med. Coll., 1949; M.P.H., Harvard U., 1959. Intern, resident internal medicine S.I. Pub. Health Hosp., 1949-50, 53-56; asso. dir. Framingham (Mass.) Heart Study, Nat. Heart and Lung Inst., 1950-53, 56-67, dir., 1967-79; cons. Framingham Union Hosp., Cushing Hosp.; asso. medicine Boston U. Med. Sch.; lectr. preventive medicine Harvard U. Med. Sch.; prof. medicine, head sect. epidemiology and preventive medicine Boston U. Med. Center; med. dir. USPHS, 1949—. Contbr. med. jours.; mem. editorial bd.: Am. Heart Jour. Served with AUS, 1943-49. Recipient Gairdner Found. award, 1976; Einthoven award Leiden U., Netherlands, 1974; Francis medal U. Mich. Med. Sch., 1975; Polish Copernicus award, 1977; Dana award, 1972; Soc. Prospective Medicine award, 1979. Fellow Am. Coll. Cardiology, Am. Coll. Epidemiology; mem. Am. Heart Assn. (fellow council epidemiology, former chmn. council epidemiology), Assn. Commd. Officers USPHS, Alpha Omega Alpha. Democrat. Jewish. Subspecialties: Epidemiology; Cardiology. Current work: Cardivascular epidemiology; preventive cardiology. Home: 30 Eliot St South Natick MA 01760 Office: Boston U Med Center 80 E Concord St Boston MA 02118

KANNINEN, MELVIN FRED, research, lab. adminstr.; b. Ely, Minn., Jan. 31, 1935; s. Fred George and Mary Francesca K.; m. Jean Elaine Kutcher, Sept. 7, 1957; children: Elaine Mary, Barbara Joan. B.S., U. Min., 1957, M.S., 1959; Ph.D., Stanford U., 1966. Nuclear engr. Gen. Electric Co., Richland, Wash., 1959-63; research leader Battelle, Columbus Labs., Columbus, Ohio, 1966—. Contbr. numerous articles to tech. jours.; Co-author: Advanced Fracture Mechanics, 1983. Mem. ASME, ASTM, Soc. Exptl. Stress Analysis, Sigma Xi. Subspecialties: Fracture mechanics; Solid mechanics. Current work: Fracture mechanics, structural integrity, residual stress analysis. Home: 1864 Tamarack Ct S Columbus OH 43229 Office: 505 King Ave Columbus OH 43201

KANTOWITZ, BARRY HOWARD, psychology educator; b. N.Y.C., Aug. 25, 1943. Ph.D., U. Wis.-Madison, 1969. Asst. prof. Purdue U., West Lafayette, Ind., 1969-72, assoc. prof., 1972-79, prof. psychol. scis. and indsl. engring., 1979—; sr. lectr. ergonomics U. Trondheim (Norway), Norwegian Inst. Tech., 1976-77. Author: Experimental Psychology, 1978, Methods in Experimental Psychology, 1981, Human Factors, 1983; editor: Human Information Procedure, 1974. NIMH research grantee, 1972-78; NASA research grantee, 1983—. Fellow Am. Psychol. Assn. Subspecialties: Cognition; Human factors engineering. Current work: Human attention and information processing and implications of theoretical models for design of human-machine system interfaces. Office: Purdue U Dept Psychol Scis West Lafayette IN 47907

KANTROWITZ, ADRIAN, surgeon, educator; b. N.Y.C., Oct. 4, 1918; s. Bernard Abraham and Rose (Esserman) K.; m. Jean Rosensaft, Nov. 25, 1948; children—Niki, Lisa, Allen. A.B., N.Y.U., 1940; M.D., L.I. Coll. Medicine, 1943; postgrad. physiology, Western Res. U., 1950. Diplomate: Am. Bd. Surgery, Am. Bd. Thoracic Surgery. Gen. rotating intern Jewish Hosp. Bklyn., 1944; asst. resident, then resident surgery Mt. Sinai Hosp., N.Y.C., 1947; asst. resident Montefiore Hosp., N.Y.C., 1948, asst. resident pathology, 1949, fellow cardiovascular research group, 1949, chief resident surgery, 1950, adj. surg. service, 1951-55; USPHS fellow cardiovascular research, dept. physiology Western Res. U., 1951-52, teaching fellow physiology, 1951-52; instr. surgery N.Y. Med. Coll., 1952-55; cons. surgeon Good Samaritan Hosp., Suffern, N.Y., 1954-55; asst. prof. surgery State U. N.Y. Coll. Medicine, 1955-56, asso. prof. surgery, 1957-64, prof., 1964-70; dir. cardiovascular surgery Maimonides Med. Center, Bklyn., 1955-64, dir. surgery, 1964-70; chmn. dept. surgery Sinai Hosp. Detroit, 1970—; prof. surgery Wayne State U. Sch. Medicine, 1970—. Contbr. articles profl. jours. Served from 1st lt. to capt., M.C. AUS, 1944-46. Recipient H.L. Moses prize to Montefiore Alumnus for outstanding research accomplishment, 1949; 1st prize sci. exhibit Conv. N.Y. State Med. Soc., 1952; Gold Plate award Am. Acad. Achievement, 1966; Max Berg award for outstanding achievement in prolonging human life, 1966; Theodore and Susan B. Cummings humanitarian award Am. Coll. Cardiology, 1967. Fellow N.Y. Acad. Sci., A.C.S.; mem. Internat. Soc. Angiology, Am. Soc. Artificial Internal Organs (pres. 1968-69), N.Y. County Med. Soc., Harvey Soc., N.Y. Soc. Thoracic Surgery, N.Y. Soc. Cardiovascular Surgery, Am. Heart Assn., Am. Physiol. Soc., Am. Coll. Cardiology, Am. Coll. Chest Physicians, Bklyn. Thoracic Surgery Soc. (pres. 1967-68), Pan Am. Med. Assn., Soaring Soc. Am., Am. Ski Assn. Subspecialties: Cardiac surgery; Artificial organs. Current work: Research in area of a practical partial artificial heart. Pub. pioneer motion pictures taken inside living heart, 1950; contbr. to devel. pump- oxygenators for human heart surgey; pioneer devel. mech., artificial hearts; performed 1st permanent partial mech. heart surgery in humans, 1966; 1st use phase-shift intra-aortic balloon pump in patient in cardiogenic shock; 1st human heart transplant in U.S., Dec. 1967. Home: 70 Gallogly Rd Pontiac MI 48053 Office: 6767 W Outer Dr Detroit MI 48253

KANTROWITZ, ARTHUR, physicist; b. N.Y.C., Oct. 20, 1913; s. Bernard A. and Rose (Esserman) K.; m. Rosalind Joseph, Sept. 12, 1943 (div.); children: Barbara, Lore, Andrea; m. Lee Stuart, Dec. 25, 1980. B.S., Columbia U., 1934, M.A., 1936, Ph.D., 1947; Dr.Engring. (hon.), Mont. Coll. Mineral Sci. and Tech., 1975, D.Sc., N.J. Inst

Tech., 1981. Physicist NACA, 1935-46; prof. aero. engring. and engring. physics Cornell U., 1946-56; dir. Avco-Everett Research Lab., Everett, Mass., 1955-72, chmn., chief exec. officer, 1972-78; sr. v.p., dir. Avco Corp., 1956-79; prof. Thayer Sch. Engring., Dartmouth Coll., 1979—; vis. lectr. Harvard, 1952; Fulbright and Guggenheim fellow Cambridge and Manchester univs., 1954; fellow Sch. Advanced Study, Mass. Inst. Tech., 1957, vis. inst. prof., 1957—; Messenger lectr. Cornell U., 1978; hon. prof. Huazhong Inst. Tech., Wuhan, China, 1980; mem. Presdl. Adv. Group on Anticipated Advances in Sci. and Tech. (head task force on sci. ct.), 1975-76; mem. tech. adv. bd. U.S. Dept. Commerce, 1974-77; mem. adv. panel NOVA, Sta. WGBH-TV; mem. bd. overseers Center for Naval Analyses; adv. council Israel-U.S. Binational Indsl. Research and Devel. Found.; Hon. trustee, past mem. mech. engring. adv. com. U. Rochester; mem. adv. council dept. aero. and mech. scis. Princeton U., 1959-77; mem. engring. adv. council Stanford U., 1966-82; mem. adv. bd. engring. Rensselaer Poly. Inst., 1982—. Contbr. articles to profl. jours. Recipient Kayan medal Columbia U., 1973, Faraday Meml. medal, 1983, Theodore Roosevelt medal of honor for Distinguished Service in Sci. Fellow Am. Acad. Arts and Scis., Am. Phys. Soc., AAAS, AIAA, Am. Astronautical Soc.; mem. Internat. Acad. Astronautics, Nat. Acad. Scis., Nat. Acad. Engring., Am. Inst. Physics, Sigma Xi. Subspecialties: Fluid mechanics; High energy laser. Current work: Interaction of science and technolgy with society. Home: 24 Pinewood Village West Lebanon NH 03784

KANUNGO, RABINDRA NATH, management educator; b. Cuttack, Orissa, India, July 11, 1935; emigrated to U.S., 1965; s. Kshirode C. and Mahindra (Mohanty) K.; m. Minati Das, Nov. 26, 1957; children: Siddhartha, Nachiketa. B.A. with honors, Utkal U., 1953; M.A., Patna U., 1955; Ph.D., McGill U., 1962. Lectr. Ravenshaw Coll., Orissa, India, 1955-60; asst. prof. Indian Inst. Mgmt., Calcutta, 1962-63, Indian Inst.Tech., Bombay, 1964-65; assoc. prof. Dalhousie U., N.S., 1965-68; prof. mgmt. McGill U., Montreal, 1969—. Co-author: (with Dutta) Memory and Affect, 1975; author: Biculturalism and Management, 1980, Work Alienation, 1982; guest editor: spl. issue Internat. Rev. of Applied Psychology, vol. 30, no. 1, 1981. Recipient sr. faculty award Seagram Found., 1971; best paper award Canadian Assn. Adminstrv. Sci., 1981; recipient various grants. Fellow Canadian Psychol. Assn.; mem. Am. Psychol. Assn., Internat. Assn. Applied Psychology, India-Can. Assn. (pres. 1979-80), Nat. Assn. Canadians of Origins in India (v.p. 1979-80). Subspecialties: Social psychology; Organizational psychology. Current work: Cross cultural research on work motivation and work alienation. Home: 11 Greenfield Rd Montreal PQ Canada H9G 2L3 Office: Faculty of Mgmt McGill U 1001 Sherbrooke St W Montreal PQ Canada H3A 1G5

KANWAL, RAM PRAKASH, mathematics educator; b. Jhang, Punjab, India, July 4, 1924; came to U.S., 1954; s. Ishar Dass and Bhagwan Bai (Nandi) Chawla K.; m. Vimla Kuman, June 17, 1954; children: Neeru Kiran, Neeraj Kumar. B.A., Punjab U., 1945, M.A., 1948; Ph.D., Ind. U., 1957. Asst. prof. U. Wis.-Madison, 1957-59; assoc. prof. math. Pa. State U., 1959-62, prof., 1962—. Author: Linear Integral Equations, 1971, Generalized Functions, 1983; Contbr. numerous articles to profl. jours. Mem. Allahabad Math. Soc., Soc. Indsl. and Applied Math. Subspecialties: Applied mathematics; Integral and differential equations. Current work: Generalized functions, boundary value problems, integral equations, wave propagation, and scattering problems. Home: 1253 Smithfield St State College PA 16801 Office: Pa State U McAllister Hall University Park PA 16802

KAO, CHARLES KUEN, elec. engr.; b. Shanghai, China, Nov. 4, 1933; s. Chun-Hsien and Tsing-Fong K.; m. May Wan Wong, Sept. 19, 1959; children—Simon M. T., Amanda M.C. B.Sc. in Elec. Engring, U. London, 1957, Ph.D., 1965. Devel. engr. Standard Telephones & Cables, Ltd., London, 1957-60; prin. research engr. Standard Telecommunications Lab., Ltd., Harlow, Eng., 1960-70; prof. electronics, chmn. dept. Chinese U. Hong Kong, 1970-74; chief scientist ITT Electro Optical Products div. ITT, Roanoke, Va., 1974-81, v.p., dir. engring., 1981—. Contbr. articles to profl. jours. Recipient Morey award Am. Ceramic Soc., 1976; Stewart Ballantine medal Franklin Inst., 1977; Rank prize Rank Trust Funds, 1978; Morris Liebmann Meml. award IEEE, 1978; L.M. Ericsson Internat. prize, 1979; Gold medal Armed Forces Communications and Electronics Assn., 1980. Fellow IEEE, Inst. Elec. Engrs. (U.K.); mem. Optical Soc. Am. Subspecialty: Fiber optics. Patentee in field 24019

KAO, FA-TEN, geneticist; b. Hankow, China, Apr. 20, 1934; s. Ling-mai and Hang-seng (Feng) K.; m. Betty Chia-mai Tang, Dec. 17, 1960; 1 son, Alan S. B.S., Nat. Taiwan U., 1955; Ph.D., U. Minn., 1964. Instr., Nat. Cancer Inst. fellow dept. biophysics U. Colo. Med. Ctr., Denver, 1965-67, asst. prof., 1967-70, assoc. prof., 1970-81, prof., 1981—; sr. fellow Eleanor Roosevelt Inst. Cancer Research, Denver, 1965—; vis. scientist Sir Wiliam Dunn Sch. Pathology, U. Oxford, Eng., 1973-74. NIH fellow, 1965-67; Eleanor Roosevelt Internat. Cancer fellow/Internat. Union Against Cancer, 1973-74. Mem. Genetics Soc. Am., Am. Soc. Cell Biology, Am. Soc. Human Genetics, Am. Assn. Cancer Research, Tissue Culture Assn., AAAS, Sigma Xi. Subspecialties: Genome organization; Genetics and genetic engineering (biology). Current work: Researcher genetic studies of somatic mammalian cells, somatic cell and molecular genetic analysis of the human genome, mapping of human genes in the human genome, recombinant DNA and genetic engineering studies in mammalian cells. Home: 305 Leyden St Denver CO 80220 Office: Eleanor Roosevelt Inst for Cancer Research B 129 4200 E 9th Ave B129 Denver CO 80262

KAO, RACE L., medical educator; b. Chunking, Sechun, China, Dec. 1, 1943; came to U.S., 1967; s. Yu-Ho and Tsing (Tsou) K.; m. Lidia Wei Liu, Aug. 18, 1969; children: Elizabeth C., Grace W. B.S., Nat. Taiwan U., 1965; M.S., U. Ill., 1971, Ph.D., 1972. Research assoc. dept. animal sci. U. Ill., Urbana, 1972; research assoc. dept. physiology Pa. State U.-Hershey, 1972-75, asst. prof., 1976-77; dir. cardiothoracic research U. Tex. Med. Branch-Galveston, 1977-82, asst. prof. surgery, physiology, 1977-82; asst. prof. dept. surgery Washington U., St. Louis, 1982—. Author: Cardiac Adaptation, 1977. Pres. U. Tex. Chinese Assn., 1981. Served to lt. U.S. Army, 1965-66. NIH grantee, 1982—; Washington U. grantee, 1982—; Am. Heart Assn. grantee, 1983—. Mem. Am. Physiol. Soc., Internat. Soc. for Heart Research, Nat. Soc. Med. Research, Nutrition Today Soc. Subspecialties: Cardiac surgery; Physiology (medicine). Current work: Myocardial metabolism under ischemic, anoxic and hypertrophying conditions; protection of myocardium during cardiopulmonary by-pass; regulation of isolated cardiomyocyte metabolism. Office: Dept Surgery Washington U 4960 Audubon St Saint Louis MO 63110 Home: 812 Haverton Dr Saint Louis MO 63141

KAO, WINSTON WHEI-YANG, biochemist, educator; b. Tainan, Taiwan, Mar. 3, 1941; came to U.S., 1970; s. Bin-Wu and Pao-Tsu (Sun) K.; m. Candace Whei-Cheng, May 10, 1970; 2 sons, Edward Chung-Peng, Charles Chung-L. B.S., Nat. Taiwan U., 1966, M.S., 1970; Ph.D., U. Pa., 1974. Teaching asst. Nat. Taiwan U., 1967-68; research and teaching specialist III Rutgers U. Med. Sch., Piscataway, N.J., 1974-76; research asst. prof. U. Pitts., 1976-78, asst. prof., 1978-82; assoc. prof., dir. ophthal. research U. Cin., 1982—. Served to 2nd lt. AUS, 1966-67. Mem. Assn. Research in Vision and Ophthalmology, Am. Soc. Biol. Chemists, Chinese Soc. Biochemists. Subspecialties: Biochemistry (medicine); Cell biology (medicine). Current work: regulation of collagen gene expression in ocular tissues. Home: 10343 Peachtree Ln Cincinnati OH 45242 Office: Dept Ophthalmolog U Cincinnati Coll Medicine Cincinnati OH 45267

KAPER, HANS GERARD, mathematician, researcher; b. Alkmaar, Netherlands, June 10, 1936; came to U.S., 1969; s. Gerrit and Trijntje (Reine) K.; m. Hillegonde J. Van Biesen, July 17, 1962; children: Tasso J., Bertrand P. Ph.D., Rijksuniversiteit, Groningen, Netherlands, 1965. Sr. mathematician Argonne (Ill.) Nat. Lab., 1969—. Author: Mathematical Theory of Transport Processes in Gases, 1972, Spectral Methods in Linear Transport Theory, 1982. Pres. Downers Grove Concert Assn., 1978-82. Mem. Am. Math. Soc., Soc. Indsl. and Applied Math., Wiskundig Genootschap. Subspecialty: Applied mathematics. Current work: Applied analysis. Home: 731 59th St Downers Grove IL 60516 Office: Argonne Nat Lab Math/Comp Sci Di 9700 S Cass Ave Argonne IL 60439

KAPLAN, ALAN MARC, immunologist, educator; b. Bklyn., Dec. 10, 1940; s. Albert J. and Esther (Warshaw) K.; m. Eva Ruzickova, Mar. 18, 1972; 1 dau., Ali Michelle. B.S., Tufts U., 1963; Ph.D., Purdue U., 1969. Postdoctoral fellow U. Toronto Sch. Medicine, 1969-72; asst. prof. depts. microbiology and surgery Med. Coll. Va., Richmond, 1972-75, assoc. prof., 1975-79, prof., 1979-82; assoc. dir. research Med. Coll. Va./Va. Commonwealth U. Cancer Center, Richmond, 1980-82, dep. chmn. dept. microbiology, 1981-82; prof., chmn. dept. med. microbiology and immunology U. Ky. Sch. Medicine, Lexington, 1982—; mem. exptl. immunology study sect. NIH, 1978-82. Named Outstanding Grad. Faculty Mem. Med. Coll. Va., 1978; Med. Research Council Can. postdoctoral fellow, 1969-72. Mem. Am. Assn. Immunologists, Am. Soc. Microbiology, Reticuloendothelial Soc. Explt. Biology and Medicine, Am. Assn. Cancer Research, N.Y. Acad. Scis. Democrat. Jewish. Subspecialties: Immunobiology and immunology; Immunogenetics. Current work: Cellular immunology, tumor immunology, macrophage immunology, hybridoma technology. Home: 3434 Brandon Dr Lexington KY 40502 Office: U Ky Med Center 800 Rose St MS 409 Lexington KY 40356-0084

KAPLAN, BARRY BERNARD, cell and molecular biologist, psychiatry and anatomy educator; b. Bronx, N.Y., Sept. 28, 1946; s. Paul A. and Ralphian (Kantor) K.; m. Annie G. Gioio, Aug. 30, 1970. A.B., Hofstra U., 1968, M.S., 1969; Ph.D., Cornell U., 1974. Research assoc. Gerontology Inst., U. So. Calif., Los Angeles, 1974-76; instr. dept. cell biology and anatomy Cornell U. Med. Coll., N.Y.C., 1976-77, asst. prof., 1977-82, assoc. prof., 1983-84; assoc. prof. depts. psychiatry and anatomy U. Pitts. Sch. Medicine, 1984—. Contbr. articles to profl. publs. Served to capt. USAR, 1968-76. Mem. Am. Soc. Neurosci., Internat. Soc. Neurochemistry, Am. Soc. Neurochemistry, Am. Soc. Cell Biology. Subspecialties: Molecular biology; Neurobiology. Current work: Control of gene expression in brain and cells of neuroectodermal origin. Office: Western Psychiatric Inst and Clinic U Pittsburgh Pitsburgh PA 15213

KAPLAN, EUGENE HERBERT, biology educator; b. N.Y.C., June 26, 1932; s. Jacob and Lea (Gerstler) K.; m. Breena Lubow, Aug. 25, 1957; children: Julie, Susan. B.S., Bklyn. Coll., 1954; M.A., Hofstra U., 1957; Ph.D., NYU, 1963. Tchr. biology Sewanhaka/Floral Park (N.Y.) High Sch., 1957-58; instr. to full prof. biology Hofstra U., Hempstead, N.Y., 1958—, dir. Marine Lab., St. Ann's Bay, Jamaica, W.I., 1982—; expert in sci. teaching UNESCO, Tel Aviv, 1968-71. Author: Problem Solving in Biology, 1968, 3rd edit. 1983, Experiences in Life Science, 1969, 2nd edit. 1976, Peterson Field Guide to Coral; Reefs of Florida and the Caribbean, 1982. Subspecialties: Ecology; Taxonomy. Current work: Coral reef ecology; Macrobrachium culture; taxonomy of zooxanthellae. Home: 148 W Waterview St Northport NY 11768 Office: Dept Biology Hofstra U 1000 Fulton Ave Hempstead NY 11550

KAPLAN, HARRIET GEVIRTZ, research scientist; b. N.Y.C., Aug. 7, 1927; d. Louis and Helen (Fischer) Gevirtz; m.; children: Linda, Lucie. B.A., Hunter Coll., 1948; M.S., N.Y.U., 1966, Ph.D., 1968. Research scientist N.Y. Inst. Basic Research, S.I., 1967—; vis. asst. prof. SUNY-Stony Brook, 1972-73. Mem. AAAS, Internat. Soc. Developmental Psychobiology, Soc. Behavorial Teratology. Subspecialties: Epilepsy; Behavioral psychology. Current work: Epilepsy, development of behavior. Office: 1050 Forest Hill Rd Staten Island NY 10314

KAPLAN, HENRY SEYMOUR, educator, radiologist; b. Chgo., Apr. 24, 1918; s. Nathan M. and Sarah (Brilliant) K.; m. Leah Hope Lebeson, June 21, 1942; children—Ann Sharon, Paul Allen. B.S., U. Chgo., 1938, Sc.D. (hon.), 1969; M.D., Rush Med. Coll., 1940; M.S. in Radiology, U. Minn., 1944; Sc.D. (hon.), Hahnemann Med. Coll., 1973. Intern Michael Reese Hosp., Chgo., 1940-41, resident radiation therapy and tumor clinic, 1941-42; tng. fellow Nat. Cancer Inst., 1943-44; instr. radiology Yale Med. Sch., 1944-45, asst. prof., 1945-47; radiologist Nat. Cancer Inst., 1947-48; prof. radiology, chmn. dept. Sch. Medicine, Stanford U., 1948-72, dir. biophysics lab., 1957-64, Maureen (Lyles D'Ambrogio prof. oncology and dir. cancer biol. research lab.), 1972—; mem. sci. adv. com. St. Jude's Children's Hosp., Memphis, 1970-74; chmn. sci. adv. com. Sharett Inst. Oncology, Hadassah Med. Center-Hebrew U., Jerusalem, 1977—; mem. com. radiology NRC, 1950-56; gastrointestinal cancer com. Nat. Cancer Inst.; mem. panel path. effects radiation Nat. Acad. Sci.-NRC; adv. com. biology Oak Ridge Nat. Lab., 1969-75; subcom. radiation carcinogenesis (chmn. 1957-58), commn. on research Internat. Union Against Cancer; nat. adv. cancer council Nat. Cancer Inst., USPHS, 1959-63; advisor Cancer Council Calif., 1959-62, bd. sci. advisors div. cancer treatment, 1975-79; nat. panel cons. on cancer U.S. Senate, 1970-71; mem. council analysis and projection Am. Cancer Soc., 1972-76. Author: (with S.J. Robinson) Congenital Heart Disease: An Illustrated Diagnostic Approach, 1954, 2d edit, (with H. L. Abrams and S. J. Robinson) Congenital Heart Disease: An Illustrated Diagnostic Approach, 1965, Angiocardiographic Interpretation in Congenital Heart Diseases, (with H.L. Abrams), 1955, Hodgkin's Disease, 1972, 2d edit., 1980; editor: (with P.J. Tsuchitani) Cancer in China, 1978, (with R. Levy) Malignant Lymphomas, 1978, (with S.A. Rosenberg) Advances in Malignant Lymphomas, 1981; editorial adv. bd.: Cancer, Proc. Nat. Acad. Sci, 1973-79, Current Topics Radiation Research Quar, 1971-77; editorial com.: Ann. Rev. Nuclear Sci, 1966-70; adv. editor: Internat. Jour. Radiation Oncology, Biology, Physics. Bd. govs. Weizmann Inst. Sci., Rehovoth, Israel, 1974—, Ben Gurion U., Beersheba, Israel, 1974—. Decorated Légion d'Honneur, France; Order of Merit, Italy; Shahbanou award, Iran; recipient Lila Motley Cancer Found. award; Atoms for Peace award, 1969; Modern Medicine award for distinguished achievement, 1968; Lucy W. James award James Ewing Soc., 1971; R.R. de Villiers award Leukemia Soc., 1971; Commonwealth Fund fellow, vis. scientist NIH, 1954-55; David A. Karnofsky Meml. award Am. Soc. Clin. Oncology, 1971; Nat. award Am. Cancer Soc., 1972; Laureat, Prix Griffuel for cancer research, France, 1975; G.H.A. Clowes Meml. award Am. Assn. Cancer Research, 1976; Erskine lectr. Radiol. Soc. N.Am., 1976; gold medal Am. Soc. Therapeutic Radiologists, 1977; medal of Honor Danish Cancer Soc., 1978; Ungerman-Lubin award Taubman Found., 1978; Lila Gruber Meml. award Am. Acad. Dermatology, 1978; Kettering prize Gen. Motors Cancer Research Found., 1979; Prentis award Mich. Cancer Found., 1980; Walker prize Royal Coll. Surgeons, 1981. Fellow Am. Coll. Radiology (chmn. commn. on cancer, bd. chancellors 1970-73, Gold medal 1981), Royal Coll. Radiologists (hon.); mem. Am. Soc. Exptl. Pathology, Assn. Univ. Radiologists (pres. 1954-55, Gold medal 1979), Radiol. Soc. N.Am., AAAS, Fedn. Am. Scientists, Soc. Exptl. Biology and Medicine, Radiation Research Soc. (pres. 1956-57), Western Soc. Clin. Research, Am. Assn. Cancer Research (dir. 1954-56, 64-67, pres. 1966-67), Am. Soc. Therapeutic Radiologists (pres. 1966-67, chmn. bd. dirs. 1967-68), Internat. Club Therapeutic Radiologists, Am. Soc. Biol. Chemists, Am. Acad. Arts and Sci., Harvey Soc. N.Y., Internat. Assn. Radiation Research (pres. 1974-79), Am. Roentgen Ray Soc., Am. Soc. Clin. Oncology, Nat. Acad. Scis., Inst. Medicine, Acad. Medicine Brazil (hon.), Swedish Soc. Med. Radiology (hon.), Acad. des Sciences, Institut de France (fgn. assoc.). Subspecialty: Radiology. Home: 631 Cabrillo Ave Stanford CA 94305

KAPLAN, IRVING EUGENE, consulting human ecologist, theoretical cosmologist; b. Phila., May 1, 1926; s. Abraham and Bertha (Posner) K.; m. Harriet Bromberg, Sept. 29, 1951; children: Addie Eve, Meryl Denise. B.S., L.I.U., 1950; M.A., New Sch. Social Research, 1953; Ph.D., U.S. Internat. U., 1971. Lic. psychologist, Calif. Depth interviewer Dr. Ernst Dichter, N.Y.C., 1950-52; employment interviewer N.Y. State Employment Service, N.Y.C., 1952-56; researcher, program dir. U.S. Naval Personnel Research Activity, San Diego, 1957-66; psychologist, human ecologist, cons., San Diego, 1966-80, human ecologist, theoretical cosmologist, writer, cons., 1980—; cons. Ctr. for Study of Democratic Instns., Santa Barbara, Calif., 1964-78; UN Law of the Sea Programme, N.Y.C., 1970-77, RIO Found., Rotterdam, Netherlands, 1980-82. Contbr. articles to profl. jours. Served with USNR, 1944-46. Mem. Am. Psychol. Assn., Am. Geophys. Union, AAAS, Internat. Ocean Inst. Subspecialties: Space chemistry; Cosmology. Current work: Climatology and its impact on mankind; concepts and designs for coping with climatic change; a phenomenological cosmology and physics; an ether theory compatible with modern physics. Office: Internat Inst for Integrative Tech 3121 Beech St San Diego CA 92102

KAPLAN, JOEL HOWARD, indsl. scientist; b. N.Y.C., Apr. 6, 1941; s. Sidney and Mary (Cohen) K.; m. Frances Bromberg, Dec. 29, 1963 (div.); children: Karen, Deborah, Jeffrey. B.S., CCNY, 1962; Ph.D., Johns Hopkins U., 1967. Postdoctoral fellow McArdle Lab. for Cancer Research, 1967-69; staff scientist Gen. Electric Corp. Research and Devel., Schenectady, 1969—. Contbr. articles to profl. jours. Nat. Cancer Inst. fellow, 1967-69. Mem. Am. Assn. Immunologists, AAAS, Sigma Xi. Subspecialties: Bioinstrumentation; Immunology (medicine). Current work: Cell electrophoresis, cell-mediated immunity, membrane biology, tumor immunology, cancer immunodiagnosis. Office: Gen Electric Co Bldg K 1 Schenectady NY 12301

KAPLAN, KENNETH CHARLES, physician; b. N.Y.C., Apr. 8, 1937; s. Bernard I. and Helen R. (Gardner) K.; m. Rebecca C. Hall, Oct. 20, 1979; children: Jessica, Jeremy, Miriam, Jonah E. A.B., Dartmouth Coll., 1958; M.D., NYU, 1962. Diplomate: Am. Bd. Internal Medicine, Am. Bd. Cardiovascular Diseases. Intern, resident Bellevue Hosp., N.Y.C.; resident Univ. Hosp., N.Y.C.; sr. attending physician Phelps Meml. Hosp., North Tarrytown, N.Y., 1969—; assoc. dir. electrocardiology, 1974—, dir. coronary care unit, 1974—; cons. in cardiology Peakskill Community Hosp., 1970—, Ossining Correctional Facility, 1969—. Bd. dirs., pres. area 9 Profl. Standards Rev. Orgn., Westchester and Putnam Counties, 1976—, sec. exec. com., Purchase, N.Y., 1980—. Served to lt. comdr. USPHS, 1963-68. Fellow ACP, Am. Coll. Cardiology; mem. Am. Heart Assn. (fellow council on clin. cardiology), Am. Fedn. for Clin. Research. Subspecialties: Cardiology; Internal medicine. Home: Teatown Rd Croton-on-Hudson NY 10520 Office: 100 S Highland Ave Ossining NY 10562

KAPLAN, MAUREEN FLYNN, analyst, researcher; b. S.I., Sept. 16, 1951; d. Hugh and Barbara (Kovacs) Flynn; m. Michael Kaplan, July 2, 1972; 1 dau., Jessica. B.A., U. Chgo., 1972; 1M.A., Brandeis U., 1974, Ph.D., 1978. Analyst, mem. tech. staff Analytic Scis. Corp., Reading, Mass., 1977—. Author: The Origin and Distribution of Tell el Yahudiyeh Ware, 1981, Archaeological Data as a Basis for Repository Marker Design, 1982. W.F. Albright fellow Am. Schs. Oriental Research, 1975-76; recipient A. Sacher award Brandeis U., 1975-76. Mem. Am. Schs. Oriental Research, Am. Nuclear Soc., Am. Statis. Assn., Archaeol. Inst. Am. Subspecialties: Nuclear fission; Statistics. Current work: Systems analysis of nuclear waste repository behavior (nuclear waste management); also mathematical and chemical analysis of archaeological artifacts. Office: TASC 1 Jacob Way Reading MA 01867

KAPLAN, RICHARD STEPHEN, research physician, educator; b. Pitts., Aug. 24, 1945; s. Simon H. and Virginia L. (Mann) K.; m. Laurelynn M. Smith, Aug. 21, 1970. B.A., U. Pitts., 1966; M.D., U. Miami, 1970. Diplomate: Am. Bd. Internal Medicine, Am. Bd. Med. Onocology. Clin. assoc. Nat. Cancer Inst., NIH, Balt. Cancer Research Center, 1971-73, sr. investigator, 1979-81; asst. prof. U. Miami and Comprehensive Cancer Center of Fla., 1975-79; assoc. prof. U. Md. Cancer Center, 1981—. Contbr. chpts. to books, articles to jours. Served with USPHS, 1971-73, 79-81. Fellow ACP; mem. Am. Assn. Cancer Research, Am. Soc. Clin. Oncology, N.Y. Acad. Scis., AAAS. Subspecialties: Cancer research (medicine); Oncology. Current work: Clinical trials of investigational drugs and treatments in lymphomas and brain tumors. Home: 3806 Greenway Baltimore MD 21218 Office: 22 S Greene St Room S9D05 Baltimore MD 21201

KAPLAN, ROBERT MALCOLM, psychology educator, psychologist; b. San Diego, Oct. 26, 1947; s. Oscar Joel and Rose (Zankan) K.; m. Catherine J. Atkins, July 10, 1977. A.B., San Diego State U., 1969; M.A., U. Calif.-Riverside, 1970, Ph.D., 1972. Sr. research assoc. Am. Inst. Research, Palo Alto, Calif., 1972-73; asst. prof. to assoc. prof. psychology U. Calif.-LaJolla, 1973—; asst. prof. to prof. San Diego State U., 1975—, dir., 1982—; mem. study sect. Nat. Ctr. Health Services Research, Washington, 1981—; dir. Advanced Study Inst. NATO, Maratea, Italy, 1981, Advanced Research Workshop, 1983; cons. in field. Author: Psychological Testing, 1982, Social Psychology: Basic and Applied, 1982, Psychology: Personal and Social Adjustment, 1977; contbr. chpts. to books and articles to profl. jours. NIMH grantee, 1977; Nat. Heart Lung & Blood grantee, 1979-82; Nat. Inst. Arthritis, Metabolic and Digestive Diseases grantee, 1982; Am. Diabetes Assn. grantee, 1982. Fellow Am. Psychol. Assn.; mem. AAAS (pres. Pacific div. 1980-82), Am. Diabetes Assn., Am. Pub. Health Assn. Democrat. Jewish. Subspecialties: Behavioral psychology; Epidemiology. Current work: Behavioral medicine, health services research, health outcome measurement. Office: Ctr for Behavioral Medicine San Diego State University San Diego CA 92182

KAPLAN, STANLEY ALBERT, pharm. research co. exec.; b. N.Y.C., Sept. 23, 1938; s. Martin and Sara (Meszel) K.; m. Lois E., Sept. 11, 1960; children: Lisa, Michelle, Martin. B.S., Columbia U., 1959, M.S., 1961; Ph.D. in Pharm. Chemistry, U. Calif., San Francisco, 1965. Postdoctoral fellow NIH, St. Mary's Hosp. Med. Sch., London, Eng., 1965-66; sr. biochemist dept. clin. pharmacology Hoffmann-LaRoche, Inc., Nutley, N.J., 1966-70, research group chief, 1971-73, asst. dir. dept. biochemistry and drug metabolism, 1964-65, assoc. dir., 1976-78, dir. dept. pharmacokinetics and biopharms., 1979. Editorial bd.: Jour. Biopharmacoutics and Drug Disposition; Contbr. numerous articles to profl.jours. Fellow Acad. Pharm. Scis., Am Coll. Clin. Pharmacology; mem. Am. Pharm. Assn., Acad. Pharm. Scis., N.Y. Acad. Scis., AAAS (leader pharm. scis. sect. 1983—), Am. Soc. Pharmacology and Exptl. Therapeutics. Lodge: B'nai B'rith. Subspecialties: Pharmacokinetics; Biochemistry (biology). Current work: Physiol. dispostion of drugs, pharamacodynamics. Home: 24 Erli St Wayne NJ 07470 Office: Hoffmann LaRoche Inc Nutley NJ 07110

KAPLANSKY, IRVING, educator, mathematician; b. Toronto, Ont., Can., Mar. 22, 1917; came to U.S., 1940, naturalized, 1955; s. Samuel and Anna (Zuckerman) K.; m. Rachelle Brenner, Mar. 16, 1951; children—Steven, Daniel, Lucille. B.A., U. Toronto, 1938, M.A., 1939; Ph.D., Harvard, 1941; LL.D. (hon.), Queen's U., 1969. Instr. math. Harvard, 1941-44; mem. faculty U. Chgo., 1945—, prof. math., 1956—, chmn. dept., 1962-67, George Herbert Mead Distinguished Service prof. math., 1969—; Mem. exec. com. div. math. NRC, 1959-62. Author books, tech. papers. Mem. Nat. Acad. Scis. Home: 5825 S Dorchester Ave Chicago IL 60637

KAPP, JOHN PAUL, neurosurgeon, educator; b. Galax, Va., Feb. 22, 1938; s. Paul Homer and Katherine (Vass) K.; m. Emily Lurleese Evans, June 23, 1961; children—Paul Hardin, Emily Camille. M.D., Duke U., 1963, B.S., 1966, Ph.D., 1967. Diplomate: Am. Bd. Neurol. Surgery. Asst. prof. neurosurgery U. Tenn., Memphis, 1971-72; attending neurosurgeon Bapt. Hosp., Memphis, 1972, Bay Med. Ctr., Panama City, Fla., 1972-80; assoc. prof. neurosurgery U. Miss. Med. Ctr., Jackson, 1980-83, prof. neurosurgery, 1983—; pres. Coast Research Found., Panama City, 1978—. Author: The Cerebral Venous System and Its Disorders, 1984. Recipient Research award Am. Acad. Neurol. Surgery, 1967; NIH fellow, 1965-67. Mem. Am. Heart Assn. (stroke council), S.W. Oncology Group (brain com., stem cell essay com.), Am. Assn. Neurol. Surgeons, ACS, Phi Beta Kappa, Sigma Xi, Alpha Omega Alpha. Democrat. Methodist. Subspecialties: Neurosurgery; Cancer research (medicine). Current work: Drug Sensitivity testing in brain tumor chemotherapy. Venous disease in the brain. Home: 1933 Petit Bois N Jackson MS 39211 Office: U Miss Med Ctr 2500 N State St Jackson MS 39216

KAPP, JOSEPH ALEXANDER, materials engineer, researcher, educator; b. Troy, N.Y., July 22, 1953; s. Fred G. and Lourdes A. (Duncan) K.; m. Nancy A. Murphy, Oct. 10, 1981. Student, Clarkson Coll, 1971-73, Hudson Valley Community Coll., 1973; B.S.M.E., Union Coll., 1975, M.S.E., 1977; Ph.D., Rensselaer Poly. Inst., 1982. Process engr. processes unit Watervliet (N.Y.) Arsenal, 1975-78; research materials engr. materials engring. sect. Benet Weapons Lab., Watervliet, 1978—; instr. in materials engring. Hudson Valley Community Coll., 1981—; adj. assoc. prof. Union Coll., 1982—. Contbr. articles to profl. jours. Mem. ASME, ASTM, Soc. Exptl. Stress Analysis. Roman Catholic. Subspecialties: Fracture mechanics; Materials (engineering). Current work: Basic reseraach interests in mechanical behavior of materials, fatigue and fracture, including environmental effects; applied research interests in thick walled pressure vessel technology. Home: 25 Elmhurst Ave Rensselaer NY 12144 Office: Research Br Benet Weapons Lab Watervliet NY 12189

KAPPAS, ATTALLLAH, med scientist, educator, cons.; b. Union City, N.J., Nov. 4, 1926; m. Katharine Bingham Hull, Oct. 26, 1963; children: Peter, Michael, Nicholas. A.B., Columbia U., 1947; M.D. with honors, U. Chgo., 1950. Diplomate: Am. Bd. Internal Medicine. Intern, resident in medicine Kings County Hosp., Bklyn., 1950-51, Peter Bent Brigham Hosp., Boston, 1954-56; from fellow to assoc. Sloan Kettering Inst., N.Y.C., 1951-53, 56-57; asst. prof., assoc. prof. medicine U. Chgo. Med Sch., 1957-67; Guggenheim fellow, 1966-67; assoc. prof. Rockefeller U., N.Y.C., 1967-71, prof., 1971—, Sherman Fairchild prof., 1981—; physician-in-chief Rockefeller U. Hosp., 1974. Contbr. over 250 articles on clin. sci., pharmacology, biochemistry, nutrition, human physiology and endocrinology to sci. jours. Bd. dirs. Vis. Nurse Service N.Y. Served with U.S. Army, 1945-46. Recipient Disting. Service award in med. sci. U. Chgo., 1975; spl. award in clin. pharmacology Burroughs Wellcome Fund, 1973; Commonwealth Fund fellow, 1961. Mem. Assn. Am. Physicians, Am. Soc. Clin. Investigation, Am. Soc. Pharmacology and Exptl. Therapeutics (award 1978), Interurban Clin. Club, Practitioners Soc. N.Y. Greek Orthodox. Clubs: University (N.Y.C); Cosmos (Washington). Subspecialties: Pharmacology; Biochemistry (medicine). Current work: Biochemical and clinical pharmacology and toxicology; endocrinology; metabolism; porphyrins; heme biology, cytochrome P450 drug/carcinogen biotransformations. Office: Rockefeller U Hosp 1230 York Ave New York NY 10021

KAPRAL, FRANK ALBERT, medical microbiologist, researcher, educator; b. Phila., Mar. 12, 1928; s. John and Erna Louise (Melching) K.; m. Marina Garay, Nov. 22, 1951; children: Frederick, Gloria, Robert. B.S., Phila. Coll. Pharmacy and Scis., 1952; Ph.D., U. Pa., 1956. Instr. microbiology U. Pa., 1956-58, assoc. in microbiology, 1958-62; assoc. microbiologist Phila. Gen. Hosp., 1962-64, chief microbiology, 1964-66; assoc. prof. med. microbiology Ohio State U., 1966-69, prof. med. microbiology and immunology, 1969—; cons. in field. Contbr. numerous articles to profl. jours. Served with U.S. Army, 1946-47. NIH grantee, 1957—; Proctor & Gamble Co. grantee, 1981, 82. Fellow Infectious Diseases Soc. Am., Am. Acad. Microbiology; mem. Am. Soc. Microbiology, Am. Assn. Immunologists, Soc. Exptl. Biology and Medicine, AAAS, N.Y. Acad. Scis., Am. Soc. Cell Biology. Roman Catholic. Subspecialties: Microbiology; Immunobiology and immunology. Current work: Staphylococcal host-parasite interactions; staphylococcal toxins; staphylococcal diseases; role of lipids in host defense mechanisms. Patentee implant chamber. Home: 873 Clubview Blvd Worthington OH 43085 Office: Ohio State U 5065 Graves Hall Columbus OH 43210

KAPRAL, RAYMOND EDWARD, chemistry educator; b. Swoyersville, Pa., Mar. 21, 1942; m., 1964. B.S., King's Coll., Pa., 1964; Ph.D. in Chemistry, Princeton U., 1967. Research assoc. in chemistry Princeton U., 1967, MIT, 1968-69; asst. prof. U. Toronto, Ont., Can., 1969-74, assoc. prof., 1974-80, prof. chemistry, 1980—. Recipient Noranda award Chem. Inst. Can., 1981. Mem. Am. Phys. Soc. Subspecialty: Physical chemistry. Office: Dept Chemistry Univ Toronto Toronto ON Canada M5S 1A3

KAPUR, KAILASH CHANDER, industrial engineering educator, consultant; b. Rawalpindi, Pakistan, Aug. 17, 1941; came to U.S., 1965; s. Gobind Ram and Vidya Vanti (Khanna) K.; m. Geraldine Palmer, May 15, 1969; children: Anjali Joy, Jay Palmer. B.S.M.E., Delhi U., 1965; M. Tech., Indian Inst. Tech., 1965; M.S., U. Calif.-Berkeley, 1967, Ph.D., 1969. Registered profl. engr., Mich. Sr. research engr. Gen. Motors Research Labs., Warren, Mich., 1969-70; mem. faculty Wayne State U., Detroit, 1970—, prof., 1980—; vis. scholar Ford Motor Co., Dearborn, Mich., summer 1973; vis. prof. U. Waterloo, Ont., Can., 1977-78; sr. reliability engr. TACOM, U.S. Army, summer 1978; cons. Author: Reliability in Engineering Design, 1977, also articles.; Assoc. editor: Jour. Reliability and Safety, 1982—. Grantee Gen. Motors Corp., 1974-77, U.S. Army, 1978, U.S. Dept. Transp., 1980-82. Mem. Inst. Indsl. Engrs. (assoc. editor 1980—), IEEE (sr.), Ops. Research Soc. Am. (sr.). Subspecialties: Industrial engineering; Operations research (engineering). Current work: Research in area of reliability engineering, design reliability, transportation systems, quality control and optimization. Home: 17291 Jeanette Southfield MI 48075 Office: Wayne State U 640 Putnam Detroit MI 48202

KARABATSOS, GERASIMOS JOHN, educator, chemist; b. Chomatada, Greece, May 17, 1932; came to U.S., 1950, naturalized, 1963; s. John P. and Athena (Papadopoulou) K.; m. Marianna Marris, Dec. 16, 1956; children—Lelena, Yanna, Jason, Byron. B.A. magna cum laude, Adelphi Coll., 1954; M.A., Harvard, 1956, Ph.D., 1959. Asst. prof. Mich. State U., East Lansing, 1959-63, asso. prof., 1963-65, prof. chemistry, 1966—, chmn. dept., 1975—; NSF sr. postdoctoral fellow U. Calif., Berkeley, 1965-66; Ford Found. vis. prof. San Marcos U., Peru, 1967; sci. dir. Greek Atomic Energy Commn., 1974—. Editor: Advances in Alicylic Chemistry, 1966; Contbr. articles profl. jours. Sloan Found. fellow, 1962-66; Recipient Sigma Xi Jr. Research award, 1970, Am. Chem. Soc. award in petroleum chemistry, 1971; Distinguished Faculty award Mich. State U., 1971. Mem. Greek Acad. of Athens (corr.), Am. Chem. Soc., Chem. Soc. London, Sigma Xi. Subspecialties: Nuclear magnetic resonance (chemistry); Organic chemistry. Current work: NMR, isotope effects; stereochemistry; stereochemistry of some biochemical reactions. Home: 1623 Old Mill Rd East Lansing MI 48823

KARAGUEUZIAN, HRAYR SEVAG, pharmacologist; b. Damascus, Syria, June 30, 1946; came to U.S., 1970, naturalized, 1977; s. Anania Dirkran and Sona (Bedrossian) K.; m. Lena Vahan Demirjian, Feb. 20, 1982; 1 child: Saro. B.Sc., Damascus U., 1969; M.Sc., Columbia U., 1972, M.Phil., 1975, Ph.D., 1978. Research assoc. U. Paris, 1978-79; postdoctoral fellow U. Calif., San Francisco, 1979-80; dir., asst. prof. Cedars-Sinai Med. Ctr., Los Angeles, 1980—. Contbr. articles to profl. jours. Los Angeles Heart Assn. grantee, 1980-82; Cino-Del-Duca research award, Paris, 1978. Mem. Cardiac Electrophysiology Assn. (sec.), Western Pharmacology Soc., Am. Fedn. Clin. Research. Democrat. Christian Orthodox. Subspecialties: Physiology (biology); Pharmacology. Current work: Study of the mechanism and site of origin of cardiac arrhythmias associated with ishcemic heart disease and determination of mode of action of drugs. Home: 12601 Miranda St North Hollywood CA 91607 Office: Cedars Sinai Med Center 8700 Beverly Blvd Los Angeles CA 91607

KARAM, RATIB ABRAHAM, nuclear engring. educator, cons.; b. Miniara, Akkar, Lebanon, Mar. 8, 1934; came to U.S., 1953; s. Abraham and Matilda (Farah) K.; m. Bobbie M. Epting, June 9, 1960 (dec. Aug. 1971); children: Ratib A., Lllemya Rachel; m. Silvia Maria Gutierrez, Aug. 14, 1975. B. Chem. Engring., U. Fla., 1958, M.S., 1960, Ph.D., 1963. Nuclear engr. Argonne (Ill.) Nat. Lab., 1963-72; prof. Ga. Inst. Tech., Atlanta, 1972—. Contbr. numerous articles on nuclear engring. to profl. jours.; editor: Advanced Reactors, 1974, Environmental Impact of Nuclear Power Plants, 1975, Energy and the Environment, 1976, Societal Cost of Energy Alternatives, 1983. Fellow Am. Nuclear Soc.; mem. Am. Phys. Soc., Sigma Xi. Subspecialties: Nuclear fission; Nuclear engineering. Current work: Alternate reactor concepts, neutron transport. Office: Ga Inst Tech Sch Nuclear Engring Atlanta GA 30332

KARANTINOS, ANDREW E., mathematics educator; b. Langa, Kastoria, Greece, June 24, 1935; came to U.S., 1951, naturalized, 1956; s. Elias and Vasiliki (Eliades) K.; m. Effie Stergios, June 26, 1960; children: Nick, Chris. B.S., Morningside Coll, Sioux City, Iowa, 1959; M.S., SUNY-Buffalo, 1963; Ph.D., U. S.D., 1973; postgrad., Mich. State U., 1980, U. Wis.-Madison, Rutgers U. Tchr. Paullina (Iowa) High Sch., 1959-61, chmn. math. dept., 1959-61; tchr. Central High Sch., Sioux City, 1961-65, chmn. math. dept., 1961-65; prof. math. U. S.D., Vermillion, 1965—; Chmn. math contest U. S.D., 1965—. Author: Math. Sci. Newsletter, 1981-83. Recipient Outstanding Prof. award U. S.D. Faculty, 1979-80. Mem. Am. Hellenic Ednl. Progressive Assn. (dist. gov. 1976, supreme gov. 1979-80, chmn. edn. found. 1981-83), Math. Assn. Am., Nat. Council Tchrs. Math., Acad. Sci., Pi Mu Epsilon, Sigma Pi Sigma, Phi Delta Kappa. Greek Orthodox. Club: Math. and Computer Sci. (adv. 1970—). Lodges: Lions.; Masons. Subspecialties: Statistics. Current work: Mathematics education; theory of learning mathematics; mathematics curriculum. Home: 1019 Crestview St Vermillion SD 57069 Office: Univ SD Dakota Hall 315 Vermillion SD 57069

KARAVOLAS, HARRY J(OHN), biochemist; b. Peabody, Mass., Feb. 21, 1936; s. John Louis and Maria (Kayavas) K.; m. Barbara A. Katsaras, Aug. 26, 1962; 1 son, Christian Mark. B.S., Mass. Coll. Pharmacy, 1957, M.S. (Am. Found. Pharm. Edn. fellow), 1959; Ph.D. (USPHS fellow), St. Louis U., 1963; postgrad., Harvard U., 1963-66. Research fellow in biol. chemistry Harvard U. Med. Sch., 1963-66, research asso., instr. biol. chemistry, 1966-68, vis. lectr. biol. chemistry, 1975; tutor biochemical scis. Harvard Coll., 1966-68; asst. prof. physiol. chemistry and endocrinology U. Wis., Madison, 1968-72, mem. endocrinology-reproductive physiology program, 1968—, asso. dir., 1974—, asso. prof. physiol. chemistry, 1972-75, prof., chmn. dept., 1975—; sect. head neuroendocrinology Waisman Center on Mental Retardation and Human Devel., 1972—; mem. study sect. biochem. endocrinology NIH, 1979—. Editorial bd.: Endocrinology, 1974-78; bd. reviewers: Federation Proceedings, 1972-77; contbr. sci. articles to profl. jours. Recipient Borden award; Merck award; Rexall award, 1957; Amoco Distinguished Teaching award U. Wis., 1977; Ford Found. research grantee, 1970; NICHD research career devel. awardee, 1972-75; NIH research grantee, 1972. Mem. Am. Assn. Biol. Chemists, Endocrine Soc., Soc. Neuroscience, AAAS, Sigma Xi. Subspecialty: Biochemistry (biology). Home: 2 Regis Circle Madison WI 53711 Office: Univ Wisconsin 591 Med Scis Bldg 1215 Linden Dr Madison WI 53706

KARGER, BARRY LLOYD, chemistry educator and administrator; b. Boston, Apr. 2, 1939; m., 1961. B.S., MIT, 1960; Ph.D. in Analytical Chemistry, Cornell U., 1963. Asst. prof. to assoc. prof. Northeastern U., Boston, 1963-72, prof. chemistry, 1972—; dir. Inst. Chem. Analysis, Applied and Forensic Sci., 1973—; sci. adv. FDA, 1973-76; cons. Tech. Distriments Corp., 1977—. Research grantee NIH, 1969—, Office Naval Research and NIH, 1969-74; recipient Gulf Research award, 1971; Steven Dal Nogare Meml. award Delaware Valley Chromotog. Form, 1975; Tewett Meml. medal USSR, 1980. Mem. AAAS, Am. Chem. Soc. (Chromatography award 1982), N.Y. Acad. Scis., Sigma Xi. Subspecialty: Analytical chemistry. Office: Inst Chem Analysis Northeastern Univ Boston MA 02115

KARIM, AZIZ, research scientist; b. Tanzania, Aug. 20, 1939; s. Karim Nasser and Zera; m. Roshan, Sept. 8, 1970; children: Arif, Navine. B. Pharm., U. London, 1964, Ph.D., 1967. Analytical biochemist Burroughs Wellcome & Co., Eng., 1961-62; research asso. Sch. Pharmacy, U. Wis., Madison, 1967-69; research investigator biol. research Searle Labs., Skokie, Ill., 1969-71, sr. research investigator biochemistry, 1971-73, group leader drug metabolism, 1973-75, research fellow, 1976-79, assoc. dir. clin. pharmacology and pharmacokinetics, 1979-81; dir. clin. bioavailability and pharmacokinetics G.D. Searle & Co., Skokie, 1981—. Contbr. articles to profl. jours. Mem. Pharm. Soc. Gt. Britain, Am. Soc. Clin. Pharmacology, Am. Soc. Exptl. Pharmacology and Therapeutics. Subspecialties: Pharmacokinetics; Medicinal chemistry. Current work: Research on disposition, metabolism and elimination of drugs. Patentee in field. Home: 8457 Madison Dr Niles IL 60648 Office: PO Box 5110 Chicago IL 60680

KARLE, ISABELLA LUGOSKI, scientist; b. Detroit, Dec. 2, 1921; m., 1942; 3 children. B.S., U. Mich., 1941, M.S., 1942, Ph.D. in Phys. Chemistry (Rackham fellow), 1942-43, 1944; D.Sc. (hon.), 1976, 79. Asso. chemist U. Chgo., 1944; instr. U. Mich., 1944-46; physicist U.S. Naval Research Lab., Navy Dept., Washington, 1946-59, head X-ray sect., 1959—; Mem. U.S. Nat. Com. on Crystallography; adv. bd. Office Chemistry and Chem. Tech., NRC; mem. bd. on internat. orgns. and programs Nat. Acad. Scis., 1980-83; mem. exec. com. Am. Peptide Symposium, 1975—. Mem. editorial bd.: Polymers, 1975-81, Internat. Jour. Peptide and Protein Research, 1981—. Recipient Superior Civilian Service award Navy Dept., 1965, Ann. Achievement award Soc. Women Engrs., 1968; Hillebrand award, 1970; Fed. Woman's award, 1973; Dexter Conrad award Office of Naval Research, 1980. Mem. Nat. Acad. Scis., Am. Phys. Soc., Am. Biophys. Soc., Am. Crystallographic Assn. (pres. 1976), Am. Chem. Soc. (Garvan award 1976). Subspecialties: X-ray crystallography; Biophysics (biology). Current work: Establishing the molecular formula and geometry of biologically important substances; establishing precise conformations of oligopeptides,solving crystal structures of complex molecules. Research in application of electron and x-ray diffraction to structure problems in chemistry and biology. Office: Naval Research Lab Laboratory for Structure of Matter Code 6030 Washington DC 20375

KARLE, JEAN MARIANNE, research bioorganic chemist; b. Washington, Nov. 14, 1950; d. Jerome and Isabella L. K. B.S. in Chemistry, U. Mich., 1976. USPHS fellow Nat. Inst. Arthritis, Diabetes and Digestive and Kidney Diseases, NIH, Bethesda, Md., 1976-78; Sr. staff fellow Nat. Cancer Inst., 1978—. Author papers, reports in field. Mem. Am. Chem. Soc., Am. Assn. Cancer Research, Internat. Soc. Xenobiotics. Subspecialties: Cancer research (medicine); Analytical chemistry. Current work: Mode of action of chemotherapeutic agents, design new anti-cancer drugs, tissue culture, high pressure liquid chromatography, enzymology.

KARLE, JEROME, research physicist; b. N.Y.C., June 18, 1918; married, 1942; 3 children. B.S., CCNY, 1937; A.M., Harvard U., 1938; M.S., U. Mich., 1942, Ph.D. in Phys. Chemistry, 1943. Research asso. Manhattan project, Chgo., 1943-44, U.S. Navy Project, Mich., 1944-46; head electron diffraction sect. Naval Research Lab., Washington, 1946-58, head diffraction br., 1958, now head lab. for structure matter; mem. NRC, 1954-56, 67-75, 78—; chmn. U.S. Nat. Com. for Crystallography, 1973-75. Fellow Am. Phys. Soc.; mem. Am. Chem. Soc., Crystallograph. Assn. (treas. 1950-52, pres. 1972), Internat. Union Crystallography (exec. com. 1978—, pres. 1981—), Am. Math. Soc., AAAS, Nat. Acad. Scis. Subspecialties: Crystallography; Diffraction Physics. Current work: Theory for interpretation of multiple wavelength anomalous dispersion experiments on crystals with objective of enhanced facility in macromolecular structure determination. Research in structure atoms, molecules, crystals, solid surfaces. Office: US Naval Research Lab Lab for Structure of Matter Code 6030 Washington DC 20375 There is too much administration of everything creative. It distorts our society and its character. The solution is to select competent, well-qualified people and give them freedom and support to pursue their creative gifts.

KARLINER, JOEL SAMUEL, cardiologist, medical educator; b. N.Y.C., Nov. 1, 1936; s. William and Elsie (Levy) K.; m. Adela Bernard, Mar. 22, 1959; children: Joshua, Rachel, Leah. A.B., Columbia U., 1958, M.D., 1962. Diplomate: Am. Bd. Internal Medicine; specialty bd. Cardiovascular Disease. Intern Bronx Mcpl. Hosp., 1962-63, resident in internal medicine, 1965-68; asst. prof. medicine U. Calif.-San Diego, 1971-75, assoc. prof. medicine, 1975-81, prof. medicine, 1981, U. Calif.-San Francisco, 1981—; Chief cardiology sect. VA MC, San Francisco. Editor: Coronary Care, 1981. Served as capt. USAR, 1963-65. NIH Fogarty sr. internat. fellow, 1978-79. Fellow Am. Coll. Cardiology, ACP, Council Clin. Cardiology, Am. Heart. Assn.; mem. Western Soc. Clin. Investigation, Western Assn. Physicians. Subspecialties: Cardiology; Molecular pharmacology. Office: VA Med Ctr 4150 Clement St San Francisco CA 94121

KARLSON, RONALD HENRY, marine ecologist, educator; b. Coalinga, Calif., Oct. 13, 1947; s. William Henry and Esther (Gregg) Nichols K.; m. Susan Ray, Oct. 8, 1977; 1 son, Jame Henry. B.A., Pomona Coll., 1969; M.A., Duke U., 1972, Ph.D., 1975. Posdoctoral fellow Johns Hopkins U., 1976-78; asst. prof. Sch. Life and Health Scis., U. Del., 1978—. NSF grantee, 1976-78, 83—. Mem. Ecol. Soc. Am., Am. Soc. Zoologists, AAAS. Subspecialties: Ecology; Species interaction. Current work: The role of biological interactions and physical disturbances in determining community structure, competitive strategies and life history parameters. Office: School of Life and Health Sciences University of Delaware Newark DE 19711

KARN, ROBERT CAMERON, mammalian geneticist, researcher; b. Berwyn, Ill., Mar. 12, 1945; s. Joseph E. and Barbara G. K.; m. Marianne Karn, Aug. 27, 1966; children: Colin Edward, Evan Cameron. B.A., Ind. U., 1967, M.A. in Zoology, 1970, Ph.D., 1972. Postdoctoral fellow dept. med. genetics Ind. U. Sch. Medicine, Indpls., 1972-74, instr., 1974-76, asst. prof., 1976-81, assoc. prof. med. genetics, 1981—; vis. faculty U. Aarhus, Denmark. Contbr. articles to sci. publs. Served to maj. USAR, 1968—. USPHS career devel. awardee, 1977-82. Mem. Genetics Soc. Am., Am. Soc. Human Genetics, Am. Inst. Biol. Scis., Am. Soc. Biol. Chemistry, Sigma Xi. Subspecialties: Genetics and genetic engineering (medicine); Gene actions. Current work: Human and mouse salivary proteins; a-amylases; regulation of gene expression. Office: Dept Med Genetics Ind Univ Sch Medicine 1100 W Michigan St Indianapolis IN 46223

KARNOSKY, DAVID FRANK, forest geneticist, cons.; b. Rhinelander, Wis., Oct. 12, 1949; s. Frank and Verna (Forsman) K.; m. Sheryl Lee Bennett, Sept. 12, 1970; children: David Brian, Jason Robert. B.S., U. Wis., Madison, 1971, M.S., 1972, Ph.D., 1975. Forest geneticist N.Y. Bot. Garden, Cary Arboretum, Millbrook, N.Y., 1975-82; dir. Inst. Urban Horticulture, Millbrook, 1982-83; prof., dir. Center for Intensive Forestry in No. Regions, also dir. Forest Biotech. Group, Mich. Tech U., Houghton, 1983—; adj. assoc. prof. Coll. Environ. Sci. and Forestry SUNY-Syracuse; cons. forestry, forest biotech. and tree improvement. Contbr. articles on forest genetics and tissue culture to profl. jours. Mem. Air Pollution Control Assn., Soc. Am. Foresters, Am. Phytopathol Soc., Tissue Culture Assn. Subspecialties: Environmental toxicology; Plant cell and tissue culture. Current work: Tree improvement, micropropogation of trees, air pollution effects on trees. Home: Route 1 Box 78A Chassell MI 49916 Office: Dept Forestry Mich Tech U Houghton MI 49931

KARNOVSKY, MANFRED L., biological chemistry educator; b. Johannesburg, South Africa, Dec. 14, 1918; m., 1952. B.S., U. Witwatersrand, 1941, 1942; M.S., 1943; Ph.D. in Organic Chemistry, U. Cape Town, 1947. Jr. lectr. in chemistry U. Witwatersrand, 1941-42; chief chemist and insp. Brit. Ministry Aircraft Prodn., South Africa, 1942-43; asst. U. Cape Town, 1944-47; research assoc. to assoc. Harvard Med. Sch., Boston, 1950-51, asst. prof. to prof., 1952-65, Harold T. White prof. biol. chemistry, 1965—; chmn. dept., 1969-73. Research fellow U. Wis., 1947-48; recipient Lederle Med. Faculty award, 1955-58, 2d award Glycerine Producers Assn., 1953, Gold medal Reticuloendothelial Soc., 1966. Fellow Am. Acad. Arts and Sci.; mem. Am. Soc. Biol. Chemists, Histochem. Soc., Am. Chem. Soc., Am. Soc. Cell Biologists. Subspecialty: Biochemistry (biology). Office: Dept Biol Chemistry Harvard Med Sch Boston MA 02115

KARP, JUDITH ESTHER, oncologist, educator; b. San Diego, July 15, 1946; d. Louis Moses and Bella Sarah (Perlman) K.; m. Stanley Howard Freedman, Sept. 21, 1975. B.A., Mills Coll., 1966; M.D., Stanford U., 1971. Diplomate: Am. Bd. Internal Medicine. Hybrid intern-resident Stanford U. Hosps., 1971-72; asst. resident Johns Hopkins Hosp., Balt., 1972-73; fellow in oncology Johns Hopkins U. Sch.Medicine, Balt., 1973-75, instr. oncology and medicine, 1975-78, asst. prof., 1978—. Contbr. articles to profl. jours. Am. Cancer Soc. fellow, 1976-79; recipient Resolution of Commendation Ho. of Dels., State of Md., 1982. Mem. Internat. Soc. for Exptl. Hematology, Am. Soc. Hematology, Am. Soc. Clin. Oncology, Cell Kinetic Soc. (chmn. nominating com. 1981-82), Phi Beta Kappa. Democrat. Jewish. Subspecialties: Chemotherapy; Cell study oncology. Current work: Timed sequential chemotherapy of acute leukemia, based on cell kinetics. Home: 15 Farmhouse Ct Baltimore MD 21208 Office: Johns Hopkins Oncology Ctr 600 N Wolfe St Baltimore MD 21208

KARP, RICHARD DALE, immunologist, educator; b. Mpls., June 19, 1943. B.A., U. Minn., 1965, M.S., 1968, Ph.D., 1972. Predoctoral fellow U. Minn., 1966-72; Celeste Durand Rogers postdoctoral fellow UCLA, 1973-75; asst. prof. immunology U. Cin., 1975-81, assoc. prof., 1981—. Contbr. articles to profl. jours., chpts. in books. NIH research grantee, 1978-81; NSF research grantee, 1984—. Mem. Am. Soc. Microbiology, Am. Assn. Immunologists, Am. Soc. Zoologists, N.Y. Acad. Scis., Sigma Xi. Subspecialties: Immunobiology and immunology; Microbiology. Current work: Phylogenetic development of immune response, immune response to solid tumor growth and nature of tumor-host interactions. Office: Dept of Biological Science University of Cincinnati Cincinnati OH 45221

KARP, WARREN BILL, dental educator, researcher; b. N.Y.C., Feb. 12, 1944; s. David and Nettie Emma (Schneiderman) K.; m. Nancy Ruth Sullivan, Apr. 4, 1968 (dec. Aug. 1970); m. Nancy Virginia Blanchard, Jan. 4, 1976; children: Heather, Michael. B.S., Pace U., 1965; Ph.D., Ohio State U., 1970; D.M.D., Med. Coll. Ga., 1977. Instr. Med. Coll. Ga., 1971-73, asst. prof. pediatrics, 1973-79, assoc. prof., 1979—; dir. Clin. Perinatal Lab., Augusta, Ga., 1977—. Contbr. numerous articles on perinatal biochemistry to profl. jours. Recipient EPA award, 1971; Biomed. Research Support award, 1980; Mercury and Heart award Ga. Heart Assn., 1981. Mem. AAAS, Am. Chem. Soc., N.Y. Acad. Sci., Internat. Assn. Dental Research, Sigma Xi (sec. 1980-83). Subspecialties: Biochemistry (medicine); Pediatrics. Current work: Bilirubin metabolism; fetal and placental metabolism; environmental contaminants; drug-albumin binding. Home: 1031 Brookwood Ave Augusta GA 30909 Office: Med Coll Ga Pediatrics Augusta GA 30912

KARPEL, RICHARD LESLIE, biochemist; b. N.Y.C., May 31, 1944; s. Louis and Mollie (Schaffer) K.; m. Madeline Blatt, Jan. 8, 1947; 1 dau., Emily Miriam. Student, Cooper Union, 1961-63; B.A., Queens Coll., CUNY, Flushing, 1965; Ph.D., Brandeis U., 1970. Postdoctoral fellow, research assoc. dept. biochem. scis. Princeton U., 1970-76; asst. prof. chemistry U. Md., Catonsville, 1976-81, assoc. prof., 1981—. Contbr. biochem. articles to profl. jours. NIH grantee, 1977-81; Am. Cancer Soc. grantee, 1977-79; Am. Heart Assn. grantee, 1979-80; NRC sr. assoc., 1982-83. Mem. AAAS, Am. Chem. Soc., Sigma Xi. Subspecialties: Biochemistry (biology); Biophysical chemistry. Current work: Protein-nucleic acid interactions; helix-destabilizing proteins; metal ion-nucleic acid interactions. Office: Dept Chemistry U Md Balt County Catonsville MD 21228

KARPIAK, STEPHEN EDWARD, JR., biomed. research scientist, educator; b. Hartford, Conn., Aug. 13, 1947; s. Stephen Edward and Olga Mary (Yanenko) K. B.A. in Psychology, Coll. Holy Cross, 1969; M.A. in Exptl. Psychology, Fordham U., 1971, Ph.D., 1972. Predoctoral fellow in neurology Columbia U. Coll. Physicians and Surgeons, 1969-72, postdoctoral fellow in neurology and psychiatry, 1972-74, asst. prof. psychiatry, 1975—; sr. research scientist div neurosci. N.Y. State Psychiat. Inst., N.Y.C., 1974—; cons. WHO, UN, 1979-81. Contbr. numerous articles and chpts to profl. jours. Mem. AAAS, Soc. Neurosci. Subspecialties: Neuroimmunology; Neurochemistry. Current work: Immunological mechanisms in epilepsy; function of membrane components in behavior; biopsychology; epilepsy; immunology; neuroimmunology. Office: NY State Psychiat Inst 722 W 168th St New York NY 10032

KARPICKE, JOHN ARTHUR, systems engineer; b. Saginaw, Mich., Nov. 26, 1945; s. Herbert A. and Eleanor (Stafford) K.; m. Susan G. Denyes, Aug. 9, 1950; 1 son: Jeffrey Denyes. B.S., Mich. State U., 1972; Ph.D., Ind. U., 1976. Postdoctoral fellow psychobiology research group Fla. State U., Tallahassee, 1976-77; asst. prof. Valparaiso (Ind.) U., 1977-81; mem. tech. staff Bell Labs., Indpls., 1981-83, Am. Bell, Inc., 1983—. Contbr. articles to profl. jours. NIH fellow, 1976; NIMH grantee, 1979. Mem. Am. Psychol. Assn., N.Y. Acad. Scis. Subspecialties: Cognition; Learning. Current work: Telecommunications systems engineering; home information systems; home communication systems; synthetic speech quality/applications; human factors engineering. Office: Engring Design and Devel Am Bell Inc 6612 E 75th St Indianapolis IN 46220

KARR, JAMES RICHARD, ecologist, educator; b. Shelby, Ohio, Dec. 26, 1943; s. Rodney Joll and Marjorie Ladonna (Copel) K.; m. Kathleen Ann Reynolds, Mar. 23, 1963 (div. 1983); children: Elizabeth Ann, Eric Leigh. B.Sc., Iowa State U., 1965; M.Sc., U. Ill., 1967, Ph.D., 1970. Postdoctoral fellow Princeton U., 1970-71; postdoctoral research assoc. Smithsonian Instn., Balboa, C.Z., 1971-72; asst. prof. biology Purdue U., West Lafayette, Ind., 1972-75; prof. ecology U. Ill., Urbana, 1975—; mem. evaluation panels NSF, 1975, 76, Energy Research and Devel. Adminstrn., 1976, Instream Flow Group, Office Biol. Services, 1978; condr. workshops U.S. EPA, 1979, 80, U.S. Fish and Wildlife Service, 1979; cons. OAS, 1980. Contbr. articles, papers, monographs, revs. to profl. jours. Recipient numerous research grants. Fellow Am. Ornithologists Union, AAAS; mem. Ecol. Soc. Am. (mem. council 1982—), Am. Soc. Naturalists, Cooper Ornithol. Soc., Wilson Ornithol. Soc. (mem. exec. council 1976-79), Assn. for Tropical Biology, Internat. Soc. for Tropical Ecology, Wildlife Soc., Am. Inst. Biol. Scis., Am. Fisheries Soc., Sigma Xi. Subspecialties: Theoretical ecology; Resource management. Current work: Community ecology of birds and stream fishes, application of ecological principles to management of natural resource systems. Office: Dept Ecology Ethology and Evolution U Ill 606 E Healey St Champaign IL 61820

KARROLL, JOSEPH E., research company executive, consultant; b. N.Y.C., Mar. 24, 1922; s. Morris and Lillian (Lerman) K.; m. Zena Muriel Gelernter, Apr. 20, 1947 (div. Feb. 1979); children: Jonathan, Jaimee, Jodee. B.S., NYU, 1950, M.A., 1951; postgrad., 1954. Registered rep. Shearson Am. Express, Los Angeles, 1978-79; pres. Alternative Research Corp., Los Angeles, 1976—; v.p. Crescent Corp., Las Vegas, Nev., 1981—, Karroll Weller & Co., Inc., Los Angeles, 1982—, Karwell Corp., Phoenix, 1982—; pres. Digital Research Corp., Reno, Nev., 1982—; dir. Nat. Safety Inst., Los Angeles, 1979—. Served with U.S. Army, 1942-45; PTO. Mem. Am. Psychol. Assn. Subspecialties: Operations research (engineering); Systems engineering. Current work: research, consulting, management. Home: 15445 Ventura Blvd Suite 10-311 Sherman Oaks CA 91413 Office: Karwell Corp PO Box 15586 Phoenix AZ 85060

KARSHMER, ARTHUR ISRAEL, computer science educator; b. New Brunswick, N.J., June 14, 1940; s. Nathan and Leona Elizabeth (Kitay) K.; m. Judith Foster, Jan. 19, 1973; children: Elana, Avi. B.A., Rutgers U., 1964; M.S., U. Mass., Amherst, 1974, Ph.D., 1978; postgrad., Hebrew U., Jerusalem, 1979-80. Pvt. practice data processing and systems programming cons., New Brunswick, 1967-72; asst. prof. dept. computer sci. N. Mex. State U., University Park, 1978-79, assoc. prof. dept. computer sci., 1980—; postdoctoral research fellow Hebrew U., Jerusalem, 1979-80. Mem. editorial bd.: Sistm. Quar, 1979—, Cognition and Brain Theory, 1980—. Mem. IEEE Computer Soc., Assn. Computing Machinery. Subspecialties: Computer architecture; Operating systems. Current work: Distributed computing systems. Home: 1756 Pomona St Las Cruces NM 88001 Office: Computer Sci Dept N Mex State U Las Cruces NM 88003

KARTHA, MUKUND KRISHNA, radiology educator; b. Pattancaud, India, July 31, 1936; s. Krishnan K. and Ammalu (Kunjamma) Namboodiri; m. Mally Karunkara Kartha, Jan. 31, 1963; Children: Vyas, Neela. B.S., Kerala U., 1958; M.S., Sugar U., 1961; Ph.D., U. Western Ont., Can., 1969. Diplomate: Am. Bd. Radiology. Sci. officer India AEC, 1961-64; Cancer research fellow Ont. Cancer Found., 1964-68; chief med. physicist Ohio State U. Hosp., 1968—; asst. prof. Ohio State U., 1968-73, assoc. prof. radiology, 1973—, assoc. prof. allied med. professions, 1973—; cons. in field. Contbr. numerous articles to profl. jours. Mem. Upper Arlington Civic Assn., Aradhana Com. Ohio, Friends of India. Served in Indian Navy. Grantee in field. Mem. Am. Assn. Phys. Medicine, Radiol. Soc. N.Am., Radiol. Research Assn., Am. Assn. Therapeutic Radiology, Am. Coll. Radiology. Hindu. Subspecialties: Radiology; Oncology. Current work: Physical, chemical, biological and medical effects of ionizing radiation.

KASAHARA, AKIRA, meteorologist; b. Tokyo, Oct. 11, 1926; m., 1952. B.S., U. Tokyo, 1948, M.S., 1950, D.Sc., 1954. Asst. Geophys. Inst., U. Tokyo, 1948-53, research assoc., 1953-54; research assoc. oceanography and meteorology Tex. A&M Coll., 1954-56; research assoc. meteorology U. Chgo., 1956-62; research scientist Courant Inst. Math. Sci., NYU, 1962-63; program scientist Nat. Ctr. Atmospheric Research, Boulder, Colo., 1963-73, sr. scientist, 1973—; Affiliate prof. dept. meteorology Tex. A&M U., 1967-70; vis. lectr. Inst. Meteorology, U. Stockholm, 1971-72; adj. prof. dept. meteorology U. Utah, 1979—. Assoc. editor: Jour. Applied Meteorology, 1967-72. Fellow Am. Meteorol. Soc.; mem. Am. Geophys. Union, Meteorol. Soc. Japan (award 1961). Subspecialty: Meteorology. Office: Nat Center for Atmospheric Research PO Box 3000 Boulder CO 80307

KASAI, GEORGE JOJI, microbiologist, virologist, consultant, researcher; b. Los Angeles, Apr. 8, 1917; s. Araji and Kichino (Miura) K.; m. Tama Katako, July 7, 1946; children: Margaret L. Kasai Freeberg, Elizabeth J. Kasai Collins, Patricia J. A.A., Los Angeles Jr. Coll., 1936; S.B., UCLA, 1942; S.M., U. Chgo., 1945, Ph.D., 1952. Research asst., bacteriologist dept. microbiology U. Chgo., 1945-52; spl. fellow Zoller Dental Clinic, 1950, research assoc. in oral bacteriology, 1952-64, research asso. cholera research, asst. prof. univ., 1964-69; supervisor microbiologist in bacteriology Camp Zama, Sagami-Ono, Japan, 1969-75; immunology subsect Brooke Army Med. Center, Ft. Sam Houston, Tex., 1975—; Cons. Swiss Serum and Vaccine, Berne, 1967, WHO Cholera Panel, 1967, John Hopkins Hosp., 1967; Joint investigator, epidemiology Vibrio parahemolyticus, Indonesia, 1972-73. Contbr. articles to profl. jours. Recipient Spl. recognition Swiss Serum and Vaccine Inst., Berne, 1967, Commendation award Japanese Ground Self Def. Forces, Tokyo, 1975, Outstanding award Dept. Army, 1973, 80; Sigma Xi fellow, 1967; AAAS fellow, 1967. Mem. Am. Soc. Microbiology, Soc. Gen. Microbiology, AAAS, N.Y. Acad. Sci., Sigma Xi. Subspecialties: Virology (medicine); Microbiology. Current work: Viral serology; enzyme immuno assay; serology of infectious diseases; clinical virology (viral identification); electron microscopy. Home: 4807 El Gusto San Antonio Tx 78233 Office: Bldg 2830 Room 257 Fort Sam Houston TX 78234

KASARSKIS, EDWARD JOSEPH, physician, neurochem. researcher; b. Chgo., Oct. 9, 1946; s. Edward J. and Valeria T. (Krauchunas) K.; m. Mary Lenroot, Aug. 30, 1969; children: Andrew J., Peter E., Larisa J. B.A., Coll. St. Thomas, St. Paul, 1968; Ph.D., U. Wis., 1975, M.D., 1974. Diplomate: Am. Bd. Psychiatry and Neurology, North Bd. Med. Examiners. Med. intern U. Wis. Ctr. Health Scis., Madison, 1974-75, resident in medicine, 1974—76; resident in neurology U. Va. Hosp., Charlottsville, 1976-79; asst. prof. neurology La. State U., Shreveport, 1979-80; staff neurologist VA Med Ctr., Shreveport, 1979-80; asst. prof.neurology U. Ky., Lexington, 1980—; staff neurologist VA Med. Ctr., Lexington, 1980—. NIH grantee, 1968-74; grantee VA Research Service, 1979, 81, Distilled Spirits Council of U.S., 1981. Mem. AMA, Am. Acad. Neurology, Soc. Neurosci., AAAS, Am. Soc. Neurochemistry, Internat. Soc. Neurochemistry, Sigma Xi. Subspecialties: Neurology; Neurochemistry. Current work: Clinical neurology; vitamin and trace metal nutrition of the developing and mature central nervous system; regulation of cerebrospinal fluid pressure. Home: 305 Glendover Rd Lexington KY 40503 Office: Dept Neurology 800 Rose St Lexington KY 40536

KASCIC, MICHAEL JOSEPH, JR., mathematical consultant; b. Jersey City, Feb. 23, 1941; m. Barbara Listowski, Aug. 18, 1962; children: Paul, Renee, Eric. B.S., St. Joseph U., Phila., 1962; M.S., NYU, 1964; Ph.D., UCLA, 1967. Computing engr. N.Am. Aviation, Los Angeles, 1963-67; J.W. Young instr. Dartmouth Coll., Hanover, N.H., 1967-69; asst. prof. math. Stevens Inst., Hoboken, N.J., 1969-74; software cons. Boole & Babbage, N.Y.C., 1974-76; math. cons. Control Data Corp., Mpls., 1976—. Mem. Am. Math. Soc., Soc. Indsl. and Applied Math. (lectr. 1981—), Sigma Xi. Roman Catholic. Subspecialties: Numerical analysis. Current work: Interested in the relationships among functional analysis, numerical analysis, and algorithm development for vector processors. Office: Control Data Corp 8100 34th Ave S Minneapolis MN 55440

KASDON, S. CHARLES, physician, educator, cancer researcher; b. N.Y.C., Dec. 19, 1912; s. David S. and Sarah C. (Mirkin) K.; m. Muriel E. Cohen, Dec. 25, 1943; chldren: David, Madeline A., Louisa Kasdon Ellias. B.S., Yale U., 1933; M.S., 1934; M.D., 1938. Diplomate: Am. Bd. Ob-Gyn. Mem. faculty Tufts U. Med. Sch., Boston, 1941-, assoc. prof. ob-gyn, 1959—; pres. Med. Research Found., Boston, 1951; past chief staff Booth Meml. Hosp., Brookline, Mass.; past chief ob-gyn staff Waltham (Mass.) Hosp.; mem. malpractice tribunal Mass. Superior Ct. Author papers in field. Served to lt. comdr. USNR, 1944-47. Mem. Am. Assn. Cancer Research, Am. Endocrine Soc., Am. Soc. Cytology. Clubs: Yale, St. Botolph (Boston). Subspecialties: Obstetrics and gynecology; Gynecological oncology. Current work: Cancer of female, endocrinology. Office: 127 Bay State Rd Boston MA 02215

KASEL, JULIUS ALBERT, virologist, educator; b. Homestead, Pa., Dec. 7, 1923; s. Julius and Anna (Kudis) K.; m. Nan Elizabeth Cleghon, Sept. 29, 1950; children: Gary Lee, John Foster, Patricia Joyce. B.S., U. Pitts., 1949; M.S., Georgetown U., 1958, Ph.D., 1960. Head med. virology sect. Nat. Inst. Allergy and Infectious Diseases, 1950-72; prof. microbiology and immunology Baylor Coll. Medicine, Houston, 1972—. Served with USPHS, 1952-45; Served with USPHA, 1958-72. Decorated Air medal; recipient meritorious medal for research USPHS. Fellow Am. Acad. Microbiologists; fellow Infectious Diseases Soc. Am.; mem. Am. Assn. Immunologists, Soc. Exptl. Biology and Medicine. Roman Catholic. Subspecialties: Virology (biology); Immunology (agriculture). Current work: Immunology of respiratory virus infections in man. Home: 1926 Country Club Dr Sugarland TX 77478 Office: Baylor Coll Houston TX 77030

KASHA, KENNETH JOHN, crop scientist, educator; b. Lacome, Alta., Can., May 6, 1933; s. John Clarence and Mary Jennette (Proudfoot) K.; m. Marion Eileen Lenz, Aug. 14, 1958; children: Lorelei Marion, David John. B.Sc., U. Alta., 1957, M.Sc., 1958; Ph.D., U. Minn., 1962. Teaching asst./research fellow U. Minn., Mpls., 1960-62; research scientist Agr. Can., Ottawa Research Sta., 1962-66; asst. prof. crop sci. dept. U. Guelph, Ont., 1966-69, assoc. prof., 1969-74, prof., 1974—; cons. CIBA-Geigy Seeds, 1974-81. Contbr. articles to profl. jours. U. Alta. research scholar; JE Olson prize in botany and lst class standing prize, 1957; Guelph Sigma Xi award for excellence in research, 1974; Agr. Inst. Can. Grindley Medal, 1977. Mem. Genetics Soc. Can., Genetics Soc. Am., Am. Soc. Agronomy, Internat. Plant Cell and Tissue Culture Assn., Sigma Xi. Subspecialties: Plant genetics; Plant cell and tissue culture. Current work: Plant cytogenetics, genome organization, gene gransfer, chromosome elimination, barley, wheat, cell and tissue culture, linkage mapping. Home: 21 Glenburnie Dr Guelph ON Canada N1E 4C4 Office: Crop Sci Dept Univ Guelph Guelph ON Canada N1G 2W1

KASHMIRI, SYED V.S., molecular biologist, educator; b. Lucknow, India, July 5, 1937; came to U.S., 1964, naturalized, 1980; s. Syed M. and Shandar (Begum) Kazim; m. Rafia Mehdi, Oct. 23, 1969; 1 son, Syed Tabish. Ph.D., Duke U., 1968. Research assoc. Rockefeller U., N.Y.C., 1968-72, U. Md. Sch. Medicine, Balt., 1972-73, Wistar Inst., Phila., 1973-74; sr. scientist Litton Bionetics, Kensington, Md., 1974-77; research assoc. Johns Hopkins U. Sch. Medicine, Balt., 1977-79; asst. prof. molecular biology U. Pa. Sch. Vet. Medicine, Kennett Square, 1979—. Contbr. articles to profl. jours. Mem. Genetics Soc. Am., Am. Soc. Microbiology, Am. Soc. Virology. Subspecialties: Molecular biology; Oncology. Current work: Molecular mechanism of leukemogenesis induced by retroviruses. Model system under study is bovine leukosis induced by bovine leukemia virus. Have molecularly cloned the virus. Studying the genome organization of the virus. Home: 16 Glen View Ln Downingtown PA 19335 Office: Leukemia Studies Unit New Bolton Center Kennett Square PA 19348

KASIMIS, BASIL SPIROS, physician, educator; b. Athens, Greece, June 19, 1946; s. Spiros Dimitrios and Theoni (Stefanides) K.; m. Katherine Markantonis Rukensteiner, July 10, 1975; children: Anne-Theoni, Elizabeth-Jennifer. M.D., Nat. U. Athens, 1970, Dr. Med. Scis., 1974. Intern, South Balt. Gen. Hosp., 1974-75, jr. asst. med. resident, 1975-76, chief resident, 1976-77; fellow in med. oncology Boston U. Hosp., 1977-79, fellow in hematology, 1979-80; asst. prof. medicine U. Calif., Irvine, 1980—; sr. staff physician Long Beach (Calif.) VA Med. Center, 1980—; attending physician U. Calif.-Irvine Med. Center, 1981—. Contbr. articles to profl. jours. Served with Greek Armed Forces, 1970-72. Mem. AMA, Am. Soc. Clin. Oncology, ACP, Am. Soc. Internal Medicine. Greek Orthodox. Subspecialties: Oncology; Hematology. Current work: Prostate cancer chemotherapy, intra-arterial hepatic infusions of chemotherapy. Office: Long Beach VA Med Center Long Beach CA 90822

KASLICK, RALPH SIDNEY, dentist, educator; b. Bklyn., Oct. 17, 1935; s. John J. and Dorothy K.; m. Jessica Hellinger, Oct. 24, 1976. A.B., Columbia U., 1956, D.D.S., 1959, cert. in periodontology, 1962. Instr. Fairleigh Dickinson U. Sch. Dentistry, Hackensack, N.J., 1965-67, asst. prof., 1967-70, asso. prof., 1970-74, prof., 1974—, asst. dean for acad. affairs, 1973-75, acting dean, 1975-76, dean, 1976—; cons. in field. Contbr. chpts. to textbooks, articles to profl. jours. Served to capt. U.S. Army, 1962-64. Recipient Stanley S. Bergen award for contbn. to dental edn. Seton Hall U., 1982. Fellow Am. Coll. Dentists, N.Y. Acad. Dentistry; mem. Council Deans Am. Assn. Dental Schs., Internat. Assn. Dental Research (past pres. N.J. sect.), Am. Acad. Periodontology, ADA, Sigma Xi, Omicron Kappa Upsilon. Subspecialties: Periodontics; Preventive dentistry. Current work: Role of host and genetic factors in rapidly progressive periodontal diseases, especially in young adults; examination of crevicular gingival fluid and defective in vivo leukocyte migration and metabolism for diagnostic and preventive purposes. Office: 110 Fuller Pl Hackensack NJ 07601

KASS, EDWARD HAROLD, physician; b. N.Y.C., Dec. 20, 1917; s. Hyman A. and Ann (Selvansky) K.; m. Fae Golden, 1943 (dec. 1973); children: Robert, James, Nancy; m. Amalie Moses Hecht, 1975; stepchildren: Anne, Robert, Thomas, Jonathan, Peter. A.B. with high distinction, U. Ky., 1939, M.S., 1941; Ph.D., U. Wis., 1943; M.D., U. Calif., 1947; M.A. (hon.), Harvard U., 1958, D.Sc., U. Ky., 1962. Diplomate: Am. Bd. Pathology, Am. Bd. Microbiology, Am. Bd. Preventive Medicine. Grad. asst., instr. bacteriology U. Ky., 1939-41; research asst., instr. U. Wis. Med. Sch., 1941-43, immunologist dept. phys. chemistry, 1944, grad. asst. dept. pathology, 1944-45; intern Boston City Hosp., 1947-48, resident, 1948-49; research fellow Thorndike Meml. Lab., 1949-52; sr. fellow in virus diseases NRC, 1949-52; instr. in medicine Harvard Med. Sch., 1951-52, asso. in medicine, 1952-55, asst. prof. medicine, 1955-58, asso. prof. bacteriology and immunology, 1958-62, asso. prof. medicine, 1968-69; asst. physician Thorndike Meml. Lab., 1951-58; asso. dir. bacteriology Mallory Inst. Pathology, 1957-63; dir. Channing lab., dept. med. microbiology Boston City Hosp., 1963-77; dir. Channing lab. and physician Peter Bent Brigham Hosp., Boston, 1977—; prof. medicine Harvard U., 1969-73, William Ellery Channing prof. medicine, 1973—; Macy Faculty Scholar Oxford U., 1974-75; vis. prof. Hebrew U.-Hadassah Med. Sch., Jerusalem, 1974; vis. prof. community medicine St. Thomas Hosp., London, 1982-83; vis. prof. med. microbiology Royal Free Hosp., London, 1982-83; lejctr. London Sch. Hygiene and Tropical Medicine, London; cons. in field. Author 8 books; editor: Jour. Infectious Diseases, 1968-79, Revs. Infectious Diseases, 1979—; mem. editorial bds. profl. jours.; contbr. articles to med. jours. Recipient Public Service award NASA; spl. award Nat. Heart, Lung and Blood Inst.; Pioneer in Antibiotic Therapy award. Fellow Am. Coll. Epidemiology, ACP (Rosenthal award), Coll. Am. Pathologists, Am. Heart Assn., N.Y. Acad. Scis., Am. Coll. Epidemiology, Royal Soc. Medicine (London), Royal Coll. Physicians (London); fellow Infectious Diseases Soc. Am. (sec. 1962-68, pres. 1970, Bristol award 1980); mem. Internat. Epidemiol. Assn. (treas. 1977-81), Internat. Congress for Infectious Disease (pres. 1983—), Mass. Soc. Pathologists, New Eng. Soc. Pathologists, Soc. Exptl. Biology and Medicine, Am. Acad. Arts and Scis., AAAS, Am. Epidemiol. Soc., Am. Fedn. Clin. Research, Pan Am. Infectious Disease Soc., AMA, Am. Pub. Health Assn., Am. Soc. Clin. Investigation, Am. Soc. Microbiology, Soc. Epidemiol. Research, Am. Soc. Nephrology, Am. Thoracic Soc., Assn. Am. Physicians, Infectious Disease Soc. Mex. (hon.), Phi Beta Kappa, Sigma Xi, Alpha Omega

Alpha. Jewish. Club: Harvard (Boston, N.Y.C.). Subspecialties: Infectious diseases; Epidemiology. Current work: Infectious disease, epidemiology. Home: Todd Pond Rd Lincoln MA 01773 Office: 180 Longwood Ave Boston MA 02115

KASS, LAWRENCE, physician, educator, researcher; b. Toledo, Sept. 30, 1938. A.B., U. Mich.-Ann Arbor, 1960; M.S., U. Chgo, 1964, M.D., 1964. Diplomate: Am. Bd. Internal Medicine, Am. Bd. Pathology. Intern Peter Bent Brigham Hosp., Boston, 1964-65, asst. resident in internal medicine, 1965-66; sr. asst. resident Univ. Hosp. Cleve., 1966-68, demonstrator in medicine, 1967-68, fellow in hematology, 1967-68; asst. in anatomy div. biol. scis. U. Chgo., 1961-62, research assoc. medicine, 1968-70; asst. prof. internal medicine U. Mich.-Ann Arbor, 1970-73, assoc. prof., 1973—; prof. medicine and pathology Case Western Res. U., 1978—, prof. hematopathology, 1978—; dir. hematopathology Cleve. Met. Gen. Hosp., 1978—; cons. in medicine VA Hosp., Ann Arbor, 1974—; mem. research career selection rev. com. VA, Washington, 1976—; editorial cons. Williams and Wilkens Pubs., Balt., 1974—. Author: Bone Marrow Interpretation, 1973, 2d edit., 1979, Monocytes, Monocytosis and Monocytic Leukemia, 1973, Refractory Anemia, 1975, Pernicious Anemia, 1976, Preleukemic Disorders, 1979, Leukemia: Cytology and Cytochemistry, 1982; contbr. numerous articles to profl. jours. Served to maj. M.C. U.S. Army, 1968-70. Recipient Merck award, 1964; C.V. Mosby award, 1964; Internat. Giovanni Di Guglielmo prize of Giovanni DiGuglielmo Found. and Accademia Nazionale Dei Lincei, Rome, 1976. Fellow ACP, Coll. Am. Pathologists; mem. AAAS, Am. Soc. Hematology, Am. Fedn. Clin. Research, Am. Soc. Clin. Oncology, Soc. Exptl. Biology and Medicine, Central Soc. Clin. Research, Phi Beta Kappa, Phi Eta Sigma, Alpha Omega Alpha. Subspecialties: Hematology; Chemotherapy. Current work: Pathophysiology, cytology and cytochemistry of blood cells, leukemia and preleukemic disorders. Office: Case Western Res U Sch Medicine Cleve Met Hosp 3395 Scranton Rd Cleveland OH 44109

KASS, ROBERT S., physiology educator, researcher; b. N.Y.C., June 13, 1946; s. Philip and Katherine (Grossman) K.; m. Susan Scampole, Aug. 28, 1982. B.Sc., U. Ill., 1968; M.Sc., U. Mich., 1969, Ph.D., 1972. Postdoctoral fellow U. Mich. - Ann Arbor, 1972-73; research assoc. in physiology Yale U., New Haven, 1973-77; Asst. prof. physiology U. Rochester, N.Y., 1977-82, assoc. prof., 1982—. Mem. Biophys. Soc., Am. Heart Assn. (sci. council), Phi Beta Kappa. Democrat. Subspecialties: Physiology (medicine); Biophysics (physics). Current work: Understanding molecular mechanisms of ionic permeation in membranes of excitable cells. Office: Dept Physiology U Rochester 601 Elmwood Ave Rochester NY 14642

KASSAN, STUART S., rheumatologist; b. White Plains, N.Y., Nov. 19, 1946; s. Robert J. and Rosalind (Suchin) K.; m. N. Gail Karosh, Apr. 4, 1971. B.A., Case Western Res. U., 1968; M.D., George Washington U., 1972. Diplomate: Am. Bd. Internal Medicine, 1975 (subcert. in rheumatology). Intern, resident Grady Meml. Hosp., Atlanta, 1972-74; clin. assoc. NIH, Bethesda, Md., 1974-76; fellow in rheumatic diseases Hosp. for Spl. Surgery and N.Y. Hosp., Cornell Med. Ctr., N.Y.C., 1976-78; asst. clin. prof. medicine U. Colo. Health Scis. Ctr., Denver, 1978—; attending physician Colo. Gen. Hosp., Denver, 1978—; head rheumatology clinic VA Hosp., Denver, 1978-80. Contbr. chpts. to books, articles in field to profl. jours. Served as surgeon USPHS, 1976-78; NIH. N.Y. Arthritis Found. fellow, grantee, 1977; George Washington U. Med. Sch. alumni scholar, 1982. Fellow ACP; mem. Am. Fedn. Clin. Research, Harvey Soc., Am. Rheumatism Assn. Jewish. Subspecialties: Rheumatology; Internal medicine. Current work: Sjogren's syndrome, central nervous system; lupus erythematosus polymyaglia rheumatica; cyclic nucleotides in the central nervous system in systamic lupus erythematosus; clinical teaching; clinical research. Home: 8101 E Dartmouth Denver CO 80231 Office: Colo Arthritis Associates 4200 W Conejos Pl Denver CO 80204

KASSNER, JAMES LYLE, JR., physicist, researcher; b. Tuscaloosa, Ala., May 1, 1931; s. James Lyle and Esther Evelyn (Fisler) K.; m. Wanda Jane Hulsart, Aug. 11, 1956; children: James Dwight, Linda Jane, Christine Carol, Peter Colan, Keven Charles. B.S. in Chemistry, U. Ala., 1952, M.S. in Physics, 1953, Ph.D., 1957. Physicist Union Carbide, Oak Ridge, summer 1954; asst. prof. U. Mo.-Rolla, 1956-59, assoc. prof., 1959-66, prof., 1966—, research assoc., 1966-67, sr. investigator, dir. cloud physics, 1967-68, dir., 1968—; cons. McDonnel Douglas Corp., 1971-72, Dexter Hulsart Co., Inc., 1964-69; mem. adv. panel NASA, 1976; mem. nucleation com. Internat. Commn. on Cloud Physics, 1971-75; mem. Internat. Commn. on Atmospheric Electricity IV Ions, Aerosol and Radiation, 1971-75; mem. cloud physics com. Nat. Ctr. for Atmospheric Research, 1973-74; regional lectr. Sigma Xi, 1973-74. Contbr. articles to profl. jours. Scoutmaster Boy Scouts Am., 1952-64. Fellow Am. Phys. Soc.; mem. Am. Assn. Physics Tchrs., Am. Meteorol. Soc., Am. Geophys. Union. Baptist. Subspecialties: Meteorology; Surface chemistry. Current work: Atmospheric condensation, nucleation, ions, cloud physics, simulation of cloud processes. Home: Route 4 Box 127 College Hills Rolla MO 65401 Office: U MO 109 Norwood Rolla MO 65401

KASTIN, ABBA JEREMIAH, physician, researcher, educator; b. Cleve., Dec. 24, 1934; s. Isadore I. and Ruth (Urdang) K. A.B., Harvard U., 1956; M.D., 1960; M.D. hon. doctorate, Universidad Nacional Fed. Vil. Lima, Peru, 1980. Intern Vanderbilt U. Hosp., 1961-61, resident, 1961-62; clin. assoc. NIH, Bethesda, Md., 1962-64; clin. investigator VA Med. Ctr., New Orleans, 1965-68, chief endocirnology, 1968—; asst. prof. medicine Tulane U., 1966-71, assoc. prof., 1971-74, prof., 1974—; mem. grad. faculty U. New Orleans, 1976—. Contbr. numerous articles, chpts. to profl. publs.; Viola player, Civic Symphony Orch. New Orleans, 1968—. Trustee Jewish Fedn. Greater New Orleans, 1982-85. Recipient Edward T. tyler Fertility award Internt. Soc. Fertility, 1975, Copernicus medal Med. Faculty Cracow, poland, 1979, William S. Middleton award VA, 1982; named to Disting. Alumni Hall of Fame Cleve. Heights High Sch., 1981. Hon. mem. Civilian Soc. Endorcrinilogy, Phillippine Soc. Endocrinology and Metabolishm, Peruvian Ob-Gyn. Soc., Peruvian Endocrine Soc., Polish Endocrine Soc. Subspecialties: Neuroendocrinology; Neurophysiology. Current work: Investigation of brain peptides, including their natural occurrence, alterations under physiologic and pharmacologic manipulation, and effects on central nervous system. Office: VA Med Ctr-Tulane U Sch Medicine 1601 Perdido St New Orleans LA 70146

KASTL, ALBERT JOSEPH, psychologist; b. N.Y.C., June 25, 1939; s. Albert and Emmy (Kannengeisser) K.; m. Donna Old, June 9, 1977; children: Alison Joy. B.A., CUNY, 1961; M.A., Yale U., 1963, Ph.D., 1965. Lic. psychologist, Calif. Staff psychologist Pacific Med. Ctr., San Francisco, 1968-70; pvt. practice psychology, Santa Rosa, Calif., 1972—; cons. N. Bay Regional Ctr., Santa Rosa, 1974—. Author: Journey Back-Escaping the Drug Trap, 1975. Served to capt. U.S. Army, 1965-67. NSF fellow, 1961-62, 62-63; VA trainee, 1963-64, 64-65; NIMH fellow, 1966-67. Mem. Am. Psychol. Assn., Redwood Psychol. Soc. (pres. 1978-80). Democrat. Mem. Assembly of God Ch. Current work: Learning disability, epilepsy, rehabilitation with brain injured individuals, forensic psychology. Home: 1650 Timber Hill Rd Santa Rosa CA 95401 Office: 114 Sotoyome St Santa Rosa CA 95405

KASTL, PETER ROBERT, ophthalmologist, educator; b. Alexandria, La., July 25, 1949; s. William H. and Elizabeth (Carlton) K.; m. Susan Glanville, Dec. 31, 1974; children: Rebecca Ruth, Rachel Elizabeth. B.S., Centenary Coll., 1971; M.D., Tulane U., 1974, Ph.D., 1978. Lic. physician, La., diplomate: Am. Bd. Ophthalmology. Instr. Tulane U., New Orleans, 1980-81, asst. prof., 1981—, head contact lens sect., 1981—. Contbr. articles to sci. jours. NIH fellow, 1975; grantee, 1980-83. Felow Am. Acad. Ophthalmology; mem. Contact Lens Assn. Ophthalmologists, AMA, La. State Med. Soc., Orleans Parish Med. Soc., La. Ophthalmology Assn. Subspecialties: Ophthalmology; Biochemistry (medicine). Current work: Tear analysis, corneal transplantation, corneal metabolism. Home: 3023 Metairie Ct Metairie LA 70002 Office: 1430 Tulane Ave New Orleans LA 70112

KASTOR, JOHN ALFRED, cardiologist, educator; b. N.Y.C., Sept. 15, 1931; s. Alfred Bernard and Ellen Voigt Bentley; m. Mae Belle Eisenberg, July 4, 1954; children—Elizabeth Mae, Anne Sarah, Peter John. B.A., U. Pa., 1953; M.D., N.Y. U., 1962. With NBC, N.Y.C., 1956-58; intern, asst. resident in medicine Bellevue Hosp., N.Y.C., 1962-64; chief resident physician N.Y. U. Hosp., N.Y.C., 1964-65; clin. and research fellow in medicine Mass. Gen. Hosp., Boston, 1965-68, clin. asst. and asst. in medicine, 1968-69; instr. in medicine Harvard Med. Sch., 1968-69; dir. med. intensive care unit Hosp. U. Pa., Phila., 1969-72, asso. chief cardiovascular sect., 1972-77, chief, 1977-81; prof. medicine U. Pa. Sch. Medicine, Phila., 1976—. Editor: Internat. Jour. Cardiology; Contbr. numerous articles on cardiac electrophysiology and gen. cardiology to med. jour. Served with U.S. Army, 1953-55. Fellow Am. Coll. Cardiology, A.C.P., Council Clin. Cardiology Am. Heart Assn., Coll. Physicians Phila.; mem. Am. Fedn. Clin. Research, Am. Heart Assn. (gov. Southeastern Pa. chpt. 1975-81), Assn. Univ. Cardiologists, Venezuelan Soc. Internal Medicine, Alpha Omega Alpha. Subspecialties: Cardiology; Internal medicine. Current work: Clinical cardiac electrophysiology and electrocardiography. Home: 1001 Westview St Philadelphia PA 19119 Office: 3400 Spruce St Philadelphia PA 19104

KASVINSKY, PETER JOHN, biochemistry educator; b. Bridgeport, Conn., Dec. 7, 1942; s. Joseph Stephen and Irene Jenny (Kedves) K.; m. Elaine Joyce Amormino, Apr. 5, 1974. B.S., Bucknell U., 1964; Ph.D., U. Vt., 1970. Instr. biochemistry Wayne State U. Sch. Medicine, Detroit, 1972-74, U. Alta. (Can.), Edmonton, 1977-79, profl. asst. biochemistry, 1974-79; asst. prof. biochemistry Marshall U. Sch. Medicine, Huntington, W.Va., 1979-82, assoc. prof., 1982—; mem. grad. faulty W.Va. U., Morgantown, 1980—. Contbr. articles in biochemistry to profl. jours. Mem. Huntington Galleries, 1980—. NIH grantee, 1981. Mem. Am. Chem. Soc., AAAS, Can. Biochem. Soc., N.Y. Acad. Scis., Sigma Xi. Democrat. Roman Catholic. Subspecialties: Biochemistry (medicine); Biochemistry (biology). Current work: Regulation of enzymes of glycogen metabolism; enzymology of phosphorylation, dephosphorylation of proteins; structure, function relationships in proteins. Office: Marshall U Sch Medicine Dept Biochemistry Huntington WV 25704

KATHREN, RONALD LAWRENCE, health physicist, educator; b. Windsor, Ont., Can., June 6, 1937; s. Ben and Sally (Forman) K.; m. Susan Krafft, Dec. 24, 1964; children: SallyBeth, Daniel. B.S., UCLA, 1957; M.S., U. Pitts., 1962. Diplomate: Am. Bd. Health Physics; registered profl. engr., Calif., Am. Acad. Environ. Engrs. Supervisory health physicst Mare Island Naval Shipyard, Vallejo, Calif., 1959-61; health physicist U. Calif. Radiation Lab., Livermore, 1962-67; sr. research scientist, sect. mgr. Battelle Pacific N.W. Labs., Richland, Wash., 1967-72; staff scientist, 1978—; corp. health physicist Portland Gen. Electric Co., Oreg., 1972-78; asst. prof. and program coordinator radiol. scis. Joint Center for Grad. Study, U. Wash., Richland, 1978—; health physicist Reed Coll., 1973-78; cons. in field; adj. prof. Oreg. State Div. of Continuing Edn., 1972-77; Mem. Oreg. State Radiation adv. com., 1977-78; mem. noise adv. com. Oreg. Dept. Environ. Quality, 1976. Editor: Health Physics Jour, 1975-80; adv. editorial bd.: Handbook of Radiation Protection and Measurement, 1978—; Contbr. articles to profl. jours.; author: Reactor Safety, 1976, Reactor Physics, 1977, Radiation Protection for Technicians, 1978, Health Physics: A Backward Glance, 1980, Radiobiology in Medicine, 1982. Recipient Elda Anderson award Health Physics Soc., 1977; USPHS fellow, 1961-62. Mem. Health Physics Soc., Am. Acad. Environ. Engrs., AAAS, Am. Assn. Physicists in Medicine, Am. Nuclear Soc. Subspecialties: Health physics; Environmental engineering. Current work: Research in health physics, radiological measurement, history of science. Office: Battelle Blvd Richland WA 99352

KATONA, PETER GEZA, biomedical engineer, educator; b. Budapest, Hungary, June 25, 1937; came to U.S., 1956, naturalized, 1962; s. Stephan and Irene (Renner) K.; m. Jaroslava Blanar, Aug. 27, 1966; children—Catherine Iris, Andrew George. B.S. in Elec. Engring, U. Mich., 1960; S.M. in Elec. Engring. (Sloan fellow, 1960-62), M.I.T., 1962, Sc.D., 1965. Asst. prof. elec. engring. M.I.T., 1965-69; assoc. prof. biomed. engring. Case Western Res. U., Cleve., 1969-78, prof., 1978—, chmn. dept., 1980—. Editorial bd.: American Jour. Physiology, 1975-81; contbr. articles on cardio-respiratory control to profl. jours. Mem. Am. Physiol. Soc., Biomed. Engring. Soc. (bd. dirs. 1977-80, pres. elect 1983-84), IEEE, Am. Soc. Engring. Edn. Subspecialty: Biomedical engineering. Home: 2886 Courtland Blvd Shaker Heights OH 44122 Office: Biomed Engring Dept Case Western Res Univ Cleveland OH 44106

KATSANIS, THEODORE, aerospace research engrineer; b. North Weymouth, Mass., July 17, 1925; s. Nichos I. and Viola Mae (Plummer) K.; m. Pauline L. Hackett, Sept. 27, 1952; children: Linda, Kimberly, Jason. B.S., St. Louis U., 1948; M.S., U. Wash.-Seattle, 1962; Ph.D., Case Inst. Tech., 1967. Design engr. Boeing Co., Seattle, 1948-52; design engr. Smith-Berger, Seattle, 1952-62; research engr. NASA, Cleve., 1963—. Author computer codes. Recipient Manly Meml. award Soc. Automotive Engrs., 1974. Assoc. fellow AIAA; mem. ASME, Sigma Xi. Presbyterian. Club: Noffa (Cleve.) (pres. 1980). Subspecialties: Aerospace engineering and technology; Fluid mechanics. Home: 16495 Parklawn Ave Middleburg Heights OH 44130 Office: NASA Lewis Research Ctr 21000 Brookpark Rd Cleveland OH 44135

KATZ, DAVID HARVEY, immunologist, research institute executive; b. Richmond, Va., Feb. 17, 1943; m., 1963. A.B., U. Va., 1963; M.D., Duke U., 1968. Med. house officer Johns Hopkins Hosp., 1968-69; staff assoc. in immunology NIH, 1969-71; instr. to assoc. prof. immunology and pathology Narvard Med. Sch., 1971-76; chmn. and mem. immunology staff Scripps Clin. and Research Found., 1976-81; chief exec. officer and pres. Quidel, 1981—; pres. and dir. Med. Biol. Inst., La Jolla, Calif., 1981—; mem. adv. com. cancer ctrs. Nat. Cancer Inst., 1972-74; mem. allergy and immunol. study sect. NIH, 1977; mem. human cell biol. adv. panel NSF, 1977-78. Mem. Am. Assn. Immunologists, Am. Soc. Clin. Investigation, AAAS, Am. Assn. Pathologists, Am. Fedn. Clin. Research. Subspecialty: Immunology (medicine). Office: Dept Immunol Med Biol Inst 11077 N Torrey Pines Rd La Jolla CA 92037

KATZ, DONALD L., chemical engineering consultant, author; b. nr. Jackson, Mich., Aug. 1, 1907; m. L. Maxine Crull, 1932 (dec. Mar. 1965); children: Marvin L., Linda M. Katz Cantrell; m. Elizabeth Harwood Correll, Nov. 26, 1965. B.S. in Chem. Engring, U. Mich., 1931, Ph.D., 1933. Research engr. Phillips Petroleum Co., Bartlesville, Okla., 1933-36; mem. faculty U. Mich., Ann Arbor, 1936-77, prof. former chmn. dept. chem. and metall. engring., Alfred Holmes White Univ. prof. chem. engring., from 1966, prof. chem. engring. emeritus, 1977—; cons. engr. on petroleum tech. and underground storage of natural gas to numerous cos. and govtl. orgns. organizer adv. com. on hazardous materials for U.S. Coast Guard, Nat. Acad. Sci., 1964, chmn., 1964-72; mem. task force on info. system World Energy Conf., 1970-71; mem. sci. adv. com. U.S. Coast Guard, 1972-75; chmn. new com. on hazardous material for U.S. Coast Guard, Nat. Acad. Sci.-Nat. Acad. Engring., 1977-79; chmn. com. on air quality and power plant emissions Nat. Acad. Engring.-Nat. Acad. Sci.-NRC, 1974-75; mem. tech. assessment of pollution control adv. com. EPA, 1976-79; lectr. South China Inst. Tech., Ckekeang U., 1982. Co-author: Compressed Air Storage; author: The Settling of Waterloo-Schnackenberg-Katz Families (award of merit Mich. Hist. Soc. 1979); contbr. over 285 articles on heat transfer, fluid dynamics and use of computers in engring. edn. to profl. jours. Former mem., pres. Ann Arbor Bd. Edn.; pres. Ann Arbor Council of Chs., 1944-45; former chmn. ofcl. bd., lay leader 1st United Meth. Ch., Ann Arbor. Recipient Hanlon award Natural Gasoline Assn. Am., 1950, Disting. Faculty Achievement award U. Mich., 1964, Disting. Pub. Service award U.S. Coast Guard, 1972, Gas Industry Research award Am. Gas Assn., 1977, Nat. Medal of Science, 1983; named Mich. Engr. of Year, 1959; Donald L. Katz lectureship in chem. engring. established at U. Mich., 1971. Fellow Am. Inst. Chem. Engrs. (former mem. council, pres. 1959, Founders award 1964, Warren K. Lewis award 1967, Walker award 1968, Anthony F. Lucas god medal 1979), AAAS; mem. Am. Chem. Soc. (E.V. Murphree award indsl. and engring. chemistry div. 1975), ASME, AIME (Mineral Industry's Edn. award 1970), Soc. Petroleum Engrs. of AIME (disting. lectr. 1961-62, John Franklin Carll award 1964), Am. Soc. for Engring. Edn., Am. Assn. Petroleum Geologists, Nat. Acad. Engring., Phi Lambda Upsilon (hon.). Subspecialty: Chemical engineering. Home: 2011 Washtenaw Ann Arbor MI 48104 Office: Dow Bldg Dept Chem Engring U Mich Ann Arbor MI 48109

KATZ, ELI JOEL, researcher; b. Bklyn., Jan. 12, 1937; s. William Harry and Helen (Green) K.; m. Barbara S. Blackman, May 2, 1938; children: Aviva, Judah, Hillel. B.S.M.E., Bklyn. Poly. Inst., 1957; M.S., Pa. State U., 1959; Ph.D., Johns Hopkins U., 1962. Lectr. Hebrew U., Jerusalem, 1963-65; research specialist Electric Boat div. Gen. Dynamics Corp., 1965-66; asst. scientist Woods Hole Oceanographic Instn., 1966-69; assoc. scientist, 1970-78; sr. lectr. Tel Aviv U., 1969-70; sr. research assoc. Lamont-Doherty Geol. Obs., Columbia U., 1979—; cons. Exxon, Israel Oceaographic and Limnological Co. Contbr. articles to profl. jours. NSF grantee, 1972, 73, 74, 78, 80, 81, 82. Subspecialty: Oceanography. Current work: Circulation of the Atlantic Ocean. Study of the response of the tropical Atlantic to the annual wind stress cycle. Office: Lamont-Doherty Geol Obs Palisades NY 10964

KATZ, FRANCES R., corn processing company executive; b. LeRoy, Ill., Aug. 16, 1937; d. Raymond C. and Emma L. (Scott) Rippy; m. Terence Belshaw, Aug. 1959; 1 son, Andrew; m. Allan S. Katz, May 2, 1982. B.S., Ind. Central U., 1960; M.B.A., U. Chgo., 1980. Food technologist Durkee Foods, Chgo., 1960-68; sect. leader CFS Continental, Chgo., 1968-72; v.p. research corn processing div. Am. Maize Products Co., Hammond, Ind., 1978—. Editor, Putman Publishing Co., Chgo., 1972-78; Exec. editor: Am. Assn. Cereal Chemists. Mem. Inst. Food Technologists, Am. Chem. Soc., Research Dirs. Chgo. Subspecialties: Carbohydrate chemistry; Food science and technology. Current work: Direct research on carbohydrates, chemicals from corn. Office: 1110 Indianapolis Blvd Hammond IN 46206

KATZ, HYMAN BERNARD, mfg. co. exec.; b. Czenstochowa, Poland, Apr. 13, 1945; came to U.S., 1946, naturalized, 1952; s. Israel and Sally (Hertzberg) K.; m. Rose Gisele Gastwirth, Nov. 22, 1969; children: Samuel, Sharon. B.Sc., Poly. Inst. N.Y., 1966; M.Sc., Purdue U., 1968; Ph.D. in Chemistry, Brandeis U., 1973. Group leader quality control Damon corp./Ortho Diagnostics Instruments subs. Johnson & Johnson, Westwood, Mass., 1973-74; quality assurance mgr. Pall Biomed. Products Corp., Glen Cove, N.Y., 1975-76, mgr. quality assurance and regulatory affairs, 1976-82, corp. dir. quality assurance and regulatory affairs, 1982—. Contbr. articles to profl. jours. Recipient Interchem. Found. award, 1965-66; USPHS research grantee, 1971-72. Mem. Am. Soc. for Quality Control (dep. chmn. gen. tech. council, McDermond award 1978), Assn. for Advancement of Med. Instrumentation (standards council), Parenteral Drug Assn. (regulatory affairs com.), Cosmetic Toiletry and Fragrance Assn. (quality assurance com.), Am. Chem. Soc., Health Industry Mfrs. Assn. Subspecialties: Analytical chemistry; Industrial engineering. Current work: Contamination control, particulate analysis, fluid clarification, sterile products, statistical sampling, operations research, information systems. Office: 30 Sea Cliff Ave Glen Cove NY 11542

KATZ, J. LAWRENCE, educator, biophysicist, biomedical engineer; b. Bklyn., Dec. 18, 1927; s. Frank and Rose (Eidenberg) K.; m. Gertrude Seidman, June 17, 1950; children: Robyn Laurie, Andrea Lee, Talbot Michael. B.S. in Physics, Poly. Inst. Bklyn., 1950, M.S., 1951, Ph.D., 1957. Teaching fellow physics, research fellow, instr. math. Poly. Inst. Bklyn., 1950-56; mem. faculty Rensselaer Poly. Inst., Troy, N.Y., 1956—, prof. physics, 1967—, prof. biophysics and biomed. engring., 1971—, chmn. dept. biomed. engring., 1982, dir. Center for Biomed. Engring., 1974—; summer research asso. Wright Aero. Co., 1956; Summer research asso. Knolls Atomic Power Lab., Schenectady, 1957; hon. research asst. crystallography Univ. Coll., London, 1959-60; vis. prof. biomed. engring., oral biology U. Miami, Fla., 1969-70; vis. prof. biomechanics Chendgu U. Sci. and Tech., China; vis. scientist program Am. Assn. Physics Tchrs.-Am. Inst. Physics, 1970-72; vis. prof. orthopaedics and rehab. U. Miami, summer 1974; cons. in field, 1950—; dir. Bioanalytical Labs., Inc., Troy; adj. prof. orthopaedics Albany Med. Coll., 1972-77, prof. surgery, 1977—; vis. prof. dept. orthopedic surgery Harvard U. Med. Sch., 1978; vis. biophysicist Children's Hosp., Boston, 1978; E. Leon Watkins vis. prof. Wichita (Kans.) State U., 1978; vis. prof. biophysics and biomed. engring. Inst. de Fisica e Quimica de São Carlos, U. São Paulo, Nov. 1978; mem. engring. biology and medicine tng. com. NIH, 1968-71; mem. U.S. Standards Inst. Com. N44; chmn. subcom. diagnostic radiology; mem. VA sci. rev. and evaluation bd. for rehab. engring. research and devel., 1981-83. Editor: (with Robert Plonsey) Marcel Dekker Series on Biomedical Engineering and Instrumentation; contbr. papers to profl. lit., chpts. to books. Mem. organizing com. Black Arts Council, 1969-70; mem. exec. com. Schenectady County Liberal party, 1963—; chmn. 4th jud. dist. nominating conv. Liberal party, 1967-68; Liberal party candidate for U.S. Congress, 1968; committeeman Liberal party, 1968-71; asst. chmn. Schenectady County Liberal party, 1969-71; nat. bd. dirs. Ams. for Democratic Action, 1977—, N.Y. state bd. dirs., 1978—; sponsor tri-city div. United Negro Coll. Fund, 1967-68; mem. Schenectady Light Opera Co., 1964—. Served with USNR, 1946-48. NSF sci. faculty fellow, 1959-60; Guggenheim fellow, 1977-78. Fellow Am. Phys. Soc.; mem. Am. Crystallographic Assn. (chmn. crystal data com.), AIME (chmn. dental med. tech. com.), AAUP (pres. Rensselaer chpt. 1974-75), Biophys. Soc., Orthopaedic Research Soc. (mem. program com. 1973-75, chmn. 1975, exec. com. 1975), Internat. Assn. Dental Research, Am. Soc. Engring. Edn. (chmn. biomed. engring. div. 1978-79, chmn.

elect and program chmn. 1977—), Soc. Biomaterials (v.p. 1977-78, pres. 1978-79), Biomed. Engring. Soc. (dir. 1981-84, pres. elect 1982-83, pres. 1983-84), ASTM (chmn. composites subcom. of com. med. implants and devices), IEEE (chmn. composites subcom. com. med. implants and devices), Fedn. Am. Scientists, Sigma Xi, Sigma Pi Sigma. Jewish (trustee temple 1962-63, 63-64, mem. social action com. 1968-69). Subspecialties: Biomedical engineering; Biophysics (physics). Patentee pretensioned prosthetic device for skeletal joints. Home: 838 Maxwell Dr Schenectady NY 12309 Office: Rensselaer Poly Inst Troy NY 12181

KATZ, JAY, physician, educator; b. Zwickau, Germany, Oct. 20, 1922; came to U.S., 1940, naturalized, 1945; s. Paul and Dora (Ungar) K.; m. Esta Mae Zorn, Sept. 13, 1952; children: Sally Jean, Daniel Franklin, Amy Susan. B.A., U. Vt., 1944; M.D., Harvard U., 1949. Intern Mt. Sinai Hosp., N.Y.C., 1949-50; resident Northport (N.Y.) VA Hosp., 1950-51, Yale U., 1953-55, instr. psychiatry, New Haven, 1955-57, asst. prof., 1957-58, asst. prof. psychiatry and law, 1958-60, asso. prof. law, asso. clin. prof. psychiatry, 1960-67, adj. prof. law and psychiatry, 1967-79, prof., 1979—; tng. and supervising psychiatrist Western New Eng. Inst. for Psychoanalysis, 1972—; cons. to asst. sec. health and sci. affairs HEW, 1972-73, mem. artificial heart assessment panel, 1972-73. Author: (with Joseph Goldstein) The Family and the Law, 1964, (with Joseph Goldstein and Alan M. Dershowitz) Psychoanalysis, Psychiatry and Law, 1967, Experimentation with Human Beings, 1972, (with Alexander M. Capron) Catastrophic Diseases—Who Decides What?, 1975. Bd. dirs. Family Service of New Haven. Served to capt. M.C. USAF, 1951-53. John Simon Guggenheim Meml. Found. fellow, 1981. Fellow ACP (William C. Menninger award 1983), Am. Psychiat. Assn. (Isaac Ray award 1975), Am. Orthopsychiat. Assn., Am. Coll. Psychiatry, Center for Advanced Psychoanalytic Studies; mem. Inst. Medicine, Nat. Acad. of Scis., Group for Advancement of Psychiatry, Am. Psychoanalytic Assn. Jewish. Subspecialty: Psychiatry. Home: 27 Inwood Rd Woodbridge CT 06525 Office: 127 Wall St New Haven CT 06520

KATZ, JEREMY MILTON, research psychologist; b. Mpls., Dec. 2, 1942; s. Manual and Ruth Leah (Benjamin) K.; m. Robin G. Guern, Dec. 17, 1969 (div. 1979); children: Stephanie Susanna, Philip Benjamin, Wendy Louise; m. Debra A. Swailes, Mar. 1, 1980; 2 daus., Rachel Lynn, Jennifer Doreen. B.A., Carleton Coll., 1964; Ph.D., U. Mich., 1973. Research psychologist Milw. County Mental Health Complex, 1973—; asst. prof. psychiatry Med. Coll. of Wis., Milw., 1978—; consortium dir. Internat. Consortium for the Study of Neurol. and Psychol. Reactions to Cardiac Surgery, Milw., 1981—; symposium dir. 2d Internat. Symposium on Psychopath. and Neurol. Dysfunctions following open heart surgery, 1979-80. Editor: Psychopathological and Neurological Dysfunction Following Open Heart Surgery, 1982. Mem. Am. Psychol. Assn. Jewish. Current work: Development of an international organization and multicenter international study on neurological and psychological reactions to cardiac surgery. Home: 11934 W North Ave #2 Milwaukee WI 53226 Office: Milw County Mental Health Complex 9191 Watertown Plank Rd Milwaukee WI 53226

KATZ, JONATHAN ISAAC, educator, cons.; b. N.Y.C., Jan. 5, 1951; s. Sol and Rebecca Leah (Samuels) K.; m. Lilly Margret Canel, Mar. 15, 1982. A.B., Cornell U., 1970, A.M., 1971, Ph.D., 1973. Mem. Inst. Advanced Study, Princeton, N.J., 1973-76; assoc. prof. UCLA, 1976-81, Washington U., St. Louis, 1981—; cons. in field. Contbr. articles to profl. jours. Grantee NSF, NASA, Sloan Found. Mem. Am. Phys. Soc., Am. Astron. Soc. Subspecialties: Theoretical astrophysics; High energy astrophysics. Current work: Compact objects, accretion, binary stars, hardened structures. Office: Dept Physics Washington U Saint Louis MO 63130

KATZ, JOSEPH JACOB, chemist, educator; b. Apr. 19, 1912; s. Abraham and Stella (Asnin) K.; m. Celia S. Weiner, Oct. 1, 1944; children—Anna, Elizabeth, Mary, Abram. B.Sc., Wayne U., 1932; Ph.D., U. Chgo., 1942. Research asso. chemistry U. Chgo., 1942-43, asso. chemist metall. lab., 1943-45; sr. chemist Argonne (Ill.) Nat. Lab., 1945—; Tech. adviser U.S. delegation UN Conf. on Peaceful Uses Atomic Energy, Geneva, Switzerland, 1955; chmn. AAAS Gordon Research Conf. on Inorganic Chemistry, 1953-54. Am. editor: Jour. Inorganic and Nuclear Chemistry, 1955—. Recipient Distinguished Alumnus award Wayne U., 1955, Profl. Achievement award U. Chgo. Alumni Assn.; Guggenheim fellow, 1956-57. Mem. Am. Chem. Soc. (award for nuclear applications in chemistry 1961, sec.-treas. div. phys. chemistry 1966-76), Nat. Acad. Scis., Am. Nuclear Soc., Phi Beta Kappa, Sigma Xi. Subspecialty: Physical chemistry. Current work: Chorophyll function in photosynthesis; laser photochemistry of chlorophyll; Synthesis of Clorophyll model systems. Home: 1700 E 56th St Chicago IL 60637 Office: 9700 S Cass Ave Argonne IL 60439

KATZ, JULIAN, gastroenterologist, educator; b. N.Y.C., Apr. 3, 1937; s. Abraham M. and Fay (Sher) K.; m. Sheila Moriber, Aug. 18, 1963; children—Jonathan Peter, Sara Katherine. A.B., Columbia U., 1958; M.D., U. Chgo., 1962. Diplomate: Am. Bd. Internal Medicine. Intern U. Chgo. Hosps., 1962-63; resident in medicine Duke U., 1963-65; fellow in gastroenterology Yale U., 1965-67; practice medicine specializing in gastroenterology, internal medicine, Phila., 1969—; prof. medicine, lectr. in physiology and biochemistry Med. Coll. Pa., 1970—, also lectr. local and nat. groups; chief clin. gastroenterology Med. Coll. Pa. Editor profl. jours.; Contbr. articles to profl. jours. and books. Served with USN, 1967-69. Fellow ACP; mem. Am. Soc. Gastrointestinal Endoscopy, Am. Soc. Study Liver Disease, Am. Gastroenterological Assn., others. Subspecialty: Gastroenterology. Current work: Immunologic aspects of gastroenterology, research in nutrition and endoscopy clinical pharmacology. Home: 701 Dodds Ln Gladwyne PA 19035 Office: Gastrointestinal Specialists 555 City Ave Bala Cynwyd PA 19004

KATZ, MICHAEL, pediatrician, educator; b. Lwow, Poland, Feb. 13, 1928; came to U.S., 1946, naturalized, 1951; s. Edward and Rita (Gluzman) K. A.B., U. Pa., 1949, postgrad. (Harrison fellow), 1950-51; M.D., SUNY, Bklyn., 1956; M.S., Columbia U. Sch. Public Health, 1968. Intern UCLA Med. Center, 1956-57; resident Presbyterian Hosp. (Babies Hosp.), N.Y.C., 1960-62, dir. pediatric service, 1977—; hon. lectr. in pediatrics Makerere U. Coll., Kampala, Uganda, 1963-64; instr. in pediatrics Columbia U., 1964-65; prof. tropical medicine Sch. Public Health, 1971—; prof. pediatrics Coll. Physicians and Surgeons, 1972-77, Reuben S. Carpentier prof., chmn. dept. pediatrics, 1977—; asso. mem. Wistar Inst., Phila., 1965-71; asst. prof. pediatrics U. Pa., 1966-71; cons. WHO Regional Offices, Guatemala, Venezuela, Egypt, Yemen; mem. U.S. del. 32d World Health Assembly, Geneva, 1979; cons. UNICEF, N.Y.C. and Tokyo., USAID, Egypt, 1982. Contbr. articles to profl. jours.; author: (with others) Parasitic Diseases, 1982; editor: (with Volker ter Meulen) Slow Virus Infections of the Central Nervous System, 1977, Med. Microbiology and Immunology, 1975—; editorial bd.: Pediatric Infectious Diseases, 1981—; also co-editor manuals. Served to lt. M.C. USNR, 1957-59. NIH grantee, 1968-76; WHO grantee, 1972-76. Fellow Infectious Diseases Soc. Am., AAAS, Am. Acad. Pediatrics; mem. Soc. Pediatric Research, Am. Pediatric Soc., Harvey Soc., Am. Soc. Microbiology, Deutsche Gesellschaft fur Neuropathologie und Neuroanatomie E.V. (corr.), Am. Soc. Tropical Medicine and Hygiene, N.Y. Soc. Tropical Medicine (pres. 1976-77), Royal Soc. Tropical Medicine and Hygiene (London), Inst. Medicine of Nat. Acad. Scis., Sigma Xi. Subspecialties: Pediatrics; Tropical medicine. Current work: Malnutrition-infection complex; latent viral infection. Home: 930 Fifth Ave New York NY 10021 Office: Coll Physicians and Surgeons Columbia U 630 W 168th St New York NY 10032

KATZ, MURRAY ALAN, medical educator, researcher; b. Albuquerque, June 15, 1941; s. David Nathan and Frances Kathryn (Segal) K.; m. Sali, Apr. 24, 1964; children: Mason Jeremy, Aaron Seth. B.A., Johns Hopkins U., 1963, M.D., 1966. Diplomate: Nat. Bd. Med. Examiners, Am. Bd. Internal Medicine. Intern Johns Hopkins Osler Med. Service, Balt., 1966-67; resident Johns Hopkins Hosp., 1967-68, U. Tex. S.W. Med. Sch., Dallas, 1968-70; asst. prof. internal medicine Temple U. Sch. Medicine, Phila., 1971-74; staff physician and nephrologist U. Ariz. and Tucson VA Med. Ctr., 1974—; prof. internal medicine U. Ariz. Health Sci. Ctr., Tucson, 1981—; assoc. chief staff for research VA Med. Ctr., Tucson, 1982—. Author: Calculus for the Life Sciences, 1976; contbr. articles to med. jours.; assoc. editor: Microvascular Research, Boston, 1982—; mem. sci. adv. bd.: Western Jour. Medicine, Seattle, 1983—. Bd. dirs. Dialysis Found. of So. Ariz., Tucson, 1976, Tucson Hebrew Acad., 1978. NIH grantee, 1971-74; merit review VA Med. Ctr., 1974—; clin. investigator, 1976-79. Fellow ACP; mem. Am. Physiol. Soc., Am. Soc. Nephrology, Microcirculatory Soc. (program com. 1982-83), Western Soc. Clin. Investigation (sec.-treas 1983-86), Am. Fedn. Clin. Research (pres. Western sect. 1980). Jewish. Subspecialties: Physiology (biology); Nephrology. Current work: Microcirculation, exchange of water and solute across capillaries. Office: Research Service Tucson VA Med Ctr Tucson AZ 85723

KATZ, PHYLLIS ALBERTS, psychologist; b. Apr. 9, 1938. A.B. summa cum laude in Psychology, Syracuse U., 1957; Ph.D. in Development and Clin. Psychology, Yale U., 1961. Clin. intern West Haven (Conn.) VA Hosp., 1959-60; clin. trainee Clifford Beers Child Guidance Clinic, New Haven, 1960-61; instr. psychology So. Conn. State Coll., 1960-61, Queens Coll., N.Y.C., 1962-63; asst. prof. psychology NYU, 1963-67, assoc. prof., 1967-69, CUNY, 1969-72, chmn. developmental psychology sect., 1969-75, prof., 1973-76; dir. Inst. Research on Social Problems, Boulder, Colo., 1975—; vis. research assoc. U. Colo., Boulder, 1975-76; research cons. Behavioral Research Inst., Boulder, 1975-76; cons. NIH, NIMH, Pa. Dept. Edn., TV stas. in N.Y. and Calif., Childpark, Inc., HEW, others. Guest cons. editor: Psychol. Reports, Perceptual and Motor Skills; editor: Sex Roles: A Jour. of Research, 1976—; developmental psychology sect. Jour. Supplement Abstract Service, Am. Psychol. Assn., 1974-77; mem. editorial bd.: Child Devel, 1975-77, Archives Sexual Behavior, 1975—; contbr. articles to profl. jours., chpts. to books. USPHS trainee, 1956-59; Boie fellow, 1960-61; NYU grantee, 1963-66; grantee CUNY, Nat. Inst. Child Health and Human Devel., NIMH. Fellow Am. Psychol. Assn.; mem. Am. Ednl. Research Assn. (program com. 1974), Soc. Research in Child Devel., Soc. Psychol. Study of Social Issues (council 1979—), Southeastern Psychol. Assn., AAUP, AAAS, Assn. Women in Sci., Sigma Xi. Subspecialty: Developmental psychology. Address: 1035 Pearl St 5th Floor Boulder CO 80302

KATZ, THOMAS JOSEPH, chemistry educator; b. Prague, Czechoslovakia, Mar, 21, 1936; s. Francis and Ida (Jungmann) K.; m. Meta Oehmsen, Dec. 27, 1963; 1 son, Joshua. B.A., U. Wis., 1956; M.A., Harvard U., 1957, Ph.D., 1959. Instr. Columbia U., N.Y.C., 1959-61, asst. prof., 1961-64, assoc. prof., 1964-68, prof., 1968—. Contbr. numerous articles on chemistry to profl. jours. Alfred P. Sloan Found. fellow, 1962-66; John Simon Guggenheim Meml. Found. fellow, 1967-68. Mem. Am. Chem. Soc., Royal Soc. of Chemistry (U.K.). Subspecialties: Organic chemistry; Catalysis chemistry. Current work: Organometallic chemistry, catalysis by metals, organic synthesis,organometallic chemistry. Home: 445 Riverside Dr Apt 82 New York NY 10027 Office: Dept Chemistry Columbia U New York NY 10027

KATZ, WILLIAM, materials scientist, educator; b. Dayton, Ohio, Dec. 10, 1953; s. Murray Louis and Anita Jean (Feller) K.; m. Patricia Valerie Powers, Aug. 12,1979. B.A., Earlham Coll., 1975; M.S. in Analytical Chemistry, U. Ill, Urbana, 1977, Ph.D., 1979. Research chemist Exxon Research and Devel. Lab., Baton Rouge, 1979-80; materials scientist Gen. Electric Corp. Research and Devel. Center, Schenectady, 1980—; adj. prof. SUNY, Albany. Contbr. numerous articles to profl. jours. Mem. Am. Inst. Physics, Am. Chem. Soc., Am. Vacuum Soc., Phi Lambda Upsilon. Subspecialties: Electronic materials; Surface chemistry. Current work: Materials characterization, surface analysis, sputtering and ionization phenomenon. Home: 46 Fredericks Rd Scotia NY 12302 Office: Gen Electric Co Bldg K-1 Room 2C27 PO Box 8 Schenectady NY 12301

KATZENELLENBOGEN, BENITA S(CHULMAN), physiologist, educator; b. N.Y.C., Apr. 11, 1945; d. Max and Miriam (Grossman) Schulman; m. John Albert Katzenellenbogen, May 10, 1944; children: Deborah Joyce, Rachel Adria. B.A. summa cum laude, Bklyn. Coll., 1965; M.A. in Biololgy, Harvard U., 1966; Ph.D. in Biology (NIH fellow), Harvard U., 1970. NIH postdoctoral research fellow in endocrinology U. Ill, Urbana, 1970-71, asst. prof. physiology Coll. Medicine and dept. physiology and biophysics, 1971-76, asso. prof., 1976-82, prof., 1982—; vis. prof. U. Calif., San Francisco, 1976-77; mem. NIH Endocrinology Study Sect., 1979-83. Contbr. numerous articles, chpts., papers to profl. jours., books, procs.; editorial bd.: Endocrinology, 1979—, Jour. Steroid Biochemistry, 1980—, Am. Jour. Physiology, Cell Physiology, 1981—. Recipient Young Scholar recognition award AAUW, 1981. Mem. Endocrine Soc. (publs. com.), Am. Physiol. Soc., Soc. Study Reprodn., Phi Beta Kappa. Subspecialties: Receptors; Cancer research (medicine). Current work: Hormonal and antihormonal regulation of reproductive tissues and tumor growth; mechanism of action of estrogens and antiestrogens and other reproductive sex steroid hormones in hormone-sensitive target tissues. Home: 501 W Washington St Urbana IL 61801 Office: U Ill 524 Burrill Hall Urbana IL 61801

KATZUNG, BERTRAM GEORGE, pharmacology educator; b. N.Y.C., June 11, 1932; s. Ewald and Kate (Soeldner) K.; m. Alice Velma Camp, Mar. 1, 1957; children: Katharine, Brian. B.A., Syracuse U., 1953; M.D., SUNY-Syracuse, 1957; Ph.D., U. Calif.-San Francisco, 1962. Lic. physician, Calif. Asst. prof. pharmacology U. Calif.-San Francisco, 1962-66, assoc. prof., 1966-71, prof., 1971—. Contbr. numerous articles to profl. jours. Markle Found. scholar, 1966-71; NIH research grantee, 1968—. Mem. AAAS, Am. Soc. Pharmacology and Exptl. Therapeutics, Biophys. Soc., Soc. Gen. Physiologists. Subspecialties: Pharmacology; Membrane biology. Current work: Cardiac electrophysiology and pharmacology. Office: U Calif Pharmacology Dept 1210S San Francisco CA 94143

KAU, SEN T., pharmacologist; b. Taiwan, Dec. 25, 1942; s. Shing Young and Yi Lan (Liau) K.; m. Praphaisri Lee, Dec. 23, 1971; children: Eric, Ryan. B.S. in Pharmacy, Nat. Taiwan U., 1967; Ph.D. in Pharmacology, Vanderbilt U., 1974. Research asso. Vanderbilt U., Nashville, 1974-75; research fellow pharmcology Cornell U. Med. Coll., N.Y.C., 1975-77; biologist ICI Ltd., Eng., 1977-78; sr. research pharmacologist ICI Ams. Ltd., Wilmington, Del., 1978-80, prin. pharmacologist, 1980-81, sect. mgr. biomed. research dept., 1981—. Contbr. articles to profl. jours. NIH grantee, 1971-75; NSF fellow, 1975-77. Mem. Am. Soc. Pharmacology and Exptl. Therapeutics, Am. Soc. Nephrology, Brit. Pharmacol. Soc., N.Y. Acad. Scis., Physiol. Soc. Phila., Am. Physiol. Soc. Subspecialties: Pharmacology; renal physiology. Current work: Renal biochemistry, physiology, pharmacology and toxicology. Office: ICI Americas Inc New Murphy Rd and Concord Pike Wilmington DE 19897

KAUFMAN, DONALD BARRY, physician, educator, researcher; b. Los Angeles, Aug. 5, 1937; s. Samuel Sheldon and Shirley Ruth (Bornstein) K.; m. Elizabeth Anne Hutchinson, June 28, 1964; children: Matthew J., Tamara S. A.B., UCLA, 1959; M.D., U. Calif., San Francisco, 1963. Diplomate: Am. Bd. Pediatrics, Am. Bd. Pediatric Nephrology. Resident U. So. Calif. and Los Angeles Children's Hosp., 1963-65, 67-68; fellow Los Angeles Children's Hosp., 1968, UCLA, 1969-71; asst. prof. pediatrics/human devel. Mich. State U., East Lansing, 1971-75, assoc. prof., 1976-82, prof., 1982—, chief immunology, 1982—. Served to capt. USAF, 1965-67. USPHS fellow, 1968-71. Mem. N.Y. Acad. Scis., Am. Acad. Scis., Am. Soc. Pediatric Nephrology, Soc. Pediatric Research, Am. Fedn. Clin. Research, Am. Assn. Immunologists. Subspecialties: Immunology (medicine); Nephrology. Current work: Cellular immunology; autoimmunity; immunodeficiency. Office: Mich State U East Lansing MI 48824

KAUFMAN, EDWIN H., JR., mathematics educator; b. Decatur, Ill., June 23, 1943; s. Edwin H. and Ada Rose (Parsinger) K. B.A., Millikin U., 1965; M.S., U. Ill.-Urbana, 1967, Ph.D., 1970. Vis. asst. prof. Mich. State U., East Lansing, 1970-72; asst. prof. math. Central Mich. U., Mt. Pleasant, 1972-74, assoc. prof., 1974-78, prof., 1978—. Co-author articles in field. Recipient Scovill award Millikin U., 1964, Univ. fellow, 1965-66; NSF trainee, 1966-70. Mem. Am. Math. Soc., Math. Assn. Am., Soc. Indsl. and Applied Math., Mensa, Sigma Xi. Republican. Current work: Approximation theory; numerical analysis. Office: Central Mich U Dept Math Mount Pleasant MI 48858

KAUFMAN, FREDERICK, chemist, educator; b. Vienna, Austria, Sept. 13, 1919; came to U.S., 1940, naturalized, 1946; s. Erwin and Else (Pollack) K.; m. Klari Simonyi, Nov. 2, 1951; 1 son, Michael Stephen. Student, Vienna Technische Hochschule, 1937-38; Ph.D., Johns Hopkins, 1948. With Ballistic Research Labs., Aberdeen (Md.) Proving Ground, U.S. Army, 1948-64, chief phys. chemistry sect., 1951-60, chief chem. physics br., 1960-64; lectr. Johns Hopkins U., Balt., 1948-64; prof. chemistry U. Pitts., 1964—, chmn. dept., 1977-80, univ. prof. chemistry, 1980—, dir. space research coordination center, 1975—; mem. advisory bd. office of chemistry tech. Nat. Acad. Scis., 1975-77; cons. to govt. agys. and industry. Editorial bd.: Jour. Chem. Physics, 1971-74, Jour. Photochemistry, 1972—, Internat. Jour. Chem. Kinetics, 1976—, Jour. Phys. Chemistry, 1980—. Recipient Rockefeller Pub. Service award Woodrow Wilson Sch., Princeton, 1955; Kent award Ballistic Research Labs., 1958; Research and Devel. award U.S. Army, 1962. Fellow Am. Phys. Soc.; fellow AAAS; mem. Am. Chem. Soc. (Pitts. award 1977), Chem. Soc. Gt. Britain, Nat. Acad. Sci., Combustion Inst. (v.p. 1978-82, pres. 1982—), AAAS, Nat. Acad. Scis., Phi Beta Kappa, Sigma Xi. Subspecialty: Physical chemistry. Home: 5854 Aylesboro Ave Pittsburgh PA 15217 Office: Dept Chemistry U Pitts Pittsburgh PA 15260

KAUFMAN, JOYCE JACOBSON, chemist, researcher, educator; b. N.Y.C., June 21, 1929; d. Abraham and Sarah (Seldin) Deutch; m. Stanley Kaufman, Dec. 26, 1948; 1 dau., Jan Caryl. B.S. with honors, Johns Hopkins U., 1949, M.A., 1959, Ph.D., 1960; D.E.S. in Thoeretical Physics, Sorbonne, Paris, 1963. Analytical research chemist U.S. Army Chem. Center, 1949-52; mem. chemistry research staff Johns Hopkins U., Balt., 1952-60, head quantum chemistry group Research Inst. Advanced Studies, 1962-69, prin. research scientist dept. chemistry Research Inst. Advanced Studies, 1969—; assoc. prof. dept. anesthesiology Sch. Medicine (Johns Hopkins U.), 1969—, div. plastic surgery dept. surgery Sch. Medicine, 1977—; cons. in field. Contbr. numerous articles to profl. jours. Recipient Garvan medal Am. Chem. Soc., 1974; Md. Chemist award, 1974. Fellow Am. Phys. Soc., Am. Inst. Chemists; mem. Am. Chem. Soc., Am. Soc. Pharmacology and Exptl. Therapeutics, Internat. Soc. Quantum Biology, N.Y. Acad. Scis., Am. Soc. Anesthesiology, Am. Coll. Toxicology, AAUP. Subspecialties: Theoretical chemistry; Physical chemistry. Current work: Quantum chemistry, experimental physical chemistry, chemical physics.

KAUFMAN, LEO, microbiologist; b. N.Y.C., Jan. 20, 1930; s. David and Dora (Stalbow) K.; m. Renee Ellen Dreyfus, Jan. 19, 1952; children: Jennifer, Gary, Steven. B.S., Bklyn. Coll. of CCNY, 1952; M.S., U. Ky., 1955, Ph.D., 1958. Diplomate: Am. Bd. Microbiology, Am. Acad. Microbiology. Instr. U. Ky., 1958-59; med. research microbiologist, bateriology div. (Communicable Disease Center), 1959-60; with div. mycotic diseases Centers for Disease Control, Atlanta, 1960—, chief fungus immunology br., 1967—; adj. assoc. prof. Sch. Public Health, U. N.C.; adj. prof. Ga. State U., 1976, Emory U., 1982—. Contbr. numerous articles to profl. jours. Recipient Disting. Service award HEW, 1980; Kimble methodology research award Conf. Pub. Health, 1974. Mem. Am. assn. Immunologists, Am. Soc. Microbiology, Sigma Zi. Subspecialties: Immunobiology and immunology; Medical mycology. Current work: Provision of tests for serodiagnosis of systemic mycotic infections. Office: Centers for Disease Control Atlanta GA 30333

KAUFMAN, LINDA, computer scientist; b. Fall River, Mass., Mar. 20, 1947; d. David and Rose (Lepes) K.; m. Fred Grabiner, Sept. 20, 1981. S.B., Brown U., 1969; M.S., Stanford U., 1971, Ph.D., 1973. Asst. prof. U. Colo., Boulder, 1973-76; mem. tech. staff Bell Labs., Murray Hill, N.J., 1976—. Contbr. articles to profl. jours. NSF grantee, 1976. Mem. Assn. for Computing Machinery, Soc. Indsl. and Applied Math. Subspecialty: Numerical analysis. Current work: Function minimization, numerical lunear algebra. Home: 45 Amherst Pl Livingston NJ 07039 Office: Bell Labs Room 2C461 600 Mountain Ave Murray Hill NJ 07974

KAUFMAN, MICHELE, astronomer, educator; b. Flushing, N.Y., Sept. 7, 1940; d. David and Rose (Lepes) K.; m. Steven James Rallis, June 22, 1970. A.B., Harvard U., 1962, Ph.D., 1968. Asst. prof. Brown U., Providence, 1968-71; research assoc. SUNY-Stony Brook, 1970-72; vis. assoc. prof. U. Notre Dame, Ind., 1975-76; vis. asst. prof. Swarthmore (Pa.) Coll., 1976-77; lectr. Ohio State U., Columbus, 1977—. Contbr. articles in field to profl. jours. Mem. Am. Astron. Soc., Internat. Astron. Union. Subspecialty: Theoretical astrophysics. Current work: Star formation and galactic evolution, research in astronomy and physics teaching. Office: Physics Dept Ohio State U Columbus OH 43210

KAUFMAN, PETER BISHOP, biologist, plant physiologist, educator; b. San Francisco, Feb. 25, 1928; s. Earle Francis and Gwendolyn Bishop (Morris) K.; m. Hazel Snyder, Apr. 5, 1958; children: Linda Myrl, Ryan, Laura Irene. B.Sc., Cornell U., 1949; Ph.D. in Botany, U. Calif.-Davis, 1954, Ohio State U., summer 1955. Research investigator Calif. Rice Growers Assn., 1951-53; instr. dept. botany U. Mich., Ann Arbor, 1956, asst. prof., 1957-62, assoc, prof., 1962-72, prof., 1972—; agrl. research scientist Shell Devel. Co., Modesto, Calif., summer 1959; vis. NSF research scholar Lund

(Sweden) U., 1964-65; chmn. Mich. Natural Areas Council, 1972-74; vis. prof. depts. chemistry and molecular, cellular and developmental biology U. Colo., Boulder, 1973, 74; vis. research scientist Prairie Regional Lab., NRC Can., Saskatoon, Sask., 1973; author TV series House Botanist, 1977; vis. research scientist U. Calgary, Alta., Can., 1979, Mich. State U., East Lansing, 1980; hon. staff mem. Faculty Agr., Nagoya (Japan) U.; vis research scientist Rice Research Inst., Los Banos, Philippines; speaker numerous symposia and confs. Editorial bd.: Plant Physilogy, 1976—; author: Laboratory Experiments in Plant Physiology, 1975, Plants, People and Environment, 1979; also numerous articles. Served with U.S. Army, 1954-56. Fellow AAAS; mem. Mich. Bot. Club (pres. 1963-64), Bot. Soc. Am. (chmn. edn. com.; vice-chmn, developmental sect. 1975-77), N.Y. Acad. Scis., Am. Soc. Plant Physiologists, Scandinavian Soc. Plant Physiologists, Internat. Soc. Plant Morphologists, Soc. Developmental Biology, Am. Soc. Electron Microscopy, Tissue Culture Assn., Am. Inst. Biol. Scis., Nature Conservancy, Internat. Assn. Plant Tissue Culture, Sigma Xi. Democrat. Presbyterian. Subspecialties: Plant physiology (biology); Gravitational biology. Current work: Plant hormones, gravity responses in plants, silicification in plants, growth responses of plants in space shuttle and space stations, rice seed proteins.

KAUFMAN, RAYMOND, electronics corporation executive, mathematics educator; b. N.Y.C., Aug. 30, 1917; s. Samuel and Mary (Jagorda) K.; m. Regina May Malina, Nov. 14, 1942; children: Stephen, Adina, Judy. B.S., CCNY, 1942; M.S., NYU, 1946, Ph.D., 1950. Asst. physicist Farrand Optical Co., N.Y.C., 1944-47; instr. physics CCNY, 1947-50; physicist Farrand Optical Co., N.Y.C., 1950-57; dir. research Del Electronics Corp., Mt. Vernon, N.Y., 1957-76, v.p., 1957-76, pres., chief exec. officer, 1976—, chief exec. officer, 1982—; adj. assoc. prof. CCNY, 1964—; dir. Sherman Dean Fund, N.Y.C., Z-Seven Fund, Inc. Mem. Am. Phys. Soc., Sigma Xi, Sigma Pi Sigma. Subspecialties: Electronics; Programming languages. Home: Pinewood Shores#10 Route 39 Sherman CT 06784 Office: 250 E Sandford Blvd Mount Vernon NY 10550

KAUFMANN, PETER G., neurophysiologist, educator, researcher; b. Lauenburg, Germany, Feb. 5, 1942; s. Gustav W. and Genoveva (Mazeika) K. B.S., Loyola U., Chgo., 1964, M.A., 1966; Ph.D., U. Chgo., 1971. Asst. prof. psychology Emory and Henry Coll., Emory, Va., 1970-72; postdoctoral fellow Duke U., Durham, N.C., 1972-75; asst. prof. depts. surgery and physiology F.G. Hall Environ. Lab., 1975—. Contbr. articles to profl. jours. Mem. Soc. Neurosci, Undersea Med. Soc. Subspecialties: Neurophysiology; Hyperbaric Medicine. Current work: Influence of inert gases on neuronal activity; inert gas narcosis; hyperbaric neurophysiology; hyperbaric oxygenation. Office: Duke U Med Center Box 3823 Durham NC 27710

KAUFMANN, YORAM, clinical psychologist, educator; b. Tel-Aviv, Israel, Oct. 3, 1939; s. Herbert Zvi and Lina (Rochlin) (Botton)) K.; m. Rise A. Jacobson, Oct. 8, 1972. B.A., Hebrew U., Jerusalem, 1962, Bar-Ilan U., Ramat-Gan, Israel, 1967; Ph.D., NYU, 1972. Staff psychologist Lincoln Inst., N.Y.C., 1968-72; analyst C.G. Jung Tng. Ctr., N.Y.C., 1972—, tng. analyst, 1972—, tchr., 1972—, supr., 1972—; Chmn. program com. C.G. Jung Found., 1978-81, bd. dirs., 1978—. Mem. Am. Psychol. Assn., N.J. Psychol. Assn., N.Y. State Psychol. Assn., Internat. Assn. for Analytical Psychology, N.Y. Assn. for Analytical Psychology. Lodge: B'nai B'rith. Current work: The archetypal basis, as expressed in symbols and images, of psychic functioning. Home and Office: 159 Deer Trail North Ramsey NJ 07446 Office: 257 Central Park W Suite 3C New York NY 10024

KAUKER, MICHAEL LAJOS, pharmacology educator, researcher; b. Szerecseny, Hungary, Jan. 24, 1935; s. Lajos and Terez (Weisz) K.; m. Judith E. Jager, Aug. 6, 1941; children: Michael Lajos, Irene, Robert, Christopher. Student, Sch. Vet. Medicine, Budapest, 1953-56; Ph.D. in Pharmacology, U. Ala.-Birmingham, 1967. NIH postdoctoral trainee U. N.C.-Chapel Hill, 1967-69; asst. prof. pharmacology U. Tenn., 1969-73, assoc. prof., 1973—; cons. to pharm. mfrs. Contbr. articles and abstracts to physiology and pharmacology jours. Served with AUS, 1958-60, 61-62. NIH predoctoral fellow, 1962-67. Mem. Am. Soc. Pharmacology and Exptl. Therapeutics, Internat. Soc. Nephrology, Am. Soc. Nephrology, Sigma Xi, Soc. Exptl. Biology and Medicine. Roman Catholic. Club: Holly Hills Country (Cordova, Tenn.). Subspecialties: Pharmacology; Physiology (medicine). Current work: Hypertension, prostaglandins, renal kinins, renal function, body fluid and electrolyte balance, hormonal and local hormonal effects on the kidney, renal micropuncture. Home: 5334 Chickasaw Rd Memphis TN 38119 Office: 800 Madison Ave Memphis TN 38163

KAUL, MAHARAJ KRISHEN, consulting engineering company executive; b. Srinagar, Kashmir, India, Nov. 11, 1940; came to U.S., 1965; s. Kashi Nath and Shobhawati (Rangroo) K.; m. Girja Zutshi, Apr. 26, 1969; 1 dau., Aparna. B.S., Punjab U., 1962; M.S., SUNY-Stony Brook, 1967; Ph.D., U. Calif.-Berkeley, 1972. Lectr. Thapar Engrng. Coll., Patiala, India, 1962-65; sr. engr. EDS Nuclear, Inc., San Francisco, 1972-77; cons. engr. Quadrex Corp., Campbell, Calif., 1977-79; v.p. Encon Inc., San Jose, Calif., 1979—. Co-author: Structural Analysis and Design of Nuclear Plant Facilities, 1976; Contbr. articles to profl. jours. Mem. ASCE (com. on probabilistic methods 1978-82), Am. Nuclear Soc. Subspecialties: Theoretical and applied mechanics; Numerical analysis. Current work: Mathematical methods and computer applications in engineering mechanics and structural engineering. Home: 43670 Vista Del Mar Fremont CA 94539 Office: Encon Inc 1885 The Alameda San Jose CA 95126

KAUL, PUSHKAR NATH, pharmacologist, researcher; b. Srinagar, Kashmir, India, June 29, 1933; s. Radha Kishmen and Prabha Wati (Dhar) K.; m. Leela, Oct. 15, 1961; children: Meena, Venita, Sonya. Ph.D., U. Calif., San Francisco, 1960. Vis. scientist U. Melbourne, Australia, 1960-61; group leader in pharmacology Farbwerke Hoechst A.G., W. Ger., 1965-68; assoc. prof. pharmacology, research medicine and pediatrics U. Okla., 1968-75, prof. pharmacodynamics and toxicology, 1976-82; prof., chmn. pharmacology Morehouse Sch. Medicine, 1982—; cons. VA Hosp., Children's Hosp., Oklahoma City, NIMH, FDA, NSF, drug industry. Contbr. numerous articles to profl. jours. Recipient Nat. Research Service award NIMH, 1975; U. Okla. Alumni Research award, 1968, 69; Ebert Prize cert. Am. Pharm. Assn., 1962. Mem. Am. Soc. Pharmacology and Exptl. Therapeutics, Internat. Soc. Biochemistry and Pharmacology, Royal Australian Chem. Inst. Subspecialties: Pharmacology; Psychopharmacology. Current work: marine pharmacology, clinical psychopharmacology, drug metabolism. Office: The Morehouse Sch Medicine 720 Westview Dr SW Atlanta GA 30314

KAUTZ, DAVID JOHNATHAN, mechanical design engineer, consultant; b. Youngstown, Ohio, Feb. 28, 1954; s. Henry George and Barbara Jean (Hollingsworth) K.; m. Judith Anne Larkin, Jan. 8, 1977; 1 dau. Branden Alexis. B.S.M.E., SUNY-Buffalo, 1976. Registered profl. engr., N.Y. Field engr. BASIC Constrn. Co., Richmond, Va., 1976-77; project engr. TAM Ceramics (subs. NL Industries-Indsl. Chem. Devel.), Buffalo, 1978-80; project design engr. Battery Disposal Tech., Clarence, N.Y., 1980-81; sr. project engr. Buffalo Color Corp., 1981—; cons. failure analysis. Served to lt. USNR, 1981—. Mem. ASME (assoc., chmn. nat. legal consequences com. 1981-83), Am. Soc. Metals, Am. Soc. Naval Engrs. (assoc.). Club: Eastern Hills Racquet (Buffalo). Subspecialties: Mechanical engineering; Metallurgy. Current work: Neutralization of hazardous, contaminated equipment; hazardous material neutralization; strategic material reclamation. Patentee method, equipment to neutralize high-energy-density batteries. Office: 340 Elk St Buffalo NY 14240

KAUTZMANN, WILLIAM ELWOOD, space and communication engineer; b. Lancaster, Pa., May 28, 1918; s. William and Sarah Ann (Reeder) K.; m. Gladys Marie Weber, Feb. 15, 1942; children: Susan Marie and Pamela Diane (twins). B.S. in Chemistry, Franklin & Marshall Coll., 1943; B.S.M.E., Drexel U., Wash. U., 1951. Registered profl. engr., Ala. Asst. estimating engr. Westinghouse Electric, Phila., 1941-43, asst. to engrng. supt., St. Louis, 1943-46; engrng. planner/design McDonald Douglas, St. Louis, 1946-49; indsl. and mfg. engr. Boeing Vertol & Boeing Space Ctr., Phila., 1951-58; Space engr., Kent/Seattle, 1962-70; electronics and space engr. RCA Service Co., 1958-62, Cherry Hill, N.J., NASA S.F.C., Goddard, Md., 1973-78; space and communications engr. Gen. Electric Co., Springfield, Va., 1978—; space engr. Fairchild Industries, Germantown, Md., 1970-72. Mem. Needwood Civic Assn., Rockville, Md., 1973. Mem. AIAA. Rep. Luth.-Meth. Club: Tri-Boro Chorus (treas.-sec. 1958-62). Subspecialties: Aerospace engineering and technology; Satellite studies. Current work: spacecraft studies; satellite development and control; design of voyage spacecraft and propulsion systems; special testing and control systems. Patentee packaging improvements; helicopter manufacturing; jet aircraft; space Apollo mission cryogenics control and flow management; electronic and video strobe blade track in helicopter manufacturing. Home: 9695 Ironmaster Dr Burke VA 22015 Office: General Electric Co 6501 Loisdale Ct Springfield VA 22150

KAUZMANN, WALTER JOSEPH, chemistry educator; b. Mt. Vernon, N.Y., Aug. 18, 1916; s. Albert and Julia Maria (Kahle) K.; m. Elizabeth Alice Flagler, Apr. 1, 1951; children: Charles Peter, Eric Flagler, Katherine Elizabeth Julia. B.A., Cornell U., 1937; Ph.D., Princeton U., 1940. Westinghouse research fellow Westinghouse Mfg. Co., E. Pittsburgh, Pa., 1940-42; mem. staff Explosives Research Lab., Bruceton, Pa., 1942-44, Los Alamos Lab., 1944-46; asst. prof. Princeton U., 1946-51, asso. prof., 1951-60, prof. chemistry, 1960-82, chmn. dept., 1964-68, David B. Jones prof. chemistry, 1963-82, chmn. biochem. sci. dept., 1980-81; vis. scientist Atlantic Research Lab., NRC Can., 1983; vis. lectr. Kyoto U., 1974; vis. prof. U. Ibadan, 1975. Author: Quantum Chemistry, 1957, Kinetic Theory of Gases, 1966, Thermal Properties of Matter, 1967, (with D. Eisenberg) Structure and Properties of Water, 1969. Jr. fellow Soc. Fellows, Harvard, 1942; Guggenheim fellow, 1957, 74-75; Recipient Linderstrom-Lang medal, 1966. Fellow Am. Acad. Arts and Scis.; mem. Nat. Acad. Scis., Am. Soc. Biol. Chemists, Am. Chem. Soc., Am. Phys. Soc., A.A.A.S., Fedn. Am. Scientists, Sigma Xi. Subspecialties: Geochemistry; Physical chemistry. Current work: Physical chemistry of molten silicates, water and aqueous solutions. Home: 4 Newlin Rd Princeton NJ 08540 Frick Chem Lab Princeton NJ 08540

KAVIPURAPU, KAVI *See also* **KAVIPURAPU, KRISHNA M.**

KAVIPURAPU, KRISHNA M. (KAVI KAVIPURAPU), computer science educator, researcher; b. Telaprolu, India, Apr. 22, 1952; s. Rangarad and Kousalya (Gudipaty) K.; m. Lisa M. McDonald, Dec. 28, 1981. B.E., Indian Inst. Sci., 1975; M.S., So. Meth. U., 1977, Ph.D., 1980. Grad. teaching and research asst. So. Meth. U., 1976-80; asst. prof. dept. computer sci. U. So. La., 1980-82, U. Tex.-Arlington, 1982—. Contbr. articles to profl. jours. Indian Inst. Sci. merit scholar awardee, 1972-75; Indian Nat. Merit Scholarship awardee, 1972; recipient F. E. Terman award So. Meth. U., 1979; Tex. Instruments grantee, 1981-82. Mem. IEEE, Assn. for Computing Machinery. Subspecialties: Computer architecture; Software engineering. Current work: New and innovative high-level computer architecture; software development to develop a distributed operating system; model ultral-reliable fault tolerant computers using data flow models. Home: 712 Lincoln Green Apt 2103 Arlington TX 76011 Office: U Tex Dept Computer Sci Engring PO Box 19015 Arlington TX 76019

KAWASE, MAKOTO, research horticulturist, educator; b. Zuizan, Korea, May 20, 1926; came to U.S., 1956, naturalized, 1971; s. Sanemasa and Shigeko K.; m. Nobuko Matsui, Oct 28, 1931; children: Yutaka, Yuriko, Mary Ko. B.A., U. Tokyo, 1951, M.A., 1954; M.S., U. Minn., 1958; Ph.D., Cornell U., 1960. Postdoctoral fellow Purdue U., West Lafayette, Ind., 1960-62; research scientist Can. Dept. Agr., Morden, Man., 1962-66; assoc. prof. Ohio Agrl. Research and Devel. Ctr., 1966-70, prof., 1970—; cons. in field. Contbr. numerous articles on horticulture to profl. jours. NSF grantee, 1970; NATO vis. prof., 1974. Mem. Am. Soc. Hort. Sci. (Alex Laurie award 1962, 67), Am. Soc. Plant Physiologist, Internat. Plant Propagators Soc., Scandinavian Soc. Plant Physiology, AAAS, Bot. Soc. Am., Phi Kappa Phi, Phi Alpha Xi, Sigma Xi. Presbyterian. Subspecialties: Plant physiology (agriculture); Plant growth. Current work: Plant propagation, rooting substance, waterlogging plant physiology, containerization of field grown plants. Home: 902 Ridgecrest Dr Wooster OH 44691 Office: Dept Horticulture Ohio Agr Research and Devel Ctr Wooster OH 44691

KAY, KENNETH GEORGE, chemist, educator; b. N.Y.C., Oct. 3, 1943; s. Leslie L. and Anne (Feuerstein) K.; m. Katarina Brody, June 16, 1968; children: Victoria, Elizabeth, Jennifer. B.S., Poly. Inst. Bklyn., 1965, M.S., 1965; Ph.D., Johns Hopkins U., 1970. NRC postdoctoral fellow U. Chgo., 1970-71; asst., then assoc. prof. Kans. Sate U., Manhattan, 1971-80, prof. chemistry, 1980—; vis. assoc. prof. Tel-Aviv (Israel) U., 1979-80; vis. prof. U. Toronto, 1982-83. Contbr. numerous articles to profl. jours. Grantee NSF, 1972, 75, 78, Petroleum Research Fund, 1973, Research Corp., 1972. Mem. Am. Chem. Soc., Am. Phys. Soc., AAUP, Phi Beta Kappa, Sigma Xi, Phi Lambda Upsilon. Subspecialty: Theoretical chemistry. Current work: Theoretical studies of chemical reaction dynamics, especially unimolecular reactions; theory of intramolecular vibrational energy transfer. Office: Dept Chemistry Kans State U Manhattan KS 66506

KAY, MARGUERITE M. BOYLE, physician, scientist, educator; b. Washington, May 13, 1947; s. Murray and Ann Margot (Boyle) K. A.B. summa cum laude, U. Calif.-Berkeley, 1970; M.D., U. Calif.-San Francisco, 1975. Staff fellow Gerontology Research Ctr., NIH, 1974; chief high resolution membrane lab. NIH, 1975-77; intern, resident Wadsworth VA Med. Ctr., UCLA, 1977-79, fellow in geriatric medicine, 1979-81, chief lab. molecular and clin. immunology, 1977-81; dir. electron microscopy facility, 1977-81; practice medicine specializing in internal medicine, Los Angeles, 1977-81; assoc. chief of staff for research Olin E. Teague Vets. Ctr., Temple, Tex., 1981—; dir. div. geriatric medicine Tex. A&M U. Med Sch, Temple, 1981—. Editor: Biological Sciences, 1981; mem. editorial bd.: Mechanism of Aging and Devel, 1979; contbr. articles to profl. jours. Served with USPHS, 1975-77. Mem. Am. Soc. Cell Biology, Gerontological Soc., Am. Geriatrics Soc., Am. Assn. Immunologists, Am. Soc. Hematology. Subspecialties: Internal medicine; Gerontology. Current work: Molecular aging; senescent cell antigen; immune restoration, molecular and cell biology. Home: 4209 Eagle Rd Temple TX 76501 Office: Olin E Teague Veterans Ctr Temple TX 76501

KAY, ROBERT LEO, chemistry educator; b. Hamilton, Ont., Can., Dec. 13, 1924; s. Norman Robert and Elizabeth (Blatz) K.; m. Ann Donata Morrow, Sept. 20, 1952; children: David Robert (dec.), Theresa Ann, Joanne Frances, Robert Leo. B.A., U. Toronto, 1949, M.A., 1950, Ph.D., 1952. With Rockefeller Inst. for Med. Research, N.Y.C., 1952-56; asst. prof. chemistry Brown U., Providence, 1956-63; sr. fellow Mellon Inst., Pitts., 1963-67; prof. chemistry Carnegie-Mellon U., Pitts., 1967—, acting dir. Center Spl. Studies, 1973-74, head dept. chemistry, 1974-83; Mem. Council Gordon Research Confs., 1974-78. Contbr. articles to sci. jours.; Editor: Jour. Solution Chemistry, 1971—; editorial bd.: Jour. Phys. Chemistry, 1973-81. Served with Canadian Army, 1943-46. Merck of Can. Postdoctoral fellow, 1952; research grantee Research Corp., 1957-59, NSF, 1959-60, 72—, AEC, 1959-63, Office Saline Water, U.S. Dept. Interior, 1963-70, NIH, 1972-74. Mem. Biophys. Soc., AAAS, Am. Chem. Soc., Pitts. Chemists Club. Subspecialties: Physical chemistry; Solution chemistry. Current work: The study of eledtrolytes and surfactant solutions at high temperature and pressure. Properties investigated are conductance, transference, dielectric constants, densities, electrophoresis, and dielectric relaxation. Home: 221 Parkway Dr Pittsburgh PA 15228

KAY, ROBERT WOODBURY, geology educator, researcher; b. N.Y.C., Jan. 21, 1943; s. George Marshall and Inez Margaret (Clark) K.; m. Suzanne Elizabeth Mahlburg, June 4, 1975; children: Jennifer Elizabeth, Alexander Marshall. A.B., Brown U., 1964; Ph.D., Columbia U., 1970. Asst. prof. Columbia U., N.Y.C., 1970-75; asst. research geologist UCLA, 1975-76; asst. prof. Cornell U., Ithaca, 1976-81, assoc. prof., 1981—; vis. assoc. Calif. Inst. Tech., Pasadena, 1982-83. Contbr.: Science Yr, Chgo., 1982—. Fellow Geol. Soc. Am.; mem. Geochem. Soc., Am. Geophys. Union, Internat. Assn. Geochemistry and Cosmochemistry. Subspecialties: Geochemistry; Petrology. Current work: Trace element and petrological constraints on the origin of igneous rocks, especially those from the Aleutian Islands. Home: 102 Genung Circle Ithaca NY 14850 Office: Cornell U Kimball Hall Ithaca NY 14853

KAYAR, SUSAN R., physiology researcher; b. Highland, Ill., May 17, 1953; d. Sedat Arif and Ruth Annalea (Houseman) K. B.S., U. Miami, Fla., 1974, Ph.D., 1978. Research asst. Everglades Nat. Park, Homestead, Fla., 1978-79; project biologist Connell Metcalf and Eddy, Miami, Fla., 1979; research assoc. U. Colo., Boulder, 1979-81, U. Colo. Med. Sch., Denver, 1981—. Contbr. articles to profl. jours. Active Denver Botanic Garden, Denver Mus. Natural History, Denver Mus. Art. Nat. Merit scholar, 1971-74; Maytag fellow, 1974-77; NIH postdoctoral fellow, 1982—. Mem. AAAS, Am. Physiol. Soc., Am. Soc. Zoologists, Sigma Xi. Subspecialty: Comparative physiology. Current work: Comparative respiratory physiology, specializing in physiological adaptations to hypoxia. Office: Dept Physiology C24 U Colo Health Scis Ct 4200 E 9th Ave Denve CO 80262

KAYE, STANLEY MARTIN, physicist; b. N.Y.C., June 5, 1952; s. Irving and Rachel (Schmeltzer) K.; m. Donna Marlene Stein, Oct. 25, 1981. B.A., Hamilton Coll., 1974; M.S., U. Wash., 1976; Ph.D., UCLA, 1979. Staff physicist Lockheed Research Labs., Palo Alto, Calif., 1979-80, Plasma Physics Lab., Princeton (N.J.) U., 1980—; adj. prof. physics Rider Coll., Lawrenceville, N.J., 1982—. Mem. Am. Phys. Soc., Am. Geophys. Union, AIAA, Union Concerned Scientists, N.Y. Acad. Scis., Sigma Xi. Democrat. Jewish. Subspecialties: Nuclear fusion; Plasma physics. Current work: Research in tokamak physics. Office: Plasma Physics Lab Princeton U Princeton NJ 08544

KAZAKS, PETER A., physics educator; b. Feb. 22, 1940; U.S., 1962; s. Alexander and Sigrida (Meijers) K.; m. Alexandra G. Hazen, Sept. 8, 1968; children: Julia, Emily, Karl, Kristopher. B.Sc. with honors, McGill U., 1962; M.A., Yale U., 1963; Ph.D., U. Calif.-Davis, 1968. Research assoc. Ohio U., Athens, 1968-70; asst. prof. St. Lawrence U., Canton, Ohio, 1970-73; assoc. prof., chmn. div. New Coll. of U. South Fla., Sarasota, 1973—. Contbr. articles to sci. jours. NSF grantee. Mem. Am. Phys. Soc. Subspecialties: Nuclear physics; Theoretical physics. Current work: Reaction and scattering theory, low and medium energy nuclear physics. Office: New College 5700 N Tamiami Trail Sarasota FL 33580

KAZARIAN, LEON EDWARD, biomechanics research scientist; b. Norwalk, Conn. B.S.A.A.E., Northrop Inst. Tech., 1966; D.Eng. in Orthopedic Biomechanics, Karolinska Inst., Sweden, 1972. Research scientist in biomechanics Aerospace Med. Research Labs., Wright Patterson AFB, Ohio, 1967—. Recipient Liljencrantz award, 1979. Mem. Orthopedic Research Soc., Aerospace Med. Assn. Subspecialty: Biomechanics. Office: Aerospace Med Research Labs Wright Patterson AFB OH 45433

KAZEK, GREGORY JOSEPH, applied physicist; b. Cleve., Oct. 26, 1947; s. Stanley and Irene Martha (Lake) K.; m. Deanne Joan, Sept. 3, 1981; children: Gregory Joseph, Geoffrey S, Rachel E. B.S. in Physics, Case Inst. Tech., 1969, M.S., John Carroll U., 1971; Ph.D. in Elec. Engring, Case Western Res. U., 1974. Engr. Lamp Bus. Group, Gen. Electric Co., Cleve., 1969-72, sr. engr., 1972-75, research physicist, 1975-79; sr. project engr. Reliance Electric Co., Euclid, Ohio, 1979-83; advanced research and devel. engr. GTE Corp., Salem, Mass., 1983—. Contbr. numberous articles to profl. jours. Mem. Am. Phys. Soc., IEEE, Assn. Iron and Steel Engrs. Democrat. Subspecialties: Electrical engineering; Plasma engineering. Current work: Energy conversion applied to product and process design; project mgmt. and engring. of digital systems to control materials processing for all aspects of arc lamp technology. Patentee radiation dominated gaseous condrs. applied as lighting sources and circuit components; numerous world-wide installations of digital systems integrating peripherals of great variety. Office: 60 Boston St Salem MA 01970

KAZI, ABDUL HALIM, nuclear engrng. adminstr., radiation effects researcher; b. Kreuzlingen, Switzerland, Jan. 12, 1935; came to U.S., 1954; s. Abdul Hamid and Zubaidah (Shutt) K.; m. Patricia Stewart Hughes, June 3, 1959; children: Aaron, Ethan. B.S. with distinction, American U., Cairo, 1954; M.S. in Physics, Rensselaer Poly. Inst., 1956; S.M. in Nuclear Engring, MIT, 1959, Ph.D., 1961. Mem. staff Gen. Atomic Co., La Jolla, Calif., 1961-63; mgr. advanced reactor analysis sect. United Nuclear Co., White Plains, N.Y., 1963-66; chief army pulse radiation facility ops. material testing directorate Aberdeen Proving Ground, Md., 1966—. Contbr. numerous articles on nuclear engrng. to profl. jours. Recipient 4 commendations U.S. Army Aberdeen Proving Ground, 1971, 74, 80, 81. Mem. Am. Nuclear Soc. Subspecialties: Nuclear engineering; Radiation effects. Current work: Design, startup, operation and utilization of radiation test facilities and nuclear environment simulators, experimental radiation transport shielding and dosimetry. Patentee (with G. A. Sofer) fast breeder nuclear reactor, 1968. Office: STEAP-MT-R Aberdeen Proving Ground MD 21005

KAZIMI, MUJID S., nuclear engr., educator; b. Jerusalem, Nov. 20, 1947; s. Suleiman I. and Fiknat (Nusseibeh) K.; m. Nazik D. Denny, 1973; children: Marwan, Yasmeen, Omar. B.S., U. Alexandria, Egypt, 1969; M.S., MIT, 1971, Ph.D., 1973. Sr. engr. Westinghouse Electric Corp., Madison, Pa., 1973-74; assoc. engr. Brookhaven Nat. Lab., Upton, N.Y., 1974-76; asst. prof. nuclear engring. MIT, Cambridge, 1976-79, assoc. prof., 1979—. Mem. Am. Nuclear Soc., Am. Inst.

Chem. Engring., ASME. Subspecialties: Nuclear engineering; Nuclear fusion. Current work: Nuclear reactor design and safety analysis. Heat transfer and fluid flow. Office: MIT 77 Massachusetts Ave Cambridge MA 02139

KEDES, LAURENCE HERBERT, molecular biologist, physician, educator; b. Hartford, Conn., July 19, 1937; s. Samuel Fly and Rosalynn (Epstein) K.; m. Shirley Gail Beck, June 15, 1958; children: Dean Hamilton, Maureen Jennifer, Todd Russell. Student, Wesleyan U., Middletown, Conn., 1955-58; B.S., Stanford U., 1961, M.D., 1962. Diplomate: Am. Bd. Internal Medicine. Intern Univ. Hosp., Pitts., 1962-64; resident in internal medicine Peter Bent Brigham Hosp., Boston, 1966-67; prof. Stanford U., 1970—; v.p., sr. scientist IntelliGenetics, Inc., 1981-82, chmn. bd., 1982—; mem. staff VA Hosp., Palo Alto, Calif. Served as surgeon USPHS, 1964-66. Mem. Am. Soc. Clin. Investigation, Am. Soc. Developmental Biology. Subspecialties: Genetics and genetic engineering (biology); Artificial intelligence. Current work: Genetic engineering, artificial intelligence. Office: VA Hosp Miranda Ave Palo Alto CA 94305

KEEFER, DONALD ASHBY, neuroendocrinologist; b. Balt., Jan. 8, 1946; s. William Hobart and Florence (Roughton) K.; m. Mandy Hutson, Aug. 3, 1966; children: Deborah, Mark. B.A., Western Md. Coll., 1968; Ph.D., U.N.C., 1974. Neurobiology fellow U.N.C., Chapel Hill, 1974; guest scientist Max-Planck Inst. for Brain Research, Frankfurt, W. Ger., 1975-76; asst. prof. anatomy U.Va., Charlottesville, 1976-82; assoc. prof., 1982-83; assoc. prof., chmn. dept. biology LoLoya Coll.-Balt., 1983—. Subspecialties: Neuroendocrinology; Receptors. Current work: Teaching, modulation of estrogen receptor levels and cellular processing. Office: Loyola Coll Dept Biology Baltimore MD 21210

KEEHN, NEIL FRANCIS, aerospace company executive, systems engineer, strategic analyst; b. Massillon, Ohio, Oct. 24, 1948; s. Russell Earl and Mary Leona (Danner) M. B.S., Ariz. State U., 1970, M.S., 1970. Mem. tech. staff Tech. Services Corp., Santa Monica, Calif., 1972-74; Hughes Aircraft, El Segundo, Calif., 1974-77; assoc. program mgr. TRW, Inc., Redondo Beach, Calif., 1977-79; mgr. advanced concepts, mil. systems div. Sci. Applications, Inc., El Segundo, 1979-80; pres. Strategic Systems Scis., Santa Monica., Calif., 1980—. Recipient Bd. Govs. citation IEEE Aerospace and Electronic Systems Soc., 1977; Ariz. State U, scholar, 1967. Mem. IEEE (vice chmn. Aerospace Def. Systems Panel 1972-76, chmn. 1976-79, chmn. membership aerospace def. systems com. 1974, Winter Conv. Outstanding Service citation 1974), AIAA, U.S. Strategic Inst. Democrat. Roman Catholic. Subspecialties: National security space systems; Systems engineering. Current work: Professional activities centered on advanced planning for national security space systems; major portion of work concentrated in development of advanced strategic/systems concepts relating to exploitation of space to enhance national security; several of these efforts have resulted in presentations throughout defense establishment and at the White House. Inventor in real time processing of pulse compression waveforms (Hughes award 1977). Home: 2603 3d St Santa Monica CA 94005 Office: Strategic Systems Scis 2603 3d St Santa Monica CA 90405

KEEHN, PHILIP MOSES, chemist, educator, cons.; b. Bklyn., Mar. 22, 1943; s. Louis and Frances (Mamches) K.; m. Lillian Brody, June 19, 1966; 3 children. B.A., Yeshiva U., 1964; M.A., Yale U., 1967, Ph.D., 1969. Research assoc. Harvard U., 1969-71; asst. prof. chemistry Brandeis U., 1971-78, assoc. prof., 1978—; vis. assoc. prof. Weizmann Inst. Sci., Rehovot, Israel, 1979-80; cons. in field. Contbr. numerous articles to chem. jours. Recipient Sir Isaac Wolfson award Brandeis U., 1979-80; Tchr.-Scholar award Camille and Henry Dreyfus Found., 1979-84; Alfred Bader Found. award, 1980. Mem. Am. Chem. Soc. Subspecialties: Organic chemistry; Laser-induced chemistry. Current work: Organic synthesis of strained rings and theoretically interesting molecules; synthetic methods; photooxidatiion; pure and applie laser chemistry. Office: Dept Chemistry Brandeis U Waltham MA 02254

KEEHN, ROBERT JOHN, biostatistician; b. Rochester, N.Y., Jan. 1, 1922; s. John Adam and Roberta (Holliday) K.; m. Evelyn Catherine Hesse, Sept. 7, 1946; children: Gordon, Alan, Carl, Joy (adopted), Mark. Statistician, N.Y. State Dept. Health, 1948-53; dir. office vital stats. Conn. Dept. Health, 1953-58; profl. assoc. Med. Follow-up Agy., Nat. Acad. Scis., NRC, Washington, 1958—; statistician Atomic Bomb Casualty Commn., Hiroshima, Japan, 1963-66, 71-73. Contbr. articles to sci. jours. Served with U.S. Army, 1942-45. Mem. Am. Pub. Health Assn., Soc. Epidemiologic Research, Royal Soc. Health, Soc. Clin. Trials. Subspecialties: Statistics; Epidemiology. Current work: Follow-up and case-control studies using the U.S. veteran populations. Office: Medical Follow-Up Agency National Research Council 2101 Constitution Ave NW Washington DC 20418

KEELER, ELAINE KATHLEEN, bio-organic chemist, educational planner; b. Passaic, N.J., Mar. 26, 1941; d. Paul E. and Ruth M. (McVey) K. B.A., Coll. St. Elizabeth, Convent Station, N.J., 1966; M.S., Purdue U., 1970; Ph.D., NYU, 1977. Tchr. chemistry Acad. of St. Elizabeth, 1964-72; instr. chemistry NYU, 1972-79; asst. prof. chemistry Coll. St. Elizabeth, 1977-80, asst. dean of studies, 1979-80; asst. prof. chemistry Drew U., 1980-81; dir. applications for nuclear magnetic resonance in medicine Fonar Corp., Melville, N.Y., 1981—. Subspecialties: Nuclear magnetic resonance (biotechnology); Biochemistry (medicine). Current work: Clinical applications of nuclear magnetic resonance in medicine, educational activities related to NMR in medicine, training of physicians and technicians, product development for NMR in clinical use. Office: 110 Marcus Dr Melville NY 11747

KEELEY, JON EDWARD, biologist, educator; b. Chula Vista, Calif., Aug. 11, 1949; s. Edward L. and Elouise V. (Hite) K.; m. Sterling C. Keeley, June 23, 1973. B.S., San Diego State U., 1971, M.S., 1973; Ph.D., U. Ga., 1977. Assoc. prof. biology Occidental Coll., Los Angeles, 1977—. Contbr. articles on population and physiol. ecology. Mem. Soc. for Study of Evolution, Ecol. Soc. Am., Am. Assn. Plant Physiologists, Am. Soc. Naturalists, Bot. Soc. Am., Calif. Bot. Soc. Subspecialties: Plant physiology (biology); Population biology. Current work: Physiological and population ecology. Home: 4773 Baltimore St Los Angeles CA 90042 Office: Dept Biology Occidental Coll Los Angeles CA 90041

KEELEY, STERLING CARTER, botanist, museum curator, educator, cons.; b. San Francisco, Oct. 23, 1948; d. John Frederick and Star (Steel) Carter; m. Jon Edward Keeley, June 23, 1973. A.B., Stanford U., 1970; M.S. in Biology, San Diego State U., 1973; Ph.D. in Botany, U. Ga., 1977. Lectr. U. Ga., Athens, 1976-77, Calif. State U., Northridge, 1977, Long Beach, 1978-79; research assoc.-curator Mus. Natural History Los Angeles, 1978—; asst. prof. biology Whittier (Calif.) Coll., 1979—; reviewer NSF; cons. congl. program low-coast shoreline erosion, 1978-80, Port of Los Angeles, 1978-80. Reviewer: Systematic Botany; editorial bd.: Madrono; contbr. articles to sci. jours. NSF grantee, 1974-76, 79-81, 82-85; Whittier Coll. faculty research grantee, 1980—. Mem. Soc. Study of Evolution, Am. Soc. Plant Taxonomists, AAAS, Inst. Ecology U. Ga., Bot. Soc. Am., Assn. So. Calif. Botanists, Calif. Native Plant Soc., Sigma Xi. Subspecialties: Systematics; Ecology. Current work: Systematics of neotropical Vernonia; germination of herb species of California chaparral and desert communities. Office: Whittier Coll Whittier CA 90608

KEELING, CHARLES DAVID, oceanography educator, researcher; b. Scranton, Pa., Apr. 20, 1928; m., 1954; 5 children. B.A., U. Ill., 1948; Ph.D. in Chemistry, Northwestern U., 1953. Fellow in geochemistry Calif. Inst. Tech., 1953-56; asst. research chemist Scripps Inst. Oceanography, U. Calif.-San Diego, 1956-60, from assoc. research chemist to assoc. prof., 1960-68, prof. oceanography, 1968—; guest prof. oceanography U. Heidelberg, Germany, 1969-70; mem. commn. atmospheric chemistry and global pollution Internat. Assn. Meteorol. and Atmospheric Physics, 1967—; mem. Panel on Energy and Climate, Nat. Acad. Sci., 1974-77, Interim Sci. Directorate Carbon Dioxide Research Program, Dept. Energy, 1977-80; guest prof. Physikaliches Inst., U. Bern, Switzerland, 1979-80. Recipient Second Half Century award Am. Meteorol. Soc., 1980. Mem. AAAS, Am. Geophys. Union. Subspecialty: Geochemistry. Office: Scripps Inst Oceanography U Calif-San Diego La Jolla CA 92093

KEEN, NOEL THOMAS, plant pathologist, educator; b. Marshalltown, Iowa, Aug. 13, 1940; s. Walter T. and Evelyn Mae (Mayo) K.; m. Esther M., Apr. 6, 1974. B.S., Iowa State U., 1963, M.S., 1965; Ph.D., U. Wis., 1968. Asst. prof. plant pathology U. Calif., Riverside, 1968-72, asso. prof., 1972-78, prof., 1978—, chmn. dept., 1983—. Contbr. articles in field to profl. jours. Mem. AAAS, Am. Soc. Plant Physiologists, Am. Phytopath. Soc., Am. Inst. Biol. Scis., Sigma Xi, Gamma Sigma Delta. Subspecialties: Plant pathology; Biochemistry (biology). Current work: Research in mechanisms of disease resistance in plants. Home: 5617 Via Junipero Serra Riverside CA 92506 Office: University of California Dept Plant Pathology Riverside CA 92521

KEENER, JAMES PAUL, mathematics educator; b. Lancaster, Pa., Nov. 26, 1946; s. Paul Koendig and Helen (Gibble) K.; m. Kristine Antoinette Vaclav, June 29, 1968; children: Samantha, Justin. B.S., Case Inst. Tech., 1968; M.S., Calif. Inst. Tech., 1969, Ph.D., 1972. Asst. prof. math U. Ariz., Tucson, 1972-78; assoc. prof. U. Utah, Salt Lake City, 1977-82; prof., 1982—. Contbr. articles to profl. jours. McGill U. research scholar, 1974-75; research fellow U. Heidelberg, W.Ger., 1981; NSF grantee, 1975-82; Gardner fellow U. Utah, 1981. Mem. Soc. Indsl. and Applied Math. Subspecialty: Applied mathematics. Current work: Mathematical biology. Office: Dept Math U Utah Salt Lake City UT 84112

KEESLING, JAMES EDGAR, mathematics educator; b. Indpls., June 26, 1942; s. Fred Edgar and Martha Belle (Grimes) K.; m. Marian Ellen Calley, Jan. 26, 1963; children: James Edgar, Marian Esther, Timothy Carl, Ruth Emily. B.S.I.E., U. Miami, Fla., 1964, M.S. in Math, 1966, Ph.D., 1968. Vis. lectr. U. Ga., Athens, 1976-77; asst. prof. U. Fla., Gainesville, 1967-71, assoc. prof., 1971-75, prof. math., 1975—. Contbr. articles to profl. jours. Mem. Am. Math. Soc., Math. Assn. Am., Soc. for Indsl. and Applied Math., Phi Kappa Phi, Tau Beta Pi, Omicron Delta Kappa. Subspecialties: Applied mathematics; Operations research (mathematics). Current work: Research in topology and applied mathematics. Office: U Fla Dept Math Gainesville FL 32611 Home: 710 NE 6 St Gainesville FL 32601

KEFALIDES, NICHOLAS ALEXANDER, biochemistry educator, physician; b. Alexandroupolis, Greece, Jan. 17, 1927; came to U.S., 1947; s. Athanasios N. and Alexandra (Aematidou) K.; m. Eugenia Georgia Kutsunis, Nov. 24, 1949; children: Alexandra, Patricia, Paul. B.A., Augustana Coll., 1951; M.S., U. Ill.-Chgo., 1956, M.D., 1956, Ph.D., 1964; M.A. (hon.), U. Pa., 1971. Asst. prof. medicine U. Ill. Sch. Medicine, Chgo., 1964-65, assoc. prof., 69-70, U. Chgo., 1965-69, U. Pa. Sch. Medicine, Phila., 1970-74, assoc. prof. biochemistry, 1973-75, prof. medicine, 1974—, prof. biochemistry and biophysics, 1975—, dir., 1972—. Contbr. numerous articles on chemistry and metabolism of connective tissues to profl. jours.; author: Biology and Chemistry of Basement Membranes, 1978. Recipient Borden award in medicine Borden Co., 1956; Guggenheim fellow, 1977. Mem. Am. Soc. Biol. Chemists, Am. Soc. for Clin. Investigation, Am. Chem. Soc., Am. Soc. for Cell Biology, Am. Assn. Pathologists. Subspecialties: Internal medicine; Biochemistry (medicine). Current work: Chemistry and metabolism of connective tissue and basement membranes, metabolism of vascular cells. Office: Connective Tissue Research Inst U Pa 3624 Market St Philadelphia PA 19104

KEGAN, DANIEL L., organization and management consultant, educator; b. Chgo., Mar. 3, 1944; s. Albert I. and Esther O. K.; m. Cynthia L. Scott, Dec. 1982. B.S.E.U., Swarthmore Coll., 1965; M.S. in Organizational Behavior, Northwestern U., 1969, Ph.D., 1971, student law, 1981-84. Lic. psychologist, Mass., Calif.; engr. in tng., Pa. Dir. instnl. research Hampshire Coll., 1973-77; mem. faculty U. Calif.-Berkeley, 1979-81, Wright Inst., Berkley, 1979-81, Pepperdine U., 1980-81, Ill. Inst. Tech., 1982; sr. ptnr. Elan Assocs., Emeryville, Calif., 1979—; mem. steering com. N.E. Assn. Instl. Research, 1975-77; mem. Congl. Clearinghouse on the Future, 1978-80; steering com. Am. Soc. Tng. and Devel. Consultants SIG, 1979-80. Contbr. articles profl. jours.; jour. referee. Alumni interviewer Swarthmore Coll., 1971—; fin. com. chmn. Watergate Community Assn., Emeryville, Calif., 1981; mem. affirmative action com. Hampshire Coll., 1974-77; cons. Community Tng. and Devel. Project, San Francisco. Walter P. Murphy fellow Northwestern U., 1965-66. Mem. Am. Psychol. Assn., ABA, Orgn. Devel. Network, Assn. Media Psychologists, Bay Area OD Network, Sigma Xi, Sigma Tau, Tau Beta Pi, Alpha Pi Mu. Subspecialties: Organizational psychology; Social psychology. Current work: Organizational psychology and law, management of research and development laboratories, interpersonal and organizational change, small group behavior, legal intellectual property, corporate and employee rights and responsibilities, trust. Home: 525 Hinman Ave Evanston IL 60202 Office: Elan Associates 7 Captain Dr 503 Emeryville CA 94608

KEGEL, GUNTER HEINRICH REINHARD, physics educator, researcher; b. Herborn, Hessen, Germany, June 16, 1929; came to U.S., 1956; s. Wilhelm Ottmar and Gertrud Marie Caroline (Schuler) K.; m. Brita Inga Maria Ahlnas, Sept. 7, 1957; children: Thomas Marcus, Ann Christina. B.S. in Physics, Faculdade Nacional de Filosofia (Universidade do Brasil), 1952, Ph.D., MIT, 1961. Engr. Inst. Nacional Tecnologia, Rio de Janeiro, 1952-56; research asst. MIT, Cambridge, 1958-61; prof. physics Cath. U., Rio de Janeiro, 1961-64; prof. U. Lowell, Mass., 1964—; cons. Controls for Radiation, Cambridge, Mass., 1964, 65, Millipore Corp., Bedford, Mass., 1966; vis. MIT, 1964-66. Mem. Am. Phys. Soc., Electrochem. Soc., Am. Nuclear Soc., IEEE, Acad. Sci. in Rio de Janeiro, N.Y. Acad. Scis. Subspecialties: Nuclear physics; Analytical chemistry. Current work: Neutron scattering on actinides, proton induced x-ray spectroscopy, Rutherford backscattering spectroscopy, neutron induced radiation damage. Office: U Lowell 1 University Ave Lowell MA 01854

KEICHER, WILLIAM EUGENE, elec. engr.; b. Pitts., Dec. 28, 1947; s. William John and Gina Rina (Magrini) K.; m. Barbara Marie Gurgacz, Aug. 12, 1972; children: Lisa Anne, Kathy Marie, William Michael. B.S., Carnegie-Mellon U., 1969, M.S., 1970, Ph.D. in Elec. Engring, 1974. Aerospace intern NASA Manned Spacecraft Ctr., Houston, 1969; lab. asst. Kodak Research Lab., Rochester, N.Y., 1970; sr. elec. engr. CBS Labs., Stamford, Conn., 1973-75; mem. tech. staff M.I.T. Lincoln Lab., Lexington, Mass., 1975—. Contbr. articles to profl. jours. Served to 1st lt. U.S. Army, 1975. NDEA fellow, 1969. Mem. IEEE (sr. mem.), Optical Soc. Am., Soc. Photo-Optical Instrumentation Engrs., Assn. Old Crows, Eta Kappa Nu. Roman Catholic. Subspecialties: Electronics; Optical systems design. Current work: Infrared technology, millimeter wave technology, electro-optic systems, laser technology, sensor-technology. Home: 6 Winn Valley Dr Burlington MA 01803 Office: MIT Lincoln Lab Lexington MA 02173

KEIGWIN, LLOYD DENSLOW, marine geologist; b. N.Y.C., Feb. 26, 1947; s. Lloyd D. and Patricia (Perry) K. A.B., Brown U., Providence, R.I., 1969; M.S., U. R.I., 1976, Ph.D., 1979. Research assoc. Grad. Sch. Oceanography, U. R.I., 1979-80; asst. scientist Woods Hole (Mass.) Oceanographic Inst., 1980—. Contbr. articles to profl. jours. Served to lt. USNR, 1969—. Mem. Geol. Soc. Am., Am. Geophys. Union, AAAS. Subspecialties: Geology; Oceanography. Current work: History of climate and ocean circulation determined from $^{18}O/^{16}O$ and $^{13}C/^{12}C$ studies of $CaCO_3$ in deep sea cores. Office: Woods Hole Oceanographic Inst Woods Hole MA 02543 Home: 20 Burtonwood Ave RFD 1 Buzzards Bay MA 02532

KEIL, ALFRED ADOLF HEINRICH, engineering educator; b. Konradswaldau, Germany, May 1, 1913; came to U.S., 1947, naturalized, 1954; s. Kurt Alfred and Marie (Berger) K.; m. Ursula Leppelt, Oct. 15, 1943; children: Michael G., Juergen G. Dr. nat. sc., U. Breslau, Germany, 1939. Research asst. U. Breslau, 1939-40; research asso. Chem.-Phys. Research Establishment, Kiel, Germany, 1940- 45; chief scientist underwater explosive research div. Norfolk Naval Shipyard, Portsmouth, Va., 1947-59; tech. dir. structural mech. lab. David Taylor Model Basin, 1959-63, tech. dir. basin, 1963-66; prof., head dept. naval architecture and marine engring. MIT, 1966-71, dean Sch. Engring., 1971-77, Ford prof. engring., 1977-78, prof. emeritus, 1978—; Mem. Nat. Adv. Com. on Oceans and Atmosphere, 1977-79. Contbr. articles profl. jours. Served with German Army, 1939-40. Recipient Civilian Distinguished Service award Navy Dept., 1963; Gibbs Bros. gold medal for naval architecture Nat. Acad. Scis., 1967. Mem. Nat. Acad. Engring., Am. Soc. Naval Engrs. (Gold Medal award 1964), Verein Deutscher Ingenieure (corr. mem.), Soc. Naval Architects and Marine Engrs., Marine Tech. Soc. (Lockheed award 1979). Subspecialties: Systems engineering; Ocean engineering. Current work: Development of ocean uses; advancing engineering education. Home: 39 Hillside Terr Belmont MA 02178 Office: Mass Inst Tech Cambridge MA 02139

KEIL, DAVID JOHN, plant toxonomist, educator, cons.; b. Elmhurst, Ill., Dec. 13, 1946; s. John Bell and Clara Elizabeth (Thomas) K. B.Sc., Ariz. State U., 1968, M.Sc., 1970; Ph.D., Ohio State U., 1973. Lectr. Ohio State U., 1973; vis. asst. prof. Grand Valley State Colls., 1973-74, Franklin Coll., 1975; postdoctoral researcher Ariz. State U., 1975-76; lectr. Calif. Poly. State U., 1976-78, asst. prof. biology, 1978-80, assoc. prof., 1980—; dir. Robert F. Hoover Herbarium; environ. cons. Contbr. articles to profl. jours.; editor (with D. R. Walters), A key to selected families and other taxa of plants native, naturalized or cultivated in California, 1978. Served to capt. USAR, 1968-77. Recipient Disting. Teaching award Calif. Poly. State U., 1980. Mem. Am. Soc. Plant Taxonomists, Bot. Soc. Am., Calif. Bot. Soc., Calif. Native Plant Soc., Internat. Assn. Plant Taxonomy, Sociedad Botánica de México, Soc. Systematic Zoology, Torrey Bot. Club, Western Soc. Naturalists, Sigma Xi. Subspecialties: Systematics; Taxonomy. Current work: Systematics, biogeography and evolution of Pectis (Compositae). Field work in U.S., Mex, Central Am. Office: Biol Sics Dept Calif Poly U San Luis Obispo CA 93407

KEIL, KLAUS, geology educator, consultant; b. Hamburg, Ger., Nov. 15, 1934; s. Walter and Elsbeth K.; m. Rosemarie, Mar. 30, 1961; children: Kathrin R., Mark K. M.S., Schiller U., Jena, Ger., 1958; Ph.D., Gutenberg U., Mainz, W.Ger., 1961. Research assoc. Mineral. Inst., Jena, 1958-60, MaxPlanck-Inst. Chemistry, Mainz, 1961, U. Calif.-San Diego, 1961-63; research scientist Ames Research Center NASA, Moffett Field, Calif., 1964-68; prof. geology, dir. Inst. Meteoritics, U. N.Mex., Albuquerque, 1968—; cons. Sandia Labs., others. Contbr. over 350 articles to sci. jours. Recipient Apollo Achievement award NASA, 1970; George P. Merrill medal Nat. Acad. Scis., 1970; Exceptional Sci. Achievement medal NASA, 1977; Regents Meritorious Service medal U. N.Mex., 1983; numerous others. Fellow Meteoritical Soc., AAAS, Mineral. Soc. Am.; mem. Am. Geophys. Union, German Mineral. Soc., others. Subspecialties: Cosmochemistry; Meteoritics. Current work: Origin of the solar system and evolution of the planets based on studies of meteorites, lunar samples, Mars (Viking) terrestrial volcanology, electron beam microanalysis. Office: Dept Geology U N Mex Albuquerque NM 87131

KEISER, BERNHARD EDWARD, consulting engineer; b. Richmond Heights, Mo., Nov. 14, 1928; s. Bernhard and Helen Barbara Julia (Buerkle) K.; m. Florence Evelyn Koenig, Dec. 17, 1929; children: Sandra, Carol, Nancy, Linda, Paul. B.S.E.E., Washington U., St. Louis, 1950, D.Sc.E.E., 1953. Registered profl. engr., Va., Md., D.C. Adminstr., advanced system planning RCA, Moorestown, N.Y., 1967-69; v.p., tech. dir. Page Communications Engrs., Washington, 1969-70; dir. advanced engring. Atlantic Research Corp., Alexandria, Va., 1971-72; dir. analysis Fairchild Space and Electronics Co., Germantown, Md., 1972-75; pres. Keiser Engring., Inc., Vienna, Va., 1975—. Author: EMI Control in Aerospace Systems, 1979; author: Principles of Electromagnetic Compatibility, 1979. RCA fellow, 1951-53; NRC fellow, 1951-53. Fellow IEEE (chmn. No. Va. sect. 1980-81), Washington Acad. Scis., Radio Club Am. Republican. Lutheran. Subspecialties: Electronics; Telecommunications. Current work: Feasibility studies in microwave communication systems, communication satellite transmission and wideband cable technology. Patentee delay line time compressor and expander.

KEITH, DAVID ALEXANDER, oral and maxillofacial surgery educator, researcher; b. Chelmsford, Essex, Eng., Aug. 28, 1944; came to U.S., 1974; s. Kenneth Alexander and Phyllis (Bullock) K.; m. Barbara Anne Lewis, Jan. 10, 1976; children: Sean, Lisa. L.D.S., R.C.S., Royal Coll. Surgeons, Eng., 1966, F.D.S., R.C.S., 1970; B.D.S., Guys Hosp. Dental Sch., London, 1966. Sr. registrar Kings Coll. Hosp., London, 1971-74; instr. oral surgery Harvard U., 1973-78, asst. prof. oral and maxillofacial surgery, 1978—; research assoc. Children's Hosp. Med. Ctr., Boston, 1977—, asst. in oral surgery, 1978-82, Mass. Gen. Hosp., Boston, 1982; cons. Mass. Mental Health Ctr. Mem. Brit. Assn. Oral Surgeons (recipient award 1973), Internat. Assn. Dental Research, Am. Dental Assn., Internat. Assn. Oral Surgeons, Harvard Odontological Soc., Soc. Cranofacial Genetics. Episcopalian. Subspecialties: Oral and maxillofacial surgery; Oral biology. Current work: Craniofacial development; temporomandibular joint; anticonvulsant bone disease; connective tissue biochemistry. Home: 70 Clifton Ave Marblehead MA 01945 Office: Children's Hosp Med Ctr Lab Human Biochemistry 300 Longwood Ave Boston MA 02115

KEITH, THEO GORDON, JR., mechanical engineering educator; b. Cleve., July 2, 1939; s. Theo Gordon and Dorothy (Meech) K.; m. Sandra Jean Finzel, Aug.20, 1960; children: Robin Lynne, Nicole Heather. B.Mech. Engring., Fenn Coll., 1964; M.S. in Mech. Engring, U. Md., 1968, Ph.D., 1971. Mech. engr. Naval Ship Research and Devel. Ctr., Annapolis, Md., 1964-71; prof. mech. engring. U., Toledo,

1971—, dept. chmn., 1977—; prin. investigator NASA grants Lewis Research Ctr., Cleve., 1977-81, 79—, 80—. Recipient ASTM student award, 1963, outstanding tchr. award U. Toledo, 1977. Mem. ASME, Am. Soc. Engring. Edn., AIAA, Soc. Automotive Engrs. (Ralph Teetor award 1978), Sigma Xi. Democrat. Lutheran. Subspecialties: Mechanical engineering; Wind power. Current work: Computational fluid dynamics - numerical modeling of wind turbines, combustors, iced-helicopter blades. Laser doppler anemometer - supersonic flow measurements. Home: 3866 LaPlante Rd Monclova OH 43542 Office: Dept Mech Engring U Toledo 2801 W Bancroft St Toledo OH 43606

KELLAR, KENNETH JON, pharmacologist, educator; b. Balt., Feb. 13, 1944; s. Joseph Aaron and Faye E. (Terkowitz) K.; m. Elizabeth Kaenzig, Sept. 24, 1972; children: Joshua Aaron, Amanda Marin. B.S., John Hopkins U., 1966; Ph.D., Ohio State U., 1974. Postdoctoral fellow NASA Ames Research Center, 1974-76; assoc. prof. dept. pharmacology Georgetown U. Sch. Medicine, Washington, 1976—. Contbr. articles to sci. jours. Mem. Am. Soc. Pharmacology and Exptl. Therapeutics, Soc. for Neurosci. Subspecialties: Molecular pharmacology; Neuropharmacology. Current work: Neuropharmacology. Office: Dept Pharmacology Georgetown U Sch Medicine Washington DC 20007

KELLAR, LUCIA AMES, neuropsychologist, educator, consultant; b. Montclair, N.J., Sept. 23, 1945; d. Curtis Bradbury and Mary (Ames) Poor. A.B., Mt. Holyoke Coll., 1968; M.A., Columbia U., 1974, M.Phil., 1975, Ph.D., 1976. Lic. psychologist, N.Y. Research assoc. neurol. unit Boston State Hosp., 1969-72; vis. asst. prof. Columbia U., 1976-78; instr. NYU Med. Ctr., 1976-78, research asst. prof. neurology, 1978—; sr. neuropsychologist Inst. Research in Behavioral Neurosci., N.Y.C., 1980—; psychologist Bellevue Geriatric Clinic, N.Y.C., 1981—; co-dir. Neuropsychology Clinic, VA Med. Ctr., N.Y.C., 1980—. Author: (with Locke and Caplan) A Study in Neurolinguistics, 1973. Nat. Research Service awardee NIH, 1976-78; NIH grantee, 1978-80. Mem. Acad. Aphasia, Internat. Neuropsychol. Soc., N.Y. Neuropsychology Group. Democrat. Episcopalian. Subspecialties: Neuropsychology; Cognition. Current work: Representation of cognitive functions in the brain, hemispheric asymmetries; aphasia. Home: 405 W 23d St 7K New York NY 10011 Office: NYU Med Ctr 550 1st Ave New York NY 10016

KELLEHER, WILLIAM JOSEPH, pharmacognosy educator; b. Hartford, Conn., July 18, 1929; s. Richard Francis and Julia (Bogash) K. B.S., U. Conn., 1951, M.S., 1953; Ph.D., U. Wis.-Madison, 1960. Asst. prof. U. Conn., Storrs, 1960-66, assoc. prof., 1967-69, prof., 1970—, asst. dean, 1976-81, acting dean, 1981; guest prof. U. Frieburg, W.Ger., 1970-71, 77. Contbr. articles on pharmacognosy to profl. jours.; assoc. editor: Lloydia, 1971-76. Served to 1st lt. USMC, 1953-55. Mem. Am. Soc. Pharmacognosy (pres. 1973-74), Am. Chem. Soc., Biochem. Soc., Am. Assn. Colls. Pharmacy. Democrat. Roman Catholic. Subspecialties: Biochemistry (medicine); Microbiology. Current work: Microbial and plant cell culture, biosynthesis of pharmacological agents. Home: 840 Wormwood Hill Rd Storrs CT 06268 Office: U Conn PO Box U-92 Storrs CT 06268

KELLEN, JOHN ANDREW, biochemist; b. Vienna, Austria, July 18, 1928; emigrated to Can., 1968, naturalized, 1972; s. Charles F. and Josepha (Kellen) Koch; m. Marta Hornakova, Jan. 17, 1959; children: Charles, John. M.D., U. Bratislava, Czechoslovakia, 1952; Ph.D., U. Brno, 1963. From intern to sr. resident in medicine and endocrinology Levoca (Czechoslovakia) Hosp., 1951-56; head biochemistry dept. Levoca and Dun Streda, Czechoslovakia, 1956-63; research fellow Research Inst. Hygiene, Bratislava, 1963-66; lectr. Inst. Postgrad. Med. Edn., U. Bratislava, 1964-68; head dept. biochemistry Cancer Inst. Bratislava, 1966-68; mem. faculty U. Toronto Faculty Medicine, 1968—; prof. clin. biochemistry, 1980—; mem. staff dept. clin. biochemistry Sunnybrook Med. Ctr., Toronto, 1968—. Med. editor: Modern Medicine Can, 1969—; Author papers in field. Mem. Can. Fedn. Biol. Soc., N.Y. Acad. Scis. Subspecialties: Cancer research (medicine); Immunology (medicine). Current work: Tumor markers in clinical practice, ectopic hormones and polypeptides in tumors, tumor immunology.

KELLER, JAIME, theoret. chemist, physicist; b. Mexico, D.F., Mexico, Nov. 10, 1936; s. Arturo and Rosario (Torres) K.; m. Cristina Perez, Apr. 29, 1943; children: Cristina, Alejandro, Roberto. B.S. in Chem. Engring, Universidad Nacional Autonoma de Mexico, 1950; Ph.D. in Physics, U. Bristol, 1971. Registered Profl. Engr., Mexico. Project engr. Industrial Quimica Pensalt, Mexico, 1959-61; tech. dir. Derivados Macroquimicos, Mexico, 1961; lectr. Universidad Nacional Autonoma de Mexico, 1961-72, prof., 1972-76, heat theoret. chemistry dept., 1974-76, prof. physics, 1976—, mem. acad. council. Contbr. numerous articles to sci. jours. Mem. Am. Phys. Soc., Societa Italiana di Fisica, European Phys. Soc., Internat. Soc. Quantum Biology, Sociedad Quimica de Mexico, Sociedad Mexicana de Fisica, Academia de la Investigacion Cientifica de Mexico, Hydrogen Energy Soc. Roman Catholic. Subspecialties: Condensed matter physics; Theoretical chemistry. Current work: The chemistry of condensed matter physics specially metals in the liquid, amorphous and crystalline state and the fundamental theory behind chemistry and condensed matter physics; appplications to actual technological problems. Developer chem. industry processes. Home: 64 Fuente de la Juventud Mexico DFMexico 11000 Office: Faculty de Quimica Ciudad Universitaria Universidad Nacional Autonoma de Mexico Mexico DFMexico

KELLER, JOHN CHARLES, dental educator, researcher; b. Rockville Centre, N.Y., Aug. 11, 1952; s. Richard Frederick and Margaret (Davidson) K.; m. Gail Ann Riggle, Sept. 26, 1981. B.S., U. Ill.-Urbana, 1974; M.S., Northwestern U., 1978, Ph.D. in Biol. Materials, 1982. Asst. prof. biophys. dentistry Med. U. S.C., 1982—; cons. Greemark, Inc., Chgo., 1977-82, Young & Assocs., Charleston, S.C., 1982—. Contbr. articles profl. jours. Biomed. research support grantee Med. U. S.C., 1982. Mem. Soc. Biomaterials, Am. Assn. Dental Research (1st place award for research 1980); fellow Acad. Dental Materials; Mem. Sigma Xi (2d place award 1982). Subspecialties: Biomaterials; Toxicology (medicine). Current work: Dental and medical materials in terms of biocompatibility properties, structure and functional properties of bone cements. Office: Medical U SC 171 Ashley Ave Charleston SC 29425

KELLER, JOHN RANDALL, microbiology educator; b. Ogdensburg, N.Y., Dec. 14, 1925; s. Lloyd Marley and Helen Eugenia (Roe) K.; m. Reita R. Radway, Aug. 6, 1960. B.S., Cornell U., 1947, Ph.D., 1952. Assoc. prof. microbiology Seton Hall U., 1960—. Recipient Leonard H. Vaughn award, 1950. Mem. Am. Soc. Microbiology, Mycol. Soc. Am., Sigma Xi. Presbyterian. Club: Cosmopolitan (Montclair, N.J.). Subspecialties: Microbiology; Virology (biology). Current work: Agrobacterium-fungus interaction, camp effect on fungal sporulation. Office: Dept of Biology Seton Hall U South Orange NJ 07079

KELLER, PATRICIA J., dental educator, researcher; b. Detroit, Nov. 16, 1923. B.S., U. Detroit, 1945; Ph.D. in Biochemistry, Washington U., St. Louis, 1953. USPHS fellow U. Wash., Seattle, 1954-55, Research assoc. biochemistry, 1955-56, instr., 1956-57, research asst. prof., 1957-62, assoc. prof., 1962-67, assoc. dean, 1974-77, prof. oral biology, 1967—, chmn. oral biology, 1979—; USPHS fellow Washington U., 1953-54; vis. fellow Inst. Marine Biochemistry, Aberdeen, Scotland, 1978-79. Mem. AAAS, Am. Soc. Biol. Chemistry, Am. Soc. Cell Biology, Am. Chem. Soc., Internat. Assn. Dental Research. Subspecialties: Oral biology; Biochemistry (biology). Office: Dept Oral Biolog U Wash Sch Dentistr Seattle WA 98195

KELLER, ROBERT MARION, computer science educator, researcher; b. St. Louis, June 12, 1944; m., 1967. B.S., Washington U., St. Louis, 1966, M.S., 1968; Ph.D. in Elec. Engring. and Computer Sci, U. Calif.-Berkeley, 1970. Asst. prof. elec. engring. Princeton U., 1970-76; assoc. prof. U. Utah, Salt Lake City, 1976-81, prof. computer sci., 1981—. Mem. Assn. Computing Machinery, IEEE. Subspecialty: Computer engineering. Office: U Utah Dept Computer Sci 3160 MEB Salt Lake City UT 84112

KELLERMANN, KENNETH IRWIN, radio astronomer; b. N.Y.C., July 1, 1937; m., 1967; 1 child. S.B., MIT, 1959; Ph.D. in Physics and Astronomy, Calif. Inst. Tech., 1963. Research scientist Radiophys. Lab., Commonwealth Sci. and Indsl. Research Orgn., 1963-65; from asst. scientist to assoc. scientist Nat. Radio Astronomy Obs., Green Bank, W.Va., 1965-69, scientist, 1969-77, asst. dir., 1977, sr. scientist, 1978—; lectr. Leiden U., 1967; research assoc. Calif. Inst. Tech., 1969; adj. prof. U. Ariz., 1970-73; dir. Max Planck Inst. Radio Astronomy, 1977-79. NSF fellow, 1965-66. Mem. Nat. Acad. Sci. (B.A. Gould prize 1973), Am. Astron. Soc. (Helen B. Warner prize 1971), Am. Acad. Arts and Sci. (Rumford prize 1970), Inter at. As ron. Union, Inter. Radio Sci. Union. Subspecialty: Radio and microwave astronomy. Office: Nat Radio Astron Obs PO Box 2 Green Bank WV 24944

KELLEY, ALBERT JOSEPH, management executive; b. Boston, July 27, 1924; s. Albert Joseph and Josephine (Sullivan) K.; m. Virginia Marie Riley, June 7, 1945; children: Mark, Shaun, David. B.S., U.S. Naval Acad., 1945, MIT, 1948; Sc.D. in Instrumentation and Control Engring., MIT, 1956; postgrad., U.S. Naval Postgrad. Sch., 1953-54. Commd. ensign USN, 1945, advanced through grades to comdr.; fire control officer U.S.S. Rochester, 1946-47; carrier squadron pilot, electronics officer USN Carrier Air Group 2, 1950-51; exptl. test pilot, project dir. U.S. Naval Air Test Center, Patuxent River, Md., 1951-53; asst. head guided missile guidance br. Bur. Weapons, 1956-58, project dir. Eagle missile system Bur. Weapons, 1958- 60; program mgr. Agena launch vehicle NASA, 1960-61, dir. electronics and control, 1961-64; dep. dir. Electronics Research Center, Cambridge, Mass., 1964-67; dean Sch. Mgmt., Boston Coll., 1967-77; pres. Arthur D. Little Program Systems Mgmt. Co., 1977—; cons. Dept. Def.; dir. LFE Corp., State St. Bank and Trust, State St. Fin Corp., Perini Corp., Nat. Space Inst., C.S. Draper Lab. Author: Venture Capital, A Guidebook for New Enterprises, New Dimensions of Project Management. Mem. space applications bd. NRC. Recipient NASA Exceptional Service medal, 1967. Fellow AIAA, IEEE; mem. Internat. Acad. Astronautics, Sigma Xi, Tau Beta Pi, Eta Kappa Nu, Sigma Gamma Tau, Beta Gamma Sigma. Subspecialties: Systems engineering; Aerospace engineering and technology. Current work: Systems engring. and tech. cons. tollarge, complex projects in communications, command and control FAA, defense, internat. telecommunications, major construction projects, aerospace and energy. Home: 351 Atherton St Milton MA 02187 Office: Arthur D Little Inc Acorn Park Cambridge MA 02140

KELLEY, HENRY JOSEPH, aerospace research engr., cons., educator; b. N.Y.C, Feb. 8, 1926; s. Bernard Joseph and Margaret Elizabeth (McKillop) K.; m. Maureen Grace Youngkin, Oct. 26, 1958; children: Henry Bernard, Maureen Elizabeth Anne. B.Aero.E., N.Y. U., 1948, M.S. in Math, 1951, Sc.D. Aero. E., 1958. Asst. chief research Grumman Aerospace Corp., Bethpage, N.Y., 1948-63; v.p. Analytical Mechs. Assocs., Inc., Jericho, N.Y., 1963-78; prof. aerospace engring. Va. Poly. Inst., Blacksburg, 1978—; pres; Optimization Inc., Blacksburg, 1978—. Contbr. articles to profl. jours. Recipient Pendray award AIAA, 1979. Fellow AIAA; mem. IEEE, Am. Astron. Soc., Soc. Indsl. and Applied Math, Internat. Fedn. Automatic Control (chmn. math. of control com.). Clubs: Historic Sports-Car, Southeast Vintage Racing Assn. Subspecialties: Aerospace engineering and technology; Applied mathematics. Current work: Dynamics, control, guidance, systems optimization; aerospace applications in these areas. Home: 29 High Meadow Dr Blacksburg VA 24060 Office: Dept Aero Engring Va Poly Inst Blacksburg VA 24061

KELLEY, MARK ELBRIDGE, III, solar engr.; b. Cambridge, Mass., June 11, 1948; s. Mark Elbridge and Adelaide (True) K.; m. Josephine Seymour Carothers, July 24, 1982. B.A., Bowdoin Coll., 1970, Northeastern U., 1978. Researcher U.S. Army Materials Lab., Watertown, Mass., 1974-78; energy systems engr. Acorn Structures, Inc., Concord, Mass., 1978—, head solar dept., 1979—. Served with USCG, 1970-74. Mem. Mass. Bay Solar Energy Assn. (chmn.), Am. Solar Energy Assn., New Eng. Solar Energy Assn., ASHRAE, ASME. Subspecialties: Solar energy; Mechanical engineering. Current work: Solar energy, residential active and passive solar heating. Office: PO Box 250 Concord MA 01742

KELLEY, MICHAEL STEPHEN, computer cons.; b. Chgo., Feb. 8, 1949; s. Kenneth G. and Alice S. (Wisowaty) K.; m. Kathryn L. Ratiu, Apr. 24, 1982. B.S. in Psychology, U. Ill., 1971. Various tech. positions and employers assoc. with Illiac IV Computer, 1969-77; dir. software devel. Microcomputer Systems Corp., Sunnyvale, Calif., 1977-80; pres. Symbionics, San Jose, Calif., 1980—. Mem. Assn. Computing Machinery. Subspecialties: Operating systems; Graphics, image processing, and pattern recognition. Current work: Developing graphics processing hardware and software. Developing fax subsystems for teleconferencing. Operating systems. Home and Office: 684 Royal Glen Dr San Jose CA 95133

KELLEY, NEIL DAVIS, environm. scientist; b. Clayton, Mo., Jan. 8, 1942; s. Davis Franklin and Louise Minnie (Zager) K.; m. Jean Irish, Jan. 14, 1967 (div. June 1977). B.S., St. Louis U., 1963; M.S., Pa. State U., 1968. Staff meteorologist Meteorology Research Inc., Altadena, Calif., 1963-67; project supr. Exxon Research & Engring. Co., Linden, N.J., 1967; instr. Pa. State U., University Park, 1967-72; group chief Nat. Center Atmospheric Research, Boulder, Colo., 1972-77; prin. scientist Solar Energy Research Inst., Golden, Colo., 1977—; tech. reviewer Am. Wind Energy Assn., Washington, 1982—. Recipient Spl. Achievement award Nat. Center, Atmospheric Research, 1974, Outstanding Achievement award Solar Energy Research Inst., 1982. Mem. Instrument Soc. Am. (sr.), Am. Meteorol. Soc., AIAA, AAAS, Sigma Xi. Subspecialties: Micrometeorology; Wind power. Current work: Environmental compatibility research of wind energy conversion systems; acoustical, electromagnetic, turbulence-induced fatigue. Office: Solar Energy Research Inst 1617 Cole Blvd Golden CO 80401 Home: 605 S 42d St Boulder CO 80303

KELLEY, PATRICIA HAGELIN, geology educator; b. Cleve., Dec. 8, 1953; d. Daniel Warn and Virginia Louise (Morgan) Hagelin; m. Jonathan Robert Kelley, June 18, 1977; 1 son, Timothy Daniel. B.A., Coll. of Wooster, 1975; A.M., Harvard U., 1977, Ph.D., 1979. Instr. geology New Eng. Coll., Henniker, N.H., 1979; asst. prof. geology U. Miss., University, 1979—; Tutor Laubach Literacy Internat., Olive Branch, Miss., 1981—. Contbr. articles to profl. jours. U. Miss. Faculty grantee, 1982-83. Mem. Paleontol. Soc., Geol. Soc. Am., AAAS, Miss. Acad. Scis., Phi Beta Kappa, Sigma Xi. Democrat. Presbyterian. Subspecialties: Paleobiology; Paleoecology. Current work: Evolutionary patterns of Miocene molluscs; Tertiary naticid gastropod predation; Cretaceous Gulf Coastal Plain biostratigraphy. Home: 3850 Bethel Rd Olive Branch MS 38654 Office: Geology and Geol Engring U Miss University MS 38677

KELLEY, PAUL LEON, physicist; b. Phila., Dec. 8, 1934; s. Henry Paul and Valerie Clementine (Courtin) K.; m. Patricia Louise Pieretti, June 14, 1958; children: Matthew William, Diana Reuss. B.A., Rutgers U., 1956; M.S., Cornell U., 1959; Ph.D., MIT, 1962. Mem. staff MIT Lincoln Lab., Lexington, Mass., 1962-68, asst. group leader, 1969-71, assoc. group leader, 1971—; lectr. MIT, Cambridge, Mass., 1966-67; vis. lectr. U. Calif., Berkeley, 1968-69; cons. mem. Working Group D (Lasers) of Office of Dir. of Def. Research and Engring.'s Adv. Group on Electron Devices, 1973—. Contbr. numerous articles on quantum electronics to profl. jours.; editor: (with B. Lax and P.E. Tannenwald) Physics of Quantum Electronics, 1966; editor series: (with P.F. Liao) Quantum Electronics: Principles and Applications, 1976—; editor: Optics Letters, 1984—. Fellow Am. Phys. Soc., Optical Soc. Am. (dir. and chmn. tech. council 1982-83); mem. AAAS, Sigma Xi. Subspecialties: Spectroscopy; Condensed matter physics. Current work: Laser spectroscopy, nonlinear optics, remote sensing. Co-patentee Method and Apparatus for Compression Optical Laser Pulses. Office: PO Box 73 Lexington MA 02173

KELLING, CLAYTON LYNN, virologist, educator; b. Killdeer, N.D., Mar. 26, 1946; s. Ervin L. and Aurora J. K.; m. Nancy Carol Amundson, Oct. 5, 1974; children: Sarah Elizabeth, Jessica Lynn. B.S., N.D. State U., 1968, M.S., 1971, Ph.D., 1975. Technologist dept. vet. sci. N.D. State U., 1968-76; asst. prof. vet. sci. U. Nebr., Lincoln, 1976-82, assoc. prof., 1982—. Contbr. articles on infectious diseases of livestock to profl. jours. Mem. Am. Soc. Microbiology, Research workers in Animal Disease, Am. Assn. Vet. Lab. Diagnosticians, Sigma Xi, Alpha Zeta, Gamma Sigma Delta. Republican. Lutheran. Subspecialty: Virology (veterinary medicine). Current work: Research on viral diseases of livestock. Home: RFD 8 Lincoln NE 68506 Office: Dept Vet Sci U Nebr Lincoln Ne 68583

KELLNER, RICHARD GEORGE, computer scientist, consultant; b. Cleve., July 10, 1943; s. George Ernest and Wanda Julia (Lapinski) K.; m. Charlene Ann Zajc, June 26, 1965; children: Michael Richard, David George. B.S. in Math, Case Inst. Tech., 1965, M.S., Stanford U., 1968, Ph.D., 1969. Staff mem. Los Alamos (N.Mex.) Sci. Lab., 1969-79; dir. software devel. KMP Computer Systems, Los Alamos, 1979—; co-owner Computer-Aided Communications, Los Alamos, 1982—; cons., 1979—. Mem. IEEE, AAAS, Assn. for Computing Machinery. Republican. Subspecialties: Database systems; Distributed systems and networks. Current work: Computer-aided communications, multi-processor computer systems. Developer Cable Star TV Computer program, 1982; co-developer Common Graphics System, computer program, 1979. Home: 4496 Ridgeway Dr Los Alamos NM 87544 Office: KMP Computer Systems Inc 703 Central Ave Los Alamos NM 87544

KELLOGG, HERBERT HUMPHREY, educator, metallurgist; b. N.Y.C., Feb. 24, 1920; s. Herbert H. and Gladys (Falding) K.; m. Jeanette Halstead, July 20, 1940; children—Thomas Bartlett, Jane Falding, David Humphrey, Elizabeth Ann. B.S., Columbia, 1941, M.S., 1943. Asst. prof. mineral preparation Pa. State U., State Coll., 1942-46; faculty Columbia, N.Y.C., 1946—, Stanley-Thompson prof. chem. metallurgy, 1968—; Chmn. titanium adv. com. Office Def. Mblzn., 1954-58. Research; contbr. numerous articles to publs. Recipient Best Paper award extractive metals div. Am. Inst. Mining., Metall. and Petroleum Engrs.; James Douglas Gold medal Am. Inst. Mining, Metall. and Petroleum Engrs., 1973. Fellow Am. Inst. Mining, Metall. and Petroleum Engrs. (chmn. extractive metallurgy div. 1958), Metall. Soc., Instn. Mining and Metallurgy (London); mem. Am. Chem. Soc., Nat. Acad. Engrs., Sigma Xi, Tau Beta Pi. Subspecialties: Metallurgical engineering; Thermodynamics. Current work: Thermodynamic behavior of liquid slags, alloys and sulfide solutions (mattes) in metal production; energy utilization in metal production. Home: Closter Rd Palisades NY 10964 Office: Columbia New York City NY 10027

KELLOGG, RALPH HENDERSON, physiologist; b. New London, Conn., June 7, 1920; s. Edwin Henry and Constance Louise (Henderson) K. B.A., U. Rochester, N.Y., 1940, M.D., 1943; Ph.D., Harvard U., 1953. Intern, Univ. Hosps., Cleve., 1944; investigator physiology U.S. Naval Med. Research Inst., Bethesda, Md., 1946; teaching fellow physiology Harvard Med. Sch., Boston, 1946-47, instr., 1947-53; asst. prof. physiology U. Calif. - Berkeley, 1953-58, 1958-59, asso. prof., 1959-65, prof., 1965—, lectr. history of health scis., 1978—, acting chmn. physiology, 1966-70; mem. physiology study sect. NIH, 1966-70; physiology test com. Nat. Bd. Med. Examiners, 1966-73, chmn., 1969-73; com. respiration nomenclature Internat. Union Physiol. Scis., 1970-77, commn. respiratory physiology, 1975-81; editorial com. U. Calif. Press, 1972-76. Physiology editor: Stedman's Med. Dictionary, 1972—; joint editorial bd.: Am. Jour. Physiology and Jour. Applied Physiology, 1962-66; editorial bd.: Jour. Applied Physiology, 1977-79; contbr. articles to profl. publs.; contbg. author books on physiology of saline and urea diuresis, respiration, high-altitude acclimatization, history of physiology. Served with M.C., USNR, 1943-46. Sr. research fellow Harvard U., 1962-63; vis. fellow Corpus Christi Coll. Univ. Lab. Physiology, Oxford (Eng.) U., 1970-71; vis. scientist Laboratoire de Physiologie Respiratoire, Centre National de la Recherche Scientifique, Strasbourg, France, 1977; NIH research grantee, 1962-76. Mem. AAAS, Am. Physiol. Soc., AAUP, Am. Assn. History of Medicine, History of Sci. Soc., Phi Beta Kappa, Sigma Xi, Alpha Omega Alpha. Clubs: Roxburghe, Harvard. Subspecialties: Physiology (medicine). Current work: Professional research publications on respiratory physiology, high altitude acclimatization, history of physiology. Teaching organ system physiology to medical students, Doctor of Pharmacy students, and P.h.D. students. Home: 601 Noriega St San Francisco CA 94122 Office: Dept Physiology U Calif San Francisco CA 94143

KELLOGG, ROBERT LEA, naval officer, researcher; b. Los Angeles, Nov. 21, 1945; s. Clyde William and Evelyn Lea (Schwartz) K.; m. Judith Ann Stadnyk, Dec. 26, 1970; children: Glenn Bryan, Laurel Leann. B.A., U. So. Calif., 1967; M.S., Case Inst. Tech., 1970, Naval Postgrad. Sch., 1980, Ph.D., 1981. Commd. U.S. Navy, 1969, advanced through grades to lt comdr., 1978; with Naval Security Group Command, Washington, 1981—; tchr. extension courses Pepperdine U., 1972-74, N.H. Coll., 1975-78. Mem. Am. Astron. Soc., AAAS. Republican. Episcopalian. Subspecialties: Aerospace engineering and technology; Satellite studies. Current work: Ionospheric physics, satellite communication, communication theory, magnetospheric research.

KELLOGG, WILLIAM WELCH, meteorologist; b. New York Mills, N.Y., Feb. 14, 1917; s. Frederick S. and Elizabeth (Walcott) K.; m. Elizabeth Thorson, Feb. 14, 1942; children: Karl S., Judith Liebert, Joseph W., Jane K. Holien, Thomas W. B.A., Yale U., 1939; M.A., U. Calif. at Los Angeles, 1942, Ph.D., 1949. With Inst. Geophysics, U. Calif. at Los Angeles, 1946-52, asst. prof., 1950-52; with Rand Corp., Santa Monica, Calif., 1947-64, head planetary scis. dept., 1959-64; asso. dir. Nat. Center Atmospheric Research, Boulder, Colo., also dir.

lab. atmospheric scis., 1964-73, sr. scientist, 1973—; Mem. earth satellite panel IGY, 1956-59; mem. space sci. bd. Nat. Acad. Scis., 1959-68, mem. com. meteorol. aspects of effects of atomic radiation, 1956-58, mem. com. atmospheric scis., 1966-72, mem. polar research bd., 1972-77; mem. Rocket and Satellite Research Panel, 1957-62; mem. adv. group supporting tech. for operational meteorol. satellites NASA-NOAA, 1964-72; rapporteur meteorology of high atmosphere, commn. aerology World Meteorol. Orgn., 1965-71; chmn. internat. commn. meteorology upper atmosphere Internat. Union Geodesy and Geophysics, 1960-67, mem., 1967-75; mem. internat. com. climate Internat. Assn. Meteorology and Atmospheric Physics, 1978—; mem. sci. adv. bd. USAF, 1956-65; chmn. meteorol. satellite com. Advanced Research Projects Agy., 1958-59; mem. panel on environment President's Sci. Adv. Com., 1968-72; mem. space program adv. council NASA, 1976-77; chmn. meteorol. adv. com. EPA, 1970-74, mem. nat. air quality criteria adv. com., 1975-76, air pollution transport and transformation adv. com., 1976-78; mem. council on carbon dioxide environ. assessment Dept. Energy, 1976-78; adv. to sec. gen. on World Climate Program, World Meteorol. Orgn., 1978-79; dir. research Naval Environ. Prediction Research Facility, Monterey, Calif., 1983-84. Served as pilot-weather officer USAAF, 1941-46. Co-recipient spl. award pioneering work in planning meteorol. satellite Am. Meteorol. Soc., 1961; recipient Risseca award contbn. human relations in scis. Jewish War Vets. U.S.A., 1962-63; Exceptional Civilian Service award Dept. Air Force, 1966. Fellow Am. Geophys. Union (pres. meteorol. sect. 1972-74), Am. Meteorol. Soc. (council 1960-63, pres. 1973-74); mem. AAAS (chmn. atmospheric and hydrospheric sect. 1984), Sigma Xi. Club: Cosmos (Washington). Subspecialties: Meteorology; Climatology. Current work: Research on meterology, dynamics and turbulence of upper atmosphere, use rockets and satellites for atmospheric research; prediction radioactive fallout and dispersal; applications of infrared techniques; atmospheres of Mars and Venus; theory of climate and causes of climate change. Research on meteorology, dynamics and turbulence of upper atmosphere, use rockets and satellites for atmospheric research; prediction radioactive fallout and dispersal; applications of infrared techniques; atmospheres of Mars and Venus; theory of climate and causes of climate change. Home: 445 College Ave Boulder CO 80307 Office: Nat Center Atmospheric Research Boulder CO 80307 If there is anything that generally characterizes a gratifying and successful career in science, it is the challenge of diversity. The really important problems of the universe, and especially of society, involve several disciplines, and we are compelled to work at these discipline interfaces. Pigeon holes are for pigeons, not scientists

KELLY, CATHERINE LOUISE, mech. engr.; b. Norfolk, Va., Mar. 8, 1958; d. and Carol W. (Wurst) K. B.S.M.E., U. Colo., 1980; postgrad. in mech. engring, U. Minn., Mpls., 1981—. Registered engr.-in-tng., Colo., 1980. Prodn. engr., Honeywell, Hopkins and New Brighton, Minn., 1980—. Mem. ASME, Minn. Soc. Profl. Engrs., Nat. Soc. Profl. Engrs. Democrat. Roman Catholic. Subspecialty: Mechanical engineering. Current work: Development program concerning tool design and try-out, production processes, schedule, cost reductions, improved product function. Home: 1601 N Innsbruck Dr apt 304 Fridley MN 55432 Office: Honeywell TCAAP Bldg 103 New Brighton MN 55112

KELLY, HUGH P., physics educator, researcher; b. Boston, Sept. 3, 1931; m., 1955; 6 children. A.B., Harvard U., 1953; M.S., UCLA, 1954; Ph.D. in Physics, U. Calif.-Berkeley, 1963. Physicist U. Calif.-Berkeley, 1962-63; research physicist U. Calif.-San Diego, 1963-64, research asst. prof., 1964-65, lectr., 1965; from asst. prof. to assoc. prof. and assoc. dean U. Va., Charlottesville, 1965-70, prof., 1970-77, Commonwealth prof. physics, 1977—, now also chmn. dept. physics. Fellow Am. Phys. Soc. Subspecialty: Theoretical physics. Office: Dept Physics U Va Charlottesville VA 22903

KELLY, JEFFREY JOHN, chemistry educator; b. Portland, Oreg., Nov. 2, 1942; s. John Lloyd and Lucile (Brainard) K.; m. Katherine Amy Schmidt, June 11, 1966; children: Heidi Welton, Lisa, Shawna. B.S. in Chemistry, Harvey Mudd Coll., 1964, Ph.D., U. Calif.-Berkeley, 1964. Research assoc. Lawrence Radiation Labs., Berkeley, Calif., 1968; asst. prof. chemistry Reed Coll., Portland, Oreg., 1968-72; mem. faculty Evergreen State Coll., Olympia, Wash., 1972—; vis. prof. chemistry Harvey Mudd Coll., Claremont, Calif., 1980-81. Mem. AAAS. Subspecialties: Biophysical chemistry; Photosynthesis. Current work: Infrared and nmr spectroscopy of biological molecules.

KELLY, KEVIN ANTHONY, research institute executive, engineering educator; b. Croydon, Eng., Mar. 29, 1945; came to U.S., 1949; s. James Gerard and Anne (Donahue) K.; m. Rita Gail Freeman, Feb. 14, 1981. B.S., Notre Dame U., 1967; M.S., Yale U., 1968, Ohio State U., 1974, 76. Sr. research assoc. Coll. Engring., Ohio State U., Columbus, 1976-77; assoc. dir. Nat. Regulatory Research Inst., Columbus, 1977—; cons. Pub. Utilities Control Authority, Hartford, Conn., 1976. Mem. Am. Nuclear Soc. Subspecialties: Nuclear engineering; Materials. Current work: Public utility policy research, electric utility economics, gas utility pricing policy, nuclear power cost studies. Office: Nat Regulatory Research Inst 2130 Neil Ave Columbus OH 43210

KELLY, PATRICK JOSEPH, neurosurgeon; b. Lackawanna, N.Y., Sept. 19, 1941; s. Joseph P. and Mary (Connor) K.; m. Carol Huey, Nov. 20, 1981; children: Patrick J., Michael. Student, U. Mich., 1959-62; M.D., SUNY, Buffalo, 1967. Diplomate: Am. Bd. Neurol. Surgery. Intern Phila. Naval Hosp., 1966-67; resident in neurosurgery Northwestern U., Chgo., 1970-72, U. Tex., Galveston, 1972-74; resident Hopital St. Anne, Paris, 1977, Hopital Foch, 1977, Western Gen. Hosp., Edinburgh, Scotland, 1977; practice medicine specializing in neurosurgery, Buffalo, 1979—; attending neurosurgeon Erie County Med. Center, Millard Fillmore Hosp., VA Med. Center, Buffalo Gen. Hosp., Mercy Hosp., Our Lady of Victory Hosp.; instr. U. Tex., Galveston, 1974-75, asst. prof., 1975-78, assoc. prof., 1978-79, SUNY, Buffalo, 1979—; chief neurosurgery Sisters of Charity Hosp., 1980—. Contbr. articles to profl. jours. Served to lt. cmdr. USN, 1967-70. Am. Cancer Soc. grantee, 1973; William P. Van Wagenen fellow Am. Assn. Neurol. Surgeons, 1976; Northwestern U. grantee, 1971-72; U. Tex. grantee, 1973-74; So. Med. Assn. grantee, 1975-76; NINCDS grantee, 1977—. Mem. Am. Assn. Neurol. Surgeons, AMA, Am. Soc. Laser Medicine and Surgery, Am. and World Soc. Sterotactic and Functional Neurosurgery, Congress Neurol. Surgeons, Galveston County Med. Soc., Houston Neurol. Soc., Internat. Assn. Study Pain, N.Y. State Neurosurg. Soc., Rocky Mountain Neurosurg. Soc., Singleton Surg. Soc., Soc. Neurosci. Tex. Med. Assn., Erie County Med. Soc., N.Y. State Neurosurg. Soc., Sigma Xi. Subspecialties: Neurosurgery; Laser surgery. Current work: Sterotactic laser neurosurgery and research. Office: 2121 Main St #308 Buffalo NY 14214

KELMAN, ARTHUR, educator, plant pathologist; b. Providence, Dec. 11, 1918; s. Philip and Minnie (Kollin) K.; m. Helen Moore Parker, June 22, 1949; 1 son, Philip Joseph. B.S., U. R.I., 1941, D.Sc. (hon.), 1977; M.S., N.C. State U., 1946; Ph.D., 1949; postgrad., U. Wis., 1947-48. Faculty N.C. State U., Raleigh, 1948-65, prof., 1957-65, W.N. Reynolds distinguished prof. plant pathology, 1961-65; chmn. dept. plant pathology U. Wis., Madison, 1965-75, L.R. Jones disting. prof., 1975—, prof. bacteriology, 1977—; vis. investigator Rockefeller Inst., 1953-54; vis. lectr. Am. Inst. Biol. Sci., 1961-62; chmn. div. biol. sci. Assembly Life Sci. NRC, 1980-82. Author: The Bacterial Wilt Caused by Pseudomonas solanacearum, 1953. Chmn. div. biol. scis. NRC, 1979-82; chmn. sect. applied biology Nat. Acad. Scis., 1981—. Served with AUS, 1942-45. NSF sr. postdoctoral fellow Cambridge (Eng.) U., 1971-72. Fellow Am. Phytopath. Soc. (chmn. sourcebook com., councilor-at-large, v.p. 1965-66, pres. 1966-67), AAAS; mem. Internat. Soc. Plant Pathology (v.p. 1968-73, pres. 1973-78), Nat. Acad. Scis. (chmn. sect. applied biology 1981-83), Am. Acad. Arts and Scis., Soc. Gen. Microbiology, Am. Soc. Microbiology, Am. Inst. Biol. Sci., Sigma Xi, Alpha Zeta, Gamma Sigma Delta, Phi Kappa Phi, Phi Sigma, Xi Sigma Pi. Subspecialties: Plant pathology; Microbiology. Current work: Bacterial doseases of plants; mechanisms of pathogenesis; tissue maceration; post harvest pathology; calcium nutrition and soft rot resistance. Home: 234 Carillon Dr Madison WI 53705

KELMAN, BRUCE JERRY, toxicologist, researcher; b. Chgo., July 1, 1947; s. LeRoy Rayfield and Louise (Rosen) K.; m. Jacqueline Anne Clark, Feb. 5, 1972; children: Aaron Wayne, Diantha Renee, Coreyanne Louise. B.S., U. Ill.-Urbana, 1969, M.S., 1971, Ph.D., 1975. Diplomate: Am. Bd. Toxicology. Research asst. Coll. Vet. Medicine, U. Ill.-Urbana, 1969-74; postdoctoral research assoc. Comparative Animal Research Lab., Oak Ridge, 1974-76, asst. prof., 1976-78; sr. research scientist devel. toxicology sect. biology dept. Battelle Pacific N.W. Labs., Richland, Wash., 1979-80, assoc. mgr., 1980-81, mgr., 1981-83, assoc. mgr. biology, biology and chemistry dept., 1983—. Contbr. numeous articles on toxicology to profl. jours. Recipient Award of Merit Northwest Sect. Soc. for Exptl. Biology and Medicine, 1979. Subspecialties: Toxicology (medicine); Teratology. Current work: Scientific Management, research in transplacental movements and prenatal effects of toxic materials, metabolism of toxic materials. Office: Battelle NW PO Box 999 Richland WA 99352

KELSEN, DAVID PAUL, physician, educator; b. Phila., Mar. 24, 1947; s. Henry and Hilda K.; m. Suzanne Joy Shrager, Sept. 1947; children: Benjamin, Judith, Tamar, Johnathan. B.A., Temple U., 1968; M.D., Hahnemann Med. Coll., 1972. Diplomate: Am. Bd. Internal Medicine. Intern Temple U. Hosp., Phila., 1972-73, resident in medicine, 1973-75, chief resident in medicine, 1975-76; research fellow Sloan-Kettering Inst., N.Y.C., 1976-78; fellow in med. oncology Meml. Sloan-Kettering Cancer Ctr., 1976-78, clin. asst. attending physician, 1978-78, asst. attending physician, 1980-81, assoc. attending physician, 1982—, research assoc., 1980—; asst. prof. medicine Cornell U. Med. Coll., 1978-83, assoc. prof. clin. medicine, 1983—. Mem. Commonwealth of Pa. Adv. Bd., 1973-76. Landrum-Karnofsky scholar, 1978-80. Fellow ACP; mem. Am. Soc. Clin. Oncology, Am. Assn. Cancer Research, N.Y. Acad. Scis., Internat. Assn. Study of Lung Cancer, N.Y. State Soc. Internal Medicine. Democrat. Jewish. Subspecialties: Cancer research (medicine); Chemotherapy. Current work: Clinical research in chemotherapy of solid tumors. Office: 1257 York Ave New York NY 10021

KELSH, DENNIS J., chemist, educator; b. Valley City, N.D., Dec. 24, 1936; s. Vincent J. and Evelyn M. (Denning) K.; m. Ginger V. Huhn, June 10, 1961; children: Bridget, Hilary, James. B.A.-B.S. in Chemistry, St. John's U., Collegeville, Minn., 1958; Ph.D., Iowa State U., 1962. Faculty Gonzaga U., Spokane, Wash., 1962—, prof. chemistry, 1972—; cons. research chemist Spokane Research Ctr., U.S. Bur. Mines, 1965—; assoc. research scientist N.Y.U., N.Y.C., 1966-67; spl. asst. to dean grad. sch. Wash. State U., Pullman, 1974-75. Contbr. articles to profl. publs. Bd. dirs. Southside Schs., 1978-81, pres., 1979-80. Mem. Am. Chem. Soc. (councilor Inland Empire sect. 1976—). Roman Catholic. Subspecialties: Surface chemistry; Coal. Current work: Sludge dewatering by electroosmosis; hydrogen production by electrolysis. Office: Gonzaga U Spokane WA 99258

KELSOE, LYNDA CAROL, aerospace engineer, computer scientist; b. Birmingham, Ala., Apr. 5, 1943; d. Johnny Willard Simmons and Marjorie Nanette (Wallace) Jones; m. Michael Lawson Pierson, Dec. 20, 1968 (div. Oct. 1978); m. Neal Marshall Kelsoe, July 18, 1981. B.A. in English, U. Montevallo, 1966; B.S. in Econs, 1966, Stevens Inst. Tech., 1970; M.A. in Lit, U. Houston, 1977, U. Houston, 1979. Programmer Bell Labs., Whippany, N.J., 1970-74; programmer/analyst IBM, Houston, 1974-77; sr. programmer/analyst Lockheed Co., Houston, 1977-78; sr. systems analyst Computer Scis. Corp., Houston, 1978-82; sr. staff analyst Jefferson Assocs., Houston, 1982; sr. aerospace scientist Intermetrics, Houston, 1982—. Author: Heraldry—Study of Coats of Arms, 1969. Councilwoman El Lago City Govt., Seabrook, Tex., 1983—. Mem. Nat. Mgmt. Assn., Assn. for Computing Machinery, Nat. Assn. Female Execs., Am. Businesswoman's Assn. Subspecialties: Aerospace engineering and technology; Software engineering. Current work: Manage a research and technology group working on the space shuttle and related projects. Home: 206 Yacht Club Ln Seabrook TX 77586 Office: Intermetrics 17625 El Camino Houston TX 77058

KEMELHOR, ROBERT ELIAS, laboratory executive; b. N.Y.C., May 19, 1919; s. Louis and Rebecca (Edelson) K.; m. Shirley Tennen, June 28, 1947; children: Judith Ellen Bielecki, Joel Martin, Barry Alan. B.S. in Mech. Engring, George Washington U., 1949. Design engr. Bur. Aeros. and Ordnance, Washington, 1940-53; chief engr. McLean Devel. Lab., Copiague, L.I., N.Y., 1953-57; dir. research and devel. Pesco div. Borg Warner Corp., Cleve., 1957-59; program mgr. Applied Physics Lab., Johns Hopkins U., Laurel, Md., 1959-82, br. supr., 1982—; cons. TRW, Cleve., 1963-64, Cleve. Pneumatic, Washington, 1959-60, Allied Research Assn., Concord, Mass., 1966. Served with USN, 1943-46. Fellow AIAA (assoc.); mem. Soc. Mfg. Engrs. (sr.). Subspecialties: Mechanical engineering; Ocean engineering. Current work: Design and fabrication of spacecraft and underwater sensor devices, supervisor and manager of above articles. Patentee in field. Home: 6200 Redwing Ct Bethesda MD 20817 Office: Applied Physics Lab Johns Hopkins U Johns Hopkins Rd Laurel MD 20707

KEMP, L(OUIS) FRANKLIN, JR., research scientist; b. N.Y.C., Mar. 19, 1940; s. Louis Franklin and Louise (Nunn) K.; m. Verena Henry, Sept. 5, 1964; children: Kaaren, Charles. B.S. Aero. Engring, Princeton U., 1962; M.S. in Math, Poly. Inst. N.Y., 1965, Ph.D., 1967. Engr. Grumman, Bethpage, N.Y., 1962-63; asst. prof. math. Poly. Inst. N.Y., Bklyn., 1967-69; research assoc. Amoco Prodn., Tulsa, 1969—; adj. prof. U. Tulsa, 1978-82. Contbr. articles, poem, to profl. jours. NASA trainee, 1964-67. Mem. Soc. Indsl. and Applied Math., Math. Assn. Am., Sigma Xi. Repulican. Congregationalist. Subspecialties: Applied mathematics; Statistics. Current work: Dipmeter correlation; rational approximation; queueing networks. Home: 5334 S 74 E Ave Tulsa OK 74145 Office: Amoco Prodn Co Research Box 591 Tulsa OK 74102

KEMP, MARWIN KING, chemist, oil company scientist; b. Strong, Ark., Nov. 23, 1942; s. Elbert L. and Elvie R. (King) K.; m. Linda Jean Shoemaker, Jan. 9, 1976; children: Kirk, Heather. B.S., U. Ark., 1964; M.S., U. Ill., 1966, Ph.D., 1968. Assoc. prof. U. Tulsa, 1968-81; sr. research scientist Amoco Prodn. Co., Tulsa, 1982—. Author: Physical Chemistry - A Step-By-Step Approach, 1979. Mem. Am. Chem. Soc., AAAS, Phi Beta Kappa, Tau Beta Pi. Subspecialties: Geochemistry; Physical chemistry. Current work: Petroleum geochemistry. Office: PO Box 591 Tulsa OK 74102

KEMP, WALTER MICHAEL, biologist; b. Big Spring, Tex., Aug. 26, 1944; s. Walter L. and Mary V. (Womack) K.; m.; children: Wiliam R., Brady S. B.S.E., Abilene Christian Coll., 1966; Ph.D., Tulane U., 1970. Asst. prof. Abilene (Tex.) Christian Coll., 1970-75; asst. prof. biology Tex. A&M U., College Station, 1975-78, assoc. prof., 1978-82, prof., 1982—; mem. study sect. NIH, 1981—. Contbr. articles to sci. jours. NIH grantee, also Edna McConnell Clark Found.; recipient Teaching Excellence award Tex. A&M U. Coll. Sci., 1981. Mem. AAAS, Am. Assn. Immunologists, Am. Soc. Parasitologists, Am. Soc. Tropical Medicine and Hygiene, Royal Soc. Tropical Medicine and Hygiene. Subspecialties: Parasitology; Immunobiology and immunology. Current work: Parasite immune escape mechanisms; schistosomiasis; surface receptors; immunoglobulins; complement system; antigen mimicry. Office: Tex A&M U College Station TX 77843

KEMPE, CHARLES HENRY, pediatrics and microbiology educator, pediatrician, researcher; b. Breslau, Germany, Apr. 6, 1922; m., 1949; 5 children. A.B., U. Calif., 1942, M.D., 1945. Intern in pediatrics U. Calif., 1945-46, from instr. to asst. prof., 1949-56; asst. Sch. Medicine, Yale U., 1948-49; prof. pediatrics and microbiology, chmn. dept. pediatrics U. Colo. Med. Center, Denver, 1956—; Fleischner Fund fellow Children's Hosp., 1946; lectr. U. Calif., 1949-50; cons. in field; mem. smallpox com. Commn. Immunization Armed Forces Epidemiol. Bd., 1953—; Fulbright prof. Inst. Superiore Sanita, Italy, 1955-56; vis. prof. Pasteur Inst., Paris, 1963-64; head battered children team U. Colo., Denver; dir. Nat. Center for Prevention and Treatment Child Abuse and Neglect. Recipient Mead Johnson award Am. Acad. Pediatrics, 1959. Mem. Inst. Medicine of Nat. Acad. Sci., Am. Pediatric Soc., Am. Soc. Pediatric Research, Am. Public Health Assn., Am. Assn. Immunology. Subspecialty: Pediatrics. Office: Dept Pediatrics U Colo Med Center Denver CO 80262

KEMPE, LLOYD LUTE, engr., educator; b. Pueblo, Colo., Nov. 26, 1911; s. Henry Edwin and Ida Augusta (Pittelkow) K.; m. Barbara Jean Bell, June 27, 1938; 1 dau., Marion Louise (Mrs. Steven Sanford Palmer). B.S. in Chem. Engring, U. Minn., 1932, M.S., 1938, Ph.D., 1948. Registered profl. engr., Minn., Mich. Research asst. in soils U. Minn., 1934-35, research asso., 1940-41, asst. in chem. engring., 1946-48; asst. san. engr. Minn. Dept. Health, 1935-40; instr. bacteriology U. Mich., Ann Arbor, 1948-49, asst. prof., 1949-50, asst. prof. chem. engring. and bacteriology, 1952-55, asso. prof., 1955-58, prof., 1958-60, prof. chem. engring. and san. engring., 1960-64, prof. chem. engring., 1964-67, prof. chem. engring. and microbiology, 1967—; asst. prof. food tech. U. Ill., 1950-52. Mem. editorial bd.: Biotech. and Bioengring, 1959-70, Applied Microbiology, 1964—, Food Tech, 1967-69, Jour. Food Sci, 1967-69. Mem. adv. com. on food irradiation Am. Inst. Biol. Scis./AEC; adv. com. on microbiology of foods Nat. Acad. Scis./NRC; adv. com. on botulism hazards HEW/FDA; adv. com. on mil. environ. research Nat. Acad. Scis./NRC. Served to col. AUS, 1941-45. Decorated Bronze Star. Mem. Am. Inst. C.E., Am. Chem. Soc., Am. Soc. Microbiology, Inst. Food Technologists, A.A.A.S., Am. Acad. Environ. Engrs., Water Pollution Control Fedn., Soc. Indsl. Microbiology, Sigma Xi, Phi Lambda Upsilon, Tau Beta Pi, Alpha Chi Sigma. Club: Mason. Subspecialties: Chemical engineering; Enzyme technology. Home: 3020 Exmoor St Ann Arbor MI 48104 Office: Dept Chem Engring U Mich Ann Arbor MI 48104

KEMPE, ROBERT ARON, scientific instrument manufacturing executive; b. Mpls., Mar. 6, 1922; s. Walter A. and Madge (Stoker) K.; m. Virginia Lou Wiseman, June 21, 1946; children: Mark A., Katherine A. B.Chem. Engring., U. Minn., 1943; postgrad. in metallurgy and bus. adminstrn, Case Western Res. U., 1946-49. Div. sales mgr. TRW, Inc., Cleve., 1943-53; v.p. Metalphoto Corp., Cleve., 1954-63, pres., 1963-71, Allied Decals, Inc. (affiliate Metalphoto Corp.), 1963-68; v.p., treas. Horizons Research, Inc., Cleve., 1970-71; pres. Reuter-Stokes, Inc., Cleve., 1971—; Vice pres. Miles Ahead, Inc., 1978—. Contbr. articles profl. jours. Served to lt. (j.g.) USN, 1944-46; PTO. Mem. Am. Nuclear Soc. (exec. officer No. Ohio sect.), Chemists Club N.Y.C. Club: Country of Hudson (Ohio). Subspecialties: Nuclear fission; Integrated pest management. Current work: General manager of high technology business, conversion of innovations and good technical ideas into profitable realities. Patentee in field. Home: 242 Streetsboro St Hudson OH 44236 Office: 18530 S Miles Pkwy Cleveland OH 44128

KEMPF, GARY WILLIAM, mech. engr.; b. Ann Arbor, Mich., Jan. 22, 1940; s. Theodore B. and Cora B. (Shafer) K.; m. Lettie L. Staples, Aug. 8, 1964; children: Karl L., Marcus B., Hans W. B.S.M.E., U. Mich., 1963. Registered profl. engr., Mich., Ohio. Project engr. John G. Hoad & Assocs., Ypsilanti, Mich., 1965-71; assoc., chief mech. engr. Ayres, Lewis, Norris & May, Inc., Ann Arbor, 1971-78; pres. Kempf Engring. and Research, Inc., Milan, Mich., 1978—. Scoutmaster Boy Scouts Am., 1980-82. Served to 1st lt. U.S. Army, 1963-65. Mem. ASME, Instrument Soc. Am., Am. Wind Energy Assn. Lutheran. Lodge: Masons. Subspecialties: Wind power; Mechanical engineering. Current work: Research in 1 to 10 kilowatt wind energy systems and components; manufacturer of wood blades, feathering blade holders for wind systems. Office: PO Box 84 Milan MI 48160

KEMPSON, STEPHEN ALLAN, physiology educator; b. Walsall, Staffordshire, Eng., July 2, 1948; came to U.S., 1975; s. Sydney and Edith (Atkins) K.; m. Deirdre M. Pankhurst, Aug. 4, 1973; children: Natalie A., Allan J. B.A., Lancaster U., 1970; M.Sc., Warwick U., 1971; Ph.D., London U., 1975. Research fellow U. Rochester, N.Y., 1975-77; research assoc. Mayo Clinic, Rochester, 1977-80; asst. prof. U. Pitts., 1980-82; asst. prof. dept physiology Ind. U., Indpls., 1982—. Minn. Heart Assn. fellow, 1978-80; Health Research and Services Found. research grantee, 1981-82; NIH research grantee, 1982-85. Mem. Am. Physiol. Soc., Endocrine Soc., Am. Soc. Nephrology, Am. Fedn. Clin. Research. Subspecialties: Membrane biology; Nephrology. Current work: Teaching and research, research interest is transport by kidney cell membranes.

KEMPTHORNE, OSCAR, statistician, educator; b. St. Tudy, Cornwall, Eng., Jan. 31, 1919; came to U.S., 1947; s. James Thomas and Emily Frances (Cobeldick) K.; m. Valda Minna Scales, June 10, 1949; children: V. Jill, W. Joan, Peter J. B.A., Cambridge (Eng.) U., 1940, M.A., 1943, Sc.D., 1960. Statistician Rothamsted Expt. Sta., Harpenden, Eng., 1941-46; assoc. prof. stats. Iowa State U., Ames, 1947-51, prof., 1951—, Disting. prof., 1964—. Author: The Design and Analysis of Experiments, 1952, Introduction of Genetic Statistics, 1957, Probability Statistics and Data Analysis, 1971. Fellow Am. Statis. Inst., Royal Statis. Soc., Biometric Soc., Am. Soc. Naturalists., Internat. Statis. Inst. Subspecialty: Statistics. Current work: Teaching graduate courses in statistics, directing doctoral candidates, research in statistics, author of books. Home: 2020 Ashmore Dr Ames IA 50010 Office: Iowa State U Ames IA 50011

KENDAL, ALAN PHILIPS, virologist; b. London, Apr. 27, 1945; s. Michael and Cynthia (Greenwood) K. B.S. in Biochemistry with honors, Univ. Coll. London, 1966, Ph.D. in Virology, 1968. Research asst. Univ. Coll. Hosp. Med. Sch., London, 1969; research assoc. U. Mich. Sch. Pub. Health, Ann Arbor, 1970-73; asst. prof. U. Md., Balt., 1973-75; supervisory research chemist Ctr. for Disease Control, Atlanta, 1975-81, chief influenza br., 1981—; dir. WHO Collaborating Ctr. for Influenza Reference and Research; chmn. orthomyscovirus subgroup Internat. Com. Taxonomy of Viruses, 1980—. Mem. Am. Soc. Microbiology, Am. Soc. Virology, Soc. Gen. Microbiology. Jewish. Subspecialties: Virology (veterinary medicine); Epidemiology. Current work: Evolution, variation, epidemiology and control of influenza viruses. Designed and constructed portable hemodialysis system; pres. non-profit corp. Portadial to assist end stage renal disease patients. Office: 1600 Clifton Rd Bldg 7 Rm 112 Atlanta GA 30333

KENDALL, ERNEST TERRY, cons. co. exec., lectr.; b. N.Y.C., Oct. 21, 1932; s. James Ernest and Aida (Kessler) K.; m. Eleanor Lois Kendall, Sept. 19, 1963; children: Sara Lois, Katherine Aida. B.S.M.E., Columbia U., 1960; M.A. in Econs, Boston U., 1970, Ph.D., 1974. Assoc. scientist research and devel. div. Avco, 1959-63; scientist Space Scis., Inc., 1963-66; program mgr. Space Systems div. Raytheon Corp., 1966-70; adj. assoc. prof. Boston U., 1973-74; project mgmt. engr., sr. economist Dept. Transp., 1974-77; dir. Boston office Nat. Econ. Research Assocs., Inc., 1977-78; pres. Commonwealth Research Group, Inc., Boston, 1978—; vis. lectr. Babson Coll. Author: Factors Affecting the Commercialization of Electric and Hybrid Vehicles, 1980, Risks Involved in Commercialization of Multiple Hearth Furnaces, 1981; contbr. articles profl. jours. Served in USAF, 1952-56. Mem. Am. Econs. Assn., Soc. Automotive Engrs., Omicron Delta Phi. Subspecialties: Energy-risk analysis; Water supply and wastewater treatment. Current work: Programmatic risks (tech. and econ.), associated with commercialization of innovative tech. in fields of energy, transp. and waste disposal. Office: 230 Beacon St Boston MA 02116

KENDALL, HENRY WAY, physics educator, researcher; b. Boston, Dec. 9, 1926. B.A., Amherst Coll., 1950, D.Sc. (hon.), 1975; Ph.D. in Nuclear Physics, MIT, 1955. NSF fellow MIT, Cambridge, 1954-56, from asst. prof. to assoc. prof., 1961-67, prof. physics, 1967—; research assoc. High Energy Lab., Stanford U., 1956-57, lectr. in physics, 1957-58, asst. prof., 1958-61. Mem. Am. Phys. Soc. Subspecialty: Particle physics. Office: Dept Physics MIT Cambridge MA 02139

KENDALL, PHILIP CHARLES, psychology educator; b. Bklyn., Mar. 2, 1950; s. Charles Edward and Stella E. (Pizzimenti) K.; m. Sue Harris, Aug. 24, 1974; 1 son, Mark Philip. B.S., Old Dominion U., 1972; M.S., Va. Commonwealth U., 1974, Ph.D., 1977. Lic. cons. psychologist, Minn. Asst. prof. psychology U. Minn., 1977-80, assoc. prof., 1980-83, prof., 1983—; conf. presenter; textbook cons. pub. cos. Author and/or editor numerous books including:: (with Norton-Ford) Clinical Psychology: Scientific and Professional Dimensions, 1982, (with Butcher) Handbook of Research Methods in Clinical Psychology, 1982, (with Franks, Wilson and Brownell) Annual Review of Behavior Therapy: Theory and Practice, Vol. 8, 1982; Author and/or editor numerous books including: Advances in Cognitive-Behavioral Research and Therapy, Vol. I, 1982; Author and/or editor numerous books including:: (with Hollon) Assessment Strategies for Cognitive-Behavioral Interventions, 1981, Cognitive-Behavioral Interventions: Theory, Research and Procedures, 1979; assoc. editor: Behavior Therapy, 1980-82, Cognitive Therapy and Research, 1981—; editorial bd.: Behavior Modification, Behavioral Psychotherapy; editorial cons. 15 profl. jours; contbr. numerous articles profl. jours. Bd. dirs. Reuben Lindh Learning Ctr., 1978-80; mem. Met. Airport Disaster Crisis Intervention Program, 1979—; med. psychology com. Palo Alto (Calif.) VA Hosp., 1976-77; investigator Calif. Bd. Med. Quality Assurance, 1977. Fellow Ctr. Advanced Study in Behavioral Scis., Stanford, Calif., 1980-81; grantee NIMH, U. Minn.; recipient Leadership and Service award Va. Commonwealth U., 1974. Mem. Am. Psychol. Assn., Assn. Advancement of Behavior Therapy, Soc. Research in Psychotherapy. Subspecialties: Behavioral psychology; Cognition. Current work: Evaluation of child psychotherapy. Home: 134 Arthur Ave SE Minneapolis MN 55414 Office: Dept Psychology Univ of Minn 75 E River Rd Minneapolis MN 55455

KENDIG, JOAN JOHNSTON, biologist, educator; b. Derby, Conn., May 1, 1939; d. Frank and Agnes (Kerr) Johnston; m. Roscoe B. Kendig, Sept. 9, 1964; children: Scott, Leslie. B.A., Smith Coll, 1960; Ph.D. (Woodrow Wilson fellow, NSF fellow), Stanford U., 1966. Acting asst. prof. dept. biol. scis. Stanford, Calif., 1967, research assoc. dept. anesthesia, 1967-71, asst. prof. biology dept. anesthesia, 1971-77, assoc. prof., 1977—; Mellon faculty fellow, 1976. Contbr. articles on biology in anesthesia to profl. jours. Mem. Am. Soc. for Pharmacology and Exptl. Therpeutics, Undersea Med. Soc., Biophys. Soc., Soc. for Neurosci., Assn. Univ. Anesthetists. Subspecialties: Neurobiology; Physiology (biology). Current work: Cellular neuropharmacology of anesthetic drugs, inhalation agents, local anesthetics, barbiturates and anti-arrhythmic agents, cellular basis of the high pressure nervous syndrome, interactions between hyperbaric pressure and anesthetic agents.

KENDRICK, JAMES BLAIR, JR., university official, research scientist; b. Lafayette, Ind., Oct. 21, 1920; s. James Blair and Violet (McDonald) K.; m. Evelyn May Henle, May 17, 1942; children: Janet Blair, Douglas Henle. B.A., U. Calif. at Berkeley, 1942; Ph.D., U. Wis., 1947. Mem. staff, faculty U. Calif. at, Riverside, 1947-68, prof. plant pathology and plant pathologist, 1961-68, chmn. dept., 1963-68; v.p. agrl. scis. U. Calif., 1968-77, v.p. agr. and univ. services, 1977—; dir. agrl. expt. sta., 1973-80, dir. coop. extension, 1975-80; Participant 10th Internat. Bot. Congress, Edinburgh, Scotland, 1964; mem. Calif. Bd. Food and Agr., 1968—, U.S. Agrl. Research Policy Adv. Com., 1976; Mem. governing bd. Agrl. Research Inst., 1974-76. Contbr. articles to profl. jours. Bd. dirs. Guide Dogs for Blind, San Rafael, Calif., 1983—. Served with AUS, 1944-46. NSF sr. postdoctoral fellow U. Cambridge (Eng.) and Rothamsted (Eng.) Exptl. Sta., 1961-62. Fellow AAAS (chmn. sect. O 1978); mem. Am. Phytopath. Soc. (editorial bd. jour. 1965-68, councilor at large 1968-70), Internat. Soc. Plant Pathology (council 1968-73), Am. Inst. Biol. Scis., Calif. C. of C. (agrl. com. 1968—), Nat. Assn. State Univs. and Land Grants Colls. (chmn. div. agr. 1972-73, exec. com. 1974-76), Western Assn. State Agrl. Expt. Sta. Dirs. (chmn. 1975), Phi Beta Kappa, Sigma Xi. Congregationalist. Club: Commonwealth of Calif. (San Francisco). Subspecialty: Agricultural research administration. Spl. research diseases vegetable crops. Home: 615 Spruce St Berkeley CA 94707

KENEALY, MICHAEL DOUGLAS, animal nutrition educator; b. Council Bluffs, Iowa, May 7, 1947; s. Aloysius Joseph and Lois Elaine (Jensen) K.; m. Carol Anne Bothwell, May 31, 1969; children: Sean Aloysius, Shannon Jo. B.S., Iowa State U., 1969, Ph.D., 1974. Nutritionist Dr. Macdonald's Inc., Ft. Dodge, Iowa, 1974-75; asst. prof. animal nutrition Iowa State U., 1975-79, assoc. prof. animal sci., 1979—. Co-author: (with George Brant) Introductory Animal Science, 1982; contbr. articles to sci. jours. Served with AUS, 1974. Mem. Am. Dairy Sci. Assn., Am. Soc. Animal Sci., Sigma Xi, Gamma Sigma Delta. Republican. Presbyterian. Club: Cyclone Corvettes. Subspecialties: Animal nutrition; Animal physiology. Current work: Animal nutrition, animal production, silage production; research in silage production. Home: 1116 Garner Circle Ames IA 50010 Office: Iowa State U 123 Kildee Hall Ames IA 50011

KENESHEA, FRANCIS JOSEPH, chemist, consultant; b. Providence, June 25, 1921; s. Francis Joseph and Antonetta (Polselli) K.; m. Hilda Irene Orsini, Nov. 30, 1944; children: Ellen, Jane. B.S., U. R.I., 1943, M.S., 1948; Ph.D., U. N. Mex., 1951. Instr. chemistry U. R.I., 1946-47; sr. research engr. N. Am. Aviation, Downey, Calif., 1951-55; sr. chemist SRI Internat., Menlo Park, Calif., 1955-71; cons. in organic chemistry, Palo Alto, Calif., 1971-74; sr. cons. Quadrex Corp., Campbell, Calif., 1974—; instr. chemistry Foothill Coll., 1973-74. Contbr. articles to profl. jours. Served to lt. j.g. USNR, 1944-46. Research Corp. fellow, 1948. Fellow AAAS; mem. Am. Chem. Soc., Am. Nuclear Soc. (exec. com. local sect. 1979-82). Democrat. Subspecialties: Nuclear fission; High temperature chemistry. Current work: Light water nuclear reactor chemistry; waste management (radioactive); risk assessment related to long-term management of radioactive waste. Office: Quadrex Corp 1700 Dell Ave Campbell CA 95008

KENIG, M(ARVIN) JERRY, mechanical engineer, cons., researcher, educator; b. Phila., Sept. 20, 1936; s. Abraham and Lillian Irene (Augenstein) K.; m. Rochelle Iris, Aug. 23, 1959; children: Neil Steven, Melissa Helene. B.S.M.E. with honors, Drexel U., 1959, M.S.M.E., 1962; M.A., Princeton U., 1963, Ph.D., 1965. Registered profl. engr., Pa., Mich. Engr. trainee Naval Air Exptl. Sta., Phila. Naval Base, 1955-56; engr. trainee Naval Air Material Ctr., 1957; mech. engr. medium turbines Westinghouse Electric Corp., Lester, Pa., 1958-59; instr. in mech. engring. Drexel Inst. Tech., 1960-62; various teaching and research assistantships Princeton U., 1962-65, postdoctoral fellow, 1965; asst. prof. mech. engring. Drexel U., 1965-68, assoc. prof., 1968-82, prof., 1982—, acting assoc. dean engring., 1974, assoc. dean, 1973-74, assoc. dean engring., 1974, assoc. dir., 1974, acting dean, 1974, asst. to pres., 1974-82; prof. and chmn. dept. mech. engring. Western Mich. U., Kalamazoo, 1983—; cons. to industry and law firms. Contbr. articles to profl. jours. Served to 1st lt C.E. USAR, 1959-60. United Engrs. and Constructors, Inc. preceptorship, 1963-64; Ford found. fellow, 1962-65; NSF grantee, 1963-65. Mem. ASME (chmn. Phila. com. applied mechanics 1969-71), Am. Soc. Engring. Educators, AAAS, Am. Def. Preparedness Assn., Am. Acad. Mechanics, Sigma Xi, Pi Tau Sigma (corr. sec. 1958-59), Tau Beta Pi, Phi Kappa Phi. Subspecialties: Theoretical and applied mechanics; Materials (engineering). Current work: Stress analysis, machine design, applied mechanics, elasticity, plasticity. Office: Western Mich U Kalamazoo MI 49008

KENNEALEY, GERARD, oncologist; b. Boston, Mar. 15, 1946; s. Thomas J. K.; m. Kathleen M. O'Connor, May 30, 1970; children: Gregory, Peter, Brendan, Douglas. B.S cum laude, Boston Coll., 1966; M.S., Yale U., 1970. Intern Yale U., 1970-71, resident, 1971-72, 74-75, oncology trainee, 1975-77; med. oncologist Waterbury (Conn.) Hematology-Oncology, 1977—; asst. clin. prof. Yale U., 1979—; chmn. tumor com. St. Mary's Hosp., Waterbury, 1980—. Contbr. chpts. to books. V.p. Am. Cancer Soc., Waterbury, 1981-83. Served with U.S. Navy, 1972-74. Mem. Am. Soc. Clin. Oncology, Waterbury Med. Assn., Tribury Jaycees (treas. 1979). Subspecialties: Chemotherapy; Hematology. Current work: Practice medical oncology-hematology; clinical research in melanoma and other tumors in conjunction with Yale University. Office: Waterbury Hematology-Oncology 1201 W Main St Waterbury CT 06508

KENNEDY, BYRL JAMES, physician, oncologist; b. Plainview, Minn., June 24, 1921; s. Arthur Sylvester and Anna Margaret (Fassbender) K.; m. Margaret Bradford Hood, Oct. 21, 1950; children: Sharon Lynn, James Bradford, Scott Douglas, Grant Preston. B.S., B.A., U. Minn., 1943, B.M., 1945, M.D., 1946; M.Sc., McGill Med. Sch., 1951. Diplomate: Am. Bd. Internal Medicine. Intern Mass. Gen. Hosp., 1945-46, asst. resident, 1946, resident, 1951-52, research fellow in medicine, 1947-49; research fellow McGill Med. Sch., 1949-50, Cornell Med. Sch.-N.Y. Hosp., 1951; asst. prof. U. Minn., 1952-57, assoc. prof., 1957-67, prof., 1967—, Masonic prof. oncology, 1970—; Vice pres. Presbyn. Homes of Minn.; Lucius Littauer fellow, 1947, Damon Runyan Clin. fellow, 1949-51. Contbr. articles to profl. jours. Recipient Am. Cancer Soc. Nat. award, 1975. Mem. AMA, Am. Soc. Clin. Oncology, Am. Assn. Cancer Research, Am. Soc. Hematology, Central Soc., Am. Fedn. Clin. Research, ACP, Am. Assn. Cancer Edn., Alpha Omega Alpha. Presbyterian. Clubs: Town and Country (St. Paul), Campus (Mpls.). Subspecialties: Oncology; Cancer research (medicine). Current work: Curent work: Medical oncology. Home: 1949 E River Rd Minneapolis MN 55414 Office: Box 286 Univ Hosps Minneapolis MN 55455

KENNEDY, CAROL TYLER, research chemist; b. Columbia, S.C., Oct. 8, 1939; d. Charles Raymond and Agnes Lovelace (Myers) K. B.S. in Organic Chemistry, U. Miami, 1960, M.S., 1962, Ph.D., Fla. State U., 1965. Chem. technician Atlas Chem. Co., Miami, 1970-76, unit supr., 1976-77; mgr. tech. info. Key Pharmaceuticals, Inc., Miami, 1977-82, research mgr. for organic chemistry, 1982—; cons. to various chem. cos.; adj. instr. chemistry Dade County Community Coll., 1983—. Contbr. articles to chem. jours. Bd. dirs. Miami chpt. Am. Cancer Soc., 1976-80; trustee Fla. State U. Mem. Am. Chem. Soc., Am. Soc. Organic Chemistry, Phi Beta Kappa, Sigma Xi. Republican. Presbyterian. Subspecialty: Organic chemistry. Home: Werik Cts 159 Madeira Coral Gables FL 33134

KENNEDY, CHARLES, pediatrician, neurologist, educator, researcher; b. Buffalo, Aug. 27, 1920; s. Charles Morehouse and Florence Louise (Chandler) K.; m. Evelyn Clarke, Mar. 9, 1946; children: Allen C., Jacqueline C., Carol M.; m. Eulsum Ko, Aug. 27, 1968. B.A., Princeton, U., 1942; M.D., U. Rochester, 1945. Diplomate: Nat. Bd. Med. Examiners, Am. Bd. Pediatrics, Am. Bd. Psychiatry and Neurology. Intern New Haven Hosp., 1945-46; resident Childrens Hosp., Buffalo, 1948-51; fellow dept. physiology and pharmacology Grad. Sch. Medicine, U. Pa., Phila., 1951-53, asst. prof. pediatrics, 1958-62, asso. prof., 1962-67; resident in neurology Hosp. of U. Pa., 1953-54; fellow in neurology Columbia-Presbyn. Med. Center, N.Y.C., 1957-58; guest worker Lab. Cerebral Metabolism, NIHM, 1967—, sr research scientist, 1980—; prof. pediatrics and neurology Georgetown U. Sch. Medicine, Washington, 1971—, chief div. child neurology dept. pediatrics, 1971—. Contbr. numerous articles to profl. jours. Served to lt. (j.g.) USNR, 1946-48. Life Ins. Med. Research fellow, 1951-53. Mem. Am. Pediatric Soc., Am. Neurol. Assn., Child Neurology Soc., Soc. for Neurosci., Am. Soc. for Neurochemistry. Subspecialties: Neurology; Brain energy metabolism. Current work: Contributions to cerebral circulation and metabolism, energy metabolism of developing brain, funcional pathways of vision. Office: Dept Pediatrics Georgetown U Sch Medicine 3800 Reservoir Rd Washington DC 20007

KENNEDY, CLIVE DALE, psychologist; b. Los Angeles, Mar. 23, 1953; s. Argustus Kennedy and Gussie Theodore (Johnson) Ford; m. Lynda Dianne Jones, June 24, 1978; 1 son, Marc Antony. A.A., Los Angeles City Coll., 1972; B.A., U. Calif.-Santa Barbara, 1975; Ph.D., U. Wash., 1981. Lic. psychologist, Calif., Tex. Mental health technician VA, Los Angeles, 1971-72; peer counselor U. Calif.-Santa Barbara, 1974-75; teaching/research asst. U. Wash., Seattle, 1976-78; clin. psychologist Titus Harris Clinic, Galveston, Tex., 1979-82; clin. asst. prof. U. Tex., Galveston, 1982; clin. psychologist Ortho Indsl. Med. Ctr., Los Angeles, 1982—; cons. York Home for Boys, Pasadena, 1982—. Contbr. articles to profl. jours. Bd. dir. Operation PUSH, Galveston, 1981-82; Pres. bd. ushers Shiloh A.M.E. Ch., Galveston, 1981, 82, UCLA Med. Aux., 1971-72. UCLA Med. Aux. scholar, 1972; Am. Psychol. Assn. fellow, 1975-76; NIMH fellow, 1975; U. Calif.-Santa Barbara Letters and Sci. scholar, 1974. Mem. Am. Psychol. Assn., Assn. Black Psychologists., Alpha Phi Alpha. Subspecialties: Health services research; Behavioral psychology. Current work: Behavioral and personality factors in hypertension and other med. problems, behavioral factors in children with learning disabilities. Office: 3800 S Figueroa St Los Angeles CA 90037

KENNEDY, DONALD, educator; b. N.Y.C., Aug. 18, 1931; s. William Dorsey and Barbara (Bean) K.; m. Jeanne Dewey, June 11, 1953; children—Laura Page, Julia Hale. A.B., Harvard, 1952, A.M., 1954, Ph.D., 1956. Mem. faculty Syracuse U., 1956-60; mem. faculty Stanford, 1960-77, prof. biol. scis., 1965-77, chmn. dept., 1965-72; sr. cons. Office Sci. and Tech. Policy, Exec. Office of Pres., 1970, commr. FDA, 1977-79; v.p., provost Stanford U., 1979-80, pres., 1980—; Bd. overseers Harvard, 1970-76. Author: (with W. H. Telfer) The Biology of Organisms, 1965; also articles.; Editor: The Living Cell, 1966, From Cell to Organism, 1967; editorial bd.: Jour. Exptl. Zoology, 1965-71, Jour. Comparative Physiology, 1965-76, Jour. Neurophysiology, 1969-75, Science, 1973-77. Fellow Am. Acad. Arts and Scis., AAAS; mem. Nat. Acad. Scis., Am. Physiol. Soc., Soc. Gen. Physiologists, Am. Soc. Zoologists, Soc. Exptl. Biology (U.K.). Subspecialty: Comparative neurobiology. Current work: Science policy, academic administration. Home: 623 Miranda Ave Stanford CA 94305 Office: Office of President Stanford U Stanford CA 94305

KENNEDY, DUNCAN TILLY, neuroscientist, medical educator; b. Bklyn., May 13, 1930; s. John Love and Elizabeth Ralston Campbell (MacKenzie) K.; m. Emma Lou Hanna, Sept. 3, 1955; children: John Robert, Hanna Lou, Elsbeth Love. B.S., Columbia U., 1955; A.M., Stanford U., 1964; Ph.D., Wayne State U., 1966. Postdoctoral fellow in neurophysiology U. Wis.-Madison, 1970-72; phys. therapist Kings Daus. Hosp., Ashland, Ky., 1955-57, Marmet (W.Va.) Hsp., 1958-60; Nat. Found. Infantile Paralysis teaching fellow Stanford (Calif.) U., 1960-62, instr., 1962; NIH fellow dept. anatomy Wayne State U. Sch. Medicine, Detroit, 1962-66, instr. divs. phys. and occupational therapy, 1964-69, asst. prof. dept. anatomy, 1967-70, 72-75; asst. prof. dept. physiology and health sci. Ball State U., Muncie, Ind., 1975-78, assoc. prof., 1978—; asst. dir. Muncie Ctr. for Med. Edn., Ind. U. Sch. Medicine at Ball State U., 1980—; adj. asst. prof. anatomy Ind. U. Sch. Medicine, 1977-83, adj. assoc. prof., 1983—; cons. Delaware County (Ind.) Dep. Coroner's Office, 1979. Contbr. articles to sci. jours. NIH fellow, 1962-66, 70-72; recipent Faculty Research award Wayne State U., 1974. Mem. AAAS, Detroit Physiol. Soc., Midwest Anatomists Assn., Soc. Neurosci., AAUP, Am. Assn. Anatomists, Ind. Acad. Sci., Sigma Xi, Sigma Zeta. Democrat. Presbyterian. Subspecialties: Neurophysiology; Anatomy and embryology. Home: 1428 W Gilbert Muncie IN 47303 Office: Ball State U Muncie IN 47306

KENNEDY, EUGENE PATRICK, educator; b. Chgo., Sept. 4, 1919; s. Michael and Catherine (Frawley) K.; m. Adelaide Majewski, Oct. 27, 1943; children—Lisa Kennedy Helprin, Sheila, Katherine. B.Sc., De Paul U., 1941; Ph.D. (Nutrition Found. fellow), U. Chgo., 1949; Sc.D. (hon.), U. Chgo., 1977, A.M., Harvard, 1960. Research chemist chem. research dept. Armour & Co., 1941-47; postdoctoral fellow Am. Cancer Soc., U. Calif. at Berkeley, 1949-50; with Ben May Lab. Cancer Research, dept. biochemistry U. Chgo., 1950-56, prof. biochemistry, 1956-60; sr. postdoctoral fellow NSF, Oxford (Eng.) U., 1959-60; Hamilton Kuhn prof. biol. chemistry Harvard Med. Sch., 1960—, head dept., 1960-65; Macy scholar Cambridge, 1976. Recipient Glycerine research award, 1955; Am. Oil Chemist Soc. Lipid Research award, 1970; Gairdner Found. award, 1976; Ledlie prize, 1976. Mem. Am. Chem. Soc. (Paul Lewis award 1958), Nat. Acad. Sci., Am. Soc. Biol. Chemists (pres. 1970-71), Am. Acad. Arts and Scis. Subspecialties: Biochemistry (biology); Membrane biology. Current work: Membrane function; transport in bacterial systems; biosynthesis of membrane liquids. Home: 63 Buckminster Rd Brookline MA 02146 Office: Dept Biol Chemistry Harvard Med Sch Boston MA 02115

KENNEDY, FRANK SCOTT, biochemistry educator; b. Washington, Oct. 16, 1944; s. Frank Scott and Margaret (Baker) K.; m. Karen Schmit, Sept. 26, 1980; children: by previous marriage: Frank Scott, Suzanne Carter. B.S., Washington and Lee U., 1966; Ph.D., U. Ill., 1969. Postdoctoral fellow Oxford U., Eng., 1970-71, Harvard U. Med. Sch., Boston, 1971-76; asst. prof. La. State U. Med. Sch., Shreveport, 1976-78, assoc. prof., 1978—; cons. White Chem. Co., Shreveport, 1977-80; asst. sec. A. D. Kennedy Co., Okmulgee, Okla., 1978—; v.p. F.S.K. Co., Shreveport, 1981—. Bd. dirs. Urban League, Shreveport, 1979; vestryman Holy Cross Episcopal Ch., Shreveport, 1979-82, 1973—. Med. Research Council of Eng. fellow, 1970; NSF fellow, 1970; NIH fellow, 1971; Stiles Research Fund grantee, 1980. Mem. Am. Chem. Soc. (sec. 1979-80). Subspecialties: Biochemistry (medicine); Nutrition (medicine). Current work: Role of trace metals in health and disease processes and how these derive from their basic inorganic chemistry. Office: La State U Med Sch PO Box 33932 Shreveport LA 71130

KENNEDY, JOHN FISHER, engineering educator; b. Farmington, N.Mex., Dec. 17, 1933; s. Angus John and Edith Wilma (Fisher) K.; m. Nancy Kay Grogan, Nov. 21, 1959; children: Suzanne Marie, Sean Grogan, Brian Matthew Fisher, Karen Lynn. B.S. in Civil Engring., U. Notre Dame, 1955; M.S., Calif. Inst. Tech., 1956, Ph.D., 1960. Research fellow Calif. Inst. Tech., Pasadena, 1960-61; asst. prof. MIT, Cambridge, 1961-64, asso. prof., 1964-66; dir. Iowa Inst. Hydraulic Research; prof. fluid mechanics U. Iowa, Iowa City, 1966—; chmn. div. energy engring. U. Iowa, 1974-76; Fulbright scholar, vis. prof. U. Karlsruhe, Germany, 1972-73; Erskine fellow U. Canterbury, Christchurch, N.Z., 1976; cons. to govt. agys., indsl. firms, engring. cons. offices, 1960—; vis. asso. in hydraulics Calif. Inst. Tech., 1977; ASCE Hunter Rouse lectr., 1981. Served to 2d lt. C.E., U.S. Army, 1957. Recipient J.C. Stevens award ASCE, 1959; W.L. Huber Research prize, 1964; Karl Emil Hilgard Hydraulic prize, 1974, 78; Engring. Honor award U. Notre Dame, 1978; Corning Glass Works fellow, 1959-60. Mem. Nat. Acad. Engring., ASCE, ASME, Am. Soc. Engring. Edn., Internat. Assn. Hydraulic Research (mem. council 1972-76, v.p. 1976-80, pres. 1981—), Sigma Xi, Chi Epsilon (hon.), Tau Beta Pi. Roman Catholic. Subspecialties: Surface water hydrology; Fluid mechanics. Current work: River mechanics and management; ice and arctic engineering; hydraulic structures. Home: 2 Ashwood Dr Iowa City IA 52240 Half of being a good sculptor is knowing when to stop carving.

KENNEDY, JOHN ROBERT, zoology educator, researcher; b. Cleve., July 17, 1937; s. John R. and Marcella H. (Martin) K.; m. Rosa Lea Kennedy, Oct. 21, 1938; children: Michael E., James M. B.S., U. Mich.-Ann Arbor, 1959, M.S., 1961; Ph.D., U. Iowa-Iowa City, 1964. Asst. prof. anatomy Bowman Gray Sch. Medicine, 1964-69; assoc. prof. zoology U. Tenn., 1969-77, prof., 1977—. Mem. Am. Assn. Anatomists, Am. Soc. Cell Biology, Electron Microscopic Soc. Am., Am. Soc. Protozoology. Subspecialties: Cell biology; Cell and tissue culture. Current work: Research on structure and function of ciliated epithelium; quantitation of ciliary beat with computer assisted spectral analysis; digital image processing in electron microscopy. Home: 7105 Shadyland Dr Knoxville TN 37919 Office: University of Tennessee Department Zoology Knoxville TN 37996

KENNEDY, JOSEPH LANE, biology educator; b. Hinesville, Ga., Feb. 5, 1946; s. E.B. and Gladys (Lane) K.; m. Dine Dixon, Dec. 18, 1966; children: John, Jim. B.S., North Ga. Coll., 1964-68; M.S., Clemson U., 1970; Ph.D., Utah State U., 1976. Mem. faculty dept. biology North Ga. Coll., Dahlonega, 1974-75; dir. Pyramid Lake Study W.F. Sigler and Assocs., Reno, 1975-79; mem. faculty dept. biol. sci. Western Mont. Coll., Dillon, 1979—, chmn. natural heritage, 1980—; cons. Pyramid Lake Tribal Enterprise, Reno, 1979-81, Collville Tribe, Omak, Wash., 1980-81. Editor, author: Pyramid Lake Ecological Study, 1978; contbr. articles to profl. jours. Served to capt. U.S. Army, 1969-71. Mem. Wildlife Soc., Mont. Ednl. Assn. (pres. chpt. 1980-81), Am. Fisheries Soc., Mont. Acad. Sci. Democrat. Subspecialties: Resource management; Ecosystems analysis. Current work: Studying interaction of users and fishery resources; also characteristics of managers and success of management program; designing and conducting creative environmental education program. Home: PO Box 1043 Dillon MT 59725 Office: Dept Biol Sci Western Mont Coll Dillon MT 59725

KENNEDY, LINDA MANN, neurobiologist, researcher, educator; b. Malden, Mass., July 29, 1939; d. Alfred William and Etta May (Maglue) Mann; m. Richard Dearman Kennedy, Apr. 15,1961; children: Pamela Lea, Ruth Alexander. Diploma in nursing, New Eng. Deaconess Hosp., Boston, 1959; A.B. (New Eng. Psychol. Assn. fellow), Simmons Coll., 1975; Ph.D. (Danforth fellow 1975-79, NSF dissertation grantee 1979), Harvard U., 1980; cert. in bus. mgmt, Harvard U., 1982. Registered nurse, Mass. Staff nurse Lahey Clinic, Boston, 1959-61, and various hosps., Mass. and Ga., 1962-72; teaching asst. Simmons Coll., 1972-75, spl. instr., 1982-83; vis. fellow Cornell U., 1977-81, lectr., 1978-79; NIH postdoctoral research fellow Worcester Found. Exptl. Biology, Shrewsbury, Mass., 1980-83; lectr. Clark U., 1983, research asst. prof., 1983—; lectr. Clark U., 1983—; cons. Gen. Goods Corp., 1980-81. Contbr. articles for profl. jours. Mem. Framingham (Mass.) Conservation Council, NOW, Soc. Neurosci., Assn. Chemoreception Scis., Eastern Psychol. Assn., AAAS, N.Y. Acad. Scis., Assn. Women in Sci., Soc. for Values in Higher Edn. Subspecialties: Neurobiology; Sensory processes. Current work: Physiol. mechanisms for transduction in taste receptor cell membranes; cellular and molecular processes in chemoreception; action of taste-altering drugs. Office: Dept Biology Clark U Worcester MA 01610

KENNEDY, MICHAEL CRAIG, neurobiologist, educator, researcher; b. Buffalo, Dec. 5, 1946; s. Daniel Francis and Katherine Kinsella (Lawsing) K.; m. Dorothy Anne French, Sept. 16, 1967; children: David Shawn, Matthew Eric, Catherine Megan. B.A., William Marsh Rice U., Houston, 1968; M.S., U. Rochester, N.Y., 1971, Ph.D., 1974. Fellow dept. cell biology N.Y.U. Med. Coll., 1974-76, asst. prof. biology, 1976-81; assoc. prof. dept. anatomy Hahnemann U. Sch. Medicine, Phila., 1981—; cons. coll. div. Harper & Row, Pubs., Inc. Contbr. articles in field to profl. jours. NIH fellow, 1974-76; NIH grantee, 1978-81; NSF grantee, 1981-82. Mem. Am. Assn. Anatomists, Cajal Club, Soc. Neurosciences. Democrat. Roman Catholic. Subspecialties: Comparative neurobiology; Anatomy and embryology. Current work: Neurobiological studies of auditory communication; neuroanatomical and developmental studies of auditory system; nerve regeneration. Office: Dept Anatomy MS 408 Hahnemann U Sch Medicine Broad and Vine Sts Philadelphia PA 19102

KENNEDY, MICHAEL FRANCIS, federal government laboratory nuclear engineer, consultant; b. Buffalo, Nov. 24, 1948; s. Daniel Joseph and Marie Alice (Clabeaux) K.; m. Linda Marie Stoeckl, June 6, 1970; children: Erin, Liam, Coleen. B.S., Canisius Coll., 1970; M.S., U. Va.-Charlottesville, 1973, Ph.D., 1978. Registered profl. engr., Conn. Sr. engr. Combustion Engring. Co., Windsor, Conn., 1973-76, prin. engr., 1977-79; group leader Argonne (Ill.) Nat. Lab., 1979—; mgr. engring. services Internat. Tech. Services, Naperville, 1981—. Mem. Am. Nuclear Soc., Sigma Xi, Tau Beta Pi. Subspecialty: Nuclear engineering. Current work: Application of computer codes to the analysis of nuclear power plant safety issues. Home: 340 N Wright St Naperville IL 60540 Office: Argonne Nat Lab 9700 S Cass Ave Argonne IL 60439

KENNEDY, ROBERT ALAN, horticulturist, educator, researcher; b. Benson, Minn., Sept. 29, 1946; s. William Henry and Mary Rose (Pothen) K.; m. Lonnie M. Eisenreich, Aug. 3, 1968; children: Caleb John, Alex E. B.S., U. Minn., 1968; Ph.D., U. Calif.-Berkeley, 1974. Asst. prof. botany U. Iowa, Iowa City, 1974-78; assoc. prof. horticulture Wash. State U., Pullman, 1979-82, prof., 1983—. Contbr. articles on plant physiology to profl. jours. Served with U.S. Army, 1969-71. NSF grantee, 1975—; U.S. Dept. Agr. grantee, 1979—. Subspecialty: Plant physiology (biology). Current work: Plant physiology, carbon metabolism, photosynthesis, stress physiology. Office: Dept Horticulture Wash State U Pullman WA 99164

KENNEDY, SAMUEL IAN T., molecular biologist, researcher; b. Glasgow, Scotland, Mar. 6, 1943; came to U.S., 1977; s. William Russell and Fiona (May) K.; m. Margaret Isobel Metcalf, Jan. 4, 1968. B.Sc. with honours, U. Glasgow, 1967, M.Sc., 1968; Ph.D., U. Reading, Eng., 1970. Lectr. div. biol. sci. U. Warwick, Eng., 1972-77; assoc. prof. biology U. Calif., San Diego, 1977-80; cons. in genetic engring., 1980-82; chmn. dept. molecular biology Med. Biology Inst.; dir. research QUIDEL, La Jolla, Calif., 1980—. Assoc. editor: Virology; contbr. articles on virology to sci. jours. Med. Research Council Britain research grantee, 1975-77; NIH grantee, 1977-80; NSF grantee, 1977-80. Mem. Am. Soc. Microbiology, Soc. Microbiology. Subspecialties: Virology (biology); Genetics and genetic engineering (biology). Current work: The structure and expression of genetic material; research in areas of animal virology, cell biology, immunology, and genetic engineering. Office: 11077 N Torrey Pines Rd La Jolla CA 92037

KENNEDY, WILLIAM JO, statistics educator; b. Blackwell, Okla., July 20, 1936; m. Carole Faye Benton, Aug. 8, 1961; 1 dau., Eve. B.S., Okla. State U., 1959, M.S., 1960; Ph.D., Iowa State U., 1969. Faculty Midwestern U., Wichita Falls, Tex., 1961-64; faculty Iowa State U., Ames, 1965—; now prof. statistics. Iowa State U.-Ames. Author: Statistical Computing, 1980. Fellow Am. Statis. Assn.; mem. Soc. Indsl. and Applied Math., Inst. Math. Statistics. Subspecialties: Statistics; Algorithms. Current work: Statistical computing methods and algorithms for using digital computers to support statistical data analysis. Home: 416 Hilltop St Ames IA 50010 Office: Iowa State U 117 Snedecor Ames IA 50011

KENNEDY, WILLIAM ROBERT, neurology educator, neurologist; b. Chgo., Nov. 2, 1927; m., 1957; 5 children. B.S., U. Ill., 1951; M.S., U. Wis., 1952; M.D., Marquette U., 1958. From asst. prof. to assoc. prof. Med. Center, U. Minn.-Mpls., 1964-71, prof. neurology, 1971—; now also dir. Neuromuscular Lab.; fellow in internal medicine Mayo Clinic, 1959-60, fellow in neurology, 1960-64. Mem. Am. Acad. Neurology, Am. Neurol. Assn., Am. Electroencephalography Soc., Am. Assn. Electromyography and Electradiagnosis (past pres.). Subspecialty: Neurology. Office: U Minn Hosp Box 187 Minneapolis MN 55455

KENNELL, E. EDISON, III, engr.; b. Seattle, July 16, 1945; s. E. Edison and Helen Elizabeth (Schweitzer) K.; m. Marilyn Alexis Stewart, July 1, 1947. B.S.M..E., U. Wash., 1968. Tool designer Boeing Aircraft, 1968-69; machine designer Crown Zellerbach, 1969-782; engr. Clean Energy Products, Seattle, 1972—. Mem. Internat. Solar Energy Soc., Am. Wind Energy Assn. Subspecialties: Wind power; Micro-hydro electric power. Current work: Researching innovative wind energy conversion systems. Office: 3534 Bagley N Seattle WA 98103

KENNETT, JAMES PETER, oceanography educator, micropaleontologist, paleoecologist; b. Wellington, N.Z., Sept. 3, 1940; m., 1964; 2 children. B.Sc., N.Z. U., 1962; Ph.D. in Geology with honors, Victoria U., 1965, D.Sc., 1976. Sci. officer N.Z. Oceanographic Inst., 1965-66; NSF fellow in micropaleontology Allan Hancock Found., U. So. Calif., 1966-68; asst. prof. Fla. State U., 1968-70; assoc. prof. Grad. Sch. Oceanography, U. R.I., Kingston, 1970-75, prof., 1974—; mem. adv. com. Antarctic Deep-Sea Drilling. Recipient McKay Hammer award Geol. Soc. N.Z., 1968. Mem. AAAS, Geol. Soc. Am., Am. Assn. Petroleum Geology, Soc. Econ. Paleontology and Mineralogy, Internat. Quaternary Assn. Subspecialty: Paleoecology. Office: Grad Sch Oceanography U RI Kingston RI 02881

KENNETT, ROGER HOWARD, geneticist, educator; b. Lakewood, N.J., Dec. 27, 1940; s. R. Howard and Henrietta (Truex) K.; m. Carol Lundberg, June 8, 1966; children: Edward, David, Timothy. A.B., Eastern Coll.; Ph.D. in Biomed. Scis, Princeton U., 1969. Postdoctoral fellow U. Calif., San Diego, 1970-71; research officer genetics labs. dept. biochemistry Oxford (Eng.) U., 1972-76; asst. prof. U. Pa. Sch. Medicine, 1976-79; assoc. prof. human genetics, 1979—; dir Human Genetics Cell Center, 1976—. Editor: (with others) Monoclonal Antibodies and Hybridomas: A New Dimension in Biological Analyses, 1980. Mem. Am. Assn. Immunologists, Genetics Soc. Am., AAAS. Subspecialties: Genetics and genetic engineering (medicine); Cancer research (medicine). Current work: Molecular basis of human genetic diseases and of human oncogenesis. Office: Dept Human Genetics Med Labs 196 U Pa Sch Medicine Philadelphia PA 19104

KENNEY, DENNIS RAYMOND, soil biochemist, educator, researcher; b. Osceola, Iowa, July 2, 1937; s. Paul Nelson and Evelyn (Beck) K.; m. Betty Ann Goodhue, June 20, 1959; children: Marcia Ann, Susan Beth. B.S., Iowa State U.-Ames, 1959; M.S., U. Wis.-Madison, 1961; Ph.D., Iowa State U.-Ames, 1965. Research asst. soils dept. U. Wis.-Madison, 1959-61, asst. prof., 1966-69, assoc. prof., 1969-74, prof., chmn. dept., 1979—; research assoc. agronomy dept. Iowa State U., Ames, 1961-65, postdoctoral fellow, 1965-66; research fellow dept. sci. and indsl. research Grasslands, Palmerston, New Zealand, 1976-77. Fellow Am. Soc. Agonomy (soil sci. award 1982), Soil Sci. Soc. Am. Lodge: Kiwanis. Subspecialties: Soil chemistry; Resource conservation. Current work: Reactions and fate of nitrogen in soils, plants and waters environmentally acceptable waste application to soils. Office: Dept Soil Sci 1525 Observatory Dr Madison WI 53706

KENNEY, FRANCIS THOMAS, biochemist, educator; b. Springfield, Mass., Mar. 16, 1928; s. Edward Michael and Mary Agnes (Byrnes) K.; m. Rose Ann Rescigno, Aug. 25, 1951; children: Jeffrey, Ellen. B.S., St. Michael's Coll., Winooski, Vt., 1950; M.S., U. Notre Dame, 1953; Ph.D., Johns Hopkins U., 1957. Research assoc. Cornell U., 1957-59; sr. staff scientist Oak Ridge Nat. Lab., 1959—; prof. biomed. scis. U. Tenn., 1970—; cons. Am. Cancer Soc. Contbr. numerous articles to profl. jours. Served with U.S. Navy, 1946-48. Recipient Claude Bernard medal U. Montreal, 1969; Disting. Alumnus Award St. Michael's Coll., 1976. Fellow AAAS; mem. Am. Soc. Biol. Chemists, Sigma Xi. Subspecialties: Biochemistry (biology); Gene actions. Current work: Molecular mechanisms of regulation of gene expression by hormones and during differentiation; dysfunction of gene expression in malignant cells. Home: 919 W Outer Dr Oak Ridge TN 37830 Office: Biology Division Oak Ridge National Laboratory PO Box Y Oak Ridge TN 37830

KENNY, ALEXANDER DONOVAN, pharmacologist, endocrinologist, educator, research institute director; b. London, Mar. 4, 1925; U.S., 1947, naturalized, 1956; s. Alexander and Alice Astley (Barton) K.; m. Dorothy Marie LeTang, Aug. 19, 1950; children: Alexander Leo, Mary Alice Kenny Sinton, Virginia Ann Kenny Drawe, Peter Donovan. B.Sc., U. London, 1945, D.Sc., 1982; Ph.D. (fellow), Athenaeum of Ohio, 1950. Sr. chemist Univ. Coll. Hosp., London, 1950-51; chief Chemistry Lab., Mass. Gen. Hosp., Boston, 1951; asst. Sch. Dental Medicine, Harvard U., 1952-54; instr., 1954-55, assoc. Med. Sch., 1955-59, assoc. prof. Med. Sch., 1959-65; prof. W.Va. U. Med. Ctr., 1965-67, U. Mo. Med. Ctr., 1967-74; U. Tex. Med. Br., Galveston, 1974-75; prof.pharmacology Tex. Tech U. Health Scis. Ctr., Lubbock, 1976—, chmn. dept. pharmacology, 1976—; dir. Tarbox Parkinson's Disease Inst., 1976—. Author: Intestinal Absorption of Calcium and its Regulation, 1981; contbr. numerous articles to sci. jours. Mem. Amarillo Catholic Diocesan Sch. Bd., 1980—, v.p., 1981-82, pres., 1982-83; pres. Thomas More Cath. High Sch. Bd., 1981-83. USPHS spl. fellow, London, 1967-68. Mem. Am. Chem. Soc., Am. Inst. Nutrition, Am. Soc. Bone and Mineral Research, Am. Soc. Pharmacology and Exptl. Therapeutics, Biochem. Soc. (U.K.), Brit. Pharm. Soc., Endocrine Soc., Soc. Endocrinology (U.K.), Soc. Exptl. Biology and Medicine, Bone and Tooth Soc. (U.K.), Sigma Xi. Roman Catholic. Clubs: University City, Serra (Lubbock) (v.p. 1981-82). Subspecialties: Pharmacology; Endocrinology. Current work: Endocrine aspects of calcium and bone metabolism; calcemic hormones; parathyroid hormone; Vitamin D; calcitonin. Home: 6606 Oxford Ave Lubbock TX 79413 Office: Dept Pharmacology Tex Tech U Health Scis Ctr Lubbock TX 79430

KENNY, GEORGE EDWARD, pathobiologist, educator; b. Dickinson, N.D., Sept. 23, 1930; s. Frank Stanley and Anna Marie (Kelsch) K.; m. Mary Pearson, Aug. 23, 1958; children: Francis W., Michael B., Maureen P., John A., Edward G. B.S., Fordham U., 1952; M.S., U. N.D., 1957; Ph.D., U. Min., 1961. Research instr. U. Wash., Seattle, 1961-63, asst. prof., 1963-67, assoc. prof., 1967-71, prof., 1971—, chmn. dept. pathobiology, 1970—. Mem. editorial bd.: Jour. Clin. Microbiology, Infection and Immunity; contbr. articles to profl. jours. Pres. Archdiocesan Sch. Bd. Cath. Archdiocese of Seattle, 1979-82. Served with U.S. Army, 1953-55. Recipient Kimble Methodology award Am. Pub. Health Assn., 1971, Disting. Alumnus award U. N.D., 1983; NIH grantee. Mem. Am. Soc. Microbiology, Infectious Diseases Soc. Am., Am. Assn. Immunologists, Soc. Exptl. Biology and Medicine, Tissue Culture Assn., Internat. Orgn. Mycoplasmologists. Subspecialties: Immunogenetics; Microbiology (medicine). Current work: Antigenic structure of pathogenic microorganisms and determination of human immune response to these agents in the course of disease; agents studies: mycoplasma, Legionaire's Disease agent, respiratory viruses of humans. Patentee in field. Home: 1504 37th Ave Seattle WA 98122 Office: Dept Pathobiology SC-38 U. Wash Seattle WA 98195

KENT, JOHN FRANKLIN, zoology educator; b. Franklin, Ind., Apr. 30, 1921; s. Robert Homer and Almyra (Huckelberry) K.; m. Mary Ruth McConnell, Feb. 28, 1942; children: Jane F., Caroline M., Sarah A. A.B., Franklin Coll., 1941; Ph.D., Cornell U., 1949. Instr. anatomy U. Mich.-Ann Arbor, 1949-53, asst. prof., 1953-57; prof. zoology Conn. Coll., 1947-79, Lucretia Allyn prof. zoology, 1979—. Served to master sgt. USAAF, 1942-45; CBI. Mem. Am. Assn. Anatomists, AAAS, AAUP, Sigma Xi, Phi Kappa. Phi. Subspecialties: Cytology and histology; Anatomy and embryology. Current work: Radiation biology, ultramiscroscopic tissue structure. Home: 4 North Ridge Rd New London CT 06320 Office: Dept of Zoology Connecticut College 270 Mohegan Ave New London CT 06320

KEPLER, RAYMOND GLEN, physicist; b. Long Beach, Calif., Sept. 10, 1928; s. Glen Raymond and Erma (Larsen) K.; m. Carol F. Flint, Apr. 19, 1953; children: Julianne, Linda, Russell, David. B.S., Stanford U., 1950; M.S., U. Calif.-Berkeley, 1955, Ph.D., 1957. Mem. tech. staff Central Research Dept., Dupont, Wilmington, Del., 1957-64; div. supr. Sandia Nat. Labs., Albuquerque, 1964-69, dept. mgr., 1969—. Contbr. articles to profl. jours. Fellow Am. Phys. Soc.; mem. AAAS. Democrat. Subspecialties: Condensed matter physics; Polymer physics. Current work: Physics of organic solids; excitons, electrons and holes in organic molecular crystals; electronic properties of polymers; piezoelectricity and pyroelectricity in polymers. Home: 9004 Bellehaven Ave NE Albuquerque NM 87112 Office: Sandia Nat Labs Albuquerque NM 87185

KEPLINGER, MORENO L., toxicologist; b. Ulysses, Kans., May 25, 1929; s. Wilbur and Mary (Pudge) K.; m. Barbara Ann Evans, July 22, 1967; children: Kerry, Steven, Robert, Cory, Janice. B.S., U. Kans., 1951, M.S., 1952; Ph.D., Northwestern U., 1956. Diplomate: Am. Bd. Toxicology. Asst. prof. U. Miami Med. Sch., Coral Gables, Fla., 1956-60, 64-68; toxicologist Hercules Med. Dept., Wilmington, Del., 1960-64; mgr. toxicology Ind. Biotest Labs., Northbrook, Ill, 1968-77; cons. toxicologist, Deerfield, Ill., 1978—; pres. M.L. Keplinger, Inc., 1978—. Contbr. articles and book chpts. to profl. lit. Fulbright scholar, 1952-53; Parke-Davis fellow in pharmacology, 1953-56. Mem. Am. Soc. Toxicology, European Soc. Toxicology, Am. Soc. Pharmacology and Exptl. Therapeutics, Am. Indsl. Hygiene Assn. Lodge: Masons. Subspecialties: Toxicology (medicine); Pharmacology. Current work: Industrial, environmental and general toxicology including research in cancer, teratology genetics and neurotoxicology. Home and office: 221 Park Ln Deerfield Il 60015

KEPPELMANN, FRANK ALFRED, plant and insect specialist; b. Phila., July 12, 1921; s. Alfred Julius and Julia Teresa (Gaul) K.; m.; children: Edward, Julie. B.A., U. Va., 1943; M.S., Col. State U., 1969. Plant and insect specialist Plant Industry div. Colo. Dept. Agr., Denver, 1972—. Served with USAAF, 1943-46. Mem. Am. Assn. Bot. Gardens and Arboreta, Am. Hort. Soc., Denver Bot. Garden, Bklyn. Bot. Garden, Am. Soc. Hort. Sci. Subspecialties: Integrated pest management; Plant pathology. Home: 9224 W 86th Pl Arvada CO 80005

KERFOTT, WILLIAM BUCHANAN, JR., research corporation, executive, consultant; b. S.I., N.Y., Mar. 13, 1944; s. William Buchanan Kerfoot and Marguerite (Myers) Baumgartel; m. Patricia Hoffmann, Aug. 21, 1965; children: Christopher Alexander, Kerry Ann. B.A. in Entomology, U. Kans., 1966; Ph.D. in Biology, Harvard U., 1970. Asst. scientist Woods Hole (Mass.) Ocenaographic Instn., 1970-75; dir. Environ. Mgmt. Inst. div., Environ. Devices Corp., Marion, Mass., 1975-78; pres. K-V Assocs., Inc., Falmouth, Mass., 1978—. Bd. dirs. dir. Assn. for Preservation of Cape Cod., Orleans, Mass., 1975—; mem. water resources council Cape Cod Planning and Econ. Devel. Commn., Barnstable, Mass., 1978—; citizens adv. com. Dept. Environ. Quality Engring., Boston, 1979—. IR 100 Indsl. Research awardee, 1977; NSF postdoctorate award, 1970; hon. Woodrow Wilson fellow Woodrow Wilson Found., 1966. Mem. Nat. Water Well Assn., ASTM, Film Soc. Woods Hold (treas. 1974-75). Episcopalian. Subspecialties: Water supply and wastewater treatment; Ground water hydrology. Current work: Development of groundwater treatment and control technology for subsurface engineering. Inventor Laminate membrane, septic leachate detector, groundwater flowmeter. Office: K-V Assos Inc 281 Main St Falmouth MA 02540 Home: 49 Ransom Rd Falmouth MA 02540

KERMAN, ARTHUR KENT, physicist; b. Montreal, Que., Can., May 3, 1929; m. Enid Kerman, Dec. 21, 1952; children—Ben, Daniel, Elisabeth, Melissa, James. B.Sc., McGill U., 1950; Ph. D., M.I.T., 1953. NRC fellow Calif. Inst. Tech., 1953-54; NRC (Can.) research fellow U. Copenhagen, 1954-55; research fellow Inst. Theoretical Physics, Copenhagen, 1955-56; mem. faculty dept. physics M.I.T., Cambridge, Mass., 1956—, prof., 1964—, dir., 1976—; Guggenheim fellow and vis. prof. U. Paris, 1961-62; vis. prof. SUNY, Stony Brook, 1970-71; adj. prof. Bklyn. Coll., 1971-75; cons. Argonne Nat. Lab., Brookhaven Nat. Lab., Lawrence Berkeley Lab., Lawrence Livermore Lab., Los Alamos Sci. Lab., Nat. Bur. Standards, Oak Ridge Nat. Lab.; chmn. vis. com. Lawrence Berkeley Lab., 1981; chmn. physics div. Argonne Nat. Lab., 1972; tech. reviewer nuclear theory subprogram Dept. Energy, 1980; mem. U.S. Nuclear Physics Delegation with People's Republic of China, 1979; mem. numerous panels Nat. Acad. Sci.; mem. Pres.'s Sci. and Acad. Adv. Com. for Lawrence Livermore Lab. and Los Alamos Sci. Lab., 1981; mem. theory div. Los Alamos Sci. Lab., 1977—; mem. vis. com. Bartol Research Found., 1965-68. Asso. editor: Revs. of Modern Physics, 1968-71. Fellow Am. Phys. Soc. (exec. com. div. nuclear physics 1970-72, publs. com., mem. Tom W. Bonner prize com. 1982), Am. Acad. Arts and Scis., N.Y. Acad. Scis. Subspecialty: Theoretical physics. Office: Dept Physics Mass Inst Tech Cambridge MA 02139

KERN, HAROLD LLOYD, biochemist, educator; b. Holyoke, Mass., July 31, 1921; s. Max and Dorothy (Heller) K.; m. Mary Elizabeth Meade, June 1949 (div. Nov. 1957); 1 son, Michael; m. Lydia Gomes, Oct. 22, 1958; children: Daniel, Joan. B.S., U. Mich., 1942; Sc.D., Johns Hopkins U., 1953. Instr. Harvard U., 1953-60; research biochemist U. Buffalo, 1960-63; asst. prof- ophthalmology Albert Einstein Coll. Medicine, Bronx, N.Y., 1963-70, assoc. prof., 1970—; mem. visual sci. study sect. Nat. Eye Inst., NIH, 1974-78. Author: Biochemistry of the Eye, 1966, Current Topics in Eye Research, vol. 1, 1979. Served in USN, 1944-46; PTO. Mem. Assn. Research in Vision and Ophthalmology. Unitarian. Subspecialties: Biochemistry (biology); Membrane biology. Current work: Analysis of transport of nutrients across cellular membranes of mammalian ocular lens. Home: 922 Merillon Ave Westbury NY 11590 Office: 1300 Morris Park Ave Bronx NY 10461

KERR, DONALD MACLEAN, JR., physicist; b. Phila., Apr. 8, 1939; s. Donald MacLean and Harriet (Fell) K.; m. Alison Richards Kyle, June 10, 1961; 1 dau., Margot Kyle. B.E.E. (Nat. Merit scholar), Cornell U., 1963, M.S., 1964, Ph.D. (Ford Found. fellow, 1964-65, James Clerk Maxwell fellow 1965-66), 1966. Staff Los Alamos Nat. Lab., 1966-76, group leader, 1971-72, asst. div. leader, 1972-73, asst. to dir., 1973-75, alt. div. leader, 1975-76; dep. mgr. Nev. ops. office Dept. Energy, Las Vegas, 1976-77, acting asst. sec. def. programs, Washington, 1978, dep. asst. sec. def. programs, 1977-79, dep. asst. sec. energy tech., 1979; dir Los Alamos Nat. Lab., 1979—; mem. Navajo Sci. Com., 1974-77; mem. sci. adv. panel U.S. Army, 1975-78; mem. engring. adv. bd. U. Nev.-Las Vegas, 1976-78, chmn. com. research and devel. Internat. Energy Agy., 1979; mem. nat. security adv. council SRI Internat., 1980—; mem. adv. bd. U. Alaska Geophys. Inst., 1980—; Mem. sci. adv. group Joint Strategic Planning Staff, 1981—; Mem. Naval Research Adv. Com., 1982—; adv. bd. Georgetown U. Center Strategic and Internat. Studies, 1981—; Mem. corp. Charles Stark Draper Lab., 1982—. Mem. AAAS, Am. Phys. Soc., Am. Geophys. Union, Southwestern Assn. Indian Affairs, Sigma Xi, Tau Beta Pi, Eta Kappa Nu. Club: Cosmos (Washington). Subspecialties: Physics research and development administration; Energy research and development administration. Research, publs. on plasma physics, microwave electronics, ionospheric physics, energy and nat. security policy. Office: Los Alamos Sci Lab Los Alamos NM 87545

KERR, WILLIAM, nuclear engineering educator; b. Sawyer, Kans., Aug. 19, 1919; s. William and Maria Louise (Gill) K.; m. Ruth Duncan, Apr. 28, 1945; children: William Duncan, John Gill, Scott Winston. B.S. in Elec. Engring, U. Tenn., 1942, M.S., 1947; Ph.D., U. Mich., 1954. Instr., then asst. prof. U. Tenn., 1942-44, 46-48; mem. faculty U. Mich., Ann Arbor, 1948—, prof. nuclear engring., 1958—, chmn. dept., 1961-74, acting dir. Mich. Meml.-Phoenix Project, 1961-65, dir., 1965—, dir. Office Energy Research, 1977—, project supr. AID Nuclear Energy Project, 1956-65; Cons. Atomic Power Devel. Assos., 1954—, Argonne Nat. Lab., Colo. Commn. on Higher Edn.; chmn. nuclear engring. edn. com. Asso. Midwest Univs., 1961-62, pres., 1966-67, bd. dirs., 1965-67; trustee Argonne Univs. Assn., 1965-71; adv. com. reactor safeguards Nuclear Regulatory Commn., 1972—. Mem. Am. Soc. Engring. Edn., IEEE, Am. Nuclear Soc., Sigma Xi, Eta Kappa Nu, Phi Kappa Phi, Tau Beta Pi. Subspecialties: Nuclear engineering; Nuclear fission. Current work: Reactor safety analysis; radiation shielding; energy production, distribution and consumption. Home: 2009 Hall St Ann Arbor MI 48104

KERR, WILLIAM CLAYTON, physicist, educator; b. Steubenville, Ohio, Mar. 8, 1940; s. George Hamilton and Agnes (Dye) K.; m. Sandria Neidus, June 9, 1963; children: Tamara, Elizabeth. B.A., Coll. of Wooster, 1962; Ph.D., Cornell U., 1967. Research assoc. Chalmers U. Tech., Gothenburg, Sweden, 1967-68, Argonne (Ill.) Nat. Lab., 1968-70; asst. prof. physics Wake Forest U., Winston-Salem, N.C., 1970-75, assoc. prof., 1975-83, prof., 1983—; research leave at U. Paris, 1976-77; vis. scientist Los Alamos Nat. Lab., summer 1981. Contbr. articles to profl. jours. Recipient Excellence in Teaching award Wake Forest U., 1976; Research Corp. grantee, 1975. Mem. Am. Phys. Soc., Am. Assn. Physics Tchrs., AAUP. Subspecialties: Condensed matter physics; Statistical physics. Current work: Theory of nonlinear, low-dimensional systems. Computer simulation of nonlinear systems of interest in statistical physics, including sine-Gordon chain, structural phase transitions, low-dimensional magnetism. Office: Dept Physics Box 7507 Winston-Salem NC 27109

KERRI, KENNETH DONALD, Civil engineering educator; b. Napa, Calif., Apr. 25, 1934; s. Kenneth Ruthven and Eunice (Beck) K.; m. Judith Reeves, Aug. 22, 1958; children: Christopher, Kathleen. B.S.C.E., Oreg. State U., 1956, Ph.D., 1965; M.S., U. Calif. - Berkeley, 1959. Registered profl. engr., Calif., diplomate: Am. Acad. Environ. Engrs. San. engr. USPHS, San Francisco, 1956-58; faculty Calif. State U. - Sacramento, 1959—, prof. civil engring., 1968—; pres. Nat. Environ. Tng. Assocs., Valparaiso, Ind., 1979-80, Assn. of Bds. of Certification, Ames, Iowa, 1983. Editor/author: Operation of Wastewater Treatment Plants, 1980, Operation and Maintenance of Wastewater Collection Systems, 1982. Served with USPHS, 1956-58. Named Trainer of Yr., Nat. Environ. Tng. Assoc., 1982; recipient Pres.'s Service award Assn. Bds. of Certification, 1982; Collection System award Water Pollution Control Fedn., 1977; Faculty Research award Sacramento State Coll., 1969. Fellow ASCE; mem. Water Pollution Control Fedn., Am. Water Works Assn. Subspecialties: Civil engineering; Water supply and wastewater treatment. Home: 5839 Shepard Ave Sacramento CA 95819 Office: Calif State Univ 6000 J St Sacramento CA 95819

KERSTETTER, REX EUGENE, biology educator, researcher; b. Ashland, Kans., Nov. 22, 1938; s. Roy Everett and Blanche Elizabeth (Sailor) K.; m. Elizabeth Sue Edwards, June 5, 1960; children: Kelvin Tod, Derek Edward. B.S., Ft. Hays Kans. State Coll., 1960, M.S., 1963; Ph.D. (NASA trainee), Fla. State U, 1967. Instr. biology dept. biol. scis. Fla. State U., Tallahassee, 1966-67; prof. biology Furman U., Greenville, S.C., 1967—; NSF summer research fellow Purdue U., 1968; Eli Lilly vis. scholar Duke U., 1976. Author: The Ecosphere: Organisms, Habitats, and Disturbances, 1974. Mem. S.C. Acad. Sci., AAAS, Bot. Soc. Am., Am. Soc. Plant Physiologists. Methodist. Subspecialties: Plant physiology (biology); Ecology. Current work: Plant hormone physiology, plant proteases. Home: 16 Zelma Dr Greenville SC 29609 Office: Dept Biology Furman U Greenville SC 29613

KERWIN, LARKIN, physics educator; b. Quebec, Que., Can., June 22, 1924; s. Timothy and Catherine (Lonergan) K.; m. Maria G. Turcot, June 10, 1950; children: Lupita, Alan, Larkin, Terrence, Rosa Maria, Gregory, Timothy, Guillermina. B.S., St. Francis Xavier U., 1944, LL.D., 1970; M.S., M.I.T., 1946; D.Sc., Laval U., 1949, U. B.C., 1973, McGill U., 1974, Meml. U. Nfld., 1978, U. Ottawa, 1981, Royal Mil. Coll., Kingston, 1982, LL.D., U. Toronto, 1973, U. Alta., 1983, Dalhousie U., 1983; D.L., Concordia U., 1976; D.C.L. (hon.), Bishop's U., 1978. Asst. prof. physics Laval U., 1948-51, assoc. prof., 1951-56, prof., 1956—, chmn. dept., 1961-67, vice-dean faculty scis., 1967-69, vice-rector, 1969-72, rector, 1972-77; v.p. Natural Scis. and Engring. Council Can., 1978-80; pres. Nat. Research Council Can., 1980—; dir. Cape Breton Devel Corp.; research physicist Geotech. Corp., Cambridge, 1945-46. Author: Atomic Physics, an Introduction, 1963; also articles. Decorated Lt. Gov.'s medal, 1941; Gov. Gen. medal, 1944; Pariseau medal, 1965; Centenary medal, 1967; Prix David, 1951; Jubilee medal, 1977; knight comdr. Order Holy Sepulchre Jerusalem; also knight grand cross; officer Order of Can.; also companion; Laval Alumni medal, 1978; Gold Medal Can. Council Profl. Engrs., 1982. Fellow Royal Soc. Can. (pres. 1976), Royal Soc. Arts, AAAS, Am. Inst. Physics; mem. Internat. Union Pure and Applied Physics (sec.-gen.), Assn. Canadienne-Francaise pour l'avancement des sciences, Assn. Univs. and Colls. Can. (pres. 1974-75), Am. Phys. Soc., Corp. Profl. Engrs. Quebec, Sociedad Mexicana Fisica, Canadian Assn. Physicists (pres. 1955, medal 1969), Sigma Xi. Club: Cercle Universitaire (Quebec). Subspecialty: Atomic and molecular physics. Current work: Mars spectrometry. Home: 2166 Parc Bourbonniere Sillery PQ G1T 1B4 Canada

KESLER, CLYDE ERVIN, engineering educator; b. Dewey, Ill., May 7, 1922; s. Roy Francis and Helen (Deffenbaugh) K.; m. Mary Anne Kirk, July 20, 1947; children: Philip Roy, David Clyde. B.S. in Civil Engring., U. Ill., 1943, M.S. in Structural Engring., 1946. Engr. aide I.C.R.R., Champaign, Ill., 1946-47; faculty U. Ill., Urbana, 1947—, prof. mechanics and civil engring., 1962—; pres. Am. Concrete Inst., Detroit, 1967-68; cons. concrete and reinforced concrete problems to pvt. cos., govt. agys., 1949—. Author: (with Taylor, Corten and Wetenkamp) Mechanical Behavior of Solids, 1959; also articles. Served to maj. C.E., U.S. Army, 1943-46. Recipient Alfred E. Lindau award Am. Concrete Inst., 1970, Halliburton Engring. Edn. Leadership award U. Ill., 1982. Fellow ASCE; mem. Am. Soc. Engring. Edn., ASTM (Sanford E. Thompson award 1958), Wire Reinforcing Inst. (hon.), Hwy. Research Bd. (assoc.), Am. Concrete Inst. (hon.), Nat. Acad. Engring., Chi Epsilon, Tau Beta Pi, Phi Kappa Phi, Sigma Xi. Subspecialties: Civil engineering; Materials (engineering). Current work: Research covers the behavior of concrete under various loadings and environments with current emphasis on fracture and fiber reinforcement. Home: RFD 3 Box 314 Champaign IL 61820 Office: Newmark Lab Civil Engring 208 N Romine St U Ill Urbana IL 61821

KESLER, DARREL JOE, animal scientist, educator; b. Portland, Ind., Sept. 21, 1949; s. David Gordon and Lucille Marie (Bullock) K.; m. Cheryl Scaletta, May 26, 1973; children: Cheralyn Elizabeth, Darrel Phillip II. B.S., Purdue U., 1971, M.S., 1974; Ph.D., U. Mo., 1977. Teaching asst. Purdue U., 1971-74; research asst. dairy husbandry U. Mo., 1974-77; asst. prof. dept. animal sci. U. Ill., 1977-81, assoc. prof., 1981—. Contbr. articles profl. jours. Recipient Animal Sci. award for teaching and counseling U. Ill. Dept. Animal Sci., 1982. Mem. Am. Soc. Animal Sci. (Outstanding Young Scientist award 1983), Am. Dairy Sci. Assn., AAAS, Nat. Assn. Colls. and Tchrs. of Agr., Soc. Study of Reprodn., Soc. Theriogenology, Sigma Xi, Gamma Sigma Delta. Presbyterian. Subspecialty: Reproductive biology. Current work: Hypothalamic-pituitary-ovarian-uterine interactions, physiology and biochemistry of reproduction and lactation. Home: 111 Willard St Urbana IL 61801 Office: 101 Animal Genetics Lab 1301 W Lorado Taft Dr Urbana IL 61801

KESSEL, DAVID, biochemistry educator; b. Monroe, Mich., Jan. 8, 1931; s. Harry and Gertrude (Herman) K.; m. Elizabeth Sykes, Sept. 8, 1979. B.S. in Chemistry, M.I.T., 1952; Ph.D. in Biol. Chemistry, U. Mich., 1959. Research fellow in pharmacology Harvard U. Sch. Medicine, 1959-63, research assoc. in pathology, 1963-67; assoc. prof. pharmacology U. Rochester, N.Y., 1967-74; prof. medicine and pharmacology Wayne State U., Detroit, 1974—. Contbr. articles to profl. jours. Mem. AAAS, Am. Soc. Biol. Chemists, Am. Soc. for Photobiology, Biochem. Soc., N.Y. Acad. Scis. Subspecialties: Cancer research (medicine); Biophysics (biology). Current work: Mode of action of anti-tumor agents; porphyrin photosensitization; circumvention of drug resistance.

KESSEN, WILLIAM, educator, psychologist; b. Key West, Fla., Jan. 18, 1925; s. Herman Lowry and Maria Angela (Lord) K.; m. Marion Lord, June 10, 1950; children: Judith, Deborah, Anne, Peter Christopher, Andrew Lord, John Michael. B.S., U. Fla., 1948; Sc.M., Brown U., 1950; Ph.D., Yale U., 1952. Postdoctoral fellow Child Study Center, Yale U., 1952-54, faculty depts. psychology and pediatrics, 1954-76, Eugene Higgins prof. psychology, 1976—, chmn. dept. psychology, 1977-80, prof. pediatrics, 1978—, acting univ. sec., 1980-81; mem. intellective processes research com. Social Sci. Research Council, 1959-63, chmn., 1961-63. Author: (with G. Mandler) The Language of Psychology, 1959, The Child, 1965, Childhood in China, 1975, (with M.H. Bornstein) Psychological Development from Infancy, 1979; editor: Mussen's Handbook of Child Psychology, vol. 3, 1983; contbr. articles to profl. jours. Mem. Carnegie Council on Children, 1973-77. Fellow Center Advanced Study Behavioral Sciences, 1959-60; Guggenheim fellow, 1970-71. Fellow AAAS, Am. Psychol. Assn. (pres. div. 7 1979-80); mem. Soc. Research Child Devel., Soc. Exptl. Psychologists. Subspecialties: Developmental psychology; Cognition. Current work: Visual perception in human infants; history of children. Home: 30 Halstead Ln Branford CT 06405 Office: Dept Psychology Yale U Box 11A Yale Sta New Haven CT 06520

KESSINGER, MARGARET ANNE, medical oncologist, educator; b. Beckley, W.Va., June 4, 1941; d. Clisby Theadore and Margaret Anne (Ellison) K.; m. Loyd I.E. Wegner, Nov. 27; 1971. M.A., W.Va. U., Morgantown, 1963, M.D., 1967. Cert. Am. Bd. Internal Medicine, Am. Bd. Med. Oncology. Intern U. Nebr. Med. Ctr., Omaha, 1967-68, resident in internal medicine, 1968-70, fellow in med. oncology, 1970-72, asst. prof. internal medicine, 1972-77, assoc. prof., 1977—. Fellow ACP; mem. Am. Soc. Clin. Oncology, Am. Assn. Cancer Research, Am. Fedn. Clin. Research, Sigma Xi. Republican. Methodist. Club: 99s (Omaha). Subspecialties: Chemotherapy; Cancer research (medicine). Current work: Development of chemotherapy treatment programs for various solid tumors. Office: U Nebraska Medical Center 42d St and Dewey Ave Omaha NE 68105 Home: Route 1 Scribner NE 68057

KESSLER, BERNARD V., physicist, opticist; b. N.Y.C., June 27, 1928; s. Samuel Maxwell and Esther V. (Silverman) K.; m. Goldie-Beth Kolodny, Nov. 22, 1957; children: Ellen, Susan. B.A., N.Y.U., 1959; M.S., Cath. U. Am., 1969, Ph.D. in Physics, 1971. Research physicist in optics Naval Ordnance Lab./Naval Surface Weapons Center, Silver Spring, Md., 1959—. Contbr. articles to profl. jours. Served with USAAF, 1946-47. Mem. Optical Soc. Am., Am. Assn. Physics Tchrs., Am. Phys. Soc., Philos. Soc. Washington. Subspecialties: Optical image processing; Infrared surveillance. Current work: Infrared surveillance: Cloud/sea clutter, systems, lasers, rangefinders, optical image processing, cloud modeling. Patentee in optics. Home: 2-D Woodland Way Greenbelt MD 20770 Office: R42 White Oak Naval Surface Weapons Center Silver Spring MD 20910

KESSLER, DONALD JOE, space scientist; b. Houston, Jan. 30, 1940; s. Joseph Valentine and Mazie Inez (Doegen) K.; m. Mary Frances Lawrence, Jan. 26, 1963 (div. 1969); m. Mary Susan Cain, Dec. 31, 1969 (div. 1978). Student, Lamar State Coll., 1961-62; B.S. in Physics, U. Houston, 1965, postgrad., 1966. Student trainee physics NASA Johnson Space Ctr., Houston, 1962-65, aerospace technologist meteroid studies, 1965-69, aerospace technologist flight mission ops., 1969-74, aerospace technologist aeronomy, 1974-78, aerospace technologist flight mechanics, 1978-79, space scientist, 1979—; program chmn. space debris Com. Sci. Research, Paris, France, 1982—. Contbr. articles to profl. jours. Served with U.S. Army, 1958-61. Recipient Superior Performance award NASA, 1977, Quality Increase award, 1980. Mem. Sigma Pi Sigma, Phi Kappa Phi. Club: Lunarfins (Houston). Subspecialty: Orbital debris studies.. Current work: Direct research with objectives to define the orbital debris environment; to predict the effects of this environment on spacecraft and discover preferred techniques to minimize the hazard of orbital debris to spacecraft. Office: NASA Johnson Space Ctr NASA Rd 1 Houston TX 77058 Home: 10822 Kirktown Houston TX 77089

KESSLER, EDWIN, meteorologist, nat. lab. adminstr.; b. N.Y.C., Dec. 2, 1928; s. Edwin and Marie Rosa (Weil) K.; m. Lottie Catherine Menger; children: Austin Rainier, Thomas Russell. A.B., Columbia Coll., 1950; S.M., MIT, 1952, Sc.D., 1957. Research scientist USAF Cambridge Research Labs., Bedford, Mass., 1954-61; dir. atmospheric physics div. Travelers Research Center, Hartford, Conn., 1961-64; dir. Nat. Severe Storms Lab., Norman, Okla., 1964—; adj. prof. U. Okla., 1964—; vis. prof. M.I.T., 1975-76. Past asso. editor: Jour. Applied Meteorology; Contbr. articles to profl. jours. Served with U.S. Army, 1947-48. Recipient award for outstanding authorship NOAA, 1971. Fellow Am. Meteorol. Soc. (past pres. Greater Boston br., nat. councilor 1966-69, past mem. com. on severe local storms, past chmn. com. on weather radar, cert. cons. meteorologist), AAAS; mem. Royal Meteorol. Soc. (fgn.), Weather Modification Assn., Am. Geophys. Union, Sigma Xi. Subspecialty: Meteorology. Current work: Editor of 3-volume work on thunderstorms, published in 1981-82. Office: 1313 Halley Circle Norman OK 73069

KESSLER, GEORGE WILLIAM, mfg. co. cons.; b. St. Louis, Mar. 1, 1908; s. William Henry and Blanche M. (Pougher) K.; m. Alice Mae Maxwell, July 28, 1951; children—Judith Ann Green, William Clarkson. B.S. in Mech. Engring, U. Ill., 1930. With Babcock & Wilcox Co., Barberton, Ohio, 1930—, v.p. power generation group, 1961—. Contbr. articles to profl. jours. Fellow ASME; mem. Am. Standards Assn. (dir.), Soc. Naval Architects and Marine Engrs. (Joseph H. Linard award 1949), Am. Soc. Naval Engrs., Franklin Inst., Nat. Acad. Engring., Welding Research Council, Tau Beta Pi, Phi Eta Sigma, Sigma Tau, Pi Tau Sigma, Alpha Sigma Phi. Clubs: Propeller, Cornell (N.Y.C.). Subspecialties: Mechanical engineering; Fuels. Patentee in field. Home: 720 Williams Dr Winter Park FL 32789

KESSLER, ROBERT, industrial research lab executive, researcher; b. Farnborough, U.K., Dec. 27, 1939; came to U.S., 1954; s. Zoltan and Ilona (Elefant) K; m. Carol Joyce Sowers, Dec. 22, 1969; 1 dau., Rebecca Elizabeth. B.S., M.I.T., 1960, M.S., 1961; Ph.D., Stanford U., 1968. Vis. asst. prof. U.S. Naval Postgrad. Sch., Monterey, Calif., 1968-69; research assoc. Stanford U., Palo Alto, Calif., 1969-72; sr. scientist Avco Everett Research Lab., Everett, Mass., 1972-81, v.p., 1981—. Contbr. articles to profl. publs. Assoc. fellow AIAA; mem. ASME, Sigma Xi. Subspecialties: Plasma engineering; Coal. Current work: MHD power generation, coal processing technologies: coal gasification, coal pyrolysis, coal combustion. Office: Avco Everett Research Lab 2385 Revere Beach Pkwy Everett MA 02149

KESSLER, ROBERT EVANS, microbiologist, consultant, researcher; b. Englewood, N.J., Mar. 5, 1951; s. Evans Rodney and Avis (Ryerson) K.; m. Michele Marie Santangelo, Aug. 10, 1974; children: Ryan, Adam. B.S. with departmental honors, Ursinus Coll., 1973; Ph.D., Temple U., 1978. Postdoctoral fellow Rockefeller U., 1978-79; asst. research scientist Sch. Dentistry, U. Mich., 1979—. Temple U. fellow, 1976. Mem. Am. Soc. Microbiology, Lancefield Soc., Internat. Assn. Dental Research, Harvey Soc., Sigma Xi. Subspecialties: Microbiology (medicine); Molecular biology. Current work: Microbial cell surface biochemistry, immunochemistry; interactions of microbial cell surfaces with host tissues including adhesion, cell-cell interactions; biological activities of lipoteichoic acids. Office: Sch Dentistry U Mich Ann Arbor MI 48109

KESTENBAUM, AMI, researcher; b. Petach Tikva, Israel, June 7, 1941; came to U.S., 1958, naturalized, 1963; s. I. K. and Regina (Lewit) K.; m. Sylvia W. Kestenbaum, Sept. 5, 1971; children: Robyn, Michael. B.E.E., CCNY, 1963; M.S. in E.E, Poly. Inst. N.Y., 1965, Ph.D., 1969. Mem. research staff Western Electric Engring. Research Center, Princeton, N.J., 1969—. Contbr. articles to profl. jours. Mem. Material Research Soc., IEEE, Laser Inst. Am. Subspecialties: Laser applications; Microchip technology (engineering). Current work: Laser applications to electronic technology. Patentee in field.

KESTIN, JOSEPH, mechanical engineer, educator; b. Warsaw, Poland, Sept. 18, 1913; came to U.S., 1952, naturalized, 1960; s. Paul and Leah K.; m. Alicja Wanda Drabienko, Mar. 12, 1949; 1 dau., Anita Susan. Dipl. Ing., Engring., U. Warsaw, 1937; Ph.D., Imperial Coll., London, 1945; M.A. ad eundem, Brown U., 1955; D.Sc., U. London, 1966. Sr. lectr. dept. mech. engring. Polish U. Coll., London, 1944-46, dept. head, 1947-52; prof. engring., dir. Center for Energy Studies, Brown U., Providence, 1952—; vis. prof. Imperial Coll., London, 1958, Summer Sch. in Jablonna, Warsaw Polish Acad. Scis., 1973; professeur associe U. Paris, 1966, Université Claude Bernard (Lyon II) and Ecole Centrale de Lyon, 1974; Fulbright lectr. Instituto Superior Tecnico, Lisbon, 1972; spl. lectr. Norges Tekniske, Hogskole, Trondheim, Norway, 1963, 71; lectr. Nobel Com. Berzelius Symposium, 1979; spl. adv. on engring. edn. to Chancellor of U. Tehran, Iran, 1968; chmn. NRC Evaluation Panel for Office of Standard Ref. Data of Nat. Bur. Standards, 1976-80; mem. Evaluation Panel for Nat. Measurement Lab. of Nat. Bur. Standards, 1978-80, Numerical Data Adv. Bd., Nat. Acad. Scis., 1976-80; cons. Nat. Bur. Standards, NATO, Rand Corp.; Mem. vis. com. U. Va., Charlottesville, 1964; mem. exec. com. Nat. Bur. Standards Evaluation Panels, 1974-78. Author 3 books on thermodynamics; editor-in-chief: Dept. Energy Sourcebook on Production of Electricity from Geothermal Energy; tech. editor: Jour. Applied Mechanics, 1956-71; mem. editorial bd.: Internat. Jour. Heat and Mass Transfer, 1961-71, Heat Transfer-Soviet Research, 1968—, Heat Transfer-Japanese Research, 1972—, Mechanics Research Communications, 1973—, Jour. Non-Equilibrium Thermodynamics, 1976—, Revue Generale de Thermique, 1975, Physica A, 1978—, Internat. Jour. Thermophysics, 1979—, Jour. Chem. and Engring. Data, 1980—; contbr. articles to profl. jours. Fellow Inst. Mech. Engrs. (London) (Water Arbitration prize 1949), ASME (task group on energy 1974-76, applied mechanics div. 1967-78, chmn. 1978, nat. nominating com. 1976-78, Centennial medals for research achievements and disting. service, James Harry Potter Gold medal 1981); mem. Am. Soc. Engring. Edn. (chmn. Curtis W. McGraw Research award com. 1976-78), Internat. Assn. Properties of Steam (U.S. del. exec. com. 1954—, chief of del. 1972—, pres. 1974-76), Internat. Union Pure and Applied Chemistry (chmn. subcom. transport properties 1981—), U.S. Acad. Engring., Sigma Xi (pres. Brown U. chpt. 1979—), Tau Beta Pi. Clubs: Univ., Faculty Brown U. (Providence). Subspecialties: Mechanical engineering. Current work: Energy research administration. Home: 140 Woodbury St Providence RI 02906 Office: Brown U Providence RI 02912

KETOLA, HENRY GEORGE, research nutritionist; b. Ithaca, N.Y., June 13, 1943; s. Henry and Viena Alve K.; m. Sandra Brookhouse, June 19, 1965; children: Tanya Brookhouse, Deborah Alve. B.S., Cornell U., 1965, M.S., 1967; Ph.D., 1973; postgrad., Colo. State U.-Fort Collins, 1967-68. NIH trainee Cornell U., Ithaca, N.Y., 1965-67; teaching asst. zoology Colo. State U., Fort Collins, 1967-68; research asst. fisheries Cornell U., Ithaca, N.Y., 1968-69, NIH trainee, 1969-73; research specialist Vet. Virus Research Inst., Ithaca, NY., 1973; research physiologist, nutritionist Tunison Lab. Fish Nutrition, Cortland, N.Y., 1973—; adj. asst. prof. Cornell U., Ithaca, N.Y., 1976—. Mem. subcom. NRC, Washington, 1981—. Recipient Citation U.S. Fish and Wildlife Service, 1974. Mem. Am. Inst. Nutrition, Am. Fisheries Soc., Am. Soc. Animal Sci., Poultry Sci. Assn., Sigma Xi. Republican. Subspecialties: Nutrition (biology); Physiology (biology). Current work: Nutritional aspects of cataracts, proteins, amino acids, minerals and vitamins in fishes, physiology of digestion in fishes. Home: Box 230 B Route 90 King Ferry NY 13081 Office: Tunison Lab Fish Nutrition US Dept Interior 28 Gracie Rd Cortland NY 13045

KETTLEWELL, NEIL MACKEWAN, neuroscience educator, researcher; b. Evanston, Ill., May 27, 1938; s. George Edward and Barbara Sidney (Kidde) K.; m. Phyllis Ann Miller, Jan. 30, 1965 (div.

Sept. 1976); 1 son, Brant Regnar; m. Toni Ann Gianoulias, June 2, 1978. B.S., Kent State U., 1962; M.A., U. Mich., 1965, Ph.D., 1969. Research asst. in psychology U. Mich., 1963-69, programmer, 1966-69, systems analyst time scheduling office, 1967-69; lectr. U. Mont., 1969-70, asst. prof. psychology, 1970-75, assoc. prof., 1976—; cons. in field. Served with USAR, 1958-66. U. Mich. Presdl. scholar, 1964. Mem. Soc. Neurosci., N.Y. Acad. Scis., Pi Mu Epsilon, Psi Chi, Phi Eta Sigma. Subspecialties: Neuropsychology; Molecular biology. Current work: Ultrastructural synaptic changes in brain as result of experience. Home: 172 Fairway Dr Missoula MT 59803 Office: Dept Psychology U Mont Missoula MT 59801

KETTRICK-MARX, MARY ALICE, microbiologist; b. Hazleton, Pa., Sept. 29, 1945; d. James P. and Malvena (Bahrt) Kettrick; m. James John Marx, Jr., Dec. 17, 1944; children: Jonathyn Andrew, Christopher James. B.S., Coll. Misericordia, 1967; M.S., Villanova U., 1969; Ph.D., W.Va. U. Med. Sch., 1973. Research assoc. biology dept. Villanova (Pa.) U., 1967-69, instr., 1969; teaching asst. microbiology dept. W.Va. U. Med. Center, Morgantown, 1971-73; supr. microbiology Marshfield (Wis.) Med. Center Labs., 1973—. Mem. Am. Soc. Microbiology, Clin. Lab. Mgmt. Assn., Sigma Xi. Subspecialties: Microbiology (medicine); Virology (medicine). Home: M225 Marsh Ln Marshfield WI 54449 Office: Dept Microbiology Marshfield Med Center Lab Marshfield WI 54449

KETY, SEYMOUR S(OLOMON), physiologist, psychobiologist, emeritus educator; b. Phila., Aug. 25, 1915; s. Louis and Ethel (Snyderman) K.; m. Josephine R. Gross, June 18, 1940; children: Lawrence Philip, Roberta Frances. A.B., U. Pa., 1936, M.D., 1940, Sc.D. (hon.), 1965, Loyola U., 1969, U. Ill., 1981, M.D., U. Copenhagen, 1979. NRC fellow Harvard, 1943-44; from instr. to asst. prof. pharmacology U. Pa. Sch. Medicine, 1943-48; prof. clin. physiology Grad. Sch. Medicine, 1948-51; scientific dir. Nat. Insts. Mental Health and Neurol. Diseases and Blindness, 1951-56; chief Lab. Clin. Sci., NIMH, 1956-61, 62-67; Henry Phipps prof., dir. dept. psychiatry Johns Hopkins Sch. Medicine, 1961-62; prof. psychiatry Harvard Med. Sch., 1967-83, prof. neurosci. in psychiatry, 1983, prof. emeritus, 1983—; dir. psychiat. research labs. Mass. Gen. Hosp., Boston, 1967-77, Mailman Research Center (McLean Hosp.), Belmont, Mass., 1977—; Thomas Dent Mütter lectr., 1951, Eastman lectr., 1957, NIH lectr., 1960, Thomas William Salmon lectr., 1961, Alvarenga Prize lectr., 1961; Acad. lectr. Am. Psychiat. Assn., 1961; Saul Korey lectr., 1964, James Arthur lectr., 1966, 3d Mental Health Research Fund. lectr., London, 1965, Benjamin Musser lectr., 1970, Edward Mapother lectr., London, 1974, George Bishop lectr., 1975, Harvey lectr., 1975, Grass Found. lectr., 1975; Henry Maudsley lectr. Editor-in-chief: Jour. Psychiat. Research, 1959-82; contbr. sci. articles to profl. publs. Organizing com. Internat. Neuro-chem. Symposia, 1952-60; sci. advisory com. Mass. Gen. Hosp., 1956-60; dir. Found. Fund Research in Psychiatry, 1962-65; assoc. Neurosci. Research Found., 1962—; trustee Rockefeller U., 1976—. Recipient Theobold Smith award AAAS, 1949, Max Weinstein award, 1954; Distinguished Service award HEW., 1958; Stanley Dean award, 1962; McAlpin award Nat. Assn. for Mental Health, 1972; Intra-Sci. award, 1975; William C. Menninger award A.C.P., 1976; Fromm-Reichman award, 1978; Founds. Fund award, 1979; Passano award, 1980. Disting. fellow Am. Psychiat. Assn. (Disting. Service award 1980); hon. fellow Royal Coll. Psychiatrists; mem. Nat. Acad. Scis. (Kovalenko award 1973), Am. Acad. Arts and Scis., Am. Philos. Soc., Assn. Research Nervous and Mental Disease (trustee, pres. 1965, 80, Research Achievement award 1980), Am. Psychopath. Assn. (pres. 1965, Paul Hoch award 1973), Soc. for Psychiat. Research, Am. Soc. Clin. Investigation, Am. Soc. Pharmacology and Exptl. Therapeutics, Soc. for Neurosci. (Grass Found. award 1975), Phi Beta Kappa, Sigma Xi, Alpha Omega Alpha. Subspecialties: Psychophysiology; Psychopharmacology. Current work: Cerebral circulation and metabolism; psychiatric genetics; biological psychiatry. Office: Mailman Research Center McLean Hosp Belmont MA 02178

KEVAN, PETER GRAHAM, biologist, educator, researcher; b. Edinburgh, Scotland, June 17, 1944; s. Douglas Kieth and Kathleen Edith (Luckin) K; m.; children: Colin Douglas, Kathleen Hannah. B.Sc. in Zoology with honors, McGill U., Montreal, Que., Can., 1965; Ph.D. in Entomology, U. Alta. (Can.), Edmonton, 1970. Nat. coordinator IBP-Conservation, Can., Edmonton, 1969-70; contract biologist Can. Wildlife Service, Inuvik, N.W.T., 1970-71; postdoctoral fellow Can. Agr., Ottawa, Ont., 1971-72; project mgr. Meml. U., St. John's, Nfld., Can., 1972-75; asst. prof. U. Colo., 1975-82; asso. prof. environ. biology U. Guelph, Ont., 1982—. Contbr. numerous articles to profl. jours. NSF grantee; Nat. Research and Engring. Research Council grantee, 1982; also various other grants. Mem. Am. Bot. Soc., Brit. Ecol. Soc., AAAS, Inst. Arctic and Alpine Research. Subspecialties: Ecology; Evolutionary biology. Current work: Pollination: palaeontology, ecology, botany, zoology, evolution, co-evolution. Home: Rural Route 3 Aberfoyle Guelph ON Canada N1H 6H9 Office: Environ. Biology U Guelph Guelph ON Canada N1G 2W1

KEWMAN, DONALD GLENN, clinical psychologist, researcher, educator; b. San Francisco, Nov. 6, 1949. B.A., Stanford U., 1971; Ph.D., U. Tex.-Austin, 1977. Lic. clin. psychologist. Instr. U. Mich., Ann Arbor, 1977-81, asst. prof. dept. phys. medicine and rehab., 1981—, dir. Psychosocial Vocat. Div., 1977—; spl. cons. St. Joseph's Mercy Hosp., Ann Arbor, 1980—. Editor: Exploring Madness, 1979. Mem. Am. Psychol. Assn., Assn. Advancement Behavior Therapy, Mich. Psychol. Assn. Democrat. Quaker. Subspecialties: Behavioral psychology; Physical medicine and rehabilitation. Current work: Behavioral medicine, health psychology, rehabilitation. Home: 1510 Franklin St Ann Arbor MI 48103 Office: Dept Phys Medicine and Rehab Univ Mich Hosps Ann Arbor MI 48109

KEYES, ROBERT WILLIAM, physicist; b. Chgo., Dec. 2, 1921; s. Lee P. and Katherine K.; m. Sophie Skadorwa, June 4, 1966; children—Andrew, Claire. B.S., U. Chgo., 1942, M.S., 1949, Ph.D., 1953. With Argonne Nat. Lab., 1946-50; staff mem. Westinghouse Research Lab., Pitts., 1953-60; mem. research staff IBM Research Lab., Yorktown Heights, N.Y., 1960—; vis. physicist Am. Phys. Soc. Vis. Indsl. Physicists Program, 1974-75, 77; vice chmn. Gordon Conf. on High Pressure Physics, 1970; chmn. Gordon Conf. on Chemistry and Physics of Microstructure Fabrication, 1976, Nat. Materials Adv. Bd. (ad hoc com. on ion implantation as a new surface treatment tech.), 1978; mem. Nat. Acad. Scis.-NAE-NRC evaluation panel Nat. Bur. Standards, 1970-73; cons. physics survey com., mem. statis. data panel Nat. Acad. Sci.-NRC Council Physics Survey Com., 1972; mem. data and info. panel Nat. Acad. Sci.-NRC Com. on Survey of Materials Sci. and Engring., 1974. Editor: Revs. Modern Physics, 1976—; corr.: Comments on Solid State Physics, 1970—. Served with USN, 1944-46. Recipient Outstanding Contbn. award IBM, 1963. Fellow Am. Phys. Soc., IEEE (chmn. subcom. cultural and sci. relations 1976, mem. del. to USSR 1975, W.R.G. Baker prize 1976); mem. Nat. Acad. Engring. Asso. Subspecialties: Condensed matter physics; Computer engineering. Current work: Heavily-doped semiconductors; physical limits in information processing. Office: IBM PO Box 218 Yorktown Heights NY 10598

KEYFITZ, BARBARA LEE, mathematician, educator; b. Ottawa, Ont., Can., Nov. 7, 1944. B.Sc., U. Toronto, 1966; M.S., Courant Inst., NYU, 1968, Ph.D., 1970. Asst. prof. Columbia U. Sch. Engring. and Applied Sci., 1970-76; lectr. dept. mech. and aerospace engring. Princeton U., 1977-79; vis. mem. Institut de Mathematiques et Sciences Physiques, U. Nice, France, 1980, Math. Research Inst., U. Warwick, Eng., 1980; asst. prof. Ariz. State U., Tempe, 1979-81, assoc. prof., 1981-83, U. Houston, 1983—; vis. assoc. prof. Duke U., 1981, U. Calif.-Berkeley, 1982. Mem. Am. Math. Soc., Soc. Indsl. and Applied Math., Math. Assn. Am. Subspecialty: Applied mathematics. Current work: Nonlinear partial differential equations and fluid dynamics; supersonic and transonic gas dynamics; hyperbolic conservation laws; numerical analysis; mathematical problems in combustion theory, elasticity and magnetohydrodynamics; applications of singularity theory to chemical reaction problems. Address: Dept Math U Houston University Park Houston TX 77004

KEYL, MILTON JACK, physiology educator; b. Decatur, Ill., Mar. 5, 1924; s. Norman Joseph and Anna Marie (Bagenski) K.; m. Audrey Fay Shearer, Aug. 30, 1948 (dec. 1978); children: Mark David, Karen Lynn. Student, Valparaiso U., 1943-44; B.S., U. Cin., 1947, M.S., 1948, Ph.D., 1957. Asst. pharmacologist William S Merrell Co., Cin., 1948-52; asst. prof. physiology U. Okla., Oklahoma City, 1957-60, assoc. prof., 1960-66, prof., 1966—; mem. rev. com. NIH, Bethesda, 1969-71. Contbr. articles to profl. jours. HIH grantee, 1959-75; U.S. Navy grantee, 1975; Am. Heart Assn grantee, 1978. Mem. Am. Physiol. Soc., Am. Soc. Lymphology, Am. Soc. Nephrology, N.Y. Acad. Sci., Sigma Xi. Republican. Lutheran. Subspecialties: Physiology (medicine); Nephrology. Current work: Teaching, research in renal and hormonal regulation of body fluids. Home: 3605 Sun Valley Dr Oklahoma City OK 73110 Office: U Okla Medical Center Physiology Box 26901 Oklahoma City OK 73190

KEYS, CHRISTOPHER BENNETT, psychologist, educator; b. N.Y.C., Mar. 18, 1946; s. William Walters and Margaret (Forman) K.; m. Elizabeth Jaffer, Sept. 14, 1969; children: Benjamin, Daniel. B.A., Oberlin Coll., 1968; M.A., U. Cin., 1971, Ph.D., 1973. Lectr. U. Cin., 1972; instr. to asst. prof. U. Ill. - Chgo., 1972-78, dir., 1973-78, assoc. prof. dept. psychology, 1978—; vis. research assoc. U. Oreg., Eugene, 1979-80; adj. faculty Union Grad. Sch., Cin., 1982—; cons. Ill. Inst. Develop. Disabilities, Chgo., 1979—, Nat. Inst. Corrections, Boulder, Colo., 1978-79, Cook County Sheriff, Maywood, Ill., 1973—, Nat. Inst. Edn., Washington, 1980. Contbr. chpts. to books, articles and tech. papers to profl. jours. Recipient Applied Research award Progressive Architecture mag., 1982; Pub. Service award Cook County, 1974; Disting. Service award AIA, 1980. Mem. Am. Psychol. Assn. (regional coordinator div. 27 1982—), Midwestern Psychol. Assn., Assn. for Advancement of Psychology, Eco-Community Psychology Interest Group (mem. steering com. 1979—), Orgn. Devel. Network. Democrat. Episcopalian. Subspecialties: Social psychology; Community psychology. Current work: Organization change and development in social institutions and community organizations; primary prevention of psychopathology; environmental psychology; group processes. Home: 533 N Cuyler Ave Oak Park IL 60302 Office: Univ Illinois at Chicago Box 4348 Chicago IL 60680

KEYWORTH, GEORGE A., physicist; b. Boston, Nov. 30, 1939; s. Robert Allen and Leontine (Briggs) K.; m. Polly Lauterbach, July 28, 1962; children: Deidre, George III. B.S., Yale U., 1963; Ph.D., Duke U., 1968; D.Sc. (hon.), Rensselaer Poly. Inst., 1982. Mem. staff Los Alamos Sci. Lab., 1968-78; head exptl. physics div., 1978-81; sci. advisor to the Pres. White House, Washington, 1981—. Recipient Chmn.'s award Am. Assn. Engring. Socs., 1982. Fellow Am. Phys. Soc., AAAS; mem. Sigma Xi. Club: Cosmos (Washington). Subspecialty: Nuclear physics. Office: Executive Office of the President Washington DC 20500

KHADDURI, FARID MAJID, mech. engr.; b. Baghdad, Iraq, Aug. 10, 1945; came to U.S., 1947, naturalized, 1954; s. Majid and Majdia (Dawaff) K.; m. Alicia Basiliko, Nov. 18, 1973; children: Alexandra, Justine. B.A. in Physics, Amherst Coll., 1967; M.S. in Engring. Sci, George Washington U., 1976. Analytical engr. Atlantic Research Corp., Gainesville, Va., 1968-72, sr. design engr., 1972-73; chief engr. Trident I Missile Post Boost Control System, 1973-78, program mgr., 1979-81, Trident II Missile Post Boost Control System, 1981—. Mem. Am. Phys. Soc., ASME. Republican. Club: Kenwood Country (Bethesda, Md.). Subspecialties: Aerospace engineering and technology; Solid rocket propulsion. Current work: Program management solid rocket propulsion. Home: 5526 Westbard Ave Bethesda MD 20816 Office: 7511 Wellington Rd Gainesville VA 22065

KHAIRALLAH, PHILIP ASAD, pharmacologist, medical researcher, priest; b. Bklyn., Feb. 3, 1928; s. Amin Asad and Laurice (Macksoud) K.; m. Margaret Ann Howard, Oct. 19, 1963; children: Mona, Tatyana, Marc, Cyril. B.S., Am. U. Beirut, 1947; M.D., Columbia U., 1951; M.A. in Religious Studies, John Carroll U., 1981. Established investigator Am. Heart Assn., 1958-673, staff div. research, 1961-71; chmn. dept. cardiovascular research Cleve. Clinic, 1971—; adj. prof. biology Cleve. State U., 1974—; pastor St. Cyril of Alexandria Melkite Greek Catholic Mission, 1982—. Contbr. numerous articles to profl. publs. Trustee Northeastern Ohio affiliate Am. Heart Assn. Served to capt. M.C., USNR, 1954-80. Mem. AMA, Am. Soc. Physiology, Am. Soc. Pharmacology, Am. Soc. Nephrology. Democrat. Subspecialties: Molecular pharmacology; Cardiology. Current work: Experimental hypertension, peptide hormone pharmacology. Receptor site studies. Cardiac hypertrophy. Medical and bioethics. Home: 16180 Lucky Bell Ln Chagrin Falls OH 44022 Office: 9500 Euclid Ave Cleveland OH 44106

KHALAF, KAMEL TOMA, biology educator; b. Mosul, Iraq, Sept. 12, 1922; came to U.S., 1964; s. Toma and Terma (Alias) K.; m. Layla Hadi Haddo, June, 1958; children: Suhad, Ramiz, Samir. B.S., High Tchrs. Coll., Baghdad, Iraq, 1945; M.S., U. Okla., 1950, Ph.D., 1952. Mem. faculty High Tchrs. Coll., Baghdad, 1953-62; research prof. Iraq Natural History Inst., Baghdad, 1962-63; asst. prof. to prof. biology Loyola U., New Orleans, 1964—. Author books in field; contbr. articles to profl. jours. Mem. Fla. Entomol. Soc., Ga. Entomol. Soc., Lepidopterists Soc., Phi Sigma. Roman Catholic. Subspecialties: Taxonomy; Behavioral ecology. Current work: Micromorphology of arthropods integument. Office: Dept Biol Sci Loyola U 6363 Saint Charles Ave New Orleans LA 70118

KHALIFAH, RAJA GABRIEL, biochemist, educator; b. Tripoli, Lebanon, May 5, 1942; came to U.S., 1962, naturalized, 1983; s. Gabriel and Mona (Faris) K.; m. Lilla Ilona Csonka, July 31, 1971; children: Peter Gabriel, Anthony Paul. B.S. in Chemistry, Am. U. Beirut, 1962, Ph.D., Princeton U., 1969. Postdoctoral fellow Harvard U., Cambridge, Mass., 1968-70; research assoc. Stanford (Calif.) U., 1970-73; asst. prof. Chemistry U. Va., Charlottesville, 1973-79; research chemist VA Med. Ctr., Kansas City, Mo., 1979—; adj. asst. prof. biochemistry Kansas U. Med. Sch., Kans., 1979—. U. Va. NIH grantee, 1975-79; Kans. U. NIH grantee, 1979-82; Kansas City VA Med. Ctr. VA grantee, 1980-83. Mem. Am. Chem. Soc., Am. Soc. Biol. Chemists. Subspecialties: Biophysical chemistry; Nuclear magnetic resonance (chemistry). Current work: Biophysical chemistry; structure and function of proteins and enzymes; enzyme mechanisms; application of nuclear magnetic resonance methods to biological systems; chemical modification of enzyme active sites; labeling of proteins and enzymes with stable isotopes; spectroscopy; kinetics; computers. Office: VA Medical Center 4801 Linwood Blvd Kansas City MO 64128

KHALIL, HATEM MOHAMED, computer science educator, consultant; b. Cairo, Egypt, July 2, 1935; came to U.S., 1962; s. Mohamed Khalil and Khedeina F. (El-Din) Midan. B.S. Tech. with honors, U. Manchester, Eng., 1962; M.S., Rice U., 1964; Ph.D., U. Del., 1967; F.B.C.S., London U., 1972; M.I.S.T., Inst. Sci. and Tech., Manchester, Eng., 1962. Research fellow Rice U., 1962-64, U. Del., 1966; computer analyst Brown & Root, Inc., Houston, 1964-65; research analyst, asst. prof. U. Del., 1968-75, assoc. prof. computer sci., 1976—, chmn. dept., 1978-81; cons. U.S. Treasury Dept., Saudi Arabia, 1983—, Goldey Beacon Coll., Wilmington, Del., 1980—, Riyadh U., 1981, Am. U. Cairo, 1975-76; research prof. Inst. Statis. Studies and Research, Cairo, 1976-77; mem. adv. bd. Goldey Beacon Coll., 1982—. Author: General Computer Science, 1976, Concepts of Computer Science, 1976, Foundation of Computer Science, 1980, Fundamentals of Computation, 1983; assoc. editor: SIGNUM News, Assn. Computing Machinery, 1976-80. Fulbright scholar, Jordan, 1982-83; U. Del. research grantee, 1975-77. Mem. Assn. Computing Machinery, AAUP, Sigma Xi. Moslem. Subspecialties: Software engineering. Current work: Teaching, research, consulting. Home: PO Box 296 Newark DE 19715 Office: Dept Computer and Info Sciences U Del Newark DE 19711

KHAN, AMANULLAH, physician; b. Jullundhar, India, Mar. 2, 1940; came to U.S., 1964, naturalized, 1977; s. Ahmad Ali Khan and Qamar-un-Nisa Khan; m. Fran Elise Austin, Dec. 9, 1972; children: Roxanna, Sabrina, Amanda. M.D., King Edward Med. Coll., Lahore, Pakistan, 1963; Ph.D., Baylor U., 1968. Fellow Wadley Insts. Molecular Medicine, Dallas, 1966, chief research fellow, 1969-70, chmn. dept. immunotherapy, 1970—; adj. prof. North Tex. State U., Denton, Tex. Woman's U., Denton. Contbr. articles to profl. jours. Mem. Dallas County Med. Soc., Tex. Med. Assn., Am. Soc. Hematology, Internat. Soc. Exptl. Hematology, Am. Assn. Cancer Research, Am. Assn. Immunology and Allergy, Am. Soc. Clin. Oncology, AMA, Internat. Soc. Hematology, Am. Coll. Allergists, ACP, Am. Assn. Immunologists. Subspecialties: Cancer research (medicine); Immunology (medicine). Current work: Research of immunology, hematology, oncology. Interest in monoclonal antibodies, lymphokines, interferon and related lymphocyte products. Home: 4035 High Summit Dallas TX 75234 Office: 9000 Harry Hines Blvd Dallas TX 75235

KHAN, FAIZ MOHAMMAD, med. physicist; b. Multan, Pakistan, Nov. 1, 1938; came to U.S., 1963, naturalized, 1974; s. Mohammad Nawaz and Fatima (Niazi) K.; m. Kathleen Joyce Boer, Feb. 19, 1966; children: Sarah, Yasmine, Rachel. B.Sc., Emerson Coll., Multan, 1957; M.Sc., Govt. Coll., Lahore, Pakistan, 1959; Ph.D., U. Minn., 1969. Diplomate: in Therapeutic Radiol. Physics, Am. Bd. Radiology. Instr. therapeutic radiology U. Minn., Mpls., 1968-69, asst. prof., 1969-74, asso. prof., 1974-79, prof., 1979—, dir. radiation physics sect., 1974—; cons. physicist VA Hosp., Mpls. Contbr. articles to sci. jours. Fulbright scholar, 1963-65. Mem. Am. Assn. Physicists in Medicine, Sigma Xi. Subspecialties: Radiology; Biophysics (physics). Current work: Physics of therapeutic radiology, radiation dosimetry, treatment planning of photon and electron therapy. Home: 1744 Lake St Saint Paul MN 55113 Office: Therapeutic Radiology Dept U Minn Hospitals Minneapolis MN 55455

KHAN, MOHAMMAD ASAD, geophysicist, educator; b. Aima, Lahore, Pakistan, Aug. 13, 1940; came to U.S., 1964, naturalized, 1975; s. Ghulam Qadir and Hajira (Karim) K.; m. Tahera Pathan, Jan. 4, 1974; 1 dau., Shehzi Samira. B.S., U. Punjab, Lahore, Pakistan, 1957, M.S., 1963; postgrad., Harvard U., 1964-65; Ph.D. (East West Center scholar), U. Hawaii, 1967. Forecaster Pakistan Meteorol. Dept., Lahore, 1958-63; lectr. in geophysics U. Punjab, 1963-64; asst. prof. geophysics and geodesy U. Hawaii, 1967-71, asso. prof., 1971-74, prof., 1974—; geophysicist, geodesist Hawaii Inst. Geophysics, 1967—; NSF and NASA fellow Summer Inst. Dynamical Astronomy at Mass. Inst. Tech., 1968-69; sr. vis. scientist geodynamics Goddard Space Flight Center NASA, Greenbelt, Md., 1972-74; sr. scientist Computer Scis. Corp., Silver Spring, Md., 1974-76, sr. cons., 1976-77; diplomatic minister/adviser Resource Survey and Devel. Pakistan, 1974-76; sr. resident asso. Nat. Acad. Scis., 1972-74; leader Am. Asian Studies and Contemporary Social Problems Seminar Series, Honolulu, 1968-69. Contbr. articles to profl. publs. Chmn. East and West: A Perspective for the 80's; mem. Hawaii Environ. Council, 1979—, chmn. exec. com., 1979—, vice chmn., 1981—; chmn. Pakistan Relief Fund, Honolulu, 1971. Fellow Explorers Club; mem. Geol. Soc. U. Punjab (pres. 1962-63), Am. Geophys. Union, Pakistan Assn. Advancement Sci., Am. Geol. Inst., Am. Geophys. Union, East West Center Alumni Assn. (dir. 1976—), Internat. Alumni of East West Center (exec. com., chmn. 1977-80). Subspecialties: Geophysics. Current work: Geophysical geodetic, oceanographic and geodynamical applications of satellites, geodynamics, planetary interiors, global tectonics, global correlations, core-mantle boundary problems, gravity and isotasy, satellite altimetry, geodesy, earth models, geophysical exploration; ocean dynamics, hydrography. Research in geophys., geodetic and oceanographic applications of satellites, geodynamics, planetary interiors, global tectonics, global correlations, core-mantle boundary problems, gravity, isostasy, satellite altimetry, geodesy, earth models, geophys. exploration, charting and hydrography, ocean dynamics. Office: Hawaii Inst Geophysics U Hawaii 2525 Correa Rd Honolulu HI 96822 Most men stand the test of adversity quite well, but if you really want to test the character of a man, give him power.

KHAN, RUDOLPH A., plant pathologist; b. Guyana, Oct. 9, 1934; came to U.S., 1963, naturalized, 1976; s. Haroon J.K. and Iris K. (Nissan) K.; m. Angela Ramlall, Aug. 12, 1958; children: Kenneth F., Annette T. B.Sc., Calif. State Poly U., 1967; M.S. in Plant Pathology, Iowa State U., 1969. Asst. plant pathologist Govt. of Guyana, 1960; research asso. dept. plant pathology U. Calif., Riverside, 1969—. Contbr. articles to profl. jours. Democrat. Roman Catholic. Subspecialties: Agricultural economics; Plant pathology. Current work: Diseases of sorghum, cotton, alfalfa, sugar beets, beans; milo root rot of southern California; verticillium wilt of cotton; phytophthora root rot of alfalfa and sugar veet; yeast spot of black-eye beans.

KHAN, WINSTON, mathematics and physical science educator; b. Port-of-Spain, Trinidad, Mar. 12, 1934; s. Amarnath and Safeeran K.; m. Joan Acklima Aziz, Dec. 22, 1961; children: Alima, Selina, Shereeza, Winston, Alim. B.Sc., London U., 1956, M.Sc., 1958; Dip.Mat.Phys., Birmingham U., 1961, Ph.D., 1964. Asst. lectr. math. London U., 1958-59; asst. prof., dir. math. U. West Indies, Trinidad, 1964-69, U. P.R., Cavey, 1969-72, assoc. prof., 1972-74, prof. physics, Mayaguez, 1974—. Author: Turbulence Phenomena, 1972. Mem. Am. Math. Soc., Soc. Indsl. and Applied Math., Am. Phys. Soc., Internat. Math. Modelling, Smithsonian Instn. Subspecialties: Applied mathematics; Physics of fluids. Current work: Fluid mechanics, turbulence phenomena, physics and mathematics of fluids, engineering science. Home: Calle Uroyan AD4 Mayaguez PR 00709 Office: U PR Mayaguez PR 00709

KHARE, ASHOK K., metallurgical engineer; b. Kanpur, India, Aug. 7, 1948; came to U.S., 1969; s. Bhagwan Prashad and Krishna

(Kumari) K.; m. Poornima Chand, Nov. 18, 1974. B.S.c., Agra U., India, 1964; B. Tech., Indian Inst. Tech., 1969; M.S., Stevens Inst. Tech., 1971. Devel. metallurgist Nat. Forge Co., Irvine, Pa., 1975—; lectr. in field. Contbr. articles to profl. jours. Mem. exec. com. Warren County Republican Com. Mem. Am. Soc. Metals (Presidents' sponsor award 1978-82, editor tech. book), ASME. Republican. Hindu. Clubs: Warren Art League, Warren Players. Lodges: Masons; Shriners. Subspecialties: Metallurgical engineering; Alloys. Current work: Heavy steel forgings, high temperature alloys and some nickel and copper base alloys; product and process development applications and failure analysis; formulation and establishment of standards. Developer of patent in field. Home: 5 Leslie Blvd Warren PA 16365 Office: Nat Forge Co Irvine PA 16329

KHARE, BISHUN NARAIN, physicist; b. Varanasi, India, June 27, 1933; d. Dwarka Nath and Ram (Pyari) Srivastava; m. Jyoti Rani Khare, Dec. 7, 1962; children: Reena, Archana. B.Sc., Banaras Hindu U., India, 1953, M.Sc., 1955; Ph.D., Syracuse U., 1961. Research assoc. U. Toronto, 1961-62; research assoc. SUNY, Stony Brook, 1962-64; assoc. research scientist Ont. Research Found., Can., 1964-66; physicist Smithsonian Astrophys. Obs., 1966-68; sr. research physicist Ctr. for Radiophysics and Space Research, Cornell U., 1968—; assoc. Harvard Obs., 1966-68. Contbr. numerous articles to profl. jours. Mem. Am. Phys. Soc., AAAS, Am. Astron. Soc., Am. Chem. Soc., Internat. Astron. Union, Internat. Soc. for Study of Origin of Life, Astron. Soc. India, Planetary Soc., Sigma Xi. Subspecialties: Planetary science; Space chemistry. Current work: Molecular structure and spectroscopy, applications of spectroscopic technques to the study of compounds synthesized in primitive terrestrial and contemporary planetary atmospheres by photochemical reaction, hydrogen bonding among molecules of biological interest, planetary surfaces and atmospheres, and interstellar chemistry. Office: 306 Space Scis Cornell U Ithaca NY 14853

KHATENA, JOE, psychology educator; b. Singapore, Oct. 25, 1925; came to U.S., 1966, naturalized, 1972; s. Jacob J. and Rachel (Rahmin) K.; m. Nelly Joshua, Dec. 17, 1950; children: Annette, Jacob Allan, Moshe, Serena. B.A., U. Malaya, Singapore, 1960, 1961; Ed.M., U. Singapore, 1964; Ph.D., U. Ga., 1969. Tchr, Singapore Govt., 1950-57; lectr. English, Singapore Tchrs. Coll., 1961-66; asst. prof. psychology E. Carolina U., Greenville, N.C., 1968-69; assoc. prof. Marshall U., Huntington, W.Va., 1969-72, prof., 1972-77; prof., head ednl. psychology Miss. State U., Mississippi State, 1977—. Editorial bd.: Gifted Child Quar., 1975—; asso. editor: Jour. Mental Imagery, 1981—; contbr. articles to profl. jours.; co-author: Khatena-Torrance Creative Perception Inventory, 1976; author: Creatively Gifted Child, 1978, Educational Psychology of the Gifted, 1982, Thinking Creatively with Sounds and Words, 1973; others. Recipient Book prize U. Malaya, 1957; Nat. Assn. Gifted disting. scholar, 1982. Fellow Am. Psychol. Assn.; mem. Nat. Assn. Gifted Children (pres 1977-79), Internat. Psychol. Assn., Am. Ednl. Research Assn., Phi Kappa Phi, Kappa Delta Pi. Subspecialty: Creativity giftedness. Current work: Creative imagination imagery, gifted, measures of originality and creative self-perception, social attitudes. Home: 8 Tally Ho Dr Starkville MS 39759 Office: Dept Ednl Psychology Miss State Univ Drawer EP Mississippi State MS 39762

KHORANA, HAR GOBIND, chemist, educator; b. Raipur, India, Jan. 9, 1922; s. Shri Ganpat Rai and Shrimati Krishna (Devi) K.; m. Esther Elizabeth Sibler, 1952; children: Julia, Emilie, Dave Roy. B.S., Punjab U., 1943, M.S., 1945; Ph.D., Liverpool (Eng.) U., 1948; D.Sc., U. Chgo., 1967. Head organic chemistry group B.C. Research Council, 1952-60; vis. prof. Rockefeller Inst., N.Y.C., 1958—; prof. co-dir. Inst. Enzyme Research, U. Wis., Madison, 1960-70, prof. dept. biochemistry, 1962-70, Conrad A. Elvehjem prof. life scis., 1964-70; Alfred P. Sloan prof. biology and chemistry MIT, Cambridge, 1970—; vis. prof. Stanford U., 1964; mem. adv. bd. Biopolymers. Author: Some Recent Developments in the Chemistry of Phosphate Esters of Biological Interests, 1961; Mem. editorial bd.: Jour. Am. Chem. Soc, 1963—. Recipient Merck award Chem. Inst. Can., 1958, Gold medal Profl. Inst. Pub. Service Can., 1960, Dannie-Heinneman Preiz, Göttingen, Germany, 1967, Remsen award Johns Hopkins U., 1968, Am. Chem. Soc. award for creative work in synthetic organic chemistry, 1968, Louisa Gross Horwitz prize, 1968, Lasker Found. award for basic med. research, 1968, Nobel prize in medicine, 1968; elected to Deutsche Akademie der Naturforscher Leopoldina, HalleSaale, Germany, 1968; Overseas fellow Churchill Coll., Cambridge, Eng., 1967. Fellow Chem. Inst. Can., Am. Acad. Arts and Scis.; mem. Nat. Acad. Sci. Subspecialties: Organic chemistry; Biochemistry (biology). Research and numerous publs. on chem. methods for synthesis of nuccleotides, coenzymes and nucleic acids; elucidation on the genetic code, lab. synthesis of genes, biol. membrane, light-transducing pigments. Office: Dept Biology and Chemistry MIT Cambridge MA 02139

KHOURY, GEORGE, physician, molecular biologist, government administrator; b. Pitts., Aug. 7, 1943; m., 1967; 2 children. A.B., Princeton U., 1965; M.D., Harvard U., 1969. Research assoc. Nat. Inst. Allergy and Infectious Diseases, 1970-75; sect. head, lab. chief Nat. Cancer Inst, NIH, Bethesda, Md., 1975—. Asst. editor: Jour. Virology, 1974—; assoc. editor: Cell, 1977—. Recipient Arthur S. Fleming award U.S. Govt., 1980. Mem. Am. Soc. Microbiology, Am. Soc. Virology. Subspecialty: Molecular biology. Office: NIH Bldg 41 Suite 200 Bethesda MD 20205

KIANG, DAVID TEH-MING, physician; b. Che-Kiang, China, Nov. 13, 1935; s. Chin and Yun-Hsu (Hsu) K.; m. Adeline C. King, Dec. 22, 1968; children: Karen, Bonnie, Jonathan. M.D., Nat. Def. Med. Ctr., Taipei, 1960; M.S., Columbia U., 1964; Ph.D., U. Minn., 1973. Intern Beckman Downtown Hosp., N.Y.C., 1964-65; resident Francis Delafield Hosp., Columbia-Presbyn. Med. Ctr., N.Y.C., 1966-68; practice medicine, specializing in internal medicine and med. oncology, Mpls., 1968—; faculty Sch. Medicine, U. Minn.-Mpls., 1971—, assoc. prof. medicine, 1979—; dir. Breast Cancer Research Lab., 1979—. Mem. Central Soc. Clin. Research, Am. Assn. Cancer Research, Am. Soc. Clin. Oncology. Subspecialties: Oncology; Cancer research (medicine). Current work: Breast cancer research and treatment. Address: Box 168 Univ Minn Med Sch Minneapolis MN 55455

KIANG, ROBERT L., research engineer; b. Chungking, China, Nov. 30, 1939; came to U.S., 1962; s. Johnson C. Kiang and C. C. Huang. B.S., Nat. Taiwan U., 1961; M.S., Stanford U., 1964, Ph.D., 1969. Research engr. Stanford U. Research Inst., 1963-73; sr. research engr. SRI Internat., Menlo Park, Calif., 1973—. Mem. ASME, Am. Nuclear Soc. Subspecialties: Nuclear fission; Fusion. Current work: Nuclear safety, thermal hydraulics, wind power, inertial confined fusion. Home: 1100 Sharon Park Dr #2 Menlo Park CA 94025 Office: SRI Internat 333 Ravenswood Ave Menlo Park CA 94025

KIANG, YUN-TZU, genetics educator; b. Taiwan, Feb. 1, 1932; m. Ming, May 12, 1957; children: Wailey, Waisen, Phine. B.S., Taiwan Normal U., 1956; M.S., Ohio State U., 1962; Ph.D., U. Calif.-Berkeley, 1970. Asst. prof. genetics U. N.H., Durham, 1970-75, assoc. prof., 1975-83, prof., 1983—, chmn. genetics program, 1980-83, summer faculty fellow, 1971-72. Contbr. articles to profl. jours., also monograph on plant isozymes. Central Univ. Research Funds grantee, 1981-82; N.H. Hwy. Dept. grantee, 1972-74; U.S. Dept. Agr. grantee, 1982—. Mem. Genetics Soc. Am., Am. Soc. Naturalists, Soc. Study of Evolution, Am. Soc. Agronomy. Subspecialties: Genetics and genetic engineering (biology); Evolutionary biology. Current work: Researcher in population, ecological and evolutionary genetics; plant breeding and plant tissue culture. Office: Dept Plant Sci Univ NH Durham NH 03824

KICLITER, ERNEST EARL, JR., neuroanatomist, educator; b. Ft. Pierce, Fla.; s. Ernest Earl and Betty Lloyd (Winn) K.; m. Veronica Pelaez Herran, Oct. 7, 1967. A.B. in Psychology, U. Fla., 1968; Ph.D. in Anatomy, SUNY Upstate Med. Ctr., Syracuse, 1973. Predoctoral fellow anatomy SUNY Upstate Med. Ctr., 1969-72; postdoctoral fellow neurosurgery U. Va., Charlottesville, 1972-74; asst. prof. neuroanatomy U. Ill., Urbana, 1974-77; assoc. prof. anatomy U. P.R., San Juan, 1977—. Contbr. articles to profl. jours. Fellow Ford Found., 1967-68, NIH, 1968-74; grantee NIH, 1976—. Mem. Nat. Acad. Sci., Am. Assn. Anatomists, AAAS, Assn. Research in Vision and Ophthalmology, Soc. Neurosci (pres. P.R. chpt. 1979-80), Sigma Xi. Clubs: Garcia y Vega (St. Louis); Cajal, J.B. Johnston. Subspecialties: Comparative neurobiology; Psychophysics. Current work: Neuroanatomy of visual systems, color vision, comparative neuroanatomy. Home: Avenida Wilson 1367 Apt 203 Condado PR 00907 Office: Blvd del Valle 201 San Juan PR 00901

KIDD, ROBERT FLETCHER, social psychologist, educator; b. Pikeville, Ky., Aug. 24, 1951; s. Herbert and Kathleen Abernathy (Gaines) K.; m. Ellen F. Chayet, Oct. 16, 1982. B.A., Vanderbilt U., Nashville, 1973; M.A., U. Wis.-Madison, 1975, Ph.D., 1977; Dipl., U. Cambridge, Eng., 1976. Asst. prof. Boston U., 1977—; cons. Police Found., Washington, 1977, ABT Assocs., Camridge, Mass., 1980; adv. editor Springer-Verlag, N.Y.C., 1979—. Editor: New Directions in Attribution Research, 3 vols, 1976, 78, 81, Advances on Applied Social Psychology, 3 vols, 1980, 83. Grantee NSF, MIMH. Mem. Am. Psychol. Assn., Soc. Exptl. Social Psychologists, Eastern Psychol. Assn. Subspecialties: Social psychology; Cognition. Current work: Decision making, information seeking. Home: 73 Marlboro St Belmont MA 02178 Office: Dept Psychology Boston Univ 64 Cummington St Boston MA 02215

KIDWELL, ALBERT LAWS, geological researcher; b. Auxvasse, Mo., Jan. 1, 1919; s. Albert Lewis and Josephine (Laws) K.; m. Marian Viola Rankin, June 16, 1943; children: William Albert, Betty Jo Kidwell Evans, Patricia Alice Kidwell Lown, Thomas Paul. B.S., Mo. Sch. Mines, U. Mo.-Rolla, 1940; M.S., Washington U., St. Louis, 1942; Ph.D., U. Chgo., 1949. Geologist Mo. Geol. Survey, Rolla, 1944-47; research geologist Carter Oil Co., Tulsa, 1950-58; research assoc. Jersey Prodn. Research Co., Tulsa, 1958-64; sr. research assoc. Esso Prodn. Research Co., Houston, 1965-80, Exxon Minerals Co., 1981-82, Exxon Prodn. Research Co., 1982—. Contbr. articles on mineralogy, econ. geology and geo chemistry to profl. jours. AEC fellow U. Chgo., 1948-49. Fellow Geol. Soc. Am., Mineral. Soc. Am.; mem. Am. Assn. Petroleum Geologists, Soc. Econ. Geologists. Republican. Methodist. Club: Tulsa Geol. Soc. (pres. 1964). Subspecialties: Sedimentology; Mineralogy. Current work: Research on effects of diagenetic processes on reservoir properties of sandstones. Home: 14403 Carolcrest Houston TX 77079 Office: Exxon Prodn Research Co PO Box 2189 Houston TX 77001

KIDWELL, WILLIAM ROBERT, research biologist; b. La Follette, Tenn., Sept. 2, 1936; s. John Simon and Hazel Jane (Plemons) K.; m. Etta Bernice Widener; children: Shari Lynn, Joy Elizabeth. B.A. in Chemistry, Berea (Ky.) Coll.; Ph.D. in Biochemistry, Washington U., St. Louis. Postdoctoral fellow U. Wis.; staff fellow Nat. Cancer Inst., Bethesda, Md.; research biologist, now chief cell cycle regulation sect. Lab. Pathophys.; sec. treas. Cellco, Inc. Contbr. articles in field to profl. jours. Served with USAF, 1955-59. Mem. Am. Assn. Cancer Research, Am. Assn Cell Biology, AAAS. Subspecialties: Cell biology (medicine); Developmental biology. Current work: Control of growth of normal and tumor cells; hormone effects on mammary cell extracellular matrixproduction; role of extracellular matrix in tumor cell growth. Patentee hollow fiber cell culture apparatus and methodology. Home: 10905 Lowell Ct Ijamsville MD 21754 Office: Nat Cancer Inst Bldg 10 Bethesda MD 20205

KIECHLE, FREDERICK LEONARD, pathologist, researcher; b. Indpls., Mar. 26, 1946; s. Frederick Leonard and Bertha Mae (Fackler) K.; m. Janet B. Green, June 21, 1975; children: Elizabeth Heather, Rachel Bourk. B.A., Evansville Coll., 1968; Ph.D., Ind. U., 1973, M.D., 1975. Diplomate: Am. Bd. Pathology. Resident in pathology William Beaumont Hosp., Royal Oak, Mich., 1975-79; fellow clin. chemistry, postdoctoral research fellow Barnes Hosp., St. Louis, 1979-80; asst. prof. pathology U. Pa., Phila., 1980—; dir. STAT Lab., Hosp. Univ. Pa., Phila., 1980—, asst. dir. div. lab. medicine, 1982—. Hartford Found. fellow, 1982—; undergrad. awardee Am. Chem. Soc., 1967. Fellow Am. Soc. Clin. Pathology; mem. Am. Assn. Pathologists, Am. Assn. Clin. Chemistry, Am. Diabetes Assn., Am. Fedn. Clin. Research. Presbyterian. Subspecialties: Pathology (medicine); Biochemistry (biology). Current work: Clinical chemistry, insulin action, second messenger of insulin, insulin and phospholipid metabolish, phospholipid methylation, ^{31}P-NMR. Office: Hosp of Univ Pa 3600 Spruce St/Gl Philadelphia PA 19104

KIEFF, ELLIOTT, physician, educator; b. Phila., Feb. 2, 1943; s. Irving and Florence (Prussell) K.; m. Jacqueline, June 27, 1965; children. B.A., U. Pa., 1963; M.D., Johns Hopkins U., 1966; Ph.D. in Microbiology, U. Chgo., 1971. Asst prof. microbiology and medicine U. Chgo., 1971-74, assoc. prof., 1974-76, prof., 1977—, chief infectious disease, 1971—; Hartford vis. prof. Washington U., St. Louis, 1979; mem. adv. bd. for microbiology Am. Cancer Soc. Assoc. editor: Jour. of Infectious Disease, 1978—, Virology, 1979-81; mem. editorial bd.: Jour. Virology, 1980—. Recipient Faculty Research award Am. Cancer Soc., 1979—; Ann Langer Cancer Research award, 1982. Mem. Am. Soc. Clin. Investigation, Infectious Disease Soc. Am. Subspecialties: Infectious diseases; Virology (medicine). Current work: Molecular biology, genetics, biology of herpes viruses. Office: 910 E 58th St Chicago IL 60037

KIEFHABER, NIKOLAUS JOSEF, software engineer, researcher; b. Munich, W. Ger., July 29, 1954; came to U.S., 1977; s. Robert Paul and Gisela (Heidtmann) K.; m. Sarah Hildebrandt, Oct. 18, 1980. Vordiplom Physics, U. Regensburg, Ger., 1977; M.S. in Computer Sci, U. Colo., 1980. Mem. sr. tech. staff Precision Visuals Inc., Boulder, Colo., 1980-82, dir. research, 1982—. Mem. Assn. for Computing Machinery. Subspecialties: Graphics, image processing, and pattern recognition; Software engineering. Current work: Graphics standards, graphics software, graphics hardware. Office: Precision Visuals Inc 6260 Lookout Rd Boulder CO 80301

KIEL, ROBERT ALLEN, periodontist, researcher, educator; b. Hartford, Conn., Mar. 31, 1954; s. Jerome George and Gloria Suzanne (Snyder) K.; m. Nancy Turner, Aug. 15, 1976. B.S., Tufts U., 1975; D.M.D., U. Conn., 1975-79, M.Dental Sci., 1981. Vis. asst. prof. U. Bern, Switzerland, 1981-82; asst. prof. periodontics U. Conn.-Farmington, 1982—; pvt. practice dentistry specializing in periodontics, Vernon, Conn., 1982—. NIH research tng. grantee, 1980-81. Mem. Am. Acad. Periodontics, ADA, Hartford Dental Soc. Democrat. Jewish. Subspecialty: Periodontics. Current work: Microbiology of periodontal diseases and computerized identification of bacteria. Home: 42 Minster Brook Dr Simsbury CT 06070 Office: U Conn Health Center 263 Farmington Ave Farmington CT 06032

KIELY, MICHAEL LAWRENCE, anatomist, researcher; b. Springfield, Ill., June 17, 1938; s. Lawrence W. and Pauline D. (Goodrich) K.; m. Myra J. Vansach, Apr. 15, 1967; children: Michelle, Jeanette. A.A., Springfield Jr. Coll., 1958; B.S., Lewis U., 1960; M.S., Loyola U., Chgo., 1964, Ph.D., 1967. Research assoc. Stritch Sch. Medicine, Chgo., 1966-67; assoc. prof. Loyola U. Sch. Dentistry, Maywood, Ill., 1967—. Editorial bd.: Jour. Dental Research, Washington, 1977—. Nat. Inst. Dental Research predoctoral fellow, 1962-67. Mem. Am. Assn. Anatomists, Internat. Assn. Dental Research, Am. Assn. Dental Research, Sigma Xi. Lodge: KB. Subspecialties: Oral biology; Anatomy and embryology. Current work: Cell kinetics of dental tissues, mechanisms of tooth eruption, influence of endocrines on dental tissues. Office: Loyola U Sch Dentistry 2160 S 1st Ave Maywood IL 60153 Home: 607 Arrowhead St Carol Stream IL 60187

KIEREIN, JOHN, aerospace exec.; b. South Bend, Ind., Nov. 29, 1936; s. Leo John and Norma (Shoop) K.; m. Serena Ruth Reighard, Apr. 15, 1963; children: Kathryn, Rebecca, Pamela. B.S. in Physics, Notre Dame U., 1959; M.B.A., Ind. U., 1961. Research adminstr. Picatinny Arsenal, Dover, N.J., 1961-63; engring. adminstr. McDonnell Douglas, St. Louis, 1963-68; with Skylab payloads Martin Marietta, Denver, 1968-75; with shuttle payloads Rockwell Internat., Downey, Calif., 1975-77; dep. program mgr. Ball Aerospace, Boulder, Colo., 1977—. Author: Kamikaze Notrump, 1979; co-editor: Skylab's Astronomy and Space Science Experiments, 1979; author tech. reports. Recipient Skylab awards NASA and Martin Marietta, 1974. Mem. Planetary Soc., AIAA, Astron. Soc. Pacific, Am. Astron Soc., Assn. Gravity Research (pres. 1982—, annual award 1964, 68, 82), AAAS. Clubs: Vid ID (Boulder) (pres. 1982); Am. Contract Bridge League.). Subspecialties: Cosmology; Radio and microwave astronomy. Current work: Chemical releases to provide windows in ionosphere for elf astronomy. Compton effect interpretation of red shift. Gravity. Home: 4377 Carter Trail Boulder CO 80301 Office: Ball Aerospace Systems Div PO Box 1062 Boulder CO 80306

KIER-SCHROEDER, ANN B., comparative pathologist; b. Littlefield, Tex., June 26, 1949; d. Robert Merlin and Martha Ann (Bond) Yarbrough; m. Friedhelm Schroeder, Dec. 9, 1978. B.A., U. Tex., 1971; B.S., Tex. A&M U., 1973, D.V.M., 1974; Ph.D., U. Mo.-Columbia, 1979. Diplomate: Am. Coll. Lab. Animal Medicine. NIH postdoctoral fellow in lab. animal medicine and comparative pathology U. Mo.-Columbia, 1976-79, asst. prof. vet. pathology, 1979—, supr. histopathology lab., 1980—. Author book chpt.; Contbr. articles to profl. jours. Served to capt. Vet. Corps U.S. Army, 1977-82. Mem. Am. Coll. Lab. Animal Medicine, Am. Assn. Lab. Animal Sci., AVMA, Reticuloendothelial Soc., Sigma Xi, Phi Zeta, Mortar Board. Subspecialties: Pathology (veterinary medicine); Immunology (medicine). Current work: Chemotaxis (neutrophils and macrophages), chemiluminescence, immunopathology, diagnostic laboratory animal pathology, animal models for human health related diseases. Home: RD 2 Carter School Rd Columbia MO 65201 Office: Dept Vet Pathology VMDL U Mo Columbia MO 65211

KIERSZENBAUM, FELIPE, microbiologist, educator; b. Cordoba, Argentina, Jan. 1, 1940; s. Nachman and Ester (Niestenpower) K.; m. Delia Beatriz Budzko, Dec. 10, 1964; children: Martin, Marina. Pharmacist U. Buenos Aires, Argentina, 1961, Biochemist, 1963, Ph.D., 1966, Asst. prof., 1971-74; sr. research scientist Wellcome Research Labs., Eng., 1974-75; research staff Yale U. Sch. Medicine, New Haven, 1975-77; assoc. prof. Mich. State U., East Lansing, 1977-82, prof. microbiology and pub. health, 1982—. Contbr. articles to profl. jours. Mem. Am. Assn. Immunologists, Am. Soc. Microbiology, Reticuloendothelial Soc., AAAS, Soc. Protozoologists, Am. Soc. Tropical Medicine and Hygiene, Am. Soc. Parasitologists. Subspecialties: Immunology (medicine); Parasitology. Current work: The mechanisms of host defense against infection with Trypanosoma cruzi; cellular immunology. Office: Mich State U Dept Microbiology East Lansing MI 48824

KIESLER, CHARLES ADOLPHUS, psychologist, association executive; b. St. Louis, Aug. 14, 1934; m. (div.); children—Tina, Thomas, Eric, Kevin. B.A., Mich. State U., 1958, M.A., 1960; Ph.D. (NIMH fellow), Stanford U., 1963. Asst. prof. psychology Ohio State U., Columbus, 1963-64; asst. prof. psychology Yale U., New Haven, 1964-66, asso. prof., 1966-70; prof., chmn. psychology U. Kans., Lawrence, 1970-75; exec. officer Am. Psychol. Assn., Washington, 1975-79; Walter Van Dyke Bingham prof. psychology Carnegie Mellon U., Pitts., 1979—, head psychology, 1980-82, acting dean, 1981-82, dean Coll. Humanities and Social Scis., 1983—. Author: (with B.E. Collins and N. Miller) Attitude Change: A Critical Analysis of Theoretical Approaches, 1969, (with S.B. Kiesler) Conformity, 1969, The Psychology of Commitment: Experiments Linking Behavior to Belief, 1971, (with N. Cummings and G. Vanden Bos) Psychology and National Health Insurance: A Sourcebook, 1979. Served with Security Service USAF, 1952-56. Fellow Am. Psychol. Assn., AAAS; mem. AAUP, Eastern Psychol. Assn., Soc. Exptl. Social Psychology, Midwestern Psychol. Assn., Assn. for Advancement of Psychology, Psychonomic Soc., Council Applied Social Research, Sigma Xi, Psi Chi, Phi Kappa Phi. Subspecialties: Social psychology; Cognition. Current work: Mental health policy; systems research related to health and mental health. Office: Office of Dean Coll Humanities and Social Scis Carnegie Mellon U Pittsburgh PA 15213

KIESLING, RICHARD LORIN, plant pathologist, educator; b. Rockford, Ill., Nov. 20, 1922; s. Earl Leon and Edith Eugenia (Gorball) K.; m. Frances Mae Groth, June 22, 1947; children: Faye, Gregory, Frances, Richard. B.Sc., U. Wis., 1949, M.S., 1951, Ph.D., 1952. Asst. prof. Mich. State U., 1952-57, assoc. prof., 1957-60; prof. plant pathology, chmn. dept. plant pathology N.D. State U., 1960—. Served with Signal Corps U.S. Army, 1942-46; ETO. Mem. Am. Phytopath. Soc., Am. Inst. Biol. Scis., Sigma Xi. Methodist. Lodge: Lions. Subspecialties: Plant pathology; Genetics and genetic engineering (agriculture). Current work: Research on genetics of barley and barley smut interactions; genetics of virulence and aggressiveness in Ustilago hordei; genetics of resistance in barley to Ustilago hordei, Ustilago nigra, and Ustilago nuda. Office: Dept Plant Pathology ND State U PO Box 5012 Fargo ND 58105

KIKUCHI, CHIHIRO, nuclear engineer, educator; b. Seattle, Sept. 26, 1914; s. Naoki and Mitsue (Ichinomiya) K.; m. Grace Keiko Fujii, June 9, 1946; children: Naomi, Carl, Gary. B.S. in Physics, U. Wash., 1939, Ph.D., 1944; M.A., U. Cin., 1943. Mem. faculty Haverford (Pa.) Coll., 1943-44, Mich. State U., East Lansing, 1944-53; assoc. prof. physics U.S. Naval Research Lab., Washington, 1953-55; mem. faculty dept. nuclear engring. U. Mich., Ann Arbor, 1955—, prof. nuclear engring., 1982—; tech. specialist IAEA, 1964, Brookhaven Nat. Lab., 1951-52; vis. prof. Kyoto (Japan) U., 1976-77; cons., Sao Paulo, Brazil, 1976-77. Mem. Am. Phys. Soc., Am. Nuclear Soc. Democrat. Congregationalist. Subspecialties: Nuclear fission; Magnetic physics. Current work: Nuclear power and energy alternatives. Office: Dept Nuclear Engrin U Mich Ann Arbor MI 48109

KIKUCHI, NOBORU, mechanical engineering educator, researcher; b. Kamioka, Shenboku-gun, Akita, Japan, Feb. 4, 1951; came to U.S., 1974, naturalized, 1978; d. Masao and Toyo (Kaneko) Saito; m. Nanae Kikuchi, May 1, 1974. B.E. in Civil Engring, Tokyo Inst. Tech., 1974; M.S. in Engring. Mechanics, U. Tex.-Austin, 1975, Ph.D., 1977. Research assoc. U. Tex.-Austin, 1977-79, asst. prof., 1979-80; asst. prof. mech. engring. U. Mich., Ann Arbor, 1980—. Author: Contract Problems in Elasticity, 1983. Recipient Outstanding Jr. Faculty award Arco Oil and Gas Co., 1981; Incentive award Japan Soc. Civil Engrs., 1981. Mem. ASME, Am. Soc. Civil Engrs., Soc. Engring. Sci. Subspecialties: Numerical analysis; Solid mechanics. Current work: Development of finite element codes for nonlinear mechanics and their numerical analysis; especially contact-impact friction problems for elasticity, elasto-plasticity and visco-plasticity. Home: 1076 Island Dr Ct 106 Ann Arbor MI 48105 Office: U Mich Dep Mechanical Engring 550 E University Ann Arbor MI 48109

KILBOURNE, EDWIN DENNIS, virologist, educator; b. Buffalo, July 10, 1920; s. Edwin I. and Elizabeth (Alward) K.; m. Joy Schmid, Dec. 20, 1952; children: Edwin Michael, Richard Schmid, Christopher Norton, Paul Alward. A.B., Cornell U., 1942, M.D., 1944. Asst. Rockefeller Inst., 1948-51; mem. faculty Tulane U., 1951-55, Cornell U. Med. Coll., N.Y.C., 1955-68, prof. pub. health, dir. div. virus research, 1961-68; prof., chmn. dept. microbiology Mt. Sinai Sch. Medicine, City U. New York, 1968—. Author: (with Wilson G. Smillie) Human Ecology and Public Health, 4th edit, 1968; Editor: The Influenza Viruses and Influenza, 1975. Mem. Health Research Council N.Y.C., 1968-75. Recipient R.E. Dyer Lectureship award NIH, 1973, Borden award Assn. Am. Med. Colls., 1974, Dowling Lectureship award, 1976, Thomas Francis Lectureship award, 1976; Harvey Lectureship award, 1978; award of distinction Cornell U. Med. Alumni Assn., 1979; academy medal N.Y. Acad. Medicine. Fellow N.Y. Acad. Scis.; mem. Nat. Acad. Sci., Harvey Soc., So. Soc. Clin. Research, Central Soc. Clin. Research (emeritus), AAAS, Am. Assn. Immunologists, Am. Acad. Microbiology, Soc. Exptl. Biology and Medicine, Am. Soc. Clin. Investigation (emeritus), N.Y. Acad. Medicine, Am. Pub. Health Assn., Assn. Am. Physicians, Am. Soc. Microbiology, Infectious Diseases Soc. Am. Subspecialties: Microbiology (medicine); Virology (medicine). Current work: Lifetime research and specialization in infectious diseases, including the clinical epidemiological and molecular biological aspects of human viruses, especially influenza; viral genetics, vaccine development. Research and publs. on hormonal influences, genetic studies and exptl. transmission of viruses, recombinant virus vaccines especially influenza. Home: 446 Hillcrest Rd Ridgewood NJ 07450 Office: City U New York Mt Sinai Sch Medicine Dept Microbiology Fifth Ave at 100th St New York NY 10029

KILBOURNE, EDWIN MICHAEL, epidemiologist; b. New Orleans, Oct. 1, 1953; s. Edwin Dennis and Joy Magdalena (Schmid) K.; m. Barbara Williams, Nov. 26, 1982. A.B., Cornell U., 1974, M.D., 1978. Lic. physician, Ala. Intern U. Ala. Hosps., 1978-79, resident in medicine, 1979-80, 82-83; epidemic intelligence service officer Cts. Disease Control, Atlanta, 1980-82; chief epidemiology and investigations sect. Center Environ Health, 1983—. Contbr. articles on physiology and environ. epidemiology to profl. jours. Served as sr. asst. surgeon USPHS, 1980-82. Recipient Alexander D. Langmuir award Ctrs. Disease Control, 1982. Subspecialties: Epidemiology; Internal medicine. Current work: Epidemiology of new diseases and of physical and chemical environmental hazards. Office: Special Studies Branch Centers for Disease Control Atlanta GA 30333

KILBURN, KAYE HATCH, physician, educator; b. Logan, Utah, Sept. 20, 1931; d. H. Parley and Winona (Hatch) K.; m. Gerrie Griffin, June 7, 1954; children: Ann Louise, Scott Kaye, Jean Marie. B.S., U. Utah, 1951, M.D., 1954. Diplomate: Am. Bd. Internal Medicine, 1963. Intern/resident Cleve. Univ. Hosp., 1954-55, U. Utah Hosp., Salt Lake City, 1955-57; fellow Duke Univ. Hosp., Durham, N.C., 1957-58, Brompton Hosp., U. London, 1960-61; asst. clin. prof. medicine U. Colo., Denver, 1958-60; asst. prof. medicine Washington U. Sch. Medicine, St. Louis, 1960-62; assoc. prof. to prof. Duke U. Sch. Medicine, Durham, N.C., 1963-73; prof. medicine, chief pulmonary div. U. Mo. Sch. Medicine, Columbia, 1973-77; prof. medicine, community medicine Mt. Sinai Sch. Medicine, CUNY, 1977-80; Ralph Edgington prof. medicine U. So. Calif. Sch. Medicine, 1980—; mem. research commn. Nat. Cystic Fibrosis Found., Atlanta, 1975-80; mem. adv. com. byssinosis NRC, Washington, 1977-81; bd. sci. counselors cancer prevention and control Nat. Cancer Inst., Washington, 1981—. Contbr. 140 articles to sci jours. Served to capt. M.C. U.S. Army, 1958-60. Trudeau Soc. teaching fellow, 1958; Nat. Heart and Lung Inst. research fellow, London, 1960-61; recipient Nat. Inst. Environ. Health Scis. Career Devel. research award, 1968-73. Mem. Am. Physiol. Soc., Am. Assn. Pathologists, Am. Soc. Cellular Biol., Am. Thoracic Soc., AAAS, Internat. Epidemiology Assn., Am. Heart Assn., Sigma Xi. Democrat. Unitarian. Subspecialties: Pulmonary medicine; Environmental toxicology. Current work: Pulmonary disease; inflammatory processes, experimental pathology; birth defects of lung; emphysema, environmental and occupational diseases, byssinosis, asbestosis, occupational asthma and chronic bronchitis. Office: U So Calif Sch Medicine 2025 Zonal Ave Los Angeles CA 90033

KILBY, JACK ST. CLAIR, inventor; b. Jefferson City, Mo., Nov. 8, 1923; s. Hubert St. Clair and Vina (Freitag) K.; m. Barbara Annegers, June 27, 1948; children: Ann, Janet Lee. B.S. in Elec. Engring, U. Ill., 1947; M.S., U. Wis., 1950. Program mgr. Globe-Union, Inc., Milw., 1948-58; asst. v.p. Tex. Instruments, Inc., Dallas, 1958-70; self-employed inventor, Dallas, 1970—; disting. prof. elec. engring. Tex. A & M U., 1978—; inventor monolithic integrated circuit, others; cons. to govt. and industry. Served with AUS, 1943-45. Recipient Nat. Medal Sci., 1969; Ballentine medal Franklin Inst., 1967; Distinguished Alumni award U. Ill., 1974; named to Holley medal ASME, 1982, Nat. Inventors Hall of Fame U.S. Patent Office, 1981. Fellow IEEE (Sarnoff medal 1966, Brunetti award 1978); mem. Nat. Acad. Engring. (Zworkin medal 1975). Subspecialty: Microchip technology (engineering). Current work: Integrated circuit technology; solar energy. Home: 7723 Midbury St Dallas TX 75230 Office: 5924 Royal Ln Suite 150 Dallas TX 75230

KILGORE, WENDELL WARREN, toxicology educator, toxicologist, consultant; b. Greenfield, Mo., June 21, 1929; s. Kingdon and Faye Rose (Jarrett) K.; m. Janet E. Kilgore, July 27, 1952; children: Steven, Eric, Nancy. Ph.D., U. Calif., 1958. Microbiologist Stanford Research Inst., Palo Alto, Calif., 1959-60; from asst. prof. to prof. U. Calif.-Davis, 1960—. Contbr. articles to profl. jours. Served with U.S. Army, 1951-54. Mem. Am. Soc. Microbiology, AAAS, Am. Chem. Soc., Soc. Toxicol., Sigma Xi. Subspecialties: Toxicology (agriculture); Toxicology (medicine). Current work: General toxicology; human exposure to pesticides; toxicity of pesticides. Home: 1303 Cedar Pl Davis CA 95616 Office: Dept Environmental Toxicology Univ California Davis CA 95616

KILLIAN, BARBARA GERMAIN, aeronautical engineer; b. San Diego, Nov. 25, 1935; d. John Koschka and Martha (Stachwick) K.; m. W. Patrick Crowley, Aug. 11, 1957; m. Lawrence Seymour Germain, June 5, 1975. B.A., San Diego State U., 1957; postgrad., Stanford U., 1960-61, U. Calif.-Berkeley, 1963-64, U. N.Mex., 1981-83. Mem. staff Lawrence Livermore (Calif.) Lab., 1958-71, group leader, 1971-76; mem. staff Los Alamos Nat. Lab., N.Mex., 1976-79, asst. div. leader, 1979—; cons. Sci. Applications, Inc., La Jolla, Calif., 1976-80. Contbr. articles to profl. jours. Dept. of Energy rep. to UN Conf. of Com. on Disarmament, 1978, 79. Mem. Seismol. Soc. Am., Am. Geophys. Union, Am. Nuclear Soc. (sec.-treas. div. alternative energy techs. 1975-76, vice chmn. 1976-77, chmn. 1977-78). Subspecialties: Theoretical computer science; Aeronautical engineering. Current work: Shock wave and stress wave propagation-numerical modeling experimental and theoretical work with shock waves and shock tubes. Home: Box 494 Los Alamos NM 87544 Office: Los Alamos Nat Lab Box 1663 MS D 446 Los Alamos NM 87545

KILLIAN, GRANT ARAM, psychologist, educator; b. Cambridge, Mass., Nov. 3, 1949; s. Leo Gary and Ann (Mazmanian) K. B.A. in Philosophy, New Coll., 1972; M.A. in Social Sci, U. Chgo., 1975; Ph.D. in Ednl. Psychology, U. Chgo., 1981. Lic. psychologist, Fla. Teaching asst. psychology and philosophy New Coll., Sarasota, Fla., 1972, individual therapist, 1973-74, group therapist, 1973-74; behavioral modification therapist Walter Fernald State Hosp., Waltham, Mass., 1973; occupational and recreational therapist Sarasota (Fla.) Palms Hosp., 1973-74; research psychologist U. Chgo., 1974-76, 1976-77; psychologist Ill. State Psychiat. Inst., Chgo., 1977-80, psychology intern, 1980-81; psychologist St. Elizabeth's Hosp., Washington, 1981-82; asst. prof. Sch. Profl. Psychology, Nova U., Ft. Lauderdale, Fla., 1982—; psycho-diagnostic assessor U. Chgo., 1974-78; family therapist Community Mental Health Ctr., Washington, 1981-82. Contbr. articles to profl. jours. Recipient Noyes Found. award U. Chgo., 1974-76, 77-78, Nat. Research Service award, 1976-77; NIMH fellow, 1980-81, 81-82. Mem. Am. Psychol. Assn. Democrat. Current work: Cognition and psychopharmacology; depression and dementia; intelligence testing and interpretation. Home: 2202 Cypress Bend Dr S Pompano Beach FL 33060 Office: Nova U Sch Profl Psychology 3301 College Ave Fort Lauderdale FL 33314

KILLIAN, JAMES R., JR., former college president; b. Blacksburg, S.C., July 24, 1904; s. James Robert and Jeannette (Rhyne) K.; m. Elizabeth Parks, Aug. 21, 1929; children: Carolyn (Mrs. Paul Staley), Rhyne Meredith. Student, Trinity Coll. Duke U., 1921-23, LL.D. (hon.), 1949; B.S., Mass. Inst. Tech., 1926; Sc.D. (hon.), Middlebury Coll., 1945; hon. degrees, Bates Coll., 1950, U. Havana, Cuba, 1953, Notre Dame U., 1954, Lowell Tech. Inst., 1954, Columbia, Coll. Wooster, Oberlin Coll., 1958, U. Akron, 1959, Worcester Poly. Inst., 1960, U. Me., 1963; LL.D., Union Coll., 1947, Bowdoin Coll., Northeastern U., 1949, Boston U., Harvard, 1950, Williams Coll., Lehigh U., U. Pa., 1951, U. Chattanooga, 1954, Tufts U., 1955, U. Cal., Amherst Coll., 1956, Coll. William and Mary, 1957, Brandeis U., 1958, Johns Hopkins, N.Y. U., 1959, Providence Coll., Temple U., 1960, U. S.C., 1961, Meadville Theol. Sch., 1962; D.Applied Sci., U. Montreal, 1958; D.Eng., Drexel Inst. Tech., 1948, U. Ill., 1960, U. Mass., 1961; Ed.D., R.I. Coll., 1962; H.H.D., Rollins Coll., 1964; D.P.S., Detroit Inst. Tech., 1972. Asst. mng. editor Technology Rev., 1926-27, mng. editor, 1927-30, editor, 1930-39; exec. asst. to pres. Mass. Inst. Tech., 1939-43, exec. v.p., 1943-45, v.p., 1945-48, pres., 1948-59, chmn. corp., 1959-71, hon. chmn., 1971-79; dir. Polaroid Corp., IBM, 1959-62, Gen. Motors Corp., 1959-75, Cabot Corp., 1963-75, AT & T, 1963-77, Ingersoll-Rand Co., 1971-76. Chmn. Carnegie Commn. on Ednl. TV, 1965-67; bd. dirs. Corp. for Pub. Broadcasting, 1968-75, chmn., 1973-74; Mem. Pres. Communication Policy Bd., 1950-51, President's Com. on Mgmt., 1950-52; mem. sci. adv. com. ODM, 1951-57; chmn. Army Sci. Adv. Panel, 1951-56, Pres.' Bd. Cons. on Fgn. Intelligence Activities, 1956-57; spl. asst. to Pres. U.S. for sci. and tech., 1957-59; chmn. Pres.' Sci. Adv. Com., 1957-59, mem., 1957-61, cons., 1961-73; pres. Atoms for Peace Awards, 1955-58, 59-69; mem. Adv. Council on State Depts. Edn., U.S. Office Edn., 1965-68; chmn. President's Fgn. Intelligence Adv. Bd., 1961-63; mem. gen. adv. bd. U.S. Arms Control and Disarmament Agy., 1969-74; Bd. visitors U.S. Naval Acad., 1953-55; moderator Am. Unitarian Assn., 1960-61; Trustee Nutrition Found., 1954-70, Washington U., 1966-70, Mt. Holyoke Coll., 1962-72, Alfred P. Sloan Found., 1954-77, Boston Mus. Fine Arts, Mitre Corp.; chmn., 1967-69; bd. dirs. Nat. Merit Scholarship Corp., 1960-63, Winston Churchill Found. U.S. Ltd.; mem. Mass. Bd. Edn., 1962-65. Recipient President's Certificate of Merit, 1948, Certificate of Appreciation Dept. of Army, 1953; Exceptional Civilian Service award Dept. of Army; Pub. Welfare medal Nat. Acad. Scis., 1957; George Foster Peabody Broadcasting Spl. Edn. awards, 1968, 76; decorated Croix d'officer, Legion of Honor France, 1957; Hoover medal, 1963; Sylvanus Thayer award U.S. Mil. Acad., 1978; Vannevar Bush award Nat. Sci. Bd., 1980; others. Fellow Am. Acad. Arts and Scis.; mem. Nat. Acad. Engring., Am. Soc. Engring. Edn. (hon.), Sigma Chi, Phi Beta Kappa (hon.), Tau Beta Pi (hon.). Clubs: St. Botolph (Boston); Century, University (N.Y.C.). Subspecialty: Engineering administration. Address: care Mass Inst Technology Cambridge MA 02139

KILLINGER, DENNIS KARL, research physicist; b. Boone, Iowa, Sept. 23, 1945; s. Karl H. and Evelyn (Johnson) K.; m. Rose L. Egger, June 15, 1969; children: Laura, Robert. B.A., U. Iowa, 1967; M.A., DePauw U., 1969; Ph.D. in Physics, U. Mich., 1978. Research physicist Naval Avionics Facility, Indpls., 1968-74; research assoc. physics U. Mich., 1974-78; quantum electronics staff and program mgr. laser remote sensing Lincoln Lab., M.I.T., 1979—; Conf. chmn. Workshop on Optical and Laser Remote Sensing, Monterey, 1982; program chmn. Optical Soc. Am. Topical Meeting on Remote Probing of the Atmosphere, Lake Tahoe, 1983. Contbr. articles to profl. jours. Mem. Am. Phys. Soc., Optical Soc. Am. Current work: Physics of new optical and laser sources, quantum electronics and nonlinear optical techniques with applications toward laser remote sensing. Office: MIT Lincoln Lab Lexington MA 02173

KILMAN, JAMES WILLIAM, surgeon, educator; b. Terre Haute, Ind., Jan. 22, 1931; s. Arthur and Irene (Piker) K.; m. Priscilla Margaret Jackson, June 20, 1968; children: James William, Julia Anne, Jennifer Irene. B.S., Ind. State U., 1956; M.D., Ind. U., 1960. Intern Ind.U. Med. Ctr., Indpls., 1960-61; resident surgery Ind.U. Med. Center, 1961-66, asst. prof., 1966-69, assoc. prof., 1969-73; prof. surgery Ohio State U. Coll. Medicine, 1973-; chmn. dept. thoracic surgery Children's Hosp.; attending surgeon Univ. Hosp., Columbus, Ohio; attending staff Children's Hosp., Columbus, pres. staff, 1978; attending staff Grant Hosp., Riverside Hosp.; cons. surgeon VA Hosp., Dayton; pres. Columbus Acad. Medicine, 1977. Trustee Central Ohio Heart Assn., Acad. Medicine Edn. Found., Children's Hosp., 1978—. Served with USNR, 1951-55. USPHS Cardiovascular fellow, 1963-64. Mem. Columbus Surg. Soc. (pres. 1973-74), Columbus Acad. Medicine (council 1971-73), Am. Surg. Assn., Soc. U. Surgeons, Am. Assn. Thoracic Surgery, Am., Central, Western surg. assns., Soc. Vascular Surgery, Internat. Cardiovascular Soc., Internat. Soc. Surgeons, Chest Club, Cardiovascular Surgery Club, Sigma Xi, Alpha Omega Alpha. Subspecialty: Cardiac surgery. Current work: Clinical cardiac surgery for adults and children. Research in vascular heart disease, pericardial disease and congenital heart anomalies. Research, articles infant cardiopulmonary bypass and surgery for congenital heart lesions. Home: 4231 Jackson Pike Grove City OH 43123 Office: 410 W 10th Ave Columbus OH 43210

KILMANN, RALPH HERMAN, educator; b. N.Y.C., Oct. 5, 1946; s. Martin H. and Lilli (Loeb) K.; m. Audrey Ann Sabol, July 7, 1977; children: Christopher Martin, Catherine Mary. B.S., Carnegie-Mellon U., 1970, M.S., 1970; C.Phil., UCLA, 1971, Ph.D., 1972. Research asst. Carnegie-Mellon U., 1969-70; teaching asso. UCLA, 1970-72; instr. U. Pitts., 1972, asst. prof., 1972-75, assoc. prof., 1975-79, prof. bus. adminstrn., 1979—; pres. Orgnl. Design Cons., Inc., Pitts., 1975—; coordinator Orgnl. Studies Group, U. Pitts., 1981—. Author: Social Systems Design, 1977, (with I. I. Mitroff) Methodological Approaches, 1978. Recipient 1st Prize Inst. Mgmt. Sci., 1976. Mem. Acad. Mgmt., Inst. Mgmt. Sci., Am. Psychol. Assn. Subspecialties: Organizational sciences; Social psychology. Current work: Developing social sci. tech. to define/solve complex orgnl. problems. Developer computer-based design MAPS Design Tech., 1975; co-developer personality assessment: The Conflict Mode Instrument, 1974. Home: 110 Weir Dr Pittsburgh PA 15215 Office: Univ Pitts Grad Sch Bus Pittsburgh PA 15260

KIM, BYUNG SUK, microbiology educator; b. Yosu, Korea, Mar. 20, 1942; s. Young Taik and Gwee Yup Jung; m. Oak Cho Kim, Apr. 19, 1967; children: Peggy, Charles. B.S., Seoul Nat. U., 1967; M.S., Va. State U., 1969; Ph.D., U. Ill., 1973. Sr. research technologist U. Chgo., 1969-70; sr. staff assoc. Inst. Cancer Research, Columbia U., 1973-74; asst. prof. Northwestern U., Chgo., 1976-81, assoc. prof., 1981—. Contbr. articles to profl. jours. Mem. Am. Soc. Microbiology, Am. Assn. Immunologists. Subspecialties: Immunobiology and immunology; Cellular engineering. Current work: Regulation of antibody synthesis and tumor immunity. Home: 4453 Main St Skokie IL 60076 Office: 303 E Chicago Ave Chicago IL 60611

KIM, CHONG-KYUN, botanist, plant physiologist, educator, researcher; b. Taejon, Korea, July 28, 1936; s. Young-Bae and Kum-Joo (Chang) K.; m. Do-Jeung Choi, Nov. 19, 1966; children: Sun-Hyoung, Ji-Hyoung, Won-Tai. B.S., Korea U., 1962, M.S., 1964; Ph.D., Brigham Young U., 1979. Research asst., lectr. Seoul Nat. U., 1964; instr. Kong-ju Nat. Tchrs. Coll., 1971-74, asst. prof., 1974-79, assoc. prof., 1979—; vis. research prof. Brigham Young U., Provo, Utah, 1982. Author: The Alginophytes from the Korean Coast, 1971, Modern Plant Physiology (Korean), 1974, Fruitbody Formation and Growth of Basidiomycetes, 1975; contbr. articles on botany to profl. jours. Served with Korean Army, 1957-60. Mem. Bot. Soc. Korea, Microbiol. Soc. of Korea, Mycological Soc. Korea. Mormon. Subspecialty: Plant physiology (biology). Current work: Membrane physiology, mechanism of salt tolerance of halophytes. Home: 9-307 Samho Garden Mension Apt Banpo-Dong Kangnam-Ku Seoul Korea 135 Office: 401 WIDB Dept Botany Brigham Young U Provo UT 84602

KIM, GEUNG-HO, statistician; b. Seoul, South Korea, July 25, 1945; came to U.S., 1969, naturalized, 1980; s. Ki-duk and Duk-bo (Lim) K.; m. Jae Oak Lee, Aug. 19, 1972; children: Jeannie, Benjamin. B.A. in Bus. Administrn, Seoul Nat. U., 1967; M.S. in Stats, Iowa State U., 1971, Ph.D., 1978. Software cons. Iowa State U., Ames, 1971-77, research asst., 1977-78; asst. prof. stats. SUNY, Amherst, 1978-80; statistician U.S. Dept. Commerce, Highlands, N.J., 1980—, cons. EEO matters, 1981—. Reviewer: Math Revs, 1980—; contbr. articles to profl. jours. Fellow Royal Statis. Soc.; mem. Inst. Math. Stats., Am. Statis. Assn. Subspecialties: Statistics; Environmental engineering. Current work: Data analysis and model building in terms of multivariate statistical and/or stochastic process techniques pertinent to the marine environmental monitoring and related problems. Home: 120 Riveredge Rd Tinton Falls NJ 07724 Office: US Department Commerce Sandy Hook Laboratory Highlands NJ 07732

KIM, HUN, pathologist; b. Seoul, Korea, Apr. 9, 1942; s. Seung-Tai and Keum-Ja (Cho) K.; m. Hee-jung Kim, May 22, 1966; children: Roy, Gina. M.D., Seoul Nat. U., 1966. Diplomate: Am. Bd. Pathology. Resident in Pathology U. Chgo. Hosps. and Clinics, 1968-71; fellow, asst. prof. Stanford U. Med. Ctr., 1971-75; pathologist City of Hope Nat. Med. Ctr., Duarte, Calif., 1975-81, Hoag Meml. Hosp.-Presbyn., Newport Beach, Calif., 1981—. Mem. AAAS, Am. Assn. Cancer Research, Am. Soc. Clin. Pathology, Internat. Acad. Pathology, Am. Assn. Pathologists, and Bacteriologists, Coll. Am. Pathologists, Am. Soc. Clin. Oncology. Subspecialty: Pathology (medicine). Current work: Clinicopathologic studies on malignant lymphomas. Office: 301 Newport Blvd Newport Beach CA 92663 Home: 1405 Rancho Rd Arcadia CA 91006

KIM, HYUN YOUNG, veterinary pathologist; b. Seoul, Korea, Sept. 12, 1939; came to U.S., 1970, naturalized, 1976; s. Kyung S. and Yoon H. K.; m. Duck Jo Kim, Dec. 25, 1967; children: Daniel, Eugene. B.A., Korean Union Coll.; D.V.M., Seoul Nat. U., 1962, M.P.H., 1969; M.S., U. Ga., 1973; D.D., I.B.I.S., 1982. Lic. vet. practitioner, Pa., Mass. Asst. prof. Korean Union Coll., 1967-70; grad. research asst. U. Ga., 1971-72; research investigator U. Pa., 1973-74; vet. pathologist Pa. State Vet. Labs., Harrisburg, 1974—. Elder Market Square Presbyterian Ch., Harrisburg. UN/FAO fellow, 1969. Mem. Pa. Vet. Med. Assn., Tissue Culture Assn., Am. Assn. Vet. Lab. Diagnosticians, Korean-Am. Vet. Soc. (pres.-elect 1982). Subspecialties: Pathology (veterinary medicine); Virology (veterinary medicine). Current work: Development of laboratory diagnosis of viral infection in animals; immunofluorescent antibody tissue cultrue system, tissue culture viral-serum neutralization test or ELISA test are employed. Office: PO Box 1430 Harrisburg PA 17105

KIM, JAE HO, clin. investigator, radiologist; b. Daegu, Korea, Dec. 17, 1935; came to U.S., 1959, naturalized, 1971; s. Sa Yeup and Moo Sun (Yoo) K.; m. Johni Kim, Sept. 14, 1963; children: Alberta, Lena. M.D., Kungpook Nat. U., Daegu, 1959; Ph.D., U. Iowa, 1963. Intern Montefiore Hosp., N.Y.C., 1968-69; resident in radiology Meml. Hosp., N.Y.C., 1969-72; research assoc. Sloan-Kettering Inst. Cancer Research, 1963-66, assoc., 1966-68; asst. prof. radiology Cornell U. Med. Coll., N.Y.C., 1968-72, assoc. prof., 1974-80, prof., 1980—. Mem. AMA, Radiation Research Soc., Radiol. Soc. N.Am., Am. Radium Soc., Am. Assn. Cancer Research. Subspecialty: Cancer research (medicine). Current work: Radiobiology, radiotherapy, hyperthermia. Office: 1275 York Ave R101

KIM, JAMES C(HIN) S(OO), pathologist; b. Seoul, Korea, Mar. 2, 1937; came to U.S., 1961; s. Sung Ho and Wun Sun K.; m. Hejung Pi, Nov. 14, 1964; children: Christine, Richard. D.V.M., Seoul Nat. U., 1960, M.P.H., 1962; M.S., Kans. State U., 1965; Sc.D., Johns Hopkins U., 1972. Pathologist and clin. prof. pathology Eastman Kodak Co. and U. Rochester, N.Y., 1979—; head, assoc. prof. pathology, dept. pathology Tulane U. Delta Primate Research Ctr., Covington, La., 1977-79; assoc. prof. pathology Mich. State U., East Lansing, 1973-77; research pathologist Oak Ridge Nat. Lab., 1972-73. Essayist on nutrition (Essayist Award 1980); editor, author: Zoo Medicine, 1980; assoc. editor: Jour. Applied Nutrition, 1980—. Pres. Korean Vet. Soc. Am., 1981-83. NIH grantee, 1974-77; recipient Acad. award Med. Sch. Korea, 1982. Fellow Royal Soc. Health, Internat. Acad. Pathology; mem. AVMA, Am. Assn. Pathologists. Democrat. Presbyterian. Subspecialties: Pathology (medicine); Pathology (veterinary medicine). Current work: Toxicologic pathology, determining Kodak chemicals by use of laboratory animals. Office: Eastman Kodak Co Rochester NY 14150

KIM, JIN KYUNG, medical researcher; b. Seoul, Korea, Dec. 13, 1939; came to U.S., 1969, naturalized, 1979; d. Yong Jo and Chun Ki

(Jung) K. B.S., Ewha Woman's U., Korea, 1961; M.S., Mich. State U., 1972; Ph.D., U. Colo., 1975. Research asst. Yonsei U., Seoul, 1961-69, Mich. State U., East Lansing, 1970-72; research assoc. U. Colo., Boulder, 1972-75; postdoctoral fellow Mayo Clinic, Rochester, Minn., 1975-78; asst. prof. medicine U. Colo. Health Sci. Ctr., Denver, 1978—. Minn. Heart Assn. fellow, 1976-77. Mem. Am. Fedn. for Clin. Research, Am. Soc. for Nephrology, Sigma Xi, Rho Chi. Subspecialties: Nephrology; Biochemistry (medicine). Current work: Cellular action of vasopressin using isolated nephrons from normal and diseased kidney. Home: 7032 E 4th Ave Denver CO 80220 Office: U Colo Health Sci Center 4200 E 9th Ave Denver CO 80262

KIM, KWANG-SHIN, microbiology educator; b. Seoul, Korea, Nov. 15, 1937; s. Daw-Woo and Sung-Duk K.; m. Bu-Choon Chung, June 16, 1943; children: Edwin Chung-Mynng, Andrew Ki-Nyung. B.S., Nat. Seoul U., 1959; M.S., Rutgers U., 1963, Ph.D., 1967. Research asst. Rutgers U., New Brunswick, N.J., 1965-66, research assoc., 1966-67; asst. research scientist NYU Med. Ctr., N.Y.C., 1967-68, assoc. research scientist, 1968-71, asst. prof. microbiology, 1971-76, assoc. prof., 1976—. Contbr. articles to profl. jours. Andrew W. Mellon Found. grantee, 1975-76; Merck Co. Found. grantee, 1974. Mem. AAAS, N.Y. Acad. Scis., Am. Soc. Microbiology, Sigma Xi. Subspecialties: Microbiology (medicine); Microscopy. Current work: Ultrastructure of bacteria and its relationship with animal cell-membranes; investigation of comparative ultrastructural study of bacteria derived from leprous human and leprosy infected armadillos. Home: 462 Barbara Ave Wyckoff NJ 07481 Office: Dept Microbiology NYU Med Ctr 550 1st Ave New York NY 10016

KIM, KYEKYOON, electrical engineering educator, researcher, consultant; b. Seoul, Korea, Oct. 5, 1941; came to U.S., 1966, naturalized, 1977; s. Chung-Hee and Jung-Sook (Park) K.; m. Jung Ja Kim, May 17, 1969; children: Caroline, David. B.S., Seoul Nat. U., 1966; M.S., Cornell U., 1968, Ph.D., 1971. Research asst. Cornell U., 1966-71, postdoctoral research fellow, 1971-72; research assoc. U. Ill.-Urbana, 1972-76, asst. prof. elec. engring., 1976-81, assoc. prof., 1981—, acting dir. Fusion Tech. Lab., 1976-80, dir. charged particle research lab., 1980—; cons. in field. Contbr. numerous articles to profl. publs. Served with Korean Inf., 1962-63. Mem. Am. Phys. Soc., IEEE, Am. Vacuum Soc. (chmn. com. on fueling plasma devices 1981—). Subspecialties: Plasma engineering; Nuclear fusion. Current work: High-power laser research and development; nuclear fusion fuel pellet research and development; electromagnetic particle accelerator; electrohydrodynamics; plasma engineering; monodisperse microparticle generation. Home: 2406 Boudreau Dr Urbana IL 61801 Office: Dept Elec Engring U Ill 1406 W Green St Urbana IL 61801

KIM, KYO SOOL, program manager, researcher; b. Kang Nung, Korea, Sept. 10, 1942; came to U.S., 1968; s. Sung Y. and Chung J. (Choi) K.; m. H. Choo Hwang, Oct. 10, 1970; children: Dennis H., Paul W. B.S., Seoul Nat. U., 1968; Ph.D., Brown U., 1974. Instr., Seoul (S. Korea) Nat. U., 1968; postdoctoral research assoc. Brown U., Providence, 1973-74; radiation analysis group mgr. United Engring. and Constrn., Phila., 1974-79; program mgr. U.S. Nuclear Regulatory Commn., Washington, 1979—. Author: (in Korean) Waste Management, 1984. Fellow Innotech Corp., Providence, 1972-74. Mem. Am. Nuclear Soc., Am. Ceramic Soc., Korean Nuclear Soc., Materials Research Soc. (steering com. 1981-82). Subspecialties: Nuclear engineering; Water supply and wastewater treatment. Current work: Nuclear waste treatment technology; waste disposal technology. Home: 9236 Quick Fox Columbia MD 21045 Office: US Nuclear Regulatory Commn Washington DC 20555

KIM, MYUNGHWAN, electrical engineering educator; b. Seoul, Feb. 8, 1932; s. Sunghak and Kyunghi (Jeon) K.; m. Youngsook Susan Hyun, June 6, 1959; children: Eugene A., Erwin Y., Edward G., Julie G. B.S., U. Ala.-Tuscaloosa, 1958; M.Engring., Yale U., 1959, Ph.D., 1962. Elec. engr. TVA, Chattanooga, 1958-59; prof. elec. engring. Cornell U., Ithaca, N.Y., 1962—; NRC sr. postdoctoral assoc. Jet Propulsion Lab., Pasadena, Calif., 1968-69; vis. Assoc. Calif. Inst. Tech., Pasadena, 1969; vis. prof. Korea Advanced Inst. Sci. and Tech. and Korea U., Seoul, 1982. Contbr. articles to profl. jours. Served to 1st lt. Republic of Korea Army, 1951-54. NIH fellow, 1970. Mem. IEEE, N.Y. Acad. Scis., Korea Inst. Elec. Engrs., Korea Inst. Electronic Engrs. Subspecialties: Bioinstrumentation; Computer engineering. Current work: Research on design and fabrication of biomedical multielectrode arrays of chemical sensors by integrated circuit technology and research on computer architecture and system design. Office: Phillips Hall Cornell Univ Ithaca NY 14853

KIM, RHYN HYUN, engineering educator, consultant; b. Seoul, Korea, Feb. 4, 1936; came to U.S., 1959, naturalized, 1971; s. Wonsik and Yoshim (Chung) K.; m. Songhae O., Sept. 24, 1966; children: Juli, Allis, Steven. B.S. in Mech. Engring., Seoul Nat. U., 1958; B.S. in Mech. Engring, Mich. State U., 1960; M.S. in Engring, Mich. State U., 1961, Ph.D., 1965. Registered profl. engr., N.C. Asst. prof. mech. engring. U. N.C., Charlotte, 1965-71, assoc. prof., 1971-83, prof., 1983—; staff engr. Office of Air Quality Planning and Standards EPA, 1976-77; mech. engr. Office of Research and Devel. Indsl. Environ. Research Lab., EPA, N.C., 1977-78; cons. in field. Contbr. articles on engring. research to profl. jours. Recipient U.N.C. at Charlotte Found. awards, 1966, 69, 70, 71; NASA-ASEE summer research fellow, 1974; recipient Gas Research Inst., and Tex. A&M U. Research award, 1981. Mem. ASME, ASHRAE. Presbyterian. Subspecialties: Mechanical engineering; Solar energy. Current work: Solar energy, thermal energy conversion, software engineering optical signal processing. Home: 2726 Wamath Dr Charlotte NC 28210 Office: U N C Engring Sci Dept UNCC Station Charlotte NC 28223

KIM, S. PETER, psychiatry educator, researcher; b. Seoul, Korea, Oct. 8, 1939; s. Chongsoon and Soonbok (Rim) K.; m. O. Mary Lee, Mar. 30, 1963; children: John, Kathy. C.P.M., Seoul Nat. U., 1959; M.D., 1963; M.M.S., 1967. Cert. specialist Am. Bd. Psychiatry, Neurology. Instr. psychiatry NYU, 1972-74, clin. asst prof., 1974-76, asst. prof., 1976-79, assoc. prof., 1979—, assoc. attending physician, 1979—; dir. Ctr. Transcultural Devel. Study, N.Y. U.; cons. liaision in pediatrics N.Y. U.-Bellevue Med. Ctr. Author: Transcultural Adoption, 1981. Bd. dirs. Korean Community Service Met. N.Y., 1982—, Asian Am. Community Mental Health Ctr., N.Y., 1981—. Served with Army Republic of Korea, 1963-67. Fellow Am. Acad. Child Psychiatry (chmn. com. transcultural child), Am. Psychiat. Assn., Am Soc. Social Psychiatry; mem. AAAS, Soc. Adolesant Psychiatry. Roman Catholic. Subspecialties: Psychopharmacology; Trans/cross-cultural medical psychology. Current work: Trans-/cross-cultural psychiatry, psychosomatic medicine and pediatrics, social psychiatry, stress and emotional-physical illness. Home: 7 Pine Terrace Bronxville NY 10708 Office: NYU Sch Medicine 550 First Ave NY NY 10016

KIM, YONG SU, nuclear engineer, reactor physicist; b. Haiju, Korea, Aug. 29, 1929; came to U.S., 1954; s. Yu Taik KiM and Hyun Duk (Park) K.; m. Young Soon, Feb. 7, 1958; children: Margaret, Joseph, Elizabeth. B.S., U. Wis., 1958; M.S., MIT, 1961; Ph.D., Cath. U. Am., 1970. Registered profl. engr., Md., Calif. Nuclear engr. Internuclear Co., Clayton, Md., 1961-62; sr. exec. engr. NUS Corp., Gaithersburg, Md., 1963—. Mem. Am. Nuclear Soc. Subspecialties: Nuclear fission; Nuclear engineering. Current work: Nuclear power consulting, nuclear engineering. Home: 210 Hillsboro Dr Silver Spring MD 20902 Office: NUS Corp 910 Clopper Rd Gaithersburg MD 20878

KIM, YONG-KI, physicist; b. Seoul, Korea, Feb. 20, 1932; came to U.S., 1959, naturalized, 1975; s. Hwanchae and Chongnyol (Chong) K.; m. Young Hee, Dec. 15, 1963; children: Edward, Charlotte. B.Sc., Seoul Nat. U., 1957; M.Sc., U. Del., 1961; Ph.D., U. Chgo., 1966. Physicist Research asso Argonne (Ill.) Nat. Lab., 1966-68, asst. prof, 1968-70, physicist, 1970-79, sr. physicist, 1979—. Contbr. numerous articles to profl. jours. Mem. Am. Phys. Soc., Radiation Research Soc. Subspecialty: Atomic and molecular physics. Current work: Theory of atomic structure, collisions. Office: Argonne Nat Lab 9700 Cass Ave Argonne IL 60439

KIM, YOON BERM, immunologist, educator, cancer researcher; b. Sainchang, Soon Chun, Korea, Apr. 25, 1929; came to U.S., 1959, naturalized, 1975; s. Sang Sun and Yang Rang (Lee) K.; m. Soon Cha Kim, Feb. 23, 1959; children: John, Jean, Paul. M.D., Seoul (Korea) U., 1958; Ph.D., U. Minn., 1965. Intern Seoul Nat. U. Hosp., 1958-59; mem. faculty U. Minn. Sch. Medicine, Mpls., 1959-73, assoc. prof. microbiology, 1970-73; prof. immunology Cornell U. Grad. Sch. Med. Scis., N.Y.C., 1973—, chmn. immunology unit, 1980-82; mem., head Lab. Ontogeny of Immune System, Sloan-Kettering Inst. Cancer Research, Rye, N.Y., 1973—; chmn. credentials review and fellowship subcom. Cornell U. Grad. Sch. Med. Sci., 1976-80; mem. sci. adv. com. Internat. Symposium Gnotobiology, Ulm, Ger. and Tokyo, 1977-81; mem. Lobund Adv. Bd., U. Notre Dame, Ind., 1977—. Contbr. numerous articles to profl. publs. Recipient Research Career Devel. award USPHS, 1968-73, grantee, 1960—; Am. Cancer Soc. grantee, 1979—. Mem. Am. Assn. Immunologists, Am. Soc. Microbiology, Am. Assn. Pathologists, Assn. Gnotobiotics (bd. dirs. 1975-79, pres. 1979-80), Reticuloendothelial Soc., Internat. Soc. Dwvel. Comparative Immunology, Korean Med. Assn. Am. (sci. edn. com. 1975—, chmn. 1980), N.Y. Acad. Scis., AAAS. Harvey Soc., Sigma Xi. Subspecialties: Immunobiology and immunology; Cancer research (medicine). Current work: Ontogeny and regulation of immune system including T/B lymphocytes, NK/K cells and monocytes/macrophages; immunochemistry and biology of bacterial toxins; gnotobiotics and host-parasite relationships. Home: 6 Wilson Ridge Rd Darien CT 06820 Office: Sloan-Kettering Inst Cancer Research 145 Boston Post Rd Rye NY 10580

KIM, YOUNG HWA, physicist, researcher; b. Seoul, Korea, May 8, 1940; came to U.S., 1970; s. Jin Hee and Yong Sook (Choi) K.; m. Myong Yon Park, Apr. 19, 1969; children: Grace Sung, Steven Sung, Julie. B.S., Korean Mil. Acad., Seoul, 1963, Seoul Nat. U., 1967; M.S. in Physics, U. Houston, 1971, Ph.D., UCLA, 1980. Teaching fellow U. Houston, 1970-71; teaching assoc. UCLA, 1975-78, research asst., 1978-80; research assoc. U. Ill.-Urbana, 1980-81, vis. asst. prof., 1981-82; sr. physicist 3M Co Central Research Lab., St. Paul, 1982—. Contbg. author: Journal de physique, 1982, Macromolecules, 1983; contbr. articles to profl. jours. Served to capt. Korean Army, 1963-68. Mem. Am. Phys. Soc. Methodist. Subspecialties: Condensed matter physics; Polymer physics. Current work: Piezo- and pyroelectricity of polymers, electronic states of polymers, polymer diffusion, nematic polymers, organic conductors. Office: 3M Co Central Research Lab 208-1 3M Center Saint Paul MN 55144 Home: 7465 Columbia Ct Woodbury MN 55125

KIM, YUNG DAI, senior scientist; b. Seoul, Korea, Mar. 24, 1936; came to U.S., 1957, naturalized, 1973; s. Ik Soo and Jung Hui (Juhn) K.; m. Young Sook, June 17, 1967; children: Jean, Sue. Ph.D., U. Minn., Mpls., 1968. Vis. scientist Kettering Lab., Yellow Springs, Ohio, 1968-69; NIH postdoctoral fellow Northwestern U., Evanston, Ill., 1969-71; NIH research fellow U. Pa., Phila., 1971-73; sr. scientist Worthington Biochem. Corp., Freehold, N.J., 1973-74, Abbott Labs., North Chicago, Ill., 1974—. Contbr. articles to profl. jours. Mem. Am. Assn. Immunologists, Am. Chem. Soc., Sigma Xi, Phi Lambda Upsilon. Subspecialties: Immunology (medicine); Cancer research (medicine). Current work: Immunochemistry, protein chemistry, enzymology, cancer diagnostics research, clin. diagnostic tests. Patentee in field. Home: 75 N Rolling Ridge Lindenhurst IL 60046 Office: Abbott Labs D-90C North Chicago IL 60064

KIMBALL, EDWARD SAUL, immunologist; b. N.Y.C., May 23, 1947; s. Benjamin J. and Miriam (Ginsberg) K.; m. Ann Quimby, July 31, 1971. B.S., CCNY, 1968; M.S., Northeastern U., 1971; Ph.D., U. Pa., 1977. Jr. research chemist Collaborative Research, Waltham, Mass., 1971-73; staff fellow Lab. Immunogenetics NIH, 1977-81, sr. staff biol. response modifiers program, 1981—. Contbr. articles to profl. jours. Mem. Am. Assn. Immunologists. Subspecialties: Biochemistry (biology); Immunobiology and immunology. Current work: Biochemistry and immunochemistry of tumor antigens, growth factors, cellular receptors. Office: Frederick Cancer Research Ctr Bldg 567 Frederick MD 21701

KIMBERLY, ROBERT P., physician; b. New Haven, July 29, 1946; s. John T. and Beatrice (Branch) K.; m. Susan Alesbury, June 1972; children: Christopher, Taylor, Sarah, Michael, Thomas. A.B., Princeton U., 1968; B.A., M.A., Oxford U., 1970; M.D., Harvard U., 1973. Diplomate: Am. Bd. Internal Medicine. Asst. prof. Cornell Med. Coll., N.Y.C., 1979—. Served to lt. comdr. USPHS, 1975-77. Rhodes scholar, 1968-70; Arthritis Found. fellow, 1977-80; recipient Tchr.-Scientist award Andrew Mellon Found., 1980. Mem. ACP Am. Fedn. for Clin. Research, Am. Rheumatism Assn. Subspecialty: Immunology (medicine). Current work: Immunology and rheumatology. Office: Hosp for Spl Surgery 535 E 70th St New York NY 10021

KIMBLE, KENNETH ALAN, plant pathologist, cons.; b. Hanford, Calif., Apr. 2, 1924; s. Elmer Lewis and Helen Clark K.; m. Janet Roberta Park, June 15, 1951; children: Susan P., Ellen, Lewis P., Sarah, Mark. B.S., U. Calif.-Davis, 1953, M.S., 1957. Staff research assoc. dept. plant pathology U. Calif.-Davis, until 1982; plant pathologist-breeder vegetables Moran Seed Inc., El Macero, Calif., 1982—; cons. in field. Contbr. articles on plant pathology to profl. jours. Recipient Oustanding Performance awards U. Calif.-Davis, 1966, 71. Mem. Am. Phytopath. Soc., Sigma Xi. Republican. Club: Davis Aquatic Masters. Subspecialties: Plant pathology; Plant virology. Current work: Development of vegetable varieties resistant to disease, disease resistant vegetables to plant pathogenic bacterial fungi and viruses. Office: Moran Seed Inc PO Box 2508 El Macero CA 95616

KIMELBERG, HAROLD KEITH, neurobiologist, educator; b. Hertford, Eng., Dec. 5, 1941; came to U.S., 1963; s. Morris and Sarah (Cohen) K.; m. Pamela Cheryl Ahrens, July 14, 1966; children: David, Michael. B.Sc., Kings Coll., U. London, 1963; Ph.D., SUNY-Buffalo, 1968. Research assoc. U. Pa. Med. Sch., Phila., 1968-69; sr. cancer research scientist Roswell Park Meml. Inst., Buffalo, 1969-74; assoc. prof. div. neurosurgery Albany (N.Y.) Med. Coll., 1974-80, prof., 1980—; reviewer numerous scholarly jours.; mem. rev. bds. NIH. Contbr. numerous articles and revs. to research jours., also chpts. to books. Mem. Am. Soc. for Neurochemistry, Soc. for Neurosci., Am. Soc. Biol. Chemists. Subspecialties: Neurobiology; Membrane biology. Current work: Functions and roles of astroglial cells in the mammalian central nervous system concentrating on their ion transport and interactions with transmitters, principally using cell cultures. Home: 11 Candlewood Ln Delmar NY 12054 Office: Div Neurosurgery Albany Med Coll Albany NY 12208

KIMES, ALANE SUSAN, biochemist; b. N.J., Aug. 23, 1946; d. Carlyle LeRoy and Justine Louise (Lausser) Petuck; m. George John Kimes, Sept. 7, 1968. B.A., Eastern Coll., St. Davids, Pa., 1968; Ph.D., U. Kans., Kansas City, 1977. Postdoctoral trainee Kans. U. Med. Ctr., Kansas City, 1977-79; postdoctoral trainee Gerontology Research Center, NIA, NIH, Balt., 1980—. NIH fellow, 1980—. Mem. Soc. for Neurosci., AAAS, Am. Chem. Soc. Subspecialties: Neurochemistry; Neurobiology. Current work: Membrane turnover, CNS function, blood brain barrier; neurotransmitters. Research on lipid uptake and membrane turnover in the CNS. Office: Gerontology Research Center NIA NIH Baltimore City Hosp Baltimore MD 21224

KIMLIN, MARY JAYNE, chemistry educator; b. Cresson, PA., July 4, 1924; s. Clarence John and Ursula Monica (Monahan) K. A.S., Mt. Aloysius Jr. Coll., 1944; B.S., St. Francis Coll., 1948; M.S., Pa. State U., 1958, Ph.D., 1969. Lab. asst. E.I. duPont, Wilmington, Del., 1944-46; instr. St. Francis Coll., 1948-58; grad. asst. Ohio State U., 1951-52, Pa. State U., 1960-61; asst. prof. chemistry St. Francis Coll., 1958-63, assoc. prof., 1963-70, prof., 1970—. NSF fellow. Mem. Am. Chem. Soc., AAAS, Sigma Xi, Iota Sigma Pi, Sigma Delta Epsilon. Democrat. Roman Catholic. Clubs: Cath. Daughters Am., Summit Country. Subspecialties: Analytical chemistry; Physical chemistry. Current work: Teaching of analytical chemistry, physical chemistry and general chemistry. Office: Saint Francis College Loretto PA 15940

KIMME, ERNEST GODFREY, mathematician; b. Long Beach, Calif., June 7, 1929; s. Ernest Godfrey and Lura Elizabeth (Dake) K.; m. Carolyn McComas Smith/Rice, Aug. 29, 1952 (div. May 1975); children: Ernest G., Elizabeth E., Karl F.; m. Jeanne Bolen, Dec. 20, 1978. B.A. magna cum laude, Pomona Coll., 1952; M.A., U. Minn., 1954, Ph.D., 1955. Mem. grad. faculty math. Oreg. State U., Corvallis, 1955-57; mem. tech. staff Bell Telephone Labs., Murray Hill, N.J., 1957-65; head applied scis. Collins Radio Co., Newport Beach, Calif., 1965-72; research engr. Northrop Electronics, Hawthorne, Calif., 1972-74; sr. staff engr. Interstate Electronics, Anaheim, Calif., 1974-79; dir. advanced systems Gould Navcomm Systems, El Monte, Calif., 1979-82; pres. Cobit, Inc., Costa Mesa, Calif., 1982—. Mem. Soc. Indsl. and Applied Math., Am. Math. Soc., IEEE (chmn. Saddleback sect. 1977-78). Republican. Club: Old Crows. Subspecialties: Probability; Applied mathematics. Current work: Communications technology, especially redundant time-bandwidth systems and signalling, signal processing and analysis. Home: 301 Starfire St Anaheim CA 92807 Office: Cobit Inc 227 N Sunset St Industry CA 91744

KIMMEL, ELLEN BISHOP, psychology educator; b. Knoxville, Tenn., Sept. 16, 1939; d. Archer W. and Mary Ellen (Baker) B.; m.; children: Elinor, Ann, Jean, Tracy. B.A., U. Tenn., 1961; M.A., U. Fla., 1963, Ph.D., 1965. Asst. prof. Ohio U., 1965-68; research assoc. dept. ednl. psychology U. South Fla., Tampa, 1969-71; research assoc. div. acad. services summer 1970, asst. prof., 1971-72, dir. div. univ. studies, 1972-73, assoc. prof., 1972-75; prof. and dir. Women and Adminstrn. Inst., 1975—; lectr., cons. in field. Contbr. over 75 articles to sci. and profl. jours. Mem. Hillsborough Democratic Exec. Com. Recipient Diana award NOW, 1975; Outstanding Prof. U.South Fla., 1978; U. Tenn. scholar, 1958, 59, 60; Gov.'s award for Outstanding Service to State of Fla., 1975. Fellow Am. Psychol. Assn. (council); mem. Southeastern Psychol. Assn. (past pres.), AAUP, Am. Ednl. Research Assn., Assn. Women in Sci., Am. Personnel and Guidance Assn., Omicron Delta Kappa, LWV, NOW. Club: Athena. Subspecialties: Behavioral psychology; Status of women. Current work: Involved with status of women: sex equity, women in adminstration, etc.; training women for leadership positions, networking, biofeedback, personal awareness and preparing women for application of these skills in education and industry. Address: U South Fla Fowler Ave FAO 268 Tampa FL 33620

KIMMICH, GEORGE ARTHUR, biochemist, researcher, educator; b. Cortland, N.Y., Dec. 8, 1941; s. John George and Evelyn (Wheeler) K.; m. Marian Rice, Aug. 3, 1963; children: Lisa, Kathy. B.S., Cornell U., 1963; M.S., U. Wis.-Madison, 1965; Ph.D., U. Pa., 1968. Postdoctoral fellow U. Rochester, N.Y., 1968-70, asst. prof. biophysics, 1970-75, assoc. prof. boophysics, 1975-82, prof. biochemistry and radiation biology and biophysics, 1983—, assoc. chmn. dept. biochemistry, 1983—; vis. lectr. biochemistry Manchester (Eng.) Inst. Sci. and Tech., 1975-76. Contbr. research articles to publs. Mem. Am. Soc. Biol. Chemists, Am. Physiol. Soc. Subspecialties: Membrane biology; Biochemistry (biology). Current work: Energetics of intestinal transport systems for organic solutes, regulation of intestinal ion transport. Home: 30 Gateway Rd Rochester NY 14624 Office: Dept Biochemistry Sch Medicine and Dentistry Univ Rochester Rochester NY 14642

KIMMONS, GEORGE HARVEY, govt. ofcl.; b. Oxford, Miss., Jan. 22, 1919; s. William Garl and Doris (Lester) K.; m. Margaret Bowman, June 27, 1947; 1 son, George Harvey. B.S., U. Miss., 1941. Registered profl. engr., Tenn. With TVA, Knoxville, Tenn., 1941—, asst. dir. constrn., 1964-65, dir. constrn., 1966, now mgr. engring. design and constrn.; Mem. U.S. Com. Large Dams.; Bd. dirs. Atomic Indsl. Forum, 1978. Served to lt. (j.g.) USNR, 1943-46. Named to Hall of Fame U. Miss., 1978; recipient Silver Knight award Nat. Mgmt. Assn., 1978. Mem. Nat. Acad. Engrs. Presbyterian. Subspecialties: Civil engineering; Nuclear engineering. Home: Route 3 Williams Rd Concord TN 37720 Office: TVA W12A9 400 Commerce Ave Knoxville TN 37902

KIMURA, DOREEN, psychology educator, researcher; b. Winnipeg, Man., Can. B.A., McGill U., 1956; M.A., 1957; Ph.D., 1961. Registered psychologist, Ont., Can. Lectr. Sir George Williams U., 1960-61; research assoc. UCLA Med. Center, 1962-63; Geigy fellow, research assoc. Neurochirurgische Klinik, Zurich, Switzerland, 1963-64; research assoc. McMaster U., 1964-67; assoc. prof. dept. psychology U. Western Ont., London, 1967-74, prof., 1974—; supr. Clin. Neuropsychology, Univ. Hosp., London, 1975—; research assoc. Ont. Mental Health Found., 1973-81. Fellow Am. Psychol. Assn., Can. Psychol. Assn.; mem. Internat. Neuropsychol. Symposium, Acad. Aphasia (gov. 1969-72). Subspecialties: Neuropsychology; Psychobiology. Current work: Functional brain asymmetry in man; neuromotor mechanisms in communication; sex differences in brain organization. Office: Dept Psychology U Western Ont London ON Canada N6A 5C2

KIMURA, EUGENE TATSURU, toxicologist; b. Sheridan, Wyo., Sept. 19, 1922; m. Grace Watanabe, Feb. 12, 1950; children: Kathryn, Eugenie, Alan. B.S., U. Nebr., 1944, M.S., 1946; Ph.D. in Pharmacology, U. Chgo., 1948. Pharmacologist Nepera Chem. Co., Yonkers, N.Y., 1949-55; with Abbott Labs., North Chicago, Ill., 1955—, staff positions corp. research and devel., 1972-76, sr. toxicologist div. drug safety evaluation, 1976—, asso. research fellow, 1980—. Contbr. articles to sci. jours. Mem. Am. Soc. Pharmacology and Exptl. Therapeutics, Soc. Exptl. Biology and Medicine, Soc. Toxicology, Am. Chem. Soc., Rho Chi, Sigma Xi. Subspecialties: Toxicology (medicine); Pharmacology. Current work: Subchronic and chronic toxicology; carcinogenicity testing; pharmacology (anti-inflammatory; mediators; general). Patentee in field. Office: Abbott Labs 14th and Sheridn North Chicago IL 60064

KINCAID, JOHN FRANKLIN, scientist; b. Blackwell, Mo., Feb. 27, 1912; s. John Randall and Rose (Rich) K.; m. Nancy Virginia Ange, June 28, 1938 (dec.); children: James Randall, John Peter, Thomas Franklin (dec.); m. Marguerite Belair Hull, Oct. 30, 1971. A.B., Central Coll., Fayette, Mo., 1934; M.A., George Washington U., 1936; Ph.D., Princeton, 1938. Instr. chemistry U. Rochester, 1938-42; div. head explosives research lab. Carnegie Inst. Tech., 1942-45; research scientist Gen. Electric Co., 1945-46; head high pressure research dept. Rohm & Haas Co., 1946-49, research supr., 1949-58; head gen. sci. br., adv. research project div. Inst. Def. Analysis, 1958-59, dep. dir. advanced research projects div., 1959-60, dir. research and engring. support div., 1960-62; ind. cons., 1962-63; v.p. research and devel. Internat. Minerals and Chem. Corp., 1963-67; asst. sec. Dept. Commerce, 1967-69; cons., 1969-71; sr. scientist Applied Physics Lab., Johns Hopkins, 1971—. Author articles. Recipient Naval Ordnance Devel. award, 1945, Presdl. cert. merit, 1948. Mem. AIAA (Wyld Propulsion award 1981), AAAS, Am. Def. Preparedness Assn. Clubs: Cosmos (Washington); Aviation (Princeton, N.J.). Subspecialty: Applied physics. Inventor or co-inventor mil. and indsl. processes and products. Home: 2111 Jefferson Davis Hwy Arlington VA 22202

KINCAID, STEVEN ALAN, veterinary anatomist; b. Indpls., July 6, 1943; s. Robert Edmond and Frances Eliza (Randall) K.; m. Carol Ruth Williams, Sept. 5, 1965 (div. Jan. 1975); children: Amy Elizabeth, Jeremy Brent; m. Nancy Gail Hibbard, June 19, 1977. B.S., Purdue U., 1965, D.V.M., 1969, M.S., 1971, Ph.D., 1977. Asst. prof. Coll. Vet. Medicine, U. Tenn., Knoxville, 1977-82; assoc. prof. anatomy Sch. Vet. Medicine, Purdue U., West Lafayette, Ind., 1982—. Recipient Norden Disting. Teaching award, 1979, 82. Mem. AVMA, Am. Assn. Anatomists, Am. Assn. Vet. Anatomists, Sigma Xi, Phi Zeta. Republican. Methodist. Subspecialties: Veterinary anatomy; Anatomy and embryology. Current work: Pathobiology of bone and hyaline cartilage. Home: 3778 Laramie Dr Lafayette IN 47905 Office: Purdue U Dept Anatom Sch Vet Medicine Lynn Hall West Lafayette IN 47907

KINCAIDE, WILLIAM CHARLES, high technology company executive, mechanical engineer; b. Newberry, Mich., Dec. 19, 1936; s. William Glenister and Helen Mae (Smith) K.; m. Patricia Ann Neuville, June 23, 1957; children: Renae Ann, Stacy Lee. B.S. in Mech. Engring, Mich. Technol. U., 1959. Mgr. Apollo space suit program NASA Manned Spacecraft Center, Houston, 1959-69; mgr. new product devel. Allis-Chalmers Co., Milw., 1969-72; mgr. advanced energy programs Teledyne Energy Systems, Timonium, Md., 1972—. Mem. Internat. Assn. for Hydrogen Energy, Mich. Technol. U. Alumni Assn. Episcopalian. Lodges: Masons; Shriners. Subspecialties: Hydrogen production systems; Electrical engineering. Current work: Development of new and unique applications/equipment for emerging hydrogen production and thermoelectric technologies. Office: Teledyne Energy Systems 110 W Timonium Rd Timonium MD 21093

KIND, PHYLLIS DAWN, immunologist, educator; b. Sidney, Mont., July 31, 1933; d. Dan E. and Margaret A. (Erickson) K. B.A. in Bacteriology, U. Mont., 1955; M.S. in Microbiology, U. Mich., 1956; Ph.D. in Immunology, U. Mich., 1960. Postdoctoral fellow dept. dermatology U. Mich., Ann Arbor, 1960-63; instr. dept. pathology U. Colo. Med. Ctr., Denver, 1963-65, asst. prof., 1965-71; research microbiologist Nat. Cancer Inst., NIH, 1971-74; assoc. prof. dept. microbiology George Washington U. Med. Ctr., 1974-79, prof., 1979—, assoc. dir. Tissue Typing Lab., 1978—; ad hoc mem. immunol. scis. study sect. NIH, 1982; mem. grad. fellowship evaluation panel NRC, 1978-80, chmn., 1982. Contbr. articles to profl. jours. NSF fellow, 1955-59; NIH spl. fellow, 1963-64; NIH grantee, 1964-71, 75-78, 82—. Mem. AAAS, Am. Soc. Microbiology, Am. Assn. Immunologists, Soc. Exptl. Biology and Medicine, Assn. Women in Sci., Am. Assn. Clin. Histocompatibility Testing, Sigma Xi, Phi Sigma, Phi Kappa Phi. Subspecialty: Immunology (medicine). Current work: Regulation of immune response; regulation of interferon production. Office: 2300 Eye St NW 727 Microbiology Washington DC 20037

KINDT, THOMAS JAMES, immunologist, educator; b. Cin., May 18, 1939; s. James Michael and Barbara Katherine (Mayer) K.; m. Marie Robinson, Sept. 4, 1964; children: Rachel, James. A.B., Thomas More Coll., 1963; Ph.D., U. Ill.-Urbana, 1967. Ad. assoc. prof. dept. medicine Cornell U. Med. Coll., N.Y.C., 1973-78; assoc. prof. Rockefeller U., N.Y.C., 1973-77, acting head lab. immunology and immunochemistry, 1975-77; chief Lab. Immunogenetics Nat. Inst. Allergy and Infectious Disease NIH, Bethesda, Md., 1977—; adj. prof. Georgetown U. Sch. Medicine and Dentistry, Washington, 1981—. Served with USN, 1957-59. Am. Heart Assn. Investigator, 1970-75; NIH fellow, 1967-70. Mem. Am. Assn. Immunologists, Am. Soc. Biol. Chemists, Harvey Soc., Sigma Xi. Subspecialties: Immunogenetics; Biochemistry (biology). Current work: Research on molecules and genes that are important in immune functions, immunogenetics and biochemistry of histo-compatibility antigens and immunoglobulins. Office: Bldg 5 Room Bl-04 Bethesda MD 20205

KINDWALL, ERIC POST, physician, hyperbaric medicine educator; b. N.Y.C., Jan. 17, 1934; s. Josef Alfred and Anna Linnea (Post) K.; m. Betsy Fernald, Sept. 12, 1964; children: Kristina, Alexander; m. Marilyn Laurie MacArthur, Aug. 5, 1978. B.A., U. Wis., 1956; M.D., Yale U. Sch. Medicine, 1960. Rotating intern U. Va. Hosp., Charlottesville, 1961-62; resident in psychiatry Mass. Mental Health Ctr.-Harvard U., Boston, 1962-65; asst. dir. U.S. Navy Sch. Submarine Medicine, New London, Conn., 1967-69; asst. clin. prof. pharmacology Med. Coll. Wis., Milw., 1969—; dir. hyperbaric medicine St. Luke's Hosp., Milw., 1969—; cons. diving and hyperbaric medicine Indonesian Navy, 1981-82; hyperbaric cons. Republic of China Navy, 1976—; cons. WHO, Geneva, Switzerland, 1981-82. Contbr. articles to profl. jours.; Editor: Hyperbaric Oxygen Rev. Jour, 1979—. Mem. Undersea Med. Soc. (pres. 1981-82), Am. Occupational Med. Assn., Aerospace Med. Assn., AAAS. Subspecialty: Hyperbaric Medicine. Current work: Clinical hyperbaric medicine research and practice; research on the decompression of deep sea divers and compressed air caisson workers. Home: 13020 Oriole Ln Brookfield WI 53005 Office: St Luke's Hosp 2900 W Oklahoma Ave Milwaukee WI 53215

KING, CARY JUDSON, III, chemical engineer, educator; b. Ft. Monmouth, N.J., Sept. 27, 1934; s. Cary Judson and Mary Margaret (Forbes) K., Jr.; m. Jeanne Antoinette Yorke, June 22, 1957; children: Mary Elizabeth, Cary Judson IV, Catherine Jeanne. B. Engring., Yale, 1956; S.M., Mass. Inst. Tech., 1958, Sc.D., 1960. Asst. prof. chem. engring. Mass. Inst. Tech., Cambridge, 1959-63; dir. Bayway Sta. Sch. Chem. Engring. Practice, Linden, N.J., 1959-61; asst. prof. chem. engring. U. Calif. at Berkeley, 1963-66, asso. prof., 1966-69, prof., 1969—, vice chmn. dept. chem. engring., 1967-72, chmn., 1972-81, dean Coll. Chemistry, 1981—; cons. Procter & Gamble Co., 1969—, CPC Internat., 1982—. Author: Separation Processes, 1971, 80, Freeze Drying of Foods, 1971; Contbr. numerous articles to profl. jours. Active Boy Scouts Am., 1947—; pres. Kensington Community Council, 1972-73, dir., 1970-73. Named Inst. Lectr. Am. Inst. Chem. Engrs., 1973, Food, Pharm. and Bioengring. Div. award, 1975, William H. Walker award, 1976. Mem. Nat. Acad. Engring., Am. Inst. Chem. Engrs., Inst. Food Tech., Am. Soc. Engring. Edn. (George Westinghouse award 1978), Am. Chem. Soc., AAAS. Subspecialties: Chemical engineering. Current work: Seperation processes, including dehydration and concentration, extraction, adsorption, distillation, spray drying and freeze drying. Patentee in field. Home: 7 Kensington Ct Kensington CA 94707 Office: Coll Chemistry U Calif Berkeley CA 94720

KING, DAVID KYLE, physician; b. Logan, W.Va., July 31, 1941; s. Kyle and Dorothy (Wagoner) K.; m. Victoria Dean, June 11, 1966; children: Liesl Ann, David Link, Bran Michael. B.A., Morris Harvey Coll., 1964; M.D., W.Va. U., 1968. Diplomate: Am. Bd. Internal Medicine. Faculty assoc. M.D. Anderson Hosp., Houston, 1973-75; pvt. practice internal medicine, Phoenix, 1975—. Editor: Samaritan Medicine, 1981—. Served to maj. U.S. Army, 1971-73. Mem. ACP, Alpha Omega Alpha. Subspecialties: Oncology; Cancer research (medicine). Current work: Clinical cancer research. Primary patient cancer care. Office: Internists Oncologists Ltd 926 E McDowell Rd Phoenix AZ 85006

KING, DAVID QUIMBY, aerospace engineer; b. Summit, N.J., Sept. 5, 1953; s. William Henry and Jane (Gurnee) K. B.S., Rutgers U., 1976; M.A., Princeton U., 1978, Ph.D., 1982. Sr. engr. Jet Propulsion Lab., Pasadena, Calif., 1981—. Contbr. articles to profl. jours. Guggenheim fellow Princeton U., 1976. Mem. AIAA, Tau Beta Pi, Pi Tau Sigma. Subspecialties: Electric propulsion; Plasma physics. Current work: Experimental and analytical plasma physics related to electric propulsion for spacecraft. Home: 3131 Montrose Ave Apt 20 La Crescenta CA 91214 Office: Calif Inst Tech Jet Propulsion Lab 4800 Oak Grove Dr Pasadena CA 91109

KING, DENNIS R., clinical psychologist; b. Cin., July 22, 1947; s. Granville P. and Loretta (Tirey) K. B.A., U. Cin., 1969; M.S. in Clin, Psychology, Calif. State U.-San Diego, 1973. Lic. psychol. examiner, Tenn., diplomate: Internat. Acad. Prof. Counseling and Psychotherapy. Psychol. examiner Hiwassee Mental Health Ctr., Cleveland, Tenn., 1973-77, programs dir., 1977-78, exec. dir., 1978-80; biomed. dir. Cleveland Pain Clinic, 1980-82, Pain Mgmt. Services, Chattanooga, 1982—, Health Mgmt. Services, Cleveland, 1980—; instr. Cleveland State Community Coll., 1975—; exam. cons. Disability Determinations sect. Social Security, Cleveland, 1974—; dir. Behavorial Research Inst., Cleveland, 1981—. Contbr. articles to profl. jours. Bd. dirs. Child Shelter, Inc., Cleveland, 1974-78; mem. policy council Headstart S.E., Tenn., 1982. Tenn. Dept. Human Services grantee, 1980, 82. Fellow Internat. Council Sex. Edn.; mem. Am. Psychol. Assn., Tenn. Psychol. Assn., Biofeedback Soc. Am., Am. Assn. Sex Educators, Counselors and Therapists, Council for Exceptional Children. Republican. Baptist. Subspecialties: Pain Management; Biofeedback. Current work: Treatment of chronic pain through biofeedback, visual imagery, hypnosis and relaxation training. Treat low back pain, tension and migraine headache, arthritis, muscle spasm. Child abuse treatment. Home: 750 Beech Circle NW Cleveland TN 37311 Office: Health Mgmt Services 2850 Westside Dr Suite I PO Box 2965 Cleveland TN 37311

KING, FREDERICK ALEXANDER, neuroscientist, educator; b. Paterson, N.J., Oct. 3, 1925; s. James Aloysius and Louise Bisset (Gallant) K.; children: Alexander Karell, Elizabeth Gallant. A.B., Stanford, 1953; A.M. (John Carrol Fulton scholar 1953-55), Johns Hopkins, 1955, Ph.D., 1956. Instr. psychology Johns Hopkins U., 1954-56; asst. prof. psychiatry Ohio State U., 1957-59; mem. faculty Coll. Medicine, U. Fla., 1959-78, asst., then asso. prof., then prof. neurosurgery, 1965-69, prof., chmn. dept. neurosci., 1969-78; dir./co-dir. Center Neurobiol. Scis., 1964-78; dir. Yerkes Primate Center, prof. anatomy, asso. dean Sch. Medicine Emory U., Atlanta, 1978—; Mem. adv. com. Primate Research Centers, NIH, 1969-73; mem. psychobiology adv. panel, biol. and med. scis. div. NSF, 1963-67, cons. med. and biol. scis., 1967-70; mem. research scientist devel. rev. com. NIMH, 1969-70, 75-78, chmn. com. for coordination and communication for dirs. in biol. research tng. programs, 1972—; sec.-treas. Fla. Anat. Bd., 1969-71; vice chmn. bd. sci. advisers Yerkes Regional Primate Research Center, 1974-78; mem. brain scis. com. NRC-Nat. Acad. Scis., 1974-78. Gen. editor: Handbook of Behavioral Neurobiology, 5 vols, 1972—; Contbr. articles to profl. jours. Served with USNR, 1943-46, 51. Research fellow NIH, 1955-56; spl. fellow NIMH, Inst. Physiology, U. Pisa (Italy) Faculty Medicine, 1961-62. Mem. Internat. Neuropsychology Soc. (sec.-treas.), Soc. Neurosci. (chmn. com. on edn.), Am. Psychol. Assn. (chmn. membership com. div. physiol. and comparative psychology, chmn. com. for animal research and experimentation). Subspecialties: Neuropsychology; Neurobiology. Home: 2681 Galahad Dr Atlanta GA 30329 Office: Yerkes Primate Research Center Emory U Atlanta GA 30322

KING, FREDERICK WARREN, chemistry educator; b. Sydney, Australia, Apr. 15, 1947; came to U.S., 1977; s. Francis Albert and Bernice (Petith) K. B.Sc. with honors, U. Sydney, 1969; M.Sc., U. Calgary, 1971; Ph.D., Queen's U., Kingston, Ont., Can., 1975. Research asst. U. Oxford, Eng., 1975-76; postdoctoral fellow Northwestern U., Evanston, Ill., 1977-78; research assoc. Brock U., St. Catharines, Ont., Can., 1978-79; adj. asst. prof. chemistry U. Wis.-Eau Claire, 1979—. Contbr. articles to profl. jours. Nat. Sci. Council Can. grantee, 1975. Subspecialties: Theoretical chemistry; Physical chemistry. Current work: Bounds on the electronic density; theory of optical properties, optical data analysis; local accuracy of wavefunctions; Raman scattering from adsorbates. Office: Dept Chemistry U Wis Eau Claire WI 54701

KING, IVAN ROBERT, astronomy educator; b. Far Rockaway, N.Y., June 25, 1927; s. Myram and Anne (Franzblau) K.; m. Alice Greene, Nov. 21, 1952; children: David, Lucy, Adam, Jane. A.B., Hamilton Coll., 1946; A.M., Harvard U., 1947, Ph.D., 1952. Instr. astronomy Harvard U., 1951-52; mathematician Perkin-Elmer Corp., Norwalk, Conn., 1951-52; methods analyst U.S. Dept. Def., Washington, 1954-56; with U. Ill., 1956-64; assoc. prof. astronomy U. Calif. at Berkeley, 1964-66, prof., 1966—, chmn. astronomy dept., 1967-70. Contbr. numerous articles to sci. jours. Served with USNR, 1952-54. Mem. Soc. of Fellows Harvard, 1947-51. Mem. Nat. Acad. Scis., Am. Astron. Soc. (councillor 1963-66, chmn. div. dynamical astronomy 1972-73, pres. 1978-80), AAAS (chmn. astronomy sect. 1974), Internat. Astron. Union. Subspecialty: Optical astronomy. Current work: Structure and dynamics of star clusters and galaxies. Study of stellar systems. Office: Dept Astronomy U Calif Berkeley CA 94720

KING, JAMES CLEMENT, microbiology educator; b. Three Rivers, Mich., Feb. 4, 1904; s. Henry Burr and Marie Francoise (Jeandrevin) K.; m. Frances Newborg, Mar. 18, 1932 (div. 1949); children: Geoffrey, Martha, Henry, John. B.S., Northwestern U., 1926; M.A., U. Chgo., 1928, Ph.D., 1933. Vice consul Fgn. Service U.S., Caracas, Venezuela, 1928; instr. politics Western Res., U., Cleve., 1934-35, Princeton (N.J.) U., 1935-37, Yale U., New Haven, 1937-38; asst. prof. internat. relations Syracuse (N.Y.) U., 1940-41; geneticist Cold Spring Harbor Labs., N.Y., 1949-58; assoc. prof. microbiology N.Y.U. Med. Center, N.Y.C., 1962—. Author: The Biology of Race, 1981. Fellow AAAS, N.Y. Acad. Scis.; mem. Am. Soc. Naturalists, Soc. Study Evolution, Genetics Soc. Am. Democrat. Subspecialties: Genome organization; Evolutionary biology. Current work: Investigation of the way in which the gene pool is integrated to produce through development the highest possible frequency of variants falling within the modal phenotype. Home: 115 E 9th St Apt 15B New York NY 10003 Office: NY U Med Center 550 1st Ave New York NY 10016

KING, JANET CARLSON, nutritional sciences educator, researcher; b. Red Oak, Iowa, Oct. 3, 1941; d. Paul Emil and Norma Carolina (Anderson) C.; m. Charles Talmadge King, Dec. 25, 1967; children: Charles Matthew, Samuel Knox. B.S., Iowa State U., 1963; Ph.D., U. Calif.-Berkeley, 1972. Asst. prof. U. Calif.-Berkeley, 1973-78, assoc. prof. dept. nutritional scis., 1978—, dir. metabolic unit, 1980—. Served to capt. U.S. Army, 1963-67. Mem. Am. Dietetic Assn., Am. Inst. Nutrition, AAAS. Democrat. Subspecialty: Nutrition (medicine). Current work: Nutritional needs during pregnancy and lactation, trace element metabolism, protein and energy dietary requirements. Office: Dept Nutritional Scis U Calif Berkeley CA 94720

KING, JOAN CALUDA, neuroanatomist, educator; b. New Orleans, Mar. 6, 1938; d. Anthony Nina and Anna Nina (Vulevich) C.; m. Lewis Eugene King, Mar. 6, 1971; m. Stuart Allen Tobet, Apr. 9, 1979. B.S., St. Mary's Dominican Coll., 1961; M.S., U. New Orleans, 1970; Ph.D., Tulane U., 1972. Instr. chemistry St. Mary's Dominican Coll., 1961-66; research assoc. neuroanatomy U. Iowa, Iowa City, 1973-74; postdoctoral NIH fellow Tulane U., 1974-76, research assoc., 1976-79, asst. prof. anatomy, 1979—. Contbr. articles to profl. jours. NSF grantee, 1981—. Mem. Internat. Soc. Psychoneuroendocrinology, Am. Assn. Anatomists, Soc. Neuroscience, Endocrine Soc., Kappa Delta Pi. Subspecialties: Neuroendocrinology; Immunocytochemistry. Current work: Control of LH pituitary secretion by LHRH hypothalmic neurons; electron microscopy of LHRH neurons; endocrine conditions. Office: 136 Harrison Ave Stearns 606 Boston MA 02111

KING, JONATHAN ALAN, molecular biology educator; b. Bklyn., Aug. 20, 1941; m. Jacqueline Dee. B.S. in Zoology,magna cum laude with high honors, Yale U., 1962; Ph.D., Calif. Inst. Tech., 1968. Brit. Med. Research Council postdoctoral fellow Cambridge (Eng.) U., 1970; asst. prof. MIT, Cambridge, 1971-73, assoc. prof., 1974-78, prof. molecular biology, 1979—, dir. biology electron microscope facility, 1971—. Contbr. numerous articles to sci. jours. Chmn. Nat. Jobs with Peace Campaign. Gen. Motors Nat. scholar 1958-62; Jane Coffin Shields Fund fellow, 1968-70; recipient U.S. Antarctic Service medal, 1968; Woodrow Wilson fellow, 1962-63; NIH fellow, 1963-67. Mem. Genetics Soc. Am., Am. Soc. Microbiology, Biophysics Soc., Teratology Soc., Soc. Occupational and Environ. Health, Am. Pub. Health Assn. Subspecialties: Biochemistry (biology); Gene actions. Current work: Morphogenesis and development, virus assembly, organelle assembly. Home: 114 Charles River Rd Watertown MA 02172 Office: MIT Dept Biology Cambridge MA 02139

KING, JONATHAN STANTON, scientific publication editor; b. Bristol, Tenn., Oct. 30, 1922; s. J. Stanton and Annie (Lowrey) K.; m. Betty Boyd, June 30, 1951; children: Melissa, Robert Stanton. Student, Berea (Ky.) Coll., 1940-43, 46; B.S. in Biol. Sci, U. Chgo., 1948; Ph.D. in Biochemistry, U. Tenn., 1954. Research biochemist S.E. Massengill Co., Bristol, 1948-51, 54-56; research assoc. in urology Bowman Gray Sch. Medicine, Wake Forest U., Winston-Salem, N.C., 1956-59, instr. in biochemistry, 1959-61, research asst. prof. biochemistry (surgery), 1961-65, research assoc. prof. urology (biochemistry), 1965-70; guest investigator Rockefeller U., N.Y.C., 1968-69; exec. dir. Am. Assn. Clin. Chemistry, 1971-74; exec. editor Clin. Chemistry, 1968—; mem. internat. com. Biometrica, Madrid, Spain; mem. adv. bd. LAB, Verona, Italy. Served with USAAF, 1940-44. Helen Hay Whitney fellow, 1959-62; recipient Career Devel. award NIH, 1962-67; Outstanding Contbns. to Clin. Chemistry in a Selected Area award Am. Assn. Clin. Chemistry, 1981. Mem. Soc. for Exptl. Biology and Medicine, N.Y. Acad. Sci., Am. Assn. Clin. Chemistry, Council Biology Editors, Soc. Nat. Assn. Publ., Com. of Editors of Biochemistry Jours., Sigma Xi. Presbyterian. Subspecialties: Clinical chemistry; Biochemistry (biology). Home: 3610 Winding Creek Way Winston-Salem NC 27106 Office: PO Box 5218 Winston-Salem NC 27113

KING, KENNETH ROY, aerospace executive; b. Boston, July 8, 1956; s. Kenneth Ira and Ruth (Smaglis) K. B.S., Boston U., 1978; M.B.A., U. Mass., 1980. Program mgr. aerospace Hamilton Standard, Windsor Locks, Conn., 1980—. Mem. AIAA. Subspecialties: Aerospace engineering and technology; Satellite studies. Current work: Satellite services, extravehicular mobility unit, space station studies, utilization of space. Home: 222 Williams St #323 Glastonbury CT 06033 Office: Hamilton Standard Bradley Field Rd Windsor Locks CT 06096

KING, (MARY) MARGARET, research scientist, research lab. adminstr.; b. Oklahoma City, May 26, 1946; d. James Dean and Mary Bell (Gregory) K. B.S. in Biology and Edn, Central State U., Edmond, Okla., 1969; Ph.D. in Med. Physiology and Biophysics, U. Okla., 1975. Tchr. pub. schs., Oklahoma City, 1969-71; postdoctoral fellow Biomembrane Research Lab., Okla. Med. Research Found., Oklahoma City, 1973-76, research assoc., 1976-77, staff scientist, 1977-80, asst. mem., 1980—; NIH Cardiovascular Physiology and Pharmacology Tng. grantee U. Okla. Health Scis. Ctr., Oklahoma City, 1971-73, assoc. in research biochemistry, 1977-81, asst. prof. research biochemistry and molecular biology, 1981—; asst. prof. U. Okla. Coll. Medicine, 1981—; cons., lectr. in field. Contbr. numerous sci. articles and abstracts to profl. publs. Nat. Inst. Environ. Health Scis. awardee, 1978-80; NIH grantee, 1978-84. Mem. Am Assn. for Cancer Research, Am. Inst. Nutrition, Am. Assn. Lab. Animal Scis., Soc. for Exptl. Biology and Medicine, N.Y. Acad. Scis., Sigma Xi, Alpha Chi, Kappa Delta Pi. Subspecialties: Cancer research (medicine); Pharmacology. Current work: s: Dietary fat and antioxidant influences on breast cancer. Office: 825 NE 13th St Oklahoma City OK 73104

KING, THEODORE OSCAR, toxicologist; b. Portsmouth, Ohio, May 29, 1922; s. Edward and Rose (Glassman) K.; m. Dorothy Sillman, Dec. 28, 1952; children: Jeremy E., Naomi E. B.S. in Pharmacy, U. Mich., 1943; postgrad., Columbia U., 1944-46; Ph.D. in Pharmacology (Am. Found. for Pharm. Edn. fellow), Georgetown U., 1949; J.D., U. Wyo., 1960. Registered pharmacist, Mich., N.Y., D.C., diplomate: Am. Bd. Toxicology. Pharm. control chemist William R. Warner & Co. Inc., N.Y.C., 1943-46; analytical chemist research div. Colgate-Palmolive Peet Co., Jersey City, 1946; asst. prof. pharmacology U. Wyo., 1949-55, prof., 1955-57; adj. lectr. pharmacology Rutgers U., 1958-68; chief pharmacologist Johnson & Johnson Research Found., 1957-59; dir. div. pharmacology Ortho Research Found., 1959-65; v.p., dir. research Bio/dynamics, Inc., 1966-71; dir. drug safety evaluation Pfizer Inc., Groton, Conn., 1971-76, sr. dir., 1976—; research asso. prof. toxicology and pharmacology U. Conn., 1981—; Fulbright research scholar Heymans Inst., U. Ghent, Belgium, 1955-56; WHO public health fellow, U.K., 1951. Mem. Soc. Toxicology, Am. Soc. for Pharmacology and Exptl. Therapeutics, Internat. Union Pharmacology, Am. Chem. Soc., AAAS, Sigma Xi. Subspecialties: Toxicology (medicine); Pharmacology. Current work: Preclinical safety evaluation of drugs and chemicals.

KINGERY, WILLIAM DAVID, educator, ceramist; b. N.Y.C., July 7, 1926; s. Lisle B. and Margaret (Reynolds) K.; m. Gertrude Phillips, Nov. 22, 1965; children—William David, Peter (dec.), Rebekah, Andrew. Grad., Taft Sch., Watertown, Conn., 1943; S.B., MIT, 1950, Sc.D., 1952; Dr. (hon.), Tokyo Inst. Tech., 1982. Mem. faculty MIT,

1951—, prof. ceramics, 1960—, Orton lectr., 1980; vis. prof. Ecole Poly. Fed. de Lausanne, Switzerland, 1980; Mem. materials adv. bd. Nat. Acad. Sci., 1960-68. Author: Property Measurements at High Temperatures, 1959, Introduction to Ceramics, 1960, 2d edit., 1976; Editor: Ceramic Fabrication Process, 1958, Kinetics of High-Temperature Process, 1959, Ice and Snow, 1963; editor-in-chief: Cerammgia Internat, 1976—; Contbr. profl. jours. Chmn. Marion-Bermuda Cruising Yacht Race, 1977-79; chmn. bd. trustees, 1980—; treas. East Marion Steamship Authority. Fellow Am. Ceramic Soc (Ross Coffin Purdy award 1952, John Jeppson award 1958, outstanding service award New Eng. sect. 1957, 1st Distinguished Sosman Meml. lectr. 1973, A.V. Bleininger award 1976, F.H. Norton award 1977, hon. life mem. 1983), Keramos (hon.), Am. Chem. Soc., Glaciological Soc.; mem. Nat. Acad. Engring. Subspecialties: Ceramics; Ceramic engineering. Current work: Structure and properties of ceramics, physical ceramics, inferences from ceramic artifacts. Home: Allens Point Rd Marion MA 02738 Office: 77 Massachusetts Ave Cambridge MA 02139

KINGHORN, CAROL ANN, psychologist; b. Memphis, Nov. 9, 1935; d. Warren Joseph and Ann Frances (Macy) K. B.A., SUNY-Albany, 1957; M.A., 1958; M.S., Hofstra U., 1969, Ph.D., 1973. Lic. psychologist, N.Y. Br. librarian N.Y. Inst. Tech., Old Westbury, N.Y., 1966-72; psychologist Hofstra U. Counseling Center, Hempstead, N.Y., 1973-77; pvt. practice clin. psychology, Roslyn, N.Y., 1974—; psychologist Garden City (N.Y.) pub. schs., 1978—; mem. faculty C.W. Post Coll., Greenvale, N.Y., 1974—. Mem. Am. Psychol. Assn., N.Y. State Psychol. Assn. (legal-legis. com. 1974-76), Nassau County Psychol. Assn. (chmn. community relations com. 1973-74, chmn. legal-legis. com. 1974-76), Am. Scottish Found. Subspecialty: Giftedness. Current work: Research on giftedness, private clinical practice, teaching. Home: 14 St James Pl Hempstead NY 11550 Office: Garden City Pub Schs: 56 Cathedral Ave Garden City NY 11550

KINGSBURY, DAVID WILSON, virologist; b. Jersey City, N.J., Apr. 2, 1933; s. Wilson Howard and Rose Carmen (Abarno) K.; m. Virginia Wrenne Murray, Aug. 3, 1957; children: Paul F., Virginia W., David W., Daniel J., Robert E., James A., Sarah E. B.S., Manhattan Coll., Riverdale, N.Y., 1955; M.D., Yale U., 1959. Intern and resident in pathology Yale U., 1959-61, USPHS postdoctoral research fellow, 1961-63; from research fellow to mem. St. Jude Children's Research Hosp., Memphis, 1963—; from asst. prof. to prof. U. Tenn. Ctr. for Health Scis., 1963—; cons. NIH. Mem. Am. Assn. Immunologists, Am. Soc. Microbiology, Am. Soc. Virology. Subspecialties: Virology (biology); Molecular biology. Current work: Virus gene structure and expression; basic research. Home: 2993 Rolling Woods Memphis TN 38128 Office: 332 N Lauderdale Memphis TN 38101

KINGSLAND, GRAYDON CHAPMAN, plant pathologist, educator, researcher; b. Burlington, Vt., Aug. 28, 1928; s. Arthur George and Eleanor Francis (Chapman) K.; m.; children: Graydon Chapman, David, Christopher, Karen. B.Sc. in Botany, U. Vt., 1952; M.Sc. in Plant Pathology, U. N.H., 1955, Ph.D., Pa. State U., 1958. Cert. in pesticide application S.C. Crop Pest Commn. Research asst. Conn. Agr. Expt. Sta., 1952-53; research plant pathologist United Fruit Co., Honduras, 1958-60; asst. prof. plant pathology Clemson (S.C.) U., 1960-67, assoc. prof., 1967—; project plant pathologist food research and devel. project AID/Southeastern Consortium for Internat. Devel., Seychelles Islands, 1981-82. Contbr. articles on plant pathology to sci. jours. Scoutmaster Boy Scouts Am., Clemson, 1960-72. Served with USN, 1946-48. Mem. Am. Phytopath. Soc., Assn. Southeastern Biologists, Sierra Club, Sigma Xi. Subspecialties: Plant pathology; Tropical agriculture. Current work: Cereal grains plant pathology, mycology of cereal grains, control of plant diseases, tropical agriculture, teaching plant pathology; ecology of microorganisms. Home: 209 Manley Dr Clemson SC 29631 Office: Dept Plant Pathology and Physiology Clemson U Clemson SC 29631

KINGSTON, DAVID GEORGE IAN, chemistry educator; b. London, Nov. 9, 1938; s. Charles John Ewart and Norah Blanche (Holcroft) K.; m. Beverly Hazel Mark, June 18, 1966; children: Joy Ellen, Christina Anne, Jonathan David. B.A., Cambridge (Eng.) U., 1960, Ph.D., 1963; Dip. Th., London U., 1962. Postdoctoral research assoc. MIT, 1963-64; NATO fellow Cambridge U., 1964-66; asst. prof. chemistry SUNY-Albany, 1966-71; assoc. prof. chemistry Va. Poly. Inst. and State U., 1971-77, prof. chemistry, 1977—. Assoc. editor Jour. Natural Products; Contbr. numerous articles to profl. jours. Fulbright grantee, 1963-64. Mem. Am. Chem. Soc., Royal Soc. Chemistry, Am. Soc. Pharmacognosy. Subspecialty: Organic chemistry. Current work: Biosynthesis of antibiotics; structure determination of biologically active natural products; anticancer drugs. Office: Dept Chemistry Va Poly Inst and State U Blacksburg VA 24061

KINNEY, EVLIN L., academic cardiologist; b. N.Y.C., July 15, 1945; d. Hugh Francis Cook and Mollie (Kalina) Cook Friedman; m. Robert J. Wright II, Sept. 21, 1974. M.D., SUNY-Downstate Med. Center, N.Y.C., 1972. Diplomate: Am. Bd. Internal Medicine. Intern D.C. Gen. Hosp., 1972-73, resident in internal medicine, 1974-77; asst. prof. medicine Hershey (Pa.) Med. Center, Pa. State U., 1979-82; vis. asst. prof. medicine Ind. U. Sch. Medicine, Indpls., 1980-82; asst. prof. medicine U. Miami (Fla.) Sch. Medicine, 1982—; dir. echocardiography Miami VA Med. Center, 1982—, U. Miami Hosp. and Clinics, 1983—; Clin. scientist Am. Heart Assn., 1979-82. Mem. Greater Miami Soc. Echocardiography (pres. 1983—), Am. Fedn. Clin. Research, ACP, Am. Heart Assn. Greater Miami, Assn. VA Cardiologists. Subspecialty: Cardiology. Current work: Hemodynamic data from echocardiolography; edge detection algorithms in echocardiography; Doppler echocardiography, tissue characterization from echocardiography; survival analysis. Office: Miami VA Med Center 111-A 1201 NW 16th St Miami FL 33125

KINNEY, JOHN JAMES, mathematics educator; b. Dansville, N.Y., Aug. 2, 1932; s. Ray Stephen and Evelyn (White) K.; m. Cherry Carter, July 7, 1962; 1 dau., Kaylyn. B.S., St. Lawrence U., 1954; M.A., Harvard U., 1956; M.S., U. Mich., 1958; Ph.D., Iowa State U., 1971. Instr. math. St. Lawrence U., Canton, N.Y., 1956-58, asst. prof., 1960-64; assoc. prof. SUNY-Oneonta, 1964-68; asst. prof. U. Nebr., Lincoln, 1971-74; prof. math. Rose-Hulman Inst., Terre Haute, 1974—. Dir. Camp Retupmoc, 1977—; pres. Dixie Bee Parent-Tchrs. Assn., 1979-80; v.p. Covered bridge Council, Girl Scouts U.S.A., Terre Haute, 1982—. NSF Sci. Faculty fellow, 1968; DuPont fellow, 1958. Mem. Am. Statis. Assn., IEEE Computer Soc., Assn. for Computing Machinery, Sigma Xi, Phi Beta Kappa. Lodge: Rotary. Subspecialties: Statistics; Programming languages. Current work: Probability theory and statistics; computer science as used in solution of mathematical problems. Home: 221 Highland Rd Terre Haute IN 47802 Office: Rose Hulman Inst Tech 5500 Wabash Ave Terre Haute IN 47803

KINNEY, ROBERT BRUCE, mechanical engineering educator, researcher; b. Joplin, Mo., July 20, 1937; s. William Marion and Olive Francis (Smith) K.; m. Carol Stewart, Jan. 29, 1961; children: Rodney, David, Linda. B.S., U. Calif.-Berkeley, 1959, M.S., 1961; Ph.D., U. Minn.-Mpls., 1965. Sr. research engr. United Aircraft Corp., East Hartford, Conn., 1965-68; assoc. prof. aerospace and mech. engring. U. Ariz., 1968-78, prof., 1978—, assoc. head dept. aerospace and mech. engring., 1981—, tech. dir. Solar Energy Research Facility, 1978-79; Forschungs-Stipendiat Alexander von Humboldt Stiftung, Bonn and Gottingen, W.Ger., 1976-77. Fellow AIAA (assoc.; chmn. sect. 1971-72), AAAS; mem. ASME (chmn. sect. 1972-73, assoc. tech. editor Jour. Heat Transfer 1973-76). Subspecialties: Mechanical engineering; Fluid mechanics. Current work: Numerical fluid mechanics and heat transfer; viscous unsteady flows; effects of buoyancy; eddy dynamics and drag reduction. Office: U Ariz Dept Aerospace and Mech Engring Tucson AZ 85721

KINNEY, TERRY B., JR., geneticist, government agricultural research administrator; b. Norfolk, Mass., Sept. 12, 1925; m., 1946; 2 children. B.S., U. Mass., 1955, M.S., 1956; Ph.D., U.Minn., 1963. Research asst. poultry U. Mass., 1955-56; geneticist Hubbard Farms, Inc., N.H., 1956-57; instr. poultry U. Minn., 1957-62; biometrician Dept. Agr., Md., 1963-65, research geneticist, Ind., 1965-69; asst. dir. animal sci. research div. Agrl. Research Service, 1969-72, assoc. dep. adminstr., 1972-74, asst. adminstr. livestock vet. sci., 1974-78, assoc. adminstr., 1978-80, adminstr. sci. and edn. adminstr.-agrl. research, 1980—. Mem. Poultry Sci. Assn. Subspecialties: Population biology; Animal research administration. Office: Agrl Research Service Dept Agr Room 302 Adminstrn Bldg Washington DC 20250

KINNIER, WILLIAM JAMES, biochemist, researcher; b. Balt., Dec. 23, 1950; s. Robert Joseph and Agnes Rose (Josephs) K.; m. Vickey Kissinger, Oct. 28, 1978. Ph.D., U. N.C., Chapel Hill, 1977. Post doctoral researcher U. N.C., Chapel Hill., 1977-78; mem. staff fellow NIMH, 1978-80; sr. research biochemist A.H. Robins Co., Richmond, Va., 1980—. Contbr. articles in field to profl. jours. Mem. Soc. Neurosci. Subspecialties: Neuropharmacology; Neurochemistry. Current work: Mechanism of action of receptors beta adrenergic, alpha adrenergic, dopamine, diazepam, muscarinic receptor. Office: 1211 Sherwood Ave Richmond VA 23102

KINO, GORDON STANLEY, electrical engineering educator; b. Melbourne, Australia, June 15, 1928; came to U.S., 1951, naturalized, 1967; s. William Hector and Sybil (Cohen) K.; m. Dorothy Beryl Lovelace, Oct. 30, 1955; 1 dau., Carol Ann. B.Sc. with 1st class honours in Math, London (Eng.) U., 1948, M.Sc. in Math, 1950; Ph.D. in Elec. Engring, Stanford U., 1955. Jr. scientist Mullard Research Lab., Salford, Surrey, Eng., 1947-51; research asst., then research asso. Stanford U., 1951-55, research asso., 1957-61, mem. faculty, 1961—, prof. elec. engring., 1965—; mem. tech. staff Bell Telephone Labs., 1955-57; cons. to industry, 1957—. Author: (with Kirstein, Waters) Space Charge Flow, 1968; also numerous papers on microwave tubes; electron optics, plasma physics, bulk effects in semiconductors, acoustic surface waves, acoustic imaging, non-destructive testing. Guggenheim fellow, 1967-68. Fellow IEEE, Am. Phys. Soc., Nat. Acad. Engring., AAAS. Subspecialties: Civil engineering; Fiber optics. Current work: Signal processing, acoustic imaging and microscopy, nondestructive testing, acoustic surface waves, acoustic and optical sensors, fiber optics. Inventor Kino electron gun, 1959. Home: 867 Cedro Way Stanford CA 94305

KINRA, VIKRAM K., mech. engr., educator; b. Lyallpur, India, Apr. 3, 1946; came to U.S., 1967; s. Gurmukh Ch and Gian Vati (Khurana) K.; m. Anita Malik, Feb. 28, 1976; 1 dau., Anushka Tanya. B. Tech., Indian Inst. Tech., Kanpur, 1967; M.S., Utah State U., 1968; Ph.D., Brown U., 1975. Structural engr. Northrop Corp., Los Angeles, 1969; project engr. Ostgaard and Assocs., Los Angeles, 1970-71; asst. prof. mech. engring. U. Colo., Boulder, 1975-82; assoc. prof. aerospace engring. Tex A&M U., College Station, 1982—; cons. Willows Water Dist., Denver, 1977-78, Corning Glass Works, N.Y., 1980—, Ponderosa Assocs., Louisville, Colo., 1980. Contbr. articles to profl. jours. NSF grantee, 1976-78, 78-79, 78-81; IBM Corp. grantee, 1982-83. Mem. Soc. for Exptl. Stress Analysis, Am. Soc. for Engring. Edn. (Dow Outstanding Young Faculty award 1980), Am. Acad. Mechanics, ASME, Soc. Automotive Engring. (Ralph R. Teetor Ednl. award 1982), Sigma Xi, Pi Tau Sigma. Subspecialties: Aerospace engineering and technology; Mechanical engineering. Current work: Quantitative nondestructive evaluation; wave propagation; fracture mechanics; composite materials engineering education and research. Home: 812A Navarro Dr College Station TX 77840 Office: Aerospace Engring Dept Tex A&M U College Station TX 77843

KINSELLA, JOHN EDWARD, food science educator, food chemist, biochemist; b. Wexford, Ireland, Feb. 22, 1938; m., 1965; 2 children. B.S., Nat. U. Ireland, 1961; M.S., Pa. State U., 1965, Ph.D. in Food Sci, 1967. Tchr. zoology, Latin and chemistry CKC Onitsha, Nigeria, 1961-63; asst. prof. food sci. Cornell U., Ithaca, N.Y., 1967-73, assoc. prof., 1973-77, prof. food sci. and chemistry, chmn. dept. food sci., 1977—, dir., 1980—, Liberty Hyde Bailey prof. food sci., 1981—. Recipient Borden award for research, 1976. Mem. Am. Inst. Nutrition, AAAS, Am. Chem. Soc., Am. Dairy Sci. Assn., Inst. Food Tech. Subspecialty: Food science and technology. Office: Dept Food Sci Cornell U Ithaca NY 14850

KINSMAN, DONALD MARKHAM, animal scientist, educator; b. Framingham, Mass., May 20, 1923; s. Joshua Starr and Florence Ruby (Markham) K.; m. Helen Katharine Bailey, Aug. 28, 1949; children: Elizabeth Lee Kinsman Keefe, David Bailey, Martha Jean. B.S., U. Mass., 1949; M.S., U. N.H., 1951; Ph.D., Okla. State U., 1964. Instr. U. N.H., Durham, 1949-51, U. Vt., Burlington, 1951-52; asst. prof. U. Mass., Amherst, 1952-56; prof. animal sci. U. Conn., Storrs, 1956—. Author, editor: International Meat Science Dictionary, 1978; editor: International Sausage Book, 1981; contbr. over 50 articles to profl. jours. Bd. dirs. Mansfield (Conn.) Retirement Community, 1976—. Served with USMC, 1942-46. Decorated Purple Heart; NSF teaching fellow, 1959-60. Mem. Am. Soc. Animal Sci. (Disting. Service award 1976), Am. Meat Sci. Assn. (Signal Service award 1975, Disting. Tchr. award 1981). Republican. Congregationalist. Subspecialties: Food science and technology; Animal breeding and embryo transplants. Current work: Humane slaughter, animal welfare, meat quality, animal stress. Home: 45 Moulton Rd Storrs CT 06268 Office: Dept Animal Industries U Conn Storrs CT 06268

KINTNER, LOREN DON, veterinary pathology educator; b. Bryan, Ohio, Feb. 7, 1922; s. John and Mertie (Eberly) K.; m. Treva Elizabeth Carpenter, Aug. 25, 1946; children: Susan Jane, David Lee. B.S., Manchester Coll., 1948; D.V.M., Ohio State U., 1949; M.S., U. Mo., 1952. Diplomate: Am. Coll. Vet. Pathologists. Asst. prof. vet. pathology Sch. Vet. Medicine, U. Mo.-Columbia, 1949-56, assoc. prof., 1956-63, prof., 1963—; cons. Midwest Research Inst., 1967—. Contbr. articles to profl. jours. Recipient Norden Disting. Tchr. award, 1968; Amoco Good Tchr. award, 1977. Mem. AVMA, Mo. Vet. Med. Assn., Am. Assn. Lab. Diagnosticians, Research Workers in Animal Diseases, Sigma Xi, Gamma Sigma Delta. Subspecialties: Pathology (veterinary medicine); Animal pathology. Current work: Diagnostic pathology. Home: 704 Morningside Dr Columbia MO 65201 Office: Vet Diagnostic Lab U Mo Columbia MO 65211

KINZEL, AUGUSTUS BRAUN, tech. exec.; b. N.Y.C., July 26, 1900; s. Otto and Josephine (Braun) K.; m., 1927 (div.); children—Carol (Mrs. Charles Uht) (dec.), Doris (Mrs. Richard Campbell), Augustus F., Angela (Mrs. John W. Talbot), Helen (Mrs. William Murray Hawkins, Jr.); m. Marie MacClymont, May 3, 1945 (dec. Nov. 1973). A.B., Columbia U., 1919; B.S., M.I.T., 1921, D.Metall. Engring., 1922; D.Sc., U. Nancy, France, 1933, D.hon. caucsa, 1963; D.Eng., N.Y. U., 1955; D.Sc., Clarkson Coll. Tech., 1957; D.Engr., Rensselaer Inst., 1965, Worcester Poly. Inst., 1965, U. Mich., 1967; LL.D., Queens U., 1966; D.Sc., Northwestern U., 1969, Poly. Inst. N.Y., 1981. Metallurgist Gen. Electric Co., Pittsfield, Mass., 1919-20, 22-23, Henry Disston & Sons, Inc., Phila., 1923-26; lectr., inst. extension courses in advanced metallurgy Temple U., 1925-26; research metallurgist Union Carbide & Carbon Research Labs., Inc., N.Y.C., 1926-28, group leader, 1928-31, chief metallurgist, 1931-45, v.p., 1945-48, pres., 1948-65, v.p., 1944-54; dir. research Union Carbide Corp., 1954-55, v.p., research, 1955-65; pres., chief exec. officer Salk Inst. Biol. Scis., La Jolla, Calif., 1965-67; trustee Systems Devel. Found., Palo Alto, Calif., 1961—, v.p., 1979—; past dir. Menasco Mfg. Co., Sprague Electric Co.; dir. Gen. Am. Investors Co., Inc., N.Y.C.; chief cons. in metallurgy Manhattan Dist. and AEC, 1943-65; chmn. Naval Research Adv. Com., 1954-55, mem., 1951-79; Regent's lectr. U. Calif., San Diego, 1971; chmn. adv. bd. The Energy Center, U. Calif., San Diego, 1976—; Chmn. governing council Courant Inst. Math. Scis., N.Y. U.; trustee Calif. Inst. Tech.; chmn. Engring. Found. Bd., N.Y.C., 1945-48; bd. dirs. Scripps Meml. Hosp., 1971-77; v.p., dir. Berkshire Farm Research Inst., 1950—; Howe Meml. lectr. Am. Inst. Mining and Metall. Engrs., 1952; Sauveur lectr. Am. Soc. Metals, 1952. Sr. author: Alloys of Iron and Chromium, vols. 1 and 2, 1937, 40; mem. editorial adv. bd.: Energy mag, 1977—; contbr. articles on metallurgy and engring. to tech. publs. Recipient Morehead medal Internat. Acetylene Assn., 1955; Stevens Inst. Tech. Powder Metallurgy medal award, 1959; Indsl. Research Inst. medal, 1960; James Douglas gold medal award AIME, 1960; Wisdom award of honor, 1973; Miller medal Am. Welding Soc., 1953. Fellow N.Y. Acad. Scis., Royal Soc. Arts, Am. Inst. Chemists, Metall. Soc., Am. Soc. Metals (Campbell lectr. 1947, Burgess Meml. lectr. 1956, past chmn. N.Y. sect.), Metall. Soc.; mem. Eurospace (hon.), Engrs. Joint Council (hon. pres. 1960-61), Am. Philos. Soc., Soc. Chem. Industry, Nat. Acad. Engring. (founding pres.), Nat. Acad. Scis., AIME (hon. mem., past pres.), Am. Welding Soc. (dir., Adams lectr. 1944), Internat. Inst. Med. Electronics and Biomed. Engring., ASTM, Soc. Automotive Engrs., Am. Iron and Steel Inst., Internat. Inst. Welding (v.p., elector Hall of Fame). Clubs: La Jolla (Calif.); Beach and Tennis; Cosmos (Washington); University Chemists, Racquet and Tennis (N.Y.C.). Subspecialties: Materials (engineering); Bioinstrumentation. Current work: Ceramics, metallurgy, mechanical stress, computerization. Patentee metallurgy and engring. Address: 1738 Castellana Rd La Jolla CA 92037 Early in my career I learned two behavioural guides. First, choose a field of endeavor that you enjoy. If you enjoy your work you will be successful. Second, any transaction must benefit all parties thereto. Otherwise, it just isn't worthwhile.

KINZLY, ROBERT EDWARD, engring. co. exec.; b. North Tonawanda, N.Y., July 4, 1939; s. Robert William and Ruth Elizabeth (Burgin) K.; m. Brenda Lutz, Oct. 12, 1963; children: Michael Robert, Jennifer Ann. B.A., SUNY, Buffalo, 1961; M.S., Cornell U., 1964. Engr. Calspan Corp., Buffalo, 1963-67, sect. head, 1968-69, br. head, 1970-75; pres. SCIPAR, Inc., Buffalo, 1975—; chmn., mem. working group on microdensitometry Am. Nat. Standards Inst. Mem. Optical Soc. Am., Assn. Old Crows, Phi Beta Kappa, Sigma Xi. Republican. Presbyterian. Subspecialties: Optical engineering; Systems engineering. Current work: Optical signatures, camouflage, image processing, micrometrology, computer applications. Patentee in field.

KIRBY, MARGARET LOEWY, anatomist educator; b. Ft. Smith, Ark., June 5, 1946; d. Henry M. and Margaret (Vaile) Loewy; m. Martin Rucks Kirby, May 26, 1971; children: Nathan Martin, Julia Vaile. A.B., Manhattanville Coll., Purchase, N.Y., 1968; Ph.D., U. Ark., 1972; postgrad., U. Chgo., 1975-77. Asst. prof. U. Central Ark., Little Rock, 1973-75; asst. prof. anatomy Med. Coll. Ga., Augusta, 1977-80, asso. prof., 1980—. Contbr. articles to profl. jours. Ga. Heart Assn. grantee, 1978; NIH grantee, 1979, 81. Mem. Soc. Neurosci., Am. Assn. Anatomists, So. Soc. Anatomists, N.Y. Acad. Scis., AAAS. Democrat. Subspecialties: Anatomy and embryology; Neurobiology. Current work: Effect of neurotransmitters on development. Office: Department Anatomy Medical College Georgia Augusta GA 30912

KIRCHNER, LARRY A., mechanical engineer, metallurgical engineer, consultant; b. Detroit, Mar. 4, 1943; s. Ralph T. and Lauraine (Reynolds) K. B.M.E., U. Minn., 1966, B.Met. E., 1970. Registered profl. engr.,. Product design and devel. Twin City Monorail, Mpls., 1965-69; metallurgist, quality assurance, mgr. research and devel. Advanced Materials Tech., St. Paul, 1970; corp. metallurgist Polaris, Mpls., 1972; metallurgist Kawasaki Jet-Ski, Mpls., 1975; facilities engr. Gen. Mills, Mpls., 1976, Magnetic Peripherals, 1978-79; metallurgist Litton Microwave, Mpls., 1983, cons. and contract engring., 1970—; instr. Hennepin County Area Vocat. Tech. Inst., 1973—. Mem. City-Wide Adv. Com. St. Louis Park Community Edn., 1974-80; mem. St. Louis Park Econ. Devel. Commn., 1977-78. Mem. ASME, Am. Soc. Metals. Subspecialties: Mechanical engineering; Metallurgical engineering. Current work: Industrial and consumer products design and development. Created new design for structural member for material handling equipment; created new design for gimbal on rotor aircraft. Home: 1300 Melrose Ave Minneapolis MN 55426

KIRIK, MICHAEL JOHN, mechanical engineer; b. Pitts., Apr. 15, 1938; s. Michael John and Helen Theresa (Stibrik) K.; m. Rosemary Louis, July 5, 1957; children: Michael, Barbara, David, Rosemarie, Theresa, Andrew. B.S.M.E. with honors, Carnegie Inst. Tech., 1967; M.S.M.E., Carneige-Mellon U., 1970. Technician Rockwell Mfg., Pitts., 1963-65, jr. engr., 1965-67, research engr., 1967-74; sr. research engr. Rockwell Internat. Corp., Pitts., 1974-78, supervisory research engr., flow control div., 1978—. Planning commr. Brentwood Borough (Pa.), 1981—. Named Engr. of Yr. Rockwell Internat. Corp., 1980. Mem. ASME. Democrat. Roman Catholic. Subspecialties: Mechanical engineering; Theoretical and applied mechanics. Current work: Development and application of theoretical and experimental methods for static and dynamic analyses of valve assemblies for nuclear and non-nuclear installations. Home: 3005 Steck Way Pittsburgh PA 15227 Office: Rockwell Internat Corp 400 N Lexington Ave Pittsburgh PA 15208

KIRK, BILLY EDWARD, microbiology educator; b. Robinson, Ill., May 5, 1927; s. Virgil and Frances (Apple) K.; m. Grace J. Tucker, Nov 20, 1952 (div. 1981); children: Daniel, Perry. B.S., U. Ill.-Urbana, 1949; M.Sc., Ohio State U., 1955, Ph.D., 1957. Diplomate: in pub. health and med. lab. virology Am. Bd. Med. Microbiology Microbiologist, Pittman-Moore Co., Zionsville, Ind., 1950-53. Microbiologist Eli Lilly Co., Indpls., 1957-62; instr. microbiology U. Mich., 1962-64; assoc. prof. W.Va. U., 1964—. Vol. Mountaineer council Boy Scouts Am., Morgantown, W.Va., 1972—. Served with U.S. Navy, 1945-46. Mem. Am. Soc. Microbiology, Soc. Gen. Microbiology, Am. Soc. Virology, Tissue Culture Assn., Sigma Xi. Republican. Lutheran. Subspecialties: Virology (medicine); Infectious diseases. Current work: Immunology and pathogenesis of persistent viral infections of animals. Home: 1102 Chestnut Hill Apt Morgantown WV 26505 Office: Department of Microbiolog West Virginia Medical Center Morgantown WV 26506

KIRK, JAMES ALLEN, mech. engr., educator; b. Lakewood, Ohio, Nov. 3, 1944; s. Charles James and Helen Sophie (Tulas) K.; m. Cynthia Ambler, Feb. 6, 1976; 1 dau., Heather Elizabeth. B.S.E.E., Ohio U., 1967; M.S.M.E., M.I.T., 1969, Ph.D., 1972. Registered profl. engr., Ohio, Md. Vehicle devel. engr. Ford Motor Co., Dearborn, Mich., 1966-67; research asst. M.I.T., Cambridge, 1967-72; asst. prof. mech. engring. U. Md., College Park, 1972-77, assoc. prof., 1977—; owner, dir. J.A. Kirk Cons. Co.; cons. in accident reconstrn., product liability, failure analysis. Contbr. articles to profl. jours.; author: Mechanical Measurements Laboratory Manual, 1979. Mem. Soc. Automotive Engrs. (Ralph Teetor award 1975), ASME, Am. Soc. for Metals, Am. Soc. for Engring. Edn. (Dow Outstanding Young Faculty award 1977). Subspecialties: Mechanical engineering; Automobile accident reconstruction. Current work: Mathematical techniques for vehicular accident reconstruction and vehicle dynamics (emphasis on programmable calculator and personal computer programs). Machine design and manufacturing processes. Tribology. Home: 7210 Windsor Ln Hyattsville MD 20782 Office: Dept Mech Engring U Md College Park MD 20742

KIRK, JOHN GALLATIN, geodesist; b. Wilmington, Ohio, Oct. 21, 1938; s. Charles Roger and Dorothy Evelyn (Mason) K. B.A., Amherst Coll., 1960; M.A., U. Mich., 1962, Ph.D., 1966. Jr. astronomer Kitt Peak Nat. Obs., Tucson, 1966-69; asst. prof. astronomy U. Toledo, Ohio, 1969-74; staff scientist Computer Scis. Corp., Silver Spring, Md., 1974-79; systems analyst Gen. Dynamics/Electronics, Vandenberg AFB, Calif., 1979-80; mem. profl. staff Geodynamics Corp., Santa Barbara, Calif., 1980—. Recipient of NASA Tech. Achievement award, 1981. Mem. Am. Astron. Soc., Sigma Xi. Subspecialties: Geodesy; Orbital mechanics. Current work: Preparation of mathematical representation of the Earth's gravity field. Home: 325 Palisades Dr Santa Barbara CA 93109 Office: 55 Hitchcock Way Suite 209 Santa Barbara CA 93105

KIRK, ROBERT L., aerospace co. exec.; m. (m). B.S., Purdue U. With Litton Industries, Inc., Washington and, Switzerland, 1959-67, v.p. until, 1967; v.p. def. systems avionics, then v.p. Internat. Tel. & Tel. Corp., N.Y.C., 1967-77; pres., chief exec. officer Vought Corp., Dallas, 1977—; group v.p. LTV Corp. Subspecialty: Aerospace engineering and technology. Office: PO Box 225907 Dallas TX 75265

KIRK, ROBERT WARREN, veterinarian, educator, dermatologist, author, editor; b. Stamford, Conn., May 20, 1922; s. Frank Howard and Edna Evelyn (Higgins) K.; m. Helen Margaret Grandish, Dec. 3, 1949; children: Kathryn J., Barbara A., Janet M. Kirk Hamilton. B.S. with high distinction, U. Conn., 1943; D.V.M., Cornell U., 1946. Diplomate: Am. Coll. Vet. Internal Medicine, Am. Coll. Vet. Dermatology. Pvt. practive vet. medicine, Vt., 1947, Conn., 1948-50; mem. faculty Cornell U., 1952—; prof. medicine N.Y. State Coll. Vet. Medicine, 1958—, chmn, dept. small animal medicine and surgery, 1969-77. Author: (with Muller) Small Animal Dermatology, 3d edit, 1983, (with Bistner) Handbook of Veterinary Procedures and Emergency Treatment, 3d edit, 1981; editor: Current Veterinary Therapy, 8th edit, 1983. Bd. dirs. Seeing Eye, Morristown, N.J., Soc. Prevent Cruelty to Animals; mem. diaconate First Congregational Ch. Served with U.S. Army, 1943-45; to capt. USAF, 1950-52. Named Veterinarian of Yr. in N.Y. State, 1969; NSF fellow, 1967-68; Mark Morris fellow, 1960-61; Williams fellow Sydney U., 1975. Mem. AVMA (asst. chmn, council on edn., Gained medal 1966), N.Y. State Vet. Med. Soc., Am. Animal Hosp. Assn. (Veterinarian of Yr. 1964), Am. Bd. Vet. Practitioners (pres.), Am. Coll. Vet. Dermatology (treas.). Republican. Club: Cornell Golf. Lodge: Masons. Subspecialty: Internal medicine (veterinary medicine). Current work: Pediatric medicine, small animal dermatology, antibiotic therapy, therapeutic nutrition. Home: 84 Turkey Hill Rd Ithaca NY 14850 Office: NY State Coll Vet Medicine VRT 423 Cornell U Ithaca NY 14850

KIRKIEN-RZESZOTARSKI, ALICJA MARIA, chemistry educator; d. Leszek Tadeusz and Francesca Irena (Mortkowicz) Kirkien; m. Zygmunt Marian Konasiewicz, Aug. 18, 1948; m. Waclaw Janusz Rzeszotarski, Dec. 14, 1972. M.Sc. in Chem. Engring, Polish U. Coll., London, 1951; Ph.D. in Phys. Organic Chemistry, Univ. Coll., London, 1955. Asst. prof. chemistry U. W.I., Jamaica, 1956-59, assoc. prof., 1959-61, Trinidad, 1961-65; prof., chmn. dept. chemistry Trinity Coll., Washington, 1965—. Contbr. articles in field to profl. jours. Treas. Polish Veterans Assn., U.S.A., 1979. Served with Polish Underground, WW II. Fellow Royal Inst. Chemistry; mem. Am. Chem. Soc., Chem. Soc. Great Britain, Polish Inst. Art and Scis., Phi Beta Kappa. Republican. Roman Catholic. Subspecialties: Physical chemistry; Kinetics. Current work: Kinetics in solutions, high resolution mass spectrometry, history of science. Home: 4607 Brandywine St NW Washington DC 20016 Office: Michigan Ave Washington DC 20017

KIRKMAN, HADLEY, anatomist, researcher; b. Richmond, Ind., Mar. 14, 1901; s. Madison Lee and Leila Piety (Hadley) K.; m. Gladys L. Tracy, Apr. 5, 1942; 1 dau., Tracy Leigh Kirkman-Liff. A.B., Iowa U., Iowa City, 1925; student, Bradley U., Peoria, Ill, 1923-24; M.S., U. Chgo., 1929; Ph.D., Columbia U., 1937. Acting asst. prof. zoology Ohio U., 1928-29; instr. anatomy N.Y. Med. Coll., 1929-32, Columbia U., 1934-36, Stanford (Calif.) U., 1936-38, asst. prof., 1938-43, assoc. prof., 1943-59, prof., 1949-65, active prof. emeritus, 1965—. Contbr. numerous articles to profl. jours. NSF, sr. fellow, 1957-58; USPHS, spl. fellow, 1958-59; research grantee Am. Cancer Soc., Jane Coffin Childs Meml. Fund, Yale U., USPHS, 1948-76. Fellow N.Y. Acad. Scis., AAAS; mem. Am. Assn. Anatomists, Am. Assn. Cancer Research, AAUP, Gamma Alpha, Sigma Xi. Democrat. Subspecialties: Cancer research (medicine); Cytology and histology. Current work: Endocrine factors in carcinogenesis and cancer cell growth, effect of steroidal hormones on thymus. Home: 623 Cabrillo Ave Stanford CA 94305 Office: Stanford University Room 8 Old Anatomy Bldg CA 94305

KIRKPATRICK, CHARLES HARVEY, clin. immunologist; b. Topeka, Nov. 5, 1931; s. Hazen Leon and Clarice Opal (Privott) K.; m. Janice F. Kirkpatrick, July 11, 1959; children: Heather, Michael, Brian. B.A., U. Kans., 1954, M.D., 1958. Intern U. Ill. Research and Edn. Hosp., Chgo., 1958-59; resident U. Kans. Med. Ctr., Kansas City, 1959-62; asst. prof. medicine U. Kans. Sch. Medicine, Kansas City, 1965-68, assoc. prof., 1968; sr. investigator, head div. clin. allergy and sensitivity Lab. Clin. Investigation, Nat. Inst. Allergy and Infectious Diseases, NIH, Bethesda, Md., 1968-79; head div. clin. immunology Nat. Jewish Hosp. and Research Ctr., Denver, 1979—; prof. medicine U. Colo. Sch. Medicine, 1979—. Contbr. over 100 articles to sci. jours. NIH grantee. Mem. Am. Soc. Clin. Investigation, Am. Acad. Allergy, Am. Fedn. Clin. Research, Western Assn. Physicians, AAAS, Transplant Soc. Episcopalian. Subspecialties: Allergy; Infectious diseases. Current work: Host defense mechanisms and resistance to infectious diseases; pathogenesis of immune deficiency diseases, pathogenesis of interstitial lung disease, transfer factor.

KIRKPATRICK, R. JAMES, geology educator; b. Schenectady, Dec. 31, 1946; s. Robert James and Audrey (Rech) K.; m. Susan Alice Wilson, Sept. 7, 1968; children: Gregory Robert, Geoffrey Stephen. A.B., Cornell U., 1968; Ph.D., U. Ill., 1972. Sr. research geologist Exxon Prodn. Research, Houston, 1972-73; research fellow in geophysics Harvard U., Cambridge, Mass., 1973-75; asst. research geologist Scripps Inst. Oceanography, La Jolla, Calif., 1975-78; asst. prof. geology U. Ill., Urbana, 1978-80, assoc. prof., 1980-83, prof., 1983—. Author, editor: Kinetics of Geochemical Processes, 1981; others. NSF grantee, 1977, 79, 82. Fellow Mineral. Soc. Am.; mem. Am. Geophys. Union. Subspecialties: Petrology; Geochemistry. Current work: Kinetics of nucleation and crystal growth, NMR spectroscopy of solids, melt structure, petrogenesis of basalts, physical processes in magma bodies. Office: Dept Geology 245 Natural History Bldg 1301 W Green St Urbana IL 61801

KIRKPATRICK, RONALD CRECELIUS, scientist; b. San Angelo, Tex., Oct. 4, 1937; s. Frank Brown and Vida (Crecelius) K.; m. Phyllis Abbie Furbeck, May 16, 1964; children: Abbie Elizabeth, Andrew Walter, Ann Marie. B.S.E.E., Tex. A&M U., 1959, M.S. in Physics, 1963; Ph.D. in Astronomy, U. Tex., 1969. Instrumentation research engr. NASA, Ames, Moffett Field, Calif., 1959-61; research engr. Southwest Research Inst., San Antonio, 1962-64; NRC postdoctoral fellow NASA, Goddard Space Flight Center, 1969-71; asst. prof. Tex. A&M U., 1971-72; staff mem. Los Alamos Nat. Lab., 1973—. Contbr. articles to profl. jours. NRC fellow, 1969. Mem. Am. Astron. Soc., Internat. Astron. Union. Baptist. Subspecialties: Theoretical astrophysics; Fusion. Current work: Dynamic models of planetary nebulae, micro-fusion. Office: MSB 220 LANL Los Alamos NM 87545

KIRSCH, DONALD, R., molecular biologist; b. Newark, Apr. 28, 1950; s. Nathaniel and Ruth M. (Wortzel) K.; m. Deana M. Kirsch, June 30, 1974. A.B., Rutgers Coll., 1972; A.M., Princeton U., 1974, Ph.D., 1978. Instr. dept. pharmacology Rutgers Med. Sch., New Burnswick, N.J., 1978-81, adj. asst. prof. dept. biochemistry, 1982—; research investigator Squibb Inst. Med. Research, Princeton, 1981—. Condr. research; contbr. writings to profl. publs. in field. Coll. Medicine and Dentistry N.J. Found. grantee, 1979. Mem. AAAS, Genetics Soc. Am., Am. Soc. for Cell Biology. Subspecialties: Molecular biology; Genetics and genetic engineering (biology). Current work: Application of genetic engineering to the study of antimicrobial agents. Office: Squibb Inst for Med Research PO Box 4000 Princeton NJ 08540

KIRSCH, LAWRENCE EDWARD, physicist, computer scientist; b. Newark, Feb. 24, 1938. A.B., Columbia U., 1960; M.S., Rutgers U., 1962, Ph.D., 1965. Postdoctoral research assoc. Columbia U., 1964-66; asst. prof. computer sci. Brandeis U., Waltham, Mass., 1966-70, assoc. prof., 1971-80, prof., 1981—; dir. Feldberg Computer Ctr., 1970—. Mem. Am. Phys. Soc., Assn. Computing Machinery, IEEE. Subspecialties: Computer architecture; Particle physics. Current work: Experiments in Hyperon spectroscopy; design of digital systems. Office: Brandeis U Felderg Computer Ctr Waltham MA 02154

KIRSCHBAUM, THOMAS HARRY, obstetrician and gynecologist; b. Mpls., Apr. 22, 1929; s. Murray M. and Ella A. (Anderberg) K.; m.; children—Steven, Kristin. B.A., U. Minn., 1949, B.S., 1951, M.D., 1953. Intern U. Minn. Hosps., Mpls., 1953-54; resident in obstetrics and gynecology U. Minn. Hosp., 1956-59; asst. prof. U. Utah Med. Sch., 1959-64; asso. prof., then prof. U. Calif. Med. Sch., Los Angeles, 1964-71; prof. obstetrics and gynecology, chmn. dept. Mich. State U. Med. Sch., East Lansing, 1971—; cons. RAND Corp., 1964-71; spl. expert Reproductive Service Br. Center Population Research NICHD, 1983—. Co-editor: Seminars in Perinatology, 1977—. Served with USNR, 1954-56. Mem. Am. Coll. Obstetricians and Gynecologists, Perinatal Research Soc., Soc. Gynecol. Investigation, Am. Gynecol. Soc., Residency Rev. Com. Obstetrics and Gynecology, Central, Pacific Coast obstet. and gynecol. socs., Phi Beta Kappa, Alpha Omega Alpha. Subspecialty: Obstetrics and gynecology. Home: 1204 Academic Way Haslett MI 48840 Office: NICHD-CPR/RSB Landow Bldg 7910 Woodmont Ave Bethesda MD 20205

KIRSCHENBAUM, DONALD MONROE, biochemistry educator; b. Bklyn., June 14, 1927; s. Nathan and Ethel Adelaide (Rothkopf) K.; m. Roslyn Schwartz, June 17, 1951; children: Lisa, Sheila. B.S., CCNY, 1950; M.S., Columbia U., 1952, Ph.D., 1956. Lectr. chemistry CCNY, 1953-56; instr. biochemistry SUNY Downstate Med. Ctr., Bklyn., 1956-57, asst. prof., 1957-63, assoc. prof., 1963—; cons. in field. Editor, compiler: Atlas of Protein Spectra in Ultraviolet and Visible Regions, vol, 1972, vol. 2, 1974, vol. 3, 1983. Served with U.S. Army, 1946-47. SUNY scholar, 1976. Mem. Am. Soc. Biol. Chemists, Harvey Soc., Sigma Xi (sec. 1973-75, pres. 1975-77). Subspecialties: Biochemistry (medicine); Data compilations. Office: Dept Biochemistry Box 8 Coll Medicine SUNY Downstate Med Center 450 Clarkson Ave Brooklyn NY 11203

KIRSCHENBAUM, LOUIS JEAN, chemist, educator, researcher; b. Washington, Apr. 17, 1943; s. Abraham Isaac and Ruth (Kraut) K.; m. Susan J. Schulman, Aug. 30, 1964; children: Jay, Cynthia. B.S. in Chemistry, Howard U., 1965; M.S., Brandeis U., 1967, Ph.D., 1968. Lectr. in chemistry Brandeis U., 1968-69; Nat. Acad. Scis.-NRC postdoctoral fellow U.S. Naval Ordnance Lab., 1969-70; asst. prof. chemistry U. R.I., 1970-76, assoc. prof., 1976-83, prof., 1983—; vis. prof. Ben Gurion U. of Negev, Beer Sheva, 1978-79. Author: (with E. Grunwald) Introduction to Quantitative Chemical Analysis, 1972. Vice pres. Jewish Fedn. R.I.; mem. Jewish Community Council South County, R.I. Mem. Am. Chem. Soc., Israel Chem. Soc., Phi Beta Kappa, Sigma Xi, Beta Kappa Chi. Club: Tavern Hall (Kingston, R.I.). Subspecialties: Inorganic chemistry; Kinetics. Current work: Preparation, characterization and reactivity of transition metals in unusual oxidation states; kinetics; electrochemistry; electron paramagnetic resonance. Home: 8 South Rd Kingston RI 02881 Office: U RI Dept Chemistry Kingston RI 02881

KIRSCHENBAUM, NEAL, psychologist; b. Bklyn., Feb. 19, 1950; s. Ezra and Judith (Dimont) K.; m. Shanie Regensberg, June 28, 1977; children: Chayim David, Miriam Adina. B.A., Bklyn. Coll., CUNY, 1970, M.A., 1973, Advanced Profl. Cert., 1973; M.S., Yeshiva U., 1977, Ph.D., 1978. Lic. psychologist, Ill.; Lic. sch. psychologist, N.Y. Instr. Downstate Med, Sch., Bklyn., 1974-78; psychologist N.Y.C. Bd. Edn., Bklyn., 1977-78; instr. psychology Bklyn. Coll., CUNY, 1973-78; psychologist, prof. psychology Nat. Coll. Edn., Evanston, Ill., 1978—; pvt. practice clin. psychology Lincolnwood, Ill., 1978—; psychologist Assoc. Talmud Torah, Chgo., 1978-; Hillel Torah Day Sch., Skokie, Ill., 1978—, Arie Crown Day Sch., Skokie, 1979—. Contbr. chpts. to books. Mem. Am. Psychol. Assn., Phi Beta Kappa, Alpha Sigma Lambda. Subspecialties: Developmental psychology; Learning. Current work: Children's recognition of emotion; young children's ability to recognize emotion. Language learning in young children; effects of peer modeling on language. Home: 6524 N Drake St Lincolnwood IL 60645 Office: Nat Coll Edn 2840 Sheridan Rd Evanston IL 60201

KIRSHNER, ROBERT PAUL, astronomer; b. Long Branch, N.J., Aug. 15, 1949; s. D.R. and Virginia Klarman K.; m. Lucy Herman, June 15, 1970; children: Rebecca, Matthew. A.B., Harvard Coll., 1970; Ph.D., Calif. Inst. Tech., 1975. Research assoc. Kitt Peak Nat. Obs., Tucson, 1974-76; asst. prof. dept. astronomy U. Mich., Ann Arbor, 1976-80, assoc. prof., 1980-82, prof., 1982—, chmn. dept. astronomy, 1982—; dir. McGraw-Hill Obs., 1980—. Contbr. numerous articles to sci. jours. Recipient Harvard Coll. Bowdoin prize, 1970; NSF fellow, 1970; Alfred P. Sloan Found. research fellow, 1978; Mich. Soc. Fellows sr. fellow, 1980. Mem. Am. Astron. Soc., Astron. Soc. Pacific, Internat. Astron. Union. Subspecialties: Optical astronomy; Cosmology. Current work: Supernovae and the origin of the elements, supernova remnants; galaxy dynamics, large-scale structure in the universe. Home: 1210 W Liberty St Ann Arbor MI 48103 Office: Dept Astronom U Mich 834 Dennison Ann Arbor MI 48109

KIRSNER, JOSEPH BARNETT, physician, educator; b. Boston, Sept. 21, 1909; s. Harris and Ida (Waiser) K.; m. Minnie Schneider, Jan. 6, 1934; 1 son, Robert S. M.D., Tufts U., 1933; Ph.D. in Biol. Scis, U. Chgo., 1942. Intern Woodlawn Hosp., Chgo., 1933-34, resident, 1934-35; asst. in medicine U. Chgo., 1935-37, mem. faculty, 1937—, asso prof., 1946-51, prof., 1951—, Louis Block Distinguished Service prof. medicine, 1968—, chief of staff, also dep. dean for med. affairs, 1970-76; Cons. NIH, 1956-69; hon. pres. Gastrointestinal Research Found., 1961—; Mem. drug efficacy adv. com. to NRC; chmn. adv. group Nat. Commn. on Digestive Diseases, 1978; chmn. emeritus sci. adv. com. Nat. Found. Ileitis and Colitis. Author: also 650 articles. Gastrointestinal Exfoliative Cytology; Editorial bd.: Médecine et Chirurgies Digestives. Served with M.C. AUS, 1943-46; ETO; PTO. Recipient Julius Friedenwald medal disting. work gastroenterology, 1975; Horatio Alger award, 1979; hon. Gold Key for Disting. Service U. Chgo. Med. Alumni Assn., 1979. Mem. Am. Assn. Physicians, A.C.P. (master), Am. Gastroenterol. Assn. (past pres., governing bd.), Am. Gastroscopic Soc. (past pres.), Am. Soc. Clin. Investigation, Central Soc. Clin. Research, Chgo. Soc. Internal Medicine (past pres.), Inst. Medicine Chgo. Subspecialties: Gastroenterology; Internal medicine. Current work: All aspects of inflammatory bowel disease (ulcerative colitis and Crohn's disease). Research in gastrointestinal disorders, inflammatory disease of gastrointestinal tract. Home: 5805 Dorchester Ave Chicago IL 60637 We need a return to higher standards, and not to equate opportunity with a decline in quality. Striving for personal excellence and achievement is the best approach to the attainment of universal excellence and peace.

KIRSTEN, WERNER HEINRICH, pathologist, educator; b. Leipzig, Germany, Oct. 29, 1925; came to U.S., 1955, naturalized, 1960; m. Inger Nielsen, May 20, 1960; children: Christian, Olaf, Thomas. Student, Med. Sch., U. Halle, Germany, 1947-51, Med. Sch., Free U. Berlin, 1951-52; M.D., U. Frankfurt/Main, Germany, 1953. Asst. in virology Paul-Ehrlich Inst., U. Frankfurt/Main, 1953-54, resident in pathology, 1954-55; intern Englewood Hosp., Chgo., 1955-56; resident in pathology U. Chgo., 1956-57, asst. dept. pathology, 1957-58, asst. prof. pathology, 1961-64, assoc. prof. pathology, 1964-68, assoc. prof. pathology and pediatrics, 1966-68, prof. pathology and pediatrics, 1968—, chmn. dept. pathology, 1972—; cons. Argonne Nat. Lab., 1972-80; chmn. cancer spl. programs adv. com. NIH; pres. Ill. div. Am. Cancer Soc., 1978—. Editor: Malignant Transformation by Viruses, vol. 14 of Recent Results in Cancer Research series, 1966, Normal and Malignant Cell Growth, vol. 17, 1966, Current Concepts in the Management of Leukemia and Lymphoma, vol. 36, 1971; editor, translator: Macropathology (W. Sandritter et al), 1973, 78; editor: Current Topics in Pathology, 1967-71; co-editor, 1971—; editor: Cancer Research, 1970-71; assoc. editor, 1971-78. Recipient Career Devel. award USPHS, 1960-68, Nat. Cancer Inst., NIH, 1961-71. Mem. Am. Assn. Pathology and Bacteriology, Am. Assn. Exptl. Pathology, AAAS, Am. Assn. Cancer Research, Midwestern Soc. Pediatric Research, Internat. Soc. Comparative Leukemia Research (dir. 1967-70), Damon Runyon Fund for Cancer Research (dir. 1977—), Univs. Associated for Research and Edn. in Pathology (dir. 1977—), Assn. Pathology Chairmen (councillor 1978—, pres. 1982-83). Office: 950 E 59th St Chicago IL 60637

KISER, DONALD OWEN, utilities co. exec, engineer; b. Woodbury, N.J.; s. Owen E. and Mildred (Brungard) K.; m. Hazel M. Kuhlthau, Mar. 25, 1949; children: Beverley Kiser Maul, Phyllis, John. B.S. in Elec. Engring, Rutgers U., 1952, M.S., Harvard U., 1953. Vice pres., gen. mgr. Western div. GTE-SSG, Mountain View, Calif., 1972-78; sr. v.p. Sylvania Systems Group, Waltham, Mass., 1978-80, sr. v.p. human relations, Stamford, Conn., 1980-81; pres. GTE Lenkurt (subs.), San Carlos, Calif., 1981-82; v.p. engring. GTE Corp., 1982—; pres. GTE Labs., Waltham, 1982—. Exec. v.p. United Way of Santa Clara County, 1973-78; chmn., bd. dirs. Jr. Achievement, 1973-78. Served with U.S. Army, 1946-48. Mem. IEEE, Tau Beta Pi, Eta Kappa Nu. Subspecialties: Electronics; Materials (engineering). Current work: Central research program in telecommunications, computer science, electronics, integrated circuits, fiber optics, ceramics, systems engrineering. Office: 40 Sylvan Rd Waltham MA 02254

KISHIMOTO, YASUO, biochemical researcher, educator, administrator; b. Osaka, Japan, Apr. 11, 1925; came to U.S., 1962; s. Yasuichi and Chiyono (Sugimura) K.; m. Miyoko Nishikawa, June 12, 1926; children: Tsutomu, Yoriko Kisimoto Collins, Takashi, Momoko Anne. B.S., Kyoto U., 1948, Ph.D., 1956. Assoc. research biochemist U. Mich., Ann Arbor, 1962-67; sr. investigator G. D. Searle & Co., Chgo., 1967-69, Eunice Kennedy Shriver Center, Waltham, Mass., 1969-76; assoc. biochemist Mass. Gen. Hosp., Boston, 1969-76; dir. biochem. research John F. Kennedy Inst., Balt., 1976—; assoc. prof. Johns Hopkins U. Sch. Medicine, Balt., 1976-80, prof., 1980—; cons. NIH study sect., 1981-82; reviewer NSF; vis. prof. Japan Soc. Advancement of Sci., 1977. Author: Methods in Neurochemistry, vol. 3, 1973, Research Methods in Neurochemistry, vol. 4, 1978, Handbook of Neurochemistry, vols. 3 and 7, 1983, The Enzymes, vol. 16, 1983; contbr. articles to profl. jours. (Moore award 1975). Research grantee NIH, NSF. Mem. Am. Soc. Biol. Chemists, Am. Soc. Neurochemistry, Internat. Soc. Neurochemistry, AAAS, Japanese Biochem. Soc. Subspecialties: Neurochemistry; Biochemistry (medicine). Current work: Brain development, mental retardation, myelin synthesis and degeneration, brain glyolipid synthesis, brain fatty acid synthesis and oxidation. Home: 4506 Roland Ave Baltimore MD 21210 Office: John F Kennedy Inst and Johns Hopkins Univ 707 N Broadway Baltimore MD 21205

KISLIUK, ROY LOUIS, biochemist, educator; b. Phila., Aug. 4, 1928; s. Max and Sue (Pogust) K; m. Ingrid Scheer, Nov. 25, 1954; children: Claudette, Michelle. B.S., Queens Coll., 1950; M.S., Yale U., 1952; Ph.D., Western Res. U., 1956. Postdoctoral fellow Oxford (Eng.) U., 1956-58; vis. scientist Nat. Inst. Arthritis and Metabolic Diseases, Bethesda, Md., 1958-60; asst. prof. pharmacology Sch. Medicine, Tufts U., Boston, 1960-64, asso. prof., 1964-72, prof. biochemistry, 1972—; program dir. for biochemistry NSF, Washington, 1972-73. Contbr. over 80 articles on anticancer agts. and folic acid metabolism to profl. jours. NSF and Nat. Cancer Inst. research grantee, 1960—. Mem. Am. Soc. Biol. Chemists, Am. Soc. Pharmacology and Exptl. Therapeutics, Am. Soc. for Cancer Research, Am. Chem. Soc., Am. Soc. for Microbiology, Common Cause, ACLU. Democrat. Jewish. Subspecialties: Biochemistry (biology); Cancer research (medicine). Current work: Folate enzymes, coenzymes and antimetabolites. Home: 65 Grasmere St Newton MA 02158 Office: 136 Harrison Ave Boston MA 02111

KISSEL, STANLEY J., child psychologist; b. Bklyn., Nov. 6, 1934; s. Nathan and Sylvia (Deutsch) K.; m. Pearl Kraus, Oct. 9, 1960; children: Steven, Neal, Elizabeth. B.B.A., CCNY, 1956; M.A., Lehigh U., 1958; Ph.D., U. Buffalo, 1961. Diplomate: Am. Bd. Profl. Psychology. Sr. clin. psychologist Rochester (N.Y.) Psychiatric Center, 1961-63; sr. clin. psychologist children and youth div. Rochester Mental Health Center, 1961-69, dir. research, 1964-69, chief psychologist, 1969—; clin. asst. U. Rochester, 1969-80; cons. dept. social services, Wayne County, 1962-64. Author: Human Resources Troubled Children, 1976, Teachers Guide Understanding, 1978, Eclectic Family Therapy, 1980, Independent Practice for Mental Health Clinician, 1983. Fellow Am. Orthopsychiat. Assn.; mem. Am. Psychol. Assn., Genesee Valley Psychol. Assn. (pres. 1976-78). Democrat. Jewish. Lodge: B'nai B'rith. Current work: Effectives of psychotherapy with troubled children, depression in children, communication in families. Office: Rochester Mental Health Center 1425 Portland Ave Rochester NY 14621

KISTNER, DAVID HAROLD, biology educator; b. Cin., July 30, 1931; s. Harold Adolf and Hilda (Gick) K.; m. Alzada A. Carlisle, Aug. 8, 1957; children—Alzada H., Kymry Marie Carlisle. A.B., U. Chgo., 1952, B.S., 1956, Ph.D., 1957. Instr. U. Rochester, 1957-59; instr., asst. prof. Calif. State U., Chico, 1959-64, asso. prof. biology, 1964-67, prof., 1967—; research asso. Field Mus. Natural History, 1967—, Atlantica Ecol. Research Sta., Salisbury, Zimbabwe, 1970—; dir. Shinner Inst. Study Interrelated Insects, 1968-75. Author: (with others) Social Insects, Vols. 1-3; editor: Sociobiology, 1975—; contbr. articles to profl. jours. Recipient Outstanding Prof. award Calif. State Univs. and Colls., Los Angeles, 1976; John Simon Guggenheim Meml. Found. fellow, 1965-66; NSF grantee, 1960—; Am. Philos. Soc. grantee, 1972. Fellow Explorers Club; mem. Entomol. Soc. Am., Pacific Coast Entomol. Soc., Kans. Entomol. Soc., AAUP, AAAS, Soc. Study of Evolution, Am. Soc. Naturalists, Am. Soc. Zoologists, Soc. Study of Systematic Zoology, Chico State Coll. Assos. (charter), Council Biology Editors. Subspecialties: Ecology; Sociobiology. Current work: Interactions of myrmecophiles and termitophiles with their social insects hosts. Origin of social parasitism. Evolution of social parasites coevolution of foreign insects in social insects societies. Research trips to Africa, Orient, Europe, S. Am., Australia, China. Home: 3 Canterbury Circle Chico CA 95926

KITCHENS, CLARENCE WESLEY, JR., government engineering manager, researcher; b. Panama City, Fla., Nov. 8, 1943; s. Clarence Wesley and Voncile (Rudolph) K.; m. Terry Lee Worsley, Dec. 26, 1966; children: Kathy Lee, Mark Wesley. B.S., Va. Poly Inst. and State U., 1966, M.S., 1968; Ph.D., N.C. State U., 1970. Registered profl. engr., Md. Instr. N.C. State U., 1968-70; aerospace engr. U.S. Army Ballistic Research Lab., Aberdeen Proving Ground, Md., 1972-77, asst. to dir, 1977-78, team leader, 1978-80, chief blast dynamics br., 1980-83, Fellow, 1980-83, chief penetration mechs. br., 1983—; bd. dirs. Aberdeen Proving Ground Fed. Credit Union, 1979—. Served to capt. U.S. Army, 1970-72. NASA trainee, 1966-67; Ford Found. fellow, 1967-69. Fellow AIAA (assoc.). Democrat. Methodist. Subspecialties: Fluid mechanics; Applied mathematics. Current work: Planning and directing research in high velocity impact dynamics, penetration mechanics and kinetic energy penetrator development in close cooperation with outside funding agencies. Inventor accelerometer, 1966. Home: 2224 Rosewood Dr Edgewood MD 21040 Office: US Army Ballistic Research Lab Terminal Ballistics Div Aberdeen Proving Ground MD 21005

KITE, JOSEPH HIRAM, JR., microbiologist, educator, researcher; b. Decatur, Ga., Nov. 11, 1926; s. Joseph Hiram and Lulie (Hatch) K.; m. Jane Pascale, Aug. 6, 1970. A.B., Emory U., 1948; M.S., U. Tenn., 1954; Ph.D., U. Mich., 1959. Med. technician in bacteriology Communicable Disease Ctr., Atlanta, 1950-51, VA Hosp., 1951-52; research assoc. U. Buffalo, 1958-59, instr., 1959-63; asst. prof. bacteriology and immunology SUNY-Buffalo, 1963-68, assoc. prof. microbiology, 1968-72, prof. microbiology, 1972—. Contbr. articles to med. jours., chpts. to med. textbooks. Served with AUS, 1945-46. Mem. Am. Assn. Immunologists, Am. Soc. Microbiology, Tissue Culture Assn., AAAS, N.Y. Acad. Scis. Methodist. Subspecialties: Immunology (medicine); Microbiology (medicine). Current work: Autoimmune diseases; teaching medical, dental and graduate students; research in mechanisms of autoimmune disease and regulation of immune response. Home: 108 Chasewood Ln East Amherst NY 14051 Office: Dept Microbiology Med Sch SUNY Buffalo NY 14214

KITHIER, KAREL, physician, researcher; b. Prague, Czechoslovakia, Dec. 6, 1930; s. Karel and Marie (Bohackova) K.; m. Viktorie Svecova, May 6, 1961; 1 son: Karel. M.D., Charles U., Prague, 1962, Ph.D., 1967. Diplomate: Med. diplomate Charles U., Prague, 1962. Resident in pediatrics J. Hradec, Czechoslovakia, 1962-64; research scientist Research Inst. Child Devel., Prague, 1967-68; research assoc. Child Research Ctr. of Mich., Detroit, 1968-71; research scientist Mich. Cancer Found., Detroit, 1972-74; assoc. prof. Wayne State U. Sch. Medicine, 1974—, assoc. prof. pathology, 1978—. Contbr. articles to profl. jours. Served with Czechoslovakian Army, 1951-53. Mem. Am. Assn. Cancer Research, Am. Assn. Immunologists, Am. Assn. Clin. Chemistry, N.Y. Acad. Scis. Subspecialties: Cancer research (medicine); Pathology (medicine). Current work: Pathology and clinical aspects of proteins, tumor-associated antigens, tumor markers, specific fetal proteins, oncofetal antigens. Office: 540 E Canfield St Room 9231 Detroit MI 48201

KITTEL, JOHN HOWARD, metallurgical engineer; b. Ritzville, Wash., Oct. 9, 1919; s. John Charles and Mary Irene (Wood) K.; m. Betty Anne Quackenbush, May 23, 1943; children: Loren Howard, Marian Elizabeth, Alan William, Jack Meredith. B.S. in Metall. Engring, Wash. State U., 1943. Aero. scientist NASA Lewis Lab., Cleve., 1943-51; sr. scientist Argonne (Ill.) Nat. Lab., 1951—; ofcl. U.S. del. UN Internat. Conf. on Peaceful Uess of Atomic Energy, Geneva, Switzerland, 1958, 64. Contbr. numerous articles on effects of irradiation on reactor materials to profl. jours. Served with USAFR, 1944-45. Recipient award for best tech. paper Inst. Environ. Scis., 1981. Fellow Am. Nuclear Soc. (cert. of merit 1965, dir. 1967-70, outstanding achievement award 1979); mem. Scientist for Engrs. for Secure Energy, Sigma Xi. Subspecialties: Nuclear fission; Metallurgical engineering. Current work: Management of nuclear waste management research and development.programs. Patentee in field. Home: 456 S Julian St Naperville IL 60540 Office: Argonne Nat Lab 9700 S Cass Ave Argonne IL 60439

KITTO, GEORGE BARRIE, biochemistry educator, researcher; b. Wellington, N.Z., July 31, 1937; came to U.S., 1962; s. George Herbert and Acushla Meta (Benjamin) K.; m. Mary Berenice Scully, June 16, 1962; children: David John, Robyn Anne, John Michael. B.S., Victoria U., 1961, M.S. with 1st class honors, 1962; Ph.D., Brandeis U., 1966. Asst. prof. U. Tex.-Austin, 1966-70, assoc. prof., 1970-77, prof., 1977—, grad. adv., 1982-84; cons. Swiss Fed. Research Inst., 1978-79, Greek Govt.-UN, 1978—. Faculty adv. YMCA, Austin, 1980—; bd. dirs. Environ. Resource and Info. Center, Austin, 1974; cons. KPFT-Pacifica Radio, Houston, 1975-77. Recipient Teaching Excellence award U. Tex., 1981; Durkee fellow, 1962-66. Fellow Chem. Soc. London; mem. Royal Soc. N.Z., AAAS, Internat. Oceanographic Found., Am. Chem. Soc., Am. Soc. Biol. Chemists, S.W. Speakers Exchange (U. Tex.). Subspecialty: Biochemistry (medicine). Current work: Biochemical and immunological study of the Tephritidae immobilized enzymes and treatment of PKU, biochemical genetics of Cultex pipiens, invertebrate hemoglobins. Office: U Tex Dept Chemistry Austin TX 78712

KIVIAT, ERIK, ecology researcher; b. N.Y.C., June 9, 1947; s. Charles and Esther (Lobensky) K.; m. Elaine M. Colandrea, June 19, 1982. B.S., Bard Coll., Annandale, N.Y., 1976; M.A., SUNY-New Paltz, 1979; postgrad., Union Experimenting Colls. and Univs., Cin., 1982—. Adj. Bard Coll., 1970-73, instr., 1973-75, asst. prof., 1975-78, dir. field sta., 1972-78, research assoc., 1978—; ecologist, bd. dirs. Hudsonia Ltd., 1981—; mem. Hudson River Valley Study Adv. Com., N.Y., 1978, Hudson River Fisheries Adv. Com., 1979-83, Estuarine Sactuary Adv. Com. N.Y., 1983—; cons. in field. Author: Hudson River East Bank Natural Areas, 1978. Mem. Ecol. Soc. Am., Soc. Wetland Scientists, Estuarine Research Fedn., Am. Ornithologists Union Torrey Bot. Club, Soc. Study Amphibians and Reptiles. Subspecialties: Ecology; Resource management. Current work: Founder-administrator, Hudsonia, alternative institute for environmental research, education, technology; research wetland ecology; nature reserve design, management; rare plant, animal conservation; recycling. Home: PO Barrytown NY 12507 Office: Hudsonia Limited Bard College Annandale NY 12504

KJAER-PEDERSEN, NIELS, nuclear energy company scientist; b. Copenhagen, June 13, 1937; U.S., 1979; s. Peter Axel and Gudrun (De Wolff) K.P.; m. Dorit Loegstrup Jensen, Dec. 8, 1962; children: Klaus, Line. M.Sc., Tech. U. Denmark, Riso Nat. Lab., Roskilde, Denmark, 1969-72, Elsinore (Denmark) Shipyard, 1972-79. Staff scientist Exxon Nuclear Co., Richland, Wash., 1979—. Mem. Am. Nuclear Soc. Subspecialties: Numerical analysis; Nuclear fission. Current work: Nuclear fuel performance modeling, systems simulation, applied calculus, data reduction and interpretation. Creator computer code WAFER. Home: 2200 Davison Ave Richland WA 99352 Office: Exxon Nuclear Co 2101 Horn Rapids Rd Richland WA 99352

KLAASSEN, CURTIS DEAN, pharmacology, educator; b. Ft. Dodge, Iowa, Nov. 23, 1942; s. Henry H. and Luwene S. (Nieman) K.; m. Cherry Jo Eichner, Sept. 30, 1964; children: Kimberly, Lisa. B.A. magna cum laude, Wartburg Coll., 1964; M.S., U. Iowa, 1966, Ph.D., 1968. Instr. U. Kans. Med. Ctr., Kansas City, 1968-70, asst. prof., 1970-74, assoc. prof, 1974-77, prof. pharmacology and toxicology, 1977—. USPHS fellow, 1965-68; von Humboldt fellow, 1978; Burroughs-Wellcome scholar, 1982—. Mem. Soc. Toxicology (Achievement award 1976), Am. Soc. Pharmacology and Exptl. Therapeutics. Lutheran. Subspecialties: Toxicology (medicine); Pharmacology. Current work: Hepatobiliary disposition of xenobiotics; toxicity of cadmium. Home: 10504 W 102 Terr Overland Park KS 66214 Office: U Kans Med Ctr Kansas City KS 66103

KLABOSH, CHARLES, research and devel. co. exec., abd cons.; b. Manchester, Conn., Mar. 11, 1920; s. Kasimeraz and Veronica (Christana) Klebojas; m. Lois Thaney Heiser, Oct. 6, 1958. B.S. in Aero. Engring, U. Mich., 1953; postgrad., U. Houston, 1962-68. Sales engr. Pratt & Whitney Aircraft Co., East Hartford, Conn., 1941-52; civilian mgr. flight test program U.S. Air Force, Wright-Patterson AFB, Ohio, 1953-55; sr. design engr. overall nuclear powerplant installations Gen. Elec. Co., Evendale, Ohio, 1955-59, 61, sr. devel. engr. gas turbine power plants Am. Airlines, Tulsa, 1960; program plans engr. advance systems NASA, Houston, 1961-66, devel. engr. lunar ops., 1967-74; owner, operator Charles Klabosh Co., Jacksonville, Fla., 1974—; cons. Served with USAAF, 1943-46; Served with Air N.G., 1946-47. Recipient numerous awards NASA, 1969-73. Mem. AIAA, AAAS, Am. Assn. Small Research Cos. Subspecialties: Aerospace engineering and technology; Artificial intelligence. Current work: Robotics, optical image, signals processing, oceanography, offshore tech., phys. chemistry, combustion processes. Inventor remote manipulator system for NASA space shuttle. Home: 4320 Gadsden Ct Jacksonville FL 32207 Office: Charles Klabosh Co 815 S Main Suite 309 Jacksonville FL 32207

KLAPPER, DAVID GARY, immunologist, medical researcher; b. N.Y.C., Mar. 15, 1944; s. Arnold and Edith (Edelstein) K.; m. Susan Rae Owens, June 18, 1972; children: Joshua Michael, Ashley Carolyn. B.S., Tulane U., 1965; M.S., U. Fla., 1970, Ph.D., 1972. Research assoc. Rockefeller U., 1972-74; instr. U. Tex. Health Sci. Ctr., Dallas, 1974-76, asst. prof., 1976-77; asst. prof. dept. microbiology and immunology U. N.C., Chapel Hill, 1983—, asst. prof., 1983—. Contbr. articles to profl. jours. NIH Am. Cancer Soc. grantee, 1977—. Mem. Am. Assn. Immunologists, Harvey Soc., Southeastern Immunology Conf. (pres. 1980), Sigma Xi, Kappa Delta Phi. Democrat. Jewish. Subspecialties: Immunology (medicine); Allergy. Current work: Structure and function of allergens and immunoglobulins; automated protein sequence analysis, hybridoma antibodies, antigenic and allergenic sites on ragweed and grass allergens. Home: 501 Sharon Rd Chapel Hill NC 27514 Office: 804 FLOB 231-H Chapel Hill NC 27514

KLAR, AMAR JIT SINGH, molecular biologist; b. India, Apr. 1, 1947; came to U.S., 1969, naturalized, 1979; s. Kehar S. and Dalip K. K.; m. Kuljit K. Grewal, June 6, 1980. Ph.D., U. Wis., 1975. Staff investigator Col Spring Harbor (N.Y.) Lab., 1978-80, sr. staff investigator, 1980-82, sr. staff scientist, 1982—; ad hoc mem. genetics study sect. NIH, 1982—. Contbr. articles to sci. jours. NIH grantee, 1978—. Mem. Genetics Soc. Am. Subspecialty: Genetics and genetic engineering (agriculture). Current work: Molecular biology; genetics and molecular approaches to study development and differentiation of cell type. Home and Office: Cold Spring Harbor Lab Cold Spring Harbor NY 11724

KLARREICH, SAMUEL HENRY, psychologist; b. Nuremburg, Ger., Oct. 10, 1947; emigrated to Can., 1949; s. Josef and Regina (reiter) K.; m. Penny Ruth Schwartz, June 8, 1969. B.A., U. toronto, 1970, M.A., 1972, Ph.D., 1975. Registered psychologist, Ont. Chief psychologist Scarborough Centenary Hosp., Toronto, 1974-80; lectr. U. Toronto, 1973-77, supr. doctoral students, 1976-80; cons. bus. and industry, Toronto, 1974-80; instr. Ont. Soc. Clin. Hypnosis, 1977-80; dir. employee assistance program Imperial Oil Ltd., Toronto, 1980—; cons. Ont. Probation Services, 1973-75. Author: Teaching Interpersonal Skills, 1975, The Employee Assistance Program and the Supervisor, 1982. Bd. dir., mem. profl. adv. bd. Freedom from Fear Found., Toronto, 1982—; bd. dirs. East Scarborough Boys Club, Toronto, 1975-76; mem. Scarborough Agy. Fedn., 1979-80. Ministry of Correctional Services grantee, 1973. Mem. Am. Psychol. Assn., Can. Psychol. Assn., Am. Orthopsychiat. Assn., Assn. for Advancement Behavioral Therapy, Am. Soc. Clin. Hypnosis, Ont. Psychol. Assn. (chmn. benefits com. 1975-78, dir. div. pvt. practive 1977-79). Subspecialties: Behavioral psychology; Cognition. Current work: Cognitive behavioral therapy in industry; cost effectiveness in using short-term psychotherapy in industry; employee productivity as influenced by counselling in industry. Home: 370 Hounslow Ave Willowdale ON TorontoCanada M2R 1H6 Office: Imperial Oil Ltd 111 St Clair Ave Toronto ON Canada M5W 1K3

KLASNER, JOHN SAMUEL, geology and geophysics educator; b. Flint, Mich., June 22, 1935; s. Henry and Ina (Clemens) K.; m. Gretchen Mary, July 17; children: Christopher, Frederick, Laura, Paul. B.S., Mich. State U., 1957, M.S., 1964; Ph.D., Mich. Tech. U., 1972. Geophys. engr. Geophys. Service, Inc., Dallas, 1958-62; geophysicist Standard Oil Co. of Calif., Western Ops. Inc., San Francisco, 1964-69; prof. geology and geophysics Western Ill. U., Macomb, 1972—; part time research positions with U.S. Geol. Survey. Contbr. articles on geology and geophysics to profl. jours.; contbr. geologic maps of spl. interest gravity anomaly maps. Served with U.S. Army, 1958-60. Mem. AAAS, Geol. Assn. Can., Geol. Soc. Am., Soc. Exploration Geophysics, Soc. Econ. Geologists. Republican. Lutheran. Club: Rotary of Macomb. Subspecialties: Geology; Geophysics. Current work: Geologic and geophysical studies in midcontinent United States directed toward assessment of economic mineral potential, regional geophysics studies of precambrian, geologic and geochemical studies of Rare II areas, Southern Illinois, geophysical and geochemical studies of archaeologic features in western Illinois. Office: Dept Geology Western Ill U Macomb IL 61455

KLAUSNER, STEVEN CHARLES, cardiologist, educator; b. N.Y.C., Dec. 19, 1941; s. Solomon and Rose (Cohen) K.; m. Jill Ann Hekelburg, Aug. 14, 1965; 1 son, Joshua Charles. B.S. in Engring., Princeton U., 1963; M.D., NYU, 1968. Diplomate: Am. Bd. Internal Medicine. Intern U. Chgo. Hosps., 1968-69, resident, 1969-70, Stanford U., 1972-73; staff cardiologist Latter-day Saints Hosp., Salt Lake City, 1976-81; chief cardiologist sect. Martinez VA Hosp., (Calif.), 1981—; asst. prof. medicine U. Utah, Salt Lake City, 1976-81; assoc. prof. U. Calif.-Davis, 1981—. Contbr. articles to profl. jours. Served to maj. AUS, 1970-72. Fellow Am. Heart Assn. (fellow council on clin. cardiology); mem. ACP, Alpha Omega Alpha. Subspecialties: Cardiology; Internal medicine. Current work: Ventricular function; cardiac imaging. Office: Martinez VA Med Center 150 Muir Rd Martinez CA 04553

KLAVINS, JANIS VILIBERTS, pathology educator; b. Rugaji, Latvia, May 6, 1921; came to U.S., 1951, naturalized, 1957; s. Janis and Ida Aline (Liepins) K.; m. Ilga Minjona Krumins, July 4, 1950; children: Ilze Mara, Lize Kristine, Janis Peteris, Filips Klavs. M.D., U. Kiel, Germany, 1948, Ph.D., 1959; diploma, Music Acad., Lubeck, Germany, 1959; diploma Music and Sci. (hon.), Latvian Cultural Found, N.Y.C., 1974. Fellow Cleve. Met. Gen. Hosp., 1952-57; assoc. prof. pathology Duke U., Durham, N.C., 1960-63, prof., 1963-65, dir., 1962-65; clin. prof. pathology Columbia U., N.Y.C., 1969-71, SUNY-Bklyn., 1965-71, prof. pathology, Stoney Brook, 1971—; pathologist-in-chief Bklyn.-Cumberland Med. Ctr., 1965-70; dir. dept. labs. Queens Med. Ctr., Jamaica, N.Y., 1970-77; chmn. pathology dept. Catholic Med. Ctr., Jamaica, 1977—. Editorial bd.: Annals Clin. & Lab. Sci, 1975—; contbr. numerous sci. articles to profl. publs. Bd. dirs. N.C. State Ballet Co., Raleigh, 1963, Symphony of U.S., N.Y.C., 1975, Flushing Meadows (N.Y.)-Corona Park Community Devel. Corp., 1975, Fedicheva Ballet Co., Glen Cove, N.Y., 1982. Recipient Honor award in Sci. Latvian Cultural Found, 1976, Honor award in Music, 1978; Singer of Yr. award Beaux Arts, Inc., N.Y.C., 1980; Scientist of Year Assn. Clin. Scientists, 1983. Fellow Assn. Clin. Scientists (pres. 1980), AAAS; mem. Soc. Urban Physicians (sec.-treas. 1970-75). Clubs: Town (Scarsdale) (edn. com. 1976-78); Fraternitas Livonica (vice sr. 1976-77). Subspecialties: Pathology (medicine); Nutrition (medicine). Current work: Biological evolution; cancer research, specifically tumor markers; ferritin synthesis; pathology of amino acid excess. Home: 5 Broadmoor Rd Scarsdale NY 10583

KLEBANOFF, PHILIP SAMUEL, scientist, researcher; b. N.Y.C., July 21, 1918; s. Morris and Celia (Solowey) K.; m. Angelyn Calvo, Dec. 23, 1950; children: Steven Michael, Susan Marian, Leonard Elliott. B.S., Bklyn. Coll., 1939; postgrad., George Washington U., 1942-43, 44-45; Dr. Engring., Hokkaido U., Sapporo, Japan, 1979. Mem. staff Nat. Bur. Standards, Gaithersburg, Md., 1941-83, asst. chief mechanics div., 1969-75, chief aerodynamics sect., 1969-75, asst. chief mechanics div., 1975-78, chief fluid mechanics sect., 1975-78, sr. scientist, 1978-83, cons., 1983—; mem. NRC/NAS U.S. Nat. Commn. Theoretical and Applied Mechanics, 1970-74; mem. indsl. profl. adv. com. Dept. Aerospace Engring. Pa. State U., 1970-75. Bd. editors: Physics of Fluids, 1970-73; contbr. articles to profl. jours. Recipient Ordnance Devel. award U.S. Navy, 1945; cert. of commendation Nat. Bur. Standards, 1968; Gold medal U.S. Dept. Commerce, Washington, 1975. Fellow AIAA, Washington Acad. Sci., Am. Phys. Soc. (chmn, exec. com. div. fluid dynamics 1969, vice chmn. 1968, recipient fluid dynamics prize 1981); mem. Philos. Soc. Washington, AAAS, Nat. Acad. Engring. Subspecialty: Fluid mechanics. Current work: Research in turbulence, boundary layers flows, flow stability and fluid mechanical measurements. Home: 6412 Tone Dr Bethesda MD 20817

KLEBANOFF, SEYMOUR JOSEPH, medical researcher, educator, physician; b. Toronto, Ont., Can., Feb. 3, 1927; came to U.S., 1957, naturalized, 1977; s. Eli Samuel and Anne (Solway) K.; m. Evelyn Norma Silver, June 3, 1951; children: Carolyn, Mark. M.D., U. Toronto, 1951; Ph.D. in Biochemistry, U. London, 1954. Intern Toronto Gen. Hosp., 1951-52; asst. prof. Rockefeller U., 1959-62; assoc. prof. medicine U. Wash., Seattle, 1962-68, prof., 1968—, head div. allergy and infectious diseases, 1976—. Author: (with R.A. Clark) The Neutrophil: Function and Clinical Disorders, 1978; contbr. numerous articles on leukocyte function to med. jours. Subspecialties: Infectious diseases; Cell biology (medicine). Current work: Mechanisms of host defense against infectious agents. Office: Dept Medicine U Wash Seattle WA 98195

KLEBER, CARL JOSEPH, dental research chemist; b. Fort Wayne, Ind., Feb. 22, 1950; s. Alvin Joseph and Irene Mary (Mudrack) K.; m. Margaret Ann Minnick, Oct. 2, 1976; children: Kathryn Irene, Sebastian Joseph. B.S. in Chemistry, Purdue U., 1972; M.S. in Dentistry, Ind. U., 1979. Research chemist Egyptian Chem. Coatings, Lafayette, Ind., 1972-73; research assoc. Preventive Dentistry Research Inst., Fort Wayne, Ind., 1973—; cons. McDennen & Co., Ltd., N.Y.C., 1981—. Mem. Internat. Assn. Dental Research, Am. Assn. Dental Research, Am. Chem. Soc., Genealogical Soc. (Allen County, Ind.), Canal Soc. Ind. Subspecialties: Cariology; Preventive dentistry. Current work: Research and development of new agents for the prevention of dental caries and periodontal disease. Patentee in field of dental materials. Home: 8331 Tewksbury Ct Fort Wayne IN 46815 Office: Indiana U School Dentistry 2101 Coliseum Blvd East Fort Wayne IN 46805

KLEBESADEL, RAY WILLIAM, astrophysicist; b. Shawano, Wis., Nov. 18, 1932; s. William George and Elsa Ann (Habeck) K.; m. Virginia Mae Erckfritz, June 13, 1959; children: William, Lawrence, Rayeann, James. M.A., U. Wis., 1960. Mem. staff U. Calif., Los Alamos Nat. Lab., 1960—. Contbr. articles to sci. jours. Served with USAF, 1951-55. Mem. Am. Astron. Soc. Republican. Lutheran. Subspecialties: Gamma ray high energy astrophysics; Space instrumentation. Current work: Experimental astrophysics, developing space instrumentation for gamma-ray burst astronomy and solar-flare physics. Home: Route 1 Box 192A Espanola NM 87532 Office: Los Alamos Nat Lab ESS-9 D436 Los Alamos NM 87545

KLEHR, EDWIN HENRY, environmental science educator, consultant, researcher; b. Shakopee, Minn., Nov. 26, 1932; s. George Michael and Mildred (Heibel) K.; m. Judith Ann Kaylor, June 15, 1957; children: Celia, Elizabeth, Mary, Kristina, Monica, Jessica. B.S. in Chemistry, St. John's U., Collegeville, Minn., 1954, Ph.D., Iowa State U., 1959. Radiochemist Ames Lab., AEC, 1954-59; asst. prof. Sienna Coll., 1959-61; asst. prof. environ. sci. U. Okla., Norman, 1961-65, assoc. prof., 1965-69, prof., 1969—; cons. in field. Contbr. numerous articles on environ. sci. and nuclear analytical techniques to profl. publs. Dept. Interior fellow, 1970; Internat. Research and Exchange Bd. fellow, 1973. Mem. Sigma Xi. Subspecialties:

Environmental chemistry; Nuclear analytical chemistry. Current work: Environmental chemistry; waste treatment; nuclear analytical techniques. Home: 306 Chautauqua Norman OK 73069 Office: U Okla 202 W Boyd Norman OK 73019

KLEIMAN, HOWARD, mathematics educator; b. N.Y.C., Apr. 15, 1929; s. Louis and Molly (Blefeld) K.; m. Edna Madge Benjamin, July 26, 1956; children: Michele, Jeffrey, Daniel. B.A., NYU, 1950, M.S., 1961; M.A., Columbia U., 1954; Ph.D., Kings Coll., U. London, 1969. Tchr. N.Y.C. Bd. Edn., 1955-56, Bur. Edn. of Physically Handicapped, N.Y.C., 1956-67; asst. prof. math. Queensborough Community Coll., CUNY, 1967-70, assoc. prof., 1970-75, prof., 1978—. Contbr. articles to profl. jours. Music scholar Met. Music Sch., N.Y.C., 1951, Bershire Music Ctr., Tanglewood, 1954. Mem. Math. Assn. Am. (vice chmn. 2-yr. coll. Met. N.Y. sect. 1971-73, treas. 1973-79), London Math. Soc., Am. Math. Soc., Soc. Indsl. and Applied Math. Subspecialty: Algorithms. Current work: A probabilistic polynomal algorithm for obtaining a hamilbur circuit in a directed graph, probabilistic polynomial algorithm, new class of diophantine equations. Home: 188 83 85th Rd Holliswood NY 11423 Office: Queensborough Community College Bayside NY 11364

KLEIN, CERRY M., researcher, consultant; b. Kansas city, Mo., Dec. 11, 1955; s. David M. and Gloria A. (Cobb) K.; m. Debra Ann Baker, May 18, 1980. B.S., N.W. Mo. State U., 1977; postgrad., U. Mo.-Kansas City, 1978; M.S., Purdue U., 1980, Ph.D., 1983. Math tutor N.W. Mo., State U., Maryville, 1974-77; high sch. tchr. Consol. Sch. Dist. #1, Kansas City, Mo., 1977-78; grad. teaching asst. Purdue U., West Lafayette, Ind., 1978-81; cons. Nissus Incorp., West Lafayette, 1981-83; grad. researcher Purdue U., West Lafyette, 1981-83; asst. prof. indsl. engring. U. Mo.-Columbia, 1984—. Active Big Bros., Maryville, 1975, 76, 77. David Ross grantee Purdue U., 1982. Mem. Operation Research Soc. Am., Soc. Indsl. and Applied Math., Am. Inst. Indsl. Engrs., Pi Mu Epsilon. Democrat. Subspecialty: Operations research (engineering). Current work: Duality in dynamic programming via conjugate functions, convolutions, and legendre transformations, submodular functions, and polymatroid intersection. Home: 1704 Vail Ct Columbia MO 65201 Office: 113 Elec Engring U Mo-Columbia Columbia MO 65211

KLEIN, DOUGLAS JAY, chemistry and physics educator, researcher; b. Portland, Oreg., Nov. 8, 1942; s. Ralph E. and Fredericka E. (Brommer) K.; m. Janet C. Goodrich, June 5, 1966; children: Steven G., Suzanne C. B.SC., Oreg. State U., 1964; M.A., U.Tex.-Austin, 1967, Ph.D., 1969. Asst. prof. physics U. Tex.-Austin, 1971-78; assoc. prof. phys. chemistry Moody Coll. Marine Sci., Tex. A&M U.-Galveston, 1979—. Mem. Am. Chem. Soc., Am. Phys. Soc. Subspecialties: Theoretical chemistry; Theoretical physics. Current work: Research, publs. in chemistry, physics. Office: Tex A&M U Galveston TX 77553

KLEIN, GEORGE DEVRIES, geologist; b. Den Haag, Netherlands, Jan. 21, 1933; came to U.S., 1947, naturalized, 1955; s. Alfred and Doris (deVries) K.; m. Chung Sook Kim Chung, May 23, 1982; children: Richard L., Roger N. B.A., Wesleyan U., 1954; M.A., U. Kans., 1957; Ph.D., Yale U., 1960. Research sedimentologist Sinclair Research Inc., 1960-61; asst. prof. geology U. Pitts., 1961-63; asst. prof. to assoc. prof. U. Pa., 1963-69; prof. U. Ill., Urbana, 1970—; vis. fellow Wolfson Coll. Oxford U., 1969; vis. prof. geology U. Calif., Berkeley, 1970; vis. prof. oceanography Oreg. State U., 1974, Seoul Nat. U., 1980; CIC vis. exchange prof. geophys. sci. U. Chgo., 1979-80; chief scientist Deep Sea Drilling Project Leg 58, 1977-78; continuing edn. lectr.; asso. Center Advanced Studies U. Ill., 1974, 83. Author: Sandstone Depositional Models for Exploration for Fossil Fuels, 2d edit, 1980, Clastic Tidal Facies, 1977, Holocene Tidal Sedimentation, 1975; asso. editor: Geol. Soc. Am. Bull, 1975-81; cons. editor: McGraw-Hill Ency. of Sci. and Yearbook, 1977—; chief cons. adv. editor: CEPCO div. Burgess Pub. Co, 1979-81; series editor: Geol. Sci. Monographs. Recipient Outstanding Paper award Jour. Sedimentary Petrology, 1970; Erasmus Haworth Disting. Alumnus award in geology U. Kans., 1980; Outstanding Geology Faculty Mem. award U. Ill. Geology Grad. Student Assn., 1983; NSF grantee. Fellow AAAS, Geol. Soc. Am., Geol. Assn. Can.; mem. Am. Geophys. Union, Am. Inst. Profl. Geologists, Soc. Econ. Paleontologists and Mineralogists, Internat. Assn. Sedimentologists, Am. Assn. Petroleum Geologists, Netherlands Geol. and Mining Soc., Sigma Xi. Subspecialties: Sedimentology; Oceanography. Current work: Sedimentology - tidal sediments, deep-ocean sediments, sediments, sedimentation processes, sandstone diagenesis, sandstone reservoirs, sedimentation patterns and plate tectonics, Sediments from continents to oceans. Office: Dept Geology Univ Ill 245 Natural History Bldg 1301 W Green St Urbana IL 61801

KLEIN, GORDON LESLIE, medical educator, researcher, consultant; b. N.Y.C., Aug. 26, 1946; s. Hyman David and Ruth Harriet (Katz) K.; m. Joann Pamela Schulz, July 1, 1973. B.A., Columbia U., 1967; M.D., Albert Einstein Coll. Medicine, 1970; postgrad., Wolfson Coll., Cambridge (Eng.) U., 1970-71; M.P.H. in Nutrition, UCLA, 1980. Diplomate: Am. Bd. Pediatrics. Intern, resident in pediatrics Stanford (Calif.) U. Med. Center, 1971-74; fellow in clin. nutrition Johns Hopkins U. Med. Sch., Balt., 1976-78; sci. dir. Nutrition Research Inst., AID, Lima, Peru, 1976-78; fellow, asst. prof. pediatrics and gastroenterology UCLA, 1978-82; dir. div. pediatric gastroenterology, asst. prof. pediatrics Tulane U. Med. Sch., New Orleans, 1982—; cons. econs. sect. Am. embassy, Lima, 1978. Peer reviewer sci. articles med. jours.; contbr. articles to profl. publs. Served to lt. comdr. USN, 1974-76. Recipient Nat. Research Service award NIH, 1979-80, Clin. Assoc. Physician award NIH, 1980-82, Schlieder Ednl. Found. fellow, 1982—. Fellow Am. Acad. Pediatrics; mem. Am. Fedn. Clin. Research, Am. Gastroenterol. Assn., Am. Soc. Bone and Mineral Research, Am. Soc. Clin. Nutrition, Delta Omega. Club: Princeton (N.Y.C.). Subspecialties: Pediatrics; Gastroenterology. Current work: Research into interrelation of nutrition and calcium metabolism, study and characterization of bone disease associated with parenteral nutrition. Office: Dept Pediatrics Tulane Univ Sch Medicine 1430 Tulane Ave New Orlean LA 70112

KLEIN, JACK S., research pharmacologist; b. Liverpool, Eng., Dec. 25, 1916; came to U.S., 1921, naturalized, 1924; m.; children: Terrence, Susann, Sandra, James. B.Sc., St. John's U., 1941; Ph.B., L.I. U., 1939; M.S., Columbia U., 1943; M.D., United Med. Coll., 1975. Intern Canoga Rehab. Hosp., Canoga Park, Calif., 1975-76; resident Cornell U. Med. Ctr., N.Y.C., 1977; dept. exptl. surgery N.Y. U. Coll. Medicine, N.Y.C., 1943-44; asst. surgeon Columbia U. Aviation Research Lab., 1944; high altitude research Dept. Hosps. City N.Y., 1944-45; pharmacy dept. S. B. Penick Drug Co., 1945-49; chief drug research Meer Corp., North Bergen, N.J., 1950—. Mem. Am. Assn. Physicists in Medicine, Am. Pharm. Assn., AAAS, Am. Chem. Soc.; mem. N.Y. Acad. Sci.; Mem. Am. Police Hall of Fame, N.J. Acad. Sci., Nat. Police and Fire Fighters Assn. Jewish. Subspecialty: Organic chemistry. Current work: Drug research and elucidation on synthetic organic structures. Home: 1810 Riverside Dr E Bradenton FL 33508 Office: Meer Corp North Bergen NJ 07047

KLEIN, JOHN SHARPLESS, mathematics educator; b. Ossining, N.Y., Sept. 9, 1922; s. Edwin Benedict and Katharine Truman (Sharpless) K.; m. Nancy Deighton Wilkins, Aug. 24, 1963; children: Jeffrey, Carolyn. B.S., Haverford Coll., 1943; M.S., M.I.T., 1949; Ph.D., U. Mich., 1959. Instr. Williams Coll., Williamstorn, Mass., 1949-51, Oberlin (Ohio) Coll., 1954-55, Case Inst. Tech., Cleve., 1955-56; asst. prof. U. R.I., Kingston, 1956-58; assoc. prof. Wilson Coll., Chambersburg, Pa., 1960-63; prof. math. Hobart and William Smith Colls., Geneva, N.Y., 1964—. Served with USNR, 1943-46. Mem. Soc. Indsl. and Applied Math., Math. Assn. Am. Democrat. Mem. Soc. of Friends. Subspecialties: Applied mathematics; Operations research (mathematics). Current work: Integral transforms. Home: 11 Norway Maple Dr Geneva NY 14456 Office: Hobart and William Smith Colls Pulteney St Geneva NY 14456

KLEIN, KEITH ALLEN, nuclear engineer, government official; b. Morgantown, W. Va., Sept. 1, 1951; s. Bernard Benjamin and Gladys Mae (Radman) K. B.S. in Elec. Engring, Cornell U., 1973; M.S. in Nuclear Engring, M.I.T., 1975. Project mgr. breeder program AEC, Washington, 1974-76; program mgr. nuclear waste U.S. Energy Research, Washington, 1976-80, staff to undersec. energy, 1980-81, dir. fuel cycle projects div., 1981—. Mem. Am. Nuclear Soc. Republican. Subspecialties: Nuclear engineering; Nuclear fission. Current work: Direct government activities for light water reactor nuclear fuel reprocessing and storage. Home: 1120 Powhatan St Alexandria VA 22314 Office: U.S. Dept Energy NE-43 Washington DC 20545

KLEIN, LAWRENCE ELLIOT, medical educator and researcher; b. Cleve., June 1, 1950; s. Jerome Allen and Lillian (Panner) K.; m. Gayle Better, Mar. 16, 1980. B.S., MIT, 1972; M.D., Johns Hopkins U., 1976. Diplomate: Nat. Bd. Med. Examiners, Am. Bd. Interna. Medicine. Fellow Johns Hopkins Med. Sch., Balt., 1980, instr. medicine, 1980-81, asst. prof., 1981—; mem. East Balt. Hypertension Task Force, Balt., 1982—. Mem. Am. Geriatrics Soc., Gerontol. Soc. Am., Soc. Research in Primary Care Internal Medicine (asst. editor jour. 1981—), Med. Decision Making Soc., Am. Fedn. Clin. Research. Subspecialties: Gerontology; Internal medicine. Current work: Compliance and adverse drug effects suffered by the elderly; dementia; consultative medicine research. Office: Johns Hopkins Hosp 600 N Wolfe St Harvey 502 Baltimore MD 21205 Home: 2605 Lightfoot Dr Baltimore MD 21209

KLEIN, MARJORIE HANSON, psychiatry educator, psychologist; b. Milw., Sept. 13, 1933; d. Norman R. and Anna (Emery) Hanson; m. William A. Klein, June 23, 1956 (div. 1969); children: Jennifer, Susan; m. Norman S. Greenfield, May 17, 1969. B.A. in Psychology, Wellesley Coll., 1955, M.A., 1957, Ph.D., Harvard U., 1964. Lic. psychologist, Wis. Lab. Asst. Wellesley Coll., 1955-57; research asst. sect. on personality NIMH, Bethesda, Md., 1957-59, postdoctoral research fellow, 1966-67; project asst. psychotherapy research group Wis. Psychiat. Research Inst., U. Wis.-Madison, 1962-64, postdoctoral research fellow, isnt., 1964-65, research assoc., 1965-66, 67-72; asst. prof. psychiatry U. Wis.-Madison, 1972-75, assoc. prof. psychiatry, 1975-80, prof. psychiatry, 1980—, assoc. prof. women's studies program, 1975-80, prof. women's studies program, 1980—; ad hoc mem. clin. projects research rev. com. NIMH, 1976, mem. spl. rev. com. for psychotherapy of depression collaborative program, 1979-80, mem. treatment, devel. and assessment research rev. com., 1979-81. Contbr. numerous articles, chpts. to profl. publs.; author profl. papers; editorial bd.: Psychiatry, 1978—, Jour. Marital and Family Therapy, 1978—, Clin, Psychology Rev, 1978—; ad hoc reviewer profl. jours. Ford Found. grantee, 1966-70; NIMH grantee, 1972—; Alcohol, Drug Abuse and Mental Health Adminstrn. grantee, 1974-76; State Wis. Prevention and Wellness Program grantee, 1979-80. Mem. Am. Psychol. Assn. (editorial bd. Div. 29 1976-79), Soc. Psychotherapy Research, Am. Psychopathol. Assn., Phi Beta Kappa, Sigma Xi. Subspecialty: Psychiatry. Current work: Depression, women's studies, computer applications. Office: U Wis Clin Sci Center Dept Psychiatry 600 Highland Ave Madison WI 53792

KLEIN, MARTIN, marine engineer; b. N.Y.C., Apr. 5, 1941; m. Diane Parenteau, June 21, 1974; children: Allen Jameson, Robin Marie. S.B. in Elec. Engring, MIT, 1962. Program mgr. EG&G, Inc., Bedford, Mass., 1962-67; pres. Klein Assocs. Inc., Salem, N.H., 1968—; participant subbottom seismic profiling survey English Channel for proposed tunnel between France and England; asst. in design graphic recorder for Cousteau "Soucoupe" diving saucer; mem. sonar exploration team Acad. Applied Sci./Loch Ness Investigation Bur.; archeol. surveys, oil exploration surveys. Named Small Bus. Person of Yr. State of N.H., 1983. Fellow Marine Tech. Soc., Explorers Club; mem. Acoustical Soc. Am., Acad. Applied Sci., Instrument Soc. Am., Inst. Navigation, IEEE, U.S. Naval Inst., Brit. Hydrographic Soc. Subspecialties: Oceanography; Ocean engineering. Current work: Development of new sophisticated high resolution sonar. Patentee apparatus for sonar and related signal texture enhancement of recording media, underwater vehicle towing and recovery apparatus, process and apparatus for compatible wet and dry paper signal recording, multi angula sector sound transmitting and receiving system. Office: Klein Assocs Inc Klein Dr Salem NH 03079

KLEIN, MICHAEL JOHN, radio astronomer; b. Ames, Iowa, Jan. 19, 1940; s. Fred Michael and Florence Marie (Graf) K.; m. Barbara D. Klein, Sept. 2, 1962; children: Kristin M., Michael John, Timothy J. B.S., Iowa State U., 1962; M.S., U.Mich., 1966, Ph.D. in Astronomy, 1968. Asst. research engr. Radio Astron. Lab. U. Mich., 1963-64; NRC-NASA resident research assoc. Jet Propulsion Lab., Calif. Inst. Tech., 1968-69, sr. scientist space div., 1969-73, mem. tech. staff, 1973-81, mem. tech. staff space sci. div., 1981—, dep. program mgr., 1981—. Mem. Am. Astron. Soc., Internat. Astron. Union, Internat. Union Radio Sci., Internat. Acad. Astron. Mem. United Ch. of Christ. Subspecialties: Radio and microwave astronomy; Planetary science. Current work: Radio astronomical research and the search for extraterrestrial intelligence, astrophysics, planetary physics, technical program management microwave and millimeter wave radio astronomy. Office: 4800 Oak Gove Dr MS 264-802 Pasadena CA 91109

KLEIN, MILTON SAMUEL, cardiologist; b. Providence, July 26, 1948; s. Carl and Helen (Steiner) K.; m. Gail Danziger, Nov. 18, 1972; children: Joshua, Stephen. B.Sc., McGill U., 1968; M.D., U. Calif.-San Diego, 1972. Diplomate: Am. Bd. Internal Medicine. Intern U. Colo., Denver, 1972-73, resident, 1973-74; fellow in cardiology Washington U., St. Louis, 1974-76; asst. prof. medicine, 1977-79; research assoc. Hammersmith Hosp., London, 1976-77; clin. asst. prof. Baylor Coll. Medicine, Houston, 1979—. Author: Medical Management, 1976. Chmn. Community Hosp. Program, Houston, 1981, Physician Edn. Com., 1982. Recipient Disting. Service award Am. Heart Assn., 1981. Fellow Am. Coll. Cardiology, ACP; mem. Am. Fedn. for Clin. Research, Houston Cardiology Soc., Houston Soc. Internal Medicine. Democrat. Jewish. Subspecialties: Cardiology; Internal medicine. Current work: Nuclear cardiology, positron emission tomography, cardiac metabolism, cardiac enzymology. Home: 2319 Glen Haven Houston TX 77030 Office: Meth Hosp 6560 Fannin St Suite 1546 Houston TX 77030

KLEIN, PETER DOUGLAS, medical educator, health research administrator; b. Elmhurst, Ill., Nov. 30, 1927; m., 1950, 68; 3 children. B.S., Antioch Coll., 1948; M.S., Wayne State U., 1950, Ph.D. in Physiol. Chemistry, 1954. From asst. biochemist to sr. biochemist Div. Biol. Med. Research, Argonne Nat. Lab., 1969-72, mem. staff, 1976-80; prof. medicine U. Chgo., 1972-80; prof. pediatrics Baylor U. Coll. Medicine, 1980—; dir. stable isotope lab. Children's Nutrition Research Center, Houston, 1980—. Mem. Am. Soc. Mass Spectrometry, Am. Assn. Study Liver Disease, Am. Soc. Biol. Chemists, Am. Gastroent. Assn., Am. Soc. Clin. Nutrition. Subspecialty: Nutrition (medicine). Office: Children's Nutrition Research Center 6608 Fannin Room 519 Houston TX 77030

KLEIN, RICHARD JOSEPH, microbiologist; b. Lugoj, Romania, Nov. 6, 1926; s. Alfred Samuel and Zelma (Neumann) K.; m. Elena Marciuca, Nov. 12, 1957; 1 child, Dorian. M.D., Med. Sch., Bucharest, 1953, D.Sc., 1967. Instr., asst. prof. Med. Sch., Bucharest, Romania, 1949-57; research scientist, sr. research scientist Cantacuzino Inst., Bucharest, 1951-72; research asst. prof., research assoc. prof. Med. Sch. NYU, N.Y.C., 1972-79, assoc. prof. microbiology, 1979—. Contbr. articles to sci. jours. Mem. Am. Soc. Microbiology, N.Y. Acad. Sci., Soc. Gen. Microbiology, Am. Soc. Virology. Subspecialties: Virology (medicine); Infectious diseases. Current work: Pathogentic mechanism of viral infections, antiviral substances, viral vaccines. Office: Dept Microbiology 550 1st Ave New York NY 10016

KLEIN, WILLIAM, J., JR., nephrologist; b. Cin., May 27, 1936; m. Sybil Cohen, 1959; 3 children. B.S., U. Cin., 1958; M.D., Columbia U., 1962; Cert. health systems mgmt., Harvard U. Bus. Sch. and Sch. Pub. Health, 1976. Diplomate: Am. Bd. Internal Medicine. Med. intern Columbia Presbyn. Med. Ctr., N.Y.C., 1962-63; physician Peace Corps, Ghana, 1963-65; vis. fellow dept. microbiology Columbia Coll. Physicians and Surgeons, N.Y.C., 1965-67; asst. resident in medicine Duke U. Med. Ctr., Durham, N.C., 1967-68, fellow in nephrology, 1968-69, assoc. dept. medicine, 1969-70; also mem. renal transplant group; asst. prof. medicine and microbiology U. Ala. Med. Ctr./Med. Sch., Birmingham, 1970-73; assoc. prof. clin. medicine U. Rochester, N.Y., 1973-77; chief ambulatory services Rochester Gen. Hosp., 1973-77; dir. N.E. Health Ctr., Rochester, 1973-77; adminstr. Myers Community Hosp., Sodus, N.Y., 1973-77; prof. medicine Coll. Human Medicine and Coll. Community Health Sci., Coll. Osteo. Medicine, Mich. State U., East Lansing, 1977-79; pvt. practice nephrology, West Reading, Pa., 1979—; med. dir. St. Joseph Hosp., Reading, 1979-1982; cons. physician Pritikin Ctr., Downingtown, Pa., 1981-82; bd. dirs., chmn. quality and standards rev. com. South Central Pa. Profl. Standards Rev. Orgn., 1979—. Contbr. articles to med. jours. Served with USPHS, 1963-65. Am. Cancer Soc. fellow, 1965-67; Mead Johnson scholar, 1967-68. Fellow ACP; mem. Am. Soc. Nephrology, Internat. Soc. Nephrology, Am. Fedn. Clin. Research, Am. Assn. Immunologists, Transplantation Soc., Phi Beta Kappa., Alpha Omega Alpha. Subspecialty: Nephrology. Address: 301 S 7th Ave West Reading PA 19611

KLEINERMAN, JEROME, pathologist, medical educator; b. Pitts., July 7, 1924; m., 1944; 3 children. B.S., U. Pitts., 1943, M.D., 1946. Diplomate: Am. Bd. Pathology. Demonstrator in pathology Sch. Medicine, Case Western Res. U., 1951-52, from instr. to assoc. prof., 1952-72, prof. pathology, 1972-80, dir. dept., 1976-80; assoc. dir. St. Luke's Hosp., 1957-64, head dept. pathol. research and clin. pathology, 1965-70, assoc, dir. med. research, 1970-80, dir. div. pathology research, 1976-80; prof. pathology, chmn. dept. pathology Mt. Sinai Med. Sch., N.Y.C., 1980—; research fellow in physiology Grad. Sch. Medicine, U. Pa., 1950-51; Am. Heart Assn. research fellow Cleve. Met. Gen. Hosp., 1952-54, asst. pathologist, 1952-62, vis. assoc. pathologist, 1962—; lectr. Sch. Medicine, U.Pitts., 1963—; attending physician in pulmonary disease VA Hosp., 1965—; cons. pathologist Saranac Lab., Trudeau Found. Mem. pathology B study sect. NIH, 1965—. Fellow Am.Soc. Clin. Pathologists; mem. Am. Soc. Exptl. Pathologists, Am. Assn. Pathologists and Bacteriologists, Am. Heart Assn., Coll. Am. Pathologists. Subspecialty: Pathology (medicine). Office: Mt Sinai Med Sch 1 Gustave L Levy Pl New York NY 10029

KLEINERT, HAROLD EARL, hand surgeon; b. Sunburst, Mont., Oct. 7, 1922; s. Amil and Christine K.; m.; children: Harold, Robert, Christine, James, Jeanne. Student, No. Mont. Coll., 1941, U. Mich., 1941-43; M.D., Temple U., 1946. Diplomate: Am. Bd. Surgery, 1955. Rotating intern Grace Hosp., Detroit, 1946-47, resident in surgery, 1949-53; instr. in surgery U. Louisville, 1954-57, asst. prof. surgery, 1957-62, asso. prof., 1962-69, clin. prof., 1969—; asso. prof. Ind. U., 1967-73, clin. prof., 1973—; nat. cons. to Surgeon Gen. U.S. Air Force, 1973—; mem. staff Jewish Hosp., Louisville, 1955—, pres., 1972; mem. staff Sts. Mary and Elizabeth Hosp., Norton-Kosair-Children's Hosp., Methodist Hosp., Baptist Hosp., Suburban Hosp., Baptist East Hosp., St. Anthony Hosp., Clark County Meml. Hosp., Floyd County Meml. Hosp., North Clark Community Hosp.; Cons. to hosps. Contbr. numerous articles to profl. pubs.; author profl. movies, videotapes. Served with USAAF, 1947-49; with Air N.G., 1953-65. Fellow A.C.S.; mem. Am. Soc. Surgery of the Hand (pres. 1976), AMA (Sci. Achievement award 1980), Am. Assn. Surgery of Trauma, Am. Soc. Plastic and Reconstructive Surgeons, Am. Acad. Orthopedic Surgeons, Am. Assn. Plastic Surgeons (Hon. Award medal 1980), Ohio Valley Soc. Plastic and Reconstructive Surgery, Am. Rheumatism Assn., So. Med. Assn., Ky. Med. Assn., So. Surg. Assn., Am. Trauma Soc., Jefferson County Med. Soc., Aerospace Med. Assn., French Soc. Surgery of Hand, Italian Soc. for Surgery of Hand, Can. Soc. for Surgery of Hand, Colombia Soc. for Surgery of Hand, Louisville Surg. Soc. (pres. 1972), Ky. Surg. Soc. Club: Rotary. Subspecialty: Microsurgery. Research in microsurgery. Inventor microsurgery instruments, tendon instruments. Office: 250 E Liberty St Louisville KY 40202

KLEINKNECHT, RONALD ARTHUR, psychology educator; b. Pasco, Wash., Feb. 10, 1942; s. William Elmer and Emma Julia (Larsen) K.; m. Sharon Stewart, Nov. 29, 1967; children: Lisa, Hans, Erich, Erica. B.A., Wash. State U., 1964, M.S., 1966, Ph.D., 1969. Lic. psychologist, Wash. Asst. prof. Western Wash. U., Bellingham, 1970-73, assoc. prof., 1973-79, prof. psychology, 1979—; affiliate prof. U. Wash. Dental Sch., Seattle, 1982; clin. lectr. U. B.C., Vancouver, 1974—. Contbr. articles to profl. jours. Mem. Am. Psychol. Assn., Assn. Advancement of Behavior Therapy. Subspecialty: Health psychology. Current work: Psychological research in fear and pain. Home: 409 15st St Bellingham WA 98225 Office: Western Wash Univ Dept Psychology Bellingham WA 98225

KLEINMAN, ARTHUR MICHAEL, psychiatrist, medical anthropologist, educator; b. N.Y.C., Mar. 11, 1941; m., 1965; 2 children. A.B., Stanford U., 1962, M.D., 1967; M.A., Harvard U., 1974. Intern in medicine New Haven Hosp., Yale U., 1967-68; resident in psychiatry Mass. Gen. Hosp., Boston, 1972-75; lectr. in anthropology Harvard U., 1974-76; clin. instr. in psychiatry Mass. Gen. Hosp. and Harvard U. Med. Sch., 1975-76; assoc. prof. and adj.

assoc. prof. U. Wash., Seattle, 1976-79, prof. psychiatry, adj. prof. anthropology, 1979—. Editor-in chief: Culture, Medicine and Psychiatry, 1976—. Recipient Wellcome medal med. anthropology Royal Anthropol. Inst., 1980; Dupont Warren fellow Harvard U. Med. Sch., 1974-75; Milton Fund fellow, 1975-76; Found. Fund Research Psychiatry fellow, 1974-76; NIMH research grantee U. Wash., 1978-79. Fellow Am. Anthrop. Assn.; mem. AAAS, Soc. Med. Anthropology, Am. Psychosomatic Soc., Inst. Medicine. Subspecialty: Medical anthropology. Office: Dept Psychiatry and Behavioral Sci U Wash Seattle WA 98195

KLEINMUNTZ, BENJAMIN, psychologist, educator; b. Cologne, Germany, Jan. 15, 1930; came to U.S., 1938, naturalized, 1944; m. Dalia Segal, July 3, 1955; children: Don N., Ira M., Oren J. B.A., Bklyn. Coll., 1952; Ph.D., U. Minn., 1958; postgrad., Yale U., 1967-68. Lic. psychologist, Pa. Psychologist U. Nebr., Lincoln, 1958-59; prof. psychology Carnegie-Mellon U., Pitts., 1959-73, U. Ill., Chgo., 1973—; mem. bd. sci. advisors Psych Systems, Balt., 1981—; research cons. London House Mgmt., Chgo., 1981—. Author: Personality Measurement, 1967, Personality and Psychological Assessment, 1982, Essentials of Abnormal Psychology, 1980; Cons. editor: Jour. Clin. and Cons. Psychology, 1981—; contbr. over 100 articles to profl. jours. Served with U.S. Army, 1952-54. Recipient Excellence in Teaching award U. Ill., Chgo., 1976, 79. Fellow Am. Psychol. Assn., Midwest Psychol. Assn., Pa. Psychol. Assn. Current work: Computer simulation of clinical judgement. Home: 819 LaCrosse Ct Wilmette IL 60091 Office: 1 Dept Psychology Univ Ill Box 4348 Chicago IL 60680

KLEINROCK, LEONARD, computer scientist; b. N.Y.C., June 13, 1934; s. Bernard and Anne (Schoenfeld) K.; m. Stella Schuler, Dec. 1, 1967; children—Nancy S., Martin C. B.S. in Elec. Engring, CCNY, 1957; M.S., MIT, 1959, Ph.D., 1963. Asst. elec. engr. Photobell Co. Inc., 1951-57; research engr. Lincoln Labs., M.I.T., 1957-63; mem. faculty UCLA, 1963—, prof. computer sci., 1970—; pres. Linkabit Corp., 1968-69, Tech. Transfer Inst., 1976—; cons. in field, prin. investigator govt. contracts. Author: Queueing Systems, Vol. I, 1975, Vol. II, 1976, Communication Nets: Stochastic Message Flow and Delay, 1964, Solutions Manual for Queueing Systems, Vol. I, 1982; also articles. Recipient Paper award ICC, 1978, Leonard G. Abraham paper award Communications Soc., 1975, Outstanding Faculty Mem. award UCLA Engring. Grad. Students Assn., 1966, Townsend Harris medal CCNY, 1982, L.M. Ericsson Prize Sweden, 1982; Guggenheim fellow, 1971-72. Fellow IEEE; mem. Nat. Acad. Engring., Ops. Research Soc. Am. (Lanchester prize 1976), Assn. Computing Machinery, Internat. Fedn. Info. Processes Systems, Amateur Athletic Union. Jewish. Subspecialty: Operations research (engineering). Office: Boelter Hall 3732 UCLA Los Angeles CA 90024

KLEINSCHUSTER, STEPHEN JOHN, animal science educator, researcher, educator; b. Bath, Pa., June 3, 1939; s. Stephen John and Elizabeth (Morro) K.; m. Karen Diane Kreutzer, June 25, 1966; children: Stephan, Luke. Student, Baylor U., 1962-63; B.S., Colo. State U., 1963, M.S., 1966; Ph.D., Oreg. State U., 1970. Research fellow Colo. State U., Ft. Collins, 1964-66; mem. faculty Oreg. State U., Corvallis, 1966-70; postdoctoral fellow U. Chgo., 1971; asst. prof. Community Coll., Denver, 1972, Met. State Coll., 1971-73; assoc. prof. Colo. State U., 1973-77; prof. Utah State U., Logan, 1977—, chmn. dept. animal, dairy and vet. scis., 1980-83; dean Coll. Life Scis. and Agr. U. N.H., Durham, 1983; dir. U. N.H. Agr. Expt. Sta.; cons. NASA, EPA, State of Utah MX Impact Rev. Group. Reviewer: Jour. Nat. Cancer Inst; contbr. articles to profl. jours. Mem. Assn. Am. Vet. Med. Colls. (dir. council of chairmen), Colo.-Wyo. Acad. Sci. (pres.), AAAS, Am. Astron. Soc., Oreg. Marine Biol. Soc., Am. Assn. Anatomists, World Assn. Vet. Anatomists, Am. Assn. Vet. Anatomists, Am. Soc. Zoologists, Utah Acad. Sci., N.Y. Acad. Scis., Am. Soc. Animal Sci., Council for Agrl. Sci. and Tech. Roman Catholic. Subspecialties: Developmental biology; Immunology (agriculture). Current work: Immunotherapy of naturally occurring tumors and cell surface antigens of morphogenesis. Office: U NH Taylor Hall Durham NH 03824

KLEITMAN, DANIEL J., mathematics educator, researcher; b. N.Y.C., Oct. 4, 1934; m., 1964; 2 children. A.B., Cornell U., 1954; A.M., Harvard U., 1955, Ph.D., 1958. NSF fellow in physics Copenhagen U., 1958-59, Harvard U., 1959-60; asst. prof. Brandeis U., 1960-66; assoc. prof. MIT, Cambridge, Mass., 1966-69, prof. math., 1969—, head dept. math., 1979—; cons. in field., 1973-81. Mng. editor: Soc. Indsl. and Applied Math. Jour. Algebraic and Discrete Methods, 1975; editor: Jour. Networks. Mem. Am. Math. Soc., Ops. Research Soc. Am., Soc. Indsl. and Applied Math., Am. Acad. Arts and Sci., N.Y. Acad. Sci. Subspecialty: Operations research (mathematics). Office: Dept Math MIT Cambridge MA 02139

KLEMA, ERNEST DONALD, engineering science educator, physicist; b. Wilson, Kans., Oct. 4, 1920; s. William Wensleau and Mary Bess (Vopat) K.; m. Virginia Clyde Carlock, May 23, 1953; children: Donald David, Catherine Marion. A.B., U. Kans., 1941, M.A., 1942; Ph.D., Rice U., 1951. Staff scientist Los Alamos Sci. Lab., 1943-46; physicist, sr. physicist, prin. physicist Oak Ridge Nat. Lab., 1950-58; assoc. prof. nuclear engring. U. Mich., Ann Arbor, 1956-58; faculty Northwestern U., Evanston, Ill., 1958-68, chmn. dept. engring. scis., 1960-66; faculty Tufts U., Medford, Mass., 1968—, prof. engring. sci., 1968—, dean coll., 1968-73, adj. prof. internat. politics Fletcher Sch. Law and Diplomacy, 1973—; chmn. neutron measurement and standards NRC, Washington, 1958-63; cons. Gen. Atomic Corp., La Jolla, Calif., 1967, Tech. Edn. Research Ctr., Cambridge, Mass., 1970, Oak Ridge Tech. Enterprises, 1972. Author: manual A Laboratory Course in Reactor Physics, 1953. Mem. exec. com. Winsor Sch., Boston, 1973-76. Fellow Am. Phys. Soc., Am. Nuclear Soc.; sr. mem. IEEE. Clubs: Princeton of N.Y. (N.Y.C.); Harbor (Seal Harbor, Maine). Subspecialties: Nuclear physics; Nuclear engineering. Current work: Production of silicon surface-barrier detectors and measurement of their energy resolution for alpha particles, heavy ions, and fission fragments; policy questions related to technological development. Patentee hydrogen purification method. Home: 53 Adams St Medford MA 02155 Office: Coll Engring Tufts U Medford MA 02155

KLEMAS, VYTAUTAS, marine studies educator, researcher, university administrator; b. Klaipeda, Lithuania, Nov. 29, 1934; m., 1960; 3 children. B.S., MIT, 1957, M.S., 1959; Ph.D. in Optical Physics, U. Brunswick, 1965. Mgr. optical physics and space exploration Space Div., Gen. Electric Co., 1959-71; assoc. prof. U. Del., Newark, 1971-80, prof. marine studies, 1980—, dir., 1975—; cons. in field; mem. commn. natural resources Nat. Acad. Sci., 1975-78, mem. ocean policy com.; mem. adv. coms. NASA, 1975—; program mgr. Scientists and Engrs. Econ. Devel., NSF, 1977-78; mem. Man and the Biosphere, UNESCO, 1977—. Recipient Achievement medal Korean Advanced Inst. Sci., 1978, Merit award India Remote Sensing Agy., 1978; Gen. Electric Co. fellow, 1963. Mem. IEEE, Assn. Am. Geographers, Am. Geophys. Union, Am. Soc. Photogrammetry. Subspecialty: Remote sensing (geoscience). Office: Coll Marine Studie U Del Newark DE 19711

KLEMENT, VACLAV, physician, educator; b. Pilsen, Czechoslovakia, May 7, 1935; m. (married). M.D., Charles U. Sch. Medicine, Prague, Czechoslovakia, 1959; Ph.D., Czechoslovak Acad. Scis., 1964. Diplomate: Am. Bd. Radiology. Resident in radiology Univ. Hosp. Pilsen, Czechoslovakia, 1959-61; part-time State Oncology Inst., Prague, 1965-66; research fellow Czechoslovakia Acad. Scis., 1964-67; vis. scientist NIH, Bethesda, Md., 1967-68; mem. faculty U. So. Calif. Sch. Medicine, Los Angeles, 1969—, assoc. prof. radiation oncology and microbiology, 1979—. Contbr. chpts. to books, articles to profl. jours. Mem. Am. Assn. Cancer Research, Internat. Assn. Comparative Research on Leukemia and Related Diseases, Am. Endocurietherapy Soc., Am. Soc. Therapeutic Radiologists, Am. Coll. Radiology. Subspecialties: Radiology; Cell study oncology. Current work: Oncology, clinical and experimental, radiation oncology. Office: Dept Radiology U So Calif Sch Medicine 2025 Zonal Los Angeles CA 90033

KLEMM, LEROY HENRY, chemistry educator; b. Maple Park, Ill., July 31, 1919; s. Henry Joseph and Anna (Reines) K.; m. Christine Jones, Dec. 27, 1945; children: Richard A., Rebecca J., Ann C. B.S. in Chemistry, U. Ill.-Urbana, 1941; M.S., U. Mich., 1943, Ph.D., 1945. Research chemist Am. Oil Co., Texas City, 1944-45; research assoc. Ohio State U., 1945-46; instr. chemistry Harvard U., 1946-47; instr. Ind. U., 1947-51, asst. prof., 1951-52, U. Oreg., Engene, 1952-56, assoc. prof., 1956-63, prof., 1963—; vis. prof. U. Cin., 1965-66, LaTrobe U., Melbourne, Australia, 1979-80; sr. assoc. U. Melbourne, 1979-80. Editorial bd.: Jour. Heterocyclic Chemistry, 1971—; contbr. chpts. to books. Guggenheim fellow, 1958-59; Fulbright fellow, 1972-73. Fellow AAAS; mem. Am. Chem. Soc. (councillor 1983—), Sigma Xi. Subspecialties: Catalysis chemistry; Organic chemistry. Current work: Organic syntheses and reactions; carbocyclic and heterocyclic compounds; catalytic, electrochemical and chromatographic methods. Office: Dept Chemistry U Oregon Eugene OR 97403 Home: 1926 Moss St Eugene OR 97403

KLEMM, WILLIAM ROBERT, brain researcher, educator; b. South Bend, Ind., July 24, 1934; s. Lincoln Wilfred and Helen (DeLong) K.; m. Doris Mewha, Aug. 27, 1957; children: Mark Dolan, Laura Margaret. D.V.M., Auburn U., 1958; Ph.D., U. Notre Dame, 1963. Assoc. prof. Iowa State U., 1963-66; prof. vet. anatomy Tex. A&M U., 1966—; cons. in field. Author: Animal Electroencephalography, 1969, Science The Brain, and Our Future, 1972; contbr. numerous articles to profl. jours.; editorial bd.: Jour. Electrophysiol. Techniques, 1978—, Psychopharmacology, 1980—. Served to lt. col. USAF, 1958. Recipient Disting. Achievement award Tex A&M U., 1979. Mem. Soc. Neurosci., Am. Physiol. Soc., Sigma Xi. Republican. Presbyterian. Subspecialties: Neurophysiology; Neuropharmacology. Current work: Research on: mechanisms of action of narcotics and alcohol; movement control systems, particularly in hypnosis and catalepsy; functions of hippocampus; nerve cell spike trains as information carriers. Office: Dept Vet Anatomy Tex A&M U College Station TX 77843

KLEMP, GEORGE OTTO, JR., management consultant in behavorial science, research psychologist; b. Phila., Mar. 1, 1946; s. George Otto and Catherine Rachel (Cridl) K. A.B., Harvard U., 1968; M.A., U. Wis.-Madison, 1969; Ph.D., Yale U., 1974. Dir. research McBer and Co., Boston, 1974-79, v.p., 1979—; mem. adj. faculty U. Chgo., 1980—. Editor, contbg. author: The Assessment of Occupational Competence, 1980. Served with U.S. Army, 1969-71. Mem. Am. Mgmt. Assn., Am. Soc. Tng. and Devel., Am. Psychol. Assn., Orgn. Devel. Network. Episcopalian. Club: Yale (Boston). Current work: Developing cognitive process models of human performance in specific occupations; measurement and prediction of individual behavior in occupations within organizations. Office: McBer and Company 137 Newbury St Boston MA 02116

KLENKE-HAMEL, KARIN EDDA, industrial organizational psychologist, educator; b. Saarbruecken, W.Ger., May 16, 1947; came to U.S., 1965, naturalized, 1983; d. Herbert and Kaethe (Peters) Klenke; m. Willem Hamel, Sept. 9, 1970; children: Katja, Max. B.S. summa cum laude, Old Dominion U., 1974, M.S. in Psychology, 1977, Ph.D., 1982. Grant coordinator Southeastern Va. Tng. Ctr., Chesapeake, 1975-77; instr. psychology Old Dominion U., 1977-80, asst. prof., 1982—; pres. Dr. Karin Hamel, Ph.D. & Assocs./Indsl./Organizational Cons., Virginia Beach, 1981—; cons. in field. Author: Human Sexuality, 1980, Exploring Human Sexuality, 1981, Psychology: Its Study and Uses, 1982. Recipient Van Haller Gilmer award Va. Poly. Inst., 1981. Mem. Acad. Mgmt., Am. Psychol. Assn., Assn. Women in Psychology, Cheiron Internat. Soc. History of Behavioral and Social Scis., Southeastern Psychol. Assn. Current work: Effects of affirmative action programs on employee attitudes and performance, applied theory-building using confirmatory factor analysis and structural equation modeling, organizational life cycles impact of technology on social fabric of organizations, work-family interdependencies. Home: 5309 Sidney Ct Virginia Beach VA 23464 Office: Department Psychology Old Dominion University Norfolk VA 23508

KLEPAC, ROBERT KARL, psychologist; b. Cleve., Feb. 28, 1943; s. Louis John and Mary Ann (Hokes) K.; m. Rosalie Louise De Franco, Aug. 7, 1965; children: Lisa, Theresa. B.S.S., John Carroll U., 1965; M.S., Kent State U., 1968, Ph.D., 1969. Lic. psychologist, Fla. Asst. prof. Western Wash. State U., Bellingham, 1969-72; assoc. prof., chmn. dept. psychology N.D. State U., Fargo, 1972-82; assoc. prof., dir. clin. tng. Fla. State U., Tallahassee, 1982—, dir. clin. tng. dept., 1982—; dir. Dental Behavior Research Clinic, Tallahassee, 1982—; cons. Kent State U., 1981—. Contbr. articles to profl. jours. NIH dental research grantee, 1978-82, 82—; biomed. research support grantee, 1982—. Mem. Internat. Assn. Dental Research, Assn. for Advancement Behavior Therapy, Internat. Soc. Study of Pain, Am. Psychol. Assn., Behavior Therapy and Research Soc. Current work: Prevention/reduction of adverse reactions to dental/medical procedures; pain phenomena, cognitive behavior therapy. Office: Fla State Univ Dept Psychology Tallahassee FL 32306

KLESIUS, PHILLIP HARRY, immunologist; b. Bryn Mawr, Pa., Mar. 1, 1938; s. Phillip M. and Mary H. K.; m. Patricia A. Wood, Oct. 28, 1968; children: Stephen, Patrick. B.S., Fla. So. Coll., 1961; M.S., Northwestern State U. La., 1963; Ph.D., U. Tex., 1966. Asst. prof. U. Tex., Austin, 1968; asst. prof. U. Ariz., 1969-72; asst. chief Ctr. for Disease Control, Ft. Collins, Colo., 1973; research leader Agrl. Research Service, U.S. Dept. Agr., Auburn, Ala., 1973—; assoc. prof. Sch. Vet. Medicine, Auburn U., 1973—; vis. prof. Sch. Vet. Medicine Tuskegee Inst.; assoc. prof. Sch. Medicine Med. U. S.C., 1975—. Contbr. articles to profl. jours. Mem. Am. Assn. Immunologists, Am. Assn. Vet. Immunologists, Am. Soc. Microbiology. Methodist. Subspecialties: Microbiology (veterinary medicine); Immunology (agriculture). Current work: Veterinary Immunology of infectious diseases. Patentee in field. Office: USDA-ARS PO 952 Auburn AL 36830

KLIBANOV, ALEXANDER MAXIM, educator; b. Moscow, July 15, 1949; U.S., 1977, naturalized, 1983; s. Maxim and Eugenia (Tomas) K.; m. Margarita Romanycheva, Apr. 21, 1972; 1 dau., Tanya. M.S., Moscow U., 1971, Ph.D., 1974. Research chemist Moscow U., 1974-77; postgrad. research chemist U. Calif., San Diego, 1978-79; asst. prof. dept. nutrition and food sci. MIT, 1979-83, assoc. prof., 1983—; cons. in field. Contbr. over 60 articles to profl. jours.; editorial bd.: Applied Biochemistry and Biotech, 1981—. E.L. Doherty Professorship M.I.T., 1981; numerous research grants. Mem. Am. Chem. Soc., Am. Soc. Microbiology. Jewish. Subspecialties: Enzyme technology; Biochemistry (biology). Current work: Enzyme Stability and stabilization, immobilized enzymes and cells, enzymes as catalysts in organic chemistry, enzymes for wastewater treatment. Home: 705 Washington St Boston MA 02135 Office: MIT Bldg 16-209 Cambridge MA 02139

KLICKA, JOHN KENNETH, cancer research scientist, physiology educator; b. Chgo., Dec. 9, 1933; s. John Joseph and Regina Margaret (Micklautz) K.; m. Dorothy Mae Chwierut, Jan. 14, 1954; children: Tom, John, Jim, Mike, Dave, Steve, Thor. B.S., No. Ill. U., 1957, M.S., 1958; Ph.D., U. Ill.-Urbana, 1962. Asst. prof. physiology U. Wis.-Oshkosh, 1962-65, assoc. prof., 1968-71, prof., 1972-79; research assoc. U. Minn., Mpls., 1980—; prof. dept. urol. surgery. U. Minn.-Mpls. Served with U.S. Army, 1954-55; Berlin. Recipient Career Devel. award Nat. Inst. Environ. Health and Sci., 1981-84; NIH fellow, 1965-67. Mem. Am. Physiol. Soc. Zoologists., Am. Assn. Cancer Research, Am. Soc. Zoologists. Subspecialties: Cancer research (medicine); Toxicology (agriculture). Current work: Estrogen metabolism and its possible relationship to chemical carcinogenesis. Home: 15017 Stevens Ave S Burnsville MN 55337 Office: VA Med Center 54th St and 48th Ave S Minneapolis MI 55417

KLIDE, ALAN MARSHALL, veterinary anesthesiologist; b. Bklyn., Aug. 25, 1939; s. Herman and Adele (Silverman) K. Student, CCNY, 1957-60, U. Ga., 1960-61; V.M.D., U. Pa., 1965. Diplomate: Am. Coll. Vet. Anesthesiologists. Postdoctoral trainee dept. anesthesia U. Pa. Sch. Medicine, Phila., 1965-67, 1965-66, instr. anesthsia, 1967-69, asst. prof., 1969-72, assoc. prof., 1972—, chief sect. anesthesia, 1972-80, sec., 1968. Author: Veterinary Acupuncture, 1977; contbr. articles to profl. jours. Recipient John E. McCoy Meml. award Wash. State U., 1978. Mem. Internat. Vet. Acupuncture Soc. (dir. 1975-81), Am. Soc. Vet. Anesthesia (chmn. program com. 1979-80), Vet. Anesthesia Soc. Am. (editorial rev. bd. 1977-78), Am. Soc. Anesthesiologists, Am. Soc. Pharmacology and Exptl. Therapeutics, Assn. Vet. Anesthetists Gt. Britain and Ireland, Am. Assn. Zoo Veterinarians, Am. Coll. Vet. Pharmacology and Therapeutics, Internat. Vet. Acupuncture Soc., Am. Soc. Regional Anesthesia, Pa. Soc. Anesthesiologists, Phi Zeta. Subspecialty: Veterinary Anesthesiology. Current work: Veterinary acupuncture, anesthesia, closed circuit, clin. pharmacology of narcotics, narcotic antagonists, narcotic agonist/antagonist analgesics. Office: U Pa Sch Vet Medicine 3850 Spruce St H1 VHUP 3006 Philadelphia PA 19104

KLIEJUNAS, JOHN THOMAS, plant pathologist; b. Sheboygan, Wis., May 4, 1943; s. Anton and Janet (Lamb) K.; m. Barbara Anderson, June 22, 1968; children: Trina Mae, Mary Margaret. B.S., Wis. State U., Stevens Point, 1965; M.F., U. Minn., 1967; Ph.D., U. Wis.-Madison, 1971. Research asso. U. Wis., Madison, 1971-72; plant pathologist U. Hawaii, Hilo, 1972-79; plant pathologist forest pest mgmt. Forest Service, U.S. Dept. Agr., San Francisco, 1979—. Contbr. articles to profl. jours. Mem. Am. Phytopath. Soc. Subspecialties: Plant pathology; Integrated pest management. Current work: Development and implementation of integrated pest management strategies for reducing losses from pests; advising on pest preventive and suppressive actions. Home: 5305 Lightwood Dr Concord CA 95421 Office: 630 Sansome St San Francisco CA 94111

KLIGER, DAVID SAUL, chemist, educator; b. Newark, Nov. 3, 1943; s. William D. and Natalie K.; m. Rache Rina, Nov. 24, 1979. B.S., Rutgers U., 1965; Ph.D., Cornell U., 1970. Postdoctoral fellow Harvard U., 1970-71; asst. prof. chemistry U. Calif.-Santa Cruz, 1971-76, assoc. prof., 1976-83, prof., 1983—; mem. molecular and cellular biophysics study sect. NIH. Research, publs. in field. Mem. Interam. Photochem. Soc., Am. Assn. Photobiology, Am. Chem. Soc., Biophys. Soc., AAAS. Subspecialties: Laser photochemistry; Spectroscopy. Current work: Laser photochemistry, molecular biophysics. Home: 328 Oxford Way Santa Cruz CA 95060 Office: Natural Sci II U Calif Santa Cruz CA 95064

KLIMAS, EDWARD JOHN, electrical engineer; b. Toronto, Ont., Can., Apr. 5, 1952; came to U.S., 1954, naturalized, 1960; s. Bruno and Alexandra (Pacevicius) K. B.E.E., Cleve. State U., 1974; M.S. in Elec. and Computer Engring, Carnegie Mellon Inst., 1976. With Reliance Electric Research Ctr., Euclid, Ohio, 1973-75; coordinator for advanced instrumentation Republic Steel Research Ctr., Independence, Ohio, 1976—. Mem. IEEE, Instrumentation Soc. Am., Optical Soc. Am., Tau Beta Pi, Eta Kappa Nu. Subspecialties: Electrical engineering; Electronics. Current work: Smart sensors; coordination of high technology instrumentation into manufacturing processes and applied research. Home: 4893 S Sedgewick Rd Lyndhurst OH 44124

KLIMEK, JOSEPH JOHN, physician; b. Wilkes-Barre, Pa., Sept. 14, 1946; s. Joseph John and Frances (Pavloski) K.; m. Jane Marie Stout, June 21, 1971; 1 son, Adam. A.B., Princeton U., 1968; M.D., Pa. State U., 1972. Diplomate: Am. Bd. Internal Medicine. Fellow in infectious disease Hartford (Conn.) Hosp., 1974-76, chief of epidemiology, 1976—, assoc. dir. infectious diseases, 1980—, assoc. dir. medicine, 1978—; asst. prof. medicine U. Conn. Sch. Medicine, Farmington, 1978—. Editor: Am. Jour. Infection Control, 1980—, Asepsis, 1979—; contbr. articles to profl. jours. Named Disting. Alumnus Pa. State U. Sch. Medicine, 1978. Fellow A.C.P.; mem. Assn. for Practitioners in Infection Control (dir. 1978-81). Subspecialties: Internal medicine; Infectious diseases. Current work: Clinical research in hospital epidemiology, hospital-acquired infections, and antibiotic pharmacokinetics. Home: 39 Woodhaven Dr Simsbury CT 06070 Office: Hartford Hosp 80 Seymour St Hartford CT 06115

KLIMSTRA, PAUL DALE, pharmacologist, researcher; b. Erie, Ill., Aug. 25, 1933; s. Paul and Jessie (Jaarsma) K.; m. Lois Arlene Kemp, Feb. 2, 1957; children: David S., Jonathan K. B.A., Augustana Coll., Rock Island, Ill., 1955; M.S., U. Iowa, 1957, Ph.D., 1959. Research investigator Searle Research and Devel., Chgo., 1959-68, sr. research investigator, 1968-70, asst. dir. chem. research, 1970-71, dir. chem. research, 1971-73, dir. preclin. research and devel., 1973-74, v.p., 1974-78, v.p. N.Am. preclin. research and devel., 1978—. Contbr. articles to profl. jours. Mem. Am. Soc. Pharmacology and Exptl. Therapeutics, N.Y. Acad. Scis., Am. Pharm. Assn., Indsl. Research Inst. (com. chmn., dir.), Pharm. Mfrs. Assn., Ill. Sci. Lecture Assn. (dir.), Sigma Xi, others. Republican. Presbyterian. Subspecialty: Organic chemistry. Current work: Management of research and development. Patentee in field. Office: 4901 Searle Pkwy Skokie IL 60077

KLINE, EDWARD SAMUEL, biochemistry educator; b. Phila., June 26, 1924; s. Morris and Bessie (Lieb) K.; m. Bernice S. Greenstein, Aug. 27, 1950; children: Andrew, Matthew. B.A., U. Pa., 1948; M.S., George Washington U., 1955, Ph.D., 1961. Bacteriologist Walter Reed Inst. Research, Washington, 1954-57; biochemist Armed Forces Inst. Pathology Washington, 1957-61; postdoctoral fellow Ind. U., Bloomington, 1961-63; asst. prof. Med. Coll. Va., Richmond, 1963-68, assoc. prof., 1968—. Contbr. articles on biochemistry to profl. jours. Served with AUS, 1943-45; ETO. Recipient merit award U.S. Govt., 1960; NIH grantee, 1964-77. Mem. AAAS, Va. Acad. Scis., Nutrition Today Soc. Subspecialty: Biochemistry (biology). Current work: Distribution and function of lactate dehydrogenase and its isozymes, role of nuclear polyphosphates. Inventor pipette dispeser. Office: Med Coll VA PO Box 614 Richmond VA 23298

KLINE, ELLIS LEE, microbiologist, researcher, consultant; b. West Palm Beach, Fla., July 12, 1941; s. George Ellis and Jean Adela (Ross) K.; m. Priscilla Alden Mackenzie, June 12, 1965; children: Heather Jean, Heidi Brooks. B.S. in edn, Greenville (Ill.) Coll., 1964; M.S., No. Ill. U., 1968; Ph.D., U. Calif.-Davis, 1972. Fellow Purdue U., 1972-74; asst. prof. Edinboro (pa.) State Coll., 1974-76, assoc. prof., 1976-78; asst. prof. Clemson U., 1978-79, assoc. prof., 1979-82; prof. microbiology, 1982—; vis. research scientist NIH, Bethesda, Md., 1980; cons. in field. Contbr. articles to profl. jours. USPHS grantee, 1968-72; Clemson U. grantee, 1979, 80. Mem. Am. Soc. Microbiology. Presbyterian. Subspecialties: Genetics and genetic engineering (biology); Microbiology. Current work: Molecular biology, recombinant DNA in procaryotic and eucaryotic systems; MGR - metabolite gene regulation. Patentee methods and materials for detection of multiple sclerosis. Home: 203 N Elm St Pendleton SC 29670 Office: Clemson University Clemson SC 29631

KLINE, NATHAN SCHELLENBERG, educator, research psychiatrist; b. Phila., Mar. 22, 1916; s. Ignatz and Florence (Schellenberg) K.; m. Margot Hess, June 29, 1942 (div. 1976); 1 dau., Marna Brill Anderson. A.B., Swarthmore Coll., 1938; postgrad., Harvard, 1938-39, New Sch. Social Research, 1940-41; M.D., N.Y. U., 1943, Princeton, 1946-47, Rutgers U., 1947-48; M.A., Clark U., 1951, Clark U., 1951-53. Diplomate: Am. Bd. Psychiatry and Neurology. Intern, resident St. Elizabeths Hosp., Washington, 1943-44; staff VA, Lyons, N.J., 1946-50; child psychiatrist Union County Mental Hygiene Soc. Clinic, 1946-47; assoc. Columbia Greystone, 1947-50, N.Y. State Brain Research Project, 1948-50; research asst. dept. neurology Columbia Coll. Physicians and Surgeons, 1948-50, research assoc., 1952-55, research assoc. dept. psychiatry, 1955-57, asst. clin. prof., 1957-69, assoc. clin. prof., 1969-73, clin. prof., 1973-80; clin. prof. psychiatry N.Y. U., 1980—; dir. research Worcester (Mass.) State Hosp., 1950-52, Rockland State Hosp. Research Center, Orangeburg, N.Y., 1952-75, Rockland Research Inst., Orangeburg, 1975—; dir. psychiat. services Bergen Pines County Hosp., Paramus, N.J., 1963-75; cons. Lenox Hill Hosp., N.Y.C., 1974—; profl. adv. com. Manhattan Soc. Mental Health, Nat. Com. Against Mental Illness; clin. adv. panel NIMH, 1957-59; temporary adviser WHO, 1957, expert adv. panel, 1973—; pres. Internat. Com. Against Mental Illness. Contbr. articles profl. publs.; contbg. editor: Excerpta Medica, 1955—; adv. editorial bd.: Internat. Jour. Social Psychiatry, 1958, Fgn. Psychiatry, 1972-74. Served to sr. asst. surgeon USPHS, 1944-46. Decorated knight Great Cross Master's Grace Sanctae Mariae Serenissimus Militaris Ordo; Commandeur Ordre Touissant-Louverture; Grande Officeur Legion d'Honneur et Merite Republic of Haiti; knight grand comdr. Liberian Humane Order of African Redemption Republic of Liberia; recipient Page One award sci. N.Y. Newspapers Guild, 1956, Adolf Meyer award Assn. Improvement Mental Health, 1956, Albert Lasker award Am. Pub. Health Assn., 1957, Henry Wisner Miller award Manhattan Soc. Mental Health, 1963, Albert Lasker clin. research award, 1964. Fellow Royal Coll. Psychiatrists (hon.), ACP, AAAS, N.Y. Acad. Medicine, Am. Psychiat. Assn. (chmn. research com. 1956-57), Royal Soc. Medicine (Eng.); found. fellow Royal Coll. Psychiatrists (Eng.); mem. AMA, Am. Psychol. Assn., Soc. Biol. Psychiatry, Soc. Exptl. Biology and Medicine, Am. Coll. Neuropsychopharmacology (pres. 1966-67), Indian Psychiat. Assn. (corr. mem.), Assn. Research Nervous and Mental Disease, Med. Soc. County N.Y., Sigma Xi; hon. mem. La Sociedad Colombian de Psiquiatria Republica de Colombia. Club: Atrium. Subspecialties: Psychopharmacology; Transcultural psychiatry. Current work: Administration of multidisciplinary research institute with staff of more than 200; consultant on psychiatry to foreign governments and universities; practice of psychiatry. Office: Rockland Research Inst Orangeburg NY 10962 also 425 E 61 St New York NY 10021 Ignorance, stupidity, laziness and a poor memory have been major factors in such success as I have achieved. Had I known the research already done, I would have been discouraged from attempting to solve the problems. Being both stupid and lazy, I managed to devise solutions which were simple and easy to utilize. The answers often appeared without effort although in retrospect I saw that they were an amalgam of half a dozen things I had read elsewhere. Fortunately I did not remember the disparate parts at that time since it might well have prevented them from all running together so successfully. Curiosity, compulsiveness and reduced sleep need also contributed.

KLINE, STEPHEN JAY, mechanical engineer; b. Los Angeles, Feb. 25, 1922; s. Eugene Field and Sheda (Lowman) K.; m. Naomi Jeffries, July 11, 1977; children: David M., Mark D., Carolyn R. B.A., Stanford U., 1943, M.S., 1949; Sc.D., M.I.T., 1952. Research analyst N. Am. Aviation, 1946-48; mem. faculty Stanford (Calif.) U., 1952—, prof. mech. engring., 1961—, chmn. thermoscis. div., 1961-73, prof. values, technology and society, 1970—; cons. Gen. Electric, Gen. Motors, United Technology, DuPont, Brown Boveri. Author: Similitude and Approximation Theory, 1965, Computation of Turbulent Boundary Layers, 1968; editor: Evaluation of Complex Turbulent Flows, 1981. Served with AUS, 1943-46. Recipient Melville medal ASME, 1959, Fluids Engring. award, 1975, Centennial award, 1980; George Stephenson medal Inst. Mech. Engrs. Britain; Bucraino medal Italian Film Soc., 1965. Fellow ASME (past chmn. fluid mechanics com. Fluids Engring. Div.); mem. Nat. Acad. Engring. Subspecialties: Mechanical engineering; Fluid mechanics. Current work: Fluid mechanids of combustion processes; turbulence modelling. Office: Dept Mech Engring Stanford U Stanford CA 94305 Since I was young, I have been intrigued with increasing understanding of physical nature and human sociotechnical systems particularly where the results are important in real life matters. It is this combination of interests that led me to a career in engineering research, and later, to extend my interest to explicity study of the interaction of technology with individual humans, social systems and ecologies.

KLINGEN, THEORDORE JAMES, chemistry educator, researcher; b. St. Louis, Oct. 7, 1931; s. Leonard J. and Margaret M. (Ehlenz) K.; m. Maura Downey, Sept. 1, 1958; children: Joseph L., Ann M. B.S., St. Louis U., 1953, M.S., 1955; Ph.D., Fla. State U., 1962. Research scientist McDonnell-Douglas Corp., St. Louis, 1962-64; asst. prof. chemistry U. Miss., 1964-67, assoc. prof., 1967-70, prof., 1970—, radiation safety officer, 1982—. Contbr. articles to profl. jours. Served to capt. USAF, 1955-57. AEC grantee, 1968-76; NSF grantee, 1972-75; Dept. Energy grantee, 1976-79. Mem. Am. Chem. Soc., Am. Phys. Soc., Am. Nuclear Soc., Sigma Xi (pres. local chpt. 1976-77). Roman Catholic. Subspecialties: Inorganic chemistry; Solid state chemistry. Current work: Radiation chemistry of the solid state; electro-chemical properties of the solid state transition metal compounds; polymer formation in the liquid crystalline state. Patentee in field. Office: U Miss Dept Chemistry University MS 38677

KLINGENSMITH, WILLIAM CLAUDE, III, nuclear medicine physician, researcher; b. Pitts., Feb. 17, 1942; s. William Claude and Marian (Dale) K.; m. Georgeanna Seegar Jones, Apr. 22, 1972; children: William Claude IV, Theodore Emory. A.B., Cornell U., 1964, M.D., 1968. Diplomate: Am. Bd. Radiology, Am. Bd. Nuclear Medicine. Intern in medicine U. Oreg., 1968-69; resident in diagnostic radiology U. Colo.-Denver, 1969-72, asst. prof. radiology, 1976-80, assoc. prof., 1980—, dir. nuclear medicine, 1979—; fellow in nuclear medicine Johns Hopkins U., 1974-76; cons. Fitzsimmons Army Hosp., Denver, 1976—. Author: Atlas of Radionuclide Hepatobiliary Imaging, 1983; contbr. chpts., numerous articles to profl. publs. Served as maj. USAF, 1972-74. NIH grantee, 1981. Mem. Am. Coll. Radiology, Am. Coll. Nuclear Physicians, Soc. Nuclear Medicine. Subspecialties: Nuclear medicine; Diagnostic radiology. Current work: Radiopharmaceutical development in nuclear medicine; data analysis in nuclear medicine. Home: 3625 S Albion St Englewood CO 80110 Office: U Colo Health Sci Center 4200 E 9th Ave Denver CO 80262

KLINGMAN, GERDA ISOLDE, biochem. pharmacologist, educator; b. Berlin, May 6, 1924; d. Norman Ellsworth and Margarete Luise (Eipel) Schultz; m. Jack Dannis Klingman, May 29, 1953; 1 dau., Karin Louise. B.S., Fordham U., 1952; Ph.D. in Pharmacology, Med. Coll. Va., 1956. Postdoctoral fellow Duke U. Med. Center, 1955-57; asso. in pharmacology dept. pharmacology and exptl. therapeutics Johns Hopkins U. Sch. Medicine, 1957-59, instr., 1959-61; instr. biochem. pharmacology SUNY-Buffalo, 1961-63, asst. prof., 1963-67, asso. prof., 1967-73, prof., 1973—; mem. grant rev. com. NIMH. Contbr. chpts. to books, articles to profl. jours. Commonwealth of Va. Dept. Health predoctoral fellow, 1952-55; USPHS/NIH research career awardee, 1962-67; NIH grantee, 1962-72, 82-83. Mem. Am. Soc. for Pharmacology and Exptl. Therapeutics. Subspecialties: Neurochemistry; Neuropharmacology. Current work: Neuropharmacology and neurochemistry of the adrenergic nervous system, immunosympathectomy (nerve growth factor antiserum), body temperature regulation, ontogenesis of the adrenergic nervous system, dopamine receptors, acute and chronic tolerance and physical dependence, acute lethal and sublethal alcoholic intoxication. Office: Dept Biochem Pharmacology SUNY Buffalo 447 Hochstetter Hall Buffalo NY 14260

KLINGMAN, JACK DENNIS, biochemistry educator; b. Johnson City, N.Y., Apr. 21, 1927; s. Lewis R. and Pearle (Dennis) K.; m. Gerda I. Schultz, 1953; 1 dau., Karin L. B.A., Syracuse U., 1951; M.S., Med. Coll. Va., 1953; Ph.D., Duke U., 1957. Teaching asst. Med. Coll. Va., Richmond, 1951-53, research assoc. in pharmacology, 1953; research asst. Duke U., Durham, N.C., 1954-57; research assoc. Johns Hopkins U., Balt., 1957-61; instr. to prof. biochemistry SUNY-Buffalo, 1961—. Assoc. editor: Preparative Biochemistry, 1975—; mem. editorial bd.: Neurotoxicology, 1980—. Served with U.S. Army, 1945-47; ETO. NIH fellow, 1974-77, 79. Mem. Am. Soc. Biol. Chemistry, Am. Soc. Neurochemistry, Am. Inst. Chemists, Am. Chem. Soc., Internat. Neurochemistry Soc., AAAS, Sigma Xi. Clubs: Nat. Wildlife, Internat. Wildlife. Subspecialties: Biochemistry (medicine); Neurochemistry. Current work: Control, metabolism in autonomic ganglia; membrane matrix metabolism; neurogrowth factors; metabolism and control of amino acid and lipid metabolism. Office: Dept Biochemistry Sch of Medicine SUNY Buffalo NY 14214

KLINMAN, CYNTHIA STONE, psychoanalyst; b. Phila., Jan. 23, 1939; d. William and Mary Stone; m. Jerome J. Klinman, June 14, 1958; children: Willann, Dara, Shani. A.B., Bryn Mawr Coll., 1960, M.A., 1962; Ph.D., U. Conn., 1964. Psychologist Allentown (Pa.) State Hosp., 1961-62; asst. prof. dept. psychology U. Conn., Storrs, 1964-65; chief psychologist Phila. Psychiat. Ctr., 1965-66; dir. psychol. services Inst. Pa. Hosp., Phila., 1966-67; cons., dir. psychiat. day hosp. Roosevelt Hosp., N.Y.C., 1967-68; pvt. practice psychoanalysis, N.Y.C., 1967—. Contbr. articles to profl. jours. Mem. Am. Psychol. Assn., William Alanson White Soc., N.Y. Psychol. Assn., Eastern Psychol. Assn., N.Y. Assn. Learning Disabled Adults and Children, Sigma Xi. Subspecialties: Behavioral psychology; Learning. Current work: Development of innovative psychotherapeutic techniques with individuals, couples, families, groups. Address: 215 E 61 St New York NY 10021

KLINMAN, JUDITH POLLOCK, chemistry educator; b. Phila., Apr. 17, 1941; d. Edward and Sylvia (Fitterman) Pollock; m. Norman Klinman, July 3, 1963 (div. June 1978); children: Andrew, Douglas. A.B., U. Pa., 1962, Ph.D., 1966. Postdoctoral fellow Weizmann Inst., Rehovot, Israel, 1966-67; postdoctoral assoc. Inst. Cancer Research, Phila., 1968-70, research assoc., 1970-72, staff scientist, 1972-78; assoc. prof. chemistry U. Calif.-Berkeley, 1978-82, prof., 1982—. Mem. editorial bd.: Jour. Biol. Chemistry, 1979—. NSF fellow, 1964; NIH fellow, 1964-66. Mem. Am. Chem. Soc. (exec. council div. biochemistry 1982—), Am. Soc. Biol. Chemists, Sigma Xi. Subspecialties: Biochemistry (biology); Biophysical chemistry. Current work: Mechanism of enzymatic catalysis and regulation. Office: Dept Chemistry U Calif Berkeley CA 94720

KLIORE, ARVYDAS JOSEPH, space scientist; b. Kaunas, Lithuania, Aug. 5, 1935; came to U.S., 1948, naturalized, 1951; s. Bronius J. and Antonia (Valaitis) K.; m. Birute Anna Ulenas, Sept. 3, 1960; children: Saule Andrea, Rima Birute. B.S.E.E., U. Ill., Urbana, 1956; M.S.E., U. Mich., 1957; Ph.D., Mich. State U., East Lansing, 1962. Engr. Armour Research Found., Chgo., 1957-59; teaching asst. Mich. State U., East Lansing, 1961-62; sr. engr. Jet Propulsion Lab., Calif. Inst. Tech., Pasadena, 1962-66, mem. tech. staff, 1966-81, research scientist, 1981—; co-chmn. Venus Internat. Reference Atmosphere Task Group, COSPAR, 1982—. Editor: The Mars Reference Atmosphere, 1982; contbr. articles to profl. jours. Recipient medal NASA, 1972, Group Achievement award, 1968, 69, 75, 80; Henry Earle Riggs fellow; Bendix fellow, 1962, 1962. Mem. Internat. Astron. Union, Internat. Com. Space Research (exec. mem.), Am. Astron. Soc., Am. Geophys. Union, AAAS, Sigma Xi. Clubs: Backa Athletic, Lithuanian American Community (Los Angeles). Subspecialties: Planetary science; Radio and microwave astronomy. Current work: Radio propagation studies planetary atmospheres, conduct radio sci. expts. with spacecraft to study planetary atmospheres and ionospheres. Home: 1475 Scenic Dr Pasadena CA 91103 Office: 4800 Oak Grove Dr 161-228 Pasadena CA 91109

KLIVINGTON, KENNETH A., med. instrument co. exec.; b. Cleve., Sept. 23, 1940; s. Albert C. and Evelyn Louise (Groom) K.; m. Marie Rose Lopez, Nov. 17, 1977; 1 son, Jason Alexis. S.B., M.I.T., 1962; M.S., Columbia U., 1964; Ph.D., Yale U., 1967. Asst. research neuroscientist U. Calif., San Diego, 1967-68; dir. research and devel. Fisher/Jackson Assoc., N.Y.C., 1968-69; program officer, adminstr. Sloan Research Fellowships, A.P. Sloan Found., N.Y.C., 1969-81; v.p. research and devel. Electrobiology, Inc., Fairfield, N.J., 1981—; vis. scientist U. Calif., San Diego, 1973. Contbr. articles to profl. jours., mags., newspapers. Mem. Soc. Neurosci., Cognitive Sci. Soc., Internat. Brain Research Soc., AAAS, Biol. Regeneration and Growth Soc. Current work: Administration of multidisciplinary research in bioelectromagnetics. Office: Electro-Biology 277 Fairfield Rd Fairfield NJ 07006

KLOCK, BENNY L., astronomer; b. Washington, Oct. 29, 1934; s. Leroy and Ertie (Crouse) K.; m. Mildred Olivia Burgess, May 10, 1976; children: Mark, Lorri, Brian. B.A., Cornell U., 1956, M.S., 1960; Ph.D., Georgetown U., 1964. Tech. asst. dir. six-inch Transit Circle Div. U.S. Naval Obs., Washington, 1965-69, dir., 1969-76, chief, 1976—; cons. astrometry. Contbr. articles in field to astronomy and high precision metrology to profl. jours. Com. chmn. Troop 493 Boy Scouts Am., Rockville, Md., 1970-73. Served to capt. USAF, 1957-59. NSF grantee, 1974-75. Mem. Internat. Astron. Union, Am. Astron. Soc. Republican. Episcopalian. Current work: High precision instrumentation for the measurement of small angles. Home: 6601 S Homestake Dr Bowie MD 20715 Office: US Naval Obs Washington DC 20390

KLOCK, JOHN CHARLES, scientist, research administrator; b. Corpus Christi, Tex., Dec. 27, 1944; s. John Charles and Ruth (Myers) K.; m. Cynthia Lavrene; children: Ernest Lorne, Cade Ross. B.S., La. State U.-Baton Rouge, 1966; M.D., Tulane U., 1970. Intern in medicine Children's Hosp. San Francisco, 1970-71; resident in medicine U. Calif.-San Francisco, 1971-73, fellow in hematology, 1973-75, assoc. dir. Cancer Research Inst. asst. prof., 1976-81; assoc. dir. Inst. Cancer Research, Pacific Med. Ctr., San Francisco, 1981—; fellow in oncology Mass. Gen. Hosp., Boston, 1975-76. Assoc. editor: Jour. Clin. Apheresis, 1981—. Fellow Leukemia Soc. Am., 1974-76; scholar, 1980-85. Mem. Am. Fedn. Clin. Research, Western Soc. Clin. Research, Am. Soc. Hematology, Internat. Assn. Leukemia and Related Diseases. Democrat. Presbyterian. Subspecialties: Hematology; Oncology. Current work: structure of blood cell antigens in normal and leukemia cells; patentee artifical antigens and preparation of universal donor plasma related to this research. Office: 2200 Webster St San Francisco CA 94115

KLOCK, PETER ILLITCH, JR., energy conservation co. exec.; b. Pitts., Feb. 24, 1940; s. Peter Illitch and Lillian Buchanan (Gresham) K.; m.; 1 dau., Holly Alexis. B.S.M.E., Calif. State U., San Jose, 1964; M.S.M.E., Stanford U., 1966. Registered profl. engr., Calif. Energy project mgr. Kaiser Engrs., Oakland, Calif., 1970-74; v.p. environ. affairs URS Corp., San Mateo, Calif., 1974-76; pres. ESC Energy Corp., San Mateo, 1976—. Mem. ASME, Assn. Energy Engrs. Subspecialty: Energy conservation. Current work: To apply cogeneration technology to commercial and industrial applications by means of innovative design and financing. Home: 494 El Granada Blvd El Granada CA 94018 Office: 1611 Borel Pl Suite 222 San Mateo CA 94402

KLOCKZIEN, VERNON GEORGE, aerospace company executive, consultant; b. Chgo., Sept. 22, 1921; s. George Anthony and Esther Marie (Otto) K.; m. Virginia Rose Mikkola, Aug. 1, 1953; children: Alice, George, Bettina, Charlene. B.S. in Aero. Engring, U. Ill., 1948, M.S., Purdue U., 1950. Research engr. Boeing Co., Seattle, 1950-53; supr. air systems Northrop Aircraft Co., Hawthorne, Calif., 1953-59; dir. thermophysics lab. Lockheed Missile & Space Co., Sunnyvale, Calif., 1959-65; with (Boeing Aerospace Co.), Seattle, 1965—, mgr. propulsion and mech. systems, 1975—. Chmn. Queen Anne Community Council, Seattle, 1969-79; vice chmn. King County Assn. Community Councils, 1972-76, Seattle Center Commn., 1976-82; trustee Seattle Mental Health, 1975-78. Served to 2d lt. USAAF, 1942-45; PTO. Assoc. fellow AIAA; mem. Acad. Polit. Sci., Assn. Am. Historians. Democrat. Subspecialties: Aerospace engineering and technology; Cryogenics. Current work: Aerospace propulsion including solid rockets, cryogenic, ramjets and electric propulsion, spacecraft thermal control, thermophysics. Home: 1635 Sunset Ave SW Seattle WA 98116 Office: Boeing Aerospace Co PO Box 3999 Seattle WA 98124

KLOSE, JULES ZEISER, physicist; b. St. Louis, Aug. 7, 1927; s. Julius Harry and Florence (Zeiser) K.; m. Evelyn Yvonne Brady, Jan. 22, 1958; children: Linda M., Jules S., Charles D., James M. A.B., Washington U., St. Louis, 1949; M.S., U. Rochester, 1953; Ph.D. in Physics, Cath. U. Am., 1958. Asst. prof. physics U.S. Naval Acad., 1953-58, assoc. prof., 1959-61; research assoc., lectr. U. Mich., 1960-61; physicist Nat. Bur. Standards, Washington, 1961—. Contbr. articles in field to profl. jours. Served with AUS, 1946-47. Mem. Am. Phys. Soc., Sigma Xi. Club: Severn Valley Racquet (Gambrills, Md.). Current work: Involved in the development and evaluation of sources of calibration in the vacuum ultraviolet to be used in space experiments. Office: Nat Bur Standards Washington DC 20234

KLOTZ, IRVING MYRON, educator, scientist; b. Chgo., Jan. 22, 1916; s. Frank and Mollie (Nasatir) K.; m. Mary Sue Hanlon, Aug. 7, 1966; children: Edward, Audie Jeanne, David. B.S., U. Chgo., 1937, Ph.D., 1940. Research asso. in chemistry Northwestern U., 1940-42, instr., 1942- 46, asst. prof., 1946-47, asso. prof., 1947-50, prof., 1950-63, Morrison prof. chemistry, 1963—; Lalor fellow Marine Biol. Lab., Woods Hole, Mass., 1947-48, corp. mem., 1947—, trustee, 1957-65. Author: Chemical Thermodynamics, 3d rev. edit., 1972, Energies in Biochemical Reactions, rev. edit., 1967; articles sci. jours. Recipient Army-Navy cert. of appreciation for wartime research, 1948. Fellow Royal Soc. Medicine, Am. Acad. Arts and Scis.; mem. Nat. Acad. Scis., Am. Soc. Biol. Chemists, Am. Chem. Soc. (Eli Lilly award 1949, Midwest award 1970), AAAS, Phi Beta Kappa, Sigma Xi, Phi Lambda Upsilon, Alpha Chi Sigma. Subspecialties: Biophysical chemistry; Catalysis chemistry. Current work: Molecular structure and function of biomacromolecules, thermodynamics. Home: 2515 Pioneer Rd Evanston IL 60201

KNACKE, ROGER FRITZ, astronomer; b. June 22, 1941; s. Theodore and Margaret L. (Kostlin) K.; m. Nancy J. Knacke, Mar. 8, 1972. A.B., U. Calif., Berkeley, 1963, Ph.D., 1969. Faculty SUNY, Stony Brook, 1971—, prof. astronomy, 1979—; postdoctoral fellow U. Calif., Santa Cruz, 1970. Contbr. articles to profl. jours. Mem. Internat. Astron. Union, Am. Astron. Soc. Subspecialties: Infrared optical astronomy; Planetary science. Current work: Planetary science, planetary atmospheres, comets, interstellar medium, interstellar dust. Office: Dept Earth and Space Sci SUNY Stony Brook NY 11794

KNAPP, DANIEL ROGER, pharmacologist, educator; b. Evansville, Ind., July 29, 1943; s. Linus John and Florence Crystal (Atherton) K.; m. Rebecca Grant, Apr. 15, 1976; 1 son, Daniel Hugh. B.A. magna cum laude, U. Evansville, 1965; Ph.D., Ind. U., 1969. Postdoctoral fellow U. Calif., Berkeley, 1969-70; asst. prof. exptl. medicine U. Cin. Coll. Medicine, 1970-72; asst. prof. pharmacology Med. U. S.C., 1972-78, assoc. prof., 1978—. Author: Handbook of Analytical Derivatization Reactions, 1979. Mem. Am. Chem. Soc., Am. Soc. Mass Spectrometry, Am. Soc. Pharmacology and Exptl. Therapeutics. Subspecialties: Mass spectrometry; Organic chemistry. Current work: Biomedical mass spectrometry; medical chemistry. Office: 171 Ashley Ave Charleston SC 29425

KNAPP, RONALD HARRISON, mechanical engineering educator; b. Tuscola, Ill., July 20, 1944; s. Robert Harrison and Bessie Margaret (Tell) K. B.S.M.E. with honors, U. Hawaii, 1967, Ph.D. in Ocean Engring, 1973; M.S.M.E., Calif. Inst. Tech., 1968. Registered profl. engr., Hawaii. Mech. engr. Naval Ocean Systems Ctr., Kaneohe, Hawaii, 1968-75; v.p. Ocean Engring. Cons., Inc., Honolulu, 1972-80; assoc. prof. dept. mech. engring. U. Hawaii, 1975—; pres. Knapp Engring., Inc., Aiea, Hawaii, 1980—. Co-author: Cable Technology Handbook; contbr. articles to profl. jours. U.S. Navy fellow, 1970-72; Alexander von Humboldt Found. fellow, 1981-82. Mem. ASME, Marine Tech. Soc., AAUP, Nat. Soc. Profl. Engrs. Methodist. Subspecialties: Mechanical engineering; Solid mechanics. Current work: Mechanical analysis and design of power and communication cables; development of computer software for the stress analysis of cables. Home: 98-1033 Kupukupu Pl Aiea HI 96701 Office: Dept Mech Engring U Hawaii 2540 Dole St Honolulu HI 96822

KNIAZEWYCZ, BOHDAN GEORGE, engineering consultant, chemical/nuclear engineer; b. Erlangen, Germany, Nov.2, 1947; came to U.S., 1950, naturalized, 1957; s. Theodor and Maria (Piorkowska) K.; m. m Mary Catherine Redig, June 21, 1980; 1 dau., Caitlin Alexandra. B.E. in Chem. Engring., Vanderbilt U., 1969; M.S. in Nuclear Engring., U. Tenn.-Knoxville, 1971. Registered profl. engr., N.C., La., Calif. Sr. engr. Carolina Power & Light Co., Raleigh, N.C., 1971-75; sr. cognizant engr. Burns & Roe, Oradell, N.J., 1975-76; div. mgr. TERA Corp., Berkeley, Calif., 1976-80, EDS Nuclear, San Francisco, 1980-82; ptnr. KLM Engring., Inc., Walnut Creek, Calif., 1982—; speaker, nat. speaking tour Am. Inst. Chem. Engrs., N.Y.C., 1976-77. Contbr. numerous articles on nuclear power, waste disposal to profl. publs. Served with C.E. USAR, 1969-77. AEC univ. trainee, 1970-71. Mem. ASME, Am. Nuclear Soc., Health Physics Soc., IEEE, Soc. Mfg. Engrs., Nat. Soc. Profl. Engrs., MENSA. Republican. Roman Catholic. Subspecialties: Nuclear engineering; Nuclear fission. Current work: Development of process technology to treat hazardous and radioactive wastes. Home: 233 San Antonio Way Walnut Creek CA 94598 Office: KLM Engring Inc 1776 Ygnacio Valley Rd Walnut Creek CA 94598

KNIEVEL, DANIEL PAUL, crop physiologist, educator, researcher; b. West Point, Nebr., Jan. 29, 1943; s. Leo William and Margaret Veronica (Ziska) K.; m. Beth Ann Snoberger, Aug. 28, 1965; children: Jason Clark, Ann Marie. B.S., U. Nebr., 1965; M.S., U. Wis.-Madison, 1967, Ph.D., 1968. Research asst. U. Wis.-Madison, 1965-68; crop physiologist U. Wyo., Laramie, 1968-72; crop sci. adv. to Argentina AID, Buenos Aires, 1972-74; crop physiologist Pa. State U., University Park, 1974—, assoc. prof., 1975—; vis. scientist USDA/ARA Grassland Soil and Water Research Lab., Temple, Tex., 1981-82. Contbr. articles on crop physiology to profl. jours.; assoc. editor: Agronomy Jour, 1978-83. Mem. Am. Soc. Agronomy, Crop Sci. Soc. Am., Am. Soc. Plant Physiologists, Council for Agrl. Sci. and Tech., Sigma Xi. Democrat. Roman Catholic. Lodge: KC. Subspecialties: Plant physiology (agriculture); Integrated systems modelling and engineering. Current work: Plant stress physiology, physiology of grain yield, crop systems model development and application. Home: 732 Edgewood Circle State College PA 16801 Office: 119 Tyson Bldg University Park PA 16802

KNIGHT, DAVID BATES, chemistry educator; b. Louisville, Sept. 23, 1939; s. Theron Turner and Mary (Bates) K.; m. Gwen Waldrop, Aug. 26, 1965; children: James Derek, Kate Elaine. B.S., U. Louisville, 1961; M.A., Duke U., 1963, Ph.D., 1966. Postdoctoral fellow Ohio State U., Columbus, 1965-67; asst. prof. chemistry U. N.C., Greensboro, 1967-72, assoc. prof., 1972-80, prof., 1980—. Contbr. articles to profl. jours. Petroleum Research fund grantee, 1979-81. Mem. Am. Chem. Soc. Subspecialties: Organic chemistry; Kinetics. Current work: Chemistry of fulvenes, kinetics of H-D exchange in hydrocarbons, hydrocarbon acidity. Office: Dept Chemistry U NC Greensboro NC 27412

KNIGHT, VERNON, medical educator; b. Osceola, Mo., Sept. 6, 1917; s. Iven Robert and Myrtle (Andrews) K.; m. Elizabeth Gordon, Sept. 21, 1946; children—Hunter, Caroline, James, John. A.B., William Jewell Coll., 1939; M.D., Harvard U., 1943. Diplomate: Am. Bd. Internal Medicine. Intern medicine Mass. Meml. Hosps., Boston, 1943; house officer N.Y. Hosp.-Cornell Med. Center, 1945-46, research fellow dept. medicine, 1946-52; asst. prof. medicine Cornell Med. Coll., 1953-54; attending physician Cornell Med. Service Bellevue Hosp. and Meml. Center for Cancer, N.Y.C., 1953-54; asso. prof. medicine Vanderbilt U. Med. Sch., 1954-59; clin. dir. Inst. Allergy and Infectious Diseases, NIH, Bethesda, Md., 1959-66; prof., chmn. dept. microbiology and immunology, prof. medicine Baylor Coll. Medicine, Houston, 1966—. Served to lt. (s.g.), M.C. USNR, 1944-46. Mem. A.C.P., Assn. Am. Physicians, Am. Soc. Clin. Investigation, Am. Clin. and Climatol. Assn., Soc. Exptl. Biology and Medicine. Subspecialty: Virology (biology). Home: 11735 Green Bay Dr Houston TX 77024 Office: 1200 Moursund Ave Houston TX 77030

KNOBIL, ERNST, physiologist; b. Berlin, Germany, Sept. 20, 1926; came to U.S., 1940, naturalized, 1945; s. Jakob and Regina (Seidmann) K.; m. Julane Hotchkiss, July 11, 1959; children: Erich Richard, Mark, Nicholas, Katharine. B.S., Cornell U., 1948, Ph.D. (Schering fellow endocrinology 1949-51), 1951; hon. doctorate, U. Bordeaux, France, 1981, Med. Coll. Wis., 1983. Asst. zoology Cornell U., 1948-49; Milton Research fellow, Harvard U., 1951-53, from instr. to asst. prof. physiology, 1953-61, John and Mary R. Markle Found. scholar med. scis., 1956-61; Richard Beatty Mellon prof. physiology, chmn. dept. U. Pitts. Sch. Medicine, 1961-81; H. Wayne Hightower prof. physiology, dean U. Tex. Med. Sch., Houston, 1981—; Bowditch lectr. Am. Physiol. Soc., 1965, Gregory Pincus Meml. lectr. Laurentian Hormone Conf., 1973; Upjohn lectr. Am. Fertility Soc., 1974; Kathleen M. Osborn Meml. lectr. U. Kans., 1974; Karl Paschkis lectr. Phila. Endocrine Soc., 1975; 1st Transatlantic lectr. Soc. Endocrinology, Gt. Brit., 1979, Lawson Wilkins Pediatric Endocrine Soc. lectr., 1980; Bard lectr. Johns Hopkins U. Sch. Medicine, 1981; Herbert M. Evans Meml. lectr. U. Calif., 1981, Am. Acad. Arts and Scis. fellow, 1981; cons. USPHS, Ford Found.; Mem. human growth and devel. study sect. NIH, 1964-66; mem. adv. council Inst. Lab. Animal Resources, NRC-Nat. Acad. Sci., 1966-69; mem. nat. sci. adv. bd. Growth, Inc., 1969; mem. liaison com. on med. edn. AMA-Am. Assn. Med. Colls., 1971-74; mem. population research com. Center for Population Research, Nat. Inst. Child Health and Human Devel., NIH, 1974-78; mem. med. adv. bd. Nat. Pituitary Agy., 1980—. Editorial bd.: Am. jour. Physiology, 1959-68, Endocrinology, 1959-75, Psychoneuroendocrinology, 1974-79, Neuroendocrinology, 1976-80; editorial com. Ann. Rev. Physiology, 1968-72; editor, 1974-77; editor-in-chief: Am. Jour. Physiology: Endocrinology and Metabolism, 1979-82. Served with U.S. Army, 1944-46. Fellow AAAS; hon. fellow Am. Assn. Obstetricians and Gynecologists, Am. Gynecol. Soc.; mem. Am. Soc. Zoologists, Soc. Exptl. Biology and Medicine, Am. Physiol. Soc. (mem. council 1969-72, pres. 1978-79), Endocrine Soc. (Ciba award 1961, council 1968-71, pres. 1976-77, Koch award 1982), Soc. Endocrinology (Gt. Britain), Internat. Soc. Research in Biology of Reprodn., Assn. Chairmen Depts. Physiology (pres. 1969), Am. Assn. Med. Colls. (adminstrv. bd. council acad. socs., exec. com.), Nat. Bd. Med. Examiners, Internat. Soc. Endocrinology (mem. exec. com. 1972—, chmn. program organizing com. 5th Congress 1976), Internat. Soc. Neuroendocrinology (chmn. exec. com. 1976-84), Soc. Study Reprodn., Japan Endocrine Soc., Internat. Soc. Research Biology Reprodn.; hon. mem. Deutsche Gesellschaft für Endokrinologie. Subspecialties: Neuroendocrinology; Physiology (biology). Spl. research pituitary gland, endocrinology reprodn.

KNOCHE, HERMAN WILLIAM, agricultural biochemistry educator; b. Stafford, Kans., Nov. 15, 1934; s. Herman William and Ollie Emeline (Keller) K.; m. Darlene K. Bowman, Feb. 5, 1955; children: Kimberly K., Christopher L. B.S., Kans. State U.-Manhattan, 1959, Ph.D., 1963. Instr. U. Nebr.-Lincoln, 1962-63, asst. prof. agrl. biochemistry, 1963-68, assoc. prof., 1968-73, prof., 1973—, head dept. agrl. biochemistry, 1974—; cons. in field. Contbr. articles to profl. jours. Mem. Nebr. Gov.'s Radiation Adv. Council, 1966-80, chmn., 1969-71. Served with U.S. Army, 1953-55. Mem. Am. Chem. Soc., Am. Soc. Biol. Chemists, Sigma Xi. Republican. Methodist. Subspecialty: Biochemistry (biology). Current work: Biochemistry of lipids, identification of natural products, plant toxins. Home: Route 2 Box 36A Ceresco NE 68017 Office: Dept Agrl Biochemistry U Nebr Lincoln NE 68583

KNODEL, ELINOR LIVINGSTON, process chemist; b. N.Y.C., Jan. 25, 1947; d. and Elinor (Findley) K. A.B., Barnard Coll., 1969; M.S., Yale U., 1972; Ph.D., U. Conn., 1976. Process chemist E.I. DuPont Co., Wilmington, Del., 1980—. Contbr. articles to profl. publs. Postdoctoral research fellow Mayo Clinic, Rochester, Minn., 1978-80, Rockefeller U., N.Y.C., 1976-77. Mem. AAAS, Soc. Neurosci., Am. Chem. Soc., Am. Assn. Clin. Chemistry, Wilmington Women in Bus. Episcopalian. Subspecialties: Enzyme technology; Clinical chemistry. Current work: Enzyme purification for clinical chemistry applications. Office: EI DuPont Co Clin Systems Div Glasgow Site Wilmington DE 19898

KNOEBEL, SUZANNE BUCKNER, physician; b. Ft. Wayne, Ind., Dec. 13, 1926; d. Doster and Marie (Lewis) Buckner. A.B., Goucher Coll., 1948; M.D., Ind. U., 1960. Diplomate: Am. Bd. Internal Medicine. Intern Ind. U. Med. Ctr., Indpls., 1960-61, resident, 1961-62; asst. prof. medicine Ind. U. Sch. Medicine, 1966-69, assoc. prof., 1969-72, prof., 1972-77, Krannert prof. medicine, 1977—, asst. dean research, 1975—; asst. chief cardiology sect. Richard L. Roudebush VA Med. Center, Indpls., 1982—; assoc. dir. Krannert Inst. Cardiology, Indpls., 1974—. Mem. Am. Fedn. Clin. Research, Assn. Univ. Cardiologists, Am. Coll. Cardiology (pres. 1982-83). Subspecialty: Cardiology. Current work: Clinical coronary heart disease, ischemia, myocardial blood flow, clinical arrhythmias and computer analysis of cardiovascular data. Office: Ind U Sch Medicine 1100 W Michigan St Indianapolis In 46223

KNOLL, ANDREW HERBERT, paleontologist, educator; b. W. Reading, Pa., Apr. 23, 1951; s. Robert Samuel and Anna Augusta (Meyer) K.; m. Marsha Craig, June 22, 1974. B.A. with highest honors, Lehigh U., 1973; A.M., Harvard U., 1974, Ph.D., 1977. Asst. prof. Oberlin (Ohio) Coll., 1977-81, assoc. prof., 1982, Harvard U., Cambridge, Mass., 1982—; mem. Com. on Planetary Biology and Chem. Evolution, U.S. Space Sci. Bd, 1981—, Am. Inst. Biol. Scis. adv. panel to NASA exobiology program, 1980—. Contbr. articles on early evolution of life on earth and evolution of land plants to profl. jours. Active choral groups Boston, Oberlin. NSF grantee, 1980-81. Mem. Bot. Soc. Am., Geol. Soc. Am., Paleontol. Soc., AAAS, Soc. Econ. Paleontol. Mineralogy, Internat. Soc. Study Origins of Life, Phi Beta Kappa, Sigma Xi. Democrat. Subspecialties: Paleontology; Evolutionary biology. Current work: Field and laboratory research on the early evolution of life and the fossil record of land plants. Office: Harvard U Herbaria 22 Divinity Ave Cambridge MA 02138

KNOLL, GLENN FREDERICK, nuclear engineering educator; b. St. Joseph, Mich., Aug. 3, 1935; s. Oswald Herman and Clara Martha (Bernthal) K.; m. Gladys Hetzner, Sept. 7, 1957; children: Thomas, John, Peter. B.S., Case Inst. Tech., 1957; M.S. in Chem. Engring., Stanford, 1959; Ph.D. in Nuclear Engring., U. Mich., 1963. Asst. research physicist U. Mich., Ann Arbor, 1960-62, asst. prof. nuclear engring., 1962-67, asso. prof., 1967-72, prof., 1972—, chmn. dept. nuclear engring., 1979—, also mem. bioengring. faculty.; Vis. scientist Institut für Angewandte Kernphysik, Kernforschungszentrum Karlsruhe, Germany, 1965-66; sr. vis. fellow dept. physics U. Surrey, Guildford, Eng., 1973; summer cons. Electric Power Research Inst., Palo Alto, Calif., 1974; cons. in field. Author: Radiation Detection and Measurement, 1979, Principles of Engineering, 1982. NSF fellow, 1958-60; Fulbright travel grantee, 1965-66; Sci. Research Council sr. fellow, 1973. Fellow Am. Nuclear Soc.; mem. Am. Assn. Engring. Edn. (Glenn Murphy award 1979); Mem. IEEE (nuclear and plasma soc., chmn. tech. com. on nuclear med. sci. 1977-79), Soc. Nuclear Medicine, Sigma Xi, Tau Beta Pi. Subspecialties: Nuclear engineering; Bioinstrumentation. Current work: Radiation detection and spectroscopy, engineering aspects in nuclear medicine and radiology. Patentee in field. Office: Dept Nuclear Engring 119 Cooley Bldg U Mich Ann Arbor MI 48109

KNOLLMAN, GIL CARL, research physicist, consultant; b. Cleve., Mar. 14, 1928; s. Paul Carl and Louise Katherine (Heidebrink) K.; m. Lorraine Tommie Gordon, Mar. 14, 1959; children: Tom, Scott, Kristi, Katrina. B.S., Ga. Inst. Tech., 1949, M.S., 1950, Ph.D., 1961; postgrad. student, Stanford U., 1965-66. Sr. research physicist Ga. Tech. Research Inst., Atlanta, 1950-62; asst. prof. math Ga. Inst. Tech., 1952-60; research scientist Lockheed Research Lab., Palo Alto, Calif., 1962-63, staff scientist, 1963-64, sr. staff scientist, 1964-66, sr. mem. research labs., 1966—; cons. Ga. Research Inst., 1962-65; Lockheed Calif. Co., Los Angeles, 1964-67, Saratoga Systems, Cupertino, Calif., 1971-76; mem. U.S. Navy Tech. Coms., Washington, 1957-67, 75—; assoc. Palo Alto (Calif.) Research Observer Program, 1965-75. Author: Meson Mass Determination, 1950, Model Dense Boson System, 1961. Mem. Jimmie Carter Presdl. Campaign Com., San Francisco, 1976, 80; Spl. Agt. U.S. Counterintelligence, 1946-47. Fellow Am. Phys. Soc., N.Y. Acad. Sci., AAAS, Acoustical Soc. Am.; mem. Am. Assn. Physics Tchrs. Republican. Subspecialties: Acoustics; Materials. Current work: Planning, directing and conducting fundamental and applied research in and development of ultrasonic methodology for materials evaluation. Consultant in fields of acoustics, fluid and solid state physics; viscoelasticity, ocean sciences, nondestructive testing, ultrasonic instrumentation and aerospace structures. Patentee ultrasonic weld inspection; adhesive bond evaluation system. Home: 705 Charleston Ct Palo Alto CA 94303 Office: Lockheed Research Lab 3251 Hanover St Palo Alto CA 94304

KNOTT, DOUGLAS RONALD, crop scientist, educator; b. New Westminster, B.C., Can., Nov. 10, 1927; s. Ronald David and Florence Emily (Keeping) K.; m. Joan Madeline Hollinshead, Sept. 2, 1950; children: Holly Ann, Heather Lynn, Ronald Kenneth, Douglas James. B.S.A. in Agr, U. B.C., 1948; M.S., U. Wis., 1949, Ph.D., 1952. Asst. prof. field. husbandry U. Sask., 1952-56, assoc. prof., 1956-65, prof., head crop sci. dept., 1965-75, prof. crop sci., 1975—. Contbr. research papers to profl. lit. Fellow Am. Soc. Agronomy, Agrl. Inst. Can., Can. Soc. Agronomy, Genetics Soc. Am., Can. Soc. Genetics, Sigma Xi. Liberal. Mem. United Ch. Subspecialties: Plant genetics; Plant pathology. Current work: Wheat cytogenetics and wheat breeding. Home: 2002 14th St E Saskatoon SK Canada S7H OB2 Office: Crop Sci Dept U Sask Saskatoon SK Canada S7N OWO

KNOTT, PETER JEFFREY, pharmacology educator; b. London, Sept. 23, 1942; s. Edward Arthur and Ivy May (Baine) K.; m. Susan Carol Panter, Mar. 16, 1968; 1 dau., Samantha. B.Sc., U. London, 1966, M.Sc., 1970, Ph.D., 1976. Lectr. Inst. Neurology, U. London, 1974-79; asst. prof. pharmacology Marshall U., 1979-81, assoc. prof., 1981—, dir. neuropharmacology program, 1982—; pharmacology cons. Huntington VA Med. Ctr. Contbr. articles to profl. jours. NIH grantee, 1981-84. Mem. Am. Soc. Pharmacology and Exptl. Theapeutics, Brit. Pharmacol. Soc., Soc. Neurosci., N.Y. Acad. Scis. Subspecialties: Pharmacology; Neuropharmacology. Current work: Study of drugs on brain function as it related to behavior, brain neurotransmitter metabolism changes studied by HPLC, changes neurotransmitter release measured by in vivo electrochemistry. Home: 155 Honeysuckle Ln Hintington WV 25701 Office: Dept of Pharmacology Marshall U Sch Medicine Huntington WV 25704

KNOWLDEN, NORMAN FRANCIS, math. statistician; b. Newburgh, N.Y., Apr. 6, 1925; s. Benjamin Walter and Mabel Leona (Lockwood) K.; m. Gloria Caren Bush; children: Erik Kristin, Ethan Konrad. B.A. (Council scholar), N.Y.U., 1946; postgrad., Columbia U., 1949-51; M.S., Stevens Inst. Tech., 1953. Research chemist Maltbie Chem. Co., Newark, 1946-47; analytical statistician Burroughs-Wellcome Co. Inc., Tuckahoe, N.Y., 1947-54; statis. group leader Am. Cyanamid Co. Lederle Labs., Pearl River, N.Y., 1954-61; mgr. corp. mgmt. scis. project Allied Chem. Corp., Morristown, N.J., 1961-68, 69-79, research assoc. in math. scis., 1979—; v.p. Telos Sci. Co., N.Y.C., 1969; mem. vis. faculty mgmt. sci. dept. Stevens Inst. Tech. Grad. Sch., 1957-72. Contbr. articles to sci. jours. Mem. Am. Statis. Assn., Am. Soc. for Quality Control (cert. quality engr. and reliability engr.; edn. com. 1982-83), Zeta Psi. Lutheran. Subspecialties: Statistics; Operations research (mathematics). Current work: Process operations analysis, epidemiology, environmental control procedures; application of mathematical science and statistics in areas of process yield and quality improvement; epidemiological studies; environmental and industrial safety controls. Home: Brockden Pl Mendham NJ 07945 Office: PO Box 1087R Morristown NJ 07960

KNOWLES, BARBARA B., microbiologist, geneticist, educator; b. N.Y.C., Feb. 27, 1937; d. Christian J. and Undine G. Bang; m.; children: Jared Appleton, Amanda Gaylord. A.B., Middlebury Coll., 1958; M.S., Ariz. State U., 1963, Ph.D., 1965. Postdoctoral trainee dept. genetics U. Calif., Berkeley, 1965-66; assoc. prof. Wistar Inst., Phila., 1967-82, prof., 1983—; assoc. prof. U. Pa. Coll. Arts and Scis., 1977—. Contbr. articles to profl. publs. Recipient USPHS research career devel. award, 1975-80. Mem. Genetics Soc. Am., Am. Soc. Human Genetics, AAAS. Subspecialties: Immunogenetics; Cell biology (medicine). Current work: Description and genetic control of normal, tumor-specific and embryonic cell surface molecules. Office: Wistar Inst 36th St and Spruce St Philadelphia PA 19104

KNOWLES, JEREMY RANDALL, chemist, educator; b. Rugby, Eng., Apr. 28, 1935; came to U.S., 1974; s. Kenneth Guy Jack Charles and Dorothy Helen (Swingler) K.; m. Jane Sheldon Davis, July 30, 1960; children: Sebastian David Guy, Julius John Sheldon, Timothy Fenton Charles. B.A., Balliol Coll., Oxford (Eng.) U., 1958; M.A., D.Phil., Christ Ch., 1961. Research fellow Calif. Inst. Tech., 1961-62; fellow Wadham Coll., Oxford U., 1962-74, univ. lectr., 1966-74; vis. prof. Yale U., 1969, 71; Sloan vis. prof. Harvard U., 1973, prof. chemistry, 1974—, Amory Houghton prof. chemistry and biochemistry, 1979—; Newton-Abraham vis. prof. Oxford U., 1983-84. Author papers, revs. bioorganic chemistry. Served as pilot officer RAF, 1953-55. Fellow Royal Soc., Chem. Soc. London, Am. Acad. Arts and Scis.; mem. Biochem. Soc. London, Am. Chem. Soc., Am. Soc. Biol. Chemists. Subspecialties: Biochemistry (biology); Organic chemistry. Home: 44 Coolidge Ave Cambridge MA 02138 Office: Dept Chemistry Harvard Univ Cambridge MA 02138

KNOWLES, LLOYD GEORGE, electrical engineering educator; b. Pottsville, Pa., Jan. 27, 1934; s. William George and Thelma May K.; m.; children: David Lloyd, Carolyn Anne, Raymond William. Sr. engr. Applied Physics Lab., Johns Hopkins U., Laurel, Md., 1961—; asst. prof. Johns Hopkins U., Balt., 1974—. Contbr. articles to tech. jours., chpts. in books. Mem. Howard County (Md.) County Council, 1974—, Howard County Planning Bd., 1969-74. Served in USN, 1955-57. Mem. Am. Assn. Physicists in Medicine, Soc. Nuclear Medicine, Soc. Photo-optical Instrumentation Engrs. Democrat. Unitarian. Clubs: Columbia (Md.) Democratic, Ellicott City (Md.) Democratic. Subspecialties: Biomedical engineering; Systems engineering. Current work: Specialist in evaluation of images in photon-limited situations, evaluating map-matching techniques in missile navigation. Home: 10850-617 Green Mountain Circle Columbia MD 21044 Office: Johns Hopkins U Applied Physics Lab Johns Hopkins Rd Laurel MD 20707

KNOWLES, RICHARD JAMES ROBERT, medical physicist, researcher, educator; b. McPherson, Kans., Aug. 2, 1943; s. Richard E. and Pauline H. (Worland) K.; m. Stephanie R. Closter, May 14, 1970; 1 dau., Guenevere Regina. B.S., St. Louis U., 1965; M.S., Cornell U., 1969; Ph.D., Poly Inst. N.Y., 1979. Cert. scientist in nuclear medicine Am. Bd. Sci. in Nuclear Medicine; cert. radiol. physicist Am. Bd. Radiology. Research asst. Wilson Synchrotron Lab., Cornell U., Ithaca, N.Y., 1967-70; research asst. Ward Reactor Lab., 1970-72; sr. technologist in nuclear medicine N.Y. Hosp., N.Y.C., 1972-77; chief med. physicist L.I. Coll. Hosp., Bklyn., 1977-81; dir. Radiation Physics Lab., Downstate Med. Center, SUNY, Bklyn., 1981-82; med. physicist N.Y. Hosp.-Cornell Med. Ctr., N.Y.C., 1982—; adj. asst. prof. physics York Coll., CUNY, Jamaica, 1982—. Contbr. articles to profl. publs. Mem. Am. Phys. Soc., Am. Assn. Physicists in Medicine, Soc. Nuclear Medicine, Health Physics Soc., N.Y. Acad. Scis. Subspecialties: Imaging technology; Graphics, image processing, and pattern recognition. Current work: Medical imaging, image processing, pattern recognition. Office: 525 E 68th St New York NY 10021

KNOWLES, STEPHEN HOWARD, astronomer; b. N.Y.C., Feb. 28, 1940; s. Howard Nesmith and Emily A. (Kent) K.; m. Joan E. Brame, June 5, 1965; children: Jennifer, Katherine. B.A., Amherst Coll., 1961; Ph.D., Yale U., 1968. Astronomer U.S. Naval Obs., Washington, 1961, U.S. Naval Research Lab., 1961-74, 76—; vis. scientist Radiophysics div. C.S.I.R., Sydney, Australia, 1974-76. Contbr. articles to profl. jours. Mem. Internat. Astron. Union, Am. Astron. Soc., Sigma Xi. Episcopalian. Subspecialties: Radio and microwave astronomy; Satellite studies. Current work: Interferometry, V.L.B.I., Astrometry, planetary detection, signal processing. Home: 12107 Harbor Dr Woodbridge VA 22192 Office: US Naval Research Lab Code 4132 Washington DC 20375

KNOX, JAMES RUSSELL, JR., biophysics educator; b. Bonne Terre, Mo., May 28, 1941; s. James Russell and Esther Verl (Vaden) K.; m. Jane Susan Levitas, Mar. 19, 1966; children: Craig, Clara. B.S. in Chemistry, U. Mo.-Rolla, 1963; Ph.D. in Phys. Chemistry, Boston U., 1967. Postdoctoral fellow Oxford (Eng.) U., 1966-69; research assoc. Yale U., New Haven, 1969-70; prof. biophysics U. Conn., Storrs, 1970—; vis. prof. Harvard U., 1977-78; cons. Hoffmann-LaRoche, Nutley, N.J., 1983—. Contbr. chpts. to books, articles to profl. jours. Organizer Mansfield (Conn.) Friends of Music, 1982—. NIH grantee, 1972—; NSF grantee, 1975-78; Merck Sharpe Dohme grantee, 1978—; Hoffman-LaRoche grantee, 1983—. Mem. Am. Chem. Soc., Am. Crystallographic Assn., Biophysical Soc., Sigma Xi. Subspecialties: Biophysical chemistry; X-ray crystallography. Current work: Structure and function of proteins and enzymes; especially those which interact with penicillin antibiotics. Office: Biochemistry and Biophysics Sect Biol Scis Group U Conn Box 136 Storrs CT 06268

KNOX, LARRY WILLIAM, paleontology educator; b. Mishawaka, Ind., Nov. 10, 1942; s. William Thompson and Burnetha Elizabeth K.; m. (m), Aug. 29, 1964; children: Christopher Ian, Jennifer Lynn. B.A. in Physics, Ind. U., 1965, M.A. in Geology, 1971, Ph.D., 1974. Research chemist Uniroyal, Inc., Mishawaka, Ind., 1965-69; assoc. prof. geology Tenn. Technol. U., Cookeville, 1974—. Contbr. articles to pubs. Recipient outstanding research award Sigma Xi, 1978. Mem. Geol. Soc. Am., Soc. Econ. Paleontologists and Mineralogists, Paleontol. Soc., Palaeontol. Assn., Internat. Paleontol. Union. Subspecialties: Paleontology; Paleoecology. Current work: Taxonomy, biostratigraphy, and paleoecology of Upper Paleozoic ostracodes; biostratigraphy of ordovician conodonts. Home: Route 6 Box 28 Cookeville TN 38501 Office: Tenn Technol Univ PO Box 5125 Cookeville TN 38505

KNUDSON, ALFRED GEORGE, JR., medical geneticist; b. Los Angeles, Aug. 9, 1922; s. Alfred George and Mary Gladys (Galvin) K.; m. Anna T. Meadows, June 20, 1977; children by previous marriage: Linda, Nancy, Dorene. B.S., Calif. Inst. Tech., 1944, Ph.D. (Guggenheim fellow), 1956; M.D., Columbia U., 1947. Chmn. dept. pediatrics City of Hope Med. Center, Duarte, Calif., 1956-62, chmn. dept. biology, 1962-66; asso. dean Health Sci. Center, SUNY, Stony Brook, 1966-69; dean Grad. Sch. Biomed. Scis., U. Tex. Health Sci. Center, Houston, 1970-76; dir. Inst. Cancer Research, Fox Chase Cancer Center, Phila., 1976—, pres., 1980—; mem. Assembly Life Scis., NRC, 1975-81. Author: Genetics and Disease, 1965. Recipient Disting. Alumni award Calif. Inst. Tech., 1978. Mem. Am. Soc. Human Genetics (pres. 1978), Assn. Am. Physicians, Am. Pediatrics Soc., Am. Assn. Cancer Research. Subspecialties: Genetics and genetic engineering (medicine); Cancer research (medicine). Research, publs. in genetics of human cancer. Office: Inst Cancer Research 7701 Burholme Ave Philadelphia PA 19111

KNUDSON, GREGORY BLAIR, geneticist, researcher, educator, army res. officer; b. Salina, Kans., Aug. 9, 1946; s. Cecil C. and Dorothy A. (Hamilton) K.; m. Kathryn H. Malloy, Oct. 21, 1972; 1 son, Todd C. B.A. in Biology, Calif. State U.-Fullerton, 1969, M.A., 1971; Ph.D. in Genetics, U. Calif.-Riverside, 1977. Commd. 2d lt. U.S. Army Res., 1970, advanced through grades to capt., 1977; research officer Biomed. Lab., Chem. Corps, Edgewood Arsenal, Md., 1970-73; res. capt. Med. Service Corps., 6252d USAR Hosp., San Bernardino, Calif., 1973-78; research geneticist U.S. Army Med. Research Inst. Infectious Diseases, Ft. Detrick, Md., 1978-81, bacterial geneticist, 1981—; mem. faculty Calif. State Coll., 1976-77; instr. recombinant DNA tech. Hood Coll., Md., 1980. Mem. Genetics Soc. Am., Am. Soc. Microbiology, AAAS, N.Y. Acad. Scis. Republican. Methodist. Subspecialties: Genetics and genetic engineering (biology); Molecular biology. Current work: Role of plasmids in pathogens; recombinant DNA technology; mechanisms of DNA repair. Office: USAMRIID Ft Detrick Frederick MD 21701

KNUTH, DONALD ERVIN, computer science educator; b. Milw., Jan. 10, 1938; s. Ervin Henry and Louise Marie (Bohning) K.; m. Nancy Jill Carter, June 24, 1961; children: John Martin, Jennifer Sierra. B.S., Case Inst. Tech., 1960, M.S., 1960; Ph.D., Calif. Inst. Tech., 1963. Asst. prof., then asso. prof. math. Calif. Inst. Tech., 1963-68; prof. computer sci. Stanford U., 1968—; guest prof. math. U. Oslo, 1972-73; cons. Burroughs Corp., 1960-68. Author: The Art of Computer Programming, vol. 1, 1968, vol. 2, 1969, vol. 3, 1973, Surreal Numbers, 1974, Mariages Stables, 1976, Tex and Metafont, 1980; Editor jours. Recipient Nat. medal of sci., 1979; Guggenheim fellow, 1972-73. Fellow Am. Acad. Arts and Scis., Brit. Computer Soc; mem. Nat. Acad. Scis., Nat. Acad. Engring., Assn. Computing Machinery (Grace Murray Hopper award 1971, Alan M. Turing award 1974), Math. Assn. Am. (Lester R. Ford award 1975), IEEE Computer Soc. (McDowell award 1980, Computer Pioneer award 1982), Am. Math. Soc., Soc. Indsl. and Applied Math., Am. Guild Organists. Lutheran. Subspecialties: Algorithms; Programming languages. Current work: Analysis of algorithims, digital typography, combinatorial mathematics. Patentee in field. Home: 1063 Vernier Pl Stanford CA 94305 Office: Computer Sci Dept Stanford Univ Stanford CA 94305

KO, HON-KIM, engineering educator; b. Hong Kong, Jan. 18, 1940; s. Ching and Mo-Ching (To) K.; m. Shui-Tze Lee, Dec. 23, 1964; children: Matilda, Cynthia. Engr. Jet Propulsion Lab., Pasadena, Calif., 1966-67; asst. prof. to prof. civil engring. U. Colo.-Boulder, 1967—. Contbr. articles to profl. jours. Grantee NSF, Dept. Def., Dept. Transp., Dept. of Interior; recipient Huber Research prize AMCE, 1979; Faculty Teaching award U. Colo., 1978; Faculty Research award, 1981. Mem. ASCE, Sigma Xi. Subspecialties: Civil engineering; Materials (engineering). Current work: Constitutive modeling of soils; centrifuge modeling of geotechnical structures; research in soil mechanics; teaching civil engineering, consulting in geotechnical engineering. Office: Box 428 Univ Colo Boulder CO 80309

KOBAYASHI, ALBERT S., mechanical engineer, educator; b. Chgo., Dec. 9, 1924; m. Elizabeth Midori, Sept. 24, 1953; children: Dori Kobayashi Ogami, Tina, Laura. B.Engring., U. Tokyo, 1947; M.S. in Mech. Engring., U. Wash., 1952; Ph.D., Ill. Inst. Tech., 1958. Tool engr. Konishiroku Photo Industry, Tokyo, 1947-50; design engr. Ill. Tool Works, Chgo., 1953-55; research engr. Armour Research Found., Ill. Inst. Tech., Chgo., 1955-58; Coll. faculty cons. Boeing Co., Seattle, 1958-76; asst. prof. mech. engring. U. Wash., Seattle, 1958-61, assoc. prof., 1961-65, prof., 1965—; cons. Mathematical Sci. N.W., 1962—; vis. scholar strength of materials lab., dept. naval architecture U. Tokyo, 1969, 77. Contbr. articles to profl. jours. Fellow ASME, Soc. Exptl. Stress Analysis (F.G. Tatnall award 1973, B.J. Lazan award, 1982); mem. Soc. Engring. Sci. Presbyterian. Subspecialties: Fracture mechanics; Theoretical and applied mechanics. Current work: Two and three dimensional fracture mechanics, stable crack growth and dynamic fracture mechanics. Office: Dept Mech Engring U Wash FU-10 Seattle WA 98195

KOBAYASHI, ROGER HIDEO, pediatrics and microbiology educator; b. Honolulu, May 21, 1947; s. Roy T. and Setsuko (Ebesugawa) K.; m. Ai Lan Doan, Jan. 21, 1974; children: Lisa S., Timothy S. B.A., U. Nebr., 1969, M.D., 1975; M.S., U. Hawaii, 1975. Pediatric resident U. So. Calif., Los Angeles, 1975-77; immunology fellow UCLA Sch. Medicine, 1977-79; asst. prof. pediatrics U. Nebr., Omaha, 1980—, asst. prof. microbiology, 1980—. Contbr. articles to med. jours. Bd. dirs. Am. Lung Assn. Nebr., 1981-82. Am. Lung Assn. grantee, 1982-83; Mead-Johnson Pharms. grantee, 1982-83; NIH grantee, 1981. Fellow Am. Acad. Pediatrics, Am. Acad. Allergy; Mem. Am. Fedn. Clin. Research, Nebr. Allergy Soc. (Pres. 1980—). Subspecialties: Pediatrics; Immunology (medicine). Current work: Neutrophil-virus interaction, newborn neutrophil function, treatment of asthma in infants and young children. Office: Dept Pediatrics U Nebr Med Center Omaha NE 68105

KOBAYASHI, SHOSHICHI, mathematician, educator; b. Kofu, Japan, Jan. 4, 1932; came to U.S., 1954; s. Kyuzo and Yoshie (Obi) K.; m. Yukiko Ashizawa, May 11, 1957; children: Sumire, Mei. B.S., U. Tokyo, 1953; postgrad., U. Paris, 1953, U. Strasbourg, 1954; Ph.D. in Math., U. Wash, 1956. Mem. Inst. Advanced Study, Princeton, N.J., 1956-58; research asso. Mass. Inst. Tech., 1958-60; asst. prof. U. B.C., Vancouver, 1960-62, U. Calif. at Berkeley, 1962-63, asso. prof., 1963-66, prof. math., 1966—, chmn. dept. math., 1978-81; vis. prof. U. Mainz, Germany, 1966, U. Bonn, 1969, Mass. Inst. Tech., 1970, U. Md., 1972, U. Tokyo, 1981. Author: (with K. Nomizu) Foundations of Differential Geometry, Vol. I, 1963, Vol. II, 1969, Hyperbolic Manifolds and Holomorphic Mappings, 1970, Transformation Groups in Differential Geometry, 1972, Differential Geometry of Curves and Surfaces, 1977; editor: Pure and Applied Mathematics, 1969—, Jour. Differential Geometry, 1973—; asso. editor: Duke Math. Jour, 1970-80. Sloan fellow, 1964-66; Guggenheim fellow, 1977-78. Mem. Am., Swiss math. socs., Canadian Math. Congress, Société Mathématique de France, Math. Soc. Japan. Subspecialty: Differential geometry. Home: 421 Michigan Ave Berkeley CA 94707

KOBE, DONALD HOLM, physics educator; b. Seattle, Jan. 13, 1934; s. Kenneth Albert and Jeneva Catherine (Holm) K. B.S., U. Texas-Austin, 1956; M.S., U. Minn., 1959; Ph.D., 1961. Vis. asst. prof. Ohio State U., Columbus, 1961-63; research assoc. Quantum Chemistry Group, Uppsala, Sweden, 1964-66; vis. asst. prof. H.C. Oersted Inst., Copenhagen, Denmark, 1966-67, Northeastern U., Boston, 1967-68; prof. No. Texas State U.-Denton, 1968—; Fulbright lectr., Taipei, Taiwan, 1963-64, Nat. Acad. Sci. lectr., Yugoslavia, 1973. Contbr. articles to profl. jours. Mem. Am. Phys. Soc., Am. Assn. Physics Tchrs., AAAS, Am. Sci. Affiliation. Subspecialties: Theoretical physics; Atomic and molecular physics. Current work: I have developed with others a manifestly gauge-invariant formulation of quantum mechanics for the interaction of electromagnetic radiation and charged matter. Home: 1704 Highland Park Rd Denton TX 76201 Office: Dept of Physic North Texas State Univ Denton TX 76203

KOBERSTEIN, JEFFREY THOMAS, chemical engineering educator, polymer consultant, researcher; b. Milw., Sept. 27, 1952; s. Calvin Matthew and Eleanor Anna (Gebhardt) K.; m. Linda Leslie Farmer, June 7, 1975; children: Meghan, Nicholas. B.S. in Chem. Engring, U. Wis.-Madison, 1974, Ph.D., U. Mass., 1979. NSF postdoctoral fellow Centre National de la Recherche Scientifique (Nat. Center Sci. Research), Strasbourg, France, 1979-80; asst. prof. dept. chem. engring. Princeton (N.J.) U., 1980—. Contbr. articles to profl. jours. Mem. Am. Chem. Soc., Am. Phys. Soc., Am. Inst. Chem. Engrs. Subspecialties: Polymers; Polymer physics. Current work: Morphology and structure-property relationships in multi-phase polymer systems; scattering techniques. Office: Dept Chem Engring Princeton Univ Princeton NJ 08544

KOBETICH, EDWARD JOHN, chemical company executive, physicist; b. Clay Center, Kans., Oct. 8, 1941; s. John Edward and Hilda Isabel (Ayre) K.; m. Jane Ella Birnay, June 11, 1977; children: Randall, Bradley, Jenice, Dawn. B.S., Kans. State U., 1965, M.S., 1967; Ph.,D., U. Nebr.-Lincoln, 1968; M.B.A., Wharton Sch., U. Pa., 1983. Research assoc. U. Nebr.-Lincoln, 1966-69, U. Bristol, Eng., 1969-70, U. Calif.-Berkeley, 1970-72; sr. research physicist DuPont, Wilmington, Del., 1972-77; mgr., prin. scientist ARCO Chem. Co., Phila., 1977-; instr. U.Calif.-Berkeley, 1971-72, St. Joseph's Coll., 1973-76, Del. Community Coll., 1975-77. Contbr. numerous articles to profl. jours., 1965-72. NASA trainee, 1967; Sci. Research Council fellow, Eng., 1969; NRC sr. assoc., 1972. Mem. Am. Chem. Soc., Am. Phys. Soc., Am. Mgmt. Assn. Subspecialties: Polymer physics; Cosmic ray high energy astrophysics. Current work: Material science, polymer science, cosmic-ray physics, astrophysics. Office: ARCO Chem 1500 Market St Philadelphia PA 19101

KOBILINSKY, LAWRENCE, immunology and forensic science, educator, forensic science consultant, researcher; b. N.Y.C., Nov. 7, 1946; s. Abraham and Sophie (Selkow) K.; m. Estelle Kartagener, June 10, 1971; 1 dau., Hayley. B.S., CCNY, 1969, M.A., 1971; Ph.D., CUNY, 1977. Research asst. Columbia Presbyn. Med. Ctr., N.Y.C., 1969-70; lectr. City Coll., CUNY, 1970-71; adj. lectr. Bklyn. Coll., 1972-74, Hunter Coll., 1974-75, John Jay Coll., 1975-77; research asst. U. Pa. Johnson Research Found., Phila., 1971; research fellow Complement and Effector Biology Lab. Sloan-Kettering Inst., 1977-79, vis. investigator, 1980-82, research assoc., 1979-80; adj. asst. prof. biology John Jay Coll., CUNY, 1977-80; instr. Cornell Grad. Sch. Med. Scis., Sloan-Kettering Div., N.Y.C., 1979-80; attending staff research scientist Animal Med. Center, N.Y.C., 1979-81; mem. biochemistry doctoral faculty CUNY Grad. Center, N.Y.C., 1981—; asst. prof. immunology and forensic sci. John Jay Coll. Criminal Justice, N.Y.C., 1980—. Contbr. articles to profl. jours. Served to 1st lt. U.S. Army, 1971-72. Mem. AAAS, Am. Chem. Soc., Am. Assn. Immunologists, Am. Acad. Forensic Sci., N.Y. Acad. Scis., N.Y. Micros. Soc. (bd. mgrs., chmn. membership com., assoc. editor newsletter), Northeastern Assn. Forensic Scientists.; mem. Sigma Xi. Subspecialties: Immunobiology and immunology; Immunology (medicine). Current work: Development of new techniques for forensic serology, teaching graduate and undergraduate students, consulting on criminal cases, research. Home: 504 Rebecca Ln Oceanside NY 11572 Office: John Jay Coll Criminal Justice 445 W 59th St New York NY 10019 Home: 504 Rebecca Ln Oceanside NY 11572

KOBLER, VIRGINIA PONDS, research and development administrator; b. Carlsbad, N. Mex., Oct. 8, 1937; d. William Edward and Magnolia Serena (Dryden) Ponds; m. Bobby Sandlin Woodruff, June 1, 1957 (div. 1972); m. Julian S. Kobler, Aug. 18, 1974; children: George Ponds, Jonathan Ponds. B.S., U. Ala., 1964, M.S., 1976, Ph.D., 1979. Research physicist U.S. Army Missile Command, Redstone Arsenal, Ala., 1967-80; mgr. research and devel. U.S. Army Ballistic Missile Def. Advanced Tech. Ctr., Huntsville, Ala., 1980—; reviewer Army Research Office, Durham, N.C., 1979— . Contbr. articles to profl. jours. Mem. IEEE (reviewer 1979—), Ops. Research Soc., Am. Phys. Soc. Democrat. Mem. Ch. of Christ. Subspecialties: Algorithms; Artificial intelligence. Current work: Research and development in data processing systems; decentralized control algorithms for distributed computer systems and command, control and communications utilizing artificial intelligence techniques. Home: 5802 Macon Dr Huntsville AL 35802 Office: US Army BMD ATC PO Box 1500 Huntsville AL 35807

KOBLICK, DANIEL CECIL, biology educator, physiology researcher; b. San Francisco, May 13, 1922; s. Morris Abramovich and Rebecca Louise (Faverman) K.; m. Vivian Mae Shaw, Sept. 1945; 1 dau., Laurinda Joan Koblick McKinlay; m. Joan Lesser, July 8, 1960; 1 dau., Rebecca Pauline. Student, San Francisco Jr. Coll., 1940-42; A.B., U. Calif.-Berkeley, 1944, Postgrad., 1946-48, 51-52; Ph.D., U. Oreg., 1957. Instr. biology U. Oreg.-Eugene, 1957-58; asst. prof. zoology U. Mo.-Columbia, 1958-59; USPHS postdoctoral fellow U. Calif.-Berkeley, 1959-60, lectr. physiology, 1960-61, asst. research physiologist, 1961-63; assoc. prof. biology Ill. Inst. Tech., Chgo., 1963—. Served with AUS, 1945-46. Fellow AAAS; mem. Biophys. Soc., Am. Physiological Soc., Soc. Gen. Physiologists. Jewish. Subspecialties: Physiology (biology); Membrane biology. Current work: Role of mechanical phenomena in osmo- and volume regulation. Office: Dept Biology Ill Inst Tech 3101 S Dearborn Chicago IL 60616

KOCAOGLU, DUNDAR F., engineering educator, consultant; b. Turkey, June 1, 1939; came to U.S., 1960; s. Irfan and Meliha (Uzay) K.; m. Alev Baysak, Sept. 17, 1968. B.S.C.E., Robert Coll., Istanbul, Turkey, 1960; M.S.C.E., Lehigh U., 1962; M.S.I.E., U. Pitts., 1972, Ph.D. in Ops. Research, 1976. Registered profl. engr., Pa. Structural engr. United Engrs., 1963-71; lectr. U. Pitts., 1973-75, vis. asst. prof., 1975-76, dir. engring. mgmt. program, 1976—; engring. and mgmt. cons., Pitts, 1971—; ednl. cons. various univs.; tech. mgmt. cons. UN, 1979-80; founding chmn. TIMS Coll. on Engring. Mgmt., 1979-81. Author: Engineering Management, 1981; editor series: Engring. Mgmt., 1981—; contbr. articles to profl. jours. Pres. Pitts. Turkish-Am. Assn., 1977-79. Served to lt., C.E. Turkish Army, 1966-68. Mem. Am. Soc. for Engring. Edn. (pres. engring. mgmt. div.), Am. Soc. for Engring. Mgmt. (regional dir.), IEEE Engring. Mgmt. Soc. (publs. dir.), Omega Rho (pres.-elect). Subspecialties: Operations research (engineering); Engineering management. Current work: Engineering management, technological innovations, resource optimization, hierarchical decisions, multicriteria decisions, participative management, project management, program evaluations. Home: 125 Highvue Dr Venetia PA 15367 Office: 1037 Benedum Hall U Pitts Pittsburgh PA 15261

KOCH, HENRY GEORGE, research entomologist; b. Mt. Holly, N.J., May 22, 1948; s. Winfield Rudolph and Eva Virginia (Moore) K.; m. Lucy Madeline Midgett, June 29, 1968; children: Marcus Winfield, Laura Ann. B.S., Okla. State U., 1971, M.S., 1974; Ph.D., N.C. State U., 1977. NSF grantee Oreg. State U., Corvallis, 1971-72; grad. research asst. Okla. State U., Stillwater, 1972-74, N.C. State U., Raleigh, 1974-77, research assoc., 1977; research entomologist Agrl. Research Service, U.S. Dept. Agr., Poteau, Okla., 1977—. Contbr. articles to profl. jours. Mem. adminstrv. bd. Meth. Ch., Poteau, 1979—. Mem. Entomol. Soc. Am., Southwestern Entomol. Soc., Council Agrl. Sci. and Tech. Subspecialties: Integrated pest management; Developmental biology. Current work: Biology and control of ticks affecting man and animals; determining key mortality factors of ticks. Office: Lone Star Tick Research Lab US Dept Agr Box 588 Poteau OK 74953 Home: 100 Wedgewood Dr Poteau OK 74953

KOCH, LEONARD JOHN, nuclear power engineer; b. Chgo., Mar. 30, 1920; s. Philip and Christine (Kauk) K.; m. Rosemarie J. Shafer, Sept. 19, 1942; 1 son, William J. B.S. in Mech. Engring, Ill. Inst. Tech., 1942; M.B.A., U. Chgo., 1968. Registered profl. engr., Ill. Dir. div. Argonne (Ill.) Nat. Lab., 1948-72; v.p. Ill. Power Co., 1972-83; cons., 1983—. Contbr. articles to profl. jours. Fellow Am. Nuclear Soc.; mem. Nat. Acad. Engring. Subspecialties: Mechanical engineering; Nuclear engineering. Current work: Nuclear power engineering; nuclear power development. Home: 6 Post Dr Decatur IL 62521

KOCH, WERNER, aerospace research scientist; b. Baden, Austria, Oct. 26, 1940; s. Karl and Genoveva (Trinker) K.; m. Marianne Lisette Wendel, Aug. 26, 1967; children: Maria Christine, Alexander Karl. Dipl. Ing. in Mech. Engring, Inst. Tech. Graz, Austria, 1964; M.Aerospace Engring., Cornell U., 1965, Ph.D., 1968. Research scientist Deutsche Forschungs-und Versuchsanstalt fuer Luft-und Raumfahrt, Aachen, W.Ger., 1968-74, Aerodynamische Versuchsanstalt, Goettingen, W.Ger., 1974—. Research publs. in field. Recipient 2d prize Henry R. Worthington European Tech. Award, Brussels, 1979; NRC sr. research assoc. NASA Langley Research Ctr., 1978-79. Mem. AIAA, Gesellschaft Fuer Angewandte, Mathematik Und Mechanik. Roman Catholic. Subspecialties: Acoustical engineering; Fluid mechanics. Current work: Duct acoustics; unsteady flow separation. Home: Tegeler Weg 39 D-34 Goettingen Federal Republic of Germany Office: DFVLR/AVA Goettingen Bunsenstrasse 10 D-34 Goettingen Federal Republic of Germany

KOCHAN, IVAN, microbiology educator; b. Tudor, Ukraine, Aug. 20, 1923; U.S., 1955, naturalized, 1961; s. Volodimir and Alexandra (Dula) K.; m. Tatiana Sawycka, June 20, 1924; children: Andrew, Mark, Christina. Cand. Med., U. Lwiw, 1944; postgrad., U. Munich, 1946-48; M.Sc., U. Man., 1955; Ph.D., Stanford U., 1958. Diplomate: Am. Bd. Microbiology. Biologist Stanford Research Inst., Menlo Park, Calif., 1956-58; research assoc. Stanford U., 1958-59; assoc. prof., chmn. dept. Baylor U., Dallas, 1959-62, prof., chmn. dept., 1962-67; prof. microbiology Miami U., Oxford, Ohio, 1967—; prof., chmn. dept. microbiology Wright State U. Sch. Medicine, Dayton, Ohio, 1974-78; mem. U.S. Japan Coop. Med. Sci. Program, 1966-75. Contbr. articles to profl. jours. NIH grantee, 1960-66; Nat. Tb Assn. grantee, 1965-67; NSF grantee, 1966-72; Am. Cancer Soc. grantee, 1975-77; VA grantee, 1978-81; recipient Research award Sigma Xi, 1978. Fellow Am. Trudeau Soc., Am. Acad. Microbiology; mem. Am. Soc. Microbiology (Ohio pres. 1975-77, councilman 1978-80), Am. Assn. Immunologists, AAAS, Sigma Xi. Republican. Greek Catholic. Subspecialties: Immunology (agriculture); Infectious diseases. Current work: Promotion of lethal bacterial infections in normal and immune animals with iron and iron binding bacterial siderophores. Home: Bull Run Dr Oxford OH 45056 Office: Miami U Oxford OH 45056

KOCHEN, MANFRED, information science educator; b. Vienna, Austria, July 4, 1928; came to U.S., 1941, naturalized, 1947; m. Paula Landerer, Aug. 15, 1954; children: David J., Mark N. B.S., MIT, 1950; M.A., Columbia U., 1951, Ph.D. in Applied Math, 1955. Asst. Spectros Lab., MIT, Cambridge, 1949-50; mathematician aeroelasticity research Biot & Arnold Co., N.Y.C., 1950-52; lectr. math. Columbia U., N.Y.C., 1952-53; mem. staff math. models in behavioral sci. Harvard U., Cambridge, Mass., 1955-56; staff mathematician Thomas J. Watson Research Ctr., IBM Corp., Yorktown Heights, N.Y., 1956-58, mem. research tech. staff, 1958-60, mgr. info. retrieval project, 1960-63; exchange vis. expert Euratom, Italy, 1963-64; prof. info. sci. and urban/regional planning, research mathematician Health Research Inst., Ann Arbor, Mich., 1965—; adj. prof. computer and bus. info. systems Sch. Bus. Adminstrn., U. Mich., 1982—; vis. prof. Rockefeller U., N.Y.C., 1980-81; cons. in field. Assoc. editor: Behavioral Science, 1968-70, Jour. Association Computing Machinery, 1972— ; managing editor: Human Systems Management, 1979—. Pres. Wise Fund, 1975—. Hon. research assoc. Harvard U., 1973-74. Fellow AAAS; mem. Am. Math. Soc., Am. Soc. Info. Sci., Fedn. Am. Scientists, Am. Phys. Soc. Subspecialties: Artificial intelligence; Learning systems. Current work: Information systems and organization of knowledge; models for information-seeking behavior, problem representation solving, cognitive learning processes; decentralization theory; science of science; social planning; artificial intelligence; decision support systems. Office: MHRI U Mich Ann Arbor MI 48109

KOCHER, CARL ALVIN, physicist; b. Seattle, Feb. 14, 1942; s. Paul Harold and Annis Adelle (Cox) K.; m. Marilyn Aleta Tennant Kocher, June 24, 1968; children: Suzanne, Paul, Scott. A.B., U. Calif.-Berkeley, 1963, Ph.D., 1967. Postdoctoral fellow Oxford (Eng.) U., 1967-68; vis. scientist M.I.T., Cambridge, 1968-69; lectr. physics Columbia U., N.Y.C., 1969-73; asst. prof. physics Oreg. State U., Corvallis, 1973-78, assoc. prof., 1978—; prin. investigator U.S. Dept. Energy, 1978—. Contbr. articles in field to profl. jours. Woodrow Wilson fellow, 1963; NSF grad. fellow, 1963-67; NSF postdoctoral fellow, 1967-69. Mem. Am. Phys. Soc., Am. Assn. Physics Tchrs., Fedn. Am. Scientists, Phi Beta Kappa, Sigma Xi. Subspecialties: Atomic and molecular physics; Spectroscopy. Current work: Experimental atomic physics; atomic collisions, radiative and ionization processes, laser spectroscopy, high Rydberg states, computer instrumentation.

KOCHI, JAY KAZUO, chemist, educator; b. Los Angeles, May 17, 1927; s. Tsuruzo and Shizuko (Moriya) K.; m. Marion Kiyono, Mar. 1, 1959; children—Sims, Ariel, Julia. Student, Cornell U., 1945; B.S., U. Calif. at Los Angeles, 1949; Ph.D., Ia. State U., 1952. Faculty Harvard, 1952-55; NIH fellow Cambridge (Eng.) U., 1956; mem. faculty Iowa State U., 1956; with Shell Devel. Co., 1957-61; mem. faculty dept. chemistry Case Western Res. U., Cleve., 1962-69, prof., 1966-69; prof. chemistry Ind. U., Bloomington, 1969-74, Earl Blough prof. chemistry, 1974—; cons. chemist, 1964—. Mem. Am. Chem. Soc., Chem. Soc. (London), Nat. Acad. Scis., Sigma Xi. Subspecialties: Organic chemistry; Inorganic chemistry. Research on mechanism of catalysis of organic reactions, organometallics, electrochemistry and photochemistry. Home: 217 S Hillsdale Dr Bloomington IN 47401

KOCHWA, SHAUL, immunochemist, educator; b. Vienna, Austria, Apr. 30, 1915; s. Joseph Saul and Anna K.; m. Hadasa Kochwa, July 9, 1940; 1 dau., Varda Kruman. M.Sc., Hebrew U. Jerusalem, 1940, Ph.D., 1949; M.P.H., Harvard U., 1953. Chief chemist Gordon Co., Tel Aviv, 1943-45; med. lab. dir. Bikur Cholim Hosp., Jerusalem, 1945-47; chief epidemiologist Israeli Army, 1947-55; assoc. dir. Rogoff Research Inst., Bellinson Hosp., Petah Tiqua, Israel, 1955-59; research assoc. Mt. Sinai Hosp., N.Y.C., 1959-66; assoc. prof. medicine Mt. Sinai Sch. Medicine, 1966-72, prof. pathology, 1972—, prof. medicine, 1979—; dir. immunochemistry lab. Mt. Sinai Hosp., 1964—. Contbr. over 170 articles to sci. jours. Served with M.C. Israeli Army, 1948-55. Mem. Am. Assn. Immunologists, Am. Chem. Soc., Am. Soc. Hematology, Harvey Soc., Internat. Soc. Internal Organs, Internat. Soc. Toxicology, N.Y. Acad. Sci. Democrat. Jewish. Subspecialties: Immunology (medicine); Hematology. Current work: Biochemistry and immunochemistry of immunoglobulins; protein-protein interaction; relation between severity of disease and protein structure. Patentee in field (2). Home: 6722 Harrow St Forest Hills NY 11375 Office: 1 Gustave Levy Pl New York NY 10029

KOCKINOS, CONSTANTIN NEOPHYTOS, mathematical scientist, researcher; b. Cairo, Egypt, Oct. 14, 1926; came to U.S., 1947; s. Dimitri and Irene (Sovrani) K.; m. Jean Freeman Lincoln, Aug. 12, 1952 (div. 1962); 1 son, Marc Demetrius. Student, Am. U., Cairo, 1944-47; B.A., U. Calif.-Berkeley, 1950, 1954, M.A., 1956; Ph.D., Stanford U., 1972. Instr. U. Calif.-Berkeley, 1958-62; sr. scientist EGG, Las Vegas, Nev., 1962-64; research mathematician Stanford Research Inst., Menlo Park, Calif., 1964-72; vis. prof. Aristotelian U., Thessaloniki, Greece, 1972-76; San Jose (Calif.) State U., 1976-79; math. scientist Lockheed Missiles & Space Div., Sunnyvale, Calif., 1979—; asst. protocol officer Greek Diplomatic Service, Cairo, Egypt, 1944-45; sec. to King George of the Hellenes; mem. U.S. Senatorial Adv. Bd., Washington. Contbr. writings to publs. Mem. Senatorial Inner Circle, 1983. Fellow Explorers Club; mem. Internat. Platform Assn., U.S. Naval Inst., Am. Math. Soc., Tensor, AIAA, Sigma Xi, Pi Mu Epsilon. Republican. Greek Orthodox. Club: Athletic. Subspecialties: Applied mathematics; Theoretical physics. Current work: Fluid mechanics; aerodynamics, adaptive grid generation; nonlinear problems; applications of modern mathematics and of geometry to engineering and physics. Patentee. Home: 2121 Creeden Ave Mountain View CA 94040 Office: Lockheed Missiles & Space Co Lockheed Way 81-10 PO Box 504 Sunnyvale CA 94086

KOCOL, HENRY, federal government health physicist; b. Chgo., July 16, 1937; s. Henry Frank and Mary Barbara (Strumidlowska) K.; m. Cleo Florence Fellers, Jan. 12, 1971; 1 son, Henry Peter. B.S., Loyola U., Chgo., 1958; M.S., Purdue U., 1961. Cert. hazard control mgr. Chemist Nat. Bur. Standards, Washington, 1961-64; Public Health Service, Las Vegas, Nev., 1964-71; chemist div. bio-effects FDA, Rockville, Md., 1971-73, radiation control officer, Phila., 1973-77, radiol. health rep., 1977-79, supr. x-ray control unit State of Wash. (on loan), Seattle, 1979-82, fed./states liaison officer, 1982—. Mem. Health Physics Soc. (councilman Cascade chpt. 1980—, mem. ann. meeting place com. Bethesda, Md. 1981-82), Conf. Radiation Control Program Dirs. (chmn. task force on radiol. data collection and analysis Franfurt, Ky. 1980-82). Subspecialties: Radiation safety; Radiation biological effects. Current work: Application of physics, biology, chemistry to the problems of radiation safety; exploration of radiation principles and perspectives to lay audiences. Office: Food and Drug Administrn 5009 FOB 909 1st Ave Seattle WA 98174

KODAMA, JIRO KENNETH, toxicologist; b. Reedley, Calif., Mar. 4, 1924; s. Jitaro and Kane (Kodama) K.; m. Aya Ashida, Oct. 27, 1951; children: Cathy, Julie, Lori, Kevin, Kelly. A.A. in Chemistry, Reedley Coll., 1949; A.B. in Biochemistry, U. Calif.-Berkeley, 1951; M.S., U. Calif.-San Francisco, 1955; Ph.D. in Pharmacology and Toxicology, U. Calif.-San Francisco, 1957. Diplomate: Am. Bd. Toxicology. Sr. pharmacologist Hazleton Labs., Falls Church, Va., 1959-63; supr. pharmacology dept. Shell Devel. Co., Modesto, Calif., 1963-68; vis. scientist Shell Research Ltd., Sittingbourne, Kent, Eng., 1969-70; staff toxicologist Shell Chem. Co., San Ramon, Calif., 1970-77, Standard Oil Co. Calif., Richmond, 1977-82; tech. liaison rep. Chevron Environ. Health Ctr., Inc., Richmond, 1982—. Contbr. articles to profl. publs. Served with U.S. Army, 1945. Mem. Soc. Toxicology, Am. Soc. Pharmacology and Exptl. Therapeutics, AAAS, Internat. Soc. Study of Xenobiotics. Subspecialties: Toxicology (medicine); Pharmacology. Current work: To provide technical liasion between corporate toxicology function and agricultural and consumer products operating companies. Patentee in field. Office: PO Box 4054 Richmond CA 94804

KODANAZ, HATICE ALTAN, neuropsychologist, educator; b. Tire, Turkey, July 20, 1930; came to U.S., 1957; d. M. Haydar and A. Sacide (Sandikoglu) K.; m. A. Aytekin, May 9, 1957; children: Ahmet, Taner. B.S., U. Istanbul, Turkey, 1955, M.A., 1956; M.S., U. Kans., 1964; Ph.D., U. Ankara, Turkey, 1976. Psychologist U. Istanbul Med. Sch., 1956-57; research coordinator U. Kans. Med. Ctr., Kansas City, 1966-70, research assoc., 1975-76, instr. dept. neurology, 1976-79, asst. prof., 1979—; founder, chief clin. psychology U. Ankara, 1970-74, cons. dept. psychiatry, 1979-82. Pres. Turkish Am. Assn., Kansas City, 1980-82; area rep. Mini-Mundo, Kansas City, 1980-83; treas. Neighborhood Homes Assn., Fairway, Kans., 1981-82. Fulbright scholar, 1957-58; Smith Mundt grantee, 1958. Mem. Kans: Psychol. Assn., Am. Pain Soc., Turkish-Am. Physicians Assn., Turkish-Am. Neuropsychiat. Assn. Subspecialties: Neuropsychology; Clinical-behavioral studies. Current work: Brain behavior relationship; dementia; psychological and cultural aspects of headaches; neuropsychological evaluation and diagnosis. Office: Kans Univ Med Center 39th and Rainbow Kansas City KS 66103

KOELLE, GEORGE BRAMPTON, university pharmacologist, educator; b. Phila., Oct. 8, 1918; s. Frederick Christian and Emily Mary (Brampton) K.; m. Winifred Jean Angenent, Feb. 6, 1954; children: Peter Brampton, William Angenent, Jonathan Stuart. B.Sc., Phila. Coll. Pharmacy and Sci., 1939, D.Sc. (hon.), 1965; Ph.D., U. Pa., 1946; M.D., Johns Hopkins, 1950; Dr. Med. (hon.), U. Zurich, Switzerland, 1972. Bio-assayist LaWall & Harrisson, 1939-42; asst. prof. pharmacology Coll. Phys. and Surg., Columbia U., 1950-52; prof. pharmacology Grad. Sch. Medicine, U. Pa., 1952-59, chmn. dept. physiology and pharmacology, dean, 1957-59, chmn. dept. pharmacology, 1959-81, disting. prof., 1981—; spl. lectr. U. London, 1961; vis. lectr. U. Brazil, 1962, Polish Acad. Scis., 1979; vis. prof., Guggenheim fellow U. Lausanne, 1963-64; vis. prof. pharmacology, chmn. dept. Pahlavi U., Shiraz, Iran, 1969-70; cons. McNeil Labs., 1951-66, Phila. Gen. Hosp., 1953—, Valley Forge Army Hosp., 1954-71, Army Chem. Corps, 1956-60, Phila. Naval Hosp., 1957—; vis. lectr. pharmacology Phila. Coll. Pharmacy and Sci., 1955-57; vis. prof. Mahidol U., Bangkok, Thailand, 1978; Mem. pharmacology study sect. USPHS, 1958-62, chmn. pharmacology study sect., 1965-68; sec. gen. Internat. Union of Pharmacology, 1966-69, v.p., 1969-72; mem. bd. sci. Counselors Nat. Heart Inst., NIH, USPHS, 1960-64, mem. nat. adv. neurol. diseases and stroke council, 1970-75. Asso. editor: Remington's Practice of Pharmacy, 1951; mem. editorial bd.: Pharmacol. Revs, 1955-63; chmn., 1959-62; hon. editorial adv. bd.: Biochem. Pharmacology, 1958-72; editorial adv. bd.: Internat. Jour. Neurosci, 1970—; editorial com.: Ann. Rev. Pharmacology, 1959-65; editorial bd.: Internat. Ency. Pharmacology and Therapeutics; editor: Cholinesterases and Anticholinestrerase Agents, 1963; Contbr. articles on pharmacology to profl. jours. Trustee Phila. Coll. Pharmacy and Sci.; bd. mgrs. Wistar Inst. Served to 1st lt. Med. Adminstrn. Corps AUS, 1942-46. Recipient Abel prize in pharmacology Am. Soc. Pharmacology and Exptl. Therapeutics, 1950; Travel award XVIIIth Internat. Physiol. Congress, Copenhagen, Federated Socs., 1950; Borden undergrad. research award, 1950. Fellow A.A.A.S. (v.p. 1971), N.Y. Acad. Scis.; mem. Am. Soc. Pharmacology and Exptl. Therapeutics (pres. 1965-66, plenary lectr. 1981), Nat. Acad. Scis., Histochem. Soc., Harvey Soc., Soc. Biol. Psychiatry, John Morgan Soc., Sydenham Coterie, Sons Copper Beeches, Brit. Pharmacol. Soc., Biol. Soc. Chile (hon.), Internat. Neurochem. Soc., Soc. for Neurosci., Pharmacol. Soc. Peru (hon.), Pharmacol. Soc. Japan (hon.), Sigma Xi, Alpha Omega Alpha. Subspecialties: Pharmacology; Neurobiology. Current work: Cholinesterases; anticholinesterases; electron and light microscopic histochemistry ; neuotrophic factors. Home: 205 College Ave Swarthmore PA 19081 Office: U Pa Philadelphia PA 19104

KOENEMAN, JAMES BRYANT, mechanical engineer; b. Graceville, Minn., Nov. 24, 1936; s. Egmund Alfred and Luverne Althea (Bryant) K.; m. Mary Ann Endecavageh, 1964; children: Edward, Paul, Brian. B.S. in M.E, U. Minn., 1959; M.S., Case Inst. Tech., 1966, Ph.D., 1970. Reactor engr. Argonne Nat. Lab., Idaho Falls, Idaho, 1959-60, U.S. AEC, Argonne, Ill., 1960-64; mem. tech. staff Bell Telephone Labs., Columbus, Ohio, 1970-74; mgr. bioengring. dept. Lord Corp., Erie, Pa., 1974-81; pres. Paulson Med. Devices, Inc., Erie, 1981—; dir. research Shriners Hosp. for Crippled Children, Erie, 1978-81. Contbr. articles to profl. jours. Active Boy Scouts Am., Indian Guides. Mem. ASME (named Man of the Yr. Erie chpt. 1982), Orthopedic Research Soc., Soc. Biomaterials, ASTM. Subspecialties: Mechanical engineering; Biomedical engineering. Current work: Founder of new venture in orthopedic devices. Patentee in field. Home: 1415 Drake Dr Erie PA 16505

KOENIG, JACK LEONARD, educator; b. Cody, Nebr., Feb. 12, 1933; s. John and Lucille May K.; m. Jeanus Brosz, July 5, 1953; children—John, Robert, Stanley, Lori. B.A., Yankton Coll., 1955; M.S., U. Nebr., 1957, Ph.D., 1959. Research chemist plastics dept. DuPont, Wilmington, Del., 1959-63; asst. prof. polymer sci. div. Case Inst. Tech., Cleve., 1963-66; asso. prof. dept. macromolecular sci. Case Western Res. U., Cleve., 1966-70, prof., 1970—; dir. Molecular Spectroscopy Lab., 1974—; co-dir. Materials Research Lab., 1974—; program dir. solid state chemistry and polymer sci. div. materials research NSF, Washington, 1972-73. Author: Chemical Microstructure of Polymer Chains, 1980, (with others) Introduction to Vibrational Spectroscopy With Applications to Polymers, 1981; contbr. articles to profl. jours. Served with U.S. Army, 1953-55. Grantee NSF, Army Research Office, NASA, Naval Research Office. Subspecialties: Polymer chemistry; Surface chemistry. Current work: Sturctural characterization of polymeric materials. Home: 15503 Dale Rd Chagrin Falls OH 44022 Office: Case Western Res U Cleveland OH 44106

KOENIG, JAMES BENNETT, geothermal energy cons., writer, lectr.; b. N.Y.C., Nov. 25, 1932; s. Philip Edward and Lorraine Rose (Woldar) K.; m. Anne Mariner Jennings, June 6, 1964; children: Laura Bethune, Andrea Croft, Cassandra Gregory. B.S. in Geology, Bklyn. Coll., 1954, M.A., Ind. U., 1956; postgrad., U.S. Navy Postgrad. Sch., Monterey, Calif., 1957-58, U. Nev., 1963-65. Registered geologist, Calif. Geologist U.S. Geol. Survey, Minn., 1955-56; jr. geologist Calif. Div. Mines, San Francisco, 1956-57; successively, asst., assoc., sr., supervising geologist Calif. Div. Mines and Geology, San Franciso and Sacramento, 1960-63, 65-72; cons. geologist, Berkeley, Calif., 1972-74; pres. GeothermEx, Inc., Richmond, Calif., 1974—; lectr. symposia; writer popular press and jours. Served to lt. USNR, 1957-60. Fellow Geol. Soc. Am.; mem. Goethermal Resources Council (dir.), Am. Geophys. Union, Internat. Assn. Volcanology. Jewish. Subspecialties: Geothermal power; Geology. Current work: Evaluation of geothermal energy prospects in field, design and mgmt. of exploration and drilling, calculation of reserves, assistance to elec. utilities in evaluating geothermal options, econs. of geothermal energy for UN, World Bank and pvt. clients. Home: 901 Mendocino Ave Berkeley CA 94707 Office: 5221 Central Ave Richmond CA 94804

KOENIG, LOUIS, sci. researcher; b. Yonkers, N.Y., Oct. 11, 1911; s. Henry and Gertrude Marie (Holste) K.; m. Janie Ray Shofner, 1952; children: Livy, Charlou, Arthur Shofner, Friederich Christian Peter. B.S., N.Y. U., 1932, Ph.D., 1936. Chemist Solvay Process Co., Syracuse, N.Y., 1936-45; chemist Lithaloys Corp., N.Y.C., 1945-46; chief research br. AEC, N.Y.C., 1946-57; chmn. dept. chemistry and chem. engring. Armour Research Found., Chgo., 1947-49; asst. dir. research Stanford Research Inst., Palo Alto, Calif., 1950-51; v.p. S.W. Research Instr., San Antonio, 1951-56; pres. Louis Koenig Research, San Antonio, 1956—; adv. bd. to Scs. of Interior on Saline Water Conversion Program; adv. council to USPHS on Advanced Waste Treatment Program. Sect. editor: Chem. Abstracts, 1944—; contbr. numerous articles to profl. jours.; syndicated newspaper columnist. Recipient best paper award resources div. Am. Water Works Assn., 1960, publs. award, 1967. Mem. Am. Chem. Soc., Am. Inst. Chem. Engrs., Am. Inst. Chemists, Am. Water Resources Assn. (W.R. Boggess award 1975), Geophys. Union, Nat. Water Well Assn., Sigma Xi. Republican. Subspecialties: Chemical engineering; Water supply and wastewater treatment. Current work: Processes and economics in water supply and pollution control, research management. Home and Office: Route 10 Box 108 San Antonio TX 78258

KOENIG, MICHAEL EDWARD DAVISON, information science educator; b. Rochester, N.Y., Nov. 1, 1941; s. Claremont Judson and Mary Fletcher (Davison) K.; m. Luciana Marulli; children: Christopher Wells Bowen, Davison Packard Koenig. B.A. in Psychology, Yale U., 1963; M.A. in L.S, U. Chgo., 1968; cert. of advanced grad. study, U. Chgo., 1969; M.B.A. in Math. Methods and Computers, 1970; Ph.D. in Info. Sci, Drexel U., 1981. Mgr. info. services Pfizer, Inc., Groton, Conn., 1970-74, N.Y.C., 1970-74; dir. devel. Inst. for Sci. Info., Phila., 1974-78; v.p. ops. Swets N.Am., Berwyn, Pa., 1978-80; assoc. prof. info. sci. Sch. Library Service, Columbia U., N.Y.C., 1980—. Contbr. articles to profl. jours. Served to lt. (j.g.) USN, 1963-65. Mem. Am. Soc. Info. Scis., ALA, AAAS. Clubs: Yale (N.Y.); Elizabethan, Mory's. Subspecialties: Information systems (information science); Database systems. Current work: Productivity and information use; bibliometrics; communication of sci-tech information; information systems. Office: School of Library Service Columbia University New York NY 10027

KOERNER, ERNEST LEE, technical company executive, environmental engineer; b. Cleve., Mar. 17, 1931; s. Ernest L. and Mary Bridget (McGovern) K.; m. Barbara Jean Payne, Aug. 8, 1953; children: Susan, Patrick, Karen, Sally, Joan, Mary. B.Ch.E., U. Dayton, 1953; M.S. in Chem. Engring, Iowa State U., 1955, Ph.D., 1956. Registered profl. engr., Okla. Research sect. leader Union Carbide Metals, Inc., Niagara Falls, N.Y., 1957-59; research specialist Monsanto Co., St. Louis, 1959-67; sr. research group leader Kerr-McGee Corp., Oklahoma City, 1967-70; pres. Tech. Research & Devel., Inc., Oklahoma City, 1970-83; owner, pres. Techrad Inc., Oklahoma City, 1983—. Contbr. articles on metal recovery to profl. jours. Mem. Am. Inst. Chem. Engrs., AIME, Nat. Soc. Profl. Engrs., Okla. Soc. Profl. Engrs., Am. Inst. Plant Engrs., Nat. Assn. Environ. Profls. Subspecialties: Water supply and wastewater treatment; Chemical engineering. Current work: Conversion of H-3-PO-4- to detergent-grade STP, recovery of vanadium from ferrophosphorus, wastewater pollution control, extractive metallurgy. Holder 14 patents in metal recovery processes, phosphoric acid purification. Home: 12721 Saint Andrews Terr Oklahoma City OK 73120 Office: 4619 N Santa Fe St Oklahoma City OK 73118

KOESTNER, ADALBERT, pathologist, scientist, educator; b. Hatzfeld, Roumania, Sept. 10, 1920; came to U.S., 1955, naturalized, 1960; s. Johann and Gertraut (Gruber) K.; m. Adelaide Wilma Wacker Koestner, Jan. 20, 1951; children: George, Alexander (dec.), Rosemarie Kathrin. D.V.M., U. Munich, Ger., 1951; M.Sc., Ohio State U., 1957, Ph.D., 1959. Diplomate: Am. Coll. Vet. Pathologist. Gen. practice vet. medicine, Untergriesbach, Ger., 1951-55; instr. Ohio State U., 1955-59, asst. prof., 1959-61, assoc. prof., 1961-64, prof., 1964-81, acting chmn., 1971-72, chmn. vet. pathobiology, 1972-81; prof., chmn. dept. pathology Mich. State U., East Lansing, 1981—. Contbr. chpts. to books and articles to profl. jours. Mem. AVMA (Gaines award 1979), Am. Coll. Vet. Pathologists, Internat. Acad. Pathology, Am. Assn. Pathologists, Am. Assn. Neuropathologists (Weil award 1971), Soc. Neurosci., Am. Assn. Cancer Research. Subspecialties: Pathology (medicine); Comparative neurobiology. Current work: Two major areas of current research deal with carcinogenesis of tumors of the nervous system and animal models for demyelinating diseases such asmultiple sclerosis. Home: 2578 Woodhill Dr Okemos MI 48864 Office: Michigan State University E Fee Hall East Lansing MI 48824

KOGOMA, TOKIO, molecular genetics educator, researcher; b. Tokyo, Dec. 5, 1939; U.S., 1968; s. Kunizo and Kiwa (Kogoma) Yoshimura; m. Fusae Kogane, Dec. 2, 1966; 1 child: Takeshi. B.S., Chiba (Japan) U., 1963; M., U. Tokyo, 1965, Ph.D., 1968. Lic. pharmacist, Japan. Instr. U. Tokyo, 1968; research assoc. Kans. State U., 1968-70, U. Utah, Salt lake City, 1970-71, asst. research prof., 1972-74; asst. prof. molecular genetics U. N.Mex., Albuquerque, 1974-80, assoc. prof., 1980—; vis. prof. Poly. U., Copenhagen, 1982; adj. assoc. prof. cell biology U. N.Mex., 1983—. Contbr. articles to profl. jours. NIH grantee, 1975, 78, 82; NSF grantee, 1977, 79. Mem. Am. Soc. Microbiology, Sigma Xi. Subspecialty: Genetics and genetic engineering (biology). Current work: Bacteria, regulation, DNA replication, mutants, plasmids, restriction endonucleases, gene expression, genetic recombination. Office: U N Mex Albuquerque NM 87131

KOHEL, RUSSELL JAMES, geneticist; b. Omaha, Nov. 30, 1934; s. James and Mary Florence (Nelson) Kl; m. Joyce Mae Harlan, Aug. 11, 1957; children: James Michael, David Russell, Kathryn Anne. B.S., Iowa State U., 1956; M.S., Purdue U., 1958, Ph.D., 1959. Research geneticist cotton genetics research U.S. Dept. Agr., College Station, Tex., 1959—. Contbr. articles to profl. jours. Served with USAR, 1960-68. Recipient Cotton Genetics Research award Joint Cotton Breeding Com., 1970. Fellow Am. Soc. Agronomy; mem. Crop Sci. Soc. Am., Genetics Soc. Am., Am. Genetics Assn., Am. Soc. Plant Physiologists. Subspecialties: Plant genetics; Gene actions. Current work: Research in basic genetics, seed quality, cotton. Office: PO Drawer DN College Station TX 77841

KOHL, STEVE, physician, researcher; b. N.Y.C., Aug. 13, 1945; s. Moses J. and Dorothy (Weisenfeld) K.; m. Sybil Janice Kohl, June 11, 1967; 1 dau., Gwynne Odette. B.S. magna cum laude, CCNY, 1966; M.D., Columbia U., 1970. Diplomate: Nat. Bd. med. Examiners, 1973, Am. Bd. Pediatrics, 1976; lic. physician, N.Y., Ga., Tex. Trainee in ob-gyn Columbia Presbyn. Med. Ctr., N.Y.C., 1970-71, vis. fellow, 1971, research assoc. pediatrics, 1971; intern in pediatrics Babies Hosp., Children's Med. and Surg. Ctr., N.Y.C., 1972, resident, 1972-74; fellow in pediatric infectious disease and immunology Emory U. Sch. Medicine, Atlanta, 1974-76; asst. prof. infectious diseases, dept. pediatrics M.D. Anderson Hosp. and Tumor Inst., U. Tex. Cancer Ctr., Houston, 1976-81, assoc. prof., 1981—; asst. prof. pediatrics, program infectious diseases and clin. microbiology U. Tex. Med. Sch., Houston, 1976-79, assoc. prof., 1979—, assoc. prof. pediatrics, program immunology, 1981—; attending Hermann Hosp., Houston; pediatric courtesy staff Meml. Hosp. System, Houston; mem. infection control com. Elks Aidemore Hosp., Atlanta, 1974-76, Hermann Hosp., Houston, 1978—; mem. adv. bd. Houston chpt. Nat. Found. March of Dimes, 1978-79. Contbr. articles to profl. jours. Recipient John H. Freeman Outstanding Teaching award U. Tex., 1981; NIH grantee, 1978—; James Donovan scholar, 1966; N.Y. State Med. Regents scholar, 1966-70; Nat. Cancer Inst. fellow, 1975-76. Fellow Infectious Diseases Soc. Am.; mem. Soc. Pediatric Research, Am. Soc. Microbiology, Am. Assn. Immunologists, Am. Acad. Pediatrics, Am. Fedn. Clin. Research, So. Soc. Pediatric Research, N.Y. Acad. Sci., Houston Pediatric Soc., Harris County Med. Soc., Houston Infectious Disease Soc., Alpha Omega Alpha, Phi Beta Kappa, Sierra Club. Subspecialties: Pediatrics; Infectious diseases. Current work: Immunology of herpes simplex virus infection; cellular cytotoxicity. Office: PO Box 20708 Houston TX 77025

KOHLER, PETER FRANCIS, physician, researcher, educator; b. Milw., Apr. 14, 1935; s. David and Marguerite (McCulley) K.; m. R. Christa Eckert, June 29, 1964; children: Michael D., Nicholas P., Andre E. A.B., Princeton U., 1957; M.D., Columbia U., 1961. Intern U. Minn. Hosps., Mpls., 1961-62, in internal medicine, 1962-64; clin. fellow, then research fellow Scripps Clinic, La Jolla, Calif., 1965-67; asst. prof. medicine U. Colo. Med. Sch.-Denver, 1967-71, assoc. prof., 1972-76, prof., 1977; co-chmn. Am. Bd. Allergy and Immunology, Phila., 1982-83. Recipient Research Career Devel. award NIH, 1969-73. Fellow Am. Soc. Clin. Investigation, Am. Acad. Allergy, A.C.P.; mem. Am. Assn. Immunologists. Republican. Roman Catholic. Subspecialties: Allergy; Immunology (medicine). Current work: Immune response in hepatitis B virus infection; familial systemic lupus erythematosias and genetics of human immune system. Office: Colo Med Center 4200 E 9th Ave Denver C0 80262

KOHLI, JAI DEV, pharmacologist, educator; b. Jullundur City, India, Dec. 27, 1918; came to U.S., 1975; s. Kripa R. and Maya D (Chadda) K.; m. Pushpa Sahney, Nov. 28, 1946; children: Atul, Tanuj,. Vandana. M.S., U. Chgo., 1952; Ph.D., U. Man., Can., 1965. Diplomate: Dip. Med., India, 1940. Fellow Indian Council Med. Research, 1942-50; sci. officer Drug Research Inst., India, 1951-61; asst. prof. Faculty Medicine, U. Man., 1961-65; scientist Food and Drug Directorate Can., 1965-70; asst. dir. Indsl. Toxicology Research, India, 1970-75; assoc. research prof. U. Chgo., 1975-79, research prof. pharmacology, 1979—. Contbr. articles to profl. jours. Recipient Drewery award U. Man., 1964; Fulbright fellow, 1951-52; Wellcome Research fellow, 1961-65. Mem. Am. Soc. Pharmacology and Exptl. Therapeutics, Pharm. Soc. Can., N.Y. Acad. Scis., Sigma Xi. Subspecialties: Pharmacology; Cardiovascular pharmacology. Current work: Autonomic pharmacology, cardiovascular pharmacology drug receptor interactions, Dopamine agonists and antagonists. Office: 914 E 58th St Room 414 Chicago IL 60607

KOHN, ALAN JACOBS, zoology educator; b. New Haven, July 16, 1931; s. Curtis I. and Harriet M. (Jacobs) K.; m. Marian S. Adachi, Aug. 28, 1959; children: Lizabeth, Nancy, Diane, Stephen. A.B., Princeton U., 1953; Ph.D., Yale U., 1957. Asst. prof. Fla. State U., Tallahassee, 1958-61; asst. prof. U. Wash., Seattle, 1961-63, assoc. prof., 1963-67, prof. zoology, 1967—. Contbr. articles in field to profl. jours. NRC sr. postdoctoral research assoc., 1967; John Simon Guggenheim Meml. fellow, 1974; NSF research grantee, 1959—. Fellow AAAS; mem. Am. Soc. Zoologist (treas. 1971-74), Am. Malacological Union (pres. 1982-83). Subspecialties: Ecology; Systematics. Current work: Marine invertebrate zoology and ecology, functional morphology, biology and paleobiology of marine molluscs.

Home: 18300 Ridgefield Rd NW Seattle WA 98177 Office: Dept Zoology U Wash Seattle WA 98177

KOHN, DONALD WILLIAM, pediatric dentistry educator; b. Hackensack, N.J., Apr. 11, 1951; s. Sidney Irving and Adele (Schwartz) K.; m. Candice Lee Johnson, June 12, 1977; 1 son, Michael Alexander. Student, Carnegie-Mellon U., 1969-70, Northwestern U., 1970-72; D.D.S., Emory U., 1976; postdoctoral cert., U. Pa., 1979. Diplomate: Am. Bd. Pedodontics. Postdoctoral fellow U. Pa., Phila., 1977-79; clin. instr. Fairleigh Dickinson U., Hackensack, N.J., 1976-77; asst. clin. prof. Yale U. Sch. Medicine, New Haven, 1979—; chief pediatric dentistry Yale-New Haven Hosp., New Haven, 1979—, acting chmn. dentistry/oral surgery, 1982—, dir. tng. program in dentistry for handicapped, 1980—. Assoc. editor: Jour. Dentistry for Children, 1980—. Fellow Am. Soc. Dentistry for Children (early career merit award 1976), Am. Acad. Pedodontics; mem. Internat. Assn. Dental Research, Acad. Dentistry for Handicapped, Am. Cleft Palate Assn. Subspecialties: Dental growth and development; Pediatric dentistry. Current work: Facial growth, dismorphology, syndromology, oral manifestations of systemic disease. Home: 35 Bartlett Dr Madison CT 06443 Office: Yale New Haven Hosp 20 York St New Haven CT 06504

KOHN, ERWIN, physical scientist, science educator; b. Vienna, Austria, Aug. 23, 1923; came to U.S., 1939; s. Julius G. and Laura (Deutsch) K.; m. Henrietta Lucille Klapman, Nov. 12, 1949; children: Joseph Howard, Michael David, Daniel Reuben, Benjamin, Samuel Lee, Susan Marlene Kohn Shay. B.S., U. Ill., 1948; M.S., U. Notre Dame, 1950; Ph.D., U. Tex.-Austin, 1955. Research scientist Monsanto Co., Texas City, Tex., 1955-66; assoc. prof. Southwestern Okla. State U., Weatherford, 1966-68, N.D. State U., Fargo, 1968-72; sr. project scientist Mason & Hanger, Amarillo, Tex., 1972—; editorial bd. Jour. Liquid, 1978-81. Contbr. articles to sci. jours. NSF grantee, 1969; Humble Oil Co. research fellow, 1951-55. Fellow Am. Inst. Chemists; mem. Am. Chem. Soc. (officer including chmn. 1974-75, nat. councilor 1975—), Am. Phys. Soc., Soc. Plastics Engrs. (sr.), Sigma Xi (grant-in-aid 1967). Subspecialties: Polymer chemistry; Analytical chemistry. Current work: Physical characterization of polymers; size exclusion chromatography polymer functional group analysis by UV, IR, NMR; applications of computers to analytical methods; surface analysis including scanning electron microscopy, microprobe analysis. Liquid chromatography; kinetics and mechanisms; structure of molecules; kinetic isotope effects. Home: 3613 Nebraska Amarillo TX 79109 Office: Mason & Hanger Devel Div PO Box 30020 Amarillo TX 79177

KOHN, KURT WILLIAM, biochemical pharmacology research administrator; b. Vienna, Austria, Sept. 14, 1930; came to U.S., 1938, naturalized, 1944; s. Siegfried and Sara (Margulies) K.; m. Elaine Kay Mogels, Feb. 5, 1956; children: Philip, Julia. A.B., Harvard U., 1952, Ph.D., 1966; M.D., Columbia U., 1956. Intern Mt. Sinai Hosp., N.Y.C., 1956-57; clin. assoc. Nat. Cancer Inst., Bethesda, 1967-60, sr. investigator, 1961-67, lab. dir., 1968—. Assoc. editor: Cancer Research, 1980, Jour. Nat. Cancer Inst, 1980—; contbr. numerous articles to profl. jours. Served to capt. USPHS, 1957—. Mem. Am. Assn. Cancer Research, N.Y. Acad. Sci., AAAS. Subspecialty: Molecular pharmacology. Current work: Direct research group studying DNA-binding drugs that have potential as anti-cancer agents; interested in mechanims of action at the cellular level. Office: National Institutes Health Bldg 37 Room 5D17 Bethesda MD 20205 Home: 11519 Gainsborough Rd Potomac MD 20854

KOHN, ROBERT M., physician; b. Syracuse, N.Y., Feb. 4, 1922; s. Maurice G. and Anne (Horwich) K.; m. Anita Firestone, June 29, 1947; children: Karen Kohn Bradley, Jay, Marjorie. A.B., Union Coll., 1943; M.D., Albany Med. Coll., 1946. Diplomate: Am. Bd. Internal Medicine. From instr. to assoc. prof. medicine SUNY Sch. Medicine, Buffalo, 1952-72, clin. prof., 1972—; pvt. practice internal medicine, Buffalo, 1952—; med. dir. Ind. Health Assn. Buffalo, 1979—. Editor: Cardiac Rehab. Quar, 1967-82; contbr. articles to profl. jours. Bd. dirs. Research and Planning Council, Buffalo, 1968-71. Served to capt., M.C.T. U.S. Army, 1947-49. Fellow A.C.P., Am. Coll. Cardiology, Am. Heart Assn. (council clin. cardiology); mem. Am. Soc. Internal Medicine (pres. Socio-econ. research found. 1974-77), N.Y. State Soc. Internal Medicine (pres. 1970-71), Phi Beta Kappa, Simga Xi. Subspecialties: Cardiology; Health services research. Current work: Research in cardiac rehabilitation; research clinical trials. Principal investigation: Coronary drug project, betablockade heart attack trial, persantin-aspirin reinfarction studies, dilleiazam prevention infarction trial. Home: 287 Deerhurst Park Blvd Buffalo NY 14223 Office: 50 High St Suite 1104 Buffalo NY 14203

KOHN, WALTER, educator, physicist; b. Vienna, Austria, Mar. 9, 1923. B.A., U. Toronto, Ont., Can., 1945, M.A., 1946, LL.D. (hon.), 1967; Ph.D. in Physics, Harvard U., 1948; Docteur es Sciences honoris causa, U. Paris, 1980; D.Sc. (hon.), Brandeis U., 1981, Ph.D., Hebrew U. Jerusalem, 1981. Indsl. physicist Sutton Horsley Co., Can., 1941-43; geophysicist Koulomzine, Que., 1944-46; instr. physics Harvard U., 1948-50; asst. prof. Carnegie Inst. Tech., 1950-53, asso. prof., 1953-57, prof., 1957-60; prof. physics U. Calif., San Diego, 1960-79, chmn. dept., 1961-63; dir. Inst. for Theoretical Physics, U. Calif., Santa Barbara, 1979—. Recipient Oliver Buckley prize, 1961, Davisson-Germer prize, 1977; NRC fellow, 1951; NSF fellow, 1958; Guggenheim fellow, 1963; NSF sr. postdoctoral fellow, 1967. Fellow AAAS, Am. Phys. Soc. (counselor-at-large 1968-72), Am. Acad. Arts and Scis.; mem. Nat. Acad. Scis. Research on electron theory of solids and solid surfaces. Subspecialties: Condensed matter physics; Theoretical physics. Office: Inst for Theoretical Physics U Calif Santa Barbara CA 93110

KOIDE, SAMUEL SABURO, medical researcher; b. Honolulu, Oct. 6, 1923; s. Sukeichi and Hideko (Dai) K.; m. Sumi Mary Mitsudo, Nov. 29, 1960; children: Sumi Lynn, Mark Kenji, Eric Akira. B.S., U. Hawaii, 1945; M.D., Northwestern U., 1953; M.S., 1954; Ph.D., 1960. Diplomate: Am. Bd. Internal Medicine. Intern Cook County Hosp., Chgo., 1953-54; resident VA Research Hosp., Chgo., 1956-57, Wesley Meml. Hosp., 1957-58; assoc. Sloan Kettering Inst., N.Y.C., 1960-65; asst. dir. Population Council Rockefeller U., N.Y.C., 1965—. Served to 1st. lt. U.S. Army, 1945-47. Fellow A.C.P. Subspecialties: Internal medicine; Biochemistry (medicine). Current work: Reproductive biology, gamete research, hormones, cell differentiation; basic and applied research in biomedicine. Home: 134 Lefurgy Ave Dobbs Ferry NY 10522 Office: Population Council 1230 York Ave New York NY 10021

KOIZUMI, KIYOMI, physiology educator, researcher; b. Kobe, Japan, Sept. 4, 1924; d. Hidekichi and Hatsue (Tatsuzawa) K.; m. Morimichi Watanabe, Sept. 21, 1954; 1 son, Tsugumichi Watanabe. M.D., Tokyo Women's Med. Coll., 1947; M.S. in Physiol. Chemistry, Wayne State U., 1951; Ph.D. in Physiology, Kobe U.Coll. Medicine, Japan, 1957. Intern Saiseikai Hosp., Tokyo, 1947-48; research fellow Med. Research Inst., Tokyo, 1948-49; instr. dept. physiology SUNY Downstate Med. Ctr., Bklyn., 1951-57, asst. prof., 1957-63, assoc. prof., 1963-70, prof., 1970—; vis. lectr. physiology Kobe U. Coll. Medicine, 1960-61; hon. research fellow U. Aberdeen, Scotland, 1962; vis. scientist Physiol. Inst., U. Heidelberg, Ger., 1971; NSF/U.S.-Japan Coop. Program vis. scientist Tokyo Met. Inst. Gerontology, 1976; Fogarty Internat. fellow vis. scientist, exptl. research dept. Semmelweis Med. U., Budapest, Hungary, 1979-80. Contbr. articles to sci. jours.; assoc. editor: Jour. Autonomic Nervous System, 1979—. Wayne State U. scholar, 1949-51; recipient medal of Honor Semmelweis Med. U., 1979; NIH fellow, 1979-80; grantee NIH, 1956—; NSF, 1974—. Fellow N.Y. Acad. Sci.; mem. Am. Physiol. Soc., Soc. Neurosci., Internat. Brain Research Orgn., Harvey Soc., AAAS, Physiol. Soc. Japan, Sigma Xi. Presbyterian. Subspecialties: Neurophysiology; Comparative physiology. Current work: Teaching medical students; research in cardiovascular physiology, the autonomic nervous system, neuroendocrinology, the hypothalamic functions. Office: 450 Clarkson Ave Brooklyn NY 11203

KOKOROPOULOS, PANOS, civil engr.; b. Thessaloniki, Greece, Aug. 10, 1927; came to U.S., 1958, naturalized, 1965; s. Constantine and Mary (Carvonides) K.; m. Carolyn A. Curran, Mar. 26, 1960; children: Mary, Constantine, George. B.S. in Chemistry, U. Thessaloniki, 1955, M.S., U. Dayton, 1964; Ph.D., U. Akron, 1972. Research chemist U. Dayton Research Inst., 1963-65; asst. prof. chem. tech. U. Akron, 1965-71, dir., 1965-69, mgr. acad. systems and programming, 1969-71, research asso. dept. civil engring., 1971-72; prof. civil engring. So. Ill. U., Edwardsville, 1973—; cons. in field; tech. reviewer Appropriate Tech. Program, Dept. Energy, 1980—, Engring. Edn. Author: (with A. Fatemi, A. Amirie) Political Economy of the Middle East, 1970; also articles. Asst. dist. commr. Greek Boy Scouts, 1945-56; mem. troop com. Cahokia Mounds council Boy Scouts Am., 1976-79. Served to 2d lt. Greek Army, 1950-52. Fulbright-Smith-Mundt grantee, 1959-60; Guggenheim grantee, 1959-60; Ford Found. grantee, 1959-60. Mem. Solar Energy Soc., Am. Chem. Soc., ASCE, ASTM, Am. Soc. Engring. Edn. (exec. bd. Ind.-Ill. sect. 1980—, program chmn. and chmn.-elect div. environ. engring. 1982—). Democrat. Greek Orthodox. Subspecialties: Environmental engineering; Solar energy. Current work: Treatment of wastes and resource recovery, energy from wastes, energy auditing, solar energy, math. modeling, domestic applications, retrofitting. Home: 414 W Union Edwardsville IL 62025 Office: So Ill U Box 65 Edwardsville IL 62026

KOLAR, OSCAR CLINTON, physicist; b. South Gate, Calif., Sept. 26, 1928; s. Oscar Clinton and Elizabeth (Deeds) K.; m. Rose Marilyn Markley, Jan. 31, 1953 (div. 1971); children: Elizabeth Louise, John Clinton, Walter Markley; m. Ingeborg Anna Brautigam, Feb. 2, 1982. B.A., UCLA, 1949; Ph.D., U. Calif.- Berkeley, 1955. Registered profl. eng., Calif. Sr. physicist Lawrence Livermore (Calif.) Nat. Lab., 1955-64; group leader, 1965-70, asst. div. leader, 1967-68, head criticality safety office, 1980—; v.p. Atomic Labs., Inc., Berkeley, 1954-60. Contbr. articles to profl. jours. Mem. Am. Nuclear Soc., Am. Phys. Soc., AAAS, Am. Assn. Physics Tchrs., Sigma Xi. Democrat. Subspecialties: Nuclear physics; Nuclear engineering. Current work: Nuclear physics, nuclear reactor physics, treaty verification, nuclear criticality safety. Home: 620 Zircon Way Livermore CA 94550 Office: Lawrence Livermore Nat Lab PO Box 808 Livermore CA 94550

KOLB, CHARLES EUGENE, physical chemist, researcher; b. Cumberland, Md., May 21, 1945; s. Charles Eugene and Doris Helen (McFarl) K.; m. Susan Foote, Aug. 19, 1965; children: Craig E., Amy C. S.B. in Chemistry, MIT, 1967; M.A. in Phys. Chemistry, Princeton U., 1968, Ph.D., 1971. Sr. research scientist Aerodyne Research Inc., Billerica, Mass., 1971-75, prin. research scientist, 1975—, dir.Center for Chem. and Environ. Physics, 1977-79, tech. dir.Applied Sci. div., 1979-80, v.p., dir. Applied Scis. div., 1981—; research assoc. in atmospheric chemistry div. applied sci. Harvard U., 1976—; research affiliate Spectroscopy Lab., M.I.T., 1981—; mem. adv. bd. Regional Laser Center MIT, 1981—. Contbr. 35 articles to sci. jours. Mem. Am. Chem. Soc., Am. Phys. Soc., Optical Soc. Am., AAAS, Combustion Inst. Subspecialties: Physical chemistry; Atmospheric chemistry. Current work: Chemical kinetics and spectroscopy of combustion systems, planetary atmospheres, environmental problems, and gas laser systems; laser and spectroscopic measurements of trace species. Home: 51 Woodmere Dr Sudbury MA 01776 Office: 45 Manning Rd Billerica MA 01821

KOLBECK, RALPH CARL, cardiovascular physiologist; b. Wausau, Wis., Sept. 2, 1944; s. John and Elma (Kersten) K.; m. Donna Jean Belling, July 16, 1966; children: Lisa Jean, John Carl. B.A., U. Minn., 1966, Ph.D., 1970. Teaching asst. U. Minn., Mpls., 1966-67, lectr., 1968-73; instr. physiology and medicine Med. Coll. Ga., Augusta, 1973-76, asst. prof., 1976-80, assoc. prof., 1980—, dir. myocardial research, 1973—, dir., 1980, 1980—. Contbr. articles to profl. jours. Den leader Boy Scouts Am., Augusta, 1979—; sec. Sci. and Engring. Fair, Augusta, 1976 . Am. Heart Assn. grantee, 1977-82, 1974—, NIH grantee, 1974-82; fellow, 1970-73. Mem. Am. Physiol. Soc., Am. Fedn. Clin. Research, Am. Heart Assn. (chmn. Ga. research com. 1980—), Ga. Heart Assn., Soc. Exptl. Biology and Medicine. Subspecialties: Physiology (medicine); Cardiology. Current work: Calcium metabolism in heart and smooth muscle. Patentee in field. Office: Dept Medicine Med Coll Ga Augusta GA 30912 Home: 3235 Winding Wood Pl Augusta GA 30907

KOLFF, WILLEM JOHAN, surgeon, educator; b. Leiden, Holland, Feb. 14, 1911; came to U.S., 1950, naturalized, 1956; s. Jacob and Adriana (de Jonge) K.; m. Janke C. Huidekoper, Sept 4, 1937; children: Jacob, Adriana P., Albert C., Cornelis A., Gualtherus C.M. Student, U. Leiden Med. Sch., 1930-38; M.D. summa cum laude, U. Groningen, 1946, U. Turin, Italy, 1969, Rostock (Germany) U., 1975, U. Bologna, Italy, 1977, D.Sc., Allegheny Coll., Meadville, Pa., 1960, Tulane U., 1975, CUNY, 1982, Temple U., 1983, U. Utah, 1983. Internist, head med. dept. Mcpl. Hosp., Kampen, Holland; staff research div. Cleve. Clinic Found., 1950-67; privaat docent, dept. medicine U. Leiden, Nether-Bunts Ednl. Inst., Cleve., 1950-67, head dept. artificial organs, 1958-67; prof. surgery, head div. artificial organs dept. surgery U. Utah Coll. Medicine, Salt Lake City, 1967—, Disting. prof. surgery, 1979—, prof. internal medicine, 1981—; research prof. engring. Inst. Biomed. Engring., 1967—. Decorated commandeur Orde Van Oranje, Netherlands, 1970; Orden de Mayo al Merito en el Grado de Gran Official, Argentina, 1974; recipient Landsteiner medal for establishment blood banks during war in Holland Netherlands Red Cross, 1942; Cameron prize U. Edinburgh (Scotland), 1964; 5,000 award Gairdner Found., 1966; Valentine award N.Y. Acad. Medicine, 1969; 1st Gold medal Netherlands Surg. Soc., 1970; Leo Harvey prize Technion, Israel, 1972; Sr. U.S. Scientist award Alexander Von Humboldt Found., 1978; Austrian Gewerbeverein's Wilhelm-Exner award, 1980. Mem. AMA, AAUP, Am. Physiol. Soc., Soc. Exptl. Biology and Medicine, AAAS, N.Y. Acad. Scis., Am. Soc. Artificial Internal Organs, Nat. Kidney Found., European Dialysis and Transplant Assn., ACP, Austrian Soc. Nephrology (hon.), Academia Nacional de Medicine (Colombia) (hon.), Rotarian. Subspecialty: Artificial organs. Devel. artificial kidney for clin. use, 1943; oxygenator, 1956. Address: Div of Artificial Organs Univ of Utah Medical Center Bldg 535 Salt Lake City UT 84112

KOLLAR, EDWARD JAMES, developmental embryologist, educator; b. Forest City, Pa., Mar. 3, 1934; s. I. J. Kollar and Mary Ann (Zaverl) K.; m. Catherine Ann Tobin, Feb. 23, 1963; children: Michelle, Elizabeth, Rachael, Brian, Rebecca. B.S., U. Scranton, Pa., 1955; M.S., Syracuse U., 1959, Ph.D., 1963. Instr. U. Chgo., 1963-66, asst. prof. biology, 1966-69, asst. prof anatomy, 1969-71; assoc. prof. U. Conn. Health Ctr., Farmington, 1971-76, prof., 1976-. Editor: Archives Oral Biology, 1982. Recipient Quantrell Teaching Award U. Chgo., 1968. Mem. ADA (cons. nat. bd. examiners 1978-83), Internat. Assn. Dental Research (Isaac Schour Research award 1981, pres. cranio-facial group 1983), Soc. Devel. Biology, Internat. Soc. Differentiation, Intern. Soc. Devel. Biology, Am. Assn. Anatomists, Tissue Culture Assn., Sigma Xi. Democrat. Roman Catholic. Subspecialties: Anatomy and embryology; Cell and tissue culture. Current work: Experimental analysis of the tissue interactions involved in the development of skin and teeth with emphasis on the etiology of birth defects. Home: 1710 Boulevard West Hartford Ct 06107 Office: U Conn Health Ctr Dept Oral Biology Farmington Ct 06032

KOLLER, LOREN D., veterinary pathologist, immunotoxicologist, researcher; b. Pomeroy, Wash., June 16, 1940; s. Edwin E. and Doris K. (Shelton) K.; m. Kathleen N. Ringness, Sept. 7, 1963; children: Susan, Michael, Christopher. D.V.M., Wash. State U., Pullman, 1965; M.S., U. Wis., Madison, 1969, Ph.D., 1971. Lic. veterinarian, Oreg., Wash., Idaho. Pvt. vet. practice, Corvallis, Oreg., 1965-66; with Nat. Inst. Environ. Health Scis., Research Triangle Park, N.C., 1971-72, Oreg. State U., Corvallis, 1972-78, WOI Regional Program in Vet. Medicine, U. Idaho, Moscow, 1978—; cons. EPA, Life Systems, Inc.; mem. grant rev. panel of sci. rev. panel health research EPA. Editorial bd.: Fundamental Applied Toxicology, 1982—; cons. editor: Jour. Reticuloendothelial Soc, 1982—; contbr. articles to sci. jours. Served to capt. M.C., AUS, 1966-68. Grantee NIH, EPA, Dept. Agr., Dow Chem. Soc., Merck, Sharp & Dohme, Pacific N.W. Region Commn. Mem. AVMA, Idaho Vet. Med. Assn., Soc. Toxicol. Pathology, Soc. Toxicology, Am. Coll. Vet. Toxicologists, Reticuloendothelial Soc., Conf. Research Workers in Animal Disease, Assn. Vet. Med. Colls., Am. Assn. Vet. Immunologists. Baptist. Subspecialties: Cancer research (veterinary medicine); Pathology (veterinary medicine). Current work: Study of pathological, immunological, toxicological and oncogenic effects of drugs and chemicals in laboratory animals. Home: 808 Park Dr Moscow ID 83843 Office: Vet Medicine U Idaho Moscow ID 83843

KOLLMAN, PETER ANDREW, chemistry educator; b. Iowa City, July 24, 1944; s. Eric C. and Gusti (Binstok) K.; m. Jean Furnish, Aug. 1, 1970; children: Sarah, Eli. B.A., Grinnell Coll., 1966; Ph.D., Princeton U., 1970. NATO postdoctoral asst. Cambridge (Eng.) U., 1970-71; asst. prof. chemistry U. Calif.-San Francisco, 1971-76, assoc. prof., 1976-80, prof., 1980—. Contbr. numerous articles on chemistry and pharm. chemistry to profl. jours.; mem. adv. editorial bd.: Jour. Med. Chemistry, 1980—. NIH Career Devel. award, 1974-79; NSF fellow, 1966; Woodrow Wilson fellow (hon.), 1966. Mem. Am. Chem. Soc., Am. Phys. Soc., Sigma Xi, Phi Beta Kappa. Democrat. Jewish. Subspecialties: Biophysical chemistry; Theoretical chemistry. Current work: Computer modeling of molecular interactions involving molecules of biological interest. Home: 150 Rivoli San Francisco CA 94117 Office: U Calif Sch Pharmacy San Francisco CA 94143

KOLMER, LEE ROY, coll. dean; b. Waterloo, Ill., Jan. 4, 1928; s. Arthur Francis and Carmelita Frances (Vogt) K.; m. W. Jean O'Brien, Apr. 19, 1952; children—Diane, James, John. B.S., So. Ill. U. at Carbondale, 1952; M.S., Iowa State U., 1952, Ph.D., 1954. Asst. prof. So. Ill. U. at Carbondale, 1954-55; prof. Iowa State U., Ames, 1956-67; asst. dean Univ. Extension, 1967-71; dean Coll. Agr., 1973—; dir. Coop. Extension Service, Ore. State U., Corvallis, 1971-73. Served with U.S. Army, 1946-48. Mem. Am. Agrl. Econs. Assn. Subspecialty: Agricultural economics. Current work: Administration, teaching, research. Home: 4118 Phoenix St Ames IA 50010

KOLODNY, ABRAHAM LEWIS, rheumatologist, clinical researcher; b. Norfolk, Va., July 2, 1917; s. William and Jennie (Eisenberg) K.; m. Mildred A. Fiske, Aug. 10, 1942; children: William (dec.), David, Suki McCormick, Douglas, Peggy. M.D., U. Va., 1941. Intern South Balt. Gen. Hosp., 1941-42; resident in rheumatology Ashburn Army Hosp., McKinney, Tex., 1945-46; rheumatologist N. Charles Gen. Hosp., Balt., 1946—, Franklin Square Hosp., 1946—, pres. staff, 1981-83; lectr. Essex Community Coll., Balt.; mem. Md. Gov.'s Task Force for Arthritis, 1981-83; cons. to drug coms. Author: Comprehensive Approach Therapy Pain, 1961; contbr. chpts. to books, articles to profl. jours. Bd. dirs. Md. chpt. Arthritis Found., Md. chpt. Lupus Found. Served to maj. U.S. Army, 1942-46. Decorated Bronze Star; recipient Arthritis Found. Disting. Service award, 1981, 82. Fellow Royal Soc. Health, Internat. Acad. Law and Sci., Am. Geriatric Soc., Internat. Coll. Angiology; mem. Am. Fedn. Clin. Research, Am. Rheumatism Assn., Md. Soc. Rheumatic Diseases, Baltimore County Med. Soc., Med. and Chirurg. Faculty Md., AMA, So. Med. Soc., Am. Assn. Med. Writers, Am. Soc. Clin. Pharmacology and Therapeutic. Clubs: Towson (Md.); Hopkins (Balt.). Subspecialties: Internal medicine; Immunopharmacology. Current work: Clinical pharmacology, analgesics, anti-rheumatics. Home: PO Box 5900 Baltimore MD 21208 Office: Franklin Square Medical Arts Bldg Franklin Square Dr Baltimore MD 21237

KOLSTAD, GEORGE ANDREW, nuclear agency administrator; b. Elmira, N.Y., Dec. 10, 1919; s. Charles Andrew and Rose (Haesloop) K.; m. Christine Joyce Stillman, July 22, 1944; children: Charles Durgin, Martha Rae, Peter Kenneth. B.S. magna cum laude, Bates Coll., 1943; postgrad., Wesleyan U., Harvard U., 1944-45, MIT, 1944-45; Ph.D. in Physics, Yale U., 1948. Research assoc. Harvard U., Cambridge, Mass., 1944-45; instr. Yale U., New Haven, 1948-50; with AEC (now ERDA), 1950—, chief physics, math. br. div. research, 1952-60, asst. dir. research physics, math., 1960-73, sr. physicist div. phys. research, Washington, 1973—, chmn. geosci. working group, 1975-76; mem. Internat. Nuclear Data Com., 1968-73, chmn., 1970-71; mem. European-Am. Nuclear Data Com., 1960-73; guest scientist Univ. Inst. Theoretical Physics, Copenhagen, Denmark, 1956-57. Trustee Laytonsville Elem. Sch.; bd. dirs. Sandy Spring (Md.) Friends Sch. Recipient spl. award AEC, 1961; Sheffield-Loomis fellow Yale, 1947-48. Fellow Am. Phys. Soc.; mem. AAAS, Am. Geophys. Union, Phi Beta Kappa, Sigma Xi (treas. chpt. 1975-76). Club: Cosmos (Washington). Home: 7920 Brink Rd Oak Hill Laytonsville MD 20760 Office: ERDA Div Physical Research Washington DC 20545

KOMANDURI, AYYANGAR M., med. physicist, radiol. physics cons.; b. Pedapadu, India, Jan. 13, 1940; s. Rangachary and Pattamma (Satluri) K.; m. Vijayalakshmi, May 25, 1970; children: Chaitanya, Saranga. B.S., Andhra U., 1958, M.S., 1960, Ph.D. in Nuclear Physics, 1965. Cert. radiol. physics Am. Bd. Radiology, 1978. Research fellow Mass. Gen. Hosp., Boston, 1966-69; sci. officer Directorate of Radiol. Protection, Bhabha Atomic Research Center, Bombay, India, 1970-74; research fellow Thomas Jefferson U. Hosp., Phila., 1975, asst. prof. med. physics, 1976-81, assoc. prof. med. physics, dept. radiation therapy and nuclear medicine, 1981—; researcher, cons. in field; computer programmer. Contbr. sci. articles to profl. jours. and confs. Recipient prizes for exhibits Soc. Nuclear Medicine, 1968, 81, 80. Mem. Am. Assn. Physicists in Medicine. Subspecialties: Medical physics; Scientific Computer Programming. Current work: Radiation therapy physics; nuclear medicine physics. Office: Radiation Therapy Dept Thomas Jefferson Univ Hosp Philadelphia PA 19107

KOMANICKY, PAVEL, medical educator, researcher; b. Habura, Czechoslovakia, June 28, 1943; came to U.S., 1969; s. Stefan and

Maria e6(Milanova) K. M.D., Šafarik U., Košice, Czechoslovakia, 1966. Diplomate: Am. Bd. Internal Medicine. Intern Univ. Hosp., Košice, 1965-66; research fellow, assoc. Czechoslovak Acad. Scis., 1966-69; resident in medicine Carney Hosp., Boston U., 1970-73; research fellow Univ. Hosp., Boston U., 1973-76, staff physician, 1976-81; asst. prof. medicine Tex. Tech. U. Health Scis. Ctr., Amarillo, 1981—; instr. medicine Boston U., 1975-81; attending physician Univ. Hosp., Boston, 1976-81, Boston City Hosp., 1976-81, Amarillo Med. Ctr., 1981—. Research grantee Univ. Hosp., Boston U., 1976, 78, Tex. Tech. U., 1982. Fellow ACP; mem. AMA (Physician Recognition award 1975, 78, 81), Am. Heart Assn., Endocrine Soc., Am. Fedn. Clin. Research. Subspecialties: Endocrinology; Internal medicine. Current work: Role of humoral factors in etiology of hypertension; role of adrenal hormones in hypertension; pathophysiology of hypertension; study of receptors for adrenal hormones; pathophysiology of cardiomegaly. Home: 6040 Belpree Dr C225 Amarillo TX 79106 Office: Texas Tech University Health Sciences Center 1400 Wallace Blvd Amarillo TX 79106

KOMECHAK, MARILYN GILBERT, psychologist; b. Wabash, Ind., Aug. 28, 1936; d. Russell and Evelyn (Snyder) Gilbert; m. George J. Komechak, Aug. 23, 1936; children: Kimberly, Gilbert. B.S., Purdue U., 1958, Tex. Christian U., 1966, M.Ed., 1968; Ph.D., North Tex. State U., 1975. Staff Child Study Ctr., Diagnostic Clinic, Ft. Worth, 1968-74; assoc. dir. behavioral studies Sch. Community Service, North Tex. State U., 1974-77; pvt. practice psychology, Ft. Worth, 1977—; adj. prof. Tex. Christian U., Ft. Worth, 1977-79, U. Tex.-Arlington, 1977-80; cons. in field; mem. bus. women's bd. Sanger-Harris Dept Stores. Author: Getting Yourself Together, 1982; contbr. articles to profl. jours. Bd. dirs. Jon Pierce, Inc., Ft. Worth, 1980-82; adv. bd. Trinity Valley Mental Health/Mental Retardation, 1980, Mental Health Assn. Tarrant County, 1980; active Easter Seal Soc., 1977. Mem. Am. Psychol. Assn., Tarrant County Psychol. Assn. (sec. 1980), Am. Soc. Clin. Hypnosis, Tex. Psychol. Assn. Episcopalian. Subspecialty: Psychotherapy. Current work: Psychohistorial research in life problems and hypnosis. Home: 8109 Rush St Fort Worth TX 76116 Office: 5280 Trail Lake Dr Fort Worth TX 76133

KOMM, DEAN ALBERT, plant pathologist; b. Pitts., Jan. 15, 1948; s. Albert H. and Edith L. (Lenord) K.; m. Mary Andrea Conlon, July 12, 1980. B.S., Point Park Coll., 1972; M.S., U. Mass., 1974; Ph.D., Purdue U., 1979. Extension plant pathologist-tobacco Va. Poly. Inst. and State U., Blacksburg, 1979—. Club: Exchange. Lodge: Lions. Subspecialty: Plant pathology. Current work: Soil-borne disease of tobacco.

KOMMEDAHL, THOR, educator; b. Mpls., Apr. 1, 1920; s. Thorbjorn and Martha (Blegen) K.; m. Faye Lillian Kommedahl, Aug. 5, 1951; children: Kris Alan, Siri Lynn, Lori Anne. Student, Bethel Coll., St. Paul, 1938-39; B.S., U. Minn., 1945, M.S., 1947, Ph.D., 1951. Instr. U. Minn., St. Paul, 1946-51; asst. prof. Ohio Agrl. Research and Devel. Ctr., Wooster, 1951-53, Ohio State U., Columbus, 1952-53, U. Minn., St. Paul, 1953-57; assoc. prof., 1957-63, prof. dept. plant pathology, 1963—; cons. botanist and taxonomist Minn. Dept. Agr., 1954-60. Sr. editor: Challenging Problems in Plant Health, 1983; editor-in-chief: Phytopathology, 1964-67; cons. editor: McGraw-Hill Ency. Sci. and Tech, 1972-78; editor: Internat. Congress Plant Protection, 2 vols, 1981; Contbr. articles to profl. jours. Fulbright grantee, Iceland, 1968; Guggenheim fellow, Australia, 1961-62. Fellow AAAS, Am. Phytopathology Soc.; mem. Weed Sci. Soc. Am. (award of excellence 1966), Am. Inst. Biol. Scis., Bot. Soc. Am., Council Biology Editors, Internat. Soc. Plant Pathology, Mycological Soc. Am., Soc. Scholarly Publ., N.Y. Acad. Sci. Baptist. Subspecialties: Plant pathology; Biological control. Current work: Biol. control of plant disease by application of antagonistic microorganisms to seeds, ecology of root disease fungi. Home: 1840 W Roselawn Ave Falcon Heights MN 55113 Office: Stakman Hall Plant Pathology 1519 Gortner Ave St Paul MN 55108

KOMOREK, MICHAEL, JR., health physicist. A.S., Genesee Community Coll., 1974; B.A., SUNY Buffalo, 1977; postgrad., 1980. Research asst. SUNY - Buffalo, 1976-78; health physicist Sch. Medicine and VA Hosp., 1978-82, clin. instr., 1979—; sr. health physicist Alara Mgmt. Co., Elma, N.Y., 1980—; cons. VA Hosp., 1980—, Wyoming County Community Hosp., 1980—. Contbr. articles to profl. jours. Mem. Health Physics Soc. (pres. chpt. 1981-82), Am. Assn. Physicists in Medicine, Nuclear Medicine Soc., Laser Inst. Am. (assoc.). Subspecialty: Nuclear physics. Office: Alara Mgmt Co 2125 Transit Rd Elma NY 14059

KONDO, YOJI, astrophysicist, educator; b. Hitachi, Japan, May 26, 1933; came to U.S., 1960, naturalized, 1968; s. Tsuneo and Hama (Yamada) K.; m. Ursula Tuetermann, Sept. 10, 1965; children: Beatrice, Cynthia, Angela. B.A., Tokyo U. Fgn. Studies, 1958; M.A., U. Pa., 1963, Ph.D., 1965. Asst. prof. U. Pa., Phila., 1965; NAS-NRC fellow NASA Goddard Space Flight Ctr., Greenbelt, Md., 1965-68; astronomer, chief astrophysics sect. Johnson Space Ctr., 1968-77; astrophysicist Goddard Space Flight Ctr., 1978—, project scientist Internat. Ultraviolet Explorer satellite, 1982—; adj. assoc. prof. U. Okla., 1971-72, prof., 1972-77, U. Houston, 1974-77, U. Pa., 1978—. Editor: Earth & Extraterestrial Sci., 1974-79; editor: Comments on Astrophysics, 1979—; contbr. articles to profl. jours.; editor: X-Ray Binaries, 1976, Advances in UV Astronomy, 1982. Recipient Cert. of Commendation, NASA Johnson Space Ctr., 1975. Fellow AAAS; mem. Am. Astron. Soc., Internat. Astron. Union (acting pres. Commn. on Astronomy from Space 1982, v.p. 1982-85), U.S. Judo Assn. Subspecialties: Ultraviolet high energy astrophysics; Satellite studies. Current work: Research of evolutional processes in interacting binary stars, study of interstellar medium, study of energy genration mechanism in BL Lacertae objects and quasars. Office: NASA Goddard Spaceflight Ctr Greenbelt MD 20771

KONG, ERIC SIU-WAI, research scientist, consultant; b. Hong Kong, Jan. 14, 1953; U.S., 1970; s. Woon-Man and Chau-Mui (Mo) K.; m. Susanna May-Man Lee, June 24, 1974. B.A., U. Calif.-Berkeley, 1974; M.Sc., Renssalaer Poly. Inst., 1976, Ph.D., 1978. Teaching asst. Rensselaer Poly. Inst., 1974-76, research asst., 1976-78; research assoc. Va. Poly. Inst., 1978-79; research scientist NASA Ames Research Ctr., Moffett Field, Calif., 1979-83; mem. tech. staff Sandia Nat. Labs., Livermore, Calif., 1983—; vis. scholar Stanford U., 1980-83. Contbr. articles to profl. jours. NASA grantee, 1979-83; Japan Soc. Promotion of Sci. fellow, 1982. Fellow Am. Inst. Chemists, N.Y. Acad. Scis.; Mem. Japan Soc. Polymer Sci., Am. Phys. Soc., Am. Chem. Soc.; mem. Soc. Plastics Engrs. (chmn. chpt.). Club: San Jose Toastmasters. Subspecialties: Polymer chemistry; Composite materials. Current work: Physical aging and its effects on reliability and durability of graphite/epoxy composites. Office: Sandia Nat Labs Exploratory Chemistry Div 8315 Livermore CA 94550

KONG, JIN AU, elec. engr., educator; b. Kiangu, China, Dec. 27, 1942; came to U.S., 1965, naturalized, 1975; s. Chin Hwu and Shiu C. (Chao) K.; m. Wen Yuan Yu, June 27, 1970; children: Shing David. Ph.D., Syracuse U., 1968. Postdoctoral research engr. Syracuse (N.Y.) U., 1968-69; mem. faculty M.I.T., 1969—, now prof. elec. engring.; cons. N.Y. Port Authority, 1971, Lunar Sci Inst., 1972, Army Engring. Topographical Lab., 1979, Lincoln Lab., 1979-82, Hughes Aircraft Co., 1981; interregional advisor UN Dept. Tech. Coop. for Devel., 1978-80. Author: Theory of Electromagnetic Waves, 1975; editor: Research Topics in Electromagnetic Wave Theory, 1981; contbr. articles to profl. jours. Vinton Hayes postdoctoral fellow, 1969. Mem. Am. Geophys. Union, Am. Phys. Soc., IEEE, Internat. Union Radio Sci., Optical Soc. Am., Sigma Xi, Phi Tau Phi, Tau Beta Pi. Subspecialties: Electrical engineering; Remote sensing (geoscience). Current work: Remote sensing and electromagnetic wave theory. Home: 72 Hillcrest Ave Lexington MA 02173 Office: 36-383 MIT Cambridge MA 02139

KONIGSBERG, JAN, energy planner and administrator, researcher, educator; b. Boston, Jan. 31, 1948; s. Charles and Frances (Benchimol) K.; m. Christy Cooper, Dec. 23, 1979. B.A., Reed Coll., 1969; M.A., U. Mont., 1975. Field researcher Bur. Govt. Research, U. Mont., Missoula, 1976; instr. Chapman Coll., Anchorage, 1976; solar coordinator Mont. Energy Office, Helena, 1977; energy planning coordinator Mont. Dept. Natural Resources and Conservation, Helena, 1978-79, bur. chief planning and analysis bur. energy div., 1979-83; ptnr. Kuntz, Konigsberg & Hadley, energy devel. services, 1983—; agy. liaison Hydro Mgmt. Ind., 1983—. Author: Montana Renewable Energy Handbook; 980. Active Nature Conservancy, Mont. Environ. Info. Ctr., Trout, Unltd., ACLU. Mem. N.W. Electric Power Planning Council (sci. and statis. adv. com.). Subspecialty: Other energy policy research. Current work: Regional energy legislation, facility siting, resource acquisition planning, conservation potential analysis, renewable resource assessment. Office: PO Box 268 Helena MT 59624

KONITS, PHILIP H., physician; b. N.Y.C., Oct. 27, 1949; s. Irving and Yetta (Berkoff) K.; m. Cindy Gail Bernstein, Junell, 1975; children: Jonathan, Adina. B.A., Adelphi U., 1971; M.D., U. Rochester, 1976. Diplomate: Am. Bd. Internal Medicine. Resident in medicine Rochester (N.Y.) Gen. Hosp., 1975-78; clin. assoc., pub. health officer Nat. Cancer Inst., Balt., 1978-81; chief oncology Luth. Hosp., Balt., 1981—. Contbr. articles to profl. jours. Mem. A.C.P., Am. Soc. Clin. Oncology. Subspecialties: Oncology; Chemotherapy. Current work: New drug development in cancer treatment. Screening and community cancer care. Office: 730 Ashburton St Baltimore MD 21216

KONOWALOW, DANIEL DIMITRI, chemistry educator; b. Cleve., Apr. 28, 1929; s. Dimitry and Mary (Ehnatt) K.; m. Marcy Ellen Rosenkrantz, July 24, 1978. B.Sc., Ohio State U., 1953; Ph.D., U. Wis.-Madison, 1961. Chemist E.I. duPont de Nemours, Wilmington, Del., 1960-62; asst. dir. U. Wis. Theoretical Chemistry Inst., Madison, 1962-65; asst. prof. chemistry SUNY-Binghamton, 1965-70, assoc. prof., 1970-80, prof., 1980—; cons. to industry. Contbr. numerous articles on atomic and molecular interactions to profl. jours. Served with Chem. Corps U.S. Army, 1953-55. Subspecialties: Theoretical chemistry; Atomic and molecular physics. Current work: Molecular structure and spectra; lasers; intermolecular forces; research on electronic structure and spectra of excited states of small molecules; long-range interactions of molecular fragments. Home: Box 473A Roberts Rd RD 1 Binghamton NY 13903 Office: SUNY Binghamton NY 13901

KONRAD, MICHAEL, biochemist, research administrator; b. San Diego, Dec. 20, 1936; s. Edmund George and Ann (Akers) K.; m. Carol Guze, 1959 (div. 1968); children: Robin, Hans; m. Page Wood, Nov. 14, 1969; 1 dau., Michele. B.S., Calif. Inst. Tech., 1958; Ph.D., U. Calif.-Berkeley, 1964. Postdoctoral fellow Harvard U., 1966-68; asst. prof. UCLA, 1968-75; staff scientist U. Calif.-Berkeley, 1975-80; sr. scientist CETUS Corp., Berkeley, 1980—. Mem. Am. Soc. Biol. Chemists. Subspecialties: Genetics and genetic engineering (medicine); Biochemistry (biology). Current work: Interferon, insulin, cell biology. Office: CETUS Corp 1400 53d St Emeryville CA 94608

KONZAK, CALVIN FRANCIS, agronomist, educator; b. Devils Lake, N.D., Oct. 17, 1924; s. Peter Henry and Mary Ann (Dion) K.; m. Mary M. Coe, June 1948; children: Kenneth, Gary; m. Margarethe C. Reiter, Dec. 17, 1967. B.S., N.D. State U., 1948; Ph.D., Cornell U., 1952. Assoc. geneticist Brookhaven Nat. Lab., Upton, N.Y., 1951-58; mem. facuty Wash. State U., Pullman, 1958—, prof. agronomy and genetics, 1966—, agronomist, 1966—; cons. Contbr. articles to sci. jours. USPHS fellow, 1965-66. Fellow Am. Soc. Agronomy; mem. Crop Sci. Soc. Am., Am. Genetic Assn., AAAS, Am. Inst. Biol. Sci., Sigma Xi. Congregationalist. Lodge: Elks. Subspecialties: Plant genetics; Information systems, storage, and retrieval (computer science). Current work: Bread wheat, durum wheat and oat breeding; wheat genetics; mutations in wheat, oats and other crops; information management; electronic systems for data collection and processing; improvement of plant breeding research technology. Home: NE 1725 Wheatland Dr Pullman WA 99163 Office: Wash State U Pullman WA 99164

KOO, PETER HUNG-KWAN, immuno-biochemistry educator, researcher; b. Shanghai, China, Aug. 10, 1940; came to U.S., 1959, naturalized, 1975; s. Yung-Foo and Sun-Wa (Ko) K.; m. Somchit Alice Hotapichayawivat, Dec. 23, 1967; children: David G., Christopher G. B.A., U. Wash., 1964; Ph.D., U. Md., 1970. Research assoc. Johns Hopkins U., 1970-74, asst. prof., 1975-77; staff fellow NIH, Balt., 1974-75; asst. prof. microbiology and immunology N.E. Ohio U. Coll. Medicine, 1977-83, assoc. prof., 1983—; lectr. in immunology, biochemistry, microbiology, cancer biology at various univs., 1974—; bd. dirs., mem. profl. edn. com. Portage County (Ohio) chpt. Am. Cancer Soc., 1982—. Contbr. articles to profl. publs., 1970—. Deacon First Christian Ch., Kent, Ohio, 1978-81, 82—, mem. fin. com., 1981—. NIH grantee, 1978-82; Am. Cancer Soc. Ohio grantee, 1978, 82; Cystic Fibrosis Care Fund grantee, 1979, 82; United Way Health Found. grantee, 1982; MEFCOM Fund grantee, 1982—. Mem. Am. Assn. Immunologists, Johns Hopkins Immunology Council, Johns Hopkins Med. and Surg. Assn., N.Y. Acad. Scis., Am. Chem. Soc., AAAS, Sigma Xi. Subspecialties: Immunobiology and immunology; Cancer research (medicine). Current work: Cancer immunology and bilogy: roles of lymphokines, cytokines and interferon inducers in tumor-host interactions; natural immune defense systems; mechanism of spontaneous cancer regression; structure and function of alpha-2-macro globulin and its interactions with hormones (particularly nerve growth factor). Office: NE Ohio U Coll Medicine SR 44 Rootstown OH 44272

KOONTZ, HAROLD VIVIEN, biology educator, researcher; b. Pendleton, Oreg., Sept. 9, 1928; s. and Marjorie (McMonies) K.; m. Roberta Ellen Foote, June 25, 1965; children: Malia Sue Koontz Nosbisch, Lisa Lu Koontz Higham. B.S., Oreg. State U., 1952; Ph.D., Wash. State U., 1957. Teaching asst. Wash. State U., 1952-57; lectr., jr. agronomist U. Calif.-Davis, 1957-60; biol. scientist Hanford Atomic products, Richland, Wash., 1960-61; prof. biology U. Conn.-Storrs, 1961—; biol. scientist IAEA, Taiwan, 1968-69. Pres. Hemlock Point Dist., Coventry, Conn., 1970-80. Served with U.S. Army, 1946-48. Mem. Am. Soc. Plant Physiology, Am. Soc. Hort. Sci., Internat. Soc. Soilless Culture, Internat. Soc. Horticulture, Sigma Xi. Republican. Subspecialties: Hydroponics; Plant physiology (agriculture). Current work: Hydroponics; controlled environment plant growth; uptake and translocation of minerals; biological photography. Home: 99 Hemlock Ln Coventry Ct 06238 Office: U Conn Biology U-42 Storrs CT 06268

KOPECKO, DENNIS J., research microbiologist, consultant; b. Ironwood, Mich., Jan. 14, 1947; s. Norbert Robert and Dorothy Eileen (La Chapelle) K.; m. Patricia Guerry, Dec. 10, 1977. B.S. in Biology (U.S. Army scholar, 1966-68), Va. Mil. Inst., 1968; Ph.D. in Microbiology (Va. State Council for Higher Edn. Scholar, 1969-72), Med. Coll. Va., 1972. Predoctoral fellow dept. microbiology Med. Coll. Va., Richmond, 1968-72; postdoctoral fellow dept. medicine Stanford (Calif.) U. Med. Sch., 1972-76, C.F. Aaron fellow, 1972-73; sr. research microbiologist Dept. Def., Walter Reed Army Inst. Research, Washington, 1979—; cons., acting sci. dir. Genetic Research Corp., Columbia, Md.; vis. lectr. genetics Georgetown U., Uniformed Services U., NIH; vis. lectr. in microbial genetics Va. Commonwealth U., Richmond, 1979-80; continuing edn. lectr. Am. Coll. Vet. Microbiologists, Chgo., 1979; invited lectr. symposium on Mil. Vet. Medicine, Washington, 1980-82; vis. lectr. U. Mich., Ann Arbor, 1981; coorganizer ann. Mid-Atlantic Regional Conf. on Extrachromosomal Genetic Elements, 1977—; co-organizer, instr. grad. microbiol. course NIH, Bethesda, Md., 1980-842; exec. sec. Walter Reed Instl. Biosafety Com., 1981—. Co-editor: Progress in Molecular and Subcellular Biology; mem. editorial bd.: Jour. Bacteriology; contbr. articles to profl. publs. in field, papers to profl. confs., U.S., W. Ger., Switzerland, Can., Czechoslovakia, Dominican Republic, Japan, India. Served to capt. U.S. Army, 1976-79. Va. Heart Assn. fellow, summers 1967-68; Bank Am.-Giannini Med. Research Found. fellow, 1973-75; recipient honors Alpha Sigma Chi, 1971-72. Mem. Genetics Soc. Am., Am. Soc. Microbiology, Mid-Atlantic Regional Extrachromosomal Genetic Elements Group, Fed. Exec. and Profl. Assn., Sigma Xi. Roman Catholic. Subspecialties: Genetics and genetic engineering (medicine); Molecular biology. Current work: Molecular genetic analysis of enteric bacterial pathogens; molecular genetics of bacteria, genetic engineering of bacterial vaccine strains, genetic manipulations of bacteria. Patentee oral vaccine for immunization against enteric disease, method for the rapid detection of typhoid fever bacteria. Home: 4601 Flower Valley Dr Rockville MD 20853 Office: Walter Reed Army Inst Research Washington DC 20307

KOPELMAN, RICHARD ERIC, management educator, consultant; b. N.Y.C., May 31, 1943; s. Seymour Harold and Leona Libby (Quint) K.; m. Carol Fran Fialkov, June 7, 1970; children: Joshua Marc, Michael Adam. B.S. in Econs, U. Pa., 1965, M.B.A., 1967; D.B.A., Harvard U., 1974. Diplomate: Accredited personnel diplomate. Instr. mgmt. Baruch Coll., N.Y.C., 1973-74, asst. prof. mgmt., 1974-77, assoc. prof. mgmt., 1978-80, prof. mgmt., 1981—; dir. Three Dimensional Circuits, Plainview, N.Y.; dir., sec. Aleph Null Corp., Carle Place, N.Y., 1981—; mgmt. cons. staff Ctr. for Mgmt. of Profl. and Sci. Work, Provo, Utah, 1980—; mgmt. cons. group Deutsch, Shea & Evans, Inc., N.Y.C., 1976-81; cons. Forest Engring. Research Inst., 1976-77, Bendix Corp., Southfield, Mich., 1977-79. Bd. dirs. Day Care Council Nassau County, Inc., Hempstead, N.Y., 1979-82; tech. reader Recs. for the Blind, Inc., N.Y.C., 1976-77; cons. Vol. Urban Cons. Group, Inc., N.Y.C., 1975-78. William B. Harding fellow Harvard U., 1972-73; Yoder-Heneman personnel awardee Am. Soc. Personnel Adminstrn., 1982. Mem. Acad. of Mgmt., Am. Arbitration Assn. (panelist), Am. Psychol. Assn., Am. Inst. Decision Scis., Am. Compensation Assn. Subspecialty: Industrial and organizational psychology. Current work: Theoretical and applied research on work motivation; surveyed demonstrated effectiveness of ten prominent behavioral science techniques used to improve productivity; originated return on effort version of expectancy theory; developed model and scales to measure work, family and interrole conflicts. Home: 65 Colgate Rd Great Neck NY 11023 Office: Baruch Coll 17 Lexington Ave New York NY 10010

KOPF, RUDOLPH WILLIAM, geologist; b. Munich, Bavaria, Germany, Sept. 10, 1922; came to U.S., 1924, naturalized, 1934; s. Max Joseph and Anna Mary (Meir) K.; m. Doris Merilyn Hickling, Oct. 7, 1949; children: Eric Carter, Wendy Shawn Kopf Caufield. B.A., U. Buffalo, 1950, M.A., 1952. Registered geologist, Calif. Oceanographer U.S. Naval Hydrographic Office, Washington, 1952; geologist U.S. Geol. Survey, Washington, 1952-55, Menlo Park, Calif., 1961-80; geol. engr. AEC, Grand Junction, Colo., 1955-61; cons. geologist, Grass Valley, Calif., 1982—; lexicographer U.S. Geol. Survey, 1952-55. Author U.S. Geol. Survey publs. Served with USNR, 1942-45; PTO. Recipient award of Merit Dept. of Interior Safety Council, 1969. Fellow Geol. Soc. Am., mem., Am. Geophys. Union, Peninsula Geol. Soc. Clubs: Geologic Div. Retirees, Empire Mine Park Assn. Subspecialties: Tectonics; Sedimentology. Current work: Development of hydrotectonic hypothesis to reinterpret origin of fault breccia, breccia pipes and dikes, diatremes, cryptoexplosion structures, and mud volcanoes. Home: 129 E Empire St Grass Valley CA 95945

KOPLIN, JAMES RAY, wildlife educator, consultant, researcher; b. Monte Vista, Colo., June 9, 1934; s. Earl and May (Williams) Walker; m. Phyllis Davenport, Mar. 17, 1956; children: Lynda, Scott, Tracey. B.S., U. Mont.-Missoula, 1959, M.S., 1962; Ph.D. in Zoology, Colo. State U., 1967. Asst. prof. biology SUNY, 1965-67; asst. prof. wildlife Humboldt State U., Arcata, Calif., 1967-71, assoc. prof., 1971-76, prof., 1976—, dir. grad. studies, 1969-74, chmn. dept. wildlife, 1974-77; vis. prof. zoology U. Mont. Biol. Sta., 1966, 67, 68; interagy personnel act wildlife biologist U.S. Forest Service, 1980-81. Contbr. articles to profl. jours. Served with U.S. Navy, 1952-55. Nat. Geog. Soc. grantee, 1969-71; N.Y. Zool. Soc. grantee, 1969-70; U.S. Forest Service grantee, 1970-71; U.S. Fish and Wildlife Service grantee, 1970-71; Calif. Dept. Fish and Game grantee, 1970-71; NSF grantee, 1972. Mem. Am. Ornithologists Union, Cooper Ornithol. Soc., Wilson Ornithol. Soc., Am. Soc. Mammalogists, Ecol. Soc. Democrat. Subspecialties: Ecology; Species interaction. Current work: Inter- and intra-specific interactions among birds to prey; predator-prey interactions between birds of prey and their prey. Home: 1641 Hyland St Bayside CA 95524 Office: Department of Wildlife Humboldt State University Arcata CA 95521

KOPP, OTTO CHARLES, geology educator, researcher; b. Bklyn., July 22, 1929; s. Frank H. and Hattie (Gruhn) K.; m. Helen E. Shotkowski, Sept. 4, 1954; children: Michael, Patricia, Mary, Paul. B.S., Notre Dame U., 1951; M.A., Columbia U., 1955, Ph.D., 1958. Assoc. prof. U. Tenn., Knoxville, 1963-68, prof. geology, 1968—; cons. Oak Ridge Nat. Lab., 1959-77, adj. research participant, 1977—; cons. Am. Smelting and Refining, Knoxville, 1970—. Author: Introduction to Physical Geology, 1980. Hon. curator Children's Mus., Knoxville, Tenn., 1975—. Served with U.S. Army, 1951-54. Recipient Centennial of Sci. award Notre Dame U., 1965; named Disting. Prof. Am. Fedn. Mineral. Socs., 1976. Fellow Geol. Soc. Am., Mineral. Soc. Am.; mem. Nat. Assn. Geology Tchrs. (exec. sec. 1981—), AIME, Sigma Xi. Subspecialties: Mineralogy; X-ray crystallography. Current work: Application of cathodoluminescence to petrologic problems, economic mineralogy. Office: Dept Geol Scis U Tenn Knoxville Tn 37996-1410

KOPP, ROGER ALAN, physicist, researcher; b. Detroit, Feb. 17, 1940; s. Walter John and Dena Blanche (Hill) K.; m. Joyce Elizabeth Schrage, June 23, 1962; children: Gregory, Lorene, Duane. B.S. in Astronomy, U. Mich., 1961; M.A., Harvard U., 1963, Ph.D. in Astronomy, 1968. Staff scientist Nat. Ctr. Atmospheric Researcher, Boulder, Colo., 1966-76; vis. scientist Max Planck Inst. Physics and Astrophysics, Munich, W.Ger., 1979-80, 71-72; physicist Los Alamos Nat. Lab., 1976—; vis. scientist Osservatorio Astrofisico di Arcetri, Florence, Italy, 1982. Contbr. articles to profl. jours. Parker scholar,

1963. Mem. Internat. Astron. Union, Am. Astron. Soc., Am. Geophys. Union. Subspecialties: Laser fusion; Solar physics. Current work: Inertial confinement fusion; theoretical physics and target design for CO-2- lasers and heavy ion drivers; solar physics MHD flows in the solar atmosphere; solar wind expansion theory. Home: 1104 Paseo Barranca Santa Fe NM 87501 Office: Los Alamos National Lab MS E531 Los Alamos NM 87501

KOPP, STEPHEN JAMES, physiologist, biophysicist, educator, researcher; b. Panama Canal Zone, Panama, Mar. 28, 1951; s. Joseph Blair and Josephine Marie (Scutella) K.; m. Jane Ellen Schade, June 17, 1972; children: Elizabeth Marie, Adam Christopher. B.S., U. Notre Dame, 1973; Ph.D., U. Ill.-Chgo., 1976. Teaching asst. U. Ill. Med. Ctr., Chgo., 1973-76; postdoctoral fellow St. Louis U. Med. Ctr., 1976-77, U. Ill. Med. Ctr., 1977-79; asst. prof. physiology Chgo. Coll Osteo. Medicine, 1979-82, assoc. prof., 1982—; guest speaker Philipps U., Marburg, W.Ger., 1982; cons. in field. Contbr. articles to profl. jours. Granite City Steel Research fellow, 1975-76; NIH grantee, 1978-79. Mem. Am. Physiol. Soc., Biophysical Soc., Soc. Environ. Geochemistry and Health, Assn. Research in Vision and Ophthalmology. Unitarian. Subspecialties: Nuclear magnetic resonance (biotechnology); Physiology (medicine). Current work: Biomedical applications of phosphorous-31 nuclear magnetic resonance spectroscopy to the study of chemical processes within living organs and delineation of pathologic mechanisms. Office: Chicago Coll Osteopathic Medicine 5200 S Ellis Ave Chicago IL 60615

KOPPLIN, JULIUS OTTO, elec. engr.; b. Appleton, Wis., Feb. 6, 1925; s. Julius O. and Renata A. (Peters) K.; m. Lola Mae Boldt, Sept. 16, 1950; children—William J., John D., Mary Susan, James R. B.S.E.E., U. Wis., 1949; M.S.E.E., Purdue U., 1954, Ph.D., 1958. Corrosion engr. No. Ind. Public Service Co., 1949-53; asst. prof. elec. engring. U. Ill., 1958-61, asso. prof., 1961-68; prof., chmn. dept. elec. engring. U. Tex., El Paso, 1968-75, Iowa State U., Ames, 1975—. Contbr. articles to profl. jours. Served with USAAC, 1943-45. Decorated Air medal, Purple Heart. Mem. IEEE, Am. Soc. Engring. Edn., Sigma Xi, Eta Kappa Nu, Sigma Pi Sigma. Subspecialties: Electrical engineering; Computer engineering. Home: 241 Trail Ridge Rd Ames IA 50010 Office: Elec Engring Dept Iowa State U Ames IA 50011

KOPROWSKI, HILARY, medical scientist; b. Warsaw, Poland, Dec. 5, 1916; came to U.S., 1944, naturalized, 1950; s. Paul and Sarah (Berland) K.; m. Dr. Irena Grasberg, July 14, 1938; children: Claude Eugene, Christopher Dorian. B.A., Mikolaj Rej Gymnasium of Luth. Congregation, Warsaw, 1933; M.D., U. Warsaw, 1939; grad., Warsaw Conservatory Music and Santa Cecilia Acad., Rome; Doctor honoris causa, Widener Coll., Phila., Ludwig-Maximilian U., Ger., U. Helsinki, U. Uppsala (Sweden). Research asst. dept. exptl. and gen. pathology U. Warsaw, 1936-39; staff Yellow Fever Research Service, Rio de Janeiro, 1940-44; staff research div. Am. Cyanamid Co., 1944-46; asst. dir. viral and rickettsial research Lederle Lab., Pearl River, N.Y., 1946-57; dir. Wistar Inst., Phila., 1957—; prof. microbiology Faculty Arts and Scis., U. Pa., 1957—; Wistar prof. research medicine U. Pa., 1957—; cons. WHO, Nat. Cancer Inst., NIH, USPHS, 1962-70. Co-editor: Current Topics in Microbiology and Immunology, 1965—. Decorated Commandeur Ordre du Mérite pour la Recherche et l'Invention; Chevalier Order Royal De Lion, Belgium; recipient Alvarenga prize. Coll. Physicians Phila., 1959; Alfred Jurzykowski Found Polish Millenium prize, 1966; Felix Wankel Tierschutz prize, 1979; Alexander Von Humboldt Sr. U.S. Scientist award; Fulbright Scholar Max Planck Inst. für Verhaltensphysiologie, Seewiesen, Germany, 1971. Fellow N.Y. Acad. Medicine, Phila. Coll. Physicians; mem. Am. Acad. Arts and Scis., Nat. Acad. Scis., N.Y. Acad. Scis. (pres. 1959, trustee 1960-72). Subspecialties: Immunobiology and immunology; Virology (biology). Research cell biology, virology and immunology; vaccine against poliomyelitis, hog cholera, rabies. Home: 334 Fairhill Road Wynnewood PA 19096 Office: Wistar Inst 36th and Spruce Streets Philadelphia PA 19104

KOPSTEIN, FELIX FRIEDRICH, systems research scientist, consultant, editor; b. Vienna, Austria, Nov. 4, 1924; came to U.S., 1939; s. Ernst and Paula (Schwarz) K.; m. Ralli Hercmark, Dec., 21, 1946; children: Alice C. Kopstein Selfridge, David M. A.B., U. Pa., 1949, A.M., 1951; Ph.D., U. Ill., 1960. Research psychologist Human Factors Ops. Research, Washington, 1950-61; dir. A-I Systems Div. Burroughs Corp., Detroit, 1961-64; chmn. inst. research group Ednl. Testing Service, Princeton, N.J., 1964-66; sr. staff scientist Human Resources Research Orgn., Alexandria, Va., 1966-73; pres., chief scientist Inst. for Psycho-Logic, Wayne, Pa., 1973-79; assoc. Applied Psychol. Service, Wayne, 1979—; sec. Am. Soc. for Cybernetics, Washington, 1972-73, v.p. tech., 1974-75; sci. editor Ednl. Tech. Publs., Englewood Cliffs, N.J., 1970-76. Author: Handbook of Computer Administered Instruction, 1973. Served with USAAF, 1943-46. NSF travel grantee to USSR, 1971. Mem. Am. Psychol. Assn., Internat. Assn. for Applied Psychology, Am. Soc. for Cybernetics, Ednl. Research Assn. Democrat. Jewish. Subspecialties: Cognition; Artificial intelligence. Current work: Human and artificial intelligence interactions in learning, instruction, and problem solving. Home: 595 Barton Ln Strafford PA 19087 Office: Applied Psychological Services Science Center Wayne PA 19087

KORABIK, KAREN SUE, psychologist; b. Chgo., May 2, 1949; s. Michael John and Helen (Jarecki) K.; m. Russell L. Nekorchuk, Aug. 23, 1969 (div. 1974). A.B., St. Louis U., 1971, M.S., 1973, Ph.D., 1975. Instr. U. Mo.-Rolla, 1974, St. Louis U., 1974-76; asst. prof. psychology U. Guelph, Ont., Can., 1976—; evaluator Supt. Pub. Instrn., Springfield, Ill., 1973, Rural Devel. Research Project, Guelph, Ont., 1980-83. Research Adv. Bd. grantee, 1977; Can. Council grantee, 1977; Social Scis. and Humanities Research Council grantee, 1980, 81. Mem. Can. Psychol. Assn. (coordinator sect. on program evaluation), Am. Psychol. Assn., Assn. Women in Psychology, Can. Evaluation Soc., Midwestern Psychol. Assn. Roman Catholic. Subspecialties: Social psychology; Program evaluation. Current work: Research on leadership styles, program evaluation. Home: 23 Woodlawn Rd E Guelph ON Canada NIH 7G6 Office: Dept Psychology U Guelph Guelph ON Canada N1G 2W1

KORF, RICHARD PAUL, mycology educator; b. Bronxville, N.Y., May 28, 1925; s. Frederick and Evelyn Frederick (Krug) K.; m. Sarah Ellen Gifft, Mar. 27, 1954; m. Kumiko Tachibaba, June 27, 1959; children: Noni, Mia, Ian Frederick, Mario Takechi. B.Sc., Cornell U., 1946, Ph.D., 1950. Asst. prof. plant pathology Cornell U., Ithaca, N.Y., 1951-55, assoc. prof., 1955-61, prof. mycology, dir. plant pathology herbarium, 1961—, prof. botany, 1982—; dir. Exe Island Biol. Sta., Portland, Ont., Can., 1973—. Mng. editor: Mycotaxon, 1974—; contbr. articles to sci. jours. Vice chmn. N.Y. State Liberal Party, 1967-68. NSF systematics program research grantee, 1959-79. Mem. Mycol. Soc. Am., Internat. Assn. Plant Taxonomists, Am. Soc. Plant Taxonomists, Bot. Soc. Am., Mycol. Soc. Britain, Mycol. Soc. France, Mycol. Soc. Japan. Republican. Mem. Universal Life Ch. Subspecialties: Mycology; Taxonomy. Current work: Taxonomy of discomycetes, botanical nomenclature, microanatomy, code of nomenclature. Home: 316 Richard Pl Ithaca NY 14850 Office: Dept Plant Pathology Cornell U Ithaca NY 14853

KORI, SHASHIDHAR HALAPPA, neuro-oncologist, educator, researcher; b. Shimoga, India, June 18, 1949; came to U.S., 1974, naturalized, 1980; s. Halappa M. and Rudramma K.; m. Shylaja Rajashekarappa, Nov. 27, 1977; 1 son, Ajay. M.B., B.S., Mysore (India) U., 1971, M.D., 1974. Diplomate: Am. Bd. Psychiatry and Neurology. Intern, Govt. Hosp., Mangalore, India, 1970-71, resident in medicine, 1971-74, VA Hosp., N.Y.C., 1974-75; resident in neurology St. Vincent's Hosp., N.Y.C., 1975-78; fellow in neuro-oncology Meml. Sloan Kettering Cancer Ctr., N.Y.C., 1978-80, pain fellow, 1978-80; postdoctoral fellow in immunobiology Sloan Kettering INst., N.Y.C., 1979-80; asst. prof. neurology Case Western Res. U., 1980—; chief neuro-oncology div. Univ. Hosps., Cleve., 1980—, dir. Neuro-Oncology Clinic, 1980—, dir. Pain Clinic, 1980—; cons. in field. Contbr. articles, chpt. to profl. publs. Recipient Nat. Research Service award USPHS, 1979; USPHS postdoctoral grantee, 1979; Am. Cancer Soc. grantee, 1981. Mem. Am. Acad. Neurology, Am. Soc. Neurol. Investigations, Am. Soc. Neuroscis., AMA, Am. Soc. Internal Medicine, Internat. Assn. Study Pain, Am. Pain Soc., Eastern Pain Assn. Hindu. Subspecialties: Cancer research (medicine); Neuroimmunology. Current work: New treatment for brain tumors; biology of brain tumors; immune responses to herpes. Inventor approach to treating malignant brain tumors. Office: Univ Hosps 2064 Adelbert Rd Cleveland OH 44106

KORMENDY, JOHN, astronomer; b. Graz, Austria, June 13, 1948; emigrated to Canada, 1951, naturalized, 1961; s. John and Margarete Ludmilla (Winkler) K. B.Sc. with honors in Astronomy, U. Toronto, 1970; Ph.D., Calif. Inst. Tech., 1976. Parisot postdoctoral fellow dept astronomy U. Calif.-Berkeley, 1976-78; sr. vis. fellow Inst. Astronomy, Cambridge, Eng., 1980, 78; postdoctoral fellow Kitt Peak Nat. Obs., Tucson, 1978-79; staff mem. Dominion Astrophys. Obs., Victoria, B.C., 1979—. Contbr. articles to profl. jours. Recipient Gold medal Royal Astron. Soc. Can., 1970. Mem. Am. Astron. Soc., Astron. Soc. Pacific, Can. Astron. Soc., Internat. Astron. Union. Subspecialty: Optical astronomy. Current work: Structure and dynamics of galaxies and star clusters. Office: 5071 W Saanich Rd Victoria BCCanada V8X 4M6

KORN, EDWARD DAVID, biochemist, government research administrator; b. Phila., Aug. 3, 1928; s. Joel and Carrie (Goldman) K.; m. Muriel Evelyn Fisher, June 23, 1950; children: Elizabeth Gail, Sarah Harris. A.B., U. Pa., 1949, Ph.D., 1954. Biochemist Nat. Heart, Lung, and Blood Inst., NIH, Bethesda, Md., 1954—, chief Lab. Cell Biology, 1975—; dep. sci. dir. Nat. Heart, Lung, and Blood Inst. NIH, 1982—. Contbr. numerous articles to profl. jours.; assoc. editor: Jour. Biol. Chemistry, 1977—. Served with USPHS, 1954-56. Mem. Am. Soc. Biol. Chemists, Biophysics Soc., Am. Soc. Cell Biology. Subspecialties: Biochemistry (biology); Biochemistry (medicine). Current work: Biochemistry of cell motility. Office: NIH Bldg 3 Room B1-22 Bethesda MD 20205

KORN, JOSEPH H., physician; b. Augsburg, Germany, Jan. 31, 1947; s. Leo and Rose (Mann) K.; m. Paulette M. Jeremias, June 26, 1971; children: Naomi, Jerald, Joshua. B.S., CCNY, 1968; M.D., Columbia U., 1972. Diplomate: Am. Bd. Internal Medicine. Intern, resident U. N.C., Chapel Hill, 1972-75; fellow, instr., then asst. prof. medicine and immunology Med. U. S.C., Charleston, 1975-78; asst. prof. U. Conn., 1978—; assoc. chief staff for research VA Med. Ctr., Newington, Conn., 1982—. Contbr. chpt. to book; articles to profl. jours. VA awardee, 1979-81. Subspecialties: Immunology (medicine); Cell biology (medicine). Current work: Immunology of connective tissue. Office: 263 Farmington Ave Farmington CT 06032 Home: 1910 Asylum Ave West Hartford CT 06117

KORNBERG, ARTHUR, biochemist; b. N.Y.C., Mar. 3, 1918; s. Joseph and Lena (Katz) K.; B.S. (N.Y. State scholar), Coll. City of N.Y., 1937, LL.D., 1960; M.D. (Buswell scholar), U. Rochester, 1941, D.Sc., 1962; L.H.D., Yeshiva U., 1963; D.Sc., U. Pa., U. Notre Dame, 1965, Washington U., 1968, Princeton U., 1970, Colby Coll., 1970, M.D. (h.c.), U. Barcelona (Italy), 1970; m. Sylvy R. Levy, Nov. 21, 1943; children: Roger, Thomas Bill, Kenneth Andrew. Intern in medicine Strong Meml. Hosp., Rochester, N.Y., 1941-42; commd. officer, USPHS, 1942, advanced through grades to med. dir., 1951; mem. staff NIH, Bethesda, Md., 1942-52, nutrition sect., div. physiology, 1942-45, chief sect. enzymes and metabolism Nat. Inst. Arthritis and Metabolic Diseases, 1947-52; guest research worker depts. chemistry and pharmacology coll. medicine, N.Y. U., 1946, dept. biol. chemistry med. sch. Washington U., 1947, dept. plant biochemistry U. Calif., 1951; prof., head dept. microbiology, med. sch. Washington U., St. Louis, 1953-59; prof., biochemistry Stanford U. Sch. Medicine, 1959—, chmn. dept., 1959-69. Mem. sci. adv. bd. Mass. Gen. Hosp., 1964-67; bd. govs. Weizmann Inst., Israel. Served lt. (j.g.), med. officer USCGR, 1942. Recipient Paul-Lewis award in enzyme chemistry, 1951; co-recipient of Nobel prize in medicine, 1959; Max Berg award prolonging human life, 1968, Sci. Achievement award AMA, 1968, Lucy Worthmam James award James Ewing Soc., 1968, Borden award Am. Assn. Med. Colls., 1968. Mem. Am. Soc. Biol. Chemists (pres. 1965), Am. Chem. Soc., Harvey Soc., Am. Acad. Arts and Scis., Royal Soc., Nat. Acad. Scis. (mem. council 1963-66), Am. Philos. Soc., Phi Beta Kappa, Sigma Xi, Alpha Omega Alpha. Subspecialty: Biochemistry (medicine). Contbr. sci. articles to profl. jours. Office: Dept Biochemistry Stanford Univ Sch Medicine Stanford CA 94305

KORNBERG, THOMAS BILL, biochemist, educator; b. Washington, Nov. 10, 1948; s. Arthur and Sylvy K. B.A., Columbia U., 1970, Ph.D., 1973; student, Juilliard Sch. Music, 1965-69. Assoc. prof. biochemistry U. Calif.-San Francisco. Subspecialties: Developmental biology; Molecular biology. Office: Dept Biochemistry and Biophysics Univ Calif San Francisco CA 94143

KORNER, ANNELIESE FRIEDE, developmental psychologist, researcher; b. Munich, W. Ger.; came to U.S., 1938, naturalized, 1943; d. Leopold and Jenny (Deutsch) Friedsam; m. Ija N. Korner, 1941 (dec.); m. Sumner L. Kalman, Oct. 19, 1952; 1 dau., Susan S. Diploma, Inst. J.J. Rousseau, Geneva, 1938; M.A., Tchrs. Coll., Columbia, U., 1940, Ph.D., 1948; postgrad, San Francisco Psychoanalytic Inst., 1950-61. Lic. psychologist, Calif. Instr. dept. psychiatry U. Chgo. Med. Sch., 1943-48; chief psychologist Mt. Zion Psychiat. Clinic, San Francisco, 1948-61; cons. San Mateo (Calif.) Mental Health Services, 1962-70; research assoc. dept. psychiatry and behavior sci. Stanford U., 1964-70, sr. scientist, 1970-74, adj. prof., 1974-82, research prof. psychiatry and behavioral scis., 1982—. Author: Hostility in Young Children, 1949; consulting editor: Monograph of Soc. for Research in Child Devel, 1971-73, Infant Behavior and Development, 1981-83; contbr. numerous articles to profl. jours. NIMH grantee; Grant Found. grantee; Boys Town grantee; HEW grantee. Fellow Am. Psychol. Assn., Am. Orthopsychiat. Assn.; mem. Soc. Research in Child Devel., Perinatal Research Soc. Subspecialties: Developmental psychology; Psychobiology. Current work: Developmental studies of preterm and full-term newborns; individual and sex differences in human newborns; effects of vestibular-proprioceptive stimulation on the behavior and development of newborns. Patentee infant waterbed, method for treating premature infants. Home: 2299 Tasso St Palo Alto CA 94301 Office: Stanford University School of Medicine Stanford CA 94305

KORNETSKY, CONAN, psychologist; b. Portland, Maine, Feb. 9, 1926; s. Alex and Ida (Rosenberg) K.; m. MarciaKornetsky, June 5, 1949; children: David, Lisa. B.A., U. Maine, 1948; M.S., U. Ky., 1951, Ph.D., 1952. Asst. research psychologist Drug Addiction Research Center, Lexington, Ky., 1949-52; research scientist NIMH, Bethesda, 1952-59; assoc. research prof. pharmacology Boston U. Med. Sch., 1959-63; prof., div. psychiatry and dept. pharmacology, 1963—; cons. Nat. Inst. Drug Abuse, NIMH, VA, FDA. Author: Pharmacology: Drugs Affecting Behavior, 1976; contbr. articles to profl. jours.; editorial bd.: Jour. Pharmacology and Explt. Therapeutics, 1974—, Jour. Abnormal Psychology, 1980—, Psychopharmacology, 1964—; U.S. editor, 1970-74; editorial bd.: Internat. Rev. Neurobiology, 1975—. Served with AUS, 1944-45. USPHS sr. fellow, 1959-60. Fellow Am. Coll. Neuropsychopharmacology, Am. Psychol. Assn.; mem. Am. Soc. Pharmacology and Exptl. Therapeutics, Soc. Neuroscis., Sigma Xi. Subspecialties: Physiological psychology; Neuropharmacology. Current work: Neurobehavioral basis of rewarding effects of abuse substances, e.g., heroin, cocaine, amphetamine, brain-stimulation reward, pain, animal models of mental illness. Home: 7 Rumford Rd Lexington MA 02173 Office: 80 E Concord St L-602 Boston MA 02118

KORNFELD, STUART ARTHUR, medical researcher; b. St. Louis, Oct. 4, 1936; m., 1959; 3 children. A.B., Dartmouth Coll., 1958; M.D., Washington U., St. Louis, 1962. From asst. prof. to assoc. prof. Sch. Medicine Washington U., 1966-72, prof. medicine and biochemistry, 1966—; mem. cell biology study sect. NIH, 1974-77. Assoc. editor: Jour. Clin. Inst, 1977-81; editor, 1981-82. Mem. Am. Soc. Hematology, Am. Soc. Biol. Chemists, Am. Soc. Clin. Investigation, Nat. Acad. Sci., Inst. Medicine. Subspecialty: Hematology. Office: Dept Medicine Washington U. Sch Medicine Saint Louis MO 63130

KORNGUTH, STEVEN EDWARD, physiol. chemist; b. N.Y.C., Dec. 1, 1935; s. Eugene I. and Helen (Pardes) K.; m. Margaret Livens, Aug. 29, 1958; children—Ingrid, David. B.A., Columbia Coll., 1957; M.S., U. Wis., 1959, Ph.D., 1961. Mem. staff N.Y. State Psychiat. Inst., N.Y.C., 1961-63; asst. prof. neurology, physiol. chemistry U. Wis., Madison, 1963-68, asso. prof., 1968-72, prof., 1972—; program dir. neurobiology sect BNS, NSF, Washington, 1981—. NIH research tng. grantee in neurochemistry, 1968-73. Mem. Am. Soc. Biol. Chemists, Am. Neurosci. Soc., Internat. Brain Research Orgn. Subspecialty: Neurobiology. Home: 5702 Hempstead Rd Madison WI 53711

KORNHAUSER, EDWARD THEODORE, engineering educator; b. Louisville, Ky., June 30, 1925; s. Sidney Issac and Anna (Marshall) K.; m. Patricia Hill, Nov. 2, 1945 (div. 1978); children: John, Alan, Anne.; m. Jincy Willett, Aug. 21, 1978. B.E.E., Cornell U., 1945; S.M., Harvard U., 1947, Ph.D., 1949; M.A. (hon.), Brown U., 1957. Instr. dept. applied physics Harvard U., Cambridge, Mass., 1949-51; asst. prof. div. engring. Brown U., Providence, 1951-56, assoc. prof., 1956-63, prof., 1963—; vis. lectr. U. Bristol, Eng., 1959-60; acad. visitor Oxford (Eng.) U., 1967, Imperial Coll., London, 1974; vis. prof. U. Calif.-San Diego, 1981-82. Contbr. articles to profl. jours. Pres. Barrington (R.I.) Citizens League, 1965. Served to lt. USNR, 1943-46. NSF, NATO fellow, 1959-60. Mem. IEEE, Sigma Xi, Tau Beta Pi, Eta Kappa Nu. Subspecialty: Electrical engineering. Home: 18 Young Orchard Ave Providence RI 02906 Office: Div Engrin Brown U Providence RI 02912

KORNSTEIN, EDWARD, tech. co. exec.; b. N.Y.C., Sept. 7, 1929; s. Max and Margit (Stahl) K.; m. Marion Beatrice Stein, Dec. 20, 1958; children: Sandra P., Martin R. B.A., NYU, 1951; M.S., Drexel U., 1954. Engr., optics RCA, Camden, N.J., 1951-57, group leader optical physics, Burlington, Mass., 1960-66, mgr. optical physics, 1966-70; cons. optics Phys. Research Lab., Boston U., 1958, Boston, 1959-60; v.p. OPTEL Corp., Princeton, 1970-72; pres. Kortron, Princeton, 1972-78; v.p. Object Recognition Systems, Princeton, 1978—; cons. electro-optics. Contbr. articles to profl., trade jours. Active Boy Scouts Am., 1973—; mem. Princeton C. of C. NSF travel grantee, 1959; Optical Soc. Am. travel grantee, 1959. Mem. Soc. Info. Display, Optical Soc. Am., IEEE, Soc. Motion Picture and TV Engrs. Subspecialties: Optical engineering; Optical image processing. Current work: Application of electro-optical techniques to automatic inspection systems, robotics and optical data processing; electro-optical displays. Patentee in field.

KORWIN, PAUL, cons. mech. engr.; b. Cracow, Poland, Jan. 5, 1914; came to U.S., 1956, naturalized, 1962; m. Yala Meisels, Oct. 22, 1949; children: Danielle, Robert. M.E., State Engring. Coll., Cracow, Poland, 1934. Registered Profl. Engr., Mass. With Chem. Constrn. Corp., N.Y.C., 1956-57; with Stone & Webster Engring. Co., Boston, 1967-68, Lummus Engring., Bloomfield, N.Y., 1968-74, Heater Consulting Services, Flushing, N.Y., 1974—; cons. engr. Mem. ASME, Am. Inst. Chem. Engrs. Subspecialties: Mechanical engineering; High-temperature materials. Current work: Hydrogen steam-hydrocarbon reformers; hydrocarbon high temperature cracking heaters. Patentee in field. Address: 150-09-77th Ave Flushing NY 11367

KORYTNYK, WALTER *See also* **KORYTNYK, WSEWOLOD**

KORYTNYK, WSEWOLOD (WALTER KORYTNYK), medicinal chemist, educator; b. Caslav, Czechoslovakia, Apr. 21, 1929; came to U.S., 1958, naturalized, 1965; s. Luka and Valentyna (Makarenko) K.; m. Olena Kozak, Sept. 1, 1957; children: Natalie, Christine, Peter. B.Sc., U. Adelaide, Australia, 1953, 1954, Ph.D. in Carbohydrate Chemistry, 1957, D.Sc. in Medicinal Chemistry, 1973. Postdoctoral research fellow U. Adelaide, 1957-58, Purdue U., Lafayette, Ind., 1958-59; research chemist U.S. Dept. Agr., Pasadena, Calif., 1959-60; research prof. Niagara U., Buffalo, 1976—; asst. research prof. SUNY, Buffalo, 1968-72, assoc. research prof., 1972-81, research prof., dir. med. chemistry program, 1981—; sr. cancer research scientist, dept. exptl. therapeutics Roswell Park Meml. Inst., Buffalo, 1960-68, 76-80, assoc. cancer research scientist, 1968-76, cancer research scientist G35, 1980—; research assoc. U. Calif., Berkeley, 1967-68. Contbr. articles to profl. jours. and books. Am. Cyanamid fellow, 1957; Sci. Exchange visitor to Poland, 1970; USPHS grantee, 1965-82; other grants. Mem. AAAS, Am. Chem. Soc., Chem. Soc. (London), N.Y. Acad. Sci., Am. Soc. Biol. Chemists, Am. Assn. Cancer Research, Shevchenko Sci. Soc. Ukrainian Catholic. Subspecialties: Medicinal chemistry; Chemotherapy. Current work: Cell surfaces of cancer cells; carbohydrate chemistry, biochemistry and pharmacology; conformational analysis; nuclear magnetic resonance spectroscopy; medicinal chemistry of Vitamin B6; anticancer agents; rational design of drugs.

KOSCHIER, FRANCIS JOSEPH, III, toxicologist; b. N.Y.C., June 16, 1950; s. Francis Joseph and Mary Frances (Schaefer) K. A.B., Bard Coll., 1972; postgrad., Dartmouth Coll. Grad. Sch., 1972-74; Ph.D., U. Miss., 1976. Diplomate: Am. Bd. Toxicology. Research asst. prof. SUNY-Buffalo, USPHS grantee, 1976-79; toxicologist Food and Drug Research Labs., Waverly, N.Y., 1979-80; staff toxicologist Am. Cyanamid Co., Wayne, N.J., 1980-83; mgr. toxicology, safety, health and ecology dept. CIBA-GEIGY Corp., 1983—. Contbr. articles on toxicology to profl. jours. Mem. Soc. Toxicology, Am. Soc. Pharmacology and Exptl. Therapeutics, Phi Kappa Phi. Roman Catholic. Subspecialties: Toxicology (medicine); Environmental toxicology. Current work: Industrial toxicology. Office: CIBA-GEIGY Corp. Ardsley NY 10502

KOSERSKY, DONALD S., pharmacologist, consultant; b. Waterbury, Conn., Oct. 16, 1932; s. William and Sally (Hanken) K.; m. Marcelle S. Manet; 1 dau., Nicole E. M.S. in Pharmacology, U. Conn., 1968, Ph.D., U. of Pacific, 1971. Research assoc. dept. pharmacology U. N.C. Med. Sch., Chapel Hill, 1971-73; assoc. prof. pharmacology Northeastern U., Boston, 1973-80; assoc. prof. pharmacology and physiology, coordinator grad. programs Mass. Coll. Pharmacy and Allied Health Scis., Boston, 1980—; pharmacol. research cons. Contbr. articles to profl. jours. Mem. Am. Soc. Pharmacology and Exptl. Therapeutics, Fedn. Am. Socs. Exptl. Biology, Neurosci. Soc. Subspecialties: Pharmacology; Neuropharmacology. Current work: Pharmacology of tolerance and dependence to narcotics and other drugs of abuse. Office: Mass Coll Pharmacy and Allied Health Sciences 179 Longwood Ave Boston MA 02115

KOSHLAND, DANIEL EDWARD, JR., educator, biochemist; b. N.Y.C., Mar. 30, 1920; s. Daniel Edward and Eleanor (Haas) K.; m. Marian Elliott, May 25, 1945; children: Ellen, Phyllis, James, Gail, Douglas. B.S., U. Calif., Berkeley, 1941; Ph.D., U. Chgo., 1949. Chemist Shell Chem. Co., Martinez, 1941-42; research asso. Manhattan Dist. U. Chgo., 1942-44; group leader Oak Ridge Nat. Labs., 1944-46; postdoctoral fellow Harvard, 1949-51; staff Brookhaven Nat. Lab., Upton, N.Y., 1951-65; affiliate Rockefeller Inst., N.Y.C., 1958-65; prof. biochemistry U. Calif. at Berkeley, 1965—, chmn. dept., 1973-78; Leo Marion lectr. Nat. Research Council Can., 1972; Harvey lectr., 1969; fellow All Souls, Oxford U., 1972; Walker Ames lectr. U. Wash., 1964; Carter Wallace lectr. Princeton U., 1970, Phi Beta Kappa lectr., 1976; John Edsall lectr. Harvard U., 1980. Author: Bacterial Chemotaxis as A Model Behavioral System, 1980; mem. editorial bds.: jour. Accounts Chem. Research; editor: Procs. Nat. Acad. Scis, 1980—. Recipient T. Duckett Jones award Helen Hay Whitney Found., 1977; Guggenheim fellow, 1972. Mem. Nat. Acad. Scis., Am. Chem. Soc. (Edgar Fahs Smith award 1979, Pauling award 1979), Am. Soc. Biol. Chemists (pres.), Am. Acad. Arts and Scis. (council), Academy Forum (chmn.), Japanese Biochem. Soc. (hon.). Subspecialties: Biophysical chemistry; Enzyme technology. Current work: Enzymology of behavior, regulatory processes. Home: 3991 Happy Valley Rd Lafayette CA 94549 Office: Biochemistry Dept U Calif Berkeley CA 94720

KOSHLAND, MARIAN ELLIOTT, immunologist, educator; b. New Haven, Oct. 25, 1921; d. Waller Watkins and Margaret Ann (Smith) Elliott; m. Daniel Edward Koshland, Jr., May 25, 1945; children—Ellen R., Phyllis A., James M., Gail F., Douglas E. B.A., Vassar Coll., 1942, M.S., 1943; Ph.D., U. Chgo., 1949. Research asst. Manhattan Dist. Atomic Bomb Project, 1945-46; fellow dept. bacteriology Harvard Med. Sch., 1949-51; asso. bacteriologist biology dept. Brookhaven Nat. Lab., 1952-62, bacteriologist, 1963-65; asso. research immunologist virus lab. U. Calif., Berkeley, 1965-69, lectr. dept. molecular biology, 1966-70, prof. dept. microbiology and immunology, 1970—, chmn. dept., 1982—; mem. Nat. Sci. Bd., 1976-82; mem. adv. com. to dir. NIH, 1972-75. Contbr. articles to profl. jours. Mem. Nat. Acad. Scis., Am. Acad. Microbiology, Am. Assn. Immunologists (pres. 1982-1983), Am. Soc. Biol. Chemists, Phi Beta Kappa, Sigma Xi. Subspecialty: Immunobiology and immunology. Office: Dept Microbiology and Immunology U Calif Berkeley CA 94720

KOSKELO, MARKKU JUHANI, software engineer; b. Turku, Finland, Feb. 11, 1951; came to U.S., 1982; s. Tauno Juhani and Tuija Inkeri (Malmio) K.; m. Pirkko Tellervo (Willberg), Oct. 14, 1972; children: Ilkka Juhani, Antti Ilmari. M.Sc., Helsinki (Finland) U. Tech., 1976, D.Tech., 1981; postgrad., U. Toronto, 1977-78. Asst. Helsinki U. Tech., Espoo, 1975-78, lab. engr., 1978-81, acting assoc. prof., 1981-82; sr. software engr. Canberra Industries, Meriden, Conn., 1982, group leader sci. software devel., 1983, software project mgr., 1983—; assoc. CERN, Geneva, 1977. Contbr. articles to profl. jours. Mem. Finnish Nuclear Soc., Am. Nuclear Soc. Subspecialties: Nuclear engineering; Algorithms. Current work: Computerized gamma spectrum analysis, special nuclear measurement systems. Office: Canberra Industries Inc 45 Gracey Ave Meriden CT 06450

KOSOW, DAVID PHILLIP, biochemist, research administrator; b. Jersey City, Mar. 15, 1936; s. Zeal and Esther (Solomon) K.; m. Jean E. Herendeen, June 28, 1958; children: Lisa D., Pamela G. B.S., Antioch Coll., Yellow Springs, Ohio, 1958; M.S., Va. Poly. Inst., Blacksburg, 1960, Ph.D., 1962. Asst. prof. biochemistry Va. Poly. Inst., 1962-63; neurochemist Phila. Gen. Hosp., 1965-66; research assoc. Inst. Cancer Research, Phila., 1966-73; research scientist ARC, Bethesda, Md., 1973-81, asst. dir., 1981—. Am. Cancer Soc. fellow, 1963-65; NIH fellow, 1980. Mem. Am. Soc. Biol. Chemists, Internat. Soc. Thrombosis and Hemostasis, Am. Chem. Soc., Sigma Xi. Subspecialties: Biochemistry (medicine); Hematology. Current work: Kinetic mechanism of coagulation factors; development of plasma derivatives for clinical use; protein-ligand interactions. Office: ARC 9312 Old Georgetown Rd Bethesda MD 20814

KOSS, LEOPOLD G., physician; b. Danzig, Poland, Oct. 2, 1920; came to U.S., 1947, naturalized, 1952; s. Abram and Rose (Merenholc) K.; m. Lydia Palla; children: Michael S., Andrew C., Richard P. M.D., U. Berne, Switzerland, 1946. Intern, Lincoln Hosp., N.Y.C., 1947-48; tng. pathology, St. Gallen, Switzerland, 1946-47, Kings County Hosp., Bklyn., 1949-51; instr. pathology L.I. Coll. Medicine, 1949-51; mem. staff Meml. Hosp. Cancer and Allied Diseases, N.Y.C., 1952-70, attending pathologist, 1961-70, chief cytology service, 1961-70; pathologist-in-chief Sinai Hosp. Balt., 1970-73; prof., chmn. dept. pathology Montefiore Hosp., Med. Center Albert Einstein Coll. Medicine, 1973—; asso. mem. Sloan-Kettering Inst. Cancer Research, N.Y.C., 1957-70; asso. prof. pathology Sloan-Kettering div. Postgrad. Sch. Med. Scis., Cornell U., 1957-70; prof. pathology Jefferson Med. Coll., Phila., 1970-73; clin. prof. pathology U. Md. Med. Sch., 1971-73; vis. pathologist James Ewing Hosp., N.Y.C., 1952-60; cons. pathologist N.Y. State Dept. Health, Hosp. Spl. Surgery, N.Y.C., Walter Reed Army Med. Center, Nassau County Med. Ctr. Author: Diagnostic Cytology and Its Histopathologic Bases, 3d edit., 1979, Tumors of the Urinary Bladder, 1975; editor: Advances in Clinical Cytology, 1981; also monographs, chpts. and articles. Served to maj. M.C., AUS, 1955-57. Recipient Wien award Papanicolaou Cancer Inst., 1963, Alfred P. Sloan award cancer research, 1964; hon. prof. pathology Severance Med. Coll., Seoul, Korea, 1956. Fellow Am. Soc. Clin. Pathology, Coll. Am. Pathologists, Internat. Acad. Cytology (Goldblatt award 1962); mem. Am. Soc. Path. Bacteriologists, James Ewing Soc., AMA, Am. Soc. Cytology (pres. 1962, Papanicolaou award 1966), Internat. Acad. Pathology; corr. mem. Royal Acad. Medicine Spain; hon. mem. Brit. Soc. Clin. Cytology, Korean Med. Assn., Mex., Argentinian socs. cytology, Japanese Soc. Pathology, Polish Soc. Pathology, Peruvian Soc. Obstetrics and Gynecology. Subspecialties: Pathology (medicine); Cytology and histology. Current work: Computer image analysis of cancer cells; flow cytometry in cancer; papillomaviruses in cancer; bladder cancer. Office: 111 E 210th St Bronx NY 10467

KOSS, MICHAEL CAMPBELL, pharmacology educator, researcher; b. Ann Arbor, Mich., Sept. 24, 1940; s. Frank and Elizabeth (Campbell) K. B.A., N.Y.U., 1966; Ph.D. in Pharmacology, Columbia U., 1971. Trainee Columbia U., 1966-67, faculty fellow, 1967-71, postdoctoral fellow, 1971; asst. prof. pharmacology Coll. Medicine, U. Okla., Oklahoma City, 1971-75, assoc. prof., 1975-81, prof., 1981—, research cons. ophthalmology, 1973-75, asst. prof. pharmacology, 1973-75, adj. asst. prof. psychiatry and behavioral scis., 1975-79, adj. assoc. prof., 1979-80, adj. asst. prof. ophthalmology, 1975-83, 1983—. Contbr. articles to sci. jours. Mem. Am. Soc. Pharmacology and Exptl. Therapeutics, Soc. Neurosci., Assn. Research in Vision and Ophthalmology, Sigma Xi. Subspecialties: Pharmacology; Neuropharmacology. Home: 2038 NW 31st St Oklahoma City OK 73118 Office: U Okla Med Center PO Box 235901 Oklahoma City OK 73190

KOSSLYN, STEPHEN MICHAEL, cognitive science educator, consultant; b. Santa Monica, Calif., Nov. 30, 1948; s. S. Duke and Rhoda (Rosenberg) K.; m. Robin Sue Rosenberg, Mar. 28, 1982. B.A. summa cum laude, UCLA, 1970; Ph.D., Stanford U., 1974. Asst. prof. Johns Hopkins U., Balt., 1974-77, prof. dept. psychology, 1982—; assoc. prof. Harvard U., Cambridge, Mass., 1977-81, Brandeis U., Waltham, Mass., 1981-82; cons. Cons. Statistician, Inc., Boston, 1979-82. Author: Image and Mind, 1980, Ghosts in the Mind and Machine, 1983; editor: (with J. Anderson) Tutorials in Learning and Memory, 1983. NIMH research career devel. awardee, 1980; recipient grants including NSF, NIMH, Office Naval Research. Mem. Am. Psychol. Assn. (Boyd R. McCandless award 1977), Cognitive Sci. Soc., Psychonomic Soc., Soc. for Research in Child Devel., AAAS, Phi Beta Kappa. Subspecialties: Cognition; Neuropsychology. Current work: Computer models of visual processing, role of analogue images in problem solving, neuropsychological studies of visual processing, cognitive engineering of computer graphics systems. Office: Dept. Psychology Johns Hopkins Univ 34th and Charles Sts Baltimore MD 21218

KOSTIUK, THEODOR, space scientist; b. Plauen, Germany, Aug. 12, 1944; s. Hryhory and Raisa (Butko) K.; m. Alexandra Marie Dobransky, July 11, 1970. B.S., CCNY, 1966; Ph.D. in Physics, Syracuse U., 1973. Nat. Acad. Scis.-NRC resident research assoc. NASA/Goddard Space Flight Center, Greenbelt, Md., 1973-74, space scientist infrared and radio astronomy br., 1974, head molecular astrophysics sect., 1983. Contbr. articles to profl. jours. Recipient Simon Sonkin medal CCNY, 1966. Mem. Am. Phys. Soc., Optical Soc. Am., Inst. Physics, Am. Geophys. Union, Am. Astron. Soc., Soc. Photo-Optical Instrumentation Engrs., Ukrainian Acad. Arts and Scis. in the U.S. Ukrainian Orthodox. Subspecialties: Infrared optical astronomy; Infrared spectroscopy. Current work: High resolution infrared studies of planetary atmospheres, stars and interstellar medium; infrared astronomy, stratospheric research, infrared heterodyne spectroscopy, diode laser spectroscopy, molecular spectroscopy, infrared upconversion. Office: NASA/Goddard Space Flight Center Code 693 Greenbelt MD 20771

KOSTRZEWA, RICHARD MICHAEL, pharmacologist; b. Trenton, N.J., July 22, 1943; s. John Walter and Lottie (Wnuk) K.; m. Florence Palmer, Sept. 4, 1965; children: Theresa, Richard, Joseph, Maria, Krystyna, Thomas John, John Palmer, Francis. B.S., Phila. Coll. Pharmacy and Sci., 1965, M.S., 1967; Ph.D., U. Pa., 1971. Research pharmacologist VA, New Orleans, 1971-75; adj. asst. prof. Tulane U. Med. Center, New Orleans, 1975-76; grad. faculty U. New Orleans, 1975-77; asst. prof. La. State U. Med. Ctr., 1975-78; assoc. prof. Quillen-Dishner Coll. Medicine, East Tenn. State U., Johnson City, 1978—. Contbr. articles to profl. jours. Mem. exec. com. Appalachian chpt. March of Dimes, 1980-82; mem. Parish Sch. Bd., 1979-82, council, 1979-80. Recipient East Tenn. State U. Found. research award, 1981. Mem. Fedn. Am. Socs. Exptl. Biology, Am. Soc. Pharmacology and Exptl. Therapeutics, Soc. Neurosci., Histochem. Soc., AAAS, Sigma Xi. Subspecialties: Regeneration. Current work: Determination of mechanisms associated with development regeneration and sprouting of noradrenergic fibers in the brain

KOSTYO, JACK LAWRENCE, educator; b. Elyria, Ohio, Oct. 1, 1931; s. Louis and Matilda (Thomasko) K.; m. Shirlianne Guth, June 10, 1953; children—Cecile A., Louis C. A.B., Oberlin Coll., 1953; Ph.D., Cornell U., 1957; M.D. (hon.), U. Göteborg, 1978. NRC fellow Harvard Med. Sch., Boston, 1957-59; asst. prof., then prof. physiology Duke, 1959-68; prof., chmn. dept. physiology Emory U., Atlanta, 1968-79, U. Mich. Med. Sch., Ann Arbor, 1979—; mem. endocrinology study sect. NIH, USPHS, 1967-71; mem. physiology test com. Nat. Bd. Med. Examiners, 1974-77. Editor-in-chief: Endocrinology, 1978-82; contbr. articles to profl. jours. Mem. adv. bd. Searle Scholars. Recipient Lederle Med. Faculty award, 1961; Ernst Oppenheimer Meml. award Endocrine Soc., 1969. Mem. Endocrine Soc. (mem. editorial bd., council), Am. Physiol. Soc. (mem. editorial bd., chmn. standing com. on edn., mem. council), Soc. for Exptl. Biology and Medicine (editorial bd.), Internat. Union Physiol. Scis. (commn. on med. edn.), Assn. Chmn. Depts. Physiology (pres. 1979, council), Sigma Xi. Subspecialty: Endocrinology. Current work: Research on the structure-function relationships of pituitary growth hormone. Home: 8 Eastbury Ct Ann Arbor MI 48105

KOTAS, ROBERT VINCENT, research institute director, medical consultant; b. Buffalo, Nov. 26, 1938; s. Vincent John and Regina Agnes (Hadynka) K.; m. Ilona Rae Fielding, Mar. 2, 1968; children: Nicole, Timothy, Robert A., Rebecca. B.S., Canisius Coll., 1959; M.D., U. Buffalo, 1963. Diplomate: Am. Bd. Pediatrics. Intern Buffalo Children's Hosp., 1963-64; resident Johns Hopkins Hosp., Balt., 1964-66; neonatology fellow Johns Hopkins Med. Sch., 1968-69; research assoc. McGill U. Med. Sch., Montreal, Que., Can., 1969-70; dir. neonatology U. Okla. Med. Sch., Oklahoma City, 1970-72; dir. physiology Warren Research Ctr., Tulsa, 1972-76, dir., 1976—, William and Natalie Warren Med. Inst., Tulsa, 1977—; pulmonary cons. Tulsa Pediatric Edn. Trust, 1972—; cons. perinatologist, attending physician Eastern Okla. Regional Perinatal Ctr., 1972—; guest scientist Nat. Inst. Child Health and Human Devel., Bethesda, Md., 1975-77, cons., 1979—; clin. assoc. prof. pediatrics U. Okla. Health Scis. Ctr., 1975-77; interim dir. neonatal services St. Francis Hosp., Tulsa, 1977, interim chmn. neonatology div., 1977, instl. rev. bd., 1977—; clin. prof. U. Okla., Tulsa Med. Coll., 1977—, chmn. research com., 1977—, pediatric dept. adv. council, 1978—; assoc. prof. U. Tex. Health Sci. Ctr., San Antonio, 1983—; cons. Nat. Heart and Lung Inst., NIH, Bethesda, 1975-78, Am. Lung Assn. Hosp. Respiratory Care Rev. Team Program, 1976—, Nat. Inst. Child Health and Human Devel., Bethesda, 1979—; sect. cons. human embryology and devel. study HHS, NIH, Bethesda, 1981—; reviewer, site visitor, mem. coms. profl. projects in field. Contbr. sects. to books in field, articles to profl. publs. Served to capt. USAF, 1966-68. Recipient award best physician written book Am. Med. Writers Assn., 1980. Fellow Am. Acad. Pediatrics, Am. Thoracic Soc., Soc. for Pediatric Research, Am. Physiol. Soc., Soc. for Gynecologic Investigation; mem. Johns Hopkins Med. and Surg. Assn., So. Soc. Pediatric Research, Central Okla. Pediatric Soc., AAAS, Am. Coll. Obstetricians and Gynecologists, Soc. Exptl. Biology and Medicine. Roman Catholic. Club: Tulsa Computer Soc. (v.p. 1979-80). Subspecialties: Pulmonary medicine; Neonatology. Current work: Pulmonary surfactant, neonatal respiratory disease therapy, lung physiology and morphology, diabetes, histocompatibility linked disorders, tumor stem cell drug susceptibility, computer information processing, pediatric research. Office: Warren Med Inst 6465 S Yale Ave Tulsa OK 74136

KOTTKE, FREDERIC JAMES, physician; b. Hayfield, Minn., May 26, 1917; s. George G. and Harriet Mae (Davidson) K.; m. Astrid Marie Erling, May 27, 1939; children—Jane, James, Mary, Thomas. B.S., U. Minn., 1939, M.S. in Physiology, 1941, Ph.D., 1944, M.D., 1945. Diplomate: Am. Bd. Phys. Medicine and Rehab. (mem. 1956-59, chmn. 1963-69). Lab. asst. in physiology U. Minn. Med. Sch., 1939-40, instr., 1941-44, Baruch fellow in phys. medicine, 1946-47, asst. prof. phys. medicine, 1947-49, asso. prof., 1949-53, prof., 1953—, head dept. phys. medicine and rehab., 1952-82, resident in phys. medicine U. Minn. Hosps., 1946-47; cons. Mpls. VA Hosp., 1952—; mem. Minn. Bd. Health, 1964-67; mem. med. adv. bd. Office Vocat. Rehab., 1961-67; mem. med. research studies sect., 1961-63; bd. dirs. Kenny Rehab. Found., 1960-79; pres. Minn. Phys. Therapy Bd. Examiners, 1951-61; bd. dirs. Am. Rehab. Found. (Sister Kenny Inst.), 1964—, sec., 1964; bd. dirs. Interstudy, 1973—; John Stanley Coulter Meml. lectr., 1968, Walter J. Zeiter Meml. lectr., 1973; Disting. lectr. P.R. Med. Assn. Sect. on Phys. Medicine and Rehab., San Juan, 1973; Sidney Licht Meml. lectr. U. Pa., 1980, Ohio State U., 1981. Editorial bd.: Archives Phys. Medicine and Rehab, 1952-71, Modern Medicine, 1955—. Recipient citation Pres.'s Com. on Employment of Physically Handicapped, 1959, award of merit Rehab. Inst. Montreal, Que., Can., 1970. Mem. Am. Acad. Phys. Medicine and Rehab. (dir. 1973—, pres. 1978), Am. Assn. Lab. Animal Sci., Am. Congress Rehab. Medicine, Am. Heart Assn. (councils clin. cardiology and cerebrovascular disease), AMA, Am. Rehab. Found., Hennepin County (Minn.) Med. Soc., Internat. Rehab. Medicine Assn. (council), Internat. Soc. Rehab. Disabled, Minn. Acad. Sci., Minn. Heart Assn., Minn. Med. Alumni Assn., Minn. Med. Found., Minn. Physiatric Soc., Minn. State Med. Assn., Nat. Rehab. Assn., Am. Congress Phys. Medicine and Rehab. (v.p. 1954-58, pres. 1960, Disting. Service key 1961), Internat. Fedn. Phys. Medicine and Rehab. (chmn. edn. com. 1976—), Sigma Xi; hon. mem. Sociedad Colombiana de Medicina Fisica y Rehabilitacion, Mexican Acad. Surgery, Academia Brazileira de Medicina de Rehabilitacion, Sociedad Venezolana de Medicina Fisica y Rehabilitacion, Dutch Soc. Phys. Medicine and Rehab. Mem. Democratic Farm Labor Party. Lutheran. Subspecialty: Physical medicine and rehabilitation. Research on rehab., therapeutic exercise, spinal cord injury, cardiac problems, poliomyelitis. Home: 2741 Drew Ave S Minneapolis MN 55416 Office: U Minn Hosps 420 Delaware St SE Minneapolis MN 55455

KOTTLOWSKI, FRANK EDWARD, geologist, state official; b. Indpls., Apr. 11, 1921; s. Frank Charles and Adella Maria (Markworth) K.; m. Florence Jean Chrisco, Sept. 15, 1945; children: Karen Harvey, Janet Jenkins, Dianna Schoderbek. Student, Butler U., 1939-42; A.B., Ind. U., 1947, M.A., 1949, Ph.D., 1951. Grad. fellow Ind. U. Bloomington, 1947-51; geologic party chief Ind. Geol. Survey, Bloomington, 1948-50; instr. geology Ind. U., 1950; econ. geologist N.Mex. Bur. Mines, Socorro, 1951—, dir. div. mineral resources, 1974—; adj. prof. geosci. dept. N.Mex. Tech. Coll., Socorro, 1970—; geologic cons. Sandia Corp., 1965-72; chmn. N.Mex. Mines Safety Adv. Bd., Santa Fe, 1974—. Author: Measuring Stratigraphic Sections, 1955, Paleozoic and Mesozoic Strata on Southwestern New Mexico, 1963, (with others) Strippable Low-Sulfur Coal in San Juan Basin, 1971, Coal Resources of the Americas, 1978. Sec. Socorro County Democratic Com., 1964-68; chmn. Socorro Planning Commn., 1960-68, 71-75. Served to 1st lt. USAAF, 1942-45. Fellow Geol. Soc. Am. (exec. com.), AAAS; mem. Am. Assn. Petroleum Geologists (disting. service award 1951, editor 1971-75), Soc. Econ. Geologists, Am. Commn. on Stratigraphic Nomenclature (chmn. 1968-70). Lutheran. Subspecialties: Geology; Coal. Current work: Mineral resources, particularly energy resources, coal, uranium, geothermal, petroleum; stratigraphy; areal geology; strategic minerals. Office: N Mex Bur Mines and Mineral Resources Campus Station Socorro NM 87801 Home: 703 Sunset Dr Socoroo NM 87801

KOTTMAN, ROY MILTON, college dean; b. Thornton, Iowa, Dec. 22, 1916; s. William D. and Millie J. (Christensen) K.; m. Wanda Lorraine Moorman, Dec. 31, 1941; children: Gary Roy, Robert William, Wayne David, Janet Kay. B.S. in Agr, Iowa State U., 1941, Ph.D., 1952; M.S. in Genetics, U. Wis., 1948; LL.D. (hon.), Coll. Wooster, 1972. Asst. prof. animal husbandry Iowa State U., 1946-47; grad. research asst. U. Wis., 1947-48; mem. faculty Iowa State U., 1949-58, prof. animal husbandry, asso. dean agr., 1954-58; dean Coll. Agr., Forestry and Home Econs.; dir. Agrl. Expt. Sta., W.Va. U., 1958-60; dean Coll. Agr. and Home Econs., Ohio State U.; also dir. Ohio Agrl. Research and Devel. Center, 1960—; dir. Coop. Extension Service, 1964-82; acting assoc. dir. Nev. Agr. Expt. Sta., 1982-83; dir. Swift Ind. Packing Co.; mem. exec. com. sci. adv. bd. DNA Plant Tech. Corp. Mem. Ohio Soil and Water Conservation Commn., 1960-82; mem. Central Ohio Water Advisory Council, 1976-82; bd. dirs. Ohio 4-H Found., 1964-82, Farm Film Found., 1973-80; mem. Agr. Higher Edn. Projects Com., 1975-80, Friends NACAA Scholarship Com., 1976-80, Ohio Agrl. Mus. Com., 1977-82, Gov.'s Task Force on Gasohol, 1979-80; bd. dirs. Farm Found., 1978—; v.p. Agrl. Research Inst., 1980-81. Recipient FFA degree Am. Farmer, 1977; named to Ohio Agr. Hall of Fame, 1983. Mem. Exec. Order Ohio Commodores, Sigma Xi, Gamma Sigma Delta, Alpha Gamma Sigma (hon.), Alpha Zeta, Phi Kappa Phi, Pi Kappa Phi (future policy com. 1976-80), Phi Zeta (hon.), Alpha Gamma Rho (hon.). Presbyterian. Clubs: Rotary (hon.), Nat. Dairy Shrine.). Subspecialties: Animal breeding and embryo transplants; Genetics and genetic engineering (agriculture). Current work: Agricultural research administration. Home: 1375 Kirkley Rd Columbus OH 43210 I believe in the goodness of people and in their desire for acceptance and respect. It is parents rather than children who are the major source of teenage discontent and crime. A concentrated effort by the media, especially TV, to arouse the consciousness of parents to the mental as well as physical abuse being inflicted on their children would go far toward reducing truancy and lack of interest in learning. It would also decrease drug abuse and the costs of law enforcement. As a nation we must become more concerned with parental delinquency if we are to come to grips with juvenile delinquency.

KOUCKY, FRANK LOUIS, geologist, educator; b. Chgo., June 24, 1927; s. Frank Louis and Ella (Harshman) K.; m. Virginia Ruhl, Sept. 10, 1949; children: Frank, David, Walter, Jonathan. Ph.B., U. Chgo., 1949, M.S., 1953, Ph.D., 1956. Instr. U. Ill., Chgo., 1949-55; asst. prof. geology Mont. Sch. Mines, Butte, 1955-58, U. Ill.-Urbana, 1958-60; assoc. prof. U. Cin., 1960-72; prof. geology Coll. Wooster, Ohio, 1972—; geologist Middle East excavations. Contbr. articles on geology and archeology to profl. jours. Served with U.S. Army, 1945-47. Danforth fellow, 1967—; Bucher fellow, Swansea, Wales, 1968; research assoc. M.I.T., 1978. Fellow Geol. Soc. Am., AAAS, Ohio Acad. Sci.; mem. Soc. Econ. Geologists, Geochem. Soc., Mineral Soc. Am., Can. Mineral Assn., Phi Gamma Delta. Subspecialties: Archaeogeology; Metallurgy. Current work: Archeogeology, acnient mining and metallurgy, research in archeogeology of Middle West archeological sites. Home: 122 W Easton Rd Burbank OH 44224 Office: Dept Geolog Coll Wooster Wooster OH 44691

KOUGH, ROBERT HAMILTON, hematologist; b. Harrisburg, Pa., Feb. 19, 1921; s. Harry Milton and Olive (Smith) K.; m. Nancy Jane

Trunnell, June 18, 1943; 1 dau., Elizabeth Trunnell Beiler. B.S., Pa. State U., 1942; M.D., U. Pa., 1945. Diplomate: Am. Bd. Internal Medicine. Intern Hosp. of U. Pa., Phila., 1945-46, resident in medicine, 1955-57, fellow in hematology, 1957-58; asst. instr. dept. pharmacology U. Pa. Med. Sch., Phila., 1949-50, instr., 1950, assoc., 1951; mem. med. staff Carlisle (Pa.) Hosp., 1952-55; dir. hematology/oncology Geisinger Med. Center, Danville, Pa., 1958—. Author: Anemias Case Studies, 1981. Bd. dirs. Pa. Med. Care Found., Le Moyne, 1980—, Capital Blue Cross, Harrisburg, 1966-; corp. mem. Pa. Blue Shield, 1972-. Served to lt. (j.g.) U.S. Navy, 1946-49. Fellow A.C.P.; mem. Pa. Soc. Internal Medicine (pres. 1979-80), Am. Soc. Hematology, Am. Soc. Clin. Oncology, Eastern Coop. Oncology Group, Phi Beta Kappa, Alpha Omega Alpha, Phi Eta Sigma, Alpha Epsilon Delta, Sigma Pi Sigma, Phi Kappa Phi. Republican. Lutheran. Subspecialty: Hematology. Current work: Hematologic neoplasms—clinic trials. Home: Red Oak Dr Route 7 Box 77 Danville PA 17821 Office: Geisinger Med Center N Academy Ave Danville PA 17822

KOUSSA, HAROLD ALAN, nuclear engineer; b. Central Falls, R.I., June 20, 1947; s. Harold Albert and June Joann (John) K.; m. Marsha Lynn Heidenis, Dec. 1, 1973. B.S. in Engring. Sci., U. R.I., 1969; M.B.A., U. Hartford, 1975; M.S. in Engring. Sci., Rensselaer Poly. Inst., 1977. Reactor engring. asst. Conn. Yankee Atomic Power Co., East Hampton, 1969-75, reactor engr., 1975-77; staff nuclear engr. Am. Nuclear Insurers, Farmington, Conn., 1977-79, sr staff nuclear engr., 1979-81, prin. engr., 1981-82, ops. mgr., 1982—. Mem. Republican Town Com., East Hampton, 1981; mem. Water Pollution Control Authority, East Hampton, 1982, East Hampton Charter Revision Com., 1982. Served as ensign USNR. Mem. Am. Nuclear Soc., ASME, Assn. M.B.A. Execs. Republican. Congregationalist. Club: U. R.I. Fast Break (Kingston, R.I.). Lodge: Masons. Subspecialties: Nuclear engineering; Nuclear fission. Current work: Manage activities of engineering staff in regard to nuclear safety inspections of commercial nuclear power reactors and related projects. Home: 73 Childs Rd East Hampton CT 06424 Office: Am Nuclear Insurers 270 Farmington Ave Farmington CT 06032

KOUTNIK, DARYL LEE, bot. systematist, bot. cons.; b. Burbank, Calif., Dec. 22, 1951; s. Robert James and Alta Maria (Raville) K. B.A. in Biology and Math, Calif. State U., Northridge, 1977; M.S. in Botany, U. Calif., Davis, 1981; Ph.D. (Chubb Found. Scholar, Pacific Tropical Bot. Garden grantee), U. Calif., Davis, 1982. Research asst. U. Calif., Davis, 1979-80, teaching asst., 1980-82; sci. officer Bolus Herbarium, U. Capetown, South Africa, 1982—; cons. endangered species and environ. impact. Contbr. articles to profl. jours. Mem. Bot. Soc. Am., Am. Soc. Plant Taxonomists, Soc. Study of Evolution, Bot. Soc. South Africa, Sigma Xi. Democrat. Subspecialties: Systematics; Morphology. Current work: Systematics of the euphorbieae (euphorbiaceae) with special reference to the genera Euphorbia and Chamaesyce. Home: 6922 Hesperia Ave Reseada CA 91335 Office: Bolus Herbarium U Capetown Rondebosch South Africa 7700

KOUTS, HERBERT JOHN CECIL, physicist; b. Bisbee, Ariz., Dec. 18, 1919; s. Oliver Allen and Lillian (Niemeyer) K.; m. Hertha Pretorius, Feb. 2, 1942; children: Anne Elizabeth, Catherine Jennifer; m. Barbara Stokes, Mar. 27, 1974; stepchildren: Francis Spitzer, Michael Spitzer, Daniel Spitzer. B.S., La. State U., 1941, M.S., 1946; Ph.D., Princeton U., 1952. With Brookhaven Nat. Lab., Upton, L.I., N.Y., 1950-73, 77—, sr. scientist, asso. div. head, 1958-73, chmn. dept. nuclear energy, 1977—; dir. div. reactor safety research AEC, Washington, 1973-75; dir. Office Nuclear Regulatory Research, U.S. Nuclear Regulatory Commn., Washington, 1975-76; mem. advisory com. reactor physics AEC, 1956-63, mem. adv. com. reactor safeguards, 1962-66; mem. European Am. Advisory Com. for Reactor Physics to European Nuclear Energy Agency, 1962-68. Served with USAAF, 1942-45. Recipient E. O. Lawrence award AEC, 1963, Disting. Service award, 1975; Disting. Service award NRC, 1976. Mem. Am. Nuclear Soc. (Theos Thompson award in nuclear reactor safety 1983), Fedn. Am. Scientists, Center Moriches Audubon Soc., Nat. Acad. Engring. Subspecialties: Nuclear engineering; Nuclear fission. Current work: Nuclear reactor safety, nuclear reactor development, neutron physics, energy economy. Home: 249 S Country Rd Brookhaven NY 11719 Office: Brookhaven Nat Lab Upton NY 11973

KOVAC, PAVOL, research chemist; b. Trencin, Czechoslovakia, Dec. 6, 1938; came to U.S., 1981; s. Vojtech and Judita (Trebitsch) K.; m. Eva Gross, Nov. 24, 1962; 1 son, Roman. M.A. in Chemistry, Slovak Tech. U., 1962; Ph.D. in Organic Chemistry, Slovak Acad. Scis., 1967. Sr. research chemist Slovak Acad. Scis., Bratislava, Czechoslovakia, 1963-81, postdoctoral fellow dept. biochemistry Purdue U., Lafayette, Ind., 1967-68; prodn. chemist and group leader Bachem Inc., Torrance, Calif., 1981-82; vis. scientist NIH, Bethesda, Md., 1982—. Contbr. numerous articles on chemistry to profl. jours.; mem. editorial bd.: Jour. Carbohydrate Chemistry, 1982—; mem. editorial adv. bd.: Jour. Carbohydrates, Nucleosides, Nucleotides, 1974-81. Mem. Am. Chem. Soc. Subspecialties: Organic chemistry; Synthetic chemistry. Current work: Synthesis of oligosaccharides and oligodeoxy nucleotides. Patentee in field. Office: NIH Rockville Pike Bldg 4 Room 210 Bethesda MD 20205

KOVACH, JOSEPH K., ethologist, research and evaluation administrator; b. Godollo, Hungary, Feb. 1, 1929; came to U.S., 1957, naturalized, 1962; s. Akos K. and Karolin (Kovats) K.; m. Magdalene Hamel, Mar. 10, 1975; children: Ian, Tobias, Ilsabe, Christopher. A.B., Elmhurst (Ill.) Coll., 1959; Ph.D., U. Chgo., 1963. Russian-Hungarian interpreter Bur. Tech. and Sci. Trans., Budapest, 1953-56; research asst. U. Chgo., 1959-63; NSF fellow U. Stockholm, 1963-64; research scientist U. Ill., 1964-66; research scientist, dir. animal behavior lab. Menninger Found., Topeka, Kans., 1966—, dir. research and evaluation, 1972—, dir. research tng., 1981—; mem. peer rev. com. Biopsychology study sect. NIH, Bethesda, Md., 1981—. Author: (with Gardner Murphy) Historical Introduction to Modern Psychology, 1972; mem. editorial bd.: Applied Animal Ethology, 1975—; cons. editor: Jour. Comparative Psychology, 1982—; contbr. articles to profl. jours. Nat. Inst. Child Health and Human Devel. grantee, 1972; recipient Research Career Devel. award NIH, 1970-80; Research Career Scientist award NIMH, 1980—. Mem. Am. Psychol. Assn., Animal Behavir Soc., AAAS, Internat. Soc. Devel. Psychobiology. Subspecialties: Ethology; Gene actions. Current work: Behavior—genetics; early development and perceptual learning. Office: The Menninger Foundation PO Box 829 Topeka KS 66601 Home: 3122 Westover Rd Topeka KS 66604

KOVACH, JULIUS LOUIS, consulting company executive, consultant; b. Hungary, Aug. 24, 1934; s. Antal and Irma (Mihacz) K.; m.; 1 son, Paul Elek. Diploma in Engring, Inst. Chem. Tech., Hungary, 1955. Scientist Columbia Cellulose Co., 1957-59; sr. scientist Barnebey Cheney Co., 1959-63, dir. research and devel., 1963-68; v.p., dir. research and devel. N.Am. Carbon, Inc., 1968-72; pres. Nuclear Cons. Services, Inc., Columbus, Ohio, 1972—; dir. U.S. Activated Carbon Co.; lectr. Harvard U.; chmn. Group of Experts on Air Cleaning in Accident Situations of OECD. Contbr. articles profl. jours., chpts. to books. Maj. CAP. Mem. Am. Nuclear Soc., ASTM. Subspecialties: Gas cleaning systems; Nuclear engineering. Current work: Adsorption, catalysis, nuclear safety, surface science. Patentee in field. Home: 2948 Brookdown Dr Worthington OH 43085 Office: PO Box 29151 Columbus OH 43229

KOVACH, PAUL JOSEPH, emergency planner; b. Colver, Pa., Feb. 7, 1956; s. Paul P. and Shirley (Jackson) K. B.S., St. Francis Coll., Loretto, Pa., 1977; M.S., U. Cin., 1979. Radiol. engr. Rockwell Internat., Hanford, Wash., 1979-81; staff engr. Gen. Physics Corp., Columbia, Md., 1981-82; emergency planner GPU Nuclear, Middletown, Pa., 1982—. Mem. N.Y. Acad. Scis., Health Physics Soc. Subspecialty: Applied health physics. Current work: Contingency planning to minimize environmental and social impact of incidents at nuclear power plants. Office: GPU Nuclear-Three Mile Island PO Box 480 Middletown PA 17057

KOVACS, CHARLES JEFFREY, radiologist, educator; b. Fairfield, Conn., Apr. 7, 1941; s. Charles Kalman and Anne Elizabeth K. B.S., Siena Coll., Loudonville, N.Y., 1963; Ph.D., St. John's U., Jamaica, N.Y., 1969. USPHS fellow Brookhaven Nat. Lab., Upton, N.Y., 1969-71; instr. Hahnemann Med. Coll., Phila., 1971-72; asst. prof. radiology U. Va. Med. Sch., 1972-76; sr. scientist cancer research unit Allgheny Gen. Hosp., Pitts., 1976-81; assoc. prof. radiology Bowman Gray Med. Sch., Wake Forest U., Winston-Salem, N.C., 1981—, also dir. radiation oncology labs.; lectr. Am. Cancer Soc., 1973-76. Author articles in field. Research fellow NSF, 1965-69; recipient Visitor's and President's award U. Va. Med. Sch., 1978. Mem. Am. Assn. Cancer Research, Cell Kinetic Soc., Radiation Research Soc., AAAS, Southeastern Assn. Cancer Research. Democrat. Roman Catholic. Subspecialties: Cancer research (medicine); Cell study oncology. Current work: Experimental combined radiation and chemotherapeutics, host-tumor interactions, regulators of normal tissue cell proliferation, modulators of homeostasis in disease states, clinical studies drug and radiation toxicity. Home: 5760 Remington Dr Winston-Salem NC 27103 Office: Dept Radiology Bowman Gray Med Sch 300 S Hawthorne Rd Winston-Salem NC 27103

KOVACS, GYULA, engring. educator, robotics and fiber optics cons.; b. Nagykutas, Hungary, Apr. 22, 1941; came to U.S., 1973; s. Imre and Anna (Gorza) K.; m. Judy G. Kovacs, Dec. 4, 1976; 1 dau., Nicola Ashley. B.Sc., U. Manchester, Eng., 1965, M.Sc., 1971; M.S., Okla. State U., 1977. Registered profl. engr. Inst. Mech. Engring. (Eng.). Engring. trainee, tool maker, Eng., 1957-65; research and devl. engr. U. Manchester, 1965-76; design engr. Fenix & Scissons, Inc., Tulsa, 1976-82; assoc. prof. engring., program dir. U. Ark., Little Rock, 1982—; cons., researcher. Author: Guidelines for the Application of Robots and Automated Processes, 1982. Mem. Am. Inst. Indsl. Engrs., ASME, Am. Soc. Engring. Edn., Soc. Mfg. Engrs., Soc. Die Casting Engrs. Subspecialties: Fiber optics; Mechanical engineering. Current work: Robotics, optical signal processing, automation. Home: 1624 S Taylor St Little Rock AR 72204 Office: U Ark Sch Engring Technology Little Rock AR 72204

KOVEN, BERNARD J., physician; b. N.Y.C., Feb. 17,1927; s. Nathan S. and Helen R. (Greenberg) K.; m. Martha R. Shapiro, Dec. 20, 1959; children: Joanne, Carolyn A. A.B., Syracuse U., 1949; M.D., SUNY, Syracuse, 1953. Diplomate: Am. Bd. Internal Medicine. Research fellow in chemotherapy Sloan Kettering Inst., 1957-58; clin. instr. medicine SUNY, Syracuse, 1958; instr. medicine Cornell U. Med. Coll., N.Y.C., 1959-61; asst. coordinator cancer teaching Seaton Hall Coll. Medicine and Dentistry, 1959-60; asst. prof. medicine N.J. Coll. Medicine and Dentistry, 1960-71, assoc. coordinator cancer teaching, 1960—, assoc. prof. clin. medicine, 1971—; clin. instr. medicine Cornell U. Med. Coll., 1971—; intern Johns Hopkins Hosp., Balt., 1953-54; resident Upstate Med. Center, Syracuse, 1954-55; resident medicine Kings County Hosp., Bklyn., 1955-57; clin. asst. OPD and wards Meml. Hosp., 1958-60, clin. asst. physician, 1960-66; clin. asst. vis. physician Kings County Hosp., Bklyn., 1958; asst. vis. physician James Ewing Hosp., 1958-68; asst. attending physician Jersey City Med. Center, 1960-67, cons. neoplastic disease, 1967-75; clin. asst. physician Hackensack Hosp., 1962-64; cons. St. Elizabeth Host., N.J., 1962-72, St. Mary's Hosp., Passaic, 1962-74, Holy Name Hosp., Teaneck, N.J., 1966-80, Englewood Hosp., 1966—, East Orange VA Hosp., 1966-75, Newark City Hosp., 1966-67; asst. chief diagnosis and admitting clinic Meml. Hosp., N.Y.C., 1963-66; chief dept. oncology Holy Name Hosp., Teaneck, 1980—, attending in medicine, 1980—; pres. Meml. Oncology Assos., Englewood, N.J., 1972—; cons. Contbr. articles to profl. jours. Served with USAAF, 1945-46. Recipient Am. Cancer Soc. Nat. Honors citation, 1969; Physician of Yr. award, 1979. Fellow ACP; mem. Am. Cancer Soc. (Bergen County bd. mgrs. 1964), Am. Fedn. Clin. Research, Am. Soc. Clin. Oncology, Bergen County Med. Soc., Oncology Soc. N.J. (pres.), Med. Soc. N.J., Johns Hopkins Med and Surg. Assn., N.J. Acad. Medicine, Pan Am. Med. Assn., N.Y. Cancer Soc., Am. Assn. Cancer Edn., Am. Assn. Cancer Research, Planned Parenthood Assn. (Bergen County bd. advisors 1974-78). Subspecialties: Chemotherapy; Cancer research (medicine). Current work: Medical oncology, carcinogenesis, med. cons. Address: 163 Engle St Apt 4A Englewood NJ 07631

KOWAL, CHARLES THOMAS, astronomer; b. Buffalo, Nov. 8, 1940; s. Charles Joseph and Rose (Myszkowiak) K.; m. Maria Antonietta Ruffino, Oct. 17, 1968; 1 dau., Loretta. B.A., U. So. Calif., 1963. Research asst. Mt. Wilson and Palomar obs.'s, 1961-63, Calif. Inst. Tech., Pasadena, 1963-65, 66-75, U. Hawaii, 1965-66; asso. scientist Calif. Inst. Tech., 1976-78, scientist, 1978-81, mem. profl. staff, 1981—; staff asso. Hale Obs., 1979-80; lectr. in field. Recipient James Craig Watson award Nat. Acad. Scis., 1979. Mem. Am. Astron. Soc., Internat. Astron. Union. Subspecialties: Optical astronomy. Current work: Searching for, and studying, asteroids and comets. Investigating the motion of Neptune. Discovered bright supernova, 1972, 13th satellite of Jupiter, 1974, large planetoid between orbits of Saturn and Uranus, 1977, also asteroids and comets; recovered lost comets and asteroids. Office: Dept Astrophysics Calif Inst Tech Pasadena CA 91125

KOWALSKI, CONRAD JOHN, research chemist; b. Chgo., July 9, 1947; s. John and Cecelia K.; m. Marcia Paige, June 8, 1968; children: John, Matthew. S.B., MIT, 1968; M.S., Calif. Inst. Tech., 1971, Ph.D., 1974. NIH postdoctoral fellow Columbia U., N.Y.C., 1974-76; asst. prof. U. Notre Dame, Ind., 1976-82; asst. dir. synthetic chemistry Smith Kline Beckman, Phila., 1982—. Founder, editor: Synthetic Pathways, 1980. 3M Corp. Young faculty fellow, 1981; NSF group travel grantee, 1982; grantee NIH, NSF, U.S. Dept. Agr., 1979-82. Mem. Am. Chem. Soc., AAAS, Sigma Xi, Phi Lambda Upsilon. Roman Catholic. Subspecialties: Organic chemistry; Synthetic chemistry. Current work: Synthesis of natural products and pharmaceuticals and the development of new synthetic methods, especially involving alpha keto dianions and alkynolate anions. Office: Smith Kline Beckman 1500 Spring Garden St PO Box 7929 Philadelphia PA 19101

KOWALYSHYN, THEODORE JACOB, physician; b. Northampton, Pa., Dec. 12, 1935; s. Stephen and Anna (Kuzyk) K.; m. Mary Ann West, Aug. 19. 1967; children: Alexander West, Andrew Jacob. B.S., Lehigh U., 1957; M.D., Hahnemann Med Coll., 1966. Diplomate: Am. Bd. Internal Medicine. Intern St. Luke's Hosp., Bethlehem, Pa., 1966-67, resident in internal medicine, 1967-70; fellow in hematology U. Cin. Med. Center, 1970-72; staff physician Pocono Hosp., East Stroudsburg, Pa., 1972—; asst. clin. prof. medicine Med. Coll. Pa., Phila., 1980—; sr. aviation med. examiner FAA, Oklahoma City, 1981—. Bd. dirs. Eastern Pa. Health Care Found., Allentown, 1977-80, Pocono Hosp., 1978-81. Served with U.S. Army, 1958-60. Mem. A.C.P., Am. Soc. Internal Medicine, Pa. Med. Soc., AMA. Subspecialties: Internal medicine; Hematology. Home: 714 Sarah St Stroudsburg PA 18360 Office: Med Assocs of Monroe Country 239 E Brown St East Stroudsburg PA 18301

KOZLOFF, LLOYD M., univ. dean, scientist; b. Chgo., Oct. 15, 1923; s. Joseph and Rose (Hollowbow) K.; m. Judith Bonnie, June 16, 1947; children: James, Daniel, Joseph, Sarah. B.S., U. Chgo., 1943, Ph.D. in Biochemistry, 1948. Asst. prof. U. Chgo., 1949-58, assoc. prof., 1958-61, prof., 1961-64; prof. microbiology, chmn. dept. microbiology U. Colo., 1964-80, assoc. dean faculty affairs, 1976-80; dean Grad. div. U. Calif.-San Francisco, 1981—, prof., 1981—. Research numerous publs. in field; contbr. chpts. to books. Bd. dirs. Proctor Found., San Francisco, 1982—. Served with USN, 1944-46. Lederle fellow, 1954. Fellow AAAS (hon.); mem. Am. Soc. Biol. Chemistry, N.Y. Acad. Scis., Am. Chem. Soc. Subspecialties: Virology (biology); Population biology. Current work: Viral morphogenesis: use of folate and folate enzymes in virus assembly; biogenic origin of ice: ice nucleating site in bacteria. Home: 1750 Grant Ave San Francisco CA 94133 Office: U Calif S-140 GradDiv San Francisco CA 94143

KOZLOWSKI, THEODORE THOMAS, botany educator; b. Buffalo, May 21, 1917; s. Theodore and Helen (Zamiara) K.; m. Maude Peters, June 29, 1954. B.S., Syracuse U., 1939; M.A., Duke U., 1941, Ph.D., 1947; postgrad., MIT, 1942-43; D.Sc. honoris causa, U. Catholique de Louvain, Belgium, 1978. Asst. prof. botany U. Mass., 1947-48, asso. prof., 1948-50, prof., head dept. botany, 1950-58; prof. forestry U. Wis., 1958-72, A.J. Riker prof., 1972—, chmn. dept., 1961-64, dir. biotron lab., 1977—; cons. NSF, Stanford Research Inst., Nat. Park Service, FAO, Oak Ridge Nat. Lab., Malaysian Govt., Mont. Univ. System, Internat. Found. for Sci., Academic Press, Time-Life Books, various comml. firms; vis. biologist Am. Inst. Biol. Scis., 1969-72; vis. scientist Soc. Am. Foresters, 1963-71; vis. prof. U. Pa., 1954; George Lamb lectr. U. Nebr., 1974; George S. Long lectr. U. Wash., 1978; Rapporteur World Consultation on Tree Improvement, 1963. Author: (with P.J. Kramer) Physiology of Trees, 1960, Physiology of Woody Plants, 1979, Water Metabolism in Plants, 1964, Growth and Development of Trees, 2 vols., 1971, Tree Growth and Environmental Stresses, 1979; editor: Tree Growth, 1962, Water Deficits and Plant Growth, 7 vols., 1968-83, Seed Biology, 3 vols., 1971, Shedding of Plant Parts, 1973, (with G.C. Marks) Ectomycorrhizae, 1973, (with C.E. Ahlgren) Fire and Ecosystems, 1974, (with J.B. Mudd) Responses of Plants to Air Pollution, 1975, (with P. de T. Alvim) Ecophysiology of Tropical Crops, 1977, (with T.W. Tibbitts) Controlled Environment Guidelines for Plant Research, 1979; editorial bd.: Forest Sci., Ecology, BioSci.; asso. editor: Can. Jour. Forest Research, Am. Midland Naturalist; editor: (book series) Physiol.-Ecology. Served to capt. USAAF, 1942-46. Sr. Fulbright research scholar Oxford (Eng.) U., 1964-65; recipient Author's award Internat. Shade Tree Conf., 1971. Mem. Am., Scandinavian socs. plant physiologists, Bot. Soc. Am., Ecol. Soc. Am., Soc. Am. Foresters (Barrington Moore biol. research award 1974), Internat. Soc. Arboriculture (Arboricultural research award 1976), Societas Forestalis Fenniae (Finland) (hon.), Societas Botanicopum Poloniae (hon.), Am. Inst. Biol. Scis., Phi Beta Kappa, Sigma Xi, Phi Kappa Phi, Phi Sigma. Subspecialty: Plant physiology (agriculture). Home: 10 S Rock Rd Madison WI 53705

KOZMAN, THEODORE ALBERT, mech. engineering official, superconductivity researcher; b. Long Beach, Calif., Nov. 18, 1946; s. Albert Henery and Iola Sivels (Curren) K; m. Katharine Davis, Mar. 21, 1970; children: Austin Jon, Kenneth Scot. B.S. in Aerospace Engring, U. So. Calif., 1967, M.S., 1969; Ph.D. in Engring. Mechanics, U. Tenn.-Knoxville, 1973. Linear accelerator operator AEC, U. So. Calif., 1966-67; Douglas Aircraft Co. asst. U. So. Calif., 1967-69; assoc. engr./scientist Douglas Aircraft Co., 1967-69; research asst. U. Tex. Space Inst., Tullahoma, 1969-70; teaching asst. U. Tenn-Knoxville, 1970-72; mem. profl staff Ctr. Naval Analyses, 1973-74; mech. engr. Lawrence Livermore (Calif.) Nat. Lab., 1974-77, assoc. project mgr. for mgmt. systems, 1977-80, assoc. project mgr. magnet systems, 1980—. Papers and presentations in field. Treas. Livermore Cultural Arts Council. NASA summer trainee, 1971; NASA fellow, 1971-72. Mem. Am. Nuclear Soc. (Outstanding Tech. Achievement award 1982). Democrat. Presbyterian. Subspecialties: Mechanical engineering; Nuclear fusion. Current work: Fusion research; superconductivity; built and tested world's largest superconducting magnet. Office: PO Box 808 MZL635 Livermore CA 94550

KRACHMAN, HOWARD ELLIS, research and development executive; b. Phila., June 12, 1938; s. Albert and Sarah (Linetsky) K.; m. Betty Gurtoff, Feb. 23, 1974; children: Adam, Gower Alexis. B.S. in Mech. Engring, Drexel U., 1961, M.S., U. So. Calif., 1964; diploma, Von Karman Ins., Brussels, 1963. Assoc. Douglas Co., Santa Monica, Calif., 1961-63; mem. tech. staff TRW, Redondo Beach, Calif., 1963-70; dir. engring. Developmental Scis., Industry, Calif., 1970—. Mem. AIAA, ASME, Assn. Unmanned Vehicle Systems. Subspecialties: Aerospace engineering and technology; Energy management. Current work: Remote piloted vehicles. Home: 2120 San Pasqual Pasadena CA 91107 Office: Devel Scis 15757 E Valley Dr City of Industry CA 91744

KRAHWINKEL, DELBERT JACOB, JR., veterinary surgeon, educator; b. Owensboro, Ky., Aug. 12, 1942; s. Delbert J. and Mary Edith (Kirkendoll) K.; m. Lyda Nave, July 6, 1962; children: Kelly Lynn, Kevin Edward. D.V.M., Auburn U., 1966; M.S., Mich. State U., 1973. Diplomate: Am. Coll. Vet. Anesthesiologists, Am. Coll. Vet. Surgeons. Pvt. practice vet. medicine, Danville, Ill., 1966-67; asst. prof. surgery and anesthesiology Mich. State U., East Lansing, 1969-75; prof. surgery Sch. Vet. Medicine, U. Tenn., Knoxville, 1975—, head dept. urban practice, 1982—. Contbr. writings to sci. jour. and books. Served to capt. USAF, 1967-69. Recipient Pres.'s award Auburn U., 1966; Norden outstanding tchr. award Mich. State U., 1973; Upjohn award. Mem. AVMA, Tenn. Vet. Med. Assn., Am. Animal Hosp. Assn., Am. Assn. Vet. Clinicians. Baptist. Club: South Knox Ruritan (pres. 1979). Subspecialties: Surgery (veterinary medicine); Veterinary anesthesiology. Home: 7009 Neubert Springs Rd Knoxville TN 37920 Office: PO Box 1071 Knoxville TN 37917

KRAIG, RICHARD PAUL, neurologist, neuroscientist, educator; b. Chgo., Feb. 1, 1949; s. Harry J. and Adelaide (Farnaus) K.; m. Marcia P. Stachura, June 13, 1971; 2 daus., Jenna L., Marisa A. B.A. in Chemistry, Cornell Coll., Mt. Vernon, Iowa, 1971; Ph.D. in Physiology and Biophysics, U. Iowa, 1976; M.D., NYU, 1978. Diplomate: Nat. Bd. Med. Examiners. Grad. research asst. U. Iowa, 1974-76; intern in medicine U. Chgo. Hosp., 1978-79; resident in neurology Cornell U. Med. Coll., N.Y.C., 1979-82, clin. assoc. in neurology, 1979-81, instr., 1981-82, asst. prof. neurology, 1982—. Contbr. numerous articles on micro ion electrodes and their use in study of nervous system physiology and pathophysiology to sci. jours. Grass Found. fellow, 1975; Rockefeller Bros. clin. scholar, 1982-83. Mem. Soc. for Neurosci., Am. Acad. Neurology. Subspecialties: Neurology; Neurophysiology. Current work: Microphysiology and biochemistry of brain and brain cell microenvironment in the study of stroke; ion microsensors.

KRAKAUER, RANDALL SHELDON, physician; b. N.Y.C., Apr. 25, 1949; s. Henry Robert and Violet (Tallmadge) K.; m. Marcia Sue Katcher, June 15, 1969; children: Meryl Lucille, Ari Martin, Barak Lee. B.S., Rensselaer Poly. Inst., 1972; M.D., Albany Med. Coll., 1972. Diplomate: Am. Bd. Internal Medicine and Rheumatology. Med. intern, resident U. Minn. Hosps., 1972-74; clin. assoc. immunology NIH, 1974-76; fellow rheumatology Mass. Gen. Hsop. Harvard U. Med. Sch., Boston, 1976-77; head sect. clin. immunology Cleve. Clinic Found., 1978-83; mem. exec. com. Study Group for Lupus Nephritis, 1981—; chmn. med. adv. bd. Ohio Lupus Found. Served to lt. comdr. USPHS, 1974-76. Fellow ACP; mem. Am. Soc. Clin. Pharmacology and Therapeutics (chmn. immunotherapy sect.), Am. Rheumatism Assn., Am. Assn. Immunologists, Am. Fedn. Clin. Research, Central Soc. Clin. Research. Subspecialty: Rheumatology. Current work: Pathogenesis of and immunotherapy for autoimmune disease. Office: 900 W Main St Freehold NJ 07728

KRAKOFF, IRWIN HAROLD, physician, educator; b. Columbus, Ohio, July 20, 1923; s. Morris Joseph and Frieda K.; m. Miriam Shocket, Sept. 1, 1946; children—Peter Alan, Charles Edward, Ellen Miriam. B.A., Ohio State U., 1943, M.D., 1947. Intern Mt. Sinai Hosp., Cleve., 1947-48, resident, 1948-50, Boston City Hosp., 1950-51; attending physician, asso. chmn. dept medicine, head clin. chemotherapy and pharmacology lab. Meml. Sloan-Kettering Cancer Center, N.Y.C., 1953-76; prof. medicine Cornell U., 1955-76; prof. medicine and pharmacology, dir. Vt. Regional Cancer Center, U. Vt., 1976—. Contbr. numerous articles to profl. jours., also chpts. to books. Served with USN, 1943-46, 51-53. Recipient Alfred P. Sloan cancer research award, 1965. Mem. Am. Assn. Cancer Research, Am. Soc. Clin. Oncology, A.C.P., Am. Fedn. Clin. Research, Am. Soc. Pharmacology and Exptl. Therapeutics. Subspecialty: Cancer research (medicine). Address: 1 S Prospect St Burlington VT 05401

KRALL, RONALD LEE, pharmaceutical clinical researcher, neuropharmacologist; b. Balt., June 24, 1947; s. Melvin and Vivian (Lowy) K.; m. Susan Jane Doerner, Nov. 22, 1975; 2 sons, Joshua Andrew, Benjamin Eric. B.A., Swarthmore Coll., 1969; M.D., U. Pitts., 1973. Diplomate: Am. Bd. Neurology and Psychiatry. Intern Los Angeles County Harbor Gen. Hosp., 1973-74; staff assoc. Epilepsy br. NIH, Bethesda, Md., 1974-77; resident, fellow U. Rochester, N.Y., 1977-80, asst. prof., 1980-83; assoc. dir. clin. research Lorex Pharms., Skokie, Ill., 1983—. Recipient Commendation Medal USPHS, 1977. Mem. Epilepsy Found. Am., Am. Acad. Neurology, Am. Epilepsy Soc., Am. Soc. Clin. Pharmacology, Sigma Xi. Subspecialties: Neurology; Neuropharmacology. Current work: interest: clinical research in neuropharmacology; mechanism of action of neuroactive drugs, especially antiepileptic drugs. Office: Lorex Pharmaceuticals 5200 Old Orchard Rd Skokie IL 60077

KRAMAN, STEVE SETH, physician, educator; b. Chgo., Aug. 30, 1944; s. Julius and Ruth (Glassner) K.; m. Lillian Virginia Casanova, May 29, 1972; children: Theresa, Pilar, Laura. B.S., U. P.R., 1968, M.D., 1973. Diplomate: Am. Bd. Internal Medicine. Intern, resident Brookdale Med. Ctr., 1973-76, fellow, 1977-78, Queens Hosp. Ctr., 1976-77; asst. prof. medicine U. Ky., 1978—; staff physician VA Med. Ctr., Lexington, 1978—. NIH grantee, 1980. Fellow Am. Coll. Chest Physicians; mem. Am. Thoracic Soc., Am. Fedn. Clin. Research, Am. Physiol. Soc., Bioelectromagnetics Soc. (assoc.). Democrat. Jewish. Subspecialties: Pulmonary medicine; Physiology (medicine). Current work: Characterization and interpretation of respiratory sounds and the determination of their mechanisms of production and sites of origin. Office: VA Med Ctr 111-H Lexington KY 40511

KRAMARSKY, BERNHARD, cell biologist; b. Hamburg, Germany, Dec. 27, 1924; s. Felix and Gutta (Nachemson) K.; m. Marion Bienes, Apr. 1946; m. Lea DellaRiccia, Apr. 20, 1951; children: Esther Winter, Jonathan Felix. B.Sc., Cornell U., 1950; D.Sc., U. Florence, Italy, 1963. Instr. microbiology U. So. Calif. Med. Sch., 1964-66; research asst., electron microscopist Albert Einstein Med. Center, Phila., 1966-67; assoc. mem., electron microscopist Inst. Med. Research, Camden, N.J., 1967-74; supervisory electron microscopist, cell biologist Cell Sci. Lab., Electro-Nucleonics, Inc., Silver Spring, Md., 1974—. Served with AUS, 1943-45. Mem. Am. Assn. Cancer Research, Am. Soc. Microbiology, Chesapeake Soc. Electron Microscopy. Subspecialties: Cell biology; Virology (biology). Current work: Ultrastructural studies on receptors, endocytosis and cell-substrate adherence. Home: 11313 Baroque Rd Silver Spring MD 20901 Office: 12050 Tech Rd Silver Spring MD 20904

KRAMER, BARNETT SHELDON, medical oncologist, researcher, educator; b. Balt., July 28, 1948; s. Mervin and Muriel (Woolf) K.; m. Ruth Solomon, June 25, 1972; 1 son, Jeremy Zachary. M.D., U. Md., 1973. Diplomate: Am. Bd. Internal Medicine. Intern Barnes Hosp., St. Louis, 1973-74, resident, 1974-75; clin. assoc. Nat. Cancer Inst., Bethesda, Md., 1975-78; asst. prof. med. oncology U. Fla., Gainesville, 1978-83, assoc. prof., 1983—. Contbr. chpts. to books and articles to profl. jours. Served with USPHS, 1975-78. Mem. Am. Soc. Clin. Oncology, Southeastern Cancer Study Group. Democrat. Jewish. Subspecialties: Internal medicine; Cancer research (medicine). Current work: Infection in cancer patients; currently involved in studies of prophylaxis and therapy of infections in cancer patients; also studies of leukemia, lung cancer. Home: 2707 SW 3d Pl Gainesville FL 32607 Office: U Fla Sch Medicine JH Miller Health Ctr Box J277 Gainesville FL 32610

KRAMER, BRUCE MICHAEL, mech. engr., cons., telecommunications co. exec., educator; b. N.Y.C., July 23, 1949; s. Morris and Ruth S. (Soloway) K. S.B., S.M., M.I.T., 1972, Ph.D., 1979. Chmn. bd. Zoom Telephonics, Inc., Boston, 1976—; asst. prof. mech. engring. M.I.T., 1979—; cons. mfg. Contbr. numerous articles on metal cutting theory to profl. jours. Mem. ASME (Blackall award 1982), Am. Soc. Metals, AAAS, Sigma Xi, Tau Beta Pi, Pi Tau Sigma. Subspecialties: Materials processing; Materials (engineering). Current work: Precision machining; new tool materials; devel. mfg. sci. base. Patentee hafnium carbide coatings. Home: 130 Bowdoin St Apt 1401: Boston MA 02108 Office: 77 Massachusetts Av Room 35-234 Cambridge MA 02139

KRAMER, PAUL JACKSON, plant physiologist, educator; b. Brookville, Ind., May 8, 1904; s. LeRoy and Minnie (Jackson) K.; m. Edith Vance, June 24, 1931; children: Jean, Richard V. A.B., Miami U., Ohio, 1926, D.Litt., 1966; Ph.D., Ohio State U., 1931, D.Sc. (hon.), 1972, U. N.C., 1966; Dr. h.c., U. Paris VII, 1975. Mem. Faculty Duke U., Durham, N.C., 1931—, James B. Duke prof., 1954-74, emeritus, 1974—; program dir. NSF, 1960-61; vis. investigator Calif. Inst. Tech., 1953; vis. lectr. Cornell U., spring 1976; vis. prof. U. Tex., fall 1976; Walker Ames vis. prof. U. Wash., 1977. Author: Plant and Soil Water Relationships, 1949, Physiology of Trees, 1960, Plant and Soil Water Relationships: A Modern Synthesis, 1969, Physiology of Woody Plants, 1979, Water Relations of Plants, 1983, A Collection of Lectures in Tree Physiology, 1982; contbr. articles to profl. jours. Recipient Soc. Am. Foresters award for achievement in biology, 1961; AEC grantee, 1949-71; NSF grantee, 1955—. Mem. AAAS, Am. Inst. Biol. Scis. (pres. 1964, Disting. Service award 1977), Am. Soc. Plant Physiologists (pres. 1945), Bot. Soc. Am. (award of merit 1956, pres. 1964), Nat. Acad. Scis., Am. Philos. Soc., Phi Beta Kappa, Sigma Xi. Republican. Methodist. Club: Cosmos. Subspecialties: Plant physiology (agriculture); Physiology of plant stress. Current work: Effects of chilling and water deficits on physiological processes and growth of plants. Home: 23 Stoneridge Cir Durham NC 27705 Office: Dept Botany Duke University Durham NC 27706 Home: 23 Stoneridge Cir Durham NC 27705

KRAMER, REX WILLIARD, JR., electric utility company executive; b. Los Angeles, June 22, 1934; s. Rex W. and Ruth R. (Roseberry) K.; m. Karen Blanchard, May 16, 1958; children: Timothy, E. Cecil. A.B., Stanford U., 1956; M.A., Catholic U., 1978. Commd. ensign U.S. Navy, 1956, advanced through grades to comdr., 1978; naval officer, San Diego, 1956-58, submarine officer, New London, Conn., 1958-64, nuclear engring. officer, San Diego, 1964-71, nuclear submarine comdr., Pearl Harbor, Hawaii, 1971-75, ops. analyst, Washington, 1975-78, ret., 1978; licensing supr. Ariz. Pub. Service Co., Wintersburg, Ariz., 1978. Mem. Am. Nuclear Soc., Atomic Indsl. Forum. Republican. Episcopalian. Lodge: Kiwanis. Subspecialties: Nuclear fission; Nuclear engineering. Current work: Licensing supervisor at Palo Verde nuclear generating station, responsible for regulatory aspects of plant operations. Home: 1144 Oro Ulsta PO Box 2357 Litchfield Park AZ 85340 Office: Ariz Pub Service Co Sta 6075 PO Box 21666 Phoenix AZ 85036

KRAMER, STEPHEN LEONARD, physicist; b. Phila., July 22, 1943; s. Berthold Conrad and Clara Elizabeth (Detky) K.; m. Jean Karen Hotchkiss, Sept. 4, 1965; children: Robyn, Keith. B.S. in Physics, Drexel Inst. Tech., Phila., 1966, M.S., Purdue U., 1968, Ph.D., 1971. Grad. research asst. Purdue U., 1968-71; research asst. Argonne Nat. Lab., Ill., 1971-74, asst. physicist, 1974-80, physicist, 1981. Recipient George Tautfest award Purdue U., 1971, Lark Horovitz award, 1971. Mem. Am. Phys. Soc., Am. Assn. Physicists in Medicine, Sigma Xi. Lutheran. Subspecialties: Particle physics; Bioinstrumentation. Current work: Accelerator/particle physics, application of accelerators to medicine.

KRAMER, STEVEN DAVID, physicist; b. Lakewood, N.J., Aug. 27, 1948; s. George and Pearl (Mohel) K. A.B., Cornell U., 1970; A.M., Harvard U., 1971, PH.D., 1976. Research asst. Harvard U., 1972-76; staff scientist Oak Ridge (Tenn.) Nat. Lab., 1976—; cons. Atom Scis., Inc. Contbr. articles to profl. jours. NSF fellow, 1970-74; Dept. Energy grantee. Mem. Am. Phys. Soc., Optical Soc. Am., Archaelol. Inst. Am. (exec. com. E. Tenn.). Subspecialties: Spectroscopy; Atomic and molecular physics. Current work: Nonlinear optics, laser spectroscopy, one atom detection, resonance ionization spectroscopy. Home: 226 Countryside Circle Knoxville TN 37923 Office: PO Box X 5500 Bldg Oak Ridge TN 37830

KRASNER, PAUL R., psychologist; b. N.Y.C., May 10, 1951; s. Ernest and Elissa (Abrams) K.; m. Trudi R. Klarman, June 17, 1973; children: Lori A., Ami A. B.A., N.Y.U., 1973; M.A., Bklyn. Coll., 1975, Adelphi U., 1976, Ph.D., 1979. Lic. psychologist, N.C.; cert. biofeedback. Health Service Providers Psychology. Mem. psychol. staff dept. human resources State of N.C., Butner, 1978-81; dir. psychology Thoms. Rehab. Hosp., Asheville, N.C., 1981—; instr. N.Y. Inst. Tech., 1975-78. Contbr. articles in field to profl. jours. Mem. Am. Psychol. Assn., S.E. Psychol. Assn., N.C. Psychol. Assn., Nat. Acad. Neuropsychology, Biofeedback So. Am. Subspecialties: Neuropsychology; Biofeedback. Current work: Neuropsychology developmental disabilities, biofeedback, rehabilitation and medical psychology. Home: 4 Maybury Pl Arden NC 28704 Office: Thoms Rehab Hosp 1 Rotary Dr Ashville NC 28803

KRASNY, HARVEY CHARLES, research scientist; b. High Point, N.C., July 27, 1945; s. Morris Theodore and Elizabeth (Nurkin) K.; m. Maria Cristina Ramirez, Apr. 9, 1979; 1 dau., Pamela Marie. B.S., Lynchburg Coll., 1967; M.S., U. N.C., 1969, Ph.D., 1976. Sr. research scientist Burroughs Wellcome Co., Research Triangle Park, N.C., 1969—. Contbr. articles on antiviral and cancer chemotherapy to sci. jours. Mem. Am. Soc. Clin. Pharmacology and Therapeutics, Am. Soc. Pharmacology and Exptl. Therapeutics, Soc. Toxicology, N.Y. Acad. Sci., Sigma Xi. Subspecialties: Pharmacokinetics; Biochemistry (medicine). Current work: Drug dispositon, pharmacokinetics and biochemistry of nucleic acid antagonist. Home: Sharon Heights Apts 16-H Chapel Hill NC 27514 Office: Dept Exptl Therapy Burroughs Wellcome Co Research Trianagle Park NC 27709

KRASS, ALVIN, psychologist, test development company executive; b. Bklyn., Sept. 14, 1928; s. Nathan M. and Nora (Feigels) K.; m. Suzanne Myra Freiwirth, Sept. 5, 1954; children: Peter, Adam, Michael. B.A., Bklyn. Coll., 1951; M.A., N.Y. U., 1952, Ph.D., 1965. Lic. psychologist, N.J. Staff psychologist Brisbane Child Treatment Center, Allaire, N.J., 1955-58; chief psychologist Monmouth Med. Center, Long Branch, N.J., 1958-62; pvt. practice psychology, Monmouth County, N.J., 1955—; pres. Key Edn., Inc., Shrewsbury, N.J., 1958—; cons. Monmouth County Parks System, 1977—, N.J. Div. Vocat. Rehab., Red Bank, 1980-82. Author: Mechanisms of the Mind, 1972, also vocat. and learning potential tests. Served with U.S. Army, 1952-54. Recipient Founders' Day award N.Y.U., 1965. Mem. Am. Psychol. Assn., N.J. Psychol. Assn., Am. Acad. Psychotherapists, Monmouth-Ocean County Psychol. Assn. (pres. 1978-79). Jewish. Subspecialties: Behavioral psychology; Learning. Current work: Computer integrated test system development for vocational assessment in special populations (handicapped, socially disadvantaged, special needs). Patentee computer integrated vocat. testing devices. Home: 205 Holland Rd Holmdel NJ 07733 Office: Key Edn Inc 673 Broad St Shrewsbury NJ 07701

KRASSNER, JERRY, research scientist. B.S., SUNY, Stony Brook, 1974; M.A., U. Rochester, 1976, postgrad., 1976. Research asst. dept. physics and astronomy U. Rochester, N.Y., 1975-77; research scientist Grumman Aerospace Corp., Bethpage, N.Y., 1977. Contbr. articles to profl. jours. Mem. Am. Astron. Soc., Astron. Soc. Pacific, Soc. Photo-optical Instrumentation Engrs., N.Y. Astron. Assn. Lodge: KC. Subspecialties: Infrared optical astronomy; Aerospace engineering and technology. Current work: Researcher in infrared technology, remote sensing, infrared and radio astronomy. Office: Grumman Aerospace Corp Plant 26 Bethpage NY 11714

KRATZ, LAWRENCE JOHN, mathematician, educator; b. Detroit, Oct. 7, 1943; s. Lawrence John and Bertha (Fecteau) K.; m. Catherine Alice Foster, Oct. 10, 1970; children: Luke, John, Anne. B.A., Xavier U., Cin., 1963; M.A., U. Wis., 1966; Ph.D., U. Utah, 1975. Instr. Idaho State U., 1966-70; asst. prof. math U. Ky., Lexington, 1975-76; asst. prof. Idaho State U., Pocatello, 1976-81, assoc. prof., 1981—; engr. EG & G, Inc., Idaho Falls, 1978-79; dir. Student Sci. Tng. Program, NSF, Pocatello, 1981. Mem. Soc. Indsl. and Applied Math, Assn. Computing Machinery, Am. Math. Soc. Subspecialties: Numerical analysis; Graphics, image processing, and pattern recognition. Current work: Multi-dimensional quadrature; efficient algorithms in computer graphics. Office: Dept Math Idaho State U Pocatello ID 83201 Home: 1414 Ammon St Pocatello ID 83201

KRATZ, RICHARD L., JR., project engineer; b. Abington, Pa., Aug. 30, 1959; s. Richard L. and Eleanor (Schlegel) K.; m. Michelle M., June 26, 1982. A.S., Spring Garden Coll., B.S., 1982. Draftsman North Wales Water Authority, Pa., 1976-78, William Raudenbush, Inc., Lansdale, Pa., 1979; with Am. Energy Corp., Lansdale, Pa., 1979—, now project engr. Subspecialties: Fuels and sources; Combustion processes. Current work: Design of industrial and institutional solid waste to energy recovery systems. Office: 1100 Sumneytown Park Lansdale Pa 19446

KRATZER, REINHOLD HERMANN, chemist; b. Kaaden, Czechoslovakia, Nov. 14, 1928; s. Reinhold R. and Johanna M. (Maehner) K. Dr.rer.nat. in Inorganic Chemistry, U. Munich, 1960. Research asst. U. So. Calif., 1960-62; research chemist Naval Ordnance Lab., Corona, Calif., 1962-64; sr. scientist MHD Research, Inc., Hercules Powder Co., 1964-66; mgr. chem. research Marquardt Corp., 1966-70; mgr. chemistry dept. Ultrasystems, Inc., Irvine, Calif., 1970—. Contbr. articles profl. jours. Mem. Am. Chem. Soc., Chem. Soc. London, German Chem. Soc., AAAS, N.Y. Acad. Scis. Subspecialties: Polymer chemistry; Fuels. Current work: Mine drainage; coal and oil shale; geothermal brine; perfluorinated ethers; high viscosity index fluids; fluid-seal interactions; flammability and combustion toxicology; boron nitride. Patentee in field. Home: 17 Shooting Star St Irvine CA 92714 Office: 2400 Michelson Dr Irvine CA 92715

KRAUS, JULIAN (JOE) DAVID, instrumentation and control systems engineer; b. Grand Rapids, Mich., Oct. 10, 1927; s. Leopold and Minna (Driesen) K.; m. Ruth Maxey Bierma (div. 1973); 1 dau., Gail Marie; m. Jo Ann Zilpha Innes, Sept. 3, 1976. B.S. in Aero. Engring, Aero. U., Chgo., 1948. Registered profl. control systems engr., Calif. Field engr. Fla. Power Corp., Crystal River, 1972-74, Fischback-Lord, Richland, Wash., 1975; research engr. Gen. Electric, San Jose, Calif., 1974-75; instrumentation engr. Kaiser, Palo Alto, Calif., 1975-76, Westinghouse, Richland, 1977-81, Burns & Roe, 1981. Mem. Am. Nuclear Soc., Instrument Soc. Am. Club: Am. Contract Bridge (div. Richland chpt. 1980-82). Subspecialties: Systems engineering; Robotics. Current work: Design control systems for automated nuclear processing and refueling. Home: Route 2 Box 2320A Benton City WA 99320

KRAUS, MARJORIE PATT, biophysicist, researcher, cons.; b. Granville, Mass., Mar. 29, 1913; d. Hermann G. and Mary A. (Wackerbarth) Patt; m. Philip B. Kraus, June 28, 1940; children: Patricia, John, Robert, Deborah, Betsy, Kathryn. Ph.B., Brown U., 1933; A.M. (Royall Victor fellow), Stanford U., 1936. Resident, tutor Sch. Nursing, Springfield (Mass.) Hosp., 1934; bacteriologist St. Lukes Hosp., N.Y.C., 1937-39; research assoc. in biophysics Columbia U., 1936-39; chem. research asst. to v.p. Nat. Oil Products, Harrison, N.J., 1939-41; instr. in chemistry U. Del., 1965, research assoc. dept. civil engring., 1965-67, sr. research assoc. in radiation chemistry, 1969-76, program dir. on blue-green algal studies, 1971-76; guest research assoc. in biophysics Pa. State U., 1976; research dir. Algal Reasearch Ctr., Landenberg, Pa., 1976—; expert witness on aquatic virus for litigation over EPA permits. Contbr numerous articles, including some on cyanophages of blue-green algae, virus problems in sludge disposal and water renovation, molecular biochem. approach to aquatic toxicology, to profl. jours. Workshop leader Chester County Soil Conservation. Grad. Women in Sci. Gerry fellow, 1975-76. Mem. Am. Chem. Soc., Phycological Soc. Am., Radiation Reseaarch Soc., Am. Soc. Photobiology, Am. Soc. Microbiology, ASTM (Virology Task Force), Phi Beta Kappa, Sigma Xi. Subspecialties: Biophysical chemistry; Water supply and wastewater treatment. Current work: Environmental virology, photosynthesis, nitrogen fixation, genetic engineering water quality and wastewater treatment, resource conservation, environmental toxicology. Home and Office: 317 London Tract Landenberg PA 19350

KRAUS, WILLIAM ARNOLD, psychologist; b. Grand Rapids, Mich., Oct. 13, 1943; s. William Arnold and Alice (Riegels) K.; m. Kathleen Anderson, Aug. 6, 1966; children: Jamie and Julie (twins). B.A., Alma Coll., 1965; M.A., U. Iowa, 1967; Ph.D., Ohio U., 1970. Staff dir. U. Mass, Amherst, 1970-74; asst. prof. U. Hartford, West Hartford, Conn., 1974-80; corp. cons. Gen. Electric, Fairfield, Conn., 1980—. Contbr. articles to profl. jours.; author: Collaboration in Organizations, 1980 (award for outstanding acad. book ALA 1981); editor: Jour. Applied Behavioral Sci, 1978. Fellow Am. Psychol. Assn.; mem. Organizational Devel. Network, Am. Mgmt. Assn., Acad. Mgmt. Subspecialties: Social psychology; Behavioral psychology. Current work: New technology implementation and organizations of the future. Home: 1501 Ridge Rd North Haven CT 06473 Office: 3135 Easton Turnpike Fairfield CT 06431

KRAUSE, CHARLES JOSEPH, otolaryngologist; b. Des Moines, Apr. 21, 1937; s. William H. and Ruby I. (Hitz) K.; m. Barbara Ann Steelman, June 14, 1962; children—Sharon, John, Ann. B.A., State U. Iowa, 1959, M.D., 1962. Diplomate: Am. Bd. Otolaryngology. Intern Phila. Gen. Hosp., 1962-63; resident in surgery U. Iowa, 1965-66, resident in otolaryngology, 1966-69; fellow dept. plastic surgery Marien Hosp., Stuttgart, W. Ger., 1970; asst. prof. otolaryngology U. Iowa, 1969-72, asso. prof., 1972-75, vice chmn. dept. otolaryngology, 1973-77, prof., 1975-77; prof., chmn. dept. otolaryngology U. Mich. Med. Sch., Ann Arbor, 1977—. Author book in field; contbr. chpts. to books, articles to profl. jours. Served to capt. USAF, 1963-65. Fellow Am. Soc. Head and Neck Surgery (Council 1980-83, chmn. research com. 1980-83); Mem. AMA, Am. Acad. Ophthalmology and Otolaryngology, Am. Acad. Facial Plastic and Reconstructive Surgery (regional v.p. 1977-80, chmn. research com. 1977-80, pres. 1981-82), A.C.S. (adv. council otolaryngology 1979-83), Assn. Head and Neck Oncologists Gt. Britain (corr. mem.), Am. Assn. Cosmetic Surgeons, Assn. Research in Otolaryngology, Washtenaw County Med. Soc. (exec. com. 1979-82), Mich. State Med. Soc., Mich. Otolaryngol. Soc., Assn. Acad. Depts. Otolaryngology, Soc. Univ. Otolaryngologists, Walter P. Work Soc., Am. Cancer Soc. (med. adv. com. Washtenaw County unit), Am. Laryngol., Rhinol. and Otol. Soc., Am. Laryngol. Assn., Centurions of Deafness Research Found. Republican. Presbyterian. Subspecialties: Otorhinolaryngology; Cancer research (medicine). Current work: Cancer of the head and neck; rehabilitation of head and neck cancer patients; plastic and reconstructive surgery of the head and neck; tumor immunology. Home: 3100 Hunting Valley Dr Ann Arbor MI 48104 Office: Dept Otolaryngology U Mich Hosp Ann Arbor MI 48109

KRAUSE, ELIOT, geneticist, educator; b. Bronx, N.Y., June 7, 1938; s. Charles and Helen (Melamud) K.; m. Judy, July 5, 1959; children: Steven, Ira, Shari. B.S., Cornell U., 1960; M.S., Purdue U., 1963, Ph.D., 1968. Grad. research asst. Purdue U., 1960-65; instr. Seton Hall U., 1965-68, asst. prof. biology, 1968—; dir. NSF-sponsored short course for secondary tchrs. biology, 1973. Mem. Genetics Soc. Am., AAAS, Am. Genetic Assn., Biometrics Soc., Sigma Xi. Democrat. Jewish. Lodge: KP. Subspecialties: Genetics and genetic engineering (biology); Statistics. Current work: Cytogenetics: use of sister chromatid exchange as a measure of low doses of potential mutagenic agts. Office: Dept Biology Setaon Hall U South Orange NJ 07079

KRAUSE, IRVIN, cons., researcher, mech. engr.; b. N.Y.C., July 18, 1932; s. Saul David and Anna (Lipstein) K.; m. Cecile Ronni, Sept. 6, 1953; children: Sherry Krause-Mazza, Paul m., Mark B., Ricki. B.M.E., CUNY, 1954; M.S. in Mech. Engring, Columbia U., 1955; Eng.Sc.D. (NSF scholar), N.Y. U., 1960. Research scientist Am. Standard Corp. Research Center, Piscataway, N.J., 1960-63; asso. prof. mech. engring., dir. Material Sci. Labs., Fairleigh Dickinson U.,

Teaneck, N.J., 1963-67; mgr. engring. div. indsl. products Singer Co., Somerville, N.J., 1967-70, mgr. automated equipment, 1970-74; dir. engring. Acushnet Co., New Bedford, Mass., 1974-78; mgr. mfg. tech. Arthur D. Little Inc., Cambridge, Mass., 1978; lectr. seminars. Contbr. articles on kinematics, material sci., automation to profl. jours. Mem. ASME, Soc. Mfg. Engrs., Sigma Xi. Subspecialties: Mechanical engineering; Computer-aided design. Current work: Material manufacture and processing; computer automation of design and mfg. activities and facilities; computer integrated mfg. Patentee automated edge guide system for sewing machine. Home: 1 Seaward Ln South Dartmouth MA 02748 Office: 20 Acorn Park Cambridge MA 02140

KRAUSE, RICHARD MICHAEL, immunologist, educator, government official; b. Marietta, Ohio, Jan. 4, 1925; s. Ellis L. and Jennie (Waterman) K. B.A., Marietta Coll., 1947, D.Sc. (hon.), 1978; M.D. Case Western Res. U, 1952; D.Sc. (hon.), U. Rochester, 1979, Med. Coll. Ohio, Toledo, 1981. Research fellow dept. preventive medicine Case Western Res. U., 1950-51; intern Ward Med. Service, Barnes Hosp., St. Louis, 1952-53, asst. resident, 1953-54; asst. physician to hosp. Rockefeller Inst., 1954-57, asst. prof., asso. physician to hosp., 1957-61, asso. prof., asso. physician to hosp., 1961-62; prof. epidemiology Sch. Medicine, Washington U., St. Louis, 1962-66, asso. prof. medicine, 1962-65, prof. medicine, 1965-66; asso. prof., physician to hosp. Rockefeller U., 1966-68, prof., sr. physician, 1968-75, dir., 1974-75, Nat. Inst. Allergy and Infectious Diseases, NIH, HEW, Bethesda, Md., 1975—; USPHS surgeon, 1975-77, asst. surgeon gen., 1977—; Bd. dirs. Mo.-St. Louis Heart Assn., 1962-66, mem. research com., 1963-66; mem. exec. com. council on rheumatic fever and congenital heart disease Am. Heart Assn., 1963-66, chmn. council research study com., 1963-66, mem. assn. research com., 1963-66, mem. policy com., 1966-70; mem. commn. streptococcal and staphylococcal diseases U.S. Armed Forces Epidemiol. Bd., 1963-72, dep. dir., 1968-72; bd. dirs. N.Y. Heart Assn., 1967-73, chmn. adv. council on research, 1969-71, mem. dirs. council, 1973-75; cons., mem. coccal expert com. WHO, 1967—; mem. steering com. Biomed. Sci. Scientific Working Group, WHO, 1978; mem. infectious disease adv. com. Nat. Inst. Allergy and Infectious Disease, NIH, 1970-74; bd. dirs. Royal Soc. Medicine Found., Inc., 1971-77, treas., 1973-75; bd. dirs. Allergy and Asthma Found. Am., 1976-77, Lupus Found. Am., 1977—. Asso. editor: Jour. Immunology, 1963-71; sect. editor: Viral and Microbial Immunology, 1974-75; editor: Jour. Exptl. Medicine, 1973-75; adv. editor, 1976—; mem. editorial bd.: Bacteriological Revs, 1969-73, Infection and Immunity, 1970-78, Immunochemistry, 1973, Clin. Immunology and Immunopathology, 1976, 1978—; Contbr. numerous articles to profl. jours. Served with U.S. Army, 1944-46. Decorated Gumhuria medal, Egypt; recipient Disting. Service medal HEW, 1979; C. William O'Neal Disting. Am. Service award. Mem. U.S. Nat. Acad. Scis., Inst. Medicine, Assn. Am. Physicians, Am. Acad. Allergy, Am. Soc. Biol. Chemists, Am. Soc. Clin. Investigation, Am. Assn. Immunologists, Am. Soc. Microbiology, Harvey Soc., Am. Venereal Diseases Soc., Am. Coll. Allergists, AAAS, Infectious Diseases Soc. Am., Royal Soc. Medicine, Am. Rheumatism Assn., Practitioner's Soc. N.Y., Am. Thoracic Soc., Am. Epidemiol. Soc. Clubs: Century Assn. (N.Y.C.); Cosmos (Washington). Subspecialty: Immunogenetics. Current work: Research on pathogenesis and epidemiology of streptoccal diseases; immunochemical studies on streptoccal antigens; immunogenetics; recognition of rabbit antibodies with molecular uniformity, genetics of immune response. Research on pathogenesis and epidemiology of streptococcal diseases; immunochem. studies on streptococcal antigens; immunogenetics; recognition of rabbit antibodies with molecular uniformity, genetics of immune response. Home: 10 West Dr Bethesda MD 20814 Office: NIH/NIAID Public Health Service Bethesda MD 20205

KRAUSE, SONJA, phys. chemist, educator; b. St. Gall, Switzerland, Aug. 10, 1933; came to U.S., 1939, naturalized, 1947; d. Friedrich and Rita (Maas) K.; m. Walter W. Goodwin, Nov. 27, 1970. B.S. in Chemistry, Rensselaer Poly. Inst., 1954; Ph.D. in Phys. Chemistry, U. Calif., Berkeley, 1957. Organic chemist USPHS Communicable Disease Center, Savannah, Ga., summers 1954-55; sr. phys. chemist Rohm & Haas Co., Phila., 1957-64; vol., mem. univ. faculties Peace Corps, Nigeria, Ethiopia, 1964-66; vis. asst. prof. U. So. Calif., 1966-67; asst. prof. phys. chemistry Rensselaer Poly. Inst., 1967-72, asso. prof, 1972-78, prof, 1978. Author: (with others) Chemistry of the Environment, 1978; contbr. numerous articles to profl. jours.; editor: Molecular Electro-Optics, 1981. NIH career devel. grantee, 1975-80; NSF grantee, 1969; Am. Chem. Soc. Petroleum Research Fund grantee, 1970-72. Fellow Am. Phys. Soc.; mem. Am. Chem. Soc., Biophys. Soc., AAAS, N.Y. Acad. Sci., Sigma Xi. Subspecialties: Polymer chemistry; Biophysical chemistry. Current work: Block copolymers; polymer-polymer mixtures; transient electric virefringence of macromolecular solutions; muscle proteins. Office: Dept Chemistry Rensselaer Poly Inst Troy NY 12181

KRAUSHAAR, PHILIP FREDERICK, JR., geophysicist, particle physicist; b. Kalamazoo, Sept. 23, 1952; s. Philip Frederick and Theresa Florence (Carpentier) K.; m. Sandra Lee DeHollander, June 22, 1974; children: Kerry Jo, Megan Kristine. B.A., Kalamazoo Coll., 1974; M.S., U. Mich., 1977, Ph.D., 1982. Geophysicist Shell Oil Co., Houston, 1982. Mem. Sault Ste. Marie Tribe of Chippewa Indians. Mem. Am. Phys. Soc., Soc. Exploration Geophysicists. Subspecialties: Particle physics; Geophysics. Current work: Hadronic interaction in particle physics; seismic data acquisition and processing. Home: 2619 Heathergold Houston TX 77084 Office: Shell Oil Co PO Box 911 Houston TX 77001

KRAUSHAAR, WILLIAM LESTER, educator, physicist; b. Newark, Apr. 1, 1920; s. Lester A. and Helen (Osterhoudt) K.; m. Margaret Freidinger, Feb. 27, 1943 (div. 1980); children—Mark Jourdan, Susan, Andrew Woolman; m. Elizabeth D. Rodgers, Aug. 9, 1980. B.S., Lafayette Coll., 1942; Ph.D., Cornell U., 1949. Physicist Nat. Bur. Standards, Washington, 1942-45; asso. prof. physics Mass. Inst. Tech., Cambridge, 1956-62, prof., 1962-65; prof. physics U. Wis.-Madison, 1965—. Author: (with Uno Ingard) Introduction to Mechanics, Matter and Waves, 1960. Fellow Am. Phys. Soc., Am. Astron. Soc., Internat. Astron. Union, Am. Acad. Arts and Scis., Nat. Acad. Sci. Subspecialty: High energy astrophysics. Research and publs. on astrophysics; study of cosmic X and gamma radiation

KRAUSS, ALAN ROBERT, physicist; b. Chgo., Oct. 3, 1943; s. Paul and Shirley (Shapiro) K.; m. Julie Emelie Rosado, Aug. 28, 1965; 1 dau., Susan. B.S. in Physics (Nat. Merit scholar), U. Chgo., 1965; M.S., Purdue U., 1968, Ph.D., 1972. Research assoc. U. Chgo., 1972-74; physicist Argonne Nat. Lab., Ill., 1974—. Contbr. chpts. to books, articles to profl. publs. Hon. 2d lt. Ill. N.G. Mem. Am. Phys. Soc., Am. Vacuum Soc. (publicity chmn. fusion tech. div.), Sigma Xi, Sigma Pi Sigma. Club: Downers Grove (Ill.) Camera. Subspecialties: Nuclear fusion; Surface chemistry. Current work: Surface physics, sputtering, secondary ion emission, interaction of energetic particles with solid surfaces, fusion-related materials problems. Inventor. Office: Argonne Nat Lab Bldg 200 Argonne IL 60439

KRAUSS, HERBERT H., psychologist; b. Phila., June 13, 1940; s. Leon and Ethel K.; m. Beatrice Joty Osgood, Aug. 26, 1965; children: Michael Conal, Daniel Avram. B.S., Pa. State U., 1961, M.S., 1962; Ph.D., Northwestern U., 1966. Asst. prof. psychiatry U. Kans. Med. Center, Kansas City, 1966-67, Ohio State U. Sch. Medicine, Columbus, 1967-69; assoc. prof. psychology U. Ga., Athens, 1969-71; prof. psychology Hunter Coll., N.Y.C., 1981; cons. Exec. Health Exam., N.Y.C., 1980; Brownlee, Dolan, Stein, N.Y.C., 1981. Co-author: Living with Anxiety and Depression, 1974; co-editor: Survival or Suicide, 1970; author articles. Mem. Am. Psychol. Assn., N.Y. Acad. Sci., N.Y. Psychol. Assn. (named Outstanding Tchr. 1972), Sigma Xi. Subspecialties: Behavioral psychology; Social psychology. Current work: Stress management; self control; person-societal interactions. Home: 5 Willow St Irvington NY 10533 Office: Hunter Coll 695 Park Ave New York NY 10021

KRAUSZ, JOSEPH PHILIP, plant pathologist; b. Flushing, N.Y., May 31, 1948; s. Joseph M. and Elenor M. (Lonergan) K.; m. Cheryl A. Parker, Aug. 14, 1971; 1 son, Thomas. B.A., SUNY-New Paltz, 1971; M.S., Cornell U., 1973, Ph.D., 1976. Research fellow Internat. Center Tropical Agr., Cali, Colombia, 1974-75; sr. research plant pathologist H.J. Heinz Co., Cleveland, Miss., 1975-76; research plant pathologist FMC Cor., Davis, Calif., 1977-78; asso. prof. plant pathology Clemson U., Florence, S.C., 1978—. Contbr. articles in field to profl. jours. Served with USNR, 1966-68. Mem. Am Phytopath. Soc., Sigma Xi, Phi Kappa Phi. Roman Catholic. Subspecialty: Plant pathology. Current work: Conduct research and ednl. activities to develop, improve and disseminate disease control practices for tobacco, corn and turfgrass. Office: Clemson University Box 5809 Florence SC 29502

KRAUSZ, STEPHEN, financial management executive, physiologist; b. Salford, Lancaster, Eng., Aug. 4, 1950; came to U.S., 1957, naturalized, 1967; s. Ernest and Anna (Slomson) K.; m. Rae Vicki Sigman, May 8, 1948; children: Joseph, Dora. B.Sc. in Biology, Bklyn. Coll., 1971; M.Sc. in Physiology, Hebrew U., Jerusalem, 1973, Ph.D., 1977. Asst. lectr. Hebrew U., 1971-76, instr., 1976-77; research fellow UCLA, 1977-78; asst. prof. physiologt Howard U., 1978-82; dist. mgr. A.L. Williams Co., Atlanta, 1982. Contbr. articles in field of physiology to profl. jours. Am. Heart Assn. fellow, 1977; recipient Master's award Hebrew U., 1972. Mem. AAAS, Am. Physiol. Soc., Sigma Xi. Democrat. Jewish. Subspecialties: Physiology (biology); Comparative physiology. Current work: Not currently active, major interests include: physiological responses to high temperatures in conscious animals-respiratory, blood chemistry; regulation of airflow in upper airways. Home and Office: 2897 S Xanadu Wa Auror Co 80014

KRAUTHAMER, GEORGE MICHAEL, neuroscientist; b. Germany, Sept. 14, 1926; s. Michael and Ellen (Muller) K. Ph.D., N.Y. U., 1959. Postdoctoral fellow, research scientist U. Paris, 1960-67; asst. prof. Coll. Physicians and Surgeons, Columbia U., 1967-69; assoc. prof. dept. anatomy Rutgers Med. Sch., 1969-79, prof., 1979. Contbr. articles to sci. jours. Served with AUS, 1944-47. Mem. Soc. Neurosci., Am. Physiol. Soc., Internat. Soc. Study of Pain, Am. EEG Soc., Internat. Brain Research Orgn. Subspecialties: Neurophysiology; Neuroanatomy. Current work: Organization of mammalian central nervous system. Office: Dept Anatomy Rutgers Med Sch Piscataway NJ 08854

KRAVITZ, DAVID ALBERT, social psychology educator; b. N.Y.C., Jan. 26, 1953; s. Boris and Lucile Loveland (Colvin) K. B.A., Carleton Coll., 1974; A.M., U. Ill., 1978, Ph. D., 1980. Vis. asst. prof. U. Ill., Urbana, 1980-81; asst. prof. social psychology U. Ky., Lexington, 1981—. Contbr. research articles to jours. Subspecialty: Social psychology. Current work: Small group interaction in mixed motive situations, particularly coalition formation and bargaining; cooperative group interaction. Office: Dept Psychology Univ Kentucky 115 Kastle Hall Lexington KY 40506

KRAVITZ, DAVID WILLIAM, electronic engineer, mathematician; b. N.Y.C., Apr. 26, 1956; s. Sam and Frances Adeline (Liebowitz) K. B.S. in Math. with highest honors, Rutgers U., 1977; M.A. in Math. Sci, Johns Hopkins U., 1978; M.S. in Elec. Engring, U. So. Calif., 1980, Ph.D., 1982. Teaching asst. Johns Hopkins U., Balt., 1977-78; research asst. Chesapeake Bay Inst., Balt., summer 1978; teaching and research asst. U. So. Calif., Los Angeles, 1978-82; fellow Woods Hole Oceanographic Instn. (Mass.), summer 1976-77; mem. tech. staff Hughes Aircraft Co., El Segundo, Calif., summers 1979-81; electronic engr. Dept. Def., Linthicum, Md., 1982. Contbr. articles to profl. jours. George H. Cook scholar, 1976-77. Mem. Math. Assn. Am., IEEE, Johns Hopkins Alumni Assn., Rutgers U. Alumni Assn., U. So. Calif. Engring. Alumni Assn., Pi Mu Epsilon. Democrat. Jewish. Clubs: Howard County Striders (Columbia, Md.); Tompkins Karate Assn. Subspecialties: Electronics; Applied mathematics. Current work: Cryptographic systems design. Home: 5656 Stevens Forest Rd Apt 148 Columbia MD 21045 21090

KRAVITZ, JOSEPH HENRY, geologist, administrator; b. Nanticoke, Pa., Aug. 14, 1935; s. Joseph Henry and Julia Gertrude (Zimniski) K.; m. Prudence Ann Bullock, Nov. 27, 1965; children: Joseph Henry, Jonathan James. B.S., Syracuse U., 1957; M.S., George Washington U., 1975, M.Ph., 1977, Ph.D., 1983. Cert. profl. geologist; registered geologist, Calif. Research asst. Yale U., 1961-64; research geologist Lamont-Doherty Geol. Observatory, Palisades, N.Y., 1964-65, oceanographer, head Geol. Lab., 1965-77; head Oceanographic Labs., Naval Oceanographic Office, Washington, 1977-78; sr. geologist Outer Continental Shelf Program, NOAA, Boulder, Colo., 1978-80; sr. scientist Office of Marine Pollution Assessment, Rockville, Md., 1981; pres. Exploration Assocs. Ltd., Bethesda, Md., 1981; research assoc. Inst. Arctic and Alpine Research, U. Colo., 1981. Author, co-author numerous sci. publs. Served with USAF, 1957-60. Fellow Geol. Soc. Am., Arctic Inst. N. Am.; mem. Soc. Economic Paleontologists and Mineralogists, Explorers Club. Republican. Roman Catholic. Subspecialty: Sedimentology. Current work: Sediments and sediment processes in high latitutde glacial marine environments; geotechnical properties of sea floor sediments in the Arctic. Office: Nat Oceanic and Atmospheric Adminstrn 6010 Executive Blvd Rockville MD 20852

KREBS, EDWIN GERHARD, biochemist, pharmacologist, educator; b. Lansing, Iowa, June 6, 1918; s. William Carl and Louisa Helena (Stegeman) K.; m. Virginia Frech, Mar. 10, 1945; children: Sally, Robert, Martha. B.A., U. Ill., 1940; M.D., Washington U., 1943. Asst. prof. biochemistry U. Wash., Seattle, 1948-52, asso. prof., 1952-57, prof., 1957-68, chmn. dept. pharmacology, 1977-83; prof., chmn. dept. biol. chemistry U. Calif., Davis, 1968-77; investigator Howard Hughes Med. Inst., 1977—. Served with USN, 1945-46. Recipient Alumni citation Washington U., 1972, Disting. Lectureship award Internat. Soc. Endocrinology, 1972, ann. award Gairdner Found., 1978; Guggenheim fellow, 1966. Mem. Nat. Acad. Scis., Am. Acad. Arts and Scis., Am. Soc. Biol. Chemists, AAAS, Am. Chem. Soc., Sigma Xi, Phi Beta Kappa, Phi Kappa Phi, Alpha Omega Alpha. Subspecialty: Pharmacology. Home: 1153 21st Ave E Seattle WA 98112 Office: Dept Pharmacology Sch Medicine U Wash Seattle WA 98195

KREBS, HELMUT WALDEMAR GRAF VON THORN, research scientist; b. Memel, Germany, June 2, 1942; s. Samuel Graf von Thorn and Else Gräfin (von Keyserling) K.; m. Carol Washington-Bennett, Dec. 24, 1963 (div.); m. Jean Miles Lauder, Nov. 24, 1974. B.A., Sir George Williams U., 1967; M.A., McGill U., 1968, Ph.D., 1971. Research assoc. dept psychology McGill U., Montreal, 1971; Can. Med. Research Council postdoctoral fellow St. Elizabeths Hosp., NIMH, Washington, 1971-73, Fogarty Internat. vis. fellow, 1973-74; ind. writer-in-residence Mansfield Center, Conn., 1974-78, Chapel Hill, N.C., 1982—; research asst. prof. anatomy U. N.C. Sch. Medicine, Chapel Hill, 1981—; cons. in field. Contbr. articles to profl. jours. Miles fellow, 1974—. Mem. Soc. for Neurosci., N.C. Soc. for Neurosci., Internat. Soc. for Developmental Neurosci., AAAS. Subspecialties: Neurobiology; Regeneration. Current work: Neuroscience; developmental neurobiology; immunocytochemistry; experimental creative writing; germanistic; theory of modern 20th century novel; the concept of relativity theory in creative writing; philology. Home: 106 Marion Way Robin's Wood Chapel Hill NC 27514 Office: 330 Swing Bldg Dept Anatomy U NC Med Sch Chapel Hill NC 27514

KREIDL, TOBIAS JOACHIM, astronomer, educator, consultant; b. Rochester, N.Y., May 6, 1954; s. Norbert Joachim and Melanie (Schreiber) K. Ph.D. in Astronomy, U. Vienna, Austria, 1979. Research assoc. Ruhr U., Bochum, W. Ger., 1979-80; astronomer Lowell Obs., Flagstaff, Ariz., 1980; part-time instr. computer sci. No. Ariz. U. Contbr. articles to publs. Mem. Am. Astron. Soc., Internat. Astron. Union, AAAS. Subspecialties: Optical astronomy; Graphics, image processing, and pattern recognition. Current work: Image processing systems, analysis of two-dimensional astron. images, lunar occulations, some planetary image analysis. Home and Office: PO Box 1269 Flagstaff AZ 86002

KREIER, JULIUS PETER, microbiologist, educator; b. Phila., Nov. 30 1926; s. George John and Anna Amelia (Necker) K.; m. Ruth Leader Casten, July 9, 1955; children: Rachel E., Jesse G. Student, Temple U., 1946-49; V.M.D., U. Pa., 1953; M.Sc., U. Ill., 1959, Ph.D., 1962. Veterinarian U.S. Dept. Agr., Vera Cruz, Mex., 1953-55, Md., 1955-56; instr. Coll. Vet. Medicine, U. Ill., Champaign-Urbana, 1956-63; from asst. prof. to prof. dept. microbiology Ohio State U., Columbus, 1963—; cons. agrl. scis. sect. Rockefeller Found. Author: Parasitic Protozoa, vols. I-IV, 1978, Malaria, vols. I-III, 1980. NIH postdoctoral fellow, 1961-63; Fulbright awardee, Montevideo, Uruguay, 1977. Mem. Am. Assn. Immunologists, AVMA, Am. Soc. Parasitologists, Am. Soc. Tropical Medicine and Hygiene. Current work: Host-parasite interaction; immunology; protozoology. Home: 257 E 18th Ave Columbus OH 43201 Office: Dept Microbiology Ohio State U Columbus OH 43210

KREIFELDT, JOHN GENE, engineering design educator, consultant; b. Manistee, Mich., Oct. 7, 1934; s. Chester Edward and Bernadine (Janicki) K.; m. Winifred Ilse Strock, June 15, 1963. Ph.d., Case Western Res. U., 1969. Asst. prof. Tufts U., Medford, Mass., 1969-75, assoc. prof., 1975-78, prof. engring. design, 1978; cons. govt., industry, hosps., legal firms, 1970; v.p. Applied Ergonomics Corp., Winchester, Mass., 1975-80, ptnr., 1980. Contbr. articles to profl.. publs. Served with USNR, 1954-57. NRC postdoctoral fellow, 1973; grantee NASA, 1973-82, NIH, 1974-77. Mem. Human Factors Soc. (chmn. Consumer Products Tech. Group), Am. Soc. Engring. Edn. Subspecialties: Human factors engineering; Biomedical engineering. Current work: Consumer product design, human-computer interface design, nuclear control room designs, computer control, expert witness for law firms. Patentee (Frank Oppenheimer Award 1972, 75). Office: Dept Engring Design Tufts Univ Medford MA 02155 Home: 16 Prospect St Winchester MA 01890

KREITH, FRANK, researcher and cons.; b. Vienna, Austria, Dec. 15, 1922; came to U.S., 1940, naturalized, 1945; s. Fred and Elsa (Klug) K.; m. (married), Sept. 21, 1951; children: Michael, Marcia, Judith. B.S., U.Calif., Berkeley, 1945; M.S., UCLA, 1949; Sc.D., U. Paris, 1965. Registered profl. engr., Colo. Research engr. Calif. Inst. Tech., Pasadena, 1945-49; Guggenheim fellow Princeton U., 1949-50; asst. prof. U. Calif., Berkeley, 1951-53; asso. prof. Lehigh U., Bethlehem, Pa., 1953-59; prof. U. Colo., Boulder, 1959-77; chief solar thermal research Solar Energy Research Inst., Golden, Colo., 1977-83, sr. research fellow, 1983—. Contbr. numerous articles to sci. jours. Recipient Robinson award Lehigh U., 1948; NATO fellow, 1975; Fulbright grantee, 1964, 65. Fellow ASME (Heat Transfer Meml. award 1972, Worcester Reed Warner medal 1981); mem. AAUP, Am. Soc. Engring. Edn., AIAA, AAAS, Internat. Solar Energy Soc., Instrument Soc. Am. (pres. Lehigh Valley sect. 1955-56), Sigma Xi (nat. lectr. 1980-81), Tau Beta Pi, Pi Tau Sigma. Subspecialties: Solar energy; Ocean thermal energy conversion. Current work: Heat transfer, solar energy. Address: Solar Energy Research Inst 1617 Cole Blvd Golden CO 80401

KREJCI, ROBERT HENRY, aerospace engineer; b. Shenadoah, Iowa, Nov. 15, 1943; s. Henry and Marie (Josephine) K.; m. Carolyn Ruth Meyer, Aug. 19, 1967; children: Christopher Scott, Ryan David. B.S., Iowa State U., 1967, M.Engring., 1971. Assoc. engr. McDonnell-Douglas Corp., St. Louis, 1963-67; flight controls engr. Collins Radio, Cedar Rapids, Iowa, 1968; dept. mgr. advanced tech. programs Wasatch div. Thiokol Corp., Brigham City, Utah, 1978—. Served to capt. USAF, 1968-78. Mem. AIAA. Subspecialty: Aerospace engineering and technology. Current work: Manages advanced technology developments related to solid rocket propulsion. Home: 885 N 300 E Brigham City UT 84302 Office: Thiokol Corp Wasatch Div PO Box 524 Brigham City UT 84302

KRELL, MITCHELL, computer science educator; b. Hattiesburg, Miss., Feb. 23, 1955; s. George and Selma Shirley (Pevsner) K.; m. Lynn Michelle Carter, May 11, 1980; children: Christopher Gabriel, Matthew Reid. B.S., U. So. Miss., 1978, M.S., 1980. Instr. computer sci. dept. U. So. Miss., Hattiesburg, 1982—; graphics systems analyst Computer Scis. Corp., NASA-NATO Space Tech. Labs., Miss., 1980-82. Mem. Assn. Computing Machinery (pres. 1979-80). Jewish. Subspecialties: Graphics, image processing, and pattern recognition; Programming languages. Current work: Computer aided design and manufacturing tools, such as an interactive three dimensional graphics system. Home: 2003 Fuller St Hattiesburg MS 39401 Office: U So Miss Computer Sci Dept 728 N Hill Dr Hattiesburg MS 39406-5106

KRELL, ROBERT DONALD, pharmacologist; b. Toledo, Dec. 2, 1943; s. Robert William and Helen Margaret (Zink) K.; m. Rebecca Marie Larkins, May 14, 1966; children: Melanie Lynn, Matthew Robert. B.S. in Pharmacy, U. Toledo, 1966; Ph.D. in Pharmacology, Ohio State U., 1972. NIH postdoctoral fellow Sch. Hygiene and Public Health, Johns Hopkins U., Balt., 1972-73; asso. sr. investigator Smith Kline & French Labs., 1973-76, sr. investigator dept. pharmacology, 1976-81; sect. mg. pulmonary pharmacology Stuart Pharms. div. ICI Ams. Inc., Wilmington, Del., 1981—. Contbr. articles to sci. lit.; also reviewer various jours. Am. Found. Pharm. Edn. fellow, 1969-70. Mem. Ohio Acad. Sci., Phila. Physiol. Soc. (counsellor 1980-81, sec. 1981-83, v.p. 1983—), AAAS, Soc. Neurosci., Am. Thoracic Soc., Nat. Soc. Med. Research, Fedn. Am. Scientists, Am. Acad. Allergy, Am. Soc. Pharmacology and Exptl. Therapeutics, Pulmonary Research Group, N.Y. Acad. Scis., Sigma Xi, Rho Chi. Republican. Club: Lake Naomi (Pocono Pines, Pa.). Subspecialties: Pharmacology; Allergy. Current work: Pharmacology of allergic diseases. Patentee (with others) imidodisulfamide derivatives, 1981. Home: 818 Nathan Hale Dr West Chester PA 19380 Office: Stuart Pharm Wilmington DE 19897

KREMERS, JACK ALAN, architect, educator; b. Grand Rapids, Mich., Feb. 13, 1940; s. Albert William and Alice (DeHaan) K.; m. Doris Elaine, Sept. 8, 1961; children: Scott, Stephen, Susan. B.Arch., U. Mich., 1964, M.Arch., 1966. Cert. Nat. Council Archtl. Registration Bds. Asso. McMillen/Palmer Architects, Grand Rapids, Mich., 1966-67, Marvin DeWinter Assos. (Architects), Grand Rapids, 1967-69; prin. Jack Alan Kremers Architect and Assos., Munroe Falls, Ohio, 1969—; asst. prof. architecture and environ. design Kent State U., 1969-73, asso. prof., 1973-80, prof., 1980—; cons. energy design and analysis. Mem. Munroe Falls Planning Commn. Mem. ASHRAE, Ohio Solar Energy Assn., Am. Solar Energy Soc. Mem. Christian Ref. Ch. Subspecialties: Solar energy; Environmental engineering. Current work: Energy conservation programming design, architecture; alternative energy planning; design process. Office: Kent State U 304 Taylor Hall Kent OH 44242

KREMKAU, FREDERICK W., medical researcher, educator; b. Mechanicsburg, Pa., Apr. 30, 1940; s. Ward Joseph and Alice Melda (Wineberg) K.; m. Lillian Ruth Beasley, Sept. 2, 1967; 1 son, Jonathan Stephen. B.E.E., Cornell U., 1963; M.S., U. Rochester, 1969, Ph.D., 1972. Registered profl. engr., N.C., Conn. Research instr. medicine Bowman Gray Sch. Medicine, Wake Forest U., Winston-Salem, N.C., 1971-74, research asst. prof., 1974-80, assoc. prof., 1980-81; assoc. prof. diagnostic radiology Yale U. Sch. Medicine, 1981—. Author: Diagnostic Ultrasound: Physical Principles and Exercises, 1980, 84; also articles. Served to lt. U.S. Navy, 1963-67. NDEA fellow, 1968-69. Mem. Am. Coll. Radiology, Radiol. Soc. N.Am.; Fellow Acoustical Soc. Am.; Mem. IEEE (sr.), N.Y. Acad. Scis., AAAS, Am. Inst. Ultrasound in Medicine (dir., Presdl. recognition award 1981), Am. Assn. Physicists in Medicine, Sigma Xi. Subspecialties: Imaging technology; Acoustical engineering. Current work: Ultrasonic molecular absorption mechanisms; acoustic properties of tissues; biological effects of ultrasound; safety of diagnostic ultrasound. Home: 570 Nut Plains Rd Guilford CT 06437 Office: Dept. Diagnostic Radiology Yale U Sch Medicine New Haven CT 06510

KRENITSKY, THOMAS ANTHONY, pharmaceutical researcher; b. Throop, Pa., Sept. 13, 1938; s. Michael and Josephine Gertrude (Rozaieski) K. B.S., U. Scranton, 1959; Ph.D., Cornell U., 1963. Postdoctoral fellow Sloan-Kettering Inst., Walker Lab., Rye, N.Y., 1963-64; research assoc. dept. biochemistry Yale U. Med. Sch., New Haven, 1964-66; sr. research biochemist Wellcome Research Lab., Tuckahoe, N.Y., 1966-68, head enzymology sect, 1968-83, dept. head exptl. therapy, Research Triangle Park, N.C., 1983—; adj. assoc. prof. dept. biochemistry Sch. Medicine U. N.C., Chapel Hill, 1976—. Mem. Am. Soc. Biol. Chemists, Am. Chem. Soc. Subspecialties: Medicinal chemistry; Enzyme technology. Current work: Enzymes as synthetic catalysts, nucleoside analogs, comparative biochemistry. Inventor synthesis of ribosides using bacterial phosphorylase, 1982. Home: 106 Laurel Hill Rd Chapel Hill NC 27514 Office: Wellcome Research Labs 3030 Cornwallis Rd Research Triangle Park NC 27709

KRESS, GERARD CLAYTON, JR., educational psychologist, university administrator; b. Buffalo, July 10, 1934; s. Gerard C. and Eleanor A. (Rupp) K.; m. Suzanne A. Raloff, May 4, 1957 (div. Mar. 1982); children: Timothy, Peter, Jennifer. A.B., U. Rochester, 1956; Ph.D., SUNY-Buffalo, 1962. Research scientist Am. Insts. for Research, Pitts., 1962-68; asst. prof. U. Pitts., 1968-71; dir. ednl research Harvard U. Sch. Dental medicine, Boston, 1971—; assoc. dean for admissions, 1981—; cons. in field. Author: Managing Problem Behavior in the Classroom, 1969; contbr. numerous articles to profl. publs. Served with USAR, 1957-60. Office of Edn. grantee, 1965, 67; Health Resources Adminstrn. grantee, 1974, 75; W. K. Kellogg Found. grantee, 1980. Mem. Am. Psychol. Assn., Am. Ednl. Research Assn., Behavioral Scientist in Dental Research (pres. 1979-80), Omicron Kappa Upsilon (hon.). Subspecialties: Behavioral psychology; Dental health policy and education research. Current work: Selection and education of dental students; dental quality assurance methods; survey research. Home: 65 Joyce Kilmer Rd Boston MA 02132 Office: Harvard U Sch Dental Medicine 188 Longwood Ave Boston MA 02115

KRESS, LANCE WHITAKER, plant pathologist; b. Camp Le Jeune, N.C., Sept. 2, 1945; s. Roy Alfred and Doris (Parker) K.; m. Diane Rae Bickel, Sept. 13, 1969; children: Nicole, Kerri, Nathan. B.S., Pa. State U., 1968, M.S., 1972; Ph.D., Va. Poly. Inst., 1978. Jr. research aide Pa. State U., 1972-73; lab technician Va. Poly. Inst. and State U., 1974-75, research assoc., 1975-80; asst. ecologist Argonne Nat. Lab., Ill., 1980—; cons. on writing air pollution criteria documents. Contbr. articles, abstracts to profl. publs. Served with USN Air Res., 1963-70. Mem. Am. Phytopath. Soc., Air Pollution Control Assn. Subspecialties: Plant pathology; Environmental toxicology. Current work: Investigating effects of air pollutants on yield of soybean, corn, wheat, sorghum, others. Home: Route 1 Regan Rd Mokena IL 60448 Office: Argonne Nat Lab ER/203 9700 S Cass Ave Argonne IL 60439

KRESSEL, HENRY, electronic company executive; b. Vienna, Jan. 24, 1934; U.S., 1946, naturalized, 1955; s. Aaron and Hudi (Zauderer) K.; m. Bertha Horowitz, Sept. 16, 1956; children—Aron, Kim. B.S. magna cum laude, Yeshiva U., 1955; M.S., Harvard U., 1956; M.B.A., U. Pa., 1959, Ph.D. (David Sarnoff fellow), 1965. Engr. Solid State div. RCA, 1959-61, engring. leader, 1961-63, 65-66; mem. tech. staff RCA David Sarnoff Research Center, 1966-70, head semicondr. device research, 1970-78, dir. materials research lab., 1978-79, staff v.p. solid state research, Princeton, N.J., 1979-83; sr. v.p. E.M. Warburg, Pincus & Co., N.Y.C., 1983—; regents' lectr. U. Calif., San Diego, 1978-79; bd. dirs. Yeshiva U. Research Inst., 1979-84; cons. solar energy U.S. ERDA, 1975. Author: Semiconductor Lasers and Heterojunction LED's, 1977; editor: Characterization of Epitaxial Semiconductor Films, 1976, Semiconductor Devices for Optical Communication, 1980; asso. editor: IEEE Jour. Quantum Electronics, 1978-81; Contbr. numerous articles to sci. jours. Served with Fin. Corps U.S. Army, 1959. Recipient David Sarnoff Research award RCA, 1974, Bevel award Yeshiva U., 1980. Fellow IEEE (pres. Quantum Electronics and Applications Soc. 1978-79), Am. Phys. Soc.; mem. AIME, Nat. Acad. Engring. Subspecialties: Superconductors; Microchip technology (engineering). Current work: Management of electronic research. Patentee in field. Home: 529 Riverside Dr Elizabeth NJ 07208 Office: E M Warburg Pincus & Co 277 Park Ave New York NY 10172

KRESSEL, HERBERT Y., radiologist, educator; b. Bklyn., Nov. 20, 1947; s. Isidore and Mildred (Schindelheim) K.; m. Shirley Lancer, Aug. 26, 1969; children: Mark, David. B.A., Brandeis U., 1968; M.D., U. So. Calif., 1972. Clin. instr. U. Calif.-San Francisco, 1976-77; asst. prof. radiology U. Pa., Phila., 1977-80, assoc. prof., 1980—. NIH fellow, 1974-76. Subspecialties: Diagnostic radiology; Nuclear magnetic resonance (biotechnology). Current work: Clinical evaluation of nuclear magnetic imaging, identifying areas of clinical application, evaluating these in the context of existing technology. Gastrointestinal radiology including conventional radiology and CT scan. Office: Hosp of U Pa 3400 Spruce St Philadelphia PA 19104

KRETSINGER, ROBERT H., molecular biology educator, researcher; b. Denver, Mar. 20, 1937. B.A. in Chemistry, U. Colo., 1958; Ph.D. in Biophysics, MIT, 1964. Postdoctoral fellow Lab. Molecular Biology, Cambridge, Eng., 1964-67; mem. faculty U. Va., Charlottesville, 1967—, prof. molecular biology, 1975—, chmn. dept. biology, 1979—; chmn. space biology and medicine com. Space Sci. Bd., 1981-83; dir. multiwire area x-ray diffractometer facility. Contbr. numerous articles to sci. jours. Mem. Am. Crystallographic Assn. Subspecialties: Molecular biology; Cell biology. Current work: Structure, evolution and function of calcium modulated proteins and function of calcium as a second messenger. Home: 406 Key West Dr Charlottesville VA 22903 Office: U Va Charlottesville VA 22901

KREUTZER, RICHARD DAVID, genetics educator; b. Evergreen Park, Ill., June 23, 1936; s. Arthur and Elsie (Hofmann) K.; m. Patricia Jo Cain, Feb. 7, 1970; children: Kimberly Ann, Tamara Erin. B.S., U. Ill., 1960, M.D., 1966, Ph.D., 1966. Instr. U. Ill. Urbana, 1967-69; asst. prof. genetics Youngstown (Ohio) State U., 1969-74, assoc. prof., 1974-80, prof., 1980—; chief vector biology Gorgas Meml. Lab., Panama City, Panama, 1977-79; cons. Tri-State Labs., Austintown, Ohio, 1975—, Walter Reed Army Inst. Research, WHO. Served with U.S. Army, 1959-61. WHO grantee, 1980; U.S. Army grantee, 1983. Mem. Am. Soc. Tropical Medicine and Hygiene, Am. Genetic Assn., Am. Mosquito Control Assn., Entomol. Soc. Am., Genetics Soc. Am., Sigma Xi, Phi Sigma, Chi Gamma Iota. Lutheran. Lodge: Masons. Subspecialties: Evolutionary biology; Parasitology. Current work: Identification and characterization of protozoan parasites and vectors of disease by isozyme electrophoresis and genetic analysis. Office: Dept Biology Youngstown State U 410 Wick St Youngstown OH 44555 Home: 6164 Acatello Pl Poland OH 44514

KREVANS, JULIUS RICHARD, university chancellor, physician; b. N.Y.C., May 1, 1924; s. Sol and Anita (Makovetsky) K.; m. Patricia N. Abrams, May 28, 1950; children: Nita, Julius R., Rachel, Sarah, Nora Kate. B.S. Arts and Scis, N.Y. U., 1943, M.D., 1946. Diplomate: Am. Bd. Internal Med. Intern, then resident Johns Hopkins Med. Sch. Hosp., mem. faculty, until 1970, dean acad. affairs, 1969-70; physician in chief Balt. City Hosp., 1963-69; prof. medicine U. Calif. at San Francisco, 1970—, dean Sch. Medicine, 1971-82, chancellor. Contbr. articles on hematology, internal med. profl. jours. Served with M.C. AUS, 1948-50. Mem. A.C.P., Assn. Am. Physicians. Subspecialty: Educational administration. Office: U Calif San Francisco CA 94143

KRICKA, HANNA HALYNA, pharmaceutical company executive; b. Czestochowa, Poland, Jan. 1, 1939; came to U.S., 1950, naturalized, 1957; d. Leonid and Helena (Sachnovska) Kryckyj. B.S., Drexel Inst. Tech., 1962; M.S. in Info. Scis, Drexel U., 1971, Columbia U., 1981. Lit. scientist Merck Sharp & Dohme Research Labs., 1963-64; sr. lit. scientist Merrell-Nat., 1964-69; registration and info. scientist E.R. Squibb & Co. Inc., 1970; assoc. dir. clin. info., sr. info. scientist Hoechst Pharms. Inc., 1971-73; dir. sci. info. Bristol Myers Internat., 1973—. Mem. Am. Chem. Soc., N.Y. Acad. Scis., Am. Soc. for Info. Sci., Assn. for Computing Machinery, Drug Info. Assn., Am. Mgmt. Assn., Nat. Ukrainian Engrs. Soc. Am. (exec. bd.), Assoc. Info. Mgrs. Republican. Byzantine Catholic. Club: Columbia (N.Y.C.). Subspecialties: Information systems, storage, and retrieval (computer science); Information systems (information science). Current work: Build customized data bases for information storage and retrieval in response to global regulatory and registrational requirements in the pharmaceutical industry. Home: 215 E 80th St Penthouse J New York NY 10021 Office: 345 Park Ave New York NY 10154

KRIEG, NOEL ROGER, microbiologist, educator; b. Waterbury, Conn., Jan. 11, 1934; s. Julius Albert and Helen Mathilda (Svenson) K. B.A., U. Conn., 1955; M.S., U. Md., 1957, Ph.D., 1960. Asst. prof. microbiology Va. Poly. Inst. and State U., Blacksburg, 1960-64, assoc. prof., 1964-72, prof., 1972-82, Alumni disting. prof., 1982—, chmn., 1982-83. Editor: Bergey's Manual of Systematic Bacteriology, vol. I, 1983; editorial bd.: Manual of Methods for Gen. Bacteriology, 1980; contbr. articles to profl. jours. Trustee Bergey's Manual Trust, 1976—; bd. dirs. Am. Type Culture Collection, Rockville, Md., 1981—. Recipient Carski Disting. Teaching award Am. Soc. Microbiology, 1978; W. E. Wine Teaching award Va. Poly. Inst., 1967; Acad. Teaching Excellence award, 1981. Mem. Am. Soc. Mircobiology (chmn. gen. microbiology div. 1981), Soc. Gen. Microbiology, AAAS, AAUP, Sigma Xi. Subspecialty: Microbiology. Current work: Microbial systematics; taxonomy of spirilla, nitrogen-fixing spirilla; microbial physiology; microaerophilic spirilla. Office: Dept Biology Va Poly Inst and State U Blacksburg VA 24061 Home: 209 Reynolds St Apt 4 Blacksburg VA 24060

KRIEGER, INGEBORG, pediatrics educator, researcher; b. Hattenbach, Germany, Aug. 25, 1927; came to U.S., 1954, naturalized, 1957; d. Karl and Margarete (Ullrich) Eisenberg; m. Harvery A. Krieger, May 30, 1952 (div. 1975); children: Jacqueline, Suzanne. B.S., Philosoph-Theol. Hochschule, Bamberg, W.Ger., 1947; M.D., U. Zurich, Switzerland, 1953. Diplomate: Am. Bd. Pediatrics. Intern Glens Falls (N.Y.) Hosp., 1954-55; resident Harper Hosp., Detroit, 1955-56, Detroit Receiving Hosp., 1956-57; research assoc., instr. Wayne State U., Detroit, 1961-64, asst. prof., 1964-70, assoc. prof., 1970-76, prof., 1976—; dir. nutrition and metabolism div. Children's Hosp. of Mich., 1977—, Clinic for Genetic, Metabolic and Devel. Disorders Wayne State U., 1980—; cons. Child Research Centers, HEW. Author: Pediatric Disorders of Feeding, Nutrition and Metabolism, 1982; contbr. articles to profl. jours. Mem. and chmn. Mich. Genetics Program, 1974-82. NIH grantee, 1964-70. Mem. Am. Pediatric Soc., Soc. for Pediatric Research, Am. Inst. Nutrition, Am. Soc. for Clin. Nutrition, Lawson Wilkins Endocrine Soc., Am. Soc. Human Genetics, Midwest Soc. for Pediatric Research. Club: Gt. Lakes yacht (St. Clair Shore, Mich.). Subspecialties: Nutrition (medicine); Genetics and genetic engineering (medicine). Current work: Genetic Disorders of amino acid metabolism, nutrition research (zinc metabolism), growth disorders. Home: 1950 Tuckaway Bloomfield Hills MI 48013 Office: Childrens Hosp of Mich 3901 Beaubien St Detroit MI 48201

KRIEGER, JOHN NEWTON, urologist, educator; b. Phila., May 3, 1948; m. Monica Schoelch, July 20, 1972. A.B., Princeton U., 1970; M.D., Cornell U., 1974. Fellow Cornell U., N.Y.C., 1974-79, instr., 1979-80; instr. surgery (urology) U. Va., Charlottesville, 1980-82; asst. prof. urology U. Wash., Seattle, 1982—. Mem. Am. Urol. Assn. (1st prize N.Y. sect. essay contest 1976, scholar 1980-82). Subspecialties: Urology; Microbiology (medicine). Current work: Genitourinary tract infections, post-operation wound infections, genitourinary oncology. Office: Dept Utology RL-10 U Wash Seattle WA 98195

KRIEGSMAN, WILLIAM EDWIN, engineer, consultant; b. N.Y.C., Feb. 22, 1932; s. Edwin and Edna (Eising) K.; m. Kay Harris, May 12, 1966; children: William Edwin Jr., Katharine E. A.B. in Chemistry, U. Rochester, 1953; M. Engring. Adminstrn., George Washington U., 1964. Staff asst. The White House, Washington, 1969-71; mgr. Arthur D. Little, Inc., Washington, 1971-73, v.p., 1975-80; commr. U.S. AEC, Washington, 1973-75; v.p. Booz, Allen & Hamilton, Bethesda, Md., 1980-82; pres. Mesa Cons. Group Inc., Arlington, Va., 1982—, also dir.; mem. sci. adv. bd. Nat. Security Agy., Fort Meade, Md., 1979—. Mem. nat. steering com. Bush for Pres., Washington, 1980. Served to lt. USN, 1953-57. Am. Polit. Sci. Assn. grantee, 1966-67. Mem. Am. Nuclear Soc., AAAS, Am. Def. Preparedness Assn. Republican. Jewish. Club: Kenwood (Bethesda). Subspecialties: Inorganic chemistry; Cryptography and data security. Current work: Management and technical services to the defense and intelligence communities. Home: 5615 Ridgefield Rd Bethesda MD 20816 Office: Mesa Cons Group Inc PO Box 12086 Arlington VA 22209

KRIEGSMANN, GREGORY ANTHONY, mathematician, educator, consultant; b. Chgo., Sept. 20, 1946; s. John M. and Jean (Gage) K.; m. Barbara Lynch, Feb. 22, 1969; children: Karl, James. B.S., Marquette U., 1969; M.S., UCLA, 1970, 1972, Ph.D., 1974. Instr. math NYU, 1974-76; mem. tech. staff Hughes Aircraft Co., Canoga Park, Calif., 1976-77; asst. prof. math U. Nebr., Lincoln, 1977-80; assoc. prof. engring. sci. and applied math Northwestern U., Evanston, Ill., 1980—; cons. Real Time Engring. Co., Oak Brook, Ill., 1980—, NUSC, New London, Conn., 1982—, Exxon Engring. Co., Linden, N.J., 1982—. Mem. Soc. Indsl. and Applied Math. Subspecialties: Applied mathematics; Numerical analysis. Current work: Applied mathematics; asymptolic methods; differential equations; bifurcation theory; wave propagation; acoustics; electromagnetics, elasticity, numerical analysis, scientific computing. Office: Dept Engring Sci and Applied Math Northwestern U Evanston IL 60201

KRIGMAN, MARTIN ROSS, pathology educator, researcher; b. N.Y.C., Sept. 4, 1933; s. Harry and Anne G. (Geichman) K.; m. Ruth Johanna Wolff, June 8, 1958; children: Judith Deborah, Hannah Rachael, Sarah Beth. B.A., Columbia Coll., 1954; M.D., Cornell Med. Sch., 1958. Diplomate: Am. Bd. Pathology. Asst. prof. dept. pathology Yale U., New Haven, 1954-66; prof. pathology Sch. Medicine U. N.C., Chapel Hill, 1966—; mem. study sect. Health Effects EPA, Washington, 1981—, Nat. Inst. Environ. Health Scis., NIH, Research Triangle Park, N.C., 1980—. Program project prin. investigator Nat. Inst. Environ. Health Scis., NIH, 1978—. Served to capt. M.C. U.S. Army, 1964-66. Mem. Am. Assn. Pathologists, Am. Assn. Cell Biologists, Internat. Brain Orgn., Soc. Neurosci., Am. Assn. Neuropathologists (com. chmn.). Democrat. Jewish. Subspecialties: Pathology (medicine); Neuropathology. Current work: Neurotoxicology of heavy metals, developmental neurobiology, morphometry of nervous system, and neuropathology. Home: 31 Mount Bolus Rd Chapel Hill NC 27514 Office: Dept Pathology Med Sch U NC Bldg 228H Chapel Hill ND 27514

KRIGSVOLD, DALE THOMAS, plant pathologist; b. Stony Brook Twp., Minn., June 21, 1937; s. Alvin Ingvold and Florence Carol (Johnson) K.; m. Marsha Marie Ward, Jan. 16, 1982. B.S. magna cum laude, Old Dominion U., 1973; Ph.D. in Plant Pathology, Va. Poly. Inst. and State U., 1979. Assoc. plant pathologist, div. tropical research United Fruit Co., La Lima, Honduras, 1979-82. Contbr. articles to profl. jours. Served to cpl. USMC, 1958-62; to sgt. U.S. Army, 1965-70. Decorated Bronze Star. Mem. Am. Phytopath. Soc., Colegio de Profesionales en Ciencias Agricolas de Honduras, Assn. for Cooperation in Banana Research in the Caribbean and Tropical Am. Republican. Lutheran. Subspecialties: Plant pathology; Microbiology. Current work: Control of post-harvest diseases of perishable food crops; root ecology; ecology of soilborne plant pathogens.

KRIKORIAN, ABRAHAM DER, plant physiologist-biochemist, educator, researcher; b. Worcester, Mass., May 5, 1937; s. Abraham Der and Tarquohie Tashjian K. B.S., Mass. Coll. Pharmacy, 1959; Ph.D., Cornell U., 1965. Assoc. prof. biology SUNY-Stony Brook, 1971-81, assoc. prof. biochemistry, 1981—. Contbr. numerous articles to profl. jours.; Western Hemisphere editor: Annals of Botany, 1976-82; mem. editorial bd.: Jour. Ethnopharmacology, 1979—; plant sci. book rev. cons.: Quar. Rev. Biology, 1979—. Recipient Cosmos Achievement award NASA, 1975, 81. Mem. Beneficial Plant Assn. (adv. bd. 1979—), Soc. Econ. Bontany (mem. council 1975-80, v.p. 1981-82, pres. 1982-83), AAAS, Am. Soc. Pharmacognosy, Bot. Soc. Am., Am. Soc. Plant Physiologists, Scandanavian Soc. Plant Physiology, Internat. Soc. Plant Morphologists, Soc. Developmental Biology, Internat. Assn. Plant Tissue Culture. Subspecialties: Plant cell and tissue culture; Plant physiology (agriculture). Current work: Clonal stability; totipotency of higher plant cells in terms of morphogenesis and biochemical competence. Office: Dept Biochemistry SUNY Stony Brook NY 11794

KRIKORIAN, OSCAR HAROLD, chemist; b. Fresno, Calif., Nov. 22, 1930; s. Hagop Bedros and Aghavnie (Mardirosian) K.; m. Marilyn Ann Kooyumjian, July 18, 1953; children: Deborah, Cheryl Krikorian Scolari. B.S. in Chemistry, Calif. State U.-Fresno, 1952, Ph.D., U.Calif.-Berkeley, 1955. Chemist Lawrence Livermore (Calif.) Nat. Lab., 1955—. Contbr. over 60 sci. and tech. articles to profl. publs. Fellow Am. Inst. Chemists; mem. Am. Chem. Soc. Subspecialties: High temperature chemistry; Thermodynamics. Current work: Development of processes and technology for applications to magnetic fusion and hydrogen production. Patentee in field. Office: Lawrence Livermore Nat Lab PO Box 808 L-369 Livermore CA 94550

KRIMIGIS, STAMATIOS MIKE, physicist, researcher, consultant; b. Chios, Greece, Sept. 10, 1938; s. Michael and Angeliki (Tsetseris) K.; m. Evangelia Kantas, Feb. 11, 1968; children: Michael, John. B.S., U. Minn., 1961; M.S., U. Iowa, 1963, Ph.D., 1965. Research assoc. and asst. prof. physics U. Iowa, Iowa City, 1965-68; supr. space physics sect. Applied Physics Lab., Johns Hopkins U., Balt., 1968-74, supr. space physics and instrumentation group, 1974-81, chief scientist space dept., 1980—; mem. Space Sci. Bd., Nat. Acad. Scis. NRC, 1983—; cons.; Mem. steering com. space sci. working group Assn. Am. Univs., 1982—. Contbr. over 130 articles to sci. jours.; author books on solar, interplanetary and magnetospheric plasma physics, cosmic rays, magnetospheres of Jupiter and Saturn. Recipient Exceptional Sci. Achievement medal NASA, 1981. Fellow Am. Geophys. Union; mem. Am. Phys. Soc., AAAS. Greek Orthodox. Subspecialties: Space plasma physics; Planetary science. Current work: Physics of Earth and planetary magnetospheres, the interplanetary medium, and the sun. Home: 613 Cobblestone Ct Silver Spring MD 20904 Office: Applied Physics Lab Johns Hopkins U Laurel MD 20707

KRIMM, SAMUEL, physicist, educator; b. Morristown, N.J., Oct. 19, 1925; s. Irving and Ethel (Stein) K.; m. Marilyn Marcy Neveloff, June 26, 1949; children: David Robert, Daniel Joseph. B.S., Poly. Inst. Bklyn., 1947; M.A., Princeton U., 1949, Ph.D., 1950. Postdoctoral fellow U. Mich., Ann Arbor, 1950-52, mem. faculty, 1952—, prof. physics, 1963—, chmn. biophysics research div., 1976—, asso. dean research, 1972-75; cons. to industry. Author papers on vibrational spectroscopy, x-ray diffraction studies of natural and synthetic polymers. Served with USNR, 1944-46. Recipient Humboldt award, 1983; Textile Research Inst. fellow, 1947-50; NSF sr. postdoctoral fellow, 1962-63; sr. fellow U. Mich. Soc. Fellows, 1971-76. Fellow Am. Phys. Soc. (High Polymer Physics prize 1977, chmn. div. biol. physics 1979, div. councilor 1981, exec. com. 1983); mem. AAAS, Am. Chem. Soc., Am. Crystallographic Assn., Biophys. Soc., N.Y. Acad. Scis. Subspecialty: Biophysics (physics). Address: Dept Physics Univ Mich Ann Arbor MI 48109

KRINSKY, JEFFREY ALAN, systems engineer, educator; b. N.Y.C., Sept. 9, 1955; s. Dan and Phyllis (Chinsky) K. B.S. with honors, Fla. Inst. Tech., 1976, M.S. in Physics, 1979. Electronics design engr. Telephonics, Huntington, N.Y., 1977; project engr. Martin Marietta, Kennedy Space Center, Fla., 1979-80; sr. systems engineer Harris Corp., Melbourne, Fla., 1980—; adj. prof. math., computer sci., physics and mech. engring. Fla. Inst. Tech., Melbourne, 1979—. Mem. AIAA, Am. Optical Soc., Am. Inst. Physics, Laser Inst. Am., Sigma Pi

Sigma. Subspecialties: Integrated circuits; Laser efficiency. Current work: Using computers to control experiments, collect data, and perform sophisticated data reduction in experiments that would be impractical to do by hand. Office: Harris Corp PO Box 37 Melbourne FL 32901

KRIPPNER, STANLEY CURTIS, psychologist, educator; b. Edgerton, Wis., Oct. 4, 1932; s. Carroll Porter and Ruth Genevieve (Volenberg) K.; m. Lelie Anne Harris, June 25, 1966; children: Caron, Robert. B.S., U. Wis., 1954; M.A., Northwestern U., 1957, Ph.D., 1961; Ph.D. (hon.), U. Humanistic Studies, 1982. Dir. Child Study Ctr., Kent (Ohio) State U., 1961-64; dir. Dream Lab., Maimonides Med. Ctr., Bklyn., 1964-73; faculty Saybrook Inst., San Francisco, 1973—, dean of faculty, 1980—; vis. prof. U. P.R., 1972, Sonoma State U., 1972-73, U. Life Scis., Bogotá, Colombia, 1974. Author: Song of the Siren, 1975, Human Possibilities, 1980, (with M. Ullman) Dream Telepathy, 1973, (with A. Villoldo) The Realms of Healing, 1976; editor: Advances in Parapsychol. Research, 1977, 78, 82. Mem. adv. bd. A.R.E. Clinic, 1970, Central Premonitions Registry, 1968, Found. for Mind Research, 1967; bd. dirs. Acad. Religion and Psychical Research, Washington, 1972, Nat. Found. for Gifted and Creative Children, Warwick, R.I., 1970. Recipient award for service to youth YMCA, 1959; citation of merit Nat.Assn. Gifted Children, 1972; cert. of recognition U.S. Office Edn., 1976. Fellow Am. Soc. Clin. Hypnosis; mem. Am. Soc. Psychical Research, Am. Acad. Social and Polit. Scis, Nat. Assn. Gifted Children (dir. 1965—), Nat. Assn. Creative Children and Adults (dir. 1975—), AAAS, Am. Ednl. Research Assn., Am. Personnel and Guidance Assn., Assn. for Transpersonal Psychology, Assn. for Psychophysiol. Study of Sleep, Biofeedback Soc. Am., Council for Exceptional Children, Internat. Soc. Gen. Semantics, Menninger Found., Nat. Soc. for Study of Edn., Soc. for Study of Religion, Soc. for Clin. and Exptl. Hypnosis, Soc. for Study of Sex, World Future Soc., InterAm. Psychol. Assn., Parapsychol. Assn. (pres. 1982-83), Assn. for Humanistic Psychology (pres. 1974-75), Am. Psychol. Assn. (pres. div. 32 1980-81). Presbyterian. Current work: Study of parapsychological phenomena in relation to altered states of consciousness; study of possible parapsychological aspects of unorthodox healing. Home: 79 Woodland Rd Fairfax CA 94930 Office: 1772 Vallejo St San Francisco CA 94123

KRISCIUNAS, KEVIN, astronomer, computer programmer, educator; b. Chgo., Sept. 12, 1953; s. Alfonse Bruno Krisciunas and Ruth (Rudys) Abbate; m. E. Carmen Torres, Apr. 8, 1978. B.S. in Astronomy and Physics with deptl. distinction, U. Ill., 1974; postgrad. in Astronomy and Astrophysics, U. Calif., Santa Cruz, 1974-75; M.A. L.S. and Statistics, U. Chgo., 1976. Software writer, flier on board Kuiper Airborne Obs., Informatics, Inc., Palo Alto, Calif., 1977-82; instr. astonomy West Valley Coll., Saratoga, Calif., 1978-82; computer programmer, tech. support mem. U.K. Infrared Telescope Unit Royal Obs., Edinburgh, Hilo, Hawaii, 1982—. Contbr. articles in astronomy jours.; translator, Pulkovo Obs., 1978, The History of Astronomy from Herschel to Hertzsprung, 1982. Mem. Am. Astron. Soc. (hist. astronomy div.). Subspecialties: Infrared optical astronomy; History of astronomy. Current work: Real-time data acquisition and analysis for astronomy research, interfacing computers with telescopes to maximize efficiency of telescope usage; history 19th century astronomy especially Russian. Home: 15087 Herring Ave San Jose CA 95124 Office: UKIRT 900 Leilani St Hilo HI 96720

KRISHAN, AWTAR, cancer researcher; b. Srinagar, Kashmir, India, Oct. 11, 1937; s. Prem Nath Ganju and Arundati Sahib; m. Sarla Ganju, Aug. 22, 1965; children: Aruna, Ameeta, Neil. Ph.D., Punjab U., 1963, U. Western Ont., 1965. Chief cell kinetics and lab. Sidney Farber Cancer Center, Boston, 1966-77; prof. oncology and chief div. cytokinetics Comprehensive Cancer Center, U. Miami, Fla., 1977-81, prof., sci. dir., 1981—. Contbr. chpts to books, articles to profl. jours. Recipient Collip research medal U. Western Ont., 1965; Nat. Cancer Inst. grantee, 1973-82. Mem. Cell Kinetics Soc. Am. (governing council), Am. Assn. Cancer Research. Democrat. Hindu. Subspecialties: Cancer research (medicine); Chemotherapy. Current work: growth of cancer cells and effect of chemotherapy on normal and tumor cells. Office: PO Box 016960 R-71 Cancer Center Miami FL 33101 Home: 14300 SW 73d Ct Miami FL 33153

KRISHNA, GOLLAPUDI GOPAL, nephrologist, internist, educator; b. Anakapally, Andhra, India, Aug. 1, 1952; came to U.S., 1975; s. Gollapudi Gopala and Lalithamba (Gandikota) K. M.B.B.S., Andhra Med. Coll., Visakhapattanam, 1974. Rotating intern King George Hosp., Visakhapatanam, 1973-74; flexible intern Norwegian Am. Hosp., Chgo., 1975-76; resident in internal medicine Bergen Pines County Hosp., Paramus, N.J., 1976-78, St. Joseph Hosp., Paterson, N.J., 1978-79; fellow in nephrology Emory U., 1979-80, UCLA, 1980-81, asst. prof. medicine, 1981-83; attending physician UCLA Center for Health Scis., 1981-83; asst. prof. medicine Temple U. Sch. Medicine, Phila., 1983—; staff physician Wadsworth VA Med. Center, Los Angeles, 1982-83; attending physician Temple U. Hosp., Phila., 1983—. Contbr. articles, chpt., abstracts to profl. publs. Nat. Kidney Found. So. Calif. fellow, 1982. Mem. Am. Soc. Nephrology, Internat. Soc. Nephrology, Am. Fedn. Clin. Research, Nat. Kidney Found. So. Calif. Subspecialties: Nephrology; Internal medicine. Current work: Mechanism of extracellular fluid volume regulation in normal and disease states; mechanisms of potassium adaptions in uremia; role of dopamine in regulating sodium excretion; pathogenesis of hypertension in uremia; role of volume in generation and maintenance of metabolic alkalosis. Office: Div Nephrology Temple U Health Sci Center 3401 N. Broad St Philadelphia PA 19140

KRISHNA, NEPALLI RAMA, biophysicist, educator; b. Masulipatam, India, Nov. 20, 1945; U.S., 1971; s. Nepalli Gopala Krishna and Nepalli Jayaprada Potharaju Murthy; m. Leela Krishna Davuluri, Mar. 9, 1974; 1 son, Praveen Srinivas Nepalli. B.Sc. with honors, Andhra U., India, 1965, M.Sc., 1966; Ph.D., Indian Inst. Tech., 1972. Postdoctoral fellow Ga. Inst. Tech., Atlanta, 1972-74; research assoc. U. Alta. (Can.), Edmonton, 1974-76; assoc. scientist Cancer Center, U. Ala., Birmingham, 1976—, asst. prof. physics, 1977—, asst. prof. biochemistry, 1979—; cons. Ortho Pharm. Corp., Raritan, N.J., 1981—. Contbr. articles to profl. jours. Recipient Rao Meml. prize Andhra U., 1966, Metcalfe Gold metal, 1966; Scholar award Leukemia Soc. Am., 1982. Mem. Am. Phys. Soc., Biophys. Soc., Smithsonian Instn. Subspecialties: Biophysics (biology); Nuclear magnetic resonance (chemistry). Current work: Structure-function studies of peptide hormones, proteins and nucleic acids; immunoregulatory peptides; neurotoxins; biomolecular conformational studies by nuclear magnetic resonance spectroscopy; development of NMR techniques; protein dynamics. Office: Dept Biochemistry U Ala in Birmingham Univ Sta CHSB-Room B31 Birmingham AL 35294

KRISHNAMURTHY, LAKSHMINARAYANAN, engineering science researcher; b. Kumbakonam, Madras, India, Oct. 23, 1941; came to U.S., 1967, naturalized, 1979; s. Arupati Krishnaswami and Kalyani (Rajagopalan) Lakshminarayanan; m. Visalam, Sept. 12, 1976. B.Engring., U. Madras, 1962; M.Engring., Indian Inst. Sci., Bangalore, India, 1964; Ph.D., U. Calif.-San Diego, La Jolla, 1972. Postgrad. research engr. U. Calif.-San Diego, 1972-73; staff scientist Duvvuri Research Assocs., Chula Vista, Calif., 1973-74; mem. research staff Princeton U., 1975-76; research assoc. Purdue U., West Lafayette, Ind., 1977-78; research engr. U. Dayton Research Inst., 1978-82, sr. research engr., 1982—. Contbr. articles revs. to profl. jours.; reviewer: Applied Mechanics Revs, 1973—. Govt. India post-matriculation merit scholar, 1956-62; Indian Inst. Sci. postgrad. fellow, 1962-64; Council Sci. and Indsl. Research sr. research fellow, 1964-65; Earle C. Anthony Fellowship grantee, 1967-68. Mem. AIAA; mem. ASME; Mem. Combustion Inst., Planetary Soc., Soc. Indsl. and Applied Math., Sigma Xi. Subspecialties: Combustion processes; Fluid mechanics. Current work: Theoretical analysis and numerical computation of nonreacting and reacting turbulent flowfields in combustors. Office: U Dayton Research Inst KL461 Dayton OH 45469

KRISHNAMURTHY, MUTHUSAMY, civil engineer, civil engring educator; b. Srivilliputtur, Tamilnadu, India, May 4, 1945; came to U.S., 1971; s. Muthusamy and Muthupackiam K.; m. Jayasree K., Feb. 20, 1978; 1 son, Arun. B.E., U. Madras, India, 1967; M.Sc., Indian Inst. Sci., Bangalore, 1971; M.E., U. Fla., 1972, Ph.D., 1975. Registered profl. engr., Md., Fla. Sr. engr. PRC Toups, Rockville, Md., 1976-78; Greenhorne & O'Mara, Greenbelt, Md., 1978-79; head water resources mgmt. dept. Reynolds Smith & Hills, Orlando, Fla., 1979—. Mem. ASCE, Marine Tech. Soc., Internat. Assn. Hydraulic Research. Subspecialties: Civil engineering; Surface water hydrology. Current work: Water resources management, sediment related work; beach erosion.

KRISHNAMURTHY, RAVINDRAN, computer scientist, researcher; b. Madurai, Tamil Nadu, India, Nov. 28, 1951; came to U.S., 1975; s. V. and Kamala K.; m. Usha Narayanswamy, Nov. 1, 1981. B.Engring. with honors, Birla Inst. Tech. and Sci., 1974; M.S., U. S.C., 1977; Ph.D., U. Tex.-Austin, 1982. Data base engr. USDATA, Austin, Tex., 1979-81; mem. research staff IBM, Yorktown Heights, N.Y., 1981—. Mem. Assn. Computing Machinery, IEEE Computer Soc. Subspecialties: Database systems; Distributed systems and networks. Current work: Concurrency control, database machine, query processing, office automation. Home: PO Box 194 Millwood NY 10546 Office: T J Watson Research IBM PO Box 218 Yorktown Heights NY 10598

KRISHNAMURTI, TIRUVALAN N., meteorology educator; b. Madras, India, Jan. 10, 1932; s. Tiruvalam J. and Meena Natarajan; m. Ruby, Oct. 23, 1934. B.Sc. with honors, Delhi U., 1951; M.S., Andhra U., 1953; Ph.D., U. Chgo., 1959. Asst. prof. UCLA, 1960-66; mem. faculty Fla. State U., Tallahassee, 1967—, now prof. meteorology; cons. Douglas Aircraft Co., Aeroject Gen., U.S. Dept. Commerce. Contbr. meteorol. articles to profl. publs. NSF grantee, 1981. Fellow Royal Meteorol. Soc.; mem. Am. Meteorol. Soc. (First Half Century award 1972), Am. Geophys. Union, Swedish Geophys. Union, Japan Meteorol. Soc. Subspecialty: Meteorology. Current work: Tropical meteorology. Home: 3014 S Shore St Tallahassee FL 32312 Office: Fla State Univ 423 Bldg Tallahassee FL 32306

KRISHNAN, LEELA, physician, physicist; b. Bangalore, India, 1943; came to U.S., 1967, naturalized, 1975; d. M. and Saroja Kothandaram; m. Engikolai C. Krishnan, 1965. B.S. in Physics, U. Mysore, India, 1964; M.S. in Physics, U. Mo., 1970; postgrad., U. Tex., Houston, 1975-76, M.D., 1979. Research asst. M.D. Anderson Hosp. and Tumor Inst., U. Tex., Houston, 1975-76; research fellow Mid-Am. Cancer Center Program, U. Kans. Med. Center, Kansas City, summers 1975-76; research assoc. U. Tex., Galveston, 1976-79; resident in radiation oncology U. Kansas Med. Center, 1979-83. Contbr. articles to profl. jours. Recipient Nat. Merit Award, India, 1963-64; Am. Cancer Soc. fellow, 1982-83. Mem. Am. Assn. Physicists in Medicine, Am. Soc. Therapeutic Radiology. Subspecialty: Radiation oncology. Current work: Oncology, radiation oncology, radiation biology and radiation physics.

KRISHNAN, PALANIAPPA, research agricultural engineer; b. Kanadukathan, India, Apr. 25, 1953; came to U.S., 1974; s. Lakshmanan and Umayal (Thenappan) K.; m. Chitra Palaniappan, June 18, 1980; 1 son, Prashanth. B.Tech. with honors, Indian Inst. Tech., 1974; M.S., U. Hawaii, 1976; Ph.D., U. Ill., 1979. Research assoc. dept. agrl. engring. U. Ill.-Urbana, 1979-80; research agrl. engr., dept. agrl. engring. Oreg. State U., Corvallis, 1980—. Hunter fellow, 1977-78. Mem. Am. Soc. Agrl. Engrs., Soc. Agronomy, Nat. Soc. Profl. Engrs., Sigma Xi, Gamma Sigma Delta. Club: Oreg. State U. Table Tennis (Corvallis). Subspecialty: Agricultural engineering. Current work: Seed conditioning research. Magnetic fluid aided separation of contaminants from crop seeds and electrostatic separation of seed mixtures. Home: 3930 NW Witham Hill Dr 23C Corvallis OR 97330 Office: Agrl Engring Dept Oreg State U Corvallis OR 97331

KRISHNAN, PALLASSANA NARAYANIER, chemistry, educator; b. Edapal, India, Nov. 1, 1941; came to U.S., 1962; s. Pallassana Subramainam Narayanan and C.P. Ananda Lakshmi; m. Padma Srinivas, Sept. 2, 1971; 1 child, Rajesh. B.S., Victoria Coll., India, 1962; Ph.D., Temple U., 1968. Research assoc. U. Ariz., Tucson, 1968-69, So. Ill. U., Carbondale, 1969-70; prof. Coppin State Coll., Balt., 1970—; cons. Nat. Bur. Standards, 1978-79, Naval Research Lab., 1983. Vice pres. Kerala Cultural Assn., Washington Balt. and Md., 1982. NSF research grantee, 1976-78. Mem. Am. Chem. Soc. Subspecialties: Physical chemistry; Spectroscopy. Current work: Raman spectroscopy; optical fibers; lasers; high temperature spectroscopy. Home: 1 Dakin Ct Baltimore MD 21234 Office: Coppin State College 2500 W North Ave Baltimore MD 21216

KRISNAMOORTHY, MUKKAI SUBRAMANIAM, computer science educator; b. Nagerkoil, Madras, India, Jan. 20, 1948; came to U.S., 1979; s. Mukkai Subramaniam and Parvathi Rajam. M.S., Indian Inst. Tech., Kanpur, 1971, Ph.D., 1976. Asst. prof. Indian Inst. Tech., Kanpur, 1977-80; asst. prof. computer sci. Rensselaer Poly. Inst., Troy, N.Y., 1980—. Mem. IEEE, Assn. for Computing Machinery, Am. Math. Soc., Soc. Indsl. and Applied Math. Subspecialties: Algorithms; Foundations of computer science. Current work: Design and analysis of algorithms for different models of computation.

KRISST-KRISCIOKAITIS, RAYMOND JOHN, physicist; b. Barzdai, Lithuania, May 29, 1937; came to U.S., 1949, naturalized, 1953; s. Frank and Leonora (Paukstys) Krisciokaitis; m. Ilze Krisst, Mar. 24, 1962; children: Rima, Abraham, Lara. A.B., U.Conn., 1958; Ph.D., Mich. State U., 1965. Research fellow in physics Harvard U., 1965-70; research affiliate in physics M.I.T., Cambridge, 1965-70; vis. scientist DESY, Hamburg, W.Ger., 1970-73; prin. physicist nuclear power dept. Combustion Engring., Inc., Windsor, Conn., 1974—; faculty Harvard U., 1967-70; condr. seminors. Producer, host: community TV show Viewpoint, 1980—; Contbr. articles to profl. jours. Pres. West Hartford Parent-Tchr. Council, 1980-82. Served to 1st lt. U.S. Army, 1958-66. Mem. Am. Phys. Soc., Am. Nuclear Soc., Am. Assn. Physics Tchrs., AAAS, Sigma Xi, Sigma Pi Sigma. Democrat. Subspecialties: Particle physics; Nuclear physics. Current work: Research in reactor physics and solar energy. Patentee in field. Office: 1000 Prospect Hill Rd: Windsor CT 06095 Home: 93 Meadowbrook Rd West Hartford CT 06107

KRISTIAN, JEROME, astronomer; b. Milw., June 5, 1933; s. Michael and Alma K.; m. Mary J. Kristian, Aug. 27, 1955; 1 son, John Michael. A.B., Shimer Coll., Mount Carroll, Ill., 1953; M.S. in Physics, U. Chgo., 1956, Ph.D., 1962. Physicist Internat. Sch. Nuclear Sci. and Engring., Argonne Nat. Lab., Ill., 1957-59; vis. lectr. physics and math. Center for Relativity, U. Tex., Austin, 1962-64; asst. prof. astronomy U. Wis., Madison, 1964-67; astonomer Mt. Wilson and Las Campanas Obs., 1967—. Mem. Am. Astron. Soc., Internat. Astron. Union, Astron. Soc. Pacific. Subspecialties: Optical astronomy; Cosmology. Current work: Extragalactic astronomy and cosmology; astron. image processing and datareduction; space astronomy. Office: 813 Santa Barbara St Pasadena CA 91101

KRISTOL, DAVID SOL, chemistry educator; b. Bklyn., June 4, 1938; s. Isidore and Yetta (Eisenberg) K.; m. Beverly Dee Newman, Oct. 5, 1975; children: Rachel, Joshua. B.S., Bklyn. Coll., 1958; M.S., NYU, 1966, Ph.D., 1969. Research asst. Jewish Hosp. Bklyn., 1958-63; instr. N.J. Inst. Tech., Newark, 1966-69, asst. prof. chemistry, 1969-76, assoc. prof., 1976-79, prof., 1979—. Mem. Am. Chem. Soc. (editor newsletter Monmouth Sect. 1978—), AAAS, N.Y. Acad. Scis., Sigma Xi. Lodge: B'nai B'rith. Subspecialties: Organic chemistry; Biomedical engineering. Current work: Organic synthesis, structure-ractivity relationships; clinical blood rheology; enzyme kinetics. Office: NJ Inst Tech 323 High St Newark NJ 07102

KRISTY, THOMAS WAYNE, manufacturing company executive; b. Fresno, Calif., Dec. 28, 1953; d. John Vernon and Sylvia M. (Steitz) K. Founder T.E.F. Mfg., Fresno, 1976, pres., 1976—; cons. solar system design. Subspecialties: Organic chemistry; Solar energy. Current work: Commercial development of solar systems. Office: 1550 N Clark St Fresno CA 93703

KRITCHEVSKY, DAVID, educator, biochemist; b. Kharkov, Russia, Jan. 25, 1920; came to U.S., 1923, naturalized, 1929; s. Jacob and Leah (Kritchevsky) K.; m. Evelyn Sholtes, Dec. 21, 1947; children—Barbara Ann, Janice Eileen, Stephen Bennett. B.S., U. Chgo., 1939, M.S., 1942; Ph.D., Northwestern U., 1948. Chemist Ninol Labs., Chgo., 1939-46; postdoctoral fellow Fed. Inst. Tech., Zurich, Switzerland, 1948-49; biochemist Radiation Lab., U. Calif. at Berkeley, 1950-52, Lederle Lab., Pearl River, N.Y., 1952-57, Wistar Inst., Phila., 1957—; prof. biochemistry Sch. Vet. Medicine, U. Pa., Phila., 1965—, 1970—, chmn. grad. group molecular biology, 1972—; Mem. USPHS study sect. Nat. Heart Inst., 1964-68, 72-76; chmn. research com. Spl. Dairy Industry Bd., 1963-70; mem. food and nutrition bd. Nat. Acad. Sci., 1976-82. Author: Cholesterol, 1958, also numerous articles.; Editor: (with G. Litwack) Actions of Hormones on Molecular Processes, 1964; co-editor: (with R. Paoletti) Advances in Lipid Research, 1963—, (with P. Nair) The Bile Acids, 1971; Western Hemisphere editor Atherosclerosis. Recipient Research Career award Nat. Heart Inst., 1962, award Am. Coll. Nutrition, 1978. Mem. Am. Inst. Nutrition (Borden award 1974, pres. 1979), Am. Soc. Biol. Chemists, AAAS, Am. Chem. Soc. (award Phila. sect. 1977), Soc. Exptl. Biology and Medicine (pres.-elect 1983—), Arteriosclerosis Council, Am. Heart Assn., Am. Soc. Oil Chemists (chmn. methods com. 1963-64), Internat. Soc. Fat Research. Subspecialties: Biochemistry (biology); Nutrition (biology). Current work: Liquid metabolism; experimental atherosclerosis: cholesterol metabolism; nutrition; aging. Research on role vehicle when cholesterol and fat produces atherosclerosis in rabbits, effects saturated and unsaturated fat, deposition orally administered cholesterol in aorta man and rabbit. Pioneered use radioactive cholesterol for metabolic expts. Home: 136 Lee Circle Bryn Mawr PA 19010 Office: Wistar Inst 36th and Spruce Sts Philadelphia PA 19104

KRIVIT, WILLIAM, pediatrician; b. Jersey City, Nov. 28, 1925; s. Samuel O. and Bertha (Weiss) K.; m. Chyrrel Heaton, Aug. 27, 1951; children—Robert, Daniel, Michael, Kim. M.D., Tulane U., 1948; Ph.D., U. Minn., 1958. Intern Charity Hosp., New Orleans, 1948-49; resident in pediatrics U. Utah, 1949-50; instr. pediatrics U. Minn., 1954-58, asst. prof., 1958-60, asso. prof., 1960-62, prof., 1962—, head dept., 1979—. Editorial bd.: Am. Jour Pediatric Hematology and Oncology, 1979, Pediatrics Jour, 1980—; contbr. numerous articles profl. jours. Served with M.C. USAF, World War II. Mem. Am. Acad. Pediatrics, Am. Soc. Hematology, Northwestern Pediatric Soc., Soc. Pediatric Research, Midwest Soc. Pediatric Research, Soc. Exptl. Biology and Medicine, Central Soc. Clin. Research, Am. Assn. Cancer Research, Am. Pediatric Soc., Am. Soc. Clin. Oncology, Am. Soc. Pediatric Hematology and Oncology (pres.), Phi Beta Kappa. Jewish. Club: Variety. Subspecialty: Pediatrics. Home: 1252 Ingerson Rd Saint Paul MN 55112 Office: Pediatrics Box 391 Mayo U of Minnesota Minneapolis MN 55455

KRIVOY, WILLIAM AARON, pharmacologist; b. Newark, Jan. 2, 1928; s. Samuel and Rose (Hirschenhorn) K. B.S. in Chemistry and Biology, Georgetown U., 1948; M.S. in Physiology, George Washington U., 1949; Ph.D. in Pharmacology, George Washington U., 1953. Pharmacologist Chem. Corps Med. Labs., Army Chem. Ctr., Edgewood, Md., 1950-54; USPHS postdoctoral research fellow dept. pharmacology U. Pa. Sch. Medicine, Phila., 1954-55; dept. pharmacology U. Edinburgh (Scotland) Sch. Medicine, 1955-57; instr. pharmacology Tulane U. Sch. Medicine, 1957-59; asst. prof. pharmacology Baylor U. Coll. Medicine, Houston, 1959-63, assoc. prof., 1963-68; pharmacologist Nat. Inst. Drug Abuse Addiction Research Ctr., Lexington, Ky., 1968—; vis. prof. pharmacology U. Tex. Dental Br., Houston, 1969—. Contbr. numerous chpts. and articles to profl. publs. Mem. Am. Coll. Neuropsychopharmacology, Am. Soc. Clin. Pharmacology and Therapeutics, Am. Soc. Pharmacology and Exptl. Therapeutics, Biophys. Soc., Brit. Pharmacol. Soc., Internat. Soc. Biochem. Pharmacology, Internat. Soc. Psychoneuroendocrinology, N.Y. Acad. Scis., Sociedade Brasileria de Farmacologia e de Terapeutica Experimental, Soc. Exptl. Biology and Medicine, Tex. Acad. Sci., Sigma Xi. Subspecialties: Neuropharmacology; Pharmacology. Current work: Neuropharmacology: research on actions of opioids and peptides on nervous system. Home: 3100 Kirklevington Dr Unit 3 Lexington KY 40502 Office: Nat Inst Drug Abuse Addiction Research Ctr PO Box 12390 Lexington KY 40583

KROGMANN, DAVID WILLIAM, biochemistry educator; b. Washington, Oct. 21, 1931; s. Rudolph Francis and Cecilia Mary (O'Dea) K.; m. Loretta Isadora Kurek, June 14, 1958; children: Michele E., Patricia A., Paul D. A.B., Cath. U., 1953; Ph.D., Johns Hopkins U., 1958. Postdoctoral research assoc. Johns Hopkins U., 1958, U. Chgo., 1959-60; asst. prof. chemistry Wayne State U., 1960-63, assoc. prof., 1963-66; program dir. molecular biology NSF, 1966-67; prof. biochemistry Purdue U., 1967—. NSF research grantee. Mem. Am. Soc. Biol. Chemists, Am. Soc. Plant Physiologists. Roman Catholic. Subspecialties: Biochemistry (biology); Photosynthesis. Current work: Structure of proteins in the photosynthetic membrane. Home: 301 Chippewa St West Lafayette IN 47906 Office: Dept of Biochemistry Purdue University West Lafayette IN 47907

KROHMER, JACK STEWART, radiological physicist, educator; b. Cleve., Nov. 7, 1921; s. Jacob and Violet Isabel (Armstrong) K.; m. Doris Elaine Lyman, Sept. 14, 1946; chlidren: Karen Elise Krohmer Carr, Jack Lyman, Candace Lynn Krohmer Blanchard. B.S., Western Res. U., Cleve., 1943, M.A., 1947; Ph.D., U. Tex., 1961. Diplomate: Am. Bd. Radiology, Am. Bd. Health Physics. Assoc. physics Western Res. U., Clev., 1947-57; prof. radiol. physics U. Tex. Southwestern Med. Sch., Dallas, 1957-63; chmn. dept. physics Roswell Park Meml. Inst., Buffalo, 1963-66; chief radiol. physics Geisinger Med. Center,

1966-72; mem. Radiol. Assocs. of Erie, Pa., 1972-79; prof. radiology and radiation oncology, dir. div. radiol. physics Wayne State U., Detroit, 1979—; cons. in uses of radiation in treatment and diagnosis of disease and in biol. effects of radiation; pres. J.S. Krohmer, Ph.D., Inc.; Trustee Am. Bd. Radiology. Contbr. articles to profl. jours. Served to 1st lt. Chem. Warfare Service U.S. Army, 1943-46. Fellow Am. Coll. Radiology; mem. Am. Assn. Physicists in Medicine (pres. 1974-75), Radiation Research Soc., Radiol. Soc. N.Am., Health Physics Soc., Am. Roentgen Ray Soc., Mich. Radiol. Soc. Republican. Presbyterian. Clubs: Grosse Pointe (Mich.) Yacht; Ill. Athletic (Chgo.). Subspecialties: Imaging technology; Biophysics (biology). Current work: Radiationimaging; low level effects of radiation, radiological shielding research, teaching and service functions in all aspects of the uses of radiation in the treatment and diagnosis of disease. Home: 1182 Hawthorne Rd Grosse Pointe Woods MI 48236 Office: Univ Health Center 4201 St Antoine Detroit MI 48201

KROLIK, JULIAN HENRY, astrophysicist; b. Detroit, Apr. 4, 1950; s. Henry A. and Bessie P. (Pearlman) K. S.B., M.I.T., 1971; Ph.D. in Physics, U. Calif., Berkeley, 1977. Mem. Inst. Advanced Study, Princeton, N.J., 1977-79; postdoctoral scientist M.I.T., Cambridge, 1979-81; research assoc. Harvard-Smithsonian Center for Astrophysics, Cambridge, Mass., 1981—. Mem. Am. Astron. Soc. Subspecialty: Theoretical astrophysics. Current work: Active galactic nuclei, molecular clouds, compact X-ray sources.

KROLL, JOHN ERNEST, mathematician, educator, researcher; b. Los Angeles, Aug. 15, 1940; s. Herman Reinhold and Isabella (Whitlock) K.; m. Elizabeth Westphal, July 14, 1973; children: Karen, Kathy Ann, Steven. B.S., UCLA, 1963, M.S., 1966; Ph.D., Yale U., 1973. Postdoctoral fellow Nova Oceanographic Lab., Ft. Lauderdale, Fla., 1972-75; instr. in applied math. MIT, 1975-76; assoc. prof. math. Old Dominion U., Norfolk, Va., 1976—. Contbr. articles to profl. jours. NSF Research grantee, 1980-82. Mem. Soc. Indsl. and Applied Math., Am. Geophys. Union, Va. Acad. Sci. Presbyterian. Subspecialties: Applied mathematics; Oceanography. Current work: Investigation of the ocean mixed layer and the surface generation of inertial oscillations. Home: 133 Sir Oliver Rd Norfolk VA 23505 Office: Dept Math Scis Old Dominion U Norfolk VA 23508

KROLL, NORMAN MYLES K., physicist, educator; b. Tulsa, Apr. 6, 1922; s. Cornelius and Grace (Aaronson) K.; m. Sally Sharlot, Mar. 15, 1945; children: Linda Ruth, Cynthia Anne, Heather Roma, Ira Joseph. Student, Rice Inst., 1938-40; A.B., Columbia U., 1942, A.M., 1943, Ph.D., 1948. Mem. faculty of Columbia, 1942-62, successively asst. in physics, asst. prof., asso. prof., prof., 1954-62, sci. staff Radiation Lab., 1943-62; prof. physics U. Calif. at San Diego, 1962—, chmn. physics dept., 1963-65, 83—; mem. Inst. Advanced Study, 1948-49; vis. scientist Brookhaven Nat. Lab., summers 1952-55. Bd. editors: Jour. Math. Physics, 1966-68; Contbr. articles to profl. jours.; Adv. com.: Physics Today, 1974-76. NRC fellow, 1948-50; Guggenheim fellow, 1955-56; Fulbright scholar U. Rome, 1955-56; NSF sr. postdoctoral fellow, 1965-66. Fellow Am. Phys. Soc.; mem. Nat. Acad. Scis., Phi Beta Kappa, Sigma Xi. Subspecialties: Theoretical physics; Particle physics. Current work: Magnetic monopoles, free electron lasers. Home: 2457 Calle del Oro La Jolla CA 92037

KROMER, LAWRENCE FREDERICK, neurobiologist, educator; b. Sandusky, Ohio, Sept. 1, 1950; s. Rolland Frederick and Geraldine Susan (Bolish) K.; m. Tanya Marie Sandros, Apr. 2, 1977. B.A., U. Chgo., 1972, Ph.D., 1977. NIH fellow in histology U. Lund, Sweden, 1977-79; asst. research neuroscientist dept. neuroscis. U. Calif.-San Diego, 1979-81; asst. prof. anatomy and neurobiology U. Vt., Burlington, 1981—. Alfred P. Sloan Found. fellow, 1980—. Mem. Internat. Soc. for Developmental Neurosci., Am. Assn. Anatomists, AAAS, Soc. for Neurosci. Subspecialties: Regeneration; Neurodeveleopment. Current work: Teaching and research to analyze development and regeneration in the mammalian central nervous system using the intracephalic transplantation technique which allows the transplantation of embryonic neural tissue into the brain of adult and neonatal rodents. Office: U Vt Dept Anatomy and Neurobiology Given Bldg Burlington VT 05405

KRONBERG, PHILIPP PAUL, astronomy and astrophysics educator, consultant; b. Toronto, Ont., Can., Sept. 16, 1939; s. Philipp and Jean Stewart (Davidson) K.; m. Roberta Beatrice Secord, Aug. 3, 1963; children: Paul Andrew, Martin Thomas, Michael Philipp Robert. B.Sc. in Engring. Physics, Queen's U., Kingston, Ont., 1961, M.Sc., 1963; Ph.D., U. Manchester, Eng., 1967. Lectr. physics U. Manchester and Jodrell Bank, U.K., 1966-68; asst. prof. astronomy U. Toronto, 1968-73, assoc. prof., 1973-78, prof., 1978—; sr. von Humboldt fellow and guest scientist Max-Planck Inst. Radioastronomie, Bonn, W. Ger., 1975-77; mem. NRC Assoc. Commn. on Astronomy, 1971-74, mem. grants com., 1974-78; mem., adv. com. for VLA Project in N.Mex. U.S. Nat. Radio Astronomy Obs., 1978-82, chmn., 1980, Assoc. U., Inc.; vis. com. Nat. Radio Astronomy Obs., 1981-82; von Humboldt fellow Max Planck Inst., 1980; mem. U. Toronto Research Bd., 1982—. Contbr. numerous articles on astronomy and astrophysics to profl. jours. Alexander von Humboldt fellow, 1975-77. Mem. Am. Astron. Soc., Can. Astron. Soc., Internat. Astron. Union. Anglican. Club: Toronto Sailing and Canoe. Subspecialties: Radio and microwave astronomy; High energy astrophysics. Current work: Galactic and extragalactic astrophysics research, radio astronomy, consulting, research in astrophysics. Home: 33 Boarhill Dr Toronto (Agincourt) ON Canada M1S 2L9 Office: Dept Astronomy U Toronoto 60 St George St Toronto ON Canada M5S 1A7

KRONFOL, NOUHAD O., physician; b. Beirut, Lebanon, July 29, 1950; came to U.S., 1977; s. O.M. and S.O. K. B.Sc., Am. U. Beirut, 1971, M.D., 1975. Diplomate: Am. Bd. Internal Medicine; Am. Bd. Nephrology. Resident Am. U. Beirut, 1975-77; fellow Med. Coll. Va., Richmond, 1977-80, instr. medicine, 1980-81, asst. prof., 1981—. Mem. ACP, Am. Fedn. for Clin. Research, Am. Soc. Nephrology, Richmond Acad. Medicine, Med. Soc. Va., Alpha Omega Alpha. Subspecialties: Internal medicine; Nephrology. Current work: Molecular transport; kidney disease. Office: Med Coll Va 11th and Marshall Sts Richmond VA 23298

KROTHAPALLI, RADHA KRISHNA, internist; b. Amarthalur, India, May 4, 1951; came to U.S., 1975; s. Raghavaiah and Lakshmi Narasamma (Parvathaneni) K.; m. Shirley Marie Hunt, Jan. 24, 1981. M.B.B.S., Guntur Med. Coll., India, 1973. Diplomate: Am. Bd. Internal Medicine, Am. Bd. Nephrology. Rotating intern Mo. Bapt. Hosp., St. Louis, 1976-77; resident in medicine Montgomery (Ala.) Internal Medicine Residency Program, 1977-80; fellow in nephrology Baylor Coll. Medicine, Houston, 1980-83; practice medicine specializing in nephrology, Montgomery, Ala. Author: Pseudonomas Peritonitis, 1982; contbr. chpts. to books, articles to profl. jours. Nat. Kidney Found. fellow, 1982-83. Mem. ACP, Am. Soc. Nephrology, Internat. Soc. Nephrology, Am. Fedn. for Clin. Research, Am. Soc. Artificial Internal Organs, N.Y. Acad. Scis., Nat. Kidney Found. Subspecialties: Internal medicine; Nephrology. Current work: Clinical research in nephrology; basic renal physiology research in field of water transport in the cortical collecting tubule. Home: 2746 Baldwin Brook Dr Montgomery AL 36116 Office: 303 S Ripley Montgomery AL 36197

KROWN, SUSAN ELLEN, physician; b. Bronx, N.Y., Sept. 8, 1946; d. Frederick B. and Paula (Hauser) K.; m. Roger Edward Pitt, May 18, 1980; 1 dau., Catherine Krown Pitt. A.B., Barnard Coll., 1967; M.D., SUNY, Bklyn., 1971. Diplomate: Am. Bd. Internal Medicine (subspecialty in med. oncology). Intern, resident in internal medicine Mt. Sinai Hosp., N.Y.C., 1971-74; fellow in med. oncology and clin. immunology Meml. Sloan-Kettering, N.Y.C., 1974-77; research asso. Sloan-Kettering Inst. for Cancer Research, N.Y.C., 1977—; clin. asst. physician, asst. attending physician Meml. Hosp., N.Y.C., 1977-81, assoc. attending physician, 1982—; asst. prof. medicine Cornell U. Med. Coll., 1977-83, assoc. prof. clin. medicine, 1983—. Contbr. chpts. to books, articles to profl. jours. Am. Cancer Soc. fellow, 1978-81; Nat. Cancer Inst. grantee, 1979, 80, 81, 82. Mem. Am. Soc. Clin. Oncology, Am. Assn. Cancer Research. Subspecialties: Cancer research (medicine); Immunology (medicine). Current work: Use of biological response modifiers, particularly interferons, in cancer treatment and their role in regulating immune responses. Office: 1275 York Ave New York NY 10021

KRUG, HARRY EVERISTUS PETER, JR., nuclear engineer; b. Kearney, N.J., Aug. 1, 1932; s. Harry Everistus and Helen (Miliski) K.; m. Madonna Eileen Martin, Nov. 23, 1977 (div. Mar. 1982); children: by previous marriage: Kirk Stanley, Karen Helen, Lynne Allison. B.S., U.S. Mcht. Marine Acad., 1955; M.Nuclear Engring., NYU, 1961. Registered profl. engr., Calif. Fellow engr. Atomic dept. Westinghouse, Pitts., 1961-68; v.p., gen. mgr. NCI, Pitts., 1969-70; nuclear engr. Exxon Nuclear Co., Richland, Wash., 1971; industry mgr. Control Data Corp., Mpls., 1972-73; supr. nuclear engring. Ill. Power Co., Decatur, 1974-75; nuclear engr. U.S. Nuclear Regulatory Commn., Atlanta, 1975—, expert witness, Bethesda, Md., 1974-82. Contbr. articles to profl. jours. Served to lt. USNR, 1956-58. Mem. Am. Nuclear Soc. (pres. Midwest chpt. 1972). Roman Catholic. Subspecialties: Nuclear engineering; Artificial intelligence. Current work: Application of artificial intelligence to the operations and design of nuclear power stations and other complex systems. Home: 3030 Old Decatur Rd Apt B315 Atlanta GA 30305 Office: US Nuclear Regulatory Commission 101 Marietta St Suite 2900 Atlanta GA 30308

KRUG, SAMUEL EDWARD, psychologist, consultant; b. Chgo., Nov. 15, 1943; s. Samuel Edward and Evelyn (LaVelle) K.; m. Marion E. Besch, June 29, 1968; children: Mark, Michael, David, Timothy. A.B., Holy Cross Coll., 1965; M.A., U. Ill.-Champaign, 1968, Ph.D., 1971. Lic. psychologist, Ill., Ariz. Research asst. U. Ill.-Champaign, 1965-70, teaching asst., 1968-69, vis. prof., 1982-83; research scientist Inst. for Personality/Ability Testing, Champaign, 1966-75, exec. dir., 1976—; pres. Metri Tech, Inc., Champaign., 1982—; mem. editorial policy bd. Multivariate Exptl. Clin. Research, Wichita State U., 1976—; cons. USCG, Nat. Computer Systems, BASF-Wyandotte Corp., others. Author books, articles, psychol. tests, programs for computer interpretation of psychol. tests. Fellow Am. Psychol. Assn., Soc. for Personality Assessment; mem. Nat. Council on Measurement in Edn. Roman Catholic. Subspecialty: Measurement/personality assessment. Current work: Development and refinement of computer-based technologies for interpretation and analysis of psychological test data; computer-based training and education. Home: 2208 Galen Dr Champaign IL 61820

KRUGER, CHARLES HERMAN, JR., mechanical engineer; b. Oklahoma City, Oct. 4, 1934; s. Charles H. and Flora K.; m. Nora Nininger, Sept. 10, 1977; children—Sarah, Charles Herman III, Elizabeth, Ellen. S.B., M.I.T., 1956, Ph.D., 1960; D.I.C., Imperial Coll., London, 1957. Asst. prof. M.I.T., Cambridge, 1960; research scientist Lockheed Research Labs., 1960-62; prof. mech. engring. Stanford (Calif.) U., 1962—, chmn. dept. mech. engring., 1982—; vis. prof. Harvard U., 1968-69, Princeton U., 1979-80; mem. Environ. Studies Bd. Nat. Acad. Scis.; mem. hearing bd. Bay Area Air Quality Mgmt. Dist., 1969-83. Co-author: Physical Gas Dynamics, 1965, Partially Ionized Gases, 1973, On the Prevention of Significant Deteriorization of Air Quality, 1981; asso. editor: AIAA Jour, 1968-71; contbr. numerous articles to profl. jours. NSF sr. postodoctoral fellow, 1968-69. Mem. AIAA (medal, award 1979), Combustion Inst., ASME, Am. Phys. Soc., Air Pollution Control Assn., Engring. Aspects of Magnetohydrodynamics. Subspecialties: Combustion processes; Plasma engineering. Current work: Combustion; partially ionized plasma; air pollution; magneto hydrodynamics; diagnostics. Office: Dept Mech Engring Stanford U Stanford CA 94505

KRUGER, LAWRENCE, anatomy educator; b. New Brunswick, N.J., Aug. 15, 1929; s. Jacob C. and Kate M. (Newman) K.; m. Virginia Findlay, sept. 30, 1961; children: Erika, Paula. Ph.D., Yale U., 1954. Post-doctoral fellow Johns Hopkins U., 1955-58; post-doctoral fellow College de France, Paris, 1958, Oxford (Eng.) U., 1958-59; asst. prof. anatomy UCLA, 1959-62, assoc. prof., 1962-66, prof., 1966—; Wellcome vis. prof. Albany Med. Coll., 1981; mem. study sect. NIH, Bethesda, Md., 1974—. Editor-in-chief: Somatosensory Research, 1982—; mem. editorial bd. various profl. jours. Recipient Lederle Med. Faculty award Am. Cyanamid, 1964-67; Fogarty scholar NIH, 1977. Subspecialties: Anatomy and embryology; Neurology. Office: Dept Anatomy UCLA Center for Health Scis Los Angeles CA 90024

KRUGMAN, SAUL, physician, educator; b. N.Y.C., Apr. 7, 1911; s. Louis and Rachel (Cohen) K.; m. Sylvia Stern, Feb. 18, 1940; children—Richard David, Carol Lynn. Student, Ohio State U., 1929-32; M.D., Med. Coll. Va., 1939. Intern, then resident Cumberland, Willard Parker and Bellevue hosps., N.Y.C., 1939-41, 46-48; teaching and med. research N.Y. U.-Bellevue Med. Center, 1948—, asso. prof. pediatrics, 1954-60, prof., 1960—, chmn. dept., 1960-75; dir. pediatric service Bellevue Hosp., 1960-75, Univ. Hosp., 1960-75; Mem. Commn. Viral Infections, 1960-72; mem. nat. adv. council Nat. Inst. Allergy and Infectious Diseases, 1965-69, chmn. infectious disease adv. com., 1971-73; mem. com. on viral hepatitis NRC, 1973-76; chmn. panel on viral and rickettsial vaccines Bur. Biologies, FDA, 1973-79. Co-author: Infectious Diseases of Children, 1981; contbr. articles on infectious diseases to med. jours. NIH research fellow, 1948-50. Fellow Am. Acad. Pediatrics; mem. Am. Pediatric Soc. (pres. 1972-73), Soc. Pediatric Research, N.Y. Acad. Medicine (chmn. pediatric sect. 1960-61), Am. Epidemiol. Soc., Harvey Soc., Assn. Am. Physicians, Nat. Acad. Scis. Subspecialties: Pediatrics; Infectious diseases. Home: 300 E 33d St New York NY 10016 Office: 550 1st Ave New York NY 10016

KRULL, IRA STANLEY, chemist, researcher, educator, cons.; b. N.Y.C., Oct. 21, 1940; s. Arthur and Anne (Nadelman) K.; m. Erica Krull, Mar. 1, 1973; 1 son, Marc Arthur. B.S. cum laude, CCNY, 1962; M.S. (teaching fellow, NIH predoctoral fellow), N.Y. U., 1966; Ph.D. (NIH postdoctoral fellow), N.Y. U., 1968. Fellow Weizmann Inst. Sci., Israel, 1970-73; asst. scientist Boyce Thompson Inst., Yonkers, N.Y., 1973-77, Thermo Electron Corp., Waltham, Mass., 1977-79; sr. scientist Inst. Chem. Analysis, Applications and Forensic Sci., Northeastern U., Boston, 1979—; tchr. Center Profl. Advancement, N.J.; cons. FDA, pvt. indsl. firms, Center Profl. Advancement. Contbr. articles and revs. to profl. jours. Union Carbide postdoctoral fellow, 1968-70, U. Wis., 1967-68; recipient Founders Day award N.Y. U., 1968. Mem. Am. Chem. Soc., Chem. Soc. London, Assn. Ofcl. Analytical Chemists, Am. Assn. Cancer Research, Sigma Xi, Phi Lambda Upsilon. Subspecialties: Analytical chemistry; Organic chemistry. Current work: Development of improved trace assays for organic/inorganic materials, environ. trace analysis, analytical instrumentation development, application of analytical instrumentation for trace analysis, analytical toxicology. Office: Northeastern U 360 Huntington Ave Boston MA 02115

KRUMHANSL, JAMES ARTHUR, physicist, educator, Industrial consultant; b. Cleve., Aug. 21919; s. James and Marcella (Kelly) K.; m. Barbara Dean Schminck, Dec. 26, 1944 (div. 1983); children: James Lee, Carol Lynne, Peter Allen.; m. Marilyn Cupp Dahl, Feb. 19, 1983. B.S. in Elec. Engring, U. Dayton, 1939; M.S., Case Inst. Tech., 1940, D.Sc. (hon.), 1980; Ph.D. in Physics, Cornell U., 1943. Instr. Cornell U., 1943-44; physicist Stromberg-Carlson Co., 1944-46; mem. faculty Brown U., 1946-48, asso. prof., 1947-48; asst. prof., then asso. prof. Cornell U., 1948-55; asst. dir. research Nat. Carbon Co., 1955-57, asso. dir. research, 1957-58; prof. physics Cornell U., 1959—, Horace White prof., 1980; dir. Lab. Atomic and Solid State Physics, 1960-64; adj. prof. U. Pa., 1979; fellow Los Alamos Lab., 1980; asst. dir. for math., phys. sci. and engring. NSF, 1977-79; cons. to industry, 1946—; dir. Allied Chem. Corp.; Adv. com for AEC, Dept. Def., Nat. Acad. Sci., 1956—; vis. fellow All Souls Coll., Oxford U., 1977, Gonville and Caius Coll., Cambridge U., 1983, Royal Soc. London, 1983. Editor: Jour. Applied Physics, 1957-60; asso. editor: Solid State Communications, 1963—, Rev. Modern Physics, 1968-73; editor: Phys. Rev. Letters, 1974—, physics Oxford U. Press; Contbr. articles to profl. jours. Guggenheim fellow, 1959-60; NSF sr. postdoctoral fellow Oxford U., 1966-67. Fellow Am. Phys. Soc. (chmn. div. solid state physics 1968, councillor 1970-74), AAAS, Am. Inst. Physics (governing bd. 1973—); mem. AAUP, Am. Assn. Physics Tchrs., Sigma Xi, Phi Kappa Phi. Republican. Presbyn. Club: Ithaca Yacht. Subspecialties: Condensed matter physics; Applied mathematics. Current work: Conducting polymers; molecular biophysics of proteins and DNA. Home: 12 D Strawberry Hill Ithaca NY 14850

KRUPP, MARCUS ABRAHAM, physician; b. El Paso, Tex., Feb. 12, 1913; s. Maurice and Esther (Siegel) K.; m. Muriel McClure, Aug. 9, 1941 (dec. Oct. 1954); children: Michael, David (dec.), Peter, Sara; m. Donna Goodheart Millen, Feb. 28, 1958. A.B., Stanford U., 1934, M.D., 1939. Diplomate: Am. Bd. Internal Medicine. Intern Stanford U. Hosp., Calif., 1938-39, resident in internal medicine, 1939-42; chief clin. pathology VA Hosp., San Francisco, 1946-50; dir. Palo Alto Med. Research Found., Calif., 1950—; dir. labs. Palo Alto Med. Clinic, 1950-80; asst. clin. prof. medicine Stanford U., 1946-56, asso. clin. prof., 1956-65, clin. prof., 1965—; mem. med. tech. adv. com. Public Employees Retirement System Calif., 1972—. Editor: (with Milton Chatton) Current Diagnosis and Treatment, ann., 1971-83, (with others) Physicians Handbook, 7th-20th edits., 1981. Vice pres. bd. dirs. Calif. Heart Assn., 1974-75; pres. bd. trustees Channing House, Palo Alto. Served to capt. U.S. Army, 1942-46. Fellow ACP; mem. Western Soc. Clin. Research, Calif. Acad. Medicine (pres. 1966), Pacific Interurban Clin. Club (pres. 1977), AAAS, AMA, N.Y. Acad. Scis., Assn. Ind. Research Insts. (pres. 1966-67), Phi Beta Kappa, Alpha Omega Alpha. Subspecialty: Internal medicine. Current work: Nephrology; endocrinology, physiology. Home: 195 Ramoso Rd Portola Valley CA 94025 Office: 860 Bryant St Palo Alto CA 94301

KRUPP, PATRICIA POWERS, medical educator; b. N.Y.C., Jan. 29, 1938; d. James Joseph and Winona (Smyzer) Powers; m.; children: Christina Marie, Andrea Janine, Elizabeth Ann. Ph.D., Hahnemann Med. Coll., 1970. Mem. faculty U. Vt. Coll. Medicine, 1972—, assoc. prof. anatomy and neurobiology, 1978—. Contbr. sci. articles to profl. publs. Recipient Alumna of Yr. award Hahnemann Med. Coll., 1982. Mem. Endocrine Soc., Am. Assn. Anatomists, Internat. Hibernation Soc., Am. Women in Sci., AAUP. Clubs: Surf (Miami Beach); Sleepy Hollow Country (Westchester); Burlington Country. Subspecialties: Microscopy; Anatomy and embryology. Current work: Experimental modification of thyroid gland structure and function; effects of endogenous stimulation with TSH, diet and drugs. Home: 7 Briar Ln Essex Junction VT 05452 Office: Vt Coll Medicine Burlington VT 05405

KRUSBERG, LORIN RONALD, plant nematologist, educator; b. East Dover, Vt., July 18, 1932; s. John and Bertha Leontine (Bernstein) K.; m. Betty Kirk, Nov. 23, 1935; children: John Lorin, Michael Charles, Susan Joan. B.S., U. Del.-Newark, 1954; M.S., N.C. State U., 1956, Ph.D., 1959. Asst. prof. botany dept. U. Md., College Park, 1960-65, assoc. prof., 1965-70, prof. plant pathology, 1970—. Mem. Soc. Nematologists, Am. Phytopath. Soc., Helminthological Soc. Washington, Sigma Xi. Democrat. Methodist. Club: Garden (Beltsville, Md.). Subspecialty: Plant pathology. Current work: Biology and control of plant parasitic nematodes.

KRUSE, DAVID HAROLD, utility co. exec.; b. Portland, Oreg., May 25, 1941; s. Harold David and Naomi Sophi (Pfaff) K.; m. Barbara Lynn Sutherland, June 12, 1965; children: Kevin David, Brian Lynn. B.S., Portland State U., 1963; M.E. in Aero. Engring, Iowa Stae U., 1966, Ph.D., 1967. Cert. data processor. Sr. aero. engr. Gen. Dynamics, Ft. Worth, 1967-75, project engr., 1975-76; with Portland (Oreg.) Gen. Electric Co., 1976—, br. mgr. fin. planning, 1981, mgr. engring. and computational systems, 1982—; instr. Tarrant Coll., Ft. Worth, 1974-76. Pres. Zero Population Growth, Fort Worth, 1973; chmn. budget com. Gladstone (Oreg.) Sch. Dist., 1983, computer com., 1982. NDEA fellow, 1964. Mem. IEEE, Assn. Computing Machinery, Soc. Cert. Data Processors. Republican. Baptist. Clubs: Trails, Sierra. Subspecialties: Software engineering; Graphics, image processing, and pattern recognition. Current work: Automatted mapping/facilities management development with graphics and database applications. Home: 17350 Crownview Dr Gladstone OR 97027 Office: Portland Gen Electric 121 SW Salem St Portland OR 97204

KRUSKAL, MARTIN DAVID, mathematical physicist; b. N.Y.C., Sept. 28, 1925; m., 1950; 3 children. B.S., U. Chgo., 1945; M.S., NYU, 1948, Ph.D. in Math, 1952. Asst., instr. dept. math. NYU, 1946-51; research scientist Plasma Physics Lab., Princeton U., 1951—, sr. research assoc., 1959—, prof. astrophys. sci., 1961—, prof. math., 1981—; cons. Los Alamos Sci. Lab., 1953-59, radiation lab. U. Calif., 1954-57, Oak Ridge Nat. Lab., 1955-58, 63—, Radio Corp. Am., 1960-62, IBM Corp., 1963—; lectr. in field. Recipient Dannie Heineman prize in math. physics, 1983; NSF sr. fellow, 1959-60. Mem. Am. Math. Soc., Math. Assn. Am., Am. Phys. Soc. Subspecialty: Applied mathematics. Office: Dept Astrophysics Princeton U Princeton NJ 08540

KRYDER, MARK HOWARD, electrical engineering educator; b. Portland, Oreg., Oct. 7, 1943; s. DeVann Halley and Eleanor (Evenson) K.; m. Sandra Lee Curtis, June 19, 1965; children: Christa Marie, Matthew Curtis. B.S.E.E., Stanford U., 1965; M.S.E.E., Calif. Inst. Tech., 1966, Ph.D., 1970. Research fellow Calif. Inst. Tech., Pasadena, 1969-71; vs. scientist U. Regensburg, West Germany, 1971-73; research staff mem., mgr. exploratory bubble devices IBM, T. J. Watson Research Center, Yorktown Heights, N.Y., 1973-78; prof. elec. engring. Carnegie-Mellon U., Pitts., 1978—; mem. adv. com. Magnetism and Magnetic Materials Conf. Grantee NSF, Air Force Office Sci. Research and Industry, NASA, Rome Air Devel. Center. Mem. IEEE (mem. adminstrv. com. magnetics soc.), Am. Phys. Soc. Subspecialty: Applied magnetics. Current work: Magnetic recording technology, magnetic bubble technology. Patentee in field. Office: Elec Engring Dept Carnegie Mellon U Pittsburgh PA 15213

KRYNICKI, VICTOR EDWARD, psychologist; b. Jersey City, N.J., Feb. 16, 1949; s. Edward and Helen A. (Tomkiel) K.; m. Bernee Lynn Koch, June 14, 1970; children: Eric Victor, Philip Bernard. B.A., Yale U., 1969; M.A., SUNY-Albany, 1972; Ph.D., Columbia U., 1976. Lic. psychologist, N.Y. Asst. research scientist Inst. for Basic Research, S.I., N.Y., 1972-73; psychologist Queens Children's Hosp., Bellerose, N.Y., 1976-78, program evaluation specialist, 1979—; assoc. prof. N.Y. Med. Coll., Valhalla, N.Y., 1978-79; dir. Behavioral Counseling Ctr., Whitestone, N.Y., 1982—. Mem. Am. Psychol. Assn., Internat. Neuropsychol. Assn. Subspecialties: Neuropsychology; Information systems, storage, and retrieval (computer science). Current work: Application of microcomputer technology to psychological and neuropsychological testing and assessments. Home: 28 Hollis Ln Croton-On-Hudson NY 10520 Office: Behavioral Counseling Ctr 162-30 Powells Cove Blvd Whitestone NY 11357

KSIR, CHARLES JOSEPH, JR., psychology educator; b. Albuquerque, May 19, 1945; s. Charles Joseph and Mary Annette (Pickens) K.; m. Sandra Susan Hoos, July 29, 1967; 1 dau., Amy Elizabeth. B.A., U. Tex., 1967; Ph.D., Ind. U., 1971. Postdoctoral fellow Worcester Found., Shrewsbury, Mass., 1971-72; asst. prof. psychology U. Wyo., Laramie, 1972-76, assoc. prof., 1976-80, prof., 1980—, acting assoc. v.p. acad. affairs, 1980-81; vis. scientist Salk Inst., La Jolla, Calif., 1981. Contbr. articles to profl. jours. NIMH grantee, 1973, 75; Am. Heart Assn. grantee, 1981; Borroughs-Wellcome Fund travel grantee, 1982. Mem. Soc. for Neurosci., Behavioral Pharmacology Soc., Psychonomic Soc., AAAS, Sigma Xi. Subspecialties: Neuropharmacology; Behavioral psychology. Current work: Behavioral neuropharmacology of amphetamine and neuropeptides, tobacco dependence. Home: 2068 N 17th St Laramie WY 82070 Office: U Wyo Box 3415 Laramie WY 82071

KUBITSCHEK, HERBERT ERNEST, biophysicist; b. Oak Park, Ill., June 9, 1920; s. Ernst M. and Anna A. (Nebel) K.; m. Jenny G., June 26, 1943; children: Carolyn, Craig, Warren, Wendy. Ph.D., U. Ill., 1949. Postdoctoral fellow Inst. Radiobiology and Biophysics, U. Chgo., 1949-51; assoc. physicist Argonne (Ill.) Nat. Lab., 1951, sr. biophysicist, 1970, genetics group leader, 1972—; adj. prof. No. Ill. U. Author: Introduction to Research with Continuous Cultures, 1970. USPHS fellow, 1949-51; Minna-James Heineman fellow, 1974. Mem. AAAS, Am. Assn. Physics Tchrs., Am. Soc. Microbiology, Am. Soc. Photobiology, Biophys. Soc., Environ. Mutagen Soc., Genetics Soc. Am., Sigma Xi. Subspecialty: Cell and tissue culture. Current work: Cell growth, mutagenesis. Office: Argonne Nat Lab 9700 S Cass Ave Argonne IL 60439

KUBY, STEPHEN ALLEN, biochemist; b. Jersey City, Aug. 5, 1925; s. Meyer and Bella (Chase) K.; m. Josette Marie Gerome, July 17, 1962. A.B., N.Y. U., 1948; M.S., U. Wis., 1951, Ph.D., 1953. Research asso. Inst. Enzyme Research, U. Wis., 1954-55, asst. prof. biochemistry, 1956-63; USPHS fellow Med. Nobel Inst., Stockholm, 1955-56; asso. prof. biochemistry, asso. research prof. medicine Lab. Study of Hereditary and Metabolic Disorders, U. Utah, 1963-69, prof. biochemistry, research prof. medicine, 1969—; Mem. physiol. chemistry study sect. Research Grants div. NIH, 1968-72. Contbr. articles to profl. jours. Served with AUS, 1943-46. USPHS fellow Johnson Found. Med. Phys., U. Pa., 1953-54. Mem. Am. Soc. Biol. Chemists, Am. Chem. Soc., AAAS, N.Y. Acad. Sci., Sigma Xi, Phi Beta Kappa, Phi Lambda Upsilon. Subspecialties: Biochemistry (biology); Biochemistry (medicine). Current work: Enzyme chemistry; protein chemistry and biochemistry of inherited and metabolic disorders. Office: University of Utah Research Park Salt Lake City UT 84108

KUC, JOSEPH, plant pathologist; b. N.Y.C., Nov. 24, 1929; s. Peter and Helen (Dubec) K.; m. Ruth Helen Shaffit, June 1, 1975; children: Paul David, Rebecca Ruth, Miriam Abigail. B.S., Purdue U., 1951, M.S., 1953, Ph.D., 1955. Asst. prof. dept. biochemistry Purdue U., West Lafayette, Ind., 1955-59, asso. prof., 1959-63, prof., 1963-74; prof. dept. plant pathology U. Ky., Lexington, 1974—, disting. alumni prof., 1978. Contbr. articles to profl. jours. Recipient Campbell award Am. Phytopathol. Soc., 1976; research award U. Ky. Research Found., 1977; Fulbright fellow, 1960-61, 66-67; Alexander von Humboldt Found. awardee, 1980. Fellow Am. Phytopathol. Soc.; mem. Am. Chem. Soc., Am. Soc. Plant Physiologists, Am. Inst. Chemists, Phytochem. Soc., Japanese Soc. Plant Physiologists, Sigma Xi, Phi Lambda Upsilon, Alpha Zeta. Subspecialties: Plant pathology; Biochemistry (biology). Current work: Biochemistry of disease resistance in plants, plant biochemistry. Home: 141 Louisiana Ave Lexington KY 40502 Office: Dept Plant Pathology U Ky Lexington KY 40546

KUC, ROMAN BASIL, electrical engineering educator, consultant; b. Ulm, W. Ger., June 24, 1946; came to U.S., 1950, naturalized, 1955; s. Alex and Irene K.; m. Robin Elizabeth Mercier, July 10, 1981. B.S. in Elec. Engring, Ill. Inst. Tech., 1968; Ph.D., Columbia U., 1977. Mem. tech. staff Bell Labs., Whippany, N.J., 1968-74, Murray Hill, N.J., 1974-75; research asst. Columbia U., N.Y.C., 1975-77, research assoc., 1977-79; asst. prof. dept. elec. engring. Yale U., New Haven, 1979-83, assoc. prof., 1983—; cons. digital signal processing. Contbr. articles to profl. jours.; assoc. editor: Jour. Cardiovascular Ultrasonography, 1982—. Mem. IEEE, Am. Inst. Ultrasound in Medicine, N.Y. Acad. Scis. Subspecialties: Electrical engineering; Biomedical engineering. Current work: Research in ultrasound tissue characterization, digital signal processing.

KUCHERLAPATI, RAJU S., molecular biologist, geneticist; b. Kakinada, India, Jan. 18, 1943; s. Raju and Varahalu (Rudraraju) K. B.S., Andhra U., India, 1962, M.S., 1962; Ph.D., U. Ill., Urbana, 1972. Research fellow Yale U., New Haven, 1972-75; asst. prof. Princeton (N.J.) U., 1975-82; prof. Ctr. Genetics, U. Ill. Med. Center, Chgo., 1982—. Contbr. articles to sci. jours. Am. Cancer Soc. grantee, 1976-78, 78-80; NSF grantee, 1976-82; NIH grantee, 1980—. Mem. AAAS, Genetics Soc. Am. Subspecialties: Genetics and genetic engineering (medicine); Cell and tissue culture. Current work: Molecular biology of animal cells. Office: 808 S Wood St Chicago IL 60612

KUDRYK, VAL, research mgr.; b. Chipman, Alta., Can., Mar. 2, 1924; came to U.S., 1948, naturalized, 1954; s. John and Nancy K.; m. Martha Cavola, Dec. 1, 1951; children: Val Lawrence, John Anthony, Bruce Timothy. B.A.Sc., U. Alta., Edmonton, 1946; M.A.Sc. (Sheritt-Gordon fellow), U. B.C., Vancouver, 1948; Ph.D. (William Campbell fellow), Columbia U., 1953. Registered profl. engr., N.Y. Project engr. Chemico, N.Y.C., 1950-56; asst. v.p. Nichols Engring., N.Y.C., 1956-59; v.p. Accurate Specialties, N.Y.C., 1959-61; cons., Closter, N.J., 1961-66; mgr. metallurgy dept. Lummus Engring., Bloomfield, N.J., 1966-68; mgr. Central Research dept. Asarco, Inc., South Plainfield, N.J., 1968—. Mem. Closter Bd. Health, 1954-62, Closter City Council, 1969-71, Bergen County Sewer Authority, 1972-77. Mem. Am. Inst. Mech. Engrs., ASTM, Sigma Xi. Subspecialty: Metallurgy. Current work: Mng. corporate research in non ferrous metals. Home: 12 Henmar Dr Closter NJ 07624 Office: Central Research Dept Asarco Inc 901 Oak Tree Rd South Plainfield NJ 07080

KUEHN, THOMAS HOWARD, mechanical engineering educator, thermal science researcher; b. Mpls., Sept. 7, 1949; s. Jerome Henry and Ruth (Sprung) K.; m. Linda Ann Rust, Nov. 28, 1980. B.M.E., U. Minn., 1971, M.S., 1973, Ph.D., 1976. Asst. prof. Iowa State U., Ames, 1976-81, assoc. prof. mech. engring., 1981-83; assoc. prof. U. Minn., Mpls., 1983—; cons. Arkae Devel., Inc., Ames, 1980-81, Engelbrecht & Griffin Architects, Des Moines, 1979-80, Air Conditioning, Inc., Marshalltown, Iowa, 1977-78. Served to 1st lt. U.S. Army, 1973-74. NSF grantee, 1978, 80. Mem. Am. Underground Space Assn., ASME (assoc.), AIAA (assoc.), ASHRAE (assoc., student avd. Iowa State U. 1979-82), Ames Folk Dancers Club (pres. 1977-79). Subspecialties: Environmental engineering; Mechanical engineering. Current work: Heating, ventilating, air conditioning of buildings; natural convection heat transfer; solar thermal energy conversion and utilization. Home: 745 D Griffin Ct Mahtomedi MN 55115 Office: Mech Engring Dept U Minn Minneapolis MN 55455

KUEI, CHIH-CHUNG, computer-aided design programmer, computer graphics analyst; b. Taipei, Taiwan, Republic of China, June 30, 1950; came to U.S., 1978; s. S.Y. and C.H. (Chi) K.; m. Kady Tan, Feb. 28, 1952; 1 child, Fan-Chin. B.A. in Math, Chung-Yuan U., 1972; M.S. in Computer Sci, Stevens Inst. Tech., 1980. Instr. Math. Chung-Yu Jr. Coll., Keelung, Taiwan, 1974-78; research asst. Stevens Inst. Tech., Hoboken, N.J., 1979-80; graphics programmer Computer Sharing Services, Inc., Oakland, Calif., 1980-82; CAD programmer TRICAD, Inc., Milpitas, Calif., 1982—. Mem. Assn. Computing Machinery, IEEE Computer Soc. Subspecialty: Graphics, image processing, and pattern recognition. Current work: Computer-aided design system software development for architecture, engineering, construction appplications. Home: 267 Edwin Way Hayward CA 94544 Office: TRICAD Inc 1655 McCarthy Blvd Milpitas CA 95035

KUEISHIONG, TU, electrical engineer; b. Tainan, Taiwan, Feb. 21, 1951; s. Wen-tai and Chin-Kwan (Hsiao) T. B.S., Taiwan U., 1972; A.M., Harvard U., 1975, Ph.D., 1981. Mem. tech. staff ITT/ATC, Shelton, Conn., 1981-83; resident visitor Bell Labs., Holmdel, N.J., 1983—; adj. prof. U. New Haven, 1982—, U. Bridgeport, 1982—. Served as lt. Taiwan Army, 1972-74. Harvard U. scholar, 1974. Mem. IEEE, Assn. Computing Machinery, Soc. Indsl. and Applied Math. Subspecialties: Distributed systems and networks; Systems engineering. Current work: Computer networks and distributed processing; digital switches, robotics applications. Home: 30 Strathmore Gardens Aberdeen NJ 07747 Office: Bell Labs Crawford Corner Rd Holmdel NJ 07733

KUEMMERLE, NANCY BENTON STEVENS, research microbiologist; b. Marshall, Minn., Mar. 6, 1948; d. Ralph Brookmeyer and Barbara Annette (Burton) Stevens; m. Edgar W. Kuemmerle, Jr., May 15, 1971. B.S. in Chemistry, U. Iowa, 1970; M.S., U. Tex. Health Scis. Ctr., Houston, 1975; postgrad., Ill. State U., 1972-73, U. Tenn.-Oak Ridge Grad. Sch. Biomed. Scis., 1978—. Microbiology technician Ill. State U., Normal, 1972-73; microbiology technologist U. Ill. Med. Center, Peoria, 1973-74; sr. research asst. U. Tenn.-ERDA Comparative Animal Research Lab., Oak Ridge, 1974-76; research assoc. Biology Div. Oak Ridge Nat. Lab., 1976—. Rosalie B. Hite fellow in cancer chemotherapy M.D. Anderson Hosp. and Tumor Inst., Houston, 1971-72. Mem. AAAS, Am. Chem. Soc., Am. Soc. Microbiology, Assn. Women in Sci. Subspecialties: Molecular biology; Gene actions. Current work: Studies of mechanisms and enzymology of DNA repair, replication and recombination. Home: PO Box 126 Seymour TN 37865 Office: Biology Div Oak Ridge Nat Lab PO Box Y Oak Ridge TN 37830

KUESEL, THOMAS ROBERT, engring. co. exec., civil engr.; b. Richmond Hill, N.Y., July 30, 1926; s. Henry N. and Marie D. (Butt) K.; m. Lucia Elodia Fisher, Jan. 31, 1959; children—Robert Livingston, William Baldwin. B.Engring. with highest honors, Yale U., 1946, M.Engring., 1947. With Parsons, Brinckerhoff, Quade & Douglas, 1947—, project mgr., San Francisco, 1967-68, partner, sr. v.p., dir., N.Y.C., 1968—; asst. mgr. engring. Parsons Brinckerhoff-Tudor-Bechtel, San Francisco, 1963-67; vice chmn. OECD Tunneling Conf., Washington, 1970; mem. U.S. Nat. Com. on Tunneling Tech., 1972-74. Contbr. 30 articles to profl. jours.; designer: over 100 bridges, 60 tunnels, numerous other structures in 36 states and 20 fgn. countries, including Newport Suspension Bridge, R.I., 1959-63, NORAD Combat Ops. Center, Colorado Springs, Colo., 1962, San Francisco Bay Area Rapid Transit System, 1963-68, Hampton Roads (Va.) Bridge-Tunnel, 1969-77, Ft. McHenry Tunnel, Balt., 1978-82, Hood Canal Bridge, Washington, 1980-82, subways, Boston, N.Y., Balt., Washington, Atlanta, Pitts., Caracas. Fellow ASCE, Am. Cons. Engrs. Council; mem. Nat. Acad. Engring., Internat. Assn. Bridge, Structural Engring., Brit. Tunnelling Soc., Yale Sci. and Engring. Assn., The Moles, Sigma Xi, Tau Beta Pi. Clubs: Yale (N.Y.C.); Wee Burn (Darien, Conn.). Subspecialty: Civil engineering. Current work: Tunnel design and construction, bridge design and costruction, engineering and construction management, rapid transit systems, major project development. Office: One Penn Plaza 250 W 34th St New York NY 10119

KUFFNER, ROY JOSEPH, chemistry educator; b. N.Y.C., Mar. 15, 1922; s. Joseph and Mae (Fernholz) K.; m. Noel Worrell, Oct. 11, 1947; children: Karl, Leslie, Nicola, Greg; m. Marilyn Adair Langer, July 7, 1978. B.S., Coll. Ozarks, 1944; postgrad., Vanderbilt U., 1949-53. Asst. prof. Emory U., 1953-54; prof. chemistry Lowell (Mass.) U., 1956-67; prof. Chgo. State U., 1967-70; prof., chmn. dept. chemistry Alverno Coll., Milw., 1970—; vis. prof. U. Wis.-Milw., 1981-82. Served with M.C. U.S. Army, 1946-49. Mem. Am. Chem. Soc., AAAS, AAUP, Sigma Xi. Subspecialties: Physical chemistry; Surface chemistry. Current work: Rate of decay of surface tension at air/solution interface of solutions of slightly soluble solutes. Home: 2782 S 60th St Apt 4 Milwaukee WI 53219 Office: 3401 S 39th S Suite 226 Milwaukee WI 53215

KUFTINEC, MLADEN MATIJA, dentistry educator, researcher; b. Zagreb, Yugoslavia, Apr. 18, 1943; came to U.S., 1965, naturalized, 1972; s. Matija and Jelena (Pevec) K.; m. Ljiljana Fucijas, Oct. 26, 1968; 1 dau., Sandra Tatjana. D. Med. Stomatology magna cum laude, U. Sarajevo, Yugoslavia, 1965; cert. in orthodontics, Harvard U. Sch. Dental Medicine—Forsyth Dental Ctr., Boston, 1968; D.M.D., Harvard U.Sch. Dental Medicine, 1972; diploma summer program in pub. health nutrition, Inst. Nutrition of Central Am. and Panama, Guatemala City, Guatemala, 1969; ScD., M.I.T., 1971. Research fellow in orthodontics Harvard U. Sch. Dental Medicine, 1965-68; clin. fellow in orthodontics Forsyth Dental Center, 1966-68, staff assoc., 1972; oral sci. postdoctoral trainee M.I.T., 1968-71, research assoc., instr., 1971-72; assoc. prof. orthodontics Va. Commonwealth U.-Med. Coll. of Va. Sch. Dentistry Basic Sci. and Grad. Faculty, 1972-76, dir. orthodontic dept., 1975-76; prof. orthodontics, chmn dept. orthodontics, dir. grad. program in orthodontics U. Louisville, 1976—, mem., 1976—; trainee in public health nutrition Inst. Nutrition of Central Am. and Panama, 1969; cons. Nat. Bd. Dental Examiners, 1977—; cons. nat. bd. ADA, 1976—; bd. dirs. regional team for treatment craniofacial anomalies, Louisville, 1976—; mem. staff Kosair Hosp., Louisville, 1976—; cons. and mem. cranio-facial anomalies team Crippled Children Commn., Ky. Dept. Human Resources, 1976—; pres. Louisville sect. Internat. Assn. Dental Research/Am. Assn. Dental Research, 1978-80. Contbr. articles to profl. jours. Recipient Spl. Research award Congress of Students in Medicine and Stomatology, Yugoslavia, 1965. Mem. Internat. Assn. Dental Research (Edward H. Hatton Research award 1968, pres. Richmond chpt. 1974-75), Am. Assn. Dental Research (co-chmn. session 1977), Nutrition Today Soc., Am. Assn. Orthodontists, So. Soc. Orthodontists, Va. Orthodontic Soc., Ky. Orthodontic Soc., Sigma Xi., Omicron Kappa Upsilon. Roman Catholic. Subspecialties: Orthodontics; Nutrition (biology). Current work: Development of oro-facial structures; clinical aspects of orthodontics; function and morphology interaction. Home: 5706 Apache Rd Louisville KY 40207 Office: U Louisville Dept Orthodontics 501 S Preston Louisville KY 40292

KUGEL, HENRY W., research physicist; b. Buffalo, Dec. 2, 1940; s. Walter M. and Eugenia M. K.; m. Sharon R. Kugel; 2 children. B.S. in physics, Canisius Coll., 1962; Ph.D. in Nuclear Physics, U. Notre Dame, 1967. Research assoc. U. Wis.-Madison, 1968-70; research fellow Rutgers U./Bell Labs., New Brunswick, N.J., 1970-72; asst. prof. physics Rutgers U., 1972-78; research physicist Princeton U., 1978—, physicist-in-charge PDX neutral beam ops., 1981—. Mem. Am. Phys. Soc., AAAS, Sigma Xi. Subspecialties: Plasma physics; Nuclear fusion. Current work: Coordinates PDX tokamak neutral beam operations, computerization, and diagnostics. Neutral beam science. Office: Princeton U Plasma Physics Lab PO Box 451 Princeton NJ 08544

KUGEL, JEFFREY AURIEL, physician, educator; b. Norristown, Pa., Nov. 8, 1945; s. Julius Dennis and Marjorie Lincoln (Mills) K.; m. Cheryl Ann Ashe, Dec. 30, 1967; children: Gregory Bartlett, Victoria Amanda. B.S. in Biology, Washington and Lee U., 1967; Ph.D. in Physiology, Fla. State U., 1976; M.D., W.Va. U., 1980. Intern Vanderbilt U. Hosp., Nashville, 1980-81, resident in diagnostic radiology, 1981—; instr. Vanderbilt Sonography Sch., 1981—. Served with AUS, 1968-72. Mem. AMA, Radiol. Soc. N. Am., Sigma Xi. Subspecialties: Diagnostic radiology; Imaging technology. Current work: Digital image processing, physiologic medical imaging, residence in diagnostic radiology.

KUH, ERNEST SHIN-JEN, engineering educator; b. Peiking, China, Oct. 2, 1928; s. Zone S. and Tsia (Chu) K.; m. Bettine Chow, Aug. 4, 1957; children: Anthony, Theodore. B.S., U. Mich., 1948; M.S., M.I.T., 1950; Ph.D., Stanford U., 1952. Mem. tech. staff Bell Telephone Labs., Murray Hill, N.J., 1952-56; prof. elec. engring. U. Calif-Berkeley, 1956—, chmn. elec. engring. and computer scis., 1968-72, dean, 1973-80; cons. IBM Research Lab., San Jose, Calif., 1957-62. Co-author: Principles of Circuit Synthesis, 1959, Theory of Linear Active Networks, 1967, Basic Circuit Theory, 1969; contbr. over 70 tech. articles to profl. publs. Recipient Lamme medal Am. Soc. Engring. Edn., 1981. Fellow IEEE (Edn. medal 1981); mem. Nat. Acad. Engring., Academia Sinica. Subspecialty: Electrical engineering. Current work: Microelectronics and computer-aided design. Office: Univ Calif Elec Engring and Computer Scis Dept 231 Cory Hall Berkeley CA 94720

KUHAR, MICHAEL J., neuropharmacologist, educator; b. Scranton, Pa., Mar. 10, 1944; s. Michael S. and Josephine P. (Malaker) K.; m. Joan Barenburg, June 8, 1969; children: David, Kate. Ph.D., Johns Hopkins U., 1970. Prof. neurosci., pharmacology and psychiatry Johns Hopkins U. Sch. Medicine, Balt., 1980—; also cons. Mem. Soc. for Neuosci., Am. Coll. Neuropsychopharmacology, Am. Soc. for Pharmacology and Exptl. Therapeutics, Internat. Brain Research Orgn., Neurochem. Soc. Subspecialty: Neuropharmacology. Current work: Drug and neurohormone receptors; brain function; receptor mapping. Office: Johns Hopkins U Sch Medicine Baltimore MD 21205

KUHI, LEONARD VELLO, astronomer, educator, university provost; b. Hamilton, Ont., Can., Oct. 22, 1936; came to U.S., 1965, naturalized, 1980; s. John and Sinaida (Rose) K.; m. Patricia Suzanne Brown, Sept. 3, 1960; children: Alison Diane, Christopher Paul. B.A.Sc., U. Toronto, 1958; Ph. D., U. Calif. - Berkeley, 1963. Postdoctoral fellow Calif. Inst. Tech. and Mt. Wilson and Palomar obs., 1963-65; asst. prof. astronomy to prof. U. Calif.-Berkeley, 1965—, chmn. dept. astronomy, 1975-76, dean phys. scis., 1976-82, provost Coll. Letters and Sci., 1983—; vis. prof. U. Colo., Boulder, 1969, Coll. de France and Inst. d'Astrophysique, Paris, 1972-73, U. Heidelberg, 1978, 80-81; dir. Assn. Univs. for Research in Astronomy, 1978—; councilor Space Telescope Sci. Inst., 1980—. Contbr. numerous articles to profl. jours. Recipient Alexander von Humboldt sr. U.S. scientist award Humboldt Found., 1980-81; research grantee NSF, 1967—, NASA, 1978. Fellow AAAS; mem. Am. Astron. Soc., Astron. Soc. Pacific (pres. 1978-80), Royal Astron. Soc. Can., Internat. Astron. Union. Democrat. Subspecialties: Optical astronomy; Ultraviolet high energy astrophysics. Current work: Stellar spectroscopy, stellar atmospheres, star formation, early stellar evolution. Office: Astronomy Dept U Calif Berkeley CA 94720

KUHN, HOWARD ARTHUR, consulting metal engineer; b. Pitts., Dec. 6, 1940; s. Howard E. and Selma (Schulze) K.; m. Beverly A. Burke, Dec. 23, 1961; children: Amy, Jeffrey, David, Stephen. B.S., Carnegie Mellon U., 1962, M.S., 1963, Ph.D., 1966. Registered profl. engr. Pa. Mem. faculty Drexel U., Phila., 1966-74, prof. materials engring, 1966-74; prof. metall. engring., prof. mech. engring. U. Pitts., 1975—; pres. Deformation Control Tech., Pitts., 1980—. Author: Powder Metallurgy Processing, 1978; contbr. articles to profl. jours. Mem. citizens adv. com. Babcock Sch. Dist., 1977-80; pres. Richland Athletic Assn., 1978-81. Mem. Am. Soc. Metals, ASME, Am. Powder Metal Inst. Republican. Methodist. Subspecialties: powder metallurgy; Materials processing. Current work: Metalworking, powder metallurgy, computer-aided design and computer-aided manufacturing. Patentee process and apparatus for forging gears from powdered metals. Home: 5408 Peach Dr Gibsonia PA 15044 Office: 231 Benedum Hall Pittsburgh PA 15261

KUHN, LESLIE A., physician, cardiologist; b. South Falls, N.Y., May 10, 1924; s. I. Russel and Mary (Rosenberg) K.; m. Edna Berk, Sept. 16,. 1950; children: Amy B., Karen H.B. Student, Harvard U., 1942-44; M.D., SUNY-Bklyn., 1948. Diplomate: Am. Bd. Internal Medicine; Subcert. in cardiovascular disease. Attending cardiologist, clin. prof. medicine Mt. Sinai Med. Sch., N.Y.C., 1972—; cons. to CCUs, 1978—; cons. cardiologist VA Hosp., Bronx, N.Y., 1976—. Served to capt. M.C. U.S. Army, 1954-55. Fellow Am. Coll. Cardiology, Am. Coll. Chest Physicians, ACP., N.Y. Cardiol Soc. (pres. 1979-80). Subspecialty: Cardiology. Current work: Experimental myocardial infarction with shock-circulatory support. Office: Mt Sinai Med Ctr 1176 Fifth Ave New York NY 10029

KUHN, WILLIAM RICHARD, atmospheric science educator; b. Columbus, Ohio, May 7, 1937; s. Marvin Jacob and Esther M. (Hartranft) K.; m. Dorothy K. Kuhn; children: Jeffrey Richard, Timothy Scott, Tracy Lynn. Ph.D., U. Colo., 1967. Mem. faculty dept. atmospheric and oceanic sci. U. Mich., Ann Arbor, 1967—, chmn. dept., 1980—. Mem. Am. Geophys. Union, Am. Astron. Soc. Subspecialties: Climatology; Aeronomy. Current work: Climate of early earth, influence of water vapor and carbon dioxide on earth's temperature, evolution of Mars atmosphere. Home: 2961 Dexter Rd Ann Arbor MI 48103 Office: Dept Atmospheric and Oceanic Sci U Mich Space Physics Bldg Ann Arbor MI 48109

KUIPER, THOMAS BERNARDUS HENRICUS, astronomer; b. Amersfoort, Netherlands, July 14, 1945; s. Antonius Henricus and

Elisabeth Henrika Wilhelmina (Nieuwendijk) K.; m. Eva Nelida Rodgiquez, June 13, 1970. B.Sc. in Physics, Loyola Coll., Montreal, Que., Can., 1966; Ph.D. in Astronomy, U. Md., College Park, 1973. Grad. asst. U. Md., 1966-73; NRC resident research assoc. Jet Propulsion Lab., Pasadena, 1973-75, sr. scientist, 1975-77, mem. tech. staff, 1977—. Contbr. articles to profl. jours. Served with Royal Can. Navy Res., 1962-67. Recipient Dorval Sch. Commn. Bursary, 1962-66. Mem. Am. Astron. Soc., Can. Astron. Soc., Internat. Astron. Union, Internat. Union Radio Sci. Subspecialty: Radio and microwave astronomy. Current work: Studies of the composition and kinematics of interstellar and circumstellar gas; devel. of space instruments and missions in far Infrared astronomy; cosmic evolution. Office: Jet Propulsion Lab T-1166 Pasadena CA 91109

KUIST, CHARLES HOWARD, chemical company executive; b. West Cheser, Pa., Apr. 24, 1931; s. Howard T. and Leone I. (Marquis) K.; m. Mary E. Seebode, Apr. 9, 1955; children: Ellen D., Pamela A., Jennifer E., Timothy H. A.B., Lafayette Coll., 1953. Chemist Extrax Co., Bklyn., 1953-55; asst. instr. chemistry Newark Coll. Engring., 1955-59; mgr. research Nat. Starch & Chem. Corp., Plainfield, N.J., 1959-74; sr. v.p. Chomerics, Woburn, Mass., 1974—. Mem. Mendham (N.J.) Bd. Edn., 1970-74; v.p. Morris County Sch. Bds. Assn., Morristown N.J., 1972-74; mem. Ednl. Service Commn., Morristown, 1973; legis. del. N.J. Sch. Bds. Assn., Trenton, N.J., 1972-74. NSF summer fellow radiation project U. Notre Dame, 1959; Newark Coll. Engring. summer research fellow, 1958. Fellow Am. Inst. Chemists; mem. AAAS, Am. Chem. Soc., Am. Phys. Soc., ASTM. Subspecialties: Physical chemistry; Composite materials. Current work: Physical chemistry of composite materials, chemistry and physics of interfaces, synthesis of electrically conductive composite materials, photo-chemistry and radiation chemistry of polymers. Patentee in field. Office: Chomerics 77 Dragon Ct Woburn MA 01888 Home: Box 437 Byfield MA 01922

KUKIN, IRA, chem. co. exec.; b. N.Y.C., Apr. 4, 1924; s. William and Clara (Wachtel) K.; m. Doris Liener, June 19, 1954; children: Marrick Lee, Lori Sue, Jonathan Liener. B.S., CCNY, 1945; M.A., Harvard U., 1950, Ph.D., 1951. Dir. fuels research Gulf Research and Devel. Co., 1951-57; dir. research Sonneborn Chem. and Refinery Corp., 1958-63; founder, pres. Appollo Technologies, Inc., Whippany, N.J., 1963—. Contbr. articles to profl. jours. Trustee Yeshiva U. Served with U.S. Army, 1946-47. Recipient 1st place award Pollution Engring. Product Advancement Competition, 1977. Mem. Nat. Assn. Corrosion Engrs., Nat. Petroleum Refiners Assn., Am. Chem. Soc., ASME. Jewish. Subspecialties: Combustion processes; Resource management. Current work: Chemical system for energy conservation and air-waterpollution control; improving performance of electrostatic precipitators; dedusing systems for coals and minerals. Patentee in field. Office: 1 Appolo Dr Whippany NJ 07981

KUKULKA, CARL GEORGE, neurophsiologist, educator; b. Olean, N.Y., Mar. 3, 1949; s. Edward Karl and Mary Ann K.; m. Nancy Landfear, Sept. 11, 1971; 1 dau., Erica. B.S., Ithaca Coll., 1971; Ph.D. (scholar), Med. Coll. Va., Va. Commonwealth U., 1979. Lic. phys. therapist, N.Y. State, Iowa. Vis. asst. fellow John B. Pierce Found., New Haven, 1979-81; asst. prof. dept. phys. therapy Ithaca (N.Y.) Coll., 1981-82, U. Iowa Med. Coll., Iowa City, 1982—. Contbr. articles and abstracts to sci. jours. Muscular Dystrophy Assn. fellow, 1979-81. Mem. Soc. for Neuroscis., N.Y. Acad. Scis., Sigma Xi. Subspecialty: Neurophysiology. Current work: Human motor control, neural control of skeletal muscle. Office: 2600 Steindler Bldg Iowa City IA 52242

KULCINSKI, GERALD LAVERN, nuclear engineering educator; b. La Crosse, Wis., Oct. 27, 1939; s. Harold Franklin and June (Kramer) K.; m. Janet Noreen Berg, Nov. 25, 1961; children: Kathryn, Brian, Karen. B.S. in Chem. Engring, U. Wis., 1961, M.S. in Nuclear Engring, 1962, Ph.D., 1965. Researcher Los Alamos Sci. Lab., 1963; sr. research scientist Battelle N.W. Lab., Richland, Wash., 1965-69, tech. group leader, 1969-71; lectr. Center for Grad. Study, Richland, 1969-71; asso. prof. nuclear engring. U. Wis., Madison, 1972-74, prof., 1974—, dir. fusion engring. program, 1973-74, 79—; mem. fusion materials coordinating com. Dept. Energy; mem. INTOR Design Team, Vienna, 1978-81. Editor: Proc. 2d Topical Meeting on Fusion Tech, 1976; assoc. editor Nuclear Engring. and Design/Fusion, 1983—; Contbr. numerous articles on materials for fission and fusion reactors to profl. jours. Recipient Acad. award Big 10 Conf., 1961, Curtis McGraw research award Am. Soc. Engring. Edn., 1978; AEC grantee, 1961-62, 64-65. Fellow Am. Nuclear Soc. (treas. Richland sect. 1971-72, program chmn. 2d topical meeting on fusion tech. 1976, Disting. Achievement award 1980); mem. Am. Soc. Engring. Edn., Red Triangles (pres. Madison chpt. 1976—). Subspecialties: Nuclear fission; Nuclear fusion. Home: 6013 Greentree Rd Madison WI 53711 Office: 1500 Johnson Dr Madison WI 53706

KULCZYCKI, ANTHONY, JR., medical researcher; b. Easton, Pa., Dec. 17, 1944; s. Anthony and Mae (Yaworski) K.; m. Judith Mary Brokaw, May 31, 1969; children: Alexander, Amy-Elizabeth. A.B., Princeton U., 1966; M.D., Harvard U., 1970. Diplomate: Am. Bd. Internal Medicine, Am. Bd. Allergy and Immunology. Intern, resident Buffalo Gen. and E.J. Meyer hosps., 1970-72; research assoc. NIAMDD, NIH, 1972-74; NIH research fellow, 1974-76; asst. prof. Washington U., St. Louis, 1977-82, assoc. prof., 1982—; assoc. investigator Howard Hughes Med. Inst., 1978—. Contbr. articles to profl. jours. Served as lt. comdr. USPHS, 1972-74. Recipient J.D. Lane award. Mem. Am Assn. Immunologists, Am. Acad. Allergy, Collegium Internationale Allergologicum, Ethical Soc. St. Louis. Clubs: Princeton, Harvard (St. Louis). Subspecialties: Immunology (medicine); Allergy. Current work: Allergic diseases and other immunologic diseases. Office: Box 8122 660 S Euclid Saint Louis MO 63110

KULHAWY, FRED HOWARD, civil engineering educator; b. Topeka, Sept. 8, 1943; s. Fred and Gloria Katherine (Hahn) K.; m. Gloria Ianna, Sept. 4, 1966. B.S. in Civil Engring, Newark Coll. Engring., 1964, M.S., 1966; Ph.D., U. Calif.- Berkeley, 1969. Registered profl. engr., N.Y., N.J., Pa., Calif. Asst. instr. Newark Coll. Engring., 1964-66; soils engr. Storch Engrs., East Orange, N.J., 1966; research asst./jr. research specialist U. Calif.- Berkeley, 1966-67, 67-69; assoc. Raamot Assoc., Syracuse, N.Y., 1969-71; asst. to assoc. prof. Syracuse U., 1969-73, 73-76; assoc. prof. Cornell U., Ithaca, N.Y., 1976-81, prof. civil engring., 1981—; cons. in field. Co-author: Numerical Methods in Geotechnical Engineering, 1977; editor: Recent Developments in Geotechnical Engineering for Hydro Projects, 1981; contbr. articles to profl. jours. Fellow Geol. Soc. Am.; mem. ASCE (pres. 1974-75, Edmund Friedman young engr. award 1974, Walter L. Huber civil engring. research prize 1982), Am. Soc. Engring. Edn., ASTM, Assn. Engring. Geologists, Internat. Assn. Engring. Geology, Internat. Soc. Rock Mechanics, Internat. Soc. Soil Mechanics and Found. Engring., Transp. Research Bd., U.S. Com. on Large Dams, Underground Tech. Research Council, Chi Epsilon. Subspecialties: Civil engineering; Engineering geology. Current work: Educator, researcher, and consultant in geotechnical engineering with current emphasis in foundations, rock mechanics, dams, underground structures and computer applications. Home: 113 Orchard St Ithaca NY 14850 Office: Cornell U Hollister Hall Ithaca NY 14853

KULIK, MARTIN MICHAEL, research plant pathologist; b. N.Y.C., Apr. 20, 1932. B.S., Cornell U., 1954; M.S., La. State U., 1956, Ph.D., 1959. Commd. 1st lt. Crops Div. U.S. Army Chem. Corps, Ft. Detrick, Md., 1959-61; plant pathologist Seed Br. Agrl. Mktg. Service, U.S. Dept. Agr., 1961-63; research plant pathologist Seed Research Lab. Agrl. Research Service, 1963—. Contbr. articles to profl. jours. Served with U.S. Army, 1959-61. Mem. Am. Phytopath. Soc., Am. Soc. Agronomy, Mycol. Soc. Am., Can. Phytopath. Soc. Subspecialty: Plant pathology. Current work: Seedborne fungi and diseases. Office: US Dept Agr Seed Research Lab B-006 BARC-West Beltsville MD 20705

KULKARNI, PRASAD SHRIKRISHNA, eye research educator; b. India, May 22, 1943; came to U.S., 1966, naturalized, 1980; s. Shrikrishna M. and Kamal S. (Kulkarni) K. M.S., SUNY Downstate Med. Ctr., N.Y.C., 1971, Ph.D., 1974. Teaching asst. Downstate Med. Ctr., 1968-74; fellow dept. pharmacology Washington U., St. Louis, 1974-76; research assoc. eye research div. Columbia U., N.Y.C., 1976-78; asst. prof. ocular pharmacology Columbua U., 1978—. Contbr. articles to profl. jours. Nat. Inst. for Eye Research grantee, 1979—. Mem. Am. Soc. for Pharmacology and Exptl. Therapeutics, Assn. for Research in vision and Ophthalmology, Internat. Soc. for Eye Research. Subspecialties: Pharmacology; Ophthalmology. Current work: Mediators of ocular inflammaton, development and mechanism of action of anti-inflammatory drugs, pharmacology of coronary blood vessels.

KUMAR, GANESH NARAYANAN, polymer chemist, research administrator; b. Madras, India, Oct. 4, 1948; came to U.S., 1970; d. Ganapathy and Sundara (Iyer) Narayanan; m. Prema Kumar, June 19, 1975; children: Bharat, Ramya. B.S., U. Madras, 1970; M.S., Clarkson Coll. Tech., Potsdam, N.Y., 1972; Ph.D., Case Western Res. U., 1975. Tech. specialist Xerox Corp., Webster, N.Y., 1974-78; sr. research scientist Johnson & Johnson Dental Products, East Windsor, N.J., 1978-80, mgr. polymer research, 1980-81, dir. polymer research, 1981-83; dir. polymer sci. Vistakon, Inc. subs. Johnson & Johnson, Jacksonville, Fla., 1983—; mem. adj. faculty U. Rochester, 1977. Recipient Gold medal U.Madras, 1970; research fellow Case Western Res. U., 1971-74. Mem. Am. Chem. Soc., Am. Phys. Soc., N.Am. Thermal Analysis Soc., Internat. Assn. Dental Research, Sigma Xi. Subspecialties: Polymer chemistry; Polymer engineering. Current work: Polymer physics and chemistry. Patentee solvent for cellulose. Office: Vistakon Inc PO Box 10157 Jacksonville FL 32247

KUMAR, MUNEENDRA, research geodesist, consultant; b. Dehra Dun, Uttar Pradesh, India, Aug. 24, 1931; came to U.S., 1969, naturalized, 1976; s. Banmali and Chandra (Kanta) Krishna; m. Devi Mulchandani, Feb. 28, 1976; 1 dau., Seema Malini. B.Sc., Agra (India) U., 1951, M.Sc., 1953; M.S., Ohio State U., Columbus, 1972, Ph.D., 1976. Asst. prof. Agra U., 1953-54; superintending surveyor Survey of India, Dehra Dun, 1955-66; tech. adv. to dir. Mil. Survey of India, Delhi, 1966-68; to dir. Aero. Survey India, New Delhi, 1968-69; geodetic engr. Matz, Childs & Assocs., Rockville, Md., 1969-70; grad. research assoc. Ohio State U., Columbus, 1971-76; research geodesist Nat. Geodetic Survey, Rockville, Md., 1976-81, Def. Mapping Agy., Washington, 1981—. Author: Vector Analysis, 1954, also numerous tech. papers. Served to maj. C.E. Indian Army, 1966-68. Recipient Gold medals Agra U., 1951; Outstanding Service awards NOAA, 1981. Mem. Indian Inst. Surveyors, Am. Geophys. Union, Am. Congress on Surveying and Mapping, Marine Tech. Soc., Am. Soc. Photogrammetry. Hindu. Current work: Inertial positioning, gravity gradiometry, ocean positioning, Doppler control, geodesy teaching/consultancy. Home: 10625 Wayridge Dr Gaithersburg MD 20879

KUMAR, NIRMAL, mech. engr., cons.; b. Darabello, Sindh, India, Apr. 14, 1941; came to U.S., 1975, naturalized, 1982; s. Shyam and Leelawati (Kewal) Sunder; m. Leela, May 21, 1970; children: Leena, Naresh. B.Sc. in Engring, Ranchi (India) U., 1964. Engr. Heavy Engring. Corp., Ranchi, 1964-66, sr. engr., 1966-75, EIMCO Process Equipment Co. (subs. Baker Internat.), Salt Lake City, 1976—; cons. gears, gear drives, mechanisms. Mem. ASME. Lodge: Lions. Subspecialty: Mechanical engineering. Current work: Gear tech.: design, manufacture and testing. Home: 4260 S 1400 E Salt Lake City UT 84117 Office: PO Box 300 Salt Lake City UT 84110

KUMAR, PANGANAMALA RAMANA, mathematics educator, researcher; b. Nagpur, India, Apr. 21, 1952; came to U.S., 1973; s. P.B. and P. Kamala (Rayudu) Murthy; m. P. Jayashree, Jan. 22, 1982. B.Tech., Indian Inst. Tech., Madras, 1973; M.S., Washington U., St. Louis, 1975, D.Sc., 1977. Research asst. Washington U., St. Louis, 1973-77; asst. prof. U. Md. Baltimore County, Balt., 1977-82, assoc. prof., 1982—. Mem. IEEE (assoc. editor transactions on automatic control 1982—, chmn, stochastic control com., mem. info. dissemination com. Control Systems Soc. 1982—), Soc. Indsl. and Applied Math., Am. Math. Soc. Subspecialties: Electrical engineering; Applied mathematics. Current work: Control systems theory, mathematical system theory. Office: Dept Math U Md Baltimore County 5401 Wilkens Ave Baltimore MD 21228 Home: 304 L Cedar Run Pl Baltimore MD 21228

KUMAR, PRASANNA K., physicist; b. Bangalore, India, Oct. 23, 1937; came to U.S., 1965, naturalized, 1980; d. Krishnamurthy Seshappa and Nagarathna (Krishnamurthy) Vastare; m. Savitri Kumar, Aug. 22, 1971; children—Monisha Anjali, Pratima Valli. B.Sc., Mysore U., Bangalore, India, 1956, 1958, M.Sc., 1959; M.A., Temple U., Phila., 1967, Ph. D., 1973. Physicist U. Calif., Lawrence Berkeley Lab., Berkeley, 1974-75; asst. prof. physics Spring Garden Coll., Phila., 1975-76; research assoc. physics Drexel U., Phila., 1976-78, adj. assoc. prof. physics, 1978-80; asst. prof. med. physics U. Pa. Med. Sch., Phila., 1980—; cons. physicist. Contbr. articles to profl. jours. Mem. Am. Inst. Physics, Am. Assn. Physicists in Medicine, Sigma Xi, Sigma Pi Sigma. Hindu. Subspecialties: Nuclear physics; Cancer research (medicine). Current work: Medical research involving powerful (mega volt) X-rays in the treatment of cancer by radiotherapy techniques. Home: 55 Lakeview Dr Cherry Hill NJ 08003 Office: Dept. Radiation Therapy Hosp U Pa 3400 Spruce St Philadelphia PA 19104

KUMAR, VIJAYA BUDDHIRAJU, microbiologist, educator; b. India, May 2, 1945; s. Rajabhushana Rao Buddhiraju and Manorama (Sankara) Rao; m. Vijaya Mysore Lakshmi, Oct. 26, 1971; children: Mamokiran, Chakradhar. B.S., Osmania U., Hyderabad, India, 1963, M.S., 1965, Ph.D., 1972. Postdoctoral fellow Washington U., St. Louis, 1972-74, research coordinator, 1974-75, research asst. prof. infectious disease div., 1975—. Mem. Assn. Mycologists Am., Am. Soc. Microbiology. Subspecialties: Biochemistry (medicine); Microbiology (medicine). Current work: Evaluation of molecular events accompnaying dimorphism in microbes; dimorphism genetic-variations in strains of fungi, mechanism of fungal infection. Home: 2725 Creekmont Ln Saint Louis MO 63125 Office: Infectious Disease Div Washington U Sch Medicine Box 8051 Saint Louise MO 63110

KUMBARACI, NURAN MELEK, physiologist, educator; b. Istanbul, Turkey, Apr. 3, 1944; d. Celal and Aysel K.; m. Francis T. Jones, Jan. 1981. B.S., Robert Coll., Turkey, 1966; M.S., Columbia U., 1973, M.A., 1975, M.Phil., 1976, Ph.D., 1977. Research and devel. chemist Eczacibasi Pharm. Co., Istanbul, 1966-71; research asst. chem. engring. Columbia U., N.Y.C., 1971-72, predoctoral fellow dept. physiology, 1972-77, postdoctoral fellow, 1977-79; asst. prof. chemistry Stevens Inst. Tech., Hoboken, N.J., 1979—. Contbr. articles to profl. pubIs. NIH fellow, 1977-79; USPHS fellow, 1972-77; Esso scholar, 1971-72. Mem. Am. Physiol. Soc., Soc. Neurosci., N.Y. Acad. Scis., N.J. Acad. Scis., Sigma Xi, Phi Lambda Upsilon. Subspecialties: Neurophysiology; Comparative physiology. Current work: Teaching and research at college. Research on testing effects of neurotoxic agents on acetylcholinesterase activity and synaptic transmission. Analysis of factors which control skeletal muscle contraction. Office: Dept Chemistry and Chem Engring Stevens Inst Tech Hoboken NJ 07030 Home: 692 Stewart St Ridgefield NJ 07657

KUNCL, RALPH WILLIAM, physician, educator; b. Glendale, Calif., July 15, 1948; s. William J. and Lois (Mears) K.; m. Bonnie Sugar, June 21, 1975; children: Parker R.S., Margaux A. S. A.B. magna cum laude, Occidental Coll., 1970; Ph.D., U. Chgo., 1975, M.D., 1977. Diplomate: Am. Bd. Psychiatry and Neurology. Resident in medicine U. Chgo., 1977-78, in neurology, 1978-80; Muscular Dystrophy Assn. neuromuscular fellow Johns Hopkins U., 1980-83, asst. prof., 1983—. Contbr. articles to profl. jours. Haines scholar, 1967; Med. Scientist tng. program fellow, 1970-77. Mem. Soc. Neurosci., Am. Acad. Neurology, Am. Soc. Neurol. Invstigation, N.Y. Acad. Scis., Sigma Xi, Psi Chi. Subspecialty: Neurology. Current work: Muscular dystrophy.

KUNDU, MUKUL R., astronomer, educator; b. Calcutta, India, Feb. 10, 1930; s. Makhan L. and Monoroma K.; m. Ranu X. Paul, Sept. 9, 1958; children: Krishna, Rina, Sanjit. B.S., Calcutta U., 1949, M.S., 1951; D.Sc., U. Paris, 1957. Research assoc. Paris Obs., 1956-58; sr. research fellow Nat. Phys. Lab., Delhi, India, 1958-59; research assoc. U. Mich., 1959-62; assoc. prof. Cornell U., Ithaca, N.Y., 1962-65, Tata Inst. Found. Research, Bombay, 1965-68; prof. astronomy U. Md., College Park, 1968—, dir. astronomy, 1978—. Contbr. articles to profl. jours.; author: Solar Radio Astronomy, 1965; editor: Radio Physics of the Sun, 1980. Recipient Anantha Kr. Meml. prize Calcutta U., 1958, Krishna Lal de Gold medal, 1959; NRC fellow, 1968-74; Humboldt Found. U.S. Sr. Scientist awardee, 1978; Smithsonian Instn. awardee, 1980, 82. Mem. Am. Astron. Soc., Internat. Astron. Union, Internat. Radio Sci. Union, IEEE, Astron. Soc. India. Subspecialties: Radio and microwave astronomy; Solar physics. Home: 9013 Gettysburg Ln College Park MD 20740 Office: Astronomy Program U Md College Park Md 20742

KUNG, CHING, molecular biologist, geneticist, educator, researcher; b. Canton, China, Apr. 30, 1939; came to U.S., 1964, naturalized, 1971; s. Man and Shing-Wan (Yuen) K.; m. Joan R. Rajala, July 20, 1965; children: Andrew, Julia, Cleo. Ph.D., U. Pa., 1968. Postdoctoral research Ind. U., 1968-70, UCLA, 1970-71; asst. prof., assoc. prof. U. Calif.- Santa Barbara, 1972-74; assoc. prof. molecular biology and genetics U. Wis.- Madison, 1974-77, prof., 1977—; mem. genetics study sect. NIH, 1982—. Editor: Jour. Neurogenetics; contbr. articles and abstracts to sci. jours. Romnes research fellow U. Wis., 1977. Mem. AAAS, Am. Soc. Cell Biology. Subspecialties: Membrane biology; Neurobiology. Current work: Genetic dissection of the excitable membrane of Paramecium; uses of genetic, electrophysiological and biochemical methods to study membrane bioelectric functions.

KUNG, PATRICK CHUNG-SHU, biotechnology exec.; b. Nanjing, China, July 10, 1947; s. Tao and Yuing (Li) K.; m. Rita W. Wu, Feb. ll, 1980; 1 dau. Julia. B.S., FuJen U., Taiwan, 1968; Ph.D., U. Calif.- Berkeley, 1974. Research fellow M.I.T., 1974-77; staff scientist DuPont Co., Wilmington, Del., 1977-78; sr. research fellow Ortho Pharm. Co., Raritan, N.J., 1978-81; v.p. research Centocor, Malvern, Pa., 1982—; vis. prof. Columbia U., N.Y.C., 1981—. Contbr. articles to profl. jours. Recipient Philip Hoffman award Johnson & Johnson Co., 1979. Mem. Am. Assn. Immunologists, N.Y. Acad. Scis., Sigma Xi. Roman Catholic. Subspecialties: Immunology (medicine); Transplantation. Office: Centocor 244 Great Valley Pkwy Malvern PA 19355

KUNG, SHAIN-DOW, plant molecular biologist; b. Shandong, China., Mar. 14, 1935; s. Chao-tzen and Chih (Zhu) K.; m. Helen C. C. Fu, Sept. 5, 1964; children: Grace, David, Andrew. Ph.D., U. Toronto, 1968. Research fellow Hosp. for Sick Children, Toronto, 1968-70; biologist UCLA, 1971-74; asst. prof. biology U. Md. Balt. County, Balt., 1974-77; assoc. prof., 1977-82, prof., 1982—, acting chmn., 1982—. Contbr. chpts. to books, articles to profl. jours. Recipient Philip Morris award for Disting. Achievement in tobacco sci., 1979; disting. scholar Nat. Acad. Sci., 1981; Fulbright awardee, 1982-83; NSF, NIH grantee. Mem. Am. Soc. Plant Physiologists, AAAS. Subspecialties: Genetics and genetic engineering (biology); Molecular biology. Current work: Studying the genetics, evolution, structure and function of RuBPCase and the organization, structure, evolution, expression of higher plant chloroplast genome using the recombinant DNA technology. Office: Dept Biol Scis U Md Baltimore County Baltimore MD 21228

KUNIN, ISAAK A., mechanical engineering educator; b. Kharkow, USSR, Sept. 11, 1924; s. Abraham L. and Sarra B. (Rosetshtein) K.; m. Inessa M. Dvoskina; 1 son, Boris. M.S., Poly. Inst., Leningrad, USSR, 1952, Ph.D., 1958; D.Scis., Inst. Thermophysics, Acad. Sci. USSR, 1968. Engr. turbogenerator factory, USSR, 1952-56; researcher Mining Inst., Novosibirsk, USSR, 1956-63; prof., chmn. dept. physics Inst. Thermophysics, Acad. Sci., Novosibirsk, 1963-74; prof., chmn. dept. math. Electrotech. Inst., Novosibirsk, 1974-79; prof. dept. mech. engring. U. Houston, 1979—. Author 6 books in field; contbr. articles to profl. jours. Recipient award of excellence Halliburton Edn. Found., 1982. Mem. ASME, Acad. Mechanics, Soc. Indsl. and Applied Math. Subspecialties: Solid mechanics; Applied mathematics. Current work: Continuum mechanics, elasticity, media with microstructure, linear and nonlinear waves, defects in solids, dislocations, quantization, lie groups and their representations, gauge theories, boundary value problems. Patentee in field. Home: 8219 Twin Tree Houston TX 77071 Office: Dept Mech Engring U Houston TX 77004

KUNISCH, KARL, mathematics educator; b. Linz, Austria, Sept. 16, 1952; s. Karl Wilhelm August and Johanna Elisabeth (Neuhauser) K.; m. Brigitte Almhofer, Aug. 27, 1976; children: Katharina, Elisabeth. M.A., Northwestern U., 1975; M.Sc., Technische Universitat Graz, Austria, 1975, Ph.D., 1978. Asst. prof. Technische Universitat Graz, 1976-80, universitats dozent, assoc. prof., 1980—; vis. asst. prof. Brown U., 1979-80; vis. assoc. prof. U. Okla., 1982-83. Recipient Koerner award Koerner Found., Vienna, 1979; Max Kade stipend Max Kade Found., 1982; Fulbright scholar, Vienna and N.Y.C., 1980. Mem. Oesterreichische Mathematische Gesellschaft, Am. Math. Soc., Soc. Indsl. and Applied Math., Italian Math. Soc. Roman Catholic. Subspecialties: Applied mathematics; Numerical analysis. Current work: Parameter identification and estimation, differential equations, applied functional analysis. Address: Institut fur Mathematik Technische Universitat Graz Graz Austria.

KUNKEL, HENRY GEORGE, immunologist; b. N.Y.C., Sept. 9, 1916; s. Louis O. and Johanna C. K.; m. Betty Jean Martens, Jan. 8, 1949; children—Louis M., Henry G., Ellen L. A.B., Princeton U., 1938; M.D., Johns Hopkins U., 1942, Uppsala U., Sweden. Intern Bellevue Hosp., N.Y.C., 1942-43; mem. faculty dept. immunology Rockefeller U., N.Y.C., 1946—, Abby Rockefeller Mauzé prof., 1976—; adj. prof. medicine Cornell U. Med. Sch., 1971—; bd.

councilors Arthritis Inst., NIH, Sloan Kettering Inst., Brookhaven Labs., Cancer Center Columbia Med. Sch. Editor: Advances in Immunology, Jour. Exptl. Medicine. Served with USNR, 1945-46. Recipient Lasker award; Gairdner award; Hazen award; Waterford award; Pasteur medal; Avery-Landsteiner award; Kovalenko medal; medal N.Y. Acad. Medicine. Mem. Nat. Acad. Scis., Assn. Am. Physicians, Am. Soc. Clin. Investigation (pres. 1963), Am. Assn. Immunologists (pres. 1975), Harvey Soc. (lectr. 1965). Clubs: Princeton, Interurban Clin. Subspecialty: Immunology (medicine). Home: 2 Sutton Pl S New York NY 10022 Office: Rockefeller Univ New York NY 10021

KUNTZ, JAMES EUGENE, forest pathologist, educator, farm administrator; b. Leipsic, Ohio, Aug. 14, 1919; s. Edward Charles and Bessie Douds (Sherrard) K.; m. Helen Louise, Dec. 28,1942; children: James, David, Patricia. B.A., Ohio Wesleyan U., 1941; M.S., U. Wis.-Madison, 1942, Ph.D., 1945. Lab. asst. Ohio Wesleyan U., 1939-41; grad. research asst. U. Wis.-Madison, 1941-45, asst. prof. to prof. plant pathology, 1946—; pathologist and breeder Wis. Seed Co., 1945-46. Contbr. articles on forest pathology to profl. jours. Mem. AAAS, Am. Forestry Assn., Internat. Soc. Arboriculture, Phi Beta Kappa, Sigma Xi, Omega Delta Kappa, Phi Mu Alpha. Subspecialties: Plant pathology; Integrated pest management. Current work: Woodland and urban forestry—disease control. Office: Dept Plant Pathology Russell Labs U Wis 1630 Linden Dr Madison WI 53706

KUNTZ, ROBERT ROY, chemistry educator, researcher; b. Barry, Ill., Apr. 10, 1937; s. John Henry and Beatrice D. (Borrowman) K.; m. Joan R. Baumgartner, July 19, 1959; children: Deborah, Kenneth. B.A., Culver-Stockton Coll., Canton, Mo., 1959; M.S., Carnegie Inst. Tech., 1962, Ph.D., 1963. Asst. prof. chemistry U. Mo., Columbia, 1962-67, assoc. prof., 1967-71, prof., 1971—; assoc. program dir. NSF, Washington, 1973-74. Contbr. articles to profl. jours. Mem. Am. Chem. Soc., Internat. Photochemi. Soc., Am. Soc. Photobiology. Subspecialties: Photochemistry; Physical chemistry. Current work: Photochemistry and photophysics of indoles. Office: U Mo 123 Chemistry Columbia MO 65211

KUNTZMAN, RONALD GROVER, pharmaceutical researcher; b. Bklyn, Sept. 17, 1933; s. Herman and Fanny (Brand) K.; m. Bernice Russman, May 29, 1955; children: Fred, Gary. B.S., Bklyn. Coll., 1955; M.S., George Washington U., 1957; Ph.D. in Biochemistry, 1962. Biochemist lab. chem. pharmacology Nat. Heart Inst., NIH, Bethesda, Md., 1955-62; sr. biochemist Wellcome Research Labs., Burroughs Wellcome & Co. (U.S.A.) Inc., Tuckahoe, N.Y., 1962-66, dep. head biochem. pharmacology dept., 1967-70; asso. dir. dept. biochemistry and drug metabolism Hoffmann-La Roche Inc., Nutley, N.J., 1970-71, asso. dir. biol. research, 1972-73, dir. therapeutics research, 1973-79, asst. v.p., 1974-81, dir. pharm. research and devel., 1980-81, v.p. pharm. research and devel., 1981—. Mem. editorial bd.: Biochem. Pharmacology, 1966-68, Jour. Pharmacology and Exptl. Therapeutics, 1968-75, Neuropharmacology, 1970—, Xenobiotica, 1970—, Archives Biochemistry and Biophysics, 1971-78, Life Scis., 1973-78; contbr. numerous articles to sci. jours. Mem. Am. Soc. Pharmacology and Exptl. Therapeutics (John Jacob Abel award 1969, nominating com. 1972, exec. com. div. drug metabolism 1973-76, chmn. nominating com. div. drug metabolism 1977, chmn. div. 1978-81, sec.-treas. 1981-83, mem. council 1981-83), Am. Soc. Biol. Chemists, Am. Coll. Neuropsychopharmacology, Soc. Toxicology, AAAS, Sigma XI. Subspecialties: Pharmacology; Biochemistry (medicine). Current work: Directs development of novel therapeutics; metabolism of drugs, steroids and carcinogens; biogenic amine metabolism; pharmocokinetics. Home: 12 Augustine Ave Ardsley NY 10502 Office: 340 Kingsland St Nutley NJ 07110

KUNZ, THOMAS HENRY, biology educator; b. Kansas City, Mo., June 11, 1938; s. William Henry and Edna Johanna (Dornfield) K.; m. Margaret Louise Brown, Dec. 27, 1962; children: Pamela Lyn, David Thomas. B.S., Central Mo. State U., 1961, M.S., 1962; M.A., Drake U., 1968; Ph.D., U. Kans.-Lawrence, 1971. Instr. biology Shawnee Mission (Kans.) Schs., 1962-67; asst. prof. biology Boston U., 1971-77, assoc. prof., 1977—, dir. grad. studies, 1978-81, assoc. chmn. biology, 1981. Editor: Ecology of Bats, 1982; contbr. over 40 articles in field to profl. jours. Adv. bd. Sch. for Field Studies, Boston, 1982. NSF grantee, 1973-81; EPA contract, 1979. Mem. Am. Soc. Mammalogists, Ecol. Soc. Am., AAAS, Sigma Xi. Congregationalist. Subspecialties: Ecology; Behavioral ecology. Current work: Research on the behavorial and physiological ecology of temperate and tropical bats, life history strategies, energy allocation, social behavior, feeding ecology. Office: Boston U 2 Cummington St Boston MA 02215

KUNZLER, JOHN EUGENE, physicist; b. Willard, Utah, Apr. 25, 1923; s. John Jacob and Freida (Meier) K.; m. Lois McDonald, Dec. 29, 1950; children—Carol Kunzler Blaine, Marilyn, Bonnie, Kim Kunzler Tomeo. B.S. in Chem. Engring, U. Utah; Ph.D., U. Calif., Berkeley. With Bell Telephone Labs., Inc., Murray Hill, N.J., 1952—, dir. electronic materials lab., 1969-73, dir. electronic materials and device lab., 1973-79, dir. electronic materials, processes and devices lab., 1979—. Contbr. articles to profl. jours. Recipient John Price Wetherill medal Franklin Inst., 1964; Internat. prize for new materials Am. Phys. Soc., 1979; Kamerlingh Onnes medal, 1979. Fellow Am. Phys. Soc., AAAS; mem. Am. Chem. Soc., AAAS, Nat. Acad. Engring., Sigma Xi, Tau Beta Pi, Alpha Chi Sigma. Subspecialty: Physical chemistry. Patentee in field. Home: Route 2 Box 130 Port Murray NJ 07865 Office: Bell Labs 600 Mountain Ave Murray Hill NJ 07974

KUO, CHO-CHOU, pathobiologist; b. Matow, Taiwan, Sept. 12, 1934; s. Tsai-Chiang and Mei (Chen) K.; m. Margaret Y.H. Lee, May 4, 1938; 1 dau., Lee Wen. M.D., Nat. Taiwan U., 1960; Ph.D., U. Wash., 1970. Resident in pediatrics Nat. Taiwan U. Hosp., Taipei, 1962-65; chief resident, 1965-66; research fellow U.S. Naval Med. Research unit 2, Taipei, 1966; sr. fellow dep. preventive medicine U. Wash., 1967-70; instr. pathobiology, 1970, asst. prof., 1971-76, assoc. prof., 1976-80, prof. pathobiology, 1980—. Served to 2d lt. Chinese Army, 1960-61. Mem. Am. Assn. Immunologists, Am. Soc. Microbiology, Am. Public Heath Assn., Am. Coll. Preventive Medicine, Am. Venereal Disease Assn. Subspecialties: Microbiology; Immunobiology and immunology. Current work: Chlamydia trachomatis infections; microbiology and immunopathogenesis; laboratory diagnosis, epidemiology and treatment and prevention. Office: Dept Pathobiology Univ Wash Seattle WA 98195

KUO, JYH-FA, pharmacologist, biochemist, educator; b. Taiwan, May 19, 1933; s. Shine-Fu and Mong (Huang) K.; m. Alexandra W. H. Lou, June 22, 1965; children: Calvin, Frances. B.S., Nat. Taiwan U., 1957; M.S., S.D. State U., 1961; Ph. D., U. Ill., 1964; M.D. (hon.), Linkoping (Sweden) U., 1980. Research biochemist Lederle Labs., Am. Cyanamid Co., Pearl River, N.Y., 1964-68; asst. prof. pharmacology Yale U. Sch. Medicine, New Haven, 1968-71; assoc. prof., 1971-72; assoc. prof. pharmacology Emory U. Sch. Medicine, Atlanta, 1972-76; prof., 1976—; mem. study sect. USPHS, NIH, 1982—. Contbr. articles to profl. jours. USPHS research career devel. awardee, 1971-76; USPHS grantee, 1970—; vis. scientist fellow Swedish Med. Research Council, 1979. Mem. AAAS, Am. Soc. Biol. Chemists, Am. Soc. Pharmacology and Exptl. Therapeutics. Subspecialties: Molecular pharmacology; Biochemistry (medicine). Current work: Mechanisms of actions of calcium and cyclicnucleotides (intracellular messengers) and their roles in pathophysiology of cardiovascular system, brain, endocrine system and in cancer. Home: 2978 Greenbrook Way NE Atlanta GA 30345 Office: Dept Pharmacology Emory Univ Sch Medicine Atlanta GA 30322

KUO, PETER TE, internist, cardiologist, medical educator; b. Foochow, Fukien, China, Mar. 21, 1916; came to U.S., 1946; s. Lan-Son Kuo and So-Chen (Liu) K.; m. Nancy N. Huang, Dec. 25, 1949; children: Lawrence, Kathy. M.D., St. John's U., Shanghai, China, 1937, B.Sc., 1938; M.Med.Sc., u. Pa., 1949, D.Sc., 1950. Instr. to asst. prof. medicine St. John's U. Med. Sch., Shanghai, 1939-46; instr. to prof. medicine U. Pa. Sch. Medicine, Phila., 1952-72; staff mem. cardiovascular-pulmonary div. U. Pa. Hosp., 1952-72, dir. arteriosclerosis-lipid research lab., 1965-72; prof. medicine, chief div. cardiovascular diseases U. Medicine and Dentistry of N.J.-Rutgers Med. Sch., Piscataway, 1973-82; prof. medicine, dir. atherosclerosis research Rutgers Med. Sch., New Brunswick, 1982—; cons. internal medicine VA Hosp., Phila. Gen. Hosp., both Phila., 1968-72; cons. in medicine and cardiology Med. Ctr. at Princeton, Muhlenberg Hosp., St. Peter's Hosp., Middlesex Gen. Hosp., all N.J., 1973—; cons. Nat. Heart-Lung Inst. Health Services, Mental Health Administrn. Author books and jours in cardiology; cons. editor: cardiovascular disease sect. Practice of Medicine (Tice); mem. editorial bd.: Chest, Angiology. Recipient Disting. Service award Am. Heart Assn., 1973, 75, 81; Sci. Achievement award Chinese Am. Med. Soc., 1974; Career Devel. awardee Nat. Heart-Lung Inst., Bethesda, 1956-61; research grantee Nat. Heart Lung Blood Inst., 1968—; others. Fellow ACP, Gerontology Soc., Arteriosclerosis Council of Am. Heart Assn., Am. Coll. Cardiology, Am. Coll. Angiology, Am. Coll. Chest Physicians; mem. Assn. Univ. Cardiologists, Am. Inst. Nutrition, AAAS, AMA, Am. Fedn. Clin. Research, Am. Soc. Clin. Nutrition, Alpha Omega Alpha. Subspecialties: Cardiology; Preventive medicine. Current work: Atherosclerosis: experimental, clinical; hyperlipidemia and atherosclerosis; prevention and control of coronary heart disease. Home: 72 N Ross Hall Blvd Piscataway NJ 08854 Office: Rutgers Med Sch Acad Health Sci CN19 New Brunswick NJ 08903

KUPERSMITH, JOEL, physician, researcher, educator; b. N.Y.C., Nov. 26, 1939; s. Charles Douglas and Sally (Schulz) K.; m. Judith Rose Friedman, June 15, 1969; children: David Z., Rebecca J., Adam J. B.S., Union Coll., Schenectady, 1960; M.D, N.Y. Med. Coll., 1964. Diplomate: Am. Bd. Internal Medicine, Sub-Bd. Cardiovascular Disease. Intern Kaiser Found. Hosp., San Francisco, 1964-65; resident and chief resident in internal medicine N.Y. Med. Coll., 1967-70; fellow in cardiology Beth Israel Hosp.-Harvard U. Med. Sch., Boston, 1970-72; research assoc. in pharmacology, asst. physician in medicine Columbia-Presbyn. Med. Center, N.Y.C., 1972-74, asst. prof. medicine, 1974-78, dir. electrocardiography and clin electrophysiology, 1975-77, chief, 1977; chief clin. pharmacology Mt. Sinai Sch. Medicine, N.Y.C., 1978—, assoc. prof. medicine and pharmacology, 1979—, chief Arrythmia Clinic, 1979—. Contbr. numerous articles to sci jours. Served with M.C. USN, 1965-67. NIH grantee, 1978—; N.Y. Heart Assn. grantee, 1978-80; Hearst Found. grantee, 1979-82. Mem. Am. Soc. Clin. Investigation, Am. Soc. Pharmacology and Exptl. Therapeutics, Am. Soc. Clin. Pharmacology and Therapeutics, Am. Fedn. Clin. Research. Subspecialties: Cardiology; Pharmacology. Current work: Cellular and clinical effects of antirrhythmic drugs; ion sensitive microelectrodes, cellular electrophysiology, antirrhythmic drugs, clinic electrophysiology, electrical cardiac mapping, cardiac arrhythmias. Home: 16 Courseview Rd Bronxville NY 10708 Office: Mt Sinai Med Center One Gustave L Levy Pl New York NY 10029

KUPKE, DONALD WALTER, biochemistry educator; b. Omaha, Mar. 16, 1922; s. George Julius and Rose (Rottmann) K.; m., June 25, 1949; children: Karen, Mark, Heidi, Lise, Mical-Jean. A.B., Valparaiso (Ind.) U., 1947; M.S., Stanford U., 1949, Ph.D., 1952. Staff mem. Carnegie Inst., Stanford, Calif., 1955-56; asst. prof. biochemistry U. Va., Charltotesville, 1957-63, assoc. prof., 1963-66, prof., 1966—, chmn, dept., 1964-66. Contbr. to books. Organizer Fair Housing Program, Charlottesville, 1968-69. Served to lt. USNR, 1942-46. NRC fellow, 1952-53, 53-54. Mem. Am. Soc. Biol. Chemists, Biophys. Soc., Am. Chem. Soc., AAAS, Sigma Xi. (pres. U. Va. Chpt. 1969-70, recipient research prize 1978). Lutheran. Subspecialties: Biochemistry (biology); Biophysics (biology). Current work: Development of magnetic suspension to measurements of density, viscosity and osmotic pressure with high accuracy on small volumes; development of theory for preferential interaction in protein solutions by density. Home: Keswick VA 22947 Office: U Va Dept Biochemistry Jordan Hall Charlottesville VA 22908

KUPST, MARY JO, psychologist, researcher; b. Chgo., Oct. 4, 1945; d. George Eugene and Winifred Mary (Hughes) K.; m. Alfred P. Stresen-Reuter, Jr., Aug. 21, 1977. B.S., Loyola U., Chgo., 1967, M.A., 1969, Ph.D., 1972. Lic. psychologist, Ill. Postdoctoral fellow U. Ill, Med. Center, Chgo., 1972—; research psychologist assoc. prof. in psychiatry and pediatrics Northwestern U. Med. Sch., 1981—. Editor: (with J.L. Schulman) The Child with Cancer, 1980; Contbr. articles to profl. jours. Mem. Am. Psychol. Assn., Ill. Psychol. Assn. Subspecialty: Psychological aspects of illness. Current work: Research in coping with physical illness (especially leukemia) in children. Home: 2779 N Kenmore Ave Chicago IL 60614 Office: Dept Child Psychiatry Children's Meml Hosp 2300 Children's Plaza Chicago IL 60614

KURAMITSU, HOWARD KIKUO, microbiologist, educator, researcher; b. Los Angeles, Oct. 18, 1936; s. Richard T. and Shirley (Tanaka) K.; m. Le Kim Chua, July 19, 1970; children: Kristine, Tracy. B.S., UCLA, 1957, Ph.D., 1962. Research biochemist UCLA, 1961-62; postdoctoral fellow Harvard U., 1962-63; research assoc. U. So. Calif., 1963-67; prof. microbiology Northwestern U., 1967—; cons. Naval Dental Research Inst., Great Lakes, Ill. NIH grantee, 1980—. Mem. Am. Soc. Biol. Chemists, Am. Soc. Microbiology, Internat. Assn. Dental Research. Subspecialties: Microbiology (medicine); Genetics and genetic engineering (medicine). Current work: Determining the molecular basis for the pathogenicity of oral microorganisms by biochemical and genetic approaches. Home: 2137 Washington Ave Wilmette IL 60091 Office: Northwestern U Med Sch 303 E Chicago Ave Chicago IL 60611

KURLAND, GEORGE STANLEY, cardiologist; b. Boston, Nov. 7, 1919; s. Harry Ellis and Julia (Klebenov) K.; m. Bernice Johnson, Nov. 30, 1952. A.B., Harvard U., 1940, M.D., 1943. Intern Beth Israel Hosp., Boston, 1943-44, resident, 1944-46, cardiac fellow, 1948-49; cardiologist, 1949—; assoc prof. medicine Harvard U., Boston. Author: (with Charm) Blood Flow and Microcirculation, 1974. Served as capt. USMC, 1946-48. Mem. Am. Heart Assn. (pres. Mass. Chpt. 1982—), Am. Thyroid Assn. Subspecialty: Cardiology. Home: 450 Dedham St Newton Center MA Office: Beth Israel Hosp 330 Brooklin Ave Boston MA 02215

KURLAND, LEONARD TERRY, research physician, educator; b. Balt., Dec. 24, 1921; s. Ellis M. and Sarah (Shein) K. B.A., Johns Hopkins, 1942, Dr.P.H., 1951; M.D. (Gold medal), U. Md., 1945; M.P.H. cum laude, Harvard, 1948. Intern U. Md., 1945-46; with USPHS, 1946-64; assigned NIMH, NIH, 1948-55, Nat. Inst. Neurol. Disease and Blindness, 1955-64, chief epidemiology br., 1955-64; fellow neurology Mayo Clinic, Rochester, Minn., 1952-53, research asst. in neurology and med. statistics, 1953-55, prof., chmn. dept. med. statistics and epidemiology, 1964—; prof. epidemiology Mayo Grad. and Mayo Sch. Medicine, 1964—; clin. asst. prof. neurology Georgetown U., 1957-60; clin. prof. neurology Howard U., 1960-64; cons. NIH, FDA, WHO, Nat. Acad. Scis.; mem. geochemistry and health subcom. NRC, 1975-80, chmn., 1978-80. Sr. author: The Epidemiology of Neurologic and Sense Organ Disorders, 1973; Co-editor: Motor Neurone Disease, 1969; Contbr. articles to med. jours. Served with AUS, 1943-45. Fellow Am. Pub. Health Assn.; mem. Am. Acad. Neurology (past chmn. neuroepidemiology sect.), Am. Neurol. Assn., Am. Epidemiologic Soc. (pres. 1974), Internat. Epidemiologic Assn., Japanese Clin. Soc. Neurology (exec. council), Nat. Multiple Sclerosis Soc. (Gold Sci. award 1966, research review panel). Subspecialties: Epidemiology; Neurology. Research on epidemiology, med. record systems, genetics of diseases of nervous system and cancer. Office: 200 1st St SW Rochester MN 55901

KURLYCHEK, ROBERT THOMAS, neuropsychologist; b. Orange, N.J., Feb. 15, 1946; s. John Louis and Mary T. (Coen) K.; m.; 1 son, Aaron. A.B., Seton Hall U., 1968; M.S., U. Oreg., 1970, Ph.D., 1977. Lic. psychologist, Oreg., diplomate: Dipolmate Am. Acad. Behavioral Medicine. Cons. psychologist, Eugene, 1977—; clin. neuropsychologist Sacred Heart Hosp., Eugene, 1980—; psychologist Lane County Corrections, Eugene, 1977-80. Assoc. editor: Corrective and Social Psychiatry, 1978—; contbr. articles to profl. jours. Kappa Sigma Found. grantee, 1976. Mem. Lane County Psychologists Assn. (sec.-treas.), Am. Psychol. Assn., Nat. Acad. Neuropsychologists, Internat. Neurol. Soc., Internat. Council of Sex Edn. and Parenthood (fellow), British Psychol. Soc. Subspecialty: Neuropsychology. Current work: Research the contribution microcomputers can play in reversing cognitive deficits sustanined in brain injury. Office: 99 W 10th St Suite 323 Eugene OR 97401

KUROSKY, ALEXANDER, human genetics and biochemistry educator, researcher; b. Windsor, Ont., Can., Sept. 12, 1938; came to U.S., July 1972; s. Peter and Stella K.; m. Anna Kinik, May 18, 1963; children: Lisa Kathryn, Tanya Kristine, Stephanie Ann. B.S., U. B.C., 1965, M.S., U. Toronto, 1969, Ph.D., 1972. Research technician Can. Dept. Agr., Harrow, Ont., 1959-64, Vancouver, B.C., 1959-64; research and devel. chemist Can. Breweries, Ltd., Toronto, 1965-67; asst. prof. U. Tex. Med. Br., Galveston, 1975-78, assoc. prof., 1978-82, prof., 1982—. Recipient Disting. Teaching award Grad. Sch. Biomed. Scis., U. Tex. Med. Br., 1981. Mem. Am. Soc. Biol. Chemists, Am. Chem. Soc., Can. Biochem. Soc., AAAS. Club: Shetland Sheepdog of Houston. Subspecialties: Biochemistry (medicine); Genetics and genetic engineering (medicine). Current work: Biochem research. Office: Dept Human Biol Chemistry and Genetics U Tex Med Br 603 Basic Sci Bldg Galveston TX 77550 Home: 6605 Golfcrest Dr Galveston TX 77551

KURTIN, WILLIAM EUGENE, chemistry educator; b. Houston, Feb. 2, 1943; s. Claude A. and Anne E. (Portele) K.; m. Sandra J. Bradshaw, Dec. 20, 1969; children: Lezlee, Quentin, Sabra, Leannah. B.A., U. St. Thomas, Houston, 1965; Ph.D., Tex. Tech. U., 1969. Postdoctoral assoc. Baylor Med. Sch., Houston, 1969-70; mem. faculty dept. chemistry Trinity U., San Antonio, 1970—, prof., 1983—. NIH fellow, 1969. Mem. Am. Chem. Soc., Am. Soc. Photobiology, AAUP. Subspecialties: Biophysical chemistry; Photochemistry. Current work: Pathogenesis of gallstone disease; factors affecting bile pigment solubility; methods for analysis of biliary constituents. Office: Dept Chemistry Trinity U 715 Stadium Dr San Antonio TX 78284

KURTZ, DAVID WILLIAMS, chemist, educator; b. Altoona, Pa., July 27, 1942; s. Paul and Beth Estelle (Williams) K.; m. Saddie Mai Pun King, Aug. 26, 1972; children: Stella Anne, Kimberly C. M., Eleanor J. B.S., Houghton Coll., 1964; Ph.D. (NDEA fellow), Syracuse U., 1972. Research assoc. U. Wis., 1971-73; asst. prof. chemistry Ohio No. U., 1973-76, assoc. prof., 1977—; Petroleum Research fellow Iowa State U., 1982. Contbr. articles on photochemistry to profl. jours. Recipient Disting. Faculty award Phi Eta Sigma, 1976. Mem. Am. Chem. Soc., Sigma Xi. Presbyterian. Subspecialties: Organic chemistry; Photochemistry. Current work: Relationships between excited state structure and photochemical relaxation mechanisms. Office: Meyer Hall Ohio No U Ada OH 45810

KURTZ, PERRY JAMES, research toxicologist, educator; b. Staten Island, N.Y., Aug. 13, 1948; s. Arthur Joseph and Isabel (Perry) K.; m. Adore Marie Flynn, Aug. 28, 1971; children: Robert Brian. B.S., Fordham U., 1970; M.S., Syracuse U., 1973, Ph.D., 1975. Instr. in psychology Syracuse U., 1974-75; research psychologist behavioral toxicologist U.S. Army Environ, Hygiene Agy., 1975-78; prin. research toxicologist Battelle Meml. Inst., Columbus, Ohio, 1978-79, assoc. mgr. toxicology and pharmacology, 1979-80, sr. research toxicologist, 1980—; adj. asst. prof. pharmacology Sch. Medicine, Ohio State U., 1980—; mem. short course faculty Wayne State U., 1980—; prin. investigator Nat. Cancer Inst. contract, 1978—; study dir. Nat. Inst. Environ. Health Sci, contract, 1978—. Research publs., presentations in field gen. toxicology, neurobehavioral toxicology and psychopharmacology. Served to capt. U.S. Army, 1975-78. N.Y. State Bd. Regents scholar, 1966-70. Mem. Soc. Toxicology, Am. Psychol. Assn., Eastern Psychol. Assn., Am. Coll. Toxicology, Sigma Xi. Subspecialties: Neurobehavioral toxicology; Toxicology (medicine). Current work: Laboratory research on design and execution of behavioral tests in laboratory animals in order to study toxic effects of drugs and enviromental chemicals on the nervous system. Office: Battelle Meml Inst 505 King Ave Columbus OH 43201

KURTZ, THEODORE STEPHEN, psychoanalyst, consultant; b. N.Y.C., Apr. 25, 1944; s. Maxwell Arthur and Reba Evelyn (Rosenberg) K.; m. Maritza J. Zurita, Sept. 12, 1975. A.B., Boston U., 1964; M.A., N.Y. U., 1966; tng. psychoanalysis, N.Y. Soc. Freudian Psychologists, N.Y.C., 1968-74. Diplomate: Am. Inst. Counseling and Psychotherapy. Pvt. practice psychoanalysis, Cold Spring Harbor, N.Y., 1966—, tchr.-coordinator classes for emotionally disabled, Northport, N.Y., 1966-70; prin. Woodward Mental Health Center, Freeport, N.Y., 1970-74; asst. prof. C.W. Post Coll., L.I. U., 1974-81; cons. to industry, 1975—. Contbr. articles to profl. jours. Fellow Am. Orthopsychiat. Assn.; mem. Acad. Psychologists in Marriage, Sex, and Family Therapy (clin.), Am. Assn. Marriage and Family Therapists (clin.), Nassau County Psychol. Assn. (exec. bd. 1978), Am. Acad. Psychotherapists, Am. Psychol. Asn., Soc. Psychoanalytic Psychotherapy. Jewish. Subspecialty: Developmental psychology. Current work: Theory and modification of psychoanalytic technique; application of psychoanalytic theory to industry; organizational and group dynamics theory; research on causes, cults, and terrorism. Home: Willow Brook Rd PO Box 529 Cold Spring Harbor NY 11724 Office: 145 E 74th St New York NY 10021

KURTZMAN, RALPH HAROLD, JR., research mycologist; b. Mpls., Feb. 21, 1933; s. Ralph Harold and Susie Marie (Elwell) K.; m. Nancy Virginia Leussler, Aug. 27, 1955; children: Steven Paul, Sue. B.S., U. Minn., 1955; M.S., U. Wis., 1958; Ph.D., 1959. Asst. Prof. plant pathology U. R.I., Kingston, 1959-62; asst. Prof. biology U. Minn., Morris, 1962-65; biochemist Western Regional Research Labs., U.S. Dept. Agr., Albany, Calif., 1965—; guest scientist Tech. Research Center Finland, 1980; inst. mushroom cultivation U. Calif. Extension, Berkeley, 1981. Contbr. chpts. to books, articles to profl. jours. Chmn.

camp com. Berkeley YMCA, 1970-80. Mem. Am. Chem. Soc., Mycological SSoc. Am., Am. Soc. Plant Physiologists, Am. Mushroom Inst., Mushroom Growers Assn. Gt. Britain, Calif. Native Plant Soc. (treas. 1971-72). Subspecialties: Mycology; Plant physiology (agriculture). Current work: Biology and cultivation of mushrooms, atlatoxin in corn. Patentee in field. Home: 445 Vassar Ave Berkeley CA 94708 Office: 800 Buchanan St Berkeley CA 94720

KURZ, JOSEPH L., chemistry educator; b. St. Louis County, Mo., Dec. 13, 1933; s. Joseph W. and Elsa (Lilliard) K.; m. Linda J. Cross, June 13, 1968. A.B., Washington U., St. Louis, 1955, Ph.D., 1958. Research fellow Harvard U., Cambridge, Mass., 1958-60; research chemist and sr. chemist Esso Research and Engring. Co., Linden, N.J., 1960-64; asst. prof. Washington U., St. Louis, 1964-67, assoc. prof., 1967-73, prof., 1973—. Contbr. articles to profl. jours. Mem. Am. Chem. Soc., AAAS, AAUP. Subspecialties: Kinetics; Thermodynamics. Current work: Kinetic and equilibrium isotope effects, transition state structures research and teaching. Office: Dept Chemistry Washington U Saint Louis MO 63130

KUSCH, POLYKARP, physicist, educator; b. Blankenburg, Germany, Jan. 26, 1911; came to U.S., 1912, naturalized, 1923; s. John Matthias and Henrietta (van der Haas) K.; m. Edith Starr McRoberts, Aug. 12, 1935 (dec. 1959); children—Kathryn, Judith, Sara; m. Betty Jane Pezzoni, 1960; children—Diana, Maria. B.S., Case Inst. Tech., 1931, D.Sc., 1956; M.S., U. Ill., 1933, Ph.D., 1936, D.Sc. (hon.), 1961, Ohio State U., 1959, Colby Coll., 1961, Gustavus Adolphus Coll., St. Peter, Minn., 1962, Yeshiva U., 1976, Coll. of Incarnate Word, 1980, Columbia U., 1983. Engaged as teaching asst. U. Ill., 1931-36; research asst. U. Minn., 1936-37; instr. Columbia U., 1937-41, assoc. prof. physics, 1946-49, prof., 1949-72, chmn. dept. physics, 1949-52, 60-63, acad. v.p. and provost, 1969-72; engr. Westinghouse, 1941-42; research asso. Columbia U., 1942-44; mem. tech. staff Bell Telephone Labs., 1944-46; prof. physics U. Tex.-Dallas, 1972—, Eugene McDermott prof., 1974-80, Regental prof., 1980-82, Regental prof. emeritus, 1982—. Recipient Nobel prize in physics, 1955, Ill. Achievement award U. Ill., 1975; Fellow; Center for Advanced Study in Behavioral Sciences, 1964-65. Fellow Am. Phys. Soc., A.A.A.S.; mem. Am. Acad. Arts and Scis., Am. Philos. Soc., Nat. Acad. Scis. Democrat. Subspecialty: Atomic and molecular physics. Research in atomic and molecular beams and optical molecular spectroscopy. Office: Univ Tex-Dallas PO Box 688 Richardson TX 75080

KUSCHNER, MARVIN, physician, educator, dean; b. N.Y.C., Aug. 13, 1919; s. Julius and Sadye (Marans) K.; m. Kathryn Pancoe, Dec. 19, 1948; 1 son, Jason. A.B., N.Y.U., 1939, M.D., 1943. Diplomate: Am. Bd. Pathology, 1949. Asst. pathologist in charge 1st div. of Bellevue Hosp., N.Y.C., 1949-54, acting dir. pathology, 1954-55, dir. pathology, 1955-70, Univ. Hosp., 1968-70; chmn. dept. pathology Sch. Medicine SUNY-Stony Brook, 1970-83, prof. pathology, 1970—, dean, 1972—; Mem. exec. com. Assoc. Univs., Inc. Contbr. articles to profl. jours. Served to capt. AUS, 1944-46. Mem. Am. Soc. Clin. Pathologists, Internat. Acad. Pathology, N.Y. Path. Soc., N.Y. Pathology Profs. Council, Suffolk County Pathology Soc., Phi Beta Kappa, Sigma Xi, Alpha Omega Alpha. Subspecialties: Pathology (medicine); Environmental toxicology. Current work: Pumonary toxicology. Home: 64 E Gate Dr Huntington NY 11743 Office: Sch Medicine Health Scis Center SUNY Stony Brook NY 11794

KUSERK, FRANK THOMAS, biology educator; b. Phila., Mar. 26, 1951; s. Frank Joseph and Jane Barbara (Homka) K.; m. Evelyn Roma Hundley, Jan. 4, 1975; 1 dau., Claire Frances. B.S. in Biology, U. Notre Dame, 1973, Ph.D., U. Del., 1978. Asst. prof. biology Moravian Coll., Bethlehem, Pa., 1977—; research assoc. Acad. Natural Scis. Phila., 1981-82. NSF grantee, 1981-82. Mem. Ecol. Soc. Am., AAAS, Pa. Acad. Scis., Sigma Xi. Subspecialties: Ecology; Microbiology. Current work: Microbial ecology; the role of microorganisms in carbon cycling; evolutionary relationships of microorangisms. Home: 1826 Pennland Ct Landsdale PA 19446 Office: Dept Biology Moravian Coll Bethlehem PA 18018

KUSHICK, JOSEPH N., chemist, educator; b. N.Y.C., July 18, 1948; s. Max and Civia (Turbbiner) K.; m. Marilyn N. Massler, June 14, 1970; 1 son, Rafael. A.B., Columbia U., 1969, Ph.D., 1975. Research asso. U. Chgo., 1974-76, NSF postdoctoral fellow, 1976; asst. prof. chemistry Amherst Coll, 1976-82, asso. prof., 1982—; vis. scholar Harvard U., 1979-80; vis. scientist Universite de Paris-Sud, 1981. Contbr. articles to profl. jours. Amherst Coll. Trustee faculty fellow, 1979; Camille and Henry Dreyfus Tchr.-scholar Dreyfus Found., 1980; NIH grantee, 1982—; Petroleum Research Fund grantee, 1978—. Mem. Am. Chemn. Soc., Am. Phys. Soc., Biophys. Soc. Jewish. Subspecialties: Theoretical chemistry; Biophysical chemistry. Current work: Computer simulation of biochem. systems; computer simulation; stat. mechanics; polypeptide and protein conformation. Office: Dept Chemistry Amherst Coll Amherst MA 01002

KUSHNER, HARVEY DAVID, research and systems engineering firm executive; b. N.Y.C., Dec. 28, 1930; s. Morris and Hilda (Zweibel) K.; m. Roe Rehert, Jan. 14, 1951; children: Gantt A., Todd R., Lesley K. B.S. in Engring, Johns Hopkins U., 1951. Assoc. engr. Bur. Ships, U.S. Navy, Washington, 1951-52; mem. tech. staff Melpar, Inc., Falls Church, Va., 1953-54; staff and mgmt. positions ORI, Inc. (formerly Ops. Research, Inc.), Silver Spring, Md., 1955—, exec. v.p., 1963-68, pres, 1969—, chmn. bd., 1977—; v.p. Reliance Group, Inc., N.Y.C., 1971-77; pres. Disclosure, Inc., Silver Spring, 1972-77; cons. Applied Physics Lab., Johns Hopkins U., 1960-64, Com. on Undersea Warfare, Nat. Acad. Sci., 1963-64; trustee, mem. exec. com. Nat. Security Indsl. Assn., Washington, 1963—; dir. Profl. Services Council, Washington, 1974—. Mem. Montgomery County Econ. Adv. Council, Rockville, Md., 1981—, chmn., 1983—; campaign chmn. Montgomery County United Way, Rockville, 1980, exec. bd., 1981—; chmn. Montgomery County Commn. on Higher Edn. in High Tech., 1983-84; mem. (Md.) Gov's High Tech. Roundtable, 1983—. Fellow N.Y. Acad. Scis.; mem. IEEE (sr.), Nat. Space Club (bd. govs.), ASME, AIAA, AAAS, Ops. Research Soc. Am., Armed Forces Communications and Electronics Assn., Inst. Mgmt. Scis., Navy League of U.S. Club: Cosmos (Washington). Subspecialties: Operations research (engineering); Systems engineering. Current work: Operations research and system engineering. Office: ORI Inc 1400 Spring St Silver Spring MD 20910

KUSHNER, IRVING, rheumatologist, immunobiologist, educator; b. N.Y.C., Jan. 16, 1929; s. Boris and Rose (Klosner) K.; m. Enid Pearl Lupeson, Jan. 2, 1955; children: Ellen Ruth, Philip Seth, David Micah. B.A., Columbia U., 1950; M.D., Washington U., St. Louis, 1954. Diplomate: Am. Bd. Internal Medicine and subsplty. in rheumatology. Intern Yale-New Haven Hosp., 1954-55; clin. assoc. clin. center NIH, Bethesda, Md., 1955-57; asst. resident Harvard U. Med. Services, Boston City Hosp., 1957-58; instr. to assoc. prof. medicine Case Western Res. U., 1958-73, prof. medicine, 1974—. Contbr. 115 publs. to profl. jours.; editor books. Served with USPHS, 1957-58. Recipient Nat. Vol. Service citation Arthritis Found. Mem. Am. Rheumatism Assn., Am. Assn. Immunologists, N.Y. Acad. Sci., AAAS, ACP, Central Soc. Clin. Research. Subspecialties: Internal medicine; Immunobiology and immunology. Current work: Studies of the regulation of biosynthesis of C-reactive protein. Office: 3395 Scranton Rd Cleveland OH 44109 Home: 22149 Rye Rd Shaker Heights OH 44122

KUSHNER, SIDNEY RALPH, geneticist; b. N.Y.C., Dec. 14, 1943; s. Joseph B. and Dora (Cohn) K.; m. Deena Dash, June 12, 1969; children: Aaron, Ze'eva. B.A., Oberlin Coll., 1965; Ph.D., Brandeis U., 1970. Postdoctoral research fellow U. Calif., Berkeley, 1970-71, Stanford U. Med. Sch., calif., 1971-73; asst. prof. biochemistry U. Ga., Athens, 1973-76, asso. prof. genetics and biochemistry, 1976-82, prof., 1982—; cons. NIH, Dept. Energy. Asso. editor: Gene, 1980—; contbr. articles to profl. jours. NIH research career devel. awardee, 1975-80. Mem. Genetics Soc. Am., Am. Soc. Biol. Chemists, Am. Soc. Microbiology, AAAS. Subspecialties: Genetics and genetic engineering (biology); Biochemistry (biology). Current work: Researcher in analysis genetic recombination and DNA repair, control of gene expression, genetic engring. Office: Dept Genetics Univ Ga Athens GA 30602

KUSHWAHA, RAMPRATAP SINGH, nutritional biochemist; b. Gugrapur, U.P., India, July 11, 1943; came to U.S., 1970, naturalized, 1982; s. Ram Singh and Ramdevi (Bais) K.; m. Pushpa Chauhan, June 4, 1964; children: Vivek, Anita, Alok. B.Sc., Agra U., 1962, M.Sc., 1964; Ph.D., Wash. State U., 1973. Research instr. medicine U. Wash., Seattle, 1977-81, research asst. prof. medicine, 1981-82; adj. assoc. prof. pathology U. Tex. Health Sci. Ctr., San Antonio, 1982—; assoc. scientist S.W. Found. Research and Edn., San Antonio, 1982—. Fellow Am. Heart Assn.; mem. Am. Inst. Nutrition, N.Y. Acad Scis. Subspecialties: Biochemistry (medicine); Nutrition (medicine). Current work: Lipoprotein metabolism and atherosclerosis, dietary hyperlipoproteinemia, genetic dyslipoproteinemias. Home: 4515 Spotted Oak Woods San Antonio TX 78249 Office: SW Found Research and Edn PO Box 28174 San Antonio TX 78284

KUSY, ROBERT P, biomedical educator, consultant; b. Worcester, Mass., Oct. 19, 1947; s. Stanley J. and Mary B. (Rutkiewicz) K.; m. Gisela Bauer, June 27, 1969; children: Kimberly, Kevin. B.S., Worcester Poly. Inst., 1969; M.S., Drexel U., 1972, Ph.D., 1972. Research tech. Vellumoid Gasket Co., Worcester, 1965-69; research assoc. U. N.C., Chapel Hill, 1972-74, asst. prof., 1974-79, assoc. prof., 1979—; cons. Contbr. chpts. to books, articles to profl. jours. Served as 1st lt. U.S. Army, 1973. Recipient NIH research award, 1977-80, 77-82. Mem. Am. Soc. Metals, Am. Chem. Soc., Internat. Metall. Soc., Am. Assn. Dental Research, Soc. Biomaterials. Roman Catholic. Subspecialties: Materials (engineering); Polymers. Current work: Research involved with the fracture toughness and morphology of glassy polymers, the use of acrylic based materials in dentistry and medicine, and the characterization and modification of ultra-high molecular weight polyethylene (particularly as it is used in biomedical applications). Home: 113 Cynthia Dr Chapel Hill NC 27514 Office: U NC Dental Research Ctr Bldg 210H Chapel Hil NC 27514

KUTCHER, MARK JAY, dental educator, researcher; b. Phila., Aug. 1, 1944; s. Edward and Edith Louise (Kornberg) K.; m. Brenda Ayers, Feb. 7, 1982; 1 dau., Christina. B.A., Temple U., 1966, D.D.S., 1970; M.S. in Oral Medicine, Ind. U., 1977. Cert. oral-maxillofacial radiologist. Instr. Temple U. Dental Sch., Phila., 1972-74; resident Vanderbilt U. Med. Sch., Nashville, 1974-75; Ind. U. Dental Sch., Indpls., 1975-77; assoc. prof. U. Md. Dental Sch., Balt., 1977—, dir. oral medicine clinic, 1978-81, emergency dental clinic, 1977-81. Author: Oral Medicine for the General Practitioner. Instr. Am. Heart Asn., Balt., 1978—. Served to capt. USAF, 1970-72. Syntex Drug Corp. grantee, 1981; Md. Cancer Soc. grantee, 1980. Mem. Am. Assn. Dental Schs., Am. Assn. Dental Research, Am Acad. Oral Pathology, Am. Acad. Oral Radiology, Orgn. of Tchrs. of Oral Diagnosis, Washington Area Radiology Study Club. Current work: Teaching undergraduate dental students in areas of oral diagnosis, physical evaluation and oral medicine, applied oral medicine. Office: U Md Dental chS 666 W Baltimore St Baltimore MD 21201

KUTINA, JAN, educator; b. Prague, Czechoslovakia, July 23, 1924; came to U.S., 1969, naturalized, 1980; s. Jan and Amalie (Tauberova) K.; m. Irena Kutinova, Apr. 10, 1950; children: Irene, Jan. PhMr., Charles U., 1948, RNDr., 1949, C.Sc., 1956, Docent, 1954. Asso. prof. geochemistry Charles U., Prague, 1954-68; vis. prof. econ. geology Lehigh U., Bethlehem, Pa., 1968-69; research scientist Geol. Survey of Can., Ottawa, Ont., 1969-70; sr. research scientist Am. U., Washington, 1977-79, 80, research prof., geologist, 1980—; cons. geologist Bethlehem Steel Corp., 1974-75, UN, 1970-74, W.A. Bowes, Inc., 1976—. Chief editor: Global Tectonics and Metallogeny, 1978—. Mem. Internat. Assn. on Genesis of Ore Deposits (sec. gen. 1964-69), Am. Geol. Soc., others. Club: e de Mineralogia (hon. mem. Brazil). Subspecialties: Geochemistry; Mineralogy. Current work: Global tectonics/metallogeny; geochemistry & mineralogy of ore deposits. Office: Dept Chemistry Am Univ Washington DC 20016

KUTLIK, ROY LESTER, mechanical engineer; b. San Francisco, Apr. 19, 1955; s. Henry Andrew and Barbara Ann (Cook) K. B.S. in Mech. Engring./Material Sci. And Engring, U. Calif., Berkeley, 1978. Registrered profl. engr., Calif. Instrument and control engr. Pacific Gas & Electric, San Francisco, 1978-81; systems engr., instrumentation and control Standard Oil Calif., San Francisco, 1981—. Mem. ASME, Instrument Soc. Am., IEEE Computer Soc., Assn. for Computing Machinery., Pi Kappa Phi. Democrat. Roman Catholic. Subspecialties: Software engineering; Systems engineering. Current work: Applications of computer systems to industrial process control. Home: 135 Cypress Ave San Bruno CA 94066

KUTNER, MARC LESLIE, educator; b. Bklyn., Aug. 20, 1947; s. Edwin and Belle (Goldstein) K.; m.; children: Eric R., Jeffrey M. A.B. cum laude, Princeton U., 1968; M.A., Columbia U., 1970, Ph.D., 1972. Nat. Acad. Sci.-NRC research assoc. NASA Goddard Inst. Space Studies, 1972-74; lectr. physics Columbia U., N.Y.C., 1974-75; asst. prof. physics Rensselaer Ply. Inst., Troy, N.Y., 1975-78, assoc. prof., 1979—; mem. Nat. Radio Astronomy Obs. Users' Com., 1978-81; sci. adv., bd. dirs. Dudley Obs., 1981—. Author: (with J.M. Pesachoff) University Astronomy, 1978, Invitation to Physics, 1981; contbr.: articles to astron. jours. Invitation to Physics. NSF grad fellow, 1969-72; NSF grantee, 1976, 79, 81; Research corp. grantee, 1977; Aerospace Corp. grantee, 1976; Dudley Obs. grantee, 1980. Mem. Am. Astron. Soc., Am. Phys. Soc., Internat. Astron. Union, Sigma Xi. Subspecialty: Radio and microwave astronomy. Current work: Radio observations of interstellar molecules to study star formation and galactic structure. Office: Physics Dep Rensselaer Poly Inst Troy NY 12181

KUTTLER, JAMES ROBERT, mathematician; b. Burlington, Iowa, Aug. 3, 1941; s. Wilbur Stewart and Almeda May (McNally) K.; m. Evelyn Marie Ridgley, June 26, 1963; children: John, Robert, Laura. B.A., Rice U., 1962; M.A., U. Md., 1964, Ph.D., 1967. Mathematician Johns Hopkins Applied Physics Lab., Laurel, Ma., 1967—. Mem. Soc. Indsl. and Applied Math, Am. Math. Soc. Subspecialty: Applied mathematics. Current work: Eigenvalues of elliptic partial differential equations. Home: 7004 Nightingale Terr Lanhan MD 20706 Office: John Hopkins U Applied Physics Lab Johns Hopkins Rd Laurel MD 20707

KUTZSCHER, EDGAR WALTER, cons. physicist; b. Leipzig, Germany, Mar. 21, 1906; s. Arno Fritz and Maria Helene K.; m. Edith Hildgard Wagner, Nov. 22, 1919; children: Detlef, Bernd. Ph.D., U. Berlin, 1931; D.Sc., Inst. Tech. Berlin, 1935; D.Eng. (hon.), U. Hannover, 1963. Research and teaching asst. U. Berlin, Inst. Tech. Berlin, 1930-33; physicist German War Dept., 1933-37; research physicist, then asst. prof. Inst. Tech. Berlin, 1936-46; dir. research Electroacustic Co., Kiel, Germany, 1937-45; dir. Univ. Ext., Flensburg, W.Ger., 1945-47; physicist U.S. Navy, 1947-51, Santa Barbara (Calif.) Research Ctr., 1951-53; dept. mgr. Lockheed Aircraft Co., Burbank, Calif., 1954-72; cons. physicist, 1972—. Contbr. articles to profl. jours. Recipient Todt prize Govt. Ger., 1944. Mem. Optical Soc. Am. Presbyterian. Subspecialty: Infrared detectors and technology. Current work: Research and development of infrared sensors and devices. Patentee in field. Home: 15450 Briarwood Dr Sherman Oaks CA 91403

KUYATT, CHRIS E. (ERNIE EARL KUYATT), physicist, research ctr. adminstr.; b. Grand Island, Nebr., Nov. 30, 1930; s. Christian A. and Rosalie L. (Repp) K.; m. Patricia Lou Peirce, Sept. 18, 1949; children: Chris S., Brian, Alan, Bruce. B.S., U. Nebr., 1952, M.S., 1953, Ph.D. in Physics, 1960. Research assoc. U. Nebr., 1959-60; physicist electron physics sect. Nat. Bur. Standards, Washington, 1960-69, chief electron and optical physics sect., 1970-73, chief surface and electron physics sect., 1973-78, chief radiation physics div., 1978-79, dir., 1979—; mem. interagy. radiation research com. Dept. Commerce-Nat. Bur. Standards; Nat. Bur. Standards liaison to Nat. Council on Radiation Protection and Measurements. Contbr. articles to profl. jours. Concertmaster Rockville Mcpl. Band, 1972—. Recipient Silver medal Dept. Commerce, 1964. Fellow Am. Phys. Soc.; mem. AAAS, Philos. Soc. Washington, Phi Beta Kappa, Sigma Xi. Subspecialties: Atomic and molecular physics; Electron and ion optics. Current work: Electron optics, electron scattering, experimental atomic physics; direction of broad programs in the areas of atomic, nuclear, plasma and optical radiation. Home: 2904 Hardy Ave Wheaton MD 20902 Office: Nat Bur Standards C229 RADP Washington DC 20234

KUYATT, ERNIE EARL *See also* **KUYATT, CHRIS E.**

KUZNESOF, PAUL MARTIN, chemistry educator; b. Bronx, N.Y., Aug. 13, 1941; s. Benjamin and Betty (Gordon) K.; m. Elizabeth Anne Parks, Apr. 19, 1969 (div. 1977); 1 son, Adam Aeschylus; m. Laura Marie McDonald, Sept. 18, 1982. Sc.B., Brown U., 1963; Ph.D., Northwestern U., 1967. Asst. prof. San Francisco State U., 1969-70; assoc. prof. Universidade Estadual de Campinas, Brazil, 1970-75; lectr. U. Mich., Ann Arbor, 1975-76; vis. assoc. prof. Trinity Coll., Hartford, Conn., 1976-78; research assoc. Case Western Res. U., Cleve., 1978-79; assoc. prof. chemistry Agnes Scott Coll., Decatur, Ga., 1979-83. Judge DeKalb County Sci. Fair, Decatur, 1980-82; soccer commr. Decatur-DeKalb YMCA, 1981-82; actor Blackfriars Prodns., Agnes Scott Coll., 1980-82; active Boy Scouts Am., 1980-82. Grantee NSF; Research Corp.; Fundacao de Amparo a Pesquisa do Estado de Sao Paulo. Mem. Am. Chem. Soc. (treas. Ga. sect. 1982-83), Quantum Chemistry Program Exchange, Sigma Xi. Subspecialties: Inorganic chemistry; Solid state chemistry. Current work: Synthesis and characterization of new low-dimensional materials and electroactive polymers. Patentee in field.

KVENVOLDEN, KEITH ARTHUR, geochemist; b. Cheyenne, Wyo., July 16, 1930; s. Owen Arthur and Agnes (Bergstrom) K.; m. Mary Ann Lawrence, Nov. 7, 1959; children: Joan A., Jon W. B. Geophysics, Colo. Sch. Mines, 1952; M.S. in Geology, Stanford U., 1958, Ph.D., 1961. Registered gologist, Calif. Sr. research technologist Mobil Oil Corp., Dallas, 1961-66; research scientist NASA-Ames Research Ctr, Moffett Field, Calif., 1966-72, chief chem. evolution br., 1972-74, chief planetary biology div., 1974-75; geologist U.S. Geol. Survey, Menlo Park, Calif., 1975—; cons. prof. geology Stanford U., 1967—; adj. prof. Calif. State U., Hayward, 1978—; chmn. JOIDES adv. panel on organic geochemistry, La Jolla, Calif., 1974-80. Assoc. editor: Geochemica et Cosmochemica Acta, 1972-75; Editor: Geochemistry and the Origin of Life, 1974, Geochemistry of Organic Molecules, 1980; contbr. articles to profl. jours. Served with U.S. Army, 1954-56. Recipient award for best paper in organic geochemistry Organic Geochemistry Div. of Geochem. Soc., 1971; Apollo Achievement award NASA, 1970. Fellow Geol. Soc. Am., AAAS, Explorers Club; mem. Am. Assn. Petroleum Geologists, Soc. Econ. Paleontologists and Mineralogists, Am. Geophys. Union, Am. Chem. Soc., Geochem. Soc. Subspecialties: Organic geochemistry; Geology. Current work: Research in marine organic geochemistry, geochemical prospecting, geochemistry of hydrocarbon gases, animo acid geochronology, petroleum geochemistry; geochemistry of gas hydrates. Patentee in field. Home: 2433 Emerson St Palo Alto CA 94301 Office: US Geol Survey 345 Middlefield Rd Menlo Park CA 94025

KVIST, TAGE NIELSEN, anatomy educator; b. Copenhagen, Denmark, Jan. 17, 1942; came to U.S., 1969; s. Kai and Alma (Nielsen) K.; m. Sharon Lea Armstrong, May 8, 1965; children: Lea-Ann, Lisa Joy, Charlene Tia. B.Sc., U. B.C., Can., 1966, M.SC., 1969; Ph.D., U. Pa., 1973. Teaching fellow U. Pa., Phila., 1969-73, research assoc., 1973-76; lectr. Rosemont (Pa.) Coll., 1972; chief neurosurgery research Joseph Stokes Jr. Research Inst., Phila., 1973-76; asst. prof. anatomy Phila. Coll. Osteo. Medicine, 1976-80, assoc. prof., 1980—; cons. in field. Contbr. articles to profl. jours. Treas. Garfield Park PTO, Willingboro, N.J., 1979-81, pres., 1981-82. Recipient Humanitarian award V.J. Sarte Nat. Hydrocephalus Research Found., 1976; March of Dimes grantee, 1973-75; NIH grantee, 1979-83; Spina Bifida grantee, 1982-83. Mem. Soc. Devel. Biology, Teratology Soc., Humanity Gifts Registry, Am. Assn. Anatomists, Spina Bifida Assn. Am. (del. 1981—), Spina Bifida Assn. Delaware Valley (dir. 1978—, pres. 1981—). Club: Willingboro Atheltic. Subspecialties: Developmental biology; Neurobiology. Current work: Development anomalies, teratology. Home: 32 Globe Ln Willingboro NJ 08046 Office: Phila Coll Osteo Med Dept Anatomy 4150 City Ave Philadelphia PA 19131

KWAN, KING CHIU, chemist; b. Hong Kong, Jan. 14, 1936; s. Cho Yiu and Wai Fan (Chow) K. B.S., U. Mich., 1956, M.S., 1958, Ph.D., 1962. Research chemist R.P. Scherer Corp., Detroit, 1962-63; research assoc. Merck Sharp & Dohme Research Labs., West Point, Pa., 1964-66, unit head, 1966-69, pharmacokinetics specialist, 1969-70, sr. research fellow, 1970-76, sr. investigator, 1976-79, sr. dir., 1979-81, exec. dir. drug metabolism, 1981—; lectr. U. Mich., 1963-64; mem. pharmacology study sect. NIH, 1980-83. Mem. editorial adv. bd.: Jour. Pharmacokinetics and Biopharmaceutics, 1975—, Pharm. Research, 1982—, Pharmacy Internat, 1982—. Subspecialty: Pharmacokinetics. Current work: Pharmacokinetics, drug metabolism, drug delivery systems. Office: Merck Sharp & Dohme Research Labs West Point PA 19486

KWIK, ROBERT JULIUS, engineering consultant; b. Newark, Jan. 6, 1936; s. Julius and Freida Rose (Schilling) K.; m. Jean Agnes Chown, Sept. 12, 1958; children: Kenneth Lawrence, Karen Elizabeth, Jeanne Frieda. M.E., Stevens Inst. Tech., 1958; M.S., Calif. Inst. Tech., 1959; B.D., Princeton Theol. Sem., 1962, Th.M., 1966; Ph.D., U. Pa., 1974. Ordained minister Presbyterian Ch., 1962; pastor 1st Presbyn. Ch., Steelton, Pa., 1962-65; instr. math. U. Vt., Burlington, 1965-55; prof. physics Coll. de Libamba, Makak, Cameroon, 1967-70; engr. Gibbs & Hill, Inc., N.Y.C., 1974-82, Stone & Webster Engring. Co., 1982—. Contbr. articles to profl. jours. Mem. Soc. for Risk Analysis, Am. Nuclear Soc., ASME, AAAS. Democrat. Subspecialties: Nuclear engineering; Mechanical engineering. Current work: Application of probabilistic risk assessment fo improvement of nuclear plant reliability. Home: 50 Terrace Ave Nutley NJ 07110 Office: Stone and Webster Engring Co 250 W 34th St New York NY 10019

KWIRAM, ALVIN L., physical chemist; b. Riverhills, Man., Can., Apr. 28, 1937; came to U.S., 1954; s. Rudolf and Wilhelmina A. (Bilske) K.; m. Verla Rae Michel, Aug. 9, 1964; children: Andrew Brandt, Sidney Marguerite. B.S. in Chemistry; B.A. in Physics, Walla Walla (Wash.) Coll., 1958; Ph.D. in Chemistry, Calif. Inst. Tech., 1963. Alfred A. Noyes instr. Calif. Inst. Tech., Pasadena, 1962-63; research asso. physics dept. Stanford (Calif.) U., 1963-64; instr. chemistry Harvard U., Cambridge, Mass., 1964-67, lectr., 1967-70; asso. prof. chemistry U. Wash., Seattle, 1970-75, prof., 1975—, chmn. dept. chemistry, 1977—. Contbr. numerous articles to sci. jours. Co-founder, 1st pres. Assn. Adventist Forums, 1967-72; chmn. bd. editors, co-editor quar. jour. Spectrum, 1975-77. Recipient Eastman-Kodak Sci. award, 1962; Woodrow Wilson fellow, 1958; Alfred P. Sloan fellow, 1968-70; Guggenheim Meml. Found. fellow, 1977-78. Mem. Am. Phys. Soc., Am. Chem. Soc., Council Chem. Research (dir. 1980—, chmn. 1982-83), Sigma Xi. Subspecialties: Physical chemistry; Biophysical chemistry. Current work: Magnetic resonance in the solid state: crystals, glasses, matrix isolated species. Optical detection of magnetic resonance in photo-excited states of aromatic molecules and biomolecules. Home: 5639 NE Keswick Dr Seattle WA 98105 Office: Dept Chemistry Univ Washington Seattle WA 98195

KWOK, MUNSON ARTHUR, research scientist; b. San Francisco., Apr. 28, 1941; s. Loy Gun and Pearl Janet (Lew) K.; m. Suellen Cheng, Aug. 10, 1977. B.S., Stanford U., 1962, M.S., 1963, Ph.D., 1967. NSF postdoctoral fellow Stanford U., 1967-68; mem. tech. staff Aerospace Corp., El Segundo, Calif., 1968-76, staff scientist, 1976-77, research scientist, 1977—. Contbr. articles to profl. jours. Pres. Aerospace Asian Am. Assn., 1982-83. Mem. Am. Phys. Soc., AIAA, ASME, Chinese Engrs. So. Calif. (v.p. 1981-82), Chinese Hist. Soc. So. Calif. (v.p., sec. 1977-79), Phi Beta Kappa, Tau Beta Pi. Subspecialties: Chemical lasers; Kinetics. Current work: Gas lasers, chemical lasers, chemical kinetics, kinetic theory, plasmas, high temperature gasdynamics, optical element metrology. Inventor, patentee in field. Office: The Aerospace Corp PO Box 92957 M5-747 Los Angeles CA 90009

KWOK, SUN, astronomer; b. Hong Kong, Sept. 5, 1949; s. Chuen-Poon and Pui-Ling (Chan) K.; m. Shiu-Tseng Emily, June 16, 1973; children: Roberta Wing-Yue, Kelly Wing-Hang. B.Sc., McMaster U., 1970; M.S., U. Minn., 1972; Ph. D., 1974. Postdoctoral fellow dept. physics U.B.C., 1974-76; asst. prof. dept. physics U. Minn., 1976-77; research assoc. Center Research in Exptl. Space Sci., York U., Toronto, Ont., Can., 1977-78, Herzberg Inst. Astrophysics, NRC, Ottawa, Ont., 1978-83; asst. prof. physics U. Calgary, (Alta.), 1983—. Contbr. articles to profl. jours. Mem. Internat. Astron. Union, Am. Astron. Soc., Can. Astron. Soc. Subspecialties: Theoretical astrophysics; Radio and microwave astronomy. Current work: Stellar winds, planetary nebulae, novae, interstellar molecules, stellar evolution. Home: 139 Edgeland Rd N Calgary AB Canada T3A 2Y3 Office: Dept Physics U Calgary Calgary AB Canada T2N 1N4

KYDD, GEORGE HERMAN, physiologist; b. Eagle Rock, Va., Aug. 20, 1920; s. George Herman and Nellie Glare (Marshall) K.; m. Mary Louise Penman, Apr. 15, 1944; children: Brenda, Jean, George Herman, Richard Adrian. B.S., W. Va. State Coll.-Institute, 1942; M.S., Ohio State U.-Columbus, 1950, Ph.D., 1955. Research physiologist Aviation Med. Acceleration Lab., Warminster, Pa., 1955-70; with U.S. Naval Air. Devel. Ctr., Warminster, 1971—, phys. sci. adminstr., 1977-81, research physiologist, 1981—. Served to 1st lt. U.S. Army, 1942-46. Recipient 1st ann. Aerospace award Nat. Med. Assn., 1970. Fellow Aerospace Med. Assn. (mem. sci. program com. 1965); mem. Am. Physiol. Soc., N.Y. Acad. Scis., Aerospace Physiol. Soc. (Fred Hitchcock award 1972), Sigma Xi. Unitarian. Subspecialties: Integrated systems modelling and engineering; Biomedical engineering. Current work: Physiological regulation in temperature extremes by integrating experimental results into models that elucidate systems and enable one to predict the behavior of the system in untested conditions. Home: 6631 Boyer St Philadelphia PA 19119 Office: US Naval Air Devel Center Warminster PA 18974

KYDD, PAUL HARRIMAN, research company executive; b. New Haven, Nov. 25, 1930; s. David M. and Beatrice (Ells) K.; m. Priscilla Clisham, Dec. 15, 1956; children: David M., Andrew H. A.B. in Chemistry, Princeton U., 1952; Ph.D., Harvard U., 1956, postgrad., 1956-57, 60-61. With Gen. Electric Research Lab., Schenectady, 1957, 66; vis. lectr. Harvard U., Cambridge, Mass., 1959-60; mgr. chem. processes br. Gen. Electric Corp. Research and Devel., 1966-75; v.p. tech., dir. Research and Devel. Ctr., Hydrocarbon Research, Inc., Trenton, N.J., 1975—. Contbr. articles to profl. jours. Mem. Am. Inst. Chem. Engrs. (chmn. Central Jersey sect. 1980-81), Mercer County C. of C., Phi Beta Kappa, Sigma Xi. Subspecialties: Fuels; Chemical engineering. Current work: Application of new technology. Patentee in field. Home: 32 Woodlane Rd Lawrenceville NJ 08648 Office: PO Box 6047 Lawrenceville NJ 08648

KYLAFIS, NIKOLAOS DIMITRIOU, astrophysicist; b. Nea-Avorani, Trihonidos, Greece, Jan. 1, 1949; came to U.S., 1972; s. Dimitrios K. and Alexandra G. (Katsampas) K.; m. Ekaterini G. Tsoni, Jan. 24, 1957. Ph.D., U. Ill., Urbana-Champaign, 1978. Research assoc U. Ill., Urbana-Champaign, 1979; research fellow Calif. Inst. Tech., Pasadena, 1979-81; mem. Inst. Advanced Study, Princeton, 1981—. Contbr. articles to profl. jours. Served with Greek Army, 1978. U. Patras (Greece) nat. scholar, 1967-71. Mem. Am. Astron. Soc. Subspecialties: 1-ray high energy astrophysics; Radio and microwave astronomy. Current work: Compact X-ray sources, interstellar medium, astrophysical plasma, transfer of radiation, applied atomic and molecular physics. Office: Inst Advanced Study Sch Natural Scis Princeton NJ 08540

KYLE, ROBERT ARTHUR, physician, educator; b. Bottineau, N.D., Mar. 17, 1928; s. Arthur Nichol and Mabel Caroline (Crandall) K.; m. Charlene Mae Showlater, Sept. 11, 1954; children: John, Mary, Barbara, Jean. B.S., U. N.D., 1948; M.D., Northwestern U., 1952; M.S., U. Minn., 1959. Diplomate: Am. Bd. Internal Medicine. Intern Evanston (Ill.) Hosp., 1952-53; fellow internal medicine Mayo Grad. Sch., 1953-55, 57-59, William. H. Donner Prof. medicine and lab. medicine Med. Sch., 1961—; cons. Mayo Clinic, 1961—; research fellow Tufts U., 1960-61; vis. prof. U. Sherbrooke, P.Q., Queen's U., U. Calif., U. B.C., U. Ariz., U. Fla. Author: Monoclonal Gammonpathies, 1976, Medicine and Stamps Volume I, 1970, Volume II 1980; contbr. chpts. to books and articles in field to profl. jours. Pres. Folwell PTA, Rochester, Minn.; chmn. troop com. Gamehaven Council Boy Scouts Am.; bd. trustees First Presbyterian Ch., also elder. Served with USAF, 1955-57. Mem. AMA, Am. Soc. Hematology, Am. Fedn. Clin. Research, N.Y. Acad. Sci., Am. Soc. Clin. Oncology, AAAS, Minn. State Med. Assn., Am. Assn. Cancer Research, Minn. Soc. Internal Medicine, Internat. Soc. Hematology, A.C.P., Sigma Xi, Phi Beta Kappa, Pi Kappa Epsilon. Republican. Subspecialties: Hematology; Immunology (medicine). Current work: Multiple myeloma and related disorders; identification of monoclonal gammopathies in lab., prospective chemotherapeutic programs of multiple myeloma, amyloidosis and macroglobulinemia. Home: 1207 6th St SW Rochester MN 55901 Office: 200 1st St SW Rochester MN 55901

KYRALA, ALI, astrogeophysicist, educator; b. N.Y.C., Dec. 14, 1921; s. George Abu-Ali and Mildred Frances (Walsh) Kheirallah; m. Judith Anne Wood, Dec. 16, 1966; children: Lawrence Benali, Cadmus Kamal, Andrea Abla. B.Sc., MIT, 1947; M.Sc., Stanford U., 1948; M.S., Harvard U., 1957; Dr.Sc., U. Vienna, 1960. Instr. math. U. Mass., Amherst, 1951-53; math. physicist Lessell & Assocs., Boston, 1953-58, Goodyear Aerospace, Litchfield Park, Ariz., 1958-60; staff scientist Motorola Semiconductor, Phoenix, 1960-62; Fulbright prof. U. Alexandria, Egypt, 1963-64; prof. math. Am. U. Beirut, 1968-70, UPM, Dhahran, Saudi Arabia, 1975-77; prof. physics Ariz. State U., Tempe, 1964—; cons. Author: Applied Functions of a Complex Variable, 1972, Theoretical Physics, 1967. NASA faculty fellow, 1968; NSF faculty fellow, 1965; Am. Cancer Soc. scholar, 1973; CNES scholar, 1980; NASA-ASEE fellow, 1981-82. Mem. Am. Phys. Soc., Am. Math. Soc., Am. Astron. Soc., European Phys. Soc., Assn. Computing Machinery, IEEE, Brit. Interplanetary Soc., Sigma Xi. Muslim. Subspecialties: Theoretical astrophysics; Geophysics. Current work: Space plasma under high voltage, planetary astrophysics, relativity, statistical physics via Markov chains, theory of galactic spirals, explanation of persistence of great red spot of jupiter. Home: 2309 S Cottonwood St Tempe AZ 85282 Office: Dept Physics Ariz State Univ Tempe AZ 85287

KYRALA, GEORGE AMINE, physicist; b. Bhamdoun, Lebanon, Apr. 20, 1946; s. Amine Asaad and Moura (Habib) Khayrallah; m. Trish Mylet, Nov. 18, 1973; children: Michaelene, Kamal. B.S., Am. U. Beirut, 1967; M. Phil, Yale U., 1969, Ph.D., 1974. Fellow physics Joint Inst. Lab. Astrophysics, U. Colo., 1974-76; research asso. Optical Sci. Center, 1976-78; research fellow lasers and molecular spectroscopy U. Ariz., 1976-78; staff mem., project leader Los Alamos Nat. Lab., 1978—; vis. faculty, cons. AL-HAZEN Research Center, Baghdad, Iraq, 1975. Contbr. articles to profl. jours. Recipient Michael Chiha prize, 1964; Rockefeller fellow, 1967; Gibbs fellow, 1967-68. Mem. Am. Phys. Soc., Arab Phys. Soc. Subspecialties: Fusion; Laser fusion. Current work: Inertial confinement fusion and x-ray lasers. Office: Box 1663 MS-E526 Los Alamos MN 87545

LAANE, JAAN, chemistry educator; b. Paide, Estonia, June 20, 1942; s. Robert Friedrich and Linda (Treufeldt) L.; m. Tiiu Virkhaus, Sept. 14, 1966; children: Christina, Lisa. B.S. (Agnes Sloan Larsen scholar, Sloan scholar, James scholar), U. Ill., 1964; Ph.D. (Woodrow Wilson fellow, NSF fellow), M.I.T., 1967. Asst. prof. Tufts U., Medford, Mass., 1967-68; asst. prof. Tex. A&M U., College Station, 1968-72, assoc. prof., 1972-76, prof., chmn. div. phys. chemistry, 1976—; vis. scientist Los Alamos Sci. Labs., summers 1964,65,67; vis. prof. U. Bayreuth, W.Ger., 1979-80, summer 1981. Contbr. articles to profl. jours. NSF fellow, 1964-67; recipient Alexander von Humboldt U.S. Sr. Scientist award, 1979, Kendall award, 1964, Kodak award, 1967. Mem. Am. Chem. Soc., Am. Phys. Soc., Soc. Applied Spectroscopy, Coblentz Soc. Subspecialties: Physical chemistry; Spectroscopy. Current work: Infrared and Raman spectroscopy, nitrogen oxides, research on vibrational spectroscopy. Office: Dept Chemistry Tex A&M U College Station TX 77843 Home: 1906 Comal Circle College Station TX 77840

LABANICK, GEORGE MICHAEL, biology educator, researcher; b. Passaic, N.J., Sept. 27, 1950; s. Michael and Ilean (Meyer) L.; m. Deborah Ann Dilks, July 28, 1979; 1 son, Thomas. B.S., Coll. William and Mary, 1972; M.A., Ind. State U., 1974; Ph.D., So. Ill. U., 1978. Vis. asst. prof. Emory (Va.) and Henry Coll., 1978-79; asst. prof. biology U. S.C., Spartanburg, 1979—. Merit badge counselor Palmetto council Boy Scouts Am., 1982—. Highlands Biol. Sta. grantee, 1975, 76, 77; Sigma Xi grantee, 1977; U. S.C. Research Fund grantee, 1980; recipient Richard E. Blackwelder award So. Ill. U., 1978. Mem. Herpetologists' League, Soc. Study of Amphibians and Reptiles, Am. Soc. Ichthyologists and Herpetologists, Soc. Study of Evolution, Sigma Xi. Methodist. Subspecialties: Evolutionary biology; Population biology. Current work: Mimicry and other anti-predator mechanisms; population biology of plethodantid salamanders. Home: Route 4 Box 165 Inman SC 29349 Office: U SC Spartanburg SC 29303

LABARBERA, ANDREW RICHARD, educator; b. Teaneck, N.J., Oct. 6, 1948; s. Mario Richard and Georgine (Mart) LaB. B.S. cum laude, Iona Coll., 1970; M.Phil., Columbia U., 1974, M.A., 1974, Ph.D., 1975. Instr. dept. biology Iona coll., New Rochelle, N.Y., 1970; staff asso. Center for Reprodn. Sci., Columbia U., N.Y.C., 1975-77; research fellow Mayo Grad. Sch. Medicine, Rochester, Minn., 1977-80; asst. prof. physiology Northwestern U., Chgo., 1980—; dir. RIA Lab., CEMN, 1980—. Contbr. articles to profl. jours. USPHS-NIH predoctoral traineeship, 1971-75; Population Council grantee, 1972-75; Northwestern U. grantee, 1980-81, 81; NIH grantee, 1982—; recipient New Investigator award Am. Diabetes Assn., 1981-82. Mem. AAAS, Am. Inst. Biol. Scis., Am. Physiol. Soc., Am. Soc. Zoologists, Chgo. Assn. Reproductive Endocrinologists, Endocrine Soc., Soc.Exptl. Biology and Medicine, Soc. for Study of Reprodn., Tissue Culture Assn, Beta Beta Beta. Subspecialties: Physiology (biology); Receptors. Current work: Mechanism of action of glycoprotein hormones; ovarian cellular physiology; cell membrane biology; follitropin receptor-adenylyl cyclase interactions; stimulus-secretion coupling. Home: 708 W Wellington Ave #3 Chicago IL 60657 Office: Center for Endocrinology Metabolism and Nutrition Northwestern Univ Med Sch 303 E Chicago Ave Chicago IL 60611

L'ABATE, LUCIANO, psychology educator, psychologist; b. Briudisi, Italy, Sept. 19, 1928; came to U.S., 1948, naturalized, 1961; s. Giovanni and Alma (Zaccaro) L'A.; m. Bess Lukas, Aug. 31, 1958, 1958; children: E. Leila, John W. B.A., Tabor Coll., 1950; M.A., Wichita U., 1953; Ph.D., Duke U., 1956. Diplomate: Am. Bd. Profl. Psychology. Asst. prof. Wash. U., St. Louis, 1959-64; assoc. prof. Emory U., 1964-65; prof. psychology Ga. State U., 1965—; cons. Cross Keys, 1979—; cons. pub. cos. Author: Principles of Clinical Psychology, 1964, Understanding and Helping the Individual in the Family, 1976, Enrichment, 1977, Family Psychology, 1983; co-author: Teaching the Exceptional Child, 1975, How to Avoid Divorce, 1981, Handbook of Marital Interventions, 1983; contbr. numerous articles to profl. jours., chpts. to books; editor books, monograph. Pres. Amberwood Civic Assn., 1974-81; bd. dirs. Twin Lakes Community Center, 1976. USPHS Postdoctoral fellowship, 1950-59. Fellow Am. Psychol. Assn.; mem. Am. Orthopsychiat. Assn., Am. Family Therapy Assn., Am. Assn. Marriage and Family Therapists. Subspecialty: Family psychology. Current work: Family evaluation; prevention and intervention with couples and families. Home: 2079 Debonair Dr Atlanta GA 30345 Office: Ga State U University Plaza Atlanta GA 30303

LABELLE, EDWARD FRANCIS, biochemist, educator; b. Worcester, Mass., Aug. 11, 1948; s. Edward F. and Viola M. (Trudel) LaB.; m. Constance M. Reichmann, Aug. 19, 1972; children: Devon Nicole, Ross Edward. A.B., Coll. Holy Cross, 1970, M.S., 1970; Ph.D., U. Mich., 1974. Postdoctoral fellow Cornell U., Ithaca, N.Y., 1974-76; asst. prof. Western Ill. U., Macomb, 1976-78, U. Tex. Med. Br., Galveston, 1978—. Research grantee NIH, 1978—, Muscular Dystrophy Found., 1979-81, Am. Diabetes Assn., 1979-80. Mem. Am. Soc. Biol. Chemistry, Am Chem. Soc. Democrat. Subspecialties: Membrane biology; Biochemistry (biology). Current work: Isolation and characterization of protein responsible for amiloride-inhibited sodium transport across rabbit kidney cell membranes. Home: 5301 Tanglebriar Dickinson TX 77539 Office: Univ Tex Med Branch Div Biochemistry Galveston TX 77550

LABEN, ROBERT COCHRANE, animal scientist, educator; b. Darien Center, N.Y., Nov. 16, 1920; s. Victor L. and Ruth (Cochrane) L.; m. Dorothy Lobb, Nov. 29, 1946; children: John V., Robert J., Elizabeth J. Laben Cunningham, Catherine L. B.S., Cornell U., 1942; M.S., Okla. State U., 1947; Ph.D., U. Mo., 1950. Research and teaching asst. Okla. State U., 1946-47, U. Mo., 1947-50; instr. animal husbandry U. Calif., Davis, 1950-52, asst. prof., 1952-58, assoc. prof., 1958-64, prof., 1964—, prof. animal sci., 1964—, asst. assoc. animal husbandman expt. sta., 1950-64, geneticist, 1964—, dir., 1964-69, master advisor in animal sci., 1970—, vice chmn. dept. animal sci., 1977-82. Contbr. articles on dairy breeding, genetics and husbandry to profl. jours. Elder, deacon Davis Community Presbyn. Ch.; hunter safety vol. instr. Calif. Fish and Game Dept., 1962—; com. mem. Boy Scouts Am., 1955-56. Served from 2d lt. to capt. U.S. Army, 1942-46. Decorated Bronze Star, Purple Heart with oak leaf cluster; recipient Outstanding Advisor award U. Calif. at Davis Coll. Agr., 1982. Mem. Am. Dairy Sci. Assn., Am. Soc. Animal Sci., Biometric Soc., Am. Genetic Assn., Sigma Xi, Alpha Zeta. Republican. Subspecialties: Animal genetics; Animal production. Current work: Animal breeding and genetics, with emphasis on productive and reproductive traits of dairy cattle, milk yield and composition, lactation stress. Home: 502 Oak Ave Davis CA 95616 Office: Dept Animal Sci U Calif Davis CA 95616

LABOV, JAY BRIAN, biology educator; b. Phila., Sept. 19, 1950; s. Irving and Lillian (Guss) L.; m. Jeri E. Estra, June 1, 1975; children: Adan D., Rachel C. B.S., U. Miami, Fla., 1972; M.S., U. R. I., 1974, Ph. D., 1979. Instr. biology Washington and Lee U., Lexington, Va., 1978-79; asst. prof. biology Colby Coll., Waterville, Maine, 1979—; vis. investigator Jackson Lab., Bar Harbor, Maine, 1981; textbook rev. cons. MacMillan Pub. Co., N.Y.C., 1982—. Contbr. articles to profl. jours.; author: Vertebrate Biology: A Laboratory and Field Manual, 1976. Participant Waterville Community Chorus, 1979—. NSF grantee, 1981-82; Andrew Mellon Found. grantee, 1980-82; Sigma Xi grantee, 1980-81. Mem. Animal Behavior Soc., Am. Soc. Mammalogists, Am. Soc. Zoologists, AAAS, Sigma Xi. Subspecialties: Sociobiology; Ethology. Current work: Factors influencing infanticidal behavior toward unrelated offspring; evolution of pregnancy blocking in rodents; role of social interactions on physiology of animals. Office: Dept Biology Colby Coll Waterville ME 04901 Home: 63 Johnson Heights Waterville ME 04901

LABUDDE, ROBERT ARTHUR, software company excutive; b. Flint, Mich., May 28, 1947; s. Verne M. and Adeline (Essa) LaB.; m. Constance M. Miller, Apr. 11, 1969; children: Philip V., Zina M. B.S., U. Mich., 1969; Ph.D., U. Wis., 1973. Lectr. in computer sci. U. Wis.-Madison, 1973; asst. scientist Math. Research Ctr., Madison, 1973-74; instr. applied math. MIT, Cambridge, Mass., 1974-75; asst. prof. Old Dominion U., Norfolk, Va., 1976-79; pres. Least Cost Formulations, Ltd., Virginia Beach, Va., 1979—; exec. v.p. LaBudde Engring. Corp., Westlake Village, Calif., 1983—; cons. IBM Corp., Atlanta, 1975-78, Allied Corp., Morristown, N.J., 1977-78, Burroughs Corp., Westlake Village, Calif., 1980-82, Optical Coating Lab., Santa Rosa, Calif., 1982-83. Contbr. articles to profl. jours. Mem. Assn. for Computing Machinery, Soc. Indsl. and Applied Math., AAAS, Sigma Xi, Phi Beta Kappa, Phi Kappa Phi, Phi Lambda Upsilon. Subspecialties: Mathematical software; Laser data storage and reproduction. Current work: Application of scientific analysis and software to manufacturing; technical consulting in the development of state-of-the-art computer products. Office: Least Cost Formulations Ltd 821 Hialeah Dr Virginia Beach VA 23464

LABUZA, THEODORE PETER, food scientist, educator, cons.; b. Perth Amboy, N.J., Nov. 10, 1940; s. Theodore and Cathrine Julia (Stycheck) L. S.B., M.I.T., 1962, Ph.D., 1965. Asst. prof. food sci. and nutrition M.I.T., 1965-68, assoc. prof., 1968-71, U. Minn., St. Paul, 1971-73, prof., 1973—; cons. to food cos.; tchr. short courses on phys. chemistry of foods, food law. Author: books including Shelf Life Testing of Foods, 1982; contbr. numerous articles to profl., popular publs. Recipient Teaching award M.I.T., 1970, cert. merit NASA, 1973, 74, 76. Fellow Inst. Food Tech. (Samuel Cate Prescott Research award 1972, William V. Crusas Teaching award 1979); mem. Am. Chem. Soc., Am. Inst. Chem. Engring., Am. Soc. Agrl. Engring., Am. Assn. Cereal Chemistry, Sigma Xi, Gamma Sigma Delta, Phi Kappa Phi. Subspecialties: Food science and technology; Nutrition (biology). Current work: Physical chemistry of food stability; water activity; kinetics; nutritional stability; shelf life; computer modeling. Patentee in field. Home: 1870 Stowe Arden Hills MN 55112 Office: Dept Food Sci and Nutrition U Minn Saint Paul MN 55108

LA CELLE, PAUL LOUIS, physician, educator; b. Syracuse, N.Y., July 4, 1929; s. George Clarke and Marguerite Ellen (Waggoner) La C; m. June Dukeshire, May 23, 1953; children: Andrea Jean, Peter Theodore, Kristina Marie, Erik Clarke. A.B., Houghton Coll., 1951; M.D., U. Rochester, 1959. Intern U. Rochester Med. Center-Strong Meml. Hosp., 1959-60, resident, 1960-62; asst. prof. medicine U. Rochester, 1967-70, asso. prof., 1970-74, prof., 1974—, chmn. dept. radiation biology and biophysics, 1977—; cons. to govt. Mem. Gates-Chili Sch. Bd., Rochester, 1964-72; trustee Houghton Coll., 1976—. Served to lt. (j.g.) USNR, 1952-55. NIH spl. fellow, 1965-66; recipient von Humboldt Sr. Scientist award, 1982-83. Mem. Biophys. Soc., Microcirculation Soc., European Microcirculation Soc., Am. Soc. Hematology, AAAS, Sigma Xi, Alpha Omega Alpha. Subspecialties: Biophysics (biology); Hematology. Current work: Research: Investigation of rheologic properties of normal and pathologic blood cells in the microcirculation. Research in biophysics of blood cells, physiology of microcirculation. Office: Box RBB Med Center 601 Elmwood Ave Rochester NY 14642

LACHANCE, LEO EMERY, geneticist; b. Brunswick, Maine, Mar. 1, 1931; s. Emery and Edwidge (Pouliot) LaC.; m. P. Joan Favreau, Aug. 6, 1955; children: Lois Anne, Marc Hunter, Matthew S. B.S., U. Maine, 1953; M.S., N.C. State U., 1955, Ph.D., 1958. Research asso. biology Brookhaven Nat. Lab., Upton, N.Y., 1958-60; research geneticist Entomology Research div. USDA, Kerville and Mission, Tex., 1960-63; research leader, radiation biology and insect genetics sect. Metabolism and Radiation Research Lab., USDA, Fargo, N.D., 1964-69, 71-76, lab. dir., 1977-82; nat. tech. adv. insect genetics, research geneticist Agrl. Research Service, 1982—; head pest control sect. FAO/IAEA, Vienna, 1969-71. Contbr. articles to profl. jours. Served with USAR, 1953-56. NIH fellow, 1956-58; NRC Nat. Acad. Scis. travel grantee, 1962; vis. lectr. Purdue U., 1972. Fellow AAAS; mem. Genetics Soc. Am., Entomol. Soc. Am., Am. Inst. Biol. Scis., Fedn. Am. Scientists, Sigma Xi. Subspecialties: Genetics and genetic engineering (agriculture); Cell biology. Current work: Researcher in sterile male technique of insect control, radiation biology, insect reproduction, hybrid sterility, genetics of insect populations, cytology. Home: 77 21st Ave N Fargo ND 58102 Office: Metabolism and Radiation Research Lab PO Box 5674 State Univ Sta Fargo ND 58105

LACHENBRUCH, ARTHUR HEROLD, research geophysicist; b. New Rochelle, N.Y., Dec. 7, 1925; s. Milton Clevel and Leah (Herold) L.; m. Edith Bennett, Sept. 7, 1950; children: Roger, Charles, Barbara. B.A., Johns Hopkins U., 1950; M.A., Harvard U., 1954, Ph.D., 1958. Registered geophysicist and geologist, Calif. Research geophysicist U.S. Geol. Survey, 1951—; vis. prof. Dartmouth Coll., 1963; mem. numerous adv. coms. and panels. Contbr. articles to sci. jours. Mem. Los Altos Hills (Calif.) Planning Commn., 1968—. Served in USAAF, 1943-46. Recipient Spl. Act award U.S. Geol. Survey, 1970, Meritorious Service award, 1972; Disting. Service award U.S. Dept. Interior, 1978. Fellow am. Geophys. Union, AAAAS, Royal Astron. Soc., Geol. Soc. Am. (Kirk Bryan award 1963), Arctic Inst. N.Am., mem. Nat. Acad. Sci. Subspecialties: Tectonics; Geophysics. Current work: Solid-earth geophysics; terrestrial heat flow; tectonophysics; permafrost. Office: 345 Middlefield Rd Menlo Park CA 94025

LACHER, MIRIAM BROWNER, neuropsychologist; b. N.Y.C., Dec. 30, 1942; d. Philip and Ruth F. (Rabinowitz) B.; m. Maury Lacher, Aug. 17, 1963. A.B., Cornell U., 1963; Ph.D., U. Mich., 1970; postgrad. in neuropsychology, Columbia U., 1981. Lic. psychologist, N.Y., N.J. Asst. prof. psychology Carleton Coll., Northfield, Minn., 1970-77; vis. research assoc. U. Calif.-Berkeley, 1976-77; vis. lectr. Vassar Coll., Poughkeepsie, N.Y., 1978-79; assoc. in neuropsychology Columbia-Presbyn Med. Ctr., N.Y.C., 1980-81; postdoctoral trainee in rehab. medicine NYU Med. Ctr., N.Y.C., 1981-82; coordinator cognitive remediation Children's Specialized Hosp., Westfield-Mountainside, N.J., 1982—; cons. in neuropsychology N.Y. State Psychiat. Inst., N.Y.C., 1981. Author articles in field. Woodrow Wilson fellow, 1963; USPHS trainee in exptl. psychology U. Mich., 1964-65, in devel. psychology, U. Mich., 1968-70. Mem. Am. Psychol. Assn., AAAS, Internat. Neuropsychol. Soc., N.Y. Neuropsychol. Group, Soc. Research in Child Devel., Eastern Psychol. Assn. Subspecialties: Neuropsychology; Developmental psychology. Current work: Rehabilitation techniques in adolescents with head traumas. Home: 37 Alda Dr Poughkeepsie NY 12603 Office: Children's Specialized Hosp New Providence Rd Westfield-Mountainside NJ 07091

LACHMAN, ROY, psychology educator; b. Bklyn., Nov. 26, 1927; s. Morris and Lilly (Ladina) L.; m. Janet L. Miner, Sept. 1, 1971; 1 dau., Dana Clare. B.S., CUNY, 1955; Ph. D., NYU, 1960. Asst. prof. U. Hawaii, Hilo, 1959-60, Hollins Coll., Roanoke, Va., 1961-63; prof. psychology SUNY-Buffalo, 1963-71; prof. U. Kansas.-Lawrence, 1971-74; prof. psychology U. Houston, 1974—, dir. grad. studies, 1974-76. Author: Cognitive Psychology and Information Processing, 1979; editor: Information Technology in the 1980s, 1982. NSF, NIMH, NIA grantee, 1959—. Fellow Am. Psychol. Assn.; mem. Philosophy of Sci. Assn., Psychonomic Soc., Cognitive Sci. Soc. Subspecialties: Cognition; Artificial intelligence. Current work: Retraining the structurally unemployed in information technology. Home: 5511 Newcastle Bellaire TX 77401 Office: Dept Psychology U Houston Houston TX 77004

LACHTER, GERALD DAVID, psychology educator; b. Bronx, N.Y., May 3, 1941; s. Lazar and Leah (Weisberg) L.; m. Abbie Jane Lowenstein, May 19, 1973; 2 daus. B.A., C.W. Post Coll., 1964; M.A., Columbia U., 1966; Ph.D., CUNY, 1970. Lectr. Queens Coll., CUNY, Flushing, 1969-70; asst. prof. C. W. Post Coll, Greenvale, N.Y., 1970-74, assoc. prof., 1974-82, prof. psychology, 1982—; cons. psychologist Suffolk Child Devel. Ctr., Smithtown, N.Y., 1981—, Suffolk Assn. Retarded Children, Commack, N.Y., 1976-81. Author: Behavioral Objectives Workbook, 1974; Contbr. articles in field to profl. jours. Mem. Psychonomic Soc., Am. Psychol. Assn., Assn. Behavior Analysis, N.Y. Acad. Scis., Sigma Xi. Democrat. Jewish. Subspecialties: Behavioral psychology; Learning. Current work: Research in reinforcement schedules, stimulus control and reinforcement theory. Office: CW Post Coll Greenvale NY 11548

LACK, DOROTHEA Z., clinical psychologist; b. Winthrop, Mass. B.S., Boston U., 1968; M.A., Syracuse U., 1971, Ph.D., 1975. Staff psychologist Lynn (Mass.) Community Health and Counseling Ctr., 1974-75; dir. pain program and psychology Lourdes Hosp., Binghamton, N.Y., 1975-82; dir. pain team Kaiser Permanente Med. Ctr., Hayward, Calif., 1982—. NIH grantee, 1972. Mem. Am. Psychol. Assn., Am. Pain Soc. Subspecialties: Behavioral psychology; Learning. Current work: Pain management, pain and women, multidisciplinary approach, development of pain management program for treatment of patients with chronic pain in a health maintenance organization, women with pain, multifamily group therapy. Home: PO Box 142 Mount Eden CA 94557 Office: Kaiser Permanente Med Ctr 27400 Herperian Blvd Hayward CA 94557

LACKEY, LAURENCE E(VAN), academic administrator, geologist, computer scientist, consultant; b. Los Angeles, Aug. 13 1947. B.S. in Geology, Principia Coll., 1969, Ph.D., U. Mich., 1974. Asst. prof. Mich. State U., 1974-75; asst. prof. geology, dir. Tenn. Earthquake Info. Ctr., Memphis State U., 1975-79; chmn. dept. earth scis. Principia Coll., 1979—; cons. geology, 1975—, computers, 1982—. Mem. Am. Geophys. Union, AAAS, Geol. Soc. Am., Am. Assn. Petroleum Geologists, Am. Quaternary Assn. Subspecialty: Geology. Current work: Geology. Office: Dept Earth Sci Principia Coll Elsah IL 62068

LACKO, ANDRAS GYORGY, biochemistry educator; b. Budapest, Hungary, Nov.10, 1936; came to U.S., 1963; s. Jeno and Klara (Mezei) L.; children by previous marriage: Annette, Marianne, Peter, Joshua. B.S.A., U. B.C., Vancouver, 1961, M.Sc., 1963; Ph.D., U. Wash., 1968. Postdoctoral fellow Albert Einstein Coll. Medicine, Bronx, N.Y., 1968-69; asst. mem. Albert Einstein Med. Ctr., Phila., 1969-71; research asst. prof. Temple U. Med. Sch., Phila., 1971-75; assoc. prof. Tex. Coll. Osteo. Medicine, Ft. Worth, 1975-83, prof., 1983—; assoc. dir. biomed. research Ctr. for Studies in Aging, N. Tex. State U., Denton, 1976—. Contbr. articles in field to profl. jours. Mem. Am. Soc. Biol. Chemists. Subspecialties: Biochemistry (medicine); Biochemistry (biology). Current work: Plasma lipid and lipoprotein metabolism. Home: 612 Woodford St Denton TX 76201 Office: Tex Coll Osteo Medicine Camp Bowie and Montgomery Fort Worth TX 76201

LACKS, SANFORD ABRAHAM, geneticist; b. N.Y.C., Jan. 28, 1934; s. Charles Jonas and Goldie Rose (Dranoff) L.; m. Elaine Rose Norris, Nov. 22, 1959; children: Jennifer, Daniel, Julia. B.S., Union Coll., 1955; Ph.D., Rockefeller U., 1960. Instr., Harvard U., Cambridge, 1960-61; guest investigator Pasteur Inst., Paris, 1957-58, Hebrew U., Jerusalem, 1970-71; geneticist Brookhaven Nat. Lab., Upton, N.Y., 1961—. Contbr. articles to profl. jours. NIH grantee, 1978—, 81—. Mem. Am. Soc. Biol. Chemists, Am. Soc. Microbiology, Genetics Soc. Am. Subspecialties: Genetics and genetic engineering (biology); Biochemistry (biology). Current work: Microbial genetics; recombinant DNA, bacterial transformation, DNA repair; biochemistry, deoxyribonucleases. Office: Biology Dept Brookhaven Nat Lab Upton NY 11973

LA CLAIRE, JOHN WILLARD, II, botanist, educator; b. Utica, N.Y., July 1, 1951; s. John Willard and Mary Magdelene (Smith) La C. B.S. in Biology, Cornell U., Ithaca, N.Y., 1973; M.A., U. South Fla., 1975; Ph.D. in Botany, U. Calif.-Berkeley, 1979. Asst. prof. botany U. Tex.-Austin, 1979—. Contbr. articles to sci. jours. NSF grantee, 1982-84. Mem. AAAS, Bot. Soc. Am., Brit. Phycol. Soc., Internat. Phycol. Soc., Phycol. Soc. Am., Sigma Xi. Subspecialties: Cell biology; Marine Phycology. Current work: Plant cell motility phenomena; cellular wound healing utilizing giant algal cells as model systems; ultrastructure/electron microscopy of marine algae (cell division, morphogenesis, protoplasts). Office: Dept Botany U Texas Austin TX 78712

LACOURSE, WILLIAM CARL, ceramic engineering educator; b. Schenectady, N.Y., June 19, 1943; s. Ludger J. and Dorothea M. (Hall) LaC.; m. Patricia M. Clarke, Sept. 3, 1966; children: Brian Clarke, Elisa Mae. B.S., SUNY-Stony Brook, 1966, M.S., 1967; Ph.D., Rensselaer Poly. Inst., 1970. Research assoc. U.S. Naval Research Lab., Washington, 1970; asst. prof. glass sci. Alfred (N.Y.) U., 1970-77, assoc. prof., 1977—. Mem. Am. Ceramic Soc., Nat. Inst. Ceramic Engrs., Am. Metallographic Soc., Ceramic Ednl. Council, Soc. Glass Tech. Republican. Roman Catholic. Subspecialties: Ceramics; Materials (engineering). Current work: Research in sol-gel processing of glass, strengthening mechanisms, glass fibers, secondary ion mass spectrometry, ion scattering spectrometry, x-ray photoelectron spectrometry, gas glass interactions. Home: 15 Reynolds St Alfred NY 14802 Office: Alfred Univ Dept Glass Science Alfred NY 14802

LACY, GEORGE HOLCOMBE, plant pathologist; b. Washington, Nov. 13, 1943; s. Robert Willimgham and Wilma Pauline (Knox) L.; m. Ruth Ingeborg Lundberg, June 6, 1964; children: Yaxpal, Kyla. B.S., Calif. State U.-Long Beach, 1966, M.S., 1971; Ph.D. in Plant Pathology, U. Calif.-Riverside, 1975. Postdoctoral assoc. dept. plant pathology U. Wis.-Madison, 1975-77; asst. plant pathologist, dept. plant pathology and botany Conn. Agrl. Expt. Sta., New Haven, 1977-80; assoc. prof. plant pathology, dept. plant pathology and physiology Va. Poly. Inst. and State U., Blacksburg, 1980—; cons. Innotech Corp., Trumbull, Conn., 1978—, Allied Chem. Corp., Solvay, N.Y., 1982—, U.S. Army C.E., Vicksburg, Miss. Editor: Phytopathogenic Prokaryotes, 1982; mem. editorial bd.: Plant Molecular Biology, 1981—. NIH fellow, 1975. Mem. Am. Phytopath. Soc. (chmn. bacteriology com. 1981-82), Am. Soc. Microbiology. Subspecialties: Plant pathology; Genetics and genetic engineering (agriculture). Current work: Genetic basis for pathogenesis caused by plant pathogenic prokaryotes. Mechanisms for pathogenesis. Host responses to pathogenesis. Office: Dept Plant Pathology and Physiology Va Poly Inst and State U Blacksburg VA 24061

LADA, CHARLES JOSEPH, astronomer; b. Webster, Mass., Mar. 18, 1949; s. Joseph and Rita Anne (Holewa) L. B.A. with distinction, Boston U., 1971; M.A., Harvard U., 1973, Ph.D., 1975. Fellow Center for Astrophysics, Harvard-Smithsonian Obs., 1975-77, Harvard Coll. Obs., 1977-78; Bart J. Bok fellow U. Ariz., Tucson, 1978-80, asst. prof., 1980—. Contbr. articles to profl. jours. Alfred P. Sloan Found. fellow, 1981; recipient Bart J. Bok prize for astrophysics Harvard U., 1982. Mem. Internat. Astron. Union, Am. Astron. Soc. Subspecialties: Radio and microwave astronomy; Infrared optical astronomy. Current work: Studies of star formation of early stellar evolution; studies of interstellar medium in galaxies. Address: Steward Obs Univ Ariz Tucson AZ 85721

LADANYI, BRANKA MARIA, chemistry educator; b. Zagreb, Yugoslavia, Sept. 7, 1947; came to U.S., naturalized, 1975; d. Branko and Nevenka (Zilic) L.; m. Marshall Fixman, Dec. 7, 1974. B.S., McGill U., 1969; M.Phil., Yale U., 1971, Ph.D., 1973. Vis. prof. chemistry U. Ill.-Urbana, 1974; postdoctoral research assoc. Yale U., New Haven, 1974-77, research assoc., 1977-79; asst. prof. chemistry Colo. State U., Ft. Collins, 1979—. Contbr. articles to profl. jours., U.S., Can. Active NOW. Sloan Found. fellow, 1978—; NSF grantee, 1980-83; Am. Chem. Soc. Petroleum Research Fund grantee, 1979-82. Mem. Am. Chem. Soc., Am. Phys. Soc., Sigma Xi. Subspecialties: Physical chemistry; Statistical mechanics. Current work: Molecular theory of liquids; theory of light scattering in fluids and polymer solutions; theory of solvent effects on chemical reactions. Office: Dept Chemistry Colo State U Fort Collins CO 80523 Home: 1100 E Pitkin St Fort Collins CO 80524

LADD, GARY WAYNE, psychologist, researcher, educator; b. Buffalo, Apr. 28, 1950; s. Herbert G. and Marion V. (Johnson) L.; m. Joan A. Jurich, Aug. 6, 1981. B.A., Grove City (Pa.) Coll., 1972; M.A., Alfred U., 1974; Ed.D., U. Rochester, N.Y., 1979. Lic. psychologist, N.Y. Research assoc. U. Rochester, 1978-79; asst. prof. child devel. and psychology Purdue U., West Lafayette, Ind., 1979—; acting head Child and Family Research Inst., 1981-82. Contbr. articles to profl. jours. Mem. Am. Psychol. Assn., Soc. Research in Child Devel., Sigma Xi. Subspecialty: Developmental psychology. Current work: Research on childhood social difficulties and their remediation, relationships between early pathology and later development. Office: Dept Child Devel and Family Studies Purdue U Gates St West Lafayette IN 47906

LADDU, ATUL RAMCHANDRA, clin. researcher, physician; b. Poona, India, Aug. 23, 1940; came to U.S., 1968, naturalized, 1972; s. Ramchandra Dnyneshwar and Sarojini (Ramchandra) L.; m. Jayashree Atul, June 19, 1965; children: Prashanta Atul, Abhay Atul. M.B., B.S., G.R. Med. Coll., Gwalior, India, 1962; M.D., Delhi U., 1967. Instr. pharmacology Maulana Azad Med. Coll., New Delhi, India, 1963-68; postdoctoral fellow dept. pharmacology Med. Coll. Wis., Milw., 1968-71; instr., asst. prof., 1971-73; group leader cardiovascular div. Lederle Labs. div. Am. Cyanamid Co., Pearl River, N.Y., 1973-75; sr. clin. investigation assoc. CIBA-GEIGY Corp., Summit, N.J., 1975-76; asst. med. dir. Ives Labs. Inc. div. Am. Home Products Inc., N.Y.C., 1976, assoc. med. dir., 1976-78, project leader, 1978-82; dir. clin. research Am. Critical Care, McGaw Park, Ill., 1982—. Mem. editorial bd.: Jour. Clin. Pharmacology, Am. Heart Jour; Contbr. articles to profl. jours. Fellow Am. Coll. Clin. Pharmacology, Am. Soc. Clin. Pharmacology and Therapeutics, Am. Soc. for Pharmacology and Exptl. Therapeutics, Am. Coll. Cardiology; mem. Am. Fedn. Clin. Research, N.Y. Acad. Scis., Soc. for Exptl. Biology and Medicine, Internat. Study Group for Research in Cardiac Metabolism, Indian Coll. Allergy and Applied Immunology, Sigma Xi. Subspecialties: Cardiology; Internal medicine. Current work: Clinical research in broadest sense, cardiovascular drugs in particular. Office: 1600 Waukegan Rd McGaw Park IL 60085

LADENSON, JACK HERMAN, clinical chemist, researcher, educator; b. Phila., Apr. 8, 1942; s. Paul J. and Lillian B. (Vinolar) L.; m. Ruth E. Carroll, June 23, 1968; children: Michele, Jeff. B.S., Pa. State U., 1964; Ph.D., U. Md., 1971. Asst. dir. clin. chemistry Barnes Hosp., St. Louis, 1972-76, co-dir. clin. chemistry, 1976-79, dir. clin. chemistry, 1980—; asst. prof. pathology Washington U. Sch. Medicine, St. Louis, 1972-79, assoc. prof., 1979—; bd. dirs. Am. Bd. Clin. Chemistry, 1979—; v.p. Commn. on Accreditation in Clin. Chemistry, 1981—. Mem. Am. Assn. Clin. Chemistry (dir. 1981-83), Acad. Clin. Lab. Physicians and Scientists, Endocrine Soc. Subspecialty: Clinical chemistry. Current work: Factors controlling partitioning of electrolytes in biological fluids; use of monoclonal antibodies for diagnosis. Office: Div Lab Medicin Washington U Sch Medicine Box 8118 Saint Louis MO 6311

LADISCH, STEPHAN, pediatrician, researcher, educator; b. Garmisch-Partenkirchen, W.Ger., July 18, 1947; came to U.S., 1948, naturalized, 1959; s. Rolf Karl and Brigitte (Gareis) L.; m. b. Brigitte Bidault, May 22, 1974; children: Gwenola, Virginie. B.S. in Chemistry, U. Pa., 1969, M.D., 1973. Diplomate: Am. Bd. Pediatrics. Intern Children's Hosp. Med. Ctr., Boston, 1973-74, resident, 1974-75; clin. assoc. pediatric oncology br. Nat. Cancer Inst., Bethesda, Md., 1975-77, investigator, 1977-78; asst. prof. pediatrics, sr. mem. human immunobiology group UCLA Sch. Medicine, 1978-82, assoc. prof., 1982—. Contbr. articles to profl. jours. NSF grantee, 1968-69, 72; Von L. Meyer travel fellow, 1975; NIH grantee, 1980—; Research career devel. awardee, 82—; Leukemia Soc. Am. scholar, 1982—. Mem. Am. Assn. Immunologists, Am. Soc. Hematology, Western Soc. for Pediatric Research, Am. Fedn. for Clin. Research, Soc. for Pediatric Research, AAAS, Phi Beta Kappa, Alpha Chi Sigma, Phi Lambda Upsilon. Subspecialties: Immunobiology and immunology; Cancer research (medicine). Current work: Modulation of immune responses by lipids. Office: Dept Pediatrics UCLA Sch Medicine Los Angeles CA 90024 Home: 650 Via de la Paz Pacific Palisades CA 90272

LAFFERTY, JAMES MARTIN, physicist; b. Battle Creek, Mich., Apr. 27, 1916; s. James V. and Ida M. (Martin) L.; m. Eleanor J. Currie, June 27, 1942; children: Martin C., Ronald J., Douglas J., Lawrence E. Student, Western Mich. U., 1934-37; B.S. in Engring. Physics, U. Mich., 1939; M.S. in Physics, U. Mich., 1940; Ph.D. in Elec. Engring, U. Mich., 1946. Physicist Eastman Kodak Research Lab., Rochester, N.Y., 1939; physicist Gen. Electric Research Lab., Schenectady, 1940, 42—, mgr. power electronics lab., 1972-81; with Carnegie Instn., Washington, 1941-42; Pres. Internat. Union Vacuum Sci. Technique and Applications, 1980-83. Editor, contbg. author: Scientific Foundations of Vacuum Technique (Dushman), 1962; editor: Vacuum Arcs, Theory and Applications, 1980; asso. editor: Jour. Vacuum Sci. and Tech, 1966-69; Editorial bd.: Internat. Jour. Electronics, 1968—; Contbr. articles to profl. jours. Mem. greater consistory Ref. Ch.; trustee Schenectady Museum, 1967-73, sec., 1971-72, pres., 1972-73. Recipient Devel. award Bur. Naval Ordnance, 1946; Distinguished Alumnus citation U. Mich., 1953; IR-100 award, 1968. Fellow Am. Phys. Soc., IEEE (Lamme medal 1979), AAAS (hon. life); mem. Am. Vacuum Soc. (dir. 1962-70, sec. 1965-67, pres. 1968-69), U.S. Power Squadrons (comdr. Lake George squadron 1975-76, comdr. Dist. 2 1981-82), Sigma Xi, Phi Kappa Phi, Iota Sigma, Tau Beta Pi. Subspecialties: Electrical engineering; Electronics. Current work: Electric vehicles. Patentee in field; inventor lanthanum boride cathode, 1950, hot cathode magnetron ionization gauge, 1961, triggered vacuum gap, 1966. Home: 1202 Hedgewood Ln Schenectady NY 12309 Office: PO Box 43 Schenectady NY 12301

LAFFLER, THOMAS G., geneticist; b. Detroit, May 10, 1946; s. Anthony Henry and Helen Viola (Hoenig) L.; m. Delfina Patta, Sept. 14, 1968; children: Bridget Barbara, Mary Josephine, John Anthony. S.B., MIT, 1968; Ph.D., U. Wash., 1974. Postdoctoral fellow U. Wis.-Madison, 1974-80; asst. prof. microbiology Northwestern U. Med. Sch., 1980—. Contbr. articles to profl. jours. Am. Cancer Soc. grantee, 1981-82; NIH grantee, 1982—. Mem. Genetics Soc. Am. (morale com.), Sigma Xi. Subspecialties: Genetics and genetic engineering (biology); Cell biology. Current work: Regulation of mitotic cycle. Office: Dept Microbiology and Immunolog Northwestern U Med Sch Chicago IL 60611

LAFUSE, WILLIAM PERRY, biochemist; b. Liberty, Ind., Nov. 1, 1950; s. Ross H. and Ruth (Smith) L. B.S. with honors, Purdue U., 1973; Ph.D., Johns Hopkins U., 1978. Postdoctoral fellow dept. immunology Mayo Clinic, Rochester, Minn., 1978-82, research assoc., 1982—; asst. prof. Mayo Clinic Med. Sch., 1981—. Postdoctoral fellow Am. Cancer Soc., 1979; grantee, 1982. Mem. Am. Assn. Immunologists. Subspecialties: Immunobiology and immunology; Immunocytochemistry. Current work: Biochemical analysis of gene products of the major histocompatibility complex of the mouse with emphasis on the structure, genetics and function of the immune response associated antigens. Home: 1203 4th Ave SW Apt 5 Rochester MN 55901 Office: Dept Immunology Mayo Clinic Rochester MN 55905

LAGARIAS, JEFFREY CLARK, mathematician; b. Pitts., Nov. 16, 1949; s. John Samuel and Virginia Jane (Clark) L. B.S., M.I.T., 1972, M.S., 1972, Ph.D., 1974. Mem. tech. staff Bell Labs., Murray Hill, N.J., 1974—; vis. asst. prof. math. dept. U. Md., College Park, 1978-79. Mem. Am. Math. Soc., Math. Assn. Am., Soc. Indsl. and Applied Math. Subspecialties: Number theory; Algorithms. Current work: Computational complexity theory, number theory, combinatorial mathematics, operations research. Home: 456 Mountain Ave Gillette NJ 07933 Office: Room 2C-370 Bell Labs 600 Mountain Ave Murray Hill NJ 07974

LAGO, BARBARA DRAKE, microbial geneticist; b. Wheeling, W.Va., Dec. 18, 1930; d. Estes C. and Margaret H. Drake; m. James Lago, Jan. 8, 1968. B.A., U. Calif.-Santa Barbara, 1952; Ph.D., Stanford U., 1959. Asst. prof. San Jose (Calif.) State Coll., 1961-62; research assoc. U. Calif.-Berkeley, 1962-63; asst. prof. U. Calif.-Davis, 1964-65; with Research Labs., Merck & Co., Inc., 1965—, assoc. dir. developmental microbiology, 1974-80, dir. microbial genetics and physiology, 1980-81, sr. dir. fermentation microbiology, 1981—. Mem. Am. Soc. Microbiology, Soc. Indsl. Microbiology, Genetics Soc. Am., Soc. Gen. Microbiology. Subspecialties: Genetics and genetic engineering (biology); Microbiology. Office: PO Box 2000 Rahway NJ 07065

LAGO, PAUL KEITH, zoology educator; b. Worthington, Minn., June 24, 1947; s. Darrel Paul and Shirley Arlene (Voorhees) L.; m. Barbara Ann Kent, Sept. 13, 1969; 1 son Eric. B.A., Bemidji State U., 1969, M.A., 1971; Ph.D., N.D. State U., Fargo, 1977. Asst. prof. biology U. Miss., University, 1976-82, assoc. prof., 1982—. Author: The Phytophagious Scarabaeidae and Troginae of North Dakota, 1979. Mem. Coleopterists Soc., Entomol. Soc. Am., Am. Entomol. Soc., N.Am. Benthological Soc., Inland Bird Banding Assn., Sigma Xi. Baptist. Subspecialties: Taxonomy; Ecology. Current work: Taxonomy and ecology of Coleoptera, Neuroptera and Trichoptera. Home: 1512 White Oak Ln Oxford MS 38655 Office: Dept Biolog U Miss University MS 38677

LAGREGA, MICHAEL DENNY, civil engineering educator, consultant; b. Yonkers, N.Y., July 19, 1944; s. Michael Edward and Toy E. (Denny) LaG.; m. Mary Anne Nezelek, Nov. 1, 1970; children: Diane, David. B.E. in Civil Engring, Manhattan Coll., 1966; M.S. in San. Engring, Syracuse U., 1971, Ph.d., 1972. Registered profl. engr., Pa. and 5 other states. Project engr./project mgr. O'Brien & Gere Engrs., Syracuse, N.Y., 1966-72; asst. prof. civil engring. Drexel U., 1972-74; assoc. prof. Bucknell U., 1974—; prin. cons. Roy F. Weston, Inc., West Chester, Pa., 1981—; mem. hazardous waste facilities planning adv. com. Pa. Dept. Environ. Resources, 1981—; chmn. planning com. 1983 Mid-Atlantic Indsl. Waste Conf.; mem. state coordinating com. Office Hazardous/Toxic Waste Mgmt., Pa. State U., 1982—; program reviewer NSF. Contbr. articles to profl. publs.; editor procs. 9th, 12th and 15th Mid-Atlantic Indsl. Waste confs., 1977, 80, 83. Mem. Water Pollution Control Fedn. (chmn. indsl. waste symposia 1982—), Water Pollution Control Assn. Pa. (chmn. indsl. waste 1979-81, Research award 1978), Internat. Assn. Water Pollution Research, ASCE, Assn. Environ. Engring. Profs. Republican. Roman Catholic. Subspecialty: Environmental engineering. Current work: Hazardous waste management; industrial wastes; water pollution control; solid waste management. Office: Dept Civil Engring Bucknell U Lewisburg PA 17837 Home: 159 Jean Blvd Lewisburg PA 17837

LAGUNAS-SOLAR, MANUEL CLAUDIO, radiochemist; b. Valparaiso, Chile, Dec. 23, 1941; came to came to U.S., 1970; s. Manuel L. and Alejandrina Solar; m. Patricia Isabel Lagunas-Solar, Dec. 27, 1968; children: Claudio, Rodrigo. Licenciate in Chemistry and Edn. summa cum laude, Catholic U., Valparaiso, 1968; M.S. magna cum laude, U. P.R., 1970; Ph. D. in Chemistry, U. Calif.-Davis, 1974. Asst. prof. chemistry Catholic U., Valparaiso, 1965-69; teaching asst. U. P.R., Mayagyez, 1968-69, instr. phys. chemistry, 1969-70; research asst. P.R. Nuclear Ctr., Mayagyez, 1968-70, U. Calif.-Davis, 1970-73, asst. research radiochemist, 1973—, chief radioisotope program, 1973—; cons. in field; dir. Picowave Processors. Co-patentee: continuous flow radioactive iodine prodn. Mem. Am. Chem. Soc., Soc. Nuclear Medicine, Am. Nuclear Soc., Sigma Xi. Roman Catholic. Subspecialty: Physical chemistry. Current work: Development of medical radionuclides and radiopharmaceuticals for diagnostic and/or therapeutic nuclear medicine; radiation chemistry. Home: 2721 Hatteras Pl Davis CA 95616 Office: Crocker Nuclear Lab U Calif Davis CA 95616

LAHERU, KEN LIEM, structural engineer, researcher; b. Kuningan, West Java, Indonesia, Aug. 23, 1935; came to U.S., 1969; m. Hilda Djaladhi, Dec. 16, 1963; children: Joshua, Daniel, Suenya. Diploma in Engring, Rhine-Westfalia Tech. U. Aachen, W.Ger., 1960; Ph.D., U. Utah, 1973. Research assoc. Tech. U. Berlin, West Berlin, Ger., 1960-62; univ. lectr. Inst. Tech., Bandung, Indonesia, 1962-69; lectr. Air Force Acad., Bandung, 1963-66; rocket scientist Nat. Inst. for Aero. and Space, Bandung, 1963-67; research assoc. U. Utah, Salt Lake City, 1973-74; assoc. scientist, solid rocket structural engr. Morton Thiokol, Brigham City, Utah, 1974—; pvt. practice cons. wind energy exploration and distillation project, Bandung, 1965-68. Contbr. articles to profl. publs., Ger., U.S. Com. chmn. Lake Bonneville council Boy Scouts Am. troop 322, Brigham city, 1980—. Ministry of Edn. scholar, Indonesia, 1954-60. Mem. Soc. Engring. Sci, Joint Army-Navy-NASA-Air Force Interagy. Propulsion Com. Mem. Christian Ref. Ch. Club: Tennis/Ski (Brigham City). Subspecialties: Aerospace engineering and technology; Theoretical and applied mechanics. Current work: Material stress-strain relation, material strength, failure and aging characteristics; layered composite material bond. Home: 851 South Law Dr Brigham City UT 84302 Office: Morton Thiokol PO Box 524 Brigham City UT 84302

LAHITA, ROBERT GEORGE, immunologist, researcher; b. Elizabeth, N.J., Dec. 30, 1945; s. George Michael and Pauline Marcella (Kropaczek) L.; m. Terry Irene Barr, May 6, 1972; children: Jason, Eric. B.S., St. Peter's Coll., Jersey City, 1967; M.D., Jefferson Med. Coll., 1973; Ph.D., Thomas Jefferson U., 1973. Asst. prof. immunology Rockefeller U., N.Y.C., 1979—. Mem. AMA, Harvey Soc., Am. Rheumatology Assn., N.Y. Rheumatology Assn. (exec. bd.), Sigma Xi. Subspecialties: Immunobiology and immunology; Immunopharmacology. Current work: Sex steroids, immune regulation, autoimmune disease. Home: 500 E 63d St New York NY 10021 Office: Rockefeller U 1230 York Ave New York NY 10021

LAHOTI, GOVERDHAN DAS, mech. engr., researcher; b. Jaipur, India, May 4, 1948; came to U.S., 1969, naturalized, 1976; s. Ram Kumar and Rukmani Devi (Maloo) L.; m. Sarala Devi, Feb. 17, 1975; 1 dau., Parul. B.Engring., Burdwan (India) U., 1969; M.S., U. Calif., Berkeley, 1970, Ph.D. in Mech. Engring, 1973. Cons. metalworking tech., Columbus, Ohio, 1973-74; sr. scientist Battelle-Columbus, 1974-81; sr. research specialist Timken Research, Canton, Ohio, 1982—. Contbr. numerous articles to profl. jours. Mem. ASME, Am. Soc. Metals, N. Am. Metalworking Research Inst. Subspecialties: Mechanical engineering; Solid mechanics. Current work: Computer-aided modeling of metalworking processes, optimization of mfg. processes, metalworking equipment and processes (forging, rolling, extrusion, etc). Home: 5373 Frank Ave NW North Canton OH 44720 Office: 1835 Dueber Ave Canton OH 44706

LAI, CARL MINGTAN, plant pathologist; b. Taiwan, Dec. 11, 1934; came to U.S., 1962; s. Sui and Tou L.; m. Martha L., Jan. 10, 1939; children: Mark, James. Ph.D., U. Calif., Berkeley, 1966. Plant pathologist, plant disease clinic lab. Calif. Dept. Food & Agr., Sacramento, 1969—. Contbr. articles to profl. jours. Mem. Am. Phytopath. Soc. Subspecialty: Plant pathology. Current work: Research in plant diseases incited by bacteria and fungi. Office: Calif Dept Food and Agr 1220 N St Sacramento CA 95804

LAI, FONG MAO, pharmacologist, educator; b. Yuan-Ling, Taiwan, Aug. 17, 1942, came to U.S., 1971, naturalized, 1980; s. Chang H. and Lee (Kuo) L.; m. Yeh O. Chen, Oct. 2, 1970; children: Jimmy C., Kenny L. B.S., Taipei Med. Coll., 1966; M.S., Nat. Taiwan U., 1969; postdoctoral fellow, Roche Inst. Molecular Biology; Ph.D., Med. Coll. Va., 1974. Instr. China Med. Coll., Taichung, Taiwan, 1969-71; pharmacologist Lederle Labs., Pearl River, N.Y., 1976-78, sr. pharmacologist, group leader, 1978—. Contbr. articles to profl. jours. Recipient Am. Cyanamid Sci. Achievement award, 1979. Mem. Am. Soc. Pharmacology, and Exptl. Therapeutics. Subspecialties: Pharmacology; Cellular pharmacology. Current work: Mechanism of hypertension development and how the antihypertensive agents will normalize the high blood pressure. Office: Lederle Lab Middletown Rd Pearl River NY 10965 Home: 7 Maiden Ln New City NY 10956

LAI, YIH-LOONG, physiologist; b. Tung-Shih, Taiwan, May 25, 1938; came to U.S., 1968; s. Li and Yi (Huang) L.; m. Suh-Mei Deng, Dec. 9, 1964; children: Paul C., Si C., Judy J. B.S., Taiwan Normal U., Taipei, 1963; M.S., Nat. Taiwan U., Taipei, 1966; Ph.D., U. Kans., 1972. NIH fellow Mayo Clinic and Found., Rochester, Minn., 1972-76; instr. Mayo Med. Sch., Rochester, 1975-76; asst. mem. Virginia Mason Research Ctr., Seattle, 1976-81, assoc. mem., 1981—. Contbr. sci. articles to profl. publs. NIH grantee, 1976—. Mem. Am. Physiol. Soc., Am. Thoracic Soc., Sigma Xi. Subspecialties: Physiology (medicine); Physiology (biology). Current work: Main work interest is pulmonary physiology-physiological changes of the lungs during prolonged CO_2 retention and during asthmatic episodes. Home: 9720 20th Ave NE Seattle WA 98115 Office: Virginia Mason Research Ctr 1000 Seneca St Seattle WA 98101

LAI, YING-SAN, valve company executive; b. Chutung, Taiwan, Sept. 9, 1937; came to U.S., 1962; s. Jung-Lai and Ching-Mei (Chien) L.; m. Nancy Pui-Chin Tom, Feb. 26, 1966; children: Nolan, Ormond, Lynna. B.S. in Mech. Engring, Nat. Taiwan U., 1960; M.S., U. Iowa, 1963; Ph.D., Northwestern U., 1973. Registered profl. engr., Ill., La., Wash. Design engr. CBI Industries, Oak Brook, Ill., 1963-69, stress analyst, 1972-73; chief engr. Dresser Industries, Alexandria, La., 1973—. Contbr. tech. papers to profl. lit. Mem. ASME, Am. Nuclear Soc. Subspecialty: Mechanical engineering. Current work: Research and development on pressure relief valves and line valves for industrial applications. Home: 4806 Westgarden Blvd Alexandria LA 71301 Office: Dresser Industries PO Box 1430 Alexandria LA 71301

LAING, FREDERICK MITCHELL, botanist; b. Barre, Vt., Nov. 29, 1919; s. George F. and Margaret (Horne) L.; m. Barbara Bixby; children: Wendy Laing Brown, John M. B.S. in Edn, U. Vt., 1951; M.S., 1953. Mem. faculty U. Vt., Burlington, 1953—, now research assoc. prof. botany and plant physiology. Served to 1st lt. U.S. Army, 1941-45. Mem. Bot. Soc. Am., Am. Soc. Plant. Physiology, AAAS, Forest Products Research Soc., Can. Soc. Plant Physiol., Council Agrl. Sci. and Tech. Subspecialties: Biomass (agriculture); Plant physiology (agriculture). Current work: Plant physiology and development of sugar maple. Intensive culture of native hardwoods for energy.

LAIPIS, PHILIP JAMES, biochem. geneticist; b. Charleston, S.C., Apr. 20, 1944; s. James and Athena (Maltezos) L.; m. Sandra Anne, Oct. 24, 1970; children: Sara, Martha. B.S. in Chemistry, Calif. Inst. Tech., 1966; Ph.D. in Genetics, Stanford U., 1972. Postdoctoral fellow Princeton U., 1972-74; asst. prof. biochemistry and molecular biology U. Fla., 1974-80, asso. prof., 1980—. Mem. AAAS, Am. Soc. Microbiology, Sigma Xi. Subspecialties: Genetics and genetic engineering (biology); Evolutionary biology. Current work: Replication in vertebrate mictochondria, devel. processes affecting mitochondrial inheritance. Office: Health Center U Fla Box J-245 JHM Gainesville FL 32610

LAIRD, JAMES DOUGLAS, psychology educator; b. N.Y.C., Oct. 11, 1937; s. Morton Armstrong and Lavina Helen and (Strattan) L.; m. Nancy Meaker, June 17, 1960; children: Jennifer, Sarah, Nicholas. B.A., Middlebury Coll., 1962; Ph.D., U. Rochester, 1967. Mem. faculty Clark U., Worcester, Mass., 1966—, prof. psychology, 1982—; cons. Holy Cross Coll., 1980-82, U. Mass. Med. Sch., 1982—, others. Contbr. articles to profl. jours. Mem. Am. Psychol. Assn., Eastern Psychol. Assn. Subspecialties: Social psychology; Cognition. Current work: Self-knowledge, emotion, performance measurement, placeboes. Home: 22 Armington Ln Holden MA 01520 Office: Dept Psychology Clark U Worcester MA 01610

LAJTHA, ABEL, neurochemist; b. Budapest, Hungary, Sept. 22, 1922; came to U.S., naturalized, 1954; s. Laszlo and Rozsa (Hollos) L.; m. Marie Snyder, Nov. 25, 1953; children: Terry, Kathryn. Ph.D., U. Budapest, 1945. Asst. prof. Biochem. Inst., U. Budapest, 1945-47; fellow Stazione Zoologica, Naples, Italy, 1947-48, Royal Instn. Gt. Britain, 1948-49; asst. prof. Inst. Muscle Research, Woods Hole, Mass., 1949-50; research asso., asst. prof. Columbia U. Coll. Phys. & Surgeons., N.Y.C., 1950-66; sucessively research scientist, sr. research scientist, asso. research scientist N.Y. State Psychiat. Inst., N.Y.C., 1950-63; prin. research scientist N.Y. State Research Inst. Neurochemistry and Drug Addiction, N.Y.C., 1963-66; research prof. psychiatry N.Y. U. Sch. Medicine, 1971—; dir. Center Neurochemistry, Ward's Island, N.Y., 1966—. Editor: Brain Barrier Systems, 1968, Handbook of Neurochemistry, 1969-72, 2d ed., 1982—, Protein Metabolism of the Nervous System, 1969, Transport Phenomena of the Nervous System, 1976, Mechanisms of Protein Synthesis in the Brain, 1977, Clinical Neurology and Neurochemistry, 1980; editor in chief: Neurochem. Research, 1975—; editor: Jour. Neurochemistry, 1962-71, Brain Research, 1966-77, Jour. Neurobiology, 1968-77, Internat. Rev. Neurobiology, 1965—, Advances in Experimental Medicine and Biology, 1967—, Biological Psychiatry, 1969—, Internat. Jour. Neurosci, 1970-74, Perspectives in the Brain Sciences, 1972—, Revs. of Neurosci, 1976—, Jour. Neurosci. Research, 1976—. Mem. Am. Acad. Neurology, Am. Soc. Biol. Chemists, Biochem. Soc., AAAs, Internat. Brain Research Orgn., Internat. Soc. Neurochemistry, Am. Soc. Neurochemistry, Internat. Soc. Psychoneuroendocrinology, Soc. Neurosci., N.Y. Acad. Sci., Am. Soc. Cell Biology, Soc. Biol. Psychiatry, Collegium Internationale Neurospychologicum, Am. Chem. Soc., Italian Soc. Neurology, Sigma Xi. Subspecialty: Neurochemistry. Current work: Brain protein metabolism, breakdown, regeneration, peptide metabolism, devel. Home: 75 Orlando Ardsley NY 10502 Office: Center for Neurochemistry Ward's Island NY 10035

LAKE, BRUCE MENO, applied physicist, site mixed concrete company executive; b. Los Angeles, Nov. 22, 1941; s. Meno Truman and Jean Ivy (Hancock) L. B.S.E., Princeton U., 1963; M.S., Calif. Inst. Tech., 1964, Ph.D., 1969. Mem. tech. staff advanced instrumentation dept. TRW Systems Group, Redondo Beach, Calif., 1969-73; head exptl. sect. fluid mechanics dept. Space & Def. Systems Groups, 1973-81, asst. mgr., 1977-81, mgr., 1981—. Contbr. articles to profl. jours. Ford Found. fellow, 1964-65. Mem. Am. Phys. Soc., AIAA. Democrat. Subspecialties: Fluid mechanics; Oceanography. Current work: Emphasis of department activities is on experimental, analytical and numerical work in modeling and measurement of ocean surface and subsurface phenomena. Office: TRW Space and Technology Group One Space Park Redondo Beach CA 90278

LAKE, CHARLES RAYMOND, psychiatrist, pharmacologist, educator; b. Nashville, July 6, 1943; s. Charles Raymond and Vera (Shute) L., m. Susan Frances de la Houssaye, Aug. 12, 1967; children: Reagan Anne, Craig Anne. B.S. in Zoology, Tulane U., 1965, M.S. in Biology, 1966, Ph.D. in Physiology and Pharmacology, 1971, M.D., 1972. Resident in psychiatry Duke U. Med. Ctr., Durham, N.C., 1972-74; research assoc. lab. clin. sci. NIMH, Bethesda, Md., 1974-77; staff psychiatrist sect. exptl. therapeutics, 1977-79; prof. psychiatry and pharmacology Uniformed Services U. Health Scis. Sch. Medicine, Bethesda, 1979—. Contbr. articles to med. jours. Served with USPHS, 1974-79. Mem. Am. Psychiat. Assn., Soc. Neuroscience, Soc. Biol. Psychiatry, Am. Acad. Clin. Psychiatry, Internat. Soc. Hypertension, Am. Coll. Neuropsychopharmacology, Endocrine Soc. Oxford-Cambridge Univ. (London). Subspecialties: Psychopharmacology; Neuropharmacology. Current work: Research in biogenic amine metabolism, amphetamine "look-alikes," and "over-the-counter" speed. Home: 5614 Overlea Rd Bethesda MD 20816 Office: 4301 Jones Bridge Rd Bethesda MD 20814

LAKE, JAMES ALAN, nuclear engineer; b. Gary, Ind., Dec. 16, 1943; s. James H. Lake and Margaret (Dow) Kulcsar; m. Sharon Marie Hall, Jan. 17, 1970; children: James David, Matthew Alan. B.A., Hanover Coll., 1965; M.A., Miami U. of Ohio, 1967; M.S., Ga. Inst. Tech., 1969, Ph.D., 1972. Grad. fellow Ga. Inst. Tech., Atlanta, 1967-72; sr. engr. Westinghouse Nuclear Fuels Div., Monroeville, Pa., 1973, Westinghouse Advanced Reactors Div., Madison, Pa., 1973-80, fellow engr., 1980—; mem. Cross Sect. Evaluation Working Group, U.S. Dept. Energy, Brookhaven, N.Y., 1981—. Chmn. New Stanton Planning Commn., 1981—. Mem. Am. Nuclear Soc. Presbyterian. Subspecialties: Nuclear engineering; Nuclear physics. Current work: Liquid metal fast breeder reactor nuclear design, critical experiments reactor physics analysis and application, waste storage criticality evaluation. Home: 106 Cortland Dr New Stanton PA 15672 Office: Westinghouse Advanced Reactor Div Madison PA 15663

LAKES, RODERIC STEPHEN, biomedical and mechanical engineering educator; b. N.Y.C., Aug. 10, 1948; s. Eric A. and Dorothy E. (Hollweg) L.; m. Diana M. Lakes, Aug. 14, 1971. Student, Columbia U., summers 1964-65; B.S. in Physics, Rensselaer Poly. Inst., 1969, Ph.D., 1975; postgrad., U. Md., 1969-70. Research assoc. in engring. and applied sci. Yale U., 1975-77; asst. prof. physics Tuskegee Inst., 1977-78; asst. prof. biomed. and mech. engring. U. Iowa, 1968-82, assoc. prof., 1982—, dir., 1980—; vis. asst. prof. biomed, engring Rensselaer Poly. Inst., summer 1978, vis. assoc. prof., summer 1982. Contbr. articles on bone biomechanics and bioelectricity to profl. jours. NIH grantee, 1979-82; NIH Biomed. Research Support grantee, 1979, 80. Mem. Am. Phys. Soc., ASME, Orthopaedic Research Soc., AAAS, Sigma Xi. Subspecialties: Biomedical engineering; Solid mechanics. Current work: Bone biomechanics, biophysics, bioelectricity, experimental generalized continuum mechanics. Inventor device for remote detection of stress waves in bone. Office: Coll Engring U Iowa Iowa City IA 52242

LAKIN, JAMES DENNIS, physician, educator; b. Harvey, Ill., Oct. 4, 1945; s. Ora Austin and Annie Petronella (Johnson) L.; m. Sally A. Stuteville, July 11, 1972; children: Margaret, Matthew. B.S.M., Northwestern U., 1968, Ph. D., 1968, M.D., 1969. Diplomate: Am. Bd. Internal Medicine, Am. Bd. Allergy and Immunology. Intern, resident in internal medicine U. Mich., 1970-72, fellow in immunology allergy, 1973; dir. Allergy Research Lab., Naval Med. Research Inst., Bethesda, Md., 1974-75; lab.dir., attending physician Okla. Allery Clinic, U. Okla., 1975—; assoc. prof., 1976—; dir. adolescent allergy clinic Children's Meml. Hosp., Oklahoma City, 1975—. Author: Allergic Diseases Diagnosis and Management, 1981; contbr. articles to profl. jours. Bd. dirs. Okla. Med. Research Found. Served with USN, 1974-75. NIH grantee, 1965-69. Fellow ACP, Am. Acad. Allergy, Am. Coll. Chest Physicians. Republican. Lutheran. Subspecialties: Allergy; Immunology (medicine). Current work: Diagnosis and management of allergic airways disease. Office: Okla Allergy Clinic PO Box 26827 Oklahoma City OK 73126

LAKOWICZ, JOSEPH RAYMOND, biochemistry educator, researcher; b. Phila., Mar. 15, 1948; s. Joseph Raymond and Frances (Kaczmarek) L.; m. Susan Margaret Kozempei, Jan. 3, 1970; children: Rebecca, Deborah. Student, Drexel Inst. Tech., summer 1968, Temple U., summer 1969; B.A. in Chemistry, LaSalle Coll., 1970; M.S. in Biochemistry, U. Ill.-Urbana, 1971, Ph.D., 1973. NSF fellow U. Ill., 1970-73; postdoctoral fellow Oxford (Eng.) U., 1973-74; asst. prof. U. Minn., St. Paul, 1975-80; assoc. prof. biochemistry U. Md., 1980—. Author: Principles of Fluorescen Spectrosocpy, 1983. Mem. Biophys. Soc., Am. Soc. Biol. Chemistry, Am. Chem. Soc. Subspecialties: Biophysical chemistry; Physical chemistry. Current work: Studies of macromolecules using fluorenscence spectrosocpy. Office: U Md Dept Biochemistry 660 W Redwood St Baltimore MD 21201

LAKS, MICHAEL MILTON, cardiologist; b. Cleve., July 25, 1928; s. Alexander and Helen (Klein) L.; m. Sandra Beller, June 13, 1959; children: Helaina, Alexander. B.A., UCLA, 1951; M.D., U. So. Calif., 1956. Diplomate: Am. Bd. Internal Medicine (cardiovascular disease). Intern Cedars of Lebanon Hosp., Los Angeles, 1956-57, resident, 1957-59, chief med. resident, 1959-60, research fellow in medicine, 1960-61, spl. NIH cardiac fellow, 1964-65, asst. dir. dept. medicine, 1961-64, dir. dept., 1964-65, physician in charge cardiovascular research lab., 1965-71; research assoc. Cedars-Sinai Med. Ctr., Los Angeles, 1962-69, sr. research scientist, 1969—; dir. heart sta. and cardiovascular research lab. Harbor-UCLA Med. Ctr., 1971—, assoc. chief div. cardiology, 1975—; prof. medicine UCLA, 1975—; vis. internist UCLA Hosp. and Clinics, 1975—; cons. Wadsworth VA Hosp., Hewlett Packard Corp.; lectr. Contbr. 200 articles and abstracts to profl. jours. Served with USN, 1948-53. Am. Chem. Soc. scholar, 1947; recipient prize for sci. paper ACP, 1961. Fellow Am. Coll. Cardiology, ACP, Am. Geriatrics Soc. (founding), Am. Coll. Chest Physicians, Council Clin. Cardiology of Am. Heart Assn., Royal Soc. Medicine; mem. AMA, Am. Fedn. Clin. Research, Am. Heart Assn., Am. Soc. Nephrology, AAAS, Am. Physiol. Soc., Pan Am. Med. Assn., Western Soc. Clin. Research, Los Angeles County Heart Assn., Calif. Med. Assn., Nat. Assn. Para-Cardiac Specialists (chmn. med. adv. com.), Internat. Study Research in Cardiac Metabolism, Am. Inst. Biol. Scis., Calif. Thoracic Soc., N.Y. Acad. Scis., Western Assn. Physicians, Phi Beta Kappa, Alpha Omega Alpha, Phi Kappa Phi. Subspecialty: Cardiology. Current work: Cardiovascular research - ventricular failure and hypertrophy; computerized ECG. Home: 1939 Edgemont Los Angeles CA 90027 Office: 1000 W Carson St Torrance CA 90509

LALA, PEEYUSH KANTI, research adminstr., educator; b. Chittagong, Bangladesh, Nov. 1, 1934; s. Sudhangshu Bimal and Nanibala (Chaudhuri) L.; m. Arati Royburman, July 7, 1962; children: Probal, Prasun. M.B.B.S., Calcutta Med. Coll., 1957; postgrad., Saha Inst. Nuclear Physics, 1959-61; Ph.D., Calcutta U., 1961; M.D., 1962. Diplomate: Indian Med. Assn. Resident in medicine Calcutta Med. Coll., 1958-59, demonstrator of pathology, 1959-62; resident research assoc. Argonne Nat. Lab., Ill., 1963-64; research biologist U. Calif.-San Francisco, 1964-66; research assoc. Chalk River Nuclear Labs., Ont., Can., 1967-68; asst. prof. anatomy McGill U., 1962-72, assoc. prof., 1972-77, prof., 1977—; vis. prof. U. Melbourne, Australia, 1977-78; vis. scientist Walter and Eliza Hall Inst. Med. Research, 1977-78. Contbr. articles to profl. jours. Fulbright scholar, 1962; research grantee Med. Research Council Can., Nat. Cancer Inst., Can. Research Soc., Nat. Cancer Inst., U.S.). Mem. Can. Assn. Anatomists, Can. Soc. Cell Biology, Can. Assn. Immunologists, Am. Assn. Cancer Research, Cell Kinetics Soc., Reticuloendothelial Soc., Internat. Soc. Exptl. Hematology, Am. Soc. Immunology, Am. Soc. Reproductive Immumology. Subspecialties: Immunobiology and immunology; Reproductive biology. Current work: Immunology of tumor-host and feto-maternal relationships. Office: 3640 University St Montreal PQ Canada H3A 2B2

LALLEY, PETER MICHAEL, physiologist, educator; b. Scranton, Pa., Jan. 21, 1940; s. Paul Francis and Mary (Nolan) L.; m. Ruth Ann Pethick, June 22, 1963; children: Maureen, Christopher, Angela, Jacqueline. B.S. in Pharmacy, Phila. Coll. Pharmacy and Sci., 1963, M.S., 1965, Ph.D. in Pharmacology, 1970. Postdoctoral fellow U. Pitts. Sch. Medicine, 1970-73, asst. prof. pharmacology, 1973-75, U. Fla., Gainesville, 1975-76; asst. prof. physiology U. Wis.-Madison, 1976-80, assoc. prof., 1980—. Reviewer for jours. NIH grantee, 1971-73, 82-83, 82—; Am. Heart Assn. grantee, 1975-78. Mem. Physiol. Soc., Am. Heart Assn., Soc. for Neurosci., AAAS, Am. Pharm. Assn., Sigma Xi. Subspecialties: Neuropharmacology; Neurophysiology. Current work: Control of cardiovascular and respiratory neurons by neurotransmitters. Home: 5518 Meadowood Dr Madison WI 53711 Office: Dept Physiology U Wis Med Sci Ctr 1300 University Ave Madison WI 53706

LAM, LEO K., research scientist, engr.; b. Hong Kong, Sept. 12, 1946; came to U.S., 1969; s. K. K. and S.B. (Kong) L. B.Sc. in Physics, math, Hong Kong U., 1969, M.A., Columbia U., 1970, Ph.D., 1975. Research assoc. Joint Inst. Lab. Astrophysics, Nat. Bur. Standards, Boulder, Colo., 1975-77; research asst. prof. dept. physics U. Mo., Rolla, 1977-79; research asst. prof. U. So. Calif., 1979-81; engr. specialist Litton Guidance & Control Systems, Woodland Hills, Calif., 1981—. Contbr. articles to profl. jours. Mem. Optical Soc. Am., Am. Phys. Soc., Sigma Xi. Subspecialties: Atomic and molecular physics; Spectroscopy. Current work: Nuclear magnetic resonance gyroscope, optical phase conjugation, non-cinzar optics, optical pumping. Office: 5500 Canoga Ave MS 25/30 Woodland Hills CA 91365

LAMAR, CARLTON HINE, anatomist, veterinarian, educator; b. Lebanon, Ind., Feb. 20, 1940; s. Clifford Carl and Ruth Eleanor (Hine) L.; m. Karman Lee Bracken, Aug. 16, 1964; children: Kimberly, Kristine. B.S., Purdue U., 1962, D.V.M., 1966, M.S., 1971, Ph.D., 1974. Asst. prof. vet. anatomy Purdue U., West Lafayette, Ind., 1974-81, assoc. prof., 1981—. Mem. AVMA, Am. Assn. Anatomists, Tissue Culture Assn. Subspecialties: Anatomy and embryology; Tissue culture. Current work: Development of in vitro systems to replace animal research models. Home: 1820 Greenbrier Ave West Lafayette IN 47906 Office: Sch Vet Medicine Purdue U West Lafayette IN 47907

LA MARCHE, VALMORE CHARLES, JR., tree-ring researcher, educator; b. Hurley, Wis., Aug. 27, 1937; s. Valmore Charles and Sylvia (Maki) La M.; m. Sally Jo Guinn, July 9, 1957 (div. June 1979); children: Valmore Charles III, Joan Elizabeth, Daniel Albert, Rebecca Minette. B.A., U. Calif.-Berkeley, 1960; M.A., Harvard U., 1962, Ph.D., 1964. Hydrologist U.S. Geol. Survey, Menlo Park, Calif., 1964-67; research assoc. U. Ariz., 1967-69, assoc. prof. dendrochronology, 1969-74, prof., 1974. Co-editor: Climate From Tree Rings, 1982; mem. editorial bd.: Quaternary Research, 1982-87. Fellow Geol. Soc. Am.; mem. Am. Quaternary Assn. (councilor 1982-86), Ecol. Soc. Am., Tree-Ring Soc., Internat. Union for Quaternary Research. Democrat. Subspecialties: Dendrochronology, Climatology. Current work: Application of tree rings to problems in geology, climatology, ecology and hydrology in western North America, Asia and Southern hemisphere. Home: 4069 N Fremont Ave Tucson AZ 85719 Office: Lab Tree-Ring Research U Ariz Tucson AZ 85721

LAMARRE, PIERRE, structural engineer; b. Montreal, Que., Can., Oct. 20, 1935; s. Emile and Blanche (Gognon) L.; m. Suzanne Dubois, Sept. 22, 1937; 1 son, Andre. B.Sc.A. in Civil Engring, Laval U., 1956; M.Sc., MIT, 1967. Constrn. engr. A. Janin Co., Montreal, 1958-59, Lamarre Constrn. Co., Jonquiere, Que., 1960-64; structural designer Lavalin Internat. Inc., Montreal, 1965-68, chief designer large structures, 1969-73, chief engr. dept. pub. works, 1974-77, dir. engring., 1978-82, corp. v.p. quality and systems, 1982—. Mem. Que. Order Engrs., Permanent Internat. Assn. Navigation Congress, Internat. Congresses on Large Dams, Internat. Bridge, Tunnel and Turnpike Assn., Internat. Assn. Hydraulic Research, Internat. Assn. Bridge Structural Engrs. Subspecialties: Civil engineering; Hydroelectric engineering. Current work: Conceptual design of large and unusual structures; responsible for the quality assurance program implementation and improvement for the Lavalin Group and updating corporate engineering standards. Office: 1130 W Sherbrooke St Suite 1389 Montreal PQ Canada H3A 2R5

LAMB, DONALD QUINCY, physicist, researcher; b. Manhattan, Kans., June 30, 1945; s. Donald Quincy and Helen Letson (Keithley) L.; m. Linda Gilkerson, Sept. 23, 1978; 1 son, Michael. B.A., Rice U., Houston, 1967; M.Sc., U. Liverpool, Eng., 1969; Ph.D., U. Rochester, N.Y., 1974. Research asst. prof. physics U. Ill., 1973-75, asst. prof., 1975-77, assoc. prof., 1977-79, prof., 1979-80; vis. assoc. prof. M.I.T., 1978-79; vis. scientist dept. astronomy Harvard U., 1978-80; physicist Harvard-Smithsonian Center for Astrophysics, Cambridge, Mass., 1980—. Editor: (with D. Sugimoto, D. Schramm) Fundamental Problems in Theory of Stellar Evolution, 1981; contbr. articles to profl. jours. Woodrow Wilson fellow, 1967; NSF fellow, 1971-73; Guggenheim fellow, 1978-79; Nat. Merit scholar, 1963-67; Marshall scholar, 1967-69. Mem. Am. Phys. Soc, Am. Astron. Soc., Royal Astron. Soc. (London), Inst. Physics (London), European Phys. Soc., Internat. Astron. Union. Subspecialties: Theoretical physics; Theoretical astrophysics. Current work: Stellar evolution, especially evolution of degenerate dwarfs and neutron stars, x-ray astronomy, esp x-ray emission from compact stars.

LAMB, WILLIS EUGENE, JR., physicist, educator; b. Los Angeles, July 12, 1913; s. Willis Eugene and Marie Helen (Metcalf) L.; m. Ursula Schaefer, June 5, 1939. B.S., U. Calif., 1934, Ph.D., 1938; D.Sc., U. Pa., 1953, Gustavus Adolphus Coll., 1975; M.A., Oxford (Eng.) U., 1956, Yale, 1961; L.H.D., Yeshiva U., 1965. Mem. faculty Columbia, 1938-52, prof. physics, 1948-52, Stanford, 1951-56; Wykeham prof. physics and fellow New Coll., Oxford U., 1956-62; Henry Ford 2d prof. physics Yale, 1962-72, J. Willard Gibbs prof. physics, 1972-74; prof. physics and optical scis. U. Ariz., Tucson, 1974—; Morris Loeb lectr. Harvard, 1953-54; cons. Philips Labs., Bell Telephone Labs., Perkin-Elmer, NASA; Vis. com. Brookhaven Nat. Lab. Recipient (with Dr. Polycarp Kusch) Nobel prize in physics, 1955; Rumford premium Am. Acad. Arts and Scis., 1953; Research Corp. award, 1955; Guggenheim fellow, 1960-61; recipient Yeshiva award, 1962. Fellow Am. Phys. Soc., N.Y. Acad. Scis.; hon. fellow Inst. Physics and Phys. Soc. (Guthrie lectr. 1958), Royal Soc. Edinburgh (fgn. mem.); mem. Nat. Acad. Scis., Phi Beta Kappa, Sigma Xi. Subspecialty: Theoretical physics. Office: Dept of Physics U Ariz Tucson AZ 85721

LAMBE, THOMAS WILLIAM, civil engineer, educator; b. Raleigh, N.C., Nov. 28, 1920; s. Claude Milton and Mary (Habel) L.; m. Catharine Canby Cadbury, Sept. 13, 1947; children—Philip Cadbury, Virginia Habel, Richard Lee, Robert Henry, Susan Elizabeth. B.S., N.C. State Coll., 1942; S.M., MIT, 1944, Sc.D., 1948. With Standard Oil Co. Calif., 1944, Dames & Moore, 1945; mem. faculty MIT, 1945-81, prof. civil engring., 1958-69, Edmund K. Turner prof., 1969-81, Edmund K. Turner prof. emeritus, 1981—, head soil mechanics div., 1958-69, 72-81, head geotech. div., 1972-81; cons. geotech. engring., 1945—, NASA, 1965—; Rankine lectr., 1973, Kapp Meml. lectr., 1980. Author books, numerous articles. Recipient Desmond Fitzgerald medal Boston Soc. Civil Engring., 1954, 56, Disting. Edngring. Alumnus award N.C. State U., 1982. Fellow ASCE (Collingswood prize 1951, Wellington prize 1961, Norman medal 1964, Terzaghi lectr. 1970, Terzaghi award 1975); mem. Internat. Soc. Soil Mechanics and Found. Engring., Nat. Acad. Engring. Subspecialty: Civil engineering. Home: 40 Elm St East Pepperell MA 01437 Office: 77 Massachusetts Ave Cambridge MA 02139

LAMBERGER, PAUL HENRY, research and development supervisor; b. Pitts., Sept. 28, 1937; s. Edward H. and Mary Elizabeth (Shane) L.; m. Carol Galloway, Sept. 30, 1961; children: Susan, David. B.A. in Chemistry, Wooster (Ohio) Coll., 1959; B.S. in Chem. Engring, U. Dayton, 1979. Registered profl. nuclear engr., Calif. Chemist Monsanto Research Corp., Miamisburg, Ohio, 1959-77, group leader, 1977—. Author and editor: document Tritium Control Technology, 1972. Chmn. tech. com. Miami Valley Lung Assn., 1975—. Recipient achievement award Monsanto Research Corp., 1980. Mem. Am. Chem. Soc., Am. Nuclear Soc. Presbyterian (elder). Subspecialties: Chemical engineering; Inorganic chemistry. Current work: Processing, recovery, enrichment of tritium; effluent control, handling technology and waste management related to tritium; management of radioactive processing facilities. Home: 320 Brydon Rd Dayton OH 45419 Office: Monsanto Research Corp PO Box 32 Miamisburg OH 45342

LAMBERT, JOHN VINCENT, astronomer; b. Pittsburg, Kans., July 28, 1945; s. Jack Leeper and Beatrice Cecilia (Holub) L.; m. Joan Bollinger, Dec. 30, 1967; 1 dau., Vanya Marie. B.S., Kans. State U., Manhattan, 1967; M.S., Air Force Inst. Tech., 1971, N. Mex. State U., 1982, postgrad., 1982—. Commd. 2d lt. U.S. Air Force, 1967, advanced through grades to capt., 1967-78; physicist, Wright-Patterson AFB, Ohio, 1967-71, Cloudcroft, N. Mex., 1971-76, project mgr., Griffis AFB, N.Y., 1976-78, astronomer, cons., Alamogordo, N. Mex., 1979-. Mem. Am. Astron. Soc., Sigma Xi. Roman Catholic. Subspecialties: Planetary science; Satellite studies. Current work: Non-imaging techniques for shape/surface property determination; computer simulation of physical processes; applications of advanced electo-optical sensors; automated sensor systems. Office: PO Box 731 Alamogordo NM 88311

LAMBERTS, ROBERT LEWIS, optical engr., researcher; b. Fremont, Mich., Sept. 8, 1926; s. Lambertus and Anna (Dick) L.; m. Margaret Elizabeth Van Mouwerik, Aug. 1, 1951; children: Ruth Lamberts DuMont, Margaret Lamberts Bendroth, Nancy Lamberts Black, William, Robert, Peter. A.B., Calvin Coll., Grand Rapids, Mich., 1949; M.S. in Physics, U. Mich., 1951; Ph.D. in Optics, U. Rochester, 1969. With research labs. Eastman Kodak Co., Rochester, N.Y., 1951—, sr. research assoc., 1981—. Contbr. articles to sci. jours. Served with USN, 1944-46. Recipient best paper award Soc. Motion Picture and TV Engrs., 1962. Fellow Optical Soc. Am.; mem. Soc. Photog. Scientists and Engrs. (Charles E. Ives award 1966), Soc. Photog. Instrumentation Engrs. Subspecialties: Optical engineering; Holography. Current work: Photographic image structure. Patentee in field. Home: 236 Henderson Dr Penfield NY 14526 Office: Research Labs Eastman Kodak Co Rochester NY 14650

LAMBETH, VICTOR NEAL, horticulturist, educator; b. Sarcoxie, Mo., July 5, 1920; s. Odus Huston and Carrie (Belle) L.; m. Sarah Katherine Smarr, May 24, 1946; children: Victoria Kay, Debra Jean. B.S. in Agr, U. Mo.-Columbia, 1942, M.S. in Horticulture, 1948, Ph.D., 1950. Asst. prof. horticulture U. Mo., 1950-51, assoc. prof., 1951-59, prof., 1959—. Contbr. articles to profl. jours. Served to lt. USN, 1943-46; PTO. Recipient awards for teaching; NSF grantee, 1982. Mem. Am. Soc. Hort. Sci., Am. Soc. Plant Physiologists, Sigma Xi, Gamma Sigma Delta. Methodist. Subspecialties: Plant genetics; Plant physiology (agriculture). Current work: Tomato genetics and breeding, plant growth media. Patentee in field. Home: 1327 Lambeth Dr Columbia MO 65202 Office: Dept Horticulture U Mo 1-40 Agr Bldg Columbia MO 65211

LAMBIRD, PERRY ALBERT, pathologist; b. Reno, Nev., Feb. 7, 1939; s. C. David and Florence (Knowlton) L.; m. Mona Sue Salyer, July 30, 1960; children: Allison Thayer, Jennifer Salyer, Elizabeth Gard, Susannah Johnson. B.A., Stanford U., 1958; M.D., Johns Hopkins U., 1962; M.B.A., Okla. City U., 1973. Diplomate: Am. Bd. Pathology. Fellow in internal medicine Johns Hopkins Hosp., Balt., 1962-63, resident pathologist, 1965-68, chief resident, 1968-69; med. cons. USPHS, Washington, 1963-65; pathologist Med. Arts Lab., Oklahoma City, 1969—, Presbyn. Hosp., South Community Hosp., 1974—, Nat. Cancer Inst., 1974-81; propr. Lambird Mgmt. Cons. Service, Oklahoma City, 1974—; cons. in field. Reviewer: Jour. Am. Med. Assn, 1983—; contbr. articles to profl. jours. Pres. Okla. Symphony Orch., 1974-75, Ballet, Okla., 1978-79; del. Republican Nat. Conv., 1976. Served to lt. comdr. USPHS, 1963-65. Recipient Exec. Leadership award Oklahoma City U., 1976; Physician's Recognition award AMA, 1969-83. Fellow Am. Soc. Clin. Pathologists, Coll. Am. Pathologists, Internat. Coll. Surgeons; mem. AMA (h. of dels.), Okla. Med. Assn. (ho. of dels., trustee), Okla. County Med. Soc. (pres.), Okla. Soc. Cytopathology (pres.), Am. Pathology Found. (pres.), Okla. Found. for Peer Rev. (dir.), Arthur Purdy Stout Soc. Surg. Pathologists, Am. Assn. Pathologists, Okla. Assn. Pathologists (pres.), So. Med. Assn., N.Y. Acad. Sci., Am. Soc. Cytology, Olser Soc., Okla. City Clin. Soc., Phi Beta Kappa, Alpha Omega Alpha. Republican. Methodist. Subspecialties: Pathology (medicine); Health services research. Current work: Applied anatomic and clinical pathology including management and administration, quality control, cost benefit analysis and efficiency measures in clinical medicine. Home: 419 Northwest 14th St Oklahoma City OK 73103 Office: Med Arts Lab 100 Pasteur 1111 N Lee St Oklahoma City OK 73103

LAMBUTH, ALAN LETCHER, chemist; b. Seattle, Jan. 5, 1923; s. Benjamin Letcher and Olive Serena (Schram) L.; m. Susan Jane DeMelt, Aug. 19, 1944; children: Wendy Lambuth Trudgian, Peter, John, Douglas. Student, U. Wash., 1940-42; Student, Yale U., 1942-43; student, U. Santa Clara, 1944; B.S. in Chemistry, U. Wash., 1947. Research chemist Monsanto Co., Seattle, 1947-57, research leader 3 groups, 1958-69; product devel. mgr. Boise Cascade Corp., Boise, Idaho, 1970-75, research and devel. mgr., 1976-81, mgr. product and process devel., 1981—, also cons. overseas wood products cos.; guest lectr., reviewer research proposals and papers U.S. Dept. Agr. Forest Products Lab., NSF, univs. Publs. in polymer, adhesives and wood structural design fields.; patentee in field. Served with Signal Corps U.S. Army, 1943-46. Recipient Monsanto award for research innovation, 1969. Mem. Forest Products Research Soc. (pres.), Am. Chem. Soc., Am. Inst. Timber Constrn., Am. Plywood Assn., ASTM, Nat. Forest Products Assn., Nat. Paint and Coatings Assn., Western Wood Products Assn., Internat. Union Forestry Research Orgns. Republican. Congregationalist. Clubs: Boise Racquet, Hillcrest Country. Subspecialties: Polymer chemistry; Biomaterials. Current work: Innovative uses for wood, especially underutilized species and waste materials; biomass as a chem. resource; wood treatment or modification for improved performance; new polymers and adhesive concepts; wood structural design; new converting and manufacturing processes. Patents. Home: 7240 Cascade Dr Boise ID 83704 Office: Boise Cascade Corp 220 S 3d St Boise ID 83702

LAMEY, HOWARD ARTHUR, plant pathologist; b. Bloomington, Ind., Dec. 20, 1929; s. Carl Arthur and Mary Lucile (Seaman) L.; m. Cynthia Joan Huenink, Aug. 22, 1956; children: Timothy, Thaddeus, Linda, Suzan, Laura. B.A., Ohio Wesleyan U., 1951; Ph.D., U. Wis.-Madison, 1954. Project assoc. U. Wis., Madison, 1954-55, 57-58; research plant pathologist U.S. Dept. Agr., Camaguey, Cuba, 1958-59, Baton Rouge, La., 1960-69; plant pathologist Internat. Inst. Tropical Agr., Ibadan, Nigeria, 1969-71; project mgr. food and agr. orgn. UN, Seoul, Korea, 1971-75; extension plant pathologist N.D. State U., Fargo, 1977—. Contbr. numerous articles to profl. jours. Patron Fargo-Moorhead Civic Opera, 1981-83. Served with AUS, 1955-57. Fellow AAAS; mem. Am. Phytopath. Soc., AAAS., Mycol. Soc. Am., N.Y. Acad. Sci., Phi Beta Kappa, Sigma Xi, Gamma Sigma Delta. Subspecialty: Plant pathology. Current work: Diseases of cereals, row crops, vegetables, fruits and ornamentals. Home: 1517 8th St N Fargo ND 58102 Office: Dept Plant Pathology ND State U Fargo ND 58105

LAMM, (AUGUST) UNO, electrical engineering consultant, writer; b. Goteborg, Sweden, May 22, 1904; s. Fredrik H. and Aino M. (Wijkander) L.; m.; children: Majken, Martin, Anita. Grad., Royal Inst. Tech., 1927; D.Sc., 1943. With ASEA AB, Sweden, 1928-69, mgr. rectifier dept., 1929-39, head static converter, high voltage DC transmission and high voltage switchgear depts., 1939-55, dir. new atomic power dept., 1955-59, electrotech. dir., 1959-69; mem. exec. com., 1959-61; chmn. steering com. ASEA-Gen. Electric Joint Venture contract for Pacific Intertie High Voltage Direct Current Transmission project, 1965-69; pvt. practice elec. engring. cons., Hillsborough, Calif., 1970—. Author: The Transductor, 1943, Livsmiljo i forandring, 1980; contbr. numerous articles to profl. jours. Served with Swedish Army, 1927-28. Recipient Howard N. Potts award Franklin Inst., 1981. Fellow IEEE (dir. 1967-68, Lamme medal 1965); mem. Swedish Acad. Engring. Scis. (Gold medal 1939), Royal Acad. Sci. Sweden, Nat. Acad. Engring., Am. Soc. Swedish Engrs. (John Ericson Gold medal 1961). Subspecialties: Electrical engineering; Systems engineering. Current work: Static electric power converters, high voltage power transmission. Patentee in field.

LAMM, MICHAEL EMANUEL, pathologist, immunologist; b. Bklyn., May 19, 1934; s. Stanley S. and Rose (Lieberman) L.; m. Ruth Audrey Kumin, Dec. 16, 1961; children: Jocelyn, Margaret. Student, Amherst Coll., 1951-54; M.D., U. Rochester, 1959; M.S., Western Res. U., 1962. Diplomate: Am. Bd. Pathology. Resident pathology U. Hosp., Cleve., 1959-62; research assoc. NIH, Bethesda, Md., 1962-64; asst. prof. pathology N.Y.U. Sch. Medicine, N.Y.C., 1964-68, assoc. prof., 1968-73, prof., 1973-81; prof., chmn. pathology Case Western Res. U., Univ. Hosps., Cleve., 1981—; mem. cancer spl. program adv. com. Nat. Cancer Inst., Bethesda, Md., 1976-79; mem. sci. adv. com. Damon Runyon-Walter Winchell Cancer Fund, N.Y.C., 1978-82. Editorial bd.: Procs. Soc. Exptl. Biology and Medicine, 1973-82, Molecular Immunology, 1979-82, Jour. Immunological Methodology, 1980—, Jour. Immunology, 1981—, Am. Jour. Pathology, 1982—; contbr. articles in field to profl. jours. Named Career Scientist Health Research Council City N.Y., 1966-75; NIH research grantee, 1965—. Mem. Am. Assn. Pathologists, Am. Assn. Immunologists, Am. Soc. Biol. Chemists, N.Y. Acad. Scis., Coll. Am. Pathologists, Internat. Acad. Pathology, Soc. Exptl. Biol. Medicine. Subspecialties: Immunobiology and immunology; Pathology (medicine). Current work: Mucosal immunity, immunopathology of renal disease. Office: Inst Pathology Case Western Res U 2085 Adelbert Rd Cleveland OH 44106

LAMON, EDDIE WILLIAM, immunologist; b. Yuba City, Calif., Aug. 30, 1939; s. James Hilyer and Annie Louise (Hannah) L.; m. Bodil I.M. Lidin, June 17, 1973; children: Eddie Wiliam, Cynthia Ann Bently, Leif Christopher. B.S., U. North Ala., 1961; M.D., U. Ala., 1969; D.Sc. with highest honors, Karolinska Inst., Stockholm, 1974. Diplomate: Am. Bd. Med. Examiners, 1970. Asst. biologist So. Research Inst., Birmingham, Ala., 1964-65; intern, resident in surgery U. Ala., Birmingham, 1969-71, asst. prof. surgery, assoc. scientist, cancer research and tng ctr., 1974-75, assoc. prof. surgery, scientist, 1975-79, asst. prof. microbiology, 1974-77, assoc. prof. microbiology, 1977-82, prof. microbiology, 1982—, prof. surgery, sr. scientist, 1979—; chief tumor immunology research Birmingham VA Hosp., 1974—; guest investigator in tumor immunology dept. tumor biology Karolinska Inst., Stockholm, 1971-73. Served with USNR, 1961-63. Nat. Cancer Inst. career devel. awardee, 1975-80; Josiah Macy, Jr. Found. faculty scholar awardee, 1980-81. Mem. Am. Assn. Immunologists, Am. Assn. Pathologists, Am. Assn. Cancer Research, Am. Soc. Microbiologists, AAUP, Sigma Xi. Subspecialties: Cancer research (medicine); Immunology (medicine). Current work: Characterization of immune responses to virus induced tumors; studies of antibody-dependent cell-mediated cytotoxicity; studies of immune complex receptors on lymphocytes. Home: 3569 Hampshire Dr Birmingham AL 35223 Office: Dept Surgery U Ala School Medicine Birmingham AL 35294

LA MOTTA, ENRIQUE JAIME, environmental engineering educator, consultant; b. Guayaquil, Ecuador, July 8, 1940; s. Ricardo and Eloisa (Diaz) La M.; m. Maria Elena Rueda, June 11, 1971; children: Lorena, David, Ivan. Civil engr., U. Central del Ecuador, 1965; M.S.S.E., U. N.C., 1969, Ph.D., 1974. Sanitary engr. IEOS, Quito, Ecuador, 1965-71; assoc. prof. U. Mass., Amherst, 1976-80; asst. prof. U. Miami, Coral Gables, Fla., 1974-76; prof. prin. Esc. Poli. Nac., Quito, Ecuador, 1980—; cons., Quito, 1980—. Contbr. numerous articles to profl. jours. Recipient Bunker award U.N.C., 1969. Mem. Am. Soc. Civil Engrs., Am. Water Works Assn., Water Pollution Control Fed., Assn. Environmental Engring. Profs., Internat. Assn. Water Pollution Research, Sigma Xi. Subspecialty: Water supply and wastewater treatment. Current work: Interest: Kinetics of Biological processes, Eutrophication, Biological films, organics removal. Home: Pesantesco 100 Y Manosca Quito Ecuador Office: Escuela Politecnica Nacional Facultad de Ingenieria Civil Quito Ecuador

LAMPE, FREDERICK WALTER, chemist, educator; b. Chgo., Jan. 5, 1927; s. Joseph Dell and Christine Wood (Phillips) L.; m. Eleanor Frances Coffin, Mar. 26, 1949; children: Joan Dell Wakeling, Kathy Lee Wakeling, Erik Steven, Beth Ann, Kristina Jean. B.S., Mich. State Coll., 1950; A.M., Columbia U., 1951, Ph.D., 1953. Research chemist Humble Oil and Refining Co., Baytown, Tex., 1953-56, sr. research chemist, 1956-60; assoc. prof. chemistry Pa. State U., 1960-65, prof., 1965—, head dept. chemistry, 1983—; sci. cons. Author: (with W. R. Allcock) Contemporary Polymer Chemistry, 1981; contbr. numerous articles to sci. jours. Served with USN, 1944-46. Recipient U.S.Sr. Scientist award Alexander Von Humboldt Found., 1973; NSF sr. postdoctoral fellow, 1966-67. Fellow Am. Phys. Soc.; mem. Am. Chem. Soc., Am. Soc. Mass Spectrometry, Inter-Am. Photochem. Soc., Sigma Xi. Methodist. Subspecialties: Kinetics; Laser photochemistry. Current work: Chem. kinetics of free radical and ionic reactions in gas-phase; multiphoton infrared laser induced chemistry; vacuum ultraviolet photochemistry; mass spectrometry. Patentee in field. Home: 542 Ridge Ave State College PA 16801 Office: Pa State U 152 Davey Lab University Park PA 16802

LAMPERT, SEYMOUR, mechanical engineer, educator; b. Bklyn., Mar. 5, 1920; s. Max and Esther (Bakst) L.; m. Shirley Ruth (Axelrod), Mar. 21, 1948; children: Rachel Beth, David Aaron, Martin Daniel. B.S., Ga. Tech. Inst., 1943; M.S., Calif. Inst. Tech., 1947, Ph.D., 1954. Instr. Ga. Tech. Inst., Atlanta, 1943-44; aero. research scientist Ames Lab., Calif Inst. Tech., 1951-54; chief engr. Odin Assos., Pasadena, Calif., 1956-57; dept. mgr. Ford Aeronutronic, Newport Beach, Calif., 1957-62; dir. advanced systems research N.Am. Aviation Space Div., 1962-67; v.p. SAI, Long Beach, Calif., 1967-71; prof. dir. solar research dept. mech. engring. U. So. Calif., Los Angeles, 1975—; dir. Davato Corp., Energy Fair Found. Editor-in-chief: Solar Scis., 1982—; contbr. articles to profl. jours. Served to lt. USNR, 1944-54. Mem. Internat. Solar Energy Soc. Subspecialties: Solar energy; Fluid mechanics. Current work: Solar research and development; educational programs in solar sciences, photovoltaics. Patentee in field. Office: U So Calif Los Angeles CA 90089

LAMPERTI, ALBERT ANTHONY, anatomist, educator; b. Bronx, N.Y., Oct. 24, 1947; s. Dante J. and Helen M. (Russo) L. Ph.D., U. Cin., 1973. Instr. anatomy U. Cin., 1973-75, asst. prof., 1975-80; assoc. prof. Temple U., Phila., 1980—. NIH grantee, 1978-81. Mem. Internat. Soc. Neuroendocrinology, AAAS, Neurosci. Soc., Soc. Study Reprodn., Am. Assn. Anatomists, Endocrine Soc. Current work: Teaching reproductive neuroendocrinology. Office: Temple Universit Dept Anatomy 3420 N Broad St Philadelphi PA 19140

LAMSTER, IRA BARRY, dental educator, periodontist; b. N.Y.C., Mar. 6, 1950; s. Nathan and Mollie (Garber) L.; m. Gail Maxine Marcovitz, Aug. 28, 1971; children: Rachel Amy, Stephanie Ann. B.A., Queens Coll., CUNY, 1971; S.M., U. Chgo., 1972; D.D.S., SUNY-Stony Brook, 1977; M.M.Sc., Harvard U., 1980. Cert. in periodontology Harvard Sch. Dental Medicine. Assoc. dir. Oral Health Research Ctr., Fairleigh Dickinson U., Hackensack, N.J., 1981—, asst. prof. periodontics and oral medicine, 1980-83, assoc. prof., 1983—; scientist Oral Health Research Ctr., 1980—. Contbr. articles to profl. jours. Recipient Young Investigator award NIH, 1982; research grantee Exxon, 1980; Warner-Lambert, 1980; Lever Bros., 1981. Mem. Am. Dental Assn., Am. Acad. Periodontology, Internat. Assn. Dental Research. Subspecialties: Periodontics; Immunobiology and immunology. Current work: Inflammatory and immune function as related to pathophysiology of periodontal disease. Office: Fairleigh Dickinson U 110 Fuller Pl Hackensack NJ 07601 Home: 902 Garrison Ave Hackensack NJ 07606

LANCASTER, FREDERICK WILFRID, information science educator; b. Stanley, Durham, Eng., Sept. 4, 1933; m. Cesaria Xolpe, Aug. 21, 1961; children: Miriam, Owen, Jude, Aaron. Assoc., Newcastle-upon-Tyne Sch. Librarianship, Eng., 1954. Sr. asst.

Newcastle-upon-Tyne Pub. Libraries, Eng., 1953-57; asst. info. officer Tube Investments Ltd., Birmingham, Eng., 1957-59; sr. librarian sci. and tech. Akron (Ohio) Pub. Library, 1959-60; tech. librarian Babcock & Wilcox Co., Barberton, Ohio, 1960-62; sr. research asst. ASLIB, London, Eng., 1962; resident cons., head systems evaluation group Herner & Co., Washington, 1964-65; info. systems specialist Nat. Library Medicine, Bethesda, Md., 1965-68; dir. info. retrieval services Westat Research Inc., Bethesda, 1969-70; assoc. prof. library sci. Grad. Sch. Library & Info. Sci., U. Ill., Urbana, 1970-72, prof., 1972—, dir. biomed. librarianship, 1970-72; cons. in field; vis. lectr. numerous univs.; lectr. numerous profl. confs.; reviewer NSF, 1974—. Author: Information Retrieval Systems: Characteristics, Testing and Evaluation, 1968, 2d. edit., 1979, Toward Paperless Information Systems, 1978, The Measurement and Evaluation of Library Services, 1977, Vocabulary Control for Information Retrieval, 1972; co-author: (with Laura Drasgow and Ellen Marks) The Impact of a Paperless Society on the Research Library of the Future, A Report to the National Science Foundation, 1980, (with S. Herner, F.W. Lancaster) An Evaluation of the Goddard Space Flight Center Library, 1979, (with E.G. Fayen) Information Retrieval On-Line, 1973; mem. editorial bd. numerous jours.; editor numerous books; contbr. chpts. to books, revs. and articles to profl. jours. Mem. Am. Soc. Info. Sci., Library Assn. (U.K.), Phi Kappa Phi. Office: Grad Sch Library & Info Sci U Illinois Urbana IL 61801

LANCASTER, JACK R., JR., biochemist, educator; b. Memphis, Tenn., Aug. 27, 1948; s. Jack R. and Shirley (Gade) L.; m. Meredith Alden, July 28, 1978; children: Tanya, Madeline. B.S. in Chemistry, U. Tenn.-Martin, 1970, Ph.D. in Biochemistry, 1974. Research assoc. Cornell U., Ithaca, N.Y., 1974-76, Duke U., Durham, N.C., 1976-80; assoc. prof. biochemistry Utah State U., Logan, 1980—; Established investigator Am. Heart Assn., 1983-88. Mem. AAAS, Am. Chem. Soc., AAUP, Biophys. Soc. Subspecialties: Biochemistry (medicine); Biophysical chemistry. Current work: Biomembranes and bioenergetics; biological electron transport; bioinorganic chemistry. Office: Utah State Univ Biochemistry and Chemistry Dept UMC 03 Logan UT 84322

LANCMAN, HENRY, physicist, educator; b. Warsaw, Poland, Mar. 19, 1932; came to U.S., 1966; s. Morris and Hannah (List) L.; m. Ina Serf, Dec. 28, 1963; children: Steven, Anna. M.S., Moscow U., USSR, 1958; Ph.D., Polish Acad. Sci., Warsaw, 1966. Asst. to sr. asst. Inst. Nuclear Research, Polish Acad. Sci., Warsaw, 1958-66; research assoc. Columbia U., N.Y.C., 1967-68; from instr. to assoc. prof. Bklyn. Coll., 1968-78, prof., 1978—; vis. prof. Rijksuniversiteit, Utrecht, Holland, 1974-75. Contbr. articles to sci. publs. Recipient CUNY Research awards, 1972-73, 76-81, U.S. Dept. Energy research contract, 1979-83, NSF research grant, 1977. Mem. Am. Phys. Soc., Sigma Xi. Subspecialty: Nuclear physics. Current work: Higher order effects in beta decay; nuclear resonance fluorescence; photofission. Office: Brooklyn College Dept Physics Bedford Ave and Ave H Brooklyn NY 11210

LAND, EDWIN HERBERT, physicist, inventor; b. Bridgeport, Conn., May 7, 1909; s. Harry M. and Martha F. L; m. Helen Maislen, 1929; children—Jennifer, Valerie. Ed., Norwich Acad.; student, Harvard, Sc.D. (hon.), 1957, Tufts Coll., 1947, Poly. Inst. Bklyn., 1952, Colby Coll., 1955, Northeastern U., 1959, Carnegie Inst. Tech., 1964, Yale U., 1966, Columbia U., 1967, Loyola U., 1970, N.Y. U., 1973; LL.D., Bates Coll., 1953, Wash. U., 1966, U. Mass., 1967, Brandeis U., 1980; L.H.D., Williams Coll., 1968. Founder Polaroid Corp., Cambridge, Mass., 1937, now chmn. bd., cons. dir. basic research; William James lectr. psychology Harvard, 1966-67, Morris Loeb lectr. physics, 1974; vis. Inst. prof. Mass. Inst. Tech., 1956—; Mem. Pres.'s Sci. Adv. Com., 1957-59, cons.-at-large, 1960-73; mem. Pres.'s Com. Nat. Medal of Sci., 1969-72, Carnegie Commn. on Ednl. TV, 1966-67, Nat. Commn. on Tech., Automation and Econ. Progress, 1964-66. Trustee Ford Found., 1967-75. Recipient numerous awards including; Presdl. Medal of Freedom, 1963; Nat. Medal of Sci., 1967; Hood medal Royal Photog. Soc., 1935; Cresson medal, 1937; Potts medal, 1956; Vermilye medal Franklin Inst., Phila., 1974; John Scott medal and award, 1938; Rumford medal Am. Acad. Arts and Scis., 1945; Holley medal Am. Soc. M.E., 1948; Duddell medal Brit. Phys. Soc., 1949; Progress medal Soc. Photog. Engrs., 1955; Albert A. Michelson award Case Inst. Tech., 1966; Kulturpreis Photog. Soc. Germany, 1967; Perkin medal Soc. Chem. Industry, 1974; Proctor award, 1963; Photographic Sci. and Engring. Jour. award, 1971; Progress medal Photog. Soc. Am., 1960; Kosar Meml. award, 1973; Golden Soc. medal Photog. Soc. Vienna, 1961; Interkamera award, 1973; Cosmos Club award, 1970; NAM award, 1940, 66; Jefferson medal N.J. Patent Law Assn., 1960; Indsl. Research medal, 1965; Diesel medal in gold, 1966; Named to Nat. Inventors Hall of Fame, 1977. Fellow Nat. Acad. Scis., Photog. Soc. Am., Am. Acad. Arts and Scis. (pres. 1951-53), Royal Photog. Soc. Gt. Britain (Progress medal 1957), Royal Micros. Soc. (hon.), Soc. Photog. Scientists and Engrs. (hon., Lieven-Gevaert medal 1976); mem. N.Y. Acad. Scis., Nat. Acad. Engring. (Founders medal 1972), Optical Soc. Am. (hon. mem. 1972, dir. 1950-51, Frederick Ives medal 1967, Dudley Wright prize 1980), Royal Instn. Gt. Britain (hon.), Am. Philos. Soc., German Photog. Soc. (hon.), Soc. Photog. Sci. and Tech. Japan (hon.), IEEE (hon.), Sigma Xi. Subspecialty: Photographic science. During coll. years invented 1st light-polarizer in form of an extensive synthetic sheet; developed a sequence of subsequent polarizers, theories and applications of polarized light, including automobile headlight system, 3 dimensional pictures, camera filters; during World War II developed optical, other systems for mil. use; created cameras, films that give instantaneous dry photographs in black and white and color; proposed Retinex Theory for mechanism of color perception and designed series of supporting expts. Home: 163 Brattle St Cambridge MA 02138 Office: 730 Main St Cambridge MA 02139

LAND, IVAN MARSHALL, research investigator, educator; b. Wilkes-Barre, Pa., Oct. 26, 1947; s. Arthur and Dena (Iskowitz) L.; m. Ellen Fox, Mar. 25, 1973; 1 dau., Erika Rachel. B.S., U. Pitts., 1969; M.S., Temple U., 1975, Ph.D., 1980. Postdoctoral research assoc. Tex. Tech. U. Med. Sch., Lubbock, 1980-82; instr., research investigator Med. Coll. Wis., Milw., 1982—. Contbg. author books in field. Med. Coll. Wis. grantee, 1982; NIH tng. fellow, 1980; univ. fellow Temple U., 1974-77. Mem. Am. Physiol. Soc., Soc. for Neurosci., Sigma Xi, Alpha Epsilon Delta, Phi Eta Sigma. Subspecialties: Physiology (medicine); Gastroenterology. Current work: Investigation of neural control of gastrointestinal tract. Office: VA Med Ctr Surg Research 151 5000 National Ave Wood WI 53193 Home: 1742 N Hi-Mount Blvd Milwaukee WI 53208

LANDA, BETH KIM, biostatistician, research phychologist; b. N.Y.C., Oct. 18, 1954; d. Jay Myron and Ruth (Kaplan) L.; m. Steven A. Balicer, Dec. 10, 1977 (div. Nov. 1982). B.A. in Psychology cum laude, Hofstra U., 1975, M.A. in Applied Psychol. Research with distinction, 1977, Ph.D., 1980. Research assoc. Hofstra U., 1977-78, assoc. faculty mem., 1980—; biostatistician L.I. Jewish-Hillside Med. Ctr., Glen Oaks, N.Y., 1978—; adj. faculty mem. L.I. U., 1982. Mem. Am. Psychol. Assn., Assn. Women in Computing, Psi Chi (pres. Hofstra chpt. 1974-75). Democrat. Jewish. Subspecialties: Developmental psychology; Statistics in psychology. Current work: Application of new statistical methods and models to ongoing medical research; current projects include development of diagnostic tool in appendicitis, psychological assessment of cardiology patients with ventricular arrhythmia. Home: 105 Dogwood Ave Roslyn Harbor NY 611576 Office: LI Jewish-Hillside Med Center Glen Oaks NY 11004

LANDAUER, ROLF WILLIAM, physicist; b. Stuttgart, Germany, Feb. 4, 1927; came to U.S., 1938, naturalized, 1944; s. Karl and Anna (Dannhauser) L.; m. Muriel Jussim, Feb. 26, 1950; children—Karen, Carl, Thomas. S.B., Harvard U., 1945, A.M., 1947, Ph.D., 1950. Solid state physicist Lewis Lab., NACA (now NASA), Cleve., 1950-52; with IBM Research (and antecedent groups), 1952—, asst. dir. research, Yorktown Heights, N.Y., 1966-69, IBM fellow, 1969—. Contbr. articles on solid state theory, computing devices, statis. mechanics of computational process to profl. jours.; asso. editor: Rev. Modern Physics, 1973-76. Served with USNR, 1945-46. Fellow IEEE, Am. Phys. Soc.; mem. Nat. Acad. Engring. Subspecialties: Solid state physics; Computer engineering. Initiated IBM programs leading to injection laser, large scale integration. Office: IBM Research Center PO Box 218 Yorktown Heights NY 10598

LANDECKER, PETER BRUCE, physicist; b. N.Y.C., Oct. 1, 1942; s. Louis and Mildred (Nesson) L.; m. Elizabeth Jane Moore, June 18, 1966. A.B., Columbia U., 1963; Ph.D., Cornell U., 1967. Instr. physics Cornell U., 1967-68; asst. research physicist U. Calif., Irvine, 1968-70; research assoc. Columbia U., 1970-74, cons., N.Y.C., 1974-75; mem. tech. staff Aerospace Corp., El Segundo, Calif., 1974-82, cons., 1982—; sr. staff engr. Space and Communications Group, Hughes Aircraft Co., El Segundo, 1982—; instr. physics El Camino Coll, 1977. N.Y. State teaching fellow, 1963-65; NSF Summer Inst. fellow, 1959; N.Y. State Regents scholar, 1959-63; N.Y. State scholar, 1959-63; NASA trainee fellow, 1965-67. Mem. Am. Phys. Soc., Internat. Astron. Union, Am. Astron. Soc. Unitarian-Universalist. Club: AEA Scuba (El Segundo). Subspecialties: Satellite studies; Solar physics. Current work: Design of meteorol., satellites, X-ray spectroscopy studies of solar flares. Mem. Beach Cities Symphony Orch., Redondo Beach, Calif. Home: 1736 Nelson Ave Manhattan Beach CA 90266 Office: Hughes Space and Communications Group Bldg S41 Mail Stop B322 PO Box 92919 Airport Sta Los Angeles CA 90009

LANDERS, JACK MAXAM, manufacturing engineering educator; b. Latour, Mo., Oct. 22, 1935; s. L. Loyd and Margret (Maxam) L.; m. Mary L., Aug. 1, 1954; children: Garry, Cheryl, Jacklyn. B.S.Ed., Central Mo. State U., 1964; M.S. Ed., 1966; Ed. D., U. Mo., 1972. Self-employed constrn. work. Tchr. Rolla (Mo.) Pub. Schs., 1964-66; asst. prof. mfg. and constrn. Central Mo. State U., 1966-71, assoc. prof., 1975-80, prof., 1980, chmn, dept. mfg. and constrn., 1981—. Author: Construction: Materials, Methods and Careers, 1976, Home Maintenance, 1982. Chmn Warrensburg (Mo.) Planning and Zoning. Mem. Soc. Mfg. Engrs. Lodge: Kiwanis. Subspecialty: Industrial engineering. Home: 26 Timberline Dr Warrensburg MO 64093 Office: Central Mo State U G-79 Warrensburg MO 64093

LANDERS, ROY ESLYN, JR., therapeutic radiol. physicist; b. Milledgeville, Ga., Aug. 4, 1944; s. Roy Eslyn and Mildred McNair (Jones) L.; m. Carol Hill, Sept. 13, 1969; children: Kimberly Diane, James Jefferson. B.S. in Physics, Ga. Inst. Tech., 1966, M.S., 1968, Ph.D., 1974. Diplomate: Am. Bd. Radiology, 1980. Instr. Devry Inst. Tech., Atlanta, 1973-75, chmn. basic studies, 1975; cons. math. and computer models Varian Assocs., Atlanta, 1974-76; therapeutic radiol. physicist, cons. Sarasota Oncology Center, Fla., 1976—. Contbr. sci. articles to profl. publs. NDEA fellow, 1969. Mem. Am. Assn. Physicists in Medicine, Health Physics Soc., Sigma Xi, Sigma Pi Sigma. Subspecialty: Therapeutic Radiological Physics. Current work: Physics of therapeutic radiology; computer software development.

LANDES, JOHN DAVID, research scientist; b. Sellersville, Pa., June 28, 1942; s. Christian T. and Mabel (hertzler) L.; m. Anne Ruth, June 20, 1964; children: Jennifer, Kristina, Rebecca, David. B.S., Lehigh U., 1964, M.S., 1965, Ph.D., 1970. Registered profl. engr., Pa. Research assoc. Pratt & Whitney Aircraft, North Haven, Conn., 1965-66; research and teaching asst. Lehigh U., Bethlehem, Pa., 1966-70; sr. engr. Westinghouse Research and Devel. Ctr., Pitts., 1970-76, fellow engr., 1976-78, adv. scientist, 1978—. Editor: Elastic Plastic Fracture, 1979. Mem. ASTM (Irwin medal 1980), Am. Soc. Metals, Soc. Engring. Scis. Methodist. Subspecialties: Fracture mechanics; Solid mechanics. Current work: Research in the fracture of metals, ductile fracture, environmental effects, structural evaluation. Home: 4441 Marywood Dr Monroeville PA 15146 Office: Westinghouse Research and Devel Ctr 1310 Beulah Rd Pittsburgh PA 15235

LANDGREBE, DAVID ALLEN, elec. engr.; b. Huntingburg, Ind., Apr. 12, 1934; s. Albert E. and Sarah A. L.; m. Margaret Ann Swank, June 7, 1959; children—James David, Carole Ann, Mary Jane. B.S. in Elec. Engring, Purdue U., 1956, M.S., 1958, Ph.D., 1962. Mem. tech. staff Bell Telephone Labs., Murray Hill, N.J., 1956; electronics engr. Interstate Electronics Corp., Anaheim, Calif., 1958, 59, 62; mem. faculty Purdue U., West Lafayette, Ind., 1962—, dir. lab. for applications of remote sensing, 1969-81, prof. elec. engring., 1970—, asso. dean engring., 1981—; research scientist Douglas Aircraft Co., Newport Beach, Calif., 1964; dir. Univ. Space Research Assn., 1975-78. Author: (with others) Remote Sensing: The Quantitative Approach, 1978. Recipient medal for exceptional sci. achievement NASA, 1973. Fellow IEEE; mem. Am. Soc. for Engring. Edn., AAAS, Am. Soc. Photogrammetry, Sigma Xi, Tau Beta Pi, Eta Kappa Nu. Club: Rotary. Subspecialties: Graphics, image processing, and pattern recognition; Information systems (information science). Current work: Information representation, image processing and pattern recognition. Office: Purdue U Dept Elec Engring West Lafayette IN 47907

LANDGREN, JOHN JEFFREY, research scientist; b. St. Paul, Nov. 16, 1947; s. Ralph William and Dorothy Florence (Rediske) L.; m. Karen Lee Durciansky, Dec. 17, 1977; children: Erik John, David William. B.S. in Math, U. Minn., 1969, M.S., 1971, Ph.D., 1976. Asst. prof. math. Ga. Inst. Tech., Atlanta, 1976-78, U. Tenn., Chattanooga, 1978-80; research scientist II Ga. Tech. Engring. Expt. Sta., Atlanta, 1980—. Contbr. articles to profl. jours. NSF grantee, 1969-71. Mem. Soc. for Indsl. and Applied Math., Assn. Old Crows, Sigma Xi, Tau Beta Pi. Lutheran. Subspecialties: Systems engineering; Applied mathematics. Current work: Analytical studies in advanced radar countermeasures. Office: Ga Tech Engring Expt Sta Systems Engring Lab Atlanta GA 30332

LANDING, BENJAMIN HARRISON, laboratory director, pathology educator; b. Buffalo, Sept. 11, 1920; s. Benjamin Harrison and Margaret C. (Crohen) L.; m. Dorothy Jean Hallas, Jan. 2, 1973; children: Benjamin, Susan, William, David; m.; 1 stepson, William R. Shankle. A.B., Harvard U., 1942, M.D., 1945. Diplomate: Am. Bd. Pathology. Intern Children's Hosp., Boston, 1945-46; resident Children's Hosp., Free Hosp. for Women, Boston Lying-In Hosp., Boston, 1948-50; asst. pathologist Children's Med. Ctr., Boston, 1950-53; instr. pathology Harvard Med. Sch., Boston, 1950-53; directing pathologist Cin. Children's Hosp., 1953-61; asst. prof., assoc. prof. pathology U. Cin. Coll. Medicine, 1953-61; pathologist-in-chief Children's Hosp. Los Angeles, 1961—; prof. pathology and pediatrics U. So. Calif. Sch. Medicine, Los Angeles, 1961—. Author: Tumors of Cardiovascular System, 1956; contbr. writings in field to publs. Vice pres., pres., bd. Burbank (Calif.) Unitarian Fellowship, 1962-73; bd. dirs., pres. Pacific S.W. Dist. Unitarian Universalist Assn., 1966-70. Served to capt. M.C. U.S. Army, 1946-48. Mem. Am. Assn. Pathologists, Internat. Acad. Pathology, Histochemistry Soc., Pediatric Pathology Club (pres. 1975), Teratology Soc.; hon. mem. Pathology Soc. Peru, Pediatric Soc. Peru, Acad. Pediatrics South Korea, Med. Assn. Costa Rica, Pathology Soc. Chile. Democrat. Subspecialties: Pathology (medicine); Evolutionary biology. Current work: Mechanisms in gene/phenotype relations; quantitative and morphometric analysis of organ and tissue structure in genetic diseases. Office: Children's Hosp Los Angeles 4650 Sunset Blvd Los Angeles CA 90027 Home: 4513 Deanwood Dr Woodland Hills CA 91364

LANDIS, ROBERT CLARENCE, scientist, government official; b. Trenton, N.J., Nov. 4, 1941; s. C.W. and Florence (Generaux) L.; m. Charlotte A. Files, Feb. 24, 1968; children: Christopher, Geoffrey. B.S., Pa. State U., 1963; M.S., Tex. A&M U., 1966. Oceanographer Navy Oceanographic Office, Washington, 1963-68; project leader The Mitre Corp., McLean, Va., 1968-72; planning officer NOAA, Rockville, Md., 1972-74, dep. chief program evaluation, 1976-81; asst. sec. intergovt. oceanograpic commn. UNESCO, Paris, France, 1974-76; chief service evaluation br. Nat. Weather Service, Silver Spring, Md., 1981—. Co-author: Technology Assessment Methodology, 1972. Recipient Outstanding Employee award NOAA, 1974, 79, 80, 81, 82. Subspecialties: Meteorology; Oceanography. Current work: Program development marine weather and oceanographic prediction services. Home: 9116 Gue Rd Damascus MD 20872 Office: Nat Weather Service 8060 13th St Silver Spring MD 20910

LANDIS, WAYNE G., research scientist; b. Washington, Jan. 20, 1952; s. James G. and Harriet E. L. B.A. cum laude with honors in Biology (Hankins scholar), Wake Forest U., 1974; M.A. in Biology, Ind. U., 1978; Ph.D. in Zoology, Ind. U., 1979. Environ. and health scientist Franklin Research Ctr., Silver Spring, Md., 1979-81; research scientist ecology sect., toxicology br. research div. Chem. Research and Devel. Ctr., Aberdeen Proving Grounds, Md., 1982—. Contbr. articles on ecology, population genetics and evolution of Paramecium, revs. of toxicol. properties of monohaloacetic acids and secondary and tertiary amines to sci. jours. NSF grantee, 1972. Mem. Genetics Soc. Am., Soc. for Study Evolution, Soc. Protozoologists, Soc. Environ. Toxicology and Chemistry, Ecol. Soc. Am. Subspecialties: Environmental toxicology; Ecology. Current work: Aquatic toxicology, population biology, community ecology, risk analysis, breeding systems and evolution of protozoa, population genetics of protozoa and invertebrates, genetic engineering of enzymes for biodegradation. Office: Ecology Sect Toxicology B Research Div Chem Research and Devel Aberdeen Proving Grounds MD 21010

LANDO, JEROME B., macromolecular scientist, educator; b. Bklyn., May 23, 1932; s. Irving and Ruth (Schwartz) L.; m. Geula, Dec. 2, 1962; children: Jeffrey, Daniel, Avital. A.B. in Chemistry, Cornell U., 1953; Ph.D. in Phys. Chemistry, Poly. Inst. Bklyn., 1963. Chemist, Camille Dreyfus Lab., Research Triangle Inst., Durham, N.C., 1963-65; asst. prof. polymer sci. and engring. Case Inst. Tech., Cleve., 1963-65; assoc. prof. macromolecular sci. Case Western Res. U., Cleve., 1965-68, prof., 1974—, chmn. dept. macromolecular sci., 1978—; vis. prof. U. Mainz, W.Ger., 1974. Author: (with S. Maron) Fundamentals of Physical Chemistry, 1974; contbr. articles to profl. jours. Alexander von Humboldt sr Am. scientist awardee, 1974. Fellow Am. Phys. Soc.; mem. Am. Chem. Soc., Am. Crystallographic Assn., Sigma Xi. Subspecialties: Polymers; Polymer physics. Current work: Solid state reactions, polymer crystal structure, electrical properties of polymers. Home: 21925 Byron Rd E Shaker Heights OH 44122 Office: Dept Macromolecular Sci Case Western Res U Cleveland OH 44106

LANDOLPH, JOSEPH RICHARD, JR., microbiology and pathology educator, cancer researcher; b. Upper Darby, Pa., Nov. 9, 1948; s. Joseph Richard and Ada Nolia (Welch) L.; m. Alice Kaufman, Jan. 19, 1980; 1 son, Joseph Richard III. B.S. in Chemistry, Drexel U., 1971, Ph.D., U. Calif.-Berkeley, 1976. Postdoctoral fellow U. So. Calif., 1977-80, asst. prof. research pathology, Los Angeles, 1981-82, asst. prof. microbiology and pathology, 1982—. Contbr. articles to profl. jours. Served as 2d lt. U.S. Army, 1976-77. Am. Cancer Soc. fellow, 1977-79; recipient NSF undergrad. sci. award, 1970-71; Merck award Drexel U., 1971. Mem. Am. Chem. Soc., Am. Soc. Cell Biology, Environ. Muteger Soc., Am. Assn. Cancer Research, Tissue Culture Assn. Republican. Methodist. Subspecialties: Oncology; Integrated systems modelling and engineering. Current work: Mechanisms of chemical carcinogenesis and mutagenesis; molecular biology of chemical carcinogenesis; mechanisms of carcinogenesis caused by carcinogenic metal salts. Home: 1009 E Mendocino St Altadena CA 91001 Office: Cancer Research Lab Dept Microbiology U So Calif Cancer Ctr 1303 N Mission Rd Los Angeles CA 90033

LANDON, LEROY JAMES, chemistry educator, researcher; b. Hibbing, Minn., Jan. 30, 1940; s. Martin Phillip and Susan Shaw (Blakeslee) L.; m. Eleanor P. Richards, May 3, 1970; children: Carolyn, Michael. B.A., Brown U., 1961; Ph.D. in chemistry, U. Wis., 1966. Postdoctoral assoc. U. Tex., Austin, 1966-67, asst. prof. chemistry, 1967-70; asst. prof. UCLA, 1970-73, assoc. prof., 1973-79, prof. chemistry, 1980—; cons. EPA, Washington, 1981—. Grantee NSF, 1981-83. Mem. Am. Chem. Soc., Internat. Soc. Electrochemistry, Electrochem. Soc. (sec. 1977-79). Lodge: K.C. Subspecialties: Analytical chemistry; Biochemistry (biology). Home: Werik Apt 3614 Atlantic Ave Long Beach CA 90807

LANDSBERG, HELMUT E(RICH), meteorologist; b. Frankfurt/Main, Ger., Feb. 9, 1906; came to U.S., 1934, naturalized, 1938; s. Georg Julius and Klara (Zedner) L.; m. A. Frances Simpson; 1 son, Bruce S. Ph.D., U. Frankfurt, 1930. DCert. cons. meteorologist Am. Meteorol. Soc. 1960. Supr. Taunus Obs., 1931-34; asst. prof. geophysics Pa. State U., State College, 1934-41; assoc. prof. meteorology U. Chgo., 1941-43; exec. dir. com. geophysics Research and Devel. Bd., 1946-51; dir. geophys. research Air Force Cambridge Research Ctr., 1951-54; dir. Office Climatology U.S. Weather Bur., Washington, 1954-65, Environ. Data Service, Environ. Sci. Services Adminstrn., 1965-66; vis. prof. Inst. for Fluid Dynamics and Applied Math., U. Md., College Park, 1964-67, research prof., 1967-76, acting dir., 1974-76, prof. emeritus, 1976; ops. analyst, cons. USAAF, 1942-45; mem. Nat. Adv. Com. on Oceans and Atmosphere, 1975-77; trustee Univ. Corp. for Atmopheric Research, 1968-72. Author: Physical Climatology, 1941, Weather and Health, 1969, The Urban Climate, 1981; editor: Advances in Geophysics, 1951-76; editor-in-chief: World Survey of Climatology; contbr. articles on meteorology to profl. jours. Recipient Internat. Meteorol. Orgn. prize, 1979; William F. Peterson Found. medal, 1982; Dept. Commerce Exceptional Service award, 1960. Fellow AAAS (v.p. sect. E 1972), Am. Acad. Arts and Scis., Am. Geophys. Union (pres. meteorology sect. 1956-59, v.p. 1966-68, pres. 1968-70, Bowie medal 1978), Am. Meteorol. Soc. (councilor 1952-60, v.p. 1962-63, Bioclimatology award 1964, C.F. Brooks award 1972, Cleveland Abbe award 1983), Washington Acad. Sci.; mem. Am. Inst. for Med. Climatology (pres. 1967-81, bd. dirs. 1981), N.Y. Acad. Scis. (hon.), World Meteorol. Orgn. (pres. commn. for climatology 1969-78), Nat Acad. Engring., Ger. Meteorol. Soc. (Wegener medal 1980), Sigma Xi, Sigma Pi Sigma, Sigma Gamma Epsilon. Clubs: Explorers, Cosmos (Washington). Subspecialties: Climatology; Micrometeorology. Current work: Climate trends and fluctuations, anthropogenic influences on climate. Office: Univ Md College Park MD 20742

LANDSBERG, LEWIS, internist, endocinologist, educator; b. N.Y.C., Nov. 23, 1938; s. Mortimer and Florence (Schulte) L.; m. Jill Warren, June 14, 1964; children: Alison, Judd. A.B., Williams Coll., 1960; M.D., Yale U., 1964. Diplomate: Am. Bd. Internal Medicine. Intern Yale New Haven Hosp., 1964-65, resident internal medicine, 1965-66, 68-70; instr. Yale U. Sch. Medicine, New Haven, 1969-70, asst. prof. medicine, 1970-72, Harvard U. Med. Sch., Boston, 1972-75, assoc. prof., 1977—. Contbr. articles to profl. publs. Served with USPHS, 1966-68; NIH grantee, 1977—. Mem. Am. Fedn. Clin. Research (chmn. eastern sect. 1979), Endocrine Soc., Am. Physiol. Soc., Am. Soc. Clin. Investigation. Subspecialties: Prosthetics; Endocrinology. Current work: Catecholamines and sympathoadrenal system. Office: Beth Israel Hosp 330 Brookline Ave Boston MA 02215

LANDSTREET, JOHN DARLINGTON, astrophysicist, educator; b. Phila., Mar. 13, 1940; s. Barent French and Louise (Darlington) L.; m. Elaine Davis, Feb. 18, 1972 (div. 1983); 1 son, David Roger. B.A., Reed Coll., 1962; M.A., Columbia U., 1963, Ph.D., 1966. Instr. physics Mt. Holyoke Coll., South Hadley, Mass., 1965-66, asst. prof., 1966-67; research assoc. astronomy Columbia U., N.Y.C., 1967-70, asst. prof., 1970; asst. prof. astronomy U. Western Ont., London, Can., 1970-72, assoc. prof., 1972-76, prof. astronomy, 1976—; mem. Nat. Scis. and Engring. Research Council Can. grant selection com. for space and astronomy, 1980-83, chmn., 1982-83; mem. Sci. Adv. Council, Can.-France-Hawaii Telescope Corp., 1980-83, chmn., 1982-83; mem. sci. adv. com. Can. Centre for Space Sci., 1981-83. Contbr. profl. papers in field to publs. Ann. research grantee Nat. Scis. and Engring. Research Council Can., 1970—. Mem. Am. Astron. Soc., Can. Astron. Soc., Internat. Astron. Union, Royal Astron. Soc. Clubs: University Club of London (Ont.); Trollope Hunt, Bayfield Yacht. Subspecialty: Optical astronomy. Current work: Measurement of magnetic fields in stars; measurement of polarization in Mira variables and other complex objects; development new astronomical measuring instruments. Office: Dept Astronom Univ Western Ont London ON Canada N6A 3K7 Home: 1280 Webster St Apt 405 London ON Canada N5V 3P4

LANDWEBER, LAWRENCE HUGH, computer science educator; b. Bklyn.; s. Hyman and Rose (Borowsky) L.; m. Jean R. Lazarov, July 3, 1966; 1 son, Michael. B.s. in Math, Bklyn. Coll., 1963; M.S., Purdue U., 1965, Ph.D. in Computer Sci, 1967. Asst. prof. computer sci. U. Wis.-Madison, 1967-70, assoc. prof., 1970-76, prof., 1976—, chmn. dept. computer sci., 1978-81; cons. computer networks and edn.; chmn. mgmt. and policy coms. Computer Sci. Network, 1981-83. NSF fellow, 1964-66. Mem. Assn. Computing Machinery (chmn. spl. interest group on automata and compatability theory 1981-83), Phi Beta Kappa, Sigma Xi. Subspecialties: Distributed systems and networks; Theoretical computer science. Current work: Computer networks, electronic mail, theory of computing. Office: 1210 W Dayton St Madison WI 53706

LANDWEBER, LOUIS, educator; b. N.Y.C., Jan. 8, 1912; s. Joseph and Lena (Rosenbush) L.; m. Mae Herschfeld, Apr. 7, 1935; children—Peter Steven, Victor Allen. B.S., CCNY, 1932; M.A., George Washington U., 1935; Ph.D., U. Md., 1951. Physicist, head hydrodynamic div. David Taylor Model Basin, Washington, 1932-54; prof. U. Iowa, research engr. Inst. Hydraulic Research, 1954—; cons. Westinghouse and Davidson Lab. of Stevens Inst. Tech., NRC, referee jours. in field of fluid mechanics. Author: (with others) Advanced Mechanics of Fluids, 1959; editor of: translation Theory of Ship Motions (Blagoveshchensky), 1962, Theory of Ship Waves and Wave Resistance (A.A. Kostyukov), 1968; contbr. articles to profl. jours., chpt. to book. Recipient Ward medal and Beldon prize CCNY, 1932, Worthington prize, 1977. Fellow Japan Soc. Promotion of Sci., Soc. Naval Architects and Marine Engrs. (Davidson medal 1978), Am. Acad. Mechanics; mem. Assn. Technique Maritime et Aeronautique (U.S. corr. mem.), Nat. Acad. Engring., Phi Beta Kappa, Sigma Xi, Sigma Pi Sigma, Tau Beta Pi. Subspecialties: Fluid mechanics; Theoretical and applied mechanics. Current work: Mathematical and numerical predictions of flow about and forces acting on ship forms (ship hydrodynamics). Home: 323 Post Rd Iowa City IA 52240 Office: Inst Hydraulic Research Iowa City IA 52242

LANE, BERNARD PAUL, pathologist, educator; b. Bklyn., June 27, 1938; s. Jack R. and Rose L. (Weiss) L.; m. Dorothy S. Lane, Aug. 5, 1962; children: Erika, Andrew, Matthew. A.B., Brown U., 1959; M.D., N.Y.U., 1963. Diplomate: Am. Bd. Pathology. Asst. prof. N.Y.U., 1963-69, assoc. prof., 1969-71, SUNY-Stony Brook, 1971-76, prof., 1976—. Contbr. articles in field to profl. jours. Bd. dirs. Am. Cancer Soc., L.I., N.Y., 1979—, v.p., 1980-82. Served with USAF, 1962—. NIH fellow, 1963-65; NIH grantee, 1973-82. Mem. N.Y. Acad. Sci., Am. Assn. Cancer Research, Am. Assn. Pathology, Internat. Acad. Pathology, Am. Soc. Cell Biology, Am. Soc. Clin. Pathology, Coll. Am. Pathology. Subspecialty: Pathology (medicine). Current work: Carcinogenesis, cancer biology. Address: 6 Intervale Rd Setauket NY 11733

LANE, DANIEL MCNEEL, hematologist, biochemist; b. Fort Sam Houston, Tex., Jan. 25, 1936; s. Samuel Hartman and Mary Maverick (McNeel) L.; m. Carolyn Ann Sprueill, Nov. 28, 1958; children: Linda Ann, Daniel M. Jr., Maury S., Oleta K. M.D., U. Tex.-Dallas, 1961; M.S., U. Tenn., 1967; Ph.D., U. Okla., 1973. Head pediatric hematology/oncology U. Okla. Med. Ctr., Oklahoma City, 1966-70; research fellow Okla. Med. Research Found., Oklahoma City, 1969-72; Head pediatric hematology/oncology Tulane Med. Sch., New Orleans, 1972-73; head hematology/oncology Oklahoma City Clinic, 1973-79, Presbyn. Hosp., Oklahoma City, 1979—; dirs. Poplar Pike Realtors, Memphis, 1978—; pres. Samuel Maverick Interests, Inc., Oklahoma City., 1976—. Fin. chmn Dunlap for Congress, 1976; head Physicians for Gov. Nigh, 1978; Democratic candidate for Congress, 5th Dist., 1982. USPHS fellow, 1964-66; spl. research fellow Nat. Heart-Lung Inst., 1969-72. Fellow Am. Acad. Pediatics; mem. Am. Soc. Clin. Oncology, Am. Soc. Hematology, AMA. Democrat. Episcopalian. Subspecialties: Oncology; Biochemistry (medicine). Current work: Plasma lipids and apolipoproteins; clinical hematology and oncology (chemotherapy); pediatrics (consultative). Home: 1504 Guilford Ln Oklahoma City OK 73120 Office: 711 Stanton Young Blvd #604 Oklahoma City OK 73104

LANE, DENNIS DEL, engineering educator; b. Peoria, Ill., Feb. 16, 1950; s. Delford Lel and Rowena (Sayler) L.; m. Kristine Lae Bell, June 1, 1969; children: Thomas Del, Jeffrey Todd, Thad Matthew, Teresa Beth. B.S., U. Ill., 1972, M.S., 1973, Ph.D., 1976. Registered profl. engr., Kans. Research asst. U. Ill., Urbana, 1974-76; asst. prof. dept. civil engring. U. Kans., Lawrence, 1976-81, assoc. prof., 1981—; cons. Hershberger Law Office, Wichita, 1977—, State of Kans., Topeka, 1977—, Midwest Research Inst., Kansas City, Mo., 1978—. Contbr. chpts. to books, articles to profl. jours. EPA grantee, 1978—; State of Kans. grantee, 1978-82; Office Water Resources Tech.-Dept. Interior grantee, 1979—. Mem. ASCE (assoc.), Air Pollution Control Assn., Midwest Air Pollution Control Assn. (program chmn. 1979-80), Am. Assn. Engring. Educators, Phi Kappa Phi. Republican. Baptist. Subspecialties: Environmental engineering; Ecosystems analysis. Current work: Aerosol science—air pollution particle sizing equipment design. Pesticide transport—the mechanics of overland flow and its effects on pollution entrainment. Ecological systems—the effects of SO2 on insect populations. Home: 700 Mississippi St Lawrence KS 66044 Office: Dept Civil Engring U Kans Lawrence KS 66045

LANE, FRANK BENJAMIN, physician; b. Little Rock, Jan. 23, 1940; s. Frank B. and Elizabeth (Stancil) L.; m. Gail Anna Ingalls, 1968 (div.); m. Debra Ann Crum, Aug. 17, 1979. B.S., U. Md., 1961; M.D., Temple U., 1965. Diplomate: Am. Bd. Internal Medicine. Fellow dept. hematology U. Wash. Sch. Medicine, Seattle, 1971-74; assoc. prof. medicine U. South Fla. Sch. Medicine, Tampa, 1974-76; practice medicine specializing in hematology 1976—; chief dept medicine Tampa Gen. Hosp., 1982. Bd. dirs. Leukemia/Blood Disease Found., Tampa, 1975-81. Served to maj. U.S. Army, 1969-71. Fellow Internat. Soc. Hematology; mem. Am. Soc. Hematology, Am. Soc. Clin. Oncology, Fla. Soc. Clin. Oncology (dir. 1978-80), ACP. Subspecialties: Chemotherapy; Hematology. Current work: BCG treatment of malignant melanoma, also treatment of immune disorders with immune suppression. Bone marrow transplants. Home: 4003 Muriel Pl Tampa FL 33614 Office: 2919 Swann Ave Suite 402 Tampa FL 33609

LANE, GEORGE ASHEL, chemical researcher, conservationist; b. Norman, Okla., May 9, 1930; s. George Henry and Clarinda Burroughs (Murphy) L.; m. Patricia Graves, June 21, 1952. A.B., Grinnell Coll., 1952; Ph.D., Northwestern U., 1955. Research assoc. Dow Chem. Co., Midland, Mich., 1955—. Author: Solar Heat Storage: Latent Heat Materials, 1983; contbr. articles sci. jours. Recipient IR-100 award, 1980; named Conservationist of Yr. Trout Unltd., 1980, Mem. of Yr. Mich. Trout Unltd., 1981. Mem. Am. Chem. Soc., AAAS, Internat. Solar Energy Soc., Am. Solar Energy Soc., Sigma Xi. Presbyterian. Subspecialties: Physical chemistry; Inorganic chemistry. Current work: Solar energy; phase equilibrium; research on heat storage using latent heat materials. Holder numerous U.S. and fgn. patents. Home: 3802 Wintergreen Dr Midland MI 48640 Office: Dow Chemical Co 1776 Bldg Midland MI 48640

LANE, JOSEPH ROBERT, metallurgical engineer; b. Chgo., Mar. 3, 1917; m., 1949. B.S., U. Ill., 1943; Sc.D. in Metallurgy, MIT, 1950. Metallurgist Metall. Lab. U. Chgo., 1943-45; research asst. MIT, Cambridge, 1945-50; br. head Naval Research Lab., Washington, 1950-55; staff metallurgist Nat. Material Adv. Bd., Nat. Acad. Sci., 1955—. Fellow Am. Soc. Metals; mem. Soc. Mfg. Engrs., AIME, Soc. for Advancement Material and Process Engring. Subspecialties: Metallurgy; Metallurgical engineering. Current work: NDE, rapid solidification, materials properties data management. Office: Nat Materials Adv Bd National Research Council 2101 Constitution Ave NW Washington DC 20418

LANE, ROGER LEE, zoology educator, researcher; b. Mt. Carmel, Ill., July 4, 1945; s. Andy Lee and Flora (Baldridge) L.; m. Paulette Hruban, July 17, 1968; children: Leigh, Brooke, Taylor. B.S., U. Nebr., Lincoln, 1968, M.S., 1971, Ph.D., 1974. Lectr. in biology John F. Kennedy Coll., 1973-75; asst. prof. Kent State U., 1975-80, assoc. prof., 1980—; mem. bd. adjudicators Kerala U., Trivandrum, India, 1977-78. Author: Histology for the General Biology Laboratory, 1980. Vice-pres. Ashtabula County Animal Protective League, 1978—. Recipient Kent State U. Teaching Devel. award, 1980. Mem. Am. Malacol. Union, Am. Micros. Soc., Am. Soc. Zoologists, Crustacean Soc., Ohio Acad. Sci. Democrat. Roman Catholic. Subspecialties: Histology and Histochemistry of Invertebrates; Morphology. Current work: Morphology, histology and histochemistry of various systems of crustaceans and molluscs. Home: 2706 Burlingham Dr Ashtabula OH 44004 Office: Kent State U 3325 W 13th St Ashtabula OH 44004

LANE, STANLEY EARL, biochemist, researcher; b. Cleve., Aug. 18, 1943; s. Myron Alfred and Ruth (Schneider) L.; m. Cheryl Kushner Lane, Mar. 25, 1946; children: Rana, Amy, Amanda. B.A., Ohio State U., 1966; Ph.D., U. S.D., 1971; M.B.A., Tenn. State U., 1982. Postdoctoral fellow Oak Ridge Nat. Lab., 1971-74; instr. dept. medicine SUNY Downstate Med. Ctr., Bklyn., 1974-76; asst. prof. Meharry Med. Coll., 1976—. Contbr. articles to profl. jours. Damon Runyon Meml. Fund postdoctoral fellow, 1971-74; NSF research grantee, 1977-79; NIH research grantee, 1977—. Mem. Sigma Xi. Jewish. Subspecialties: Biochemistry (medicine); Molecular pharmacology. Current work: Hepatic involvement of alcohol and barbiturate drugs; affect of these drugs on the regulation of heme biosynthesis and cytochrome P-450. Home: 7456 Harness Dr Nashville TN 37221 Office: Meharry Med Coll 1005 18th Ave N Nashville TN 37208

LANG, CONRAD MARVIN, chemist, educator; b. Chgo., July 1, 1939; s. Arne Conrad and Myrtle Oliva (Erickson) L.; m. Louise June Swanson, June 17, 1961; children: Kevin Alan, Kurtis Erik, Kenneth Marvin. B.S., Elmhurst (Ill.) Coll., 1961; M.S., U. Wis.-Madison, 1964; Ph.D., U. Wyo., 1970. Instr. U. Wis.-Stevens Point, 1964-66, asst. prof., 1966-70, assoc. prof., 1970-78, prof., 1978—; W.B. King vis. prof. Iowa State U., 1976-77; cons. in field. Contbr. articles to profl. jours. Grantee in field. Mem. Am. Chem. Soc. (Outstanding Service award Wis. sect.). Subspecialties: Physical chemistry; Electron spin resonance. Current work: Electron spin resonance application to spin labeled fluids. Office: Dept Chemistry U Wis Stevens Point WI 54481 Home: 3015 Cherry St Stevens Point WI 54481

LANG, PEARON GORDON, dermatologist, chemosurgeon; b. Washington, Nov. 9, 1943; s. Pearon Gordon and Mary Emma (Cupp) L.; m. Jean B.Wilson, June 4, 1972; children: Sara, David. B.A., U. Va., 1966, M.D., 1970. Diplomate: Am. Bd. Dermatology, Am. Bd. Dermatopathology. Intern Barnes Hosp., St. Louis, 1970-71; resident in dermatology U. Mich. Med. Ctr., Ann Arbor, 1971-74; fellow dept. dermatology George Washington U. Med. Ctr., Washington, 1980; Brittingham fellow in chemosurgery U. Wis. Madison, 1980; practice medicine, Columbia, S.C., 1974-77; clin. instr. dermatology Med. U. S.C., Charleston, 1974-77, asst. prof., 1980—; asst. prof. dermatology Emory U., Atlanta, 1977-80. Mem. steering com. Dockstreet Theatre, Charleston, 1982—. Fellow Am. Acad. Dermatology, Am. Soc. Dermatopathology; mem. Am. Soc. Dermatol. Surgery, Am. Coll. Chemosurgery (assoc.), Soc. for Investigative Dermatology. Presbyterian. Subspecialty: Dermatology. Current work: Research in skin cancer as well as general dermatology; microscopic controlled excision skin cancer. Home: 716 Angus Ct Mount Pleasant SC 29464 Office: Med U SC 171 Ashley Ave Charleston SC 29425

LANG, ROBERT, medical educator, physican; b. Albany, N.Y., Feb. 1, 1944; s. Michael Spencer and Esther (Best) L.; children: Julia Koren, Jodie Alison, Jason MacLean, Simon Garrett. A.B., Rutgers Coll., 1965; M.D., Albany Med. Coll., 1969. Asst. prof. U. Pa., Phila., 1975-76; asst. prof. Yale U. Sch. Medicine, New Haven, 1976-81, assoc. prof., 1981—; assoc. dir. Clin. Research Ctr., New Haven, 1977-80. Mem. Community Youth Devel. Com., Guilford, Conn., 1982-83. Served with USN, 1973-75. Mem. Am. Fedn. Clin. Research, Am. Soc. Bone and Mineral Research, So. Health and Human Values, Alpha Omega Alpha. Democrat. Jewish. Subspecialties: Endocrinology; Internal medicine. Current work: Humanistic aspects of medical education, metabolic bone diseases, disorders of calcium metabolism. Home: 1300 Boston Post Rd #404 Guilford CT 06437 Office: Yale U Sch Medicine 333 Cedar St New Haven CT 06510

LANG, WILLIAM WARNER, physicist; b. Boston, Aug. 9, 1926; s. William Warner and Lilla Gertrude (Wheeler) L.; m. Asta Ingard, Aug. 31, 1954; 1 son, Robert. B.S., Iowa State U., 1946, Ph.D., 1958; M.S., M.I.T., 1949. Acoustical engr. Bolt Beranek and Newman, Inc., Cambridge, Mass., 1949-51; instr. in physics U.S. Naval Postgrad. Sch., Monterey, Calif., 1951-55; cons. engr. E.I. du Pont de Nemours & Co., Wilmington, Del., 1955-57; mem. research staff M.I.T., 1958; physicist IBM, Poughkeepsie, N.Y., 1958—, program mgr. acoustics tech., 1977—. Editor: Designing for Noise Control, 1978. Pres. Noise Control Found., Poughkeepsie, 1975—; adj. prof. physics Vassar Coll., 1978—; chmn. working group Internat. Orgn. Standardization, 1969—; tech. com. 29 Internat. Electrotech. Commn., 1975—. Served with USN, 1944-47, 52. Decorated Meritorious Service medal. Fellow Audio Engring. Soc., IEEE (Group on Audio and Electroacoustics Achievement award 1970, dir. 1970-71), Acoustical Soc. Am.; mem. Nat. Acad. Engring., Inst. Noise Control Engring. (pres. 1978), AAAS. Episcopalian. Club: Rotary (pres. local club 1975-76). Subspecialties: Acoustical engineering; Acoustics. Current work: Noise control engineering. Home: 29 Hornbeck Ridge Poughkeepsie NY 12603 Office: IBM Acoustics Lab C18/704 PO Box 390 Poughkeepsie NY 12602

LANGACKER, PAUL GEORGE, physics educator; b. Evanston, Ill., July 14, 1946; s. George Rollo and Florence Lorraine (Hinesley) L.; m. Irmgard Sieker, June 25, 1983. B.A., MIT, 1968; M.A., U. Calif.-Berkeley, 1969, Ph.D., 1972; M.A. (hon.), U. Pa., 1982. Research assoc. Rockefeller U., N.Y.C., 1972-74, U. Pa., Phila., 1974-75, asst. prof., 1975-81, assoc. prof. dept. physics, 1981—; chmn. 4th Workshop on Grand Unification, Phila., 1983—. Mem. Am. Phys. Soc. Subspecialties: Particle physics; Cosmology. Current work: Theoretical elementary particle physics: weak interactions; grand unification; cosmology. Home: 4646 Larchwood Ave Philadelphia PA 19143 Office: Dept Physics Univ Pa Philadelphia PA 19104

LANGE, CHRISTOPHER STEPHEN, radiobiology educator, researcher; b. Chgo., Feb. 11, 1940; s. Oscar R. and Irene A. (Oderfeld) L.; m. Kathleen Gale Johnson, June 1964 (div. 1971); 1 dau., Tamara Alice Merry; m. Eleanor Esther Gitlin, Sept. 29, 1973; 1 son, Theodore Oskar. B.S. in Physics, MIT, 1961; D.Phil., U. Oxford, Eng., 1968. Research officer, then sr. research officer Paterson Labs., Christie Hosp. and Holt Radium Inst., Manchester, Eng., 1962-69; asst. prof. radiation biology and biophysics, asst. prof. radiology U. Rochester (N.Y.) Sch. Medicine and Dentistry, 1969-80; prof., dir. radiobiology div. SUNY Downstate Med. Ctr., Bklyn., 1980—; vis. prof. Chemistry U. Calif.-San Diego, 1975-76; cons. SUNY-Buffalo, U. Tenn. Ctr. Health Sci., FDA. Author: The Mechanism of the Radiation Effect at the Cellular Level, 1968; contbr. chpts. to books, articles to profl. jours. Democratic committeeman, Brighton, N.Y., 1972-75 Democratic committeeman, Rochester, 1973-75; chmn., mem. coms. Religious Soc. of Friends, Rochester, 1971-80. Recipient cert. of gratitude Pres. U. Hirosaki, Japan, 1979; NIH research career devel. awardee, 1972-77; others. Mem. Assn. Radiation Research, Radiation Research Soc., Biophys. Soc., Am. Soc. Cell Biology, N.Y. Acad. Scis., Sigma Xi, Omicron Delta Epsilon. Democrat. Subspecialties: Biophysics (biology); Genome organization. Current work: Cellular basis of radiation lethality and ageing; electrochemical control of differentiation; molecular basis of cellular radiation lethality; DNA damage and repair mechanisms; organization of DNA in the mammalian chromosome; biophysics of DNA sedimentation and viscoelastometry. Office: 450 Clarkson Ave Brooklyn NY 11203

LANGE, EUGENE ALBERT, metallurgical engineer, cons.; b. Stevens Point, Wis., Oct. 22, 1923; s. Albert Gustav and Linda (Thalacker) L.; m. Lois June, Feb. 8, 1951. B.S. in Chem. Engring, U. Wis., 1945, M.S. in Metall. Engring, 1951. Registered profl. engr., Wis., 1951. Research metallurgist U. Wis., Madison, 1951-53; research engr. Gray Iron Research Inst., Columbus, Ohio, 1953-56; supervising research metallurgist Naval Research Lab., Washington, 1956-80; cons. metall. engring., Washington, 1980—; tchr. short courses on structural integrity tech. Union Coll., Schenectady, U. Wis., Madison. Contbr. numerous articles to profl. jours. Recipient Research Publ. award Naval Research Lab., 1968, Outstanding Performance award USN, 1969, 70. Mem. Am. Soc. Metals, ASME (cert. appreciation 1975), ASTM, Am. Welding Soc. Club: Bolling Air Force Officers. Subspecialties: Fracture mechanics; Metallurgical engineering. Current work: Engring. applications of fracture mechanics tech.; dynamic fracture mechanics; structural integrity; fracture-safe design; criteria for fracture toughness. Patentee dynamic tear tes; developed double cantiliver split-pin displacement gauge, high-strength, non-magnetic alloy for steel castings, fluidity test. Home and Office: 5101 River Rd Bethesda MD 20816

LANGE, ROBERT CARL, diagnostic imaging educator; b. Stoneham, Mass., Aug. 26, 1935; s. Carl Bert and Anne Delores (Kelliher) L.; m. Mary Ann Groeniger, June 27, 1959 (div. June 1979); children: Louis M., Christopher E.; m. Bunny J. Lambert, July 3, 1982. B.S., Northeastern U., 1957; Ph.D., MIT, 1962. Group leader physics Mound Lab., Miamisburg, OH, 1962-69; tech. dir. nuclear medicine Yale-New Haven Hosp., New Haven, 1969—; assoc. prof. diagnostic imaging Yale U. Sch. Medicine, New Haven, 1969—; cons. to various hosps. Author: Nuclear Medicine for Technicians, 1972. Mem. Am. Chem. Soc., Soc. Nuclear Medicine, Am. Phys. Soc., AAAS, N.Y. Acad. Scis. Subspecialties: Nuclear medicine; Nuclear magnetic resonance (biotechnology). Current work: Computer applications in nuclear medicine, nuclear magnetic resonance imaging. Home: 5 Hughes Pl New Haven CT 06511 Office: Yale U Sch Medicine 333 Cedar St New Haven CT 06510

LANGER, ROBERT SAMUEL, biochem. engring. educator, researcher; b. Albany, N.Y., Aug. 29, 1948; s. Robert Samuel and Mary (Swartz) L.; m. Gerda Constance Endemann, June 27, 1982. B.S., Cornell U., 1970; Sc.D., M.I.T., 1974. Research assoc. Children's Hosp., Boston, 1974-77; asst. prof. M.I.T., Cambridge, 1977-81, assoc. prof., 1981—; cons. Merck, Ventrex, Genentech, IMC, 1981—. Contbr. over 80 articles to profl. jours. Recipient Compton award M.I.T., 1973, Outstanding Tchr. award, 1982, Poitras Devel. chair, 1982. Mem. Am. Chem. Soc., Biomed. Engring. Soc., Controlled Release Soc. (bd. govs.), Am. Soc. Artificial Internal Organs, Am. Inst. Chem. Engring. Subspecialties: Biomedical engineering; Biomaterials. Current work: Drug delivery systems, medical applications of enzyme technology, biomaterials, tumor neovascularization. Patentee in field. Home: 46 Greenville St Somerville MA 02143 Office: MIT Dept Biochem Engring Cambridge MA 02139

LANGER, WILLIAM DAVID, physicist; b. N.Y.C., Sept. 28, 1942; s. Eli and Anne (Meislik) L. B.S., N.Y. U., 1964; M.S., Yale U., 1965, Ph.D., 1968. Asso. fellow astrophysics Goddard Inst. Space Studies, NASA, 1968-70; NSF-NATO fellow in physics Niels Bohr Inst., Copenhagen, 1970-71; research asst. prof. astrophysics N.Y. U., 1971-76; asst. prof. U. Pa., 1976-78; assoc. prof. physics and astronomy U. Mass., 1978-80; cons. radio astronomer Bell Telephone Labs., 1976-82; research physicist Princeton (N.J.) Plasma Physics Lab., 1980—. Mem. Am. Phys. Soc., Am. Astron. Soc., Internat. Astron. Union, AAAS. Subspecialties: Plasma physics; Radio and microwave astronomy. Current work: Plasma transport in divertors and edge of tokomaks; atomic and molecular processes in plasmas; dynamical evolution of interstellar clouds, star formation. Office: Princeton U PO Box 451 Princeton NJ 08544

LANGFORD, LARKIN HEMBREE, research farm superintendent; b. Stockton, Mo., Oct. 24, 1919; s. A. Elmer and Effie Jewel (Hembree) L.; m. Eleanor W. Young, Dec. 13, 1947; children: H. Dale, Don A., Sharon K., Pamela J. B.S., U. Mo.-Columbia, 1942, M.A. in Agr, 1947. Instr. San Carlos (Ariz.) Vets. Sch., 1947-49; animal husbandman N.D. Agrl. Coll., Dickinson, 1949-64; supt. North Mo. Ctr., U. Mo., Spickard, 1964—, also assoc. prof. Author articles. Served with field arty. U.S. Army, 1942-46; maj. Res. ret. Decorated Bronze Star, Purple Heart. Mem. Nat. Cattlemen's Assn. Republican. Mem. Christian Ch. (Disciples of Christ). Lodges: Masons; Rotary. Subspecialties: Animal nutrition; Animal physiology. Current work: Supervise research in beef cattle breeding and feeding.

LANGFORD, WILLIAM FINLAY, mathematics educator, researcher; b. Thunder Bay, Ont., Can., Sept. 11, 1943; s. William Everett and Mary Pearl (Finlay) L.; m. Grace Ann Cooper, Aug. 1, 1970; children: Cathena Dionne, Anne Elisabeth. B.S. with honors, Queen's U., Kingston, Ont., 1966; Ph.D., Calif. Inst. Tech., Pasadena, 1971. Ass. prof. McGill U., Montreal, Que., assoc. prof., 1978-82; assoc. prof. math. U. Guelph, Ont., 1982—; research visitor Inst. de Math. et Scis. Physiques, Nice, France, 1979-80; U. Victoria, B.C., Can., summer 1978. Nat. Scis. and Engring. Council Can. grantee, 1971-82; Que.-France travel grantee, Nice, France, 1979-80. Mem. Soc. Indsl. and Applied Math., Can. Math soc., Can. Applied Math. Soc., Am. Math. Soc. Mem. United Ch. of Canada. Subspecialty: Applied mathematics. Current work: Numerical analysis of bifurcation problems; classification and unfoldings of degenerate bifurcations. Office: U Guelph Dept Math & Stats Guelph ON Canada NIG 2W1 Home: 959 Gordon St Guelph ON Canada NIH 6H9

LANGHOFF, PETER WOLFGANG, physicist, theoretical chemist, educator, researcher; b. N.Y.C., Jan. 19, 1937; s. Joachim and Frieda A. (Damm) L.; m. Judith Dianna Perrotta, June 30, 1962; children: Lisa M., Kristen D., Allison K. B.S. in Physics, Hofstra U., 1958, Ph.D., SUNY, Buffalo, 1965. Research physicist Cornell Aero. Labs., Buffalo, 1962-65; research fellow Harvard U., 1967-69; asst. prof. chemistry Ind. U. Bloomington, 1969-73, assoc. prof., 1973-77, prof., 1977—, chmn. program in chem. physics, 1977—; cons. MIT Lincoln Lab., 1967-69, Lawrence Livermore (Calif.) Lab., 1977-83; Inst. fellow in physics Brandeis U., 1961; vis. research scientist dept. chemistry Harvard U., 1970, 1971, Stanford U., 1975, Stanford-Ames faculty fellow dept. aeros. and astronautics, 1976; vis. fellow Joint Inst. for Lab. Astrophysics, Nat. Bur. Standards and U. Colo., Boulder, 1976-77; sr. NRC resident research asso. NASA Ames Research Center, Mountain View, Calif., 1978-79; vis. research scientist Max Planck Inst. for Physics and Astrophysics, Munich, W.Ger., 1980—; vis. fellow theoretical chemistry U. Sydney, Australia, 1981; professeur associé U. Paris, Orsay Cedex, France, 1981; vis. prof. U. B.C., 1983. Contbr. numerous articles to sci. jours. Served to capt. U.S. Army, 1965-67. Decorated Army Commendation medal; NSF grantee, 1980—; Petroleum Research Found. grantee, 1971—; NASA grantee, 1978—. Mem. Am. Phys. Soc. Subspecialties: Theoretical chemistry; Atomic and molecular physics. Current work: Studies of atomic and molecular photoexcitation and ionization. Office: Dept Chemistry Ind U Bloomington IN 47405

LANGLANDS, ROBERT PHELAN, mathematician; b. New Westminster, Can., Oct. 6, 1936; came to U.S., 1960; s. Robert and Kathleen (Phelan) L.; m. Charlotte Lorraine Cheverie, Aug. 13, 1956; children: William, Sarah, Robert, Thomasin. B.A., U. B.C., 1957, M.A., 1958; Ph.D., Yale U., 1960. From instr. to asso. prof. Princeton (N.J.) U., 1960-67; prof. math. Yale U., New Haven, 1968-72, Inst. Advanced Study, Princeton, 1972—. Editor: Annals of Math., 1979—; author: Euler Products, 1971, (with H. Jacquet) Automorphic Forms on GL(2), 1970, On the Functional Equations Satisfied by Eisenstein Series, 1976, Base Change for GL(2), 1980. Recipient Wilbur Lucius Cross medal Yale U., 1975. Fellow Royal Soc. London, Royal Soc. Can.; mem. Am. Math. Soc. (Cole prize 1982), Can. Math. Soc. Subspecialty: Parasitology. Current work: Automorphic forms, group representations, number theory. Office: Inst Advanced Study Princeton NJ 08540

LANGLER, RICHARD FRANCIS, chemistry educator; b. Sarnia, Ont., Can., Oct. 24, 1945; came to U.S., 1981; s. Richard Frederick and Edith Alberta (Richardson) L.; m. Jeanette Reid, Dec. 11, 1982. B.Sc., Dalhousie U., Halifax, N.S., Can., 1967, Ph.D., 1975; M.Sc., U. N.B., Fredericton, 1969. Vis. lectr. U. Toronto, 1974-75; postdoctoral fellow Dalhousie U., 1975-77, spl. lectr., 1977-79; asst. prof. chemistry Mt. Allison U., Sackville, N.B., Can., 1979-80; asst. prof. St. Mary's U., Halifax, N.S., 1980-81, Fla. Inst. Tech., Melbourne, 1981—. Mem. Chem. Inst. Can., Am. Chem. Soc., Am. Sci. Glassblowers Soc. Subspecialties: Organic chemistry; Synthetic chemistry. Current work: Regiochemistry of oxidation reactions of organosulfur compounds. Office: Fla Inst Tech 150 W University Blvd Melbourne FL 32901

LANGLEY, RICHARD HOWARD, chemistry educator, researcher; b. Richmond, Ind., Aug. 27, 1948; s. Horace Wilson and Norma Virginia (Stewart) L.; m. Maureen Barbara O'Reilly, June 2, 1979; m. Mary Patricia Gaston, June 12, 1971 (div. Dec. 1975). B.S., Miami U., Oxford, Ohio, 1971, 1973, M.S., 1976; Ph.D., U. Nebr., 1977. Postdoctoral fellow Ariz. State U., Tempe, 1977-79; asst. prof. U. Wis.-River Falls, 1979-82, Stephen F. Austin State U., Nacogdoches, Tex., 1982—. Contbr. articles in field to profl. jours. U. Wis.-River Falls faculty research grantee, 1979, 80; State Wis. Instructional Improvement grantee, 1981. Mem. Am. Chem. Soc., Mineral. Soc. Am., AAAS, Soc. Coll. Sci. Tchrs., Sigma Xi. Republican. Mem. Ch. of Christ. Subspecialties: Solid state chemistry; Inorganic chemistry. Current work: Synthesis of fluoride containing solids, synthesis of unusual oxidation state metals, chemical education. Home: 502 Shady Ln Nacogdoches TX 75961 Office: Stephen F Austin State U Box 13006 SFA Sta Nacogdoches TX 75962

LANGNER, CARL GOTTLIEB, mechanical and research engineer; b. Brownsville, Tex., Nov. 30, 1938; s. Charles August Gottlieb and Marie (Dallmeyer) L.; m. Peggy Ann Griffis, Oct. 12, 1968; children: Margaret Ann, Marilee Suzanne. B.S. in Mech. Engring, U. Tex.-Austin, 1960, M.S., Rice U., 1963; Ph.D. in Engring. Mechanics, U. Fla., 1973. Registered profl. engr., Tex. Research engr. S.W. Research Inst., San Antonio, 1962-65; staff research engr. Shell Devel. Co., Houston, 1966-70, 73—. NSF scholar Johns Hopkins U., 1965-66. Mem. ASME. Subspecialties: Solid mechanics; Mechanical engineering. Current work: Development of technology to construct and maintain pipelines and flowline bundles in the deep ocean. Patentee in field. Home: 6702 Spring Leaf Dr Spring TX 77379 Office: PO Box 1380 Houston TX 77001

LANGRIDGE, ROBERT, educator, scientist; b. Essex, Eng., Oct. 26, 1933; came to U.S., 1957; s. Charles and Winifred (Lister) L.; m. Ruth Gottlieb, June 26, 1960; children: Elizabeth, Catherine, Suzanne. B.Sc. in Physics (1st class honours), U. London, Eng., 1954, Ph.D. in Crystallography, 1957. Vis. research fellow biophysics Yale, 1957-59; research asso. biophysics M.I.T., 1959-61; research asso. pathology Children's Cancer Research Found., Boston; research asso. biophysics, lectr. biophysics, also tutor biochem. scis. Harvard, 1961-66; research asso. Project MAC, Lab. for Computer Sci., M.I.T., 1964-66; prof. biophysics and info. scis. U. Chgo., 1966-68; prof. chemistry and biochem. scis. Princeton, 1968-76; prof. pharm. chemistry, biochemistry and biophysics, med. info. scis., dir. Computer Graphics Lab., U. Calif., San Francisco, 1976—; mem. organizing com. 1969 Internat. Biophysics Congress; mem. computer and biomath. research study sect. NIH, USPHS, 1968-72, chmn., 1975-77; mem. vis. com. biology dept. Brookhaven Nat. Lab., 1977-80, mem. adv. com. neutron diffraction, biology dept., 1980—; cons. NIH, NSF, Nat. Acad. Sci., AEC, ERDA.; sr. research collaborator heuristic programming project Stanford U., 1979—, vis. prof. dept. computer sci., 1983-84. Mem. editorial bd.: Handbook of Biochemistry and Molecular Biology, 1973—; Contbr. articles on molecular biology, x-ray diffraction and computer graphics to profl. jours. Guggenheim fellow, 1983-84. Mem. Am. Soc. Biol. Chemists, Am. Chem. Soc., Biophys. Soc. (editorial bd. 1970-73, council 1971-74), Assn. Computing Machinery, Sigma Xi. Subspecialties: Molecular biology; Graphics, image processing, and pattern recognition. Office: 926 Med Scis Bldg U Calif San Francisco CA 94143

LANGSTON, MICHAEL ALLEN, computer science educator; b. Glen Rose, Tex., Apr. 21, 1950; s. Allen Deliphus and Wanda Hepsi (Gray) L.; m. Ina Marie Stedham, May 18, 1975; 1 son, Glen. B.S., Tex. A&M U., 1972; Ph.D., 1981; M.S., Syracuse U., 1975. Instr. overseas div. U. Md., Stuttgart, W.Ger., 1975-78; lectr. Tex. A&M U., 1978-81; asst. prof. computer sci. Wash. State U., Pullman, 1981—; reviewer NSF, 1981—. Contbr. articles to profl. jours. Served to capt. U.S. Army, 1972-78. Gulf Found. fellow, 1979. Mem. Assn. Computing Machinery (reviewer 1981—), Soc. Indsl. and Applied Math. (referee 1982—), IEEE, Ops. Research Soc. Am. (referee 1981—), Phi Eta Sigma, Phi Kappa Phi, Pi Mu Epsilon, Upsilon Pi Epsilon. Subspecialties: Algorithms; Theoretical computer science. Current work: Design and analysis of computer algorithms, combinatorial optimization, operations research, computational complexity, scheduling and allocation techniques. Home: NW 135 Thomas Pullman WA 99163 Office: Dept Computer Sci Wash State U Pullman WA 99164

LANGWAY, CHESTER CHARLES, JR., geology educator; b. Worcester, Mass., Aug. 15, 1929; s. Chester Charles and Agnes Hedwig (McLush) L.; m.; children: Nancy, Mary, JoAnn, Thomas. A.B., Boston U., 1955, M.A., 1956; Ph.D., U. Mich.-Ann Arbor, 1965. Research geologist-glaciologist U.S.A. SIPRE, Wilmette, Ill., 1956-59, U.S.A. CRREL, Hanover, N.H., 1961-65, chief snow and ice br., research div., 1966-77; prof., chmn. dept. geol. sci. SUNY-Buffalo, Amherst, 1975—; cons. Ecology and Environment, Inc., Buffalo, 1981. Served with U.S. Army, 1946-48; Served with USAF, 1948-52. Recipient cert. of achievement U.S.A. CRREL, 1978. Fellow Geol. Soc. Am., Arctic Inst. N.Am.; mem. Internat. Glaciological Soc., Am. Geophys. Union, AAAS, Collegium Disting. Alumni Boston U., Phi Kappa Phi. Club: Cosmos (Washington). Lodge: Elks. Subspecialties: Geology; Climatology. Current work: coordinating and conducting joint international glaciological research program on Greenland Ice Sheet; shallow, intermediate and deep ice core acquisition, investigations involving physical, mechanical, and chem. analysis of glacier snow and ice core samples for paleoclimatic, geochemical and geophysical records. Office: Dept Geol Scis SUNY-Buffalo 4240 Ridge Lea Rd Amherst NY 14226

LANING, DAVID BRUCE, nuclear engineer, consultant; b. Milw., Sept. 18, 1953; s. Bruce Ivo and Patricia Jane (O'Brien) L.; m. Ana Maria Sarosiek, Aug. 5, 1978; children: George, Peter-David. B.S. with honors in Engring. Physics, U. Wis.-Madison, 1975; S.M. in Nuclear Engring., MIT, 1978, Sc.D., 1980, S.M. in Mgmt., 1983. Research and devel. project mgr. C.S. Draper Lab., Cambridge, Mass., 1980-83; mgr. new product devel. GA Technologies Inc., San Diego, 1983—; cons. DBL Diagnostics Co., Cambridge. Eugene Wigner fellow, 1979; NIH fellow, 1977; Whitaker Health Sci. fellow, 1978; C.S. Draper fellow, 1982. Mem. Am. Phys. Soc., Am. Nuclear Soc., IEEE, AAAS. Roman Catholic. Club: Rotary. Subspecialties: Nuclear fission; CAT scan. Current work: Nuclear reactor instrumentation design and biomedical instrumentation design. Office: GA Technologies 10955 John Jay Hopkins Dr San Diego CA 92138

LANKS, KARL WILLIAM, pathology educator; b. Phila., Nov. 1, 1942; s. Gustav W. and Elizabeth E. (Reutschler) L.; m. Nena W. Chin, Apr. 16, 1979; children: Paul, Charles, Cristina, Claire, Belinda. B.S., Pa. State U., 1963; M.D., Temple U., 1967; Ph.D., Columbia U., 1971. Diplomate: Am. Bd. Pathology. Research fellow dept. chemistry Harvard U., Cambridge, Mass., 1968-70; instr. dept. pathology Columbia U., N.Y.C., 1970-72; asst. prof. pathology SUNY-Bklyn., 1975-78, assoc. prof., 1978—. Contbr.; articles to profl. jours. Served as maj. U.S. Army, 1972-74. Mem. Am. Soc. Cell Biology, Am. Soc. Exptl. Pathology, Biophys. Soc. Am., Soc. Biol. Chemists. Subspecialties: Pathology (medicine); Cell biology (medicine). Current work: Teaching and research in the field of experimental pathology. Office: SUNY-Bklyn Downstate Med Ctr 450 Clarkson Ave Box 25 Brooklyn NY 11203

LANMAN, ROBERT CHARLES, pharmacology/toxicology educator, consultant; b. Bemidji, Minn., Oct. 2, 1930; s. Thomas Bradford and Inga Othelia (Engen) L.; m. Dorothy Ann Desnoyers, Dec. 16, 1937; children: Michael Bradford, Dianne Marie, Douglas Robert, Krista Ann. B.S. in Pharmacy, U. Minn., 1956, Ph.D. in Pharmacology, 1967. Pharmacologist NIH, Bethesda, Md., 1962-66; asst. prof. pharmacology U. Mo.-Kansas City, 1966-72, assoc. prof., 1972-81, prof., 1981—; cons. Cancer Research Ctr., Columbia, Mo., 1978-80, Marion Labs., Inc., Kansas City, Mo., 1980—. Served with USN, 1948-52. Named Prof. of Yr. U. Mo-Kansas City, 1970, 74; Amoco teaching award, 1974; recipient cert. of appreciation Rho Chi, 1982. Mem. Am. Soc. Pharmacology and Exptl Therapeutics, Am. Assn. Colls. Pharmacy, Am. Pharm. Assn., Acad. Pharm. Scis., Fedn. Am. Socs. Exptl. Biology, N.Y. Acad. Scis., Kappa Psi. Subspecialties: Pharmacology; Toxicology (medicine). Current work: Kinetics of the absorption, distribution, metabolism and excretion of therapeutic and toxic substances. Home: 8202 W 72nd St Overland Park KS 66204 Office: Univ Mo-Kansas City 2411 Holmes St M3-123 Kansas City MO 64108

LANNING, HOWARD HUGH, astronomer; b. El Centro, Calif., May 26, 1946; s. James Clyde and Ethel Mary (Malan) L.; m. Sheryl Marie Falgout. A.A. in Astronomy, Imperial Valley Jr. Coll., 1966, M.S., San Diego State U., 1969, 1974. Research asst. Hale Obs., Pasadena, Calif., 1970-72; grad. teaching asst. San Diego State U., 1972-73; research asst. NASA Ames Research Center at San Diego State U., 1973-74; night asst., astronomer Mt. Wilson Obs., 1974—; sci. writer Calipatria Herald, Mt. Wilson Obs. Assn. Contbr. articles to profl. jours. Mem. Am. Astron. Soc., Royal Astron. Soc., Astron. Soc. Pacific. Subspecialty: Optical astronomy. Current work: Observation and discovery of variable stars including photometric and spectroscopic studies of white dwarf binaries, cataclysmic variables and novae. Office: Mt Wilson Obs Mount Wilson CA 91023

LANOUX, SIGRED, chemistry educator; b. New Orleans, Nov. 1, 1931; s. Nelson J. and Lucille (Speyrer) L.; m. Emily A. Broussard, Jan. 9, 1954; children: Yvonne Marie, Jeannine Marie. B.S., U. Southwestern La., Lafayette., 1957; Ph.D., Tulane U., 1962. Research assoc. U. Ill., Urbana, 1961-62; research chemist E.I. Dupont de Nemours, Chattanooga, 1962-66; asst. prof. U. Southwestern La., 1966-72, assoc. prof., 1972-74, dept. head, 1972—, prof., 1974—. Contbr. articles to chemistry jours. Bd. dirs. Cathedral-Carmel Sch., 1972-75; v.p. Bayou Council Girl Scouts U.S.A., 1974-76, pres., 1976-81. Served to sgt. U.S. Army, 1951-53. Mem. Am. Chem. Soc., La. Acad. Sci. (chmn. chemistry sect. 1968-70, dir. phys. sci. div. 1968-72, pres. 1972-73), Am. Inst. Chemists. Democrat. Roman Catholic. Subspecialty: Inorganic chemistry. Current work: Synthesis of novel phosphazenes. Patentee polyamide stabilizers, fabric flame retardants. Home: 104 Ridgewood St Lafayette LA 70506 Office: U Southwestern La PO Box 44370 Lafayette LA 70504

LANPHERE, MARVIN ALDER, geologist; b. Spokane, Wash., Sept. 29, 1933; s. P. Dale and Elsie Christine (Alder) L.; m. Joyce Elvia Brown, July 1, 1961; children: Christine, Darcy, Andrew. B.S., Mont. Sch. Mines, 1955; M.S., Calif. Inst. Tech., 1956, Ph.D., 1962. Registered geologist, Calif. Research fellow Calif. Inst. Tech., Pasadena, 1961-62; geologist U.S. Geol. Survey, Menlo Park, Calif., 1962-67, 69-75, dept. asst. chief geologist, Washington, 1967-69, geologist, Menlo Park, 1976—; vis. fellow Australian Nat. U., Canberra, 1975-76. Author: K-AR Dating, 1969. NSF fellow, 1958-61; Penrose grantee Geol. Soc. Am., 1958. Fellow Geol. Soc. Am.; mem. Am. Geophys. Union. Republican. Presbyterian. Subspecialties: Geochemistry; Geology. Current work: Geochronology (isotopic age measurements) and isotope tracer studies applied to petrology. Office: US Geol Survey Mail Stop 37 345 Middlefield Rd Menlo Park CA 94025 Home: 1036 Oakland Ave Menlo Park CA 94025

LANSDELL, HERBERT CHARLES, neurosci. adminstr.; b. Montreal, Que., Can., Dec. 22, 1922; came to U.S., 1954, naturalized, 1961; s. Archibald and Emma Maude (Leonard) L.; m. Judith Purnell, Oct. 5, 1963 (sep.); children: Grant, Bret. B.Sc., Sir George Williams Evening Coll., Montreal, 1944; Ph.D., McGill U., 1950. Asst. prof. McGill U., Montreal, 1949-50; def. service sci. officer Def. Research Med. Lab., Toronto, Ont., Can., 1950-54; asst. prof. U. Buffalo, 1954-58; research psychologist Nat. Inst. Neurol. and Communicative Disorders and Stroke, Bethesda, Md., 1958-70, health scientist adminstr., 1970—. Editor: Physiol. and Comparative Psychology Newsletter, 1979—; contbr. articles on human brain function, including sex differences to profl. jours. Served with Royal Can. Navy, 1944-45. Fellow Am. Psychol. Assn.; mem. Acad. Aphasia, AAAS, Eastern Psychol. Assn., Internat. Brain Research Orgn., Internat. Neuropsychol. Soc., Psychometric Soc., Soc. for Neurosci., Soc. for Philosophy and Psychology, Sigma Xi. Democrat. Subspecialties: Neuropsychology; Physiological psychology. Current work: Administration of grants and contracts in neuroscience; analysis and publication of neuropsychological data. Designer more legible numbers design. Office: NIH Room 916 Fed Bldg Bethesda MD 20205

LANTZ, KEITH ALLEN, computer scientist, educator; b. Chehalis, Wash., Aug. 21, 1952; s. Earle W. and Rubye (Johnson) L.; m. Karen C. Ciccariello., Mar. 20, 1982; 1 dau., Tiffany. B.S., Wash. State U. (1975), Pullman; M.S., U. Rochester, 1977, Ph.D., 1980. Teaching and research asst. Wash. State U., 1974-75; systems programmer SRI Internat., Menlo Park, Calif., 1977, cons., 1979-80; teaching and research asst. U. Rochester, 1975-79, research assoc., 1979-80; asst. prof. computer sci. Stanford U., 1980—; cons. IBM, Xerox, Olivetti, 1979—; mem. Internet Research Group, 1979—. Phi Kappa Phi fellow, 1975-76. Mem. Assn. Computing Machinery, IEEE. Subspecialties: Distributed systems and networks; Operating systems. Current work: Architecture and implementation of distributed operating systems; network graphics; user interfaces; partitionable computer systems. Office: Computer Systems Lab Stanford Univ Stanford CA 94305

LANZANO, PAOLO, physical scientist, researcher; b. Cairo, Nov. 29, 1923; came to U.S., 1949, naturalized, 1955; s. Giuseppe and Anna (Battaglia) L.; m. Bernadine Clare Law, Feb. 2, 1957. B.S., U. Rome, 1944, Ph.D., 1947. Assoc. prof. St. Louis U., 1949-56; design specialist Douglas Aircraft, El Segundo, Calif., 1956-58; sr. staff mem. Space Tech. Lab., El Segundo, 1958-61; space scientist N.Am. Aviation, Downey, Calif., 1961-72; sr. research scientist Naval Research Lab., Wshington, 1972—. Author: Deformations of an Elastic Earth, 1982; contbr. numerous articles to profl. jours. Fellow AIAA (assoc.); mem. Am. Geophys. Union, Soc. Indsl. and Applied Math., Am. Math. Soc., Sigma Xi. Republican. Roman Catholic. Subspecialties: Geophysics; Astronautics. Current work: Geodynamics, earth tides, ocean tides, polar motion, celestial mechanics. Home: 8614 Pappas Way Annandale VA 22003 Office: Naval Research Lab Code 5110 Washington DC 20375

LANZEROTTI, LOUIS J., physicist, educator; b. Carlinville, Ill., Apr. 16, 1938; m. Mary Yvonne DeWolf, June 19, 1965; children: Mary, Louis. B.S., U. Ill., 1965; A.M., Harvard U., 1963, Ph.D., 1965. Physicist, Bell Telephone Labs., Murray Hill, N.J., 1965—; adj. prof. U. Fla., Gainesville, 1978—; mem. adv. bd. Geophys. Inst., U. Alaska, Fairbanks, Mac Planck Institut fur Aeronomie W.Ger. Co-author: Particle Diffusion in the Radiation Belts; co-editor: Solar System Plasma Physics, Upper Atmosphere Research in Antarctica; author articles. Mem. Harding Twp. Sch. Bd., 1982—. Fellow Am. Phys. Soc., AAAS; mem. Am. Geophys. Union, Assn. Geomagnetism and Geoelectricity Japan, IEEE. Subspecialties: Satellite studies; Geophysics. Current work: Magnetospheres of earth and planets; interplanetary medium; planetary environments; geomagnetic depth sounding. Office: 600 Mountain Ave Murray Hill NJ 07974

LANZKOWSKY, PHILIP, physician, educator; b. Cape Town, South Africa, Mar. 17, 1932; came to U.S., 1965, naturalized, 1974; m. Rhona Chiat, Dec. 4, 1965; children: Shelley, David Roy, Leora, Jonathan, Marc. M.B., Ch.B., U. Cape Town, 1954, M.D., 1959. Diplomate: Child Health Eng. Intern Groote Schuur Hosp., U. Cape Town, 1955-56; resident Red Cross Children's Hosp., U. Cape Town, 1957-60; fellow in pediatric hematology Duke U., 1961-62, U. Utah, Salt Lake City, 1962-63; dir. pediatric hematology N.Y. Hosp.-Cornell U. Med. Ctr., 1965-70; asst. prof., assoc. prof. pediatrics Cornell U. Med. Sch., 1965-70; chmn. pediatrics, chief pediatric hematology-oncology L.I. Jewish-Hillside Med. Ctr., New Hyde Park, N.Y., 1970—; chief of staff Children's Hosp., 1983—; prof. pediatrics SUNY Med. Sch., Stony Brook, 1970—; cons. pediatrics Nassau County Med. Ctr., East Meadows, N.Y., 1970—, Cath. Med. Ctr., Queens, N.Y., Peninsula Hosp. Ctr., Far Rockaway, N.Y., Jamaica, N.Y. Hosp., Queens, St. John's Episcopal Hosp., Far Rockaway, Hillcrest Hosp., Queens, Health Ins. Plan N.Y. Author: Pediatric Hematology-Oncology: A Treatise for the Clinician, 1980, Pediatric Oncology: A Treatise for the Clinician, 1983; contbr. articles to med. jours. Mem. med. adv. bd. L.I. chpt. Leukemia Soc. Am., Met. chpt. Hemophilia Soc. Am., St. Mary's Hosp. for Children, Bayside, N.Y., L.I. chpt., Nat. Found. Sudden Infant Death. Cecil John Adams Meml. traveling fellow, 1960; Hill-Pattison-Struthers bursar, 1960; Nutrition Found. grantee, 1968. Fellow Royal Coll. Physicians Edinburgh, Am. Acad. Pediatrics; mem. Soc. Pediatric Research, Am. Pediatric Soc., Harvey Soc., Am. Hematology Soc., Am. Soc. Clin. Oncology, Am. Assn. Cancer Research, Am. Council Emigres in the Professions. Subspecialties: Pediatrics; Hematology. Home: 159 W Shore Rd Great

Neck NY 11024 Office: LI Jewish-Hillside Med Ctr New Hyde Park NY 11040

LANZKRON, ROLF WOLFGANG, electrical engineer, manufacturing company executive; b. Hamburg, West Germany, Dec. 9, 1929; m. Amy Virginia Yarri, Mar 3, 1961; children: Paul, Sophie and Lisa. B.S. in Elec. Engring, Milw. Sch. Engring., 1953, Ph.D., U. Wis.-Madison, 1955, 1956. Chief systems integration br. Martin-Marietta Co., Balt., 1960-62; div. chief Apollo program NASA, Houston, 1962-68; program mgr. Computer Display Channel, Raytheon Corp., Sudbury, Mass., 1968-73, 1973-78, mgr. graphic systems, 1978-79, program mgr. Washington Post/Rayedit, 1979-80, mgr. graphic systems, 1980-82, mgr. radar displays, 1982, dep. dir. air traffic control, 1973—. Mem. Am. Rocket Soc., Inst. Radio Engring. Subspecialty: Electrical engineering. Current work: Advanced technology, computers, radar display and graphics. Home: 35 Gardner Rd Brookline MA 02146 Office: Raytheon Co 528 Boston Post Rd Box 4-1-650 Sudbury MA 01776

LAPATOVICH, WALTER PETER, physicist; b. Shenandoah, Pa., July 12, 1953; s. Walter Joseph and Martha Theresa L.; m. Penny Louise, Apr. 24, 1982; 1 dau., Erika. S.B., M.I.T., 1975, Ph.D., 1980. Research asst. dept. physics and research lab. electronics M.I.T., Cambridge, 1975-80, teaching asst. dept. physics, 1974-75; mem. tech. staff GTE Labs., Inc., Waltham, Mass., 1980—. Contbr. writings to profl. publs. in field. Mem. Am. Phys. Soc. (div. Electron and Atomic Physics), Sigma Xi. Subspecialties: Atomic and molecular physics; Spectroscopy. Current work: Molecular spectroscopy of plasmas and vapors, plasma discharges, spectroscopy, molecular discharges. Office: GTE Labs Inc 40 Sylvan Rd Waltham MA 02254

LAPICKI, GREGORY, physicist, educator; b. Warsaw, Poland, Feb. 14, 1945; came to U.S., 1967; s. Andrzej and Zofia (Chrzaszczewska) L.; m. Carin Alton; 1 son, Jeremy. Magister Fizyki, Warsaw U., 1967; Ph.D. in Physics, NYU, 1975. Postdoctoral trainee dept. physics NYU, N.Y.C., 1975-77, research scientist Radiation and Solid State Lab., 1977-78; vis. asst. prof. dept. physics Tex. A&M U., College Station, 1979-80; asst. prof. dept. chemistry and physics Northwestern State U., Natchitoches, La., 1980-81; asso. prof. dept. physics East Carolina U., Greenville, N.C., 1981—; participant in research, nuclear div. Oak Ridge Nat. Lab., 1981—. Contbr. 18 articles to profl. jours. Nat. Bur. Standards grantee, 1982-84. Mem. Am. Phys. Soc., Sigma Xi. Subspecialties: Atomic and molecular physics; Theoretical physics. Current work: Inner shell ionization; stopping power problems; theory of penetration of charged particles through matter. Office: Dept Physics East Carolina U Greenville NC 17834

LAPIDUS, MICHEL LAURENT, mathematics educator; b. Casablanca, Morocco, July 4, 1956; came to U.S., 1979; s. Serge and Myriam Gisele (Benathar) L.; m. Odile Ioos, July 5, 1980. M.S., U. Paris, France, 1977; Diplome d'Etudes Apponfondies, 1978; Doctorate Pure Math, 1980. Research assoc. Inst. math. U. Paris, 1978-80; Georges Lurcy fellow U. Calif.-Berkeley, 1979-80; asst. prof. math. U. So. Calif., 1980-82, asst. prof., 1982—. Author: Domaine de Dependance, 1978; Contbr. articles to profl. jours. Mem. Am. Math. Soc., Math. Assn. Am., French Math. Soc. Current work: Linear and nonlinear functional analysis, semigroups of operators, Trotter-Lie formula; Schrodinger Equation, Feynman Integral. Office: U So Calif Dept Math DRB 306 Los Angeles CA 90089

LAPIETRA, JOSEPH RICHARD, chemistry educator; b. N.Y.C., July 20, 1932; s. Richard and Columbia (Sannino) LaP.; m. Barbara Louise Weldele, Aug. 7, 1976. B.A., Marist Coll., 1954; Ph.D., Catholic U., 1961. Tchr. St. Helena High Sch., Bronx, N.Y., 1954-56; mem. faculty Cath. U., Washington, 1960-61, Marist Coll., Poughkeepsie, N.Y., 1961—, prof. chemistry, 1976—, acad. dean, 1969-75. Bd. dirs. Rehab. Programs, Inc., Poughkeepsie, 1969-75, pres., 1973-74. Subspecialties: Physical chemistry. Current work: History and philosophy of science. Office: Dept Chemistry Marist Coll Poughkeepsie NY 12601

LAPLANCHE, LAURINE ANNA, chemistry educator; b. N.Y.C., July 4, 1938; d. Samuel J. and Bertha A. (Hagen) LaP. B.S., U. Md., 1959; Ph.D. in Phys. Chemistry, Mich. State U., 1963. Asst. prof. physics W.Va. State Coll., 1964-65; asst. prof. chemistry No. Ill. U., DeKalb, 1965-72, assoc. prof., 1972—. Contbr. articles to sci. jours. Mem. Am. Chem. Soc., Sigma Xi, Phi Kappa Phi. Subspecialties: Nuclear magnetic resonance (chemistry); Molecular pharmacology. Current work: Teaching and research in physical chemistry, specifically in nuclear magnetic resonance spectroscopy. Office: No Ill U DeKalb IL 60115

LAPORTE, DANIEL D'ARCY, physicist; b. Oakland, Calif., Jan. 26, 1934; s. Rollo Collett and Erma Jean A. (Todd) LaP.; m. Carole M. Morrison, June 22, 1957; children: Kristine, Brock, Shareen. B.S., Lewis and Clark Coll., 1956. With Naval Weapons Ctr., Pasadena, Calif., 1956-57; mem. tech. staff Jet Propulsion Lab., Pasadena, 1957-71, supr.IR Instruments Group, 1961-68, staff physicist, 1968-71; mem. tech. staff Santa Barbara Research Ctr., Goleta, Calif., 1971—, staff physicist, 1971-82, sr. staff physicist, advanced applications, electro-optical instrumentation product line, 1982—; lectr. Santa Barbara Sch. Dist. Gifted and Talented Program; bd. dirs. Santa Barbara Intra-Sch. Sci. Fair; co-investigator Mars atmospheric water vapor detection and mapping expt. Viking Mars Mission, NASA, 1967-77. Named to Outstanding Young Men Am. U.S. Jaycees, 1969; NASA grantee, 1967. Mem. Optical Soc. Am., Soc. Applied Spectroscopy. Republican. Mem. United Churches of Christ. Club: Santa Barbara Sailing. Subspecialties: Remote sensing (atmospheric science); Infrared spectroscopy. Current work: Design of optical instruments for measurements from spacecraft. Office: 75 Coromar Dr Goleta CA 93117

LARAGH, JOHN HENRY, physician, scientist, educator; b. Yonkers, N.Y., Nov. 18, 1924; s. Harry Joseph and Grace Catherine (Coyne) L.; m. Adonia Kennedy, Apr. 28, 1949; children—John Henry, Peter Christian, Robert Sealey; m. Jean E. Sealey, Sept. 22, 1974; children—John Henry, Peter Christian. M.D., Cornell U., 1948. Diplomate: Am. Bd. Internal Medicine. Intern medicine Presbyn. Hosp., N.Y.C., 1948-49, asst. resident, 1949-50; fellow cardiology, trainee Nat.. Heart Inst., 1950-51; research fellow N.Y. Heart Assn., 1951-52; asst. physician Presbyn. Hosp., 1950-54, asst. attending, 1954-61, asso. attending, 1961-69, attending physician, 1969—, pres. elect med. bd., 1972-74; dir. cardiology Delafield Hosp., N.Y.C., 1954-55; mem. faculty Columbia Coll. Phys. and Surg., 1950—, prof. clin. medicine, 1967—, spokesman exec. com. faculty council, 1971-73; Hilda Altschul Master prof. medicine, dir. Cardiovascular Center and Hypertension Center N.Y. Hosp.-Cornell Med. Center, 1975—, chief Cardiology div., 1976—; dir. Hypertension Center and Nephrology div. Columbia-Presbyn. Med. Center, 1971; cons. USPHS, 1964—. Editor-in-chief: Cardiovascular Reviews and Reports; Editor: Hypertension Manual, 1974, Topics in Hypertension, 1980; Editorial bd.: Am. Heart Jour. Mem. policy adv. bd. detection and follow-up program Nat. Heart and Lung Inst., 1971, bd. sci. counselors, 1972-78; Vice chmn. bd. trustees for profl., scientific affairs Presbyn. Hosp., 1974-75; chmn. U.S.A.-USSR Joint Program in Hypertension, 1977. Served with AUS, 1943-46. Recipient Stouffer prize med. research, 1969. Fellow A.C.P., Am. Coll. Cardiology; mem. Am. Heart Assn. (chmn. council high blood pressure research 1968-72), Am. Soc. Clin. Investigation, Assn. Am. Physicians., Am. Soc. Contemporary Medicine and Surgery (adv. bd.), Internat. Soc. Hypertension (sci. council), Harvey Soc., Kappa Sigma, Nu Sigma Nu, Alpha Omega Alpha. Clubs: Winged Foot (Mamaroneck, N.Y.); Shinnecock Hills Golf (Southampton). Subspecialty: Physiology (medicine). Research on hormones and electrolyte metabolism and renal physiology, on mechanisms of edema formation and on the causes and treatment of human high blood pressure. Home: 435 E 70th St New York NY 10021 Office: NY Hosp-Cornell Med Center 525 E 68th St New York NY 10021 In my scientific research, a great resource has been the ability to look at everyday clinical phenomena from a different point of view to develop new ideas and principles about human physiology. These perceptions enable experiments and hypotheses for quantum leaps forward in creation of new knowledge and redirection of medical thinking.

LARDY, HENRY ARNOLD, biological sciences educator; b. Roslyn, S.D., Aug. 19, 1917; s. Nick and Elizabeth (Gebetsreiter) L.; m. Annrita Dresselhuys, Jan. 21, 1943; children: Nick, Diana, Jeffrey, Michael. B.S., S.D. State U., 1939, D.Sc. (hon.), 1979; M.S., U. Wis., 1941, Ph.D., 1943. Mem. faculty U. Wis.-Madison, 1945—, asso. prof. biochemistry, 1947-50, prof., 1950—, Vilas prof. biol. scis., 1966—. Co-editor: The Enzymes. Pres. Citizens vs. McCarthy, 1952. Recipient Paul-Lewis award enzyme chemistry Am. Chem. Soc., 1949, Neuberg medal Am. Soc. European Chemists, 1956, award in arg. Wolf Found., 1981, Nat. award for agrl. excellence, 1982. Mem. Nat. Acad. Sci., Am. Chem. Soc. (chmn. biol. div. 1958), Am. Soc. Biol. Chemists (pres. 1964), Am. Philos. Soc., Golden Retriever Club Am. (pres. 1962), Am. Acad. Arts and Scis., Endocrine Soc., Japanese Biochem. Soc. (hon.). Subspecialties: Behaviorism; Behaviorism. Current work: Sperm metabolism and functions, regulation of gluconeogenecis of hormones. Home: Thorstrand Rd Madison WI 53705

LARGE, H. LEE, JR., pathologist; b. Rocky Mount, N.C., Mar. 1, 1918; s. H. Lee and Nellie Pearl (Brockwell) L.; m. Iris Christine Haynes, Jan. 3, 1941. B.S. in medicine, U. N.C., 1939; M.D., Vanderbilt U., 1942. Cert. pathology and clin. pathology. Intern Vanderbilt Univ. Hosp., Nashville, 1942-43; resident in pathology Med. Coll. Va., Richmond, 1946-47, Charlotte (N.C.) Meml. Hosp., 1947-49; asst. pathologist Charlotte Meml. Hosp., 1949-52; cons. Charlotte ENT Hosp., 1972—; pathologist Presbyn. Hosp., Charlotte, 1952—. Bd. dirs. Am. Cancer Soc. (past pres.), ARC. Served as capt. AUS, 1943-46. Fellow Coll. Am. Pathologists (ho. of dels.), Am. Soc. Clin. Pathologists, Am. Assn. Pathologists and Bacteriologists, Internat. Acad. Pathologists; mem. Am. Soc. Cytology, N.C. Bd. Med. Examiners (pres. 1967, 71). Presbyterian. Lodge: Rotary. Subspecialties: Pathology (medicine); Cytology and histology. Current work: Surgical pathology and cytology. Home: 7443 Windyrush Rd Matthews NC 28105 Office: Presbyn Hosp PO Box 33549 Charlotte NC 28233

LARGEN, JOHN WILLIAM, JR., clinical neuropsychologist; b. Johnson City, Tenn., Nov. 5, 1950; s. John William and Mary (Fouraker) L.; m. Janet Burns, July 25, 1975. B.A., Rutgers Coll., 1973; M.A., Montclair State Coll., 1975; Ph.D., U. Houston, 1980. Cert. in psychology. Clin. neuropsychologist Tex. Research Inst. Mental Scis., Houston, 1980—; pvt. practice neuropsychology, Houston, 1982—; sci. adv. com. Alzheimer's Disease and Related Disorders Assn. Houston, 1979—. Contbr. chpts. to books, articles to jours. Mem. Am. Psychol. Assn., Internat. Neuropsychol. Soc., Gerontol. Soc. Am., Biofeedback Soc. Am. (awards 1978, 79, 80), Biofeedback Soc. Tex. (pres. 1983), Biofeedback Soc. Harris County (pres. 1979-80). Methodist. Subspecialties: Neuropsychology; Biofeedback. Current work: Research, clinical neuropsychology of dementing disorders and schizophrenia, computed tomography and cerebral blood flow research. Home: 3519 Cave Springs Kingwood TX 77339 Office: Texas Research Inst of Mental Scis 1300 Moursund Houston TX 77030

LARGMAN, KENNETH, strategic analyst, strategic defense analysis company executive; b. Phila., Apr. 7, 1949; s. Franklin Spencer and Roselynd Marjorie (Golden) L.; m. Suzanna Forest, Nov. 7, 1970 (div. Nov. 1978); 1 dau., Jezra. Student, SUNY-Old Westbury, 1969-70. Ind. strategic analyst, 1970-77; chmn., chief exec. officer World Security Council, San Francisco, 1980—. Author: research documents Space Peacekeeping, 1978, Preventing Nuclear Conflict: An International Beam Weaponry Agreement, 1979, Space Weaponry: Effects on the International Balance of Power and the Prevention of Nuclear War, 1981, Preventing Nuclear War: Coordinating U.S. and Soviet Space Defense Against Nuclear Attack, 1982; 14 vols. on threats, vulnerabilities and safeguards of coordinated U.S./Soviet Space def. system, 1983. Mem. Air Force Assn., Am. Astron. Soc., AIAA, World Affairs Council. Club: Commonwealth. Subspecialties: Satellite studies; Military space systems. Current work: Use of military space systems to prevent nuclear war; design of space-based laser defense system able to destroy nuclear missiles and bomber attacks, yet leaving opposing nations unable to attack each other or Earth targets with system. Office: World Security Council 275C World Trade Center San Francisco CA 94111

LARIMER, FRANK WILLIAM, geneticist; b. Mt. Pleasant, Mich., Feb. 26, 1948; s. Lee Weldon and Virginia J. (Bell) L.; m. Constance Ann Zendel, Aug. 19, 1972. B.A., Albion Coll., 1971; M.S., Fla. State U., 1973, Ph.D., 1975. Staff scientist biology div. Oak Ridge Nat. Lab., 1976—; instr. U. Tenn.-Oak Ridge Grad. Sch. Biomed. Scis., 1977—. Nuclear Research fellow Fla. State U., 1971; NIH trainee, 1973-75. Mem. Genetics Soc. Am., Am. Soc. Microbiology, Environ. Mutagen Soc. Subspecialties: Genetics and genetic engineering (biology); Molecular biology. Current work: Genetic control of mutagenesis and DNA repair; isolation of eukaryotic DNA repair genes; mechanisms of chemical mutagenesis; repair of chemical lesions in DNA; organization and expression of the ARO1 cluster-gene of yeast; genetic toxicology. Office: Oak Ridge Nat Lab Box Y Oak Ridge TN 37830

LARKIN, EDWARD CHARLES, internal medicine and hematology educator; b. Waltham, Mass., Aug. 7, 1937; s. John Joseph and Mary Elizabeth (McKoan) L.; m. Sally Stillinger, June 19, 1965; children: Michael, Lucy. A.B., Harvard U., 1959; M.D., Yale U., 1963. Diplomate: Am. Bd. Internal Medicine. Intern New Eng. Med. Ctr., Boston, 1963-64, resident in internal medicine, 1964-66; prof. medicine and pathology U. Calif., Davis, 1981—. Served to maj. USAF, 1968-70. Fellow ACP; mem. Am. Soc. Hematology, Internat. Soc. Hematology, Western Soc. Clin. Investigation. Subspecialty: Genetics and genetic engineering (biology). Home: 2560 San Miguel Dr Walnut Creek CA 94596

LARKIN, JOHN MONTAGUE, microbiology educator; b. Phila., Apr. 7, 1936; s. J. Walter and Mary (Montaque) L.; m. Brenda Wynn McDonald, Aug. 18, 1962; children: Laura, Jennifer. B.S., Ariz. State U., 1961, M.S., 1963; Ph.D., Wash. State U., 1967. Asst. prof. microbiology La. State U., Baton Rouge, 1967-71, assoc. prof., 1971-79, prof., 1979—. Served with USMC, 1959. HEW grantee, 1969-71; NSF grantee, 1976-79; U.S. Geol. Survey, 1978-81. Mem. Am. Soc. Microbiology (chmn. gen. microbiology 1969-80, councillor 1980—). Republican. Club: Sci. (pres. 1980-81). Subspecialties: Microbiology; Physiology (biology). Current work: Biology of large sulfur-oxidizing bacteria. Office: Dept Microbiology La State Univ Baton Rouge LA 70803

LAROCCA, JOSEPH PAUL, educator; b. La Junta, Colo., July 5, 1920; s. Vito Michael and Mary (Parlapiano) LaR.; m. Blair Camak, Apr. 19, 1947; children—Carl Anthony, Mary Blair, Charlotte Ann, Elizabeth. B.S., U. Colo., 1942; M.S. (Nat. Formulary fellow), U. N.C., 1944; Ph.D., U. Md., 1948. Research chemist U.S. Naval Research Lab., Washington, 1947-49; asso. prof. medicinal chemistry U. Ga., 1949-50, prof., 1950—, head dept., 1968—; lectr. Oak Ridge Asso. Univs. Mobile Radioisotope Program, 1965, 66, cons. physician manpower USPHS, 1967. Author research papers, reports. Bd. dirs. Am. Assn. Colls. Pharmacy, 1974-76; Chmn. bd. Clark County unit Am. Cancer Soc., 1965-66, pres., 1964-65; chmn. bd. Clark County Community Chest, 1963-64, pres., 1962-63; pres. Athens Assn. Retarded Children, 1960-61. Served with AUS, 1944-46. Gustauus A. Pfeiffer Meml. Research fellow, 1964-65; Orins Research participant, 1964, 65, 66. Fellow A.A.A.S.; mem. Am. Pharm. Soc., Am. Chem. Soc., Ga. Pharm. Assn., Ga. Acad. Sci., Sigma Xi, Phi Kappa Phi, Rho Chi, Phi Delta Chi. Club: Rotarian. Subspecialty: Medicinal chemistry. Home: 115 Fortson Circle Athens GA 30606

LARRABEE, MARTIN GLOVER, educator; b. Boston, Jan. 25, 1910; s. Ralph Clinton and Ada Perkins (Miller) L.; m. Sylvia Kimball, Sept. 10, 1932 (div. 1944); 1 son, Benjamin Larrabee Scherer; m. Barbara Belcher, Mar. 25, 1944; 1 son, David Belcher Larrabee. B.A., Harvard, 1932; Ph.D., U. Pa., 1937; M.D. (hon.), U. Lausanne, Switzerland, 1974. Research asst., fellow U. Pa., Phila., 1934-40, asso. to asso. prof., 1941-49; asst. prof. physiology Cornell U. Med. Coll., N.Y.C., 1940-41; asso. prof., Johns Hopkins, Balt., 1949-63, prof. biophysics, 1963—. Contbr. articles to scientific jours. Mem. Am. Physiol. Soc., Biophys. Soc., Am. Soc. Neurochemistry, Internat. Neurochem. Soc., Nat. Acad. Scis., Soc. for Neurosci. (treas. 1970-75), Physiol. Soc. (asso., Eng.), Phi Beta Kappa. Clubs: Appalachian Mountain, Sierra, Mountain of Md. Subspecialties: Neurophysiology; Neurochemistry. Current work: Research on circulatory, respiratory and nervous systems of animals, especially on synaptic and metabolic mechanisms in sympathetic ganglia. Research on circulatory, respiratory and nervous systems of animals, especially on synaptic and metabolic mechanisms in sympathetic ganglia, 1934—; wartime research on oxygen lack, decompression sickness, nerve injury, infrared viewing devices, 1941-45. Home: 4227 Long Green Rd Glen Arm MD 21057 Office: Biophysics Dept Johns Hopkins Univ Baltimore MD 21218

LARRICK, JAMES WILLIAM, physician, medical researcher; b. Denver, Jan. 4, 1950; s. William Franklin and Louise (Gottschalk) L. B.A. magna cum laude, Colo. Coll., 1972; Ph.D., Duke U., 1979, M.D., 1980. Intern in internal medicine Stanford U. Med. Center, 1980-81, Marie Stauffer Sigall Found. fellow, 1981-82; research scientist, project leader Cetus Immune Research Labs., Palo Alto, Calif., 1982—. Contbr. articles, chpts. to profl. publs. Boettcher Found. fellow, 1968-70; Watson Found. fellow, 1972-73; NIH Med. Scientist Tng. Program fellow, 1973-80. Mem. AAAS, N.Y. Acad. Sci., Am. Soc. Clin. Research, Physicians for Social Responsibility. Subspecialties: Cancer research (medicine); Cellular engineering. Current work: Characterization and cloning of human monoclonal antibodies and lymphokines for therapy of cancer, immunological diseases, hepatitis, cytomegalovirus, gram negative sepsis and other infectious diseases. Home: Star Route Box 48 Woodside CA 94062 Office: Cetus Immune Research Labs 3400 Bayshore W Palo Alto CA 94303

LARRIMORE, JAMES ABBOTT, nuclear engineer; b. San Francisco, Aug. 18, 1934; s. James and Dorothy Elisabeth (Martin) L.; m. Irene Maria Kalbfleisch, Mar. 6, 1965; children: Mark Joseph, Corinne Marilou. B.Eng. Physics, Cornell U., 1957; Ph.D. in Nuclear Engring, M.I.T., 1963. Dir. M.I.T. Engring. Practice Sch., Oak Ridge, 1960-62; asst. prof. nuclear engring. M.I.T., Boston, 1960-63; mem. vis. com. nuclear engring. dept., 1976-78; asst. to head reactor physics dept. Euratom Research Center, Ispra, Italy, 1963-69; br. chief, dept. head Gen. Atomic Co., San Diego, 1969-76, sr. staff adviser, 1979-80; sr. officer div. nuclear power and reactors IAEA, Vienna, Austria, 1976-78; dir. internat. programs GA Techs. Inc., San Diego, 1980—. Ford Found. fellow, 1962-63. Mem. Am. Nuclear Soc., Sigma Xi. Republican. Subspecialties: Nuclear fission; Nuclear engineering. Current work: International program development for high temperature gas-cooled nuclear reactor research and development and commercialization. Home: 14044 Rue San Remo Del Mar CA 92014 Office: GA Technologies Inc PO Box 85608 San Diego CA 92138

LARSEN, AUSTIN ELLIS, veterinarian; b. Provo, Utah, Nov. 1, 1923; s. Ariel Ellis and Vera Alice (Austin) L.; m. Jean Woodward, Sept. 8, 1946 (div. 1974); children: Reed W., Nick Dean, Brent Austin, Donald Allen; m. Joan Ensign Brown, Aug. 20, 1974. D.V.M., Wash. State U., 1949; Ph.D., U. Utah, 1969. Pvt. practice vet. medicine, Salt Lake City, 1949-68; cons. Schering Corp., 1963—; non-med. cons. VA, 1968—; cons. veterinarian Fur Breeders Agrl. Coop., 1968—; assoc. prof. cellular, viral and molecular biology U. Utah, Salt Lake City, 1968—, dir. vivarium, 1968—. Served to sgt. U.S. Army, 1943-46. Mem. AVMA, Intermountain Vet. Assn., Utah Vet. Med. Assn., Fur Farmers Research Inst., Am. Soc. Microbiologists, Am. Soc. for Exptl. Pathology, Am. Soc. Lab. Animal Practitioners, Am. Assn. for Lab. Animal Sci., Sigma Xi. Mormon. Subspecialties: Laboratory Animal Medicine; Microbiology (veterinary medicine). Current work: Laboratory animal science; animal welfare information; slow virus diseases.

LARSEN, EDWARD WILLIAM, mathematician; b. Flushing, N.Y., Nov. 12, 1944; s. James Millan and Marjorie L.; m. Juliette Mary Mainieri, Mar. 23, 1974. B.S. in Math, Rensselaer Poly. Inst., 1966, Ph.D., 1971. Asst. prof. math. N.Y. U., N.Y.C., 1971-76; assoc. prof. math. U. Del., Newark, 1976-77; mem. staff Los Alamos Nat. Lab., 1977—. Mem. editorial bd.: SIAM Jour. Applied Math, 1976-83, Transport Theory and Statis. Physics, 1976—. Mem. Am. Nuclear Soc., Soc. Indsl. and Applied Math. Subspecialties: Applied mathematics; Nuclear fission. Current work: Neutral and charged particle transport theory; numerical methods and acceleration of iterative methods for solving transport problems; asymptotic soluations of transport problems. Home: 155 Piedra Loop Los Alamos NM 87544 Office: Los Alamos Nat Lab PO Box 1663 MS-B226 Los Alamos NM 87434

LARSEN, KENNETH DAVID, medical research scientist; b. Woodburn, Oreg., Oct. 10, 1947; s. Carl Sherman and Doris Lorraine (Lewis) L.; m. Nora Cheng, June 13, 1980; 1 dau., Karina. B.A., Oreg. State U., 1970; M.S., U. N.D., 1973; Ph.D., Emory U., 1976; M.D., U. Miami, 1983. Research assoc. Rockefeller U., N.Y.C., 1977-79, asst. prof., 1979-80; asst. prof. dept. physiology U. Pitts. Sch. Medicine, 1980-81; resident in anesthesiology U. Pa., Phila., 1983—. Contbr. articles to profl. jours. NIH fellow, 1977. Mem. Soc. Neurosci. Subspecialty: Neurophysiology. Current work: Motor system neurophysiology, neurophysiology of mammalian motor cortex, cerebellum and basal ganglia.

LARSEN, KENNETH MARTIN, mathematics educator; b. Ogden, Utah, June 26, 1927; m. Merlee May Smith, Nov. 10, 1955; children—Debra, Joseph, Rebecca, Paul, John, Peter, James, Mark, Rachel. B.A., U. Utah, 1950; M.A., Brigham Young U., 1956; Ph.D., UCLA, 1964.

Prof. math Brigham Young U., Provo, Utah, 1960—. Mem. Am. Math. Soc., Am. Phys. Soc. Republican. Mem. Ch. Jesus Christ of Latter-day Saints. Subspecialties: Applied mathematics; Numerical analysis. Co patentee confinement of high temperature plasmas. Home: 2270 N 300 E Provo UT 84604 Office: Dept Math Brigham Young U Provo UT 84602

LARSON, BRUCE L., biochemistry educator; b. Mpls., June 24, 1927; m., 1954; 3 children. B.S., U. Minn., 1948, Ph.D. in Biochemistry, 1951. Mem. faculty U. Minn., 1948-51; from instr. to assoc. prof. U. Ill., 1951-66, prof. biol. chemistry, 1966—, mem. staff nutrition sci. faculty, 1972—; Fulbright lectr., 1965. Mem. AAAS, Am. Chem. Soc. (award 1966), Am. Soc. Biol. Chemists, Am. Dairy Sci. Assn. Subspecialty: Biochemistry (biology). Office: U Ill Dept Dairy Sci 315 Animal Scis Lab Urbana IL 61801

LARSON, DANIEL JOHN, physicist, educator; b. Mpls., Nov. 8, 1944; s. Edwin W. and Verva (Johnson) L.; m. Beverly Jean Hill, Aug. 13, 1966; children: Emily Jean, Andrew John. B.A. summa cum laude, St. Olaf Coll., 1966; M.A. in Physics, Harvard U., 1967, Ph.D., 1971. Asst. prof. physics Harvard U., 1970-75, assoc. prof., 1975-78; assoc. prof. physics U. Va., 1978—. Contbr. articles to profl. jours. Woodrow Wilson fellow (hon.), 1966; NSF grad. fellow, 1966-70; Nat. Bur. Standards Precision Measurements grantee, 1975-78. Mem. Am. Phys. Soc. (mem. program com. div. electron and atomic physics 1980). Subspecialty: Atomic and molecular physics. Current work: Laser spectroscopy, trapped ions, negative ions; precision measurements, microwave spectroscopy. Home: 505 Ivy Farm Dr Charlottesville VA 22901 Office: Dept Physics U Va Charlottesville VA 22901

LARSON, DENNIS L., agrl. engr. educator; b. Mason City, Iowa, Feb. 3, 1940; s. Vernon C. and Adelaide L. (Wamstad) L.; m. Cheryl A. Larson, June 1, 1963; children: Scott, Kristine, Steven, Kathryn. B.S., Iowa State U., Ames, 1963; M.S., U. Ill., Urbana, 1964; Ph.D., Purdue U., Lafayette, Ind., 1971. Registered profl. engr., Ill, Calif., Ariz. Product engr. John Deere Planter Works, Moline, Ill, 1966-68; asst. prof., tech. advisor in Colombia U. Nebr., Lincoln, 1970-73; food engr. Mich. State U., East Lansing, 1973; engr. Chrysler Def. Products, Detroit, 1973; asst. prof. agrl. engring. U. Ariz., Tucson, 1973-79, assoc. prof., 1979—. Served to 1st lt. AUS, 1964-66. Mem. Am. Soc. Agr. Engrs., Am. Soc. Engring. Edn., Am. Solar Energy Soc., Sigma Xi. Subspecialty: Agricultural engineering. Current work: Energy use in agriculture and energy alternatives, use of solar energy to drive irrigation pumps, agricultural systems analysis.

LARSON, DONALD CLAYTON, physicist, educator, cons.; b. Wadena, Minn., Jan. 29, 1934; s. Clyde Melvin and Selma L. Engen; m. Susan Dunnet, July 17, 1960; children: Tor, Jon, Erika. B.S., U. Wash., 1956; M.S., Harvard U., 1957, Ph.D., 1962. Asst. prof. U. Va., 1962-67; assoc. prof. physics Drexel U., Phila., 1967-83, prof. phsics, 1983—; cons. Mem. Am. Phys. Soc., ASTM, Internat. Solar Energy Soc., Phi Beta Kappa, Sigma Xi, Tau Beta Pi. Democrat. Subspecialties: Condensed matter physics; Semiconductors. Current work: Mirror-boosted solar collectors, field techniques for insulation thermal testing, optical waveguides, semiconductors, optical and electrical properties of films, solar energy, thermal insulation. Office: Dept of Physics and Atmospheric Sci Drexel U Philadelphia PA 19104

LARSON, JOHN LEONARD, applications engineer; b. Champaign, Ill., Feb. 6, 1949; s. Rudolph G. and Bernice M. L.; m. Emmylou Studier, Sept. 2, 1972. B.S. with distinction, U. Ill.-Urbana, 1971, M.S., 1974, Ph.D., 1978. Sr. systems analyst Burroughs Corp., Paoli, Pa., 1978-82; sr. applications engr. Cray Research, Inc., Chippewa Falls, Wis., 1982—; conf. participant. Contbr. articles profl. jours., chpts. in book. Mem. Assn. Computing Machinery, Soc. Indsl. and Applied Mathematicians, IEEE, Am. Math. Soc. Club: Ambler Stamp (v.p.). Subspecialties: Mathematical software; Numerical analysis. Current work: Numerical software development for pipelined and parallel supercomputers, automatic roundoff error analysis. Office: Cray Research Inc Chippewa Falls WI 54729

LARSON, LARRY GALE, electronics/aerospace co. exec.; b. New Effington, S.D., Aug. 26, 1931; s. and Dorothy (Dybdahl) L.; m. Delores J., Nov. 4, 1950; children: Larry Gale, Gregory A., Dawn Marie Larson Blair. B.E.E., U. Minn., 1959; postgrad. in corp. fin, U. Calif., 1959-61. Group supr. Lockhead Missiles & Space Co., Sunnyvale, Calif., 1960-62; sect. mgr. Honeywell, Inc., Mpls., 1962-68; v.p. Honeywell Electro-Optics Ops., Lexington, Mass., 1979—; dir. programs DALMO Victor Co., Belmont, Calif., 1968-69; pres., chief exec. officer EOCOM Corp., Tustin, Calif., 1969-79. Contbr. numerous articles to tech. and trade assn. jours. Mem. Am. Security Council, 1972—, Am. Def. Preparedness Assn., 1973—; chmn. Nat. UN Day Com., 1977. Served with USAF, 1950-54. Recipient cert. of appreciation 1976 Pollution Engring. Congress. Mem. Am. Electronics Assn., Assn. U.S. Army, Soc. Photo-Optical Instrumentation Engrs., IEEE, U. Minn. Alumni. Republican. Subspecialties: Electronics; Information systems (information science). Home: 2 McDonald Circle Andover MA 01810 Office: 2 Forbes Rd Lexington MA 02173

LARSON, (LEONARD) ALLAN, biochemistry researcher; b. Takoma Park, Md., Nov. 7, 1953; s. Leonard Allan and Edna Ruth (Naugle) L. B.S., U. Md., 1975, M.S., 1977; Ph.D., U. Calif.-Berkeley, 1982. Student asst. USPHS Office for Civil Rights, HEW, 1971-75; teaching asst. U. Md., 1975-77; teaching assoc. U. Calif.-Berkeley, 1978-82, research specialist dept. biochemistry, 1982—; researcher Mus. Vertebrate Zoology, 1978—. Contbr. articles to profl. jours. Mem. Genetics Soc. Am., Soc. Study Evolution, Soc. Systematic Zoology. Subspecialties: Evolutionary biology; Genetics and genetic engineering (biology). Current work: Molecular and morphological evolution in vertebrates: protein evolution, genome evolution, morphological evolution, phylogenetic reconstruction, amphibian systematics, electrophoresis, evolutionary theory, population genetics.

LARSON, PHILIP RODNEY, research plant physiologist; b. N. Branch, Minn., Nov. 26, 1923; s. Eric Gunnar and Anna Ruth (Fagerstrom) L.; m. Yvonne Evelyn Sybrant, Sept. 3, 1948; children: Cynthia Marie, Paula Rae. B.S., U. Minn., 1949; M.S., 1952; Ph.D., Yale U., 1957. Research forester Southeastern Forest Exptl. Sta., Lake City, Fla., 1952-56; plant physiologist North Central Forest Exptl. Sta., Rhinelander, Wis., 1957—, leader pioneering research unit. Contbr. numerous articles to profl. jours. Served to lt. USNR, 1942-46. Recipient Disting. Service award U.S. Dept. Agr., 1975; N.Y. Bot. Garden Research award, 1977. Fellow Internat. Acad. Wood Sci.; mem. Soc. Am. Foresters (Barrington Moore award 1975), Bot. Soc. Am., Am. Soc. Plant Physiologists. Subspecialties: Plant physiology (agriculture); Plant growth. Current work: Physiology of wood formation, structure and function of the vascular system of trees, relation of primary to secondary vasculature in all aerial organs of trees.

LARSON, RICHARD BONDO, astronomy educator; b. Toronto, Ont., Can., Jan. 15, 1941; came to U.S., 1963; s. Carl Johan and Elsie (Bondo) L. B.Sc., U. Toronto, 1962, M.A., 1963; Ph.D., Calif. Inst. Tech., 1968. Asst. prof. astronomy Yale U., New Haven, 1968-73, assoc. prof., 1973-75, prof., 1975—, chmn. dept. astronomy, 1981—. Contbr. articles to profl. jours. Mem. Am. Astron. Soc., Royal Astron. Soc., Internat. Astron. Union, Sigma Xi. Subspecialty: Theoretical astrophysics. Current work: Research on formation of stars, galaxies. Address: Dept Astronomy Yale U PO Box 6666 New Haven CT 06511

LARSON, RICHARD GUSTAVUS, mathematics educator, consultant; b. Pitts., May 16, 1940; s. Gustavus and Anna Lisa (Nelson) L. B.S., U. Pa., 1961; M.S., U. Chgo., 1962, Ph.D., 1965. Instr. M.I.T., 1965-67; asst. prof. math. U. Ill.-Chgo., 1967-70, assoc. prof., 1970-77, prof., 1977—. Contbr. articles to profl. jours. Mem. Am. Math. Soc., Soc. Indsl. and Applied Math., Assn. Computing Machinery, IEEE Computer Soc. Subspecialties: Mathematical software; Theoretical computer science. Current work: Applications of computers to mathematical problems; algorithms for prime factorization. Office: Dept Math Statistics and Computer Sci U Ill-Chgo PO Box 4348 Chicago IL 60680

LARSON, STEVEN MARK, physician, researcher; b. Tacoma, Wash., Nov. 30, 1941; s. Louis Edward Emanuel and Evelyn Agusta (Peterson) L.; m. Elaine L. Williamson, June 14, 1964; children: Justine L., Nathan P. B.A., U. Wash., 1963, M.D., 1968. Diplomate: Am. Bd. Internal Medicine, Am. Bd. Nuclear Medicine. Intern Virginia Mason Hosp., Seattle, 1968-69, resident in medicine, 1969-70; asst. prof. radiology Johns Hopkins Med. Inst., Balt., 1972-75; assoc. prof. med. pathology U. Oreg., 1975-76; assoc. prof. medicine, radiology U. Wash. Med. Sch., Seattle, 1976-81, prof. medicine, 1981—; chief nuclear medicine VA Hosp., Portland, Oreg., 1975-76, asst. chief nuclear medicine, Seattle, 1976-81, chief nuclear medicine, 1981—. Contbr. numerous articles to profl. jours., chpts. to books. Served with USNR, USPHS, 1970-72. Fellow Am. Coll. Nuclear Physicians (regent 1975-76); mem. Soc. Nuclear Medicine (trustee 1976-80). Presbyterian. Subspecialties: Internal medicine; Nuclear medicine. Current work: Diagnostic and therapy of cancer with radioactive drugs. Home: 6048 Princeton Ave NE Seattle WA 98115 Office: VA Med Ctr 4435 Beacon Ave S Seattle WA 98108

LARSON, VIVIAN M., virologist, researcher; b. Erie, N.D., Oct. 3, 1931; d. Orlando C. and Alice M. (Port) L. B.S., N.D. State Coll., 1953; M.P.H., U. Mich., 1958; Ph.D., 1963. Cert. med. technologist Am. Soc. Clin. Pathology. Bacteriologist Detroit Dept. Health Labs., 1953-59; research fellow Merck Sharp & Dohme Research Labs., West Point, Pa., 1963-69, Sr. research fellow, 1969-71, dir. Herpes virus research, 1980—; dir. NIH Virus Lab., 1971-80. Contbr. articles on bacteriology, virology, cancer immunology and viral vaccine devel. to sci. jours. Mem. Am. Soc. for Microbiology, Am. Acad. Microbiology, AAAS, N.Y. Acad. Scis., Phi Kappa Phi, Delta Omega. Lutheran. Subspecialties: Virology (medicine); Infectious diseases. Current work: Research on the development of herpes virus vaccines for human use by genetic engineering. Patentee Herpes subunit vaccine. Home: 362 Park Dr Harleysville PA 19438 Office: Merck Sharp & Dohme Research Labs West Point PA 19486

LARUE, DAVID KNIGHT, geology educator; b. Oakland, Calif., Jan. 5, 1953; s. Gerald Alexander and Lois Pearl (Knight) L. B.S., U. So. Calif., 1974; M.S., Northwestern U., 1976, Ph.D., 1979. Lectr. dept. geology Northwestern U., Evanston, Ill., 1979-80; asst. prof. dept. geology Stanford U., 1980—. NSF grantee, 1979, 80; Petroleum Research Found. grantee, 1980. Mem. Geol. Soc. Am., Soc. Econ. Mineralogists and Paleontologists. Subspecialties: Sedimentology; Tectonics. Current work: Precambrian plate tectonics; geology of Barbados; accretionary complex; turbidites, folding, strain, facies analysis. Office: Stanford U Dept Geology Stanford CA 94305 Home: 51 Pearce Mitchell Pl Stanford CA 94305

LASATER, ERIC MARTIN, neurophysiologist; b. Stuttgart, W.Ger., Jan. 6, 1953; s. Gene Martin and Naomi Ruth (Krahn) L.; m. Jill Ann Smith, Aug. 6, 1977. B.S., Colo. State U., 1975; M.S., U. Calif., Davis, 1977; Ph.D., U. Tex. Med. Br., Galveston, 1980. Teaching asst. U. Calif., Davis, 1975-77; research and teaching fellow U. Tex. Med. Br., 1977-80; research assoc. Harvard U., Cambridge, Mass., 1980—. Contbr. articles to profl. jours. Recipient Sigma Xi award for excellence in research, 1979. Mem. Soc. for Neurosci., Assn. for Research in Vision and Ophthalmology, AAAS, Sigma Xi. Subspecialties: Neurophysiology; Biophysics (biology). Current work: Neurophysiology and biophysics of information processing by the visual system. Office: Harvard U Bio Labs 16 Divinity Ave Cambridge MA 02138

LASKIN, OSCAR LARRY, clinical pharmacology educator, virologist; b. Phila., Sept. 11, 1951; s. Bernard and Blanche (Friedman) L.; m. Christine Ann Goril, Apr. 4, 1981. A.B. summa cum laude, Temple U., 1972, M.D. with honors, 1976. Diplomate: Am. Bd. Internal Medicine. Intern Johns Hopkins Hosp., Balt., 1976-77, resident in medicine, 1977-79, fellow in medicine, 1979-82, fellow in pharmacology, 1981-82; asst. prof. clin. pharmacology Cornell U. Med. Coll., N.Y.C., 1982—, asst. prof. pharmacology and medicine, 1982—; asst. attending physician N.Y. Hosp., 1982—. Contbr. articles to profl. jours. NIH fellow, 1981; clin. scholar Rockefeller Bros. Fund, 1982; Hartford Found. fellow, 1983; recipient research prize Am. Heart Assn., 1975. Fellow ACP; mem. Am. Fedn. Clin. Research, Am. Soc. Microbiology, Alpha Omega Alpha. Republican. Jewish. Subspecialties: Internal medicine; Pharmacology. Current work: Clinical pharmacology of antiviral drugs, rapid viral diagnosis and therapy of viral diseases especially herpesviruses. Home: 450 E 63d St Apt 5H New York NY 10021 Office: Cornell Med Coll 1300 York Ave New York NY 10021

LASKOWSKI, SHARON J., computer scientist; b. Bristol, Conn., Apr. 15, 1954; d. Anthony and Irene (Piascik) L.; m. Joseph F. Jaja, June 5, 1982. B.S., Trinity Coll., Hartford, Conn., 1975; M.S., Yale U., 1977, M.Phil., 1978, Ph.D., 1980. Asst. prof. computer sci. Pa. State U., University Park, 1979-83; system engr. Mitre Corp., McLean, Va., 1983—. NSF grantee, 1980. Mem. Assn. Computing Machinery. Subspecialties: Algorithms; Theoretical computer science. Current work: Artificial intelligence, expert systems, arithmetic complexity, analysis of algorithms. Office: Mitre Corp 1820 Dolley Madison Blvd McLean VA 22102

LASLEY, STEPHEN MICHAEL, research scientist, toxicologist; b. Louisville, Feb. 12, 1950; s. Hubert Elmer and Doris Jean (Epperson) L.; m. Carolyn Jean Culver, Aug. 3, 1974; 1 son, Michael Paul. M. Engring., U. Louisville, 1973, Ph.D. in Psychology, 1979. Engring. intern Formica Corp., Cin., 1970; student engr. Pub. Service Co. Ind., New Albany, 1971; research asst. dept. psychology U. Louisville, 1973-79; postdoctoral fellow dept. environ. health U. Cin., 1979-82; with dept. pharmacology Tex. Coll. Osteo. Medicine, Ft. Worth, 1983—. Contbr. articles to profl. jours. Recipient Nat. Research Service award Nat. Inst. Environ. Health Scis., 1979-82. Mem. AAAS, Soc. Neurosci. Democrat. Roman Catholic. Subspecialties: Neurochemistry; Environmental toxicology. Current work: Effects of environmental lead exposure on CNS neurotransmitters in rats, regional brain neurotransmitter activity in rats as result of lead exposure and the interaction of psychoactive agents. Home: 2813 Centre Ct Fort Worth TX 76116 Office: Dept Pharmacology Tex Coll Osteo Medicine Camp Bowie at Montgomery Fort Worth TX 76107

LASRY, JEAN-CLAUDE MAURICE, psychologist, educator; b. Casablanca, Morocco, July 21, 1937; emigrated to Can., 1957; s. Marcel and Violette (Leon) L.; m. Gaby Benbaruk, Oct. 12, 1969; children: Eytan, Arielle. B.A., U. Montreal, Que., 1965, M.A., 1966, Ph.D., 1968. Lectr. U. Montreal, 1966-68, asst. prof., 1968-75, assoc. prof., 1975-81, prof. psychology, 1981—; research assoc. Jewish Gen. Hosp., Montreal, 1969—. Editor: quar. Cross-Cultural Psychol Bull, 1980—; contbr. chpts. to books, articles to profl. jours. Founding pres. Ecole Maimonide, Montreal, 1969-71, pres., 1976-80; pres. Assn. Sepharade Francophone, Montreal, 1972-74; officer Can. Jewish Congress, 1975—. Can. NRC scholar, 1964-68; recipient Barkoff Leadership award Allied Jewish Community, Montreal, 1972. Mem. Can. Psychol. Assn., Que. Corp. Psychologists (award 1966), Am. Psychol. Assn., Internat. Assn. for Cross-Cultural Psychology, Que. Family Therapy Assn. (founding pres. 1979-82). Subspecialties: Social psychology; Psychotherapy. Current work: Cross-cultural psychology; mental health; immigrants' adaptation and identity; family structure and power. Home: 406 Ellerton Ave Mount Royal PQ Canada H3P 1E4 Office: Jewish Gen Hosp 433 Cote Ste Catherine Montreal PQ Canada H3T 1F2

LASSLO, ANDREW, medicinal chemist, educator; b. Mukacevo, Czechoslovakia, Aug. 24, 1922; came to U.S., 1946, naturalized, 1951; s. Vojtech Laszlo and Terezie (Herskovicova) L.; m. Wilma Ellen Reynolds, July 9, 1955; 1 dau., Millicent Andrea. M.S., U. Ill., 1948, Ph.D., 1952, M.L.S., 1961. Research chemist organic chems. div. Monsanto Chem. Co., St. Louis, 1952-54; asst. prof. pharmacology, div. basic health scis. Emory U., 1954-60; prof., chmn. dept. medicinal chemistry Coll. Pharmacy, U. Tenn. Center for Health Scis., 1960—; cons. Geschickter Fund for Med. Research Inc., 1961-62; dir. postgrad. tng. program sci. librarians USPHS, 1966-72; chmn. edn. com. Druf Info. Assn., 1966-68; dir. postgrad. tng. program organic medicinal chemistry for chemists FDA, 1971; exec. com. adv. council S.E. Regional Med. Library Program, Nat. Library of Medicine, 1969-71; chmn. regional med. library program com. Med. Library Assn., 1971-72; mem. pres.'s faculty adv. council U. Tenn. System, 1970-72; chmn. energy authority U. Tenn. Center for Health Scis., 1975-77, chmn. council departmental chmn., 1977, 81; chmn. Internat. Symposium on Contemporary Trends in Tng. Phamacologists, Helsinki, 1975. Producer, moderator: TV and radio series Health Care Perspectives, 1976-78; Editor: Surface Chemistry and Dental Integuments, 1973, Blood Platelet Function and Medicinal Chemistry, 1984; Contbr. numerous articles in sci. and profl. jours.; Mem. editorial bd.: Jour. Medicinal and Pharm. Chemistry, 1961, U. Tenn. Press, 1974-77; Composer: piano Synthesis in C Minor, 1968. Served to capt. M.S.C. U.S. Army Res., 1953-62. Recipient Sigma Xi Research prize, 1949, Honor Scroll Tenn. Inst. Chemists, 1976; Americanism medal D.A.R., 1976; U. Ill. fellow, 1950-51; Geschickter Fund Med. Research grantee, 1959-65; USPHS Research and Tng. grantee, 1958-64, 66-72, 82—; NSF research grantee, 1964-66; Pfeiffer Research Found. grantee, 1981—. Fellow AAAS, Am. Inst. Chemists (nat. councillor for Tenn. 1969-70); mem. ALA (life), Am. Chem. Soc. (sr.), Am. Pharm. Assn., Am. Soc. Pharmacology and Exptl. Therapeutics (chmn. subcom. pre-and postdoctoral tng. 1974-78, exec. com. ednl. and profl. affairs 1974-78), Sigma Xi (pres. elect U. Tenn. Center for Health Sci. chpt. 1975-76, pres. 1976-77), Beta Phi Mu, Rho Chi. Methodist (past trustee, ofcl. and adminstrv. bd., chmn. com. edn. and social concerns). Clubs: Health Scis. Center Faculty Club (charter bd. dirs. 1966-68, exec. com. 1967-68. Subspecialties: Medicinal chemistry; Molecular pharmacology. Current work: Relationships between chemical constitution, physicochemical characteristics and biodynamic response; development of novel antithrombotic agents; science information and library resources. Inventor, patentee. Home: 5479 Timmons Ave Memphis TN 38119 Office: 26 S Dunlap St Memphis TN 38163 Of all the pleasures a human being can savor, none exceeds the satisfaction of a genuine sense of accomplishment. It undergirds all elements of creative living and surmounts vicissitudes exceeding conventional human endurance.

LAST, JEROLD ALAN, medicine and biological chemistry educator, researcher; b. N.Y.C., June 5, 1940; s. Herbert and Florence (Lieberman) L.; m. Sandra Mizes, Jan. 26, 1969 (div. Jan. 1974); 1 son, Andrew; m. Elaine Zimelis, June 1, 1975; children: Matthew, Michael. B.S., U. Wis.-Madison, 1959, M.S., 1961; Ph.D., Ohio State U., 1965. Sr. research microbiologist Squibb, Inc., New Brunswick, N.J., 1967-69; mng. editor Nat. Acad. Scis., Washington, 1970-73; research assoc. Harvard U., Cambridge, Mass., 1973-76; assoc. prof. U. Calif. Med. Sch., Davis, 1976—. Editor: Methods in Molecular Biology Series, 9 Vols, 1967-74; contbr. articles to profl. jours. Fulbright fellow, 1983. Mem. Soc. Toxicology (Frank R. Blood award 1980), Am.Thoracic Soc., Am. Soc. Biol. Chemistry. Subspecialties: Biochemistry (medicine); Pulmonary medicine. Current work: Research on mechanisms of lung fibrosis, lung collagen structure and function, air pollution toxicology, health effects of air pollution. Patentee in field. Home: 510 Hubble St Davis CA 95616 Office: U Calif IM/Pulmonary Medicine Davis CA 95616

LASTER, RICHARD, food co. exec.; b. Vienna, Austria, Nov. 19, 1923; naturalized, 1944; s. Alan and Caroline L.; m. Liselotte Schneider, Oct. 17, 1948; children—Susan (Mrs. Franklin Rubenstein), Thomas. Student, U. Wash., 1941-42; B.Ch.E. cum laude, Poly. Inst. Bklyn., 1943; postgrad., Stevens Inst. Tech., 1945-47. With Gen. Foods Corp., 1944—, corp. research and devel., Hoboken, N.J., 1944-58, ops. mgr., Woburn, Mass., 1958-64, mgr. research devel., White Plains, N.Y., 1964-67, corp. mgr. quality assurance, White Plains, 1967-68, ops. mgr., 1968-69, exec. v.p., 1969-71, pres., 1971-73, corp. group v.p., White Plains, 1973-74, exec. v.p., 1974—, also dir., pres. research, devel. and food-away-from-home, 1975—; dir. Firestone Tire & Rubber Co. Contbr. articles on food sci. to profl. publs. Mem. sch. bd., Chappaqua, N.Y., 1971-73, pres., 1973-74; chmn., mem. bd., 1st v.p. United Way of Westchester, 1978; chmn. adv. com. Poly. Inst. Westchester, 1977; bd. dirs. Am. Health Found.; trustee Nutrition Found. Inc., Poly. Inst. N.Y.; mem. coll. council SUNY, Purchase. Recipient Distinguished Alumnus award; elected fellow Poly. Inst. N.Y. Mem. Nat. Acad. Scis., Am. Inst. Chem. Engrs. (Food and Bioengring. award 1972), Am. Chem. Soc., Am. Inst. Chemists, AAAS, Tau Beta Pi, Phi Lambda Upsilon. Clubs: Birchwood Tennis; Woodstock (Vt.); Country. Subspecialties: Chemical engineering; Food science and technology. Current work: Managing an agricultural biotechnology company from start-up situation into a going business. Patentee in field. Home: 23 Round Hill Rd Chappaqua NY 10514 Office: 250 North St White Plains NY 10625

LATA, GENE FREDERICK, biochemist; b. N.Y.C., May 17, 1922; s. Louis F. and Tessie (Greco) L.; m.; children: Paul, Matthew, Thomas, Catherine, Mary. B.S., CCNY, 1942; M.S., U. Ill., 1948, Ph.D., 1950. Lab. asst. CCNY, 1941-42; jr. chemist Gen. Foods Central Research Labs., 1942, 1946-47; teaching asst. U. Ill.-Champaign, 1947, research fellow, 1947-50, instr., 1950-53, asst. prof., 1953-63; assoc. prof. U. Iowa, Iowa City, 1963—, traveling fellow, 1965-66; vis. lectr. Scientist Harvard U., 1965-69. Contbr. articles on biochemistry to profl. jours. State del. dist. conv. Democratic Party, 1982. Served with AUS, 1942-46. Mem. Am. Soc. Biol. Chemistry, Am. Chem. Soc., AAAS, N.Y. Acad. Scis., Iowa Acad. Sci., Sigma Xi, Phi Sigma, Phi Lambda Upsilon. Roman Catholic. Subspecialties: Biochemistry (biology); Receptors. Current work: Physical chemical character of steroid solutions, steroid-protein binding, inter-relationships of retinal angiogensis and steroid hormones. Office: Dept Biochemistry U Iowa Iowa City IA 52242

LATHAM, ALLEN, JR., manufacturing company executive; b. Norwich, Conn., May 23, 1908; s. Allen and Caroline (Walker) L.; m. Ruth Nichols, Nov. 11, 1933; children: W. Nichols, Harriet (Mrs. William S. Robinson), David W., Thomas W. B.S. in Mech. Engring, MIT, 1930, Sloan fellow, 1936. Devel. engr. E.I. duPont, Belle, W.Va., 1930-35; engr., treas. Polaroid Corp., Cambridge, Mass., 1936-41; engr., v.p. Arthur D. Little, Cambridge, 1941-66; pres. Cryogenic Tech., Waltham, Mass., 1966-71; chmn. bd. Haemonetics, Braintree, Mass., 1971—, also dir.; corporator Eliot Savs. Bank, Boston. Named Engr. of Yr. Socs. New Eng. Engring., 1970. Mem. ASME, Am. Inst. Chem. Engrs., Instrument Soc. Am., AAAS, Nat. Acad. Engring. Club: Country (Brookline, Mass.). Subspecialties: Mechanical engineering; Biomedical engineering. Current work: Development of centrifugal separation systems with disposable fluid disposable fluid pathways for therapeutic application in human blood disorders. Patentee in blood processing equipment and processes. Home: 66 Malcolm Rd Jamaica Plain MA 02130 Office: 400 Wood Rd Braintree MA 02184

LATHAM, ELEANOR EARTHROWL, neuropsychologist; b. Enfield, Conn., Jan. 12, 1924; d. Francis Henry and Ruth (Harris) Earthrowl; m. Vaughan Milton Latham, July 20, 1946; children: Rebecca, Carol, Jennifer, Vaughan Milton. B.A., Vassar Coll., 1945; M.A., Smith Coll., 1947, Clark U., 1974, Ed.D., 1979. Music supr. Worcester (Mass.) Pub. Schs., 1962-65, counselor, 1966-74, psychologist, 1975-82; pvt. clin. practice, Worcester, 1981—; neuropsychologist St. Vincent Hosp., Worcester, 1982—, Hahnemann Hosp., 1982—; assoc. in pediatrics U. Mass. Med. Sch., 1982—; cons. sch. systems, 1981—. Trustee Performing Arts Sch., Worcester, 1980—; bd. dirs. Worcester Childrens Friend Soc., 1970-76, 83—. Mem. Am. Psychol. Assn., Council Anthropology and Edn., Am. Ednl. Research Assn., Internat. Neurospsychol. Soc. Republican. Unitarian. Subspecialties: Neuropsychology; Developmental psychology. Current work: Effects of neurological disease upon cognitive functioning and behavior; clinically planning educationally for rehabilitation and learning; neuromuscular disease, Duchenne muscular dystrophy; psychosocial issues in families of boys with Duchenne muscular dystrophy. Home: 59 Berwick St Worcester MA 01602

LATHAM, GARY PHILLIP, industrial and organizational psychologist, researcher; b. Halifax, N.S., Can., Nov. 16, 1945; came to U.S., 1967, naturalized, 1973; s. William Phillip and Hope Elizabeth (Marshall) L.; m. Sharon Lee Bridgwater, Sept. 6, 1969; children: Bryan Phillip, Brandon William. B.A., Dalhousie U., 1967; M.S., Ga. Inst. Tech., 1969; Ph.D., U. Akron, 1974. Staff psychologist Am. Pulpwood Assn., Atlanta, 1969-71; mgr. human resource research Weyerhaeuser Co., Tacoma, 1973-76; research assoc. U. Wash., 1976—; pres. G. P. Latham Inc., Seattle, 1980—; human resource cons. Weyerhaeuser Co., 1976—; Scott Paper Co., Everett, Wash., 1979—. Author: Increasing Productivity Through Performance Appraisal, 1981, Developing and Training Human Resources in Organizations, 1981, Goal Setting: A Motivational Technique that Works, 1983; mem. editorial bd.: Jour. Applied Psychology, 1982—, Acad. Mgmt. Jour, 1982—, Jour. Organizational Behavior Mgmt, 1978—. Bd. dirs. St. James Sch., Seattle, 1982—. Office Naval Research grantee, 1979-82. Fellow Am. Psychol. Assn. (chmn. membership com. 1977-78); mem. Can. Psychol. Assn. (chmn. indsl. organl. psychology 1975-76), Acad. Mgmt., Soc. Organizational Behavior. Subspecialties: Industrial-organizational psychology; Behavioral psychology. Current work: Motivation, quality of work life; selection-staffing; performance appraisal. Home: 15260 Maplewild Ave SW Seattle WA 98166 Office: U Wash Psychology Dept Seattle WA 98195

LATOZA, KENNETH CHARLES, numerical analyst. B.S., Valparaiso U., 1975; postgrad., N.C. State U., 1975-76; M.S., Northwestern U., 1978. Tech. programmer Nalco Chem. Corp., Chgo., 1976-77; research engr. Ford Motor Co., Dearborn, Mich., 1979—. Mem. Nat. Found. Ileitis and Colitis, N.Y.C. Mem. Soc. Indsl. and Applied Math., Sigma Pi Sigma. Subspecialties: Numerical analysis; Theoretical and applied mechanics. Current work: Develop new vehicle dynamic models, kinematic and dynamic analysis of non-linear tuned absorbers. Office: Ford Motor Co Rm 102 Engring Computer Ctr 20000 Rotunda Dr Dearborn MI 48121

LATTA, JOHN NEAL, research scientist; b. Ottumwa, Iowa, Apr. 11, 1944; s. Oris N. and Edna M. (Hanna) L.; m. Karen N. Jamison, June 10, 1966; children: J. Neal, Mark R. B.E.S., Brigham Young U., 1966; M.S.E.E., U. Kans., 1969; Ph.D. in Elec. Engring., 1971. Mem. tech. staff RCA Labs., Princeton, N.J., 1967; Bell Telephone Labs., Murray Hill, N.J., 1969; sr. engr. U. Mich., Ann Arbor, 1969-77; sr. scientist Sci. Applications, Inc., Falls Church, Va., 1977-83; pres. Adroit Systems, Inc., Alexandria, Va., 1983—; cons. holographic optics. Mem. IEEE (sr. mem.), Soc. Automotive Engring., Assn. Computing Machinery, Optical Soc. Am., Soc. Photo-Optical Instrumentation Engrs., Pattern Recognition Soc., Am. Underground Space Assn., Air Force Assn., Sigma Xi. Subspecialties: Holography; Graphics, image processing, and pattern recognition. Current work: Design of large image processing systems using distributed network concepts. Home: PO Box 1297 Arlington VA 22210 Office: PO Box 6547 Alexandria VA 22306

LAU, JARK C(HONG), research associate, consultant; b. Singapore, Oct. 18. 1935; came to U.S., 1975; s. Chon-Poh and Wan-Quen (Lok); m. Alice Wong, Mar. 26, 1961 (dec. 1980); children: Yung R., Ming S. Fellowship diploma, Royal Melbourne (Australia) Inst. Tech., 1956; A.E., Calif. Inst. Tech., 1964, M.S., 1963; Ph.D., Inst. Sound and Vibration Research, U. Southampton, Eng., 1971. Sr. lectr. Singapore Poly., 1958-71; assoc. prof. U. Singapore, 1971-75; tech. cons. Lockheed-Ga. Co., Marietta, Ga., 1975-80; sr. research assoc. Kimberly-Clark Corp., Neenah, Wis., 1980—, Roswell, Ga., 1983—, in-house cons. U.S. and fgn., 1980—. Contbr. articles to profl. publs. Scoutmaster Boy Scouts Assn., Australia, 1955-58. Assoc. fellow AIAA. Seventh-day Adventist. Subspecialties: Fluid mechanics; Aeronautical engineering. Current work: Research to try to characterize structure of shear layer turbulence, development of better techniques of fiber-web formation, consultation service to many mills in the corporation. Inventor. Home: 2525 Powdes Ridge Roswell GA 30076 Office: Kimberly-Clark Corp 1400 Holcomb Bridge Rd Roswell GA 30076

LAUB, LEONARD JOSEPH, physicist, business executive; b. Chgo., July 27, 1946; s. Walter and Beatrice R. (Kaye) L.; m. Bonnie Lynn Kohl, Jan. 9, 1981; 1 stepson, Marc Schwartz. Student, U. Chgo., 1963-64; B.Sc., Ill. Inst. Tech., 1970; postgrad., Northwestern U., 171-72. Sect. mgr. electro-optics, research dept. Zenith Radio Corp., Chgo., 1965-76; mgr. optical storage tech. Xerox Electro-Optical Systems, Pasadena, Calif., 1976-78; tech. dir. , dir. product planning Star Systems div. Exxon Enterprises, Pasadena, 1978-81; pres. Vision Three, Inc., Pasadena, 1981—. Contbr. articles to profl. jours. Mem. Optical Soc. Am., IEEE, Assn. for Computing Machinery, Soc. Photog. Scientists and Engrs., Soc. Photo-optical Instrumentation Engring., Audio Engring. Soc. Subspecialties: Laser data storage and reproduction; Distributed systems and networks. Current work: Optical information storage, information-finding systems, man-machine interaction, design of visual and audio delivery systems, coding theory. Patentee signal processing, laser-based metrology, optical storage. Home and Office: 1400 Arroyo View Dr Pasadena CA 91103

LAUBENTHAL, NEIL DAVID, nuclear plant engr., naval officer; b. Mobile, Ala., May 21, 1954; s. Joseph Bruce and Jane Evelyn (Frederick) L.; m. Constance Demery, Nov. 20, 1976. B.S. in Mech. Engring, U. Miami, 1976; student, Nuclear Power Sch., Orlando, Fla., 1976-77. Commd. ensign U.S. Navy, 1976, advanced through grades to lt., 1980; weapons officer U.S.S. Grayling, Charleston, S.C., 1977-81, engr. officer, 1981-82; asst. prof. naval sci. ROTC, Auburn U., 1982—. Decorated Navy Expeditionary medal, Navy Achievement medal with gold star. Mem. ASME, U.S. Naval Inst., Tau Beta Pi, Pi Tau Sigma. Roman Catholic. Subspecialty: Nuclear fission. Home: 117 Carter St Auburn AL 36830 Office: Auburn U ROTC Auburn AL 36830

LAUBSCHER, ROY EDWARD, astronomer; b. Cleve., May 13, 1944; s. William John and Florence (Huge) L.; m. Mayumi Aono, Dec. 31, 1980. B.S., Case Inst. Tech., 1966; Ph.D. in Astronomy, Yale U., 1980. Analyst Computer Scis. Corp., Silver Spring, Md., 1973-78; sr. analyst Gen. Dynamics Corp., Vandenberg AFB, Calif., 1978-80; mem. prof. staff Geodynamics Corp., Santa Barbara, Calif., 1980—; astronomer Astronomisches Rechen-Institut, Heidelberg, W.Ger., 1970-72. Contbr. articles to profl. jours. Mem. Am. Astron. Soc., German Astron. Soc., Sigma Xi. Subspecialties: Satellite studies; General relativity. Current work: Analysis of satellite geodetic data. Home: 728 W Micheltorena St Santa Barbara CA 93101

LAUDENSLAGER, MARK LEROY, physiological psychologist, educator; b. Charlotte, N.C., May 13, 1947; s. Mark LeRoy and Rosemary (Baskerville) L.; m. Carol Ann Kemler, May 26, 1945; 1 son, Eric D. A.B., U. N.C., 1969; Ph.D., U. Calif.-Santa Barbara, 1975. Lectr., teaching asst. U. Calif.-Santa Barbara, 1972-73, research asst., 1973-75, research psychologist, 1978-80; postdoctoral fellow U. Calif. Scripps Inst., San Diego, 1975-78, U. Colo. Health Sci. Ctr., Denver, 1980-81; researcher, instr. Denver U., 1981—. Contbr. numerous articles to profl. jours. NIMH fellow, 1970, 1975, 76, 80. Mem. Am. Physiol. Soc., Soc. Neurosci., Am. Ornithologists Union, Sigma Xi. Subspecialties: Physiological psychology; Neuroimmunology. Current work: Brain-behavior relationships, comparative animal behavior, stress and immune functioning, neuroendocrinology. Office: Denver U Dept Psychology Denver CO 80208

LAUDER, JEAN MILES, neurobiologist, educator; b. Haverhill, Mass., June 29, 1945; d. Robert William and Frances (Miles) L.; m. Ralph Nicholson, June 5, 1965 (div.); m. Helmut Krebs, Nov. 28, 1974. B.A., U. Maine, 1967; Ph.D., Purdue U., 1972. Staff fellow NIMH, Washington, 1972-74; research asso. U. Conn., Storrs, 1974-76, asst. prof. in residence, 1976-78; asso. prof. anatomy U. N.C. Sch. Medicine, Chapel Hill, 1978—. Contbr. articles to profl. publs. Recipient Career devel. award NIH, 1980-85. Mem. Internat. Soc. Devel. Neurosci., Soc. Neurosci., N.C. Soc. Neurosci., AAAS, Internat. Soc. Psychoneuroendocrinology, Am. Soc. Cell Biology. Subspecialties: Neurobiology; Developmental biology. Current work: Research in area of development of brain, in particular neurotransmitters and hormones, including control of brain development by these humoral substances. Office: U NC Sch Medicine Dept Anatomy 111 Swing Bldg 217H Chapel Hill NC 27514

LAUDISE, ROBERT ALFRED, research chemist; b. Amsterdam, N.Y., Sept. 2, 1930; s. Anthony Thomas and Harriette Elizabeth (O'Neil) L.; m. Joyce Elizabeth DeSilvia, Aug. 24, 1957; children: Thomas Michael, Margaret Joyce, John David, Mary Elizabeth, Edward Robert. B.S. in Chemistry, Union Coll., Schenectady, N.Y., 1952; Ph.D. in Chemistry (A.D. Little fellow), M.I.T., 1956. Mem. Tech. staff Bell Telephone Labs., Murray Hill, N.J., 1956-60, head crystal chemistry research dept., 1960-72, asst. dir. materials research lab., 1972-74, dir. materials research lab., 1974-77, dir. phys. and inorganic chemistry research lab., 1977—; vis. prof. U. Aix, Marseilles, France, 1971, Hebrew U., Jerusalem, 1972, Shandong U., China, 1980; cons. Pres.'s Sci. Com., 1960-64; adv. com. Nat. Bur. Standards, 1970-78; solid state scis. com. NRC, 1977-81; adv. com. NASA, 1977—. Author: The Growth of Single Crystals, 1970; editor: Jour. Crystal Growth, 1978—; contbr. articles to sci. jours. Recipient Sawyer award, 1974. Fellow Am. Mineral. Soc.; mem. Nat. Acad. Engring., Internat. Orgn. Crystal Growth (pres., award 1981), Am. Assn. Crystal Growth (pres. 1971-77), Am. Chem. Soc., Am. Ceramic Soc., AAAS, Electrochem. Soc., IEEE, Am. Crystall. Soc., Kappa Sigma. Roman Catholic. Clubs: Twin Lakes (Pa.); Sailing. Subspecialties: Electronic materials; Solid state chemistry. Current work: Crystal growth of electronic materials; connection between chemical bonding, structure and useful electronic properties. Patentee in field. Home: 65 Lenape Ln Berkeley Heights NJ 07922 Office: Bell Telephone Labs 600 Mountain Ave Murray Hill NJ 07974

LAUENROTH, WILLIAM KARL, systems ecologist, educator; b. Carthage, Mo., July 31, 1945; s. Reinhold and Elaine (Frediani) L.; m. Jacqueline E. Willey, Dec. 17, 1967. B.S., Humboldt U., 1968; M.S. in Plant Ecology, N.D. State U., 1970; Ph.D. in Range Ecology, Colo. State U., 1973. Postdoctoral fellow Natural Resource Ecology Lab., Colo. State U., Ft. Collins, 1973-75, research ecologist, 1973—, assoc. prof. range sci. dept., 1981—. Contbr. articles to sci. jours. Grantee EPA, Elec. Power Research Inst., NSF. Mem. Bot. Soc. Am., Soc. Range Mgmt., Ecol. Soc. Am., AAAS. Subspecialties: Ecosystems analysis; Resource management. Current work: Integrated systems, ecosystems analysis, simulation modeling, ecological theory. Home: 1049 Summit View Dr Fort Collins CO 80524 Office: Colo State U Room 202a NR Bldg Fort Collins CO 80523

LAUER, DENNIS ERROL, mathematics educator; b. Salina, Kans., June 19, 1938; s. Marvin LaVerne Lauer and Awyn DeGracia (Awyn) Lauer G.; m. Lillie Jean Rehlander, May 15, 1976; children: Ryan, Jeanine, Christian. B.S. in Physics, U. Kans., Lawrence, 1960, M.S., 1963; M.S. in Math, Purdue U., 1966. Instr. Ill. Coll., Jacksonville, 1964-66; asst. prof. math. Purdue U., Westville, Ind., 1968—. Mem. Am. Math. Assn., AAUP. Subspecialties: Operations research (mathematics); Theoretical physics. Home: 1706 Peachtree Dr Valparaiso IN 46383 Office: Purdue U N Central Campus Westville IN 46383

LAUER, FLORIAN I., horticultural science educator, plant breeder; b. Richmond, Minn., Sept. 13, 1928; s. Christian and Barbara (Wilms) L.; m. Mary Ann, Aug. 21, 1954; children: Sarah, Sheila, Maureen, Paul. B.S., U. Minn., 1951, Ph.D., 1956. Research fellow U. Minn., Mpls., 1956-60, asst. prof., 1962-62, assoc. prof., 1962-66, prof. hort. sci., 1966—. Mem. Crop Sci. Soc. Am., Potato Assn. Am. Subspecialties: Genetics and genetic engineering (biology); Plant growth. Current work: Genetics, plant breeding, horticulture; unreduced gametes; insect resistance, disease resistance. Office: Dept Hort Sci U Minn Saint Paul MN 55113

LAUFMAN, LESLIE RODGERS, physician, researcher, educator; b. Pitts., Dec. 13, 1946; d. Marshall C. and Ruth (Petrauskas) Rodgers; m. Harry B. Laufman, Apr. 25, 1970. B.A., Ohio Wesleyan U., 1968; M.D., U. Pitts., 1972. Intern Montefiore Hosp., Pitts., 1972-73, resident in internal medicine, 1973-74; fellow hematology-oncology Ohio State U., Columbus, 1974-76; staff physician Grant Hosp., Columbus, 1976—; prin. investigator SWOG Satellite grantee, Columbus, 1977—; clin. asst. prof. Ohio State U., 1977—; dir. med. oncology Grant Hosp., 1977—; mem. gyn. com. S.W. Oncology Group, Kansas City, Kans., 1982—; mem. Nat. Surg. Adjuvant Breast Project, Pitts., 1980—. Mem. Am. Med. Women's Assn., Am. Soc. Clin. Oncology. Subspecialties: Hematology; Chemotherapy. Current work: Therapy of cancer of ovary; mitomycin C toxicity. Office: Grant Hosp 323 E Town St Columbus OH 43215

LAUGHLIN, MAURICE HAROLD, physiologist, educator; b. Fairfield, Iowa, Feb. 12, 1948; s. Charles M. and Charolete O. (Stark) L.; m. Linda S. Harmon, Mar. 28, 1970; children: Daniel, Michael, David. B.A., Simpson Coll., 1970; Ph.D., U. Iowa, 1974. Postdoctoral fellow U. Iowa, Iowa City, 1970-74; capt., research aerospace physiologist USAF Sch. Medicine, Brooks AFB, Tex., 1976-80; asst. prof. Oral Roberts U. Sch. Medicine, Tulsa, Okla., 1980—. Contbr. articles to sci. publs. NIH research grantee, 1980. Fellow Am. Heart Assn.; assoc. fellow Aerospace Med. Assn.; mem. Am. Physiol. Soc., Am. Coll. Sports Medicine, Microcirculatory Soc. Democrat. Methodist. Subspecialties: Physiology (medicine); Gravitational biology. Current work: interactions of exercise on coronary and skeletal muscle blood flows in health and disease; cardiovascular effects on acceleration stress. Home: 2938 E 77th St Tulsa OK 74136 Office: Physiology Dept Oral Roberts U School Medicine 7777 S Lewis Tulsa OK 74171

LAURENDEAU, NORMAND MAURICE, mechanical engineering educator; b. Lewiston, Maine, Aug. 16, 1944; s. Maurice Paul and Lydia Alma (Roy) L.; m. Marlene Odette Carlos, Feb. 14, 1972; children: Andre, Jules. B.S., U. Notre Dame, 1966; M.S., Princeton U., 1968; Ph.D., U. Calif.-Berkeley, 1972. Mem. faculty dept. mech. engring. Purdue U., West Lafayette, Ind., 1972—, prof., 1981—, dir. Coal Research Ctr., 1980—. Mem. Am. Chem. Soc., Combustion Inst., ASME, Optical Soc. Am. Subspecialties: Combustion processes; Spectroscopy. Current work: Coal combustion and gasification, combustion kinetics, laser-induced fluorescence. Office: Dept Mech Engring Purdue U West Lafayette IN 47907 Home: 1507 Central St Lafayette IN 47905

LAURENSON, ROBERT MARK, engineer; b. Pitts., Oct. 25, 1938; s. Robert Mark and Mildred Othelia (Frandsen) L.; m. Alice Ann Scroggins, Aug. 26, 1961; children: Susan Elizabeth, Shari Lynn. Student, Drury Coll., 1956-58; B.S., Mo. Sch. Mines, 1961; M.S.E., U. Mich., 1962; Ph.D., Ga. Inst. Tech., 1968. Registered profl. engr., Mo. Dynamics engr. McDonnell Douglas Corp., St. Louis, 1962-64, sr. dynamics engr., 1968-71, group engr., 1971-74, staff engr., 1974-75, tech. specialist, 1975-78, sr. tech. specialist, 1978-81, sect. chief, 1981—; lectr. St. Louis U., 1969-71; adj. assoc. prof. U. Mo.-Rolla, 1980—; participant various confs. and symposia. Contbr. articles, reviewer papers for profl. jours.; editor: 1982 Advances in Aerospace Structures. NASA tng. grantee. Mem. AIAA (gen. chmn. conf. 1981), ASME (chmn. exec. com. aerospace div. 1983-84), Sigma Xi, Pi Tau Sigma, Tau Beta Pi, Phi Kappa Phi, Sigma Phi Epsilon. Subspecialties: Mechanical engineering; Aerospace engineering and technology. Current work: Dynamic analysis of missile and space vehicle structures. Home: 349 Beaver Lake Dr Saint Charles MO 63301 Office: PO Box 516 Saint Louis MO 63166

LAURIE, JOHN ANDREW, oncologist; b. Stirling, Scotland, Feb. 15, 1947; came to U.S., 1974; s. Andrew and Agnes Margaret (Watson) L.; m. Theresa J. Schneider, Dec. 13, 1969; children: Alison, Pamela, Gillian, Ian. Sc.B., Edinburgh U., 1968, M.B., 1971. Diplomate: Am. Bd. Internal Medicine (Medical Oncology). Intern and resident Edinburgh (Scotland) U., 1971-74; resident in internal medicine Mt. Sinai Hosp., Milw., 1974-75, oncology fellow, 1975-77; oncologist Grand Forks (N.D.) Clinic, 1977—; exec. com. N. Central Cancer Treatment Group, Rochester, Minn., 1978—. Nat. Cancer Inst. grantee, 1980. Mem. ACP, Am. Soc. Clin. Oncology. Subspecialties: Chemotherapy; Cancer research (medicine). Current work: Clinical research in oncology. Office: Grand Forks Clinic 1000 S Columbia Rd Grand Forks ND 58201

LAUTER, CARL BURTON, medical educator, medical program director; b. Detroit, Dec. 30, 1939; s. Reuben David and Sadie (Kaplowitz) L.; m. Jain Beth Mogill, Dec. 21, 1975; children: Shira Lynn, Rebekah Dana, Jonathan Norton. B.A. in Chemistry, Wayne State U., 1962, M.D., 1965. Diplomate: Am. Bd. Internal Medicine, Am. Bd. Allergy and Immunology, Am. Bd. Infectious Diseases. Intern Henry Ford Hosp., Detroit, 1965-66; resident Wayne State U./ Detroit Receiving Hosp., 1966-67, Hosp. U. Pa., 1969-70; fellow in infectious diseases Wayne State U. Addiliated Hosps., 1970-73; fellow in allergy and immunology Henry Ford Hosp., 1979-80; dir. med. edn., chief infectious disease sect. Harper-Grace Hosp., Detroit, 1973-79; assoc. prof. medicine Wayne State U. Sch. Medicine, Detroit, 1978—; vice chief med. services William Beaumont Hosp., Royal Oak, Mich., 1979-82, chief med. services, 1982—; chmn. William Beaumont Hosp., Infection Controll Com., 1980—; cons. Harper-Grace, Hutzel and Sinai hosps., Children's Hosp., Detroit, 1973—. Contbr. articles to profl. jours. Served to capt. USAF, 1967-69. Fellow ACP, Am. Coll. Clin. Pharmacy, Infectious Disease Soc. Am.; mem. Am. Coll. Allergy and Clin. Immunology, Am. Soc. Microbiology, AMA, AAAS, Am. Fedn. Clin. Research, Phi Beta Kappa, Alpha Omega Alpha. Jewish. Subspecialties: Allergy; Infectious diseases. Current work: Toxic shock syndrome - clinical aspects and pathophysiology; antiviral chemotherapy. Home: 3168 Woodland Ridge West Bloomfield MI 48033 Office: William Beaumont Hosp 3601 W 13 Mile Rd Royal Oak MI 48072

LAUTERBUR, PAUL C(HRISTIAN), chemistry educator, researcher; b. Sidney, Ohio, May 6, 1929; s. Edward Joseph and Gertrude Frieda (Wagner) L.; m.; children: Daniel James, Sharon Lynn. B.S., Case Inst. Tech., 1951; Ph.D., U. Pitts., 1962. Research asst. Mellon Inst., Pitts., 1951-53, research fellow, 1955-63; assoc. prof. chemistry SUNY, Stony Brook, 1963-69, prof. chemistry, 1969—, also research prof. radiology; cons. in field. Contbr. numerous articles to profl. jours.; editor-in-chief: Magnetic Resonance in Medicine, 1982—. Served with U.S. Army, 1953-55. Fellow Am. Phys. Soc. (Biol. Physics award 1983); mem. Am. Chem. Soc. (Johannes S. Buck-Willis R. Whitney award Eastern N.Y. Sect. 1976), AAAS, Soc. Magnetic Resonance in Medicine (pres. 1981-83, Gold medal 1982). Subspecialties: Nuclear magnetic resonance (chemistry); Nuclear magnetic resonance (biotechnology). Current work: Magnetic resonance in medicine and biology.

LAUTZENHEISER, CLARENCE ERIC, research and development institute administrator, metallurgist; b. Lincoln, Nebr., May 21, 1921; s. Delbert L. and Emma (Semler) L.; m. Audrey Aileen Bayliss, Jan. 1, 1948; children: Clarence Eric, John Andrew, Ann. B.S., M.I.T., 1952. Registered profl. engr., Tex., 1961. Flight instr. Moore's Flying Service, Corpus Christi, Tex., 1940-42, Cal-Aero Flight Acad., Ontario, Calif., 1942-43; owner, operator Gulf Coast Flying Service, Kingsville, Tex., 1946-49; with Dow Chem. Co., Freeport, Tex., 1952-62, research engr., 1952-54, maintenance specialist, 1954-62; with S.W. Research Inst., San Antonio, 1962—, dir., 1971-74, v.p. quality assurance systems and engring. div., 1974—; cons. in field. Contbr. articles to profl. jours. Served to 1st lt. USAAF, 1943-46; to lt. col. USAAFR, 1946-81. Fellow Am. Soc. Nondestructive Testing; mem. Am. Welding Soc., Am. Soc. Metals, Nat. Assn. Corrosion Engrs., ASME (Centennial award 1980). Subspecialties: Metallurgical engineering; Systems engineering. Current work: Engineering problem solutions through nondestructive testing technologies; structured integrity assessments using nondestructive testing technologies; materials evaluations utilizing destructive and nondestructive testing technologies. Home: Route 1 Box 58 Medina TX 78055 Office: 6220 Culebra Rd San Antonio TX 78238

LAUX, DAVID CHARLES, microbiologist, immunologist, educator; b. Sarver, Pa., Jan. 1, 1945; s. Charles L. and Margaret (Klein) L.; m. Sara E. Pollen; 1 son, Benjamin. B.A., Washington and Jefferson Coll., 1966; M.S., Miami U., Oxford, Ohio, 1968; Ph.D., U. Ariz., 1971. Postdoctoral fellow Pa. State U., 1971-73; faculty U. R.I., 1973—, assoc. prof., 1978—. Contbr. articles to profl. jours. NIH grantee, 1973—. Mem. Am. Assn. Cancer Research. Subspecialties: Immunobiology and immunology; Microbiology. Current work: Immunobiology and microbiology. Office: 318 Morrill Microbiology U RI Kingston RI 02881

LAVENDER, JOHN FRANCIS, research virologist; b. Chgo., Nov. 16, 1929; s. John Francis and Elsie M. (Walker) L.; m. Bessie Fern Davis, June 21, 1969; stepchildren: Michael, Patricia, Penny. B.A., Drake U., 1951; M.S., U. Ill., 1953; Ph.D., UCLA Med. Sch., 1963. Microbiologist U.S. Army Chem. Corps, 1955-57; instr. UCLA Med. Sch., 1961-63; sr. virologist Eli Lilly & Co., Indpls., 1964-71, research virologist, 1972—. Author sci. articles. Served in U.S. Army, 1953-55. Mem. N.Y. Acad. Scis., Am. Soc. Microbiology, Sigmi Xi, Beta Beta Beta. Subspecialties: Virology (medicine); Microbiology (medicine). Current work: Vaccine research for canine distemper and rabies; virus chemotherapy for herpes simplex viruses 1 and 2. Patentee in field. Home: 543 W Dr Woodruff Pl Indianapolis IN 46201 Office: Lilly Research Labs ML 539 Indianpolis IN 46285

LAVIETES, MARC HARRY, medical educator, physician; b. New Haven, July 7, 1941; s. Paul Harold and Ruth (Sweedler) L.; m. Beverly Faye Blatt, Aug. 13, 1966; children: Bryan Ross, Jonathan David. B.A., Yale U., 1963; M.D., Case-Western Res. U., 1969. Intern Harlem Hosp. Ctr., N.Y.C., 1969-70; resident in medicine, fellow Bellevue Hosp., N.Y.C., 1971-73, Columbia-Presbyn. Med. Ctr., 1973-75; asst. prof. U. Medicine and Dentistry N.J., Newark, 1975-82, assoc. prof. clin. medicine, 1982—. Contbr.: Diagnostic Aspects and Management of Asthma, 1981. N.J. Thoracic Soc. research grantee, 1982-83. Mem. Am. Thoracic Soc., Am. Fedn. Clin. Research. Subspecialties: Internal medicine; Pulmonary medicine. Current work: Inter-relationships between respiratory regulation and respiratory mechanics in asthmatic subjects. Office: Coll Hosp Pulmonary Div I 354 100 Bergen St Newark NJ 07103

LAVIGNE, ROBERT JAMES, entomology educator; b. Herkimer, N.Y., May 30, 1930; s. Robert James and Dorothea (Eckerson) L.; m. Judith Jane Watson, Aug. 14, 1976; children: Jay Wayne, Michelle Renee, Todd Jeffery, Cathi Thone. B.A., Am. Internat. Coll., Springfield, Mass., 1952; M.S., U. Mass., 1958, Ph.D., 1961. Instr. U. Mass., 1957-59; asst. prof. entomology U. Wyo., Laramie, 1959-65, assoc. prof., 1965-71, prof., 1971—; cons. Mine Reclamation Cons., Laramie, 1976-79, Ecology Econs., Ft. Collins, Colo., 1973-75. Co-author: Rangeland Entomology, 1974; editor: Jour. Econ. Entomology, 1982—. Mem. Entomol. Soc. Am., Entomol. Soc. Washington, Pan-Am. Acridological Soc., Soc. Range Mgmt., Kans. Entomol. Soc., Phi Kappa Phi, Alpha Chi. Lodge: Moose. Subspecialties: Integrated pest management; Ethology. Current work: Biocontrol of weeds, chemical and cultural control of forest insects, ethology and taxonomy of predatory flies. Home: 1323 Mill St Laramie WY 82071 Office: Dept Entomology U Wyo Ivinson St Laramie WY 82071

LAW, PETER KOI, neurophysiologist, educator; b. Chengsha, China, Feb. 25, 1946; s. Lawrence C.K. and Josephine S.L. (Wong) L.; m. Rosalie Y.Y. Cheng, Apr. 21, 1973; children: Dawn, Peter. B.S. with honors, McGill U., 1968; M.Sc., U. Toronto, 1969, Ph.D., 1972. Postdoctoral fellow McMaster U., Hamilton, Ont., 1972-75; asst. prof. neurology Vanderbilt U. Sch. Medicine, Nashville, 1975-79; sr. researcher Jerry Lewis Neuromuscular Diseases Research Center, Nashville, 1975-79; assoc. prof. neurology, physiology/biophysics U. Tenn. Center for Health Scis., Memphis, 1979—; also research, cons. Contbr. articles and abstracts to profl. lit. Grantee univs, founds. and muscular dystrophy orgns. Mem. Am. Physiol. Soc., Soc. Neuroscis., N.Y. Acad. Sics., AAAS, Internat. Soc. Developmental and Comparative Immunology, Memphis Acad. Neurology, Am. Assn. Electromyography and Electrodiagnosis. Roman Catholic. Subspecialties: Neurophysiology; Genetics and genetic engineering (medicine). Current work: Treating muscle weakness of hereditary myopathies by genetic complementation. Developing new transplant techniques to facilitate regeneration and fusion of normal donor myoblasts with myogenic cells of dystrophic hosts. Home: 6992 Corsica Dr Germantown TN 38138 Office: 956 Court Ave 2A05 Memphis TN 38163

LAWLOR, EVELYN DAVIS, psychologist; b. Hartford, Conn.; d. Frank Wilbert and Adelaide (Byrd) Davis; m. Ignatius E. Lawlor, Aug. 20; 1960; m. (dec. 1978). R.N., Lincoln Sch. Nursing, 1930; B.S., N.Y.U., 1940, M.A., 1944; M.S.S., New Sch. Social Research, 1953; Ph.D., Yeshiva U., 1960. Registered psychologist, N.Y., diplomate: Am. Bd. Profl. Psychology, Internat. Acad. profl Counseling and Psychotherapy. Asst. prof. Yeshiva U., N.Y.C., 1959-61; adj. asst. prof. Queens coll, 1965-74; clin. psychologist B.A.R.O. Clinic, Bklyn., 1954-57; ednl./psychol. cons. Title I Dist. 8, N.Y.C., 1970-71; sch. psychologist Bur. Child Guidance, Bd. Edn., N.Y.C., 1960-70; staff therapist Postgrad. Center for Mental Health, N.Y.C., 1962-76; supr. Postgrad. W., N.Y.C., 1978-80; cons. in field; conductor seminars in field. Recipient Achievement awards Phi Delta Kappa, 1958, 59, 60, 75, 79, 80; Service award P.G.C. for Mental Health, 1972; Grace Humanitarian award Grace Bapt. Ch., 1977; others. Fellow Internat. Council Psychologists; mem. Am. Psychol. Ass., N.Y. Vocational Guidance Assn., Am. Assn. Vocat. Guidance, N.Y. Psychol. Assn., Am. Personnel and Guidance Assn., Am. Group Psychotherapy Assn., Soc. for Clin. and Exptl. Hypnosis, others. Democrat. Presbyterian. Clubs: League Women Voters, Manhood Found., others. Current work: Hypnosis with symptom of epileploid states - defense stress mgmt. and relaxation; burn out; transcendental meditation; autogenic tng. Address: 706 S. 6th Ave Mount Vernon NY 10550

LAWRASON, F. DOUGLAS, physician, drug company executive; b. St. Paul, July 30, 1919; s. Joseph F. and Clara (Mueller) L.; m. Elaine J. Wilson, Mar. 18, 1944; children—Peter D., Jock D., Susan E. Student, U. Chgo., 1937-40; B.A., U. Minn., 1941, M.A., 1944; M.D., 1944. Intern, resident, fellow Yale Sch. Medicine and New Haven Hosp., 1944-49; instr., asst. prof. medicine Yale, 1949-53; profl. asso. NRC, 1950-53; asst. prof. medicine, asst. dean Sch. Medicine, U. N.C., 1953-55; prof. medicine, provost U. Ark.; dean U. Ark. Sch. Medicine, 1955-61; exec. dir. med. research Merck Sharp & Dohme Research Labs., West Point, Pa., 1961-66, v.p. med. research, 1966-69; prof. medicine, dean acad. affairs U. Tex. Southwestern Med. Sch., Dallas, 1969-72, dean, 1972-73; sr. v.p. sci. affairs Schering-Plough Corp., 1973-75, pres. research div., 1975-81, sr. v.p. sci., 1981—, also corporate dir. Served with M.C. USNR, 1946-47. Mem. N.Y. Acad. Sci., AAAS, Fedn. for Clin. Research, Nat. Soc. Med. Research (dir. 1981—), Am. Heart Assn., Acad. Hematology, Am. Coll. Cardiology, Sigma Xi. Subspecialties: Internal medicine; Hematology. Current work: Physician (internal medicine, hematology, oncology, chemotherapy) and executive in sciences, mainly in new biotechnology (molecular biology, genetic engineering, etc.). Home: 53 Spring Valley Rd Convent NJ 07961

LAWRENCE, CHRISTOPHER WILLIAM, biologist, educator; b. London, Oct. 2, 1934; s. William John Cooper and Edna Lesley (Yates) L.; m. Judiana Vorster, Aug. 5, 1961; children: Nicholas Ralph, Andrew Giles, Peter John. B.Sc., Univ. Coll. Wales, Aberystwyth, 1956; Ph.D., U. Birmingham, Eng., 1959. Sci. officer U.K. Atomic Energy Authority, Harwell, Eng., 1959-61, sr. sci. officer, 1961-69; assoc. prof. U. Rochester (N.Y.), 1969-82, prof., 1982—. Subspecialties: Molecular biology; Genetics and genetic engineering (biology). Current work: Molecular biology of mutagenesis in the lower eukaryote Saccharonmyces cerevisiae (baker's yeast). Home: 119 Village Ln Rochester NY 14610 Office: U Rochester Med Sch 260 Crittenden Blvd Rochester NY 14642

LAWRENCE, DALE NOLAN, medical research immunologist, educator; b. Covington, Ky., Feb. 24, 1944; s. Hershiel Steger and Mildred Lucille (Smalley) L.; m. Trude Sellin Lowenbach, Aug. 4, 1973; children: Hans, Elizabeth. Student in chemistry, U. Cin., 1962-65; M.D., Duke U., 1969; postgrad. in immunogenetics, U. Tex., 1972-73, in pub. health, Emory U., 1979—. Diplomate: Am. Bd. Internal Medicine, also Sub-Bd. Infectious Disease. Intern U. Tex. Health Sci. Ctr., 1969-70, resident in medicine, 1970-71, fellow in infectious disease, 1972-73; commd. lt. comdr. USPHS, 1973, advanced through grades to comdr., 1975; epidemic intelligence service officer (U.S. Ctrs. for Diseases Control), 1973-75, med. officer, Atlanta, 1975-77, 79-81, chief, 1981—, med. officer in charge epidemiology of blood product exposure, 1982—; vis. scientist Genetics Lab., U. Oxford, Eng., 1978-79; mem. med. adv. bd. Ctrs. for Disease Control; med. officer NSF Genetics Research Program, Amazonas, Brazil, 1976; clin. prof. epidemiology U. Miami, 1981—; founder, sponsor Windsor Scholarship, Cin. Scholarship Found., 1982—. Named outstanding resident physician clin. tchr. dept. medicine U. Tex. Health Scis. Ctr., San Antonio, 1973; Dean W. C. Davison scholar, 1969. Mem. Am. Soc. Human Genetics, Soc. Epidemiol. Research, AAAS, Am. Coll. Epidemiology, Am. Assn. Clin. Histocompatibility Testing, Am. Fedn. Clin. Research, Sigma Xi, Omicron Delta Kappa. Republican. Subspecialties: Epidemiology; Infectious diseases. Current work: Genetics of human immune response and disease susceptibility; epidemiology of blood product exposure, Acquired Immune Deficiency Syndrome Task Force, Centers for Disease Control. Office: Centers for Disease Control Atlanta GA 30333

LAWRENCE, DAVID A., immunologist; b. Paterson, N.J., Jan. 9, 1945; s. J. Arthur and Doris Helen (Graf) L.; m. Georgia Lawrence, Dec. 29, 1967. A.B., Rutgers U., 1966; M.S., Boston Coll., 1968; Ph.D., 1971. Research fellow Scripps Clinic and Research Found., LaJolla, Calif., 1971-74; mem. faculty dept. microbiology and immunology Albany (N.Y.) Med. Coll., 1974—, assoc. prof., 1977—. Sinsheimer Trust awardee, 1976-78; NIH grantee, 1976—; Am. Cancer Soc. grantee, 1976-78. Mem. Am. Assn. Immunologists, Am. Soc. Microbiologists, N.Y. Acad. Sci. Subspecialties: Immunobiology and immunology; Immunotoxicology. Current work: Immunotoxicology of heavy metals and chemical carcinogens; reactivities of T-lymphocyte subsets. Office: Dept Microbiology and Immunology Albany Med Coll Albany NY 12208

LAWRENCE, ERNEST CLINTON, physician, educator; b. Gilmer, Tex., Apr. 19, 1948; s. Price Winton and Mattie Florene (Welsh) L.; m. Cheryl Renee Craig, Dec. 28, 1969; children: Bryan Wayne, David Clinton, Craig Andrew. B.A., U. Tex.-Austin, 1973; M.D., U. Tex. Southwestern Med. Sch., Dallas, 1973. Diplomate: Am. Bd. Internal Medicine, Am. Bd. Allergy and Immunology. Intern Moffit Hosp., U. Calif., San Francisco, 1973-74; resident in medicine U. Calif. Affiliated Hosps., San Francisco, 1974-75; clin. assoc., investigator Metabolism Br., Nat. Cancer Inst., NIH, Bethesda, Md., 1975-78; pulmonary fellow Baylor Coll. Medicine, Houston, 1978-79, asst. prof. medicine, 1979—; dir. Rockwell Keough Pulmonary Immunology Lab. of Meth. Hosp., Houston, 1979—. Contbr. articles to profl. jours. Served to lt. comdr. USPHS, 1975-78. Tex. Merit scholar, 1971, 72; recipient student research award Southwestern Med. Sch., 1973; Pfizer Labs. award, 1973; Clin. Assoc. Physician award Gen. Clin. Research Ctr., Baylor Coll. Medicine, 1980-82. Mem. ACP, Am. Thoracic Soc., Am. Lung Assn., Am. Fedn. Clin. Research, Am. Assn. Immunologists, Harris County Med. Soc., Tex. Med. Assn. Methodist. Subspecialties: Pulmonary medicine; Immunology (medicine). Current work: Immune mechanism in normal human lungs and in inflammatory lung diseases; research and treatment of patients with inflammatory or fibrotic lung diseases, such as asbestosis, sarcoidosis, idiopathic pulmonary fibrosis and miscellaneous lung disease. Home: 10738 Valley Hills Houston TX 77071 Office: 6565 Fannin St Mail Sta F501 Houston TX 77030

LAWRENCE, ERNEST GREY, JR., plant pathologist; b. Charlotte, N.C., Oct. 22, 1952; s. Ernest Grey and Opal Pauline (Wright) L. B.S., Greensboro Coll., 1975; M.S., Clemson U., 1977, Ph.D., 1979. Research assoc. Va. Poly. Inst. and State U., Blacksburg, 1980-82; field research rep. Stauffer Chem. Co., Winston-Salem, N.C., 1982—. Contbr. articles to profl. jours. Mem. Am. Phytopath. Soc. Subspecialty: Plant pathology. Current work: Field testing of experimental pesticides. Office: 911 Motor Rd Winston-Salem NC 27105

LAWRENCE, GARY WRIGHT, plant pathologist, technical supervisor; b. Charlotte, N.C., June 24, 1954; s. Ernest Grey and Opal Pauline (Wright) L.; m. Amanda Minix, Nov. 26, 1954. B.A., Greensboro Coll., 1976; M.S., N.C. State U., 1979; Ph.D., La. State U., 1983. Grad. research asst. dept. plant pathology N.C. State U., 1976-78; grad. research asst. dept. plant pathology La. State U., 1979-82; tech. supr. Pennwalt Corp. (AgChem. Div.), Baton Rouge, 1982—. Contbr. articles to tech. jours. Mem. Am. Phytopath. Soc., Soc. Nematologists, Gamma Sigma Delta, Beta Iota Omega. Baptist. Subspecialties: Plant pathology; Integrated systems modelling and engineering. Current work: Vegetable crop disease; nematode management; assaying resistance-breaking races of nematodes; nematode population dynamics; pesticide research on agrinomic crops.

LAWRENCE, LEO ALBERT, nuclear scientist; b. Oakland, Calif., June 21, 1938; s. Albert and Teresa (Weise) L.; m. Terry M. Gulliksen, Aug. 22, 1964; children: John, Craig, Mark, Lisa. B.S. in Physics, Humboldt State U., 1964; M.S., U. Denver, 1967. Research scientist Batelle Northwest Labs., Richland, Wash., 1967-70; research scientist Westinghouse Hanford Co., Richland, 1970—, fellow scientist, 1979—. Contbr. many articles, papers to profl. lit. Bd. dirs. Wash. Jr. Golf Assn., Tacoma, 1982. NSF trainee, 1966. Mem. Am. Nuclear Soc. (materials sci. program chmn. 1982—), Sigma Phi Sigma (alumni). Roman Catholic. Club: Tri-City Country (Kennewick, Wash.). Subspecialties: Nuclear fission; Nuclear reactor fuels. Current work: In-reactor thermo-mechanical performance of mixed uranium-plutonium oxide fast breeder reactor fuels including advanced designs, fuel pin chemistry and fuel-cladding compatability. Home: 713 N Reed Pl Kennewick WA 99336 Office: Westinghouse Hanford Co PO Box 1970 W/E-1 Richland WA 99351

LAWRENCE, MERLE, med. educator; b. Remsen, N.Y., Dec. 26, 1915; s. George William and Alice Rutherford (Bowne) L.; m. Roberta Ashby Taylor Harper, Aug. 8, 1942; children—Linda Alice Lawrence Nolt, Roberta Harper Lawrence Henderson, James Bowne. A.B., Princeton, 1938, M.A., 1940, Ph.D., 1941. NRC fellow Johns Hopkins Hosp., 1941; asst. prof. psychology, Princeton, 1946-50, asso. prof., 1950-52; asso. research Lempert Inst. Otology, N.Y.C., 1946-52; asso. prof. dept. otolaryngology U. Mich. Med. Sch., 1952-57, prof. otolaryngology, 1957—; research asso. Inst. Indsl. Health, 1952—; prof. psychology U. Mich. Coll. Lit. Sci. and Arts, 1957—; dir. Kresge Hearing Research Inst., 1961—; mem. Nat. Adv. Neurol. and Communicative Disorders and Stroke Council, 1976-79; mem. communicative disorders research tng. com. Nat. Inst. Neurol. Diseases and Blindness, 1961-65; mem. communicative scis. study sect. div. research grants NIH, 1965-69, chmn., 1967-69, mem. communicative disorders rev. com., 1972-76. Served as naval aviator USNR, 1941-46, 50-51; PTO; Served as naval aviator USNR, Korean conflict; PTO. Decorated Purple Heart, Air medal with nine gold stars; recipient Sec. Navy Commendation, award merit Am. Acad. Ophthalmology and Otolaryngology, 1965, Assn. Research in Otolaryngology, 1979, Am. Otol. Soc., 1967, Distinguished Service award Princeton Class of 1938.; NRC fellow, 1941. Fellow Otosclerosis Study Group, Am. Laryngol., Rhinolog. and Otolaryngol. Soc.; mem. AAAS, Acoustical Soc. Am., Mich. Acoustical Soc. (pres. 1956), Am. Acad. Ophthalmology and Otolaryngology, Am. Otological Soc., Collegium Oto-Rhino-Laryngologicum Amicitiae Sacrum, Soc. U. Laryngologists, Assn. Research Otolaryngology, Am. Auditory Soc. (council 1978-82), Am. Tinnitus Assn., Quarter Century Wireless Assn. Clubs: Rotary (pres. 1978-79), Centurion, Mich. Masters Swim, Ann Arbor Racquet. Subspecialties: Neurophysiology; Sensory processes. Current work: Auditory physiology especially micro circulation of blood and lymph in the ear. Home: 2029 Vinewood Blvd Ann Arbor MI 48104

LAWS, D(ONALD) R(ICHARD), experimental psychologist, consultant; b. St. Louis, Nov. 7, 1934; s. Claude and Gertrude (Crow) L. B.A., U. Mo.-Columbia, 1959; M.A., So. Ill. U.-Carbondale, 1964, Ph.D., 1969. Med. research assoc. Behavior Research Lab., Anna, Ill., 1966-68; asst. prof. ednl. psychology So. Ill. U., 1968-69; exptl. psychologist Atascadero (Calif.) State Hosp., 1970—; cons. behavioral assessment, program devel., equipment design; v.p. Pacific Profl. Assocs., Sherman Oaks, Calif., 1980—. Contbr. chpts., articles to profl. publs.; reviewer for various profl. jours., 1973—. Served with U.S. Army, 1955-57. Mem. Am. Psycol. Assn., Assn. Advancement Behavior Therapy, Brit. Assn. Behavioral Psychotherapy. Subspecialties: Behavioral psychology; Physiological psychology. Current work: Design, development, implementation of behavioral assessment; behavior therapy and cognitive behavior therapy for sexual disorders, especially sexual deviance; human sexual behavior. Home: 350 Bernardo Morro Bay CA 93442 Office: Sexual Behavior Lab Atascadero State Hosp 10333 El Camino Real Atascadero CA 93423

LAWSON, ALFRED JAMES, radiation biologist, oncologist; b. Welland, Ont., Can., May 2, 1950; came to U.S., 1953, naturalized, 1963; s. Walter Samuel and Mary (Waddell) L.; m. Margaret Mary Heurkens, Aug. 18, 1973; 1 son, Ian. B.A., SUNY-Buffalo, 1972, Ph.D., 1977. Post-doctoral fellow U. Iowa, Iowa City, 1976-78; asst. prof. radiation oncology U. Ala. in Birmingham, 1978—, asst. prof. pathology, 1981, dir. histology facility, 1979—, assoc. dir. radiation facility, 1979—, adir. div. radiation biology, 1981—. Contbr. articles to profl. jours. Nat. Cancer Inst. grantee, 1979—. Mem. Radiation Research Soc., N.Y. Acad. Sci., Ala. Soc. Radiation Oncologists, Ala. Soc. Histotech., Sigma Xi. Subspecialties: Radiology; Pathology (medicine). Current work: Combined modality treatment of cancer, radiation pathology of normal tissues, gastrointestinal, bone marrow transplant, nuclear magnetic resonance spectroscopy. Offfice: U Ala Dept Radiation Oncology PO Box 441 Volker Hall Birmingham AL 35294

LAWSON, EDWARD EARLE, pediatrician, educator, researcher; b. Winston-Salem, N.C., Aug. 6, 1946; s. Robert Barrett and Elsie Chatterton (Earle) L.; m. Rebecca Newhall Fitts, June 21, 1969; children: Katherine Tabor, Robert Barrett. B.A., Harvard U., 1968; M.D., Northwestern U., 1972. Intern, resident Children's Hosp. Med. Ctr., Boston, 1972-75; research fellow Harvard U. Med. Sch., 1975-78; asst. prof. pediatrics U. N.C.-Chapel Hill, 1978-82, assoc. prof., 1982—. Recipient Research Career Devel. award NIH, 1982, Sidney Farber Research Grant award United Cerebral Palsy, 1983. Fellow Am. Acad. Pediatrics, Am. Thoracic Soc.; mem. Am. Physiol. Soc., Soc. Pediatric Research. Subspecialties: Pediatrics; Physiology (biology). Current work: Central neural control of breathing in newborns. Office: U NC Dept Pediatrics Chapel Hill NC 27514

LAWSON, JAMES EDWARD, JR., geophysicist; b. Lebanon, Mo., Dec. 27, 1938; s. James Edward and Edwina Alma (Buley) L.; m. Carole June Horton, Aug. 22, 1981; 1 dau., Adrianna Vashti. B.S., U. Tulsa, 1965, B.A., 1965, M.S., 1968, Ph.D., 1972. Geophysicist Earth Scis. Obs., U. Okla., 1970-78, adj. prof., 1978—; chief geophysicist Okla. Geol. Survey, 1978—. Author: Earthquake Map of Oklahoma, 1979. Vol. cons. Okla. Red Cross Blood Ctr. Reference Lab., 1968—. Mem. Seismological Soc. Am. (life), Am. Geophys. Union (life), Internat. Soc. Blood Transfusion, Sigma Xi (life). Subspecialty: Geophysics. Current work: Earthquake seismology. Home: Stonebluff Route 2 Box 85-C Haskell OK 74436 Office: Okla Geophysical Obs Box 8 Leonard OK 74043

LAWTON, EMIL ABRAHAM, chemist; b. Detroit, Oct. 12; s. Irvin and Jennie (Belkin) L.; m. Renee Burk, Dec. 7, 1967; children: Gil M., Ron D.; m. Cynthia Ann Frumhoff, Mar. 12, 1976. B.S., Wayne State U., 1946; Ph.D., Purdue U., 1952. Chemist Nat. Bur. Standards, Washington, 1952-53; project leader, prin. scientist Battelle Meml. Inst., Columbus, Ohio, 1953-57; from mgr. to program mgr. Rocketdyne, Canoga Park, Calif., 1957-72; sect. head. Thiokol Chem. Corp., Santa Monica, Calif., 1972-75; v.p. research Tech. Mgmt. Cons., Woodland Hills, Calif., 1976-77; mgr. advanced programs Shock Hydrodynamics, North Hollywood, Calif., 1977—. Contbr. articles to profl. jours. Served to lt. (j.g.) USNR, 1943-46. Mem. Am. Chem. Soc., Am. Solar Energy Soc., Sigma Xi, Phi Lambda Upsilon. Club: Sierra. Subspecialties: Polymer chemistry; Inorganic chemistry. Current work: Polymeric absorbants, pulse combustion, synthesis, propellants, fuels, inorganic chemistry. Patentee in field. Office: 4716 Vineland Ave North Hollywood CA 91602 Home: 13025 Hesby St Sherman Oaks CA 91423

LAX, MELVIN DAVID, mathematics educator; b. Boston, March 20, 1947; s. Martin and Mildred (Berman) L. B.S., Rensselaer Poly. Inst., 1969, M.S., 1971, Ph.D., 1974. Lectr., So. Ill. U., Carbondale, 1974-76; vis. asst. prof. Okla. State U., Stillwater, 1976-77; asst. prof. Calif. State U., Long Beach, 1977-81, assoc. prof., 1981—; panel rev. Zentralblatt fur Mathematik und ihre Grenzgebiete, Berlin, 1977—. Contbr. articles to books, profl. jours. Mem. Am. Math. Soc., Soc. Indsl. and Applied Math., Pi Mu Epsilon. Subspecialty: Applied mathematics. Current work: Approximate solution of random differential equations and random integral equations. Office: Calif State U Dept Math Long Beach CA 90840

LAX, PETER DAVID, mathematician; b. Budapest, Hungary, May 1, 1926; came to U.S., 1941, naturalized, 1944; s. Henry and Klara (Kornfeld) L.; m. Anneli Cahn, 1948; children: John, James D. B.A., N.Y. U., 1947, Ph.D., 1949; D.Sc. (hon.), Kent State U., 1976; D. honoris causa, U. Paris, 1979. Asst. prof. N.Y. U., 1949-57, prof., 1957—; dir. Courant Inst. Math. Scis., 1972-80. Author: (with Ralph Phillips) Scattering Theory, 1967, Scattering Theory for Automorphic Functions, 1976, (with A. Lax and S.Z. Burstein) Calculus with applications and computing), 1976, Hyperbolic Systems of Conservation Laws and the Mathematical Theory of Shock Waves, 1973. Mem. Pres.'s Com. on Nat. Medal of Sci., 1976, Nat. Sci. Bd., 1980—. Served with AUS, 1944-46. Recipient Semmelweis medal Semmelweis Med. Soc., 1975. Mem. Am. Math. Soc. (pres. 1979-80, Norbert Wiener prize), Nat. Acad. Scis. (applied math. and numerical analysis award 1983), Am. Acad. Scis., Math. Assn. Am. (bd. govs., Chauvenet prize), Soc. Indsl. and Applied Math.; fgn. asso. Académie des Scis. (France). Subspecialties: Numerical analysis; Applied mathematics. Home: 300 Central Park W New York NY 10024 Office: 251 Mercer St New York NY 10012

LAXER, CARY, computer science educator; b. Bklyn., July 16, 1955; s. Stanley and Harriet (Greenbaum) L. B.A., N.Y.U., 1976; Ph.D., Duke U., 1980. Research asst. prof. computer engring. Duke U., Durham, N.C., 1980-81; asst. prof. computer sci. and elec. engring. Rose-Hulman Inst. Tech., Terre Haute, Ind., 1981—. Mem. Big Bros., Durham County Social Services, Durham, N.C., 1976-81, Big Bros./ Sisters Vigo County, Terre Haute, Ind., 1981—; religious sch. tchr. United Hebrew Congregation, Terre Haute, 1982—. Mem. IEEE, Assn. Computing Machinery, Biomed. Engring. Soc., Am. Soc. Engring. Edn. (sect. treas. Ind./Ill. sect. 1982—), AAUP. Jewish. Subspecialties: Education biomedical applications; Biomedical engineering. Current work: Computer analysis of cardiac electrical potentials to determine presence of myocardial disease. Home: 4330 S 5th St Apt 1 Terre Haute IN 47802 Office: Rose-Hulman Inst Tech 5500 Wabash Ave Terre Haute IN 47803

LAYMAN, DONALD K(EITH), nutritional biochemist; b. Kewanee, Ill., Feb. 15, 1950; s. Milan Henry and Pearl (Gleich) L.; m. Gayle S. Garrison, Aug. 4, 1973; 1 son, Michael. B.S., Ill. State U., 1972, M.S., 1974; Ph.D., U. Minn.-St. Paul, 1977. Asst. prof. nutrition U. Ill., Urbana, 1978—; nutrition cons. Regional Office of Edn., State of Ill., 1979-80. Research grantee Child Health and Human Devel., NIH, 1982. Mem. Am. Inst. Nutrition, N.Am. Soc. Study of Obesity, Ill. Nutrition Com., Sigma Xi. Lutheran. Subspecialties: Nutrition (medicine); Biochemistry (medicine). Current work: Protein and energy nutrition, controls of protein synthesis, obesity, muscle cell growth, exercise and nutrition. Home: 2007 Winchester Dr Champaign IL 61820 Office: Dept Foods and Nutrition Univ Ill 905 S Goodwin Urbana IL 61801

LAYMON, STEPHEN ALAN, mechanical engineer; b. San Francisco, Dec. 18, 1946; s. Charles Richard and Evelyn Alice (Lust) L.; m. Nancy Elizabeth Roth, July 14, 1979; 1 dau., Ashley Elizabeth. B.S. in Engring. Sci, Purdue U., 1968; M.S. in Engring. Mechanics, Ill. Inst. Tech., 1971. Registered profl. engr., Ind. Project engr. FMC, Corp., Tipton, Ind., 1972-73; mgr. turbocharger devel. Schwitzer div. Household Internat., Indpls., 1973—. Served with U.S. Army, 1969-71. Mem. ASME, ASTM, Am. Soc. Metals. Subspecialties: Solid mechanics; Materials. Current work: Manage product development and qualification testing. Home: 5544 Meckes Ln Indianapolis IN 46237 Office: Schwitzer div Household Internat 1125 Brookside Ave Indianapolis IN 46202

LAYTON, RICHARD GARY, physics educator, researcher; b. Salt Lake City, Dec. 24, 1935; s. Lynn Cornell and and Leone Gardiner (Gedge) L.; m. Susan Emily Brinkman, Dec. 27, 1963; children: Catherine Louise, Paul Richard, Spencer Lee. B.A., U. Utah, 1960, M.A., 1963; Ph.D., Utah State U., 1965. Asst. research physicist, electrodynamic lab. Utah State U., Logan, 1963-65; asst. prof. physics SUNY-Fredonia, 1965-68, assoc. prof., 1968-69; assoc. prof. physics No. Ariz. U., Flagstaff, 1969—; research assoc. Lowell Obs., Flagstaff, summer 1970, 71; sci. collaborator Grand Canyon Nat. Park, Ariz., 1971-75. Contbr. articles to sci. jours. Mem. exec. bd. Grand Canyon Council Boy Scouts Am., 1973. Mem. Am. Phys. Soc., Am. Assn. Physics Tchrs. (award 1971), Am. Meteorol. Soc., Internat. Assn. Colloid and Interface Scientists, Sigma Xi. Republican. Mem. Ch. Jesus Christ of Latter-day Saints. Subspecialties: Surface chemistry; Cloud physics. Current work: Ice nucleation on inorganic and organic (including bacteria) substrates and related effects in cloud and precipitation physics as well as in frost damage to plants and its prevention. Patentee in field. Home: 655 E David Dr Flagstaff AZ 86001 Office: Dept Physics No Ariz U Flagstaff AZ 86011

LAZARCHICK, JOHN, physician, medical educator; b. Pottsville, Pa., Nov. 1, 1942; s. John and Ann (Peshock) L.; m. Lynda L. Cabashinsky, 1960; 1 son, Jeffrey. B.A., Lafayette Coll., 1964; M.D., Jefferson Med. Coll., 1968. Diplomate: Am. Bd. Internal Medicine. Resident Mayo Clinic, Rochester, Minn., 1969-71; fellow Mass. Gen. Hosp., Boston, 1971-72, U. Conn. Health Ctr., Farmington, 1975-77, asst. prof., 1977-79; asst. prof. div. hematology Med. U. S.C., Charleston, 1979-82, assoc. prof., 1982—, assoc. dir. hematology-hemostasis labs., 1982—; cons. S.C. Med. Consortium. Contbr. articles to profl. jours. Served to lt. comdr. USN, 1972-75. Nat. Hemophilia Found. grantee, 1975-77; NIH grantee, 1980-83. Mem. Am. Soc. Hematology, Am. Fedn. Clin. Research, AAAS, Assn. Clin. Scientists. Republican. Roman Catholic. Subspecialties: Internal medicine; Hematology. Current work: Study of the mechanism of action of factor VIII Antibodies. Home: 798 Creekside Dr Mount Pleasant SC 29464 Office: Med U SC 171 Ashley Ave Charleston SC 29403

LAZARETH, OTTO WILLIAM, physicist; b. Bklyn., Sept. 16, 1938; s. Otto William and Marie (Moller) L.; m. Victoria McLane, June 1, 1977. B.S. in Physics, Wagner Coll., 1961, M.S., Queens coll., CUNY, 1967, Ph.D., 1973. Computer programmer Brookhaven Nat. Lab., Upton, N.Y., 1967-73, physicist, 1974—. Served with U.S. Army, 1960-63. Mem. Am. Phys. Soc., Am. Nuclear Soc. Subspecialties: Nuclear fusion; Materials. Current work: Radiation effects in solids, computer modeling of atomic and nuclear systems. Office: Brookhaven Nat Lab Bldg 510 B Upton NY 11973

LAZARUS, ALLAN K., chemist, educator; b. Bangor, Maine, May 20, 1931; s. Julius E. and Ruth (Hecht) L.; m. Gloria Berkowitz, Dec. 24, 1957; children: Carol R., Martin C., Warren R. B.A., N.Y.U., 1952; M.S., 1955; Ph.D., 1957. Group leader Cities Service Research & Devel. Co., Cranbury, N.J., 1957-59; chemist FMC Corp., Princeton, 1959-65, Esso Research & Engring. Co., Linden, N.J., 1965-66; group leader Tenneco Chems. Inc., Piscataway, N.J., 1966-71; asst. prof. chemistry Trenton (N.J.) State Coll., 1972—. Contbr. articles to profl. jours. Allied Chem. and Dye Corp. fellow, 1956. Mem. Am. Chem. Soc., Phi Beta Kappa, Sigma Xi, Phi Lambda Upsilon. Subspecialties: Organic chemistry; Synthetic lubricants. Current work: Synthesis of high-molecular-weight monomeric esters of potential utility as base fluids for synthetic lubricating oils and greases. Patentee in field. Home: 149 Lawr-Penn Rd Lawrenceville NJ 08648 Office: Dept Chemistry Trenton State Coll CN 550 Trenton NJ 08625

LAZARUS, ARNOLD ALLAN, clinical psychologist; b. Johannesburg, South Africa, Jan. 27, 1932; came to U.S., 1966, naturalized, 1976; s. Benjamin and Rachel Leah (Mosselson) L.; m. Daphne Ann Kessel, June 10, 1956; children: Linda Sue, Clifford Neil. B.A. with honors, U. Witwatersrand, Johannesburg, 1956, M.A., 1957, Ph.D., 1960. Diplomate: Am. Bd. Profl. Psychology. Pvt. practice clin. psychology, South Africa, 1959-63; vis. asst. prof. dept. psychology Stanford U., 1963-64; prof. psychology Temple U. Med. Sch., 1967-70; dir. clin. tng. Yale U., 1970-72, disting. prof. psychology Rutgers U., New Brunswick, N.J., 1972—; exec. dir., founder Multimodal Therapy Inst., Kingston, N.J. and N.Y.C., 1976—; pvt. practice psychotherapeutics, Princeton, N.J., cons. in field. Author: 8 books including Behavior Therapy and Beyond, 1971, Multimodal Behavior Therapy, 1976; 9 books including The Practice of Multimodal Therapy, 1981; editorial bd. profl. jours.; contbr. articles to profl. jours. Recipient Disting. Service award Am. Bd. Profl. Psychology. Fellow Am. Psychol. Assn.; mem. Nat. Acads. of Practice, Am. Acad. Psychotherapy, Assn. Advancement Behavior Therapy (past pres.), Assn. Advancement Psychotherapy., Internat. Acad. Profl. Counseling and Psychotherapy (diplomate). Subspecialties: Behavioral psychology; Cognition. Current work: Discovering rapid and durable methods of psychotherapy in the traeatment of anxiety, depression and marital family discord. Home: 56 Herrontown Circle Princeton NJ 08540 Office: Psychology Bldg Busch Campus Rutgers Univ New Brunswick NJ 08903 "Whatever modicum of success" I may have achieved is probably due, in large part, to my view of parity as a way of life. I am committed to the notion that there are no superior human beings — we are all different, indeed unique, but equal. While some people possess superior skills and abilities, this does not make them superior human beings. To respect others for their exceptional capacities, but never to deify them, enables one to learn from others instead of envying them and denigrating oneself. This egalitarian view transforms acquisitiveness, power, and agression into love, intimacy, and productive activity.

LAZERSON, JACK, physician, researcher; b. N.Y.C., Jan. 9, 1936; s. Mayer and Jennie (Gerson) L.; m. Eleanor Marie Ianniccari, Mar. 17, 1962 (div. 1973); children: Deborah Ann, Darlene Ann, Donna Ailene, David Aaron; m. Cheryl Justine Lello, Nov. 17, 1979; 1 son, Samuel Aaron. B.A., NYU, 1957; M.D., Chgo. Med. Sch., 1961. Sr. fellow biochemistry U. Wash. Sch. Medicine, Seattle, 1966-69, instr. pediatrics, 1966-69; asst. prof. pediatrics Stanford U. Sch. Medicine, Palo Alto, Calif., 1969-72, U. So. Calif. Sch. Medicine, 1972-76; assoc. prof. pediatrics Med. Coll. Wis., Milw., 1976-79; prof. pediatrics/ pathology U. Calif.-Davis Sch. Medicine, 1979—; med. dir. Great Lakes Hemophilia Found., Milw., 1972-76. Contbr. articles to profl. jours. Served to capt. USAF, 1964-66. Named Disting. Alumnus, Chgo. Med. Sch., 1980. Democrat. Jewish. Subspecialties: Pediatrics; Hematology. Current work: Biochemistry of hemostasis, clinical evaluation of hemostatic disorders of childhood. Home: 205-2 Selby Ranch Rd Sacramento CA 95825 Office: U Calif Sch Medicine 4301 X St Sacramento CA 95817

LAZO, JOHN STEPHEN, pharmacology educator, scientist; b. Phila., Dec. 15, 1948; s. John and Mildred Doris (Popowich) L.; m. Jacqui Lynne Fiske, Oct. 12, 1974; 1 dau., Jacquelyn Kristina Fiske. A.B. in Chemistry, Johns Hopkins U., 1971; Ph.D. in Pharmacology, U. Mich., 1976. USPHS-NIH predoctoral research trainee dept. pharmacology U. Mich., Ann Arbor, 1971-76; Am. Cancer Soc. postdoctoral fellow, dept. pharmacology Yale U. Sch. Medicine, New Haven, 1976-78, asst. prof., 1978-83, assoc. prof., 1983—. Author: (with A.C. Startorelli and J.R. Bertino) Molceular Actions and Targets for Cancer Chemotherapeutic Agents, 1981, (with P.S. Dannies and J. W. Kozarich) Pharmacology: Pretest Self-Assessment and Review; contbr. numerous articles on pharmacology to profl. jours. NIH fellow, 1978. Mem. N.Y. Acad. Scis., Am. Soc. for Pharmacology and Exptl. Therapeutics, AAAS, Am. Assn. Cancer Research, Tissue Culture Assn. Subspecialties: Pharmacology; Cancer research (medicine). Current work: Pharmacology of anticancer agents, cancer, drugs, biochemical, treatment. Home: 262 Stonehedge Ln Guilford CT 06437 Office: 333 Cedar St New Haven CT 06510

LAZOWSKA, EDWARD DELANO, computer scientist educator; b. Washington, Aug. 3, 1950. A.B., Brown U., 1971; M.S., U. Toronto, 1974, Ph.D., 1977. Assoc. prof. dept. computer sci. U. Wash., Seattle. Subspecialties: Distributed systems and networks; Operating systems. Current work: Computer Systems: modelling and analysis, design and implementation, distributed systems. Office: Computer Science FR-3 University of Washington Seattle WA 98195

LAZZARI, EUGENE PAUL, biochemistry educator; b. Archibald, Pa., Mar. 12, 1931; m. Martha N. Neumann, June 23, 1973; children: Victoria Ennis, Katherine Ennis. B.S., U. Scranton, 1953; M.A., Williams Coll., 1955; Ph.D., Iowa State U., 1961. Research assoc. Brookhaven Nat. Labs., Upton, N.Y., 1961-62, City of Hope Med. Ctr., Duarte, Calif., 1962-63; mem. faculty Dental Br., U. Tex., Houston, 1963—, prof. biochemistry, 1974—. Editor: Dental Biochemistry, 1968, Experimental Aspects of Oral Biochemistry, 1983. NASA summer faculty fellow, 1969. Mem. Internat. Assn. for Dental Research, Am. Chem. Soc., AAUP, Sigma Xi. Democrat. Subspecialties: Biochemistry (biology); Organic chemistry. Current work: Amino acid derivatization, peptide and protein isolation and sequence determination. Office: Texas Dental Branch Tex Med Ctr Houston TX 77025

LEA, JEAN HEDRICK, diagnostic medical sonographer; b. Midland, Tex., Apr. 7, 1949; d. Ralph Randolph and Anne Hedrick L.; m. Dan Spencer Spitz, Apr. 7, 1979. B.A., U. Tex., Austin, 1970; M.P.H., U. Okla., 1980. Pub. health nutritionist U. Miss., Jackson, 1972-74; sonographer, supr. U. Rochester, N.Y., 1974-76, assoc. faculty, 1976-77; asst. prof. diagnostic sonography U. Okla. Coll. Allied Health, Oklahoma City, 1977-80, assoc. prof., 1980—; mem. task force on safety, efficacy and use diagnostic ultrasound in obstets. Nat. Inst. Child Health Human Devel., 1983. Columnist: Med. Ultrasound, 1982—. Mem. Soc. Diagnostic Med. Sonographers (chmn edn. com. 1978-81, mem. legis. task force 1980—), Am. Inst. Ultrasound Medicine. Democrat. Methodist. Subspecialties: Imaging technology; Perinatal diagnosis and therapy. Current work: Obstetrical ultrasound, gynecology. Home: 10920 Woodbridge Rd Oklahoma City OK 73132 Office: Okla Coll Allied Health PO Box 26901 Oklahoma City OK 73107

LEACH, DAVID GOHEEN, geneticist, researcher, author; b. New Bethlehem, Pa., Jan. 18, 1913; s. Andrew Steelman and Nellie (Goheen) L.; m.; children: Robin, David Goheen, Brian. B.S., Coll. Wooster, Ohio, 1934, D.Sc., 1966. Engaged in research and bybridization of ornamental plants, 1953—, subsidiary research, nutrition pathology, entomology, physidogy and propagation of ericaceous plants, 1953—, research, toxicology of ericaceous plants, 1966-68, 80-83; dir. expt. sta. for hybridization of woody plants, Brookville, Pa., then Madison, Ohio, 1972—, extensive overseas plant exploration; lectr., U.S., overseas.; chmn. Internat. Conf. Chemotaxonomy, N.Y. Bot. Garden, 1978; co-chmn. Internat. Conf. Taxonomy, Royal Bot. Garden, Edinburgh, Scotland, 1982. Author: Rhododendrons of the World, 1961, Hybrids and Hybridizers, 1978; contbr. numerous articles to profl. jours. and encys. Bd. dirs. Western Res. Fine Arts Assn., East Suburban Concerts, Met. Outdoor Family Ctr., YMCA, all Northeastern Ohio. Gold medal of honor Garden Club Am., 1969; Cert. of Merit, Pa. Hort. Soc., 1970; Jackson Dawson Meml. medal Mass. Hort. Soc., 1972; Research award Nurserymen's Assn., 1980. Mem. Am. Soc. Hort. Sci., Am. Hort. Soc. (Recognition award 1974), Am. Assn. Bot. Gardens and Arboreta, Am. Rhododendron Soc. (Gold medal 1965), Internat. Plant Propagators Soc., Royal Hort. Soc. (Loder Gold Cup award 1969), Australian Rhododendron Soc., others. Subspecialty: Plant genetics. Current work: Research in gamma radiation effects, induced mutations, cytoplasmic inheritance, pigment inheritance and other characteristics for improved hybrids of ornamental plants.

LEACH, FRANKLIN ROLLIN, biochemistry educator; b. Gorman, Tex., Apr. 2, 1933; s. Frank Rollin and Jewel (Casey) L.; m. Anna Belle Coke, Feb. 27, 1970; 6 children; 3 stepchildren. Student, Cicso Jr. Coll, 1950-51; B.A. summa cum laude, Hardin-Simmons U., 1953; Ph.D., U. Tex.-Austin, 1957. Nat. research fellow in med sci. U. Calif.-Berkeley, 1957-59; research fellow in biology Calif. Inst. Tech., 1965-66; asst. prof. Okla. State U., 1960-63, assoc. prof., 1963-68, prof., 1968—. Contbr. articles to profl. jours. Rosalie B. Hite Cancer fellow U. Tex.; NRC fellow, 1957-58; Research Career Devel. award NIH, 1962-72. Mem. Am. Soc. Biol. Chemists, Am. Chem. Soc., Am. Soc. Microbiology. Club: Stillwater Flying (pres. 1968—). Subspecialties: Biochemistry (biology); Microbiology. Current work: Analytical biochemistry, enzymology, bacterial transformation. Home: 1101 N Lincoln Stillwater OK 74075 Office: Dept Biochemistry Okla State U Stillwater OK 74078

LEACH, JEANETTE, geneticist; b. Corvallis, Oreg., May 26, 1953; d. Charles Morley and Dorothy Jean (Cort) L. Student, Reed Coll., 1971-73; B.Sc., Oreg. State U., 1976; Ph.D., Cornell U., 1981. Postdoctoral research assoc. dept. botany U. B.C. (Can.), Vancouver, 1982—. Mem. Am. Phytopath. Soc., AAAS, Genetics Soc. Am., Sierra Club. Subspecialties: Gene actions; Plant pathology. Current work: Genetics of plant-pathogen interaction. Office: Dept Botany U BC Vancouver BC Canada V6T 2B1

LEACOCK, ROBERT ARTHUR, physics educator; b. Detroit, Oct. 3, 1935; s. Robert Cowles and Kathleen Frances (Maguire) L.; m. Jean Mackenzie Searles, Dec. 2, 1961; children: Nina Kathleen, Nicole Irene, Alexis Catherine. B.S., U. Mich., 1957, M.S., 1960, Ph.D., 1963. Instr. U. Mich., Ann Arbor, 1963-64; vis. scientist CERN, Geneva, 1964-65; postdoctoral assoc. Ames (Iowa) Lab., 1965-67; asst. prof. Iowa State U., Ames, 1967-71, assoc. prof. physics, 1971—. Contbr. articles to profl. jours. Corning Glass Works fellow U. Mich., 1962-63; Am.-Swiss Found. Sci. Exchange fellow CERN, 1964-65. Mem. Am. Phys. Soc., Sigma Xi. Subspecialties: Particle physics; Theoretical physics. Current work: Construction of new quantum mechanical formalisms, application to phenomena of particle physics, especially hadronic physics. Home: 2323 Donald St Ames IA 50010 Office: Dept Physics Iowa State U Ames IA 50010

LEADON, BERNARD MATTHEW, engineering educator, consultant; b. Farmington, Minn., Nov. 29, 1917; s. Bernard Matthew and Ruth Cecilia (McDonnell) L.; m. Ann Teresa Sweetser, Sept. 21, 1946; children: Bernard Matthew, Christopher, Mary, Thomas, Mark, Monica, Michael, Margaret, Catherine, Francis. B.S., Coll. St. Thomas, St. Paul, 1938; M.S., U. Minn., 1942, Ph.D., 1955. Registered profl. engr., Fla. Engr.'s aide Pacific Gas & Electric Co., San Francisco, 1941; instr. U. Minn., Mpls., 1942-43, lectr.-scientist, 1946-57; sr. aerodynamicist Cornell Aero Lab., Buffalo, 1943-46; sr. staff scientist Gen. Dynamics Corp., San Diego, 1957-64; prof. engring. sci. U. Fla., Gainesville, 1964—; cons. to various groups including U.S. Air Force, Pratt & Whitney Aircraft Group. Contbr. chpts. to tech. books, articles to profl. jours.; editor: Procs. 3d Nat. Conf. Wind Engring. Research, 1978. NATO fellow, Eng., 1973. Assoc. fellow AIAA; mem. Am. Phys. Soc., Sigma Xi. Roman Catholic. Subspecialties: Aerospace engineering and technology; Theoretical and applied mechanics. Current work: Full scale measurements of wind pressures on tall buildings, film cooling of turbine blades. Home: 412 N E 13th Ave Gainesville FL 32601 Office: Dept Engring Sci U Fla Gainesville FL 32611

LEAF, ALEXANDER, physician, educator; b. Yokohama, Japan, Apr. 10, 1920; came to U.S., 1922, naturalized, 1936; s. Aaron L. and Dora (Hural) L.; m. Barbara Louise Kincaid, Oct. 1943; children—Caroline Joan, Rebecca Louise, Tamara Jean. B.S., U. Wash., 1940; M.D., U. Mich., 1943; M.A., Harvard, 1961. Intern Mass. Gen. Hosp., Boston, 1943-44, mem. staff, 1949—, physician-in-chief, 1966-81; resident Mayo Found., Rochester, Minn., 1944-45; research fellow U. Mich., 1947-49; practice internal medicine, Boston, 1949—; faculty Med. Sch., Harvard, 1949—, Jackson prof. clin. medicine, 1966-81, Ridley Watts prof. preventive medicine, 1980—, chmn. dept. preventive medicine and clin. epidemiology, 1980—. Served to capt. M.C. AUS, 1945-46. Recipient Outstanding Achievement award U. Minn., 1964; vis. fellow Balliol Coll., Oxford, 1971-72; Guggenheim fellow, 1971-72. Fellow Am. Acad. Arts and Scis.; mem. Am. Soc. Clin. Investigation (past pres.), Am. Physiol. Soc., Biophys. Soc., Assn. Am. Physicians, Nat. Acad. Sci., Inst. Medicine, A.C.P. (master). Subspecialties: Internal medicine; Nephrology. Current work: Renal physiology and ion transport; cardiovascular risk factor reduction. Home: 1 Curtis Circle Winchester MA 01890 Office: Mass Gen Hosp Boston MA 02114

LEAHY, DENIS ALAN, scientist; b. Taber, Alta., Can., June 13, 1952; s. George Denis and Frances Carmen (Morton) L.; m. Judith Susan Lorenz, July 4, 1981. B.S. in Physics, U. Waterloo, Ont., Can., 1975; M.S. in Physics (Nat. Scis. and Engring. Research Council Can. scholar 1975-79), 1976, Ph.D., 1980. Student research asst. NRC, Ottawa, Ont., Can., 1975; resident research assoc. Marshall Space Flight Ctr., Ala., 1980-82; mem. faculty dept. physics U. Calgary, Can., 1982—. Contbr. articles to profl. publs. Mem. Am. Astron. Soc. Subspecialties: 1-ray high energy astrophysics; Theoretical astrophysics. Current work: X-ray data analysis and interpretation in terms of theoretical models, computer programming. Office: Dept Physics U Calgary Calgary Canada T2N1N4

LEAKE, DONALD LEWIS, oral and maxillofacial surgeon; b. Cleveland, Okla., Nov. 6, 1931; s. Walter Wilson and Martha Lee (Crow) L.; m. Rosemary Dobson, Aug. 20, 1964; children: John Andrew Dobson, Elizabeth, Catherine. A.B., U. So. Calif., 1953, M.A., 1957; D.M.D., Harvard U., 1962; M.D., Stanford U., 1969. Intern Mass. Gen. Hosp., Boston, 1962-63, resident, 1963-64; postdoctoral fellow Harvard U., 1964-66; practice medicine specializing in oral and maxillofacial surgery; asso. prof. oral and maxillofacial surgery Harbor-UCLA Med. Center, Torrance, 1970-74, dental dir., chief oral and maxillofacial surgery, 1970—, prof., 1974—; asso. dir. UCLA Dental Research Inst., 1979-82, dir., 1982—; cons. to hosps.; dental dir. coastal health services region, Los Angeles County, 1974-81. Contbr. articles to med. jours. Recipient 1st prize with greatest distinction for oboe and chamber music Brussels Royal Conservatory Music, Belgium, 1956. Fellow ACS; mem. Internat. Assn. Dental Research, So. Calif. Soc. Oral and Maxillofacial Surgeons, Internat. Assn. Oral Surgeons, AAAS, Soc. for Biomaterials, ASTM, European

Assn. Maxillofacial Surgeons, Brit. Assn. Oral Surgeons, Internationale Gesellschaft fur Kiefer-Gesichts-Chirurgie, Phi Beta Kappa, Phi Kappa Phi. Club: Harvard of Boston. Subspecialties: Oral and maxillofacial surgery; Biomaterials. Current work: Biomarerials for reconstructive surgery. Home: 2 Crest Rd W Rolling Hills CA 90274 Office: Harbor-UCLA Med Center 1000 W Carson St Torrance CA 90509

LEAMNSON, ROBERT NEAL, biology educator, researcher, cons.; b. Evansville, Ind., July 16, 1932; s. Robert Clinton and Mary Catherine (Farley) L. B.S., U. Notre Dame, 1955; M.S., 1965; Ph.D., U. Ill., 1973. Tchr. secondary schs., Ind., Ohio, 1955-65; research project chemist CTS Research, West Lafayette, Ind., 1966-67; asst. chemist Ill. State Geol. Survey, Urbana, 1967-69; asst. prof. research Wistar Inst., Phila., 1974-77; asst. prof. biology Southeastern Mass. U., 1978—; researcher, cons. Children's Hosp., Boston, Brown U. Contbr. articles to profl. publs. Mem. AAAS, Am. Soc. Microbiology, N.Y. Acad. Sci. Subspecialties: Virology (biology); Molecular biology. Current work: Molecular biology of viral proteins and nucleic acids; isoelectric points of viral proteins; functional sequences of nucleic acids of retroviruses. Office: Southeastern Mass U Dept Biology North Dartmouth MA 02747

LEAN, ERIC G., electrical engineer; b. Fukien, China, Jan. 11, 1938; m. Alice T. Kiang; children: Eugenia, Angela. B.S.C., Cheng-Kung U., Taiwan, 1959; M.S., U. Wash., 1963; Ph.D., Stanford U., 1967. With IBM Research Ctr., Yorktown Heights, N.Y., 1967—, mgr. acoustic physics, 1968-74, mgr. optical solid state tech., 1974-79, mgr. printing tech., 1979-81, mgr. output device technologies, 1981—. Contbr. chpts to books, articles to profl. jours; mem. editorial bd.: Applied Physics Letters, 1975-78; patentee in field. Mem. IEEE (sr. mem.), Optical Soc. Am., Sigma Xi. Subspecialty: Civil engineering. Current work: Device, material and application of printer and display technologies. Patentee in field. Office: PO Box 218 Yorktown Heights NY 10598 Home: 10 Berrybrook Ln Chappaqua NY 10514

LEAR, GEORGE EMORY, engineer; b. Hollidays Cove, W. Va., July 17, 1927; s. Emory Kiel and Kathryn (Hoffman) L.; m.; children: Amy, Mary, Georgianna, Dorothy. B.S., U.S. Mil. Acad., 1950; M.S. in Civil Engring, Northwestern U., 1959, Catholic U. Am., Washington, 1970; cert., U.S. Command and Staff Coll., Ft. Leavenworth, Kans., 1962, Ft. Lee, Va., 1971. Registered profl. engr., D.C. Commd. 2d lt. C.E. U.S. Army, 1950, advanced through grades to col., 1969; ret., 1970; project mgr. nuclear reactor regulation Nat. Regulatory Commn., Bethesda, Md., 1970-74, chief operating reactors br., 1974-77, chief environ. specialists br., 1977-79, chief hydrologic and geotech. engring., 1979-83, chief structures and geotech. engring., 1983—; instr. to asst. prof. dept. math. U.S. Mil. Acad., 1962-65. Vol. coach U.S. Naval Acad. Sailing Squadron, Annapolis, Md., 1980—. Decorated Meritorious Service medal (2) U.S. Army, 1969, 70. Mem. Am. Nuclear Soc., ASCE, Balt.-Washington Health Physics Soc. Club: Annapolis Naval Sailing Assn. Subspecialties: Civil engineering; Nuclear engineering. Current work: Management of structural and geotechnical engineers in the review of nuclear power plant and other nuclear facility siting requirements. Office: Nuclear Regulatory Commn 7920 Norfolk Ave Bethesda MD 20815

LEARY, JOSEPH ALOYSIUS I, physical chemist; b. N.Y.C., Nov. 22, 1919; s. James W. and Florence (Robinson) L.; m. Theresa Grabowski, Feb. 22, 1943; children: Lynne T. Leary Hull, Jeffry J. B.S. in Chem. Engring., Newark Coll. Engring., 1943; Ph.D. in Chemistry, U. N.Mex., 1956. Research scientist Los Alamos Sci. Lab., 1944-56, group leader plutonium R & D, 1956-74; dir. nuclear nonproliferation U.S. Dept. Energy, Washington, 1979-81, dir. fuels mgmt. and safeguards-nuclear energy, 1981—; adj. prof. U. N.Mex., 1959-61. Author: Advanced LMFBR Fuels, 1977; contbr. articles to profl. jours. Served with U.S. Army, 1944-43. Mem. Am Nuclear Soc., Tau Beta Pi, Sigma Xi. Club: Cosmos (Washington). Subspecialties: High temperature chemistry; Nuclear engineering. Patentee in field. Office: US Dept Energy Ne-44 Washington DC 20545

LEARY, MARK RICHARD, psychology educator, researcher; b. Morgantown, W. Va., Nov. 29, 1954; s. Edward and Eleanor (Durrett) L.; m. Wendy H. McColskey, Jan. 2, 1982. B.A., W.Va. Wesleyan Coll., 1976; M.A., U. Fla., Gainesville, 1978, Ph.D., 1980. Research fellow U. Fla., Gainesville, 1977-80; asst. prof. psychology Denison U., Granville, Ohio, 1980-83; asst. prof. ednl. psychology U. Tex., Austin, 1983—. Author: Understanding Social Anxiety, 1983; contbr. articles to profl. jours. NSF fellow, 1977-80. Mem. Am. Psychol. Assn., Soc. Personality and Social Psychology, Southwestern Psychol. Assn., Phi Beta Kappa, Sigma Xi. Subspecialty: Social psychology. Current work: Research on self-presentation, social anxiety and distorted hindsight. Office: Dept Ednl Psychology U Tex Austin TX 78712

LEARY, RICHARD LEE, geologist, paleobotanist, museum curator; b. Portsmouth, Va., Sept. 19, 1936; s. Wilbur Talmadge and Mary Katherine (Lee) L.; m. Eleanor Marie Riehl, June 18, 1961; children: Seth Richard, Sara Marie. B.S. in Geology (Inst. scholar), Va. Poly. Inst., 1959; M.S. in Geology (Univ. fellow), U. Mich., 1961, Ph.D., U. Ill., 1980. Summer field asst. Calif. Oil Co., Colo., 1959; field asst. Mobil Oil Co., N. Mex., 1960; asst. Kelsey Mus., Ann Arbor, 1960-61; mus. apprentice Newark Mus., 1961-62; curator geology Ill. State Mus., Springfield, 1962—; mem. environ. geology faculty Sangamon State U., 1973-78; mem. faculty The Clearing, Ellison Bay, Wis., alt. summers 1975—. Contbr numerous articles to sci. and non-tech. jours., papers to profl. confs.; research in paleobotany, geology. Deacon, Westminster Presbyn. Ch., Springfield. Ill. State Acad. Sci. grantee, 1975; NSF grantee, 1981. Mem. Geol. Soc. Am., Bot. Soc. Am., Internat Orgn. Paleobotany, Ill. State Acad. Sci. (v.p. 1975-76, pres. 1976-77), Sigma Xi. Subspecialties: Paleobiology; Paleoecology. Current work: Carboniferous (Pennsylvanian/Mississippian) non-swamp paleoenvironments; paleobotany, paleoecology, sedimentation, paleotopography. Discoverer new species of fossil plants, new genus and species of fossil scorpion. Office: Ill State Mus Springfield IL 62706

LEATH, KENNETH THOMAS, research plant pathologist; b. Providence, Apr. 29, 1931; s. Thomas and Elizabeth (Wootten) L.; m. Marie Lorraine, Aug. 6, 1955; children: Kenneth, Steven, Kevin, Maria. B.S., U. R.I., 1959; M.S., U. Minn., 1966, Ph.D., 1966. Research technician Coop. Cereal Rust Lab., St. Paul, 1959-66; research plant pathologist Pasture Research Lab., U.S. Dept. Agr., University Park, Pa., 1966—, adj. prof., 1966—. Contbr. articles in field to profl. jours. Bd. dirs. Pa. Forage and Grassland Council, 1976-82. Served with USN, 1951-55. Mem. Internat. Soc. Plant Pathology, Am. Phytopath. Soc., Am. Soc. Aghronomyt, Am. Forage and Grassland Soc., Pa. Forage and Grassland Soc. Lodge: Elks. Subspecialties: Plant pathology; Microbiology. Current work: Researcher in forage legume diseases. Office: US Dept Agriculture Pasture Research Lab University Park PA 16802

LEATH, PAUL LARRY, university administrator, physicist, researcher; b. Moberly, Mo., Jan. 9, 1941; s. James L. and Naomia A. (Burton) L.; m. Rosemary Rippel, June 2, 1962; children: Steven E., Kimberly L. A.S., Moberly Jr. Coll., 1960; B.S., U. Mo.-Columbia, 1961, M.S., 1963, Ph.D., 1966. Research officer Oxford (Eng.) U., 1966-67; asst. prof. Rutgers U., New Brunswick, N.J., 1967-71, assoc. prof., 1971-78, prof. physics, 1978—, assoc. provost sci. and engring., 1978—; bd. dirs. Research and Devel. Council N.J., 1980-82; cons. Oak Ridge nat. Lab., 1975-79. Author: (with R.J. Elliott, J.A. Krumhansl) Theory and Properties of Randomly Disordered Crystals and Related Physical Systems; contbr. articles to profl. jours. Town councilman Millstone (N.J.) Borough Council, 1978—. Sci. Research Council U.K. sr. vis. fellow Oxford U., 1972-73, 81. Mem. Am. Phys. Soc., AAAS, Inst. Physics (Britain), N.Y. Acad. Scis., Sigma Xi. Democrat. Methodist. Subspecialties: Theoretical physics; Condensed matter physics. Current work: Properties of alloys, mixed crystals, and disordered systems, percolation phenomena and critical behavior. Office: Rutgers U Serin Physics Labs Piscataway NJ 08854

LEATHERS, CHESTER RAY, microbiologist, educator; b. Claremont, Ill., May 15, 1929; s. Herman Walter and Edna Pearl (Shaw) L.; m.; children: Mark, Brent, Michelle, Lisa. B.S., Eastern Ill. U.; M.S., Ph.D., U. Mich. Teaching fellow U. Mich., Ann Arbor, 1951-55; research mycologist U.S. BioWar Labs., Frederick, Md., 1955-57; asst. prof. microbiology Ariz. State U., Tempe, from 1957, now assoc. prof. Contbr. articles to profl. jours. Served with Chem. Corps U.S. Army, 1955-57. Fellow AAAS; mem. Mycological Soc. Am., Ariz. Acad. Sci. (pres. 1961). Subspecialties: Microbiology (medicine); Plant pathology. Current work: Human pathogenic fungi; plant pathogenic fungi; epidemiology of human diseases. Office: Dept Botany/ Microbiology Ariz State U Tempe AZ 85281

LEAVITT, FRED I., psychology educator; b. Bklyn., Dec. 12, 1940; s. Ezekiel and Goldie (Erdwein) L.; m. Diane Lee Bright, Aug. 30, 1964; children: Jessica, Melaine. B.A., Eastern N.Mex. U., 1963; Ph.D., U. Mich., 1968. Postdoctoral fellow U. Calif.-Berkeley, 1968-69; vis. lectr. Williams Coll., Williamstown, Mass., 1969-70; prof. psychology Calif. State U., Hayward, 1970—; analyst, cons. Helix Orgn., Berkeley, Calif., 1978. Author: Drugs and Behavior, 1974, 2d edit., 1982. Subspecialty: Psychopharmacology. Current work: Determinants of placebo response, methodology in research, relationships between philosophy of science and scientific practices and between philosophy and psychological problems. Home: 1293 Trestle Glen Oakland CA 94610 Office: Psychology Dept Calif State U Hayward CA 94542

LEAVITT, RICHARD DELANO, physician, cancer researcher; b. Chgo., Aug. 23, 1945; s. Jerome S. and Frieda Shanks L.; m. Susan M. Siegel, May 8, 1966; children: Toby Karen, Michael Baird. S.B., U. Chgo., 1967; M.D., U. Ill.-Chgo., 1971. Cert. Am. Bd. Internal Medicine (Hematology and Oncology). Research asst. Mt. Sinai Hosp., Chgo., 1968-70, Institut de Pathologie Cellulair, Paris, 1970-71; intern George Washington Hosp., Washington, 1971-72; oncology fellow Nat. Cancer Inst., Balt., 1972-75; hematology fellow Sydney (Australia) Hosp., 1975-76; resident in medicine Johns Hopkins Hosp., 1976-78, instr. medicine, 1978-80; asst. prof. medicine and oncology U. Md. Cancer Ctr., Balt., 1980—. Mem. editorial bd.: Jour. Biol. Response Modifiers, 1982—. Served with USPHS, 1972-75. Rogers Found. scholar, 1978-80; Nat. Leukemia Assn. fellow, 1981—. Mem. Am. Soc. Clin. Oncology, Am. Fedn. Clin. Research, ACP, Cancer and Acute Leukemia Group B. Subspecialties: Cancer research (medicine); Immunology (medicine).

LEAVITT, ROBERT LAVERNE, engineer; b. Shell, Wyo., June 23, 1934; s. Paul Barclay and Ellen Ann (Wilkerson) L.; m. Rosalen Joy Wark, Dec. 6, 1959; children: Gerald Roy, Russell Alan, Thomas Robert. B.A., Calif. State U.-Fullerton, 1978. Registered profl. engr., Calif. Prodn. supr., engr. Beckman Instrument Inc., Fullerton, Calif., 1969-71, inspection supr, quality assurance supr., 1971-78; dir. quality assurance med. products div. Gould, Inc., Oxnard, Calif., 1978—. Treas. Brea Little League, 1974. Served with U.S. Army, 1957-59. Mem. Am. Soc. Quality Control Regulatory Affairs Profl. Soc. Republican. Subspecialties: Biomedical engineering; Electronics. Current work: Coordination of product reliability programs, direction of planning and execution of division quality assurance, reliability and control programs. Office: 1900 Williams Dr Oxnard CA 93030

LEBENTHAL, EMANUEL, pediatrician, educator; b. Tel Aviv, Israel, Apr. 12, 1936; came to U.S., 1970; s. Abraham and Frumit L.; m. Hannah Leichtung, June 6, 1962; children: Avi, Yael, Tamar, Michal, David. M.D., Hebrew U., Jerusalem, Israel, 1964. Resident in pediatrics Hadassah Med. Ctr., Jerusalem, 1965, Beilinson Med. Ctr., Tel-Aviv U., 1966-68; fellow in pediatric gastroenterology Stanford (Calif.) Sch. Medicine, 1970-72; asst. in medicine Children's Hosp., Boston, 1972-74, assoc. in medicine, 1974-76; assoc. prof. pediatrics SUNY-Buffalo, 1976-80, prof. pediatrics, 1980—; chief Div. Gastroenterology Children's Hosp., Buffalo, 1976—. Editor: Textbook of Gastroenterology and Nutrition in Infancy; editor: Jour. Pediatric Gastroenterology and Nutrition, 1981; mem. editorial bd.: Am. Jour. Gastroenterology. Served to capt. Israeli Paratroopers, 1954-57. Mem. Soc. Pediatric Research, N.Am. Soc. Pediatric Gastroenterology, Am. Gastroenterology Assn., Am. Soc. Clin. Nutrition, Am. Inst. Nutrition. Jewish. Subspecialties: Gastroenterology; Pediatrics. Current work: Gastroenterology, pediatrics. Home: 332 Forestview Dr Williamsville NY 14221 Office: Children's Hosp Buffalo 219 Bryant St Buffalo NY 14222

LEBER, RALPH ERIC, energy researcher, adminstr.; b. Seattle, Nov. 30,1949; s. Ralph Theodore and Ann Elisa (Ellsworth) L.; m. Lori Marie Ramonas, Apr. 28, 1979. Student, Lycee de Maths et Sciences, Lyon, France, 1966-67; B.A., Reed Coll., Portland, Oreg., 1972; M.S., Yale U., 1973, M.Ph., 1974, Ph.D. in Chemistry, 1975. Sr. reactor operator, supr. Reed Coll. Nuclear Reactor Facility, 1969-72; summer trainee Pacific N.W. Nat. Labs., 1970, Brookhaven Nat. Lab., 1971; research assoc., teaching asst. Yale U., 1972-75; postdoctoral research fellow Lawrence Berkeley Lab., 1975-77; AAAS Congressional Sci. fellow U.S. Congress, 1977-78; dir. energy research, dir. Demonstration of Energy-Efficient Devels., asst. dir. tech. services Am. Public Power Assn., Washington, 1978—; exec. com. Fuel Cell Users Group; supervisory com. Intgrated Energy Systems Group; synthetic fuels rev. panel U.S. Synthetic Fuels Corp. Editor: Public Power mag, 1978—, DEED Digest, 1980—; contbr. articles tech. jours. Mem. U.S. del. U.S./USSR exchange fields of housing and other constrn.; Advisor Washington Internships for Students in Engring. Program; judge Washington Area High Sch. Sci. Fairs; nat. chmn. Reed Coll. Alumni Fund Drive; bd. dirs. Spring Lake Condominium Assn., Bethesda, Md.; steering and ann. fund coms. Reed Coll. Bd. Trustees. Recipient Richard Wolfgang prize Yale U. Chemistry Dept., 1976; nat. finalist White House Fellowship, 1975. Mem. AAAS, Am. Chem. Soc. (Coryell Undergrad. award 1972), Am. Nuclear Soc. (award of merit 1972), Am. Phys. Soc., Am. Wind Energy Assn., ASME, Internat. Cogeneration Soc., Internat. Dist. Heating Assn., Phi Beta Kappa, Sigma Xi. Clubs: Wash. State Soc., Yale (Washington). Subspecialties: Fuels and sources; Systems engineering. Current work: Research and application of energy sci., technologies, and techniques to community energy programs, evaluation of related tech., environ., fin., operational and instnl. questions. Office: 2301 M St NW Washington DC 20037

LEBHERZ, HERBERT G(ROVER), biochemistry educator, researcher; b. San Francisco, July 27, 1941; s. Herbert A. and Edvige (Bionda) K. A.A., City Coll. San Francisco, 1961; B.A., San Francisco State U., 1964, M.A., 1966; Ph.D., U. Wash., 1970. Research assoc. Swiss Fed. Inst. Tech., Zurich, 1971-72, sr. research assoc., 1972-74; cancer research scientist Roswell Park Meml. Inst., Buffalo, 1975; assoc. prof. chemistry San Diego State U., 1976-80, prof., 1980—. Contbr. numerous articles to profl. jours. NIH grantee, 1976—; Muscular Dystrophy Assn. grantee, 1977—; Am. Heart Assn. grantee, 1981. Mem. AAAS, Am. Soc. Biol. Chemists. Subspecialties: Biochemistry (biology); Biochemistry (medicine). Current work: Regulation of gene expression in normal and abnormal cells; protein evolution. Home: 5637 Baja Dr San Diego CA 92115 Office: San Diego State U San Diego CA 92182

LEBLANC, ROBERT BRUCE, chemist, research co. exec.; b. Alexandria, La., Jan. 28, 1925; s. Moreland Paul and Carmen Mary (Haydel) LeB.; m. Barbara Ann Sanders, Oct. 11, 1968. B.S. in Chemistry, Loyola U., 1947; M.S., Tuland U., 1949, Ph.D., 1950. Diplomate: Am. Bd. Bioanalysts, 1976. Asst. prof. chemistry Tex. A&M U., College Station, 1950-52; group leader research Dow Chem. Co., Freeport, Tex., 1952-63, sect. head, Midland, Mich., 1963-67; mgr. textile chem. devel. Ashland Chem. Co., Charlotte, N.C., 1967-68; research mgr. Nat. Cotton Council, Memphis, 1968-70; pres. LeBlanc Research Corp., North Kingstown, R.I., 1970—. Editor 6 books; contbr. articles to profl. jours. Served to lt. USNR, 1943-46. Republican. Roman Catholic. Lodge: Rotary. Subspecialties: Polymer chemistry; Clinical chemistry. Current work: Textile chemistry, flammability and flame retardance of textiles. Patentee in field. Home: 99 Main St Wickford RI 02852 Office: 6172 Post Rd North Kingstown RI 02852

LEBLANC, ROGER MAURICE, biophysics educator, researcher; b. Trois-Rivières, Que., Can., Jan. 5, 1942; s. Henri and Rita (Moreau) L.; m. Micheline Veillette, June 26, 1965; children: Nancy, Hugues, Daniel. B.Sc. in Chemistry, U. Laval, 1964, D.Sc. in Phys. Chemistry, 1968. Postdoctoral fellow Davy Faraday Research Lab., Royal Instn., London, 1968-70; mem. faculty U. Que., Trois-Rivières, 1970—, prof. phys. chemistry, 1970—, chmn. chemistry-biology dept., 1971-75, dir. biophysics research group, 1978-81; chmn. Photobiophysics Research Ctr., 1981—; vis. prof. INRS-Energie, 1972—, U. Sherbrooke, 1977—; vis. scientist U. Liège, Belgium, 1977, 78, 79. Contbr. articles to sci. jours.; reviewer: Revue Canadienne de Biologie. Recipient prix du Govt. France, 1962-63; prize Allied Chem. Corp., 1962-63; Sir William Price award, 1963-65; Noranda prize Chem. Inst. Can., 1982; others. Mem. Assn. Canadienne Française pour l'Avancement des Sciences (Vincent prize 1978), Chem. Inst. Can., Am. Chem. Soc., Biophys. Soc., N.Am. Photochem. Soc., Brit. Photobiology Soc., Brit. Biophys. Soc., Am. Soc. Photobiology, European Photochem. Assn. Subspecialties: Biophysics (biology); Surface chemistry. Current work: Photovoltaic properties of chloroplast pigments; optical properties of chlorophylls in artificial biological membranes; electrical properties of the mixtures phospholipids and retinals in bilayer models; interactions phospholipids and rhodopsin.

LEBLOND, CHARLES PHILIPPS, educator; b. Lille, France, Feb. 5, 1910; s. Oscar and Jeanne (Desmarchelier) L.; m. Gertrude Elinor Sternschuss, Oct. 22, 1936; children: Philippe Louis, Pierre Francis, Paul Noel, Marie-Pascale Murphy. M.D., U. Paris, 1934; Ph.D., U. Montreal, 1942; Dr.Sc., Sorbonne U., 1945; D.Sc. (hon.), McGill U., 1982. Rockefeller fellow Yale U., 1935; in charge biology sect. Lab. Synthese Atomique, Paris, 1937; lectr. history McGill U., Montreal, Que., Can., 1941-46, assoc. prof. anatomy, 1946-48, prof., 1948—, chmn. dept., 1957-74. Contbr articles in field to profl. jours. Served with Free French Army, 1944-45. Named Prix Saintour French Acad., 1935; NATO prof. U. Louvain, Belgium, 1959; recipient award Gairdner Found., 1965; Biology Prize Province of Que., 1968. Fellow Royal Soc. Can., Royal Soc. London.; Mem. Am. Assn. Anatomist (exec. 1954-58, pres. 1962-63), Am. Soc. Cell Biology (mem. council 1968-71), Can. Assn. Anatomists (pres. 1965), Histochem. Soc. (pres. 1956). Club: Faculty (Montreal, Que.). Subspecialties: Microscopy; Cell biology (medicine). Current work: The work done on the microscopic localization of biological substances by radioautography has added dynamic features to classical histology.

LEBO, ROGER VAN, molecular biologist, geneticist; b. Pottsville, Pa., Mar. 1, 1948; s. Stanley E. and Irene M. (Bush) L.; m. Susan Thelma Southard, Apr. 3, 1972; children: Frank, Paul. A.B., Chaffey Coll., Alta Loma, Calif., 1968; B.S., Pa. State U., 1970; Ph.D., Duke U., 1974. Diplomate: Am. Bd. Med. Genetics. NIH trainee Duke U., 1970-74; postdoctoral fellow U. Calif.-San Francisco, 1974-76, asst. research biochemist, 1976—; assoc. Howard Hughes Med. Inst., San Francisco, 1979—; prin. project investigator gene mapping by chromosome sorting, 1979. Recipient research award Duke U., 1974. Mem. Am. Soc. Human Genetics, Soc. Analytical Cytometry, Am. Genetic Assn., Am. Fedn. Clin. Research. Republican. Subspecialties: Genetics and genetic engineering (medicine); Cell biology (medicine). Current work: Human genetic disease, further dissection of human genome, gene expression using chromosome-specific recombinant DNA libraries. Inventor optical bench of dual laser chromosome sorter. Home: 538 Miramar Ave San Francisco CA 94112 Office: University of California 1556 Health Sciences E San Francisco CA 94143

LEBOEUF, CECILE MARIE, solar energy researcher; b. Schenectady, Apr. 19, 1955; d. Maurice Bernard and Eleanor Alma (Schmitt) L. B.S. in Mech. Engring., U. Mass., 1977; M.S. in Mech. Engring. (grad. research asst.), Colo. State U., 1979. Registered profl. engr., Colo. Technician, Gen. Electric Co., Schenectady, 1974-76; staff engr. Solar Energy Research Inst., Golden, Colo., 1979—. Contbr. articles to conf. procs. Recipient Outstanding Achievement award Solar Energy Research Inst., 1981. Mem. Internat. Solar Energy Soc., Boulder Solar Energy Soc., Tau Beta Pi. Subspecialties: Solar energy; Mechanical engineering. Current work: Systems engineering of solar thermal energy concepts including concept development, analytical and computer modeling and system design; recent work has involved analysis of solar pond performance for various applications. Office: 1617 Cole Blvd Golden CO 80401

LEBOFSKY, LARRY ALLEN, planetary astronomer; b. Bklyn., Aug. 31, 1947; s. Harry and Clara (Goodman) L.; m. Marcia J. Lebofsky, July 15, 1973; m. Nancy R. Lebofsky, May 9, 1980; 1 dau., Miranda B. B.S., Calif. Inst. Tech., 1969; Ph.D., MIT, 1974. Cons. Sci. Applications, Inc., 1975; NRC resident research assoc. Jet Propulsion Lab., 1975-77; cons., 1981—; research assoc. Lunar and Planetary Lab., U. Ariz., 1977-80, sr. research assoc., 1980-82, research fellow, 1982-83, assoc. research scientist, 1983—; cons. Infrared Astron. Satellite Sci. Team, 1981—. Contbr. chpts. to books and articles to profl. jours. Mem. Am. Astron. Soc., Internat. Astron. Union, Sigma Xi. Democrat. Jewish. Subspecialties: Planetary science; Infrared optical astronomy. Current work: Visual and infrared telescopic and satellite observations of asteroids, satellites and planets. Home: 2333 E 7th St Tucson AZ 85719 Office: Lunar and Planetary Lab Ariz Tucson AZ 85721

LEBOW, MICHAEL DAVID, psychologist, researcher, therapist, educator; b. Detroit, June 24, 1941; emigrated to Can., 1974; s. William LeBow and Mildred (Rosenman) Boudreau; m. Barbara Louise Ernest, Aug. 20, 1969; children: William Michael, Matthew David. B.A., UCLA, 1964; M.A., U. Utah, 1967, Ph.D., 1969. Asst. prof. psychology U. Man. (Can.), Winnipeg, 1969-72, assoc. prof., 1974-79, prof., 1979—; asst. prof. Dartmouth Coll. Med. Sch., 1972-74.

Author: books, including Weight Control: The Behavioral Strategies, 1981, Child Obesity: A New Frontier of Behavior Therapy, 1983; contbr. articles to profl. jours., author profl. papers. Mem. Am. Psychol. Assn., Assn. Advancement Behavior Therapy, Nat. Health Service Providers in Psychology, Soc. Behavioral Medicine, Sigma Xi, Phi Kappa Phi. Subspecialty: Behavioral psychology. Current work: Short and long-term impact of eating and activity control technologies on the weight and fatness of adults, adolescents, and children. Home: 548 Manchester Blvd S Winnipeg MB Canada R3T 1N8 Office: U Man Winnipeg MB Canada R3T 2N2

LEBOWITZ, JOEL LOUIS, physicist, educator; b. Taceva, May 10 1930; U.S., 1946, naturalized, 1951; m. Estelle Mandelbaum, June 21, 1953. B.S., Bklyn. Coll., 1952; M.S., Syracuse U., 1955; Ph.D., 1956; Ph.D. hon. doctorate, Ecole Polytechnique Federale, Lausanne, Switzerland, 1977. NSF postdoctoral fellow Yale, 1956-57; mem. faculty Stevens Inst. Tech., 1957-59, Yeshiva U., N.Y.C., 1959-77, prof. physics, 1965-77; acting chmn. dept. physics Belfer Grad. Sch. Sci., 1964-67, chmn. dept., 1967-76; prof. math. and physics, dir. Center for Mathematical Scis., Rutgers U., New Brunswick, N.J., 1977—, George William Hill prof. math., 1980—. Editor: Jour. Statis. Physics, 1975—, Studies in Statis. Mechanics, 1973—, Com. Math. Physics, 1973—; contbr. articles to profl. jours. Guggenheim fellow, 1976-77. Fellow Am. Phys. Soc., AAAS; mem. Nat. Acad. Scis., N.Y. Acad. Scis. (pres. 1979), AAUP, Am. Math. Soc., Phi Beta Kappa, Sigma Xi. Subspecialties: Probability; Statistical physics. Office: Dept Math Hill Center Busch Campus Rutgers U New Brunswick NJ 08903

LEBOWITZ, MICHAEL DAVID, medical educator; b. N.Y.C., Dec. 21, 1939; s. Harry and Rachel (Dick) L.; m. Joyce Schmidt, Dec. 9, 1960; children: Jon, Kira, Debra. A.B., U. Calif.-Berkeley, 1961, M.A., 1965; Ph.D. in Preventive Medicine, U. Wash., 1969, U. Wash., 1971. Research assoc. dept. environ. health U. Wash., Seattle, 1967-71; asst. prof. pulmonary disease sect. U. Ariz. Health Sci. Ctr., Tucson, 1971-75, assoc. prof., 1975-80, prof., 1980—; vis. fellow Cardiothoracic Inst., U. London, 1978-79; advisor environ. health scis. WHO, Geneva, 1979—; cons. Ariz. Dept. Health Services, Phoenix, 1972—, U.S. Indian Health Services, Tucson, 1973-76; mem., chmn. Pima County (Ariz.) Air Quality Adv. Commn., 1975-78, Ariz. Chest Symposium Com., Tucson, 1979—. Author: Research Methodology, 1976; author monographs. Grantee EPA, NIH, FDA. Fellow Am. Coll. Chest Physicians; mem. Soc. for Epidemiol. Research, Biometrics Soc., Am. Thoracic Soc., Internat. Epidemiol. Soc., Am. Epidemiol. Soc., Ariz. Thoracic Soc., Ariz.-Nev. Acad. Sci. Democrat. Subspecialties: Epidemiology; Pulmonary medicine. Current work: Epidemiology of airway obstructive diseases, environmental health and occupational health, biostatistics. Office: Div Respiratory Scis U Ariz Health Scis Ctr Tucson AZ 85724

LECAR, MYRON, astrophysicist, educator; b. Bklyn., Apr. 10, 1930; s. Joshua Irving and Rachel (Stoun) L. S.B., M.I.T., 1951; M.S., Case Inst. Tech., 1953; Ph.D., Yale U., 1963. Physicist NASA Goddard Space Flight Ctr., 1958-61, NASA Inst. for Space Studies, 1961-65; astrophysicist Harvard-Smithsonian Ctr. for Astrophysics, Cambridge, Mass., 1965—; lectr. Harvard U., 1965—; lectr. Yale U., 1963-65; vis. prof. Hebrew U., Jerusalem, Israel, 1966-67. Editor: Gravitational N-Body Problem, 1970. Served to lt. j.g. USN, 1954-58. Prin. investigator, grantee to found first astron. obs. in Israel, 1971. Mem. AAAS, Am. Astron. Soc., Internat. Astron. Union, Royal Astron. Soc. (U.K.). Democrat. Jewish. Subspecialties: Theoretical astrophysics; Cosmology. Current work: Origin and evolution of galaxies, origin of solar system. Office: 60 Garden St Cambridge MA 02138

LECHNER, JOSEPH HADRIAN, chemistry educator; b. Boston, Nov. 13, 1951; s. Hadrian Blair and Mary Margaret (Sumner) L.; m. Catherine Elizabeth Lechner Cook, June 13, 1981. B.S. in Chemistry, Roberts Wesleyan Coll., 1972; Ph.D., U. Iowa, 1977. Research fellow Northwestern U., 1978-79; asst. prof. dept. chemistry Mt. Vernon (Ohio) Nazarene Coll., 1979—. Clarinetist Knox Symphony, Mt. Vernon, 1981—; scoutmaster Boy Scouts Am., North, Chili, N.Y., 1970-72. NIH grantee, 1978. Mem. Am. Chem. Soc., Am. Sci. Affiliation, AAUP, Sigma Xi. Subspecialties: Biochemistry (medicine); Oral biology. Current work: Biochemical research on structure and self-assembly of the protein and mineral components of mammalian connective tissues. Office: Mount Vernon Nazarene Coll Martinsburg Rd Mount Vernon OH 43050

LECHNER, NORBERT MANFRED, architect, educator, researcher; b. Pahl, W.Ger., Apr. 19, 1944; came to U.S., 1952, naturalized, 1962; s. Karl and Leni L.; m. Judith Victoria Nosty, Mar. 31, 1967; children: Walden, Ethan. B.S. in Architecture cum laude, CCNY, 1971; M.S. in Archtl. Tech, Columbia U., 1973. Registered architect, Ala. Customer engr. Nat. Cash Register Co., N.Y.C., 1962-65; designer Port of N.Y. and N.J. Authority, N.Y.C., 1970-72; mem. faculty Auburn U., 1974—, asst. prof. bldg. sci., 1974-82, asso. prof., 1982—; cons. in field. Democrat. Unitarian. Subspecialty: Solar energy. Current work: Design of integrating sun simulator to test phys. models of bldg. to determine their responsiveness to direct solar energy. Home: 719 Mercer Circle Auburn AL 36830 Office: Bldg Sci Dept Auburn U Auburn AL 36849

LEDBETTER, CARL SCOTIUS, JR., mathematician; b. Fort Riley, Kans., May 27, 1949; s. Carl Scotius and Ruth Slocum (Weymouth) L.; m. Charlene Hudson, June 8, 1968 (div. July 1974); m. Elizabeth Robin McClard, Oct. 17, 1978; children: Noah McClard, Megan Elizabeth. B.S., U. Redlands, 1971; M.A., Brandeis U., 1975; Ph.D., Clark U., 1977. Instr. math. Clark U., Worcester, Mass., 1975-77; asst. prof. math. Wellesley (Mass.) Coll., 1977-79; dean acad. planning Sonoma State U., Rohnert Park, Calif., 1979-80, exec. asst. to pres., 1980-81; exec. dir. Calif. Seismic Safety Commn., Van Nuys, 1981; sr. scientist IBM, Los Angeles, 1981-82; dir. acad. computing, Stamford, Conn., 1982—; cons., 1981—; dir. Directors Research Ltd., N.Y.C.; ednl. cons. St. Johns Coll., Santa Fe, 1979-81, Ariz. State U., Tempe, 1979-81. Author: Differential Equations with Linear Algebra and Analysis, 1971, Sheaf Representations and First Order Conditions, 1977, The Mitigation Preparedness Tableau: An Instrument for Risk Assessment and Public Policy Analysis, 1981. Mem. steering com. Gov.'s Task Force on Earthquake Preparedness, Calif., 1980-82; dir. job opportunity benefits United Way, Apple Valley, Calif., 1982-83. Served with USAR, 1967-73. Named Outstanding Tchr. of Yr. Clark U., 1975-76; NSF fellow, 1971-72; Ariz. State U. fellow, 1972-74; Brandeis U. fellow, 1974-75. Mem. Am. Math Soc., Math. Assn. Am., Soc. Indsl. and Applied Math., AAAS, Earthquake Engring. Research Soc. Am., Inst. Mgmt. Scis., Consortium for Math. and Its Applications, ACM, IEEE, Phi Beta Kappa, Omicron Delta Kappa. Democrat. Mem. Ch. of Christ. Subspecialties: Algorithms; Operations research (mathematics). Current work: Operations research; linear programming; optimization problems in mathematics, computer science and management; academic computing. Office: PO Box 10244 Stamford CT 06904 Home: 583 Topstone Rd New Fairfield CT 06810

LEDBETTER, MARY LEE STEWART, cell biologist; b. Monterrey, Nuevo Leon, Mex., Aug. 30, 1944; d. William Sheldon and Maria Rosalind (Markham) Stewart; m. Steven John Ledbetter, Sept. 10, 1966; children: William John, Joanna Marie. B.A., Pomona Coll., 1966; Ph.D., Rockefeller U., 1972. Research assoc. dept. microbiology Dartmouth Med. Sch., Hanover, N.H., 1972-75, instr., 1975-77; research assoc. in psychiatry, 1977-78, research asst. prof. biochemistry, 1978-80; asst. prof. biology Coll. Holy Cross, Worcester, Mass., 1980—. Contbr. articles to sci. jours. Damon Runyon-Walter Winchell Cancer Research Fund fellow, 1972-73; Leukemia Soc. Am. fellow, 1973-75, 77; Nat. Cancer Inst. fellow, 1975-77. Mem. Am. Soc. Cell Biology, AAAS, Genetics Soc. Am., Assn. Women in Sci., Sigma Xi. Democrat. Subspecialties: Cell and tissue culture; Membrane biology. Current work: Regulation of growth and development as exhibited in cultured mammalian cells; control of cell communication; control of protein degradation. Office: Dept Biology Coll Holy Cross Worcester MA 01610

LEDDICK, GEORGE RUSSELL, psychology educator, therapist; b. Newman, Calif., Apr. 6, 1948; s. Kenneth L. and Ann (Karl) L. B.A., DePauw U., 1970; M.A., Fisk U., 1977; Ph.D., Purdue U., 1980. Instr. Purdue U., 1979-80; prof. counseling psychology and ednl. psychology Ind. U., Ft. Wayne, 1980—; pres. Ind. Specialists in Group Work, Ft. Wayne, 1982-84. Cons. editor: Jour. for Specialists in Group Work, 1982-85, The Clinical Supervisor jour, 1982-85; Contbr. chpts. to books, ency., articles to profl. jours. Fellow Internat. Council Parenthood Edn.; mem. Am. Assn. Counseling and Devel., Am. Psychol. Assn., Am. Assn. Marriage and Family Therapy, Assn. Specialists in Group Work, Assn. Counselor Edn. and Supervision, Orgnl. Devel. Network, Phi Delta Kappa. Subspecialty: Counseling Psychology. Current work: Psychotherapy supervision; marriage and family therapy; group work; organizational development; microcomputer applications. Office: Indiana-Purdue U 2101 Coliseum Blvd E Fort Wayne IN 46815

LEDEEN, ROBERT WAGNER, neurochemist, educator; b. Denver, Aug. 19, 1928; s. Hyman and Olga (Wagner) L.; m. Lydia Rosen Hailparn, July 2, 1982. B.S., U. Calif., Berkeley, 1949; Ph.D., Oreg. State U., 1953. Postdoctoral fellow in chemistry U. Chgo., 1953-54; research asso. in chemistry Mt. Sinai Hosp., N.Y.C., 1956-59; research fellow Albert Einstein Coll. Medicine, Bronx, N.Y., 1959, asst. prof., 1963-69, asso. prof., 1969-75, prof., 1975—. Contbr. articles on brain glycoconjugates to profl. jours.; mem. editorial bd.: Jour. Neurochemistry. Mem. neurol. scis. study sect. NIH. Served with U.S. Army, 1954-56. NIH grantee, 1963-81; Nat. Multiple Sclerosis Soc. grantee, 1967-74. Mem. Internat. Soc. Neurochemistry, Am. Soc. Neurochemistry, Am. Chem. Soc., Am. Soc. Biol. Chemists, N.Y. Acad. Sci., Am. Oil Chemists Soc. Jewish. Subspecialty: Neurochemistry. Current work: Research on gangliosides and other glycolipids of nervous system; biochemistry of myelin. Home: 8 Donald Ct Wayne NJ 07470 Office: 1300 Morris Park Ave Bronx NY 10461

LEDERBERG, JOSHUA, univ. pres., geneticist; b. Montclair, N.J., May 23, 1925; s. Zwi Hirsch and Esther (Goldenbaum) L.; B.A., Columbia, 1944; Ph.D., Yale, 1947, Sc.D. (hon.), 1960; Sc.D. (hon.), U. Wis., 1967, Columbia U., 1967, Yeshiva U., 1970, Mt. Sinai Coll. Medicine, 1979, Rutgers U., 1981; M.D. (hon.), U. Turin, 1969; Litt.D. (hon.), Jewish Theol. Sem., 1979; LL.D. (hon.), U. Pa., 1979; m. Marguerite S. Kirsch, Apr. 5, 1968; children—David Kirsch, Anne. With U. Wis., 1947-58; prof. genetics Stanford Sch. Medicine, 1959-78; pres. Rockefeller U., N.Y.C., 1978—; mem. bd. sci. advisers Cetus Corp., Berkeley, Calif.; mem. study sects. NSF, coms. NASA; mem. U.S. Def. Sci. Bd.; cons. ACDA; chmn. Pres.'s Cancer Panel, 1979-81; dir. Inst. Sci. Info. Inc., Phila., Ann. Revs., Inc., Palo Alto, Calif.; mem. adv. com. med. research WHO, 1971-76. Bd. dirs. Natural Resources Def. Council, N.Y.C., Chem. Industry Inst. Toxicology, N.C.; mem. Pres.'s Panel on Mental Retardation, 1961-62; mem. nat. mental health adv. council NIMH, 1967-71; Recipient Nobel prize in physiology and medicine for research genetics of bacteria, 1958. Mem. Nat. Acad. Scis., Royal Soc. London (fgn.). Subspecialties: Genetics and genetic engineering (biology); Artificial intelligence. Current work: Strategies of biomedical research; inculcating and sustaining scientific creativity. Office: Office of Pres Rockefeller U 1230 York Ave New York NY 10021

LEDERER, CHARLES MICHAEL, nuclear chemist, adminstr.; b. Chgo., June 6, 1938; s. Philip Charles and Jane (Bernheimer) L.; m. Claudette Evenson, Feb. 26, 1970; children: Laura Jane, Mark Edward. A.B., Harvard U., 1960; Ph.D., U. Calif.-Berkeley, 1963. Staff sr. scientist U. Calif.-Berkeley Lawrence Lab., 1963-78, dept. head for info. and data analysis, 1978-80, dep. dir., 1980—. Co-author: Table of Isotopes, 6th edit., 1966, 7th edit., 1978; contbr. articles to profl. jours. Nat. Merit scholar, 1956; recipient Harvard U. Book award, 1959. Mem. Am. Phys. Soc., AAAS, Am. Nuclear Soc., Phi Beta Kappa, Sigma Xi. Democrat. Subspecialties: Nuclear physics; Energy use in buildings. Home: 3040 Buena Vista Way Berkeley CA 94708 Office: Universitywide Energy Research Group U Calif Bldg T-9 Berkeley CA 94720

LEDERMAN, LEON MAX, physicist, educator; b. N.Y.C., July 15, 1922; s. Morris and Minna (Rosenberg) L.; m. Florence Gordon, Sept. 19, 1945; children—Rena S., Jesse A., Heidi R. B.S., Coll. City N.Y., 1943; A.M., Columbia, 1948, Ph.D., 1951. Asso. in physics Columbia, N.Y.C., 1951, asst. prof., 1952-54, asso. prof., 1954-58, prof., 1958—, Eugene Higgins prof. physics, 1973—; dir. Fermi Nat. Accelerator Lab., Batavia, Ill., 1979—; Dir. Nevis Labs., Irvington, N.Y., 1960-67, 69—; guest scientist Brookhaven Nat. Labs., 1955—; cons., nat. accelerator lab. European Orgn. for Nuclear Research (CERN), 1970—; mem. high energy physics adv. panel AEC, 1966-70; mem. adv. com. to div. math. and phys. scis. NSF, 1970-72. Contbr. articles to profl. jours. Served to 2d lt. Signal Corps, AUS, 1943-46. Recipient Nat. Medal of Sci., 1965, Wolf prize, 1983; NSF fellow, 1967; Guggenheim fellow, 1958-59; Ford Found. fellow European Center For Nuclear Research, Geneva, 1958-59; recipient Townsend Harris medal City U. N.Y., 1973; Elliot Cresson medal Franklyn Inst., 1976. Fellow A.A.A.S., Am. Phys. Soc.; mem. Italian Phys. Soc., Nat. Acad. Sci. Subspecialty: Nuclear physics. Home: 34 Overlook Rd Dobbs Ferry NY 10522

LEDINKO, NADA, biologist, educator; b. Girard, Ohio, Dec. 16, 1924; d. John Daniel and Zora Maria (Valencic) L. Ph.d., Yale U., 1952. Scientist Pub. Health Research Inst., N.Y.C., 1956-62; spl. fellow Nat. Cancer Inst., Cold Spring Harbor Lab. and Salk Inst., 1963-65; chief virologist Putnam Meml. Hosp., Research Inst., Bennington, Vt., Putnam, 1965-71; prof. biology U. Akron, 1971—. Contbr artcles to profl. jours. Mem. Am. Soc. Microbiology, Am. Assn. Cancer Research, Tissue Culture Assn., Phi Beta Kappa, Sigma Xi, Phi Kappa Phi. Subspecialties: Molecular biology; Cell and tissue culture. Current work: Molecular biology of cancer cells; biochemistry of transferred cell phenotype. Office: Dept Biology U Akron Akron OH 44325

LEDNICKY, RAYMOND ANTHONY, mechanical engineer; b. Manila, Philippines, Sept. 15, 1923; s. Victor Eugene and Maria (Valero) L.; m. Marilyn Janice Gemmell, Aug. 10, 1948; children: Richard Allen, Jay Alfred. B.S. in Mech. Engring, U. Kans., 1952. Sales engr. E.J. Nell Co., Manila, 1952-59; design engr., product engr. Western Electric Co., Lee's Summit, Mo., 1960-74; owner, research and devel. engr., Lee's Summit, 1974—. Served with USAF, 1943-46. Mem. ASME, Soc. Mfg. Engrs. Republican. Roman Catholic. Subspecialties: Polymer engineering; Mechanical engineering. Current work: Research and development of molding techniques to produce prototypes for market surveys. Patentee in field. Address: 207 N Walnut St Lee's Summit MO 64063

LEDOUX, CHRIS BOB, forest engineering educator, researcher; b. Taos, N.Mex., Apr. 9, 1950; s. Julian H. and Estella T. LeD.; m. Stephanie H. Riley, Dec. 23, 1978. B.S., U. Idaho, 1973; M.S., Oreg. State U., 1975, Ph.D., 1984. Supervisory engr. Forest Service, U.S. Dept. Agr., Whitebird, Idaho, 1972-74; engring. researcher Oreg. State U., 1973-74; logging engr. Bur. Land Mgmt., Medford, Oreg., 1974-76; mem. faculty Stephen F. Austin State U., 1976-78; prof. forest engring. Oreg. State U., Corvallis, 1978—; cons. U.S. Forest Service, Portland, Oreg., 1978—; cons. to forestry industry in Pacific N.W., 1978—. Author computer models. Mem. Tex. Surveyors Assn., Soc. Am. Foresters, Xi Sigma Pi. Subspecialties: Operations research (engineering); Systems engineering. Current work: Statistics, operations research, computer modeling, systems analysis. Office: Dept Forest Engring Oreg State U Corvallis OR 97331

LEE, BERNARD SHING-SHU, research institute executive, chemical engineer; b. Nanking, China, Dec. 14, 1934; came to U.S., 1949, naturalized, 1956; s. Wei-Kuo and Pei-Fen (Tang) L.; m. Pauline Pan, Sept. 7, 1963; children: Karen, Lesley, Tania. Student, U. Va., 1952-54; B.Ch.E., Poly. Inst. Bklyn., 1956, D.Ch.E., 1960. Registered profl. engr., N.Y., Ill. Mem. staff Arthur D. Little, Inc., Cambridge, Mass., 1960-65; successively supr., mgr., asst. dir., asso. dir., dir. coal gasification research Inst. Gas Tech., Chgo., 1965-75, v.p. process research, 1976, exec. v.p., 1977, pres., 1978—, trustee, 1978—; dir. GDC, Inc., Hycrude Corp. Contbr. articles to profl. jours. Westinghouse fellow, 1957; recipient Personal Achievement award Chem. Engring. mag., 1978; Disting. Alumnus award Poly. Inst. N.Y., 1980. Fellow Am. Inst. Chem. Engrs. (33d Ann. Inst. lectr. 1981); mem. Am. Chem. Soc., AIME, AAAS, Am. Gas Assn. Methodist. Subspecialty: Energy research management. Home: 6900 N Kilpatrick Ave Lincolnwood IL 60646 Office: 3424 S State St Chicago IL 60616

LEE, BERT GENTRY, aerospace and media executive; b. N.Y.C., Mar. 29, 1942; s. Harrell Estes and Peggy (Harding) L.; m. Catharine Sue Cooper, Oct. 5, 1975; children: Cooper Gentry, Austin Myles. B.A. summa cum laude, U. Tex., Austin, 1963; M.S., MIT, 1964; postgrad., U. Glasgow, Scotland, 1964-65. Staff engr. Martin Marietta Aerospace Co., Denver, 1965-72; mgr. Viking mission ops. and design, Denver, 1972-74; dir. Viking sci. analysis and mission planning, Pasadena, Calif., 1974-76; sect. mgr. mission design Jet Propulsion Lab., Pasadena, 1976-78; mgr. mission ops. and engring. Project Galileo, 1978-80, chief engr., 1981-83; series mgr. Cosmos for PBS, 1978-80; v.p., gen. mgr. Carl Sagan Prodns., Inc., sci. for media, 1976-83; lectr. sci., space and the future, 1975—. Contbr. profl. jours. Woodrow Wilson fellow, 1963; Marshall fellow, 1964; recipient Exceptional Sci. Achievement medal NASA, 1977. Mem. Phi Beta Kappa. Subspecialties: Software engineering; Aerospace engineering and technology. Current work: System engineering for large, high technology systems. Home: 5441 Rock Castle Rd La Canada CA 91011 Office: 4800 Oak Grove Dr Pasadena CA 91103

LEE, CHENG-CHUN, pharmacologist, toxicologist, researcher; b. Chiangtu, Chiangso, China, May 24, 1922; m. Janice Y. C. Wang, Feb. 9, 1959; children: James P., Ray W. D.V.M., Nat. Central U., China, 1945, M.S., 1948; M.S., Mich. State U., 1950, Ph.D., 1952. Teaching asst. Nat. Central U. Nanking, China, 1945-48; research pharmacologist Lilly Research Lab., Indpls., 1952-63; sr., prin. pharmacologist Midwest Research Inst., Kansas City, Mo., 1963-67, head pharmacology and toxicology, 1967-76, asst. dir., assoc. dir., dept. dir. biol. sci. div., 1976-79; sr. sci. adv. health and environ. rev. div. EPA, Washington, 1979—; lectr. Sch. Pharmacy U. MO.-Kansas City, 1965-66; dept. pharmacology Kans. U. Med. Sch., 1966-79; professorial lectr. dept. pharmacology George Washington Med. Sch., Washington, 1981—; cons. Sci. Working Group on Chemotherapy of Malaria WHO, 1980, 81. Contbr. numerous articles on pharmacology and toxicology to profl. jours. Recipient Prin. Adv. award Midwest Research Inst., 1979; recipient Bronze medal EPA, 1979, Award for Spl. Achievement and Contbn., 1981. Mem. Am. Physiol. Soc., Am. Soc. for Pharmacology and Exptl. Therapeutics, Soc. Toxicology, N.Y. Acad. Scis., Am. Soc. for Exptl. Biology and Medicine, Chinese Physiol. Soc., Chinese Animal Husbandry and Vet Med. Assn., Am. Coll. Toxicology. Subspecialties: Pharmacology; Toxicology (medicine). Current work: Drug metabolish and safe evaluation, health effects of chemicals, toxicology, pharmacology, hazard assessment. Home: 1351 Snow Meadow Ln McLean VA 22102 Office: HERD (TS-796)/OTS/EPA 401 M St SW Washington DC 20460

LEE, CHI OK, physiology educator; b. Tanyang, Choongbuk, Korea, June 8, 1939; came to U.S., 1968; s. Yoon Hee and Kyung Duk (Song) L.; m. Kwanghee Kin Kim, Sept. 20, 1969; children: Albert S., Daniel S. B.S., Seoul Nat. U., 1965, M.S., 1967; Ph.D., Ind. U., 1973. Research asst. Atomic Energy Research Inst., Seoul, 1967-68; postdoctoral fellow U. Chgo., 1973-76; asst. prof. Cornell U. Med. Coll., N.Y.C., 1976-81, assoc. prof., 1981—. Recipient L.N. Katz Basic Sci. Research prize Am. Heart Assn., 1974; established investigator, 1976. Mem. Am. Physiol. Soc. (mem. editorial bd. 1981-84), Biophys. Soc., Am. Heart Assn., N.Y. Acad. Scis., Korean Scientists and Engrs. Assn. Am. (pres. N.Y. chpt. 1980-81). Subspecialties: Physiology (medicine); Biophysics (biology). Current work: Electrophysiology and contractility of heart muscle cells, ion transport, mechanism of cardiac glycoside action, application of ion-selective microelectrodes, teaching physiology. Home: 1161 York Ave New York NY 10021 Office: Dept Physiology Cornell U Med Coll 1300 York Ave New York NY 10021

LEE, CHING TSUNG, physicist, educator; b. Taiwan, July 1, 1937; s. Sing Fu and You (Chou) L.; m. Su-Hwa, Sept. 2, 1967; children: Lily, Perry. B.S., Nat. Taiwan U., Taipei, 1962; Ph.D., Rice U., 1967. Postdoctoral fellow Tex. A&M U., 1967-68; research assoc. Rice U., Houston, 1968-69; assoc. prof. Ala. A&M U., Normal, 1969-73, prof. physics, 1973—; researcher in field. Mem. Am. Phys. Soc. Subspecialties: X-ray lasers; Condensed matter physics. Current work: Research on superradiance and X-ray laser. Office: Physics Dept Ala A&M Univ Normal AL 35762

LEE, CHING Y., geneticist, researcher, educator; b. Peking, China, Sept. 15, 1935; d. Ting P. and Wei H. (Wong) L.; 1 son, Norman. B.S. in Genetics, Taiwan Chung-Hsing U., 1957; M.S. in Microbial Genetics, Bishop's U., Que., Can., 1972. Asst. prof. genetics Taiwan Chung-Hsing U., 1963-67; sr. microbial geneticist Sch. Medicine, U. Ottawa, Ont., Can., 1970-73; research scientist Ill. State Psychiat. Inst., Chgo., 1976—; cons. Contbg. author: Biological Markers in Psychiatry and Neurology, 1981. Mem. Am. Genetics Soc., N.Y. Acad. Scis. Republican. Subspecialties: Neurochemistry; Genetics and genetic engineering (medicine). Current work: Neurochemical and genetic approach to schizophrenia. Home: 12600 Roma Rd Palos Park IL 60464 Office: 1601 W Taylor St Room 222W Chicago IL 60612

LEE, CHING-LI, research chemist, educator; b. Chia-Yi, Taiwan, Mar. 30, 1942; came to U.S., 1970; s. Yuan-Chi and Liung-Chu (Chang) L.; m. Ming-Lea Liao, Jan. 6, 1972; children: Thomas, George, Jenny. B.S., Chung-Hsing U., Taiwan, 1969; M.S., Wayne State U., 1972-, Ph.D., 1975. Research assoc. dept. immunology Mayo Clinic, Rochester, Minn., 1975-77; research scientist Roswell Park

Meml. Inst., Buffalo, 1977-79, sr. research scientist, 1979—; asst. prof. research SUNY-Buffalo, 1981—. Mem. Am. Assn. Immunology, Am. Assn. Cancer Research, Am. Chem. Soc. Subspecialties: Immunology (medicine); Cancer research (medicine). Current work: isolation and characterization of human tumor antigens (or enzymes) and their clinical application; studying antigenic structure of globular proteins. Office: Roswell Park Memorial Institute 666 Elm St Buffalo NY 14263

LEE, CHUNG, med. researcher, educator; b. Shanghai, China, Sept. 18, 1936; s. Ho Chaung and Hwy Chi (Pao) L.; m. Daphne Chin-Quee, Aug. 21, 1966; children: Michael, Traci. B.S., Nat. Taiwan U., 1959; M.S., W.Va. U., 1966; Ph.D., 1969. USPHS postdoctoral fellow Albany (N.Y.) Med. Coll., 1969-71; assoc. biochemistry and obgyn Northwestern U. Med. Sch., Chgo., 1971-74, asst. prof., 1974-78, assoc. prof., 1979—, dir. urology research labs.; assoc. research dept. Evanston (Ill.) Hosp.; lectr. Cook County Grad. Sch. Medicine, Chgo.; cons. Abbott Labs. Contbr. chpts. to books, articles to profl. jours. NIH fellow, 1969-71; Am. Cancer Soc. grantee, 1973-74; NIH grantee, 1974—. Mem. Am. Physiol. Soc., Endocrine Soc., Am. Assn. Cancer Research, Am. Soc. Andrology, Soc. Study Reproduction. Subspecialties: Physiology (biology); Receptors. Current work: Laboratory research toward a better understanding in the physiology and pathology of the prostate. Home: 3406 Seine Ct Hazel Crest IL 60429 Office: 303 E Chicago Ave Chicago IL 60611

LEE, DAVID, nephrologist; b. Shanghai, China, June 26, 1941; came to U.S., 1969; m. Paulette W. Wang, June 17, 1972. M.B., B.S., U. Hong Kong, 1964. Diplomate: Am. Bd. Internal Medicine, Am. Bd. Nephrology. Intern U. Singapore Med. Sch., 1964-65; sr. house officer, registrar Royal Free Hosp., Royal Postgrad. Med. Sch., London, 1966-69; fellow in nephrology UCLA Med. Ctr., 1969-71, dir. dialysis program, 1971-76, dir. renal transplant program, 1977-82; chief div. nephrology UCLA-Sepulveda VA Med. Ctr., 1981—. Contbr. articles to profl. jours. Chmn. sci. adv. council Nat. Kidney Found., So. Calif., 1982; mem. rev. com. for state regional dialysis ctrs. Calif. Dept. Health, 1978-80. NIH grantee; VA grantee, 1977—. Mem. Royal Coll. Physicians, Am. Soc. Nephrology, Internat. Soc. Nephrology, Western Soc. Clin. Investigation, Transplantation Soc. Democrat. Lodge: Rotary. Subspecialties: Internal medicine; Nephrology. Current work: Nephrology, metabolic disorders with special emphasis on mineral metabolism, hypertension and kidney transplantation. Home: 10373 Summer Holly Circle Los Angeles CA 90077 Office: Sepulveda VA Med Ctr 16111 Plummer St Sepulveda CA 91343

LEE, E. BRUCE, system scientist, educator; b. Brainerd, Minn., Feb. 1, 1932; s. Ernest R. and Hazel (Bruce) L.; m. Judith D., Sept. 4, 1954; children: Brian, Kevin, Timothy, Joel, Cara, Elizabeth. A.A., Brainerd State Coll., 1952; B.S. in Mech. Engring, U. N.D., 1955, M.S., 1956; Ph.D., U. Minn., 1960. Instr. U. N.D., 1955-56; sr. scientist Honeywell, Mpls., 1956-63; assoc. prof., prof., head dept. elec. engring. U. Minn., Mpls., 1963—; vis. scientist Research Inst. Advanced Studies, Balt., 1961; vis. prof. Calif. Inst. Tech., Pasadena, 1967-68, Tech. U. Warsaw, Poland, 1974, 80; sr. fellow Sci. Research Council Eng., Warwick, Warwickshire, 1968; assoc. dir. Ctr. Control Sci. and Dynamics, Mpls., 1981—. Author: Foundations of Optimal Control Theory, 1967. Recipient Nat. Gold Medal award ASME. Fellow IEEE; mem. Soc. Indsl. and Applied Math., Pi Tau Sigma. Subspecialties: Distributed systems and networks; Algorithms. Current work: Development of theory of interacting systems with delays, two dimensional systems and applications to circuit design and image processing. Home: 1705 Innsbruck Pkwy Minneapolis 55421 Office: Univ Minn 123 Church St Minneapolis MN 55455

LEE, EDWARD YUE SHING, computer systems engineer, cons. data base management systems; b. Hong Kong, Aug. 1, 1937; came to U.S., 1957, naturalized, 1971; s. George Arthur and King Won (Lo) L.; m. Esther K. C. Ng, July 3, 1963; children: Philip S.W., Christina K. Y. B.S., Portland State U., 1961; M.S., Purdue U., 1963, Ph.D., 1971. Physicist, engr. Tektronix, Inc., Beaverton, Oreg., 1963-66; research asst. Purdue U., 1966-71; prin. investigator U.S. Army Constrn. Research Lab., Champaign, Ill., 1971-74; mem. tech. staff (Jet Propulsion Lab.), Pasadena, Calif., 1974-77; staff engr. Def. System Group, TRW, Redondo Beach, Calif., 1977-80, sr. project engr., 1980—, lectr. database system, 1979—. Bd. dirs. Chevy Chase Estate Home Owner Assn., 1976-78. Mem. Assn. Computing Machinery (recipient outstanding mem. award 1978-79, chmn. Los Angeles chpt. 1981-82), IEEE Computer Soc., Am. Phys. Soc. Subspecialties: Distributed systems and networks; Database systems. Current work: Research and development on time critical distributed data processing system; distributed database management systems and information systems; storage and retrieval. Office: TRW Def and System Group One Space Park Dr Redondo Beach CA 90278 Home: 30675 Via La Cresta Rancho Palos Verdes CA 90274

LEE, GARRETT, physician, researcher; b. San Francisco, June 23, 1946; s. Frederick B. and Josephine (Woo) L. B.A. in Genetics, U. Calif.-Berkeley, 1968; M.D., U. Calif.-Davis, 1972. Med. intern Duke U. Med. Ctr., 1972; med. resident U. Calif.-Davis, 1973, cardiac fellow, 1974, asst. prof. medicine, 1975, dir., 1977-83, Cardiovascular Laser Research Lab., Cedars Med. Ctr., Miami, Fla., 1983—. Contbr. articles to profl. jours. Recipient Med. Student Research award U. Calif.-Davis, 1972. Fellow Am. Coll. Clin. Pharmacology; mem. Am. Heart Assn. (dir. Golden Empire chpt.), Alpha Omega Alpha. Subspecialties: Cardiology; Laser medicine. Current work: Use of laser in the treatment of cardiovascular diseases, cardiovascular laser research, cardiologist, cardiac catheterization. Office: 1295 NW 14th St Cedars South-Suite K Miami FL 33125

LEE, GRIFF CALICUTT, civil engr.; b. Jackson, Miss., Aug. 17, 1926; s. Griff and Lyda (Higgs) L.; m. Eugenia Humphreys, July 29, 1950; children—Griff Calicutt III, Robert H., Carol E. B.E., Tulane U., 1948; M.S., Rice U., 1951. Civil engr. Humble Oil & Refining Co., New Orleans, Houston, 1948-54; design engr. J. Ray McDermott & Co., Inc., New Orleans, 1954-66, chief engr., 1966-75, group v.p., 1975-80, v.p., group exec., 1980—; mem. vis. com. dept. civil engring. U. Tex.; mem. adv. bd. Tulane Coll. Engring.; mem. marine bd. NRC. Contbr. articles to profl. jours. Served with USN, 1944-46. Mem. Nat. Acad. Engring., Am. Bur. Shipping, Am. Concrete Inst., ASCE, Am. Welding Soc., Soc. Petroleum Engrs., Am. Petroleum Inst., Law of Sea Inst. Presbyterian. Clubs: International House, Petroleum of New Orleans, Vista Shores Country. Subspecialty: Civil engineering. Home: 6353 Carlson Dr New Orleans LA 70122 Office: 1010 Common St New Orleans LA 70112

LEE, HARRY WILLIAM, forest engineering educator, researcher; b. Bonners Ferry, Idaho, Sept. 26, 1938; s. Robert Edward and Laura Rhoda (Knowles) L.; m. Evelyn Louise Robinson, Aug. 7, 1962; children: Larissa Dawn, Branda Carol. B.S. in Chem. Engring, U. Idaho, 1972, M.S., 1977, Ph.D. in Agrl Engring, 1982. Asst. county planner, Latah County, Moscow, Idaho, 1972-74; regional planner State Idaho, Moscow, 1974-79; research assoc. Agrl. Engring. U. Idaho, Moscow., 1979-80, instr. forest products., 1980-82. Author: Latah County Flood Plain Analysis, 1973, Flood Insurance Brochure, 1976. Planning Commn. Latah County, 1980; study com. Sch. Facilities, Moscow, 1978. Served with U.S. Army, 1961-63. Recipient outstanding civil engring. student ASCE, 1972. Mem. Nat. Audubon Soc., Am. Soc. Agrl. Engrs. Republican. Subspecialties: Resource conservation; Civil engineering. Current work: Primary interests are in soil and water conservation applied to forest harvesting schedules; forest harvesting equipment research and development. Home: 1017 N Almon Moscow ID 83843 Office: Dept Forest Products U Idaho Moscow ID 83843

LEE, HENRY JOUNG, biological chemist, educator; b. Seoul, Korea, Nov. 17, 1941; s. Hun-Sang and Chung-ok (Kim) L.; m. Hyojo Sue, July 11, 1969; children: Lois, Angie, Jenny. B.S., Seoul Nat. U., 1964, M.S., 1966; Ph.D., Okla. State U., 1971. Research assoc. Mt. Sinai Sch. Medicine, N.Y.C., 1971-73; asst. prof. Fla. A&M U., Tallahassee, 1973-79, assoc, prof., 1979-82, prof., 1982—, head, 1982—; vis. scientist Rockefeller U., N.Y.C., 1979; NSF grant reviewer. Editor: Progress in Research and Clinical Application of Corticosteroids, 1982. NIH grantee, 1978, 79; recipient Fla. A&M U. research award, 1981, service award, 1982. Mem. Am. Soc. Biol. Chemistry, Am. Chem. Soc., Sigma Xi. Democrat. Baptist. Subspecialty: Medicinal chemistry. Current work: chemical synthesis and evaluation of new cortiosteroids for the development of new anti-inflammatory steroids without side effects. Home: 2521 Harriman Circle Tallahassee FL 32312 Office: Fla A&M U Tallahassee FL 32307

LEE, HIKYU, computer scientist; b. Seoul, Korea, Dec. 30, 1951; came to U.S., 1975. B.S. in Electronics Engring, Seoul Nat. U., 1973; M.S. in Computer Sci, Princeton U., 1976, M.A., 1977, Ph.D., 1980. Asst. prof. dept. elec. engring. and computer sci. Northwestern U., Evanston, Ill., 1979-81; mem. tech. staff Bell Labs., Murray Hill, N.J., 1981—. Mem. IEEE Computer Soc., Assn. Computing Machinery, Korean Scientists and Engrs. in Am. Subspecialties: Distributed systems and networks; Operating systems. Current work: Distributed Unix systems. Home: 69 Possum Way Murray Hill NJ 07974 Office: Bell Labs 600 Mountain Ave Murray Hill NJ 07974

LEE, INSU PETER, pharmacologist; b. Kaesong, Korea, Dec. 25, 1935; came to U.S., 1956, naturalized, 1965; s. C.S. and J.S. L.; m. Chong W. Lee, Mar. 11, 1962; children: Deborah, Frederick, Michelle, Jacqueline. B.A., Pacific Luth. U., 1959; Ph.D., U. Wash., 1969. Pharmacologist Nat. Cancer Inst., Bethesda, Md., 1969-72; sr. staff fellow Nat. Inst. Environ. Health Sci., Research Triangle Park, N.C., 1972-74, pharmacologist, 1974—. Contbr. numerous articles to U.S. and internat. sci. jours. Am. Cancer Soc. research grantee Sch. Medicine, U. Wash., 1968-69; NIH predoctoral trainee, 1968-69; fellow Japan Soc. Promotion Sci., 1981. Mem. Am. Soc. Microbiology, AAAS, Soc. Toxicology, European Soc. Toxicology, Am. Soc. Pharmacology and Exptl. Therapeutics, Internat. Union Toxicology, Sigma Xi. Republican. Subspecialties: Cellular pharmacology; Toxicology (medicine). Current work: Reproductive and developmental toxicology, enzyme induction, gene expression, PAH metabolism, DNA repair, mutagenesis and carcinogenesis.

LEE, JAMES CHING, biochemistry educator, researcher; b. Shanghai, China, Dec. 16, 1941; came to U.S., 1963, naturalized, 1976; s. Winston Tai-Kong and Annie (Lau) L.; m. Lucy Ling-York, June 14, 1969; children: Genevieve Ching-Wen, Amanda Ching-Man. A.B., Hope Coll., 1966; Ph.D., Case Western Res. U., 1971. Postdoctoral fellow Brandeis U., 1971-76; asst. prof. biochemistry St. Louis U., 1976-80, assoc. prof., 1980-83, prof., 1983—; cons. NIH, Washington, 1982—. Recipient Godfrey award Hope Coll., 1966; Katzman award St. Louis U., 1978. Mem. Biophys. Soc., Am. Soc. Biol. Chemists, Am. Chem. Soc. Subspecialties: Biophysical chemistry; Biochemistry (biology). Current work: Thermodynamics of protein interactions. Office: St Louis U 1402 S Grand Blvd Saint Louis MO 63104

LEE, JAMES TRAVIS, JR., surgeon; b. Wichita Falls, Tex., Apr. 20, 1943; s. James Travis and Mary Ann (Walker) L. B.A., U. Tex., 1964; M.S., U. Ill., 1966; Ph.D., 1968; M.D., U. Minn., 1975. Diplomate: Am. Bd. Surgery. Chemist Phillips Petroleum Co., Bartlesville, Okla., 1964; sr. chemist 3M Co., St. Paul, 1968-72; research fellow U. Minn., Mpls., 1975-81, asst. prof. surgery, 1981—; staff surgeon VA Hosp., Mpls., 1981—. Mem. Am. Fedn. Clin. Research, Sigma Xi, Alpha Chi Sigma, Phi Lambda Upsilon, Phi Kappa Phi. Republican. Club: Rattlesnake Lodge. Subspecialties: Surgery; Organic chemistry. Current work: Surgical infections in compromised hosts. Home: 2085 Dotte Dr White Bear Lake MN 55110 Office: Mpls VA Hosp Surg-012 Minneapolis MN 55417

LEE, JEN-SHIH, biomedical engineering educator; b. Chukang, Kwangtung, China, Aug. 22, 1940; came to U.S., 1962, naturalized, 1975; s. Yee-I and Yao-tzu (Lai) L.; m. Lian-pin Ma, June 11, 1966; children: Lionel, Grace, Albert Lee. B.S. in Mech. Engring, Nat. Taiwan U., 1961; M.S., Calif. Inst. Tech., 1963, Ph.D., 1966. Research fellow Calif. Inst. Tech., Pasadena, 1966; asst. research engr. U. Calif. San Diego, La Jolla, Calif., 1966-69; asst. prof. U. Va., Charlottesville, 1969-74, assoc. prof. div. biomed. engring., 1974-83, prof. biomed. engring., 1983—; cons. Inst. Physiology, U. Graz, Austria, 1979—. Assoc. editor: Annals Biomed. Engring, 1982; contbr. articles to profl. jours. Anthony scholar Calif. Inst. Tech., 1964-66; advanced research fellow San Diego County Heart Assn., 1966-69; recipient Research Career Devel. award Nat. Heart, Lung and Blood Inst., 1974-80. Mem. Am. Physiol. Soc., Biomed. Engring. Soc., Am. Microcirculatory Soc., AAAS, ASME. Subspecialties: Biomedical engineering; Theoretical and applied mechanics. Current work: Mechanics and mass transport of microvascular flow, mixing in high frequency ventilation, and transcapillary fluid movement and the measurement of the blood and plasma density. Office: U Va Med Center Div Biomed Engring PO Box 377 Charlottesville VA 22908

LEE, JOHN CHAESEUNG, nuclear engineer, educator; b. Seoul, Korea, July 29, 1941; came to U.S., 1965, naturalized, 1976; s. Kwan Hee and Chin Bae (Kim) L.; m. Theresa Sungock Lee, June 26, 1971; 1 dau., Nina. B.S., Seoul Nat. U., 1963; Ph.D., U. Calif.-Berkeley, 1969. Sr. engr. Westinghouse Electric Corp., Pitts., 1969-73, Gen. Electric Co., San Jose, Calif., 1973-74; asst. prof. nuclear engring. U. Mich., Ann Arbor, 1974-78, assoc. prof., 1978-81, prof., 1981—; cons. adv. com. on reactor safeguards U.S. Nuclear Regulatory Commn., Washington, 1975—; cons. Los Alamos Nat. Lab., 1977—. Contbr. articles and papers to profl. jours. Recipient Disting. Service award U. Mich. Class of 1938E, 1979. Mem. Am. Nuclear Soc., AAAS. Episcopalian. Subspecialties: Nuclear fission; Nuclear engineering. Current work: Nuclear reactor theory, reactor core physics and design analysis, reactor kinetics, fuel cycle analysis, reactor safety analysis, power plant simulation and control. Home: 2128 Georgetown Blvd Ann Arbor MI 48105 Office: Dept Nuclear Engring Univ Michigan Ann Arbor MI 48109

LEE, JOHN FRANCIS, management consulting company executive; b. Boston, Sept. 19, 1918; s. Michael Francis and Catherine Mary (Arrigal) L.; m. Helene Zinka Comes, May 15, 1946 (div. 1972); children: Anne-marie Lee Dorman, Robert Paul, Virginia Louise Lee Linden, Jacqueline. S.B., The Citadel, Charleston, S.C., 1947; S.M., Harvard U., 1948; Sc.D., U. London, 1968; Litt.D. (hon.), U. Malaga, Spain, 1972. Registered profl. engr., D.C., Maine. Asst. prof. engring. U. Maine, Orono, 1948-50, assoc. prof., 1950-52; Broughton prof. engring. N.C. State U., 1952-61; pres. SUNY-Stony Brook, 1961-62; spl. adviser NSF, Washington, 1962; pres., chief exec. officer Internat. Devel. Services, Inc., Washington, 1962-71, Promotorco (S.A.), Luxembourg, 1971-79, Intercontinental Mgmt. Consultants, Inc., Torrance, Calif., 1979—; participant White House Conf. on Internat. Coop., Washington, 1965; personal rank of ambassador Intergovtl. Com. European Migration-Argentina Negotiations, Washington, Buenos Aires, 1968. Author: Theory & Design of Steam & Gas Turbines, 1961, Thermodynamics, 1962, Statistical Thermodynamics, 1973; contbr. numerous articles to profl. jours. Served to maj. U.S. Army, 1941-45. Named Ambassador of Goodwill, State of N.C., 1961; decorated Order So. Cross Govt. of Brazil, Order Bernardo O'Higgins Govt. of Chile, Chevalier Legiond'Honeur Govt. of France. Mem. IEEE, AIAA, Optical Soc. Am., Internat. Soc. Hybrid Microelectronics, Soc. Photo-Optical Instrumentation Engrs., Am. Fgn. Service Assn. Unitarian. Clubs: Cosmos, Internat. (Washington); Jockey (Paris). Subspecialties: Thermodynamics; Statistical mechanics. Current work: Communication theory; ballistic missile defense; high technology management. Home: 6702 Los Verdes Dr NUM2 Rancho Palos Verdes CA 90274 Office: Intercontinental Management Consultants Inc 3838 Carson St #110 Torrance CA 90503

LEE, KOTIK KAI, laser scientist; b. Chungking, China, May 30, 1941; came to U.S., 1967, naturalized, 1975; s. Shi Shien and Wen Ru (Hsia) L.; m. Lydia S.M. Ruo, Sept. 8, 1967; children: Jennifer M., Peter H. B .Sc., Chung Yuan Coll., 1964; M.Sc., U. Ottawa, 1967; Ph.D., Syracuse U., 1972. Instr. physics Syracuse (N.Y.) U., 1972-73; asst. prof. physics Rio Grande (Ohio) Coll., 1973-74; vis. prof. math. U. Ottawa, Ont., Can., 1974-76; scientist U. Rochester, N.Y., 1977-78, research scientist, applied mathematician, 1979—; mem. tech. staff Santa Barbara Research Ctr., Goleta, Calif., 1978-79. Contbr. many articles to profl. jours.; reviewer various jours. Can. NRC grad. studentship, 1966-67; research grantee, 1975-78. Mem. Am. Math. Soc. (reviewer Math. Rev. 1973—), Am. Phys. Soc. (referee Phys. Rev. 1974—), Can. Assn. Physicists, Soc. Indsl. and Applied Math., Internat. Soc. Gen. Relativity and Gravitation (referee Jour. 1980—). Roman Catholic. Subspecialties: Theoretical physics; Laser physics. Current work: Laser physics, laser plasma interactions, laser fusion, materials science, differential geometry and differential equations, applied mathematics, relativity and gravitation, other areas of theoretical physics. Home: 18 Winding Brook Dr Fairport NY 14450 Office: Lab for Laser Energetics U Rochester 250 E River Rd Rochester NY 14623

LEE, KWANG-SUN, physician, researcher, educator; b. Seoul, Korea, Aug. 30, 1941; s. Wha-Kyung and Kum-Bok (Chung) L.; m. On-Sook Lee, Sept. 9, 1967; children: Iris J.W., Jennifer J.W. M.D., Seoul Nat. U., 1965. Diplomate: Am. Bd. Pediatrics. Intern Lincoln Hosp., Bronx, N.Y., 1967-68, resident in pediatrics, 1968-71; fellow in neonatology Albert Einstein Coll. Medicine, Bronx, 1971-73, instr. in pediatrics, 1973-75, attending neonatologist, 1973-80, asst. prof. pediatrics, 1975-79, assoc. prof., 1979-80; assoc. prof. pediatrics Pritzker Sch. Medicine, U. Chgo., 1980—, dir. neonatology, 1980—, co-adminstr., 1980—. Contbr. articles to profl. jours. Mem. adv. bd. Chgo. Chpt. March of Dimes, 1981—. Recipient Clin. Investigator award Nat. Inst. Arthritis, Metabolism and Digestive Diseases, 1976-79; Irma T. Hirschl Career Scientist award, 1978-80; Nat. Inst. Child Health and Human Devel. grantee. Mem. Am. Acad. Pediatrics, Soc. Pediatric Research. Subspecialties: Pediatrics; Neonatology. Current work: Bilirubin metabolism and perinatal epidemiology; medical care, education, research, and organization for the improvement of intensive newborn care. Home: 5510 S Kimbark Ave Chicago IL 60637 Office: Box 325 Pediatrics Univ Chicago Chicago IL 60637

LEE, KYU TAIK, pathologist, educator; b. Taegu, Korea, Sept. 17, 1921; came to U.S., 1960, naturalized, 1965; s. Chai Yung and Chung Bai (Kim) L.; m. Sook Kyung Lee, Oct. 10, 1944; children: Chongme Lee, Chongho Lee. M.D., Severance Union Med. Coll., 1943; Ph.D., Washington U., 1956. Prof., chmn. medicine Kyung Pook U. Med. Sch., Taegu, 1956-60; assoc. prof. pathology Albany (N.Y.) Med. Coll., 1961-66, prof. pathology, 1966—, assoc. dean grad. studies, 1976-80; mem. com. on pathology NRC, 1969-72; heart/lung research Rev. com. NIH, 1979—. Mng. editor: Exptl. and Molecular Pathology, 1969—; contbr. articles to profl. jours. NIH-Nat. Heart, Lung and Blood Inst. grantee, 1956—. Fellow Council on Arterosclerosis, AMA; mem. Am. Soc. Exptl. Pathology, Internat. Acad. Pathology, Council Biology Editors, Korean Med. Assn. Am. (pres. 1979-80), Korean Cardiology Soc. (pres. 1956-60), Sigma Xi, Phi Beta Kappa. Club: Schuyler Meadows Country. Subspecialties: Pathology (medicine); Nutrition (medicine). Current work: Atherosclerosis research in experimental animals/humans. Home: 114 Myton Ln Menands NY 12204 Office: Albany Med Coll 47 New Scotland Ave Albany NY 12208

LEE, LAWRENCE HWA-NI, aerospace and mechanical engineering educator, consultant; b. Shanghai, China, Jan. 5, 1923; came to U.S., 1947, naturalized, 1953; s. Ven-Foo and Soo-Lien (Wu) L.; m. Lydia Shui-Yen Shen, Dec. 18, 1948; children: Lynn Lawrence. B.S. in Civil Engring, La Universitato Utopia, Shanghai, 1945, M.S., U. Minn.-Mpls., 1947, Ph.D., 1950. Registered profl. engr., Ind. Instr. U. Minn.-Mpls., 1949-50; asst. prof. U. Notre Dame, ind., 1950-56, assoc. prof., 1956-60, prof. aerospace and mech. engring., 1960—; cons. Ind. State Bd. Registration Profll Engr., Indpls., 1953-63, Bendix Corp., South Bend, Ind., 1953-73, Gen. Motors Corp., Santa Barbara, Calif., 1960-64, 73, Rock Island Arsenal, Ill., 1976. Editor: Developments in Mechanics, Vol. 6, 1972; contbr. articles to prof. jours. Recipient Structural Mechanics Research award Office Naval Research - AIAA, 1972. Fellow ASCE (com. chmn. 1970-72); mem. Soc. Engring. Sci. (dir. 1981-84), ASME, Am. Soc. Engring. Edn., Am. Acad. Mechanics. Subspecialties: Solid mechanics; Theoretical and applied mechanics. Current work: Educator in aerospace and mechanical engineering, research interests in the field of nonlinear mechanics. Home: 17939 Edgewood Walk South Bend IN 46635 Office: U Notre Dame Notre Dame IN 46556

LEE, LIENG-HUANG, chemist, researcher, consultant; b. Fuzhou, Fujian, China, Nov. 6, 1924; came to U.S., 1952, naturalized, 1964; s. Bin and Da-Mei (Chang) L.; m. Chiu-Bin Wu, Feb. 8, 1949; children: Muriel Ei-hui Payne; Daniel Chia-Sen, Robert Min-sen, Grace Mei-hai. B.S. in Chemistry, Amoy (Xiamen) U., Xiamen, China, 1947, M.S., Case Inst. Tech., 1954, Ph.D., 1955. Chemist, Taiwan, 1947-52; lectr. Tunghai U., Taiwan, 1956-57; vis. prof. Taiwain NOrmal U., 1957-58; cons. Union Indsl. Research Inst., Hsin-Chu, Taiwan, 1957-58; research chemist Dow Chem. Co., Midland, Mich., 1958-63; sr. research chemist, 1963-68; sr. scientist Xerox Corp., Rochester, N.Y., 1968—; invited lectr. Chinese Acad. Scis., Beijing, 1979; Disting. scholar Nat. Acad. Scis. and Chinese Acad. Scis., 1983. Contbr. numerous articles to profl. jours.; editor: Adhesion, Friction and Wear, 6 books, 1973-81; editorial bd.: Jour. Adhesion, 1971—, Jour. Chinese Polymer Sci, 1982. Recipient Mabery prize Case Inst., 1954; Firestone fellow, 1952-58; Lubrizol fellow, 1954-55; Am. Chem. Soc. Petroleum Research Fund grantee, 1975, 76, 79, 83. Fellow Am. Inst. Chemists; mem. Am. Phys. Soc., Am. Chem. Soc. (chmn. div. organic coatings and plastics chemistry 1976), Sigma Xi (pres. Midland br. 1965). Subspecialties: Polymer chemistry; Surface chemistry. Current work: Adhesion, friction and wear of polymers, and electrophotography. Patentee in field. Office: Xerox Corp Webster Research Center Webster NY 14580 Home: 796 John Glenn Blvd Webster NY 14580

LEE, LIHSYNG STANFORD, cancer researcher, geneticist, engr.; b. China, Oct. 28, 1945; came to U.S., 1969, naturalized, 1980; s. Honping and Kuorung (Shea) L; m. Alice S.F. Chang, Sept. 8, 1974; children: Jenny, Oriana. B.S. (Ten-yu-tan scholar), Nat. Taiwan U., 1968; M.Phil. (fellow), Yale U., 1971, M.S., 1972, Ph.D., 1974. Postdoctoral fellow Roswell Park Meml. Inst., Buffalo, 1974-76; staff assoc. Columbia U. Coll. Physicians and Surgeons, N.Y.C., 1976-79; mem. staff Gen. Electric Research Center, Schenectady, 1979—. Contbr. numerous articles to sci. jours. Served to 2d lt. Taiwan Army, 1968. NIH grantee, 1976-79. Mem. Am. Soc. Pharmacology and Exptl. Therapeutics, Am. Soc. Cell Biology, Am. Soc. Microbiology, Am. Assn. Cancer Research, Am. Coll. Toxicology. Subspecialties: Genetics and genetic engineering (biology); Cancer research (medicine). Current work: Cell culture, biotechnology, hybridoma, monoclonal antibody, recombinant DNA, gene cloning, cellular pharmacology, biochemistry molecular biology, membrane biology, immunoradio-technology, computer technology. Home: 1269 Hempstead Rd Schenectday NY 12309 Office: K1-3B33 General Electric Research Center Schenectady NY 12301

LEE, LONG C., electrical engineering educator; b. Taiwan, Oct. 19, 1940; s. Chin L. and Wang (Wen) L.; m. Laura M. Cheng., Dec. 1, 1967; children: Gloria, Thomas. B.S., Taiwan Normal U., 1964; A.M., U. So. Calif., 1967, Ph.D., 1971. Research assoc. U. So. Calif., 1971-77; sr. physicist SRI Internat., Menlo Park, Calif., 1977-81; prof. elec. and computer engring. San Diego State U., 1982—. Contbr. articles to profl. jours. NASA grantee, 1979—; NSF grantee, 1980—; Air Force Office of Research grantee, 1980—. Mem. Am. Phys. Soc., Interam. Photochemistry Soc., Am. Geophys. Union. Subspecialties: Atomic and molecular physics; Spectroscopy. Current work: Photoabsoprtion, photonization, and photodissociation processes of molecules. Office: Dept Elec and Computer Engring San Diego State U San Diego CA 92182

LEE, LUCY FANG, research biochemist; b. Fukien, China, Nov. 3, 1931; d. Paul and Theresa Fang; m. Joseph J. Lee, Oct. 2, 1954; children: Rebecca, Yvonne. Ph.D., Mich. State U., 1967. Teaching asst. Mich. State U., 1963-67, U. Md., 1965-67; postdoctoral fellow U. Chgo., 1967-68; research chemist Dept. Agr. Regional Poultry Research Lab., East Lansing, Mich., 1968—; cons. poultry disease, viral oncology. Contbr. numerous articles on Marek's disease to profl. jours. Mem. Am. Soc. Biol. Chemists, Am. Assn. Immunologists, Am. Soc. Microbiology. Roman Catholic. Subspecialties: Cancer research (veterinary medicine); Virology (veterinary medicine). Current work: Research on basic mechanism on virology and immunology of Marek's disease in chickens; use of monoclonal antibodies via hybridoma tech. Home: 954 Whittier Dr East Lansing MI 48823 Office: 3606 E Mount Hope East Lansing MI 48823

LEE, M. HOWARD, physics educator; b. Pusan, Korea, May 21, 1937; came to U.S., 1953; s. Sang-Hun and Hojung (Park) L.; m. Margaret F. Kendig, Feb. 26, 1967; 1 dau., Jennifer Katharine. B.S., U. Pa., 1959, Ph.D., 1967. Postdoctoral fellow U. Alta. (Can.), Edmonton, 1967-69; staff physicist MIT, Cambridge, Mass., 1969-73; prof. physics U. Ga., Athens, 1973—. Contbr. articles to profl. jours. Recipient Michael award U. Ga., 1976; Fulbright-Hays sr. research scholar, 1978-79. Mem. Am. Phys. Soc., Biophys. Soc. Subspecialties: Statistical physics; Theoretical physics. Current work: Study of time-dependent behavior of quantum many-body systems. Development of the generalized Langevin equation by the method of recurrence relations. Home: 275 Sandstone Dr Athens GA 30605 Office: Dept Physics U GA Athens GA 30602

LEE, MARGARET *See also* **LEE, YEU-TSU N.**

LEE, MARTIN ALAN, space physicist; b. Bromley, Eng., Oct. 9, 1945; came to U.S., 1948, naturalized, 1955; s. Erastus H. and Shirley (Wilson) L. B.S. in Physics, Stanford U., 1966, Ph.D., U. Chgo., 1971. Research assoc. Max-Planck-Inst. Extraterrestrial Physics, Munich, W.Ger., 1971-73, U. Chgo., 1973-74; asst. prof. physics Washington U., St. Louis, 1974-79; sr. research scientist U. N.H., Durham, 1979—. Contbr. articles to profl. jours. NASA grantee; recipient Mark Perry Geller prize U. Chgo., 1971; NATO postdoctoral fellow, 1971. Member. Am. Geophys. Union, Am. Astron. Soc. Subspecialties: Theoretical astrophysics; Solar physics. Current work: Theoretical plasma astrophysics: energetic particle transport, shock acceleration of energetic particles, plasma waves, cosmic ray physics. Home: N River Rd RFD 1 Newmarket NH 03857 Office: Space Sci Ctr DeMereitt Hall U NH Durham NH 03824

LEE, MICKEY MITCHELL, psychological and educational consultant, researcher; b. Erie, Pa., Dec. 20, 1950; s. Mickey Mitchell and Anita L. B.S., Slippery Rock State Coll., 1973; Ed.S. and M.Ed., Edinboro State Coll., 1975; Ph.D., U. Ala., 1977. Cert. ednl. psychology researcher and sch. psychologist. Post-doctoral intern N.W. Regional Edn. Lab., Portland, Oreg., 1978, research assoc., Juneau, Alaska, 1978-79; ednl. psychologist Auburn U., Montgomery, Ala., 1977-78; psychol. and ednl. cons. La. State U. Sch. Dentistry, New Orleans, 1979—; adj. faculty U. New Orleans, 1980-81; cons. Geotrud Bapber Center, Erie, Pa., 1974, Alaska Dept. Edn., Juneau, 1978, La. State U. Sch. Nursing, 1980, La. State U. Med. Center, New Orleans, 1982. Counselor Crisis Line, New Orleans, 1982. Mem. Am. Assn. Cancer Edn., Internat. Assn. Dental Research, Am. Assn. Dental Schs., Am. Ednl. Research Assn., Mid-South Ednl. Research Assn. Russian Orthodox. Current work: Survey design and analysis; prediction of academic success from non-cognitive variables; evaluation studies. Home: 2719 Ursuline (side) New Orleans LA 70119 Office: La State U Sch Dentistry 1100 Florida Ave New Orleans LA 70119

LEE, RICHARD FRANK, plant pathologist; b. Smith Center, Kansas., Mar. 19, 1945; s. Frank P. and Irene Adele (McNall) L.; m. Janet Loree Shellito, July 22, 1967; children: Allan Scott, Lori Renee, Jason Todd. B.S., Ft. Hays Kans. State U., 1968; M.S., Kans. State U., 1973, Ph.D., 1977. Grad. research asst. dept. plant pathology Kans. State U., Manhattan, 1971-72, research asst., 1972-77; research plant pathologist dept. plant pathology U. Calif., Davis, 1977-78; asst. plant pathologist U. Fla. Agr. Research and Edn. Center., Lake Alfred, 1978—. Contbr. articles to profl. jours. Served with U.S. Army, 1968-70. Mem. Am. Phytopath. Soc., Fla. Hort. Soc., Sigma Xi. Subspecialties: Plant pathology; Plant virology. Current work: Plant viruses, identification and characterization of rickettsialike bacteria, citrus virology. Office: 700 Experiment Sta Rd Lake Alfred Fl 33850

LEE, RICHARD M, psychologist; b. Long Beach, N.Y., May 6, 1938; m. Julia Ann Kerstetter, Jan. 30, 1965. B.S., U. Mich., 1959; Ph.D., U. Md., 1966. Lic. psychologist, Mich. USPHS fellow U. Md., College Park, 1963; research assoc. Henry Ford Hosp., Detroit, 1966-67, div. chief, 1967-73; lab. dir. E.B. Ford Inst., Detroit, 1971-78; pvt. practice psychology, Bloomfield Hills, Mich., 1978—; adj. assoc. prof. psychology Wayne State U., Detroit, 1975—; reviewer/cons. NSF, Washington, 1970-78, Science, 1965-75. Contbr. articles in field to profl. lit. Research grantee NIMH, 1969, 78, Mich. Heart Assn., 1973, 76. Mem. Biofeedback Soc. Mich. (pres. 1982—), Am. Psychol. Assn., Biofeedback Soc. Am., Soc. for Neurosci. Subspecialties: Psychophysiology; Psychobiology. Current work: Behavioral science methods for blood pressure control; discrimination (estimation) of blood pressure, self-regulation methods. Inventor, researcher blood pressure tracking system. Home: 3885 Lone Pine West Bloomfield MI 48033 Office: 1575 Woodward Ave Bloomfield Hills MI 48013

LEE, ROBERT BONGKYU, nuclear corporation executive, nuclear engineer; b. Seoul, Korea, Oct. 18, 1940; came to U.S., 1965, naturalized, 1973; s. Byong Y. and Dal K. (Lee) L.; m. Young Ja Kim, May 6, 1967; children: Catherine, Lucille. B.S., Han Yang U., Seoul, 1962; M.S., Iowa State U., 1966, Ph.D., 1968; M.B.A., Fairleigh Dickinson U., Rutherford, N.J., 1982. Research assoc. Iowa State U., Ames, 1968-71, Pa. State U., University Park, 1971-73; mgr. nuclear fuels GPU Nuclear Corp., Parsippany, N.J., 1973—. Subspecialties: Nuclear engineering; Nuclear fission. Current work: Nuclear fuels management; nuclear analysis; safety analysis. Home: RD 8 Glenn Rd Flemington NJ 08822 Office: GPU Nuclear Corp 100 Interpace Pkwy Parsippany NJ 07054

LEE, ROBERT WINGATE, geneticist, educator; b. Boston, June 11, 1942; s. Richard Henry and Marcella Isabelle (MacKenzie) L. B.S., U. Mass., 1964, M.A., 1966; Ph.D., SUNY, Stony Brook, 1971. Postdoctoral fellow Duke U., Durham, N.C., 1971-73; asst. prof. dept. biology Dalhousie U., Halifax, N.S., Can., 1973-77, assoc. prof., 1977—. Contbr. articles to profl. jours. Mem. Genetics Soc. Can., Genetics Soc. Am., Nova Scotian Inst. Sci., Plant Molecular Biology Assn. Club: Armdale Yacht (Halifax). Subspecialties: Plant genetics; Genome organization. Current work: Molecular genetics of plants. Home: 5685 Inglis St Halifax NS Canada B3H 1K2 Office: Dept Biology Dalhousie U Halifax NS Canada B3H 4J1

LEE, ROY YUEWING, computer scientist; b. Hong Kong, Mar. 12, 1942; came to U.S., 1960, naturalized, 1964; s. George and King (Wan) L.; m. Louene Hickcox, Oct. 1, 1963; children: Arthur, Benji, Timothy, Anthony. B.S., Oreg. State U., 1963; M.S., U. Wis., 1964, Ph.D., 1967. Postdoctoral researcher U. Minn., Mpls., 1967-68; mem. tech. staff Sandia Labs., Albuquerque, 1968-69, Livermore, Calif., 1969-75, supr., 1975—; Publs. chmn. Compcon, spring 1981—. Office Naval Research/NRC grantee, 1968. Mem. Math. Assn. Am., Am. Computing Machinery Assn. Subspecialties: Graphics, image processing, and pattern recognition; Software engineering. Current work: Computer graphics software packages, software engring. and methodology; distributed systems and networks; intelligent terminals, data communication. Office: PO Box 969 Livermore CA 94550

LEE, SANBOH, device physicist, materials scientist; b. Taiwan, July 2, 1948; came to U.S., 1975; s. Ching-Shiang and Shioh-Yeh (Chang) L. Ph.D., U. Rochester, 1980. Research engr. Taiwan Power Research Labs., 1972-75; mem. tech. staff Xerox Webster (N.Y.) Research Ctr., 1980—. Contbr. articles to profl. jours. Mem. Am. Physic. Soc., AIME, Am. Metal Soc. Subspecialties: Electronic materials; Semiconductors. Current work: Applied research in semiconductor materials. Computer modelling in device physics. Home: 208 Peakview Dr Rochester NY 14467 Office: 800 Phillips Rd Webster NY 14580

LEE, SANG HE, tumor biologist; b. Seoul, Korea, Mar. 7, 1946; s. Sei Woo and Hong Tai (Kim) L. Ph.D., U. Houston, 1976. Postdoctoral fellow dept. pharmacology Yale U. Sch. Medicine, 1976-79, research assoc., 1979-80; sr. scientist dept. pharmacology Genetech, Inc., South San Francisco, Calif., 1980—. Contbr. articles to profl. jours. Robert A. Welch Found. fellow, 1973-76; Damon-Runyon-Walter Winchell Cancer Fund fellow, 1977-79. Mem. Am. Assn. Cancer Research. Subspecialty: Immunobiology and immunology. Current work: Tumor biology, immunology and chemotherapy. Home: 1035 San Carlos Ave El Granada CA 94018 Office: 460 Point San Bruno Blvd South San Francisco CA 94018

LEE, SANG HOON, medical oncologist, internist, hematologist; b. Daegu, Korea, July 22, 1946; came to U.S., 1973; s. Moo-Chul and Ryang (Jang) L.; m. Young Ae Kang, June 9, 1973; children: Soojung, Jean, Eusung,. M.D., Yonsei U., Seoul, Korea, 1970. Diplomate: Am. Bd. Internal Medicine, Am. Bd. Med. Oncology. Intern U. Wis. Hosp., Madison, 1973-74, radiotherapy resident, 1974-75; med. resident U. Miss. Hosps., Jackson, 1975-78; hematology/oncology fellow U. Calif. Med. Ctr., Irvine, 1978-79; med. oncology fellow U. So. Calif.-Los Angeles County Hosp., Los Angeles, 1979-80; practice medicine, specializing in med. oncology, Torrance, Calif., 1980—. Served to capt. Republic of Korea Army, 1970-73. Mem. Am. Soc. Clin. Oncology, ACP, AMA. Subspecialties: Chemotherapy; Internal medicine. Office: 4201 Torrance Blvd Suite 780 Torrance CA 90503

LEE, SANG MOON, management science educator; b. Seoul, Korea, Apr. 1, 1939; s. Chang W. and D.S. (Bahng) L; m. Laura L. Moncrief, Apr. 20, 1968; children: Tosca M., Amy L. B.A., Seoul Nat. U., 1961; M.B.A., Miami U., Oxford, Ohio, 1963; Ph.D., U. Ga., 1968. Prof. Va. Poly. Inst. and State U., Blacksburg, 1968-76; disting. prof. and dept. chmn. U. Nebr., Lincoln, 1976—; cons. City Pub. Service, San Antonio, 1970-74, City of Lincoln, 1977-79, Omaha Pub. Power, 1978-80. Author: Goal Programming, 1972, Decision Science, (1975) Management Science, 1983; others. Recipient Outstanding Research award U. Nebr., 1980, Disting. Teaching award, 1982. Fellow Am. Inst. Decision Sci. (pres. 1983—, disting. Service award); mem. Acad. Mgmt., Inst. Mgmt. Sci. Democrat. Subspecialties: Operations research (engineering); Mathematical software. Current work: Application of operations research to decision making in organizations, with special emphasis on multiple objectives for productivity improvement. Office: Univ Nebr 210 CBA Lincoln NE 68588 Home: 7245 N Hampton Rd Lincoln NE 68506

LEE, SHUISHIH SAGE, pathologist, electron microscopist; b. Soochow, Kiang-su, China, Jan. 5, 1948; came to U.S., 1972; s. We-Ping Wilson and Min-Chen (Sun) Chang; m. Chung-Seng Lee, Mar. 31, 1973; children: Yvonne Claire, Michael Chung. M.D., Nat. Taiwan U., 1972; Ph.D., U. Rochester, 1976. Resident in pathology Strong Meml. Hosp., Rochester, N.Y., 1976-78, Northwestern Meml. Hosp., Chgo., 1978-79; pathologist and dir. cytology and electron-microscopy Parkview Meml. Hosp., Fort Wayne, Ind., 1979—. Contbr. numerous articles to med. jours. Fellow Am. Soc. Clin. Pathologists, Coll. Am. Pathologists; mem. Am. Assn. Pathologists, Internat. Acad. Pathology, N.Y. Acad. Scis., Am. Soc. Cytology, Electron Microscopic Soc. Am., AMA, Ind. Med. Assn., Ind. Assn. Pathologists, Buckeye Soc. Cytology. Subspecialties: Pathology (medicine); Cytology and histology. Current work: Tumor marker study. Home: 5728 Prophet's Pass Fort Wayne IN 46825 Office: Fort Wayne Med Lab 500 Med Center Bldg Fort Wayne IN 46802

LEE, SI GAPH, obstetrics and gynecology educator, reproductive endocrinologist; b. Seoul, Korea, Jan. 2, 1937; came to U.S., 1965; s. Chang Ran and Sun-Ho L.; m. Esther M Bang, Aug. 29, 1967; children: Susie, Daniel. B.S., Yonsei U., Seoul, 1958, M.D., 1962. Diplomate: Am Coll. Obstetricians and Gynecologists, Am. Bd. Reproductive Endocrinology. Intern Fitkin Meml. Hosp., N.J., 1965-66; resident in ob-gyn Temple U. Hosp., 1966-69; asst. prof. ob.-gyn Temple U. Sch. Medicine, 1971-73; assoc. prof. ob-gyn Sch. Medicine, Sioux Falls, 1974-78, prof. ob-gyn, dir. reproductive endocrinology, 1979—. Contbr. articles to profl. jours. Fellow Am. Coll. Obstetricians and Gynecologists; mem. Endocrine Soc., Am. Fertility Soc., Am. Fedn. Clin. Research. Subspecialties: Obstetrics and gynecology; Reproductive endocrinology. Current work: Peptides and steroid hormone of the hypothalmic pituitary ovarian feedback system. Home: 2800 Stonehedge Ln Sioux Falls SD 57103 Office: Dept of Obstetrics and Gynecology University of South Dakota 2701 S Spring Ave Sioux Falls SD 57105

LEE, SOO IK, neurology educator; b. Yesan, Korea, May 15, 1932; came to U.S., 1974; s. Joo Hyun and Onyo (Chun) L.; m. Ock Kim, Apr. 17, 1962; children: Kyusang, Jean, Sunny Young. M.D., Yonsei U., 1958, D.M.Sc., 1974. Diplomate: Am. Bd. Psychiatry and Neurology, Korean Bd. Internal Medicine; cert. in electroencephalography. Intern, resident in internal medicine Yonsei U. Severance Hosp., Seoul, 1959-64; resident in neurology, fellow U. Va. Hosp., Charlottesville, 1964-68; fellow Mayo Grad. Sch., Rochester, Minn., 1968-69; from asst. prof. to prof. Yonsei U., 1969-74, chmn. neurology, dir. EEG Lab., 1970-74; assoc. prof. neurology U. Va., 1974-80, prof., 1980—; dir. EEG Lab. U. Va. Hosp., 1974—; cons. VA Hosp., Salem, Va., 1974—. Author: Epilepsy Case Studies, 1981; contbr. articles to profl. jours. Mem. Am. Acad. Neurology, Am. Epilepsy Soc., Am. EEG Soc., Va. Neurol. Soc., AMA. Presbyterian. Subspecialty: Neurology. Current work: Neurology, electroencephalogreaphy, evoked potentials. Home: 2504 Woodhurst Rd Charlottesville VA 22901 Office: VA Med Ctr Jefferson Ave Charlottesville VA 22908

LEE, TED CHOONG KIL, research scientist; b. Seoul, Korea, Dec. 3, 1940; came to U.S., 1965; s. Chong H. and Soon Ye (Kim) L.; m. Sook J. Moon, Sept. 15, 1966; children: Shirley, Charles, Michelle. B.S., Korea U., Seoul, 1965; Ph.D., Okla. State U., Stillwater, 1971. Research assoc. Rockefeller U., 1971-73; asst. prof. Howard U., 1973-78; vis. asst. prof. Cornell U., Ithaca, N.Y., 1978-81; sr. research scientist Revlon Health Care Group, Tuckahoe, N.Y., 1981—. Served with Korean Army, 1961-63. NIH grantee, 1973-78. Mem. Am. Soc. Biol. Chemists, Am. Chem. Soc. Subspecialties: Biochemistry (biology); Biophysical chemistry. Current work: Mechanism of hemoglobin biosynthesis, antisickling agents, isolation of plasma proteins by novel methods. Home: 69 Jennifer Ct Yorktown Heights NY 10598 1 Scarsdale Rd Tuckahoe NY 10707

LEE, THOMAS A, plant pathologist; b. Gorman, Tex., Dec. 24, 1945; s. Thomas A. and Lela L.; m. Barbara Lee, June 18, 1966; children: Chad, Clay. B.A., Tex. A&M U., 1968, M.S., 1970, Ph.D., 1973. Extension plant pathologist Tex. Agrl. Extension Service, 1973—. Mem. Am. Phytopath. Assn., Am. Peanut Research and Edn. Soc. Baptist. Subspecialties: Plant pathology; Plant physiology (agriculture). Current work: New fungicides and nematicide application techniques. Home: 100 Sandra Palmer Stephanville TX 76401 Office: Box 1177 Stephenville TX 76401

LEE, THOMAS HENRY, electrical engineering educator, consultant; b. Shanghai, China, May 11, 1923; s. Y.C. and N.T. (Ho) L.; m. Kim Ping, June 12, 1948; children: William, Thomas Henry, Richard. B.S., Nat. Chiao Tung U., 1946; M.S., Union Coll., 1950; Ph.D., Rensselaer Poly. Inst., 1954. With Gen. Electric Co., 1948-80, mgr. lab. ops., 1967-71, mgr. tech. resources, 1967-71, mgr. strategic planning, 1974-77, staff exec. strategic planning, 1977-78, staff exec., chief technologist, 1978-80; vis. research scientist MIT, 1979-80, prof. elec. engring., 1980—, assoc. dir. energy lab., Philip Sporn prof. energy processing, 1982—, dir. elec. power systems engring. lab., 1982—; exec. com. U.S. Nat. com. CIGRE Internat. Conf. Large High Voltage Electric Systems; lectr. Author: Physics and Engineering of High Power Switching Devices, 1975; contbr. articles to profl. jours. Recipient ann. achievement award Chinese Inst. Engr., 1962; achievement award Chinese Engrs. and Scientists Assn. So. Calif., 1976; meritorious award Power Engring. Soc., 1976; managerial awards Gen. Electric co., 1955-58. Fellow IEEE (pres. Power Engring. Soc.); mem. Nat. Acad. Engring. (founding mem.), Conn. Acad. Sci. and Engring. (chmn.), Am. Phys. Soc., Am. Vacuum Soc., Sigma Xi. Subspecialties: Electrical engineering; Energy systems. Current work: Energy systems; energy technology and policy; electric power systems engineering, physical electronics; technology assessment and planning. Patentee in field. Home: 33 Chestnut St Boston MA 02108 Office: Massachusetts Institute of Technology Room 10-172 Cambridge MA 02139

LEE, TONY JER-FU, pharmacologist, educator; b. Hualien, Taiwan, Nov. 10, 1942; s. Huo-Yen and Wan L.; m. Mei-shya Su, June 24, 1978; children: Jonathan, Cheryl. B.S., Taipei (Taiwan) Med. Coll., 1967; Ph.D., W.Va. U., 1973. Postdoctoral fellow UCLA, 1973-75; asst. prof. pharmacology So. Ill. U. Sch. Medicine, Springfield, 1975-80, assoc. prof., 1980—. Contbr. articles to profl. jours. Am. Heart Assn. grantee, 1976—; NIH grantee, 1981—. Mem. Am. Soc. Pharmacology and Exptl. Therapeutics, Electron Microscopy Soc. Am., Soc. Neurosci. Subspecialties: Pharmacology; Microscopy. Current work: Pharmacology and morphology of blood vessels, cerebral vasodilator and constrictor transmitters, neurogenic control of cerebral and peripheral blood vessel tone in hypertension; immunocytochemistry. Home: 61 W Hazel Dell Springfield IL 62707 Office: Dept Pharmacology So Ill Univ Sch Medicine Springfield IL 62702

LEE, TZUO-CHANG, electro-optics specialist; b. Nanchang, Kiangsi, China, Jan. 24, 1936. Ph.D. in Elec. Engring, Stanford U., 1964. With Honeywell Tech. Ctr., Bloomington, Minn., 1965—, group leader optics research, 1977-80, dept. mgr. electro-optics, 1980—. Recipient Sweatt award Honeywell, 1976. Mem. IEEE, Optical Soc. Am. Subspecialties: Optical signal processing; Semiconductors. Current work: Management of high technology areas including optical computing, optoelectronics—solid state LEDs and detectors, and integrated circuits based on advanced semiconductors.

LEE, VAN MING, engineering consultant, educator; b. Shanghai, China, Feb. 17, 1946; came to U.S., 1968; s. Yang Chung and Tsoi Wei (Woo) L.; m. Elizabeth Son-Nuu Tang, June 1, 1973; 1 son, Calvin. B.S., Nat. Taiwan Chung-Hsing U., 1968; M.S., Kans. State U., 1971; Ph.D., NYU, 1978. Cons. Consultants & Designers Inc., Hartford, Conn., 1972-73; environ. engr. Stone & Webster Engring. Corp., Boston, 1973-76; mgr. modeling Equitable Environ. Health, Woodbury, N.Y., 1976-79; project mgr. Parsons Brinckerhoff, N.Y.C., 1979—; adj. asst. prof. CUNY, N.Y.C., 1980-81, SUNY-Old Westbury, 1982-83; asst. prof. computer sci. N.Y. Inst. Tech., 1983—. Mem. ASTM, ASCE (task force, com. 1981—), Acoustical Soc. Am., Soc. Indsl. and Applied Math., Am. Soc. Photogrammetry. Subspecialties: Programming languages; Integrated systems modelling and engineering. Current work: Application of asymptotic statistics to engineering problems; mathematical modeling and computer simulation; scientific programming and computer graphics. Office: NY Inst Tech Old Westbury NY 11568

LEE, WILLIAM WAI LIM, energy, environmental consultant; b. Shanghai, China, Aug. 6, 1948; came to U.S., 1965; s. Frank H. Yao. and Jean (Holt) L. B.S.E., Tulane U, 1969; M.S.E., U. Mich., 1970; S.M.C.E., MIT, 1976, Sc.D., 1977. Cert. civil engr., Calif., N.Y., Pa. Project engr. County Sanitation Dists, Los Angeles, 1970-72; project engr. Woodward-Clyde Cons., San Francisco, 1977-79; asst. prof. dept. civil and urban engring. U. Pa., 1979-82; project dir. R.F. Weston Inc., Rockville, Md., 1982—. Author: Decisions in Marine Mining, 1979; author: (with M.S. Baram and D.B. Rice) Marine Mining of the Continental Shelf, 1978; contbr. articles in field to profl.

jours. Mem. AAAS, ASCE, Soc. Risk Analysis, Ops. Research in Soc. Am., Inst. Mgmt. Scis. Am. Baptist. Subspecialties: Resource management; Operations research (engineering). Current work: Use of quantitative analysis to provide insights into natural resources management and environmental quality problems. Home: 1 Mirrasou Ln Gaithersburg MD 20878

LEE, WYLIE IN-WEI, biomedical engineer; b. Tainan, Taiwan, Aug. 18, 1941; s. Tien-chi and Joan (Huang) L.; m. Nancy Kuo, Jan. 29, 1966; children: Marilyn, Jennifer, Arthur. B.Sc., Taiwan Normal U., 1963; Ph.D., U. Mass., Amherst, 1971. Postdoctoral fellow Manchester (Eng.) U., 1971-72; bioengring. fellow U. Wash., Seattle, 1975-77, asst. prof., 1977-81, assoc. prof., 1981—; cons. WESTMED, Seattle, 1982—; Poalyta Co., Taipei, 1982—, Syva Co., Palo Alto, Calif., 1980-81. Pres. Taiwanese Am. Assn., Seattle, 1976. Recipient grants NIH, 1978, 83; Burroughs Wellcome Fund, 1982. Mem. N. Am. Taiwanese Profs. Assn. (dir. 1982-83), Am. Phys. Soc., Biophys. Soc., Am. Fertility Soc. Subspecialties: Biomedical engineering; Laser medicine. Current work: Biomedical applications of dynamic laser scattering with emphasis on reprodutive biology and characterization of biomaterials; development of new clinical instruments using high technology such as microcomputers, lasers optical fibers and graphic display. Patentee cilioscope, laser spermometer. Office: Univ Wash Center Bioengineering Fl-20 Seattle WA 98195

LEE, YEU-TSU N. (MARGARET LEE), surgeon; b. Sian, Shensi, China, Mar. 18, 1936; d. Kiang-Piao Nee and Lien-Luan Soong; m. Vin-Jang Thomas Lee, Dec. 29, 1962; 1 son, Maxwell Ming-Dao. Student, Nat. Taiwan U., 1953-55; B.A., U. S.D., 1957; M.D. cum laude, Harvard U., 1961. Intern, resident in gen. surgery U. Mich. Hosp., Ann Arbor, 1961-64; sr., chief resident in gen. surgery U. Mo. Hosp., Columbia, 1964-66; post-residency clin. fellow, then cons. surgeon Ellis Fischel State Cancer Hosp., Columbia, 1966-72; asst. prof. UCLA, 1972-73; assoc. prof. surgery U. So. Calif., Los Angeles, 1973—. Author book; contbr. articles to profl. jours. Recipient Service to Humanity award United Chinese-Am. League, 1974; Nat. Cancer Inst. grantee, 1981—. Fellow ACS; mem. Am. Assn. Cancer Research, AMA, Am. Soc. Clin. Oncology, Los Angeles County Med. Assn., Los Angeles Surg. Soc., Soc. Surg. Oncology, Orgn. Chinese-Am. Women. Democrat. Roman Catholic. Subspecialties: Surgery; Oncology. Current work: Surgical diagnosis, treatment and clinical research on cancer. Home: 1935 Micheltorena St Los Angeles CA 90039 Office: 1200 N State St Los Angeles CA 90033

LEE, YUAN TSEH, chemist, educator, cons.; b. Hsinchu, Taiwan, China; s. Tsefan and Pei (Tasi) L.; m. Bernice Wu, June 28, 1963; children: Ted, Sidney, Charlotte. B.S., Nat. Taiwan U., 1959; M.S., Nat. Tsinghue U., 1961; Ph.D., U. Calif.-Berkeley, 1965. Asst. prof. to prof. chemistry U. chgo., 1968-74; then prof. U. Calif., Berkeley, 1974—, also prin. investigator. Contbr. numerous articles on chem. physics to profl. jours. Recipient Ernest O. Lawrence award Dept. Energy, 1981; Alfred P. Sloan fellow, 1969-71; Camille and Henry Dreyfus Found. tchr. scholar grantee, 1971-74; John Simon Guggenheim fellow, 1976-77. Fellow Am. Phys. Soc., Am. Acad. Arts and Scis.; mem. Am. Chem. Soc., AAAS, Nat. Acad. Scis. Subspecialties: Physical chemistry; Kinetics. Current work: Dynamics of elem. chem. reactions and laser photochemistry by crossed molecular beams methods. Office: U Calif Lawrence Berkeley Lab 70A-4414 Berkeley CA 94720

LEECH, J(AMES) NATHAN, nuclear licensing engineer; b. Marion, Ind., Jan. 14, 1949; s. Ralph James and Betty Ruth (Mitchell) L.; m. Mary Ellen Geibel, June 4, 1977; 1 son, James Nathan. B.S. in Aero. Engring, Purdue U., 1972; student, Naval Nuclear Power Sch., 1974; M.B.A., Eastern Mich. U., 1983. Engring. asst. NASA, Houston, 1968-71; nuclear licensing engr. Consumers Power Co., Jackson, Mich., 1978—. Served to lt. USN, 1972-78. Mem. Am. Nuclear Soc., Am. Soc. Quality Control, Am. Soc. Naval Engrs. Republican. Subspecialties: Nuclear fission; Nuclear power plant quality assurance. Current work: Activities related to obtaining operating license for nuclear power plants under construction. Home: 74 W Olcott Lake Jackson MI 49201 Office: Concumers Power Co 1945 W Parnall Rd Jackson MI 49201

LEED, PETER LEWIS, project engineer, consultant; b. N.Y.C., July 1, 1952; s. Seymour and June Blossom (Nagin) L.; m. Denise Quesnel, Oct. 17, 1976; children: Alison Julia, Andrew Gregory. B.S. in Environ. Sci, Tufts U., 1973, Columbia U., 1976; M.B.A. candidate, Babson Coll. Profl. engr., N.J., Mass. Tech. service engr. Enviro-Clear, Somerville, N.J., 1976-79; project engr. Bird Machine, South Walpole, Mass., 1980—; instr. Central New Eng. Coll. Tech. Mem. ASME, Inst. Noise Control Engrs. Subspecialties: Mechanical engineering; Acoustical engineering. Current work: Computer aided engineering, automated design of structures. Home: 5 Pocumtuck Norfolk MA 02056

LEELAMMA, SRINIVASA G., mathematics educator; b. Mysore, India, May 11, 1936; came to U.S., 1966; s. Srinivasa and Alamelamma Gopalachar. B.Sc., Osmania U., Hyderabad, India, 1955, M.Sc., 1957; Ph.D., Marathwada U., Aurangabad, India, 1966. Lectr. Math. Women's Coll., Kurnool, India, 1958-65; asst. prof. U. R.I., 1966-68; assoc. prof. SUNY-Geneseo, 1968-73, prof. math., 1973—. Contbr. articles to profl. jours. Mem. Am. Math. Soc., Math. Assn. Am., Soc. Indsl. and Applied Math. Democrat. Hindu. Current work: Research in qualitative study of differential equations. Office: Dept Math SUNY College Geneseo NY 14454

LEELING, JERRY L., pharmacologist, researcher; b. Ottumwa, Iowa, July 11, 1936; s. Harold A. and Dorcas L.; m. Mary Kilfoil, Apr. 27, 1957; children: James, Susan, Stephanie. B.S., Parsons Coll., Iowa, 1959; M.S., U. Iowa, 1962, Ph.D., 1964. Research scientist Miles Labs, Inc., Elkhart, Ind., 1964-70, sr. research scientist, 1970-73, sect. head, 1973—. Contbr. articles to profl. jours. Mem. Am. Chem. Soc., AAAS, Soc. Toxicology, Am. Soc. Pharmacology and Exptl. Therapeutics. Subspecialties: Pharmacology; Pharmacokinetics. Current work: Drug metabolism; pharmacokinetics. Home: 54576 Briarwood Dr Elkhart IN 46514 Office: PO Box 40 Elkhart IN 46515

LEEMAN, SUSAN EPSTEIN, physiologist, educator; b. Chgo., May 9, 1930; d. Samuel and Dora (Gubernokoff) Epstein; m.; children: Eve, Jennifer, Raphael. B.A., Goucher Coll., 1951; M.A., Radcliffe Coll., 1954, Ph.D., 1958. Instr. Harvard U., 1958-59, adj. asst. prof., 1966-68, asst. research prof., 1968-71; asst. prof. Lab. Human Reproduction and Reproductive Biology, Harvard U., Boston, 1972-73, asso. prof., 1973-80; postdoctoral fellow Brandeis U., 1959-62, sr. research asso., 1962-66; prof. physiology U. Mass. Med. Center, Worcester, 1980—; mem. endocrinology study sect. div. research grants NIH. Contbr. articles to profl. jours.; Mem. editorial bd.: Jour. Neuroscience. USPHS fellow, 1954-58; NIH grantee, 1959-62, 63-67, 68-72; recipient Astwood award Endocrine Soc., 1981; Alberta Heritage for med. research Vis. Prof. award, 1981; Van Dyke award Coll. Physicians and Surgeons Columbia U., 1982. Mem. Am. Physiol. Soc., Endocrine Soc. (council), Soc. Neuroscience, Sigma Xi. Subspecialties: Neurophysiology; Neuroendocrinology. Current work: Neuropeptides, biochemistry and physiology. Patentee in field. Home: 139 Park St Newton MA 02158 Office: Department Physiology University Massachusetts Medical Center 55 Lake Ave North Worcester MA 01605

LEES, ALISTAIR JOHN, chemistry educator; b. Preston, Eng., July 12, 1955; came to U.S., 1979; s. Peter and Enid Elizabeth (Stobbs) L.; m. Margaret Vivien Gribben, Aug. 3, 1979. B.Sc., U. Newcastle-upon-Tyne, Eng., 1976, Ph.D., 1979. Research assoc. U. So. Calif., Los Angeles, 1979-81; asst. prof. chemistry SUNY, Binghamton, 1981—; cons. Anitec Image Corp., Binghamton, 1983—. Contbr. articles to profl. jours. Am. Chem. Soc.-Petroleum Research Fund grantee, 1982; Research Corp. grantee, 1982. Mem. Am. Chem. Soc. Subspecialties: Inorganic chemistry; Laser photochemistry. Current work: Physical inorganic chemistry, photochemistry and photophysics of transition metal complexes, excited-state spectroscopy. Home: 511 Old Lane Rd Binghamton NY 13903 Office: Dept Chemistr SUNY Binghamton NY 13901

LEES, LESTER, aeronautical engineering educator; b. N.Y.C., Nov. 8, 1920; s. Harry and Dorothy (Innenberg) L.; m. Constance Louise Morton, Aug. 30, 1941; 1 son, David Grayson. B.S., M.I.T., 1940, M.S., 1941. Research fellow, instr. math. Calif. Inst. Tech., 1942-44, asso. prof. aeros., 1953-55, prof., 1955-74, dir. environ. quality lab., 1970-74, prof. environ. engring. and aeronautics, mem. sr. staff environ. quality lab., 1974—; aero. engr. Nat. Adv. Com for Aeros., Langley Field, Va., 1944-46; asst. prof. aero. engring. Princeton U., 1946-48, assoc. prof., 1948-53; cons. TRW, 1953—, Aerospace Corp., 1960-65. Contbr. articles to profl. jours. Fellow Am. Acad. Arts and Scis., AIAA (hon.); mem. Nat. Acad. Engring. Democrat. Jewish. Subspecialties: Ecology; Theoretical ecology. Current work: Effects of carbon dioxide emission from man-made sources on earth's temperature and climate. Home: 1911 N Pepper Dr Altadena CA 91001 Office: Calif Inst Tech 1201 E California Blvd Pasadena CA 91125

LEES, MARJORIE BERMAN, biomedical researcher; b. N.Y.C., Mar. 17, 1923; d. Isadore I. and Ruth (Rogal) Berman; m. Sidney Lees, Sept. 17, 1946; children: David, Andrew, Eliot. Ph.D., Radcliffe Coll.-Harvard U., 1951. Asst. and assoc. biochemist McLean Hosp., 1955-62; assoc. biochemist and biochemist, 1966-76; sr. research assoc. in pharmacology Dartmouth Coll. Med. Sch., 1962-66; sr. research assoc., mem. faculty Harvard U. Med. Sch., Boston, 1975—; biochemist E.K. Shriver Center, Mass. Gen. Hosp., Boston, 1976—; mem. nat. adv. council Nat. Inst. Neurol. and Communicative Disorders and Stroke, 1979-82; grant reviewer for NIH and NSF. Contbr. numerous articles to sci. jours.; mem. editorial bd.: Jour. Neurochemistry. Named to Hunter Coll. Hall of Fame, 1982; NIH grantee, 1962—. Mem. Am. Soc. Biol. Chemists, Internat. Soc. for Neurochemistry, Am. Soc. for Neurochemistry (treas. 1975-81, mem., pres. 1982-84); Mem. Soc. for Neurosci., Am. Soc. Neuropathologist, N.Y. Acad. Scis. Subspecialties: Neurochemistry; Neuroimmunology. Current work: Neurochemical research as applied to problems of mental retardation; myelin and demyelinating disorders. Home: 50 Eliot Memorial Rd Newton MA 02158 Office: EK Shriver Center 200 Trapelo Rd Waltham MA 02254

LEEVY, CARROLL MOTON, medical educator, hepatology researcher; b. Columbia, S.C., Oct. 13, 1920; s. Isaac S. and Mary (Kirkl) L.; m. Ruth S. Barboza, Feb. 4, 1956; children—Carroll Barboza, Maria Secora. A.B., Fisk U., 1941; M.D., U. Mich., 1944; Sc.D., N.J. Inst. Tech., 1973; D.Hum., Fisk U., 1981. Intern Jersey City Med. Center, 1944-45, resident, 1945-48, dir. clin. investigation, 1947-57; fellow Banting-Best Inst., U. Toronto, 1953; research assoc. Harvard Med. Sch., 1959; assoc. prof. Coll. Medicine and Dentistry of N.J., 1960-64, prof., 1964; dir. div. hepatology and nutrition N.J. Med. Sch., 1959-75, acting chmn. dept. medicine, 1966-68, chief of medicine, 1968-71; physician-in-chief Coll. Hosp.; dir. Liver Center, Coll. Medicine and Dentistry N.J., 1983—; chief of medicine VA Hosp., East Orange, N.J., 1966-71; cons. NIH, 1965—, FDA, 1970-80; Alcohol and Nutrition Found., 1970—, Am. Liver Found., 1979—; mem. expert com. on chronic liver disease WHO, 1978. Author: Practical Diagnosis and Treatment of Liver Disease, 1957, Evaluation of Liver Function in Clinical Practice, 1965, 2d edit., 1974, Liver Regeneration in Man, 1973, The Liver and Its Diseases, 1973, Diseases of the Liver and Biliary Tract, 1977, Guidelines for Detection of Drug and Chemical-Induced Hepatotoxicity, 1979, Alcohol and the Digestive Tract, 1981; Contbr. numerous articles to med., sci. jours. Served with USNR, 1954-56. Mem. Am. Assn. for Study of Liver Diseases (pres. 1967-68, chmn. steering com. 1968-74), Internat. Assn. for Study of Liver (pres. 1970-74, chmn. criteria com. 1972—), Am. Gastroenterol. Assn. (mem. edn. and tng. com. 1967-71), ACP (mem. publs. com. 1969-74), AMA (vice-chmn., chmn. program com. sect. on gastroenterology 1971-74), Assn. Am. Physicians, Soc. Exptl. Biology and Medicine, Am. Soc. Clin. Nutrition, Am. Inst. Nutrition, AAAS Nat. Med. Assn., Am. Fedn. Clin. Research, Sigma Pi Phi. Subspecialty: Hepatology. Current work: Portal hypertension; liver regenerational; immunologic reactivity in liver disease; drug-induced liver injury; fibrogenesis; hepatocellular cancer. Home: 35 Robert Dr Short Hills NJ 07078 Office: New Jersey Med Sch Coll Medicine and Dentistry of NJ 100 Bergen St Newark NJ 07103

LEFEBVRE, ARTHUR HENRY, mechanical engineer; b. Long Eaton, Eng., Mar. 14, 1923; came to U.S., 1976; s. Henri and May (Brown) L.; m. Elizabeth Marcella Betts, Dec. 20, 1952; children: David Ivan, Paul Henry, Anne Marie. B.Sc., Nottingham U., 1946; Ph.D., Imperial Coll., London, 1952, D.Sc., 1975. Combustion engr. Rolls Royce, Derby, Eng., 1952-61; prof. aircraft propulsion Cranfield Inst. Tech., Eng., 1961-71, prof., head Sch. Mech. Engring., 1971-76; prof., head Sch. Mech. Engring., Purdue U., West Lafayette, Ind., 1976-80, Reilly prof. combustion engring., 1980—; cons. on combustion to various cos., Britain, Sweden and U.S.A.; mem. propulsion and energetics panel Adv. Group Aero. Research and Devel., 1972-76. Contbr. tech. articles to profl. jours.; author Gas Turbine Combustion, 1983. Fellow Royal Aero. Soc., Instn. Mech. Engrs., Royal Soc. Arts. Subspecialties: Combustion processes; Fuels. Current work: Combustion engineering, fuel atomization, evaporation, flame stabilization, ignition, flame propagation, combustion efficiency, gas turbine combustion. Patentee combustion equipment. Home: 1741 Redwood Ln Lafayette IN 47905 Office: Sch Mech Engring Purdue U West Lafayette IN 47907

LEFEBVRE, RICHARD CRAIG, clinical psychologist, researcher; b. Bethesda, Md., Jan. 18, 1954; s. Eugene Francis and Gladys (Engblom) L.; m. Jacqueline Mary Kling, Aug. 9, 1975. B.A., Roanoke Coll., 1974; M.S., North Tex. State U., 1979, Ph.D., 1981. Lic. clin. psychologist, Va. Behavioral medicine fellow U. Va. Med. Center, 1980-81; psychologist, dir. tng. U. Va. Counseling Center, 1981—; pvt. practice cons. psychology, Charlottesville, 1981—; cons. Nat. Heart, Lung and Blood Inst., Bethesda, 1982—. Contbr. chpt., articles to profl. publs. VA merit rev. grantee, 1980. Mem. Am. Psychol. Assn., Eastern Psychol. Assn. Episcopalian. Subspecialties: Health Psychology; Biofeedback. Current work: Stress research; health psychology; behavioral cardiology; stress management; biofeedback training in clinical and health psychology; cognitive-behavioral psychotherapies. Office: U Va 204 University Way Charlottesville VA 22901

LEFFLER, CHARLES WILLIAM, physiology educator; b. Cleve., May 21, 1947; s. William Bain and Marjorie Adele (Smith) L.; m. Robin Burke, Aug. 23, 1968; 1 dau., Noelle Burke. Student, DePauw U., 1965-68; B.S., U. Miami, 1969; M.S., U. Fla.-Gainesville, 1971, Ph.D., 1974. Postdoctoral fellow U. Fla., Gainesville, 1974-76; asst. prof. respiratory physiology U. Louisville, 1976-77; asst. prof. U. Tenn., 1977-81, assoc. prof., 1981—; established investigator Am. Heart Assn., 1982. Mem. Am. Physiol. Soc., Soc. Exptl. Biology and Medicine, Cardiopulmonary Council Am. Heart Assn., Am. Soc. Zoologists, Sigma Xi, Phi Beta Kappa. Subspecialties: Physiology (medicine); Physiology (biology). Current work: Control of vascular resistance by autocoids; perinatal circulatory transition, pulmonary circulation, pregnancy-cardiovascular control, prostanoids. Office: U Tenn Sch Medicine 894 Union Ave NA426 Memphis TN 38163

LEFFLER, JOHN WARREN, ecology educator; b. Reading, Pa., Mar. 1, 1949; s. Lee Roy and Louise (Romig) L.; m. Lynda Trzaska, June 5, 1971; 1 son, Andrew J. B.S., Albright Coll., 1971; Ph.D., U. Ga., 1977. Research asst. Woods Hole Oceanographic Inst., Mass., 1972; instr., research assoc. U.Ga., Athens, 1975-78; asst. prof. Ferrum Coll., Va., 1978-82, assoc. prof. environ. studies, 1982—; mem. ISEP adv. com. NSF, 1981; external reviewer U.S. EPA, 1980; proposal reviewer NSF, 1975—, research equipment com., 1979-81; prin. investigator Contract Battelle Meml. Inst., Columbus, Ohio, 1979-81; proposal reviewer NOAA, 1980-81. Contbr. articles to profl. jours. Served with USAR, 1971-77. NSF fellow, 1971-74; Woods Hole Oceanographic Inst. fellow, 1970. Mem. Ecol. Soc. Am., Am. Inst. Biol. Scis., AAAS, Assn. Southeastern Biologists. Subspecialties: Ecology; Environmental toxicology. Current work: Ecosystem theory, analysis, structure, function; microcosms in environmental research; microcosms as test systems for environmental toxicology. Home: Route 4 Box 312B Rocky Mount VA 24151 Office: Ferrum Coll Ferrum VA 24088

LEFKOWITZ, ISSAI, physics educator; b. N.Y.C., Mar. 13, 1926; s. Leo and Rose (Ossofsky) L.; m. Libby Cohen, June 22, 1952; children: Neil, Phillip. B.A. in Physics, Bklyn. Coll., 1959, Ph.D., Cambridge (Eng.) U., 1964. Dir. research and devel. Gulton Industries, Metuchen, N.J., 1954-60; chief solid state Frankford Arsenal, Phila., 1964-71; pres., chief exec. officer Princeton Materials Sci., N.J., 1971-75; prof. dept. physics U. N.C., Chapel Hill, 1975—; grants officer U.S. Army Research Office, Research Triangle Park, N.C., 1975—; vis. scientist NSF, Washington, 1971-79; cons. Dept. Energy, Washington, 1976-80. Co-editor: Ferroelectric Letters. Fellow EURATOM; mem. Am. Phys. Soc., The Phys. Soc. (U.K.), Internat. Union Pure and Applied Physics (chmn. sect. 1979—). Subspecialties: Condensed matter physics; Superconductors. Current work: Ferroelectrics; high temperature superconductivity; diagnostic cancer techniques. Office: Dept Physics UNC Chapel Hill 27514 Home: 67 Burris Pl Chapel Hill NC 27514

LEFKOWITZ, ROBERT JOSEPH, medical educator; b. N.Y.C., Apr. 15, 1943; s. Max and Rose (Levine) L.; m. Arna Susan, June 11, 1963; children: David, Larry, Cheryl, Mara, Joshua. B.A. in Chemistry cum laude, Columbia U., 1962, M.D., 1966. Diplomate: Am. Bd. Internal Medicine. Intern Columbia-Presbyn. Med. Center, 1966-67; asst. resident in medicine, 1967-68; clin. and research assoc. NIH, 1968-70; sr. resident Mass. Gen. Hosp., Boston, 1970-71, clin. and research fellow cardiology, 1971-73; teaching fellow Harvard U. Med. Sch., 1971-73; assoc. prof. medicine, asst. prof. biochemistry Duke U. Med. Ctr., 1973-77, prof. medicine, 1977-76; investigator Howard Hughes Med. Inst., 1976-82; James B. Duke prof. medicine Duke U. Med. Ctr., 1982—. Author: (with L.T. Williams) Receptor Binding Studies in Adrenergic Pharmacology, 1978, (with others) Mammalian Biochemistry, 1983; editor: Receptor Regulation, 1981; editorial bd.: Molecular Pharmacology, 1979—, Jour. Cardiovascular Pharmacology, 1978—, Membrane Biochemistry, 1978—, Life Scis, 1978—, Archives of Biochemistry and Biophysics, 1979-82, Jour. Receptor Research, 1979—, Jour. Clin. Investigation, 1979—, Jour. Biol. Chemistry, 1982—, Jour. Hypertension, 1982—; contbr. articles to profl. jours. Served to lt. comdr. USPHS, 1968-70. Recipient Roche prize, 1965, Janeway prize, 1966, Young Scientist award Passano Found., 1978, Young Investigator award Am. Coll. Clin. Pharmacology, 1979, George W. Thorn award Howard Hughes Med. Inst., 1979, Gordon Wilson medal Am. Clin. and Climatol. Assn., 1982. Mem. Am. Fedn. Clin. Research (nat. council 1978-83, sec.-treas. 1980-81), Am. Soc. Biol. Chemists, Am. Soc. Clin. Investigation (counselor), Am. Heart Assn., Am. Soc. Pharmacology and Exptl. Therapeutics (John J. Abel award 1978), Endocrine Soc. (Ernst Oppenheimer Meml. award 1982), Assn. Am. Physicians, Phi Beta Kappa, Alpha Omega Alpha. Subspecialties: Biochemistry (biology); Biochemistry (medicine). Current work: Hormone and drug receptors. Home: 3539 Hamstead Ct Durham NC 27707 Office: Duke University Medical Center Box 3821 Durham NC 27710

LEGAN, HARRY LEWIS, orthodontic educator, researcher; b. St. Paul, Sept. 16, 1948; s. Leo Theodore and Charlotee June (Most) L.; m. Robertanne Turner, Mar. 27, 1982. B.S., U. Minn-Mpls., 1969, B.A., 1969, D.D.S., 1973; orthodontic cert., U. Conn., 1977. Asst. prof. surgery U. Tex. Southwestern Med. Sch., Dallas, 1977—, dir. orthodontics, 1977—; guest lectr. Baylor Coll. Dallas, 1980—; research cons. U. Conn.-Farmington, 1982-85; orthodontic cons. Craniofacial Deformities Team, Dallas, 1978—. Contbr. chpts., articles to profl. publs. NIH fellow, 1974-77; U. Tex.-Dallas grantee, 1978. Mem. Am. Assn. Orthodontists, ADA, Southwestern Soc. Orthodontists, Cleft Palate Assn., Internat. Assn. Dental Research, Sadi Fontaine Acad. (hon.), Sociedad Colombiana de Ortodoncia (hon.), Alpha Omega (pres. club 1972-73). Subspecialty: Orthodontics. Current work: Clinical and histologic evaluation of temporomandibular joint; planning and stability of maxillofacial surgical procedures; histochemical characterization of head and neck musculature in patients with craniofacial deformities; biomechanics. Office: U Tex Southwestern Med Sch 5323 Harry Hines Blvd Dallas TX 75235

LEGLER, JOHN MARSHALL, biology educator, anatomy educator; b. Mpls., Sept. 9, 1930; s. Fredick W. and Helen (Hertig) L.; m. Avis J. Johnson, Dec. 22, 1952; children: Austin F., Edward P., Gretchen T., Allison K. Student, U. Kiel, U. Heidelberg, W. Ger., 1950-51; B.A., Gustavus Adolphus Coll., 1953; Ph.D., U. Kans., 1959. Asst. instr. U. Kans., Lawrence, 1969; asst. curator herpetology Mus. Natural History, 1955-59; faculty U. Utah, Salt Lake City, 1959—, prof. biology, 1969—; curator herpetology Mus. Zoology, 1959—; research assoc. Los Angeles County Mus., Los Angeles, 1975—, Gorgas Meml. Lab., Panama, 1964-68; vis. prof. U. New Eng., Armidale, New South Wales, Australia, 1972—. Contbr. articles to profl. jours. Recipient Disting. Teaching award U. Utah, 1967. Mem. Am. Soc. Ichthyologists and Herpetologists (gov. 1967-71), Herpetologists League (pres. 1968-70), Brit. Herpetological Soc., Sigma Xi. Subspecialties: Evolutionary biology; Ecology. Current work: The biology of turtles, taxonomy, ecology and behavior; research involves exploratory work and collecting in poorly known regions. Office: Dept Biology U Utah Salt Lake City UT 84112

LEHMAN, THOMAS JOSEPH ANSORGE, rheumatologist, immunology researcher; b. Boston, Jan. 15, 1949; s. Benjamin Joseph and Nell (Zamkin) L.; m. Karen Ann Teters, May 22, 1982. B.A., U. Calif.-Berkeley, 1970; M.D., Jefferson Med. Coll., Phila., 1974. Resident in pediatrics Children's Hosp. Los Angeles, 1974-77, fellow in rheumatology, 1977-79; med. staff NIH, Bethesda, Md., 1981—. Author: Behavior Therapy of Children, 1983. Served with M.C. USN, 1979-81. Recipient E. Harol Hinman prize Jefferson Med. Coll., 1974. Mem. Arthritis Found, Am. Acad. Pediatrics, Am. Fedn. Clin. Research, Phi Beta Kappa. Subspecialties: Pediatrics; Infectious diseases. Current work: Animal models of juvenile rheumatoid arthritis, pathogenesis, immunology and genetics; epidemiologic studies of systemic lupus erythematosis. Office: ARB: NIADDK NIH 9000 Rockville Pike Bethesda MD 20205

LEHMAN, WILLIAM JEFFREY, physiology educator; b. N.Y.C., June 20, 1945; s. Karl and Lisa (Frank) L.; m. Diana Carol Martin, Oct. 6, 1982. B.S., SUNY-Stony Brook, 1966; Ph.D., Princeton U., 1969. Postdoctoral fellow Brandeis, Waltham, Mass., 1969-72; higher sci. officer Oxford U., Eng., 1973; guest scientist Max Planck Inst., Heidelberg, W.Ger., 1975; asst. prof. Boston U., 1973-78, assoc. prof., 1978—. Contbr. articles to profl. jours. N.Y. State Regents scholar, 1962; Charles Grosvenor Osgood fellow, 1968; Am. Heart Assn. investigator, 1982; Underwood fellow, 1973. Mem. Biophys. Soc., Biochem. Soc. (U.K.), Soc. Gen. Physiologists, Am. Heart Assn. Subspecialties: Physiology (biology); Molecular biology. Current work: Investigating the molecular nature of the regulation of muscular contraction. Office: Boston U Sch Medicine 80 E Concord St Boston MA 02118

LEHMANN, ERICH LEO, statistics educator; b. Strasbourg, France, Nov. 20, 1917; came to U.S., 1940, naturalized, 1945; s. Julius and Alma Rosa (Schuster) L.; m. Juliet Popper Shaffer; children: Stephen, Barbara, Fia. M.A., U. Calif. at Berkeley, 1943, Ph.D., 1946. Asst. dept. math. U. Calif. at Berkeley, 1942-43, asso., 1943-46, instr., 1946-47, asst. prof., 1947-51, asso. prof., 1951-54, prof., 1954-55, prof. dept. statistics, 1955—, chmn. dept. statistics, 1973-76; vis. asso. prof. Columbia, 1950-51, Stanford, 1951-52; vis. lectr. Princeton, 1951. Author: Testing Statistical Hypotheses, 1959, Basic Concepts of Probability and Statistics, 1964, 2d edit, (with J.L. Hodges, Jr.) Basic Concepts of Probability and Statistics, 1970, Nonparametrics: Statistical Methods Based on Ranks, 1975, Theory of Point Estimation, 1983. Guggenheim fellow, 1955, 66, 79; Miller research prof., 1962-63, 72-73. Fellow Inst. Math. Statistics, Am. Statis. Assn.; mem. Internat. Statis. Inst., Am. Acad. Arts and Scis., Nat. Acad. Scis. Subspecialty: Statistics. Current work: Theory of estimation, comparison of experiments, multiple comparison problems. Office: Dept Statistics U Calif Berkeley CA 94720

LEHMKUHL, HOWARD DUANE, microbiologist; b. Wahoo, Nebr., Nov. 9, 1943; s. Howard Preston and Violet Mae (Carlson) L.; m. Lois Jean Lehmkuhl, May 27, 1972; children: Aaron D., Eric J., Melissa A. B.A., Nebr. Wesleyan U., 1967; M.S., U. Nebr., 1972; Ph.D., Iowa State U., 1978. Research technician dept. vet. sci. U. Nebr., 1970-72; research assoc. Vet. Med. Research Inst., Ames, Iowa, 1972-77; microbiologist Nat. Animal Disease Ctr., Ames, 1977—. Contbr. articles in field to profl. jours. Cub master Mid-Iowa Council Boy Scouts Am.; mem. bd. edn. St. Cecillas Parish, Ames. Served with U.S. Army, 1968-70. Decorated Purple Heart, Bronze Star. Mem. Am. Soc. Microbiology, Am. Soc. Virology, Conf. Research Workers in Animal Disease, Gamma Sigma Delta, Sigma Xi. Lutheran. Subspecialties: Virology (veterinary medicine); Animal virology. Current work: Sheep respiratory tract disease; isolating, characterizing and identifying microbial agents from diseased sheep lungs, studying transmission and pathogenesis of diseases of the respiratory tract of sheep. Home: 908 Arizona Ave Ames IA 50010 Office: PO Box 70 Ames IA 50010

LEHMKUHL, L(LOYD) DON, clinical neurophysiologist, researcher; b. Lodgepole, Nebr., Jan. 2, 1930; s. Lloyd Henry and Lillian Fay (Hafer) L.; m. Carol Dill, Oct. 3, 1953; children: Pamela Kay, Sandra Sue, Linda Dee. B.S., U. Nebr.-Lincoln, 1953; M.S., U. Iowa, 1948, Ph.D., 1959, cert. phys. therapy, 1956. Instr. U. Iowa Coll. Medicine, 1959-60; asst. prof. Western Res. U. Coll. Medicine, 1960-68; assoc. prof. Case Western Res. U. Coll. Medicine, 1968-70; asst. dir. dept. allied med. professions and services, staff mem. for Council on Med. Edn., AMA, Chgo., 1970-76; clin. neurophysiologist Inst. Rehab. and Research, Houston, 1976—; research cons. Met. Gen. Hosp., Cleve., 1966-70. Author: Clinics in Physical Therapy: Electrotherapy, 1981, (with L.K. Smith) Brunnstrom's Clinical Kinesiology, 4th edit, 1983. Elder Braeburn Presbyterian Ch., Houston, 1979—; leader sr. high youth group, 1980—. Served to lt. (j.g.) USNR, 1953-55; Atlantic. Recipient Excellence in Teaching award Grad. Class Phys. therapy, Case Western Res. U., 1969, Appreciation award Joint Rev. Com., Rochester, Minn., 1976, Am. Occupational Therapy Assn., New Orleans, 1977, Staff Dept. Phys. Therapy, Inst. Rehab. and Research, 1979, Lucy Blair Service award Am. Phys. Therapy Assn., 1983. Mem. Am. Phys. Therapy Assn. (editorial bd. 1982—), Am. Physiol. Soc., Soc. Behavioral Kinesiology (pres. 1977-81), AAAS, Sigma Xi. Republican. Club: Post Oak Prominaders (Houston) (dir. 1980-81). Subspecialties: Neurophysiology; Physical medicine and rehabilitation. Current work: Using electrophysiological recording techiques to map transmission of nerve impulses through the nervous system of patients with injuries of brain or spinal cord to assist in localizing injury and planning therapy. Home: 3015 Winslow Houston TX 77025 Office: Inst Rehab and Research 1333 Moursund Ave Houston TX 77030 Home: 3015 Winslow Houston TX 77025

LEHNINGER, ALBERT LESTER, biochemistry educator; b. Bridgeport, Conn., Feb. 17, 1917; s. Wally and Selma (Heymer) L.; m. Janet Wilson, Mar. 12, 1942; children: James Wilson, Erika L. Whitmore. B.A., Wesleyan U., 1939; M.S., U. Wis., 1940, Ph.D., 1942. Instr. dept. phys. chemistry U. Wis.-Madison, 1942-45; asst. prof. dept. biochemistry U. Chgo., 1945-49; mem. council biol. scis. Pritzker Sch. Medicine, 1977—; vis. prof. U. Frankfurt, W.Ger., 1951; DeLamar prof., dir. dept. physiol. chemistry Johns Hopkins U., Balt., 1952-78, Univ. prof. med. sci., 1977—; vis. prof. Guy's Hosp., London, 1963; bd. visitors Cornell U. Med. Coll., N.Y.C., 1981-82. Author: The Mitochondrion, 1964, Bioenergetics, 1965, 72, Biochemistry, 1970, 75, Short Course in Biochemistry, 1973, Principles of Biochemistry, 1982; contbr. articles to profl. jours. Mem. Inst. Medicine, Nat. Acad. Sci., Am. Soc. Biol. Chemists, Am. Philos. Soc. (v.p.), Am. Acad. Arts Scis., Am. Soc. Cell Biology, Biochemistry Soc., Biophys. Soc., Phi Beta Kappa, Sigma Xi. Clubs: Green Spring, Hamilton Street, Gibson Island. Subspecialties: Biochemistry (biology); Biophysical chemistry. Current work: Bioenergetics of normal and cancer cells; mitochondrial activities; biochemistry of calcification. Home: 15020 Tanyard Rd Sparks MD 21152 Office: Johns Hopkins U 725 N Wolfe St Baltimore MD 21205

LEHRER, PAUL MICHAEL, psychology educator; b. N.Y.C., Aug. 30, 1941; s. Samuel and Ethel (Lubin) L.; m. Phyllis Alpert, June 13, 1965; children: Jeffrey, Suzanne. A.B., Columbia U., 1963; Ph.D., Harvard U., 1969. Lic. psychologist, N.J. Instr. psychiatry Tufts U. Med. Sch., Boston, 1968-70; asst. prof. psychology Rutgers U., New Brunswick, N.J., 1970-72; asst. prof. psychiatry U. Medicine and Dentistry N.J.-Rutgers Med. Sch., Piscataway, 1972-78, assoc. prof., 1978—; 1st v.p. SERV Inc., Trenton, N.J., 1979—. Editor: Handbook of Relaxation and Stress Reduction Techniques, 1983. Roche Inc. research grantee, 1975-76. Mem. Am. Psychol. Assn., Biofeedback Soc. Am., Biofeedback Soc. N.J. (pres. 1979), Soc. Psychophysiol. Research, Assn. Advancement of Behavior Therapy. Democrat. Jewish. Current work: Behavioral medicine; the psychological and physiological effects of stress reduction techniques. Office: Dept Psychiatry Rutgers Med Sch Piscataway NJ 08854

LEHRER, SAMUEL B., clinical immunologist, educator; b. New Britain, Conn., Apr. 1, 1943; s. Charles Rudy and Nettie (Fleischer) L.; m. Gila Ashinazi, June 20, 1971; children: Rudy, Mark, Sandra. B.S., Upsala (N.J.) Coll., 1966; Ph. D., Temple U., 1971. Postdoctoral fellow Scripps Clin. Research Found., LaJolla, Calif., 1971-75; asst. prof. medicine Tulane Med. Sch., New Orleans, 1975-79, assoc. prof., 1979—, adj. prof. microbiology and immunology, 1980—. Nat. Inst. Allergy and Infectious Diseases awardee, 1978-81; Am. Lung Assn. grantee, 1978-80. Mem. Am. Soc. Microbiology, Am. Assn. Immunology, Soc. Exptl. Biology and Medicine, Am. Thoracic Soc., AAAS, Collegium Internat. Allergologicum. Republican. Jewish. Subspecialty: Immunobiology and immunology. Current work: Isolation and identification of allergens, IgE biosynthesis. Home: 142 Brockenbrough Ct Metairie LA 70005 Office: Tulane U Med Sch 1700 Perdido St New Orleans LA 70112

LEIBHARDT, EDWARD, optics mfg. co. exec., research cons.; b. New Rome, Wis., Oct. 13, 1919; s. Stephan and Roza (Jilling) L.; m. Maidi Wiebe, June 3, 1961; children: Barbara, Leslie. B.A., Northwestern U., 1954; Ph.D. in Astronomy, 1959. Engraver R.R. Donnelly and SonsCo., Chgo., 1937-43; ptnr. Liebhardt Bros., Maywood, Ill., 1943-46; prin. Leibhardt Engring., Maywood, 1946-51; pres. Diffraction Products Inc., Woodstock, Ill., 1951—; cons. optics research and devel. Mem. Optical Soc. Am., Optical Soc. Chgo., Soc. Applied Spectroscopy, Soc. Photo-Optical Instrumentation Engrs., Physics Club Chgo., Sigma Xi. Subspecialties: Diffraction Gratings; Holography. Current work: Diffraction grating ruling and holography; spectroscopy and photometry research in astronomy. Home: 9416 W Bull Valley Rd Village of Bull Valley IL 60098 Office: PO Box 645 Woodstock IL 60098

LEIBMAN, KENNETH CHARLES, biochemical pharmacologist, educator; b. N.Y.C., Aug. 7, 1923; s. Charles Joseph and Laura Adelaide (Hillman) L.; m.; children: Gregory K., Melissa H. B.S., Poly. Inst. Bklyn., 1943; M.Sc., Ohio State U., 1948; Ph.D., NYU, 1954. Chemist Nat. Lead Co., 1943-44; grad. asst. Ohio State U., 1946-48, NYU, 1949-53; research assoc. U. Wis., 1953-55; specialist U.S. Tech. Coop. Mission to India, 1955-56; instr. U. Fla., 1956-57, asst. prof. pharmacology, 1957-62, assoc. prof., 1962-68, prof., 1968—. Contbr. numerous articles to profl. publs.; editor-in-chief: Drug Metabolism amd Disposition, 1972-82. Mem. Am. Soc. Pharmacology and Exptl. Therapeutics. Quaker. Subspecialties: Molecular pharmacology; Drug Metabolism. Current work: Metabolism of drugs and toxicants; enzymic mechanisms and regulation of drug metabolism; biochemical toxicology. Office: Dept Pharmacology U Fla Med Sch Gainesville FL 32610

LEIBOVIC, K. NICHOLAS, neuroscientist, educator; b. Lithuania, June 14, 1921; s. Joseph A. and Chassia (Michailova) L.; m. Marianne Karpf, Aug. 4, 1944; children: David A., Stephen J. B. Engring., Trinity Coll., Cambridge (Eng.) U., 1943; B.Sc. (hon.) in Math, London U., 1952. Sect. leader. math. cons. Brit. Oxygen Co., London, 1956-60; sr. mathematician Westinghouse Research, Pitts., 1960-63; asst. dir., sect. chmn. Ctr. Theoretical Biology, SUNY, Buffalo, 1964-74, prof. dept. biophys. scis., 1965—; cons. NIH, others. Author: Nervous Systems Theory, 1972; editor: Information Processing in the Nervous System, 1969; asso. editor, referee various sci. journ.; contbr. articles to profl. publs. NIH grantee, 1966-70, 82—; NSF grantee, 1966; NRC grantee, 1971, 75. Mem. AAAS, Biophys. Soc., Brit. Computer Soc., N.Y. Acad. Scis., Ops. Research Soc., Soc. Indsl. and Applied Math., Soc. for Neurosci., Assn. for Research in Vision and Opthalmology. Jewish. Subspecialties: Neurobiology; Biophysics (biology). Current work: Informaton processing in nervous system at cellular and systems levels: at former, transduction and transmission of signals, at latter, properties of convergent and divergent pathways. Home: 105 High Park Blvd Buffalo NY 14226 Office: Dept Biophysics Cary Hall SUNY Buffalo NY 14214

LEIBOWITZ, MICHAEL JONATHAN, molecular biologist, physician, educator; b. Bklyn., May 14, 1945; s. Harry and Helen (Oser) L.; m. Sharon Marilyn Stein, June 5, 1966; children: Amy Susan, Brian Sheldon. A.B. in Chemistry, Columbia U., 1966; Ph.D. in Molecular Biology, Albert Einstein Coll. Medicine, Yeshiva U., 1972, M.D., 1973. Intern, asst. in medicine Barnes Hosp., Washington U., St. Louis, 1974-74; pharmacology research assoc. Nat. Inst. Gen. Med. Scis., Bethesda, Md., 1974-76; guest worker Lab. Biochem. Pharmacology, Nat. Inst. Arthritis and Metabolic and Digestive Diseases, NIH, Bethesda, 1974-76; sr. staff fellow, 1976-77; asst. prof. microbiology U. Medicine Dentistry N.J.-Rutgers Med. Sch., Piscataway, 1977-82, assoc. prof., 1982—; mem. grad. faculty microbiology Rutgers U., New Brunswick, N.J., 1977—; mem. instl. biosafety com. Schering Corp.; Participant Human Resources File, Manalapan-Englishtown (N.J.) Regional Schs. Contbr. articles to profl. jours. Served as surgeon with USPHS, 1974-76. March of Dimes Birth Defects Found. Basil O'Connor scholar, 1978-81; Alexandrine and Alexander L. Sinsheimer scholar, 1980-83; grantee NIH, Am. Cancer Soc. Mem. AAAS, Am. Chem. Soc., Genetics Soc. Am., Am. Soc. Microbiology, N.Y. Acad. Scis., U.S. Fedn. Culture Collections, N.Y. Zool. Soc., Phi Beta Kappa, Alpha Omega Alpha. Jewish. Subspecialties: Genetics and genetic engineering (biology); Virology (biology). Current work: Host-virus interactions study to develop anti-viral agents. Home: 3 Baron Ct Manalapan NJ 07726 Office: Dept Microbiology UMDNJ Rutgers Med School Piscataway NJ 08854

LEIDERSDORF, CRAIG B., consulting company executive; b. Mineola, N.Y., Mar. 12, 1950; s. Carl Bernard and Shirley Katherine (Brecht) Rawson L. B.S., Standord U., 1972; M.S., U. Calif.-Berkeley, 1975. Capt., Cutter Abraxas, Caribbean and South Pacific, 1972-74; research asst. U. Calif.-Berkeley, 1975; engr. Swan Wooster Engring., Vancouver, B.C., Can., 1976-77, Tetra Tech, Inc., Pasadena, Calif., 1977-80; sr. v.p. Tekmarine, Inc., Sierra Madre, Calif., 1980—, dir., 1981—. Mem. ASCE, Marine Tech. Soc., Phi Beta Kappa, Tau Beta Pi. Subspecialties: Civil engineering; Coastal engineering. Current work: Scientific research and engineering applications relating to coastal processes and nearshore oceanography, with particular emphasis on the Arctic. Patentee in field. Office: Tekmarine Inc 37 Auburn Ave Sierra Madre CA 91024

LEIDHEISER, HENRY, JR., chemistry educator, researcher; b. Union City, N.J., Apr. 18, 1920; s. Henry and Margaret (Steinel) L.; m. Virginia Townsend, Feb. 21, 1944; children: Margaret LeBaron, Henry. B.S. in Chemistry, U. Va., 1941, M.S. in Phys. Chemistry, 1944, Ph.D., 1946. Research assoc. U. Va., 1946-49; lab. mgr. Va. Inst. Sci. Research, Richmond, 1949-60, dir., chief exec. officer, 1960-68; prof. chemistry Lehigh U., Bethlehem, Pa., 1968—, dir., 1968—; mem. various coms. Nat. Acad. Scis.; bd. dirs. Petroleum Research Fund; cons. industry and govt. Contbr. articles to profl. jours. Recipient numerous awards including Silver medal Am. Electroplaters Soc., 1978, Arch T. Colwell award Soc. Automotive Engrs., 1979, Willis Rodney Whitney award Nat. Assn. Corrosion Engrs., 1983. Mem. Am. Chem. Soc., Electrochem. Soc., Nat. Assn. Corrosion Engrs. Republican. Presbyterian. Club: Saucon Valley Country (Bethlehem). Subspecialties: Surface chemistry; Corrosion. Home: RD 7 Pleasant Dr Bethlehem PA 18015 Office: Sinclair Lab 7 Lehigh Univ Bethlehem PA 18015

LEIFMAN, LEV JACOB, mathematician, translator; b. Kiev, Ukraine, Apr. 12, 1929; came to U.S., 1979; s. Jacob Lev and Nina Boris (Tsyrlin) L.; m. Miriam Israel Eidelson, May 16 1962; children: Jacob, Tatyana. M.Sc., Kiev U., 1952; Ph.D., Moscow U., 1962. Research prof. Ukrainian Inst. Trade, Kiev, 1959-63; assoc. prof. Novosibirsk U., USSR, 1963-67; research prof. USSR Acad. Scis., Novosibirsk, 1963-70; assoc. prof. Inst. Civil Engrs., Novosibirsk, 1971-73, Haifa (Israel) U., 1974-78; translation editor Am. Math. Soc., Providence, 1979—; head and sci. supr. lab. exptl. programming Inst. Automated Control Systems, Novosibirsk, 1964-69; faculty R.I. Coll, 1982—. Author: Netzplantechnik bei begrenzten Ressourcen, 1968, Modelling of Private Consumption, 1972; editor: Network Planning under Restraints on Resources, 1971; series Modelling of Control Processes, 1967-73, Theory of Probability and Math. Stats, 1978—, Vestnik of the Leningrad U, 1979, Procs. of Steklov Inst. Math, 1983—. Mem. Internat. Assn. Cybernetics, Am. Math. Soc., Soc. Inds. and Applied Math. Jewish. Subspecialties: Algorithms; Operations research (mathematics). Current work: Optimization - development and analysis of methods and algorithms; mathematical models and methods in operations research and economics; mathematical models of translation between natural languages and their implementation on computers. Home: 467 Pleasant St Pawtucket RI 02860 Office: PO Box 6248 Providence RI 02940

LEIGH, EGBERT GILES, JR., biologist; b. Richmond, Va., July 27, 1940; s. Egbert Giles and Lucinda Lee (Kinsolving) L.; m. Elizabeth Murray Hodgson, Mar. 21, 1968; children: John Murray, Mary Bruce. A.B., Princeton U., 1962; Ph.D., Yale U., 1966. Asst. prof. biology Princeton (N.J.) U., 1966-72; biologist Smithsonian Tropical Research Inst., Balboa, Panama, 1969—. Author: Adaptation & Diversity, 1971; sr. co-editor: Ecology of a Tropical Forest, 1982. Mem. Am. Soc. Naturalists, Ecol. Soc. Am., Brit. Ecol. Soc., Paleontol. Research Instn., Rocky Mountain Biol. Lab. Subspecialties: Evolutionary biology; Theoretical ecology. Office: Smithsonian Tropical Research Inst APO Miami FL 34002

LEIGH, RICHARD JOHN, neurologist, educator, researcher; b. Stoke-on-Trent, Eng., Dec. 28, 1946; s. Richard and Leila (Mason) L.; m. Diana Jo Willmott, Aug. 28, 1971; 2 children. M.B.B.S., U. Newcastle-upon-Tyne, 1970, M.D., 1975; postgrad., Cornell Med. Coll., 1975-77, Johns Hopkins U., 1977—. Med. resident, Newcastle-upon-Tyne, 1970-73, neurology research fellow, 1973-75; resident in neurology Cornell Med. Coll., N.Y.C., 1975-77; neurology/opthalmology fellow Johns Hopkins U. Med. Sch., Balt., 1977-79, asst. prof. neurology and ophthalmology, 1979-83; assoc. prof. neurology Case Western Res. U., Cleve., 1983—. Author monograph. Recipient Frank R. Ford award Johns Hopkins Hosp., 1982. Mem. Am. Acad. Neurology, Soc. Neurosci. Subspecialties: Neurophysiology; Neurology. Current work: Neural control of eye movements; disorders of ocular motility. Office: Dept Neurology University Hosps Cleveland OH 44106

LEIGH, THOMAS FRANCIS, entomology educator, researcher; b. Loma Linda, Calif., Mar. 6, 1923; s. Herbert C. and Sarah S. (Robinson) L.; m. Nina Anatol Eremin, Nov. 26, 1954; children: Michael Andrew, Nicholas Jonathan. B.S., U. Calif.-Berkeley, 1949, Ph.D., 1956. Cert. Am. Registry of Profl. Entomologists. Supervising entomologist Westley Pest Control Assn., Calif., 1949-50; from asst. to assoc. entomologist U. Ark., Fayetteville, 1953-58; from asst. entomologist to entomologist U. Calif.-Davis, 1958—; cons. Rockefeller Found., 1970, IAEA, Senegal, 1977, FAO, Syria, 1981, INIA, Mex., various times. Contbr. over 100 articles to sci. and tech. jours. Active Boy Scouts Am. Served in U.S. Army, 1943. Grantee NIH, 1965-68, Rockefeller Found., 1971-76, NSF, 1974-78, U.S. Dept. Agr., 1965-67, Calif. cotton industry, 1959—. Mem. Entomol. Soc. Am. (br. pres. 1980-81), Am. Registry Profl. Entomologists (governing council 1978-81, pres. 1983), Calif. Assn. Registry Profl. Entomologists (founding pres. 1978-79). Episcopalian. Lodge: Rotary. Subspecialties: Integrated pest management; Ecology. Current work: Biology and ecology of the insects, mites and spiders of the cotton agroecosystem and management of pest species; plant resistance to insects and spider mites. Home: 242 Pine St Shafter CA 92363 Office: 17053 Shafter Ave Shafter CA 93263

LEIGHTON, ROBERT BENJAMIN, physicist, educator; b. Detroit, Sept. 10, 1919; s. George B. and Olga (Homrig) L.; m. Alice M. Winger, July 31, 1943 (div. 1974); children—Ralph, Alan; m. Margaret L. Lauritsen, Jan. 7, 1977. B.S., Calif. Inst. Tech., 1941, M.S., 1944, Ph.D., 1947. Asst. prof. Calif. Inst. Tech., Pasadena, 1949-53, asso. prof., 1953-59, prof. physics, 1959—, chmn. div. physics, math., astronomy, 1970-75; Prin. investigator TV Expt. Mariner 4, 1964, Mariners 6 and 7, 1969; co-investigator TV Expt. Mariner 9, 1971. Author: Principles of Modern Physics, 1959, (with others) The Feynman Lectures on Physics, 1964. Mem. Am. Phys. Soc., Am. Astron. Soc., Nat. Acad. Sci., Am. Acad. Arts and Scis. Subspecialties: Cosmic ray high energy astrophysics; Infrared optical astronomy. Research in cosmic rays, solar physics, space astronomy, infrared astronomy, telescope design. Office: Calif Inst Tech Pasadena CA 91125

LEIPHOLZ, HORST HERMANN EDUARD, civil engineer, educator, researcher; b. Plonhofen, Germany, Sept. 26, 1919; s. Ernst and Marta (Wohlfeil) L.; m. Ursula Schlag, May 9, 1942; children: Barbara, Gunthara. Diplom, U. Stuttgart, Germany, 1958, Dr.-Ing., 1959. Asst. U. Stuttgart, 1958-62, docent, 1962-63; prof. U. Karlsruhe, Germany, 1963-69, U. Waterloo, Ont., Can., 1969—, chmn. dept. civil engring., 1982—, chmn. solid mechanics div., 1971-77, assoc. dean grad. studies, 1976-81. Author: Theory of Elasticity, 1974, Direct Variation Methods, 1977, Stability of Elastic Systems, 1980, others.; Contbr. articles to profl. jours. Recipient Disting. Tchr. award U. Waterloo, 1976. Fellow Engring. Inst. Can., Am. Acad. Mechanics; mem. Can. Soc. Mech. Engring. (chmn. research and devel. div. 1978-80), German Soc. Engring. Math. and Mechs. (exec. council 1978-81). Subspecialties: Theoretical and applied mechanics; Applied mathematics. Current work: Stability and control of structures. Home: 401 Warrington Dr Waterloo ON Canada N2L 2P7 Office: Dept Civil Engring Univ Waterloo Waterloo On Canada N2L 3G1

LEIS, JONATHAN PETER, biochemistry educator, researcher; b. Bklyn., Aug. 17, 1944; s. Morris and Beatrice (Sperber) L.; m. Susan Schoenfeld, Dec. 20, 1970; children: Benjamin, Betsy. B.A. in Chemistry, Hofstra U., 1965; Ph.D. in Biochemistry, Cornell U., 1970. Postdoctoral fellow Albert Einstein Coll. Medicine, 1970-73; asst. prof.

Duke U. Med. Ctr., 1974-79; assoc. prof. biochemistry Case Western Res. U. Med. Sch., 1979—; ad hoc reviewer NSF, Leukemia Soc.; mem. grant rev. panel Am. Cancer Soc. (Ohio Div.), 1981—. Contbr. numerous articles to profl. publs.; reviewer various sci. jours. Cubmaster Greater Cleve. council Boy Scouts Am., 1981-83. Recipient Research Career Devel. award Nat. Cancer Inst., NIH, 1974-79; Hofstra U. scholar, 1962-65; NIH trainee, 1965-70; Damon Runyon cancer research fellow, 1971-73; Leukemia Soc. scholar, 1974; NSF grantee, 1976-82; Am. Cancer Soc. grantee, 1974-83; Nat. Cancer Inst. grantee, 1974-83. Mem. Am. Soc. Biol. Chemists. Subspecialties: Biochemistry (medicine); Virology (medicine). Current work: Mechanisms of replication of retroviruses in cells; regulation of viral mechanisms of transcription, integration, transposition, RNA processing and virion assembly. Home: 3628 Palmerston Rd Shaker Heights OH 44122 Office: Case Western Res U Sch Medicine 2119 Abington Rd Cleveland OH 44106

LEISER, ANDREW TWOHY, environ. horticulturist, educator; b. Kettle Falls, Wash., May 13, 1923; s. Oliver Edwin and Frances Ida (Mann) L.; m. Vera Ruth Hornaday, Aug. 1947 (dec. 1975); children: Christine, Mark, Kathleen; m. Shirley June Dana, Feb. 1978; 4 stepchildren. B.A., Wash. State U., 1947, M.S., 1949; Ph.D., UCLA, 1959. Asst. prof. Purdue U., West Lafayette, Ind., 1957-61; horticulturist W.R. Grace & Co., Clarksville, Md., 1961-64; prof. dept. environ. horticulture U. Calif., Davis., 1964—. Author: (with Elizabeth McClintock) An Annotated Checklist of Woody Ornamental Plants of California, Oregon, and Washington, 1979, (with Donald H. Gray) Biotechnical Slope Protection and Erosion Control, 1982. Served with U.S. Army, 1942-46. Recipient Calif. State Assembly award, 1971; Alex Laurie award Am. Soc. Hort. Sci., 1974; Bell award Mental Health Assn. Yolo County, 1977. Mem. Am. Soc. Hort. Sci., Bot. Soc. Am., Royal Hort. Soc. (London), Internat. Soc. Arboriculture, Internat. Plant Propagators Soc., Calif. Hort. Soc., Calif. Assn. Nurserymen (Research award 1971, Edn. award 1980), Am. Rhododendron Soc., Ind. Assn. Nurserymen, Am. Assn. Nurserymen, Sigma Xi, Phi Kappa Phi, Alpha Zeta, Pi Alpha Xi. Episcopalian. Subspecialties: Ornamental horticulture; Taxonomy. Current work: Plant introduction and evaluation, revegetation, plant propagation, taxonomy of woody ornamental plants, plant adaptation. Home: 939 Pecan Pl Davis CA 95616 Office: Dept Environ Horticulture U Calif Davis CA 95616

LEISMAN, GILBERT ARTHUR, biology educator; b. Washington, May 12, 1924; s. Arthur Gustave and Agnes Pauline (Schrohsein) L.; m. Marie Katherine Andresen, June 21, 1952. B.S., U. Wis., 1949; M.S., U. Minn., 1952, Ph.D., 1955. Asst. prof. biology Emporia State U., 1955-59, assoc. prof., 1959-64, prof., 1964—. Contbr. numerous articles to sci. jours. Served with USAAF, 1942-45. Decorated DFC, Air medal.; NSF Research grantee, 1958, 62, 63. Mem. Kans. Acad. Sci. (past pres.), Botan. Soc. Am. (past chmn. paleobot. sect.), Am. Inst. Biol. Sci., Internat. Orgn. Paleobotany, AAAS, Am. Legion. Democrat. Subspecialty: Paleobiology. Current work: Paleobotany of the carboniferous age, coal balls, education. Home: 1501 Sherwood Way Emporia KS 66801 Office: 1200 Commercial St Emporia KS 66801

LEITE, RICHARD JOSEPH, aeronautical engineer, consultant; b. Fremont, Ohio, Mar. 8, 1923; s. Carl Albert and Marie Margaret (Dolweck) L.; m. Barbara M. Higgins, Nov. 19, 1955; children: Mark Richard, Jeffrey Howard, Mary Elizabeth. B.Naval Sci. magna cum laude, U. Notre Dame, 1945, B.S. in Aero, engring, cum laude, 1947; M.S.E. in Aero. Engring, U. Mich., 1948, Ph.D., 1956. Research assoc. U. Mich., 1948-56; sr. engr. Booz-Allen Applied Research Inc., Dayton, Ohio, 1956-58; research engr. U. Mich., 1958-71; sr. staff engr. Bendix Corp., Inc., Ann Arbor, Mich., 1971-72; sr. scientist KMS fusion, Inc., Ann Arbor, 1972-77; sr. staff engr. TRW Systems, Inc., Redondo Beach, Calif., 1977—; cons. in field. Served as ensign USNR, 1943-46; PTO. Mem. Am. Vacuum Soc., Sigma Xi, Phi Kappa Phi. Roman Catholic. Subspecialties: Aerospace engineering and technology; Planetary atmospheres. Current work: Develop miniature sensors and instruments for planetary atmosphere probe exploration that are compatible with modern microprocessor controlled electronic control and data systems. Patentee pyrochem. processes for decomposition of water. Home: 6742 Abbotswood Dr Rancho Palos Verdes CA 90274 Office: TRW Systems Inc One Space Park Redondo Beach CA 90278

LEITH, EMMETT NORMAN, educator, electrical engineer; b. Detroit, Mar. 12, 1927; s. Albert Donald and Dorothy Marie (Emmett) L.; m. Lois June Neswold, Feb. 17, 1956; children: Kim Ellen, Pam Elizabeth. B.S., Wayne State U., 1950, M.S., 1952, Ph.D., 1978. Mem. research staff U. Mich., 1952—, prof. elec. engring., 1968—; cons. several indsl. corps. Contbr. books, profl. jours. Served with USNR, 1945-46. Recipient Gordon Meml. award S.P.I.E., 1965; citation Am. Soc. Mag. Photographers, 1966; Achievement award U.S. Camera and Travel mag., 1967; Excellence of Paper award Soc. Motion Picture and TV Engrs., 1967; Daedalion award, 1968; Stuart Ballantine medal Franklin Inst., 1969; Distinguished Faculty Achievement award U. Mich., 1973; Alumni award Wayne State U., 1974; cited by Nobel Prize Commn. for contbns. to holography, 1971; Holley medal ASME, 1976; named Man of Year Indsl. Research mag., 1966; Nat. medal of Sci., 1979; Russel lecture award U. Mich., 1981; recipient Dennis Gabor medal Soc. Photo-Instrumentation Engrs., 1983. Fellow Optical Soc. Am. (Wood medal 1975), IEEE (Liebmann award 1967, Inventor of Year award 1976); mem. Nat. Acad. Engring., Sigma Xi, Sigma Pi Sigma. Subspecialties: Holography; Optical image processing. Current work: White light optical processing and interferometry; optical processing of images. Patentee in field. First demonstrated (with colleague) capability of holography to form high-quality 3-dimensional image. Home: 51325 Murray Hill Canton MI 48187 Office: Univ Mich Inst Sci and Tech PO Box 618 Ann Arbor MI 48107

LEITMANN, GEORGE, mechanical engineer; b. Vienna, Austria, May 24, 1925; s. Josef and Stella (Fischer) L.; m. Nancy Lloyd, Jan. 28, 1955; children: Josef Lloyd, Elaine Michelle. B.S., Columbia U., 1949, M.A., 1950; Ph.D., U. Calif., Berkeley, 1956. Physicist, head aeroballistics sect. U.S. Naval Ordnance Sta., China Lake, 1950-57; mem. faculty U. Calif., Berkeley, 1957—, prof. engring. sci., 1963—, asso. dean grad. affairs, 1981—; cons. to aerospace industry and govt. Author: An Introduction to Optimal Control, 1966, Quantitati ve and Qualitative Games, 1969, The Calculus of Variations and Optimal Control, 1981, others; contbr. articles to profl. jours. Served with AUS, 1944-46; ETO. Decorated Croix de Guerre, France; recipient Pendray Aerospace Lit. award, 1979; Von Humboldt U.S. sr. scientist, 1980; Levy medal, 1981; Miller Research prof., 1966. Mem. Acad. Sci. Bologna, Internat. Acad. Astronautics (corr.), Nat. Acad. Engring. Subspecialty: Mechanical engineering. Office: Coll Engring U Calif Berkeley CA 94720

LEIVE, LORETTA, biochemist, microbiologist, researcher; b. N.Y.C., Apr. 12, 1936; d. James and Ann Carol (Gordon) Lambert; m.; 1 dau., Cynthia M. A.B., Barnard Coll., 1956; Ph.D. in Bacteriology and Immunology, Harvard U., 1963. With Nat. Inst. Arthritis, Diabetes, Digestive and Kidney Diseases, NIH, Bethesda, Md., 1963—, sect. chief, 1983—; speaker numerous confs. and symposia. Editor: Bacterial Membranes and Walls, 1973; contbr. articles and abstracts to profl. jours. NIH fellow, 1962-63; NSF fellow, 1956-58. Mem. Am. Soc. Microbiology, Am. Soc. Biol. Chemists (conf. chmn., book editor). Subspecialties: Microbiology; Biochemistry (biology). Current work: Outer membrane of gram-negative bacteria; macrophage membranes and phagocytosis; (bacterial outer membrane; macrophages; phagocytosis). Home: 6817 Sorrell St McLean VA 22101 Office: NIH Bldg 4 Room 116 Bethesda MD 20205

LEKOUDIS, SPIRO G(EORGE), aerospace engineering educator, researcher; b. Komotini, Rodopi, Greece, July 13, 1949; came to U.S., 1972; s. George P. and Loukia (Exarchou) L.; m. Helen Miliotis, Nov. 8, 1979; 1 son, George. Diploma in Mech. Engring, Nat. Tech. U., Athens, Greece, 1972; M.S. in Engring. Sci, Va. Tech., 1973; Ph.D. in Engring. Sci. and Mechanics, Va. Inst. Tech., 1977. Research asst. Va. Poly. Inst., Blacksburg, 1972-76; cons. Lockheed Ga., Marietta, 1977-79; asst. prof. aerospace engring. Ga. Inst. Tech., Atlanta, 1980-82, assoc. prof., 1982—; cons. Lockheed Ga., 1980—, Gen. Electric, 1983—. Contbr. articles to profl. jours. Mem. AIAA, Am. Phys. Soc. Subspecialties: Aeronautical engineering; Algorithms. Current work: Research in fluid mechanics of interest in vehicle design. Home: 115 Bonnie Ln Atlanta GA 30228 Office: Aerospace Engring Georgia Inst Tech Atlanta GA 30332

LELI, DANO ANTHONY, clinical neuropsychologist; b. S.I., July 22, 1947; s. Peter and Viola (Troisi) L.; m. Patricia Gaile Percy, June 1, 1972; 1 son, david Anthony. B.A., U. Central Fla., 1972, M.S., 1974; Ph.D., U. South Fla., 1978. Lic. psychologist, Ala. Postdoctoral fellow in clin. neuropsychology depts. neurology and psychiatry U. Ala.-Birmingham, 1978-79, research asst. prof. neurology, 1979—, asst. prof. psychology, 1980—, asst. prof. psychiatry, 1981—, co-chmn. med. psychology coordinating com., 1982—, adminstr., 1979—; cons. to Alzheimer's Support Group, Center for Aging, U. Ala.-Birmingham, 1981—. Presenter TV program on migraine headache, 1981; Research publs. in field. Served with USAF, 1968-70; Iceland. Mem. Am. Psychol. Assn., Southeastern Psychol. Assn., Internat. Neuropsychol. Assn., Ala. Acad. Neurology and Psychiatry, Ala. Psychol. Assn. Democrat. Subspecialties: Neuropsychology; Clinical psychology. Current work: Clinical neuropsychology; brain functions; assessment and treatment psychological difficulties resulting from neurological disease; regional cerebral blood flow. Office: Dept Neurology U Ala Birmingham AL 35294

LELLINGER, DAVID BRUCE, botanist, curator; b. Chgo., Jan. 24, 1937; s. Nicholoas Francis and Rose (Kreicker) L.; m. Linda Mae Kuhles, June 15, 1963; children: Richard, Anne. A.B., U. Ill., 1958; M.S., U. Mich., 1960, Ph.D., 1965. Assoc. curator U.S. Nat. Herbarium, Smithsonian Instn., Washington, 1963—. Editor: Am. Fern. Journ, 1966—; assoc. editor: Pteridologia, 1979—. Nat. Geog. Soc. grantee, 1971; Smithsonian Research Found. grantee, 1971. Mem. Brit. Pterid Soc., Internat. Soc. Plant Taxonomists, Am. Fern. Soc. Subspecialty: Taxonomy. Current work: Taxonomy of ferns and fern-allies, especially new world tropics. Address: U S Nat Herbarium N4B-166 Smithsonian Instn Washington DC 20560

LEMAY, MOIRA KATHLEEN, engineering psychology educator, researcher; b. N.Y.C., Apr. 12, 1934; d. Bernard Howard and Kathleen (Sullivan) Fitzpatrick; m. Joseph A. LeMay, June 14, 1958; children: Valerie, Joseph. B.S., Queens Coll., 1956; M.S., Pa State U., 1960, Ph.D., 1970. Engring. psychologist U.S. Naval Research Lab., Washington, 1960-62, ITT FEd. Labs., Nutely, N.J., 1962-64; instr. Manhattanville Coll., 1964-69; asst. porf. Skidmore Coll., 1969-70; dir. indsl./organizational psychology program Montclair State Coll., 1969—; faculty fellow Wright-Patterson AFB, Ohio, 1978; mem. N.J. Panel Sci. Advs., 1980—; vis. scientist Jet Propulsion Lab., Calif. Inst. Tech., 1982-83. Contbr. articles on man-machine interface to profl. publs. Committeeperson Bergen County (N.J.) Democratic party, 1980—. USAF Office Sci. Research grantee, 1979-80. Mem. Humas Factors Soc., Am. Psychol. Assn., AAAS. Subspecialties: Human factors engineering; Human-computer interaction. Current work: Human factors in interactive computer systems; perception in computer graphic displays. Home: 1023 Hillcrest Rd Ridgewood NJ 07450 Office: Montclair State Coll Upper Montclair NJ 07043

LEMBACH, KENNETH JAMES, biochemist, researcher; b. Rochester, N.Y., June 16, 1939; s. Charles Walter and Dorothy Frances (Reynolds) L.; m. Regis Catherine Mann, Aug. 21, 1965; children: Lara, Aimee. B.S., MIT, 1961; Ph.D., U. Pa., Phila., 1966. Research assoc. MIT, Cambridge, Mass., 1968-69; asst. prof. biochemistry Vanderbilt U., Nashville, 1969-76, assoc. prof., 1976-79; sr. research biochemist Cutter Labs., Berkeley, Calif., 1979-80, sect. head plasma product research, 1980-82, prin. scientist II, 1982—. Contbr. articles in field to profl. jours. USPHS research grantee, 1971-79. Mem. Am. Soc. Biol. Chemists, Am. Soc. Cell Biology, Tissue Culture Assn., AAAS, Sigma Xi. Democrat. Roman Catholic. Club: Sierra (San Francisco). Subspecialties: Biochemistry (medicine); Cell and tissue culture. Current work: Development of ethical pharmaceuticals via identification and isolation from biological sources, including plasma, cell culture and R-DNA technology. Home: 662 Park Hill Rd Danville CA 94526 Office: Cutter Labs 4th and Baker Sts Berkeley CA 94710

LEMISH, JOHN, economic geology educator, consultant; b. Rome, N.Y., July 4, 1921; s. Adam Gregorovich and Anastasia Pavlovna (Kiselevich) L.; m. Jane Louise Thompson, June 19, 1946 (dec. Oct. 1973); children: Jeffrey Thompson, Judith Anastasia, Jocelyn Anne, Jennifer Elizabeth, Julia Karen. B.S., U. Mich., 1947, M.S., 1948, Ph.D., 1955. Lab. technician Gen. Cable Corp., Rome, N.Y., 1939-41; exploration geologist, U.S. Geol. Survey (East Tintic Dist.), Utah, 1949-51; teaching fellow U. Mich., Ann Arbor, 1951-53, instr., 1954-55; prof. geology Iowa State U., Ames, 1955—; mem. coms. on performance of concrete NRC Hwy. Research Bd., Washington, 1958-69; chmn. translations and publ. com. Am. Geol. Inst., Washington, 1966-80; dir. rehab. project State Mining Bd. Iowa, Des Moines, 1969; chmn. State Mining Bd., 1964-73; mem. steering com. Iowa Coal Project, Iowa State U., Ames, 1974-77. Author: (with J.T. Lemish) Jeff Carson, Young Geologist, 1960, Mineral Deposits of Iowa, 1969. Served to 1st lt. USAAF, 1942-45. NSF grad. fellow, 1952-53. Fellow Geol. Soc. Am.; mem. Soc. Econ. Geologists, Am. Assn. Petroleum Geologists, AIME, Geochem. Soc. Democrat. Presbyterian. Subspecialty: Geology. Current work: Teaching, research in coal geology of Iowa, geology of deep coal in Iowa, geology of Forest City basin, limestone exploration. Patentee in field. Home: 536 Forest Glen Ames IA 50010 Office: Dept Earth Scis Iowa State U Ames IA 50011

LEMONS, THOMAS MILLARD, consulting electrical engineer; b. Indpls., Sept. 15, 1934; s. Kenneth Eugene and Katharine (Fillmore) L.; m. Priscilla Moore, July 18, 1959; children: Elizabeth Bellnap, Katharine Fillmore. B.E.E., Purdue U., 1956. Registered profl. engr., Mass. Ptnr. Audio-Lite, Indpls., 1949-56; applications engr. Sylvania Lighting div. GTE, Salem and Danvers, Mass., 1956-70, mgr. applications engring., 1968-70; pres. TLA Lighting Cons., Inc., Salem, Mass., 1970—; instr. Chamberlayne Jr. Coll., Boston, 1978—. Pres. Salem Council Chs., 1964-65; pres. Marblehead Arts Festival, 1964, 68, 69. Served with AUS, 1957. Fellow Illuminating Engring. Soc., U.S. Inst. Theatre Tech.; mem. Internat. Assn. Lighting Designers, Nat. Soc. Profl. Engrs., Internat. Commn. on Illumination, Soc. Motion Picture and TV Engrs., Assn. Energy Engring. Episcopalian. Current work: Illumination systems; lighting product and installation designs. Home: 14 Orne St Marblehead MA 01945 Office: 72 Loring Ave Salem MA 01970

LENARD, JOHN, physiology and biophysics educator; b. Vienna, Austria, May 17, 1937; came to U.S., 1943, naturalized, 1948; s. George and Frances (Perten) L.; m. Nancy Roberta Stevenson, Oct. 5, 1973; children: Eric, Keith, Karen, Steven. B.A., Cornell U., 1958, Ph.D., 1964. Post-doctoral fellow U. Calif., San Diego, 1964-67; asst. prof. biochemistry Albert Einstein Coll. Medicine, 1967-68; assoc. Sloan Kettering Inst., 1968-72; asst. prof. biochemistry Cornell U., 1968-72; adj. asst. prof. biology Hunter Coll., 1970-73; assoc. prof. physiology Coll. Medicine and Dentistry N.J.-Rutgers Med. Sch., 1973-76, prof. physiology and biophysics, 1976—, asst. chmn. dept., 1976—, asst. dean, 1980-82, assoc. dean, 1982—; mem. molecular biology adv. panel NSF, 1978-79, mem. cell biology panel, 1982—. Contbr. articles in field to profl. jours. NIH grantee, 1980—. Mem. Am. Soc. Biol. Chemists, Biophysical Soc., Am. Soc. Cell Biology, Am. Soc. Microbiology, Harvey Soc. Subspecialties: Biochemistry (medicine); Cell biology (medicine). Current work: Membranes, esp viral envelopes, viral entry and assembly, lysosomal structure and function. Home: 444 Harrison Ave Highland Park NJ 08904 Office: Department Physiology and Biophysics Univ Medicine and Dentistry NJ Rutgers Medical Sch Piscataway NJ 08854

LENCHNER, NATHANIEL HERBERT, dentist, consultant; b. N.Y.C., Aug. 28, 1923; s. Edward and Jennie (Reizes) L.; m. Florence Smith; children: Jonathan, Michael, Debra. B.A., N.Y. U., 1943, D.D.S., 1950. Instr. N.Y. U. Coll. Dentistry, 1950-55; dental cons. Whaledent, Internat., N.Y.C., 1977—; pvt. practice dentistry, Forest Hills, N.Y., 1950—; asst. clin. prof. Columbia U. Sch. Dentistry, 1974-80, lectr., 1980—; adj. assoc. prof. biomed. engring. U. Miami Sch. Engring. and Architecture, 1981—; mem. admissions com. NYU Coll. Dentistry, 1982—. Assoc. editor: Jour. Prosthetic Dentistry, 1974—. Fellow Northeastern Gnathological Soc. (pres. 1966-70), Greater N.Y. Acad. Prosthodontics, Acad. Gen. Dentistry; mem. Am. Prosthodontic Soc., ADA. Clubs: Lake Success Golf, Lake Success Tennis (Lake Success, N.Y.); Fountains Country (Lake Worth, Fla.). Subspecialties: Prosthodontics; Biomedical engineering. Current work: General practice dentistry with strong emphasis in prosthodontics; research in biomedical devices, i.e. electrosurgery. Home: 6 Bridle Path Ln Lake Success NY 11020 Office: PC 104-20 Queens Blvd Forest Hills NY 11375

LENNARD, PAUL ROSS, motor systems neurobiologist, educator, researcher; b. N.Y.C., May 12, 1949; s. Murray P. and Ellen (Adelman) L.; m. Susan Buder, Jun 10, 1978. Ph.D., Washington U., St. Louis, 1975. Research assoc. Washington U., 1976; USPHS postdoctoral fellow Stanford U., 1977-79; asst. prof. biology Emory U., 1979—; mem. spl. study sect. NIH, 1980; reviewer NSF, Sci., Jour. Neurophysiology. McCandless grantee, 1980; USPHS grantee, 1981-84. Mem. Soc. for Neurosci. (mem. program com., councillor Atlanta area chpt.), Am. Soc. Zoologists. Subspecialties: Neurobiology; Neurophysiology. Current work: Analysis of neural control of locomotion: CNS pattern generation, afferent modulation supraspinal control, and interlimb coordination; development of computer-based motor-pattern recognition systems. Office: Emory U Dept Biology Atlanta GA 30322

LENNON, VANDA ALICE, neuroimmunologist; b. Sydney, Australia, Aug. 1, 1943; came to U.S., 1972; s. Norman John and Claire Florence (Williams) L.; m. Edward Howard Lambert, Dec. 23, 1975. M.B., B.S., U. Sydney, 1966; Ph.D., U. Melbourne, 1973. Intern, resident in medicine Montreal (Que., Can.) Gen. Hosp., 1966-68; postdoctoral fellow Walter and Eliza Hall Inst. Med. Research, Melbourne, Australia, 1968-71; research assoc., asst. prof. Salk Inst. Biol. Studies, San Diego, 1972-78; adj. assoc. prof. neuroscis. U. Calif., San Diego, 1977; assoc. prof. neurology Mayo Med. Sch., Rochester, Minn., 1978-82, asso. prof. immunology, 1979-82, prof. immunology and neurology, 1982—. Contbr. articles to profl. jours. Nat. Health Med. Research Council postgrad. scholar, Australia, 1969-71; Nat. Multiple Schlerosis Soc. postdoctoral fellow, 1972-73. Mem. Am. Assn. Immunologists, Am. Acad. Neurology, Am. Soc. Neurochemistry, Am. Assn. Neuropathologists. Subspecialty: Neuroimmunology. Current work: Acetylcholine receptor, myasthenic syndromes, demyelinating diseases, autoimmunity, muscle development, myelin development. Home: 202 14th St NE Rochester MN 55901 Office: Mayo Clinic 828 Guggenheim Bldg Rochester MN 55905

LENSCHOW, DONALD HENRY, meteorologist; b. La Crosse, Wis., July 17, 1938; s. Henry John and Margaret (Marsch) L.; m. Janette Len Chin, Aug. 28, 1964; children: Christine, Audrey. B.S. in Elec. Engring, U. Wis., 1960, M.S. in Meteorology, 1962, Ph.D., 1966. With Nat. Center for Atmospheric Research, Boulder, Colo., 1965—, sr. scientist, 1979—; adj. prof. meteorology Colo. State U., Ft. Collins, 1972—. Contbr. over 50 articles on atmospherics to profl. publs. Mem. Am. Meteorol. Soc. Subspecialties: Micrometeorology; Atmospheric chemistry. Current work: Understanding structure of atmospheric boundary layer; airplane measurements of boundary layer structure; fate of trace reactive species in boundary layer. Home: 95 Pawnee Dr Boulder CO 80303 Office: Nat Center Atmospheric Research PO Box 3000 Boulder CO 80307

LENT, ROBERT WILLIAM, counseling psychologist; b. Bklyn., Apr. 1, 1953; s. Jack Harvey and Gladys (Unger) L. B.A., SUNY-Albany, 1975; M.A., Ohio State U., 1977, Ph.D., 1979. Teaching assoc. Ohio State U., 1976-77, psychology intern, 1978-79, Mpls. VA Hosp., 1977-78; asst. prof. Student Counseling Bur., U. Minn.-Mpls., 1979—. Co-editor: Handbook of Counseling Psychology, 1984; behavioral sci. sect. editor: Jour. Minn. Acad. Scis, 1982—; Contbr. articles to profl. jours. Vol. counselor Walk-In Counseling Center, Mpls., 1977-78. Recipient Ohio State U. Fellowship award, 1975-79. Mem. Am. Psychol. Assn. (program editor div. counseling psychology 1981—), Minn. Psychol. Assn., Phi Beta Kappa. Subspecialty: Behavioral psychology. Current work: Teaching and supervision of counseling, research, consultation, counseling; comparison of anxiety reduction methods; study of placebo control methodology, study of counselor supervision techniques. Office: Student Counseling Bur U Minn 101 Eddy Hall Minneapolis MN 55455

LENTSCH, JACK WAYNE, nuclear engineer, health physicist; b. Corpus Christi, Tex., Feb. 29, 1944; s. William J. and Rebyl J. (Galloway) L.; m. Cheryl K. Weissbeck, Aug. 28, 1965; children: Tracy, Janice. B.S., Oreg. State U., 1965, M.S., 1966; Ph.D., N.Y.U.,

1972. Cert. health physicist. Nuclear engr. Bechtel Power Corp., San Francisco, 1971-72, Portland Gen. Electric, Oreg., 1972-76; supr. radiol. engring., 1976-79, mgr. licensing & analysis, 1979—. Mem. Am. Nuclear Soc. (dir., vice-chmn., chmn. 1979—), Health Physics Soc. Republican. Catholic. Subspecialties: Nuclear engineering; Nuclear fission. Current work: Nuclear power plant licensing and analysis; health physics. Home: 13780 SW 114th Ave Tigard OR 97223 97204

LEON, MYRON A., immunologist, educator; b. Troy, N.Y., July 13, 1926; s. Samuel and Sadye (Lasky) L.; m. Janet E., June 7, 1948; children: Miriam, David. B.S., Columbia U., 1950, Ph.D., 1965. Assoc. surg. research St. Luke's Hosp., Cleve., 1955-61; adj. asst. prof. microbiology Western Res. U., Cleve., 1955-64; assoc. head pathology research St. Luke's Hosp., Cleve., 1964-74; adj. prof. pathology Western Res. U., 1973-75; prof. immunology and microbiology Wayne State U. Med. Sch., Detroit, 1975—, acting chmn. dept. immunology and microbiology, 1979-80, 82—. Pres. Holly Hill Farms Homeowners Assn., 1982—. Served with USN. Mem. AAAS, Am. Assn. Immunologists, N.Y. Acad. Scis., Sigma Xi. Subspecialty: Immunobiology and immunology. Current work: Control of response to polysaccharides in humans and mice; T-cell reactivity of rheumatoid arthritis patients; regulation of B-cell differentiation. Office: Wayne State U Med Sch 540 E Canfield St Detroit MI 48201

LEON, SHALOM A., biochemist; b. Jerusalem, Israel, Apr. 7, 1935; came to U.S., 1965, naturalized, 1974; s. Albert S. and Bertha L.; m. Ofra Leon, July 5, 1962; children: Avner, Avital, Iris. M.Sc., Hebrew U., Jerusalem, 1959, Ph.D., 1964. Sr. research asst. pharmacology Hebrew U., 1960-65; postdoctoral research fellow Ind. U., 1965-67; asst. mem. research labs. Albert Einstein Med. Center, Phila., 1968-70, mem. dept. nuclear medicine, 1970—; clin. assoc. prof. radiobiology Temple U. Med. Sch., Phila., 1978—. Author in field. Fellow Am. Cancer Soc., 1980-82. Mem. Radiation Research Soc., N.Y. Acad. Scis., Am. Assn. Cancer Research, Am. Soc. Immunologists, Am. Chem. Soc., Am. Rheumatism Assn., AAAS. Subspecialties: Cancer research (medicine); Immunology (medicine). Current work: Effects of ionizing radiation on DNA and radioprotection, DNA circulation in cancer, rheumatoid diseases, radioimmunotherapy for malignancy. Patentee in field. Office: Albert Einstein Med Center Philadelphia PA 19141

LEON, STEVEN JOEL, mathematics educator, research consultant; b. Detroit, Sept. 16, 1943; s. Rudolph and Florence (White) L.; m. Judith Ellen Russ, Dec. 17, 1972. B.S., Mich. State U., East Lansing, 1965, M.S., 1966, Ph.D., 1971. Grad. asst. Mich. State U., 1967-70; asst. prof. math. Weber State Coll., Ogden, Utah, 1970-75; assoc. prof., 1975-79; assoc. prof. math. Southeastern Mass. U., North Dartmouth, 1979—; research cons. Lockheed Ga. Co., Marietta, summer 1981; math. coordinator NSF project, Ogden, 1976-77. Author: Linear Algebra with Applications, 1980. Mem. Soc. Indsl. and Applied Math., Am. Math. Soc., Math. Assn. Am., Assn. Computing Machinery, Spl. Interest Group on Numerical Mathematics. Subspecialties: Numerical analysis; Applied mathematics. Current work: Numerical analysis, linear algebra, numerical linear algebra, mathematical modelling, scientific and statistical computing. Home: 37 Idlewood Ave North Dartmouth MA 02747 Office: Southeastern Mass U North Dartmouth MA 02747

LEONARD, ANTHONY, fluid dynamics researcher; b. Rock Island, Ill., June 2, 1938; s. Charles Anthony and Gertrude Catherine (Hoffman) L.; m. Gretchen Ann Ver Husen, July 23, 1960; children: Kerrin, Jeffrey. B.S. in Mech. Engring, Calif. Inst. Tech., 1959; M.S., Stanford U., 1960, Ph.D., 1963. Mem. tech. staff Rand Corp., Santa Monica, Calif., 1963-66; asst. prof. mech. engring. Stanford (Calif.) U., 1966-72, assoc. prof., 1972-73; research scientist NASA Ames, Moffett Field, Calif., 1973—; Cons. Nuclear Energy div. Gen. Electric, San Jose, Calif., 1966—, Calspan, Buffalo, 1970-73. Recipient Edward Teller award Am. Nuclear Soc., 1963. Mem. Am. Phys. Soc., AIAA, Soc. Indsl. and Applied Math. Subspecialties: Fluid mechanics; Applied mathematics. Current work: Computational fluid dynamics, turbulent flows, computational physics, applied mathematics. Home: 1450 Marlborough Ct Los Altos CA 94022 Office: NASA Ames Research Center MS 202 A-1 Moffett Field CA 94035

LEONARD, BENJAMIN FRANKLIN, III, geologist; b. Dobbs Ferry, N.Y., May 12, 1921; s. Benjamin Franklin and Florence J. (Smith) L.; m. Eleanor Vandewater, Mar. 18, 1950; children: Ruth L. O'Neil, William C. B.S., Hamilton Coll., 1942; M.A., Princeton U., 1947, Ph.D., 1951. Geologic field aide Geol. Survey Nfld., 1942; geologist U.S. Geol. Survey, Denver, 1943—; vis. prof. geology Colo. Sch. Mines, Golden, 1967-68. Editor: (with A. E. J. Engel and H. L. James) Buddington Volume, Geological Society America, 1962; editor: Internat. Platinum Symposium Soc. Econ. Geologist, 1976; assoc. editor: Canadian Mineralogist, 1976-78; editorial bd.: Eana. Geology, 1972-73. Fellow Geol. Soc. Am., Mineral. Soc. Am.; mem. Internat. Assn. Genesis Ore Deposits (officer paragenetic commn. 1972-82), Internat. Commn. Ore Microscopy (v.p. 1982—), Mineral. Assn. Can., Soc. Econ. Geologists (councilor 1976-79), Soc. Geology Applied to Mineral Deposits, Colo. Sci. Soc. (pres. 1956), Sigma Xi, Phi Beta Kappa. Subspecialties: Mineralogy; Ore or mineral deposits. Current work: Geology and ore deposits of central Idaho, ore minerals and alteration products, geochemical and biogeochemical exploration. Home: 2907 Sunset Dr Golden CO 80401 Office: US Geol Survey Mail Stop 905 Box 25046 Federal Center Denver CO 80225

LEONARD, ELLEN MARIE, physicist; b. N.Y.C., Nov. 28, 1944; d. Eberhard F. and Ann M. (Bazzone) Dullberg; m. Tom A. Leonard, Nov. 24, 1965 (div. 1974); m. John L. Kammerdiener, June 28, 1975; children: Susan A., Michael J. B.S., U. Mich., 1966, M.S., 1968, Ph.D., 1973. Mem. staff Los Alamos Nat. Lab., 1973-80, program mgr., 1980—. Mem. IEEE (mem. environ. quality com. 1982-83), Am. Nuclear Soc., Am. Phys. Soc., AAAS. Subspecialties: Fuels; Fusion. Current work: Program manager for fossil energy program. Home: 1910 Spruce St Los Alamos NM 87544 Office: Los Alamos Nat Lab Los Alamos NM 87545

LEONARD, MYER SAMUEL, oral and maxillofacial surgeon, educator; b. Newcastle, Eng., July 3, 1937; came to U.S., 1976; s. Solomon Lazarus and Ada (Vilenski) L.; m. Mary Patricia (Dalton), Feb. 25, 1969; children: Chantal, Charlotte and Ralph (twins). License in dental surgery, Royal Coll. Surgeons, Edinburgh, Scotland, 1961; B.Dental Surgery, U. Durham, Eng., 1962. Registrar oral and maxillofacial surgery East Anglea Hosp., Ipswich, Eng., 1971-72; sr. registrar Royal Dental Hosp., St. George's Hosp., St. Thomas Hosp., London, 1972-76; vis. prof. U. Mich., Ann Arbor, 1975-76; assoc. prof. U. Minn., Mpls., 1976—. Contbr. articles to profl. jours. NIH grantee, 1981-84; Royal Dental Hosp. grantee, 1974. Fellow Royal Coll. Surgeons, Am. Assn. Oral and Maxillofacial Surgeons; mem. Brit. Assn. Oral Surgeons, ADA. Subspecialties: Oral and maxillofacial surgery; Graphics, image processing, and pattern recognition. Current work: Computer graphics related to craniodfacial morphology; analysis of acoustic emissions from human and animal joints; computers in medical education. Inventor surg. instruments. Home: 1340 Fairlawn Way Golden Valley MN 55416 Office: Dept Oral & Maxillofacial Surgery U Minn 515 Delaware SE Minneapolis MN 55455

LEONARD, NELSON JORDAN, chemistry educator; b. Newark, Sept. 1, 1916; s. Harvey Nelson and Olga Pauline (Jordan) L.; m. Louise Cornelie Vermey, May 10, 1947; children: Kenneth Jan, Marcia Louise, James Nelson, David Anthony. B.S. in Chemistry, Lehigh U., 1937, Sc.D., 1963; B.Sc., Oxford (Eng.) U., 1940, D.Sc., 1983; Ph.D., Columbia U., 1942; D.h.c., Adam Mickiewicz U., Poland, 1980. Fellow and asst. chemistry U. Ill., Urbana, 1942-43, instr., 1943-44, asso., 1944-45, 46-47, asst. prof., 1947-49, asso. prof., 1949-52, prof. organic chemistry, 1952-68, head div. organic chemistry, 1954-63, prof. chemistry, also mem. Center for Advanced Study, 1968—, prof. biochemistry, 1973—; investigator antimalarial program Com. Med. Research OSRD, 1944-46; sci. cons. and spl. investigator Field Intelligence Agy. Tech., U.S. Army and Dept. Commerce, 1945-46; mem. Can. NRC, summer 1950; Swiss-Am. Found. lectr., 1953, 70; vis. lectr. UCLA, summer 1953; Reilly lectr. U. Notre Dame, 1962; Stieglitz lectr. Chgo. sect. Am. Chem. Soc., 1962; Robert A. Welch Found. lectr., 1964; disting. vis. lectr. U. Calif.-Davis, 1975; vis. lectr. Polish Acad. Scis., 1976; B.R. Baker Meml. lectr. U. Calif., Santa Barbara, 1976; Ritter Meml. lectr. Miami U., Oxford, Ohio; Werner E. Bachman Meml. lectr. U. Mich., Ann Arbor, 1977; vis. prof. Japan Soc. Promotion of Sci., 1978; Arapahoe lectr. U. Colo., 1979; mem. program com. in basic scis. Arthur P. Sloan, Jr. Found., 1961-66; Philips lectr. Haverford Coll., 1971; Baker lectr., Groningen, Netherlands, 1972; FMC lectr. Princeton U., 1973; plenary lectr. Laaxer Chemistry Conf., Laax, Switzerland, 1980, 82; Calbiochem-Behring Corp. U. Calif.-San Diego Found. lectr., 1981; Watkins vis. prof. Wichita State U. (Kans.), 1982; Ida Beam Disting. vis. prof. U. Iowa, 1983; mem. adv. com. Searle Scholars program Chgo. Community Trust, 1982—; ednl. adv. bd. Guggenheim Found., 1969—, mem. com. of selection, 1977—. Editor: Organic Syntheses, 1951-58; mem. adv. bd., 1958—; bd. dirs., 1969—; v.p., 1976-80; pres., 1980—; editorial bd.: Jour. Organic Chemistry, 1957-61, Jour. Am. Chem. Soc, 1960-72; adv. bd.: Biochemistry, 1973-78; contbr. articles to profl. jours. Recipient Am. Chem. Soc. award, 1963; medal Synthetic Organic Chem. Mfrs., 1970; Rockefeller Found. fellow, 1950; Guggenheim Meml. fellow, 1959, 67. Fellow Am. Acad. Arts and Scis.; mem. Polish Acad. Scis. (fgn.), Nat. Acad. Scis., Ill. Acad. Sci. (hon.), Gesellschaft Deutscher Chemiker, Am. Chem. Soc. (Edgar Fahs Smith award and lectureship Phila. sect. 1975, Centennial lectr. 1976, Roger Adams award 1981), Am. Soc. Biol. Chemists, AAAS, Chem. Soc. London, Swiss Chem. Soc., Am.-Can. Soc. Plant Physiologists, Am. Soc. Photobiology, Inter-Am. Photochem. Soc., Phi Beta Kappa, Phi Lambda Upsilon (hon.), Tau Beta Pi, Alpha Chi Sigma. Subspecialty: Organic chemistry. Home: 606 W Indiana Ave Urbana IL 61801

LEONE, IDA A., plant pathologist; b. Elizabeth, N.J.; d. Joseph and Josephine (Aprigliano) L. B.S., Douglass Coll., 1944; M.S., Rugters U., 1946. Research asst. Rutgers U., 1946-50, research assoc., 1950-58, asst. research specialist, 1958-70, assoc. research specialist, 1970-76, research prof. plant pathology, 1976—; Bd. mgrs. Central Jersey Regional Air Pollution Control Agy., 1970-81; mem. Elizabeth (N.J.) Mayor's Ad Hoc Com. on Air Pollution, 1965-66. Contbr. numerous articles to profl. jours., chpts. to books. Bd. dirs. Delaware Valley Citizens Council for Clean Air, Union City Tb and Respiratory Disease Assn. EPA grantee, 1975-81. Mem. Am. Soc. Plant Physiologists, Am. Phytopathol. Soc., Air Pollution Control Assn., N.J. Acad. Sci., N.Y. Acad. Scis., Indian Assn. Air Pollution Control. Roman Catholic. Subspecialties: Plant physiology (agriculture); Environmental toxicology. Current work: Air pollution effects on vegetation, re-vegetation of former refuse landfills. Home: 876 Rayhon Terrace Rahway NJ 07065 Office: Dept Plan Pathology Cook Coll PO Box 231 New Brunswick NJ 08903

LEONG, STANLEY PUI-LOCK, physician, surgeon, researcher; b. Shanghai, China, Sept. 8, 1948; came to U.S., 1967; s. Joseph Yuk-Bor and Maria Yau-Ying (Ling) L.; m. Elizabeth T. Leong, Dec. 7, 1974. B.S. in Biology cum laude, Tulane U., 1971, M.D. and M.S. in Immunology and Microbiology, 1974. Lic. physician, La., Mass., Calif., Md. Intern Charity Hosp., New Orleans, 1974-75; fellow surgery and oncology Tulane U., 1975-76; research assoc. Boston U., 1976-78; mem. Hubert H. Humphrey Cancer Research Ctr., 1976-78; resident internal medicine New Eng. Deaconess Hosp., Boston, 1978-79, sr. research assoc. Cancer Research Inst., 1978-79; clin. fellow Harvard U., 1978-79; resident in gen. surgery U. Calif., Orange, 1979-83; research assoc. div. surgery City of Hope Nat. Med Ctr., Duarte, Calif., 1979-82; clin. assoc. surgery br. Nat. Cancer Inst., NIH, Bethesda, Md., 1983—. Contbr. articles to profl. jours. Greater New Orleans Cancer Assn. fellow, 1971; Damon Runyon-Walter Winchell Cancer Fund Oncology fellow, 1981-82. Mem. AMA, Mass. Med. Soc., Am. Fedn. Clinic Research, Am. Assn. Cancer Research., AAAS, N.Y. Acad. Sci. Roman Catholic. Subspecialties: Immunobiology and immunology; Cancer research (medicine). Current work: Tumor immunology; melanoma and sarcoma antigens; cytohistochemistry, immunofluorescence; monoclonal antibody; fluorescence activated cell sorter; general surgery. Office: Surgery Br Bldg 10 Rm 2B42 Nat Cancer Inst NIH Bethesda MD 20205

LEOPOLD, ALDO STARKER, wildlife biologist; b. Burlington, Iowa, Oct. 22, 1913; s. Aldo and Estella (Bergere) L.; m. Elizabeth Weiskotten, Aug. 7, 1938; children—Frederic Starker, Sarah Pendleton. B.S., U. Wis., 1936; postgrad., Yale U. Sch. Forestry, 1936-37; Ph.D. in Zoology, U. Calif., Berkeley, 1944. Wildlife biologist Mo. Conservation Commn., 1939-44, Pan Am. Union, Mex., 1944-46; mem. faculty U. Calif., Berkeley, 1946—, prof. zoology and forestry, 1967-78, emeritus, 1978—; adv. bd. Nat. Parks, 1972-76. Author: Game Birds and Mammals of California, 1951, Wildlife in Alaska, 1953, Wildlife of Mexico, 1959, The Desert, 1961, The California Quail, 1977, also tech. papers. Recipient Conservation award Dept. Interior, 1964; Guggenheim fellow, 1947-48. Mem. Nat. Acad. Sci., Nat. Acad. Arts and Sci., Am. Inst. Biol. Scis. (Disting. Service award 1980), Wildlife Soc., Wilderness Soc., Marine Mammal Commn., Calif. Acad. Scis. (pres. 1959-71), Sierra Club. Club: Bohemian (San Francisco). Subspecialties: Ecology; Resource management. Current work: Wildlife ecology and conservation. Home: 712 The Alameda Berkeley CA 94707 Office: Dept Forestry U Calif Berkeley CA 94720

LEOPOLD, ESTELLA BERGERE, botanist, educator; b. Madison, Wis., Jan. 8, 1927; d. Aldo and Estella (Bergere) L. Ph.B., U. Wis., 1948; M.S., U. Calif., Berkeley, 1950; Ph.D., Yale, 1955. Asst. physiology and embryology Genetics Expt. Sta., Smith Coll., 1951-52; asst. biologist, then asst. animal ecology Yale U., 1952-54; botanist, paleontology and stratigraphy br. U.S. Geol. Survey, 1955-76; adj. prof. biology U. Colo., 1967-76; dir. Quaternary Research Center; prof. botany and forest research U. Wash., Seattle, 1976—. Co-recipient Conservationist of Year award Nat. Wildlife Fedn., 1969; award Keep Colo. Beautiful Inc., 1976; travel grantee NSF, Spain, 1957, NSF, Poland, 1961, 1977. Fellow AAAS; mem. Nat. Acad. Scis. (environ. studies bd. 1976-79), Ecol. Soc. Am. (climate research bd. 1979—), Internat. Quaternary Assn. U.S. Nat. Com., Am. Quaternary Assn. (pres.-elect 1980), Bot. Soc. Am., Inst. Ecology (dir.). Subspecialties: Ecology; Botany. Home: 5608 17th St NE Seattle WA 98105 Office: Quaternary Research Center Univ Wash Seattle WA 98195

LEOPOLD, LUNA BERGERE, geology educator; b. Albuquerque, Oct. 8, 1915; s. Aldo and Estella (Bergere) L.; m. Barbara Beck Nelson, 1973; children: Bruce Carl, Madelyn Dennette. B.S., U. Wis., 1936, D.Sc. (hon.), 1980; M.S., UCLA, 1944; Ph.D., Harvard, 1950; D.Geography (hon.), U. Ottawa, 1969, D.Sc., Iowa Wesleyan Coll., 1971, St. Andrews U., 1981. With Soil Conservation Service, 1938-41, U.S. Engrs. Office, 1941-42, U.S. Bur. Reclamation, 1946; head meterorologist Pineapple Research Inst. of Hawaii, 1946-49; hydraulic engr. U.S. Geol. Survey, 1950-71, chief hydrologist, 1957-66, sr. research hydrologist, 1966-71; prof. geology U. Calif. at Berkeley, 1973—. Author: (with Thomas Maddock, Jr.) The Flood Control Controversy, 1954, Fluvial Processes in Geomorphology, 1964, Water, 1974, (with Thomas Dunne) Water in Environmental Planning, 1978; also tech. papers. Served as capt. air weather service USAAF, 1942-46. Recipient Distinguished Service award Dept. of Interior, 1958; Kirk Bryan award Geol. Soc. Am., 1958; Veth medal Royal Netherlands Geog. Soc., 1963; Cullum Geog. medal Am. Geog. Soc., 1968; Rockefeller Pub. Service award, 1971; Busk medal Royal Geog. Soc., 1983. Mem. ASCE, Nat. Acad. Scis. (Warren prize), Geol. Soc. Am. (pres. 1972), Am. Geophys. Union, Am. Acad. Arts and Scis., Am. Philos. Soc., Sigma Xi, Tau Beta Pi, Phi Kappa Phi, Chi Epsilon. Club: Cosmos (Washington). Subspecialty: Hydrology. Home: 400 Vermont Ave Berkeley CA 94707

LEOPOLD, ROGER ALLEN, entomologist; b. Redwood Falls, Minn., Mar. 23, 1937; s. Alvin Jacob and Hazel Marie (Wheaton) L.; m.; children: Darin, Derek. B.A., Concordia Coll., Moorhead, Minn., 1962; Ph.D., Mont. State U., 1967. Research entomologist Agrl. Research Service, U.S. Dept. Agr., Fargo, N.D., 1967—; adj. prof. N.D. State U., Fargo, 1968—, Concordia Coll., Moorhead, 1978—. Contbr. articles to profl. jours. NDEA fellow, 1962; NSF grantee, 1970. Mem. Entomol. Soc. Am., Am. Soc. Zoology, N.D. Acad. Sci., Sigma Xi. Lodge: Elks. Subspecialties: Reproductive biology; Developmental biology. Current work: Reproductive physiology and development of insects; cryogenic storage of insect germplasm. Office: US Dept Agr Metabolism and Radiation Research Lab PO Box 5674 Fargo ND 58105

LEOSCHKE, WILLIAM LEROY, chemistry educator; b. Lockport, N.Y., May 2, 1927; s. William Christian and Hildegarde (Miller) L.; m. Marjorie Ann Kovell, June 16, 1956; children: Peter Conrad, Anna Elizabeth. B.S., Valparaiso U., 1950; M.S., U. Wis.-Madison, 1952, Ph.D., 1954. Research assoc. U. Wis.-Madison, 1954-59; faculty Valparaiso (Ind.) U., 1959—, prof. chemistry, 1968—; cons. research Nat. Mink Foods, New Holstein, Wis., 1955—. Chmn. area com. Am. Friends Service Com., Valparaiso, 1972-77; pres. Porter County Human Relations Council, Valparaiso, 1962-65. Served with USN, 1945-46. Mem. Am. Chem. Soc., AAUP (chpt. pres. 1965-67), Sigma Xi. Democrat. Lutheran. Subspecialties: Animal nutrition; Animal physiology. Current work: Experimental work on the nutrition and physiology of mink leading to development of pellets for practical mink nutrition, studies of urinary calculi and digestion physiology of mink. Home: 278W 100 S Valparaiso IN 46383 Office: Valparaiso U Valparaiso IN 46383

LEPLEY, ARTHUR RAY, chemist; b. Peoria, Ill., Nov. 1, 1933; s. Ray and Maud Alverta L.; m. (div.); 4 children. A.B., Bradley U., 1954; M.S., U. Chgo., 1956, Ph.D., 1958. Research assoc. organic chemistry U. Munich, 1958-59, U. Chgo., 1959-60; asst. prof. chemistry SUNY-Stony Brook, 1960-65; assoc. prof. Marshall U., Huntington, W. Va., 1965-68, prof. chemistry, 1968—; vis. prof. U. Utah, 1969-71; guest worker Lab. Chem. Physics, Nat. Inst. Arthritis, Metabolism and Digestive Disorders, 1975-76. NSF postdoctoral fellow, 1958-59; USPHS postdoctoral fellow, 1960. Mem. AAAS, Am. Chem. Soc., Royal Soc. Chemistry, Am. Inst. Chemistry, Sigma Xi. Subspecialties: Kinetics; Nuclear magnetic resonance (chemistry). Office: Marshall U Dept Chemistry Huntington WV 25701

LEPP, HENRY, geology educator and consultant; b. Halbstadt, Russia, Mar. 4, 1922; s. David Abraham and Marie (Willms) L.; m. Maxine Marie Foster, Sept. 15, 1952; children: Kathleen, Stephen, David, Tamara. B.Sc., U. Sask., Can., 1944; Ph.D., U. Minn., 1954. Geologist COMINCO Ltd., Yellowknife, N.W.T., 1944-46, N.W. Byrne Inc., Yellowknife, 1946-48, Alcan Alumnium, Conakry, Guinea, 1948-50, Freeport Sulphur, N.Y.C., 1954; prof. geology U. Minn.-Duluth, 1954-65, Macalester Coll., St. Paul, 1964—; cons. Que. Cartier, Montreal, 1955-58, Roberts Mining Co., Duluth, 1960-63. Author: Dynamic Earth, 1973; editor: Geochemistry of Iron, 1975. Fellow Geol. Soc. Am.; mem. Soc. Econ. Geologists, Sigma Xi. Subspecialties: Geology; Geochemistry. Home: 3042 Sandy Hook Dr Saint Paul MN 55113 Office: Dept Geology Macalester Coll Saint Paul MN 55105

LERCH, IRVING ABRAM, med. physicist; b. Chgo., June 29, 1938; s. Abraham and Rissel (Lutwak) L.; m. Sharon Lerch, Feb. 24, 1963. B.S., U.S. Mil. Acad., 1960; M.S., U. Chgo., 1966, Ph.D., 1969. Research assoc. U. Chgo., 1969-73; first officer IAEA, Vienna, Austria, 1973-76; prof. N.Y. U. Sch. Medicine, N.Y.C., 1976—; tech. asst. expert cons. IAEA, WHO, others. Contbr. articles to profl. jours.; Books and publs. editor: Med. Physics, 1980-82; sci. editor: Biomedical Dosimetry, 1975, Physics, Dosimetry and Biomedical Aspects of Californium, 1976. Served to 1st lt. U.S. Army, 1960-63. USPHS fellow, 1964-69; Nat. Cancer Inst.-USPHS grantee, 1972-73, 69-73; others. Mem. Am. Phys. Soc., Am. Assn. Physicists in Medicine, Radiation Research Soc., Radiol. and Med. Physics Soc. N.Y. Subspecialties: Cancer research (medicine); Radiology. Current work: Radiation dosimetry standardization in use of ionizing radiations in detection and treatment of cancer. Home: 146 W 74th St Apt 3 New York NY 10023 Office: 566 First Ave New York NY 10016

LERMAN, ABRAHAM, geological sciences educator. M.Sc., Hebrew U., Jerusalem, 1960; Ph.D., Harvard U., 1964. Sr. scientist isotope research Weizmann Inst. Sci., Rehovot, Israel, 1966-69; research scientist in chem. limnology Can. Centre for Inland Waters, Burlington, Ont., 1969-71; assoc. prof. geol. sci. Northwestern U., Evanston, Ill., 1971-75, prof., 1975—; vis. prof. Swiss Fed. Poly. Inst., Zurich, 1976-77, U. Karlsruhe, Germany, 1979, 81; cons. Battelle-PNL, Richland, Wash., 1980, Rockwell Internat., Richland, 1981—, Dept. Energy, 1982. Author: Goechemical Processes, 1979; editor: Lakes: Chemistry, Geology, Physics, 1978; contbr. articles to profl. jours. Guggenheim fellow, 1976. Fellow Geol. Soc. Am.; mem. Am. Chem. Soc., AAAS, Am. Geophys. Union, Geochem. Soc., Sigma Xi. Subspecialties: Geochemistry; Environmental engineering. Current work: Transport in water in sediment environments, global geochemical cycles, disposal and migration of nuclear and chemical wastes, geochemistry of lakes, rivers. Office: Dept Geol Sci Northwestern U Evanston IL 60201

LERMAN, SIDNEY, ophthalmology educator; b. nr. Montreal, Que., Can., Oct. 6, 1927; s. Aaron and Rachel L.; m. Marilyn F. Frank, Apr. 14, 1957; children: Lora Rachel, Mark Jonas. B.Sc., McGill U., 1948, M.D.C.M., 1952; M.S. in Biochemistry, U. Rochester, 1961. Diplomate: Am. Bd. Ophthalmology. Intern Montreal Gen. Hosp., 1952, resident, 1953-55; successively instr., asst. prof., assoc. prof. and dir. ophthalmology research U. Rochester, 1957-68; prof. ophthalmology and biochemistry, dir. dept. ophthamol. research McGill U., 1968-75; prof. ophthalmology, adj. prof. chemistry Emory U. and Ga. Inst. Tech., 1975—. Author: Glaucoma—Chemical Mechanical, 1961, Cataract Therapy, 1964, Basic Ophthalmology,

1966, Radiant Energy and the Eye, 1980. Fellow Am. Acad. Ophthalmology, Oxford Ophthalmol. Congress, ACS; mem. Am. Soc. Biol. Chemists, Am. Chem. Soc., Am. Soc. Photobiology, Internat. Soc. Eye Research, Assn. Research in Vision and Ophthalmology. Subspecialties: Biophysical chemistry; Photochemistry. Current work: Photobiology, photochemistry, pharmacology and toxicology, biologic effects of radiation, biophysical research, optical spectroscopy, NMR spectroscopy. Patentee in field. Home: 1648 Musket Ridge Rd NW Atlanta GA 30327 Office: Emory University 1708 Haygood Dr NE Atlanta GA 30322

LERNER, LEON MAURICE, biochemistry educator; b. Chgo., Feb. 2, 1938; s. Sidney and Yetta (Weiner) L.; m. Helen Jane Abrams, Aug. 23, 1959; children: Linda, Marcia, Gary. B.S., Ill. Inst. Tech., 1959, M.S., 1961; Ph.D., U. Ill.-Chgo., 1964. Postdoctoral researcher U. Ill. Med. Ctr., Chgo., 1964-65; instr. SUNY Downstate Med. Ctr.-Bklyn., 1965-67, asst. prof., 1967-73, assoc. prof., 1973-80, prof. biochemistry, 1980—. Contbr. articles to profl. jours. Nat. Cancer Inst.-NIH grantee, 1971-80; Am. Cancer Soc. grantee, 1967-69. Mem. Am. Chem. Soc., AAAS, Sigma Xi. Subspecialties: Biochemistry (medicine); Organic chemistry. Current work: Carbohydrate chemistry, nucleic acid chemistry, antimetabolites. Office: SUNY Downstate Med Center 450 Clarkson Ave Brooklyn NY 11203

LERNER, LEONARD JOSEPH, endocrinologist, educator; b. Roselle, N.J., Sept. 26, 1922; s. Hyman and Esther Celia (Honig) L. B.S., Rutgers U., 1943, A.B., 1951, M.S., 1953, Ph.D., 1954; postgrad., N.Y.U., 1951. Registered pharmacist, N.J., Ohio, Calif., N.Y. Research asst. Bur. Biol. Research, Rutgers U., 1952-54; sect. head William S. Merrell Co., Cin., 1954-58; sect. head endocrinology Squibb Inst. Med. Research, New Brunswick, N.J., 1958-65; dir. dept. endocrinology, 1965-70, Lepetit Research Lab., Gruppo Lepetit, Milan, Italy, 1971-77; research prof. Jefferson Med. Coll., Phila., 1977—; adj. prof. Hahnemann Med. Coll., 1971—; cons., lectr. in field. Contbr. numerous articles to profl. jours. Served with AUS, 1944-46. Grantee pharm. co., various agys. Mem. Endocrine Soc., AAAS, Am. Physiol. Soc., N.Y. Acad. Sci., Am. Assn. Cancer Research, Am. Fertility Soc., Soc. Study Reprodn., Soc. Exptl. Biology and Medicine, Internat. Study Group Steroid Hormones, Sigma Xi. Subspecialties: Endocrinology; Reproductive biology. Current work: Reproductive endocrinology, endocrine pharmacology. Patentee in field. Home: K-8 Windsor Castle Apts Cranbury NJ 08512 Office: 1025 Walnut St Philadelphia PA 19107 Home: K-8 Windsor Castle Apts Cranbury NJ 08512

LERNER, RICHARD ALAN, physician, researcher; b. Chgo., Aug. 26, 1938. Student, Northwestern U., 1956-59; B.A., Stanford U., M.D., 1964. Intern Palo Alto Stanford Hosp., 1964-65; research fellow dept. exptl. pathology Scripps Clinic and Research Found., La Jolla, Calif., 1965-68, assoc., 1970-72, assoc. mem., 1972-74, mem. dept. immunopathology, 1974-77; mem. dept. cellular and developmental immunology Research Inst. Scripps Clinic, 1977-82, head com. for study of molecular genetics, 1980-82, chmn. dept. molecular biology, 1982—; chmn. Internat. Symposium on Molecular Basis of Cell-Cell Interaction, 1977-80; cons. spl. virus cancer program Nat. Cancer Inst. Mem. editorial bd.: Jour. Virology. Recipient Career Devel. award NIH-AID, 1970; award Parke, Davis, 1978. Fellow ACS (mem. fellowship screening com. Calif. div.); mem. Am. Soc. Virology (charter), Biophys. Soc., N.Y. Acad. Scis., Am. Soc. Microbiology, Am. Soc. Exptl. Pathology, Am. Assn. Immunologists, Am. Soc. Nephrology, Alpha Omega Alpha, Phi Eta Sigma. Subspecialty: Molecular biology. Home: 7750 E Roseland La Jolla CA 92037 Office: Research Inst Scripps Clinic 10666 N Torrey Pines Rd La Jolla CA 92037

LERNER, STEPHEN ALEXANDER, microbiology educator; b. Chgo., Oct. 4, 1938; s. David G. and Florence (Trace) L.; m. Ronna Bergman, June 6, 1963; children: Deborah, Daniel, Susan. A.B., Harvard U., 1959, M.D., 1963. Med. intern Peter Bent Brigham Hosp., Boston, 1963-64, resident, 1964-65; research assoc. NIH, Bethesda, Md., 1965-68; postdoctoral fellow Stanford Biochemistry Dept., Palo Alto, Calif., 1968-71; asst. prof. dept. medicine U. Chgo., 1971-78, assoc. prof., 1978—. Editor: Aminogly Coside Ototoxicity, 1981; editorial bd.: Antimicrobial Agents and Chemotherapy, 1981—. Recipient Borden Undergrad. award Harvard U., 1963. Fellow Infectious Diseases Soc. Am; mem. Am. Soc. Microbiology, Phi Beta Kappa, Alpha Omega Alpha. Democrat. Jewish. Subspecialties: Microbiology; Pharmacology. Current work: Genetic and physiologic machanisms of bacterial resistance to antibiotics, antibiotic pharmacology. Home: 4918 S Kimbark Ave Chicago IL 60615 Office: U Chgo Sch Medicine 950 E 59th St Chicago IL 60637

LEROUX, PIERRE J.A., mathematics educator; b. Quebec, Que., Can., Aug. 18, 1942; s. Guy and Gabrielle (Dufresne) L.; m. Madeleine Loubert, Oct. 3, 1964; children: Marie-Claude, Sophie. B.A., Coll. Ste Marie, 1961; B.Sc., U. Montreal, 1964, M.Sc., 1966, Ph.D., 1970. Instr. U. Montreal, 1965-70; invited prof. U. de Nantes, France, 1972, Inst. Nat. Statis, Econ. and Applications, Rabat, Morocco, 1975; prof. math. U. Que., Montreal, 1971—, dir. module de math., 1976-78. Author: Algebre lineaire, une approche matricielle, 1982, revised edit., 1983. Pres. Fedn. Que. Canot-kayak-camping, Montreal, 1972-76. Can. NRC fellow, 1970-71; FCAC grantee, 1980-83; CRSNG grantee, 1982-85. Mem. Assn. Math. Que. (council 1980—), Soc. Math. Can., Am. Math. Soc., Soc. Indsl. and Applied Math. Subspecialties: Linear algebra; Combinatorics. Current work: Research in enumerative combinatorics, combinatorial theories for generating functions with applications to classical analysis, for instance, orthogonal polynomials; applications of linear algebra, mobius inversion and mobius categories to combinatorics. Office: U Que Dept Math CP 8888 succ A Montreal PQ Canada H3C 3P8

LERSTEN, NELS RONALD, botany educator; b. Chgo., Aug. 6, 1932; s. Anders Einar and Elvira Maria (Bloom) L.; m. Patricia Ann Brady (dec. June 13, 1958); children: Samuel, Andrew, Julie. B.S., U. Chgo., 1958, M.S., 1960; Ph.D., U. Calif.-Berkeley, 1963. Asst. prof. botany Iowa State U., 1963-66, assoc. prof., 1967-70, prof., 1970—. Contbr. articles to profl. jours., chpts. in books. Served with USCG, 1952-55; Korea. NSF grantee, 1959, 1971-74, 78-81. Mem. Am. Soc. Plant Taxonomists, Bot. Soc. Am., Internat. Assn. Plant Taxonomists, Sigma Xi. Subspecialties: Morphology; Systematics. Current work: Anatomy of flowering plant reproduction, plant anatomy applied to taxonomy flowering plants, plant reproduction, plant anatomy, systematic anatomy. Home: 1913 Roosevelt Ave Ames IA 50010 Office: Dept Botany Iowa State Ames IA 50011

LESCHACK, LEONARD ALBERT, research company executive; b. N.Y.C., Mar. 6, 1935; s. David and Selma (Kaminsky) LeS.; m. Lorraine L., Mar. 3, 1962; children: Christopher E., Adam A. B.S. in Petroleum Geology, Rensselaer Poly. Inst., 1957; diploma in oceanography, Grad. Sch. U.S. Dept. Agr., 1962; postgrad. in geophysics, U. Wis-Madison, 1963-64. Cert. profl. geol. scientist. Geophys. trainee Shell Oil Co., Houston, 1957; asst. seismologist Byrd Sta. Traverse Party, U.S. Nat. Com. Internat. Geophys. Yr., Antarctica, 1957-59; oceanographer Naval Oceanographic Office, Suitland, Md., 1964-65; polar regions project officer EXPO-67, Montreal, Que., Can., 1965-66; pres. LeSchack Assocs., Ltd., Long Key, Fla., 1967—; U.S. rep. Argentine Antarctic Expdn., 1962-63; participant 2d Internat. Permafrost Conf., Siberia, 1973; pres. Trident Arctic Exploration Ltd., Montreal, 1979—. Contbr. chpts. to books, articles to profl. jours. Served to lt. (j.g.) USN, 1959-63; to capt., 1980-81. Decorated Legion of Merit; recipient Antarctica Service medal Nat. Acad. Scis.-NRC, 1966; grantee Arctic Inst. N.Am., 1963. Mem. Am. Geophys. Union, Soc. Exploration Geophysicists, Am. Inst. Profl. Geologists. Subspecialties: Geophysics; Remote sensing (geoscience). Current work: Collection and analysis of Arctic sea ice data using remote sensing techniques; developing data collection techniques with manned submersibles; exploration for geothermal resources. Patentee graphical data digitizer.

LESKO, ROBERT JOSEPH, mgmt. and tech. cons.; b. Homestead, Pa., Sept. 24, 1942; s. Joseph and Irene Teresa (Anderson) L.; m. Kathleen Menzie, Aug. 7, 1965; children: Mark Joseph, Robert Anderson. B.S. in Mech. Engring, U. Notre Dame, 1964; M.B.A., U. Pa. Wharton Sch., 1967; postgrad., Georgetown U. Law Center, 1966-67; M.F.A., Catholic U. Am., 1979. Mem. tech. staff Computer Scis. Corp., Falls Church, Va., 1967-68; pres. Centaur Mgmt. Cons., Inc., Washington, 1968-72; exec. v.p. Med. Aid Tng. Schs. Inc., Silver Spring, Md., 1972-74; pres. Software Architecture and Engring., Inc., Arlington, Va., 1974-82; v.p. Applied Mgmt. Sci., Inc., Silver Spring, Md., 1982—; mem. faculty Grad. Sch. Bus., George Washington U., 1969-70, Am. U., 1970-71; dir. Knowledge Engring., Inc., Arlington. Bd. dors Oxfam-Am., Washington, 1970-72. Mem. Assn. Energy Engrs. Subspecialty: Information systems, storage, and retrieval (computer science). Current work: Info. systems, energy data systems, advanced data processing systems, strategic planning.

LESKO, STEPHEN ALBERT, biophysics educator; b. Cassandra, Pa., Dec. 30, 1931; s. Stephen and Elizabeth (Krumenaker) L.; m. Shu-Uin Yang, Sept. 4, 1981. B.S., Indiana U. Pa., 1959; Ph.D., U. Md., 1965. Instr., Johns Hopkins U., Balt., 1965-69, research assoc., 1969-74, asst. prof., 1974-82, assoc. prof., 1982—. Served as cpl. U.S. Army, 1952-54; Korea. Mem. Am. Chem. Soc., AAAS, Am. Assn. Cancer Research, Biophys. Soc. Roman Catholic. Subspecialties: Biochemistry (medicine); Environmental toxicology. Current work: Oxygen toxicity, chemical and radiation carcinogenesis, nucleic chemistry and biology. Home: 7406 Knollwood Rd Baltimore MD 21204 Office: 615 N Wolfe St Baltimore MD 21205

LESLIE, JOHN FRANKLIN, research microbiologist; b. Dallas, July 2, 1953; s. Frank R. and Peggy J. (Shelton) L.; m. Ingelin Lono, Jan. 10, 1976; children: Timothy Franklin, Inger Joyce. B.A. in Biology, U. Dallas, 1975; M.S. in Genetics, U. Wis., 1977, Ph.D., 1979. Univ. fellow, then NIH trainee Lab. Genetics, U. Wis., Madison., 1975-79; postdoctoral research affiliate dept. biol. sci. Stanford (Calif.) U., 1979-81; research microbiologist corp. research and devel. Internat. Mineral and Chem. Corp., Terre Haute, Ind., 1981—; tech. adv. Inst. Christian Resources, San Jose, Calif. Contbr. articles to profl. jours. Mem. Genetics Soc. Am., Mycol. Soc. Am., Am. Soc. Microbiology, Soc. Gen. Microbiology, Brit. Mycol. Soc., AAAS, Sigma Xi. Presbyterian. Subspecialties: Genetics and genetic engineering (agriculture); Microbiology. Current work: Fungal genetics in model and plant pathogenic systems. Genetic enhancement of microbial metabolite production using classical and recombinant DNA techniques. Dissection of fungal development and population structure using chromosome rearrangements and classical genetics. Empirical and theoretical Population genetics. Home: 2851 S Brown Ave Terre Haute IN 47802 Office: Corp Research and Devel Internat Mineral & Chem Corp PO Box 207 Terre Haute IN 47808

LESSER, RONALD PETER, neurologist; b. Los Angeles, Jan. 17, 1946; m. Sara Elizabeth Roesler; 2 children. B.A. in Anthropology, Pomona Coll, 1966; M.D., U. So. Calif., 1970. Diplomate: Nat. Bd. Med. Examiners, Am. Bd. Psychiatry and Neurology (Psychiatry), 1978, Am. Bd. Psychiatry and Neurology (Neurology), 1980, Am. Bd. Qualification in EEG, 1981. Intern in pediatrics Mayo Grad. Sch. Medicine, Rochester, Minn., 1970-71; resident in psychiatry N.Y. State Psychiat. Inst, Columbia-Presbyn. Med. Ctr., N.Y.C., 1973-76; research fellow neurophysiology and EEG Neurol. Inst., Columbia-Presbyn. Med. Ctr., 1975-76, resident in neurology, 1976-79; spl. fellow EEG Cleve. Clinic Found., 1979, staff neurologist, dir. clin. neurophysiology service and adult seizure clinic, 1980—; sr. asst. surgeon, community health officer USPHS Indian Hosp., Rapid City, S.D., 1971-73. Contbr. articles to profl. jours. Recipient William Kirk Meml. prize in Anthropology Pomona Coll., 1966. Mem. Am. Acad. Neurology, Am. EEG Soc. (chmn. guidelines com.), Am. Epilepsy Soc., Am. Psychiat. Assn., Central Assn. Electroencephalographers. Subspecialty: Neurology. Office: 9500 Euclid Ave Cleveland OH 44106

LESTER, DAVID, biochemist, educator, cons.; b. New Haven, Jan. 22, 1916; s. Asher and Esther (Rubin) L.; m. Ruth Weiss, Sept. 18, 1938; children: Anne Deborah Lester Schager, James Matthew. B.S., Yale U., 1936, Ph.D., 1940. Diplomate: Am. Bd. Clin. Chemistry, 1958. Mem. faculty Lab. Applied Physiology, Yale U., 1940-62; prof. biochemistry Rutgers U., New Brunswick, N.J., 1962—; dir. grad. program in pharmacology Rutgers U./U. Medicine and Dentistry N.J., 1980-82; sci. dir. Nat. Alcohol Research Center, 1978-83. Author book in field; contbr. chpts to books, articles to profl. jours.; asso. editor: Jour. Studies on Alcohol, 1950—. Rep. Borough of Princeton; mem. instl. biosafety com. Princeton U. Mem. Am. Chem. Soc., N.Y. Acad. Scis., Am. Soc. Pharmacology and Exptl. Therapeutics, AAAS, Research Soc. Alcoholism. Subspecialties: Biochemistry (medicine); Pharmacology. Current work: Animal models alcoholism, predictive factors alcoholism. Home: 29 Forester Dr Princeton NJ 08540 Office: Center Alcohol Studies Rutgers Univ New Brunswick NJ 08903

LESTER, DAVID BRENT, nuclear engineer, consultant; b. Bunkie, La., Mar. 18, 1941; s. James Ezra and Alice Helena (Shubert) L.; m. Linda Marie Begg, Apr. 22, 1967; children: Glenwood Alice, David Brent, Brady Clay. B.S. in Nuclear Sci. and Engring., U.S. Naval Acad., 1964. Commd. ensign U.S. Navy, 1964, advanced through grades to lt., 1968; ret., 1970; engr. La. Power & Light Co., New Orleans, 1970-78, project mgr. nuclear plant, 1979, mgr. nuclear plant, 1980-82, asst. to v.p. nuclear ops., 1982-83; prin. Ins. Cons. Services, Inc., Metairie, La., 1983—. Mem. Am. Nuclear Soc., La. Nuclear Soc. (chmn. 1979), U.S. Naval Inst. Republican. Roman Catholic. Subspecialties: Nuclear fission; Nuclear engineering. Current work: Management of contruction, engineering, start-up and operation of nuclear power plants. Office: La Power & Light Co PO Box B Lillona LA 70066

LESTER, JOHN BERNARD, astronomer, educator; b. San Diego, Mar. 11, 1945; s. Bernard Edward and Margaret (Miller) L.; m. Rose Ann Patterson, July 1, 1972; children: Catherine, Margaret. Student, San Diego State U., 1963-65; B.A., Northwestern U., 1967; M.S., U. Chgo., 1969, Ph.D., 1972. Research assoc. Smithsonian Astrophys. Obs., Cambridge, Mass., 1972-76; asst. prof. astronomy U. Toronto, Ont., Can., 1976-81, assoc. prof., 1981—. Contbr.: articles to Astrophys. Jour. Mem. Am. Astron. Soc., Internat. Astron. Union, Astron. Soc. Pacific. Subspecialties: Optical astronomy; Stellar composition. Current work: Stellar atmospheres, chemical compositiopns, ultraviolet astronomy, computer models. Home: 2581 Barcella Crescent Mississauga ON Canada L5K 1E5 Office: Erindale Coll Mississauga ON Canada L5L 1C6

LESTER, RICHARD KEITH, engineering educator, consultant; b. Leeds, Eng., Jan. 3, 1954; came to U.S., 1974; s. Bernard and Mona (Smuckler) L.; m. Anne Elizabeth Columbia, Oct. 20, 1979. Sc.B. in Chem. Engring, Imperial Coll., 1974; Ph.D. in Nuclear Engring, M.I.T., 1979. Vis. fellow Rockefeller Found., N.Y.C., 1977-78; instr. M.I.T., Cambridge, 1978-79, asst. prof. nuclear engring., 1979-80, Edgerton asst. prof. nuclear engring., 1980-82, assoc. prof. nuclear engring., 1982—; cons. several U.S. cos. Kennedy scholar, 1974-76. Mem. Am. Nuclear Soc., AAAS. Subspecialty: Nuclear engineering. Current work: Nuclear chemical engineering, nuclear waste management, international nuclear relations, energy policy analysis. Office: Dept Nuclear Engring MIT 77 Massachusetts Ave Cambridge MA 02138

LESTREL, PETE ERNEST, educator, research anthropologist; b. Quito, Ecuador, Feb. 19, 1938; came to U.S., 1948, naturalized, 1954; s. Hans and Berta (Schwab) L.; m. Dagmar Centa Kowalzyk, Apr. 20, 1968; children: Nicole, Valerie. A.B., UCLA, 1964, M.A., 1966, Ph.D., 1975. Engr. N.Am. Aviation, Los Angeles, 1962-65; instr. Santa Monica (Calif.) Coll., 1967-73; asst. prof. anthropology Case Western Res. U., Cleve., 1973-75, cons. dept. anatomy, 1974-75, asst. prof. UCLA Sch. Dentistry, 1977-80, assoc. prof., 1981—; research anthropologist VA Med. Center, Sepulveda, Calif., 1976—. Editorial bd.: Human Biology Jour, 1980-83; contbr. articles to profl. jours., chpts. in books. Fellow Human Biology Council; mem. Am. Assn. Phys. Anthropologists, Am. Assn. Dental Research, Internat. Assn. for Dental Research. Democrat. Subspecialties: Dental growth and development; Morphometrics. Current work: Quantitative description of complex morphological forms frequently encountered in biology and medicine; computer modelling of growth and devel. and evolutionary processes. Home: 7327 De Celis Pl Van Nuys CA 91406 Office: 16111 Plummer Ave Sepulveda CA 91343

LETO, SALVATORE, laboratory director, andrologist; b. Borgetto, Sicily, Oct. 28, 1937; came to U.S., 1946; s. Antonino and Elisabetta (Armato) L.; m. Margaret A. Smith, Sept. 12, 1964 (div. July 1970); children: Anthony L., Gerald A.; m. Evelyn H. Brady, Dec. 28, 1973. B.S. in Chemistry, CCNY, 1961; Ph.D. in Biology, Georgetown U., 1967. Cert. clin. lab. supr., N.Y. Staff fellow Nat. Inst. Child Health and Human Devel., NIH, Balt., 1967-71; lab. supr. IDANT Corp., N.Y.C., 1971-72, lab. dir., Balt., 1972-73, Washington Fertility Study Ctr., 1973—; cons., 1980—. Author: Clinical Advances in Andrology, vol. 8, 1982, Male Reproduction and Fertility, 1983; contbr. articles to sci. jours. Served in U.S. Army, 1961-63. Mem. Am. Fertility Soc., Am. Soc. Andrology, Am. Physiol. Soc., AAAS, Am. Assn. Tissue Banks, Pan Am. Congress Andrology, Sigma Xi. Democrat. Roman Catholic. Subspecialties: Andrology; Reproductive biology. Current work: Andrology, reproductive biology, cryobiology (human sperm cryo-preservation), immuno-infertility, endocrinology of reproduction. Home: 20 Sparrow Hill Ct Baltimore MD 21228 Office: 2600 Virginia Ave Suite 500 Washington DC 20037

LEUHRS, DEAN CARL, chemistry educator; b. Fremont, Nebr., Apr. 20, 1939; s. Glen C. and Louise A. (Haebler) L.; m. Joan Marie Moege, June 14, 1969; 1 dau., Janine. B.S., Mich. State U., 1961; Ph.D., U. Kans., 1965. Asst. prof. Mich. Technol. U., Houston, 1965-70, assoc. prof. chemistry, 1970—. NSF fellow, 1961-64; NIH fellow, 1964-65. Mem. Am. Chem. Soc., Am. Pomological Soc. Baptist. Subspecialty: Inorganic chemistry. Current work: Reactions in nonaqueous solvents, kinetics of multidentate substitution reactions, synthesis of one-dimensional electrically conducting solids. Patentee in field. Office: Dept Chemistry Mich Technol U Houghton MI 49931

LEUNG, BENJAMIN SHUET-KIN, medical educator, researcher; b. Hong Kong, June 30, 1938; s. Yun-Pui and Kan-Yau (Lee) L.; m. Helen Tsan-Fu Hsu, Oct. 19, 1964; children: Kay, Titus, Steven. B.S., Seattle Pacific Coll., 1963; Ph.D., Colo. State U., 1969; postgrad., Vanderbilt U., 1971. Affiliate mem. dept. biochemistry, dir. research, dir. Clin. Research Ctr. Lab., dir. Steroid Receptor Lab. U. Oreg., Portland, 1971-76, asst. prof. dept. surgery, 1971-74, assoc. prof., 1974-76; sr. research scientist Cedars-Sinai Med. Ctr., Los Angeles, 1976-78; assoc. oncologist Med. Sch. UCLA, 1976-78; assoc. prof. dept. ob-gyn U. Minn., Mpls., 1978—; ad hoc cons. NIH, Nat. Cancer Inst., NSF, 1974—; lectr. in field. Reviewer-referee various profl. publs.; editor: Hormonal Regulation of Experimental Breast Tumors, 1981; contbr. articles to sci. jours. Chmn. Dad's Club, Bridlemile Sch., Portland, Oreg., 1975-76; pres. bd. dirs. Ch. in Mpls., 1978-83. NIH fellow, 1966-68, 69-71; Ford Found. fellow, 1969-71; numerous research grants. Mem. Am. Soc. Biol. Chemists, Endocrine Soc., Soc. Gynecologic Investigation, AAAS, Minn. Ob-Gyn Soc. Subspecialties: Cell and tissue culture; Receptors. Current work: Mechanism of steroid hormone action; role of hormones in mammary and gynecologic tumors. Home: 6076 Olinger Blvd Edina MN 55436 Office: University Minn Dept Ob-Gyn Box 395 Mayo 420 Delaware St Minneapolis MN 55455

LEUNG, CHRISTOPHER CHUNG-KIT, anatomist, educator, researcher; b. Hong Kong, Jan. 3, 1939; U.S., 1960, naturalized, 1975; s. Nai Kuen and Sau Wah (Chan) L.; m. Stella Tang, May 11, 1970; children: Jacquelyn, Therese. Ph.D., Jefferson Med. Coll., 1969. Instr. Jefferson Med. Coll., 1969-72, asst. prof., 1972-74, U. Kans. Sch. Medicine, 1974-79; assoc. prof. anatomy La. State U. Sch. Medicine, 1979—. Contbr. numerous articles to profl. jours. NIH grantee, 1978—. Mem. Am. Assn. Anatomists, Soc. Developmental Biology, Am. Soc. Cell Biology, Am. Assn. Immunologists, Soc. Reproductive Biology. Democrat. Roman Catholic. Subspecialties: Anatomy and embryology; Immunology (medicine). Current work: Embryology, teratology, immunology, pathology. Office: La State U Sch Medicine Shreveport LA 71130

LEUNG, CHUN MING, astrophysicist, educator; b. Hong Kong, June 5, 1946; U.S., 1966; s. Yeuk F. and Sing W. (Chiu) L.; m. Anna C.W. Shing, Dec. 22, 1973; 1 son, Jonathan J.F. B.S., Western Mich. U., 1969; Ph.D. in Astronomy, U. Calif., Berkeley, 1975. Research assoc. Nat. Radio Astronomy Obs., Charlottesville, Va., 1974-76, asst. scientist, 1976-78; asst. prof. physics Rensselaer Poly. Inst., Troy, N.Y., 1978—; vis. staff mem. Los Alamos Nat. Lab. Contbr. articles to profl. jours. Recipient Dudley award Dudley Obs., Schenectady, 1982. Mem. Internat. Astron. Union, Am. Astron. Soc., Sigma Xi. Subspecialties: Theoretical astrophysics; Radio and microwave astronomy. Current work: Theoretical astrophysics; interstellar matter; star formation; radiative transfer; numerical fluid dynamics; interstellar chemistry; numerical methods. Office: Dept Physics Rensselaer Poly Inst Troy NY 12181

LEUNG, WING HAI, chemistry educator, researcher; b. Hong Kong, July 29, 1937; U.S., 1969; s. Ju-Dug and Ping-Fan (Yu) L.; m. Lai-Yin Kwan, Aug. 15, 1965; children: Kar-Woo, Kar-Hong, Kar-Peck. B.Sc., U. Hong Kong, 1963; M.S., U. Miami, 1970, Ph.D., 1974. Research assoc. SUNY-Buffalo, 1974-76; sr. chemist GAF Corp., Binghamton, N.Y., 1976-77; scientist Clinton (Iowa) Corn Corp., 1977-78; asst. prof. chemistry Hampton (Va.) Inst., 1978-82, assoc. prof., 1982—; mem. U.S. Congl. Adv. Bd., Washington, 1982-83. Contbr. articles to profl. jours. NASA grantee, 1981. Mem. Am. Chem. Soc., Am. Geophys. Union, N.Y. Acad. Sci., Sigma Xi. Subspecialties: Surface chemistry; Physical chemistry. Current work: Surface phenomena, kinetic studies of crystal growth and the interaction of solid/solution interfaces such as iron oxide and other suspended particles. Office: Department of Chemistry Hampton Institute Hampton VA 23668

LEUZE, MICHAEL REX, computer science educator and researcher; b. Oak Ridge, June 18, 1951; s. Rex Ernest and Ruth Imogene (Morris) L. B.S., U. Tenn.-Knoxville, 1973, M.S., 1975; Ph.D., Duke U., 1981. Research asst. prof. Duke U., Durham, N.C., 1981-82; asst. prof. Vanderbilt U., Nashville, 1982—. Mem. Assn. Computing Machinery, Soc. Indsl. and Applied Math., Sigma Xi. Baptist. Subspecialties: Numerical analysis; Computer architecture. Current work: Numerical methods, parallel algorithms and architectures. Office: Vanderbilt U Computer Sci Dept Box 1679 Sta 8 Nashville TN 37235

LEUZE, REX ERNEST, chemical engineer; b. Sabetha, Kans., Mar. 7, 1922; s. Ernest Jacob and Madelyn (Newman) L.; m. Ruth Imogene Morris, June 19, 1948; children: Michael Rex, Robert Morris, Thomas Ernest. B.S., Kans. State U.- Manhattan, 1944; M.S., U. Tenn., 1956. Analytical chemist Monsanto Chem. Co., Ill., 1944-45; radiochem. analyst Clinton Lab., Oak Ridge (Tenn.) Nat. Lab., 1945-46, process devel. engr., 1946-63, asst. sect. head, 1963-76, expt. engr., sect. head, 1976-81, pilot plant sect. head, 1981—. Served with U.S. Army, 1945-46. Fellow Am. Inst. Chemists; mem. Am. Chem. Soc., Am. Nuclear Soc., Sigma Xi, Phi Kappa Phi, Phi Lambda Upsilon, Sigma Tau. Baptist. Subspecialties: Chemical engineering; Nuclear fission. Current work: Transuranium element process development and separations, nuclear fuel cycles, radiochemical processing, preparation of hydrous oxide sols and gels, solvent extraction, ion exchange. Patentee in field. Home: 517 W 5th Ave Lenoir City TN 37771 Office: Oak Ridge Nat Lab PO Box X Oak Ridge TN 37830

LEV, MAURICE, pathologist; b. St. Joseph, Mo., Nov. 13, 1908; s. Benjamin and Rose L.; m. Lesley Beswick, Sept., 1947; children: Benita J., Peter B. B.S., N.Y.U., 1930; M.D., Creighton U., 1934; M.A. in Philosophy, Northwestern U., 1966; H.H.D., DePaul U., 1981. Diplomate: Am. Bd. Pathology. Intern Michael Reese Hosp., Chgo., 1937-38, resident in pathology, 1938-40; instr. pathology U. Ill., Chgo., 1939-46, asst. prof., 1947-48, assoc. prof., 1948-51, lectr., 1963—; pathologist Chgo. State Hosp., 1940-42; asst. prof. Creighton U., 1946-47; assoc. prof. U. Miami, Coral Gables, Fla., 1951-56, prof., 1956-57; pathologist, dir. research labs. Mt. Sinai Hosp. Greater Miami, 1951-57; prof. pathology Northwestern U., 1957-77, prof. emeritus, 1977—; dir. Congenital Heart Disease Research and Tng. Ctr., Chgo. Heart Assn., Hektoen Inst. Med. Research, 1957-82, career investigator and educator, 1966—; professorial lectr. U. Chgo., 1959—; cons. cardiovascular pathology Children's Meml. Hosp., Chgo., 1957—; lectr. pathology Chgo. Med. Sch. U. Health Scis., 1970—; lectr. Loyola U., Maywood, Ill., 1971—; Disting. prof. pediatrics Rush Med. Coll., Chgo., 1974—, Disting. prof. dept. internal medicine, 1975—, Disting. prof. dept. pathology, 1977—; lectr. Cook County Grad. Sch. Medicine, Chgo., 1977—; dir. Clin. Lab. Deborah Heart & Lung Ctr., Browns Mills, N.J., 1982—. Contbr. articles to profl. jours. Served to lt. col. U.S. Army, 1942-46. Recipient Alumni Assn. Michael Reese Hosp. & Med. Ctr. Disting. Alumnus award, 1978; Creighton U. Alumni Achievement award, 1979; named to City of Chgo. Sr. Citizens Hall of Fame, 1980; Brennamen award, 1983. Fellow AMA, Am. Coll. Cardiology (v.p. 1963-64), Am. Soc. Clin. Pathologists, Coll. Am. Pathologists, N.Y. Acad. Sci., Am. Coll. Chest Physicians, Inst. Medicine Chgo., Am. Heart Assn. (award of merit 1977, research achievement award 1980); mem. Am. Soc. Pathologists and Bacteriologists, Am. Assn. Anatomists, Histochem. Soc., Gerontol. Soc., Chgo. Path. Soc. (v.p. 1966-67, pres. 1967-68), Ill. Path. Soc., Midwest Soc. Electron Microscopists, Nashville Cardiovascular Soc., Sigma Xi, Phi Delta Epsilon. Subspecialties: Health services research; Pathology (medicine). Current work: Congenital heart and conduction system. Office: Deborah Heart & Lung Center Trenton Rd Browns Mills NJ 08015

LEVANDER, ORVILLE ARVID, research chemist, consultant; b. Waukegan, Ill., Apr. 6, 1940; s. Oscar Arvid and Emelia (Sahlsten) L.; m. Ruth Novelli, Aug. 10, 1981. B.A., Cornell U., 1961; M.S., U. Wis.-Madison, 1963, Ph.D., 1965. Postdoctoral fellow Columbia U., N.Y.C., 1965-66; research assoc. Harvard U., Boston, 1966-67; research chemist FDA, Washington, 1967-69, U.S. Dept. Agr., Beltsville, Md., 1969—. Editor: Micronutrient Interactions, 1980. Cons. WHO, Geneva, 1979—, NRC, Washington, 1976—. William Evans vis. fellow U. Otago, 1982. Mem. Am. Inst. Nutrition, Am. Chem. Soc., AAAS. Subspecialty: Nutrition (medicine). Current work: Selenium, vitamin E, trace element nutrition, heavy metal toxicology, lead, mercury, cadmium, prostaglandins, drug-nutrient interactions. Office: US Dept Agr Human Nutrition Center Beltsville MD 20705

LEVANT, RONALD F., psychology educator and researcher; b. Los Angeles, Oct. 26, 1942; s. Harry G. and Wilma I. (Adler) L.; m. Joyce Gilbert, Dec. 9, 1962 (div. 1966); 1 dau., Caren Elizabeth; m. Carol A. Beaudoin, Aug. 4, 1981. A.B., U. Calif.-Berkeley, 1964, 1969; postgrad., U. Calif.-San Francisco, 1965-67; Ed.D. in Clin. Psychology and Pub. Practice, Harvard U., 1973. Asst. dir. Robert W. White Sch., Boston, 1972-73, dir., 1973-74; assoc. prof. dept. psychology Boston State Coll., 1974-75; assoc. in edn. Harvard U., 1974-75; asst. prof. counseling psychology Boston U., 1975—; treatment team leader, inpatient unit Human Resource Inst. Boston, 1974-75, cons. psychologist, organizational cons., 1974—; cons. evaluative research Roxbury (Mass.) Children's Services, Inc., 1978-80. Contbr. articles to profl. jours. USPHS fellow, 1965, 66; Mass. Dept. Mental Health grantee, 1970; Boston U. grantee, 1981. Mem. Am. Psychol. Assn., Eastern Psychol. Assn., New Eng. Psychol. Assn., Mass. Psychol. Assn. (bd. profl. affairs 1977-81, dir. 1980-82, legis. com. 1982—), Internat. Assn. Applied Psychology, Soc. Psychotherapy Research, Am. Orthopsychiat. Assn., Acad. Psychologists in Marital and Family Therapy (v.p. Am. Bd. Family Psychology 1980—), Am. Family Therapy Assn., Nat. Council Family Relations, Am. Personnel and Guidance Assn. (marriage and family counseling com. 1982—). Democrat. Jewish. Club: Greater Boston Track. Subspecialty: Counseling psychology. Current work: Family dynamics and therapy; preventive/developmental programs for families based on social skill training. Parenting and parent-child relationships (particular reference to father-daughter relationship); new roles for men in the family. Development of the marriage over the life cycle. Office: Boston U 605 Commonwealth Ave Boston MA 02215

LEVARY, REUVEN ROBERT, management sciences educator, researcher; b. Bucurest, Romania, Jan. 6, 1944; came to U.S., 1975, naturalized, 1983; s. Jacob and Carola (Fisher) L.; m. Martha, Merritt, Dec. 16, 1978; 1 dau., Sarah. B.Sc., Technion, Haifa, Israel, 1969, M.Sc., 1972; M.S., Case Western Res. U., 1976, Ph.D., 1978. Teaching asst. Technion, 1969-72; grad. asst. Case Western Res. U., 1975-77; asst. prof. mgmt. scis. St. Louis U., 1978-81, assoc. prof., 1981—; cons. in field. Contbr. articles to profl. jours. Case Western Res. U. travel grantee, 1977; St. Louis U. grantee, 1979. Mem. IEEE, Ops. Research Soc. Am., Inst. Mgmt. Scis., Omega Rho. Subspecialties: Operations research (engineering); Resource management. Current work: Applied optimization, computer simulation, production management, distribution systems, energy modeling. Office: Saint Louis U 3674 Lindell Blvd Saint Louis MO 63108

LEVEEN, HARRY (HENRY LEVEEN), surgery educator, biomedical engineer; b. Woodhaven, N.Y., Aug. 10, 1916; s. Edward Phillip and Anna LeV.; m. Jeanette L. Rubricius; children: Robert Frederick, Eric G. B.A., Princeton U., 1936; M.D., NYU, 1940; M.S., U. Chgo., 1947; Ph.D., Loyola U., Chgo., 1952. Intern Queens Gen. Hosp., 1940-42, resident, 1943-44, Meadowbrook Hosp., Hempstead, N.Y., 1942-43, U. Chgo., 1945-46; cons. Kings County Hosp., N.Y.C., 1956-79, Downstate Med. Center, 1956-79; chief surgery VA Hosp., Bklyn., 1956-79; prof. surgery SUNY, N.Y.C., 1960-79, Med. U. S.C., Charleston, 1979— Fellow ACS; mem. Internat. Cardiovascular Soc., Am. Surg. Assn., Soc. Vascular Surgery. Subspecialties: Cancer research (medicine); Chemotherapy. Current work: Radiofrequency thermotherapy for treatment of cancer. Inventor shunt for ascites. Home: 321 Confederate Circle Charleston SC 29425 Office: Med U SC 171 Ashley Ave Charleston SC 29425

LEVEEN, HENRY *See also* **LEVEEN, HARRY**

LEVENSON, MILTON, chemical engineer; b. St. Paul, Jan. 4, 1923; s. Harry and Fanny M. L.; m. Mary Beth Novick, Aug. 27, 1950; children: James L., Barbara G., Richard A., Scott D., Janet L. B.Ch.E., U. Minn., 1943. Jr. engr. Houdaille-Hershey Corp., Decatur, Ill., 1944; research engr. Oak Ridge Nat. Lab., 1944-48; with Argonne (Ill.) Nat. Lab., 1948-73, assoc. lab. dir., 1973; dir. nuclear power div. Electric Power Research Inst., Palo Alto, Calif., 1973-80; cons. to pres. Bechtel Power Corp., San Francisco, 1981—; lectr. in field. Contbr. articles to profl. jours., chpts. to books. Served with C.E. U.S. Army, 1944-46. Fellow Am. Inst. Chem. Engrs. (Robert E. Wilson award 1975), Am. Nuclear Soc. (pres. 1983-84); mem. AAAS, Nat. Acad. Engring. Subspecialty: Nuclear engineering. Patentee in field. Office: 50 Beale St San Francisco CA 94119

LEVENSPIEL, OCTAVE, chemical engineering educator; b. Shanghai, China, July 6, 1926; s. Abraham and Elizabeth (Greenhouse) L.; m. Mary Jo Smiley, July 13, 1952; children: Bekki, Barney, Morris. B.S., U. Calif.-Berkeley, 1947; M.S., Oreg. State U., 1949, Ph.D., 1952. Registered profl. engr., Oreg. Asst. prof. Oreg. State U., Corvallis, 1952-54, prof. chem. engring., 1969—; asst. prof. Bucknell U., Lewisburg, Pa., 1954-56, assoc. prof., 1956-58, Ill. Inst. Tech., Chgo., 1958-62, prof., 1962-68. Author: Chemical Reaction Engineering, 1962, rev. edit., 1972, (with D. Kunii) Fluidization Engineering, 1969, Chemical Reactor Omnibook, 1979. Mem. Am. Inst. Chem. Engrs. (Wilhelm award 1979), Am. Chem. Soc. Subspecialty: Chemical engineering. Current work: Design of chemical reactors, modelling of fluidized bed reactors and combustors, development of novel gas/solid and solid/solid contactors, magnetic valves, and distributor/downcomers.

LEVENTHAL, BRIGID GRAY, research physician; b. London, Eng., Aug. 31, 1935; came to U.S., 1940, naturalized, 1954; d. Hugh Joseph and Barbara Theordora (Church) Gray; m. Carl M. Leventhal, Feb. 4, 1962; children: George Leon, Sarah Elizabeth, Dinah Susan, James Gray. B.A. with highest honors, UCLA, 1955; M.D., Harvard U., 1960. Lic. Physician, Mass. Md., diplomate: Am. Bd. Pediatrics (Sub-cert. in pediatric Nematology/oncology). Intern Mass. Gen. Hosp., Boston, 1960-61, jr. asst. resident in pediatrics, 1961-62; sr. asst. resident in pediatrics Boston City Hosp., 1962-63; USPHS trainee St. Elizabeth's Hosp., Brighton, Mass., 1963-64; postdoctoral fellow leukemia service Nat. Cancer Inst., Bethesda, Md., 1964-65; sr. investigator, 1965-73, head chemoimmunotherapy sect. pediatric oncology br., 1973-76; dir. div. pediatric oncology Johns Hopkins U., Balt., 1976—, assoc. prof. oncology, 1976—, assoc. prof. pediatrics, 1976—; mem. oncology and pediatric staffs Johns Hopkins Hosp., 1976—; mem. bd. sci. counselors div. cancer treatment Nat. Cancer Inst.; sci. adv. bd. St. Jude Hosp. Judge Westinghouse Sci. Talent Search. Editorial bd.: Jour. Nat. Cancer Inst, 1974-76, Leukemia Research, 1977—, Jour. Biol. Response Modifiers, 1981—; contbr. numerous articles and abstracts to profl. jours., chpts. in books. Recipient Fed. Woman's award, 1974; Outstanding Career Woman award Nat. Council Women, 1979; Edward A. Dickson Alumnus of Yr. Achievement award UCLA, 1982. Mem. Am. Fedn. Clin. Research, Am. Soc. Hematology, Am. Assn. Cancer Research, Am. Soc. Clin. Oncology (sec.-treas. 1976-79, dir. 1980-83), Transplantation Soc., AAAS, Internat. Soc. Exptl. Hematology, Soc. Pediatric Research, Am. Soc. Clin. Investigation, Am. Pediatric Soc., Phi Beta Kappa, Alpha Omega Alpha. Democrat. Jewish. Club: Variety (Balt.). Subspecialties: Cancer research (medicine); Pediatrics. Current work: Clinical research in pediatric cancer patients with emphasis on interaction of drugs and immune system. Home: 9254 Old Annapolis Rd Columbia MD 21045 Office: Oncology Ctr 3 12 Johns Hopkins Hospital Baltimore MD 21205

LEVENTHAL, GERALD SEYMOUR, psychologist; b. Bklyn., Apr. 29, 1936; s. Elias Leventhal and Mildred (Turetsky) Leventhal R.; m. Miriam Levin, July 9, 1961 (div. 1980); children: Dulcie, Jeffrey. B.A., Queens Coll., 1957; Ph.D., Duke U., 1962. NIMH postdoctoral fellow Columbia U., N.Y.C., 1963-64; acting asst. prof. psychology UCLA, 1963-64; from asst. to assoc. prof. psychology N.C. State U., Raleigh, 1964-70; prof. psychology Wayne State U., Detroit, 1970-79; NIMH fellow in psychology Yale U. Sch. Medicine, New Haven, 1979-80, Mental Health Services fellow, 1980-81; project dir. Community Mental Health Center, St. Claire's Hosp., Denville, N.J., 1981—; cons. quality assurance and info. systems Greystone Park State Psychiat. Hosp., Morris Plains, N.J., 1981—, Mental Health Hosps. div. N.J. Dept. Human Services, Trenton, N.J., 1981—. Contbr. chpts. to books. NSF grantee, 1965-70, 71-79. Fellow Am. Psychol. Assn.; mem. Soc. Exptl. Social Psychology (mem. dissertation prize com. 1972-76, chmn. 1974-75), Eastern Psychol. Assn. Jewish. Subspecialties: Social psychology; Information systems (information science). Current work: I direct project to introduce computer-assisted information systems throughout the New Jersey State Psychiatric Hospital system; and in key community mental health centers that serve the same population of patients; while directing the organization, development, implementation and training of phases, I am also gathering evaluation data and doing research. Home: 10 Roosevelt Pl Montclair NJ 07042 Office: Community Mental Health Center Research and Evaluation Dept St Clare's Hosp Morris Ave Denville NJ 07834

LEVENTHAL, MARVIN, research physicist; b. N.Y.C., Dec. 4, 1937; s. Jerome and Helen (Treppel) L.; m. Alice Judith, Apr. 16, 1961; children: Liza, and Tama (twins). B.S. in Physics, CCNY, 1958; Ph.D. in Physics (Owens-Ill. fellow, 1961-64), Brown U., 1964. Research asso. Yale U., New Haven, 1964-66, asst. prof. physics, 1966-68; mem. tech. staff Bell Labs., Murray Hill, N.J., 1968—. Contbr. articles to profl. publs. Mem. Am. Phys. Soc., Am. Astron. Soc., Sigma Xi. Jewish. Subspecialties: Gamma ray high energy astrophysics; Atomic and molecular physics. Current work: Gamma-ray line astronomy, positronium physics; balloon borne gamma-ray line astronomy with germanium detectors. Home: 28 Sunset Dr Summit NJ 07901 Office: Room 1E-349 Bell Labs Murray Hill NJ 07974

LEVENTHAL, STEPHEN HENRY, rsearch mathematician, reservoir engineer; b. N.Y.C., Apr. 2, 1949; s. Louis and Sylvia (Miller) L.; m. Ellen Sue Warach, Aug. 29, 1971; children: Daniel Scott, Seth Andrew. B.A. in Math, Rutgers U., 1969, M.A., U. Md., 1971, Ph.D., 1973. Mathematician Naval Surface Weapons Center, Silver Spring, Md., 1973-77; research mathematician Gulf Research and Devel. Co., Pitts., 1977-79, sr. research mathematician, 1979-81, supr. math. research, 1981-83, dir. reservoir simulation, 1983—. Contbr. articles to profl. jours. Organizer New Leadership of Israel Bonds, Pitts., 1982. NDEA fellow U. Md., 1969-72; NRC Postdoctoral fellow, 1973-74. Mem. Soc. Indsl. and Applied Math., Soc. Petroleum Engrs., Pi Mu Epsilon. Subspecialties: Numerical analysis; Petroleum engineering. Current work: Research into new and improved numerical methods for partial differential equations, with principle field of application being petroleum engineering. Home: 7236 Del Rey Houston TX 77071 Office: Gulf Research and Devel Co PO Box 37048 Houston TX 77236

LEVER, ALFRED BEVERLEY PHILIP, chemist, educator; b. London, Feb. 21, 1936; s. Reginald Walter and Rose Anne (Verber) L.; m. Bernice Roth, July 19, 1963. B.S., Imperial Coll. Sci. and Tech., 1957, Ph.D., 1960. Assoc. prof. dept. chemistry York U., Downsview, Ont., Can., 1967-72, prof. chemistry, 1972—; grad. program dir., 1968-76; acting chmn. Chem. Inst. Can. Inorganic div., 1971-72; vis. research assoc. Dell'Universita di Firenze, Italy, 1973, U. Sydney, Australia, 1972; cons. Scintrex Ltd., Can. Contbr. numerous articles to profl. publs. Japan Soc. Promotion Sci. fellow, 1984. Fellow Am. Electrochem. Soc.; mem. Chem. Soc. London, Am. Chem. Soc., Chem. Inst. Can. (Alcan lectr. award 1981). Subspecialties: Biophysical chemistry; Photochemistry. Current work: Chemistry; biophysical chemistry; photochemistry, laser, synthetic chemistry. Home: 79 Denham Dr Thornhill Ont Canada Office: York University 4700 Keele St Downsview Ont Canada M3J 1P3

LEVERETT, DENNIS H., dental educator, researcher; b. Cleve., June 22, 1931; s. Everett T. and Mary (Snow) L.; m. Joyce E. Hazard; children: Timothy, Amanda; m.; children by previous marriage: David, Lise, Terese, Stephen, Christopher. D.D.S., Ohio State U., 1956; M.P.H., Harvard U., 1968. Diplomate: Am. Bd. Dental Pub. Health. Dental officer USPHS, 1956-60; pvt. practice dentistry, New Orleans, 1960-66; pub. health dentist N.Mex. Dept. Health, Santa Fe, 1966-67; research fellow Harvard U. Sch. Dental Medicine, Boston, 1967-69; exec. dir. Center for Community Dental Health, Portland, Maine, 1969-73; dept. chmn. Eastman Dental Center, Rochester, N.Y., 1973—; mem. profl. adv. council Portland Health Dept., 1969-75; bd. dirs. Westside Health Dept., Rochester, 1975-79; cons. N.Y. State Dept. Health, 1980—. Mem. editorial adv. bd.: Jour. Dental Research, 1977—; contbr. numerous articles to profl. publs., chpts. to textbrooks. Nat. Inst. Dental Research grantee, also pvt. industry grants. Mem. Am. Pub. Health Assn. (sec. dental health sect. 1979-82), Am. Assn. Dental Research (councilor 1981—), Internat. Assn. Dental Research, European Orgn. Caries Research A (sr.). Subspecialties: Cariology; Preventive dentistry. Current work: Epidemiology, clinical trials of caries preventive agents, postdoctoral training in research methodology. Home: 51 Bellevue Dr Rochester NY 14620 Office: Eastman Dental Center 625 Elmwood Ave Rochester NY 14620

LEVICH, BENJAMIN GREGORY, physicist; b. Charkov, USSR, Mar. 30, 1917; s. Gregory and Evgeny (Atlasnez) L.; m. Tanya Rubinstein, Nov. 4, 1943; children—Alexander, Evgeny. M.Physics, Kharkov U., USSR, 1937; Ph.D., Moscow Pedagogical Inst., 1940; D.Sc. (hon.), Hebrew U. Jerusalem, Boston U., Carnegie-Mellon U. Sr. researcher, head theoretical dept. Inst. Electrochemistry, USSR Acad. Scis., 1940-58; prof. phys. chemistry Kazan U., 1943-54; prof. theoretical physics, head dept. Moscow Inst. Physics and Engring., 1954-65; head dept. chem. mechanics Moscow U., 1964-72; prof. Tel-Aviv U., 1978; Albert Einstein prof. sci. City U. N.Y., 1979—; dep. of sec. gen. Com. on Atomic Energy, USSR, 1946-50. Editorial bd.: Energy Conversion, Physico-Chem. Hydrodynamics; Author: Physico-Chemical Hydrodynamics, 1952, English edit., 1962, Statistical Physics, 1949, 53, Theoretical Physics. An Advanced Text, 2 vols, 1962, rev. edit., 1971, English rev. edit., 1972-73, Spanish rev. edit., 1975-78, also articles. Recipient Mendeleev prize in Chemistry, 1960; Palladium medal Am. Electrochem. Soc., 1973; medal Brit. Chem. Soc., 1981. Hon. mem. Imperial Coll., London.; Corr. mem. Acad. Scis. USSR (expelled 1979); mem. Norwegian Acad. Scis. and Letters (fgn.), N.Y. Acad. Scis. (hon.). Office: City College Inst Applied Chem Physics New York NY 10031

LEVIN, ALAN EDWARD, research engineer; b. Balt., May 17, 1953; s. Armand Herbert and Shula Hannas (Ziff) L. S.B.M.E., MIT, 1975, Sc.D., 1980. Registered profl. engr., Tenn. Devel. staff mem. Oak Ridge Nat. Lab., 1980—. Mem. Am. Nuclear Soc. (mem. program com. thermal-hydraulics div. 1981—), Tenn. Soc. Profl. Engrs., Nat. Soc. Profl. Engrs., Am. Inst. Chem. Engrs. (mem. nuclear heat transfer com. 1981—), Wine Soc. East Tenn. (dir. 1981—), Les Amis du Vin, Sigma Xi, Pi Tau Sigma, Tau Beta Pi. Subspecialties: Nuclear fission; Heat transfer and fluid dynamics. Current work: Liquid metal fast breeder reactor safety, sodium boiling behavior, two-phase flow medeling, flow instabilities in boiling systems statisical data analysis, experimental design and planning. Office: Oak Ridge Nat Lab PO Box Y Bldg 9108 Ms-2 Oak Ridge TN 37830

LEVIN, BARBARA CHERNOV, government research executive, toxicologist; b. Providence, May 5, 1939; d. Edward and Lillian (Mills) Chernov; m. Ira William Levin, June 18, 1961; children: David Michael, Jordan James. A.B., Brown U., 1961; postgrad., U. Wash., 1961-62; Ph.D., Georgetown U., 1973. Research asst. dept. endocrinology Johns Hopkins U. Sch. Medicine, 1962-63; teaching asst. dept. biology Georgetown U., 1968-73; postdoctoral fellow Lab. Molecular Biology, Nat. Inst. Neurol. and Communication Disorders and Stroke, NIH, Bethesda, Md., 1973-75, staff fellow, 1975-78; research biologist Ctr. for Fire Research, Nat. Bur. Standards, Washington, 1978-82, head fire toxicology research, 1982—; del. U.S. tech. adv. group on toxic hazards in fire Internat. Standards Orgn. Author Nat. Bur. Standards publs.; contbr. articles to profl. jours. Recipient outstanding performance award Nat. Bur. Standards, 1979, 82, sustained superior performance award, 1982. Mem. Soc. Toxicology, Am. Coll. Toxicology, Am. Chem. Soc., ASTM, Am. Soc. Microbiology, Sigma Xi. Subspecialties: Toxicology (medicine); Genetics and genetic engineering (biology). Current work: Toxicology of combustion products; combustion; toxicity; bioassay; carbon monoxide; hydrogen cyanide; test methods; inhalation; major fires. Office: Center for Fire Research Nat Bur Standards Bldg 224 Room A363 Washington DC 20234

LEVIN, GEORGE BENJAMIN, waste management program executive; b. Chgo., July 10, 1941; s. William I. and Lillian Gerber L.; m. Imogene Vanderpool, Nov. 30, 1974; 1 son, William Ralph. B.S., U. Mich.-Ann Arbor, 1963; Ph.D., Calif. Inst. Tech., 1974. Nuclear engr. Stone & Webster Engring. Corp., Boston, 1974-77; program mgr. United Nuclear Industries, Richland, Wash., 1977-79, EG & G Idaho, Idaho Falls, 1979—. Served to lt. USN, 1963-69. Mem. Am. Nuclear Soc., Canadian Nuclear Soc. Subspecialties: Nuclear engineering; Theoretical chemistry. Current work: The Safe and efficient management of radioactive wastes. Home: 2720 Ross St Idaho Falls ID 83401 Office: EG & G Idaho Inc PO Box 1625 Idaho Falls ID 83415

LEVIN, HARVEY STEVEN, neuropsychologist, educator; b. N.Y.C., Dec. 12, 1946; s. Nathan and Mary (Weinberg) L.; m. Ellen Margaret Halliczer, June 23, 1968; 1 son, Marc. B.A., CCNY, 1967; M.A., U. Iowa, 1971, Ph.D., 1972. Lic. psychologist, Tex. Postdoctoral research fellow U. Iowa Hosps., 1972-73; asst. prof. neurosurgery U. Tex. Med. Br., Galveston, 1974-79, assoc. prof., 1979-83, prof., 1983; cons. neuropsychologist U.S. Army Hosp., Landstuhl, W.Ger., 1981.

Author: Neurobehavioral Consequences of Closed Head Injury, 1982; cons. editor: Cortex, 1981—. NSF fellow, 1970. Fellow Am. Psychol. Assn.; mem. Soc. Neurosci., Internat. Neuropsychol. Soc., Assn. Nervous and Mental Disease, Acad. Neurology, N.Y. Acad. Sci., Nat. Head Injury Found. (cons.). Subspecialties: Neuropsychology; Neuropharmacology. Current work: Recovery from traumatic brain injury; neuropsychology and pharmacology of memory disorders; neuropsychological correlates of aging and dementia. Home: 10 Quintana Dr Galveston TX 77551 Office: U Tex Med Br Div Neurosurgery E17 Galveston TX 77550

LEVIN, JOSHUA ZEV, computer programmer; b. Cambridge, Mass., Feb. 5, 1949; s. Herschel Levin and Betty Louise (Tennenbaum) Zimmerman; m. Susan Evelyn Goldsmith, Feb. 20, 1982. B.A. in Physics, Queens Coll., 1971; M.S.E.E., NYU, 1974; Ph.D., Rensselaer Poly. Inst., 1980. Sr. engr. Boeing Computer Systems, Seattle, 1979-80; sr. software engr. Gen. Dynamics, E.D.S.C., Groton, Conn., 1980-81; sr. applied mathematician Aydin Controls, Ft. Washington, Pa., 1981—; cons. Mfg. Data Systems, Inc., Gen. Electric. Contbr. articles to profl. jours. Mem. Assn. for Computing machinery, IEEE, Phi Beta Kappa. Democrat. Jewish. Subspecialties: Graphics, image processing, and pattern recognition; Mathematical software. Current work: Devel. of algorithms in algebraic geometry for application to computer graphics and computer-aided design. Home: Bldg 2 Townhouse 2 English Village Apts North Wales PA 1945 Office: 401 Commerce Dr Fort Washington PA 19034

LEVIN, MICHAEL HOWARD, environmental researcher; b. N.Y.C., Sept. 25, 1936; s. Irving and Bess (Ruderman) L.; m.; 1 dau., Eleanor Marie. B.S., U. Vt., 1958; M.S., Rutgers U., 1960, Ph.D., 1964. Cert. sr. ecologist, 1981. Research assoc. N.Y. Bot. Garden, 1964; asst. prof. U. Notre Dame, 1964-66, U. Man., 1966-68, U. Pa., 1968-73; mem. Delaware River Basin Commn., 1977-78; pres., dir. research Environ. Research Assocs., Inc., Villanova, Pa., 1970—; dir. Ambric Environ. Sci., Inc., 1980—; adj. prof. Del. State Coll., 1980. Contbr. articles to profl. jours. Served to 1st lt. Med. Service Corps U.S. Army, 1960-61. Fellow AAAS; mem. Ecol. Soc. Am., Am. Soc. Plant Taxonomy, Sigma Xi, Alpha Zeta. Subspecialties: Ecology; Systematics. Current work: Research director of environmental sciences involved in testing of air, land, water and specialty in wood technology and materials testing. Address: 490 Darby-Paoli Rd Villanova PA 19085

LEVIN, ROBERT DAVID, oncologist, hematologist; b. Memphis, Dec. 10, 1943; s. Jack and Sarah (Stagman) L.; m. Pamela Joy Gilford, June 11, 1976; children: Nickolai, Douglas. B.S. in Chemistry, Calif. Inst. Tech., 1965; M.D., U. Chgo., 1969. Diplomate: Am. Bd. Internal Medicine. Intern Gen. Rose Hosp., Denver, 1969-70; resident in medicine Northwestern U. Hosp., Chgo., 1972-74, fellow in hematology and oncology, 1974-77; asst. chief Div. Hematology and Oncology Mt. Sinai Hosp., Chgo., 1977—; mem. Physicians Edn. program Am. Cancer Soc., Chgo. Contbr. articles to sci. publs. Served to capt. USAF, 1970-72. Mem. AMA, Am. Fedn. Clin. Research, Am. Soc. Hematology, Eastern Oncology Group. Jewish. Subspecialties: Oncology; Cancer research (medicine). Current work: Studying basic approaches to immunotherapy of lung and bladder cancer; some transfer factor work; coagulation abnormalities of diabetes mellitus. Office: Mt Sinai Hosp 15th and California Chicago IL 60608

LEVIN, ROBERT JOHN, physician, educator; b. N.Y.C., Dec. 29, 1934; s. Benjamin Bernard and Ruth Florence (Schwartz) L.; m.; children: John Graham, Elizabeth Hurt. Student, Duke U., 1951-54; M.D. with distinction, George Washington U., 1958. Med. house officer Peter Bent Brigham Hosp., Boston, 1958-59, asst. resident in medicine, 1959-60; clin. assoc. Nat. Heart Inst., Bethesda, Md., 1960-62, investigator, 1963-64; chief med. resident VA Hosp., West Haven, Conn., 1962-63, clin. investigator, 1964-66, attending physician, 1966—; instr. medicine Yale U., 1962-63, instr. medicine and pharmacology, 1964-65, asst. prof. medicine and pharmacology, 1965-68, chief sect. clin. pharmacology, 1966-74, assoc. prof., 1968-73, prof. medicine, lectr. pharmacology, 1973—, dir. physician's assoc. program, 1973-75; clin. asst. medicine Yale-New Haven Med. Ctr., 1964-65, asst. attending physician, 1965-68, attending physician, 1968—; mem. Conn. Adv. Com. on Foods and Drugs, 1967—, sec., 1969-71, chmn., 1971-73; mem. com. myocardial infarction Nat. Heart and Lung Inst., 1969-72, mem. lipid metabolism adv. com., 1977-79; mem. blue ribbon panel on behavior modification drugs for sch. age children HEW, 1971; mem. div. med. scis. NRC, 1971-74; mem. rev. com. drug abuse research ctr. NIMH, 1972; cons. Nat. Commn. for Protection of Human Subjects of Biomed. and Behavioral Research, 1974-78; chmn. award com. Nellie Westerman Prize for Research in Ethics, 1975-79. Author: Ethics and Regulation of Clinical Research, 1981; hon. advisor editorial bd.: Biochem. Pharmacology, 1968—; assoc. editor, 1969-74; editor: Clin. Research, 1971-76; mem. editorial bd.: Internat. Ency. Pharmacology and Therapeutics, 1974—, Forum on Medicine, 1977-80; editor: IRB: A Rev. of Human Subjects Research, 1978—; mem. editorial adv. bd.: Jour. Family Practice, 1975—; contr. articles in field to profl. jours. Bd. dirs. Medicine in the Pub. Interest, Inc., 1976—, Pub. Responsibility in Medicine and Research, 1981—; vice chmn. Commn. on Fed. Drug Approval Process, 1981-82. Served with USPHS, 1960-64. Multiple Research grantee. Fellow ACP, Am. Coll. Cardiology, Hastings Center Inst. of Soc. Ethics and Life Scis.; mem. Am. Soc. Clin. Investigation, Am. Soc. Clin. Pharmacology and Therapeutics (dir.), Am. Fedn. Clin. Research (nat. council 1967-76, exec. com. 1971-76), Am. Soc. Pharmacology and Exptl. Therapeutics (exec. com. 1974-77), AAAS, Histamine Club, Sigma Xi, Alpha Omega Alpha. Subspecialties: Internal medicine; Pharmacology. Current work: Ethics of medicine and research. Office: Sch Medicine Yale U 333 Cedar St New Haven CT 06510

LEVIN, STEPHEN MICHAEL, orthopedic surgeon, researcher; b. Toronto, Ont., Can., July 10, 1932; s. Benjamin and Anabelle (Glasman) L.; m. Patricia R. Rollo, May 27, 1976; children: Leslie, Mindy, Jody, Erin. B.S., CCNY, 1954; M.D., SUNY-Bklyn., 1958; postgrad., U. Pa., 1965-66. Diplomate: Am. Bd. Orthopedic Surgery. Intern Robert Packer Hosp., Sayre, Pa., 1958-59; resident in orthopedic surgery Nat. Orthopedic Hosp., Arlington, Va., 1963-67; pvt. practice medicine, specializing in orthopedic surgery, Alexandria, Va., 1967—, researcher in biomechanics, 1976—; instr. Howard U., Washington; sec. N.Am. Acad. Manipulative Medicine. Contbr. articles to profl. jours. Served to capt. U.S. Army, 1959-62. Fellow ACS, Am. Coll. Orthopedic Surgeons; mem. Eastern Orthopedic Assn. Subspecialties: Orthopedics; Biomedical engineering. Current work: Back pain and biomechanics. Office: 5021 Seminary Rd Suite 125 Alexandria VA 22311

LEVIN, WILLIAM COHN, physician, university president; b. Waco, Tex., Mar. 2, 1917; s. Samuel P. and Jeanette (Cohn) L.; m. Edna Seinsheimer, June 23, 1941; children: Gerry Lee Levin Hornstein, Carol Lynn. B.A., U. Tex., 1938, M.D., 1941; M.D. (hon.), U. Montpellier, 1980. Diplomate: Am. Bd. Internal Medicine. Intern Michael Reese Hosp., Chgo., 1941-42; resident John Sealy Hosp., Galveston, Tex., 1942-44; mem. staff U. Tex. Med. Br. Hosps., Galveston, 1944—, asso. prof. internal medicine, 1948-65, prof., 1965—; now Warmoth prof. hematology; pres. U. Tex. Med. Br., 1974—; past chmn., past mem. cancer clin. investigation rev. com. Nat. Cancer Inst. Exec. com., mem. nat. bd. Union Am. Hebrew Congregations; trustee Houston-Galveston Psychoanalytic Found., 1975—, Menil Found., 1976—. Recipient Nicholas and Katherine Leone award for adminstrv. excellence, 1977; decorated Palmes Académiques, France). Fellow A.C.P., Internat. Soc. Hematology; mem. Am. Fedn. Clin. Research, Central Soc. Clin. Research, Am. Soc. Hematology, Phi Beta Kappa, Sigma Xi, Alpha Omega Alpha. Subspecialties: Hematology; Oncology. Home: 1301 Harbor View Dr Galveston TX 77550 Office: 301 University Bldg Suite 646 Adminstrn Bldg Galveston TX 77550

LEVINE, GARY M., gastroenterologist; b. N.Y.C., Oct. 29, 1942; s. Jacob L. and Ruth (Rosenberg) L.; m. Dianne L. Sholinsky, July 2, 1966; children: Ari, David. B.A., Queens Coll., 1964; M.D., NYU, 1968. Intern Duke Hosp., Durham, N.C., 1968-69, resident, 1969-70; gastroenterological fellow U. Pa. Hosp., Phila., 1972-74, attending physician, 1974-82; chief gastroenterology VA Med. Ctr., Phila., 1978-82; head div. gastroenterology Albert Einstein Med. Ctr., Phila., 1982—. Contbr. articles to profl. jours. Served to maj. U.S. Army, 1970-72. Jonas Salk scholar, N.Y.C., 1964. Fellow ACP, Am. Coll. Gastroenterology; mem. Am. Gastroenterological Assn., Am. Fedn. Clin. Research, Am. Physiol. Soc., Phi Beta Kappa, Alpha Omega Alpha (prize NYU 1968). Democrat. Jewish. Subspecialties: Gastroenterology; Nutrition (biology). Current work: Intestinal adaptation; investigation into role of luminal nutrients and other factors in regulating intestinal structure and function; clinical nutrition; investigation of effects of enteral and parenteral nutrition on the gastrointestinal tract. Office: Albert Einstein Med Ctr York & Tabor Rds Philadelphia PA 19141

LEVINE, GILBERT, engineering educator; b. Teaneck, N.J., Apr. 12, 1927; s. Sidney and Edith (Friedl) L.; m. Ilma S. Levine, Aug. 30, 1950; children: Susan A., Ruth E. B.S., Cornell U., 1949, Ph.D., 1952. Asst. prof. agrl. engring. Cornell U., Ithaca, N.Y., 1952-57, assoc. prof., 1957-65, prof., 1965—, dir., 1974-76, also dir.; vis. assoc. prof. irrigation U. Calif.-Davis, 1957-58; sr. cons. Ministry of Pub. Works, Venezuela, 1976-77, Ministry of Environment, 1977—; cons. AID, Ford Found. Contbr. articles to profl. jours. Served with AUS, 1945-47. Mem. Am. Soc. Agrl. Engring., Am. Soc. Engring. Edn., Soil Conservation Soc. Am., ASCE, U.S. Com. Irrigation and Drainage. Subspecialties: Agricultural engineering; Integrated systems modelling and engineering. Current work: Improvement in the management of tropical irrigation systems research, technical assistance and consulting in relation to tropical irrigation system problems. Office: Cornell Univ 468 Hollister Hall Ithaca NY 14853

LEVINE, JERRY DAVID, nuclear engineer; b. Mount Vernon, N.Y., June 27, 1952; s. Stanley Irwin and Edith (Souberman) L.; m. Ronnie Elaine Freedman, July 31, 1977. B.S. in Physics and Earth and Space Sci, SUNY-Stony Brook, 1974; M.S. in Nuclear Engring, Poly. Inst. N.Y., 1976. Asst. engr. Ebasco Services, Inc., N.Y.C., 1976-77, assoc. engr., Princeton, N.J., 1977-78, engr., 1978-80; sr. engr. Envirosphere Co. div., 1980—. Mem. Am. Nuclear Soc. Subspecialties: Fusion; Nuclear engineering. Current work: Involved in nuclear safety studies for the Tokamak Fusion Test Reactor since 1976. Also involved in licensing of first mined geologic repository for high level nuclear wastes since 1982. Home: Princeton Arms North Apt 166 Cranbury NJ 08512 Office: Envirosphere Co div Ebasco Services Inc James Forrestal Campus Princeton NJ 08544

LEVINE, JOHN MYRON, psychology educator; b. Des Moines, Nov. 28, 1942; s. Ben W. and Bessie R. (Spiwak) L.; m. Jan Wortman, July 16, 1965; children: Jeffrey, Andrew. B.A., Northwestern U., 1965; M.S., U. Wis.-Madison, 1967, Ph.D., 1969. Asst. prof. U. Pitt., 1969-75, assoc. prof., 1975—. Editor: Teacher and Student Perceptions, 1983. NSF grantee, 1981. Mem. Am. Psychol. Assn., Soc. Exptl. Social Psychology, Am. Ednl. Research Assn. Subspecialty: Social psychology. Current work: Group Dynamics, socialization in groups, social comparison, social psychology of education. Home: 291 Orchard Dr Pittsburgh PA 15228 Office: U Pitts Pittsburgh PA 15260

LEVINE, MARVIN, psychology educator, researcher; b. Bklyn., Mar. 16, 1928; s. Louis and Dora (Chaifetz) L.; m. Matilda E. Cascio, July 14, 1951; children: Laurie, Todd. B.A., Columbia U., 1950; M.A., Harvard U., 1952; Ph.D., U. Wis.-Madison, 1959. Asst. prof. Ind. U.-Bloomington, 1959-65; prof. psychology SUNY-Stony Brook, 1965—. Author: A Cognitive Theory of Learning, 1976. NSF fellow Brussels, 1961-62; NIMH grantee, 1963-75; NSF grantee, 1978-82. Fellow Am. Psychol. Assn.; mem. Psychonomic Soc. Subspecialties: Cognition; Behavioral psychology. Current work: Theory of problem-solving, especially role of imagery and strategies. Home: 5 Hawks Nest Rd Stony Brook NY 11790 Office: Dept Psychology SUNY Stony Brook NY 11794

LEVINE, MELVIN MORDECAI, nuclear engineer, researcher; b. Richmond, Va., Nov. 20, 1925; s. Samuel J. and Leah (Want) L.; m. Lilo Guggenheim, Dec. 31, 1951; children: Susan Fox, Wendy Levine, David. B.S., MIT, Cambridge, Mass., 1946; Ph.D., U. Va., 1955. Instr. Pa. State U., Hazelton, 1946-48; physicist Babcock and Wilcox Co., Lynchburg, Va., 1955-59, Brookhaven Nat. Lab., Upton, N.Y., 1959—. Served as ensign U.S. Navy, 1944-46. Mem. Am. Nuclear Soc. Subspecialties: Nuclear engineering; Nuclear fission. Current work: Safety of nuclear power reactors, including theralhydraulics and reactor physics. Home: 16 Cove Lane Port Jefferson NY 11777 Office: Brookhaven Nat Lab Bldg 130 Upton NY 11973

LEVINE, MICHAEL STEVEN, brain researcher, educator; b. Bklyn, Sept. 22, 1944; s. David and Rose (Katz) L. B.A., Queens Coll., 1966; Ph.D., U. Rochester, 1970. Postdoctoral fellow UCLA, 1970-72, research assoc. dept. psychiatry, 1972-76, asst. prof., 1976-79, assoc. prof., 1979—; cons. Hereditary Disease Found., Los Angeles, 1975. Mem. Am. Psychol. Assn., Soc. Neurosci., Internat. Brain Research Orgn., Sigma Chi, Phi Beta Kappa, Psi Chi. Democrat. Jewish. Subspecialties: Neurophysiology; Psychobiology. Current work: Developmental research on basic neurobiological mechanisms of maturation and aging. Office: Dept Psychiatry UCLA Los Angeles CA 90024

LEVINE, RANDOLPH HERBERT, computer co. exec.; b. Denver, Nov. 20, 1946; s. Harold and Muriel Faye (Sachs) L.; m. Sarah Loewenberg, June 21, 1970; children: Seth Jason, Johanna Beth. A.B., U. Calif.-Berkeley, 1968; M.A., Harvard U., 1969, Ph.D., 1972. Vis. scientist Nat. Ctr. Atmospheric Research, Boulder, Colo., 1972-74; research assoc., lectr. astronomy Harvard Coll. Obs., Cambridge, Mass., 1974-81; sr. staff scientist, dir. computing AER, Inc., Cambridge, 1981-82; unit mgr. Ednl. Services, Digital Equipment Corp., Maynard, Mass., 1982—; cons. Univs. Space Research assn., Aerospace Corp., NASA; tchr. astronomy Harvard Coll., 1974-81. Mem. Assn. Computing Machinery, IEEE Computer Soc., Am. Phys. Soc., Am. Astron. Soc., Am. Geophys. Union, Internat. Astron. Union. Subspecialties: Computer Education; Solar physics. Current work: Research and development in computer based education; research in solar physics. Home: 50 Carver Rd Newton Highlands MA 02161 Office: 12 Crosby Dr Bedford MA 01730

LEVINE, SAMUEL HAROLD, physicist; b. Hazlehurst, Ga., Nov. 30, 1925; s. Abraham and Rebecca (Starr) L.; m. Trudy Foner, Aug. 14, 1955; children: Renee, Lisa, Suzanne. B.S., Va. Poly. Inst. and State U., 1947; M.S., U. Ill., 1948; Ph.D., U. Pitts., 1954. Instr. Va. Poly. Inst. and State U., 1949-50; mgr. Westinghouse Bettis Atomic Plant, Pitts., 1954-59; reactor physicist Gen. Atomic Co., San Diego, 1959-61; group head Rocketydne Co., Canoga Park, Calif., 1961-62; head nuclear sci. lab. Northrop Space Lab., Hawthorne, Calif., 1962-68; prof. physics Pa. State U., University Park, 1968—; dir. Breazeale Nuclear Reactor, 1968—. Served with USAAF, 1943-44. Recipient Invention award NASA, 1973; Westinghouse fellow, 1953; Lady Davis fellow Technion, Haifa, Israel, 1976. Mem. Am. Nuclear Soc., Am. Phys. Soc., Sigma Xi, Phi Kappa Phi. Subspecialties: Nuclear engineering; Nuclear fission. Current work: In-core fuel management; experimental reactor physics; neutron and beta radiation dosimetry. Home: 528 E Hamilton Ave State College PA 16801 Office: Breazeale Nuclear Pa State U University Park PA 16802

LEVINE, SEYMOUR, psychology in psychiatry educator, researcher; b. Bklyn., Jan. 23, 1925; s. Joseph and Rose (Raines) L.; m. Barbara Lou McWilliams, Feb. 19, 1949; children: Robert Thomson, Leslie Ingrid, Alicia Margaret. B.A., U. Denver, 1948; M.A., N.Y. U., 1950, Ph.D., 1952. Asst. prof. Boston U., 1952-53; asst. prof. Ohio State U., 1956-60; assoc. prof. psychiatry and behavioral Scis. Stanford U. Sch. Medicine, 1962-69, prof., 1969—, dir. Stanford Primate Facility, 1976—, dir. biol. scis. research tng. program, 1971—; cons. Found. Human Devel., Univ. Coll., Dublin, Ireland, 1973—. Editor: Hormones and Behavior, 1972, Psychobiology of Stress, 1978, Coping and Health, 1980. Served with U.S. Army, 1942-45; ETO. Recipient Hoffheimer Research award, 1961; Career Devel. award NIMH, 1962; Research Scientist award, 1967—. Fellow Am. Psychol. Assn., AAAS.; mem. Internat. Soc. Develop. Psychobiology (dir. 1969, pres. 1975-76), Internat. Soc. Psychoneuroendocrinology, Am. Soc. Primatologists, Internat. Primatological Soc. Subspecialties: Psychobiology; Neuroendocrinology. Current work: Hormones and behavior; developmental psychobiology; stress and health. Home: 927 Valdez Pl Stanford CA 94305 Office: Dept Psychiatry and Behavioral Scis Stanford U Sch Medicine Stanford CA 94305

LEVINSKAS, GEORGE JOSEPH, toxicologist; b. Tariffville, Conn., July 8, 1924; s. Joseph John and Frances Julia (Eurkunas) L.; m. Ruth Irene Hublitz, Dec. 28, 1946; children: Robert John, Nancy Jane Levinskas Armstrong, Edward Joseph. A.B., Wesleyan U., Middletown, Conn., 1949; Ph.D., U. Rochester, 1953. Diplomate: Am. Bd. Toxicology, Acad. Toxicological Scis. Research assoc. U. Rochester Atomic Energy Project, 1952; research assoc., lectr., asst. prof. applied toxicology Grad. Sch. Pub. Health, U. Pitts., 1953-58; research pharmacologist, chief indsl. toxicologist, dir. Environ. Health Lab. central med. dept. Am. Cyanamid Co., Stamford, Conn., also Princeton, N.J., 1958-71; dir. environ. assessment and toxicology, dept. medicine and environ. health Monsanto Co., St. Louis, 1971—. Served to 2d lt., inf. U.S. Army, 1943-46. NRC fellow, 1949-52; recipient Graham prize for excellence in natural sci. Wesleyan U., 1949. Mem. Am. Chem. Soc., Am. Indsl. Hygiene Assn., Soc. Toxicology (charter mem.), Environ. Mutagen Soc., Am. Soc. Pharmacology and Exptl. Therapeutics, N.Y. Acad. Scis., AAAS, Phi Beta Kappa, Sigma Xi. Subspecialties: Toxicology (medicine); Environmental toxicology. Current work: Pharmacology and toxicology of industrial and agricultural chemicals and food additives; chemistry of bone mineral. Office: 800 N Lindbergh Blvd Saint Louis MO 63167

LEVINSKY, NORMAN GEORGE, physician, educator; b. Boston, Apr. 27, 1929; s. Harry and Gertrude (Kipperman) L.; m. Elena Sartori, June 17, 1956; children—Harold, Andrew, Nancy. A.B. summa cum laude, Harvard U., 1950, M.D. cum laude, 1954. Diplomate: Am. Bd. Internal Medicine. Intern Beth Israel Hosp., Boston, 1954-55, resident, 1955-56; commd. med. officer USPHS, 1956; clin. asso. Nat. Heart Inst., Bethesda, Md., 1956-58; NIH fellow Boston U. Med. Center, 1958-60; practice medicine, specializing in internal medicine and nephrology, Boston, 1960—; chief of medicine Boston City Hosp., 1968-72; physician-in-chief, dir. Evans dept. clin. research Univ. Hosp., Boston, 1972—; asst. prof., asso. prof. medicine Boston U., 1960-68, Wesselhoeft prof., 1968-72; Wade prof. medicine, 1972—; Mem. drug efficacy panel NRC.; mem. nephrology test com.-Am. Bd. Internal Medicine, 1971-76. Editor: (with R.W. Wilkins) Medicine: Essentials of Clinical Practice, 1978, 2d edit., 1983; Contbr. chpts. to books, sci. articles to med. jours. Mem. AAAS, Am. Fedn. Clin. Research, Am. Soc. Clin. Investigation, Assn. Am. Physicians, Am. Physiol. Soc., Am. Soc. Nephrology, Phi Beta Kappa, Alpha Omega Alpha. Club: Interurban. Subspecialties: Nephrology; Physiology (biology). Current work: Renal KalliKrein systems; acute renal failure. Home: 20 Kenwood Ave Newton MA 02159 Office: 75 E Newton St Boston MA 02118

LEVINSON, MARK, mechanical engineering educator; b. Bklyn., June 12, 1929; s. Samuel Eleazar and Rose (Tartakow) L.; m. Suzanne Josephson, Dec. 27, 1953; children: Robert M., Madeline J. B.Aero. Engring., Poly. Inst. Bklyn., 1951, M.S., 1960; Ph.D., Calif. Inst. Tech., 1964. Registered profl. engr., Ont. Asst. prof. mech. engring. Oreg. State U., Corvallis, 1959-61; assoc. prof. mech. engring. Clarkson Coll., Potsdam, N.Y., 1964-66; assoc. prof. mechanics W.Va. U., Morgantown, 1966-67; assoc. prof. mech. engring. McMaster U., Hamilton, Ont., Can., 1967-71, prof. engring. mechanics, 1971-80; A. O. Willey prof. mech. engring. U. Maine, Orono, 1980—. Contbr. articles to profl. jours. Founding mem. Heritage Hamilton (Ont., Can.), v.p., 1973-78; bd. govs. Ont. Inst. Studies in Edn., Toronto, 1978-80; acad. mem. Council Ont. Univs., 1977-79. Served with U.S. Army, 1952-54. Fellow AAAS, AIAA; mem. ASME, Soc. Indsl. and Applied Math., Soc. History Tech. Subspecialties: Solid mechanics; Applied mathematics. Current work: Refined models for thick elastic plates and elastic foundations. Office: Dept Mech Engring U Maine Boardman Hall Orono ME 04469

LEVINTHAL, CHARLES FREDERICK, research psychologist, educator; b. Cin., July 6, 1945; s. Sam and Mildred Caroline (Greenburg) L.; m. Beth Ellen Kuby, Dec. 16, 1973; children: David, Brian. A.B., U. Cin., 1967; M.A. in Psychology, U. Mich, 1968, Ph.D., 1971. Asst. prof. psychology Hofstra U., 1971-78, assoc. prof., 1978—; dir. Ph.D. program in applied research and evaluation, 1978—. Author: The Physiological Approach in Psychology, 1979, Introduction to Physiological Psychology, 2d edit, 1983; contbr. articles to profl. jours. Served with USAR, 1968-74. Woodrow Wilson fellow, 1967; NSF Found. grad. fellow, 1967-71. Mem. Am. Psychol. Assn., Midwestern Psychol. Assn., Soc. Psychophysiol. Research, Soc. Neurosci., Phi Beta Kappa. Subspecialties: Neuropsychology; Physiological psychology. Current work: Research on physiological bases for cognitive activity in the brain, hemispheric specialization in brain, electrophysiology. Home: 9 Royal Oak Dr Huntington NY 11743 Office: Hofstra Hempstead NY 11550

LEVINTHAL, ELLIOTT CHARLES, physicist; b. Bklyn., Apr. 13, 1922; s. Fred and Rose (Raiben) L.; m. Rhoda Arons, June 4, 1944; children—David, Judith, Michael, Daniel. B.A., Columbia Coll., 1942; M.S., Mass. Inst. Tech., 1943; Ph.D., Stanford U., 1949. Project engr. Sperry Gyroscope Co., N.Y.C., 1943-46; research asso. nuclear physics Stanford U., 1946-48; research physicist Varian Assos., Palo Alto, Calif., 1949-50, research dir., bd. dirs., 1950-52; chief engr. Century Electronics, Palo Alto, 1952-53; pres. Levinthal Electronics, Palo Alto, 1953-61; sr. scientist, dir. Instrumentation Research Lab., dept. genetics Stanford U. Sch. Medicine, 1961-74, asso. dean for research affairs, 1970-73, adj. prof. genetics, 1974-80; dir. Instrumentation Research Lab., 1974-80; dir. def. scis. office Def. Advanced Projects Agy., Dept. Def., Arlington, Va., 1980—; mem. NASA Adv. Council; cons. HEW. Recipient NASA Public Service medal, 1977. Mem. AAAS, Am. Phys. Soc., IEEE, Optical Soc. Am., Biomed. Engring. Soc., Sigma Xi. Democrat. Jewish. Subspecialties: Biomedical engineering; Nuclear physics. Home: 700 New Hampshire Ave NW Washington DC 20037 Office: 1400 Wilson Blvd Arlington VA 22209

LEVINTHAL, MARK, geneticist, researcher, educator; b. Bklyn., Mar. 3, 1941; s. Louis and Bertha (Nissenbaum) L.; m. Maxine Kassirla, Dec. 22, 1962; children: Peter, Sarita. B.S., Bklyn. Coll., 1962; Ph.D., Brandeis U., 1966. Postdoctoral fellow Johns Hopkins U., Balt., 1966-68; staff fellow lab. molecular biology Nat. Inst. Arthritis, Metabolic and Digestive Diseases, NIH, Bethesda, Md., 1968-72; assoc. prof. biol. sci. Purdue U., West Lafayette, Ind., 1972—. Contbr. articles to profl. jours. Mem. steering com. New Moblzn. to End the War, 1969-71; mem. Vietnam moratorium com. NIH/NIMH, 1969-72. Recipient Biology prize Blkyn. Coll., 1958; NIH fellow, 1966, 66-68; grantee, 1972-74; NSF grantee, 1974-77. Mem. Am. Soc. Microbiology, AAAS, Genetics Soc. Am., AAUP, Italian Molecular Biology Soc., ACLU, Sigma Xi, Darwin Soc., Soc. for Study of Evolution. Mem. Peace and Freedom Party. Zen Buddhist. Club: Brooklyn (Lafayette). Subspecialties: Gene actions; Evolutionary biology. Current work: Experimental molecular evolution of prokaryotes. Genetic engineering; molecular evolution; prokaryotes; genetics evolution; metabolic evolution.

LEVITAN, ALEXANDER ALLEN, physician, educator; b. Boston, Oct. 19, 1939; s. Sacha and Gertrude L.; m. Lucy Kerr Albree, Jan. 14, 1967; children: Lara, Denise, Karen. A.B., Cornell U., 1959; M.D., U. Rochester, 1959-63; M.P.H., U. Minn., 1969. Diplomate: Am. Bd. Internal Medicine, also Sub-Bd. Med. Oncology, Am. Bd. Med. Hypnosis. Intern Vanderbilt U. Hosp., Nashville, 1963-64; resident in medicine Harvard U. med. service Boston City Hosp., 1964-65; clin. assoc. Nat. Cancer Inst., Bethesda, Md., 1965-67; med. fellow U. Minn., 1967-69; v.p. Mpls. Med. Specialists, Mpls., 1969—; chief medicine Unity Hosp., Fridley, Minn., 1975; vice chmn. Edn. and Research Found., Am. Soc. Clin. Hypnosis, Des Plaines, Ill., 1980—. Reviewer: Am. Jour. Clin. Hypnosis, 1983. Bd. dirs. Research and Edn. Found. Health Central Inc., Mpls., 1973-74, Long Lake Improvement Assn., New Brighton, Minn., 1980—; mktg. com. Unity Hosp., Fridley, Minn., 1981—. Served to lt. comdr. USPHS, 1965-67. Fellow ACP, Am. Soc. Clin. Hypnosis; mem. Am. Bd. Med. Hypnosis (dir. 1982—), Am. Soc. Clin. Oncology, Minn. Soc. Internal Medicine, Am. Fedn. Clin. Research. Republican. Presbyterian. Subspecialties: Internal medicine; Oncology. Current work: Clinical oncology chemotherapy protocol trials, applications of hypnosis to medical oncology. Home: 2051 Long Lake Rd New Brighton MN 55112 Office: Minneapolis Med Specialists 509 Osborn Rd NE Suite 115 Minneapolis MN 55432

LEVITAN, HERBERT, neuroscience educator, researcher; b. Bklyn., Apr. 25, 1939; s. Meyer and Lena (Kohl) L.; m. Karen Brounstein, Aug. 23, 1964; children: James, Danielle. Student, Bklyn. Coll., 1956-58; B.E.E., Cornell U., 1962, Ph.D., 1965. Sr. staff assoc. Lab. Neurobiology Nat. Inst. Child Health and Human Devel., Bethesda, Md., 1970-72; neurophysiologist Lab. Neuroscience, Nat. Inst. Aging, Balt., 1979-82; assoc. prof. dept. zoology U. Md., 1972-83, prof. dept. zoology, 1983—. Contbr. articles in field to profl. jours. NIH fellow, 1965-67, 68-70, 70. Mem. Soc. Neuroscience, Am. Physiology Soc., Am. Soc. Cell Biology, Soc. Gen. Physiologist, AAUP, Sigma Xi. Subspecialties: Neurobiology; Neuropharmacology. Current work: Biophysical mechanisms underlying the effects of drugs on excitable and inexcitable cells. Home: 212 Dale Dr Silver Spring MD 20910 Office: Department Zoology University Maryland College Park MD 20742

LEVITIN, LEV BEROVICH, engineering educator; b. Moscow, Sept. 25, 1935; U.S., 1981; s. Ber L. and Tzetzilia (Gushansky) L.; m. Yulia Shmukler, 1959 (div. 1970); 1 son, Boris. M.Sc., Moscow U., 1960; Ph.D., Acad. Scis. of USSR, 1969. Sr. research scientist Inst. Info. Transmission Problems, Moscow Acad. Scis., 1961-73; sr. lectr. Tel-Aviv U., 1974-80; vis. prof. Bielefeld U., W. Ger., 1980-81, Syracuse (N.Y.) U., 1981-82; prof. engring. Boston U., 1982—; vis. scientist Heinrich-Hertz Inst., Berlin, 1980, Institut für Optoelektronik, Oberpfaffenhofen, W.Ger., 1981; cons. Vishay Israel, Ltd., Tel-Aviv, 1979. Editor: Principles of Cybernetics (in Russian), 1967; contbr. articles sci. jours. Mem. Popov Sci. and Engring. Soc., IEEE, Israel Statis. Assn., AAUP, Am. Math. Soc., Assn. Computing Machinery, Soc. Indsl. and Applied Math., AAAS. Subspecialties: Applied mathematics; Computer engineering. Current work: Information theory and its applications, physical information theory, optical communication, quantum theory of measurements, physics of computation, coding theory, automata theory, statistical physics, computational complexity, VLSI testing. Office: Boston University College of Engineering 110 Cummington St Boston MA 02215

LEVITSKY, DAVID AARON, educator; b. Phila., Oct. 15, 1942; s. Isadore and Reba (Beruowitz) L.; m.; children: Steven, Sandy, Susan. B.A., Rutgers U., 1964, M.S., 1966, Ph.D., 1968. Asst. prof. psychology Cornell U., Ithaca, N.Y., 1970-75, assoc. prof., 1975—; cons. Allied Chem., N.Y.C., 1969-70; mem. study panel food and nutrition bd. Nat. Acad. Sci., 1979—. Author: Malnutrition, Learning, Environment, 1979. Pres. Ithaca Assn. Jewish Studies, 1980-82. Recipient New Leadership award Nutrition Found., 1970; Career Devel. award NIH, 1970-75; Best Paper award Internat. Soc. Developmental Psychobiology, 1975. Mem. Am. Psychol. Assn., Am. Inst. Nutrition, Eastern Psychol. Assn., Psychonomics Soc. Democrat. Jewish. Subspecialties: Psychobiology; Nutrition (medicine). Current work: Relationship between nutrition, biochemistry-physiology, brain function and behavior and energetic component in regulation of body weight. Home: 327 Winthrop Dr Apt 5 Ithaca NY 14850 Office: Cornell Univ Savage Hall Ithaca NY 14853

LEVITT, MORTON, pharmacologist; b. N.Y.C., Jan. 4, 1929; s. Abe S. and Nellie (Glass) L.; m. Renee Rosenberg, Jan. 5, 1952; children: Ilene, Steven, David. B.S., CCNY, 1951, Fordham U., 1957; M.S., George Washington U., 1959; Ph.D., Howard U., 1966. Research asst. George Washington U., Washington, 1957-59; pharmacologist Nat. Heart Inst., 1962-67; research biologist Sterling Winthrop, Rensselaer, N.Y., 1967-70; sr. research fellow dept. psychiatry Columbia U., N.Y.C., 1970—, N.Y. State Psychiat. Inst., 1970—. Contbr. articles on pharmacology to profl. jours. Served with USMC, 1951-53. Mem. Am. Soc. for Pharmacology and Exptl. Therapeutics, N.Y. Acad. Scis., Sigma Xi. Subspecialties: Molecular pharmacology; Neurochemistry. Current work: Genetics of mental illness, hypertension, advenergic mechanisms, catacholamines, biochemistry of behavior.

LEVITZ, MORTIMER, medical educator; b. N.Y.C., May 11, 1921; s. Hyman and Ida (Goldberg) L; m. Catherine Blum, June 1, 1947; children: Ellen Maud, Stuart Michael. B.S., CCNY, 1941; M.A., Columbia U., 1947, Ph.D., 1951. Research assoc. Columbia U., N.Y.C., 1951-52; asst. prof. NYU Med. Ctr., N.Y.C., 1952-56, assoc. prof., 1959-67, prof. obstetrics and gynecology, 1967—; mem. endocrine study sect. NIH, 1973-75, chmn. subcom. 3, 1983—. Contbr. articles to profl. jours.; editorial bd.: Jour. Clin. Endocrinology and Metabolism, 1962-74. Served with U.S. Army, 1944-46. Recipient NIH Research Career Devel. award, 1962-72; Nat. Cancer Inst. grantee, 1953—. Fellow N.Y. Acad. Sci.; mem. Am. Soc. Biol. Chemists, Endocrine Soc., Am. Chem. Soc., Soc. Gynecol. Investigation. Subspecialties: Biochemistry (biology); Endocrinology. Current work: Mechanism of steroid hormone action in uterus and breast and steroid metabolism in normal and pathological breast. Office: N Y U Sch Medicine 550 1st Ave New York NY 10016

LEVITZKY, MICHAEL GORDON, physiology educator; b. Elizabeth, N.J., Jan. 3, 1947; s. Edward and Shirley (Worfman) L.; m. Ellen M. DeRonde, June 27, 1969. B.A., U. Pa., 1969; Ph.D., Albany Med. Coll. of Union U., 1975. Instr. physiology Albany (N.Y.) Med. Coll., 1974-75; asst. prof. physiology La. State U. Med. Center, New Orleans, 1975-80, assoc. prof. physiology, 1980—. Author: Pulmonary Physiology, 1982. Research grantee NIH, 1978-81, 82—; Young Investigator Pulmonary Research grantee, 1976-78; George Bel grantee Am. Heart Assn., 1981-82. Mem. Am Physiol. Soc., Am. Thoracic Soc., N.Y. Acad. Scis., Soc. Exptl. Biology and Medicine, Sigma Xi. Subspecialties: Physiology (medicine); Pulmonary medicine. Current work: Cardiopulmonary physiology, hypoxic pulmonary vasoconstriction; research pulmonary circulation. Office: Dept Physiology La State Univ Med Center Sch Medicine 1901 Perdido St New Orleans LA 70112-1393

LEVNER, MARK HENRY, microbiologist; b. Milw., Jan. 1, 1941; s. Sidney Aaron and Faye (Bindler) L.; m. Abigail Straus, Nov. 16, 1969; children: Adam Harris, Ethan Straus; m.; 1 son by previous marriage, Geoffrey Michael. B.S. in Physics, U. Wis.-Madison, 1963, M.S., U. Ill.-Urbana, 1965; Ph.D. in Biophysics, U. Chgo., 1971. Postdoctoral assoc. Inst. Cancer Research, Fox Chase, Pa., 1971-73; research assoc. Haverford (Pa.) Coll., 1973; sr. microbiologist Wyeth Labs., Radnor, Pa., 1973-83, supr. molecular biology unit, 1983—. Contbr. articles to profl. publs.; patentee processes for enhancing prodn. of enterotoxin. NIH postdoctoral research fellow, 1971-72, 72-73. Subspecialties: Genetics and genetic engineering (biology); Microbiology (medicine). Current work: Genetic, biochemical studies of bacterial toxins; vaccine development. Patentee processes for enhancing prodn. of enterotoxin. Office: Wyeth Labs PO Box 8299 Philadelphia PA 19101

LEVY, DAVID EDWARD, neurologist, researcher, educator; b. Washington, May 10, 1941; s. Maurice W. and Cele (Blue) L.; m. Ellen K., Jan. 8, 1967. A.B. cum laude, Harvard U., 1963, M.D., 1968. Diplomate: Am. Bd. Inernal Medicine, Am. Bd. Psychiatry and Neurology. Intern in medicine N.Y. Hosp., N.Y.C., 1968-69, resident in medicine, 1969-70, resident in neurology, 1970-73; instr. in neurology Cornell U., N.Y.C., 1973-75, asst. prof. neurology, 1975-80, assoc. prof., 1980—; practice medicine specializing in neurology; mem. staff N.Y. Hosp., N.Y.C.; established investigator Am. Heart Assn., 1978-83; grant reviewer Nat. Inst. Neurol. and Communicative Diseases and Stroke. Contbr. numerous articles, abstracts to profl. jours.; editorial bd.: Stroke. Community rep. N.Y.C. Mayor's com. on subway constrn. Robert Wood Johnson Found. grantee, 1981-84. Mem. Am. Acad. Neurology, Am. Neurol. Assn., ACP, Soc. Neurosci., Stroke Council of Am. Heart Assn., Phi Beta Kappa. Subspecialties: Neurology; Database systems. Current work: Mechanisms and prevention of ischemic brain damage, establishment of computer database systems usable for determining prognosis from major illnesses. Office: Dept Neurology (A-569 Cornell U Med Col) 1300 York Ave New York NY 10021

LEVY, DEBORAH LOUISE, research scientist, neurobiologist, psychologist, educator; b. Mpls., Nov. 3, 1950; d. Walter Julius and Hilma Bernice (Cohn) L. B.A., U. Chgo., 1972, Ph.D., 1976. Lic. psychologist, Kans., Ill. Postdoctoral fellow Menninger Found., Topeka, 1977-79; asst. prof. U. Chgo., 1979—; research scientist Ill. State Psychiat. Inst., Chgo., 1979—. Recipient Karl Menninger Sci. Day prize Menninger Alumni Assn., 1979, research scientist devel. award NIMH, 1981; Scottish Rite research grantee, 1980. Mem. AAAS, N.Y. Acad. Scis., Am. Psychol. Assn. Jewish. Subspecialties: Neurology; Neuropharmacology. Current work: Indices of biological vulnerability to psychosis; neurobiological markers, oculomotor and vestibular system assessment; structure-activity relationships. Office: Ill State Psychiatric Inst 1601 W Taylor St Chicago IL 60612

LEVY, ELINOR M., immunologist; b. N.Y.C., Mar. 18, 1942; d. Louis S. and Tillie (Shepeten) Miller; m. Charles Joseph Levy, July 11, 1962; children: Benjamin Michael, Rebecca Ruth. B.A. in Chemistry, Brandeis U., 1963; Ph.D. in Biophysics, Emory U., 1972. Postdoctoral fellow dept. pathology U. B.C., Vancouver, 1973-75; research assoc. Boston U. Sch. Medicine, 1975-76, instr. dept. microbiology, 1976-79; asst. prof. Div. Med. and Dental Scis. Boston U. Grad. Sch., 1979—. Contbr. articles to profl. jours. Mem. Am. Assn. Immunologists, Phi Beta Kappa. Subspecialties: Immunology (medicine); Immunobiology and immunology. Current work: Immunoregulation by suppressor cells particularly in trauma and cancer patients and in Sr treated mice; isolation of lymphocyte subpopulations using electrophoresis and counter current distribution. Office: Boston U 80 E Concord St Boston MA 02118

LEVY, EUGENE HOWARD, astrophysicist, planetary physicist, researcher, cons., educator; b. N.Y.C., May 6, 1944; s. Isaac Philip and Anita Harriette (Guttman) L.; m. Margaret Rader, Oct. 13, 1967; children: Roger Philip, Jonathan Saul, Benjaminm Howard. A.B., Rutgers U., 1966; Ph.D. in Physics (NASA grad. fellow), U. Chgo., 1971. Postdoctoral fellow Ctr. Theoretical Physics, U. Md., 1971-73; asst. prof. physics and astrophysics Bartol Research Found., Swarthmore, Pa., 1973-75; asst. prof. planetary sci. U. Ariz., Tucson, 1975-78, assoc. prof., 1978—, assoc. head dept. planetary scis., faculty of applied math., 1981—; researcher; nat. sci. policy advisor; chmn. com. on planetary and lunar exploration Nat. Acad Scis., 1978-82, mem. space sci. bd., 1978-82; mem. NASA Solar-System Exploration Com., 1980-82; cons. Contbr. articles profl. jours. Ctr. for Theoretical Physics fellow, 1971-73; NSF and NASA research grantee, 1975—. Mem. Am. Phys. Soc., Am. Astron. Soc., Am. Geophys. Union, Internat. Astron. Union, Phi Beta Kappa, Sigma Xi. Subspecialties: Theoretical astrophysics; Planetology. Current work: Theoretical astrophysics, solar system physics, planetary physics, magnetohydrodynamics, plasma physics, solar physics. Office: Lunar and Planetary Lab U Ariz Tucson AZ 85721

LEVY, HILTON BERTRAM, medical researcher, administrator; b. N.Y.C., Sept. 21, 1916; s. Harry and Dorothy (Edelman) L.; m. Nettie Zack, Jan. 18, 1942; children: Charles, Harriet. B.S., CCNY, 1935; M.A., Columbia U., 1936; Ph.D., Poly. Inst. N.Y., 1946. Biochemist Overly Found., N.Y.C., 1946-52; head sect. molecular virology Nat. Inst. Allergy and Infectious Diseases, Bethesda, Md., 1952—. Editor: Biochemistry of Viruses, 1969, Interferon, 1970; contbr. numerous articles to profl. jours. Recipient Outstanding Performance award HEW, 1970. Fellow AAAS; mem. Am. Soc. Biol. Chemists, Am. Assn. Immunologists, Soc. Exptl. Biology and Medicine, Infectious Disease Soc., Soc. de Microbiologie de France, Am. Soc. Microbiologists. Subspecialties: Biochemistry (medicine); Virology (medicine). Current work: Virus replication, mechanism action of interferon, development of interferon inducers, immune modulation by IFN inducers, clincal studies with interferon inducers in virus diseases, cancer and neurologic diseases. Office: NIH Frederick MD 21701 Home: 9400 Linden Ave Bethesda MD 20814

LEVY, JOHN STUART, dentist, clinical consultant; b. New Haven, Dec. 26, 1946; s. Morton Julian and Pearl Ruth (Brodes) L.; m. Beverly Eileen Eden, Nov. 28, 1971; 1 dau., Perri Melissa. B.S., George Washington U., 1968; D.D.S., Georgetown U., 1976. Registered dentist. Postdoctoral fellow Yale U., New Haven, 1976-78; clin. instr. Georgetown U., Washington, 1976—; pres. Levy-D.D.S., P.C., New Haven, 1976—. Contbr. articles to profl. jours. Bd. dirs. Camp Laurelwood, Madison, Conn., 1979-82; bd. dirs., sec. med. bd. Jewish Home for Aged, New Haven, 1978—, bd. dirs. Jewish Fedn., New Haven; cabinet, mem. United Jewish Appeal Nat. Young Leadership, N.Y.C., 1980—. Served with U.S. Army, 1968-71. Recipient Joseph Borkowski award for professionalism Georgetown U., 1976; Nat. Service award NIH, 1977, 78. Mem. Internat. Assn. Dental Research, ADA. Jewish. Subspecialty: Cariology. Current work: Flouride-dental enamel studies, biomaterials related to dental applications. Home: 147 Kohary Dr New Haven CT 06515 Office: John S Levy DDS PC 52 Trumbell St New Haven CT 06510

LEVY, NELSON LOUIS, immunologist, pharmaceutical company executive; b. Somerville, N.J., June 19, 1941; s. Myron and Sylvia (Cohen) L.; m. Louisa Stiles; children: Michael, Andrew; m.; children from previous marriage: Scott, Erik, Jonathan. B.A., B.S., Yale U., 1963; M.D., Columbia U., 1967; Ph.D., Duke U., 1973. Diplomate: Am. Bd. Allergy and Immunology. Asst. prof. immunology Duke U., Durham, N.C., 1973-76, assoc. prof., 1976-80, prof., 1980-81; dir. biol. research Abbott Labs., North Chicago, Ill., 1981-82, v.p. pharm. discovery, 1982—; cons. NIH, 1973—, Gen. Motors Cancer Research Found., 1981—. Assoc. editor: Jour. Immunology, 1975—; contbr. articles to profl. jours. Served with USPHS, 1968-70. Mem. Am. Assn. Immunologists, Am. Assn. Cancer Research, Alpha Omega Alpha, Phi Gamma Delta. Club: Yale (Chgo.). Subspecialties: Immunology (medicine); Neurobiology. Current work: Directing multidisciplinary program to find new therapeutics entities in the areas of neuroscience, immunology, cardiovascular disease and infectious disease. Office: Abbott Labs Abbott Park North Chicago IL 60064

LEVY, PETER MICHAEL, physics educator; b. Frankfurt, Ger., Jan. 10, 1936; came to U.S., 1940; s. Alfred and Alice (Wolf) L.; m. Darline Gay Shapiro, Oct. 9, 1965; children: Erik Jacques Pierre, Serge Jacques Francois. B.M.E. summa cum laude, CCNY, 1958; M.A., Harvard U., 1960, Ph.D., 1963. Chargé de recherche Lab. Lovis Neel, Grenoble, France, 1963-64; research assoc. dept. physics U. Pa., Phila., 1964-66; asst. prof. dept. applied sci. Yale U., New Haven, 1966-70; assoc. prof. physics NYU, N.Y.C., 1970-75, prof., 1975—, chmn. dept., 1976-82. Contbr. 70 articles to profl. jours. Recipient Fulbright-Hays award, 1975-76; NSF exchange scientists award, 1975-76; médaille de Vermeil Soc. d'Encouragement au Progrès, Paris, 1978. Fellow N.Y. Acad. Sci. (bd. govs., v.p. phys. scis.); mem. Am. Phys. Soc. Subspecialties: Condensed matter physics; Magnetic physics. Current work: Magnetic properties of solids. Phase transisitions in magnetic materials; orbital effects in magnetic alloys and intermetallic compounds. Office: Dept Physics NYU 4 Washington Place New York NY 10003

LEVY, ROBERT ISAAC, physician, university official; b. Bronx, N.Y., May 3, 1937; s. George Gerson and Sarah (Levinson) L.; m. Ellen Marie Feis, 1958; children: Andrew, Joanne, Karen, Patricia. B.A. with high honors and distinction, Cornell U., 1957; M.D. cum laude, Yale U., 1961. Intern, then asst. resident in medicine Yale-New Haven Med. Center, 1961-63; clin. asso. molecular diseases Nat. Heart and Lung Inst., Bethesda, Md., 1963-66; chief resident Nat. Heart, Lung and Blood Inst., Bethesda, 1965-66, attending physician molecular disease br., 1965-80, head sect. lipoproteins, 1966-80, dep. clin. dir. inst., 1968-69, chief clin. services molecular diseases br., 1969-73, chief lipid metabolism br., 1970-74, dir. div. heart and vascular diseases, 1973-75, dir. inst., 1975-81; v.p. health scis., dean Sch. Medicine; v.p. health scis., dean Tufts U., Boston, 1981—, prof. medicine, 1981-83; v.p. health scis. Columbia U., N.Y.C., 1983—, prof. medicine, 1983—; attending physician Georgetown U. med. div. D.C. Gen. Hosp., 1966-68; spl. cons. anti-lipid drugs FDA. Editor: Jour. Lipid Research, 1972-80, Circulation, 1974-76, Am. Heart jour, 1980—; contbr. articles to profl. jours. Served as surgeon USPHS, 1963-66. Recipient Kees Thesis prize Yale U., 1961; Arthur S. Flemming award, 1975; Superior Service award HEW, 1975; Research award and Van Slyke award Am. Soc. Clin. Chemists, 1980. Mem. Am. Heart Assn. (mem.-at-large exec. com. council basic sci., mem. exec. council on atherosclerosis), Am. Inst. Nutrition, Am. Fedn. Clin. Research, N.Y. Acad. Scis., Am. Soc. Clin. Nutrition, Am. Soc. Clin. Investigation, Am. Coll. Cardiology, Inst. Medicine of Nat. Acad. Scis., Am. Soc. Clin. Pharmacology and Therapeutics, Assn. Am. Physicians, Phi Beta Kappa, Sigma Xi, Alpha Omega Alpha, Alpha Epsilon Delta, Phi Kappa Phi. Subspecialties: Cardiology; Internal medicine. Current work: Lipoproteins; cholesterol; preventive cardiology; nutrition; atherosclerosis; disordrs of lipid transport and lipid metabolism. Office: Columbia U Coll Physicians and Surgeons 630 W 168th St New York NY 10032

LEVY, STEPHEN RAYMOND, consultant, research, development and products company executive; b. Everett, Mass., May 4, 1940; s. Robert George and Lillian (Berfield) L.; m. Sandra Helen Rosen, Aug. 26, 1961; children: Phillip, Susan. B.B.A., U. Mass., 1962. Pres., chief exec. officer, dir. Bolt Beranek & Newman, Inc., Cambridge, Mass. Served with AUS, 1963-66. Decorated Army Commendation medal. Mem. Assn. Computing Machinery, Fin. Execs. Inst. Subspecialty: Computer company administration.. Home: 175 Commonwealth Ave Boston MA 02116 Office: 10 Moulton St Cambridge MA 02138

LEWELLEN, ROBERT THOMAS, research geneticist; b. Nyssa, Oreg., Apr. 27, 1940; s. John and Frances M. (Klinkenberg) L.; m. Priscilla E. Stark, Sept. 15, 1962. B.S., Oreg. State U., 1962; Ph.D., Mont. State U., 1966. Agronomist Mont. State U., Bozeman, 1965-66; geneticist USDA-ARS, Salinas, Calif., 1966—. Contbr. articles to profl. jours. Mem. Am. Soc. Agronomy, Am. Soc. Crop Sci., AAAS, Am. Soc. Phytopathology, Sigma Xi. Subspecialties: Genetics and genetic engineering (agriculture); Integrated pest management. Current work: Germplasm development, plant breeding methodology, plant breeding research, genetics of sugarbeet, plant pathology, agronomy, integrated pest management. Office: PO Box 5098 Salinas CA 93915

LEWIN, BENJAMIN, molecular biologist, editor, writer; b. Eng.; s. Sherry and Ann L.; m. Ann; children: Nicholas Sheridan, Jonathan Asher. B.A., Cambridge (Eng.) U., 1967, M.A., 1970, Ph.D., 1976; M.Sc., U. London, 1968. Tutorial fellow U. Sussex, Eng., 1969-70; editor Nature New Biology, 1970-71; vis. scientist Nat. Cancer Inst., NIH, Bethesda, Md., 1972-73. Editor: Cell mag, M.I.T., 1974—; Author: Gene Expression 1: Bacterial Genomes, 1974, Gene Expression 3: Plasmids and Phages, 1977, Gene Expression 2: Eucaryotic Chromosomes, 1980, Genes, 1983. Subspecialties: Gene actions; Genome organization. Current work: Communication in molecular biology. Office: Cell Offices 292 Main St Cambridge MA 02142

LEWIN, MARK HENRY, psychologist; b. N.Y.C., May 4, 1935; s. Leonard Lemuel and Amelia (Schwartz) L.; m. Eleanor Sue Mattes, June 23, 1957; children: David Samuel, Deborah Francine. B.A., Queens Coll., Flushing, N.Y., 1956; M.A., U. Wis.-Madison, 1959, Ph.D., 1961. Psychologist Rohrer, Hibler and Replogle, Rochester, N.Y., 1967-68; pvt. practice, Rochester, 1968-75; dir. Upstate Psychol. Service Center, Rochester, 1975—; exec. officer Am. Bd. Profl. Psychology, 1970-77; dir. George Heisel Corp., Rochester, 1980—. Author: Establishing and Maintaining A Successful Professional Practice, 1976; Contbr. articles to profl. jours. Bd. dirs. Rochester Area Hillel Found., Rochester, 1968—, pres., 1979-82; bd. dirs. Jewish Home and Infirmary, Rochester, 1976—; chmn. evaluation com United Way, 1982. Served to maj. U.S. Army, 1959-67. Recipient Disting. Profl. Achievement award Genesee Valley Psychol. Assn., 1979; Disting. Profl. Contbn. award Am. Bd. Profl. Psychology, 1978. Mem. Am. Psychol. Assn., Genesee Valley Psychol. Assn. (pres. 1972-74), Rochester Area Assn. Clin. Psychologists. Democrat. Subspecialties: Behavioral psychology; Neuropsychology. Current work: Use of psychological and behavioral science knowledge to better understand human behavior in order to engender more productive problem solving. Home: 64 Palmerston Rd Rochester NY 14618 Office: Upstate Psychol Service Center 756 E Main St Rochester NY 14605 Home: 64 Palmerston Rd Rochester NY 14618

LEWIN, WALTER H(ENDRIK) G(USTAV), physicist, educator; b. The Hague, Netherlands, Jan. 29, 1936; came to U.S., 1966; s. Walter S. and Pieternella J. (Van der Tang) L.; m. Huibertha G. van Teunenbroek; children: Pauline, Emanuel, Emma, Yakob; m. Ewa M.M. Basinska. Ph.D. in Physics, U. Delft (Netherlands), 1965. Instr. Libanon Lyceum, Rotterdam and U. Delft, 1960-66; asst. prof. physics MIT, 1966-68, asso. prof., 1968-74, prof., 1974—. Recipient medal for exceptional sci. achievement NASA, 1978. Mem. Am. Phys. Soc., Am. Astron. Soc., Internat. Astron. Union. Subspecialty: 1-ray high energy astrophysics. Research and publs. in high energy astrophysics, x-ray astronomy; collaboration with artists. Office: MIT 37-627 Cambridge MA 02139

LEWINE, RICHARD RALPH JEAN, clinical psychologist, consultant; b. Chartres, France, Jan. 29, 1947; came to U.S., 1950, naturalized, 1965; s. Morris and Paulette Suzanne (Barrier) L.; m. Janice Dorothy Florin, June 15, 1969 (div. 1981). B.A., Harvard U., 1969; Ph.D., U. Pa., 1975. Postdoctoral fellow NIMH, Amherst, Mass., 1975-78; research assoc. U. Denver, 1978-80; asst. prof. U. Ill.-Champaign, 1980-81; assoc. prof. U. Chgo. Pritzker Sch. Medicine, 1981—; assoc. dir. Mental Health Clin. Research Ctr., Chgo., 1981—. Author: Schizophrenia: Symptoms, Causes, Treatment, 1979, The Caring Family Living with Chronic Mental Illness, 1981; cons. editor: Jour. Social and Clin. Psychology, 1982. MacArthur Found. research grantee, 1982-85; Grant Found. grantee, 1979; NIMH research grantee, 1982. Mem. Am. Psychol. Assn., Am. Psychopath. Assn., Soc. Life History Research. Clubs: Harvard, Flute Soc. (Chgo.). Subspecialty: Psychology research. Current work: Conduct research into the phenomenology, demographics, and psychosocial aspects of schizophrenia, with a special focus on the problems of diagnosis. Office: Lab Biol Psychiatr Ill State Psychiat Inst 1601 W Taylor St Chicago IL 60612

LEWIS, ALAN JAMES, pharmacologist, research exec.; b. Newport, Wales, Sept. 29, 1945; came to U.S., 1979; s. William Tyssul and Elizabeth Ella (Deers) L.; m. Judith Ann Royle, Sept. 14, 1947; children: Nina Francis, Huw Gareth, Victoria Elizabeth. B.S. in Physiology and Biochemistry, Southampton (Eng.) U., 1967; Ph.D., U. Wales, 1970. Postdoctoral fellow research dept. biomed. scis. U. Guelph, Can., 1970-72, Yale U. Lung Research Center, New Haven, 1972-73; group leader Organon Labs., Glasgow, Scotland, 1973-79; assoc. dir. exptl. therapeutics Wyeth Labs., Inc., Phila., 1979—. Contbr. articles to profl. publs. and jours. Mem. Brit. Pharmacol. Soc., Brit. Immunological Soc., Am. Pharmacol. Soc., Reticuloendothelial Soc. Subspecialties: Immunogenetics; Allergy. Current work: Pharmacological modulation of autoimmune diseases and allergic diseases. Home: 1041 Shearwater Dr Audubon PA 19407 Office: Wyeth Labs Inc PO Box 8299 Philadelphia PA 19101

LEWIS, ALLEN ROGERS, biology educator; b. Ithaca, N.Y., Aug. 11, 1947; s. Norman F. and Edith M. (Kelsey) L.; m. Laurie J. Irvine, Oct. 18, 1969; children: Jessica, Kenneth. B.S., Cornell U., 1969; M.S., U. Del., 1971, U. Rochester, 1977, Ph.D., 1979. Marine extension specialist U. Del., Lewes, 1971-74; asst. prof. biology U. P.R., Mayaguez, 1978-81, assoc. prof., 1981—. Editor: Caribbean Jour. Sci, 1982—; contbr. articles to profl. jours. Mem. Ecol. Soc. Am., AAAS, Soc. for Study of Evolution, Assn. Tropical Biology. Subspecialties: Behavioral ecology; Population biology. Current work: Studies in behavior and ecology of Puerto Rican birds and reptiles, emphasis on habitat utilization and social organization. Home: W-17 Sierra Cayey Alturas de Algarrobo Mayaguez PR 00708 Office: Dept Biology U PR Mayaguez PR 00708

LEWIS, ANDREW MORRIS, JR., virologist, researcher; b. Cheriton, Va., Nov. 28, 1934; s. Andrew M. and Wilsye H. L.; m. Gladys R. Shorrock, June 8, 1960; children: T. Reid, Andrew M. III. B.A., Duke U., 1956, M.D., 1961. Intern in pediatrics Duke U. Med. Ctr., 1961-62, resident in pediatrics, 1962-63; research virologist USPHS Nat. Inst. Allergy and Infectious Diseases, NIH, Bethesda, Md., 1963—, Lab. Molecular Microbiology, 1981—. Served with USPHS, 1963—. Decorated Commendation medal USPHS. Mem. AAAS, Am. Soc. Microbiology, Am. Assn. Immunologists. Subspecialties: Virology (biology); Cell study oncology. Current work: Mechanisms of viral carcinogenesis; biology of DNA virus transformed mammalian cells; replication of Ad2-SV40 hybrids. Office: Bldg 5 Rm B1-32 NIH Bethesda MD 20205

LEWIS, ARCHIE JEFFERSON, III, horticulture educator; b. McCormick, S.C., Dec. 4, 1945; s. Archie Jefferson and Frances Margaret (Schumpert) L.; m. Judith Elizabeth Sheppard, Dec. 22, 1973; children: Andrew Pickens, Frances Elizabeth. B.S., Clemson U., 1967, M.S., 1970, Ph.D., 1975. Instr. hort. Clemson (S.C.) U., 1972-76; asst. prof. horticulture Va. Poly. Inst. and State U., Blacksburg, 1976—. Served with USAF, 1970-72. Recipient Outstanding Tchr. award horticulture dept. Clemson U., 1976. Mem. Am. Soc. Hort. Sci., Sigma Xi, Alpha Zeta, Gamma Sigma Delta, Pi Alpha Xi. Methodist. Subspecialties: Plant physiology (agriculture); Plant growth. Current work: Basic and applied research in production, growth and development, acclimitization and display life of floricultural and greenhouse crops with emphasis on growth modifying chemicals. Office: Dept Horticulture Va Poly Inst and State U Blacksburg VA 24061

LEWIS, BRIAN MURRAY, radio astronomer; b. Oxford, Eng., June 20, 1943; s. Brian Clive and Christina (Murray) L.; m. Dianne Nell Barnett, May 27, 1967; children: Rupert Murray, Penelope Anne. Ph.D., Australian Nat. U., 1970. With Jodrell Bank, Manchester, Eng., 1969-71; dir. Carter Obs., Wellington, N.Z., 1973-81; research assoc. Arecibo (P.R.) Obs., 1982—. Contbr. articles to sci. jours. Mem. Royal Astron. Soc., Am. Astron. Soc., Australian Astron. Soc. Subspecialties: Radio and microwave astronomy; Theoretical astrophysics. Current work: 21 cm observations of external galaxies, structure of clusters of galaxies, missing mass, neutrino mass. Home: 120-122 L St Ramey Aguadilla PR 06006 Office: Box 995 Arecibo PR 06120

LEWIS, DAVID KENNETH, chemistry educator, researcher; b. Poughkeepsie, N.Y., Feb. 11, 1943; s. Emery Othello and Bradleigh (Bowen) L.; m. Nancy Louise Bertke, Aug. 22, 1964; children: David, Nina, Carl. A.B. in Chemistry, Amherst Coll, 1964; Ph.D. in Phys. Chmeistry, Cornell U., 1970. Research asst. dept. vegatable crops Cornell U., 1959, research asst. chemistry, 1964-69; research asst. chemistry and atmospheric sci. Arthur D. Little, Inc., 1961-64; asst. prof. chemistry Colgate U., 1969-76, asso. prof., 1976-80, prof., 1980—; vis. sr. research fellow U. Colo. NOAA, 1977-78; cons. in field. Contbr. articles in field to profl. jours. Bd. dirs. Madison County (N.Y.) Assn. Retarded Citizens, 1971—. Grantee in field. Mem. Am. Chem. Soc., Council on Undergrad. Research (dir. 1980—), Sigma Xi. Subspecialties: Kinetics; Atmospheric chemistry. Current work: Rates and mechanisms of energy transfer and chemical reactions in gases at high temperature; examining differences in energy flow and reaction pathway between thermally induced and laser induced reactions of small molecules. Office: B-3 McGregory Hall Colgate University Hamilton NY 13346

LEWIS, DAVID SLOAN, JR., aircraft company executive; b. North Augusta, S.C., July 6, 1917; s. David S. and Reuben (Walton) L.; m. Dorothy Sharpe, Dec. 20, 1941; children: Susan, David Sloan III, Robert, Andrew. Student, U. S.C., 1934-37; B.S. in Aero. Engring, Ga. Inst. Tech., 1939. Aerodynamicist Glenn L. Martin Co., Balt., 1939-46; chief aerodynamics McDonnell Aircraft Corp., St. Louis, 1946-52, chief preliminary design, 1952-55, mgr. sales, 1955-56, mgr. projects, 1956-57, v.p. project mgmt., 1957-59, sr. v.p. ops., 1960-61, exec. v.p., 1961-62, pres., 1962-67, McDonnell-Douglas Co.; also chmn. Douglas Aircraft Co. div., 1967-70; chmn., chief exec. officer Gen. Dynamics Corp., 1970—; dir. Ralston Purina Co., St. Louis, BankAm. Corp., San Francisco, Mead Corp. Alderman, Ferguson, Mo., 1951-54; Trustee Washington U., St. Louis. Fellow AIAA. Episcopalian. Subspecialty: Aeronautical engineering. Office: Pierre Laclede Center Saint Louis MO 63105

LEWIS, DONALD HOWARD, microbiologist, educator; b. Stamford, Tex., May 31, 1936; s. Jessie H. and Myrtle M. (Randolph) L.; m. Carolyn Kay Conoley, Feb. 1, 1960; children: Donald Howard, John M., David E. (dec.), James M. B.A., U. Tex., Austin, 1959; M.A., S.W. Tex. State U., San Marcos, 1964; Ph.D., Tex. A&M U., College Station, 1967. Mem. faculty Tex. A&M U., 1967—, prof., acting head dept. vet. microbiology, 1979—. Contbr. articles to sci. jours.; composer religious selections. Served with AUS, 1959-62. Recipient 11 research grants, 1975—. Mem. Am. Fisheries Soc., Am. Soc. Microbiology, Tex. Acad. Scis., Internat. Assn. Aquatic Animal Medicine, Sigma Xi, Gamma Sigma Delta. Democrat. Presbyterian. Subspecialty: Microbiology. Current work: Disease processes in aquatic animals, microbial ecology, immunology. Inventor technique to control biofouling, vaccine for aeromonas hydrophilic. Home: 1205 Winding Rd College Station TX 77843 Office: Dept Vet Microbiology Texas A&M U College Station TX 77843

LEWIS, EDWARD B., biology educator; b. Wilkes-Barre, Pa., May 20, 1918; s. Edward B. and Laura (Histed) L.; m. Pamela Harrah, Sept. 26, 1946; children: Hugh, Glenn(dec.), Keith. B.A., U. Minn., 1939; Ph.D., Calif. Inst. Tech., 1942. Instr. biology Calif. Inst. Tech., Pasadena, Calif., 1946-48, asst. prof., 1948-49, assoc. prof., 1949-56, prof. biology, 1956—. Editor: Genetics and Evolution, 1961. Mem. Nat. Adv. Com. on Radiation, 1958-61. Served to capt. USAAF, 1942-46. Rockefeller fellow, 1948-49. Mem. Genetics Soc. Am. (sec. 1962-64, pres. 1967—), Nat. Acad. Scis., AAAS, Am. Soc. Human Genetics, Am. Soc. Naturalists, Am. Adad. Arts and Scis. Subspecialties: Evolutionary biology; Radiation biology. Home: 805 Winthrop Rd San Marino CA 91108 Office: Calif Inst Tech Pasadena CA 91109

LEWIS, GWYNNE DAVID, agricultural educator; b. Hackensack, N.J.; s. Gwynne Trevor and Dorothy H. (DeWitt) L.; m. Mary Alice Pfotzer, June 25, 1960; 1 son, Trevor; m. Gary Anne Dunn, Dec. 5, 1970. B.S., Rutgers U., 1951; M.S., Purdue U., 1953; Ph.D., Cornell U., 1958. Asst. prof. plant pathology Rutgers U., New Brunswick, N.J., 1958-64, assoc. prof., 1964-70, prof., 1970—. Contbr. over 200 articles to profl. jours. Mem. Colts Neck (N.J.) Shade Tree Commn., 1972—, chmn., 1976—. Recipient Bronze medal Am. Rhododendron Soc., 1975. Mem. Am. Phytopath. Soc., Soc. Nematologists, Sigma Xi. Mem. Christian Ch. Subspecialties: Plant pathology; Integrated pest management. Current work: Chemical, biological and genetic control of plant diseases. Home: 52 Glenwood Rd Colts Neck NJ 07722 Office: Rutgers U New Brunswick NJ 08903

LEWIS, HERSCHEL PAUL, neurosurgeon; b. Niagara Falls, N.Y., May 22, 1934; s. Herschel Paul and Florence Margarite (Hanna) L.; m. Barbara Lou Jager, Aug. 11, 1956; children: Michael, Susan; m. Linda Sue Lewis, Oct. 20, 1979; children: Rebecca, Katie, Christy. A.B., Hamilton Coll.; M.D., N.Y. Med. Coll. Diplomate: Am. Bd. Neurol. Surgery. Intern, Resident U. Cin.; NIH fellow in head injury, 1966-68, pvt. practice medicine specializing in neurosurgery, Cin.; co-dir. neurosurgery Bethesda Hosp., Cin., mem. exec. com.; asst. prof. U. Cin.; dir. neurosurgery Mercy Hosp., Fairfield, Ohio, St. Luke's Hosp., Ft. Thomas, Ky.; pres. Neurol. Surgeons Assocs. of Cin. Contbr. articles to profl. jours. Mem. Central Ohio River Valley Health Planning Assn. Served to maj. USAR. Mem. Am. Assn. Neurol. Surgeons, Congress Neurol. Surgeons, Cin. Med. Soc., Ky. Med. Soc., Brain Club Cin., ACS. Methodist. Club: Bankers of Cin. Subspecialties: Neurosurgery; Microsurgery. Current work: Cerebral circulatory pathophysiology, head injury, cerebral blood flow, laser surgery, ultrasound diagnosis. Home: 9300 Cunningham Rd Cincinnati OH 45243 Office: 10496 Montgomery Rd NUM110 Cincinnati OH 45242

LEWIS, JAMES ALEXANDER, neurobiologist, educator; b. Phila., Mar. 7, 1946; s. George Campbell and Elizabeth Glenn (Zipf) L.; m. Anne Sun Wah Kuan, Dec. 27, 1975; children: Diane, Cynthia. B.S. in Chemistry, M.I.T., 1964-68; Ph.D. in Biochemistry, U. Calif.-Berkeley, 1972. Postdoctoral fellow M.R.C. Lab. Molecular Biology, Cambridge, Eng., 1972-74; asst. prof. Columbia U., N.Y.C., 1974-81; research asst. prof. U. Pitts., 1981-82; asst. prof. U. Mo., Columbia, 1982—. Contbr. articles to profl. jours. NSF fellow, 1968-72; Am. Cancer Soc. fellow, 1972-74. Mem. Soc. Neuroscience. Subspecialties: Neurobiology; Gene actions. Current work: Genetic and pharmacological characterization of drug-resistant mutants of the nematode Caenorhabditis elegans that appear acetylcholine receptor-deficient; synthesis of a radioactive receptor ligand and demonstration through binding assays of mutant receptor deficiency. Office: Division Biological Sciences University Missouri Columbia MO 65211

LEWIS, JAMES BRYSON, maintenance engineer; b. Wyandotte, Mich., July 27, 1950; s. Jack Bryson and Gladys Irene (Forsyth) L.; m. Anna M. Welke, Feb. 28, 1976 (div. July 1981); 1 son, Phillip. B.S., U. Mich.-Ann Arbor, 1972. Registered profl. engr., Mich. Field engr. Gen. Electric Co., Chgo., 1972-76, startup mgr., Platteville, Colo., 1976-77, field engr.-nuclear, Oak Brook, Ill., 1977-78, service supr., 1978-79; sr. engr. Consumers Power Co., Jackson, Mich., 1979-81, gen. supr. mech. maintenance, 1981—. Mem. Am. Nuclear Soc. Subspecialties: Mechanical engineering; Fracture mechanics. Current work: Failure analysis of steam turbine generators and related mechanical equipment. Office: Consumers Power Co 1945 W Parnall Rd Jackson MI 49201

LEWIS, JOHN ALLEN, mathematician; b. Detroit, Jan. 21, 1923; s. Arthur Laighton and Norma (Allen) L.; m., June 21, 1948; children: Julia Gay, Jonathan Allen, Celia Chapin. B.S., Worcester Poly. Inst., 1944; M.S., Brown U., 1948, Ph.D., 1950. Research physicist Corning Glass Co., N.Y., 1950-51; mem. tech. staff Bell Labs., Murray Hill, N.J., 1951—. Served to lt. (j.g.) USNR, 1943-45. Mem. Am. Math. Soc., Soc. Indsl. and Applied Math, Math Assn. Am., Soc. Rheology. Democrat. Subspecialties: Applied mathematics; Fluid mechanics. Current work: Heat transfer, semiconductor fabrication, satellite attitude control, electronic materials processing. Home: 109 Maple St Summit NJ 07901 Office: Bell Labs Rm 7B228 Murray Hill NJ 07901

LEWIS, JOHN SIMPSON, JR., planetary sciences educator, consultant; b. Trenton, June 27, 1941; s. John Simpson and Elsie Dinsmore (Vandenbergh) L.; m. Ruth Margaret Adams, Aug. 1, 1964; children: John Vandenbergh, Margaret Tanner, Christopher Franklin, Katherine Rose, Elizabeth Adams, Peter Mandeville. A.B., Princeton U., 1962; M.A., Dartmouth Coll., 1964; Ph.D., U. Calif.-San Diego, 1968. Asst. prof. geochemistry and chemistry MIT, 1968-72, assoc. prof., 1972-80, prof. planetary sci., 1980-82, adj. prof., 1982-83; vis. prof. planetary sci. Calif. Inst. Tech., 1973; prof. planetary sci. U. Ariz., Tucson, 1982—; cons. Avco Corp., Martin Marietta Aerospace, Dynatrend, Inc., 1969-80; founder, pres. J.S. Lewis Assoc., Inc., 1980—. Author: (with R.G. Prinn) Origin of Planets and Evolution of their Atmospheres, 1983; translator: (with Ruth A. Lewis) Cosmic Evolution: Patience dans L'azur (Hubert Reeves), 1983. Recipient James B. Macelwayne award Am. Geophys. Union, 1976; NASA Exceptional Sci. Achievement medal, 1983. Mem. AAAS, Am Chem. Soc., Am Astron. Soc. (div. planetary scis.), N.Y. Acad. Scis. Mormon. Subspecialties: Planetary science; Planetary atmospheres. Current work: Origin and evolution of planets and their atmospheres, applications of chemistry to space science, relationships between earth, other planets and meteorites. Home: 5010 W Sweetwater Dr Tucson AZ 85745 Office: Dept Planetary Scis U Ariz Tucson AZ 85721

LEWIS, LINDON L., atomic physicist, researcher; b. Galesburg, Ill., Oct. 13, 1950; s. Everett LeRoy and Norma Beth (Famulener) L.; m. Merilee Anne Schultheiss, May 20, 1972; 1 son, Kenneth Edward. B.A. in Physics, Knox Coll., 1972, Ph.D., U. Wash., Seattle, 1978. Research assoc. U. Wash., Seattle, 1978-79; physicist Nat. Bur. Standards, Boulder, Colo., 1979—, project leader cesium clock research. Contbr. articles to profl. publs., papers to confs., U.S., France. Mem. Am. Physics Soc., AAAS, Phi Beta Kappa. Subspecialty: Atomic and molecular physics. Current work: Diode lasers; atomic spectroscopy; optical pumping; atomic frequency standards. Office: 325 Broadway 524 11 Boulder CO 80303

LEWIS, LISA, clinical psychologist, neuropsychologist; b. Canonsburg, Pa., Aug. 19, 1952; d. Fredrick John and Audre (Riggs) L. B.S., Pa. State U., 1973; M.A., Conn. Coll., 1976; Ph.D., Miami U., Oxford, Ohio, 1979. Psychology trainee Cin. VA Hosp., 1977-78; predoctoral intern U. Fla. Med. Sch., 1978-79; postdoctoral fellow Menninger Found., Topeka, 1979-81, staff psychologist, 1981—, instr., 1981—; Mem. adv. bd. Kans. Head Injury Found., Kansas City, 1982—. Contbr. articles, chpts. to profl. publs. Mem. Am. Psychol. Assn., Kans. Psychol. Assn., Phi Beta Kappa, Sigma Xi. Subspecialties: Clinical psychology; Neuropsychology. Current work: Brain-behavior relationships; psychiatric diagnosis. Office: Menninger Found Bldg 8 Box 829 Topeka KS 66601

LEWIS, MARY ANITA, research administrator; b. Dublin, Ga., Sept. 21, 1950; s. F.H. and Barbara A. (Brandt) L. B.A., U. Ga., 1972, M.S., 1974, Ph.D., 1976. Chief tng. and evaluation researcher Civil Aeromed. Inst., FAA, Oklahoma City, 1976-79; mgr. applied behavioral research PPG Industries, Inc., Pitts., 1979—; adj. assoc. prof. psychology U. Okla., Norman, 1978-79. Contbr. articles to profl. jours. Recipient A.S. Edwards award U. Ga. Psychology Dept., 1976. Mem. Am. Psychol. Assn., AAAS, Soc. Indsl. and Organizational Psychology. Subspecialty: Applied behavioral research. Current work: Organizational assessment, organizational change, ergonomics, management assessment/development, environment and behavior, psychometrics, organizational strategy-structure-climate. Home: 2794 Milford Dr Bethal Park PA 15102 Office: PPG Industries Inc 1 Gateway Ctr Pittsburgh PA 15222

LEWIS, MICHAEL, research psychologist, educator; b. N.Y.C., Jan. 10, 1937; s. Bernard W. and Leah (Cohen) L.; m. Rhoda Rosenzweig, Aug. 18, 1962; children: Benjamin, Felicia. B.A. with honors, U. Pa., 1958, M.A., 1960, Ph.D., 1962. Asst. prof. psychology Antioch Coll., 1962-65, assoc. prof., 1965-68; vis. assoc. prof. Harvard U., 1965; sr. investigator psychology Fels Research Inst., Yellow Springs, Ohio, 1966-68; research psychologist Ctr. Psychol. Studies, Ednl. Testing Service, Princeton, N.J., 1968-71; dir. infant lab. Ednl. Testing Service, 1971-78, sr. research psychologist, 1971-82; dir. Inst. Study Exceptional Children, 1977-82; adj. prof. dept. pediatrics, U. Medicine and Dentistry N.J., Piscataway, 1980-82, prof. psychology, 1979-80, prof. pediatrics and psychiatry, 1982—; cons. to founds.; adj. prof. psychology Temple U., 1979-80; sr. research scientist pediatric service Roosevelt Hosp.-St. Luke's Med. Ctr., N.Y.C., 1977-82; vis. prof. CUNY, 1973-78; clin. prof. pediatric psychology Columbia U., 1972—. Author and/or editor: (with Hale) numerous books including Attention and Cognitive Development, 1979, (with Brooks-Gunn) Social Cognition and the Acquisition of Self, 1979, (with Rosenblum) The Child and Its Family: The Genesis of Behavior, Vol. 2, 1979, The Uncommon Child: The Genesis of Behavior, Vol. 3, 1981; cons. editor, cons. reader to profl. jours.: The Uncommon Child: The Genesis of Behavior, Vol. 3. Fellow Am. Psychol. Assn., N.Y. Acad. Scis., AAAS; mem. Soc. Research in Child Devel., Ohio Acad. Sci., Psychologists Interested in Advancement Psychotherapy, Soc. Research in Psychophysiology, Sigma Xi. Subspecialties: Developmental psychology; Neonatology. Current work: Early development of socio-economic abilities and cognitive-communicative competence in normal and dysfunctional children. Home: 95 Linwood Circle Princeton NJ 08540 Office: Rutgers Medical School Academic Health Science Center CN19 New Brunswick NJ 08903

LEWIS, MICHAEL DOLAN, mechanical engineer; b. Houston, Nov. 7, 1952; s. Henry John and Dolores Marie (Barkis) L.; m. Roberta Francis, Oct. 22, 1977; children: Matthew Dolan, Kelli Christine. B.S.M.E., Tex. A&M U., 1975; M.B.A., U. Houston, 1982. Registered profl. engr., Tex. Asst. design engr. subsea drilling equipment Nat. Supply Co., Houston, 1975-77, asso. design engr. standard wellhead, 1977-79, design engr. standard wellhead, 1979-81, product engr. standard wellhead, 1981-82, product supr. standard wellhead, 1982—; v.p. Cornell Ranches Corp., 1976—. Mem. Meadow Brook Civic Club, 1978—. Model for Mgmt. award Armco Inc., 1982; Quality Plus award, 1982. Mem. ASME, Tex. Soc. Profl. Engrs., Tex. A&M U. Former Students Assn. (century). Republican. Roman Catholic. Club: WCS Mgmt. (Houston) (treas.). Subspecialties: Mechanical engineering; Metallurgical engineering. Current work: Design and devel. of wellhead equipment for critical applications. Inventor metal seal, flow control choke with studded inlet, high temperature casing packoff, hydraulic casing jack.

LEWIS, PHILIP M., electrical engineer, computer scientist; b. N.Y.C., May 30, 1931; m., 1953; 2 children. B.E.E., Rensselaer Poly. Inst., 1952; S.M., MIT, 1954, Sc.D., 1956. From instr. to asst. prof. elec. engring. MIT, 1954-59; mem. tech. staff Gen. Electric Research and Devel. Ctr., Schenectady, 1959-69, cons. automation theory and software design, 1959, mgr., 1969-78, mgr. computer sci. br., 1978—; cons. Epsco, Inc., 1955, Lincoln Labs., MIT, 1955-56, Hycon Eastern Inc., 1956-57, Sanders Assocs., Inc., 1958; adj. prof. Rensselaer Poly. Inst., 1960—. Mng. editor: Jour Computer Soc., Indsl. and Applied Math, 1971—. Coolidge fellow Gen. Electric Research and Devel. Corp., 1977—. Fellow Soc. Indsl. and Applied Math.; mem. Assn. Computing Machinery, IEEE. Subspecialty: Theoretical computer science. Office: Computer Sci Branch Gen Electric Research and Devel Ctr PO Box 8 Schenectady NY 12309

LEWIS, RALPH JAY, III, human resources mgmt. educator, cons.; b. Balt., Sept. 25, 1942; s. Ralph Jay and Ruth E. (Schmeltz) L. B.S. in Indsl. Engring. and Mgmt. Scis, Northwestern U., 1966; M.S. in Adminstrn, U. Calif.-Irvine, 1968; Ph.D. in Socio-Tech. Systems, UCLA, 1974. Research analyst Chgo. Area Expressway Surveillance Project, 1963-64, Gen. Am. Transp. Corp. (research div.), Chgo., 1965-66; assoc. prof. human resources mgmt. Calif. State U.-Long Beach, 1972—; cons. Rand Corp., Santa Monica, Calif., 1964-74, Air Can., Montreal, 1972-73; adv. Project Quest, Los Angeles, 1969-71, Maxisystems, Santa Anna, Calif. Co-author: Studies in the Quality of Life, 1972; author instructional CAI program system, 1972-80. Mem. AAAS, Am. Psychol. Assn., Assn. Humanistic Psychology, World Future Soc. Subspecialties: Social psychology; Distributed systems and networks. Current work: Methods to study the interactions among complex technical and industrial systems, advanced scientific developments, managerial and organizational systems and individual and interpersonal impacts to better cope with turbulence. Home: 3749 Oakfield Dr Sherman Oaks CA 91423 Office: Dept Mgmt and Human Resources Mgmt Calif State U Long Beach 1250 Bellflower Blvd Long Beach CA 90840

LEWIS, RANDOLPH VANCE, biochemistry educator; b. Powell, Wyo., Apr. 8, 1950; s. Jack F. and Evelyn J. (Vonburg) L.; m. Lorrie Dale, May 28, 1972; 1 son, Brian. B.S., Calif. Inst. Tech., Pasadena, 1972; M.S., U. Calif.-San Diego, 1974, Ph.D., 1978. Postdoctoral fellow Roche Inst. Molecular Biology, Nutley, N.J., 1978-80; asst. prof. U. Wyo., Laramie, 1980—; cons. Baker Chem. Co., Phillipsburg, N.J., 1980—, Cetus Corp., Berkeley, Calif., 1980, Syntro Corp., San Diego. Contbr. articles on biochemistry to profl. jours. Nat. Collegiate Athletic Assn. fellow, 1972; Alfred P. Sloan Found. research fellow, 1983-85; NIH grantee, 1981-84; Research Corp. award, 1981; Wyo. Heart Assn. grantee, 1982. Mem. Am. Chem. Soc. (sect. treas. 1982—), AAAS, N.Y. Acad. Scis. Republican. Subspecialties: Biochemistry (medicine); Neuroendocrinology. Current work: Research on the biosynthesis of bioactive peptides particularly in the adrenal medulla. Home: 1067 Arapaho Laramie WY 82070 Office: U Wyo PO Box 3944 Univ Sta Laramie WY 82071

LEWIS, ROBERT, JR., research plant pathologist; b. Shaw, Miss., Feb. 28, 1945; s. Robert and Cora Lee (Foster) L.; m. Mattie Ducking, Dec. 22, 1968; children: Eldrin, Zanetta, Robert. B.S., Jackson (Miss.) State U., 1969; Ph.D. in Plant Pathology, Tex. A&M U., 1976. Research plant pathologist U.S. Forest Service, Stoneville, Miss., 1976—. Contbr. articles to profl. jours. Mem. Delta Area council Boy Scouts Am. Recipient Outstanding Achievement award So. Forest Insect Workshop, 1981. Mem. Am. Phytopath. Soc., Miss. Acad. Sci., Sigma Xi. Baptist. Subspecialty: Plant pathology. Current work: Detection, evaluation and control of economically important tree diseases. Home: 344 Kentucky Greenville MS 38701 Office: PO Box 227 Stoneville MS 38776

LEWIS, TERENCE DAVID, physician, educator; b. Cooranbong, New South Wales, Australia, Nov. 6, 1943; came to U.S., 1978; s. Oswald O. and Dorothy G. (Adderton) L.; m. Valerie Jean Daley, Jan. 4, 1966; children: Michelle, Antony, Jane, Katy. M.B.B.S., U. Sydney, 1967. Diplomate: Am. Bd. Internal Medicine. Intern Repatriation Hosp., Concord, New South Wales, 1967-68; resident Royal Perth Hosp., Australia, 1968-69, Repatriation Hosp., Heidelberg, Victoria, Australia, 1969-72; asst. physician clin. research unit Alfred Hosp., Melbourne Hosp., Australia, 1972-74; asst. prof. McMaster U., Hamilton, Ont., Can., 1974-78; asst. prof. dept. gastroenterology Loma Linda (Calif.) U., 1978-81, assoc. prof., 1981—. Contbr. articles to profl. jours. Named Clin. Investigator of Yr. W.E. McPherson Soc., 1980. Fellow ACP; mem. Am. Fedn. Clin. Research, Am. Gastroent. Assn., Sigma Xi. Subspecialties: Gastroenterology; Internal medicine. Current work: Human gastrointestinal motility; gastrointestinal smooth muscle physiology and pharmacology; gastrointestinal diseases. Home: 11627 Cedar Way Loma Linda CA 92354 Office: Loma Linda U Med Center 11234 Anderson St Loma Linda CA 92350

LEWIS, URBAN JAMES, biochemist, researcher; b. Flagstaff, Ariz., Apr. 28, 1923; s. Urban James and Mary Mildred (Rozen) L.; m. Loraine Joyce Chambers, July 16, 1950; children: Geoffrey P., Wayne K. B.A., San Diego State Coll., 1948; M.S., U. Wis.-Madison, 1950, Ph.D., 1952. Research assoc. Am. Meat Inst., U. Chgo., 1953-54; research assoc. Merck, Sharp & Dohme, Rahway, N.J., 1954-61; mem. Scripps Clinic and Research Found., La Jolla, Calif., 1961-82, Whittier Inst. for Diabetes and Endocrinology, La Jolla, 1982—; fellow Med. Nobel Inst., Stockholm, Sweden, 1952-53; mem. med. adv. bd. Nat. Pituitary Agy., Balt., 1972-75; mem. endocrinology study sect. NIH, Washington, 1978-82; cons. Calbiochem-Behring, La Jolla, Serono Labs., Braintree, Md. Contbr. articles to profl. jours. Served with AUS, 1943-46; PTO. Recipient Babcock award U. Wis., 1951; WHO Travel fellow, 1970; grantee NIH, Am. Cancer Soc., Am. Diabetes Soc., 1961—. Mem. Am. Soc. Biol. Chemists, Endocrine Soc. Democrat. Unitarian. Subspecialties: Biochemistry (medicine); Endocrinology. Home: 5733 Skylark Pl La Jolla CA 92037 Office: Whittier Inst Diabetes and Endocrinology 9894 Genesee Ave La Jolla CA 92037

LEWIS, VICTOR LAMAR, plastic surgery educator; b. Evanston, Ill., Sept. 22, 1942; s. Victor L. and Anne (Ward) L.; m. Jayne Martin, Dec. 2, 1972; children: Torrey, Michael. B.A., Yale U., 1964; M.D., Northwestern U., 1968. Diplomate: Am. Bd. Surgery, Am. Bd. Plastic Surgery. Intern Hosp. of U. Pa., Phila., 1968-69; resident in surgery La. State U. Service Charity Hosp. of La., New Orleans, 1969-73; asst. instr. La. State U., 1972-73; resident in plastic surgery Northwestern U., Chgo., 1975-77, instr., 1976-77, assoc. in surgery, 1977-78, asst. prof. surgery, 1978—. Served to lt. comdr. USNR, 1973-75. Plastic Surg. Edn. Found. grantee, 1979. Fellow ACS; mem. Chgo. Surg. Soc., Chgo. Plastic Surgery Soc., Plastic Surgery Research Council, Soc. Surgery of Trauma. Anglican Catholic. Subspecialty: Surgery. Current work: Wound healing, increased length of tissue survival, healing of membranous bone. Home: 837 W Fullerton St Chicago IL 60614 Office: Northwestern University Medical School 303 E Chicago Ave Chicago IL 60611

LEWIS-PINKE, ELLEN RUTH, neurobiologist; b. N.Y.C., May 11, 1954; d. Martin Herman and Joan Sheila (Klauber) Lewis; m. James Richard Pinke, July 22, 1982. B.A., U. Rochester, 1976; Ph.D., U. Calif., Irvine, 1981. Postdoctoral researcher McLean Hosp., Belmont, Mass., 1981-82; postdoctoral assoc Yale Med. Sch., New Haven, 1982—. Mem. AAAS, Soc. for Neurosci., Nat. Med. Research Council. Subspecialties: Neurobiology; Regeneration. Current work: Development and regeneration in nervous system. Intracerebral implants. Office: Dept Neurosurgery Yale-New Haven Med Center New Haven CT 06520

LEY, RICHARD WAYNE, mathematician, software consultant; b. Harrisburg, Pa., Aug. 31, 1947; s. Robert Emil and June Doris (Weaver) L. A.B. in Math, U. Calif.-Berkeley, 1969, M.S., Stanford U., 1970. Mathematician Dept. Def., Ft. Meade, Md., 1974-81, 82—; assoc. mgr. BDM Corp., Los Angeles, 1981-82; owner, mgr. Minicomputer Software Projects, Silver Spring, Md., 1982—; ptnr. K & L Software, Silver Spring, Md., 1983—. Pres. Citizens for Greenbelt Assn., 1980. Served to lt. USNR, 1971-73. NSF fellow, 1969; recipient Spl. Achievement award Dept. Def., 1981. Mem. Math. Assn. Am., Soc. for Indsl. and Applied Math., Ops. Research Soc. Am. Republican. Roman Catholic. Subspecialties: Mathematical software; Probability. Current work: Conducting computer-aided research in speech and digital signal processing; writing microcomputer applications software; performing computer systems analysis and simulation testing. Home: 11425 Oak Leaf Dr Silver Spring MD 20901

LEY, TIMOTHY JAMES, physician, researcher; b. Buffalo Center, Iowa, June 17, 1953; s. William Dean and Clara Ruth (Odl) L. B.A., Drake U., Des Moines, Iowa, 1971-74; M.D., Washington U., St. Louis, 1978. Diplomate: Am. Bd. Internal Medicine. Intern Mass. Gen. Hosp., Boston, 1978-79, jr. resident, 1979-80; clin. assoc. Nat. Heart, Lung and Blood Inst., NIH, Bethesda, Md., 1980—. Contbr. articles on the structure and function of human globin genes, the treatment of B-thalassemia, and manipulation of globin gene function in vivo to profl. jours. Mem. Phi Beta Kappa, Alpha Omega Alpha. Presbyterian. Subspecialties: Genetics and genetic engineering (biology); Hematology. Current work: Therapeutic manipulation of human globin genes; hematologist; genetic engineer. Office: NIH Bldg 10-ACRF Room 7C103 Bethesda MD 20205

LI, CHOH HAO, biochemist, endocrinologist; b. Canton, China, Apr. 21, 1913; came to U.S., 1935, naturalized, 1955; s. Kan-chi Li and Mewching Tsui; m. Annie Lu, Oct. 1, 1938; children: Wei-i Li, Ann-si Li, Eva Li. B.S., U. Nanking, 1933; Ph.D., U. Calif.-Berkeley, 1938; M.D. (hon.), Cath. U. Chile, 1962; LL.D., Chinese U., Hong Kong, 1970; D.Sc., U. Pacific, 1971, Marquette U., 1971, St. Peter's Coll., 1971; hon. doctor, Uppsala U., 1977; D.Sc., U. San Francisco, 1978, L.I. U., 1981, U. Colo., 1981, Med. Coll. Pa., 1982. Instr. chemistry U. Nanking, 1933-35; research asso. U. Calif.-Berkeley, 1938-44, asst. prof. exptl. biology, 1944-47, asso. prof., 1947-49; prof. biochemistry, prof. exptl. endocrinology dir. Hormone Research Lab., Berkeley and San Francisco, 1950—; mem. acad. adv. bd. Chinese U., Hong Kong, 1963—; adv. bd. Chem. Research Center, Nat. Taiwan U., 1964—; vis. scientist Children's Cancer Research Found., Boston, 1955, 63-73; co-chmn. Internat. Symposium on Growth Hormone, Milan, Italy, 1967, Internat. Symposium on Protein and Polypeptide Hormones, Leige, Belgium, 1968; hon. pres. Internat. Symposium on Gonadotropins, Bangalore, India, 1973; chmn. Internat. Symposium on Proteins, Taipei, 1978, Internat. Symposium Growth Hormones and other Biologically Active Peptides, Milan, Italy, 1979; vis. prof. U. Montreal, 1948, Nat. Taiwan U., 1958, Chinese U. Hong Kong, 1967, Marquette U., 1973. Co-editor: Perspectives in the Biochemistry of Large Molecules, Supplement 1, 1962; sect; editor: Chem. Abstracts, 1960-63; co-asso. editor: Internat. Jour. Peptide and Protein Research, 1969-76, editor-in-chief, 1976—; mem. editorial adv. bd.: Family Health, 1969-81, Biopolymers, 1979—; editorial bd.: Current Topics in Exptl. Endocrinology, 1969—, Archives Biochem. Biophysics, 1979—; editor: Hormonal Proteins and Peptides, 1973—, Versatility of Proteins, 1979. Recipient numerous honors, including Ciba award in endocrinology, 1947, Amory prize Am. Acad. Arts and Scis., 1955, Albert Lasker award for basic med. research, 1962, Golden Plate award Am. Acad. Achievement, 1964, Univ. medal, Liege, Belgium, 1968; Sci. Achievement award AMA, 1970; Nat. award Am. Cancer Soc., 1971; Nicholas Andry award Assn. Bone and Joint Surgeons, 1972; Lewis prize Am. Philos. Soc., 1977; Nichols medal Am. Chem. Soc., 1979; Sci. award Academia Santa Chicra, Genoa, 1979; Koch award Endocrine Soc., 1981; Heyrovsky Gold medal Czech Acad. Sci., 1982; Harvey lectr., 1951; Faculty Research lectr. U. Calif., San Francisco, 1962-63; Evan lectr., 1976; Pres. Marcos lectr., Manila, 1967; Lasker award lectr. Salk Inst., 1969; Nord lectr. Fordham U., 1972; Geschwind lectr. U. Calif., Davis, 1980; Grattarola lectr. U. Milan, 1981; Guggenheim fellow, 1948. Fellow Am. Acad. Arts and Scis., AAAS, N.Y. Acad. Sci., Am. Inst. Chemists; mem. Am. Chem. Soc. (Calif. sect. award 1951), Am. Soc. Biol. Chemists, Endocrine Soc., Biochem. Soc. London, Soc. Exptl. Biol. Medicine; hon. mem. Harvey Soc., Argentina Soc., Endocrinol. Metabolism Biol. Soc. Chile, Academia Sinica (Republic of China; Hu Shih meml. lectr. 1967), Nat. Acad. Scis.; fgn. mem. Chilean Acad. Scis., Israel Biochem. Soc. Subspecialties: Endocrinology; Biochemistry (biology). Home: 901 Arlington Ave Berkeley CA 94707

LI, JONATHAN J., pharmacologist, educator; b. N.Y.C., June 22, 1939; s. Wah O. and Theresa M. Lee; m. Sara Antonia, Aug. 28, 1942; children: Christopher, Stephanie. A.B., Brown U., 1962; postgrad., Columbia U., 1963-64, Stanford U., 1965-67; Ph.D., SUNY-Syracuse, 1972. Research fellow dept biol. chemistry Harvard U., 1971-74; asst. dir. spl. diagnostic. and treatment unit endocrine sec. VA Med. Ctr., Mpls., 1974-76, research scientist, 1976—; asst. prof. dept. urologic surgery U. Minn., Mpls., 1980—; ad hoc mem. clin. cancer program project rev. com. NIH, HHS, 1983; con-founder, co-chmn. Gordon Research Conf. on Hormonal Carcinogenesis, 1985. Contbr. numerous articles in field to profl. jours. VA med. research grantee, 1974—; Nat. Cancer Inst. grantee, 1977—. Mem. Assn. VA Scientists (nat. sec. 1980—), Am. Soc. Biol. Chemists, Am Assn. Cancer Research, Endocrine Soc., Histochemical Soc., N.Y. Acad Sci., AAAS. Subspecialties: Cancer research (medicine); Receptors. Current work: Estrogen and chemical carcinogenesis, sex steroid receptors, mechanism of action of steroid hormones; metabolism of natural and synthetic steroids, inhibitors of carcinogenesis, mixed function oxidases, toxicology and drug metabolism. Office: 54th St and 48th Ave S Med Research Labs Bldg 49 VA Med Ctr Minneapolis Mn 55417

LI, JOSEPH K(WOK) K(WONG), molecular biologist, researcher, biotech. analyst; b. Hong Kong, Jan. 13, 1940; came to U.S., 1963, naturalized, 1983; s. Kan and Wai Ching (Chan) L.; m. Livia K., June 28, 1970; children: Karen K., Brenda K. Ph.D. in Microbiology, UCLA, 1975. Research asst. UCLA, 1970-74; postdoctoral fellow Duke U., 1975-77, med. research asso., 1977-80; mgr. immunology and virology Becton Dickinson Research Ctr., Research Triangle Park, N.C., 1980-82; research assoc. prof. U. N.C., Chapel Hill, 1982—; cons. in field. Mem. Am. Soc. Microbiology, Duke U. Comprehensive Cancer Center, N.Y. Acad. Sci. Subspecialties: Virology (biology); Species interaction. Current work: Hybridoma, recombinant DNA and synthetic polypeptide synthehsis; in vivo imaging; drugtargeting; DNA tumor virus and retrovirus research. Mem. All Star and All Conf. Team, So. Calif. Soccer Conf., Nat. Coll. Athletic Assn., 1964-66, capt., 1965-66. Office: Dept Pharmacology U NC Chapel Hill NC 27514

LI, LI-HSIENG, biochemist; b. Peiping, China, Dec. 31, 1933; came to U.S., 1959, naturalized, 1971; s. T.C. and S.F. (Wang) L.; m. Grace P. Ho, Mar. 23, 1960; children: Diana A., Robert G. B.S., Nat. Taiwan U., 1955; M.S., Va. Poly. Inst. and State U., 1962, Ph.D., 1964. Research assoc. Ind. U., Bloomington, 1964-65; research assoc. The Upjohn Co., Kalamazoo, 1965-73, sr. research scientist, 1973—; mem. exptl. therapeutics study sect. NIH, Bethesda, Md., 1978-80. Chmn. pub. edn. com. Am. Cancer Soc., Kalamazoo, 1977-79, pub. issue com., 1981—; chmn. pub. edn. com. Mich. Div., 1981—, bd. dirs., exec. com., 1981—. Nat. Taiwan Sugar Corp. scholar, 1953-55; NIH grantee, 1962-64; U.S. Dept. Agr. fellow, 1969. Mem. Am. Assn. Cancer Research, Am. Soc. Biol. Chemists, Am. Chem. Soc., Phi Sigma, Phi Lambda Upsilon, Sigma Xi, Phi Kappa Phi. Subspecialties: Cancer research (medicine); Biochemistry (medicine). Current work: Cancer research in areas of in vitro screening research and development; biochemical mechanism of action of antitumor agents and chemoimmuno-therapy of cancer. Office: The Upjohn Co 301 Henrietta St Kalamazoo MI 49001

LI, MING CHIANG, physicist; b. Ningpo, China, June 18, 1935; s. Yung Fu and Shih Heng (Fei) L.; m. Betty Wang, Jan. 30, 1965. B.S., Peking U., 1958; Ph.D., U. Md., 1965. Mem. Inst. Advanced Study, Princeton, 1965-67; asst. prof. Va. Poly. Inst. and State U., Blacksburg, 1967-72, assoc. prof., 1972—. Mem. Am. Phys. Soc., Sigma Xi. Subspecialties: Theoretical physics; Atomic and molecular physics. Current work: Laser and high precision spectroscopy; light scattering devel. interferometry; measurement with coherent beams.

LI, PAUL HSIANG, plant physiology educator; b. Shantung, China, May 4, 1933; came to U.S., 1958; s. T.S. and Y.Y. (Chang) L.; m. Paulina Wang, Aug. 17, 1963; children: John, Jennifer. B.S., Nat. Chung-Shing U., Taiwan, 1956; Ph.D., Oreg. State U., 1963. Prof. plant physiology U. Minn.-St. Paul, 1977—; vis. prof. Internat. Potato Ctr., Lima, Peru, 1973-74, Japan Soc. Promotion of Sci., 1976, Chinese Ministry of Agr., 1980. Editor: Plant Cold Hardiness and Freezing Stress, Vol. 1, 1978, Vol. 2, 1982. Mem. Am. Soc. Hort. Sci. (Dow Chem. award 1965, Alex Lauris award 1966), Am. Soc. Plant Physiologists, Crop Sci. Soc. Am., Potato Assn. Am., Soc. Cryobiology. Roman Catholic. Subspecialties: Plant physiology (agriculture); Plant physiology (biology). Current work: Environmental stress physiology with emphasis on temperature stress of crop plants. Home: 3489 Milton St Saint Paul MN 55112 Office: University of Minnesota Saint Paul MN 55108

LI, TINGYE, electrical engineer; b. Nanking, China, July 7, 1931; came to U.S., 1953, naturalized, 1963; s. Chao and Lily Wei-peng (Sie) L.; m. Edith Hsiu-hwei Wu, June 9, 1956; children: Deborah Chunroh, Kathryn Dairoh. B.Sc. in Elec. Engring, U. Witwatersrand, South Africa, 1953; M.S., Northwestern U., Evanston, Ill., 1955, Ph.D., 1958. Mem. tech. staff Bell Labs., Holmdel, N.J., 1957-67, dept. head repeater techniques research dept., 1967-76, lightwave media research dept., 1976—. Assoc. editor: Jour. on Lightwave Tech., 1983—; Editorial bd.: Procs. IEEE, 1974-83; contbr. articles on microwave antennas and propagation, lasers, coherent optics, optical communications, optical-fiber transmission to sci. jours., chpts. in books. Recipient Alumni Merit award Northwestern U., 1981. Fellow IEEE (W.R.G. Baker prize 1975, David Sarnoff award 1979), Optical Soc. Am. (chmn. optical communications tech. group 1979-80); mem. AAAS, Chinese Inst. Engrs. U.S.A. (dir. 1974-78, achievement award 1978), Nat. Acad. Engring., Sigma Xi, Eta Kappa Nu, Phi Tau Phi. Club: F.F. Fraternity. Subspecialties: Fiber optics; Electrical engineering. Current work: Optical-fiber communication; optical fiber fabrication; transmission properties; semiconductor lasers and detectors. Patentee in field. Office: Bell Laboratories Box 400 Holmdel NJ 07733

LI, VICTOR ON-KWOK, electrical engineering educator; b. Hong Kong, Oct. 11, 1954; U.S., 1973; s. Chia-Luen and Wai-Ying (Chan) L.; m. Regina Yui-Kwan Wai, Aug. 14, 1977. B.S., MIT, 1977, M.S., 1979, Elec. Engr., 1980, Sc.D., 1981. Cons. Pub. Systems Evaluation, Inc., Cambridge, Mass., 1977-80; teaching asst. MIT Cambridge, 1977-78, research asst., 1979-81; asst. prof. elec. engring. U. So. Calif., Los Angeles, 1981—; founding mem. Communication Scis. Inst., 1982—; chief systems cons. Optima Systems, Inc., Woodland Hills, Calif., 1983—; joint services elec. program researcher U.S. Dept. Def., 1982—. Contbr. numerous articles to profl. publs. N.Y.C. Urban fellow, 1975; NSF grantee, 1982—. Mem. IEEE, Ops. Research Soc. Am., Assn. Computing Machinery, Eta Kappa Nu, Tau Beta Pi. Club: MIT of So. Calif. (Los Angeles). Subspecialties: Database systems; Distributed systems and networks. Current work: Performance modeling; distributed databases; communication networks. Home: 3485 Ashwood Ave Los Angeles CA 90066 Office: Elec Engring Dept U So Calif Los Angeles CA 90089

LIANG, CHANG-SENG, cardiologist, medical researcher; b. Fukien, China, Jan. 6, 1941; came to U.S., 1966, naturalized, 1976; s. You-rang and Mu-lan L.; m. Betty P. Wang, July 29, 1967; children: Marilyn, Marybeth, Michelle. M.D., Nat. Taiwan U., 1965; Ph.D., Boston U., 1971. Diplomate: Am. Bd. Internal Medicine. Intern in medicine Brookdale Hosp. Center, Bklyn., 1966-67; resident in anesthesiology Univ. Hosp., Boston, 1967-68, resident in medicine, 1970-71, research fellow, 1971-74; fellow in cardiology, instr. medicine Boston U. and Boston City Hosp., 1973-74; asst. prof. pharmacology Boston U., 1973-78, asst. prof. medicine, 1974-78, assoc. prof. medicine and pharmacology, 1978-82; assoc. prof. medicine U. Rochester, N.Y., 1982—; mem. cardiovascular and pulmonary study sect. NIH, 1981—. Contbr. articles to profl. jours. NIH spl. fellow, 1971-72. Mem. Am. Physiol. Soc., Am. Fedn. Clin. Research, Am. Soc. Pharmacology and Exptl. Therapeutics, Am. Soc. Clin. Investigation, Am. Heart Assn., Am. Coll. Cardiology. Subspecialties: Cardiology; Pharmacology. Current work: Circulatory regulation; cardiovascular pharmacology; clinical pharmacology; coronary circulation; sympathetic nervous system pharmacology. Office: U Rochester Med Center Box 679 Rochester NY 14642

LIAO, PAUL F(OO-HUNG), physicist; b. Phila., Nov. 10, 1944; s. Tseng Wu and Tung Mei (Lin) L.; m. Karen Ann Pravetz, Aug. 31, 1968; children: Teresa, Joanna. B.S. in Physics, M.I.T., 1966, Ph.D., Columbia U., 1973. Research assoc. Columbia U. Radiation Lab., 1972-73; mem. tech. staff Bell Labs., Holmdel, N.J., 1973-80, head dept. quantum electronics research, 1980-83; div. mgr. physics and optical scis. research CSO, 1983—; mem. tech. program com. Internat. Quantum Electronics Conf., 1980, tech. program co-chmn. N.Am. subcom., 1982; mem. tech. program com. Conf. on Lasers and Electro Optics, 1981-82, chmn. subcom. on laser spectroscopy, nonlinear optics and phase conjugation, 1983. Contbr. numerous articles to profl. jours. and; editor: (with P.L. Kelley) Quantum Electronics series, 1981; mem. adv. bd.: CRC Handbook of Laser Science and Technology, 1982—. Mem. Am. Phys. Soc. (rep. to Joint Council on Quantum Electronics), Optical Soc. Am., IEEE. (mem. adminstrv. bd. quantitative elec. application Soc.). Methodist. Clubs: Fair Haven (N.J.) Sailing (first officer 1980-81, chief officer 1982. Subspecialties: Spectroscopy; Condensed matter physics. Current work: Quantum electronics, solid state physics, laser spectroscopy. Patentee in nonlinear optics, quantum electronics and laser spectroscopy. Office: Bell Labs Holmdel NJ 07733

LIAO, SHUEN-KUEI, immunology educator; b. Morioka, Japan, June 27, 1940; s. Lung-Sheng and Fa-Mei Liao (Hsu) L.; m. Mary

Elizabeth Rumble, Sept. 23, 1972; children: May-Lynn, Nelson G-Y. B.Sc., Tunghai U., 1964; Ph.D., McMaster U., 1971. Demonstrator, fellow, histology U. Toronto, Ont., Can., 1970-73; profl. asst. Ont. Cancer Found., Hamilton, 1973-74, research assoc., 1976—; lectr. to asst. prof. McMaster U., Hamilton, 1974-80, assoc. prof. pathology, pediatrics and immunology, 1980—; mem. staff lab. medicine Henderson Gen. Hosp., Hamilton, 1974—. Contbr. numerous sci. articles to profl. publs. Danish Cancer Soc. fellow, 1970; Med. Research Council Can. grantee, 1976—; Ont. Cancer Treatment and Research Found. grantee, 1978—; Nat. cancer Inst. Can. grantee, 1981—. Mem. AAAS, Am. Assn. Cancer Research, Can. Soc. Immunology, Cancer Soc. Cell Biology, Internat. Pigment Cell. Soc., N.Y. Acad. Sci. Subspecialties: Cancer research (medicine); Immunobiology and immunology. Current work: Work includes cancer immunology, cell biology, monoclonal antibodies, somatic genetics, and gene cloning. Office: McMaster Univ Pathology Dept Hamilton ON Canada L8N 3Z5

LIAO, SHUTSUNG, biochemist, educator; b. Tainan, Taiwan, Jan. 1, 1931; came to U.S., 1956, naturalized, 1970; s. Chi-Chun and Chin-Shen L.; m. Shuching Kuo, Mar. 19, 1960; children: Jane Tzufen, May Tzuming. Ph.D., U. Chgo., 1961. Asst. prof. U. Chgo., 1964-69, assoc. prof., 1969-71, prof. biochemistry, 1972—. Contbr. articles to profl. jours. NIH research grantee, 1965—; Am. Cancer Soc. research grantee, 1975—. Mem. Am. Soc. Biol. Chemists, Endocrine Soc., Internat. Soc. Biochem. Endocrinolgy. Subspecialties: Biochemistry (biology); Endocrinology. Current work: Mechanism of action of hormones, regulation of biosynthesis of proteins and nucleic acids. Home: 5632 S Woodlawn Ave Chicago IL 60637 Office: U Chgo 950 E 59th St Chicago IL 60637

LIAO, TA-HSIU, chemistry educator; b. Taipei, Taiwan, Feb. 22, 1942; came to U.S., 1965, naturalized, 1978; s. Shing-Tsu and A-Men (Shieh) L.; m. Tsui-Hsing, Liao/Hou, Feb. 6, 1973; children: Cinderella, Richard. B.S., Nat. Taiwan U., 1964; Ph.D., UCLA, 1969. Research assoc. Rockefeller U., N.Y.C., 1970-73, asst. prof., 1973-74, Okla. State U., Stillwater, 1974-76, assoc. prof., 1976—. NIH grantee, 1976. Mem. Am. Soc. Biol. Chemists, Am. Chem. Soc., AAAS. Subspecialties: Biochemistry (medicine); Biochemistry (biology). Current work: Structure-function relationships of proteins and peptides. Office: Okla State U Dept Biochemistry Stillwater OK 74078

LIAW, CHARLES *See also* **LIAW, HAW-MING**

LIAW, HAW-MING (CHARLES LIAW), physicist; b. Hau-lien, Taiwan, Jan. 20, 1942; s. Tse-wei and Chu-mei (Wang) L.; m. Sandra Jane Gabler. B.S., Taiwan Nat. Normal U., 1965; M.S. in Physics, U. Lowell, Mass., 1970; Ph.D. in Acoustic Biophysics, Boston U., 1979. Instr. U. Lowell, 1967-69; tchr. Bedford (Mass.) High Sch., 1969-71; vis. prof. Taiwan Nat. Normal U., Taipei, 1971-72; postdoctoral fellow U. Calif.-San Francisco, U. Calif.-San Diego, 1979-80; research scientist Lockheed-Calif., Burbank, 1981-82, Hughes Aircraft Co., El Segundo, Calif., 1982—. Contbr. articles to profl. jours. Subspecialties: Acoustics; Electro-optical systems. Current work: Acoustic and electro-optical sensor technology and systems; signal processing with computers. Office: Hughes Aircraft Co El Segundo CA 90245

LIBBEY, LEONARD MORTON, food science educator, consultant; b. Boston, Apr. 17, 1930; s. Leonard Frank and Eleanor (Jones) Miller) L.; m. Janet May Young, Aug. 15, 1971. B.V.A., U. Mass., Amherst, 1953; M.S., U. Wis.-Madison, 1953-54; Ph.D., Wash. State U.-Pullman, 1961. Research asst. U. Wis.-Madison, 1953-54, Wash. State U., Pullman, 1955-61; research assoc. Oreg. State U., Corvallis, 1962-68, assoc. prof., 1969-80, prof. dept. food sci., 1981—; cons. U.S. Army QMC, Natick, Mass., 1970-74. Editor: (with others) The Chemistry and Physiology of Flavors, 1967; contbr. articles to profl. jours. Grantee NSF, 1977-82, Nat. Cancer Inst., 1978—. Mem. Am. Chem. Soc., Inst. Food Technologists, Am. Soc. Mass Spectrometry, AAAS, Am. Oil Chemists Soc., U.S. Chess Fedn., N.Am. Truffling Soc. Republican. Subspecialties: Food science and technology; Mass spectrometry. Current work: Application of gas chromatography/mass spectrometry to flavor chemistry, carcinogen analysis, pheromone analysis with insects. Home: 905 NW 30th St Corvallis OR 97330 Office: Dept Food Sci Oreg State U Corvallis OR 97331

LIBBY, PAUL ROBERT, research biochemist; b. Torrington, Conn., Sept. 2, 1934; s. David and Lillian (Kaminsky) L.; m. Barbara Sue Rosenblum, Dec. 28, 1958; children: Matthew Jay, Kenneth Alan, Bruce Peter. B.S., Yale U., 1956; Ph.D., U. Chgo., 1962. Cancer research scientist dept. breast surgery Roswell Park Meml Inst., Buffalo, 1963-64, sr. cancer research scientist, 1964-72, assoc. cancer research scientist, 1972-77, assoc. cancer research scientist dept. exptl. therapeutics, 1977—; assoc. prof. dept. pharmacology SUNY, Buffalo; prof. dept. biology Niagara U. Mem. Am. Chem. Soc., AAAS, Am. Assn. Cancer Research, Am. Soc. Cell Biology, Am. Soc. Biol. Chemistry, Endocrine Soc., Sigma Xi. Democrat. Jewish. Subspecialties: Biochemistry (biology); Cell biology. Current work: Biochemistry of gene activation; biochemistry of tumor promotion. Office: Roswell Park Meml Inst Buffalo NY 14263

LIBBY, VIBEKE, integrated circuit development scientist; b. Copenhagen, July 29, 1950; U.S., 1980, naturalized, 1983; d. Preben Viggo and Lis (Sorensen) Eider; m. Stephen B. Libby, Apr. 27, 1980. Candidatus scientiarum degree in physics and chem, Niels Bohr Inst., U. Copenhagen, 1981. Assoc. scientist Stone & Webster Engring Corp., Boston, 1981-82; sr. scientist Raytheon Research Div., Lexington, Mass., 1982—. Contbr. articles in nuclear engring. to publs. Bd. dirs. Women for Energy, Boston, 1981-82. Mem. Am. Nuclear Soc. Subspecialties: Condensed matter physics; Integrated circuits. Current work: Semiconductor research: development of integrated circuit processing technology; computer-aided design; ternary logic; nuclear engineering; accident analysis, fission product transport and low level waste disposal. Home: 159 Emeline St Providence RI 02906 Office: Raytheon Research Div 131 Spring St Lexington MA 02173

LIBCKE, JOHN HANSON, pathologist; b. Detroit, Aug. 25, 1935; s. John William and Martha Ingebore (Hanson) L.; m. Jean Louise McIver, Aug. 16, 1957; children: Robert, Martha, Jane. A.B., Albion Coll., 1956; B.S., Wayne State U., 1959, M.D., 1959. Diplomate: Am. Bd. Pathology, 1964. Intern Womack Army Hosp., Ft. Bragg, N.C., 1959-60; resident Walter Reed Gen. Hosp., Washington, 1960-64; chief lab. services 121st Evacuation Hosp., Korea, 1964-65; chief anat. pathology Letterman Gen Hosp., San Francisco, 1965-69; asst. chief pathology Pontiac (Mich.) Gen. Hosp., 1969-73, dir. labs., 1973—. Served to lt. col. AUS, 1958-69. Mem. Mich. Assn. Blood Banks (pres. 1979-80, Mich. Soc. Pathologists trustee 1979-82). Republican. Presbyterian. Club: Detroit Yacht (fleet surgeon 1981-82). Subspecialty: Pathology (medicine). Current work: Surgical pathology; cytopathology. Office: Pontiac Gneral Hospital Pontiac MI 48053

LIBERATORE, MATTHEW JOHN, operations research educator, researcher; b. Phila., June 23, 1950; s. Matthew Albert and Phyllis Rose (Saltarelli) L.; m. Mary Jane Cunningham, Oct. 18, 1975; children: Kathryn, Michelle. B.A., U. Pa., 1972, M.S., 1973, Ph.D., 1976. Mgr. ops. research FMC Corp., Phila., 1976-78, mgr. research and devel. planning and evaluation, Princeton, N.J., 1978-80; asst. prof. mgmt. Temple U., Phila., 1980—; cons. FMC Corp., Phila. and Princeton, N.J., 1980-81. Mem. Inst. Mgmt. Sci., Ops. Research Soc. Am. Republican. Roman Catholic. Subspecialties: Operations research (mathematics); Operations research (engineering). Current work: Application and development of operations research methods in technology management and planning. Office: Temple U 13th and Montgomery Aves Philadelphia PA 19122

LIBURDY, ROBERT PETER, immunotoxicology educator; b. Detroit, Oct. 23, 1947; s. Amo and Deloris (Verna) L.; m. Nora L. Burgess, June 21, 1975. B.Sc., Cornell U., 1969; Ph. D., Brown U., 1975; M.B.A., U. Colo.-Boulder, 1981. Asst. prof. environ. toxicology NYU Med. Ctr., N.Y.C., 1981—. Served to capt. USAF, 1975-81. Mem. Am. Assn. Immunologists, Am. Soc. Biol. Chemists, Soc. Toxicology, Radiation Research Soc. Subspecialties: Immunotoxicology; Biochemistry (medicine). Current work: Cell-Mediated immunology; lymphoyte membrane; microwave radiation; nonionizing radiation. Invented flour-stat assay for human Ig, 1979, electronic resonance chromatography, 1980, cell-membrane dielectric breakdown assay, 1981. Home: 1245 Park Ave Apt 16A New York NY 10028 Office: NY Univ Med Ctr 550 First Ave New York NY 10016

LICHSTEIN, EDGAR, cardiologist; b. N.Y.C., Nov. 27, 1936; s. Joseph and Ruth (Weisner) L.; m. Marilyn Dorf, June 19, 1966; children: Adam, Amy. B.A., Columbia U., 1957; M.D., SUNY-Downstate Med. Ctr., 1961. Diplomate: Am. Bd. Internal Medicine (subspecialty cardiovascular disease). Intern Lenox Hill Hosp., N.Y.C.; resident in Medicine N.Y.U. Med. Center; dir. cardiology Mt. Sinai Hosp. Services, City Hosp. Ctr., Elmhurst, N.Y., 1968-76; dir. div. cardiology Maimonides Med. Ctr., Bklyn., 1976—; chmn. heart info. com. N.Y. Heart Assn., 1979-83, chmn. council on pub. edn., 1983—. Author: Hemodynamics Reference File, 1970; contbr. articles to profl. jours. Mem. New Rochelle (N.Y.) Bd. Edn., 1976-81. Served to capt. USAF, 1966-68. Nat. Heart and Lung Inst. grantee, 1978-82. Fellow Am. Coll. Cardiology, ACP, Am. Coll. Chest Physicians, Am. Council Clin. Cardiology, N.Y. Cardiologic Soc. Jewish. Subspecialty: Cardiology. Current work: Clinical research in the field of cardiovascular medicine.

LICHTENFELD, KAREN MOSS, medical oncologist; b. Phila., June 23, 1947; s. Jack N. and Miriam (Soifer) M. B.A., U. Pa., 1968; M.D., Jefferson Med. Coll., 1972. Diplomate: Am. Bd. Internal Medicine, Am. Bd. Med. Oncology. Intern Sinai Hosp., Balt., 1972-73; resident U. Md. Hosp., 1973-75; fellow Balt. Cancer Research Ctr., 1975-78, staff assoc. clin. br., 1975-77, investigator, 1977-78; practice medicine specializing in med. oncology, Balt., 1978—; mem. staff dept. medicine Sinai Hosp., Balt., 1978, Baltimore County Gen. Hosp., 1979, N. Charles Gen. Hosp., 1982; asst. prof. medicine U. Md. Sch. Medicine, Balt., 1977—. Mem. ACP, Am. Women's Med. Assn., Am. Soc. Clin. Oncology, Md. Soc. Internal Medicine, Baltimore City Med. Soc., H. A. Hare Honor Med. Soc., J. F. Gibbon Jr. Surg. Soc. Subspecialty: Chemotherapy. Office: 2435 W Belvedere Ave Suite 15 Baltimore MD 21215

LICHTENWALNER, DIANE MARIE, biochemist, researcher; b. Somerville, N.J., Dec. 17, 1952; d. Benedict Casimir and Agnes Helen (Gelzinis) Marshall; m. Mark Richard Lichtenwalner, July 20, 1974; g61 son, Daniel Benedict. B.A., Lafayette Coll., 1974; Ph.D., Temple U., 1978. Postdoctoral researcher Temple U., Phila., 1978-79, postdoctoral fellow, 1979—. Contbr. articles to profl. jours. Mem. Am. Soc. Microbiology, Am. Fedn. Clin. Research. Subspecialties: Biochemistry (medicine); Molecular pharmacology. Current work: Biomedical research.

LICHTER, ROBERT LOUIS, chemistry educator, researcher, administrator; b. Cambridge, Mass., Oct. 26, 1941. A.B., Harvard U., 1962; Ph.D., U. Wis.-Madison, 1967. Postdoctoral fellow Tech. U. Braunschweig, Ger., 1967-68, Calif. Inst. Tech., Pasadena, 1968-70; asst. prof. Hunter Coll., N.Y.C., 1970-75, assoc. prof., 1975-80, prof. chemistry, 1980-83, chmn. dept., 1977-81; regional grants dir. Research Corp., N.Y.C., 1983—. Co-author: 15-N NMR Spectroscopy, 1979, 13-C NMR Spectroscopy, 1980. NRC Travel awardee, 1975, 77; NSF awardee, 1981. Mem. Am. Chem. Soc., Royal Soc. Chemistry, Soc. Applied Spectrscopy. Subspecialty: Organic chemistry. Current work: 15-N and -13-C NMR spectroscopy, applications to organic structure elucidation. Office: 405 Lexington Ave New York NY 10174

LICHTMAN, MARSHALL A., physician, educator; b. N.Y.C., June 23, 1934; s. Samuel and Vera L.; m. Alice Jo Maisel, June 23, 1957; children: Susan, Joanne, Pamela. A.B., Cornell U., 1955; M.D., U. Buffalo, 1960. Resident in medicine Strong Meml. Hosp., Rochester, N.Y., 1960-63, chief resident, instr. medicine, 1965-66; USPHS postdoctoral research assoc. Sch. Public Health U. N.C., 1963-65; sr. instr., research trainee in hematology U. Rochester Sch. Medicine, 1966-67, asst. prof. medicine, spl. postdoctoral research fellow in hematology, 1968-70, assoc. prof. medicine and radiation biology, biophysics, 1971-74, prof. medicine and radiation biology and biophysics, 1974—, chief hematology unit dept. medicine, 1975—, assoc. dean acad. affairs and research, 1979, sr. assoc. dean, 1980—. Editor: Abnormalities of Granulocytes and Monocytes, 1975, Hematology for Practitioners, 1978, Hematology and Oncology, 1980; co-editor: Hermatology, 3d edit, 1983; contbr. articles to profl jours. Leukemia Soc. Am. scholar, 1969-74; USPHS grantee, 1971—. Fellow ACP; mem. Am. Fedn. Clin. Research, AAAS, Am. Soc. Hematology, Internat. Soc. Hematology, N.Y. Acad. Scis., Am. Soc. Clin. Investigation, Assn. Am. Physicians, Am. Assn. Cancer Research, Am. Physiol. Soc., Reticuloendothelial Soc., Soc. Exptl. Biology and Medicine, Am. Soc. Cell Biology. Subspecialties: Hematology; Physiology (medicine). Current work: Research interests include physiology and biochemistry of blood cells, hemoporisis and leukemia. Home: 138 Roby Rd Rochester NY 14618 Office: U Rochester Sch Medicine 601 Elmwood Ave Rochester NY 14642

LICHTMAN, ROBERT MARK, psychologist; b. Bklyn., Dec. 7, 1937; s. Jack and Matilda (Rubel) L.; m. Florence Pearl Greenstein, Dec. 14, 1958 (div. Sept. 1980); children: Ira Mark, Melissa Joy, Stewart Gordon, Jennifer Beth; m. Virginia Karla Witt, May 1, 1982. A.B. in Psychology, SUNY, N.Y.C., 1976; B.A., New Sch. Social Research, 1978, M.A., 1980; Ph.D. in Psychology, Columbia Pacific U., Calif., 1981. Cert. in behavior therapy L.I.U. Psychotherapist Nassau Hosp., Mineola, N.Y., 1977-82, cons. psychotherapist, 1982—; psychotherapist Nassau Pain & Stress Ctr., L.I., 1980—; psychologist Creedmoor Psychiatric Clinic, N.Y.C., 1982—. Mem. Psychologists for Social Responsibility, Washington, 1982. Served with U.S. Army, 1955-58. Mem. Am. Psychol. Assn. (cert. exptl. psychotherapist), Psychologists in Marital, Sex & Family Therapy, Assn. for Advancement Behavior Theraphy, Am. Assn. Artist Therapists, Mensa. Subspecialties: Behavioral psychology; Cognition. Current work: Currently working with schizophrenic population in state hospital and in the area of marital, sex and fmaily counseling in private practice; consultant to cardiac patients in stress management. Home: 48-19 38th St Long Island City NY 11101 Office: The Nassau Pain & Stress Center 222 Station Plaza N Mineola N.Y. 11501

LICHTON, IRA JAY, nutritionist, physiologist; b. Chgo., Sept. 18, 1928; s. Alex Sander and Elizabeth (Sollo) L.; m. Marilyn Mendel, Nov. 5, 1949; 1 son, Alex. Ph.D., U. Chgo., 1947; B.S., U. Ill., 1950, M.S., 1951, Ph.D., 1954. Research assoc. ob-gyn U. Chgo., 1954-56; research fellow cardiovascular Michael Reese Hosp., Chgo., 1956-58; instr. physiology Stanford (Calif.) U., 1958-62; assoc. prof. nutrition U. Hawaii, Honolulu, 1962-68, prof. human nutrition, 1968—; vis. colleague ob-gyn Bonn U., Venusberg, W.Ger., 1959; vis. prof. nutrition U. Md., College Park, 1976. Pres. Temple Emanu-el, Honolulu, 1971-73. Am. Heart Assn. research fellow, 1954-56. Fellow AAAS; mem. Am. Physiol. Soc., Soc. Study of Reproduction (charter), N.Y. Acad. Scis., Sigma Xi. Subspecialties: Physiology (medicine); Nutrition (biology). Current work: Appetite control, anthropometric study of child growth, fluid and electrolyte metabolism, diet and blood pressure. Office: Univ Hawaii Manor 1800 East-West Rd Honolulu HI 96822

LICHTWARDT, ROBERT WILLIAM, botany educator; b. Rio de Janeiro, Brazil, Nov. 27, 1924; s. Henry Herman and Ruth (Moyer) L.; m. Elizabeth Ann Thomas, Jan. 27, 1951; children: Ruth Elizabeth, Robert Thomas. A.B., Oberlin Coll, 1949; M.S., U. Ill., 1951, Ph.D., 1954. Research assoc. Iowa State U., Ames, 1955-57; asst. prof. dept. botany U. Kans., Lawrence, 1957-60, assoc. prof., 1960-65, prof., 1965—, chmn. dept., 1981—. Editor-in-chief: Mycologia, 1965-70. NSF postdoctoral fellow, 1954-55; NSF sr. postdoctoral fellow, 1963-64. Mem. Myco. Soc. Am. (hon. life mem., pres. 1971-72), Bot. Soc. Am. (chmn. microbiol. sect. 1975-76), Brit. Myco. Soc., Japan Myco. Soc. Subspecialty: Mycology. Current work: Fungi that parasitize or are symbiotically associated with arthropods, especially trichomycetes; medical mycology, especially histoplasmosis. Home: 2131 Terrace Rd Lawrence KS 66044 Office: University of Kansas Department of Botany Lawrence KS 66045

LIDDLE, GRANT WINDER, physician, educator; b. American Fork, Utah, June 27, 1921; s. Parley H. and Elizabeth (Winder) L.; m. Victoria Ragin; children—Kathryn (Mrs. James C. Wallwork), Annette (Mrs. Dennis Weight), Rodger, Robert, Patricia. B.S., U. Utah, 1943; M.D., U. Calif. at San Francisco, 1948. Intern in internal medicine U. Calif. Hosp., 1948-49, asst. resident, 1949-51; research fellow metabolic unit U. Calif. at San Francisco, 1951-53; instr. medicine U. Calif. Sch. Medicine, San Francisco, 1953; surgeon sect. clin. endocrinology Nat. Heart Inst., 1953-56; chief endocrine service Vanderbilt U. Sch. Medicine, 1956—, asso. prof. medicine, 1956-61, prof., 1961—, Harrie Branscomb disting. prof., 1979-80, chmn. dept. medicine, 1968—; Mem. endocrinology study sect. USPHS, 1958-62, chmn. diabetes and metabolism tng. grants com., 1963-66; mem. Nat. Adv. Arthritis and Metabolic Diseases Council, 1967-71; investigator selection com. VA, 1970-74. Contbr. articles to profl. jours. Served with AUS, 1943-46. Recipient Sir Henry Hallett Dale medal Soc. Endocrinology, Gt. Britain, 1973; Earl Sutherland prize for achievement in research, 1979. Master A.C.P. (John Phillips Meml. award 1977); mem. Assn. Am. Physicians, Endocrine Soc. (council 1962-65, 70-75, pres. 1973-74, Upjohn award 1962, Distinguished Leadership award 1971), Am. Soc. Clin. Investigation (sec.-treas. 1963-66, pres. 1966-67), So. Soc. Clin. Investigation (council 1963-64, 66-67, pres. 1965-66, Founder's medal 1977), Internat. Soc. Endocrinology (chmn. exec. com. 1968-76, pres. 1976-80), Assn. Profs. Med. (pres. 1977-78). Subspecialty: Endocrinology. Home: 770 Norwood Dr Nashville TN 37204

LIDE, DAVID REYNOLDS, JR., physicist, sci. adminstr.; b. Gainesville, Ga., May 25, 1928; s. David Reynolds and Kate (Simmons) L.; m. Mary Ruth Lomer, Nov. 5, 1955; children: David A., Vanessa G., James H., Quentin R. B.S., Carnegie Inst. Tech., 1949; M.A., Harvard U., 1951, Ph.D., 1952. Research physicist Nat. Bur. Standards, 1954-63, chief molecular spectroscopy sect., 1963-69; chief Office Standard Reference Data, 1969—. Author research papers on microwave and infrared spectroscopy, high temperature sci., infrared lasers and related areas; also articles. Recipient Samuel Wesley Stratton award Nat. Bur. Standards, 1968; Gold medal Dept. Commerce, 1968. Mem. Am. Phys. Soc., Internat. Council Sci. Unions. (sec. gen., com. on data for sci. and tech.), Internat. Union Pure and Applied Chemistry (pres. phys. chemistry div.), Am. Chem. Soc., AAAS. Subspecialties: Information systems (information science); Atomic and molecular physics. Current work: Direct program for developing scientific and engineering data bases with aid of computer and telecommunications technology. Home: 4604 Tournay Rd Bethesda MD 20816 Office: Nat Bur Standards Washington DC 20234

LIEB, MARGARET, microbiology educator; b. Bronxville, N.Y., Nov. 28, 1923; d. Werner and Margaret L. B.A., Smith Coll., 1945; M.A., Ind. U., 1946; Ph.D., Columbia U., 1950. Postdoctoral fellow Calif. Inst. Tech., Pasadena, 1950-53, Pasteur Inst., Paris, 1953-54, Radium Inst., 1954-55; asst. prof. biology Brandeis U., Waltham, Mass., 1955-60; assoc. prof. microbiology U. So. Calif., Los Angeles, 1960-67, prof, 1967—; program dir. genetic biology NSF, Washington, 1972-73, mem. genetic biology panel, 1976-79. Mem. editorial bd.: Gene. Nat. Found. for Infantile Paralysis fellow, 1953-54; career devel. award, 1962-72. Fellow AAAS; mem. Am. Soc. for Microbiology (chmn. virology div. 1978), Genetics Soc. Am., Phi Beta Kappa. Subspecialties: Chronobiology; Gene actions. Current work: Mutation and recombination in bacteriophage lambda, mismatch repair, gene regulation. Office: Univ So Calif Sch Medicine 2020 Zonal Ave Los Angeles CA 90033

LIEBENBERG, DONALD HENRY, physicist, science administrator, researcher; b. Madison, Wis., July 10, 1932; s. Rex Lionel and Made (Sachtjen) L.; m. Norma Malmanger, Sept. 7, 1957; children: Karl Henry, Kira Jean. B.S., U. Wis., M.S., Ph.D. Mem. staff Los Alamos Nat. Lab., N.Mex., 1961—; program dir. solar terrestrial research NSF, 1967-68, program dir. low temperature physics, 1981—. Contbr. articles to profl. jours. Pres. Los Alamos Concerts Assn., 1972-77, Los Alamos Sinfonietta, 1971, Miner's Candle Homeowners Assn., Breckenridge, Colo., 1979. Boris A. Bahkmeteff fellow, 1959. Mem. AAAS, Am. Astron. Soc., Am. Phys. Soc., Am. Geophys. Union. Club: Cosmos (Washington). Subspecialties: Condensed matter physics; Low temperature physics. Current work: High pressure materials properties studies especially with the light isotopes; superfluid helium film studies; solar corona studies (High temperature plasma); magneto-optical studies. Office: Nat Sci Found 1800 G St NW Washington DC 20550

LIEBERHERR, KARL JOSUA, computer science educator; b. Wattwil, Switzerland, Aug. 27, 1948; came to U.S., 1977; s. Karl and Berta (Weder) L.; m. Ruth Sylvia Kuber, Sept. 3, 1976; children: Andrea, Eva. Dr.Sc.Math., Swiss Fed. Inst. Tech., 1977. Research and teaching asst. Swiss Fed. Inst. Tech., Zurich, 1973-77, vis. prof., 1982-83; vis. asst. prof. Fla. State U., Tallahassee, 1977-78, U. N.Mex., Albuquerque, 1978-79; asst. prof. computer sci. Princeton U., 1978—; cons. Swiss Air Force, Bern, 1973-75, Crypto AG, Zug, Switzerland, 1981—. Contbr. articles to profl. jours. NSF grantee, 1980, 82; Swiss NSF grantee, 1982. Mem. Assn. Computing Machinery, Soc. Indsl. and Applied Math., IEEE. Subspecialties: Algorithms; Computer-aided design. Current work: Hardware description languages, VLSI algorithms, silicon compilers, combinatorial optimization, cryptography. Home: 95 Hartley Ave Princeton NJ 08540 Office: Dept EECS Princeton Univ Princeton NJ 08544

LIEBERMAN, EDWARD MARVIN, physiologist, researcher, educator, consultant; b. Lowell, Mass., Feb. 10, 1938; s. Irving and Nellie (Silverman) L.; m. Harriet Handman, June 12, 1960; children: Lynn R., Dana B., Kurt M. B.S., Tufts U., 1959; M.S., U. Mass., 1961; Ph.D., U. Fla. Coll. Medicine, 1965. Swedish Med. Research Council fellow U. Uppsala, Sweden, 1966-68; asst. prof. Bowman Gray Sch. Medicine, 1968-72, assoc. prof., 1972-76; assoc. prof. physiology Sch. Medicine, East Carolina U., 1976-78, prof., 1978—; bd. dirs. N.C. Heart Assn., Chapel Hill; extramural research reviewer NSF. Contbr. articles to profl. jours. NIH grantee, 1967-68, 70-74, 76-78; NSF grantee, 1978-81, 82-85; U.S. Army Research Office grantee, 1982-85. Mem. Biophys. Soc., Soc. Neurosci., Am. Heart Assn., N.Y. Acad. Sci., Am. Physiol. Soc. Subspecialties: Neurophysiology; Biophysics (biology). Current work: Electrophysiology of glia, neuron-glia interactions, membrane transport of nerve and muscle, cellular metabolism. Office: Dept Physiology East Carolina U Sch Medicine Greenville NC 27834

LIEBERMAN, GERALD JACOB, mathematician; b. Phila., July 11, 1948; s. Samuel V. and Sophie (Serinsky) L.; m. Diana Ziegler, June 14, 1970; children: Barry, Eliza. B.A., Wesleyan U., Middletown, Conn., 1968; M.S., U. Rochester, 1971, Ph.D., 1973. Asst. prof. Colby Coll., Waterville, Maine, 1973-76; staff research scientist Gen. Motors Research Labs., Warren, Mich., 1976—. Nat. Merit scholar, 1964; NSF trainee, 1968-72. Mem. Am. Math. Soc., Math. Assn. Am., Ops. Research Soc. Am., Pub. Choice Soc., U.S. Wayfarer Assn. (commodore 1982—). Subspecialty: Applied mathematics. Current work: Building mathematical models to quantitatively describe and analyze the interface between General Motors and society as an aid to corporate policy-making. Home: 15956 Lauderdale Dr Birmingham MI 48009 Office: Gen Motors Research Labs Warren MI 48090

LIEBERMAN, MICHAEL MERRIL, microbiologist, army officer; b. Chgo., June 10, 1944; s. Charles and Anne (Oberlander) L.; div.; 1 son, Ted G. B.S. in Biochemistry, U. Chgo., 1966, Ph.D. in Microbiology, 1969; grad., Command and Gen. Staff Coll., 1979. Postdoctoral research assoc. NASA-Ames Research Ctr., Moffett Field, Calif., 1969-71; sr. research microbiologist Cutter Labs., Inc., Berkeley, Calif., 1971-75; commd. capt. U.S. Army, 1976; sr. research microbiologist dept. clin. investigation Brooke Army Med. Ctr., Ft. Sam Houston, Tex., 1976-83; chief microbiology service, dept. clin. investigation Tripler Army Med. Ctr., Hawaii, 1983—. Contbr. articles to profl. jours. Mem. Am. Soc. Microbiology, Am. Assn. Immunologists. Subspecialties: Infectious diseases; Microbiology (medicine). Current work: Research in microbiology and immunology, specifically the development of a vaccine against the gram-negative bacterium, Pseudomonas aeruginosa, consisting of ribosomes prepared from the organism. Home: 3130 Ala Ilima St Apt 24D Honolulu HI 96818 Office: Tripler Army Med Center Dept Clin Investigation Tripler AMC HI 96859

LIEBERMAN, PHILIP, video game company executive; b. Trenton, N.J., Feb. 8, 1955; s. Mendel and Bela (Jerzey) L.; m. Doris Mary Loring, July 29, 1980. A.A., San Francisco City Coll., 1976; B.A. in Physics, San Francisco State U., 1980. Technician, Kevex Corp., Foster City, Calif., 1972-73, field engr., 1977-78; prin. owner Lieberman & Assocs., Mill Valley, Calif., 1978-81; v.p., prin. Pacific Novelty & Mfg. Inc., Marina Del Rey, Calif., 1981-83; prin. Lieberman & Assocs., Los Angeles, 1983—; instr. computer sci., and engring. UCLA, 1982—. Editor books in computer field. Mem. IEEE, Assn. Computing Machinery, Spl. Interest Group on Graphics. Subspecialties: Graphics, image processing, and pattern recognition; Artificial intelligence. Current work: Design of video games with graphics, sound and strategy. Developer video games.

LIEBERMAN, ROBERT PERRY, radiology educator; b. Buffalo, July 6, 1947; s. Samuel Lawrence and Erma Ada (Perry) L. B.S., U. Chgo., 1969; M.S., NYU, 1975, M.D., 1975, Ph.D., 1975. Diplomate: Nat. Bd. Med. Examiners, Am. Bd. Radiology. Intern Cleve. Met. Gen. Hosp., 1975-76; resident in radiology NYU Med. Ctr., 1976-79; instr. U. Wis. Hosp., Madison, 1979-80, asst. prof., 1980-83; assoc. prof. U. Nebr. Med. Center, Omaha, 1983—. NIH fellow, 1969-75. Mem. Radiol. Soc. N.Am., Am. Roentgen Ray Soc. Subspecialty: Diagnostic radiology. Current work: Needle biopsy, interventional radiology of the urinary tract. Office: U Wis Hosp 600 Highland Ave Madison WI 53792

LIEBERMAN, SEYMOUR, educator; b. N.Y.C., Dec. 1, 1916; s. Samuel D. and Sadie (Levin) L.; m. Sandra Spar, June 5, 1944; 1 son, Paul B. B.S., Bklyn. Coll., 1936; M.S., U. Ill., 1937; Ph.D. (Rockefeller scholar 1939-41), Stanford U., 1941; Traveling fellow, U. Basle, Switzerland, Eidgenoess. Tech. Hochschule, Zurich, Switzerland, 1946-47. Chemist Schering Corp., 1938-39; spl. research asso. Harvard, 1941-45; asso. mem. Sloan-Kettering Inst., 1945-50; mem. faculty Columbia Coll. Phys. and Surg., N.Y.C., 1950—, prof. biochemistry, 1962—; pres. Inst. Health Scis., St. Luke's Roosevelt Hosp. Center, 1981—; Mem. subcom. human applications radioactive materials N.Y.C. Health Dept.; Pfizer traveling fellow McGill U., 1968; Syntex lectr. Mexican Endocrine Soc., 1970; mem. Am. Cancer Soc. panel steroids, 1945-49, 1949-50, 1957-60; mem. endocrine study sect. NIH, 1959-63, chmn., 1963-65, mem. gen. clin. research centers, 1967-71; Mem. med. adv. com. Population Council, 1961-73; mem. endocrinology panel Cancer Chemotherapy Nat. Service Center, 1958-62; cons. WHO human reprodn. unit, 1972-74, Ford Found., 1974-77. Editor: Jour. Clin. Endocrinology and Metabolism, 1963-67; editorial bd., 1958-63, 68-70, Jour. Biol. Chemistry, 1975-80; contbr. articles to profl. jours. Recipient Distinguished Alumnus award Bklyn. Coll., 1971. Fellow N.Y. Acad. Scis., Nat. Acad. Scis.; mem. Am. Soc. Biol. Chemists, Am. Chem. Soc., Internat. Soc. Endocrinology (U.S. del. central com.), Endocrine Soc. (Ciba award 1952, Koch award 1970, council 1970-73, pres. 1974-75), Harvey Soc. Asso. Subspecialties: Biochemistry (medicine); Endocrinology. Current work: Steroid hormone biochemistry, endocrine biochemistry. Home: 32-22 163d St Flushing NY 11358 Office: 630 W 168th St New York NY 10032

LIEBIG, WILLIAM JOHN, medical manufacturing company exec.; b. Huntingdon, Pa., Mar. 24, 1923; s. William A. and Gertrude L. (Schierz) L.; m. Suzanne V. King, Nov. 16, 1978; 1 dau.: Barbara. B.S., Juniata Coll., 1943; M.S., Phila. Coll. Textiles, 1949; M.B.A., NYU, 1951. Div. mgr. Susquehanna Mills, Inc., N.Y.C., 1949-54; gen. mgr. Meadox Weaving Co., Waldwick, N.J., 1954-55; pres. Dormeyer Sales Corp., Haledon, N.J., 1955-60; div. sales mgr. Webcor, Inc., Mt. Vernon, N.Y., 1955-60, Camfield, Inc., Mt. Vernon, 1955-60; pres., chief exec. officer Meadox Meds., Inc., Oakland, N.J., 1961—; v.p., dir. Huntingdon Throwing Mills, Mifflinburg, Pa., 1967—; pres., dir. Meadox Prostetics Ltd., Eng., 1978—; dir. N.Am. Bank, Phoenix; chmn. Liebig Found. Contbr. articles to profl. jours. Pres. Harrington Park (N.J.) Bd. Edn., 1957-65; chmn. Tri-County Com. on I287 Alignment, 1965-73. Served with USAAF, 1942-45. Decorated Air medal with 3 oak leaf clusters, D.F.C., Presdl. citation with oak leaf cluster. Mem. Assn. Advancement Med. Instrumentation, Health Industry Mfg. Assn., Asia Soc., Alumni Assn. NYU, Phi Beta Kappa. Lodge: Masons. Subspecialties: Biomedical engineering. Office: 103 Bauer Dr Oakland NJ 07436

LIEBMAN, JEFFREY MARK, research scientist; b. Milw., Nov. 7, 1946; s. Albert and Elaine Shirley (Smuckler) L.; m. Anita H. Liebman, June 22, 1974; children: Cynthia, Jacquelyn. B.A. cum laude, Oberlin (Ohio) Coll., 1968; Ph.D. in Psychology (NSF predoctoral fellow), UCLA, 1973. Postdoctoral fellow U. Calif.-San Diego, 1973-76; mgr. CIBA-GEIGY, Summit, N.J., 1976—. Contbr. articles to profl. jours. Mem. Soc. Neurosci., Am. Soc. Pharmacology and Exptl. Therapeutics, Sigma Xi. Club: Sierra. Subspecialties: Neuropharmacology; Physiological psychology. Current work: Behavioral pharmacology, mechanisms of drug effects on animal behavior, development of psychotherapeutic drugs. Office: 556 Morris Ave Summit NJ 07901

LIEBMAN, JOEL FREDRIC, chemistry educator; b. Bklyn., [illegible] Murray and Lucille H. (Spitz) L.; m. Deborah Van Vechten, June 12, 1970. B.S., Bklyn. Coll., 1964; M.A., Princeton U., 1967, Ph.D., 1970. NATO postdoctoral fellow U. Cambridge, Eng., 1971; NRC-Nat. Bur. Standards postdoctoral research assoc. Nat. Bur. Standards, 1972; asst. prof. chemistry U. Md.-Baltimore County, Catonsville, 1972-77, assoc. prof., 1977-82, prof., 1982—. Contbr. articles to profl. jours. Recipient Outstanding Young Tchr. award U. Md.-Baltimore County, 1981; NSF trainee and fellow, 1967, 68-70; German Acad. Exchange Service faculty fellow, 1976; U. Md. Faculty research fellow, 1977. Mem. Am. Chem. Soc., Am. Phys. Soc., Sigma Xi. Subspecialties: Theoretical chemistry; Organic chemistry. Current work: Chemical bonding, organic thermochemistry, ion energetics, noble gas and fluorine compounds, molecular geometry, mathematical and formal theory. Home: 5107 Edmondson Ave Baltimore MD 21229 Office: U Md Baltimore County 5401 Wilkens Ave Catonsville MD 21228

LIEBMAN, JON CHARLES, civil engineer, educator; b. Cin., Sept. 10, 1934; s. J. Charles and Joan (Heineman) L.; m. Judith Rae Stenzel, Dec. 27, 1958; children: Christopher Brian, Rebecca Anne, Michael Jon. B.S., U. Colo., Boulder, 1956; M.S., Cornell U., Ithaca, N.Y., 1963, Ph.D., 1965. Asst. prof., then asso. prof. Johns Hopkins U., Balt., 1965-72; prof. civil engring. U. Ill., Urbana, 1972—, head dept., 1978—. Served from ensign to lt. USN, 1956-61. Mem. Am. Soc. Engring. Edn. (Western Electric Fund award 1969), ASCE, Assn. Environ. Engring. Profs. (dir. 1980-82), Ops. Research Soc. Am. Subspecialties: Environmental engineering; Operations research (engineering). Office: Newmark Lab 208 N Romine St Urbana IL 61801

LIEBOW, CHARLES, physiologist, educator, dentist; b. Bklyn., June 17, 1944; s. Raymond and Ruth (Slavitsky) L.; m. Roslyn Lee Raskin, July 4, 1968; children: Bradley, Adam, Lisa. A.B., Univ. Coll., N.Y. U., 1966; D.M.D., Harvard U., Boston, 1970; Ph.D., U. Calif.-San Francisco, h61973. Dental research trainee Boston outpatient VA Harvard U. Grad. Sch., Boston, 1970-71; postdoctoral fellow U. Calif.-San Francisco, 1971-73; asst. prof. physiology Cornell U. Med. Sch., N.Y.C., 1973-80; assoc. sci. dir. Nat. Pancreatic Cancer Project, New Orleans, 1980—; assoc. prof. surgery and physiology La. State U. Med. and Dental Sch., New Orleans, 1980—; research reviewer Bd. Regents La., New Orleans, 1981. Contbr. articles to profl. publs. Cystic Fibrosis Found. fellow, 1973; grantee, 1976; Andrew Mellon Found. fellow, 1975; NIH grantee, 1976, 78; Schlieder Found. grantee, 1982. Mem. Am. Physiol. Assn., Am. Pancreatic Assn., Internat. Assn. Dental Research, N.Y. Acad. Sci. Jewish. Subspecialties: Physiology (medicine); Gastroenterology. Current work: Research on non-exocytic mechanisms of enzyme secretion from pancreas. 70112 Home: 4017 Martinique Ave Kenner LA 70062

LIEBOWITZ, HAROLD, aeronautical engineering educator, university dean; b. Bklyn., June 25, 1924; s. Samuel and Sarah (Kaplan) L.; m. Marilyn Iris Lampert, June 24, 1951; children: Alisa Lynn, Jay, Jill Denice. B. in Aero. Engring., Poly. Inst. Bklyn., 1944, M., 1946, D., 1948. With Office Naval Research, 1948-60; asst. dean Grad. Sch., exec. dir. engring. expt. sta. U. Colo. at Boulder, 1960-61; also vis. prof. aero. engring.; head structural mechs. br. Office Naval Research, 1961-68, dir. Navy programs in solid propellant mechanics, 1962-68; dean Sch. Engring. and Applied Sci., George Washington U., Washington, 1968—; research prof. Cath. U., Washington, 1962-68; Cons. Office Naval Research, 1960—, Pratt & Whitney Aircraft Co., 1981-82, Pergamon Press, 1968—, Acad. Press, 1968—; mem. Israeli-Am. Materials Adv. Group, 1970—; sci. adviser Congl. Ad Hoc Com. on Environ. Quality, 1969—; co-dir. Joint Inst. for Advancement Flight Scis. NASA-Langley Research Center, Hampton, Va., 1971—. Editor: Advanced Treatise on Fracture, 7 vols., 1969-72; founder, editor-in-chief: Internat. Jour. Computers and Structures, 1971—; founder, editor: Internat. Jour. Engring. Fracture Mechanics, 1968—; contbr. articles to profl. jours. Recipient Outstanding award Office Naval Research, 1961, Research Accomplishment Superior award, 1961; Superior Civilian Service award USN, 1965, 67; Commendation Outstanding Contbns. sec. navy, 1966. Fellow AAAS, AIAA, Am. Soc. Metals; mem. Soc. Engring. Scis. (past pres.), Sci. Research Soc. Am., Am. Technion Soc. (dir.), Am. Acad. Mechanics (founder, pres.), Internat. Coop. Fracture Inst. (founder, v.p.), Engrs. Council for Profl. Devel., Nat. Acad. Engring. (home sec.), Sigma Xi, Tau Beta Pi, Sigma Gamma Tau, Omega Rho, Pi Tau Sigma, Sigma Tau. Subspecialties: Fracture mechanics; Aeronautical engineering. Current work: Aeronautical and structures research. Home: 9112 LeVelle Dr Chevy Chase MD 20815 Office: 725 23d St NW Washington DC 20052

LIECHTI, CHARLES A., electrical engineer, physicist, corporation executive; b. Mar. 12, 1937. Ph.D. in Elec. Engring, Swiss Fed. Inst. Tech., 1967, Diploma in Physics. Now head dept. device physics Hewlett-Packard Labs., Hewlett-Packard Co., Palo Alto, Calif.; nat. lectr. Microwave Theory and Techniques Soc., 1979. Fellow IEEE (several outstanding paper awards). Subspecialty: Electrical engineering. Office: Device Physics Dept Hewlett-Packard Co 1501 Page Mill Rd Palo Alto CA 94304

LIECHTY, RICHARD DALE, surgery educator, researcher; b. Lake Geneva, Wis., Oct. 20, 1925; s. Ernest A. and Margaret (Demerath) L.; m. Valerie Grunow; children: Robert Mark, Valerie Ann, Richard Cameron. B.A., Yale U., 1950; M.D., Northwestern U., Chgo., 1954. Diplomate: Am. Bd. Surgery. Intern U. Mich., Ann Arbor, 1954-55, resident in surgery, 1955-59; asst. prof. to assoc. prof. U. Iowa, 1959-71; assoc. prof. to prof. surgery U. Colo., Denver, 1971—. Author, editor: Synopsis of Surgery, 1968. Bd. dirs. Project Hope, 1968. Served with USNR, 1944-46; PTO. Recipient teaching award U. Iowa, 1970, U. Colo. Kaiser Permanente, 1974. Fellow ACS; mem. Western Surg. Assn. (sec.), Am. Assn. Endocrine Surgeons, Denver Acad. Surgery. Republican. Presbyterian. Subspecialties: Endocrinology; Surgery. Current work: Endocrine clinical research, thyroid and parathyroid diseases, surgical treatment of obesity. Office: Dept of Surgery University of Colorado Medical Center 4200 E 9th Ave Denver CO 80220

LIEDL, GERALD LEROY, materials engineering educator, researcher, consultant; b. Fergus Falls, Minn., Mar. 2, 1933; s. Max A. and Leeta C. (Spencer) L.; m. Carol E. Lee, June 1, 1957; children: Barbara, Janice. B.S., Met.E., Purdue U., 1955, Ph.D., Met.E., 1960. Asst. prof. materials engring. Purdue U., West Lafayette, Ind., 1959-65, assoc. prof., 1965-73, asst. head, 1973-78, head, prof., 1978—; mem. tech. staff Bell Telephone Labs., 1964. Contbr. articles to profl. jours. Served with U.S. Army, 1960—. Recipient outstanding undergrad. teaching award Purdue U., 1971, 76; Incentive award Alumni Found., 1976. Mem. AIME, Am. Soc. Metals, Am Soc. for Engring. Educators. Presbyterian. Lodge: Rotary. Subspecialties: Materials; Metallurgy. Current work: X-ray diffraction, electron microscopy, phase transformations. Home: 2830 Henderson St West Lafayette IN 47906 Office: Purdue U Materials Engring CMET Bldg West Lafayette IN 47907

LIEF, LAURENCE HOWARD, physician; b. St. Louis, Nov. 12, 1947; s. Milton and Rose (Modest) L.; m. Patricia Ross Cleary, June 8, 1980. A.B., Princeton U., 1970; M.D., Cornell U., 1974. Diplomate: Am. Bd. Internal Medicine (Cardiovascular Diseases). Intern and resident Georgetown U. Hosp., Washington, 1974-77; fellow in cardiology U. Calif.-San Diego, 1977-79; instr. U. So. Calif., Los Angeles County Med. Ctr., Los Angeles, 1979-80; asst. clin. prof. medicine U. Calif.-San Francisco, 1980—; attending cardiologist Mt. Zion Hosp., San Francisco, 1980—. Fellow Am. Coll. Cardiology, Am. Heart Assn. (council clin. cardiology), Am. Fedn. Clin. Research; mem. ACP. Subspecialty: Cardiology. Current work: Ventricular function in ischemic, valvular, myopathic heart disease. Office: Mt Zion Hosp PO Box 7921 San Francisco CA 94120

LIENTZ, BENNET PRICE, information systems educator; b. Hollywood, Calif., Oct. 24, 1942; s. B. Price and Josephine (Palen) L.; m. Martha B. Benson, Aug. 29, 1964; children: Bennet, Andrew, Charles. B.A., Claremont Men's Coll., 1964; M.S., U. Wash., 1966, Ph.D., 1968. Systems analyst, mgr. System Devel. Corp., Santa Monica, Calif., 1968-70; asst. prof. info. systems U. So. Calif., Los Angeles, from 1970, assoc. prof., to 1974; prof. info. systems UCLA Grad. Sch. Mgmt., 1974—; cons. Atlantic Richfield, Security Pacific Bank, Aerojet Gen., Trans. Tech. Author: Systems in Action, 1978, Software Maintenance Management, 1980, Distributed Systems, 1981; contbr. articles to profl. jours. Office of Naval Research grantee, 1974—. Mem. Ops. Research Soc., Assn. Computing Machinery, Am. Statis. Assn., Inst. Math. Stats. Club: Los Angeles Atheletic. Subspecialties: Information systems (information science); Distributed systems and networks. Current work: Software maintenance, distributed systems, microcomputers. Office: U Calif Grad Sch Mgmt 405 Hilgard St Los Angeles CA 90024

LIEPA, GEORGE ULDIS, nutrition educator; b. Oldenburg, Ger., Oct. 4, 1946; came to U.S., 1949, naturalized, 1962; s. Oskars and Anna (Baldonis) L.; m. Candice Harr, July 7, 1979; 1 dau., Arianne. B.A., Drake U., Des Moines, 1968, M.A., 1970; Ph.D., Iowa State U.-Ames, 1976. Asst. to dean Drake U., 1968-70; teaching asst. Iowa State U., 1970-74; fellow U. Tex. Heath Sci. Ctr., San Antonio, 1976-79; asst. prof. nutrition Tex. Woman's U., 1979—; cons. Anderson Clayton Foods, Dallas. Contbr. articles to profl. jours. Served to maj. Tex. Army N.G., 1976-83. Mem. Am. Inst. Nutrition, Gerontol. Soc., Am. Oil Chemists Soc. (sec. protein sect. 1981-83), Sigma Xi, Phi Delta Gamma. Presbyterian. Club: Univ. (pres. 1982-83). Subspecialty: Nutrition (medicine). Current work: Effects of nutrition on human health and disease. Home: 2303 Mercedes St Denton TX 76201 Office: Texas Woman's U Denton TX 76204

LIESKE, JAY HENRY, astronomer; b. Warsaw, Indi, June 22, 1941; s. Henry Louis and Marguerite Virginia Ida (Jones) L.; m. Sarah Adeline Crichlon, May 18, 1965; children: Jay Henry, Camilla, Stephan. B.S., Valparaiso U., 1963; M.S., Yale U., 1964, Ph.D., 1968. Research asst. Yale U., 1965-68; sr. research engr. Jet Propulsion Lab., Pasadena, Calif., 1968-70, mem. tech. staff, 1970-80, sr. research scientist, 1980—; instr. West Coast U., 1969-73. Author: Theory of Motion of Jupiter's Galilean Satellites, 1977, Expressions for the Precession Quantities Based Upon the IAU System of astronomical Constants, 1977. Active Boy Scouts Am. Recipient NASA medal for Exceptional Sci. Achievement, 1977, NASA Achievement award for Planetary Ephemeris Devel., 1979, NASA Achievement award for Satellite Ephemeris Devel., 1981; Alexander von Humboldt award for Sr. Am. Scientists, 1980. Mem. Am. Astronom. Soc., Internat. Astronom. Union, Sigma Xi. Democrat. Lutheran. Subspecialties: Satellite studies; Relativity and gravitation. Current work: Development of state-of-the-art planet and satellite ephemerides; development of reference coordinate systems; analysis of observations of galilean satellites for ephemeris improvement. Office: Jet Propulsion Lab 4800 Oak Grove Dr Pasadena CA 91103

LIGHT, ALAN RAY, neurophysiologist, educator; b. Tulsa, Mar. 22, 1950; s. Raymond Edmund and Zenobia Rose (Ogle) L.; m. Kathleen C., Dec. 22, 1972. B.A., Hamilton Coll., 1972; Ph.D., SUNY-Syracuse, 1976. Research asst. SUNY-Syracuse, 1972-76; State of N.Y. Research Found. fellow U.N.C., Chapel Hill, 1976-77, research instr. dept. physiology, 1977-79, research asst.prof., 1979—. Mem. Soc. for Neurosci. Subspecialty: Neurobiology. Current work: Researcher in pain systems and descending control of pain. Office: U NC Dept Physiology 51 Med Research Bldg 2064 Chapel Hill NC 27514

LIGHTMAN, ALLAN JOEL, research physicist, optical diagnostics designer; b. Toronto, Ont., Can., Sept. 2, 1942; came to U.S., 1971; m. Nona Podlisker, June 30, 1963; children: Dalia, Arik. B.A.Sci., U. Toronto, 1963, M.A., 1965; Ph.D., Weizmann Inst., Rehovot, Israel, 1971. Asst. prof. elec. engring. Wayne State U., 1971-78; research physicist U. Dayton, 1978—. Contbr. articles to profl. jours. Mem. Optical Soc. Am., IEEE, AIAA. Subspecialties: Laser-optical diagnostics; Combustion processes. Current work: Combustion study using laser diagnostics-LDA, RAMAN, visualization; computer controlled experimentation with distributed processing; optical data analysis, holography, interferometry. Patentee real-time optical data processor for coherent radar. Office: U Dayton Research Institute KL 461 Dayton OH 45469

LIGLER, FRANCES SMITH, immunologist; b. Louisville, June 11, 1951; d. George Frederick and Mary Hagan Smith; m. George Todd Ligler, Aug. 18, 1972; children: Amy Elizabeth, Adam George. B.S., Furman U., Greenville, S.C., 1972; D.Phil., Oxford (Eng.) U., 1977. Postdoctoral fellow U. Tex. Med. Ctr., San Antonio, 1975-76; asst. instr. Southwestern Med. Sch., Dallas, 1976-78, instr., 1978-80; primary scientist E.I. DuPont de Nemours and Co., Wilmington, Del., 1980—. Contbr. articles to profl. jours. Mem. Am. Assn. Immunologists, Wilmington Women in Bus. Presbyterian. Subspecialties: Immunology (agriculture); Cancer research (medicine). Current work: Analysis of functional potential of human leukemia and lymphoma cells by cytofluorimetric deterinations of surface markers, in vitro differentiation, response to chemotherapy. Office: DuPont 500 S Ridgeway Ave Glenolden PA 19036

LIGLER, GEORGE TODD, computer graphics co. exec.; b. Gary, Ind., Oct. 4, 1949; s. George Edward and Audrey (Anderle) L.; m. Frances Hart Smith, Aug. 19, 1972; 1 dau., Amy. B.S., Furman U., 1971; M.Sc., Oxford (Eng.) U., 1973, D.Phil., 1975. Asst. prof. U. Tex.-San Antonio, 1975-76; research mgr. Tex. Instruments, Dallas, 1976-80; dep. gen. mgr. Burroughs Corp., Paoli, Pa., 1980-82; pres. Aydin Controls, Ft. Washington, Pa., 1982—; cons. in field; panel chmn. Air Force Studies Bd., Washington, 1981. Rhodes scholar, 1971; Woodrow Wilson fellow, 1971; named to Furman U. Hall of Fame, 1979. Mem. IEEE, Assn. Computing Machinery, Am. Assn. Rhodes Scholars, Am. Trust for Wolfson Coll. (v.p., trustee). Republican. Presbyterian. Subspecialties: Graphics, image processing, and pattern recognition; Software engineering. Current work: Design, marketing and manufacturing of computer graphics workstations, display generators, high resolution color monitors. Home: 606 Kilburn Rd Wilmington DE 19803 Office: 401 Commerce Dr Fort Washington PA 19034

LIGON, JAMES TEDDIE, agricultural engineer; b. Easley, S.C., Feb. 20, 1936; s. Henry Grace and Gracia (Carson) L.; m. Martha Nelle Craig, June 11, 1958; children: Melissa Grace, James Mark, Polly Claire. B.S., Clemson U., 1957; M.S., Iowa State U., 1959, Ph.D., 1961. Registered profl. engr., S.C. Asst. prof. U. Ky., Lexington, 1961-66; assoc. prof. Clemson (S.C.) U., 1966-71, prof. agrl. engring., 1971—; dir. Water Resources Research Inst., 1975-78. Mem. Am. Soc. Agrl. Engrs., Am. Geophys. Union. Prebyterian. Subspecialties: Agricultural engineering; Resource conservation. Current work: Soil drainage, subsurface irrigation, soil water plant relationships. Office: Agrl Engring Dept Clemson U Clemson SC 29631

LIH, MARSHALL MIN-SHING, chemical engineer, consultant, researcher; b. Nanking, China, Sept. 15, 1936; s. Kun-Hou and Marion Worsang (Liang) L.; m. June Tsun-Ping Young, Apr. 22, 1962; children: Matthew, Andrew, Angela. B.S., Nat. Taiwan U., 1958; M.S., Wis.-Madison, 1960, Ph.D., 1962. Research engr. E.I. duPont de Nemours & Co., Inc., Buffalo, 1962-64; mem. faculty Cath. U. Am., Washington, 1964-74; program dir. thermodynamics and mass transfer NSF, Washington, 1973-76, sect. head engring. chemistry and energetics, 1976-79, div. dir. chem. and process engring., 1979—; sr. research scientist Nat. Biomed. Research Found., 1966-73; vis. prof. and chmn. chem. engring. Nat. Taiwan U., 1970-71; cons. to industry; mem. adv. com. govt. agys; lectr. Author: Transport Phenomena in Medicine and Biology, 1975, also articles. Bd. dirs. Sino-Am. Cultural Soc., Washington, 1976-79. B.F. Ruth Meml. Symposium lectr. Iowa State U.; Ralph B. Derr Meml. lectr. Bucknell U.; Dupont Disting. vis. scholar Va. Poly. Inst. and State U. Mem. Am. Inst. Chem. Engrs. (chmn. Nat. Capital sect. 1968-69), Sigma Xi, Phi Lambda Upsilon. Subspecialties: Chemical engineering; Biomedical engineering. Current work: Catalysis, biomedical engineering, color technology and operations research. Office: 1800 G St NW Suite 1126 Washington DC 20550

LIKENS, GENE ELDEN, ecologist; b. Pierceton, Ind., Jan. 6, 1935; s. Colonel Benjamin and Josephine (Garner) L.; m.; children: Kathy, Gregory, Leslie. B.S., Manchester (Ind.) Coll., 1957, D.Sc. (hon.), 1979; M.S., U. Wis., 1959, Ph.D., 1962. Asst. zoology Manchester Coll., 1955-57; grad. teaching asst. U. Wis., 1957-59, vis. lectr., 1963; instr. zoology Dartmouth Coll., 1961, instr. biol. scis., 1963, asst. prof., then assoc. prof., 1963-69; mem. faculty Cornell U., 1969—, prof. ecology, 1972—, Charles A. Alexander prof. biol. scis., 1983—; dir. Inst. Ecosystem Studies, 1983—; v.p. N.Y. Bot. Garden, 1983—; vis. prof. (Center Advanced Research); also dept. environ. scis. U. Va., Charlottesville, 1978-79; lectr. Williams Summer Inst. Coll. Tchrs., 1966, 67, Drew Summer Inst. Coll. Tchrs., 1968, Cornell U. Alumni Assn., 1978; Paul C. Lemon ecology lectr. SUNY, Albany, 1978; chmn. New Eng. div. task force conservation aquatic ecosystems U.S. Internat. Biol. Program, 1966-67; vis. assoc. ecologist Brookhaven Nat. Lab., 1968; C.P. Snow lectr. Ithaca Coll., 1979; Robert S. Campbell lectr. U. Mo., 1980; Disting. Ecologist lectr. N.C. State U., 1980; cons. in field, mem. numerous govt. and sci. panels, participant numerous confs. Author books and papers in field. Recipient Conservation award Am. Motors Corp., 1969; 75th anniversary award U.S. Forest Service, 1980; Disting. Achievement award Lab. Biomed. and Environ. Studies, UCLA, 1982; NATO sr. fellow, 1969; Guggenheim fellow, 1972-73; grantee NSF, EPA. Fellow AAAS; mem. Nat. Acad. Scis., Ecol. Soc. Am. (chmn. study com. 1971-74, v.p. 1978-79, pres. 1981-82), Am. Soc. Limnology and Oceanography (pres. 1976-77, First G.E. Hutchinson award for excellence in research 1982), Am. Acad. Arts and Scis., Am. Polar Soc., Explorers Club, Freshwater Biol. Assn., Internat. Assn. Gt. Lakes Research, Internat. Water Resources Assn. (charter), Soc. Internat. Limnologiae (nat. rep.), Sigma Xi, Gamma Alpha, Phi Sigma. Methodist. Subspecialties: Ecology; Ecosystems analysis. Current work: Biogeochemistry, acid rain, ecosystem analyses, watershed studies. Office: Inst Ecosystem Studies NY Bot Garden Cary Arboretum Box AB Millbrook NY 12545

LILJE, KARL DAVID, mech. engr., educator, cons.; b. Scranton, Pa., Aug. 10, 1935; s. Ralph Waldo and Helen Rhoda (Ball) L.; m. Ann Marie Victoria, Nov. 6, 1960; children: Anneliese, Erik. B.S.M.E., Pa. State U., 1957; M.S.M.E., N.Y.U., 1960. Registered profl. engr., Ohio. Instr. in mech. engring. N.Y.U., 1957-60; mfg. engr. Westinghouse Electric Corp., Columbus, Ohio, 1960-63; opto-mech. engr. Kollmorgen Corp., Northampton, Mass., 1963-65; sr. opto-mech. engr. Eastman Kodak Co., Rochester, N.Y., 1965-71; sr.mech. engr. Cogar Corp., Hopewell Junction, N.Y., 1971-72; cons. mech. engring., Sharon, Conn., 1972-76; sr. opto-mech. engr. Perkin Elmer Corp., Danbury, Conn., 1976-81; assoc. prof. engring. tech. Calif. Poly. State U., San Luis Obispo, 1981—; cons. opto-mechanics Perkin Elmer, 1981—. Mem. ASME, Am. Soc. Elec. Engrs. Subspecialties: Mechanical engineering; Opto-mechanics consulting. Current work: Teaching applied machine design, mechanisms, dynamics, statics, descriptive geometry, sr. design projects. Inventor safety interlock for high-speed centrifuges; co-inventor tape controlled memory clip sorting machine. Home: 251 Del Mar Ct San Luis Obispo CA 93401 Office: Dept Engring Tech Calif Poly State U San Luis Obispo CA 93407

LILLEGRAVEN, JASON ARTHUR, geology educator; b. Mankato, Minn., Oct. 11, 1938; s. Arthur Oscar and Agnes Mae (Eaton) L.; m. children: Brita Anna, Ture Andrew. A.A., Los Angeles Harbor Coll., 1959; B.A., Calif. State U., 1962; M.S., S.D. Sch. Mines & Tech., 1964; Ph.D., U. Kans., 1968. Faculty San Diego State U., 1969-75; faculty U. Wyo., Laramie, 1976—, prof. geology, 1979, prof. zoology, 1983—; program dir. systematic biology program NSF, Washington, 1977-78. Contbr. articles to profl. jours. Mem. Am. Soc. Mammalogists, Geol. Soc. Am., Paleontol. Soc., Soc. Vertebrate Paleontology, Soc. Study of Evolution. Subspecialties: Systematics; Paleobiology. Current work: Researcher in evolutionary history of Mesozoic and early Cenozoic mammals as indicated by the fossil record and comparative reproductive biology of living species. Office: Dept Geology and Geophysics U Wyo PO Box 3006 University Station Laramie WY 82071

LILLER, MARTHA H(AZEN), astronomer; b. Cambridge, Mass., July, 1931; d. Harold L. and Katherine P. (Salisbury) Hazen; m. William Liller (div.); children: John Avery, Hilary Webb. A.B., Mt. Holyoke Coll., 1953; M.A., U. Mich., 1955, Ph.D., 1958. Instr. U. Mich., summer 1957, research assoc., lectr., 1959-60; instr. Mt. Holyoke Coll., 1957-59; lectr. Wellesley Coll., 1961-63, 66-67; research fellow Harvard U., 1957-59, 60-69; adj. assoc. prof. Boston U., 1979; curator astron. photographs Harvard Coll. Obs., Harvard U., Cambridge, Mass., 1969—; lectr. Harvard U., 1983—. Contbr. numerous articles to profl. jours. Mem. Am. Astron. Soc., Internat. Astron. Union, Phi Beta Kappa, Sigma Xi. Subspecialty: Optical astronomy. Current work: Research in globular star clusters, variable stars, stellar evolution. Office: Harvard Coll Observatory 60 Garden S Cambridge MA 02138

LILLER, WILLIAM, astronomer; b. Phila., Apr. 1, 1927; s. Carroll and Catherine (Dellinger) L.; m. Lorraine Dundas Brown, Apr. 4, 1951; 1 dau., Tamara Kay; m. Martha Hazen, June 20, 1959; children: John Avery, Hilary Webb. A.B., Harvard U., 1949; M.A., U. Mich., 1951, Ph.D., 1953. Instr. to assoc. prof. U. Mich., Ann Arbor, 1953-59; Robert Wheeler Willson prof. applied astronomy Harvard U., Cambridge, Mass., 1960-82; sr. research astronomer Inst. Isaac Newton, Santiago, Chile, 1982—. Contbr. articles to profl. jours. Served with USN, 1945-46. Grantee U. Mich., 1955-60, ONR, 1958-63, NSF, 1963-82, Harvard U., 1973-79. Mem. Am. Astron. Soc., Internat. Astron. Union, AAAS, Am. Acad. Arts and Scis., Astron. Soc. Pacific. Clubs: Instituto to Pro Musica, Vina del Mar. Subspecialties: Optical astronomy; 1-ray high energy astrophysics. Current work: X-ray sources, globular clusters, planets, comets.

LILLESAND, THOMAS MARTIN, remote sensing educator, researcher; b. Laurium, Mich., Oct. 1, 1946; s. Walter J. and Gladys L. (Johnston) L.; m. Theresa Ann Hefmeister, Aug. 31, 1968; children: Mark Thomas, Kari Lea, Michael Thomas. B.S., U. Wis., 1969, M.S., 1970, Ph.D., 1973. Mem. faculty Sch. Environ. and Resource Engring., SUNY-Syracuse Coll. Environ. Sci. and Forestry, 1973-78; mem. faculty dept. forest resources and dept. civil and mineral engring. U. Minn., Mpls., 1978-82; prof. environ. studies, forestry and civil and environ. engring. U. Wis., Madison, 1982—, dir., 1982—; mem. (Land Remote Sensing Satellity Adv. Com.), 1982—. Author: (with Ralph W. Kiefer) Remote Sensing and Image Interpretation, 1979. Mem. ASCE, Am. Soc. Photogrammetry (Alan Gordon Meml. award 1980), Remote Sensing Soc. U.K., Internat. Soc. Photogrammetry and Remote Sensing. Roman Catholic. Subspecialty: Remote sensing (geoscience). Current work: Broad interests in the application of remote sensing in agriculture, forestry, water resources, environmental monitoring, and land use analysis. Home: 321 Cheyenne Trail Madison WI 53705 Office: 1225 W Dayton St Room 1231 Madison WI 53706

LILLYWHITE, MALCOLM ALDEN, thermophysical scientist, educator; b. Washington, June 7, 1940; s. Benjamin Alden and Leah (Plowman) L.; m. Lynda Knobloch, May 25, 1979. B.S. in Physics, Coll. William and Mary, 1963; Ph.D., U. Calif., 1970. Registered thermophys. scientist. Physicist NASA-Langly Space Ctr., Hampton, Va., 1958-62; sr. physicist Slumberger Oil Co., Washington, 1964-67; chief engr., supr. Thermophysics Lab., Martin Marietta Corp., Denver, 1967-72; dir. Domestic Tech. Inst., Denver, 1973-80; pres. Domestic Tech. Internat., Inc., Denver, 1982—; prof. tech. mgmt. Grad. Sch. Internat. Studies, U. Denver, 1982—. Author: Solar Simulation Technology, 1967, Thermophysical Properties Handbook, 1968, Passive Solar Greenhouse Design, 1983; contbr. articles to profl. publs. Bd. dirs. Nat. Gasohol Comm., Lincoln, Nebr., 1978-81; chmn. passive solar tax credits com. Internat. Solar Energy Soc., Denver, 1980—. Recipient George Washington Meml. Engring. award Va. Acad. Scis., 1959. Mem. AIAA, IEEE, ASTM, Internat. Solar Energy Soc., Inst. Environ. Scis. Subspecialties: Food science and technology; Solar energy. Current work: Third World food and energy technology development; small scale industry decentralized food and energy production; rural techno-economic development; technical manpower development and training; renewable energy technology. Office: Domestic Technology International Inc 6726 S Happy Hill Rd Evergreen CO 80439

LIM, DANIEL V., microbiology educator; b. Houston, Apr. 15, 1948; s. Don H. and Lucy (Toy) L.; m. Carol Sue Lee, Sept. 2, 1973. B.A., Rice U., 1970; Ph.D., Tex. A&M U., 1973. Postdoctoral fellow Baylor Coll. Medicine, Houston, 1973-76; asst. prof. microbiology U. South Fla., Tampa, 1976-81, assoc. prof., 1981—, interim chmn. dept. biology, 1983—; cons. GIBCO Labs., 1982—, The Conservancy, 1980—, Pharmacia Diagnostics. Trustee, treas. Suncoast Assn. Chinese-Ams., Inc., Tampa, 1982—. Robert A. Welch Found. fellow, 1971-73, 73-76. Mem. Inter-Am. Soc. Chemotherapy (v.p. 1982—, trustee), Am. Soc. Microbiology (policy com. chmn. southeastern br. 1981-83, Carski Found. Disting. Teaching award com. 1983-86), Am. Pub. Health Assn., Internat. Soc. Chemotherapy (mem. council 1983—), AAAS, Sigma Xi. Subspecialties: Microbiology (medicine); Ecosystems analysis. Current work: Development of rapid techniques for identification of pathogenic bacteria; studies of virulence mechanisms in pathogenic bacteria, specifically Neisseria gonorrhoeae, Streptococcus agalactiae and Vibrio cholerae; environmental microbiology. Office: Biology Dep U South Fla Tampa FL 33620

LIN, CHAR-LUNG CHARLES, statistical programmer; b. Gee-Lung, Taiwan, Feb. 26, 1951; came to U.S., 1976; s. Sheng and Ma-Jen (Chen) L.; m. Nancy N. Chu, Aug. 17, 1979 (dec. May 6, 1982); 1 dau., Melissa. M.S., Memphis State U., 1978; Ph.D., Iowa State U., 1982. Statis. programmer BMDP Statis. Software, Los Angeles, 1982—. Mem. Am. Statis. Assn. Subspecialties: Statistics; Mathematical software. Current work: Computation related problems in statistics. Office: BMDP Statis Software 1964 Westwood Blvd Suite 202 Los Angeles CA 90025

LIN, CHINLON, electronic engr., optical scientist; b. Taiwan, Jan 19, 1945; came to U.S., 1968; s. Bing-Chuan and Shiao-Chi (Tsang) L.; m. Helen C. L., Aug. 10, 1969. B.S., Taiwan U., Taipei, 1967; M.S., U. Ill., 1970; Ph.D., U. Calif.-Berkeley, 1973. Mem. tech. staff Bell Labs., Holmdel, N.J., 1974—; research asst. Electronics Research Labs. U. Calif., 1970-73. Contbr. articles to profl. jours. Recipient Electronics Premium award Inst. Elec. Engrs. London, 1980. Mem. IEEE, Optical Soc. Am. (topical advisor on fiber and integrated optics). Subspecialties: Fiber optics; Laser technology. Current work: Optical electronics and fiber optics; optical fiber communication, laser technology, electro-optical engring. Patentee in field. Office: Bell Labs Holmdel NJ 07733

LIN, CHUN CHIA, research physicist, educator; b. Canton, China, Mar. 7, 1930; s. Yue Hang Lam and Kin Ng. B.S., U. Calif.-Berkeley, 1951; M.A., 1952; Ph.D., Harvard U., 1955. Asst. prof. physics U. Okla., Norman, 1955-59; assoc. prof. physics, 1959-63, prof. physics, 1963-68, U. Wis., Madison, 1968—; cons., univ. retainee Tex. Instruments Inc., 1960-68; cons. Sandia Labs., 1976-81; sec. Gaseous Electronics Conf., 1972-73. Contbr.: sci. research articles to publs. including Jour. Chemical Physics, Phys. Rev. Sloan Found. fellow, 1962-66; research grantee NSF and Air Force Office Sci. Research. Fellow Am. Phys. Soc. (sec. div. electron and atomic physics 1974-77). Subspecialties: Atomic and molecular physics; Condensed matter physics. Current work: Atomic and molecular collision processes; radiation of atoms and molecules excited by electron impact and laser irradiation; electronic energy band theory of crystalline solids, impurity atoms in solids, amorphous solids. Home: 1652 Monroe St Apt C Madison WI 53711 Office: Dept Physics Univ Wis Madison WI 53706

LIN, DOROTHY SUNG, physician; b. Detroit, Mar. 28, 1925; d. M.N. and Tso-Yen (Liu) Sung; m. Chao-Ming Lin, Apr. 4, 1954; 1 dau., Sharyn. B.Sc., Foochow Christian U., 1948; M.D., Shantung Med. Coll., Tsinan, China, 1953. Diplomate: Am. Bd. Radiology, Am. Bd. Nuclear Medicine. Intern Shantung Med. Coll. Hosp., 1953, resident in diagnostic radiology, 1953-56, Phila. Gen. Hosp., 1976-77; fellow in nuclear medicine Hosp. of U. Pa., Phila., 1977-78; attending physician Thomas Jefferson U. Hosp., Phila., 1978-80, U. Miss. Med. Center, Jackson, 1980-83, VA Med. Center, 1980—. Recipient Physicians Recognition award AMA, 1980-83. Mem. Soc. Nuclear Medicine, Miss. Med. Assn., Central Med. Soc. Miss. Subspecialties: Diagnostic radiology; Nuclear medicine. Office: VA Med Center 1500 E Woodrow Wilson Dr Jackson MS 39216

LIN, HAYASHI *See also* **LIN, PEI-JAN PAUL**

LIN, HEH-SEN, elec. engr., biomed. product researcher; b. China, Jan. 3, 1942; came to U.S., 1966, naturalized, 1975; s. Chao-Yuan and Shu-Ying (Shea) L.; m. Alice Ko-Chien Ho, June 8, 1968; children: Andrew Li-Shing, David Li-Wen. Ph.D. in Elec. Engring, Case-Western Res. U., 1968. Project mgr. LFE Corp., Waltham, Mass., 1970-75; sr. staff mem. Medtronic Inc., Mpls., 1975-77; elec. engring. mgr. Edwards Pacemaker div. Am. Hosp. Supply Corp., Los Angeles, 1977-80; mgr. ELA Med. Co., Paris, 1980-82; founder, pres. Lin & Co., Los Angeles, 1982—; assoc. prof. elec. engring. dept. U. Minn., Mpls., 1975-76; cons. on Chinese trade and biomed. products. Contbr. articles to profl. jours. Mem. Pres. Reagan's Task Force. Served as lt. Chinese Air Force, 1965-66. NSF scholar, 1970-73. Mem. IEEE. Subspecialty: Biomedical engineering. Current work: Implantable pacemakers in cardiology field; pacemaker, cardiology, EKG,biomaterials. Patentee pacemaker field. Office: Lin & Co Los Angeles CA

LIN, HSIU-SAN, cell biologist, radiation oncologist; b. Nagoya, Japan, Mar. 15, 1935; came to U.S., 1962, naturalized, 1976; s. Mao-Sung and Tao L.; m. Su-Chiung Chen, Sept. 22, 1962; children: Kenneth, Bertha, Michael. M.D., Nat. Taiwan U., 1960; Ph.D., U. Chgo., 1968. Cert. Am. Bd. Radiology, 1982. Intern Cook County Hosp., Chgo., 1962-63; resident in internal medicine, 1963-64; resident in therapeutic radiology Mallinckrodt Inst. Radiology, St. Louis, 1979-81; asst. prof. radiology Washington U., St. Louis, 1971-76; assoc. prof., 1976—; vis. scientist U. Oxford, Eng., 1977-78. Cons. editor: Jour. Reticuloendothelial Soc, 1982—; contbr. articles to profl. jours. Recipient Research Career Devel. award NIH, 1974-79. Mem. Am. Soc. Microbiology, Am. Assn. Cancer Research, Reticuloendothelial Soc. Subspecialties: Cell biology (medicine); Radiology. Current work: Differentiation of mononuclear phagocytes. Office: 510 S Kingshighway Saint Louis MO 63110

LIN, LOUIS MIN-TSU, dental educator; b. Tainan, China, Mar. 14, 1939; came to U.S., 1969, naturalized, 1978; s. Chin-Shian and Luan-Tsu (Chen) L.; m. Betty C. Huang, Apr. 23, 1970; 1 son, John Jeffy. B.D.S., Chung Shan Med. and Dental Coll., Taiwan, 1964; Ph.D. in Pathology, U. Okla., 1972; D.M.D., U. Medicine and Dentistry N.J.-Newark, 1976. Diplomate: Am. Bd. Endodontics. Resident in oral pathology La. State U.-New Orleans, 1972-74; chief resident in endodontics U. Conn., Farmington, 1976-79; asst. prof. dentistry Fairleigh Dickinson U., Hackensack, N.J., 1979-80; assoc. prof. dentistry U. Medicine and Dentistry N.J.-Newark, 1983—. La. State U. fellow, 1972; Found. U. Medicine and Dentistry N.J. grantee, 1981. Mem. Internat. Assn. for Dental Research, ADA, Am. Acad. Oral Pathology, Am. Assn. Endodontists. Subspecialties: Endodontics; Oral pathology. Current work: Pathogenesis of pulpal-periapical tissue complex; tissue response to viable and killed bacteria. Home: 16 Cherry Ln Parsippany NJ 07054 Office: U Medicine and Dentistry NJ 100 Bergen St Newark NJ 07103

LIN, MING-CHANG, research chemist, educator; b. Hsinpu, Hsinchu, Taiwan, Oct. 24, 1936; came to U.S., 1967, naturalized, 1975; s. Fushin and Tao May (Hsu) L.; m. Juh-Huey Chern, June 26, 1966; children: Karen, Linus H., Ellena J. B.Sc., Taiwan Normal U., Taipei, 1959; Ph.D., U. Ottawa, 1966. Postdoctural research fellow U. Ottawa, Ont., Can., 1965-67; postdoctural research assoc. Cornell U., Ithaca, N.Y., 1967-69; research chemist Naval Research Lab., Washington, 1970-74; supervisory research chemist, head chem. kinetics sect., 1974-82, Sr. Scientist for chem. kinetics, 1982—; adj. prof. chemistry Cath. U. Am., Washington, 1981—. Contbr. over 120 articles on reaction kinetics, chem. lasers, combustion and planetary atmosphere chemistry and applications of lasers to studying chem. kinetics to profl. jours. Served as 2d lt. Taiwan ROTC, 1960-62. Recipient Hillebrand prize Chem. Soc. Washington, 1975; Navy Civilian Meritorious Service award, 1979; Humboldt award, 1982; Guggenheim fellow, 1982. Fellow Washington Acad. Scis. (Phys. Scis. award 1976); mem. Am. Chem. Soc., Combustion Inst., N. Am. Taiwanese Profs. Assn., Sigma Xi (Pure Sci. award Naval Research Lab. chpt. 1978). Club: Taiwanese Assn. Am. Subspecialties: Kinetics; Laser-induced chemistry. Current work: Lasers and other modern diagnostic tools are used to study kinetics and mechanisms of homogeneous and heterogeneous (catalytic) chemical reactions. Office: Code 6100 Naval Research Lab Washington DC 20375

LIN, PEI-JAN PAUL (HAYASHI LIN), diagnostic radiol. physicist, cons. radiol. physics; b. Taipei, Taiwan, China, Aug. 25, 1946; s. Jintoku and Rie Hayashi; m. Keiko M. Lin, May 11, 1946; children: Rika, Rina. B.S. in Physics, Rikoy U., Tokyo, 1969, M.S., De Paul U., 1974; Ph.D. in Sci, U. Tsukuba, Ibaraki, Japan, 1981. Cert. diagnostic radiol. physics Am. Bd. Radiology, 1977. Asst. med. physicist, dept. therapeutic radiology Rush-Presbyterian-St. Luke's Hosp., Chgo., 1971-73; instr. dept. radiology Northwestern U. Med. Sch., Chgo., 1973-76; assoc., 1976-77, asst. prof., 1977-82, assoc. prof., 1982—; radiol. cons. various hosps.; staff physicist, dept. diagnostic radiology affiliated profl. staff Northwestern Meml. Hosp. Contbr. papers to profl. publs. Spl. research grantee Philips Med. Systems, Inc., 1982. Mem. Am. Coll. Radiology, Am. Assn. Physicists in Medicine, Soc. Photo-Optical Instrumentation Engrs., Radiol. Soc. N.Am., Soc. Radiol. Engring. Subspecialties: Imaging technology; Biomedical engineering. Current work: Imaging properties in diagnostic radiology; information transfer and mass data storage for radiology; physics and engring. of diagnostic radiol. imaging equipment evaluation, performance evaluation and testing of radiological imaging equipment, including Computed Tomography, Digital Subtraction Angiography and nuclear magnetic resonance imaging equipment. Office: Dept Radiology Northwestern Univ Med Sch 303 E Chicago Ave Room 2-307 Chicago IL 60611

LIN, ROBERT PEICHUNG, physicist; b. China, Jan. 24, 1942; s. Tung Hua L. B.S., Calif. Inst. Tech., 1962; Ph.D., U. Calif.-Berkeley, 1967. Asst. research physicist U. Calif.-Berkeley, 1967-74, assoc., 1974-79, research physicist, 1979—, sr. fellow, 1980—. Contbr. articles to profl. jours. Mem. Am. Astron. Soc., Am. Geophys. Union. Subspecialties: Solar physics; High energy astrophysics. Current work: Energetic solar flare particles; X-ray and gamma ray spectroscopy of the sun and cosmic sources. Office: Space Scis Lab U Calif-Berkeley Berkeley CA 94720

LIN, SHEN, mathematician; b. Amoy, Fukien, China, Feb. 4, 1931; s. Chio-Shih and Shui-Hsian (Wang) L.; m. Jih-Jie Chang, Oct. 23, 1971; m. Mona Lo, Nov. 12, 1956 (div. 1964); children: John, David, Robert. B.S. summa cum laude, U. Philippines, 1951; M.A., Ohio State U., 1953, Ph.D., 1963. Instr. Ohio State U., Columbus, 1956-59, lectr., research assoc., 1962-63; asst. prof. Ohio U., Athens, 1959-62; vis. lectr. Princeton (N.J.) U., 1972; mem. tech. staff Bell Labs., Murray Hill, N.J., 1963—; cons. AT&T Long Lines, Bedminster, N.J., 1976—. Named Disting. mem. tech. staff Bell Labs., 1982. Mem. Am. Math. Soc., Math. Assn. Am., Soc. Indsl. and Applied Math. Subspecialties: Mathematical software; Algorithms. Current work: Design of algorithms and mathematical software to perform optimization of telecommunication network. Home: 159 Southgate Rd Murray Hill NJ 07974 Office: Bell Labs Inc 600 Mountain Ave Murray Hill NJ 07974

LIN, SHENG HSIEN, chemist, educator, research, cons.; b. Taiwan, Sept. 17, 1937; came to U.S., 1962, naturalized, 1971; s. Ching-Po and Li-Mei (Chow) L.; m. Pearl, Aug. 30, 1970; 1 son, Huie. B.S., Nat. Taiwan U., Taipei, 1959, M.S., 1961; Ph.D., U. Utah, 1964. Postdoctoral fellow Columbia U., 1964-65; asst. prof. chemistry Ariz. State U., 1965-68, assoc. prof., 1968-72, prof., 1972—; vis. prof. U. Cambridge, Eng., 1972-73, Tech. U. Munich, W.Ger., 1978-80; invited lectr. Academia Sinica, 1980-81, Nuclear Research Ctr., Strasbourg, France, 1982. Co-author: books, including Basic Chemical Kinetics, 1980, Multi-photon Spectroscopy of Molecules, 1983; editor: Radiationless Transitions, 1980; Contbr. articles to profl. jours. Served to 2d lt. Chinese Air Force, 1961-62. Recipient U.S. Sr. Scientist award Alexander von Humboldt Found., 1979-80, Disting. Research award Ariz. State U., 1983-84; Alfred P. Sloan fellow, 1967-71; Guggenheim fellow, 1971-73. Mem. Am. Chem. Soc., Am. Photochem. Soc., Am. Soc. Photobiology, AAAS, Soc. Columbia Chemists, Sigma Xi, Phi Lambda Upsilon, Phi Kappa Phi. Subspecialties: Physical chemistry; Photochemistry. Current work: Multi-photon processes, time-resolved x-ray scattering, ion spattering, solid state chemistry. Home: 1915 E Calle de Caballos Tempe AZ 85284 Office: Dept Chemistry Ariz State U Tempe AZ 85281

LIN, SHIH-CHIA CHEN, research scientist; b. Ka-Shing, Chekiang, China, Nov. 3, 1917; came to U.S., 1948; d. Tse-kung and Malon (Fong) C.; m. Teh Ping Lin, Sept. 17, 1948; children: Florence Jean, Henry John. B.S. in Chemistry, Central U. Chungking, China, 1940; M.S. in Oceanography, UCLA, 1951; Ph.D. in Biochemistry, U. Calif.-Berkeley, 1954. Jr. research pharmacologist U. Calif.-San Francisco, 1960-61, asst. research pharmacologist, 1961-67, 73—; biochem. pharmacologist SRI Internat., Menlo Park, Calif., 1967-68. Mem. Am. Soc. Pharmacology and Exptl. Therapeutics, Internat. Soc. Study of Zenobiotics, Western Pharmacol. Soc., Sigma Xi, Iota Sigma Pi. Subspecialties: Neurochemistry; Neuropharmacology. Current work: Cation transport at the nerve endings, interaction of CA2 -ATPase and the release of neurotransmitter, relation of cation transport to the development of tolerance and physical dependence to CNS depressant agent. Office: Univ Calif Dept Pharmacology Parnassus San Francisco CA 94143

LIN, TSUNG-MIN, research scientist; b. Chefooa Shangtung Province, China, Oct. 8, 1916; s. Chiu-Poo and Shu-Ying (You) L.; m. Hsia Lin, July 7, 1943; children: Abraham Tau-Tse, Dora Tao-Loo. B.S., Nat. Tsing Hun U., 1933; M.S., U. Ill., 1952; Ph.D., 1954. Asst. in physiology Tsing Hun U., 1939-41; instr. Peking Union Med. Coll., 1948-51; asst. prof. clin. sci. U. Ill., 1954-56; sr. pharmacologist Lilly Research Labs., Indpls., 1956-60; research scientist, 1960-64, research assoc., 1964—; del. Internat. Physiol. Congress, 1959; U.S. del. Congrss Pharmacology, 1961. China Med. Bd. fellow, 1951-54. Mem. Am. Physiol. Soc., Am. Pharmacol. Soc., Am. Gastroen. Assn., Am. Pancreatic Assn., Parietal Cell Club. Subspecialties: Physiology (medicine); Pharmacology. Current work: Histmaine, anti-histamines, gastrointestinal hormones, Glucagon on gastric, pancreatic bile secretions; pancreatic polypeptide and somatostain on gastrointestinal functions. Office: 307 E McCarty St Indianapolis IN 46285

LIN, TU, physician; b. Fukien, China, Jan. 18, 1941; came to U.S, 1967, naturalized, 1971; s. Tao-Sheng and Chien-An (Chang) L.; m. Pai-Li Lin, July 1, 1967; children: Vivian, Alexander, Margaret. M.D., Nat. Taiwan U., 1966. Diplomate: Am. Bd. Internal Medicine. Intern Episcopal Hosp., Phila., 1967-68; resident Berkshire Med. Ctr., Pittsfield, Mass., 1968-70; research fellow Brown U., Providence, 1971-73; chief endocrine sect. VA Hosp., Columbia, S.C., 1975—; asst. prof. U. S.C., Columbia, 1977-80, assoc. prof., 1980—. Contbr. articles to profl. jours. Recipient Disting. Investigator award U. S.C., 1981. Fellow Am. Coll. Medicine; mem. Endocrine Soc., Am. Soc. Andrology, Am. Fedn. Clin. Research, N.Y. Acad. Sci., So. Soc. Clin. Investigation. Subspecialties: Endocrinology; Cell biology (medicine). Current work: Effect of aging on Leydig cell function; cell membrane receptors, adenylate cyclase activity, cyclic AMP formation, protein Kinase activity, protein phosphorylation, Steroidogenesis, estrogen receptors. Home: 46 Eastbranch Ct Columbia SC 29204 Office: William Jennings Bryan Dorn VA Hosp Columbia SC 29201

LIN, TUNG-HUA, engineering and applied science educator; b. Chungking, China, May 26, 1911; came to U.S., 1934, naturalized, 1953; s. Yao Ching and Yu (Kuo) L.; m. Susan Chiang, Mar. 19, 1939; children: Rita P. Chiou, Robert P., James P. B.S., Chiaotung U., China, 1933; M.S., M.I.T., 1936; D.Sc., U. Mich., 1953. Prof. aero. engring. Tsing-Hua U., China, 1937-39; chief engr. Chinese Aircraft Mfg. Factory, Nanchuen, Sichuen, 1939-45; officer Chinese Tech. Mission in Eng., London, 1936-49; assoc. prof. aero. engring. U. Detroit, 1949-55; prof. engring. and applied sci. UCLA, 1955—; cons. Rockwell Internat. Corp., Los Angeles, 1964-70, ARA, Inc., West Covina, Calif., 1965—. Author: Theory of Inelastic Structures, 1968; contbr. articles to tech., sci. jours. Tsinghua U. Fellowship awardee to study in U.S., 1933. Fellow ASME; mem. Chinese Am. Engrs. and Scientists in So. Calif. (pres. 1964), Chinese Inst. Engrs. in U.S., So. Calif. Chinese Engrs. and Scientists, Am. Acad. Mechanics, Phi Tan Phi. Subspecialties: Solid mechanics; Mechanical engineering. Current work: Derivation of polycrystal stress-strain relations from those of single crystals beyond elastic range and at elevated temperatures; fatigue crack in initiation in metals. Office: UCLA Hilgard Ave Los Angeles CA 90024 Home: 906 Las Pulgas Rd Pacific Palisades CA 90272

LIN, WINSTON T., educator, researcher; b. Taiwan, Oct. 16, 1944; came to U.S., in 1970, naturalized, 1977; s. Chang C. and Shian S. (Young) L.; m. Wendy W. Szu, Nov. 6, 1967; 1 son, Paul C. B.A., Nat. Taiwan U., Taipei, 1966; M.A., Northwestern U.-Evanston, Ill., 1972, Ph.D., 1976. Research asst. fellow Academia Sinica, Taipei, Taiwan, 1967-70; sr. economist John Deere & Co., Moline, Ill., 1974-75; asst. prof. SUNY-Buffalo, 1975-80, assoc. prof. mgmt. sci. and fin., 1980—; mem. Orgn. Prin. Investigators, Buffalo, 1983-84. Author: Applied Econometrics for Management, 1986; editor: Readings in Mathematical Finance, 1985; assoc. editor: Advances in Modeling and Simulation, 1983. Recipient award Health Research Council, 1982-83; U. Buffalo Found. grantee, 1978-79; Research Found. SUNY grantee-in-aid, 981-83. Mem. Ops. Research Soc. Am., Am. Statis. Assn., Inst. Mgmt. Scis., Am. Fin. Assn., Western Fin. Assn., Econometric Soc. Clubs: Chinese, Formosan (Buffalo) (com. 1982-83). Subspecialties: Statistics; Operations research (mathematics). Current work: Modeling and decision systems, optimal control optimization, statistical analysis with computers, analytic methods of planning, corporate finance. Office: SUNY-Buffalo Main St Buffalo NY 14214

LIN, YING-MING, chemistry educator; b. Kaohsiung, Taiwan, Nov. 18, 1937; came to U.S., 1968, naturalized, 1981; s. Er and Yu-Hsou (Yang) L.; m. Fu Mei Wang, Mar. 9, 1966; children: Hwei Tzer, Rosalind Hweimei. B.S. in Pharmacy, Nat. Taiwan U., 1960; Ph.D., U. Tenn., 1973. Pharmacist Kaohsiung City Hosp., 1962-68; teaching asst. U. Tenn., Memphis, 1969; fellow, research assoc. St. Jude Children's Research Hosp., Memphis, 1970-77; staff assoc. Nat. Cancer Inst., Bethesda, Md., 1976-77; asst. prof. chemistry Tenn. State U., Nashville, 1977-81, assoc. prof., 1981—. Contbr. chpts. to books, articles to profl. jours. NIH grantee, 1979-82, 82—. Mem. AAAS. Subspecialties: Biochemistry (biology); Biochemistry (medicine). Current work: Enzymes and effectors of cyclic nucleotide metabolism; calmodulin; calcium in biological regulation. Office: Tenn State U 3500 Centennial Blvd Nashville TN 37203

LIN, YUE JEE, geneticist, educator; b. China, Oct. 8, 1945; s. Yung C. and Rye M. (Chen) L.; m. Chiu Y. Lin, June 29, 1972; children: Jeffrey, Sarah. B.S., Nat. Taiwan U., 1967; M.S., Ohio State U., 1972; Ph.D., 1976. Research asst. Taiwan Agrl. Research Inst., 1968-69; research asst. Nat. Taiwan U., 1969-70; teaching assoc. Ohio State U., Columbus, 1970-76; asst. prof. St. John's U., Jamaica, N.Y., 1976-82; assoc. prof. genetics, 1982—. Contbr. articles to sci. jours. Mem. Am. Soc. Cell Biology, Genetics Soc. Am., Am. Genetic Assn., Sigma Xi. Subspecialty: Genome organization. Current work: Cytogenetics of complex heterozygotes, of polyploids; cytogenetic effects of mutagens and environmental chemicals. Office: St John's U Jamaica NY 11439

LINCOLN, CHARLES EBENEZER, cognitive psychologist; b. N.Y.C., Dec. 1, 1951; s. John K. and Helen E. (Peck) L.; m. June E. Carlson, Aug. 14, 1982. B.S., U. Ill., Urbana, 1974; Ph.D., Rice U., 1980. Mem. tech. staff Bell Labs., Piscataway, N.J., 1974-75, Naperville, Ill., 1980—; research assoc. U. Tex. Health Sci. Ctr., Houston, 1979-80; statis. cons. Baylor Coll. Medicine, 1977-79. Contbr. articles to profl. jours. Mem. Am. Psychol. Assn. Subspecialties: Cognition; Human factors engineering. Current work: Human factors of person-machine interfaces, especially speech interfaces. Office: Bell Labs Naperville-Wheaton Rd Naperville IL 60566 Home: 27W675 Elm Dr West Chicago IL 60185

LINCOLN, DAVID ERWIN, biology educator, researcher; b. Detroit, Oct. 8, 1944; s. Robert Thomas and Mildred (Johns) L.; m. Patricia Grace Yates, Aug. 20, 1970. B.A., Kalamazoo Coll., 1971; Ph.D., U. Calif.-Santa Cruz, 1978. Research asst. A.M. Todd Co., Kalamazoo, Mich., 1962-73, U. Calif.-Santa Cruz, 1973-78; postdoctoral research affiliate Stanford (Calif.) U., 1978-80, teaching fellow, 1980; asst. prof. U. S.C., Columbia, 1980—. Contbr. articles to profl. jours. Served with U.S. Army, 1966-69. Mem. Ecol. Soc. Am., Bot. Soc. Am., AAAS, Phytochem. Soc. N.Am., Assn. Southeastern Biologists. Subspecialties: Species interaction; Plant physiology (biology). Current work: Evolution and ecology of carbon allocation to leaf chemicals which deter herbivores. Home: 212 S Waccamaw Ave Columbia SC 29205 Office: Dept Biology U SC Columbia SC 29208

LINCOLN, WALTER BUTLER, JR., ocean engineer, consultant; b. Phila., July 15, 1941; s. Walter B. and Virginia (Callahan) L.; m. Sharon Platner, Oct. 13, 1979; 1 dau., Amelia Adams. B.S., U. N.C., 1963; Ocean Engr., M.I.T., 1975; M.B.A., Rensselaer Poly. Inst., 1982. Registered profl. engr., N.H. Ops. research analyst applied physics lab. Johns Hopkins U., Silver Spring, Md., 1969-71; research asst. M.I.T., Cambridge, 1971-75; ocean engr. USCG Research and Devel. Center, Groton, Conn., 1975-79; prin. ocean engr. Sanders Assocs., Inc., Nashua, N.H., 1979—, head ocean engring. analysis group, 1982—. Mem. Planning Bd., Town of Brookline, N.H., 1982—. Served to lt. comdr. USN, 1963-70. Mem. Soc. Naval Architects and Marine Engrs., Marine Tech. Soc. (exec. bd. New Eng. sect. 1980—). Clubs: Nat. Assn. Underwater Instrs., Am. Schooner Assn. Subspecialties: Systems engineering; Algorithms. Current work: Integrated systems modelling and engineering of deep ocean acoustic systems for U.S. Navy, development of algorithms for simulation of hydromechanics of ocean systems. Home: Mason Rd Brookline NH 03033 Office: Sanders Assocs Inc 95 Canal St Nashua NH 03061

LIND, HENRIK OLAV, computer scientist; b. Stockholm, Sweden, Jan. 5, 1949; s. Osvald and Nina (Noor) L.; m. Laura Alice Bein, Mar. 19, 1977; 1 son, Jesse Alexander. B.S. in Physics, Carnegie-Mellon U., 1970; M.S.E.E., 1976; postgrad., U. Wash., 1982—. Sr. programmer Kentron, Cambridge, Mass., 1973-74; programmer, analyst, 1976-78; computer scientist Bolt, Beranek & Newman, Inc., Cambridge, 1978-82. Subspecialties: Distributed systems and networks; Artificial intelligence. Current work: System architecture, formal semantics. Office: Dept Math U Wash Seattle WA 98195

LIND, V. GORDON, physics educator; b. Brigham City, Utah, Feb. 12, 1935; s. Vance Otto and Vida (Hansen) L.; m. Linda Sue Wilson, Dec. 18, 1964; children: Bretton, Mark, Kimara, Justin, Cherise, Zachary, Rixa, Vanessa, Marilyse, Lisette. B.S., Utah State U., 1959; M.S., U. Wis., 1961, Ph.D., 1964. Research assoc. Mich. State U., 1964; asst. prof. dept. physics Utah State U., Logan, 1964-68, assoc. prof., 1968-75, prof., 1975—, head dept., 1981—. NSF grantee, 1967-71, 1980-83. Mem. Am. Phys. Soc., Phi Kappa Phi, Sigma Pi Sigma. Republican. Mormon. Subspecialties: Particle physics; Nuclear physics. Current work: Intermediate energy nuclear physics, cosmology, astrophysics. Home: 250 East 1st North Smithfield UT 84335 Office: Dept Physics Utah State U UMC 41 Logan UT 84322

LINDAHL, LASSE ALLAN, biology educator; b. Copenhagen, Sept. 9, 1944; U.S., 1973. M.Sc., U. Copenhagen, 1969, Ph.D., 1973. Postdoctoral fellow U.Wis.-Madison, 1973-76; asst. prof. U. Aarhus, Denmark, 1976-78; asst. prof. biology U. Rochester, N.Y., 1978—. Contbr. articles to profl. jours. Mem. Danish Biochem. Soc. (sec. 1971-73), Am. Soc. Microbiology. Subspecialties: Molecular biology; Gene actions. Current work: Regulation of ribosome synthesis in Escherichia coli. Office: Dept Biology U Rochester Rochester NY 14627

LINDAU, INGOLF EVERT, physicist, educator; b. Vaxjo, Sweden, Oct. 4, 1942; came to U.S., 1971; s. Ture Gustav Verner and Siri Syster (Johansson) L.; m. Inge-Britt Elisabeth Lof, Apr. 26, 1980. Civil Engr., Chalmers U. Tech., Gothenburg, Sweden, 1968, Tech. lic., 1970, Ph.D., 1971. Research asst. Chalmers U. Tech., 1968-71; research scientist Varian Assocs., Palo Alto, Calif., 1971-72; research assoc. Stanford U., 1972-74; adj. prof. physics, 1974-81, prof., 1981—; assoc. dir. Synchrotron Radiation Lab., 1980—. Contbr. chpts. to books and articles in field to profl. jours. Served with Swedish Army, 1963-64. Mem. Am. Phys. Soc., Am. Vacuum Soc., Am. Chem. Soc., AAAS. Subspecialties: Condensed matter physics; Atomic and molecular physics. Current work: Electron spectroscopy studies of surfaces and interfaces using synchrtoron radiation. Home: 135 Peter Coutts Stanford CA 94305 Office: SEL Stanford U Stanford CA 94305

LINDAUER, MAURICE WILLIAM, chemist, educator, cons.; b. Millstadt, Ill., Sept. 25, 1924; s. Herbert Johann and Pearl (Maserang) L.; m. Janie Ruth Shiver, Feb. 14, 1946; children: Jane E. Lindauer Elder, Rosemary Lindauer Brannen, Maurice Jack. A.B., Washington U., St. Louis, 1949; A.M., 1952; M.Ed., Harvard U., 1962; Ph.D. (Oak Ridge Inst. fellow 1964; NSF fellow 1964-65, Oak Ridge Assn. fellow 1968), Fla. State U., 1970. Research chemist Mallinckrodt Chem. Works, St. Louis, 1952-55; research engr. Am. Zinc Co., Monsanto, Ill., 1955-56; research chemist Allied Chem. & Dye Corp., Hopewell, Va., 1956-57; asst. prof. chemistry Valdosta State Coll., 1957-63, assoc. prof., 1963-71; prof., 1971—, head dept. chemistry, 1982—; participant NSF Summer Programs, 1959-63; NSF Acad. Yr. Inst., 1962. Contbr. articles to profl. jours. Served with USN, 1943-46. AEC grantee, 1959, 61, 63, 67; NEDA equipment grantee, 1966. Mem. Am. Chem. Soc., Sigma Xi. Club: Harvard (Atlanta). Subspecialties: Analytical chemistry; Physical chemistry. Current work: History of Chemistry; thermodynamics; chemistry education. Home: 1401 Miramar St Valdosta GA 31601 Office: Dept Chemistry Valdosta State Coll Valdosta GA 31698

LINDEMANN, RICHARD HENRY, geology educator; b. Lansing, Mich., Sept 12, 1950; s. Otto Hans and Carol (Deering) L.; m. Faith E. Grey, Nov. 10, 1979. B.S. in Geology, SUNY-Oneonta, 1972, M.S., Rensselear Poly. Inst., 1974, Ph.D., 1980. Asst., lectr. Skidmore Coll., 1976-80, asst. prof. geology, 1980—. Mem. Paleontol. Soc., Soc. Econ. Paleontologists and Mineralogists, Paleontol. Research Instn., Internat. Paleontol. Assn., History of Earth Scis. Soc. Subspecialties: Paleoecology; Sedimentology. Current work: Biostratigraphy and paleocology of Styliolina fissurella (hall); paleoecology, sedimentary petrology, and diagenesis of Onondaga limestone. Home: RD 1 Box 122 Greenfield Center NY 12833 Office: Dept Geolog Skidmore Coll Saratoga Springs NY 12866

LINDEN, HENRY ROBERT, chemical engineering research executive; b. Vienna, Austria, Feb. 21, 1922; came to U.S., 1939, naturalized, 1945; s. Fred and Edith (Lermer) L.; m. Natalie Govedarica, 1967; children by previous marriage: Robert, Debra. B.S., Ga. Inst. Tech., 1944; M.Chem. Engring., Poly. Inst. N.Y., 1947; Ph.D., Ill. Inst. Tech., 1952. Chem. engr. Socony Vacuum Labs., 1944-47; with Inst. of Gas Tech., 1947-78, various research mgmt. positions, 1947-61, dir., 1961-69, exec. v.p., dir., 1969-74, pres., trustee, 1974-78; research asso. prof. Ill. Inst. Tech., 1954-62, adj. prof., 1962-78, research prof. chem. engring., prof. gas engring., 1978—; chief operating officer Gas Devels. Corp., 1965-73, chief exec. officer, 1973-78; also dir.; pres., dir. Gas Research Inst., 1976—; dir. Sonat Inc., So. Natural Gas Co., Reynolds Metals Co., UGI Corp.; mem. energy research adv. bd. Dept. Energy. Author tech. articles. Recipient award of merit, operating sect. Am. Gas Assn., Disting. Service award Am. Gas Assn., Gas Industry Research award, 1982; Walton Clark Medal Franklin Inst.; Bunsen-Pettenkofer-Ehrentafel medal Deutscher Verein des Gas- und Wasserfaches.; named to IIT Hall of Fame, 1982. Fellow Am. Inst. Chem. Engrs., Inst. of Fuel; mem. Am. Chem. Soc. (recipient H.H. Storch award, chmn. div. fuel chemistry 1967, councilor 1969-77), Nat. Acad. Engring. Subspecialties: Fuels; Chemical engineering. Holder U.S. and fgn. patents in fuel tech. Home: 1515 N Astor St Chicago IL 60610 Office: 8600 W Bryn Mawr Ave Chicago IL 60631

LINDENBLAD, IRVING WERNER, astronomer; b. Port Jefferson, N.Y., July 31, 1929; s. Nils Erik and Elsie Christine (Lawson) L.; m. Ann Bolling Terry, Dec. 21, 1958; children: Irving Werner, Nils Bolling. A.B., Wesleyan U., Middletown, Conn., 1950; M.Div., Colgate Rochester Div. Sch., 1956; postgrad., Harvard U., 1956; M.A., George Washington U., 1963. Ordained to ministry Baptist Ch., 1956; minister Savannah (N.Y.) Congl. Ch., 1954-55, Market St. Bapt. Ch., Harrisburg, Pa., 1957, Montowese Bapt. Ch., North Haven, Conn., 1961-62; astronomer U.S. Naval Obs., Washington, 1953, 58-60, 63—. Contbr. articles to profl. jours. Founder, pres. local chpt. Nat. Fedn. Fed. Employees, 1967-69, Arlington County Civic Fedn., 1970-71. Served with AUS, 1951-53. Recipient Sustained Superior Performance award U.S. Navy, 1979. Fellow Royal Astron. Soc.; mem. N.Y. Acad. Sci., Am. Astron. Soc., Am. Geophys. Union. Subspecialties: Geodetic astronomy; Optical astronomy. Current work: Operation of largest photographic zenith tube in world. Home: 4735 Arlington Blvd Arlington VA 22203 Office: US Naval Obs Washington DC 20390

LINDENMEYER, PAUL HENRY, materials scientist, consultant; b. Bucyrus, Ohio, May 4, 1921; s. Richard A. and Pauline (Datwyler) L.; m. Carol V. McCartney, Jan. 18, 1944; children: Thomas H., Paul H., Peter E. B.S., Bowling Green State U., 1944; Ph.D., Ohio State U., 1951. Mgr. pioneering research Union Carbide Corp., Chgo., 1951-59; mgr. fiber sci. Chemstrand Research Ctr., Durham, N.C., 1959-69; head materials lab. Boeing Sci. Research Labs., Seattle, 1969-73; program dir. NSF, Washington, 1973-75; sr. prin. scientist Boeing Aerospace Co., Seattle, 1975—; pres. Dynamic Materials, Inc., Seattle, 1982—; vis. scientist materials dept. M.I.T., 1966-67; vis. scientist Fritz Haber Institut, Berlin, 1975-76; cons. numerous cos., U.S. Europe, 1972—. Contbr. articles to sci. jours.; editor: Supramolecular Structure in Fibers, 1967. Pres. Palos Heights (Ill.) Sch. Bd., 1958. Served with AUS, 1942-46. Recipient U.S. Sr. Scientist award Alexander v. Humboldt-Stifung, Bonn, Ger., 1975. Fellow Am. Phys. Soc. (chmn. div. polymer physics 1967-71), Am. Inst. Chemists; mem. Am. Chem. Soc., Fiber Soc. (bd. govs. 1969-77), Soc. Aerospace Materials and Process Engring., AIAA. Subspecialties: Thermodynamics; Composite materials. Current work: Application of the theory of dissipative structures to the solidification of materials using microprosser and real time data reduction technology to measure, accelerate and ultimately control the properties of materials that involve the dissipation of energy. Home: 165 Lee St Seattle WA 98109

LINDER, SOLOMON LEON, physicist, educator; b. Bklyn., Mar. 13, 1929; s. Aaron and Miriam Sabena (Stern) L.; m. Barbara Sue German, Nov. 29, 1953; children: Aaron, David, Burton. B.S. in Physics, Rutgers U., 1950; Ph.D. in Physics (NSF fellow), Washington U., St. Louis, 1955. Mem. tech. staff Bell Telephone Labs., Whippany, N.J., 1955-62; sr. group engr. McDonnell Douglas Astronautics Co., St. Louis, 1962-67, 72—, Titusville, Fla., 1967-71, tech. specialist, St. Louis, 1977—; part-time instr. physics and math. Fairleigh Dickinson U., Madison, N.J., 1959-62, Washington U., 1963-67, 76—, U. Central Fla., Orlando, 1970-71, So. Ill. U., Edwardsville, 1973-74. Contbr. articles to tech. jours. Mem. IEEE (sr.), Optical Soc. Am. Subspecialties: Guidance systems; Electro-optical systems. Current work: Applied research and development of optical and electro-optical systems. Patentee in optics and electro-optics fields. Home: 14751 Coeur D'Alene Ct Chesterfield MO 63017 Office: McDonnell Douglas Astronautics Co PO Box 516 Saint Louis MD 63166

LINDLEY, BARRY DREW, medical educator, researcher; b. Orleans, Ind., Jan. 25, 1939; s. Paul Lemuel and Martha (Drew) L.; m. Sondra Lee Patterson, June 20, 1964 (div. 1980); children: Theodore, Matthew, Sarah; m. Marian Elizabeth Price, Apr. 24, 1982. B.A., DePauw U., 1960; Ph.D., Western Res. U., 1964. Asst. prof. physiology Western Res. U., 1965-68, assoc. prof., 1968—; vis. scientist Cambridge U., 1972, Duke U. Marine Lab., 1973. Contbr. numerous articles to profl. jours. Recipient Lederle Med. Faculty award Lederle Found., 1967-70, research career devel. award USPHS, 1971-76. Mem. Am. Physiol. Soc., Biophysical Soc., Soc. Gen. Physiologists, Phi Beta Kappa. Democrat. Presbyterian. Subspecialties: Biophysics (biology); Physiology (medicine). Current work: Muscle physiology and biophysics, denervation, membrane transport; tempature jump methods, irreversible thermodynamics.

LINDQUIST, ANDERS GUNNAR, mathematics educator; b. Lund, Sweden, Nov. 21, 1942; came to U.S., 1972; s. Gunnar David and Gudrun Katarina (Dahl) L.; m. Kerstin Birgitta Rickander, Jan. 7, 1966; children: Johan, Martin. Civ. Ing., Royal Inst. Tech., Sweden, 1967, Tekn.Lic., 1968, Tekn.Dr., 1972. Docent Royal Inst. Tech., Stockholm, 1972; Vis. asst. prof. U. Fla., Gainesville, 1972-73; assoc. prof. research Brown U., Providence, 1973; assoc. prof. math. U. Ky., Lexington, 1974-80, prof., 1980—, Royal Inst. Tech., Stockholm, j71982. NSF grantee, 1974—; Air Force Office Sci. Research grantee, 1977—. Mem. Soc. Indsl. and Applied Math., IEEE. Subspecialties: Applied mathematics; Optimization and Systems Theory. Current work: Control theory, systems theory, stochastic processes, estimation theory. Office: Dept Math Univ Ky Patterson Office Tower Lexington KY 40506

LINDQUIST, DAVID GREGORY, biology educator; b. Chgo., Feb. 14, 1946; s. Donald Raefield and Ruth Margaret (Rogers) L.; m. Donna Marie Bishop, Dec. 22, 1973; children: Gregory David, William James. B.A., UCLA, 1968; M.A., Calif. State U.-Hayward, 1972; Ph.D., U. Ariz., 1975. Research technician Moss Landing (Calif.) Marine Labs., 1969-71; teaching asst. U. Ariz., Tucson, 1971-74; asst. prof. biology U. N.C., Wilmington, 1975-81, assoc. prof., 1981—; adj. prof. East Carolina U., Greenville, N.C., 1977-79, research assoc., 1981-83; cons. Coastal Zone Resources Corp., Wilmington, 1977-80. Editor: Developments in Environmental Biology of Fishes, 1983; contbr. articles to profl. jours. Mem. Sci. Support Coordination Team, Wilmington, 1982; judge New Hanover County Sci. Fair, Wilmington, 1982. U.S. Dept. Interior grantee, 1978-82. Mem. Am. Soc. Ichthyologists and Herpetologists (sec.-treas. Southeastern sect. 1981, v.p. 1982), N.C. Acad. Sci. (chmn. zoology 1983), Assn. Southeastern Biologists, Southeastern Fishes Council. Democrat. Presbyterian. Subspecialties: Behavioral ecology; Species interaction. Current work: Ichythyology: Undersea investigation of reef fishes; endangered and endemic freshwater fishes of Southeastern U.S. Home: 2511 Sidbury Rd Wilmington NC 28405 Office: Dept Biol Scis U NC PO Box 3725 Wilmington NC 28406

LINDQUIST, OIVA HERBERT, space program official; b. Superior, Wis., Feb. 8, 1926; s. Karl August and Hannah Ahro L.; m. Ruth I. Bechtel, June 11, 1948; children: Dan, Jim. B.S. in Physics, Iowa State U., 1948, M.S., 1949; M.S. in Biophysics, U. Minn., 1952. Advanced programs mgr. Honeywell, Mpls., 1977-80, SASS program mgr., 1980-82, mgr. space programs, 1982—, mem. program mgmt. council, 1981—. Originator Apollo flight director-attitude indicator, 1963; patentee integrated engine performance indicator, energy rate indicator. Served with USNR, 1943-46. Mem. AIAA. Republican. Lutheran. Subspecialties: Human factors engineering; Infrared sensor systems. Current work: Smart spectrally agile infrared sensors; man-machine system development. Home: 7501 11th Ave S Minneapolis MN 55423 Office: Honeywell Inc 2600 Ridgway Pkwy Minneapolis MN 55440

LINDQUIST, SUSAN LEE, biology educator, researcher; b. Chgo., June 5, 1949; d. Iver John and Eleanor (Maggio) L.; m. John David McKenzie, Aug. 1, 1971 (div. Aug. 1980). B.S., U. Ill., 1971; Ph.D., Harvard U., 1976. Teaching fellow Harvard U., Cambridge, Mass., 1971, 72; Am. Cancer Soc. postdoctoral fellow U. Chgo., 1976-78, prof. biology, 1978—. Andrew Mellon Found. fellow, 1979-80; grantee NIH, 1978—, NSF, 1978. Mem. AAAS, Genetics Soc., Soc. Cell Biology. Subspecialties: Cell biology; Genetics and genetic engineering (medicine). Current work: Regulation of gene expression, transcriptional and translational control mechanisms, regulation and function of the heat shock response. Home: 5461 S Ingleside Chicago IL 60615 Office: 1103 E 57th St Chicago IL 60637

LINDSAY, WILLIAM FRANCIS, project engineer scientist, researcher; b. Marinette, Wis., July 19, 1926; s. Francis William and Mae Sylvia (Scherer) L.; m. Beverly Ann, Apr. 16, 1953; children: Ann, David, Thomas, Michael, Barbara, Maureen, Carole, Janet, Diane. B.S., Marquette U., Milw., 1950; postgrad., Ill. Inst. Tech., 1951-52, U. Calif.-Livermore, 1959-60. Sr. technician Argonne Nat. Lab., 1950-52; research engr. Allis Chalmers Mfg. Co., Milw., 1953-55; staff engr. Boeing Co., Seattle, 1955-56; staff physicist U. Calif., Livermore, 1956-61, EGG, Santa Barbara, Calif. and Albuquerque, 1961-70; sr. engr. KOA Inc., Albuquerque, 1970-77; engr. scientist Sci. Applications, Inc., Albuquerque, 1977—. Served with USN, 1944-46. Mem. Am. Phys. Soc., Am. Nuclear Soc., Soc. Am. Mil. Engrs., Sigma Xi, Sigma Pi Sigma. Republican. Roman Catholic. Subspecialties: Nuclear physics; Particle physics. Current work: Directed energy beams, radiation effects in materials, systems engineering. Patentee solid state detector. Home: 641 Stagecoach Rd SE Albuquerque NM 87123 Office: Sci Applications Inc 505 Marquette St NW Albuquerque NM 87102

LINDSAY-HARTZ, JANICE, psychologist, educator; b. Sioux City, Iowa, Dec. 4, 1947; d. Charles Douglas and Margaret Gertrude (Barbour) Lindsay; m. Steven Edward Marshal Hartz, June 12, 1976. B.Sc., Brown U., 1970; M.A., 1975; Ph.D., 1980. Lic. psychologist, Fla. Research asst. U. Pa., Phila., 1970-72; tchr. Germantown Friends Sch., Phila., 1971-73; psychotherapist Bklyn. Ctr. for Psychotherapy, Bklyn., 1978-79; staff psychologist Psychiatric Assocs. of Grant Center Hosp., Miami, Fla., 1980-82; psychologist, family therapist in pvt. practice, Miami, 1981—; asst. prof. Nova U. Sch. Profl. Psychology, Ft. Lauderdale, 1982—. Contbr. articles to profl. jours. NSF fellow, 1974-78; Clark U. scholar, 1973-74; Elijah Benjamin Andrews scholar, 1968, 69. Mem. Am. Psychol. Assn., Dade County Psychol. Assn., S. Fla. Soc. Adolescent Psychiatry, Mental Health Assn., Sigma Xi. Subspecialties: Social psychology; Phenomenological pshchology. Current work: Phenomenological/interview studies of emotions and of survivor experiences. Home: 5880 SW 81st St South Miami FL 33143 Office: 3301 College Ave Fort Lauderdale FL 33314 also: 5901 SW 74th St Suite 208 South Miami FL 33143

LINDSLEY, DAVID FORD, neurophysiology educator; b. Cleve., May 18, 1936; s. Donald Benjamin and Ellen (Ford) L.; m. Elizabeth McBride, Aug. 20, 1960; children: Eric, Karen, Victoria. A.B., Stanford U., 1957; Ph.D., UCLA, 1961. USPHS-USSR exchange fellow dept. physiology Moscow State U., 1961-62; USPHS postdoctoral fellow dept. psychology U. Cambridge, 1962-63; asst. prof. dept. physiology Stanford Med. Sch., 1963-67; assoc. prof. dept. physiology U. So. Calif. Med. Sch., Los Angeles, 1967—. Recipient Lederle Med. Faculty award Stanford Med. Sch., 1964-67; Max-Planck Inst. Psychiatry Guggenheim fellow, 1974-75. Mem. Am. Physiol. Soc., Am. Assn. Anatomists, Soc. Neurosci., Internat. Brain Research Orgn., AAAS. Republican. Subspecialty: Neurophysiology. Current work: Central nervous system neurophysiology and behavior. Home: 517 11th St Santa Monica CA 90402 Office: U So Calif Med Sch Dept Physiology 2025 Zonal Ave Los Angeles CA 90033

LINDSLEY, DONALD BENJAMIN, physiological psychologist, educator; b. Brownhelm, Ohio, Dec. 23, 1907; s. Benjamin Kent and Mattie Elizabeth (Jenne) L.; m. Ellen Ford, Aug. 16, 1933; children: David Ford, Margaret, Robert Kent, Sara Ellen. A.B., Wittenberg Coll. (now U.), 1929, D.Sc., 1959; A.M., U. Iowa, 1930, Ph.D., 1932; Sc.D., Brown U., 1958, Trinity Coll., Hartford, Conn., 1965; D.Sc., Loyola U. Chgo., 1969; Ph.D. (hon.), Johannes Gutenberg U., Mainz, W.Ger., 1977. Instr. psychology U. Ill., 1932-33; NRC fellow Harvard U. Med. Sch., 1933-35; research asso. Western Res. U. Med. Sch., 1935-38; asst. prof. psychology Brown U.; also dir. psychol. and neurophysiol. lab. Bradley Hosp., 1938-46; dir. war research project on radar operation Yale, OSRD, Nat. Def. Research Com., Camp Murphy and Boca Raton AFB, Fla., 1943-45; prof. psychology Northwestern U., 1946-51; prof. psychology, physiology, psychiatry and pediatrics UCLA, 1951-77, prof. emeritus, 1977—, mem. Brain Research Inst., 1951—, chmn. dept. psychology Brain Research Inst., 1959-62; William James lectr. Harvard, 1958; Univ. Research lectr. U. Calif. at Los Angeles, 1959; Phillips lectr. Haverford Coll., 1961; Walter B. Pillsbury lectr. psychology Cornell U., 1963; vis. lectr. Kansas State U., 1966, Tex. A & M U., 1980; Mem. sci. adv. bd. USAF, 1947-49; undersea warfare com. NRC, 1951-64; cons. NSF, 1952-54; mem. mental health study sect. NIMH, 1953-57; neurol. study sect. Nat. Inst. Neurol. Diseases and Blindness, 1958-62; cons. Guggenheim Found., 1963-70, mem. ednl. adv. bd., 1970-78; chmn. behavioral scis. tng. com. Nat. Inst. Gen. Med. Scis., 1966-69; mem. behavioral biology adv. panel AIBS-NASA; mem. space sci. bd. Nat. Acad. Scis., 1967-70; mem. com. space medicine, 1969-71; mem. Calif. Legis. Assembly Sci. and Tech. Council, 1970-71. Cons. editor: Jour. Exptl. Psychology, 1947-68, Jour. Comparative and Physiol. Psychology, 1952-62, Jour. Personality, 1958-62; editorial bd.: Internat. Jour. Physiol. and Behav, 1965—, Exptl. Brain Research, 1965-76, Developmental Psychobiology, 1968-82; Contbr. numerous articles on physiol. psychology, neurosci., brain and behavior to sci. jours., also numerous chpts. in books. Trustee Grass Found., 1958—. Awarded Presdl. Cert. of Merit (for war work), 1948; Guggenheim fellow, Europe, 1959; Distinguished Sci. Achievement award Calif. Psychol. Assn., 1977. Mem. Nat. Acad. Scis. (chmn. com. long duration missions in space 1967-72, mem. space sci. board), Am. Psychol. Assn. (Distinguished Sci. Contbn. award 1959), Am. Physiol. Soc., Soc. Exptl. Psychologists, Am. Electroencephalographic Soc. (pres. 1964-65, hon. mem. 1980—), AAAS (v.p. 1954, chmn. sect. J 1977), Midwest Psychol. Assn. (pres. 1952), Am. Acad. Cerebral Palsy, Western Soc. Electroencephalography (hon. mem. with great distinction, pres. 1957), Western Psychol. Assn. (pres. 1959-60), Am. Acad. Arts and Scis., Internat. Brain Research Orgn. (treas. 1967-71), Soc. Neuroscis. (Donald B. Lindsley prize in behavioral neurosci. established in his name), Sigma Xi, Alpha Omega Alpha, Gamma Alpha, Phi Gamma Delta. Conglist. Subspecialties: Neurophysiology; Physiological psychology. Current work: Electrophysiological studies of brain and behavior in emotion attention, perception, learning and information processing; studies of spinal cord and brain stem reflexes; electromyographic stuides of neuromuscular disorders; electroencephalographic studies of behavior disorders in children and adults. Home: 471 23d St Santa Monica CA 90402 Office: Dept Psychology U Calif Los Angeles CA 90024

LINDZEY, GARDNER, educator, psychologist; b. Wilmington, Del., Nov. 27, 1920; s. James and Marguerite (Shotwell) L.; m. Andrea Lewis, Nov. 28, 1944; children—Jeffrey, Leslie, Gardner, David, Jonathan. A.B., Pa. State U., 1943, M.S., 1945; Ph.D., Harvard U., 1949. Research analyst OSRD, 1944-45; instr. psychology Pa. State U., 1945; head guidance center, lectr., psychology Western Res. U., 1945-46; teaching fellow Harvard U., 1946-47, research fellow, 1947-49, research asso., asst. prof., 1949-53, lectr., chmn. psychol. clinic staff, 1953-56, prof., chmn. dept., 1972-73; prof. psychology Syracuse U., 1956-57, U. Minn., 1957-64, U. Tex., 1964-72, chmn., 1964-68, v.p. acad. affairs, 1968-70, v.p. ad interim, 1971, v.p., dean Grad. Studies, prof. psychology, 1973-75; dir. Center for Advanced Study in Behavioral Scis., Stanford, Calif., 1975—; Mem. psychopharmacology study sect. NIMH, 1958-62, mem. program-project com., 1963-67, mem. adv. com. on extramural research, 1968-71; mem. com. faculty research fellowships Social Sci. Research Council, 1960-63, bd. dirs., 1962-76, mem. com. problems and policy, 1963-70, 72-76, chmn., 1965-70, mem. exec. com., 1970-75, chmn., 1971-75, mem. com. genetics and behavior, 1961-67, chmn., 1961-65; mem. com. biol. bases social behavior, 1967—, mem. com. work and personality in middle years, 1972-77; mem. sociology and social psychology panel NSF, 1965-68, mem. spl. commn. social scis., 1968-69, mem. advisory com. research, 1974—, mem. Waterman award com., 1976—; mem. exec. com., assembly behavioral and social sci. Nat. Acad. Sci.-NRC, 1970—, mem. com. life sci. and pub. policy, 1968-74, mem. panel nat. needs for biomed. and behavioral research personnel, 1974—; mem. com. social sci. in NSF, 1975—; mem. Inst. Medicine, 1975—; mem. com. on drug abuse Office Sci. and Tech., 1962-63; mem. Presdl. Com. Nat. Medal Sci., 1966-69; bd. dirs. Found.'s Fund Research in Psychiatry, 1967-70, Am. Psychol. Found., 1968-76, v.p., 1971-73, pres., 1974-76. Author: (with Hall) Theories of Personality, 1957, 70, 78, (with Allport and Vernon) Study of Values, 1951, 60, Projective Techniques and Cross-Cultural Research, 1961, (with J.C. Loehlin and J.N. Spuhler) Race Differences in Intelligence, 1975, (with C.S. Hall and R.F. Thompson) Psychology, 1975; also articles.; Editor: Handbook of Social Psychology, Vols. 1 and 2, 1954, Vols. 1-5, 1969, Assessment of Human Motives, 1958, Contemporary Psychology, 1967-73, History of Psychology in Autobiography, Vol. 6, 1974; asso. editor: Psychol. Abstracts, 1960-62, Ency. Social Scis, 1962-67; co-editor: Century Psychology Series, 1960-74, Theories of Personality; Primary Sources and Research, 1965, History of Psychology in Autobiography, Vol. V, 1968, Behavioral Genetics: Methods and Research, 1969, Contributions to Behavior-Genetic Analysis, 1970. Fellow Center Advanced Study Behavioral Scis., Stanford, 1955-56, 63-64, 71-72, Inst. Medicine, 1975—. Fellow Am. Psychol. Assn. (dir. 1962-68, 70-74, mem. publs. bd. 1956-59, 70-73, chmn. 1958-59, mem. council of reps. 1959-67, 68-74, pres. div. social and personality psychology 1963-64, mem. policy and planning 1975, 78; 1964-66, pres. assn. 1966-67, mem. council of editors 1968-73, chmn. com. sci. award 1968-69, pres. div. gen. psychology 1970-71), Am. Acad. Arts and Scis., Am. Philos. Soc., AAAS, Am. Sociol. Assn.; mem. Am. Eugenics Soc. (dir. 1962-70), Soc. Social Biology (dir. 1972—, pres. 1978—), Am. Psychol. Assn. (dir. ins. trust 1973—), Univs. Research Assn. (dir. 1968-71), Assn. Univ. for Research Astronomy (dir. 1973-75). Club: Cosmos. Subspecialty: Social psychology. Current work: Personality, behavior genetics, academic adminstration. Home: 890 Robb Rd Palo Alto CA 94306

LINE, ROLAND F., plant pathologist; b. Winona, Minn., Jan. 11, 1934. B.S., U. Minn., 1956, M.S., 1959, Ph.D., 1963. Research plant pathologist Regional Cereal Disease Research Lab., U.S. Dept. Agr., Pullman, Wash., 1968—. Subspecialty: Plant pathology. Current work: Research on cereal diseases, especially control of rusts of wheat. Office: US Dept Agr Regional Cereal Disease Research Lab 361 Johnson Hall Washington State U Pullman WA 99164

LINEBACK, DAVID R., food scientist; b. Russellville, Ind., June 7, 1934; m., 1956; 3 children. B.S., Purdue U., 1956; Ph.D. in Organic Chemistry, Ohio State U., 1962. Research chemist Monsanto Chem. Co., 1956-57; fellow U. Alta., 1962-64; from instr. to asst. prof. biochemistry U. Nebr., Lincoln, 1964-69; from assoc. prof. to prof. grain sci. and industry Kans. State U., 1969-76; prof. food sci. and head dept. Pa. State U., 1976-80, N.C. State U., 1980—. Mem. Am. Assn. Cereal Chemists, Inst. Food Technologists, Am. Chem. Soc. Subspecialty: Food science and technology. Office: Dept Food Sci NC State U Raleigh NC 27650

LINEHAN, MARSHA M., psychology educator; b. Tulsa, Okla., May 5, 1943. B.S. cum laude, Loyola U., Chgo., 1968, M.A., 1970, Ph.D., 1971; postgrad (fellow), SUNY-Stony Brook, 1972-73. Lectr. psychology Loyola U., Chgo., 1969-71, adj. asst prof., 1973, 75; adj. asst. prof. State U. Coll., Buffalo, 1972; asst. prof. Catholic U. Am., 1973-77, dir., 1974-77; asst. prof. U. Wash., Seattle, 1977—, adj. asst.prof. psychiatry and behavioral scis., 1981—, dir., 1977-79, mem. adj. med. staff, 1981—; 1pvt. practice psychology, Seattle, 1974—; cons. in field. Assoc. editor: Behavior Therapy; editorial bd.: Behavior Modification, Behavioral Assessment, Cognitive Therapy and Research; contbr. numerous articles to profl. jours. NDEA fellow, 1968-71; grantee in field. Mem. Am. Psychol. Assn., Am. Assn. Suicidology, Assn. Advancement of Behavior Therapy, Soc. Psychotherapy Research, Western Psychol. Assn., Wash. Psychol. Assn. Subspecialty: Behavioral psychology. Current work: Behavior therapy with suicidal individuals. Office: Dept Psychology U Wash Seattle WA 98195

LINFIELD, ROGER PAUL, research astronomer; b. Newton, Mass., Nov. 12, 1954; s. Paul Robert and Lois Caroline (Tewksbury) L. B.S. in Astronomy and Astrophysics, Mich. State U., 1975; Ph.D. in Astronomy (NSF fellow), Calif. Inst. Tech., 1981. Postdoctoral research astronomer U. Calif.-Berkeley, 1982—. Mem. Am. Astron. Soc. Subspecialty: Radio and microwave astronomy. Current work: Extragalactic radio sources. Office: 264-781 Jet Propulsion Lab Pasadena CA 91109

LING, FREDERICK FONGSUN, educator; b. Tsingtao, China, Jan. 2, 1927; came to U.S., 1947, naturalized, 1962; s. Frank Fengchi and Helen (Wong) L.; m. Linda Kwok, Feb. 20, 1954; children—Erica Helen, Alfred Frank, Arthur Theodoric. B.S. in Civil Engring, St. John's U., Shanghai, 1947, Bucknell U., 1949; M.S., Carnegie-Mellon U., 1951, Sc.D., 1954. Engr. Kwan, Chu & Young (cons. engrs.), Shanghai, 1947, Loftus Engring. Corp., Pitts., 1951; asst. prof. math. Carnegie-Mellon U., 1954-56; asst. prof. Rensselaer Poly. Inst., Troy, N.Y., 1956-58, asso. prof., 1958-63, prof., 1963-73, William Howard Hart prof. rational and tech. mechanics, 1973—, chmn. dept. mechanics, 1967-74, chmn. dept. mech. engring., aero. engring. and mechanics, 1974—; vis. prof. U. Leeds, Eng., 1970-71; Cons. S.W. Research Inst., Gen. Electric Co., Mitre Corp., Alco Products, Inc., Mech. Tech., Inc., Wear Scis., Inc. Contbr. articles to profl. jours. NSF sr. fellow, 1970. Fellow AAAS, Am. Soc. Lubrication Engrs. (Nat. award 1977), ASME (v.p. research 1977-81, gov. 1981-83), Am. Acad. Mechanics; mem. Nat. Acad. Engring., N.Y. Acad. Scis., Soc. Engring. Sci., Am. Phys. Soc., Gordon Research Conf. on Friction, Lubrication and Wear (chmn. 1970), Sigma Xi, Tau Beta Pi, Pi Mu Epsilon, Pi Tau Sigma. Subspecialty: Mechanical engineering. Home: 30 Mellon Ave Troy NY 12180

LING, JOSEPH TSO-TI, mining and manufacturing company executive, environmental engineer; b. Peking, China, June 10, 1919; came to U.S., 1948, naturalized, 1963; s. Ping Sun and Chong Hung (Lee) L.; m. Rose Hsu, Feb. 1, 1944; children: Lois Ling Weber, Rosa-Mai Ling Ahlgren, Louis, Lorraine. B.C.E., Hangchow Christian Coll., Shanghai, China, 1944; M.S. in Civil Engring, U. Minn., 1950; Ph.D. in San. Engring, U. Minn., 1952. Registered profl. engr., Minn., Ala., N.J., Okla., W.Va., N.Y., Ill., Ind., Pa., Mich. Civil engr. Nanking-Shanghai R.R. System, 1944-47; research asst. san. engring. U. Minn., 1948-52; sr. staff san. engr. Gen. Mills, Inc., Mpls., 1953-55; dir. dept. san. engring. research Ministry Municipal Constrn., Peking, 1956-57; prof. civil engring. Bapt. U., Hong Kong, 1958-59; head dept. water and san. engring. Minn. Mining & Mfg. Co., St. Paul, 1960-66, mgr. environ. and civil engring., 1967-70, dir. environ. engring. and pollution control, 1970-74, v.p. environ. engring. and pollution control, 1975—; Adv. mem. on air pollution Minn. Bd. Health, 1964-66; mem. Minn. Gov.'s Adv. Com. on Air Resources, 1966-67; mem. adv. panel on environ. pollution U.S. C. of C., 1966-71; mem. chem. indsl. com., adv. to Ohio River Valley Water Sanitation Commn., 1962-76; mem. environ. quality panel Electronic Industries Assn., 1971-80; mem. environ. quality com. NAM, 1965—; mem. Pres.'s Adv. Bd. on Air Quality, 1974-75; vice chmn. environ. com. U.S. Bur. and Industry Adv. Com. to OECD, 1975—; mem. adv. subcom. on environ., health and safety regulations Pres.'s Domestic Policy Rev. of Indsl. Innovation, 1978-79; adv. panel indsl. innovation and health, safety and environ. regulation Office Tech. Assessment of U.S. Congress, 1978—; exec. com. engring. assembly NRC, 1977-80; also environ. studies bd. Commn. Natural Resources, 1977—; mem. staff services subcom. of environ. com. Bus. Roundtable, 1975—; mem. environ. com. U.S. Council, Internat. C. of C., 1978—; adv. com. on research applications policy NSF, 1976-80. Contbr. articles to profl. jours. Trustee Belwin Outdoor Lab., St. Paul; bd. dirs. Fresh Water Biol. Found., Northwest Area Found., St. Paul Area YMCA, Midwest China Study Center, Minn. Environ. Sci. Found. Woodrow Wilson Sr. fellow, 1975—. Fellow ASCE; mem. Nat. Acad. Engring., Am. Acad. Environ. Engrs. (chmn. examination update com. 1981—), Minn. Assn. Commerce and Industry, Am. Water Works Assn., Air Pollution Control Assn. (dir.), Chem. Mfg. Assn., Water Pollution Control Fedn. Club: Rainbow (Mpls.). Subspecialties: Environmental engineering; Ecosystems analysis. Current work: Resource conservation oriented technologies (clean-low/non waste technologies) and promotion of preventive concept through control of pollution at the sources. Home: 2090 Arcade St Saint Paul MN 55109 Office: Minn Mining & Mfg Co Box 33331 900 Bush Ave Saint Paul MN 55133 There must be a goal in any stage of one's life. It must be high enough to offer a challenge. One should not wait for the opportunity but create the opportunity to achieve that goal.

LING, ROBERT F., statistics educator; b. Hong Kong, Apr. 21, 1939; U.S., 1957, naturalized, 1960. B.A., Berea (Ky.) Coll., 1961, M.A., U. Tenn., 1963; M.Phil., Yale U., 1968, Ph.D., 1971. Asst. prof. stats. U. Chgo., 1970-75; assoc. prof. math. scis. Clemson (S.C.) U., 1975-77, prof., 1977—; vis. prof. Vanderbilt U., Nashville, 1982, U Chgo., 1983; vis. lectr. Com. of Pres. of Statis. Socs., 1984-86. Co-author: Exploring Statistics with IDA, 1979, IDA? A User's Guide, 1981, Conversational Statistics, 1982; assoc. editor: Am. Statis. Assn, 1977—; mem. editorial bd.: Jour. of Classification, 1983. Office of Naval Research grantee, 1973—. Mem. Am. Statis. Assn., Internat. Assn. for Statis. Computing, Mensa. Subspecialties: Statistics; Mathematical software. Current work: Applied statistics and data analysis; cluster analysis and classification; numerical approximations of statistical distributions; statistical computing. Home: 102 Brookwood Dr Clemson SC 29631 Office: Dept Math Scis Clemson U Clemson SC 29631

LINGREN, RONALD HAL, psychology educator; b. Gowrie, Iowa, June 26, 1935; s. Herbert Nathanial and Zula Melissa (Bolton) L.; m. Bernetta Kilpatrick, Nov. 28, 1981; children by previous marriage: Scott Allen, Kristin Lee. B.S., Iowa State U., 1960; M.A., U. Iowa, 1961, Ph.D., 1965. Lic. psychologist, Wis. Asst. dir. Pine Sch. Research Project, U. Iowa, Iowa City, 1962-64, cons. child psychiatry, 1963-65; prof., dir. Ctr. Behavioral Studies U. Wis.-Milw., 1969-74, prof., 1965—; state rep. Wis. State Assembly, Madison, 1975-81; cons. Willo Glen Acad., 1981—, St. Aemelian Child Care Center, Milw., 1968-76, Juneau Acad., 1970-74; dir. Community Children's Ctr., Menomenee Falls, 1982—. Contbr. articles to profl. jours. Served with U.S. Army, 1953-55. Recipient legis. citation Wis. State Assembly, 1981; Disting. Service award U. Wis.-Milw., 1981. Mem. Wis. Psychol. Assn., Wis. Sch. Psychologists Assn., Nat. Assn. Sch. Psychologists, Wis. Council Assns. for Pupil Services. Subspecialties: Behavioral psychology. Current work: Education and minority groups; educational and psychological measurement; political persuasion and decision making; behavior disorders in children and adolescents. Address: W149 N8301 Norman Dr Menomonee Falls WI 53051

LINK, MICHAEL PAUL, pediatric hematologist/oncologist; b. Cleve., Jan. 3, 1949; s. J. Alexander and Betty Irene (Lewis) L. A.B., Columbia U., 1970; M.D., Stanford U., 1974. Intern Children's Hosp. Med. Ctr., Boston, 1974-75, resident, 1975-76; faculty Stanford U., 1979—, asst. prof. pediatrics, 1979—. Mem. Phi Beta Kappa, Alpha Omega Alpha. Subspecialties: Pediatrics; Chemotherapy. Current work: Care of children with malignant disease; research and treatment malignancies in children. Office: 520 Willow Rd Palo Alto CA 94304

LINN, D. WAYNE, biology educator; b. Estherville, Iowa, Oct. 9, 1929; s. Ward Ira and Mary Belle (Warrington) L.; m. Fae Arlene Anderson; children: Jennifer Kathleen, Jay Gregory, Douglas David.

B.A., Mankato State U., 1952; M.S., Oreg. State U., 1955; Ph.D., Utah State U., 1963. Research chemist Mayo Clinic, Rochester, Minn., 1952-53; fisheries research biologist U. Wash., Seattle, 1955-58; research assoc. Utah State U., Logan, 1958-62; asst. prof. biology Dakota Wesleyan U., Mitchell, S.D., 1962-64; fisheries officer Peace Corps, Lilongwe, Malawi, 1973-75; asst. and assoc. prof. biology So. Oreg. State Coll., Ashland, 1964-73, prof., 1975—, chmn. dept., 1969-73; advisor 208 Water Quality Program, Jackson County, Oreg., 1976-80; cons. N.W. Biol. Cons., Ashland, 1980—. Author lab. manual; contbr. articles to profl. jours. Pres. Jackson County Vector Control Bd., 1980—. Served with U.S. Army, 1947-49. AEC fellow, 1959; USPHS fellow, 1960; Harvard U. fellow, 1973. Fellow Am. Sci. Affiliation (exec. bd. Oreg. chpt. 1970-73), Gideons (local and state officer), Sigma Xi, Phi Kappa Phi, Phi Sigma, Xi Sigma Pi. Mem. Ch. of Christ. Subspecialties: Ecology; Resource conservation. Current work: Research in water behavior, characteristics, components and quality. Home: 899 Hillview Dr Ashland OR 97520 Office: So Oreg State Coll Siskiyou Blvd Ashland OR 97520

LINN, JOHN, computer scientist; b. Ithaca, N.Y., Aug. 8, 1955; s. John Gaywood and Eleanor Morrison (Ringer) L.; m. Eve F.W. Eve Wahrsager, May 15, 1982. B.A., Hampshire Coll., Amherst, Mass., 1977. Programming analyst Data Gen. Corp., Westboro, Mass., 1977-78; computer scientist research staff Bolt Beranek and Newman, Cambridge, Mass., 1978—. Mem. Assn. Computing Machinery. Subspecialties: Cryptography and data security; Distributed systems and networks. Current work: Techniques and architectures to support reliable and secure communication in computer networks; monitoring and control of networks. Office: Bolt Beranek and Newman 10 Moulton St Cambridge MA 02238

LINNEMANN, ROGER EDWARD, radiologist, radiation mgmt. co. exec., cons., educator; b. St. Cloud, Minn., Jan. 12, 1931; s. Martin C. and Esther C. L.; m.; children: Thomas M., Kathryn E., Roger Edward, Kurt F., Nickolas J. B.A. cum laude, U. Minn., Mpls., 1952, B.S., M.D., 1956. Diplomate: Am. Bd. Radiology, Am. Bd. Nuclear Medicine. Intern Walter Reed Army Hosp., Washington, 1956-57; resident in radiology, 1962-65, research asst. dept. radiobiology, 1961-62; commd. capt., M.C. U.S. Army, 1957, advanced through grades to lt. col., 1968; gen. med. officer U.S. Army Europe, Berlin, W. Ger., 1957-61; comdg. officer Nuclear Medicine Research Dept., Europe; radiol. health cons. U.S. Army Europe, Landstuhl, W. Ger., 1962-68; ret., 1968; asst. prof. radiology U. Minn., 1968; nuclear medicine cons. Phila. Electric Co., 1968-69; pres. Radiation Mgmt. Corp., Phila., 1969-82; vice chmn. bd., 1982—; clin. assoc. prof. radiology U. Pa., 1974—; vis. assoc. prof. Northwestern U., 1977—. Decorated Legion of Merit; NRC scholar, 1969. Mem. AMA (Physician's Recognition award 1971, 77), Am. Coll. Radiology, Am. Nuclear Soc., Am. Public Health Assn., Indsl. Med. Assn., Radiol. Soc. N.Am., Soc. Nuclear Medicine, Med. Liaison Officer's Network, Edison Electric Inst., Atomic Indsl. Forum, Am. Coll. Nuclear Physicians, Am. Assn. Physicists in Medicine, Indsl. Med. Assn., Bavarian Am. Radiol. Soc., Assn. Medicine and Security (Spain) (hon.). Clubs: Union League (Phila.); Aronimink Golf (Newtown Square, Pa.). Subspecialty: Nuclear medicine. Current work: Medical response to radiation injuries. Office: 3508 Market St Philadelphia PA 19104

LINSCOTT, WILLIAM DEAN, immunologist; b. Bakersfield, Calif., Apr. 23, 1930; s. Mark Ruskin and Wilma (Wells) L.; m. Nancy C. Carmody, Aug. 20, 1955; children: Kevin Russell. Ph.D., UCLA, 1960. Postdoctoral fellow Howard Hughes Med. Inst., Miami, 1960-62, Scripps Clinic and Research Found., LaJolla, Calif., 1962-64; prof. dept. microbiology U. Calif.-San Francisco, 1964—. Author: pub. Linscott's Directory of Immunological and Biological Reagents, 1979-81, 82-83, 83-85. Bd. dirs. Mendocino (Calif.) Folklore Camp, 1970-83. Grantee NIH, 1965-68, Am. Cancer Soc., 1968-70, NSF, 1977-80, U. Calif. Com. Cancer Research, 1981-83. Mem. Am. Assn. Immunologists. Subspecialty: Immunobiology and immunology. Current work: Role of serum complement system in immune induction and in cancer immunology. Office: Dept Microbiology U Calif San Francisco CA 94143

LINSKY, JEFFREY LAWRENCE, astronomer, educator; b. Buffalo, June 27, 1941; s. Max and Rose (Cheplowitz) L.; m. Lois Fleischer, Jan 14, 1941; children: Joel, Samara. B.S. in Physics, MIT, 1963; M.A. in Astronomy, Harvard U., 1965, Ph.D., 1968. Astronomer Nat. Bur. Standards, Boulder, Colo., 1969—; postdoctoral research assoc. U. Colo., Boulder, 1968-69, lectr., 1969-74, assoc. prof. adj., 1974-78, prof. adj., 1978—. Mem. editorial bd.: Solar Physics; contbr. articles to profl. jours. Smithsonian Soc. fellow, 1967-68; recipient numerous NASA grants. Fellow Joint Inst. Lab. Astrophysics; mem. Internat. Astronom. Union, Am. Astronom. Soc. Subspecialties: Ultraviolet high energy astrophysics; Theoretical astrophysics. Current work: The study of the atmospheres, coronae and winds of stars, using x-ray, ultraviolet, optical and radio techniques; stellar flares; solar physics; radiative transfer theory. Home: 1645 Bear Mountain Dr Boulder CO 80303 Office: JILA University of Colorada Boulder CO 80309

LINSLEY, JOHN (DAVID), astrophysicist, educator; b. Mpls., Mar. 12, 1925; s. James Adolphus and Martha Carolina (Wennerholm) L.; m. Betty Howard, Aug. 4, 1945; m. Lora Nalbandian, Apr. 26, 1955; m. Paola Bianca Quargnali, Oct. 26, 1966; children: Joanna Marie, Amina Martha, Alexander Piero. B.Physics with high distinction, U. Minn., 1946; Ph.D., 1952. Teaching asst. physics U. Minn., 1947-49, research asst., 1950; asst. prof. physics U. Va., 1951, M.I.T., 1954-57, research assoc., 1958-70; research prof. physics astronomy U. N.Mex., Albuquerque, 1971—; v.p. Tex. Center for Advancement of Sci. and Tech., 1982—; mem. faculty internat. Sch. Cosmic-Ray Astrophysics Majorana Centre, Erice, Sicily, 1978, 82; lectr. in field. Contbr. articles in field to profl. jours. Served with U.S. Army, 1944-46. Recipient Premio Internazionale San Valentino d'Oro City of Terni, Italy, 1981. Fellow Am. Phys. Soc.; m. Am. Astron. Soc., Internat. Astron. Union, AAAS, N.Y. Acad. Sci., Sigma Xi. Subspecialties: Cosmic ray high energy astrophysics; Particle physics. Current work: Exptl. and theoretical study of the origin, nature and interactions of cosmic rays; leading U.S. expert on highest energy cosmic rays and extensive air showers; inventor of the ionization loss Cerenkov radiation method of measuring cosmic ray charge composition.

LINSLEY, RAY KEYES, civil engr.; b. Hartford, Conn., Jan. 13, 1917; s. Ray Keyes and Flora Madelaine (Ladd) L.; m. Anne Virginia Cutler, Nov. 26, 1937; children—Dianne, Stephen, Alan, Brian. B.S., Worcester Poly. Inst., 1937, D.Engring. (hon.), 1979; D.Sc., U. Pacific, 1973. Engr. TVA, 1937-40; engr. U.S. Weather Bur., Washington, 1941, Sacramento, 1942-44, chief hydrology, Washington, 1945-50; asso. prof. Stanford, 1950-55, prof., head dept. civil engring., 1956-67, asso. dean, 1956-58, dir. project engring.-econ. planning, 1962-71, prof. emeritus, 1975—; Fulbright prof. Imperial Coll., London, 1957-58; v.p. Carroll Bradberry & Assos., 1959-67; pres. Hydrocomp Internat., 1967-78; chmn. Hydrocomp, Inc., 1972-78; pres. Linsley Kraeger Assos. Ltd., 1979—; cons. engr.; with Office Sci. and Tech., Washington, 1964-65. Author: Applied Hydrology, 1949, Elements of Hydraulic Engineering, 1955, Hydrology for Engineers, 1958, 3d edit., 1982, Water-Resources Engineering, 3d edit, 1979. Commr. U.S. Nat. Water Commn., 1968-73. Recipient Meritorious Service award Dept. Commerce, 1949. Fellow Am. Geophys. Union (pres. hydrology sect. 1956-59), ASCE (Collingwood prize 1943, Julian Hinds award 1978); mem. Nat. Acad. Engring., Am. Meteorol. Soc., Nat. Soc. Profl. Engrs., Venezuelan Soc. Hydraulic Engrs. (hon.), Japan Soc. Civil Engrs. (hon.), Sigma Xi, Tau Beta Pi. Subspecialty: Surface water hydrology. Home: 280 Swanton Blvd Santa Cruz CA 95060 Office: 527 Bayview Dr Aptos CA 95003

LINT, THOMAS FRANKLIN, immunologist; b. Pitts., Dec. 22, 1946; s. Franklin E. and Martha E. (Dugan) L. B.S., U. Dayton, 1968; M.S., Case Western Res. U., 1970; Ph.D., Tulane U., 1973. Diplomate: Am. Bd. Med. Lab. Immunology. Postdoctoral fellow Rush-Presbyterian St. Luke's Med. Ctr., Chgo., 1973-75, asst. prof. immunology, 1975-80; assoc. prof., 1980—; vis. fellow Cambridge (Eng.) U. Tumor Immunity Group, 1977; vis. prof. U. Notre Dame, 1982-84; cons. NIH, 1979, 80. Contbr. articles, chpts. to profl. publs. Recipient Career Devel. award Schweppe Found., 1976-79; Research Career Devel. award NIH Nat. Cancer Inst., 1979-84; Chgo./Am. Heart Assn. grantee, 1976-80; Leukemia Research Found. grantee, 1978-79; NIH Nat. Cancer Inst. grantee, 1979-87. Mem. Am. Assn. Immunologists, Chgo. Assn. Immunologists (sec.-treas. 1980-82, chmn. 1982-83), Reticuloendothelial Soc., Ill. Assn. Cancer Research. Subspecialties: Immunology (medicine); Cancer research (medicine). Current work: Interaction of humoral immune factors with tumor cells; immune deficiency research into control humoral immune effector mechanisms; analysis of genetic deficiencies of same mechanism; biochemistry of phagocytic cells. Office: Dept Immunology Microbiology 1753 W Congress Chicago IL 60612

LINTHICUM, DARWIN SCOTT, immunology educator; b. San Gabriel, Calif., Dec. 11, 1951; s. Ernest Lee and June Loraine (Robertson) L. B.A., U. Calif.-San Diego, 1973, Ph.D., 1976. Postdoctoral fellow Walter and Eliza Hall Inst. Med. Research, Melbourne, Australia, 1977-79; asst. prof. microbiology and neurology U. So. Calif., 1979-82; assoc. prof. pathology U. Tex., Houston, 1982—. Contbr. articles to profl. jours. Kermit E. Osserman fellow, 1977; Nat. Multiple Sclerosis Soc. fellow, 1977-79; award Leukemia Soc. Am., 1979—. Mem. Am. Assn. immunologists. Subspecialties: Immunology (medicine); Neuroimmunology. Current work: Multiple sclerosis, neuroimmunology diseases. Office: U Tex Health Sci Center Houston TX 77025

LINVILL, JOHN GRIMES, engineering educator; b. Kansas City, Mo., Aug. 8, 1919; s. Thomas G. and Emma (Crayne) L.; m. Marjorie Webber, Dec. 28, 1943; children: Gregory Thomas, Candace Sue. A.B., William Jewell Coll., 1941; S.B., Mass. Inst. Tech., 1943, S.M., 1945, Sc.D., 1949; Dr. Applied Sci., U. Louvain, Belgium, 1966. Asst. prof. elec. engring. Mass. Inst. Tech., 1949-51; mem. tech. staff Bell Telephone Labs., 1951-55; asso. prof. elec. engring. Stanford U., 1955-57, prof., dir. solid-state electronics lab., 1957-64, prof., chmn. dept. elec. engring., 1964-80, dir. Center for Integrated Systems, 1980—; co-founder, dir. Telesensory Systems, Inc.; dir. Spectra-Physics, Inc., Cromemco, Inc., Am. Microsystems, Inc. Author: (with J.F. Gibbons) Transistors and Active Circuits, 1961, Models of Transistors and Diodes, 1963. Recipient citation for achievement William Jewell Coll., 1963, Medal of Achievement Am. Electronics Assn., 1983. Fellow IEEE (Edn. medal 1976); mem. Nat. Acad. Engring. (John Scott medal 1980), Am. Acad. Arts and Scis. Subspecialty: Integrated circuits. Inventor Optacon, reading aid for blind. Home: 30 Holden Ct Portola Valley CA 94025 Office: Dept Elec Engring Stanford U Stanford CA 94305

LINVILLE, MALCOLM EUGENE, JR., psychology educator; b. Kansas City, Mo., Mar. 30, 1939; s. Malcolm Eugene and Maebell (Reimert) L.; m.; children: Douglas, Deborah. B.A. in Psychology, Kansas City U., 1957, M.A. in Edn. and Psychology, 1963; Ph.D. in Counseling, U. Mo.-Kansas City, 1974. Lic. psychologist, Mo.; cert. Nat. Bd. Cert. Counselors. Intern Walter Reed Hosp., Los Angeles, 1959; tchr. Consol. Dist. No. 1, Hickmon Mills, Mo., 1962-64; psychologist Prairie Sch. Dist., Prairie Village, Kans., 1964-69; prof. psychology U. Mo.-Kansas City, 1969—; pvt. practice psychology, 1978; cons. Kansas City Sch. Dist., Kansas City Urban Affairs Dept., Menorah Med. Ctr. Speech and Hearing Dept. Editor: The Counseling Interviewer, 1979-82; contbg. editor: Jour. Sch. Psychology, 1975—; contbr. articles to profl. jours. Bd. dirs. New Sch. Human Edn., 1973-79; pres. Booster Club, Kansas City, 1976-79; chmn. Human Resources Corp., Kansas City, 1977. Served to sgt. AUS, 1959-62. Danforth assoc., 1980-84; named Outstanding Tchr. AMOCO Found., Kansas City, 1977. Mem. Am. Psychol. Assn., Am. Personnel and Guidance Assn., Assn. Tchr. Educators, Mo. Guidance Assn. (v.p. 1981-2), Phi Delta Kappa. Subspecialty: Counseling psychology. Current work: Research dealing with family therapy, especially in outcome studies. Office: University of Missouri 365 Education Bldg Kansas City MO 64110

LINZ, ARTHUR, physicist, researcher; b. Barcelona, Spain, Jan. 30, 1926; s. Arthur and Dorothy (Warnock) L. B.S.E.E., Brown U., 1946; M.S. in Physics, U. N.C., 1950, Ph.D., 1952. Physicist Nat. Lead. Co., South Amboy, N.J., 1952-58; sr. research assoc. dept. elec. engring. and computer scis. M.I.T., 1958—; participant summer faculty research program Naval Research Lab., Washington, summers 1980-81. Contbr. articles to profl. jours. Served with USN, 1943-45. Mem. Am. Phys. Soc., Am. Inst. Physics, Optical Soc. Am., Am. Assn. Crystal Growth, AAAS, Fedn. Am. Scientists, Sigma Xi. Subspecialty: Laser research. Current work: Electro-optical materials; new solid state tunable laser materials; application of computer control techniques to crystal growth of high optical quality laser crystals; tailored band gap oxides for solar photolysis. Patentee in field. Office: Room 13-3154 MIT Cambridge MA 02139

LIOU, JUHN G., geology educator; b. Miaoli, Taiwan, China, Dec. 28, 1939; came to U.S., 1965; s. A-wen and Zhi-mei (Chiu) L.; m. Hsiu-Yin, Jan. 28, 1965; children: Gary Chi-Zen, Grace F., Joyce M. B.S., Nat. Taiwan U., 1963; postgrad., Inst. Geology, Taiwan, 1963-65; Ph.D., UCLA, 1970. Research asst. Inst. Geology, Nat. Taiwan U., Taipei, 1963-65; dept. geology UCLA, 1965-70; postdoctoral research assoc. NASA, Houston, 1970-72; asst. prof. dept. geology Stanford U., 1972-76, assoc. prof., 1976—; mem. panel NSF Earth Sci. Div., Washington, 1981-84. Author: Zeolite Stability, 1969, 72, East-Taiwan Ophiolite, 1977. Served with U.S. Army, 1972-73. Guggenheim fellow, 1978-79; NSF research grantee, 1972-83. Fellow Mineral. Soc. Am. (award 1977), Geol. Soc. Am.; mem. Am. Geophys. Union. Subspecialties: Geochemistry; Petrology. Current work: Synthesis and stability of minerals under hydrothermal conditions; rock-water interactions in geothermal systems; paragenesis of minerals in metamorphic rocks. Home: 640 Georgia Ave Palo Alto CA 94306 Office: Stanford U Dept Geology Stanford CA 94305

LIOU, KUO-NAN, atmospheric scientist, educator; b. Taiwan, China, Nov. 16, 1943; came to U.S., 1966, naturalized, 1976; s. and S. Y. (Yang) L.; m. Agnes L. Y., Aug. 4, 1968; 1 dau., Julia C. C. B.S., Taiwan U., Taipei, 1965; M.S., N.Y., U., 1968; Ph.D., 1970. Research asst. N.Y. U., 1966-70; research assoc. Goddard Inst. Space Studies, N.Y.C., 1970-72; asst. prof. U. Wash., 1972-74; assoc. prof. U. Utah, 1975-79; prof., 1980—; dir. grad. studies in meteorology, 1981—. Author: An Introduction to Atmospheric Radiation, 1980; contbr. numerous articles, chpts. to profl. publs. Recipient Founders Day award N.Y. U., 1971; David P. Gardner Fellow award U. Utah, 1978. Mem. Am. Meteorol. Soc., Am. Geophys. Union, Optical Soc. Am., AAAS, N.Y. Acad. Scis. Subspecialties: Meteorology; Remote sensing (atmospheric science). Current work: Radiation and climate modelling, remote sensing from satellites, light scattering by nonspherical particles. Office: Dept Meteorology U Utah Salt Lake City UT 84112

LIPICKY, RAYMOND JOHN, physician, clinical pharmacologist, government official; b. Cleve., May 3, 1933; s. John and Margret Karolina (Szantos) L.; m. Janet Lee Eisenhut, Mar. 8, 1958 (div.); children: Laura Lee, Josh Wesley; m.; m. d, Freda Sanders Jacobsen, Jan. 20, 1983. A.B., Ohio U., 1955; M.D., U. Cin., 1960. Intern Barnes Hosp., St. Louis, 1960-61; resident in medicine, 1961-62, Strong Meml. Hosp., Rochester, N.Y., 1964-65, trainee in cardiology, 1965-66; asst. prof. pharmacology and medicine U. Cin., 1966-71, assoc. prof., 1971-72, prof. pharmacology, 1972-79, prof. medicine, 1973-79, clin. prof. Medicine, 1979—, dir. div. clin. pharmacology, 1973-79; med. officer FDA, Rockville, Md., 1979—, acting dir. cardiorenal drug products div., 1982—; vis. scientist Lab. Biophysics, NIH, Bethesda, Md., 1979-83. Contbr. articles to profl. publs. USPHS fellow, 1957-58; Nat. Acad. Sci. fellow, Hungary, 1976; NIH grantee, 1967-80; Muscular Dystrophy Assn. grantee, 1975-80. Mem. Am. Physiol. Soc., Soc. Neurosci., Am. Soc. Pharmacology and Therapeutics, Biophys. Soc. Subspecialties: Internal medicine; Biophysics (physics). Current work: Drug effects on electrically excitable membranes; regulator; research scientist. Home: 2051 Apricot Ln Gaithersburg MD 20878 Office: 5600 Fishers Ln Room 16-B-45 Rockville MD 20857

LIPKA, STEPHEN ERIK, product development official, computer scientist; b. Bklyn., Dec. 23, 1946; s. Martin Justin and Matilda (Liebowitz) L. B.S. in Physics, Poly. Inst. Bklyn., 1968, M.S. in Elec. Engring, 1970; Ph.D. in Computer Sci, SUNY-Stony Brook, 1975. Teaching asst. Poly. Inst. Bklyn., 1968, SUNY-Stony Brook, 1970-75; system cons. Softech, Inc., Waltham, Mass., 1975-80; dir. database products devel. Computer Corp. Am., Cambridge, Mass., 1980—; adj. staff Coll. Engring., Boston U., 1981, Met. Coll., 1978; adj. faculty Old Dominion U., Norfolk, Va., 1976. N.Y. State Scholar, 1964-68. Mem. Assn. Computing Machinery, IEEE Computing Soc., Alpha Epsilon Pi (chpt. pres. 1968). Subspecialties: Database systems; Software engineering. Current work: Manager development of relational-like database management system product. Home: 118 Commonwealth Ave Boston MA 02116 Office: Computer Corp Am 4 Cambridge Ctr Cambridge MA 02142

LIPKIN, BERNICE SACKS, health scientist adminstr., researcher, computer scientist; b. Boston, Dec. 21, 1927; d. Milton and Esther (Berchuck) S.; m. Lewis Edward Lipkin, Sept. 14, 1952; children: Joel Arthur, Libbe Ann. B.S., Northeastern U., 1949; M.A., Boston U., 1950; Ph.D., Columbia U., 1961. NIH trainee in neurology dept. neurosurgery Mt. Sinai Hosp., N.Y.C., 1955-56; United Cerebral Palsy fellow in neurology dept. neurosurgery Columbia-Presbyn. Med. Center, 1961-62; research assoc. Walter Reed Lab. Sensory Psychology, 1962-63; research/devel. scientist life scis. group MELPAR, Falls Church, Va., 1963-64, Office Research and Devel., Directorate Sci. and Tech., CIA, 1964-70; scientist computer sci. dept. U. Md., College Park, 1971-72; exec. sec. div. research grants NIH, Bethesda, Md., 1972—. Author: (with A. Rosenfeld) Picture Processing and Psycho pictorics, 1970. Mem. Am. Psychol. Assn., Optical Soc. Am., IEEE, AAAS, Assn. for Computing Machinery, Soc. for Neurosci., Assn. for Research in Vision and Ophthalmology, Sigma Xi. Jewish. Subspecialties: Information systems, storage, and retrieval (computer science); Graphics, image processing, and pattern recognition. Current work: Methods of shaping text data into informational statements and quantifiable summary information that can be modelled and/or analyzed statistically; the automated office, using the Matext system to develop customized programs for personalized mail, labels, rosters, budgets, assignment lists. Office: NIH Westwood Bldg Bethesda MD 20205

LIPKIN, GEORGE, physician; b. N.Y.C., Dec. 31, 1930; s. Samuel and Celia (Greenfield) L.; m. Sari Berger, June 16, 1957; children: Michael David, Lisa Susan. A.B., Columbia Coll., 1952; M.D., SUNY Downstate Med. Ctr., 1955. Diplomate: Am. Bd. Dermatology, 1961. Intern Montefiore Hosp., Bronx, N.Y., 1955-56; resident in dermatology N.Y. U. Med. Center, 1957-58, 59-61; mem. faculty dept. dermatology N.Y.U. Sch. Medicine, 1961—, assoc. prof., 1967-74, prof. dermatology, 1974—; dir. Berger Found. Cancer Research; vis. scientist U. Zurich, 1972-73. Contbr. chpts. to books, articles to profl. jours. Served to capt. M.C. U.S. Army, 1957-59. Mem. Am. Acad. Dermatology, Soc. Investigative Dermatology, Am. Assn. Cancer Research, Harvey Soc., AAAS, Fedn. Am. Scientists, Union Concerned Scientist. Subspecialties: Dermatology; Cancer research (medicine). Current work: Cancer research, emphasis on study of growth inhibitors, their identification, purification, application to control of tumor growth. Home: 61 Virginia Ave Clifton NJ 07012 Office: 530 1st Ave New York NY 10016

LIPKOWITZ, KENNETH BARRY, chemistry educator; b. N.Y.C., Apr. 1, 1950; s. George and Phyllis (Soloway) L.; m. Raima Marzee Larter, June 12, 1977; children: Nathan, Benjamin. B.S., SUNY-Geneseo, 1972; Ph.D., Mont. State U., 1975. Postdoctoral assoc. Ohio State U., Columbus, 1976-77; asst. prof. chemistry Ind. U.-Purdue U., Indpls., 1977-81, assoc. prof., 1982—; vis. prof. Princeton (N.J.) U., 1981-82. Contbr. articles to profl. jours. Mem. Am. Chem. Soc., Quantum Chemistry Program Exchange, Ind. Acad. Sci. Democrat. Jewish. Club: Bushville (N.Y.) Athletic (co-founder). Subspecialties: Organic chemistry; Theoretical chemistry. Current work: Physical organic chemistry and theoretical medicinal chemistry. Office: Dept Chemistry Ind U-Purdue U PO Box 647 Indianapolis IN 46223 Home: 5431-D Rue Manet Indianapolis IN 46220

LIPMAN, PETER WALDMAN, geologist; b. N.Y.C., Apr. 21, 1935; s. Howard W. and Jean (Hertzberg) L.; m. Beverly J. Showalter, June 17, 1962; children: Ben, Tim. B.S., Yale U., 1958; M.S., Stanford U., 1962, Ph.D., 1962. Geologist Bear Creek Mining Co., 1957; geologist, field asst. U.S. Geol. Survey, Denver, 1958, Menlo Park, Calif., 1959, Denver, 1962—; adj. vis. scientist program Am. Geophys. Union, Washington, 1970-71. Editor: 1980 Eruptions, Mt. St. Helens, 1981; assoc. editor: Bull. Geol. Soc. Am, 1981—, Geology, 1976-78. Bd. dirs. Friends of Contemporary Art, Denver, 1966—. NSF fellow, Tokyo, 1964-65; U.S. Geol. Survey Mendenhall lectureship, 1982. Fellow Geol. Soc. Am., Mineral. Soc. Am.; mem. Am. Geophys. Union (sec.), Colo. Sci. Soc. (hon.), Internat. Assn. Volcanology (exec. com.). Subspecialties: Petrology. Current work: Volcanism-a record of igneous processes within the earth, relations between volcanism and ore deposits, volcanic hazards and sci. responsibilities. Home: 26596

Columbine Glen Golden CO 80401 Office: US Geol Survey MS 913 Federal Center Denver CO 80225

LIPMANN, FRITZ (ALBERT), biochemist; b. Koenigsberg, Germany, June 12, 1899; came to U.S., 1939, naturalized, 1944; s. Leopold and Gertrud (Lachmanski) L.; m. Elfreda M. Hall, June 23, 1931; 1 son, Stephen. Student, U. Koenigsberg, 1917-22, U. Munich, 1919; M.D., U. Berlin, 1924, Ph.D., 1928; M.D. (hon.), U. Marseilles, 1947, U. Copenhagen, 1972, M.A., Harvard, 1949, D.Sc., U. Chgo., 1953, U. Paris, 1966, Harvard, 1967, Rockefeller U., 1971; L.H.D., Brandeis U., 1959, Yeshiva U., 1964. Research asst. Prof. Meyerhof's Lab., Kaiser Wilhelm Inst., Berlin and Heidelberg, 1927-30; Dr. A Fischer's Lab., Berlin, 1930-31; Rockefeller fellow Rockefeller Inst. Med. Research, N.Y.C., 1931-32; research asso. Biol. Inst. Carlsberg Found., Copenhagen, Denmark, 1932-39; dept. biochemistry Med. Sch. Cornell U., 1939-41; research chemist, head biochem. research lab. Mass. Gen. Hosp., Boston, 1941-57; prof. biol. chemistry Med. Sch. Harvard, 1949-57; prof. Rockefeller U., 1957—. Author: Wanderings of a Biochemist, 1971; sci. papers. Recipient Carl Neuberg medal, 1948; Mead Johnson & Co. award for outstanding work on Vitamin B-complex, 1948; Nobel prize for medicine and physiology, 1953; Nat. Medal Sci., 1966. Fellow N.Y. Acad. Sci., Danish Royal Acad. Scis.; fgn. mem. Royal Soc.; mem. Nat. Acad. Scis., Am. Chem. Soc., Am. Soc. Microbiology, Biochem. Soc., A.A.A.S., Am. Soc. Biol. Chemists, Harvey Soc., Am. Philos. Soc. Subspecialties: Biochemistry (biology); Microbiology. Current work: Phospho-proteinkinase, malignant transformation, sulfo proteinkinase. Home: 201 E 17th St New York City NY 10003 also RD 2 Box 347 Rhinebeck NY 12572 Office: Rockefeller University New York City NY 10021

LIPPINCOTT, JAMES ANDREW, biochemist, educator; b. Cumberland County, Ill., Sept. 13, 1930; s. Marion Andrew and Esther Oral (Meeker) L.; m. Barbara Sue Barnes, June 2, 1956; children: Jeanne Marie, Thomas Russell, John James. A.B., Earlham Coll., 1954; A.M., Washington U., St. Louis, 1956; Ph.D., 1958. Lectr., research assoc. Washington U., 1958-59; postdoctoral fellow Centre National de la Recherche Scientifique, Gif-sur-Yvette, France, 1959-60; asst. prof. biol. scis. Northwestern U., Evanston, ILL., 1960-66, assoc. prof., 1966-73, prof., 1973-81, prof. biochemistry, molecular biology and cell biology, 1981—, assoc. dean biol. scis., 1980-83. Contbr. articles to profl. jours. Recipient Tanner-Shaughnessy merit award Ill. Soc. for Microbiology, 1981; Jane Coffin Childs fellow, 1959-60. Mem. Am. Soc. Biol. Chemists, Am. Soc. Microbiology, Am. Soc. Plant Physiologists, Am. Soc. Phytopathology, Bot. Soc. Am., Scandinavian Soc. Plant Physiologists, Japanese Soc. Plant Physiologists. Subspecialties: Plant physiology (biology); Microbiology. Current work: Plant tumorigenesis and development. Office: Dept Biochemistry Molecular Biology and Cell Biology Northwestern U Evanston IL 60201

LIPPINCOTT, SARAH LEE, astronomer; b. Phila., Oct. 26, 1920; d. George E. and Sarah (Evans) L.; m. Dave Garroway. Student, Swarthmore Coll., 1938-39, M.A., 1950; B.A., U. Pa., 1942; D.Sc., Villanova U., 1973. Research asst. Sproul Obs., Swarthmore (Pa.) Coll., 1941-50, research asso., 1951-72, dir., 1972-81, prof., 1977-81, prof. and dir. emeritus, 1981—, research astronomer, 1981—. Author: (with Joseph M. Joseph) Point to the Stars, 1963, 3d edit., 1977, (with Laurence Lafore) Philadelphia, the Unexpected City, 1965; Contbr. articles to profl. jours. Mem. Savoy Opera Co., Phila., 1947—; bd. mgrs. Societe de Bienfaisance de Philadelphie, 1966-69. Recipient achievement award Kappa Kappa Gamma, 1966; Distinguished Daus. of Pa. award, 1976; Fulbright fellow, Paris, 1953-54. Mem. Rittenhouse Astronom. Soc., Am. Astronom. Soc. (lectr. 1961—), Internat. Astron. Union (v.p. commn. 26 1970-73, pres. 1973-76), Sigma Xi (nat. lectr. 1971-73). Subspecialties: Optical astronomy. Current work: Study of stars in vicinity of the sun: distances, stellar mases; search for low mass unseen companions, plants, other solar systems. Home: 507 Cedar Ln Swarthmore PA 19081 Office: Sproul Observatory Swarthmore Coll Swarthmore PA 19081

LIPPMANN, DAVID ZANGWILL, chemistry educator; b. Houston, July 6, 1925; s. Oscar Ulrich and Sadie (Buchwald) L.; m. Sylvia Jane Neustein, Dec. 25, 1969. B.Sc., U. Tex.-Austin, 1947, M.A., 1949; Ph.D., U. Calif.-Berkeley, 1953. Researcher in industry, 1953-63; asst. prof. to prof. chemistry. Southwest Tex. State U., San Marcos, 1963—. Contbr. articles to profl. jours. Served with U.S. Army. Decorated Combat Infantryman's Badge, Purple Heart. Mem. Am. Chem. Soc., AAAS, Sigma Xi. Jewish. Subspecialties: Kinetics; Physical chemistry. Current work: Kinetics of condensation of gases; theoretical calculations of kinetics. Home: 2308 Tower Dr Austin TX 78703 Office: Dept Chemistry SW Tex State Univ San Marcos TX 78666

LIPPS, JERE HENRY, geologist, educator; b. Aug. 28, 1939; s. Henry John and Margaret R. L.; m. Karen Elizabeth Loeblich, June 25, 1964 (div. 1970); m. Susannah McClintock, Sept. 28, 1978; 1 son, Jeremy Christian. B.S., UCLA, 1962, Ph.D., 1966. Research geologist Chevron Research Co., La Habra, Calif., 1963, UCLA, 1966-67; from asst. to assoc. prof. U. Calif.-Davis, 1967-75, prof., 1975—, chmn. dept. geology, 1972, 79—, dir. Inst. Ecology, 1971; participating scientist Deep Sea Drilling Project, 1969-71, U.S. Antarctic Research Program, 1971-80; research assoc. Los Angeles County Mus. Natural History. Author: Controversies in Earth Science, 1975; author numerous articles for sci. publs. Lipps Island named in honor, 1978; recipient NSF Antarctic medal, 1979. Fellow AAAS, Cushman Found. Am. Research (v.p. 1982-83, dir.); mem. Paleontol. Soc., Soc. Systematic Zoology, Soc. Am. Naturalists, Explorers Club. Subspecialties: Paleobiology; Evolutionary biology. Current work: Evolutionary paleobiology of marine organisms and their relationship to earth history. Office: Dept Geology U Calif Davis CA 95616

LIPSCHUTZ, MICHAEL ELAZAR, chemistry educator; b. Phila., May 24, 1937; s. Maurice and Anna (Kaplan) L.; m. Linda Jane Lowenthal, June 21, 1959; children: Joshua Henry, Mark David, Jonathan Mayer. B.S., Pa. State U., 1958; S.M., U. Chgo., 1960, Ph.D., 1962. Gastdocent U. Bern, Switzerland, 1964-65; asst. prof. chemistry Purdue U., West Lafayette, Ind., 1965-68, assoc. prod., 1968-73, prof., 1973—; vis. assoc. prof. Tel Aviv (Israel) U., 1971-72; cons. NASA, 1973-75, 77-79, Lunar and Planetary Inst., 1980—. Assoc. editor: Proc. 11th Lunar and Planetary Sci. Conf., 3 vols, 1980; contbr. articles to profl. jours. NSF-NATO fellow, 1964-65; Fulbright fellow, 1971-72; NASA grantee; NSF grantee. Fellow Meteoritical Soc. (treas. 1978—); mem. Am. Chem. Soc., AAAS, Geochem. Soc., Am. Geophys. Union. Democrat. Jewish. Subspecialties: Space chemistry; High temperature chemistry. Current work: Geo- and cosmochemistry; trace and ultra-trace element analysis; high temperature and high pressure chemistry; solid state chemistry; cosmic ray-induced nuclear reactions. Office: Dept Chemistry Purdue U West Lafayette IN 47907 Home: 2900 Henderson Ave West Lafayette IN 47906

LIPSCOMB, WILLIAM NUNN, JR., educator, phys. chemist; b. Cleve., Dec. 9, 1919; s. William Nunn and Edna Patterson (Porter) L.; m. Mary Adele Sargent, May 20, 1944; children—Dorothy Jean, James Sargent. B.S., U. Ky., 1941, D.Sc. (hon.), 1963; Ph.D., Calif. Inst. Tech., 1946; Dr.h.c., U. Munich, 1976; D.Sc. (hon.), L.I. U., 1977, Rutgers U., 1979, Gustavus Adolphus Coll., 1980, Marietta Coll., 1981. Phys. chemist OSRD, 1942-46; with U. Minn., 1946-59, successively asst. prof., asso. prof. and acting chief phys. chemistry div., prof. and chief phys. chemistry div., 1954-59; prof. chemistry Harvard U., 1959-71, Abbott and James Lawrence prof., 1971—, chmn. dept. chemistry, 1962-65; Mem. U.S.A. Nat. Com. for Crystallography, 1954-59, 60-63, 65-67; chmn. program com. Fourth Internat. Congress of Crystallography, Montreal, 1957. Author: The Boron Hydrides, 1963, (with G.R. Eaton) NMR Studies of Boron Hydrides and Related Compounds, 1969; Asso. editor: Jour. Chemical Physics, 1955-57; contbr. articles to profl. jours.; Clarinetist, mem., Amateur Chamber Music Players. Guggenheim fellow Oxford U., Eng., 1954-55, Cambridge U., Eng., 1972-73; NSF sr. postdoctoral fellow, 1965-66; Overseas fellow Churchill Coll., Cambridge, Eng., 1966, 73; Robert Welch Found. lectr., 1966, 71; Howard U. distinguished lecture series, 1966; George Fisher Baker lectr. Cornell U., 1969; centenary lectr. Chem. Soc., London, 1972; lectr. Weizmann Inst., Rehovoth, Israel, 1974; Evans award lectr. Ohio State U., 1974; Gilbert Newton Lewis Meml. lectr. U. Calif., Berkeley, 1974; also lectureships Mich. State U., 1975, U. Iowa, 1975, Ill. Inst. Tech., 1976, numerous others; also speaker confs.; Recipient Harrison Howe award in Chemistry, 1958; Distinguished Alumni Centennial award U. Ky., 1965; Distinguished Service in advancement inorganic chemistry Am. Chem. Soc., 1968; George Ledlie prize Harvard, 1971; Nobel prize in chemistry, 1976; Disting. Alumni award Calif. Inst. Tech., 1977; sr. U.S. scientist award Alexander von Humboldt-Stiftung, 1979; award lecture Internat. Acad. Quantum Molecular Sci., 1980. Fellow Am. Acad. Arts and Scis., Am. Phys. Soc.; mem. Am. Chem. Soc. (Peter Debye award phys. chemistry 1973, chmn. Minn. sect. 1949-50), Am. Crystallographic Assn. (pres. 1955), Nat. Acad. Sci., Netherlands Acad. Arts and Scis. (fgn.), Math. Assn. Bioinorganic Scientists (hon.), Academie Europeenne des Sciences, des Arts et des Lettres, Phi Beta Kappa, Sigma Xi, Alpha Chi Sigma, Phi Lambda Upsilon, Sigma Pi Sigma, Phi Mu Epsilon. Subspecialty: Physical chemistry. Office: Dept Chemistry Harvard U Cambridge MA 02138

LIPTAY, JOHN STEPHEN, electrical engineer, computer designer; b. N.Y.C., Nov. 28, 1940; s. Imre J. and Annette K. L.; m. Lynne M. Oakland, Dec. 30, 1972; children: Thomas, Steven. B.E.E., Rensselaer Poly. Inst., 1962; M.E.E., 1966. With IBM, Poughkeepsie, N.Y., 1965—, adv. engr., 1972-77, sr. engr., 1977—. Mem. IEEE, Assn. Computing Machinery. Subspecialties: Computer engineering; Computer architecture. Current work: Internal organization and logic design of large computers. Patentee in computer design. Home: RD 3 Box 284 Rhinebeck NY 12572 Office: IBM Corp Poughkeepsie NY 12602

LIPTON, ALLAN, physician; b. N.Y.C., Dec. 29, 1938; s. Murray and Rutb L.; m. Nancy Whitcomb; children: Samuel, Joshua A., Sarah Jan. A.B., Amherst Coll., 1959; M.D., N.Y.U., 1963. Intern Bellevue Hosp., N.Y.C., 1963-64; resident, 1964-65; fellow in hematology-oncology Meml. Hosp., 1967-69; mem. staff Hershey (Pa.) Med. Center, 1971—; chief div. oncology, 1972—, prof. medicine, 1979—. Served with USPHS, 1965-67. Mem. Am. Soc. Clin. Oncology, Am. Assn Cancer Research. Subspecialties: Cell biology; Chemotherapy. Current work: Cancer research and treatment, cell growth. Address: Hershey Med Center Hershey PA 17033

LIPTON, HOWARD LEE, neurologist; b. Detroit, Aug. 19, 1939; s. Joseph and Pearl (Sherman) L.; m. Pam Rosen, Mar. 30, 1967; children: Joseph Robert, Robert Mead, Chad Martin. B.S., U. Nebr., 1962; M.D., 1964. Diplomate: Dipolmate Am. Bd. Neurology and Psychiatry. Intern, Ohio State U. Hosp., Columbus, 1967-70; resident in neurology Johns Hopkins Hosp., Balt.; NIH spl. fellow in neurovirology, 1970-72; asst. prof. neurology Northwestern U. Med. Sch., Chgo., 1972-77; assoc. prof., 1977-81, Ernest J. and Hattie H. Magerstadt prof. neurology, 1981—. Served to capt. AUS, 1965-67. Recipient NIH research career devel. award, 1977-82; Josiah Macy Jr. Found. scholar, 1980-81. Mem. Am. Acad. Neurology, Am. Neurol. Assn., Am. Assn. Immunologists, Alpha Omega Alpha. Subspecialties: Neurology; Virology (medicine). Current work: Multiple sclerosis, neurovirology. Office: 303 E Chicago Ave Chicago IL 60611

LIPTON, JAMES ABBOTT, federal health executive, dental educator, researcher; b. N.Y.C., July 24, 1946; s. Benjamin M. and Ann (Rappaport) L.; m. Jill Friedman, Oct. 8, 1978. B.S., CCNY, CUNY, 1967; D.D.S., Columbia U., 1971, M.Phil., 1974, Ph.D., 1980. Trainee USPHS, N.Y.C., 1971-74, dental cons., 1976-77, commd. dental officer, 1976—; chief Nat. Health Service Corps, Region II, N.Y.C., 1977—; asst. clin. prof. Dental Sch. Columbia U., N.Y.C., 1976-83; asst. prof. Sch. Pub. Health Columbia U., N.Y.C., 1983—. Contbr. articles to profl. jours. Recipient commd. officer plaque USPHS, 1980. Mem. ADA, Am. Pub. Health Assn., Internat. Assn. Dental Research, Internat. Assn. for Study of Pain, Am. Sociol. Assn. Jewish. Subspecialties: Facial pain; Medical sociology. Current work: Psychosocial and linguistic aspects of facial pain patients; sociocultural dimensions of health and illness behavior. Office: USPHS 26 Federal Plaza Room 3304 New York NY 11278

LIS, MARTIN, endocrinologist, research lab. executive, educator; b. Praha, Czechoslovakia, Jan. 6, 1935; emigrated to Can., 1968, naturalized, 1973; s. Jan and Marta (Gaigher) L.; m. Marie Simkova, Sept. 10, 1960; children: Martina, Andrew. D.V.M., Sch. Vet. Medicine, Brno, Czechoslovakia, 1960; Ph.D., Czechoslovak Acad. Sci., 1967, McGill U., 1972. Fellow Czechoslovak Acad. Scis., 1960-68, C.N.R.Z. Lab. Physiology of Lactation, Jouy-en-Josas, France, 1965-66, Inst. Exptl. Medicine and Surgery, U. Montreal, Que., Can., 1968-69; fellow Clin. Research Inst., Montreal, 1970-73, sr. scientist, 1973-75, dir. Lab Comparative Endocrinology, 1976-83; dir. Nat. Wildlife Research Ctr., Environment Can., Ottawa, Ont., 1983—; assoc. prof. dept. medicine U. Montreal, 1976-83. Contbr. articles to profl. jours. Recipient ann. award Czechoslovak Acad. Scis., 1966; Clarke Inst. Psychiatry, Toronto, Ont., Can., 1977. Mem. Endocrine Soc., Am. Assn. Cancer Research, Can. Soc. Endocrinology and Metabolism, Tissue Culture Assn., Am. Assn. Lab. Animal Sci., N.Y. Acad. Scis., AAAS, Club de Recherche Clinique de Quebec. Subspecialties: Endocrinology; Cell biology (medicine). Current work: Peptide hormones, monoclonal antibodies, endocrine tumors, tumor markers, wildlife toxicology and pathology research management. Office: Environment Can Nat Wildlife Research Ctr Ottawa ON Canada K1A 0E7

LISAK, ROBERT PHILIP., physician, educator; b. Bklyn., Mar. 17, 1941; s. Irving Arthur and Sylvia Lillian (Kadish) L.; m. Deena Freda Penchansky, Aug. 2, 1964; children: Ilene Ann, Michael. A.B. cum laude with highest honors in History, NYU, 1961; M.D., Columbia U., 1965. Diplomate: Nat. Bd. Med. Examiners, Am. Bd. Neurology. Intern Montefiore Hosp. and Med. ctr., Bronx, N.Y., 1965-66; research assoc. NIMH, Bethesda, Md., 1966-68; resident in medicine Bronx Mcpl. Hosp. Ctr., 1968-69; resident in neurology Hosp. U. Pa., Phila., 1969-72; asst. prof. neurology U. Pa. Sch. Medicine, 1972-76; assoc. prof., 1976-80, prof., 1980—; hon. vis. fellow Univ. Coll. London, 1978-79, Inst. Neurology, London, 1978-79. Co-author: Myasthenia Gravis, 1982; contbr. articles to sci. jours., chpts. to books. Served as sr. asst. surgeon USPHS, 1966-68. Recipient Helen M. Jones award NYU, 1961; Founders Day award, 1961; USPHS-Nat. Inst. Neurol. Communicative Disease and Stroke tchr.-investigator award, 1972-77; Fulbright-Hays scholar, 1978-79. Fellow Am. Acad. Neurology; mem. AAAS, Am. Neurol. Assn., N.Y. Acad. Scis., Am. Assn. Immunologists, Am. Fedn. Clin. Research, Soc. Neurosci., Med. Soc. London, Phila. Neurol. Soc. Subspecialties: Neuroimmunology; Neurology. Current work: Neuroimmunology: study of experimental and human disorders of possible immunopathogenesis, including multiple sclerosis Guillian-Barre Syndrome, myasthenia gravis; neurology: clinical and teaching; cell and tissue culture; neurobiology and immunology of cultured oligodendrocytes and Schwann cells. Office: U Pa Sch Medicine 3400 Spruce St Philadelphia PA 19104

LISCHER, LUDWIG FREDERICK, cons., former utility company executive; b. Darmstadt, Germany, Mar. 1, 1915; came to U.S., 1923, naturalized, 1933; s. Ludwig J. and Paula (Stahlecker) L.; m. Helen Lucille Rentz, Oct. 1, 1938; 1 dau., Linda Sue. B.S. in Elec. Engring, Purdue U., 1937, D.Eng. (hon.), 1976. Registered profl. engr., Ill. With Commonwealth Edison Co., Chgo., 1937-80, v.p. charge engring., research and tech. activities, 1964-80; ret., 1980; cons. and mem. adv. com. to engring. tech. div. Oak Ridge Nat. Lab., 1980—; cons. energy and electric utility fields; mem. various tech. adv. coms. to fed. agys. and Edison Electric Inst.; chmn. research adv. com. Electric Power Research Inst. Contbr. articles to profl. jours. Bd. dirs. Chgo. Engring. and Sci. Center; trustee Ill. Inst. Tech. Served to lt. col. AUS and USAAF, 1941-45. Fellow IEEE; mem. ASME, Nat. Acad. Engring., Am. Nuclear Soc., Tau Beta Pi, Eta Kappa Nu. Subspecialties: Electrical engineering; Nuclear engineering. Current work: Energy conversion and electric power generation and transmission. Home: 124 W Willow St Lombard IL 60148

LISCHNER, HAROLD WILL, pediatric immunologist, educator; b. San Diego, Feb. 4, 1925; s. Hyman and Frances (Rieger) L.; m. Kyong Ok Kim, Dec. 25, 1958; children: Neal Yong, Ray Hwan, David Ho, Ben Woo. B.S. in Chemistry, UCLA, 1947, M.S., 1950; M.D., U. Chgo., 1952. Diplomate: Am. Bd. Pediatrics, Am. Bd. Med. Lab. Immunology. Rotating intern U. Utah Hosp., 1952-53; intern in pediatrics U. Minn. Hosps., 1953-54; physicia, dir. Houses for Korea Village Devel. Project, 1954-56; vis. instr. in pediatrics and pathology Soo Do and Young Sae Med. Colls., Seoul, Korea, 1956; resident in pediatrics U. Calif. Med. Center, Los Angeles, 1957-58, chief resident in pediatrics, 1958-59; instr. U. Mo., 1959-60, asst. prof. pediatrics, 1960-61; asst. prof. U. Pa., 1961-64; dir. pediatric outpatient dept. U. Pa., Hosp., 1963-64; spl. fellow in hematology and immunology St. Christopher's Hosp. for Children, Phila., 1964-66, chief immunology sect., 1968—, dir. Immunology Lab., 1970-83, chief allergy, immunology, and rheumatology sect., 1983—; prof. pediatrics Temple U., 1974, prof. microbiology/immunology, 1976—. Contbr. articles to profl. jours. NIH grantee, 1962-74, 76-77, 80-83; NSF grantee, 1975-76. Mem. Physicians for Social Responsibility, Soc. Social Responsibility in Sci., ACLU, Med. Soc. for Prevention War., Am. Assn. Immunologists, Transplantation Soc., Reticuloendothelial Soc., Am. Acad. Pediatrics, Am. Soc. Microbiologists, Internat. Soc. Exptl. Hematology, Internat. Soc. Immunopharmacology. Subspecialty: Immunology (medicine). Home: 20 Hamilton Circle Philadelphia PA 19130 Office: Saint Christopher's Hosp for Children 5th and Lehigh Ave Philadelphia PA 19133

LISH, PAUL MERRILL, scientific consultant, pharmacist, researcher; b. McCammon, Idaho, Feb. 2, 1921; s. Marion M. and Viola (Davis) L.; m. Irene Howe, Nov. 17, 1960 (dec.); children: Michael, Marilyn, Gloria, Carol; m. Grete Enderweit, Oct. 27, 1978. B.S. in Pharmacy, U. Idaho, 1949; M.S., U. Nebr., 1951; Ph.D. in Pharmacology, St. Louis U. Med. Sch., 1955. Pharmacologist, dept. head, v.p. biol. scis. Mead Johnson Co., 1955-69; dir. research and devel. Chromalloy Am. Corp., 1969-78; sci. advisor on pharmacology, and regulatory affairs toxicology Ill. Inst. Tech., Chgo., 1978—. Author. Served as 1st petty officer USN, 1943-46. Recipient Mead Johnson Presdl. award for meritorious research, 1960. Mem. Am. Soc. Pharmacology and Exptl. Therapeutics, Soc. Exptl. Biology and Medicine, Sigma Xi, Rho Chi. Mormon. Subspecialties: Pharmacology; Toxicology (medicine). Current work: Testing toxicity of munitions; FDA liaison work in new drug development. Patentee in field. Office: Ill Inst Tech Research Inst 10 W 35th St Chicago IL 60616

LISHAK, ROBERT STEPHEN, zoology educator; b. Perth Amboy, N.J., Aug. 15, 1949; s. Stephen and Anna (Petenko) L.; m. Betty J. Toth, Sept. 10, 1971. B.S., Seton Hall U., 1971; Ph.D., Ohio State U., 1976. Mem. faculty dept. zoology Auburn (Ala.) U., 1976—. Contbr. articles to profl. publs. Recipient Biology award Seton Hall U., 1971; Teaching award Ohio State U., 1974. Mem. Am. Soc. Mammalogists, Wildlife Soc., Ala. Acad. Sci., Sigma Xi. Subspecialty: Ethology. Current work: Behavior of rodents; specifically, animal sounds and communication. Office: Auburn U Auburn AL 36849

LISTER, RICHARD MALCOLM , plant virologist; b. Sheffield, Eng., Nov. 14,1928; s. Francis William and Frances May (Scotrick) L.; m. Jean Isabel Mills, May 19, 1953; children: Susan, Christina, Rosalind, John. B.Sc., Sheffield (Eng.) U., 1949; Dip. Agrl. Sci., Cambridge (Eng.) U., 1951; Dip. Trop. Agr., Imperial Coll. Trop. Agr., Trinidad, W.I., 1952; Ph.D., St. Andrews (Scotland) U., 1964. Scientist West African Cocoa Research Inst., Ghana and Nigeria, 1952-56; princ. scientist Scottish Hort. Research Inst., Invergowrie, 1956-66; prof. dept. botany and plant pathology Purdue U., West Lafayette, Ind., 1966—. Contbr. chpts. to books, articles to profl. jours. NSF grantee; U.S. Dept. Agr. grantee; Ind. Crop Improvement Soc. grantee; Am. Phytopath. Soc., fellow, 1973. Mem. Assn. Applied Biologists, Am. Phytopath. Soc., AAUP. Subspecialties: Plant virology; Plant pathology. Current work: Mission-oriented virological problems of crop plants. Office: Dept Botany and Plant Pathology Purdue U West Lafayette IN 47907

LIT, ALFRED, psychologist; b. N.Y.C., Nov. 24, 1914; s. Zachary Oscar and Elsie (Jaro) L.; m. Imogene Speegle, Jan. 27, 1947. B.S., Columbia U., 1938; M.A., 1943; Ph.D., 1948. Lic. optometrist, N.Y., psychologist, N.Y. Lectr. optometry Columbia U., N.Y.C., 1946-48, asst. and assoc. prof., 1949-56; staff, research project dept. psychology and U.S. Office Naval Research, 1949-56; research psychologist Vision Research Labs., U. Mich., Ann Arbor, 1956-59; head human factors engring. staff Bendix Systems Div., Ann Arbor, 1959-61; prof. psychology, prof. elec. scis. and systems engring., prof. engring. biophysics program and prof. molecular sci. doctoral program So. Ill. U., Carbondale, 1961—; mem. com. on vision NRC, Nat. Acad. Sci., Washington, 1965—. Contbr. articles to profl. jours.; sci. referee: Am. Jour. Optometry, 1975—. Served to 1st lt. USAAF, 1943-46. Recipient Kaplan Research award Sigma Xi, 1971; Ill. Dept. Mental Health grantee, 1962-65; NSF grantee, 1965-69, 76-78; USPHS grantee, 1967—. Fellow Am. Psychol. Assn., Am. Acad. Optometry, Optical Soc. Am., Psychonomic Soc., AAAS, N.Y. Acad. Sci., Soc. Engring.

Psychology; mem. Human Factors Soc., Assn. for Research in Vision and Ophthalmology, Eastern Psychol. Assn., Midwestern Psychol. Assn., N.Y. County Optometric Soc. (pres. 1954-55). Subspecialties: Sensory processes; Optometry. Current work: Psychophysiological and electrophysiological studies on spatio-temporal factors in monocular and binocular perception, the visual latent period, and their applications to clinical practice. Home: Route 1 PO Box 164A Murphysboro IL 62966 Office: Dept Psychology So Ill Univ Carbondale IL 62901

LITCHFIELD, JOHN HYLAND, microbiologist; b. Scituate, Mass., Feb. 13, 1929; s. Frank Albert and Alma (Hyl) L.; m. Dianne Chappell, Apr. 15, 1966; 1 son, Robert Chappell. S.B., MIT, 1950; M.S., U. Ill., 1954, Ph.D., 1956. Research chief chemist Searle Food Corp., Hollywood, Fla., 1950-51; research scientist Swift & Co., Chgo., 1956-57; asst. prof. Ill. Inst. Tech., Chgo., 1957-60; research leader Battelle Meml. Inst., Columbus, Ohio, 1960—; adj. assoc. prof. human nutrition and food mgmt. Ohio State U., 1977-78. Contbr. articles to profl. jours. Served to 1st lt. AUS, 1951-53; ETO. Recipient Disting. Inventor award Battelle Meml. Inst., 1977. Fellow AAAS, Am. Acad. Microbiology, Inst. Food Technologists (Outstanding Service award 1980), Am. Inst. Chemists, Am. Pub. Health Assn., Royal Soc. Health; mem. Soc. Indsl. Microbiology (pres. 1971, Charles Porter award 1977). Subspecialties: Food science and technology; Microbiology. Current work: Applied and industrial microbiology, food science and technology. Patentee in field. Home: 255 Bryant Ave Worthington OH 43085 Office: Battelle Memorial Institute 505 King Ave Columbus OH 43201

LITLE, PATRICK ALAN, Psychologist; b. Pomona, Calif., Nov. 14, 1946; s. Ralph and Doris Elizabeth (Little) L.; m. Veronica James, June 16, 1979. B.Music, U. Redlands, Calif., 1968; M.A., Calif. State U.-Long Beach, 1982; postgrad., U. Del., 1979—. Registered music therapist, Calif., Va. Music therapy intern Lanterman State Hosp., Pomona, Calif., 1977; music tchr. Pomona City Schs., 1977; music therapist Coll. Hosp., Cerritos, Calif., 1977-78; therapist trainee Community Psychology Clinic, Long Beach, Calif., 1978-79; psychology trainee VA, Perry Point, Md., 1980-82; psychology intern VA Outpatient Clinic, Los Angeles, 1982-83; psychology technician VA Med. Ctr., Coatesville, Pa.; psychology cons. Wilmington (Del.) City Schs., 1980-81. Served with USN, 1969-72. Calif. State scholar, 1964-68, 69. Mem. Am. Psychol. Assn. (assoc.), Nat. Assn. Music Therapy, Psychomusicology Soc., Phi Kappa Phi, Phi Mu Alpha Sinfonia (chpt. pres. 1967-68). Democrat. Baptist. Subspecialties: Behavioral psychology; Neuropsychology. Current work: Psychology of music: psychophysiological aspects of musical rhythms, role of personality factors and musical preferences in psychophysiological responses to music; hypnosis. Home: 908 Kenilworth Ave Newark DE 19711 Office: VA Med Ctr Coatesville PA 19320

LITLE, PATRICK ALAN, music therapist; b. Pomona, Calif., Nov. 14, 1946; s. Ralph and Doris Elizabeth (Little) L.; m. Veronica James, June 16, 1979. B.Music, U. Redlands, 1968; M.A., Calif. State U.-Long Beach, 1982; postgrad., U. Del., 1979—. Registered music therapist, Calif., Va. Music therapy intern Lanterman State Hosp., Pomona, Calif., 1977; music tchr. Pomona City Schs., 1977; music therapist Coll. Hosp., Cerritos, Calif., 1977-78; therapist trainee Community Psychology Clinic, Long Beach, Calif., 1978-79; psychology trainee VA, Perry Point, Md., 1980-82; psychology intern VA Outpatient Clinic, Los Angeles, 1982-83; psychology cons. Wilmington (Del.) City Schs., 1980-81. Served with USN, 1969-72. Calif. State scholar, 1964-68, 69. Mem. Am. Psychol. Assn. (assoc.), Nat. Assn. Music Therapy, Psychomusicology Soc., Phi Kappa Phi, Phi Mu Alpha Sinfonia (chpt. pres. 1967-68). Democrat. Baptist. Subspecialties: Behavioral psychology; Neuropsychology. Home: 13702-H Via del Palma Whittier CA 90602

LITOV, RICHARD EMIL, nutritionist, researcher; b. N.Y.C., Sept. 28, 1953; s. Tzvetan A. and Alice V. L. B.S., U. Calif., Davis, 1975, Ph.D., 1980. Scientist Mead Johnson & Co., Evansville, Ind., 1980—. Recipient Edward Frank Kraft award U. Calif., Davis, 1972. Mem. Am. Inst. Nutrition, Am. Coll. Sports Medicine, Inst. Food Technologists, Sigma Xi. Methodist. Subspecialties: Nutrition (biology); Biochemistry (biology). Current work: Trace mineral bioavailability in infant and adult nutrition. Office: Nutritional Sci Dept Mead Johnson & Co Evansville IN 47721

LITOVCHENKO, VLADIMIR ALEXEI, research chemist; b. Makhachkala, USSR, Nov. 15, 1934; came to U.S., 1981; s. Alexei Z. and Alexandra P. (Shepetkova) L.; m. Irina A. Poletaieva, July 28, 1957; m. Olga Aroseva., Nov. 10, 1978; children: Ioann, Anastasia-Christine. B.Sc. in Physics, U. Moscow, 1956; M.Sc. in Theoretical Physics, 1958; Ph.D., People's Friendship U., Moscow, 1967. Predoctoral fellow Lebedev Inst. Physics, USSR Acad. Sci., 1958-61; asst., then assoc. prof. dept. applied math. Faculty of Sci., People's Friendship U., 1961-73; chief pharmacopoeia dept. Inst. antibiotics, Moscow, 1973-74; adj. assoc. prof. dept. quantum theory Faculty Physics U. Moscow, 1971-77, assoc. prof. dept. chemistry, 1977-81; vis. asst. prof. dept. math. and research assoc. dept. chemistry U. Okla., Norman, 1982—. Contbr. articles to Soviet and U.S. profl. publs. Mem. Am. Phys. Soc. Subspecialties: Theoretical physics; Physical chemistry. Current work: Theoretical physics: interaction of polarized radiation with chiral media, statistical physics of chiral media, theoretical solid state physics. Physical chemistry: solubilization and micellization study by fluorescence, Raman and infrared spectroscopy. Optics: Polarization of fluorescence, synchrotron radiation. Office: 620 Parrington Oval Room 208 Norman OK 73019

LITROWNIK, ALAN JAY, educator; b. Los Angeles, June 25, 1945; s. Irving and Mildred Mae (Rosin) L.; m. Hollis Merle Glazer, Aug. 20, 1967; children: Allison Brook, Jordan Michael. B.A., UCLA, 1967; M.A., U. Ill., 1969, Ph.D., 1971. Asst. prof. psychology San Diego State U., 1971-75, assoc. prof., 1975-78, prof., 1978-81, prof., chmn. dept. psychology, 1981—; cons. San Diego County Dept. Edn., 1975-81, Jay Nolan Ctrs., 1981-82; exec. bd. Ctr. for Behavioral Medicine, San Deigo, 1982—. Contbr. articles to profl. jours. U.S. Office Edn. grantee, 1975-78, 80-81. Mem. Am. Psychol. Assn., Assn. for Advancement Behavior Therapy. Democrat. Jewish. Subspecialties: Behavioral psychology; Cognition. Current work: Application of social learning and information processing approaches to understanding and treatment of disturbed and/or handicapped children. Office: Dept Psychology San Diego State Univ 5300 Campanile Dr San Diego CA 92182

LITSTER, JAMES DAVID, physics educator; b. Toronto, Ont., Can., June 19, 1938; came to U.S., 1966; s. James Creighton and Gladys May (Byers) L.; m. Cheryl Schmidt, June 26, 1965; children: Robin, Heather. B.S., McMaster U., 1961; Ph.D., M.I.T., 1965. Asst. prof. physics M.I.T., Cambridge, 1966-71, assoc. prof., 1971-75, prof., 1975—; lectr. physics Harvard U. Med. Sch., Boston, 1974—; cons. N.Y. State Edn. Dept., 1978, NSF, Washington, 1977-81; vis. scientist Riso (Denmark) Nat. Lab., 1978. John Simon Guggenheim Meml. Found. fellow, 1971. Fellow Am. Phys. Soc.; mem. AAAS. Subspecialties: Condensed matter physics; Statistical physics. Current work: Light scattering and x-ray scattering studies of condensed matter physics. Office: Mass Inst Tech Cambridge MA 02139

LITTERST, CHARLES LAWRENCE, toxicologist; b. Cleve., Aug. 9, 1944; s. Robert Leroy and Olga Roesche (Porter) L.; m. Gretchen Nyland, 1966; children: Michael David, Stephen Charles. B.S., Purdue U., 1966; M.S., U. Wis., 1968, Ph.D., 1970. Pharmacologist FDA, Washington, 1970-72; toxicologist Nat. Cancer Inst., NIH, Bethesda, Md., 1972—; toxicology cons. U.S. Surgeon Gen. Office, Washington, 1977—; instr. Found. for Advanced Edn. in Scis., NIH, 1974—. Mem. Soc. Toxicology, Am. Soc. Pharmacology and Exptl. Therapeutics, N.Y. Acad. Scis., Sigma Xi, Phi Kappa Phi. Presbyterian. Subspecialties: Toxicology (medicine); Chemotherapy. Current work: Animal and human toxicity of antineoplastic chemotherapeutic agents; antidotes; mechanisms of action. Office: Nat Cancer Inst NIH Bldg 37 Rm 6D28 Bethesda MD 20205

LITTLE, ALAN BRIAN, obstetrician, gynecologist, educator; b. Montreal, Que., Can., Mar. 11, 1925; emigrated to U.S., 1951, naturalized, 1959; s. Herbert Melville and Mary Lizette (Campbell) L.; m. Nancy Alison Campbell, Aug. 20, 1949; children: Michael C. (dec.), Susan MacF. and Deborah MacF. (twins), Catherine E., Jane A., Mary L. B.A., McGill U., 1948, M.D., C.M., 1950. Intern Montreal Gen. Hosp., 1950-51; resident Boston Lying-in and Free Hosp. for Women, 1951-55, asst. obstetrician, asso. obstetrician and gynecologist, 1955-65; teaching fellow, asst. prof. Harvard Med. Sch., 1952-65; prof. ob-gyn, then Arthur H. Bill prof. ob-gyn, dir. dept. reproductive biology Case Western Res. U. Sch. Medicine, Cleve., 1965-82; prof. chmn. ob-gyn McGill U. Faculty of Medicine, Montreal, Que., Canada, 1982—; dir. dept. ob-gyn Univ. Hosps., Cleve., to 1982; mem. com. population research Nat. Inst. Child Health and Human Devel. Author: (with B. Tenney) Clinical Obstetrics, 1962; editor: (with others) Gynecology and Obstetrics-Health Care for Women, 1975; contbr. articles profl. jours. Served with RCAF, 1943-45. Fellow Am., Royal colls. surgeons, Am. Coll. Obstetricians and Gynecologists; mem. AMA, Endocrine Soc., Am. Gynecol. Soc., Am., Central assns. ob-gyn, Assn. Profs. Ob-Gyn, Soc. Gynecol. Investigation. Subspecialty: Obstetrics and gynecology. Office: 687 Pine Ave W Montreal PQ H3A 1A1 Canada

LITTLE, ELBERT LUTHER, JR., botanist, dendrologist; b. Ft. Smith, Ark., Oct. 15, 1907; s. Elbert Luther and Josephine (Conner) L.; m. Ruby Rema Rice, Aug. 14, 1943; children: Gordon Rice, Melvin Weaver, Alice Conner. B.A., U. Okla., 1927, B.S., 1932; M.S., U. Chgo., 1929, Ph.D., 1929; postgrad., U. Mich., 1927, Utah State U., 1928. Asst. prof. biology Southwestern Okla. State U., 1930-33; asst. to assoc. forest ecologist Forest Service, U.S. Dept. Agr., Tucson, 1934-41; dendrologist Forest Service, Washington, 1942-67, chief dendrologist, 1967-76; botanist Fgn. Econ. Adminstrn., Bogota, Colombia, 1943-45; prodn. specialist U.S. Comml. Co., Mexico City, 1945; cons. FAO, Cost Rica, 1964-65, 67, Ecuador, 1965, 75, Nicaragua, 1971; collaborator dept. botany Nat. Mus. Natural History, Smithsonian Inst., Washington, 1965-76, research asso., 1976—; vis. prof. Universidad de Los Andes, Merida, Venezuela, 1953-54, 1960, Va. Poly. Inst. and State U., 1966-67, U. D.C., 1979; forester Okla. Forestry Div., 1930, 77-78. Author: Checklist of U.S. Trees, 1979, Atlas of U.S. Trees, 1971-81, Trees of Puerto Rico and the Virgin Islands, 1964-74, Alaska Trees and Shrubs, 1972, Arboles Commes de Esmeraldas, 1969, Audubon Society Field Guide to North American Trees, Eastern Region and Western Region, 1980, Forest Trees of Oklahoma, 1981; Contbr. articles to profl. jours. Recipient Superior Service award U.S. Dept. Agr., 1960, Disting. Service award, 1973, Outstanding award, 1975; Disting. Service award Am. Forestry Assn., 1981; Profl. Achievement award U. Chgo., 1982. Fellow Soc. Am. Foresters, Washington Acad. Sci., Okla. Acad. Sci., AAAS; mem. Bot. Soc. Am., Am. Soc. Plant Taxonomists, Ecol. Soc. Am., Internat. Soc. Plant Taxonomy, Am. Inst. Biol. Sci., Internat. Soc. Tropical Foresters, Sociedad Botanica de Mexico, Explorers club. Subspecialties: Taxonomy; Ecology. Current work: Research on trees of U.S. and tropical America; preparation of references for tree identification. Home: 924 20th St S Arlington VA 22202 Office: Dept Botan Smithsonian Inst Washington DC 20560

LITTLE, HENRY NELSON, biochemistry educator; b. Portland, Maine, Oct. 17, 1920; s. George T. and Bertha H. (Nelson) L.; m. Isabelle Billings, June 20, 1948; children: Nathan, Joyce, Dana, Karen. B.S., Cornell U., 1942; M.S., U. Wis.-Madison, 1946, Ph.D., 1948. Research assoc. U. Chgo., 1948-49; asst. prof. biology Johns Hopkins U., Balt., 1949-51; assoc. prof. chemistry U. Mass., Amherst, 1951-56, prof., 1956-66, prof. biochemistry, 1966—. Served in U.S.M.C., 1942-44; PTO. Mem. Am. Chem. Soc., Am. Soc. Biol. Chemists, AAAS, Am. Soc. Plant Physiologists, Sigma Xi. Subspecialty: Biochemistry (biology). Current work: The mechanism and regulation of the biosynthesis of tetrapyrroles and hemoproteins. Office: U Ma-s Amherst MA 01002

LITTLE, ROBERT COLBY, physiologist, educator; b. Norwalk, Ohio, June 2, 1920; s. Edwin Robert and Eleanor Thresher (Colby) L.; m. Claire Campbell Means, Jan. 20, 1945; children—William C., Edwin C. A.B., Denison U., 1942; M.D., Western Res. U., 1944, M.S., 1948. Intern Grace Hosp., Detroit, 1944-45; USPHS postdoctoral research fellow Western Res. U., 1948-49; resident internal medicine Crile VA Hosp., Cleve., 1949-50; asst. prof. physiology, then asso. prof. physiology and medicine U. Tenn. Sch. Medicine, 1950-54; research participant Oak Ridge Inst. Nuclear Studies, 1952; dir. clin. research Mead Johnson & Co., 1954-57; lectr. medicine U. Louisville, 1955-57; dir. cardio pulmonary labs. Scott Sherwood and Brindley Found., Temple, Tex., 1957-59; prof. physiology, asst. prof. medicine Seton Hall Coll. Medicine and Dentistry, 1957-64; prof. physiology, chmn. dept., also asst. prof. medicine Ohio State U. Sch. Medicine, 1964-73; prof. physiology, chmn. dept., prof. medicine Med. Coll. Ga. Sch. Medicine, Augusta, 1973—; Cons. in field. Author: Physiology of the Heart and Circulation, 1977, 2d edit., 1981; editor: Physiology of Atrial Pacemakers and Conductive Tissues, 1980; contbr. articles to profl. jours. Served to capt. M.C. AUS, 1945-47. Mem. Am. Physiol. Soc., So. Soc. Clin. Investigation, Am. Heart Assn., Assn. Chmn. Depts. Physiology, Soc. Exptl. Biology and Medicine, Am. Fedn. Clin. Research, Sigma Xi, Sigma Chi, Alpha Kappa Kappa. Subspecialties: Biomedical engineering; Theoretical and applied mechanics. Current work: Biomechanics studies of musculoskeletal system and tissue biomechanics. Home: 44 Plantation Hills Dr Evans GA 30809

LITTLE, ROGER GEORGE, photovoltaic research co. exec; b. Adams, Mass., Sept. 28, 1940; m. Caro Lyn; children: Mark, Jean, Michelle. B.A. in Physics, Colgate U., 1962; Ph.D., M.I.T., 1964. Research physicist Ion Physics Corp., Burlington, Mass., 1964-68; founder, pres. Spire Corp., Bedford, Mass., 1968—; mem. solar energy panel Energy Research Adv. Bd., 1982. Contbr. articles to profl. jours. Mem. Am. Nuclear Soc., Am. Phys. Soc., ASTM, Electrochem. Soc., Materials Research Soc., Optical Soc. Am. Subspecialty: Solar energy. Current work: Developing photovoltaic manufacturing equipment and process lines. Coordinating near-term solar cell and module equipment engineering, mid-term process development of high-efficiency cells, and far-term research in thin film materials for high-efficiency, low cost modules. Patentee in field. Home: 228 Dudley Rd Bedford MA 01730 Office: Patriots Park Bedford MA 01730

LITTLEFIELD, JOHN WALLEY, geneticist, educator, pediatrician; b. Providence, Dec. 3, 1925; s. Ivory and Mary Russell (Walley) L.; m. Elizabeth Legge, Nov. 11, 1950; children—Peter P., John W., Elizabeth L. M.D., Harvard U., 1947. Diplomate: Am. Bd. Internal Medicine. Intern Mass. Gen. Hosp., Boston, 1947-48, resident in medicine, 1948-50, staff, 1956-74, chief genetics unit children's service, 1966-73; asso. in medicine Harvard U. Med. Sch., 1956-62, asst. prof. medicine, 1962-66, asst. prof. pediatrics, 1966-69, prof. pediatrics, 1970-73; prof., chmn. dept. pediatrics Johns Hopkins U. Sch. Medicine, Balt., 1974—; pediatrician-in-chief Johns Hopkins U. Hosp., 1974—. Author: Variation, Senescence and Neoplasia in Cultured Somatic Cells, 1976. Served with USNR, 1952-54. Guggenheim fellow, 1965-66. Mem. Am. Acad. Arts and Scis., Nat. Acad. Scis., Am. Soc. Biol. Chemists, Am. Soc. Clin. Investigation, Tissue Culture Assn., Soc. Pediatric Research, Am. Soc. Human Genetics, Am. Pediatric Soc., Am. Acad. Pediatrics, Assn. Am. Physicians, Phi Beta Kappa, Alpha Omega Alpha. Subspecialties: Genetics and genetic engineering (medicine); Pediatrics. Current work: Human and medical genetics, cell culture, molecular biology. Home: 304 Golf Course Rd Owings Mills MD 21117 Office: Children's Medical and Surgical Center 2-116 Johns Hopkins Hosp Baltimore MD 21205

LITTLETON, JESSE TALBOT, III, radiology educator; b. Corning, N.Y., Apr. 27, 1917; s. Jesse Talbot and Bessie (Cook) L.; m. Martha Louise Morrow, Apr. 17, 1943; children: Christine, Joanne, James, Robert, Denise. Student, Emory (Va.) and Henry Coll., 1934-35, Johns Hopkins U., 1935-39; M.D., Syracuse U., 1939-43. Lic. physician N.Y., Pa., Fla., Ala.; cert. Am. Bd. Radiology. Intern Buffalo Gen. Hosp., 1943; resident in medicine, surgery, radiology Robert Packer Hosp., Sayre, Pa., 1946-51, assoc. radiologist, 1951-53, chmn. dept. radiology, 1953-76; prof. radiology U. S. Ala., Mobile, 1976—; cons. in field. Author textbooks (2); contbr. chpts. to books and 80 articles to profl. jours., sci. exhibits to profl. confs. Served to capt. U.S. Army MC, 1944-46; Pacific. Fellow Am. Coll. Radiology; mem. AMA, N.Y. Acad. Scis., Am. Roetgen Soc., Ala. Acad. Radiology, Med. Assn. Ala., Sigma Xi, Alpha Omega Alpha. Republican. Methodist. Club: Country of Mobile (Ala.). Subspecialty: Radiology. Current work: Conventional tomography, physical principles, equipment development and testing and clinical applications; transportation and radiology of the acutely ill and traumatized patient; angiography, development of equipment for sectional radiographic anatomy. Home: 5504 Churchill Downs Ave Theodore AL 36582 Office: U S Ala Med Ctr 2451 Fillingim St Mobile AL 36617

LITTLETON, JOHN EDWARD, educator, astrophysicist; b. Ballston Spa, N.Y., July 28, 1943; s. Clarence Eugene and Anna Lucy (Seeley) L.; m. Barbara Anita Rhein, July 24, 1976. B.S. in Engring. Physics, Cornell U., 1965; Ph.D. in Astrophysics, U. Rochester, 1972. Research assoc. Belfer Grad. Sch. Sci., Yeshiva U., N.Y.C., 1972-73; research fellow Harvard Coll. Obs., Cambridge, Mass., 1973-75; asst. prof. physics W.Va. U., Morgantown, 1975-81, assoc. prof. physics, 1981—. Contbr.: research articles to sci. publs. including Astrophys. Jour. Mem. Am. Astron. Soc., AAAS, Internat. Astron. Union, Am. Assn. Physics Tchrs., Sigma Xi. Subspecialties: Theoretical astrophysics; Plasma physics. Current work: Astrophysical fluid mechanics and astrophysical plasma processes; atomic and molecular processes in atmospheres of cool stars. Home: 1432 Dorsey Ave Morgantown WV 26505 Office: Dept Physics WVa U PO Box 6023 Morgantown WV 26506

LITTMAN, MICHAEL GEIST, research scientist, educator; b. Washington, Mar. 29, 1950; s. Maxwell Leonard and Mildred (Geist) L.; m. L. Marion Katz Littman, Aug. 29, 1971. B.A., Brandeis U., 1972; Ph.D., M.I.T., 1977. Postdoctoral assoc. M.I.T., 1977-79; asst. prof. dept. mech. and aero. engring. Princeton U., 1979—. Contbr. articles in field to profl. jours. Co-recipient Alfred Rheinstein Faculty award Princeton U., 1981. Mem. Am. Phys. Soc., Optical Soc. Am., Phi Beta Kappa, Sigma Xi. Jewish. Subspecialties: Atomic and molecular physics; Spectroscopy. Current work: Effects of external fields on excited atoms; stark effect, spectroscopy of alkali atoms, atomic beams, pulsed tunable lasers. Office: Princeton U D418 EQ Princeton NJ 08544

LITTON, COLUMBUS C., research agronomist; b. Shoopman, Ky., Nov. 21, 1921; s. John Sherman and Mary Ann (Blevins) L.; m. Neva Sue Miller, Dec. 25, 1954; 1 son, Thomas Christopher. B.S. in Agr, U. Ky., 1949, M.S., 1952. Asst. prof. U. Ky., 1952-53; agt. Agrl. Research Service, U.S. Dept. Agr., U. Ky., 1953-55; research agronomist West Ky. Substation, Princeton, 1955-60, Agrl. Research Service, U. Ky., Lexington, 1960—. Contbr. articles to profl. jours. Served with U.S. Army, 1942-46. Mem. Am. Phytopath. Soc. Republican. Baptist. Subspecialty: Plant pathology. Current work: Breeding for disease resistance in burley and dark air and dark fire cured tobaccos. Home: 2024 Bellefonte Dr Lexington KY 40503 Office: Dept Plant Pathology U Ky Lexington KY 40546

LITTON, STEPHEN FREDERICK, orthodontist, anatomist, educator; b. Bklyn., Jan. 8, 1943; s. Murray A. and Eda (Schwartz) L.; m. Bonnie Tarnoff, July 4, 1965; children: Jeremy, Jonathan. B.A., U. Minn., 1965; B.S. with distinction, 1965; D.D.S., 1967; Ph.D., 1972. Research fellow, teaching asst. U. Minn., 1966-72, asst. prof. dept. anatomy, 1972—; cons. orthodontics Pilot City Health Ctr., Mpls., 1972-73; dir. orthodontics Children's Health Ctr., Mpls., 1972-83; chief dentistry Mt. Sinai Hosp., Mpls., 1980—; cons. orthodontics State of Minn., St. Paul, 1979—. Named Tchr. of Year Sch. Dentistry, U. Minn., 1979. Mem. Internat. Assn. Dental Research (W.H. Crawford award Minn. sect. 1965), ADA, Minn. Dental Assn., Mpls. Dist. Dental Soc., Am. Assn. Orthodontics, Minn. Soc. Orthodontists (award 1967), Midwestern Soc. Orthodontists, Omicron Kappa Upsilon, Alpha Omega (scholarship 1967, pres. 1966-67). Subspecialties: Orthodontics; Anatomy and embryology. Current work: collagen connective tissue, orthodontic tooth movement. Home: 1850 Kelly Dr Golden Valley MN 55427 Office: Univ Minnesota Dept Anatomy Minneapolis MN 55455

LITWHILER, DANIEL W., mathematician; b. Ringtown, Pa., Feb. 28, 1942; s. Daniel W. and Dorothy (Lynch) L.; m. Margaret Pendergast, Apr. 30, 1966; children: Daniel, Christopher, Kevin, Heather. B.S., Fla. State U., 1963, M.S., 1965; Ph.D., U. Okla., 1977. Commd. 2d lt. U.S. Air Force, 1965, advanced through grades to lt. col., 1980; chief ops. research div. dept. math (U.S. Air Force Acad.), Colorado Springs, Colo., 1972-82, dep. head dept. math. scis., 1983—, issues and policy analyst, 1982-83. Colo. liaison to U.S. Olympic Com., U.S. Baseball Fedn., 1978-79. Recipient Significant Achievement cert. Sec. of Air Force, 1983; Joint Service commendation Fifth Air Force, 1972; Air Force commendation, 1970, 82; Meritorious Service award U.S. Air Force, 1972, 80, 83, U.S. Air Force Acad., 1980, 1983. Mem. Ops. Research Soc. Am., Air Force Assn. (Silver medal Area C 1964). Republican. Roman Catholic. Subspecialties: Operations research (mathematics); Industrial engineering. Current work: Location theory.

Home: 5215 Bluestem Dr Colorado Springs CO 80917 Office: Dept Math USAF Acad Colorado Springs CO 80840

LITYNSKI, DANIEL MITCHELL, army officer, elec. engr., educator; b. Amsterdam, N.Y., Mar. 13, 1943; s. Mitchell Peter and Stella Agnes (Pawlowski) L.; m. Dianne Helene Miller, Dec. 28, 1963; children: Laura Ann, James Mitchell, John Thomas. B.S. in Physics, Rensselaer Poly. Inst., 1965, Ph.D., 1978; M.S. in Optics, U. Rochester, 1971. Commd. 2d lt. U.S. Army, 1965, advanced through grades to lt. col., 1980; comdr., Vietnam, 1967, Aberdeen Proving Ground, Md., 1967-68, research physicist, 1969, materiel officer, Vietnam, 1971-72, optical physicist, 1972-73, research officer, asst. prof. dept. physics, West Point, N.Y., 1974-78, assoc. prof. dept. elec. engring., 1980—, bn. exec. officer, Giessen, W.Ger., 1978-80. Contbr. articles to profl. jours. Decorated Bronze Star (3), Meritorious Service medal, Army Commendation medal. Mem. IEEE, Optical Soc. Am., Laser Inst. Am., Sigma Xi. Roman Catholic. Subspecialties: Laser research; Optical signal processing. Current work: Electromagnetic radiation propagation and interaction; coherent pulse propagation in active, guided media; electro-optics and integrated optics. Home: 292B Lee Rd West Point NY 10996 Office: Dept Elec Engring US Mil Acad West Point NY 10996

LITZENBERG, DAVID P., nuclear engineer, business, executive; b. Llanarch, Pa., Aug. 5, 1924; s. Charles C. and Elisabeth (Parker) L.; m. Pat Vickers, Sept. 1, 1979; m. Jeanne Carpenter, Feb. 27, 1947 (dec. Aug. 1977). B.S., U. Pa., 1946; M.B.A., Harvard U., 1952; Ph.D. in Nuclear Sci, Calif. Inst. Tech., 1958. Registered profl. engr., Ill. Chief engr. Zenith Engring. Corp., Phila., 1958-62; pres. Chempump Corp., Phila., 1958-65; chmn. bd. Powerdyne Corp., Phila., 1965—, Scottsdale, Ariz., 1965—; owner Litzenberg Engring. Soc., Phila., 1965—; chmn. bd. Pioneer Industries, Inc., Hayden Corp.; dir. L.A. Mitchell Ltd., Gt. Britain, N.K.K.K. Ltd., Japan, Heinze Fabrikat, Germay. Mem. Regional Econ. Council, 1957-62; chmn. United Fund, 1965, Am. Assn. Grad. Schs., 1969. Served to col. U.S. Army, 1947-52. Recipient Thomas awrad Am. Inst. Mgmt., 1965; Cutler award Inernat. Mktg. Soc., 1962. Mem. Am. Nuclear Soc. (chmn.), Am. Ordnance Assn. (life), Soc. Am. Mil Engrs. (dir.), Navy League (dir.), Franklin Inst., Pa. C. of C., Pa. Mfrs. Assn., Am. Nuclear Soc., Am. Inst. Mgmt. Subspecialties: Industrial engineering; Mechanical engineering. Patentee process equipment (3). Home: 8638 N 84th Pl Scottsdale AZ 85258

LIU, CHANG YU, mechanical engineer; b. Potin, Hopei, China, June 21, 1935; s. Tien-fu and Cherish (Pei-chang) L.; m. Wu peng yun, Aug. 13, 1968; children: Zeh-chen, Zeh-Wen. B.S., Chinese Naval Coll. Tech., Tsoying, 1958; M.S., Colo. State U., 1965, Ph.D., 1967. Design engr. Chinese 1st Naval Shipyard, Tsoying, 1959-60; instr. Chinese Naval Acad., Tsoying, 1961-63; assoc. prof. mech. engring. and naval architecture Nat. Taiwan U., Taipei, 1967-75; assoc. prof. Singapore U., 1975-78; prof. mech. engring. Unicamp, Campinas, Brazil, 1978—. Author: Principles of Naval Architecture, 1974. Ministry Edn. Taiwan scholar, 1976. Mem. AIAA, Aeros. and Astronautics Soc. Republic China. Subspecialties: Fluid mechanics; Wind power. Current work: Instrumentations, theoretical and experimental study of boundary layer, heat transfer and energy storage; flow through porous medium. Patentee semibalanced rudder, twin-wire resistance probe manometer. Home: Rua Luiz Vicentin 144 Barao Geraldo Campinas Sao Paulo Brazil 13100 Office: DEM FEC Unicamp Barao Geraldo Campinas Sao Paulo Brazil 13100

LIU, CHEN-HUEI, aerospace engineer, researcher; b. Hsing Ling, Kwangtung, China, June 29, 1941; came to U.S., 1966, naturalized, 1975; s. Hsiang-Hsien and Lin (Teng) L.; m. Grace Chi-Kun, Feb. 9, 1968; children: Edwin, Gene, Eric. B.S., Cheng Kung U., Tainan, Taiwan, 1963; M.S., NYU, 1968, Ph.D., 1971. Sr. analyst Computer Sci. Corp., Hampton, Va., 1972-73; research scientist NRC, NASA-Langley, Hampton, Va., 1973-75, aerospace engr., 1975—. Mem. Am. Phys. Soc., ASME, AIAA. Subspecialties: Fluid mechanics; Aeronautical engineering. Current work: Computational fluid dynamics, aerodynamics, vortex flows. Home: 128 Macaulay Rd Williamsburg VA 23185 Office: NASA Langley Research Center Mail Stop 252 Hampton VA 23665

LIU, CHING-TONG, research physiologist, educator; b. Kiangsu, China, Oct. 19, 1931; s. Lien Yi and Su Ju (Ku) L.; m. In-May Hsin, Feb. 28, 1970; children: Rex, Grace, Jeannette, Christine. M.S., U. Tenn., Memphis, 1959, Ph.D., 1963. Assoc. research biologist Sterling-Winthrop Research Inst., Renssalaer, N.Y., 1965-66; asst. prof. physiology Baylor Coll. Medicine, Houston, 1966-73, adj. prof., 1979—; research physiologist U S Army Med. Research Inst. Infectious Diseases, Ft. Detrick, Md., 1973—. Contbr. numerous articles to profl. jours. Mem. Soc. Exptl. Biology and Medicine, Am. Physiol. Soc., Am. Soc. Pharmacology and Exptl. Therapeutics, Am. Soc. Nephrology. Subspecialties: Physiology (medicine); Pharmacology. Current work: Dynamic functional changes and systematically integrated responses to certain viral infections in animals. Home: 7915 W 7th St Frederick MD 21701 Office: Med Div USAMRIID Fort Detrick Frederick MD 21701

LIU, EDWIN H., plant biochemist, educator; b. Honolulu, Apr. 11, 1942; s. Edward F. and Margaret Y. (Yuen) L.; m. Jeanne M., June 5, 1965. A.B., Johns Hopkins U., 1964; Ph.D., Mich. State U., 1971. Asst. prof. biology U. S.C., 1972—; asst. ecologist Savannah River Ecology Lab., Aiken, S.C., 1978—. Sea Grant grantee, 1978-81. Mem. Am. Chem. Soc., Am. Soc. Plant Physiologists. Subspecialties: Biochemistry (biology); Ecology. Current work: Biochemical adaptations of natural populations to environmental stress. Home: 407 Key Ave New Ellenton SC 29809 Office: Savannah River Ecology Lab PO Drawer E Aiken SC 29801

LIU, HAN-SHOU, space scientist, researcher; b. Hunan, China, Mar. 9, 1930; came to U.S., 1960, naturalized, 1972; s. Yu-Tin and Chun-Chen (Yeng) L.; m. Sun-Ling Yang Liu, May 2, 1957; children: Michael Fu-Yen, Peter Fu-Tze. Ph.D., Cornell U., 1963. Research asst. Cornell U., 1962-63; research assoc. Nat. Acad. Sci., Washington, 1963-65; scientist NASA Goddard Space Flight Center, Greenbelt, Md., 1965—; Pres. Mei-Hwa Chinese Sch., 1980-81. Contbr. articles to profl. jours. Fellow AAAS; mem. Am. Astron. Soc., Am. Geophys. Union, Planetary Soc., AIAA. Subspecialties: Tectonics; Geophysics. Current work: Dynamics of the earth and planets. Home: 2301 Laurelwood Terr Silver Spring MD 20904 Office: Code 921 NASA Goddard Space Flight Center Greenbelt MD 20771

LIU, HUA-KUANG, elec. engr., educator; b. Kueilin, Kwangsi, China, Sept. 2, 1939; s. Shui-chien and Cheng-Hsia (Fang) L.; m. Shao-Fen Liu; children: Tien-Wen, Ren-Wen. B.S.E.E., Nat. Taiwan U., 1962; M.S.E.E., U.Iowa, 1965; Ph.D., Johns Hopkins U., 1969. Jr. instr. Johns Hopkins U., 1965-69; mem. faculty dept. elec. engring. U. Ala., 1969—; prof., 1977—; vis. assoc. prof. Stanford U., 1975-76; vis. scientist M.I.T., 1982-83; cons. to industry U.S. Army Missile Command. Whitehead fellow, 1965-69; NSF grantee, 1970-73, 75-77. Mem. Optical Soc. Am., IEEE, Sigma Xi. Subspecialties: Optical engineering; 3emiconductors. Current work: Optical image processing, holography and printing. Patentee in field. Office: Dept Elec Engring U Ala University AL 35486

LIU, JAMES CHI-WING, med. physicist; b. Hong Kong, Jan. 6, 1950; U.S., 1969, naturalized, 1982; s. Chin Wan and Ying Mui (Wong) L. B.S., U. Mo., Columbia, 1973, M.S., 1974; Ph.D., 1981. Diplomate: Am. Bd. Radiology. Radiol. physicist Midwest Center of Radiol. Physics, Pitts., 1976-77; asst. med. physicist U. Mo. Med. Center, Columbia, 1977-81; med. physicist City of Hope, Duarte, Calif., 1981—. Mem. Am. Coll. Radiology, Am. Soc. Therapeutic Radiologists, Am. Assn. Physicists in Medicine. Roman Catholic. Club: Hills. Subspecialty: Medical physics. Current work: Radiation therapy dosimetry and diagnostic imaging. Office: 1500 E Duarte Rd Duarte CA 91010

LIU, KWOK-ON ELISHA, physicist, researcher; b. China, Jan. 19, 1946. Diploma in Physics, Hong Kong Bapt. Coll., 1969, M.Sc., Fla. Inst. Tech., 1975, Ph.D., U. Houston, 1982. Lab. dir. Hong Kong Bapt. Coll., 1969-73; teaching asst. Fla. Inst. Tech., Melbourne, 1973-75; research and teaching asst. physics dept. U. Houston, 1976—. Subspecialties: Laser fusion; Microchip technology (materials science). Current work: Investigation on dense plasma focus x-ray source for x-ray lithography. Developer plasma x-ray source for x-ray lithography. Office: 7400 Plum Creek #1605 Houston TX 77012

LIU, LEROY FONG, molecular biologist, educator; b. Taiwan, July 28, 1949; came to U.S., 1973; s. Shao-yao and Wen-hwa L.; m. Angela A. Yang, July 28, 1974; children: Christina E., Elaine E. Ph.D., U. Calif.-Berkeley, 1977. Postdoctoral fellow Harvard U., 1977-78, U. Calif.-San Francisco, 1978-80; asst. prof. Johns Hopkins Med. Sch., 1980—. Contbr. articles to profl. jours. Searle scholar, 1981. Mem. AAAS. Subspecialties: Molecular biology; Gene actions. Current work: I am interested in studies of protein-nucleic acids interactions. Currently, I have been investigating the biological roles of DNA topoisomerases. Home: 2211 Sulgrave Ave Baltimore MD 21209 Office: 725 N Wolfe St Baltimore MD 21205

LIU, MAW-SHUNG, physiology educator, dentist; b. Tai-Chung, Taiwan, China, Feb, 2, 1940; came to U.S., 1968; s. Chau-Tan and Shien (Wang) L.; m. Min-Chau Chang, Sept. 15, 1966; 1 child, Chien-Ye. D.D.S., Kaohsiung Med. Coll., Taiwan, 1964; M.Sc., U. Ky., 1970; Ph.D., U. Ottawa, 1976. Intern in pathology U. Ky. Med. Ctr., Lexington, 1968-69; instr. physiology La. State U. Sch. Medicine, New Orleans, 1974-76, asst. prof., 1976-78; assoc. prof. physiology Bowman Gray Sch. Medicine, Wake Forest U., Winston-Salem, N.C., 1978-82; prof. physiology St. Louis U. Sch. Medicine, 1982—; prin. investigator research lab. NIH, 1977—. Guest reviewer: Jour. Gerontology. Mem. Am. Physiol. Soc., Shock Soc., Internat. Soc Heart Research. Subspecialties: Physiology (medicine); Biochemistry (medicine). Current work: Myocardial and hepatic metabolism in endotoxin shock. Office: Dept Physiology St Louis U Sch Medicine 1402 S Grand Blvd St Louis MO 63104

LIU, PAUL CHI, oceanographer; b. Chefoo, Shuntung, China, June 18, 1935; came to U.S., 1959, naturalized, 1972; s. Joseph T. C. and Agatha I. M. (Wang) L.; m. Teresa Sheau-mei Wang, Jan. 30, 1965; 1 dau., Christina P. B.S., Nat. Taiwan U., 1956; M.S., Va. Poly. Inst., 1961; Ph.D., U. Mich.-Ann Arbor, 1977. Research phys. scientist U.S. Lake Survey, U.S. Army C.E., Detroit, 1965-71; research phys. scientist Nat. Ocean Survey, NOAA, Detroit, 1971-74, oceanographer, Ann Arbor, Mich., 1974—; vis. scholar U. Mich., 1978—. NOAA fellow, 1971-72. Mem. Am. Acad. Mechanics, Am. Geophys. Union, Am. Meteorol. Soc., ASCE, Soc. Indsl. and Applied Math., Sigma Xi, Phi Kappa Phi. Roman Catholic. Subspecialties: Ocean engineering; Fluid mechanics. Current work: Wind-generated waves, coastal engineering, physical oceanography. Home: 2939 Renfrew St Ann Arbor MI 48105 Office: Great Lakes Environ Research Lab NOAA 2300 Washtenaw Ave Ann Arbor MI 48104

LIU, TEH-YUNG, biochemist, researcher; b. Tainan, Taiwan, China, May 24, 1932; came to U.S., 1956; s. Chu-An and Chi-Li (Wu) L.; m. Sue C. Chen, Aug. 30, 1961; children: Cornelia, J. Rebecca, Daniel H. B.S., Nat. Taiwan U., Taipei, 1955; Ph.D., U. Pitts., 1961. Research assoc. Rockefeller U., 1961-65, asst. prof. protein biochemistry, 1965-67; biochemist Brookhaven Nat. Lab., Upton, N.Y., 1967-73; sect. chief NIH, Bethesda, Md., 1973-74; dep. dir. div. bacterial products bur. biologics FDA, Bethesda, 1974-79; dir. div. biochemistry and biophysics Nat. Ctr. for Drug and Biologics, Bethesda, 1979—. Contbr. numerous articles to profl. jours. Mem. Am. Soc. Biol. Chemists, Phi Lambda Upsilon. Subspecialties: Biochemistry (biology); Molecular biology. Current work: Structure, function, and biosynthesis of macromolecules; human c-reactive protein, limulus coagulation system, bacterial polysaccharides. Office: Division Biochemistry and Biophysics Office of Biologics National Ctr for Drug and Biologics Room 516 Bldg 29 8800 Rockville Pike Bethesda MD 20205

LIU, WING KAM, engineering educator; b. Hong Kong, May 19, 1952. B.S. in Engring. Sci. with highest honors, U. Ill.-Chgo., 1976; M.S. in Civil Engring, Calif. Inst. Tech., 1977, Ph.D., 1980. Research asst. dept. material engring. U. Ill.-Chgo.-Rush-Presbyn.-St. Luke Hosp., 1974-76; research asst. div. engring. and applied sci. Calif. Inst. Tech., 1976-80; asst. prof. dept. mech. and nuclear engring. Northwestern U., Evanston, Ill., 1980-83, assoc. prof., 1982—; cons. Hughes, Inc.; cons. in reactor analysis and safety Argonne Nat. Lab.; cons. U.S. Ballistic Research Lab., Aberdeen Proving Ground, Md. Contbr. articles to profl. jours.; reviewer: Jour. Nuclear engring. and Design. Recipient Melville medal ASME, 1979; Ralph R. Teetor Ednl. award, 1983; grantee NSF, NASA. Mem. ASCE, ASME, Am. Acad. Mechanics, Instn. Civil Engrs. Gt. Britain, Soc. Automotive Engrs., Phi Kappa Phi. Subspecialties: Solid mechanics; Fluid mechanics. Current work: Finite elements, computer aided engineering, fluid-structure interaction, thermal-fluid-solid interactions. Office: Dept Mech Engring Northwestern Univ Evanston IL 60201

LIU, YUNG SHENG, scientist; b. Anhwei, China, Sept. 23, 1944; came to U.S., 1967, naturalized, 1982; s. Hsing Chi and Li Wen (Wang) L.; m. Ming Lee, Jan 20, 1945; children: Alan, Jenny. B.S., Nat. Taiwan U., 1966; Ph.D., Cornell U., 1972. Research asst. Cornell U., 1968-72, teaching asst., 1968; physicist Gen. Electric Research Center, Schenectady, 1972—; vis. scientist UCLA, 1969; cons. in field; mem. U.S. Congressional Adv. Bd., 1982—. Contbr. articles to profl. jours. AVCO fellow, 1970; USAF grantee, 1975—; recipient Gen. Electric Outstanding Achievement award, 1977, Publ. award, 1982. Mem. Am. Phys. Soc., Optical Soc. Am., IEEE, Sigma Xi. Subspecialties: Laser Optics; Microelectronics. Current work: Laser physics and nonlinear optical devices; high power solid state lasers; laser processing of semiconductors, microelectronics. Home: 101 Woodhaven Dr Scotia NY 12302 Office: KWB-1307 PO Box 8 Schenectady NY 12345

LIU, YUNG YUAN, materials/energy research engineer; b. Taipei, Taiwan, Mar. 20, 1950; came to U.S., 1973, naturalized, 1982; s. Kan C. and Mon W. (Chou) L.; m. Teresa L. Ngai, Jan. 4, 1975; children: Sharon H.Y., Alvin H.L. B.S., Nat. Tsing-Hua U., Taiwan, 1971; M.A., MIT, 1976, Sc.D., 1978. Staff engr. Entropy Ltd., Lincoln, Mass., 1977-78; asst. nuclear engr. Argonne (Ill.) Nat. Lab., 1978-81, nuclear engr., 1981—; mem. life code com. U.S. Dept. Energy, 1978-81, mem. fuel performance evaluation task force, 1978-81; cons. Los Alamos Nat. Lab., 1982—. Contbr. articles to profl. jours.; editor-in-chief: Free Chinese Monthly, Cambridge, Mass., 1973-75. Mem. Am. Nuclear Soc., Am. Soc. Metals. Baptist. Subspecialties: Nuclear fission; Nuclear fusion. Current work: Materials related energy technology, high temperature materials behavior; radiation damage; physics of deformation and fracture; computer simulation of materials behavior. Home: 2043 Sunnydale St Woodridge IL 60517 Office: Materials Sci and Tech Argonne Nat Lab 9700 S Cass Ave Argonne IL 60439

LIU, YUNG-PIN, biochemist; b. Tao-Yuan, Taiwan, Aug. 26, 1937; came to U.S., 1963, naturalized, 1975; s. A-Yen and Kan-Mei (Hsu) L.; m. Mei-Shiang Sung, July 10, 1967; 1 dau., Bo-May. B.S., Chung-Yuan U., 1960; M.S., Lowell Tech. U., 1967; Ph.D., Baylor U., 1970. Teaching asst. Lowell Tech. U., Lowell, Mass., 1965-67; research asst. Wadley Inst. Molecular Medicine, Dallas, 1967-69; postdoctoral fellow Yale U., New Haven, 1969-72; research assoc. St. Jude Children's Hosp., Memphis, 1972-75; asst. prof. U. Tenn., Memphis, 1973-75; cancer expert NIH, Bethesda, Md., 1975-82; research biochemist Walter Reed Army Inst. Research, Washington, 1982—. Contbg. author: Cyclic Nucleotides in Disease, 1975, Ocular Pathology Update, 1980, Molecular Mechanics Photoreceptor Transduction, 1981, Structure of the Eye, 1982, Pharmacologic Principles Cancer Treatment, 1982. Named Outstanding Supr. NIH, 1979; recipient Outstanding Achievement award Fight for Sight, Inc., 1980. Mem. Am. Soc. Biol. Chemists, Chinese Biochem. Soc. Subspecialties: Biochemistry (biology); Hematology. Current work: Isolation, identification, and characterizations of plasminogen activator and plasminogen inhibitor in the human promyelocytic leukemia cell line. Home: 9723 Singleton Dr Bethesda MD 20817 Office: Walter Reed Army Inst Research 14th St and Dahlia Washington DC 30207

LIVE, DAVID HARRIS, chemist; b. Phila., Apr. 3, 1946; s. Israel and Anna (Harris) L. B.S., U. Pa., 1967; Ph.D., Calif. Inst. Tech., 1974. Research assoc. Rockefeller U., N.Y.C., 1974-78, asst. prof. dept. phys. biochemistry, 1978—; cons. Jet Propulsion Lab., Calif. Inst. Tech., 1975—. Mem. Am. Chem. Soc., AAAS, N.Y. Acad. Sci. Subspecialties: Biophysical chemistry; Polymer chemistry. Office: Rockefeller U 1230 York Ave New York NY 10021

LIVINGSTON, CLARK H., plant pathologist; b. Eau Claire, Wis., Nov. 25, 1920; s. George Wallace and Helen Louise (Holcomb) L.; m. Ann Jean Garney, June 9, 1947; children: Kay Ann, Thomas Clark. B.S., Colo A&M U., 1951, M.S., 1953; Ph.D., U. Minn., 1966. Mem. faculty Colo. State U., Ft. Collins, 1955—, now plant pathologist, prof. botany plant pathology. Contbr. articles in field to profl. jours. Active Boy Scouts Am. Served with U.S. Army, 1943-46. Mem. Am. Phytopath. Soc., Potato Assn Am. Republican. Subspecialties: Plant pathology; Plant virology. Current work: Potato, bean and tree viruses and their related diseases. Home: 3008 Shore Rd Fort Collins CO 80524 Office: 113 Potato Virus Research Lab Fort Collins CO 80523

LIVINGSTON, DAVID MORSE, physician, molecular biologist; b. Cambridge, Mass., March 29, 1941; s. Arthur Joshua and Phyllis Freda (Kanters) L.; m. Jacqueline Sue Gutman, June 23, 1963; children: Catherine Ellen, Julie. A.B., Harvard U., 1961; M.D. Magna Cum Laude, Tufts U., 1965. Diplomate: Am. Bd. Internal Medicine. Intern, resident in medicine Peter Bent Brigham Hosp., Boston, 1965-67; research fellow NIH, Bethesda, Md., 1967-69, scientist, 1971-73; research fellow in biochemistry, fellow Helen Hay Whitney Found. Harvard Med. Sch., Boston, 1969-71, asst. prof., 1973-1976, assoc. prof., 1976-82, prof. medicine, 1982—; mem. virology study sect. NIH, 1979-83. Mem. editorial bd.: Virology, Jour. Virology, 1976-79; mem. editorial bd.: Cell, 1976-79, Archives of Virology, 1977-79. Served with USPHS, 1967-69, 1971-73. Mem. Am. Soc. Biol. Chemists, Am. Soc. Clin. Investigation, Am. Soc. Microbiology. Subspecialties: Molecular biology; Cancer research (medicine). Current work: molecular biology of tumor viruses, molecular genetics and biology of blood clotting proteins, cancer medicine. Office: Dana Farber Cancer Inst Harvard Med Sch Boston MA 02115

LIVINGSTON, JOHN DAVID, computer consultant; b. Jefferson City, Mo., Apr. 16, 1928; s. Joseph Hall and Laura Mae (Terrill) L.; m. Rachel Ann Jennings, Dec. 27, 1952; children: David Hall, Jennie Ann, Francis Joseph. B.S. in Indsl. Edn., Ind. State U., 1954; postgrad., Purdue U., 1954. With U.S. Bur. Census, Suitland, Md., 1960-64; mgr. computer test facility Bunker Ramo, Ft. Huachuca, Ariz., 1964-65; mgr. computation Lab. Electronic Assocs., Inc., Princeton, N.J., 1965-68; computer ctr. mgmt. expert UN, Bucharest, Romania, 1969-70, prof. info. tech., Turin, Italy, 1970-79; UN chief tech. advisor Egyptian Fgn. Ministry, Cairo, 1979—. Author: What is a Computer?, 1971, Computer Operations Management, 1969. Served in U.S. Army, 1945-49. Fellow Inst. Prodn. Control, Visible Record Soc.; mem. IEEE, Assn. Computing Machinery, Data Processing Mgmt. Assn., Nat. Micrographics Assn. Subspecialties: Information systems, storage, and retrieval (computer science); Information systems (information science). Current work: Computer assisted instruction research; current information retrieval and management systems; special interests computers in aid to developing countries. Home: Rt 6 Box 171 Mountain Home AZ 72653 Office: UN-Cairo PO Box 20-Grand Central New York NY 10163

LIVINGSTON, WILLIAM CHARLES, astronomer; b. Santa Ana, Calif., Sept. 13, 1927; s. William Charles and Ada Markley (Parvin) L.; m. Dorothy Wingate Newell, June 22, 1957; children: Ann, Peter. Student, Los Angeles City Coll, 1946-49; A.B., UCLA, 19521; Ph.D., U. Calif.-Berkeley, 1959. Observing asst. Mt. Wilson Obs., 1950-52, 57-58; jr. astronomer Kitt Peak Nat. Obs., Tucson, 1959-60; asst. astronomer, 1960-65, asso. astronomer, 1965-69, astronomer/tenure, 1969—, researcher meterol. optics. Served with USAF, 1945-46. Mem. Am. Astron. Soc., Internat. Astron. Union (pres. com. 9 instruments), Astron. Soc. India. Subspecialties: Optical astronomy; Solar physics. Current work: Solar instruments and telescopes, solar magnetic fields, solar eclipses, observations of solar spectrum, solar variability.

LIVOLSI, VIRGINIA ANNE, pathologist, educator; b. N.Y.C., July 29, 1943; d. Epifanio and Mary A. (LaPorta) LiV. B.S., Coll. Mt. St. Vincent, 1965; M.D., Columbia U., 1969. Diplomate: Am. Bd. Pathology. Intern Columbia-Presbyn. Med. Center, N.Y.C., 1969-70; fellow in surg. pathology Nat. Cancer Inst.CPMC, 1970-74; asst. prof. Yale U., New Haven, 1974-79, assoc. prof., 1979-83; prof. pathology, U. Pa., Phila., 1983—; chief surg. pathology U. Pa. Hosp. Author: Practical Clinical Cytology, 1979; editor: Thyroiditis, 1981. Fellow Internat. Acad. Pathology, Am. Soc. Clin. Pathology; mem. Internat. Acad. Cytology. Subspecialty: Pathology (medicine). Current work: Gynecologic pathology; endocrine pathology; thyroid disease.

LLAURADO, JOSEP G., nuclear medicine physician, scientist; b. Barcelona (Catalonia), Spain, Feb. 6, 1927; s. Jose and Rosa (Llaurado) Garcia; m. Deirdre Mooney, Nov. 9, 1966; children: Raymund, Wilfred, Mireya; m. Catherine D. Entwistle, June 28, 1958 (dec.); children: Thadd, Oleg, Montserrat. B.S., B.A., Balmes Inst., Barcelona, 1944; M.D., Barcelona U., 1950; Ph.D., 1960; M.Sc. Biomed. Engrng., Drexel U., 1963. Diplomate: Am. Bd. Nuclear Medicine. Resident Royal Postgrad Sch. Medicine, Hammersmith Hosp., London, 1952-54; fellow M.D. Anderson Hosp. and Tumor Inst., Houston, 1957-58, U. Utah Med. Coll., Salt Lake City, 1958-59; asst. prof. U. Otago Dunedin, N.Z., 1954-57; sr. endocrinologist Pfizer Med. Research Lab., Groton, Conn., 1959-60; assoc. prof. U. Pa., 1963-67; prof. Med. Coll. Wis., Milw., 1970-82, Marquette U., 1967-82; clin. dir. nuclear medicine service VA Med. Ctr., Milw., 1977-82; chief nuclear medicine service VA Hosp., Loma Linda, Calif., 1983—; prof. dept. radiation scis. Loma Linda U. Sch. Medicine, 1983—; U.S. rep. symposium on dynamic studies with radioisotopes in clin. medicine and research IAEA, Rotterdam, 1970, Knoxville, 1974. Contbr. numerous articles to profl. jours. Merit badge counselor Boy Scouts Am., 1972—; pres. Hales Corners (Wis.) Hist. Soc., 1981-83. Recipient Commendation cert. Boy Scouts Am., 1980. Fellow Am. Coll. Nutrition; mem. Soc. Nuclear Medicine (computer and acad. councils), IEEE (sr.), Biomed. Engrng. Soc. (charter), Am. Physiol. Soc., Am. Soc. Pharmacology and Exptl. Therapeutics, Soc. Math. Biology (founding), Endocrine Soc., Royal Soc. Health, Soc. Catalana de Biologia. Roman Catholic. Subspecialties: Nuclear medicine; Biomedical engineering. Current work: Cardiac nuclear medicine; nuclear magnetic resonance; computer applications to biomedicine; compartmental analysis. Office: VA Hosp 115 Loma Linda CA 92357

LLINAS, RODOLFO RIASCOS, medical educator, researcher, physician; b. Bogotá, Colombia, Dec. 16, 1934; U.S., 1960, naturalized, 1973; s. Jorge Enrique and Bertha (Riascos) L.; m. Gillian Kimber, Dec. 24, 1965; children: Rafael Hugo, Alexander Jorge. B.S., Gimnasio Moderno, Bogotá, 1952; M.D., Universidad Javeriana, Bogotá, 1958; Ph.D., Australian Nat. U., Canberra, 1965. Rotating intern Hosp. San José, Bogotá, 1959; postdoctoral research fellow Stanley Cobb Lab., Mass. Gen. Hosp., Harvard U. Med. Sch., Boston, 1959-61; NIH researchfellow in physiology U.Minn., Mpls., 1961-63, assoc. prof. physiology, 1965-66; assoc. mem. AMA Inst. Biomed. Research, Chgo., 1966-68, mem., 1970, head neurobiology unit, 1967-70; assoc. prof. neurology and psychiatry Northwestern U., 1967-71; professorial lectr. in pharmacology U. Ill. Coll. Medicine-Chgo., 1967-68, clin. prof. pharmacology, 1968-72; prof. physiology, head neurobiology div. U. Iowa Sch. Medicine, 1970-76; prof. physiology and biophysics, chmn. dept. physiology and biophysics NYU Med. Sch., 1976—; guest prof. physiology Wayne State U. Sch. Medicine, 1967-74; professorial lectr. College de France, Paris, 1979; mem. neurol. sci. research tng. com. Nat. Insts. Neurol. Diseases and Stroke NIH, 1971-73; assoc. neuroscis. research program MIT, 1974-83; mem. neurology study sect. div. research grants NIH, 1974-78; mem. panel on basic neurosci. research Nat. Acad. Sci. Task Force, 1978; mem. sci. adv. bd. for basic research Max-Planck Inst. Psychiatry, Munich, W.Ger., 1979-83. Authors: (with Hubbard and Quastel) Electrophysiological Analysis of Synaptic Transmission, 1969; editor: Neurobiology of Cerebellar Evolution and Development, 1969, (with W. Precht) Frog Neurobiology: A Handbook, 1976; chief editor: Neurosci, 1974—; editorial bd.: Jour. Neurobiology, 1980—, Pfluegers Archiv, 1981—, Jour. Theoretical Neurobiology, 1981—. Recipient John C. Krantz award U. Md. Sch. Medicine, 1976. Mem. Soc. Neurosci. (council 1974-78), Am. Physiol. Soc. (Bowditch lectr. 1973, mem. task force on neurophysiology 1974), Am. Soc. Cell Biology, Biophys. Soc., Harvey Soc., Internat. Brain Research Orgn. (mem. U.S. nat. com. 1978-81, acting chmn. com. 1982, chmn. com. 1983-84), N.Y. Acad. Scis., Tensor Soc., Alpha Omega Alpha. Subspecialties: Neurophysiology; Biophysics (biology). Current work: Synaptic transmission in vertebrates and invertebares; single cell electrophysiology; neuronal nets; cerebellar function; mathematical modelling. Office: NYU Med Ctr 550 1st Ave New York NY 10016

LLORCA, ARTHUR LEE, psychologist; b. Chgo., Sept. 21, 1940; s. Arthur Edward and Zelia (Collebrusco) L.; m. Rita Mary Pedretti, Oct. 31, 1964 (div. Dec. 1979); children: Raymond, Michael, Robert; m. Ann Elise Aucker, May 7, 1980. M.A., U. N.C., Greenville, 1980; M.T.I.D., N.C. State U., Raleigh, 1975. Staff psychologist Sandhills Ctr., Pinehurst, N.C., 1981—; adj. faculty Campbell U., Buies Creek, N.C., 1976—; cons. N.C. Headstart Program, Raleigh, 1982—. Served to majr., inf. U.S. Army, 1960-78. Mem. Am. Psychol. Assn., Internat. Soc. Polit. Psychology, Pinehurst Clinic Group. Roman Catholic. Subspecialties: Clinical psychology. Current work: International relations/human behavior. Home: Box 433 Pinehurst NC 28374 Office: Sandhills Center Drawer 639 Pinehurst NC 28374

LLOYD, EVAN ELLIOTT MORGAN, engineering company exec.; b. New Haven, Oct. 16, 1945; s. David P.C. and Kathleen Mansfield (Elliott) L.; m. Cathy Ann Disque, Sept. 5, 1965; children: Steven Kenneth Elliott, Melissa Kim. A.A.S., Capitol Inst. Tech., 1968, B.S. E.T., 1968. Customer engr. IBM Corp., Arlington, Va., 1966-68; instr. Capitol Inst. Tech., Kensington, Md., 1968-69; sr. engr. Link div. Singer Co., Silver Spring, Md., 1969-75, mgr.bus. devel., 1976-80; group mgr. Quadrex Corp., Campbell, Calif., 1980-82; v.p. Interfacts, Inc., San Jose, Calif., 1982—. Recipient Ann. Research and Devel. award Singer Co., 1975. Mem. Am. Assn. Engrng. Edn., IEEE, Am. Nuclear Soc. (standards com. 1980—). Subspecialties: Distributed systems and networks; Theoretical computer science. Current work: Directing development of new power generation simulation systems and training data base systems for power utility applications. Office: Interfacts Inc 7174 Santa Teresa Blvd San Jose CA 95139

LLOYD, JAMES EDWARD, entomology educator; b. Oneida, N.Y., Jan. 17, 1933; s. Harry Austin and Ann Lucille (Lynch); m. Dorothy June Pafka, Aug. 16, 1958; children: Robert Stanley, Kyle Anne. B.S., SUNY-Fredonia, 1960; M.A., U. Mich., 1962; Ph.D., Cornell U., 1966. Sci. tchr. Dunkirk (N.Y.) pub. schs., 1960; prof. U. Fla., Gainesville, 1966—. Contbr. sci. articles to profl. jours. Served with USN, 1951-55. Recipient Disting. Alumni award SUNY, Fredonia, 1982. Mem. AAAS, Fla. Entomol. Soc., Coleopterists Soc., Am. Naturalists Soc., Animal Behavior Soc. Subspecialties: Evolutionary biology; Ethology. Current work: Systematics and behavior of fireflies (lampyridae), evolution of insect communication systems, insect behavioral ecology. Home: 915 NW 40th Terr Gainesville FL 32605 Office: Dept Entomology U Fla Gainesville FL 32611

LLOYD, JOHN TRACY, psychologist, consultant; b. Astoria, Oreg., Dec. 20, 1946; s. Fred. H. and Alice Rose (Baker) L.; m. Susan Lynn Quackenbush, Aug. 20, 1970. B.A., U. Oreg., 1969; M.S., U. Idaho, 1972, Ph.D., 1977. Lic. psychologist, Wash. Instr. U. Idaho, Moscow, 1972-73, 75-77; research faculty Whitman Coll., Walla Walla, Wash., 1973-75; psychologist OUI: Rehab. Ctr., Lewiston, Idaho, 1976-77; asst. prof. Wash. State U., Pullman, 1977-82; psychologist Dolorology Assocs., Spokane, Wash., 1982—; cons. psychologist Eptom Soc., Pullman, 1976-82, supt. pub. instrn., Olympia, Wash., 1979-82; text rev. cons. Random House, Inc., N.Y.C., 1981-82; field supr./cons. Psychol. Corp., N.Y.C., 1981. Contbr. articles to profl. jours. Mem. Am. Psychol Assn., Wash. State Psychol. Assn., Western Assn. for Counselor Edn. and Supervision, Southeastern Wash. Counselor Edn. Consortium (co-dir. 1981). Subspecialties: Cognition; Neuropsychology. Current work: Closed head brain trauma and its effects; learning disability and neuroeducation. Home: South 1249 Wall Spokane WA 99204

LO, ALOYSIUS KOU-FANG, research scientist; b. Nanking, China, Oct. 24, 1936; emigrated to U.S., 1962; s. Ta-kuan and Mai (Chow) L.; m. Lisa Kwangping, Feb. 20, 1965. B.Sc., Nat. Taiwan U., 1961; M.Sc., U. Detroit, 1964; M.S., U. Mich., 1966; Ph.D., U. Toronto, 1970. Teaching fellow U. Detroit, 1962-64; research engr. AVCO Everett (Mass.) Research Labs., 1965; teaching fellow U. Mich., Ann Arbor, 1966; research assoc. U. Toronto, 1970-72; research scientist Environment Can., Toronto, 1972—. Contbr. articles to profl. jours. Served as 2d lt. Chinese Army, 1961-62. Mem. AIAA, Profl. Engrs. Ont., Can. Meteorol. Soc. Subspecialties: Aeronautical engineering; Planetary atmospheres. Current work: Dynamics and turbulence structures of atmospheric boundary layers and long-range transport of air pollutions. Engineering studies on alternate (renewable) energies. Home: 144 Kingsdale Ave Willowdale ON Canada M2N 3W9 Office: Atmospheric Environment Services 4905 Dufferin St Toronto ON Canada M3H 5T4

LO, KWOK-YUNG, radio astronomer, educator; b. Nanking, China, Oct. 19, 1947; came to U.S., 1965, naturalized, 1980; m. Helen Chen, Jan. 1, 1973; children: Jan Hsin, Pei Hsin. S.B., M.I.T., 1969; Ph.D., 1974. Research fellow radio astronomy Calif. Inst. Tech., Pasadena, 1974-76, sr. research fellow radio astronomy, 1978-80, asst. prof., 1980—; Miller fellow basic research in sci. U. Calif.-Berkeley, 1976-78, asst. research astronomer, 1978. Contbr. articles in field to profl. jours. Mem. Internat. Astron. Union, Internat. Union Radio Sci., Am. Astron. Soc. Subspecialties: Radio and microwave astronomy; Infrared optical astronomy. Current work: The galatic center, star formation, intergalactic hydrogen, dwarf galaxies, very long baseline interferometry. Office: Dept Astronom Calif Inst Tech 105-24 Pasadena CA 91125

LOACH, KENNETH WILLIAM, chemist; b. Portsmouth, Eng., Sept. 5, 1934; s. William A.F. and Josephine (Wilkes) L.; m. Sandra K. Miller, June 11, 1966; children: Matthew W., Catherine S. B.Sc., U. Auckland, N.Z., 1956, M.Sc., 1958; Ph.D., U. Wash., 1969. Analytical chemist Dept. Agr., Hamilton, N.Z., 1958-60, C.S.I.R.O., Canberra, Australia, 1960-63; asst. prof. chemistry SUNY,-Plattsburgh, 1969-77, assoc. prof., 1977—, assoc. prof. computer sci., 1978—. Mem. Am. Chem. Soc., AAAS, Assn. Computing Machinery, Fedn. Am. Scientists. Democrat. Roman Catholic. Subspecialties: Analytical chemistry; Information systems, storage, and retrieval (computer science). Current work: Analytical chemical theory; trace analysis; information science; computerized data analysis; computer-assisted instruction. Office: Dept Chemistry SUNY Plattsburgh NY 12901

LOATMAN, ROBERT BRUCE, computer scientist; b. Washington, Aug. 23, 1945; s. Paul John and Miriam Joyce (Barna) L.; m. Carol Anne Chalmers, June 6, 1969; children: Thomas, Cynthia, Ryan, Michael. B.A., Fordham U., 1967, M.A., 1972, Ph.D., 1976. Programmer Gen. Electric Co., Schenectady, 1968-69; instr. math. Fordham U., Bronx, N.Y., 1972-73, Georgetown U., Washington, 1973-76; info. systems engr. Mitre Corp., McLean, Va., 1978; sr. assoc. Killalea Assocs., Inc., Alexandria, Va., 1976-80; computer scientist Planning Research Corp., McLean, Va., 1980—; cons. Phonic Ear, Palo Alto, Calif., 1980; NSF trainee, 1970-73. Recipient Medals of Achievement Math. Assn. Am., 1963, Rensselaer Poly. Inst., 1963; NSF research grantee, 1965. Mem. Am. Assn. Artificial Intelligence, Am. Math. Soc., Soc. Indsl. and Applied Math., Assn. Computing Machinery, Spl. Interest Group in Artificial Intelligence, AAAS. Roman Catholic. Subspecialty: Artificial intelligence. Current work: Knowledge representation, knowledge-based systems, expert systems, knowledge acquisition, heuristics, knowledge engineering, natural language interface. Home: 9903 Snowbound Ct Vienna VA 22180 Office: Research and Devel Div Planning Research Corp 1500 Planning Research Dr McLean VA 22102

LOBO, PETER ISAAC, physician, educator; b. Kabale, Uganda, Apr. 11, 1943; s. Leonard Luciano and Carmen Isabelle L.; m. Monica Lobo, May 5, 1971; children: Toinette Carmen, Ingrid Elizabeth, Leonard Jason. M.B.Ch.B., Makerere U., Kampala, Uganda, 1966. Diplomate: Am. Bd. Internal Medicine, Am. Bd. Nephrology. Resident in internal medicine and nephrology U. Va. Sch. Medicine, Charlottesville, 1971-76, instr., 1976-77, asst. prof., 1977-81, assoc. prof. medicine, 1981—, dr. tissue typing lab., 1978—. Contbr. articles to sci. jours. Fellow ACP; mem. Am. Assn. Immunologists, Am. Histocompatibility Soc., Internat. Transplantation Soc., So. Soc. Clin. Investigatio. Roman Catholic. Subspecialties: Nephrology; Transplantation. Current work: Function of human lymphocyte subsets, immune-regulation, studies on formation and role of alloantibodies pre- and post-transplantation. Home: 348 Keywest Dr Charlottesville VA 22901 Office: U Va Med Center Box 133 Charlottesville VA 22908

LOCASCIO, JOSEPH JASPER, psychologist; b. Chgo., Mar. 15, 1950; s. Dominic and Angela (Lipari) L. B.S., Loyola U., Chgo., 1972; M.A., U. Kans., 1974; Ph.D., Northwestern U., 1982. Statistician Chgo. Dept. Health div. Mental Health, 1975-76; coordinator research and evaluation, 1976-79; instr. Northwestern U., Evanston, 1982; postdoctoral researcher dept. psychology U. Ill., Urbana, 1982—. Mem. Am. Psychol. Assn., Am. Statis. Assn. Subspecialties: Evaluation research; Quantitative/statistical methodology. Current work: Evaluation research methodology and quantitative/statistical methods especially as applies to mental health and education. Home: 108 E Stoughton St Apt 7 Champaign IL 61820 Office: Dept Psychology Univ Ill Champaign IL 61820

LOCHER, PAUL JOHN, psychology educator; b. Allentown, Pa., Sept. 27, 1941; s. Paul Otto and Teresa (Bedics) L. Ph.D., Temple U., 1972. Assoc. prof. Montclair State Coll., Upper Montclair, N.J., 1972—. NSF grantee, 1967. Mem. Am. Psychol. Assn., Jean Piaget Soc., Eastern Psychol. Assn., Psychonomic Soc. Subspecialty: Cognition. Current work: Using several techniques to record eye movements, relation between visual encoding strategies and higher cognitive processes. Home: 586 Upper Mountain Ave Upper Montclair NJ 07043 Office: Montclair State Coll Upper Montclair NJ 07043

LOCKE, BEN Z(ION), government science administrator, epidemiologist; b. N.Y.C., Sept. 8, 1921; s. Samuel and Gussie (Radinofsky) L.; m. Frances Bellin, Oct. 25, 1947; children: Meryl, Lauren, Robert, Kenneth. B.A., Bklyn. Coll., 1947; M.S.P.H., Columbia U., 1950. Jr. statistician N.Y. State Dept. Health, 1946-49, prin. statistician, 1950-56; chief Consultation Sect., NIMH, Bethesda, Md., 1956-66; dir. research and service evaluation program, assoc. prof. psychiatry Temple U. Med. Sch., 1966-67; asst. chief Center for Epidemiologic Studies, NIMH, Rockville, Md., 1967-75, chief, 1975—; cons. in field. Co-author: Mental Disorders/Suicide, 1972; co-editor series: Monographs in Psychosocial Epidemiology, vols. 1-3, 1980-82. Served with U.S. Army, 1942-46. Decorated Purple Heart.; Recipient award for meritorious service Alcohol, Drug Abuse and Mental Health Adminstrs., 1982. Fellow Am. Coll. Epidemiology, Am. Public Health Assn.; mem. Soc. Epidemiologic Research, Am. Statis. Assn., AAAS. Jewish. Subspecialties: Epidemiology; Statistics. Current work: Through epidemiologic research, elucidate the etiology of specific mental disorders allowing for programs of intervention and prevention. Home: 11803 Saddlerock Rd Silver Spring MD 20902 Office: 5600 Fishers Ln Room 18-105 Rockville MD 20857

LOCKE, BILL J., psychology educator, consultant; b. Erick, Okla., Feb. 5, 1936; s. Walter A. and Ruby G. (Davis) L.; m. Dorothy P Hardin, Aug. 11, 1957; children: Jon David, James Mitchell. B.A., Okla. U., 1959; M.A., Okla. State U., 1960, Ph.D., 1962; postgrad., U. Kans.-Lawrence, 1962-63. Lic. psychologist, Tex. Coordinator research tng. Kans. U. Bur. Child Research, Lawrence, 1963-66; dir. research tng. Parsons (Kans.) Research Ctr., 1967-68; assoc. prof. psychology Tex. Tech. U., Lubbock, 1969-72, prof., 1972—, clin. psychology cons.; bd. dirs. Lubbock Mental Health Assn., 1969-75. Contbr. articles to sci., profl. jours., 1962—. Coach Lubbock YMCA Athletics for Youth, 1970-75. NIMH fellow, 1959-62; Nat. Inst. Neurol. Diseases and Blindness fellow, 1963. Fellow Am. Assn. Mental Deficiency; mem. Am. Psychol. Assn., Lubbock Assn. Psychologists (pres. 1974). Democrat. Baptist. Subspecialty: Behavioral psychology. Current work: Research into etiology, consequences and management of mentally retarded people's problems. Office: Dept Psychology Tex Tech U PO Box 4100 Lubbock TX 79409

LOCKHART, JAMES ARTHUR, botanist, educator; b. Grand Rapids, Mich., June 7, 1926; s. Arthur John and Ruth Laura (Eyles) L.; m. Donna Margaret; m. Joan Conway, June 6, 1943; 1 dau., Joan Elizabeth. B.S., Mich. State Coll., M.S., 1952; Ph.D. (NSF fellow), UCLA, 1954. NSF postdoctoral fellow U. Pa., Phila., 1954-55; postdoctoral fellow Calif. Inst. Tech., 1955-60; assoc. plant physiologist U. Hawaii, 1960-65; prof. botany U. Mass., Amherst, 1965—. Research publs. in field. Served to sgt. U.S. Army, 1944-46, 50-51. NSF grantee, 1960-72. Mem. AAAS, Am. Soc. Plant Physiologists, Bot. Soc. Am., Am. Inst. Biol. Scis., VFW, Sigma Xi. Subspecialties: Plant physiology (biology); Population biology. Current work: Analyses of integration of plant functions to define alternative survival strategies as functions of different environments. Home: 294 Puffton Village Amherst MA 01002 Office: Dept Botany U Mass Amherst MA 01003

LOCKHEAD, GREGORY ROGER, psychology educator; b. Boston, Aug. 8, 1931; s. John Roger and Ester Mae (Bixby) L.; m. Jeanne Marie Hutchinson, June 9, 1957; children: Diane, Elaine, Dana John. B.S., Tufts U., 1958; Ph.D., Johns Hopkins U., 1965. Research staff mem. IBM Research, Katona, N.Y., 1958-61; asst. prof., assoc. prof., prof. psychology Duke U., Durham, N.C., 1965—; research assoc. U. Calif., Berkeley, 1971-72; vis. prof. Stanford (Calif.) U., 1971-72; research prof. Oxford (Eng.) U., 1980-81. Contbr. chpts. to books, articles to profl. jours. Recipient research awards NSF, NIH, 1965-83. Fellow Am. Psychol. Assn.; mem. Psychonomic Soc. Democrat. Subspecialties: Psychophysics; Cognition. Current work: Perception, sensory processes, cognitive mechanisms, psychophysical transfer functions, man-machine systems. In general: How does man know his world. Home: 101 Emerald Circle Durham NC 27713 Office: Dept Psychology Duke Univ Durham NC 27706

LOCKRIDGE, OKSANA, research scientist; b. Sept. 4, 1941; d. Gregor and Svetlana (Dolgoruky) Maslivec; m.; 1 dau., Katya. B.A., Smith Coll., 1963; Ph.D., Northwestern U., 1971. Postdoctoral scholar U. Mich., Ann Arbor, 1970-74, research assoc., 1974-82, asst. research scientist, 1982—. Research grantee NIH, 1979-82, Dept. Def., 1982—. Mem. Am. Soc. Biol. Chemists Assn. Women in Sci., Am. Chem. Soc. Subspecialties: Biochemistry (biology); Molecular pharmacology. Current work: Structure and amino acid sequence of human serum cholinesterase. Home: 1508 Morton Ann Arbor MI 48104 Office: Pharmacology Dept Med Science I Univ Mich Med School Ann Arbor MI 48109

LOCKWOOD, JOHN LEBARON, plant pathology and botany educator; b. Ann Arbor, Mich., May 28, 1924; s. George LeBaron and Mary Bonita (Leininger) L.; m. Jean Elizabeth Springborg, Mar. 21, 1959; children: James LeBaron, Laura Ann. B.A., Mich. State Coll., 1948, M.S., 1950; Ph.D., U. Wis.-Madison, 1953. Mem. faculty Mich. State U., East Lansing, 1955—, prof. plant pathology and botany, 1967—. Served with U.S. Army, 1943-46. Fellow Am. Phytopath. Soc., Sigma Xi. Subspecialties: Plant pathology; Microbiology. Current work: Ecology of fungi in soil; soybean diseases; root diseases; biological control. Home: 1931 Yuma Trail Okemos MI 48864 Office: Mich State Univ Botany and Plant Pathology Dept East Lansing MI 48824

LODGE, GEORGE TOWNSEND, educator; b. Kent, Ohio, Nov. 2, 1907; s. Edward B. and Martha (Townsend) L.; m. Edith Bennett, Nov. 19, 1927; children: Ann, David. B.A., Oberlin Coll., 1929; M.A., Ohio State U., 1932; Ph.D., Case Western Res. U., 1940; postgrad., Columbia U., 1944. Diplomate: Am. Bd. Profl. Psychology. Chief, clin. psychology training unit VA, San Francisco, 1946-48, Letterman Army Hosp., 1948-51, U.S. VA (various locations), 1951-60; dir. human factors research U.S. Naval Safety Center, Norfolk, Va., 1960-67; prof. psychology Old Dominion U., Norfolk, 1967-74; prof. psychiatry and behavioral scis. Eastern Va. Med. Sch., 1975—. Mem. Mental Health and Mental Retardation Profl. adv. bd. State of Va., Richmond, 1970-78. Served to comdr. USN, 1941-46. Fellow AAAS, Am. Psychol. Assn.; mem. Human Factors Soc., Internat. Assn. Applied Psychology, Sigma Xi. Subspecialties: Human factors engineering; Gerontology. Current work: Consulations in human factors as related to problems of aging. Address: 3100 Shore Dr Apt 1230 Virginia Beach VA 23451

LODISH, HARVEY FRANKLIN, cell biologist, educator; b. Cleve., Nov. 16, 1941; s. Nathan H. and Sylvia B. (Friedman) L.; m. Pamela Chentow, Dec. 29, 1963; children: Heidi, Martin, Stephanie. A.B., Kenyon Coll, 1962; D.Sch. (hon.), 1982; Ph.D., Rockefeller U., 1966. Postdoctoral fellow MRC Lab. Molecular Biology, Cambridge, Eng., 1966-68; asst. prof. biology M.I.T., 1968-71, assoc. prof., 1971-76, prof., 1976—; cons. scientist Children's Hosp., Boston, 1979—; mem. sci. bd. Damon Biotech Inc., Med. Found., Am. Cancer Soc. Contbr. articles to profl.jours. Am Cancer Soc. fellow, 1966-68; Guggenheim fellow, 1977-78. Mem. Am. Chem. Soc., Am. Soc. Biol. Chemists, AAAS, Am. Soc. Microbiology, Am. Soc. Cell Biologists, Phi Beta Kappa. Subspecialties: Cell biology; Biochemistry (biology). Current work: Research on cell biology; synthesis and function of membcrane proteins; regulation of gene expression. Consultant biotechnology.

LODWICK, GWILYM SAVAGE, radiologist, educator; b. Mystic, Iowa, Aug. 30, 1917; s. Gwylim S. and Lucy A. (Fuller) L.; m. Maria Antonia De Brito Barata; children by previous marriage—Gwilym Savage III, Philip Galligan, Malcolm Kerr, Terry Ann. Student, Drake U., 1934-35; B.S., State U. Iowa, 1942, M.D., 1943. Resident pathology State U. Iowa, 1947-48, resident radiology, 1948-50; fellow, sr. fellow radiologic and orthopedic pathology Armed Forces Inst. Pathology, 1951; asst., then asso. prof. State U. Iowa Med. Sch., 1951-56; prof. radiology, chmn. dept. U. Mo. at Columbia Med. Sch., 1956-78, research prof. radiology, 1978—, interim chmn. dept. radiology, 1980-81, chmn. dept. radiology, 1981-83, prof. bioengrng., 1969-83, acting dean, 1959, assoc. dean, 1959-64; assoc. radiologist Mass. Gen. Hosp., 1983—; vis. prof. dept. radiology Harvard Med. Sch., 1983—; cons. in field; vis. prof. Keio U. Sch. Medicine, Tokyo, 1974; chmn. sci. program com. Internat. Conf. on Med. Info., Amsterdam, 1983;

trustee Am. Registry Radiologic Technologists, 1961-69, pres., 1964-65, 68-69; mem. radiology tng. com. Nat. Inst. Gen. Med. Scis., NIH, 1966-70; com. radiology Nat. Acad. Scis.-NRC, 1970-75; chmn. com. computers Am. Coll. Radiology, 1965, Internat. Commn. Radiol. Edn. and Information, 1969—; cons. to health care tech. div. Nat. Center for Health Services, Research and Devel., 1971—; dir. Mid-Am. Bone Tumor Diagnostic Center and Registry, 1971—; adv. com. mem. NIH Biomed. Image Processing Grant Jet Propulsion Lab., 1969-73; nat. chmn. MUMPS Users Group, 1973-75; mem. radiation study sect. div. research grants NIH, 1976-79, mem. study sect. on diagnostic radiology and nuclear medicine div. research grants, 1979—, chmn. study sect. on diagnostic radiology div. research grants, 1980-82. Author: also articles. others; Adv. editorial bd.: Radiology, 1965—, Current/Clin. Practice, 1972—; mem. editorial bd.: Jour. Med. Systems, 1976—, Radiol. Sci. Update div. Biomedia, Inc, 1975—; mem. cons. editorial bd.: Skeletal Radiology, 1977—, Contemporary Diagnostic Radiology, 1978-80. Served to maj. AUS, 1943-46. Decorated Sakari Mustakallio medal, Finland; named Most Disting. Alumnus in Radiology State U. Ia. Centennial, 1970; recipient Sigma Xi Research award U. Mo., Columbia, 1972, Gold medal XIII Internat. Conf. Radiology, Madrid, 1973. Fellow A.M.A. (radiology rev. bd. council med. edn., council rep. on residency rev. com. for radiology 1969-74); mem. Am. Coll. Radiology, European Soc. Radiology (European steering com. on computers), Radiol. Soc. N.Am. (3d v.p. 1974-75, chmn. ad hoc com. representing assoc. scis. 1979—), Salutis Unitas (chmn. assoc. scis. com. 1981—), Assn. U. Radiologists, Mo. Radiol. Soc. (1st pres. 1961-62), Alpha Omega Alpha; hon. mem. Portuguese Soc. Radiology and Nuclear Medicine, Tex. Radiol. Soc., Ind. Roentgen Soc., Phila. Roentgen Ray Soc., Finnish Radiol. Soc. (hon.). Subspecialties: Diagnostic radiology; Imaging technology. Current work: Modeling and analysis; bone images, automated image analysis; medical decision making; radiology information systems.

LOEB, GERALD ELI, neurophysiologist, physician, biomedical engineering consultant; b. New Brunswick, N.J., June 26, 1948; s. Louis C. and Ruth (Blumberg) L.; m. Sandra Kay Loeb, Sept. 8, 1968; 1 son, Jason Eliot. B.A., Johns Hopkins U., 1969, M.D., 1972. Lic. physician, Md. Guest research assoc. U. Utah Artificial Eye Project, 1971; resident in surgery U. Ariz., 1972-73; med. officer, permanent sr. investigator Lab. Neural Control, NIH, Bethesda, Md., 1973—; guest researcher dept. otolaryngology U. Calif., San Francisco, 1979—; pres. Biomed Concepts, Inc., Clarksburg, Md., 1981—. Contbr. articles to sci. jours. Vice pres. Clarksburg Community Assn., 1979-80. Seeing Eye fellow, 1969-72. Mem. Soc. Neurosci., Biomed. Engring. Soc., Phi Beta Kappa. Subspecialties: Neurophysiology; Artificial organs. Current work: Information processing in the nervous system; sensorimotor neurophysiology, neural prosthetics, biomedical engineering. Home: 15730 Comus Rd Clarksburg MD 20871 Office: NIH Rm 36-5A29 Bethesda MD 20205

LOEB, VIRGIL, JR., physician; b. St. Louis, Sept. 21, 1921; s. Virgil and Therese (Meltzer) L.; m. Lenore Harlow, Sept. 8, 1950; children: Katherine Loeb Barksdale, Elizabeth Loeb McCane, David, Mark. Student, Swarthmore Coll., 1938-41; M.D., Washington U., St. Louis, 1944. Diplomate: Am. Bd. Internal Medicine. Intern Barnes and Jewish hosps., St. Louis, 1944-45; resident in internal medicine, research fellow in hematology Washington U. and Barnes Hosp., 1947-52; med. faculty, 1951—, prof. clin. medicine, 1978—, clin. research and cons. practice med. oncology, hematology, 1956—; dir. Central Diagnostic Labs., Barnes Hosp., 1952-68; staff numerous hosps., 1951—; cons. Nat. Cancer Inst., 1966—, chmn. cancer clin. investigation rev. com., 1966-69, mem. diagnostic research adv. com., 1972-75, bd. sci. counselors, 1983—; mem. nat. adv. coms. Am. Cancer Soc., 1969—. Contbg. author books; contbr. articles to profl. publs. Mem. oncology merit rev. bd. VA, 1971-75; bd. dirs. Nat. and Mo. div. Am. Cancer Soc., pres. Mo. div., 1983-85; bd. dirs. St. Louis Blue Cross and Blue Shield; trustee John Burroughs Sch., 1966-69. Served with M.C. AUS, 1945-47. Fellow ACP; mem. Central Soc. Clin. Research, Inst. Medicine of Nat. Acad. Sci., Am. Assn. Cancer Research, Am. Soc. Clin., Oncology, Am. Fedn. Clin. Research, Internat. Soc. Hematology, Am. Soc. Hematology, St. Louis Soc. Internal Medicine (pres. 1974), Sigma Xi, Alpha Omega Alpha. Subspecialties: Chemotherapy; Hematology. Current work: Clinical trials in cancer therapeutic research. Home: 24 Deerfield Rd Saint Louis MO 63124 Office: 3103 Queeny Tower Barnes Hosp Plaza Saint Louis MO 63110

LOEB, WILLIAM A., mechanical engineer; b. N.Y.C., Sept. 7, 1924; s. Samuel L. and Ethel (Kossman) L.; m. Marion Conrad, May 26, 1971; children: David, Jonathan, Suzanne. B.S., MIT, 1945, M.S., 1947. Registered profl. engr., Mass., N.Y., Alaska. Research engr. Delaval Steam Turbine Co., Trenton, N.J., 1947-51; v.p. engring. United Nuclear Corp., White Plains, N.Y., 1951-65; pres. Iso Nuclear Corp., N.Y.C., 1965-67; v.p. research Tech. Investors Corp., N.Y.C., 1967-71; dir. spl. projects Combustion Engring., Windsor Locks, Conn., 1971-75; pres. West Stockbridge Enterprises, Inc., Mass., 1975—; cons. U.S. Dept. Energy, Washington, 1976-78. Served to lt. (j.g.) USN, 1942-46; PTO. Mem. ASME, Nat. Soc. Profl. Engrs., Am. Congress Surveying and Mapping, Am. Nuclear Soc., Internat. Solar Energy Soc., Mass. Soc. Profl. Engrs., Sigma Xi, Tau Beta Pi. Subspecialties: Solar energy; Systems engineering. Current work: Small hydroelectric plants, computer-based facility control systems, economic small energy systems. Home: Austerlitz Rd West Stockbridge MA 01266 Office: West Stockbridge Enterprises Inc Box 100 West Stockbridge MA 01266

LOEPPKY, JACK ALBERT, physiologist, researcher; b. Saskatoon, Sask., Can., Jan. 14, 1944; came to U.S., 1967; s. George and Sarah (Martens) L.; m. Janet Sue By, Nov. 22, 1974; 1 son, Kristopher. B.A. with distinction, U. Sask., 1966; M.S., U. N.Mex., 1969, Ph.D., 1973; postgrad., Colo. State U., 1969-70. Instr. U. Sask., Saskatoon, 1966-67; technician Lovelace Med. Found., Albuquerque, 1966-69, research assoc., 1970-75, assoc. scientist, 1975—, head dept. cardiopulmonary physiology, 1981—; head respiratory technologist Wellington (N.Z.) Hosp., 1975; adj. asst. prof. U. N.Mex., Albuquerque, 1982—. Editor: Oxygen Transport to Human Tissues, 1982; contbr. articles to profl. jours. Scholar, 1962-66. Mem. Am. Physiol. Soc. Club: N.Mex. Mountain (Albuquerque). Subspecialties: Physiology (medicine); Gravitational biology. Current work: Pulmonary gas exchange, exercise physiology, cardiopulmonary effects of gravitational stress. Office: Research Div Lovelace Med Found 5400 Gibson Blvd SE Albuquerque NM 87108

LOESCHER, WAYNE HAROLD, physiology educator; b. Lima, Ohio, Nov. 6, 1942; s. Harold and Helen Elizabeth (Miller) L.; m. Judith Elizabeth Hunt, Aug. 26, 1967; 1 son, Karl Benjamin. B.A., Miami U., Oxford, Ohio, 1964, M.S., 1967; Ph.D., Iowa State U., 1972. Instr. Millersville (Pa.) State Coll., 1966-67; postdoctoral research assoc. agronomy Iowa State U., Ames, 1971-73; plant physiologist Los Angeles Arboretum, Arcadia, Calif., 1973-75; asst. prof. Wash. State U., Pullman, 1975-80, assoc. prof., 1980—; cons. in field. Contbr. articles to profl. jours. Mem. AAAS, Am. Soc. Plant Physiologists, Am. Soc. Hort. Sci., Bot. Soc. Am., Internat. Assn. Plant Tissue Culture. Subspecialties: Plant physiology (biology); Plant cell and tissue culture. Current work: Regulation of carbohydrate metabolism, use of tissue cultures as model systems for study of plant metabolism. Office: Dept Horticulture Wash State Univ Pullman WA 99164

LOEWENSTEIN, WERNER RANDOLPH, physiologist, biophysicist; b. Spangenberg, Germany, Feb. 14. 1926; came to U.S., 1957, naturalized, 1965; s. Siegfried and Adele (Muller) von Loewenstein; m. Birgit Rose, Oct. 7, 1971; children: Claudia, Patricia, Harriett, Stewart. B.S., U. Chile, 1945, Ph.D., 1950. Instr. physiology U. Chile, Santiago, 1951-53, asso. prof., 1955-57; fellow in residence Wilmer Inst., Johns Hopkins U., Balt., 1953-54; research zoologist U. Calif., Los Angeles, 1954-55; asst. prof. physiology Columbia U. Coll. Physicians and Surgeons, N.Y.C., 1957-59, asso. prof., 1959-66, prof., dir. cell physics lab., 1966-71; prof. physiology and biophysics, chmn. dept. U. Miami (Fla.) Sch. Medicine, 1971—; Block lectr. U. Chgo., 1960; lectr. Royal Swedish Acad. Sci., 1966; Max Planck lectr., 1967, Fulbright disting. prof., 1970, USSR Acad. Sci. lectr., Leningrad, 1975; mem. Pres.'s Biomed. Research Adv. Panel, 1975-77, USAF Sci. Adv. Panel, 1982—. Author, editor several books; editor: Biochimica et Biophysica Acta, 1967-74; editor-in-chief: Jour. Membrane Biology, 1969—; editor: Handbook of Sensory Physiology, 12 vols., 1971-77; contbr. numerous articles on membrane biophysics, physiology of intercellular communication, and neurophysiology to profl. jours. Kellogg internat. fellow in physiology, 1953-55; Commonwealth Fund internat. fellow, 1967; NSF, NIH research grantee. Fellow N.Y. Acad. Scis.; mem. Am. Physiol. Soc., Biophys. Soc., Marine Biol. Lab. Woods Hole (corp. mem.). Clubs: Woods Hole Yacht, Coconut Grove Sailing, Royal Kev Biscayne Tennis and Racquet. Subspecialties: Biophysics (biology); Neurobiology. Current work: Intercellular communication. Home: 1090 Mariner Dr Key Biscayne FL 33149 Office: Dept Physiology and Biophysics Univ Miami Sch Medicine PO Box 520875 Miami FL 33152

LOEWY, ROBERT GUSTAV, engineering educator, consulting engineer; b. Phila., Feb. 12, 1926; s. Samuel N. and Esther (Silverstein) L.; m. Lila Myrna Spinner, Jan. 16, 1955; children—David G., Esther Elizabeth, Joanne Victoria, Raymond M. B.Aero. Engring., Rensselaer Poly. Inst., 1947; M.S., MIT, 1948; Ph.D., U. Pa., 1962. Sr. vibrations engr. Martin Co., Balt., 1948-49; asso. research engr. Cornell Aero. Lab., Buffalo, 1949-52; staff stress engr. Piasecki Helicopter Co., Morton, Pa., 1952-53; prin. engr. Cornell Aero. Lab., 1953-55; chief dynamics engr., also chief tech. engr. Vertol div. Boeing Co., 1955-62; faculty U. Rochester, 1962-74, prof. mech. and aerospace scis., 1965-74; dir. Space Sci. Center, 1966-71; dean Coll. Engring. and Applied Sci., 1967-74; v.p., provost Rensselaer Poly. Inst., 1974-78; Inst. prof. Rensselaer Poly Inst., 1978—; chief scientist USAF, 1965-66; cons. govt. and industry, 1959—; Mem. aircraft panel Pres.'s Sci. Adv. Council, 1968-72; mem. Air Force Sci. Adv. Bd., 1966-75, 78—, vice chmn., 1971, chmn., 1972-75, chmn. aero. systems adv. group, 1978—; mem. Post Office Research and Engring. Adv. Council, 1966-68; mem. research and tech. adv. com. on aeros. NASA, 1970-71, mem. research and tech. adv. council, 1976-77, chmn. aero. adv. com., 1978—; mem. aerospace engring. bd. NRC, 1972-78; mem. naval studies bd. Nat. Acad. Scis., 1979-82; chmn. tech. adv. com. FAA, 1976-77. Contbr. articles to profl. jours. Served with USNR, 1944-46. Gotshall-Powell scholar Rensselaer Poly. Inst.; USAF Exceptional Civilian Service awards, 1966, 75. Hon. fellow Am. Helicopter Soc. (tech. dir. 1963-64); fellow AIAA (Lawrence Sperry award 1958), AAAS; mem. Am. Soc. Engring. Edn., Nat. Acad. Engring., Sigma Xi, Sigma Gamma Tau, Tau Beta Pi. Subspecialties: Aeronautical engineering; Aerospace engineering and technology. Current work: Structural dynamics, unsteady aerodynamics, aeroelasticity. Home: Dutch Village Menands NY 12204 Looking back, I was fortunate to have known somehow, from an early age, that I would be an aeronautical engineer. That profession, through industry, research and education, has provided challenge, satisfaction and valued associations.

LOFGREN, GARY ERNEST, federal government space center geologist; b. Los Angeles, Apr. 17, 1941; m. Kenneth Gordon and Mildred Edith (Johnson) L.; m. Ellen Much, June 22, 1965. B.S., Stanford U., 1963, Ph.D., 1969; M.A., Dartmouth Coll., 1965. Space geoscientist NASA Johnson Space Ctr., Houston, 1968—; adj. prof. geology U. Houston, 1976—; team leader Basaltic Volcanism Study Project, Lunar and Planetary Inst., Houston, 1976-81. Author/editor: Basaltic Volcanism, 1981; contbr. articles to profl. jours.; editor: Lunar Sci. Conf. Procs, 1968; assoc. editor: Geophys. Research Letters, 1972-76. Recipient Superior Achievement award NASA Johnson Space Ctr., 1978; NDEA fellow, 1965-68. Fellow Mineral. Soc. Am., Geol. Soc. Am. (Spl. commendation 1973); mem. Internat. Union Geol. Socs. (com. on igneous rocks 1976-80), Am. Geophys. Union, Am. Assn. Crystal Growth, Houston Philos. Soc., Sigma Xi. Subspecialties: Petrology; Geochemistry. Current work: Nucleation and crystallization in natural rock forming silicate melts and synthetic analogs; experimental igneous petrology. Home: 214 Bay Colony Circle La Porte TX 77571 Office: NASA Johnson Space Ctr Mail Code SN4 Houston TX 77058

LOFQUIST, LLOYD HENRY, educator; b. Mpls., June 11, 1917; s. Frederick G. and Hilma B. (Lundin) L.; m. Lillian Mary Holm, Nov. 1, 1941; children—Mary Lillian, Mark Frederick. B.A., U. Minn., 1940, M.A., 1941, Ph.D., 1955. Chief psychology service Mpls. VA Hosp., 1948-56; asso. prof. psychology dept. U. Minn., Mpls., 1956-60, prof. psychology, 1960—, asso. dean social scis., 1967-69, asst. v.p., acad. adminstrn., 1969-73, chmn. dept. psychology, 1975—; chmn. Minn. Bd. Examiners of Psychologist, 1960-65; cons. VA, 1957-79; prin. investigator various HEW research and tng. grants. Author: Vocational Counseling with the Physically Handicapped, 1957, Egyptian transl, 1961, Psychological Research and Rehabilitation, 1960, Problems in Vocational Counseling, 1961, Adjustment to Work, 1969; mem. editorial bds.: Rehab. Counseling Bull, 1975-80, Jour. Vocat. Behavior, 1970-75. Served to capt. U.S. Army, 1942-46. Decorated Bronze Star. Fellow Am. Psychol. Assn.; mem. Am. Personnel and Guidance Assn. (research award 1967), Am. Rehab. Counseling Assn. (pres. 1968, research awards 1960, 65). Democrat. Lutheran. Subspecialties: Behavioral psychology; vocational psychology. Current work: Adjustment to work.

LOFTFIELD, ROBERT BERNARD, biochemistry educator, researcher; b. Detroit, Dec. 15, 1919; s. Sigurd and Katherine Maria (Roller) L.; m. Ella Bradford, Aug. 24, 1946; children: Lore, Eric, Linda, Norman, Bjorn, Curtis, Katherine, Earl, Allison, Ella-Kari. B.S., Harvard U., 1941, M.A., 1942, Ph.D., 1946. Research assoc. physics MIT, 1946-48; research assoc. medicine Mass. Gen. Hosp., 1948-64; tutor biochemistry sci. Harvard U., 1950-64, asst. prof. biochemistry, 1959-64; prof., chmn. biochemistry U. N.Mex., Albuquerque, 1964—. Contbr. articles to profl. jours. Served with U.S. Army, 1945-46. Recipient Warren Triennial prize, 1956; Damon Runyon fellow, 1952-53; Guggenheim fellow, 1961-62; NIH fellow, 1971-72; Fulbright fellow, 1977; NATO fellow, 1983. Subspecialties: Biochemistry (medicine); Enzyme technology. Current work: Mechanisms of enzymic rate enhancement and specificity, synthesis of protein, aminoacyl tRNA; biochemistry of burn and trauma. Home: 707 Fairway St NW Albuquerque NM 87107 Office: Department Biochemistry University of New Mexico Albuquerque NM 87131

LOFTUS, ELIZABETH F., psychology educator; b. Los Angeles, Oct. 16, 1944; d. Sidney A. and Rebecca (Breskin) Fishman; m. Geoffrey R. Loftus, June 30, 1968. B.A., UCLA, 1966; M.A., Stanford U., 1967, Ph.D., 1970; D.Sc. (hon.), Miami U., Oxford, Ohio, 1982. Asst. prof. New Sch. Social Research, N.Y.C., 1970-73; asst. prof. to prof. psychology U. Wash., Seattle, 1973—; cons. Gen. Services Adminstrn., FTC, Dept. of Justice, numerous corps. Author: Eyewitness Testimony, 1979 (Am. Psychol. Found. Disting. Contbn. award 1980), Memory, 1980; 7 other books. Fellow Ctr. Advanced Study in Behavioral Scis., 1978-79. Fellow Am. Psychol. Assn. (council reps.); mem. Western Psychol. Assn. (pres. 1983), Am. Psychology-Law Soc. (bd. dirs.), Psychonomic Soc. Subspecialties: Cognition; Learning. Current work: Experimental research on human memory. Home: 1221 22d Ave E Seattle WA 98112 Office: Psychology Dept University of Washington Seattle WA 98195

LOGAN, TERRY JAMES, soil chemistry educator; b. Georgetown, Guyana, Feb. 6, 1943; came to U.S., 1962, naturalized, 1970; s. David A. and Elsa E. (Gonsalves) L.; m. Marilyn I. Winters, Feb. 1964 (div. 1972); 1 son, David.; m. Cynthia A. Hawk, June 9, 1973; children: Noelle, Jennifer. B.S., Calif. Poly. State U., 1966; M.S., Ohio State U., 1969, Ph.D., 1971. Prof. soil chemistry Ohio State U., Columbus, 1971—. Contbr. articles to profl. jours. Mem. Am. Soc. Agronomy, Soil Sci. Soc. Am., Soil Conservation Soc. Am., Internat. Assn. Gt. Lakes Research. Subspecialties: Soil chemistry; Inorganic chemistry. Current work: Nutrient losses from land. Disposal of wastes on land. Soil erosion and conservation. Office: Dept Agronomy Ohio State U 2021 Coffey Rd Columbus OH 43210

LOH, PHILIP CHOO-SENG, virologist, educator, reseracher, cons.; b. Singapore, Sept. 14, 1925; U.S., 1947, naturalized, 1958; s. Poon Lip and Soh Choo (Teo) L.; m. Susie S. H., Feb. 5, 1955; children: Valerie K.H., Rhonda K.H. B.S., Morningside Coll., 1950; M.Sc., State U. Iowa, 1953; M.P.H., U. Mich., 1954; Ph.D. (Horace Rackham fellow), 1958. Asst. prof. U. Mich., 1961; assoc. prof. virology U. Hawaii, 1961-65; prof., 1965—; vis. prof. U. Bristol, Eng., 1975. Contbr. numerous articles to profl. jours. Recipient Regent's Excellence in research award U. Hawaii, 1965; USPHS spl. research fellow, 1967; Am. Cancer Soc. Eleanor Roosevelt Internat. Cancer fellow, 1975. Fellow AAAS; mem. Am. Soc. Microbiology, Am. Assn. Immunology, Soc. Exptl. Biology and Medicine, Tissue Culture Assn., AAAS, Sigma Xi. Subspecialties: Virology (medicine); Water supply and wastewater treatment. Current work: Viral replication and inhibitors, pathobiology of viral diseases, environ, virology.

LOH, YOKE PENG, neurobiologist; b. Singapore, July 27, 1947; s. Poon Lip and Sue Heng L. B.Sc., Univ. Coll. Dublin, 1969; Ph.D., U. Pa., 1973. Vis. fellow Lab. Devel. Neurobiology, NIH, Bethesda, Md., 1973-76, sr. staff fellow, 1976-79; research chemist, 1979-83; chief sect. cellular neurobiology Lab. Neurochemistry and Neuroimmunology, NIH, 1983—. Contbr. articles to profl. jours. Fogarty Internat. Postdoctoral fellow, 1973-76; recipient Deutscher Akademischer Austauschdienst award, 1978. Mem. Soc. Neurosci. Neurochemistry Soc., N.Y. Acad. Sci., Biochem. Soc. Gt. Britain. Subspecialties: Neurobiology; Neuroendocrinology. Current work: Biosynthesis of neuropeptides and their role in neuronal functions; neuropeptide biosynthesis; nerve regeneration and protein synthesis. Office: NIH Bldg 36 Rm 2A-21 Bethesda MD 20205

LOHMAN, GUY MARING, computer scientist; b. St. Louis, July 16, 1949; s. Ira H. and Louise M. L. B.A., Pomona Coll., 1971; M.S., Cornell U., 1975, Ph.D., 1977. Teaching asst. Sch. Ops. Research, Cornell U., 1972-75; assoc. ops. research analyst Atlantic Richfield, Los Angeles, summers 1973, 74; database adminstr. Cornell Law Assn., Cornell U., 1975-76; sr. engr. Jet Propulsion Lab., Pasadena, Calif., 1976-79, acting group supr., 1977-79; mem. tech. staff End-to-End Info. Systems Tech. Devel. Group, 1979-81; group supr. Advanced DMBS Engring. Group, 1981-82; mem. research staff IBM Research Lab., San Jose, Calif., 1982—; instr. safety/systems mgmt. U. So. Calif., 1978-79. Contbr. articles to and referee for profl. jours. Served to capt. USAR, 1972. Cornell U. fellow, 1975-76. Mem. Assn. Computing Machinery (spl. interests group on mgmt. data 1973, vice chmn. San Gabriel Valley chpt. 1979-81, publicity chmn. 1981-82), IEEE Computer Soc. (internat. publicity chmn. 1982 Internat. Conf. on Very Large Data Bases), Sierra Club, Phi Beta Kappa, Sigma Xi. Clubs: Jet Propulsion Lab. Ski (pres. 1979-80), Jet Propulsion Lab. Hiking.). Subspecialties: Database systems; Distributed systems and networks. Current work: Distributed data base management systems; query optimization, novel applications (remotely-sensed geophysical data, CAD/CAM), backup/recovery, modelling. Office: IBM Research Lab K55-281 5600 Cattle Rd San Jose CA 95193

LOHR, KRISTINE MARIE, medical educator, physician; b. Buffalo, Oct. 29, 1949, s. Leonard J. and Lucille E. (Roger) L. B.A., Canisius Coll., 1971; M.D., U. Rochester, 1975. Diplomate: Am. Bd. Internal Medicine. Med. intern, clin. instr. Ohio State U. Hosps., Columbus, 1975-78; research fellow Duke U. Med. Center, Durham, N.C., 1978-81; asst. prof. medicine Med. Coll. Wis., Milw., 1981—; staff physician VA Med. Center, Wood, Wis., 1981—. Recipient Merit Review award VA, 1982; Nat. Eye Inst. grantee, 1982; Wis. Arthritis Found. grantee, 1982, 83. Mem. Am. Fedn. Clin. Research, ACP, Am. Med. Women's Assn., Reticuloendothelial Soc., Am. Rheumatism Assn., AAAS. Subspecialties: Internal medicine; Rheumatology. Current work: Phagocyte chemotaxis, inflammation, membrane receptors, membrane fluidity. Home: 2249 N 114th St Wauwatosa WI 53226 Office: Med Coll Wis 8700 W Wisconsin Ave Milwaukee WI 53226

LOHSE, DAVID JOHN, physicist; b. N.Y.C., Sept. 14, 1952; s. Edward and Mildred Edna (Hofmeister) L.; m. Maria I.M. Garcia, Sept. 2, 1978. B.S. in Physics and Computer Sci, Mich. State U., 1974; Ph.D., U. Ill., 1978. Research assoc. U. Ill., 1974-78; research engr. Exxon Chem. Co., Linden, N.J., 1980—. NRC fellow, 1978-80. Mem. Am. Phys. Soc., Am. Chem. Soc. Lutheran. Subspecialty: Polymer physics. Current work: Theoretical statistical mechanics of polymer solutions and polymer blends; use of small angle neutron scattering on polymer solutions and blends. Home: 345 Evanston Dr East Windsor NJ 08520 Office: Exxon Chemical Company PO Box 45 Linden NJ 07036

LOKEN, MERLE KENNETH, radiologist, educator; b. Hudson, S.D., Jan. 21, 1924; s. Albert R. and Emma T. (Lunder) L.; m. Fern Mae Buhler, June 1, 1947; children—Allen, Evonne, Gwenda, Lynette, Warren. B.A., Augustana Coll., Sioux Falls, S.D., 1946; B.S., Mass. Inst. Tech., 1948, M.S., 1949; Ph.D., U. Minn., 1956, M.D., 1962. Diplomate: Am. Bd. Nuclear Medicine. Asst. prof. Augustana Coll., 1949-50; mem. faculty U. Minn. Sch. Medicine, 1953—, prof. radiology, 1969—, dir. div. nuclear medicine, 1963—, faculty rep. to Big 10 Intercollegiate Conf., 1974-81. Contbr. numerous sci. papers. Served with USNR, 1943-46, 51-52. Fellow Am. Coll. Radiology; mem. Soc. Nuclear Medicine (pres. central chpt. 1972-74, nat. treas. 1972-75, trustee 1976-80, nat. pres. 1983-84), Minn. Radiol. Soc., Radiol. Soc. N.Am., Sigma Xi. Republican. Lutheran. Subspecialties: Nuclear medicine; Nuclear medicine. Current work: Development and evaluation of techniques utilizing radioactive materials in the practice of medicine. Home: 16 Park Ln Minneapolis MN 55416

LOKENSGARD, JERROLD PAUL, chemistry educator and researcher; b. Saskatoon, Sask., Can., July 30, 1940; came to U.S., 1943; s. Bernhard Oliver and Eleanor Ruth (Bensen) L.; m. Elizabeth Ann Hopkins, Aug. 14, 1965; children: Michael John, Ann-Marie.

B.A., Luther Coll., 1962; M.A., U. Wis., 1964; Ph.D., 1967. NIH postdoctoral fellow Iowa State U., Ames, 1967; asst. prof. Lawrence U., Appleton, Wis., 1967-76, assoc. prof. chemistry, 1976—; research assoc. U. Toronto, Ont., Can., 1973-74; vis. assoc. prof. Cornell U., Ithaca, N.Y., 1980-81. Contbr. articles to profl. jours. Danforth grad. fellow, 1962; NSF grad. fellow, 1962, 63. Mem. Am. Chem. Soc. (local sect. chmn. 1970), AAAS, AAUP. Lutheran. Subspecialties: Organic chemistry; Synthetic chemistry. Current work: Organic synthesis, insect defensive substances, strained hydrocarbons. Office: Lawrence Univ PO Box 599 Appleton WI 54912

LOLLI, PETER PATRICK, psychologist; b. Providence, Sept. 2, 1950; s. Peter and Anne (Lambiase) L. B.A., Yale U., 1972; M.A., Duke U., 1975, Ph.D., 1982. Lic. psychologist, N.C. Psychol. cons. N.W. Region Childfind, North Smithfield, R.I., 1976; staff psychologist Developmental Evaluation Ctr., Greensboro, N.C., 1976-77; staff psychologist, evaluation team leader, 1977-83, pvt. practice psychology, 1983—. Chmn. exec. bd. Epilepsy Assn. Greater Greensboro, 1981-82. Mem. Am. Psychol. Assn., Soc. Pediatric Psychology, N.C. Psychol. Assn. Subspecialties: Developmental psychology; Neuropsychology. Current work: Neuropsychology of learning disabilities; emotional problems of developmentally disabled children. Home: 2409 Dent St Greensboro NC 27408 Office: 2311 W Cone Blvd Suite 143 Greensboro NC 27408

LOMBARD, R. ERIC, animal morphologist, educator; b. Bklyn., May 16, 1943; s. Richard Jose and Jacquline (Sears) L.; m. Mary Maxwell, June 24, 1967; children: Johanna Kaye, Stefan Andrew. B.A., Hanover Coll., 1965; Ph.D., U. Chgo., 1970. Postdoctoral fellow U. Calif., Berkeley, 1970-71, U. So. Calif., Los Angeles, 1971-72; asst. prof. anatomy U. Chgo., 1972-78, assoc. prof., 1978—; research assoc. Field Mus. Natural History, Chgo., 1980—. Contbr. chpts. to books, articles to profl. jours. NSF grantee, 1976-79, 80-83. Mem. AAAS, Am. Soc. Zoologists, Am. Soc. Ichthyologists and Herpetologists, Soc. Study Evolution, Sigma Xi. Club: Nat. Model R.R. Assn. Subspecialties: Morphology; Evolutionary biology. Current work: Origin and evolution of adaptive novelties; functional morphology of vertebrates; systematics of fossil and living amphibians; the interaction of development and evolution. Office: Dept Anatomy 1025 E 57th St Chicago IL 60637

LOMBARDI, GABRIEL G., physicist; b. Buenos Aires, Argentina, Sept. 5, 1954; s. Richard R. and Nelly A. (Bassi) L.; m. Cathy Ann, Mar. 19, 1977; 1 dau., Abigail Anne. B.A., U. Chgo., 1975; Ph.D., Harvard U., 1981. Teaching fellow Harvard U., Cambridge, Mass., 1977-79, research asst., 1976-80; physicist Nat. Bur. Standards, Washington, 1980-82, TRW-Inc., Redondo Beach, Calif., 1982—. Contbr. articles in field to profl. jours. NSF grad. fellow, 1975; NRC postdoctotal assoc., 1980. Mem. Am. Phys. Soc., Optical Soc. Am. Subspecialties: Atomic and molecular physics; Spectroscopy. Current work: Laser spectroscopy of atomic vapors and beams and or plasmas, pulsed discharges, optical interferometry. Office: TRW One Space Park #R1-1196 Redondo Beach CA 90278

LOMEN, DAVID ORLANDO, mathematics educator, consultant; b. Decorah, Iowa, May 11, 1937; s. Erlin Rueben and Ellen Dorthea (Jensen) L.; m. Constance Sylvia Trecek, Dec. 25, 1961; 1 dau., Catherine Ellen. B.A., Luther Coll., 1959; M.S., Iowa State U., 1962, Ph.D., 1964. Design specialist Gen. Dynamics/Convair, San Diego, 1963-66; prof. math. U. Ariz., 1966—; vis. scientist Cambridge U., 1972; research scientist ICW, Wageningen, Netherlands, 1978; vis. prof. U. Oslo, 1980; cons. to industry. Contbr. articles to math. jours. Bd. dirs. Tucson Cystic Fibrosis Found., 1974-80. Recipient Creative Teaching award Univ. Found., Tucson, 1978; sr. scientist NTNF of Norway, Oslo, 1980. Mem. Am. Math. Soc., Soc. Indsl. and Applied Math., Soil Sci. Soc. Am., Am. Geophys. Union. Republican. Lutheran. Club: Norsemen Fedn. of Tucson (treas.). Subspecialties: Applied mathematics; Hydrology. Current work: Research in soil physics and other areas applied mathematics; development of computer-aided instruction materials in mathematics. Office: Dept Mathematics University of Arizona Tucson AZ 85721

LONDON, EDYTHE DANICK, neuropharmacologist; b. Rome, Italy, Sept. 14, 1948. B.S. in Zoology, George Washington U., 1969; M.S. in Biol. Scis, Towson State U., 1973; Ph.D. in Pharmacology, U. Md., 1976. NIMH postdoctoral fellow div. psychopharmacology dept. pharmacology and exptl. therapeutics Johns Hopkins U. Sch. Medicine, Balt., 1976-78; staff fellow Lab. of Neuroscis. Gerontology Research Center Nat. Inst. in Aging, Balt. City Hosps., Balt., 1979-80, pharmacologist, 1981-82, Gerontology Research Center, Nat. Inst. Aging, NIH, 1982, NIDA Addiction Research, 1982—. Contbr. articles on neuropharmacology, neurochemistry to prof. jours. Mem. Am. Soc. for Pharmacology and Exptl. Therapeutics, Am. Soc. for Neurochemistry, Soc. for Neurosci, AAAS, Am. Aging Assn., Internat. Soc. fo Cerebral Blood Flow and Metabolism, Am. Assn. for Lab. Animal Sci. Subspecialties: Neurochemistry; Neuropharmacology. Current work: Neuropharmacology, neurochemistry of development and aging. Office: Gerontology Research Center Nat Inst on Aging Balt City Hosps Baltimore MD 21224

LONDON, IRVING MYER, physician, educator; b. Malden, Mass., July 24, 1918; s. Jacob A. and Rose (Goldstein) L.; m. Huguette Piedzicki, Feb. 27, 1955; children: Robert L.J., David T. A.B., Harvard U., 1939, M.D., 1943; D.Sc., U. Chgo., 1966. Sheldon Traveling fellow Harvard U., 1939-41, Delamar research fellow med. sch., 1940-41; intern Presbyn. Hosp., N.Y.C., 1943, asst. resident, 1946-47, asst. physician, 1946-52, asso. attending physician, 1954-55; Rockefeller fellow in medicine Coll. Physicians and Surgeons, Columbia U., 1946-47; instr. Columbia U., 1947-49; asso. in medicine Coll. Phys. and Surg., Columbia U., 1949-51; asst. prof. Coll. Phys. and Surg., Columbia, 1951-54, asso. prof., 1954-55; prof., chmn. dept. medicine Albert Einstein Coll. Medicine, N.Y.C., 1955-70, vis. prof. medicine, 1970—; prof. biology M.I.T., 1969—; vis. prof. medicine Harvard Med. Sch., 1969-72; dir. div. health scis. and tech. Harvard and M.I.T., 1969—, prof. medicine, 1972—, Grover M. Hermann prof. health scis. and tech., 1977—; dir. Whitaker Coll. Health Scis., Tech. and Mgmt., M.I.T., 1978-83; dir. med. service Bronx Mcpl. Hosp. Center, 1955-70; Roger Morris lectr. U. Cin., 1958; Stuart McGuire lectr. Med. Coll. Va., 1960; Phi Delta Epsilon lectr. U. Colo., 1962, Harvey lectr., 1961; Jacobaeus lectr., Stockholm, Sweden, 1964; vis. scientist Pasteur Inst., Paris, 1962-63; Commonwealth Fund fellow, 1962-63; Alpha Omega Alpha lectr. Yale, Boston U., Columbia, SUNY Downstate Med. Center, U. Chgo.; Harry L. Alexander vis. prof. Washington U., St. Louis, 1968; Alpha Omega Alpha vis. prof. Johns Hopkins U., 1970; Eugene A. Stead Jr. vis. lectr. Duke Med. Center, 1970; Cons. to Surgeon Gen., AUS, 1957-60; chmn. metabolism study sect. USPHS, 1961-63; Med. fellowship bd. Nat. Acad. Scis., NRC, 1955-64; mem. bd. sci. cons. Sloan Kettering Inst., 1960-72; bd. sci. counselors Nat. Heart Inst., 1964-68; exec. com. Health Research Council, City N.Y., 1958-63; mem. sci. adv. council Pub. Health Research Inst., N.Y.C., 1958-63; mem. adv. com. to dir. NIH, 1966-70, nat. cancer adv. bd., 1972-76; chmn. research group Nat. Commn. on Arthritis, 1975-76; chmn. adv. com. Div. Health Scis., Inst. Medicine, 1979-82; mem. Bd. Sci. Counselors, NIH, 1979-83. Asso. editor: Jour. Clin. Investigation, 1952-57; mem. editorial bd.: Am. Jour. Medicine, 1965-79. Served as capt. AUS, 1944-46. Recipient Theobald Smith award in med. scis. AAAS, 1953. Fellow Am. Acad. Arts and Scis.; mem. Am. Soc. Biol. Chemists, Am. Soc. Clin. Investigation (pres. 1963-64), Nat. Acad. Scis. (mem. bd. medicine 1967-70, mem. exec. com. Inst. Medicine 1970-72), Internat. Soc. Hematology, Soc. Exptl. Biology and Medicine, Assn. Am. Physicians, Nat. Acad. Scis., Phi Beta Kappa, Alpha Omega Alpha. Subspecialties: Hematology; Internal medicine. Current work: Research, teaching, academic medicine. Office: Harvard-Mass Inst Tech Div Health Scis and Tech 77 Massachusetts Ave Cambridge MA 02139

LONDON, RAY WILLIAM, clinical, consulting and medical psychologist, researcher; b. Burley, Idaho, May 29, 1943; s. Loo Richard and Maycelle Jerry (Moore) L. A.S., Weber State Coll., 1965, B.Sc., 1967; M. S.W., U. So. Calif., 1973, Ph.D., 1976. Diplomate: Am. Bd. Psychol. Hypnosis, Am. Acad. Behavioral Medicine, Am. Bd. Psychotherapy, Internat. Bd. Medicine and Psychology. Congl. asst. U.S. Ho. of Reps., 1964-65; research assoc. Bus. Advs., Inc., Ogden, Utah, 1965-67; dir. counseling and consultation services Meaning Found., Riverside, Calif., 1966-69; mental health and social service liaison San Bernardino County (Calif.) Social Services, 1968-72; clin. trainee VA Outpatient Clinic, Los Angeles, 1971-72, Children's Hosp., 1972-73, clin. fellow, 1973-74; clin. trainee Reiss Davis Child Study Ctr., Los Angeles, 1973-74, Los Angeles County-U. So. Calif. Med. Center, 1973; psychotherapist Benjamin Rush Neuropsychiat. Ctr., Orange, Calif., 1973-75; clin. psychology postdoctoral trainee Orange County (Calif.) Mental Health, 1976-77; postdoctoral fellow U. Caif.-Irvine Coll. Medicine, 1978; now clin. faculty; clin. psychologist Orange Police Dept., 1974-80; pvt. practice consultation and psychotherapy, Santa Ana, Calif., 1974—; cons. to public schs., hosps., bus., nationally and internationally, 1973—; pres. bd. govs. Human Factor Programs, Ltd., 1976—; pres. Internat. Bd. Medicine and Psychology, 1980—; mem. faculty Internat. Congress of Hypnosis and Psychosomatic Medicine, Soc. Clin. and Exptl. Hypnosis, Internat. Adv. Bd. European Congress Hypnosis and Psyhcosomatic Medicine; research assoc. Nat. Commn. for Protection of Human Subjects of Biomed. and Behavioral Research, 1976; fellow Inst. for Social Scientists on Neurobiology and Mental Illness, 1978. Editor: Internat. Bull. Medicine and Psychology, 1980; assoc. editor: Australian Jour. Clin. Hypnotherapy and Hypnosis, 1980; editorial cons.: Internat. Jour. Clin. and Exptl. Hypnosis, 1981; pub.: London Behavioral Medicine Assessment, 1982; producer: TV series Being Human, 1980. Recipient Congl. recognition U.S. Ho. of Reps., 1978; named scholar laureate Erickson Advanced Inst., 1980. Fellow Internat. Acad. Medicine and Psychology (dir. 1981—), Soc. Clin. Social Work (dir. 1979-80), Royal Soc. Health; mem. Acad. Psychosomatic Medicine, Am. Assn. Social Psychiatry, Am. Assn. Marriage and Family Therapy, Am. Psychology-Law Soc., Am. Psychol. Assn., CAlif. Psychol. Assn., Am. Group Psychotherapy Assn., Am. Orthopsychiat. Assn., Am. Soc. Clin. Hypnosis, Am. Soc. Psychosomatic Dentistry and Medicine, Biofeedback Soc. Am., Internat. Soc. Hypnosis, N.Y. Acad. Sci., Soc. Behavioral Medicine, Soc. Clin. and Exptl. Hypnosis, World Assn. Social Psychiatry, Phi Delta Kappa, Delta Sigma Rho, Tau Kappa Alpha, Pi Rho Phi, Lambda Iota Tau. Club: Toastmasters. Subspecialties: Psychophysiology; Cognition. Current work: Medical psychology; behavioral medicine; biofeedback and hypnosis; research and clinical application of behavioral, cognitive, psychophysiology, cancer, nephrology, immunology, pain, pediatrics, phobias, neuropsychology and cardiology problems. Office: 1125 E 17th St Suite E-211 Santa Ana CA 92701

LONDON, WILLIAM THOMAS, research physician; b. N.Y.C., Mar. 11, 1932; s. William Wolf and Lillian (Mann) L.; m. Linda Jane Greenman, June 23,1968; children: Barbara, Katharine, Emily, Nancy. B.A., Oberlin Coll., 1953; M.D., Cornell U., 1957. Postdoctoral research fellow in endocrinology Sloan-Kettering Inst., N.Y.C., 1960-62; sr. surgeon USPHS NIH, Bethesda, Md., 1962-66; research epidemiologist Inst. Arthritis ands Metaboloic Disease, 1962-66; research physician Inst. Cancer Research, Phila., 1966-67; sr. research physician, 1978—; adj. prof. U. Pa. Sch. Medicine, 1978—. Contbr. articles to profl. jours. Bd. dirs. Cheltenham Twp. Adult Sch. Nat. Cancer Inst. grantee, 1966—; Nat. Inst. Arthritis, Metabolic and Digestive Diseases grantee, 1978-81. Fellow ACS; mem. AMA, ACP, Endocrine Soc., Am. Assn. Cancer Research, Am. Soc. Preventive Oncology, Soc. Epidemiol. Research. Democrat. Jewish. Subspecialties: Cancer research (medicine); Epidemiology. Current work: Hepatitis B virus, acute hepatitis, chronic hepatitis, cirrhosis. Office: 7701 Burholme Ave Philadelphia PA 19111

LONG, BILLY WAYNE, medical educator, gastroenterologist; b. Tupelo, Miss., Apr. 5, 1948; s. Bonner W. and Lorene (Moore) L.; m. Rebecca Merritt Holmes, June 17, 1972; children: Matthew, Scott. B.A., David Lipscomb Coll., 1969; M.D., U. Miss., 1973. Diplomate: Am. Bd. Internal Medicine. Intern U. Miss. Med. Ctr., 1973-74, resident, 1974-75; clin. assoc. NIH, Bethesda, Md., 1975-77; fellow in gastroenterology U. Pa., Phila., 1977-79, asst. prof., 1979-81; clin. asst. prof. U. Miss., Jackson, 1981—; pvt. practice medicine, specializing in gastroenterology, Jackson, 1981—. Contbr. chpts. to books, articles to sci. jours. Served to lt. comdr. USPHS, 1975-77. Recipient RAGS award VA, 1978, Merit award, 1980; Young Investigator award NIH, 1980. Fellow ACP, Am. Coll. Gastroenterology; mem. Am. Gastroent. Assn., Soc. Gastrointestinal Endoscopy, Am. Fedn. Clin. Research, Miss. First, Jackson C. of C. Mem. Ch. of Christ (deacon). Subspecialties: Gastroenterology; Internal medicine. Current work: Gastrointestinal hormones; pancreatic exocrine secretion; intestinal epithelial cell absorption and secretion; clinical gastroenterology. Home: 755 Gillespie Pl Jackson MS 39202 Office: Woodrow Wilson 500-B E Jackson MS 39216

LONG, DANIEL RUSSEL, physics educator, researcher; b. Redding, Calif., June 9, 1938; s. Russel F. and Jane (Lemm) L.; m. Linda Kay Welp, Aug. 20, 1961 (div. 1982); children: Tracy Jane, Michael Daniel. B.A., U. Calif.-Berkeley, 1960; Ph.D., U. Wash., 1967. Prof. physics Eastern Wash. U., Cheney, 1967—. Sloane Found. fellow, 1977. Mem. Am. Phys. Soc., Sigma Xi. Republican. Subspecialty: Relativity and gravitation. Current work: Study of the mass separation dependence of the gravitation constant for small mass separations. Home: 119 W 6th St Cheney WA 99004 Office: Physics Dept Eastern Wash U Cheney WA 99004

LONG, DAVID AINSWORTH, environmental engineering educator, consultant; b. Rainback, Iowa, Aug. 17, 1933; s. Wilbur McKinnis and Lora Myrtle (Ainsworth) L.; m. Marilyn Jane Norris, Aug. 8, 1954; children: Jeffrey, James and Joan (triplets), Thomas. B.S.C.E., State U. Iowa, 1957, M.S., 1959; Ph.D., Pa. State U., 1971. Registered profl. engr., Iowa, Pa., diplomate: Am. Acad. Environ. Engrs. Research assoc. State U. Iowa, Iowa City, 1957-59; drainage basin engr. Wis. Commn. Water Pollution, Wisconsin Rapids, 1959-61; project engr. Baxter & Woodman, Inc., Crystal Lake, Ill., 1961-64; prof. environ. engring. Pa. State U., University Park, 1964—; cons. Gilbert Assocs., Reading, Pa., 1978-80, Lyco Systems, Williamsport, Pa., 1979—, Sweetland Engring., State College, Pa., 1977—. Chmn. State Coll. Boro Water Authority, State College, Pa., 1977—; commr. Juniata Valley council Boy Scouts Am., State College, 1972—. Mem. Water Pollution Control Fedn. (bd. dirs. 1976-79, Philip F. Morgan award 1973, Arthur Sidney Bedell award 1975), ASCE, Am. Water Works Assn., Assn. Environ. Engring. Profs. Republican. Episcopalian. Subspecialty: Water supply and wastewater treatment. Current work: Operation and management of water and wastewater treatment utilities. Home: 1009 Metz Ave State College Pa 16801 Office: Pa State U 212 Sackett Bldg University Park PA 16802

LONG, FRANKLIN A., emeritus chemistry educator; b. Great Falls, Mont., July 27, 1910; s. F.A. and Ethel (Beck) L.; m. Marion Thomas, 1937; children: Franklin, Elizabeth. A.B., U. Mont., 1931, M.A., 1932; Ph.D., U. Calif., 1935. Instr. chemistry U. Calif., 1935, U. Chgo., 1936; instr. chemistry Cornell U., 1937-38, prof., 1939-79, prof. emeritus, 1979—, chmn. dept., 1950-60, v.p. research and advanced studies, 1963-69, Henry Luce prof. sci. and society, 1969-79, dir. program on sci., tech. and society, 1969-73, dir. peace studies program, 1976-79; Dir. United Tech. Corp., Exxon Corp., 1969-81, cons., 1970-82; Mem. President's Sci. Adv. Com., 1961-66; asst. dir. U.S. Arms Control and Disarmament Agy., 1962-73; dir. Arms Control Assn., 1971-76; mem. adv. com. for planning and instnl. affairs NSF, chmn., 1973-74, mem. adv. panel for policy research and analysis, 1976-80; co-chmn. Am. Pugwash Steering Com., 1974-79; mem. U.S.-India subcom. for ednl. and cultural affairs, 1974-82, co-chmn., 1977-82; bd. dirs. Associated Univs., Inc., 1947-74, hon. bd. dirs., 1975—; bd. dirs. Council Sci. and Tech. for Devel., 1977—, Albert Einstein Peace Prize Found., 1979—. Mem. editorial bd.: Am. Scientist, 1974-81; Contbr. articles on chemistry, sci. policy and pub. affairs and arms control and disarmament to books, jours., encys. and reference works. Trustee Cornell U., 1956-57, Alfred P. Sloan Found., 1970—, Fund for Peace, 1980—. Guggenheim fellow, 1970. Fellow Am. Acad. Arts and Scis. (v.p. 1976-80); mem. Nat. Acad. Scis., AAAS, Council on Fgn. Relations, Am. Chem. Soc. Subspecialties: Physical chemistry. Current work: Policies for hazardous chemical wastes; military technology; Science and thehnology for development. Home: 429 Warren Rd Ithaca NY 14850

LONG, GARY JOHN, chemist, educator; b. Binghamton, N.Y., Dec. 3, 1941; s. Clifford J. and Margaret B. (Goodnow) L.; m. Audrey A. Ristway, Aug. 24, 1963; 1 son, Jeffrey R. B.S., Carnegie-Mellon U., 1964; Ph.D., Syracuse U., 1968. Prof. chemistry U. Mo.-Rolla, 1968—. Contbr. numerous articles on chemistry to profl. jours. Fellow Royal Soc. Chemistry, Am. Chem. Soc., Sigma Xi. Subspecialties: Inorganic chemistry; Solid state chemistry. Current work: Solid state and physical inorganic chemisry, Mossbauer effect spectroscopy, magnetic susceptibilities, X-ray and neutron diffraction, synthetic inorganic chemistry. Office: Dept Chemistry U Mo-Rolla Rolla MO 65401

LONG, GEORGE L., biochemistry researcher, molecular biologist; b. Atkin, Minn., Dec. 20, 1943; s. Walter and Patricia (Minehart) L.; m. Sharon L. King, June 21, 1967; children: Regan, Marnie. B.S., Pacific Lutheran U., 1966; Ph.D., Brandeis U., 1971. Asst. prof. chemistry Pomona Coll., Claremont, Calif., 1973-80; NIH sr. fellow U. Wash., Seattle, 1980-82; sr. scientist Lilly Research Labs., Indpls., 1982—. NIH fellow, 1980-82. Mem. Am. Chem. Soc., Am. Soc. Biol. Chemists, AAAS, Sigma Xi. Lutheran. Subspecialties: Genetics and genetic engineering (biology); Molecular biology. Current work: Cloning, structure determination and expression of eukaryotic genes. Office: 307 E McCarty St Indianapolis IN 46285

LONG, JEROME RUDISILL, physics educator; b. Lafayette, Ia., May 17, 1935; s. Hollis Moody and Joyce (Rudisill) L.; m. Sandra Tardo, May 27, 1962; children: Christopher, Jeremy. B.S., U. Southwestern La., 1956; M.S., La. State U., 1958, Ph.D., 1965. Postdoctoral fellow U. Pa., Phila., 1965-67; vis. prof. physics Simon Fraser U., Burnaby, B.C., Can., 1978-79; asst. prof. physics Va. Poly. Inst., Blacksburg, 1967-72, assoc. prof., 1972—. Mem. Am. Phys. Soc., AAUP. Episcopalian. Subspecialty: Condensed matter physics. Current work: Superconducting materials; magnetic materials; transport properties; magnetic properties. Home: 606 Woodland Dr NW Blacksburg VA 24060 Office: Physics Dept Va Poly Inst Blacksburg VA 24061

LONG, JOHN PAUL, pharmacologist, educator; b. Albia, Iowa, Oct. 4, 1926; s. John Edward and Bessie May L.; m. Marilyn Joy Stookesberry, June 11, 1950; children: Jeff, John, Jane. B.S., U. Iowa, 1950, M.S., 1952, Ph.D., 1954. Research scientist Sterling Winthrop Co., Albany, N.Y., 1954-56; asst. prof. U. Iowa, Iowa City, 1956-58, asso. prof., 1958-63, prof., 1963-70, 1970—, head dept. pharmacology, 1970-83. Author 268 research publs. in field. Served with U.S. Army, 1945-46. Recipient Abel award Am. Pharm. Assn., 1958; Ebert award Pharmacology Soc., 1962. Mem. Am. Soc. Pharm. Exptl. Therapy, Soc. Exptl. Biol. Medicine. Republican. Subspecialties: Neuropharmacology; Molecular pharmacology. Current work: Structure-activity relationships of dopamine receptor agonist. Home: 1817 Kathlin Dr Iowa City IA 52240 Office: Dept Pharmacology Coll Medicine Iowa City IA 52242

LONG, LEONARD MICHAEL, engineering consultant; b. New Orleans, July 6, 1955; s. William Anthony and Joyce (Eiserloh) L. B.S. in Mech. Engring, La. State U., Baton Rouge, 1977, M.S., 1979. Project/maintenance engr. Vulcan Materials Co., Geismar, La., 1979-80; mech. engr. Beard Engring., Inc., Baton Rouge, 1980—. Contbr. articles to profl. jours. Mem. ASME (treas. Baton Rouge chpt.). Republican. Roman Catholic. Subspecialty: Mechanical engineering. Current work: Innovative technology for treatment of industrial wastes. Home: 3030 Congress Blvd Number 69 Baton Rouge LA 70808

LONG, LYLE NORMAN, aeronautical engineer; b. Fergus Falls, Minn., Apr. 7, 1954; s. Norman Laverne and Shirley Ann (Leeman) L.; m. Laura Jean Greuel, July 11, 1981. B.M.E. with distinction, U. Minn., 1976; M.S., Stanford U., 1978; D.Sc., George Washington U., 1983. Engr. Donaldson Co., Inc., Mpls., 1974-76; research asst. Stanford (Calif.) U./NASA-Ames, 1977-78; research assoc. Joint Inst. for Advancement Flight Scis., NASA-Langley Research Center, Hampton, Va., 1979-83; sr. aero. engr. Lockheed-Calif. Co., Burbank, Calif., 1983—. Mem. AIAA (chmn. Langley student br. 1978-80), Am. Soc. Engring. Edn., Soc. Indsl. and Applied Math., Pi Tau Sigma, Audubon Soc. Club: Sierra (San Francisco). Subspecialties: Aeronautical engineering; Fluid mechanics. Current work: Hydrodynamical stability of fluids, aerodynamics of rotating blades using an acoustic formulation in the time domain, computational aerodynamics. Home: 11739 Darlington Ave Los Angeles CA 90049 Office: George Washington University/NASA-Langley Research Center MS 461 Hampton VA 23665

LONG, ROBERT RADCLIFFE, educator; b. Glen Ridge, N.J., Oct. 24, 1919; s. Clarence D. and Gertrude (Cooper) L.; m. Cristina Nersing, 1962; children: John Radcliffe, Robert W. A.B. in Econs, Princeton, 1941; M.S. in Meteorology, U. Chgo., 1949, Ph.D., 1950. Meteorologist U.S. Weather Bur., Paris, France, 1946-47; asst. prof. Johns Hopkins, 1951-56, asso. prof., 1956-59, prof. fluid mechanics, 1959—, dir. hydrodynamics lab., 1951—. Author: also articles in field. Engineering Science Mechanics; Editor: Tellus, 1966—, Jour. Geophys. Research, 1968—. Served from aviation cadet to capt. USAAF, 1942-45. Mem. Am. Meteorol. Soc. Subspecialties: Meteorology; Fluid mechanics. Current work: Fluid mechanics with application to meteorological and oceanographic systems. Home: 802 Beaverbank Circle Towson MD 21204 Office: Johns Hopkins U Baltimore MD 21218

LONG, WILLIAM HENRY, III, entomology and zoology educator, agricultural consultant; b. Decatur, Ala., Sept. 20, 1928; s. William Henry and Sara Lapsley (Liston) L.; m. Janice Helon Rogers, Sept. 11, 1953; children: Janice Faye Williams, Nancy Anne, Daniel Henry. B.A., U. Tenn., 1952; M.S., N.C. State U., 1954; Ph.D., Iowa State U., 1957. Asst. prof. dept. entomology La. State U., Baton Rouge, 1957-59, assoc. prof., 1959-64, prof., 1964-65; prof. dept. biology Nicholls State U., Thibodaux, La., 1965—; agrl. cons., pres. Long Pest Mgmt., Inc., Thibodaux, 1965—; expert, cons. UN FAO, Cairo, 1973-74, UN, IAEA, Pirracicaba, Brazil, 1975-77. Contbr. articles in field to profl. jours. Served with USAF, 1946-49; PTO. Mem. Entomol. Soc. Am., Nat. Alliance Ind. Crop Cons., La. Entomol. Soc. (pres. 1970), La. Agrl. Cons. Assn. (pres. 1977). Democrat. Presbyterian. Subspecialty: Integrated pest management. Current work: Agricultural consultant, professor of entomology and zoology, pursuing research in the management of sugarcane and soybean crops. Home: RFD 1 Box 548D Thibodaux LA 70301 Office: Nicholls State U Hwy 1 S Thibodaux LA 70301

LONGENECKER, GESINA (LOUISE) LIZANA, pharmacology researcher, medical educator; b. New Orleans, June 25, 1945; d. Florian Joseph and Shirley Louise (White) Lizana; m. Herbert Eugene Longenecker, June 12, 1965; children: Lani Louise, Herbert Eugene, Aimee Lee. B.S. with honors in Chemistry, Tulane U., 1965; Ph.D. in Pharmacology, Cornell U. Grad Sch. Med. Scis., N.Y.C., 1971. Postdoctoral fellow in pharmacology Cornell U. Grad. Sch. Med. Scis., 1971-72; postdoctoral fellow in biochemistry U. South Ala. Coll. Medicine, 1972-74, instr. pharmacology, 1974-75, asst. prof. pharmacology, 1975-81, assoc. prof., 1981—; mem. environ. biology rev. panel EPA, 1981—; mem. study sects. on use perfluorochems., sickle cell disease, coronamy vasospasm NIH; cons. Gulf States div. ARC, 1975—. Contbr. numerous articles, abstracts to profl. publs.; trainee editor: Clin. Pharmacology and Therapeutics, 1971-72; editor: Jour. Electrophysiol. Techniques, 1974-79. U. South Ala. Intramural grantee, 1975-78; Am. Lung Assn. grantee, 1978-80; NIH grantee, 1980-83; Council Tobacco Research—U.S.A., Inc. grantee, 1981-84; Upjohn Co. grantee, 1981-82. Mem. Neurosci. Soc., Soc. Exptl. Biology and Medicine, N.Y. Acad. Scis., Southeastern Pharmacology Soc., AAAS. Subspecialties: Pharmacology; Hematology. Current work: Platelet-endothelial interactions; prostaglandins; free radicals. Home: 3728 Claridge Rd S Mobile AL 36608 Office: Pharmacology-MSB Coll Medicin U South Al Mobile AL 36688

LONGMORE, WILLIAM JOSEPH, biochemist, educator; b. La Jolla, Calif., Oct. 7, 1931; m. Martha L. Baxter, Oct. 12, 1953; children: David, Brian, Timothy, Christopher. A.B., U. Calif.-Berkeley, 1957; Ph.D., U. Kans., 1961. Postdoctoral fellow Scripps Clinic and Research Found., La Jolla, 1961-63, assoc., 1963-66; asst. prof. St. Louis U. Sch. Med., 1966-69, assoc. prof., 1969-73, prof., 1973—, interim chmn. dept. biochemistry, 1982—; vis. prof. State U., Netherlands, 1977-78; cons. USPHS, NIH, Bethesda, Md., 1974—. Served with USAF, 1951-53. USPHS postdoctoral fellow, 1961-63; research career award, 1966-75; sr. fellow, 1977-78. Mem. AAAS, Am. Chem. Soc., Am. Soc. Biol. Chemists, Am. Physiol. Soc., Sigma Xi. Subspecialty: Biochemistry (medicine). Current work: Pulmonary biochemistry related to the lung surfactant system, metabolic regulation, lipid biochemistry. Home: 517 Beaucaire Dr Saint Louis MO 63122 Office: Dept Biochemistry Saint Louis U 1402 S Grand St Saint Louis MO 63104

LONGNECKER, DANIEL SIDNEY, pathologist, educator; b. Omaha, June 8, 1931; s. Walter Winfield and Hope Aline (Ranney) L.; m. Louise Elizabeth Miller, June 22, 1952; children: Matthew Paul, Daniel Alan, Jan Aline, Thomas Winfield. A.B., U. Iowa, 1954; M.D., 1956; M.S., 1962. Diplomate: Am. Bd. Pathology. Asst. prof., then assoc. prof. pathology U. Iowa, 1962-69; assoc. prof. St. Louis U. Med. Sch., 1969-82; pathology Dartmouth Coll. Med. Sch., 1972—; mem. Nat. Adv. Envron. Health Scis. Council, 1978-81. Author papers in field. Served to lt USNR. Research grantee NIH, 1974—. Mem. Am. Assn. Cancer Research, Am. Assn. Pathologists, Am. Pacreatic Assn., Soc. Exptl. Biology and Medicine, Common Cause. Subspecialties: Pathology (medicine); Cancer research (medicine). Current work: Experimental carcinogenesis, carcinoma of the pancreas. Office: Dept Pathology Dartmouth Med Sch Hanover NH 03755

LONGROY, ALLAN LEROY, chemistry educator; b. Flint, Mich., May 28, 1936; s. Arthur and Cornelia Martha (Schipanski) L.; m. Ruth Ann Patton, Dec. 26, 1955; children: Kathryn, Darla, Julie. A.S., Flint Jr. Coll., 1956; A.B., Flint Coll., 1958; M.S. in Chemistry, U. Mich., 1961, Ph.D., 1963. Teaching fellow U. Mich., 1958-62; postdoctoral research fellow Brandeis U., 1962-64; asst. prof. chemistry Ind. U.-Ft. Wayne, 1964-67; asst. prof. Purdue U.-Ft. Wayne, 1967-68, assoc. prof., 1968—. Fellow Union Carbide Corp., 1962; faculty research grantee Ind. U., 1966. Mem. Am. Chem. Soc. (treas. Northeastern Ind. sect. 1982—). Subspecialty: Organic chemistry. Current work: Organic reaction mechanisms, chemistry demonstrations. Home: 4566 Haffner Dr Fort Wayne IN 46815 Office: Indiana University-Purdue University 2101 Coliseum Blvd E Fort Wayne IN 46805

LONGWELL, JOHN PLOEGER, chemical engineering educator; b. Denver, Apr. 27, 1918; s. John Stalker and Martha Dorothea (Ploeger) L.; m. Marion Reed Valleau, Dec. 11, 1945; children: Martha Reed, Elizabeth Ann, John Dorney. B.S. in Mech. Engring, U. Calif., Berkeley, 1940; Sc.D. in Chem. Engring, M.I.T., 1943. With Exxon Research & Engring. Co., Linden, N.J., 1943-77; dir. Exxon Research & Engring. Co. Central Basic Research Lab., 1960-69; sr. sci. adv. Exxon Research & Engring. Co. (Central Basic Research Lab.), 1969-77; prof. chem. engring. M.I.T., Cambridge, 1977—; chmn. NRC Com. on Advanced Energy Storage Systems, 1979. Contbr. articles to profl. jours. Recipient Sir Alfred Egerton medal for contbns. to combustion Nat. Acad. Engring., 1976. Mem. Am. Chem. Soc., Combustion Inst. (past pres.), Am. Inst. Chem. Engrs. (award 1979), Sigma Xi, Tau Beta Pi. Republican. Subspecialties: Combustion processes; Fuels. Current work: High temperature fuel conversion processes-combustion, gasification pyrolysis; control of pollutants from combustion. Patentee in field. Home: 22 Follen St Cambridge MA 02138 Office: Mass Inst Tech Room 66-350 Cambridge MA 02139

LONNGREN, KARL ERIK, electrical and computer engineer, educator; b. Milw., Aug. 8, 1938; s. Bruno Leonard and Edith Irene (Osterlund) L.; m. Vicki Mason, Feb. 16, 1963; children: Sondra Lyn, Jon Erik. B.S., U. Wis.-Madison, 1960, M.S., 1962, Ph.D., 1964. Asst. prof. elec. and computer engring. U. Iowa, Iowa City, 1965-67, assoc. prof., 1967-72, prof., 1972—; vis. scientist, Sweden, 1964, Japan, 1972, 81, Denmark 1982, Oak Ridge, 1967-69, Los Alamos, 1979-80, Can., 1971, Wis., 1976-77. Contbr. numerous articles to profl. jours.; co-editor: Solitons in Action, 1978. Fellow Am. Phys. Soc., IEEE. Subspecialties: Electrical engineering; Plasma physics. Current work: Research in nonlinear waves, solitons and shocks, nonlinear mathematics with emphasis in plasma physics. Office: Elec and Computer Engring Dept U Iowa Iowa City IA 52242

LONSDALE, HAROLD KENNETH, high tech. co. exec.; b. Westfield, N.J., Jan. 19, 1932; s. Harold K. and Julia (Papandrea) L.; m. Constance Kerr, June 20, 1953; children: Karen Anne, Harold Kenneth. B.S. in Chemistry, Rutgers U., 1953; Ph.D. in Phys. Chemistry, Pa. State U., 1957. Staff mem. Gen. Atomic Co., San Diego, 1959-70; prin. scientist Alza Corp., Palo Alto, Calif., 1970-72; vis. scientist Max Planck Inst. for Biophysics, Frankfurt, W.Ger., 1973-74, Weizmann Inst. Sci., Rehovot, Israel, 1973-74; pres. Bend Research, Inc., Oreg., 1975—. Editor: Reverse Osmosis Membrane Research, 1972; contbr. articles to profl. jours. Served to 1st lt. USAF, 1957-59. Named Small Bus. Entrepreneur of Yr. Oreg. Bus. Mag., 1982. Mem. Am. Chem. Soc. Subspecialty: Physical chemistry. Current work: Founder, pres. high technology co.; membrance science and technology; water desalination and purification by membrane processes; controlled release products; gas separations. Patentee in field. Office: Bend Research Inc 64550 Research Rd Bend OR 97701

LONTZ, JOHN FRANK, biophysicist, educator; b. Bridgeport, Pa., Oct. 14, 1909; s. Frank Jacob and Mary Anna (Weinczyk) L.; m. Alexandra Elizabeth Gorzelska; children: Robert Jan., John F. B.S., Rutgers U., 1931; A.M., Temple U., Phila., 1933; Ph.D., Yale U., 1936; L.H.D., Holy Family Coll., 1971. Research chemist E. I. du Pont de Nemours, Wilmington, Del., 1936-41, research supr., 1950-60, research assoc., 1960-72; prof. biophysics Holy Family Coll., Torresdale, Pa., 1959-72; adj. prof. Temple U. Sch. Dentistry, Phila., 1972—; assoc. dir. maxillofacial clinic VA Med. Ctr., Wilmington, 1972—; cons. UN Indsl. Devel. Orgn., 1962-72, NIH, 1964-72. Bd. dirs., pres. Cath. Social Services, Wilmington, 1954-62. Served to lt. col. AUS, 1941-46; to col. USAR, 1960-70. Mem. Am. Chem. Soc., Am. Phys. Soc., AAAS, Soc. Biomaterials, Tissue Culture Assn., ASTM, Res. Officers Assn. Roman Catholic. Club: Athenaeum (Wilmington). Subspecialties: Polymer physics; Biophysics (biology). Current work: Polymer structure, molecular viscoelasticity, biological cell activity, toxicity and tumorigenicity, safe and effective prosthetics. Patentee plant growth stimulants, new structures, Teflon. Home: 515 Eskridge Dr Wilmington DE 19809 Office: VA Med Ctr 1601 Kirkwood Hwy Wilmington DE 19805

LOO, TI LI, pharmacologist, educator, researcher; b. Changsha, China; m. Marie Lee; Children: Michale, Agnes, Jonathan. B.Sc. (scholar), Tsing Hua U., Peking, 1940; Ph.D. (Sino-Brit. scholar), Oxford U., 1947. Research assoc. Christ Hosp. Inst., Cin., 1951-54; supervisory chemist Nat. Cancer Inst., Bethesda, Md., 1955-73; pharmacologist, prof. therapeutics M.D. Anderson Hops., U. Tex., Houston, 1973-80; chief pharmacology br., 1975—, Ashbel Smith prof. therapeutics, 1970—, chief pharmacology br., 1980—. Mem. Am. Soc. Clin. Oncology, Am. Soc. Pharms. and Exptl. Therapeutics, Am. Assn. Clin. Research, Am. Cancer Soc., others. Roman Catholic. Subspecialties: Pharmacology; Chemotherapy. Office: 6723 Bertner Ave Houston TX 77030

LOO, YEN-HOONG, biochemist; b. Honolulu, Dec. 19, 1914; s. Goon and Sun (Luke) L. B.A., Barnard Coll., 1937; M.S., U. Mich., 1938, Ph.D., 1943. Postdoctoral fellow U. Tex., Austin, 1943-44; research biochemist U. Ill., Urbana, 1944-51; biochemist Nat. Heart Inst., Bethesda, Md., 1951-52; research biochemist Eli Lilly & Co., Indpls., 1952-68; research scientist N.Y. State Inst. Basic Research in Devel. Disabilities, S.I., 1968—. Author numerous articles, sci. papers and book chpts. U. Mich. Barbour scholar, 1942-43; NIH vis. fellow, Babraham, Eng., 1966-67; Eli Lilly and Co. grantee, 1969-70; NIH grantee, 1976—. Fellow AAAS; mem. Am. Soc. Biol. Chemists, Am. Soc. Neurochemistry, N.Y. Acad. Scis. Subspecialties: Neurochemistry; Biochemistry (medicine). Current work: Biochemistry of the developing brain; synaptic development; biochemical mechanism of brain dysfunction in classical phenylketonuria and in maternal phenylketonuria. Office: Inst Basic Research in Developmental Disabilities 1050 Forest Hill Rd Staten Island NY 10314

LOOFT, DONALD JOHN, electronics co. exec.; b. Lakota, Iowa, Jan. 29, 1924; s. Edward Gustav and Louise Wilhemina (Knoner) L.; m. Helen Elena Turkish, Apr. 12, 1946; children: Nikki Lynn Lowry, Christel Louise, David Edward. Student, U. Dubuque, 1941-43; student, U. Wyo, 1943; A.B. in Econs-Physics, George Washington U.; M.S. in Engring. Mgmt, George Washington U., 1967-72. With Electro Tech. Lab., U.S. Army Mobility Equipment Research and Devel. Ctr., Ft. Belvoir, Va., 1946-69; dir., 1967-69, Night Vision Lab., Ft. Belvoir, 1969-75; dep. dir. Def. Advanced Research Projects Agy., Arlington, Va., 1975-78; cons., 1978-79; v.p., gen. mgr. Magnavox Govt. and Indsl. Electronics Co., Mahwah, N.J., 1978—; cons. Dept. Def. Contbr. articles to profl. jours. Served to 1st lt. AUS, 1942-46. Recipient mdeal for leadership Ft. Belvoir Comdg. Gen., 1962; Meritorious Service medal dept. Def., 1967, 72, 75, 78. Mem. IEEE, Soc. Am. Mil. Engrs. Subspecialties: Optical engineering; Optical image processing. Current work: Management in electro-optical civil and defense industry. Office: 46 Industrial Ave Mahwah NJ 07430

LOOK, A. THOMAS, pediatric hematologist, leukemia researcher; b. Midland, Mich., Dec. 5, 1948; s. Alfred Thomas and Betty (Elzi) L.; m. Karen Daly; children: Paul, Anne. B.S., U. Mich., 1971, M.D., 1975. Diplomate: Am. Bd. Pediatrics, Sub-Bd. Pediatric Hematology. Intern U. Mich., Ann Arbor, 1975-76, resident in pediatrics, 1976-77; fellow in Pediatric hematology St. Jude Hosp., Memphis, 1977-79, research assoc., 1979-81, asst. mem., 1981—. Fellow Am. Acad. Pediatrics; mem. Am. Assn. Cancer Research, Am. Assn. Clin. Oncology, Soc. Analytical Cytology, Am. Soc. Hematology. Subspecialty: Cancer research (medicine). Current work: Cellular biology of human acute leukemia using flourescence flow cytometry. Office: St Jude Children's Hosp Memphis TN 38101

LOOK, DWIGHT CHESTER, JR., mechanical engineering educator, researcher; b. Smith Center, Kans., Aug. 25, 1938; s. Dwight C. and Margery Rae (Bash) L.; m. Patricia A. Wellbaum, June 4, 1960; children: Dwight, Douglas. B.A., Central Coll., Fayette, Mo., 1960; M.S., U. Nebr., 1962; Ph.D., U. Okla., 1969. Teaching asst. dept. physics U. Nebr., Lincoln, 1960-63; aerosystems engr. Ft. Worth div. Gen. Dynamics, 1963-67; instr. evening coll. Tex. Christian U., Ft. Worth, 1967; spl. instr. mech. engring. U. Okla., Norman, 1969; from asst. prof. to prof. mech. engring. U. Mo.-Rolla, 1969—. Author: Thermodynamics, 1982; contbr. articles to profl. jours. NDEA fellow U. Okla., 1967-69; recipient Ralph R. Teetor award Soc. Automotive Engrs., 1978. Mem. ASME, AIAA, Am. Soc. Engring. End., Mo. Acad. Sci. Subspecialty: Mechanical engineering. Current work: Research project dealing with two-dimensional electromagnetic scattering. Office: Univ Mo Rolla 204 Mech Engring Bldg Rolla MO 65401

LOOMIS, EDMOND CHARLES, parasitologist; b. San Francisco, June 25, 1921; s. Lenton Charles and Elizabeth Mary (Wheal) L.; m. Alice Mae Lee, Nov. 27, 1948; children: Michael, Patrick E., Terrance E., Leeanne. B.A., U. Calif.-Berkeley, 1947, Ph.D., 1959. Assoc. vector control specialist Calif. Dept. Pub. Health, 1948-51, sr. vector control specialist, 1952-58; malaria advisor AID, Jamaica and Indonesia, 1959-61; extension parasitologist, extension program dir. animal and avian scis., forestry, marine and wildlife scis. and vet. medicine U. Calif.-Davis, 1962—; cons. parasitologist Commonwealth Sci. Industry and Sci. Research Orgn., Australia, 1969-70; parasitologist Ivan-Watkins Dow, Ltd., N.Z., 1974-75. Co-author: Parasites of Livestock and Poultry In California, 1982, Ticks of California, 1983; contbr. chpts. numerous articles to sci. publs., also to books. Served with USN, 1942-46; PTO. Mem. Entomol. Soc. Am. (chmn. vet. medicine 1979, disting. award in entomology 1979, disting. award in extension edn. 1981), Am. Soc. Parasitologists, Am. Soc. Vet. Parasitologists, Am. Mosquito Control Assn., Internat. Soc. Acarology, Calif. Vet. Med. Assn. Democrat. Roman Catholic. Subspecialties: Parasitology; Microbiology (veterinary medicine). Current work: Biology and control of parasites affecting livestock and poultry; taxonomy and biology of acarina. Office: Extension Vet Medicine U Calif Davis CA 95616

LOOMIS, HAROLD GEORGE, mathematics educator, researcher; b. Erie, Pa., Apr. 22, 1925; s. Harold Lloyd and Mildred Elsie (Backstrom) L.; m. Phyllis Goodrich Wright, Nov. 26, 1947 (div. 1972); children: Thomas P., Richard W., Nancy A., Mary S.; m. Robin Uluwehi Fern, Feb. 15, 1972. B.S., Stanford U., 1950; M.A., Pa. State U., 1952, Ph.D., 1957. Scientist HRB Singer, State College, Pa., 1952-55; instr. Pa. State U., State College, 1955-57; asst. prof. Amherst (Mass.) Coll., 1957-62, U. Hawaii, Honolulu, 1963-66, prof., 1981—; scientist NOAA, Honolulu, 1966-81; sec. tsunami commn. Internat. Union Geodesy and Geophysics, 1979—. Contbr. chpt. to book, articles to profl. jours. Served with USMC, 1942-46. Mem. Math. Assn. Am., Soc. Indsl. and Applied Math. Democrat. Unitarian. Subspecialties: Oceanography; Applied mathematics. Current work: Generation, propagation and terminal effects of tsunamis. Numerical hydrodynamics of long waves. Home: 1125 B 9th Ave Honolulu HI 96816 Office: Univ Hawaii 2540 Dole St Honolulu HI 96822

LOOMIS, MARY ELIZABETH SNUGGS, management information systems educator, information systems consultant; b. Chgo., Nov. 7, 1948; d. John Francis and Marion (Gunness) Snuggs; m. Timothy Patrick Loomis, Dec. 8, 1974; 1 dau., Allison Catherine. B.S., Purdue U., 1969; M.S., UCLA, 1972, Ph.D., 1975. Postgrad. research fellow IBM Research Lab., San Jose, Calif., 1975, ind. cons., Tucson, 1976—; asst. prof. mgmt. info. systems U. Ariz., Tucson, 1975-82, assoc. prof., 1982-83; database cons. D. Appleton Co., Inc., Manhattan Beach, Calif., 1982—, v.p., 1983—; rep. Codasyl Data Definition Lang. Com., 1981-82. Author: Data Management and File Processing, 1983, Data Communications, 1983. Mem. Assn. Computing Machinery, IEEE Computer Soc., Soc. Info. Systems. Subspecialties: Database systems; Distributed systems and networks. Current work: Development of database management system technology for distributed computer networks. Office: Dept Mgmt Info Systems Univ Arizona BPA Tucson AZ 85721

LOOMIS, STEPHEN HENRY, zoology educator; b. Flint, Mich., Oct. 3, 1952; s. Robert Henry and Evangeline (Horton) L.; m. Jeanne Rose Ohrtman, Oct. 18, 1981. B.S., U. Calif.-Davis, 1974, Ph.D., 1979. Research assoc. Rice U., Houston, 1979-80; asst. prof. dept. zoology Conn. Coll., New London, 1980—. Mem. Am. Soc. Zoologists, AAAS, Am. Mus. Natural History. Democrat. Lutheran. Subspecialties: Physiology (biology); Biochemistry (biology). Current work: Anhydrobiosis in nematodes, freezing tolerance in pulmonate gastropods, physiol. ecology of Crepidula fornicata. Home: 23 River Ridge Rd New London CT 06320 Office: Conn Coll Dept Zoology New London CT 06320

LOOMIS, TED ALBERT, physician, toxicologist; b. Spokane, Apr. 24, 1917; s. George and Sadie (Turner) L.; m. Marion Adams, Aug. 19, 1923; children: Bonnie, Becky. B.S., U. Wash., 1939; M.S., U. Buffalo, 1941; Ph.D., 1943; M.D., Yale U., 1946. Prof. pharmacology U. Wash., Seattle, 1947—. Contbr. numerous articles to sci. jours. Mem. Soc. Pharmacology and Exptl. Therapeutics, Soc. Toxicology, Soc. Exptl. Biology and Medicine. Subspecialty: Toxicology (medicine). Current work: Mechanisms of chemical induced toxicity; clinical toxicology. Address: Dept Pharmacology U Wash Sch Medicine Mail Stop SJ-30 Seattle WA 98195

LOONEY, PAUL BRYAN, well logging services company manager; b. San Deigo, Jan. 28, 1951; s. Francis Leroy and Cecelia Christina (McCartney) L.; m. Sarah Ellen Vickers, Jan. 22, 1982. B.S., Pa. State U.-State College, 1980. Ordinary seaman Woods Hole Oceanographic Inst., Mass., 1971-72; disc jockey St. Francis Coll., Loretto, Pa., 1974-76; warehouseman/driver A & M Beverage, Inc., Cresson, Pa., 1973-76; liquor store clk. Pa. Liquor Control Bd., State College, Pa., 1977-80; unit mgr. The Analysts/Schlumberger, Houston, Ventura, Calif., 1980—. Mem. Animal Behavior Soc., Inst. Diving, Nature Conservancy, Cousteau Soc., Oceanic Soc., Marine Tech. Soc. Subspecialties: Fuels; Applied mathematics. Current work: Helping to locate and identify oil zones through computer analysis of drilling data, maintaining safe rig operations to prevent environmental damage from oil well blow outs, also monitoring while drilling activities for instantaneous evaluation. Office: The Analysts/Schlumberger 4951 Olivas Park Dr Ventura CA 93003

LOOSEN, PETER THOMAS, psychiatrist, researcher, educator; b. Freiburg, Germany, Mar. 19, 1944; s. Otto Willi and Olga Maria (Hiener) L.; m. Laura Jeanne d'Angelo, Apr. 10, 1982. M.D., Ludwig Maximilian-U., Munich, W.Ger., 1970; Dr.med., 1974. Diplomate: German Bds. Psychiatry and Psychoanalysis, Am. Bd. Psychiatry and Neurology. Resident in psychiatry Psychiat. Hosp., U. Munich, 1972-74; psychiat. research fellow, resident dept. psychiatry U. N.C. Sch. Medicine, Chapel Hill, 1975-79, assoc. prof. psychiatry, dir. clin. research unit, 1979-83; prof., dir. affective disorders clin. Duke U. Med. Ctr., 1983—. Contbr. articles to profl. jours. Mem. Am. Psychiat. Assn., Internat. Soc. Psychoneuroendocrinology, N.Y. Acad. Scis., Collegium Internationale Neuropsychopharmacologicum, Am. Coll. Neuropsychopharmacology, Soc. Biol. Psychiatry, Soc. Neuroscience. Subspecialties: Psychiatry; Neuroendocrinology. Current work: Psychoneuroendocrinology. Office: Department Psychiatry Box 3857 Duke U Med Ctr Durham NC 27710

LOPATIN, DENNIS EDWARD, immunologist; b. Chgo., Oct. 26, 1948; s. Leonard Harold and Cynthia (Shifrin) L.; m. Marie S. Ludmer, June 6, 1971 (div. 1983); 1 son, Jeremy G.; m. Constance M. McLeod, July 24, 1983. B.S., U. Ill, Urbana, 1970; M.S., 1972; Ph.D. in Microbiology, 1974. Postdoctoral fellow Northwestern U. Med. Sch., Chgo., 1974-75; research scientist U. Mich., Ann Arbor, 1976—; assoc. prof., 1982—. Contbr. articles to sci. jours. Mem. Am. Assn. Immunologist, Am. Soc. Microbiology, Internat. Assn. Dental Research, Sigma Xi. Subspecialties: Immunology (medicine); Periodontics. Current work: Evaluation of the influence of microorganisms on immunoregulatory mechanisms and the influence of immunoregulatory mechanisms on response to certain microorganisms. Office: U Mich Ann Arbor MI 48109

LOPER, JOHN CAREY, microbiologist, educator; b. Hadley, Pa., June 21, 1931; s. Clark M. and Anna B. (Carey) L.; m. Dorothy L. Moredock, Dec. 23, 1956; children: John T., Robert D., Christopher L. B.A. cum laude, Western Md. Coll., 1952; M.S., Emory U., 1953; Ph.D., Johns Hopkins U., 1960. Instr., sr. instr., asst. prof. pharmacology St. Louis U. Sch. Medicine, 1960-63; asst. prof. microbiology and molecular genetics U. Cin. Coll. Medicine, 1963-66, assoc. prof., 1966-74, prof., 1974—, prof. environ. health, 1979—. Contbr. articles to sci. jours. Served with Med. Service Corps U.S. Army, 1954-56. NIH spl. research fellow, 1970-71; vis. fellow in genetics Australian Nat. U., Canberra, 1970-71. Mem. Genetics Soc. Am., Environ. Mutagen Soc., Am. Soc. for Microbiology, AAAS, Fedn. Am. Scientists, N.Y. Acad. Scis., AAUP. Subspecialties: Gene actions; Environmental toxicology. Current work: Environmental mutagenesis, mutagens and carcinogens in drinking water, gene engineering of yeast for degradation of hazardous waste. Home: 6315 Parkman Pl Cincinnati OH 45213 Office: Dept Microbiology and Molecular Genetics U Cin Coll Medicine Cincinnati OH 45267

LOPEZ, HECTOR, physicist, researcher; b. El Paso, Tex., Sept. 6, 1947; s. Isidro E. and Victoria C. L. B.S., U. Tex.-El Paso, 1970; M. Engring., U. Va., 1976. Radiol. physics cons. Physics Control, Inc., 1976-77; with Bur. Radiol. Health, FDA, USPHS, 1977—. Contbr. articles to profl. jours. Served with USAF, 1970-74. Recipient Commendation award USAF, 1974; USPHS unit commendation, 1981. Mem. Am. Assn. Physicists in Medicine, Am. Inst. Ultrasound in Medicine. Democrat. Roman Catholic. Subspecialties: Acoustics; Biomedical engineering. Current work: Ultrasound image analysis; ultrasound signal processing and analysis. Patentee in field. Office: Nat Ctr for Devices and Radiol Health HFX-210 5600 Fishers Ln Rockville MD 20857

LOPEZ, LARRY MARKELL, pharmacist; b. Louisville, May 29, 1946; s. Clarence Earl and Jean Doris (Markel) L; m. Patricia Catherine Garrett, Dec. 28, 1969; children: Laura Michell, Lisa Marie. B.S., U. Fla., 1969, Pharm.D., 1979. Registered pharmacist Fla., Ky. Community pharmacist Charles R. Walgreen & Co., West Palm Beach, Fla., 1968-70, Jack Eckerd Corp., Clearwater, Fla., 1970-77; asst. prof. pharmacy and medicine U. Fla., Gainesville, 1979—; Clin. pharmacy cons. Family Practice Med. Group, 1982—. Schering Corp. grantee, 1980; Am. Soc. Hosp. Pharmacy grantee, 1981; Miles Labs. grantee, 1982. Mem. Am. Coll. Clin. Pharmacy, Fla. Soc. Hosp. Pharmacists, Alachua County Assn. Pharmacists. Subspecialties: Family practice; Pharmacokinetics. Current work: Teaching rational drug therapy of hypertension and rational dosing of drugs using principles of pharmacokinetics. Home: 1630 NW 16th Ter Gainesville FL 32605 Office: U Fla College of Pharmacy Box J-4 Gainesville FL 32610

LÓPEZ-MARJANO, VINCENT, nuclear medicine administrator, educator; b. Madrid, Apr. 3, 1921; U.S., 1956; s. Bernardo and María (Majano) López); m. Jadwiga Jaworska, Mar. 4, 1952; children: Denise, Paul Vincent. B.A., B.S., U. Madrid, 1939, M.D., 1945, Ph.D., 1951; M.D., U. Costa Rica, 1952. Diplomate: Am. Bd. Nuclear Medicine, specialty bds. Pulmonary Disease, Spain and Costa Rica, Chest Surgery, Costa Rica. Intern Gen. Hosp. Madrid, 1945-46, resident, 1946-49; chief pulmonary function lab. Mt. Wilson (Md.) State Hosp., 1956-61, VA Hosp., Balt., 1961-70; asst. in medicine Johns Hopkins Med. Instns., Balt., 1961-65, instr. radiology, 1965-66, asst. prof. environ. medicine, 1968-70; chief pulmonary physiology lab. VA Hosp., Balt., 1964-70, chief tng. sect. nuclear medicine, Hines, Ill., 1970-74; clin. assoc. prof. medicine Loyola Stritch Sch. Medicine, Maywood, Ill., 1970-74; dir. nuclear medicine Gottlieb Meml. Hosp., Melrose Park, Ill., 1973-77; chmn. nuclear medicine div. Cook County Hosp., Chgo., 1977—, assoc. chmn. dept. radiology, 1979—; vis. prof. medicine San Marcos Med. Sch., Lima, Peru, 1966; vis. prof. Universidad Nacional de México, 1973, Med. Sch., Mérida, Yucatán, 1974; vis. scientist Nat. Acad. Scis., U.S., 1981; clin. assoc. prof. medicine Chgo. Med. Sch., 1973—. Contbr. chpts. to books, articles to profl. publs.; editor: profl. jours. including Revista de Bioloía y Medicina Nuclear, 1970, El Torax (Uruguay), Respiration, 1971, Jour. Nuclear Medicine and Allied Sciences, 1982. Mem. Am. Fedn. Clin. Research, Am. Physiol. Soc., Am. Soc. Nuclear Medicine, Johns Hopkins Med. Soc., Peruvian Soc. Nuclear Medicine, Mex. Soc. Nuclear Medicine (hon.), Brit. Nuclear Medicine Soc., Spanish Nuclear Medicine Soc., Colombian Nuclear Medicine Soc. Quaker. Subspecialties: Nuclear medicine; Pulmonary medicine. Current work: Regional lung function especially in sarcoidosis and silicosis. Home: 3100 N Sheridan Rd Chicago IL 60657 Office: Cook County Hosp 1825 W Harrison St Chicago IL 60612

LOPEZ-OVEJERO, JORGE ANDRES, physician, researcher, educator, consultant; b. San Pedro, Jujuy, Argentina, Oct. 16, 1938; came to U.S., 1966; s. Andres Isidoro Lopez-Ovejero and Delia Murguiondo; m. Linda Marie Krieger; 1 son, Andres. M.D., U. Buenos Aires, Argentina, 1965. Jr. resident in medicine Research Inst., U. Buenos Aires Sch. Medicine, 1965-66, research fellow dept. nephrology, 1966; intern Elmhurst (N.Y.) div. Mt. Sinai Hosp. Services, 1967-68; resident in medicine Manhattan VA Hosp., N.Y.C., 1969-70, Coney Island Hosp.-Maimonides Downstate Sch. Medicine, Bklyn., 1970-71, chief resident in medicine, 1971-72; attending physician Columbia U., Presbyterian Hosp., N.Y.C., 1972-73; asst. attending physician N.Y. Hosp., Cornell Med. Center, N.Y.C., 1973—; instr. in medicine Cornell U., N.Y.C., 1973-77, asst. prof. medicine, 1977—; cons. reviewer med. jours. Contbr. articles to profl. jours. Recipient Virginia Nash award Virginia Nash Fund, 1971; N.Y. Heart Assn. research fellow Columbia U., 1972-73. Mem. Am. Fedn. Clin. Research. Roman Catholic. Subspecialties: Internal medicine. Current work: Clinical research/treatment of hypertension; nutritional factors in the development and treatment of various chronic diseases. Office: 525 E 68th St Room k 400 New York NY 10021

LORAN, MURIEL RIVIAN, pharmacologist, consultant, researcher; b. N.Y.C., Feb. 6, 1925; d. Morris and Anna Frances (Gavrin) L. B.S. cum laude, L.I. U., 1947; M.S., Phila. Coll. Pharm. and Sci., 1948; Ph.D., Ohio State U., 1951. Lic. pharmacist, N.Y., D.C. Asst. prof. pharmacy U. Mont., Missoula, 1951-54; asst. research pharmacologist, asst. prof. pharmacology U. Calif.-San Francisco, 1954-55, asst. research pharmacologist, lectr. in pharmacology, 1955-57, asst. research pharmacologist, div. gastroenterology, 1957-59, assoc. research pharmacologist, 1959-66; research pharmacologist Children's Hosp. and Adult Med. Center, San Francisco, 1967-71; prof. cell biology and pathology U. Tel-Aviv Sch. Medicine, Tel Hashomer, Israel, 1968-69; dir. pharmacy Good Samaritan Hosp., Suffern, N.Y., 1971-72; research and quality control supr., dept. pharmacy services Montefiore Hosp. and Med. Center, Bronx, N.Y., 1973-80; dir. S M Consultants, N.Y.C., 1980—. Contbr. articles to publs., papers to profl. confs., U.S.; contbr., Japan, Netherlands. Am. Found. for Pharm. Edn. fellow, 1947-50; recipient Humanitarian award Epilepsy Soc. for Social Service, 1979, 81. Fellow Am. Inst. Chemists; mem. Am. Physiol. Soc., Acad. Pharm. Scis., Am. Chem. Soc., N.Y. Acad. Scis., Am. Pharm. Assn., Controlled Release Soc., Rho Chi. Jewish. Subspecialties: Pharmacology; Pharmacokinetics. Current work: Formulation of controlled release pharmaceuticals, anticonvulsants, tranquilizers and other products with a narrow therapeutic index. Work covers formulation, scale-up, clinical and animal studies and pharmocokinetics. Background in polymer chemistry contributes to understanding of dynamics of bonding. Patentee oral suspension of Phenytoin. Office: S M Consultants 200 Cabrini Blvd New York NY 10033

LORANCE, ELMER DONALD, chemistry educator; b. Tupelo, Okla., Jan. 18, 1940; s. Elmer Dewey and Imogene (Triplett) L.; m. Phyllis Ilene Miller, Aug. 30, 1969; children: Edward Donald, Jonathan Andrew. B.A., Okla. State U., 1962; M.S., Kans. State U., 1967; Ph.D., Okla. U., 1977. Asst. prof. chemistry So. Calif. Coll., Costa Mesa, 1970-73, assoc. prof., 1973-80, prof., 1980—. Contbr. articles in chemistry to profl. jours. Mem. Am. Chem. Soc., AAAS, Phi Lambda Upsilon, Phi Theta Kappa. Republican. Subspecialties: Organic chemistry; Biochemistry (biology). Current work: Research in chemotaxonomical studies of populations of desert plants at different elevations. Home: 3101 S Sycamore St Santa Ana CA 92707 Office: Southern California College Dept Chemistry Costa Mesa CA 92626

LORD, FREDERIC MATHER, psychologist; b. Hanover, N.H., Nov. 12, 1912; s. Frederic Pomeroy and Lydia Jeannette (Mather) L. A.B., Dartmouth Coll., 1936; M.A., U. Minn., 1943; Ph.D., Princeton U., 1951. Research assoc. Ednl. Testing Service, Princeton, N.J., 1949-60, sr. research psychologist, chmn. psychometric research group, 1960-76, disting. research scientist, chmn. psychometric research group, 1976—; vis. lectr. Princeton U., 1959-71; Brittingham vis. prof. U. Wis., 1963-64. Author: (with M.R. Novick) Statistical Theories of Mental Test Scores, 1968, Applications of Item Response Theory to Practical Testing Problems, 1980; mem. edit. council: Psychometrika, 1953—; chmn. edit. council, 1973-80. Fellow Am. Psychol. Assn. (pres. div. 5 1975-76), Am. Statis. Assn., Inst. Math. Stats.; mem. Psychometric Soc. (pres. 1958-59), Am. Ednl. Research Assn. Current work: Mental test theory and applications. Office: Educational Testing Service Rosedale Rd Princeton NJ 08541

LORENZ, EDWARD NORTON, meteorologist; b. West Hartford, Conn., May 23, 1917; m., 1948; 3 children. A.B., Dartmouth Coll., 1938; M.A., Harvard U., 1940; S.M., MIT, 1943. Asst. meteorology MIT, 1946-48, mem. staff, 1948-54; from asst. prof. to assoc. prof. meteorology, 1955-62, prof., 1962—, formerly dept. chmn.; vis. assoc. prof., UCLA, 1954-55. Recipient Crafoord prize, 1983. Mem. Nat. Acad. Sci., Am. Math. Soc., Am. Meteorol. Soc. Subspecialty: Meteorology. Office: Dept Meteorology MIT Cambridge MA 02139

LORENZEN, ROBERT T., agricultural engineering educator; b. New Leipzig, N.D., Feb. 16, 1917; s. Theodore and Hattie (Marek) L.; m. Mary K. Junkman, Feb. 6, 1954. B.S. in Agr, N.D. State U., 1943, U. Wis., 1954; M.S. in Agrl. Engring, U. Calif., Davis, 1957. Registered profl. engr., N.Y. Constrn. foreman Nat. Park Service, Mt. Rainier, Wash., 1936-39; constrn. supr. U. Wis., Madison, 1946-54; instr. U. Calif., Davis, 1954-56; asst.prof. agrl. engring. Colo. State U., Ft. Collins, 1956-59; from asst. prof. to prof. agrl. engring. Cornell U., Ithaca, N.Y., 1959-82, prof. emeritus, 1982—; design engr. Potlatch Forest, Lewiston, Idaho, 1978-79. Served to lt., inf. U.S. Army, 1943-46; ETO. Mem. Am. Soc. Agrl. Engring., Forest Products Research Soc., ASTM, ASHRAE, Sigma Xi, Chi Epsilon. Republican. Lutheran. Subspecialties: Agricultural engineering; Solar energy. Current work: Farm structures and production systems design. Office: Cornell U Riley-Robb Hall Ithaca NY 14853

LORENZETTI, OLE JOHN, pharmaceutical company research administrator; b. Chgo., Oct. 25, 1936; s. Natale and Quintilia (Bertochinni) L.; m. Lorna Joyce Bailey, June 20, 1962; children: Elizabeth Anne, Maria Anne, Dario Natale. B.S., U. Ill.-Chgo., 1958; M.S., Ohio State U., 1962, Ph.D. in Pharmacology-Toxicology, 1965. Sr. scientist, head pharmacology Miles Labs., Elkhart, Ind., 1965-69; head pharmacology and toxicology Alcon Labs., Ft. Worth, 1969-72, mgr. biol. sci., 1972-79; assoc. dir. research and devel. Owen Labs. div., 1979-82, dir. research and devel., 1982—; exec. v.p. Profl. Cons., Inc., Ft. Worth, 1978—. Patentee in field. Pres. Rotary Found., Ft. Worth, 1979. Fellow Acad. Pharm. Sci., Am. Acad. Clin. Toxicology, Soc. Cosmetic Chemists (local chmn. 1975-77); mem. Am. Acad. Ophthalmology (lectr.), Am. Acad. Dermatology (lectr.), Sigma Xi, Rho Chi. Republican. Roman Catholic. Subspecialties: Ophthalmology; Dermatology. Current work: Development of drug therapy for ophthalmic and dermatological disease. Home: 1945 Bekeley Pl Fort Worth TX 76110 Office: Alcon Labs Inc 6201 S Freeway Fort Worth TX 76101

LORENZINI, PAUL GILBERT, nuclear engineer; b. Portland, Oreg., Apr. 16, 1942; s. Gilbert Henry and Viola P. (Gates) L.; m. Janet Grace Jesperson, Aug. 19, 1967; children: Christy, Michael. B.S., U.S. Mcht. Marine Acad., 1964; Ph.D., Oreg. State U., 1969; J.D., Loyola U., Los Angeles, 1975. Cert. nuclear engr., Calif. Project engr. Atomics Internat., Canoga Park, Calif., 1973-74, program mgr., 1974-76; assoc. Tooze, Kerr, et al., Portland, 1976-79; asst. chief counsel Rockwell Hanford Ops., Richland, Wash., 1979, dir. health, safety, environment, 1979-82, asst. gen. mgr., 1982—. Founder, pres. Washington Voice Energy, Oreg. Voice Energy. Mem. Am. Nuclear Soc. (dir. Oreg. sect. 1978-79, fuel cycle/waste mgmt. exec. com. 1981—, rules and by laws com. 1982—), Health Physics Soc., Calif. Bar. Assn., Oreg. Bar. Assn. Republican. Presbyterian. Subspecialties: Nuclear fission; Nuclear engineering. Home: 121 W Orchard Way Richland WA 99352 Office: Rockwell Hanford Ops PO Box 800 Richland WA 99352

LORING, ARTHUR PAUL, geology educator, consultant; b. Bklyn., May 22, 1936; s. Alex N. and Mildred (Arkawy) Lifschutz; m. Carol Lynn Lotterer, Aug. 10, 1963; children: Wendy, Karen, David. A.B., Columbia U., 1958; M.S., Pa. State U., 1961; Ph.D., NYU, 1966. Teaching asst. Pa. State U., 1959-61; lectr.-instr. Bklyn. Coll., 1961-67; asst. prof. Upsala Coll., 1967; asst. prof. geology York Coll., CUNY, 1967-72, assoc. prof., 1972—; adj. prof. Bklyn. Coll., 1980, 82; cons. geology, Glen Cove, N.Y., 1975—. Contbr. articles in geology to profl. jours. Fellow Geol. Soc. Am.; mem. Am. Soc. Photogrammetry and Remote Sensing, Assn. Engring. Geologists (research mem.). Subspecialties: Remote sensing (geoscience); Ground water hydrology. Current work: Application of remote sensing techniques in search of ground water in crystalline terrains. Home: 8 David Ct Glen Cove NY 11542 Office: York Coll CUNY Jamaica NY 11451

LORTIE, JOHN WILLIAM, solar mktg. specialist, research corp. execl, solar cons., phys. sci. researcher; b. Chgo., July 11, 1920; s. William Arthur and Alice Marie (McNamee) L.; m. Mary Elaine Sullivan, Sept. 21, 1946; children: Colleen, Kevin, Timothy. Student in radar sci, Ill. Inst. Tech., 1940-42, in edn, U. Ala., 1976. Radar technician Western Electric Co., 1946-50; pres. Wm. A. Lortie & Sons, Westchester, Ill., 1950-65, Monark Instant Homes, Ocean Springs, Miss., 1965-75; exec. officer Energy Research Corp., Mobile, Ala., 1975-82, dir. research, 1980-82; head dept. solar tech. Carver State Tech. Coll., 1976-81; pres. Essential Solar Products, Mobile, 1980—; solar cons. Served in U.S. Army, 1942-46. Mem. Ala. Solar Energy Assn. (sec.-treas.), Ala. Solar Industries Assn. (dir.), Internat. Solar Energy Soc., Nat. Assn. Solar Contractors. Republican. Roman Catholic. Subspecialties: Combustion processes; Fuels and sources. Current work: Patent research; laser guidance systems.

LORTON, LEWIS, dental researcher, army officer; b. N.Y.C., Nov. 3, 1939; s. Fred and Rose (Engel) L.; m. Jacqueline Carol Andor, Aug. 4, 1982; children: Elizabeth, Michael S., Mark, Michael B., Erin. B.A., Brandeis U., 1960; D.D.S., U. Pa., 1964; M.S. in Dentistry, Ind. U., 1978; cert. in fixed prosthodontics, Fitzsimmons Army Med. Center, 1973. Commd. 2d lt. U.S. Army, 1966, advanced through grades to col., 1983; clinic chief dental activity, Hanau, W. Ger., 1968-71; chief fixed prosthetics Dental Activity, Ft. Carson, Colo., 1973-76; research officer, research coordinator U.S. Army Inst. Dental Research, Washington, 1976-82, chief bioengring. br., San Francisco, 1982—. Mem. editorial bd.: Jour. of Endodontics, 1981—. Mem. ADA, Am. Assn. Dental Research, Internat. Assn. Dental Research. Subspecialties: Biomaterials; Prosthodontics. Current work: Research in maxillofacial injury and shielding, development of dental field equipment, portable battery-operated X-ray, identification problems in post mortem, prediction of dental emergencies. Home: 1302 Kobbe Ave San Francisco CA 94129 Office: US Army Institute of Dental Research Letterman Army Institute of Research San Francisco CA 94129

LORUSSO, PAUL DAVID, mechanical engineering administrator; b. Somerville, Mass., Mar. 2, 1954; s. Raphael and Concetta (Mangano) L.; m. Janet Rae Pendleton, July 17, 1982. B.S.M.E., Northeastern U., 1976; M.B.A., Babson Coll., 1983. Engr. shock and vibration testing and system design Barry Wright Corp., Watertown, Mass., 1972-76; with Digital Equipment Corp., Maynard, Mass., 1976—, supr. engring., 1978-81, mgr. engring., 1981—. Recipient 2d place award Plastic's Better Way Competition, 1979. Evangelical. Subspecialty: Mechanical engineering. Current work: Electronics packaging of small systems; mini and personal computers. Home: 993 Massachusetts Ave Apt 228 Arlington MA 02174 Office: Digital Equipment Corp 146 Main St MLO6B-2/E66 Maynard MA 01754

LOSS, FRANK J., mechanical engineer; b. Homestead, Pa., May 14, 1936. B.S., Carnegie-Mellon U., 1958, M.S., 1959, Ph.D., 1961. Registered profl. engr., Calif. Engr. Westinghouse Bettis Atomic Power Lab., Pitts., 1961-62; head mechanics of materials br. U.S. Naval Research Lab., Washington, 1964-82; tech. dir. Materials Engring. Assocs., Lanham, Md., 1982—. Contbr. numerous articles on engring. to profl. jours. Served to 1st lt., C.E. U.S. Army, 1962-64. Mem. ASTM, ASME. Subspecialties: Fracture mechanics; Materials (engineering). Current work: Director of research and engineering, structural integrity technology development, fracture mechanics assessment of structural steels, corrosion fatigue, assessment of nuclear structural steels, radiation embrittlement, failure analysis. Home: 10318 Royal Rd Silver Spring MD 20903 Office: 9700-B Palmer Hwy Lanham MD 20706

LOSTROH, ARDIS JUNE, physiology educator; b. Malcolm, Nebr., Dec. 21, 1925; d. Louis Henry and Huldah (Larson) L.; m. Maurice Edward Krahl, Feb. 4, 1967. A.B. with distinction, U. Nebr.-Lincoln, 1950, M.A., 1952; Ph.D., U. Calif.-Berkeley, 1956. Postdoctoral fellow USPHS, U. Chgo., 1959-61; asst. prof. exptl. endocrinology U. Calif.-San Francisco, 1961-65, assoc. prof., 1965-77; vis. assoc. prof. physiology Stanford U., 1970-77; prof. dept. physiology Med. Coll. Va., Richmond, 1977-78; cons., 1978—. Nat. Found. Infantile Paralysis Postdoctoral fellow, Paris, 1957-58; USPHS Postdoctoral fellow, 1959-61. Mem. Am. Phyiol. Soc., Phi Beta Kappa, Sigma Xi, Alpha Lambda Delta. Republican. Lutheran. Club: Catalina Racquet and Swim (Tucson). Subspecialties: Endocrinology; Cell and tissue culture. Current work: Mammalian models; cellular-organ functions: regulation by hormones, ions, etc.; gonadal-pituitary hormones; control of reproduction; pituitary hormones: general metabolic effects; pituitary hormones and insulin: growth/diabetes. Home: 2783 W Casas Circle Tucson AZ 85741

LOTH, JOHN LODEWYK, aerospace engr., educator; b. The Hague, Netherlands, Sept. 14, 1933; s. Julius Edward and Fyna Johanna (Van de Berg) L.; m. Harriet H. Huffman; children by previous marriage: Eric, Frank, Marianne. B.A., U. Toronto, 1957, M.A. Sc., 1958, Ph.D. in Mech. Engring, 1962. Research fellow Centre National de Recheche Scientifique, Paris, 1958-59; lectr. U. Toronto, 1961-62; asst. prof. aero. engring. U. Ill., Urbana, 1962-67; prof. aerospace engring. W.Va. U., Morgantown, 1967—; pres. Dynamic Flow, Inc. (Engring. Cons.), 1972—. Contbr. articles to profl. jours. Mem. AIAA, Am. Soc. Engring. Edn., Combustion Inst., Profl. Engrs. Soc. W. Va., Sigma Xi, Sigma Gamma Tau. Subspecialties: Aerospace engineering and technology; Combustion processes. Current work: Vertical axis wind turbine optimization and aerodynamic automatic speed control; powered high lift wing development, designed and tested the first circulation controlled stol aircraft. Home: PO Box 4094 Morgantown WV 26505 Office: Engring Sci Bldg W Va U Morgantown WV 26506

LOTLIKAR, PRABHAKAR DATTARAM, biochemist, educator; b. Shirali, Karnatak, India, May 21, 1928; came to U.S., 1955, naturalized, 1966; s. Dattaram Vithalrao and Laxmibai Dattaram (Kamath) L.; m. Faye Lun Chin, June 17, 1960; 1 son, Jeffrey. B.S., Bombay U., 1950, M.S., 1954; Ph.D., Oreg. State U., 1960. Asst. chemist Raptakos Brett & Co., Bombay, 1950-55; postdoctoral fellow, then project assoc. McArdle Lab. Cancer Research, U. Wis.-Madison, 1960-65, instr., 1965-66; research instr. biochemistry Fels Research Inst., Temple U. Med. Sch., Phila., 1967-68, asst. prof. biochemistry, 1968-75, assoc. prof., 1975—. Recipient USPHS Career Devel. award, 1969-73; Nat. Cancer Inst. grantee, 1968—. Mem. Am. Chem. Soc., AAAS, Am. Assn. Cancer Research, Am. Soc. Biol. Chemists., Soc. Exptl. Biology and Medicine, Biochem. Soc., N.Y. Acad. Scis. Subspecialties: Cancer research (medicine); Oncology. Current work: Mechanisms of chemical carcinogenesis. Home: 1042 Randolph Dr Yardley PA 19067 Office: 3420 N Broad St Philadelphia PA 19140

LOTTI, VICTOR JOSEPH, pharmacologist, pharm. co. exec., researcher; b. Trenton, Jan. 6, 1938; s. Joseph Anthony and Elizabeth (Persiano) L.; m. JoAnna Pitcock, Sept. 7, 1960; m. Barbara M., Nov. 17, 1976; children: Lynn E., Victor J., Lisa A. B.S. in Pharmacy, U. Conn., 1959; M.S. in Pharmacology, U. Mo., 1961; Ph.D. in Pharmacology (NIH fellow), UCLA, 1965. Registered pharmacist, Mo. With Merck & Co., 1967—, sr. dir. research coordination, Rahway, N.J., 1975-77, sr. dir. pharmacology, Chibret, France, 1977-79, dir. microbial pharmacometrics, West Point, Pa., 1980—. Contbr. numerous articles, abstracts, chpts. to profl. publs. NIH postdoctoral fellow, 1966. Mem. Am. Soc. Pharmacology and Exptl. Therapeutics, Am. Chem. Soc., Western Pharmacol. Soc., Soc. Neuroscis., Assn. Research Vision and Ophthalmology. Roman Catholic. Subspecialties: Ophthalmology; Neuropharmacology. Current work: Ocular pharmacology; glaucoma; ocular and CNS neurotransmitters; ocular and CNS receptor mechanism; drug mechanisms of action; microbial natural product interaction with mammalian enzymes and receptors. Patentee composition and method of treatment of dopamine deficiency, methods of treating hypertension. Home: 214 Brookside Circle Harleysville PA 19438 Office: Merck Sharp and Dohme West Point PA 19486

LOUGHMAN, BARBARA ELLEN EVERS, research immunologist; b. Frankfort, Ind., Oct. 26, 1940; d. Jimmie and Ruth (Hoyer) Evers; m. Terry B. Loughman, June 28, 1962; children: Lance Evers, Chad Elliott. B.S., U. Ill., 1962; Ph.D., U. Notre Dame, 1972. With Ames Research Labs., Miles Labs., Inc., Elkhart, Ind., 1962-72; NIH staff fellow Gerontology Research Center, Balt., 1972-74; research scientist hypersensitivity diseases research Upjohn Co., Kalamazoo, 1974-78, sr. research scientist, 1978-79, reserach head, 1979—; mem. study sect. NIH, 1982—. Contbr. articles to sci. jours. Mem. AAAS, Am. Assn. Immunologists, Assn. Gnotobiology. Club: Altrusa (Kalamazoo). Subspecialties: Immunobiology and immunology; Immunopharmacology. Current work: Research administration; cellular immunology; somatic cell hybridization; clinical pharmacology of biologicals; immune pharmacology. Patentee in field. Home: 7881 Cayhill Richland MI 49083 Office: 301 Henrietta St Kalamazoo MI 49001

LOUI, MICHAEL CONRAD, electrical engineering educator; b. Phila., June 1, 1955. B.S., Yale U., 1975; S.M., MIT, 1977, Ph.D., 1980. Vis. asst. prof. elec. engring., vis. research asst. prof. Coordinated Sci. Lab., U. Ill., Urbana-Champaign, 1981-82; asst. prof. elec. engring., research asst. prof., 1982—. Fannie and John Hertz Found. grad.

fellow, 1975-80. Mem. Assn. Computing Machinery, IEEE, Math. Assn. Am., Soc. Indsl. and Applied Math., Sigma Xi, Phi Beta Kappa, Tau Beta Pi. Subspecialty: Theoretical computer science. Current work: Analysis of algorithms, automata, computational complexity, data structures, distributed computation. Office: Coordinated Sci Lab U Ill 1101 W Springfield Ave Urbana IL 61801

LOUIS, JOHN, hematologist; b. Chgo., June 21, 1924; s. Demitrios John and Artemis (Halkiopolous) L.; m. Priscilla Humay, July 5, 1975; children: Priscilla Artemis, Demitrios John. B.S., U. Ill., 1948, M.S., 1950, M.D., 1950. Diplomate: Am. Bd. Internal Medicine, Am. Bd. Pathology. Instr., research assoc. U. Ill., 1951-65; asst. prof. medicine Loyola U., Chgo., 1965-70; pvt. practice medicine, 1970-75; prof. medicine VA Med. Center North Chgo./Chgo. Med. Sch., 1975—; cons. in field. Contbr. articles to profl. jours. Served with AUS, 1942-44. NIH grantee, 1958-65, 60-65. Fellow ACP; mem. Am. Soc. Clin. Oncology, Am. Assn. Cancer Research, Am. Soc. Hematology, Internat. Soc. Hematology. Greek Orthodox. Subspecialties: Hematology; Chemotherapy. Current work: Clinical pharmacology. Address: 347 Circle Ln Lake Forest Il 60045

LOUIS, THOMAS MICHAEL, anatomy educator, researcher, consultant; b. Pensacola, Fla., Dec. 27, 1944; s. Russell George and Gertrude (McGuire) L.; m. Mary Patricia Lilliquist, Dec. 27, 1969 (div. 1981); children: Elizabeth Ione, Michael Patrick. B.S. in Biology, Va. Poly. Inst., 1968, M.S. in Zoology, 1970; Ph.D., Mich. State U., 1975; postdoctoral fellow, Oxford (Eng.) U., 1975-76. Lalor fellow Oxford (Eng.) U., 1975-76; vis. prof. Pasteur Inst., Paris, summer 1976; asst. prof. East Carolina Med. Sch., Greenville, N.C., 1976-79, assoc. prof. anatomy dept., 1979—; ptnr. Eastern Anatomy Cons., Greenville, 1981—. Author: Prostaglandan and Control of Estrous Cycle, 1973 (Richard Hoyto research prize). Mem. Am. Physiol. Soc., Am. Anatomical Soc., Soc. for Study of Reprodn., Endocrine Soc., Sigma Xi. Roman Catholic. Lodge: KC. Subspecialties: Reproductive biology (medicine); Maternal and fetal medicine. Current work: Fetal and neonatal biology and reproductive biology. Office: Anatomy Dept Sch Medicine East Carolina U Greenville NC 27834 Home: 5 Edgewood Greenville NC 27834

LOURENCO, RUY VALENTIM, physician, educator; b. Lisbon, Portugal, Mar. 25, 1929; came to U.S., 1959, naturalized, 1966; s. Raul Velentim and Maria Amalia (Gomez-Rosa) L.; m. Susan Jane Lowenthal, Jan. 18, 1960; children: Peter Edward, Margaret Philippa. B.S., U. Lisbon, 1946, M.D., 1951. Intern Lisbon U. Hosp., 1953-55; instr. medicine U. Lisbon, 1956-59; fellow in medicine Columbia-Presbyn. Med. Center, N.Y.C., 1959-63; asst. prof. N.J. Coll. Medicine, Jersey City, 1963-66; assoc. prof., 1966-67; assoc prof. medicine and physiology U. Ill., Chgo., 1967-69, prof. medicine, 1969—, head dept. medicine, 1977—, Foley prof. medicine, 1978—; attending physician U. Ill. Hosp., 1967—, VA Hosp., Chgo., 1967—; dir. respiratory physiology lab. Cook County Hosp., 1967-69, chmn. dept. pulmonary medicine, 1969-70, attending physician, 1969—. Fellow AASS, Am. Coll. Chest Physicians (pres. Ill. chpt. 1974-75), ACP, Am. Fedn. Clin. Research; mem. Am. Heart Assn., Am. Physiol. Soc., Am. Soc. Clin. Investigation, Am. Thoracic Soc., Soc. for Exptl. Biology and Medicine, Assn. Am. Physicians, Assn. Profs. Medicine, Central Soc. Clin. Research (councillor 1973-77), Sigma Xi, Alpha Omega Alpha (faculty mem.), Phi Kappa Phi. Subspecialties: Internal medicine; Pulmonary medicine. Current work: Pulmonary diseases, respiratory physiology and biochemistry. Home: 1000 N Lake Shore Dr Chicago IL 60611 Office: Department of Medicine University of Illinois Hospital 840 S Wood St Chicago IL 60612

LOUX, JOSEPH J., toxicologist; b. Pittston, Pa., Aug. 27, 1936; s. Joseph J. and Eleanor A. (Mazaitis) L. B.S., U. Scranton, 1958; M.S., W.Va. U., 1962. Sr. scientist Smith Kline & French, 1969-71; biomed. mgr. Franklin Inst. Research Labs., Phila., 1971; sr. scientist Cooper Labs., Cedar Knolls, N.J., 1972-75, sect. head gen. pharmacology, ophthalmic pharmacology and toxicology, 1976, assoc. dir. ophthalmic research, 1976-77; group leader pharmacology/toxicology Carter Products Research, Cranbury, N.J., 1977—; lectr. Ctr. Profl. Advancement, New Brunswick, N.J. Contbr. articles to profl. jours. Mem. Soc. Comparative Ophthalmology (steering com.), N.Y. Acad. Scis., AAAS, Physiol. Soc. Phila., Acad. Pharm. Scis., Am. Pharm. Assn., Am. Chem. Soc., Am. Soc. Pharmacology and Exptl. Therapeutics, Am. Coll. Toxicology, Soc. Cosmetic Chemists, Internat. Union Pharmacology, Soc. Toxicology. Roman Catholic. Subspecialties: Toxicology (medicine); Pharmacology. Current work: Product efficacy and safety, pharmacodynamics and therapeutics; ocular toxicology, dermatology, dental biochemistry. Office: Carter Products Research Half Acre Rd Cranbury NJ 08512

LOVE, WARNER EDWARDS, biophysics educator; b. Phila., Dec. 1, 1922; s. J. Warner E. and Elizabeth (Ford) L.; m. Lois Jane Hosbach, Dec. 26, 1945; children: Rebecca Edwards, Michael Warner. B.A. in Zoology, Swarthmore Coll., 1946; Ph.D. in Physiology, U. Pa., 1951. Fellow, then asso. biophysics U. Pa., 1951-55; research fellow, then research asso. physics Inst. Cancer Research, Fox Chase, Pa., 1955-57; faculty Johns Hopkins, 1957—, prof. biophysics, 1965—, dept. chmn., 1971-74, 80-83. Served with Am. Field Service, 1943-45; Served with AUS, 1945-46. Mentioned in despatches. Subspecialty: Biophysics (physics). Research on x-ray crystal structure analysis of biol macro-molecules, particularly hemoglobins. Home: 1419 Eutaw Pl Baltimore MD 21217

LOVELAND, KATHERINE ANNE, psychologist, educator; b. Atlanta, June 30, 1955; d. Edward Henry and Olga Helen (Christie) L.; m. James Voorhis Temple, June 3, 1979. B.A., U. Va., 1975; Ph.D., Cornell U., 1979. Vis. asst. prof. Rice U., Houston, 1979-82, adj. asst. prof., 1982—; develop. psychologist Tex. Research Inst. Mental Scis., Houston, 1982—. Nat. Inst. Neurol. and Communicative Disorders and Stroke new investigator research awardee, 1982-84. Mem. Soc. for Research in Child Devel., Am. Psychol. Assn., AAAS, N.Y. Acad. Scis., Phi Beta Kappa. Subspecialties: Developmental psychology; Cognition. Current work: Psychology of human development; language, perceptual development, especially in abnormal populations. Home: 15342 Falmouth Ave Houston TX 77084 Office: Tex Research Inst of Mental Scis 1300 Moursund St Houston TX 77030

LOVELL, RICHARD ARLINGTON, biochemist; b. Kentland, Ind., Aug. 4, 1930; s. Floyd Arlington and Eva Marie (Reinhart) L.; m. Joanne Therese Walsh, Dec. 27, 1965; children: Luke Richard, Maria Therese, Jude Joseph, John Floyd, Andrew Patrick, Alice Marie. B.S., Xavier U., Cin., 1952, M.S., 1953; Ph.D., McGill U., Montreal, Que., Can., 1963. Postdoctoral fellow U. Wis.-Madison, 1964-65, Yale U., 1965-66; instr. psychiatry U. Chgo., 1966-69, asst. prof., 1969-75; mgr. biochemistry Ciba-Geigy Corp., Summit, N.J., 1975—. Mem. Am. Chem. Soc., N.Y. Acad. Scis., Am. Epilepsy Soc., Internat. Soc. Neurochemistry, Soc. Neurosci., AAAS. Subspecialties: Biochemistry (biology); Neurochemistry. Current work: Neurochemistry, Neuropharmacology of mental illness. Office: Ciba-Geigy Corp 556 Morris Ave Summit NJ 07901 Home: 479 Snyder Ave Berkley Heights NJ 07922

LOVELL, RICHARD THOMAS, fisheries scientist and educator; b. Lockesburg, Ark., Feb. 21, 1934; s. Thomas Nesbitt and Henry Etta (Campbell) L.; m. Ganata Nettles, Dec. 21, 1963; children: Thomas Alan, Richard Graves. B.S., Okla. State U.-Stillwater, 1956, M.S., 1957; Ph.D., La. State U., 1963. Asst. prof., assoc. prof. La. State U., Baton Rouge, 1963-69; assoc. prof. to prof. fisheries dept. Auburn U., Ala., 1969—. Author: Manual for Fish Nutrition Studies and Fish Feed Analysis, 1975, 81; assoc. editor: Transactions Am. Fisheries Soc, 1978-80; columnist: Aquaculture, 1975—. Chmn. dist. com. Boy Scouts Am., Opelika-Auburn, Ala., 1981-82, scoutmaster, Stillwater, Okla., 1958-60; chmn. council on ministries First United Methodist Ch., Opelika, Ala., 1975-80; vice-chmn. ofcl. bd., 1982—. Named Profl. Scientist So. Assn. Agrl. Scientists, 1975. Fellow Am. Inst. Chemists; mem. Am. Fisheries Soc., Am. Inst. Nutrition, Catfish Farmers Am. (chmn. research 1976-81, disting. service award 1979, 80), Alpha Zeta, Phi Kappa Phi. Methodist. Subspecialties: Animal nutrition; Food science and technology. Current work: Research feeding and nutrition of fish, off-flavors in fish. Home: 622 Terracewood Dr Opelika AL 36801 Office: Fisheries Dept Auburn U Auburn AL 36830

LOVEN, ANDREW WITHERSPOON, engring. co. exec.; b. Crossnore, N.C., Jan. 31, 1935; s. Andrew W. and Annie Laura (Crowell) L.; m. Elizabeth Joann De Groot, June 20, 1959; children: Laura Elizabeth, James Edward. B.S., Maryville (Tenn.) Coll., 1957; Ph.D. (NSF fellow), U. N.C., Chapel Hill, 1962. Diplomate: Am. Acad. Environ. Engrs.; P.E., Va., Ga., S.C., N.C., Md. Fla. Research assoc. U. N.C., 1962-63; sr. research chemist Westvaco Corp., Charleston, S.C., 1963-66, Covington, Va., 1963-66, mgr. carbon devel., 1966-71; mgr. Westvaco Cons. Service, 1968-71; project mgr. and mgr. product devel. Engring.-Sci., Inc., Washington and Atlanta, 1971-74, v.p., regional mgr., 1974-80, sr. v.p. engring., 1980-81, group v.p. and dir., 1981—. Mem. Am. Inst. Chem. Engrs., Water Pollution Control Fedn., Am. Pub. Works Assn., ASTM. Subspecialties: Environmental engineering; Chemical engineering. Current work: All phases of environmental science and engineering. Patentee and publs. in environ sci. and engring. fields. Home: 1941 Huntington Hall Ct Atlanta GA 30338 Office: Engring Sci Inc 57 Executive Park S Suite 590 Atlanta GA 30329

LOVETT, PAUL SCOTT, biol scientist, educator, researcher, cons.; b. Phila., Dec. 14, 1940; s. Paul Joseph and Kathleen Jean (Mulhern) L.; m. Patricia Elene, Jan. 5, 1982; children: Beth A., Mark S. B.S. in Biology, Delaware Valley Coll., 1964; Ph.D. in Microbiology, Sch. Medicine, Temple U., 1968. Postdoctoral fellow Scripps Clinic and Research Found., La Jolla, Calif., 1968-70; asst. prof. dept. biol. scis. U. Md. Baltimore County, Catonsville, 1970-74, assoc. prof., 1974-78, prof. biol. scis., 1978—; cons., researcher in field. Mem. Am. Soc. Microbiology. Subspecialties: Genetics and genetic engineering (biology); Molecular biology. Current work: Gene cloning and expression in Bacillus subtilis. Office: 5401 Wilkens Ave Catonsville MD 21228

LOW, BOON-CHYE, physicist; b. Singapore, Feb. 13, 1946; came to U.S., 1968; s. Kuei-Huat and Ah-Tow (Tee) Lau; m. Daphne Nai-Ling Yip, Mar. 31, 1971; 1 son, Yi-Kai. B.Sc., U. London, 1968; M.S., U. Chgo., 1969, Ph.D., 1972. Research assoc. U. Chgo., 1972-73; research fellow High Altitude Obs., Boulder, Colo., 1973-74; scientist, 1981—; engaged in shipbldg. bus., Singapore, 1975-78; vis. scientist U. Tokyo, U. Chgo., 1978-79; Nat. Acad. Sci. sr. research assoc. NASA Marshall Space Flight Ctr., Huntsville, Ala., 1980-81. Contbr. articles to profl. jours. Japan Soc. for Promotion of Sci. fellow, 1978-79. Mem. Am. Phys. Soc., Am. Astron. Soc. Subspecialties: Theoretical astrophysics; Solar physics. Current work: Astrophysical hydromagnetics and plasma physics, theoretical research. Office: High Altitude Obs NCAR PO Box 3000 Boulder CO 80307

LOW, FRANCIS EUGENE, physics educator; b. N.Y.C., Oct. 27, 1921; s. Bela and Eugenia (Ingerman) L.; m. Natalie Sadigur, June 25, 1948; children—Julie, Peter, Margaret. B.S., Harvard, 1942; M.A., Columbia, 1947, Ph.D., 1949. Mem. Inst. Advanced Study, 1950-52; asst. prof. U. Ill., Urbana, 1952-55, asso. prof., 1955-56; prof. physics MIT, 1957-67, Karl Taylor Compton prof., 1968—, dir. Center for Theoretical Physics, 1973-76, dir. Lab. for Nuclear Scis., 1979-80, provost, 1980—; cons. in field. Contbr. articles to profl. jours. Served with USAAF, 1942-43; Served with AUS, 1944-46. Mem. Nat. Acad. Scis., Phys. Soc. (chmn. div. particles and fields 1974, councillor-at-large 1979—), Fedn. Am. Scientists (mem. nat. council 1973-77), ACLU, Am. Acad. Arts and Scis., Union Concerned Scientists (chmn. 1969). Subspecialty: Theoretical physics. Home: 28 Adams St Belmont MA 02178 Office: Office of Provost Room 3-208 Mass Inst Tech Cambridge MA 02139

LOW, FRANK JAMES, physicist, educator; b. Mobile, Nov. 23, 1933; s. Albert S. and Flora (Woodruff) L.; m. Edith Estella Morgan, Sep. 8, 1956; children—Valerie Ann, Beverly Ellen, Eric David. B.S., Yale, 1955; M.A., Rice U., 1957, Ph.D., 1959. Mem. tech. staff Tex. Instruments, Inc., Dallas, 1959-62; asso. scientist Nat. Radio Astronomy Obs., Green Bank, W.Va., 1962-65; research prof. U. Ariz., Tucson, 1965—; prof. space sci. Rice U., Houston, 1966-71, adj. prof., 1971-79; pres. Infrared Labs., Inc., Tucson, 1967—. Bd. dirs. Tucson East YMCA-YWCA, 1973-75. Recipient H.A. Wilson award Rice U., 1959, Spl. founders prize Tex. Instruments Found., 1976. Mem. Am. Astron. Soc. (Helen B. Warner prize 1968), Am. Phys. Soc., Nat. Acad. Scis., Sigma Xi. Subspecialty: Infrared optical astronomy. Current work: Infrared astronomy; space astronomy, infrared instrumentation. Home: 4940 Calle Barril Tucson AZ 85718 Office: Univ Ariz Tucson AZ 85721

LOW, GEORGE MICHAEL, college president; b. Vienna, Austria, June 10, 1926; s. Arthur and Gertrude (Burger) L.; m. Mary R. McNamara, Sept. 3, 1949; children: Mark S., Diane E., G. David, John M., Nancy A. B.S. in Aero. Engring, Rensselaer Poly. Inst., 1948, M.S., 1950, Eng.D. (hon.), 1969; Sc.D., U. Fla., 1969; Eng.D. (hon.), Lehigh U., 1979, LL.D., Hartwick Coll., 1981; L.H.D., Villanova U., 1982. With NASA (and predecessor), 1949—, dep. assoc. adminstr. manned space flight, 1963-64, dep. dir. Manned Spacecraft Center, Houston, 1964-67, mgr. Apollo Spacecraft program, 1967-69, dep. adminstr. Apollo Spacecraft program, Washington, 1969-76; pres. Rensselaer Poly. Inst., Troy, N.Y., 1976—; dir. Gen. Electric Co. Trustee Com. Econ. Devel., Hartford Grad. Center. Recipient Outstanding Leadership medal NASA, 1962, 2 Distinguished Service medals, 1969; Space Flight award Am. Astronautical Soc., 1968; Arthur S. Flemming award U.S. Jr. C. of C., 1963; Paul T. Johns trophy Arnold Air Soc., 1969; Astronautics Engr. award Nat. Space Club, 1970; Robert H. Goddard Meml. trophy, 1973; Nat. Civil Service League Career Service award, 1973; Rockefeller Pub. Service award for adminstrn., 1974; Cross of Honor Sci. and Art 1st class, Austria, 1980; NASA Disting. Service medal, 1981; Citizen Laureate award Univ. Found. Albany, 1983; nat. engring. award Am. Assn. Engring. Sci., 1983. Fellow AIAA (hon. fellow, Louis W. Hill Space Transp. award 1969); mem. Assn. Colls. and Univs. State of N.Y. (v.p.), Nat. Acad. Engring. (Founders award 1978, Fellow, Founders award 1978, mem., council, chmn. com. on sci. engring. and pub. policy). Subspecialty: Aerospace engineering and technology. Office: Office of Pres Rensselaer Poly Inst Troy NY 12181

LOW, KENNETH BROOKS, JR., molecular geneticist; b. New Rochelle, N.Y., Jan. 19, 1936; s. Kenneth Brooks and Elizabeth (Hammond) L.; m. Elise Langworthy, Aug. 21, 1960; children: Kennan, David. A.B., Amherst Coll., 1958; M.S., U. Pa., 1960, Ph.D., 1965. Postdoctoral fellow N.Y.U. Med. Center, 1966-68; mem. faculty Yale U. Sch. Medicine, 1968—, asst. prof., 1968-73, assoc. prof., 1973-78, sr. scientist, 1978—. Contbr. articles to profl. jours.; Mem. editorial bd.: Jour. of Bacteriology, 1973-78. Mem. Am. Soc. Microbiology. Subspecialties: Gene actions; Genetics and genetic engineering (medicine). Current work: Mechanisms of genetic recombination, DNA repair and gene transfer in microorganisms. Office: Radiobiol Labs Yale U 333 Cedar St New Haven CT 06510

LOW, PHILIP STEWART, chemistry educator; b. Ames, Iowa, Aug. 8, 1947; s. Philip Funk and Mayda Matilda (Stewart) L.; m. Joan B. Foord, Dec. 19, 1969; children: Philip, Tara, Emily, Justin. B.S. cum laude, Brigham Young U., 1971; Ph.D., U. Calif.-San Diego, 1975. Postdoctoral fellow U. Mass., Amherst, 1975-76; asst. prof. Purdue U., West Lafayette, Ind., 1976-82, assoc. prof. chemistry, 1982—; cons. Monsanto Co., St. Louis, 1983—. Contbr. articles to profl. jours. NIH grantee, 1977-83. Mem. Am. Chem. Soc., Am. Soc. Biol. Chemists, Red Cell Club, Sigma Xi. Mormon. Subspecialties: Biochemistry (biology); Membrane biology. Current work: Biological membrane structure probed by differential scanning calorimetry, fluorescence and NMR spectroscopy; currently investigating red blood cell and chloroplast membranes. Office: Dept Chemistr Purdue U West Lafayette IN 47907

LOW, TERESA LINGCHUN KAO, biochemistry researcher, educator; b. Hankow, Hupei, China, Feb. 17, 1941; came to U.S., 1964, naturalized, 1971; d. Ling-mei and Hang-seng (Fan) Kao; m. Chow-Eng Low, June 26, 1966; children: Huan-Cheh, Cecilia, Jasmine. B.S., Tunghai U., Taiwan, 1962; M.S., Tex. Woman's U., 1966; Ph.D., U. Tex., Austin, 1970. Sci. research specialist La. State U., Baton Rouge, 1970-71; research assoc. Ind. U., Bloomington, 1972-75; postdoctoral fellow U. Tex. Med. Br., Galveston, 1976-77, instr., 1977-78; asst. prof. George Washington U., Washington, 1978-81, assoc. prof. biochemistry, 1981—. Patentee in field. NIH research grantee, 1982-84. Mem. Am. Center Chinese Med. Scis. (sec. 1980-82), Chinese Med. and Health Assn., N.Y. Acad. Scis., Sigma Xi. Subspecialty: Immunobiology and immunology. Current work: The chemical and biological characterization of thymosins, a family of hormones derived from the thymus gland and demonstrated to have potent immunopotentiating properties. Home: 9105 Bramble Pl Annandale VA 22003 Office: Dept Biochemistry George Washington U Med Center 2300 I St NW Washington DC 20037

LOW, WALTER CHENEY, neurophysiologist, researcher; b. Madera, Calif., May 11, 1950; s. George Chen and Linda Quan (Gong) L. B.S., U. Calif., Santa Barbara, 1972; M.S., U. Mich., 1974, Ph.D. (NIH fellow), 1979. NSF fellow Cambridge U., U.K., 1979-80; AGAN fellow U. Vt., Burlington, 1980—. Contbr. articles to profl. jours. Recipient Nat. Research Service award Nat. Inst. Neurol. and Communicative Disorders and Stroke, 1979, Nat. Heart, Lung and Blood Inst., 1981—. Mem. AAAS, Soc. Neurosci., N.Y. Acad. Scis., Sigma Xi, Eta Kappa Nu. Subspecialties: Neurophysiology; Neural transplantation. Current work: Research in structure and function of central nervous system, and use of neural transplants in recovery of function. Office: Dept Physiology and Biophysics U Vt Burlington VT 05405

LOWE, IRENE POSNER, biologist; b. Bklyn, Oct. 23, 1926; d. Leo Lazar and Helen Rose (Brootzkoos) Posner; m. Irving Jack Lowe, June 14, 1953; children: Marc Leo, Margo Lucy. B.S., George Washington U., 1945, M.S., 1949; Ph.D., Washington U., 1954. Faculty psychiatry dept. Washington U. Med. Sch., St. Louis, 1953-58; faculty anatomy dept. U. Minn. Med. Sch., Mpls., 1959-62; Nat. Research Service Award research assoc. dept. pharmacology U. Pitts. Sch. Medicine, 1979—. Woods Hole Predoctoral fellow, 1950. Mem. AAAS, Soc. Neuroscis., Soc. Biol. Chemists, Sigma Xi. Subspecialties: Developmental biology; Immunocytochemistry. Current work: Developmental biology of neuropeptides in spinal cord relating to spinal injury. Office: Dept Pharmacology U Pitts Med Sch Pittsburgh PA 15261

LOWE, JAMES EDWARD, cardiovascular surgeon and researcher; b. Brunswick, Ga., Dec. 27, 1946; s. Gordon E. and Elsie Muriel (Foster) L.; m. Lorraine Elizabeth Sassone, Sept. 7, 1969; children: Summer Elizabeth, Natalie Susanne. B.A., Stanford U., 1969; M.D., UCLA, 1973. Diplomate: Am. Bd. Surgery. Resident in surgery Duke U., Durham, N.C., 1973-81, NIH surg. scholar, 1975-81, research fellow in pathology, 1975-77, asst. prof. surgery and pathology, 1981—; asst. dir. cardiovascular surgery Durham VA Med. Center, 1981—; established investigator Am. Heart Assn., 1981—. Mem. Alpha Omega Alpha. Subspecialty: Cardiac surgery. Current work: Pathogenesis of reversible and irreversible ischemic myocardial injury (metabolic, ultrastructural and electrophysiological alterations). Home: 2712 Circle Dr Durham NC 27705 Office: Dept Surger Duke Univ Med Center Erwin Rd Durham NC 27710

LOWE, NICHOLAS JAMES, med. educator, dermatologist; b. Wolverhampton, Eng., July 10, 1944; came to U. S., 1975; s. Betty and Chitty L.; m. Pamela Susan Gee, Aug. 10, 1968; children: Nichola, Phillipa. M.B., Ch.B., Med. Sch., Liverpool U., Eng., 1968, M.D., 1977. Diplomate: Am. Bd. Dermatology, 1980. Surg. intern David Lewis Hosp., Liverpool, 1968-69; med. intern Walton Hosp., Liverpool, 1969; dep. indsl. health med. officer H.M. Dockyard, Chatham, Eng., 1969-70; resident in internal medicine Royal Navy Hosp., Haslar, Hampshire, Eng., 1970-73, specialist in medicine, 1973; med. officer H.M. ships Fox and Fawn, W.I., 1973; mem. staff dermatology dept. U. Southampton, Eng., 1973; sr. registrar dept. dermatology Liverpool Royal Infirmary, 1974-77; tutor U. Liverpool, 1974-77; research fellow in dermatology Scripps Clinic and Research Found., La Jolla, Calif.; instr. dermatology U. Calif.-San Diego, 1975-76; asst. prof. dermatology U. Wis.-Madison, 1977; asst. prof. medicine/dermatology, dir. dermatology clinics UCLA Sch. Medicine, 1978-79, dir. phototherapy and psoriasis treatment center, 1978-79, asst. prof., 1979-80, assoc. prof., 1980—, acting dir. combined, 1981; Cons. Westwood Pharms., Johnson and Johnson, Owen Pharms., 1978-82, Neutrogena Dermatologics, Nelson Research and Devel., Allergan-Herbert Pharms., Hoffman La Roche Pharms., 1980—, Centre de Internationale Rechecherches Dermatologique, Vallbone, France, 1981-. Contbr. numerous articles on dermatology to profl. jours.; co-editor: (with Howard Maibach) Animal Models in Dermatology: Dermatopharmacology and Dermatotoxicology, 1983. Served as surgeon Royal Navy, 1969-73; Eng. Named Outstanding Clin. Tchr. of Yr. UCLA (div. dermatology), 1979; NIH grantee, 1979-83; Nat. Cancer Inst. grantee, 1981-85. Fellow ACP, Am. Acad. Dermatology (Bronze award 1977, Gold award 1979, Bronze award 1982); mem. Brit. Med. Assn., Royal Coll. Physicians, Brit. Assn. Dermatology (investigative group), Royal Soc. Medicine (dermatology), Wis. Dermatol. Soc., Am. Fedn. Clin. Research, Soc. Investigative Dermatology, Dermatology Found., Los Angeles Dermatol. Assn., Am. Soc. Photobiology, Western Soc. Clin. Research, Am. Assn. Cancer Research, Pacific Dermatol. Assn., Los Angeles County Med. Assn., Nat. Psoriasis Found. (med. adv. bd.). Subspecialties: Dermatology; Oncology. Current work: Researcher ultraviolet light and epidermal polyamine and DNA synthesis; ultra violet carcinogenesis; bioassay of psoralens; psoriasis therapy. Office: UCLA Sch Medicine Div Dermatology 52-121 CHS Los Angeles CA 90024 Home: 531 12th St Santa Monica CA 90402

LOWE, PHILLIP A(RNOLD), oil company executive, energy consultant; b. Salt Lake City, Feb. 25, 1939; s. Ellis Edward and Elinor (Havus) L.; m.; 1 dau., Susan. B.S., U. Utah, 1961; M.S., U. R.I., 1964; Ph.D., Carnegie-Mellon U., 1968. Registered profl. engr., Pa. Sr. engr. Westinghouse Electric Co., Pitts., 1964-70; mgr. prodn. devel. Combustion Engring., Windsor, Conn., 1970-74; dep. asst. ins. gen. Dept. Energy, Washington, 1974-80; dir. advanced tech. NUS Corp., Gaithersburg, Md., 1980—; chief exec. officer B & P Energy Inc., Potomac, Md., 1982—, also dir. Editor: Jour. Fluid Mechanics, 1964-66. Served to lt. USN, 1962-64. Recipient Superior award Dept. Energy, 1979. Mem. ASME (dir. chpt. 1972-74), Am. Petroleum Engrs., Am. Nuclear Soc. Subspecialties: Fuels; Mechanical engineering. Current work: Consulting on use of coal and synthetic fuels; production of oil and gas. Home: 11316 Rouen Dr Potomac MD 20854 Office: B & P Energy Inc 11316 Rouen Dr Potomac MD 20854

LOWE, SUNNY KEN, plant pathologist; b. 1930. B.S., U. Hawaii, 1955; M.S., U. Calif.-Davis, 1962. Research assoc. plant pathology U. Calif.-Davis, 1957—. Contbr. articles to profl. jours. Mem. Am. Phytopath. Soc. Subspecialty: Plant pathology. Current work: Stone fruit virus and mycoplasma diseases, thermal and chemo-therapy. Office: Plant Pathology Dept U Calif Davis CA 95616

LOWE, TIMOTHY JOE, management educator; b. Marshalltown, Iowa, May 5, 1942; s. Raymond C. and Barbara (Kulisek) L.; m. Marsha Schupbach, Feb. 26, 1967; children: Marc W., Carrie A. B.S. in Indsl. Engring., Iowa State U., 1965, M.Engring., 1967; Ph.D., Northwestern U., 1973. Process engr. Enjay Chem. Co., Baton Rouge, 1965-67, 69-70; asst. prof. U. Fla., Gainesville, 1973-77, assoc. prof., 1977-78; assoc. prof. mgmt. Purdue U., West Lafayette, Ind., 1978-83, prof., 1983—, dir. research and Ph.D. programs Sch. Mgmt., 1982—. Assoc. editor: Jour. Ops. Research, 1980-82; departmental editor: Jour. Inst. Indsl. Engrs., 1980—. Served to 1st lt. U.S. Army, 1967-69. Mem. Inst. Mgmt. Scis., Ops. Research Soc. Am., Inst. Indsl. Engrs., Math. Programming Soc. Democrat. Presbyterian. Subspecialties: Operations research (engineering); Industrial engineering. Current work: Mathematical programming; optimization; facility design and analysis. Home: 3000 Wilshire Ave West Lafayette IN 47906 Office: Krannert Grad Sch Mgmt Purdue U West Lafayette IN 47907

LOWENGRUB, MORTON, educator; b. Newark, Mar. 31, 1935; s. Samuel and Anna (Grossman) L.; m. Carol Wendy Sterns, Aug. 20, 1961; children—John, Wendy, Paul. B.A., N.Y. U., 1956; M.S., Calif. Inst. Tech., 1958; Ph.D., Duke U., 1961. Asst. prof. N.C. State U., Raleigh, 1961-62, Wesleyan U., Middletown, Conn., 1963-67; asso. prof. math. Ind. U., Bloomington, 1967-71, prof., 1971—, chmn. dept., 1977-80. Author: Crack Problems in the Classical Theory of Elasticity, 1969, Topics in Calculus, 2d edits, 1970, 75. Leverhulme fellow, Gt. Britain, 1962-63; NSF postdoctoral fellow, 1966-67; Sci. Research Council fellow, Gt. Britain, 1973-74. Mem. Am. Math. Soc., Soc. Indsl. and Applied Math., Math. Assn. Am. Subspecialty: Applied mathematics. Current work: Mixed boundary value problems arising in mathematical theory of elasticity and applications to fracture mechanic. Office: Dept Math Ind U Bloomington IN 47401

LOWENSOHN, HOWARD STANLEY, research physiologist, educator; b. Columbus, Ohio, Jan. 23, 1931; s. Howard Stanley and Grace Marion (Jones) L.; m. Martha Jane Mitchell, Oct. 3, 1953; 1 son, Thomas Andrew. B.S., Franklin and Marshall Coll., 1956; M.S., U. So. Calif., 1962; Ph.D., U. Md., 1972. Chemist Brewer & Co., Worcester, Mass., 1956-59; lab. technician U. So. Calif., Los Angeles, 1959-61; lab. technologist R.I. Hosp., Providence, 1962-63; research physiologist Walter Reed Army Inst. Research, Washington, 1963—; cons. Johns Hopkins U., Balt., 1977—; research advisor NRC, Washington, 1979—; adj. assoc. prof. Uniformed Services U. Health Scis., Bethesda, Md., 1980—. Treas. Cub Scouts, Silver Spring, Md., 1971; curriculum advisor Woodlin Elem. Sch., Silver Spring, 1973-74; trustee Grace Ch. Day Sch., Silver Spring, 1980-81. Served with USN, 1951-55. Recipient Letter Commendation Walter Reed Army Med. Ctr., 1965, 76, 77. Mem. Am. Physiol. Soc., Am. Heart Soc., AAAS, N.Y. Acad. Scis., Phi Sigma. Episcopalian. Subspecialties: Physiology (medicine); Pharmacology. Current work: Effects of drugs upon myocardial function and coronary blood flow at rest and during exercise, coronary circulation and myocardial hyperfunction, bile dynamics and drugs. Inventor soluble phenobarbital tablet for injection, 1957. Office: Dept Pharmacolog Walter Reed Army Inst Research Walter Reed Army Med Ctr Washington DC 20307

LOWENSTAM, HEINZ A., paleoecologist; b. Siemianowitz, Germany, Oct. 9, 1912; s. Curt and Frieda Berta (Sternberg) L.; m.; children—Ruth L. Golstein, Michael D., Steven A. Student, U. Frankfurt, 1932, U. Munich, 1933-37; Dr. h.c. (hon.), U. Munich, 1981; Ph.D., U. Chgo., 1939. Curator of paleontology Ill. State Mus., Springfield, 1939-42; asso. geologist Ill. State Geol. Survey, Urbana, 1942-48; research asso. in biochemistry, U. Chgo., 1948-50, asso. prof. dept. geology, 1950-52; prof. paleocology dept. geology and planetary scis. Calif. Inst. Tech., Pasadena, 1952—. Mem. editorial bd.: Jour. Sedimentary Petrology, 1956-76, Jour. Paleogeography, Palaeo Climatology and Palaeo Ecology, 1965—. Research grantee Shell Oil Co., Office of Naval Research, NSF. Fellow Am. Acad. Arts and Scis., AAAS, Geol. Sci. Am.; mem. Nat. Acad. Scis., Paleontol. Soc., Soc. Study of Evolution, Am. Assn. Petroleum Geologists. Jewish. Subspecialty: Paleoecology. Office: 1201 E California Blvd Pasadena CA 91125

LOWERY, CAROL ROTTER, psychology educator, psychotherapist, family therapist; b. St. Louis, Nov. 3, 1948; d. Milton J. and Mary L. (Cantrell) L.; m. Wayne Echols, Sept. 4, 1970 (div. 1972); m. John Rotter, Aug. 14, 1982; stepchildren: Tanya, Jeremy. B.A., St. Louis U., 1970; M.A., U. Kans., 1975, Ph.D., 1977. Lic. clin. psychology, Ky. Coordinator children's services Bluegrass Comprehensive Care Ctr., Frankfort, Ky., 1977-78; asst. prof. psychology U. Ky., 1978—; cons. Westinghouse, Atlanta. Mem. Lexington Children's Services Com., 1979—. NSF grantee, 1980. Mem. Am. Psychol. Assn., Am. Psychology Law Soc. Subspecialty: Family therapy. Current work: Effects of divorce on family members; post-divorce adjustment. Home: 1712 Liberty Rd Lexington KY 40505 Office: Department Psychology 0044 University of Kentucky Lexington KY 40506

LOWN, BERNARD, cardiologist, educator; b. Utena, Lithuania, June 7, 1921; s. Nisson and Bella L.; m. Louise C., Dec. 29, 1946; children: Anne, Frederick, Naomi. B.S. summa cum laude, U. Maine, 1942; M.D., Johns Hopkins U., 1945. Asst. in pathology Yale U. Med. Sch., New Haven (Conn.) Hosp., 1945-46; intern Jewish Hosp., N.Y.C., 1947-48; asst. resident in medicine Montefiore Hosp., N.Y.C., 1948-50; research fellow cardiology Peter Bent Brigham Hosp., Boston, 1950-53, asst. in medicine, 1955-56; dir. Levine Cardiovascular Research Lab., 1956-58, jr. assoc., 1956-62, research assoc., 1958-59, assoc., 1962-63, sr. assoc., 1963-70; dir. Levine Coronary Care Unit, 1965-74, physician, 1970—; asst. in medicine Harvard U. Med. Sch., Boston, 1955-58; asst. prof. medicine dept. nutrition Harvard Sch. Pub. Health, Boston, 1961-67, assoc. prof. cardiology, 1967-74, prof., 1974—, dir. cardiovascular research lab. dept. nutrition, 1961—; coordinator U.S.-USSR coop. study, sudden death, 1973-81; cons. in cardiology Beth Israel Hosp., Boston, 1963—, Newton-Wellesley Hosp., 1963—, Children's Hosp., 1971—. Co-author: (with Samuel A. Levine) Current Advances in Digitalis Therapy, 1954, (with Harold D. Levine) Atrial Arrhythmias, Digitalis and Potassium, 1958; contbr. chpts. to books, lectrs. to profl. confs. and articles to profl. jours.; mem. editorial bds. of numerous profl. jours. Fellow Am. Coll. Cardiology; mem. Am. Soc. Clin. Investigation, Am. Heart Assn., Assn. Am. Physicians, AAAS, Am. Fedn. Clin. Research, British Cardiac Soc., Cardiac Soc. Australia and New Zealand, Swiss Soc. Cardiology, Belgian Royal Acad. Medicine, Physicians for Social Responsibility (founder, first pres. 1960-70), U.S.-China Physicians Friendship Assn. (pres. 1974-78), NIH, Internat. Physicians for Prevention of Nuclear War (pres. 1980—), Am. Council Sci. and Health, Phi Beta Kappa, Alpha Omega Alpha. Subspecialty: Cardiology. Current work: Sudden cardiac death, cardiac arrhythmias, higher nervous activity. Office: Harvard Sch Pub Health 665 Huntington Ave Boston MA 02115

LOWN, JAMES WILLIAM, chemistry educator, researcher, consultant; b. Blyth, Eng., Dec. 19, 1934; emigrated to Can., 1961, naturalized, 1969; s. James William and Dinah (Pears) L.; m. Elizabeth M. Beatty, Aug. 18, 1962; children: Andrew James, Peter William. B.Sc., Imperial Coll., London, 1956; Ph.d., D.I.C., 1960. Research assoc. Walter Reed Research Inst., Washington, 1962-63; asst. prof. chemistry U. Alta. (Can.), Edmonton, 1964-68, assoc. prof., 1968-74, prof., 1974—, Alta. Heritage Found. for Med. Research prof.-in-residence, 1982—; regional dir. Nat. Found. Cancer Research; numerous invited lectures, U.S., France, Germany, Japan, Italy, Eng., Argentina. Research numerous publs. in field. Mem. Am. Chem. Soc., Royal Soc. Chemistry, Internat. Soc. Heterocyclic Chemistry, Sigma Xi. Anglican. Subspecialties: Organic chemistry; Cancer research (medicine). Current work: Primary research interest is study of molecular mechanism of action of clinically useful anticancer agents.

LOWNDES, HERBERT EDWARD, pharmacology educator, consultant; b. Barrie, Ont., Can., July 12, 1943; s. Herbert and Amy B. (Eddy) L.; m. Helen Marion Hasselfield, July 16, 1966; children: Gregory William, Eleanor Jean, Paul Edward. B.A., U. Sask., 1964, cert., 1968, M.Sc., 1970; Ph.D., Cornell U., 1972. Postdoctoral fellow U. Western Ont., 1972-73; asst. prof. Coll. Medicine and Dentistry N.J., Newark, 1973-77, assoc. prof., 1977-81, prof. pharmacology, 1981—; cons. neurotoxicology to corps. and fed. agys. Served to lt. Royal Can. Navy, 1961-66. Grantee USPHS, N.J. Med. Sch., Pharm. Mfrs. Assn., ALS Soc. Am., others. Mem. Soc. Toxicology, Sci. Neurosci., Am. Soc. Pharmacology and Exptl. Therapeutics, N.Y. Acad. Sci., Pharm. Soc. Can., Am. Assn. Neuropathologists, AAAS, others. Subspecialties: Neurobiology; Toxicology (medicine). Current work: Effects of toxic chemicals on the nervous system. Neurotoxicology employing neurophysiological, neuromorphological and neurochemical techniques. Home: 18 Fairview Dr East Hanover NJ 07936 Office: 100 Bergen St Newark NJ 07103

LOWRIGHT, RICHARD HENRY, geology educator, consultant on ground water and petroleum; b. Bethlehem, Pa., Aug. 31, 1940; s. Raymond Charles and Marion Helen (Shaeffer) L.; m. Margaret Mary Brockbank, June 26, 1966. A.A., Hershey Jr. Coll., 1960; A.B., Franklin and Marshall Coll., 1962; Ph.D., Pa. State U., 1971. Tchr. Horace Greeley High Sch., Chappaqua, N.Y., 1964-66; mem. faculty Susquehanna U., Selinsgrove, Pa., 1971—, assoc. prof. geology, 1978—; cons. Fletcher-Lowright & Assocs., Selinsgrove, 1979—. Contbr. articles to profl. jours. Mem. Soc. Econ. Paleontologists and Mineralogists, Yellowstone-Bighorn Research Assn. Subspecialty: Sedimentology. Current work: Hydraulic equivalence studies as indication of depositional processes in modern and ancient sands. Home: 106 Maple St Selinsgrove PA 17870 Office: Dept Geology Susquehanna U Selinsgrove PA 17870

LOWRY, WALLACE DEAN, geology educator; b. Medford, Oreg., Oct. 5, 1917; s. Bert Bebe and Neva May (Britten) L.; m. Marian Dorothea Wyckoff, Nov. 20, 1942; 1 son, Robert Edward. B.S., Oreg. State U., 1939, M.S., 1940; Ph.D., U. Rochester, 1943. Geologist Oreg. Dept. Geology and Mineral Industries, Portland, 1942-47, Texaco, Inc., Taft, Calif., 1947-49; prof. geology Va. Poly. Inst. and State U., Blacksburg, 1949-82, prof. emeritus, 1982—. Fellow Geol. Soc. Am.; mem. Am. Assn. Petroleum Geologists, Va. Acad. Sci. Club: Torch. Subspecialties: Geology; Tectonics. Current work: Arizona source of Poway Conglomerate (late Eocene) of San Diego area and the limited offset of the San Andreas Fault. Home: 607 Rose Ave SW Blacksburg VA 24060 Office: Dept Geol Sci Va Poly Inst and State U Blacksburg VA 24061

LOWY, DAVID CHARLES, health physicist; b. N.Y.C., Dec. 15, 1955; s. Bertram Alan and Laura (Friedman) L. B.S., Union Coll., 1977. Seismic observer Geosource, Denver, 1977; environ. physicist Impace Ltd., Denver, 1978-80, Camp Dresser & McKee Inc., 1980-81; health physicist Mobil Oil Corp., Denver, 1981—. Mem. AAAS, Health Physics Soc., Am. Nuclear Soc. Subspecialty: Health physics. Current work: Hydrological and environmental affairs. Office: Mobil Oil Corporation PO Box 5444 Denver CO 80217

LOWY, DOUGLAS RONALD, physician, researcher; b. N.Y.C., May 25, 1942; s. Milton M. and Frances (Siegel) L.; m. Lyndia, Dec. 15, 1968; children: Stephanie, Diane, Matthew. A.B., Amherst Coll., 1964; M.D., NYU, 1968. Diplomate: Am. Bd. Internal Medicine, Am. Bd. Dermatology. Intern in internal medicine Stanford U. Med. Center, 1968-69, resident in internal medicine, 1969-70; resident in dermatology Yale-New Haven Med. Center, 1973-75; commd. lt. comdr. USPHS, 1970, advanced through grades to med. dir., 1980; research assoc. Lab. Viral Diseases, Nat. Inst. Allergy and Infectious Diseases, Bethesda, Md., 1970-73; sr. investigator dermatology br. Nat. Cancer Inst., Bethesda, 1975-82, dep. chief dermatology br., 1982-83, chief lab. cellular oncology, 1983. Editorial bd.: Jour. Virology, 1980—, Archives of Virology, 1983—, Jour. Investigative Dermatology, 1981—. Mem. Am. Acad. Dermatology, Soc. Investigative Dermatology, Am. Soc. Microbiology, AAAS. Subspecialties: Virology (medicine); Genetics and genetic engineering (medicine). Current work: Retroviruses, papillomaviruses, oncogenes, dermatology. Office: Bldg 37 Rm 1B-26 Nat Cancer Inst Bethesda MD 20205

LOXLEY, THOMAS EDWARD, energy consultant, educator; b. Beaver, Pa., Jan. 20, 1940; s. Raymond Bernard and Stella Ann (Eisert) L.; m. Mary Cynthia Miller, Oct. 16, 1976 (div. 1979). B.S. in Engring. Sci, Case Inst. Tech., 1961. Cert. mech. engr., U.S. Civil Service Comm. Mech. engr. U.S. Naval Weapons Lab., Dahlgren, Va., 1961-65, U.S. Army Watervliet (N.Y.) Arsenal, 1965-68; system engr. Internat. Hydrodynamics Co., Vancouver, B.C., Can., 1968-69; pres. Manned Submersible System Co., Beaver, Pa., 1969-71; mech. engr. U.S. Naval Surface Weapons Ctr., Dahlgren, 1971-75; asst. prof. Va. Poly. Inst. and State U., 1975-78; pvt. practice energy consulting, Shepherdstown, W.Va., 1978—. Author: Progress in Solar Energy, 1982; editor: pub. newsletter Inverted Cave Constrn. System News, 1982—. Mem. Am. Solar Energy Soc., Marine Tech. Soc., ASHRAE. Democrat. Lodges: Kiwanis; Rotary. Subspecialties: Geothermal power; Solar energy. Current work: Applied systems research teaming ultra law grade solar and geothermal resources with heat pumps for super-efficient and cost effective space heating and cooling. Patentee electronic instrumentation system. Home and Office: PO Box 299 Steamboat Run Shepherdstown WV 25443

LOYALKA, SUDARSHAN K., nuclear engineering educator; b. Pilani, India, Apr. 11, 1943; came to U.S., 1964; s. Brijmohan and Draupadi Devi L.; m. Nirja Awasthi; children: Pranav, Prashant, Shashwat. B.E., U. Rajasthan, India, 1964; M.S., Stanford U., 1965, Ph.D., 1967. Registered profl. engr., Mo. Asst. prof. nuclear engring. U. Mo., Columbia, 1967-72, assoc. prof., 1972-76, prof., 1976—. Recipient Chancellor's award U. Mo., Columbia, 1982. Fellow Am. Phys. Soc.; mem. Am. Nuclear Soc., Am. Chem. Soc., Sigma Xi. Subspecialty: Nuclear engineering. Current work: Kinetic theory of gases; aerosol mechanics; neutron transport theory; nuclear reactor thermal hydraulics and safety. Office: U Mo-Columbia 0039 Engring Columbia MO 65210 Home: 2701 Raintree Ct Columbia MO 65201

LOZMAN, JEFFREY, orthopaedic surgeon; b. Bklyn., May 26, 1947; s. Teddy and Laura (Jabush) L.; m. Nancy Lamara, Jan. 31, 1977; children: Joshua, Rebecca. B.S., Fairleigh Dickinson U., 1969; M.D., Albany Med. Coll., 1972. Intern Albany Med. Ctr. Hosp., 1972-73, jr. asst. resident, 1972-76, chief resident, 1976-77, asst. instr., 1973-77, asst. clin. prof., 1979—, dir. psychomotor skills, 1982—; research fellow Shock trauma unit Albany Med. Coll., 1973-74; ASIF fellow, Bern, Switzerland, 1979-80, 80-81. Contbr. articles to profl. jours. NIH grantee, 1980—. Mem. Am. Thoracic Soc., Am. Inst. Clin. Research, N.Y. Soc. Medicine, Am. Acad. Orthopedic Surgeons. Subspecialty: Orthopedics. Current work: Primary bone healing and management of multiple traumas. Address: 1 Executive Park Dr Albany NY 12203

LU, ALLEN AN-HUA, hydrologist; b. Hankow, Hubei, China, Dec. 29, 1934; came to U.S., 1964, naturalized, 1973; s. Shu-Chih and Hsiu Wen (Chu) L.; m. Caroline An-Chun Lu/Li, June 18, 1972; children: Janice Hsin-Min, Irene Ya-Min. B.S., Chinese Naval Acad., 1958; M.S., Tsin-Hua U., 1964; Ph.D., Rensselaer Poly. Inst., 1969. Asst. prof. Siena Coll., Loudonville, N.Y., 1969-76; research scientist N.Y. State Dept. Health, Albany, 1976-78, sr. research scientist, 1978-79; staff scientist Rockwell Hanford Operation, Richland, Wash., 1979-81, staff hydrologist, 1981—. Mem. Am. Physics Soc., Am. Nuclear Soc., Sigma Pi Sigma. Subspecialties: Hydrology; Numerical analysis. Current work: Develop and apply computer models for predicting the gaseous and liquid radioactive waste transport and assessing the impact on the environment. Home: 2237 Carriage Ave Richland WA 99352 Office: Rockwell Hanford Operation PO Box 800 Richland WA 99352

LU, CHUN CHIAN, physicist; b. Taiwan, China, May 15, 1938; came to U.S., 1964, naturalized, 1975; s. Wen-Lai and Gue-Mei (Yeh) L.; m. Ann S.C. Liu, Dec. 14, 1968; 1 child, Ja. B.S., Nat. Taiwan U., 1962; Ph.D., U. Tenn., 1969. Postdoctoral fellow Oak Ridge Nat. Lab., 1969-71, Ctr. Theoretical Studies, Coral Gables, Fla., 1971-73; engr. Fla. Power & Light Co., Miami, 1973-78; sr. engr. Westinghouse Research/Devel. Ctr., Pitts., 1978-80; princ. engr. Racal Milgo Inc., Miami, 1980—; v.p. Sci. Programming, Inc. Contbr. articles to profl. jours. Mem. Am. Phys. Soc. Subspecialties: Atomic and molecular physics; Distributed systems and networks. Current work: High speed communication networks, digital sigmal processing; symmetry principles in matter, microprocessor applications. Office: 8600 NW 41 St Miami FL 33166

LU, JOHN KUEW-HSIUNG, ob-gyn educator, endocrinologist; b. Miaoli, Taiwan, Sept. 16, 1937; came to U.S., 1967, naturalized, 1980; s. En-Gie and Jan-Mei (Wu) L.; m. Marianne Mann Wang, Dec. 29, 1969; children: Judith Maria, John Lawrence. B.Sc., Nat. Taiwan Normal U., 1961; M.Sc., Nat. Taiwan U., 1967; Ph.D., Mich. State U., 1972. Postdoctoral fellow U. Pitts., 1972-74; research assoc. Mich. State U., East Lansing, 1974-75; asst. prof. U. Calif.-San Diego, La Jolla, 1975-77, UCLA, 1977-83, assoc. prof. ob-gyn and anatomy, 1982—. Author: Progress in Prolactin Physiology, 1978, Parkinson's Disease II, 1978, Dynamics of Ovarian Function, 1981, Neuroendocrinology of Aging, 1983. Recipient Disting. Research award Sigma Xi, 1970; Nat. Cancer Inst. research grantee, 1977; Nat. Inst. Aging research grantee, 1980. Mem. Soc. Gynecol. Investigation, Am. Physiol. Soc., Endocrine Soc., Soc. Study Reproduction, N.Y. Acad. Scis., Sigma Xi. Subspecialties: Neuroendocrinology; Reproductive biology. Current work: Developmental changes in ovarian and neuroendocrine function during reproductive aging, relationship between ovarian steroids and neuroendocrine regulation of gonadotropin and prolactin secretion. Home: 1129 Iliff St Pacific Palisades CA 90272 Office: Dept Ob-Gyn Sch Medicine UCLA 405 Hilgard Ave Los Angeles CA 90024

LU, KAU U., mathematics educator; b. Canton, China, July 10, 1939; came to U.S., 1963; s. Shuk-to and Shon (Haung) L.; m. Huey Mei Lee, Sept. 12, 1968; 1 dau., Pamela. B.S. E.E., Nat. Taiwan U., 1961; Ph.D. in Math, Calif. Inst. Tech., 1968. Asst. prof. Calif. State U.-Long Beach, 1968-75, assoc. prof., 1975-79, prof. math., 1979—; cons. Tridea Electronics, El Monte, Calif., 1969-70; research assoc. U. Calif.-Berkeley, 1981. Contbr. articles to publs. in field. Mem. Am. Math. Soc., Pacific Astronomy Soc., Soc. for Indsl. and Applied Math. Democrat. Subspecialty: Applied mathematics. Current work: Applied mathematics, mathematical analysis, astronomy, astrophysics, solar physics, analytic number theory. Office: Calif State Univ Dept Math Long Beach CA 90840

LU, KUO HWA, health science educator; b. Antung, Liaoning, China, Jan. 7, 1923; came to U.S., 1947, naturalized, 1960; s. Pei San and Ei Chen (Sun) L.; m. Catherine Collins, Oct. 13, 1943; children: Randolph Steven, Conway Kent, Nancy Lynn, Cliff Bertrand. B.S., Nat Central U., 1945; M.S., U. Minn., 1948, Ph.D., 1951. Postdoctoral fellow U. Minn., Mpls., 1952-56; asst. prof. Utah State U., Logan, 1956-58, assoc. prof., 1958-60, Oreg. Health Sci. U., Portland, 1960-63, prof., 1963-66, 67—; cons. Nuclear Medicine, AEC, UCLA, 1962-70, Tempo, Gen. Electric, 1977-78, Proctor & Gamble, 1976—, Tinitis Registry, 1979—. NIH fellow, 1966-67; NIDR fellow, 1975-78; Proctor & Gamble fellow, 1976—. Mem. Biometric Soc., Math. Assn. Am., Am. Statis. Assn., Internat. Assn. Dental Research, Sigma Xi. Subspecialties: Statistics; Mathematics applied to medicine. Current work: Clinical evaluation of theraputicals, computer simulation of disease processes. Home: 11780 SW Terra Linda Beaverton OR 97005 Office: Ore Health Sci Univ 611 SW Campus Dr Portland OR 97201

LU, PONZY, molecular biologist, educator; b. Shanghai, China, Oct. 7, 1942; came to U.S., 1949, naturalized, 1963; s. Abraham and Beth (Chou) L.; m. Heidi Fahl, Jan. 13, 1975; 1 dau., Kristina. B.S., Calif. Inst. Tech., 1964; Ph.D., M.I.T., 1970. Postdoctoral fellow Arthritis Found., Max Plauck Inst., Goettingen, W.G., U. Geneva, 1970-73; asst. prof. dept. chemistry U. Pa., Phila., 1973-78, assoc. prof., 1978-82, prof., 1982—; mem. study sect. NIH, 1982—. Recipient Career Devel. award NIH, 1977-82. Mem. Am. Soc. Biol. Chemists, Biophys. Soc., Sigma Xi. Subspecialties: Molecular biology; Biophysical chemistry. Current work: Gene regulation studied by NMR spectroscopy, use of modern spectroscopic and genetic methods to look at the molecules involved in gene regulation. Office: Dept Chemistry U Pa Philadelphia PA 19104

LUBAR, JOEL FREDRIC, psychologist, clinical psycholophysiologist, educator; b. Washington, Nov. 16, 1938; s. Raymond and Barbara Frances (Pollak) L.; m. Judith Ostrovsky, June 18, 1961; children: Sandra Gita, Edward Justin. B.S. (scholar), U. Chgo., 1960, Ph.D., 1963; Sr. postdoctoral fellow, UCLA, 1976. Asst. prof. psychology U. Rochester, N.Y., 1963-67; assoc. prof. U. Tenn., Knoxville, 1967-71, prof., 1971—; cons. VA; pvt. practice physiol. psychology and biofeedback Southeastern Biofeedback Inst.; Bd. dirs., mem. profl. adv. bd. Knox Area Epilepsy Found., 1978—; research dir. Chileda Inst. Ednl. Devel., LaCrosse, Wis., 1976—; profl. adv. bd. Epilepsy Found. Am., 1979-83. Author: Biological Foundations of Behavior, 1969, A Primer of Physiological Psychology, 1971, (with Isaacson, Schmaltz and Douglas) Study Guide to Accompany a Primer of Physiological Psychology, 1971, A First Reader in Physiological Psychology, 1972, Biological Foundations of Behavior, 1974, Brain and Behavior, 1975, (with W.M. Deering) Behavioral Approaches to Neurology, 1981; regional editor: Physiology and Behavior, 1970—; contbr. articles to profl. jours. NIMH grantee, 1964-75; NSF grantee, 1975-76. Fellow N.Y. Acad. Scis.; mem. Am. Psychol. Assn., AAAS, AAUP, Southeastern Psychol. Assn., Soc. Neurosci., European Brain Research Orgn., Biofeedback Soc. Am., Biofeedback Soc. Tenn. (pres. 1977-78), Sigma Xi. Subspecialties: Neuropsychology; Physiological psychology. Current work: Biofeedback research in regard to epilepsy, learning disorders and psychophysiol. disorders. Office: Dept Psychology U Tenn Knowville TN 37916

LUBAROFF, DAVID MARTIN, immunologist; b. Phila., Feb. 1, 1938; s. Albert and Mary (Kahn) L.; m. Martha Ida, June 25, 1961; children: Saul, Scott, Matthew. B.S., Phila. Coll. Pharmacy and Sci., 1961; M.S., Georgetown U., 1963; Ph.D., Yale U., 1967. Postdoctoral fellow dept med. genetics U. Pa., Phila., 1967-69, assoc. depts. pathology and med. genetics, 1969-70, asst. prof. depts. med. genetics and pathology, 1970-73; asst. prof. depts. urology and microbiology U. Iowa, Iowa City, 1973-77, assoc. prof., 1977-82, prof., 1982—; research immunologist VA Hosp., Iowa City, 1973—; mem. U. Iowa Cancer Ctr. Mem. editorial bd.: The Prostate; contbr. articles to profl. jours. Trustee Agudas Achim Synagogue, Iowa City, 1975—, pres., 1981—. Nat. Cancer Inst. grantee; VA grantee. Mem. AAAS, Am. Assn. Immunologists, Internat. Transplantation Soc., Internat. Soc. Preventive Oncology. Democrat. Subspecialties: Immunobiology and immunology; Cancer research (medicine). Current work: Studying subpopulations of lymphocytes in cellular immunologic responsed using membrane antigens to identify the populations; studying the role immune responses to tumor antigens play in regression of metastatic cancer. Office: Dept Urology U Iowa Iowa City IA 52242

LUBATTI, HENRY JOSEPH, physics educator; b. Oakland, Calif., Mar. 16, 1937; s. John and Pauline (Massimino) L.; m. Catherine Jean Berth Ledoux, June 29, 1968; children: Karen E., Henry J., Stephen J.C. M.S., U. Ill., 1963; A.A., U. Calif.-Berkeley, 1957, A.B., 1960, Ph.D., 1966. Research assoc. U. Paris, Orsay, France, 1966-68; asst. prof. physics M.I.T., Cambridge, 1968-69; mem. faculty U. Wash., Seattle, 1969—, prof. physics, 1974—, sci. dir., 1969—. Contbr. over 50 articles to sci. publs. Alfred P. Sloan fellow, 1971-75. Fellow Am. Phys. Soc.; mem. Sigma Xi, Tau Beta Pi. Subspecialty: Particle physics. Current work: Experimental: rare decays of K-mesons; deep inelastic lepton scattering (neutrinos and muons). Office: Physics Dept Univ Wash FM-15 Seattle WA 98195

LUBELL, MICHAEL STEPHEN, educator; b. N.Y.C., Mar. 25, 1943; s. Richard M. and Lillian (Arnoff) L.; m. Ellen Bloom, June 29, 1969; 1 dau., Karina Bloom. B.A., Columbia U., 1963; M.S., Yale U., 1965, Ph.D., 1969. Instr. physics Yale U., New Haven, 1971-72, asst. prof., 1972-77, assoc. prof., 1977-80; assoc. prof. physics CCNY, 1980-82, prof., 1983—; NRC steering com. for Army Basic Research, 1980—; sci./tech. adv. to Senator Chistopher J. Dodd, 1980—; mem. exec. com. Internat. Conf. Physics of Elec. and Atomic Collisions, 1983—. Contbr. articles to profl. jours. NSF fellow, 1965-67; AEC fellow, 1969-70; Sloan fellow, 1979—; NSF grantee, 1974—; Dept. of Energy grantee, 1980—; Office Naval Research grantee, 1983—. Mem. Am. Phys. Soc. (panel on pub. affairs 1982—), AAAS, N.Y. Acad. Sci. Democrat. Subspecialties: Atomic and molecular physics; Particle physics. Current work: Experimental studies of parity violation effects in neutral weak currents, spin-dependence in electron-hydrogen scattering, spin structure of nucleons, laser cooling of atomic beams. Office: Convent Ave at 138th St New York NY 10031 Home: 171 Bayberry Ln Westport CT 06880

LUBINIECKI, ANTHONY STANLEY, microbiologist; b. Greensburg, Pa., Oct. 4, 1946; s. Stanley Anthony and Helen Marie L.; m. Robin Lea Brudowsky, June 8, 1968; 1 son, Gregory. B.S., Carnegie-Mellon U., 1968; Sc.D. in Microbiology, U. Pitts., 1972. Research asst. U. Pitts., 1971-72, asst. research prof., 1972-74; prin. scientist Meloy Labs., Inc., Springfield, Va., 1974-78, mng. dir., 1979-80; tech. dir. biol. products Flow Labs., Inc., McLean, Va., 1980-82; mgr. cell culture ops. Genentech Inc., South San Francisco, 1982-83, dir. cell culture R&D, 1983—. Contbr. articles to profl. jours. Mem. Reston Community Assn., 1976—. NIAID/NIH grantee, 1973-74, 74-82; other grants. Mem. Am. Soc. Microbiology (chmn. indsl. microbiology com.), Am. Assn. Immunologists, N.Y. Acad. Scis., Soc. Gen. Microbiology, Tissue Culture Assn., Soc. Exptl. Biology and Medicine. Roman Catholic. Subspecialties: Genetics and genetic engineering (medicine); Cell biology (medicine). Current work: Development and production of human biological products for clinical trials; large-scale cell culture; cell culture studies of human cancer risk and genetics. Office: Genentech Inc 460 Point San Bruno Blvd South San Francisco CA 94080

LUBLIN, FRED DAVID, neurologist, researcher, educator; b. Phila., Sept 28, 1946; s. P. Paul and Sara (Raynes) L.; m. Barbara H. Swartz, June 11, 1969; children: Alex, Derek. A.B. magna cum laude, Temple U., 1968; M.D. summa cum laude, Jefferson Med. Coll., 1972. Diplomate: Am. Bd. Med. Examiners, Am. Bd. Psychiatry and Neurology. Intern Albert Einstein Med. Ctr., Bronx, N.Y., 1972-73; resident in neurology N.Y. Hosp.-Cornell Med. Ctr., N.Y.C., 1973-76; instr. neurology Cornell Med. Ctr., 1975-76, Thomas Jefferson U., Phila., 1976-78, asst. prof. neurology and biochemistry, 1978-82, assoc. prof. neurology and biochemistry, 1982—; attending neurologist Thomas Jefferson U. Hosp., 1976—; cons. neurologist Wilmington and Coatesville VA Hosps., 1978—; dir. Kensington Hosp., 1980—. Recipient Tchr.-Investigator award NIH, 1978; Nat. Multiple Sclerosis Soc. research grantee, 1981. Mem. Am. Assn. Neurology, Soc. Neurosci., N.Y. Acad. Scis., AAAS, Assn. Research in Nervous and Mental Diseases, Sigma Xi, Alpha Omega Alpha. Subspecialties: Neuroimmunology; Neurology. Current work: Neuroimmunologic research utilizing animal models of multiple sclerosis and other demyelinating disorders, studies of lymphocyte activity and immunomodulation in neuroimmunologic disorders. Office: Thomas Jefferson University Medical College 1025 Walnut St Philadelphia PA 19107

LUCANTONI, DAVID MICHAEL, operations research scientist; b. Balt., Aug. 31, 1954; s. Vincent Michael and Marie (Fitzpatrick) L. B.S. in Math, Towson State U., 1976; M.S. in Stats, U. Del., 1978; Ph.D. in Ops. Research, U. Del., 1981. Mem. tech. staff Bell Labs., Holmdel, N.J., 1981—. Author: An Algorithmic Analysis of A Communication Model, 1983; contbr. articles to sci. jours. Mem. Ops. Research Soc. Am., Math. Assn. Am. Subspecialties: Operations research (mathematics); Probability. Current work: Algorithmic methods in stochastic modelling; performance analysis of computer systems and data communication networks. Home: 37 Victoria Dr Eatontown NJ 07724 Office: Bell Laboratories Crawfords Corner Rd Holmdel NJ 07733

LUCAS, GLENN EUGENE, nuclear engineering educator, educator, consultant; b. Los Angeles, Mar. 8, 1951; s. Glenn Edwin and Mary Lorraine (Shaw) L.; m. Mary Susan Ricketts, Sept. 2, 1972; children: Kelly, Ryan, Shannon. B.S., U. Calif., Santa Barbara, 1973; M.S., MIT, 1975, Sc.D., 1977. Engr. Exxon Nuclear Co., Inc., Richland, Wash., 1978; asst. prof. U. Calif., Santa Barbara, 1978—; cons. engring., Santa Barbara, 1978—. Author: Effects of Anisotropy and Irradiation on the Creep Behavior of Zircaloy-2, 1978; co-author: Creep of Zirconium Alloys in Nuclear Reactors, 1983. Bd. dirs., v.p. Santa Barbarans for Rational Energy Policy, 1978—. Mem. Am. Nuclear Soc. (significant contbn. award 1982), AAAS, Am. Soc. for Metals, Sigma Xi. Subspecialties: Nuclear engineering; Metallurgy. Current work: Mechanical metallurgy and environmental effects on structural alloys for advanced energy systems. Home: 618 Andamar Way Goleta CA 93117 Office: Dept Chem and Nuclear Engring U Calif Santa Barbara CA 93106

LUCAS, MYRON CRAN, geneticist; b. Cin., Nov. 15, 1946; s. Ralph Frank and Dallace Fair (Coberly) L. B.S., Lewis and Clark Coll., 1969; Ph.D., Wash. State U., 1974. Postdoctoral research assoc. botany U. Ill., 1973-75, U. Ga., 1975-76; adj. asst. prof. biol. scis. Fla. State U., 1977; postodoctoral research asso. U. Idaho, Moscow, 1977-78; assoc. prof. biol. scis. La. State U., Shreveport, 1978—. La. State U. grantee, 1981; Magale Found. grantee, 1982; NDEA fellow, 1969-71. Mem. Am. Soc. Microbiology, Genetics Soc. Am., AAAS, N.Y. Acad. Sci., Sigma Xi. Subspecialty: Gene actions. Current work: Researcher in gene regulation and devel. in fungi; structure and function of messenger RNA. Office: Dept Biol Scis La State U Shreveport LA 71115

LUCAS, RICHARD JOHN, mathematician, educator; b. Chgo., June 26, 1948. B.S., U. Ill.-Chgo., 1971, M.S., 1973, Ph.D., 1979. Asst. prof. Ill. Inst. Tech., 1979-82; asst. prof. math. scis. Loyola U., Chgo., 1982—; research assoc. Office Naval Research, 1982. Research publs. in applied math., wave propagation, plasma physics. Mem. Am. Phys. Soc., Soc. Indsl. and Applied Math. Subspecialty: Applied mathematics. Current work: Research in applied mathematics, wave propagation, plasma physics. Office: Loyola U Dept Math Scis Chicago IL 60626

LUCAS, ROBERT ALAN, mechanical engineer, educator; b. Allentown, Pa., June 13, 1935; s. Robert Horace and Helen Marguerite (Koons) L.; m. Joanne Adele Wetherhold, June 15, 1957; children: Michael John, Elizabeth Adele, Leslie Anne, Marya Courtney. B.S., Lehigh U., 1957, M.S., 1959, Ph.D., 1964. Design engr. Air Products and Chem. Co., Trexlertown, Pa., 1957-58; mem. faculty Lehigh U., Bethlehem, Pa., 1959—, asst. prof. mech. engring., 1964-69, assoc. prof., 1969—; postdoctoral research assoc. Naval Research Lab., Washington, 1965-66. Subspecialty: Mechanical engineering. Current work: Computer aided engineering; design of mechanical systems. Home: 2604 Walnut St Allentown PA 18104 Office: Dept Mech Engring Lehigh Bethlehem PA 18015

LUCAS, WILLIAM JOHN, botanist, educator; b. Adelaide, S. Australia, Feb. 23, 1945; came to U.S., 1977; s. Robert Bruce and Thelma Rose (Packer) L.; m. Diana Kristine Gorny, Dec. 17, 1966; children: Jessica Kristine, Judith Margret, Joanna Ruth. B.S. with honors, U. Adelaide, 1971, Ph.D. in Plant Physiology, 1975. Postdoctoral fellow and research assoc. dept. botany U. Toronto, Ont., Can., 1975-77; asst. prof. botany U. Calif., Davis, 1979-80, assoc. prof., 1980-83, prof., 1983—, dir. lab. research program. Contbr. articles on plant physiology and plant biophysics to profl. jours. NSF grantee, 1978-86. Mem. Am. Soc. Plant Physiologists, Bot. Soc. Am., Can. Soc. Plant Physiologists, Soc. for Exptl. Biology. Subspecialties: Plant physiology (biology); Biophysics (biology). Current work: Regulation of plant membrane transport systems with special emphasis on plasmalemma transport, photosynthetic assimilation of exogenous HCO3 by Chara Corallina, K influx into corn roots-mechanisms and site of entry, phloem physiology. Home: 1001 Deodara Ct Davis CA 95616 Office: Dept Botany U Calif Davis CA 95616

LUCAS, WILLIAM RAY, official NASA; b. Newbern, Tenn., Mar. 1, 1922; m., 1948; 3 children. B.S., Memphis State U., 1943; M.S., Vanderbilt U., 1950, Ph.D. in Chem. Metallurgy, 1952; L.H.D. (hon.), Mobile Coll., 1977, D.Sc., Southeastern Inst. Tech., 1980, U. Ala., Huntsville, 1981. Instr. chemistry Memphis State U., 1946-48; chemist guided missile devel. div. Redstone Arsenal, 1952-54, chief chem. sect., 1954-55; chief engr. material sect. Army Ballistic Missile Agy., 1955-56, chief engr. material br., 1956-60; with Marshall Space Flight Center, NASA, 1960—, material div., 1963-66, dir. propulsion and vehicle engring. lab., 1966-68, dir. program devel., 1968-71, dep. dir., 1971-74, dir., 1974—. Served to lt. USNR, 1943-46. Recipient Exceptional Sci. Achievement medal NASA, 1964, 2 Exceptional Service medals, 1969, Distinguished Service medal, 1972; Presdl. rank disting. exec., 1980; Roger W. Jones award for outstanding exec. leadership Am. U., 1981; Space award for outstanding contbns. in field of space VFW, 1983. Fellow Am. Soc. Metals, Am. Astronautical Soc., AIAA (Oberth award 1965, Holger N. Toftoy award 1976); mem. Nat. Acad. Engring., Am. Chem. Soc., Sigma Xi. Subspecialties: Aerospace engineering and technology; Metallurgy. Current work: Research and development of space launch vehicles and spacecraft. Research in materials engring. metallurgy, inorganic chemistry, environ. effects on materials, especially space environ. effects. Office: NASA Marshall Space Flight Center AL 35812

LUCATORTO, THOMAS BENJAMIN, research physicist; b. N.Y.C., May 9, 1937; s. Benjamin and Faye (Mautone) L.; m. Kathleen Gross, Aug. 5, 1966 (div. Feb. 1979); children: Theresa, Rachael; m. Linda Bryson, Apr. 7, 1979. B.S. in Physics, CCNY, 1960; M.S., Columbia U., 1964, Ph.D., 1968. Postdoctoral fellow Columbia U., 1968-69; physicist Nat. Bur. Standards, Washington, 1969—. Contbr. articles to profl. jours. Recipient Silver medal Dept. Commerce, 1981. Mem. Am. Phys. Soc. Subspecialties: Atomic and molecular physics; Plasma physics. Current work: Shell collapse in atoms, atomic autoionization, atomic effects in solids, resonant laser-vapor interactions. Home: 3600 Van Ness St Washington DC 20008 Office: Nat Bur Standards A251-Physics Washington DC 20234

LUCE, JAMES KENT, physician, research oncologist; b. Le Mars, Iowa, May 11, 1922; s. Ernie LeRoy and Gladys (Knapp) L.; m. Joan Larpenter, Mar. 31, 1967; children: Holly, Laura, Douglas, Gregory; m. Candace Ann Myers, Feb. 14, 1981. A.B. in Geology (James Monroe MacDonald scholar), U. Calif.-Berkeley, 1948; M.D., Yale U., 1952. Intern Tripler Army Hosp., Honolulu, 1952-53; resident in internal medicine VA Hosp., Iowa City, 1958-60, U. Calif.-San Francisco, 1962-63; asst. prof. medicine U. Tex.-M.D. Anderson Hosp., Houston, 1966-71; dir. clin. research and med. dir. Mountain States Tumor Inst., Boise, Idaho, 1971-77; dir. div. community and clin. activities West Coast Cancer Fedn., San Francisco, 1977-78; dir. clin. oncology Adria Labs. Inc., Dublin, Ohio, 1978—. Contbr. articles to profl. jours. Mem. bd. dirs. Thomas A. Dooley Found., N.Y.C., 1964—. Served with USNR, 1943-45; with USAF, 1952-54. Mem. Am. Assn. Cancer Research, Am. Soc. Clin. Oncology. Democrat. Subspecialty: Chemotherapy. Current work: Clinical research in medical oncology; development of new or improved cancer chemotherapeutic agents. Home: 343 W Main St Plain City OH 43064 Office: 5000 Post Rd Dublin OH 43017

LUCE, ROBERT DUNCAN, educator; b. Scranton, Pa., May 16, 1925; s. Robert R. and Ruth Lillian (Downer) L.; m. Gay Gaer, June 5, 1950 (div. 1967); m. Cynthia Newby, Oct. 2, 1967 (div. 1977); 1 dau., Aurora Newby. B.S. in A.E, M.I.T., 1945, Ph.D. in Math, 1950; M.S. (hon.), Harvard U., 1976. Asst. prof. sociology and math stats. Columbia U., N.Y.C., 1953-57; lectr. social relations Harvard U., Cambridge, 1957-59, Alfred N. Whitehead prof. psychology, 1976-81, prof. psychology, 1981—; prof. U. Pa., Phila., 1959-69; vis. prof. Inst. for Advanced Study, Princeton, N.J., 1969-72; prof. social scis. U. Calif-Irvine, 1972-75; chmn. sect. psychology Nat. Acad. Scis., Washington, 1980-83. Author: (with H. Raiffa) Games and Decisions, 1957, Individual Choice Behavior, 1959; co-editor: Handbook Math. Psychology, 1963, 65 (3 vols); author: (with D.H. Krantz, P. Suppes, A. Tversky) Foundations of Measurement, 1971. Chmn. Assembly of Behavioral and Social Sci. NRC, 1976-79. Served with USNR, 1942-45. Guggenheim fellow, 1980-81; NSF fellow, 1966-67. Fellow AAAS, Am. Psychol. Assn. (Disting. Sci. Contbrs. award 1971); mem. Psychonomic Soc., Soc. for Math. Psychology (pres. 1979), Psychometric Soc. (pres. 1976), Am. Math. Soc., Am. Acad. Arts and Scis., Nat. Acad. Scis. Current work: Axiomatic measurement (conjoint, non-associative, utility), reaction time models, psychophysical models, choice behavior. Home: 33 Washington St Cambridge MA 02140 Office: Harvard U William James Hall Cambridge MA 02138

LUCE, WILLIAM GLENN, animal science educator; b. Beaver Dam, Ky., Mar. 21, 1936; s. William Horton and Annie Elizabeth (Shultz) L.; m. Dorcas Linda Ward, Dec. 20, 1957; m. Nancy Ebey Ballard, Nov. 24, 1970; children: William Glenn, Bryan Ward. B.S. in Animal Husbandry, U. Ky., 1958; M.S. in Animal Sci, U. Nebr., 1964; Ph.D. in Animal Nutrition, U. Nebr., 1965. Cert. animal scientist Am. Registry Cert. Animal Scientists. Co-mgr. Kroger Co., Louisville, 1958-62; research asst. U. Nebr., 1962-65; extension animal scientist U. Ga., 1965-68; prof. dept. animal sci., extension livestock specialist Okla. State U., 1968—. Served with USAR, 1958—. Recipient Outstanding Service award Okla. Swine Breeders Assn., 1977; Tyler award Okla. State U., 1980. Mem. Am. Soc. Animal Sci., Sigma Xi, Sigma Phi Epsilon. Democrat. Methodist. Subspecialty: Animal nutrition. Current work: Swine nutrition and management; swine extension specialist for State Oklahoma; brood sow nutrition and baby pig nutrition research. Home: 2817 W 17th Ave Stillwater OK 74074 Office: Animal Sci Dept Okla State U Stillwater OK 74078

LUCEY, ROBERT FRANCIS, educator; b. Worcester, Mass., Mar. 13, 1926; s. Cornelius Joseph and Mary (Shea) L.; m. Marie E. Lemire, Sept. 27, 1952; children—Robert Francis, Eileen, Jeanne-Marie, Mary, James, Elizabeth, Cornelius J., II, Brian. B.Vocat. Agr., U. Mass., 1950; M.S., U. Md., 1954; Ph.D., Mich. State U., 1958. Asst. prof. agronomy U. N.H., Durham, 1958-61; asst. prof. field crops Cornell U., 1961-67, asso. prof., 1967-70, prof., 1970—, chmn. dept. agronomy, 1975—. Recipient N.Y. Farmers award, 1968; highest achievement certificate Epsilon Sigma Phi, 1968. Mem. Am. Soc. Agronomy, Crop Sci. Soc. Am., N.Y. Acad. Scis., AAAS, Am. Grassland Council. Subspecialty: Agronomy. Home: RD 2 Ithaca NY 14850 Office: Dept Agronomy Cornell U Ithaca NY 14850

LUCHINS, EDITH HIRSCH, mathematics educator; b. Breziny, Poland, Dec. 21, 1921; came to U.S., 1928; d. Max and Leah (Kravetsky) Hirsch; m. Abraham S. Luchins, Oct. 10. 1942; children: David, Daniel, Jeremy, Anne, Joseph. B.A. cum laude in Math, Bklyn. Coll., 1942; M.S., NYU, 1944; Ph.D., U. Oreg., 1957. Instr. dept. math. Bklyn. Coll., 1944-46, 48; research asst. applied math. project NYU, 1946; lectr. dept. math. U. Miami, Coral Gables, Fla., 1959-60, assoc. prof, 1960-62; assoc. prof. dept. math. scis. Rensselaer Poly. Inst., Troy, N.Y., 1962-70, prof., 1970—. Author: (with A.S. Luchins) Revisiting Wertheimer's Seminars, 1978, Rigidity of Behavior, 1959, Logical Foundations of Mathematics for Behavioral Scientists, 1965. Bd. dirs. Union of Orthodox Congregations Am., N.U.C., 1975—; chmn. Renssalaer faculty council, 1982-83. Danforth assoc., 1980—. Fellow AAAS; mem. Am. Math. Soc., Math. Assn. Am., Assn. Women Math., Phi Beta Kappa, Pi Mu Epsilon. Subspecialties: Mathematics and psychology; Learning. Current work: Sex differences in math and computer education. Home: 53 Fordham Ct Albany NY 12209 Office: Dept Math Scis Rensselaer Poly Inst 110 8th St Troy NY 12181

LUCIER, RONALD DAVID, atomic electric company systems engineer; b. Ware, Mass., Feb. 27, 1956; s. Leonard Joseph and Mary (Moriarty) L.; m. Cheryl Letitia Tracht, June 24, 1978; 1 son, Brian. B.S. in Mech. Engring., Worcester Poly. Inst., 1979. Physicist Yankee Atomic Electric Co., Framingham, Mass., 1977-81, systems engr., 1981—, public speaker, 1981—. Vol. firefighter West Brookfield (Mass.) Fire Dept., 1975—; mem. West Brookfield Local Assessment Com. Hazardous Waste, 1981—. Mem. Am. Nuclear Soc. Subspecialties: Nuclear engineering; Systems engineering. Current work: Nuclear safety related engineering of systems, power plant performance analysis. Home: RFD 1 Box 220 West Brookfield MA 01585 Office: Yankee Atomic Electric Co 1671 Worcester Rd Framingham MA 01701

LUCK, DENNIS NOEL, biology educator, researcher; b. Durban, South Africa, Dec. 8, 1939; came to U.S., 1969; s. Peter Burvill and Eva (Taylor) L.; m. Joan Burchall, Jan. 18, 1969; 1 son, Roy Burvill. B.Sc., U. Natal, South Africa, 1961, M.Sc., 1963; D.Phil., U. Oxford, Eng., 1966. Lectr. dept. biochemistry U. Natal, 1966-69; vis. asst. prof. biochemistry Baylor U. Coll. Medicine, Houston, 1969-70; asst. prof. zoology U. Tex.-Austin, 1970-72; asst. prof. biology Oberlin (Ohio) Coll., 1972-75, assoc. prof., 1975-82, prof., 1982—; cons., research biochemist Gilford Instrument Labs., Oberlin, 1980-82; fgn. expert

Shanxi Agrl. U., Taigu, People's Republic of China, summer 1982. Contbr. articles to research jours. NSF grantee, 1975-80; Eleanor Roosevelt Internat. Cancer fellow, 1978-79. Mem. Biochem. Soc. (London), Am. Soc. Cell Biology, Am. Soc. Developmental Biology, AAAS, Sigma Xi (chpt. pres. 1982-83). Subspecialties: Biochemistry (biology); Molecular biology. Current work: Biochemical endocrinology; nucleic acid metabolism. Teaching-biochemistry; molecular biology, molecular genetics, genetic engineering. Home: 240 Oak St Oberlin OH 44074 Office: Oberlin Coll Oberlin OH 44074

LUCKETT, LARRY WAYNE, medical physicist, army officer; b. San Antonio, Oct. 8, 1948; s. Robert Miles and Muriel Mae (Stearns) L.; m. Judith Ann Blodgett, June 2, 1972. B.S. cum laude, Trinity U., San Antonio, 1971; M.S., Tex. A&M U., College Station, 1973; postgrad., Rensselaer Poly. Inst., 1980—. Cert. Am. Bd. Health Physics. Commd. 1st lt. M.S.C. U.S. Army, 1973, advanced through grades to maj., 1983; cons. physicist U.S. Army Environ. Hygiene Agy., Aberdeen, Md., 1973-76; radiation protection officer Tripler Army Med. Center, Honolulu, 1977-80; now with dept. radiology Walter Reed Army Med. Center, Washington. Chmn. subcom. on environ. factors and cancer Am. Cancer Soc., Honolulu, 1978-80. Mem. Health Physics Soc., Am. Assn. Physicists in Medicine, Soc. Nuclear Medicine, Am. Nuclear Soc. (assoc.), Tau Beta Pi, Alpha Nu Sigma, Sigma Pi Sigma. Baptist. Subspecialties: Nuclear medicine; Imaging technology. Current work: Quantitative nuclear image analysis; computed tomography; population exposures to environmental radiation sources. Home: 17224 Hobble Bush Ct Rockville MD 20855 Office: Dept Radiology Walter Reed Army Med Center Washington DC 20307

LUCKY, ROBERT WENDELL, electrical engineer; b. Pitts., Jan. 9, 1936; s. Clyde Arthur and Grace Katherine (Luck) L.; m. Joan Miriam Jackson, Aug. 19, 1961; children: David William, Karen Joan. B.S. in Elec. Engring, Purdue U., 1957, M.S., 1959, Ph.D., 1961. With Bell Telephone Labs., Holmdel, N.J., 1961—, supr. signal theory, 1964-65, head dept. advanced data communications, 1965-76, dir. Electronic and Computer Systems Research Lab., 1977-81, exec. dir. research Communications Scis. Div., 1982—; Mem. engring. adv. bd. Purdue U., 1973-75; mem. USAF Sci. Adv. Bd., 1979—, vice chmn., 1983—; mem. adv. com. NSF, 1983—. Author: Principles of Data Communication, 1968; Editor: Proc. of IEEE, 1974-76, Computer Communication, 1975; cons. editor: Plenum Press, 1979—. Named Distinguished Alumnus Purdue U., 1969. Fellow IEEE (v.p. 1978-79, exec. v.p. 1980—, publs. bd. 1970-76, bd. govs. info. theory group 1969-74); mem. Nat. Acad. Engring., Communications Soc. (pres. 1977—, Achievement award 1975), Eta Kappa Nu (nat. dir. 1974-76). Subspecialty: Computer engineering. Current work: Telecommunications systems; computer communications; computer science. Patentee in field. Home: 238 Kemp Ave Fair Haven NJ 07701 Office: Room 7B-220 Bell Telephone Labs Murray Hill NJ 07974

LUCOT, JAMES BERNARD, pharmacologist, educator; b. McKeesport, Pa., Aug. 17, 1951; s. Joseph Bernard and Florence Patricia (Greger) L.; m. Loretta Lahr, May 10, 1975; m. Katherine Ann Lucot, May 22, 1982. B.S. magna cum laude, U. Pitts., 1973; Ph.D., U. N.C.-Chapel Hill, 1977. Grad. research asst. neurobiology U. N.C., 1976-77; postdoctoral fellow U. Chgo., 1977-80; asst. prof. depts. pharmacology and toxicology, psychiatry Wright State U., Dayton, Ohio, 1980—; researcher; community resource person for info. on central nervous system pharmacology. Contbr. articles to profl. jours. Recipient Wright State U.seed grant, Dayton, Ohio, 1980—. Mem. AAAS, Soc. Neurosci., N.Y. Acad. Scis., Soc. Stimulus Properties of Drugs. Democrat. Subspecialties: Psychopharmacology; Neuropharmacology. Current work: Categorization of serotonin receptors and serotonin antagonists, mechanism of action of antidepressants, mechanism of motion sickness. Office: Wright State U Dept Pharmacology and Toxicology Sch Medicine Dayton OH 45435

LUDERER, ALBERT AUGUST, medical research scientist; b. Jersey City, May 23, 1948; s. Albert Frank and Hilda Marie (Koenig) L.; m. Margaret Annetta Diaz, June 27, 1970; children: Hilary Faye, Albert William. A.B., Drew U., 1970; M.S., Rutgers U., 1973, Ph.D., 1975. With Corning Glass Works, N.Y., 1976—, sr. research scientist and tech. dir. cancer research div., 1979-83, mgr. biomed. research health and sci. group, 1983—; Co-chmn. 3d Internat. Conf. Diagnostic Immunology, Henniker, N.H., 1981. Contbr. numerous sci. articles to profl. publs. Postdoctoral fellow Jefferson Med. Coll. Pa., Phila., 1974-76. Mem. Am. Assn. Immunologists, N.Am. Hyperthermia Group, Am. Genetic Assn., N.Y. Acad. Sci., Tissue Culture Assn. Subspecialties: Cancer research (medicine); Immunology (medicine). Current work: Early diagnosis and therapy of cancer. Application of hybridoma and DNA technology of the diagnosis and therapy of neoplastic diseases. Office: Corning Glass Works Sullivan Research Park FR-64 Corning NY 14831

LUDLOW, CHRISTY LESLIE, speech pathologist; b. Montreal, Que., Sept. 7, 1944; d. Forester Wilcox and Margaret Helen (Sweet) Leslie; m. Gregory Ludlow, Sept. 7, 1968. B.Sc., McGill U., 1965, M.Sc., 1967; Ph.D., NYU, 1973. Staff speech pathologist, research asst. NYU Med. Ctr., 1967-70; Walter A. Anderson fellow, 1970-72; project mgr. Am. Speech and Hearing Assn., Washington, 1973-74; research speech pathologist, communicative disorders program Nat. Inst. Neurol. and Communicative Disorders and Stroke, NIH, Bethesda, Md., 1974—; cons. Vietnam head injury study Walter Reed Army Med. Ctr., 1980—. Assoc. editor: Jour. Speech and Hearing Research, 1981—. Recipient Dir.'s award NIH, 1977. Fellow Internat. Acad. Research in Learning Disabilities, Am. Speech and Hearing Assn. (chmn. sci. affairs com. 1978-82); mem. Internat. Assn. Research in Otolanyngology, Internat. Neuropsychol. Soc., Acad. Aphasia, N.Y. Acad. Sci., Am. Speech Lang. Hearing Assn., Am. Assn. Phonetic Sci. Subspecialties: Neuropsychology; Otorhinolaryngology. Current work: Speech science, neurolinguistics, brain injury and language function. Home: 8801 Garfield St Bethesda MD 20817 Office: 7550 Wisconsin Ave Bethesda MD 20205

LUDLUM, DAVID BLODGETT, medical scientist, clinical pharmacologist; b. N.Y.C., Sept. 30, 1929; s. C. Daniel and Elsie B. (Blodgett) L.; m. Carlene L. Dyke, Dec. 23, 1952; children: Valerie Jean Ludlum Wright, Kenneth David. B.A., Cornell U., 1951; Ph.D., U. Wis., 1954; M.D., NYU, 1962. Research scientist DuPont Co., Wilmington, Del., 1954-58; intern Bellevue Hosp., N.Y.C., 1962-63; asst. prof. pharmacology Yale U. Sch. Medicine, New Haven, 1963-68; assoc. prof. pharmacology U. Md. Sch. Medicine, Balt., 1968-70, prof., 1970-76; prof., chmn. dept. pharmacology Albany (N.Y.) Med. Coll, 1976-80, prof. pharmacology and medicine, dir. oncology research, 1980—; cons. indsl. exposure; adj. prof. chemistry Rensselaer Poly. Inst., Troy, N.Y., 1977—; vis. prof. oncology Johns Hopkins U., 1973-76; vis. prof. Courtauld Inst., London, summer, 1970. Contbr. articles to profl. jours.; assoc. editor: Cancer Research, 1980—. WARF fellow, 1951-52; Am. Heart Assn. fellow, 1960-62; NIH Research Career Devel. awardee, 1968; Markle scholar, 1967; NIH grantee, 1963—; Am. Cancer Soc. grantee, 1981-82. Mem. Am. Soc. Pharmacology and Exptl. Therapeutics, Am. Soc. Clin. Pharmacology and Therapeutics, Am. Assn. Cancer Research, Am. Soc. Biol. Chemists, Am. Chem. Soc., Phi Beta Kappa, Phi Kappa Phi, Sigma Xi, Alpha Omega Alpha. Subspecialties: Pharmacology; Cancer research (medicine). Current work: Pharmacology of antineoplastic agents, clinical pharmacology, molecular pharmacology, mutagenesis and carcinogenesis. Home: 24 Linda Ct Delmar NY 12054 Office: Albany Med Coll 47 New Scotland Ave Albany NY 12208

LUDUENA, RICHARD FROILAN, biochemist, educator, researcher; b. San Francisco, Feb. 9, 1946; s. Froilan Pindaro and Sara (Lafert) L.; m. Linda Lucille Gleason, June 27, 1981. B.A., Harvard Coll., 1967; Ph.D., Stanford U., 1973. Postdoctoral fellow Stanford U. Sch. Medicine, Calif., 1973-76; asst. prof. U. Tex. Health Sci. Ctr., San Antonio, 1976-82, assoc. prof., 1982—. Jane Coffin Childs Meml. Fund postdoctoral fellow, 1973-75; Yamagiwa-Yoshida travel grantee Internat. Union Against Cancer, Paris, 1979; Nat. Inst. Gen. Med. Scis. research grantee, 1979—; Nat. Cancer Inst. research grantee, 1979—; Robert Welch Found. research grantee, 1978—. Mem. Am. Soc. Cell Biology, Internat. Soc. Neurochemistry, Internat. Soc. Evolutionary Protistology, Am. Soc. Biol. Chemists. Subspecialty: Biochemistry (medicine). Current work: Structure and evolution of tubulin regulation of microtubule assembly. Office: Dept Biochemistry U Tex Health Sci Ctr 7703 Floyd Curl Dr San Antonio TX 78284

LUDWICK, JOHN CALVIN, JR., geologist, educator, researcher; b. Berkeley, Calif., Apr. 25, 1922; s. John Calvin and Margretta; ; s. John Calvin and Rouse L.; m. Norah O'Donnell Sutherland, Oct. 24, 1957. A.B., UCLA, 1947, M.S., 1949, Ph.D., 1951. Sr. research geologist Gulf Research and Devel. Co., Tex. and Pa., 1950-68; Samuel L. and Fay M. Slover prof. oceanography Old Dominion U., Norfolk, Va., 1968—, dir., 1968-80; mem. task force to study future disposal of dredged material Commonwealth of Va., 1976-78; commr. Adv. Comm. on Environ. Quality, 1982—. Served with AUS, 1942-46; PTO. Fellow Geol. Soc. Am.; mem. Am. Shore and Beach Preservation Assn. Subspecialties: Oceanography; Sedimentology. Current work: Mechanics of marine sediment transport; shoaling in estuaries; beach processes; analysis of tidal currents. Office: Dept Oceanography Old Dominion U Norfolk VA 23508

LUDWIG, JAMES J., medical research institute executive; b. N.Y.C., July 7, 1925; s. Jesse J. and Constance (Allman) L.; m. Judith Chase, May 15, 1954; m. Eileen Denari, Aug. 8, 1970; children: Jessica, Nicole, Lisa. B.A., Yale U., 1947. Vice pres. Saks Fifth Ave., San Francisco, 1950-80; pres., chief exec. officer Med. Research Inst. San Francisco at Pacific Med. Center, 1981—; chmn. Century Oil & Gas Co., Denver, 1977-79. Vice pres. San Francisco Ballet Assn.; bd. dirs. San Francisco Zool. Soc., Hamlin Sch.; pres. Uptown Parking Corp., San Francisco Mcpl. Ry. Improvement, San Francisco Mcpl. Ry. Improvement Corp. Clubs: Burlingame Country (Hillsborough, Calif.); Family (San Francisco); Menlo Circus (Atherton). Subspecialty: Medical research administration. Current work: Administration of medical research in fields of heart, arthritis, vision, cancer, behavior, biomedical engineering. Office: Med Research Inst of San Francisco Pacific Med Ctr 2000 Webster St San Francisco CA 94115

LUDWIG, JOHN HOWARD, environ. engr.; b. Burlington, Vt., Mar. 7, 1913; s. Rudolf Frederick and Emily Henrietta (Sikora) L.; m. Gilda Mary Silva, Nov. 9, 1946; children—Howard Russell, Robert William. B.S. in Engring, U. Calif., Berkeley, 1934, M.S., U. Colo., 1941, Harvard, 1956, Sc.D., 1958. Registered profl. engr., Calif., Oreg. Engr. Bur. Reclamation, Denver, 1936-39, C.E., Portland, 1939-43, Sacramento, 1949-51; engr. water pollution USPHS, Washington, 1951-55; dir. research and devel. EPA, Washington, 1955-69; asst. commr. Nat. Air Pollution Control Adminstrn., 1969-70, asso. commr., 1970-72; cons. environ. engring. to fgn. and internat. agys., 1972—; U.S. del. Econ. Commn. for Europe, Geneva, 1969-72, Orgn. for Econ. Co-op. and Devel., Paris, 1969-72, NATO Com. on Challenges of Modern Soc., Brussels, 1969-72. Contbr. articles to profl. jours. Mem. Santa Barbara Environ. Quality Bd., 1973-75. Served to capt. USAAF, 1943-46. Recipient Commendation medal HEW, 1963, Superior Service award, 1967; Gold medal for Exceptional Service EPA, 1971; Gordon Fair award Am. Acad. Environ. Engrs., 1973; named Distinguished Engring. Alumnus U. Colo., 1976. Mem. Nat. Acad. Engrs., Cosmos Club, ASCE, Am. Meteorol. Soc., Am. Acad. Environ. Engrs., Phi Beta Kappa, Sigma Xi, Tau Beta Pi, Chi Epsilon, Delta Omega. Subspecialties: Environmental engineering; Water supply and wastewater treatment. Current work: Consulting in administrative and technical aspects of air pollution control programs. Home and Office: 43 Alston Pl Santa Barbara CA 93108

LUE, LOUIS PING-SION, entomologist; b. Taipei, Taiwan, May 16, 1938; s. I-Lo and Ar-chow (Chiang) L.; m. Martha Ching-Shiow Hwang, Dec. 20, 1969; 1 son, Andrew Luh-Ming. B.S., Taiwan U., 1963; M.S., Auburn U., 1970, Ph.D., 1974. Teaching asst. dept. anatomy Taiwan U. Med. Coll., Taipei, 1964-67; research asso. U. Ark., 1974-76, Purdue U., 1976-78; research assoc. La. State U., 1978-79; asst. prof., research scientist Va. State U., Petersburg, 1979—. Contbr. articles on determination of boll weevil chromosome number to sci. jours. Mem. Entomol. Soc. Am., Am. Chem. Soc., Genetics Soc. Am., AAAS; mem. Va. Acad. Sci.; Mem. Sigma Xi. Subspecialties: Toxicology (agriculture); Genetics and genetic engineering (agriculture). Current work: Effects of pesticides on insects and plants; chromosome abnormality caused by pesticide treatment, pesticide evaluation and pesticide residue analysis. Office: Va State U Box 271 Petersburg VA 23803

LUERSSEN, FRANK WONSON, steel company executive; b. Reading, Pa., Aug. 14, 1927; s. George V. and Mary Ann (Swoyer) L.; m. Joan M. Schlosser, June 17, 1950; children: Thomas, Mary Ellen, Catherine, Susan, Ann. B.S. in Physics, Pa., State U., 1950; M.S. in Metall. Engring, Lehigh U., 1951. Metallurgist research and devel. div. Inland Steel Co., East Chicago, Ind., 1952-54, mgr. various positions, 1954-64, mgr. research, 1964-68, v.p. research, 1968-77, v.p. steel mfg., 1977-78, pres., 1978—, chmn., 1983—; dir. Continental Ill. Corp. Author various articles on steelmaking tech. Trustee Calumet Coll., Whiting, Ind., 1972-80, Northwestern U., 1980—; v.p., dir. Munster (Ind.) Med. Research Found., 1972—; trustee, sec., treas. Munster Sch. Bd., 1957-66. Served with USNR, 1945-47. Fellow Am. Soc. Metals, Nat. Acad. Engring.; mem. Am. Inst. Mining, Metall. and Petroleum Engrs., Am. Iron and Steel Inst., Metals Soc. Gt. Britain. Subspecialty: Metallurgical engineering. Home: 8226 Parkview Ave Munster IN 46321 Office: 30 W Monroe St Chicago IL 60603

LUGAR, ROBERT MYERS, nuclear waste disposal engineer; b. Hershey, Pa., Dec. 18, 1956; s. Gilbert L. and Irene (Myers) L. B.S., Slippery Rock State Coll., 1978; M.Environ. Pollution Control, Pa. State U., 1981. Tech. sales rep., project engr. Apollo Technologies, Inc., Whippany, N.J., 1978-80, Dallas, 1978-80; grad. asst. Ctr. Air Environment Studies, Pa. State U., University Park, 1980-82, civil engring. dept., 1980-82; application engr. Hittman Nuclear & Devel. Corp., Columbia, Md., 1982—. Mem. Am. Nuclear Soc., Air Pollution Control Assn., Water Pollution Control Assn. Subspecialty: Environmental engineering. Current work: Low-level radwaste management; cement solidification, high integrity containers, low-level burial site, liquid radwaste treatment, volume reduction, incineration, environmental impact and siting of hazardous and low-level radioactive waste landfills. Office: Hittman Nuclear and Devel Corp 9151 Rumsey Rd Columbia MD 21045

LUHBY, A. LEONARD, medical educator; b. N.Y.C., Dec. 21, 1916; s. Barnet and Lottie L.; m. Sara K.; 1 dau., Tami. A.B., Columbia U., 1938; M.D., N.Y.U., 1943. Diplomate: Am. Bd. Pediatrics, Am. Bd. Nutrition. Fellow in hematology Children's Hosp., Boston, 1945-47, fellow in hematology and immunology, Columbus, Ohio, 1948-50; instr. pediatrics N.Y. Med. Coll., N.Y.C., 1950-54, from asst. prof. to prof., 1955—, chief sect. hematology, oncology and nutrition, dept. pediatrics, 1954—; pres. Am. Bd. Nutrition, 1976-80. Recipient disting. service award Cooley's Anemia Found.; NIH grantee. Fellow or mem. Am. Soc. Clin. Nutrition, Am. Physiol. Soc., Am. Inst. Nutrition, Am. Assn. Cancer Research. Subspecialties: Cancer research (medicine); Pediatrics. Current work: Role of nutrients in prevention, adjuvant therapy and cure of selected malignant tumors. Home: 2794 Webb Ave Bronx NY 10468 Office: NY Med Coll Valhalla NY 10595

LUI, MAY SO-YING, biochemist, educator; b. Hong Kong, Dec. 25, 1950; came to U.S., 1969, naturalized, 1982; d. David and Gek Tian (Li) Sim; m. Alec Yen Nein Lui, Dec. 31, 1971; 1 dau., Natalie Shau Bie. B.Sc., U. Minn., 1973; Ph.D., Med. Coll. Wis., 1977. Teaching asst. dept. biochemistry U. Minn., Mpls., 1973; research asst. Med. Coll. Wis., Milw., 1973-77; research assoc. Lab. Exptl. Oncology, Ind. U. Sch. Medicine, Indpls., 1977-79, lectr., 1979-82, asst. prof., 1982—. Contbr. articles to profl. jours. Recipient prize as best biochemistry student Med. Coll. Wis., 1975. Mem. Am. Assn. for Cancer Research. Subspecialties: Cancer research (medicine); Biochemistry (medicine). Current work: Cancer chemotherapy, cancer biochemistry, nucleotide metabolism, biochemical pharmacology of anticancer drugs. Office: 1100 W Michigan St Indianapolis IN 46223

LUINE, VICTORIA NALL, neurobiologist, researcher, educator; b. Pine Bluff, Ark., Apr. 22, 1945; d. Vincent and Peggy Lynn (Nall) L.; m. David B. Russell, May 21, 1976; 1 son, Richard David. B.S., Allegheny Coll., 1967; Ph.D., SUNY, Buffalo, 1971. Research assoc. Rockefeller U., N.Y.C., from 1972, now asst. prof. dept. neurobiology. Contbr. articles to profl. jours. Mem. AAAS, Soc. for Neurosci. Subspecialties: Neurochemistry; Neuroendocrinology. Current work: Role of neurotransmitters, neurotramsmitter related enzymes and receptors in the mediation by gonadal hormones of sexual behavior and gonadotropin secretion. Office: Rockefeller U New York NY 10021

LUKAS, JEFFREY HILTON, psychologist; b. Flushing, N.Y., Oct. 8, 1944; s. Arne Victor and Mary Eleanor (Light) L.; m. Lynda Forrest Lukas, Aug. 21, 1971; children: Brennen Thomas, Ryan Jeffrey. B.A., Syracuse U., 1967; M.A., U. Del., 1975, Ph.D., 1976. Research psychologist U.S. Army Human Engring Lab., Aberdeen Proving Ground, Md., 1967-72; team leader, behavioral research directorate U.S. Army Human Engring. Lab., 1975—; adj. prof. U. Del., Newark. Contbr. articles to profl. jours. Bd. dirs. Am. Cancer Soc. Harford County, 1981—; v.p., bd. dirs. Community Assn. Mem. AAAS, Am. Psychol. Assn., Soc. Neurosci., N.Y. Acad. Sci., Soc. Psychophysiol. Research. Subspecialties: Physiological psychology; Sensory processes. Current work: Selective attention: arousal effects on central nervous system and behavior; human and animal brain research utilizing single units, evoked potentials and EEG analyses, brain behavior relationships. Home: 2517 Fairway Dr Bel Air MD 21014 Office: US Army Human Engring Lab Aberdeen Proving Ground MD 21005

LUKAS, SCOTT EDWARD, research pharmacologist, drug abuse educator; b. Pitts., May 2, 1953; s. Thomas Edward and Carol Lee (Fox) L.; m. Karin Hawkinson, Aug. 30, 1975; 1 dau., Lillian Ruth. B.S. in Biology, Pa. State U., 1975; Ph.D. in Pharmacology, U. Md., 1979. Lab. scientist dept. pharmacology and toxicology U. Md., Balt., 1977-79, instr. pharmacology and neurosci., 1978-79; postdoctoral fellow dept. psychiatry and behavioral sci. Johns Hopkins U., Balt., 1979-82, asst. prof., 1982—, NIDA Addiction Research Ctr., 1982—; lectr. on problems of drug abuse. Contbr. articles to profl. jours.; illustrator for: sci. publs. Time Life Books and. Emerson fellow, 1978-79; Am. Soc. for Pharmacology and Exptl. Therapeutics travel grantee, 1979; Neurosci. tng. grantee, 1980-81; Nat. Research Service grantee, 1981-83. Mem. Am. Soc. for Pharmacology and Exptl. Therapeutics, Soc. for Neurosci., AAAS, Behavioral Pharmacology Soc. Subspecialties: Pharmacology; Neurobiology. Current work: Electroencephalographic and behavioral correlates of drug abuse. Effects of drugs of abuse on hearing and vision and these relationships to reinforcing efficacy using self-administration techniques in animals and humans. Drug dependence and abuse liability of opiates, sedatives, stimulants and hallucinogens. Office: NIDA Addiction Research Ctr care Baltimore City Hosps PO Box 5180 Baltimore MD 21224

LUKASIK, STEPHEN JOSEPH, aerospace company executive; b. S.I., N.Y., Mar. 19, 1931; s. Stephen Joseph and Mildred F. (Tynan) L.; m. Marilyn B. Trappiel, Jan. 31, 1953; children: Carol J., Gregory C., Elizabeth A., Jeffrey P.; m. Virginia Dogan Armstrong, Feb. 11, 1983; stepchildren: Elizabeth L. Armstrong, Alan D. Armstrong. B.S., Rensselaer Poly. Inst., 1951; M.S., MIT, 1953, Ph.D., 1956. With Advanced Research Projects Agy., Dept. Def., Washington, 1966-74; v.p. Xerox Corp., Rochester, N.Y., 1974-76; chief scientist, sr. v.p. Rand Corp., Santa Monica, Calif., 1977-79; chief scientist FCC, Washington, 1979-82; v.p. Northrop Corp., Palos Verdes, Calif., 1982—; mgr. Northrop Research and Tech. Ctr., Palos Verdes, Calif., 1982—; cons. to govt. Trustee Stevens Inst. Tech.; mem. adv. com. Stanford U. Computer Center; assoc. editor The Info. Soc. Author: Some Relations between Technology and Arms Control, 1983, The 1979 WARC - Its Results and Subsequent FCC Actions, 1981, Military Research and Development, 1977, Technology, Transfer and National Security, 1976. Served to capt. USAR, 1951-66. Recipient Disting. Civilian Service medal Sec. Def., 1973, 74. Mem. Am. Phys. Soc., AAAS. Club: Cosmos (Washington). Subspecialties: Fluid Dynamics; Information systems (information science). Current work: Direct corporate research laboratory in areas of microelectronics, image processing and manufacturing productivity. Office: One Research Park Palos Verdes CA 90274

LUKERT, MICHAEL THOMAS, geology educator, consultant; b. Kansas City, Mo., June 28, 1937; s. George Gordon and Bonita Ellen (Gardener) L.; m. Emmalou Rittmeyer, Aug. 20, 1961; children: Claudia, George, Bonita. B.S., U. Ill., 1960; M.S., No. Ill. U., 1962; Ph.D., Case-Western Res. U., 1973. Asst. geologist Ill. Geol. Survey, Urbana, summer 1964; prof. geology Edinboro (Pa.) State Col., 1967—; adj. prof. Thiel Coll., Greenville, Pa., 1975-76, Mercyhurst Coll., Erie, Pa., 1979, 81; geologist Va. Div. Mineral Resources, Charlottesville, summers 1970-74, 76-77; cons. LTL Assocs., Edinboro, 1979—. Contbr. numerous articles to sci. publs. Mem. Geol. Soc. Am., Geochem. Soc. Republican. Methodist. Subspecialties: Petrology. Current work: Geology of Virginia Blue Ridge and Piedmont, geochronology, environmental geology. Office: Dept Geoscis Edinboro State College Edinboro PA 16444

LUKEZIC, FELIX LEE, plant pathologist, educator; b. Florence, Colo., May 27, 1933; s. Felix and A. and Irene (Wands) L.; m. Arlene Claire, Aug. 18, 1955; children: Bret, Craig, Susan. B.S., Colo. State U., 1956, M.S., 1958; Ph.D., U. Calif.-Davis, 1963. Lab. technician II U. Calif.-Davis, 1958-63; asst. plant pathologist United Fruit Co., LaLima, Honduras, 1963-65; asst. prof. plant pathology Pa. State U., University Park, 1965-70, assoc. prof., 1970-75, prof., 1975—. Contbr. articles in field to profl. jours. Danforth Found. fellow, 1969. Mem.

Am. Phytopath. Soc., Am. Soc. Plant Physiologists. Democrat. Methodist. Club: Nittany Valley Sailing. Subspecialties: Plant pathology; Microbiology. Current work: Researcher in physiological interaction between the host and pathogens. Home: 531 Hillside Ave State College PA 16801 Office: Pa State U Dept Plant Pathology 211 Buckhout Lab University Park PA 16802

LUKIN, MARVIN, chemist, educator, researcher; b. Cleve., Feb. 12, 1928; s. Adolph and Yetta (Babkow) L.; m. Judith, Jan. 1, 1962; children: Jonathan, Joshua. B.S., Ohio U., 1949; M.S. (Research Corp. fellow), Western Res. U., 1954, Ph.D., 1956. Research fellow Mellon Inst., Pitts., 1956-57; postdoctoral fellow Albert Einstein Coll. Medicine, N.Y.C., 1957-61; research fellow Cleve. Clinic, 1963-65; postdoctoral fellow Case Western Res. U., 1966-67; asst. prof. chemistry Youngstown State U., 1967-75, assoc. prof., 1975-83, prof., 1983—. Contbr. articles to profl. jours. Mem. Am. Chem. Soc., Royal Soc. Chemistry, Sigma Xi. Subspecialties: Biochemistry (biology); Organic chemistry. Current work: Peptide synthesis; enzyme mechanisms; organic synthesis; teaching biochemistry.

LUM, LAWRENCE GEORGE, immunologist; b. Sacramento, Calif., June 28, 1947; s. George Wai and Nellie (Quan) L.; m. Carol Ann Lee, June 17, 1973; children: Hillary Eileen, Amanda Renee. B.S., U. Redlands, 1969; M.D., U. Calif., San Francisco, 1973. Diplomate: Am. Bd. Pediatrics. Intern pediatrics dept. pediatrics U. Calif., San Francisco, 1973-74, resident pediatrics, 1974-75, U. Colo., Denver, 1975-76; clin. asso., fellow immunology, cellular immunology sect., metabolism br. Nat. Cancer Inst., NIH, Bethesda, Md., 1976-78, expert cons., 1978-79; staff mem. dept. pediatrics Prince Georges County Hosp., 1978-79; mem. dept. pediatrics So. Md. Hosp., 1978-79; asst. prof. pediatrics dept. pediatrics U. Wash., Seattle, 1979—; asst. mem. Fred Hutchinson Cancer Research Center, Seattle, 1979—; mem. staff Children's Orthopedic Hosp. Med. Center, Seattle, 1979—. Contbr. articles in field to profl. jours. Recipient awards Bank Am., 1965, Optimist Club, 1965, Outstanding Alumnus award U. Redlands.; Am. Bapt. nat. scholar, 1965-69. Mem. Am. Assn. Immunologists, Am. Fedn. Clin. Research, Phi Delta Epsilon. Subspecialties: Immunology (medicine); Transplantation. Current work: Human T cell subpopulations and the regulation of antibody prodn. in bone marrow transplant recipients and immunodeficiency disorders. Office: Fred Hutchinson Cancer Research Center Room 785 1124 Columbia St Seattle WA 98104

LUMB, RALPH FRANCIS, physical chemist, consultant; b. Worcester, Mass., May 27, 1921; s. Irving E. and Clara L. (Bunker) Clapp; m. Phyllis M. Wetherbee, Feb. 11, 1941; Sandra Lumb Eddy, Joy Lumb Gardner, Cheryl Lumb Drescher, Randall W., Richard I., Linda Lumb Warner, Steven R., Bonnie S. Lumb Lillis. A.B., Clark U., 1947, Ph.D. in Phys. Chemistry, 1951. Cert. safeguards specialist. Chief physics and chemistry br. AEC, Washington, 1951-56; v.p. Quantum, Wallingford, Conn., 1956-60; dir. Western N.Y. Nuclear Research Center; SUNY, Buffalo, 1960-68; pres. Advanced Tech. Cons. Corp., Wallingford, 1968-71, NASAC, Inc., Reston, Va., 1971—. Editor: Management of Nuclear Materials, 1960. Mem. Gov.'s Com. on Western N.Y. Devel., 1962-64; chmn. N.Y. State Adv. Com. on Radiation Utilization, 1962-64. Served with USAF, 1943-45. Fellow Am. Inst. Chemists, AAAS; mem. Inst. Nuclear Materials Mgmt., Am. Nuclear Soc. Subspecialties: Fuels and sources; Nuclear engineering. Current work: Nuclear fuel consulting; measurement and control of nuclear materials; development and evaluation of systems for measurement and control of nuclear materials and for quality assurance for nuclear fuel. Patentee in field. Office: 1850 Samuel Morse Dr Reston VA 22090

LUMENG, LAWRENCE, physician, biochemist, researcher, nutritionist; b. Manila, Philippines, Aug. 10, 1939; came to U.S., 1958; s. Lu and Lucia (Lim) Meng; m. Pauline J-P Young, Nov. 26, 1966; children: Carey N-K, Emily P-Y. A.B., Ind. U., 1960; M.D., Ind. U.-Indpls., 1964, M.S., 1969. Intern U. Chgo. Hosps., 1964-65; resident, fellow Ind. Univ. Hosps., Indpls., 1965-69; asst. prof. Ind. U., Indpls., 1971-74, assoc. prof., 1974-79, prof., 1979—; chief gastroenterology VA Med. Ctr., 1979—; vis. prof. U. So. Calif., Los Angeles, 1977. Author research papers; editor of workshop book. Served to maj. M.C. U.S. Army, 1969-71. VA grantee. Fellow ACP; mem. Am. Soc. Biol. Chemists, Am. Soc. Clin. Investigation, Am. Inst. Nutrition, Am. Assn. Study of Liver Diseases. Subspecialties: Gastroenterology; Biochemistry (biology). Current work: Alcohol metabolism, alcoholic liver disease, alcoholism, pyridoxine metabolism, mitochondrial metabolism, hepatic metabolism. Office: Ind U Sch Medicine 545 Barnhill DR Indianapolis IN 46223

LUMLEY, JOHN LEASK, physicist, educator; b. Detroit, Nov. 4, 1930; s. Charles S. and Jane Anderson Campbell (Leask) L.; m. Jane French, June 20, 1953; children: Katherine Leask, Jennifer French, John Christopher. B.A., Harvard, 1952; M.S. in Engring, Johns Hopkins, 1954, Ph.D., 1957. Postdoctoral fellow Johns Hopkins, 1957-59; mem. faculty Pa. State U., 1959-77, prof. aerospace engring., 1963-74, Evan Pugh prof. aerospace engring., 1974-77; Willis H. Carrier prof. engring. Cornell U., 1977—; prof. d'echange U. d'Aix-Marseille, France, 1966-67; Fulbright sr. lectr. U. Liege; vis. prof. U. Louvain-La-Neuve, Belgium; Guggenheim fellow U. Provence (Ecole Centrale de Lyon), France, 1973-74. Author: (with H.A. Panofsky) Structure of Atmospheric Turbulence, 1964, Stochastic Tools for Turbulence, 1970, (with H. Tennekes) A First Course in Turbulence, 1971; also articles.; Tech. editor: Statistical Fluid Mechanics, 1971, 75, Variability of the Oceans, 1977; assoc. editor: Physics of Fluids, 1971-73, Ann. Rev. of Fluid Mechanics, 1976—; chmn. tech. editorial bd.: Izvestiya: Atmospheric and Oceanic Physics, 1971—; editorial bd.: Fluid Mechanics: Soviet Research, 1972—; Prin.: films Deformation of Continuous Media, 1963, Eulerian and Lagrangian Frames in Fluid Mechanics, 1968. Recipient medallion U. Liege, Belgium, 1971. Fellow Am. Acad. Arts and Scis., Am. Acad. Mechanics, Am. Phys. Soc. (exec. com. div. fluid dynamics 1972-75, 81-84, chmn. exec. com. div. fluid dynamics 1982); mem. Soc. Natural Philosophy, N.Y. Acad. Scis., AAAS, Am. Geophys. Union, AIAA (fluid and plasmadynamics award 1982), Johns Hopkins Soc. Scholars (charter), Sigma Xi. Subspecialties: Fluid mechanics; Micrometeorology. Current work: Turbulence modeling in geophysical flows and in technology and biology. Home: 743 Snyder Hill Rd Ithaca NY 14850 Office: 238 Upson Hall Cornell U Ithaca NY 14853

LUMSDEN, ROBERT DOUGLAS, plant pathologist; b. Washington, June 21, 1938; s. George Napier and Mary Louise (Shropshire) L.; m. Valerie Brook, June 11, 1960; children: Douglas Robert, Thomas Brook. B.S., N.C. State U., 1961, M.S., 1963; Ph.D., Cornell U., 1966. Research plant pathologist Soilborne Diseases Lab., Beltsville (Md.) Agrl. Research Ctr., U.S. Dept. Agr., 1966—. Mem. Phytopath. Soc., Mycological Soc., Am. Can. Phytopath. Soc., Mexican Phytopath. Soc. Subspecialties: Plant pathology; Microbiology. Current work: Ecology of soilborne plant pathogens and antagonistic microorganisms, biological control and integrated control of plant diseases. Office: 262 Biosci Bldg Beltsville Agrl Research Ctr Beltsville MD 20705

LUNCHICK, MYRON EDWIN, assn. exec., mech. engr.; b. Boston, Apr. 23, 1926; s. H.B. and E.M. (Trager) L.; m. Ethel Louise Swankin, Oct. 10, 1953; children: James, Elizabeth, Laura, Jacqueline. B.S. in Civil Engring, Worcester Poly. Inst., 1948; M.S. in Hydraulic Engring, La. State U., 1949; Ph.D. in Theoretical and Applied Mechanics, U. Ill., 1952. Head design evaluation br. David Taylor Model Basin, Carderock, Md., 1955-61; chief applied mechanics Pneumo Dynamics Corp., Bethesda, Md., 1961-63; head structures br. Surface Effect Ships Project Office, Carderock, 1966-68; cons. engr., 1963-66, 66-68; pres. Sea Co., Bethesda, 1972-81; assoc. dir. N.E. Field Service Office, ASME, Danbury, Conn., 1982—; mem. faculty U. Ill., 1949-51, U. Md., 1957-61, George Washington U., 1970-71. Contbr. articles to profl. jours. Served to sgt. U.S. Army, 1944-46. Mem. ASME (Tech. Achievement award Washington sect. 1975, Centennial medallion 1981), Marine Tech. Soc. Republican. Club: Cosmos (Washington). Subspecialties: Mechanical engineering; Fluid mechanics. Current work: Structural integrity of submersibles and high performance ships; design of hyperbaric test facilities; manufacturing technology for advanced ship structures. Home: 143 Fillow St Norwalk CT 06850 Office: Lake Ave Extension Danbury CT 06810

LUND, CHARLES EDWARD, stress engineer; b. Fremont, Mich., Apr. 4, 1946; s. Henry Knute and Dorothy Carol (Beatty) L.; m. Sandra Leaha Ruff, July 30, 1967; 1 dau., Deanna Marie. B.S.E., U. Mich., 1968, M.S.E., 1970; postgrad., Stanford U., 1972-75. Assoc. engr. McDonnell-Douglas Astron. Co., Culver City, Calif., 1968-69; structures engr. Lockheed Missiles & Space Co., Sunnyvale, Calif., 1970-77; stress engr. II NWL Control Systems, Kalamazoo, 1977—. Treas. Angling Rd. Sch. Parent Tchr. Orgn., Portage, Mich., 1981-82, chmn. carnival, 1981-83. Mem. AIAA. Club: Mood Makers Band (dir. 1979—). Subspecialties: Aeronautical engineering; Mechanical engineering. Current work: Finite element analysis. Patentee digital plotter pen caddy. Home: PO Box 3335 Kalamazoo MI 49003 Office: NWL Control Systems 2220 Palmer Ave Kalamazoo MI 49001

LUND, JOHN WILLIAM, civil engineering educator; b. Berkeley, Calif., July 7, 1936; s. John Jorgensen and Lydia Marie (Olsen) L.; m. Jacqueline Lee Urling, June 19, 1962; 1 son, John David. B.S. in Civil Engring, U. Colo., 1958, Ph.D., 1967; M.E. in Transp. Engring, U. Calif., 1962. Registered civil engr., Calif., Oreg. Asst. hwy. mgr. Calif. Div. Hwys., Bishop, 1958-59, 62-64; asst. prof. civil engring. U. Alaska, Fairbanks, 1964-65; assoc. prof. to prof. civil engring. Oreg. Inst. Tech., Klamath Falls, 1967-82, assoc. dean, head dept. engring. tech., 1980-82; vis. prof. civil engring. Oreg. State U., Corvallis, 1982—; assoc. dir. Geo.-Heat Ctr., Klamath Falls, 1974-82; ptnr. LLC Goethermal Cons., Klamath Falls, 1976-82; sr. engr. Eliot Allen & Assocs., Salem, Oreg., 1981—. Author: Road Design Handbook, 1971; editor: Direct Utilization of Geothermal Energy, 1979; proc. profl. conf.; editorial bd.: Geothermics Jour, 1982—. Chmn. bd. Klamath Falls Transit Authority, 1970-73; mem. adv. com. U.S. Dept. Energy, Geothermal Div., 1976-80; mem. adv. bd. U.S. Bur. Land Mgmt., 1978-82. Served to lt. U.S. Army, 1959-60. Mem. ASCE, Am. Soc. Engring. Edn., Geothermal Resources Council (v.p. 1980—, Best Paper award 1981), Sierra Club (chpt. pres. 1967-70), Sigma Tau, Chi Epsilon. Democrat. Lutheran. Club: 20-30. Subspecialties: Geothermal power; Civil engineering. Current work: Geotechnical engineering, soil mechanics and highway materials; transportation engineering geology and geothermal energy. Home: 302 S Monmouth Ave 101 Monmouth OR 97361 Office: Dept Civil Engring Tech Oreg Inst Tech Klamath Falls OR 97601

LUND, MELVIN ROBERT, dental educator; b. Siren, Wis., Oct. 17, 1922; s. Alexander and Emelia (Hamson) L.; m. Margaret Elizabeth Reith, Nov. 10, 1946; children: Mark, Kristine, Kelly. D.M.D., U. Oreg., 1946; M.S., U. Mich., 1954. Mem. restorative faculty Loma Linda (Calif.) U. Sch. Dentistry, 1954-56, chmn. restorative faculty, 1956-69, mem. faculty, 1969-71; chmn. operative dentistry Ind. U. Sch. Dentistry, Indpls., 1971—. Author: Operative Dentistry, 4th edit, 1982; contbr. articles to profl. jours. Served to capt. U.S. Army, 1946-48; Korea. Mem. Internat. Assn. Dental Research, Am. Acad. Restorative Dentistry, Am. Assn. Dental Schs. Republican. Seventh-day Adventist. Lodge: Lions Internat. Subspecialties: Dental materials; operative dentistry. Current work: By controlling the technical and manipulative procedures for purpose of observing effects on the restorative material. Office: Ind U Sch Dentistry 1121 W Michigan St Indianapolis IN 46202

LUND, RICHARD, biology educator; b. N.Y.C, Sept. 17, 1939; s. Gissur Olav and Tikvah (Rabinowitz) L.; m. Wendy Laurie Glickman, May 7, 1978; children by previous marriage: Barbara, Freya, Jessica. B.S., U. Mich., 1961, M.S., 1963; Ph.D., Columbia U., 1968. Curatorial asst. Carnegie Mus., Pitts., 1965-68; asst. prof. U. Pitts., 1968-74; asst. prof. biology Adelphi U., Garden City, N.Y., 1974-76, assoc. prof., 1976-81, prof., 1981—; research assoc. Carnegie Mus., 1968—, Am. Mus. Natural History, N.Y.C., 1981—. NSF grantee, 1974, 77, 79, 82. Mem. Soc. Vertebrate Paleontology, Paleontol. Soc., Ecol. Soc. Am., Am. Soc. Ichthyologists and Herpetologists. Subspecialties: Paleobiology; Paleoecology. Current work: Evolutionary morphology of Carboniferous fishes. Home: 349 Kensington Rd S Garden City NY 11530 Office: Dept Biolog Adelphi U South Ave Garden City NY 11530

LUNDBLAD, ROGER LAUREN, biochemist, educator; b. San Francisco, Oct. 31, 1939; s. Lauren Alfred and Doris Ruth (Peterson) L.; m. Susan Taylor, Oct. 15, 1966; children: Christina Susan, Cynthia Karin. B.S., Pacific Luth. U., 1961; Ph.D., U. Wash., 1965. Research assoc. biochemistry U. Wash., 1965-66, Rockefeller U., 1966-68; asst. prof. to prof. depts. pathology, biochemistry and periodontics U. N.C., Chapel Hill, 1968—; assoc. dir. adminstrn. Dental Research Center, U. N.C., 1968—; assoc. dir. program devel. Center Thrombosis and Hemostasis, U. N.C., 1968—. Contbr. articles to profl. jours. Mem. Am. Heart Assn., Am. Soc. Biol. Chemists, Am. Chem. Soc., AAAS, Internat. Soc. Thrombosis and Hemostatis, Sigma Xi. Democrat. Lutheran. Subspecialties: Biochemistry (biology); Genetics and genetic engineering (biology). Current work: Thrombosis and hemostasis, regulation of gene expression in eukaryotic systems, recombinant DNA. Home: 638 Wellington Dr Chapel Hill NC 27514 Office: Dental Research Center 210H U NC Chapel Hill NC 27514

LUNDE, PETER J., Solar engineering consulting company executive, educator; b. N.Y.C., June 8, 1931; s. Otto Helmer and Grace Elizabeth (Smith) L.; m. Dorothy Marie Tomaszewski, Nov. 26, 1957; children: Jennifer, Katharine, Timothy. B.S., Pa. State U., 1953, M.S., 1960, Ph.D., 1962. Registered profl. engr., Conn. Research engr. Chevron Research Co., Richmond, Calif., 1961-63; sr. research engr. Carrier Corp., Syracuse, N.Y., 1963-67; head process analysis Space Systems Life Support, Hamilton Standard div. United Tech., Windsor Locks, Conn., 1967-73; sr. research scientist Ctr. Environment and Man, Hartford, 1973-77; cons., pres. New Energy Resources, Simsbury, Conn., 1977—. Contbr. articles to profl. jours.; author: Solar Thermal Engineering, 1980. Served wth USN, 1953-56. Mem. ASHRAE, Am. Chem. Soc., Internat. Solar Energy Soc. Episcopalian. Subspecialties: Solar energy; Mathematical software. Current work: Solar energy system performance prediction and economic evaluation. Patentee in field. Address: 4 Daniel Ln West Simsbury CT 06092

LUNDGREN, JAMES REINHOLD, civil engr.; b. Vancouver, B.C., Can., Jan. 11, 1945; s. Nels Reinhold and Agnes May (Fulton) L.; m. Angela Andrian Plaza, Nov. 24, 1973; children: Steven, Douglas, Mary. B.A.Sc., U. B.C., 1968; M.S., U. Ill., 1970. Registered profl. engr., Que. Project dir. Can. Nat. Rys., Edmonton, Atla., Montreal, 1971-73, sr. project engr., Montreal, 1973-76; mgr. Fast Project Assn. Am. R.R.s, Pueblo, Colo., 1976-77, mgr. track research div., Chgo., 1977-78, dir. research and test ops., Pueblo, 1978-82; exec. dir. Transp. Test Center, Pueblo, 1982-83, asst. v.p., 1983—. Contbr. articles to profl. jours. Mem. ASME; mem. Locomotive Maintenance Officers Assn.; Mem. Am. Ry. Engring. Assn., Roadmasters and Maintenance of Way Assn., Car Dept. Officers Assn., Ry. Fuel and Operating Officers Assn. Presbyterian. Subspecialties: Civil engineering; Mechanical engineering. Current work: Management of rail transportation related hardware research and test activities. Office: PO Box 11130 Pueblo CO 81001

LUNDIN, BRUCE THEODORE, energy tech. adminstr., cons.; b. Alameda, Calif., Dec. 28, 1919; s. Oscar L. and Elizabeth E. (Erickson) L.; m. Barbara B. Bliss, July 27, 1946; children: Dianne, Robert, Nancy; m. Jean A. Oberlin, Mar. 20, 1982. B.S. in Mech. Engring, U. Calif.-Berkeley, 1942; D. Sc. (hon.), U. Toledo, 1975. Assoc. dir. NASA Lewis Research Center, Cleve., 1961-68, dep. assoc. adminstr., 1968-70, dir., 1970-77; cons. EPRI, others; staff dir. Pres.'s Commn. on Accident at Three-Mile Island. Contbr. articles on aircraft propulsion to profl. jours. Pres. Westhore Unitarian Ch.; trustee Southwest Gen. Hosp. Recipient NASA medal for Outstanding Leadership, 1965, Public Service medal, 1971, 79; Disting. Service medal, 1971, 77; named Nat. Space Club Engr. of Yr., 1976. Fellow AIAA, AAAS, Royal Aero. Soc. Subspecialties: Aerospace engineering and technology; Nuclear engineering. Current work: Safety advisory board, recovery of Three Mile Island.

LUNDIN, ROBERT WILLIAM, psychologist; b. Chgo., Apr. 28, 1920; s. Adolph Eugene and Agnes (King) L.; m. Margaret Waitt, Aug. 8, 1952; children: Sara Jane, Robert King. A.B., DePauw U., 1942; M.A., Ind. U., 1943, Ph.D., 1947. Asst. prof. Denison U., Granville, Ohio, 1947-49; assoc. prof. Hamilton Coll., Clinton, N.Y., 1949-64; prof. U. of South, Sewanee, Tenn., 1964—, Kenan prof., 1981—. Author: Psychol. Music, 1967, Personality: Behavioral Analysis, 1974, Personality: Experimental Approach, 1961, Theories and Systems of Psychology, 1979. Fellow Am. Psychol. Assn. (sec.-treas. 1969-76); mem. Southeastern Psychol. Assn., Sigma Xi. Club: Ecce Quo Bonum. Subspecialties: Behavioral psychology; Learning. Current work: Experimental psychologist: psychological theory, personality studies, psychology of music, history of psychology. Home: Greens View Rd Sewanee TN 37375 Office: Univ Of South Sewanee TN 37375

LUNN, ANTHONY C., polymer scientist; b. Huddersfield, Yorkshire, Eng., Sept. 25, 1946; came to U.S., 1967; s. Norman S. and Alice M. (Crowther) L.; m. Phyllis M. Teitelbaum, June 4, 1972. M.A., Cambridge (Eng.) U., 1967; M.S., Harvard U., 1968; S.D., MIT, 1972. Research scientist DuPont Co., Wilmington, Del., 1969; research assoc. MIT, 1972-73; project leader Am. Cyanamid Co., Stamford, Conn., 1973-78, Bound Brook, N.J., 1978-81; sect. mgr. suture research Ethicon Inc., Somerville, N.J., 1981—. Contbr. articles on polymer sci. to profl. jours. Frank Knox fellow, 1967; Fulbright grantee, 1967; Phillips fellow, 1970; DuPont fellow, 1968. Mem. Am. Chem. Soc., Am. Phys. Soc., Sigma Xi. Democrat. Subspecialty: Polymers. Current work: Synthetic surgical sutures; polymer and fiber phsyics. Patentee in field. Office: Ethicon Inc Route 22 Somerville NJ 08876 Home: 47 Hawthorne Ave Princeton NJ 08540

LUNT, OWEN RAYNAL, educator, biologist; b. El Paso, Tex., Apr. 8, 1921; s. Owen and Velma (Jackson) L.; m. Helen Hickman, Aug. 8, 1953; children—David, Carol, Janet. B.A. in Chemistry, 1947; Ph.D. in Agronomy, 1951. Mem. faculty UCLA, 1951—, prof. plant nutrition, 1964-72, prof. biology, 1972—, acting chmn. dept. biophysics, 1965-70; dir. Lab. Biomed. and Environ. Scis., 1968—. Served with USN, 1944-46. Fellow Am. Soc. Agronomy, Soil Sci. Soc. Am.; mem. Am. Soc. Plant Physiologists, Internat. Soc. Soil Sci., AAAS, Am. Nuclear Soc. (Los Angeles chpt.), Sigma Xi. Subspecialties: Soil chemistry; Plant physiology (biology). Current work: Research administration; soil chemistry, fertility and managemant. Research in soil chemistry, fertility, plant physiology. Home: 1200 Roberto Ln Los Angeles CA 90024 Office: 900 Veteran Ave Los Angeles CA 90024 I was reared in a cheerful, harmonious family on a farm. During childhood, we were poor. We all had chores and the entire household was enthusiastic about work. The family was very generous with others who were less fortunate. Unwaivering allegiance to high ethical and moral standards was expected. The whole family was active in the Mormon church. Neither parent had finished high school, but they read extensively. From an early age, I understood the family would support me in securing any educational objectives. In retrospect, I believe I had one of the best of starts.

LUPASH, LAWRENCE O(VIDIU), computer analyst, researcher; b. Bucharest, Romania, May 29, 1942; came to U.S., 1980; s. Ovidiu N. and Stefania (Lebu) L.; m. Corina Constantineanu, Dec. 26, 1975. M.S., Poly. Inst. Bucharest, 1965, Ph.D., 1972. Project engr. Inst. for Automation, Bucharest, 1965-68, sr. engr., 1971-72; researcher Romanian Acad. Scis., Bucharest, 1968-71; sr. researcher/lectr. U. Bucharest, 1972-79; sr. analyst Intermetrics, Inc., Huntington Beach, Calif., 1980—; asst. prof. Poly. Inst. Bucharest, 1966-68, 71-72; vis. lectr. U. Tirana, Albania, 1973. Author: (with V. Ionescu) Numerical Techniques in System Theory, 2 vols, 1974; Contbr. numerous articles to profl. publs. Recipient Republican award Poly. Inst. Bucharest, 1962; Romanian Acad. Scis. grantee, 1968; Case Western Res. U. grantee, 1969. Mem. IEEE, Soc. Indsl. and Applied Math., Orange County Philatelic Soc., Am. Philatelic Soc. Greek Orthodox. Subspecialties: Mathematical software; Computer-aided design. Current work: Numerical methods in system theory, mathematical software, computer-aided design of control systems, numerical analysis, simulations. Office: Intermetrics Inc 5392 Bolsa Ave Huntington Beach CA 92649

LUPIANI, DONALD ANTHONY, psychologist; b. N.Y.C., June 7, 1946; s. Louis and Josephine (Boccia) L.; m. Linda Moyik, June 1970; 1 dau., Jennifer. B.A., Iona Coll., 1968; M.A., Columbia U., 1971, Ph.D., 1973. Diplomate: Am. Bd. Profl. Psychology, Am. Acad. Behavioral Medicine, Am. Bd. Profl. Psychotherapy. Postdoctoral Behavior Therapy Inst., White Plains, N.Y., 1974-76; Clin. intern Psychiat. Service Center, White Plains, 1974-76; adj. assoc. prof. Iona Coll., New Rochelle, N.Y., 1973—; psychologist Kaliski Sch., Bronx, 1979-80; clin. assoc. Tchrs. Coll. Columbia U., 1975—; psychologist Franciscans, N.Y.C., 1979—; dir. psychology and spl. edn. services Riverdale Country Sch., Bronx, 1973—. NIMH fellow, 1968-73. Fellow Am. Orthopsychiat.Assn., Am. Coll. Psycology; mem. Am. Psychol. Assn., Westchester Psychol. Assn., N.Y. State Psychol. Assn., Nat. Assn. Sch. Psychologists; Mem. Am. Orthopsychiatry Assn. Roman Catholic. Subspecialty: Behavioral psychology. Current work: Diagnosis and treatment of psychological disorders and learning difficulties. Home: 227 Square Mile Rd Yonkers NY 10701 Office: Riverdale Country Sch W 253d St and Fieldston Rd Riverdale NY 10471

LUPO, MICHAEL VINCENT, aeronautical engineer; b. St. Louis, May 6, 1952; s. Vincent Joseph and JoAnn Marie (Macke) L.; m. Marilyn Kay Kern, June 5, 1976; 1 dau., Maria Christine. B.S. in Aero. Engring, U. Mo., Rolla, 1975, M.S. in Engring. Mgmt, 1982. Gen.

engr. U.S. Army, Texarkana, Tex., 1975-76, St. Louis, 1976—. Mem. AIAA (sec. Rolla br. 1975), Planetary Soc. (founding mem.). Roman Catholic. Subspecialties: Reliability and maintainability; Satellite studies. Current work: Development of space technology; research in space and lunar enterprise management; work in reliability and maintainability of aircraft. Home: 9332 Talbot St Saint Louis MO 63123 Office: US Army Aviation Research and Devel Command 4300 Goodfellow St Saint Louis MO 63120

LUPULESCU, AUREL PETER, medical educator, researcher, physician; b. Manastiur, Banat, Romania, Jan. 1, 1923; came to U.S., 1967, naturalized, 1973; s. Peter Vichentiu and Maria Ann (Dragan) L. M.D. magna cum laude, Sch. Medicine, Bucharest, Romania, 1950; Ph.D. in Biology, Faculty of Scis., U. Windsor, Ont., Can. Diplomate: Am. Bd. Internal Medicine. Chief Lab. Investigations, Inst. Endocrinology, Bucharest, 1950-67; research assoc. SUNY Downstate Med. Ctr., 1968-69; asst. prof. medicine Wayne State U., 1969-72; assoc. prof., 1973—; vis. prof. Inst. Med Pathology, Rome, 1967; cons. VA Hosp., Allen Park, Mich, 1971-73. Author: Steroid Hormones, 1958, Ultrastructure of Thyroid Gland, 1968, Hormones and Carcinogenesis, 1983; contbr. chpts., numerous articles to profl. publs. Mem. N.Y. Acad. Sci., AMA, Am. Soc. Cell Biology, Soc. Exptl. Biology and Medicine, AAAS. Republican. Subspecialties: Endocrinology; Cancer research (medicine). Current work: Hormones and tumor biology; studies regarding role of hormones in carcinogenesis. Office: Wayne State U Sch Medicine 540 E Canfield Ave Detroit MI 48021

LURAIN, JOHN ROBERT, III, gynecologic oncologist; b. Princeton, Ill., Oct. 27, 1946; s. John Robert and Elizabeth (Grampp) L.; m. Nell Lee Snavely, June 14, 1969; 1 dau., Alice Elizabeth. A.B., Oberlin Coll., 1968; M.D., U. N.C., 1972. Diplomate: Am. Bd. Ob-Gyn, Sub-Bd. Gynecology Oncology. Resident in ob-gyn U. Pitts., Magee Womens Hosp., 1972-75; fellow in gynecologic oncology Roswell Park Meml. Inst., Buffalo, 1977-79; assoc. dir. gynecologic oncology Northwestern U. Med. Sch., Prentice Womens Hosp., Chgo., 1979—, assoc. dir., 1979—. Contbr. chpts. to books, articles to med. jours. Served as lt. comdr., M.C. USNR, 1975-77. Galloway fellow, 1975; Am. Cancer Soc. fellow, 1977-79, 80-83. Fellow Am. Coll. Ob-Gyn, Am. Soc. Colposcopy and Cervical Pathology, Am. Soc. Clin. Oncology, Central Assn. Ob-Gyn, Chgo. Gynecol. Soc.; mem. Assn. Chgo. Gynecologic Oncologists (sec.-treas. 1981-83), Soc. Gynecologic Oncologists, Ill. Assn. Cancer Research. Subspecialties: Gynecological oncology; Oncology. Current work: Gynecologic cancer treatment and research; trophoblastic disease (choriocarcinoma) treatment and research. Home: 1212 Forest Ave Oak Park IL 60302 Office: Northwestern U Med Sch Prentice Womens Hosp 333 E Superior St Chicago IL 60611

LURIA, SALVADOR EDWARD, biologist; b. Turin, Italy, Aug. 13, 1912; U.S., 1940, naturalized, 1947; s. David and Ester (Sacerdote) L.; m. Zella Hurwitz, Apr. 18, 1945; 1 son, Daniel. M.D., U. Turin, 1935. Research fellow Curie Lab., Inst. of Radium, Paris, 1938-40; research asst. surg. bacteriology Columbia U., 1940-42; successively instr., asst. prof., asso. prof. bacteriology Ind. U., 1943-50; prof. bacteriology U. Ill., 1950-59; prof. microbiology M.I.T., 1959-64, Sedgwick prof. biology, 1964—, Inst. prof., 1970—, dir. center cancer research, 1972—; non-resident fellow Salk Inst. Biol. Studies, 1965—; lectr. biophysics U. Colo., 1950; Jesup lectr. zoology Columbia U., 1950; Nieuwland lectr. biology U. Notre Dame, 1959; Dyer lectr. NIH, 1963; with OSRD, Carnegie Instn., Washington, 1945-46. Asso. editor: Jour. Bacteriology, 1950-55; editor: Virology, 1955—; sect. editor: Biol. Abstracts, 1958-62; editorial bd.: Exptl. Cell Research Jour, 1948—; adv. bd.: Jour. Molecular Biology, 1958-64; hon. editorial adv. bd.: Jour. Photochemistry and Photobiology, 1961—. Guggenheim fellow Vanderbilt U. and Princeton U., 1942-43, Pasteur Inst., Paris, 1963-64; Co-recipient Nobel prize for medicine, 1969. Mem. Am. Philos. Soc., Am. Soc. Microbiology (pres. 1967-68), Nat. Acad. Scis., Am. Acad. Arts and Scis., AAAS, Soc. Gen. Microbiology, Genetics Soc. Am., AAUP, Sigma Xi. Subspecialty: Microbiology. Home: 48 Peacock Farm Rd Lexington MA 02173 Office: Dept Biology MIT Cambridge MA 02139

LURIA, SAUL MARTIN, research psychologist, educator; b. Athol, Mass., Dec. 24, 1929; s. Maurice and Florence (Shefts) L.; m. Honi Surnamer, Apr. 7, 1963; children: David, Steven. B.S., U. Richmond, 1949; M.S., U. Pa., 1951, Ph.D., 1955. Research psychologist Naval Submarine Med. Research Lab., Groton, Conn., 1957—; instr. U. New Haven, 1971-82. Served with U.S. Army, 1954-57. Subspecialties: Psychophysics; Behavioral psychology. Current work: Vision research. Home: 35 Beacon Hill Dr Waterford CT 06385 Office: Naval Submarine Med Research Lab Submarine Base Groton CT 06349

LURQUIN, PAUL FRANCIS, genetics educator; b. Brussels, Oct. 13, 1942; s. Leon L. and Josephine (Vander Est) L.; m. Denise F. Devoghel, June 26, 1965; children: Beatrice, Michael. B.S., U. Brussels, 1962, M.S., 1964, Ph.D., 1970. Head research group Belgian Nuclear Ctr. Mol, 1970-80; lectr. U. Liège, Belgium, 1973-80; asst. prof. genetics Wash. State U.-Pullman, 1980-81, assoc. prof., 1981—; cons. Univ. Genetics, Norwalk, Conn., 1983—. Editor: Genetic Engineering in Eukaryotes, 1983. Recipient Stas Prize, Belgian Royal Acad., 1970; Fulbright fellow, 1976; NATO fellow, 1976. Mem. Belgian Biochem. Soc., Plant Molecular Biology Assn., Am. Soc. Plant Physiologists, Internat. Cell Research Orgn. Subspecialties: Genetics and genetic engineering (biology); Plant cell and tissue culture. Current work: Developed a liposome system for introducing recombinant DNA into plant protoplasts. Address: Wash State Univ Dept Genetics Pullman WA 99164

LUSKIN, ALLAN TESSLER, immunologist, educator; b. Chgo., Jan. 21, 1943; s. Bert L. and Ruth (Katz) L.; m. Diana I. Luskin, Dec. 6, 1968 (div. Dec. 1982); 1 son, Joshua Dante. B.S. with honors, U. Ill.-Chgo., 1964, M.D., 1968. Diplomate: Am. Bd. Internal Medicine. Intern/resident U. Ill., Chgo., 1968-72, chief resident in medicine, 1973-74; fellow in allergy and clin. immunology Grant Inst. Allergy and Immunology, Chgo., 1974-75; asst. prof. Rush Med. Ctr., Chgo., 1976-81, assoc. prof., 1982—, dir. clin. services, 1976—, dir. tng. program, 1979—. Served to maj., m.c. USMC, 1970-72. Recipient Phoenix award Rush Med. Ctr., 1979, 81. Mem. Chgo. Assn. Immunologists, Am. Acad. Allergy, Lupus Erythematosus Soc. Ill., Assn. Animal Allergic Vets., Alpha Omega Alpha. Subspecialties: Immunology (medicine); Allergy. Current work: Psychoneuroimmunology; mechanisms of mediator release; acute phase response in immunologic disease. Office: Rush Med Center 1725 W Harrison Chicago IL 60612

LUSTICK, SHELDON, educator; b. Syracuse, N.Y., Aug. 16, 1934; s. Alex and Dorothy (Rosoff) L; m. Denise Duschenes, Mar. 22, 1970; children: Danielle, Erica. B.S., San Fernando Valley Coll., 1963; M.S., Syracuse U., 1965; Ph.D., UCLA, 1968. Teaching asst. Syracuse U., 1963-65, UCLA, 1965-66, instr., 1966-68; asst. prof. zoology Ohio State U., Columbus, 1968-71, assoc. prof., 1971-75, prof., 1975—, dir. grad. program in environ. biology, 1978—. Editor: Behavior and Physiology: The Cost of Survival in Vertebrates, 1983; contbr. articles to profl. jours. Served with U.S. Army, 1957-59. NSF grantee, 1976—; Air Force Office Sci. Research grantee, 1972-77; Dept. Interior grantee, 1968-78. Mem. Ecol. Soc. Am., Cooper Ornithol. Soc., Am. Ornithol. Soc., Sigma Xi. Subspecialties: Animal physiology; Phyaiological ecology. Current work: How animals adapt to stressful environ. conditions both behaviorally and physiologically. Home: 6939 Riverside Dr Powell OH 43065 Office: Ohio State U 1735 Neil Ave Columbus OH 43210

LUTAS, ELIZABETH M., physician, educator; b. N.Y.C., Oct. 2, 1951; d. Michael and Maria (Dano) L. B.A., N.Y.U., 1972, M.D. with honors, 1976. Diplomate: Am. Bd. Internal Medicine, Nat. Bd. Med. Examiners. Intern N.Y.U. Med. Center, N.Y.C., 1976-77, resident, 1977-79, instr. medicine, 1977-81, fellow in cardiology, 1979-81; research fellow in cardiology, instr. medicine N.Y. Hosp-Cornell U. Med. Center, N.Y.C., 1981-82, asst. prof. internal medicine, 1982—. Mem. ACP, Am. Coll. Cardiology, Am. Fedn. Clin. Research. Subspecialties: Cardiology; Internal medicine. Current work: Extensive research on hypertension, mitral valve prolapse. Office: NY Hosp-Cornell Med Ctr 525 E 68th St New York NY 10021

LUTH, WILLIAM CLAIR, geoscience administrator, researcher; b. Winterset, Iowa, June 28, 1934; s. William Henry Luth and Ora Anna (Klingaman) Sorenson; m. Betty Lou Heubrock, Aug. 23, 1953; children: Linda Diane, Robert William, Sharon Jean. B.A., U. Iowa, 1958; M.S., 1960; Ph.D., Pa. State U., 1963. Research assoc. Pa. State U., 1963-65; asst. prof. MIT, 1965-68; assoc. prof. geology Stanford U., 1968-77, prof., 1977-79; geoscientist Dept. Energy, 1976-78; div. supr. Sandia Nat. Labs., Albuquerque, 1979-82, mgr. geoscis. dept., 1982—; Alfred P. Sloan research fellow MIT, 1966-68; cons. Council on Environ. Quality, Washington, 1978; mem. vis. staff Los Alamos Labs., 1978. Contbr. chpts. to books, articles to profl. jours. Served to cpl. U.S. Army, 1953-56. Fellow Geol. Soc. Am., Mineral. Soc. Am.; mem. Am. Geophys. Union, Geochem. Soc., Sigma Xi. Republican. Subspecialties: Geochemistry; Petrology. Current work: Crystallization of the igneous rocks, experimental petrology, radioactive waste disposal, geochemical aspects of in situ fossil fuel utilization. Home: 1600 La Cabra Dr SE Albuquerque NM 87123 Office: Sandia Nat Labs Geosci Dept 1540 Albuquerque NM 87185

LUTHER, EDWARD TURNER, geologist; b. Nashville, Feb. 11, 1928; s. Eligah T. and Margaret M. (McCall) L.; m. Patricia Ann Worthy, Sept. 17, 1955; children: Margaret, Daniel. B.A., Vanderbilt U., 1950, M.S., 1951. Student geologist Tenn. Div. Geology, Nashville, 1950-51, geologist, 1951-52, sr. geologist, 1952-57, prin. geologist, 1958-61, asst. dir., 1961—; fuels engr. TVA, Chattanooga, 1957-58; cons. Environ. Systems Corp., Knoxville, Tenn., 1980—; instr. geology U. Tenn., Nashville, 1955-57, 70-80; cons. goal geology, Nashville, 1963-80. Author: Our Restless Earth, 1977. Served with USN, 1946-48. Fellow Geol. Soc. Am. (div. sec. 1973-79), Tenn. Acad. Sci. (sect. chmn.), Sigma Xi. Methodist. Subspecialties: Geology; Sedimentology. Current work: Supervise research program of state geological survey, primarily geologic mapping, compiling bulletin on stratigraphic usage in Tennessee, research on basement configuration in Tenn. Home: 838 Summerly Dr Nashville TN 37209 Office: Tenn Div Geology 701 Broadway Nashville TN 37203

LUTTGES, MARVIN WAYEN, aerospace engineering and bioengineering educator, consultant; b. Chico, Calif., Feb. 3, 1941; s. Clarence Adolph and Pearl Mary (Grieve) L.; m. Nancy Jo Sullivan, Aug. 21, 1969. B.S., U. Oreg., 1962; Ph.D., U. Calif.-Davis, 1968; postgrad. (USPHS fellow), Northwestern U., 1969. Grad. asst. U. Oreg., 1962-64, U. Calif.-Irvine, 1964-68; prof. aerospace engring., bioengring. U. Colo., Boulder, 1969—; cons. in field of acoustics; acting dir. EPA Region VIII Noise Tech. Assistance Ctr. Acoustical designer for various bldgs. including, Navajo Skill Ctr., Windowrock, Ariz.; contbr. numerous articles to profl. jours. Recipient Research award Aerospace Engring. Sci., 1980. Mem. AAAS, Soc. Neurosci., Sigma Xi, Tau Beta Pi. Subspecialties: Biomedical engineering; Space medicine. Current work: Biomedical engineering; motor and sensory systems; unsteady fluid dynamics experimentation; nervous system degeneration and regeneration. Home: 4323 N 30th St Boulder CO 80301 Office: University of Colorado Aerospace Engineering Sciences 429 Boulder CO 80309

LUTTINGER, DANIEL ALAN, neuropsychophramacologist; b. N.Y.C., Apr. 21, 1953; s. Philip and Selma (Marks) L. Ph.D. in Pharmacology and Physiology, U. Chgo. Postdoctoral research fellow U. N.C., Chapel Hill. Contbr. articles to profl. jours. NINH fellow. Mem. Soc. for Neurosci. Subspecialties: Neuropharmacology; Behavioral psychology. Current work: Neuropeptides and behavior. Office: U NC Sch Medicine Biol Sci Research Center 220H Chapel Hill NC 27514

LUTTINGER, JOAQUIN MAZDAK, physicist, educator; b. N.Y.C., Dec. 2, 1923; s. Paul and Shirley (Levy) L.; m. Abigail Thomas, Sept. 17, 1970; 1 dau., Catherine; stepchildren—Sarah, Jennifer, Ralph. S.B., Mass. Inst. Tech., 1944, Ph.D., 1947. Swiss-Am. exchange fellow Swiss Fed. Inst. Tech., Zurich, 1947-48, Nat. Research fellow, 1948-49; Jewitt fellow Inst. for Advanced Study, Princeton, N.J., 1949-50; asst. prof. physics U. Wis., 1950-51, asso. prof., 1951-53, U. Mich., Ann Arbor, 1953-57; prof. U. Pa., Phila., 1958-60, Columbia U., N.Y.C., 1960—, chmn. dept. physics, 1977—; cons. in field. Contbr. numerous articles to profl. jours. Served with Signal Corps U.S. Army, 1944-46. NRC fellow Ecole Normale Superieure, Paris, 1957-58, Rockefeller U., 1967-68; Guggenheim fellow Rockefeller U., 1975-76. Fellow Am. Phys. Soc.; mem. Am. Acad. Arts and Scis., Nat. Acad. Scis. Subspecialty: Theoretical physics. Home: 29 Claremont Ave New York NY 10027 Office: Dept Physics Columbia U New York NY 10027

LUTTON, LEWIS MONTFORT, zoology educator; b. Cin., July 14, 1945; s. Edwin Scott and Virginia (Melchior) L.; m. Marianne Hendow, Nov. 26, 1982; children by previous marriage: Wolf, Bram. B.A., Swarthmore Coll., 1968; Ph.D., Cornell U., 1976. Instr. Allegheny Coll., Meadville, Pa., 1974-76, asst. prof., 1976-80; asst. prof. zoology Mercyhurst Coll., Erie, Pa., 1980-83, assoc. prof., 1983—, chmn. dept. biology, 1983—, dir. sci depts., 1983—, chmn. pre-health com., 1980-83, dir. summer program for high sch. studies, 1980-83, Egan Scholar advisor, 1983—. Pres. Unitarian Universalist Ch., Meadville, Pa., 1979; coordinator Unitarian Universalist Summer Inst., Kenyon, Ohio, 1980-82; bd. dirs. Crawford County Drug and Alcohol Assn., Meadville, 1978-79, Family Service Assn., Erie, 1981-83. NSF grantee, 1979. Mem. Am. Soc. Zoologists, Am. Soc. Mammalogists, N.Y. Acad. Scis., Animal Behavior Soc., AAAS, Phi Kappa Phi. Democrat. Subspecialties: Physiology (biology); Ethology. Current work: Biological effects of marijuana. Home: 1930 W 35th St Erie PA 16508 Office: Dept Biology Mercyhurst Coll Erie PA 16546

LUTZ, BARRY LAFEAN, research astronomer, educator; b. Windsor, Pa., Jan. 2, 1944; s. Ray D. and Nina C. (Bull) L.; m. Karen L. Witman, Sept. 3, 1966 (div.); m. Mary Susanna Maxwell, July 25, 1981. B.S. magna cum laude, Lebanon Valley Coll., 1965; A.M., Princeton U., 1965, Ph.D., 1968. Postdoctoral fellow Nat. Research Council Can., Ottawa, 1968-70; research astronomer Lick Obs., U. Calif.-Santa Cruz, 1970-71; sr. research assoc., adj. asst. prof., then adj. asso. prof. SUNY-Stony Brook, 1971-77, adj. asso. prof., 1977-81; astronomer Lowell Obs., Flagstaff, Ariz., 1977—; adj. assoc. prof. physics Ariz. State U., 1981-83, adj. prof., 1983—. Contbr. articles to profl. publs. Democratic town committeeman, 1972-78; nat. del. Dem. Conv., 1974. NSF grantee, 1972, 74, 76, 77, 78, 80; NASA grantee, 1977, 78, 79, 80, 81, 82. Mem. Am. Astron. Soc., Astron. Soc. Pacific, Internat. Astron. Union, Sigma Xi. Subspecialties: Infrared optical astronomy; Planetary science. Home: 3330 Gillenwater Dr Flagstaff AZ 86001 Office: Lowell Obs Box 1269 Flagstaff AZ 86002

LUTZ, FRANK BROBSON, JR., internist; b. Millville, N.J., Aug. 22, 1950; s. Frank Brobson and Nell (Smith) L. B.A., Vanderbilt U., 1969; M.D., Tulane U., 1974, M.P.H., 1975. Diplomate: Am. Bd. Internal Medicine. Intern Charity Hosp., New Orleans, 1974-75, resident in medicine, 1975-77, fellow Tulane U., 1977-78; dir. New Orleans Health Dept., 1983. Recipient writing award Med. Econs., 1980. Fellow ACP; mem. So. Med. Assn., Internat. Union Against Venereal Diseases, Am. Fedn. Clin. Research, Med. Soc. Study of Venereal Diseases. Democrat. Presbyterian. Club: Chinese Shar-Pei. Subspecialty: Internal medicine. Current work: Sexually transmitted diseases, gerontology, antibiotic clinical research. Home: 1022 Dumaine St New Orleans LA 70116 Office: New Orleans Health Dept 1545 Tulane Ave New Orleans LA 70112

LUTZ, PETER LOUIS, physiology educator; b. Glasgow, Scotland, Sept. 29, 1939; came to U.S., 1971. B.S. with honors, Glasgow U., 1964, Ph.D., 1970. Lectr. Ife U., Ibadem, Nigeria, 1964-66; asst. prof. Duke U., Durham, N.C., 1971-72; lectr. Bath U., Eng., 1972-76; assoc. prof. Marine Sch., U. Miami, Fla., 1976-82, prof. and chmn. biology marine sch., 1983—. Contbr. numerous articles to profl. jours. Mem. Am. Physiol. Soc., Am. Soc. Zoologists, World Mariculture Soc., Soc. Exptl. Biology. Subspecialties: Physiology (biology); Animal physiology. Current work: Anoxic brain metabolism, physiology of diving animals, growth and metabolism of aquaculture species. Home: 561 Satinwood Dr Miami FL 33149 Office: Rosenstiel Sch Marine Sci U Miami 4600 Rickenbacker Causeway Miami FL 33149

LUTZ, RICHARD ARTHUR, marine biologist, educator; b. N.Y.C., June 8, 1949; s. F. Arthur and Alice Eva (Campbell) L.; m. Sarah Hurlburt, Apr. 18, 1981. B.A., U. Va., 1971; Ph.D., U. Maine, 1975. Research asst. U. Maine, Walpole, 1971-75, research assoc., 1975-77; postdoctoral fellow Yale U., New Haven, 1977-79; asst. prof. Rutgers U., New Brunswick, N.J., 1979—. Editor: Mussel Culture and Harvest, 1980, Skeletal Growth of Aquatic Organisms, 1980. Served with USN, 1971-72. Mem. Nat. Shellfisheries Assn. (pres. 1983-84, Thurlow C. Nelson award 1972), AAAS, World Mariculture Soc., Am. Soc. Zoologists, Sigma Xi. Presbyterian. Subspecialties: Deep-sea biology; Ecology. Current work: Biology of deep-sea hydrothermal vents; molluscan aquaculture; molluscan shell structure; invertebrate larval ecology; molluscan larval biology. Home: 52 Main St PO Box 215 Bloomsbury NJ 08804 Office: Rutgers U Dept Oyster Culture Nelson Biol Labs PO Box 1059 Piscataway NJ 08854

LUTZ, THOMAS EDWARD, astronomer; b. Teaneck, N.J., Nov. 20, 1940; s. Harry Joseph and Mary Agnes (Farrell) L.; m. Julie Haynes, July 8, 1967; children: Melissa, Clea. B.M.E., Manhattan Coll., Riverdale, N.Y., 1962; M.S., U. Ill.-Urbana, 1965, Ph.D., 1969. Asst. prof. astronomy Wash. State U., 1969-75, assoc. prof., 1975-81, prof., dir. program in astronomy, 1981—. Contbr. articles to profl. jours. Mem. Am. Astron. Soc., Internat. Astron. Union, Royal Astron. Soc., Astron. Soc. Pacific. Subspecialty: Optical astronomy. Current work: Application of statistical methods to astronomy. Home: NE 1200 McGee Way Pullman WA 99163 Office: Program in Astronomy Washington State U Pullman WA 99164

LUTZKER, JOHN R., psychology educator; b. Burlingame, Calif., Oct. 28, 1947; s. Abe and Anne (Solomon) L.; m. Sandra Zabel, Dec. 23, 1966; children: Dov, Tov. B.A., San Francisco State U., 1969, M.A., 1970; Ph.D., U. Kans., 1973. Asst. neurology Mass. Gen. Hosp., 1972-73; asst. prof. U. Pacific, Stockton, Calif., 1973-78; assoc. prof. coordinator behavior analysis and therapy program So. Ill. U., Carbondale, 1978—. Author: Behavior Change, 1981; editor: The Behavior Therapist, 1981—. Ill. Dept. Pub. Aid grantee, 1980-83. Mem. Am. Psychol. Assn. (exec. com, 1981—). Subspecialty: Behavioral psychology. Current work: Ecobehavioral approach to the treatment and prevention of child abuse and neglect. Office: Behavior Analysis and Therapy Program So Ill U Carbondale IL 62901

LUUS, REIN, chemical engineer, educator; b. Tartu, Estonia, Mar. 8, 1939; emigrated to Can., 1949, naturalized, 1955; s. Edgar and Aili (Prakson) L.; m. Taina Hilkka Inkeri Jaakola, June 17, 1973; 1 son, Brian Markus. B.A., M.A. in Sci, U. Toronto, 1962; A.M., Princeton U., 1963, Ph.D., 1964. Fellow in chem. engring. Princeton U., 1964-65; asst. prof. chem. engring. U. Toronto, Ont., 1965-68, asso. prof., 1968-74, prof., 1974—; dir. Chem. Engring. Research Cons. Ltd., Toronto, Ont., Can., 1966—; cons. in field.; Nat. Research Council Can. sr. fellow Steel Co. Canada, 1972-73. Author: (with L. Lapidus) Optimal Control of Engineering Processes, 1967. Recipient E.W. R. Steacie prize NRC, Can., 1976. Fellow Chem. Inst. Can.; mem. Can. Soc. for Chem. Engring. (ERCO award 1980), Assn. Profl. Engrs. Ont., Sigma Xi. Lutheran. Club: Hart House Squash. Subspecialties: Chemical engineering; Operations research (engineering). Current work: Optimization of engineering processes,suboptimal control, model reduction and parameter estimation. Home: 3 Terrington Ct Don Mills ON M3B 2J9 Canada Office: Dept Chemical Engineering University Toronto Toronto ON M5S 1A4 Canada Through conscientious and continuous effort, the intellectual ability of a person can be substantially improved. A positive attitude is just as important as the ability to tackle a challenging problem. Life is beautiful if lived in an appreciative manner.

LUYKX, PETER VAN OOSTERZEE, biology educator; b. Detroit, Dec. 14, 1937; s. Henrik Maria and Barbara (McFadden) L.; m. Nancy Palmer, Aug. 20, 1960; children: Aurolyn, Diana; m. Jeannette King., Nov. 22, 1978. A.B., Harvard U., 1959; Ph.D., U. Calif.-Berkeley, 1964. Asst. prof. dept. zoology U. Minn., Mpls., 1964-67; asst. prof. dept. biology U. Miami, Fla., 1967-72, assoc. prof., 1972-82, prof., 1982—. Author: Cellular Mechanisms of Chromosome Distribution, 1970; contbr. articles to profl. jours. NIH predoctoral fellow, 1961-64; research grantee, 1964-67, 68-73; NSF grantee, 1978-80, 82—. Mem. AAAS, Genetics Soc. Am., Internat. Union Study of Social Insects, Am. Soc. Cell Biology. Subspecialties: Evolutionary biology; Genome organization. Current work: Sex-chromosome evolution, population cytogenetics, enzyme surveys, behavior and social organization of termites. Office: Univ Miami Dept Biology Coral Gables FL 33124 Home: 2140 SW 80th Ct Miami FL 33155

LVOVSKY, EDWARD ABRAHAM, physician, researcher; b. Jassy, Romania, Jan. 17, 1937; came to U.S., 1974, naturalized, 1980; s. Abracham Boruch and Beila (Partugeis) L.; m. Valentina Geogia Deomichieva, Oct. 17, 1969; children: Boris, Paul. M.D., Leningrad (USSR) Med. Sch., 1961; Ph.D., Radiology Inst., Leningrad, 1967. Practice medicine, Inta, USSR, 1961-63, Leningrad, 1964; research scientist Radiology Inst., 1967-74; vis. scientist NIH, Bethesda, Md., 1974-79; house staff Md. U. Hosp., 1980, George Washington U. Hosp., Washington, 1981-83; asst. prof. radiation medicine Georgetown U., Washington, 1983—. Contbr. chpts. to books and articles to profl. jours. Fogarty fellow, 1974-76. Mem. Am. Soc. Therapeutic Radiologists, AMA, Am. Tissue Culture Assn. Democrat. Jewish. Subspecialties: Immunopharmacology; Cancer research (medicine). Current work: Synthesis and biological effects of interferon inducers; patient care, reseach in field of interferon. Home: 1100 W Side Dr Gaithersburg MD 20878

LWOWSKI, WALTER W., research chemist, educator; b. Garmisch-Partenkirchen, Bavaria, Germany, Dec. 28, 1928; came to U.S., 1959; s. Hans A. F. and Anna (Hanstein) L. Dipl. Chem, U. Heidelberg, 1954; Dr. rer.nat, 1955. Postdoctoral fellow UCLA, 1955-57; jr. faculty U. Heidelberg, Germany, 1957-59; postdoctoral fellow Harvard U., Cambridge, Mass., 1959-60; asst. prof. chemistry Yale U., New Haven, 1960-66; research prof. chemistry N.Mex. State U., Las Cruces, 1966—; dir. Boehringer Mannheim Corp., Indpls., 1980—. Author, editor: Nitrenes, 1970. Fellow AAAS, N.Y. Acad. Scis.; mem. Am. Chem. Soc., Royal Soc. Chemistry, Sigma Xi. Subspecialties: Organic chemistry; Photochemistry. Current work: Chemistry of highly reactive intermediates, nitrenes, heterocyclic compounds, photochemistry, instrumental analysis. Home: 905 Conway Ave Condo 20 Las Cruces NM 88005 Office: Chemistry Dept NM State U Box 3-C Las Cruces NM 88003

LYDA, STUART DAVISSON, plant pathology educator; b. Bridger, Mont., June 6, 1930; s. Clyde Davisson and Olive Josephine (Brottem) L.; m. JoAnne Lucile Koeneke, Apr. 17, 1953; children: Harriette, Thomas, Sonja, Karen, Timothy. B.S. in Agr, Mont. State U., 1956, M.S. in Botany, 1958; Ph.D., U. Calif.-Davis, 1962. Asst. prof. plant pathology U. Nev., 1962-66, assoc. prof., 1966-67; assoc. prof. plant pathology Tex. A&M U., College Station, 1967-77, prof., 1977—. Served with USMC, 1948-52. Mem. Am. Phytopathol. Soc., Mycol. Soc. Am. Republican. Subspecialties: Plant pathology; Microbiology. Current work: Ecology and Physiology of soil-borne plant pathogens. Home: PO Box 9308 College Station TX 77840 Office: Department Plant Science Texas A&M University College Station TX 77843

LYGREE, DAVID GERALD, chemistry educator, university administrator; b. Minot, N.D., Aug. 10, 1942; s. C. Gerald and Esther R. (Fossom) L.; m. Laurae Y. Johnson, Aug. 20, 1966; children: Jedd, Lindsay. B.A., Concordia Coll., Moorhead, Minn., 1964; Ph.D., U. N.D., 1968. Postdoctoral fellow Case Western Res. U., 1968-70; asst. prof. chemistry Central Wash. U., Ellensburg, 1970-73, assoc. prof., 1973-79, prof., 1979—; asst. dean, 1980—. Author: Life Manipulation, 1979. Am. Cancer Soc. fellow, 1968; Research Corp. grantee, 1970. Mem. AAAS, Am. Chem. Soc., Sigma Xi. Subspecialty: Biochemistry (biology). Current work: My current interest is in the biology of aging, and I am preparing to write a book on this subject for the non-specialist. Home: 805 B St Ellensburg WA 98926 Office: Central Washington University Dept Chemistry Ellensburg WA 98926

LYKOS, PETER GEORGE, educator, scientist; b. Chgo., Jan. 22, 1927; s. George Peter and Theodora (Psimoulis) L.; m. Marie Nina Shumicki, July 2, 1950; children—George, Kristina, Andrew. B.S., Northwestern U., 1950; Ph.D., Carnegie Inst. Tech., 1955. Prof. chemistry Ill. Inst. Tech., 1955—, dir. computation center and computer sci. dept., 1963-71; cons. solid state sci. Argonne Nat. Lab., 1958-66; cons. radiation therapy Michael Reese Hosp., Chgo., 1966-71; head computer impact on soc. NSF, 1971-73; Pres. Four Pi, Inc., Oak Park, Ill., 1966-80; chmn. com. on computers in chemistry, chemistry div. Nat. Acad. Scis.-NRC; dir. Assn. Media-based Continuing Engring. Edn. Author tech. articles. Bd. dirs. Interactive Instrnl. TV, 1976-78, Oak Park Public Library, 1976-81; trustee Beacon Unitarian Ch. Served with USNR, 1944-46. Mem. Am. Chem. Soc. (chmn. edn. com. Chgo., co-dir. Operation Interface I, chmn. div. computers in chemistry, mem. com. on profl. tng., adv. bd. Chemistry and Engring. News), Assn. Computing Machinery (chmn. Chgo., nat. lectr., chmn. spl. interest group on computers and soc., liaison to Am. Fedn. Info. Processing Socs. 1976-78, chmn. com. on computers and public policy), Sigma Xi, Alpha Chi Sigma, Phi Sigma. Subspecialties: Theoretical chemistry. Current work: Proselytizing chemists to use computers. Home: 316 N Ridgeland Ave Oak Park IL 60302 Office: Ill Inst Tech Chicago IL 60616

LYKOUDIS, PAUL S., educator; b. Preveza, Greece, Dec. 3, 1926; came to U.S., 1953, naturalized, 1964; s. Savvas Paul and Loukia (Miliaressis) L.; m. Maria Komis, Nov. 26, 1953; 1 son, Michael. Mech. and Elec. Engring. degree, Nat. Tech. U. Athens, Greece, 1950; M.S. in Mech. Engring, Purdue U., 1954, Ph.D., 1956. Mem. faculty Purdue U., 1956—, prof. aeros., astronautics and engring. scis., 1960-73, prof., head sch. nuclear engring., 1973—; dir. Aerospace Scis. Lab., 1968-73; Vis. prof. aero. engring. Cornell U., 1960-61; cons. RAND Corp., 1960—, Argonne Nat. Lab., 1974—. Contbr. numerous papers in field of heat transfer, fluid mechanics, magneto-fluid-mechanics, astrophysics, and fluid mechanics of physiol. systems. Grantee NSF, 1960—. Asso. fellow Am. Inst. Aeros. and Astronautics (former asso. editor jour.); mem. Am. Phys. Soc., Am. Astron. Soc., Am. Nuclear Soc., Sigma Xi. Subspecialty: Nuclear engineering. Home: 222 E Navajo West Lafayette IN 47906 Office: Sch Nuclear Engring Purdue U Lafayette IN 47907

LYLE, EVERETT SAMUEL, JR., soil science educator researcher; b. Dyersburg, Tenn., Mar. 17, 1927; s. Everett Samuel and Helen Louise (Dodson) L.; m. Nancy Evelyn Davis, Aug. 30, 1947; children: Everett Samuel, Donald Benson. B.S. in Forestry, U. Ga., 1951; M.F., Duke U., 1952; Ph.D., Auburn U., 1969. Registered forester, cert. profl. soil scientist, Ala. Staff asst. Woodlands div. Union Camp Paper Corp., Savannah, Ga., 1952-57; assoc. prof. dept. forestry Auburn (Ala.) U., 1957—; cons. in field. Mem. Ala. Surface Mining Commn., 1976. Served with USN, 1944-46. Research grantee EPA, 1977, Ala. Surface Mining Reclamation Council, 1973, Drummond Coal Co., 1980. Mem. Soc. Am. Foresters, Am. Soc. Agronomy, Soil Sci. Soc. Am. Presbyterian. Subspecialties: Soil chemistry; Plant growth. Current work: Land reclamation and forest soils. Home: Route 5 Box 361 Opelika AL 36801 Office: Dept Forestry Auburn Univ Auburn AL 36849

LYLES, DOUGLAS SCOTT, virology educator and researcher; b. Jackson, Miss., July 12, 1950; s. William Scott and Elva (Hall) L.; m. Mary Fennell, June 3, 1972; children: Jessica, Graham. B.A. in Biochemistry, U. Pa., 1972, Ph.D., U. Miss. Med. Ctr., Jackson, 1975. Postdoctoral fellow Rockefeller U., N.Y.C., 1975-78; asst. prof. Bowman Gray Sch. Medicine Dept. Microbiology and Immunology, Winston-Salem, N.C., 1978—. Contbg. author sects. to publs. in field. NIH postdoctoral fellow, 1975-78. Mem. Am. Soc. Virology, Am. Soc. Microbiology, Phi Beta Kappa. Subspecialties: Virology (biology); Membrane biology. Current work: Penetration of enveloped ciruses into host cells. Assembly of viral proteins into cell membranes. Home: 634 Summit St Winston-Salem NC 27101 Office: Dept Microbiology and Immujology Bowman Gray Sch Medicine Winston-Salem NC 27103

LYMAN, GARY HERBERT, cell biologist, cancer researcher, educator; b. Buffalo, Feb. 24, 1946; s. Leonard Samuel and Beatrice Louise L.; m. Carolyn Gertrude Zalewski, Nov. 21, 1979; children by previous marriage: Stephen Leonard, Christopher Henry, Robert Dean. B.A., SUNY-Buffalo, 1968, M.D., 1972; M.P.H., Harvard U., 1982. Diplomate: Am. Bd. Internal Medicine (med. oncology, hematology). Resident in medicine U. N.C.-Chapel Hill, 1972-74; fellow in oncology Roswell Park Meml. Inst., Buffalo, 1974-77; research instr. medicine SUNY Med. Sch.-Buffalo, 1974-77; mem. faculty U. South Fla. Coll. Medicine, Tampa, 1977—, assoc. prof. medicine, 1980—, dir. div. med. oncology, 1979—. Co-author: Cancer Chemotherapy Therapeutics Agents: Handbook of Clinical Data, 2d edit, 1982; Contbr. articles to profl. jours., chpts. to books. Spl. fellow Leukemia Soc. Am., 1976-77; postdoctoral fellow biostats. Harvard U., 1981-82; spl. clin. fellow Roswell Park Meml. Inst., Buffalo, 1975-76. Fellow ACP; mem. Physicians for Social Responsibility. Subspecialties: Oncology; Epidemiology. Current work: Cancer clinical trials, biostatistics, epidemiology, clinical decision analysis. Office: 12901 N 30th St Tampa FL 33612

LYMAN, JOHN, psychology and engineering educator; b. Santa Barbara, Calif., May 29, 1921; s. Oren Lee and Clara Augusta (Young) L. A.B. in Psychology and Math., UCLA, 1943, M.S., 1950, Ph.D. in Psychology, 1951. Research technician Lockheed Aircraft Corp., Burbank, Calif., 1940-43, mathematician, 1943-44; with dept. psychology UCLA, 1947—, asso. prof., 1957-63, prof., 1963—, from instr. to assoc. prof. Sch. Engring. and Applied Sci., 1950-63, prof. Sch. Engring. and Applied Sci., 1963—, chmn. engring. systems dept., head Biotech. Lab.; research engr. Inst. Traffic and Transp., 1967-73; vis. prof. bioengring. Technol. Inst., Delft, Netherlands, 1965; spl. cons. Nat. Acad. Scis., Washington, 1973; cons. VA, Los Angeles, 1962-66, 67—, NIH, 1963-66, 68-73, med. devices div. FDA, 1976—, Perceptronics, Inc., Woodland Hills, Calif., 1978-79, other agys. and cos. Author chpts. in books, articles in profl. jours.; editor in field. Served to lt. (j.g.) U.S. Navy, 1944-46. Recipient numerous fellowships and grants. Fellow Am. Psychol. Assn., Soc. Engring. Psychologists, AAAS, Human Factors Soc. (Paul Fitts award 1971, pres. 1967-68); mem. Biomed. Engring. Soc. (pres. 1980-81), IEEE, Am. Soc. Engring. Edn., Sigma Xi, Tau Beta Pi. Subspecialties: Human factors engineering; Biomedical engineering. Office: UCLA 7619 Boelter Hall U Calif Los Angeles CA 90025

LYMANGROVER, JOHN ROBERT, physiologist, educator; b. Ft. Wayne, Ind., July 24, 1944; s. Robert D. and Mary Rose (Farley) L. B.S., Xavier U., 1966; M.S., U. Ky., 1968; Ph.D., U. Cin., 1972. Postdoctoral fellow Med. Coll. Ohio, Toledo, 1972-75; asst. prof. physiology Tulane U., New Orleans, 1972-80, adj. assoc. prof., 1980—; asst. prof. Bowman Gray Med. Sch., Winston Salem, N.C., 1980—; cons. NIH, 1979-82. Contbr. articles to profl. jours. U. Cin. fellow, 1968-72; Grantee Schleider Edn. Found., 1976-79, Am. Heart Assn., 1978-82, Electric Power Research Inst., 1978-82. Mem. Endocrine Soc., Bioelectromagnetics Soc., Am. Heart Assn., AAAS, Sigma Xi. Democrat. Roman Catholic. Subspecialties: Physiology (medicine); Endocrinology. Current work: Control of pituitary and adrenal hormone secretion and electric field interaction with endocrine tissue. Home: 2441 Lomond St Winston Salem NC 27107 Office: Bowman Gray Med Sch 300 S Hawthorne St Winston Salem NC 27103

LYNCH, CAROL BECKER, biology educator, researcher evolutionary genetics; b. St. Albans, N.Y., Dec. 3, 1942; d. Milton Taylor and Catherine Margaret (Kupsh) Becker; m. George Robert Lynch, Aug. 19, 1967. A.B., Mt. Holyoke Coll., 1964; M.A., U. Mich., 1965; Ph.D., U. Iowa, 1971. Research assoc. U. Iowa, Iowa City, 1971; postdoctoral fellow U. Colo., Boulder, 1972-73; asst. prof. Wesleyan U., Middletown, Conn., 1973-78, assoc. prof., 1979—. Mem. Behavior Genetics Assn., Genetics Soc. Am., Soc. Study Evolution, Am. Soc. Naturalists, Am. Behavior Soc., AAAS. Subspecialties: Evolutionary biology; Animal genetics. Current work: Mechanisms of evolutionary adaptation; studies of genetic correlations in traits involved in thermoregulation and growth. Office: Dept Biology Wesleyan U Middletown CT 06457

LYNCH, DENIS PATRICK, pathologist, educator; b. Kansas City, Kans., Oct. 5, 1951; s. Patrick Edward and Helen Mary (Dragastin) L.; m. Monica Colosimo, June 29, 1973; children: Sydney Alexis, Shannon Meredith. D.D.S., U. Calif.-San Francisco, 1976; cert. oral pathology, U. Ala.-Birmingham, 1978, postgrad., 1977—. Sr. intern in oral medicine U. Calif.-San Francisco, 1975-76; resident in anatomic pathology U. Ala.-Birmingham, 1976-77, chief resident in oral pathology, 1977-78, NIH-Nat. Inst. Dental Research postdoctoral fellow, 1977-80; asst. prof. pathology U. Tex. Dental Br., Houston, 1981—. Editorial bd.: Jour. Dental Edn, 1978-81; cons. editor: Dental Student Mag, 1981—. Tchr. Confrat. of Christian Doctrine, Birmingham, Ala., 1978-81 Tchr. Confrat. of Christian Doctrine, Houston, 1981—. U. Calif.-San Francisco Pres.'s scholar, 1972; Regent's scholar, 1973-76; named Outstanding Instr. U. Ala.-Birmingham, 1977. Mem. Am. Assn. Dental Schs. (v.p. and mem. exec. com. 1977-79, Recognition award 1981), Tex. Assn. Advs. for Health Professions, Internat. Assn. Dental Research, Houston Soc. Clin. Pathology, Omicron Kappa Upsilon. Roman Catholic. Subspecialties: Oral pathology; Pathology (medicine). Current work: Immunopathology and treatment of mucocutaneous disease; nutritional effects on immune response; secretory immunology and dental caries. Home: 5711 Claridge Houston TX 77096 Office: U Tex Dental Br PO Box 20068 Houston TX 77225

LYNCH, JOHN AUGUST, analytical chemist, educator; b. S.I., Jan. 29, 1947; s. John A. and Eleanor Mary (Nitsche) L.; m. Sandra May Fekety, Aug. 28, 1971 (div. 1980); m. Mary Lynn Howard, Jan. 17, 1983. B.S., St. Peter's Coll., 1970; Ph.D. (NSF fellow 1971-73, USPHS fellow 1973-75), Pa. State U., 1970-76. NSF research grantee St. Peter's Coll., 1969-70; teaching asst. Pa. State U., 1970-71; asst. prof. chemistry U. Tenn., Chattanooga, 1975-80, assoc. prof., 1980—. Pres. Chattanooga Wine Tasters, 1981-82. Mem. Am. Chem. Soc. (chmn. Chattanooga sect. 1978-79, sect. Disting. Service award 1981). Subspecialties: Analytical chemistry; Kinetics. Current work: Investigation of chlorine addition to alkenes; use of non-linear regression analysis to discover degree of unsaturation of oils; computer interfacing. Home: 900 Mountain Creek Rd Apt T-405 Chattanooga TN 37405 Office: Dept Chemistry U Tenn Chattanooga TN 37402

LYNCH, MERVIN DEAN, psychology educator, psychologist; b. Grundy Ctr., Iowa, Nov. 29, 1933; s. Bert Leroy and Anna Margaret (Henningsen) L.; m. Vivian Reba Jasper, Dec. 31, 1961; children: Daphid Alexander, Maryl Lee. B.S., Iowa State U., 1956; M.S., U. Wis.-Madison, 1960, Ph.D., 1963. Lic. psychologist, Mass. Research analyst Klau, VanPeiterson, Dunlap, Milw., 1956-57; lectr. communications research Ind. U., 1962-63; asst. prof. U. Mo., 1963-67; assoc. prof. ednl. research Boston-Bouve Coll. Human Devel. Professions, Northeastern U., 1968-74, prof., 1974—; Social Scis. Research Council postdoctoral fellow dept. stats. Harvard U., 1968-69; evaluator Tchrs. Corps, 1979-80; cons. New Eng. Med. Regional Library Consortium. Author: Elements of Statistical Inference for Education and Psychology, 1976, Self Concept: Some Recent Advances in Theory and Research, 1981. Mem. Am. Ednl. Research Assn., Eastern Ednl. Research Assn., Am. Psychol. Assn., Northeast Ednl. Research Assn., Nat. Reading Conf., Lambda Chi Alpha. Unitarian. Club: Minuteman Feline Fanciers (Lexington, Mass.) (pres. 1976-77). Subspecialties: Social psychology; Developmental psychology. Current work: Research interests in the study of self-concept and creativity, assessment of children's behavior and studies of writing style; teaching interests are in statistics, research methods and communication. Home: 27-1 Arlington Rd Woburn MA 01801 Office: Northeastern U 102 Fenway St Boston MA 02115

LYNCH, PETER ROBIN, physiology educator, research radiologist; b. Phila., July 18, 1927; s. Harold Vincent and Elsa (Richter) L.; m. Lynda A. Roller, Dec. 26, 1952 (div. 1978); children: Christopher R., Jonathan D., Elizabeth A.; m. Eileen P. Thomas, June 20, 1978; stepchildren: Cathy Hodgson, Beth Hodgson. B.S., U. Miami, Fla., 1950; M.S., Sch. Medicine Temple U., 1954; Ph.D., Temple U., 1958. Cert. diving instr., CPR instr. Instr. physiology Sch. Temple U., Phila., 1958-60, asst. prof. physiology, 1960-66, assoc. prof. physiology, 1966-70, prof. physiology, 1970—, prof. radiology, 1970—; adj. prof. Druckkammer Laboratorium, Zurich, Switzerland, 1977-78. Contbr. articles to sci. jours. Served with USN, 1944-46. Mem. Am. Physiol. Soc., Phila. Physiol. Soc. (pres. 1982-83), Undersea Med. Soc., N.Am. Soc. for Cardiac Radiology, Am. Littoral Soc. (founder, council 1980). Quaker. Club: Submariners (Ambler, Pa.) (pres. 1972-73). Subspecialties: Physiology (medicine); Imaging technology. Current work: Cardiovascular physiology, experimental radiology and underwater medicine. Home: 1316 Zachary Rd Roslyn PA 19001 Office: Temple U Sch Medicine 3420 N Broad St Philadelphia PA 19140

LYNCH, ROBERT MICHAEL, statistics educator; b. Bklyn., May 30, 1944; s. John Patrick and Emily Marie (Matson) L.; m. Terry Lynn Bell, Dec. 13, 1969; children: Christopher Patrick, Cary Amanda. B.S., SUNY-Brockport, 1966; Ph.D., U. No. Colo., 1971. Asst. prof. stats. Eastern Ill. U., Charleston, 1971-72; prof. stats. U. No. Colo., Greeley, 1973—; vis. prof. stats. Coll. of V.I., St. Thomas, summer 1982; researcher Oak Ridge Associated Univs., 1982. Editor: Topics in Management Information Systems, 1973, Bayesian Decision Analysis, 1975, Techniques of Business Research, 1974; cons. editor: Jour. Exptl. Edn, 1978—; reviewer: Jour. Computing Revs, 1973—; editorial collaborator: Current Index to Stats, 1980—. Exec. Office of Gov. of Ill. fellow, 1972-73; Fulbright scholar, Thailand, 1978-79. Fellow Royal Statis. Soc. London; mem. Assn. Computing Machinery, Am. Inst. Decision Scis. Subspecialties: Statistics; Information systems (information science). Current work: Professor of statistics and information systems; statistical computing, statistical methods, Data Base Management Systems. Home: 2213 27th Ave Greeley CO 80631 Office: U No Colo Greeley CO 80639

LYNCH, SEAN ROBORG, hematology educator, researcher; b. Johannesburg, South Africa, Mar. 27, 1938; came to U.S., 1977; s. John Rudolph and Dagny (Sondergaard) L.; m. Alison Marah Storrar, Apr. 30, 1966; children: John, Heather. M.B., B.Ch., U. Witwatersrand, 1961, M.D., 1970. Diplomate: Am. Bd. Internal Medicine. Specialist physician U. Witwatersrand, Johannesburg, 1972-77; assoc. prof. hematology U. Kans. Med. Ctr., Kansas City, 1977-81, prof., 1981—. Contbr. articles to profl. jours. USPHS postdoctoral fellow, 1969. Fellow ACP; mem. Am. Soc. Hematology, Central Soc. Clin. Research, Am. Fedn. Clin. Research. Subspecialties: Hematology; Nutrition (medicine). Current work: Hematology, iron metabolism, iron nutrition. Office: U Kans Med Ctr Rainbow and 39th Sts Kansas City KS 66103

LYNDS-CHERRY, PATRICIA GAIL, psychologist; b. Woodlake, Calif., Feb. 7, 1950; d. Edgar David and Frances Jean (Eberle) Lynds; m. Albert Lee Cherry, Nov. 13, 1982. B.A., Calif. State U., 1972; M.A., U. Nebr., 1975, Ph.D., 1977; postgrad., U. Calif.-Davis, 1978-81. Project coordinator U. Calif.-Davis, 1979-80; psychologist Sacramento County Office Edn., 1980-81, Kings County Supt. of Schs., Hanford, Calif., 1981—. Contbr. articles to profl. jours. Chmn. Kings County Child Abuse Com., 1982. Maude Hammond Fling fellow U. Nebr., 1973-74. Mem. Am. Psychol. Assn., Calif. Assn. Sch. Psychologists and Psychometrists, Assn. Women in Sci., Women in Neursci. Club: VA-25 Officers Wives. Subspecialties: School psychology; Psychobiology. Current work: Child abuse prevention and intervention, genetic bases of learning disabilities, spl. edn. for bilingual/bicultural students. Office: Kings County Supt of Schools Government Center Hanford CA 93230

LYNESS, JAMES N., applied mathematician, research scientist. B.A., Oxford U., 1954, M.A., 1955, D.Phil., 1957. Lectr. U. New South Wales, Sydney, Australia, 1960-66; applied mathematician Argonne (Ill.) Nat. Lab., 1967—. Subspecialties: Mathematical software; Numerical analysis. Current work: Scientific research in numerical analysis and numerical computer software evaluation. Office: Argonne Nat Lab MCS Argonne IL 60439

LYNN, GEORGE LESLIE, psychologist; b. N.Y.C., Oct. 4, 1947; s. Fred H. and Nellie K. (Riszcerles) L. B.A., Hunter Coll., 1971; M.A., Columbia U., 1972; Psy.D., Ill. Sch. Profl. Psychology, 1981. Diplomate: Am. Acad. Behavioral Medicine.; lic. psychologist, N.Y.; Cert. psychotherapist. Counselor Bellevue Hosp., N.Y.C., 1973-74; psychologist South Beach Psychiat. Ctr., S.I., 1974-75; psychotherapist L.I. Consultation Ctr., Rego Park, 1975-76; counselor Northwestern U. Med. Ctr., Chgo., 1977-78; psychology intern Manhattan Psychiat. Ctr., Ward's Island, 1978-79; supr., psychotherapist Community Guidance Service, N.Y.C., 1975—; supr., faculty Postgrad. Ctr. for Metnal Health, N.Y.C., 1980—; supr. Inst. for Human Identitiy, 1982; faculty New Sch. for Social Research, 1982—; psychologist Metro. Hosp.-N.Y. Med. Coll., 1982—. Contbr. articles to profl. jours.; commentator, WOR-Radio, Nighttalk Show, N.Y.C., 1980—. Cons. psychologist Lenox Hill Neighborhood Assn., N.Y.C., 1972-73; bd. govs. Commn. on Consumer and Patient Rights, N.Y.C., 1979—. L.I. Cons. Center fellow, 1973-75. Fellow Am. Orthopsychiatric Assn., Am. Inst. Psychotherapy & Psychoanalysis; mem. Acad. Psychosomatic Medicine, Am. Acad. Psychotherapists, Am. Group Psychotherapy Assn., Am. Psychol. Assn., AAAS, N.Y. Acad. Sci., Soc. Pediatric Psychology, Assn. for Birth Psychology, Assn. Applied Psychoanalysis. Current work: Research activities as relates to psychotherapy, psychosomatic di. Inventor Draw-A-Group Test (personality assessment), 1981, Psychosomatic Therapy (psychol. treatment), 1975. Address: 165 E 89th St #3D New York NY 10028

LYNN, WALTER ROYAL, civil engineering educator, university administrator; b. N.Y.C., Oct. 1, 1928; s. Norman and Gussie (Gdalin) L.; m. Barbara Lee Campbell, June 3, 1960; children: Michael Drew. B.S., U. Miami, 1950; M.S., U. N.C., 1955; Ph.D., Northwestern U., 1963. Registered profl. engr., N.Y. State; registered land surveyor, Fla. Land surveyor Ehly Constrn. Co., Miami, Fla., 1950-51; chief party Rader Engring. Co., Miami, 1951; supt. sewage treatment, lectr. civil engring. U. Miami, 1951-53, asst. prof. mech. engring., 1954-55; asst. prof. civil engring., 1955-57, research asst. prof. marine lab., 1957-58, asso. prof. civil and indsl. engring., 1959-61; dir. research Ralph B. Carter Co., 1957-58; asso. prof. san. engring. Cornell U., Ithaca, N.Y., 1961-64, prof. civil and environ. engring., 1964—, dir. Center Environ. Quality Mgmt., 1966-76, dir. Sch. Civil and Environ. Engring., 1970-78, dir. Program on Sci., Tech. and Society, 1980—, adj. prof. pub. health, 1971—, trustee, 1980—; mem. spl. adv. commn. solid wastes NRC, 1968-76; com. to rev. Washington met. water supply study Nat. Acad. Engring., 1976-84, chmn., 1980-84; chmn. Bd. Water Sci. and Tech. NRC, 1982-85; cons. WHO, 1969—; chmn., 1980—; cons. Rockefeller Found., 1976—, SEARO, 1978. Editor: (with A. Charnes) Mathematical Analysis of Decision Problems in Ecology, 1975; asso. editor: Jour. Ops. Research, 1968-76, Jour. Environ. Econs. and Mgmt, 1972—; Contbr.: chpt. to Human Ecology and Public Health, 1969; author articles. Chmn. Ithaca Mayor's Citizens Adv. Com., 1964-65, Ithaca Urban Renewal Agy., 1965-68; trustee Cornell U., 1980—; bd. dirs. Cornell Research Found., 1978—. Served with AUS, 1946-48. Fellow ASCE; mem. AAAS, Inst. Mgmt. Sci., Ops. Research Soc. Am.,

Nat. Acad. Engrs. Mex. (corr.), Sigma Xi, Phi Kappa Phi. Subspecialties: Civil engineering; Operations research (engineering). Current work: Analysis of university /industrial research. Home: 102 Iroquois Pl Ithaca NY 14850

LYNTS, GEORGE WILLARD, petroleum geologist; b. Edgerton, Wis., July 26, 1936; s. George Edward and Eleda May (Stoke) L.; m. Joan Leigh McCullough, Aug. 29, 1959; children: Sharolyn, Margaret. B.S., U. Wis., 1959, M.S., 1961, Ph.D., 1964. Instr. U. Wis., Madison, 1962-63; fellow Lamont-Doherty Geol. Obs., Columbia U., Palisades, N.Y., 1964-65; prof. geology Duke U., Durham, N.C., 1965-82; geologist Cities Service Co., Tulsa, 1982—; cons. Bechtel Corp. Los Angeles, 1971-72, Ebasco Services, Inc., N.Y.C., 1979. Author: (with others) Miami Geological Memoir 1, 1971; contbr. articles to profl. jours. Served with USNR, 1955-56. Fellow NSF, summer 1962, USPHS, 1964-65. Mem. Am. Assn. Petroleum Geologists, Am. Geophys. Union, Internat. Paleontol. Assn., Cushman Found. for Foramiferal Research, AAAS. Subspecialties: Geology; Offshore technology. Current work: Development and application of new exploration technology primarily by the use of predictive tectonic models to frontier basins. Home: 702 Meadowood Dr Broken Arrow OK 74012 Office: Cities Service Co PO Box 3908 Tulsa OK 74102

LYON, RICHARD EVAN, polymer scientist; b. Falls Church, Va., June 4, 1951; s. Richard John and Mary Dina (Succi) L.; m. Virginia Lee O'Brien, July 1, 1978; children: Blair, Marie. B.S., U. Mass., 1975, M.S., 1981. Research asst. U. Tex.-Port Aransas, 1975-77; research chemist Foss Mfg. Co. Inc., Haverhill, Mass., 1977-79; postgrad. U. Mass., Amherst, 1979—. Mem. Am. Phys. Soc., Am. Chem. Soc. (polymer div.). Subspecialties: Polymer engineering; Thermodynamics. Current work: Thermomechanical research on polymers including structure, property, thermodynamic relationships. Home: 180 Summer St Amherst MA 01002 Office: Polymer Sci and Engring Dept U Mass Amherst MA 01003

LYON, WILLIAM SOUTHERN, JR., company administrator; b. Pulaski, Va., Jan. 25, 1922; s. William Southern and Irene (Hunter); m. Carey Helen Greer, May 13, 1946; children: Victoria Carey, William III. B.S. in Chemistry, U. Va., 1943; M.S., U. Tenn., 1967. Chemist E.I. du Pont, Belle, W.Va., 1943-44, Richland, Wash., 1944-45; lab foreman Tenn. Eastman, Oak Ridge, 1945-47; chemist Oak Ridge Nat. Lab., 1947-61, group leader, 1961-76, sect. head, 1976—; cons. Thai Atomic Energy Peace, Bangkok, Thailand, 1967-70, Internat. Atomic Energy Agy., Vienna, 1970, 76. Author: Trace Element Measurements at the Coal Fired St. Plant, 1977; editor-author: Guide to Activation Analysis, 1964; editor: Nuclear and Atomic Activation Analysis, 1976, Analytical Chemistry in Nuclear Fuel Reprocessing, 1977, Radio element analysis: Program and Problems, 1980, Analytical Chemistry in Nuclear Technology, 1982. Recipient Radiation Industry award Am. Nuclear Soc., 1980; George Hevesy medal J. Radioanalical Chemistry, Toronto, 1981. Mem. Am. Nuclear Soc. (exec. com. 1976-82, vice chmn.-elect 1982-83), Am. Chem. Soc. Episcopalian. Clubs: Fox Den Country, Concord, Tenn. Subspecialties: Analytical chemistry. Current work: Neutron activation analysis, radiochemistry, scientometrics, technical information. Home: 7007 Rockingham Dr Knoxville TN 37919 Office: Oak Ridge Nat Lab E160 4500S Oak Ridge TN 37830

LYONS, JOHN WINSHIP, chemist; b. Reading, Mass., Nov. 5, 1930; s. Louis M. and Margaret (Tolman) L.; m. Grace Hanley, Nov. 28, 1953; children: Margaret, Mary Ann, John H., Louis M. A.B., Harvard U., 1952; A.M., Washington U., St. Louis, 1963, Ph.D. in Phys. Chemistry, 1964. Various positions in research and devel. mgmt. Monsanto Co., St. Louis, 1955-73; dir. Center for Fire Research, Inst. Applied Tech., Nat. Bur. Standards, Washington, 1973-77, 1977-78, 1978—; mem. adv. com. engring. NSF, 1981—. Author: Viscosity and Flow Measurement, 1963, The Chemistry and Uses of Fire Retardants, 1970; contbr. chpts. to books, articles to profl. jours. Mem. adv. council Coll. Engring., U. Md.; bd. dirs. Nat. Fire Protection Assn.; cochmn. U.S. Japan Nat. Resources Panel of Fire Research. Served with U.S. Army, 1953-55. Recipient Presdl. Mgmt. Improvement award, 1977; Gold medal Dept. Commerce, 1977; Presdl. Rank award of Disting. Exec., 1981. Fellow AAAS, Washington Acad. Sci.; mem. Am. Chem. Soc. (chmn. St. Louis 1970), Am. Inst. Chem. Engrs. Roman Catholic. Subspecialties: Physical chemistry; Materials (engineering). Current work: Fire research, polyelectrolyte solutions, rheology. Patentee in field. Office: Nat Bur Standards Nat Engring Lab Tech B119 Washington DC 20234

LYONS, JOHN W(INSHIP), government agency administrator, chemist; b. Reading, Mass., Nov. 5, 1930. A.B. in Chemistry, Harvard U., 1952; A.M. in Phys. Chemistry, Washington U., St. Louis, 1963, Ph.D., 1964. With Monsanto Co., 1955-73, group leader, sect. mgr. research dept., inorganic chems. div., 1962-69, mgr. comml. devel., head fire safety center, 1969-73; mem. ad hoc panel on fire research Nat. Bur. Standards, Washington, 1971-73; dir. Center for Fire Research, 1973-77, Inst. Applied Tech., 1977-78, Nat. Engring. Lab., 1978—; acting dep. dir. Nat. Bur. Standards, 1983—; chmn. Products Research Com. (trust which adminstrs. fire research funds), 1974-79; bd. dirs. Nat. Fire Protection Assn., 1978—; vis. lectr. various univs.; co-chmn. U.S.-Japan Natural Resources Panel on Fire Research, 1975-78; mem. adv. com. on engring. NSF, 1981—; mem. adv. council U. Md. Coll. Engring., 1980—. Author: Viscosity and Flow Measurement, 1963, The Chemistry and Uses of Fire Retardants, 1970; contbr. numerous articles to profl. publs. Served with U.S. Army, 1953-54. Recipient Gold medal Dept. Commerce, 1977, Pres.'s Mgmt. Improvement award White House, 1978, Pres.'s Disting. Exec. Rank award, 1981. Fellow AAAS, Washington Acad. Sci.; mem. Am. Chem. Soc. (chmn. St. Louis sect. 1971-72), Am. Inst. Chem. Engrs., Sigma Xi. Subspecialty: Engineering research administration. Current work: Acting deputy director National Bureau of Standards. Office: Nat Engring Lab Nat Bur Standards Washington DC 20234

LYONS, KENNETH PAUL, nuclear medicine physician and, educator; b. Worcester, Mass., Nov. 12, 1938; s. William Patrick and Jean (Mattson) L.; m. Joanna Harris, Sept. 5, 1975; children: James, Kathleen, Kevin, Michael. B.S., Loyola U., Los Angeles, 1961; M.D., Creighton U., 1965. Diplomate: Am. Bd. Nuclear Medicine. Intern St. Joseph Hosp., Omaha, 1965-66; resident in internal medicine Harbor Gen. Hosp., Los Angeles, 1968-71; instr., fellow in nuclear medicine UCLA, 1971-73; asst. prof. radiol. scis. U. Calif.-Irvine, 1973-81, assoc. prof., 1981—; chief nuclear medicine Long Beach VA Med. Ctr., 1973—; dept. assoc. dir. 1982; prin. investigator VA, 1981-82. Author: Cardiovascular Nuclear Medicine, 1981; editor: Atlas of Nuclear Medicine, 1981, Jour. Clin. Nuclear Medicine, 1975—; contbr. over 100 articles to med. jours. Served to capt. USAF, 1966-68. Research fellow USPHS, 1962-66. Mem. Soc. Nuclear Medicine (sec./treas. So. Calif. chpt. 1982-83), Am. Coll. Nuclear Physicians. Republican. Roman Catholic. Subspecialties: Nuclear medicine; Internal medicine. Current work: Detection and assessment of coronary artery disease; development of miniature avalanche radiation detectors. Office: Long Beach VA Med Ctr 5901 E 7th St Long Beach CA 90822

LYONS, MICHAEL JOSEPH, microbiologist, virologist, educator; b. Cork, Ireland, Sept. 16, 1930; came to U.S., 1960, naturalized, 1973; s. Michael J. and Margaret M. (Hopkins) L.; m. Yvonne J. T. Barnett, Sept. 7, 1960; children: Fiona, Conor, Patricia, Desmond. B.S., Nat. U. Ireland, 1953, M.S., 1954; Ph.D., U. Glasgow, Scotland, 1959. Research assoc. Rockefeller Univ., N.Y.C., 1961-66; asst. prof. dept. microbiology Sch. Medicine, U. Pa., Phila., 1966-68; asst. prof. Cornell U. Med. Coll., N.Y.C., 1968-76; adj. assoc. prof. Lab of Bacteriology and Immunology Rockefeller U., N.Y.C.; assoc. dir. Center for Natural Scis., N.Y. Inst. Tech., Old Westbury, N.Y. Contbr. articles on microbiology/virology and chem. carcinogenesis to profl. jours. USPHS fellow, 1961-63; NIH grantee, 1969-73; Nat. Multiple Sclerosis Soc. grantee, 1978-82. Mem. Harvey Soc., Am. Soc. for Microbiology, Sigma Xi. Club: Rockefeller U. Faculty and Students (N.Y.C.). Subspecialties: Neuroimmunology; Virology (medicine). Current work: Virus induced obesity, pathogenesis of multiple sclerosis. Home: 53 Eiler Ln Irvington NY 10533 Office: Rockefeller U 1230 York Ave New York NY 10021

LYONS, WILLIAM KIMBEL, mathematician; b. Uniontown, Pa., June 2, 1953; s. William Leon and Laurel (Provance) L. B.S., Ga. Inst. Tech., 1974; M.S., Brown U., 1978, Ph.D., 1980. Instr. math. Simon's Rock Coll., Great Barrington, Mass., 1979-80; research engr. UOP Process, Des Plaines, Ill., 1980—. Mem. Am. Math. Soc., Soc. Indsl. and Applied Math, Soc. Computer Simulations, N.Y. Acad. Scis., Sigma Xi. Subspecialties: Mathematical software; Chemical engineering. Current work: Mathematical modelling and curve fitting of industrial chemical processors for use in process simulators. Co patentee Sulfolane and parex process simulators. Home: 6724 S Echo Ln Apt 2 Westmont IL 60559 Office: UOP Process Div 20 UOP Plaza Des Plaines IL 60016

LYONS-RUTH, KARLEN, research psychologist, educator; b. Sheffield, Ala., Aug. 21, 1945; d. Vernon Everett and Helen (Karlen) Lyons; m. William Atkinson Ruth, Aug. 11, 1973; children: Adrienne, Gregory. B.A., Duke U., 1967; Ph.D., Harvard U., 1974. Fellow Boston U. Med. Sch., 1974-77; instr. psychology Harvard U. Med. Sch., Cambridge, Mass., 1977—; psychologist Cambridge Hosp., 1977-80, prin. investigator social devel. project, 1981—; dir. Family Support Project, Somerville, Mass., 1980—. Research grantee NIMH, 1977, 80; recipient New Investigator award Nat. Inst. Child Health and Human Devel., 1981. Mem. Am. Psychol. Assn., Soc. for Research in Child Devel. Subspecialty: Developmental psychology. Current work: Social development, developmental psychopathology, effectiveness of prevention services for infants at risk for emotional disorder. Office: Dept Psychology Cambridge Hosp 1493 Cambridge St Cambridge MA 02139

LYTLE, JAMES MARK, electronic kits and educational products company executive, consultant; b. Pitts., Sept. 16, 1939; s. William Allen and Mary Elizabeth (Leahey) L.; m. Katherine Martha Rausch, Feb. 21, 1964; children: Mark, David, Susan. B.S., Capitol Inst. Tech., 1968. Engr. RCA Service Co., Bendix, McGraw Hill, 1967-79; design engr. Heath Co., St. Joseph, Mich., 1979-82, product line mgr., 1982—. Served with USN, 1957-67. Subspecialties: Electronics; Robotics. Current work: Defining hi-technology electronic products to be used for educational purposes. Home: 5780 Echo Ridge Stevensville MI 49127 Office: Heath Co Hilltop Rd St Joseph MI 49085

MA, BENJAMIN MINGLEE, nuclear engineering educator; b. Szeyang, Kiangsu, China, Feb. 2, 1924; s. Donald K. and Chin C. Chang; m. Phyllis Y., Dec. 28, 1956; children: Stephen S., Theodore S. B.S., Nat. Central U., Nanking; M.S., Eng. D., Stanford U.; Ph.D., Iowa State U. Registered profl. mech. engr., Ohio; profl. nuclear engr., Calif. Instr. U. Mich.-Ann Arbor, 1951-54, asst. prof. to assoc. prof., 1954-63; assoc. prof. Iowa State U., Ames, 1963-76, Fulbright prof. nuclear engring., 1977—; cons. NASA Space Propulsion Research Ctr., 1963-64, Argonne (Ill.) Nat. Lab., 1965-71, IAEA-CNEA, Argentina, 1973-74, Hanford Energy Devel. Lab., Richland, Wash., 1975-76. Author: Design of Nuclear Reactor Fuel Element, Nuclear Reactor Materials and Applications; contbr. over 150 sci. articles to profl. publs. Fulbright fellow, 1977; fellow Iowa State U. Fellow Am. Nuclear Soc.; mem. Am. Phys. Soc., ASME, Am. Soc. Engring. Edn., IEEE, Sigma Xi. Subspecialties: Nuclear engineering; Nuclear fission. Current work: Nuclear energy: nuclear fusion, nuclear fission, nuclear radioisotopic energy; pleasma stability; thermoelectric power; reactor design; fuel element design; safety analysis; safety systems. Home: 701 Ash Ave Ames IA 50010

MAACK, THOMAS MICHAEL, physiologist, medical educator; b. Insterburg, Germany, July 17, 1935; came to U.S., 1965; s. Hans H. and Kate (Maligson) M.; m. Isa Tavares, Jan. 13, 1962; children: Marisa Tavares, Marcia Tavares. M.D., Escola Paulista de Medicina, Brazil, 1961, hon. doctorate in nephrology, 1981. Intern Hosp. das Clinicas, São Paulo, 1961-62; instr. physiology Faculdade de Medicina da Universidade de Sao Paulo, 1963-64; instr. to asst. prof. physiology SUNY-Upstate Med. Ctr., Syracuse, 1965-69; mem. faculty Cornell U. Med. Coll., N.Y.C., 1969—, prof. physiology, 1976—; investigator Mt. Desert Island Biol. Lab., Salisbury, Cove, Maine, 1966-72. Mem. editorial bd.: Am. Jour. Physiology, 1971-82; contbr. sci. articles to profl. publs. Recipient Associação prize Faculdade de Medicina, Universidade de São Paulo, 1961; Am. Heart Assn. Advanced Research fellow, 1966. Mem. Am. Physiol. Soc., Am. Soc. Nephrology, Internat. Soc. Nephrology, Sociedade Brasileira Para o Progresso de Ciencia, Salt and Water Club. Subspecialties: Physiology (biology); Nephrology. Current work: Renal physiology; filtration, transport and metabolism of proteins and peptide hormones; mechanisms and regulation of fluid and electrolyte transport; isolated kidney preparation; lysosomal physiology and endocytosis in epithelial tissues and cells. Home: 1161 York Ave New York NY 10021 Office: Cornell U Med Coll Physiology Dept 1300 York Ave New York NY 10021

MAATUK, JOSEF, engineering company executive; b. Bagdad, Iraq, Mar. 30, 1942; came to U.S., 1969, naturalized, 1975; s. Nahgi and (Israj) M.; m. Debbie Lee Hecox, July 12, 1981; 1 child: Briyanna. B.S.M.E., Technion, Haifa, Israel, 1966; M.S.M.E., UCLA, 1973, Ph.D., 1976. Registered profl. engr., Calif. Pres. Seven Engring., Santa Monica, Calif., 1978—; cons. to Carl Zeiss in, Oberkochen, Germany, 1981, Daimler-Benz, Stuttgart, Germany, 1977, Ford Aerospace, Palo Alto, 1981. UCLA Chancellor's fellow, 1972-75. Mem. Soc. Automotive Engrs., Nat. Soc. Profl. Engrs., IEEE. Subspecialties: Theoretical and applied mechanics; Aerospace engineering and technology. Current work: Multi-disciplinary projects in dynamics, structures and controls. Developer computationally efficient dynamic formulation for multi-rigid and multi-flexible systems in open or closed loop configurations.

MABRY, PAUL DAVIS, JR., psychobiologist, research scientist, educator; b. Meridian, Miss., Sept. 28, 1943; s. Paul Davis and Frances Elizabeth (Thigpen) M. B.S., Millsaps Coll., Jackson, Miss., 1965; M.S., U. Miss., 1967, Ph.D., 1970. Predoctoral fellow neurosurgery U. Miss. Med. Center, 1969-70; research assoc. neurosci. and behavior program Princeton U., 1970-76; assoc. prof., chmn. psychology dept., head behavioral and phys. scis. Sacred Heart Coll., Belmont, N.C., 1976—. Contbr. articles to profl. jours. Mem. Soc. Neurosci., Am. Psychol Assn., Eastern Psychol. Assn., AAAS, Sigma Xi. Current work: Functional devel. of brain, research, education, psychobiology. Home: 6500 Garsdale Pl Charlotte NC 28210 Office: Dept Psychology Sacred Heart Coll Belmont NC 28012

MACAGNO, EDUARDO ROBERTO, neurobiologist, educator; b. San Juan, Argentina, June 13, 1943; s. Enzo Oscar and Matilde (Cusminsky) M.; m. Nancy Parilla, Oct. 1, 1974. B.A., U. Iowa, 1963; M.S., Columbia U., 1965, Ph.D. in Physics, 1968. Assoc. prof. biol. scis. Columbia U., N.Y.C., 1980—. Contbr. articles to profl. jours. Mem. Soc. Neurosci., Biophys. Soc., AAAS, Marine Biol. Lab., N.Y. Acad. Scis., Phi Beta Kappa, Sigma Xi. Subspecialties: Neurobiology; Developmental biology. Current work: Studies in formation of synaptic connections in development and regeneration of the nervous system; structure and function of visual systems. Office: Columbia U 1003B Fairchild Ctr New York NY 10027

MACARIO, ALBERTO JUAN LORENZO, research physician, immunology laboratory administrator, educator; b. Naschel, Argentina, Dec. 1, 1935; s. Alberto Carlos and Maria Elena (Giraudi) M.; m. Everly Conway, Mar. 16, 1963; children: Alex, Everly. M.D., Nat. U. Buenos Aires, 1960. Intern Ramos Mejia Hosp., Buenos Aires, 1958-60, resident, 1960, Rivadavia Hosp., Buenos Aires, 1961-62, physician, hematologist, 1962-64; fellow NRC Argentina, Buenos Aires, 1964-69; head dept. radioactive isotopes Inst. for Hematological Investigations, Nat. Acad. Medicine, Buenos Aires, 1967-69; Eleanor Roosevelt fellow dept. tumorbiology Karolinska Inst., Stockholm, 1969-71; mem. sci. staff Lab. Cell Biology NCR of Italy, Rome, 1971-73; head Lab. Immunology, Internat. Agy. Research on Cancer WHO, Lyons, France, 1973-74; research scientist Brown U., 1974-76, N.Y. State Dept. Health, Albany, 1976-79, chief hematology, 1979-81, dir. clin. and exptl. immunology sect., 1981—; reviewer grants NIH, 1976-79; supt. cytohematology segment Hematology Proficiency Testing Program State N.Y., 1976-79, dir., 1979-81; lectr. in field. Contbr. chpts. to books, articles to profl. jours. Ford Found.-Nat. Acad. Sci. fellow, 1968; grantee in field; recipient Diploma de Honor prize Nat. U. Buenos Aires, 1961; Bernardino Rivadavia prize Nat. Acad. Medicine Argentina, 1967; Ciencia e Investigacion prize Argentinian Soc. Advancement Sci., 1967. Mem. Scandinavian Soc. Immunology, Italian Assn. Immunologists, French Soc. Immunology, Am. Assn. Immunologists, Am. Soc. Microbiology. Subspecialties: Cellular engineering; Cell and tissue culture. Current work: Immunochemistry-immunobiology of morphogenesis of surface and supracellular structures; monoclonal antibodies for microbial biotechnology-bioengineering; archaebacteria, immunology, immunochemistry, ecology, biotechnology; monoclonal antibodies, hormones and receptors, biotechnology; methanogens surface and cell-wall immunochemistry and architecture; erythroblastic nest, morphogenesis, cell-cell communication, differentiation. Office: Center for Laboratories and Research New York State Department of Health Albany NY 12201

MACASAET, FRANCISCO FRIGINAL, pathologist, army officer; b. Infanta, Quezon, Philippines, Mar. 19, 1939; came to U.S., 1965, naturalized, 1979; s. Ricardo Orozco and Juliana (Friginal) M.; m. Evelyn Paralejas, Mar. 9, 1946; children: Eloise, Joel, David, Alan. A.A., Far Eastern U., Manila, 1958, M.D., 1963. Diplomate: Am. Bd. Pathology. Intern Far Eastern Univ. Hosp., 1962-63; resident Silliman U. Med. Ctr., Philippines, 1963-65; fellow in virus research Kans. U. Med. Ctr., 1965-67; research physician Silliman U. Med. Center, 1967-70; resident in pathology Cleve. Met. Gen. Hosp., 1970-73; fellow in microbiology Mayo Clinic, Rochester, Minn., 1974; asst. prof. pathology Case Western Res. U. and Met. Gen. Hosp., Cleve., 1975-78; maj. U.S. Army, 1978, advanced through grades to lt. col., 1982; clin. pathologist U.S. Army Research Inst. Infectious Diseases, Ft. Detrick, Md., 1978—. Contbr. articles to profl. jours. Recipient Spl. award Nat. Sci. Devel. Bd., Philippines, 1974. Fellow Am. Soc. Clin. Pathologists; Mem. Coll. Am. Pathologists, Am. Soc. Microbiology, N.Y. Acad. Sci. Subspecialties: Microbiology (medicine); Pathology (medicine). Current work: Clinical pathology, virology, rapid diagnosis, enzyme-linked immunospecific assay. Office: US Army Med Research Inst of Infectious Diseases Fort Detrick Frederick MD 21701

MACBRYDE, BRUCE, plant conservationist; b. St. Louis, May 21, 1941; s. Cyril M. and Anita E. (Koehler) MacB.; m. Olga S. Herrera Carvajal, Aug. 24, 1968; 1 son, Brendon Douglas. A.B., Washington U., St. Louis, 1963, Ph.D., 1970. Founder herbarium, researcher Pontificia Universidad Catolica del Ecuador, Quito, 1970-72; extramural asst. prof. St. Louis U., 1970-72; postdoctoral fellow Mo. Bot. Garden, 1970-72; research assoc. U. B.C. Bot. Garden, Vancouver, Can., 1972-75; botanist Office Endangered Species, U.S. Fish and Wildlife Service, Dept. Interior, Washington, 1975—. Co-author: Prevalent Weeds of Central America, 1975, Vascular Plants of British Columbia, 1977; co-editor: Genetics and Conservation, 1983; contbr. articles to sci. and popular jours. Bd. dirs. Biol. Inst. Tropical Am. Grantee Ga.-Pacific Corp., Gulf Oil Co. Mem. Bot. Soc. Am. (chmn. conservation com. 1976, 77), Internat. Assn. Plant Taxonomy, Nat. Acad. Scis., Sigma Xi. Presbyterian. Current work: Endangered U.S. plants; conservation in Latin America; plant taxonomy synthesis and outreach. Office: US Fish and Wildlife Service Office Endangered Species Washington DC 20240

MACCABEE, BRUCE SARGENT, research physicist; b. Rutland, Vt., May 6, 1942; s. Earle Philbrick and Bernice Frances (Sargent) M.; m. Barbara Chistine Shoeneman, Sept. 9, 1982; 1 dau., Vanessa. B.S., Worcester Poly. Inst., 1964; M.S., Am. U., 1967, Ph.D. in Physics, 1970. Research assoc. Am. U., Washington, 1970-71; cons. Compackager Corp., 1970-73, Tracor, 1970-71, Sparcom, Inc., 1971, Sci. Applications, Inc., 1973-74; research physicist Naval Surface Weapons Center, White Oak, Md., 1972—; Chmn. Fund for UFO Research. Contbr. articles to sci. jours. Recipient profl. awards, 1978, 80. Mem. Am. Phys. Soc., Optical Soc., AAAS, Fedn. Am Scientists, Cousteau Soc. Subspecialties: Laser physics; Atmospheric optics. Current work: High energy lasers, laser generation of underwater sound, atmospheric opics, optics, chemical lasers. Patentee in electronics, optics and lasers.

MACCALLUM, CRAWFORD JOHN, physicist; b. N.Y.C., May 28, 1929; s. Ian Crawford and Lucile Annette (Heath) MacC.; m. Reut Leah Ran, July 13, 1981; children: John, Bruce, Reid, Taber, Ari. B.A., Princeton U., 1951; Ph.D., U. N.Mex., 1962. Mem. tech. staff Sandia Nat. Labs., Albuquerque, 1957—. Contbr. articles to profl. jours. Fulbright fellow Ain Shams U., Cairo, Egypt, 1964-65. Mem. Am. Phys. Soc., Am. Astron. Soc. Subspecialty: Gamma ray high energy astrophysics. Current work: Gamma ray line astronomy, explosive microwave generation.

MACCOBY, ELEANOR EMMONS, psychology educator; b. Tacoma, May 15, 1917; d. H. Eugene and Viva (Johnson) Emmons; m. Nathan Maccoby, Sept. 16, 1938; children: Janice B. Maccoby Carmichael, Sarah Maccoby Hatch, Mark. Student, Reed Coll., 1934-35, 36-37; B.A., U. Wash., 1939; M.A., U. Mich., 1949, Ph.D., 1950. Study dir., div. program surveys Dept. Agr., 1942-46, Survey Research Center, U. Mich., 1946-48; lectr. social relations Harvard U., 1950-58; faculty Stanford U., 1958—, prof. psychology, 1958—, chmn. dept., 1973-76. Author: (with R.R. Sears, H. Levin) Patterns of Child Rearing, 1957, (with M. Zellner) Experiments in Primary Education, 1970, (with C.N. Jacklin) The Psychology of Sex Differences, 1974, Social Development: Psychological Growth and the Parent-Child Relationship, 1980; Editor: (with Newcomb and Hartley) Readings in Social Psychology, 1957, Development of Sex Differences, 1966;

Contbr. articles to profl. jours. Center for Advanced Study in Behavioral Sci. fellow, 1969-70. Fellow Am. Psychol. Assn. (div. pres. child psychology); mem. Soc. Research Child Devel. (gov. council 1963, pres. 1981-83), Social Sci. Research Council (chair 1983-84), Western Psychol. Assn. (pres. 1974-75). Subspecialty: Developmental psychology. Home: 729 Mayfield Ave Stanford CA 94305

MACCOY, DOUGLAS MAIDLOW, veterinary surgeon, educator; b. Washington, Aug. 15, 1947; s. Edgar Milton and Charlotte (Maidlow) MacC. B.S., Purdue U., 1969; D.V.M., U. Ga., 1973. Diplomate: Am. Coll. Vet. Surgeons. Intern N.Y. State Coll. Vet. Medicine, Cornell U., Ithaca, 1973-74, resident in surgery, 1974-76, asst. prof., 1976-82, Coll. Vet. Medicine, U. Ill., Urbana, 1982—. Contbr. articles to profl. jours. Mem. AVMA, Am. Coll. Vet. Surgeons, Vet. Cancer Soc., Am. Assn. Vet. Clinicians. Democrat. Episcopalian. Subspecialties: Surgery (veterinary medicine); Cancer research (veterinary medicine). Current work: Application of advanced surgical techniques to animal disease (reconstructive and microsurgery); fracture repair in birds; clinical treatment of animal cancer. Office: Dept Clin Scis Coll Vet Medicine U Ill Urbana IL 61801

MAC CREADY, PAUL BEATTIE, aero. engr.; b. New Haven, Sept. 29, 1925. B.S. in Physics, Yale U., 1947; M.S., Calif. Inst. Tech., 1948, Ph.D. in Aeros. cum laude, 1952. Founder, pres. Meteorology Research, Inc., 1951-70, Atmospheric Research Group, 1958-70; founder, 1971; since pres. AeroVironment, Inc., Pasadena, Calif.; cons. in field, 1951—, mem. numerous govt. tech. adv. coms. Author research papers in field. Recipient Collier trophy Nat. Aero. Assn., 1979, Edward Longsreth medal Franklin Inst., 1979, Gold Air medal Fedn. Aero. Internat., 1981, Inventor of Year award Assn. Advancement Innovation and Invention, 1981; named Engr. of Century ASME, 1980. Mem. Nat. Acad. Engring., Am. Acad. Arts and Scis., Am. Meteorol. Soc. (chmn. com. atmospheric measurements 1968-69, councilor 1971-74), AIAA (Reed Aero. award 1979). Leader team that developed Gossamer Condor (Kremer prize 1977), 1976-77, Gossamer Albatross (Kremer prize 1979) for human-powered flight across English channel, 1979, Solar Challenger, ultralight aircraft powered by solar cells, 1981. Address: AeroVironment Inc 145 Vista Ave Pasadena CA 91107

MAC DIARMID, ALAN GRAHAM, chemistry educator, researcher; b. Masterton, N.Z., Apr. 14, 1927; m., 1954; 4 children. B.Sc., U. N.Z., 1948, M.Sc., 1950; M.S., U. Wis., 1952, Ph.D. in Chemistry, 1953, Cambridge (Eng.) U., 1955. Asst. lectr. in chemistry St. Andrews U., 1955; from instr. to assoc. prof. U. Pa., Phila., 1955-64, prof. chemistry, 1964—. Sloan fellow U. Pa., 1959-63. Mem. Am. Chem. Soc., Royal Soc. Chemistry. Subspecialty: Solid state chemistry. Office: Dept Chemistry U Pa Philadelphia PA 19104

MACDONALD, G(EORGE) WAYNE, psychologist; b. Sydney, N.S., Nov. 6, 1945; s. John George and Winnifred Evelyn (MacMillan) MacD.; m. Susan Elizabeth Barrett, May 9, 1970; children: Josh, Scott, Lindsay. B.A. magna cm laude, St. Francis Xavier U., 1967; M.A., U. Windsor, 1969, Ph.D., 1974. Registered psychologist, N.S. Intern Windsor Western Hosp., 1969-72; asst. prof. St. Francis Xavier U., Antigonish, N.S., 1972-73; chief psychologist Sarnia Lambton Centre, Sarnia, Ont., 1973-76; staff psychologist I.W.K. Hosp. for Children, Halifax, N.S., 1976; dir. Dept. Psychology, 1976—; cons. in clin. neuropsychology; profl. adv. bd. Canadian Assn. for Children with Learning Disorders, 1982—. Fellow Assn. Psychologists of N.S. (pres. 1980-81); mem. Internat. Neuropsychology Soc., Am. Psychol. Assn., Canadian Psychol. Assn. Roman Catholic. Subspecialties: Neuropsychology; Behavioral psychology. Current work: Neuropsychological correlates of dietary phenylalanine restriction in older children with pheylketonuria; elucidation of neuropsychological deficits in wide variety of pediatric conditions. Home: 2541 Westmount St Halifax NS Canada B3L 3G7 Office: PO Box 3070 Halifax NS Canada B3J 3G9

MAC DONALD, GORDON JAMES FRASER, geophysicist; b. Mexico City, July 30, 1929; s. Gordon and Josephine (Bennett) MacD.; m. Marcelline Kuglen (dec.); children: Gordon James, Maureen, Michael; m. Betty Ann Kipniss; 1 son, Bruce. A.B. summa cum laude, Harvard U., 1950, A.M., 1952, Ph.D., 1954. Asst. prof. geology, geophysics Mass. Inst. Tech., 1954-55, asso. prof. geophysics, 1955-58; staff asso. geophysics lab. Carnegie Inst. Washington, 1955-63; cons. U.S. Geol. Survey, 1955-60; prof. geophysics U. Calif. at Los Angeles, 1958-68, dir. atmospheric research lab., 1960-66, asso. dir., 1960-68; v.p. research Inst. for Def. Analyses, 1966-67, exec. v.p., 1967-68, trustee, 1966-70; vice chancellor for research and grad. affairs U. Calif. at Santa Barbara, 1968-69, prof. physics and geophysics, 1968-69; mem. council on Environ. Quality, Washington, 1969-72; Henry R. Luce prof. environ. studies and policy, dir. environ. studies program Dartmouth, 1972—; trustee Mitre Corp., disting. vis. scholar, 1977-79, chief scientist, 1979—, v.p., 1983—; cons. NASA, 1960-70, mem. lunar and planetary missions bd., 1967; mem. def. sci. bd. Dept. Def., 1966-70; cons. Dept. State, 1967-70; mem. Pres.'s Sci. Adv. Com., 1965-69; adv. panel on nuclear effects Office Tech. Assessment, 1975—. Author: The Rotation of the Earth, 1960; co-author: Sound and Light Phenomena: A Study of Historical and Modern Occurrences, 1978; Contbr. articles to sci., tech. jours. Fellow Am. Acad. Arts and Scis., AAAS; mem. Am. Math. Soc., Nat. Acad. Scis. (chmn. environmental studies bd. 1970, 72-73, chmn. commn. on natural resources 1973—), Am., Royal astron. socs., Am. Mineral. Soc., Am. Geophys. Union., Geochem. Soc. Am., Geol. Soc. Am., Am. Meteorol. Soc., Am. Philos. Soc., Seismol. Soc. Am., Soc. Indsl. and Applied Math., Council Fgn. Relations, Sigma Xi. Subspecialties: Geophysics; Statistics. Address: MITRE Corp 1820 Dolly Madison Blvd McLean VA 22102

MACDONALD, JAMES ROSS, physicist, educator; b. Savannah, Ga., Feb. 27, 1923; s. John Elwood and Antonina Jones (Hansell) M.; m. Margaret Milward Taylor, Aug. 3, 1946; children: Antonina Hansell, James Ross, William Taylor. B.A., Williams Coll., 1944; S.B., Mass. Inst. Tech., 1944, S.M., 1947; D.Phil. (Rhodes scholar), Oxford (Eng.) U., 1950, D.Sc., 1967. Mem. staff Digital Computer Lab., Mass. Inst. Tech., 1946-47; physicist Armour Research Found., Chgo., 1950-52; asso. physicist Argonne Nat. Lab., 1952-53; with Tex. Instruments Inc., Dallas, 1953-74, v.p. corporate research and engring., 1968-73, v.p. corporate research and devel., 1973-74, cons., 1974—; dir. Simmonds Precision Products Inc., 1979—; William Rand Kenan Jr. prof. physics U. N.C., Chapel Hill, 1974—; adj. prof. biophysics U. Tex. Med. Sch., Dallas, 1954-74; mem. solid state scis. panel NRC, 1965-73; mem. adv. com. for sci. edn. NSF, 1971-73; mem. vis. com. physics Mass. Inst. Tech., 1971-74; mem. external adv. com. Engring. Expt. Sta., Ga. Inst. Tech., 1976-79. Contbr. articles to profl. jours. Mem. Dallas Radio Commn., 1967-71; mem. sci. adv. council Callier Hearing and Speech Center, Dallas, 1974-78; bd. dirs. League for Ednl. Advancement in Dallas, 1965-70. Fellow Am. Phys. Soc. (com. on edn. 1973-75, com. on applications of physics 1975-78), IEEE (awards 1962, 74, asso. editor Transactions of Profl. Group on Audio 1961-66, Transactions on Audio and Electroacoustics 1966-73), AAAS; mem. Nat. Acad. Engring. (exec. com. assembly of engring. 1975-78, council 1971-74), Nat. Acad. Scis. (chmn. numerical data adv. bd. 1970-74, com. on motor vehicle emissions 1971-74, chmn. com. on motor vehicle emissions 1973-74, mem. com. on satellite power systems 1979-81, mem. com. on sci., engring., and pub. policy 1981-83), Am. Inst. Physics. (governing bd. 1975-78, chmn. com. on profl. concerns 1976-78), Electrochem. Soc., Audio Engring. Soc., Phi Beta Kappa, Sigma Xi. Subspecialties: Condensed matter physics; Analytical chemistry. Current work: Analysis of the electrical response of solids and liquids, especially solid and liquid electrolytes. Patentee in field. Home: 308 Laurel Hill Rd Chapel Hill NC 27514 Office: Dept Physics and Astronomy U NC Chapel Hill NC 27514

MACDONALD, JOHN CHISHOLM, chemistry educator; b. Boston, Mar. 17 1933; s. John James and Catherine Marie MacD.; m. Patricia Butler, July 6, 1964; children: Elizabeth, Nancy. B.S., Boston Coll., 1955, M.S., 1957; Ph.D., U. Va., 1962. AEC postdoctoral fellow Pa. State U., 1962-63; sr. research chemist Monsanto Research Corp., 1963-66; asst. prof. chemistry Fairfield (Conn.) U., 1966-70, assoc. prof., 1970-75, prof., 1975—; editorial cons. Served with U.S. Army, 1957-58. NIH fellow Yale Med. Sch., 1972-73; NIH grantee, 1973-75; Swedish research fellow U. Stockholm, 1979. Mem. Am. Chem. Soc., AAAS, Sigma Xi. Subspecialties: Analytical chemistry; Graphics, image processing, and pattern recognition. Current work: Application of computers to medical data; science for society; chemistry of the new nutrition. Home: 250 Strobel Rd Trumbull CT 06611 Office: North Benson Rd Fairfield CT 06430

MACDONALD, JOHN STEPHEN, oncologist, educator; b. Bklyn., June 2, 1943; s. John Stephen and Margaret (Martin) M.; m. Mary Suzanne Stock, July 11, 1964; children: Margaret Wilson, John Stephen, Kathleen Lenore, Frederick Stock. A.B., Dartmouth Coll., 1965, B.M.S., 1967; M.D., Harvard U., 1969. Intern and resident in medicine Beth Israel Hosp., Boston, 1969-71; clin. assoc. immunology and med. oncology Nat. Cancer Inst., Bethesda, Md., 1971-74, assoc. dir. cancer therapy evaluation program, div. cancer treatment, 1979-82, med. oncologist Washington Clin., 1982—; instr., asst. prof., then assoc. prof. medicine Georgetown U., Washington, 1974-79, clin. assoc. prof., 1979—, George Washington U., 1980—. Editor-in-chief: Cancer Treatment Reports, 1979—; contbr. over 100 articles to med. jours. Bd. mgmt. YMCA, 1979—; bd. dirs. CYO, 1979—. Served with USPHS, 1971-74. Jr. faculty clin. fellow Am. Cancer Soc., 1974-76. Fellow ACP; mem. Am. Fedn. Clin. Research, Am. Soc. Clin. Oncology, Am. Assn. for Cancer Research. Republican. Roman Catholic. Subspecialties: Chemotherapy; Internal medicine. Current work: Clinical cancer treatment and research. Home: 3405 Glenmoor Dr North Chevy Chase MD 20015 Office: 7910 Woodmont Ave Bethesda MD 20205

MACDONALD, TIMOTHY LEE, chemistry educator, consultant; b. Long Beach, Calif., Mar. 12, 1948; s. Ivan L. and Patricia A. (Haley) M.; m. Deborah L. Patrick, Aug. 17, 1971; 1 dau., Kate L. B.Sc., UCLA., 1971; Ph.D., Columbia U., 1975. NIH postdoctoral fellow Stanford (Calif.) U., 1975-77; asst. prof. Vanderbilt U., Nashville, 1977-82; assoc. prof. chemistry U. Va., Charlottesville, 1982—. Contbr. in field to profl. jours. Sloan Found. fellow Vanderbilt U., 1981; recipient Young Faculty award Eli Lilly and Co., Vanderbilt U., 1981. Mem. Am. Chem. Soc., Soc. Toxicology. Subspecialties: Organic chemistry; Biochemistry (biology). Current work: Research directed at synthetic organic chemistry for new methods development and their application to total synthesis and at bioorganic chemistry of drug-receptor or enzyme binding. Patentee in field. A.P. Home: 2625 Jefferson Park Circle Charlottesville VA 22901 Office: Dept Chemistry U Va McCormick Rd Charlottesville VA 22901

MACDOUGALL, JOHN ARCHIBALD, metals co. exec., mech. engr., cons.; b. Inverness, N.S., Can., Jan 28, 1938; came to U.S., 1978; s. John Bernard and Martha Christene (Mac Lellan) MacD.; m. Anne Marguerite Donovan, Sept. 16, 1961; children: John J. B., James R. T. Dipl. Eng., Royal Mil. Coll., Kingston, Ont., Can., 1961; B.Sc. M.E., Queen's U., Kingston, 1962. Profl. engr., Ont., Que. Project engr. Alcan Aluminum Co., Oswego, N.Y., 1965-69; project engr. Midland Ross Corp., Toledo (Ohio) and Portland, Oreg., 1969-73, project mgr., 1973-75; cons. agglomeration and briquetting, Montreal (Que., Can.) and Pitts., 1975-81; mgr. engring. Inmetco div. Inco Ltd., Ellwood City, Pa., 1981—; cons. in field; mem. math. faculty SUNY-Oswego, 1965-68. Bd. dirs. Jaycees, Oswego, 1968-69; baseball coach Portland Little League, 1970-72. Served to flight lt. RCAF, 1961-65. Mem. ASME, Inst. for Briquetting and Agglomeration (dir.), Combustion Inst., Iron and Steel Soc. Roman Catholic. Subspecialties: Materials processing; Alloys. Current work: Direct reduction of iron bearing waste materials and ores, also recycling metals and associated technology. Patentee in field. Home: 5 Marquette Rd Pittsburgh PA 15229 Office: PO Box 720 Ellwood City PA 16117

MACE, MARSHALL ELLIS, plant pathologist; b. Lincolnton, N.C., Jan. 9, 1927; s. Every Jasper and Ila Zoe (Ellis) M.; m. Evelyn Marie Martin, Mar. 26, 1952; children: Cynthia Marie, Pamela Gail. B.S., Wake Forest Coll., 1954; Ph.D., N.C. State U., 1960. Research plant pathologist United Fruit Co., Norwood, Mass., 1959-65; research plant pathologist U.S. Dept. Agr., Beltsville, Md., 1965-72, College Station, Tex., 1972—. Contbr. articles to profl. jours. Served to sgt. Signal Corps U.S. Army, 1945-46. Mem. Am. Phytopath. Soc., Gamma Sigma Delta. Republican. Methodist. Subspecialties: Plant pathology; Biochemistry (biology). Current work: Histochemistry and biochemistry of diseased plants.

MAC EWEN, GEORGE DEAN, physician, medical institute executive; b. Metcalfe, Ont., Can., Nov. 10, 1927; s. George W. and Catherine (Grant) MacE.; m. Marilyn Ruth Heidelberger, May 29, 1954; children: Kathryn, Jane, Nancy, David, John. M.D., C.M., Queen's U., Kingston, Ont., Can., 1953. Diplomate: Am. Bd. Orthopaedic Surgery. Intern D.C. Gen. Hosp., Washington, 1953-54; resident in gen. surgery Emergency Hosp., Washington, 1954-55; fellow in orthopaedic surgery Campbell Clinic, Memphis, 1955-58; asst. med. dir. Alfred I. duPont Inst., Wilmington, Del., 1958-68, med. dir., 1969—, surgeon-in-chief, 1961-79; chief orthopaedic service VA Hosp., Wilmington, 1960-69, cons. in orthopaedics, 1961—, Wilmington Med. Center, 1961—, St. Francis Hosp., Wilmington, 1961—; cons. in orthopaedic surgery Surgeon Gen. U.S. Navy, 1964—, USAF Hosp., Dover, Del.; med. cons. John G. Leach Sch. and; Evan G. Shortlidge Sch. for Handicapped, Wilmington; cons. USAF Base, Dover, Del., 1961—, Pocono Med. Services, 1973-74, Walter Reed Army Hosp., Washington, 1978—; med. adv. com. Del. Systems for Exceptional Children, 1973—; assoc. prof. orthopaedic surgery Jefferson Med. Coll., Thomas Jefferson U., Phila., 1970-76, prof., 1976—; clin. prof. U. Del., Newark, 1977—; lectr. U.S. Naval Hosp., Phila., 1964—, vis. prof., Portsmouth, Va., 1967—; exec. com. Del. sch. health adv. com., 1966—; med. adv. com. VA Hosp., 1965-79; mem. U. Del. Research Found., 1976-79. Editorial bd.: Del. Med. Jour., 1966-69; contbr. articles to books, jours. Bd. trustees Wilmington Coll., 1980—; bd. dirs. Arthritis Found., 1974-77, Blood Bank of Del., 1969-78; exec. com. Blood Bank of Del., 1976-78, hon. mem., 1979—; Mem. editorial bd. of evaluation and health professions U. Del. Acad. and Profl. Programs Div. of Continuing Edn., 1977—. Fellow Am. Acad. Cerebral Palsy; mem. Acad. of Medicine (Wilmington), Am. Acad. Orthopaedic Surgeons (chmn., mem. various coms.), Am. Acad. Pediatrics (orthopaedic surg. fellow 1977—), A.C.S., AMA (ho. of dels. 1982—), Am. Orthopaedic Assn. (traveling fellow 1967, pres. 1980-81), Brit. Orthopaedic Assn. (corr.), Campbell Club (pres. 1979), Can. Orthopaedic Assn. (trustee 1969-75), Coll. of Physicians (Phila.), Cosmos Club, Del. State Med. Soc., Eastern Orthopaedic Assn., Fla. Orthopaedic Assn. (hon.), New Castle County Med. Soc. (del. 1970-72), Nat. Assn. Children's Hosps. and Related Instns. (bd. trustees 1980—), N.J. Orthopaedic Assn. (hon.), Del. Soc. Orthopedic Surgeons (charter), Orthopaedic Forum Club, Orthopaedic Research Soc., Pan Am. Orthopedic Orgn., Pediatric Orthopaedic Soc. (pres. 1972-73), Pa. Orthopaedic Soc. (hon.), Phila. Orthopaedic Soc. (pres. 1968-69), Scoliosis Research Soc. (founding, exec. com. 1969-74, pres. 1971-73), Societe de Scoliose du Quebec (hon.), Societe Internationale de Chirurgie Orthopedique et de Traumatologie (sec., treas. U.S.A. sect. 1979—), S. African Orthopaedic Assn. (hon.), Tex. Orthopaedic Assn. (hon.), Twentieth Century Orthopaedic Assn. Subspecialty: Orthopedics. Home: PO Box 86 Wilmington DE 19899 Office: PO Box 269 Wilmington DE 19899

MACFARLANE, MALCOLM DAVID, pharmacologist; b. Cambridge, Mass., Sept. 26, 1940; s. Robert Malcolm and Elizabeth Agnes (Hennessey) MacF.; m. Mary Kay Allen, Mar. 8, 1975. B.S. in Pharmacy, Northeastern U., 1962; Ph.D. in Pharmacology, Georgetown U., 1967. Instr. pharmacology Kirksville (Mo.) Coll. Osteo. Medicine, 1967-69; Assoc. prof. pharmacology U. So. Calif., 1969-74; dir. research Meyer Labs., Ft. Lauderdale, Fla., 1974-78; v.p. profl. services Glaxo Inc., Research Triangle Park, N.C., 1978-81, v.p. regulatory affairs, 1981—. Contbr. articles to numerous profl. publs. Pres Oakland Shores Condominium Assn. Kappa Psi scholar, 1961; Rho Chi scholar, 1961. Fellow Am. Coll. Clin. Pharmacology; mem. Am. Soc. Pharmacology and Exptl. Therapeutics, Am. Soc. Clin. Pharmacology, Regulatory Affairs Professionals Soc., Sigma Xi, Rho Chi, Kappa Psi, U.S. Power Squadron. Subspecialty: Pharmacology. Current work: Pharmacotherapy of aging. Office: 5 Moore Dr Research Triangle Park NC 27709

MACHADO, LESTER, dentist, researcher; b. San Jose, Calif., July 8, 1955; s. Manuel Joseph and Eleanor V. (Costigan) M. B.A., U. Calif.-Davis, 1977; D.D.S., U. Pacific, 1980. Clin. research assoc. U. Pacific Sch. Dentistry, San Francisco, 1982-83; dir. Red Cross Dental Asst. Tng. Program, Travis AFB, Calif., 1980-83. Author: Chairside Assisting in Operative Dentistry, 1983. Served as capt. USAF, 1980-83. Calif. Gov.'s scholar, 1973; recipient F. Gene Dixon award Calif. Dental Service, 1980; Clin. Excellence award U. Pacific Sch. Dentistry, 1980; Dean's award, 1980. Mem. Internat. Assn. Dental Research, Acad. Operative Dentistry, Acad. Gen. Dentistry, Am. Dental Assn., Christian Med. Soc., Theta Xi, Omicron Kappa Upsilon. Democrat. Roman Catholic. Subspecialty: Oral and maxillofacial surgery. Current work: Applications of microsurgical technique to oral and maxillofacial surgery, organic and functional disorders of the temporomandibular apparatus. Home: 2562 Walnut Blvd Apt 96 Walnut Creek CA 94598 Office: David Grant Med Center SGD Travis AFB CA 94535

MACHALSKI, ROBERT CHESTER, pharmaceutical company executive, database analyst; b. Chgo., Mar. 14, 1945; s. Chester Stanley and Genevieve Marie (Slowik) M.; m. Teodora Vera Kalbarczyk, Jan. 27, 1968; children: Cathrine, Philip, David. Student, Loyola U., Chgo., 1970-72; B.S. in Biology, Roosevelt U., 1975; M.B.A. in Data Communication Systems, Roosevelt U., 1982. Divisional quality control mgr. Sweetheart Cup Corp., Chgo., 1975-76; mgr. sect. reliability and quality engring. labs. Baxter Travenol Labs., Round Lake, Ill., 1977—. Mem. Am. Soc. Quality Control (cert. quality engr.), Soc. Plastic Engrs., Soc. Mfg. Engrs. Roman Catholic. Subspecialties: Biomedical engineering; Database systems. Current work: Converting manval engineering and metrology systems into automated computer-databse systems. Office: Baxter Travenol Labs Route 120 and Wilson Rd Round Lake IL 60073

MACHOVEC, GEORGE STEPHEN, information specialist, editor, consultant; b. Columbus, Ohio, June 10, 1952; s. Charles R. and Geraldine (Elieff) M. B.S. in Physics and Astronomy, U. Ariz., 1974, M.L.S., 1977. Solar energy librarian, coordinator computer reference service for sci. and engring. Ariz. State U. Library, Tempe, 1977—; cons. Info. Intelligence, Inc., Phoenix, 1980—. Author: Solar Energy Index, 1980, Supplement I, 1982; editor: Info. Intelligence Online Newsletter, 1981—, Info. Intelligence Online Hotline, 1982—; mng. editor: Online Libraries and Microcomputers, 1983—. Mem. Am. Soc. Info. Sci., ALA., Spl. Libraries Assn., Am. Solar Energy Soc., Internat. Solar Energy Soc., Ariz. Online User Group (chmn. 1979-81). Democrat. Subspecialties: Solar energy; Information systems (information science). Current work: Have developed computerized index to solar energy publications, involved with solar energy, information storage and retrieval, online systems, database systems. Home: PO Box 785 Tempe AZ 85281 Office: Arizona State U Library Tempe AZ 85287

MACHTIGER, HARRIET GORDON, psychotherapist; b. N.Y.C., July 27, 1927; d. Michael J. and Miriam D. (R) Gordon; m. Sidney Machtiger, Feb. 7, 1948; children: Avram Coleman, Marcia Gordon, Bennett Rand. B.A., Bklyn. Coll., 1947; diploma with distinction, U. London, 1966, Ph.D., 1974. Lic. clin. psychologist, Pa.; cert. psychoanalyst, Jungian analyst. Tchr. Phila. Pub. Schs., 1962-64; ednl. therapist Child Guidance Tng. Ctr., London, 1966-68; specialist Sch. Psychol. Service, Inner London Ednl. Authority, 1968-70; psychotherapist Paddington Day Hosp., London, 1970-71, London Ctr. Psychotherapy, 1971-74, Staunton Clinic U. Pitts., 1974-78; pvt. practice psychotherapy and psychoanalysis, Pitts., 1976—; pres. C.G. Jung Ctr., Pitts., 1975-80. Recipient Disting. Contbn. to Edn. award Commonwealth of Pa., 1962. Fellow Am. Orthopsychiat. Assn.; mem. Am. Psychol. Assn., Brit. Psychol. Soc., Inter-Regional Soc. Jungian Analysts (tng. com. 1979-81), Brit. Assn. Psychotherapists. Democrat. Jewish. Subspecialties: Psychotherapy; Developmental psychology. Current work: Psychotherapeutic techniques, particularly transference/counter transference issues; working with difficult and handicapped patients; teaching. Home: 207 Tennyson Ave Pittsburgh PA 15213 Office: 110 The Fairfax 4614 5th Ave Pittsburgh PA 15213

MACINNES, DAVID FENTON, JR., chemistry educator; b. Abington, Pa., Mar. 19, 1943; s. David F. and Kathleen (O'Neill) MacI.; m. Barbara Hardy MacInnes, June 22, 1974; children: Colin, Breanyn. B.A., Earlham Coll., 1965; M.A., Princeton U., 1968, Ph.D., 1972. Tchr. chemistry, head dept. sci. Westtown Sch., Pa., 1970-73; asst. prof. chemistry Guilford Coll., 1973-81, assoc. prof., 1982—; fellow U. Pa., 1980-81; summer research assoc. Brookhaven Nat. Lab., Upton, N.Y., 1982; adv. Toxic Waste Task Force, Greensboro, 1979-82. Contbr. articles in field to profl. jours. Mem. Am. Chem. Soc. (Dolittle award 1981), N.C. Acad. Arts and Scis. Quaker. Subspecialties: Inorganic chemistry; Polymer chemistry. Current work: Conductive polymers, organic batteries, photovoltaic cells, computers in chemical education. Office: Dept Chemistry Guilford Coll Greensboro NC 27410

MACINTYRE, WALTER MACNEIL, computer center administrator; b. Kenosha, Wis., Nov. 28, 1930; s. Alexander MacNeil and Agnes Ferguson (Weir) M.; m. Elizabeth Henderson Hamilton, Aug. 1954. Instr. in chemistry Brown U., 1956-57; in various chemistry and computing positions U. Colo., 1957-73, assoc. prof. computer sci., 1963-68, assoc. prof. chemistry, 1963-73; head Computing Lab., Nat. Inst. Med. Research, London, 1973-76; dir. univ. computing services SUNY-Buffalo, 1976-80; dir. sci. computing div. Nat. Center Atmospheric Research, Boulder, Colo., 1980—. Mem. Am. Chem.

Soc., AAAS, Assn. Computing Machinery, Brit. Computer Soc., Brit. Chem. Soc. Presbyterian. Subspecialties: X-ray crystallography; Information systems, storage, and retrieval (computer science). Current work: Development of supercomputer services. Office: Nat Center Atmospheric Research PO Box 3000 Boulder CO 80307

MACINTYRE, WILLIAM JAMES, physicist; b. Canaan, Conn., Nov. 26, 1920; s. William M. and Helen (Hoyt) MacI.; m. Patricia Nelle Grossman, Sept. 16, 1947; children: Kathleen S., Steven J. B.S., Western Res. U., 1943, M.A., 1947; M.S., Yale U., 1948, Ph.D., 1950. Prof. biophysics Sch. Medicine, Case Western Res. U., Cleve., 1949-72; staff physicist Cleve. Clinic Found., 1972—; cons. in field. Contbr. articles to profl. jours.; author: Quantitative Nuclear Cardiology, 1975. Served with U.S. Army, 1943-46. Mem. Soc. Nuclear Medicine, Am. Phys. Soc., Biophys. Soc., Am. Assn. Physicists in Medicine, Central Research Soc., Am. Heart Assn., Soc. Magnetic Resonance in Medicine, Soc. Magnetic Resonance in Imaging. Subspecialties: Nuclear medicine; Imaging technology. Current work: Emission tomography, applications of nuclear magnetic resonance to cardiac studies, cardiovascular nuclear medicine. Home: 3108 Huntington Rd Shaker Heights OH 44120 Office: 9500 Euclid Ave Cleveland OH 44106

MACK, JAMES LEWIS, Psychology educator; b. Kansas City, Mo., Nov. 25, 1936; s. Jack D. and Katherine (Barton) M.; m.; children: Andrea, Paul David. B.A., Stanford U., 1960; Ph.D., U. Minn., 1967. Lic. psychologist, Ohio. Instr. psychology U. Minn., Mpls., 1962-67; psychologist State of Minn., Circle Pines, 1965-67; sr. psychologist Norristown (Pa.) State Hosp., 1967-70; vis. asst. prof. psychology U. Pa., Phila., 1968-70; adj. assoc. prof. psychology Temple U., Phila., 1967-70; asst. prof. psychology Depts. Neurology and Psychiatry, Case Western Res. U., Sch. Medicine, Cleve., 1970—; cons. psychologist Cleve. VA Med. Ctr., 1971—. Contbr. articles to profl. jours. Alfred P. Sloan Found. scholar, 1954-58. Mem. Am. Psychol. Assn., Internat. Neuropsychol. Soc., Acad. of Aphasia. Democrat. Jewish. Subspecialty: Neuropsychology. Current work: Research in cognitive correlates of neurologic disorders; clin. evaluation of neurologic patients. Home: 2280 S Overlook Rd Cleveland Heights OH 44106 Office: Dept Neurology Case Western Res U Sch of Medicine 2065 Adelbert Rd Cleveland OH 44106

MACK, MICHAEL EDWARD, physicist; b. Poughkeepsie, N.Y., May 28, 1939; s. Edward Joseph and Anita Eleanor (Barton) M.; m. Sarah Marie McManus, July 20, 1963; children: Patrick E., Michael P., Kathleen E., Maura A. B.S. in Physics, M.I.T., 1961, Ph.D., 1967. Prin. research scientist United Technologies Research Ctr., East Hartford, Conn., 1967-73; dir. laser systems, measurements, tech. and systems group Avco Everett Research Labs., Revere Beach, Everett, Mass., 1973-81; chief scientist Eaton Ion Implantation Div., Beverly, Mass., 1981—; fellow Ctr. Advanced Engring. Studies, M.I.T., 1981. Contbr. articles to prof. jours. Mem. Manchester Harbor Adv. Com., 1978—. Mem. Optical Soc. Am., Am. Phys. Soc., Laser Inst. Am. Club: Manchester Harbor Boat. Subspecialties: Atomic and molecular physics; Spectroscopy. Current work: Wafer cooling and dosimetry in ion implantation; charge neutralization during ion implantation; high power ion sources. Home: 7 Hidden Ledge Rd Manchester MA 01944 Office: 16 Tozer Rd Beverly MA 01915

MACKAY, JOHN KELVIN, renewable energy company executive; b. Kirkland Lake, Ont., Can., Jan 4, 1938; s. John William and Ruth Arnold (McNairn) (Hesse) MacK.; m. Marie Alice Grossman, Dec. 28, 1963; children: Alexander, Tania. B.Sc., Queens U., Kingston, Ont., 1959; Ph.D., McGill U., Montreal, Que., Can., 1967. Tech. officer Imperial Chem. Industries, Runcorn, U.K., 1964-66; mgr. research and devel., dir. Occidental Petroleum Corp., various locations U.S. and Europe, 1966-77; tech. dir. Continental Group Inc., Cupertino, Calif., 1977-82; v.p. ops. United Energy Corp., Foster City, Calif., 1982—. Contbr. articles to profl. jours. Cominco fellow, 1962; Can. NRC fellow, 1959-63. Mem. Am. Chem. Soc., N.Y. Acad. Scis., AAAS, Marine Tech. Soc. Subspecialties: Solar energy; Physical chemistry. Current work: Research, development and rapid commercialization of advanced energy systems. Patentee in field. Home: 12665 Orella Ct Saratoga CA 95070 Office: United Energy Corp 420 Lincoln Centre Dr Foster City CA 94404

MACKAY, JOHN ROSS, geographer; b. Tamsui, Taiwan, Dec. 31, 1915; s. George William and Jean (Ross) M.; m. Violet Anne Meekins, Feb. 19, 1944; children—Margaret Anne, Leslie Isabel. B.A., Clark U., 1939; M.A., Boston U., 1941; Ph.D., U. Montreal, Que., Can., 1949; Docteur de l'université (hon.), Ottawa U., 1972. Asst. prof. geography McGill U., Montreal, 1946-49; prof. U. B.C., Can., Vancouver, 1949—; Disting. vis. prof. U. Wash., fall 1977. Served with Can. Army, 1941-46. Recipient Massey medal Royal Can. Geog. Soc., 1967; Centennial medal Govt. Can., 1967; Queen's Silver Jubilee medal, 1977. Mem. Can. Assn. Geographers (pres. 1953-54), Assn. Am. Geographers (pres. 1969), Royal Soc. Can. (Miller medal 1975), Arctic Inst. N. Am. (chmn. bd. govs. 1962-64), Geol. Soc. Am. Presbyterian. Subspecialty: Arctic studies. Current work: Permafrost, ground ice; freeze phenomena. Research, numerous publs. on permafrost, ground ice, ice wedges and pingos, Can. Arctic. Office: Dept Geography U BC Vancouver BC V6T 1W5 Canada

MACKAY, RAYMOND ARTHUR, chemistry educator; b. Queens, N.Y., Oct. 30. 1939; s. Theodore Henry and Helen Marie (Cusack) M.; m. Patricia Dora Graber, June 9, 1962; children: Brett Edward, Chelsea Christine; m. Mary Dilberian, Aug. 13, 1966. B.S. in Chemistry, Rensselaer Poly. Inst., 1961; Ph.D., SUNY-Stony Brook, 1966. Research assoc. Brookhaven Nat. Lab., Upton, N.Y., 1966; asst. prof. dept. chemistry Drexel U., Phila., 1969-75, assoc. prof., 1975-80, prof., 1980—. Contbr. articles to profl. jours. Served to capt., chem. corps U.S. Army, 1967-69. Recipient U.S. Army Research and Devel. award, 1980; NSF fellow, 1964-66. Mem. Am. Chem. Soc., Am. Inst. Chemists, N.Y. Acad. Sci., AAAS, AAUP, Inter-Am. Photochem. Soc., Sigma Xi, Phi Lambda Upsilon. Subspecialties: Physical chemistry. Current work: Reactions in and physical studies of microemulsions; kinetics of and infrared emission from gas-aerosol reactions. Office: Dept Chemistry Drexel U Philadelphia PA 19104

MACKE, HARRY JERRY, mechanical engineer; b. Newport, Ky., Aug. 26, 1922; s. Harry Jerome and Mildred Ruth (Rauch) M.; m. Virginia Heinlein, Apr. 1, 1948; children: Janice, Jennifer. B.S. in Mech. Engring, U. Ky., 1947, M.S., Harvard U., 1948, S.D. in Applied Mechanics, 1951. Registered profl. engr., Ohio. Instr. applied mechanics U. Ky., 1947; teaching fellow in civil engring. Harvard U., 1948-50; mech. engr. Aircraft Engine Bus. Group, Gen. Electric Co., Boston, 1951-52, Cin., 1952-71, cons. engring. mechanics, 1971-77, mgr. applied exptl. stress, 1977—; adj. assoc. prof. aerospace engring. U. Cin., 1961-66. Contbr. articles to profl. jours. Served to 1st lt. Signal Corps U.S. Army, 1943-46. Mem. ASME, Soc. Exptl. Stress Analysis, Sigma Xi, Tau Beta Pi. Subspecialties: Mechanical engineering; Theoretical and applied mechanics. Current work: Exptl. mechanics, stress analysis, photoelasticity, aircraft gas turbines. Home: 7305 Drake Rd Cincinnati OH 45243 Office: 1 Neumann Way Mail Drop G60 Cincinnati OH 45215

MACKENZIE, DAVID ROBERT, plant pathologist, educator; b. Beverly, Mass., Oct. 19, 1941; s. William Forrester and Eleanor (Robertson) Mack.; m. Doris Layton, June 29, 1963; children: Wendy Ellen, Scott Eobert, Todd Forrester. B.S., U. N.H., 1964; M.S., Pa. State U., 1967, Ph.D., 1970. Mem. field staff Rockefeller Found., Mex., Philippines, 1970-73; head dept. vegetable breeding Asian Vegetable Research and Devel. Ctr., Taiwan, 1973-74; mem. faculty dept. plant pathology Pa. State U., University Park, 1974-83, prof., 1981-83; head dept. plant pathology and crop physiology La. State U., 1983—; dir. Pa. Potato Breeding Program; cons. Nat. Crop Loss Assessment System; expert cons. FAO IPC Project. Contbr. articles to profl. jours. Mem. Am. Phytopath. Soc., Potato Assn. Am. Home: 916 Rue Crozat Baton Rouge LA 70810 Office: Dept Plant Pathology and Crop Physiology 202 Life Sci Bldg La State U Baton Rouge LA 70808

MACKENZIE, JOHN DOUGLAS, educator; b. Hong Kong, Feb. 18, 1926; U.S., 1954, naturalized, 1963; s. John and Hannah (Wong) MacK.; m. Jennifer Russell, Oct. 2, 1954; children—Timothy John, Andrea Louise, Peter Neil. B.Sc., U. London, 1952, Ph.D., 1954. Research asst., lectr. Princeton U., 1954-56; ICI fellow Cambridge (Eng.) U., 1956-57; research scientist Gen. Electric Research Center, N.Y.C., 1957-63; prof. materials sci. Rensselaer Poly. Inst., 1963-69; prof. engring. U. Calif., Los Angeles, 1969—; U.S. rep. Internat. Glass Commn., 1964-71. Author books in field (6); editor: Jour. Non-Crystalline Solids, 1968—; contbr. articles to profl. jours. Fellow Am. Ceramic Soc., Royal Inst. Chemistry; mem. Nat. Acad. Engring., Am. Phys. Soc., Electrochem. Soc., ASTM, Am. Chem. Soc., Soc. Glass Tech. Subspecialty: Ceramics. Patentee in field. Office: 6531 Boelter Hall Univ of Calif Los Angeles CA 90024

MACKENZIE, ROBERT DOUGLAS, biochemist, research scientist, safety coordinator; b. Chgo., Aug. 18, 1928; s. Vernon Gordon and Alice (Rasmussen) MacK.; m. Ruth Ann Noelcke, Sept. 6, 1952; children: R. Bruce, Barbara A., Catherine A., Deborah R. B.S., U. Cin., 1952; M.S., Mich. State U., 1954, Ph.D., 1957. Research asst. William S. Merrell Co., summer 1952; grad. asst. Mich. State U., 1953-54, 54-57, postdoctoral fellow, 1957; research biochemist William S. Merrell Co., 1957-63; sect. head Merrell Nat. Labs., 1963-81; mgr. operational safety, loss prevention and security Merrell Dow Research Ctr., Cin., 1981—; adj. assoc. prof. biochemistry U. Cin., 1967-69. Contbr. articles to profl. jours., chpt. to book. Served to cpl. AUS, 1946-48. Mem. AAAS, Am. Chem. Soc., Am. Heart Assn., Am. Soc. Pharmacology and Exptl. Therapeutics, Internat. Soc. Thrombosis and Haemostasis, Soc. Exptl. Biol9gy and Medicine, N.Y. Acad. Scis., Am. Mgmt. Assn., Sigma Xi. Club: S.W. Ohio Vintners Assn. Subspecialties: Biochemistry (medicine); Hematology. Current work: Blood coagulation, thrombosis research, laboratory safety, research administration. Home: 5532 Morrow Blackhawk Rd Morrow OH 45152 Office: 2110 E Galbraith Rd Cincinnati OH 45215

MACKERER, CARL ROBERT, oil co. exec., research scientist; b. Jersey City, Sept 2, 1940; s. Carl Joseph and Kathryn Anna M.; m. Marie Elizabeth Kempster, Dec. 3, 1946; children: Mary Ann, Carl William, Linda Marie. A.B., Rutgers U., 1963, postgrad., 1967-68; Ph.D. in Med. Biochemistry, U. Nebr., 1971. Cardiovascular pharmacologist Hoffmann-La Roche, Nutley, N.J., 1963-67; instr. biochemistry U. Nebr. Coll. Medicine, Omaha, 1971; research investigator G.D. Searle, Inc., Skokie, Ill., 1971-73, sr. research investigator, 1973-75, research scientist, 1975-77, sr. research scientist, 1978; mgr. biochem. toxicology Toxicology div. Mobil Oil Corp., Princeton, N.J., 1978—. Contbr. 76 articles to sci. jours. Mem. Am. Coll. Toxicology, Am. Soc. for Pharmacology and Exptl. Therapeutics, Biochem. Soc., Am. Chem. Soc. Subspecialties: Pharmacology; Toxicology (medicine). Current work: Manager of a section working in fields of genetic toxicology, pharmacokinetics, analytical chemistry, clinical chemistry, pharmacology attempting to determine and explain toxic effects of chemicals. Home: 5 Blue Spruce Dr Pennington NJ 08536 Office: PO Box 1029 Princeton NJ 08540

MACKEY, GEORGE WHITELAW, educator, mathematician; b. St. Louis, Feb. 1, 1916; s. William Sturges and Dorothy Frances (Allison) M.; m. Alice Willard, Dec. 9, 1960; 1 dau., Ann Sturges Mackey. B.A., Rice Inst., 1938; A.M., Harvard U., 1939, Ph.D., 1942; M.A., Oxford, 1966. Instr. math. Ill. Inst. Tech., 1942-43; faculty instr. math. Harvard, 1943-46, asst. prof., 1946-48, asso. prof., 1948-56, prof. math., 1956-69, Landon T. Clay prof. math. and theoretical sci., 1969—; vis. prof. U. Chgo., summer 1955, UCLA, summer 1959, Tata Inst. Fundamental Research, Bombay, India, 1970-71; Walker Ames vis. prof. U. Wash., summer 1961; Eastman vis. prof. Oxford, 1966-67; asso. prof. U. Paris, 1978. Author: Mathematical Foundations of Quantum Mechanics, 1963, Lectures on the Theory of Functions of a Complex Variable, 1967, Induced Representations and Quantum Mechanics, 1968, The Theory of Unitary Group Representations, 1976, Unitary Group Representations in Physics, Probability and Number Theory, 1978; Contbr. articles math. jours. Served as civilian, operational research sect. 8th Air Force, 1944; applied math. panel NDRC, 1945. Guggenheim fellow, 1949-50, 61-62, 70-71. Mem. Am. Math. Soc. (v.p. 1964-65, Steele prize 1974), Nat. Acad. Scis., Société Mathematique de France, Am. Philos. Soc., Am. Acad. Arts and Scis., Phi Beta Kappa, Sigma Xi. Subspecialty: Unitary group representations. Home: 25 Coolidge Hill Rd Cambridge MA 02138

MACKNIGHT, WILLIAM JOHN, chemistry educator, consultant; b. N.Y.C., May 5, 1936; s. William John and Margaret (Stuart) MacK.; m. Carol Marie Bernier, Aug. 19, 1967. B.S. in Chemistry, U. Rochester, 1958, M.S., 1963, Ph.D., 1964. Research assoc. Princeton (N.J.) U., 1964-65; asst. prof. chemistry U. Mass., Amherst, 1965-69, assoc. prof. chemistry, 1969-74, prof. chemistry, 1974-76, prof., head polymer sci. and engring., 1976—; cons. Exxon Research and Engring., Linden, N.J., 1967—. Author: Polymeric Sulfur and Related Polymers, 1965, Introduction to Polymer Viscoelasticity, 1972, 2d edit., 1983. Served to lt. USN, 1958-61. Fellow Am. Phys. Soc. Subspecialty: Polymer chemistry. Current work: Polymer blends, ionomers, polyurethanes. Home: 127 Sunset Ave Amherst MA 01002 Office: Polymer Sci and Engring U Mass Amherst MA 01003

MACLACHLAN, JAMES MORRILL, educator; b. Geneva, Ill., Mar. 21, 1934; s. John Andrew and Gladys (Morrill) MacL.; m. Sally Gerig, Oct. 21, 1978; children: Sheila, Carolyn, Laura. B.S., Carnegie Inst. Tech., 1956; M.B.A., Harvard U., 1971; Ph.D., U. Calif.-Berkeley, 1975. Pubr. The Tri-Town News, Sidney, N.Y., 1958-69; asst. prof. NYU, N.Y.C., 1975-79; assoc. prof. Columbia U., N.Y.C., 1980, Rensselaer Poly. Inst., Troy, N.Y., 1981—; pres. Timely Decisions, Inc., Delmar, N.Y., 1978—, Biblical Films, Inc., Delmar, 1981—; camp pres. Gideon Bible Assn., Albany, N.Y., 1981—. Author: Response latency: New Measure of Advertising, 1977. Served with U.S. Army, 1956-58. Mem. Am. Psychol. Assn., Soc. Motion Picture and TV Editors, Nat. Assn. Religious Broadcasters. Club: Harvard. Subspecialty: Cognition. Current work: Developed process for time compression of TV; developing advanced chroma-key system for TV. Home: 310 Elm Ave S Delmar NY 12054 Office: Rensselaer Poly Inst Lally Mgmt Bldg Troy NY 12181 Home: 310 Elm Ave S Delmar NY 12054

MACLANE, SAUNDERS, mathematician, educator; b. Taftville, Conn., Aug. 4, 1909; s. Donald Bradford and Winifred (Saunders) MacL.; m. Dorothy M. Jones, July 21, 1933; children: Margaret Ferguson, Cynthia M. Hay. Ph.B., Yale, 1930; A.M., U. Chgo., 1931; D.Phil., Goettingen, Germany, 1934; D.Sc. (hon.), Purdue U., 1965, Yale, 1969, Coe Coll., 1973, U. Pa., 1977; LL.D., Glasgow U., Scotland, 1971. Sterling Research fellow Yale, 1933-34; Benjamin Pierce instr. Harvard, 1934-36; instr. Cornell U., 1936-37, U. Chgo., 1937-38; asst. prof. Harvard, 1938-41, asso. prof., 1941-46, prof., 1946-47; prof. math. U. Chgo., 1947-63, chmn. dept., 1952-58, Max Mason Distinguished Service prof. of math., 1963—; exec. com. mem. Internat. Math. Union, 1954-58; research mathematician Applied Math. Group, Columbia, 1943-44, dir., 1944-45; Mem. Nat. Sci. Bd., 1974-80. Author: (with Garrett Birkhoff) Survey of Modern Algebra, 1942, Homology, 1963, (with Garrett Birkhoff) Algebra, 1967, Categories for the Working Mathematician, 1971; Editor: bull. Am. Math. Soc, 1943-46; mng. editor, 1946-47; editor: Trans. Am. Math. Soc, 1949-54; Chmn. editl. com.: Carus Math. Monographs, 1940-45; Contbr.: sci. articles on math. to Annals Math. Recipient Chauvenet prize for mathematical exposition by Math. Assn. Am., 1941; Montclair Yale Cup, 1929; Proctor prize, 1979; John Simon Guggenheim fellow, 1947-48, 72-73. Mem. Am. Math. Soc. (v.p. 1946-47, council 1939-41, pres. 1973-74), Math. Assn. Am. (v.p. 1948-49, pres. 1950-52, Distinguished Service award 1975), Nat. Acad. Sci. (council 1958-61, 69-72, v.p. 1973-81, chmn. editorial bd. procs. 1960-68), Royal Danish Acad. Scis. (fgn. mem.), Am. Philos. Soc. (mem. council 1960-63, v.p. 1968-71), Akademie der Wissenschaften (Heidelberg), Royal Soc. Edinburgh, Assn. for Symbolic Logic (exec. com. 1945-47), Am. Acad. Arts and Sci. (council 1981—), Phi Beta Kappa, Sigma Xi. Conglist. Subspecialty: Category Theory. Current work: Philosophy of science; category theory. Home: 5712 S Dorchester Ave Chicago IL 60637

MACLAY, WILLIAM NEVIN, physical chemist, mfg. co. exec.; b. Belleville, Pa., Dec. 30, 1924; s. Robert B. and Grace V. (Royer) M.; m. Betty Jane Boucher, June 4, 1949; children: Gary L., Dennis K., Rebekah L., Bonnie L., Beth S. B.S. magna cum laude, Juniata Coll, Huntingdon, Pa., 1947; Ph.D. in Phys. Chemistry, Yale U., 1950. Assoc. prof. chemistry Davis and Elkins Coll., 1950-51; research scientist B.F. Goodrich Co., 1951-59; with Koppers Co., Inc., Monroeville, Pa., 1959—, asst. mgr. research dept., 1968, v.p., mgr. research and devel., 1968-84, v.p., mgr. external research, 1984—; dir. Ceramatec Corp., Kopvenco, Inc.; Bd. dirs. Indsl. Health Found., 1974-77; mem. Materials Adv. Panel, State of Pa., 1974-77, MSF Indsl. Panel on Sci. and Tech., 1973—. Contbr. articles to profl. jours. Trustee Juniata Coll., 1971-74; chmn. Pres.'s Devel. Council, 1969; deacon, elder United Presbyn. Ch. Served with USN, 1944-46. Mem. Am. Chem. Soc. Subspecialties: Physical chemistry; Polymer chemistry. Current work: Emulsion polymerization, suspension polymerization, surface and colloid chemistry. Patentee in field. Home: 539 Greenleaf Dr Monroeville PA 15146 Office: 440 College Park Dr Monroeville PA 15146

MACLEAN, PAUL DONALD, govt. ofcl.; b. Phelps, N.Y., May 1, 1913; s. Charles Chalmers and Elizabeth (Dreyfus) MacL.; m. Alison Stokes, July 16, 1942; children—Paul, David, Alexander, James, Alison. B.A., Yale U., 1935; postgrad., U. Edinburgh, Scotland, 1935-36; M.D. cum laude, Yale U., 1940. Intern in medicine Johns Hopkins U., 1940-41; asst. resident medicine New Haven Hosp., Yale Sch. Medicine, 1941-42, research asst. pathology, 1942, asst. prof. physiology, 1949-51, asst. prof. psychiatry, physiology and neurology, 1951-53, asso. prof. physiology, 1956-57; clin. instr. medicine U. Wash. Med. Sch., Seattle, 1946-47; USPHS research fellow Harvard U. Med. Sch., also Mass. Gen. Hosp., 1947-49; dir. EEG lab. New Haven Hosp., 1951-52; asso. prof. psychiatry, physiology and neurology, attending physician Grace-New Haven Hosp., 1953-56; sr. postdoctoral fellow NSF dept. physiology U. Zurich, 1956-57; chief sect. limbic integration and behavior Lab. Neurophysiology Intramural Research, NIMH, USPHS, Dept. Health and Human Services, Bethesda, Md., 1957-71; chief lab. brain evolution and behavior Intramural Research, Bethesda, Md., 1971—. Editorial bd.: Jour. Nervous and Mental Disease. Emeritus trustee L.S.B. Leakey Found. Served to maj. M.C. AUS, 1942-46; PTO. Recipient award for disting. research Assn. for Research in Nervous and Mental Disease, 1964; Salmon lectureship award, 1966; Superior Service award HEW, 1967; Hincks Meml. lectr., Ont.; Spl. award Am. Psychopathol. Assn., 1971; G. Burroughs Mider NIH Lectureship award, 1972; Karl Spencer Lashley award Am. Philos. Soc., 1972. Mem. Am. Electroencephalographic Soc., Am. Neurol. Assn., Am. Assn. History of Medicine, Am. Physiol. Soc., Am. Soc. Naturalists, Eastern Assn. Electroencephalographers, Am. Assn. Neurol. Surgeons, Soc. Neurosci., Am. Assn. Anatomists, Sigma Xi, Alpha Omega Alpha. Subspecialties: Neurophysiology; Neuranatomy. Current work: Neruo behavioral research on role of striatal complex and limbic systems in species-typical behavior and prosematic (non-verbal) communication. Home: 9916 Logan Dr Potomac MD 20854 Office: NIMH Lab Brain Evolution and Behavior 9000 Rockville Pike Bethesda MD 20205

MACON, WILLIAM LINUS, IV, surgeon; b. St. Louis, Sept. 1, 1937; s. William L. and Hazel (Rohrer) M.; m. Mary Lynn Shilling, Oct. 24, 1944; children: William H., Christopher A., Stewart M. A.B., Princeton U., 1959; M.D., Harvard U., 1963. Diplomate: Am. Bd. Surgeons. Intern Strong Meml. Hosp., Rochester, N.Y., 1963-64, resident in surgery, 1964-65, 67-71; asst. prof. surgery Case Western U., Cleve., 1971-76; dir. surgery ICU, St. Agnes Hosp., Balt., 1977-82; head dept. Surgery St. Joseph Hosp., Towson, Md., 1983—; cons. E.M.S., Cleve., 1973-76. Served as capt. U.S. Army, 1965-67. Recipient Upjohn achievement award, 1970. Fellow ACS; mem. Soc. Critical Care Medicine, Southeastern Surg. Congress, Am. Fedn. Clin. Research, Coll. Internal Chirurgiae Digestivae, Am. Soc. Parenteral and Enteral Nutrition, Sigma Xi. Presbyterian. Club: Baltimore Country. Subspecialty: Surgery. Current work: Head department surgery, gastrointestinal surgery and intensive care medicine. Home: 805 St Georges Rd Baltimore MD 21210 Office: St Joseph Hosp 7620 York Rd Towson MD 21204

MACOSKO, CHRISTOPHER WARD, chemical engineering educator; b. Bridgeport, Conn., June 14, 1944; s. Theodore J. and Mary Jeane (Olander) M.; m. Kathleen Del Snow, Sept. 3, 1967; children: Brynne, Jed, Maren, Kase. B.Ch.E., Carnegie Mellon U., 1966; M.Sc. in Chem. Engring., Imperial Coll., London U., 1967, Ph.D., Princeton U., 1970. Asst. prof. dept. chem. engring. and materials sci. U. Minn., 1970-74, assoc. prof., 1974-79, prof., 1979—; cons. Rheometrics, Union, N.J., 1970—, 3M Co., St. Paul, 1980—. Contbr. over 100 articles to tech. jours. Mem. Soc. Rheology, Am. Chem. Soc., Soc. Plastics Engring., Am. Inst. Chem. Engrs., Brit. Soc. Rheology. Subspecialties: Chemical engineering; Polymer engineering. Current work: Polymer processing and rheology especially systems which polymerize or crosslink during processing; reaction injection molding; new rheological test methods; coating flows. Patentee in field. Office: Dept Chem Engring and Materials Sci U Minn 421 Washington Ave Minneapolis MN 55455

MACOVSKI, ALBERT, electrical engineering educator, researcher; b. N.Y.C., May 2, 1929; s. Philip and Rose (Wionogr) M.; m. Adelaide Paris, Aug. 5, 1950; children: Michael, Nancy. B.E.E., CCNY, 1950; M.E.E., Poly. Inst. Bklyn., 1953; Ph.D., Stanford U., 1968. Engr. RCA Lab., Princeton, N.J., 1950-57; assoc. prof. Poly. Inst. Bklyn., 1957-60; staff scientist Stanford Research Inst., Menlo Park, Calif., 1960-71; fellow U. Calif. Med. Center, San Francisco, 1971-72; prof. elec. engring. and radiology Stanford U., 1972—; cons., researcher med. imaging. Author: Medical Imaging Systems, 1983; contbr. numerous articles to profl. jours. Recipient RCA achievement awards, 1952, 54; NIH spl. fellow, 1972. Fellow IEEE (Zworykin award 1973, Profl. Group on BTR award 1957), Ops. Soc. Am.; mem. Am. Assn. Physicists in Medicine, Sigma Xi, Tau Beta Pi. Democrat. Jewish. Subspecialties: Biomedical engineering; Imaging technology. Current work: Med. imaging using x-ray and ultrasound, multiple energy x-ray imaging, NMR imaging. Holder more than 100 patents in field. Home: 2505 Alpine Rd Menlo Park CA 94025 Office: Stanford University Electrical Engineering Dept Durand Bldg Stanford CA 94305

MACPHERSON, HERBERT GRENFELL, nuclear engr.; b. Victorville, Calif., Nov. 2, 1911; s. Duncan William and Minnie Belle (Morrison) MacP.; m. Janet Taylor Wolfenden, June 5, 1937; children—Janet Lynne, Robert Duncan. Student, San Diego State Coll., 1928-31; A.B., U. Calif., Berkeley, 1932, Ph.D. in Physics, 1937. Research scientist Nat. Carbon Co., Cleve., 1937-50, asst. dir. research, 1950-56; dir. reactor programs Oak Ridge Nat. Lab., 1956-64, dep. dir., 1964-70; prof. nuclear engring. U. Tenn., 1970-76; acting dir. Inst. for Energy Analysis, Oak Ridge, 1974-75, cons., 1975—; cons. in field. Editor: (with Lane and Maslan) Fluid Fuel Reactor, 1958. Fellow Am. Phys. Soc., Am. Nuclear Soc.; mem. AAAS, Nat. Acad. Engrs., Phi Beta Kappa, Sigma Xi. Clubs: Cosmos, Oak Ridge Country. Subspecialties: Nuclear engineering; Nuclear fission. Current work: Promoting international research and development on molten salt reactors. Devel. nuclear grade graphite, 1941-50, molten salt reactor, 1956-65; assisted in founding Inst. Energy Analysis, 1974-75. Home: 102 Orchard Circle Oak Ridge TN 37830 Office: Oak Ridge Associated Univs PO Box 117 Oak Ridge TN 37830 I have tried to understand the nature of my own mix of talents and to find a place where its application could make a positive contribution. This led me into the complex area of the development of nuclear power, where coordination of disciplines and cooperation of participants are vital.

MAC QUEEN, ROBERT MOFFAT, solar physicist; b. Memphis, Mar. 28, 1938; s. Marion Leigh and Grace (Gilfillan) MacQ.; m. Caroline Gibbs, June 25, 1960; children: Andrew, Marjorie. B.S., Southwestern U., Memphis, 1960; Ph.D., Johns Hopkins U., 1968. Asst. prof. physics Southwestern U., Memphis, 1961-63; instr. physics and astronomy Goucher Coll., Towson, Md., 1964-66; sr. research scientist Nat. Center for Atmospheric Research, Boulder, Colo., 1967—, dir. high altitude obs., 1979—; prin. investigator NASA Apollo program, 1971-75, NASA Skylab program, 1970-76, NASA Solar Maximum Mission, 1976-79, NASA/ESA Internat. Solar Polar Mission, 1978-83; lectr. U. Colo., 1968-79, adj. prof., 1979—; mem. com. on space astronomy Nat. Acad. Scis., 1973-76, mem. com. on space physics, 1977-79; mem. Space Sci. Bd., 1983—. Recipient NASA Exceptional Sci. Achievement medal, 1974. Mem. Am. Astron. Soc. (chmn. solar physics div. 1976-78), Optical Soc. Am., Am. Assn. Physics Tchrs., Sigma Xi. Subspecialty: Solar physics. Current work: Solar Coronal processes, space instrumentation. Home: 1366 Northridge Ct Boulder CO 80302 Office: Box 3000 Nat Center for Atmospheric Research Boulder CO 80307

MACQUILLAN, ANTHONY MULLENS, microbiology educator, researcher; b. London, Feb. 3, 1928; U.S., 1958; s. Alexander Terence and Mary Theresa (O'Connor) MacQ.; m. Mary Frances MacMain, Sept. 11, 1953; children: Teresa (dec.), Jennifer, Alan, Andrea. B.S.A., U. B.C., 1956, M.Sc., 1958; Ph.D., U. Wis.-Madison, 1962. Postdoctoral fellow NRC, Ottawa, Can., 1961-63; asst. prof. microbiology U. Md., 1963-72, assoc. prof., 1972—. Contbr. articles to profl. jours. Served with RAF, 1946-48. NIH grantee, 1971-74; Office Naval Research grantee, 1982-83. Mem. Am. Soc. Microbiology, Am. Soc. Photobiology. Roman Catholic. Subspecialties: Microbiology; Genetics and genetic engineering (biology). Current work: Applicability of genetic engineering/molecular genetics to marine microorganisms using yeast and bacterial model systems; genetics and molecular genetics of photorepair in yeast. Office: University of Marylan Dept Microbiology College Park MD 20742

MACRIDES, FOTEOS, biologist, researcher; b. Stamford, Conn., Aug. 6, 1943; s. Dennis and Chryssie (Liatsos) M.; m. Patricia Mary Hayes, Sept. 6, 1969; 1 dau., Kalle Patricia. B.S. in Biology, MIT, 1965, Ph.D. in Psychobiology, 1970; postgrad., Harvard U. Med. Sch., 1965-67. Research assoc. dept psychology MIT, Cambridge, 1970-72; staff scientist Worcester Found. for Exptl. Biology, Shrewsbury, Mass., 1972-75, sr. scientist, 1976—; Mem. sensory physiology and perception adv. panel NSF. Contbr. articles to profl. jours. NSF grantee, 1970-72, 1972—; NIH grantee, 1975-83. Mem. Soc. for Neurosci., Soc. for Chemoreception Scis., European Chemoreception Research Orgn., AAAS, Sigma Xi, Alpha Chi Sigma, Delta Upsilon. Subspecialties: Neurobiology; Comparative neurobiology. Current work: Electrophysiological and neurochemical interactions between the olfactory and limbic/hypothalamic systems in the regulation of behavioral and hormonal functions. Office: WFEB 222 Maple Ave Shrewsbury MA 01545

MACSWAN, IAIN CHRISTIE, plant pathologist; b. Ocean Falls, C., Can., Apr. 15, 1921; s. John and Catherine Rintoul (Kennedy) MacS.; m. Helen Constance White, Feb. 17, 1943; children: Neil, Catherine, Margot. B.S.A., U. B.C., 1942, M.S.A., 1961. Prin. research asst. Dominion Lab. Plant Pathology, U. B.C. (Can.), Vancouver, 1946-47; asst. provincial plant pathologist B.C. Dept. Agr., Vancouver, 1947-55; extension plant pathologist Oreg. State U., Corvallis, 1955—, asst. prof., 1955-61, assoc. prof., 1961-67, prof., 1967—. Contbr. articles to profl. jours. Served with RCAF/RAF, 1942-46. Mem. Am. Phytopathol. Soc., Can. Phytopathol. Soc. Republican. Lodge: Elks. Subspecialty: Plant pathology. Current work: Fungicide testing and pear storage rots. Home: 1629 NW 14th ST Corvallis OR 97330 Office: Oreg State U Cordley Hall 1089 Corvallis OR 97331

MACTUTUS, CHARLES FRANCIS, research scientist; b. Bklyn, Aug. 9, 1953; s. Charles Clifford and Emily Ann (Kresse) M. B.S., St. Lawrence U., 1975; M.A., Kent State U., 1977, Ph.D., 1979. Postdoctoral research fellow Johns Hopkins U., 1979-81; staff fellow Lab. Behavioral and Neurol. Toxicology, Nat. Inst. Environ. Health Scis., Research Triangle Park, N.C., 1981—. Contbr. articles to profl. jours. Recipient Sigma Xi research grantee, 1977, 79, 78. Mem. AAAS, Soc. Neurosci., Internat. Soc. Developmental Psychology, Behavioral Teratology Soc., Eastern Psychol. Assn., Midwestern Psychol. Assn., Nat. Soc. Med. Research, Sigma Xi. Democrat. Roman Catholic. Subspecialties: Psychobiology; Environmental toxicology. Current work: Developmental neurotoxicology, psychology of learning and memory. Office: NIH-NIEHS PO Box 12233 Research Triangle Park NC 27709

MACURDA, DONALD BRADFORD, JR., energy consultant, researcher; b. Boston, Aug. 8, 1936; s. Donald Bradford and Eleanor (Parsons) M.; m. Evelyn Anderson, June 31, 1961 (div. 1978); children: Bruce, Kristin; m. Noma Warneken Elliot, July 11, 1981. B.S., U. Wis.-Madison, 1956, Ph.D., 1963. Prof. geology and mineralogy U. Mich.-Ann Arbor, 1963-78; sr. research specialist Exxon Prodn. Research Co., Houston, 1978-81; cons. The Energists, Houston, 1981—, Geoquest, Internat., 1982—. Author: Phylogeny of Fissiculate Blastoids, 1972, The Fissiculate Blastoids, 1983. Served with USAF, 1957-60. Fellow Geol. Soc. Am.; mem. Am. Assn. Petroleum Geologists, Paleontol. Soc., Am. Geophys. Union, Soc. Economic Paleontologists and Mineralogists, Soc. Exploration Geophysicists. Republican. Subspecialties: Geophysics; Paleobiology. Current work: Seismic stratigraphy; the evolution of sedimentary basins; Echinoderms-evolution and paleocrology; ecology of living crinoids. Home: 214 Leaflet Ln Spring TX 77373 Office: The Energists 10260 Westheimer St Suite 110 Houston TX 77042

MADDIN, ROBERT, educator; b. Hartford, Conn., Oct. 20, 1918; s. Isadore and Mae (Jacobs) M.; m. Odell E. Steinberg, July 9, 1945; children—Leslie Jeanne, Jill Andrea. B.S., Purdue U., 1942; D.Engring., Yale, 1948. Research fellow Yale, 1948-49; asst. prof. Johns Hopkins, 1949-52, asso. prof., 1952-55; vis. prof. U. Birmingham, Eng., 1954-55, Oxford (Eng.) U., 1970; prof. U. Pa., 1955-56; prof., dir. Sch. Metallurgy and Materials Sci., 1957-72, Univ. prof. metallurgy, 1973—; lectr. univs. and mfg. plants, Eng., France, Germany, Israel, India, Japan. Author: (with H. Kimura) Quench Hardening in Metals, 1971; also chpts. in books.; Editor: Creep and Recovery; Editor-in-chief: Materials Science and Engineering; Contbr. tech. articles profl. publs. Served as lt. USAAF., World War II. Named Distinguished Engring. Alumnus Purdue U., 1974. Fellow Am. Soc. Metals, Metall. Soc. of Am. Inst. Mining and Metall. Engrs. (chmn. publs. com., exec. com.); mem. Yale Engring. Assos., Yale Metall. Assos., U. Birmingham Metall. Soc. (hon. life), Md. Inst. Metals (hon. dir.), Sigma Xi. Subspecialty: Metallurgy

MADDOX, ROBERT ALAN, research scientist, program manager; b. Granite City, Ill., July 12, 1944; s. Robert Alvin and Maxine (Elledge) M.; m. Rebecca Ann Speer, Dec. 29, 1967; children: Timothy Alan, Jason Robert. B.S., Tex. A&M U., 1967; M.S., Colo. State U., 1973, Ph.D., 1981. Weather observer U.S. Weather Bur., Hazelwood, Mo., 1967; commd. 2d lt. U.S. Air Force, 1967, advanced through grades to capt., 1970, served as weather officer, until 1975; research scientist Nat. Oceanographic and Atomospheric Adminstrn., Boulder, Colo., 1976-82, program mgr., 1982—; adj. prof. Nat. Weather Service Tng. Ctr., Kansas City, Mo., 1978—. Mem. Am. Meteorol. Soc. (chpt. v.p. 1979), AIAA, Nat. Weather Assn. (award 1981), AAAS. Subspecialties: Meteorology; Synoptic meteorology. Current work: Research on understanding and prediction of mesoscale, convective storm systems in middle lattitudes. Office: NOAA/ERL Weather Research Program R 1E22 325 Broadway Boulder CO 80303

MADDUX, JAMES EUGENE, psychology educator; b. Norfolk, Va., Jan. 8, 1954; s. Charles E. and Doris F. (Cranford) M. B.A., U. Richmond, 1976; M.A., U. Ala., 1979, Ph.D., 1982. Intern in psychology U. Oreg. Med. Sch., 1980-81; asst. prof. psychology Tex. Tech. U., 1981—. Contbr. articles to profl. jours. NIMH Tng. Fellowship award, 1976-79; Tex. Tech. U. grantee, 1982. Mem. Am. Psychol. Assn., Southwestern Psychol. Assn., Southeastern Psychol. Assn., Assn. for Care of Children's Health, Soc. Pediatric Psychology. Current work: Health psychology; attitude change; parenting skills training. Office: Tex Tech U Box 4100 Lubbock TX 79409

MADER, DONALD LEWIS, forestry educator; b. Balt., Nov. 7, 1926; s. Lewis Henry and Mary Elizabeth (Whipple) M.; m. Mary Alice Williams, Nov. 11, 1950; children: James, Susan, Russell. B.S., N.Y. State Coll. Forestry, 1950; M.S., U. Wis.-Madison, 1954; Ph.D., 1956. Asst. prof. U. Mass.-Amherst, 1956-61, assoc. prof., 1961-70, prof., 1970—. Served with USN, 1945-46. AAAS fellow, 1968. Mem. Soc. Am. Foresters, Soil Sci. Soc. Am. Republican. Methodist. Current work: Forest soil and site quality, forest hydrology, urban forest soils. Home: 683 E Pleasant St Amherst MA 01002 Office: U Mass Dept Forestry and Wildlife Mgmt Amherst MA 01003

MADERSON, PAUL F.A., biologist educator, biology researcher; b. Sidcup, Kent., Eng., Dec. 19, 1938; came to U.S., 1965; s. Reginald Arthur and Margaret (Parslow) M.; m. Una Kemple, June 17, 1961; 1 dau., Siobhan. B.Sc. with honors in Zoology, Kings Coll., London, 1960; Ph.D. in Zoology, St. Mary's Hosp., London, 1962, D.Sc., London U., 1972. Asst. lectr. zoology U. Hong Kong, 1962; lectr. U. Calif., 1965-66; postdoctoral Harvard Med. Sch., 1966-68; adj. asst. prof. Boston U., 1966-68; asst. prof. biology Bklyn. Coll., 1968-71, assoc. prof., 1971-73, prof., 1973—. Mem. editorial bd.: Jour. Morphology, 1975—, Am. Zoologist, 1977-82; contbr. numerous articles to profl. jours. Fellow AAAS; mem. Am. Soc. Zoologists, Zool. Soc. London, Anatomical Soc. Gt. Brit. and Northern Ireland. Subspecialties: Developmental biology; Morphology. Current work: Cell proliferation and differentiation in skin of snakes and lizards; amniote ear development; correlation of developmental data with evolutionay problems. Home: 2715 E 23d St Brooklyn NY 11235 Office: Brooklyn Coll CUNY Brooklyn NY 11210

MADISON, DON HARVEY, physicist, educator; b. Pierre, S.D., Jan 4, 1945; s. Walter Leon and Marguerite M.; m. Lina Erika Madison, Aug. 27, 1966; children: Lisa, Kristina. B.A. with highest honors in Math., Sioux Falls (S.D.) Coll., 1967; M.S. in Physics, Fla. State U., Tallahassee, 1970, Ph.D., 1972. Admissions counselor Sioux Falls Coll, 1967-68; research assoc. U. N.C.-Chapel Hill, 1972-74; asst. prof. physics Drake U., Des Moines, 1974-77, assoc. prof., 1977-81, prof., 1981—; sci. adv. com. for profl. meetings; referee profl. jours. Contbr. articles to profl. jours, chpts. to books. Bd. dirs. Home, Inc. Served with Army N.G., 1962-70. Recipient Disting. Alumni award Sioux Falls Coll., 1982; Centennial Scholar award Drake U., 1981. Mem. Am. Phys. Soc., Am. Assn. Physics Tchrs. Baptist. Subspecialty: Atomic and molecular physics. Current work: Atomic excitation and ionization by charged particle impact, research in theoretical aspects of atomic scattering. Office: Dept Physics Drake U Des Moines IA 50311

MADURA, JAMES ANTHONY, surgery educator; b. Campbell, Ohio, June 10, 1938; s. Anthony Peter and Margaret Ethel (Sebest) M.; m. Loretta Jayne Sovak, Aug. 8, 1959; children: Debra Jean, James Anthony II, Vicki Sue. B.A., Colgate U., 1959; M.D., Western Res. U., 1963. Diplomate: Am. Bd. Surgery. Intern, resident in surgery Ohio State U., 1963-71; asst. prof. surgery Ind. U.-Indpls., 1971-76, assoc. prof., 1977-80, prof., 1980—. Contbr. articles profl. jours., chpts. in books. Chmn. ad hoc com. redistricting Washington Twp. Sch. Bd., Indpls., 1979-80; safety dir. Indpls. Youth Hockey League, 1975-77. Served to capt. U.S. Army, 1964-66; Vietnam. Fellow ACS (pres. Ind. chpt. 1982-83); mem. Central Surg. Assn., Western Surg. Assn., Soc. Surgery Alimentary Tract, Midwest Surg. Soc., Assn. Acad. Surgery, Assn. Surg. Edn., ASTM, Sigma Xi. Republican. Roman Catholic. Club: Columbia. Lodge: Elks (Indpls.). Subspecialties: Surgery; Nutrition (medicine). Current work: Education of medical students and surgical residents; nutritional treatment of critically ill patients. Home: 9525 Copley Dr Indianapolis IN 46260 Office: Dept of Surgery Indiana University Medical Center 545 Barnhill Dr Indianapolis IN 46223

MAGE, MICHAEL GORDON, immunochemist; b. N.Y.C., Aug. 17, 1934; s. Myron and Mildred (Farber) M.; m. Rose Goldman, June 12, 1955. A.B., Cornell U., 1955; D.D.S., Columbia U., 1960. Postdoctoral fellow in microbiology Columbia U., 1960-62; commd. officer U.S. Public Health Service, 1962; advanced through grades; immunologist Lab. Microbiology, Nat. Inst. Dental Research, Bethesda, Md., 1962-66; immunochemist Lab. Biochemistry, Nat. Cancer Inst., NIH, Bethesda, 1966—. Contbr. articles to profl. jours. Activist on public health aspects of energy and nuclear war. Mem. Am. Assn. Immunologists, Am. Public Health Assn., Am. Chem. Soc., AAAS, Fedn. Am. Scientists, Sigma Xi. Subspecialties: Immunobiology and immunology; Immunocytochemistry. Current work: Lymphoid cell separation; T lymphocyte differentiation; molecular mechanisms of cytotoxic T lymphocyte immunization and effector function. Office: NIH Bldg 37 Room 4C28 Bethesda MD 20205

MAGEE, DONAL F., physiologist, educator, researcher; b. Aberdeen, Scotland, June 4, 1924; came to U.S., 1968; s. Hugh Edward and M.; m.; 5 children. B.M., B.Ch., Oxford U., 1948, M.A., 1948, D.M., 1972; Ph.D. in Physiology, U. Ill., Chgo., 1951. Instr. clin. sci. U. Ill., Chgo., 1948-52; asst. prof., asso. prof., prof. pharmacology U. Wash., Seattle, 1952-65; prof. physiology Creighton U., Omaha, 1965—, chmn. dept., 1965—. Author, editor books on gastrointestinal physiology; contbr. numerous articles to sci. publs. Guggenheim fellow, 1959-60; Fogerty internat. fellow, 1978-79. Mem. Am. Physiol. Soc., Am. Gastroent. Soc., Soc. Exptl. Biology and Medicine, Brit. Med. Assn., AAUP. Subspecialties: Comparative physiology; Molecular pharmacology. Current work: Gastrointestinal physiology—stomach, pancreas, gall bladder. Office: Dept Physiology Creighton U Omaha NE 68178

MAGEE, PAUL TERRY, molecular geneticist, educator; b. Los Angeles, Oct. 26, 1937; s. John Paul and Lois Lorene M.; m. Beatrice Eve Buten, Apr. 19, 1940; children: Alexander John, Amos Hart. B.S. in Chemistry, Yale U., 1959; Ph.D. in Biochemistry, U. Calif.-Berkeley, 1964. Am. Cancer Soc. fellow CNRS, France, 1964-66; asst. prof. microbiology Yale U., New Haven, 1966-72, assoc. prof. microbiology and human genetics, 1972-75, assoc. prof. human genetics, 1975-77; prof. microbiology and pub. health, chmn. dept. Mich. State U., East Lansing, 1977—; adv. panel Genetic Biology Programs, NSF, 1979-83. Mem. AAAS, Am. Soc. Microbiology, Genetics Soc. Am., Am. Soc. Biol. Chemists. Jewish. Subspecialties: Molecular biology; Genetics and genetic engineering (biology). Current work: Molecular genetics of development in saccharomyces genetics of virulence in Candida albicans. Home: 912 Stuart Ave East Lansing MI 48823 Office: Dept Microbiology and Pub Health Michigan State University East Lansing MI 48824

MAGEE, PETER NOEL, pathologist; b. Sheffield, Yorkshire, Eng., Dec. 21, 1921; s. John Albert and Elsie May (Mould) M.; m. Ines Verspyck, Dec. 15, 1951; children—Rosemary Anne, Michael John. Student, Worksop Coll. Cambridge (Eng.) U., 1940-42; M.B., B.Chir., Univ. Coll. Hosp., London, 1945. House physician Univ. Coll. Hosp., Kent and Sussex Hosp., 1946-47; Graham scholar in pathology U. London, U. Coll. Hosp. Med. Sch., 1951-53; mem. sci. staff, toxicology research unit Med. Research Council Labs., Carshalton, Surrey, 1953-68; Philip Hill prof. exptl. biochemistry U. London at Courtauld Inst. Biochemistry, Middlesex Hosp. Med. Sch., 1967-75; dir. Fels Research Inst., Temple U. Med. Sch., Phila., 1975—. Contbr. sci. articles to profl. jours. Mem. grand council Cancer Research Campaign of Gt. Britain, 1972—. Served with Med. Br. RAF, 1947-49. Recipient Johann Georg Zimmermann Sci. prize for cancer research Medizinische Hochschule, Hannover, 1975. Mem. Brit. Med. Assn., Am. Assn. Pathologists, Am. Assn. for Cancer Research, Pathol. Soc. Gt. Britain and Ireland, Biochemical Soc. (U.K.), Soc. Toxicology, Internat. Acad. Environmental Safety. Episcopalian. Club: Athenaeum (London). Subspecialty: Cancer research (medicine). Home: 1240 Morris Rd Wynnewood PA 19096 Office: 3420 N Broad St Philadelphia PA 19140

MAGEN, MYRON SHIMIN, physician, educator; b. Bklyn., Mar. 1, 1926; s. Barney and Gertrude Beatrice (Cohen) M.; m. Ruth Sherman, July 6, 1952; children—Jed, Ned, Randy. D.O., Coll. Osteo. Medicine and Surgery, 1951; Sc.D. (hon.), U. Osteo. Medicine and Health Scis., Des Moines, 1981. Rotating intern Coll. Hosp., Des Moines, 1951-52, resident in pediatrics, 1953-54; chmn. dept. pediatrics Coll. Osteo. Medicine and Surgery, Des Moines, 1958-62, Riverside Osteo. Hosp., Trenton, Mich., 1962-68, Detroit Osteo. Hosp., 1965-67; med. dir., dir. med. edn. Zieger-Botsford Hosps., Farmington, Mich., 1968-70; dean Mich. Coll. Osteo. Medicine, Pontiac, 1968-70; dean, prof. pediatrics Mich. State U. Coll. Osteo. Medicine, East Lansing, 1970—; mem. spl. med. adv. group to chief med. dir. VA, 1973-77; mem. grad. med. edn. nat. div. com. U.S. Dept. Health and Human Services, Washington, 1978-80. Contbr. articles to profl. jours. Served with USN, 1943-45. Recipient Disting. Service award Okla. Coll. Osteo. Medicine and Survery, 1975; Founder's medal Tex. Coll. Osteo. Medicine, 1978; James Watson Disting. lectr. Ohio Osteo. Assn., 1974. Mem. Am. Assn. Colls. Osteo. Medicine (pres. 1977), Am. Osteo. Assn. (com. edn.; La. Burns lectr. 1977), Am. Coll. Osteo. Pediatrics (pres. 1965-66), Mich. Assn. Osteo. Physicians and Surgeons. Subspecialty: Pediatrics. Home: 1251 Farwood Dr East Lansing MI 48823 Office: 308 E Fee Hall Mich State Univ East Lansing MI 48824

MAGER, ARTUR, aerospace co. exec.; b. Nieglowice, Poland, Sept. 21, 1919; came to U.S., 1939, naturalized, 1944; s. Herman and Ella (Kornbluh) M.; m. Phyllis Weisman, Aug. 19, 1942; 1 dau., Ilana G. B.S., U. Mich., 1943; M.S., Case Inst. Tech., 1951; Ph.D., Calif. Inst. Tech., 1953. Registered profl. engr., Ohio. Aero. research scientist NASA Lewis Labs., Cleve., 1946-51; research scientist Marquardt Corp., Van Nuys, Calif., 1954-60; dir. Nat. Engring. Sci. Co., Pasadena, Calif., 1960-61; dir. spacecraft scis. Aerospace Corp., El Segundo, 1961-64; gen. mgr. applied mechanics div., 1964-68, v.p., gen. mgr. engring. sci. ops., 1968-78, v.p. engring. group, 1978-82, cons., 1982—; mem. BSD Re-entry Panel, 1961-63, NASA com. missile and space vehicle aerodynamics, 1963-65; mem. adv. com. AFML, 1971-72. Contbr. to profl. publs. Mem. alumni fund council Calif. Inst. Tech., 1972-74; mem. indsl. council West Coast U.; bd. councilors U. So. Calif. Sch. Engring.; mem. developmental disabilities bd. Area X, chmn., 1976-78; bd. dirs. Calif. Assn. for Retarded, 1972-74, recipient Golden Rule award, 1977; pres. Exceptional Children's Found., 1970-72. Recipient Distinguished Alumni award U. Mich., 1969. Fellow Inst. Advanced Engring., Am. Inst. Aeros. and Astronautics (chmn. Los Angeles sect. 1967-68, dir. 1975-77, pres. 1980), AAAS; mem. Gesellschaft Angewandte Mathematik Mechanik, Technion Soc., Am. Phys. Soc., ASME, Nat. Acad. Engring., Sigma Xi. Subspecialty: Aerospace engineering and technology. Home: 1353 Woodruff Ave Los Angeles CA 90024 Office: 2350 E El Segundo Blvd El Segundo CA 90245 Aerospace Corp PO Box 92957 Los Angeles CA 90009

MAGID, RONALD MICHAEL, chemistry educator; b. Bklyn., Dec. 19, 1938; s. Leo and Florence (Miller) M.; m. Sabina S. Stone, July 3, 1960 (div. 1966); children: Laurence H., Steven P.; m. Linda L. Jenny, May 10, 1969. B.S., Yale U., 1959, M.S., 1960, Ph.D., 1963. Nat. Acad. Sci./NRC postdoctoral fellow Stanford (Calif.) U., Calif., 1963-64; asst. prof. Rice U., Houston, 1964-70, U. Tenn., Knoxville, 1970-72, assoc. prof., 1972-80, prof. chemistry, 1980—. Contbr. articles in field to profl. jours. NSF grad. fellow, 1959-63; recipient Salgo Noren Teaching awards Rice U., 1968, 70; Alumni Teaching award U. Tenn., 1976. Mem. Am. Chem. Soc., Chem. Soc. (London). Subspecialty: Organic chemistry. Current work: Mechanism and stereochemistry of reactions of allylic compounds, orbital symmetry controlled reactions. Home: 7914 Gleason Rd Apt 1161 Knoxville TN 37419 Office: Dept Chemistry U Tenn Knoxville TN 37996

MAGILL, CLINT WILLIAM, genetics educator; b. Washington, Sept. 15, 1941; s. Floyd C. and Wilma (Mesman) M.; m. Jane Mary Magill, July 31, 1965; children: Anne Mary. B.S., U. Ill., 1963; Ph.D., Cornell U., 1968. Postdoctoral fellow U. Minn., 1968-69; asst. prof. genetics Tex. A&M U., College Station, 1969-75, assoc. prof., 1975—. Contbr.

articles to profl. jours. NIH grantee, 1965-68; NSF fellow, 1968-69; grantee, 1975-78. Mem. AAAS, Genetics Soc. Am., Environ. Mutagen Soc., Am. Soc. Microbiology. Subspecialties: Genetics and genetic engineering (agriculture); Plant cell and tissue culture. Current work: Researcher in biochemical genetics of eukaryotic microorganisms and plant cell culture, effects of metabolic blocks on purine metabolism in neurospora. Home: 3603 Midwest St Bryan TX 77801 Office: Dept Plant Sci Tex A& M U College Station TX 77843

MAGLICH, BOGDAN C., physicist; b. Yugoslavia, Aug. 5, 1918; came to U.S., 1956, naturalized, 1967; s. Cveta and Ivanka (Bingulac) M.; m. Elowyn Castle Westervelt, 1959 (div. 1969); children: Marko Castle, Ivanka Taylor; m. Sharon Bundy Chagnaud, 1970 (div. 1977); 1 dau., Roberta Cveta.; m. Shelia Sanders Aldrich Mosler, 1982. Diploma physics, U. Belgrade, 1951; M.S., U. Liverpool, Eng., 1955; Ph.D., MIT, 1959. Staff mem. Lawrence Berkeley Lab., 1959-62; dep. group leader Brit. group CERN European Orgn. Nuclear Research, Geneva, 1962-63, leader Swiss group, 1964-67; vis. prof., joint faculty mem. Princeton U.-U. Pa. accelerator U. Pa., 1967-69; prof. physics, prin. investigator high energy physics Rutgers U., 1969-74; chmn. Migma Inst. High Energy Fusion, 1974—; pres., chmn. Fusion Energy Corp., Princeton, N.J., 1972-81, Aneutronix Inc., 1982-83, Sci. Transfer Assocs., Inc., 1981—, United Scis. Inc., 1984—; dir. Nat. Computer Analysts, Princeton.; Resident scientist UN-ILO Seminar Econ. Devel. East Africa, Kenya, 1967; lectr. Postdoctoral Sch. Physics, Yerevan, USSR, 1965, Internat. Sch. Majorana, Italy, 1969; mem. U.S. delegation Internat. Conf. High Energy Physics, Vienna, 1968, Kiev, 1970; spl. rep. of U.S. Pres. to Yugoslavia, 1976; sci. project dir. Univ. Research Ctr., King Abdulaziz U., Saudi Arabia, 1981-82; vis. prof. elec. engring. Poly. Inst. N.Y., 1981—. Editor: Adventures in Experimental Physics. Chmn. Yugoslav-Am. Bicentennial Com., 1975-76, Sheila and Bogdan Maglich Found., 1983—. Recipient White House citation, 1961; Bourgeois d'honneur de Lens, Switzerland, 1973; UNESCO fellow, 1957-58. Fellow Am. Phys. Soc.; mem. Ripon Soc. (bd. govs.), Sigma Xi. Clubs: Nassau, Mass. Inst. Tech. (N.Y. and Princeton). Subspecialty: Nuclear fusion. Discoverer omega-meson, 1961, missing-mass spectrometer, 1964, delta-meson, 1963, g-meson, S, T, U-mesons, 1965, precetron, 1969, migma-cell, 1972, aneutronic energy process, 1982. Patentee in field. Office: PO Box 2005 Princeton NJ 08540 Office: 2118 Rockefeller Center 1270 Ave of Americas New York NY 10020

MAGNANTI, THOMAS LEE, management educator; b. Omaha, Oct. 7, 1945; s. Lee A. and Florence L. (Lindquist) M.; m. Beverly A. McVinney, June 10, 1967; 1 son, R. Randall. B.S. summa cum laude, Syracuse U., 1967; M.S. in Stats, Stanford U., 1969, Stanford U., 1971, Ph.D., 1972. Asst. prof. Sloan Sch. Mgmt., MIT, Cambridge, 1971-75, assoc. prof., 1975-79, prof., 1979—, head mgmt. sci. area, 1982—; vis. fellow CORE, Leuven, Belgium, 1976-77; vis. scientist Grad. Sch. Bus., Harvard U., Cambridge, Mass., 1980-81. Author: (with S. Bradley, A. Hax) Applied Mathematical Programming, 1977; editor in chief: Ops. Research, 1983—; co-editor: Math. Programming, 1981-83; assoc. editor: Mgmt. Sci, 1978-81, Soc. Indsl. and Applied Math. Jour. on Applied Math, 1976-81, Jour. on Algebraic and Discrete Methods, 1981—. Com. chmn. Cub Scouts, Holliston, Mass., 1981—. Mem. Ops. Research Soc. Am. (com. chmn. 1978-82), Inst. Mgmt. Scis., Soc. Indsl. and Applied Math., Math. Programming Soc. Subspecialties: Operations research (engineering); Operations research (mathematics). Current work: Research in theoretical and applied optimization, particularly network optimization, transportation and distribution planning. Home: 616 Prentice St Holliston MA 01746 Office: Sloan Sch Mgmt MIT Bldg E53-350 Cambridge MA 02139

MAGNUSON, GUSTAV DONALD, physicist; b. Chgo., Aug. 22, 1926; s. Gust and Anna (Sjostr) M.; m. Phyllis Elaine, Mar. 18, 1950; children: Randal B., Donald G., Erik J., Scott K. Ph.B., U. Chgo., 1949, B.S., 1950; M.S., U. Ill., 1952, Ph.D., 1957. Lifetime teaching cert., Calif. Sr. staff scientist Gen. Dynamics/Convair, San Diego, 1957-66; now staff scientist; research assoc. prof. U. Va., Charlottesville, 1966-69; staff scientist Gulf Gen. Atomic, San Diego, 1969-72, IRT Corp., 1972-76; instr. San Diego Community Coll, San Diego State U.; judge San Diego Sci. Fair. Contbr. papers to profl. jours.; speaker profl. confs. Served with U.S. Army, 1946-47. Mem. Am. Phys. Soc., Am. Assn. Physics Tchrs., AAAS, Sigma Xi. Republican. Roman Catholic. Club: Nat. Assn. Clock and Watch Collectors. Subspecialties: Atomic and molecular physics; Magnetic physics. Current work: Analysis and design of large superconducting magnets for magnetohydrodynamics, fusion research, isotope separation. Patentee. Home: 1755 Catalina Blvd San Diego CA 92107 Office: Gen Dynamics/Convair PO Box 80847 MZ 16-1073 San Diego CA 92138

MAGNUSON, JOHN JOSEPH, zoology educator, researcher; b. Evanston, Ill., Mar. 8, 1934; s. John J. and Florence H. (Hellstrom) M.; m. Norma Edna Domian, June 14, 1959; children: Susan Florence, Jennifer Lou. Student, No. Ill. U., 1952-53; B.S., U. Minn., 1956, M.S., 1958; Ph.D., U. B.C., 1961. Fishery research biologist Bur. Commn. Fisheries, Honolulu, 1961-67; prof. U. Wis.-Madison, 1968-75, 1976—; program dir. for ecology NSF, Washington, 1975-76; dir. Ctr. for Limnology U. Wis.-Madison, 1982—, chmn. oceanography and limnology grad. program, 1978—; chmn. fishery task force Nat. Acad. Sci., Washington, 1980—; bd. tech. experts Gt. Lakes Fishery Commn., Ann Arbor, Mich., 1976—; adv. com. Mar. Resources Research FAO, Rome, 1981—; mem. ecology com. SAB EPA, Washington, 1979-81; mem. sci. adv. bd. environ. sci. Oak Ridge (Tenn.) Nat. Lab., 1977-79; mem. ad hoc Lake Michigan Task Force Wis. DNR, Madison, 1974-75. Contbr. articles on zoology to profl. jours.; editor report, Gt. Lakes Fishery Commn., 1979; Fish Behavior in Capture and Culture Fisheries, 1980. Nat. Research Council scholar, 1959-61; recipient cons. award Fedn. Ont. Naturalists, 1981. Fellow AAAS; mem. Am. Soc. Limnology and Oceanography, Ecol. Soc. Am. (aquatic ecology pres. 1975-76), Am. Fisheries Soc. (pres. 1981-82). Current work: Distributional ecology of fishes, lakes as islands, long term ecological research on lake ecosystems, community ecology. Office: U Wis-Madison Ctr for Linmology Madison WI 53706

MAGRATH, IAN TREVOR, cell biologist, physician; b. Isle of Man, U.K., Oct. 31, 1944; s. Albert Ernest and Ivy Gladys (Myers) M.; m. Pamela Vera Green, Mar. 4, 1943; children: Samantha, James, Simon. M.B., B.S., London U., 1967. Intern Charing Cross Hosp., London, 1967-68; redisent in medicineloncology, also Royal Postgrad. Med. Sch., 1968-70; Research fellow in immunology Kennedy Inst. Rheumatology, 1970-71; hon. lectr. Inst. Diseases of the Chest, London U., 1970-71; dir. Lymphoma Treatment Ctr., Kampala, Uganda, 1971-74; sr. investigator and vis. scientist pediatric oncology Nat. Cancer Inst., Bethesda, Md., 1974—. Editor: The Influence of the Environment in the Pathogenesis of Leukemias and Lymphomas; contbr. chpts. to books, sci. articles to profl. publs. Recipient Gov.'s Clin. Gold medal Charing Cross Hosp., London, 1967; Llewellyn scholar, 1967; Cancer Research Campaign grantee, 1971-74. Mem. Am. Soc. Clin. Oncology, Am. Assn. Cancer Research, Royal Coll. Physicians. Subspecialties: Chemotherapy; Cell study oncology. Current work: Major interest is lymphoid neoplasia, especially Burkitt's lymphoma; studies involve pathogenesis, cell biology and treatment; the geography of lymphoid neoplasia. Home: 814 Patton Dr Silver Spring MD 20901 Office: NIH Bldg 10 Room 13C103 Bethesda MD 20205

MAGUIRE, MARJORIE PAQUETTE, geneticist, educator; b. Pearl River, N.Y., Sept. 2, 1925; d. Percy Carlton and Pearl Ella (Phillips) P.; m. Bassett Maguire, Jr., June 13, 1950; children: William C., David J. B.S., Cornell U., 1947, Ph.D., 1952. Research assoc. U. Tex., Austin, 1958-64, research scientist, 1964-75, assoc. prof., 1975-81, prof., 1981—. Contbr. articles to profl. jours. NIH career devel. awardee, 1965-75; grantee NSF, NIH. Mem. Genetics Soc. Am., Am. Genetic Assn., Am. Soc. Cell Biology, Bot. Soc. Am., Sigma Xi. Subspecialties: Gene actions; Cell biology. Current work: Researcher in genetic recombination, meiotic chromosome structures, muclear ultrastructure, interrelationships on synapsis, crossing over and chromosome disjunction. Home: 2702 Valley Springs Rd Austin TX 78746 Office: Dept Zoology U Tex Austin TX 78712

MAGUIRE, WILLIAM MICHAEL, experimental psychologist, educator; b. Yonkers, N.Y., Aug. 22, 1948; s. William Patrick and Dorothy Anne (Whalen) M.; m. Margaret Elizabeth Short, Feb. 11, 1973 (div. 1982); 1 son, Patrick Short. B.A., Fordham U., 1970; Ph.D., SUNY-Buffalo, 1977. Research asst. prof. SUNY-Buffalo, 1977-81, vis. asst. prof., 1982; asst. prof. psychology St. John's U., 1981—; adviser Psi Chi chpt. Mem. Am. Psychol. Assn., Eastern Psychol. Assn. Subspecialties: Cognition; Neurophysiology. Current work: Functional properties of visual association cortex; the role of attention and preattentive mechanisms in vision. Office: Saint John's University Grand Central and Utopia Pkwy Jamaica NY 11437

MAHAFFEY, MICHAEL KENT, mechanical engineer; b. Burlington, Iowa, Dec. 31, 1941; s. Kent Clay and Gertrude (Fogey) M.; m. Judith Ann Mawdsley, June 6, 1964; children: Kristin Kay, Michael Kent. B.S.N.E., Kans. State U., 1964, M.S.N.E., 1966. Research engr. devel. Battelle N.W., Richland, Wash., 1965-70; with Westinghouse Hanford Co., Richland, 1970—, mgr., 1982. Treas. troop 201 Boy Scouts Am., 1980-83; mgr. McDonalds Softball Team, Richland, Wash., 1981-82. Mem. Am. Nuclear Soc., ASME (vice-chmn. 1982, chmn. Columbia Basin sect. 1983). Republican. Methodist. Subspecialties: Nuclear fission; Nuclear engineering. Current work: Research and development associated with cleanup and radioactive waste disposal from Three Mile Island nuclear power plant. Home: 2520 Granada Ct Richland WA 99352 Office: Westinghouse Hanford Co PO Box 1970 Richland WA 99352

MAHAN, DONALD CLARENCE, animal research scientist, educator; b. East Chicago, Ind., May 28, 1938; s. Clarence and Irene M.; m. Amy Jo Osburne, Nov. 17, 1962; children: Melaine, Jean, Laurie. B.S., Purdue U., 1960, M.S., 1965; Ph.D., U. Ill., 1969. County extension agt. Purdue U., Lafayette, Ind., 1961-63; asst. prof. Ohio State U.-Columbus, 1969-73; assoc. prof. Ohio State U.-Wooster, 1973-81, prof. animal nutrition, 1981—; cons., Poland, Bulgaria, U.S. Soybean Assn., 1981, Honduras, U.S. Feed Grains Council, 1982. Contbr. articles in field to profl. jours. Officer Gideons Internat., Wayne County, Ohio, 1981-83; mem. ofcl. bd. Grace Brethren Ch., Wooster, Ohio, 1979-83, asst. Sunday Sch. supt., 1979-83. Served with U.S. Army, 1960-61. Mem. Am. Soc. Animal Sci., Am. Inst. Nutrition, Animal Nutrition Research Council, Am. Registry Cert. Animal Scientists, Sigma Xi, Gamma Sigma Delta. Subspecialties: Animal nutrition; Nutrition (biology). Current work: Swine, nutrition, selenium, calcium, phosphorus, weanling swine, reproduction, protein. Home: 2583 Taylor Ave Wooster OH 44691 Office: Ohio Agrl Research and Devel Ctr Ohio State U 110 Gerlaugh Hall Wooster OH 44691

MAHAN, HARRY CLINTON, psychologist, consulting statistician in neuropsychology; b. Ashtabula, Ohio, Mar. 14, 1909; s. Ray Noble and Jennie (Strickler) M.; m. Eleanor Gearhart, Apr. 20, 1944; 1 son, Michael G. A.B., Ohio U., 1931; M.A., Ohio State U., 1932, Ph.D., 1940. Psychologist Warren (Pa.) State Hosp., 1932-35, 1938-41; pvt. practice indsl. psychology, Wichita, Kans., 1946-50; assoc. prof. psychology U. Wichita, 1947-50; head dept. social sci. Oceanside (Calif.) Coll., 1954-57; head behavioral sci. dept. Palomar Coll., San Marcos, Calif., 1957-76; cons. neuropsychology statistician, Oceanside, Calif., 1977—; dir. Project Socrates, 1965—; 1st chmn. Calif. State Psychology Examining Com., 1957-60. Author: The Interactional Psychology of J.R. Kantor, 1968, A Primer of Interactional Psychology, 1970. Served to col. USMCR, 1943-69. Fellow Am. Psychol. Assn. (life); mem. Nat. Acad. Neuropsychologists (life), Blue Key, Phi Delta Theta. Presbyterian. Subspecialties: Statistics; Neuropsychology. Current work: Statistical consulting and computational assistance. Home: 811 Leonard Ave Oceanside CA 92054

MAHAPATRA, RAJAT KANTI, cardiologist; b. Sautia, West Bengal, India, Aug. 6, 1943; s. Naba K. and Mohamaya (Bhattacha Riva) M.; m. Dipta Panda, May 19, 1973; children: Anirban Gora, Rita. B.S., Govt. Med. Coll., India; M.D. in Internal Medicine, Post Grad. Med. Inst., India, 1971, All India Med. Inst., 1976. Bd. cert. in Internal Medicine and Cardiology, India. Fellow in hepatology Postgrad. Med. Inst., Chandigarh, India, 1971-72; registrar in cardiology All India Med. Inst., New Delhi, India, 1973-76; chief cardiology Safdar Jang Hosp., New Delhi, 1976-77; cardiology fellow Maimonides Med. Ctr., Bklyn., 1977-78, SUNY-Downstate Med. Ctr., 1978-79; fellow in hypertension U. Louisville, 1979-80; chief cardiologist USAF Carswell AFB Hosp., Ft. Worth, 1980-82, civilian cardiologists, 1983—, dir. cardiology. Served to maj. USAF, 1980-82. Recipient All India Med. Inst. Benjamin Castleman award, 1974. Fellow Am. Coll. Angiology; mem. Am. Fedn. Clin. Research. Subspecialties: Cardiology; Internal medicine. Current work: investigate work in clinical pharmacology, drug trials, non-invasive cardiology and hypertension. Home: 1200 Mesquite Trail Benbrook TX 76126 Office: Carswell AFB Hosp Ft Worth TX 76127

MAHBOUBI, EZZAT OLLAH, physician; b. Dargaz, Iran, Apr. 21, 1929; s. Abbas and Khadijah M.; m. Pouran Minou, Apr. 12, 1957; children: Arta, Artin. M.D., U. Teheran, 1961; M.P.H., UCLA, 1964; cert. in cancer epidemiology, communications sci., preventive medicine and epidemiology. Dir. Babol Med. Research Sta.; head Internat. Agy. Research on Cancer, Teheran, Iran; acting chief epidemiology Am. Health Found., N.Y.C., 1980-81; prof. epidemiology Eppley Cancer Inst., U. Nebr. Med. Ctr., Omaha, 1973—; mem. grad. faculty U. Nebr., 1976—. Contbr. numerous articles to profl. jours., chpts. to books. Mem. Am. Assn. Cancer Research, Am. Public Health Assn., Internat. Epidemiol. Assn. Republican. Subspecialties: Epidemiology; Preventive medicine. Current work: Causation of cancer; etiological factors in cancer. Office: Eppley Cancer Inst UNMC Omaha NE 68105 Home: 9933 Devonshire Dr Omaha NE 68114

MAHER, JAMES VINCENT, JR., physics educator; b. N.Y.C., Aug. 25, 1942; s. James V. and Anne C. (Canneen) M.; m. Angela B. Braunstein, Aug. 13, 1966; children: Robin, James. B.S., U. Notre Dame, 1964; M.S., Yale U., 1965, Ph.D., 1968. Postdoctoral appointee Argonne (Ill.) Nat. Lab., 1968-70; asst. prof. physics U. Pitts., 1970-74, assoc. prof., 1974-80, prof., 1980—, dir., 1979-80. Contbr. articles to profl. jours. NSF grantee, 1970-83; Research Corp. grantee, 1982-83; vis. scientist grantee Niels Bohr Inst., Copenhagen, 1978, Kerrfysisch Versneller Inst., Netherlands, 1976; Danforth fellow, 1964-68; Gen. Motors nat. scholar, 1960-64. Fellow Am. Phys. Soc.; mem. Sigma Xi. Roman Catholic. Subspecialties: Statistical physics; Condensed matter physics. Current work: Experiments to study the statistical physics of liquids and interfaces between liquids. Home: 1313 Denniston Ave Pittsburgh PA 15217 Office: Dept Physics and Astronomy Univ Pitts Pittsburgh PA 15260

MAHER, TIMOTHY JOHN, pharmacologist; b. Boston, Nov. 24, 1953; s. Robert D. and Veronica I. (Cody) M.; m. Barbara Jean Walz, Aug. 20, 1977; children: Andrew Michael, Matthew Edward. B.S. with honors, Boston State Coll., 1976; Ph.D., Mass. Coll. Pharmacy, 1980. Asst. prof. pharmacology Mass. Coll. Pharmacy, Boston, 1980—; postdoctoral fellow dept. nutrition and food sci. MIT, Cambridge, Mass., 1980—; postdoctoral research fellow Ctr. for Brain Sci. and Metabolism, Cambridge, 1981—. Contbr. articles to sci. jours. Mem. Soc. Neurosci. Roman Catholic. Subspecialties: Neuropharmacology; Neurochemistry. Current work: Investigations of the interactions between nutrient intake and central nervous system neurotransmitter synthesis, release and function. Office: 179 Longwood Ave Boston MA 02115

MAHER, VERONICA MARY, microbiologist, educator, research laboratory administrator; b. Detroit, Feb. 20, 1931; d. Henry Cornelius and Veronica Margaret (Kelly) M. B.S. in Biology, Marygrove Coll., 1951, M.S., U. Mich., 1958; Ph.D. in Molecular Biology, U. Wis., 1968. Joined Servants of the Immaculate Heart of Mary, Roman Catholic Ch., 1951; research assoc. Yale U. Sch. Medicine, 1968-69; asst. prof. Marygrove Coll., 1969-71; staff scientist Mich. Cancer Found., Detroit, 1970-76; assoc. prof. microbiology and biochemistry Mich. State U., 1976-79, prof., 1979—, co-dir., 1976—; mem. sci. adv. com. Nat. Ctr. Toxicological Research, 1978-81; mem. cancer spl. programs adv. com. Nat. Cancer Inst., NIH, 1980—. Contbr. chpts. to books, articles to profl. publs. Nat. Cancer Inst. grantee, 1972—; Dept. Energy grantee, 1977—; EPA grantee, 1978-81; Nat. Inst. Environ. Health Scis. grantee, 1977-83. Mem. Am. Assn. Cancer Research, Environ. Mutagen Soc., Am. Soc. Microbiology. Democrat. Subspecialties: Cancer research (medicine); Cell biology (medicine). Current work: Transformation of normal human cells into cancer cells by exposing them to chemicals or radiation; mechanisms of carcinogenesis. Home: 6091 Brook Haven Ln Apt 27 East Lansing MI 48823 Office: Mich State U Carcinogenesis Lab Fee Hall East Lansing MI 48824

MAHESHWAR, PREM NARAIN, radiation physicist, cons. radiation safety; b. Agra, India, May 21, 1939; s. Moti Lal and Choti Devi M.; m. Pushpa Rani, July 4, 1961; children: Archana, Mukul. B.S., Agra U., 1957, M.S., 1959, Ph.D., 1975; M.S. with distinction in Nuclear Physics, Laval U., Que., Can., 1968. Sci. officer Atomic Energy, Trombay, India, 1959-66, Laval U., 1966-70; physicist Hopital Maisonneuve, Montreal, Que., Can., 1970; radiation safety officer, instr. computer sci. McGill U., Montreal, 1972-74; chief physicist Bapt. Med. Center, Jacksonville, Fla., 1977-81, Med. Ctr. Puntagorda, Fla., 1977-81, Assn. Radiologists Inc., Charleston, W.Va., 1981—; cons. radiation safety Atomic Speed Co., 1977—. Contbr. articles in field to profl. books and pubs., U.S. and Can., papers to profl. confs., U.S., Can., France, India. Atomic Energy Can. Ltd. research grantee, 1966. Mem. Am. Assn. Physicists in Medicine, Am. Nuclear Soc. Subspecialty: Cancer research (medicine). Current work: Cancer treatment of radiation related cancer, radiation safety X-ray. Office: 3100 MacCorkale Ave B-1 Charleston WV 25321

MAHLBERG, PAUL GORDON, botanist, educator; b. Milw., Aug. 1, 1928; s. Paul Rudolph and Antoinette Marie (Heinkel) M.; m. Marilyn Margaret Waite, Aug. 14, 1954; children: Melinda Sue, Heidi Margaret. B.S., U. Wis., 1950; Ph.D., U. Calif.-Berkeley, 1958. Asst. prof. botany U. Pitts., 1958-65; prof. biology Ind. U., Bloomington, 1965—; cons. to cos. engaged in plant tissue culture to produce secondary products, rubber, alkaloids, in a culture system. Author: Laboratory Program in Plant Anatomy, 1971; editor 8 books; Contbr. articles to profl. jours. Recipient numerous grants from nat. agys. Subspecialties: Plant cell and tissue culture; Cell and tissue culture. Current work: Laticifer cell, glandular cells.

MAHLER, INGA R., molecular biologist, educator; b. Beuthen, Germany, Sept. 13, 1925; d. Julius and Margarete (Goldstein) Ring; m. Donald L. Mahler, May 28, 1950. B.A., U. Pa., 1947, M.S., 1949; Ph.D., Brandeis U., 1961. Researcher dental bacteriology Tufts U. Dental Sch., 1950-57; postdoctoral fellow dept. biochemistry Brandeis U., Waltham, Mass., 1961-63, jr. research assoc., 1963-70; sr. scientist Rosenstiel Biomed. Research Ctr., Waltham, 1970—; cons. in field. Contbr. writings to profl. publs. Active NOW, LWV, World Wildlife Fund. Career devel. grantee Nat. Health Found., 1957-59. Mem. Am. Soc. Microbiology. Democrat. Jewish. Subspecialty: Molecular biology. Current work: Cloning of bacillus subtilis genes with emphasis on sporulation genes; transl. and control of sporulation genes. Home: 36 Boulder Rd Newton Center MA 02159 Offic: Rosenstiel Research Ct Brandeis U Waltham MA 02154

MAHOWALD, ANTHONY P., genetics educator; b. Albany, Minn., Nov. 24, 1932; s. Aloys and Cecelia (Maus) M.; m. Mary Briody, Apr. 11, 1971; children: Maureen, Lisa, Michael. B.S., Spring Hill Coll., 1958; Ph.D., Johns Hopkins U., 1962. Asst. prof. Marquette U., 1966-70; asst. mem. Inst. Cancer Research, 1970-72; asso. prof. biology Ind. U., Bloomington, 1972-76, prof., 1976-82; prof., chmn. dept. genetics and anatomy Case Western Res. U., Cleve., 1982—. Editor-in-chief: Developmental Biology. Mem. Am. Soc. Cell Biology, Am. Soc. Genetics, Soc. Devel. opmental Biology, Am. Soc. Zoology. Subspecialties: Developmental biology; Gene actions. Current work: Analysis of oogenesis and early development, utilizing mutations, recombinant DNA and hybridoma approaches. Offic: Dept Genetics and Anatomy Case Western Res Cleveland OH 44106

MAI, WILLIAM FREDERICK, plant pathologist, educator; b. Greenwood, Del., July 23, 1916; s. William Frederick and Lurana (Owens) M.; m. Barbara Lee Morrell, June 2, 1941; children: Virginia Mai Austin, William Howard, Elizabeth Mai Hardy. B.S., U. Del., 1939; Ph.D., Cornell U., 1945. Asst. prof. plant pathology Cornell U., Ithaca, N.Y., 1946-49, assoc. prof., 1949-52, prof., 1952—; cons. nematode diseases of plants. Editor: Plant Parasitic Nematodes, 2 vols, 1971; co-author: Pictorial Key to Genera of Plant-Parasitic Nematodes, 1975; mem. editorial bd.: Phytopathology, 1960-63, Parasitology, 1970-77, Plant Disease Reporter, 1975-76; author articles, expt. sta. bulls. and pamphlets. Active United Fund, civic orgns. Served with USN, 1945-46. Recipient Adventurere in Research award 9th Internat. Congress Plant Protection, 1979; NSF grantee, 1960-67; grantee Nat. Acad. Sci., Internat. Potato Ctr., AID, United Brands. Mem. Am. Inst. Biol. Scis., Orgn. Tropical Am. Nematologists, Am. Phytopath. Soc. (pres. Northeastern Div. 1963), Soc. Nematologists (pres. 1969), Soc. European Nematologists, Helminthological Soc. Wash., Potato Assn. Am., Sigma Xi, Phi Kappa Phi. Lodge: Rotary. Subspecialty: Plant pathology. Home: 613 E Shore Dr Ithaca NY 14850 Office: Dept Plant Pathology Cornell U Ithaca NY 14853

MAIBACH, HOWARD I., dermatologist; b. N.Y.C., July 18, 1929; s. Jack Louis and Sidonia (Fink) M.; m. Siesel Wile, July 8, 1953; children—Lisa, Ed, Todd. A.B., Tulane U., 1950, M.D., 1955. Diplomate: Am. Bd. Dermatology. Intern William Beaumont Army Hosp., El Paso, Tex., 1955-56; resident, fellow in dermatology USPHS, Hosp. of U. Pa., 1959-61; asst. instr. U. Pa., 1958-61, lectr., 1960-61;

practice medicine specializing in dermatology U. Calif. Hosps., San Francisco, 1961—; asst. prof. dermatology U. Calif. Sch. Medicine, San Francisco, 1961-63, asso. prof., 1967-73; research asso. Cancer Research Inst., 1967—; mem. staff U. Calif.-H.C. Moffitt Hosps., 1961—; cons. Laguna Honda Hosp., 1962-66, chief dermatology service, 1963-67; cons. Letterman Gen. Hosp., Calif. Med. Facility, Vacaville, San Francisco Gen. Hosp., Sonoma State Hosp., Eldridge, Calif., Stanford Research Inst., Menlo Park, Calif., Calif. Dept. Public Health, Berkeley, VA Hosp., Research Inst. Fragrance Materials, Inc., David Grant USAF Hosp. of Travis AFB, Naval Hosp., San Diego, Wilford Hall AFB, Tex., Army Environ. Health Agy., Md.; mem. Internat. Contact Dermatitis Research Com. Editor: Animal Models in Dermatology, 1965; co-editor: Dermatotoxicology and Pharmacology, 1977, Skin Microbiology, 1981; bd. editors: Internat. Jour. Dermatology, 1974—; editorial bd.: Contact Dermatitis: Environ. Dermatology, 1974—, Clin. Toxicology, 1976—; internat. editorial bd.: Excerpta Media, 1976—; contbr. numerous articles to profl. jours. Served to capt. M.C. U.S. Army, 1955-58. Recipient awards Soc. Cosmetic Chemists, 1970, 71, 73. Fellow A.C.P.; mem. Am. Acad. Dermatology (award for essay 1961), San Francisco Dermatol. Soc. (pres. 1970-71), Pacific Dermatol. Assn., Soc. Investigative Dermatology, N.Y. Acad. Scis., Calif. Med. Assn., Am. Fedn. Clin. Research, AMA, San Francisco Med. Soc., Am. Dermatol. Assn., Internat. Soc. Tropical Dermatology, Am. Soc. Clin. Pharmacology and Therapeutics, Am. Coll. Toxicology; hon. mem. Swedish Dermatol. Soc., Am. Vet. Dermatol. Assn., Am. Acad. Vet. Dermatology, Danish Dermatol. Soc., German Dermatol. Soc. Subspecialties: Dermatology; Toxicology (medicine). Office: University of California Hospital San Francisco CA 94143

MAIBAUM, MATTHEW, behavioral science consultant, technical writer; b. Chgo., Aug. 14, 1946; s. Richard Walter and Sylvia (Kamion) M. A.B., U. Calif.-Berkeley, 1969; M.P.A., UCLA, 1973; Ph.D., Calif. Sch. Profl. Psychology, 1975, Claremont Grad. Sch., 1980. Lic. psychologist, Calif. Administrv. analyst, Los Angeles County, 1969-71, psychol. trainee, So. Calif., 1971-76; psychologist State of Calif., Los Angeles, 1976-78; pvt. practice psychol. cons., So. Calif., 1978—, freelance cons. behavioral scientist, 1969—; cons. Pico-Robertson Sr. Citizens Ctr., Los Angeles, 1973-74, Furthermore Found. Clinic, 1975—, Chabad Mental Health Programs, 1981—; instr. Redlands U., UCLA, U. Phoenix, 1976—. Author: novels Sanity, 1975, Hope is Waking Man's Dream, 1979; contbr. tech. articles to publs. Mem. Am. Psychol. Assn., So. Calif. Polit. Sci. Assn., Dramatists Guild Am., Authors Guild Am., Amity Circle of Los Angeles, Calif. State Psychol. Assn., Internat. Social Sci. Honor Soc., Sigma Xi. Subspecialties: Social psychology; Psychiatry. Current work: Organization behavior, social conflict, prejudice, intergroup relations, research methodology development, forensic psychology. Home: 826 Greentree Rd Pacific Palisades CA 92072

MAICKEL, ROGER PHILIP, pharmacologist, educator; b. Floral Park, N.Y., Sept. 8, 1933; s. Philip Vincent and Margaret Mary (Rose) M.; m. Lois Louise Pivonka, Sept. 8, 1956; children: Nancy Ellen Maickel Ward, Carolyn Sue. B.S., Manhattan (N.Y.) Coll., 1954; postgrad., Poly. Inst. Bklyn., 1954-55; M.S., Georgetown U., 1957, Ph.D., 1960. Biochemist Nat. Heart Inst., Bethesda, Md., 1955-65; asso. prof. pharmacology Ind. U., 1965-69, prof., 1969—, head sect. pharmacology med. scis. program, 1971-77; prof. pharmacology and toxicology, head dept. Sch. Pharmacy and Pharmacal Scis. Purdue U., West Lafayette, Ind., 1977-83; acting v.p. product acquisition and devel. BetaMED Pharms., Inc., Indpls., 1983—. Adv. editor: Pergamon Press, 1970—; adv. editorial bd.: Neuropharmacology, 1974—. Bd. dirs. TEAMS, Inc., 1981—. Recipient Alumni award in medicine Manhattan Coll., 1972. Fellow AAAS, Am. Coll. Neuropsychopharmacology, Am. Inst. Chemists, Royal Soc. Chemistry, Collegium Internat. de Neuro-Psychopharmacologicum; mem. Am. Chem. Soc., Am. Soc. Pharmacology and Exptl. Therapeutics, Am. Soc. Clin. Pharmacology and Therapeutics, ASTM, Soc. Forensic Toxicologists, Internat. Assn. Chiefs Police, Internat. Soc. Psyconeuroendocrinology, N.Y. Acad. Scis., Soc. Neurosci., Soc. Toxicology, Sigma Xi, Rho Chi. Subspecialty: Pharmacology. Home: 3567 Canterbury Dr Lafayette IN 47905 Office: Pharmacy Bldg Purdue U West Lafayette IN 47907 As a human being, I hope to be able to do my best in the roles of scientist, teacher, and citizen by fulfilling the academic criteria of teaching, research, and service to the utmost degree humanly possible.

MAIER, JOHN JOSEPH, computer scientist, educator; b. Utica, N.Y., Dec. 11, 1931; s. Joseph Francis and Katherine (Pryzbyc) M.; m. Jean Elizabeth Walus, May 28, 1955 (dec.); children: Douglas J., Dawn Marie, Sandra J.; m. Shirley Ann Roberts, July 27, 1977. B.A. in Physics, Utica Coll., 1959; M.S. in Elec. Engring, Syracuse U., 1975; postgrad. in elec. engring. and computer sci, Syracuse U., 1976—. Test equipment engr. Gen. Electric Co., Utica, 1958-63, engring. analyst, 1963-65; phys. scientist U.S. Air Force, Griffiss AFB, N.Y., 1965-78, computers scientist, 1978—; adj. tchr. math. Mohawk Valley Community Coll, Utica, 1960—. Served with USN, 1950-54; Korea. Mem. Optical Soc. Am., Sigma Pi Sigma. Roman Catholic. Clubs: Utica Maennerchor, Rome Polish Home. Subspecialties: Distributed systems and networks; Software engineering. Current work: Currently responsible for efforts related to the handling on nonnumerical data involving large distributed computer systems; also responsible for an exploratory development effort to develop a data base machine. Home: RD 1 Lovers Ln Oriskany NY 13424 Office: RADC/IRDT NY 13441

MAIER, LEO ROBERT, JR., mechanical engineering educator; b. Allentown, Pa., Oct. 23, 1939; s. Leo Robert and Dorothy Smith (Grim) M.; m. Marilyn Kater, Dec. 27, 1969; 1 son: Clifford John. B.S., Purdue U., 1961; M.Engring., Pa. State U., King of Prussia, 1967; Ph.D., Iowa State U., 1972. Registered profl. engr., Ohio, N.J. Sr. mech. engr. Power Generators, Inc., Trenton, N.J., 1961-69; tchr., research grad. asst. Iowa State U., Ames, 1969-72; sr. mech. design engr. United Aircraft Corp., Norwalk, Conn., 1972-73; sr. analytical engr. Atlantic Research Corp., Alexandria, Va., 1973-75; assoc. prof. mech. engring. Ohio No. U., Ada, 1975—; cons. in field, 1977—. Active United Way, 1979—, ARC, 1978—. Grantee NSF-Def. Civil Preparedness Agy.-Am. Soc. for Engring. Edn., 1976; NASA-Am. Soc. for Engring. Edn. summer faculty fellow, 1977; USAF-Am. Soc. for Engring. Edn. summer faculty fellow, 1978. Mem. ASME, Am. Soc. for Engring. Educators. Subspecialties: Mechanical engineering; Solid mechanics. Current work: Failure analysis, polymers, fracture mechanics, robotics. Office: Dept Mech Engring Ohio No U Ada OH 45810

MAIMAN, THEODORE HAROLD, physicist; b. Los Angeles, July 11, 1927. B.S. in Engring. Physics, U. Colo., 1949; M.S. in Elec. Engring, Stanford U., 1951; Ph.D. in Physics, Stanford U., 1955. Sect. head Hughes Research Labs., 1955-61; pres., founder Korad Corp., Santa Monica, Calif., 1961-68, Maiman Assos., Los Angeles, 1968-76; v.p., founder Laser Video Corp., Los Angeles, 1972-75; v.p. advanced tech. and new ventures TRW Inc. Electronics and Def. Sector, Los Angeles, 1975—. Author papers in field.; Adv. bd.: Indsl. Research mag. Served with USNR, 1945-46. Recipient award Fannie and John Hertz Found., 1966; Ballantine award Franklin Inst., 1962; award for devel. laser Aerospace Elec. Soc.-Am. Astron. Soc., 1965; named Alumni of Century U. Colo., 1976. Fellow Soc. Motion Picture and TV Engrs., Am. Phys. Soc. (Oliver E. Buckley prize 1966), Optical Soc. Am. (R.W. Wood prize 1976), Soc. Photog. and Instrumentation Engrs.; mem. Nat. Acad. Engrs., Nat. Acad. Scis., IEEE, Soc. Info. Display, Sigma Xi, Sigma Pi Sigma, Sigma Tau, Pi Mu Epsilon. Subspecialties: Graphics, image processing, and pattern recognition; Laser data storage and reproduction. Responsible devel. 1st laser.

MAIN, JAMES HAMILTON PRENTICE, oral pathologist, educator; b. Biggar, Scotland, June 7, 1933; naturalized, Canadian citizen 1976; s. George P. and Helen H. (Stark) M.; m. Patricia A. Robertson, July 28, 1961; children: Fiona G., George I. P. B.D.S., U. Edinburgh, Scotland, 1955, Ph.D., 1964. King George VI fellow Northwestern U., 1955-56; Carnegie research fellow U. Edinburgh, 1960-61, lectr. in oral pathology, 1961-66, sr. lectr., 1966-69; prof. oral pathology U. Toronto, Ont., Can., 1969—; head dept. dentistry Sunnybrook Med. Ctr., 1971—. Contbr. numerous articles, chpts. to profl. publs. Served to flight lt. RAF, 1957-60; France. Recipient Colgate Palmolive prize Internat. Assn. Dental Research, London, 1966; Clark prize Edinburgh Path. Soc., 1968. Fellow Royal Coll. Surgeons (Edinburgh), Royal Coll. Pathologists (London), Royal Coll. Dentists of Can. (pres. 1981-83); mem. Can. Dental Assn., Can. Acad. Oral Pathology (pres. 1974-75), Am. Acad. Oral Pathology. Mem. United Ch. of Canada. Club: Sleep Hollow Golf and Country (Stouville, Ont.). Subspecialties: Oral pathology; Cancer research (medicine). Office: U Toronto Faculty Dentistry Oral Pathology 124 Edward St Toronto ON Canada M5G 1G6 Home: 85 Dawlish Ave Toronto ON Canada M4N 1H2

MAIN, STEPHEN PAUL, biologist, educator, researcher; b. Iowa City, Aug. 26, 1940; s. Clarence Ervin and Margaret (Paul) M.; m. Elaine Carol Blum, Aug. 3, 1963. B.S., Valparaiso U., 1962, M.A.L.S., 1965; Ph.D., Oreg. State U., 1972. Secondary sci. tchr., Crescent City, Ill., 1962-63, Mill City, Oreg., 1965-69; asst. prof. biology Wartburg Coll., Waverly, Iowa, 1972-79, asso. prof., 1979—. Contbr. articles to profl. jours. Subspecialties: Ecology; Taxonomy. Current work: Taxonomy and ecology of benthic diatoms in rivers. Office: Wartburg College Waverly IA 50677

MAINIGI, DAIVENDER KUMAR, biochemist, research co. exec.; b. Narowal, Punjab, India; came to U.S., 1974, naturalized, 1980; m. Kusum Kumar, Jan. 7, 1969. B.S., A.M. U., Aligarh, India, 1960, M.S., 1962; Ph.D. in Biochemistry, M.S. U., Baroda, India, 1967. Asst. prof. biochemistry M.S. U., Baroda, 1964-74; postdoctoral assoc. Inst. for Cancer Research, Phila., 1974-77; instr. toxicology Albany (N.Y.) Med. Coll., 1980-81; sr. research assoc., div. nutrition scis. Cornell U., Ithaca, N.Y., 1978-80, vis. fellow, 1981-82; dir. life scis. Am. Standards Testing Bur., Inc., West Conshohocken, Pa., 1982—. Contbr. articles to profl. jours. Recipient various fellowships. Mem. Am. Assn. Cancer Research, European Assn. Cancer Research, Soc. Toxicology, Am. Chem. Soc. Subspecialties: Cancer research (medicine); Toxicology (medicine). Current work: Biochemical toxicology, nutritional, chemical carcinogenesis: promotional and inhibitory effects of dietary components, drugs, and sex hormones on different phases of carcinogenesis and toxicity. Metabolism and detoxification of xenobiotics, mechanisms of carcinogen absorption/transportation. Office: 1075 DeHaven St West Conshohocken PA 19428

MAJESTY, MELVIN SIDNEY, psychologist, former air force officer, consultant; b. New Orleans, June 6, 1928; s. Sidney Joseph and Marcella (Kieffer) M.; m. Bettye Newanda Gordon, Dec. 18, 1955; 1 dau., Diana Sue. B.A., La. State U., 1949; M.S., Western Res. U., 1951; Ph.D., Case-Western Res. U., 1967. Cert. psychologist. Commd. 2d lt. U.S. Air Force, 1951, advanced through grades to lt. col., 1976; research psychologist Human Resources Research Ctr., Denver, 1951-57, Los Angeles, 1958-64; asst. dir. tng. research Human Resources Lab., Dayton, Ohio, 1967-69; dir. faculty and profl. edn. research U.S. Air Force Acad., Colorado Springs, Colo., 1969-72; dir. pilot candidate selection program U.S. Air Force Officer Tng. Sch., San Antonio, 1972-76; ret., 1976; cons. Calif. State Personnel Bd., Sacramento, 1976—. Contbr. articles to profl. jours. Decorated Air Force Commendation medal, Meritorious Service medal. Mem. Am. Psychol. Assn., Internat. Personnel Mgmt. Assn., Western Psychol. Assn., Personnel Testing Council, Soc. Indsl. and Organizational Psychology. Current work: Personnel selection, personnel test validation, development of physical ability tests for law enforcement jobs. Patentee communication ctr., 1966. Office: Calif State Personnel Board 801 Capital Mall Sacramento CA 95814

MAJEWSKI, FRANK THOMAS, mechanical and industrial engineer, automation and robotics consultant; b. Newark, Dec. 12, 1920; s. Thomas Michael and Mary (Gleba) M.; m. Adele J. Poznanski, Nov. 26, 1950; 1 dau., Patricia A. A.A., Seton Hall U., 1956, B.S., 1958, M.B.A., 1965; D.B.A., Heed U., 1983. Lic. profl. engr., N.J., Calif. Vice pres. Pullman Kellog, Inc., Jersey City, 1955-58; exec. v.p. Astrotherm Corp., Indpls., 1958-60; pres. RMF, Inc., King of Prussia, Pa., 1960-66, dir., 1960-67; v.p. MAI Corp., N.Y.C., 1966-67; cons. in automation and robotics IBM, Boca Raton, Fla., 1967-83; mem. adv. com. W.Va. Inst. Tech., 1960-64, Seton Hall U., 1960-67. Recipient U.S. Naval Ordnance award, 1945; Outstanding Contbn. award IBM, 1968; Inventions award, 1981, 82, 83. Mem. ASME, Soc Mfg. Engrs. (cert. mfg. engr.), Am. Inst. Indsl. Engrs., Soc. Advancement Mgmt., Robot Inst. Am., Alpha Kappa Psi. Subspecialties: Mechanical engineering; Industrial engineering. Current work: Industrial robots and robotic applications, plant automation. Inventor high altitude pressure suits, 1956, wrist joint for robots, 1979. Home: 2551 NE 35th St Lighthouse Point FL 33064

MAJOR, EUGENE OLIVER, molecular virology researcher, educator; b. Chgo., Feb. 24, 1945; s. Wilfred Eugene and Irene Therese (Rongers) M.; m. Loretta T. Sieg, Oct. 10, 1965; children: Eugene E., Julie M.; m. Grace G., June 26, 1982; stepchildren: Blakely Burns, Bethany Burns. A.B., Holy Cross Coll., 1966; M.S., U. Ill. Med. Center-Chgo., 1971, Ph.D., 1973. Research asst. U. Ill. Med. Center-Chgo., 1967-73; postdoctoral fellow Tumor Virus Lab., 1973-75; asst. prof. Loyola U. Med. Sch., Maywood, Ill., 1975-80, assoc. prof., 1980-82, assoc. dean, Chgo., 1980-82; head unit on molecular virology and genetics Infectious Diseases br. Nat. Inst. Neurol., Communicative Disorders and Stroke, NIH, Bethesda, Md., 1982—; lectr. in med. virology, infectious disease curriculum; cons. genetic engring. Contbr. articles to profl. jours. Mem. Am. Soc. Microbiology, N.Y. Acad. Scis., AAAS, Tissue Culture Assn. Subspecialties: Molecular biology; Virology (biology). Current work: Molecular pathology of brain disorders caused by viruses, genetic research using recombinant DNA technology. Office: Bldg 36 Nat Inst Neurol Communicative Disorders and Stroke NIH Bethesda MD 20205

MAKAR, BOSHRA HALIM, mathematics educator; b. Sohag, Egypt, Sept. 23, 1928; came to U.S., 1966, naturalized, 1971; s. Halim and Hakima (Khair Mikhail) M.; m. Nadia E. Eissa, Jan. 1, 1960; children: Ralph, Roger. B.Sc., Cairo U., 1943-47, M.Sc., 1952, Ph.D., 1955. Mem. faculty Cairo U., 1948-65; vis. assoc. prof. Am. U., Beirut, Lebanon, 1966, Mich. Tech. U., Houghton, 1967; prof. math. St. Peter's Coll., Jersey City, 1967—. Mem. Am. Math Soc., Math. Assn. Am., AAUP. Republican. Roman Catholic. Clubs: Poetry Soc. London; United Poets Internat. (Philippines) (v.p. 1971-76). Subspecialties: Theory of functions and functional analysis; Cryptography and data security. Current work: Mathematical analysis, functions of a complex variable and functional analysis, and cryptology. Home: 410 Fairmount Ave Jersey City NJ 007306 Office: St Peter's Coll Math Dept Kennedy Blvd Jersey City NJ 07306

MAKI, TAKASHI, surgery educator, physician; b. Tokyo, Sept. 18, 1940; U.S., 1971; s. Toru and Umeko (Kato) M.; m. Teruko Okumura, Oct. 9, 1966; children: Hisako, Reiko, Keiko, Satoshi. M.D., U. Tokyo, 1966; Ph.D., SUNY-Downstate Med. Ctr., 1977. Intern U. Tokyo Hosp., 1966-67, resident, 1967-71; asst. prof. surgery SUNY-Downstate Med. Ctr., 1977-78; assoc. mem. Cancer Research Inst., New Eng. Deaconess Hosp., Boston, 1978—; asst. prof. surgery Harvard Med. Sch., 1979-83, assoc. prof. surgery, 1983—. Contbr. numerous articles to sci. jours. Whitaker Health Scis. Fund grantee, 1979-80; William F. Milton Fund grantee, 1982. Mem. Am. Assn. Immunologists, N.Y. Acad. Sci., Transplantation Soc., Am. Soc. Transplant Surgeons, Sigma Xi. Congregationalist. Clubs: Cambridge Boat, Harvard. Subspecialties: Transplantation; Cancer research (medicine). Current work: Immunological studies of specific unresponsiveness; suppressor cells in allograft system; immunological studies of spontaneous mammary tumors. Home: 17 Redwood Rd Westwood MA 02090 Office: Cancer Research Inst 185 Pligrim Rd Boston MA 02215

MAKINEN, MARVIN WILLIAM, educator; b. Chassell, Mich., Aug. 19, 1939; s. William John and Milga Katarina (Myllyla) M.; m. Michele de Groot, July 30, 1966; children: Eric William, Stephen Matthew. B.A., U. Pa., 1961; postgrad., Free U. Berlin, 1960-61; M.D., U. Pa., 1968; D.Phil., U. Oxford, Eng., 1976. Diplomate: Am. Bd. Med. Examiners. Intern dept. pathology Coll. Physicians and Surgeons, Columbia U., 1968-69; vis. fellow Lab. Molecular Biophysics, U. Oxford, 1971-74; asst. prof. dept. biophysics and theoretical biology U. Chgo., 1974-80, assoc. prof., 1980—. Contbr. articles to profl. jours. Served with USPHS, 1969-71. NIH spl. fellow, 1971-74; established investigator Am. Heart Assn., 1975-80. Mem. Am. Chem. Soc., Biophys. Soc., Am. Soc. Biol. Chemist. Subspecialties: Biophysics (biology); Photochemistry. Current work: Enzyme structure and function, physical chemistry of biological macromolecules, molecular spectroscopy. Office: Dept Biophysics and Theoretical Biology U Chicago 920 E 58th St Chicago IL 60637

MAKINODAN, TAKASHI, biologist; b. Hilo, Hawaii, Jan. 19. 1925; s. Shinsuke and Mitsuyo (Haitani) M.; m. Jane Oganeku, Dec. 19. 1954; 1 dau., Ann. B.S., U. Hawaii, 1949; M.S., U. Wis.-Maidson, 1950, Ph.D., 1953. Assoc. biologist biology div. Oak Ridge Nat. Lab., 1955-57, biologist, head immunology group, 1957-72; prof. Oak Ridge Grad. Sch. Biomed. Scis., U. Tenn., 1968-72, dir., 1968-72; chief Lab. Cellular and Comparative Physiology Gerontology Research Ctr., Nat. Inst. Aging, NIH, Balt. City Hosps., 1972-76; prof. medicine in residence UCLA, 1976—; dir. geriatric research Edn. and Clin. Ctr. VA Wadsworth Med. Ctr., Los Angeles, 1976—; adj. prof. biology U. So. Calif., 1982—; mem. microbioloby fellowship rev. com. NIH, 1967-70; mem. adv. panel regulatory biology program NSF, 1971-73; mem. study sect. immunolo. scis. NIH, 1978-82; mem. adv. com. Andrew Norman Inst. Advanced Study in Gerontology and Geriatrics, U. So. Calif., 1981-83; cons. Radiation Effects Research Found., Hiroshima, Japan., 1976—; mem. sci. council Intra-Sci. Research Found., 1980—. Contbr. articles to profl. jours. Recipient Biomed. Scis. award ethel Percy Andrus Gerontology Ctr., Los Angeles, 1976; NSF fellow, 1961-62. Mem. Am. Assn. Immunologists, AAAS, Am. Chem. Soc., Gerontological Soc., Soc. Exptl. Biology and Medicine. Subspecialties: Immunobiology and immunology; Gerontology. Current work: Aging and the immune system; late radiation effects onimmune function. Office: VA Wadsworth Med Ctr Wilshire and Sawtelle Blvds Los Angeles CA 90037

MAKMAN, MAYNARD HARLAN, biochemist, pharmacologist, educator; b. Cleve., Oct. 6, 1933; s. Saul Harold and Thelma (Singer) M.; m. Marianne Wertheim, Sept. 27, 1959; children: Judith, Lisa. B.A., Cornell U., 1955; M.D., Case-Western Res. U., 1962, Ph.D. in Pharmacology, 1962. USPHS surgeon Lab. Cellular Pharmacology, NIMH, 1962-64; asst. prof. biochemistry and molecular pharmacology Albert Einstein Coll. Medicine, 1964-70, assoc. prof., 1970-79, prof., 1979—. Contbr. numerous articles to profl. publs. Served with USPHS, 1962-64. Recipient Career Devel. award USPHS, 1966-71; USPHS spl. fellow, 1965-66. Mem. Am. Soc. Biol. Chemists, Am. Soc. Pharmacology and Exptl. Therapeutics. Subspecialties: Biochemistry (medicine); Pharmacology. Current work: Mechanisms of action of neurotransmitters, hormones and growth factors; neurobiology, receptors, catecholamines, peptide hormones, adenylate cyclase, brain, retina, growth and differentiation, aging research. Office: 1300 Morris Park Ave Bronx NY 10461

MAKOFSKE, WILLIAM JOSEPH, environ. physicist, educator; b. Bklyn., July 27, 1943; s. Harold A. and Amelia (Heinz) M.; m. Mary Frances Morris, June 10, 1967; children: David, Adam. B.S. in Physics, Pratt Inst., Bklyn., 1964; M.S. (teaching fellow, research fellow), Rutgers U., 1966; Ph.D. in Physics, Rutgers U., 1968. Postdoctoral fellow Rutgers U., 1968; research assoc., lectr. U. Minn., Mpls., 1968-71, Columbia U., 1971-74; prof. physics Ramapo Coll. N.J., Mahwah, 1974—, dir., 1975—. Contbr. articles profl. jours. Co-chmn. Town of Warwick Conservation Bd., 1978-82; mem. Westchester Scientists for Environ. Info., 1972-74. Recipient award Pratt Inst. Engring. Alumni Assn., 1964, Faculty Research grant Ramapo Coll., 1975, NSF grant, 1978. Mem. Am. Phys. Soc., Am. Assn. Physics Tchrs., Met. Solar Energy Soc., Tau Beta Pi. Subspecialties: Nuclear physics; Solar energy. Current work: Nuclear structure physics, solar greenhouse and passive solar research, design and devel. Office: 505 Ramapo Valley Rd Mahwah NJ 07430

MAKOUS, WALTER, visual scientist, educator; b. Milw., Nov. 22, 1934; s. Lawrence and Ruth Lorraine (Luehring) M.; m. Marilyn Ann Carlson, Feb. 2, 1958; children: Ann Louise, James Carl, Matthew Lloyd; m. Barbara Anne Duggins, Apr. 29, 1982. B.S., U. Wis., 1958; M.Sc., Brown U., 1961, Ph.D., 1964. Staff mem. IBM Research Center, Yorktown Heights, N.Y., 1963-66; asst. prof. psychology, lectr. physiology and biophysics U. Wash., Seattle, 1966-69; vis. scientist IBM Research Center, 1970-71; assoc. prof. to prof. psychology U. Wash., 1969-79; prof. psychology and ophthalmology, dir Center Visual Sci., U. Rochester, N.Y., 1979—; cons. Served in USNR, 1953-55. NIH and NSF grantee and fellow. Fellow Optical Soc. Am.; mem. Assn. Research in Vision and Ophthalmology, Neurosci. Soc., Psychonomic Soc. Subspecialties: Psychophysics; Sensory processes. Current work: Research in visual psychophysics. Office: Center for Visual Science U Rochester Rochester NY 14627

MALAMUD, HERBERT, physicist, med., health physicist, cons.; b. N.Y.C., June 28, 1925; s. Max and Anna (Mintzer) M.; m. m Sylvia, Oct. 27, 1951; children: Ronni Sue, Marc David, Kathi Jan. B.S. in Physics, CCNY, 1949, M.S., U. Md., 1952, Ph.D., N.Y. U., 1957; M.S. in Mgmt. Sci. and Engring, L.I. U., 1976. Diplomate: Am. Bd. Sci. in Nuclear Medicine. Various indsl. research positions, 1957-70; sr. physicist dept. nuclear medicine Queens Hosp., N.Y.C., 1970-79; tech. dir. Nuclear Assos./Victreen, Carle Place, N.Y., 1979—; cons. med. physics, 1976—; adj. prof. various conns. and univs., 1955—. Contbr. articles to profl. publs. Served with Signal Corps U.S. Army, 1943-46. Mem. Am. Phys. Soc., AAAS, Soc. Nuclear Medicine, Am. Assn. Physicists in Medicine, Health Physics Soc., Am. Inst. Ultrasound in

Medicine. Subspecialties: Imaging technology; Nuclear medicine. Current work: Physics of radiology, nuclear medicine, radiotherapy, ultrasound, med. nuclear magnetic resonance, health (radiation) physics; nuclear and atomic physics. Home: 30 Wedgewood Dr Westbury NY 11590 Office: 100 Voice Rd Carle Place NY 11514

MALAMY, MICHAEL HOWARD, molecular biology educator; b. N.Y.C., Apr. 20, 1938; s. Henry Robert and Rhoda (Resnick) M.; m. Frances E. Siegel, June 15, 1958; children: Adam Craig, Jocelyn Enid. B.A., NYU, 1958, Ph.D., 1963. Postdoctoral fellow Pasteur Inst., Paris, 1963-65, Princeton (N.J.) U., 1965-66; asst. prof. Tufts U., Boston, 1966-70, assoc. prof., 1970-76, prof., 1976—. Subspecialties: Molecular biology; Genetics and genetic engineering (biology). Current work: Mechanisms of transportation and function of transposons and insertion sequences; control of plasmid and chromosomal gene expression. Office: Tufts U 136 Harrison Ave Boston MA 02111

MALANGA, CARL JOSEPH, pharmacist, educator, researcher; b. N.Y.C., Aug. 26, 1939; s. Joseph John and Carolina J. (Graziano) M.; m. Mary Louise Villano, July 30, 1966; 1 son, Carl Joseph III. B.S. in Pharmacy, Fordham U., 1961, M.S., 1967, Ph.D., 1970. Registered pharmacist, N.Y. State. Lab. instr. Coll. Pharmacy, Fordham U., 1964-67, instr., 1967-70; asst. prof. Sch. Pharmacy, W.Va. U., Morgantown, 1970-73, assoc. prof., 1973-78, prof., 1978—, chmn. basic pharm. scis., 1978—. Contbr. articles to profl. jours. Served to 1st lt., inf. U.S. Army, 1962-64. Decorated U.S. Army Commendation medal; recipient Borden prize, 1961, Merck award, 1961, Bristol award, 1961, Outstanding Tchr. award W.Va. U., 1975-82; N.Y. State scholar, 1957-61. Mem. Am. Pharm. Assn., Am. Soc. for Pharmacology and Exptl. Therapeutics, Am. Assn. Colls. Pharmacy, Rho Chi. Democrat. Roman Catholic. Lodge: Kiwanis. Subspecialties: Pharmacology; Comparative physiology. Current work: Pharmacological control of ciliary activity and mucociliary transport. Home: 2202 Surrey Dr Morgantown WV 26505 Office: Sch Pharmacy WVa U Med Ctr Morgantown WV 26506

MALATESTA, VICTOR JULIO, clinical psychologist; b. Wilmington, Del., Apr. 3, 1951; s. Victor Julio and Gloria Marie (Faenza) M. B.A., U. Del., 1972; M.S., U. Ga., 1975, Ph.D., 1978. Alcoholism counselor State Alcoholism Services, Wilmington, Del., 1972-73; psychiat. technician Athens (Ga.) Gen. Hosp., 1974-78; psychology instr. U. Ga., Athens, 1977-80; clin./research fellow, 1978-80; psychology intern Med. U. S.C., Charleston, 1980-81, asst. prof., 1981—; cons. Fenwick Hall Hosp., John's Island, S.C., 1981—; acting dir. Psych. Assessment Ctr., Med. U. S.C., 1981-82; program rep. Nat. Council Aging, 1978-79. Contbr. articles to profl. jours. Nat. inst. Alcoholism & Alcohol Abuse fellow, 1981; U. Ga. fellow, 1978; Zimmer Research awardee, 1976. Mem. Am. Psychol. Assn., Southeastern Psychol. Assn., So. Soc. Philosophy and Psychology, Sigma Xi. Subspecialties: Behavioral psychology; Neuropsychology. Current work: Psychol. assessment, alcoholism, human sexuality, behavior therapy. Home: PO Box 740 Folly Beach SC 29439 Office: Dept Psychiatry Med U SC 171 Ashley Ave Charleston SC 29425

MALBICA, JOSEPH ORAZIO, research biochemist; b. N.Y.C., Apr. 6, 1925; s. Orazio and Venera (Strano) M.; m. Joanne Ruth Craft, July 30, 1947; children: Colleen Malbica Hewes, Kathryn Malbica Umina, Joseph O., Suzanne Malbica Miller. B.S., Bklyn. Coll., 1949; M.S., Fordham U., 1954; Ph.D., Rutgers U., 1967. Med. service rep. E.L. patch, 1949-52; research biochemist Hoffman-LaRoche, Inc., 1954-65; Hess & Clark Div. Richardson-Merrill, Inc., 1967-69; mgr. drug metabolism Stuart Pharms. div. ICI Americas Inc., Wilmington, Del., 1969—. Contbr. articles to sci. publs. Served with M.C. U.S. Army, 1943-46. Mem. Am. Chem. Soc., N.Y. Acad. Scis., Am. Soc. Pharmacology and Exptl. Therapeutics, Am. Soc. Mass Spectrometry, AAAS, Sigma Xi. Mormon. Subspecialties: Biochemistry (medicine); Analytical chemistry. Current work: Drug metabolism, pharmacokinetics, biotransformation, drug protein binding kinetics. Home: 1039 Carolyn Dr West Chester PA 19380 Office: Concord Pike and Murphy Rd Wilmington DE 19897

MALGADY, ROBERT GEORGE, educational statistics educator, consultant; b. Newark, Nov. 14, 1949; s. George J. and Marie A. (Calabrese) M. B.A., Rutgers U., New Brunswick, N.J., 1971; Ph.D., U. Tenn.-Knoxville, 1975. Asst. prof. SUNY-Brockport, 1975-78; asst. prof. ednl. stats. NYU, 1978-82, assoc. prof., 1982—; research assoc. Fordham U., 1981—; vis. investigator Meml. Sloan-Kettering Cancer Center, N.Y.C., 1981—; evaluator N.Y. State Dept. Edn., Albany, 1975-81, Div. Youth and Family Services, Trenton, 1980-81; cons. in field. Contbr. articles, chpts. to profl. publs.; author: test Vocational Adaptation Rating, 1980; adv. editor: Jour. Ednl. Psychology, 1977—; book editor, Harper & Row Pubs., 1978—, Holt, Rinehart & Winston, 1979—. SUNY-Albany fellow, 1976. Mem. Am. Psychol. Assn., Am. Ednl. Research Assn., Am. Stats. Assn., Psychonomic Soc., Phi Kappa Phi. Subspecialties: Psychometrics/quantitative psychology; Statistics. Current work: Development of psychodiagnostic tests and therapeutic modalities for Hispanics; measurement of psychological adaptation of cancer patients. Office: NY U 933 Shimkin Hall New York NY 10003

MALHOTRA, MANOHAR LAL, metallurgist, dental gold co. exec.; b. Multan, West Pakistan, Sept. 12, 1939; came to U.S., 1968, naturalized, 1977; s. Chetan Dass and Prakash Wati (Khanna) M.; m. Usha Kapoor, June 6, 1966; children: Dinesh K., Arun K. B.Sc. with honors in Physics, Delhi (India) U., 1961, M.Sc. in Physics, 1963; M.S. magna cum laude in Physics, Fairleigh Dickinson U., 1970; Ph.D. in Physics, Banaras Hindu U., 1972, U. Va., 1974. Scientist Nat. Phys. Lab., New Delhi, 1965-68; teaching cum research fellow dept. physics Fairleigh Dickinson U., Teaneck, NJ., 1968-70; NIH predoctoral research fellow in materials sci. dept. U.Va., 1970-74; NSF postdoctoral research fellow in chem. engring. SUNY, Buffalo, 1974-75; NSF postdoctoral research fellow dept. dental materials U. Mich., Ann Arbor, 1975-77; metallurgist Degussa Dental Inc., Long Island City, N.Y., 1977, dir. research, 1977—. Contbr. articles on dental research to profl. jours. Mem. Internat. Assn. for Dental Research, Am. Assn. for Dental Research, Am. Soc. Metals, Am. Def. Preparedness Assn., Internat. Precious Metals Inst., Sigma Xi. Democrat. Subspecialties: Dental materials; Alloys. Current work: Reseach and development of low-gold dental alloys with higher tarnish and corrosion resistance in the oral environment. Home: 104-07 Weside Ave Corona NY 11368 Office: Degussa Dental Inc 21-25 44th Ave Long Island City NY 11101

MALHOTRA, SHYAM KUMAR, orthodontist; b. Lahore, Pakistan, Apr. 1, 1942; came to U.S., 1970; s. Som Nath and Tara M.; m. Veena Malhotra, Nov. 20, 1969; children: Rachna, Rishi. B.Sc., U. Lucknow, India, 1960, B.D.S., 1964, M.D.S., 1968; D.D.S., Meharry Med. Coll., 1982. Lectr. K. G. Med. Coll., Lucknow, 1968-70; research assoc. Meharry Med. Coll., Nashville, 1970-73, instr., 1973-76, asst. prof., 1976—. Contbr. sci. articles to profl. jours. Mem. Am. Dental Assn., Internat. Assn. Dental Research, Am. Assn. Dental Research, Tenn. Dental Assn., Nashville Dental Soc. Subspecialties: Orthodontics; Dental growth and development. Current work: Human growth and development. Office: Meharry Med Coll Box 33-A 1005 18th Ave N Nashville TN 37208

MALICK, JEFFREY B., pharmacology researcher; b. Bklyn., Nov. 14, 1942; s. Joseph Bevan and Beatrice Lillian (Breen) M.; m. Jean Carol Rabush, Sept. 8, 1962; children: Jeffrey B., Jonathan, Jay, James, Jason. B.A., Rutgers U., 1965, M.S., 1968; Ph.D. in Psychopharmacology, NYU, 1973. Research asst. in neuropharmacology Lederle Labs., Pearl River, N.Y., 1962-67; supr. neurophysiology Union Carbide Corp., Tuxedo, N.Y., 1967-69; group leader CNS pharmacology Schering Corp., Bloomfield, N.J., 1969-74; mgr. CNS pharmacology ICI Americas, Inc., Wilmington, Del., 1974—; prof. psychopharmacology Bloomfield Coll., 1974; prof. neurophysiology Fairleigh Dickinson U., Teaneck, N.J., 1974-75; prof. neurochemistry U. Del., Newark, 1980-82. Editor: Antidepressants: Neurochemical, Behavioral and Clinical Perspectives, 1981, Endomorphins: Chemistry, Physiology, Pharmacology and Clinical Relevance, 1982, Anxiolytics: Neurochemical Behavioral and Clinical Perspectives, 1983. Fellow Internat. Soc. for Research on Aggression; mem. Am. Soc. for Pharmacology and Exptl. Therapeutics, Collegium Internationale Neuro-psychopharmacologicum, Soc. for Exptl. Biology and Medicine, Soc. for Neurosci. Republican. Methodist. Subspecialties: Neuropharmacology; Pharmacology. Current work: Neuropharmacology; psychopharmacology; nerochemical, neuroanatomical and physiological substrates of behavior and drug action. Discovery of novel therapeutic agents. Home: 2415 Heather Rd E Wilmington DE 19803 Office: Stuart Pharms Div ICI Americas Inc Concord Pike and New Murphy Rd Wilmington DE 19897

MALIK, DAVID JOSEPH, chemistry educator; b. Pittsburg, Calif., July 24, 1945; s. Joseph E. and Marguerite (Jacopetti) M. B.S., Calif. State U.-Hayward, 1968, M.S., 1969; Ph.D., U.-Calif.-San Diego, 1976. Lectr. U. Calif.-San Diego, 1976-77; research assoc. U. Ill., Urbana, 1977-80; asst. prof. chemistry Ind. U.-Purdue U., Indpls., 1980—. Contbr. articles to profl. jours. NSF grantee, 1980; Cottrell Research Corp. grantee, 1981. Mem. Am. Chem. Soc. (com. chmn. edn. sect. 1981-83), Am. Phys. Soc., Sigma Xi. Subspecialties: Theoretical chemistry; Atomic and molecular physics. Current work: Quantum mechanics, intermolecular energy transfer, potential energy surfaces. Office: Dept Chemistry Ind U-Purdue U Indianapolis IN 46224

MALIK, FAZLEY BARY, physics educator, researcher, cons.; b. Bankura, India, Aug. 16, 1934; came to U.S., 1960, naturalized, 1976; s. Malik Abdul and Feroza Bary. B.Sci. with honors, Calcutta (India) U., 1953; M.Sci., Dacca (Bangladesh) U., 1955; Dr. Rer. Nat., Gottingen (Germany) U., 1958. Research assoc. Max Planck Inst. for Physics and Astrophysics, Munich, W.Ger., 1958-60; research assoc. Princeton U., 1960-63; asst. prof. physics Yale U., 1964-68; assoc. prof. Ind. U., 1968-76, prof., 1976-80; prof., chmn. dept. physics So. Ill. U., 1980—; vis. prof. numerous univs. in, Germany, Switzerland, Turkey, and, Pakistan; cons. U.S. AEC, Swedish AEC, French AEC, Ford Found.; speaker numerous Am., Brit., German, French, Swedish, Italian, Asian phys. soc. meetings; cabinet-level adv. Bangladesh Planning Commn. Contbr. numerous articles to profl. jours. Mem. ACLU. Mem. Am. Phys. Soc., Ind. Fedn. Tchrs. (state truestee), Sigma Xi. Subspecialties: Nuclear physics; Atomic and molecular physics. Current work: Theoretical atomic and nuclear physics. Home: 200 Emerald Ln Carbondale IL 62901 Office: Dept Physics So Ill U Carbondale IL 62901

MALIK, MUJEEB REHMAN, mechanical engineer; b. Rabwah, Punjab, Pakistan, Nov. 14, 1951; came to U.S., 1975; s. Saif-Ur-Rehman and Amatul-Rashid Shaukat; m. Naila Shah, May 12, 1976; 1 child, Ahsen Mujeeb. B.S., Pakistan Engring. U., Pakistan, 1973; M.Eng., U. Toronto, 1975; Ph.D., Iowa State U., 1978. Research asst. Engring. Research Inst., Ames, Iowa, 1975-78; prin. scientist Systems and Applied Scis. Corp., Hampton, Va., 1978-81; pres. High Tech. Corp., Hampton, 1981—. Contbr. articles to profl. jours. Mem. ASME, AIAA. Subspecialties: Fluid mechanics; Mechanical engineering. Current work: Computational fluid mechanics and heat transfer; viscous drag reduction; laminar flow control; stability and transition; turbulence; energy technology. Office: High Tech Corp PO Box 7262 Hampton VA 23666

MALIN, DOUGLAS HARWELL, nuclear engineer; b. N.Y.C., Sept. 2, 1951; s. Bernard Malin and Naomi (Nocks) Lopez; m. Suzanne Cannon, Aug. 31, 1975; children: Ashley, Lauren. B.S., Ga. Inst. Tech., 1973, M.S., 1975; M.B.A., Ga. State U., 1981. Research asst. Ga. Inst. Tech., Atlanta, 1974-75; engr. Idaho Nat. Engring. Lab., Idaho Falls, Idaho, 1975-77; project engr. Nuclear Assurance Corp., Atlanta, 1978-82; fuel engr., analyst Fla. Power & Light Co., Miami, Fla., 1982—. Contbr. articles to profl. jours. Mem. Am. Nuclear Soc. Subspecialties: Nuclear engineering; Nuclear fission. Current work: Nuclear fuel cycle and management. Office: Fla Power & Light Co PO Box 529100 Miami FL 33152

MALINA, JOSEPH FRANCIS, JR., educator, civil engineering educator; b. Bklyn., Aug. 24, 1935; s. Joseph Francis and Mary (Wesolowski) M.; m. Ida Marie Klein, Jan. 9, 1965; children: Kristyn Rosemarie, Joseph Alexander, Mary Amelia, Alexandra Frances. B.C.E., Manhattan Coll., Riverdale, N.Y., 1957; M.S., U. Wis., 1959, Ph.D., 1961. Diplomate: Am. Acad. Environ. Engrs. Instr. Manhattan Coll., 1957; instr. U. Wis., 1957-58; asst. prof. U. Tex., Austin, 1961-64, assoc. prof., 1964-70, prof. civil engring., 1970—, C.W. Cook prof. environ. engring., 1980—, chmn. dept., 1976—; dir. Environmental Health Engring. Labs., 1970-76. Editor: (with R.E. Speece) Applications of Commercial Oxygen to Water and Wastewater Systems, 1973, (with B.P. Sagik) Virus Survival in Water and Wastewater Systems, 1974, (with E.F. Gloyna and E.M. Davis) Stabilization Ponds as a Wastewater Treatment Alternative, 1976. Mem. ASCE, Assn. Environ. Engring. Profs., Internat. Assn. Water Pollution Research, Nat., Tex. socs. profl. engrs., Water Pollution Control Fedn., Tex. Water Pollution Control Assn., Sigma Xi, Epsilon Sigma Pi, Chi Epsilon, Tau Beta Pi, Omicron Delta Kappa, Phi Kappa Phi. Subspecialty: Water supply and wastewater treatment. Home: 4508 Crestway Dr Austin TX 78731

MALINA, ROGER FRANK, astronomer, editor; b. Paris, July 6, 1950; U.S., 1968; s. Frank Joseph and Marjorie (Duckworth) M. B.Sc. in Physics, M.I.T., 1972; Ph.D. in Astronomy, U. Calif.-Berkeley, 1979. Research asst. Univ. Coll., London, 1979-81; asst. research astronomer U. Calif.-Berkeley, 1980—, project scientist, 1981—. Contbr. articles to profl. jours. Mem. Brit. Interplanetary Soc., Royal Astron. Soc., Am. Astron. Soc., Internat. Soc. Arts Sci. and Tech. (dir. 1982—), Sigma Xi. Subspecialty: 1-ray high energy astrophysics. Current work: Project scientist, extreme ultraviolet explorer, exec. editor Leonard Jour. Internat. Soc. Arts, Sci. and Tech. Home: 508 Connecticut St San Francisco CA 94107 Office: Space Science Lab U Calif Berkeley CA 94720

MALINDZAK, GEORGE STEVE, JR., physiology educator, cons., researcher; b. Cleve., Jan. 3, 1933; s. George Steve and Mary (Zemancik) M.; m. Marianne Beamer, June 27, 1959; children: Katherine, R. Scott, Edward G., Eric S. A.B. cum laude in Chemistry and Biology, Western Res. U., 1956; M.S. in Physiology and Biophysics, Ohio State U., 1958, Ph.D., 1961. Phys. metallurgist Thompson Co., Cleve., 1956-57; research asst. dept. physiology Ohio State U., Columbus, 1957-60, NIH fellow, 1960-61, research assoc., 1961-62; cons. adv. com. computers in research NIH, Bethesda, Md., 1961-62; instr. dept. physiology Bowman Gray Sch. Medicine, Winston-Salem, N.C., 1962-63, asst. prof., 1963-68, dir., 1964-65, assoc. prof., 1968-73; research physiologist clin. studies div. EPA, U. N.C., Chapel Hill, 1973-76; adj. assoc. prof. depts. physiology and surgery U. N.C. Sch. Medicine, 1974-76; prof., chmn. dept. physiology Northeastern Ohio U. Coll. Medicine, Rootstown, 1976—; prof. div. biol. scis. Kent (Ohio) State U., 1977—; adj. prof. elec. engring. U. Akron, Ohio, 1977—; cons. Simulators, Inc., Northbrook, Ill., 1968-69, NASA, 1967-74, Peer Rev. Corp., Health Service Adminstrn., Akron, 1976—80. Contbr. over 100 articles on physiology, pharmacology, engineering mathematics to profl. jours. Mem. research com. N.C. Heart Assn., 1968-69; research rev. com. Ohio chpt. Am. Heart Assn., 1977—; mem. research com. Akron chpt. Am. Heart Assn., 1977—; Bd. dirs. N.C. chpt. Nat. Cystic Fibrosis Found., 1969-71, High Rock Lake Assn., 1972-77; mem. spl. com. on engring. edn. N.C. State U., 1973-75; trustee, mem. exec. com. Akron chpt. Am. Heart Assn., 1979—, mem. nat. profl. edn. com., 1977-81. Served with U.S. Army, 1950-53. Decorated Purple Heart, Bronze Star; sr. research investigator award N.C. Heart Assn., 1965-70, 1970-73; NIH grantee, 1961-62, 1962-73; EPA grantee, 1973-76; Am. Heart Assn. grantee, 1977—. Mem. Am. Physiol. Soc., Am. Soc. Pharmacology and Exptl. Therapeutics, Assn. Chairmen Depts. of Physiology, IEEE (chmn. engring. in medicine and biology group Winston-Salem chpt. 1968-73, editorial rev. bd. 1979—, exec. com. N.C. sect. 1970-74), Biomed. Engring. Soc., Assn. Computing Machinery (treas. spl. interest group in biomed. computing 1973-75, 1975-77, sec.-treas. Symposium on Health Sci. Computing Dallas 1977), AAUP, Biophys. Soc., Acoustical Soc. Am., Sigma Xi, Phi Soc., Beta Beta Beta. Subspecialties: Physiology (medicine); Cardiology. Current work: Cardiovascular disease-coronary heart disease, coronary artery spasm. Office: NE Ohio Univ Coll Medicine Rootstown OH 44272

MALINOW, MANUEL RENE, cardiovascular educator, physician; b. Buenos Aires, Argentina, Feb. 27, 1920; m. Marta Arias, May 20, 1953; children: Ana, Roberto, Juan Sebastian. B.D., Buenos Aires Med. Sch., 1944. Research fellow Michael Reese Hosp., Chgo, 1945-46; intern Hosp. Salaberry, Buenos Aires, 1943-44, resident, 1944-45; chief lab. cardiovascular diseases Oreg. Regional Primate Research Ctr., Beaverton, Oreg., 1964—; prof. medicine Oreg. Health Scis. U., Portland, 1963—; med. dir. YMCA, Portland, 1974—; practice medicine, specializing in cardiology, Portland, 1982—; mem. staff St. Vincent Hosp., Portland. Mem. AAAS, Royal Soc. Medicine, Interna. Primatol. Soc., AAUP, N.Y. Acad. Scis., Am./Oreg. Heart Assn., Western Assn. Physicians, Am. Fedn. Clin. Research, Fedn. Am. Socs. Exptl. Biology (Clin. Nutrition). Subspecialties: Cardiology; Physical medicine and rehabilitation. Current work: Experimental atherosclerosis, atherosclerosis regression. Office: Oreg Regional Primate Research Center 505 NW 185th Ave Beaverton OR 97006

MALINOWSKI, EDMUND ROBERT, chemist, educator, consultant; b. Mahanoy City, Pa., Oct. 16, 1932; s. Francis and Stella (Pieczul) M.; m. Helen Devcich, Sept. 6, 1958; children: Paul, Robert. B.S., Pa. State U., 1954; M.S., Stevens Inst. Tech., 1956, Ph.D. (Robert-Crooks-Stanley fellow), 1961. Research assoc. Stevens Inst. Tech., 1960-63, asst. prof. chemistry, 1963-66, assoc. prof., 1966-70, prof., 1970—; cons. in field. Author: (with D. Howery) Factor Analysis in Chemistry, 1980; research numerous publs. in chemistry. Recipient Jess H. Davis Research award Stevens Inst. Tech., 1977. Mem. Am. Chem. Soc. Roman Catholic. Subspecialties: Physical chemistry; Analytical chemistry. Current work: Computer analysis of chem. data. Home: 49 New England Dr Lake Hiawatha NJ 07034 Office: Dept Chemistry and Chem Engring Stevens Inst Tech Hoboken NJ 07030

MALINS, DONALD CLIVE, fisheries center director; b. Lima, Peru, May 19, 1931; came to U.S., 1947, naturalized, 1953; s. Richard Henry and Mabel Madeline (Warner) M.; m. Mary Louise Leiren, Jan. 27, 1962; children: Christopher, Gregory, Timothy. B.A., U. Wash., 1953; B.S., Seattle U., 1956; Ph.D., U. Aberdeen, Scotland, 1967, D.Sc., 1976. Research chemist N.W. and Alaska Fisheries Center, Seattle, 1956-74; research prof. Seattle U., 1972—; affiliate prof. U. Wash., Seattle, 1974—; dir. environ. conservation div. N.W. and Alaska Fisheries Ctr., Seattle, 1974—; cons. UN Environ. Program, Nairobi, Kenya, 1980—. Editor: (with J.R. Sargent) Effects of Petroleum on Arctic and Subarctic Marine Environments and Organisms, Vols. I and II, 1977; editor-in-chief: Aquatic Toxicology, Seattle, 1979—. Recipient gold medal Dept Commerce, 1982, spl. achievement award, 1975, 80; superior performance award Dept. Interior, 1960, 61; Bond award cert. Am. Oil Chemists Soc., 1961. Mem. Am. Soc. Biol. Chemists, Am. Chem. Soc. Clubs: Wash. Athletic, Sand Point Golf and Country (Seattle). Subspecialties: Environmental toxicology; Biochemistry (medicine). Current work: Marine biochemistry and toxicology, studies on marine pollutants and pollutant related diseases in marine life, biochemistry of porpoise sonar. Office: Environ Conservation Div NW and Alaska Fisheries Center 2725 Montlake Blvd E Seattle WA 98112

MALITZ, SIDNEY, physician; b. N.Y.C., Apr. 20, 1923; s. Benjamin and Etta (Cohen) M. Student, N.Y. U., 1940-42, Tulane U., 1942-43; B.M., Chgo. Med. Ch., 1946, M.D., 1947. Diplomate: Am. Bd. Psychiatry and Neurology. Intern St. Mary's Hosp., Huntington, W.Va., 1946-47; sr. intern Bethesda, Hosp., Cin., 1947-48; resident N.Y. State Psychiat. Inst., N.Y.C., 1948-51, sr. research psychiatrist, 1954-56, acting prin. research psychiatrist, 1956-58, acting chief psychiat. research, chief dept. exptl. psychiatry, 1958-64, chief psychiat. research dept. exptl. psychiatry, 1964-72, dep. dir., 1972-75, acting dir., 1975-76, 81—, dep. dir., 1976-78; in charge psychiat. drug clinic Vanderbilt Clinic, Presbyn. Hosp., N.Y.C., 1956-75, asst. attending psychiatrist, 1960-66, asso. attending psychiatrist, 1966-71, attending psychiatrist, 1971—, acting dir. psychiatry service, 1975-76, 81—; asst. dept. psychiatry Coll. Physicians and Surgeons, Columbia U., N.Y.C., 1955-57, asso., 1957-59, asst. clin. prof., 1959-65, asso. prof., 1965-69, prof., 1969—, vice chmn. dept. psychiatry, 1972-75, acting chmn., 1975-76, 81—, vice chmn., 1976-78; mem. panel impartial psychiat. experts N.Y. State Supreme Ct., 1960—; mem. adv. com. subcom.; cons. U.S. Pharmacopeia; mem. adv. com. subcom. health N.Y. State Constl. Conv.; cons. div. med. scis. NRC, Washington, 1967-70; cons. Rush Found., Los Angeles, 1968—; mem. ad hoc rev. com. to select Nat. Drug Abuse Research Centers, Center Studies Narcotic and Drug Abuse, NIMH, 1972—. Contbr. numerous articles to profl. jours. Fellow Am. Psychiat. Assn. (chmn. com. biol. psychiatry 1961-62, program com. 1961-62, sec-treas. chpt. 1962-63, mem. com. research 1966-68, pres. chpt. 1969-70, chmn. Council Research and Devel. 1971-73), N.Y. Acad. Medicine, AAAS (council 1969—), Am. Coll. Neuropsychopharmacology, Collegium Internationale Neuropsychopharmacologicum, Am. Coll. Psychiatrists (archivist-historian 1978—), Royal Coll. Psychiatrists, Am. Coll. Psychoanalysts, N.Y. Soc. Clin. Psychiatry, Assn. Research Newvous and Mental Disease, N.Y. State, N.Y. County med. socs., AMA (cons. council drugs 1960—), N.Y. Acad. Scis, N.Y. Psychiat. Soc., Am. Psychopath. Assn., Soc. Biol. Psychiatry, Schilder Soc. Subspecialties: Psychopharmacology; Neuropsychology. Current work: Lateralization of brain function; regional cerebral blood flow as it relates to mental illness; brain metabolism; cognitive and affective changes with electroconvulsive therapy. Address: Coll Physicians and Surgeons Columbia U Dept Psychiatry 161 Fort Washington Ave New York NY 10032

MALKIN, RICHARD, biologist, educator; b. Chgo., Mar. 25, 1940; s. Maurice Benjamin and Phyllis (Bilkis) M.; m. Carole Jacobs, Dec. 25, 1960; children: Daniel, Karin, Jesse. B.S., Antioch Coll., 1962; Ph.D.,

U. Calif.-Berkeley, 1967. Postdoctoral fellow U Goteborg, Sweden, 1969; asst. prof. U. Calif., Berkeley, 1970-75, assoc. prof., 1975-79, prof., 1980—. Mem. Am. Soc. Plant Physiologists, Am. Soc. Biol. Chemists, Biophys. Soc., Am. Soc. Photobiology. Subspecialty: Photosynthesis. Current work: Energy-transduction in photosynthetic membranes; electron transport and primary photochemical reactions. Office: U Calif 313 Hilgard Hall Berkeley CA 94720

MALKINSON, ALVIN MAYNARD, biochemist, educator; b. Buffalo, Jan. 5, 1941; s. Irving nd and Ida (Gitin) M.; m. Lynn Ellen Reynolds, Dec. 26, 1967; children: Sabra E., Zachary D. B.A., U. Buffalo, 1963; Ph.D., Johns Hopkins U., 1968. Served with US. Peace Corps, 1968-72; postdoctoral fellow U. Leicester, Eng., 1971-72; lectr. U. Nairobi, Kenya, 1969-71; postdoctoral Yale U. Med. Sch., New Haven, 1972-74; asst. prof. U. Minn. Med. Sch., Mpls., 1975-78, U. Colo., Boulder, 1978—. Author: Hormone Action, 1975; also articles. White House summer fellow, 1964; Minn. Med. Found. grantee, 1975; NIH fellow, 1975-78; grantee, 1980—; Nat. Found.-March of Dimes Basil O'Connor fellow, 1976-78; Colo. Heart Assn. grantee, 1979. Mem. Assn. Advancement of Cancer Research, Am. Soc. Pharmacology and Exptl. Therapeutics, Rho Chi. Subspecialties: Cancer research (medicine); Biochemistry (medicine). Current work: Role of protein phosphorylation in regulation normal and neoplastic lung development in mice. Home: 3855 Broadway Boulder CO 80302 Office: U Colo Boulder CO 80309

MALKUS, WILLEM VANRENSSELER, mathematics educator; b. Bklyn., Nov. 19, 1923; s. Hubert Paul and Alida Fitzhugh (Wright) M.; m. Joanne Starr Simpson, 1948; m. Ulla Charlotte, Dec. 28, 1964; children: David S., Steven W., Karen E., Perm. Ph.D. in Physics, U. Chgo., 1950. Asst. prof. natural sci. U. Chgo., 1950-51; phys. oceanographer Woods Hole (Mass.) Oceanographic Instn., 1951-60; prof. geophysics UCLA, 1960-67, prof. geophysics and math., 1967-69; prof. applied math. MIT, Cambridge, 1969—. Contbr. articles to sci. jours. Served to lt. (j.g.) USNR, 1942-46. Guggenheim fellow U. Cambridge, (Eng.) and Stockholm, 1971-72, U. Cambridge, 1979-80; NSF fellow, 1964-65. Fellow Am. Phys. Soc., Am. Acad. Arts and Scis., Am. Geophys. Union; mem. Nat. Acad. Sci. Subspecialties: Applied mathematics; Statistical physics. Current work: Turbulence; geophysical fluid dynamics; hydromagnetics. Home: 72 Shade St Lexington MA 02173 Office: MIT Room 2-369 Cambridge MA 02139

MALLAMA, ANTHONY, astronomer, analyst; b. Cleve., Jan. 19, 1949; s. Domenic and Marie (Ippolito) M. B.A., Vanderbilt U., 1971; M.S., U. Toledo, 1975. Staff scientist Computer Scis. Corp., Silver Spring, Md., 1975—; observer Cerro Toledo Obs., Chile, 1980, 81, 82. Contbr. articles to profl. jours. Mem. Am. Astron. Soc., Phi Beta Kappa. Subspecialties: Optical astronomy; Binary stars. Current work: Design and implementation of a system to accurately measure 10,0000 stars to serve as a reference frame for NASA's space telescope. Home: 5-C Eastway Greenbelt MD 20770 Office: Computer Sci Corp 8728 Colesville Rd Silver Spring MD 20910

MALLATT, MARK EDWARD, dental researcher and educator, consulting clinical examiner; b. Gary, Ind., July 6, 1950; s. Russell Clayton and Marjorie May (Hoagl) M.; m. Kathleen Ann Quill, Aug. 21, 1976. B.S., Ind. U.-Bloomington, 1972, D.D.S., 1975. Cert. dentist, Ind. Clin. research assoc. Oral Health Research Inst., Indpls., 1975-77, assoc. dir. clin. research, 1978—; instr. Ind. U. Sch. Dentistry, 1977-78, asst. prof. preventive dentistry, 1978—; cons. clin. examiner for pvt. industry, Ind., Ohio, Tex., 1977—. Pvt. industry grantee, 1977-82. Mem. Am. Assn. Dental Research (v.p. Ind. sect. 1981-82, pres. sect. 1982-83), Am. Assn. Dental Schs., ADA, Ind. Dental Assn., Psi Omega (Achievement award 1974). Republican. Subspecialties: Preventive dentistry; Cariology. Current work: Clinical investigations in assessing the efficacy and safety of oral health products and procedures relative to: dental caries, gingivitis, plaque, calculus, pellicle, and oral soft tissue pathology. Home: 1753 Esther Ct Plainfield IN 46168 Office: Oral Health Research Inst 415 Lansing St Indianapolis IN 46202

MALLAY, JAMES FRANCIS, engineering company executive, consultant; b. Morristown, N.J., Dec. 8, 1936; s. Paul C. and Rachel R. (Jones) M.; m. Mary Lou Egert, June 12, 1960; children: Cindy, Russell. B.S.M.E., Lafayette Coll., Pa., 1959; M.S. in Nuclear Engring, MIT, 1961. Registered profl. engr., Ohio, Va. Mgr. licensing Babcock & Wilcox, Lynchburg, Va., 1971-75, mgr. components, Akron, Ohio, 1975-77, mgr. European ops., Lynchburg, 1977-79; mgr. safety assessment Nuclear Safety Analysis Ctr., Palo Alto, Calif., 1980-81; mgr. performance analysis Babcock & Wilcox, Lynchburg, 1982; gen. mgr. NUTECH Engrs., San Jose, Calif., 1982—; lectr. Lynchburg Coll., 1967-71; vis. prof. Internat. Ctr. Theoretical Physics, Trieste, Italy, 1982. Project dir.: Procedures Guide for Prohabilisitc Risk Assessment, 1980-82. Served to capt. U.S. Army, 1962-65. Mem. ASME, Am. Nuclear Soc. (chmn. standards steeering com. 1980—). Methodist. Subspecialties: Nuclear engineering; Nuclear physics. Current work: Nuclear safety; probabilistic risk assessment. Home: 16615 Oak View Circle Morgan Hill CA 95037 Office: 6835 Via del Oro San Jose CA 95119

MALLEY, ARTHUR, immunochemistry educator; b. Chgo., Jan. 7, 1931; s. Isadore and Shirley (Zoloth) M.; m. Sharon Roadman, June 10, 1961; children: Brian A., Derek S. B.A., San Francisco State U., 1953, B.S., 1957; Ph.D., Oreg. State U., 1961. Postdoctoral fellow Calif. Inst. Tech., Pasadena, 1961-63; chmn. immune disease dept. Oreg. Regional Primate Research Ctr., Beaverton, 1963—; prof. microbiology and immunology dept. Oreg. Health Scis. U., 1973—. Contbr. articles to profl. jours. Served with U.S. Army, 1953-55. Welcome Trust fellow, 1981; NIH grantee. Mem. Am. Chem. Soc., Am. Assn. Immunologists, Am. Acad. Allergy. Democrat. Club: Irvington Tennis (Portland). Subspecialties: Immunology (medicine); Cellular engineering. Current work: Regulation of immune responses using modified antigens, soluble proteins secreted from antigen-specific cells and anti-idiotypic antibodies. Patentee in field. Home: 345 NW 95th Ave Portland OR 97229 Office: 505 NW 185th Ave Beaverton OR 97006

MALLIAKOS, ASIMIOS, nuclear engineer; b. Athens, Greece, May 21, 1952; came to U.S., 1975; s. Constantine and Elisavet (Horozoglou) M.; m. Eilwen Amdall, Apr. 4, 1981. B.S., U. Thessaloniki, Greece, 1975; M.S., Poly Inst. of N.Y., 1977; Ph.D., U. Mo.-Columbia, 1980. Prin. engr. Combustion Engring., Inc., Windsor, Conn., 1980—. Mem. Am. Nuclear Soc., Conn. chpt. Am. Nuclear Soc. Club: Toast-masters Internat. (Windsor). Subspecialty: Nuclear engineering. Current work: Thermohydraulic safety analyses, startup test analyses, computer code simulation for nuclear power plants; research in improved methods in reliability analysis. Home: 19 Carriage Dr Endfield CT 06082 Office: Conbustion Engring Inc Windosr CT 06095

MALLICK, PANKAJ KUMAR, research engineer, educator; b. Hooghly, India, Feb. 14, 1946; came to U.S., 1967; s. Phani Bhusan and Anandamoyee (Nandy) M.; m. Sunanda Kundu, Aug. 7, 1970; 1 son, Samip Kumar. M.S., Ill. Inst. Tech., 1970, Ph.D. in Mech. Engring, 1973. Research assoc. Ill. Inst. Tech., Chgo., 1973-74; research engr. Am. Can Co., Batavia, Ill., 1974-75, Eagle Internat. Corp., Addison, Ill., 1975-76; sr. research scientist Ford Motor Co., Detroit, 1976-79; asst. prof. mech. engring. U. Mich.-Dearborn, 1979—. Contbr. articles to profl. jours. Nat. scholar Ministry of Edn., Govt. of India, 1960-66. Mem. ASME, Soc. Plastics Engrs., Am. Soc. Body Engrs., Sigma Xi. Subspecialties: Composite materials; Polymers. Current work: Mechanics and mechanical behavior of polymeric and composite materials; manufacturing techniques for mass production of composite materials. Patentee in field. Office: 4901 Evergreen Rd Dearborn MI 48128

MALLINGER, ALAN GARY, psychiatrist, psychopharmacologist; b. Pitts., July 29, 1947; m. Joan Ellen Mallinger, Aug. 20, 1972; children: Julie Beth, Daniel Todd. B.S., U. Pitts., 1969, M.D., 1973. Diplomate: Am. Bd. Psychiatry and Neurology. Resident in gen. psychiatry Univ. Health Center, Pitts., 1973-75, 79-81; instr. psychiatry and pharmacology U. Pitts., 1975-76, asst. prof., 1976—; attending psychiatrist Western Psychiat. Inst. and Clinic, Pitts., 1981—. Contbr. chpt. to books, articles to profl. jours. Laughlin fellow, 1981; Mead Johnson travel fellow, 1982. Mem. Am. Soc. for Clin. Pharmacology and Therapeutics, Am. Coll. Clin. Pharmacology, Soc. for Neurosci., Am. Psychosomatic Soc., AAAS, Phi Beta Kappa, Alpha Omega Alpha. Subspecialties: Psychopharmacology; Psychiatry. Current work: Research on cell membrane transport processes in relation to mood disorders and the mechanisms of action of psychotherapeutic drugs. Office: 3811 O'Hara St Pittsburgh PA 15213

MALLMANN, ALEXANDER JAMES, physicist, educator; b. Sheboygan, Wis., Dec. 8, 1937; s. Alexander Bernard and Anne Frances (Govek) M.; m. Jean Louise, Aug 10, 1968 (div.); children: James, David. B.S. in Physics, U. Wis.-Milw., 1965, M.S., 1968; Ph.D. in Materials Sci, Marquette U., 1977. Mem. faculty Milw. Sch. Engring., 1968—, prof. physics, 1979—; cons. in optics and acoustics. Contbr. articles to profl. jours. Served with USAF, 1955-59. Mem. Optical Soc. Am., Am. Assn. Physics Tchrs. (pres. Wis. sect. 1981-82), Internat. Solar Energy Soc., Physics Club Milw. (pres. 1982-83). Current work: Computer simulation of optical phenomena of the atmosphere. Home: 20250 Jeffers Dr New Berlin WI 53151 Office: Dept Physics PO Box 644 Milwaukee WI 53201

MALLORY, CLELIA WOOD, chemistry educator, researcher; b. Bklyn., Feb. 9, 1938; d. Charles Otis and Clelia (Adams) Wood; m. Frank B. Mallory, Nov. 26, 1965. A.B., Bryn Mawr Coll., 1959, M.A., 1960, Ph.D., 1963. Research assoc. Bryn Mawr Coll., 1963-77; lectr. chemistry Yale U., 1977-80, U. Pa., Phila., 1980—. Contbr. articles to profl. jours. Mem. Am. Chem. Soc., AAAS, Phila. Organic Chemists Club (chmn. 1976-77). Subspecialties: Organic chemistry; Photochemistry. Current work: Organic chemistry, photochemistry, nuclear magnetic resonance spectroscopy. Home: 321 Caversham Rd Bryn Mawr PA 19010 Office: Dept Chemistry U Pa Philadelphia PA 19104

MALLORY-BARKLEY, BARBARA ZOMMER, psychologist; b. New Haven., Conn., May 25, 1936; d. Peter and Estelle (Serba) Zommer; m. George Boudreau, Apr. 11, 1955 (div. 1969); children: Deborah Boudreau, George Boudreau, Scott Boudreau.; m. Hunter Mallory, May 25, 1972 (div. 1976). B.A., So. Conn. State Coll., New Haven, 1968; M.A., U. Conn.-Storrs, 1972; postgrad., Harvard U., 1974-76. Supr. research program Mass. Gen. Hosp., Boston, 1972-74; edn. dir. Sch. for Learning Disabled, Beacon Sch., Brookline, Mass., 1973-74; assoc. dir. clinic Eagle Hill Sch., Greenwich, Conn., 1974-76; assoc. to pres., dir. New Eng. div. Ednl. Records Bur., Wellesley, Mass., 1978—; lectr. Harvard U., 1972-73. Author: Developmental Implications of Iconio Memory, 1972; test battery Objectives per Item of Comprehensive Testing Program, 1980. State of Conn. grantee, 1972. Mem. Internat. Neuropsychology Assn., Am. Psychol. Assn., Power Squadron (asst. edn. dir. 1983-84), World Affairs Council (mem. council Forum II 1983—), Phi Delta Kappa. Republican. Episcopalian. Clubs: Harvard Faculty, Newport (R.I.) Yacht. Subspecialties: Cognition; Learning. Current work: Director diagnostics and evaluation. Home: 411 Marrett Rd Lexington MA 02173 Office: Educational Records Bureau 37 Cameron St Wellesley MA 02181

MALLOV, SAMUEL, pharmacology educator; b. N.Y.C., Apr. 19, 1919; s. Harry and Fannie (Rubin) M.; m. Charlotte Wolkov, Mar. 14, 1943; children: Joseph, Jonathan. B.S., CCNY, 1939; M.S., N.Y. U., 1941; Ph.D., Syracuse U., 1952. From instr. to prof. SUNY Upstate Med. Ctr., Syracuse, 1953—; phys. chemist Carnegie Inst. Tech., Pitts., 194548. Author research papers. Mem. Soc. Pharmacology and Exptl. Therapeutics, Internat. Soc. Heart Research, Am. Heart Assn. Subspecialties: Pharmacology; Cellular pharmacology. Current work: Biochemistry and pharmacology of heart and blood vessels. Home: 223 Dawley Rd Fayetteville NY 13066 Office: 766 Irving Ave Syracuse NY 13210

MALMGREN, LESLIE THEODORE, JR., biomed. researcher, educator; b. Brockton, Mass., Mar. 28, 1946; s. Leslie Theodore and Abbie H. (Denson) M.; m. Kathleen A. A.B., Bridgewater (Mass.) State Coll., 1968, M.A., 1972; Ph.D., Clark U., 1975. Asso. prof. otolaryngology Upstate Med Center, Syracuse, N.Y. Grantee NIH, 1971-73, 76-78, 82—, Swedish Med. Research Council, 1977-78, Hendircks Fund, 1978-80, Voice Found., 1978-79, 80-81, Deafness Research Found., 1980—. Mem. Am. Assn. Anatomists, Assn Research in Otolaryngology, Soc. Neurosci. Subspecialties: Otorhinolaryngology; Neurobiology. Current work: Histochemistry and electron microscopy of head and neck muscles and their innervation. Experimental neuropathology of head and neck muscles. Office: Otolaryngology Research Labs SUNY Upstate Med Center Weiskotten Hall Rm 89 7600 Irving Ave Syracuse NY 13210

MALONE, DANIEL RICHARD, neuropsychologist; b. Utica, N.Y., Sept. 27, 1949; s. Leo Daniel and Gloria Rita (Adorino) M.; m. Eileen Mary Sheridan, Mar. 20, 1982. B.S., Rensselaer Poly. Inst., 1972; M.S., Auburn U., 1975, Ph.D., 1978. Lic. psychologist, Tex., Md. Research worker Inst. of Psychiatry, London, 1978-79; postdoctoral fellow U. Tex. Med. Br., Galveston, 1980-81; asst. prof. U. Bielefeld, W.Ger., 1981-82; staff neuropsychologist Springfield Hosp. Ctr., Sykesville, Md., 1982—. Recipient Nat. Research Service award NIMH, 1980. Fellow Md. Psychol. Assn.; mem. Am. Psychol. Assn., Internat. Neurophychol. Soc. Subspecialty: Neuropsychology. Current work: Identification and description of the effects of brain injury and disease on higher cognitive functions, with emphasis on cognitive rehabilitation. Sensory integration in human and non-human primates. Home: 1400 Eutaw Pl Baltimore MD 21217 Office: Springfield Hosp Ctr Sykesville MD 21784

MALONE, MARVIN HERBERT, pharmacologist, educator, researcher; b. Fairbury, Nebr., Apr. 2, 1930; s. Herbert August Frederick and Elizabeth Florinda (Torrey) M.; m. Shirley Ruth Cane, Dec. 21, 1952; children: Carla Margaret, Gayla Christa. B.S. in Pharmacy, U. Nebr., 1951, M.S. in Physiology and Pharmacology, 1953; Ph.D. in Pharmacology and Pharm. Scis, 1958; postgrad., Rutgers U., 1954-55. Asst. U. Nebr., Lincoln, 1951-53, 1956-58; research ast. Squibb Inst. Med. Research, New Brunswick, N.J., 1953-56; asst. prof. U. N. Mex., Albuquerque, 1958-60; assoc. prof. U. Conn., Storrs, 1960-69; prof. physiology and pharmacology U. of Pacific, Stockton, Calif., 1969—; cons. Drug Plant Labs., U. Wash., 1960-64, Research Pathology Assoc., 1967-70, Amazon Natural Drug Co., 1967-70, ICI USA Inc., 1968-78, SISA Inst. Research, 1977—; mem. task force on plants for fertility regulation WHO Spl. Program for Research, Devel. and Research Tng. in Human Reproduction, 1982—. Author: Experiments in the Pharmaceutical Biological Sciences, 1973; editorial bd.: Jour. Natural Products: Lloydia, 1971—, Jour. Ethnopharmacology, 1978—; editor: Am. Jour. Pharm. Edn, 1974-79; contbr. articles on pharmacology to profl. jours. Recipient U. Pacific Distinction of Merit, 1980, Mead Johnson Labs. award, 1964; USPHS grantee, 1960-63, 1968-73; U.S. Army grantee, 1962-63; U. Conn. research Found. grantee, 1964-68. Fellow Am. Found. Pharm. Edn., Am. Inst. Chemists, AAAS; mem. Acad. Pharm. Scis. (sr.), Am. Soc. Pharmacology and Exptl. Therapeutics, Western Pharmacology Soc., Am. Soc. Pharmacognosy, Am. Pharm. Assn., Am. Assn. Colls. Pharmacy, Acad. Pharm. Scis., Soc. Econ. Botany, Ethnopharmacology Soc., Sigma Xi, Rho Chi, Phi Lambda Upsilon, Phi Kappa Phi. Subspecialties: Pharmacology; Toxicology (medicine). Current work: Screening and assay of natural products, pharmacology of inflammation and anti-inflammation, pharmacodynamics of psychotropic and autonomic agents, biometrics, fertility regulation. Home: 722 Bedford Rd Stockton CA 95204 Office: Physiology and Pharmacology Unit Univ Pacific Sch Pharmacy Stockton CA 95207

MALONE, MICHAEL JOSEPH, neurologist, research center administrator, educator; b. Portland, Maine, Apr. 28, 1930; s. Patrick Joseph and Margaret Marie (Ridge) M.; m. Dorothy Helen Corcoran, July 4, 1957; 1 son, Michael Patrick. B.A. cum laude, Boston Coll., 1951; M.D., Georgetown U., 1956. Diplomate: Am. Bd. Psychiatry and Neurology. Intern Receiving Hosp., Detroit, 1956-57; resident in neurology Boston U.-Boston VA Hosp., 1960-63; research assoc., assoc. dept. neurochemistry Harvard U., Cambridge, Mass., 1963-65, assoc. dept. neurology, 1965-67; prof. neurology and pediatrics George Washington U. Med. Sch., Washington, 1970-75; assoc. dir. Boston City Hosp., 1965-70; asst. prof. neurology Boston U. Med. Sch., 1965-70, prof. neurology and psychiatry, 1975—; dir. Geriatric Research Edn. Clin. Ctr., VA, Bedford, Mass., 1976—. Contbr. articles to profl. publs. Served to capt. U.S. Army, 1957-60. NIH grantee, 1963—. Mem. Am. Acad. Neuroloy, Am. Soc. Neurochemistry, Internat. Soc., Pediatric Neurology, AAAS, Neurosci. Soc. Democrat. Roman Catholic. Club: Beverly (Mass.) Yacht. Subspecialties: Neurology; Neurochemistry. Current work: Neurochemistry; membranes, lipids lipoproteins; glycosphingolipids; aging. Office: VA Hosp 200 Springs Rd Bedord MA 01730

MALONE, ROBERT EDWARD, microbiology educator; b. Los Angeles; s. Robert E. and Shirley M. (Russell) M.; m. Cheryl L. Blascow, Dec. 3, 1976. B.S., UCLA, 1970; Ph.D., U. Oreg., 1976. Postdoctoral fellow in biology U. Chgo., 1977-80; asst. prof. microbiology Loyola U. of Chgo. Stritch Sch. Medicine, Maywood, Ill., 1980—. NIH research grantee, 1980-83; recipient Career Devel. award Schweppe Found. of Chgo., 1981-84. Mem. Am. Soc. Microbiology, Genetics Soc. Am., AAAS. Club: Trout Unltd. Subspecialties: Genetics and genetic engineering (biology); Molecular biology. Current work: Molecular description of genetic recombination in eucaryotic systems, its relationship to chromosome structure and segregation. Office: Dept of Microbiology Loyola University School of Medicine Maywood IL 60153

MALONE, THOMAS BECKER, research psychologist; b. Wilmington, Del., Oct. 30, 1936; s. John L. and Wanda C. M.; m. Mary Tobin, Sept. 6, 1958; 1 son, John T. B.S., St. Joseph U., 1958; M.A., Fordham U., 1962, Ph.D., 1964. Research scientist Grumman Aerospace, Bethpage, N.Y., 1963-65; v.p., dir. human factors Matrix Corp., Alexandria, Va., 1965-71; sr. v.p. Essex Corp., Alexandria, 1971-82; exec. v.p. Carlow Assoc., Fairfax, Va., 1982—. Editor: Proceedings HFS Conv, 1973, 76. Served with U.S. Army, 1959-60. Visual research assistantship Naval Tng. Device Com., Fordham U., 1962-63. Mem. Human Factors Sco., IEEE, AIAA, Am. Nuclear Soc., Soc. Mfg. Engrs. Subspecialties: Human factors engineering; Experimental psychology. Current work: Human factors engineering, man-computer interface, cognitive process training research and development robotics, artificial intelligence. Home: 6633 Kennedy Ln Falls Church VA 22042 Office: Carlow Assos Inc 8315 Lee Hwy Fairfax VA 22031

MALONEY, DENNIS MICHAEL, psychologist, communicator; b. Portland, Oreg., Dec. 19, 1947; s. Bernard A. and Margaret (Stickel) M.; m. Karen Blase, Nov. 25, 1972 (div. Sept. 1980); 1 son, Kevin. A.B. with honors, Gonzaga U., 1969; M.A., U. Kans.-Lawrence, 1972, Ph.D., 1973. Lic. psychologist, Nebr. Dir. research and evaluation Western Carolina Center, Morganton, N.C., 1973-75; co-dir. teaching-family project Boys Town Center, Boys Town, Nebr., 1975-78, dir. tech. communications dept. youth care, 1978-82; asst. dir. Boys Town U.S.A., 1982—; cons. to various publs., 1973—. Contbr. numerous articles to profl. publs.; editor: various publs., including Teaching-Family Newsletter, 1979—. U. Kans. grantee, 1972; Law Enforcement Assistance Adminstrn. grantee, 1973-75; NIMH grantee, 1974-76; Appalachian Regional Commn. grantee, 1974-76. Mem. Am. Psychol. Assn., Nat. Teaching-Family Assn. (founding), Assn. Advancement Behavior Therapy, Internat. Assn. Bus. Communicators. Subspecialty: Behavioral psychology. Current work: Communications research; translation of technical information for lay audiences. Office: Boys Town USA Dept Boys Town NE 68010

MALONEY, JOHN PATRICK, mathematics and computer science educator, solar consultant; b. Omaha, Dec. 9, 1929; s. John Daniel and Helen Frances (Brader) M. B.S.E.E., Iowa State Coll., 1958; Ph.D. in Math, Georgetown U., 1965. Instr. Georgetown U., 1963-65; asst. prof. U. Nebr.-Lincoln, 1965-67; prof. math. and computer sci. U. Nebr.-Omaha, 1967—. Served with USAF, 1950-54; Ger. Mem. Soc. Indsl. and Applied Math., Math. Assn. Am. Subspecialties: Solar energy; Applied mathematics. Current work: Work with passive solar energy research team at U. Nebr.-Omaha. Home: 4839 Pine St Omaha NE 68016 Office: U Nebr 60th Dodge St Omaha NE 68182

MALOSH, JAMES BOYD, mechanical engineer, educator; b. Licking, Ill., July 30, 1943; s. John Andrew and Nadean (Crandall) M.; m. Sandra Sue Browne, June 22, 1963 (div. 1971); 1 son, Ronald; m. Helen Faye Kangas, Oct. 22, 1973; children: Jeffrey, Brian, Ronald, Melanie. B.S., Wayne State U., Detroit, 1966; M.S., Mich. Tech. U., Houghton, 1969; Ph.D., 1980. Registered profl. engr., Mich., Ohio, Pa. Acoustical engr. Walker Research Labs., Grass Lake, Mich., 1969-70; research engr. Lawrence Livermore Labs., Livermore, Calif., summers 1971, 72; research assoc. Mich. Tech. U., 1970-75; sr. research engr. U.S. Steel Research Labs., Monroeville, Pa., 1975-80; prin. research scientist Battelle Meml. Labs., Columbus, Ohio, 1980-81; assoc. prof. mech. engring. U. Alaska, Fairbanks, 1981—; cons. Arctic Designers, Fairbanks, 1982, Cowper and Madson Attys., 1982; prin. investigator Alaska Dept. Transp., Fairbanks, 1982. Mem. ASME, Nat. Soc. Profl. Engrs., Sigma Xi, Tau Beta Pi, Pi Tau Sigma. Subspecialties: Acoustical engineering; Theoretical and applied mechanics. Current work: Application of experimental and theoretical dynamic analysis with computers to the solution of problems in noise, vibration and pulsation control. Particular interests are in transient fluid flow, finite waves and energy system pulsation and control. Patentee automotive silencer, blast furnace stove burner. Home: 717 Chandalar Univ Alaska Fairbanks AK 99701 Office: U Alaska Fairbanks AK 99701

MALOZEMOFF, PLATO, mining executive; b. Russia, 1909. Student, U. Calif., 1931, Mont. Sch. Mines, 1932. Chmn., chief exec. officer Newmont Mining Corp., N.Y.C.; pres., dir. Resurrection Mining Co., Newmont Exploration Ltd., Carlin Gold Mining Co.; chmn. Newmont Mines Ltd., Newmont Proprietary Ltd.; chmn., dir. Idarado Mining Co.; v.p., dir. Newmont Exploration of Can., Ltd.; vice chmn., dir. Magma Copper Co.; dir. O'Okiep Copper Co., So. Peru Copper Corp., Sherritt Gordon Mines, Ltd., Atlantic Cement Co. Inc., Peabody Coal Co., Palabora Mining Co., Ltd., Tsumeb Corp., Ltd., Foote Mineral Co., Browning-Ferris Industries, Highveld Steel & Vanadium Corp., Newmont Mining Corp. Bd. dirs. Boys' Clubs Am.; v.p., trustee Am. Mus. Natural History.; trustee Carnegie Hall Soc. Subspecialty: Metallurgy. Current work: Mining engineering. Office: 200 Park Ave New Mont Mining Coap 36th Fl New York NY 10166

MALSKY, STANLEY JOSEPH, physicist; b. N.Y.C., July 15, 1925; s. Joseph and Nellie (Karpinski) M.; m. Gloria E. Gagliardi, Oct. 15, 1965; 1 son, Mark A. B.S., NYU, 1949, M.A., 1950, M.S., 1953, Ph.D., 1963. Nuclear physicist Dept. Def., 1959-64; chief physicist VA, 1954-63; adj. asso. prof., then prof. radiol. sci. Manhattan Coll., Bronx, N.Y., 1960-74; non-resident research collaborator med. div. Brookhaven Nat. Labs., Upton, N.Y., 1964-69; research prof. radiology NYU Sch. Medicine, N.Y.C., 1975-77; pres. Radiol. Physics Assn., White Plains, N.Y., 1965—. Contbr. chpts. to books. Served with U.S. Army, 1945-46. Recipient James Picker Found. award, 1963-67; Founder's Day award NYU, 1964; Leadership award Manhattan Coll., 1969. Fellow Am. Public Health Assn., AAAS, Royal Soc. Health; charter mem. Am. Assn. Physicists in Medicine, Health Physics Soc. Roman Catholic. Subspecialties: Cancer research (medicine); Biophysics (physics). Current work: Radiation doismetry; radiation exposures; med.physics applied to diagnostic radiology; nuclear medicine; radiation therapy and tng. of staffs;legal aspects. Address: PO Box 31 Elmsford NY 10523

MALVERN, DONALD, aircraft manufacturing company executive; b. Sterling, Okla., Apr. 22, 1921; s. George Michael and Anna Francesca (Elsass) M.; m. Ruth Marie Vogler, June 4, 1949; 1 son, Michael John. B.S., U. Okla., 1946. Engr. Victory Architects and Engrs., Clinton, Ohio, 1943, Douglas Aircraft Co., Santa Monica, Calif., 1943; with McDonnell Aircraft Co., St. Louis, 1946—, exec. v.p., 1973-82, pres., 1982—; v.p. McDonnell Douglas Corp., 1973—; pres. McDonnell Douglas Saudi Arabia, Inc., 1978—. Served to 1st lt. USAAF, 1943-46. Asso. fellow AIAA (Tech. Mgmt. award 1968, Reed Aeros. medal 1980); mem. Am. Def. Preparedness Assn. (pres. St. Louis chpt. 1979-80), Navy League U.S. (life), Nat. Aeros. Assn., Air Force Assn., Armed Forces Mgmt. Assn., Pi Tau Sigma, Tau Beta Pi, Tau Omega. Clubs: Bellerive Country., St. Louis. Subspecialty: Aeronautical engineering. Current work: General management. Home: 37 Baron Ct Rural Route 2 Florissant MO 63034 Office: PO Box 516 Saint Louis MO 63166

MALVERN, LAWRENCE EARL, engineering educator; b. Sterling, Okla., Sept. 14, 1916; s. George Michael and Anna Francesca (Elsass) M.; m. Marjorie Malene McCarther, Aug. 8, 1939; 1 dau., Maureen. Sc.B., Southwestern Okla. State Coll., 1937; M.A., U. Okla., Norman, 1939; Ph.D., Brown U., 1949. Asst. prof. math. and mechanics Carnegie-Mellon U., 1949-53; assoc. prof. mechanics Mich. State U., 1953-58, prof., 1958-69; prof. engring. scis. U. Fla., 1969—. Author: Introduction to the Mechanics of a Continuous Medium, 1969, Engineering Mechanics—Statics and Dynamics, 1976; assoc. editor: Jour. Applied Mechanics, 1978-85. Served to lt. (j.g.) USNR, 1944-46. Guggenheim fellow, 1959-60. Fellow Am. Acad. Mechanics; fellow ASME; mem. Soc. Engring. Sci. (dir. 1967-70), AIAA, Am. Math. Soc., Soc. Indsl. and Applied Math., Soc. Exptl. Stress Analysis, Am. Acad. Mechanics, Am. Soc. Engring. Edn., Sigma Xi, Tau Beta Pi. Subspecialties: Solid mechanics; Applied mathematics. Current work: Exptl. and analytical research in metal plasticity and composite materials mechanics; teaching grad. and undergrad. engring. mechanics. Home: 4400 NW 14th Pl Gainesville FL 32605 Office: U Fla 231 Aerospace Bldg Gainesville FL 32611

MAN, CHI-SING, mechanics educator, researcher; b. Hong Kong, Aug. 23, 1947; emigrated to Can., 1981; s. Yip and Sau-Ying (Leung) M.; m. May Lai-Ming Chan, July 5, 1973; children: Li-Xing, Yi-Heng. B.Sc. with honors, U. Hong Kong, 1968, M.Phil., 1976; Ph.D., Johns Hopkins U., 1980. Tutor in math. and physics Hong Kong Bapt. Coll., 1970-72, asst. lectr. in physics, 1972-76; postdoctoral fellow Johns Hopkins U., Balt., 1980-81; asst. prof. U. Man., Winnipeg, Can., 1981—. Editor: jour. Dousou, 1974-76. Natural Sci. and Engring. Research Council Can. grantee, 1982-84. Mem. Soc. for Natural Philosophy. Subspecialties: Theoretical and applied mechanics; Offshore technology. Current work: Subsea permafrost; in-situ pressuremeter tests; settlement of artificial islands; creep of frozen soils; foundations of continuum thermomechanics; continuum theories of phase transitions; percolation; frozen soils as multiphase mixtures. Home: 315 Dalhousie Dr Winnipeg MB Canada R3T 2Z4 Office: Dept Civil Engring U Manitoba Winnipeg MB Canada R3T 2N2

MANALIS, MELVYN SAMUEL, research physicist, educator; b. Los Angeles, Oct. 16, 1939; s. Barney M. and Kathryn (Swiler) M.; m. Marilyn Jean White, June 21, 1965; children: Andrew, Scott, Jeremy. B.A. in Math, Calif. State U.-Northridge, 1961; M.S. in Physics, U. N.H., 1964, Ph.D., U. Calif.-Santa Barbara, 1970. Instr. dept. physics Colby Coll., Waterville, Maine, 1963-64; scientists Jet Propulsion Lab., Calif. Inst. Tech., Pasadena, 1965; research scientist II U. Colo., Boulder, 1966; physicist Nat. Bur. Standards, Washington, 1967; scientist The Te Co., Santa Barbara, Calif., 1970-72; research physicist U. Calif.-Santa Barbara, 1972-79, lectr., 1975, adj. lectr. environ. studies, 1975—, research physicist, 1975—. Contbr. articles on physics to profl. jours. Cons. Santa Barbara County Bd. Suprs. Gen. Motors fellow, 1967-68; U. Calif. Sea grantee, 1977-78, 1978-79. Mem. Am. Inst. Physicists, Am. Assn. Physics Tchrs., Sigma Xi. Club: Friends of the Earth. Subspecialties: Wind power; Solar energy. Current work: Directing one of the largest wind energy assessment programs in the state of California, recent results of this work were presented at the first U.S.-China Conference on energy resources and the environment. Office: Dept Environ Studies U Calif Santa Barbara CA 93106

MANCINI, ERNEST ANTHONY, geologist, geology educator; b. Reading, Pa., Feb. 27, 1947; s. Ernest and Marian K. (Filbert) M.; m. Marilyn E. Lee, Dec. 27, 1969; children: Lisa L., Lauren N. B.S. in Biology, Albright Coll., 1969; M.S. in Geology, So. Ill. U., 1972, Ph.D., Tex. A&M U., 1974. Petroleum exploration geologist Cities Service Oil Co., Denver, 1974-76; assoc. prof. geology U. Ala.-University, 1976—; petroleum geologist Mineral Resources Inst., University, 1976-78; state geologist, oil and gas supr. Geol. Survey Ala., University, 1982—. Contbr. articles and abstracts on petroleum geology, paleoecology, biostratigraphy and paleontology to profl. jours. Mem. Soc. Econ. Paleontologists and Mineralogists (v.p. Gulf Coast sect. 1982-83), Am. Assn. Petroleum Geologists (A.I. Levorsen award in petroleum geology Gulf Coast sect. 1980, 1st best paper award 1980), N.Am. Micropaleontology Soc., Internat. Micropaleontology Soc., Ala. Geol. Soc. (pres. 1980-81), Sigma Xi, Phi Kappa Phi, Phi Sigma. Democrat. Presbyterian. Subspecialties: Paleoecology; Petroleum engineering. Current work: Major research interests are Cretaceous and Tertiary biostratigraphy and their application for petroleum with emphasis on the Gulf Coastal Plain of the United States. Home: 1503 Briarcliff Northport AL 35476 Office: Geol Survey Ala PO Drawer 420 Hackberry L University AL 35486

MANDAL, ANIL KUMAR, physician, researcher; b. West Bengal, India, Nov. 12, 1935; came to U.S., 1967, naturalized, 1978; s. Nirmal C. and Kamala B. M.; m. Pranati Ganguly, June 18, 1964; children: Aditi, Atashi. Intermediate sci., Suri Vidyasagar Coll., West Bengal, India, 1953; M.D., Calcutta Nat. Med. Coll., 1959. Diplomate: Am. Bd. Internal Medicine. Research fellow in nephrology West Suburban Hosp., Oak Park, Ill., 1962-63, U. Edinburgh, Scotland, 1968-69; med. officer med. research Inst. Postgrad. Medicine, Calcutta, India, 1963-66; registrar in medicine R.G. Med. Coll. Hosps., Calcutta, 1966-67; lectr. dept. pathology U. Edinburgh, 1968-69; research fellow in nephrology U. Ill. Hosps., Chgo., 1969-70, resident in internal medicine, 1970-71; chief resident, instr. medicine and phys. diagnosis U. Ill., Chgo., 1971-72; staff physician, asst. and assoc. prof. medicine VA Med. Ctr., U. Okla. Health Sci. Ctr., Oklahoma City, 1972-82; staff physician, prof. medicine VA Med. Center, Med. Coll. Ga., Augusta, 1982—. Author: (with James E. Wenzl) Electron Microscopy of the Kidney in Renal Disease and Hypertension - A Clinicopathological Approach, 1979; editor: (with Sven O. Bohman) The Renal Papilla and Hypertension, 1980. Recipient Outstanding Service award West Suburban Hosp., Oak Park, Ill., 1970, Physicians Recognition award AMA, 1972. Fellow ACP; mem. Internat. Soc. Nephrology, Am. Soc. Nephrology, So. Soc. Clin. Investigation, Central Soc. Clin. Research. Subspecialties: Nephrology; Pathology (medicine). Current work: The spleen and acute renal failure: an experimental model, protection is afforded against acute renal failure by splenectomy, mechanisms of protection are studied. Office: Nephrology Sect VA Med Center Augusta GA 30910

MANDEL, H(AROLD) GEORGE, pharmacologist; b. Berlin, June 6, 1924; U.S., 1937, naturalized, 1944; s. Ernest A. and Else (Crail) M.; m. Marianne Klein, July 25, 1953; children: Marcia Vivian, Audrey Lynn. B.S., Yale U., 1944, Ph.D., 1949. Lab. instr. in chemistry Yale U., 1942-44, 47-49; research asso. dept. pharmacology George Washington U., 1949-50, asst. research prof., 1950-52, asso. prof. pharmacology, 1952-58, prof., 1958—, chmn. dept. pharmacology, 1960—; Advanced Commonwealth Fund fellow Molteno Inst. Cambridge (Eng.) U., 1956; Commonwealth Fund fellow U. Auckland (N.Z.) and U. Med. Scis., Bangkok, Thailand, 1964; Am. Cancer Soc. Eleanor Roosevelt Internat. fellow Chester Beatty Research Inst. London, 1970-71; Am. Cancer Soc. scholar U. Calif., San Francisco, 1978-79; mem. cancer chemotherapy com. Internat. Union Against Cancer, 1966-73; mem. external rev. com. Howard U. Cancer Research Center, 1972—; cons. Bur. Drugs, 1975-79, EPA, 1978-82; mem. toxicology adv. com. FDA, 1975-78; mem. med. research service merit rev. bd. in alcoholism and drug dependence VA, 1975-78; mem. cancer spl. program adv. com. Nat. Cancer Inst., 1974-78, chmn., 1976-78; mem. Nat. Large Bowel Cancer Project Working Cadre, 1980—; mem. com. on toxicology NRC-Nat. Acad. Sci., 1978-82; mem. Kettering award selection com. Gen. Motors Cancer Research Found., 1979-81. Editorial bd.: Jour. Pharmacology and Exptl. Therapeutics, 1960-65; field editor, 1978—; editorial bd.: Molecular Pharmacology, 1965-69, Research Communications in Chem. Pathology, Pharmacology, 1972—, Cancer Research, 1974-76; asso. editor, 1977-81. Served with AUS, 1944-46. Recipient John J. Abel award in pharmacology Eli Lilly and Co., 1958, Distinguished Achievement award Washington Acad. Scis., 1958, Golden Apple Teaching award Student AMA, 1969. Mem. Am. Chem. Soc., Am. Soc. Biol. Chemists, Am. Soc. Pharmacology and Exptl. Therapeutics (pres. 1973-74), Am. Assn. Cancer Research, AAAS, Assn. Med. Sch. Pharmacology (pres. 1976-78), Internat. Soc. Biochem. Pharmacology, Alpha Omega Alpha, Sigma Xi. Democrat. Club: Cosmos (Washington). Subspecialty: Pharmacology. Research, numerous publs. on cancer chemotherapy, mechanism of growth inhibition, antimetabolites, drug disposition. Home: 5500 Christy Dr Bethesda MD 20816 Office: 2300 Eye St NW Washington DC 20037

MANDEL, HARVEY PHILLIP, psychologist; b. Montreal, Feb. 3, 1944; s. Bernard and Ida (forman) M.; m. Dorothy Moore, June 17, 1972. B.Sc., McGill U., 1965; M.Sc., Ill. Inst. Tech., 1968, Ph.D., 1969. Registered psychologist, Ont. Chmn. Counselling and Devel. Ctr., Downsview, Ont., 1978-82; assoc. prof. psychology York U., Downsview, 1970—; psychol. cons. Youthdale Treatment Ctrs., Toronto, 1970—, Youth Clin. Services, 1980—. Author/editor: Brief Psychotherapy, 1981. Can. Council leave fellow, 1977; Calif. State U. vis. scholar, 1977. Mem. Canadian Psychol. Assn., Am. Psychol. Assn., Ont. Psychol. Assn., Am. Orthopsychiat. Assn. Jewish. Subspecialties: Clinical psychology; Developmental psychology. Current work: Brief psychotherapy; adolescence and adulthood; underachievement. Office: York Univ 4700 Keele St Downsview ON Canada M3J 1P3 Home: 96 Wild Briarway Willowdale ON Canada M2J 2L4

MANDEL, LEONARD, physicist; m. Jeanne Elizabeth Kear, Aug. 20, 1953; children: Karen Rose, Barry Paul. B.Sc., U. London, Eng., 1947, 1948, Ph.D., 1951. Tech. officer Imperial Chem. Industries, 1951-54; lectr., sr. lectr. Imperial Coll., U. London, 1954-64; prof. physics U. Rochester, N.Y., 1964—, prof. optics, 1977-80; joint sec. Rochester Confs. on Coherence and Quantum Optics, 1966, 72, 77, 83. Editor books; asso. editor: Optic Letters, 1977-79; Contbr. numerous sci. articles to profl. jours. First recipient Max Born prize Optical Soc. Am., 1982. Fellow Am. Phys. Soc., Optical Soc. Am. (asso. editor jour. 1970-76, 82—, chmn. com. for soc. objectives and policy 1977). Subspecialties: Atomic and molecular physics; Laser optics. Current work: Research o lasers and quantum optics, both experimental and theoretical,particularly interactions between atoms and light photon statistics phase transitions in lasers. Office: Dept Physics and Astronomy U Rochester Rochester NY 14627

MANDLER, GEORGE, psychology educator; b. Vienna, Austria, June 11, 1924; came to U.S., 1940, naturalized, 1943; s. Richard and Hede (Goldschmied) M.; m. Jean Matter, Jan. 29, 1957; children: Peter C., Michael A. B.A., NYU, 1949; postgrad., U. Basel, Switzerland, 1947-48; M.S., Yale U., 1950, Ph.D., 1953. Asst. prof. and lectr. Harvard U., Cambridge, Mass., 1953-60; prof. U. Toronto, Ont., Can., 1960-65; prof. psychology, dir. Ctr. Human Info. Processing, U. Calif.-San Diego, La Jolla, 1965—; fellow Ctr. Advanced Study in Behavioral Sci., Stanford U., 1959-60; vis. prof. Oxford U., Eng., 1971-72; vis. research fellow Univ. Coll. London, 1978—. Author: Mind and Emotion, 1975, Mind, Consciousness and the Emotions, 1983, (with M. Mandler) Thinking, 1964, (with W. Kessen) The Language of Psychology, 1959. Served with U.S. Army, 1940-43. Guggenheim fellow, 1971-72; SSRC Research awardee, 1959. Fellow Am. Psychol. Assn. (div. pres. 1980); mem. Psychonomic Soc., Fedn. Behavioral, Psychol. and Cognitive Socs. (pres. 1981). Subspecialty: Cognition. Current work: Theoretical analyses of mental representation with special reference to consciousness, emotion and memory. Experimental work in the analysis of emotion, memory, including amnesia. Home: 1406 La Jolla Knoll La Jolla CA 92037 Office: Center for Human Information Processing Univ Calif San Diego La Jolla CA 92093

MANDRA, YORK T., geology educator; b. N.Y.C., Nov. 24, 1922; s. Raymond and Irene (Farruggio) M.; m. Highochi Simon, Jan. 26, 1946. B.A., U. Calif.-Berkeley, 1948, M.A., 1950; Ph.D., Stanford U., 1958. Mem. faculty dept. geology San Francisco State U., 1950—, prof., 1964—; vis. prof. U. Calif.-Santa Barbara, 1972—, Syracuse U., summer 1963, U. Maine, summer 1969; vis. scientist N.Z. Geol. Survey, Lower Hutt, 1970. Contbr. articles to profl. jours. Pres., bd. dirs. David S. Sohigan Found., 1975—. Served with USAAF, 1942-46. NSF fellow, 1959-60; grantee, 1967-77; Danforth Found. fellow, 1956-57. Fellow Geol. Soc. Am., AAAS, Calif. Acad. Scis.; mem. Nat. Assn. Geology Tchrs. (pres. Far Western sect. 1953-54, Robert Wallace Webb award 1977). Subspecialty: Micropaleontology. Current work: Fossil silico flagellates; societal problems of energy. Office: Geosciences Dept San Francisco State U San Francisco CA 94132

MANGANO, RICHARD MICHAEL, scientist; b. Mt. Vernon, N.Y., Mar. 2, 1950; s. Frank G. and Louise (Amoruso) M.; m. Patricia B. Siuta, Oct. 28, 1978. B.S. in Chemistry, Iona Coll., New Rochelle, N.Y., 1972; Ph.D. in Biochemistry, Fordham U., 1980. Instr. dept. chemistry Manhattan Coll., Riverdale, N.Y., 1977-79; faculty research assoc. Md. Psychiat. Research Center, 1979-81; dept. psychiatry U. Md., Balt., 1979-81; fellow Hoffmann-LaRoche, Inc., Nutley, N.J., 1981-82, sr. scientist dept. pharmacology, 1982—. Contbr. articles to profl. jours. Mem. Am. Chem. Soc., N.Y. Acad. Sci., Soc. Neurosci., Phi Lambda Upsilon, Sigma Xi. Subspecialties: Biochemistry (biology); Neurochemistry. Current work: Research interests in biochemistry and biochemical pharmacology include investigations of cellular metabolism, toxicity and cellular degeneration, regulation of ligand receptor interactions as pertains to drug devel.

MANGER, WILLIAM MUIR, physician, research scientist; b. Greenwich, Conn., Aug. 13, 1920; s. Julius and Lilian B. (Weissinger) M.; m. Lynn Seymour Sheppard, Jan. 14, 1942; children: William, Lilian, Shep, Charles. B.S., Yale U., 1944; M.D., Columbia U., 1946; Ph.D., Mayo Clinic, 1959. Diplomate: Am. Bd. Internal Medicine. Intern Presbyn. Hosp., N.Y.C., 1946-47, resident, 1949-50; fellow in internal medicine Mayo Clinic, 1950-57; successively instr., assoc., asst. attending physician, lectr. in medicine Columbia U. Med. Center, 1957—; dir. Manger Research Found., 1960-77; assoc. prof. clin. medicine N.Y.U. Med. Center, 1975-83; clin. prof. medicine N.Y.U. Med. Ctr., 1983—; cons. in internal medicine Southampton (L.I.) Hosp., 1972—; assoc. attending physician Bellevue Hosp., 1977—, Univ. Hosp., 1977—. Author: Chemical Quantitation of Epinephrine and Norepinephrine in Plasma, 1959, Pheochromocytoma, 1977, Catecholamines in Normal and Abnormal Cardiac Function, 1982; editor: Hormones and Hypertension, 1966; contbr. articles to profl. jours. Trustee Thyroid Found., 1979—; bd. dirs. Found. for Depression and Manic Depression, Inc., 1978—, pres., 1979—; trustee Found. for Advancement Internat. Research in Microbiology.; Trustee St. Albans Sch., Washington, 1958-64, 67-74, chmn. bd., 1967-69; trustee Buckley Sch., N.Y.C., 1975—; former deacon, elder, now trustee Fifth Ave. Presbyn. Ch., N.Y.C. Served to lt. (j.g.) USNR, 1947-49. Recipient Mayo Alumni award for meritorious research, 1955. Fellow Am. Heart Assn. (council high blood pressure research), ACP, Am. Physiol. Soc., Am. Coll. Cardiology, N.Y. Acad. Medicine (admission and edn. cons.), Acad. Psychosomatic Medicine, Am. Inst. Chemists, Royal Soc. Health, Am. Coll. Clin. Pharmacology; mem. AMA, Am. Thoracic Soc., N.Y. Acad. Scis., AAAS, Am. Chem. Soc., Am. Soc. Pharmacology and Exptl. Therapeutics, Endocrine Soc., Pan Am. Med. Assn., Soc. Exptl. Biology and Medicine, Physicians and Surgeons Alumni Soc., Soc. Alumni Presbyn. Hosp., Mayo Fellow Assn. (past pres.), Mayo Alumni Assn. (pres. elect.), Internat. Soc. Hypertension, Inter-Am. Soc. Hypertension, Am. Fedn. Clin. Research, Catecholamine Club (pres. 1981), Am. Soc. Nephrology, Med. Strollers (past exec. com.), Nat. Hypertension Assn. (founder, chmn. 1977—), Internat. Med. Council Drug Use, Sigma Xi, Nu Sigma Nu. Clubs: Yale, University, N.Y. Athletic, Meadow, Southampton Bathing Corp. Subspecialties: Cardiology; Neuroendocrinology. Current work: Hypertension; catecholamines; pheochromocytoma; endocrinology; research concerns studies on human and experimental pheochromocytoma and prevention of tumor development. Office: New York University Medical Center 400 E 34th St New York NY 10016

MANGLITZ, GEORGE RUDOLPH, entomologist, researcher; b. Washington, Aug 26, 1926; s. Adolph A. and Margaret (Bakersmith) M.; m. Marjorie Joan Welsch, Oct. 20, 1953; children: Ruth A. Manglitz Carr, Harry G., Paul A., Mary M., Joel D. B.S., U. Md., 1951, M.S., 1952; Ph.D., U. Nebr., 1962. Entomologist United Fruit Co., C.Am., 1951; entomologist Agrl. Research Service, US. Dept. Agr., Minot, N.D., 1952-54, Tifton, Ga., 1954-56, Beltsville, Md., 1956-68, research entomologist, Lincoln, Nebr., 1958—; prof. U. Nebr., 1972—; cons. Editor: Jour. Environ. Quality, 1972-75. Served in USN, 1944-46; PTO. Mem. Entomol. Soc. Am., Am. Inst. Biol. Sci., Am. Registry Profl. Entomologists (chpt. sec.-treas. 1982—), Forage Insect Research Conf. (chmn. 1977-78). Democrat. Mem. United Ch. of Christ. Club: YMCA (Lincoln). Subspecialties: Entomology; Genetics and genetic engineering (agriculture). Current work: Development of forage crop varieties that resist insects. Home: 955 N 67th St Lincoln NE 68505 Office: USDA-ARS East Campus U Nebr Lincoln NE 68583

MANGUS, MARVIN DALE, petroleum geologist; b. Altoona, Pa., Sept. 13, 1924; s. Alfred Ross and Myrna B. (Truby) M.; m. Jane Gray, Apr. 16, 1950; children: Alfred R., Donald H. B.S. in Earth Sci, Pa. State U., 1945, M.S. in Geology, 1946. Exploration geologist U.S. Geol. Survey, Alaska, 1946-58, cons. geologist, 1977-82; sr. exploration geologist Atlantic Richfield Co., Guatemala and Bolivia, 1958-60, Calgary, Alta., Can., 1960-62, Anchorage, Alaska, 1962-70, cons. geologist, 1970, 71-78, Calderwood and Mangus, Anchorage, 1970-77, Marvin D. Mangus, 1977—; exploration geologist Forest Oil, Anchorage, 1970-77, Simasko Prodn., 1977-82. Author or co-author profl. papers. Fellow Geol. Soc. Am.; mem. Am. Assn. Petroleum Geologists, Canadian Soc. Petroleum Geologists, Am. Inst. Profl. Geologists, Explorers Club. Clubs: Arts, Landscape (Washington) (pres. 1957-58). Subspecialties: Geology; Sedimentology. Current work: Consulting on the surface and subsurface geology of Northern Alaska and Arctic islands of Canada; mainly petroleum geology and exploration.

MANHOLD, JOHN HENRY, pathology educator; b. Rochester, N.Y., Aug. 20, 1919; s. John Henry and Helen Martha (Schulz) M.; m. Enriqueta Andino, Apr. 25, 1971. D.M.D., Harvard U., 1944; M.A., Washington U., 1956; B.A., U. Rochester, 1940. Instr. pathology Tufts Coll. Medicine and Dentistry, Boston, 1947-50; asst. prof. pathology Washington U., St. Louis, 1954-56; asst. prof., dept. chmn. Seton Hall Coll. Medicine and Dentistry, Jersey City, 1957-57, assoc. prof., 1958, prof., 1958-70; prof., chmn. dept. pathology Coll. Medicine and Dentistry N.J., Newark, 1970—. Author: Introduction to Psychosomatic Dentistry, 1956, Outline of Pathology, 1960, Clinical Oral Diagnosis, 1965, Gingival Tissue Respiration and the Oxygenating Agents, 1977. Served to lt. USN, 1940-46, 50-54. Recipient cert. of Recognition U. Md., 1965, U. Medicine and Dentistry of N.J., 1981. Fellow Internat. Coll. Dentistry, Am. Coll.

Dentistry; mem. Acad. Psychosomatic Medicine (pres. 1977-78), Am. Soc. Clin. Pathologists, Internat. Assn. Dental Research, Sigma Xi, Omicron Kappa Upsilon. Subspecialties: Psychophysiology; Pathology (medicine). Current work: Psychosomatic and pathology research. Home: 352 Shunpike Rd Chatham Township NJ 07928 Office: Univ Medicine and Dentistry of New Jersey 100 Bergen St Newark NJ 07103

MANKIN, DON, research cons.; b. Phila., Apr. 22, 1942; s. Harry and Eva M. B.S.E.E., Drexel Inst. Tech., 1964; M.A., Johns Hopkins U., 1966, Ph.D., 1968. Asst. prof. psychology dept. Lehigh U., 1968-75; vis. research assoc. U. So. Calif., 1975-76; assoc. prof. futures studies program U. Houston, Clear Lake City, 1976-78; assoc. prof. U. Med., 1978-80; research cons. Rand Corp., Santa Monica, Calif., 1981—, U. So. Calif., 1980—; grad. program dir. Antioch U., 1981—. Author: Toward a Post-Industrial Psychology, 1978; editor: Classic in Industrial and Organizational Psychology, 1980. Bd. dirs. Barrios Unidas. Mem. Acad. Mgmt., Am. Psychology Assn., World Future Soc. Current work: Social and organizational impacts of computers, organizational innovation, employee self-management. Home: 17 Ozone Ave Apt 9 Venice CA 90291 Office: Rand Corp 1700 Main St Santa Monica CA 90406

MANLEY, AUDREY FORBES, physician; b. Jackson, Miss., Mar. 25, 1934; d. Jesse Lee and Ora Lee (Buckhalter) Forbes; m. Albert Edward Manley, Apr. 3, 1970. A.B. with honors (tuition scholar), Spelman Coll., Atlanta, 1955; M.D. (Jesse Smith Noyes Found. scholar), Meharry Med. Coll., 1959. Diplomate: Am. Bd. Pediatrics. Intern St. Mary Mercy Hosp., Gary, Ind., 1960; from jr. to chief resident in pediatrics Cook County Children's Hosp., Chgo., 1960-62; fellow neonatology U. Ill. Research and Ednl. Hosp., Chgo., 1963-65; staff pediatrician Chgo. Bd. Health, 1963-66; practice medicine specializing in pediatrics, Chgo., 1963-66; assoc. Lawndale Neighborhood Health Center North, 1966-67; asst. med. dir., 1967-69; asst. prof. Chgo. Med. Coll., 1966-67; instr. Pritzker Sch. Medicine, U. Chgo., 1967-69; asst. dir. ambulatory pediatrics, asst. dir. pediatrics Mt. Zion Hosp. and Med. Center, San Francisco, 1969-70; med. cons. Spelman Coll., 1970-71, med. dir. family planning program, chmn. health careers adv. com., 1972-76; med. dir. Grady Meml. Hosp. Family Planning Clinic, 1972-76; with Health Services Adminstrs., Dept. Health and Human Services, 1976—; commd. officer USPHS, 1976—; chief genetic diseases services br. Office Maternal and Child Health, Bur. Community Health Services, Rockville, Md., 1978—; assoc. adminstr. clin. affairs Office of Adminstr. Health Resources and Services Adminstrn., 1983—; cons. in field.; assoc. adminstr. clin. affairs Health Resources and Services Adminstrn., USPHS, Rockville, Md., 1983—. Author numerous articles, reports in field. Trustee Spelman Coll., 1966-70. Recipient Meritorious Service award USPHS, 1981, Mary McLeod Bethune award Nat. Council Negro Women, 1979; also numerous service and achievement awards. Fellow Am. Acad. Pediatrics; mem. Nat. Inst. Medicine of Nat. Acad. Sci., Nat. Med. Assn., Am. Public Health Assn., AAUW, AAAS, Spelman Coll. Alumnae Assn., Meharry Alumni Assn., Operation Crossroads Africa Alumni Assn. Subspecialties: Pediatrics; Genetics and genetic engineering (medicine). Home: 2807 18th St NW Washington DC 20009 Office: 5600 Fishers Ln Rockville MD 20857

MANLEY, DONALD GENE, entomology educator; b. Monterey Park, Calif., Sept. 15, 1946; s. Maurice Emerson and LaVada Louise (Bryant) M.; m. Julia Ann Beaver, Feb. 15, 1969; children: Stephanie Suzanne, Christine Louise. B.A., UCLA, 1973; M.A., Calif. State U.-Long Beach, 1975; Ph.D., U. Ariz., 1978. Asst. prof. entomology Clemson (S.C.) U., 1978-82, assoc. prof., 1982—. Active Boy Scouts Am. Mem. Entomol. Soc. Am., AAUP, Entomol. Soc. S.C., Am. Registry Profl. Entomologists, Ga. Entomol. Soc., Phi Kappa Phi, Kappa Delta Pi, Alpha Zeta. Republican. Roman Catholic. Subspecialties: Integrated pest management; Taxonomy. Current work: Developed and coordinated integrated pest management program on tobacco in South Carolina; research on biology, ecology and taxonomy of velvet ants. Office: Pee Dee Expt Sta Clemson U PO Box 5809 Florence SC 29502

MANLY, PHILIP JAMES, high technology company executive, consulting health physicist; b. Cin., Apr. 12, 1944; s. Richard Samuel and Marian (LeFevre) M.; m. Jean Maron, Mar. 25, 1967; children: Charlotte, Fred, Peter, Elizabeth. B.S., MIT, 1967; M.S., Rensselaer Poly. Inst., 1971. Head Radiol. tech. div. Pearl Harbor Naval Shipyard, Honolulu, 1971-73, chief health physicist, 1973-74, head radiol. tng. div., 1974-78, pres. Gamma Corp., Wahiawa, Hawaii, 1978—. Editor: Procs. Health Physics Tng, 1979; tech. cons. videotape: Slowly Dying Embers, 1982. Mem. Health Physics Soc. (founder and 1st pres. Hawaii chpt. 1976, chmn. 13th Mid-Yr. Symposium, Honolulu 1979), Am. Nuclear Soc., Am. Pub. Health Assn. Subspecialties: Radiological Engineering; Software engineering. Current work: Appliations of radiation for solving problems in industry; computer applications for microcomputers in radiation related-activities. Home: 228 Plum St Wahiawa HI 96787 Office: Gamma Corp PO Box 430 Wahiawa HI 96786

MANN, ALFRED KENNETH, physics educator; b. N.Y.C., Sept. 4, 1920; m. Jayne Bowers, 1946; children: Stephen, Cecile, David, Brian. A.B., U. Va., 1942, M.S., 1946, Ph.D., 1947. With Manhattan Project, 1942-44; instr. physics Columbia U., N.Y.C., 1947-49; from asst. prof. to assoc. prof. physics U. Pa., Phila., 1949-57, prof. physics, 1957—; assoc. dir. Princeton-Pa. Accelerator, 1966-68; trustee Associated Univs., Inc., 1970-83; mem. subpanel on new facilities, high energy physics adv. panel, 1975; mem. physics program evaluation com. Regents of SUNY, 1974-75. Contbr. articles to profl. jours. Fulbright fellow Australian Nat. U., 1955-56; NSF fellow CERN, 1962-63; Guggenheim fellow, 1981-82. Fellow Am. Phys. Soc. (chmn. exec. com. 1983), Univs. Research Assn. (trustee 1979—), Sigma Xi, Phi Beta Kappa. Subspecialty: Particle physics. Current work: Initially in isotope separation, mass spectroscopy and molecular beams; in late 1950's began work in elementary particle physics including measurements of space-time structure of strangeness changing weak hadronic currents, and studies of rare decay modes of charged K-mesons; neutrino physics experiments at Fermi Nat. Accelerator Lab., Brookhaven Nat. Lab. Office: Dept Physic Univ Pa Philadelphia PA 19104

MANN, CHRISTIAN JOHN, geology educator, consultant; b. Junction City, Kans., Oct. 16, 1931; s. Christian J. and Vaughnie Jean (Waynick) M.; m. Diane K. Messmann, Feb. 11, 1961; children: John, Anna, Vaughn, Katrina. B.S., U. Kans., 1953, M.S., 1957; Ph.D., U. Wis.-Madison, 1961. Geologist Gulf Oil Corp., Ft. Worth, 1953, Calif. Co., Jackson, Miss., 1957-58, Chevron Oil Co., Jackson, 1961-64; sr. scientist Hazeton-Nuclear Sci. Corp., Palo Alto, Calif., 1964-65; faculty mem. U. Ill.-Urbana, 1965—, prof. geology, 1979—; ptnr. Lanman Assocs., Urbana, 1976—. Contbr. numerous papers, reports to profl. lit. Univ. coordinator Ill. Jr. Acad. Sci., Urbana, 1970-78. Served to 1st lt. AUS, 1953-55. Recipient Disting. Service award Miss. Petroleum Council, 1964. Fellow Geol. Soc. Am., AAAS (nominating com. 1980—); mem. Internat. Assn. Math. Geologists (charter, Best Paper award 1978, dep. editor 1981-83), Nat. Acad. Sci. (Nat. com. on math geology 1981-83), Am. Assn. Petroleum Geologists (ho. of dels. 1981-83), Math. Geologists of U.S. (pres. 1976-78). Subspecialties: Sedimentology; Mathematical geology. Current work: Mathematical and computer applications in geology; quantitative lithostratigraphic correlation; theoretical stratigraphy; petroleum geology; analysis of cyclicities and periodicity in geology. Office: Dept Geology Univ Ill 1301 W Green St Urban IL 61801

MANN, FREDERICK MICHAEL, nuclear science researcher; b. San Francisco, May 8, 1948; s. Leslie E. and Dorthy M. (Kirby) M. B.S., Stanford U., 1970; Ph.D., Calif. Inst. TEch., 1975. Research fellow Calif. Inst. Tech., 1975; advanced engr. Westinghouse Hanford Co., Richland, Wash., 1975-78, sr. scientist, 1978—. Contbr. numerous articles to profl. jours. Mem. Am. Phys. Soc., Am. Nuclear Soc., Phi Beta Kappa. Subspecialty: Nuclear physics. Current work: Supply nuclear data for energy-related projects, including theoretical and experimental efforts; active in Cross Section Evaluation Working Group; written computer codes to generate, evaluate, process, and use nuclear data. Home: 240 Saint Ct Richland WA 99352 Office: Westinghouse Hanford Co PO Box 1970b Richland WA 99352

MANN, ROBERT ARTHUR, biochem. co. exec.; b. Cleve., Mar. 26, 1947; s. Samuel and Grace (Gerson) M. B.A., U. Miami, 1970. Vice pres. U.S. Biochem. Corp., Cleve., 1973-77, sr. v.p., 1977—, also dir. Mem. Iron Arrow Soc., Alpha Epsilon Rho, Omega Soc. Democrat. Jewish. Subspecialties: Enzyme technology; Nutrition (biology). Current work: Biochem. applications in clin. chemistry. Office: 21000 Miles Pkwy Cleveland OH 44128

MANN, ROBERT WELLESLEY, educator, engineer; b. Bklyn., Oct. 6, 1924; s. Arthur Wellesley and Helen (Rieger) M.; m. Margaret Ida Florencourt, Sept. 4, 1950; children: Robert Wellesley, Catherine Louise. S.B., Mass. Inst. Tech., 1950, S.M., 1951, Sc.D., 1957. With Bell Telephone Labs., N.Y.C., 1942-43, 46-47; research engr. Mass. Inst. Tech., 1951-52, research supr., 1952-57, mem. faculty, 1953—, prof. mech. engring., 1963-70, Germeshausen prof., 1970-72, prof. engring., 1972-74, Whitaker prof. biomed. engring., 1974—, head systems and design div., mech. engring. dept., 1957-68, 82-83, founder, engring. projects lab., 1959-62; founder, chmn. steering com. Center Sensory Aids Evaluation and Devel., 1964—, chmn. div. health scis., tech., planning and mgmt., 1972-74, founder biomechanics and human rehab. lab., 1975; lectr. engring. in faculty of medicine Harvard U., 1973—; exec. com. program in health scis. and tech. Harvard-Mass. Inst. Tech., 1972—; research assoc. in orthopedic surgery Children's Hosp. Med. Center, 1973—; cons. in engring. sci. Mass. Gen. Hosp.; cons. in field, 1953—; mem. Commn. Engring. Edn., 1962-69; com. prosthetics research and devel. Nat. Acad. Scis., 1963-69, chmn. sensory aids subcom., 1965-68, com. skeletel system, 1969; mem. com. interplay engring. with biology and medicine Nat. Acad. Engring., 1969-73, chmn. sensory aids subcom., 1968-73; mem. com. sci. policy for medicine and health Inst. Medicine, 1973-74, 82-84; mem. com. on nat. needs for rehab. physically handicapped Nat. Acad. Scis., 1975-76; mem.-at-large confs. com. Engring. Found., 1975-81. Author numerous articles in field.; Cons. editor: Ency. Sci. and Tech.; editorial bd.: Jour. Visual Impairment and Blindness, 1976-80; assoc. editor: IEEE Trans. in Biomed. Engring, 1969-78; asso. editor: ASME Jour. Biomed. Engring, 1976-82. Pres., trustee Amanda Caroline Payson Scholarship Fund, 1965—; bd. dirs. Carroll Rehab. Center, 1967-74, pres., 1968-74; mem. corp. Perkins Sch. for Blind, 1970—, Mt. Auburn Hosp., 1972—. Served with AUS, 1943-46. Recipient Sloan award for outstanding performance, 1957; Talbert Abrams photogrammetry award, 1962; award Assn. Blind of Mass., 1969; IR-100 award for Braillemboss, 1972; Bronze Beaver award Mass. Inst. Tech., 1975; J.R. Killian faculty achievement award Mass. Inst. Tech., 1983; UCP Goldenson Research for Handicapped award, 1976; H.R. Lissner award, 1977; New Eng. award, 1979. Fellow Am. Acad. Arts and Scis., IEEE (chmn. group on engring. in biology and medicine 1974-78), AAAS; mem. Nat. Acad. Engring., Nat. Acad. Scis., Inst. Medicine of Nat. Acad. Scis., ASME (gold medal 1977), AAAS, Biomed. Engring. Soc. (dir. 1981-84), M.I.T. Alumni Assn. (pres. 1983-84), Sigma Xi (nat. lectr. 1979-81), Tau Beta Pi, Pi Tau Sigma. Roman Catholic. Subspecialties: Biomedical engineering. Current work: Human musculo-skeletal-joint biomechanics, asteoarthritis; human rehabilitaion; computer-aided surgical simulation. Patentee in field. Home: 5 Pelham Rd Lexington MA 02173 Office: 77 Massachusetts Ave Cambridge MA 02139

MANNING, IRWIN, physicist; b. Bklyn., Mar. 7, 1929; s. Louis and Nettie (Jaffee) M.; m. Amelia Ann Young, May 24, 1964; children: Emily, Sarah. B.S. in Math, M.I.T., 1951, Ph.D. in Physics, 1955. Research assoc. Syracuse U., 1955-57, U. Wis., 1957-59; research physicist Naval Research Lab., Washington, 1959—. Mem. Am. Phys. Soc. Subspecialty: Theoretical physics. Current work: Materials modification by ion implantation. Home: 1801 Courtland Rd: Alexandria VA 22306 Office: Condensed Matter and Radiation Scis Div Naval Research Lab Washington DC 20375

MANNING, JOHN WILLIAM, physiology educator; b. New Orleans, Nov. 14, 1930; s. John W. and Dorothy (Baylor) M.; m. Cynthia Satterlee, Jan. 9, 1954; children: Kathleen, Michael, Donald, Edward, Cheryl Ann, Robert. B.S., Loyola U. of South, 1951; M.S., Tulane U., 1955; Ph.D., Loyola U., Chgo., 1958. Postdoctral research fellow Emory U., Atlanta, 1958-65, asst. profl., 1964-66, assoc. prof., 1966-70, prof. physiology, 1970—; vis. scientist Karolinska Inst. Stockholm, 1963-64; vis. prof. Shinshu U., Matsumoto, Japan, 1972; co-dir. tng. in physiology NIH, 1970-75; mem. Cardiac Function br. task group for 10 yr. rev. and update of heart and vasular disease program plan, 1981—; cons. edn. officer Am. Physiol. Soc., 1976—, others. Contbr. articles to profl. jours. Mem. Am. Physiol. Soc., Soc. Neurosci., Am. Assn. Anatomists, Can. Physiol. Soc. Democrat. Roman Catholic. Club: Cajal. Subspecialties: Physiology (medicine). Current work: Central nervous system regulation of cardiovascular activity; central integration of autonomic behavior; central response and transmission of small cutaneous afferents. Home: 1406 Council Bluff Dr Atlanta GA 30345 Office: Rm 056 1648 Pierce Dr Atlanta GA 30322

MANNING, SUSAN KARP, psychology educator; b. N.Y.C.; d. Walter and Estelle (Grossman) Karp; m. Robert E. Manning, June 22, 1963 (div. Aug. 1980); m. Leonard Cohan, Jan. 1, 1981. B.A., U. Mich.-Ann Arbor, 1960, M.A., 1961; Ph.D., U. Calif.-Riverside, 1971. Tech. specialist Mitre Corp., Bedford, Mass., 1962-64; research asst. Pacific State Hosp., Pomona, Calif., 1964-65; teaching asst. U. Calif.-Riverside, 1966-67; lectr. CCNY, 1967-68; assoc. prof. psychology, lectr. Hunter Coll., 1968—; research asst. MIT, 1961-62, Harvard Med. Sch., 1964. Contbr. articles to profl. jours. Mem. Manhattan Community Bd. 6, 1979—; chmn. health and social services com., 1981—; mem. community bd. Bellevue Hosp., 1982—. NSF research grantee, 1977-80; recipient Shuster faculty award Hunter Coll., 1981. Mem. Psychonomic Soc., Eastern Psychol. Assn. Democrats. Club: Gramercy Stuyvesant Ind. Subspecialty: Cognition. Current work: Research in short-term memory, especially concerned with recall of items presented tactually, visually, and auditorily. Office: Hunter College 695 Park Ave New York NY 10021

MANOHAR, MURLI VAID, veterinary biosciences educator; b. Amritsar, Panjab, India, Oct. 3, 1947; came to U.S., 1975; s. Kewal Krishan and Kanta (Persinni) V.; m. Panna Panna, Dec. 16, 1974; children: Leslie, Patrick, Christine, Crystal. B.V..Sc., Panjab Agrl. U., 1968, Ph.D., 1973; Ph.D., U. Wis.-Madison, 1978. Asst. prof. vet. bioscis. U. Ill.-Urbana, 1979-81, assoc. prof., 1982—. Mem. Am. Physiol. soc., Am. Heart Assn. Subspecialty: Health services research. Current work: Myocardial function and perfusion. Office: U Ill Coll Vet Medicine 212 Large Animal Clinic Urbana IL 61801

MANOR, ROBERT EDWARD, physicist, cons.; b. Whitehouse, Ohio, Dec. 3, 1937; s. Clifford William and Emily Melissa (Bradford) M. B.S., U. Toledo.; M.S., Kent State U., 1969. Devel. physicist LOF Glass Co., Toledo, 1964-65; math. physicist Toledo Engring. Co., Inc., 1965-67; devel. physicist Haughton Elevator Co., Toledo, 1969-70; cons. math., devel. physicist laser-energy beam systems Energystics, Inc., Toledo, 1976-77; owner, chief physicist REM Agy., Whitehouse, Ohio, 1970—. Subspecialties: Integrated circuits; Laser fusion. Current work: Development mathematics and quantum optics for a continuous laser fusion of deuterium and tritium pellets; solid state computer with digital electronic readout system.

MANSFIELD, CARL MAJOR, radiotherapist, educator; b. Phila., Dec. 24, 1928; s. Edward and Vivian A. (Wright) M.; m. Sara Lynn, Sept. 11, 1958; children: Joel, Kara. A.B. in Chemistry, Lincoln U., 1951; postgrad., Temple U., 1952; M.D., Howard U., 1956. Diplomate: Am. Bd. Radiology. Intern Episcopal Hosp., Phila., 1956-57, resident in radiology, 1957-58, 60, 61-62; resident in radiation therapy and nuclear medicine Thomas Jefferson Med. Coll. Hosp., Phila., 1960-61, NIH fellow, 1962-63, instr. radiology, 1964-65, assoc. radiology, chief div. nuclear medicine, 1966-67, asst. prof., 1967-69, assoc. prof. radiation therapy and nuclear medicine, chief sect. ultrasound, 1970-74, prof., 1974-76; lectr. radiology U. Pa., 1967-73; prof., chmn. dept. radiation therapy U. Kans. Med. Ctr., Kansas City, 1976-83-; prof., chmn. dept. radiation therapy and nuclear medicine Jefferson Med. Coll., Phila., 1983—; vis. prof. Hahnemann Med. Coll. Hosp., 1971; chmn. dept. radiation therapy Menorah Med. Ctr., Kansas City, Mo., 1977—; mem. staff Baptist Meml Hosp., Kansas City, Mo., Bethany Med. Ctr., Kansas City, Kans., Shawnee Mission (Kans.) Med. Ctr. Contbr. numerous articles to profl. jours. Served to capt. USAF, 1958-60. NIH fellow, 1963-64; Chernicoff fellow, 1964-66; Am. Cancer Soc. fellow, 1965-68. Mem. Am. Coll. Nulcear Medicine, Am. Coll. Radiology, AMA, Am. Soc. Therapeutic Radiologists, Am. Coll. Nuclear Physicians, Am. Inst. Ultrasound in Medicine, Assn. Univ. Radiologists, Brit. Inst. Radiology, Nat. Med. Assn., Radiol. Soc. N.Am., Royal Soc. Medicine, Soc. Nuclear Medicine, N.Y. Acad. Sci., Am. Radium Soc., Am. Cancer Soc., Radiation Research Soc., Am. Soc. Clin. Oncologists. Subspecialties: Radiology; Nuclear medicine. Current work: Cancer treatment and research.

MANSFIELD, LOIS, mathematics educator; b. Portland, Maine, Jan. 2, 1941; d. Robert Carleton and Mary (Bowdish) M. B.S., U. Mich.-Ann Arbor, 1962; M.S., U. Utah-Salt Lake City, 1966, Ph.D., 1969. Vis. asst. prof. computer sci. Purdue U., West Lafayette, Ind., 1969-70; asst. prof. computer sci. U. Kans., Lawrence, 1970-74, assoc. prof., 1974-78; assoc. prof. math. N.C. State U., Raleigh, 1978-79; assoc. prof. applied math. U. Va., Charlottesville, 1979—; mem. adv. panel for computer sci. NSF, Washington, 1975-78; vis. scientist Inst. Computer Applications in Sci. and Engring., NASA, Hampton, Va., 1977, cons., 1978. NSF research grantee, 1976—. Mem. Am. Math. Soc., Soc. Indsl. and Applied Math. (editorial bd. Jour. Sci. and Statis. Computing 1979—), Assn. Computing Machinery (bd. dirs. SIGNUM 1980-83). Subspecialty: Numerical analysis. Current work: Numerical solution of partial differential equations. Home: 5800 Burnett Ln Ruckersville VA 22968 Office: U Va Thornton Hall Charlottesville VA 22901

MANSON, JOHN ALEXANDER, chemistry educator; b. Dundas, Ont., Can., Aug. 4, 1928; s. Gordon Alexander and Anna Isabel (McDonald) M.; m. Emma Natalie Dean, June 29, 1951; children: Lois Patricia Manson Pyle, John Michael Gordon, William Dean, Barbara Denise. B.Sc in Chemistry and Physics with honors, McMaster U., 1949, M.Sc., 1951, Ph.D., 1956. Assoc. research engr. Engring. Research Inst. U. Mich., Ann Arbor, 1956-67; sr. chemist Air Reduction Co., Murray Hill, N.J., 1957-59, sect. head, 1959-61, supr., 1961-66; mem. faculty Lehigh U., Bethlehem, Pa., 1966—, prof. chemistry, 1970—; dir. Polymer Lab., Materials Research Ctr., 1966—; cons. in field. Co-author: Polymer Blends and Composites, 1976, Fatigue in Engineering Plastics, 1980; contbr. sci. articles to profl. publs. Subspecialties: Polymers; Polymer chemistry. Current work: Structure and properties of polymers and composites; fracture and fatigue; polymer-concrete composites. Home: 1320 Sycamore Ave Bethlehem PA 18017 Office: Lehigh Univ Cox Lab 32 Bethlehem PA 18015

MANSON, STEVEN TRENT, physics educator; b. Bklyn., Dec. 12, 1940; s. Henry Joshua and Rosalind (Frey) M.; m. Bettye Bonds, July 17, 1944; children: Jonathan, Andrew. B.S., Rensselaer Poly. Inst., 1961; M.A., Columbia U., 1963, Ph.D., 1966. NAS-NRC postdoctoral fellow Nat. Bur. Standards, Washington, 1966-68; asst. prof. Ga. State U., Atlanta, 1968-72, assoc. prof., 1972-77, prof., 1977—; cons. Oak Ridge Nat. Lab., Argonne Nat. Lab., Batelle Meml. Inst. Editorial bd.: Jour. Electron Spectroscopy and Related Phenomena; Contbr. articles to profl. jours. Active Winnona Park Neighbors Assn., 1977—, Decatur Neighborhood Alliance, 1977—; mem. Decatur Community Goals Com., 1980—. NSF grantee, 1973—; U.S. Army Research Office grantee, 1974—; recipient Alumni Disting. Prof. award Coll. Arts and Scis., Ga. State U., 1977. Fellow Am. Phys. Soc.; mem. Inst. Physics (London), Sigma Pi Sigma. Subspecialties: Atomic and molecular physics; Theoretical physics. Current work: Theoretical studies of atomic collisions and photoabsorption. Home: 463 Kirk Rd Decatur GA 30030 Office: Dept Physics and Astronomy Ga State Univ Atlanta GA 30303

MANSOUR, EDWARD GEORGE, surgical oncologist; b. Bethlehem, Dec. 4, 1933; U.S., 1965, naturalized, 1974; s. George H. and Eugenie M. (Hamameh) M.; m. Mary Elizabeth Nolan, July 23, 1966; children: George, John, Maura, William. B.Sc., Am. U. Beirut, 1958, M.D., 1962. Diplomate: Am. Bd. Gen. Surgery. Intern. Am. U. Hosp., Beirut, 1961-62; resident Cleve. Met. Gen. Hosp., Case Western Res. U., 1965-69; physician Arabian Am. Oil Co. Hosps., Dhahran, Saudi Arabia, 1962-65; instr. surgery Case Western Res. U., Cleve., 1969-70, asst. prof, 1970-77, assoc. prof., 1977—; dir. surg. oncology Cleve. Met. Gen. Hosp., 1969—, dir. Oncology Clinic, 1975—, assoc. surgeon, 1977—; cons. surg. oncology Luth. Med. Ctr., Cleve., 1971—; pres. staff Cuyahoga County Hosps.; cons. Nat. Cancer Inst.; prin. investigator, chmn. surg com. Eastern Coop. Oncology Group; Trustee, v.p. Ohio div.. Am. Cancer Soc., bd. dirs. Cuyahoga unit. Contbr. articles to profl. jours. Mem. Acad. Medicine Cleve., Am. Assn. Cancer Research, Am. Assn. Cancer Edn., ACS, AMA, Am. Soc. Clin. Oncology, Central Surg. Assn., Cleve. Surg. Soc. (pres.), Ohio

State Med. Assn., N.E. Ohio Soc. Oncologists, Soc. Surg. Oncology, Soc. Environ. Geochemistry and Health, Soc. Surgery of Alimentary Tract, Collegium Internationale Chirurgiae Digestivae, Alpha Omega Alpha. Subspecialties: Surgery; Oncology. Current work: Adjuvant therapy in breast and colorectal cancer; prognostic value of biological markers in cancer; hepatic artery infusion. Office: 3395 Scranton Rd Cleveland OH 44109

MANSOUR, TAG ELDIN, pharmacologist; b. Belkas, Eqypt, Nov. 6, 1924; came to U.S., 1951, naturalized, 1956; s. Elsayed and Rokaya (Elzayat) M.; m. Joan Adela MacKinnon, Aug. 6, 1955; children—Suzanne, Jeanne, Dean. B.Sc., Cairo U., 1946; PH.D., U. Birmingham, Eng., 1949, D.Sc., 1974. Lectr. U. Cairo, 1950-51; Fulbright instr. physiology Howard U., Washington, 1951-52; sr. instr. pharmacology Western Res. U., 1952-54; asst. prof., assoc. prof. pharmacology La. State U. Med. Sch., New Orleans, 1954-61; assoc. prof., prof. pharmacology Stanford U. Sch. Medicine, 1961—, Donald E. Baxter prof., chmn. dept. pharmacology, 1977—; cons. USPHS, WHO, Nat. Acad. Scis.; Mem. adv. bd. Med. Sch., Kuwait U.; Heath Clarke lectr. London Sch. Hygiene and Tropical Medicine, 1981. Contrbr. sci. articles to profl. jours. Commonwealth Fund fellow, 1965; Mercy Found. scholar NIMR, 1982. Fellow AAAS; mem. Am. Soc. Pharmacology and Exptl. Therapeutics, Am. Soc Biol. Chemists, Am. Heart Assn., World Affairs Council, Sierra Club, Sigma Xi. Club: Stanford Faculty. Subspecialties: Molecular pharmacology; Biochemistry (medicine). Current work: Regulation of enzymes and effect of hormones on regulatory enzymes-biochemistry and pharmacology of parasitic worms. Office: 300 Pasteur Dr Stanford CA 94305

MANSUR, LOUIS KENNETH, science manager, researcher; b. Lowell, Mass., Apr. 18, 1944; s. Louis Francis and Adele Agnes (Abraham) M.; m. Helen Ruth Hotz, Aug. 30, 1969; children: Kendra Jane, Joanna Beth, Warren Keith. B.S. in Physics, U. Lowell, 1966; M.S. in Engring. Physics, Cornell U., 1968; Ph.D. in Materials Sci, Cornell U., 1974. Reactor engr. U.S. AEC, Washington, 1966-67, 70-71; reactor physicist assigned to Gen. Electric Co., Fayetteville, Ark., 1968-69; materials scientist Oak Ridge Nat. Lab., Tenn., 1974-82, group leader, 1983—. Editor: Phase Stability during Irradiation, 1981; contbr. chpts. to books, articles to profl. jours. Mem. Am. Nuclear Soc., Am. Soc. Metals (chmn. chpt. 1981-82), AAAS, Metall. Soc. of AIME, Sigma Xi. Subspecialties: Materials. Current work: Defect mechanisms, theory of radiation effects in metals and alloys, ion bombardment, kinetics of microstructural evolution. 37716Office Oak Ridge Nat Lab Bldg 5500 Oak Ridge TN 37830

MANSUR, OVAD MORDECAI, computer science educator, researcher; b. Jerusalem, Israel, Nov. 14, 1937; came to U.S., 1974, naturalized, 1973; s. Mordecai M. and Flora (Silas) M.; m. Chaya Tania Auraham, Jan. 2, 1963; children: Michelle, Amiel. B.A., Hebrew U., Jerusalem, 1965, M.B.A., 1969; Ph.D., Case Western Res. U., 1977. Systems analyst, projects mgr. Nat. Ins. Inst., Jerusalem, 1965-69, dir. systems analysis and programming, 1970-72; dir. econ. research and long-range planning Israel Ports Authority, Tel Aviv, 1972-74; asst. prof. computer sci. Cleve. State U., 1978—; cons. Spl. Control Systems, Cleve., 1979—. Mem. Assn. Computing Machinery (v.p. Cleve. chpt. 1981-82), Am. Soc. Info. Sci., Data Processing Mgmt. Assn. Jewish. Subspecialties: Software engineering; Information systems, storage, and retrieval (computer science). Current work: Design of information retrieval systems; software metrics; software engineering economics; management and control of software projects. Office: CIS Dept Cleve State U Cleveland OH 44115

MANTEI, KENNETH ALAN, chemistry educator; b. Culver City, Calif., Nov. 22, 1940; s. Ludwig and Olga (Deckert) M.; m. Rosalind A. Harrell, Sept. 9, 1967; children: Randall Wade, Erika Jean. B.A., Pomona Coll., 1962; Ph.D., Ind. U., 1967. Researcher UCLA, 1967-68; faculty Calif. State Coll., San Bernardino, 1968—, prof. chemistry, 1981—, chmn. dept., 1975—. Subspecialties: Physical chemistry; Programming languages. Current work: FORTH computer programming language; implementing FORTH on 68000-based microcomputers. Home: 1254 E 26th St San Bernardino CA 92404 Office: Dept Chemistry Calif State Coll 5500 State College Pkwy San Bernardino CA 92407

MANTHEY, MICHAEL JOHN, computer scientist; b. Mpls., 1944; m.; 1 dau., Solveig. B.S., Rensselaer Poly. Inst., 1966; Ph.D., SUNY, Buffalo, 1980. Cons. U.S. Navy, 1969-71; asst. prof. computer sci. U. Aarhus, Denmark, 1972-78; dir. labs., dept computer sci. SUNY, Buffalo, 1978-80; asst. prof. computer sci. U. N.Mex., Albuquerque, 1980—. Mem. Assn. for Computing Machinery, AAAS, Sigma Xi. Subspecialties: Distributed systems and networks; Foundations of computer science. Current work: Unification of quantum physics and concurrent computational concepts; non determinism, uncertainty conservation of information.

MANYAM, BALA VENKTESHA, neurologist, researcher; b. Bangalore, India, Oct. 15, 1942; d. Kolar and Swarnam Venktesier; m. Rangavalli Bala Manyam, June 10, 1970; 1 child, Shaila. M.B., B.S., Bangalore Med. Coll., 1967. Resident internal medicine Dalhousie U., 1972; resident in neurology Med. Coll. Wis. Hosp., Milw., 1972-75; Thomas Jefferson U. Hosp., Phila., 1972-75; asst. research officer Nat. Inst. Nutrition, Hyderabad, India, 1970; staff neurologist VA Med. Center, Wilmington, Del., 1975-81, asst. chief neurology service, dir. movement disorders clinic, 1981—; instr. Thomas Jefferson U., Phila., 1975-81, asst. prof. depts. neurology and pharmacology, 1981—. Contbr. articles to profl. jours., chpts. to books. Recipient Spl. Performance award VA, 1977; VA grantee, 1980—. Mem. Am. Acad. Neurology, Am. Soc. Pharmacology and Exptl. Therapeutics, World Fedn. Neurology (research com. neuroepidemiology), Soc. Neurosci., Soc. Clin. Trials, Assn. Research in Mental and Nervous Diseases, Sigma Xi. Hindu. Subspecialties: Neurology; Neuropharmacology. Current work: Biochemical and pharmacological alterations of central nervous system in movement disorders and the effect of various drugs. Home: 2312 Patwynn Rd Wilmington DE 19810 Office: VA Med Center Wilmington DE 19805

MAO, CHI CHIAN, physician; b. Canton, China, Aug. 7, 1942; s. Sung N. and Chen Y. (Choy) M.; m. Shelley Giak Shiaw Lim, Sept. 5, 1966; children: Hua-ching, Dillon, Noreen. B.S., Nat. Taiwan U., Taipei, 1965; Ph.D., U. Okla., 1971; M.D., U. Tex.-Houston, 1981. Vis. fellow NIMH, Washington, 1972-74, vis. assoc., 1975-76, sr. staff fellow, 1976-77; from assoc. prof. to prof. biology and physiology Nat. Taiwan Normal U. and China Med. Coll., Taipei, 1974-75; vis. research scientist U. Tex. Med. Sch., Houston, 1978-80; intern U. Tex. Med. Sch. at Houston and affiliated hosps., 1981-82, resident in internal medicine, 1982-83; resident Baylor Med. Coll. Affiliated Hosps.; Houston, 1983—. Contbr. articles to profl. publs. Recipient award Distilled Spirits Council U.S., 1980-81. Mem. Nat. Taiwan U. Zool. Soc. (pres. 1964-65), Am. Soc. Pharmacology and Exptl. Therapeutics, N.Y. Acad. Scis., A.C.P. Subspecialties: Internal medicine; Neuropharmacology. Current work: Neurochemistry; neuropharmacology; gastrointestinal physiology and endocrinology. Office: Univ Tex Med Sch Physiology Dept PO Box 20708 Houston TX 77025 Home: 12108 Stone West Houston TX 77035

MAPLE, TERRY LEE, psychology educator, behavioral scientist; b. Maywood, Calif., Sept. 10, 1946; s. Merrill and Evelyn May (Hayes) M.; m. Adelaide G. Petersen, June 17, 1972; children: Molly Jean, Emily May. A.B., U. of the Pacific, 1968; M.A., U. Calif-Davis, 1971, Ph.D., 1974. Asst. prof. Emory U., Atlanta, 1975-78; assoc. prof. Sch. Psychology, Ga. Inst. Tech., Atlanta, 1978—. Author: Gorilla Behavior, 1982, Orangutan Behavior, 1980, Captivity and Behavior, 1979, Aggression, Hostility and Violence, 1973. Research grantee Am. Mus. Natural History, 1979, 81, NIH, 1981, 82. Mem. Am. Psychol. Assn., Am. Soc. Primatologists, Animal Behavior Soc. Democrat. Presbyterian. Subspecialties: Psychobiology; Learning. Current work: Primate behavior; animal behavior; learning and memory; comparative psychology. Office: Sch Psychology Ga Inst Tech Atlanta GA 30332

MAPOU, ROBERT LEWIS, psychologist; b. Washington, May 28, 1955; s. Albert and Phyllis Helen (Smul) M. B.S., U. Md., 1977; M.A. Candidate, Emory U., 1978—. Research asst. U. Md., College Park, 1973-77; engr., programmer/analyst Link div. Singer Co., Silver Spring, Md., 1978; research asst. Emory U., Atlanta, 1978-81, clin. adminstrv. asst. neuropsychology lab., 1982—; biomed. engring. technician Yerkes Regional Primate Research Ctr., Atlanta, 1981—. Recipient Sklar award U. Md., 1977; Emory U. stipendee, 1978—. Mem. Am. Psychol. Assn., Southeastern Psychol. Assn., Phi Beta Kappa, Phi Kappa Phi. Democrat. Jewish. Subspecialties: Neuropsychology; Clinical psychology. Current work: Hemispheric specialization and visual processing; computer applications to psychological research, assessment and clinical intervention. Home: 1111 Clairmont Rd F3 Decatur GA 30030 Office: Emory Univ Dept Psychology Atlanta GA 30030

MAR, BRIAN WAYNE, civil engineering educator; b. Seattle, Aug. 5, 1933; s. Frank W. and Ruth L. (Dott) M.; m. Hazel Joy Woo, Dec. 17, 1955; children: Melanie, Marnie, Murray, Monte, Mimi. B.S. in Chem. Engring, U. Wash., Seattle, 1955, M.S., 1957, Ph.D., 1958, M.S.E. in Civil Engring, 1967. Research specialist Boeing Co., Seattle, 1959-67; mem. faculty U. Wash., 1967—, prof. civil engring. Inst. for Environ. Studies, 1973—, chmn. water and air resources div. Inst. for Environ. Studies, 1975-77, asso. dean Coll. Engring., 1976-81; cons. in field. Author: (with others) The Natural Environment, Waste and Control, 1973, Environmental Modeling and Decision Making-The U.S. Experience, 1976, also articles, reports. Mem. ASCE, AAAS, Am. Geophys. Union, Phi Beta Kappa, Sigma Xi, Tau Beta Pi. Subspecialty: Environmental engineering. Home: 10615 60th Ave S Seattle WA 98178 Office: Dept Civil Engring Coll Engring Univ Wash Seattle WA 98106 It is not what you know, but how you use what you know in working with others.

MARAMOROSCH, KARL, virologist, educator; b. Vienna, Austria, Jan. 16, 1915; came to U.S., 1947, naturalized, 1952; s. Jacob and Stefanie Olga (Schlesinger) M.; m. Irene Ludwinowska, Nov. 15, 1938; 1 dau., Lydia Ann. M.S. magna cum laude in Entomology, Agrl. U., Warsaw, Poland, 1938; student, Poly. U. Bucharest, Rumania, 1944-46; fellow, Bklyn. Bot. Garden, 1947-48; Ph.D. (predoctoral fellow Am. Cancer Soc. 1948-49), Columbia, 1949. Civilian internee in, Rumania, 1939-46; asst., then asso. Rockefeller Inst., N.Y.C., 1949-61; sr. entomologist Boyce Thompson Inst., Yonkers, N.Y., 1961-74, program dir. virology and insect physiology, 1962-74; prof. microbiology Waksman Inst., Rutgers U., New Brunswick, N.J., 1974—; vis. prof. agr. U. Wageningen, Netherlands, 1953, Cornell U., 1957, Rutgers U., 1967-68, Fordham U., 1973, Sapporo U., Japan, 1980; Mendel lectr. St. Peters Coll., Jersey City, 1963; virologist FAO to Philippines, 1960; Mendel lectr. Fudan U., Shanghai, 1982; cons. World-wide survey, 1963; chmn. U.S.-Japan Coop. Seminar, 1965, 74; mem. panel food and fiber Nat. Acad. Scis., 1966; cons. rice virus diseases AID-IRRI, Hyderabad, India, 1971; cons. UNDP, Bangalore, India, 1978-79; virologist FAO/UNDP, Sri Lanka, 1981; AIBS lectr., 1970-72, Found. Microbiology Nat. lectr., 1972-73, Fulbright Distingd. prof., Yugoslavia, 1972-73, 78; mem. tropical medicine and parasitology study sect. NIH, 1972-76; chmn. 1st-3d Internat. Confs. Comparative Virology, 1969, 73, 76. Author: Comparative Symptomatology of Coconut Diseases of Unknown Etiology, 1964; Editor: Biological Transmission of Disease Agents, 1962, Insect Viruses, 1968, Viruses, Vectors and Vegetation, 1969, Comparative Virology, 1971, Mycoplasma Diseases, 1973, Viruses, Evolution and Cancer, 1974, Invertebrate Immunity, 1975, Legume Diseases in the Tropics, 1975, Invertebrate Tissue Culture: Research Applications, 1976, Invertebrate Tissue Culture: Applications in Medicine, Biology and Agriculture, 1976, Aphids as Virus Vectors, 1977, Insect and Plant Viruses: An Atlas, 1977, Viruses and Environment, 1978, Practical Tissue Culture Applications, 1979, Leafhopper Vectors and Plant Disease Agents, 1979, Vectors of Plant Pathogens, 1980, Invertebrate Systems in Vitro, 1980, Vectors of Disease Agents, 1981, Methods in Virology, 1964—, Advances in Virus Research, 1972—, Archives of Virology, 1973—, Intervirology, 1973—, Advances in Cell Culture, 1979—; Editor-in-chief: Jour. N.Y. Entomol. Soc, 1972—; Asso. editor: Virology, 1964-68, 75—. Recipient Sr. Research award Lalor Found., 1957; Nat. Ciba-Geigy award in agr., 1976; Wolf prize in agr., 1980; Jurzykowski prize in biology, 1980. Fellow N.Y. Acad. Scis. (A. Cressy Morrison prize natural sci. 1951, chmn. div. microbiology 1956-60, rec. sec. 1960-61, v.p. 1962-63), Nat. Acad. Scis. India (hon.), AAAS (Campbell award 1958); mem. Harvey Soc., Growth Soc., Phytopath. Soc. (councellor), Entomol. Soc. Am., Indian, Japan, Can. phytopath. socs., Leopoldina Acad., Internat. Com. Virus Nomenclature, Electron Microscope Soc., Am. Soc. Microbiology (Waksman award 1978), Tissue Culture Assn. (pres. N.E. br. 1978-81), Sigma Xi (pres. Rugers chpt. 1978). Subspecialty: Microbiology. Home: 17 Black Birch Ln Scarsdale NY 10583 Office: Waksman Inst Microbiology Rutgers U New Brunswick NJ 08903

MARAN, JANICE WENGERD, pharmacologist; b. Balt., June 30, 1942; d. Edgar Arthur and Mildred Ilease (Laughter) Wengerd; m. Anthony J. Maran, Apr. 29, 1966. B.S., Juniata Coll., 1964; Ph.D. (NIH fellow, George D. and Grace H. Shafer fellow), Stanford U., 1974. Research asst. Stanford (Calif.) U., 1964-66, research asso., 1966-69; research scientist McNeil Labs., Ft. Washington, Pa., 1977-78; sr. scientist McNeil Pharm., Spring House, Pa., 1978—, project mgr., 1980—; NATO postdoctoral fellow in sci. U. Bristol, Eng., 1974-75; NIH postdoctoral fellow Johns Hopkins U. Med. Sch., 1975-77. Contbr. articles to sci. publs., chpts. to books. NSF award for summer study in chemistry, 1960. Mem. Am. Physiol. Soc., Biomed. Engring. Soc., Soc. for Neurosci., N.Y. Acad. Scis., AAAS, Internat. Platform Assn., Sigma Xi. Subspecialties: Neuropharmacology; Neurophysiology. Current work: Discovery of novel psychotrophic drugs and the eleucidation of the mechanism of action of novel and existing central nervous system drugs through electrophysiological techniques; project management of drug development of a cerebral vasodilator. Home: 106 Anton Rd Wynnewood PA 19096 Office: McNeil Pharmaceutical Spring House PA 19477

MARAN, STEPHEN PAUL, astronomer; b. Bklyn., Dec. 25, 1938; s. Alexander P. and Clara F. (Schoenfeld) M.; m. Sally Ann Scott, Feb. 14, 1971; children: Michael Scott, Enid Rebecca, Elissa Jean. B.S., Bklyn. Coll., 1959; M.A., U. Mich., 1961, Ph.D., 1964. Astronomer Kitt Peak Nat. Obs., Tucson, 1964-69; project scientist for orbiting solar observatories NASA-Goddard Space Flight Center, Greenbelt, Md., 1969-75, head advanced systems and ground observations br., 1970-77; mgr. operation Kohoutek, 1973-74, sr. staff scientist, 1977—; Cons. Westinghouse Research Labs., 1966; vis. lectr. U. Md., College Park, 1969-70; sr. lectr. U. Calif. at Los Angeles, 1976. Author: (with John C. Brandt) New Horizons in Astronomy, 1972, 2d edit., 1979, Arabic edit., 1979; Editor: Physics of Nonthermal Radio Sources, 1964, The Gum Nebula and Related Problems, 1971, Possible Relations Between Solar Activity and Meteorological Phenomena, 1975, New Astronomy and Space Science Reader, 1977, A Meeting with the Universe, 1981; assoc. editor: Earth, Extraterrestrial Scis, 1969-79; editor: Astrophys. Letters, 1974-77; assoc. editor, 1977—; Contbr. articles on astronomy, space to popular mags. Named Distinguished Visitor Boston U., 1970; recipient Group Achievement awards NASA, 1969, 74, Hon. Mention AAAS-Westinghouse Sci. Writing Award, 1970. Mem. Internat. Astron. Union, AAAS, Am. Astron. Soc. (Harlow Shapley vis. lectr. 1981—), Royal Astron. Soc., Am. Phys. Soc., Am. Geophys. Union, Am. Inst. Aeros. and Astronautics. Subspecialties: Ultraviolet high energy astrophysics; Optical astronomy. Current work: Observations of nebulae from space; space telescope research. Office: Code 680 NASA-Goddard Space Flight Center Greenbelt MD 20771

MARANGOS, PAUL JEROME, neurochemist, researcher; b. Bklyn., July 2, 1947; s. Peter John and Ann (Ventra) M.; m. Maia Doumato, Aug. 18, 1968; children: Peter Jerome, Victoria Ann. B.A., R.I. Coll., 1969; Ph.D., U. R.I., 1973. Postdoctoral fellow Roche Inst. Molecular Biology, Nutley, N.J., 1973-76; sr. staff fellow NIMH, Bethesda, Md., 1976-79, chief unit on neurochemistry, 1979—; adj. professorial lectr. George Washington U. Sch. Medicine. Editor: Methods in Neuroscience, 1983; Contb. articles to profl. jours. Recipient Bennett award in biol. psychiatry Soc. Biol. Psychiatry, 1980. Mem. AAAS, Am. Soc. Neurochemistyr, Soc. Neurosci. Subspecialties: Neuropharmacology; Neurochemistry. Current work: Brain specific proteins, neurotransmitters and drug receptors in brain. Mechanism of benzodiazepine action, adenosine receptors. Home: 18409 Kingshill Rd Germantown MD 20874 Office: 9000 Rockville Pike Bldg 10 Room 4S-239 Bethesda MD 20205

MARASH, STANLEY ALBERT, statistical consultant; b. N.Y.C., Dec. 18, 1938; s. Albert and Esther (Cunio) M.; m. Muriel Sylvia Sutchin, Dec. 9, 1941; children: Judith Ilene, Alan Scott. B.B.A., CCNY, 1961; M.B.A., Bernard M. Baruch Coll., 1970. Registered profl. engr., Calif. Statistician Gen. Dynamics/Electric Boat, Groton, Conn., 1961-62, Idaho Nuclear Energy Lab., Idaho Falls, 1962-63; statistician RCA Memory Products Operation, Needham, Mass., 1963-64, engr. quality assurance, 1964-65; cons. engr. (RCA Astro Electronics div.), Princeton, N.J., 1965-66; corp. mgr. quality assurance Ideal Corp., Bklyn., 1966-68; mgr. quality assurance Gen. Instrument SignaLite, Neptune, N.J., 1968; pres. Stat-A-Matrix Group, Edison, N.J., 1968—; p. Stat-A-Matrix, Internat., Inc., Edison, n.j., 1975—; trustee, mng. dir. Stat-A-Matrix Inst., 1975—; adj. prof., indsl. adv. com. Middlesex County Coll., Edison, N.J., 1971—; vis. prof. U. Sao Paulo, 1974, 75, 77, Madrid Poly U., 1975; mem. exec. standards council Am. Nat. Standards Inst. Author: Quality Assurance for the Nuclear Power Industry, 1972, Statistical Quality Control, 1972; editor: Industrial Quality Programs, 1976, Managing Quality Costs, 1975, Nuclear Quality Assurance, 1976, Auditing Nuclear Quality Assurance, 1975, (with Louis I. Korn) Reliability in Nuclear Power Generating Stations, 1974. Fellow Am. Soc. Quality Control (cert. quality engr., cert. reliability engr., Ellis R. Ott award 1981); mem. IEEE, Am. Statis. Assn., Am. Soc. Tng. and Devel., ASTM, ASME. Subspecialties: Statistics; Quality assurance. Office: 2124 Oak Tree Rd Edison NJ 08820

MARASHI, MUSA S., computer cons.; b. Nadjaf, Iraq, Oct. 27, 1947; s. Sayed-Hadi and Hakema M. B.S. in Computer Sci, Fairleigh Dickinson U., 1977, M.S., 1978. Cons. ITT, 1977-80, Merrill Lynch, 1977-80, Bell Labs, Holmdel, N.J., 1981—; part-time cons. Chem-Pro, Fairfield, N.J. Mem. Assn. Computing Machinery. Subspecialties: Database systems; Software engineering. Address: 52 Ellen Heath Dr Matawan NJ 07747

MARAVOLO, NICHOLAS C., biology educator; b. Chgo., Dec 4, 1940; s. Nicholas and Mary A. (Plonis) M. S.B., U. Chgo., 1962, S.M., 1964, Ph.D., 1966. Asst. prof. biology Lawrence U., Appleton, Wis., 1966-75, assoc. prof., 1975-83, prof., 1983—. Mem. Bot. Soc. Am., Am. Soc. Bryology and Lichenology, Am. Soc. Plant Physiology. Subspecialty: Plant growth. Current work: Developmental physiology of growth regulators in hepatics. Home: 845 E Alton St Appleton WI 54911 Office: Dept Biology Lawrence U PO Box 599 Appleton WI 54912

MARCATILI, ENRIQUE ALFREDO JOSÉ, electrical engineer; b. Cordoba, Argentina, Aug. 1, 1925; came to U.S., 1954, naturalized, 1961; s. Enrique S. and Aguinalda M.; m. Ana B. Wainfeld, Dec. 5, 1953; children: Viviana, Enrique Richard. B.S. in Aero. Engring., U. Cordoba, 1948. Elec. engr., 1949. Mem. tech. staff Bell Labs., Holmdel, 1954—, head coherent optics research dept., 1976—. Author. Fellow IEEE (co-recipient W.R.G. Baker prize 1975); mem. Nat. Acad. Engring. Subspecialties: Electronics; Microelectronics. Current work: Electromagnetism; integrated optics on semiconductors (photonics); optical communications. Patentee in field. Home: 2 Markwood Ln Rumson NJ 07760 Office: Crawford Hill Lab Holmdel NJ 07733

MARCHALONIS, JOHN JACOB, biochemistry adminstr., researcher; b. Scranton, Pa., July 22, 1940; s. John Louis and Anna Irene (Stadner) M.; m. Anne Baldwin Caldwell, Aug. 16, 1969 (div. 1977); 1 dau., Lee Celia; m. Sally Ann Sevy, May 5, 1978; 1 dau., Elizabeth Ann. A.B., Lafayette Coll., 1962; Ph.D., Rockefeller U., 1967. Postdoctoral fellow Walter and Eliza Hall Inst., Victoria, Australia, 1967-69; cons. in immunology Miriam Hosp., Providence, 1969-70; asst. prof. Brown U., Providence, 1969-70; head lab. molecular immunology Walter & Eliza Hall Inst., Victoria, 1970-76, asst. prof., 1970-73, assoc. prof., 1973-76; adj. prof. U. Pa., Phila., 1977—; head cell biology Frederick Cancer (Md.) Cancer Research Ctr., 1977-79; prof. and chmn. Med. U. S.C., Charleston, 1980—; mem. allergy and immunology study sect. NIH, Washington, 1979-83; dir. Am. Type Tissue Culture Collection, 1983-85. Author: Immunity in Evolution, 1977; editor: Comparative Immunology, 1976, The Lymphocyte, 1977, (with N. Cohen) Seof-Non-Seof Discrimination, 1980; editorial bd.: Jour. Immunogenetics, 1976—, Devel. and Comparative Immunology, 1977—, Contemporary Topics in Immunobiology, 1978—, Proc. Soc. Exptl. Biology in Medicine, 1982—; adv. bd.: Quar. Rev. Biology, 1978—; editor: (with G. W. Warr) Antibody as Tool, 1982; editor-in-chief: Cancer Biology Revs, 1980, 81, 82; contbr. numerous articles to med. jours. Frank R. Lillie fellow, 1965-70; NIH grantee, 1981-84, 1973-76, 1975-76; Am. Cancer Soc. grantee, 1980; fellow, 1977-78; Am. Heart Assn. grantee, 1974-76; Damon Runyon-Walter Winchell Cancer Fund grantee, 1975. Mem. Am. Soc. Zoologists (chmn. div. comparative immunology 1981-82), Am. Assn. Immunologists, Am. Soc. Biol. Chemists, Sigma Xi. Episcopalian. Subspecialties: Biochemistry (medicine); Immunology (medicine). Current work: Molecular basis of cellular recognition and differentiation. Home: 986 Casseque Province Mount Pleasant SC 29464 Office: Med U SC 171 Ashley Ave Charleston SC 29425

MARCHATERRE, JOHN FREDERICK, nuclear engineer; b. Hermansville, Mich., Aug. 28, 1932; s. Frederick Joseph and Louise

Marie (Boucher) M.; m. Joan Patricia Konvalinka, Sept. 14, 1957; children: Mary, Ann, Madeleine. B.S.Ch.E., Mich. Tech. U., 1954, M.S.Ch.E., 1956; M.S.Ch.E. Engr, 1960; M.B.A., U. Chgo., 1980. Registered profl. engr., Ill. Asst. engr. Argonne Nat. Lab., Ill., 1955-58, group leader, 1956-62, project mgr., 1962-69, assoc. div. dir., reactor analysis and safety div., 1969—; invited lectr. Advanced Summer Inst., Kjeller, Norway, 1959; tech. advisor City of Chgo., 1966-70; cons. Adv. Com. Reactor Safeguards, 1980—; mem. nuclear engring. edn. com. U. Chgo., 1976—. Mem. elec. adv. bd. City of Naperville, 1972-80; precinct committeeman, DuPage County, Ill., 1970-78. Mem. Am. Nuclear Soc., Sigma Xi. Republican. Roman Catholic. Subspecialties: Nuclear engineering; Chemical engineering. Current work: Heat transfer and hydrodynamics; nuclear reactor safety. Office: 9700 S Cass Ave Bldg 208 Argonne IL 60439 Home: 1209 Atlas Ln Naperville IL 60540

MARCHETTA, FRANK CARMELO, surgeon, cancer researcher, educator; b. Utica, N.Y., Apr. 28, 1920; s. Donato Anthony and Teresa Maria (Romano) M.; m. Jean Elenor Cile, Apr. 30, 1949; children: Linda Marchetta Wild, Charles, Joanne. Student, Cornell U., 1938-41; M.D., U. Buffalo, 1944. Diplomate: Am. Bd. Surgery. Intern Deaconess Hosp., Buffalo, 1944-45, resident in surgery, 1947-50, Roswell Park Meml. Inst., Buffalo, 1950-51, assoc. cancer research head and neck surgeon, 1951-54, assoc. chief cancer research head and neck surgery, 1954-56, chief cancer research head and neck surgeon sect. A, 1956-75, cons., 1975—; assoc. clin. prof. head and neck surgery Dental Sch. SUNY-Buffalo, 1958—, 1962—; cons. VA Hosp., Buffalo. Contbr. articles to sci. jours. Served to capt. M.C. AUS, 1945-47. Recipient award Roswell Park Meml. Inst., 1975. Mem. N.Y. State Med. Soc., New York County Med. Soc., ACS, Soc. Head and Neck Surgeons (sec.), Soc. Surg. Oncology, Am. Assn. for Cancer Research. Subspecialties: Surgery; Oncology. Current work: Clinical research in head and neck tumors; laboratory experiments related to problems in head and neck surgery. Home: 192 High Park Blvd Eggertsville NY 14226 Office: 2804 Main St Buffalo NY 14214

MARCHETTE, NYVEN JOHN, microbiologist, researcher, educator; b. Murphys, Calif., June 26, 1928; s. Victor Joseph and Marie Anderson (Hacker) M.; m. Elaine Marie, June 11, 1950; 1 son, David Joseph. B.A., U. Calif.-Berkeley, 1950, M.A., 1953; Ph.D., U. Utah, 1960. Postdoctoral fellow U. Calif., San Francisco, 1961-63; asst. research microbiologist, 1963-69, assoc. research microbiologist, 1969-70; assoc. prof. tropical medicine U. Hawaii, 1970-74, prof. tropical medicine and public health, 1974—. Author: Ecological Relationships and Evolution of the Rickettsiae, vols. I and II, 1982. Fogarty sr. internat. fellow John Curtin Sch. Med. Research, Canberra, Australia, 1977-78. Mem. Am. Soc. Microbiology, AAAS, Am. Soc. Trop. Medicine. Subspecialties: Microbiology (medicine); Infectious diseases. Current work: Pathogenesis, ecology, epidemiology of infectious diseases; Kawasaki disease, sexually transmitted diseases, etiology, virology, Rickettsiology, pathogenesis, epidemiology. Home: 211 Kuukama St Kailua HI 96734 Office: 3675 Kilauea Ave Honolulu HI 96816

MARCHIN, GEORGE L., microbiologist, educator; b. Kansas City, Kans., July 12, 1940; s. George Leonard and Ann Maria (Hanis) M.; m. Anne S., July 20, 1974; children: Melissa Ann, Katherine Louise, Madelaine Christine. A.B., Rockhurst Coll., 1962; Ph.D., U. Kans., 1967. Grad. teaching asst., grad. research asst. U. Kans. Med. Center, Kansas City, 1962-67; research assoc. Purdue U., West Lafayette, Ind., 1967-70; asst. prof. Kans. State U., Manhattan, 1970-77, assoc. prof. microbiology, 1978—; researcher bacteriophage T4, giardiasis; vis. scientist, Swedish exchange grantee Umea (Sweden) Universitet, 1977-78. Contbr. articles in field to profl. publs. NIH postdoctoral fellow, 1968-70; NIH research grantee, 1972-75, 75-78. Mem. AAAS, Am. Soc. Microbiology. Democrat. Roman Catholic. Club: Optimist. Lodge: Jednota Lodge. Subspecialties: Microbiology; Molecular biology. Current work: Modification of protein synthesis in Escherichia coli by bacteriophage T4; inactivation and pathogenesis of giardia lamblia. Home: Route 4 Box 43 Driftwood Estates Manhattan KS 66502 Office: Div of Biolog Kans State U Manhattan KS 66506

MARCOULLIS, GEORGE PANAYIOTIS, physician, biochemist, medical educator, biochemical science researcher; b. Limassol, Cyprus, Apr. 4, 1949; s. Panayiotis Stylianou and Aggeliki (Joannou) M.; m. Erato Kozakou, Aug. 20, 1970; 1 son, Panos. M.D., Athens U., 1974; Ph.D., Helsinki U., 1978. Lic. in medicine, Greece, Finland. Jr. researcher Athens U., 1971-73; biomed. researcher Minerva Inst., Helsinki, 1973-77; mem. staff Maria Hosp., Helsinki, 1976-77; asst. prof. medicine N.Y. Med. Coll., 1977-79; vis. asst. prof. U. Nancy (France) Faculty of Scis., 1979-80; vis. assoc. prof. Nancy Med. Sch., 1981-82; assoc. prof. medicine SUNY-Downstate Med. Center, Bklyn., 1980-83; med. resident Washington VA Hosp., 1983; chief med. resident Mt. Vernon Hosp., (N.Y.), 1983-84; fellow in hematology N.Y. Med. Coll., 1984. Author: (with others) Contemporary Issues in Clinical Nutrition, 1983, Progress in Gastroenterology, Vol. IV, 1983; Contbr. articles to profl. jours. Recipient medal award Nancy U., 1980; grantee Athens U., 1971-74, WHO, 1975-77, INSERM, 1977-79, NIH, 1979-82, VA, 1982—. Mem. Am. Fedn. Clin. Research, Soc. for Exptl. Biology and Medicine. Subspecialties: Biochemistry (biology); Hematology. Current work: Use of protein chemistry techniques to delineate the structure and role of membrane receptors and transport proteins in cellular recognition, in hormone, drug and nutrient metabolism, and in receptor-related diseases. Home: 453 E 14th St Apt 11C New York NY 10009 Office: SUNY Downstate Med Ctr and VA Med Ctr 800 Poly Pl Brooklyn NY 11209

MARCUS, ELLIOT MEYER, neurologist, educator; b. Bridgeport, Conn., Sept. 13, 1932; s. Kalman and Bessie Edith (Schefkind) M.; m. R. Nuran Turksoy, Aug. 29, 1962; 1 dau., Erin. B.A magna cum laude, Yale U., 1954; M.D., Tufts U., 1958. Diplomate: Am. Bd. Psychiatry and Neurology. Intern Yale U., New Haven, 1958-59; clin. fellow in neurology Tufts New Eng. Med. Ctr., Boston, 1959-60; fellow in neuropathology Columbia Presbyn. Hosp., N.Y.C., 1960-61; chief resident neurology Tufts New England Med. Center, Boston, 1961-62, NIH fellow in neurophysiology, 1962-63, mem. faculty, 1965-76, prof. neurology, 1975-76, sr. neurologist, 1975-76; dir. neurology service St. Vincent Hosp., Worcester, Mass., 1976—; prof. neurology U. Mass., Worcester, 1976—; lectr. Tufts U., 1976—; Bd. dirs., mem. med. adv. bd. Mass. M.S. Soc. Author: (with Curtis and Jacobson) An Introduction to the Neurosciences, 1972; contbr. articles to prefl. jours. Served to capt. M.C. AUS, 1963-65. NIH grantee, 1966-71. Mem. Mass. Neurol. Assn. (1st pres. 1979), Am. Acad. Neurology, Am. Neurol. Assn., Am. EEG Soc., Stroke Council, AMA, Soc. Neurosci. Clubs: Bristol Yacht, Yale of Boston. Subspecialties: Neurology; Neurophysiology. Current work: Coma Epilepsy, EEG, evoked potentials, neuroanatomical basis of epilepsy; role of corpus callosum in explt. epilepsy. Office: St Vincent Hosp Worcester MA 01604

MARCUS, MELVIN GERALD, physical geographer, educator; b. Seattle, Apr. 13, 1929; s. Albert Joseph and Lucille (Plumm) M.; m. (married), June 6, 1953; children: William Andrew, Annette Allen, Alison Leigh, Benjamin Logan. Student, Yale U., 1947-50; B.A., U. Miami, Coral Gables, Fla., 1956; M.A., U. Colo., 1957; Ph.D., U. Chgo., 1963. With Lab. Climatology, Thorthwaite Assocs., Centerton, N.J., 1959-60; mem. faculty Rutgers U., 1961-64, U. Mich., 1964-73, Ariz. State U., Tempe, 1974—, prof. geography, 1969—; mem. Ariz. Climate Com. Contbr. articles to profl. jours. Dir. Yosemite Inst. Served to 1st lt. USAF, 1951-55. Mem. Gov.'s Commn. on Ariz. Environment., AAAS, Assn. Am. Geographers, Am. Geog. Soc. (council), Ariz. Acad. Sci., Internat. Glaciological Soc. Subspecialties: Climatology; Glaciology. Current work: Alpine environments, climatology, glaciology.

MARCUS, RUDOLPH ARTHUR, chemist; b. Montreal, Que., Can., July 21, 1923; came to U.S., 1949, naturalized, 1958; s. Myer and Esther (Cohen) M.; m. Laura Hearne, Aug. 27, 1949; children: Alan Rudolph, Kenneth Hearne, Raymond Arthur. B.S., McGill U., 1943, Ph.D., 1946; D.Sc. (hon.), U. Chgo., 1983. Postdoctoral research asso. NRC of Can., Ottawa, 1946-49, U. N.C., 1949-51; asst. prof. Poly. Inst. Bklyn., 1951-54, asso. prof., 1954-58, prof., 1958-64, U. Ill., Urbana, 1964-78; Arthur Amos Noyes prof. chemistry Calif. Inst. Tech., Pasadena, 1978—; temp. mem. Courant Inst. Math. Scis., N.Y. U., 1960-61; trustee Gordon Research Confs., 1966-69, chmn. bd., 1968-69, mem. council, 1965-68; mem. rev. panel Argonne Nat. Lab., 1966-72, chmn., 1967-68; mem. rev. panel Brookhaven Nat. Lab., 1971-74; mem. adv. council in chemistry Princeton U., 1972-78; mem. rev. com. Radiation Lab., U. Notre Dame, 1975-80; mem. panel on atmospheric chemistry climatic impact com. Nat. Acad. Scis.-NRC, 1975-78, mem. com. kinetics of chem. reactions, 1973-77, chmn., 1975-77, mem. com. chem. scis., 1977-79; vis. com. div. chemistry and chem. engring. Calif. Inst. Tech., 1977-78; adv. council chemistry Poly. Inst. N.Y., 1977-80; adv. com. for chemistry NSF, 1977-80; vis. prof. theoretical chemistry U. Oxford, Eng., 1975-76; also professorial fellow Univ. Coll. Mem. editorial bd.: Jour. Chem. Physics, 1964-66, Ann. Rev. Phys. Chemistry, 1964-69, Jour. Phys. Chemistry, 1968-72, 80—, Accounts of Chem. Research, 1968-73, Internat. Jour. Chem. Kinetics, 1976—, Molecular Physics, 1977-80, Chem. Physics Letters, 1980—, Laser Chemistry, 1982—; contbr. articles to profl. jours. Recipient Anne Molson prize in chemistry McGill U., 1943, Alexander von Humboldt Found. Sr. U.S. Scientist award, 1976, Robinson medal Faraday div. Royal Soc. Chemistry, 1982; Alfred P. Sloan fellow, 1960-63; NSF sr. postdoctoral fellow, 1960-61; sr. Fulbright-Hays scholar, 1972. Fellow Am. Acad. Arts and Scis. (exec. com. Western sect., Co-chmn. 1981—); mem. Nat. Acad. Scis., Am. Chem. Soc. (past div. chmn., mem. exec. com., mem. adv. bd. petroleum research fund, Irving Langmuir award Chem. Physics 1978), Am. Phys. Soc. (exec. com. div. chem. physics), Electrochem. Soc., AAUP, Alpha Chi Sigma. Subspecialties: Kinetics; Theoretical chemistry. Current work: Theories of electron transfer reactions, unimolecular reactions, and intra-molecular dynamic. Home: 331 S Hill Ave Pasadena CA 91106

MARCUS, STEVE IRL, electrical engineering educator; b. St. Louis, Apr. 2, 1949; s. Herbert A. and Peggy L. (Polishuk) M.; m. Jeanne M. Wilde, June 4, 1978; 1 son, Jeremy A. B.A., Rice U., 1971; S.M., MIT, 1972, Ph.D., 1975. Registered profl. engr., Tex. Research engr. The Analytic Scis. Corp., Reading, Mass., 1973; asst. prof., U. Tex-Austin, 1975-80, assoc. prof. elec. engring., 1980—; cons. Tracor, Inc., Austin, 1977. Author: Nonlinear Estimation, 1984; regional co-mng. editor: Acta Applicandae Mathematicae, 1982—; contbr. articles in field to profl. jours. NSF fellow, 1971-74; Werner N. Pornberger Centennial teaching fellow U. Tex.-Austin, 1982. Mem. IEEE (chmn. conf. on decision and control program com. 1983, assoc. editor transactions on automatic control 1980-81), Am. Math. Soc., Soc. Indsl. and Applied Math., Ops. Research Soc. Am., Eta Kappa Nu, Tau Beta Pi. Jewish. Subspecialties: Systems engineering; Electrical engineering. Current work: Teaching and research in analysis and estimation techniques for nonlinear stochatic systems, stochastic control theory, and geometric methods in systems. Home: 1703 W 32d St Austin TX 78703 Office: Dept Elec Engring U Tex Austin TX 78712

MARCUSE, DIETRICH, physicist; b. Koenigsberg, Germany, Feb. 27, 1929; s. Richard and Gertrud (Solty) M.; m. Haide Schwarz, Sept. 18, 1934; children: Christina, Mikel. Diplom physiker, Freie Universitaet Berlin, 1954; Doktor Ingenieur, Universitaet Karlsruhe, Germany, 1962. Physicist Central Lab. Siemens und Halske, Berlin, 1954-57; mem. tech. staff Bell Labs., Holmdel, N.J., 1957—. Author 4 books; contbr. articles to profl. jours. Fellow Optical Soc. Am., IEEE (Quantum Electronics award 1981). Subspecialties: Fiber optics; Electrical engineering. Current work: Fiber optic communications theory; semiconductor lasers. Patentee in field. Office: Bell Labs Holmdel NJ 07733

MARCUVITZ, NATHAN, educator; b. Bklyn., Dec. 29, 1913; s. Samuel and Rebecca (Feiner) M.; m. Muriel Spanier, June 30, 1946; children—Andrew, Karen. B.E.E., Poly. Inst. Bklyn., 1935, M.E.E., 1941, D.E.E., 1947. Engr. RCA Labs., 1936-40; research asso. Radiation Lab., Mass. Inst. Tech., 1941-46; asst. prof. elec. engring. Poly. Inst. Bklyn., 1946-49, asso. prof., 1949-51, prof., 1951-65, dir., 1957-61, v.p. research, acting dean, 1961-63, prof. electrophysics, 1961-66, dean research, dean, 1964-65; asst. dir. def. research and engring. Dept. Def., Washington, 1963-64; prof. applied physics N.Y.U., 1966-73; prof. electrophysics Poly. Inst. N.Y., 1973—, inst. prof., 1978—; vis. prof. Harvard U., spring 1971. Author: Waveguide Handbook, Vol. 10, 1951, (with L. Felsen) Radiation and Scattering of Waves, 1973; also numerous articles. Fellow IEEE; mem. Nat. Acad. Engring., Am. Phys. Soc., Sigma Xi, Tau Beta Pi, Eta Kappa Nu. Subspecialties: Electrical engineering; Plasma physics. Home: 7 Ridge Dr E Great Neck NY 11021 Office: Poly Inst NY Grad Center Rt 110 Farmingdale NY 11735

MARCZYNSKI, THADDEUS JOHN, pharmacologist, educator; b. Poznan, Poland, Nov. 30, 1920; came to U.S., 1964, naturalized, 1973; s. John Adam and Wanda (Sielski) M.; m. Barbara Konieczna, Oct. 10, 1956; children: John T., Gregory T. M.D., U. Cracow, 1951, Ph.D. in Pharmacology, 1959. Asst., sr. asst., adj. dept. pharmacology Sch. Medicine, Cracow (Poland) U., 1954-64; asst. prof. dept. pharmacology U. Ill, Chgo., 1964-69, asso. prof., 1969-73, prof, 1973—, prof. depts. psychiatry and bioengring., 1974—. Contbr. numerous articles on physiol. and pharm. aspects of brain function, brain research, EEG and clin. neurophysiology to profl. jours. Rockefeller Found. fellow Brain Research Inst., UCLA, 1961-62; NIH research grantee, 1966-71. Mem. Soc. for Neurosci., Fedn. Am. Socs. for Exptl. Biology. Republican. Roman Catholic. Subspecialties: Neurophysiology; Neuropharmacology. Current work: Stochastic models of neuronal firing patterns; information processing in the mammalian brain; mechanisms of learning and behavior; mechanisms of positive reinforcement and learning; pharmacology of learning and behavior. Home: 1217 Hinman Ave Evanston IL 60202 Office: Dept Pharmacology U Ill Coll Medicine 835 S Wolcott Ave Chicago IL 60612

MARDER, BARRY M(ICHAEL), mathematician; b. Newark, Oct. 31, 1942; s. Seymour and Beatrice (Lehner) M.; m. Ellen Macy, June 14, 1964 (div. 1977); children: Wendy, Dinah; m. Kathy Sapp, Mar. 7, 1982. B.S., Stevens Inst. Tech., 1964; Ph.D., NYU, 1969. Staff mem. Los Alamos Sci. Lab., 1969-75; guest researcher Max Planck Institut für Plasmaphysik, Garching, W.Ger., 1975; staff mem. Sandia Labs., Albuquerque, 1976—. Mem. Am. Assn. Artificial Intelligence. Subspecialties: Applied mathematics; Artificial intelligence. Current work: Artificial intelligence. Home: 7212 Gladden Ave Albuquerque NM 87110 Office: Sandia Nat Labs Albuquerque NM 87115

MARDER, EVE E., neurobiology educator; b. N.Y.C.; d. Eric and Dorothy (Silverman) M. A.B., Brandeis U., 1969; Ph.D., U. Calif.-San Diego, 1974. Postdoctoral fellow Laboratories de Neurobiologie, Ecole Normale Superieure, Paris, 1975-78; asst. prof. biology Brandeis U., 1978—. McKnight Found. Scholar's awardee, 1978; Sloan Found. fellow, 1980-84. Mem. Soc. Neuroscis., AAAS. Subspecialties: Neurophysiology; Neuropharmacology. Current work: Mechanisms of synpatic transmission and pattern generation. Office: Dept Biology Brandeis U Waltham MA 02254

MARENGO, NORMAN PLAYSON, biology educator; b. N.Y.C., Feb. 21, 1913; s. Frederick and Edith Lorimer (Fairchild) M.; m. Mildred McAusland, Nov. 23, 1939 (dec. 1976); children: Robert F., Mary Kalb. B.S., NYU, 1936; M.A., 1939, M.S. in Biology, 1942, Ph.D., 1949. Instr. NYU, 1939-43; instr., then asst. prof. Lafayette Coll., Easton, Pa., 1948-50; asst. prof. Hofstra Coll., Hempstead, N.Y., 1950-54; tchr. sci. Central High Sch. Dist., Merrick, N.Y., 1954-55; asst. prof. biology C.W. Post Coll., L.I.U., Greenvale, N.Y., 1955-57, assoc. prof., 1957-60, prof., 1960-82, adj. prof., 1982—. Contbr. articles to profl. jours. Served to 1st lt. AUS, 1943-46. Mem. Bot. Soc. Am., Torrey Bot. Club, Am. Fern Soc., Sigma xi. Republican. Subspecialties: Cell biology; Animal genetics. Current work: Ultrastructural studies of fern spores and sporogenesis. Home: 123 Leverich St Hempstead Ny 11550 Office: C W Post Coll of LIU Greenvale NY 11548

MARGERUM, DALE WILLIAM, chemistry educator; b. St. Louis, Oct. 20, 1929; s. Donald C. and Ida Lee (Nunley) M.; m. Sonya Lora Pedersen, May 16, 1953; children: Lawrence Donald, Eric William, Richard Dale. B.A., S.E. Mo. State U., 1950; Ph.D., Iowa State U., 1955. Research chemist Ames Lab., AEC, Iowa, 1952-53; instr. Purdue U., West Lafayette, Ind., 1954-57, asst. prof., 1957-61, asso. prof., 1961-65, prof., 1965—, head dept. chemistry, 1978-83; phys. chemist, vis. scientist Max Planck Inst., 1963, 70; vis. prof. U. Kent, Canterbury, Eng., 1970; mem. med. chem. study sect. NIH, 1965-69; mem. adv. com. Research Corp., 1973-78; mem. chemistry evaluation panel Air Force Office Sci. Research, 1978-82. Cons. editor: McGraw Hill, 1962-72; editorial bd.: Jour. Coordination Chemistry, 1971-81, Analytical Chemistry, 1967-69. Recipient Grad. Research award Phi Lambda Upsilon, 1954; Sigma Xi research award Purdue U., 1973; Herbert C. McCoy research award Perdue U., 1983; NSF sr. postdoctoral fellow, 1963-64. Fellow AAAS; mem. Am. Chem. Soc. (chmn. Purdue sect. 1965-66), AAUP, Sigma Xi, Phi Lambda Upsilon. Subspecialty: Inorganic chemistry. Office: Dept Chemistry Purdue U West Lafayette IN 47907

MARGOLIASH, EMANUEL, biochemist, educator; b. Cairo, Feb. 10, 1920; s. Wolf and Bertha (Kotler) M.; m. Sima Beshkin, Aug. 22, 1945; children: Reuben, Daniel. B.A., Am. U., Beirut, 1940, M.A., 1942, M.D., 1945. Research fellow, lectr., acting head cancer research labs. Hebrew U., Jerusalem, 1945-58; research asso. U. Utah, Salt Lake City, 1958-60, McGill U., Montreal, Que., Can., 1960-62; research fellow Abbott Labs., North Chicago, Ill., 1962-69, sr. research fellow, 1969-71, head protein sect., 1962-71; prof. biochemistry and molecular biology Northwestern U., Evanston, Ill., 1971—; mem. com. on cytochrome nomenclature Internat. Union Biochemistry, 1962—; mem. adv. com. Plant Research Lab., Mich. State U./AEC, 1967-72; co-chmn. Gordon Research Conf. on Proteins, 1967. Editorial bd.: Jour. Biol. Chemistry, 1966-72, Biochem. Genetics, 1966—, Jour. Molecular Evolution, 1971—; contbr. articles and revs. to sci. jours. Rudi Lemberg fellow Australian Acad. Sci., 1981; Guggenheim fellow, 1983. Fellow Am. Acad. Arts and Scis.; mem. Nat. Acad. Scis., Biochem. Soc. (Keilin Meml. lectr. 1970), Harvey Soc. (lectr. 1970-71), Am. Soc. Biol. Chemists (publs. com. 1973-76), Am. Chem. Soc., Can. Biochem. Soc., Soc. Developmental Biology, Biophys. Soc. (exec. com. U.S. bioenergetics group 1980—), N.Y., Ill. acads. scis., Am. Soc. Naturalists, Sigma Xi (nat. lectr. 1972-73, 74-77). Subspecialties: Biochemistry (biology); Molecular biology. Current work: Structure-function relations, molecular biology and evolution of respiratory proteins; cytochrome as a model antigen. Home: 353 Madison Ave Glencoe IL 60022 Office: Dept Biochemistry Molecular Biology and Cell Biology Northwestern U Evanston IL 60201

MARGOLIN, PAUL, geneticist, microbiology educator; b. N.Y.C., Aug. 31, 1923; s. Simon and Elizabeth (Offenhenden) M.; m. Jean Spielberg, May 19, 1946. Student, U. Wash., 1940-41, NYU, 1941-42, U. Miss., 1943; B.S., NYU, 1947; Ph.D., Ind. U., 1955. Research fellow Calif. Inst. Tech., 1955-56, Edinburgh (Scotland) U., 1957-58; sr. staff investigator Cold Spring Harbor (N.Y.) Lab., 1958-66; mem., chief dept. genetics Pub. Health Research Inst. City of N.Y., Inc., 1966—; research prof. microbiology NYU Sch. Medicine, 1966—; mem. adv. com. etiology of cancer Am. Cancer Soc., 1969-73; mem. planning group for nat. meeting of biosafety com. chairpersons NIH, 1980. Contbr. numerous articles to sci. jours., books. Trustee Cold Spring Harbor Lab., 1967-68. Ind. U. fellow, 1955; Caltech fellow, 1956; NIH Postdoctoral fellow, 1957-58; NSF, NIH research grantee, 1961—; NIH Career Devel. awardee, 1961-65. Mem. Genetics Soc. Am., Am. Soc. Microbiology, AAAS, Sigma Xi. Subspecialties: Gene actions; Molecular biology. Current work: Research on regulation of gene expression; mutation mechanisms. Home: 100 Bleecker St Apt 10B New York NY 10012 Office: 455 1st Ave New York NY 10016

MARGOLIS, FRANK LEONARD, research biochemist; b. Bklyn., Jan. 21, 1938; m. Joyce Weisman; children: Jonathan, Peter, Michael. B.S., Antioch Coll., 1964; Ph.D. in Biochemistry, Columbia U., 1964. Asst. research microbiologist UCLA Sch. Medicine, 1966-69; research assoc. Roche Inst. Molecular Biology, Nutley, N.J., 1969-71, asst. mem., 1971-73, assoc. mem., 1974-80, mem., 1981—; adj. prof, grad. program in biochemistry CUNY, 1972—. Contbr. papers, chpts and abstracts to profl. lit. Mem. AAAS, Am. Soc. Biol. Chemists, Am. Soc. Pharmacology and Exptl. Therapeutics, Am. Soc. Neurochemistry, Internat. Soc. Neurochemistry, Soc. Neurosci., Sigma Xi, Phi Lambda Upsilon. Subspecialties: Biochemistry (medicine); Neurochemistry. Current work: Biochemistry of olfaction, neurochemistry, regulation of mammalian gene expression. Office: Roche Inst Molecular Biology Nutley NJ 07110

MARGOLIS, SAM AARON, biochemist, educator; b. Cambridge, Mass., Nov. 17, 1933; s. Abraham and Rose (Huberman) M.; m. Evelyn Gertrude Meltzer, Aug. 7, 1960; children: Laura Beth, Daniel Abraham. A.B., Boston U., 1955; Ph.D., 1963; M.S., U. R.I., 1957; postgrad., Yale U., 1957-59. Fellow Enzyme Inst., U. Wis., 1964-66; pharmacologist FDA, Washington, 1966-68; microbiologist NIH, 1968-70, Nat. Cancer Inst., 1970-72; research chemist Nat. Bur. Standards, Washington, 1972—; instr. pharmacology and biochemistry U.S. Dept. Agr. Contbr. articles to profl. jours. Mem. Am. Chem. Soc., Am. Soc. Microbiology, N.Y. Acad. Sci., Am. Assn. Clin. Chemistry, Sigma Xi. Subspecialties: Biochemistry (medicine); Analytical chemistry. Current work: Methods development and preparation of standards for clinical chemistry and food analysis (nutrients) using HPLC, NMR. amino acid analysis and other analytical techniques. Office: National Bureau Standards Bldg 222 Washington DC 20234

MARGOLIUS, HARRY STEPHEN, pharmacologist, physician, educator, cons.; b. Albany, N.Y., Jan. 29, 2938; s. Irving and Betty (Zweig) M.; m. Francine Rockwood, May 22, 1964; children:

Elizabeth Anne, Craig Matthew. B.S., Union U., 1959; Ph.D., Albany Med. Coll., 1963; M.D., U. Cin., 1968. Diplomate: Nat. Bd. Med. Examiners. NIH postdoctoral fellow U. Cin., 1963-66, instr., 1967; USPHS fellow NIH, Bethesda, Md., 1967; intern, med. resident, clin. fellow in medicine Harvard U. Med. Sch., Boston City Hosp., 1968-70; pharmacology research asso. Nat. Heart, Lung and Blood Inst., NIH, Bethesda, 1970-72, sr. clin. invstigator, 1972-74; asso. prof. pharmacology, asst. prof. medicine Med. U. S.C., Charleston, 1974-77, prof. pharmacology, 1977—, asso. prof. medicine, 1977-80, prof. medicine, 1980—; program dir. Gen. Clin. Research Center, 1974-81; vis. scholar dept. pharmacology Cambridge (Eng.) U., 1980-81; attending staff Med. Univ. Hosp., VA Hosp., Charleston Meml. Hosp., St. Francis Hosp.; mem. research grant rev. com. NIH, NSF, 1976-80, FDA, 1982-86. Contbr. articles to sci. jours., chpts. to books. Served to sr. surgeon USPHS, 1970-74. NIH grantee, 1975—; Burroughs Wellcome scholar, 1976-81; Pfizer lectr., 1976. Mem. AAAS, Am. Soc. Pharmacology and Exptl. Therapeutics, Am. Soc. Clin. Pharmacology and Therapeutics, Am. Fedn. Clin. Research, Am. Soc. Clin. Investigation, So. Soc. Clin. Investigation, Council High Blood Pressure Research of Am. Heart Assn., Alpha Omega Alpha, Pi Kappa Epsilon. Jewish. Club: Hobcaw Yacht. Subspecialties: Cellular pharmacology; Cell biology (medicine). Current work: Kallikreins, kinins and membrane ion transport in epithelial cells (gastrointestinal, renal). Kallikrein synthesis and activity in cells, organs and secretions of amphibians and mammals; clinical investigation in hypertension, renal, cardiovascular, and diabetic diseases. Home: 645 Molasses Ln Mount Pleasant SC 29464 Office: 171 Ashley Ave Charleston SC 29425

MARGON, BRUCE H., astronomer, educator; b. N.Y.C., Jan. 7, 1948; s. Leon and Maxine E. (Margon) Siegelbaum; m. Carolyn J., May 8, 1976; dau., Pamela. A.B., Columbia U., 1968; M.A., U. Calif.-Berkeley, 1971, Ph.D., 1973. NATO fellow Univ. Coll. London, 1973-74; mem. faculty U. Calif.-Berkeley, 1974-76, UCLA, 1976-80; chmn., prof. astronomy U. Wash., Seattle, 1980—; bd. dirs. Associated Univs. for Research in Astronomy, Inc. Contbr. research articles to profl. jours. Sloan research fellow, 1979—. Mem. Am. Astron. Soc. (Pierce prize 1981), Internat. Astron. Union, Am. Phys. Soc., Royal Astron. Soc. Subspecialties: 1-ray high energy astrophysics; Optical astronomy. Current work: X-ray and ultraviolet astronomy; optical observations of X-ray source counterparts. Office: Astronomy Dept FM-20 U Wash Seattle WA 98195

MARGRAVE, JOHN LEE, college administrator, chemist; b. Kansas City, Kans., Apr. 13, 1924; s. Orville Frank and Bernice J. (Hamilton) M.; m. Mary Lou Davis, June 11, 1950; children: David Russell, Karen Sue. B.S. in Engring. Physics, U. Kans., 1948, Ph.D. in Chemistry, 1950. AEC postdoctoral fellow U. Calif. at Berkeley, 1951-52; from instr. to prof. chemistry U. Wis., Madison, 1952-63; prof. chemistry Rice U., 1963—, chmn. dept., 1967-72, dean advanced studies and research, 1972-80, v.p., 1980—; Reilly lectr. Notre Dame, 1968; vis. distinguished prof. U. Wis., 1968, U. Iowa, 1969, U. Colo., 1975, Ga. Inst. Tech., 1978, U. Tex. at Austin, 1978, U. Utah, 1982; Seydel-Wooley lectr. Ga. Inst. Tech., 1970; Dupont lectures U. S.C., 1971; Abbott lectures U. N.D., 1972; Cyanamid lectures U. Conn., 1973; Sandia lectures U. N.Mex., 1981; NSF-Japan Joint Thermophys. Properties Symposium, 1983, cons. to govt. and industry, 1954—; pres. Mar Chem., Inc., 1970—, High Temperature Sci., Inc., 1976—; Dir. Rice Design Center, Houston Area Research Ctr.; v.p. U. Kans. Research Found., Gulf Univs. Research Consortium, Energy Research and Edn. Found., 1972-78. Contbr. profl. jours.; Contbg. editor: Characterization of High Temperature Vapors, 1967, Mass. Spectrometry in Inorganic Chemistry, 1968; editor: High Temperature Science, 1969—, Procs. XXIII and XXIV Confs. on Mass Spectrometry, 1975, 76. Served with AUS, 1943-46; capt. Res. Sloan research fellow, 1957-58; Guggenheim fellow, 1960; recipient Kiekhofer Teaching award U. Wis., 1957; IR-100 award for CFX lubricant powder, 1970; Tex. Honor Scroll award, 1978; Disting. Alumni citation U. Kans., 1981. Fellow Am. Inst. Chemists, Tex. Acad. Sci., Am. Phys. Soc.; mem. Nat. Acad. Sci., Am. Chem. Soc. (Inorganic Chemistry award 1967, Southwest Regional award 1973, Fluorine Chemistry award 1980), Am. Ceramic Soc., AAAS, Electrochem. Soc., Chem. Soc. (London), Tex. Philos. Soc., Am. Soc. Mass Spectrometry (dir.), Sigma Xi, Omicron Delta Kappa, Sigma Tau, Tau Beta Pi, Alpha Chi Sigma. Methodist. Subspecialties: High temperature chemistry; Inorganic chemistry. Current work: Chemistry under extreme condition-high temperatures, low temperatures, high pressures, high reactivity, and microgravity. Patentee in field. Home: 5012 Tangle Ln Houston TX 77056

MARGULIES, ROBERT ALLAN, naval officer, physician, educator; b. Jersey City, May 29, 1942. M.D., Coll. Medicine and Dentistry N.J.-Newark, 1969; M.P.H. in Environ. Medicine, Johns Hopkins U., 1978. Diplomate: Am. Bd. Preventive Medicine. Commd. ensign U.S. Navy, 1969, advanced through grades to capt., 1980; intern Martland Hosp.-Coll. Medicine and Dentistry N.J., Newark, 1969-70; resident in gen. surgery Nat. Naval Med. Center, Bethesda, Md., 1975-76; resident in aerospace medicine Johns Hopkins Sch. Hygiene and Public Health, 1977-78; comdg. officer Naval Submarine Med. Research Lab., Groton, Conn., 1978-81; attending physician Naval Submarine Med. Center, Naval Submarine Base, New London, Conn., 1978-81; lectr. epidemiology (environ. health) Yale U., 1979—; lectr. mil. medicine Uniformed Services U. Health Scis., 1976—, chmn. operational and emergency medicine, 1982—; cons. emergency medicine Naval Hosp., Bethesda. Mem. Assn. Mil. Surgeons U.S., Aerospace Med. Assn., Undersea Med. Soc., Soc. U.S. Naval Flight Surgeons, AIAA, Internat. Soc. Aviation Safety Investigators, Am. Coll. Emergency Physicians, Univ. Assn. Emergency Medicine, Assn. Tchrs. of Preventive Medicine, Am. Coll. Preventive Medicine (sec.-treas. 1981—). Subspecialties: Preventive medicine; Emergency Medicine. Current work: Emergency medical equipment.

MARIENFELD, CARL JOSEPH, community medicine and environ. health educator; b. Chgo., July 11, 1917; s. Frederic Carl and Maria (Pink) M.; m. Pauline Tuech, Aug. 14, 1943; children: Robert, Kenneth, Mark, Joyce. B.A., Lake Forest Coll., 1938; M.D., U. Ill., 1943; M.P.H., Johns Hopkins U., 1960. Diplomate: Am. Bd. Pediatrics. From asst. prof. to assoc. prof. U. Ill., Chgo., 1947-57; med. dir. Heart Disease Control, USPHS, Washingtgon, 1957-61; prof. preventive medicine U. Mo., Columbia, 1961—, dir. environ. health, 1966—; mem. Cancer Commn., State of Mo., 1963-73. Contbr. articles to med. jours. Served to capt. U.S. Army, 1944-46. NIH grantee, 1966-73; EPA grantee, 1977-81. Mem. AAAS, Mo. Med. Soc., Boone County Med. Soc., Soc. Epidemiology. Mem. United Ch. of Christ. Subspecialties: Epidemiology; Pediatrics. Current work: Environmental epidemiology—birth defects and cancer epidemiology—public health. Home: Route 1 Box 182 Ashland MO 65010 Office: U Mo Sinclair Farm Route 3 Columbia MO 65201

MARINO, ANDREW ANTHONY, biophysicist, lawyer; b. Phila., Jan. 12, 1941; s. Frank and Frances M.; m. Linda Lee, Aug. 14, 1965; children: Lawrence, Andrew, Christopher, Lisa. B.S. in Physics, St. Joseph's U., Phila., 1962; M.S. in Biophysics, Syracuse U., 1965, Ph.D., 1968, J.D., 1974. Bar: N.Y. bar 1975. Biophysicist VA Med. Ctr., Syracuse, N.Y., 1964-81; asst. prof. dept. orthopaedic surgery La. State U. Med. Ctr., Shreveport, 1981—; pres. Plastafil Corp. Author: (with R.O. Becker) Electromagnetism and Life, 1982; editor: Jour. Bioelectricity; contbr. articles to profl. publs. Subspecialty: Biophysics (physics). Current work: Biological effects of electromagnetic radiation bioelectricity and therapeutic aspects of electromagnetism. Home: PO Box 127 Belcher LA 71004 Office: La State U Med Center PO Box 33932 Shreveport LA 71130

MARINUS, MARTIN GERARD, pharmacology educator; b. Amsterdam, Netherlands, June 22, 1944; came to U.S., 1968, naturalized, 1977; m. Isabel Cristina Castro, Dec. 21, 1970; children: Lucinda M., Julian M., Alicia I. B.Sc., U. Otago, N.Z., 1965, Ph.D., 1968. Posdoctoral fellow Yale U. Sch. Medicine, 1968-70; vis. fellow Free U. Amsterdam, 1970-71, research assoc. Rutgers Med. Sch., 1971-74; assoc. prof. pharmacology U. Mass. Med. Sch., 1974—. Contbr. articles to profl. jours. Recipient Faculty Research award Am. Cancer Soc., 1976-81. Mem. Am. Soc. Microbioloby. Subspecialty: Genetics and genetic engineering (medicine). Current work: DNA methylation, DNA repair; mutagenesis. Home: 58 Newton St West Boylston MA 01583 Office: Dept Pharmacology U Mass Med Sch Worcester MA 01605

MARION, DANIEL FRANCIS, tree and ornamental plant pathologist; b. Washington, May 3, 1944; s. Edmund M. and Marie P. (Pate) M.; m. Dorothy K., Nov. 27, 1965; children: Daniel, Matthew, Meghan. B.Sc., U. Ga., 1966; M.Sc., Ohio U., 1968; Ph.D., U.R.I., 1974. Regional plant pesticide specialist Diamond Shamrock Corp., Cleve., 1968-70; research assoc. instr. plant disease control U. R.I., Kingston, 1970-74; assoc. prof. tree and ornamental plant protection SUNY-Canandaigua, 1974—; pvt. practice tree pathology, Canandaigua, N.Y., 1975—; regional research coordinator, cons. Mauget Co., Burbank, Calif., 1976—. Contbr. articles in field to profl. jours. Mem. Internat. Soc. Arboriculture, Internat. Phytopath. Soc., Am. Forestry Assn., Am. Phytopath. Soc., N.Y. Arborists Assn., Pesticide Assn. N.Y., Soc. Am. Foresters, Gamma Sigma Delta, Phi Sigma. Republican. Roman Catholic. Subspecialties: Plant pathology; Integrated pest management. Current work: Researcher/consultant in chemotherapy for suppression of internal disorders of trees; diseases of turfgrasses. Home: 3650 Middle Cheshire Rd Canandaigua NY 14424 Office: Dept Environ Conservation/Horticulture SUNY-CCFL Lincoln Hill Campus Canandaigua NY 14424

MARION, WAYNE RICHARD, researcher, ecology educator, consultant; b. Ithaca, N.Y., June 28, 1947; s. Maurice Verne and Doris Grace (Cross) M.; m. Marjorie Anne Durfee, June 7, 1969; children: Julie Lynn, Kristi Jill. B.S., Cornell U., 1969; M.S., Colo. State U., 1970; Ph.D., Tex. A&M U., 1974. Cert. wildlife biologist. Grad. research asst. Tex. A&M U., 1970-74; instr. Cornell U., 1974; asst. prof. U. Fla., Gainesville, 1975-80, assoc. prof., 1980—; cons. Applied Biology, Inc., Decatur, Ga., 1979—; pres. Wildlife Resources Mgmt., Gainesville, 1979—; chmn. edn. com. Morningside Nature Ctr., Gainesville, 1982—. Contbr. chpts. to books, numerous articles to profl. jours. Bd. dirs. Morningside Nature Ctr., 1982—; ruling elder Westminster Presbyterian Ch. U.S. Forest Service grantee, 1978-83; Occidental Chem. Co., grantee, 1979-81; Fla. Game and Fish Commn. grantee, 1980-82; Fla. Inst. Phosphate research grantee, 1981-83. Mem. Wildlife Soc., Sigma Xi, Xi Sigma Pi, Phi Sigma, Alpha Zeta, Phi Kappa Phi, Gamma Sigma Delta. Republican. Subspecialties: Ecology; Resource management. Current work: Currently, I am conducting research on the effects of intensive forest management and intensive phosphate mining activity on wildlife populations with emphasis upon designing and enhancing future reforested and reclaimed areas as wildlife habitats. Home: 1807 NW 39th Terr Gainesville FL 32605 Office: Sch Forest Resources and Conservation U Fla 118 Newins-Ziegler Hall Gainesville FL 32611

MARISKA, JOHN THOMAS, astrophysicist; b. Fairbanks, Alaska, Feb. 25, 1950; s. Ben F. and Genevieve K. (Thomas) M.; m. Patricia A. Wood, June 10, 1972; 1 dau., Sara. B.A., U. Colo., 1972; A.M., Harvard U., 1973, Ph.D., 1977. Nat. Acad. Sci./NRC research assoc. Naval Research Lab., Washington, 1977-79, astrophysicist, 1979—. Contbr. articles to profl. jours. Mem. Am. Astron. Soc., Am. Geophys. Union, AAAS, Internat. Astron. Union. Subspecialties: Solar physics; Ultraviolet high energy astrophysics. Current work: Physics of the outer layers of the solar atmosphere. Stellar physics. Ultraviolet observational astronomy. Home: 8313 Uxbridge Ct Springfield VA 22151 Office: Code 4175M Naval Research Lab Washington DC 20375

MARK, HANS MICHAEL, government official, physicist, engineer; b. Mannheim, Germany, June 17, 1929; came to U.S., 1940, naturalized, 1945; s. Herman Francis and Maria (Schramek) M.; m. Marion G. Thorpe, Jan. 28, 1951; children: Jane H., James P. A.B. in Physics, U. Calif. at Berkeley, 1951; Ph.D., MIT, 1954; Sc.D. (hon.), Fla. Inst. Tech., 1978, D. Eng., Poly. Inst. N.Y., 1982. Research asso. MIT, 1954-55, asst. prof., 1958-60; research physicist Lawrence Radiation Lab., U. Calif. at Livermore, 1955-58, 60-69, exptl. physics div. leader, 1960-64; asso. prof. nuclear engring. U. Calif. at Berkeley, 1960-66, prof., 1966-69, chmn. dept. nuclear engring., 1964-69; lectr. dept. applied sci. U. Calif. at Davis, 1969-73; cons. prof. engring. Stanford, 1973—; dir. NASA-Ames Research Center, 1969-77; undersec. Air Force, Washington, 1977-79, sec., 1979-81; dep. adminstr. NASA, Washington, 1981—; mem. Pres.'s Adv. Group Sci. and Tech., 1975-76. Author: (with N.T. Olson) Experiments in Modern Physics, 1966; also numerous articles.; Editor: (with S. Fernbach) Properties of Matter Under Unusual Conditions, 1969. Recipient Distinguished Service medal NASA, 1972, 77; Exceptional Civilian Service award U.S. Air Force, 1979; Disting. Public Service medal Dept. Def., 1981. Fellow Am. Phys. Soc., AIAA; mem. Nat. Acad. Engring., Am. Nuclear Soc., Am. Geophys. Union. Subspecialties: Atomic and molecular physics; Aerospace engineering and technology. Research on nuclear energy levels, nuclear reactions, applications nuclear energy for practical purposes, atomic fluorescence yields, measurement X-rays above atmosphere. Home: 2704 George Mason Pl Alexandria VA 22305 Office: NASA 400 Maryland Ave SW Washington DC 20546

MARK, JAMES EDWARD, chemistry educator; b. Wilkes-Barre, Pa., Dec. 14, 1934; s. Frank Charles and Anna (Raisch) M.; m. Eliza Pollard, Nov. 28, 1964; children: Elizabeth, Eleanor. B.S., Wilkes Coll., 1957; Ph.D. in Chemistry, U. Pa., 1962. Asst. prof. chemistry Poly. Inst. Bklyn., 1964-67; mem. faculty dept. chemistry U. Mich., Ann Arbor, 1967-77, prof., 1972-77; prof. chemistry, dir. polymer research ctr. U. Cin., 1977—; cons. Dow Corning, Midland, Mich., 1981—. Co-editor 1 book in field; contbr. articles to profl. jours. Fellow Am. Phys. Soc., N.Y. Acad. Sci.; mem. Am. Chem. Soc. (best paper award rubber div. 1975-81). Republican. Subspecialties: Polymer chemistry; Physical chemistry. Current work: Rubberlike elasticity, statistical properties of polymers, polymer-coated electrodes. Home: 5236 Oakhill Ln Cincinnati OH 45239 Office: Dept Chemistry U Cin Cincinnati OH 45221

MARK, MARVIN ROBERT, dental educator, physiologist, researcher; b. Chgo., Ot. 31, 1943; s. Albert A. and Esther (Simon) M.; m. Jennie Beth Rickerson, June 15, 1968 (div. 1980). B.S., Roosevelt U., 1970; M.S. in Physiology, U. Mich., 1973, Ph.D., 1980; D.M.D., Med. Coll. Ga., 1982. Teaching fellow in physiology U. Mich., Ann Arbor, 1970-73, in anatomy, 1973-76; asst. prof. Med. Coll. Ga., Augusta, 1977—; cons. tech. photography Lamda Research, Augusta, 1980—. Photographer: book Architechtonics of Human Cerebral Fiber Systems, 1973. NSF trainee, 1971-72; NIH grantee, 1973-76. Mem. Soc. Neurosci., Biol. Photog. Assn. Jewish. Subspecialties: Neuropharmacology; Cell biology (medicine). Current work: The trophic influence of autonomic nerves on effector cells. Inventor-coordinate plotting instrument; convertible plug; multislide composer. Home: 222 12th St SE Washington DC 20003 Office: VA Med Ctr Washington DC 20422

MARKATOS, NICOLAS-CHRISTOS GREGORY, research company executive, educator; b. Athens, Greece, Mar. 1, 1944; s. Gregory Nicolas and Katherine John M. Dipl. Eng., Nat. Tech. U., Athens, 1967, M.Sc., 1967, M.A., Athens Sch. Econs., 1970, DIC, Imperial Coll., London, 1973; Ph.D., 1974. Systems engr. Proctor & Gamble, Athens, 1969-70; research asst. Imperial Coll., London, 1973-74; group leader Concentration, Heat and Momentum Research Ltd., London, 1975-78, tech. mgr., 1978—; co-dir. Centre Math. Modelling and Process Analysis, London, 1983—; head Math. Modeling and Computer Simulation, 1983—; lectr. in field. Research numerous publs. in field. Recipient cert. of recognition for creative devel. tech. innovation NASA, 1980. Mem. Chartered Engring. Instn., Instn. Chem. Engrs., Am. Inst. Chem. Engrs., AIAA, Internat. Soc. Computer Simulation, Greek Tech. Chambers. Christian Orthodox. Subspecialties: Chemical engineering; Mathematical software. Current work: Research and teaching in fluid mechanics, heat/mass transfer and combustion; particular interest in mathematical modelling and computer simulation of turbulent fluid flows with special emphasis on flows with chemical reaction, two-phase and compressible supersonic flows; and thermophysical processes in general purpose of work is advancement of computer modelling as powerful tool to scientific research, industry and environment. Home: 21 Shottfield Ave London England SW14 8EA Office: Faculty Sci and Math Sch Math Stats Computing Thames Poly Wimbledon England SE18 6PF

MARKELONIS, GEORGE JOSEPH, JR., anatomist, educator; b. Balt., Jan. 5, 1947; s. George Joseph and Emily T. (Skabisky) M. B.S., U. Md., 1969; M.S., Villanova U., 1972; Ph.D., U. Md., 1976. Postdoctoral fellow U. Md. Sch. Medicine, 1976-78, Muscular Dystrophy Assn. postdoctoral fellow, 1979, research asst. prof. anatomy, 1979-81, asst. prof., 1981—. Contbr. articles to profl. jours. NIH grantee, 1979. Mem. Am. Soc. Neurochemistry, Soc. Neurosci., N.Y. Acad. Sci., AAAS. Subspecialties: Neurobiology; Neurochemistry. Current work: Study of trophic interactions between nerve and muscle; study of protein derived from nerves which supports maintenance of muscle in tissue culture; study of factors influencing muscle maintenance and protein turnover. Office: Dept Anatomy U Md Sch Medicine Baltimore MD 21201

MARKENSON, JOSEPH AARON, physician; b. Lincoln, Nebr., Jan. 2, 1943; s. George and Esther M.; m. Alicejane Lippner, Jan. 6, 1966; 5 children. A.B., Boston U., 1965; M.S., Rutgers U., 1966; M.D., SUNY-Bklyn., 1970. Diplomate: Am. Bd. Internal Medicine. Intern in medicine Kings County–Downstate Med. Ctr., Bklyn., 1970-71; med. resident Cornell U.–N.Y. Hosp., N.Y.C., 1973-75; fellow in rheumatology Hosp. Spl. Surgery, N.Y.C., 1975-76, asst. attending physician, 1976—, asst. scientist research dept., 1976—, asst. scientist, 1976—, dir. personnel health service, 1978—; asst. prof. medicine Cornell U. Med. Coll., N.Y.C., 1976—; asst. attending physician N.Y. Hosp., N.Y.C., 1976—; cons. dept. medicine Meml. Hosp., N.Y.C., 1976—, cons. rheumatology, 1979—; mem. pub. edn. com. Arthritis Found., 1981. Lupus Erythematosus Found. fellow, 1975-76. Fellow ACP, N.Y. Acad. Medicine; mem. Am. Assn. Immunologists, Am. Occupational Med. Assn. (mem. subcom. on hosp. personnel health service), Am. Rheumatism Assn. (mem. postgrad. clin. seminar subcom., sec. treas. N.E. regional), Am. Soc. Microbiology, Am. Fedn. Clin. Research, Soc. Exptl. Medicine and Biology, N.Y. Acad. Sci., N.Y. Rheumatism Assn. (sec.-treas.). Office: Hosp Spl Surgery 535 E 70th St New York NY 10021 Home: 435 E 70th St New York NY 10021

MARKERT, CLEMENT LAWRENCE, educator; b. Las Animas, Colo., Apr. 11, 1917; s. Edwin John and Sarah (Norman) M.; m. Margaret Rempfer, July 29, 1940; children—Alan Ray, Robert Edwin, Betsy Jean. B.A. summa cum laude, U. Colo., 1940; M.A., U. Calif. at Los Angeles, 1942; Ph.D., Johns Hopkins, 1948. Merck-NRC fellow Calif. Inst. Tech., 1948-50; asst. prof. zoology U. Mich., 1950-56, assoc. prof., 1956-57; prof. biology Johns Hopkins, 1957-65; chmn. dept. biology Yale U., 1965-71, prof. biology, 1965—, dir., 1974—; Panelist NSF, 1959-63; panelist subcom. on marine biology President's Sci. Adv. Com., 1965-66; mem. council Am. Cancer Soc., 1976-78; co-chmn. developmental biology interdisciplinary cluster President's Biomed. Research Panel, 1975; mem. com. on animal models and genetic stocks NRC, 1979—; trustee Bermuda Biol. Sta. for Research, 1959—; bd. sci. advisers La Jolla Cancer Research Found.; mem. bd. sci. advs. Jane Coffin Childs Meml. Fund for Med. Research, 1979-83; chmn. bd. trustees BIOSIS, 1981, mem., 1976—. Editor: Prentice-Hall Series in Developmental Biology, Procs. 3d Internat. Conf. Isozymes, 1974; Mng. editor: Jour. Exptl. Zoology, 1963—; editorial bd.: Archives of Biochemistry and Biophysics, 1963-81, Sci. mag, 1979—, Differentiation, 1973—, Developmental Genetics, 1979—, Cancer Research, 1982—. Served with Internat. Brigades, 1938; Spain; with Mcht. Marine, 1944-45. Mem. Am. Inst. Biol. Scis. (pres. 1965), Nat. Acad. Scis. (governing council 1977-80), Inst. Medicine, Am. Soc. Biol. Chemists, Internat. Soc. Developmental Biologists, Am. Soc. Developmental Biology (pres. 1963-64), Soc. Study of Reprodn., Am. Acad. Arts and Scis. (governing council 1980—), Am. Genetic Assn. (pres. 1980), Am. Soc. Naturalists (v.p. 1967), Am. Soc. Zoologists (pres. 1967), Genetics Soc. Am., Am. Soc. Cell Biology, Phi Beta Kappa, Sigma Xi. Subspecialties: Genetics and genetic engineering (biology); Reproductive biology. Current work: Manufacture of chimeric mammals; microsurgical alterations of the genome of mammalian eggs. Home: 64 Hartford Turnpike Hamden CT 06517

MARKHAM, CLAIRE AGNES, chemistry educator; b. New Haven, Aug. 12, 1919; s. James J. and Agnes V. (Manning) M. B.A., St. Joseph Coll., 1940; Ph.D., Cath. U. Am., 1952. Faculty fellow Chem. Biodynamica, U. Calif.-Berkeley, 1967-68, Inst. Tech., Trondheim, Norway, 1967; Under sec. Conn. Dept. Energy, Hartford, 1977-79; prof. chemistry St. Joseph Coll., West Hartford, Conn., 1952—, dean grad. div., 1979—, chmn. chemistry dept., 1960-70, mem. pres. council, 1980—; mem. planning com. Grad. Consortium, Hartford, Conn., 1976—. Editor: Basic Science Series, 1962-67; contbr. articles to profl. jours. Mem. White House Conf. on Energy Prodn., Washington, 1978; chmn. adv. bd. Conn. Environ. Mediation Ctr., Hartford, 1982—; bd. dirs. Conn. Energy Council for Tchrs., Hartford, 1982—; mem. adv. bd. Conn. Energy Round Table, Hartford, 1982—. Recipient Research awards Am. Chem. Soc., NSF, 1963-73; Sci. Edn. awards NSF, U.S. Dept. Energy, 1959-80; NSF faculty fellow, 1967-68; NSF travel grantee, 1974-76, 77. Mem. Am. Chem. Soc. (councilor 1968-70, 74-85, chmn. sect. 1971-73), AAAS, Conn. Acad. Sci. and Engring. (chmn. membership com. 1978-79), Sigma Xi. Subspecialties: Catalysis chemistry; Photochemistry. Current work: Photochemical reactions at photoconductor surfaces, chlorophyll/carotene interactions in two-phase systems. Home: 1678 Asylum Ave West Hartford CT 06117 Office: St Joseph Coll West Hartford CT 06117

MARKHAM, THOMAS LOWELL, mathematics educator; b. Apex, N.C., Jan. 2, 1939; s. Excell and Mary (Patterson) M. B.S., U. N.C.-

Chapel Hill, 1961, M.A., 1964; Ph.D., Auburn U., 1967. Asst. prof. U. N.C.-Charlotte, 1967-68; asst. prof. math. U. S.C., 1968-72, assoc. prof., 1972-82, prof., 1982—; vis. assoc. prof. Oreg. State U., 1977. Contbr. articles, numerous revs. to profl. jours. Mem. Am. Math. Soc., Math. Assn. Am., Soc. Indsl. and Applied Math., Sigma Xi, Pi Mu Epsilon, Phi Kappa Phi. Subspecialties: Linear algebra; Numerical analysis. Current work: Research on special classes of matrices and their applications to numerical problems. Home: 3005 Essex Rd Columbia SC 29204 Office: U SC Dept Math/Stats Columbia SC 29208

MARKLAND, FRANCIS SWABY, JR., biochemist, educator; b. Phila., Jan. 15, 1936; s. Francis Swaby and Willie Lawrence (Averritt) M.; m. Barbara Blake, June 27, 1959; children: Cathleen Blake, Francis Swaby. B.S., Pa. State U., 1957; Ph.D., Johns Hopkins U., 1964. Postdoctoral fellow UCLA, 1964-66, asst. prof. biochemistry, 1966-74; vis. asst. prof. U. So. Calif., 1973-74, assoc. prof., 1974-83, prof., 1983—; cons. Clin. Lab. Med. Group, Los Angeles. Contbr. numerous articles to profl. jours. Mem. Northridge Masterworks Chorale; asst. coach Pop Warner Football, Little League Baseball. Recipient NIH research career devel. award, 1968-73; grantee Am. Cancer Soc., Nat. Cancer Inst., 1979—, Nat. Heart, Lung and Blood Inst., 1978—. Mem. Am. Soc. Biol. Chemistry, Am. Chem. Soc., Am. Soc. Hematology, Internat. Soc. Toxinology, Am. Assn. Cancer Research, Endocrine Soc., Am. Heart Assn., Sigma Xi, Phi Mu Alpha, Alpha Zeta. Subspecialties: Biochemistry (medicine); Cancer research (medicine). Current work: Protein receptors for steroid hormons, snake venom coagulant enzymes. Office: USC Cancer Center Cancer Research Lab Rm 106 1303 N Mission Rd Los Angeles CA 90033

MARKLEY, JOHN LUTE, biophysicist; b. Denver, Mar. 6, 1941; s. Miles Russell and Winnifred Farrar (Lute) M.; m. Diane A. Sheehan, Aug. 9, 1975; 1 dau., Jessamyn Sheehan. B.A., Carleton Coll., 1963; Ph.D., Harvard U., 1969. Research chemist Merck Inst. Therapeutic Research, Rahway, N.J., 1966-69; postdoctoral fellow U. Calif.-Berkeley, to 1972; asst. prof. chemistry Purdue U., 1972-76, assoc. prof., 1976-81, prof., 1981—; dir. Biochem. Magnetic Resonance Lab., 1977—, chmn. biochemistry div., 1982—. Contbr. articles to profl. jours. NIH fellow, 1970-71; Fogarty fellow, 1980-81; recipient NIH Research Career Devel. award, 1976-80. Mem. AAAS, Am. Soc. Biol. Chemists, Am. Chem. Soc., Biophys. Soc., Internat. Soc. Magnetic Resonance, Soc. Magnetic Resonance in Medicine, Sigma Xi, Phi Beta Kappa. Club: Lafayette Soaring. Subspecialties: Nuclear magnetic resonance (biotechnology); Biophysical chemistry. Current work: Application of nuclear magnetic resonance spectroscopy and biotechnology to biomedical problems, protein chemistry, protein nucleic acid interactions, enzyme mechanisms, in vivo NMR. Office: Dept Chemistry Purdue U West Lafayette IN 47907

MARKLEY, KENNETH ALAN, psychologist, educator; b. Harrisburg, Pa., Sept. 30, 1933; s. Charles Donald and Gladys (Wallis) M.; m. Susan Watson, Sept. 14, 1957; children: Christopher David, Jennifer Elaine. B.A., Dickinson Coll., 1955; M.A., NYU, 1959; Hum. D. (hon.), Clarksville (Tenn.) Sch. Theology, 1976. Cert. profl. counselor, Pa. Sch. psychologist Central Dauphin Schs., Harrisburg, Pa., 1959-61; exec. dir. United Cerebral Palsy, Inc., Camp Hill, Pa., 1961-64; Eastern dir., seminars dir. Narramore Found., Rosemead, Calif., 1964—; mem. faculty psychology Dickinson Coll., 1963; cons. psychologist Harrisburg pvt. schs., 1963-64. Author: Our Speaker This Evening, 1968. Served to 1st lt. AUS, 1955-57. Mem. Am. Psychol. Assn., Pa. Psychol. Assn., Am. Personnel and Guidance Assn., Pa. Personnel and Guidance Assn. Republican. Mem. Christian and Missionary Alliance. Subspecialties: Behavioral psychology; Social psychology. Current work: Research into integration of theological concepts and behavioral dynamics. Home and Office: 104 N 26th St Camp Hill PA 17011

MARKS, DENNIS WILLIAM, educator; b. Madison, Wis., Nov. 5, 1944; s. Louis Sheppard and Helen Teresa (Oeters) M.; m. Sita Patricia Smith, Nov. 30, 1968. B.S., Fordham U., 1966; Ph.D., U. Mich., 1970. Postdoctoral fellow, asst. prof. astronomy U. Toronto, Ont., Can., 1970-71; asst. prof. physics and astronomy Valdosta (Ga.) State Coll., 1971-74, dir. Planetarium and Obs., 1973-80, assoc. prof. physics and astronomy, 1974-78, prof., 1978—; vis. prof. physics and astronomy Iowa State U., Ames, 1982-83; NASA trainee, 1966-69. Author: The Rotation of Viscous Polytropes, 1970. NSF predoctoral fellow, 1969-70; NSF internat. travel grantee, 1978; U.S. Nat. Com.-Internat. Astron. Union travel grantee, 1979; Com. Humanities in Ga. project grantee, 1980. Mem. Internat. Soc. Gen. Relativity and Gravitation, AAUP (nat. council 1978-81), Am. Astron. Soc., Ga. Acad. Sci. Subspecialties: Relativity and gravitation; Theoretical astrophysics. Current work: Relativistic thermo-hydrodynamics, energy-momentum tensor for radiation, cosmic background radiation, radiative viscosity, differential rotation of stars. Office: Valdosta State Coll 1500 N Patterson St Valdosta GA 31698

MARKS, HENRY LEWIS, research geneticist; b. Waynesboro, Va., Sept. 6, 1935; s. Charles Alexander and Mabel Elizabeth (Batten) M.; m. Shirley Anita Reasor, Dec. 27, 1959; 1 son, Daniel Steven. B.S. with honors, Va. Poly. Inst. and State U., 1958, M.S., 1959; Ph.D., U. Md., 1967. Poultry husbandman Agrl. Research Service, U.S. Dept. Agr., Beltsville, MD., 1960-65, research geneticist, 1965-67, Athens, Ga., 1967—. Contbr. to profl. jours. Coordinator, So. Regional genetics project N.E. Ga. Council, Boy Scouts Am., 1971-74; v.p. PTA., Athens, 1975-76, pres., 1976-77. Recipient Poultry Sci. Research award, 1969. Mem. World's Poultry Sci. Assn., Poultry Sci. Assn., Sigma Xi, Phi Kappa Phi. Lutheran. Lodge: Rotary. Subspecialty: Animal genetics. Current work: Long-term selection responses; genetics of mechanisms manipulated by selection; genetic parameters of resistance to mycotoxins; genetic parameters of body composition. Home: 897 Spartan Ln Athens GA 30606 Office: Ga Livestock-Poultry Bldg Athens GA 30602

MARKS, LAWRENCE EDWARD, psychologist; b. N.Y.C., Dec. 28, 1941; s. Milton and Anne (Parnes) M.; m. Joya Ellen Cazes, Dec. 24, 1963; children: Liza Robin, Laura Jill. A.B., Hunter Coll., 1962; Ph.D., Harvard U., 1965. Lectr., research asst. Harvard U., 1965-66; asst. fellow John B. Pierce Found. Lab., Yale U., New Haven, 1966-69, asso. fellow, 1969—, research asso., 1966-76, asst. prof., 1970-76, asso. prof., 1976—; cons. Inst. Artificial Organs. Author: Sensory Processes: The New Psychophysics, 1974, The Unity of the Senses, 1978; contbr. articles to profl. jours. Mem. Am. Psychol. Assn., Soc. Neuroscience, Internat. Soc. Artificial Organs, Acoustical Soc. Am., Optical Soc. Am. Subspecialties: Psychophysics; Sensory processes. Current work: Processes of hearing, vision, touch and sensory perception, development of aids for blind and deaf. Office: 290 Congress Ave New Haven CT 06519

MARKS, NEVILLE, neurochemist, educator; b. Dublin, Ireland, Apr. 10, 1930; s. Rudolf and Freda (Rapp) M.; m. Liliane Dahan, Dec. 31, 1971; children: Timothy Joseph, Lionel Victor. Ph.D., Inst. Psychiatry, U. London, 1960. Research scientist U. Mich., 1960; mem. staff N.Y. State Dept. Mental Hygiene, 1961—, prin. research scientist, 1972—; assoc. prof. psychiatry NYU, 1979—. Editor: Research Methods in Neurochemistry, Vols. 1-5, 1982, Protein Catabolism, 1976; contbr. articles to profl. jours. NIH grantee. Mem. Am. Soc. Biol. Chemists, Internat. Soc. Neurochemistry, Am. Soc. Neurochemistry, Soc. Endocrinology, Biochem. Soc. (U.K.), Council Research Scientists State of N.Y. (past chmn.). Subspecialties: Neurochemistry; Biochemistry (biology). Current work: Purification and assay of enzymes involved in protein turnover and polypeptide metabolism in CNS tissues. Home: 1 Lincoln Plaza 38N New York NY 10023 Office: Ctr for Neurochemistry Nathan S Kline Research Inst for Psychiat Research New York NY 10035

MARKS, PAUL ALAN, oncologist, cell biologist; b. N.Y.C., Aug. 16, 1926; s. Robert R. and Sarah (Bohorad) M.; m. Joan Harriet Rosen, Nov. 28, 1953; children—Andrew Robert, Elizabeth Susan, Matthew Stuart. A.B. with gen. honors, Columbia U., 1945, M.D., 1949; Doct. Biol. Sci. h.c., U. Urbino (Italy), 1982. Fellow Columbia Coll. Physicians and Surgeons, 1952-53, asso., 1955-56, mem. faculty, 1956—, dir. hematology tng., 1961-74, prof. medicine, 1967-82, dean faculty of medicine, v.p. med. affairs, 1970-73, dir., 1972-80, v.p. health scis., 1973-80, prof. human genetics and devel., 1969-82, Frode Jensen prof. medicine, 1974-80; attending physician Presbyn. Hosp., N.Y.C., 1967—; pres., chief exec. officer Meml. Sloan-Kettering Cancer Center, 1980—; attending physician Meml. Hosp. for Cancer and Allied Diseases, 1980—; mem. Sloan-Kettering Inst. for Cancer Research, 1980—; adj. prof. Rockefeller U., 1980—; vis. physician Rockefeller U. Hosp., 1980—; instr. Sch. Medicine, George Washington U., 1954-55; cons. VA Hosp., N.Y.C., 1962-66; Asso. investigator Nat. Inst. Arthritis and Metabolic Diseases, NIH, Bethesda, Md., 1953-55; mem. adv. panel hematology tng. grants program NIH, 1969-73, chmn. hematology tng. grants program, 1971-73; vis. scientist Lab. Cellular Biochemistry, Pasteur Inst., 1961-62; vis. prof. 1st di Chemica Biologica, U. Genoa, Italy, 1963; mem. adv. panel on developmental biology NSF, 1964-67; mem. Delos Conf., Athens, Greece, 1971, 72; mem. founding com. Radiation Effects Research Found., Japan, 1975; mem. Pres.'s Biomed. Research Panel, 1975-76, Pres.'s Cancer Panel, 1976-79, Pres.'s Commn. on Accident at Three Mile Island, 1979; chmn. exec. com. div. med. scis. Nat. Acad. Scis.-NRC, 1973-76; ad hoc adviser White House Conf. on Aging, 1981; adviser Leopold Schepp Found.; dir. Pfizer, Inc., Dreyfus Mut. Funds. Editor: Monographs in Human Biology, 1963; Contbr. articles to profl. jours.; Mem. editorial bd.: Blood, 1964-71; assoc. editor, 1976-77; editor-in-chief, 1978-82, Jour. Clin. Investigation, 1967-72. Trustee St. Luke's Hosp., 1970-80, Roosevelt Hosp., 1970-80, Presbyn. Hosp., 1972-80; mem. jury Albert Lasker Awards, 1974—; bd. dirs. Pub. Health Research Inst., N.Y.C., 1971-74; bd. govs. Weizmann Inst., 1976—; bd. dirs. Revson Found., 1976—; bd. sci. counselors, div. cancer treatment Nat. Cancer Inst., 1980—; mem. council div. biol. scis. and Pritzker Sch. Medicine, U. Chgo.; mem. tech. bd. Milbank Meml. Fund, 1978—; trustee Metpath Inst. Med. Edn., 1977-79. Recipient Charles Janeway prize Columbia, 1949, Joseph Mather Smith prize, 1959, Stevens Triennial prize, 1960, Swiss-Am. Found. award in med. research, 1965, Columbia U. Coll. Physicians and Surgeons Disting. Service medal, 1980; Commonwealth Fund fellow, 1961-62. Fellow Am. Acad. Arts and Scis.; mem. Inst. Medicine (mem. council 1973-76), Nat. Acad. Scis. (chmn. sect. med. genetics, hematology and oncology 1980-83, chmn. Acad. Forum Adv. Com. 1980-81), Red Cell Club (past chmn.), Am. Fedn. Clin. Research (past councillor Eastern dist.), Am. Soc. Clin. Investigation (pres. 1972-73), Am. Soc. Biol. Chemists, Am. Soc. Human Genetics (past mem. program com.), Am. Assn. Cancer Research, ACP, Am. Soc. Cell Biology, Am. Soc. Hematology (pres.-elect 1983, pres. 1984), Assn. Am. Physicians, Enzyme Club, Harvey Soc., Internat. Soc. Developmental Biologists, Interurban Clin. Club, Soc. for Study Devel. and Growth, World Soc. Ekistics. Subspecialty: Oncology. Research on biochemistry erythrocyte aging, predisposing factors to anemia, messenger RNA for hemoglobin, polyribosomes, molecular defect in thalassemia, erythroid cell differentiation and leukemia. Home: Beach Hill Rd Bridgewater CT 06752 Office: 1275 York Ave New York NY 10021

MARKS, PETER JACOB, laboratory services, company executive, environmental engineer; b. Vestonburg, Pa., Dec. 25, 1941; s. Peter and Agnes Irene (Miller) M.; m. Kay Helen, Aug. 17, 1963; children: Sean Patrick, Eric Scott. B.S. in Biology; M.S. in Environ. Sci. and Engring. With Research lab. for Analytical Methods Devel., Lancaster County (Pa.) Gen. Hosp., 1963-64; with Roy F. Weston, Inc., West Chester, Pa., 1965—, v.p. lab. services. Mem. ASTM, Water Pollution Control Fedn., Water Pollution Control Assn. Pa. Subspecialties: Environmental engineering; Analytical chemistry. Current work: Program manager on research and development project for Dept. Army to develop technologies for Army hazardous waste treatment. Home: 1332 Sherwood Dr West Chester PA 19380 Office: Roy F Weston Inc Weston Way West Chester PA 19380

MARKS, WILLIAM BYRON, research neurophysiologist; b. Montgomery, Ala., July 30, 1933; s. John S. and Ellen N. (Allison) M.; m. Anne Pattillo Foerster; 3 children. B.S., MIT, 1956; Ph.D., Johns Hopkins U., 1963. Faculty biophysics Johns Hopkins U., 1964-79; research physiologist Lab. Neurol. Control, Nat. Inst. Neurol., Communicative Diseases and Stroke, NIH, Bethesda, Md., 1974—. Subspecialty: Neurophysiology. Current work: Visual physiology, nerve-muscle physiology. Office: NIH Bldg 36 Room 5A29 Bethesda MD 20205 Home: 1615 Manchester Ln Washington DC 20011

MARKUS, LAWRENCE, mathematician, educator; b. Hibbing, Minn., Oct. 13, 1922; s. Benjamin and Ruby (Friedman) M.; m. Lois Shoemaker, Dec. 9, 1950; children: Sylvia, Andrew. B.S., U. Chgo., 1942, M.S., 1946; Ph.D., Harvard U., 1951. Instr. meteorology U. Chgo., 1942-44; research meteorologist Atomic Project, Hanford, 1944; instr. math. Harvard U., 1951-52; instr. Yale U., 1952-55; lectr. Princeton U., 1955-57; asst. prof. U. Minn., 1957-58, asso. prof., 1958-60, prof. math., 1960—, asso. chmn. dept. math., 1961-63, dir. control scis., 1964—, Regents' prof. math. and dir. Control Sci. and Dynamical Systems Center, 1980—; Leverhulme prof. control theory, dir. control theory centre U. Warwick, Eng., 1970-73, Nuffield prof. math., 1970—; regional conf. lectr. NSF, 1969; vis. prof. Yale U., Columbia U., U. Calif., U. Warsaw, 1980, Tech. Inst. Zurich, 1983, Peking U. (China), 1983; dir. conf. Internat. Centre Math., Trieste, 1974; lectr. Internat. Math. Congress, 1974, Iranian Math. Soc., 1975, Brit. Math. Soc., 1976, Japan Soc. for Promotion Sci., 1976, Royal Instn., London, 1982; vis. prof. U. Tokyo, 1976; mem. panel Internat. Congress Mathematicians, Helsinki, 1978; sr. vis. fellow Sci. Research Council, Imperial Coll., London, 1978; vis. prof. Tech. U., Denmark, 1979; mem. UNESCO sci. adv. com. Control Symposium, U. Strasbourg, France, 1980; IEEE Plenary lectr., Orlando, Fla., 1982; Mem. adv. bd. Office Naval Research, Air Force Office Sci. Research. Author: Flat Lorentz Manifolds, 1959, Flows on Homogeneous Spaces, 1963, Foundations of Optimal Control Theory, 1967, Lectures on Differentiable Dynamics, 1971, rev. edit., 1980, Generic Hamiltonian Dynamical Systems, 1974; editor: Internat. Jour. Nonlinear Mechanics, 1965-73, Jour. Control, 1963-67; contbr. articles to profl. jours. Served to lt. (j.g.) USNR, 1944-46. Recipient Research prize, Kiev, 1969; Fulbright fellow, Paris, 1950; Guggenheim fellow, Lausanne, Switzerland, 1963. Mem. Am. Math. Soc. (past mem. nat. council), Am. Geophys. Soc., Soc. Indsl. and Applied Math. (past nat. lectr.), Phi Beta Kappa, Sigma Xi. Subspecialties: Cosmology; Systems engineering. Office: Dept Math U Minn Minneapolis MN 55455

MARKWELL, MARY ANN K., research scientist; b. Cleve., Sept. 30, 1948. Ph.D. in Biochemistry, Mich. State U., 1975. NIH trainee Mich. State U., 1970-75, NIH postdoctoral fellow, 1975-76; Damon Runyon fellow UCLA, 1976-80, asst. research prof. dept. microbiology, 1980—. Contbr. articles to profl. jours. Mem. AAAS, Am. Chem. Soc., Am. Soc. Biol. Chemistry, Am. Soc. Microbiology, Assn. Women in Sci., Am. Soc. Virology, Fedn. Am. Scientists, Sigma Xi. Subspecialties: Membrane biology; Virology (biology). Current work: Virus receptors, gangliosides, glycoconjugates, membrane fusion. Office: Molecular Biology Inst UCLA Los Angeles CA 90024

MARLATT, ROBERT BRUCE, plant pathologist, researcher; b. Cleve., July 18, 1920; s. Leslie Arthur and Alice (Walsh) M.; m. Merle F. Greiner, Nov. 2, 1946; children: Randall A., Richard F., Ross E.; m. Lillian Friedman, May 5, 1979. B.S. in Chemistry, Adrian Coll., 1942; Ph.D. in Plant Pathology, U. Ariz., 1952. Plant pathologist Calif. Dept. Agr., Sacramento, 1952, U. Ariz. Research Ctr., Mesa, 1952-62, N.Z. Dept. Agr., Levin, 1962-64, U. Fla. Research Ctr., Homestead, 1964—; cons. Co-author: Foliage Plant Production, 1981; contbr. articles sci. jours. and popular mags. Served with USNR, 1942-46. Burgess fellow, 1951; NSF fellow, 1961; Dept. Agr. Collaborator, 1960. Mem. Am. Phytopath. Soc., Am. Soc. Hort. Sci., Ariz. Acad. Sci., Sierra Club. Unitarian. Subspecialty: Plant pathology. Current work: Diseases of tropical foliage plants, diseases of tropical fruits. Office: 18905 SW 280th St Homestead FL 33031

MARLER, PETER ROBERT, research biologist; b. London, Eng., Feb. 24, 1928; came to U.S., 1957, naturalized, 1963; s. Robert A. and Gertrude (Hunt) M.; m. Judith Golda Gallen, Sept. 1, 1954; children: Christopher, Catherine, Marianne. B.Sc., U. London, 1948, Ph.D. in Botany, 1952, U. Cambridge, Eng., 1954; D.Sc. hon., SUNY, 1983. Research fellow Jesus Coll., Cambridge (Eng.) U., 1954-56; from asst. prof. to prof. U. Calif.-Berkeley, 1957-66; prof. Rockefeller U., N.Y.C., 1966—, dir. Center for field Research in Ecology and Ethology, 1972—; sr. research zoologist N.Y. Zool. Soc., 1966-69; dir. Inst. Research Animal Behavior, 1969-72. Author: (with W.J. Hamilton III) Mechanisms of Animal Behavior, 1966, (with J. Vandenbergh) Social Behavior and Communication, Handbook of Behavioral Neurobiology, Vol. 3, 1979. Guggenheim fellow, 1964-65. Fellow N.Y. Zool. Soc., Am. Acad. Arts and Scis., Nat. Acad. Scis., Am. Psychol. Assn., Am. Philos. Soc.; mem. Animal Behavior Soc. (pres. 1968-69), Am., Brit. ornithologists unions, Cooper Ornithol. Soc., Am. Soc. Zoologists, Internat. Primatological Soc. Subspecialties: Ethology; Psychobiology. Current work: The ethology and psychobiology of vocal communication in birds and primates. Animal semantics. The roles of nature and nurture in behavioral development, especially vocal behavior. Cultural transmission of animal behavior. The role of learning in the development of birdsong and the analysis of its social functions. Home: Reservoir Rd Staatsburg NY 12580 Office: Rockefeller U New York NY 10021

MARON, MELVIN EARL, librarian, educator; b. Bloomfield, N.J., Jan. 23, 1924; s. Hyman and Florence (Goldman) M.; m. Dorothy Elizabeth Mastin, Aug. 16, 1948; children: Nadia, John. B.S. in Mech. Engring, U. Nebr., 1945, B.A. in Physics, 1947; Ph.D. in Philosophy, UCLA, 1951. Lectr. philosophy UCLA, 1951-52; tech. engr. IBM Corp., San Jose, Calif., 1952-55; mem. tech. staff Ramo-Wooldridge Corp., 1955-59; mem. sr. staff, computer scis. dept. RAND Corp., 1959-66; prof. library sci. U. Calif.-Berkeley, 1966—, assoc. dir., 1966—. Contbr. articles to profl. jours. Served with AUS, 1943-46. Mem. ACM, Am. Soc. Info. Sci., AAAS, Philosophy Sci. Assn. Home: 63 Ardilla Rd Orinda CA 94563 Office: Sch Library and Info Scis U Calif Berkeley CA 94720

MARON, MICHAEL BRENT, physiologist, educator, researcher; b. Long Beach, Calif., Oct. 8, 1949; s. Victor and Florence Kathryn (Perlman) M. A.A., Long Beach City Coll., 1967; B.A., U. Calif.-Santa Barbara, 1972, Ph.D., 1976. Research asst. U. Calif.-Santa Barbara, 1973-76; ad hoc instr. U. Wis.-Milw., 1977, clin. asst. prof., 1977-79; postdoctoral fellow Med. Coll. Wis., 1977-79; asst. prof. physiology Northeastern Ohio Univs. Coll. Medicine, 1979-83, assoc. prof. physiology, 1984—. Research publs. in field. Am. Lung Assn. grantee, 1980-83; Akron Dist. chpt. Am. Heart Assn. grantee, 1981-83; Nat. Heart, Lung, Blood Inst. grantee, 1983-86. Mem. Am. Physiol. Soc. (Harwood S. Belding award in environ. physiology 1976), Am. Coll. Sports Medicine, AAAS, N.Y. Acad. Scis. Subspecialty: Physiology (medicine). Current work: Regulation of lung fluid balance and pulmonary blood flow; physiology of prolonged physical work. Office: Northeastern Ohio U Coll Medicine State Route 44 Rootstown OH 44272

MARONI, GUSTAVO PRIMO, biologist, educator; b. Merlo, Argentina, Nov. 20, 1941; came to U.S., 1968; s. Victor and Iole (Bright) M.; m. Donna Farolino Kubai, Dec. 16, 1974. Licenciado in Biology, U. Buenos Aires, 1968; Ph.D., U. Wis., 1972. Research assoc. Institut fur Genetik, U. Cologne, 1974-75; asst. prof. dept. zoology U. N.C., Chapel Hill, 1975-80, assoc. prof., 1980-82, assoc. prof. dept. biology, 1982—. Contbr. articles to profl. jours. Mem. Genetics Soc. Am., AAAS. Subspecialty: Gene actions. Current work: Gene control in Drosophila; molecular and genetic analyses of gene-protein systems; alcohol dehydrogenase; metal-binding proteins. Office: Dept Biology U NC Chapel Hill NC 27514 Home: 412 Landerwood Ln Chapel Hill NC 27514

MARQUARDT, DONALD WESLEY, statistician, researcher; b. N.Y.C., Mar. 13, 1929; s. Kurt C. and Amelia P. (Moller) M.; m. Margaret E. Rittershaus, Sept. 13, 1952; children: Paul E. (dec.), Joan N. A.B., Columbia U., 1950; M.A., U. Del., 1956. Research engr./mathematician E.I. du Pont de Nemours & Co., Inc., Wilmington, Del., 1953-57, research project engr./sr. mathematician, 1957-64, cons. supr., 1964-72, cons. mgr. engring. dept., 1972—; mem. NRC eval. panel for applied math. for Nat. Bur. Standards; mem. Am. Nat. Standards Inst. com. on quality assurance, vice chmn., 1983, chmn., 1984; rep. to Internat. Standards Orgn. tech. coms. on statis. methods and quality assurance., chmn. coordinating groups. Mem. editorial bd. Communications in Stats.; assoc. editor: Jour. Bus. and Econ. Stats; contbr. articles to profl. jours. Served with U.S. Army, 1951-52. Fellow Am. Statis. Assn.; mem. AAAS, Am. Soc. for Quality Control (sr. mem., Youden prize 1974), Assn. for Computing Machinery, Soc. for Indsl. and Applied Math., Am. Inst. Chem. Engrs., Sigma Xi. Presbyterian. Subspecialties: Statistics; Algorithms. Current work: Quality mgmt. systems tech.; analysis of unequally-spaced time series for environ. and chronobiology data. Home: 1415 Athens Rd Wilmington DE 19803 Office: Engring Dept El du Pont de Nemours & Co Inc Wilmington DE 19898

MARQUARDT, WARREN WILLIAM, veterinary scientist, educator; b. Erhard, Minn., Nov. 1, 1930; s. William Gust and Freida Amelia (Neubert) M.; m. Alice Mae Overboe, Sept. 4, 1955; children: Gregory D., Sheila P., Cheryl K., Steven R. B.S., U. Minn., 1959, D.V.M., 1961, Ph.D., 1970. Research fellow Coll. Vet. Medicine, U. Minn., St. Paul, 1962-65, instr., 1965-69; assoc. prof. vet. sci. U. Md., College Park, 1969-80, prof., 1980—, dir. avian disease research sect., 1971—. Contbr. articles to profl. jours. Served as staff sgt. USAF, 1951-55. U.S. Dept. Agr. grantee. Mem. AVMA, Am. Soc. Microbiology, World Vet. Poultry Assn., Am. Assn. Avian Pathologists, Minn. Vet. Med. Assn., Sigma Xi, Phi Zeta, Gamma Sigma Delta.. Lutheran. Subspecialties: Virology (veterinary medicine); Infectious diseases. Current work: Hybridoma technology—monoclonal antibodies for the antigenic assessment of avian corona viruses; biochemistry of virus antigens; enzyme-linked immunosorbent assays in diagnostic virology. Office: VA-MD Regional Coll Vet Medicine College Park MD 20742

MARQUART, JOHN ROBERT, chemistry educator; b. Benton Harbor, Mich., Feb. 3, 1933; s. Herman B. and Irene (Mull) M. B.S., U. Ariz., 1955; M.S., U. Ill., 1961, Ph.D., 1963. Chemist Shell Devel. Co., Emeryville, Calif., 1963-68; prof. Mercer U., Macon, Ga., 1968-78; vis. prof. U. Ill., Urbana, 1978-79, 82-83; assoc. prof. chemistry Eastern Ill. U., Charleston, 1979—. Served to capt. U.S. Army, 1955-61. NSF fellow, 1970, 74, 82. Mem. Am. Chem. Soc., Ga. Acad. Sci. (chmn. chem. sect. 1978-79), Sigma Xi, Gamma Sigma Epsilon (exec. dir. archives 1975-79). Subspecialties: Physical chemistry; Analytical chemistry. Current work: High temperature fast kinetic studies. Fast analytical techniques. Home: Box 231 Route 4 Charleston IL 61920 Office: Eastern Ill U Charleston IL 61920

MARQUIS, NORMAN RONALD, pharmaceutical company executive; b. Laconia, N.H., Jan. 2, 1936; m. Alice S. Sawyer, Feb. 2, 1957 (dec. Nov. 2, 1975); children: Norma, Karen, Mark; m. Kathy R. Rhame, Apr. 1, 1979. B.A., U. N.H., 1959, M.S., 1960; Ph.D., U. Mich., 1965. Teaching/research fellow Harvard U., Boston, 1965-67; research assoc. Mead Johnson, Evansville, Ind., 1967-70, prin. investigator med. research, 1970-79, mgr. clin. devel., 1979-81, assoc. dir. pharm. div., 1981—; adj. prof. U. Evansville, 1974-80. Contbr. articles to profl. publs. Mem. Evansville Philharm. Choir/Soc., 1982—. Mem. Am. Soc. Biol. Chemists (hon.), Soc. Exptl. Biology and Medicine (hon.), Am. Heart Assn. Thrombosis Council, Soc. Non-Invasive Vascular Tech., Tri-State Critical Care Soc. Club: Evansville Kennel. Subspecialties: Biochemistry (medicine); Physiology (medicine). Current work: Medical pharmaceutical administration; drug development; clinical investigation; drug marketing consultantion. Patentee antithrombogenic process. Office: Mead Johnson Pharm Div Evansville IN 47721

MARQUIS, ROBERT EDWARD, microbiology educator, program director; b. Sarnia, Ont., Can., Jan. 21, 1934; came to U.S., 1952; s. Louis Joseph and Margaret Ellen (Mulvale) M.; m. Diana Elizabeth Mears, June 8, 1957; children: Linda, Heather, Andre. B.S., Wayne State U., 1956; M.S., U. Mich., 1958, Ph.D., 1961. Postdoctoral fellow U. Edinburgh, Scotland, 1961-63; sr. instr. U. Rochester, N.Y., 1963-65, asst. prof., 1965-70, assoc. prof., 1970-78, prof. microbiology, 1978—; sr. fellow Scripps Instn., La Jolla, Calif., 1970-71; cons. Dept. Energy, Bartlesville, Okla., 1979-82; editor Petroleum Microbiology, London, 1981-82. Contbr. articles to sci. publs. Mem. Am. Soc. Microbiology, Undersea Med. Soc., Internat. Assn. Dental Research, Soc. Gen. Microbiology, Sigma Xi. Subspecialty: Microbiology. Current work: Microbial enhancement of oil recovery, physiology and control of oral bacteria, bacterial endospores, microbial barophysiology. Home: 163 Warrington Dr Rochester NY 14618 Office: U Rochester 260 Crittenden Blvd Rochester NY 14642

MARRAZZI, MARY ANN, pharmacologist, educator; b. Ann Arbor, Mich., Dec. 22, 1945; d. Amedeo Sorrentino and Rose Florence (Netter) M. B.A., U. Minn., 1966; Ph.D., Washington U., St. Louis, 1972. NIH postdoctoral fellow dept. pharmacology Washington U. Sch. Medicine, 1972-74; vis. investigator Inst. Psychiatry, U. Mo., St. Louis, 1974; asst. prof. pharmacology Sch. Medicine, Wayne State U., Detroit, 1974-78, assoc. prof., 1978—; assoc. dept. psychiatry Harper-Grace Hosps., Detroit, 1981—. Contbr. articles to profl. jours. NSF internat. travel award, 1976. Mem. Am. Soc. Pharmacology and Exptl. Therapeutics, Soc. for Neurosci., Am. Soc. for Neurochemistry, Sigma Xi. Subspecialties: Neuropharmacology; Neurochemistry. Current work: CNS control of metabolic homeostasis and feeding, hypothalamic glucoreceptors, energy metabolism in brain, metabolic encephalopathies, anorexia nervosa, prostaglandins in brain. Home: 962 Lochmoor Grosse Pointe Woods MI 48236 Office: Dept Pharmacology Sch Medicine Wayne State U Detroit MI 48201

MARROCCO, RICHARD THOMAS, psychology educator; b. Rochester, N.Y., July 27, 1943; s. William Thomas and Audrey Janette (Grein) M.; m. Mary Ellen Manley, June 20, 1970; 1 son, Steven Thomas. B.A., UCLA, 1965; Ph.D., Ind. U., 1971. Postdoctoral fellow U. Calif.-Berkeley, 1971-73; asst. prof. U. Oreg., Eugene, 1973-79, assoc. prof. psychology, 1979—. NIH research grantee, 1974-77, 78-82; NSF research grantee, 1982-85. Mem. AAAS, Soc. Neurosci. Democrat. Subspecialties: Neurophysiology; Neurobiology. Current work: Primate visual neurophysiology. Home: 1015 Barber Dr Eugene OR 97405 Office: Dept Psychology U Oreg Eugene OR 97403

MARRS, BARRY LEE, biochemist, educator; b. Newark, Sept. 23, 1942; s. Donald Lee and Goldie (Mack) M.; m. Barbara Griswold, Aug. 20, 1966; children: Abbe Lea, Gwendolyn Griswold. B.A., Williams Coll., 1963; Ph.D., Case Western Res. U., 1968. NSF postdoctoral fellow U. Ill., 1967-69; Am. Cancer Soc. postdoctoral fellow Stanford U., 1969-71; research assoc. Ind. U., Bloomington, 1971-72; asst. prof. Sch. Medicine St. Louis U., 1972-75, assoc. prof., 1975-78, prof. biochemistry, 1978-83, program dir. cell and molecular biology tng. program, 1977-83; sr. research assoc. Exxon Research and Engring. Co., 1983—; group head Exxon, 1983—. Research grantee NIH, 1973-83, NSF, 1973-83; recipient Career Devel. award Gen. Med. Scis. Inst., NIH, 1975-80. Mem. Am. Soc. Microbiology, Am. Soc. Biol. Chemists. Subspecialties: Genetics and genetic engineering (biology); Molecular biology. Current work: Genetics of photosynthetic bacteria; regulation of membrane development; research on structure, function and development photosynthetic membranes using classical genetic and recombinant DNA and biochemical approaches. Home: 401 Jaguar Ln Bridgewater NJ 08807 Office: Corp Research Sci Route 22E Annandale NJ 08801

MARSCHER, WILLIAM DONNELLY, mech. engr.; b. Utica, N.Y., Feb. 18, 1948; s. William Ransford and Margaret Elizabeth (Donnelly) M.; m. Deborah Lynn Schmidt, May 27, 1972; children: Michael, Colleen, Megan. B.S., Cornell U., 1970, M.Engring., 1972; M.S. in Applied Mechanics, Rensselaer Poly. Inst., 1976. Engr. electronic fuel injection div. Bendix Corp., Troy, Mich., 1970-73; sr. engr. Pratt & Whitney Aircraft, East Hartford, Conn., 1973-78; project engr. Creare, Inc., Hanover, N.H., 1978-82; dir.mechanics Worthington Group/McGraw Edison, Mountainside, N.J., 1982—; instr. mechanics Lebanon (N.H.) Coll., 1979-80. Contbr. articles to profl. jours. NASA fellow, 1972; Internat. Fellow exchange student to, Lima, Peru, 1965. Mem. ASME, Am. Soc. Lubrication Engrs., ASTM, Soc. for Exptl. Stress Analysis. Democrat. Roman Catholic. Subspecialties: Theoretical and applied mechanics; Solid mechanics. Current work: Finite element analysis, friction and wear, fatigue research, rotor dynamics, pump and compressor mechanical designs. Patentee in field. Home: 10 Grace Way Morristown NJ 07960 Office: 270 Sheffield St Mountainside NJ 07092

MARSDEN, SULLIVAN SAMUEL, JR., petroleum engineering educator, consultant; b. St. Louis, June 3, 1922; s. Sullivan Samuel and Irene Margaret M.; m.; children: Sullivan F., Robert S., Mary V. Marsden Hilton, Anastasia E. B.A. in Chem. Engring, Stanford U., 1944, Ph.D. in Phys. Chemistry, 1948. Phys. chemist Tenn. Eastman Corp., Oak Ridge, 1945; with Stanford Research Inst., Menlo Park, Calif., 1947-50; asst. dir. Nat. Chem. Lab. India, Poona, 1950-53; assoc. prof. petroleum and nat. gas engring. Pa. State U., 1953-57; asso. prof. petroleum engring. Stanford (Calif.) U., 1957-62, prof., 1962—; cons. in field. Fulbright awardee U. Tokyo, 1963-64, Inst. Petroleum, Ploesti, Romania and Gubkin Inst., Moscow, 1978. Mem. Soc. Petroleum Engrs. Republican. Club: Stanford U. Faculty. Subspecialties: Petroleum engineering; Fuels. Current work: Development of use of barge-mounted processing plants for offshore and remote (Arctic) natural gas fields; subsurface hydraulic mining methods for tar sand deposits. Patentee in field. Office: Dept Petroleum Engring Stanford U Stanford CA 94305

MARSH, BENJAMIN BRUCE, meat and animal science educator; b. Petone, N.Z., Nov. 15, 1926; came to U.S., 1971, naturalized, 1982, s. Spencer Albon and Thomasina (Leeder) M.; m. Doris Lillian Johannesson, Dec. 6, 1952; children: Richard J., David B. B.Sc., Victoria U., Wellington, N.Z., 1946, M.Sc., 1947; Ph.D., Cambridge (Eng.) U., 1951. Chemist Dominion Lab., 1951-57; biochemist, dep. dir. Meat Industry Research Inst., Hamilton, N.Z., 1957-71; prof. meat and animal sci. U. Wis.-Madison, 1971—; dir. Muscle Biology Lab., 1971—. Contbr. articles to profl. jours. Mem. Am. Meat Sci. Assn. (Disting. Meats Research award 1970), Am. Soc. Animal Sci. (Research award 1978), Inst. Food Technologists. Subspecialties: Food science and technology; Biochemistry (biology). Current work: Muscle biology and meat science: early postmorten behavior and properties; muscle contraction; muscle metabolism; rigor mortis; meat quality. Office: 1805 Linden Dr Madison WI 53706

MARSH, JAMES DALTON, medical researcher, cardiologist, educator; b. Chgo., Dec. 27, 1947; s. Russell Elliot and Kathryn (Lain) M.; m. Susan Kohli, May 27, 1974; 1 son, Eric. A.B., Harvard Coll., 1970; M.D., Harvard Med. Sch., 1974. Diplomate: Am. Bd. Internal Medicine; cert. bd. cardiovascular disease. Intern Peter Bent Brigham Hosp., Boston, 1974–75, resident in cardiovascular disease, 1975-78; assoc. physician Brigham and Women's Hosp., Boston, 1979—; instr. Harvard Med. Sch., Boston, 1981-82, asst. prof. medicine, 1982—. Am. Heart Assn. grantee, 1981—; recipient NIH clin. investigator award, 1980—. Fellow Am. Coll. Cardiology, Am. Coll. Physicians; mem. Am. Fedn. Clin. Research. Subspecialties: Cellular pharmacology; Cell biology (medicine). Current work: Regulation of cardiac cellular response to hormones and neurotran. Office: Brigham and Womens Hosp 75 Francis St Boston MA 02115

MARSH, JEANNE CAY, social service adminstrn. educator; b. Madison, Wis., July 9, 1948; d. Herbert Louis and Helen Irene (Moeckly) M.; m. Steven King Shevell, Oct. 3, 1976. B.A., Mich. State U., 1969; M.S.W., U. Mich., 1972, Ph.D., 1975. Postdoctoral fellow U. Mich., Ann Arbor, 1975-77; research investigator Inst. for Social Research, 1977; asst. prof. Sch. Social Service, U. Chgo., 1978—; Cons. Hull House Assn., Chgo., 1980—; dept. child psychiatry U. Chgo., 1980—, Ill. Coalition Against Domestic Violence, 1982—. Author: Rape and the Limits of Law Reform, 1982. Mem. Ind. Voters Ill., Chgo., 1979—, Hyde Park-Kenwood Community Council, 1979—. Ford Found. fellow, 1972-73; Rackham Prize fellow, 1973-74. Mem. Am. Psychol. Assn., Nat. Assn. Social Workers, Evaluation Research Soc., Phi Beta Kappa, Mortar Board. Subspecialty: Social psychology. Current work: Social program and policy evaluation research; mental health policy research; knowledge utilization in research. Home: 5240 S Greenwood Chicago IL 60615 Office: U Chgo Sch Social Service 969 E 60th St Chicago IL 60637

MARSH, KENNETH ALBERT, industrial scientist; b. Rochford, Eng., Apr. 1, 1949; came to U.S., 1975; s. Albert Frederick Alexander and Irene Mabel (Emery) M.; m. Lucile Chapman Marsh, Aug. 28, 1976; children: Rhonda, Jessica. B.Sc. with honors, U. Canterbury, N.Z., 1969, M.Sc., 1971; Ph.D., York U., Toronto, Ont., Can., 1975. Research fellow solar astronomy Calif. Inst. Tech., 1975-82; mem. tech. staff Sci. Ctr., Rockwell Internat., Thousand Oaks, Calif., 1982—. Contbr. articles to profl. jours. Mem. Am. Astron. Soc. Subspecialties: Radio and microwave astronomy; Graphics, image processing, and pattern recognition. Current work: Image reconstruction as applied to ultrasonics for non-destructive evaluation, also image interpretation for robot vision. Office: 1049 Camino Dos Rios Thousand Oaks CA 91360

MARSH, WILLIAM MICHAEL, environmental scientist, planning cons., author; b. Marquette, Mich., Sept. 20, 1942; s. William Rol and Audrey Jane (Steinhoff) M.; m. Susan Amanda Reed, Aug. 31, 1963; children: Katherine W., Christopher R. B.S., Central Mich. U., 1964; M.A., U. Mo., 1965; postgrad. (Rackham fellow, NSF trainee), U. Mich., 1968-70. Mem. faculty geography dept. Central Mich. U., Mt. Pleasant, 1965-68; research asso. Environ. Research Inst. Mich., Ann Arbor, 1971-74; mem. faculty phys. geography dept. and environ. studies program U. Mich., Flint, 1970—, chmn. dept. resource and community sci., 1981—; planning cons. landscape architecture and archtl. projects; vis. prof. urban planning U. Mich.; vis. lectr. Mich. State U. Author: Environmental Analysis for Land Use and Site Planning, 1978, Landscape: An Introduction to Physical Geography, 1981, Landscape Planning, 1983. Mem. Am. Planning Assn., Ecol. Soc. Am., Soil Conservation Soc. Subspecialties: Resource management; Geomorphology. Current work: Water management on devel. sites; aeolian processes on coastal slopes. Home: 1602 Calumet Flint MI 48503 Office: Resource & Community Sci U Mich Flint MI 48503

MARSHAK, MARVIN LLOYD, physics educator; b. Buffalo, Mar. 11, 1946; s. Kalman and Goldie (Hait) M.; m. Anita Sue Kolman, Sept. 24, 1972; children: Rachel, Adam. A.B., Cornell U., 1967; M.S., U. Mich., 1969, Ph.D., 1970. Research assoc. U. Mich., Ann Arbor, 1970; research assoc. U. Minn., Mpls., 1970-74, asst. prof., 1974-78, assoc. prof. dept physics, 1978—; prin. investigator U.S. Dept. Energy contract in elem. particle physics, 1982; project dir. NSF grants, 1979, 81. Contbr. articles to profl. jours. Mem. Am. Phys. Soc. Subspecialties: Particle physics; Cosmic ray high energy astrophysics. Current work: Strong interaction experiments, protein decay, experiments in cosmic rays. Home: 2855 Ottawa Ave S Minneapolis MN 55416 Office: 116 Church St SE Minneapolis MN 55455

MARSHAK, ROBERT EUGENE, physicist, educator; b. N.Y.C., Oct. 11, 1916; s. Harry and Rose (Shapiro) M.; m. Ruth Florence Gup, Apr. 18, 1943; children—Ann, Stephen. A.B., Columbia U., 1936; Ph.D., Cornell U., 1939, Utkal U., India, 1977, City U. N.Y., 1979, CCNY, 1980. Instr. dept. physics U. Rochester, N.Y., 1939-43, asst. prof., 1943-46, asso. prof., 1946-49, prof., 1949-70, chmn. dept. physics and astronomy, 1950-64, Disting. Univ. prof., 1964-70; pres. CCNY, 1970-79, pres. emeritus, 1979—; Univ. Disting. prof. physics Va. Poly. Inst. and State U., Blacksburg, 1979—; lectr. Harvard Obs., summer 1940; professeur d'Echange (Guggenheim fellow) at Sorbonne, 1953-54; vis. prof. Columbia U., summer 1950, U. Mich., 1952, Tata Inst. Bombay, 1953, French Sch. for Theoretical Physics, 1954, Sch. for Theoretical Physics, Tokyo, 1965, Goteburg, 1970; guest prof. at CERN, Geneva; Ford Found. and Guggenheim fellow, 1960-61, Guggenheim fellow, 1967-68, Nobel Found. prof., Sweden, 1970; mem. Inst. Advanced Study, Princeton, spring 1948; physicist radiation lab. Mass. Inst. Tech., 1942-43, Montreal Atomic Energy project, 1943-44; dep. group leader in theoretical physics Los Alamos Sci. Lab., 1944-46; vice chmn. N.Y. State Adv. Com. on Atomic Energy, 1958; Avco vis. prof. Cornell U., 1959; chmn. vis. physics com. Brookhaven Nat. Lab., 1964-65; Niels Bohr vis. prof. Inst. Math. Sci., Madras, India, 1963; lectr. Yalta Internat. Sch., 1966, Hercig Novi Internat. Sch., 1967; head Nat. Acad. Sci. del. to Poland, 1964, to Yugoslavia, 1965; mem. Sloan Fellowship Com., 1967-73; mem. sci. council Internat. Center of Theoretical Physics, Trieste, 1967-75; Buhl vis. prof. Carnegie-Mellon U., 1968; distinguished visitor U. Tex., 1970; mem. Solvay Congress, 1967, 82, Pugwash Conf., 1967, U.S.-Japan Com. on Sci.-Cooperation, 1969-72; mem. nat. com. UNESCO, 1970-73; trustee Univ. Research Assn., 1968-70, Atoms for Peace Award, 1958-70; chmn. div. particles and fields, 1970-71; founder Rochester Confs. on High Energy Physics, 1950—; vis. com. physics Carnegie Mellon U., 1967-70; bd. dirs. Internat. Found. for Sci., 1970-75. Author: Meson Physics, 1952, (with L.I. Schiff and E.C. Nelson) Our Atomic World, 1946, (with E.C.G. Sudarshan) Elementary Particles, 1961, (with Riazuddin and C. Ryan) Theory of Weak Interactons, 1969; Assoc. editor: Phys. Rev, 1953-55; editor interscience: Tracts and Monographs in Physics and Astronomy, 1955-70; dir.: Jour. History of Ideas, 1973—. Recipient A. Cressy Morrison Astron. prize N.Y. Acad. Scis., 1940, J. Robert Oppenheimer Meml. prize. Fellow Am. Phys. Soc. (exec. com. 1968-69, council 1965-69, pres. 1983), AAAS; mem. AAUP, Fedn. Am. Scientists (chmn. 1947-48), Nat. Acad. Scis. (past chmn. adv. com. on sci. exchanges with USSR and Eastern Europe; council 1971-74), Am. Acad. Arts and Scis., Am. Philos. Soc., Council Fgn. Affairs, Internat. Union Pure and Applied Physics (past sec. commn. on high energy physics, past vice chmn. nat. commn.), Phi Beta Kappa (Scholar 1982), Sigma Xi (nat. lectr. 1969), Phi Kappa Phi. Subspecialties: Particle physics; Theoretical physics. Current work: Theory of weak interactions in particle physics, grand unified theories, structure of quarks and leptons. Researcher in theoretical particle physics. Office: Va Poly Inst and State U Dept Physics Blacksburg VA 24061

MARSHAK, ROBERT REUBEN, university dean, veterinarian; b. N.Y.C., Feb. 23, 1923; s. David and Edith (Youselovsky) M.; m. Ruth Emilie Lyons, Dec. 4, 1948; children: William Lyons, John Ball, Richard Best.; m. Margo Post Marshall, June 25, 1983. Student, U. Wis., 1940-41; D.V.M., Cornell U., 1945, U. Bern, 1968, M.A., U. Pa., 1971. Diplomate: Am. Coll. Vet. Internal Medicine (charter). Practice vet. medicine, Springfield, Vt., 1945-56; prof., chmn. dept. medicine Sch. Vet. Medicine, U. Pa., Phila., 1956-58; prof. medicine Grad. Sch. Medicine, 1957-64; chmn. dept. clin. studies Sch. Vet. Medicine, 1958-73; dir. Bovine Leukemia Research Center, 1965-73; dean Sch. Vet. Medicine, 1973—; co-dir. center on Interactions Animals and Soc., 1975-79; also mem. grad. group com. in comparative med. scis. Center on Interactions Animals and Soc.; adv. council James A. Baker Inst. and N.Y. State Coll. Vet. Medicine (both Cornell U.); adv. bd. Pa. Dept. Agr.; mem. com. on vet. medicine scis. Nat. Acad. Scis.; chmn. Gov.'s Study Group on Horse Racing Industry in Pa., 1979; mem. del. to evaluate vet. med. and research Chinese Ministry Agr. Mem. editorial bd.: Jour. of Am. Vet. Radiology Soc, 1964—; Contbr. numerous articles to sci. jours. Served with AUS, 1943-44. Fellow N.Y. Acad. Scis., Phila. Coll. Physicians; mem. John Morgan Soc. (pres. 1967-68); AAAS Am. Assn. Cancer Research, Am., Pa. vet. med. assns., Pa. Livestock Assn. (dir.), Conf. Research Workers in Animal Diseases, Am. Assn. Vet. Med. Colls., Sigma Xi, Phi Zeta. Subspecialty: Internal medicine (veterinary medicine). Home: 280 Melrose Ave Merion PA 19066 Office: U Pa Sch Vet Medicine 3800 Spruce St Philadelphia PA 19104

MARSHALL, ALAN GEORGE, chemistry and biochemistry educator; b. Bluffton, Ohio, May 26, 1944; s. Herbert Boyer and Cecile Louise (Mogil) M.; m. Marilyn Josephine Gard, June 13, 1965; children: Gwendolyn Scott, Brian George. B.A., Northwestern U., 1965; Ph.D., Stanford U., 1969. Asst. prof. chemistry U. B.C. (Can.), Vancouver, 1969-75, assoc. prof., 1975-80; prof. chemistry and biochemistry Ohio State U., Columbus, 1980—, dir., 1980—. Author: Biophysical Chemistry, 1978; author, editor: Fourier, Hadamard and Hilbert Transforms, 1982. Sloan fellow, 1976-80; NSF fellow, 1965-69. Mem. Am. Chem. Soc., Am. Phys. Soc., Biophys. Soc., AAAS, Sigma Xi. Subspecialties: Biophysical chemistry; Analytical chemistry. Current work: Analytical and biological spectroscopy; Fourier and Hilbert transform spectroscopy. Office: Dept Chemistry Ohio State U 140 W 18th Ave Columbus OH 43210

MARSHALL, FRANKLIN NICK, pharmacologist; b. Chgo., July 5, 1933; s. John Nick and Elizabeth (Hegel) M.; m. Nancy Rose Emmering, Nov. 25, 1955; 1 son: Nicholas Franklin. B.S., U. Iowa, 1957, M.S. (scholar), 1959, Ph.D., 1961. Instr. pharmacology U. Iowa, Iowa City, 1961; pharmacologist Dow Chem. Co., Indpls., 1961-66, project leader, 1966-72, group leader, 1972-73, assoc. scientist, 1973-80; dir. pharmacology Merrell Dow Pharms. Co., Indpls., 1980—. Contbr. articles to profl. jours. Mem. Am. Soc. Pharmacology and Explt. Therapeutics, Soc. Explt. Biology and Medicine, N.Y. Acad. Scis., Am. Heart Assn., AAAS, Sigma Xi. Subspecialties: Pharmacology; Cardiology. Current work: Renal, cardiovascular and autonomic pharmacology. Home: Route 2 Box 111A Thorntown IN 46071 Office: PO Box 68511 Indianapolis IN 46268

MARSHALL, GARLAND ROSS, biomedical scientist, educator; b. San Angelo, Tex., Apr. 16, 1940; s. Garland Ross and Jewel Gray M.; m. Suzanne Russell, Dec. 26, 1959; children: Stuart C., Keith W., Melissa A., Kenneth L. B.S., Calif. Inst. Tech., 1962; Ph.D., Rockefeller U., 1966. Instr. depts. physiology and biophysics and biol. chemistry Washington U., St. Louis, 1966-67, asst. prof., 1967-72, asso. prof., 1972-76, prof., 1976—; vis. prof. Massey U., Palmerston North, N.Z., 1975; pres. Tripos Assos., Inc.; mem. sci. council Nelson Research.; Am. Heart Assn. investigator, 1970-75. Mem. Am. Chem. Soc., Endocrine Soc., Biophys. Soc., Am. Phys. Soc., Am. Soc. Biol. Chemistry, Am. Soc. Pharmacology and Exptl. Therapeutics. Subspecialties: Molecular pharmacology; Graphics, image processing, and pattern recognition. Current work: Computer-aided drug design, conformation of peptides receptors, molecular graphics, peptide chemistry, conformational analysis. Office: Dept Physiology and Biophysics Washington Univ Med Sch Saint Louis MO 63110

MARSHALL, GERALD FRANCIS, optical physicist; b. Seven Kings, Essex, Eng., Feb. 26, 1929; came to U.S., 1967; s. Albert Edward and Ethelena Maria (Plaskett) M.; m. Irene Mary Stook, Oct. 11, 1958; children: Clare Marshall De Forge, Mark, Guy, Maria. B.Sc. in Physics, London U., 1952. Mktg. tech. advisor Morganite Internat. Ltd., London, 1954-59; sr. research and devel. engr. Ferranti Ltd., Edinburgh, 1959-67; dir. engring. Med. Lasers, Burlington, Mass., 1969-70; staff cons. Schiller Industries, Warren, Mich., 1971-76; dir. optical engring. Energy Conversion Devices Inc., Troy, Mich., 1976—; cons. physicist, 1967—. Contbr. chpts. to books, articles to profl. jours. Served with RAF, 1947-49. Fellow Inst. of Physics, Internat. Soc. for Optical Engring.; mem. Optical Soc. Am. Roman Catholic. Subspecialties: Optical engineering; Laser alignment. Current work: Editing and writing optical engineering, optical design of flat panel displays, of swimming goggles with prescription lenses and optical instrumentation. Patentee in optics. Office: 32400 Edward St Madison Heights MI 48071

MARSHALL, HARRY DWIGHT, machine tool co. exec.; b. Grand Rapids, Mich., Aug. 26, 1915; s. Harry Dwight and Edith Lenore (Butler) M.; m. Judy Corrigan, Dec. 26, 1942; children: Nancy Marshall Taylor, John Stephen. B.S.M.E. in Mech. and Indsl. Engring, U. Mich., 1939. Cost engr. Eastman Kodak Co., Rochester, N.Y., 1939-40; asst. to v.p. mfg. Gallmeyer & Livingston Co., Grand Rapids, 1940-52, v.p., 1952—, treas., 1959—, also dir.; dir. Mut. Home Fed.

Savs. & Loan Assn., Grand Rapids, 1964—, Mut. Ins. Co. Grand Rapids, 1964—, v.p., 1967, pres., 1968-74; bd. dirs. Employers Assn. Grand Rapids, 1956—, v.p., 1957, pres., 1958-60. Bd. dirs. Kent County (Mich.) chpt. ARC, 1960-69; elder, trustee Westminster Presbyterian Ch., Grand Rapids. Mem. ASME, Tau Beta Pi, Phi Kappa Phi. Republican. Clubs: Univ., Penninsular. Lodges: Rotary (dir. 1960-62, 69-71, v.p. 1961-62, 69-70) (Grand Rapids)). Subspecialties: Mechanical engineering; Industrial engineering. Current work: Mfg. and fin. responsibilities and decisions of machine tool mfr. Home: 1661 Fisk Rd SE Grand Rapids MI 49506 Office: 336 Straight Ave SW Grand Rapids MI 40504

MARSHALL, VINCENT DEPAUL, microbiologist; b. Washington, Apr. 5, 1943; s. Vincent P and Mary F. (Bach) M.; children: Vincent, Amy. B.S., Northeastern Okla. State U., 1965; M.S., U. Okla., 1967, Ph.D., 1970. Research asst. U. Okla., 1965-70; research assoc. U. Ill.-Urbana, 1970-73; research scientist Upjohn Co., Kalamazoo, 1973-75; research head Upjoun Co., 1975, sr. scientist, 1976—. Contbr. chpts., numerous articles to profl. publs. NIH predoctoral fellow, 1968; post-doctoral fellow, 1971. Mem. Am. Soc. Microbiology, Am. Soc. Biol. Chemists, Sigma Xi. Subspecialty: Microbiology. Current work: Antibiotic experimental chemotherapy, bioconversions, antibiotic discovery. Patentee in field. Office: Upjohn Co 300 Henrietta St Kalamazoo MI 49001

MARSHALL, WALTER LINCOLN, chemist; b. Princeton, N.J., May 6, 1925; s. Abraham L. and Edith M. (Lambert) M.; m. Jean A. Clark, Dec. 23, 1948; children: David, John, Katherine. A.B. in Chemistry, Princeton U., 1946, Ph.D., Harvard U., 1950. With Gen. Electric Co., Schenectady, N.Y., 1950—, mgr. turbine tech. lab. large steam turbine div., 1968—. Mem. Am. Chem. Soc., IEEE, AAAS. Subspecialties: Physical chemistry; Materials (engineering). Patentee in field. Office: 55 201 North Ave Schenectady NY 12345

MARSHALL, WILLIAM ROBERT, JR., educator; b. Calgary, Alta., Can., May 19, 1916; came to U.S., 1925, naturalized, 1944; s. William Robert and May (Barnum) M.; m. Dorothy Robbins, Jan. 9, 1943; children—Margaret O., Mary J., William Robert. B.S. in Chem. Engring, Armour Inst. Tech., 1938; Ph.D., U. Wis., 1941. Registered profl. engr. Chem. engr. E. I. duPont Co. Expt. Sta., 1941-47; cons. to industry, 1948—; mem. faculty U. Wis., Madison, 1948—, prof. chem. engring., 1953—, asso. dir. engring. experiment sta., 1953-66, 66-71, exec. dir., 1971-81, dean, dir. univ. research program, 1981—; Mem. engring. scis. adv. panel Nat. Sci. Found., 1957-59, chmn. adv. com. to engring. div., 1966-67, mem. sci. applications task force, 1977; chmn. chem. engring. rev. com. Argonne Nat. Lab., 1959, 66, mem., 1958-66; Chmn. Wis. Licensing Bd. Architects, Profl. Engrs., Land Surveyors and Designers; mem. Korea-U.S. joint commn. Nat. Acad. Sci., 1977—; Fourth Inst. lectr. Am. Inst. Chem. Engrs., 1952; Mem. adv. com. dept. applied sci. Brookhaven Nat. Lab., 1973-75, chmn., 1975. Co-author: Application of Differential Equations to Chemical Engineering, 1949; Author articles in field, sect. in handbook. Pres. Wis. Ballet Co., 1963-66; Trustee Argonne U. Assn., chmn. nuclear engring. edn. com., 1967-69, chmn. bd. commn. on edn.; bd. dirs. Commn. on Engring. Edn., 1965-68. Recipient William H. Walker award, 1953; Gold medal Verein Deutsches Inqineurs, W. Ger., 1975. Fellow Am. Acad. Arts and Scis., Am. Inst. Chem. Engrs. (chmn. com. on equipment testing procedures 1951-56, dir. 1956-58, Profl. Progress award 1959, pres. 1963, chmn. continuing edn. com. 1964-67, Founders award 1973, treas. 1976-80); mem. Engrs. Joint Council (v.p. 1963-65, exec. com., dir.), Assn. Midwest U.'s (pres. 1962-63, dir. 1960—), Am. Assn. Land Grant Colls. (chmn. research sect. engring. div. 1957-58), Am. Soc. Engring. Edn., Soc. for History Tech. (exec. council 1966-68), Engring. Colls. Consortium for Minorities (pres. 1975-81), Nat. Acad. Engring. (com. on interplay engring. with biology and medicine 1967—, vice chmn. 1969-70, chmn. 1970—, mem. commn. on edn. 1968-70, chmn. 1970-73, membership com. 1967-69), Phi Kappa Phi. Subspecialty: Chemical engineering. Home: 4917 Tonyawatha Trail Madison WI 53716 Office: Univ of Wisconsin Warf Office Bldg 610 Walnut St-Rm 1215 Madison WI 53706

MARSTEN, RICHARD BARRY, scientific organization executive; b. N.Y.C., Oct. 28, 1925; s. Jesse and Rosalind M.; m. Sarah Betty Jaffe, June 26, 1949; children: Michael Frederick, Jessica Claire. B.S. in Elec. Engring, M.I.T., 1946, M.S., 1946, postgrad., 1946-49; Ph.D., U. Pa., 1951. Registered profl. engr., N.J. With RCA, 1957-69, mgr. radar systems projects Missile and Surface Radar div., 1959-61, mgr. spacecraft electronics, 1961-67, chief engr. Astro Electronics div., 1967-69; dir. communications programs Office of Applications, NASA, Washington, 1969-76; dean Sch. Engring. City U. N.Y., 1975-79, prof. engring., 1979-81; mgr. space program Office of Technology Assessment, U.S. Congress, 1980-81; exec. dir. Bd. on Telecommunications and Computer Applications, Nat. Acad. Engring./NRC, Washington, 1981—; mem. broadcast panel study space applications Nat. Acad. Scis., 1967, chmn. point-to-point communications panel, 1968; mgr., dir. study on communications for social needs Pres.'s Domestic Council, 1971; mem. panel automation opportunities to health care Fed. Council Sci. and Tech., 1972-73; mem. panel on telecommunications research in U.S. and selected fgn. countries, com. on telecommunications Nat. Acad. Engring., 1973-74; chmn. telecommunications adv. panel Office of Technology Assessment, 1979-80; Adv. council Nat. Energy Found. Editor: Communications Satellite Systems Technology, 1966. Recipient Exceptional Service medal, also Group Achievement award NASA, 1974; citation White House, 1972, Internat. Women's Yr., 1975; NASA/USSR Apollo-Soyuz Space medal, 1976. Asso. fellow AIAA (chmn. nat. tech. activities com. communications 1967-69, chmn. tech. program 1st Communications Satellite Systems Conf. 1966, chmn. tech. splty. group on info. systems 1970-71); fellow IEEE (com. policy bd. 1972—, adv. EASCON 1973-75, program chmn. EASCON 1970, bd. govs. com. soc. 1977-80, chmn. com. on telecommunications policy 1978-80, cons. mem. 1980-81, mem. 1981-83, mem. steering com. U.S. Tech. Policy Conf. 1978—, mem. U.S. Tech. Policy Com. 1982—, bd. govs. aero. and elec. systems soc. 1982-85); mem. AAAS, N.Y. Acad. Scis. Current work: Satellite applications; satellite communications, broadcasting and remote sensing systems; telecommunications and information systems. Address: NAE/NRC 2100 Pennsylvania Ave NW Washington DC 20418

MARTEN, GORDON CORNELIUS, research agronomist, educator; b. Wittenberg, Wis., Sept. 14, 1935; s. Clarence George and Cora Levina (Verpoorten) M.; m. Lynette Joy Hanson, Sept. 9, 1961; 1 dau., Kimberly Joy. B.S. with highest honors, U. Wis., 1957; M.S., U. Minn., 1959, Ph.D., 1961. Research agronomist U.S. Dept. Agr.-Agrl. Research Service, St. Paul, 1961-72; asst. prof. U. Minn., St. Paul, 1962-66, assoc. prof., 1966-71, prof. dept. agronomy and plant genetics, 1971—; research leader, research agronomist Agrl. Research Service, U.S. Dept. Agr., St. Paul, 1972—; bd. dirs. Am. Forage and Grassland Council, Lexington, Ky., 1976-79; bd. govs., program chmn. XIV Internat. Grassland Congress, Lexington, 1978-81. Assoc. editor: Crop Sci, 1972-74; contbr. over 100 articles to profl. jours.; co-breeder low-alkaloid reed canarygrass, 1983. Chmn. Roseville (Minn.) Sch. Dist. Curriculum Com. Recipient Merit Cert. Am. Forage and Grassland Council, 1976; Outstanding Service awards Am. Forage and Grassland Council and U. Ky., 1981, Minn. Forage and Grassland Council, 1982. Fellow Am. Soc. Agronomy; mem. Crop Sci. Soc. Am. (dir. 1975-76, div. chmn. 1976), Biol. Club U. Minn. (pres. 1978). Lutheran. Subspecialties: Ecology; Animal nutrition. Current work: Evaluation techniques for nutritional value of forage crops, environmental and genetic influences on feed value of forages and grasslands, discovery of antiquality constituents in forage feeds for ruminant animals. Office: Dept Agronomy and Plant Genetics 1509 Gortner Ave Saint Paul MN 55108

MARTENS, FREDERICK HILBERT, nuclear engineer; b. Peoria, Ill., Dec. 15, 1921; s. Hilbert A. C. and Anna D. (Seebergen) M.; m. Carolyn Lee Arnold, Aug. 23, 1947; children: Phillip A., Christine L. B.S., U. Chgo., 1946; M.S., U. N.Mex., 1948. Cert. quality assurance engr. Nuclear engr. Argonne (Ill.) Nat. Lab., 1950—. Served with U.S. Army, 1942-45. Mem. Am. Nuclear Soc., Sigma Xi. Subspecialty: Nuclear engineering. Current work: Reactor operations. Invented fuel assay reactor, 1958. Home: 502 W Union St Plainfield IL 60544 Office: Argonne Nat Lab 9700 S Cass Ave Argonne IL 60439

MARTIGNONI, MAURO EMILIO, microbiologist, educator; b. Lugano, Switzerland, Oct. 30, 1926; came to U.S., 1956, naturalized, 1963; s. Angiolo Fedele and Lucia Albina (Rava) M.; m. Marie Louise di Suvero, May 15, 1953; children: Enrico A., Matteo L. Diploma in biol. sci, Fed. Inst. Tech., Zurich, Switzerland, 1950, Ph.D., 1956; postgrad., U. Calif., Berkeley, 1951-52. Asst. Fed. Inst. Tech., Zurich, 1950, 52; Am.-Swiss Found. for Sci. Exchange fellow Inst. Internat. Edn., 1951-52; cons. FAO, UN, Rome, 1952-53; entomologist Swiss Forest Research Inst., Zurich, 1953-56; asst. pathologist, lectr. invertebrate pathology U. Calif., Berkeley, 1956-63, assoc. prof., 1963-65; prin. research microbiologist Forestry Scis. Lab., Dept.Agr., Corvallis, Oreg., 1965-68, chief microbiologist, 1968—; prof. invertebrate pathology Oreg. State U., 1965—; co-coordinator U.S. Working Group in Microbiology U.S./USSR Joint Comm. on Sci. and Tech. Cooperation, 1977-83. Author: Laboratory Exercises in Insect Microbiology and Insect Pathology, 1961; contbr. chpts., numerous articles to profl. publs.; editorial bd.: Jour. Invertebrate Pathology, 1964-67, Current Topics in Comparative Pathobiology, 1970-73. Mem. Sierra Club, Citizens for a Clean Environment of Oreg. Recipient Silver medal and Kern award Fed. Inst. Tech., 1957, Superior Service Unit award Dept. Agr., 1977; USPHS grantee, 1958-64; Dept. Agr. grantee, 1975-80. Mem. AAAS, Am. Soc. Microbiology, Am. Soc. Virology, Entomol. Soc. Am., Schweizerische Entomologische Gesellschaft, Soc. Invertebrate Pathology (founding), Sigma Xi. Democrat. Roman Catholic. Subspecialties: Virology (biology); Animal pathology. Current work: Etiology, pathogenesis and pathophysiology of viral diseases of insects; current emphasis is in identification of insect-pathogenic viruses, biological assay and standardization of virus preparations, and safety evaluation of insect virus preparations. Office: Forestry Scis Lab 3200 Jefferson Way Corvallis OR 97331

MARTIN, BILLY R., pharmacologist, educator; b. Winston-Salem, N.C., Apr. 25. 1943; s. John Lee and Leva Montrose (King) M.; m. Jean Yarbrough, May 5, 1947; children: Zachary, Lindsay. A.B., U. N.C., 1965; Ph.D. in Pharmacology, 1974. Asst. prof. pharmacology Med. Coll. Va., Richmond., 1976-82, assoc. prof., 1982—. Contbr. chapts. to books, articles to profl. jours. Swedish Med. Research Council fellow, 1975-76; Wellcome Trust fellow, 1976. Mem. Am. Soc. Pharmacology and Exptl. Therapeutics, Va. Acad. Scis., Sigma Xi. Subspecialty: Pharmacology. Current work: Pharmacology. Office: Med Coll Va Box 613 MCV Station Richmond VA 23298

MARTIN, CHRISTOPHER MICHAEL, medical research administrator, physician; b. N.Y.C., Sept. 25, 1928; s. Christopher William and Genevieve (Grennon) M.; m. Phyllis Walsh, Oct. 16, 1954; children: Eileen Margaret, Christopher Walsh, Marianne. A.B., Harvard U., 1949, M.D., 1953. Intern Boston City Hosp., 1953-54, resident, 1956-57; research fellow Harvard Med. Sch., 1957-59; sr. asst. surgeon NIH, Bethesda, Md., 1954-56; asst. prof. medicine Seton Hall (N.J.) Coll. Medicine, 1959-63, assoc. prof., 1963-65; prof. medicine Georgetown U., 1965-70; sr. dir. med. affairs Merck Sharp & Dohme, West Point, Pa., 1970-77, exec. dir. infectious diseases, 1977—. Contbr. numerous articles to profl. jours. Served with USPHS, 1954-56. Grantee in field. Fellow Infectious Disease Soc. Am.; mem. Am. Soc. Pharmacology and Therapeutics, Am. Soc. Microbiology, Phi Beta Kappa, Alpha Omega Alpha. Roman Catholic. Subspecialties: Infectious diseases; Microbiology (medicine). Current work: Discovery and development of new antibiotics, antimicrobials and vaccines. Home: 395 Woodcrest Rd Strafford PA 19087 Office: Merck Sharp & Dohme Research Labs Sumneytown Pike West Point PA 19486

MARTIN, DAVID EDWARD, physiology educator; b. Green Bay, Wis., Oct. 1, 1939; s. Edward Henry and Lillie (Luckman) M. B.S. in Zoology, U. Wis.-Madison, 1961, M.S. in Zoology and Edn, 1963, Ph.D. in Physiology, 1970. Asst. prof. physiology Ga. State U., Atlanta, 1970-74, assoc. prof., 1974-80, prof., 1980—; collaborating scientist Yerkes Primate Center, Atlanta, 1970-79, affiliate scientist, 1980—. Author: The Marathon Footrace, 1979, The High Jump Book, 1982, La Corsa di Maratona, 1982; contbr. articles to profl. jours. Marathon chmn. Men's Olympic Devel. Com. of Athletics Congress, Indpls., 1979—; coordinator Satellite Physiologic Testing Ctr., U.S. Olympic Com. Elite Athlete Spl. Project, Ga. State U., 1982—. Named Disting. Prof. Ga. State U., 1975, 81; U.S.A. del. Internat. Olympic Acad., 1978; recipient Disting. Service award Athletics Congress, 1982. Mem. Am. Physiol. Soc., Soc. Study of Reproduction, Am. Coll. Sports Medicine, Internat. Primatol. Soc. Republican. Unitarian. Subspecialties: Physiology (medicine); Reproductive biology. Current work: Elucidation of adaptive cardiopulmonary parameters in elite distance runners; rehabilitation of sexual dysfunction in spinal cord injured males: endocrine, testicular morphological, uroneurologic correlates. Home: 510 Coventry Rd Stonehouse Decatur GA 30030 Office: Coll Health Scis Ga State U Atlanta GA 30303

MARTIN, DAVID LEE, biochemist, research scientist; b. St. Louis, May 30, 1941; s. L. Frederick and Ruth Elma (Wilson) M.; m. Sandra Bloom, Aug. 20, 1966; children: Laura Eleanor, Rachel Kirsten. B.S. (Internat. Milling Co. scholar), U. Minn., 1963; M.S., U. Wis., 1965, Ph.D. (NIH fellow), 1968. Research assoc. U. Wis., 1968; asst. prof. chemistry U. Md., 1968-72, assoc. prof., 1972-80; research chemist Armed Forces Radiobiology Research Inst., Bethesda, Md., 1976; vis. scientist, 1977-80; research scientist Ctr for Labs. and Research, N.Y. State Health Dept., Albany, 1980—; dir. Chemistry Assocs. Md. Inc.; cons. in field. Author: Molecules in Living Systems, 2d edit, 1978, Teacher's Guide to Molecules in Living Systems, 2d edit, 1978. NIH grantee., 1975—. Mem. Am. Soc. Biol. Chemists, Biochem. Soc., Soc. Neurosci., Am. Soc. Neurochemistry, N.Y. Acad. Scis., AAAS, Sigma Xi. Subspecialties: Biochemistry (biology); Neurochemistry. Current work: Neurochemistry, neurotoxicology, especially of synaptic transmission and of glial cells. Office: Ctr for Labs and Research NY State Dept Health Albany NY 12201

MARTIN, DONALD VINCENT, psychologist, psychology educator; b. Bethpage, N.Y., July 19, 1952; s. James and Rita (Coriell) M.; m. Maggie Jones, May 13, 1972; children: Paige, Sean. B.S., Roanoke Coll., 1973; M.S., Radford U., 1975; Ph.D., N. Tex. State U., 1980. Diplomate: Am. Acad. Behavioral Medicine.; Lic. psychologist, Okla. Asst. prof. psychology Southside Va. Community Coll., Keysville, Va., 1975-76; asst. prof. psychology Northeastern State U., 1980—; cons. numerous sch. dists., Okla., Tex.,. mental health and guidance Ctrs. Author: Conversations, 1981, Behavioral Medicine, 1982, Blended or Step Families, 1982; Contbr. articles to profl. jours. Coach Tahlequah (Okla.) Youth Soccer, 1981-82; coach Denton (Tex.) Youth Basketball, 1976-80. Mem. Am. Psychol. Assn., Am. Assn. Marriage and Family Therapists, Tex. Assn. Group Therapists, others. Subspecialties: Physiological psychology; Behavioral psychology. Current work: Research in areas of effects of nutritional habits (vitamin intak. Home: Route 2 Box 129C Parkhill OK 74451 Office: Psychology Dept Northeastern State U Tahlequah OK 74464

MARTIN, FREDERICK WIGHT, physicist, educator, consultant; b. Boston, Feb. 16, 1936; s. Frederick E. and Rhoda (Nichols) M.; m. Elizabeth Foltz, Apr. 24, 1965; children: Frederick N., Katharine. A.B., Princeton U., 1957; M.S., Yale U., 1958, PH.D., 1964. Research asst. Yale U., 1959-63; physicist Ion Physics Corp., Burlington, Mass., 1963-64, sr. physicist, 1964-66; amaneunsis I fysik U. Aarhus, Denmark, 1966-68; research assoc., asst. prof. physics U. Ky., Lexington, 1968-70; asst. prof. physics U. Md., College Park, 1970-78; pres. Microscope Assocs., Inc., Dedham, Mass., 1978—; vis. scientist SUNY, Stony Brook, 1975-76; Nat. Acad. Sci.-NRC sr. research assoc., 1978; adj. assoc. prof. physics Worcester Poly. Inst., 1983—. Contbr. articles to profl. publs. Mem. Am. Phys. Soc., IEEE, Sigma Xi. Subspecialties: Microelectronics; Atomic and molecular physics. Current work: Research on ion implantation and device fabrication in semiconductors; stopping of particles; principal investigator of various grants on X-ray production in single atomic collisions at high velocities; studies on use of ions rather than electrons in microscopy. Patentee in field. Office: 50 Village Ave Dedham MA 02026

MARTIN, GEORGE FRANKLIN, anatomist; b. Englewood, N.J., Feb. 20, 1937; s. George Franklin and Nellie Alamanda (Williams) M.; m. Muriel Joy Brandkamp, Aug. 13, 1960; children: George F., Laurie Lee. B.S. in Biology, Bob Jones U., 1960; postgrad., Emory U., 1956-57; M.S. in Anatomy, U. Ala., Birmingham, 1963, Ph.D., 1965. Instr. anatomy Ohio State U. Coll. Medicine, Columbus, 1965-67, asst. prof., 1967-69, assoc. prof., 1970-73, prof., 1973—; mem. neurobiology rev. group NIH. Mem. editorial bd.: Jour. Comparative Neurology, 1981—, Anat. Record, 1982—; contbr. articles to profl. jours. NIH grantee, 1967-80; NSF grantee, 1980—. Mem. Am. Assn. Anatomists, Pan Am. Assn. Anatomists, Am. Neurosci. Soc., AAAS, Internat. Soc. Developmental Neurobiology. Subspecialties: Comparative neurobiology; Anatomy and embryology. Current work: Research in developmental and comparative neuroscience, development of motor systems, development of cerebellum. Office: 1645 Neil Ave Columbus OH 43210

MARTIN, JAMES ARTHUR, aerospace engineer; b. Ft. Benning, Ga., Aug. 9, 1944; s. Paul A. and Mildred Ruby (McDowell) M.; m. Carol Sue Feather, Dec. 31, 1966; children: Fred, Heather, Andy. B.S., W.Va. U., 1966; M.S., MIT, 1967, Aero-Astronautical Engr., 1969; D.Sc., George Washington U., 1982. Aerospace engr. NASA LARC, Hampton, Va., 1966—. Author tech. reports and papers. Mem. AIAA (assoc. editor jour. 1982—, mem. liquid propulsion tech. com. 1980—), Am. Astronautics Soc. Subspecialties: Aerospace engineering and technology; Systems engineering. Current work: Earth-to-Orbit and orbit transfer vehicles; trajectory optimization; preliminary design; cost estimation. Patentee launch vehicle, 1980. Home: Rt 3 Box 1140 Gloucester VA 23061 Office: NASA LARC MS 365 Hampton VA 23665

MARTIN, JAMES CULLEN, chemistry eductor, consultant; b. Dover, Tenn., Jan. 14, 1928; s. Joseph Emmett and Myrtle Irene (Futrell) M.; m. June Echols, Aug. 25, 1951; children: Joseph, Bruce, Kendall, Steven, Christopher. B.A., Vanderbilt U., 1951, M.S., 1952; Ph.D., Harvard U., 1956. Instr. U. Ill., Urbana, 1956-59, asst. prof., 1959-62, assoc. prof., 1962-65, prof. chemistry, 1965—; cons. Dow Chem. Co., Midland, Mich., 1953-68, Eli Lilly, Indpls., 1960-62; expert witness in field, 1978—. Contbr. articles to profl. jours. Served with U.S. Army, 1946-50. Guggenheim Found. fellow, 1966; Alfred P. Sloan Found. fellow, 1962-66; recipient U.S. Sr. Scientist award; Alexander von Humboldt awardee, Germany, 1979. Fellow AAAS; mem. Am. Chem. Soc. (Buck-Whitney medal 1979, chmn. subcom. on new products and services 1977-82, chmn. organic div 1982-83, chmn. subcom. on jours. monitoring 1983), Sigma Xi, Alpha Chi Sigma, Phi Beta Kappa. Presbyterian. Subspecialties: Organic chemistry; Inorganic chemistry. Current work: Coordination chemistry of nonmetallic elements, hypervalent species, mechanisms of reactions, free-radical initiators, physical-organic chemistry. Home: 11 Lake Park Dr Champaign IL 61820 Office: 245 Roger Adams Lab U Ill 1209 W California St Urbana IL 61801

MARTIN, JAMES TILLISON, biologist; b. Bluefield, W.Va., Aug. 10, 1946; s. James T. and Ora Marie (Cross) M. B.A., W.Va. U., 1968; M.S., U. Conn., 1971; Dr. rer. nat., U. Munich, W.Ger., 1974. Postdoctoral asso. U. Utrecht, Netherlands, 1974; postdoctoral asso. U. Minn., 1975-76; asst. prof. biology Stockton State Coll., Pomona, N.J., 1976—; vis. scientist M.I.T., 1980-81. Mem. Soc. Neurosci., Animal Behavior Soc., Am. Soc. Zoologists, Internat. Soc. Psychoneuroendocrinology, Phi Beta Kappa. Subspecialties: Neuroendocrinology; Psychobiology. Current work: Endocrine influence on growth and devel. of nervous system and behavior; physiol. and behavioral aspects of domestication. Office: Biology Program Stockton State Coll Pomona NJ 08240

MARTIN, JOHN AYRES, research psychologist, educator, consultant; b. Taunton, Mass., July 18, 1952; s. Ayres and Beatrice (Almeida) M. B.A. summa cum laude, Harvard U., 1974; Ph.D. with distinction, Stanford U., 1978. Instr. pediatrics Stanford (Calif.) Med. Ctr., 1979-82; research assoc. dept. psychology Stanford U., 1978-83; asst. research psychologist U.Calif.-Berkeley, 1981—; assoc. prof. Calif. Sch. Profl. Psychology-Berkeley, 1983—; cons. Mt. Zion Med. Ctr., San Francisco, 1981—; ad hoc reviewer various profl. jours., 1975—; Vis. scientist Tavistock Inst., London, 1973. Author: Making Sense with Data, 1984; Editorial bd.: Child Devel., Jour. Child Psychology and Psychiatry; contbr.: articles to profl. jours. including Pediatrics. Grad. fellow NSF, 1974-77; research grantee William T. Grant Found., 1983—. Mem. Am. Psychol. Assn., Soc. for Research in Child Devel., Am. Orthopsychiat. Assn. Subspecialties: Developmental psychology; Social psychology. Current work: Development of interpersonal functioning; social development; longitudinal research; systems theory; stress and coping (early adolescence through adulthood); applied multivariate research methods. Home: 263 Castro St San Francisco CA 94114 Office: Calif Sch Profl Psychology 1900 Addison St Berkeley CA 94704

MARTIN, JOSE GINORIS, energy engineering educator, consultant, researcher; b. Havana, Cuba, Feb. 4, 1941; s. José and María De La Paz (Ginoris) M.; m. Dagma Faria Neto, Sept. 2, 1975; children: Victor José Faria, Cassia Inês Faria. B.S. in Engring, Miss. State U., 1964, M.S., U. Wis., 1966, Ph.D., 1970. Engr. Theory div. Lawrence Radiation Lab., Livermore, Calif., 1968; prof. Instituto Politecnico Nacional, Mexico City, 1970-75; prof., grad. coordinator dept. energy engring. U. Lowell, Mass., 1975—; vis. prof. Instituto Militar de Engenharia, Rio de Janeiro, 1972-73, U. Mexico, 1976-77; vis. faculty Coll. Architecture Ariz. State U., Tempe, 1983—; prin. investigator Nuclear Regulatory Commn. Project, Lowell, Mass., 1977-80, NSF project, 1981—; mem. internat. test and evaluation team Sandia Nat. Labs./Small Solar Power Systems, Almeria, Spain, 1982-83; Cons. Oak Ridge Nat. Lab, Los Alamos Nat. Lab., N. Mex. Editor: Procs. of Internat. Workshop on Distributed Solar Collectors; Contbr. articles to profl. publs. Bd. dirs. Unitas-Service Orgn., Lowell, 1979-82, Talented and Gifted program Lowell Schs., 1980, Lowell Manpower Bd., 1980-82. Wis. Alumni Research Found. fellow, 1965-66; AEC fellow, 1966-69; NSF grantee, 1978—. Fellow Am. Soc. Engring. Edn.; mem. Internat. Solar Energy Soc., Am. Nuclear Soc., Am. Phys. Soc., AAAS, Mexican Soc. Physics, Sigma Xi, Tau Beta Pi, Phi Kappa Phi. Subspecialties: Solar energy; Nuclear engineering. Current work: Solar thermal engineering; fusion and fission energy systems; cost of power. Patentee in field. Home: 85 Mansur St Lowell MA 01852 Office: Department of Energy Engineering University of Lowell Lowell MA 01854

MARTIN, JOSEPH PATRICK, JR., biologist, educator; b. Lynn, Mass., May 21, 1952; s. Joseph Patrick and Lorraine Albertine (Levesque) M. A.B. in Biology, Harvard U., 1973; Ph.D. in Zoology, Duke U., 1979. NSF postdoctoral fellow Duke U. Med. Center Biochemistry, 1979-80, NIH postdoctoral fellow, 1980-81; asst. prof. biology William Marsh Rice U., Houston, 1981—. Contbr. articles to profl. publs. Am. Heart Assn.-Tex. Affiliate research grantee; Robert A. Welch Found. grantee, 1982. Mem. Genetics Soc. Am., Soc. Study of Evolution, Am. Soc. Zoologists, Am. Chem. Soc. Roman Catholic. Subspecialties: Biochemistry (biology); Evolutionary biology. Current work: The intracellular production and deleterious effects of oxygen free radicals. The evolution and biological functions of superoxide dismutase. Home: 19191/2 South Blvd Houston TX 77098 Office: 6100 S Main St Houston TX 77001

MARTIN, LARRY DEAN, geology educator; b. Bartlett, Nebr., Dec. 8, 1943; s. Orval George and Nellie B. (Frey) M.; m. Jean Anne Bright, Jan. 28, 1967; children: Amanda, Mary. B.S. in Zoology, U. Nebr., 1966, M.S. in Geology, 1969; Ph.D. in Biology, U. Kans., 1973. Field party leader U. Nebr. State Mus., Lincoln, 1965-71; curator Mus. Natural History, prof. systematics and ecology and adj. prof. geology U. Kans., Lawrence, 1972—. Contbr. articles to profl. jours. NSF grantee, 1976-82; Nat. Geog. Soc. grantee, 1980-81; U. Kans. grantee, 1973-84. Mem. Tertiary-Quaternary Assn., Soc. Vertebrate Paleontology, Internat. Union Quaternary Research, Sigma Xi, Sigma Gamma Epsilon. Subspecialties: Systematics; Paleobiology. Home: 2569 Ousdahl Rd Lawrence KS 66044 Office: U Kans Dept Systematics Ecology Lawrence KS 66045

MARTIN, L(AURENCE) ROBBIN, research scientist; b. Annapolis, Md., Sept. 21, 1939; s. Lawrence H. and Ethel A. (Robbin) M. B.A., Pomona Coll., 1961; Ph.D., MIT, 1966. Noyes postdoctoral fellow Calif. Inst. Tech., 1966-68; asst. prof. chemistry U. Calif., Riverside, 1968-73; research scientist Aerospace Corp., Los Angeles, 1973—. Author: (with Gokcen) Solution Manual for Thermodynamics, 1978, 'SO2, NO, and NOx Oxidation. . ." Atmospheric Considerations, 1983; also articles. Grantee Research Corp., NSF, EPA, Am. Chem. Soc. Mem. Am. Phys. Soc., AAAS, Am. Geophys. Union, Sigma Xi. Subspecialties: Kinetics; Atmospheric chemistry. Current work: Gas phase and aqueous kinetics, thermodynamics, atmospheric chemistry; research in atmospheric chemistry. Office: Chemistry and Physics Labs Aerospace Corp Los Angeles CA 90009

MARTIN, LAWRENCE RONALD, nuclear engineer, aerospace engineer; b. Sterling, Ill., Nov. 22, 1946; s. Arthur Wilford and Dorothy Helen (Snyder) M.; m. Karen Patricia Morgan, Mar. 5, 1982; children: Mary, Benjamin. B.S., U. Kans.-Lawrence, 1972; M.S., Air Force Inst. Tech., 1980. Commd. 2d lt. U.S. Marine Corps, 1972, advanced through grades to maj., 1982; combat engr. firepower div. devel. ctr. (U.S. Marine Corps Devel. and Edn. Command), Quantico, Va., 1972—. Mem. Am. Nuclear Soc., IEEE. Roman Catholic. Subspecialties: Nuclear engineering; Laser applications. Current work: Military applications, nuclear and chemical warfare and directed energy weapons, research and development of offensive and defensive systems, including system survivability design hardening against their effects.

MARTIN, LEE DAVID, defense electronics scientist, consultant; b. Chillicothe, Mo., June 23, 1923; s. Albert Justin and Matilda (Hankins) M.; m. Betty Jeanne Holmes, Feb. 16, 1946; children: Mary Ann Martin Toohey, William Albert. B.S.E. in Chemistry, George Washington U., 1958; M.S. in Mgmt, Rensselaer Poly. Inst., 1963. Commd. 2d lt. U.S. Marine Corps, 1946, advanced through grades to lt. col., 1964, ret., 1966; mem. tech. staff Hdqrs. Marine Corps, Washington, 1953-58; mem. logistics staff First Marine Aircraft Wing, Iwakuni, Japan, 1958-59; mem. staff F8U aircraft program Marine Corps Air Sta., Beaufort, S.C., 1959-62; officer-in-charge planning office Marine Corps East Coast Logistics Staff, Albany, Ga., 1963-66; advance planner Gen. Electric Fleet Ballistic Missile Support Project, Gen. Electric Co., Pittsfield, Mass., 1966-78; mgr. tech. planning Gen. Electric Ordnance Systems, 1978—. Fellow Wildfowl Trust (U.K.); mem. Ops. Research Soc. Am., Am. Geophys. Union, Am. Phys. Soc. Republican. Presbyterian. Current work: Secondary Effects on orbital characteristics of orbital craft. Home: 165 Tower Rd Dalton MA 01226 Office: General Electric Company Ordnance Systems 100 Plastics Ave Pittsfield MA 01201

MARTIN, LEONARD JAMES, planetary astronomy researcher; b. Seattle, Mar. 10, 1930; s. Alvord Bushnell and Margaret Catherine (Ditty) M.; m. Claudia Hepke, June 18, 1954; children: Christoph, Nicholaus, Jennifer. B.A., U. Wash., 1958. Cartographer Aero. Chart and Info. Ctr., U.S. Air Force, St. Louis, 1958-63; Cartographer, lunar observer, Flagstaff, Ariz., 1963-67; planetary researcher-observer Lowell Obs., Flagstaff, 1967—. Contbr. articles to profl. jours. Bd. dirs. Flagstaff Festival of Arts, 1981-84. Served with U.S. Army, 1952-54. Mem. Am. Geophys. Union, Am. Astron. Soc. Republican. Clubs: Fairfield Continental Country, Page-Lake Powell Boating. Subspecialties: Planetary science; Planetary imaging. Current work: Computer-aided analysis and photometry of Viking Orbiter images of Mars, especially concerning limb clouds, dust clouds and possible volcanic clouds; acquisition and analysis of telescopic photography of Mars. Home: 3208 N 4th St Flagstaff AZ 86001 Office: Lowell Obs PO Box 1269 Flagstaff AZ 86002

MARTIN, LOREN GENE, medical educator and researcher; b. Danville, Ill., Oct. 31, 1942; s. Eugene Edwin and Nona Lucille (Jones) M.; 1 son, Peter L. A.B., Ind. U., 1965, Ph.D., 1969. Postdoctoral research assoc. environ. physiology Ind. U., Bloomington, 1969-70; instr., then asst. prof. Temple U. Med. Sch., Phila., 1970-73; asst.prof., assoc. prof. U. Ill. Coll. Medicine, Chgo. and Peoria, 1973-78; assoc. prof. Va. Commonwealth U., Richmond, 1978-81, Okla. Coll. Osteo. Medicine and Surgery, Tulsa, 1981—; research grant reviewer regulatory biology sect. NSF, Washington, 1982—. Contbr. articles to profl. jours. Sr. staff fellow FDA, 1981; research grantee Ind. Acad. Sci., 1968, Ill. Heart Assn., 1977, Va. Center on Aging., 1979, 80. Mem. Am. Physiol. Soc., Endocrine Soc., Internat. Assn. Heart Research, Soc. Exptl. Biology and Medicine, Sigma Xi. Subspecialty: Physiology (medicine). Current work: Physiological research: high altitude and endocrine effects upon cardiovascular system, aging and sex steroid effects on heart. Home: 321 N Santa Fe Tulsa OK 74127 Office: Okla Coll Osteo Medicine and Surgery 1111 W 17th St PO Box 2280 Tulsa OK 74101

MARTIN, MALCOLM ALAN, molecular geneticist, researcher; b. Washington, Aug. 3, 1938; s. Raymond Earl and Rose Lillian (Cohn) M.; m. Susan E. Martin, Aug. 18, 1973; children: Daniel B., David A. A.A., George Washington U., 1958; M.D., Yale U., 1962. Diplomate: Nat. Bd. Med. Examiners. Intern Strong Meml. Hosp., Rochester, N.Y., 1962-63, resident in internal medicine, 1963-64; sr. scientist Nat. Inst. Allergy and Infectious Diseases, NIH, Bethesda, Md., 1971—, chief Lab. Molecular Microbiology, 1981—; cons. in field. Contbr. articles to profl. jours. Served with USPHS, 1964-71. Subspecialties: Molecular biology; Virology (biology). Current work: Molecular biology of animal viruses and their integration into mammalian cells. Home: 6408 Crane Terr Bethesda MD 20817 Office: Bldg 5 NIH Bethesda MD 20205

MARTIN, PAUL CECIL, physicist, educator; b. Bklyn., Jan. 31, 1931; s. Harry and Helen (Salzberger) M.; m. Ann Wallace Bradley, Aug. 7, 1957; children: Peter, Stephanie, Daniel. A.B., Harvard U., 1952, Ph.D., 1954. Faculty Harvard U., 1957—, prof. physics, 1964-82; J. H. VanVleck prof. pure and applied physics; chmn. dept. physics Harvard U., 1972-75, dean div. applied scis., 1977—; vis. prof. Ecole Normale Superieure, Paris, 1963, 66, U. Paris (Orsay), 1971. Bd. editors: Jour. Math Physics, 1965-68, Annals of Physics, 1968-82, Jour. Statis. Physics, 1975-80. Bd. dirs. Assoc. Univs. for Research in Astronomy, 1979—; trustee Assoc. Univs., Inc., 1981—. NSF postdoctoral fellow, 1955; Sloan Found. fellow, 1959; Guggenheim fellow, 1966, 71. Fellow Nat. Acad. Scis., Am. Acad. Arts and Scis., AAAS, Am. Phys. Soc. (Councillor-at-large 1982-84); mem. Inst. Theoretical Physics (chmn. adv. com. 1979). Subspecialties: Condensed matter physics; Statistical physics. Current work: Phase transaction, nonequililbrium phenomena, turbulence and chaos. Home: 27 Stone Rd Belmont MA 02178 Office: Lyman Lab Physics Pierce Hall Harvard U Cambridge MA 02138

MARTIN, PAUL JOSEPH, research director, consultant; b. Hammond, Ind., May 22, 1936; s. Joseph Edward and Verna Catherine (Heidgerken) M.; m. Jeanne Therese Oubre, Sept. 10, 1960; children: Mary Kay, Barry, Craig, Colleen. B.S.E.E., U. Tex.-Austin, 1961; M.S., Drexel U., 1962; Ph.D., Case-Western Res. U., 1967. Research fellow Latter-day Saints Hosp., Salt Lake City, 1962-63; research bioengr. Tech., Inc., Dayton, Ohio, 1963-64; research assoc. Mt. Sinai Med. Ctr., Cleve., 1967-75, head bioengring. sect., 1975—; asst. prof. Case-Western Res. U., Cleve., 1967-75, assoc. prof., 1975-83, prof., 1983-; research cons. NIH, NSF, EPA, Cleve. Clinic, 1975—. Assoc. editor: Am. Jour. Physiology, 1975-80; editorial bd.: Circulation Research, 1975-82. Served with USN, 1954-57. Research grantee Am. Heart Assn., 1967-78, NIH, 1972—. Fellow Am. Heart Assn. (chmn. research study sect. 1973-76, mem. regional research com. council on circulation 1977—); mem. Am. Physiol. Soc., Biomed. Engring. Soc. (membership com. 1972—), IEEE. Clubs: Classical Guitar Soc., Western Res. Woodworkers (Cleve.). Subspecialties: Physiology (medicine); Software engineering. Current work: Dynamic and steady-state control of the heart and circulation, cardiovascular pharmacokinetics, laboratory instrumentation and computer control, computers in health care delivery. Home: 2099 Lamberton Rd Cleveland Heights OH 44118 Office: Mount Sinai Med Center University Circle Cleveland OH 44106

MARTIN, PRESLEY FRANK, biologist; b. Jeffersonville, Ind., Feb. 27, 1948; s. Presley Frank and Mary Garnett (Hartling) M.; m. Andrea Ruth Martin, June 15, 1968; children: PResley Kenneth, Benjamin Louis. Student, Dartmouth Coll., 1966-68; B.S. with honors in Biology, Ind. U., Bloomington, 1971; Ph.D. in Biology, Johns Hopkins U., 1978. Busch fellow Waksman Inst. Microbiology Rutgers U., 1980-82, research asst. prof., 1982—. Am. Cancer Soc. fellow, 1978-80. Mem. Genetics Soc. Am. Subspecialties: Genetics and genetic engineering (biology); Developmental biology. Current work: Developmental genetics of drosophila, molecular mechanisms of gene expression. Home: 29 Hickory St Metuchen NJ 08840 Office: PO Box 759 Piscataway NJ 08854

MARTIN, RICHARD LEE, research chemist; b. Garden City, Kans., Oct. 13, 1950; s. George Abraham and Della Faye (Faulkender) M. B.S. in Chemistry, Kans. State U., 1972, Ph.D., U. Calif., Berkeley, 1976. Research asso. U. Wash., Seattle, 1977; Chaim Weizman fellow, 1978; mem. staff theoret. div. Los Alamos Nat. Lab., 1979—. Contbr. numerous articles to sci. jours. Mem. Am. Chem. Soc., Am. Phys. Soc. Subspecialties: Theoretical chemistry; Physical chemistry. Current work: Quantum chemistry; ab initio electronic structure of molecules. Office: Theoretical Div MSJ 569 Los Alamos Nat Lab Los Alamos NM 87544

MARTIN, SCOTT ELMORE, food science educator; b. Wilmington, Del., Sept. 17, 1946; s. Elmore Louis and Elizabeth (Welch) M.; m. Charlotte Jean Goff; children: Reid Scott, Laine Elizabeth. B.A., Tarkio Coll., 1968; M.S., Wichita State U., 1970; Ph.D., Kans. State U., 1973. Research asst. Kans. State U., Manhattan, 1970-73; research assoc. U. Calif.-Irvine, 1973-75, U. Ill.-Urbana, 1975-77, asst. prof. dept. food sci., 1977-81, assoc. prof., 1981—; cons. Combe Labs., Rantoul, Ill., 1979—, TeePak, Inc., Danville, Ill., 1979—. Contbr.: chpt Sources of Food Spoilage, 1979, others. Mem. Am. Soc. Microbiology, Inst. Food Technologists, Sigma Xi. Presbyterian. Subspecialties: Microbiology; Cancer research (medicine). Current work: Sublethal injury to microorganisms, oxygen toxicity, selenium and carcinogenic bioactivation, cambylobacter infection, mode of action of antimicrobials, sanitation consulting. Office: Dept Food Sci Univ Ill 905 S Godowin Ave Urbana IL 61801 Home: 2506 Brett Dr Champaign IL 61821

MARTIN, SCOTT MCCLUNG, biochemist, editor; b. Charleston, W.Va., Mar. 2, 1943; s. Harry Milton and Phyllis Lee (Amos) M. B.S., Marshall U., 1965; M.S., Brigham Young U., 1968; Ph.D., Ohio State U., 1973. Asst. prof. biology Wittenberg U., Springfield, Ohio, 1973-75; postdoctoral study Ohio State U., Columbus, 1975-76; curator Bacillus Genetic Stock Center, 1978-79; postdoctoral study U. Mo., Kansas City, 1976-78, vis. asst. prof. biology, 1978; asso. editor of biochemistry Chem. Abstracts Service, Columbus, 1980—. Contbr. articles to profl. publs. Mem. Am. Soc. Microbiology, Soc. Protozoologists, Am. Inst. Biol. Scis., AAAS, Ohio Acad. Sci. Democrat. Mormon. Subspecialties: Microbiology; Genome organization. Home: 712 Harley Dr Columbus OH 43202 Office: Dept Biochemistry Chem Abstracts Service Columbus OH 43210

MARTIN, T. JOE, plant breeder/pathologist; b. Baxter Springs, Kans., Dec. 28, 1947; s. Verl Curtis and Marie (Bowers) M.; m. Lorna June Price, Nov. 5, 1966; children: Todd Curtis, Denny Joe, Jeffery Price. B.S., Pittsburg (Kans.) State U., 1970; M.S. in Plant Pathology, Kans. State U., 1971; Ph.D., Mich. State U., 1974. Asst. prof., plant pathologist Kans. State U., Fort Hays, 1974-80; asst. prof., wheat breeder, 1980-82, assoc. prof., wheat breeder, 1982—. Mem. crop Sci. Soc. Am., Am. Phytopath. Soc. Subspecialty: Plant genetics. Address: Expt Sta Kans State U Hays KS 67601

MARTIN, TERRY ZACHRY, astronomer; b. N.Y.C., Aug. 7, 1946; s. Edward Berry and Marjorie Mae (Boyd) M.; m. Marcia Baker, May 30, 1982. A.B., U. Calif., Berkeley, 1967; M.S., U. Hawaii, 1969, Ph.D., 1975. Research geophysicist UCLA, 1975-79; mem. tech. staff Jet Propulsion Lab., Pasadena, Calif., 1979—. Mem. Am. Astron. Soc., Common Cause, ACLU, Cousteau Soc., Planetary Soc. Subspecialties: Planetary science; Infrared optical astronomy. Current work: Planetary atmospheres, infrared detector tech., optics, vacuum tech. Home: 2031 Ahlin Dr La Canada Flintridge CA 91011 Office: 4800 Oak Grove Dr 11-116 Pasadena CA 91109

MARTIN, THOMAS LYLE, JR., univ. pres.; b. Memphis, Sept. 26, 1921; s. Thomas Lyle and Malvina (Rucks) M.; m. Helene Hartley, June 12, 1943; children—Michele Marie, Thomas Lyle. B.E.E., Rensselaer Poly. Inst., 1942, M.E.E., 1948, D.Eng., 1967; Ph.D., Stanford U., 1951. Prof. elec. engring. U. N.Mex., 1948-53; prof. engring. U. Ariz., 1953-63, dean engring., 1958-63, U. Fla., Gainesville, 1963-66, So. Meth. U., Dallas, 1966-74; pres. Ill. Inst. Tech., Chgo., 1974—; dir. Stewart-Warner Co., Inland Steel Co., Amsted Industries, Cherry Elec. Products Corp., Commonwealth Edison Co., Sundstrand Corp., Hyatt Internat. Author: UHF Engineering, 1950, Electronic Circuits, 1955, Physical Basis for Electrical Engineering, 1957, Strategy for Survival, 1963, Electrons And Crystals, 1970, Malice in Blunderland, 1973. Mem. Dallas-Fort Worth Regional Airport Bd., 1970-74; bd. dirs. Museum Sci. and Industry, 1975—. Served to capt. Signal Corps AUS, 1943-46. Decorated Bronze Star. Fellow IEEE; mem. Nat. Acad. Engring., Inst. Gas Tech. (dir.). Subspecialties: Electronics; 3emiconductors. Home: 990 Lake Shore Dr Apt 19C Chicago IL 60611

MARTIN, WILLIAM EUGENE, physicist; b. St. Joseph, Mo., Dec. 19, 1941; s. George Edward and Ruth (Berry) Hartnett) M.; m. Susan Dewar, Dec. 28, 1964; children: Christopher E., David L. B..S. in Physics, San Diego State U., 1969, M.S., 1970; Ph.D. in Applied Physics, U. Calif.-San Diego, 1974. Physicist Naval Electronics Lab., San Diego, 1970-75, Lawrence Livermore Nat. Lab., Livermore, Calif., 1975—. Served with USN, 1963-66. Naval Electronics Lab. Ctr. fellow, 1974; Atomic Weapons Research Establishment, Aldermaston, Eng.) fellow, 1981-82. Mem. Optical Soc. Am., Am. Nuclear Soc., AAAS, Calif. Acad. Sci. Republican. Subspecialties: Particle physics; Laser propagation physics, free electron lasers. Current work: Laser and particle beam physics, high efficiency lasers, laser fusion research, laser/particle beam propogation. Patentee in field of integrated optics, optical logic devices, 1972-77. Home: 3192 Malvasia Ct Pleasanton CA 94566 Office: Lawrence Livermore Nat Lab L-436 PO Box 808 Livermore CA 94550

MARTINDALE, COLIN EUGENE, psychology educator; b. Ft. Morgan, Col., Mar. 21, 1943; s. Roy W. and Martha (Dill) M.; m. Judith Calhoun Kopecky, Aug. 21, 1964 (div. 1973). B.A., U. Col., 1964; Ph.D., Harvard U., 1970. Asst. prof. to prof. psychology U. Maine, Orono, 1970—. Author: Romatic Progression, 1975, Cognition and Consciousness, 1981; editor: Empirical Studies of the Arts, 1982—. Mem. state exec. com. Libertarian Party of Maine, 1981. Recipient 1st prize Am. Insts. for Research, 1970. Fellow Am. Psychol. Assn.; mem. Assn. Computers and the Humanities (exec. council 1979-80), Soc. for Personology (founding), Assn. Lit. and Linguistic Computing (founding), Phi Beta Kappa. Subspecialties: Psychological Aesthetics; Psychological Hedonics. Current work: Development and testing an evlutionary theory of the hstory of art and literature; general determinants of aesthetic pleasure; creativity. Home: 89 3d St Bangor ME 04401 Office: U Maine Orono ME 04469

MARTINEZ, AUGUSTO JULIO, neuropathologist, cons.; b. St. Cruz del Sur, Cuba, Apr. 12, 1930; s. Augusto and Aurora (Avila) M.; m. Josephine Martinez, Oct. 15, 1967; children: Killeen, Bridget, Mary. M.D., U. Havana, 1959. Diplomate: Am. Bd. Pathology. Asst. prof. pathology Med. Coll. Va., 1969-70, asso. prof. pathology, 1972-76; asso. prof. pathology, asst. prof. neurology U. Tenn., 1970-72; prof. pathology U. Pitts., 1976—; neuropathologist John Gaston Hosp., West Bowl Hosp., LeBonheur Children's Hosp., Memphis, 1970-72, St. Jude Children's Research Hosp., 1970—, Presbyn.-Univ. Hosp., Pitts., 1970—. Editorial bd.: Jour. Neuropathology and Exptl. Neurology, 1980—; Contbr. articles to profl. jours. Grantee Health Research and Services Found. of Pitts., 1977-79, Muscular Dystrophy Assn., 1978-79, NIH, 1978-81. Mem. Coll. Am. Pathologists, Am. Soc. Clin. Pathologists, Am. Assn. Neuropathologists, AMA, Internat. Acad. Pathology, Am. Acad. Neurology, Am. Assn. Pathologists. Republican. Roman Catholic. Subspecialties: Pathology (medicine); Neurology. Current work: Encephalitis due to free living amebas. Brain tumors—free-living amebic infections—electron microscopy. Pathology of skeletal muscle disease—immunoperoxidase technique. Office: U Pitts Sch Medicine Pittsburgh PA 15261

MARTINEZ, IRVING RICARDO, JR., dermatologist, anatomist and electron microscopist; b. New Orleans, Apr. 30, 1935; s. I. Ricardo and Amelia (Areces) M.; m. Dolly-Dean Kimball, June 24, 1961; children: Tessa Mariana, I. Ricardo. B.S. in Biology, Loyola U., New Orleans, 1958, M.S. in Physiology, 1960; M.D., La. State U., 1965; Ph.D. in Anatomy, Boston U., 1971. Diplomate: Am. Bd. Dermatology and Dermatopathology. Instr. physiology La. State U., New Orleans, 1959-60; dermatology Boston U., 1969-70; assoc. chmn. dermatology Ochsner Clinic, New Orleans, 1971-74; head electron microscopu lab. Alton Oschsner Med. Found., New Orleans, 1972-74, asst. dir. research, 1973-74; Practice medicine specializing in dermatology, Metairie, La., 1974—; clin. assoc. prof. Tulane Med. Sch., New Orleans, 1970—, La. State U., 1970—; cons. E. Jefferson Gen. Hosp., Metairie, 1974—, Lakeside Hosp., 1974—. Author: Wound Healing, 1972; editor: Med. Sch. Newspaper, 1963-64, Yearbook, 1963-65. Bd. dirs. New Orleans Opera Assn., 1975, Delta Festival Ballet Co., New Orleans, 1977. Served to capt. M.S.C. U.S. Army, 1958-60. Recipient Russel L. Holman award La. State U. Med. Sch., 1964, Mabel Claire Elmore award, 1965; NIH grantee, 1974-76. Fellow Am. Acad. Dermatology (Bronze award 1971), ACP, Am. Soc. Dermapathology; mem. Am. Assn. Anatomists, La. Dermatol. Soc. (sec.-treas. 1972-73). Clubs: Bienville, Pendennis (New Orleans); Metairie Country. Subspecialties: Dermatology; Cell biology (medicine). Current work: Dermatology, dermatopathology, electron microscopy, research in normal and abnormal keratinization, wound healing, membrane coating granules, psoriasis, laser surgery. Office: Dermatology-Allergy-Dermatologic Surgery 3333 Kingman St Suite M Metairie LA 70002

MARTINEZ-CARRION, MARINO, biochemistry educator; b. Felix, Almeria, Spain, Dec. 2, 1936; came to U.S., 1957; s. Juan Martinez and Maria Carrion; m. Jeanne Larson, Sept. 7, 1957; children: Victoria, Juan. B.A., U. Calif.-Berkeley, 1959, M.A., 1961, Ph.D., 1964. Asst. prof. U. Notre Dame, Ind., 1965-69, assoc. prof., 1969-74, prof., 1974-77; prof., chmn. dept. biochemistry Va. Commonwealth U., Richmond, 1977—; chmn. biophys. chemistry study sect. NIH, 1977-82. Contbr. over 80 articles to sci. jours. NIH Research Career awardee, 1972-77; recipient Faculty Excellence award Va. Commonwealth U., 1982. Fellow N.Y. Acad. Scis.; mem. Am. Soc. Biol. Chemists, Pan Am. Assn. Biochem. Socs. (sec. 1981—), Am. Soc. Neurochemistry, Am. Chem. Soc., Spanish Biochem. Soc. Subspecialties: Biochemistry (biology); Biophysics (biology). Current work: Neuroreceptors, enzyme structure and function, spectroscopy of macromolecules, membrane function. Office: Va Commonwealth U Dept Biochemistry Richmond VA 23298

MARTIN-ROBINSON, KERA GAYLE, technical consultant; b. N.Y.C., Dec. 13, 1953; d. Sigmunt and Pauline Lillian (Saretsky) Schwager; m. William Edward Robinson, July 12, 1981. Systems analyst New Eng. Life Ins. Co., Boston, 1973-77; database analyst Standard Oil of Calif., San Francisco, 1977-79; database tech. rep. Nat. CSS, San Francisco, 1979-82; pres. Softstone Cons., San Francisco, 1980—; tech. cons. info. ctr. Fed Res. Bank, San Francisco, 1982-83; tech. instr. Exploratorium of San Francisco, 1982-83; sr. tech. instr., course designer Fortune Systems, Redwood City, 1983—. Mem. Assn. Computing Machinery, Nat. Computer Graphics Assn. Democrat. Subspecialties: Database systems; Graphics, image processing, and pattern recognition. Current work: Advances in database technology to enhance user acceptance. Special interests in computer graphics and animation for effective communication and in the arts; in computer aided instrn. for end-user training. Office: Fortune Systems 101 Twin Dolphin Blvd Redwood City CA 94065

MARTINS, DONALD HENRY, astronomer, astrophysicist, educator; b. Poplar Bluff, Mo., July 31, 1945; s. Otto Henry and Winifred Fanny (Bellamy) M.; m. Joyce Ann, Aug. 1, 1969; 1 dau., Winifred. B.S., U. Mo., 1967, M.S., 1969; Ph.D., U. Fla., 1974. Research assoc. NASA/NAS/NRC, Johnson Space Center, Tex., 1974-76; Houston Baptist U., 1976-77; instr. U. Houston, Clear Lake City, 1976-77, U. St. Thomas, 1976-77; asst. prof. dept. physics and astronomy U. Ga., Athens, 1977-82; asst. prof. dept. physics/chemistry/astronomy U. Alaska, Anchorage, 1982—. Contbr. articles to profl. publs. Served with U.S. Army, 1969-70. Research Corp. grantee, 1978-79. Mem. Am. Astron. Soc., Astron. Soc. Pacific, Internat. Astron. Union, Nat. Geog. Soc., Am. Orchid Soc. Subspecialties: Optical astronomy; Graphics, image processing, and pattern recognition. Current work: Surface and stellar photometry of globular clusters using image processing techniques; astron. instrumentation. Office: Dept Physics/Chemistry/Astronomy Univ Alaska Anchorage AK 99508

MARTINSON, CHARLIE ANTON, plant pathology educator, researcher, cons.; b. Orchard, Colo., Sept. 15, 1934; s. George W. and Bertha E. (Wirsing) M.; m. Kathryn A. Reichert, June 9, 1957; children: Gerald, Karen, Nancy, Brian. B.S., Colo. State U., 1957, M.S., 1959; Ph.D., Oreg. State U., 1964. Asst. prof. Cornell U., Ithaca, N.Y., 1963-68; assoc. prof. plant pathology Iowa State U., Ames, 1968—; cons. Corn Prodn. Systems, Chgo., Am. Agrl. Industries, Rosemont, Ill. Research publs. in field. Served to capt. USAR, 1957-63. Mem. Am. Phytopath. Soc., AAAS. Lutheran. Subspecialties: Plant pathology; Integrated pest management. Current work: Physiology and control of corn diseases; biol. control and mgmt. fungal pathogens; resistance and breeding of corn for resistance. Office: Iowa State U 425 Bessey Hall Ames IA 50011

MARTINSON, IDA MARIE, nurse-physiologist, educator; b. Mentor, Minn., Nov. 8, 1936; d. Oscar and Marvel (Nelson) Sather; m. Paul Varo Martinson, Mar. 31, 1962; children—Anna Marie, Peter. Diploma, St. Luke's Hosp. Sch. Nursing, 1957; B.S., U. Minn., 1960, M.N.A., 1962; Ph.D., U. Ill., Chgo., 1972. Instr. Coll. St. Scholastica and St. Luke's Sch. Nursing, 1957-58, Thornton Jr. Coll., 1967-69; lab. asst. U. Ill. at Med. Center, 1970-72; lectr. dept. physiology U. Minn., St. Paul, 1972—, asst. prof., 1972-74, asso. prof., research, 1974-77, prof., dir. research, 1977—; vis. research prof. Nat. Taiwan U. Def. Med. Center, 1981. Author: Mathematics for the Health Science Student, 1977; editor: Home Care for the Dying Child, 1976, Women in Stress, 1979; contbr. chpts. to books, articles to profl. jours. Active Am. Cancer Soc., Am. Heart Assn. Recipient Am. Bus. Press award, 1977; recipient various grants. Mem. Council Nurse Researchers, Nat. League for Nursing, Am. Acad. Nursing, Am. Nurses Assn., Inst. Medicine, Sigma Xi, Sigma Theta Tau. Lutheran. Subspecialties: Gerontology; Pediatrics. Current work: Block nurse project; home care. Office: Dept Family Health Care Nursing U Calif San Francisco CA 94135 The challenge of quality health care to all of society and the critical role nursing has to play in order to achieve this goal has motivated me throughout my professional life. The richness of talent in this country spurs me on.

MARTON, LAURENCE JAY, physician, educator; b. Bklyn., Jan. 14, 1944; s. Bernard Dov and Sylvia M.; m. Marlene Lesser, Ju 27, 1967; 1 son, Eric Nolan. A.B., Yeshiva U., 1965; M.D., Albert Einstein Coll. Medicine, 1969. Surg. intern Los Angeles County Harbor Gen. Hosp., 1969-70; resident in neurosurgery U. Calif., San Francisco, 1970-71, resident in lab. medicine, 1973-75, asst. research biochemist dept. neurosurgery, 1973-74, asst. clin. prof. depts. lab. medicine and neurosurgery, 1974-75, asst. prof., 1975-78, assoc. prof., 1978-79, prof., 1979—, asst. dir. div. clin. chemistry dept. lab. medicine, 1974-75, dir., 1975-79, acting chmn., 1978-79, chmn., 1979—. Editor: (with D.R. Morris) Polyamines in Biology and Medicine, 1981, (with P.M. Kabra) Liquid Chromatography in Clinical Analysis, 1981; contbr. articles and abstracts to sci. jours. Served as sr. asst. surgeon USPHS, 1971-73. Nat. Cancer Inst. research career devel. award, 1975-80; research grantee, 1975—; Am. Cancer Soc. research grantee, 1981-82. Mem. Am. Assn. for Cancer Research, Am. Assn. for Clin. Chemistry, AAAS, Assn. Clin. Scientists, Acad. Clin. Lab. Physicians and Scientists, Assn. Pathology Chairmen, Am. Assn. Pathologists, Calif. Med. Assn., San Francisco Med. Soc., Alpha Omega Alpha. Subspecialties: Cancer research (medicine). Current work: Polyamines and tumors; clinical applications of liquid chromatography; polyamine biosynthesis inhibitors as anti-tumor agents; application of advanced technology to patient diagnosis. Home: 69 Aloha Ave San Francisco CA 94122 Office: Dept Lab Medicine U Calif San Francisco CA 94143

MARTZ, ERIC, immunologist; b. Columbus, Ind., Apr. 30, 1940; s. Karl and Margaret Rebekah (Brown) M.; m. Phyllis Donna Ustin, June 5, 1971; children: Gabriel Ustin, Elias Samuel. A.B. in Chemistry with high honors, Oberlin Coll., 1963; Ph.D. in Biology, Johns Hopkins U., 1969. Postdoctoral fellow Princeton (N.J.) U., 1970; dept. pathology Harvard U. Med. Sch., Boston, 1970; asst. prof. dept. microbiology U. Mass., Amherst, 1974-77, assoc. prof., 1977—. Contbr. articles to sci. jours. Winner Westinghouse Nat. Sci. Talent Search, 1957; NSF fellow, 1963-68; Damon Runyon Cancer Fund postdoctoral fellow, 196-71; 1969-71; recipient Research Career Devel. award NIH, 1977-82. Mem. Am. Assn. Immunologists. Subspecialties: Immunobiology and immunology; Cell biology. Current work: Effector mechanisms in cellular immunity; mechanisms of T lymphocyte-mediated killing. Home: 48 Hunter's Hill Circle Amherst MA 01002 Office: U Mass Amherst MA 01003

MARTZ, FREDRIC ALLEN, research animal scientist; b. Whitley County, Ind., May 24, 1935; s. Joseph Clemont and Dora Belle (Barnes) M.; m. Donna Joan Wilkinson, Jan. 25, 1959; children: Erich, Kile, Connie, Kevin, Kathy. B.S., Purdue U., 1957, M.S., 1959, Ph.D., 1961. Asst. prof. U. Mo.-Columbia, 1961-66, assoc. prof., 1966-73, prof., 1973—, chmn. dept. dairy sci., 1978-82; research animal scientist Agrl. Research Service, USDA, 1982—. Contbr. articles to profl. jours. Leader 4-H Club. Recipient Outstanding Young Scientist award Gamma Sigma Delta, 1972; Cert. of Merit Am. Forage and Grassland Assn., 1978. Mem. Am. Dairy Sci. Assn., Am. Soc. Animal Sci., Am. Inst. Nutrition, Am. Forage and Grassland Assn. (dir. 1983—). Subspecialty: Animal nutrition. Current work: Mineral and fiber utilization by animals. Research in cattle, mineral availability, fiber utilization. Home: Rural Route 8 Box 81 Columbia MO 65202 Office: S142 Animal Sci Center Columbia MO 65211

MARUYAMA, YOSH, physician, educator; b. Pasadena, Calif., Apr. 30, 1930; s. Edward Yasaki and Chiyo (Sakai) M.; m. Fudeko Tsuji, July 18, 1954; children: Warren H., Nancy C., Marian M., Karen A. A.B., U. Calif.-Berkeley, 1951; M.D., U. Calif.-San Francisco, 1955. Diplomate: Am. Bd. Radiology. Intern San Francisco Hosp., 1955-56; resident Mass. Gen. Hosp., Boston, 1958-61; James Picker advanced acad. fellow Stanford U., 1962-64; asst. prof. radiology Coll. Med. Scis., U. Minn., Mpls., 1964-67, asso. prof., 1967-70, dir. div. radiotherapy, 1968-70; prof., chmn. dept. radiation medicine Coll. Medicine, U. Ky., Lexington, 1970—, dir. Radiation Cancer Ctr., 1975—; bd. dirs. Ephraim McDowell Cancer Ctr.; cons. VA Hosp., Lexington. Asso. editor: Applied Radiology; editor: New Methods in Tumor Localization, 1977; Contbr. articles to profl. jours. Served with M.C. AUS, 1956-58. Am. Cancer Soc. fellow, 1960-61. Mem. Am. Coll. Radiology (commn. on radiation therapy and patterns of care study), AAAS, Am. Radium Soc., Cell Kinetics Soc., Ky. Cancer Commn., Radiation Research Soc., Soc. Exptl. Biol. Medicine, Am. Assn. Cancer Research, Radiol Soc. N.Am., Am. Soc. Therapeutic Radiology, Soc. Chmn. Acad. Radiology Oncology Programs, Am. Assn. Immunologists, Southeastern Cancer Research Assn. (dir.), Southeastern Cancer Group, Ky. Med. Assn., Order Ky. Cols. Minn., N.Y. Acad. Scis., Japan Soc. Ky., Shodan Judoka, Kodokan Inst. (Tokyo), Phi Beta Kappa, Sigma Xi (chpt. pres.), Alpha Omega Alpha. Club: Spindletop Hall (Lexington). Subspecialties: Oncology; Cancer research (medicine). Current work: Neutronbrachytherapy; chemo-radio therapy leukemia/lymphona. Home: 1739 Lakewood Dr Lexington KY 40502

MARVIN, JOSEPH GEORGE, research engineer; b. San Jose, Calif., June 9, 1934; s. John and Mary (Thomas) M.; m. Gwendolyn Hoskins, Feb. 21, 1957; children: John, Cheryl, Gregory, Cynthia, Laura. B.S. in Mech. Engring, Santa Clara U., 1956; M.S. in Engring. Mechanics, Stanford U., 1964. Registered profl. engr., Calif. Research scientist Ames Research Ctr., NASA, Moffett Field, Calif., 1956-66, asst. chief fluid mechanics, 1966-71, chief exptl. fluid dynamics, 1971—. Contbr. chpts. to books, articles to profl. jours. Served to 1st lt. U.S. Army, 1956-57. Recipient Shuttle Aerothermo Group award NASA, 1981, Shuttle Aeroheating award, 1982. Fellow AIAA (assoc.); mem. Calif. Soc. Profl. Engrs., Tau Beta Pi. Subspecialties: Aeronautical engineering; Fluid mechanics. Current work: Fluid mechanics, turbulent flows, entry heating, facility design, computational fluid dynamics. Office: NASA Ames Research Center MS 229-1 Moffett Field CA 94035

MARWAHA, JWAHARLAL, pharmacologist, educator, researcher; b. Uganda, May 27, 1952; U.S., 1978, naturalized, 1982; s. Suraj Prakash and Kamla Rani (Kapur) M.; m. Carmen Kim Fonkalsrud, Dec. 20, 1978. B.Sc. with honours, U. London, 1974; Ph.D., U. Alta. Sch. Medicine, 1978. Instr. U. Colo. Health Scis. Center, Denver, 1978-80; research asso. Yale U. Sch. Medicine, New Haven, 1980-81; asst. prof. pharmacology and life scis. Ind. U. Sch. Medicine, Terre Haute, 1981—. Contbr. articles to sci. jours. Mem. Soc. for Neurosci., AAAS, Am. Soc. Pharmacology and Exptl. Therapeutics, Am. Physiol. Soc., Soc. Biol. Psychiatry (A.E. Bennet award 1981), N.Y. Acad. Scis., Sigma Xi. Subspecialties: Neuropharmacology; Neurology. Current work: Neurotransmitters; neuroreceptors; hormones; electrophysiology; monoamines; central nervous system. Office: 135 Holmstedt Hall Ind U Terre Haute IN 47809

MARX, DONALD HENRY, plant pathologist; b. Ocean Falls, C., Can., Oct. 3, 1936; s. Edmund N. and Francis (Gorski) M.; m. Selina Van Giesen, Dec. 21, 1957; children: Selina, Teresa, Mary, Donald, Frank. B.S., U. Ga., 1961, M.S., 1962; Ph.D., N.C. State U., 1966. With Forest Service, U.S. Dept. Agr., 1966—, chief plant pathologist, 1973—, dir. inst., 1975—; cons. Internat. Found Sci., Sweden. Contbr. numerous articles to profl. jours. Served with USMC, 1954-57. Recipient U.S. Dept. Agr., Superior Service award, 1970; Ruth Allen award, 1977; Arthur Fleming award, 1975; Barrington Moore award, 1977. Mem. Am. Phytopath. Soc., AAAS, Sigma Xi. Roman Catholic. Subspecialties: Plant growth; Resource management. Current work: Forest regeneration and reclamation; nursery tree seedling prodn. Home: 465 Little John Ct Watkinsville GA 30677 Office: Inst Mycorrhizal Research and Devel USDA Forest Service Forestry Scis Lab Carlton St Athens GA 30602

MARX, JAMES JOHN, JR., immunologist, researcher; b. Paris, Tex.; s. James John and Grace E. (Beckfeld) M.; m. Mary Alice Kettrick, Aug. 25, 1973; 2 sons, Jonathyn Andrew, Christopher James. Civ., St. Vincent Coll., Latrobe, Pa., 1966; M.S. in Immunology, W.Va. U., 1970, Ph.D., 1973. Sr. scientist Marshfield (Wis.) Med. Found., 1973—. Contbr. articles to profl. jours. Mem. nat. Explorer com. Boy Scouts Am.; mem. Wildwood Park Zool. Soc. Served with Med. Service Corps U.S. Army, 1966-68. Mem. Am. Soc. Microbiology, Am. Assn. Clin. Immunology and Allergy, Am. Lung Assn., Am. Acad. Allergy, Reticuloendothelial Soc., Wis. Lung Assn., Am. Assn. Immunologists, Sigma Xi. Roman Catholic. Subspecialties: Immunology (medicine); Chemotherapy. Current work: Definition of immunologic lung disease - mechanisms of pathogenesis; culture of human tumor cells for defining chemotherapy. Home: M225 Marsh Ln Marshfield WI 54449 Office: 510 N Saint Joseph Ave Marshfield WI 54449

MARX, MELVIN HERMAN, research psychologist; b. Bklyn., June 8, 1919; m. Kathleen Kendall, Sept. 5, 1948; children: Diana, Christine, Ellen, James. A.B., Washington U., St. Louis, 1940, M.A., 1941, Ph.D., 1943. Instr. U. Mo.-Columbia, 1944-45, asst. prof. psychology, assoc. prof., 1945-64, research prof., 1964—; research psychologist; U.S. Air Force, Lackland AFB, San Antonio, 1950-51. Editor: books, including Psychological Theory, 1951; author: Systems and Theories in Psychology, 1963; 3d edit., 1979. Recipient Research Career award NIMH, 1964—. Fellow Am. Psychol. Assn. (councillor); mem. Midwestern Psychol. Assn. (pres. 1965), AAAS, Phi Beta Kappa, Sigma Xi. Subspecialties: Behavioral psychology; Learning. Current work: Experimental psychology of learning, memory, and motivation; research in nonconscious information processing. Home: 308 Banyan St Duck Key (Marathon) FL 33050 Office: U Mo 16 McAlester Hall Columbia MO 65211

MARZELLA, LIBERO LOUIS, physician, researcher, educator; b. Giovinazzo, Italy, May 14, 1948; came to U.S., 1964; s. Frank and Rose (Turturro) M. B.S. in Biology, Loyola Coll.-Balt., 1970; M.D., U. Md., 1974; Ph.D. in Pathology, Karolinska Inst., Stockholm, 1979. Intern/resident U. Md. Hosp., Balt., 1974-76; asst. dir. research Md. Inst. Emergency Med. Services Systems, 1981—, asst. prof. pathology, 1981—. Contbr. articles to sci. publs. U. Md. Bressler Fund grantee, 1980; Am. Heart Assn. grantee, 1981; Distilled Spirits Council grantee, 1983. Mem. Am. Soc. Cell Biology, Am. Assn. Pathologists. Subspecialties: Cell biology; Family practice. Current work: Elucidate the pathophysiology of sepsis and shock; develop methods for the study of structure and function of subcellular organelles, particularly the lysosomes. Office: Dept Pathology U Md 10 S Pine St Baltimore MD 21201

MARZLUF, GEORGE AUSTIN, biochemistry educator; b. Columbus, Ohio, Sept. 29, 1935; s. Paul B. and Faun (Simmons) M.; m. Jackie B. Rice, Mar. 2, 1972; chilren: Bruce, Julie, Phillip, Glenn. Asst. prof. biology Marquette U., 1966-70; assoc. prof. biochemistry Ohio State U., 1970-75, prof., 1975—, dir. Program in Molecular, Cellular and Devel. Biology, 1980—. Mem. Am. Soc. Biol. Chemists, Genetics Soc. Am., Am. Soc. Microbiology. Subspecialties: Molecular biology; Genetics and genetic engineering (agriculture). Current work: Regulation of gene expression. Home: 3200 Rochfort Bridge Dr Columbus OH 43220 Office: 484 W 12th St Columbus OH 43210

MASCARENHAS, JOSEPH PETER, biologist, educator, researcher, consultant; b. Nairobi, Kenya, Nov. 19, 1929; came to U.S., 1957; s. Theotonio C. and Philomena Olive (D'Sousa) M.; m. Patricia Schneider, Jan. 23, 1960; children: Nonika, Shaun. B.Sc. in Agr, U. Poona, India, 1952, M.Sc., 1954; Ph.D., U. Calif.-Berkeley, 1962. Research officer Parry & Co., Thiruvalla, India, 1953-56; instr. in biology Amherst Coll., 1962-63; instr. biol. scis. Wellesley Coll., 1963-64, asst. prof. biol. scis., 1964-67; vis. asst. prof. and research assoc. dept. biology M.I.T., 1966-68; assoc. prof. biol. scis. SUNY-Albany, 1968-74, prof., 1974—; cons. in field. Contbr. numerous articles to sci. jours. Brown Hazen Fund of Research Corp. grantee, 1963-64; NSF grantee, 1965—; Research Found. SUNY grantee, 1969-78; Am. Soybean Assn. grantee, 1980-81; U.S. Dept. Agr. grantee, 1983—. Mem. Am. Soc. Plant Physiologists, Bot. Soc. Am., Soc. Developmental Biology, Am. Soc. Cell Biology, AAAS, Plant Molecular Biology Assn., Internat. Soc. Developmental Biology, Internat. Plant Tissue Culture Assn. Subspecialties: Genetics and genetic engineering (agriculture); Developmental biology. Current work: Molecular regulation of plant development; engineering of plants for heat stress tolerance. Office: Dept Biol Scis SUNY Albany NY 12222

MASH, ERIC JAY, psychology educator; b. N.Y.C., Feb. 6, 1943; s. Abe and Mary (Lifschitz) M.; m. Heather Mary Grace Henderson, Dec. 24, 1969. B.B.A., CUNY, 1964; M.A., Temple U., 1965; Ph.D., Fla. State U., 1969. Cert. psychologist, Alta. Intern in clin. psychology U. Oreg. Med. Sch., Portland, 1968-69, resident in med. psychology, 1969-70; asst. prof. psychology U. Calgary, Alta., Can., 1970-75, assoc. prof., 1975-79, prof., 1979—; vis. assoc. prof. U. Oreg. Health Sci. Ctr., Portland, 1976-77; dir. Family Study Project, Calgary, 1979—. Editor: Behavior Assessment Child Disorders, 1981, Behavior Therapy Assessment, 1976, Behavior Modification and Families, 1976; assoc. editor: Behavioral Assessment, 1982—. USPHS research fellow, 1968; retardation fellow, 1967; Can. Council leave fellow, 1976; Killam Found. fellow, 1981. Mem. Am. Psychol. Assn., Assn. Behavior Analysis, Psychol. Assn. Alta., Can. Psychol. Assn., Assn. Advancement of Behavior Therapy. Subspecialties: Developmental psychology; Behavioral psychology. Current work: Family interaction research, behavioral assessment. Home: 6415 Silver Ridge Dr NW Calgary AB Canada T3B 3T1 Office: Dept Psychology Univ Calgary Calgary AB Canada T2N 1N4

MASHIMO, PAUL AKIRA, oral biology educator, researcher; b. Osaka City, Japan, Oct. 25, 1926; came to U.S., 1966; s. Eizo and Teruko (Kimura) M.; m. Eunice Kusue, Nov. 23, 1955; children: Minoru, Hiroshi. D.D.S., Osaka Dental U., 1948; Ph.D., Kyoto (Japan) Med. U., 1955. Instr. oral surgery Osaka Dental U., 1948-50, lectr. dept. microbiology, 1950-54, asst. prof. dept. preventive dentistry, 1954-55, assoc. prof., 1955-66; asst. prof. oral biology SUNY-Buffalo, 1966-71, assoc. prof., 1971—; cons. Sunstar Inc., Osaka, IPD Co. Ltd., Shorin Pub. Co., Tokyo, Belmont-Takara Co., N.J. Translator: Psychology and Dentistry, 1977, Clinical Pharmacology in Dental Professionals, 1980. John C. Ball fellow, 1956-58; Japan Soc. fellow, 1957; Sunstar Inc. fellow, 1979; recipient Sunstar Inc. Interferon Research award, 1982. Mem. Internat. Assn. Dental Research, Am. Soc. Microbiologists, N.Y. Acad. Sci., Sigma Xi. Subspecialties: Oral biology; Periodontics. Current work: Microbiological analysis on periodontal diseases, especially in the field of anaerobic microbiology, searching the etiologic factors of the diseases. Home: 639 Cottonwood Dr Williamsville NY 14221 Office: Sch of Dentistry SUNY Room 205 Foster Hall 3435 Main St Buffalo NY 14214

MASKER, WARREN EDWARD, molecular biologist, research scientist, educator; b. Honesdale, Pa., July 8, 1943; s. Warren Russel and Eva (Chuprevich) M. B.S. in Engring. Physics, Lehigh U., 1965; Ph.D. in Physics, U. Rochester, 1970. Postdoctoral fellow dept. radiation biology and biophysics U. Rochester, N.Y., 1969-71; postdoctoral fellow dept. biol. scis. Stanford (Calif.) U., 1971-73; postdoctoral fellow dept. biochemistry Harvard Med. Sch., Boston, 1973-74; mem. research staff biology div. Oak Ridge Nat. Lab., 1975—; prof. grad. sch.biomed. sci. U. Tenn., Oak Ridge, part-time 1975—. Contbr. articles to sci. jours. Am. Cancer Soc. postdoctoral fellow, 1970; Helen Hay Whitney Found. postdoctoral fellow, 1971; USPHS research grantee, 1980. Fellow AAAS; mem. Am. Phys. Soc., Biophys. Soc., Am. Soc. Photobiology, Am. Soc. Microbiology. Subspecialties: Molecular biology; Microbiology. Current work: replication, repair, recombination DNA molecular mechanisms of DNA damage and repair, in vitro DNA replication, in vitro packaging of DNA, mutagenesis. Home: 103 Viola Rd Oak Ridge TN 37830 Office: Biology Div PO Box Y Oak Ridge Nat Lab Oak Ridge TN 37830

MASLOFF, JACQUELINE JO, computer scientist, systems analyst; b. N.Y.C., July 27, 1954; d. Phillip and Jane (Zimmerman) M.; m. Howard J. Levy, June 19, 1977 (div. 1980). B.S., B.Arch., Rensselaer Poly. Inst., 1976; postgrad., Boston U., 1976—. Architect Dooling & Siegel, Newtonville, Mass., 1976-77, Tufts U., Medford, Mass., 1978-80; instr. Applicon, Woburn, Mass., 1980-81; tech. systems analyst Foxboro (Mass.) Co., 1981—; instr. Boston U., 1981—. Author: Encylopedia of Energy, 1979, Computer Programs to Design and Monitor Energy Systems, 1978. Mem. Boston Computer Soc., Assn. for Computing Machinery (newsletter editor 1980-82, chmn. 1982—), IEEE Computer Soc., Siggraph. Democrat. Jewish. Club: Rensselaer Poly. Inst. Alumni Club Boston (pres. 1980-82). Subspecialties: Distributed systems and networks; Information systems, storage, and retrieval (computer science). Current work: User support (internal) of personal computing; installation and projects related to inhouse broadband local area networks. Home: 101 F2 Chestnut St Foxboro MA 02035 Office: Foxboro Co 38 Neponset Ave Foxboro MA 02154

MASLOW, DAVID E(ZRA), cancer researcher; b. Bklyn., July 6, 1943; s. Archie N. and Tillie A. (Karp) M.; m. Sarah T. Holstein, Nov. 26, 1972; children: Dina Meira, Aleeza Nechama, Jonathan Aaron. B.S., Bklyn. Coll., 1963; Ph.D., U. Pa., 1968. Asst. instr. U. Pa., 1967-68; cancer research scientist I dept. exptl. pathology Roswell Park Meml. Inst., Buffalo, 1968-72, research scientist II, 1972-75, research scientist IV, 1977—; asst. prof. Erie Community Coll., Buffalo, 1972-75, SUNY at Buffalo, 1978—. Contbr. articles to profl. jours. NIH fellow, 1964-68; Nat. Cancer Inst. grantee, 1972—. Mem. Am. Assn. Cancer Research, Am. Soc. Zoologists, Soc. Developmental Biology, AAAS. Subspecialties: Cell and tissue culture; Cancer research (medicine). Current work: Specificity of interactions of normal and cancer cells. Office: Roswell Park Meml Inst Buffalo NY 14263 Home: 125 Brooklane Dr Williamsville NE 14221

MASON, CHARLES PERRY, biologist; b. Newport, R.I., Aug. 12, 1932; s. Charles Perry and Bernice Marie (Passmore) M.; m. Harriet

Gale, Sept. 7, 1958; children: Grant Vause, Gale Perrie. B.S., U. R.I., 1954; M.S., U. Wis., Madison, 1958; Ph.D., Cornell U., 1961. Asst. prof., then assoc. prof. biology Hamline U., 1961-67; assoc. prof. Gustavus Adolphus Coll., 1967-81, prof., 1981—; vis. scientist Gray Freshwater Biol. Inst., U. Minn., summers 1977-81. Mem. Phycological Soc. Am., Internat. Phycological Soc., Nature Conservancy, Minn. Acad. Sci. Lutheran. Subspecialties: Ecology; Physiology (biology). Current work: Physiological ecology of algae. Home: 905 S 5th St Saint Peter MN 56082 Office: Dept Biology Gustavus Adolphus Coll Saint Peter MN 5608

MASON, CHARLES T., JR., plant taxonomy educator; b. Joliet, Ill., Mar. 26, 1918; s. Charles T. and Alphie (Longley) M.; m. Patricia Bovyer, June 26, 1943; 1 son, Charles T. B.S., U. Chgo., 1940; M.A., U. Calif.-Berkeley, 1942, Ph.D., 1949. Tchr. secondary schs., Long Beach, Calif., 1942-47; instr. U. Wis.-Madison, 1949-53; asst. prof. dept. ecology U. Ariz., Tuscon, 1953-58, assoc. prof., 1958-62, prof. plant toxonomy, 1962—. Fellow AAAS; mem. Internat. Assn. Plant Taxonomy, Am. Assn. Plant Taxonomists, Calif. Bot. Soc., Am. Inst. Biol. Scis., Sigma Xi, Phi Kappa Phi. Lodge: Masons. Subspecialties: Systematics; Taxonomy. Current work: Curator of herbarium. Office: Herbarium Dept Ecology U Ariz Tucson AZ 85721 Home: 2945 N Bear Canyon Rd Tucson AZ 85749

MASON, CURTIS L., plant pathologist; b. Daingerfield, Tex., Oct. 9, 1919; s. James L. and Sarah C. (Baker) M.; m. Mary E. Funkhouser, May 25, 1942; children: Robert L., Sarah A. (dec.). B.S., Tex A&M U., 1940, M.S., 1942; Ph.D., U. Ill., 1947. Agt., Bur. Plant Industry cotton div. U.S. Dept. Agr., 1939-42, assoc. plant pathologist, 1947-48; asst. prof. plant pathology U. Ark., 1948-52; plant pathologist (Niagara Chem. Div.), Middleport, N.Y., 1952-54, asst. sales mgr., 1954-57, mgr. tech. services, 1957-60, regional mgr., 1960-62; area mgr. Buckman Labs., Richmond, Va., 1962-70; ext. plant pathologist U. Ark. Coop. Ext. Service, Little Rock, 1971-77, pesticide coordinator, 1977—, coordinator So. region pesticide impact assessment program, 1977—. Served with USAAF, 1943-45. Mem. Am. Phytopath. Soc., Gamma Sigma Delta, Epsilon Sigma Phi. Subspecialties: Plant pathology; Integrated pest management. Current work: Training pesticide applicators; risk/benefit assessment of pesticides. Home: 4712 Hampton Rd North Little Rock AR 72116 Office: 1201 McAlmont PO Box 391 Little Rock AR 72203

MASON, DAVID DICKENSON, statistics educator; b. Abingdon, Va., Jan. 22, 1917; s. William Thomas and Eva (Dorton) M.; m. Virginia Louise Pendleton, Oct. 28, 1944; children: Marjorie F., David P. B.A., King Coll., 1936; M.S. (Acad. Merit fellow), Va. Poly. Inst., 1938; postgrad., Ohio State U., 1939-40; Ph.D., N.C. State U., 1948. Asst. soil scientist Va. Poly. Inst., 1938-39; asst. prof. soils Ohio State U., 1947-49; prin. biometrician Dept. Agr., Beltsville, Md., 1949-53; prof. stats. N.C. State U., 1953-62, prof., head dept. stats., 1962-81, emeritus prof. and head, 1981—; head Inst. Stats., 1962-81, emeritus head, 1981—; Sr. cons. United Fruit Co., Boston, 1957—. Contbr. articles to profl. jours. Instl. rep. So. Regional Edn. Bd. com. statistics, 1963—, chmn., 1973-75; Bd. dirs. Triangle Univs. Computation Center Corp., chmn., 1968-70. Served with AUS, 1941-45. Fellow Am. Statis. Assn., Am. Soc. Agronomy, Soil Sci. Soc. Am.; mem. Biometric Soc., Sigma Xi, Phi Kappa Phi, Gamma Sigma Delta. Presbyn. (elder, deacon). Club: Rotarian. Subspecialties: Statistics. Home: 4212 Arbutus Dr Raleigh NC 27612 Office: Inst Statistics Box 5457 Raleigh NC 27650

MASON, EDWARD ARCHIBALD, chem. and nuclear engr.; b. Rochester, N.Y., Aug. 9, 1924; s. Henry Archibald and Monica (Brayer) M.; m. Barbara Jean Earley, Apr. 15, 1950; children—Thomas E., Kathleen M., Paul D., Mark J., Anne M., Mary Beth. B.S., U. Rochester, 1945; M.S., Mass. Inst. Tech., 1948, Sc.D., 1950. Asst. prof. chem. engring. Mass. Inst. Tech., Cambridge, 1950-53, asso. prof. nuclear engring., 1957-63, prof., 1963-77, dept. head, 1971-75; commr. Nuclear Regulatory Commn., Washington, 1975-77; v.p. research Standard Oil Co. (Ind.), Chgo., 1977—; dir. research Ionics, Inc., Cambridge, Mass., 1953-57; sr. design engr. Oak Ridge Nat. Lab., 1957; mem. adv. com. reactor safeguards AEC, 1972-75; cons. other govt. agys., industry; dir. Cetus Corp., Commonwealth Edison Co. Contbr. articles to profl. jours. Mem. adv. com. M.I.T., U. Chgo., U. Tex., Ga. Inst. Tech.; bd. dirs. John Crerar Library. Served with USNR, 1943-46. NSF Sr. Postdoctoral fellow, 1965-66. Fellow Am. Acad. Arts and Scis., Am. Nuclear Soc., Am. Inst. Chemists; mem. Nat. Acad. Engring. (councilor 1978—), N.Y. Acad. Scis., Am. Inst. Chem. Engrs. (R.E. Wilson award in nuclear chem. engring. 1978), Western Soc. Engrs., Am. Chem. Soc., AAAS, Phi Beta Kappa, Sigma Xi, Tau Beta Pi. Clubs: Execs. (Chgo.); Hinsdale (Ill.) Golf. Subspecialty: Nuclear engineering. Patentee in field. Home: 145 Hillcrest Ave Hinsdale IL 60521 Office: 200 E Randolph Dr Chicago IL 60601

MASON, JOHN MONTGOMERY, JR., marine fisheries biologist; b. Bridgeport, Conn., July 9, 1945; s. John Montgomery and Jayne (Hutton) M.; m. Barrie Orr, June 8, 1974; 1 son, Jeremy M. B.A., DePauw U., 1967; M.S., U. R.I., 1976. Cert. fisheries scientist Am. Fisheries Soc.; cert. Am. Inst. Fishery Research Biologists. Research assoc. Woods Hole (Mass.) Oceanographic Instn., 1970-79; sr. fishery biologist Mid-Atlantic Fishery Mgmt. Council, Dover, Del., 1979-80; research scientist N.Y. State Div. Marine Resources, Stony Brook, 1980—; advisor U.S. sect. Internat. Commn. Conservation of Atlantic Tunas, 1979—. Contbr. articles to sci. jours. Sunday sch. tchr. St. Mark's Episcopal Ch., Westhampton, N.Y.; co-founder N.Y. Sport Fishing Fedn., 1982. Rector scholar DePauw U., 1963-67. Mem. Am. Fisheries Soc. (life; cert. fisheries scientist), AAAS, Am. Inst. Fishery Research Biologists, Internat. Game Fish Assn., Babylon Tuna Club (hon.), Va. Bluewater Gamefish Assn. (hon.), Atlantic Tuna Club, Nat. Coalition Marine Conservation (adv. bd. North Atlantic region 1982—), Phi Sigma. Subspecialties: Resource management; Population biology. Current work: Study of marine fish and fisheries for determination of management needs including research on catch and effort, species distribution, behavior and population structure. Home: 8 Annette Ln East Moriches NY 11940 Office: NY State Dept Environ Conservation Bldg 40 SUNY Stony Brook NY 11794

MASON, SUSAN ELIZABETH, psychology educator, reseacher; b. Glen Ridge, N.J., Sept. 10, 1952; d. Julien Jacquelin and Helen Emma (Grimm) M.; m. Howard Malcolm Reid, June 10, 1979; 1 dau., Mary Mason. B.S. in Math. and Psychology, Dickinson Coll., 1974; M.S. in Psychology, Ga. Tech. U., 1975, Ph.D., 1977. Cert. math. educator, Pa. Asst. prof. psychology Buffalo State Coll., 1977-78; asst. prof. W.Va. U., 1978-79; assoc. prof. Niagara U., 1979—; adj. prof. Buffalo State Coll., 1980—; mem. exec. bd. Network on Aging, Buffalo, 1981—; cons. in field. Contbr. chpts. to books and articles to profl. jours. SUNY Research Found. grantee, 1978; Niagra U. Research Council grantee, 1981, 82. Mem. Am. Psychol. Assn., Gerontol. Soc. Am., Eastern Psychol. Assn., AAAS, Sigma Xi. Subspecialties: Developmental psychology; Cognition. Current work: Cognitive changes in adulthood. Office: Psychology Dept Niagara University Niagara University NY 14109

MASON, W. ROY, chemistry educator; b. Charlottesville, Va., Feb. 6, 1943; s. W. Roy and Mary Ruffin (Jones) M.; m. Nanette R. Marshall, Aug. 25, 1963; children: Kristina Celeste, Katherine Ruffin, Alice Eloise. B.S., Emory U., 1963, M.S., 1966, Ph.D., 1966. Research assoc. Calif. Inst. Tech., Pasadena, 1966-67; asst. prof. chemistry No. Ill. U., DeKalb, 1967-70, assoc. prof., 1970-80, prof., 1980—; vis. research fellow H.C. Orsted Inst., Copenhagen, 1974-75. Contbr. articles to profl. jours. Petroleum Research Fund grantee, 1967, 69, 80; NSF grantee, 1980. Mem. Am. Chem. Soc. (chmn. Rock River sect. 1970-71, sec.-treas. 1980—), Sigma Xi. Episcopalian. Subspecialties: Inorganic chemistry; Physical chemistry. Current work: Magneto optics and electronic spectra of heavy metal coordination compounds. Home: 1120 Northern Ct DeKalb IL 60115 Office: Dept Chemistry No Ill U DeKalb IL 60115

MASON, WILLIAM A(LVIN), psychologist, educator; b. Mountain View, Calif., Mar. 28, 1926; s. Alvin Frank and Ruth Sabina (Erwin) M.; m. Virginia Joan Carmichael, June 27, 1948; children: Todd, Paula, Nicole, Hunter. B.A., Stanford U., 1950, M.A., 1952, Ph.D., 1954. Research assoc. U. Wis.-Madison, 1954-59, Yerkes Labs. of Primate Biology, Orange Park, Fla., 1959-63; head dept. behavioral sci. Delta Primate Ctr. Tulane U., Covington, La., 1963-71; prof. psychology, research psychologist Calif. Primate Research Ctr. U. Calif.-Davis, 1971—; cons. USPH, Washington. Contbr. chpts. to books, articles to profl. publs. in field. Served with USMC, 1944-46. USPH spl. fellow, 1963-64; various grants USPH, NSF. Fellow Animal Behavior Soc., AAAS, Am. Psychol. Assn. (pres. div. 6 1982); mem. Internat. Soc. Devel. Psychobiology (pres. 1971-72, paper award 1976), Internat. Primatol. Soc. (pres. 1976-80, 80—), Am. Soc. Primatologists. Subspecialties: Psychobiology; Developmental psychology. Current work: Social behavior, development, nonhuman primates. Home: Calif Primate Research Ct U Calif Davi CA 9561

MASOUREDIS, SERAFEIM P., physician, educator; b. Detroit, Nov. 14, 1922; s. Panagiotis George and Lemonia (Moniodis) M.; m. Marion Helen Mykytew, Oct. 4, 1943; children: Claudia, Linus. A.B., U. Mich., 1943; M.D., 1948; Ph.D. in Med. Physics, U. Calif., Berkeley, 1952. Intern San Francisco Hosp., 1952-53; research asso. Donnor Lab., Berkeley, Calif., 1953-54; resident in medicine U. Calif., San Francisco Gen. Hosp., 1954-55; asst. prof. pathology, asst. dir. central blood bank U. Pitts., 1955-58, assoc. prof., 1958-59; assoc. prof. preventive medicine U. Calif., San Francisco, 1959-62; assoc. prof. medicine, dir. Moffitt Hosp. Blood Bank, 1962-67; exec. dir. Milw. Blood Center, 1967-69; prof. medicine Marquette U. Sch. Medicine, 1967-69; prof. pathology, dir. Univ. Hosp. Blood Bank, U. Calif., San Diego, 1969—; cons. WHO, 1966-67; mem. merit rev. bd. hematology VA, 1969-75; sci. counselor div. biologics standards NIH, 1971-72; bd. dirs. Milw. Blood Center, 1969—, San Diego Blood Bank, 1970-76. Contbr. articles to profl. jours. Mem. Western Soc. Clin. Research, Am. Assn. Blood Banks, Am. Assn. Cancer Research, Am. Fedn. Clin. Research, Am. Assn. Immunologists, Am. Physiol. Soc. Human Genetics, Brit. Soc. Immunology, Central Soc. Clin. Research, Internat. Soc. Blood Transfusion, Internat. Soc. Hematology, Soc. Exptl. Biology and Medicine, Western Assn. Physicians. Greek Orthodox. Subspecialties: Hematology; Immunology (medicine). Current work: Blood group antigens, red cell membrane biochemistry, hemolytic anemias, blood group antibodies. Office: U Calif Sch Medicine San Diego La Jolla CA 92093

MASSA, DENNIS JON, research scientist, physical chemist; b. Myrtle Beach, S.C., Sept. 29, 1945; s. Amel C. and Josephine N. M.; m. Donna J. Fabry, June 4, 1966; children: Shelley, Rebecca, Elizabeth. B.A. in Chemistry, Bradley U., 1966; Ph.D. in Phys. Chemistry, U. Wis.-Madison, 1970. NSF postdoctoral fellow U. Calif.-San Diego, 1970-71; research assoc. Research Labs., Eastman Kodak Co., Rochester, N.Y., 1971—. Mem. Am. Chem. Soc., Am. Phys. Soc., Soc. Rheology. Subspecialties: Polymer chemistry; Polymer physics. Current work: Polymer physical chemistry, polymer physics, materials science, rheology. Home: 40 Deer Creek Rd Pittsford NY 14534 Office: Research Lab Eastman Kodak Co Rochester NY 14650

MASSAD, CAROLYN EMRICK, educational psychologist, psychometrician, educator, consultant; b. Cleve., Nov. 24, 1935; d. Steve George and Mary Elizabeth (Evans) Emrick; m.; 1 son, Mark Isam. B.S., Kent State U., 1956, M.Ed., 1963, Ph.D., 1967. Tchr., Havana, 1955, Solon (Ohio) Bd. Edn., 1957-67; assoc. examiner Ednl. Testing Service, Princeton, N.J., 1967-74, examiner, 1974-80, sr. examiner, group head test devel., 1980—; cons. La. Dept. Edn., Bermuda Dept. Edn. Internat. Assn. Evaluation of Ednl. Achievement, 1972-83. Author: English as a Foreign Language in 10 Countries, 1975; author, editor: Handbook of Information for Evaluation and Assessment, 1977; rev. editor: Am. Ednl. Research Jour, 1972-75. Spencer Found. postdoctoral fellow, Stockholm, 1972-73; NDEA fellow, 1961, 64. Mem. Am. Ednl. Research Assn., Am. Psychol. Assn., Nat. Council Measurement in Edn., Internat. Reading Assn. Republican. Roman Catholic. Subspecialties: Developmental psychology; Psychometrics. Current work: Development of measures for job licensure/selection; development of language skills and how they can be measured/assessed. Home: 303 Emmons Dr 5B Princeton NJ 08540 Office: Educational Testing Service Rosedale Rd Princeton NJ 08540

MASSARO, EDWARD JOSEPH, biochemist, toxicologist, research center adminstr.; b. Passaic, N.J., June 7, 1933; s. Anthony and Sarah (Topchik) M.; m. Janet Carole Elser, Nov. 23, 1953 (div. 1977); children: David Alan, Anita Diane Massaro Edwards, Paul Anthony, Steven Joseph; m. Arlene Margaret Mahood, May 31, 1978. A.B., Rutgers U., 1955; M.A., U. Tex., 1958, Ph.D. in Biochemistry, 1962. Instr. Blinn Coll., Brenham, Tex., 1956-57; research assoc. Johns Hopkins U., Balt., 1965, Yale U., 1965-68; asst. prof. biochemistry SUNY, Buffalo, 1968-71, assoc. prof., 1971-75, prof., 1975-78, research prof., 1978-79; dir. chem. carcinogenesis Mason Research Inst., Worcester, Mass., 1977-78, dir. toxicology and chem. carcinogenesis, 1978; dir. Center for Air Environment Studies, Pa. State U., 1978—, prof. toxicology, dept. vet. sci., 1978—; cons. in field. Contbr. articles to profl. jours. Bd. dirs. Environ. Clearing House, Inc., Buffalo, 1973; bd. dirs. Central Pa. Lung and Health Service Assn., 1979—, mem. exec. com. of adv. bd., 1981—, mem. budget and program com., 1981—, v.p., 1981—. USPHS fellow, 1960-62, 62-63, 63-65; fellow Rachel Carson Coll., SUNY, Buffalo, 1968-78. Mem. AAAS, Am. Assn. for Lab. Animal Sci., Am. Assn. Pathologists, Am. Coll. Toxicology, Am. Soc. Biol. Chemists, Am. Soc. Icthyologists and Herpetologists, Am. Soc. for Pharmacology and Exptl. Therapeutics, Am. Wildlife Fedn., Behavioral Teratology Soc., Internat. Assn. Bioinorganic Scientists, Internat. Assn. Water Pollution Research, Internat. Soc. for Study of Xenobiotics, Internat. Union Pharmacology, Internat. Soc. for Ectotoxicology and Environ. Safety, Nat. Soc. for Med. Research, N.Y. Acad. Scis., Soc. for Environ. Geochemistry and Health, Soc. Environ. Toxicology and Chemistry, Soc. for Prodn. of Old Fishes, Soc. Toxicology, Teratology Soc., Toxicology Forum, Sigma Xi. Subspecialties: Biochemistry (biology). Current work: Lipid peroxidation, xenobiotic metabolism, heavy metal toxicology. Antioxidant, prostaglandin, cyclic nucleotide, polyaromatic hydrocarbon metabolism; perinatal toxicology of lead. Office: Center for Air Environment Studies 226 Fenske Lab Pa State U University Park PA 16802

MASSELL, WULF FRIEDRICH, geophysicist; b. Töltz, Germany, June 2, 1943; came to U.S., 1955; s. Paul and Gisela (Below) M.; m. Georgeann Christine Kind, June 9, 1968; children: Christina, Erika. B.S., U. Minn., 1966; Ph.D., Ind. U., 1972. Mgr. labs. U.S. Antartic Research Program, McMurdo, Australia, 1966-67; asst. prof. geophysics U. Tex.-Austin, 1972-77; sr. geophysicist Amoco Products Co., Houston, 1977-80; dir. research and devel. Exploration Services div. Geosource, Inc., Houston, 1980—. Mem. Soc. Exploration Geophysicists, European Assn. Exploration Geophysicists, Seismol. Soc. Am., AAAS, Mensa. 1. Subspecialties: Geophysics; Geology. Current work: Exploration geophysics applied to the search for hydrocarbons; information theory and seismic data processing; digital graphics. Home: 4002 El James Dr Spring TX 77373 Office: Geosources Inc PO Box 36306 Houston TX 77036

MASSEY, EVAN MORGAN, coal co. exec.; b. Richmond, Va., Dec. 29, 1926; s. Evan and Gertrude Anne (Morgan) M.; m. Elizabeth Anne Herscher, July 9, 1949; children: Elizabeth Anne, Evan Morgan, John W., Susan C., Nancy Lee; m. Joan Carpenter, Oct. 12, 1963; children: Charles Taylor, Craig Lewis. B.M.E., U. Va., 1949; M.B.A., U. Richmond, 1965. With A. T. Massey Coal Co., Richmond, 1948—; organizer Royalty Smokeless Coal Co. and other cos., Clifftop, W.Va., 1950—, pres., 1972—. Mem. Republican Nat. Com., St. Joe Polit. Assistance Com. Recipient Black Bull award Pa. State Alumni, 1982. Mem. AIME, Nat. Coal Assn. (dir.), Bituminous Coal Operators Assn. (dir.), Richmond C. of C. Republican. Episcopalian. Clubs: Bull & Bear (Richmond); Anglo Belgian (London). Subspecialties: Combustion processes; Coal. Current work: Research and tech.; furthering prodn., sale and use of coal as a growing source of energy. Office: A T Massey Coal Co PO Box 26765 Richmond VA 23261

MASSEY, GAIL AUSTIN, elec. engring. educator; b. El Paso, Tex., Dec. 2, 1936; s. Albert Harley and Mary Frances (Edmundson) M.; m. Barbara Suzanne Koch, July 2, 1960. B.S., Calif. Inst. Tech., 1959; M.S., Stanford U., 1967, Ph.D., 1970. Engr. Raytheon Co., Santa Barbara, Calif., 1959-63; sr. engr. GTE Sylvania, Mountain View, Calif., 1963-72; assoc. prof. Oreg. Grad. Ctr., Beaverton, 1972-74, prof., 1974-80; prof. elec. engring. San Diego State U., 1981—; cons. in field. Contbr. chpts. to books and articles to profl. jours. Fellow Optical Soc. Am.; mem. IEEE, Acoustical Soc. Am., Soc. Photo-Optical Instrumentation Engrs. Subspecialties: Laser instrumentation; Nonlinear optics. Current work: Generation of electron beams and images using laser-excited nonlinear photoelectric emission; laser instrumentation for remote measurement of gas temperature and density. Home: 7107 Birchcreek Rd San Diego CA 92119 Office: San Diego State Dept Engring San Diego CA 92182

MASSEY, WALTER EUGENE, physicist, university official; b. Hattiesburg, Miss., Apr. 5, 1938; s. Almar Clevel and Essie (Nelson) M.; m. Shirley Streeter, Oct. 25, 1969; children: Keith Anthony, Eric Eugene. B.S., Morehouse Coll., 1958; M.A., Washington U., St. Louis, 1966, Ph.D., 1966. Physicist Argonne (Ill.) Nat. Lab., 1966-68; asst. prof. physics U. Ill., Urbana, 1968-70; asso. prof. Brown U., Providence, 1970-75, prof., dean, 1975-79; cons. Argonne Nat. Lab., dir., 1979—; v.p. for research U. Chgo., 1982—; mem. NSB, NSF, 1978—; cons. Nat. Acad. Scis., 1973-76. Contbr. articles on sci. edn. in secondary schs. and in theory of quantum fluids to profl. jours. Trustee Brown U., 1980—; bd. dirs. Urban League R.I., 1972-74. NSF fellow, 1962; NDEA fellow, 1959-60. Fellow AAAS (dir. 1981—); Mem. Am. Phys. Soc. (councillor-at-large 1980—), Am. Assn. Physics Tchrs. (Disting. Service award 1975), Sigma Xi. Subspecialties: Condensed matter physics; Theoretical physics. Home: 4950 Chicago Beach Dr Chicago IL 60615 Office: Argonne Nat Lab Argonne IL 60439

MASSIER, PAUL FERDINAND, engineering project manager, administrator; b. Pocatello, Idaho, July 22, 1923; s. John and Kate (Arki) M.; m. Miriam Alice Parks, May 1, 1948 (dec. 1975); children: Marilyn Massier Schwegler, Paulette Massier Holden; m. Dorothy Margaret Hedlund, Sept. 12, 1978. B.S. in Mech. Engring, U. Colo., 1947, M.S., MIT, 1949. Engr. insp. Pan Am. Refining Corp., Texas City, Tex., 1947-48; design engr. Maytag Co., Newton, Iowa, 1949-51; research engr. Boeing Co., Seattle, 1951-55, Jet Propulsion Lab., Pasadena, Calif., 1955-58, group supr., 1958-81, project mgr., exec. asst. research sect., 1981—. Contbr. articles to profl. publs. in field; reviewer research papers for 7 jours.; organizer, chmn. conf. sessions; mem. sci. coms.; contbr. presentations tech. symposia. Served with U.S. Army, 1943-46. Recipient Apollo Achievement award NASA, 1969, basic noise research award, 1981. Assoc. fellow AIAA; mem. Sigma Xi. Subspecialties: Acoustics; Fluid mechanics. Current work: Conducted research in noise, rocket propulsion, heat transfer, supersonic diffusers, thermally ionized gases, shock waves, supersonic nozzles, gas turbines. Office: 1000 N First Ave Arcadia CA 91006

MASSON, MICHAEL EDWARD JOSEPH, psychology educator; b. Vancouver, C., Jan. 16, 1954; s. Lawrence C. and Sanny (Kralkay) M.; m. Debra A. Robbins, Aug. 9, 1980. B.A., U. B.C., 1975; M.A., U. Colo., 1977, Ph.D., 1979. Postdoctoral fellow Carnegie-Mellon U., Pitts., 1979-80; asst. prof. psychology U. Victoria, B.C., Can., 1980—. Contbr. articles to profl. jours. Social Sci. & Humanities Research Council Can. grantee, 1981-84; Nat. Sci. & Engring. Research Council Can. postdoctoral fellow, 1979-80; grantee, 1981-84. Mem. Am. Psychol. Assn., Canadian Psychol. Assn., Cognitive Sci. Soc., Psychonomic Soc., Nat. Conf. for Use of On-Line Computers in Psychology. Subspecialties: Cognition; Learning. Current work: Conceptual and perceptual processing of text is being studied in context of rapid reading processes such as skimming, speed reading, computerized presentation of text. Office: Dept Psychology U Victoria Victoria BC Canada V8W 2Y2 Home: 1490 Garnet Rd Victoria BC Canada V8P 5L1

MASSOPUST, LEO CARL, anatomy educator, researcher; b. Milw., Nov. 12, 1920; s. Leo Carl and Lillian (Bauer) M.; m. Elizabeth Mary Crain, Sept. 18, 1943; children: Mary Kathryn, Christopher Michael. B.S., Marquette U., 1943, M.S., 1947; Ph.D., U. Colo., 1953. Instr. anatomy Westminster Coll., Fulton, Mo., 1947-48; instr. U. Colo., 1948-54; sr. scientist NIH, Bethesda, Md., 1954-58; sr. physiologist Southeast La. Hosp., Mandeville, 1958-60; researcher Cleve. Psychiat. Hosp., 1960-74; assoc. prof. St. Louis U., 1974—; reviewer NSF, Washington, 1965—. Served to lt. j.g. USN, 1943-46. Mem. Am. Anatomy Soc., Am. Physiol. Soc., Soc. Neurosci., Am. Acad. Neurology, Sigma Xi, Phi Sigma Omicron. Subspecialties: Anatomy and embryology; Neurophysiology. Current work: Neural pathway tracing via autoradiography, intracellular enzyme markers and electrophysiology. Home: 8105 Valcour Ave Saint Louis MO 63123 Office: Saint Louis U Sch Medicine 1402 S Grand Blvd Saint Louis MO 63104

MASTERS, COLIN LOUIS, medical researcher, neuropathologist; b. Perth, Australia, Feb. 5, 1947; s. Robert Otto and Gwendoline Claire (Wright) M.; m. Helen Lannah, Nov. 24, 1976; children: Seth Lucian, Mary Katherine. M.B.B.S., U. Western Australia, 1970, M.D., 1977. Med. researcher, Australia, 1970-76, 81—, U.S. and Europe, 1976-81; sr. research fellow dept. pathology U. Western Australia, 1981—. Contbr. articles on neuropathology, neurovirology and neurobiology to profl. jours. Subspecialties: Neurology; Pathology (medicine). Current work: Research in chronic degenerative brain diseases. Office: Dept Pathology U Western Australia Nedlands Western Australia 6009

MASTERS, ROBERT EDWARD LEE, neural re-education researcher, psychotherapist, human potential educator; b. St. Joseph, Mo., Jan. 4, 1927; s. Robert and Katherine (Leeper) M.; m. Jean Houston, May 8, 1965. B.A. in Philosophy, U. Mo.-Columbia, 1951; Ph.D. in Clin. Psychology, Humanistic Psychology Inst., 1974. Dir. Library of Sex Research, N.Y.C., 1962-66, Dir. Sensory Inagery Program, 1965-68; dir. research Found. for Mind Research, N.Y.C. and Pomona, N.Y., 1968—; dir. Zarathustra Project, Pomona, 1980—; co-dir. Human Capacities Tng. Program, Ramapo, N.J., 1982—; pvt. practice psychotherapy, neural re-edn.; prin. tchr. Psychophys. Method Tchr. Tng. Programs, 1980—; pres. Human Capacities Corp., Pomona, 1982—; prin. tchr. Hypnotherapist Tng., Pomona, 1982—. Author: books, including Eros and Evil, 1962, (with J. Houston) Varieties of Psychedelic Experience, 1966, Mind Games, 1972, Listening to the Body, 1978; author: Psychophysical Method Exercises Vols. I-VI, 1983. Served with USN, 1945-46; PTO. Grantee Erickson Found, 1966; Kleiner Found, 1968; Babcock Found, 1970; Doris Duke Found., 1972. Mem. Am. Psychol. Assn., N.Y. Acad. Scis., Am. Assn. Sex Educators, Counselors and Therapists, AAAS, Assn. Humanistic Psychology. Subspecialties: Gerontology; Neuropsychology. Current work: Human potentials; genius and high-level creativity; hypnosis and altered states of consciousness; neural re-education; body image; movement disinhibition; nonverbal communication; parapsychology; sexology; applications of brain research to education. Office: Foundation for Mind Research PO Box 600 Pomona NY 10970

MASTROIANNI, LUIGI, JR., physician; b. New Haven, Nov. 8, 1925; s. Luigi and Marion (Dallas) M.; m. Elaine Catherine Pierson, Nov. 4, 1957; children—John James, Anna Catherine, Robert Luigi. A.B., Yale U., 1946; M.D., Boston U., 1950, D.Sc. (hon.), 1973, M.A., U. Pa., 1970. Diplomate: Am. Bd. Obstetrics and Gynecology (subsplty. gynecol. endocrinology). Intern Met. Hosp., N.Y.C., 1950-51, resident obstetrics and gynecology, 1951-54; practice medicine specializing in obstetrics and gynecology, New Haven, 1955-61, Los Angeles, 1961-65, Phila., 1965—; fellow in fertility and endocrinology Med. Sch., Harvard and Free Hosp., 1954-55; instr. obstetrics and gynecology Yale Sch. Medicine, 1955-56, asst. prof., 1956-61; prof. obstetrics and gynecology U. Calif. at Los Angeles, 1961-65; chief obstetrics and gynecology Harbor Gen. Hosp., Torrance, Calif., 1961-65; prof., chmn. obstetrics and gynecology U. Pa., 1965—, Goodell prof. gynecology, 1967—; chief obstetrics and gynecology Woman's Hosp. of Hosp. U. Pa., 1965—; chmn. primate research centers adv. com. NIH, 1965-68; sci. adv. bd. Internat. Planned Parenthood Fedn., 1966—. Editor: Jour. Fertility and Sterility, 1969-76; contbr. articles to profl. jours. Bd. dirs. Sex Info. Ednl. Council U.S., 1968-71. Recipient Squibb prize Am. Fertility Soc., 1965; Ortho medal for contbns. to reprodn., 1966; Lindback Found. award for disting. teaching, 1969; Barren Found. medal, 1977. Fellow A.C.S., Am. Coll. Obstetricians and Gynecologists, Coll. Physicians Phila., Am. Gynecol. Soc.; mem. Am. Physiol. Soc., Soc. Exptl. Biology and Medicine, Soc. Gynecologic Investigation, Am. Fertility Soc., Pacific Coast Fertility Soc. (hon. mem.), Obstet. Soc. Phila., Endocrine Soc., Soc. Study of Reprodn. (dir. 1972—), Soc. Study of Fertility, Central Assn. Obstetricians and Gynecologists (hon.), Peruvian Obstet. and Gynecol. Soc. (hon.), Societ Gynaecologica et Obstetrica Italia (hon.), Brazilian Fertility Soc. (hon.), Sigma Xi, Alpha Omega Alpha. Roman Catholic. Subspecialty: Reproductive endocrinology. Current work: In vitro fertillization, reproductive biology-human infertility. Home: 351 Laurel Ln Haverford PA 19041 Office: 3400 Spruce St Philadelphia PA 19104

MASTRONARDI, RICHARD, mechanical engineering administrator; b. N.Y.C., Nov. 14, 1947; s. Pasquale Severio and Lucy Lillian (Spoto) M.; m. Susan Elizabeth Kessler, Apr. 25, 1970; 1 dau., Sara Elizabeth. B.S. in Aero. Engring, Rensselaer Poly. Inst., 1969; M.B.A., Northeastern U., 1975. Aeroelastic design engr. Boeing Co., Seattle, 1969-70; exptl. flight test engr. Sikorsky Helicopter, Stratford, Conn., 1970; project engr. Atkins & Merrill Inc., Ashland, Mass., 1970-75; pres. Aerowedge Inc., Medfield, Mass., 1975-81; dir. mech. engring. Am. Sci. and Engring. Inc., Cambridge, Mass., 1975—; cons. various orgns. Co-chmn. New Start Com. (Vietnam refugees), Medfield, 1980—. Mem. AIAA, Planetary Soc., NASA Get Away Spl. User Com. Subspecialties: Aerospace engineering and technology; 1-ray high energy astrophysics. Current work: Program and engineering manager for developing X-ray and higher energy high-resolution digital inspection systems for non-destructive testing. Inventor proximity rotary table and elevator for CT scanner, hollow electro-less nickle baffle tube, aerowedge truck wind deflector. Home: 55 South St Medfield MA 02052 Office: American Science and Engring Inc Fort Washington Cambridge MA 02139

MASURSKY, HAROLD, geologist; b. Ft. Wayne, Ind., Dec. 23, 1923; s. Louis and Celia (Ochstein) M.; m.; 4 children. B.S. in Geology, Yale U., 1943, M.S., 1951; D.Sc. (hon.), No. Ariz. U., 1981. With U.S. Geol. Survey, 1951—, chief astrogeologic studies br., 1967-71, chief scientist, Flagstaff, Ariz., 1971-75, sr. scientist, 1975—, lunar orbiter, 1965-67; team leader, prin. investigator TV experiment (Mariner Mars), 1971, co-investigator Apollo field geol. team Apollo 16 and 17, also mem. Apollo orbital sci. photog. team, Apollo site selection group; leader Viking landing site staff, dep. team leader orbiter visual imaging systems, 1975, mem. teams, 1977, chmn. mission ops. group Venus Pioneer Mission, 1978, co-chmn. mission operational group Galileo Mission, 1981, mission ops. leader, radar team Venus Orbiting Imaging Radar Mission, 1981—; mem. Space Sci. Adv. Com., 1978-81, solar system exploration com., 1980-83; mem. Space Sci. Bd., 1982—; v.p. intedisc com. B, COSPAR; sec. Coordinating Com. of Moon and Planets. Asso. editor: Icarus, Geophys. Rev. Letters. Served with AUS, 1943-46. Fellow Geol. Soc. Am. (asso. editor bull., pres. planetary geol. div.), AAAS, Am. Geophys. Union, Internat. Astron. Union (commns.), Am. Astron. Assn., Com. Space Research. Subspecialties: Geology; Satellite studies. Address: US Geol Survey 2255 N Gemini St Flagstaff AZ 86001 The vigorously ongoing exploration of outer space continues the exploration tradition of man. It has been exhilarating to be involved with the manned and unmanned exploration programs. Each of the nine programs so far has required new thinking and approaches to solving the technical and scientific problems. We now will encounter Uranus and Neptune and map the surface of Venus in detail with radar, as we are doing with the Earth.

MATALON, MOSHE, engring. and applied math. educator; b. Cairo, Feb. 25, 1949; U.S., 1974; s. Ernest and Alisa (Saadia) M.; m. Hedva Waizmann, Nov. 1, 1978; 1 child, Leehe. B.Sc., Tel Aviv U., 1969, M.Sc., 1976; Ph.D., Cornell U., 1977. Asst. prof. Poly. Inst. N.Y., Farmingdale, 1978-80; asst. prof. Northwestern U., 1980—. Mem. AIAA, Am. Phys. Soc., Soc. Indsl. and Applied Math., The Combustion Inst., Sigma Xi. Subspecialties: Combustion processes; Fluid mechanics. Current work: Combustion theory, fluid dynamics, applied mathematics. Home: 9006 Oleander St Morton Grove IL 60053 Office: Northwestern U Technol Inst Dept Engring Scis and Applied Math Evanston IL 60201

MATARAZZO, JOSEPH DOMINIC, psychologist; b. Caiazzo, Italy, Nov. 12, 1925; s. Nicholas and Adeline (Mastroianni) M.; m. Ruth Wood Gadbois, Mar. 26, 1949; children: Harris, Elizabeth, Sara. Student, Columbia U., 1944; B.A., Brown U., 1946; M.S., Northwestern U., 1950, Ph.D., 1952. Fellow in med. psychology Washington U. Sch. Medicine, 1950-51; instr. Washington U., 1951-53, asst. prof., 1953-55; research asso. Harvard Med. Sch., asso. psychologist Mass. Gen. Hosp., 1955-57; prof., head med. psychol. dept. Oreg. Health Scis. U., Portland, 1957—; mem. nursing research and patient care study sect., behavioral medicine study sect. NIH; spl. research com. Am. Cancer Soc.; Bd. regents Uniformed Services U. Health Scis., 1974-80. Author: Wechsler's Measurement and Appraisal of Adult Intelligence, 5th edit., 1972, (with A.N. Wiens) The Interview: Research on its Anatomy and Structure, 1972, (with Harper and Wiens) Nonverbal Communication, 1978; asso. editor: Human Orgn, 1954-66; editorial bd.: Jour. Clin. Psychology, 1962—, Psychotherapy: Theory, Research and Practice, 1964-68, Mental Hygiene, 1969-77, Jour. Continuing Edn. in Psychiatry, 1975-79; cons. editor: Contemporary Psychology, 1962-70, 80—, Jour. Community Psychology, 1974-81, Behavior Modification, 1976—, Intelligence: An Interdisciplinary Jour, 1976, Jour. Behavioral Medicine, 1977—, Profl. Psychology, 1978—, Jour. Cons. and Clin. Psychology, 1978—; editor: Psychology series Aldine Pub. Co, 1964-74; psychology editor: Williams & Wilkins Co, 1974-77; contbr. articles to psychol. jours. Served to ensign USNR, 1943-47; capt. Res. Recipient Hofheimer prize Am. Psychiat. Assn., 1962. Fellow Am. Psychol. Assn. (pres. div. health psychology 1978-79, mem. Council of Reps. 1982—), AAAS; mem. Western, Oreg. psychol. assns., Am. Assn. State Psychology Bds. (pres. 1963-64), Nat. Assn. Mental Health (dir.), Oreg. Mental Health Assn. (dir., pres. 1962-63), Internat. Council Psychologists (dir. 1972-74, pres. 1976-77), Assn. Advancement of Psychology (trustee 1980-84, chmn. bd. trustees 1983-84), Sigma Xi. Subspecialties: Psychobiology; Neuropsychology. Current work: Lifestyle risk factors in health and illness; brain-behavior parameters in cognition. Home: 1934 SW Vista Ave Portland OR 97201

MATEKER, EMIL JOSEPH, JR., geophysicist, geophys. co. exec.; b. St. Louis, Apr. 25, 1931; s. Emil Joseph and Lillian A. (Broz) M.; m. Lolita Ann Winter, Nov. 25, 1954; children: Mark Steven, Anne Marie, John David. B.S., St. Louis U., 1956, M.S., 1959, Ph.D. in Geophysics, 1964. Registered geophysicist, geologist, Calif. Geophysicist Standard Oil Co. Calif., 1957-60; instr. geophysics St. Louis U., 1960-63; asst. prof. geophysics Washington U., 1963-65, assoc. prof., 1965-69; cons. oil cos., 1962-70; mgr. geophysical research Western Geophys. Co. Am., 1969-70, v.p. research and devel., 1970-74; pres. Westrex, 1974-77, Litton Resources Systems, 1977; pres. aero service div., v.p. corp. Western Geophys. Co. Am., 1974—. Author: A Treatise on Modern Exploration Seismology, 2 vols, 1965; contbr. articles to profl. jours. Chmn. bd. dirs. St. Agnes Acad., Houston, 1977—; pres. Strake Jesuit Boosters Club, Houston, 1977-78; mem. parish council St. John Vianney Ch., Houston. Served with U.S. Army, 1952-54. Recipient Merit award St. Louis U., 1976. Mem. Soc. Exploration Geophysicists, Am. Geophys. Union, Seismol. Soc. Am., AAAS, Geophys. Soc. Houston, European Assn. Exploration Geophysicists, Sigma Xi. Roman Catholic. Subspecialties: Geophysics; Remote sensing (geoscience). Current work: Exploration geophysics, synthetic aperture radar. Home: 419 Hickory Post Houston TX 77079 Office: 8100 Westpark Dr Houston TX 77063

MATHAI, ARAK MATHAI, mathematics educator, researcher. Ph. D., U. Toronto, 1964. Prof. dept. math. McGill U., Montreal, Que., Can., 1964—. Author six books; Contbr. articles to profl. jours.; mem. editorial bds. various math. and statis. jours. Mem. Internat. Statis. Inst., Soc. Internat. and Applied Math., Statis. Sci. Assn. Can (exec. com. 1972-77), Inst. Math. Stats. Subspecialties: Statistics; Probability. Current work: Theoretical and applied statistics and probability and some aspects of analysis, information theory, population problems and model building. Office: Dept Math McGill U 805 Sherbrooke St W Montreal PQ Canada H3A 2K6

MATHER, JOHN CROMWELL, physicist; b. Roanoke, Va., Aug. 7, 1946; s. Robert Eugene and Martha Belle (Cromwell) M.; m. Jane Anne Hauser, Nov. 22, 1980. B.A. in Physics with highest honors, Swarthmore Coll., 1968; Ph.D. in Physics (NSF fellow, hon. Woodrow Wilson fellow, 1968, Fannie and John Hertz fellow 1970-74), U. Calif.-Berkeley, 1974. NAS/NRC research assoc. Goddard Inst. Space Studies, N.Y., 1974-76; astrophysicist Goddard Space Flight Center, Greenbelt, Md., 1976—; lectr. in astronomy Columbia U., N.Y.C., 1975-76. Contbr. sci. articles to profl. publs. Mem. Optical Soc. Am., Am. Phys. Soc., Am. Astron. Soc., Soc. Photo-optical Instrumentation Engrs., Sigma Xi, Phi Beta Kappa. Democrat. Unitarian. Subspecialties: Cosmology; Infrared optical astronomy. Current work: Measurement of cosmic background radiation from the Big Bang using infrared and microwave instruments on NASA Cosmic Background Explorer Satellite planned for 1987 launch. Home: 4400 Romlon St Beltsville MD 20705 Office: Code 693.2 NASA Goddard Greenbelt MD 20771

MATHES, STEPHEN JOHN, plastic and reconstructive surgeon, educator; b. New Orleans, Aug. 17, 1943; s. John Edward and Norma (Deutsch) M.; m. Jennifer Wood-Bridge, Nov. 1966; children : David, Brian, Edward. B.S., La. State U., 1964, M.D., 1968. Diplomate: Am. Bd. Surgery, Am. Bd. Plastic and Reconstructive Surgery. Intern U. Va., Charlottesville, 1968-69, resident in gen. surgery, 1969-70, Emory U., Atlanta, 1972-75, resident in plastic surgery, 1975-77; asst. prof. Washington U., St. Louis, 1977-78; assoc. prof. U. Calif.-San Francisco, 1978—. Author: Clinical Atlas of Muscle and Musculocutaneous Flaps, 1979, Clinical Applications for Muscle and Musculocutaneous Flaps, 1982; contbr. articles to profl. jours. Served as maj. U.S. Army, 1970-72. NIH grantee, 1982; recipient best paper award Am. Soc. Aesthetic Plastic Surgery, 1981; 1st prize Plastic Surgery Ednl. Found. Scholarship Contest, 1981. Fellow ACS; mem. Am. Soc. Plastic and Reconstructive Surgeons (James Barrett Brown prize 1982), Am. Burn Assn., Internat. Soc. Reconstructive Microsurgery, Plastic Surgery Research Council, Am. Soc. for Surgery of Hand. Episcopalian. Subspecialties: Microsurgery; Surgery. Current work: Clinical and experimental research in microsurgery, wound healing, wound infections and flap use in reconstructive surgery. Home: 618 Dorchester Rd San Mateo CA 94402 Office: Dept Surgery U Calif San Francisco CA 94143

MATHEWS, C(OLLIS) WELDON, physical chemist, educator; b. Troy, Ala., July 19, 1938; s. James Michael and Lottie Bae (Green) M.; m. Betty Ann Logan, June 13, 1959; children: Terri Lynn, Cynthia Elaine. B.S., U. Ala., 1960; Ph.D., Vanderbilt U., 1965. Postdoctoral fellow Vanderbilt U., 1964-65, NRC Can., Ottawa, Ont., 1965-67; asst. prof. chemistry Ohio State U., 1967-72, assoc. prof., 1972—; vis. research assoc. U. Md. and Naval Research Labs., summer 1979. Contbr. articles, cpts. to profl. publs.; editor: (with K.N. Rao) Molecular Spectroscopy: Modern Research, 1972. Fellow Optical Soc. Am.; mem. Am. Chem. Soc., Sigma Xi, Alpha Chi Sigma, Sigma Pi Sigma. Subspecialties: Physical chemistry; Spectroscopy. Current work: Molecular spectroscopy of transient species; conventional and laser spectroscopic studies of gas-phase reactive intermediates, e.g. high-temperature chemistry. Office: Dept Chemistry Ohio State U Columbus OH 43210

MATHEWS, GRANT JAMES, nuclear astrophysicist; b. Saginaw, Mich., Oct. 14, 1950; s. Kenneth Ward and Agnes Hazel M. B.S., Mich. State U., 1972; Ph.D., U. Md., 1977. Postdoctoral research assoc. U. Md., 1977, 1977; research assoc. U. Calif. (Lawrence Berkeley Lab.), Berkeley, 1977-79; research fellow Calif. Inst. Tech., Pasadena, 1979-81; physicist U. Calif. Lawrence Livermore Nat. Lab., Livermore, 1981—. Contbr. articles to research jours. Recipient Sigma Xi Research Excellence award, 1976. Mem. Am. Phys. Soc., Am. Astron. Soc., Am. Chem. Soc. Subspecialties: Nuclear astrophysics; Nuclear physics. Current work: Nuclear astrophysics, stellar evolution, galactic evolution, cosmological nucleosynthesis, cosmic-ray propagation, explosive nucleosynthesis. Office: U Calif Lawrence Livermore Nat Lab Livermore CA 94550

MATHEWS, KENNETH PINE, physician, educator; b. Schenectady, Apr. 1, 1921; s. Raymond and Marguerite Elizabeth (Pine) M.; m. Alice Jean Elliott, Jan. 26, 1952 (dec.); children: Susan Kay, Ronald Elliott, Robert Pine; m. Winona Beatrice Rosenburg, Nov. 8, 1975. A.B., U. Mich., 1941, M.D., 1943. Diplomate: Am. Bd. Internal Medicine, subsplty. bd. in allergy. Intern, asst. resident, resident in medicine Univ. Hosp., Ann Arbor, Mich., 1943-45, 48-50; mem. faculty dept. medicine U. Mich. Med. Sch., 1950—, assoc. prof. internal medicine, 1956-61, prof., 1961—, head div. allergy, 1967—, chmn. residency rev. com. for allergy and immunology; cons. Ann Arbor VA Hosp., Wayne County Gen. Hosp. Co-author: A Manual of Clinical Allergy, 2d edit, 1967; editor: Jour. Allergy and Clin. Immunology, 1968-72; contbr. numerous articles in field to profl. jours. Served to capt. M.C. AUS, 1946-48. Recipient Disting. Service award Am. Acad. Allergy, 1976. Fellow Am. Acad. Allergy (past pres.), A.C.P.; mem. Am. Assn. Immunologists, Central Soc. Clin. Research (emeritus), Am. Fedn. Clin. Research, Am. Thoracic Soc., Mich. Allergy Soc. (past pres.), Washtenaw County Med. Soc., Mich. State Med. Soc., Alpha Omega Alpha, Phi Beta Kappa. Subspecialties: Allergy; Immunopharmacology. Current work: Mechanisms of urticaria and angioedema; IgE antibodies; mechanisms of drug hypersensitivity. Home: 1145 Aberdeen Dr Ann Arbor MI 48104 Office: Box 27 Univ of Mich Med Center Ann Arbor MI 48109

MATHEWS, MAX VERNON, elec. engr.; b. Columbus, Nebr., Nov. 13, 1926; s. Lavern Buckingham and Ruth June (Vernon) M.; m. Marjorie Winslow Morse, Dec. 27, 1947; children—Vernon, Guy, Boyd. B.S., Calif. Inst. Tech., 1950; M.S., M.I.T., 1952, Sc.D., 1954. With Bell Labs., Murray Hill, N.J., 1955—, dir. acoustical and behavioral research, 1962—; Sci. adv. Inst. Research and Coordination Acoustics/Music, Paris. Author: Technology of Computer Music, 1969. Served with USNR, 1944-46. Fellow IEEE (David Sarnoff Gold Medal award 1973), Acoustical Soc. Am.; mem. Nat. Acad. Scis., Audio Engring. Soc., Psychonomic Soc., Sigma Xi. Subspecialties: Acoustical engineering; Electrical engineering. Home: 81 Oakwood Dr New Providence NJ 07974 Office: 2D-554 Mountain Ave Murray Hill NJ 07974

MATHEWS, R. MARK, psychologist, educator; b. Topeka, Kans., Apr. 12, 1953; s. Ralph C. and Dorothy M.; m. Stephanie A. Scheffler, June 15, 1976. B.G.S., U. Kans.-Lawrence, 1974, M.A., 1975, Ph.D., 1980. Research assoc. Center for Pub. Affairs, Lawrence, 1977-80; asst. prof. psychology U. Hawaii, Hilo, 1980-83; research dir. R & T Center Ind. Living, U. Kans., Lawrence, 1983—; cons. Hawaii County Community Mental Health Center, Hilo, 1981—. Author: Matching Clients and Service, 1981; Contbr. articles to profl. jours. Chmn. adv. bd. Hawaii County Community Mental Health Ctr., 1981—; mem. Hawaii Mental Health Adv. Council, 1981—. Mem. Am. Psychol. Assn., Assn. for Advancement Behavior Therapy, Assn. for Behavior Analysis. Subspecialty: Behavioral psychology. Current work: Research interests include the development and analysis of community service programs, the evaluation of educational interventions and the assessment and training of job-finding skills. Office: Bur Child Research U Kans Lawrence KS 66045

MATHEWSON, CHRISTOPHER COLVILLE, engineering geologist, educator, researcher, consultant; b. Plainfield, N.J., Aug. 12, 1941; s. George Anderson and Elsa Ray (Shrimpton) M.; m. Janet Marie Olmstead, Nov. 2, 1968; children: Heather Alexis, Glenn George Anderson. B.S. in Civil Engring, Cast Inst. Tech., 1963; M.S. in Geol. Engring, U. Ariz., 1965, Ph.D., 1971. Profl. engr., Tex., Ariz.; profl. geologist, Oreg. Asst. prof. Tex. A&M U., College Station, 1971-75, assoc. prof., 1975-81, prof., 1981—, dir. Center for Engring. Geoscis., 1982. Author: Engineering Geology, 1981. Chmn. Planning and Zoning Commn., College Station, 1973-78. Served to lt. NOAA, 1965-70. Fellow Geol. Soc. Am., Geol. Assn. Can.; mem. Assn. Engring. Geologists. (editor 1981—, C.P. Holdredge award), Am. Geophys. Union. Subspecialties: Geology; Coal. Current work: Engineering geology applied to coal development, urban planning, coastal and river processes and geomorphic processes. Office: Tex A&M Dept Geology College Station TX 77843

MATHIAS, MELVIN MERLE, nutrition educator; s. Burrell R and Mary L. (Wilfong) M.; m. Evelyn Jane Clayson, July 20, 1963; children: Michael, Beth, David. B.S., Purdue U., 1961; M.S., Cornell U., 1964, Ph.D., 1967. Research investigator U.S. Army, Denver, 1966-68; asst. prof. Colo State U., Fort Collins, 1968-76, assoc. prof., 1976-81, prof., 1981—; faculty trainee Donner Lab. U. Calif.-Berkeley, 1971. Scoutmaster Boy Scouts, Fort Collins, 1976-82. Served with U.S. Army, 1966-68. Mem. Am. Inst. Nutrition, Am. Oil Chemists Soc., AAAS, Colo. Heart Assn., Phi Kappa Phi (pres. 1982-83). Republican. Methodist. Club: Optimist (vp 1976). Subspecialties: Nutrition (medicine); Biochemistry (medicine). Current work: Role of dietary lipids in prostaglandin metabolism. Office: Colo State U Food Science and Nutrition Dept Fort Collins CO 80523 Home: 413 Tulane Dr Fort Collins CO 80525

MATHIESON, ARTHUR CURTIS, botanist, educator; b. Los Angeles, Dec. 26, 1937; s. Robert Pinkerton and Helen Bernice (Gibb) M.; m. Myla Jean Edwards, June 14, 1958; children: Cindy, Bert, Kristi. B.A., UCLA, 1960, M.S. in Botany, 1961; Ph.D. in Botany-Oceanography, U. B.C., 1965. Asst. prof. botany U. N.H., 1965-70, assoc. prof., 1970-73, prof., 1973—, dir., 1970—; cons. in field. Contbr. articles to profl. jours. Mem. N.H. Coastal Zone Mgmt. Com., 1980-82; mem. N.H. Estuarine Planning Com. NRC fellow, 1962-65; NOAA grantee. Mem. Phycological Soc. Am., Internat. Phycological Soc., New Eng. Estuarine Research Soc. Republican. Subspecialty: Ecology. Current work: Physiological ecology-floristics of seaweeds, including their mariculture and economic utilization. Office: Jackson Estuarine Lab U NH Adams Point Rt 1 Durham NH 03824

MATHIESON, BONNIE JEAN, research scientist; b. LaSalle, Ill., May 10, 1945; d. Gilbert Joseph and Dorothy May (Hamer) Lindenmier; m. Donald John Mathieson, Sept. 3, 1966; 1 son, Daniel John. B.S., U. Ill., 1967; postgrad., Stanford U., 1969-70; Ph.D., Cornell U., 1975. Research asst. Stanford U., 1967-70; postdoctoral fellow lab. microbial immunity Nat. Inst. Allergy and Infectious Diseases, NIH, Bethesda, Md., 1975-76, staff fellow, 1976-79, sr. staff fellow, 1979-82; mem. Basel (Switzerland) Inst. Immunology, 1982—. Mem. editorial bd.: Devel. and Comparative Immunology, 1982—; contbr. articles to profl. jours. Mem. AAAS, Am. Assn. Immunologists, Am. Soc. Zoologists, Found. Advanced Edn. in Scis., N.Y. Acad. Scis., Soc. Devel. and Comparative Immunology. Subspecialties: Immunobiology and immunology; Immunogenetics. Current work: Genetics of lymphocyte antigen expression. Home: 5906 Kingsford Pl Bethesda MD 20817 Office: Basel Inst Immunology Grenzacherstrasse 487 Basel Switzerland CH-4005

MATHIS, PHILIP MONROE, biology educator; b. Lowes, Ky., June 2, 1942; s. James Monroe and Genet (Reasons) M.; m. Marilyn Myron, Feb. 3, 1964; 1 dau., Lori Michelle. B.S., Murray (Ky.) State U., 1964; M.S., Middle Tenn. State U., 1964-67; Ed.S., Vanderbilt U., 1971; Ed.D., U. Ga., 1973. Tchr. sci. Illmo-Scott City (Mo.) High Sch., 1964-67; mem. faculty dept. biology Middle Tenn. State U., Murfreesboro, 1967—, now prof.; vis. scientist Tenn. Acad. Sci., 1976—; cons. Wadsworth Pub. Co., 1979—. Author: Laboratory Activities in Biology, 1982. Recipient Outstanding Service award Middle Tenn. State U., 1977; Outstanding Publ. award Edn. Digest, 1975. Fellow Sci. Assn. Tenn. (bd. dirs. 1978-80); mem. Soc. Coll. Sci. Tchrs. (editor jour. 1980—, nat. steering com. 1980—). Democrat. Club: Cannonsburg Long Rifles. Subspecialties: Species interaction; Genome organization. Current work: Evaluation of effects of aerial pollutants on corticoulous lichens; sampling and procedural innovations, karyology of freshwater fishes. Home: Route 8 Brandywood Subdiv Murfreesboro TN 37130 Office: Middle Tenn State U Box 578 Murfreesboro TN 37132

MATHRE, DONALD EUGENE, plant pathology educator; b. Frankfurt, Kans., Jan. 5, 1938; s. John E. and Anna Mae (Hurlburt) M.; m. Judith Ann, Aug. 25, 1961; children: Elizabeth Ann, Susan Jean. B.S., Iowa State U., 1960; Ph.D., U. Calif.-Davis, 1964. Asst. prof. plant pathology U. Calif.-Davis, 1964-67; asst. prof. Mont. State U., Bozeman, 1967-69, assoc. prof., 1969-72, prof., 1972—. Mem. Am. Phytopath. Soc., Crop Sci. Soc. Am. Subspecialty: Plant pathology. Current work: Researcher in soil-borne diseases of cereals and forage crops. Office: Mont State U Dept Plant Pathology Bozeman MT 59717

MATHUR, PERSHOTTAM P(RASAD), pharmacologist, researcher, educator; b. Delhi, India, Jan. 19, 1938; came to U.S., 1963, naturalized, 1976; s. Mukat Prasad and Bhatto Devi M.; m. Madhu Munni, Jan. 11, 1971; 1 dau., Sharad. B.S. in Chemistry and Biology, U. Delhi, 1957; Ph.D. in Pharmacology, U. Fla., 1968. NIH fellow in cardiology St. Luke's Hosp., N.Y.C., 1969-72; sect. head biochem. pharmacology William H. Rover Inc., Ft. Washington, Pa., 1972-76; group mgr. cardiovascular pharmacology A.H. Robins Co., Richmond, Va., 1977-80, sci. coordinator, sr. project coordinator clin. pharmacology, 1980—; adj. asst. prof. pharmacology Med. Coll. Va., 1977—. Contbr. articles to profl. jours.; reviewer for profl. jours. Fellow Am. Coll. Clin. Pharmacology; mem. Am. Soc. Pharmacology and Exptl. Therapeutics, Am. Fedn. Clin. Research, Soc. Exptl. Medicine and Biology, Va. Heart Assn., Rho Chi. Subspecialty: Pharmacology. Current work: Anti-hypertensives, anti-arrhythmics. Office: A H Robins Co 1211 Sherwood Ave Richmond VA 23220

MATHUR, RADHEY MOHAN, electrical engineering educator; b. Alwar, India, Feb. 2, 1936; s. Radhey Shyam and Phool (Kumari) M.; m. Aruna Mathur, June 30, 1943; children: Sudhanshu, Shikha. B.Sc., U. Rajasthan, Jaipur, India, 1956; B.Tech. with honors, Indian Inst. Tech., Kharaspur, 1960; Ph.D., U. Leeds, Eng., 1969. Instr. U. Man., Winnipeg, 1969-70, asst. prof. to assoc. prof., 1970-78, prof., 1978—, chmn. dept. elec. engring., 1980—. Recipient Carter prize U. Leeds, 1969; prize Inst. Engrs. India, 1962; Outreach award U. Man., 1982. Mem. IEEE (sr. mem.), Inst. Elec. Engrs. (London). Subspecialties: Wind power; Electrical engineering. Current work: Special high efficency electrical machines for wind power generation. Application of power electronics to electrical power transmission systems. Home: 39 Lafayette Bay Winnipeg MB Canada R3T 3J9 Office: Dept Elec Engring Univ Manitoba Winnipeg MB Canada R3T 2J2

MATHUR, SUBBI, immunologist, educator; b. Insein, Burma, Oct. 13, 1939; came to U.S., 1969; d. T.A. Ramaswami Pillai and Seethalakshmi Ramaswami; m. Rajesh S. Mathur, May 4, 1968; children: Veena Rani, Arun Kumar. B.A., U. Madras, India, 1957; M.A., 1959; M.S., 1960; Ph.D., 1966. Research assoc. Med. U. S.C., Charleston, 1975-81, asst. prof., 1981—. Mem. adv. panel Am. Assn. Minority Women in Science, Washington, 1982. NIH research grantee, 1981; Career Devel. awardee, 1982. Mem. Am. Soc. Microbiologists, Internat. Assn. Reproductive Immunology, Am. Assn. Immunologists, Am. Fertility Soc. Hindu. Subspecialties: Immunology (medicine); Transplantation. Current work: Reproductive immunology; immune mechanisms in infertility and endocrinological disorders; immune mechanisms between mother and fetus; immunology of chronic vaginal candidiasis; human leukocyte antigens and infertility. Office: BCIM Dept Med U SC 171 Ashley Ave Charleston SC 29425

MATIJEVIC, EGON, chemistry educator, consultant; b. Otocac, Croatia, Yugoslavia, Apr. 27, 1922; s. Gregor and Stefica (Spiegel) M.; m. Bozica Biscan, Mar. 9, 1947. Diploma in chem. engring, U. Zagreb, 1944, Ph.D. in Chemistry, 1948, Dr. Habil. in Phys. Chemistry, 1952; D.Sc. (hon.), Lehigh U., 1977. Instr. chemistry U. Zagreb, 1944-47, sr. instr. phys. chemistry, 1949-52; pvt. dozent in colloid chemistry, 1952-54, dozent in phys. and colloid chemistry, 1955-56; assoc. prof. chemistry Clarkson Coll. Tech., Potsdam, N.Y., 1960-62, assoc. dir., 1966-68, dir., 1968-81, prof. chemistry, 1962—, chmn. dept., 1981—; lectr.; vis. prof. Sci. U. Tokyo, 1979; vis. scientist U. Leningrad, 1977; vis. prof. U. Melbourne, 1976, Japan Soc. Promotion of Sci., 1973, Clarkson Coll. Tech., 1957-59; research fellow colloid sci. U. Cambridge, Eng., 1957-57. Am. regional editor: Colloid and Polymer Sci, 1966—; editor series: Surface and Colloid Science, 1964—; mem. editorial bd.: Colloids and Surfaces, 1977; asst. editor: Croatica Chemica Acta, 1953-57; adv. bd.: Jour. Colloid and Interface Sci, 1967-69; contbr. numerous articles to sci. jours., chpts. to books. Recipient Gold medal Am. Electroplaters Soc., 1976; Disting. Teaching award Clarkson Coll. Tech., 1975; Kendall award Am. Chem. Soc., 1972. Mem. Am. Chem. Soc. (chmn. No. N.Y. sect. 1962-63, councillor 1976-83), Am. Water Works Assn., Croatian Chem. Soc. (exec. bd. 1950-57), Internat. Assn. Colloid and Interface Scientists (pres.-elect), Kolloid-Gesellschaft (Germany), Chem. Soc. Japan (hon., testimonial scroll 1978), Sigma Xi. Subspecialties: Colloid chemistry; Surface chemistry. Current work: Colloids, corrosion of metals, aerosols, adhesion. Office: Clarkson College Potsdam NY 13676

MATKOWSKY, BERNARD JUDAH, applied mathematician, educator; b. N.Y.C., Aug. 19, 1939; s. Morris N. and Ethel H. M.; m. Florence Knobel, Apr. 11, 1965; children: David, Daniel, Devorah. B.S., CCNY, 1960; M.E.E., N.Y. U., 1961, M.S., 1963, Ph.D., 1966. Fellow Courant Inst. Math. Scis., N.Y. U., 1961-66; mem. faculty dept. math. Rensselaer Poly. Inst., 1966-77; prof. applied math. and math. Northwestern U., Evanston, Ill., 1977—; vis. prof. Tel Aviv U., Israel, 1972-73; vis. scientist Weizmann Inst. Sci., Israel, summer 1976, summer 1980, Tel Aviv U., summer 1980; cons. Argonne Nat. Lab., Sandia Labs. Editor: Wave Motion-An Internat. Jour, 1979—; editor: SIAM Jour. Applied Math, 1976; asso. mng. editor, 1978—; contbr. chpts. to books, articles to profl. jours. Fulbright grantee, 1972-73; Guggenheim fellow, 1982-83. Mem. Soc. Indsl. and Applied Math., Am. Math. Soc., AAAS, Combustion Inst., Conf. Bd. Math. Scis. (council, com. human rights of math. scientists), Com. Concerned Scientists, Soc. Natural Philosophy, Sigma Xi, Eta Kappa Nu. Subspecialty: Applied mathematics. Current work: Combustion theory, stochastic differential equations, bifurcation theory, nonlinear stability theory, singular pertubation theory. Home: 3704 Davis St Skokie IL 60076 Office: Technological Inst Northwestern U Evanston IL 60201

MATLIN, MARGARET WHITE, psychology educator; b. Washington, Nov. 14, 1944; d. Donald E. and Helen (Severance) White; m. Arnold H. Matlin, May 6, 1967; children: Elizabeth Anne, Sara Ellen. B.A., Stanford U., 1966; M.A., U. Mich., 1967, Ph.D., 1969. Asst. prof. psychology SUNY-Geneseo, 1971-78, assoc. prof., 1978-82, prof., 1982—, coordinator women's studies minor, 1977—. Author: The Pollyanna Principle, 1978, Human Experimental Psychology, 1979, Cognition, 1983, Perception, 1983. Co-founder Genesee Valley Citizens for Peace, 1972. Recipient Chancellor's award for excellence in teaching SUNY, 1977. Mem. Am. Psychol. Assn., Assn. Women in Psychology, Psychonomic Soc. Democrat. Subspecialty: Cognition. Current work: Cognition, experimental psychology. Home: 2290 Anderson Rd Linwood NY 14486 Office: SUNY Geneseo NY 14486

MATSAKIS, DEMETRIOS NICHOLAS, radio astronomer; b. St. Louis, June 30, 1949; s. Nicholas D. and Theodora (Pappageorge) M.; m. Cynthia L. Anderson, Oct. 14, 1977; 1 son, Nicholas D. B.S. in Physics, M.I.T., 1971, M.S., U. Calif., Berkeley, 1972, Ph.D., 1978. Nat. Sci. Commn. postdoctoral fellow Naval Research Labs., Washington, 1978-79; astronomer U.S. Naval Obs., Washington, 1980—. Contbr. articles to profl. jours. NDEA fellow, 1972-74. Mem. Internat. Astron. Union, Am. Astron. Soc., Astron. Soc. Pacific, Phi Beta Kappa, Sigma Xi. Subspecialty: Radio and microwave astronomy. Current work: Astrometry, interferometry, molecular clouds. I build receivers, do interferometry for astrometric applications, and also study molecular emission from clouds in interstellar medium. Producer documentary film Solomon's Child. Home: 319 Elm Ave Takoma Park MD 20912 Office: US Naval Obs Washington DC 20390

MATSON, JOHNNY LEE, psychology educator; b. Watseka, Ill., June 23, 1951; s. Walter Lee and Mary Lee (Purcell) M.; m. Deann Marie Dahlquist, Aug. 21, 1976; 1 dau., Meggan Lee. B.S. in Psychology, Eastern Ill. U., 1973, M.S., 1974; Ph.D., Ind. State U., 1976. Program dir. Jemison Ctr. Partlow State Sch. & Hosp., Tuscaloosa, Ala., 1977-78; asst. prof. child psychiatry and clin. psychology U. Pitts., 1978-80; assoc. prof. learning devel. and spl. edn. No. Ill. U., 1980—. Author: Handbook of Behavior Modification in the Mentally Retarded, 1981, Psychopathology in the Mentally Retarded, 1982, Methodology in Psychopharmacology, 1983. Spencer fellow Nat. Acad. Edn., 1981—. Mem. Am. Psychol. Assn. (program chmn. div. child and youth service 1981), Assn. Advancement of Behavior Therapy (co-chmn. div. mental retardation 1978—), Am. Assn. Mental Deficiency. Methodist. Subspecialty: Behavioral psychology. Current work: Assessment and treatment research in psychopathology of children and the developmentally disabled. Home: 2313 Concord Ct DeKalb IL 60115 Office: No Ill U Dept Learning Devel and Spl Edn DeKalb IL 60115

MATSUMOTO, MASAKAZU, veterinary microbiologist; b. Tokyo, Oct. 28, 1941; s. Seijo and Tsurue (Matsuda) M.; m. Nancy Elaine, July 17, 1971; children: Paul K., Mari L. B.S., U. Hokkaido, Japan, D.V.M., 1964; M.S., U. Hawaii, 1966; Ph.D., U. Calif.-Davis, 1972. Diplomate: Am. Coll. Vet. Microbiologists. Postdoctoral fellow Purdue U., West Lafayette, Ind., 1972-73; microbiologist Plum Island Animal Disease Ctr., U.S. Dept. Agr., Greenport, N.Y., 1973-75; asst. prof. Sch. Vet. Medicine, Oreg. State U., Corvallis, 1975-80, assoc. prof., 1980—; cons. Abbot Labs., 1976-78. Contbr. articles to profl. jours. Am. Cancer Soc. grantee, 1973; Agrl. Research Found. grantee, 1977; NIH grantee, 1979-81; U.S. Dept. Agr. grantee, 1979-81. Mem. AVMA, Am. Soc. Microbiology, Am. Assn. Avian Pathologists, Common Cause. Democrat. Subspecialties: Microbiology (veterinary medicine); Infectious diseases. Current work: Development of vaccines against bovine respiratory diseases and poultry diseases. Mechanisms of respiratory diseases. Aging and respiratory infections.

MATSUSHIMA, SATOSHI, astronomer, educator; b. Fukui, Japan, May 6, 1923; came to U.S., 1950, naturalized, 1963; s. Kyoyu and Hatsuye (Isaka) M.; m. Reiko Hori, Oct. 25, 1955; children—Peter Koichi, Anne Yuko. B.S., U. Kyoto, 1946; Ph.D., U. Utah, 1954; postgrad., Harvard, 1952-54; Sc.D., U. Tokyo, 1966. Asst. U. Kyoto, 1946-50; research asso. U. Pa., 1954-55; sr. research fellow Paris (France) Obs., 1956-57; research asso. U. Kiel, Germany, 1957-58; asst. prof. physics Fla. State U., 1958-60; assoc. prof. astronomy U. Iowa, 1960-67; prof. astronomy Pa. State U., University Park, 1967—, head dept. astronomy, 1976—, acting head dept. physics, 1981-82; vis. astronomer Utrecht Obs., Netherlands, 1956; sr. research fellow Calif. Inst. Tech., 1958-60; vis. prof. U. Tokyo, 1965-66, 74, Tohoku U., 1974; cons. U.S. Naval Research Lab., 1963; panel mem. Nat. Acad. Scis.-NRC, 1970-73. Contbr. articles to profl. jours. UNESCO-IAU grantee, 1956-58; German Humboldt Found. fellow, 1957-58; German Astron. Soc. grantee, 1957-58; Research Corp. grantee, 1958; U.S.-Japan Coop. Sci. program fellow, 1965-66; NSF research grantee, 1963—. Fellow Royal Astron. Soc. Eng.; mem. Am. Astron. Soc., Am. Geophys. Union, Astron. Soc. Pacific, Internat. Astron. Union, Sigma Xi. Subspecialties: Optical astronomy; Theoretical astrophysics. Current work: Theory of stellar and planetary atmospheres. Administration of research projects in stellar, extragalactic, and high energy astronomy. Home: 274 Osmond St State College PA 16801 Office: 525 Davey Lab University Park PA 16802

MATTAS, RICHARD FRANCIS, materials scientist, manager fusion program; b. Chgo., Sept. 14, 1947; s. Charles Joseph and Lilian (Sebek) M.; m. Loretta Ann Urbaczewski, June 27, 1970. B.S. in Physics, Yale U., 1969; M.S. in Metall. Engring, U. Ill., 1971, Ph.D., 1974. Research asst. U. Ill.-Urbana, 1969-74; asst. metallurgist Argonne (Ill.) Nat. Lab., 1974-79, metallurgist, 1979—. Mem. Am. Nuclear Soc., Am. Soc. for Metals, Sigma Xi. Subspecialties: Nuclear fusion; Metallurgy. Current work: Research; examination and testing of irradiated materials including metals, fuels, fusion breeding materials; technical: coordinating activites in area of nuclear analysis and materials for Fusion Engineering Design Center. Home: 5713 Middaugh Downers Grove IL 60516 Office: Argonne Nat Lab 9700 S Cass Ave Argonne IL 60439

MATTEI, JANET AKYUZ, astronomer; b. Bodrum, Turkey, Jan. 2, 1943; came to U.S., 1962; d. Baruh and Polise (Isbir) A.; m. Michael Mattei, Dec. 17, 1972. B.A., Brandeis U., 1965; Yuksek Lisans in Astronomy, Ege U., Turkey, 1970, M.S., U. Va., 1972. Tchr. physics, astronomy, phys. scis. Am. Collegiate Inst., Turkey, 1967-69; teaching asst. astronomy Ege U., Turkey, 1969-70; research asst. astronomy Maria Mitchell Obs. Nantucket, Mass., summer 1969; asst. to dir. Am. Assn. Variable Star Observers, Cambridge, Mass., 1972-73, dir., 1973—; worldwide coordinator variable star observations; cons. dwarf nova type variable stars; speaker colls., high schs. Contbr. articles to sci. jours. Recipient Wien internat. scholarship award; prin. investigator grants made to Am. Assn. Variable Star Observers from NSF, NASA Research Corp. Mem. Internat. Astron. Union, Am. Astron. Soc., Am. Assn. Variable Star Observers. Subspecialties: Optical astronomy; Variable stars. Current work: Visual and photometric studies of variable stars, particularly dwarf novae, long period variables, symbiotic variables. Office: 187 Concord Ave Cambridge MA 02138

MATTEN, LAWRENCE CHARLES, paleobotanist; b. Newark, Sept. 1, 1938; s. Bernard and Florence (Law) M.; m. Marlene Rotbart, Sept. 6, 1959; children: Sharlene, Alan, Sharon, Ronald. B.A., Rutgers U., 1959; Ph.D., Cornell U., 1965. Cert. tchr. gen. sci., biology, physics, chemistry, N.J. Gen. Sci. tchr. Woodstown (N.J.) High Sch., 1959-60; teaching asst., research asst., extension corr. Cornell U., Ithaca, N.Y., 1960-64; instr. biology SUNY, Cortland, 1964-65; asst. prof. dept. botany So. Ill. U., Carbondale, 1965-69, assoc. prof., 1969-77, prof., 1977—; cons., researcher. Contbr. articles on paleobotany to profl. jours. NSF grantee, 1981-85; Am. Philos. Soc. grantee, 1982-84. Mem. Linnean Soc. (London), Bot. Soc. Am., Paleontol. Soc., Torrey Bot. Club, Internat. Orgn. Paleobotany, Ill. Acad. Sci. Jewish. Subspecialties: Evolutionary biology; Paleobiology. Current work: Evolutionary relationships of earliest seed plants and related groups, structure of progymnosperms and Devonian ferns, floras of Ireland, development of techniques for preparing petrified remains. Office: Dept Botany So Ill U Carbondale IL 62901

MATTER, DARRYL EDGAR, psychology educator; b. Marysville, Kans., Mar. 20, 1940; s. Murray Edgar and Mary Gladys (Bodge) M.; m. Roxana Marie Warren, Sept. 1, 1980. B.S., Kans. State U., 1964, M.S., 1966; Ph.D., U. Md., 1971. Asst. prof. So. U., Baton Rouge, La., 1977, Western Carolina U., Cullowhee, N.C., 1977-78; assoc. prof. Ga. Coll., Milledgeville, 1978, W. Ga. Coll., Carrollton, 1978—. Contbr. articles in field to profl. jours. Mem. SAR, Am. Psychol. Assn., Am. Personnel and Guidance Assn., AAUP, Am. Home Econs. Assn., Phi Delta Kappa (research award 1981). Subspecialties: Developmental psychology; Educational psychology. Current work: Counseling and educational psychology, human development, instructional processes. Home: PO Box 1190 Carrollton GA 30117 Office: Dept Counseling and Ednl Psychology Edn Center W Ga Coll Carrollton GA 30118

MATTESON, DAVID ROY, psychologist, therapist, educator; b. Hornell, N.Y., Oct. 18, 1938; s. Lucious Hurd and Pearl Eleanor (James) M.; m. Sandra Lynn Sall, Aug. 26, 1962; children: Eric Nathan, Heather Michelle. B.A., Alfred U., 1960; B.Div., Colgate Rochester Div. Sch., 1964; Ph.D., Boston U., 1968. Registered psychologist, Ill.; cert. nat. health provider. Instr. Fitchburg (Mass.) State Coll., 1965-68; assoc. prof., counselor Marietta (Ohio) Coll., 1968-75; lectr. Royal Danish Sch. Edn., Copenhagen, 1973-74; clin. dir. Washington County Mental Health, Marietta, 1974-75; prof. psychology Governors State U., Park Forest South, Ill., 1975—; clin. group psychologist Gay Horizons, Chgo., 1977-80. Author: Adolescence Today: Sex Roles and the Search for Identity, 1975, Ego Identity: A Handbook for Psychosocial Research, 1984. Mem. Gay and Lesbian Task Force, Chgo., 1980—. Mem. Am. Psychol. Assn., Midwestern Psychol. Assn., ACLU. Mem. Citizens Party. Unitarian. Subspecialties: Developmental psychology; Social psychology. Current work: Adolescent identity; sex roles changes; gay men in heterosexual marriages; couple and family therapy. Home: 814 Blackhawk Dr Park Forest South IL 60466 Office: Governors State U Steunkel Rd Park Forest South IL 60466

MATTHEWS, BRIAN WESLEY, educator; b. South Australia, May 25, 1938; s. Lionel and E.L. (Harris) M.; m. Helen F. Denley, Sept. 7, 1963; 2 children. B.Sc., U. Adelaide, 1959, Hons., 1960, Ph.D. (Physics), 1964; Australian Atomic Energy Commn. fellow, 1960. Mem. staff Med. Res. Council Lab. Molecular Biology, Eng., 1963-66; vis. asso. molecular biologist NIH, Bethesda, Md., 1967-68; prof. physics and dir. Inst. Molecular Biology, U. Oreg., Eugene, 1969—. Imperial Chem. Ind. Australia and N.Z. fellow, 1962; John S. Guggenheim Meml. Found. fellow, 1977; Alfred P. Sloan fellow, 1971. Mem. Am. Crystallographic Assn., Am. Chem. Soc. Subspecialty: Molecular biology. Current work: Protein structure and function; crystallography. Address: Inst of Molecular Biology Univ Ore Eugene OR 97403

MATTHEWS, HARRY ROY, biological chemistry educator; b. Faversham, Kent., Eng., May 25, 1942; came to U.S., 1980; s. Elton Roy and Mavis (Sheppard) M.; m. Margaret Jean Wynne, Sept. 25, 1965 (div. May 1979); m. Iris Mary Sheppard, May 16, 1979. B.Sc., King's Coll. U. London, 1963, Ph.D., 1967; postgrad., U. Hong Kong, 1963-64. Sr. lectr. Portsmouth (Eng.) Poly., 1967-80; guest investigator Rockefeller U., N.Y.C., 1976-79; assoc. prof. U. Calif.- Davis, 1980—. Author: Fractionation of Nucleic Acids and Oligonucliotedes, 1976, DNA, Chromation and Chromosomes, 1981. Sci. Research Council Eng. grantee, 1970-80; Cancer Research Campaign grantee, 1976-80. Mem. Am. Soc. Biol. Chemists. Subspecialties: Biochemistry (biology); Gene actions. Current work: The role of histone and non-histone protein modifications in chromosome structure, gene expression, replication and mitosis particularly phosphorylation, acetylation and methylation. Home: 1301 Lake Blvd Davis CA 95616 Office: U Calif Sch Medicine Davis CA 95616

MATTHEWS, JOHN BRIAN, marine science educator; b. Glazebrook, Eng., Feb. 15, 1938; s. John and Evelyn (Bates) M.; m. Deborah Susan Judwin, Apr. 15, 1974 (div. 1980); 1 dau., Heather Dina; m. Nina Mary Stricker, Dec. 26, 1982. B.Sc., London U., 1960; D.I.C., Imperial Coll., 1963; Ph.D., London U., 1963. Research assoc. Imperial Coll., London, 1960-63; research assoc. U. Ariz., Tucson, 1963-66; asst. prof. U. Alaska, Fairbanks, 1966-69, assoc. prof., 1969-75, prof. marine sci., 1976—; vis. prof. Inst. Coastal Oceanography and Tides, Bidston, Eng., 1970-71; environ. fellow IBM Research Labs., Yorktown Heights, N.Y., 1973-74; research scientist Bedford Inst., Halifax, N.S., Can., 1975-76; chmn. Working Group 57, Sci. Com. on Oceanic Research, 1975—; chmn. U. Alaska Statewide Computer Adv. Com., 1980-82. Editor: Am. Geophys. Monograph Series on Coastal and Estuarine Regimes, 1978—, Marine Pollution Bull, 1980—. Founding pres. Fairbanks Light Opera Theatre, 1969-74; mem. Fairbanks Northstar Borough Library Commn., 1970-72; v.p. Alaska Assn for Arts, 1973-74. Fellow Royal Geog. Soc.; mem. Am. Geophys. Union (life), Estuarine and Brackish Water Fedn. (life), Am. Meteorol. Soc. Subspecialties: Oceanography; Mathematical software. Current work: Physical oceanography of coastal and estuarine regions. Office: Geophys Inst U Alaska Fairbanks AK 99701

MATTHEWS, KATHLEEN SHIVE, biochemistry educator, researcher; b. Austin, Tex., Aug. 30, 1945; d. William and Gwyndolyn (White) Shive; m. Randall Matthews, July 8, 1967. B.S. in Chemistry, U. Tex.-Austin, 1966; Ph.D., U. Calif.-Berkeley, 1970. Postdoctoral fellow Stanford (Calif.) U., 1970-72; asst. prof. Rice U., Houston, 1972-77, assoc. prof. biochemistry, 1977—; mem. study sect. NIH, 1980—. Contbr. articles to profl. publs. NSF fellow, 1967-70; AAUW fellow, 1970-71; Giannini Found. fellow, 1971-72; recipient Teaching award Rice U., 1977, 79. Mem. Am. Soc. Biol. Chemists, AAAS, Soc. Neurosci., Am. Chem. Soc., N.Y. Acad. Scis. Subspecialties: Biochemistry (biology); Biophysical chemistry. Current work: Protein-DNA interactions; individual metabolism. Office: Dept Biochemistry Rice Univ PO Box 1892 Houston TX 77251

MATTHEWS, LESLIE SCOTT, orthopedic surgeon, educator; b. Balt., Sept. 18, 1951; s. Warren Gamelial and Jane (Black) M.; m. Julie Ann Nolan, June 14, 1981. B.A. in Natural Sci, Johns Hopkins U., 1973; M.D., Baylor U., 1976. Intern Johns Hopkins Hosp., Balt., 1976-77, resident in surgery, 1976-78, resident in orthopedics, 1978-81, asst. prof. orthopedic surgery, 1981—; asst. chief orthopedic surgery Union Meml. Hosp., Balt., 1981—; cons. Loch Raven Veterans Hosp., Balt., 1981—, Bethesda (Md.) Naval Hosp. Trustee St. Paul's Sch. Boys, Brooklandville, Md., 1982. Recipient C. Markland Kelly award Johns

Hopkins U., 1972, 73, Barton Cup, 1973. Mem. Johns Hopkins Med. and Surg. Soc., Arthroscopy Assn. N.Am. Episcopalian. Subspecialty: Orthopedics. Current work: Arthroscopic surgical techniques, particularly involving the shoulder joint. Home: 3228 Abell Ave Baltimore MD 21218 Office: Union Memorial Hospital 201 E University Pkway Baltimore MD 21218

MATTHEWS, RICHARD HUGH, physician, research scientist; b. Providence, R.I., July 12, 1940; s. Eldon Ronald and Geneva Maud (Davidge) M.; m. Barbara Lee Cannon, Aug. 24, 1963; children: Elisabeth Victoria, David Thomas. B.S., Kalamazoo Coll., 1962; M.S., U. Mich., 1964, Ph.D., 1968; M.D., Ohio State U., 1983. Fellow McGill Cancer Research Unit, Montreal, Quebec, Can., 1967-69; asst. prof. to assoc. prof. physiol. and chemistry Ohio State U., Columbus, 1969-80, med. intern, researcher, 1980—. Mem. Am. Assn. Cancer Research, Am. Soc. Biol. Chemists, Biophys. Soc., Icsaber Soc. (sec.-treas. 1977), Ohio State U. Research Soc. (pres. 1978-79). Republican. Unitarian. Subspecialties: Cancer research (medicine); Biochemistry (medicine). Current work: multimodality therapy of metastatic colon cancer to liver; prolactin receptors and breast cancer; caffeine effects on breast diseases. Home: 385 King Ave Columbus OH 43201 Office: Ohio State U 333 W 10th Ave Columbus OH 43210

MATTHEWS, ROWENA GREEN, biological chemistry educator, research scientist; b. Cambridge, Eng., Aug. 20, 1938; d. David Ezra and Doris (Cribb) Green; m. Larry Stanford Matthews, June 18, 1960; children: Brian, Keith. B.A., Harvard U., 1960; Ph.D., U. Mich., 1969. Mem. faculty U. Mich., Ann Arbor, 1975--, assoc. prof. biol. chemistry, 1981--; mem. phys. biochemistry study sect. NIH, Washington, 1982-86; co-chmn. Conf. on Enzymes, Coenzymes and Metabolic Pathways, Gordon Research Confs., Kingston, R.I., 1983. Contbr. numerous sci. articles to profl. publs. Established investigator Am. Heart Assn., 1978-83. NIH grantee, 1978--. Mem. Am. Soc. Biol. Chemists, Am. Chem. Soc., Biophys. Soc., Phi Beta Kappa. Subspecialties: Biochemistry (medicine); Biophysics (physics). Current work: Research interests focus on the mechanisms of enzymes involved in folate metabolism, and studies on the reguation of folate metabolism in mammalian cells. Home: 1609 S University St Ann Arbor MI 48104 Office: U Mich Biophysics Research Div 2200 Bonisteel Blvd Ann Arbor MI 48109

MATTHYSSE, ANN GALE, biologist, educator; b. Chgo., Oct. 25, 1937; d. George W. and Ann (Van Nice) Gale; m. Steven W. Matthysse, Aug. 25, 1962; 1 son, Michael. A.B. in Biochem. Scis. magna cum laude, Radcliffe Coll., 1961; postgrad. in Life Scis., Rockefeller U., 1961-63; Ph.D. in Biology, Harvard U., 1967. Postdoctoral research fellow in molecular biology Calif. Inst. Tech., 1966-69; in bacterial physiology Harvard Med. Sch., 1969-70; lectr. in biology Harvard U., Cambridge, Mass., 1970-71; asst. prof. microbiology Ind. U. Sch. Medicine, Indpls., 1971-75; asst. prof. botany U. N.C., Chapel Hill, 1975-77, assoc. prof., 1977-82, assoc. prof. biology, 1982—; Allen Lectr. in phytobacteriology U. Wis., 1982. Editorial bd.: Jour. Bacteriology; contbr. articles to profl. jours. Treas., Orange County Assn. Retarded Citizens, 1976-78. Mem. Am. Soc. Microbiology, Am. Phytopath. Soc., Am. Soc. Plant Physiology, AAAS, Phi Beta Kappa. Quaker. Subspecialties: Plant pathology; Molecular biology. Current work: Molecular biology of bacterial diseases of plants, crown gall, Agrobacterium, plant pathology, bacterial attachment, plant tissue culture. Office: Coker Hall 010A Univ NC Chapel Hill NC 27514

MATTISON, DONALD ROGER, physician, researcher; b. Mpls., Apr. 28, 1944; s. Milford Zachary and Elizabeth Ruth (Davey) M.; m. Margaret Rose Libby; children: Jon Simpson, Amy Catherine. B.A., Augsburg Coll., Mpls., 1966; M.S., M.I.T., 1968; M.D., Columbia U., 1973. Resident in ob-gyn Presbyn. Hosp., N.Y.C., 1973-75, 77-78; commd. officer USPHS, 1975, advanced through grades to comdr., 1982; research assoc. developmental pharmacology br. sect. molecular toxicology Nat. Inst. Child Health and Human Devel., NIH, 1975-77; med. officer pregnancy research br., 1978—; cons. dept. ob-gyn Nat. Naval Med. Ctr., Bethesda, Reproductive Risk Assessment Group EPA. Contbr. articles to profl. jours.; mem. editorial bd.: Pediatric Pharmacology. Mem. Am. Assn. Cancer Research, AAAS, Am. Coll. Ob-Gyn, N.Y. Acad. Scis. Subspecialties: Obstetrics and gynecology; Toxicology (medicine). Current work: Reproductive toxicology, molecular mechanism; mechanisms of action of reproductive toxine, xenobiotics ovarian toxicity, oocyte destruction, ovarian metabolic processing. Office: NICHD NIH Bldg 10ACRF Room 8C313 Bethesda MD 20205 Home: 1719 Wilmart St Rockville MD 20852

MATTSON, MARGARET ELLEN, biologist, government research institute officer; b. Phila., May 13, 1947; d. John C. and Margaret M. (Brazz) M. B.A. magna cum laude, Holy Family Coll., 1969; Ph.D. in Neurobiology and Behavior, Cornell U., 1975. Health scientist/epidemiologist Enviro Control, Inc., Rockville, Md., 1976-78; program scientist Behavioral Medicine br. Nat. Heart Lung and Blood Inst., NIH, Bethesda, Md., 1978-81, project. officer for behavioral studies Clin. Trials br., 1981—; guest lectr. Found. Advanced Edn. in Scis., Bethesda, Md. Contbr. articles to profl. jours. Referral coordinator, newsletter editor Women's Center and Referral Service, Adelphi, Md., 1975-77. Recipient Outstanding Alumni award Holy Family Coll., 1981; NDEA grad. fellow Cornell U. Mem. Am. Psychol. Assn., Soc. Behavioral Medicine, Soc. Clin. Trials, Grad. Women in Sci. (pres. Washington chpt. 1982—). Subspecialties: Behavioral psychology; Epidemiology. Current work: Research and funding program development in the interdisciplinary area of biobehavioral sciences, especially applications of behavioral science to epidemiology and clinical trials. Office: Federal Bldg 7550 Wisconsin Ave Bethesda MD 20902

MATULIC, LJUBOMIR FRANCISCO, physicist; b. Potosi, Bolivia, May 8, 1923; s. Jeronimo and Vinka (Luksic) M.; m. Alice Demeure, Mar. 1, 1930; children: Paul, Miryam Matulic Keller, Katia. Licenciate, U. Chile, Santiago, 1949; M.S., Ind. U., 1963; Ph.D., U. Rochester, 1970. Prof. math. Instituto Normal Superior, La Paz, Bolivia, 1949-54; prof. math., physics Institut Leguerrier, Montreal, Que., Can., 1954-60; instr. math. Royal Mil. Coll., St. Jean, Que., 1958-60; prof. physics St. John Fisher Coll., Rochester, N.Y., 1963—. Author: (with Popovich) Algebra, 1954. NSF fellow, 1968-70. Mem. Asociacion Matematica Argentina, Am. Phys. Soc., Am. Assn. Physics Tchrs. Democrat. Roman Catholic. Subspecialty: Atomic and molecular physics. Current work: Quantum optics, short pulse propagacion in nonlinear absorbers, phase modulation of these pulses. Home: 183 High St Ext Fairport NY 14450 Office: Saint John Fisher College Rochester NY 14618

MATZ, ROBERT, physician, medical educator; b. N.Y.C., Aug. 5, 1932; s. Milton and Celia (Wachovsky) M.; m. Lita S Freed, Dec. 24, 1955 (dec. July 9, 1982); children: Jessica, Jonathan, Daniel.; m. Betty Lynne Orloff, Aug. 4, 1983. B.A., NYU, 1952, M.D., 1956. Diplomate: Am. Bd. Internal Medicine. Intern Bronx Mcpl. Hosp. Ctr., 1956-57, resident in internal medicine, 1957-60; assoc. dir. medicine Morrisania Hosp., Bronx, N.Y., 1963-76; dir. medicine North Central Bronx Hosp., 1976—; assoc. prof. medicine Albert Einstein Coll. Medicine, Bronx, 1972-80, prof. medicine, 1980—. Contbr. 50 abstracts, 30 articles to med. lit.; author 3 chpts. on diabetic coma. Served to capt. M.C. U.S. Army, 1959-61. Recipient Frederich Holden award N.Y.U. Med. Coll., 1956. Fellow ACP; mem. Am. Diabetes Assn., N.Y. Diabetes Assn. (v.p. 1978-80, pres.-elect 1980—), Am. Heart Assn., Phi Beta Kappa, Beta Lambda Sigma. Subspecialties: Endocrinology; Internal medicine. Current work: Diabetic coma; diet treatment of diabetes mellitus; teaching internal medicine and care of diabetics. Home: 32 Buena Vista Dr Hastings-on-Hudson NY 10706 Office: North Central Bronx Hosp 3424 Kossuth Ave Bronx NY 10467

MATZINGER, DALE FREDERICK, geneticist, researcher; b. Alleman, Iowa, Apr. 14, 1929; s. Albert Christian and Violet Marie (Schrumpf) M.; m. Joan Camilla Cox, June 11, 1960; children: Michael, Ellen. B.S., Iowa State U., 1950, M.S., 1951, Ph.D., 1956. Asst. statistician N.C. State U., Raleigh, 1956-57, asst. prof. stats., 1957-60, asso. prof. genetics, 1960-64, prof. genetics, 1964—. Editorial bd.: Tobacco Sci, 1964-71, J Jour. Heredity, 1981—; Contbr. articles to profl. lit. Served with U.S. Army, 1951-53. Recipient Sigma Xi research award, 1961, Gamma Sigma Delta award of merit, 1970, Philip Morris award for disting. achievement in tobacco sci., 1971; Am. Soc. Agronomy fellow, 1965. Mem. Am. Soc. Agronomy, Biometrics Soc., Genetics Soc. Am., Am. Genetic Assn., AAAS. Methodist. Subspecialties: Plant genetics; Population biology. Current work: Quantitative genetics of nuclear and cytoplasmic organelle variability in self-fertilizing plant species. Home: 3413 Doyle Rd Raleigh NC 27607 Office: Dept Genetics NC State U Raleigh NC 27650

MAUERSBERGER, KONRAD, physics educator; b. Lengefeld, Ger., Apr. 28, 1938; came to U.S., 1969, naturalized, 1972. Diploma, U. Bonn, W.Ger., 1964, Ph.D., 1968. Research assoc. dept. physics U. Bonn, 1968-69; research assoc. dept physics and astronomy U. Minn., Mpls., 1969-74, asst. prof., 1974-77, assoc. prof., 1977-82, prof., 1982—. Recipient Research award Inst. Tech., U. Minn., 1979. Mem. Am. Geophys. Union, AAAS. Subspecialties: Atmospheric chemistry; Planetary atmospheres. Current work: Atmospheric physics, chemistry and physics of stratosphere, planetary atmosphere, instrumentation to determine density and composition of atmospheres. Office: Sch Physics and Astronomy U Minn 116 Church St SE Minneapolis MN 55455

MAUGHAM, JAMES HENRY-EAMON, III, recovery company executive, environmental specialist, consultant; b. Hoboken, N.J., Sept. 16, 1947; s. James Henry and Elsa Monk (Pedersen) M.; m. Elain Marie Ziminski, Sept. 2, 1972; children: Erika Gillin, Stephanie Katherine, James Henry-Eamon, IV. B.S., Queen's Coll., 1975. Lic. master mariner. Ship crews mem., various, internat., 1962-76; marine surveyor London Rayner, Ltd., London and N.Y.C., 1970-71, Hull & Cargo Surveyors, N.Y.C., 1975-77; regional mgr. marine service Am. Marine Service, Phila., 1977-78; pres. J.H.E. Maugham & Assoc., Phila., 1978—, Resource Mgmt., Medford, N.J., 1982—; cons. USCG, Washington, 1978—, Water Quality Ins. Syn., N.Y.C., 1978—, Spill Control Assn., Washington, 1977—. Contbr. articles to publs. Pres. Hoot Owl Civic Assn., Medford; trustee White Birch Lake Assn., Medford; bd. dirs. Burlington County (N.J.) council Boy Scouts Am., Heritage Ship Guild Phila. Served with U.S. Army, 1966-69. Decorated Bronze Star with V, Purple Heart(3).; Recipient cert. of appreciation City of N.Y., 1971. Mem. Marine Tech. Soc. (chmn. Del Val chpt. 1979—), Internat. Oceanographical Found., Nat. Maritime Hist. Soc., St. Michael's Maritime Mus., Am. Soc. Marine Artists, Soc. Marine Consuls. Democrat. Roman Catholic. Subspecialties: Oceanography; Environmental engineering. Current work: Disposal and reclamation of hazardous waste and materials utilizing best available technology. Oil and hazardous materials marine spill recovery. Marine salvage. Home: 42 Christopher's Mill Rd Medford NJ 08055 Office: Resource Management Services Inc PO Box 83 Medford NJ 08055

MAUGHAN, O. EUGENE, fishery biologist; b. Weston, Idaho, Jan. 3, 1943; s. Owen Weston and Ione (Stocks) M.; m. Lu Dean Lewis, Aug. 30, 1962; children: Terry Lynn, Cindy Jean, Kimberly Dene, James Benjamin. B.S., Utah State U., 1966; M.A., Kans. U., 1968; Ph.D., Wash. State U., 1972. Fishery biologist U.S. Fish and Wildlife Service, Spokane, 1971-72, Blacksburg, Va., 1972-77, Stillwater, Okla., 1977—; leader Okla. Coop. Fishery Research Unit, 1977—. Contbr. articles to profl. jours. Recipient Commendation award Sport Fishing Inst., 1978. Mem. Am. Fisheries Soc., Soc. Herpetologists and Ichthyologists, Okla. Acad. Sci., Sigma Xi. Mormon. Subspecialties: Ecology; Theoretical ecology. Current work: Fish culture, fish ecology, reservoir management, stream and riverine management, instream flow methodology, habitat evaluation procedures. Office: Okla Coop Fishery Research Unit Okla State U Stillwater OK 74078

MAULIK, DEBABRATA, obstetrician, gynecologist, cons. perinatologist; b. Calcutta, West Bengal, India, Mar. 3, 1939; came to U.S., 1976; s. Ananta Kumar and Tara (Chakrabarti) M.; m. Shibani Mukherji, Aug. 11, 1973; children: Devika, Davesh. M.D., U. Calcutta, 1962; M.R.C.O.G., Royal Coll. Obstetrics/Gynecology, London, 1967; Ph.D., U. London, 1975. Diplomate: Am. Bd. Ob-Gyn. Intern Med. Coll. Hosps., Calcutta, 1962-63, resident, 1963-65; resident various hosps., 1965-69; lectr. Charing Cross, U. London, 1973-76; dir. maternal/fetal dept. Deaconess Hosp., Buffalo, 1976-80; asst. prof. SUNY, Buffalo, 1976-80; asst. prof. dept. ob-gyn U. Rochester (N.Y.) Sch. Medicine, 1980—; assoc. attending ob-gyn Strong Meml. Hosp., Rochester, 1982—, Deaconess Hosp., Buffalo, 1976-80; mem. nat. program com. Assn. Planned Parenthood Physicians Am., 1978-81; mem. orgn. com. 9th Trophoblast Conf., Rochester, 1982. Bd. dirs. Buffalo Council World Affairs, 1979-80. Recipient Gold medal U. Calcutta, 1962; Nuffield Found. fellow, 1970-73; U. Buffalo Found. research grantee, 1978-79; E.R. and M.S. Goode research grantee, 1981-82; NIH research grantee, 1982-85. Fellow Am. Coll. Ob-Gyn; mem. Planned Parenthood Fedn. Am. (program com. 1978—), Royal Coll. Ob-Gyn, AMA, AAAS, Indian Assn. Rochester. Hindu. Subspecialties: Maternal and fetal medicine; Perinatal diagnosis and therapy. Current work: Extracorporeal perfusion of human placenta, model system, transport kinetic and toxicologic study, development of doppler flowmetry technology and applications in medicine, computers in perinatal medicine. Developer extracorporeal perfusion of human placenta, 1975. Home: 225 Oakdale Dr Rochester NY 14618 Office: U Rochester 601 Elmwood Ave Strong Meml Hosp Rochester NY 14642

MAURER, ROBERT DISTLER, industrial physicist; b. St. Louis, July 20, 1924; s. John and Elizabeth J. (Distler) M.; m. Barbara A. Mansfield, June 9, 1951; children: Robert M., James B., Janet L. B.S., U. Ark., 1948, LL.D., 1980; Ph.D., MIT, 1951. Mem. staff MIT, 1951-52; with Corning Glass Works, N.Y., 1952—, mgr. physics research, 1963-78, research fellow, 1978—. Contbr. articles to profl. jours., chpts. to books. Served with U.S. Army, 1943-46. Recipient Indsl. Physics prize Am. Inst. Physics, 1978; L.M. Ericsson Internat. prize in telecommunications, 1979. Fellow Am. Ceramic Soc. (George W. Morey award 1976); mem. IEEE (Morris N. Liebmann award 1978), Am. Phys. Soc., Nat. Acad. Engring. Subspecialties: Materials; Fiber optics. Current work: Physical behavior of glasses. Patentee in field. Home: 6 Roche Dr Painted Post NY 14870 Office: Sullivan Park Corning Glass Works Corning NY 14830

MAURISSEN, JACQUES PAUL JEAN, toxicologist; b. Liège, Belgium, Feb. 11, 1947; came to U.S., 1973; s. Paul Laurent Gustave and Hubertine Marie Joséphine (Albert) M.; m. Ghislaine Dallemagne, July 22, 1972; 1 dau., Stéphanie. Lic. in Psychology (M.S.), U. Liège, 1970; Ph.D., M.S. in Toxicology, U. Rochester, 1979. Teaching asst. U. Liège, 1969-70; research fellow Foundation Camille Héla, Liège, 1970-71, Fonds National de la Recherche Scientifique, Rome, 1971, Accords Culturels Franco-Belges, Strasbourg, France, 1971-72, European Tng. Program in Brain & Behavior Research, Stasbourg, France, 1971-72, Svenska Institutet för Kulturellt Utbyte Med Utlandet, Gothenburg, Sweden, 1973, Belgian Am. Ednl. Found., Rochester, N.Y., 1973-74; expert witness Dept. Labor, Rochester, 1978; postdoctoral fellow U. Rochester, 1979-80, instr., 1980-82; sr. research toxicologist Dow Chem. Co., Midland, Mich., 1982—. Contbr. articles to profl. jours. Mem. Soc. Neurosci., European Soc. Toxicology, Am. Psychol. Assn., Digital Equipment Computer Users Soc., Behavioral Pharmacology Soc. Subspecialties: Toxicology (medicine); Psychophysics. Current work: Neurotoxicology. Home: 534 Woodcock Rd Midland MI 48640 Office: Dow Chem Co 607 Bldg Neurobehavioral Toxicology Midland MI 48640

MAUZERALL, DAVID CHARLES, biophysics researcher, educator; b. Sanford, Maine, July 22, 1929; s. David James and Jeannette (Morin) M.; m. Miriam I. Jacob, July 29, 1959; children: Denise, Michele. B.S., St. Michael's Coll., 1951; Ph.D., U. Chgo., 1954. Research assoc. Rockefeller U., N.Y.C., 1954-59, asst. prof., 1959-64, assoc. prof., 1964-65, 1966-69, prof., 1969—. Contbr. numerous articles to profl. jours. Guggenheim fellow, 1965-66. Subspecialties: Biophysics (biology); Laser photochemistry. Current work: Photo electron transfer reactions related to photosynthesis, photo driven proton transfer in rhodopsins. Office: Rockefeller Univ 1230 York Ave New York NY 10021

MAVROYANNIS, CONSTANTINE, theoretical physicist; b. Athens, Greece, Nov. 13, 1927; s. George and Paraskevi (Xylas) M.; m. Paraskevi Logothetis, Aug. 30, 1961; children: Maria, Irene. B.S. in Ch.E, Tech. U. Athens, Greece, 1957; Ph.D. in Phys. Chemistry, McGill U., 1961; D.Phil. in Math, Oxford U., Eng., 1963. Postdoctoral fellow NRC Can., Ottawa, Ont., 1963-64, asst. research officer, 1964-65, assoc. research officer, 1966-73, sr. research officer, 1974—. Contbr. numerous articles to sci. jours. Served with Greek Army, 1951-54. NRC Can. Student fellow, 1959-61; NATO Sci. Postdoctoral Overseas fellow, 1961-63; NRC Can. postdoctoral fellow, 1963-64. Fellow Chem. Inst. Can.; Mem. Am. Phys. Soc., Can. Assn. Physicists. Subspecialties: Condensed matter physics; Atomic and molecular physics. Current work: Many-body theory in solid state physics, collective electronic excitations, optical properties of solids, surface physics, quantum optics and quantum electronics. Home: 2121 Maywood St Ottawa ON Canada K1G 1E8 Office: NRC Can 100 Sussex Dr Ottawa ON Canada K1A OR6

MAX, CLAIRE ELLEN, physicist; b. Boston, Sept. 29, 1946; d. Louis William and Pearl (Bernstein) M.; m. Jonathan Arons, Dec. 22, 1974; 1 son, Samuel. A.B., Radcliffe Coll., 1968; Ph.D., Princeton U., 1972. Postdoctoral fellow in physics U. Calif.-Berkeley, 1972-74; physicist, laser fusion program Lawrence Livermore Lab., Livermore, Calif., 1974—; cons. on plasma physics, role of women in physics. Editor: (with J. Arons and C.F. McKee) Particle Acceleration Mechanisms in Astrophysics, 1979; author articles. Mem. physics adv. com. Nat. Sci. Found., 1972-82. Fellow Am. Phys. Soc. (exec. com. div. plasma physics 1981-82); mem. Am. Astron. Soc. (exec. com. high energy astrophys. div. 1975-76), Assn. Women in Sci., Sigma Xi, Phi Beta Kappa. Subspecialties: Plasma physics; Theoretical astrophysics. Current work: Laser-plasma interactions; plasma astrophysics. Home: 617 Grizzly Peak Blvd Berkeley CA 94708 Office: Lawrence Livermore Lab L-477 Livermore CA 94550

MAX, STEPHEN RICHARD, biochemist, educator; b. Providence, Dec. 25, 1940; s Leo and Paula (Strasberg) M.; m. Barbara H. Sohmer, Aug. 8, 1976; 1 dau., Paula E. Max-Sohmer. B.S., U. R.I., 1962, Ph.D., 1966. Guest worker NIH, 1968-70; asst. prof. biochemistry Howard U. Sch. Medicine, 1966-70; asst. prof. neurology U. Md., Balt., 1970-74, assoc. prof., 1974-81, prof., 1981—. Contbr. articles to profl. publs. Served to capt. U.S. Army, 1966-68. Research grantee NIH, Muscular Dystrophy Assn., Roche Research Found., NASA. Mem. Soc. for Neurosci., Soc. Biol. Chemistry, Am. Soc. Neurochemistry, Internat. Soc. Neurochemistry, Internat. Brain Research Orgn., Am. Chem. Soc., Am. Acad. Neurology. Subspecialties: Biochemistry (biology); Neurochemistry. Current work: Neuromuscular biochemistry; endocrinology, neurochemistry, neuroscience, regeneration, denervation. Office: Dept Neurolog U Md Sch Medicine Baltimore MD 21201

MAXSON, CARLTON JAMES, mathematician, educator; b. Cortland, N.Y., Apr. 19, 1936; s. Edward C. and Eleanor M. (Brooks) M.; m. Carol Y. Maxson, Aug. 31, 1957; children: Debra, James. B.S., SUNY, Abalny, 1958; M.S. (NSF fellow), U. Ill., 1961; Ph.D. (NSF fellow, Faculty fellow), SUNY, Buffalo, 1967. Tchr. high sch. math., Hammondsport, N.Y., 1958-62; asst. prof. to assoc. prof. SUNY, Fredonia, 1962-69; assoc. prof. to prof. Tex. A&M U., 1969-81, prof. math., assoc. dean of sci., 1981—; vis. prof. Teeside Poly., Middlebrough, Eng., 1981. Contbr. articles on algebra and discrete math. to profl. jours. Mem. Am. Math. Soc., Math. Assn. Am., AAAS, Edinburgh Math. Soc., Sigma Xi. Methodist. Subspecialty: Algebra. Current work: Algebra and discrete math., near-rings, applications of algebra to problems in computer sci.

MAXSON, STEPHEN CLARK, behavioral geneticist, educator; b. Newport, R.I., Apr. 13, 1938; s. Clark S. and Louise K. (Williams) M.; m. Kathleen N. Wilson, Aug. 28, 1965; m. Susan Irene Wolf, June 21, 1980. S.B., U. Chgo., 1960, Ph.D., 1966. Instr. biology, research assoc. in behavior genetics U. Chgo., 1966-69; asst. prof. depts. biobehavioral scis. and psychology U. Conn., Storrs, 1969-74, assoc. prof., 1974—. Fellow Internat. Soc. Research on Aggression; mem. AAUP, Soc. Neurosci., Behavior Genetics Assn., Sigma Xi. Subspecialties: Gene actions; Psychobiology. Current work: Effects of Y-Chromosome on sexually dimorphic behaviors in mammals. Genetics, development, aggression, sexual behavior, inbred mice. Office: Dept Behavioral Scis U-154 U Conn Storrs CT 06268

MAXWELL, ARTHUR EUGENE, geophysics institute director, researcher; b. Maywood, Calif., Apr. 11, 1925; s. John Henry and Nelle Irene (Arnold) M.; m. Beulah McKay, 1946 (div. 1963); children: Delle Rae, Eric Arnold, Lynn Marie; m. Rita Louise Johnson, Nov. 23, 1963; children: Brett Alan, Gregory James. B.S. with honors in Physics, N.Mex. State U., 1949; M.S. in Oceanography, Scripps Inst. Oceanography U. Calif.-San Diego, 1952, Ph.D., 1959. Head oceanographer Office Naval Research, Washington, 1955-60, head geophysics br., 1960-65; assoc. dir., dir. research Woods Hold (Mass.) Oceanographic Inst., 1967-71, provost, 1971-81; dir. Inst. for Geophysics, U. Tex.-Austin, 1982—; mem. corp. Woods Hole Oceanographic Inst., 1971-81, Marine Biol. Lab., Woods Hole, 1973-82, Mus. Sci., Boston, 1972—; bd. dirs. Palisades Geophys. Inst., Nyack, N.Y., 1982—. Editor: The Sea, Vol. IV, 1970; jour. Bolletino de Geofisica, 1968-79. Mem. Nat. Adv. com. on Oceans and Atmosphere, Washington, 1972-76; mem. Mass. Gov.'s Adv. Com. on Sci. and Tech., 1965-71, Mass. Commn. on Ocean Mgmt., 1968-71. Served with USN, 1942-46. Recipient meritorious civilian service award Office Naval Research, 1958; superior civilian service award

Asst. Sec. Navy, 1963; disting. civilian service award Sec. Navy, 1964; disting. alumni award N.Mex. State U., 1965. Fellow Am. Geophys. Union (pres. 1976-78), Marine Tech. Soc. (pres. 1981-82), Internat. Oceanographic Found.; mem. N.Y. Acad. Scis., Sigma Xi. Clubs: Cosmos (Washington); Explorers (N.Y.). Subspecialties: Geophysics; Sea floor spreading. Current work: Heat flow through ocean floor and other marine geophysical measurements. Office: Inst for Geophysics U Tex Austin TX 78712 Home: 3607 Greystone Apt 0622 Austin TX 78731

MAXWELL, DONALD LEE, nuclear engineering consultant, consulting firm executive, information management systems consultant; b. Youngstown, Ohio, Sept. 11, 1927; s. Jay and Ruth LaRue (Muller) M.; m. Margie Ann Lawton, July 19, 1948; children: Karen Ann, Sharon Lee Terry Lynn. B.S. in Sci, Ohio U., 1951. Registered profl. engr., Calif. Mem. Apollo Support Group, Gen. Electric Corp., 1962–72; mem. quality assurance staff NUS Corp., Clearwater, Fla., 1972–77; mgr. quality assurance Ill. Power Co., Decatur, 1977; cons. corp. staff Black & Veatch, Overland Park, Kans., 1977-79; dir. quality assurance Reliability Engineering - Info. Mgmt. Systems, S.W. Research Inst., San Antonio, 1979-81; sr. systems engr. Rockwell Hanford Ops., Richland, Wash., 1981-82; v.p. Project Assistance Corp., San Antonio, 1982—; Dir./developer electronic power plant info. control system and project mgmt. control system, 1980. Served to 1st lt. C.E. U.S. Army, 1946-48, 51-53; Korea. Recipient Apollo award NASA, Houston, 1968. Mem. Am. Soc. Quality Control, Am. Nuclear Soc. Republican. Lodge: Elks. Subspecialties: Information systems, storage, and retrieval (computer science); Nuclear fission. Current work: Developing large computer-assisted information management control systems for electric power companies and petrochemical industry for productivity improvement. Home and Office: 8618 Timberwilde Dr San Antonio TX 78250

MAXWELL, DONALD ROBERT, pharm. co. exec.; b. Paris, Mar. 30, 1929; U.S., 1974; s. Titus Bonner and Helen Mary Camille M.; m. Catherine Billon, Aug. 16, 1956; children: Monica, Nicholas, Christopher, Caroline, Denis, Dominic, Marie-Claire, Philip. B.A., U. Cambridge, Eng., 1952, M.A., 1956, Ph.D., 1955. Research scholar Med. Research Council, Eng., 1953-55; attaché de récherche Institut Pasteur, Paris, 1955-56; research pharmacologist May & Baker Ltd., Eng., 1957-74; dir. preclin. research Warner-Lambert, Morris Plains, N.J., 1974-77; v.p. preclin. research Warner-Lambert/Parke-Davis, Ann Arbor, Mich., 1977—. Contbr. articles to profl. jours. Mem. Am. Soc. Pharmacology and Exptl. Therapeutics, Brit. Pharmacol. Soc., Physiol. Soc., Biochem. Soc., Internat. Coll. Neuro-Psychopharmacology, Royal Soc. Medicine (Eng.); fellow Inst. Biology (Eng.). Subspecialties: Pharmacology; Neuropharmacology. Current work: Direction of drug discovery research in areas of central nervous, cardiovascular and related areas. Office: Warner-Lambert/Parke Davis 2800 Plymouth Rd Ann Arbor MI 48106

MAXWELL, JOYCE BENNETT, biology educator, textbook reviewer; b. Merced, Calif., June 18, 1941; d. Leland Grover and Thelma Mae (Freeman) Bennett; m. William Calvin Maxwell, July 29, 1962; children: Kathleen, Patrick. A.B. in Zoology, UCLA, 1963; Ph.D. in Genetics and Biochemistry, Calif. Inst. Tech., 1970. Asst. prof. biology Calif. State U.-Northridge, 1970-77, assoc. prof., 1978-82, prof., 1982—. Woodrow Wilson Found. fellow, 1963-64; Mayr Found. fellow, 1964-65. Mem. AAAS, Am. Women in Sci., Genetics Assn. Democrat. Presbyterian. Subspecialty: Gene actions. Current work: Biochemical genetics of Neurospora crassa, serine biosynthesis in Neurospora crassa, unstable genes. Home: 36 Wales St Thousand Oaks CA 91360 Office: Dept of Biology California State University Northridge CA 91330

MAY, EUGENE PINKNEY, psychologist; b. Louisville, May 1, 1931; s. Eugene Pinkney and Amanda (Baskette) M. B.A., George Peabody Coll., 1953, M.A., 1966; Ph.D., U. Ill., 1971. Lic. psychologist, Ohio, Ill. Psychologist, chief counseling psychology VA Med. Ctr., Cleve., 1971-81; psychologist VA Outpatient Clinic, Oakland Park, Fla., 1981—; cons. psychologist Cleve. Met. Gen. Hosp., 1974-81; adj. prof. Ursuline Coll., 1980-81; adj. clin. asst. prof. Case Western Res. U., 1978-81; pvt. practice psychology, Cleve., 1973-81. Contbr. articles to profl. jours. Mem. Am. Psychol. Assn., Ohio Psychol. Assn., Cleve. Psychol. Assn., Cleve. Acad. Cons. Psychologists, Soc. Psychol. Study of Social Issues, Assn. Humanistic Psychology, U. Ill. Alumni Assn. Subspecialties: Behavioral psychology; health psychology. Current work: Stress management, relaxation therapy, health-inducing habit promotion, psychotherapy, behavioral medicine. Home: 2173 NE 63d St Fort Lauderdale FL 33308 Office: 5599 N Dixie Hwy Oakland Park FL 33334

MAY, RICHARD BEARD, psychology educator; b. Seattle, Dec. 20, 1938; emigrated to went to Can., 1966; s. Louie Beard and Ruby June (Simmons) M.; m. Marjorie Stevenson, Aug. 25, 1962; children: Robert, Richard. B.A., Whitman Coll., 1961; M.A., Claremont Grad. Sch., 1963, Ph.D., 1966. Asst. prof. U. Victoria, B.C., Can., 1966-71, assoc. prof., 1971-81, prof., 1981—, asst. chmn. dept. psychology, 1974-78, 80-83; research assc. U. Oxford, Eng., 1972-73; cons. Victoria Sch. Bd., 1971-72, Dept. Human Resources, 1974-75. Contbr. articles to profl. jours. USPHS fellow, 1962-66; NRC Can. fellow, 1969-77, 72, 79. Mem. Am. Psychol. Assn., Canadian Psychol. Assn., Soc. for Research in Child Devel., Psychonomic Soc. Subspecialties: Cognition; Developmental psychology. Current work: Cognition, Concept learning, memory development. Office: Dept Psychology U Victoria Victoria BC Canada V8W 2Y2

MAY, ROBERT MCCREDIE, biology educator, research administrator; b. Sydney, Australia, Jan. 8, 1936; came to U.S., 1973; s. Henry Wilkinson and Kathleen (McCredie) M.; m. Judith Feiner, Aug. 3, 1962; 1 dau., Naomi. B.S., Sydney U., 1956; Ph.D., 1959. Research fellow Harvard U., Cambridge, Mass., 1960-61; sr. lectr. Sydney (Australia) U., 1962-64, reader in physics, 1964-69, prof. physics, personal chair, 1969-72; prof. biology Princeton (N.J.) U., 1973—. Author: Model Ecosystems, 1974, Theoretical Ecology, 1976, Biology of Infectious Diseases, 1982. Recipient Weldon medal Oxford U., 1980, Pawsey medal Australian Acad. Sci., 1967; Prather lectr. Harvard U., 1979. Fellow Royal Soc., AAAS, mem. various profl. assns. Clubs: Athenaeum, London. Subspecialties: Population biology; Applied mathematics. Current work: Basic understanding of dynamic behavior of plant and animal populations, and of community ecology. Applications to resource management and infectious diseases. Office: Princeton U Nassau St Princeton NJ 08544

MAY, RODNEY ALAN, mechanical engineer; b. Kansas City, Kans., Aug. 6, 1953; s. Eldon and Agnes (Ecker) M.; m. Teresa A. Harris, June 17, 1978. B.S.M.E., U. Kans., 1975, M.S.M.E., 1977. Mem. tech. staff applied mechanics div. Sandia Nat. Labs., Albuquerque, 1977—; adj. prof. mech. engring. U. N.Mex., 1980. Author articles on packaging and transport of radioactive material. Summerfield scholar, 1972-75; Black & Veatch scholar, 1972-76. Mem. ASME, Am. Soc. Metals, Tau Beta Pi, Pi Tau Sigma, Phi Kappa Phi. Subspecialties: Applied mechanics; Fracture mechanics. Current work: Design, testing and evaluation of containers and systems for the safe transport and storage of radioactive materials. Office: PO Box 5800 Albuquerque NM 87185

MAY, WALTER GRANT, chemical engineer; b. Saskatoon, Sask., Can., Nov. 28, 1918; came to U.S., 1946, naturalized, 1954; s. George Alfred and Abigail Almira (Robson) M.; m. Mary Louise Stockan, Sept. 26, 1945; children: John R., Douglas W., Caroline O. B.Sc., U. Sask., Saskatoon, 1939, M.Sc., 1942; Sc.D., M.I.T., 1948. Chemist British Am. Oil Co., Moose Jaw, Sask., 1939-40; asst. prof. U. Sask., 1943-46; with Exxon Research & Engring. Co., Linden, N.J., 1948-83; sr. sci. adv., 1976-83; prof. U. Ill., 1983—; with Advanced Research Projects Agy., Dept. Def., 1959-60; industry based prof. Stevens Inst. Tech., 1968-74, Rensselaer Poly. Inst., 1975-77. Recipient Process Indsl. Div. award ASME, 1972. Mem. Am. Inst. Chem. Engrs., ASME, Nat. Acad. Engring. Subspecialty: Chemical engineering. Home: 941 Cheshire Dr Champaign IL 61821 Office: Dept Chem Engring U Ill 1209 W California St Urbana IL 61801

MAYBERRY, WILLLAM ROY, microbiology educator; b. Grand Junction, Colo., Nov. 30, 19328; s. Leo Roy and Josephine Frances (Purkat) M.; m. Katy Jane Carson, Jan. 14, 1967. B.A., U. Colo., 1961; M.A., Western State Coll., 1963; Ph.D., U. Ga., 1966. Chemist Lucius Pitkin, Inc., Grand Junction, Colo., 1960-61; postdoctoral fellow U. Ga., Athens, 1966-67, U. S.D. Coll. Medicine, Vermillion, 1967-68, asst. prof. microbiology, 1968-75, assoc. prof., 1975-77; assoc. prof. microbiology East Tenn. State U. Coll. Medicine, Johnson City, 1977—. Mem. AAAS, Am. Soc. Microbiology, Am. Chem. Soc., ASTM, Internat. Orgn. Mycoplasmology. Subspecialties: Microbiology; Analytical chemistry. Current work: Lipid and carbohydrate components of microbial cell-envelopes; fatty acid analysis (cellular) of bacteria; monohydroxy and dihydroxy fatty acids of prokaryotes. Office: Dept Microbiology East Tenn State U Coll Medicine Johnson City TN 37614

MAYDEW, RANDALL CLINTON, research laboratory administrator, aerodynamics researcher; b. Lebanon, Kans., Jan. 29, 1924; s. Kermit A. and Lelia (Phifer) M.; m. Maxine Norvell, Sept. 2, 1944 (dec. May 1970); children: Jenan Louise, Randall Paul, Barbara Ann; m. Susanna Jean Glaze, Dec. 1, 1971. B.S. in Aero. Engring, U.Colo.-Boulder, 1948, M.S., 1949. Research asst. U. Colo. Engring. Expt.Sta., 1948-49; aero. research scientist NACA Ames Lab., Moffett Field, Calif., 1949-52; mem. staff dept. aerodynamics Sandia Nat. Labs., Albuquerque, 1952-57, supr. exptl. aerodynamics div., 1957-65, mgr. dept. aerodynamics, 1965—. Contbr.: numerous articles to profl. publs., and McGraw Hill Ency. Sci. and Tech, 1961, 67 and 71 edits. Served to 1st lt. AC U.S. Army, 1943-45; Saipan. Decorated D.F.C.; Air medal with 3 oak leaf clusters. Fellow AIAA (assoc.); mem. Supersonic Tunnel Assn. (sec. 1963-64, pres. 1969-70), Sigma Xi, Pi Tau Sigma. Subspecialties: Aeronautical engineering; Wind power. Current work: Aerodynamics and flight mechanics of bombs, shells, rockets, reentry vehicles, guided missiles and parachutes; vertical axis wind turbines; boundary layer stability. Home: 5305 Queens Ct NE Albuquerque NM 87109 Office: Dept Aerodynamics Sandia Nat Labs Albuquerque NM 87185

MAYEDA, KAZUTOSHI, biology educator; b. Santa Monica, Calif., June 17, 1928; s. Haruju and Kohana (Mitsuuchi) M.; m. Betty Miwako Waki, June 17, 1941; children: Karen, Kathy, Michael. B.S., U. Utah, 1957, M.S., 1958, Ph.D., 1961. Diplomate: Am. Bd. Med. Genetics. Asst. prof. biology Wayne State U., Detroit, 1961-66; assoc. prof. 1966-71, prof., 1971—; genetic cons. C.S. Mott Ctr., 1973—, dir. cytogenetics, 1978—; research assoc. Nat. Inst. Genetics, Mishima, Japan, 1970-71. Bd. govs. Mid-West dist. Japanese Am. Citizens League, Chgo., 1979-81, pres., Detroit, 1976-77. Served with U.S. Army, 1950-52. NIH grantee, 1982—; predoctoral fellow, 1958-61. Mem. Am. Soc. Human Genetics, Genetic Assn., Am., AAAS, Soc. Study of Evolution, N.Y. Acad. Sci., Sigma Xi. Subspecialties: Genetics and genetic engineering (medicine); Genetics and genetic engineering (biology). Current work: Isolation and location of human DNA; linkage analyses of human genes. Office: Dept Biol Sci Wayne State Univ Detroit MI 48013

MAYER, DAVID JONATHAN, physiologist, educator; b. Mt. Vernon, N.Y., July 18, 1942; s. Jerome Herman and Doris (Cantor) M.; m. Ursula Brigitte Fischer, Aug. 2, 1972. B.A., Hunter Coll., 1966; Ph.D., UCLA, 1971. Postdoctoral fellow UCLA, 1971-72; asst. prof. Med. Coll. Va., Richmond, 1972-75, assoc. prof., 1975-78, prof., 1978—. Contbr. over 100 articles to profl. jours. NIDA grantee, 1973—. Mem. Soc. Neurosci., Am. Physiol. Soc., Internat. Assn. Study of Pain (founding), Am. Pain Soc. (dir.). Subspecialties: Neurophysiology; Psychobiology. Current work: Neurobiology of pain; endogenous opiates and pain control systems. Home: 502 Honaker Ave Richmond VA 23226 Office: Med Coll Va Richmond VA 23298

MAYER, KLAUS, physician; b. Mayence, Germany, May 21, 1924; came to U.S., 1934; s. Stephan Karl and Caecilie Matilda (Mueller) M.; m. Vera Strasser, May 6, 1950; children—Rulon Richard, Carla Christina. B.S., Queens Coll., 1945; M.D., U. Groningen, Holland, 1950. Intern Hosp. St. Raphael, New Haven, 1950-51; resident Meml. Hosp. Cancer and Allied Diseases, N.Y.C., 1952-55; scientist Brookhaven Nat. Lab., 1951-52; mem. staff Meml. Hosp. Cancer and Allied Diseases, 1952—; practice medicine specializing in internal medicine and hematology, N.Y.C., 1956—; attending hematologist, dir. blood bank Hosp. Spl. Surgery, 1957—; dir. blood banks Meml. Hosp. for Cancer and Allied Diseases, 1966—, dir. hematology labs., 1971—, attending physician, 1972—, asso. chmn. dept. medicine, 1980—; attending physician N.Y. Hosp., 1979—; prof. clin. medicine Cornell U. Med. Coll., N.Y.C., 1979—; cons. Rockefeller U. Hosp. and Manhattan Eye and Ear Hosp., 1978—; mem. Mayor's Com. Radiation, 1968—, Bd. Examiners Blood Banks, N.Y.C., 1971-74; mem., prin. ad hoc com. to create Am. Blood Commn., 1974-75, sec.-treas., 1975-76, chmn. utilization com., 1976—; mem. N.Y. State Council on Human Blood and Transfusion Services, 1976—, chmn. com. supply and distbn. Bd. dirs. Fishers Island Health Project. Served with AUS, 1943-46; col. M.C., Res. Damon Runyon Cancer Research fellow, 1955-58. Fellow A.C.P., Am. Coll. Nuclear Medicine, Am. Soc. Clin. Pathology; mem. Am. Assn. Blood Banks (pres. 1973-74, Disting. Service award 1976), N.Y. Acad. Medicine (com. public health 1977). Subspecialties: Thermodynamics; Statistical mechanics. Current work: Thermodynamics, statistical mechanics. Research use radioactive isotopes in hematology. Home: 45 Sutton Pl S New York NY 10022 Office: 1275 York Ave New York NY 10021

MAYER, MANFRED MARTIN, microbiology educator; b. Frankfurt on Main, Germany, June 15, 1916; s. Gustav and Julie (Sommer) M.; m. Elinor S. Indenbaum, Dec. 6, 1942; children: Jonathan Marnin, David Michael, Dan Ellis, Matthew Jared. B.S., Coll. City N.Y., 1938; Ph.D., Columbia U., 1946; Dr. Med. Sci. (hon.), Johannes Gutenberg U., Mainz, Germany, 1969. Sci. staff OSRD, Columbia, 1942-45, instr. biochemistry, 1946; asst. prof. bacteriology Sch. Hygiene and Pub. Health, Johns Hopkins, 1946-48, assoc. prof. microbiology, 1948-61, prof. microbiology, 1961—; cons. USPHS, NSF, U.S. Dept. Agr., Office Naval Research. Author: (with Elvin A. Kabat) Experimental Immunochemistry, 1948, rev. edit., 1960; editorial bd.: Jour. Immunology. Recipient Kimble award methodology, 1953; Karl Landsteiner award Am. Assn. Blood Banks, 1974; Albion O. Bernstein award Med. Soc. State N.Y., 1976; Internat. award Gairdner Found., 1982. Fellow AAAS; mem. Am. Chem. Soc., Am. Assn. Immunologists (ccuncillor 1971, pres. 1976), Soc. Exptl. Biology and Medicine, Internat. Coll. Allergists, Nat. Acad. Sci., Am. Soc. Biol. Chemists, Biochem. Soc., Phi Beta Kappa, Sigma Xi. Subspecialties: Immunology (medicine); Membrane biology. Current work: Production of trans-membrane channels by complement attract; and its consequences on living cells. Home: 562 Sudbrook Ln Baltimore MD 21208

MAYER, RAMONA ANN, quality assurance ofcl., researcher; b. Algona, Iowa, May 9, 1929; d. William John and Esther Theresa (Wolf) M. B.A. in Chemistry, State U. Iowa, 1956; postgrad. in chemistry, Ohio State U., 1960. Lab. asst. Tb Hosp., Iowa City, 1952-53; library asst. State U. Iowa, Iowa City, 1954-56; info. specialist Battelle Meml. Inst., Columbus, Ohio, 1956-60, researcher, 1960-77, dir. quality assurance unit, 1977—; abstractor Chem. Abstracts Service, Columbus, 1958-79. Recipient Achievement award Nat. Aerospace Assn., 1970. Fellow Am. Inst. Chemists; mem. Am. Chem. Soc., Am. Soc. for Quality Control, ASTM, Nat. Soc. for Med. Research. Subspecialties: Toxicology (medicine); Cancer research (veterinary medicine). Current work: Quality assurance on programs dealing with toxicology, pathology, teratology, animal behavior, biochemistry, immunology, chemotherapy, ecology. Home: 4314 Chaucer Ln Columbus OH 43220 Office: 505 King Ave Columbus OH 43201

MAYER, RICHARD FREDERICK, neurologist, research neurophysiologist; b. Olean, N.Y., June 2, 1929; s. Frank W. and Rosemond F. (Bush) M.; m. Janet R. Bury, Oct. 10, 1959; children: Kathryn, Julianna, Andrea, Christopher, Randall. B.S., St. Bonaventure U., 1950; M.D., SUNY, Buffalo, 1954; postgrad., Mayo Found., U. Minn., 1955-56, Harvard U., 1956-57, 60-62, Neurology Inst., U. London, 1957-58. Diplomate: Am. Bd. Psychiatry and Neurology. Intern Boston City Hosp., 1954-55; resident in neurology Mass. Gen. Hosp., Boston, 1956-60; instr., asso. neurol. unit Harvard U., Boston City Hosp., 1962-66, asso. prof., 1966-68; prof. neurology U. Md. Hosp. and Sch. Medicine, 1968—; dir. EMG Lab., Myasthenia Gravis Clinic and Neuromuscular Service; guest researcher Lab. Neural Control, Nat. Inst. Neurol. and Communicative Disease and Stroke, NIH, Bethesda, Md.; med. adv. bd. Nat. Myasthenia Gravis Found., 1974—; mem. neurobiology merit rev. bd. VA Office, Washington. Contbr. articles profl. jours. Served to lt. USNR, 1958-60. Grantee M.S. Soc., NIH, Myasthenia Gravis Soc. Fellow Am. Acad. Neurology; mem. Am. Neurol. Assn., Soc. Neurosci., N.Y. Acad. Scis., Am. EEG Soc., Am. Assn. Neuropathology, Alpha Omega Alpha. Roman Catholic. Subspecialties: Neurology; Neurophysiology. Current work: Studies of diseases of neuromuscular system (Myasthenia gravis, muscular dystrophy), studies of control and orgn. of muscles and motor units. Office: U Md Hosp Redwood Greene Baltimore MD 21201

MAYES, LESLIE WILLIAM, technical services supervisor, educator; b. Thunder Bay, Ont., Can., Dec. 7, 1955; s. Edward Arthur and Marguerite Elizabeth (Blake) M.; m. Martha B. Laking, Sept. 6, 1975. B.S., Queen's U., Kingston, Ont., 1979. Programmer part-time dept. computing and info. sci. Queen's U., 1976-79; systems programmer Lakehead U. Computer Centre, Thunder Bay, 1979—, lectr. dept. math., 1980—. Ont. Ministry of Edn. scholar, 1974. Mem. Assn. Computing Machinery, IEEE. Mem. United Ch. Can. Clubs: Lakehead Toastmasters, Port Arthur Curling and Athletic. Subspecialties: Programming languages; Operating systems. Current work: Enhance and maintain operating systems and programming environments on VAX11/780 and DEC 2020 computers. Home: 26 Wishart Crescent Thunder Bay ON Canada P7A 6G3 Office: Lakehead U Oliver Rd Thunder Bay ON Canada P7B 5E1

MAYFIELD, DONALD LEWIS, radiation protection scientist, health physicist; b. Scottsbluff, Nebr., Jan. 20, 1938; s. Lewis Kenneth and Irene Elizabeth (Longstreth) M.; m. Jorlene Lana Schuler, Mar. 22, 1959; children: Brandon James, Christinna Lynne Mayfield Dyson, David Paul, Gail Loraine. B.S., U. Colo., Boulder, 1964; M.P.H., U. Mich., 1969. Radiation protection specialist Batelle-Northwest, Richland, Wash., 1969-70, radiation monitoring supr., 1970-72, devel. engr., 1972-74; mem. staff Health Physics Group, Los Alamos Nat. Lab., 1974-76, Environ. Surveillance Group, 1976—; cons. radiation protection. Mem. Health Physics Soc., Am. Assn. Physicists in Medicine, Am. Phys. Soc. Republican. Lutheran (Mo. Synod). Subspecialties: Radiological protection; Health physics. Current work: Radiation dosimetry, radiation protection instrumentation, radioactive waste disposal, environmental radiation surveillance, alternative fuels, solar energy. Home: 45 Puye Ct Los Alamos NM 87544 Office: Los Alamos Nat Lab Box 1663 Mail Stop K490 Los Alamos NM 87545

MAYHALL, JOHN TARKINGTON, oral anatomy educator, dental anthropologist; b. Greencastle, Ind., Apr. 7, 1937; emigrated to Can., 1970; s. Ward Davis and Esther (Tarkington) M.; m. Melinda Kay Fuller, Dec. 30, 1960. B.A., DePauw U., 1959; D.D.S., Ind. U.-Indpls., 1963; M.A., U. Chgo., 1968, Ph.D., 1976. Lic. dentist, Ont., Ind. Research fellow U. Toronto, 1971-72, assoc. dentistry, 1971-72, asst. prof. oral anatomy, 1972-75, assoc. prof., 1975-79, prof, 1979—; abstractor Oral Research Abstracts, 1968-75. Editorial adv. bd.: Jour. Dental Research, 1967-70. Served with USPHS, 1963-66. Research trainee Nat. Inst. Dental Research, Chgo., 1966-70; research fellow Can. Internat. Biol. Programme, Toronto, 1971-72. Fellow Human Biology Council; mem. Internat. Assn. Dental Research, Am. Assn. Phys. Anthropologists, Soc. Study of Human Biology, Omicron Kappa Upsilon. Subspecialties: Oral biology; Anatomy and embryology. Current work: Dental morphology, epidemiology of North American Indians and Inuit, genetics of dental morphology, growth and development, development of dentition. OfficeFaculty of Dentistry University of Toronto 124 Edward St Toronto ON Canada M5G 1G6

MAYHEW, ERIC GEORGE, cancer research scientist; b. London, June 22, 1938; s. George J. and Doris I. (Tipping) M.; m. Barbara Doe, Sept. 28, 1966; m. Karen Ann Carvana, Apr. 1, 1978; children: Miles, Ian. B.Sc., U. London, 1960, M.Sc., 1964, Ph.D., 1967. Research asst. Chester Beatty Research Inst., London, 1960-64; Cancer research scientist Roswell Park Meml. Inst., 1964—; vis. scientist Internat. Inst. Cellular and Molecular Pathology, Brussels, 1977-78; asso. research prof. SUNY, Buffalo, 1979—. Contbr. numerous articles to profl. jours. Mem. N.Y. Acad. Scis., Am. Assn. Cancer Research, AAAS. Subspecialties: Cancer research (medicine); Pharmacology. Current work: Basis of metastasis, selective delivery of drugs to tumors. Office: Dept Exptl Pathology Roswell Park Meml Inst Buffalo NY 14263

MAYNARD, SHERWOOD DAVIS, marine educator; b. Rochester, Minn., Oct. 4, 1946. B.S. with honors, U. Wash.-Seattle, 1968, M.S., 1972; Ph.D., U. Hawaii, 1982. Teaching asst. dept. oceanography U. Wash., Seattle, 1969-71; marine technician/diver Naval Arctic Research Lab., Barrow, Alaska, 1971; teaching asst. dept oceanography U. Hawaii, Honolulu, 1972-75, research asst., 1975-76, research technician, 1977-80; dir. Marine Option Program/Blue-Water Marine Lab., 1980—; cons. Legal Aid Soc., Honolulu, 1980-83, Mokauea Fishermen's Assn., 1980-82. Mem. Marine Tech. Soc. (edn. dir. 1982—), Am. Soc. Limnology and Oceanography, AAAS, Hawaii Sci. Tchrs. Assn., Hawaiian Acad. Sci. Subspecialties: Deep-sea biology; Oceanography. Current work: Marine education-secondary and undergraduate, mesopelagic and fish biology. Office: U Hawaii Marine Option Program 1000 Pope Rd Room 208 Honolulu HI 96822

MAYNE, RICHARD, anatomy and biochemistry educator; b. Oxford, Eng., May 10, 1942; came to U.S., 1968; s. Robert John and Dorothy (Emma) M.; m. Pauline Oakley, July 25, 1966; children: Rebecca Maynard, Lucy Rosina Elizabeth. B.A., U. Oxford, 1964, M.A., 1967, D.Phil., 1967. Asst. prof. anatomy U. Pa., Phila., 1972-74; asst. prof. anatomy U. Ala. in Birmingham, 1974-78, assoc. prof. anatomy, 1978-83; prof. assoc. prof. anatomy, 1983—; assoc. prof. biochemistry U. Ala. in Birmingham, 1982—. Mem.editorial bd.: Arteriosclerosis, 1981—. Established investigator Am. Heart Assn., 1978-83. Mem. Soc. Developmental Biology, Am. Soc. Cell Biology, Am. Assn. Anatomists. Subspecialties: Cell and tissue culture; Developmental biology. Current work: Monoclonal antibodies, cell culture techniques, connective tissue proteins, structure of cartilage, skeletal muscle and blood vessels. Home: 241 Big Springs Dr Birmingham AL 35216 Office: Dept Anatomy U Ala inBirmingham Univ Sta Birmingham AL 35294

MAYO, CLYDE CALVIN, organizational psychologist; b. Robstown, Tex., Feb. 2, 1940; s. Clyde Culberson and Velma (Oxford) M.; m. Jeanne Lynn McCain, Aug. 24, 1963; children: Brady Scott, Amber Camille. B.A., Rice U., 1961; B.S., U. Houston, 1964, Ph.D., 1972; M.S., Trinity U., 1966. Lic. psychologist, Tex. Mgmt. engr. LWFW, Inc., Houston, 1966-72, sr. cons., 1972-78, prin., 1978-81; ptnr. Mayo, Thompson, Bigby, Houston, 1981—; counselor Interface Counseling Ctr., Houston, 1976-79; instr. St. Thomas U., Houston, 1979—, U. Houston Downtown Sch., 1972. Author: Bi/Polar Inventory of Strengths, 1978. Coach, mgr. Meyerland Little League, 1974-78, So. Belles Softball, 1979-80, S.W. Colt Baseball, 1982—. Mem. Houston Psychol. Assn. (membership dir. 1978), Tex. Psychol. Assn., Am. Psychol. Assn., Am. Soc. Tng. and Devel., Houston Area Indsl. Orgnl. Psychologists. Club: Meyerland. Subspecialties: Industrial/organizational psychology; Behavioral psychology. Current work: Career development within organizations, executive assessment, organizational development. Home: 8723 Ferris St Houston TX 77096 Office: Mayo Thompson Bigby 3663 Briarpark St Houston TX 77042

MAYO, JOHN SULLIVAN, research and devel. co. exec.; b. Greenville, N.C., Feb. 26, 1930; s. William Louis and Mattie (Harris) M.; m. Lucille Dodgson, Apr. 1957; children—Mark Dodgson, David Thomas, Nancy Ann, Lynn Marie. B.S., N.C. State U., 1952, M.S., 1953, Ph.D., 1955. With Bell Telephone Labs., Inc., Murray Hill, N.J., 1955—, exec. dir. toll electronic switching div., 1973-75, v.p. electronics tech., 1975-79, exec. v.p. network systems, 1979—; chmn., program chmn. Internat. Solid State Circuits Conf.; bd. dirs. Nat. Engring. Consortium, Inc. Contbr. articles to profl. jours. Recipient Alexander Graham Bell award; named Outstanding Engring. Alumnus N.C. State U., 1977. Fellow IEEE; mem. Nat. Security Indsl. Assn. (chmn. study of global communications, mem. COMCAC adv. com.), Nat. Acad. Engring. Subspecialty: Electrical engineering. Current work: Technology and systems for telecommunications with emphasis on digital approaches for transmssion and switching. Patentee in field. Office: 600 Mountain Ave Murray Hill NJ 07974

MAYOCK, ROBERT LEE, internist; b. Wilkes-Barre, Pa., Jan. 19, 1917; s. John F. and Mathilde M.; m. Constance M. Peruzzi, July 2, 1949; children: Robert Lee, Stephen Philip, Holly Peruzzi. B.S., Bucknell U., 1938; M.D., U. Pa., 1942. Diplomate: Am. Bd. Internal Medicine. Intern Hosp. U. Pa., Phila., 1943-44, resident, 1944-45, chief med. resident, 1945-46, attending physician, 1946—; chief pulmonary disease sect. Phila. Gen. Hosp., 1959-72, sr. cons. pulmonary disease sect., 1972—; asst. prof. clin. medicine U. Pa., 1949-59, assoc. prof., 1959-70, prof. medicine, 1970—; mem. med. adv. com. for Tb Commonwealth of Pa., 1965-74, mem. med. adv. com. on chronic respiratory disease, 1974—, chmn. adv. com., 1981—; cons. Subsplty Bd. Pulmonary Disease Am. Bd. Internal Medicine, 1971-76. Contbr. articles in field to med. jours. Served to capt. U.S. Army, 1952-54. Fellow ACP, Am. Coll. Chest Physicians (regent 1972-79); mem. Pa. Med. Soc., Phila. County Med. Soc., Physiology Soc. Phila., Laennec Soc. Phila. (pres. 1963-64), Am. Thoracic Soc., N.Y. Acad. Scis., AMA, Am. Fedn. Clin. Research, Am. Lung Assn. (dir. Phila. and Montgomery County 1961—, pres. 1966-69, dir.-at-large 1983—), Am. Heart Assn., Pa. Lung Assn. (dir. 1976—), Sigma Xi, Alpha Omega Alpha. Clubs: Merion Cricket, Swiftwater Reserve.; Westmoreland (Wilkes Barre, Pa.). Subspecialties: Pulmonary medicine; Pharmacokinetics. Current work: Educator in pulmonary disease. Home: 244 Gypsy Ln Wynnewood PA 19096 Office: Ravdin Bldg 3d Floor Suite 1 U Pa Philadelphia PA 19104

MAYOL, PETE SYTING, biology educator; b. Negros Occidental, Philippines, Aug. 12, 1936; came to U.S., 1963; s. Abundio Y. and Restituta (Syting) M.; m. Flocer Villarente, July 11, 1964; children: Paul A., Virgil A., Jason M. B.S., U. Philippines, 1957; M.S., Okla. State U., 1965; Ph.D., Purdue U., 1968. Plant pathologist Bur. Plant Industry, Manila, 1957; lab. tech. Forest Products Research Inst., Philippines, 1957-58; instr. dept. agronomy U. Philippines, Los Banos, 1958-63; grad. research asst. dept. botany Okla. State U., Stillwater, 1963-65; grad, teaching asst. dept botany and plant pathology Purdue U., West Lafayette, Ind., 1965-66; asst. prof. dept. biol. scis. Calif. State Coll.-Stanislaus, Turlock, 1968-73, assoc. prof., 1973-81, prof., 1981—; mem. sci. rev. group NIH, Washington, 1981—. Contbr. articles to profl. jours. Entrance scholar U. Philippines, 1953; David-Ross fellow Purdue U., 1966; Danforth Found. assoc., 1980. Mem. AAAS, Calif. Acad. Scis., Filipino-Am. Assn. Stanislaus County (pres. 1976-78), U. Philippines Alumni Assn. Central Calif. (dir. 1982—), Sigma Xi, Phi Sigma. Subspecialties: Microbiology; Plant pathology. Current work: Interrelationships between soil microorganisms and plant parasitic nematodes particularly the root-knot nematodes. Home: 1620 Carleton Dr Turlock CA 95380 Office: Dept Biol Scis Calif State Coll Stanislaus 800 Monte Vista Ave Turlock CA 95380

MAYOR, HEATHER DONALD, microbiologist, educator; b. Melbourne, Australia, July 6, 1930; d. Joseph Arthur Lindsay and Elizabeth Emily (Boyd) Donald; m. Richard Blair Mayor, May 28, 1956; children: Diana Boyd, Philip Hastings. B.S., U. Melbourne, 1948, M.S., 1950; Ph.D., U. London, 1954; D.Sc., U. Melbourne, 1970. Walter and Eliza Hall postdoctoral fellow Inst. Med. Research, Melbourne, 1954-56; asst. prof. virology Harvard U. Med. Sch., Cambridge, Mass., 1956-59; assoc. prof. Baylor U. Coll. Medicine, Houston, 1963-75, prof. microbiology and immunology, 1975—; cons. NIH, AEC, U. Tex. Health Sci. Ctr., others. Contbr. over 200 articles to sci. jours. Mem. women's com. Houston Symphony Orch.; mem. Houston Contemporary Arts Mus., Young Women in the Arts, Houston Friends of Music. Recipient Disting. award Center for Interaction: Man. Sci. and Culture, 1971. Mem. Am. Assn. Immunologists, Am. Soc. Virologists, Am. Soc. Microbiology, Biophys. Soc., Electron Microscopy Soc. Am., Am. Soc. Cell Biology, Tissue Culture Assn., Sigma Xi. Episcopalian. Clubs: Tuesday Music, Doctors, Houstonian (Houston). Subspecialties: Virology (medicine); Genetics and genetic engineering (biology). Current work: Defective viruses and cancer, interference and prevention.

MAYR, ERNST, emeritus zoology educator; b. Kempten, Germany, July 5, 1904; came to U.S., 1931; s. Otto and Helene (Pusinelli) M.; m. Margarete Simon, May 4, 1935; children: Christa E., Susanne. Cand. med., U. Greifswald, 1925; Ph.D., U. Berlin, 1926, Uppsala U., Sweden, 1957, D.Sc., Yale U., 1959, U. Melbourne, 1959, Oxford U., 1966, U. Munich, 1968, U. Paris, 1974, Harvard U., 1980, Guelph U., U. Cambridge, 1982. Asst. curator zool. mus. U. Berlin, 1926-32; mem. Rothschild expdn. to Dutch New Guinea, 1928, expdn. to Mandated Ty. of New Guinea, 1928-29, Whitney Expdn., 1929-30; research asso. Am. Mus. Natural History, N.Y.C., 1931-32, asso. curator, 1932-44, curator, 1944-53; Jesup lectr. Columbia U., 1941; Alexander Agassiz prof. zoology Harvard U., 1953-75, emeritus, 1975—; dir. Harvard (Mus. Comparative Zoology), 1961-70. Author: List of New Guinea Birds, 1941, Systematics and the Origin of Species, 1942, Birds of the Southwest Pacific, 1945, Birds of the Philippines, (with Jean Delacour), 1946, Methods and Principles of Systematic Zoology, (with E. G. Linsley and R. L. Usinger), 1953, Animal Species and Evolution, 1963, Principles of Systematic Zoology, 1969, Populations, Species and Evolution, 1970, Evolution and the Diversity of Life, 1976, (with W. Provine) Evolutionary Synthesis, 1980, Biologic de l'Evolution, 1981, The Growth of Biological Thought, 1982; editor: Evolution, 1947-49. Pres. XIII Internat. Ornith. Congress, 1962. Recipient Leidy medal, 1946; Wallace Darwin medal, 1958; Brewster medal Am. Ornithologists Union, 1965; Daniel Giraud Elliot medal, 1967; Nat. Medal of Sci., 1970; Molina prize Accademia delle Scienze, Bologna, Italy, 1972; Linnean medal, 1977; Gregor Mendel medal, 1980. Fellow Linnean Soc. N.Y. (past sec. editor), Am. Ornithol. Union (pres. 1956-59), New York Zool. Soc.; mem. Am. Philos. Soc., Nat. Acad. Sci., Am. Acad. Arts and Scis., Am. Soc. Zoologists, Soc. Systematic Zoology (pres. 1966), Soc. Study Evolution (sec. 1946, pres. 1950); hon. or corr. mem. Royal Australian, Brit. ornithol. unions, Zool. Soc. London, Soc. Ornithol. France, Royal Soc. New Zealand, Bot. Gardens Indonesia, S. Africa Ornithol. Soc., Linnean Soc. London, Deutsche Akademie der Naturforsch Leopoldina., Accad. Naz. dei Lincei. Subspecialty: Evolutionary biology. Current work: Philosophy of biology, intellectual history of biology. Office: Mus Comparative Zoology Harvard U Cambridge MA 02138

MAYS, CHARLES WILLIAM, radiation physicist, radiobiologist, educator; b. Corsicana, Tex., Mar. 17, 1930; s. Charles William and Fay Lockhart M.; m. Evelyn Ekker, June 24, 1951; 3 daus., Shelby, Sharon, Susan; m. Desiree McMahon, Oct., 18, 1972; 2 sons, David, Rory. B.S. in Physics, U. Utah, 1951, Ph.D., 1958. Cert. health physicist. Mem. faculty U. Utah, Salt Lake City, 1957—, research prof. anatomy, 1975-82, adj. prof. physics, 1979—, research prof. pharmacology, 1979—; Mem. U.S. Nat. Council on Radiation Protection, Washington, 1966—; mem. internat. dose com. Internat. Commn. Radiol. Protection, Sutton, Eng., 1972—; chmn. adv. com. U.S. Transuranium Registry, Hanford, Wash., 1978-83. Editor: Some Aspects of Internal Irradiation, 1962, Delayed Effects of Bone-Seeking Radionuclides, 1969, Biological Effects of Ra-224 and Thorotrast, 1978; contbr. over 135 sci. articles to profl. publs. Served to 1st lt. U.S. Army, 1951-54; Korean War. Decorated Bronze Star; recipient Disting. Teaching award U. Utah, 1977. Mem. AAAS, Radiation Research Soc., Health Physics Soc. Subspecialties: Toxicology (medicine); Nuclear physics. Current work: Risk from radiation-induced cancer; removal of radioactivity from the body. Co-patentee improved radiation detector. Office: Univ Utah Radiobiology Dept Bldg 351 Salt Lake City UT 84112

MAZIA, DANIEL, biologist; b. Scranton, Pa., Dec. 18, 1912; s. Aaron and Bertha (Kurtz) M.; m. Gertrude Greenblatt, June 19, 1938; children—Judith Ann, Rebecca Ruth. A.B., U. Pa., 1933, Ph.D., 1937; Ph.D. h.c, U. Stockholm, 1976. National Research Council fellow in biology sci. Princeton, 1937-38; asst. prof. zoology U. Mo., 1938-41, asso. prof., 1942-47, prof., from 1947; asso. prof. zoology U. Calif., Berkeley, 1951-53; now prof.; prof. biol. scis. Stanford U., 1980—; Mem. corp. Marine Biol. Lab., Woods Hole.; mem. bd. trustees and physiology teaching staff, 1950—. Mem. editorial bd.: Exptl. Cell Research. Served as 1st lt. AAF, 1942-44; capt. aviation medicine, 1944-45. Fellow Am. Acad. Arts and Scis.; mem. Am. Soc. Cell Biology, Soc. Gen. Physiologists, Nat. Acad. Scis. Subspecialty: Cell biology. Current work: Mitosis. Address: Hopkins Marine Station (Stanford U) Pacific Grove CA 93950

MAZO, ROBERT MARC, chemistry educator, researcher; b. Bklyn, Oct. 3, 1930; s. Nathan and Rose Marian (Mazo) M.; m. Joan Ruth Spector, Sept. 5, 1954; children: Ruth Sara, Jeffrey Alan, Daniel Paul. A.B., Harvard U., 1952; M.S., Yale U., 1953, Ph.D., 1955. NSF postdoctoral fellow U. Amsterdam, Netherlands, 1955-56; research assoc. U. Chgo., 1956-58; asst. prof. Calif. Inst. Tech., 1958-62; assoc. prof. chemistry U. Oreg., 1962-65, prof., 1965—; cons. in field. Author: Statistical Mechanical Theories of Transport Process; contbr. numerous articles to profl. publs. Alfred P. Sloan fellow, 1961-65; NSF sr. postdoctoral fellow, 1968-69; Heinrich Hertz Stiftung fellow, 1981-82; Meyerhof fellow, 1982. Mem. Am. Chem. Soc., Am. Phys. Soc., AAAS, AAUP, Sigma Xi. Subspecialties: Statistical mechanics; Statistical physics. Current work: Theory of irreversible processes; applications of stat.mechanics to varied fields. Patentee in field. Home: 2460 Charnelton St Eugene OR 97405 Office: Inst Theoretical Sci U Oreg Eugene OR 97403

MAZUREK, THADDEUS JOHN, physicist, physics and astrophysics educator; b. Tarnogrod, Poland, Aug. 11, 1942; came to U.S., 1951, naturalized, 1962; s. Josef and Marianna (Pavliha) M.; m. Carolyn Bryant, Nov. 30, 1974; 1 dau.: Fabrienne T. M.S., Fordham Coll., 1966; M.A., Yeshiva U., 1968, Ph.D., 1973. Research assoc. Yeshiva U., 1973; research assoc. Harvard Coll. Obs., Cambridge, Mass., 1973-74; research scientist, lectr. U. Tex.-Austin, 1974-79; asst. prof. SUNY-Stony Brook, 1979-82; sr. physicist Mission Research Corp., Santa Barbara, Calif., 1982—; lectr., guest prof. Nordita, Copenhagen, Denmark, 1979. Contbr. articles to profl. jours. NSF Research grantee, 1978-79; Dept. Energy Research grantee, 1980-82. Mem. Am. Astron. Soc., Internat. Astron. Union, N.Y. Acad. Sci., AAAS. Democrat. Subspecialties: Theoretical physics; Theoretical astrophysics. Current work: Theory of fluid and plasma flows in nuclear explosions within the atmosphere; supernova theory, including hydrodynamics and energy transport of stellar explosions. Office: Mission Research Corp 735 State St Santa Barbara CA 93102

MCABEE, THOMAS ALLEN, psychologist, consultant; b. Spartanburg, S.C., Mar. 31, 1949; s. Thomas Walker and Doris Lee (Gillespie) McA. Student, Ga. Inst. Tech., 1967-69; B.A., Furman U., 1971; M.A., U. S.C., 1975, Ph.D., 1979. Co-dir. community problems survey Eau Claire Community Project, Columbia, S.C., 1975; instr. U.S.C., Columbia, 1976—; NSF intern S.C. State Legislature, Columbia, 1978; research dir. S.C. Legislative-Gov.'s Com. on Mental Health and Mental Retardation, Columbia, 1979-80; co-dir. Children's TV project Feelings Just Are, Columbia Area Mental Health Center, 1980—; psychologist S.C. Dept. Mental Retardation, Florence, 1982—; chmn. Primary Prevention Public Media Com., S.C. Dept. Mental Health, 1979-81; research cons. S.C. Protection and Advocacy System for Handicapped Citizens, 1980, 81; cons. numerous govtl. and pvt. agys. Recipient Palmetto Pictures Photography award U. S.C., 1977, Gabriel award, 1981; NIMH fellow, 1976-77. Mem. Am. Psychol. Assn. Subspecialties: Behavioral psychology; Community psychology. Current work: Behavior modification for the developmentally disabled; mass media-based community interventions; children's television. Home: 830-H Parker Dr Florence SC 29501 Office: SC Department Mental Retardation PO Box 3209 Florence SC 29502

MC AFEE, JERRY, retired oil company executive, chemical engineer; b. Port Arthur, Tex., Nov. 3, 1916; s. Almer McD. and Marguerite (Calfee) McA.; m. Geraldine Smith, June 21, 1940; children—Joe R., William M., Loretta M., Thomas R. B.S. in Chem Engring, U. Tex. at Austin, 1937; Sc.D. in Chem. Engring, Mass. Inst. Tech., 1940; student, Mgmt. Problems for Execs., U. Pitts., 1952. Research chem. engr. Universal Oil Products Co., Chgo., 1940-43, operating engr., 1944-45; tech. specialist Gulf Oil Corp., Port Arthur, Tex., 1945-50; successively dir. chemistry, asst. dir. research, v.p., asso. dir. research subs. Gulf Research & Devel. Co., Hamarville, Pa., 1950-55, v.p. engring. mfg. dept. of corp., 1955-60; v.p., exec. tech. advisor of corp., 1960-64, also dir. planning and econs., 1962-64; sr. v.p. Gulf Oil Corp., 1964-67, chmn. bd., chief exec. officer, 1976-81, now dir.; sr. v.p. Gulf Eastern Co., London, 1964-67; exec. v.p. Brit. Am. Oil Co., Ltd., Toronto, Ont., 1967-69; pres., chief exec. officer, dir. Gulf Oil Can. Ltd., 1969-75; adv. bd. Mellon Bank.; dir. McDonnell Douglas Corp. Bd. dirs. Am. Petroleum Inst., Aspen Inst. Humanistic Studies, MIT Corp., Pitts. Symphony Soc.; chmn. Allegheny Conf. Community Devel. Mem. Am. Inst. Chem. Engring. (v.p. 1959, pres. 1960), Nat. Acad. Engring., Am. Petroleum Inst., Am. Chem. Soc. Presbyterian. Clubs: Duquesne (Pitts.); Toronto, York (Toronto); Fox Chapel Golf; Rolling Rock (Ligonier, Pa.). Subspecialties: Chemical engineering; Fuels. Current work: Petroleum and its products. Patentee in field. Home: 4 Indian Hill Rd Pittsburgh PA 15238 Office: Gulf Oil Corp Gulf Bldg Pittsburgh PA 15219

MCALLISTER, RUSSELL GREENWAY, JR., cardiologist; b. Richmond, Va., Nov. 23, 1941; s. Russell Greenway and Kathryn Lee (Young) McA.; m. Ann Parks, Nov. 9, 1968; children: Kathryn Ann, Edward Russell. B.S., Hampden-Sydney Coll., 1963; M.D., Med. Coll. Va., 1967. Diplomate: Am. Bd. Internal Medicine. Asst. prof. medicine U. Ky. Coll. Medicine, Lexington, 1972-78, assoc. prof., 1978-82, prof., 1982—; clin. investigator VA Med. Ctr., Lexington, 1974-78, assoc. chief staff for research and devel., 1979—. Contbr. articles to profl. jours. NIH Fogarty Sr. Internat. fellow, 1978-79. Fellow , Am. Coll. Cardiology (mem. cardiovascular drugs com. 1978—), Am. Coll. Clin. Pharmacology, A.C.P.; mem. Central Soc. for Clin. Research, Fayette County Med. Soc. Republican. Presbyterian. Club: Lexington Tennis. Subspecialties: Cardiology; Pharmacology. Current work: Investigations of clinical pharmacology of cardiovascular drugs. Home: 1612 Bon Air Dr Lexington KY 40502 Office: VA Med Ctr (151) Cooper Dr Lexington KY 40511

MCANDREW, FRANCIS THOMAS, psychologist; b. Augsburg, W.Ger., Jan. 27, 1953; came to U.S., 1953; s. John Francis and Jane Ann (Tuman) McA.; m. Maryjo Ann McCarthy, July 29, 1978; children: Timothy Ned, Maura Jill. B.S. cum laude, Kings Coll., 1974; Ph.D., U. Me., 1981. Equipment mgr. Cath. Youth Center, Wilkes-Barre, Pa., 1970-71; research asst. King's Coll., Wilkes-Barre, 1972-74; teller United Penn Bank, Wilkes-Barre, 1971-74, 76; grad. asst. U. Me., Orono, 1975-79; instr. psychology Knox Coll., Galesburg, Ill., 1979-81, asst. prof., 1981—. Contbr. articles to profl. jours. U. Me. fellow, 1974-75. Mem. Eastern Psychol. Assn., N.Y. Acad. Sci., Am. Psychol. Assn., Midwestern Psychol. Assn., Internat. Soc. Human Ethology, Psi Chi. Club: Northgate Racquetball. Subspecialties: Social psychology. Current work: Environ. psychology; nonverbal communication; animal behavior; emotion; consumer psychology. Address: Knox Coll Dept Psychology PO Box 51 Galesburg IL 61401

MCAREAVY, JOHN FRANCIS, govt. ofcl.; b. Coggon, Iowa, Sept. 22, 1927; s. John B. and Kathryn (McMeel) McA.; m. Joan N. Nilles, Sept. 8, 1949; children: Susan, Kathryn, Mary, Martha, Brian, John L., Joseph, Thomas, Julie, Amy, Douglas, Molly. B.A., U. Iowa, 1951, M.S., 1955, Ph.D., 1969. Instr. math. Muscatine (Iowa) High Sch., 1952-56, Muscatine Jr. Coll., 1955-56; mathematician Ameta, Rock Island, Ill., 1956-61, dept. head math., 1961-65, dept. head pub. adminstr., 1965-80, dir., 1980—; cons. Mgmt. Services, Inc., Davenport, Iowa, 1972-74; adv. bd. Dept. Indsl. Engring., U. Iowa, 1981—; mem. I.E. Task Group St. Ambrose Coll., Davenport, 1981. Author: Analysis of Factors Affecting Achievement, 1969. Mem. Sch. Bd., Muscatine, Iowa, 1970, pres., 1971. Served with U.S. Army, 1946-48. Recipient Superior Performance award Ameta, 1959; Outstanding Achievement award Sec. of Army, 1974. Mem. Acad. Mgmt., Am. Psychol. Assn., Sigma Iota Epsilon. Roman Catholic. Lodges: KC; Elks. Subspecialties: Learning; Social psychology. Current work: Directing research program spanning sociotech. systems analysis, man/machine interfaces in decision support systems, productivity enhancement. Address: 1031 Newell Ave Muscatine IA 52761

MCARTHUR, DAVID ALEXANDER, research engineer, marketing consultant; b. Meridian, Miss., Aug. 14, 1938; s. Robert Stainton and Kathleen Adele (Sanders) McA.; m. Beverley Bogue, Sept. 16, 1961; children: Shirin R., Kara Kay, Paul D. B.S., U. Ariz., Tucson, 1960; postgrad., U. Munich, Ger., 1961; Ph.D., U. Calif.-Berkeley, 1967. Mem. tech. staff Bell Telephone Labs., Holmdel, N.J., 1967-69; with Sandia Nat. Labs., Albuquerque, 1969—, Disting. mem. tech. staff, 1983—. Contbr. articles to profl. jours. Woodrow Wilson fellow, 1962; NSF fellow, 1963-67; Nat. Merit scholar, 1956; Fulbright scholar, 1960-61. Mem. Am. Phys. Soc., Am. Nuclear Soc. Republican. Presbyterian. Subspecialties: Nuclear engineering; Nuclear fission. Current work: Nuclear reactor safety research; numerical modelling of plasmas, reactor-excited lasers, coded-aperture imaging of fission sources. Patentee reactor excited laser. Office: PO Box 5800 Albuquerque NM 87185

MCARTHUR, ELDON DURANT, research geneticist, botany and range science educator; b. Hurricane, Utah, Mar. 12, 1941; s. Eldon and Denise (Dalton) McA.; m. Virginia Johnson, Dec. 20, 1963; children: Curtis Durant, Monica, Denise, Ted Owen. A.S. with high honors, Dixie Coll., 1963; B.S. cum laude, U. Utah, 1965, M.S., 1967, Ph.D., 1970. Agrl. Research Council postdoctoral fellow U. Leeds, Eng., 1970-71; research geneticist Intermountain Forest and Range Expt. Sta., Forest Service, U.S. Dept. Agr., Ephraim, Utah, 1972-75, Provo, Utah, 1975-78, prin. research geneticist, 1978—, project leader, 1983—; adj. assoc. prof. dept. botany and range sci. Brigham Young U., 1976-78, adj. prof., 1978—. Contbr. numerous articles to profl. jours., govt. publs., chpts. to books. Recipient U.S. Dept. Agr. cert. of Merit, 1979; NDEA fellow, 1965-68; NIH fellow, 1968-70; Sigma Xi grantee, 1970-71; NSF grantee, 1981—. Mem. Soc. Range Mgmt., Soc. Study of Evolution, Bot. Soc. Am., Am. Genetic Assn., Rocky Mountain Forest Genetics Com., Sigma Xi. Mormon. Subspecialty: Plant genetics. Current work: Cytogenetics, breeding systems, evolution; improvement of Western wildland shrubs. Home: 555 N 1200 E Orem UT 84057 Office: 735 N 500 E Provo UT 84601

MCARTHUR, WILSON COOPER, engineering consultant executive, radiation protection consultant; b. Clinton, N.C., June 28, 1936; s. Charles Dixon and Margaret (McLamb) McA.; m. Robbie Louise Taylor, Aug. 23, 1959; children: Gregory, Suzette, Alexander. B.S., East Carolina U., 1965; M.S., U. N.C., 1967; Ph.D., Purdue U., 1971. Registered profl. engr.; cert. hazards control mgr. Prin. engr. Carolina Power and Light Co., Raleigh, N.C., 1971-77; v.p., gen. mgr. Hittman Corp., Columbia, Md., 1977-78; div. mgr. Tera Corp., Berkeley, Calif., 1978-80, EDS Nuclear, Walnut Creek, Calif., 1980-82; pres. KLM Engring., Inc., Walnut Creek, 1982—. Served with USAF, 1955-59. Named Outstanding physics student East Carolina Coll., 1965. Mem. Am. Nuclear Soc. (pres. N.C. chpt. 1977), Health Physics Soc. (pres.

N.C. chpt. 1976). Subspecialties: Nuclear fission; Nuclear engineering. Current work: Development of radiation protection criteria and development of techniques to process, transport and dispose of radioactive waste. Home: 1478 Ramsay Circle Walnut Creek CA 94598 Office: KLM Engring Inc 1776 Ygnacio Valley Rd Suite 200 Walnut Creek CA 94596

MCATEE, JAMES LEE, JR., chemist, educator, cons.; b. Waco, Tex., Aug. 29, 1924; s. James Lee and Edith Beatrice (Smith) McA.; m. Francis Louise Linville, Sept. 6, 1947; children: Susan Monday, James Lee, Rosamonde Slakie, Winfield Linville. B.S., Tex. A&M U., 1947; M.S., Rice U., 1949, Ph.D., 1951. Lab. technician Shell Exploration & Prodn. Research, summers 1948-50; research scientist Barold div. NL Industries, Houston, 1951-59, supvr. tech. service labs., 1954-59; mem. faculty dept. chemistry Baylor U., 1959—, asst. prof. chemistry, 1959-61, assoc. prof., 1961-69, prof., 1969—, chmn. dept., 1981—; cons. chemistry of clays to industry. Contbr. numerous articles on clays, clay minerals, Am. minerals, surface and colloid sci. to profl.jours. Deacon, elder First Presbyterian Ch., Waco. Served to lt. USAAF, 1943-45. Fellow Minerals Soc. Am.; mem. Am. Chem. Soc., Am. Crystollography Soc., Clay Minerals Soc., Tex. Electron Microscopy Soc., N.Am. Thermal Analysis Soc. Club: Kiwanis (Waco) (pres. 1970-71). Subspecialties: Physical chemistry; Surface chemistry. Current work: Study of clay minerals and organo-clay complexes; surface and rheology of clays and organo-clays extending to high pressure areas. Home: 5625 Oakview Waco TX 76710 Office: Dept Chemistry Baylor U Waco TX 76798

MCAULEY, VAN ALFON, aerospace mathematician; b. Travelers Rest, S.C., Aug. 28, 1926; s. Stephen Floyd and Emily Floree (Cox) McA. B.A., U. N.C., Chapel Hill, 1951. Mathematician Army Ballistic Missile Agy., Huntsville, Ala., 1956-59; physicist NASA, Marshall Center, Huntsville, 1960-61, research mathematician, 1962-70, mathematician, 1970-81. Served with U.S. Army, 1944-46. Recipient Apollo achievement award NASA, 1969, cost savs. award, 1973, Skylab achievement award, 1974, Outstanding Performance award, 1976. Mem. Am. Math. Soc., Soc. Indsl. and Applied Math., AAAS, N.Y. Acad. Scis., Phi Beta Kappa. Subspecialties: Aerospace engineering and technology; Numerical analysis. Current work: Documentation of methods devised for the numerical solution of heat flow equations involving both elliptic and parabolic partial differential equations. Patentee in field, of control system invention. Home: 3529 Rosedale Dr Huntsville AL 35810

MCBRIDE, ANGELA BARRON, psychiatric nurse, psychologist; b. Balt., Jan. 16, 1941; s. John Stanley and Mary Constance (Szczepanska) Barron; m. William Leon McBride, June 12, 1965; children: Catherine Alexandra, Kara Angela. B.S.N., Georgetown U., 1962; M.S.N., Yale U., 1964; Ph.D., Purdue U., 1978; D.P.S. hon., U. Cin., 1983. Instr. Yale U. Sch. Nursing, 1964-67, lectr., 1967-68, research asst., 1969-72, asst. prof., 1972-73; assoc. prof. Ind. U. Sch. Nursing, 1978-81, prof., 1981—; adj. assoc. prof. Sch. Medicine, 1980-81, adj. prof., 1981—; adj. assoc. prof. Purdue Sch. Sci., 1980-81, adj. prof., 1981—. Author: The Growth and Development of Mothers, 1973, Living with Contradictions: A Married Feminist, 1976; editor: Psychiatric Nursing and the Demand for Comprehensive Health Care, 1972; mem. adv. bd. Ann. Rev. of Nursing Research, 1981—; reviewer: Research in Nursing and Health, 1980—, Nursing Outlook, 1982—. Recipient Disting. Alumnus award Yale U. Sch. Nursing, 1978, Research and Scholarship award Am. Nurses Assn. Council of Specialists in Psychiat. and Mental Health Nursing, 1984; nat. fellow Kellogg Found., 1981—. Fellow Am. Acad. Nursing; mem. Am. Nurses Assn., Am. Psychol. Assn., Nat. Women's Health Network, Nat Council Family Relations, Sigma Xi, Sigma Theta Tau. Democrat. Roman Catholic. Subspecialties: Developmental psychology; Social psychology. Current work: Teacher-researcher with interests in women's health, adult development, psychology of parenthood, psychology of obesity. Home: 744 Cherokee Ave Lafayette IN 47905 Office: Indiana Sch Nursing 610 Barnhill Dr Indianapolis IN 46223

MCBRIDE, EARLE FRANCIS, geology educator, consultant; b. Moline, Ill., May 25, 1932; s. Earle Curtis and Elsa (Burch) McB.; m. Donna Joann, June 10, 1956; children: Deborah, Suzanne. B.A., Augustana Coll., 1954; M.A., U. Mo.-Columbia, 1956; Ph.D., Johns Hopkins U., 1960. Asst. prof. to prof. geology U. Tex., Austin, after 1959—; now Wilton E. Scott prof. geology; pres. Sandstones, Inc. Contbr. articles on sedimentary geology to profl. jours. Fellow Geol. Soc. Am.; mem. Am. Assn. Petroleum Geologists, Soc. Econ. Paleontologists and Mineralogists, Internat. Assn. Sedimentologists. Subspecialties: Sedimentology; Petrology. Current work: Origin of sandstone; geology educator; consultant. Office: Dept of Geol Scis U Tex Austin TX 78712

MCBRIDE, WILLIAM JOSEPH, JR., neurochemist, educator; b. Phila., Dec. 24, 1938; s. William Joseph and Elizabeth (Fantini) McB.; m. Dianne Burnice Stewart, July 14, 1962; children: Vicki, Richard, Rebecca, Andrew. B.A., Rutgers U., 1964; Ph.D., SUNY, Buffalo, 1968. Postdoctoral fellow in neurobiology Ind. U., Bloomington, 1968-71, asst. prof. psychiatry and biochemistry, Indpls., 1971-75, assoc. prof., 1975-79, prof., 1979—; manuscript referee Jour. Neurochemistry and Pharmacology, Biochemistry and Behavior; research proposals evaluator VA Hosp. Contbr. articles to sci. jours. Served with USAF, 1956-60. Recipient Research Scientist Career Devel. award NIMH, 1979—. Mem. Soc. for Neurosci., Am. Soc. for Neurochemistry, Research Soc. on Alcoholism, Internat. Soc. for Neurochemistry. Subspecialties: Neurochemistry; Neurobiology. Current work: Neurochemistry of alcoholism; interactions of neurotransmitter systems; amino acid transmitters and nervous system function. Home: 7530 Hoover Rd Indianapolis IN 46260 Office: Inst. Psychiatric Research Ind U Sch Medicine Indianapolis IN 46223

MCBROOM, ROBERT CHISM, criticality safety specialist; b. Magdalena, N.Mex., May 24, 1947; s. Russell Thomas and Chanda (Brown) McB.; m. Karen Marie Reczek, Aug. 15, 1971; children: Sean Russell, Scott Edward. B.S., Angelo State U., San Angelo, Tex., 1968; M.S., Ill. Inst. Tech., Chgo., 1971; Ph.D., U. Fla., Gainesville, 1977. Registered profl. Engr., Calif. Critic Gen. Atomic Co., San Diego, 1976-81; criticality safety specialist Exxon Nuclear Idaho Co., Idaho Falls, 1981—. Mem. Inst. Nuclear Materials Mgmt., Am. Nuclear Soc. Democrat. Subspecialties: Nuclear fission; Numerical analysis. Current work: Application of the body of neutron behavior data to the problems of safety and process optimization for a generalized uranium reprocessing plant. Home: 2232 Dickson Circle E Idaho Falls ID 83402 Office: Exxon Nuclear Idaho Co Inc 1955 Fremont Ave Idaho Falls ID

MCBRYDE, FELIX WEBSTER, consulting ecologist, thematic cartographer; b. Lynchburg, Va., Apr. 23, 1908; s. John McLaren and Flora O'Neall (Webster) McB.; m. Frances Van Winkle, July 23, 1934; children: Richard Webster, Sarah Elva, John McLaren. B.A., Tulane U., 1930, LL.D. (hon.), 1967; Ph.D., U. Calif.-Berkeley, 1940. Instr. geography Ohio State U., Columbus, 1937-42; geographer U.S. War Dept., Washington, 1942-45; dir. Peruvian Office, Inst. Social Anthropology Smithsonian Instn., Lima, 1945-47; geographer-cons. U.S. Bur. Census, Washington and, Latin Am., 1958-64; field dir. Gordon A. Friesen Assocs., Washington and, Guatemala, 1958-64; chief geography br. Interam. Geodetic Survey, C.Z., 1964-65; field dir. Battelle Meml. Inst., Panama and Colombia, 1965-70; dir. McBryde Ctr. for Human Ecology, Potomac, Md., 1970—; U.S. Census Mission chief 1st Nat. Census of Ecuador, 1949-51; ecologist, expert on tourism UN Devel. Programme, Jamaica, 1971; ecologist, expert on hydrology, Argentina, 1972; census geography cons. U.S. Bur. Census, Honduras, 1972; ecology cons. hydroelectric dam World Bank, Bayano River, Panama, 1973; ecology cons. on transp. Battelle-U.S. Dept. Transp., Panama, 1973; ecology cons. on mining UN, N.Y. and Cerro Colorado, Panama, 1981; dir. geog. research div. Transemantics, Inc., Washington, 1975—. Author: Solola: A Guatemalan Town, 1933; author: Cultural and Historical Geography of Southwest Guatemala, 1947; founding editor: Bull. Am. Soc. Profl. Geographers (now Profl. Geographer), 1943-45; patentee in field. Election campaign adviser to Pres. Milleda Morales Exec. Research Inc. N.Y., Honduras, 1957; election campaign adviser to Pres. Ydigoras Fuentes F.W. McBryde Assocs., Inc., Guatemala, 1957-58; mem. nat. adv. bd. Am. Security Council, Washington, 1977—. U. Colo. fellow, 1930-31; Clark U. research fellow, 1931-32; U. Calif.-Berkeley teaching fellow, 1933-35, 37; Social Sci. Research Council fellow, 1935-36; NRC fellow, 1940; Pan Am. Airways travel fellow, 1940-41; Ohio State U. Grad. Sch. research grantee, 1940-41. Mem. Ecuadorian Inst. Anthropology and Geography (hon. dir. 1951—), Assn. Am. Geographers (founding pres., sec., treas. 1943-45), Am. Congress Surveying and Mapping, Am. Geophys. Union, Am. Inst. Biol. Scis., Soc. Am. Mil. Engrs., Marine Tech. Soc., AAAS, N.Y. Acad. Scis., Explorers Club. Episcopalian. Club: U.S. Senatorial (Washington). Subspecialties: Geography, space relationship analysis; Thematic cartography, cartographic design. Current work: World map plottings on original series of equal-area map projections, special interest in all oceanic distributional data. Home: 10100 Falls Rd Potomac MD 20854 Office: McByde Center for Human Ecology 10100 Falls Rd Potomac MD 20854

MCCABE, ALIYSSA KIM, psychologist, educator, researcher; b. Pitts., May 22, 1952; d. Henry C. and Mary (Kuhr) McC.; m. Charles E. Cuneo, July 25, 1981. A.B., Oberlin Coll., 1974; M.A., U. Va., 1977, Ph.D., 1980. Asst. prof. psychology Wheaton Coll., NOrton, Mass., 1980—. Author: Developmental Psycholinguistics, 1983; contbr. articles to profl. jours. Mellon grantee, 1981, 82; Funds for Improvement Post-Secondary Edn. grantee, 1981. Mem. Am. Psychol. Assn., Soc. Research in Child Devel., Internat. Soc. Ecol. Psychology. Subspecialties: Cognition; Developmental psychology. Current work: Research on development of discourse structure in child language; research on judgement, interpretation, and memory of metaphors in context. Home: 9 Mather St Boston MA 02124 Office: Dept Psychology Wheaton Coll Norton MA 02766

MC CABE, BRIAN FRANCIS, physician; b. Detroit, June 16, 1926; s. Charles J. and Rosalie T. (Dropiewski) McC.; m. Yvonne L. Fecteau, Sept. 8, 1951; children—Brian F., Bevin E. B.S., U. Detroit, 1950; M.D., U. Mich., 1954. Intern Univ. Hosp., Ann Arbor, Mich., 1954-55; resident U. Mich. Med. Sch., 1955-59; practice medicine specializing in otolaryngology and maxillofacial surgery, Iowa City; mem. staff Iowa City VA Hosp.; prof., head dept. otolaryngology and maxillofacial surgery U. Iowa; also chmn. residency rev. com. for otolaryngology.; Dir. Bd. Examiners Otolaryngology. Co-editor: Annals of Otology, Rhinology and Laryngology. Mem. Am. Acad. Ophthalmology and Otolaryngology (sec. for otolaryngology), Am. Laryngol., Rhinol. and Otolaryngol. Soc., Am. Otol. Soc. (editor-librarian), Am. Laryngol. Soc., Otosclerosis Study Group, Galens Hon., Linn County, Am., Iowa med. socs., Collegium Oto-Rhino-Laryngologicum Amicitae sacrum, New Zealand Soc. Otolaryngology, Snipe Class Internat. Racing Assn., Nat. Amateur Yacht Racing Assn., DN Ice Yacht Racing Assn., Alpha Omega Alpha. Clubs: Centurion, Barton Boat, Hawkeye Sailing. Subspecialties: Otorhinolaryngology; Neurophysiology. Current work: Vestibular neurophysiology, cochlear electrode inplant physiology. Home: 237 Ferson St Iowa City IA 52240 Office: Univ Hosps and Clinics Iowa City IA 52240 It seems to me the greatest single attribute leading to success is the ability to transmit a thought or concept clearly and simply.

MCCAFFERTY, WILLIAM PATRICK, educator; b. Murray, Utah, Oct. 11, 1945; s. William Hugh and Beverly Ann (Trott) Coon; m. Nadine Poland, Sept. 15, 1964; children: Kirk, Maureen, Shena Lee. B.A., U. Utah, 1967, M.A., 1969; Ph.D., U. Ga., 1971. Prof. biology Dixie Coll., St. George, Utah, 1970-71; prof. entomology Purdue U., W. Lafayette, 1971—; environ. cons. Minn. Dept. Natural Resources, Mpls., 1975-79, Inst. Paper Chemistry, Appleton, Wis., 1979-82; instr. U.S. EPA Workshop, Cin., 1978; dir. Purdue Entomol. Mus., 1971—. Author: Aquatic Entomology, 1981. Recipient Parent's Award for Instructional Innovation Purdue U., 1974; NSF grantee, 1982; U.S. EPA grantee, 1977; Office Water Resources Research grantee, 1978. Fellow Ind. Acad. Sci.; mem. Entomol. Soc. Wash., N. Am. Benthological Soc., Entomol. Soc. Am. Club: Optimists. Subspecialties: Systematics; Ecosystems analysis. Current work: Systematics and ecology of aquatic insects, subaquatic behavior, video-analysis of functional morphology. Home: 306 Sylvia St West Lafayette IN 47906 Office: Dept Entomology Purdue Univ West Lafayette IN 47907

MCCALL, DAVID WARREN, chemist, researcher; b. Omaha, Nebr., Dec. 1, 1928; s. H. Byron and Grace (Cox) McC.; m. Charlotte M. Dunham, July 30, 1955; children: William Christopher, John Dunham. B.S., U. Wichita, 1950; M.S., U. Ill-Urbana, 1951, Ph.D., 1953. Mem. tech. staff Bell Labs., Murray Hill, N.J., 1953-62, dept. head, 1962-69, asst. chem. dir., 1969-73, chem. dir., 1973—. Co-author: Challenges, Needs and Opportunities, 1981; editor: Polymer Jour, 1976—. Fellow AAAS; mem. Am. Chem. Soc., Chem. Abstracts Adv. Bd., Royal Soc. Chemistry, Am. Phys. Soc. Subspecialties: Nuclear magnetic resonance (chemistry); Physical chemistry. Office: Bell Laboratories 600 Mountain Ave Murray Hill NJ 07974

MCCALL, EDWARD HUFFAKER, scientific consultant, research computer scientist; b. Oxford, Miss., Dec. 11, 1938; s. Ephriam Forrest and Mariada (Huffaker) McC.; m. Judith Irene Bohn, Aug. 20, 1960; children: Scott Edward, Cathy Ann, Douglas James. Student, U. Ill.-Urbana, 1956-59; B.S. in Chem. Engring, Iowa State U., 1960; M.S. in Computer Sci, U. Minn.-Mpls., 1970, Ph.D., 1979. Jr. engr. Panhandle Eastern Pipeline Co., Kansas City, Mo., 1958; chem. engring. researcher 3M Co., Maplewood, Minn., 1960-63, programmer, 1963-68; programmer, project mgr. Sperry Univac, Roseville, Minn., 1968-79, researcher, sci. cons., 1979—; adj. asst. prof. U. Minn.-Mpls., 1979—. Asst. scoutmaster Indianhead council Boy Scouts Am., White Bear Lake, Minn., 1974-80; football coach Mounds View Jr. Football League, New Brighton, Minn., 1979-80. Mem. Math. Programming Soc., Ops. Research Soc., Soc. Indsl. and Applied Math. Republican. Lutheran. Subspecialties: Operations research (mathematics); Numerical analysis. Current work: numerical algorithms for vector and parallel processors; Large scale mathematical programming, especially nonconvex global optimization, linear programming and mixed integer programming scientific processing applications. Home: 4710 Debra Ln Shoreview MN 55112 Office: Sperry Univac 2276 Highcrest Rd Roseville MN 55113

MCCALL, JOHN W., parasitology educator; b. Blackshear, Ga., June 3, 1941; s. Harold L. and Vera E. McC.; m.; children: Christy Ann, Deborah Elizabeth (dec.) Sharon Elaine. B.S. in Zoology, U. Ga., 1963, M.S. in Entomology, 1966, Ph.D., 1970. Research assoc. dept. pathology and parasitology Coll. Vet. Medicine U. Ga., 1970-72; research assoc. dept. parasitology, 1972-78, assoc. prof., 1978—; pres. TRS Labs., Inc., Athens, Ga., 1971—. Contbr. articles to profl. jours. Mem. Am. Soc. Parasitologists, Southeastern Soc. Parasitologists, Am. Soc. Tropical Medicine and Hygiene, Am. Heartworm Soc., Am. Assn. Vet. Parasitologists, Entomol. Soc. Am., Ga. Entomol. Soc., Assn. Southeastern Biologists, Internat. Filariasis Assn., Ga. Mosquito Control Assn., Sigma Xi, Phi Zeta, Phi Sigma. Republican. Baptist. Club: Immunology. Subspecialty: Parasitology. Current work: Filariasis animal models for filariasis; antifilarial drugs; anthelmintics cryobiology; immunity to parasitic diseases. Office: Department Parasitology College Veterinary Medicine University of Georgia Athens GA 30602

MCCALL, ROBERT BOOTH, psychologist; b. Milw., June 21, 1940; s. John I. and Blanche (Booth) McC.; m. Rozanne Allison, June 13, 1962; children: Darin Scott, Stacey Allison. A.B., DePauw U., 1962; M.A., U. Ill., 1964, Ph.D., 1965. NSF postdoctoral fellow Harvard U., Cambridge, Mass., 1965-66; asst. prof. U. N.C., Chapel Hill, 1966-68; chmn. Dept. Psychology and chief perceptual cognitivive devel. sect. Fels Research Inst., Yellow Springs, Ohio, 1968-77; asso. prof. Antioch Coll., Yellow Springs, 1968-77; sr. scientist, sci. writer Boys Town Center, Nebr., 1977—. Co-producer: news feature series Sci. for Families, 1981; author: Infants: The New Knowledge, 1980, Fundamental Statistics ofr Psychology, 1980; contbr. articles to profl. jours.; mem. 7 profl. editorial bds.; guest reviewer numerous jours. NIH grantee, 1970-77; Nat. Inst. Edn. grantee, 1973-76; NSF grantee, 1978-81; many others. Fellow Am. Psychol. Assn. (fellow, exec. com. 1977—); mem. AAAS, Soc. for Research in Child Devel., Sigma Xi. Subspecialty: Developmental psychology. Current work: Infancy, devel. of intelligence, parent-child relations; communication of behavioral research to gen. public. Office: Boys Town Center Boys Town NE 68010

MCCALMON, ROBERT THOMAS, JR., immunobiologist, researcher; b. Bremerton, Wash., May 5, 1943; s. Robert Thomas and Dorothy Jane (Miller) McC.; m. Sandra Ann Gagnon, Mar. 20, 1976; 1 son, Scott Thomas. B.A., U. Colo., 1967; M.A., Drake U., 1969; Ph.D., U. Ariz., 1973. Postdoctoral fellow Nat. Jewish Hosp. and Research Ctr., Denver, 1973-75; asst. prof. dept. surgery U. Colo. Health Scis. Ctr., Denver, 1975-79; dir. surg. immunology Denver VA Hosp., 1975-79; dir. Immunol. Assocs. Denver, 1979—, pres., 1980-82. Contbr. articles to profl. jours. Pres. bd. trustees Mile High Transplant Bank; mem. utilization com. End-Stage Renal Disease Coordinating Council. NIH grantee. Mem. Am. Assn. Clin. Histocompatibility Testing, Am. Assn. Immunologists, Am. Soc. Microbiology. Club: Denver Athletic. Subspecialty: Immunology (medicine). Current work: Biological significance of lymphocyte surface antigens. Office: 3570 E 12th Ave #200 Denver CO 80206

MCCAMMON, JAMES ANDREW, chemistry educator; b. Lafayette, Ind., Feb. 8, 1947; s. Lewis Brown and Jean (McClintock) McC.; m. Anne Woltmann, May 15, 1947. A.B. magna cum laude, Pomona Coll., Claremont, Calif., 1969; A.M., Harvard U., 1970, Ph.D., 1976. Asst. prof. chemistry U. Houston, 1978-81, M.D. Anderson prof., 1981—, chmn. phys. chemistry div., 1979—. NSF/NIH fellow, 1976-78; Sloan Found. fellow, 1980; recipient Research Career Devel. award NIH, 1980, Tchr.-scholar award Dreyfus Found., 1982. Mem. Am. Phys. Soc., AAAS, Am. Chem. Soc., Biophys. Soc. Subspecialties: Theoretical chemistry; Biophysical chemistry. Current work: Statistical mechanics of macromolecules and liquids; theory of protein structure, dynamics and function. Office: U Houston Dept Chemistry Houston TX 77004

MCCAMPBELL, STANLEY REID, physician; b. Nashville, Dec. 16, 1925; s. Basil Davis and Louise (McCall) McC.; m. Joan F. Garner, Nov. 7, 1953; children: Louise, Robert, James, Kelly. B.A., Vanderbilt U., 1949, M.D., 1952. Intern, resident Cornell/Bellevue Hosp., N.Y.C., 1952-55; fellow in cardiology Cornell Med. Ctr., N.Y.C., 1955-56, U. London, 1956-57; practice medicine, specializing in cardiology, Oklahoma City, 1957—. Contbr. articles to profl. jours. Served with USN, 1943-46. Named Phi Beta Kappa of Yr., 1981. Mem. Okla. County Med. Soc. (pres. 1970), Okla. Med. Assn. (pres. 1972), World Med. Tennis Soc. (pres. 1971—, Internat. Congress Psychosomatic and Preventive Medicine, pres. 1978—), Okla. Assn. of Phi Beta Kappa (pres. 1975), Am. Med. Tennis Assn. (pres. 1969-71). Republican. Presbyterian. Subspecialty: Cardiology. Current work: Practice of cardiology, pacemaker electronics, physical fitness, coronary artery disease. Office: 1211 N Shartel Rm 408 Oklahoma City OK 73103 Home: 6609 N Hillcrest St Oklahoma City OK 73116

MCCANN, FRANCES V., biomedical educator, research scientist, consultant; b. Manchester, Conn., Jan. 15, 1927; d. John J. and Grace Elizabeth (Tuttle) McC.; m. Elden J. Murray, Sept. 20, 1962. A.B., U. Conn., 1952, Ph.D., 1959; M.S., U. Ill., 1954; M.A. (hon.), Dartmouth Coll., 1967. Instr. physiology Dartmouth Coll., until 1959, asst. prof., 1959-61, assoc. prof., 1961-67, prof., cons. in medicine, 1967—; cons. NIH; exec. com. basic sci. Am. Heart Assn. Contbr. numerous articles profl. jours. Trustee Montshire Mus. Sci.; Investigator Am. Heart Assn., 1965-70. Mem. Soc. Gen. Physiology, Soc. Neurosci., Am. Soc. Zoologists, Marine Biol. Labs. Corp., Am. Physiol. Soc. Subspecialties: Cell study oncology; Comparative physiology. Office: Dept Physiology Dartmouth College Medical School Hanover NH 03755

MCCARLEY, ROBERT EUGENE, chemistry educator, researcher; b. Denison, Tex., Aug. 17, 1931; m., 1952; 4 children. B.S., U. Tex., 1953, Ph.D. in Chemistry, 1956. From instr. to assoc. prof. chemistry Iowa State U., 1956-70, prof., 1970—, chmn. dept. chemistry, 1976—; research assoc. Ames Lab., 1956-57, assoc. chemist, 1957-63, chemist, 1963-74, sr. chemist, 1974—. Mem. Am. Chem. Soc. Subspecialty: Inorganic chemistry. Office: Dept Chemistry Iowa State U Ames IA 50010

MCCARLEY, ROBERT WILLIAM, neurophysiologist, psychiatrist; b. Mayfield, Ky., Aug. 17, 1937; s. Robert Smith and Mary Agnes (McGill) McC.; m. Alice M. Bowen, Aug. 10, 1968; children: Robby, Scott. A.B. summa cum laude, Harvard U., 1959; postgrad., Gutenberg U., Mainz, W.Ger., 1959-60; M.D., Harvard U., 1964. Diplomate: Am. Bd. Psychiatry and Neurology, 1972. Intern Brigham and Women's Hosp., Boston, 1964-65; psychiat. resident Mass. Mental Health Ctr., Boston, 1965-68; lectr. psychiatry Simmons Coll. Social Work, Boston, 1967; instr. Harvard U. Med. Sch., Boston, 1970-75, asst. prof. psychiatry, 1975-78, assoc. prof., 1978—, co-dir. lab. neurophysiology, 1975—; co-dir. clin. research tng. program Mass. Mental Health Ctr., 1980—; cons. Mass. Rehab. Commn.; mem. rev. panel NIMH. Contbr. over 100 articles to sci. jours. Gen. Motors scholar, 1955-59; NIMH grantee, 1968—; NSF grantee, 1974—; others. Mem. AAAS, Sleep Research Soc., Soc. Neurosci., Am. Psychiat. Assn., Mass. Psychiat. Assn., Phi Beta Kappa. Democrat. Clubs: Harvard (Boston); Neighborhood (West Newton, Mass.). Subspecialties: Neurophysiology; Psychiatry. Current work: Neurophysiology of sleep; mathematical and computer modeling of sleep cycle control; computer techniques of neurophysiological cellular data description and display; topographic mapping of human electrical activity during sleep and schizophrenia. Office: 74 Fenwood Rd Boston MA 02115

MCCARROLL, WILLIAM HENRY, chemistry educator; b. Bklyn., Mar. 19, 1930; s. Joseph Allen and Elva Jeanette (Hill) McC.; m. Chantal Olivette Gleanzar, Oct. 25, 1958; children: Monique, Marthe, Marc, Marianne. B.A., U. Conn., 1953, M.S., 1955, Ph.D., 1957. Mem. tech. staff RCA Labs., Princeton, N.J., 1956-67; prof. chemistry Rider Coll., Lawrenceville, N.J., 1967—, chmn. dept., 1972-80. Mem. Lawrence Twp. Council, Lawrenceville, 1976; chmn. Lawrence Republican Orgn., Lawrenceville, 1967-69. Research Corp. grantee, 1980-81; NSF summer fellow, 1982. Mem. Am. Chem. Soc. (chmn. Trenton sect. 1977-79, 83), Sigma Xi. Republican. Presbyterian. Subspecialties: Solid state chemistry; 3emiconductors. Current work: Synthesis and characterization of transition metal oxides in reduced valence states with emphasis on molybdenum. Metal atom clusters in oxide systems. Home: 14 Monroe Ave Lawrenceville NJ 08648 Office: Rider Coll PO Box 6400 Lawrenceville NJ 08648

MCCARTER, RONALD STEPHEN, company executive; b. Hackett, Ark., Oct. 23, 1931; s. Joel Kirby and Emily Francis (Buttler) McC.; m. Estrella Gaces, Nov. 29, 1980; children by previous marriage: Gregory S., Phyllis Ann, Timothy G. B.S., Tex. A&M Coll., 1957; M.E.E., N.Y. U., 1959; postgrad., Stanford U., 1980. With Bell Telephone Labs., Whippany, N.J., 1957-70; with Teledyne Brown Engring., Huntsville, Ala., 1970-79; pres. Teledyne Ryan Aero., San Diego, 1979—. Served with U.S. Navy, 1949-53. Mem. Navy League U.S., AIAA, Am. Def. Prepardness Assn., Phi Eta Sigma, Tau Beta Pi, Eta Kappa Nu. Office: 2701 Harbor Dr San Diego CA 92138

MCCARTER, STATES MARION, plant pathology educator; b. York County, S.C., Sept. 30, 1937; s. Willie Meek and Ossie Emma (Melton) McC.; m. Martha Jane Howell, Oct. 16, 1963; children: Stephen, Elizabeth. B.S., Clemson U., 1959, M.S., 1961, Ph.D., 1965. Extension plant pathologist, nematologist Auburn (Ala.) U., 1965-66; research plant pathologist U.S. Dept. Agr., Tifton, Ga., 1966-68; asst. prof. plant pathology U. Ga., Athens, 1968-72, assoc. prof., 1972-77, prof., 1977—. Contbr. articles to profl. jours. Served to capt., Chem. Corps U.S. Army, 1963-65. Recipient Teaching awards Gamma Sigma Delta, 1978, U. Ga. Agrl. Alumni, 1978; U.S. Dept. Agr. grantee, 1980-81. Mem. Am. Phytopathol. Soc., Ga. Acad. Sci., Ga. Assn. Plant Pathologists. Republican. Methodist. Subspecialty: Plant pathology. Current work: Bacterial diseases of vegetables. Office: Dept Plant Pathology U Ga Athens GA 30602

MCCARTHY, DAVID MURRAY, physician, researcher; b. Morristown, N.J., Dec. 12, 1945; s. John Murray and Mary Francis (Apgar) McC.; m. Linda Frisa, Aug. 30, 1969; 1 dau., Alice. A.B., Yale U., 1967; M.D., Columbia U., 1971. Diplomate: Am. Bd. Internal Medicine, also Sub-Bd. Cardiology. Intern U. Chgo, Hosp., 1971-72; resident in medicine St. Luke's Hosp., N.Y.C., 1974-75; resident in cardiology Presbyn. Hosp., N.Y.C., 1975-77; research fellow in cardiology Columbia U., N.Y.C., 1977-79; asst. prof. medicine U. Pa.; dir. nuclear cardiology Hosp. of U. Pa., Phila., 1979—. Served with USPHS, 1972-74. S.E. Pa. Heart Assn. research grantee, 1982-83. Fellow Am. Coll. Cardiology; mem. Am. Fedn. Clin. Research, ACP, Soc. Nuclear Medicine. Subspecialties: Cardiology; Nuclear medicine. Current work: Nuclear imaging of cardiac structure and function, with particular interest in new methodology. Home: 1306 Knox Rd Wynnewood PA 19096 Office: 3400 Spruce St Philadelphia PA 19104

MCCARTHY, DENNIS DEAN, astronomer; b. Oil City, Pa., Sept. 22, 1942; s. William Henry and Evelyn Dorothy (Siembida) McC.; m. Diane Kay Wallingford, Sept. 17, 1966; children: Dunca Sean, Deidre Carrie. B.S., Case Inst. Tech., 1960; M.A., U. Va., 1970, Ph.D. in Astronomy, 1972. Astronomer U.S. Naval Obs., 1965-75, project leader, 1975-79, head astron. time and polar motion sect., 1979-82, chief earth orientation and spl. projects br., 1982—; vis. prof. Japan Soc. Promotion of Sci., 1979. Contbr. articles to profl. jours. Mem. Am. Astron. Soc., Am. Geophys. Union, Internat. Astron. Union. Subspecialties: Astrometry; Geodesy. Current work: Orientation and rotation of the earth; modeling and prediction of earth orientation parameters; application of modern technology to the determination of earth orientation. Home: 2432 Riviera Dr Vienna VA 22180 Office: US Naval Observatory Washington DC 20390

MC CARTHY, JOHN, meteorologist; b. New Orleans, July 3, 1942. B.A., Grinnell Coll., 1964; M.S., U. Okla., 1967; Ph.D. in Geophys. Sci., U. Chgo., 1973. Research meteorologist Weather Sci. Inc., 1966-67, Cloud Physics Lab., U. Chgo., 1967-68; asst. prof. meteorology U. Okla., 1973-78, assoc. prof., 1978-80, scientist Nat. Center Atmospheric Research, Boulder, Colo., 1980—; mem. univ. relations com. Univ. Corp. Atmospheric Research, 1974—. Mem. Am. Meteorol. Soc., AIAA, Sigma Xi. Subspecialty: Meteorology. Office: Nat Ctr Atmospheric Research PO Box 3000 Boulder CO 80307

MCCARTHY, JOHN, computer scientist, educator; b. Boston, Sept. 4, 1927; s. Patrick Joseph and Ida McC.; m.; children: Susan Joanne, Sarah Kathleen. B.S., Calif. Inst. Tech., 1948; Ph.D., Princeton U., 1951. Instr. Princeton U., 1951-53; acting asst. prof. math. Stanford U., 1953-55; asst. prof. Dartmouth Coll., 1955-58; asst. and assoc. prof. communications scis. M.I.T., Cambridge, 1958-62; prof. computer sci. Stanford U., 1962—; dir. Info. Internat. Inc. Served with AUS, 1945-46. Mem. Assn. Computing Machinery (A.M. Turing award 1971), Am. Math Soc., Am. Assn. Artificial Intelligence. (pres. 1983-84). Republican. Subspecialty: Artificial intelligence. Current work: Artificial intelligence, time-sharing computer systems, mathematical theory of computation, symbolic comuterization. Home: 846 Lathrop Dr Stanford CA 94305 Office: Dept Computer Sci Stanford U Stanford CA 94305

MC CARTHY, JOHN FRANCIS, JR., aeronautical engineer; b. Boston, Aug. 28, 1925; s. John Francis and Margaret Josephine (Bartwood) McC.; m. Camille Dian Martinez, May 4, 1968; children: Margaret I., Megan, Jamie M., Nicole E., John F. S.B., M.I.T., 1950, S.M. in Aero. Engring, 1951; Ph.D. in Aeros. and Physics, Calif. Inst. Tech., 1962. Supr. air/ground communications TWA, Rome, 1946-47; project mgr. aeroelastic and structures research lab. M.I.T., 1951-55, prof. aeros. and astronautics, 1971-78; ops. analyst Hdqrs. SAC, Offutt AFB, Nebr., 1955-59; dir., asst. chief engr. Apollo, Space div. N.Am. Aviation, Inc., Downey, Calif., 1961-66; v.p. Los Angeles div./Space div., Rockwell Internat. Corp., 1966-71; dir. and prof. M.I.T. Center for Space Research, 1974-78; dir. NASA Lewis Research Center, Cleve., 1978-82; v.p., gen. mgr. Electromech. div. Northrop Corp., 1982—; mem. Internat. Council Aero. Scis., Koln, Germany, 1978—, USAF Sci. Adv. Bd., Washington, 1970-82; mem. sci. adv. group Joint Chiefs of Staff, Joint Strategic Target Planning Staff, Offutt AFB, 1976-81; com. mem. Energy Engring. Bd., Assembly Engring., NRC, 1979; cons. in field.; cons. Office of Undersec. Def. for Research and Devel. Author numerous tech. reports. Campaign chmn. Downey Community Hosp., 1968-69; bd. govs. Nat. Space Club, Washington, 1978—; chmn. Fed. Exec. Bd., Cleve., 1979-80; mem. adv. bd. for dept. mech. engring., aero. engring. and mechanics Rensselaer Poly. Inst., 1981—. Served with USAAF, 1944-46. Recipient Apollo Achievement award NASA, 1969, Meritorious Civilian Service medal USAF, 1973, Exceptional Civilian Service medal, 1978, Disting. Service medal NASA, 1982. Fellow Am. Astronautical Soc., AIAA (dir. 1975-76), Royal Aero. Soc. (London); mem. Nat. Acad. Engring., Am. Soc. Engring. Edn. (exec. com. aerospace div. 1969-72), Am. Mgmt. Assn. (pres.'s council 1978—), Sigma Xi, Sigma Gamma Tau. Unitarian. Clubs: Cosmos, 50 of Cleve. Subspecialties: Aerospace engineering and technology; Aeronautical engineering. Current work: Engaged in research, development and manufacture of passive sensors, automatic test equiptment and aircraft components. Patentee impact landing system. Home: 19171 Via del Caballo Yorba Linda CA 92686 Office: Northrop Electro-Mech Div 500 E Orangethorpe Ave Anaheim CA 92801 Success is hard work and the guts to change careers when the going gets too easy.

MCCARTHY, JOSEPH GERALD, plastic surgeon, surgery educator, investigator; b. Lowell, Mass., Nov. 28, 1938; s. Joseph H. and Eva McC.; m. Karlan L. Sloan, June 6, 1964; children: Cara, Stephen. A.B., Harvard U., 1960; M.D., Columbia U., 1964. Diplomate: Am. Bd. Surgery, Am. Bd. Plastic Surgery. Intern Columbia-Presbyn. Med. Ctr., N.Y.C., 1964-65, resident in surgery, 1967-71; resident in plastic surgery NYU Med. Ctr., N.Y.C., 1971-73; instr. plastic surgery NYU Sch. Medicine, 1973-75, asst. prof. plastic surgery, 1975-78, assoc. prof. plastic surgery, 1978-81, Lawrence D. Bell prof. plastic surgery, 1981—; dir. Inst. Reconstructive Plastic Surgery, NYU Med. Ctr., 1981—; attending plastic surgeon Univ. Hosp.; vis. plastic surgeon, dir. service Bellevue Hosp.; attending surgeon Manhattan Eye, Ear and Throat Hosp.; assoc. attending physician VA Hosp., N.Y.C. Assoc. editor: Reconstructive Plastic Surgery, 7 vols, 1977; assoc. editor: Jour. Plastic and Reconstructive Surgery. Served to lt. comdr. USPHS, 1965-67. Am. Cancer Soc. fellow, 1969-70. Fellow ACS; mem. Am. Soc. Plastic and Reconstructive Surgeons, N.Y. Regional Soc. Plastic and Reconstructive Surgeons, Plastic Surgery Research Council, N.Y. Acad. Scis., Am. Cleft Palate Assn., Am. Assn. Plastic Surgeons. Clubs: Harvard of N.Y.C., Englewood (N.J.) Field.). Subspecialty: Surgery. Current work: Principal investigator craniofacial anomalies-etiology and treatment for NIH. Home: 150 Brayton St Englewood NJ 07631 Office: Inst Reconstructive Plastic Surgery 560 1st Ave New York NY 10016

MCCARTHY, ROBERT D., food science educator; b. Manoa, Pa., Sept. 25, 1932; s. David and Ida (Senger) McC; m. Jeanne L. Bertholf, June 16, 1956; children: Jeanne, Robert, Eda, Mary, Joseph, Jere. B.A., Pa. State U., University Park, 1954, M.S., 1956; Ph.D., U. Md.-College Park, 1958. Research assoc. U. Md., College Park, 1958; research assoc. Pa. State U., University Park, 1958-61, asst. prof., 1961-65, assoc. prof., 1965-71, prof. food sci., 1971—. Contbr. articles in field to profl. jours. Chmn. Planning Commn., Haines Twp., Pa., 1978—; pres., v.p. Our Lady of Victory Confrat. Christian Doctrine, State College, Pa., 1965-72. Mem. Am. Dairy Sci. Assn., Am. Inst. Nutrition, N.Y. Acad. Sci., Sigma Xi, Phi Kappa Phi, Gamma Sigma Delta. Republican. Lodge: KC. Subspecialties: Food science and technology; Nutrition (medicine). Current work: Quantitatively minor food factors with physiological effects, influencing cholesterol, metabolism, hypertension, etc. Home: RD Aaronsburg PA 16820 Office: Pa State U 114 Borland Lab University Park PA 16802

MCCARTY, LESLIE PAUL, pharmacologist, chemist; b. Detroit, May 30, 1925; s. Leslie Evert and Ruth Winifred (Clouse) McC.; m. Wanda Mae Brown, Sept. 8, 1948 (div. Idiv. 1974); children: Michael, Patricia, Maureen, Brian; m. Marie Jeanne Beullay, May 8, 1976. B.S., Salem Coll., 1947; M.S., Ohio State U., 1949; Ph.D., U. Wis.-Madison, 1960. Research chemist Upjohn Co., Kalamazoo, 1949-55; pharmacologist Dow Chem. Co., Midland, Mich., 1960—. Contbr. writings to profl. publs. Served with USNR, 1943-45. Mem. AAAS, Am. Soc. Pharmacology and Exptl. Therapeutics, Sigma Xi. Subspecialties: Pharmacology; Neuropharmacology. Current work: Pharmacology, toxicology. Patentee. Home: 4588 S Flajole Rd Midland MI 48640 Office: U S Area Med Dow Chemical Co Bldg 1803 Midland MI 48640

MCCARTY, MACLYN, medical scientist; b. South Bend, Ind., June 9, 1911; s. Earl Hauser and Hazel Dell (Beagle) McC.; m. Anita Alleyne Davies, June 20, 1934 (div. 1966); children: Maclyn, Richard E., Dale, Colin; m. Marjorie Steiner, Sept. 3, 1966. A.B., Stanford U., 1933; M.D., Johns Hopkins U., 1937; Sc.D., Columbia, 1976, U. Fla., 1977, Rockefeller U., 1982. House officer, asst. resident physician Johns Hopkins Hosp., 1937-40; asso. Rockefeller Inst., 1946-48, asso. mem., 1948-50, mem., 1950—, prof., 1957—, v.p., 1965-78, phys. in chief to hosp., 1961-74; research in streptococcal disease and rheumatic fever.; Cons. USPHS, NIH. Mem. distbn. com. N.Y. Community Trust, 1966-74; chmn. Health Research Council City N.Y., 1972-75; Mem. bd. trustees Helen Hay Whitney Found. Served with Naval Med. Research Unit, Rockefeller Hosp. USNR, 1942-46. Fellow medicine N.Y. U. Coll. Medicine, 1940-41; NRC fellow med. scis. Rockefeller Inst., 1941-42; Recipient Eli Lilly award in bacteriology and immunology, 1946, 1st Waterford Biomed. Research award, 1977. Mem. Am. Soc. for Clin. Investigation, Am. Assn. Immunologists, Soc. Am. Bacteriologists, Soc. for Exptl. Biology and Medicine (pres. 1973-75), Harvey Soc. (sec. 1947-50, pres. 1971-72), N.Y. Acad. Medicine, Assn. Am. Physicians, Nat. Acad. Scis., Am. Acad. Arts and Scis., N.Y. Heart Assn. (1st v.p. 1967, pres. 1969-71), Am. Philos. Soc. Subspecialties: Microbiology (medicine); Infectious diseases. Current work: Pneumococcal transformation and discovery of genetic role of DNA; streptococcal infections and pathogenesis of rheumatic fever. Home: 500 E 63d St New York NY 10021 Office: Rockefeller U 66th St and York Ave New York NY 10021

MCCAULEY, JOHN FRANCIS, astrogeologist, government research administrator; b. N.Y.C., Apr. 2, 1932; m., 1956; 3 children. B.S., Fordham U., 1953; M.A., Columbia U., 1957, Ph.D. in Geology, 1959. Coop. geologist Pa. Geol. Survey, 1956-58; asst. prof. U. S.C., 1958-62, assoc. prof., 1962-63; geologist U.S. Geol. Survey, 1963-70, chief br. astrogeol. studies, 1970—; lectr. Columbia U., 1957-58; project geologist div. geology S.C. Devel. Bd., 1958-63; cons. geology, 1958-63; co-investigator Mars Mariner TV Team, 1971. Fellow Geol. Soc. Am.; mem. Am. Astron. Soc. Subspecialties: Planetology; Geology. Address: 60 Wilson Ln Flagstaff AZ 86001

MCCLAMROCH, N. HARRIS, aerospace engineer, educator; b. Houston, Oct. 7, 1942; s. N. Harris and Dorothy (Or) McC.; m. Margaret Hobart, Aug. 10, 1963; 1 dau., Kristin. B.S., U. Tex.-Austin, 1963, M.S., 1965, Ph.D., 1967. Asst. prof. to prof. aerospace engring. U. Mich.-Ann Arbor, 1967—; research engr. Delft (Netherlands) U., 1976, Sandia Corp., Albuquerque, 1977, C.S. Draper Lab., Inc., Cambridge, Mass., 1982. Author: State Models of Dynamic Systems, 1980. Recipient Disting. Service Award U. Mich., 1971. Mem. IEEE, Soc. of Indsl. and Applied Math., AAAS. Subspecialties: Systems engineering; Robotics. Current work: Research into control of robotic manipulators, large flexible space structures. Home: 4056 Thornoaks Ann Arbor MI 48104 Office: Dept Aerospace Engring Mich Ann Arbor MI 48109

MCCLELLAN, ROGER ORVILLE, scientist, research mgr.; b. Tracy, Minn., Jan. 5, 1937; s. Orville and Gladys Lavern (Paulson) McC.; m. Kathleen Mary Dunagan, June 23, 1962; children—Eric John, Elizabeth Christine, Katherine Ruth. D.V.M. with highest honors, Wash. State U., 1960; M.Mgmt., Robert O. Anderson Grad. Sch. Mgmt., U. N.M., 1981. Diplomate: Am. Bd. Vet. Toxicology (pres. 1970-73); cert. Am. Bd. Toxicology. Research asst. dept. vet. microbiology Wash. State Univ., Pullman, 1957-60; jr. biol. scientist Biology Lab., Hanford Labs., Gen. Electric Co., Richland, Wash., 1957-58, biol. scientist, 1959-62, sr. scientist, 1963-64; sr. scientist biology dept. Pacific N.W. Labs., Battelle Meml. Inst., Richland., 1965; scientist med. research br., div. biology and medicine AEC, Washington, 1965-66; asst. dir. research, dir. fission product inhalation program Lovelace Found. for Med. Edn. and Research, Albuquerque, 1966-73; v.p., dir. research adminstrn., dir. Inhalation Toxicology Research Inst., 1973-76; dir. Inhalation Toxicology Research Inst., pres. Lovelace Biomed. and Environ. Research Inst., Albuquerque, 1976—; adj. prof. dept. radiation sci. Sch. Pharmacy, U. Ark. Med. Sch., Little Rock, 1970—; clin. asso. dept. pathology U. N.Mex. Sch. Medicine, Albuquerque, 1971—, adj. prof. dept. biology, 1973—; adj. prof. dept. vet. and comparative anatomy, pharmacology and physiology, mem. grad. faculty program vet. sci. Coll. Vet. Medicine, Wash. State U., 1980—; mem. Joint Space Nuclear Systems/Biomed. and Environ. Research Working Group, 1967-73; cons. Nat. Inst. Environ. Health Scis., NIH, 1968-71, mem. toxicology study sect., 1969-73; mem. subcom. on whole animal radiobiology and pathology Los Alamos Meson Physics Facility, 1970-75; mem. biomed. adv. com. on health effects for reactor safety study to Nuclear Regulatory Commn., 1975-76; mem. Nat. Council on Radiation Protection and Measurements, 1971—; chmn. Ad Hoc Rev. Com. on Sci. Criteria for Environ. Lead, 1977-78; mem. exec. com. sci. adv. bd. EPA, 1974-80, mem. Sci. adv. bd., 1974—, mem. environ. health com. sci. adv. bd., 1980—, chmn., 1981—; mem. Ad Hoc Clean Air Sci. Adv. Com. on Health Effects of Sulfur Oxide and Particulate Matter, 1980—, N. Am. Late Effects Group Steering Com., 1974—; mem. com. toxicology Nat. Acad. Sci. NRC, 1979—, chmn., 1980—, ad hoc mem. bd. toxicology and environ. health hazards, 1980—; mem. com. on animal models for research on aging Inst. Lab. Animal Resources, 1977-80, also chmn. subcom. on carnivores, 1977-80; mem. dose assessment adv. group Dept. Energy, 1980—; mem. health research com. Health Effects Inst., 1981—. Asso. editor: Lab. Animal Sci, 1976-80; mem. editorial bd.: Jour. Toxicology and Environ. Health, 1980—. Recipient Elda E. Anderson award Health Physics Soc., 1974. Fellow AAAS; mem. Radiation Research Soc. (chmn. fin. com. 1979—), Health Physics Soc. (program com. 1970-73, chmn. 1972), Am. Coll. Vet. Toxicologists (councilman 1968-71), AVMA (adv. bd. on vet. spltys. 1973-76), N.Mex. Vet. Med. Assn., N.Mex. Zool. Soc. (dir. 1970-72), Wash. State Vet. Med. Assn., Soc. Exptl. Biology and Medicine, Soc. Toxicology (com. on legis. assistance 1978-80, program com. 1980—), Soc. for Risk Analysis, Sigma Xi, Phi Kappa Phi, Phi Zeta, Alpha Psi. Lutheran. Subspecialties: Toxicology (medicine); Environmental toxicology. Current work: Evaluation of toxicity of emissions of energy technologies (nuclear, coal combustions and gasification, solar, conservations, automotive) with emphasis on inhalation hazards and risk to man. Home: 1111 Cuatro Cerros SE Albuquerque NM 87123 Office: PO Box 5890 Albuquerque NM 87185

MCCLENDON, JOHN HADDAWAY, plant physiologist, educator; b. Mpls., Jan. 17, 1921; s. Jesse Francis and Margaret (Stewart) McC.; m. Betty Morgan, Nov. 16, 1923; children: Susan, Lise, Natalie. B.A., U. Minn., 1942; Ph.D., U. Pa., 1951. Research assoc. Stanford (Calif.) U., 1951-52; U. Minn., Mpls., 1952-53; asst. prof. dept. agrl. biochemistry U. Del., Newark, 1952-64; assoc. prof. dept. botany U. Nebr., Lincoln, 1965—. Contbr. articles to sci. jours. Chmn. Nebr. chpt. Zero Population Growth, 1970-73; mem. Nebr. Environ. Coalition. Served to lst lt. U.S. Army, 1943-46. NSF grantee, 1960, 65. Mem. Am. Soc. Plant Physiologists, AAAS, Am. Inst. Biol. Scis., Bot. Soc. Am., Sierra Club, Wilderness Soc., Sigma Xi. Democrat. Subspecialties: Plant physiology (biology); Evolutionary biology. Current work: Photosynthesis in trees; origin of life. Home: 1970 B St Lincoln NE 68502 Office: U Nebr Lincoln NE 68588

MCCLINTOCK, BARBARA, geneticist; b. June 16, 1902. Ph.D. in Botany, Cornell U., 1927; D.Sc. (hon.), U. Rochester, U. Mo., Smith Coll., Williams Coll., Western Coll. for Women. Instr. botany Cornell U., Ithaca, N.Y., 1927-31, research assoc., 1934-36, Andrew D. White prof.-at-large, 1965—; asst. prof. U. Mo., 1936-41; mem. staff Carnegie Instn. of Washington, Cold Spring Harbor, N.Y., 1941-47, Disting. Service mem., 1967—; cons. agrl. sci. program Rockefeller Found., 1962-69. Recipient Achievement award Assn. Univ. Women, 1947, Nat. medal of Sci., 1970, MacArthur Found. prize; Rosenstiel award, 1978, Nobel prize, 1983; NRC fellow, 1931-33; Guggenheim Found. fellow, 1933-34. Mem. Nat. Acad. Scis. (Kimber genetics award 1967), Am. Philos. Soc., Am. Aad. Arts and Scis., Genetics Soc. Am. (pres. 1945), Bot. Soc. Am. (award of merit 1957), AAAS, Am. Inst. Biol. Sci., Am. Soc. Naturalists. Subspecialty: Gene actions. Office: Carnegie Instn of Washington Cold Spring Harbor NY 11724

MCCLURE, CARL KENNETH, research scientist, mech. engr.; b. Roaring Spring, Pa., Nov. 23, 1936; s. William Warren and Ida Belle (Kimmel) McC.; m. Joyce Elaine Montgomery, June 22, 1959; children: Susan Amelia, Dean Maurice, Hugh Charles. B.S.M.E., Pa. State U., 1959; Ph.D. in Engring. Mechanics, 1972; M.S.M.E., Drexel U., 1962. Registered profl. engr., Pa. Engr. Philco-Ford Corp., Phila., 1959-62; instr. Pa. State U., University Park, 1964-69; sr. research scientist Armstrong World Industries, Inc., Lancaster, Pa., 1969—. Contbr. articles to profl. jours. Docent North Mus., Franklin and Marshall Coll., Lancaster, Pa., 1979—. Served to capt. U.S. Army, 1962-68. Mem. ASME, Soc. for Exptl. Stress Analysis. Methodist. Subspecialties: Solid mechanics; Mechanical engineering. Current work: Application of solid mechanics to research directed toward the development of plastic and elastomeric products. Vibrations and impact mechanics of composite materials, and material testing techniques. Office: PO Box 3511 Lancaster PA 17604

MC CLURE, DONALD STUART, physical chemist, educator; b. Yonkers, N.Y., Aug. 27, 1920; s. Robert Hirt and Helen (Campbell) McC.; m. Laura Lee Thompson, July 9, 1949; children: Edward, Katherine, Kevin. B.Chemistry, U. Minn., 1942; Ph.D., U. Calif., Berkeley, 1948. With war research div. Columbia U., 1942-46; mem. faculty U. Calif., Berkeley, 1948-55; group leader, mem. profl. staff RCA Labs., 1955-62; prof. chemistry U. Chgo., 1962-67, Princeton U., 1967; vis. lectr. various univs.; cons. to govt. and industry. Author: Electronic Spectra of Molecules and Ions in Crystals, 1959, Some Aspects of Crystal Field Theory, 1964; also articles. Guggenheim fellow Oxford (Eng.) U., 1972-73; Humboldt fellow, 1980; recipient Irving Langmuir prize, 1979. Fellow Am. Acad. Arts and Scis., Nat. Acad. Scis.; mem. Am. Chem. Soc., Am. Phys. Soc. Subspecialties: Spectroscopy; Photochemistry. Current work: Spectroscopy and Photochemistry of inorganic and organic compounds and crystals. Home: 23 Hemlock Circle Princeton NJ 08540

MCCLURE, JAMES DOYLE, laser technologist; b. Clayton, N.Mex., June 23, 1935; s. Robert Roy McClure and Wilma (Carpenter) Frost;

m. Moyra C.T. McAdam, Oct. 26, 1957 (div. 1981); children: Karen Allyn, Robert Kevin. B.S., U. Wash., 1956, M.S., 1958; Ph.D., MIT, 1962. Sr. research scientist Boeing Sci. Research Labs., Seattle, 1962-71, Boeing Aerospace Co., 1971-81, mgr. advanced laser tech., 1981—. Contbr. articles to profl. publs. Boeing fellow MIT, 1958; DuPont fellow U. Wash., 1956. Mem. Am. Inst. Physics, AIAA. Current work: Research and development of advanced lasers for defense and communications. Office: Boeing Aerospace Co/MS8H-29 PO Box 3999 Seattle WA 98124

MCCLURE, WILLIAM ROBERT, biological sciences educator; b. Detroit, Dec. 30, 1941; s. Robert Douglas and Catherine (Coffin) McC.; m. Donna Lee DeRight, Dec. 20, 1966; children: Janice, Katherine. B.S., U. Alaska, 1966; Ph.D., U. Wis.-Madison, 1970. Postdoctoral fellow Max-Planck Inst., Gottingen, W.Ger., 1970-73; asst. prof. Harvard U., Cambridge, Mass., 1973-78, assoc. prof., 1978-81, Carnegie-Mellon U., Pitts., 1981-82, prof. biol. scis., 1982—; cons. NSF, 1981—. Contbr. research articles to profl. publs. Helen Hay Whitney Found. postdoctoral fellow, 1970-73. Mem. Am. Soc. Biol. Chemists, Am. Chem. Soc., Biophys. Soc., Am. Soc. Microbiology, AAAS. Subspecialties: Biochemistry (biology); Molecular biology. Current work: Polynucleotide enzymology; mechanism and regulation of E. Coli, RNA polymerase. Home: 716 East End Ave Pittsburgh PA 15221 Office: Carnegie-Mellon Univ 4400 Fifth Ave Pittsburgh PA 15213

MCCLUSKEY, GEORGE EADON, JR., astronomer, educator, researcher; b. Hammonton, N.J., Aug. 28, 1938; s. George Eadon and Eleanor (Cappuccio) McC.; m. Carolina Paciencia Salas, Apr. 28, 1979. Ph.D. in Astronomy, U. Pa., 1965. Faculty Lehigh U., Bethlehem, Pa., 1965—, prof. astronomy, 1976—; guest investigator Copernicus satellite, 1973-76, astron. Netherlands satellite, 1976—, internat. ultraviolet Explorer satellite, 1978—. Contbr. articles to astron. jours. NASA grantee, 1966-67, 73-76, 78—. Mem. Am. Astron. Soc., AAAS, Internat. Astron. Union, Royal Astron. Soc. Subspecialties: Theoretical astrophysics; Satellite studies. Current work: Close binary stars, ultraviolet astronomy, x-ray astronomy, cosmology, relativity. Office: Div Astronomy Dept Math Christmas-Saucon Hall Bldg 14 Lehigh U Bethlehem PA 18015

MCCOLL, JOHN DUNCAN, pharmacologist, consultant; b. London, Ont., Can., Nov. 11, 1925; s. Gordon and Mary Rosamund (Clunis) McC.; m. Patricia Amy Ridont, May 29, 1954; children: Pamela, Susan, Gordon. B.A., U. Western Ont., 1946, M.S., 1950; Ph.D., U. Toronto, 1953. Asst. dir. research Frank W. Horner Ltd., Montreal, Que., Can., 1950-69; v.p. biol. scis. Mead-Johnson Research Center, Evansville, Ind., 1969-75; v.p., dir. research Chattem Labs., Inc., Chattanooga, 1975-83; pres. McColl & Assocs., Inc., Chattanooga, 1983—. Contbr. numerous articles on pharmacology, toxicology and biochemistry to profl. jours. Served to capt. Can. Armed Forces, 1943-46. Mem. Am. Soc. Pharmacology and Exptl. Therapeutics, Soc. of Toxicology, Am. Chem. Soc., Biochem. Soc. (Gt. Brit.), Am. Soc. Clin. Pharmacology and Therapeutics, Soc. Toxicology of Can., Deutsch Pharmakologische Gesellschaft, European Soc. Study of Drug Toxicity, Pharm. Soc. Can. Episcopalian. Club: Walden (Chattanooga). Subspecialties: Pharmacology; Toxicology (medicine). Current work: Clinical pharmacology; consulting. Patentee in field. Home and Office: 901 Brynwood Dr Chattanogga TN 37415

MCCOLLOUGH, MICHAEL LEON, astronomy educator; b. Sylva, N.C., Nov. 3, 1953; s. Stribling Mancell and Vivian Hazel (Bradley) McC. B.S., Auburn U., 1975, M.S., 1981; Ph.D. candidate, Ind. U. Lab. instr. Auburn (Ala.) U., 1974-75, grad. asst., 1975-77, lab. technician, 1977-78; assoc. instr. Ind. U., Bloomington, 1978—. Mem. Sigma Pi Sigma, Am. Astron. Soc., Royal Astron. Soc., Astron. Soc. Pacific, Am. Phys. Soc., Optical Soc. Am., Am. Assn. Physics Tchrs., Soc. Physics Students. Baptist. Subspecialties: Optical astronomy; Theoretical astrophysics. Current work: Reduction of observations of active galaxies, computer modeling of supernova remnants and modeling of effects of turbulence in stellar atmospheres. Home: 988 Eigenmann Hall Bloomington IN 47401 Office: Dept Astronomy Ind Univ 319 Swain W Bloomington IN 47401

MCCOLM, DOUGLAS WOODRUFF, physics educator, researcher; b. Kisaran, Sumatra, Sept. 26, 1933; m., 1955; 3 children. B.A., Oberlin Coll., 1955; Ph.D. in Physics, Yale U., 1961. Staff physicist Lawrence Radiation Lab., U. Calif.-Berkeley, 1961-66; assoc. prof. physics U. Calif.-Davis, 1966—, now also chmn. dept. physics. Mem. Am. Phys. Soc. Subspecialty: Atomic and molecular physics. Office: Dept Physics U Calif 1000 Plum Ln Davis CA 95616

MCCOMBE, BRUCE DOUGLAS, physicist, government research administrator; b. Sanford, Maine, Mar. 2, 1938; m., 1963; 1 child. A.B., Bowdoin Coll., 1960; Ph.D. in Physics, Brown U., 1966. Research assoc. semicondr. physics NRC/Naval Research Lab., 1965-67; research physicist Naval Research Lab., Washington, 1967-70, acting head surface and transp. sect., 1970, head semicondrs. br., 1974-80, supr. electronics tech. div., 1980—; mem. ad hoc com. elec. properties NRC/Nat. Acad. Sci., 1970, mem. ad hoc panel high magnetic field research and facility, 1977-78; Naval Research Lab. sabbatical study program participant Max Planck Inst. Solid State Research, Stuttgart, Germany, 1972-73; Navy mem. tech. rev. com. Joint Service Electronics Program, Dept. Def., 1974—; mem. various program coms. internat. conf., 1974—; adj. prof. physics SUNY-Buffalo, 1978; mem. materials research adv. com. NSF, 1978—. Recipient Pure Sci. award Sci. Research Soc. Am., 1972. Fellow Am. Phys. Soc.; mem. Sigma Xi. Subspecialty: Solid state physics. Office: Electronics Tech Div Code 6800 Naval Research Lab Washington DC 20375

MCCOMBS, BARBARA LEONA, research psychologist, consultant; b. Galesburg, Ill., Nov. 30, 1942; d. Raymond Louis and Emily Marie (Schulz) Adolf; m. E. Scott Leherissey, Dec. 19, 1963 (div. Mar. 1970); m. George M. McCombs, Aug. 19, 1972 (div. June 1980); children: Heather Rae, Ryan William. B.S., Fla. State U., 1968, M.S., 1969, Ph.D., 1971. Cert. ednl. psychologist, Mo. Research asst. Computer Assisted Instrn. Ctr. Fla. State U., 1969-70, research assocc., 1970-71; chief human factors McDonnell Douglas Corp., Denver, 1971-82; sr. research psychologist Denver Research Inst., 1982-; cons. Fla. State U., 1982—. Editor: Handbook of Research on Teaching, 1982. U.S. Dept. Edn. fellow, 1968-69. Mem. Am. Psychol. Assn., Am. Ednl. Research Assn., Human Factors Soc. Lutheran. Subspecialties: Learning; Cognition. Current work: Individualized instruction, individual differences, cognitive and affective learning strategies, computer based learning, instructor roles. Home: 1050 S Monaco Pkwy Unit 66 Denver CO 80224 Office: Denver Research Institute SSRE Div U Denver Denver CO 80208

MCCOMBS, CANDACE CRAGEN, immunologist, researcher; b. Springfield, Ill., Jan. 30, 1947; d. Francis H. and Blanche (McVey) Cragen; m. Lawrence M. McCombs, Dec. 23, 1964 (div.); 1 dau., Jeanette Roxanne. B.S., A.B., Stanford U., 1969; Ph.D., U. Calif.-Berkeley, 1974. Research immunologist VA Center, Long Beach, Calif., 1978-82; asst. prof. medicine La. State U. Med. Center, New Orleans, 1982—. Contbr. articles in field to profl. jours. Recipient Research Career Devel. award NIH, 1981-86; Arthritis Found. research fellow, 1978-81. Mem. Am. Assn. Immunologists, Am. Soc. Human Genetics, Am. Heart Assn. (mem. basic sci. council). Subspecialties: Immunogenetics; Transplantation. Current work: Genetic predisposition to autoimmune disease, HLA and disease associations, a rapid MLC test for organ transplantation, regulation of immune responses. Home: Route 5 Box 169K Covington LA 70433 Office: Dept Medicine La State U Med Center 1542 Tulane Ave New Orleans LA 70112

MCCOMBS, HARRIET G., social psychologist, educator; b. Columbia, S.C., Nov. 9, 1954; d. William F. and Harriet C. McC. B.S., U. S.C., 1974; M.A., U. Nebr., 1976, Ph.D., 1978. Instr. U. Nebr., Lincoln, 1975-77; congressional intern U.S. Ho. of Reps., Washington, 1976; asst. prof. Wayne State U., Detroit, 1978-82; asst. prof. psychology Yale U., New Haven, 1983—. Editor: newsletter Sojourner, 1977-82. Evaluator State Crime Commn., Lincoln, 1976-77, Wayne County Children's Center, Detroit, 1979-82. Psychol. Assn. minority fellow, 1975-78. Am. ; Mem. Am. Psychol. Assn., Evaluation Research Soc., Sigma Xi, Sigma Gamma Rho. African Methodist. Subspecialty: Social psychology. Current work: Application of social psychological principles to the study and improvement of the mental health of individuals and communities. Office: Yale Child Study Center 333 Cedar St New Haven CT 06510

MCCOMBS, ROLLIN KOENIG, radiation oncologist; b. Denver, Aug. 17, 1919; s. Curtis and Emma Elizabeth (Koenig) McC.; m. Judy Louise Bacon, Oct. 22, 1924; children: David, Daniel, Susan, Kathleen, Michael. B.A., U. Colo., 1941, M.A., 1944; M.D., Stanford U., 1954. Diplomate: Am. Bd. Radiology, Am. Bd. Nuclear Medicine; lic. physician, Calif., Ariz. Instr. physics U. Colo., Boulder, 1942-48; intern in surgery Stanford U. Hosps., San Francisco, 1953-54; research fellow, assoc. physician Donner Lab., U. Calif., Berkeley, 1954-57; resident, staff VA Hosp., Long Beach, Calif., 1957-67; radiation oncologist Long Beach Community Hosp., 1967—; asst. clin. prof. radiology U. So. Calif., Los Angeles, 1978—; Meml. Fund lectr. Radiol. Soc. N.Am., 1956. Contbr. articles to profl. jours. Mem. Am. Assn. Physicists in Medicine, Brit. Inst. Radiology, Am. Coll. Radiology, Am. Coll. Nuclear Medicine, Am. Soc. Therapeutic Radiologists, AMA, Phi Beta Kappa, Sigma Xi, Alpha Chi Sigma, Pi Mu Epsilon, Sigma Pi Sigma. Presbyterian. Club: Frank Isaac Meml. Soc. Lodge: Masons. Subspecialties: Radiology; Nuclear medicine. Current work: Particle radiotherapy, med. physics, radiation dosimetry, ferroki.

MCCOMMAS, STEVEN ANDREW, zoologist; b. Amarillo, Tex., Sept. 7, 1949; s. Rufus Andrew and Beverly Jean (Strickl) McC. Student, Tex. Christian U., 1970, Pan Am. U., 1970; B.A., U. Tex., 1972; Ph.D., U. Houston, 1982. Research asst. U. Tex. Marine Sci. Ctr., Port Aransas, summers 1973, 74, 76; teaching asst. U. Tex., Austin, 1973-75, U. Houston, 1978, research assoc. marine sci. program, 1981—. Contbr. articles to profl. jours. Moody Found. fellow, 1979-80. Mem. Soc. Study of Evolution, Genetics Soc. Am., Soc. Systematics Zoologists, Am. Soc. Zoologists, Electrophoresis Soc., Phi Kappa Phi. Subspecialties: Evolutionary biology; Genome organization. Current work: Mechanisms of speciation, combining population genetics and quantitative genetics approaches. Office: Marine Sci Program U Houston 4700 Ave U Galveston TX 77550

MCCONAHEY, PATRICIA JANE, med. research scientist; b. Dover, Ohio, June 9, 1936; d. Wallace Veigh and Edith Elizabeth McC. B.S., Waynesburg (Pa.) Coll., 1958. Sr. Technician dept. pathology U. Pitts. Sch. Medicine, 1958-61; research asst. dept. immunopathology Scripps Clinic and Research Found., La Jolla, Calif., 1961-63, research assoc., 1964-74, asst. mem. dept. immunology, 1974—; cons. Johnson & Johnson Co., 1981—. Contbr. articles to profl. jours. Brown Hazen grantee, 1968. Mem. Am. Assn. Immunologists. Subspecialty: Immunology (medicine). Current work: Genetics influencing autoimmunity as observed in mouse model systems. Home: 8882 Caminito Primavera La Jolla CA 92137 Office: 10666 N Torrey Pines Rd La Jolla CA 92037

MCCONKEY, JOHN WILLIAM, physicist, educator; b. Portadown, Northern Ireland, Feb. 20, 1937; s. James William and Matilda Jean (Lawson) McC.; m. Maureen Greer, Aug. 31, 1963; children: Ruth, Andrew, Deborah, Christa. B.Sc., Queens U., Belfast, Northern Ireland, 1958, Ph.D., 1962. Lectr. Queen's U., 1962-70; vis. research fellow Sorbonne, Paris, 1964; prof. physics U. Windsor, Ont., Can., 1970—. Contbr. articles to profl. jours. Royal Soc. Commonwealth fellow, 1976-77; NATO sr. fellow, 1976-77; recipient numerous grants from Can. govt. and industry. Fellow Brit. Inst. Physics; Mem. Can. Assn. Physicists, Am. Phys. Soc. Baptist. Subspecialty: Atomic and molecular physics. Current work: Electron and photon interactions with atoms and molecules.

MCCONNELL, DENNIS BROOKS, horticulturist, educator; b. Waupun, Wis., Aug. 18, 1938; s. Robert Brooks and Edith Carol (Erickson) McC.; m. Ruth Ann Bickel, Nov. 28, 1964; children: Michael Brooks, Sharon Kay. B.S., U. Wis.-River Falls, 1966; M.S., U. Wis.-Madison, 1968, Ph.D., 1971. State extension foliage specialist Agrl. Research Ctr., U. Fla., Apopka, 1970-73, mem. faculty dept. ornamental horticulture, Gainesville, 1973—, asst. prof. ornamental horticulture, 1970-75, assoc. prof., 1975—. Author: (with C. A. Conover and R. W. Henley) Professional Guide to Green Plants, 1976, The Indoor Gardener's Companion, 1978; contbr. chpts., articles, abstracts to profl. publs.; editor: (with P. L. Neel and J. T. Midcap) Florida Nurserymen's Retail Sales Handbook, 1974, (with G. S. Smith and J. T. Midcap) Florida Landscape Installation Handbook), 1975, Florida Landscape Maintenance Handbook, 1975. Served with U.S. Army, 1960-63. Florist Trans-World grantee, 1975; Dept. Agr. grantee, 1978; Foliage Edn. and Research Found. grantee, 1980; Sea-Land Corp. grantee, 1982. Democrat. Methodist. Club: Gainesville Torch. Subspecialties: Plant growth; Morphology. Home: 3745 SW 3d Pl Gainesville FL 32607 Office: Dept Ornamental Horticulture 1541 HS/PP Bldg U Fla Gainesville FL 32611

MCCONNELL, HARDEN MARSDEN, chemistry educator, researcher; b. Richmond, Va., July 18, 1927; m. Sofia Glogovac, 1956; Hunter, Trevor, Jane. B.S., George Washington U., 1947; Ph.D. in Chemistry, Calif. Inst. Tech., 1951. NRC fellow in physics U. Chgo., 1950-52; research chemist Shell Devel. Co., 1952-56; from asst. prof. to prof. chemistry Calif. Inst. Tech., 1956-64; prof. Stanford U., 1964-79, Robert Eckles Swain prof. chemistry, 1979—; cons. Exxon Corp., Becton, Dickinson & Co.; Harkins lectr. U. Chgo., 1967; Falk-Plaut lectr. Columbia U., 1967; 21st Renaud Found. lectr., 1971; Debye lectr. Cornell U., 1973; Harvey lectr. Rockefeller U., 1977; A.L. Patterson lectr. Inst. Cancer Research, U. Pa., 1978; 8th Pauling lectr. Stanford U., 1981; vis. com. dept. chemistry MIT, 1980-81; vis. com. dept. biology Carnigie-Mellon U., 1983—. Mem. editorial bd.: Bio Organic Chemistry; mem. editorial bd.: Jour. Membrance Biology, Jour. Supra Molecular Structure. Recipient Alumni Achievement awrd George Washington U., 1971, Dickson Prize for Science Carnegie-Mellon U., 1982, Disting. Alumni award Calif. Inst. Tech., 1982. Fellow Am. Phys. Soc., AAAS; mem. Nat. Acad. Sci., Am. Soc. Biol. Chemists, Biophys. Soc., Am. Chem. Soc. (Pure Chemistry award 1962, Harrison Howe award 1968, Irving Langmuir award 1971), Swedish Biophys. Soc. (hon. mem.). Subspecialty: Biophysical chemistry. Home: 421 El Escarpado Stanford CA 94305 Office: Dept Chemistry Stanford U Stanford CA 94305

MCCONNELL, JOHN CHARLES, physics educator; b. Belfast, Northern Ireland, Sept. 11, 1945; s. John Douglas and Margaret Mary (Johnston) McC.; m. Winifred Elizabeth Davidson, Aug. 8, 1968; children: Deirdre Louise, Alison Joyce. B.Sc. with 1st class honors, Queen's U., Belfast, 1966, Ph.D., 1969. Research assoc. Kitt Peak Natl. Obs., Tucson, 1969-70; research fellow div. engring. and applied physics Harvard U., Cambridge, Mass., 1970-72; asst. prof. dept. physics faculty sci. York U., Downsview, Ont., Can., 1972-75, assoc. prof., 1975-80, prof. atmospheric sci., 1980—. Contbr. articles in field to profl. jours. Recipient awards Natural Scis. and Engring. Research Council, 1972—, Atmospheric Environment Service, 1975—. Mem. Am. Geophys. Union, Canadian Assn. Physicists, Am. Astron. Soc. Club: Royal Scottish Country Dance Soc. Subspecialties: Planetary science; Aeronomy. Current work: Physics and chemistry of planetary atmospheres, radiative transfer, planets, stratosphere, aurora, aeronomy. Office: Dept Earth and Atmospheric Sci York U 4700 Keele St Downsview ON Canada M3J 1P3

MCCONNELL, WILLIAM RAY, toxicologist; b. Wise, Va., Oct. 15, 1943; s. William and Rebecca Ruth (Stallard) McC.; m. Barbara Hicks, June 24, 1967. B.S., Va. Poly. Inst. and State U., 1965; M.S., Med. Coll. Va., 1973, Ph.D., 1976. Research pharmacologist So. Research Inst., Birmingham, Ala., 1976-79; sr. research toxicologist A.H. Robins Co., Inc., Richmond, Va., 1979—. Contbr. articles to profl. jours. Mem. AAAS, Am. Soc. Pharmacology and Exptl. Therapeutics. Republican. Baptist. Subspecialties: Toxicology (medicine); Pharmacology. Current work: Safety testing of pharm. drugs. Office: 1211 Sherwood Ave PO Box 26609 Richmond VA 23261

MCCOOL, PATRICK MICHAEL, plant pathologist, soil microbiologist; b. Santa Ana, Calif., Dec. 31, 1952; s. William Douglas and Ruth Marilyn (Stentz) McC.; m. Wendy Lane Kobaly, Dec. 4, 1982. B.S., U. Calif., Irvine, 1975, M.S. in Plant Pathology, 1977, Ph.D., 1981. Research asst. dept. plant pathology U. Calif., Riverside, 1976-78; research asst. Statewide Air Pollution Research Center, 1978-81, asst. research plant pathologist, 1981—, assoc. dept. plant pathology, 1981—. Contbr. articles to profl. jours. Calif. Dept. Food and Agr. grantee, 1981-82, 82—. Mem. Am. Phytopath. Soc. Presbyterian. Subspecialties: Plant pathology; Air pollution effects on plants. Current work: Interaction of air pollutants with plants, plant pathogens and soil microorganisms. Ozone, SO2, oxidants, crop loss, pollution/pathogen interactions, plant pathogens, endomycorrhizae. Office: Statewide Air Pollution Research Center U Calif Riverside Ca 92521

MC CORD, THOMAS BARD, planetary scientist, educator; b. Elverson, Pa., Jan. 18, 1939; s. Thomas M. Mc C. and Hazel V. (Bard) McCord; m. Carol Susan Bansner, Dec. 8, 1939. B.S. in Physics, Pa. State U., 1964; M.S. in Geology, Calif. Inst. Tech., 1966; Ph.D. in Astronomy and Planetary Scis., Calif. Inst. Tech., 1968. Research assoc. Calif. Inst. Tech., 1968; prof. MIT, 1968-77; asst. dir. Inst. Astronomy, U. Hawaii, Honolulu, 1976-79, prof. planetary sci., 1979—; chmn. planetary geosci. div. Hawaii Inst. Geophysics, 1979—; chmn. bd. SETS, Inc., 1980—. Contbr. 100 articles to profl. jours. Served with USAF, 1958-61. Recipient numerous grants. Fellow AAAS; mem. Am. Astron. Soc.; Mem. Am. Geophys. Union.; mem Explorers Club. Subspecialties: Planetary science; Solar energy. Current work: Origin and evolution of solar system; solar energy conversion; salt-gradient solar ponds. Home: 2328 Halehaka St Honolulu HI 96821 Office: Hawaii Inst Geophysics Univ Hawaii Honolulu HI 96822

MCCORKLE, GEORGE MASTON, molecular biologist; b. Wise, Va., Aug. 1, 1921; s. Claiborne Ross and Hazel Stewart (Webb) McC.; m. Harriet List Trees, Apr. 10, 1947 (div. 1972); children: Katharine McCorkle Snyder, Judith McCorkle Cole, James; m. Barbara Swanton Backus, Apr. 13, 1974. B.A., Yale U., 1942, M.Phil., 1973, Ph.D., 1975. With U. S. Dept. State, 1946; Charles Scribner's Sons, N.Y.C., 1947-60, v.p., 1955-60; also dir.; v.p. fin. New Am. Library, Inc., N.Y.C., 1960-67; with R.R. Bowker Co., N.Y.C., 1967-70, pres., 1968-70; research assoc. dept. biol. scis. Purdue U., 1975-79; dept. biology Yale U., 1979—. Served to lt. comdr. USNR, 1942-46, 50-52. Decorated personal commendation ribbon. Mem. Am. Soc. Microbiology, Genetics Soc. Am., Soc. Devel. Biology, Phi Beta Kappa. Subspecialties: Molecular biology; Developmental biology. Current work: Regulation of bacterial and eucaryotic gene expression; expression of the lactate dehydrogenase gene (ldh-3) in developing mouse spermatocytes. Home: 45 Mill Rock Rd Hamden CT 06511 Office: Kline Biology Tower Yale U New Haven CT 06520

MCCORMACK, MICHAEL G., consultant in science, energy and government; b. Basil, Ohio, Dec. 14, 1921; s. Henry Arthur and Nancy Jane (Jenkins) McC.; m. Margaret Higgins, June 21, 1947; children: Mark, Steven, Timothy. B.S. in Chemistry, Wash. State U., Pullman, 1948, M.S., 1949; Ph.D., Stevens Inst. Tech., 1976; LL.D., Salisbury State Coll., 1981. Research scientist Hanford AEC Facility, Richland, Wash., 1950-70; mem. 92d-96th Congresses from 4th Dist. Wash., 1970-80; individual practice as cons., Washington, 1980-82; pres. McCormack Assocs., Inc., Washington, 1982—; dir. Universal Voltronics Corp., Mt. Kisco, N.Y.; mem. Space Applications Bd., Washington, 1981—, NASA Adv. Council Task Force, 1982, Fusion Adv. Panel, Washington, 1980—. Mem. Wash. Stat Ho. of Reps., 1956-60, Wahs. State Senate, 1960-70. Served to 1st lt. U.S. Army, 1943-46; ETO. Recipient Centennial award ASME, 1980; Disting. Pub. Service award IEEE, 1980; Legis. Recognition award Nat. Soc. Profl. Engrs., 1980; named Solar Energy Man of Yr. Solar Energy Industries Assn., 1975; Triple E award Nat. Environ. Devel. Assn., 1976; Alumni Achievement award Wash. State U., 1981, one of top 100 innovators in world Technology mag., 1981. Fellow AAAS, Am. Nuclear Soc., Am. Inst. Chemists; mem. Am. Chem. Soc., Am. Legion, VFW. Democrat. Lodges: Masons; Grange. Subspecialties: Nuclear fusion; Nuclear fission. Current work: Public education in science and energy.

MCCORMICK, BARNES WARNOCK, aerospace engineering educator; b. Waycross, Ga., July 15, 1926; s. Barnes Warnock and Edwina (Brogdon) McC.; m. Emily Joan Hess, July 18, 1946; 1 dau., Cynthia Joan. B.S. in Aero. Engring, Pa. State U., 1948, M.S., 1949, Ph.D., 1954. Research asso. Pa. State U., University Park, 1949-54, asso. prof., 1954-55, prof. aero. engring., 1959—, head dept. aerospace engring., 1969—; assoc. prof., chmn. aero. dept. Wichita U., 1958-59; chief aerodynamics Vertol Helicopter Co., 1955-58; cons. to industry. Author: Aerodynamics of V/Stol Flight, 1967, Aerodynamics, Aeronautics and Flight Mechanics, 1979; Contbr. articles to profl. jours. Served with USNR, 1944-46. Recipient joint award for achievement in aerospace edn. Am. Soc. Engring. Edn.-Am. Inst. Aeros. and Astronautics, 1976. Fellow Am. Inst. Aeros. and Astronautics (assoc.); mem. ASEE, Am. Helicopter Soc. (tech.

council), Sigma Xi, Sigma Gamma Tau, Tau Beta Pi. Club: Masons. Subspecialty: Aeronautical engineering. Current work: Low-speed aerodynamics, propellers, helicopters, aircraft accident litigation, airplane spinning. Patentee in field. Home: 611 Glenn Rd State College PA 16801 Office: Pa State U Coll Engring University Park PA 16802

MCCORMICK, FERRIS ELLSWORTH, computer manufacturing company exec.; b. Bloomington, Ind., Aug. 16, 1946; s. John F. and Alpha E. (Moore) McC. B.A., Ind. U., 1968, postgrad., 1971. Compiler specialist Ind. U., Bloomington, 1969-74; mgr. compilers Harris Corp., Ft. Lauderdale, Fla., 1974-79, dir. software, 1979-82, dir. software product mktg., 1982—; instr. computer sci. Ind. U., Bloomington, 1971-73; cons. Dykema, Gossett, Spencer, Goodnow & Trigg, Detroit, 1981-82. Pres. Bloomington Symphony Orch., 1973. Mem. Assn. Computing Machinery, Am. Math. Soc., Soc. Indsl. and Applied Math., Math. Assn. Am., Am. Mgmt. Assn., Phi Beta Kappa. Republican. Subspecialties: Programming languages; Software engineering. Current work: Long-range and medium range software product planning, improved long term software reliability. Home: 775 NW 42d St Fort Lauderdale FL 33309 Office: Harris Corp Computer Systems 2101 W Cypress Creek Fort Lauderdale FL 33309

MCCORMICK, MICHAEL EDWARD, ocean engineering educator; b. Washington, Sept. 11, 1936; s. Edward Joseph and Loretta Marie (O'Donnell) McC.; m. Mary Ann Sarsfield, Sept. 7, 1963; children: Michael, Matthew, Brendan, Eamon. B.A., Am. U., 1959; M.S., Cath. U., 1961, Ph.D., 1966. Registered profl. engr., Md., Pa. Hydrodynamicist David Taylor Model Basin, Carderock, Md., 1958-61; instr. Swarthmore (Pa.) Coll., 1961-62, vis. scholar, 1976-77; asst. prof. Cath. U., Washington, 1962-65, Trinity Coll, Hartford, Conn., 1965-68; assoc. prof. U.S. Naval Acad., Annapolis, Md., 1968-74, prof. ocean engring., 1974—; cons. Pratt & Whitney/United Aircraft Corp., East Hartford, Conn., 1967-68, Chesapeake Inst. Corp., Shadyside, Md., 1969-70, U.S. ERDA/Dept. Energy, Washington, 1975-80, Gibbs & Cox, Inc., Arlington, Va., 1980-81, Johns Hopkins U./Applied Physics Lab., Laurel, Md., 1981—; U.S. mem. exec. com. Internat. Energy Agy. Author: Ocean Engineering Wave Mechanics, 1973, Ocean Wave Engineering Conversion, 1981; editor: Ocean Engineering, 1973—, Alternate Energy, 1981—, Ocean Engring, 1976—. Vice-pres. St. Mary's P.T.A., Annapolis, Md., 1969-70, pres., 1970-71; coordinator, coach St. Mary's Soccer, Annapolis, Md., 1979—; coach Naval Youth League Basketball, Annapolis, 1979-81. Mem. ASCE, ASME (intersoc. coordinator), Am. Soc. Engring. Edn. (chmn. ocean engring. div.), Marine Tech. Soc. (jour. editor), Soc. Underwater Tech. Democrat. Roman Catholic. Subspecialty: Species interaction. Patentee in field. Home: 1906 Sands Dr Annapolis MD 21401 Office: US Naval Acad Ocean Engring Annapolis MD 21402

MCCORMICK, NORMAN JOSEPH, nuclear engineering educator; b. Hays, Kan., Dec. 9, 1938; s. Clyde T. and Vera (Miller) McC.; m. Mildred Miring, Aug. 20, 1961; children: Kenneth John, Nancy Lynn. B.S.M.E., U. Ill., 1960; M.S. in Nuclear Engring., 1961; Ph.D., U. Mich., 1965. Registered profl. engr., Wash. NSF postdoctoral fellow, Ljubljana, Yugoslavia, 1965-66; asst. prof. U. Wash., Seattle, 1966-70, assoc. prof., 1970-74, prof. nuclear engring., 1975—; Nat. Acad. Scis. research fellow, Ljubljhana, 1971; scientist Science Applications, Inc., Palo Alto, Calif., 1974-75; N.Am. editor Progress in Nuclear Energy, 1980—; mem. editorial bd. Transport Theory and Statis. Physics, 1982—; cons. Hanford Engring. Devel. Lab., Richland, Wash., 1974—. Author: Reliability and Risk Analysis, 1981. Mem. Am. Nuclear Soc., Soc. Risk Analysis, Inst. Nuclear Materials Mgmt., Optical Soc. Am. Subspecialties: Nuclear engineering; Remote sensing (atmospheric science). Current work: Reliability and risk analysis, methods and nuclear/non-nuclear power applications; inverse transport solutions for radiative transfer. Patentee in field. Office: U Wash Dept Nuclear Engring BF-10 Seattle WA 98195

MCCOWN, BRENT HOWARD, horticulturist, educator, cons.; b. Chgo., Feb. 21, 1943; s. C. Y. and Francis E. (Howard) McC.; m. Deborah D. Donoghue, June 15, 1968; children: Elizabeth, Nevin James. B.S., U. Wis., 1965, M.S., 1967, Ph.D. (U. Wis. Alumni Research Found. 4-Yr. prize fellow), 1969. Asst. prof. plant physiology Inst. Arctic Biology, U. Alaska, Fairbanks, 1970-72; research coordinator Cold Regions Research and Engring. Lab., Hanover, N.H., 1970-72; asst. prof. dept. horticulture U. Wis., Madison, 1972-77, assoc. prof., 1977-82, prof., 1982—, Inst. Environ. Studies, 1972—, instr. plant prodn., 1972—; cons. on genetic engring. and biotech. applications to agr. Contbr. articles to profl. jours., chpts. to books. Served to capt. Chem. Corps U.S. Army, 1969-72. Recipient Outstanding Teaching award U. Wis., 1977. Mem. Internat. Assn. Plant Tissue Culture, Internat. Plant Propagator's Soc. (Outstanding Research Paper award 1980), Am. Soc. Hort. Sci., Am. Soc. Plant Physiology, Scandinavian Soc. Plant Physiology, Am. Soc. Plant Growth and Regulation, Wis. Acad. Arts and Sci., Sigma Xi, Phi Kappa Phi (pres. Wis. chpt.). Subspecialties: Plant cell and tissue culture; Plant physiology (agriculture). Current work: Micropropagation of crop plants, especially woody; somatic hybridization; plant growth regulators. Home: 236 E Sunset Ct Madison WI 53705 Office: Dept Horticulture U Wis Madison WI 53706

MCCOY, JAMES ERNEST, physicist, consultant; b. Glendale, Calif., May 4, 1941; s. Ernest F. and Amy M. (Cook) McC.; m. Sharon Donnelly, 1971 (div. 1974); m. Linda Black, 1963 (div. 1971); children: Laure Ann, Kenneth James. B.S., Calif. Inst. Tech., 1963; Ph.D., Rice U., 1968. Astrophysicist NASA, Houston, 1963—; cons. NW Energy, Salt lake City, 1980—; v.p., geophysics instr. Vernon Devel., Vancouver, B.C., Can., 1981—. Mem. Am. Geophys. Union, AIAA, IEEE. Subspecialties: Plasma physics; Remote sensing (geoscience). Current work: Space plasma interactions with high voltage solar arrays; coherent scattering ionospheric radar; satellite studies of solar terestrial interactions; imaging geophysical exploration radars; satellite navigation. Home: 10319 Ney St Houston TX 77034 Office: NASA-JSC Code SN3 Houston TX 77058

MCCOY, RANDOLPH EDWARD, plant pathologist; b. San Antonio, June 19, 1943; s. William Edward and Johanna (Hansen) McC.; m. Suzanne Williams, Apr. 17, 1982; children: by previous marriage: Merrie, Rebecca, William. B.S., Okla. State U., 1975, M.S., 1967; Ph.D., Cornell U., 1971. Faculty U. Fla., Fort Lauderdale, 1971—, prof. plant pathology, 1980—. Contbr. articles in field to profl. jours. Mem. Internat. Orgn. Mycoplasmology, Am. Phytopath. Soc., Fla. Hort. Soc., Palm Soc., Exptl. Aircraft Assn. Subspecialties: Plant pathology; Microbiology. Current work: Researcher in fastidious vascular pathogens, palm diseases and mycoplasmas, protozoa that infect plants. Home: 1841 SW 44th Terr Fort Lauderdale FL 33317 Office: U Fla: 3205 SW College Ave Fort Lauderdale FL 33314

MCCOY, ROGER MICHAEL, geography educator, researcher; b. Dewey, Okla., Feb. 3, 1933; m., 1955; 2 children. B.S., U. Okla., 1957; M.A., U. Colo., 1964; Ph.D. in Geography, U. Kans., 1967. Asst. prof. geography U. Ill.-Chgo., 1967-69; asst. prof. U. Ky., 1969-72; assoc. prof. U. Utah, 1972-80, prof. geography, chmn. dept. geography, 1980—; cons. remote sensing. Mem. Am. Soc. Photogrammetry (Ford-Bartlett award, Assn. Am. Geographers.). Subspecialty: Remote sensing (geoscience). Office: Dept Geography U Utah Salt Lake City UT 84112

MCCOY, SCOTT, JR., oil co. exec.; b. Indpls., Oct. 8, 1939; s. Scott and Esther Alberta (McClintock) McC.; m. Carol Joanne Fassnacht, June 24, 1964; 1 dau., Shawna Bernita. B.S., Wheaton Coll., 1961; M.S., U. Ariz., 1964. Registered geologist, Calif.; cert. profl. geologist. Geologist, biostratigrapher Phillips Petroleum Co., Bartlesville, Okla., 1964-73, Denver, 1973-77; geologist/biostratigrapher Amoco Prodn. Co., Denver, 1977-81, geol. supr., 1981—. Internat. Union Geol. Sci. travel grantee, Paris, 1980. Fellow Geol. Soc. Am.; mem. Am. Assn. Petroleum Geologists, Am. Inst. Profl. Geologists., Paleontological Soc., Soc. Econ. Paleontologists and Mineralogists, Soc. Rocky Mt. Geologists, Internat. Union Geol. Sci. (mem. subcommn. on neogene stratigraphy). Republican. Episcopalian. Subspecialties: Geology; Paleontology. Current work: Tectonics of Pacific margin; Molluscan biostratigraphy of no. Pacific; hydrocarbon exploration in no. Pacific. Office: 1670 Broadway Denver CO 80202 Home: 733 S Elkhart St Aurora CO 80012

MCCRACKEN, FRANCIS IRVIN, research plant pathologist; b. Bicknell, Ind., Aug. 19, 1935; s. Irvin E. and Mildred L. (Wineinger) McC.; m. Karen R. Jorgensen, June 4, 1938; children: Kelly J., Alan F. B.S., Ariz. State U., 1957; M.S., Okla. State U., 1959; Ph.D., Wash. State U., 1972. Instr. biology U. Idaho, Moscow, 1962-67; research plant pathologist U.S. Forest Service, Stoneville, Miss., 1967—; adj. prof. Miss. State U., State College, 1978—. Contbr. articles to profl. jours. Mem. Am. Phytopath. Soc., Sigma Xi. Methodist. Subspecialty: Plant pathology. Current work: Researcher in eitology and control of hardwood tree diseases. Home: 309 Gerald St Leland MS 38756 Office: US Forest Service PO Box 227 Stoneville MS 38776

MCCRAIN, GERALD RAY, biologist, cons., planner; b. Asheville, N.C., Aug. 21, 1950; s. Raymond and Aileen Ruth (Gambill) McC.; m. Patricia Jean Ray, June 19, 1981. B.S. in Botany, N.C. State U., 1972, M.S. in Botany and Ecology, 1975. Instr. biology U. N.C., Asheville, 1975; grounds supr. Bluebeards Castle Hotel, St. Thomas, V.I., 1975-77; biologist Bur. of Fish and Wildlife, St. Thomas, 1978; environ. specialist div. coastal zone mgmt., St. Thomas, 1979-82, eviron. cons., Houston, St. Thomas, 1982—. Contbr. articles on biology to profl. jours. Mem. Bot. Soc. Am. Subspecialties: Ecology; Resource management. Current work: Ecosystem analyses, resource planning and management, cons. air pollution, water pollution and CZM permitting, landscape planning, environment impact statement analyses. Home: 10184 Oakberry Ln Houston TX 77042

MCCRAW, RONALD KENT, psychologist; b. Houston, Dec. 6, 1947; s. Leon Frank and Lorna Mae (Bailey) McC. B.A., U. Tex.-Austin, 1970; M.A., U. Tex. Med. Br., Galveston, 1972; Ph.D., U. South Fla., 1981. Research technician U. Tex. Med. Br., Galveston, 1972-74; grad. asst. Fla. Mental Health Inst., Tampa, 1975-76; resident in clin. psychology U. Tex. Health Scis. Ctr., San Antonio, 1977-78; psychometrist Hillsborough Community Mental Health Ctr., Tampa, 1978-79; clin. psychologist USAF Hosp., Chanute AFB, Ill., 1982—; commd. capt. U.S. Air Force, 1982; Film/book reviewer Sci. Books & Films. AAAS, 1978—. Contbr. articles to profl. jours. Mem. Galveston County Cultural Arts Council, 1973; asst. coach Leaguerettes, Temple Terrace, Fla., 1979; coach Girls Softball Assn., Baytown, Tex., 1980-82; instr. ARC, Baytown, Tampa, 1980—. Fellow Am. Orthopsychiat. Assn.; mem. Am. Psychol. Assn., Am. Acad. Behavioral Medicine, Soc. Personality Assessment, Acad. Psychosomatic Medicine, Assn. for Advancement Behavior Therapy, N.Y. Acad. Sci., Sigma Xi, Psi Chi, Omicron Delta Kappa. Methodist. Lodges: Masons; Order of Demolay. Subspecialty: Behavioral psychology. Current work: Behavioral medicine, personality assessment, pregnancy and childbirth, individual and family therapy, group therapy with adolescents. Office: Mental Health Clinic USAF Hosp Chanute AFB IL 61868 Home: PSC Box 1429 Chanute AFB IL 61868

MCCRAY, RICHARD ALAN, astrophysicist, educator, university administrator; b. Los Angeles, Nov. 24, 1937; m., 1961; 2 children. B.S., Stanford U., 1959; M.A., UCLA, 1962, Ph.D. in Physics, 1967. Research fellow in physics Calif. Inst. Tech., 1967-68; asst. prof. astronomy Harvard Coll. Obs., 1968-71; assoc. prof. U. Colo., Boulder, 1971-75, prof. physics and astrophysics, 1975—, fellow, 1972—, chmn. Inst. Lab., 1981—. John Simon Guggenheim Meml. Found. fellow, 1975. Mem. Am. Astron. Soc., Internat. Astron. Union. Subspecialty: Theoretical astrophysics. Address: Joint Inst Lab Astrophysics U Colo Boulder CO 80309

MCCREA, ROBERT ALAN, neurobiology educator; b. Louisville, Dec. 27, 1948; s. Donald A. and Naomi (Eberhard) McC.; m. Janet Margaret Denomme, Mar. 24, 1972. B.A., Wayne State U., 1970; Ph.D. in Anatomy, 1976. Instr. Wayne State U., Detroit, 1977; research scientist NYU, N.Y.C., 1977-81; asst. prof. pharmacology and physiology U. Chgo., 1982—. Contbr. numerous sci. articles to profl. publs. Recipient Nat. Research Service award NIH, 1977, grantee, 1983. Mem. AAAS, Am. Assn. Anatomists, Soc. Neurosci., Internat. Union Physiol. Scis. (Travel award 1980). Democrat. Subspecialties: Neurobiology; Neurophysiology. Current work: Anatomical and physiological substrates for the brain's control of eye movements.

MC CREDIE, KENNETH BLAIR, physician, educator; b. Christchurch, N.Z., July 2, 1935; came to U.S., 1969, naturalized, 1976; s. Gordon Blair and Margaret J. (Stevenson) McCredie; m. Maria Isabel Delgado, Oct. 29, 1980; children: Wendy Jane, Anna Margaret, Jennifer Mary. Student, Canterbury U., 1953-54; M.B., Ch.B., Otago U., 1960. Intern medicine Napier Hosp., N.Z., 1961-62, resident, 1962-65; resident medicine Prince Henry Hosp., Sydney, Australia, 1965-66, sr. fellow hematology, 1966-69; project investigator dept. development therapeutics U. Tex. System Cancer Ctr., M.D. Anderson Hosp. and Tumor Inst., Houston, 1969-70, asst. internist, asst. prof. medicine, 1970-73, assoc. internist, assoc. prof. medicine, 1973-76, chief leukemia service, 1973, internist, assoc. prof. medicine, 1976, prof. medicine, internist, chief leukemia service, 1978, prof. dept. dental oncology, 1979, dep. dept. head, chief leukemia service, 1981—, asst. prof. program in gen. internal medicine Med. Sch., 1973-74, assoc. prof., 1974, assoc. prof. dept. dental oncology Health Sci. Center, Dental br., 1977, prof. medicine, internist, chief leukemia service, dep. chmn. dept. hematology of div. medicine, 1983—; prof. dept. internal medicine U. Tex. Med. Sch., Houston, 1983—; Sir James Wattie vis. prof., N.Z., 1980. Mem. editorial bd.: Exptl. Hematology, 1973-78. Vice-pres. med. and sci. affairs Leukemia Soc. Am., Inc., 1979—, chmn. med. and sci. adv. com., 1979—, v.p. Tex. Gulf Coast chpt., Houston, 1981. Served to capt. Royal N.Z. Army, 1954-65. Recipient Outstanding Service to Mankind award Tex. Gulf Coast chpt. Leukemia Soc. Am., 1981, Dr. John Kenny award Nat. chpt. Tex. Gulf Coast chpt. Leukemia Soc. Am., 1983. Fellow Royal Australasian Coll. Physicians, ACP, Philippine Coll. Physicians (hon.); mem. Royal Soc. Medicine, Haematology Soc. Australia, Am. Fedn. Clin. Research, Am. Soc. Hematology, Am. Assn. Cancer Research, AMA, Am. Soc. Clin. Oncology, Internat. Soc. Exptl. Hematology, Working Group Leukocyte Procurement Nat. Cancer Inst., Tex. Med. Assn., S.W. Cancer Chemotherapy Study Group, Harris County Med. Soc., Internat. Assn. Comparative Research on Leukemia and Related Diseases, N.Y. Acad. Scis. Subspecialties: Cancer research (medicine); Oncology. Office: 6723 Bertner Ave Houston TX 77030

MCCUBBIN, THOMAS KING, JR., physicist, educator; b. Balt., June 1, 1925; s. T. King and Isabella M. McC.; m. Mary Lamb, Jan. 27, 1951; children: Mary Louisa, Thomas, Ruth. B.E.E., U. Louisville, 1946; Ph.D., Johns Hopkins U., 1951. Mem. faculty dept. physics Pa. State U., 1957—, prof., 1964—; mem. research staff M.I.T., 1954-57, Johns Hopkins U., 1951-54; professeur etranger Universite de Dijon, France, 1974. Contbr. articles to profl. jours. Served with USN, 1944-46. NSF grantee. Fellow Am. Phys. Soc., Optical Soc. Am. (dir.-at-large 1968-70). Democrat. Episcopalian. Subspecialties: Infrared spectroscopy; Atomic and molecular physics. Current work: Infrared and Raman and fluorescence spectroscopy of molecules. Home: 909 Willard Circle State College PA 16801 Office: 104 Davey Lab University Park PA 16802

MCCULLOCH, CHRISTOPHER ALLAN, dental scientist, periodontist; b. Winnipeg, Man., Can., Mar. 26, 1951; s. Allen Whitten and Velma Gertrude (Cockerill) McC.; m. Judith Millicent Beamish, July 9, 1982. B.Sc., U. Toronto, 1972, D.D.S., 1976, Ph.D., 1982; certificate in Periodontics, Columbia U., 1978. Cert. in periodontics, Ontario. Pvt. practice in periodontics, Hamilton, Ontario, 1978-82; research assoc. faculty medicine U. Toronto, 1982—. Author: Biocompatibility of Dental Materials, 1981. Medical Research Council Can. fellow, 1980, 81, 82, 83. Fellow Royal Coll. Dentists. Anglican. Subspecialties: Oral biology; Periodontics. Current work: cell kinetics of gastro-intestinal mucosa and periodontium of mice; flow cytometry; cell sorting. Home: 9 Holloway Rd Islington Ontario Canada M9A 1E7 Office: Univ Toronto 1189 Medical Sci Bldg Ontario Canada M5S 1A8

MCCULLOCH, JOHN HATHORN, oncologic surgeon; b. Evanston, Ill., Apr. 28, 1940; s. Hathorn Waugh and Esther (Str) McC.; m. Susan Groendyke, Aug. 29, 1964; children: David, Beth; m. Carol Bugerhoff, Nov. 29, 1980. B.A., Pomona Coll., 1962; M.D., U. Ill.-Chgo., 1966. Diplomate: Am. Bd. Surgery. Intern Cook County Hosp., Chgo., 1966-67, resident in gen. surgery, 1967-69, 1970-72; NIH fellow in surg. oncology U. Ill.-Chgo., 1969-70; practice medicine specializing in oncologic surgery, Spartanburg, S.C., 1974—; tumor registry dir. Spartanburg Gen. Hosp., 1975—; attending physician S.C. State Aid Cancer Clinic, 1974—; clin. investigator Piedmont Oncologic Assocs., 1977—; clin. instr. surgery U. Ill. Med. Sch., Chgo., 1968-72, U. Tex. Med. Sch. of San Antonio, 1972-74, Med. Coll. of S.C., Charleston, 1974-77, clin. assoc. prof. surgery, 1977—. Conbr. articles on oncologic surgery to profl. jours. Bd. dirs. Am. Cancer Soc., Spartanburg, 1974—, pres., 1977-79. Served to maj. USAF, 1972-74. Recipient Merck Manual award U. Ill. Med. Sch., 1966. Mem. ACS, Internat. Soc. for Preventive Oncology, Inc., Am. Soc. Clin. Oncology, S.C., Oncology Soc., Am. Soc. Preventive Oncology, AMA, San Antonio Surg. Soc., Soc. Clin. Surgeons, S.C. Surg. Soc. Republican. Unitarian. Club: Piedmont (Spartanburg). Subspecialties: Surgery; Oncology. Current work: Clinical cancer research trials. Office: 1776 Skylyn Dr PO Box 2768 Spartanburg SC 29304

MC CUNE, WILLIAM JAMES, JR., manufacturing company executive; b. Glens Falls, N.Y., June 2, 1915; s. William James and Brunnhilde (Decker) McC.; m. Janet Waters, Apr. 19, 1940; 1 dau., Constance (Mrs. Leslie Sheppard); m. Elisabeth Johnson, Aug. 8, 1946; children—William Joseph, Heather H.D. S.B., Mass. Inst. Tech., 1937. With Polaroid Corp., Cambridge, Mass., 1939—, v.p. engring., 1954-63, v.p., asst. gen. mgr., 1963-69, exec. v.p., after 1969, pres., chief exec. officer, dir., 1980-82, chmn. bd., pres., chief exec officer, 1982—. Chmn. bd., trustee Mitre Corp.; Trustee Boston Mus. Sci., Mass. Gen. Hosp. Fellow Am. Acad. Arts and Scis.; mem. Nat. Acad. Engring. Subspecialty: Manufacturing, engineering administration. Home: Old Concord Rd Lincoln MA 01773 Office: Polaroid Corp 549 Technology Sq Cambridge MA 02139

MCCURDY, LAYTON, psychiatrist, educator; b. Florence, S.C., Aug. 20, 1935; m., 1958; 2 children. M.D., Med. U. S.C., 1960. Rotating intern Med. Coll. S.C., 1960-61; resident in psychiatry N.C. Meml. Hosp., 1961-64; asst. prof. psychiatry Emory U., 1966-68; prof. psychiatry, chmn. dept. medicine U. S.C.-Charleston, 1968—; cons. in field; psychoanalysis trainee Columbia U. Psychoanalytic Clinic, Atlanta, 1967-68; mem. research rev. com., applied research br. NIMH, 1970-74; mem. Nat. Adv. Mental Health Council, 1980-83. NIMH research fellow Maudsley Hosp., London, 1974-75. Fellow Am. Coll. Psychiatrists, Am. Psychiat. Assn., Am. Psychosomatic Soc., Assn. Acad. Psychiatry (pres. 1970-72); mem. Assn. Am. Med. Colls. Subspecialty: Psychiatry. Office: Dept Psychiatry Med U SC Charleston SC 29401

MCDANIEL, BOYCE DAWKINS, physicist, educator; b. Brevard, N.C., June 11, 1917; s. Allen Webster and Grace (Dawkins) McD.; m. Jane Chapman Grennell, Aug. 3, 1941; children: Gail P., James G. B.S., Ohio Wesleyan U., 1938; M.S., Case Inst. Tech., 1940; Ph.D., Cornell U., 1943. Staff mem. radiation lab. Mass. Inst. Tech., 1943; physicist Los Alamos Sci. Lab., 1943-46; mem. faculty Cornell U., Ithaca, N.Y., 1945—, prof., 1956—, assoc. dir. lab. nuclear studies, 1960-67, dir. lab. nuclear studies, 1967—, Floyd R. Newman prof. nuclear studies, 1977—; head accelerator sect. Nat. Accelerator Lab., Batavia, Ill., 1972; mem. high energy physics adv. panel ERDA, 1975—. Contbr. articles to profl. jours. Trustee Associated Univs., Inc., 1962-75. Vis. fellow Brookhaven Nat. Lab., 1966; Fulbright Research grantee Australian Nat. U., 1953; Guggenheim and Fulbright grantee U. Rome and Synchrotron Lab., Frascati, Italy, 1959-60. Fellow Am. Phys. Soc.; mem. Univs. Research Assn. (trustee 1971-77, 83—), Nat. Acad. Sci. Subspecialty: Particle physics. Spl. research neutron spectroscopy, gamma ray spectroscopy, high energy photoprodn. K mesons and hyperons, instrumentation for high energy physics, accelerator design and constrn. Home: 26 Woodcrest Ave Ithaca NY 14850

MCDANIEL, HUGH THOMAS, JR., veterinarian, educator; b. Berkeley, Calif., Jan. 20, 1934; s. Hugh Thomas and Agnes (Walters) McD.; m. Carolyn Leagon, Feb. 14, 1954; children: Robin McDaniel Barton, Hugh Thomas III, Walter Scott. B.S. in Zoology, N.C. State U., 1956; D.V.M., U. Ga., 1958. Diplomate: Am. Coll. Theriogenologists, 1979. Pvt. practice vet. medicine, Mayfield, Ky., 1958-65; field research asst. Hess & Clark, Ashland, Ohio, 1965-67; poultry specialist Pillsbury Co., Ft. Smith, Ark., 1967-68; ptnr. Mayfield (Ky.) Vet. Clinic, 1968-76; from instr. to asst. prof. U. Ga., Athens, 1976-82, assoc. prof. herd health mgmt., 1982—. Bd. dirs. Classic City Kennel Club, 1981-82. U.S. Dept. Agr. grantee, 1981; Ga. Dept. Agr. grantee, 1980-81. Mem. AVMA, Am. Assn. Bovine Practitioners, Am. Assn. Swine Practitioners, Soc. for Therogenology. Presbyterian. Subspecialties: Preventive medicine (veterinary medicine); Reproduction. Current work: Sarcocystis and mycobacteriosis in swine. Home: 110 Dunwoody Dr Athens GA 30605 Office: Coll Vet Medicine U Ga Athens GA 30602

MCDANIEL, KIRK COLE, animal science educator; b. Berkeley, Calif., Apr. 5, 1949; s. William J. and Ernestine W. (Wakeham) Mc.D.; m. Eileen M. McDaniel, Jan. 7, 1949; children: Corey, Casey, Thomas. B.S. in Range Mgmt, Humboldt State U.; M.S. in Range and Resources, Oreg. State U; Ph.D., Tex. A&M U. Range technician Bur. Land Mgmt.; research asst. Oreg. State U., 1972; range ecologist U.S. Forest Service, 1972-74; research assoc. Tex. A&M U.; asst. prof. dept. animal sci. N. Mex. State U., 1978—. Contbr. to profl. jours. Mem. Soc. Range Mgmt., Weed Sci. Soc. Am. Subspecialties: Resource management; Integrated pest management. Current work: Range brush and weed management; range science; range ecology; resource management. Home: 1743 Royal St Las Cruces NM 88001 Office: N Mex State U PO Box 31 Dept Animal and Range Sci Las Cruces NM 88001

MCDANIELS, DAVID KEITH, physicist, educator; b. Hoquiam, Wash., May 21, 1929; s. Forest L. and Grace L. McD.; m. Patricia R. McDaniels, Feb. 9, 1966; children: D. Douglas, Keith A., Kevin P. B.S., Wash. State U., 1951; M.S., U. Wash., Seattle, 1958, Ph.D., 1960. Physicist Hanford Atomic Products Operation, Gen. Electric Co., 1951-54; NSF research fellow Nuclear Research Centre, Saclay, France, 1962; successively asst. prof., assoc. prof., prof. physics U. Oreg., Eugene, 1963—; vis. staff mem. Los Alamos Nat. Lab., 1970-71, Uppsala U., Sweden, 1979. Author: The Sun: Our Future Energy Source, 1979; contbr. numerous articles profl. jours. Served with AUS, 1954-55. Grantee NSF, Bonneville Power Adminstr., 1963—. Mem. Am. Phys. Soc., AAAS, Am. Assn. Physics Tchrs., Internat. Solar Energy Soc. Subspecialties: Solar energy; Nuclear physics. Current work: Nuclear physics and solar energy, giant resonance studies at intermediate energies, solar radiation. Home: 2365 W 23rd St Eugene OR 97405 Office: Physics Dept U of Oregon Eugene OR 97403

MCDERMOTT, DANIEL JOSEPH, pharmaceutical corporation executive, medical researcher; b. Pitts., May 19, 1936; s. Daniel J. and Anne (Collins) McD.; m. Janet Patricia Duhow, 1979; children: Jennifer, Brian. B.S., Georgetown U., 1958; M.S., U. Minn., 1965; Ph.D., Marquette U., 1969. Postdoctoral fellow Med. Coll. Wis., Milw., 1969-72, asst. prof. physiology, 1972-73; assoc. dir. cardiovascular clin. research G.D. Searle & Co., Chgo., 1973-79; v.p. med. research Smith Labs, Inc., Northbrook, Ill., 1979—. Served to lt. USN, 1959-63; served to capt. USNR. Mem. Am. Soc. Clin. Pharmacology and Exptl. Therapeutics, AAAS, Am. Heart Assn., N.Y. Acad. Scis., Sigma Xi. Subspecialties: Pharmacology; Physiology (medicine). Current work: Clinical trials of chymopapain; randomized controlled trials to evaluate new drugs. Home: 4 Anglican Ln Lincolnshire IL 60015 Office: Smith Labs Inc 2215 Sanders Rd Northbrook IL 60062

MCDERMOTT, JOHN DONOVAN, computer science researcher, consultant; b. Glenwood, Minn., Oct. 2, 1942; s. Dominic and Maxine (Ferring) McD.; m. Margaret Shaughnessy, Dec. 30, 1967; children: Brigid, Paul, Owen, Liam. B.A., St. Louis U., 1966, M.A., 1967; Ph.D., U. Notre Dame, 1969. Asst. prof. philosophy Eisenhower Coll., Seneca Falls, N.Y., 1969-74; vis. research assoc. dept. computer sci. Carnegie-Mellon U., Pitts., 1974-75, research assoc. and lectr., 1975-78, research computer scientist and asst. dept. head, 1978-82, sr. research computer scientist and assoc. dept. head, 1982—; cons. in field. Contbr. articles to profl. jours. Mem. Assn. Computing Machinery, Am. Assn. for Artificial Intelligence. Subspecialty: Artificial intelligence. Current work: Addressing the problems involved in developing expert systems, computer programs that contain the wealth of domain, specific knowledge that human experts require in order to function effectively. Office: Dept Computer Sci Carnegie-Mellon U Pittsburgh PA 15213

MCDEVITT, HUGH O'NEILL, immunologist, educator, hospital administrator; b. Cin., 1930. M.D., Harvard U., 1955. Diplomate: Am. Bd. Internal Medicine. Intern Peter Bent Brigham Hosp., Boston, 1955-56, sr. asst. resident in medicine, 1961-62; asst. resident Bellevue Hosp., 1956-57; research fellow dept. bacteriology and immunology; Harvard U., 1959-61; USPHS spl. fellow Nat. Inst. Med. Research, Mill Hill, London, 1962-64; assoc. prof. med. immunology Stanford U., 1969-72, prof., 1972—, chief div. immunology, 1970—; physician (Univ. Hosp.), 1966—; cons. physician VA Hosp., Palo Alto, Calif., 1968—. Mem. Nat. Acad. Sci., AAAS, Am. Fedn. Clin. Research, Am. Soc. Clin. Investigation, Am. Assn. Immunologists. Subspecialty: Immunology (medicine). Office: Dept Med Microscopy D-345 Fairchild Stanford U Hosp Stanford CA 94305

MCDONAGH, PAUL FRANCIS, physiology educator, researcher; b. Norwood, Mass., Mar. 25, 1945; s. John J. and Josephine Thomas McD.; m. Jennifer L. Crockett, May 21, 1978; children: John, Denise. B.S., Worcester Poly. Inst., 1967; M.S., Columbia U., 1969; Ph.D., U. Calif.-Davis, 1977; postgrad., U. Ariz.-Tucson, 1979. Research asst. chem. engring. Columbia U., N.Y.C., 1967-68; research assoc. U. Calif.-Davis, 1974-75; postdoctoral fellow U. Ariz., Tucson, 1976-79; asst. prof. Yale U., New Haven, 1979—; mem. research com. Conn. Heart Assn., 1982-84; cons. Miles Inst. Preclin. Pharmacology, 1982. Mem. editorial bd.: Microcirculation, 1983. Served to 1st lt. U.S. Army, 1968-70. Standard Oil fellow, 1970; U. Calif. Chancellor's fellow, 1974; grantee Ariz. Heart Assn., Yale Fluid Research, NIH, Conn. Heart Assn., Charles Oshe Fund, Miles Labs. Mem. Am. Physiol. Soc., Microcirculatory Soc., Biomed Engring. Soc., N.Y. Acad. Scis. Subspecialties: Physiology (medicine); Biomedical engineering. Current work: Pathophysiology, coronary microvascular physiology, improved protection and preservation of hearts during surgery. Office: Dept Surgery Yale U Sch Medicine 333 Cedar St New Haven CT 06510

MCDONALD, CHARLES CAMERON, research administrator; b. Wyoming, Ont., Can., Aug. 5, 1926; m., 1949; 4 children. B.Sc., U. Western Ont., 1949, M.Sc., 1950; Ph.D. in Chemistry, Ill. Inst. Tech., 1954. Research fellow in chemistry Ill. Inst. Tech., 1954-55; research chemist E I du Pont de Nemours & Co., Inc., Wilmington, Del., 1955-65, research supr. central research dept., 1965-78, research mgr. central research and devel. dept., 1979. Subspecialty: Research management. Office: Central Research Dept E I du Pont de Nemours & Co Wilmington DE 19898

MCDONALD, GERAL IRVING, forest pathologist; b. Wallowa, Oreg., Dec. 31, 1935; s. Harvey Irving and Lola M. (Gorbett) McD.; m. Judy Lynne Manos, Feb. 4, 1962; children: Michele, Scott, Alyssa. B.S. in Forestry, Wash. State U., 1963; Ph.D. in Plant Pathology, 1969. Research plant pathologist Forestry Scis. Lab., U.S. Forest Service, Moscow, Idaho, 1966—. Contbr. articles to profl. jours. Served with Wash. N.G., 1954-61. Mem. Am. Phytopath. Soc., Sigma Xi. Lodge: Lions. Subspecialties: Plant genetics; Integrated pest management. Current work: Genetics of host pest systems, mathematical analysis and computer simulation of disease epidemics, development of integrated pest management. Office: 1221 S Main Moscow ID 83843

MCDONALD, KEITH LEON, theoretical physicist; b. Murray City, Utah, Apr. 20, 1923; s. Thomas Francis and Ada Pearl (Russell) McD. B.S. in Physics, U. Utah, Salt Lake City, 1950, M.S., 1951, Ph.D., 1956. With research and devel. dept. U.S. Naval Ordnance Test Sta., China Lake, Calif., 1951-52; theoretical grad. researcher U. Utah, 1954-56; mem. staff U. Calif.-Los Alamos Sci. Lab., 1956-57; researcher math. and physics U.S. Army Chem. Corps Proving Ground, Dugway, Utah, 1957-60; mem. faculty Brigham Young U., 1960-62, U. Utah, 1963; with Nat. Bur. Standards, Boulder, Colo., 1963-65, Idaho State U., 1965; theoretical researcher ESSA Environ. Research Labs., Boulder, 1966-68; cons. NOAA, Environ. Research Labs., Boulder, 1969-71; cons., researcher cosmic hydromagnetism, Salt Lake City, 1971—. Contbr. articles to profl. jours. Mem. Am. Phys. Soc., Am. Geophys. Union, Am. Assn. Physics Tchrs., Am. Astron.Soc., Optical Soc. Am., Phi Beta Kappa, Sigma Xi, Sigma Pi Sigma. Subspecialties: Theoretical physics; Theoretical astrophysics. Current work: Cosmic hydromagnetism and dynamo theory, solar-terrestrial relations, high speed streams with origin in breaches in the sun's upper magnetic toroid. Office: PO Box 2433 Salt Lake City UT 84110

MCDONALD, STANFORD LAUREL, clinical psychologist, educator; b. Lincoln, Nebr., Mar. 14, 1929; s. Laurel C. and Irene V. (Frey) McD.; m. Shirley P. Peterson, Apr. 26, 1964; children: Stacia, Jeffrey, Kathleen, Patricia. A.B., Nebr. Wesleyan U., Lincoln, 1956; M.A., U. Nebr., Omaha, 1959; Ph.D., Fielding Inst., Santa Barbara, Calif., 1974; postgrad., U. Nebr., Lincoln, 1958-60. Lic. psychologist, Ill. Supr. psychol. services SPEED Devel. Ctr., Chicago Heights, Ill., 1965-80; pres. Stanford L. McDonald, P.C., Olympia Fields, Ill., 1980—; lectr. profl. groups. Bd. dirs. Community Chest, Park Forest, Ill., 1981—, Good Shepherd Ctr., Flossmoor, Ill., 1981—, Ill. Epilepsy Assn., 1979. Served with USMC, 1950-52; Korea. Fellow Am. Orthpsychiat. Assn.; mem. Ill. Psychol. Assn., Am. Psychol. Assn., Biofeedback Soc. Ill. (pres.), Biofeedback Soc. Am., Soc. Behavioral Medicine, N.Y. Acad. Scis., Psi Chi, Phi Delta Kappa. Lodge: Kiwanis. Subspecialties: Behavioral psychology; Neuropsychology. Current work: Biofeedback research clinician; bioinstrumentation applications within behavioral medicine. Home: 255 Rich Rd Park Forest IL 60466 Office: 2555 W Lincoln Hwy Suite 203 Olympis Fields IL 60461

MCDONOUGH, THOMAS REDMOND, astrophysicist; b. Boston, Oct. 4, 1945; s. Redmond Augustus and Sophie Theresa (Stankewich) McD. B.S., M.I.T., 1966; Ph.D., Cornell U., 1973. Postdoctoral researcher Cornell U., Ithaca, N.Y., 1973-75; resident research assoc. Jet Propulsion Lab., Pasadena, 1976-77, cons., 1978-81; lectr. engring. Calif. Inst. Tech., Pasadena, 1979—; coordinator Search for Extraterrestrial Intelligence, Planetary Soc., Pasadena, 1981—; cons. Avco Embassy Pictures, Hollywood, Calif., 1981-82. Contbr. articles to profl. jours.; popular sci./sci. fiction stories pub. in Creative Computing. Recipient citation NASA, 1979. Fellow Brit. Interplanetary Soc.; mem. Internat. Astron. Union, Am. Astron. Soc., Am. Phys. Soc., AIAA, Authors Guild, Sci. Fiction Writers Am., Internat. Platform Assn. Club: Toastmasters. Subspecialties: Planetary science; The search for extraterrestrial intelligence. Current work: Spacecraft research on giant planets; search for extraterrestrial intelligence. Home: 500 S Oak Knoll 46 Pasadena CA 91101 Office: 138 78 Calif Inst Tech Pasadena CA 91125

MCDOUGAL, W(ILLIAM) SCOTT, urologist, researcher, educator; b. Grand Rapids, Mich., July 11, 1942; s. William Julian and Verna Wilma (Pasma) McD.; m. Susan I. Schroeder, June 8, 1967; 1 dau., Molly Katherine. A.B., Dartmouth coll., 1964; M.D., Cornell U. Med. Coll., 1968. Diplomate: Am. Bd. Surgery, Am. Bd. Urology. Intern Univ. Hosp., Cleve.; intern Case-Western Res. U., 1968-69, resident, 1969-75, asst. prof., 1977-78, assoc. prof., 1978-80; postdoctoral fellow Yale U. Med. Coll., 1971-72; assoc. prof. urology Dartmouth Coll., 1980—, chmn. dept. urology, 1982—; exam mem. Am. Bd. Urology, 1980—. Author: Manuel of Burns, 1978 (Am. Med. Writing award 1979), Traumatic Injuries of Genito-Urinary System, 1980. Served to maj. U.S. Army, 1975-77. NIH acad. tng. grantee, 1971-75. Fellow ACS; mem. Am. Urologic Assn., Assn. Acad. Surgery, AMA. Presbyterian. Subspecialties: Urology; Surgery. Current work: Obstructive uropathy, metabolism, nutrition. Home: PO Box B1165 Hanover NH 03755 Office: Dartmouth-Hitchcock Med Center 2 Maynard St Hanover NH 03755

MCDOUGALD, LARRY ROBERT, poultry parasitologist; b. Broken Bow, Okla., Dec. 31, 1941; s. Charlie and Ann Marie (Hall) McD.; m. Dana Lou Lovell, June 14, 1962; children: Charles, Tracey. B.S., S.E. Okla. State U., 1962, M.S., 1966; Ph.D., Kans. State U., 1969. Postdoctoral research assoc. U. Ga. Athens, 1969-71, assoc. prof. poultry parasitology, 1977—; sr. parasitologist Eli Lilly & Co., Indpls., 1971-77; cons. in field. Mem. Am. Soc. Parasitology, Am. Assn. Poultry Sci., Soc. Protozoology, Sigma Xi. Subspecialties: Parasitology; Preventive medicine (veterinary medicine). Current work: Exptl. research in coccidia, host specificity, chemotheraphy, pathology. Office: Dept Poultry Sci U Ga Athens GA 30602

MC DOWELL, CHARLES ALEXANDER, physical chemist, educator; b. Belfast, Ireland, Aug. 29, 1918; s. Charles and Mabel (McGregor) McD.; m. Christine Joan Staddart, Aug. 10, 1945; children—Karen Mary Anne, Christina Anne, Avril Jeanne. B.Sc., M.Sc., Queens U., Belfast, 1942; D.Sc., U. Liverpool, Eng., 1955. Sr. asst. lectr. Queen's U., 1941-42; sci. officer U.K. Civil Def., 1942-45; lectr. U. Liverpool, 1945-55; prof., head dept. chemistry U. B.C., Vancouver, 1955-81, Univ. prof., 1981—, Killiam sr. research fellow, 1969-70; vis. prof. Kyoto (Japan) U., 1965, 69-70; Disting. vis. prof. U. Fla., Gainesville, 1974; NRC Sr. Research fellow Cambridge (Eng.) U., 1963-64; W.U.S. cons. U. Honduras, 1975; Disting. vis. prof. U. Capetown, 1975; Frontiers of Chemistry lectr. Wayne State U., 1978; Disting. vis. lectr. Faculty of Sci., U. Calgary, 1978; chmn. Internat. Union Pure and Applied Chemistry Congress, Vancouver, 1981. Editor: Mass Spectrometry, 1963, Magnetic Resonance, 1972; contbr. articles to profl. jours. Recipient Letts Gold medal in theoretical chemistry Queen's U., 1941, Sci. medal Université de Liège, Belgium, 1955; Centennial medal Gov. of Can., 1967; Queen Elizabeth Silver Jubilee medal, 1977. Fellow Chem. Soc. London; Mem. Chem. Inst. Can. (medal 1969, Montreal medal 1982, pres. 1979), Royal Soc. Chemistry (U.K.), Royal Soc. Can., Am. Chem. Soc., Am. Mass Spectrometry Soc., Mass Spectrometry Soc. Japan, Am. Phys. Soc. Subspecialties: Nuclear magnetic resonance (chemistry); Physical chemistry. Current work: Electron spin resonance spectrometry; nuclear magnetic resonance spectroscopy; ENDOR spectrometry; electronic structures of molecules; photoionization and photoelectron spectroscopy; cryogenic studies on chemical substances. Home: 5612 McMaster Rd Vancouver BC V6T 1JB Canada

MCDOWELL, DAVID JAMISON, clinical psychologist, educator, researcher; b. Pitts., Jan. 11, 1947; s. David Emerson and Auleene Marley (Jamison) McD.; m. Nancy Annis, Jan. 13, 1973; children: Sasha Annis, Christopher Daniel. B.A., Princeton U., 1968; Ph.D., U. Maine-Orono, 1980. Lic. psychologist. Predoctoral intern Worcester (Mass.) State Hosp., 1976-77, part-time admission officer, 1979-82; instr. Coll. of the Holy Cross, Worcester, 1977-78; clin. dir. Milford (Mass.) Assistance Program, Inc., 1978-80; asst. prof. psychiatry and pediatrics U. Mass. Med. Ctr., Worcester, 1980-83; vis. lectr. Assumption Coll., Worcester, 1979-80; ptnr. Worcester County Counseling Assocs., Bolton, Mass., 1982—; dir. Adolescent Service, Intatient Psychiat. Unit, U. Mass. Med. Ctr., 1982-83; clin. dir. Newton (Mass.) Multi-Service Ctr., 1983—. Co-author: The Mental Health Industry, 1978; contbr. chpts. to books, articles to ency. State of Maine mental health fellow, 1973-75; Chemstrand Corp. nat. merit scholar, 1964-68. Mem. Am. Psychol. Assn., Mass. Psychol. Assn., Am. Orthopsychiat. Assn., Phi Kappa Phi. Subspecialty: Cognition. Current work: Cognitive and neuropsychological aspects of schizophrenia; sociology of mental health consumers and providers; treatment of adolescents; individual and family psychotherapy. Office: U Mass Med Ctr Div Psychology 55 Lake Ave N Worcester MA 01605

MCDOWELL, LEE RUSSELL, animal nutritionist, nutritional consultant, educator; b. Lyons, N.Y., Apr. 11, 1941; s. Russell Gale and Ida May (Lee) McD.; m. Lorraine Marie Worden, June 19, 1965; children: Suzannah, Joana, Teresa. A.A.S., Alfred (N.Y.) Agr. and Tech., 1961; B.S., U. Ga., 1964, M.S., 1965; Ph.D. in Animal Nutrition, Wash. State U., 1971. Vol Peace Corps, Santa Cruz, Bolivia, 1965-67; asst. prof. dept. animal sci. U. Fla., Gainesville, 1971-77, assoc. prof., 1977-82, prof., 1982—; nutritional cons. in, Latin Am. and, S.E. Asia. Co-author: Latin American Tables of Feed Composition, 1972, 74, Latin American Symposium on Mineral Research with Grazing Ruminants, 1978; contbr. chpts. to books. Mem. Am. Dairy Sci. Assn., Am. Soc. Animal Sci., Am. Inst. Nutrition, Am. Forage and Grassland Council, Council Agrl. Scis. and Tech., Asociación Latinoamericana de Producción Animal. Methodist. Subspecialty: Animal nutrition. Current work: Research in international animal agriculture: tropical feeds and mineral deficiencies and toxicities for grazing livestock. Home: 13 SW 15B Archer FL 32618 Office: U Fla 125 Animal Sci Bldg Gainesville FL 32611

MCDOWELL, ROBERT E., JR., animal scientist, educator, researcher; b. Charlotte, N.C., June 27, 1921; s. Robert E. and Grace Wilson (Bradford) McD.; m. Dorothy Gill, Dec. 8, 1945; children: Jean, Ann, Robert. B.S., N.C. State U., 1942; M.S., U. Md., 1949, Ph.D., 1955. Tchr. agrl. program VA, 1946; research investigator, program dir. USDA, Beltsville, Md., 1946-67; prof. internat. animal sci. Cornell U., Ithaca, N.Y., 1967—; cons. FAO, U.S. AID, Rockefeller Found., U. P.R., U. Venezuela, U. Dominican Republic, U.S. Peace Corps, U. Calif.-Davis, USDA; chmn. bd. trustees Internat. Livestock Ctr. for Africa; vice chmn. external evaluation panel U.S. AID; chmn. steering com. internat. agr. Cornell U. Author: Improvement Livestock Production in Warm Climates, 1972; Contbr. chpts. to books, articles to profl. jours. Served with USMC, 1942-46. Decorated Bronze Star, Meritorious Commendation medal.; Recipient Superior Service award USDA, 1962; Animal Agr. award in internat. animal agr. Am. Soc. Animal Sci., 1979. Mem. Am. Diary Sci. Assn., Am. Soc. Animal Sci., AAAS, Nat. Dairy Council, Ret. Officers Assn., Alpha Zeta. Subspecialties: Animal physiology; Animal genetics. Current work: Contributions—animals to support man in developing countries; research, training and development of small farm agriculture in developing countries. Office: Cornell U 131 Morrison Hall Ithaca NY 4853

MCDOWELL, WILLIAM JACKSON, chemist, researcher; b. McMinnville, Tenn., July 14, 1925; s. Lucien Lafayette and Flora Elesibeth (Lassiter) McD.; m. Betty Jeanne Lawson, June 7, 1951; children: William Lawson, Stephen Jackson. B.S., Tenn. Tech. U., 1951; M.S., U. Tenn.-Knoxville, 1954. Jr. chemist Union Carbide Nuclear, Oak Ridge, Tenn., 1951-52, chemist, 1954-75, group leader, sr. research chemist, 1975—. Pres. P.T.A., Knoxville, Tenn., 1969. Recipient IR-100 Indsl. Research and Devel. Mag., 1981. Mem. Am. Chem. Soc., Am. Nuclear Soc., AAAS, Tenn. Acad. Sci., Sigma Xi. Club: Carbide Camera (Oak Ridge). Subspecialties: Analytical chemistry; Inorganic chemistry. Current work: Separation science, the science of using substance selective phase transfer agents, applications, theory and mechanisms. Home: 10903 Melton View Ln Knoxville TN 37921 Office: Oak Ridge Nat Lab PO Box X Oak Ridge TN 37830

MCDUFFIE, FREDERIC CLEMENT, physician, foundation administrator; b. Lawrence, Mass., Apr. 27, 1924; m. Isabel Simpson Wiggin, May 31, 1952; children: Elisabeth Wiggin, Joan Selden, Deborah Howard, Charles Dennett. Grad., Harvard U., 1947, M.D. cum laude, 1951. Diplomate: Am. Bd. Internal Medicine. Intern Peter Bent Brigham Hosp., Boston, 1951-52, resident, 1952-53, 56-57; tng. in phys. chemistry Harvard U., 1953-56; in immunology Columbia Coll. Physicians and Surgeons, 1954-56; asst. prof. internal medicine U. Miss., Jackson, 1957-62, asst. prof. microbiology, 1957-64, asso. prof., 1964-65; cons. medicine and microbiology Mayo Clinic and Mayo Found., Rochester, Minn., 1965; asst. prof. internal medicine and microbiology Mayo Grad. Sch. Medicine, 1965-69, asso. prof., 1969-73, Mayo Med. Sch., 1973, prof. internal medicine and immunology, 1974-79; prof. medicine Emory U., Atlanta, 1979—; vis. investigator Center for Disease Control, Atlanta, 1979; sr. v.p. med. affairs Arthritis Found., Atlanta, 1979—; pres. Miss. chpt. Arthritis Found., 1962-63, bd. dirs., mem. exec. com., 1974-79, chmn. med. and sci. com., 1975-79, nat. trustee, 1978-79, chmn. nat. research com., 1978-79. Editorial bd.: Arthritis and Rheumatism, 1976—, Jour. Rheumatology, 1974—; editor: Jour. Lab. and Clin. Medicine, 1977-79; contbr. articles to profl. jours. Served with U.S. Army, 1943-45. Mem. Am. Heart Assn., Am. Assn. Immunologists, Am. Rheumatism Assn., Central Rheumatism Assn. (pres. 1973-74), Central Soc. Clin. Research (council 1977-79), Soc. Exptl. Biology and Medicine, Am. Fedn. Clin. Research, Joint Club, A.C.P., Alpha Omega Alpha. Home: 3155 Arden Rd NW Atlanta GA 30305 Office: Arthritis Found 1314 Spring St NW Atlanta GA 30309

MCELHANEY, JAMES HARRY, biomedical engr.; b. Phila., Oct. 27, 1933; s. James H. and Olga M. (Lazuk) McE.; m. Eileen M. Esbensen, Nov. 17, 1954; children—Kathy, Amy, Liza. B.S., Villanova U., 1955; M.S., U. Pa., 1960; Ph.D., W.Va. U., 1964. Registered profl. engr., Pa. Mech. engr. Philco Corp., Phila., 1955-59; asst. prof. mech. engring. Villanova (Pa.) U., 1959-62; prof. theoretical and applied mechanics W.Va. U., Morgantown, 1962-69; asso. prof. mech. engring., head biomechanics dept. Hwy. Safety Research Inst., U. Mich., 1969-74; prof. biomechanics Duke U., Durham, N.C., 1974—, dir. grad. studies, 1974-75; pres. Safety Electronics Corp.; cons. Ford Motor Co., Am. Motors Co., Chrysler Corp., Westinghouse Corp., NIH, others; chmn. com. on indsl. head protection Am. Nat. Standards Inst. Editor: mech. engring. sect. Jour. of Bioengring, 1977—; mem. editorial adv. bd.: Jour. of Biomechanics, 1967—; Contbr. articles on biomechanics to profl. jours. Served with U.S. Army Res., 1955-63. NIH grantee, 1967—. Mem. Am. Soc. Engring. Edn., ASME (chmn. bioengring. div. 1975—, asso. editor Trans. on Biomech. Engring.), Soc. for Exptl. Stress Analysis, Soc. Automotive Engrs. (chmn. anthropometric dummy study com.), Southeastern Neurol. Soc. (pres. elect), Biomedical Engring. Soc., Am. Acad. Mechanics, Sigma Xi, Pi Tau Sigma. Club: Duke Men's. Subspecialty: Biomedical engineering. Home: 3411 Cambridge Rd Durham NC 27707 Office: Biomedical Engring Dept Duke U Durham NC 27707

MCELHATTAN, GLENN RICHARD, chemistry educator; b. Knox, Pa., Dec. 3, 1934; s. Glenn D. and Florence B. McE.; m. Mary Frances Master, Feb. 11, 1956; children: Brenda, Dianne, David, Curtis. B.S., Clarion State Coll., 1956; M.S., Western Res. U., 1963; Ed.D., U. Pitts., 1973. Tchr. chemistry Rocky Grove High Sch., Franklin, Pa., 1959-68; prof. chemistry Clarlon State Coll., 1968—. Contbr. in field. Served as 1st. lt. USMC, 1956-59. U.S. Office Edn. grantee, 1971-73. Mem. Am. Chem. Soc., Nat. Sci. Tchrs. Assn., Pa. Sci. Tchrs. Assn.

Methodist. Subspecialties: Inorganic chemistry; Nutrition (medicine). Current work: Teaching general chemistry and nutrition to liberal arts and nursing students. Home: Rural Delivery Box 813 Hershey Rd Franklin PA 16323 Office: Venango Campus Clarion U Pa Oil City PA 16301

MC ELLIGOTT, JAMES GEORGE, neuroscientist; b. N.Y.C., June 20, 1938; s. James P. and Mildred C. (Biernesser) McE.; m. Sandra G. Fitzpatrick., Aug. 26, 1967; children: Seamus, Sean. B.S., Fordham Coll., 1960; M.A., Columbia U., 1963; Ph.D., McGill U., 1967. NIH postdoctoral fellow Brain Research Inst. UCLA, 1967-68, research assoc., 1968-71; assoc. prof. pharmacology Temple U., 1971—; cons. in field. Mem. Soc. Neurosci., AAAS, Sigma Xi. Subspecialties: Neuropharmacology; Neuropsychology. Current work: Neurophysiology, bioengineering, computer science.

MCELLIGOTT, SANDRA G., sci. researcher; b. Port-of-Spain, Trinidad, West Indies, Sept. 29, 1945; d. Alvin K. and Enid V. (Rampersaud) Fitzpatrick; m. James G. McElligott, Aug. 26, 1967; children: Seamus, Sean. B.A., UCLA, 1968; M.A., Temple U., 1974, Ph.D., 1978. Research asst. Temple U., Phila., 1974-77; postdoctoral fellow Monell Chem. Senses Center, Phila., 1977-79; research fellow U. Calif., Berkeley, 1979-80; research assoc. U. Pa., Phila., 1980—. Mem. Soc. Neuroscis. Subspecialties: Neurobiology; Developmental biology. Current work: Molecular mechanisms of synapse formation during development. Adhesion of pre- and post- synaptic membranes. Developed histochemical assay for galactosyltransferase. Home: 923 Jenifer Rd Horsham PA 19044 Office: Biology Dept U Pa Philadelphia PA 19104

MCELROY, DOUGLAS BOYDEN, astronomer; b. Long Beach, Calif., Aug. 29, 1952; s. Boyden Stewart and Helen Mildred (Hahn) McE.; m. Linda Jeanne Rupp, June 22, 1974; 1 dau., Wendy Lynne. B.S., Calif. Inst. Tech., 1974; M.S., Ariz. State U., 1977, Ph.D., 1981. Student research asst. Calif. Inst. Tech., Pasadena, 1972-74; teaching asst. Ariz. State U., 1975-77; summer research asst. Kitt Peak Nat. Obs., 1977, 78, research asst., 1978-81; postdoctoral research assoc. U. Minn., Mpls., 1981—. Contbr. articles to profl. jours. V.M. Slipher scholar Ariz. State U., 1981. Mem. Astron. Soc. Subspecialty: Optical astronomy. Current work: Research in galactic dynamics, globular clusters, photometry, kinematic and dynamic structure of external galaxies. Office: 116 Church St SE Minneapolis MN 55455 Home: 3809 Upton Ave S Minneapolis MN 55410

MC ELROY, WILLIAM DAVID, biochemist; b. Rogers, Tex., Jan. 22, 1917; s. William D. and Ora (Shipley) McE.; m. Nella Winch, Dec. 23, 1940 (div.); children—Mary Elizabeth, Ann Reed, Thomas Shipley, William David; m. Marlene A. DeLuca, Aug. 28, 1967; 1 son, Eric Gene. B.A., Stanford, 1939; M.A., Reed Coll., 1941; Ph.D., Princeton U., 1943; D.Sc., U. Buffalo, 1962, Mich. State U., 1970, Loyola U., Chgo., 1970, U. Notre Dame, 1975, Calif. Sch. Profl. Psychology, 1978; D.Pub. Service, Providence Coll., 1970; LL.D., U. Pitts., 1971, Johns Hopkins U., 1977. War research, com. med. research OSRD, Princeton, 1942-45; NRC fellow Stanford, 1945-46; instr. biology dept. Johns Hopkins, 1946, successively asst. and asso. prof., prof. biology, 1951-69, chmn. biology dept., 1956-69; also dir. McCollum-Pratt Inst., 1949-64; dir. NSF, Washington, 1969-71; chancellor U. Calif., San Diego, 1972-80, prof., 1980—. Author textbook.; Editor: (with Bentley Glass) Copper Metabolism, 1950, Phosphorus Metabolism, 2 vols, 1951, 52, Mechanism of Enzyme Action, 1954, Amino Acid Metabolism, 1955, The Chemical Basis of Heredity, 1957, The Chemical Basis of Development, 1959, Light and Life, 1961, Cellular Physiology and Biochemistry, 1961, (with C.P. Swanson) Foundations of Modern Biology series, 1961-64. Mem. Sch. Bd. Baltimore City, 1958-68. Recipient Barnett Cohen award in bacteriology, 1958; Rumford prize Am. Acad. Arts and Scis., 1964. Mem. Am. Inst. Biol. Scis. (pres. 1968), Am. Chem. Soc., Nat. Acad. Sci., Am. Soc. Biol. Chemists (pres. 1963-64), Soc. Gen. Physiology (pres. 1960-61), Soc. Naturalists, Soc. Zoologists, Am. Acad. Arts and Scis., Am. Soc. Bacteriologists, Am. Philos. Soc., AAAS (pres. 1976, chmn. 1977), Sigma Xi, Kappa Sigma. Subspecialties: Biochemistry (biology); Microbiology. Current work: Basic science of bioluminescence-applilcations to clinical, agriculture and industrial problems. Office: Univ Calif San Diego La Jolla CA 92093

MCENALLY, TERENCE ERNEST, JR., physics educator; b. Richmond, Va., Apr. 21, 1927; s. Terence Ernest and Ellen (Cottrell) McE.; m. Marie Gianino, June 28, 1964; children: Terence E., Deirdre. B.S., Va. Poly. Inst., 1950, M.S., 1955; Ph.D., MIT, 1966. Mem. faculty dept. physics Va. Poly. Inst., Blacksburg, 1952-55; mem. faculty dept. elec. engring. N.C. State U., Raleigh, 1961-63; instr. dept. physics MIT, Cambridge, Mass., 1966-67; mem. faculty dept. physics East Carolina U., Greenville, N.C., 1967—. Served with USNR, 1945-46. Mem. Am. Phys. Soc., Sigma Xi. (chpt. pres. 1972). Subspecialties: Magnetic physics; Crystallography. Current work: ESR of minerals. Home: 113 N Woodlawn Ave Greenville NC 27834 Office: Dept Physics East Carolina U Greenville NC 27834

MCENTEE, KENNETH, veterinary pathologist, reproductive pathology consultant, educator; b. Oakfield, N.Y., Mar. 30, 1921; s. Michael Thomas and Florence (Bobsene) McE.; m. Janet Louise Fraser, Aug. 6, 1952; children: Michael F., Margaret C. D.V.M., Cornell U., 1944; V.M.D. (hon.), Royal Vet. Coll., Stockholm, Sweden, 1975. Diplomate: Am. Coll. Vet. Pathologists, 1952. Pvt. practice vet. medicine, Newport, Vt., 1944-45; from asst. prof. to prof. N.Y. State Vet. Coll., Cornell U., Ithaca, 1947-80, assoc. dean, 1969-73; prof. reproductive pathology Coll. Vet. Medicine. U. Ill., Urbana, 1980—; dir. Internat. Registry of Reproductive Pathology. Contbr. chpts. to textbooks, articles to profl. jours. Served to capt. Vet. Corps U.S. Army, 1945-47. Recipient Borden award, 1971; Eastern Artificial Insemination Coop. Research award, 1973. Mem. AVMA, Am. Coll. Vet. Pathologists, Internat. Acad. Pathology, Am. Assn. Pathologists, U.S. Power Squadron, Sigma Xi, Phi Zeta, Phi Kappa Phi. Subspecialties: Animal pathology; Comparative Reproductive Pathology. Current work: Reproductive diseases of domestic animals. Home: 1805 S Carle Dr Urbana IL 61801 Office: Coll Vet Medicine 2001 S Lincoln Ave Urbana IL 61801

MC EVILLY, THOMAS VINCENT, seismologist; b. East Saint Louis, Ill., Sept. 2, 1934; s. Robert John and Frances Nathalie (Earnshaw) McE.; m. Dorothy K. Hopfinger, Oct. 23, 1970; children: Mary, Susan, Ann, Steven, Joseph, Adrian. B.S., St. Louis U., 1956, Ph.D., 1964. Geophysicst California Co., New Orleans, 1957-60; engring. v.p. Sprengnether Instrument Co., St. Louis, 1962-67; asst. prof. seismology U. Calif., Berkeley, 1964-68, assoc. prof., 1968-74, prof., 1974—, chmn. dept. geology and geophysics, 1976-80, asst. dir. seismographic sta., 1968—; assoc. dir., head earth sci. div. Lawrence Berkeley Lab., 1982—; cons. numerous govt. agys., geotech. cos. Contbr. numerous articles to profl. jours. Mem. Am. Geophys. Union, Royal Astron. Soc., Earthquake Engring. Research Inst., Seismol. Soc. Am. (editor bull. 1976—), Soc. Exploration Geophysicists, AAAS. Subspecialty: Geophysics. Current work: Earthquake processes and crustal structure; applied seismology and seismic instrmentation. Office: Dept Geology and Geophysics U Calif Berkeley CA 94720

MCEWEN, BRUCE SHERMAN, neuroscientist, educator; b. Ft. Collins, Colo., Mar. 17, 1938; s. George Middleton and Esther Cornelia (Lenters) McE.; m. Nancy Lu Ames, Jan. 8, 1937; children: Carolyn Ann, Sarah Louise. A.B. summa cum laude in Chemistry, Oberlin Coll., 1959; Ph.D. in Biology, Rockefeller U., 1969. USPHS postdoctoral fellow U. Goteborg, Sweden, 1964-65; asst. prof. zoology U. Minn., Mpls., 1965-66; asst. prof. dept. neurosci. Rockefeller U., N.Y.C., 1966-71, assoc. prof., 1971-81, prof., 1981—. Mem. Soc. Neurosci., Endocrine Soc., Am. Soc. Neurochemistry, Phi Beta Kappa, Sigma Xi. Subspecialties: Neurochemistry; Neuropsychology. Current work: Hormone action as the brain and behavior; cellular and molecular actions of neuroactive substances, including hormones. Office: Rockefeller U 1230 York Ave New York NY 10021

MCFADDEN, PETER WILLIAM, educator; b. Stamford, Conn., Aug. 2, 1932; s. Kenneth E. and Marie (Gleason) McF.; m. Kathleen Mary Garvey, Aug. 11, 1956; children—Peter, Kathleen, Mary. B.S. in Mech. Engring, U. Conn., 1954, M.S., 1956; Ph.D., Purdue U., 1959. Registered profl. engr., Ind., Conn. Asst. instr. U. Conn., 1954-56, dean Sch. Engring., prof. mech. engring., 1971—; mem. faculty Purdue U., 1956-71, prof. mech. engring., head, 1965-71; postdoctoral research Swiss Fed. Inst., Zurich, 1960-61; cons. to industry, 1959—. Mem. ASME, Am. Soc. Engring. Edn. Subspecialty: Mechanical engineering. Research in cryogenics, heat transfer, mass. transfer. Home: 85 Willowbrook Rd Storrs CT 06268

MCFALL, ELIZABETH, molecular biologist, educator, cons.; b. San Diego, Oct. 28, 1928; d. C. E. and Teresa (Moore) McF.; m. Andrew L. Floyd. B.S. in Chemistry, San Diego State Coll., 1950; M.A. in Econs, U. Calif., Berkeley, 1954; Ph.D. in Biochemistry, U. Calif., Berkeley, 1957. Research fellow Harvard U., 1957-60; research asso. M.I.T., Cambridge, 1960-61 62-63; NIH spl. research fellow Nat. Inst. Med. Research, London, 1961-62; prof. microbiology Sch. Medicine, N.Y. U., N.Y.C., 1963—. Editor: Jour. Bacteriology. Recipient NIH career devel. award, 1964-74. Mem. Am. Soc. Biol. Chemistry, Am. Soc. Microbiology, AAAS, Am. Acad. Microbiology, N.Y. Acad. Scis. Democrat. Subspecialties: Genetics and genetic engineering (biology); Gene actions. Current work: Molecular biology, control of gene expression. Office: Dept Microbiology Sch Medicine NY Univ New York NY 10016

MCFARLANE, HAROLD FINLEY, nuclear engineer; b. Hagerstown, Md., Apr. 23, 1945; s. Alvis Finley and Lila (Anderson) McF.; m. Mary Ellen Newberry, June 22, 1968; 1 son, Matthew Harold. B.S. magna cum laude in Physics, U. Tex.-Austin, 1967; M.S. in Engring. Sci, Calif. Inst. Tech., 1968, Ph.D., 1971. Research assoc. Calif. Inst. Tech., 1971; asst. prof. NYU, 1971-72; engr. Argonne Nat. Lab., Idaho Falls, Idaho, 1972-77, group leader, 1977-80, sect. head, 1980-82, program mgr., 1982—. Mem. Am. Nuclear Soc. (program chmn. reactor physics div. 1982), Phi Beta Kappa, Phi Kappa Phi. Club: Idaho Falls Ski. Subspecialties: Nuclear fission; Nuclear engineering. Current work: Physics of fast breeder reactors; critical experiments. Home: Lower Power Plant Rd Idaho Falls ID 83402 Office: Argonne Nat Lab PO Box 2528 Idaho Falls ID 83402

MCFEE, WILLIAM WARREN, soil scientist; b. Concord, Tenn., Jan. 8, 1935; s. Fred Thomas and Ellen Belle (Russell) McF.; m. Barbara Anella Steelman, June 23, 1957; children—Sabra Anne, Patricia Lynn, Thomas Hallie. B.S., U. Tenn., 1957; M.S., Cornell U., 1963, Ph.D., 1966. Mem. faculty Purdue U., 1965—, prof. soil sci., 1973—, dir. natural resources and environ. sci. program, 1975—; vis. prof. U. Fla., 1972-73; cons. acid rain program Electric Power Research Inst., EPA. Author articles in field, chpts. in books. Served with USAR, 1958-61. Alpha Zeta scholar, 1957; named Outstanding Agr. Tchr. Purdue U., 1972. Fellow Am. Soc. Agronomy, Soil Sci. Soc. Am.; mem. Internat. Soil Sci. Soc., Soc. Am. Foresters, Sigma Xi. Presbyterian. Subspecialties: Soil chemistry; Ecosystems analysis. Current work: Effects of atmosphere depositon on soils; properties of reclaimed minelands. Home: 709 McCormick Rd West Lafayette IN 47906 Office: Agronomy Dept LILY Purdue U West Lafayette IN 47907

MCGAGHIE, WILLIAM CRAIG, medical educator, researcher; b. Chgo., June 28, 1947; s. William and Vivian Iona (Skoglund) McG.; m. Pamela Wall, Mar. 13, 1976; children: Michael Craig, Kathleen Ann. B.A., Western Mich. U., 1969; M.A., Northwestern U., 1971, Ph.D., 1973. Lectr. ednl. psychology Northwestern U., 1973-74; asst. prof. U. Ill., 1974-78, U. N.C., Chapel Hill, 1978-82, assoc. prof., 1982—. Author: Competency-Based Circiculum Development in Medical Education, 1978; editor: Profiles in College Teaching, 1972; mem. editorial bd.: Evaluation and the Health Professions, 1981—; contbr. articles to profl jours. USPHS grantee, 1981—. Mem. Am. Psychol. Assn., Am. Ednl. Research Assn. Episcopalian. Subspecialties: Medical education; Medical psychology. Current work: Medical education research and development; behavioral medicine. Home: 112 Village Ln Chapel Hill NC 27514 27514

MCGARITY, ARTHUR EDWIN, engineering educator, energy and environmental systems researcher; b. Chgo., Apr. 2, 1951; s. Owen and Lois Wilson (Thomas) McG.; m. Jane Ziegler, June 11, 1977; children: Kate Elizabeth, Owen Carlos. B.S., Trinity U., 1973; M.S.E., Johns Hopkins U., 1978, Ph.D., 1979. Elec. engr. San Antonio Pub. Service, 1973-74; profl. asst. NSF, Washington, summer 1975; summer intern Exec. Office of Pres., Washington, summer 1975; asst. prof. engring. Swarthmore (Pa.) Coll., 1978—; scientist-in-residence Solar Energy Group, Argonne (Ill.) Nat. Lab., 1981-82. Author: Solar Heating and Cooling: An Economic Assessment, 1977. Solar Energy Research Inst. course devel. grantee, 1979; Sigma Xi summer research grantee, 1979; Argonne Nat. Lab. faculty research leave grantee, 1981-82. Mem. Inst. Mgmt. Sci., Internat. Solar Energy Soc. Democrat. Presbyterian. Subspecialties: Solar energy; Operations research (engineering). Current work: Solar energy systems, ops. research, environ. policy, research on solar heating systems design, computer simulation methods, seasonal storage, compound parabolic concentrator solar collectors. Home: 525 Elm Ave Swarthmore PA 19081 Office: Dept Engring Swarthmore Coll Swarthmore PA 19081

MCGARVEY, JOHN JAMES, computer systems firm executive; b. Yonkers, N.Y., Dec. 26, 1931; s. John Joseph and Elizabeth Marie (Flanagan) McG.; m. Patricia Ann Marsteller, May 18, 1957; children: Daniel, Sheila, Michael, John Timothy. B.S., Pa. State U., 1956; M.S., Naval Postgrad. Sch., 1966. Commd. ensign U.S. Navy, 1956, advanced through grades to comdr., 1969, ret., 1977; sr. systems designer Bay Area Rapid Transit, Oakland, Calif., 1977-78; tech. systems mgr. Am. Pres. Lines, Ltd., Oakland, 1978-79; dir. profl. services Optimum Systems, Inc., Santa Clara, Calif., 1979-80; v.p. Info. Systems Group, Inc., San Leandro, Calif., 1979—; cons. ISG Inc., 1979—; lectr. Calif. State U., Hayward, 1978—, U. Mo.-College Park, 1975-77. Mem. Assn. Computing Machinery, Assn. for System Mgmt., Naval Inst. Republican. Roman Catholic. Subspecialties: Information systems (information science); Database systems. Current work: Programming lang./software. Home: 8111 Phaeton Dr Oakland CA 94605 Office: 15200 Hesperian Blvd San Leandro CA 94578

MCGAUGHEY, ROBERT WILLIAM, zoology educator; b. Mpls., Sept. 21, 1941; s. M.W. and R.V. (Chase) McG.; m. Beverly M., July 2, 1982; 1 son, Matthew G.; stepchildren: Marck A. Sawyer, Kelli M. Sawyer, Shana B. Sawyer. B.A., Augustana Coll., Sioux Falls, S.D., 1963; M.A., U. Colo., 1965; Ph.D., Boston U., 1968. Staff scientist Worcester Found. Exptl. Biology, Shrewsbury, Mass., 1968-69; mem. faculty dept. zoology Ariz. State U., Tempe, 1971—; postdoctoral fellow Agr. Research Council unit of reproductive physiology and biochemistry, Cambridge, Eng., 1969-70. Contbr.: chpts. to books, articles to profl. jours. Bd. dirs. Planned Parenthood, Phoenix, 1972. Lalor Found. fellow, 1969-70; Nat. Inst. Child Health and Human Devel. grantee, 1974—; Whitehall Found. grantee, 1981-82. Mem. Soc. Developmental Biology. Subspecialties: Developmental biology; Reproductive biology. Current work: Regulation of mammalian oocyte maturation; cytogenetics of mammalian development. Office: Dept Zoology Ariz State U Tempe AZ 85287

MCGEAN, THOMAS JAMES, transp. cons. co. exec.; b. N.Y.C., Apr. 8, 1937; s. James T. and Lilian R. (Sargent) McG.; m. Doris L., Aug., 1962; children: Terence, Cynthia. B.M.E. cum laude, NYU, 1959; M.S.M.E., Calif. Inst. Tech., 1960. Registered profl. engr., Va., Fla. Engr. Bell Telephone Labs., Murray Hill, N.J., 1960-67; engr. Computer Sci. Corp., Falls Church, Va., 1967-69; sr. engr. Mitre Corp., McLean, Va., 1969-74; dir. transp. tech. Deleuw Cather, Washington, 1974-76; cons. transp., Annandale, Va., 1976-81; exec. v.p. N. D. Lea & Assocs., Washington, 1981—; ptnr. Lea, Elliott, McGean and Co., Washington, 1983—; assoc. professorial lectr. George Washington U.; vis. asso. prof. Howard U. Author: Urban Transportation Technology, 1976. Pres. Washington chpt. Vols. for Internat. Tech. Assistance, 1962-73. Recipient William R. Bryan medal in engring. N.Y. U., 1959, David Orr prize in mech. engring., 1959. Mem. Transp. Research Bd., Inst. Transp. Engrs., ASME, Tau Beta Pi, Pi Tau Sigma. Subspecialties: Systems engineering; Mechanical engineering. Current work: Advanced automated transit systems and cost/benefit assesment of advanced research and devel. needs. Patentee coaxial cable design. Home: 3711 Spicewood Dr Annandale VA 22003 Office: Dulles Internat Airport 600 W Service Rd Suite 32 PO Box 17030 Washington DC 20041

MC GEE, MARK GREGORY, research psychologist, author; b. Mpls., Dec. 7, 1946; s. Theodore G. and Martha A. (Khune) McGee; m. Ellen Shiplet, Aug. 8, 1977; children: Alyson, Katy, Maureen. B.A. summa cum laude, U. Minn.-Mpls., 1973, Ph.D., 1976. Teaching assoc. U. Minn.-Mpls., 1974-76; asst. prof. Tex. A&M U., 1976-80; research fellow U. Colo. Med. Sch., Denver, 1980-81; research psychologist Brain Scis. Labs., Nat. Jewish Hosp. and Research Ctr., Denver, 1981—; editorial cons. Author: Human Spatial Abilities, 1979, Introductory Psychology Reader, 1980, Introductory Psychology (textbook), 1984. Served as sgt. USMC, 1968-71; Vietnam. Recipient Lucetta O. Bissel research award U. Minn.-Mpls., 1975; NIMH postdoctoral research fellow, 1981-83. Mem. Am. Psychol. Assn., AAAS, Am. Ednl. Research Assn., Behavior Genetics Assn., Brain Behavioral Scis. Assn. Subspecialties: Developmental psychology; Psychobiology. Current work: Research on adolescent bio-cognitive development,individual differences in functional brain development and development of specific cognitive abilities; electrophysiological measurement of brain activity during reading in normal and reading disabled readers. Home: 27848 Whirlaway Trail Evergreen CO 80439 Office: Nat Jewish Hosp and Research Ctrs Brain Scis Lab 3800 E Colfax Ave Denver CO 80206

MCGHEE, GEORGE RUFUS, JR., paleobiologist, educator; b. Henderson, N.C., Sept. 25, 1951; s. George Rufus and Mary London (Cobb) McG.; m. Marae Wilcox Paschall, Nov. 25, 1971. B.Sc. with honors, N.C. State U., 1973; M.Sc., U. N.C.-Chapel Hill, 1975; Ph.D., U. Rochester, 1978. Hydrologic field asst. U.S. Geol. Survey, Raleigh, N.C., 1972-73; geprufte wissenschaftliche hilfskraft U. Tubingen, W.Ger., 1977; asst. prof. geology and ecology Rutgers U.-New Brunswick, 1978—; vis. scientist Field Mus. Natural History, Chgo., 1981; research assoc. Am. Mus. Natural History, N.Y.C., 1982—; gastdozent U. Tubingen, 1982. Assoc. editor: Paleobiology, 1982-85. NSF grantee, 1981-84; Deutsche Forschungsgemeinschaft grantee, 1982; Am. Chem. Soc. grantee, 1982-85. Mem. Internat. Palaeontol. Assn., Paleontol. Soc., Die Palaontologische Gesellschaft, Palaeontol Assn. (Eng.), Soc. Systematic Zoology. Democrat. Quaker. Subspecialties: Paleobiology; Paleoecology. Current work: Ecosystem evolution, marine paleoecology, evolutionary morphology. Office: Geol Scis Rutgers U New Brunswick NJ 08903 Home: 66 Wyckoff New Brunswick NJ 08901

MCGILL, SCOTT DOUGLAS, data processing educator, consultant; b. Meadville, Pa., Sept. 24, 1946; s. Gaylord Arthur and Margaret Arnetta (Kebert) McG.; m. Cathleen Ann Chaffin, Nov. 28, 1970; children: Kelly Meghan, Kerry Shannon. B.S. in Math, Allegheny Coll., Meadville, 1968, NYU, 1971; M.A. in Computer Systems Mgmt, U. Nebr., 1972; postgrad., U. Colo., Denver, 1974—. Systems analyst Sperry Univac, Washington, 1972-73; sr. systems analyst Colorado Springs, Colo., 1973-75; mgr. programming and design City of Colorado Springs, 1975-76, dir. data processing, 1975—; mem. honorarium faculty U. Colo., Colorado Springs, 1975—; sr. instr. Mgmt. Devel. Found., Colorado Springs, 1978; dir. Cibar Systems Inst., Colorado Springs, 1982—; data processing cons. Mem. bus. adv. council Computer Tng. for Severely Handicapped, Denver, 1982; bd. dirs. Goodwill Industries, Colorado Springs, 1982—; mem. adv. council U. So. Colo., Pueblo, 1981—; mem. bus. adv. council U. Colo., 1978—; mem. exec. council Pikes Peak council Boy Scouts Am., Colorado Springs, 1979—. Served to capt. USAF, 1968-72. Recipient Disting. Greater Colo. Service award Denver Fed. Exec. Bd., 1980; Silver Individual Service award Data Processing Mgmt. Assn., 1981; Outstanding Contbn. to Data Processing award Colo. Intergovtl. ADP Council, 1978; Outstanding Data Processing Orgn. of Yr. award Rocky Mountain Assn. Local Govt. Computer Users, 1979. Mem. Soc. for Info. Mgmt. (chmn. Rocky Mountain chpt. 1983), Data Processing Mgmt. Assn. (pres. 1978, internat. dir. 1982), IEEE, Assn. for Computing machinery (chmn. Pikes Peak chpt. 1977-78), Am. Mgmt. Assn. (program chmn. 1983). Republican. Club: Rocky Mountain (dir. 1982). Lodges: Kiwanis (1st v.p. 1982-83); Masons; Shriners; Elks. Subspecialties: Software engineering; Information systems (information science). Current work: Data processing management (motivation, productivity, system life cycle, government applications). Home: 4529 Misty Dr Colorado Springs CO 80907 Office: Mgmt Info Ctr 217 S Wahsatch Colorado Springs CO 80903

MCGILVERY, ROBERT WARREN, biochemistry educator, adminstr.; b. Coquille, Oreg., Aug. 25, 1920; s. Neil and Mary (Berry) McG.; m. Alice Marie Lusby, Nov. 1, 1943. B.S., Oreg. State U., 1941; Ph.D., U. Wis.-Madison, 1947. USPHS fellow U. Wis.-Madison, 1947-48, asst. prof., 1948-51, assoc. prof., 1951-57, U. Va., Charlottesville, 1957-62, prof., 1962—, chmn. dept. biochemistry, 1982—. Author: Biochemistry-A Functional Approach, 1970, 2d edit., 1979, 3d edit., 1983, Biochemical Concepts, 1975, Fructose 1.6 Diphosphatase, 1962. Served with USN, 1942-45. Fellow AAAS, mem., Am. Soc. Biol. Chemists, AAUP. Subspecialty: Biochemistry (medicine). Current work: Fuel metabolism, muscle metabolism, nutrition. Office: PO Box 3852 Charlottesville VA 22903

MC GINNIS, ROBERT CAMERON, univ. dean; b. Edmonton, Alta., Can., Aug. 18, 1925; s. Reginald James and Louie Mildred (Hyde) McG.; m. Alyce Lenore Wright, July 21, 1951; children—Kathryn Mae, Robert Kelly, Shauna-Marie Claretta. B.Sc., U. Alta., 1949, M.Sc., 1951; Ph.D., U. Man., 1954. Cytogeneticist Agr. Can., Winnipeg, Man., 1951-60; asso. prof. plant sci. U. Man., 1960-65, head dept., 1965-75, dean Faculty Agr., 1979—; dir. Plant Breeding Sta.,

Njoro, Kenya, 1973-75; asso. dir. Internat. Crops Research Inst. for Semi-Arid Tropics, Hyderabad, India, 1975-79; trustee Internat. Inst. Tropical Agr., Ibadan, Nigeria, 1980; mem. governing bd. Biomass Energy Inst., Winnipeg, 1980; cons. in field. Author papers in field. Chmn. bd. Ft. Garry Public Library, 1968-73; mem. bd. govs. Grace Hosp., Winnipeg, 1971-73. Recipient Centennial medal, 1967. Mem. Agrl. Inst. Can., Man. Inst. Agrologists, Genetics Soc. Can., Am. Genetics Soc., N.Y. Acad. Scis. Mem. United Ch. Can. Club: Rotary. Subspecialty: Plant genetics. Current work: Plant genetics. Home: 800 Cloutier Dr Winnipeg Canada R3V 1L2 Office: Faculty Agr Univ Man Winnipeg MB Canada R3T 2N2

MCGOOKEY, DONALD PAUL, geologist, consultant; b. Sandusky, Ohio, Sept. 19, 1928; s. Earl Jay and Rosella Ethel (Kramer) McG.; m. Doris Jean Masell, Dec. 29, 1951; children: Douglas, Daniel, Dianna, David, Donald, Doreen. B.S., Bowling Green (Ohio) State U., 1951; M.A., U. Wyo., Laramie, 1952; Ph.D., Ohio State U., 1958. Cert. petroleum geologist. Geologist Texaco, Rocky Mountain Area., 1952-68, N.Y.C., 1969-71, chief geologist, Houston, 1971-78, asst. div. mgr. exploration, Midland, Tex., 1978-79; sr. v.p. exploration Omni Exploration, Radnor, Pa., 1979-80; self-employed petroleum geologist, Midland, 1980—. Author: Cretaceous-Rocky Mountain Area, 1972, Tertiary of Gulf Coast Area, 1975; Editor: Typical Rocky Mountain Traps, 1966. Served with USN, 1946-48. Bownocker fellow Ohio State U., 1957-58. Fellow Geol. Soc. Am.; mem. Am. Assn. Petroleum Geologists (assoc. editor 1966-81, chmn. publ. com. 1975-81), West Tex. Geol. Soc., Soc. Econ. Paleontologists and Mineralogists, Soc. Ind. Profl. Earth Scientists. Republican. Episcopalian. Subspecialties: Geology; Sedimentology. Current work: Research into the development of new concepts of the deposition of sediments and resulting preservation in the geologic record. Home: 2406 Culpepper Midland TX 79701 Office: 228 Western United Life Bldg Midland TX 79701

MC GOON, DWIGHT CHARLES, surgeon, educator; b. Marengo, Iowa, Mar. 24, 1925; s. Charles Douglas and Ada Belle (Buhlman) McG.; m. Betty Lou Hall, Apr. 2, 1948; children: Michael, Susan, Betsy, Sarah. Student, Iowa State U., 1942-43, St. Ambrose Coll., Davenport, Iowa, 1943-44; M.D., Johns Hopkins U., 1948. Intern Johns Hopkins Hosp., 1948-49, resident in surgery, 1949-54; cons. in surgery Mayo Clinic, Rochester, Minn., 1957—; Stuart W. Harrington prof. surgery Mayo Med. Sch., 1975-79. Editor-in-chief: Jour. Thoracic and Cardiovascular Surgery, 1977—; editorial bd.: Circulation, 1970-76, Surgery, 1971-77, Am. Jour. Cardiology, 1969-77, Am. Heart Jour, 1969-76; contbr. numerous articles to profl. jours. Served with USN, 1943-45; with M.C. USAF, 1954-56. Fellow ACS; mem. Am. Assn. Thoracic Surgery (pres. 1983-84), Am. Coll. Cardiology (trustee 1979-83), Am. Surg. Assn., Soc. Clin. Surgery, Soc. Univ. Surgeons, Johns Hopkins Soc. Scholars, Phi Beta Kappa, Alpha Omega Alpha. Presbyterian. Subspecialty: Cardiac surgery. Home: 706 12th Ave SW Rochester MN 55901 Office: Mayo Clinic 200 1st St SW Rochester MN 55901

MCGOVREN, JAMES PATRICK, pharmacologist; b. Washington, Ind., June 12, 1947; s. Paul Francis and Elizabeth Mary (McCrisaken) McG.; m. Cecilia Marie Shaw, Nov. 28, 1968; children: Laura Elizabeth, Katie Susan. Student, U. Notre Dame, 1965-67; B.S. in Pharmacy, Purdue U., 1970; Ph.D. in Pharm. Scis, U. Ky., 1975. Research assoc. cancer research Upjohn Co., Kalamazoo, 1983—, biology research head, career research, 1983—. NSF trainee, 1970-73; Charles J. Lynn Meml. fellow Am. Found. Pharm. Edn., 1973-75. Mem. Am. Assn. Cancer Research, Am. Pharm. Assn., Acad. Pharm. Scis. Subspecialties: Cancer research (medicine); Pharmacology. Current work: Pharmacology of antitumor agents and pharmacokinetics. Home: 2239 Crimora Schoolcraft MI 49087 Office: Upjohn Co Kalamazoo MI 49001

MCGOWAN, JON GERALD, mechanical engineer, educator; b. Lockport, N.Y., May 3, 1939; s. Gerald F. and Xenia W. (Guenther) McG.; m. Suzanne Jessop, Sept. 25, 1965; children: Gerald, Edward. B.S., Carnegie Inst. Tech., 1961, Ph.D., 1965; M.S., Stanford U, 1962. Devel. Engr. E.I. DuPont de Nemours, Wilmington, Del., 1965-67; prof. mech. engring. U. Mass., Amherst, 1967—; cons., ptnr. Windpower Assocs. Contbr. articles to profl. jours. Trustee Dickenson Meml. Library, Northfield, Mass., 1975—. Mem. ASME, Internat. Solar Energy Soc., Air Pollution Control Assn., Internat. Hydrogen Energy Soc. Republican. Congregationalist. Subspecialties: Solar energy; Combustion processes. Current work: Solar energy research and consulting; wind energy systems research; combustion engineering—applied fluid mechanics and thermodynamics. Home: 134 Main St Northfield MA 01360 Office: Dept Mech Engring U Mass Amherst MA 01003

MCGRAIN, PRESTON, state government agency geologist; b. Corydon, Ind., Dec. 10, 1917; s. Albert Miller and Eva Alice (Shuck) McG.; m. Magdalene Schlotthauer, Feb. 16, 1959. A.B., Ind. U., 1940, M.A., 1942. Cert. profl. geologist, Ind. Asst. geologist Ind. Geol. Survey, Bloomington, summers 1940, 46, 47; geologist Ind. Flood Control and Water Resources Commn., Indpls., 1947-50; asst. state geologist Ky. Geol. Survey, Lexington, 1950-83; cons. geologist, summer 1941. Contbr. numerous articles, reports in field to profl. lit.; editor: Ky. Geol. Survey, 1952-64. Served to capt. C.E. AUS, 1942-46. Fellow Geol. Soc. Am., Ind. Acad. Sci.; mem. Soc. Mining Engrs., Am. Assn. Petroleum Geologists, Sigma Xi. Christian Scientist. Club: Lake Ellerslie (Lexington). Subspecialties: Economic geology; Geology. Current work: Geology of Kentucky; economic aspects of Kentucky minerals deposits. Home and Office: 1221 Providence Rd Lexington KY 40502

MCGRATH, JOSEPH FAY, research mathematician; b. Chgo., Apr. 10, 1942. B.S. in Math. and Physics, U. Albuquerque, 1964, M.S., U. Dayton, 1968, Ph.D., U. N.Mex., 1974. Research mathematician KMS Fusion, Inc., Ann Arbor, Mich., 1975—. Mem. Washtenaw County Democratic Com., Ann Arbor, 1980—. Mem. Soc. for Indsl. and Applied Math. Subspecialties: Applied mathematics; Mathematical software. Current work: Mathematics applied to basic research problems in inertial fusion that give rise to first kind integral equations, initial value problems, or boundary value problems. Home: 2936 Renfrew St Ann Arbor MI 48105 Office: KMS Fusion Inc PO Box 1567 Ann Arbor MI 48106

MCGRAW, JOHN THOMAS, astronomer; b. Gaylord, Minn., Mar. 22, 1946; s. Thomas Roger and Grace Ann (Friederichs) McG.; m. Maryann Margaret McDonough, June 23, 1973; 1 son, Thomas Patrick. B.A., St. Olaf Coll., 1968; M.A., U. Tex., 1973, Ph.D., 1977. Research scientist asst. dept. astronomy U. Tex.- Austin, 1970-77, research assoc., 1977-78; research assoc. dept. astronomy U. Cape Town, South Africa, 1975-76; research assoc. Steward Obs., U. Ariz., Tucson, 1979-81, asst. astronomer, 1981—. Contbr. articles in field to profl. jours. Vice pres. Manana Vista Homeowners Assn. NSF research grantee, 1981-83. Mem. Am. Astron. Soc., Royal Astron. Soc., Astron. Soc. Pacific, Sigma Xi, Sigma Pi Sigma. Subspecialties: Optical astronomy; Cosmology. Current work: Development of automated survey telescope to do observational cosmology. Office: Steward Obs U Ariz Tucson AZ 85721

MC GREGOR, DOUGLAS HUGH, pathologist; b. Temple, Tex., Aug. 28, 1939; s. Harleigh Heath and Joyce Ellen (Lambert) McG.; m. Mizuki Kitani, July 6, 1969; children: Michelle Sakuya, David Kenji. B.A., Duke U., 1961, M.D., 1966; postgrad., U. Edinburgh, Scotland, 1961-62. Diplomate: Am. Bd. Pathology. Intern and chief resident in pathology UCLA Med. Ctr., Los Angeles, 1966-68; sr. surgeon and lt. comdr. Atomic Bomb Casualty Commn., Hiroshima, Japan, 1968-71; chief resident in pathology Queens Med. Ctr., Honolulu, 1971-73; asst. and assoc. prof. pathology U. Kans. Med. Ctr., Kansas City, 1973-82, prof., 1982—; dir. anat. pathology VA Med. Ctr., Kansas City, Mo., 1975—. Contbr. numerous articles to profl. jours., chpts. to books. Leader YMCA Indian Princess Program, Overland Park, Kans., 1977-79, Indian Guide Program, 1978-80, Cub Scouts Am., Overland Park, 1980-82, Boy Scouts Am., Leawood, Kans., 1982. Served as lt. comdr. USPHS, 1968-71; Japan. Grantee Merck, Sharp and Dohme, 1980, NIH, 1980. Fellow Coll. Am. Pathologists, Am. Soc. Clin. Pathologists; mem. Am. Assn. Pathologists, Internat. Acad. Pathologists, Soc. Exptl. Biology and Medicine, N.Y. Acad. Scis., AAAS, Kansas City Soc. Pathologists (sec.-treas. 1982-83, pres. 1983-84). Club: Leawood Country. Subspecialties: Pathology (medicine); Microscopy. Current work: Biology and pathology of parathyroid hormone secretion; platelet-leukocyte aggregation; ultrastructure and pathobiology of neoplasms; radiation carcinogenesis; morphogenesis of atherosclerosis. Home: 9400 Lee Blvd Leawood KS 66206 Office: VA Medical Ctr 4801 Linwood Blvd Kansas City MO 64128

MCGREGOR, WHEELER KESEY, JR., physicist; b. Akron, Ohio, Apr. 20, 1929; s. Wheeler Kesey and Emma Zada (Turner) McG.; m. Frankie Marie Simons, Feb. 12, 1930. B.S. in Engring. Physics, U. Tenn., 1951, M.S. in Engring. Sci, 1961, Ph.D. in Physics, 1969. With ARO, Inc. (now Sverdrup Tech., Inc.), 1951; at Arnold Engring. Devel. Center, U.S. Air Force, Arnold AFB, Tenn., 1951—, now sr. tech specialist; sabbatical Rocket Propulsion Lab., Edwards AFB, Calif., 1977-78; adj. faculty U. Tenn. Space Inst. Contbr. articles to profl. jours. Recipient Arnold award for contbns. to aero. and astronautics Am. Rocket Soc., 1961. Mem. AIAA, Am. Phys. Soc., Air Force Assn. Subspecialties: Atomic and molecular physics; Aerospace engineering and technology. Current work: Plasma physics, combustion technology, quantitative spectroscopy. Home: Route 6 Stillwood Dr Manchester TN 37355 Office: Mail Stop 660 Arnold AF Statio TN 37389

MCGROARTY, ESTELLE JOSEPHINE, biochemistry educator, biochemical researcher; b. Lafayette, Ind., Sept. 14, 1945; d. George Phillip and Grace Elanor (Williamson) Beihl; m. Dennis Lee McGroarty, Aug. 27, 1967; children: Michael Patrick, Kathleen Erin. B.S., Purdue U., 1967, Ph.D., 1971. Lectr. biochemistry Purdue U., 1971-72; research assoc. Mich. State U., 1972-73, asst. prof., 1973-78, assoc. prof., 1978—. NASA fellow, 1967-70; Research Corp. Mich. State U. Grantee, 1981. Mem. Biophys. Soc., Am. Soc. Microbiology, AAAS, Sigma Xi. Presbyterian. Subspecialties: Biophysics (biology); Biochemistry (biology). Current work: Biochemistry and biophysical properties of membranes and extraced lipids; analysis of the structure of erythrocyte membranes, bacterial membranes and membranes from normal and transformed fibroblasts. Home: 3933 Willow Ridge Dr Holt MI 48842 Office: Department of Biochemistry Michigan State University Biochemistry Bldg East Lansing MI 48824

MCGUIRE, EUGENE JOSEPH, research physicist; b. N.Y.C., May 15, 1938; s. Andrew Joseph and Anne Teresa (Gwynne) McG.; m. Coralie Seidler, Aug. 6, 1965; children: Andrew, Rachel, David. B.E.E., Manhattan Coll., 1959; Ph.D., Cornell U., 1965. With Sandia Nat. Labs., Albuquerque, 1965-74, staff mem., 1974—, supr. laser theory div., 1982—. Contbr. articles to physics jours. Mem. Am. Phys. Soc. Subspecialties: Atomic and molecular physics; Laser fusion. Current work: Atomic physics, inner-shell ionization and decay, application of atomic physics to laser and particle-beam fusion. Home: 4083 Dietz Farm Circle NW Albuquerque NM 87107 Office: Sandia Nat Labs Kirtland AFB E Albuquerque NM 07185

MCGUIRE, JAMES HORTON, physicist, educator; b. Canandaigua, N.Y., June 7, 1942; s. Horton E. and Karolyn W. (McGuire); m. V. Jane Rasmussen, Oct. 10, 1981; children: Bruce, Brooke, Carrie, Marti. B.S., Rensselaer Poly. Inst., 1964; M.S., Northeastern U., 1966, Ph.D., 1969. Vis. scientist Hahn Meitner Inst., Berlin, 1980-81; asst. prof. Tex. A&M U., 1969-72, Kans. State U., Manhattan, 1972-76, assoc. prof. physics, 1976-83, prof., 1983—; cons. Picatinny Arsenal, Naval Surface Weapons Lab., White Oak, Md., Lawrence Livermore Lab.; summer fellow Goddard Space Flight Ctr. Contbr. articles to profl. publs. U.S. Air Force grantee, 1970-72; Dept. Energy grantee, 1974—; NSF travel grantee to USSR, 1978. Mem. Am. Phys. Soc. Subspecialties: Atomic and molecular physics; Theoretical physics. Current work: Atomic collisions, multiple ionization, electron capture, ionization. Office: Physics Dept Kans State U Manhattan KS 66506

MCGUIRE, JOHN L., pharmaceutical company research executive; b. Kittanning, Pa., Nov. 3, 1942; s. Lawrence F. and Florence (Jones) McG.; m. Pamela Hale, Aug. 2, 1969; children: Megan L., Christa H. B.S., Butler U., 1965; M.A., Princeton U., 1968, Ph.D., 1969. Assoc. scientist pharmacology dept. Ortho Pharm. Corp., Raritan, N.J., 1969-70, scientist, 1970-72, sect. head biochem. research dept., 1972-75, exec. dir. research, 1975-80, v.p. basic scis., 1980—; adj. assoc. prof. M.S. Hershey Sch. Medicine, Pa. State U., 1978—; adj. prof. Rutgers U., 1983—. Contbr. articles to profl. jours. Mem. exec. bd. Keystone Area Boy Scouts Am., 1975—, George Washington council, 1980—; trustee Hunterdon Med. Ctr., Flemington, N.J., 1978—; bd. dirs. Hunterdon County YMCA, 1982—, Hunterdon County United Way, 1983—. Mem. Am. Soc. Pharmacology and Exptl. Therapeutics, Soc. Exptl Biology and Medicine, Am. Phys. Soc., Endocrine Soc., Am. Soc. Clin. Pharmacology and Therapeutics, Am. Chem. Soc., Am. Fertility Soc., Soc. Adv. Contraception, N.J. Health Scis. Group. Club: Princeton (N.Y.C.). Subspecialties: Pharmacology; Neuroendocrinology. Current work: Management pharmaceutical research and development activies; endocrine pharmacology. Patentee in field. Home: 9 Sunnyfield Dr Whitehouse Station NJ 08889 Office: Ortho Pharm Corp Raritan NJ 08869

MCGUIRE, STEPHEN CRAIG, nuclear engineering, physics educator, consultant; b. New Orleans, Sept. 17, 1948; s. Harry Stewart and Ruth (Barsock) McG.; m. Saundra Yancy, Aug. 28, 1971; children: Carla Abena, Stephanie Niyonu. B.S., So. U., Baton Rouge, 1970; M.S., U. Rochester, 1974; Ph.D., Cornell U., 1978. Research asst. U. Rochester, N.Y., 1971-74, Cornell U., Ithaca, N.Y., 1975-78; lectr. Stanford (Calif.) U. Linear Accelerator Ctr., 1976; devel. assoc. Oak Ridge Nat. Lab., 1978-82; asst. prof. nuclear engring, physics Ala. A&M U., Normal, 1982—; cons. chem. tech. div. Oak Ridge Nat. Lab., 1982—. John McMullen fellow in nuclear sci. Cornell U., 1977; Crown Zellerbach Found. fellow, 1968. Mem. Am. Phys. Soc., Am. Nuclear Soc., AAAS, Sigma Xi. Subspecialties: Nuclear engineering; Nuclear physics. Current work: Nuclear spectroscopy, neutron physics, atomic charge exchange processes, nuclear waste management. Office: Dept Physics Ala A&M U PO Box 523 Normal AL 35762

MCGUIRE, TERRY RUSSELL, biology educator; b. Wadsworth, Ohio, June 12, 1950; s. Myron Elwood and Mildred Irene (Boyer) McG.; m. Jeannette Marie Haviland, Jan. 2, 1983. Safety engr. PerMold, Inc., Medina, Ohio, 1972; teaching asst. Sch. Life Scis., U. Ill., 1972-78; postdoctoral fellow div. genetics U. Rochester (N.Y.) Sch. Medicine and Dentistry, 1978-79; asst. prof. dept. biol. scis. Rutgers U., New Brunswick, N.J., 1979—. Author: (with J. Hirsch) Behavior-Genetic Analysis, 1982; contbr. chpts. to books, articles to profl. jours. Mem. Animal Behavior Soc., Behavior Genetics Assn., Genetics Soc. Am., N.Y. Entomol. Soc., Columbia Population Biology Group, Sigma Xi. Subspecialties: Gene actions; Behavior genetics. Current work: Behavior genetics; component analysis of the blow fly; phormia regina, conditioning, central excitatory state. Office: Dept Biol Scis Livingston Campus Rutgers U New Brunswick NJ 08903

MCHENRY, CHARLES STEVEN, biochemistry researcher, biochemistry educator; b. Indpls., Jan. 1, 1948; s. William Horace and Barbara Anne (Hayes) McH.; m. Mary Michele Herzogenrath, May 31, 1980; 1 son, Patrick Venzke. B.S., Purdue U., 1970; Ph.D., U. Calif.-Santa Barbara, 1974. Postdoctoral fellow Stanford U. Med. Sch., Palo Alto, Calif., 1974-76; asst. prof. U. Tex. Med. Sch., Houston, 1976-82, assoc. prof. biochemistry, 1982—. Contbr. articles in field to profl. jours. Cystic Fibrosis Found. postdoctoral fellow, 1974; recipient Faculty Research award Am. Cancer Soc., 1982; Research Grants Am. Cancer Soc., 1977—, NIH, 1978—, Robert Welch Found., 1981—. Mem. Am. Chem. Soc., Am. Soc. Microbiology, Am. Soc. Biol. Chemists. Subspecialties: Biochemistry (biology); Molecular biology. Current work: Enzymology, structure of multienzyme complexes, mechanism and regulation of DNA synthesis. Office: Dept Biochemistry and Molecular Biology U Tex Med Sch PO Box 20708 Houston TX 77025

MCHENRY, KEITH WELLES, JR., oil company executive; b. Champaign, Ill., Apr. 6, 1928; s. Keith Welles and Jayne (Hinton) McH.; m. Lou Petry, Aug. 23, 1952; children: John, William. B.S. in Chem. Engring, U. Ill., 1951, Ph.D., Princeton U., 1958. With Standard Oil Co. (and affiliates), 1955—; asst. project chem. engr. research and devel. dept. Amoco Oil Co., Whiting, Ind., 1955-58, group leader, 1958-67, research asso., 1967-68, asst. dir. fuels research, 1968-70, dir. process and analytical research, 1970-74, mgr. process research, Naperville, Ill., 1974-75, v.p. research and devel., 1975—; Hurd lectr. Northwestern U., 1981; mem. adv. council Catalysis Center, U. Del., Newark, 1978-83, chmn., 1981—; mem. adv. council Sch. Engring. and Applied Sci., Princeton U., 1976—, Coll. Engring., U. Ill., Chgo., 1979—; mem. U.S. Nat. Com. World Petroleum Congress, 1975-83. Trustee North Central Coll., Naperville, 1978—; chmn. area com. Jr. Achievement, 1981-83; ordained elder Presbyn. Ch., 1964. Served with U.S. Army, 1946-47. Gen. Electric fellow, DuPont fellow, 1952-54. Mem. Nat. Acad. Engring., Am. Inst. Chem. Engrs. (editorial bd. jour. 1974-78), Am. Chem. Soc., AAAS, Am. Petroleum Inst., Sigma Xi, Tau Beta Pi. Subspecialty: Chemical engineering. Office: PO Box 400 Naperville IL 60566

MCHENRY, MARTIN CHRISTOPHER, internist; b. San Francisco, Feb. 9, 1932; s. Merl and Marcella (Bricca) McH.; m. Patricia G. Hughes, Apr. 27, 1957; children: Michael, Christopher, Timothy, Mary Ann, Jeffrey, Paul, Kevin, William, Monica, Martin. Student, U. Santa Clara, 1950-53; M.D., U. Cin., 1957; M.S. in Medicine, U. Minn.-Mpls., 1966. Diplomate: Am. Bd. Internal Medicine. Rotating intern Highland Alameda County Hosp., Oakland, Calif., 1957-58; fellow in internal medicine Mayo Clinic, Rochester, Minn., 1958-61; spl. appointee in infectious diseases, 1963-64; staff physician div. infectious diseases Henry Ford Hosp., Detroit, 1964-67; staff physician Cleve. Clinic, 1967-72, head dept. infectious diseases, 1972—; asst. clin. prof. medicine Case Western Res. U., 1970-77, assoc. clin. prof., 1977—. Contbr. numerous chpts., articles to profl. publs.; co-editor First, Second and Third Cleve. Symposiums on Infectious Diseases, 1974, 75, 78; editorial adv. bd.: Med. Update, 1978-79. Chmn. manpower com. Swine Flu Program, Cleve., 1976. Named Disting. Tchr. of Yr. in Medicine Cleve. Clinic, 1972. Fellow ACP, Am. Coll Chest Physicians (chmn. com. cardiopulmonary infections 1975-77, 81-83); mem. Infectious Diseases Soc. Am., Am. Thoracic Soc., Am. Fedn. Clin. Research, Am. Soc. Microbiology, N.Y. Acad. Scis., Am. Soc. Tropical Medicine and Hygiene, Am. Soc. Clin. Pharmacology and Therapeutics (chmn. sect. infectious diseases and antimicrobial agts. 1970-77, 80—, dir. 1972-75), Am. Geriatrics Soc., AAAS, Alumnae Assn. Mayo Clinic Found., AMA, Cleve. Acad. Medicine, Ohio Soc. Internal Medicine, Am. Soc. Clin. Pathology, Royal Soc. Medicine Gt. Brit., Am. Veneral Disease Assn., So. Med. Assn. Subspecialties: Infectious diseases; Internal medicine. Current work: Cooperative study on vertebral osteomyelitis. Home: 2779 Belgrave Rd Pepper Pike OH 44124 Office: Cleve Clinic Found 9500 Euclid Ave Cleveland OH 44106

MCHONE, JAMES GREGORY, geology educator; b. Schenectady, N.Y., Aug. 5, 1949; s. and Helen (Gregory) McH.; m. Cynthia Abbott, Dec. 26, 1971; 1 dau., Sarah. B.A., U. Vt., 1971, M.Sc., 1975; Ph.D., U. N.C., 1978. Cert. geologist, Ind. Project geologist Chiasma Cons., Inc., South Portland, Maine, 1978-80; asst. prof. geology Ind. U., Indpls., 1980—. Served with USNG, 1971-77. Soc. Sigma Xi grantee, 1973; Ind. U. fellow, 1981. Mem. Geol. Soc. Am., Am. Goephys. Union, Mineral. Soc. Am., Vt. Geol. Soc., Sigma Xi. Democrat. Subspecialties: Petrology; Tectonics. Current work: Alkalic magmas and plate rifting; research area Eastern North America and the North Atlantic Ocean. Office: Geology Department Indiana University/ Purdue University 425 Agnes St Indianapolis IN 46202

MCILHENNY, HUGH MEREDITH, drug research and development executive; b. Gettysburg, Pa., Sept. 25, 1938; s. Hugh C. and Helen E. (Burgoon) McI.; m. Loretta May Mink, May 8, 1938; children: Lynn Meredith, Craig Meredith. B.S., Pa. State U., 1960; M.S., U. Mich., 1964, Ph.D., 1966. Research asst. antibiotics research and dvel. Parke-Davis & Co., Detroit, 1960-62; research investigator drug metabolism Pfizer, Inc., Groton, Conn., 1966-79, regulatory affairs liaison, 1979—. Contbr. articles to profl. jours. Mem. East Lyme (Conn.) Planning Commn., 1975-77. Mem. Am. Soc. Pharmacology and Exptl. Therapeutics, Regulatory Affairs Profl. Soc. Subspecialties: Drug research and regulation; Pharmacokinetics. Current work: Drug regulatory affairs; regulatory agency liaison; new drug investigation and registration. Home: 10 Mayfield Terr East Lyme CT 06333 Office: Eastern Point Rd Groton CT 06340

MC ILRATH, THOMAS JAMES, physics educator, researcher; b. Dowagiac, Mich., May 10, 1938; s. William Frederick and Leora May (Lewis) McIlrath; m. Valerie Hoy, June 30, 1962; children: Christine, Laura. B.S., Mich. State U., 1960; Ph.D., Princeton U., 1966. Research assoc. Harvard Coll. Obs., Harvard U., 1967-73; prof. U. Md., College Park, 1973—; physicist Nat. Bur. Standards, Gaithersburg, Md., 1974—. Contbr. articles to profl. jours. NSF fellow, 1960; Woodrow Wilson fellow, 1960; NATO fellow, 1966-67; recipient Dept. Commerce Silver medal, 1980. Mem. Optical Soc. Am., Am. Phys. Soc. Subspecialties: Spectroscopy; Atomic and molecular physics. Current work: Short wavelength optical radiation; non-liner optics, remote sensing lasers, generation of coherent short wave length radiation, interaction of intense radiation with matter, atmospheric sensing. Patentee compact diode array spectrometer. Home: 5944 Westchester Park Dr College Park MD 20742 Office: IPST Md College Park MD 20742 Home: 5944 Westchester Park Dr College Park MD 20740

MCILWAIN, CARL EDWIN, physicist, educator; b. Houston, Mar. 26, 1931; s. Glenn William and Alma Ora (Miller) McI; m. Mary Louise Hocker, Dec. 30, 1952; children: Janet Louise, Craig Ian. Ph.D., U. Iowa, 1960. Asst. prof. physics U. Iowa, 1960-62; assoc. prof. physics U. Calif.-San Diego, 1962-66, prof., 1966—; mem. Fachbeirat Inst. Extraterrestrial Physics, Max Planck Inst., Garching, Germany. Contbr. articles to scientific and research jours. Guggenheim fellow, 1968, 72; recipient Alexander von Humboldt Found. sr. U.S. scientist award, 1976. Fellow Am. Geophys. Union; mem. Am. Phys. Soc., Am. Astron. Soc. Subspecialties: Planetary science; Ultraviolet high energy astrophysics. Current work: Developing new techniques for measuring faint astronomical images and for measuring electric fields in naturally occurring plasmas; inventor multichannel photometer for low light level astronomy.

MCINERNEY, JOSEPH JOHN, nuclear physicist, researcher; b. Boston, Aug. 13, 1932; s. John Joseph and Anne Copeland (Berry) McI.; m. Claire Patricia Shine, July 2, 1955 (div. Aug. 1970); children: Joseph, Lynn, Maureen; m. Suzanne Finke, Oct. 20, 1970; children: Kathleen, John. B.S. in Mech. Engring, Northeastern U., 1960; M.S. in Nuclear Engring, Pa. State U.-State College, 1962, Ph.D., 1964; M.S. in Human Physiology, Pa. State U.-Hershey, 1980. Nuclear physicist Knolls Atomic Power Lab., Schenectady, 1964-76; NIH fellow Pa. State U. Sch. Medicine, Hershey, 1976-79, staff research scientist, 1979—; cons. Whitaker Found., Harrisburg, Pa., 1981—. Served with USNR, 1951-55. Research grantee Whitaker Found., 1978-81, Am. Heart Assn., 1979-80, NIH, 1982—. Mem. Am. Nuclear Soc. (referee 1966—), AAAS, Am. Physiol. Soc., Am. Fedn. for Clin. Research, Am. Heart Assn., Am. Assn. Physics in Medicine. Democrat. Subspecialties: Biomedical engineering; Imaging technology. Current work: Development of new imaging techniques for detection and diagnosis of cardiovascular disease. Patentee imaging device. Home: RD 6 Box 11 Hummelstown PA 17036 Office: M S Hershey Med Ctr Pa State U Sch Medicine Box 850 Hershey PA 17033

MCINNIS, DONALD OWEN, research geneticist, educator; b. Honolulu, June 18, 1951; s. Harry D. and Marjorie E. (Graber) McI.; m. Peggy Lorraine Sutton, July 21, 1979. B.A., U. Va., 1973; M.Life Scis., N.C. State U., 1975, Ph.D. in Genetics, 1978. Mem. staff screwworm research lab. U.S. Dept. Agr./Tex. A&M U., Mission, 1978-80, research geneticist, 1980-82; research geneticist tropical fruit and vegetable research lab. U.S. Dept. Agr., Honolulu, 1982—. Contbr. articles to sci. jours. Mem. Phi Beta Kappa, Sigma Xi. Subspecialties: Genetics and genetic engineering (biology); Evolutionary biology. Current work: Genetic control of insects; sterile insect release method; biological control methods; integrated control. Home: 1216 Manu Aloha St Kailua HI 96734 Office: 2727 Woodlawn Dr Honolulu HI 96804

MCINTIRE, FLOYD COTTAM, biochemist, educator, researcher; b. Price, Utah, Sept. 10, 1914; s. Brigham Franklin and Effie (Cottam) McI.; m. Helen Maurine Walseth, June 26, 1943; children: Meredith Kay, Kenneth Marvin, Gregory Franklin. A.B., Brigham Young U., 1936, M.A., 1937; Ph.D., U. Wis.-Madison, 1940. Asst. agrl. chemist N.D. Agrl. Coll., Fargo, 1941-42; research biochemist Abbott Labs., North Chicago, Ill., 1942-62, head biochemistry, 1962-70, dir. exptl. biology, 1970-74; prof. oral biology U. Colo.-Denver, 1974—. Mem. Am. Soc. Biol. Chemists, Am. Chem. Soc., Am. Soc. Microbiology, AAAS, Internat. Assn. Dental Research. Republican. Mormon. Subspecialties: Oral biology; Biochemistry (biology). Current work: Chemistry of interbacterial adherance, bacterial surface carbohydrates. Home: 7780 E Cornell Ave Denver CO 80231 Office: University of Colorado School of Dentistry 4200 E 9th Ave Denver CO 80262

MCINTOSH, ROBERT E., JR., electrical engineer, educator; b. Hartford, Conn., Jan. 19, 1940; s. Robert E. and Natalie R. (Glynn) McI.; m. Anne Marie Potvin, July 7, 1962; children: Robert E., III, Edgar J., Michael T., William P., Matthew P. B.S.E.E., Worcester Poly. Inst., 1962; S.M. in Applied Physics, Harvard U., 1964; Ph.D., U. Iowa, 1967. Mem. tech. staff Bell Telephone Labs., North Andover, Mass., 1962-65; research asst. U. Iowa, Iowa City, 1965-67; asst. prof. U. Mass., Amherst, 1967-70, assoc. prof., 1970-73, prof., 1973—; area coordinator microwave devices and systems dept. ECE; guest prof. U. Nijmegen, Netherlands, 1973-74; cons. NSF, Dept. Def. Contbr. articles to profl. jours. Active Babe Ruth Baseball. Grantee NSF, Air Force Office Sci. Research, Army Research Office, Office Naval Research, Nat. Acad. Scis. Mem. IEEE (sr.), Am. Phys. Soc., Antennas and Propagation Soc. (sec.-treas.), Sigma Xi, Tau Beta Pi, Epsilon Kappa Nu. Lodge: Rotary. Subspecialties: Applied magnetics; Electronics. Current work: Research in microwave remote sensing using radars and radiometers. Home: 138 Columbia Dr Amherst MA 01002 Office: Dept ECE U Mass Amherst MA 01003

MCINTOSH, ROBERT PATRICK, biology educator, editor; b. Milw., Sept. 24, 1920; s. Andrew Benno and Emma Minnie (Waller) McI.; m. Joan Ailene Wright, May 24, 1977; children: Toby James, Sarah, Ellen. B.S., Lawrence U., 1942; M.A., U. Wis.-Madison, 1948, Ph.D., 1950. Prof. biology Middlebury (Vt.) Coll., 1950-53; prof. biology Vassar Coll, Poughkeepsie, N.Y., 1953-58, U. Notre Dame., Ind., 1958—. Editor: Am. Midland Naturalist, Notre Dame, 1970—, Phytosociology, 1978. Served to 1st lt. U.S. Army, 1942-45; S. Pacific. NSF sr. postdoctoral fellow, Wales, 1964. Fellow AAAS; mem. Ecol. Soc. Am. (sec. 1981-85), Am. Soc. Naturalists, Brit. Ecol. Soc. Subspecialty: Ecology. Current work: Community ecology, history of ecology. Home: 1216 Hillcrest Rd South Bend IN 46617

MCINTOSH, TRACY KAHL, neuroendocrinologist, educator; b. N.Y.C., May 8, 1953; s. Ezra Albert and Roe McI. A.B., Williams Coll., 1975; Ph.D., Rutgers, U., 1980. Postdoctoral fellow Boston U. Med. Sch., 1980-82, asst. prof., 1982—; spl. scientist Boston City Hosp., 1982—; instr. U. Mass., Boston, 1982—; cons. Surg. Assocs., Boston, 1982—. Mem. Com. for Social Responsibility, Concord, Mass., 1983; mem. Crime Prevention Com., Jamaica Plain, Mass., 1982—. Recipient Hyde Scholarship award Williams Coll., 1975; NIH postdoctoral fellow, 1980-82. Mem. AAAS, Endocrine Soc., Internat. Soc. Psychoneuroendocrinology, Neuroendocrine Soc., N.Y. Acad. Sci., Phi Beta Kappa, Sigma Xi. Unitarian. Subspecialties: Neuroendocrinology; Neuroimmunology. Current work: To explore the endocrine response to stress, shock and trauma, particularly endophins and enkephalins, and relationship to development of immunosuppressive syndromes. Home: 29 Lakeville Rd Jamaica Plain MA 02130 Office: Boston U Med Center 80 E Concord St Boston MA 02118

MCINTYRE, GARY ALLEN, plant pathologist, educator; b. Portland, Oreg., July 16, 1938; s. John H. and Onie (Meihoff) McI.; m. Leone, Sept. 1, 1963; children—Paula, Laura. B.S., Oreg. State Coll., 1960, Ph.D., 1964. Faculty U. Maine, Orono, 1969-75; faculty Colo. State U., Fort Collins, 1975—, prof., chmn. dept. botany and plant pathology, 1975—. Contbr. articles in field to profl. jours. NDEA fellow, 1960-63. Mem. Am. Phytopath. Soc., Potato Assn. Am., Sigma Xi, Phi Kappa Phi, Phi Sigma. Subspecialties: Plant pathology; Integrated pest management. Current work: Researcher in skin diseases of potato. Home: 904 Cheyenne Dr Fort Collins CO 80525 Office: Colorado State Dept Botany and Plant Pathology Fort Collins CO 80523

MCINTYRE, OSWALD ROSS, physician; b. Chgo., Feb. 13, 1932; m. Jean Geary, June 5, 1957; children—Margaret Jean, Archibald Ross, Elizabeth Geary. A.B. cum laude, Dartmouth Coll., 1953, postgrad, 1953-55; M.D., Harvard U., 1957. Intern U. Pa. Hosp., 1957-58; resident in medicine Dartmouth Med. Sch. Affiliated Hosps., 1958-60; instr. medicine Dartmouth Coll., 1964-66; attending physician VA Hosp., White River Junction, Vt., 1964; asst. prof. medicine Dartmouth, 1966-69, asso. prof., 1969-75, prof., 1976—, James J. Carroll prof. oncology, 1980—, dir., 1975—; cons. in hematology, oncology; individual practice medicine specializing in hematology-oncology, Hanover, N.H., 1964—. Served with USPHS, 1961-63. Mem. Am. Fedn. Clin. Research, AAAS, Am. Soc. Hematology, N.Y. Acad. Sci., Internat. Assn. Study Lung Cancer. Subspecialties: Hematology; Oncology. Home: River Rd Lyme NH 03768 Office: Norris Cotton Cancer Center Hanover NH 03755

MCISAAC, ROBERT J(AMES), pharmacology educator, university executive; b. Bklyn., Jan. 9, 1923; s. Milton J. and June Z. (Barrus) McI.; m. Carol M. Langner, May 22, 1976; children: James, Laurie, Heidi, Heather, Gregory. B.S., Sch. Pharmacy, U. Buffalo, 1949; Ph.D. in Pharmacology, Sch. Medicine, 1954. Postdoctoral fellow Grad. Sch. Medicine, U. Pa., 1956-58; asst. prof. pharmacology U. Buffalo, 1958-64, assoc. prof., 1964-68, prof., 1968—, dir. grad. studies in pharmacology, 1969-76, acting. chmn. dept. pharmacology, 1970-71, asst. v.p. for research and grad. studies, 1981—. Contbr. articles to profl. jours. Mem. Niagara River Quality Control Com., 1981—. Served to sgt. U.S. Army, 1943-46. Spl. postdoctoral fellow Farmakologiska Institutet, U. Lund, Sweden, 1965-66; NIH grantee, 1959-62, 65-69, 70-79. Mem. Soc. Pharmacology and Exptl. Therapeutics, Soc. Neurosci., AAAS, N.Y. Acad. Sci. Methodist. Subspecialties: Cellular pharmacology; Neuropharmacology. Current work: Facilitation and inhibition of synaptic transmission in autonomic ganglia; especially the mechanism of prolonged facilitation of transmission and the role of non-nicotinic transmitters in this process.

MCKAY, KENNETH G(ARDINER), electronics executive, telecommunications consultant; b. Montreal, Que., Can., Apr. 8, 1917; s. James Gardiner and Margaret (Nicholas) McK.; m. Renee, July 25, 1942; children: Margo, Kenneth. B.Sc., McGill U., 1938, M.Sc., 1939; D.Sc., MIT, 1941; D.Eng. (hon.), Stevens Inst. Tech., 1980. Engr. NRC, Ottawa, Ont., Can., 1942-46; various positions Bell Telephone Labs., Murray Hill, N.J., 1946-65; v.p. engring. AT&T, N.Y.C., 1965-72; exec. v.p. Bell Labs., Murray Hill, 1972-80; chmn. bd. C.S. Draper Labs., Cambridge, Mass., 1980—; dir. Nat. Aviation & Tech. Fund, Nat. Telecommunications & Tech. Fund. Contbr. numerous articles to profl. jours. Mem. adv. council Sch. Engring. Stanford U., 1974—; mem. sci. and acad. adv. com. U. Calif., 1980—; trustee Stevens Inst. Tech., 1974—; bd. govs. N.Y. Coll. Osteo. Medicine, 1980—. Served in RCAF, 1942-46. Recipient Pub. Service award NASA, 1969, Pub. Service Group Achievement award, 1969. Fellow Am. Phys. Soc., IEEE; Mem. Nat. Acad. Scis., Nat. Acad. Engring., N.Y. Acad. Scis., Sigma Xi. Clubs: Weston (Conn.) Gun; St. Andrews Golf (Hastings-on-Hudson, N.Y.). Subspecialties: Telecommunications; Aerospace engineering and technology. Patentee semicondrs. and TV pick-up tubes (8). Address: 200 E 66th St New York NY 10021

MCKEARNEY, JAMES WILLIAM, pharmacologist; b. Bay Shore, N.Y., Apr. 4, 1938; s. James M. and Veronica W. (Green) McK; m. children: Erika, Diana. B.A., C.W. Post Coll., 1962; M.S., U. Pitts., 1965, Ph.D., 1966. Research fellow in pharmacology Harvard U., 1966-68, instr. psychobiology, 1968-69; staff scientist Worcester Found. Exptl. Biology, Shrewsbur, Mass., 1970-72, sr. scientist, 1972—; adj. prof. psychology Boston U. Contbr. articles to sci. jours., chpts. to books. Served with USN, 1956-59. Research grantee NIMH Nat. Inst. Drug Abuse. Mem. , Am. Soc. Pharmacology and Exptl. Therapeutics; Am. Psychol. Assn. Behavioral Pharmacology Soc., Eastern Psychol. Assn. Subspecialties: Neuropharmacology; Psychobiology. Current work: Basic animal research on mechanisms of action of psychoactive drugs. Home: 11 Orchard St Newton MA 02158 Office: Worcester Foundation Shrewsbury MA 01545

MCKEE, CHRISTOPHER FULTON, astrophysics educator; b. Washington, Sept. 6, 1942; m., 1965; 3 children. A.B., Harvard U., 1963; Ph.D. in Physics, U. Calif.-Berkeley, 1970. Physicist Lawrence Livermore Lab., U. Calif., 1969-70; research fellow in astrophysics Calif. Inst. Tech., Pasadena, 1970-71; asst. prof. astronomy Harvard U., 1971-74; asst. prof. physics U. Calif.-Berkeley, 1974-77, assoc. prof., 1977-78, prof. physics and astronomy, 1978—. Mem. Am. Astron. Soc., Am. Phys. Soc., Internat. Astron. Union. Subspecialty: Theoretical astrophysics. Office: Physics Dept U Calif Berkeley CA 94720

MCKEE, JAMES STANLEY COLTON, physics educator; b. Belfast, Northern Ireland, June 6, 1930; s. James and Dorothy (Colton) McK.; m. Christine Diane Savage, July 16, 1961; children: James Conor, Sylvia Dorothy Siobhan. B.Sc., Queen's U., Belfast, 1952, Ph.D., 1956; D.Sc., U. Birmingham, Eng., 1968. Asst. lectr. Queens U., Belfast, 1954-56; lectr. Birmingham U., 1956-64, sr. lectr., 1964-74; prof. physics, dir. Cyclotron Lab. U. Man., 1974—; mem. grant selection com. Nat. Scis. and Engring. Research Council, 1980-83. Contbr. articles to profl. jours. Fellow Inst. Physics, mem., Canadian Assn. Physicists. Liberal. Subspecialties: Nuclear physics; Solar energy. Current work: Few body problems in nuclear and particle physics; polarisation studies; solar concentrators. Home: 1443 Wellington Crescent Winnipeg MB Canada R3N OB2 Office: U Man Rm 223 Allen Bldg Winnipeg Canada R3T 2N2

MCKEE, KEITH EARL, civil engr., research center adminstr.; b. Chgo., Sept. 9, 1928; s. Charles Richard and Maude Alice McK.; m. Lorraine Marie Celichowski, Oct. 26, 1957; children: Pamela Ann, Paul Earl. B.S., Ill. Inst. Tech., 1950, M.S., 1956, Ph.D., 1962. Designer Swift & Co., Chgo., 1953-54; research engr. Armour Research Found., Chgo., 1954-63; dir. mech. design and product assurance Andrew Corp., Chgo., 1963-68; dir. engring. Ill. Inst. Tech. Research Inst., Chgo., 1968-80, dir., 1977—, also adj. prof. Author articles. Served to capt. USMC, 1950-53. Mem. ASME, ASCE, Soc. Mfg. Engrs., Numerical Control Soc., Am. Assn. Engring. Socs., Am. Def. Preparedness Assn. (pres. Chgo. chpt. 1975—), Navy League, Air Force Assn., Assn. U.S. Army. Subspecialties: Industrial engineering; Robotics. Current work: Manufacturing technology and productivity, flexible manufacturing systems, robotics, computer-aided design and manufacturing. Home: 608 Burns St Flossmoor IL 60422 Office: 10 W 35th St Chicago IL 60616

MC KEE, LEWIS WITHERSPOON, mechanical engineer; b. Portsmouth, N.H., Mar. 12, 1923; s. Andrew Irwin and Katherine (Brown) McKee; m. Sophia Kuhlmann Greenstein, June 29, 1946; m. Pat, Oct. 26, 1963; children: Lewis, Sophia, Miriam, Eugenia. B.S.M.E., MIT, 1944. Registered profl. engr., Calif. Chief product engr. Barden Corp., Danbury, Conn., 1947-73; engring. mgr. NMB Corp., Chatsworth, Calif., 1973-74; cons. mech. engring., Westlake Village, Calif., 1974-75, Thousand Oaks, Calif., 1978-80; mgr. bearing engring. W.S. Shamban & Co., Newbury Park, Calif., 1975-77; mgr. engring. Tribotech Corp., Redwood City, Calif., 1977-78; engring. specialist Aerospace Corp., El Segundo, Calif., 1980—. Served with USN, 1942-45. Recipient Project Hindsight award Dept. Def., 1966. Mem. ASME. Republican. Clubs: Westlake Yacht, Thousand Oaks Racquet. Current work: Consulting activities on bearing applications for satellites. Patentee ball bearing components and assemblies. Office: Aerospace Corp El Segundo CA 90009

MCKEE, MICHAEL GEOFFREY, clinical psychologist; b. Santa Monica, Calif., May 11, 1931; s. George Thomas and Ellen Mary Dorothea (Evans) McK.; m. Theodora Thorson, Jan. 27, 1956; children: Cory, Carol, Anna, Evan. B.A., U. Calif.-Berkeley, 1953, Ph.D., 1960. Vis. lectr. dept. psychology U. Calif.-Berkeley, 1968-69; chief of research, psychol. services staff CIA, Washington, 1960-68; sr. psychology cons., adj. faculty Case Western Res. U., Cleve., 1969—; mem. profl. staff VA, 1978—; head biofeedback sect. Cleve. Clinic Found., 1976—; cons. VA. Contbr. chpts. to books. Mem. Am. Psychol. Assn. (council rep. 1979-82), Am. Orthopsychiat. Assn., Biofeedback Soc. Am., Assn. for Advancement of Behavior Therapy, Am. Pain Soc., Am. Psychosomatic Soc., Soc. Clin. and Exptl. Hypnosis, Ohio Psychol. Assn. (pres. 1975-76, Disting. Service award 1981). Subspecialties: Biofeedback; Behavioral psychology. Current work: Psychophysiologic diagnosis and treatment via biofeedback instrumentation of stress related disorders. Home: 3171 Warrington Rd Shaker Heights OH 44120 Office: Cleveland Clinic Found 9500 Euclid Ave Cleveland OH 44106

MCKEE, ROBERT BRUCE, JR., engring. educator; b. Kalispell, Mont., Jan. 15, 1924; s. Robert B. and Myrtle I. (Flaten) McK.; m. Ann Jameson, June 22, 1949; children: Janet, James, Jane; m. Kathryn Vinyard, Oct. 8, 1970. B.S. in Mech. Engring, Mont. State U., 1948; M.S., U. Wash., 1952; Ph.D., UCLA, 1967. Registered profl. engr., Nev. Engr. Pratt & Whitney Aircraft, Hartford, Conn., 1948-50, Dow Chem. Co., Midland, Mich., 1952-56, Mastro Plastics, N.Y.C., 1956-57; prof. mech. engring. U. Nev., Reno, 1957—. Served to 2d lt. USAF, 1943-46. Mem. ASME, Nat. Soc. Profl. Engrs., ASHRAE, Internat. Solar Energy Soc. Republican. Subspecialties: Solar energy; Composite materials. Current work: Simulation of solar radiation and of passive solar structures. Home: 590 Greenstone Reno NV 89512 Office: Mech Engring U Nev Reno NV 89557

MCKELVY, JEFFREY FORRESTER, neurobiologist, educator; b. Akron, Ohio, Aug. 25, 1938; s. Jack Forrester and Beverly Brickner McK.; m. Linda Mollin, June 7, 1963 (div.); children: Kathleen Forrester, Jill Leslie; m. Mary Louise Tally, Feb. 22, 1977. B.Sc., U. Akron, 1963; Ph.D., Johns Hopkins U., 1968. Asst. prof. anatomy U. Conn. Health Center, 1971-76; assoc. prof. biochemistry U. Tex. Health Sci. Center, Dallas, 1976-79; assoc. prof. psychiatry U. Pitts. Sch. Medicine, 1979-81; prof. neurobiology and behavior SUNY, Stony Brook, N.Y., 1981—; cons. NIH, mem. neurology study sect., 1978-82; mem. NIH Task Force on Research Needs in Endocrinology, 1979—. Contbr. numerous articles to sci. jours.; Editor: Current Methods in Cellular Neurobiology, 1982. Recipient NIH Research Career Devel. award, 1977-82. Mem. Am. Soc. Biol. Chemists, Soc. Neurosci., Am. Soc. Neurochemistry, Endocrine Soc., Internat. Brain Research Orgn. Democrat. Episcopalian. Subspecialties: Neurobiology; Molecular biology. Office: Dept Neurobiology and Behavior SUNY Grad Biol Bldg Stony Brook NY 11794

MCKENZIE, ROBERT JAMES, plant physiologist, consultant; b. Warrenton, Va., Sept. 8, 1943; s. Charles Robert and Emma Kathleen (Brosius) McK.; m. Linda Rae McKenzie, Apr. 21, 1967 (div.); children: Amanda Brooke, Sean Robert. B.S., U. Wash., 1970; Ph.D., Wash. State U., 1978. Plant physiology Hawaiian Sugar Planters Assn., Aiea, 1979-82; cons. plant cell and tissue culture, genetic engring., Phila., 1982—. Contbr. articles to sci. jours. Served in U.S. Army, 1965-67; Vietnam. Mem. AAAS, Am. Bot. Soc., Am. Soc. Plant Physiologists, Am. Inst. Biol. Scis. Subspecialties: Plant cell and tissue culture; Genetics and genetic engineering (agriculture). Current work: Plant cell and tissue culture; protoplast technology; organelle transfer; organelle genetics; genetic modification; photosynthetic physiology; nitrogen transformations (nitrification and denitrificatrion). Home and Office: 2020 Walnut St 12D Philadelphia PA 19103

MCKENZIE, WILLIAM FRANK, geologist; b. Dallas, Aug. 17, 1941; s. William Karl and Joe Marie (Johnson) McK; m. Sarah Elizabeth Goodin, May 3, 1981. B.S. in Chemistry, U. Okla., 1963; M.S. In Chemistry, Colo. Sch. Mines, 1970; Cert. Geothermics, Internat. Inst. Geothermal Research, Pisa, Italy, 1973; M.A. in Geology, U. Calif.-Berkeley, 1978, Ph.D., 1980. Chemist Pan Am. Petroleum, Tulsa, 1963-64, Marathon Oil Co., Littleton, Colo., 1964-67; lectr. geology U. Md., 1974-79; geochemist Nuclear Geology Lab., Pisa, Italy, 1973-74, U.S. Geol. Survey, Menlo Park, Calif., 1974-76; geologist Chevron Resources Co., San Francisco, 1980 ; cons. in field. Fellow Geol. Soc. Am.; mem. Soc. Econ. Geologists, Am. Geophys. Union, Geochem. Soc., Mineral. Soc. Am., U. Calif. Alumni Assn. Democrat. Club: Commonwealth. Subspecialties: Geology; Geochemistry. Current work: Mineral exploration, genesis of ore deposits, active geothermal systems and thermodynamic behavior of aqueous electrolytes at high pressure and temperature. Office: PO Box 7147 San Francisco CA 94120

MC KETTA, JOHN J., JR., chemical engineering educator; b. Wyano, Pa., Oct. 17, 1915; s. John J. and Mary (Gelet) McK.; m. Helen Elisabeth Smith, Oct. 17, 1943; children: Charles William, John J. III, Robert Andrew, Mary Anne. B.S., Tri-State Coll., Angola, Ind., 1937; B.S.E., U. Mich., 1943, M.S., 1944, Ph.D., 1946; D.Eng. (hon.), Tri-State Coll., 1965, Drexel U., 1977; Sc.D., U. Toledo, 1973. Diplomate: registered profl. engr., Tex., Mich. Group leader tech. dept. Wyandotte Chem. Corp., Mich., 1937-40, asst. supt. caustic soda div., 1940-41; teaching fellow U. Mich., 1942-44, instr. chem. engring., 1944-45; faculty U. Tex., Austin, 1946—, successively asst. prof. chem. engring., asso. prof., then prof. chem. engring., 1951-52, 54—, E.P. Schoch prof. chem. engring., 1970-81, Joe C. Walter chair, 1981—; asst. dir. Tex. petroleum research com., 1951-52, 54-56, chmn. chem. engring. dept., 1950-52, 55-63, dean Coll. Engring., 1963-69; exec. vice chancellor acad. affairs U. Tex. System, 1969-70; editorial dir. Petroleum Refiner, 1952-54; pres. Chemoil Cons., Inc., 1957-73; dir. Gulf Pub. Co., AID, Inc., Dallas, Dresser Industries, Howell Corp., Houston, Commonwealth Refining Co., San Antonio, Tipperary Corp., Midland, Tex., Tesoro Petroleum Co., San Antonio, Vulcan Materials Co., Birmingham, Ala., KINARK Corp., Tulsa.; Chmn. Tex. AEC, So. Interstate Nuclear Bd.; mem. Tex. Radiation Adv. Bd., 1978—; chmn. Nat. Energy Policy Com., 1970-72, Nat. Air Quality Control Com., 1972—; mem. adv. bd. Carnegie-Mellon Inst. Research, 1978—; mem. U.S. Acid Precipitation Task Force. Author: series Advances in Petroleum Chemistry and Refining; Chmn. editorial com.: Petroleum Refiner; mem. adv. bd.: Internat. Chem. Engring. mag; editorial bd.: Ency. of Chem. Tech; exec. editor: Ency. of Chem. Processing and Design. Bd. regents Tri-State U., 1957—. Recipient Bronze plaque Am. Inst. Chem. Engrs., 1952, Charles Schwab award Am. Steel Inst., 1973; Lamme award as outstanding U.S. educator, 1976; Joe J. King Profl. Engring. Achievement award U. Tex., 1976; Gen. Dynamics Teaching Excellence award, 1979; Triple E award for contbns. to nat. issues on energy, environment and econs. Nat. Environ. Devel. Assn., 1976; Boris Pregal Sci. and Tech. award N.Y. Acad. Scis., 1978; named distinguished alumnus U. Mich Coll. Engring., 1953, Tri-State Coll., 1956; fellow Allied Chem. & Dye, 1945-46; distinguished fellow Carnegie-Mellon U., 1978. Mem. Am. Inst. Mining Engrs., Am. Chem. Soc. (chmn. Central Tex. sect. 1950),

Am. Inst. Chem. Engrs. (chmn. nat. membership com. 1955, regional exec. com., nat. dir., nat. v.p. 1961, pres. 1962, service to soc. award 1975), Am. Soc. Engring. Edn., Chem. Markets Research Assn., Am. Gas Assn. (adv. bd. chems. from gas 1954), Houston C. of C. (chmn. refining div. 1954, vice chmn. research and statistics com. 1954), Engrs. Joint Council (dir.), Engrs. Joint Countil Profl. Devel. (dir. 1963—), Nat. Acad. Engring., Sigma Xi, Chi Epsilon, Alpha Psi Omega, Tau Omega, Phi Lambda Upsilon, Phi Kappa Phi, Iota Alpha, Omega Chi Epsilon, Tau Beta Pi, Omicron Delta Kappa. Subspecialties: Chemical engineering; Fuels. Current work: High pressure thermodynamics; the energy supply and demand. Home: 5227 Tortuga Trail Austin TX 78731

MCKINNEY, FRANK KENNETH, geology educator; b. Fairfield, Ala., Apr. 13, 1943; s. Frank Gene and Edna Ruth (Pass) McK.; m. Marjorie Ann Jackson, Sept. 1, 1964; children: Michael, Benjamin, Rachel, Mariana. B.S., Old Dominion U., 1964; M.S., U. N.C., 1967, Ph.D., 1970. Instr. Appalachian State U., Boone, N.C., 1968-70, asst. prof. geology, 1970-72, assoc. prof., 1972-76, prof., 1976—; vis. prof. U. Durham, Eng., 1978; exchange scientist Nat. Acad. Scis.-Soviet Acad. Scis., Moscow, 1978, Nat. Acad. Scis.-Czech Acad. Scis., Prague, 1981, 83. Contbr. articles to profl. jours. Smithsonian Instn. fellow, 1972-73. Mem. Paleontol. Soc. (councillor 1980-81), Soc. Econ. Paleontologists and Mineralogists, Paleontol. Research Inst., Am. Soc. Zoologists, Internat. Bryozoology Assn. (councillor 1977-83), Sigma Xi. Subspecialties: Paleobiology; Evolutionary biology. Current work: Taxonomy, evolution, functional interpretation, growth models of Paleozoic Bryozoa; evolutionary patterns in history of Bryozoa, Ordovician-recent. Home: Route 1 Box 39F Boone NC 28607 Office: Appalachian State U Boone NC 28608

MCKINNEY, JOHN EDWARD, physicist; b. Altoona, Pa., Apr. 6, 1925; s. Clayton A. and Katie (Kessler) McK.; m. Ursula Guttstadt, Aug. 17, 1958. B.S., Pa. State U., 1950. Physicist Nat. Bur. Standards, Washington, 1949—, dental and med. materials group, 1978—. Author: Density and Compressibility of Liquids, 1968. Chmn. transp. com. East Bethesda Citizens Com., Bethesda, Md., 1975-78. Served with U.S. Army, 1943-45. Decorated Purple Heart, Bronze Star medal. Mem. Internat. Assn. Dental Research, Rheology Soc., Philos. Soc. Washington. Subspecialties: Polymer physics; Composite materials. Current work: Wear and physical properties of dental and medical materials, development of related instrumentation. Home: 9124 McDonald Dr Bethesda MD 20817 Office: Nat Bur Standards Washington DC 20234

MCKINNEY, JOHN PAUL, psychology educator; b. Royal Oak, Mich., Feb. 19, 1935; s. Lorenz Hastings and Marie Irene (Janisse) McK.; m. Madeleine Angele McKinney-Fermin, Oct. 13, 1962; children: Peter J., Martin F., Maureen M., Elizabeth M. A.B. cum laude, U. Detroit, 1956; M.A., Fordham U., 1958; Ph.D., Ohio State U., 1961. Research assoc. McGill U., 1963; asst. prof. Smith Coll., Northampton, Mass., 1963-66; assoc. prof. psychology Mich. State U., East Lansing, 1966—, prof. psychology, pediatrics and human devel., 1971—. Editor: Developmental Psychology: Studies in Human Development, 1970; co-author: Developmental Psychology: The Infant and Young Child, 1977, Developmental Psychology: The Adolescent and Young Adult, 1977, Developmental Psychology: The School-age Child. Mem. Am. Psychol. Assn., Soc. for Research in Child Devel., AAAS, Eastern Psychol. Assn., Midwestern Psychol. Assn. Subspecialty: Developmental psychology. Current work: Personality-social devel. in children and adolescents, teaching nad research in developmental psychology, psychotherapy. Home: 6196 Skyline Dr East Lansing MI 48823 Office: Dept Psychology Mich State Univ East Lansing MI 48824

MCKINNEY, RALPH VINCENT, JR., dental educator, academic administrator; b. Columbus, Ohio, Jan. 9, 1933; s. Ralph Vincent and Roth Rosa (Braatz) McK.; m. Mary Irene Shoemaker, July 2, 1955; children: Heather, Holly, Laurel, Rosemary. B.S., Bowling Green U., 1954; D.D.S. cum laude, Ohio State U., 1961; Ph.D., U. Rochester, 1971. Practice gen. dentistry, Rocky River, Ohio, 1961-65; clin. asst. prof. Case-western Res. U., 1961-65; postdoctoral fellow U. Rochester, 1965-70; assoc. prof. oral pathology Med. Coll. Ga., 1970-75, prof., 1975—, chmn. dept., 1979—, chmn. faculty, 1979; bd. dirs. Med. Coll. Ga. Research Inst., 1980—, sec. bd., 1980—; speaker in field. Contbr. numerous articles, abstracts to profl. publs. Bd. dirs. Central Savannah River council Girl Scouts U.S.A., 1981—. Served to 1st lt. arty. U.S. Army, 1954-57; Germany; served to col. Dental Corps. USAR, 1957—. Decorated Meritorious Service medal with 2 oak leaf clusters; recipient Disting. Service award Faculty Med. Coll. Ga., 1982; USPHS fellow, 1965-70; NIH grantee, 1971-78; pvt. industry grantee, 1978—. Mem. ADA, Internat. Assn. Dental Research (dir. 1980-81, pres. Exptl. Pathology Group 1980—), Am. Acad. Oral Pathology, Internat. Acad. Pathology, Soc. Biomaterials Research, Sigma Xi. Subspecialties: Oral pathology; Implantology. Current work: Animal and human research on endosseous dental implants; animal research on wound healing. Home: 505 Loyola Dr Augusta GA 30909 Office: Med Coll Ga Augusta GA 30912

MC KINNEY, ROSS ERWIN, civil engineering educator; b. San Antonio, Aug. 2, 1926; s. Roy Earl and Beatrice (Saylor) McK.; m. Margaret McKinney Curtis, June 21, 1952; children: Ross Erwin, Margaret E., William S., Susanne C. B.A., So. Meth. U., 1948, B.S. in Civil Engring, 1948; S.M., MIT, 1949, Sc.D., 1951. San. scientist S.W. Found. for Research and Edn., San Antonio, 1951-53; asst. prof. MIT, 1953-58, assoc. prof., 1958-60; prof. U. Kans., 1960-63, chmn. dept. civil engring., 1963-66, Parker prof. civil engring., 1966-76, N.T. Veatch prof. environ. engring., 1976—; v.p. Rolf Eliassen Assos., Winchester, Mass., 1954-60; pres. Environ. Pollution Control Services, Lawrence, Kans., 1969-73. Author: Microbiology for Sanitary Engineers, 1962; Editor: Nat. Conf. on Solid Waste Research, 1964, 2d Internat. Symposium for Waste Treatment Lagoons, 1970. Mem. Cambridge (Mass.) Water Bd., 1953-59, Lawrence-Douglas County Health Bd., 1969-76, Kans. Water Quality Adv. Council, 1965-76, Kans. Solid Waste Adv. Council, 1970-76, Kans. Environ. Adv. Bd., 1976—. Served with USNR, 1943-46. Recipient Harrison P. Eddy award, 1962, Water Pollution Control Fedn. Rudolph Hering award, 1964; U.S. Presdl. Commendation, 1971; Environ. Quality award EPA Region VII, 1979. Mem. ASCE, Am. Water Works Assn., Water Pollution Control Fedn. (Thomas R. Camp medal 1982), Am. Pub. Works Assn., Am. Chem. Soc., Am. Soc. Microbiologists, AAAS, Am. Soc. Engring. Edn., Internat. Assn. Water Pollution Research, Am. Acad. Environ. Engrs., Kans. Water Pollution Control Assn., Nat. Acad. Engring., N.Y. Acad. Sci., AAUP, Sigma Xi, Sigma Tau, Kappa Mu Epsilon, Chi Epsilon, Tau Beta Pi. Subspecialty: Water supply and wastewater treatment. Current work: Biological and chemical wastewater treatment research; aerobic and anaerobic treatment systems. Patentee water treatment process. Home: 2617 Oxford Rd Lawrence KS 66044

MCKINNEY, STANLEY JOE, actuary, educator; b. Memphis, Aug. 8, 1959; s. Sammie Joe and Amy Ruth (Sanders) McK. B.S., U. Tex.-Arlington, 1980; M.S., East Tex. State U., 1981. Instr. bus. math. East Tex. State U., Commerce, 1981; tchr. 6th grade Paris (Tex.) Ind. Schs., 1981; actuarial asst. A.S. Hansen, Inc., Dallas, 1981—; mgr. computer graphics unit, 1982—; instr. tech. calculus Eastfield Coll., Dallas, 1982—. Mem. Math. Assn. Am., Am. Math. Soc., Soc. Indsl. and Applied Math., Tex. Acad. Sci. Subspecialties: Applied mathematics; Mathematical software. Current work: Computing actuarial techniques; graphics use as an information tool; probability. Home: 5937 Milton Apt 127-SB Dallas TX 75206

MCKLVEEN, JOHN WILLIAM, nuclear environmental engineer, educator, consultant; b. Washington, May 31, 1943; s. James H. and Helen (Faust) McK.; m.; children: Tori Leigh, Holly Hearne. B.S., U.S. Naval Acad., 1965; M.E. in Nuclear Engring, U. Va., 1971, Ph.D., 1973. Cadet U.S. Naval Acad., Annapolis, Md., 1961-65, commd. ensign, 1965; advanced through grades to lt. 1965, served as nuclear submarine officer, 1965-70, ret., 1970; faculty elec. and computer engring. Coll. Engring. and Applied Scis., Ariz. State U., Tempe, 1974—; pres. Radiation and Environ. Monitoring Inc., 1980—; cons. various nat. labs. and orgns. Author: Fast Neutron Activation Analysis, 1981; contbr. articles to publs. AEC spl. fellow, 1970-73. Mem. Am. Nuclear Soc. (chmn. environ. scis. div. 1982—, 1chmn. Ariz. sect. 1980-81, nat. program com. 1981—). Subspecialties: Nuclear engineering; Radiation Measurement. Current work: Environmental radiation measurements, neutron activation analysis, fast neutron activation, liquid scintillation spectrometry, energy education. Home: 8301 S Terrace Tempe AZ 85284

MCKNIGHT, RICHARD D., nuclear engineer; b. Cin., June 30, 1944; s. U.D. and Mary Louise (Dornseif) McK.; m. Pamela Susan Bratrude, Aug. 13, 1978. B.S. in Chem. Engring, U. Cin., 1967, M.S. in Nuclear Engring, 1969, Ph.D., 1974. Nuclear engr. applied physics div. Argonne (Ill.) Nat. Lab., 1974—. Mem. Am. Nuclear Soc. Subspecialties: Nuclear fission; Nuclear engineering. Current work: fast breeder reactor design; theory and analysis of zero power reactor fast critical assembly experiments; nuclear data. Home: 1512 George St Downers Grove Il 60516 Office: Argonne Nat Lab 9700 S Cass Ave Argonne IL 60439

MCKOWN, CORA F., home economics educator; b. Atoka, Okla., Aug. 21, 1943; d. W.F. and Zelma (Minton) McK. B.S., Southeastern Okla. U., 1964; M.S., Okla. State U., 1968; Ph.D., U. Mo., 1972. Instr. U. Okla. Extension, 1965-67; asst. prof. U. Ark., 1972-77; prof., chairperson housing and consumer dept. Tex. Tech U., 1977—; cons.; energy research. Author: Texas, Earth Sheltered, 1982, Residential Energy Alternatives, 1979; author articles on energy and housing. Mem. Am. Soc. Interior Designers, Am. Assn. Housing Educators, Am. Home Econs. Assn., AAUW, LWV. Subspecialties: Solar energy; Behavioral ecology. Current work: Passive solar housing; energy consumption; earth sheltered design. Office: PO Box 4170 Lubbock TX 79409

MC KUSICK, VICTOR ALMON, physician, geneticist, educator; b. Parkman, Maine, Oct. 21, 1921; s. Carroll L. and Ethel M. (Buzzell) McK.; m. Anne Bishop, June 11, 1949; children: Carol Anne, Kenneth Andrew, Victor Wayne. Student, Tufts Coll., 1940-43; M.D., Johns Hopkins U., 1946; D.Sc., N.Y. Med. Coll., 1974, U. Maine, 1978, Tufts U., 1978; M.D. (hon.), Liverpool U., 1976; Sc.D. (hon.), U. Rochester, 1979, Meml. U. Nfld., 1979; D.Med.Sc., Med. U. S.C., 1979; D.M.Ch., U. Helsinki, 1981. Tng. in clin. medicine, lab. research Johns Hopkins U./USPHS, 1946-52; instr. medicine Johns Hopkins Sch. Medicine, 1952-54, asst. prof., 1954-57, asso. prof., 1957-60, prof. medicine, 1960—, chief div. med. genetics, dept. medicine, 1957-73, prof. epidemiology, biology, 1969-78, William Osler prof. medicine, 1978—, chmn. dept. medicine, 1973—; physician-in-chief Johns Hopkins Hosp., 1973—; mem. research adv. com. Nat. Found., 1959-78; mem. med. adv. bd. Howard Hughes Med. Inst., 1967-83; pres. Internat. Med. Congress, Ltd., 1972-78; mem. Nat. Adv. Research Resources Council, 1970-74; mem. bd. sci. advisers Roche Inst. Molecular Biology, 1967-71; trustee Jackson Lab., 1979—. Author: Heritable Disorders of Connective Tissue, 1956, 60, 66, 72, Cardiovascular Sound in Health and Disease, 1958, Medical Genetics 1958-60, 1961, Human Genetics, 1964, 69, On the X Chromosome of Man, 1964, Mendelian Inheritance in Man, 1966, 68, 71, 75, 78, 83, (with others) Genetics of Hand Malformations, 1978, Medical Genetic Studies of the Amish, 1978; editor med. textbook. Recipient John Phillips award ACP, 1972; Gairdner Internat. award, 1977; Premio Internazionale Sanremo per le Ricerche Genetiche, 1983. Fellow Am. Acad. Orthopedic Surgeons (hon.), Royal Coll. Physicians (London), AAAS; mem. Nat. Acad. Sci., Am. Philos. Soc., Am. Soc. Human Genetics (pres. 1974, Wm. A. Allan award 1977), Assn. Am. Physicians, Am. Soc. Clin. Investigation, Académie Nationale Médecine (France) (corr.), Phi Beta Kappa, Alpha Omega Alpha. Presbyterian. Clubs: West Hamilton Street, St. Andrew's Soc. Balt. Subspecialties: Gene actions; Genome organization. Current work: Delineation of genetic disorders and traits-gene mapping of human chromosomes. Home: 221 Northway Baltimore MD 21218 Office: Johns Hopkins Hosp Baltimore MD 21205

MCLAFFERTY, FRED WARREN, chemist, educator; b. Evanston, Ill., May 11, 1923; s. Joel E. and Margaret E. (Keifer) McL.; m. Elizabeth E. Curley, Feb. 5, 1948; children: Sara L., Joel E., Martha A., Samuel A., Ann E. B.S., U. Nebr., 1943, M.S., 1947; Ph.D., Cornell U., 1950. Postdoctoral fellow U. Ia., 1949-50; research chemist, div. leader Dow Chem. Co., 1950-56; dir. Eastern Research Lab., 1956-64; prof. chemistry Purdue U., 1964-68, Cornell U., 1968—; mem. chem. sci. and tech. bd. NRC. Author: Mass Spectrometry of Organic Ions, 1963, Mass Spectral Correlations, 2d edit, 1981, Interpretation of Mass Spectra, 3d edit, 1980, Tandem Mass Spectrometry, 1983, Advances in Analytical Chemistry and Instrumentation, (with C.N. Reilley), Vols. 4-7), 1967-70, Index and Bibliography of Mass Spectrometry, (with J. Pinzelik), 1967, Atlas of Mass Spectral Data, (with E. Stenhagen and S. Abrahamsson), 1969, (with E. Stenhagen and S. Abrahamsson) Registry of Mass Spectral Data, 1974; Co-editor: Archives of Mass Spectral Data, 1969-72. Served with AUS, 1943-45; ETO. Decorated Purple Heart, Combat Inf. badge, Bronze Star medal with 4 oak leaf clusters; recipient Pitts. Spectroscopy award Spectroscopy Soc. Pitts., 1975; John Simon Guggenheim fellow, 1972. Fellow N.Y. Acad. Scis., Nat. Acad. Scis.; mem. Am. Chem. Soc. (chmn. analytical chem. div. 1969, chmn. Midland sect. 1956, Northeastern sect. 1964, award chem. instrumentation 1971, award analytical chemistry 1981), Am. Soc. Mass Spectroscopy (founder, sec. 1957-58), Chem. Soc. London, A.A.A.S., Sigma Xi, Phi Lambda Upsilon, Alpha Chi Sigma. Presbyn. Subspecialty: Analytical chemistry. Current work: Mass spectrometry of molecules, computer identification of unknown spectra. Home: 110 The Parkway Ithaca NY 14850

MC LAREN, DIGBY JOHNS, geologist, educator; b. Carrickfergus, No. Ireland, Dec. 11, 1919; m. Phyllis Mary Matkin, Mar. 25, 1942; children: Ian, Patrick, Alison. Student, Queens' Coll., Cambridge U., 1938-40, B.A., 1941, M.A. (Harkness scholar), 1948; Ph.D., Mich. U., 1951; D.Sc. (hon.), U. Ottawa, 1980. Geologist Geol. Survey Can., Ottawa, Ont., 1948-80, dir. gen., 1973-80; sr. sci. advisor Can. Dept. Energy, Mines and Resources, Ottawa, 1981—; vis. prof. U. Ottawa, 1981—; 1st dir. Inst. Sedimentary and Petroleum Geology, Calgary, Alta., Can., 1967-73; pres. Commn. on Stratigraphy, Internat. Union Geol. Scis., 1972-76; apptd. 14th dir. Geol. Survey Can., 1973; chmn. bd. Internat. Geol. Correlation Program, UNESCO-IGCP, 1976-80. Contbr. memoirs, bulls., papers, geol. maps, sci. articles in field to profl. lit. Served to capt. Royal Arty. Brit. Army, 1940-46. Gold medalist (sci.) Profl. Inst. Pub. Service of Can., 1979. Fellow Royal Soc. Can., Royal Soc. London, European Union of Geoscis. (hon.); fgn. assoc. U.S. Nat. Acad. Scis, Geol. Soc. France; mem. Geol. Soc. London, Geol. Soc. Germany (hon.), Geol. Soc. Am. (pres. 1982), Paleontol. Soc. (pres. 1969), Geol. Assn. Can., Can. Soc. Petroleum Geologists. (pres. 1971). Subspecialty: Geology. Office: Dept Geology Univ Ottawa Ottawa ON K1N 6N5 Canada

MCLAUGHLIN, CALVIN S., biochemist, educator; b. St. Joseph, Mo., May 29, 1936. B.A. magna cum laude; scholar, King Coll., 1958; postgrad. (Rockefeller Bros. fellow), Yale U., 1958-59; Ph.D. in Biochemistry (Upjohn fellow, Nutrition Found. fellow, NIH fellow), M.I.T., 1964. Am. Cancer Soc. postdoctoral fellow Institut de Biologie-Physico-chimique, Paris, 1964-66; asst. prof. biochemistry U. Calif. Irvine, 1966-69, asso. prof., 1969-73, prof., 1973-78, vice-chmn. dept. biol. chemistry, 1978—; dir. Cancer Research Inst., 1981—; Gabriel Lester Meml. lectr., 1979; chemist Tenn. Eastman. Co., Kingsport, 1957-58; vis. prof. Sch. Botany, Oxford (Eng.) U., 1976, Linacre Coll., Oxford, 1980; mem. sci. adv. bd. Am. Cancer Soc., NSF, NIH; sec.-treas. Pacific Slope Biochem. Conf., 1973-76; participant profl. cons. and seminars. Editorial bd.: Jour. Bacteriology; reviewer: Molecular and Cellular Biology; contbr. articles to sci. jours. Danforth asso., 1976-82; grantee NIH, Burns Family Found., Nat. Mycology Research Center, Nat. Cancer Inst. Mem. AAAS, Genetics Soc. Am., Am. Soc. Microbiology, Am. Soc. Biol. Chemists, Mycol. Soc. Am. Subspecialties: Genetics and genetic engineering (biology); Biochemistry (biology). Office: Dept Biol Chemistr U Calif Irvine CA 92625

MCLAUGHLIN, CAROL LYNN, physiology researcher; b. Hartford, Conn., Jan. 15, 1943; d. Edward Joseph and Eunice (Robinson) McL. B.A., Oberlin Coll., 1965; M.A., West Chester State Coll., 1977; Ph.D., U. Pa., 1981. Research asst. Harvard U., 1965-70; nutritional physiologist Smith Kline & French, West Chester, 1971-74; research specialist U. Pa., Kennett Square, Pa., 1975-79, research assoc., 1981-82; postdoctoral fellow Thomas Jefferson U., 1981-82; research assoc. prof. Washington U., St. Louis, 1982—. Contbr. numerous articles to sci. jours. Mem. Am. Inst. Nutrition, Soc. Neurosci. Subspecialties: Physiological psychology; Nutrition (biology). Current work: Control of food intake by peptides in obese rodents. Office: Box 8113 Washington U 4566 Scott Ave Saint Louis MO 63110

MCLAUGHLIN, DONALD REED, chemist, educator, researcher; b. Los Angeles, Oct. 6, 1938; s. Alfred Reed and Anita Grace (Squires) McL.; m. Linda Irene, June 15, 1964. B.A., UCLA, 1960; Ph.D., U. Utah, 1965. Assoc. prof. chemistry U. N.Mex., 1965—; dir. Los Alamos Center for Grad. Studies. Mem. Am. Chem. Soc., Sigma Xi. Subspecialties: Theoretical chemistry; Laser photochemistry. Current work: Quantum chemistry; dynamics and theoretical molecular spectroscopy. Office: Dept Chemistry U NMex Albuquerque NM 87131

MCLAUGHLIN, MICHAEL RAY, plant pathologist, researcher; b. Carroll, Iowa, June 20, 1949; s. Harold Eugene and Charlene Martha (Yost) McL.; m. Cindy Lee Richardson, May 30, 1970; children: Erin, Caragh, Mark. B.S., Iowa State U., 1971, M.S., 1974; Ph.D., U. Ill., 1978. Vis. asst. prof. plant pathology and physiology Clemson (S.C.) U., 1978-79; asst. prof. entomology and plant pathology U. Tenn., Knoxville, 1979-82; research plant pathologist USDA Agr. Research Service, Mississippi State, Miss., 1982—; adj. assoc. prof. Miss. State U., Mississippi State, 1982—. Mem. Am. Photopath. Soc., Sigma Xi, Gamma Sigma Delta. Subspecialties: Plant virology; Plant pathology. Current work: Viruses and virus desease of forage legumes. Office: PO Drawer PG Mississippi State MS 39762

MCLEAN, JOHN ROBERT, clinical scientist; b. St. Thomas, Ont., Can., Apr. 15, 1926; came to U.S., 1954; s. John and Dorothy (McFarlane) McL.; m. Elizabeth Virtue, Sept. 8, 1951; children: Dorothy, Mark. Ph.D., Queen's U., Ont., 1954. Dir. neuropharmacology Parke-Davis & Co., Ann Arbor, Mich., 1972-77, assoc. dir. biochemistry, 1977-82, group leader neurol. diseases, 1982-83, asst. dir. clin. sci., 1983—. Mem. Am. Soc. Pharmacology and Exptl. Therapeutics, Am. Heart Assn. Council on Thrombosis, Am. Epilepsy Soc. Subspecialties: Neurology; Neuropharmacology. Current work: Development of drugs for the treatment of neurological diseases. Home: 1708 Covington Ann Arbor MI 48103 Office: 2800 Plymouth Rd Ann Arbor MI 48105

MCLENDON, BENNIE DERRELL, agricultural engineering educator; b. Cuthbert, Ga., Dec. 26, 1941; s. Joseph Edwin and Bronnie (Belcher) McL.; m. Frances Wright, Nov. 28, 1964; children: Jeffery Derrell, Kevin Allen. Mem. faculty U. Ga., Athens, 1965—, asst. prof. agrl. engring., 1973-82, assoc. prof., 1982—. Contbr. articles to profl. publs. NSF Sci. Faculty fellow Cornell U., 1969-71. Mem. Am. Soc. Agrl. Engrs., Am. Soc. Engring. Edn. Subspecialty: Agricultural engineering. Current work: Microprocessor-based control for agricultural production and processing. Teaching and research in the area of micro processor-based control systems. Home: 388 Cherokee Ridge Athens GA 30606 Office: Univ Ga Driftmier Engring Center Athens GA 30602

MCLENITHAN, KELLY DANIEL, theoretical physicist, numerical analyst; b. Grand Rapids, Mich., June 11, 1954; s. Earle Chester and Norma Fay (Gierloff) McL.; m. Priscilla Colleen Pflug, Aug. 7, 1976; children: Shannon Leigh, Kevin Daniel. A.B., Ill. Coll., 1976; M.S., U. Ill.-Urbana, 1978, Ph.D., 1982. Research trainee Oak Ridge Nat. Lab., summer 1975; teaching asst. U. Ill.-Urbana, 1976-78, research asst., 1978-82, research assoc., 1982; project assoc. U. Wis.-Madison, 1982—. Mem. Am. Phys. Soc., Math. Assn. Am., Soc. Indsl. and Applied Math., Phi Lambda Upsilon. Subspecialties: Theoretical physics; Applied mathematics. Current work: Quantum mechanical scattering problems; magnetohydrodynamic equilibrium and stability of confined fusion plasmas; thermonuclear reaction dynamics. Home: 3018 Jason Pl Madison WI 53719 Office: Torsatron/Stellarator Lab U Wis 1415 Johnson Dr Madison WI 53706

MCLENNAN, JEAN GLINN, virologist; b. Charlotte, N.C., Feb. 8, 1934; d. John Manning and Irene Mae (Pool) Glinn; m. William Eldon McLennan, Sept. 17, 1961. B.A., U. Pa., Phila., 1955; B.S., U. Nebr.-Omaha, 1983; postgrad., Hunter Coll., 1960-61, U. Nebr., Omaha, 1976—. Microbiologist Emory U. Sch. Medicine, 1955-57, virology, 1963-66; microbiologist Cornell U. Sch. Medicine, 1957-61, Ga. Dept. Public Health, Atlanta, 1961-62, Cutter Labs., Berkeley, Calif., 1966-75; virologist Burns-BioTec Lab., Elkhorn, Nebr., 1975—. Mem. Am. Soc. Microbiology, Assn. Vet. Microbiologists (editor newsletter 1982—). Democrat. Lutheran. Subspecialties: Animal virology; Microbiology (veterinary medicine). Current work: Quality control and statistical analysis. Home: 1442 S 163d St Omaha NE 68130 Office: PO Box 3113 Omaha NE 68103

MC LERAN, JAMES HERBERT, ednl. adminstr., dentist; b. Audubon, Iowa, Apr. 9, 1931; s. Louis D. and Alma K. (Christensen) McL.; m. Hermine Weinert Hayden, July 15, 1979; 1 son, Stephen Andrew. B.S., Simpson Coll., 1953; D.D.S., U. Iowa, 1957, M.S., 1962. Diplomate: Am. Bd. Oral and Maxillofacial Surgeons. Instr. U. Iowa, 1959-60, asst. prof., 1963-67, asso. prof., 1967-69, prof. oral surgery, 1972—, asso. dean, 1972-74, dean, 1974—; prof. and chmn. dept. oral surgery U. N.C., Chapel Hill, 1969-72. Served with Dental Corps USN, 1957-59. Recipient Finkbine Leadership award U. Iowa, 1957;

named Instr. of Yr. Jr. ADA, 1964, Outstanding Instr. award, 1965. Mem. Am. Assn. Oral and Maxillofacial Surgeons, Iowa Assn. Oral and Maxillofacial Surgeons, Internat. Assn. Dental Research, ADA, Midwestern Assn. Oral and Maxillofacial Surgeons, Am. Dental Soc. Anesthesiology, N.C. Soc. Dental Surgeons, N.C. Dental Soc., Iowa Dental Assn., Univ. Dist. Dental Soc., Johnson County Dental Soc., Am. Coll. Dentists, Internat. Coll. Dentists, Am. Assn. Dental Schs. (pres. 1978-79), Psi Omega (adv.). Republican. Club: Rotary. Subspecialty: Oral and maxillofacial surgery. Home: 6 Glendale Terr Iowa City IA 52240 Office: Coll Dentistry U Iowa Iowa City IA 52242

MCLERRAN, ARCHIE RALPH, mechanical engineer, consultant; b. Milam County, Tex., Aug. 27, 1918; s. Archie Roy and Pearl Marie (Smith) McL.; m. Inez Smith, Aug. 14, 1917; 1 dau., Marilyn McLerran Souchek. B.S. in Mech. Engring, Tex. A&M U., 1939. Registered profl. engr., Tex., La. With Ideco div. Dresser Industries, Beaumont, Tex., 1939-60, chief research and devel. engr., 1939-60; owner, prin. engr. A.R. McLerran & Assocs. (cons. engrs.), Beaumont, 1961-64; mgr. field operation, project mohole NSF, Houston, 1964-68, field ops. officer, deep sea drilling project, La Jolla, Calif., 1968-83; mgr. ops. and engring. ocean drilling program Tex. A&M U., College Station, 1983—; cons. Bd. dirs. Isla Verde Assn., Inc., Solana Beach, Calif., 1979-81, pres., 1980-81. Recipient Meritorious Service award NSF, 1978, Disting. Service award NSF, 1983. Mem. ASME, Am. Petroleum Inst., Republican. Baptist. Club: Lomas Santa Fe Country (Solana Beach). Subspecialties: Mechanical engineering; Offshore technology. Current work: Management of scientific ocean drilling program and development deep sea drilling and coring technology. Patentee in field. Office: Ocean Drilling Project Coll Geosci Tex A&M U College TX 77843

MCLINDEN, LYNN, mathematician; b. Cleve., July 27, 1943. A.B., Princeton U., 1965; Ph.D., U. Wash., 1971. Vis. mem. Math Research Ctr., U. Wis.-Madison, 1971-73; asst. prof. math. U. Ill., Urbana-Champaign, 1973-78, assoc. prof., 1978—; research fellow Ctr. for Ops. Research and Econometrics, Université Catholique de Louvain, Louvain-la-Neuve, Belgium, 1978-79; vis. mem. Math. Research Ctr., U Wis., Madison, 1983-84. NSF research grantee, 1975—. Mem. Am. Math. Soc., Soc. Indsl. and Applied Math., Ops. Research Soc. Am., Math. Programming Soc. Subspecialties: Applied mathematics; Operations research (mathematics). Current work: Optimization theory and related analysis. Office: Dept Math U Ill-Urbana-Champaign 1409 W Green St Urbana IL 61801

MCLOON, LINDA KIRSCHEN, anatomy researcher, educator; b. N.Y.C., July 28, 1953; m. Steven Charles McLoon. B.S. in Biology, SUNY, Binghamton, 1974; Ph.D. in Anatomy, U. Ill. at Med. Center, Chgo., 1979. Teaching asst. dept. anatomy U. Ill. Med. Center, 1974-79; instr. dept. anatomy Ill. Coll. Podiatric medicine, Chgo., 1975-78; research assoc. dept. biol. structure U. Wash., Seattle, 1979; research assoc. dept. ophthalmology U. Minn. Contbr. articles and abstracts to profl. jours. Mem. Soc. Neurosci., Am. Assn. Anatomists, AAAS. Subspecialties: Neurobiology; Developmental biology. Current work: Developmental neurobiology; transplantation of nervous tissue; trophic factors in development. Office: Dept Ophthalmology Box 493 U Minn 516 Delaware SE Minneapolis MN 55455

MC LUCAS, JOHN LUTHER, corp. ofcl.; b. Fayetteville, N.C., Aug. 22, 1920; s. John Luther and Viola (Conley) McL.; m. Patricia Knapp, July 27, 1946 (div. 1981); children: Pamela McLucas Byers, Susan, John C., Roderick K.; m. Harriet D. Black, Sept. 25, 1981. B.S., Davidson Coll., 1941; M.S., Tulane U., 1943; Ph.D., Pa. State U., 1950, D.Sc., 1974. Vice pres., tech. dir. Haller, Raymond & Brown, Inc., State College, Pa., 1950-57; pres. HRB-Singer, Inc., State College, 1958-62; dep. dir. research and engring. Dept. Def., 1962-64; asst. sec.-gen. for sci. affairs NATO, Paris, France, 1964-66; pres., chief exec. officer Mitre Corp., Bedford, Mass., 1966-69; under sec. of air force, 1969-73, sec. of air force, 1973-75; adminstr. FAA, 1975-77; pres. Comsat Gen. Corp., Washington, 1977-79; exec. v.p. COMSAT, 1979-80, pres. world systems div., 1980—; mem. Def. Intelligence Agy. Sci. Advisory Com., 1966-69, U.S. Air Force Sci. Advisory Bd., 1967-69, 77—, Def. Sci. Bd., 1968-69; chmn. USAF SDAG, 1979-83. Contbr. articles to tech. lit. Bd. dirs. von Karman Inst., Brussels, 1966-69; vice-chmn. bd. Wolf Trap Found. Served with USNR, 1943-46. Recipient Disting. Service award Dept. Def., 1964, 1st bronze palm, 1973, silver palm, 1975. Fellow IEEE; fellow AIAA (pres.-elect); mem. N.Y. Acad. Scis., Ops. Research Soc. Am., Am. Phys. Soc., Nat. Acad. Engring., Chief Execs. Forum, Sigma Xi, Sigma Pi Sigma. Club: Cosmos. Subspecialty: Telecommunications company management. Patentee in field. Home: 309 N. Lee St Alexandria VA 22314 Office: Comsat 950 L'Enfant Plaza Washington DC 20024

MC MAHON, CHARLES JOSEPH, JR., materials science educator; b. Phila., July 10, 1933; s. Charles Joseph and Alice (Schu) McM.; m. Helen June O'Brien, Jan. 31, 1959; children: Christine, Charles, Elise, Robert, David. B.S., U. Pa., 1955; Sc.D., MIT, 1963. Instr. metallurgy MIT, 1958-62, research asst., 1962-63; postdoctoral fellow U. Pa., 1963-64; asst. prof. dept. materials sci. and engring., 1964-68, assoc. prof., 1968-74, prof., 1974—; cons. in field. Editor: Microplasticity, 1968; Contbr. articles to profl. jours. Served with USN, 1955-58. Churchill Overseas fellow Churchill Coll., Cambridge (Eng.) U., 1973-74. Fellow Am. Soc. Metals, Nat. Acad. Engring., Inst. Metallurgists U.K.; mem. AIME, Instn. Metallurgists, Metals Soc. (U.K.), AAAS. Democrat. Roman Catholic. Subspecialty: Metallurgy. Current work: Interfacial fracture, solute segregation to surfaces and interfaces. Home: 7103 Sherman St Philadelphia PA 19119 Office: U Pa Philadelphia PA 19104

MCMAHON, DONALD HOWLAND, optical physicist; b. Buffalo, Apr. 18, 1934; s. Arthur Philmore and Elnora Winifred (Langley) McM.; m. Mable Grinnell, Apr. 24, 1934. B.S. in Physics with honors, U. Buffalo, 1957; Ph.D. in Exptl. Physics, Cornell U., 1964. Mem. research staff Sperry Research Ctr., Sudbury, Mass., 1963-71, program mgr. in optics tech., 1971-73, mgr. optics dept., 1973—. Contbr. numerous articles, on optics to profl. jours. Fellow Am. Optical Soc.; mem. Am. Phys. Soc., IEEE, Phi Beta Kappa. Subspecialties: Fiber optics; Optical image processing. Current work: Fiber optic sensors, fiber optic communication, optical pattern recognition, electro optics, acousto optics, optical info. processing. Patentee in field, mainly in optics. Home: 74 Judy Farm Rd Carlisle MA 01741 Office: 100 North Rd Sudbury MA 01776

MCMAHON, MARIBETH, mfg. co. exec., physicist; b. Bradford, Pa., June 8, 1949; d. James Harry and Josephine Rose (Sylvester) Lovette; m. Frank Joseph McMahon, Nov. 19, 1976. B.S. in Physics, Pa. State U., 1971; M.S. in Physics, Pa. State U., 1974, Ph.D., 1976. Research asst. in physics Pa. State U., 1971-76; advanced research and devel. engr. GTE Sylvania, Danvers, Mass., 1976-77; sr. physicist 3M Co., St. Paul, 1977-78, market devel. supr., 1978—; dog obedience instr. Mem. Optical Soc. Am., Sigma Xi, Sigma Pi Sigma. Subspecialties: Optical properties of materials; Applied mathematics. Current work: Optical properties of materials; product devel. of photog. film for radiography. Office: 3M Bldg 223-2SW Saint Paul MN 55144

MCMANAMON, PAUL FRANCIS, electronic engr.; b. Cleve., July 1, 1946; s. John and Catherine (Bauman) McM.; m. Laura Sarina (DeDonne), July 5, 1969; children: Joan, Steven, David., Deborah. B.S., John Carroll U., 1968; M.S., Ohio State U., 1973, Ph.D., 1977. Physicist Electronic Warfare Analyses U.S. Air Force, 1968-70, electronics engr., 1970-71, 71-73; tech. specialist Electronic Warfare Avionics, Wright Patterson AFB, Ohio, 1973-79; group leader Passive IR Sensor Group, 1979—. Mem. Am. Phys. Soc., Optical Soc. Am., Assn. Old Crows. Roman Catholic. Subspecialty: Thermal imaging. Current work: Research on focal plane array infrared sensors.

MCMANUS, MARIANNE LEE, psychologist, cons.; b. N.Y.C., July 30, 1932; d. James William and Mary (Lee) McM. B.A., Coll. New Rochelle, 1953; M.A., U. Wis., 1958, Ph.D., 1963. Lic. psychologist, Calif. Learning skills specialist U. Wis., Madison, 1956-57, clin. psychologist, lectr., 1958-68; psychology intern Wis. Diagnostic and Treatment Center, Madison, 1957-58; pvt. practice psychology, Madison, after 1960, Ames, Iowa, to 1979; asst. v.p., assoc. prof., clin. psychologist Iowa State U., Ames, 1968-79; corp. psychologist Rohrer, Hibler & Replogle, Los Angeles, 1979-83; pres., cons. psychologist McManus & Assocs., Santa Monica, Calif., 1983—; Media/tng. cons. Fortune 500 Cos., Los Angeles, 1979—. Contbr. articles to profl. jours. Am. Psychol. Assn. travel grantee, 1976; Iowa State U. grantee, 1972, 77. Mem. Am. Psychol. Assn. (chmn. award com. for excellence in cons. 1982—), Am. Mgmt. Assn., Am. Soc. for Tng. and Devel., Women in Communication, AAAS, AAUP. Democrat. Club: Los Angeles Athletic. Current work: Evaluation of candidates for key positions; appraisal of management skills and effectiveness; research on occupational stressors and effects on productivity; executive training and development; team building. Office: McManus and Assocs 757 Ocean Ave Suite 111 Santa Monica CA 90402

MCMEEKING, ROBERT MAXWELL, mechanical engineer, educator; b. Glasgow, Scotland, May 22, 1950; came to U.S., 1972; s. Robert Maxwell and Elizabeth Higginson (Craighead) McM.; m. Norah Anne Madigan, Sept. 4, 1976; children: Gavin Robert, Anne Catherine. B.Sc. in Engring. with 1st class honors, U. Glasgow, 1972; M.S., Brown U., 1974, Ph.D., 1977. Acting asst. prof. mech. engring. Stanford U., 1976-78; asst. prof. theoretical and applied mechanics U. Ill., Urbana-Champaign, 1978-82, assoc. prof., 1982—. Contbr. articles to profl. jours. NSF grantee, 1980-82, 82—; recipient George Harvey prize U. Glasgow, 1972, Goudie prize, 1972, George Russel prize, 1972. Mem. ASME, Am. Acad. Mechanics, ASTM, AAAS, Sigma Xi. Subspecialties: Theoretical and applied mechanics; Materials. Current work: Research in mechanics of ductile rupture, fracture, micromechanics of materials, plasticity, finite element methods, metal forming. Home: 2312 Barberry Dr Champaign IL 61820 Office: U Ill Dept Theoretical and Applied Mechanics 216 Talbot Lab 104 S Wright St Urbana IL 61801

MCMENAMIN, EDWARD WILLIAM, engring. exec.; b. Phila., Jan. 22, 1928; s. Edward A. and Wilhelmina (Muth) McM.; m. Catherine E. Rowland, May 1, 1954; children: James R., Robert T., Cathy E. B.S.M.E., Drexel U., Phila., 1952. Registered profl. engr., Pa. Successively draftsman, coop. student, design engr., project engr. Proctor & Schwartz, Phila., 1945-59; successively chief engr., v.p. engring. and sales Gimpel Corp., Langhorne, Pa., 1959—. Mem. ch. council, fin. sec., chmn. fin. and bldg. coms. St. John's Lutheran Ch.; adult leader Boy Scouts Am. Served with USAAF, 1946-47. Mem. ASME, Am. Soc. Non-destructive Testing. Republican. Lodges: Odd Fellows; Masons. Subspecialties: Mechanical engineering; Metallurgy. Current work: Leading designer and developer trip throttle valves for steam turbines for marine, petro, chem. industries and utilities. Home: 680 Mason Dr Warminster PA 18974 Office: 250 Woodbourne Rd PO Box 188 Langhorne PA 18974

MC MILLAN, EDWIN MATTISON, physicist, educator; b. Redondo Beach, Calif., Sept. 18, 1907; s. Edwin Harbaugh and Anna Marie (Mattison) McM.; m. Elsie Walford Blumer, June 7, 1941; children—Ann B., David M., Stephen W. B.S., Calif. Inst. Tech., 1928, M.S., 1929; Ph.D., Princeton U., 1932; D.Sc., Rensselaer Poly. Inst., 1961, Gustavus Adolphus Coll., 1963. Nat. research fellow U. Calif. at Berkeley, 1932-34, research asso., 1934-35, instr. in physics, 1935-36, asst. prof. physics, 1936-41, asso. prof., 1941-46, prof. physics, 1946-73, emeritus, 1973—; mem. staff Lawrence Radiation Lab., 1934—, asso. dir., 1954-58, dir., 1958-73; on leave for def. research at Mass. Inst. Tech. Radiation Lab., U.S. Navy Radio and Sound Lab., San Diego, and Los Alamos Sci. Lab., 1940-45; mem. gen. adv. com. AEC, 1954-58; mem. commn. high energy physics Internat. Union Pure and Applied Physics, 1960-67; mem. sci. policy com. Stanford Linear Accelerator Center, 1962-66; mem. physics adv. com. Nat. Accelerator Lab., 1967-69; chmn. 13th Internat. Conf. on High Energy Physics, 1966; guest prof. CERN, Geneva, 1974. Trustee Rand Corp., 1959-69; Bd. dirs. San Francisco Palace Arts and Scis. Found., 1968—; trustee Univs. Research Assn., 1969 74. Recipient Research Corp. Sci. award, 1951; with Glenn T. Seaborg) Nobel prize in chemistry, 1951; with Vladimir I. Veksler) Atoms for Peace award, 1963; Alumni Distinguished Service award Calif. Inst. Tech., 1966; Centennial citation U. Calif. at Berkeley, 1968; Faculty Research lectr. U. Calif. at Berkeley, 1955. Fellow Am. Acad. Arts and Scis., Am. Phys. Soc.; mem. Nat. Acad. Scis. (chmn. class I 1968-71), Am. Philos. Soc., Sigma Xi, Tau Beta Pi. Subspecialty: Nuclear physics. Address: Lawrence Berkeley Lab Berkeley CA 94720

MC MILLAN, ROBERT SCOTT, astronomer; b. Pitts., Feb. 26, 1950; s. William Robert and Mary Eunice (Amos) McM.; m. Gloria Lee, Dec. 22, 1980; 1 son, Christopher Norman. B.S., Case Inst. Tech., 1972; M.A., U. Tex., 1974, Ph.D., 1977. Research and teaching asst. U. Tex., Austin, 1972-77; research asso. astrophysics br. Space Scis. Labs., NASA Marshall Space Flight Center, Huntsville, Ala., 1977-79; sr. research asso. Lunar and Planetary Lab., U. Ariz., Tucson, 1979—. Mem. Am. Astron. Soc., Soc. Photo-optical Instrumentation Engrs., Sigma Xi, Tau Beta Pi. Subspecialties: Optical astronomy; Optical image processing. Current work: Developer of an optical spectrometer for accurately measuring Doppler shifts of stars and of an electronic camera for detecting nearby asteroids. Home: 428 E Adams St Tucson AZ 85705 Office: Univ of Arizon Space Science Bldg Tucson AZ 85721

MCMILLAN, ROBERT WALKER, physicist, researcher; b. Sylacauga, Ala., Apr. 18, 1935; s. Robert Thomas and Mary Alma (Bush) McMillen; m. Dorothea Ann Simmons, Sept. 11, 1955; children: Marisa, Robert, Natalie. B.S., Auburn U., 1957; M.S., Rollins Coll., 1966; Ph.D., U. Fla., 1974. Assoc. engr. Westinghouse Electric Corp., Balt., 1960-61; staff engr. Martin Marietta Aerospace Corp., Orlando, Fla., 1961-76; prin. research scientist Ga. Inst. Tech. Engring. Expt. Sta., Atlanta, 1976—. Contbr. articles to profl. jours. Served to 1st lt. USAF, 1958-60. Mem. Optical Soc. Am., IEEE. Democrat. Baptist. Subspecialties: Electrical engineering; Condensed matter physics. Current work: Millimeter wave systems, spectroscopy, radiometry, atmospheric physics, millimeter wave devices. Patentee in field. Home: 6332 Queen View Ln Stone Mountain GA 30087 Office: Engring Experiment Sta Ga Inst Tech Atlanta GA 30332

MCMILLIN, DAVID EDWIN, geneticist, educator; b. Pitts., Apr. 16, 1952; s. John Glenn and Dorothy Ruth (Lauterette) McM.; m. Mary Katherine McMillin, Dec. 25, 1974; children: John David, Phillip Matthew. B.S. in Biology, Ga. So. Coll., 1974; M.S. in Plant Genetics, Tex. A & M U., 1977; Ph.D. in genetics, N.C. State U., 1981. Asst. prof. dept. biol. scis. N. Tex. State U., 1981—. Contbr. articles to profl. jours. Recipient Kenneth R. Keller award, 1981. Mem. Genetics Soc. Am., AAAS, Soc. Devel. Biology, Gamma Sigma Delta. Subspecialties: Gene actions; Plant genetics. Current work: Devel. genetics, gene regulation; nuclear organelle interactions; crop improvement using choromosome mediated gene transfer. Home: 1721 Teasley Ln #271 Denton TX 76203 Office: Dept Biological Science North Texas State U Denton TX 76203

MCMILLIN, JOHN MICHAEL, endocrinology educator and administrator; b. Mpls., Jan. 16, 1940; s. John Dominic and Margaret Dagmar (Marklund) McM.; m. Joan Francis Austin, May 10, 1970; children: J. Andrew, Christy Ann. B.S., U. Minn., 1962, M.D., 1965, postgrad., 1969-73. Diplomate: Am. Bd. Internal Medicine. Intern Parkland Meml. Hosp., Dallas, 1965-66; research fellow, resident in psychiatry Mass. Gen. Hosp., Boston, 1968-69; fellow in medicine U. Minn., Mpls., 1969-71, fellow in endocrinology, 1971-73, asst. prof. medicine, 1973-76; head div. endocrinology U. S.D., Sioux Falls, 1976—; assoc. chief of staff for research devel. VA, Sioux Falls, 1976—. Author: Dissipative Structores and Spatiotemporal Organization in Biomedical Research, 1980; contbr. articles to profl. jours. Served to capt. AUS, 1966-68; Vietnam. Recipient Order of Ski Umah U. Minn., 1965; VA research grantee, 1976—; recipient Faculty Recognition award U. S.D., 1981. Mem. Am. Cancer Soc. (pres. S.D. 1977-81, 82-83, nat. med. del.), Am. Diabetes Assn. (dir. S.D.), Am. Fedn. Clin. Research, Endocrine Soc., Soc. Study of Reprodn., Internat. Soc. Chronobiology, Alpha Omega Alpha. Democrat. Episcopalian. Subspecialty: Neuroendocrinology. Current work: Control and function of growth hormone, prolactin and sex steroids. Home: 1601 Cedar Ln Sioux Falls SD 57103 Office: U South Dakota Sch of Medicine 2501 W 22d St Sioux Falls SD 57105

MCMORRIS, F(REDERICK) ARTHUR, research biologist, educator; b. Lawton, Okla., Sept. 17, 1944; s. William Arthur and Mary Alfredda (Pope) McM. A.B. with high honors, Brown U., 1966; Ph.D., Yale U., 1972. Research asoc. M.I.T., 1972-74; asst. prof. Wistar Inst., Phila., 1974—; mem. grad. groups in molecular biology, neurosci. and genetics U. Pa., 1975—. Contbr. articles to profl. jours. Bd. dirs. Spruce Hill Community Assn., 1976—, exec. v.p., 1980—; bd. dirs. Friends of Clark Park, 1978-80. NIH fellow, 1972. Mem. AAAS, Am. Soc. Cell Biology, Am. Soc. Neurochemistry, Am. Soc. Human Genetics, Soc. Devel. Biology, Soc. Neurosci., Sigma Xi. Democrat. Subspecialties: Neurochemistry; Gene actions. Current work: Gene regulation and biochemical expression in cells of the nervous system. Home: 4333 Larchwood Ave Philadelphia PA 19104 Office: Wistar Institute 36th and Spruce Sts Philadelphia PA 19104

MCMURRY, JOHN EDWARD, chemistry educator; b. N.Y.C., July 27, 1942; s. Edward and Marguerite Ann (Hotchkiss) McM.; m. Susan Elizabeth Sobuta, Sept. 4, 1964; children: Peter Michael, David Andrew. B.A., Harvard U., 1964; M.A., Columbia U., 1965, Ph.D., 1967. Asst. prof. U. Calif.-Santa Cruz, 1967-71, assoc. prof., 1971-75, prof., 1975-80; prof. chemistry Cornell U., Ithaca, N.Y., 1980—. Assoc. editor: Accounts Chem. Research, 1975—. Sloan Found. fellow, 1969-71; NIH grantee, 1975-80. Mem. Am. Chem. Soc., AAAS. Subspecialty: Organic chemistry. Current work: Research on developing new chemical reactions for use in the laboratory synthesis of organic molecules. Home: 112 Oak Hill Rd Ithaca NY 14850 Office: Baker Lab Cornell U Ithaca NY 14853

MCNALLY, JAMES RAND, JR., fusion energy cons.; b. Boston, Nov. 10, 1917; s. James R and Margaret (Turley) McN.; m. Margaret Anne McKenna, Nov. 26, 1942; children: James Rand, Peter Joseph, Mary Ellen, Francis Edward, Anne Therese, Michael Stephen, Margaret Rose. B.S. in Physics, Boston Coll., 1939; M.S., M.I.T., 1941, Ph.D. in Physics, 1943. Mem. faculty M.I.T., 1939-48; with Oak Ridge Nat. Lab., 1948-82, sr. research physicist, until 1982; fusion energy cons., 1982—. Contbr. articles to profl. jours. Trustee Dedham Pub. Library, 1945-48. Recipient Manhattan Dist. award AEC, 1946. Fellow Optical Soc. Am. (charter), AAAS; mem. Am. Phys. Soc., Am. Assn. Physics Tchrs. Republican. Roman Catholic. Club: KC. Subspecialties: Nuclear fusion; Plasma physics. Current work: Advanced fuel fusion reactor development utilizing multi-MeV accelerators, interpretation of giant red spots formed in upper atomospher by nuclear weapons tests or unusual aurora, in fusion-fission breeder prospects, in properties of nuclear dynamos fueld by fusion fuel, safety questions of large or multiple nuclear weapons explosions. Address: 103 Norman Ln Oak Ridge TN 37830

MCNAMARA, JAMES O'CONNELL, neurologist; b. Portage, Wis., Sept. 25, 1942; s. Louis V. and Lucille M. (O'Connell) McN.; m. Anne M. Niebler, Aug. 15, 1964; children—Dennis, Brigid, James O'Connell, Michael Brian. A.B., Marquette U., postgrad. Med. Sch., 1964-66; M.D., U. Mich., 1968. Diplomate: Am. bd. Psychiatry and Neurology; cert Am. Bd. Qualification in EEG. Asst. prof. medicine and neurology Duke U. Med. Ctr., 1975-80; dir. Epilepsy Ctr., Durham VA Med. Center-Duke U., 1976—, assoc. prof., 1980—, asst. prof. pharmacology, 1982—; dir. Center Advanced Study of Epilepsy, 1982—; Chmn. profl. adv. bd. Epilepsy Assn. N.C., 1978-80; mem. profl. adv. bd. Epilepsy Found. Am., 1981—. Contbr. articles to profl. publs. Served to maj. M.C. U.S. Army, 1971-73. NIH, VA research grantee. Mem. Am. Neurol. Assn., Am. Acad. Neurology, Soc. Neurosci., Am. Epilepsy Soc., Am. Soc. Pharmacology and Exptl. Therapeutics. Roman Catholic. Club: Chapel Hill Country. Subspecialties: Neurology; Neurobiology. Current work: Neurobiologic approach to epilepsy. Clinical and scientific approaches to epilepsy. Scientific disciplines utlutilized span morphology, pharmacology, biochemistry and physiology. Office: Dept Neurology Durham VA Hos Duke U Med Center 508 Fulton St Durham NC 27705

MCNAMARA, T(HOMAS) F(RANCIS), microbiologist, immunologist educator; b. Bklyn., Jan. 26, 1928; s. Thomas Francis and Georgiana Regina (Costello) McN.; m. Virginia Anne Daly, June 27, 1953; children: Thomas J. (dec.), Michael A., Mary K., Carolyn V., Georgine T. B.S., Manhattan Coll., 1949; M.A., Hofstra U., 1950; Ph.D., Catholic U. Am., 1959. Sr. scientist Eaton Labs., Norwich, N.Y., 1957-59; sr. research asso. Am. Cyanamid Co., Pearl River, N.Y., 1959-61; asso. dir. research Warner Lambert Co., Morris Plains, N.J., 1961-72; asso. prof. Sch. Dental Medicine, SUNY-Stony Brook, 1972—; cons. Warner Lambert, Morris Plains, 1972—, Richardson-Vick, N.Y.C., 1975-77, Johnson & Johnson, New Brunswick, N.J., 1980—, Am. Cyanamid, Morris Plains, 1982—. Organizer Conservative Party, Rockland County, N.Y.C., 1963-64. Served with U.S. Army, 1950-52. Recipient award Soc. Cosmetic Chemists, 1965, Presdl. citation Sigma Xi, 1979; Cath. U. fellow, 1955-57. Mem. AAAS, Internat. Assn. Dental Research (Leadership award N.J. chpt. 1974), N.Y. Acad. Sci., Sigma Xi. Republican. Roman Catholic. Subspecialties: Microbiology; Immunobiology and immunology. Current work: Microbiology of caries and periodontal disease. Patentee in biol. field. Home: PO Box 44 Port Jefferson NY 11777

MCNAUGHTON, SAMUEL JOSEPH, botanist, educator; b. Takoma Park, Md., Aug. 10, 1939; s. Frank and Ruth Ellen (Flanders) McN.; m. Margaret M. Smith, Sept. 7, 1959; children: R. Sean, Erin M. B.S., N.W. Mo. State Coll., 1961; Ph.D., U. Tex., Austin, 1964. Asst. prof. biology Portland (Oreg.) State Coll., 1964-65; USPHS postdoctoral trainee Stanford U., 1965-66; asst. prof. botany Syracuse

(N.Y.) U., 1966-73, prof., 1973—. Contbr. numerous articles on botany to profl. jours. Recipient Chancellors Citation Syracuse U., 1982; NSF grantee, 1966—. Fellow AAAS, Explorers Club; mem. Bot. Soc. Am., AAUP, Am. Soc. Naturalists, Am. Soc. Plant Physiology, Ecol. Soc. Am., Brit. Ecol. Soc., Sigma Xi (Syracuse U. faculty research award 1980). Subspecialties: Ecology; Ecosystems analysis. Current work: Ecology and ecosystems analysis. Office: 130 College Pl Syracuse NY 13210

MCNEAL, LYLE GLEN, animal scientist, educator; b. Calif., May 16, 1942; s. Darrell Glen and Elizabeth Bessie (Mista) McN.; m. Nancy Wilkie, Aug. 10, 1962; children: Tamara, Sean, Joshua, Travis, Susannah, Jenny, Ian, Ilene. B.S., Calif. Poly. State Coll., 1964; M.S., U. Nev., Reno, 1966; Ph.D., Utah State U., 1978. Asst. Mgr. Hidden Trails Ranch, Agoura, Calif., 1960-61; shepherd, dept. animal husbandry Calif. Poly. State Coll., Pomona, 1962-64; grad. research asst., div. animal sci. U. Nev., Reno, 1964-66; county agrl. extension agt. U. Nev. Coop. Extension Service, 1966-68; assoc. prof. animal sci. Calif. Poly. State U., San Luis Obispo, 1969-79; Sheep scientist U.S. Sheep Experiment Sta., DuBois, Idaho, 1974-78; dept. animal, dairy and vet. sci. Utah State U., Logan, 1979—; nat. sheep cons.; dir. Navajo Sheep Project. Active Boy Scouts Am. Recipient Disting. Teaching award Calif. Poly. State U., 1973; Cindy award Informational Film Producers Am., 1975; award of achievement Internat. Audio Visual Competition, Soc. for Tech. Competition, 1980. Mem. Am. Soc. Animal Sci., Nat. Wool Growers Assn., Am. Assn. Sheep and Goat Practitioners, Am. Polypay Sheep Assn. (hon. life). Republican. Mormon. Subspecialties: Animal breeding and embryo transplants; Animal physiology. Current work: Sheep production and management—systems and applied approaches; wool production and wool fiber biology; restoration of Navajo sheep germ plasm. Home: 85 Quarter Circle Dr Nibley UT 84321 Office: UMC 48 ADVS Dept Utah State U Logan UT 84322

MCNEESE, LEONARD EUGENE, chemical engineer, program director; b. Round Rock, Tex., May 11, 1935; s. Leonard and Rosemary (Hall) McN.; m. Mildred Gurene Allen, Aug. 29, 1954; children: Gregory Eugene, Sharon Nanette, Michael Alan. B.S. in Chem. Engring, Tex. Tech. U., 1957, M.S., U. Tenn., 1962. Research, group leader Oak Ridge (Tenn.) Nat. Lab., 1957-69, sect. head, chem. tech. div., 1969-74, dir. molten salt Reactor program, 1974-76, assoc. dir. chem. tech. div., 1976, dir. fossil energy program, 1976—. Mem. Am. Inst. Chem. Engrs., Am. Nuclear Soc., ASME. Subspecialties: Chemical engineering; Coal. Current work: Management of research and development on fossil energy related topics. Patentee (5) nuclear fuel processing. Home: 103 Morgan Rd Oak Ridge TN 37830 Office: Oak Ridge Nat Lab PO Box X Oak Ridge TN 37830

MCNEILL, THOMAS HUGH, neuroanatomist; b. Denver, Jan. 1, 1947; s. Virgil Hugh and Gloria (Tenopir) McN.; m. Florence McNeill, July 26, 1980. B.S., Colo. State U., 1971; M.S., Colo. Stae U., 1974; Ph.D., U. Rochester, 1980. USPHS trainee U. Rochester, 1975-79, NIH postdoctoral fellow, 1979-81, research assoc., 1979-81, asst. prof. neurology, 1981—. Contbr. articles to profl. jours. NIH postdoctoral fellow, 1979-81; young investigator award, 1982—; Alzheimer's Disease and Related disorders Assn. grantee, 1982-83; United Cancer Council grantee, 1982-83; Nat. Cancer Inst. grantee, 1983—; Nat. Inst. Aging grantee, 1982—. Mem. Soc. Neurosci., Am. Aging Assn., Histochem. Soc., Brit. Brain Research Assn., European Brain and Behavioral Soc., Am. Assn. Anatomists, Am. Acad. Neurology. Subspecialties: Neurobiology; Immunocytochemistry. Current work: Neurobiology of neurotransmitter and neuropeptide systems; immunocytochemistry, fluorescence histochemistry, senile dementia, aging, development, radiation sensitizers, Alzheimer's disease. Home: 97 Azalea Rd Rochester NY 14620 Office: Dept Neurology Box 673 U Rochester Rochester NY 14642

MCNEW, GEORGE LEE, research scientist; b. Alamogordo, N.M., Aug. 22, 1908; s. William Henry and Nettie Henrietta (Fry) McN.; m. Elizabeth Ann Mehlhop, May 28, 1932; 1 dau., Freda Louise. B.S., N.Mex. State U., 1930, D.Sc. (hon.), 1954; M.S., Iowa State U., 1931, Ph.D., 1935. With Iowa Agrl. Expt. Sta., 1930-35; with Rockefeller Inst. Med. Research, Princeton, N.J., 1935-39, N.Y. Agrl. Expt. Sta., 1939-43; mgr. research and devel. agrl. chems. U.S. Rubber Co., Bethany, Conn., 1943-47; prof., head botany and plant pathology dept. Iowa State U., 1947-49; mng. dir. Boyce Thompson Inst., Yonkers, N.Y., 1949-74; disting. scientist, cons. dean agr. N.M. State U., Las Cruces, 1978-83, adj. prof. plant pathology, 1983—. Contbr. numerous articles to sci. jours. Pres. N.Y. State Tuberculosis Assn., 1961. Fellow Am. Phytopath. Soc.; mem. Am. Chem. Soc., Bot. Soc. Am., Am. Soc. Plant Physiology.; MEM Torrey Botanical Club. Republican. Presbyterian. Club: Torrey Botanical. Subspecialties: Plant pathology; Plant physiology (agriculture). Current work: Pesticide development, ecology of endomycorrhizal fungi; specialization and biology of plant pathogens. Home: 1406 Georgianna Ct Las Cruces NM 88005 Office: Dept Entemology and Plant Pathology NMex State U Las Cruces NM 88003

MCNICOL, LORE ANNE, microbiology educator, researcher; b. LaJolla, Calif., Nov. 15, 1944; d. Kermit Alverson and Anne Caroline (Ferm) Long; m. David Leon McNicol, Mar. 25, 1967; 2 daus., Katharine Anne, Elizabeth Mary. Student, MIT, 1962-63, U. Calif.-Berkeley, 1964; B.A., U. Mont., 1965; Ph.D., Boston U., 1968. Research fellow dept. molecular biology and microbiology Tufts U., 1968-71; Pa. Plan scholar dept. human genetics U. Pa., 1972-74, asst. prof. microbiology, 1971-76; vis. asst. prof. biology Calif. Inst. Tech., 1976-77; asst. prof. microbiology U. Md., 1977-83; guest lectr. U. Mass., 1971-72; postdoctoral assoc. Inst. Cancer Research, 1974-76; mem. Nat. Bd. Pre-Test Cons. Com.; environ. biology study panel EPA. Contbr. articles sci. jours. Mem. Larsage Neighborhood Assn., 1971-76, Old Town Coll. Park Hist. Preservation Assn., 1978—. Recipient Nat. Service award Readers Digest Found., 1962; Phi Sigma award, 1962; NSF Undergrad. Research trainee, 1964-65; NIH postdoctoral fellow, 1968-70; NASA predoctoral fellow, 1965-68; Penn Mut. Life Ins. Fund scholar, 1972. Mem. AAAS, Am. Soc. Microbiology, AAUP, Grad. Women in Sci., Sigma Xi, Sigma Delta Epsilon. Republican. Lutheran. Club: Md. U., Harvard (Washington). Subspecialties: Molecular biology; Microbiology. Current work: Genetic engineering of a vaccine against malaria.

MCNITT, RICHARD PAUL, engineering science and mechanics educator; b. Reedsville, Pa., Aug. 1, 1935; s. Robert M. and May (Wieder) McN.; m. Jean Kearns, Jan. 27, 1957; children: Colleen, Elizabeth, Karen, Brian. B.S., Pa. State U., 1957, M.S., 1959; Ph.D., Purdue U., 1965. Asst. prof. engring. sci. and mechanics Va. Inst. Tech., 1965-68, assoc. prof., 1968-74, prof., 1974-81; prof., head engring. sci. and mechanics Pa. State U., 1981—, vice chmn. Co-editor: Environmental Degradation of Engineering Materials, 1977, Environmental Degradation of Engineering Materials in Hydrogen, 1981, Environmental Degradation of Engineering Materials in Corrosive Environments, 1981; editor: Vol. 8 Development in Theoretical and Applied Mechanics, 1976. Mem. ASEE, Soc. Engring. Sci., Am. Acad. Mechanics, Soc. Exptl. Stress Analysis, Tau Beta Pi. Republican. Presbyterian. Subspecialties: Theoretical and applied mechanics; Metallurgical engineering. Current work: Environmental degradation of materials; failure analysis and prevention; system safety; catastrophic failures in engineering; engineering education. Home: 500 E Prospect Ave State College PA 16801 Office: Pa State U 227 Hammond St University Park PA 16802

MCNULTY, PETER J., physics educator, researcher; b. N.Y.C., Aug. 2, 1941; s. Peter J. McNulty and Winifred (Bones) Molen; m. Patricia A. Arnold, Nov. 5, 1966; children: Patricia, Peter. B.S., Fordham U., 1962; Ph.D., SUNY-Buffalo, 1965. Postdoctoral fellow SUNY-Buffalo, 1965-66; sr. research assoc. Air Force Cambridge Research Labs., Hanscom AFB, Mass., 1970-71; vis. assoc. scientist Brookhaven Nat. Lab., Upton, N.Y., 1973-74; prof. physics Clarkson Coll., Potsdam, N.Y., 1966—; sr. research assoc. Air Force Geophysics Labs., Hanscom AFB, 1979-80; cons. USAF, 1973-83, BDM Corp., 1980-83, Jet Propulsion Lab., 1982-83, Gen. Electric Co., 1981-82. Mem. Am. Phys. Soc., AAAS, Radiation Research Soc., IEEE (sr.). Democrat. Roman Catholic. Subspecialties: Nuclear physics; Biophysics (physics). Current work: Radiation physics, biophysics. Home: PO Box 292 Canton NY 13676 Office: Physics Dep Clarkson Coll Potsda NY 13676

MCQUARRIE, IRVINE GRAY, neuroscientist, neurosurgeon, educator, cons.; b. Ogden, Utah, June 27, 1939; s. Irwin Bruce and Ruby Loretta (Epperson) McQ.; m. Katharine Gamble Rogers, Mar. 11, 1967 (div.); children: Michael Gray, Mollie, Morgan Elizabeth; m. Maryann Priscilla Kaminski, Aug. 14, 1980. B.S. in Biology, U. Utah, 1961; M.D., Cornell U., 1965, Ph.D., 1977. Diplomate: Am. Bd. Neurol. Surgery. Intern, asst. surgeon, surgeon N.Y. Hosp., N.Y.C., 1965-71, 72-73; research fellow dept. physiology Cornell U. Med. Coll., N.Y.C., 1971-72, 74-76, asst. prof. depts. physiology and surgery, 1976-81; vis. asst. prof. dept. anatomy Case-Western Res. U., Cleve., 1979-81, asst. prof. neurosurgery, 1981—, asst. prof. devel. genetics and antonomy, 1981—; clin. investigator in neurol. surgery VA Med. Center, Cleve., 1981—; asst. neurosurgeon Univ. Hosps. of Cleve., 1981—. Contbr. articles to sci. jours. Served to comdr., M.C. UNSR, 1973-74. Recipient Andrew W. Mellon Tchr.-Scientist award, 1977-79; NIH fellow, 1971-72, 74-76; VA career devel. award and individual research grantee, 1981—. Mem. AAAS, Soc. for Neurosci., Am. Soc. for Cell Biology, Am. Assn. Anatomists, Congress Neurol. Surgeons, Am. Assn. Neurol. Surgeons. Democrat. Presbyterian. Subspecialties: Regeneration; Neurosurgery. Current work: Mechanism of axonal regeneration in the central nervous system; biochemical investigations on the maintenance and replacement of nerve cell processes (called axons and dendrites) by complex intraneuronal transport mechanisms. Office: 2119 Abington Rd Cleveland OH 44106

MCRAE, D(AVID) SCOTT, aerospace engineering educator, researcher; b. Pinehurst, N.C., Aug. 3, 1939; s. W. S. and R. (Cook) McR.; m. Linda K. Morgan, Apr. 25, 1964; children: Alexander S, Marcus A. B.S.M.E., N.C. State U., 1961; M.S.M.E., U. Mo.-Columbia, 1966; Ph.D., Air Force Inst. Tech., 1976. Commd. 2d lt. U.S. Air Force, 1961, advanced through grades to maj., 1972; served with electronics installation, Robins AFB, GA., 1961-64, at Beale AFB, Calif., 1966-69, Wright Patterson AFB, Ohio, 1971-75, Wright Patterson AFB, 1975-77, Moffett Field, Calif., 1977-81, ret., 1981; assoc. prof. aerospace engring. N.C. State U., 1981—. Author profl. papers. Mem. AIAA (v.p. Carolina Sect. 1981—). Subspecialties: Fluid mechanics; Numerical analysis. Current work: Solution of the Navier-Stokes equations by numerical techniques; error analysis of finite difference algorithms. Office: Dept Mech and Aero Engring NC State U PO Box 5246 Raleigh NC 27650

MCSHANE, DAMIAN ANTHONY, psychologist, educator, researcher, consultant; b. Spooner, Wis., May 19, 1950; s. Kenneth George and Katherine Joyce (Waggoner) McS.; m. Anne Marie Herring, Apr. 30, 1974; children: Clear Brook, Damian Peter. B.A., Mankato State Coll., 1973; M.A., George Peabody Coll., 1976; Ph.D., Vanderbilt U., 1980. Lic. clin. psychologist, sch. psychologist, Wis.; lic. psychologist; cert. sch. psychologist, Minn. Diagnostic specialist child and family therapist Minn. Pub. Schs., Mpls., 1977-78; dir. mental health unit IHB Med. Clinic, Mpls., 1978-79; spl. edn. coordinator Lac Courte Oreilles Schs., Hayward, Wis., 1979-81; research assoc. George Peabody Coll., 1981—; postdoctoral fellow in mental health U. Wis., 1982; asst. prof. psychology Oreg. Health Scis. U., Portland, 1982—; adv. com. Am. Indian Paracounselor Tng. Program, 1977-78; mem. Mpls. Spl. Edn. Adv. Com., 1978—, SEARCH, 1970-78; bd. dirs. South Mpls. Day Care Activity Ctr., 1977-78, Health Systems Agy. Western Superior, 1981; panel of physicians Wis. Bur. Social Security Disabilities, 1981. Contbr. articles to profl. jours. Alworth scholar, 1969-70; NIMH fellow, 1973-76. Subspecialties: Developmental psychology; Cognition. Current work: Transcultural diagnosis and assessment; physiological, morphological ethnic differences relating to congition and learning. Office: Dept Psychiatry Oregon Health Scis U 3181 SW Sam Jackson Park Rd Portland OR 97201

MC SHARRY, JAMES JOHN, microbiologist, educator, researcher; b. Newark, May 28, 1942; s. John Thomas Mc S. and Marie (Weakl) McSharry; m. Mary-Martin, June 24, 1967; children: Karen, Patrick. B.S., Manhattan Coll., 1965; M.S., U. Va., 1967, Ph.D., 1970. Research asst. Manhattan Coll., 1964-65; postdoctoral fellow Rockefeller U., 1970-73; asst. prof. virology Albany Med. Coll., 1973-76, assoc. prof., 1976-83, prof., 1983—; adj. prof. Coll. St. Rose, Albany, N.Y., 1983—, SUNY-Albany, 1983—. Mem. Am. Assn. Microbiology, AAAS, Harvey Soc., Sigma Xi. Roman Catholic. Subspecialties: Virology (medicine); Microbiology (medicine). Current work: Structure-function relations of viral proteins; mode of action of antiviral drugs.

MCSHERRY, ARTHUR JAMES, consulting engineer; b. N.Y.C., Jan. 8, 1945; s. Leonard Quinten and Doris (Berley) McS.; m. Donna Gail Brager, Jan. 13, 1968. B.S. cum laude in Metall. Engring, U. Wash., 1973, M.S. in Nuclear Engring, 1974. Sr. engr. Gen. Electric Co., San Jose, Calif., 1974-81; engring. mgr. NUTECH, San Jose, 1981—. Author: Measurement of Irradiation Enhanced Creep in Nuclear Materials, 1977; contbr. chpt. to book. Served with USN, 1962-69. Katherene Marsh engring. scholar U. Wash., 1971-73. Mem. Am. Nuclear Soc., Tau Beta Pi. Subspecialties: Nuclear fission; Alloys. Current work: Application of probabilistic risk assessment methodology to nuclear power plant safety, licensing and design trade-offs. Home: 48815 Big Horn Ct Fremont CA 94539 Office: NUTECH 6835 Via Del Oro San Jose CA 95119

MCVEY, JAMES THOMAS, health physicist; b. St. Petersburg, Fla., June 20, 1946; s. James Elmer McVey and Dorothy Jean (Moore) Hickrod) McV.; m. Ruthanne Casey, Dec. 17, 1966; children: Melissa Ann, James Sean. A.A., Santa Fe Community Coll.-Fla., 1968; B.S., U. Fla., 1970, M.S., 1972. Researcher dept. medicine U. Fla., Gainesville, 1967-74; pres., cons. Environ. Radiation Mgmt., Gainesville, Fla., 1973-77; S.E. rep. Pulcir Inc., Oak Ridge, 1975-76; health physics services physicist FDA, Jefferson, Ark., 1977-80; mgr. Health Physics Systems Inc., Gainesville, 1980—; cons. corp. radiation Quadrex, San Jose, Calif., 1980-81, safety officer, 1982—. Contbr. articles in field to profl. jours. Mem. Hazardous Material Com., Gainesville, Fla., 1981; v.p. Alachua County Fair Assn. USPHS fellow, 1970. Mem. Jr. C. of C. (v.p. Gainesville), Am. Nuclear Soc., Health Physics Soc., Ark. Acad. Sci., Research Scientist Group (chmn. 1977-80). Democrat. Am. Baptist. Subspecialties: Nuclear physics; Nuclear medicine. Current work: Health physics, decontamination, radioactive waste volume reduction, personnel dosimetry and other hazardous materials handling. Home: 4117 NW 34th Pl Gainesville FL 32606 Office: Health Physics Systems Inc 1940 NW 67th Pl Gainesville FL 32606

MCWHIRTER, JAMES JEFFRIES, counselor educator, psychologist, consultant; b. Big Spring, Tex., Jan. 31, 1938; s. James Davidson and Hazel Barbara (Jeffries) McW.; m. Mary Clare Plasker, Aug. 27, 1960; children: Robert, Benedict, Anna, Mark, Paula. B.A., St. Martin's Coll., 1961; M.Ed., U. Oreg., 1964, Ph.D., 1969; M.Ed., Oreg. State U., 1965. Cert. psychologist Ariz. Tchr., counselor Central Catholic High Sch., Portland, Oreg., 1961-63, Centennial High Sch., Portland, 1963-67; psychology resident VA Hosp-U. Oreg. Med. Sch., Portland, 1969-70; asst. prof. counselor edn. Ariz. State U., 1970-72, assoc. prof., 1972-77, prof., 1977—; vis. summer prof., 10 instns., U.S., Can., Europe, 1967-83; Fulbright prof. Hacettepe U., Ankara, Turkey, 1977-78; pvt. practice psychology, Tempe, Ariz., 1970-83, cons. in field, 1970—. Author: Learning Disabled Child, 1977 (selection of 2 book clubs), Problem Solving in Families, 1983; contbr. numerous articles to profl. jours.; editor books, 1981. Mem. Am. Psychol. Assn., Am. Personnel and Guidance Assn., Assn. Counselor Edn. and Supervision, Am. Sch. Counselors Assn. Democrat. Roman Catholic. Subspecialties: Behavioral psychology; Social psychology. Current work: Small group interaction; relaxation and imagery training. Home: 110 E Fremont Dr Tempe AZ 85282 Office: Ariz State U 425I Payne Hall Tempe AZ 85287

MEAD, ALBERT RAYMOND, biologist, educator; b. San Jose, Calif., July 17, 1915; s. Lester A. and Jennie (Fiske) M.; m. Eleanor Morrow, Feb. 8, 1942; children: Ruth Mead Cruz, James Irving. B.S., U. Calif.-Berkeley, 1938; postgrad., U. Calif.-Davis, 1941; Ph.D., Cornell U., 1942. Research fellow U. Calif.-Berkeley, 1946; asst. prof., then assoc. prof. U. Ariz., 1946-52, prof., 1952—, head dept. zoology, 1956-67, coordinator marine sci. program, 1967-70, coordinator undergrad. program biol. sci., 1970-76, assoc. dean liberal arts coll., 1976-80; research assoc. Pacific Sci. Bd., Nat. Acad. Sci., 1948. Author: The Giant African Snail: A Problem in Economic Malacology, 1961, Economic Malacology with Particular Reference to Acatina Fulica, 1979; contbr. 79 papers to profl. jours. Served to capt. AUS, 1942-46. Grantee in field, 1953—. Fellow AAAS (pres. div., nat. council); mem. Am. Malacological Union (nat. pres.), numerous others. Lodge: Elks. Subspecialties: Population biology; Taxonomy. Current work: Population biology. Office: U Ariz Tucson AZ 85721 Home: 401 Sierra Vista Dr Tucson AZ 85719

MEAGHER, DONALD JOSEPH, engineering administrator, researcher; b. Albany, N.Y., May 2, 1949; s. John J. and Caroline H. (Glatz) M.; m. Linda Jean Stapleton, June 5, 1971; children: Kelly J., Donald F., Sean J. B.S. in Elec. Engring., Rensselaer Poly. Inst., Troy, N.Y., 1971, M.Eng., 1972, M.Eng. in Computer and Systems Engring., 1979, Ph.D. in Elec. Engring., 1982. Assoc. electronics engr. Cornell Aero. Lab., Buffalo, 1972-74; mem. tech. staff Argo Systems, Palo Alto, Calif., 1974-77; mgr. Interactive Computer Graphics Ctr., Rensselaer Poly. Inst., 1978-80, instr., 1980-81; dir. computer graphics devel. Phoenix Data Systems, Inc., Albany, N.Y., 1981—. Contbr. articles to profl. jours. Mem. IEEE, Assn. Computing Machinery, Sigma Xi. Republican. So. Baptist. Subspecialties: Graphics, image processing, and pattern recognition; Computer-aided design. Current work: Development of "Octree" techniques for real-time solid modeling and image generation. Applications in medical imaging, CAD/CAM, simulation, cinematography, etc. Home: 18 Lynwood Dr Loudonville NY 12211 Office: Phoenix Data Systems Inc 80 Wolf Rd Albany NY 12205

MEAL, LARIE, chemistry technology educator; b. Cin., June 15, 1939; s. George Lawrence and Dorothy Louise (Heileman) M. B.S., U. Cin., 1961, Ph.D. in Chemistry, 1966. Research chemist U.S. Indsl. Chem. Co., Cin., 1966-67; assoc. prof. chem. tech. U. Cin., 1968—; ind. cons., 1974—. Contbr. articles to profl. jours. Mem. League for Animal Welfare, Cin., 1961—, Animal Rescue Fund, 1980—. Mem. Am. Chem. Soc., AAAS, N.Y. Acad. Scis., Iota Sigma Pi. Democrat. Subspecialties: Analytical chemistry; Organic chemistry. Current work: Arson analysis: chemical analysis of fire debris. Analysis for accelerants using second derivative ultraviolet spectrophotometry. Home: 2231 Slane Ave Norwood OH 45212 Office: Dept Chem Tech U Cin 100 E Central Pkwy Cincinnati OH 45210

MEALIEA, WALLACE LAIRD, JR., dental educator, clinical psychology researcher; b. Weehawken, N.J., June 9, 1939; s. Wallace Laird and Rose (La Russo) M.; m. Linda Kay Wills, July 13, 1963; children: Wallace, Laura. B.A., Central Coll., Pella, Iowa, 1962; M.A., U. Mo., 1965, Ph.D., 1967. Asst. prof., dir. student council Earlham Coll., Richmond, Ind., 1967-69; asst. prof. U. Wis.-Madison, 1969-71; assoc. prof. Ind. State U., Terre Haute, 1971-72; assoc. prof., dir. psycho. services Dalhousie U., Halifax, N.S., Can., 1972-75; prof. basic dental scis. U. Fla., Gainesville, 1975—; psychol. cons. Coop. Edn. Services, Sheboygan, Wis., 1969-070, Atlantic Inst. Edn. and CBC, Halifax, 1973-75, VA Hosp., Gainesville, 1975—. Mem. sch. bd. Halifax Grammar Sch., 1973. Nat. Inst. Dental Research grantee, 1978; U. Fla. grantee, 1979. Mem. Am. Psychol. Assn., Assn. for Advancement Behavior Therapy, Internat. Assn. Dental Research, Southeastern Psychol. Assn. Subspecialties: Behavioral psychology; Behavioral sciences in dentistry. Current work: Behavioral therapy, parameters of behavior change, pain management, biofeedback. Home: 8209 SW 1st Pl Gainesville FL 32607 Office: Basic Dental Scis U Fla JHMHC-J-424 Gainesville FL 32610

MEANS, LARRY WILLIAMS, psychologist, educator; b. The Dalles, Oreg., Apr. 20, 1941; s. Lawrence W. and Bernice E. M.; m. Carolyn E. Means, June 15, 1963; children: Julie A., Michael J. B.S., Portland State U., 1963; M.A., U. Minn., 1966; Ph.D., Claremont Grad. Sch., 1969. Postdoctoral fellow U. Fla., 1968-70; asst. prof. psychology, then assoc. profl, then prof. East Carolina U., Greenville, N.C., 1970—, also dir. grad. program in gen. psychology. Contbr. articles to profl. jours. Mem. Soc. Neurosci., Behavioral Teratology Soc., Research Soc. on Alcoholism, Sigma Xi. Democrat. Subspecialties: Learning; Physiological psychology. Current work: Brain lesions and behavior, fetal alcohol syndrome; behavioral deficits, drugs and behavior. Home: 202 Adams Blvd Greenville NC 27834 Office: Dept Psycholog East Carolina U Greenville NC 27834

MECKEL, ALFRED HANS, dental consultant 2nd educator; b. Munich, Germany, Mar. 12, 1924; came to U.S., 1954, naturalized, 1960; s. Hermann and Elisabeth Steub M.; m. Ingeborg Ruth Schuster, Nov 24, 1951; 1 son, Timothy Stephan. De.med. dent., U. Munich, 1951; D.D.S., Northwestern U., 1957. Civilian dentist U.S. Army, Garmisch-Partenkirchen, Germany, 1949-54; research dentist Procter & Gamble Corp., Cin., 1954-57, 57—; Vis. prof. U. Autónoma de Nuevoleon, Monterrey, Mex., 1978—. Served with German Air Force, 1942-45. Mem. Internat. Assn. Dental Research (pres. cardiology group 1981-82), Fedn. Dentaire Internat. (cons. 1977-84). Subspecialty: Dental research consulting. Current work: Tooth structure, electron microscopy, clinical investigations in safety and effectiveness of oral products. Home: Route 4 Box 221A West Harrison IN 47060

MECKSTROTH, WILMA KOENIG, chemistry educator, researcher; b. Auglaize County, Ohio, Feb. 12, 1929; d. Otto John and Edna Wilhemina (Rehn) Koenig; m. Robert Rueben Meckstroth, Sept. 5,

1949 (div. 1976); children: Linda A. Koenig, David J., Jeffrey. B.S. cum laude, Ohio State U., 1949, Ph.D., 1968. Chemistry instr. Urbana (Ohio) U., 1959-62; teaching asst. Ohio State U., Columbus, 1962-63, research assoc., 1963-66, chemistry instr., 1966-67, assoc. prof. chemistry, 1968—; with research and devel. Anchor Hocking Corp., Lancaster, Ohio, 1967-68. Recipient disting. teaching award Ohio State U. Alumni Assn., Columbus, 1979-80, teaching excellence award Mfg. Chemists Assn., 1976. Mem. Am. Chem. Soc., Phi Kappa Phi, Alpha Lambda Delta. Republican. Subspecialty: Physical chemistry. Current work: Pulse radiolysis of metal carbonyls; reactivity of transient species is studied as well as temperature dependence. Office: Ohio State Univ Newark Campus University Dr Newark OH 43055

MEDINA, MIGUEL ANGEL, educator; b. Laredo, Tex., July 5, 1932; s. Napoleon and Luz (Malacara) M.; m. Johnnie Lee Word, May 17, 1963; children: Paul, Amy, Sara. B.S., St. Mary's U., 1957, M.S., 1963; Ph.D., Southwestern Med. Sch., 1967. Research chemist Amoco Oil Co., Texas City, Tex., 1957-59; chemist Sch. Aerospace Medicine, San Antonio, 1959-64, research pharmacologist, 1967-70; asst. prof. U. Tex. Health Sci. Ctr., San Antonio, 1970-74, assoc. prof., 1974-80, prof., 1980—; grant reviewer SW Found. Edn. and Research, 1980—; mem. Nat. adv. bd. research resources council NIH, 1978-82. Co-editor: Pharmacology and Toxicology of Propellant Hydrazins, 1968; contbr. to books. Served with USN, 1952-54. Fulbright fellow, 1973; UNESCO fellow, 1976. Mem. Am. Soc. Pharmacology and Exptl. Therapeutics, Am. Chem. Soc., Am. Soc. Neurochemistry; hon. mem. Peruvian Pharmacology Soc. Subspecialty: Neurochemistry. Current work: Effects of drugs on brain intermediary metabolism. Home: 803 Clarion Dr Durham NC 27705 Office: Dept Civil Engring Duke Univ Durham NC 27706

MEDINA, MIGUEL ANGEL, JR., engineering educator; b. Havana, Cuba, Dec. 9, 1946; came to U.S., 1961; m. Margarita Piedra, Aug. 28, 1976; 1 son, Miguel Angel III. B.S. in Civil Engring, U. Ala., M.S., 1972; Ph.D., U. Fla., 1976. Asst. prof. civil engring. Duke U., Durham, N.C., 1976-81, assoc. prof., 1981—, chmn. univ. research council, 1982—; cons. in field. Contbr. articles to profl. jours. Served to 1st lt. U.S. Army, 1969-71. Decorated Army Commendation medal (2); research grantee U.S. E.P.A., 1977, NSF, 1978, Dept. Interior, 1980; others. Mem. ASCE, Am. Geophys. Union, Am. Water Resources Assn., Sigma Xi, Tau Beta Pi (adviseor N.C. Gamma chpt. 1980—). Club: Duke Faculty. Subspecialties: Civil engineering; Integrated systems modelling and engineering. Current work: Research and development of hydrologic and water quality computer simulation, mathematical models of pollutant transport systems, water resources systems. Office: Dept Pharmacology U Tex Health Sci Ctr 7703 Floyd Curl Dr San Antonio TX 78284 Home: 120 W Summit San Antonio TX 78212

MEDITCH, JAMES STEPHEN, electrical engineering educator; b. Indpls., July 30, 1934; s. Vladimir Stephen and Sandra (Gogeff) M.; m. Theresa Claire Scott, Apr. 4, 1964; children: James Stephen, Sandra Anne. B.S. in Elec. Engring, Purdue U., 1956, Ph.D., 1961; S.M., M.I.T., 1957. Staff engr. Aerospace Corp., Los Angeles, 1961-65; assoc. prof. elec. engring. Northwestern U., 1966-67; mem. tech. staff Boeing Sci. Research Labs., Seattle, 1967-70; prof. U. Calif., Irvine, 1970-77; prof., chmn. dept. elec. engring. U. Wash., 1977—, also adj. prof. computer sci. dept. Author: Stochastic Optimal Linear Estimation and Control, 1969; editor: proc. IEEE, 1983—. Fellow IEEE; mem. AAAS, Am. Soc. Engring. Edn., Assn. for Computer Machinery. Subspecialties: Computer engineering; Distributed systems and networks. Current work: Computer-communication systems and networks, local area networks,distributed processing. Office: Dept Elec Engring FT-10 U Wash Seattle WA 98195

MEDLEY, DONALD MATTHIAS, psychological educator; b. Faulkton, S.D., Feb. 18, 1917; s. Thomas A. and Cecilia A. (Kellen) M.; m. Betty Ann Robertsen, Aug. 23, 1948; 1 son, Timothy Laurence. B.S., St. Thomas Coll., 1938; M.A., U. Minn., 1950, Ph.D., 1954. Instr. St. Thomas Coll., 1948-50; prof. CUNY, 1954-65; sr. research psychologist Ednl. Testing Service, Princeton, N.J., 1965-70; prof. ednl. research U. Va., 1970—; cons. univs., social systems, govt. agys. Contbr. numerous articles to profl. jours., encys. Served with AUS, 1942-46. Fellow AAAS, Am. Psychol. Assn.; mem. Am. Ednl. Research Assn., Nat. Council on Measurement in Edn., Phi Delta Kappa. Democrat. Roman Catholic. Subspecialties: Educational psychology; Learning. Current work: Research in teaching; evaluation of tchrs., educational and psychological measurement: design of educational research. Office: Sch Edn U Va 405 Emmet St Charlottesville VA 22903

MEDVE, RICHARD JOHN, biology educator; b. California, Pa., Jan. 28, 1936; s. Andrew D. and Marianne E. (Bardelli) M.; m. Mary Lee T. Everett, Nov. 26, 1958; children: Pam, Steve, Ken, Kathy, Dave. B.S., California (Pa.) State U., 1957; M.A., Kent State U., 1958; Ph.D., Ohio State U., 1969. Tchr. high sch. Willoughby-(Ohio)-Eastlake Schs., 1958-66; mem. faculty dept. biology Slippery Rock (Pa.) State Coll., 1966—, prof., 1971—; cons. Aquatic Ecology Assocs., Pitts., 1975—. Contbr. articles to profl. jours. Recipient Disting. prof. award Slippery Rock State Coll., 1981; NSF fellow, 1962-64; Western Pa Conservancy grantee, 1981-82. Mem. Nat. Assn. Biol. Tchrs., Pa. Acad. Sci., Ohio Acad. Sci., Torrey Bot. Club, Western Pa. Conservancy, Western Pa. Bot. Soc. Subspecialty: Ecology. Current work: Mycorrhizae, stripmine revegetation. 16057

MEDZIHRADSKY, JOSEPH LADISLAS, immunologist; b. Levice, Czechoslovakia, Apr. 30, 1924; s. Joseph and Gisela (Dohany) M.; m. Zdenka Helena, Apr. 8, 1967. M.D., Comenius U., Bratislava, Czechoslovakia, 1951. Asst. prof. Comenius U. Med. Sch., 1951-60; sr. scientist Cancer Inst., Bratislava, 1960-75; vis. scientist Roswell Park Meml. Inst., Buffalo, 1974-75; sr. scientist Wellcome Research Labs., Research Triangle Park, N.C., 1975—. Eleanor Roosevelt Internat. Cancer fellow Internat. Union Against Cancer, Geneva, 1973. Mem. Am. Assn. Immunologists, Internat. Soc. Immunopharmacology. Subspecialties: Immunobiology and immunology; Immunology (agriculture). Current work: Cellular immunology and drug interactions: manipulations in immune reactions against cancer.

MEEHAN, THOMAS, research biochemist; b. Youngstown, Ohio, May 16, 1942; s. Frank C. and Verona (Zanover) M.; m. Nancy Aherns, Feb. 1967; m. Susan P. Hawkes, Sept. 17, 1978. B.S., U. Akron, 1965; Ph.D., St. Louis U., 1973. Trainee Lawrence Radiation Lab., U. Calif.-Berkeley, 1973-76; biochemist Lawerence Radiation Lab., U. Calif.-Berkeley, 1976-78; research scientist Mich. Molecular Inst., Midland, 1978—; adj. prof. Case Western Res. U., 1978—; cons. in field. Contbr. articles to profl. jours. NIH grantee, 1967-73, 74-76; Nat. Cancer Inst. grantee, 1979—. Mem. AAAS, Am. Assn. Cancer Research, Am. Chem. Soc., Biophys. Soc. Subspecialties: Cancer research (medicine); Molecular biology. Current work: Molecular mechanisms of tumor initiation by chemical carcinogens. Office: 1910 W St Andrews Rd Midland MI 48640

MEEK, DEVON WALTER, chemistry educator, consultant; b. River, Ky., Feb. 24, 1936; s. Don Carlos and Willa Mae (Walters) M.; m. Violet Illene Imhof, Aug. 21, 1965; children: Brian Philip, Karen Anne. B.A., Berea Coll., 1958; M.S., U. Ill.-Urbana, 1960, Ph.D., 1961. Teaching/research asst. U. Ill.-Urbana, 1958-60, research fellow, 1960-61; asst. prof. Ohio State U., Columbus, 1961-66, assoc. prof., 1966-69, prof. chemistry, 1969—, chmn. dept., 1977-81; cons. Procter & Gamble Co., Cin., 1979—, Shepherd Chem. Co., 1976-80. Co-author: Experimental General Chemistry, 2d edit, 1974; co-editor: Catalytic Aspects of Metal Phosphine Complexes, 1982. Guggenheim fellow, 1981-82; Sci. Research Council sr. research fellow, 1974. Fellow Phi Kappa Phi, Phi Lambda Upsilon; mem. Am. Chem. Soc. (sec. inorganic div. 1983-87). Lutheran. Subspecialties: Catalysis chemistry; Inorganic chemistry. Current work: Homogeneous catalysis of O_2, CO and H_2 reactions with polyphosphine complexes of Pt, Rh, Ir and Ru; phosphorus-31 and metal nuclear magnetic resonance. Office: Ohio State U Dept Chemistry Room 120 McPherson Chem Lab 140 W 18th Ave Columbus OH 43210

MEEKINS, JOHN FRED, astrophysicist; b. Boston, Oct. 4, 1937; s. Donald Fred and Signe (Bjork) M.; m. C. Ann.Turner, Sept. 9, 1961; children: David George, Brian John. B.A., Bowdoin Coll., 1959; Ph.D., Cath. U. Am., 1973. Research physicist Naval Research Lab., Washington, 1959-79, astrophysicist, 1979—. Contbr. articles to profl. jours. Asst. scoutmaster Boy Scouts Am., 1980-82, scoutmaster, 1982—. Mem. Am. Astron. Soc., Sigma Xi. Subspecialties: High energy astrophysics; 1-ray high energy astrophysics. Current work: Study of high energy astrophys. objects, by means of their radiation, especially X-Ray emission. Home: 5674 Ravenel Ln Springfield VA 22151 Office: Code 4125 Naval Research Lab Washington DC 20375

MEEKS, MARION LITTLETON, astronomer, physicist, researcher; b. Gainesville, Ga., Oct. 1, 1923; s. Jesse Littleton and Ione (Tumlin) M.; m. Bennie Scurry Stone, May 25, 1944 (div.); children: Marshall, Fleming, Marion; m. Louise Ann Vogt, Aug. 14, 1970; children: Lita, Mariano. B.S., Ga. Inst. Tech., 1943, M.S., 1947; Ph.D., Duke U., 1950. Asst. prof., assoc. prof. physics Ga. Inst. Tech., Atlanta, 1951-59; assoc. Harvard Coll. Obs., Cambridge, Mass., 1959-61; staff MIT Lincoln Lab., Lexington, 1961-70, research staff, 1978—; head radio astronomy ops. Haystack Obs., Westford, Mass., 1970-78; adj. prof. astronomy U. Mass., Amherst, 1971-75; researcher; film maker. Editor: Radio Telescopes and Observations, 1976; author: Radar Propagation at Low Altitudes, 1982; contbr. sci. papers to publs; author computer-generated astronomy films. Served to lt. j.g. USN, 1943-46. Mem. Internat. Astron. Union, Internat. Sci. Radio Union, Am. Astron. Soc., Sigma Xi. Episcopalian. Club: Valley Pond (Lincoln). Subspecialties: Radio and microwave astronomy; Remote sensing (atmospheric science). Current work: Remote sensing from satellites, radio and microwave propagation studies, computer animation for video and films. Home: Stonehenge RFD 4 Lincoln MA 01773 Office: MIT Lincoln Lab PO Box 73 Lexington MA 02173

MEEM, JAMES LAWRENCE, JR., nuclear scientist; b. N.Y., Dec. 24, 1915; s. James Lawrence and Phyllis (Deaderick) M.; m. Buena Vista Speake, Sept. 5, 1940; children: James, John. B.S., Va. Mil. Inst., 1939; M.S., Ind. U., 1947, Ph.D., 1949. Aero. research sci. NACA, 1940-46; dir. bulk shielding reactor Oak Ridge Nat. Lab., 1950-53, in charge nuclear operation aircraft reactor expt., 1954-55; chief reactor sci. Alco Products, Inc., 1955-57; in charge startup and initial testing Army Package Power Reactor, 1957; prof. nuclear engring. U. Va., Charlottesville, 1957-81, dept. chmn., dir. reactor facility, 1957-77, prof. emeritus, 1981—; cons. U.S. Army Fgn. Sci. and Tech. Ctr., 1981—; vis. cons. nuclear fuel cycle programs Sandia Labs., Albuquerque, 1977-78; vis. staff mem. Los Alamos Sci. Lab., 1967-68; mem. U.S.-Japan Seminar Optimization of Nuclear Engring. Edn., Tokai-mura, 1973. Author: Two Group Reactor Theory, 1964. Fellow Am. Nuclear Soc. (sec. reactor ops. div. 1966-68, vice chmn. 1968-70, chmn. 1970-71, Exceptional Service award 1980); mem. Am. Phys. Soc., Am. Soc. Engring. Edn., SAR. Subspecialties: Nuclear fission; Nuclear engineering. Home: Mount Airy RFD 2 Box 118 Charlottesville VA 22901

MEFFERD, ROY BALFOUR, JR., medical research scientist, human resources technology consultant; b. Hico, Tex., Sept. 22, 1920; s. Roy B. and Delfa (Russell) M.; m. Mary Louise Key, Aug. 25, 1940; children: Marsha Ellen Mefferd Steele, Roy Scott. A.S., Tarleton State Coll., 1938; B.S., Tex. A&M U., 1940; M.S., 1940; Ph.D., U. Tex.-Austin, 1951. Dir. Metabolism Lab., S.W. Found. for Research and Edn., San Antonio, 1951-54; dir. Mental Health Research Lab., Biochem. Inst., U. Tex.-Autin, 1954-59, Psychiat. and Psychosomatic Research Lab., VA Med. Ctr., Houston, 1949-80, ret., 1980; asst. prof. physiology Baylor Coll. Medicine, Houston, 1949-51, assoc. prof., 1952-55, prof., 1956—; v.p. research Birkman & Assocs., Inc., Houston, 1972-0-; pres. Birkman-Mefferd Research, Houston, 1972-82, ret., 1982; adj. prof. psychology U. Houston, 1972-80; asj. prof. behavioral sci. U. Tex. Sch. Public Health, Houston, 1976 ; cons. to numerous med. and indsl. orgns., 1951—. Served to capt. inf. U.S. Army, 1942-47. Japan. U. Tex. Rosalie B.Hite fellow, 1949-51; S.W. Found. for Research and Edn. Damon Runyon fellow, 1952-53; numerous grants from various govt., pvt. agys., 1951—. Fellow AAAS; mem. Am. Psychol. Assn., Am. Physiol. Soc., Am. Chem. Soc., Sigma Xi. Subspecialties: Psychiatry; Human resources technology. Current work: Consulting: psychometrics, statistics, validation studies, selection, human resources analyses for industry, productivity studies, medical studies—psychophysiology, neuropsychology, research design. Home: 823 Longview Sugar Land TX 77478

MEGARGEE, EDWIN INGLEE, psychology educator; b. Plainfield, Fla., Feb. 27, 1937; s. S. Edwin and Jean (Inglee) M.; m. Sara Jill Mercer, June 27, 1980; 1 stepdau., Heather Lynn Dunham.; m.; children: by previous marriage: Elyn Jean, Edwin Inglee, Christopher John, Stephen Andrew. B.A., Amherst Coll., 1958; Ph.D., U. Calif.-Berkeley, 1964. Lic. psychologist, Fla. Clin. psychologist Alameda County Probation Dept., San Leandro, Calif., 1961-64; asst. prof. U. Tex.-Austin, 1964-67; assoc. prof. psychology Fla. State U., Tallahassee, 1967-70, prof., 1970—; pres., chmn. bd. Criminal Justice Assessment Service, Tallahassee, 1980—; cons. psychologist U.S. Secret Service, Washington, 1980—. Author: Delinquency in Three Cultures, 1969, California Psychological Inventory Handbook, 1972, Classifying Criminal Offenders, 1979, Research in Clinical Assessment, 1966, The Dynamics of Agression, 1970; contbr. articles to profl. jours. Pres. Soc. Arts & Crafts, Tallahassee, 1978-79. Fellow Am. Psychol. Ass., Soc. for Personality Assessment; mem. Am. Assn. Correctional Psychologists (pres. 1973-75), Internat. Differential Treatment Assn. (dir.), Am. Correctional Assn. (dir. 1973-75), Phi Beta Kappa, Sigma Xi. Club: Cosmos. Subspecialty: Clinical psychology. Current work: Aggression and violence; applications of psychology to the criminal justice system. Home: 3348 E Lakeshore Dr Tallahassee FL 32312 Office: Psychology Dept Fla State U Tallahassee FL 32306

MEGEL, HERBERT, toxicologist; b. Newark, Nov. 10, 1926; s. Benjamin Abraham and Anna (Geller) M.; m. Eleanor Sitzman, Jan. 20, 1951; children—Diana, Joseph. B.A., N.Y. U., 1948, M.S., 1950, Ph.D., 1954. Diplomate: Am. Bd. Toxicology. Dir. bioassay Princeton Labs. Inc., 1954-59; sr. research physiologist Boeing Co., Seattle, 1959-62; dir. biochemistry Nat. Drug. Co., Phila., 1962-70; sect. head immunology Merrell-Nat. Labs., Cin., 1970-78; sr. research toxicologist pathology-toxicology dept. Merrell Dow Pharm., Inc., Cin., 1978—. Contbr. articles and book chpts. to profl. lit. Fellow N.Y. Acad. Sci.; mem. Am. Soc. Immunologists, Am. Soc. Pharmacology and Exptl. Therapeutics, Soc. Toxicology, Phi Beta Kappa. Subspecialties: Toxicology (agriculture); Immunotoxicology. Current work: Immunopharmacology, immunotoxicology, compliance with good laboratory practice. Home: 1250 Forest Ct Cincinnati OH 45205 Office: 2110 E Galbraith Rd Cincinnati OH 45215

MEHEARG, LILLIAN ERL, psychologist; b. Parkdale, Ark., Oct. 2, 1918; d. Thomas Albert and Lillie Leuna (Massey) M.; m. George Edwin Volz, Jan. 2, 1955 (dec. 1960). B.A., Millsaps Coll., 1957; M.A., La. State U., 1958; Ph.D., U. So. Miss., 1963. Lic. psychologist, Miss. Exec. dir., psychologist Lake Charles (La.) Mental Health Center, 1959-61; clin. dir., psychologist Hammond (La.) Mental Health Center, 1962-64; from asst. prof. to prof. psychology U. So. Miss., Hattiesburg, 1964-83; pvt. practice clin. psychology, Hattiesburg, 1975—; chmn. Miss. State Bd. Psychol. Examiners, 1969-70, exec. sec., 1968-69; cons. VA Hosps. (various locations), 1967-76. Contbr. articles to profl. jours. Served with WAC, 1944-45. Mem. Am. Psychol. Assn., Biofeedback Soc. Am., Miss. Psychol. Assn., Biofeedback Soc. Miss., Phi Kappa Phi, Psi Chi. Republican. Unity Ch. Subspecialty: Behavioral psychology. Current work: Pain management, biofeedback, clinical hypnosis. Address: 3601 Morningside Dr Hattiesburg MS 39401

MEHENDALE, HARIHARA MAHADEVA, toxicologist, educator; b. Philya, India, Jan 12, 1942; came to U.S., 1963, naturalized, 1969; s. Shinginkodlu Mahadevabhat and Narmada (Mahadeva) M.; m. Rekha H. Mehendale, Mar. 5, 1947; children: Roopa, Neelesh. B.Sc., Karnatak U., Dharwar, India, 1963; M.S. in Physiology, N.C. State U., 1966, Ph.D., 1969. Diplomate: Am. Bd. Toxicology. Postdoctoral fellow in toxicology U. Ky., 1969-71; vis. fellow Nat. Inst. Environ. Health Scis., 1971-72, staff fellow, 1972-75; asst. prof. dept. pharmacology and toxicology U. Miss. Med. Center, Jackson, 1975-78, asso. prof., 1978-80, prof., 1980—, dir. toxicology tng. program, 1982-; vis. prof. forensic medicine Karolinska Inst., Stockholm, 1983-84; also cons.; mem. adv. panel toxicology study sect. NIH, 1981—. Contbr. numerous articles and chpts. to sci. lit.; editorial bd.: Fundamental and Applied Toxicology, Jour. Toxicology and Environ. Health. Pres. India Assn. of Miss. Mem. Am. Soc. Pharmacology and Exptl. Therapeutics, Soc. Toxicology, Am. Chem. Soc., Internat. Soc. Study Xenobiotics, Am. Thoracic Soc., Internat. Union Pharmacology (sect. toxicology), AAAS, Assn. Scientists of Indian Origin in Am., Indian Sci. Congress Assn., Entomol. Soc. Am., Entomol. Soc. India, Miss. Acad. Scis., Miss. Heart Assn., Sigma Xi (chpt. pres. 1982-83). Subspecialties: Toxicology (agriculture); Environmental toxicology. Current work: Environmental toxicology. Office: U Miss Med Center Dept Pharmacology/Toxicology 2500 N State St Jackson MS 39216

MEHRA, RITA VIRMANI, dentist, educator, researcher; b. Lyall-pur, Punjab, India, Mar. 1, 1938; came to U.S., 1963; s. Ram Narain and Veeran (Chawla) Virmani; m. Surinder K. Mehra, Dec. 30, 1970; children: Vivek, Rahul, Arun. B.S., Punjab U., 1958, B.D.S., 1962; D.D.S., Ind. U., 1970, M.S.B., 1966. Intern Punjab Dental Coll. and Hosp., 1962-63, Birmingham (Eng.) Dental Hosp., 1963-64; grad. asst. Ind. U. Sch. Dentistry, 1965-67, asst. prof., 1970-72; asst., dir. Johnson & Johnson Dental Products Co., East Windsor, N.J., 1974-76; Chmn. dept. fixed partial denture Fairleigh Dickinson U. Sch. Dentistry, Hackensack, N.J., 1981—; cons. Hilltop Research, East Brunswick, N.J., 1978-82, NIDR Panel, Washington, 1980, FDA, 1981-83. Recipient Gold medal Punjab U., 1962. Mem. Internat. Assn. Dental Research, Omicron Kappa Upsilon. Republican. Hindu. Subspecialties: Prosthodontics; Dental materials. Current work: Clinical research in dental materials. Home: 1441 Martine Ave Scotch Plains NJ 07076 Office: Fairleigh Dickinson U Sch Dentistry 110 Fuller Pl Hackensack NJ 07601

MEHRABADI, MORTEZA M(IRZAIE), mechanical engineer, educator, researcher; b. Tehran, Iran, July 11, 1947; came to U.S., 1972; s. Ali Mirzaie and Mehri (Alipoor) M.; m. Fatemeh Ashraf, June 6, 1975; 1 dau., Roxana. B.S.M.E., Tehran U., 1969; M.S., Tulane U., 1973, Ph.D., 1979. Research asst. dept. mech. engring. Tulane U., 1973-79, asst. prof. mech. engring., 1982—; postdoctoral fellow and lectr. dept. civil engring. Northwestern U., Evanston, Ill., 1979-82. Contbr. articles to profl. jours. Mem. Am. Acad. Mechanics, ASME, Soc. Engring. Sci., N Tau Beta Pi. Subspecialties: Theoretical and applied mechanics; Solid mechanics. Current work: Continuum mechanics, elasticity, plasticity, mechanics of granular materials. Office: Dept Mech Engring Tulane U New Orleans LA 70118

MEHTA, BIPIN MOHANLAL, microbiologist; b. Bombay, India, July 25, 1935; s. Mohanlal Jamnadas and Vimala (Harkisondas) M.; m. Varsha Bipin Mehta, May. 8, 1960. Ph.D., U. Bombay, 1963. Lectr. nutrition U. Bombay, 1963-65; research assoc., teaching asst. SUNY Downstate Med. Center, 1965 66; vis. research fellow Sloan-Kettering Inst. Cancer Research, N.Y.C., 1966-69, research assoc., 1972-74, assoc., 1974-81, asst. mem., 1982—; instr. microbiology Grad. Sch. Med. Scis., Cornell U., 1973-75, asst. prof. biology unit, 1975-80, pharmacology and exptl. therapeutics unit, 1980—; research assoc. U. Ottawa, 1969-72. Contbr. articles to profl. jours. Bd. Dirs. Westchester div. Am. Cancer Soc. Mem. Am. Soc. Microbiology, Can. Soc. Cell Biology, Soc. Gen. Microbiology (U.K.), AAAS, N.Y. Acad. Scis., Am. Assn. Cancer Research, Am. Soc. Clin. Oncology. Subspecialties: Microbiology; Pharmacokinetics. Current work: Microbial genetics, physiology and nutrition with pharmacology of anticancer agents. Office: 145 Boston Post Rd Rye NY 10580

MEHTA, JAWAHAR L., cardiologist; b. India, Aug. 10, 1946; came to U.S., 1970, naturalized, 1978; s. Mohan Lal and Ishwar (Devi) M.; m. Paulette Smedresman, Oct. 20, 1977; children: Asha, Jason. B.S. in Biology, Panjab U., Chandigarh, India, 1962, M.D., 1967. Instr. medicine SUNY-Stony Brook, 1973-75, U. Minn., Mpls., 1975-76; asst. prof. medicine U. Fla., Gainesville, 1976-80, assoc. prof., 1980—; dir. CCU VA Med. Ctr., Gainesville, 1976—. Editor: Platelets and Prostaglandins in Cardiovascular Disease, 1981; contbr. articles to profl. jours. Fellow ACP, Am. Coll. Cardiology, Am. Heart Assn. (council on circulation, council on clin. cardiology); mem. Am. Soc. Clin. Investigation, Alachua County Med. Soc. Subspecialties: Cardiology; Internal medicine. Current work: Management of unstable angina, acute myocardial infarction and congestive heart failure. Role of platelets and prostaglandins in cardiovascular disease. Home: 6604 NW 18th Ave Gainesville FL 32605 Office: Dept Medicine U Fla JHM Health Ctr Box J-277 Gainesville FL 32610

MEHTA, NOSHIR RUSTOM, periodontist, educator; b. Secunderabad, A.P., India, Mar. 5, 1945; came to U.S., 1969, naturalized, 1981; s. Rustom Cawasha and Sooni R. (Darabsha) M.; m. Dara Rosenblatt, Nov. 10, 1977; 1 dau., Larina Noshir. B.D.S., Punjab Govt. Dental Coll. and Hosp., Amritsar, India, 1967; M.D.S. in Periodontics, Govt. Dental Coll., Lucknow, India, 1969; M.S. in Peridontics, Tufts U., 1973, D.M.D., 1976. Research assoc. Dental Sch. Tufts U., 1970-71; clin. instr. 1971-73, asst. prof., 1973-75, clin. instr., 1975-77, asst. clin. prof., 1977-82, assoc. clin. prof., 1983—, co-dir. cranio-mandibular pain dysfunction ctr., 1979—; periodontal cons. USPHS Hosp., Brighton, Mass., 1974-81, Dental Service Corp Mass., Boston, 1980—, Indsl. Accident Bd. Mass., 1979—. Contbr. articles to profl. jours. Fellow Internat. Acad. Dental Studies (internat. liaison officer 1982—); mem. Mass. Periodontal Soc., Am. Acad. Periodontology, ADA, Mass. Dental Soc., Internat. Assn. Dental Research, Indian Acad. Periodontology (hon.). Zorastrian. Subspecialties: Periodontics; Cranio-mandibular cervical pain and dysfunction. Current work: Research in cranio-mandibular and cervical dysfunctions and research in clinical periodontology and dental occlusion. Office: 77 Commercial St Boston MA 02181

MEHTA, RAJENDRA G., cell biologist; b.; b. Dabhoi, India, Aug. 31, 1947; s. Govindlal H. and Arvinda G. (Vora) M.; m. Raksha R. Mehta, Feb. 23, 1976; 1 child, Sonkulp. M.Sc., Gujarat U., Ahmedabad, India, 1968; Ph.D., U. Nebr., 1974. Research asst. U. Nebr., Lincoln, 1971-74; research assoc. U. Rochester (N.Y.) Sch. Medicine and Cancer Center, 1974-76, U. Louisville Sch. Medicine, 1976-77; research scientist Ill. Inst. Tech. Research Inst., Chgo., 1978-80, sr. biochemist, 1980—. Contbr. articles to profl. jours. Mem. Am. Assn. Cancer Research. Hindu. Subspecialties: Cancer research (medicine); Receptors. Current work: Mechanism of hormone and retinoid action during normal and neoplastic differentiation of mammary gland. Home: 16537 S 76th Ave Tinley Park IL 60477 Office: 10 W 35th St Chicago IL 60616

MEIER, WILBUR LEROY, JR., educator, university dean; b. Elgin, Tex., Jan. 3, 1939; s. Wilbur Leroy and Ruby (Hall) M.; m. Judy Lee Longbotham, Aug. 30, 1958; children: Melynn, Marla, Melissa. B.S., U. Tex., 1962, M.S., 1964, Ph.D., 1967. Planning engr. Tex. Water Devel. Bd., Austin, 1962-66, cons., 1967-72; research engr. U. Tex., Austin, 1966; asst. prof. indsl. engring. Tex. A&M U., College Station, 1967-68, asso. prof., 1968-70, prof., 1970-73, asst. head dept. indsl. engring., 1972-73; prof., chmn. dept. indsl. engring. Iowa State U., Ames, 1973-74; prof., head sch. of indsl. engring. Purdue U., West Lafayette, Ind., 1974-81; dean Coll. Engring., Pa. State U., University Park, 1981—; cons. Indsl. Research Inst., St. Louis, 1979, Environments for Tomorrow, Inc., Washington, 1970—, Water Resources Engrs., Inc., Walnut Creek, Calif., 1969-70, Computer Graphics, Inc., Bryan, Tex., 1969-70, Kaiser Engrs., Oakland, Calif., 1971, Tracor, Inc., Austin, 1966-68; cons. div. planning coordination Tex. Gov.'s Office, 1969; cons. Office of Tech. Assessment, 1982—, Southeast Ctr. for Elec. Engring. Edn., 1978—. Editor: Marcel Dekker Pub. Co., 1978—; Contbr. articles to profl. jours. Named Outstanding Young Engr. of Year Tex. Soc. Profl. Engrs., 1966; USPHS fellow, 1966. Mem. Am. Inst. Indsl. Engrs. (dir. ops. research div. 1975, pres. Ind. chpt. 1976, program chmn. 1973-75, editorial bd. Transactions, publ. chmn., newsletter editor engring. economy div. 1972-73, v.p. region VIII 1977-79, exec. v.p. chpt. ops. 1981-83), Ops. Research Soc. Am., Inst. Mgmt. Scis. (v.p. S.W. chpt. 1971-72), ASCE (sec.-treas. Austin br. 1965-66, chmn. research com., tech. council water resources planning and mgmt. 1972-74), Nat., Ind. socs. profl. engrs., Am. Soc. for Engring. Edn. (chmn. indsl. engring. div. 1978-83, pres. Tex. A&M U. chpt. 1971-72), Nat. Assn. State Univ. and Land Grant Colls. (mem. engring. legis. task force 1983—), Sigma Xi, Tau Beta Pi, Alpha Pi Mu (asso. editor Cogwheel 1970-75, regional dir. 1976-77, exec. v.p. 1977-80, pres. 1980-82), Phi Kappa Phi, Chi Epsilon. Club: Rotary. Subspecialties: Industrial engineering; Operations research (engineering). Home: 596 Shadow Ln State College PA 16801 Office: Coll Engring 101 Hammond Bldg Pennsylvania State U University Park PA 16802

MEIJER, WILLEM, botanist, educator, consultant; b. The Hague, Netherlands, June 27, 1923; came to U.S., 1968; s. Fredrik Gerrit and Aaltjen (Holtzapfel) M.; m. Jennie Margurite, Aug. 12, 1942; children: George Johan Willem, Frederica. Ph.D., U. Amsterdam, Netherlands, 1951. Botany asst. U. Amsterdam, 1946-51; botanist Bot. Gardens, Bogor, Java, Indonesia, 1951-59; lectr. U. Andalas, Pajakumbuh, West Sumara, 1955-57, prof., 1957-58; forest botanist Forest Dept., Sandakan, Malaysia, 1959-68; assoc. prof. botany U. Ky., 1968—; cons., mng. ptnr. Environ. Plant Life Services. Author books and articles on flora and vegetation in Holland, Malaysia, Ky. NSF travel grantee, Indonesia, 1973, 1975, 1981; Nat. Geog. Soc. grantee, 1975, 1976, 81, 83. Mem. Am. Biol. Soc., Bot. Soc. Am., Soc. Econ. Botany, Internat. Assn. Plant Taxonomists. Subspecialties: Taxonomy; Ecology. Current work: Taxonomy of Tiliaceae-basswood family; Flora Neotropica; Flora of Cameroon; tree familits in Ceylon; Rafflesiaceae-worldwide. Office: Sch Biol Scis U Ky Lexington KY 40506

MEILLEUR, STEVEN GRANT, solar energy association exec.; b. N.Y.C., June 12, 1951; s. Charles Truman and Jane Loveridge (Burrows) M.; m. Joan Marie Bulman, May 28, 1977; 1 son, David Michael. B.A., Bucknell U., 1973. Co-dir. Resources for Community Alternatives, Santa Fe, 1975-79; exec. dir. N.Mex. Solar Energy Assn., Santa Fe, 1979—. Author works in field. Bd. dirs. Vol. Involvement Service Santa Fe. Mem. Am. Solar Energy Soc., New Eng. Solar Energy Assn., Santa Fe Human Services Coalition. Quaker. Subspecialties: Solar energy; Information systems (information science). Current work: Technology transfer, program design, solar training and information, management systems development. Office: PO Box 2004 Santa Fe NM 87501

MEINDL, JAMES DONALD, electrical engineer; b. Pitts., Apr. 20, 1933; s. Louis M. and Elizabeth F. (Steinhauser) M.; m. Frederica Ziegler, May 21, 1961; children—Peter James, Candace Ann. B.S., Carnegie Mellon U., 1955, M.S., 1956, Ph.D., 1958. Engr. Autonetics Co., Downey, Calif., 1957, Westinghouse Co., Pitts., 1958-59; head sect. microelectronics U.S. Army Electronics Command, Ft. Monmouth, N.J., 1959-62, chief br. semicondr. and microelectronics, 1962-65, dir. div. integrated electronics, 1965-67; assoc. prof. elec. engring. Stanford U., 1967-70, prof., 1970—, dir. integrated circuits lab., 1967—, dir. Electronics Labs., 1972—, dir. Center Integrated Systems, 1981—; dir. Telesensory Systems Inc., Palo Alto, Calif., 1971—; cons. to govt., industry. Author: Micropower Circuits, 1969; contbr. numerous articles to profl. publs. Served to 1st lt. AUS, 1959-61. Recipient Arthur S. Flemming Commn. award Washington Jr. C. of C., 1967; J.J. Ebers award IEEE Electron Devices Soc., 1980. Fellow IEEE (editor Jour. Solid State Circuits 1966-71, Internat. Solid-State Circuits Conf. Outstanding Paper award 1970, 75, 76, 77, 78), AAAS; mem. Nat. Acad. Engring., Electrochem. Soc., Biomed. Engring. Soc. (co-editor Annals of Biomed. Engring. 1976-80), Assn. Advancement of Med. Instrumentation, AAUP, Sigma Xi, Tau Beta Pi, Eta Kappa Nu, Phi Kappa Phi. Subspecialties: Integrated circuits; Integrated circuits. Current work: Engineering education, research and development of integrated systems. Patentee integrated circuit field. Office: 118 AEL Bldg Stanford U Stanford CA 94305

MEINEL, ADEN BAKER, optical sci. and astronomy educator; b. Pasadena, Calif., Nov. 25, 1922; s. John George and Gertrude (Baker) M.; m. Marjorie Pettit, Sept. 5, 1944; children: Carolyn, Walter, Barbara, Edward, Elaine, Mary, David. A.B., U. Calif., Berkeley, 1947, Ph.D., 1949. Asso. prof., asso. dir. Yerkes Obs., U. Chgo., 1950-57; dir. Kitt Peak Nat. Obs., 1958-61; prof. astronomy and optical scis. U. Ariz., Tucson, 1961—, chmn. dept. astronomy, 1961-65, dir., 1962-65, dir. optical sci. center, 1965-74; cons. Perkin-Elmer Corp., 1961-70, Itek Corp., 1970-75, U.S. Air Force, 1963-76, Sci. Applications, Inc., 1976-82. Author 5 books in field; contbr. chpts. to books, articles to profl. jours. Regent Calif. Luth. Coll., 1963-71. Served with USNR, 1944-46. Mem. Optical Soc. Am. (pres. 1972-73, Adolph Lomb medal 1952, Ives medal 1980), Am. Astron. Soc. (Warner prize 1954), Internat. Astron. Union (pres. Commn. 9 1973-76). Lutheran. Subspecialties: Optical system engineering; Optical astronomy. Current work: Telescope design, solar energy, atmospheric optics; development of designs for major telescopes for several countries including U.S. Home: 10121 Catalina Hwy Tucson AZ 85749 Office: U Ariz Optical Scis Center Tucson AZ 85721

MEINWALD, JERROLD, chemist, educator; b. Bklyn., Jan. 16, 1927; s. Herman and Sophie (Baskind) M.; m. Yvonne Chu, June 25, 1955 (div. 1979); children: Constance Chu, Pamela Joan; m. Charlotte Greenspan, Sept. 7, 1980. Ph.B., U. Chgo., 1947, B.S., 1948; M.A., Harvard, 1950, Ph.D., 1952. Mem. faculty Cornell U., 1952-72, 73—, Goldwin Smith prof. chemistry, 1980—; research dir. Internat. Centre Insect Physiology and Ecology, Nairobi, 1970-77; prof. chemistry U. Calif. at San Diego, 1972-73; vis. prof. Rockefeller U., 1970; Camille and Henry Dreyfus Disting. scholar Mt. Holyoke Coll., 1981, Bryn Mawr Coll., 1983; cons. to industry; mem. vis. com. chemistry Brookhaven Nat. Lab., 1969-72, chmn., 1972; mem. med. A chemistry study sect. NIH, 1963-67, chmn., 1965-67; mem. adv. bd. Petroleum Research Found., 1971-73; mem. adv. council chemistry dept. Princeton U., 1978—; mem. adv. bd. Research Corp., 1978-83; mem. adv. bd. chemistry div. NSF, 1979-83; organizing chmn. Sino-Am. Symposium on Chemistry of Natural Products, Shanghai, 1980. Bd. editors: Jour. Organic Chemistry, 1962-66, Organic Reactions, 1968-78, Organic Synthesis, 1968-72, Jour. Chem. Ecology, 1974—, Insect Sci, 1979—; contbr. articles to profl. jours. Sloan fellow, 1958-62; Guggenheim fellow, 1960-61, 76-77; NIH spl. postdoctoral fellow, 1967-68; NIH Fogarty internat. scholar, 1980—; Japan Soc. Promotion of Sci. fellow, 1983. Mem. Am. Chem. Soc. (chmn. organic div. 1969), Swiss Chem. Soc., AAAS, Nat. Acad. Sci., Am. Acad. Arts and Scis., Royal Chem. Soc., Phi Beta Kappa, Sigma Xi. Subspecialties: Organic chemistry; Species interaction. Current work: Organic chemical defense and communication mechanisms in nature; chemical ecology, isolation, characterization, synthesis and biosynthesis of natural products. Office: Baker Lab Cornell U Ithaca NY 14853

MEISELS, GERHARD GEORGE, university dean, chemist, educator; b. Vienna, Austria, May 11, 1931; came to U.S., 1951, naturalized, 1961; s. Leo and Adele Josefa Maria (Seehofer) M.; m. Sylvia Claire Knopsnider, June 28, 1958; 1 dau., Laura Germaine. Student, U. Vienna, 1949-51, 52-53; M.S., U. Notre Dame, Ind., 1952, Ph.D., 1956. Postdoctoral research asso. U. Notre Dame, 1955-56; chemist Gulf Oil Corp., Pitts., 1956-59; part time instr. Carnegie Inst. Tech., Pitts., 1956-58; chemist nuclear div. Union Carbide Corp., Tuxedo, N.Y., 1959-63, asst. group leader, 1964-65; asso. prof. U. Houston, 1965-70, prof., 1970-75, dept. chmn., 1973-75; prof., chmn. dept. chemistry U. Nebr., Lincoln, 1975-81, dean Coll. Arts and Scis., 1981-83, dean of univ., 1983—; cons. Union Carbide Corp., Gearhart-Owen Industries. Editor: spl. issue Jour. Radiation Physics and Chemistry, 1980; contbr. writings in field to profl. publs. Sec., pres. Ramsey (N.J.) Jr. C. of C., 1959-64. Fulbright fellow, Smith-Mundt fellow, 1951-52; sr. fellow Sci. Research Council, Eng., 1976. Mem. Am. Chem. Soc. (com. chmn.), Am. Soc. for Mass Spectrometry (charter, com. chmn.), Nebr. Acad. Scis., AAAS, Am. Phys. Soc., Sigma Xi. Clubs: Houston Kennel (dir. 1968-70), Cornhusker Kennel (pres., dir., del. to Am. Kennel Club 1976-81), Internat. Arabian Horse Assn.). Subspecialties: Physical chemistry; Kinetics. Current work: Unimolecular and bimolecular reactions of gaseous ions with well defined internal energies; radiation chemistry and structure of heavy ion tracks. Home: 6001 Frontier Rd Lincoln NE 68516 Office: 1223 Old Father Hall U Nebr Lincoln NE 68588

MEISER, JOHN HENRY, chemist, educator; b. Cin., Nov. 21, 1938; s. Paul M. and Mildred P. (Turck) M.; m. Enya P. Flores, Aug. 12, 1967; children: Cristina, Teresa, Katharina. B.S., Xavier U., 1961; Ph.D., U. Cin., 1966. Asst. prof. U. Dayton, Ohio, 1966-69; asst. prof. Ball State U., Muncie, Ind., 1969-74, assoc. prof., 1974-80, prof. chemistry, 1980—. Author: Physical Chemistry, 1982. Recipient Student award Am. Inst. Chemists, 1961. Mem. Am. Chem. Soc., Am. Phys. Soc., Ind. Acad. Sci. (sec. 1979-82). Subspecialties: Physical chemistry; Solid state chemistry. Current work: X-ray diffraction, solid-state chemistry, radiocarbon dating. Office: Dept Chemistry Ball State U Muncie IN 47306

MEISKE, JAY C., animal nutrition educator, researcher; b. Hartley, Iowa, June 22, 1930; s. John D. and Sadie B. (Zinn) M.; m. Donna Marie Poland, Dec. 23, 1956; children: Susan, Sally, Thomas, Mary. B.S., Iowa State U., 1952; M.S., Okla. State U., 1953; Ph.D., Mich. State U., 1957. Instr. U. Minn., St. Paul, 1957-59, asst. prof., 1959-65, assoc. prof., 1965-70, prof., 1970—. Subspecialty: Animal nutrition. Office: Dept Animal Sci U Minn 1364 N Eckles St Paul MN 55108

MEISS, DENNIS EARL, neurobiologist, educator; b. Oakland, Calif., Oct. 26, 1947; s. Albert L. and Irene D. (Riley) M.; m. Cheryl L. Petty, Dec. 18, 1971. B.S., St. Mary's Coll., Moraga, Calif., 1969; M.A., UCLA, 1971; Ph.D., U. Conn., 1976. Postdoctoral fellow Scarborough Coll., U. Toronto, 1976-79; marine program Boston U., 1979; asst. prof. physiology Clark U., Worcester, Mass., 1979—. Contbr. numerous articles to profl. publs. Bd. dirs. Greater Worcester chpt. Am. Cancer Soc., 1980—. Grantee Muscular Dystrophy Assn. Can., 1976-79, NIH, 1979-81, Grass Found., 1977, NSF, 1980—. Mem. AAAS, Am. Soc. Zoologists, Bermuda Biol. Sta., Marine B Biology Lab. (Woods Hole, Mass.), N.Y. Acad. Sci., Soc. Neurosci. Subspecialties: Neurophysiology; Neuropharmacology. Current work: Developmental neurobiology; neuropharmacology. Development and growth of neural and neuromuscular systems; mechanisms of action of barbituates. Office: Dept Biology Clark U Worcester MA 01610

MEISTER, ALTON, biochemist, educator; b. N.Y.C., June 1, 1922; s. Morris and Florence (Glickstein) M.; m. Leonora Garten, Dec. 26, 1943; children: Jonathan Howard, Kenneth Eliot. B.S., Harvard U., 1942; M.D., Cornell U., 1945. Intern, asst. resident N.Y. Hosp., N.Y.C., 1946; head clin. biochem. research sect. Nat. Cancer Inst., NIH, 1951-55; commd. officer USPHS, NIH, 1946-55; prof. biochemistry, chmn. dept. Tufts U. Sch. Medicine, 1955-67; prof., chmn. dept. biochemistry Cornell U. Med. Coll., N.Y.C., 1967—; biochemist-in-chief N.Y. Hosp., 1971—; Vis. prof. biochemistry U. Wash., 1959, U. Calif. at Berkeley, 1961; cons. USPHS, 1964-68; chmn. physiol. chemistry study sect.; cons. com. on growth NRC, 1954; cons. biochemistry study sect. USPHS, 1955-60, cons. biochemistry tng. com., 1961-63; cons. Am. Cancer Soc., 1958-61, 71-74. Author: Biochemistry of Amino Acids; Mem. editorial bd.: Jour. Biol. Chemistry, 1958-64; asso. editor, 1976—; mem. editorial bd.: Biochem. Preparations, 1957-64, Biochemistry, 1962-71, 80—, Methods in Biochem. Analysis, 1963—, Biochimica et Biophysica Acta, 1965-77, Ann. Rev. Biochemistry, 1961-65; asso. editor, 1965—; Editor: Advances in Enzymology, 1969—; Contbr. chpts. to books, articles on enzymes, amino acids and glutathione to sci. jours. Recipient Paul-Lewis award enzyme chemistry Am. Chem. Soc., 1954; chmn. U.S. com. Internat. Union Biochemistry, 1960-65, 79—. Fellow Am. Acad. Arts and Scis.; mem. Nat. Acad. Scis., Inst. of Medicine of Nat. Acad. Sci. (sr.), Biophys. Soc., Biochem. Soc. London, Chem. Soc. London, AAAS, Am. Chem. Soc. (chmn. div. biol. chemistry 1965), Am. Assn. Cancer Research, Am. Soc. Biol. Chemists (pres. 1977), Japanese Biochem. Soc. (hon.), Harvey Soc. (hon.), Sigma Xi, Alpha Omega Alpha. Subspecialty: Biochemistry (medicine). Address: 1300 York Ave New York NY 10021

MELBY, EDWARD CARLOS, JR., veterinarian; b. Burlington, Vt., Aug. 10, 1929; s. Edward C. and Dorothy H. (Folsom) M.; m. Jean Day File, Aug. 15, 1953; children—Scott E., Susan J., Jeffrey T., Richard A. Student, U. Pa., 1948-50; D.V.M., Cornell U., 1954. Diplomate: Am. Coll. Lab. Animal Medicine. Practice veterinary medicine, Middlebury, Vt., 1954-62; instr. lab. animal medicine Johns Hopkins U. Sch. Medicine, Balt., 1962-64, asst. prof., 1964-66, asso. prof., 1966-71, prof., dir. div. lab. animal medicine, 1971-74; prof. medicine, dean Coll. Veterinary Medicine, Cornell U., Ithaca, N.Y., 1974—; cons. VA, Nat. Research Council, NIH. Author: Handbook of Laboratory Animal Science, Vols. I, II, III, 1974-76. Served with USMC, 1946-48. Mem. Am., N.Y. State, So. Tier, Vt. veterinary med. assns., Am. Assn. Lab. Animal Sci., Am. Coll. Lab. Animal Medicine, Am. Assn. Accreditation Lab. Animal Care, AAAS, Phi Zeta. Subspecialties: Pathology (veterinary medicine); Laboratory animal medicine. Home: 625 Highland Rd Ithaca NY 14850

MELCHER, ANTONY HENRY, university research administrator, dentistry educator; b. Johannesburg, Transvaal, South Africa, July 1, 1927; emigrated to Can., 1969, naturalized, 1974; s. A. Robert and Anne (Lewis) M.; m. Marcia R. Marcus, Nov. 15, 1953; children: Rowena, Lindsay. B.D.S, U. Witwatersrand, Johannesburg, 1949, H.D.D., 1958, M.D.S., 1960; Ph.D., U. London, 1964. Research fellow Inst. Dental Surgery, U. London, 1962-64; Leverhulme research fellow, sr. lectr. Royal Coll. Surgeons, 1964-67, research fellow in dental sci., sr. lectr., 1967-69; prof. dentistry U. Toronto, Ont., Can., 1969—, Med. Research Council Group in Periodontal Physiology, 1973-83; chmn. grad. dept. dentistry U. Toronto, Ont., Can., 1973-83, dir. postgrad. dental edn., 1983—; vis. research worker tissue and organ culture unit Imperial Cancer Research Fund, London, 1966-69. Editor, contbg. author: Biology of Periodontium, 1969. Med. Research Council Can. grantee, 1969—. Fellow AAAS; mem. Internat. Assn. Dental Research (pres. 1982-83), Am. Soc. Cell Biology, Can. Dental Assn., Am. Acad. Periodontology (hon.), South African Soc. Periodontology (hon.). Club: Island Yacht (Toronto). Subspecialties: Microscopy; Cell biology (medicine). Current work: Structure and function of connective tissues and wound healing, particularly in supporting structures of teeth. Office: U Toronto Faculty Dentistry 124 Edward St Toronto ON Canada M5G 1G6 Home: 101 Banstock Dr Willowdale ON Canada M2K 2H7

MELCHER, JAMES RUSSELL, electrical engineering educator; b. Giard, Iowa, July 5, 1936; s. Melvin Charles and Opal Maxine (Getty) M.; m. Janet Louise Damman, June 15, 1957; children: Jennifer, Eric, Douglas. B.S. in Elec. Engring, Iowa State U., 1957, M.S. in Nuclear Engring, 1958; Ph.D., Mass. Inst. Tech., 1962. Mem. faculty Mass. Inst. Tech., 1962—, asst. prof., 1962-66, asso. prof., 1966-69, prof. elec. engring., 1969—, Stratton chair in elec. engring., 1981, dir. High Voltage Research Lab., 1980; Guggenheim fellow Churchill Coll., Cambridge, Eng., 1972; cons. to industry, 1962—. Author: Field-Coupled Surface Waves, 1963, (with H.H. Woodson) Electromechanical Dynamics, 1969, Continuum Electromechanics, 1981. Recipient first Mark Mills award Am. Nuclear Soc., 1958, Western Electric Fund award New Eng. sect. Am. Soc. Engring. Edn., 1969, Young Alumnus Recognition Ia. State U., 1971; Profl. Achievement citation Iowa State U., 1981. Fellow IEEE, Am. Soc. Engring. Edn., Am. Phys. Soc., Am. Chem. Soc., Sigma Xi, Tau Beta Pi. Subspecialties: Electrical engineering; Mechanical engineering. Research and publ. in continuum electromechanics, continuum feedback control, energy conversion, electrohydrodynamics, air-pollution control, electromech. precipitation and coalescence of particles, electromechanics of electrochem. and biol. systems. Patentee: maker films for Nat. Com. Elec. Engring. Films. Home: 29 Fairlawn Ln Lexington MA 02173 Office: Mass Inst Tech Cambridge MA 02139

MELESE D'HOSPITAL, GILBERT BERNARD, nuclear engineer; b. Paris, Oct. 17, 1926; U.S., 1949, naturalized, 1966; s. Pierre M. and Madeleine M. (Hirsch) Melese; m. Yolande M. d'Hospital, May 28, 1954; children: Francois, Patrick, Philip, Isabelle. Baccalaureat, Lycee, Toulouse, France, 1943; M.S. Sch. Aeros., U. Paris, 1949; Ph.D., Johns Hopkins U., 1954; diploma, Summer AEC Sch., U. Mich.-Ann Arbor, 1958. Registered profl. engr., Calif. Research asst. Johns Hopkins U., Balt., 1952-54; staff engr. French AEC, Saclay, 1954-57; asst. prof. Columbia U., N.Y.C., 1957-60, assoc. prof., 1960; staff mem. GA Technologies, San Diego, 1960-82; vis. prof. MIT, Cambridge, 1983—; adj. prof. Naval Postgrad. Sch., Monterey, Calif., 1983; cons. Republic Aviation, Framingdale, L.I., N.Y., 1959-60; adj. assoc. prof. Columbia U., N.Y.C., 1960-62; lectr. U. Calif.-San Diego, 1961-62. Editor: Nuclear Engineering, 1961; editorial bd.: Nuclear Engring. and Design, Amsterdam, Netherlands, 1971—. Served to lt. French Air Force, 1951-52. Fulbright scholar U.S./French Govt., 1949-50. Fellow Am. Nuclear Soc. (dir. 1972-75, exceptional service award 1980), AIAA; mem. ASME, U.S. Nat. Com. World Energy Conf. (dir. 1979-82), AAAS, Sigma Chi. Subspecialties: Nuclear fission; Nuclear engineering. Current work: Heat transfer and fluid flow in advanced fission reactors and in fusion systems. Home: 824 Hudson St Davis CA 95616

MELIUS, PAUL, chemistry educator, researcher; b. Livingston, Ill., Nov. 21, 1927; s. Louis and Tina (Contoyanis) M.; m. Vaya Gourbis, Sept. 13, 1976; m. Alice Randall Motley, Aug. 20, 1953 (div. 1968); children: Randall, Mark, Alan, Lisa. B.S., Bradley U., Peoria, Ill., 1950; M.S., U. Chgo., 1952; Ph.D., Loyola U., 1956. Research chemist Nat. Aluminate Corp., Chgo., 1952-53; research biochemist Med. Sch. Northwestern U., 1956-57; prof. chemistry Auburn U., 1957—. Author: Problem Workbook Biochemistry, 1975; mem. editorial bd.: Jour. Biosystems, 1976—; contbr. numerous articles to profl. jours. Bd. dirs. United Fund, Auburn, Ala., 1982. Served with U.S. Army, 1946-47. Mem. Am. Chem. Soc. (chmn. sect. 1965-80), Biochem. Soc. London, Am. Soc. Biol. Chemists. Subspecialties: Biochemistry (biology); Enzyme technology. Current work: Protein and enzyme chemistry, aromatic and haloorganic compound metabolism in fish and marine organisms. Office: Auburn University Department Chemistry Auburn AL 36849

MELLBERG, JAMES RICHARD, chemist, dental researcher; b. Manitowac, Wis., June 3, 1932; s. Millard Filmore and Marion Elanor (Zimmerman) Hagenson; m. Gail Maureen Loehning, Sept. 26, 1956; children: Eric, Diane, Laura. B.S., Wis. State Coll.-Oshkosh, 1955; M.S., Loyola U., Chgo., 1960. Head dental research Kendall Co., Barrington, Ill., 1958-75; sr. scientist Colgate-Palmolive Co., Piscataway, N.J., 1975—; cons. U.S. Naval Research Inst., Great Lakes, Ill., 1972—, Nat. Inst. Dental Research, Bethesda, Md., 1974-82; adj. prof. U. Ill., Chgo., 1974-75. Author: Fluoride in Preventive Dentistry, 1983. Deacon Pottersville (N.J.) Reformed Ch., 1980-83; troop com. Boy Scouts Am., Barrington, 1962-75. Served with U.S. Army, 1956-58. Recipient 1st place (7) 2d place (4) Sci. Exhibits awards ADA, 1966-83. Mem. AAAS, Am. Chem. Soc., Internat. Assn. Dental Research, Am. Assn. Dental Research. Subspecialties: Cariology; Preventive dentistry. Current work: Career research includes study of anticaries mechanisms of fluoride, development of new fluoride products and methods for clinical use; the formation and remineralization of caries lesions. Patentee (7) Fluoride for Caries Prevention, 1967-79. Home: 6 Addison Dr Pottersville NJ 07979

MELLIN, THEODORE NELSON, research scientist; b. Paterson, N.J., Dec. 24, 1937; s. Nelson F. and Helen S. (Steib) M.; m. Beverly

E. Donahue, July 4, 1959; children: Jennifer, Victoria, Abigail; m. Josephine R. Carlin, June 11, 1977. B.S., U. Vt., 1959; M.S., U. Maine, 1961; Ph.D. (David Ross fellow), Purdue U., 1964. USPHS fellow, postdoctoral trainee in steriod biochemistry Worcester Found. Exptl. Biology, 1964-66; with Merck Inst. Therapeutic Research, Rahway, N.J., 1966, research fellow dept. biochemistry, 1974—, guest lectr. Merck speakers program; reviewer endocrinology sect. Prostaglandin Jour.; presenter papers at profl. meetings. Contbr. articles and abstracts to profl. jours. Mem. Soc. Neurosci., Soc. Study of Reprodn., N.J. Soc. Neurosci., Merck Sci. Club (pres. 1969-71), Sigma Xi, Alpha Zeta. Independent. Presbyterian. Club: Old Straw Hat Ski (pres. 1975-77). Subspecialty: Neuroendocrinology. Current work: BPH, hirsutism and acne. Home: 32 Overhill Dr North Brunswick NJ 08902 Office: Merck & Co PO Box 2000 Rahway NJ 07065

MELLONIG, JAMES THOMAS, periodontist, periodontal researcher; b. Milw., Jan. 9, 1942; s. Val Martin and Ann Mary (Pugel) M.; m. Karen Susan Wojcik, Nov. 23, 1967; children: Timothy, Amy, Kimberly. A.B., Marquette U., 1964; D.D.S., 1969; M.S., George Washington U., 1973; periodontics cert., Naval Grad. Dental Sch., 1974. Diplomate: Am. Acad. Periodontology, Am. Acad. Oral Medicine. Commd. Lt. U.S. Navy, 1969, advanced through grades to comdr., 1978; clin. investigator Naval Med. Research Inst., Bethesda, Md., 1976-80; chief periodontics Naval Regional Dental Ctr., Gt. Lakes, Ill., 1980—, cons. in periodontics, 1980—; cons. in periodontics VA, North Chicago, Ill., 1980—; clin. assoc. prof. Loyola U. Dental Sch., Chgo., 1982—, U. Ill. Dental Sch., 1982—. Contbr.: chpt. to Tissue Management in Restorative Dentistry, 1982. Mem. Am. Acad. Periodontology (exec. council), Am. Acad. Oral Medicine, ADA, Am. Assn. Tissue Banks, Gt. Lakes Dental Soc. Roman Catholic. Subspecialties: Periodontics; Oral Medicine. Current work: Reconstructive periodontics, bone grafts, bone allografts. Home: 833 Braemar Dr Mundelein IL 60060 Office: US Naval Regional Dental Center Great Lakes IL 60088

MELLORS, ROBERT CHARLES, pathologist; b. Dayton, Ohio, 1916; s. Bert S. and Clementine (Steinmetz) M.; m. Jane K. Winternitz, Mar. 25, 1944; children: Alice J., Robert Charles, William K., John. Ph.D., Western Res. U., 1940; M.D., Johns Hopkins, 1944. Diplomate: Am. Bd. Pathology in path. anatomy. Intern Nat. Naval Med. Center, Bethesda, Md., 1944-45; research fellow medicine Meml. Center Cancer and Allied Diseases, N.Y.C., 1946-50, research fellow pathology, 1950-53, asst. attending pathologist, 1953-57, assoc. attending pathologist, 1957-58; sr. fellow Am. Cancer Soc., 1947-50; sr. clin. research fellow Damon Runyon Meml. Fund, 1950-53; asst. attending pathologist Meml. Hosp., N.Y.C., 1953-57, asso. attending pathologist, 1957-58; asst. attending pathologist Ewing Hosp., N.Y.C., 1953-57, assoc. attending pathologist, 1957-58; instr. biochemistry Western Res. U., 1940-42; research assoc. Poliomyelitis Research Ctr. and Dept. Epidemiology Johns Hopkins U. Sch. Hygiene, 1942-44; asst. prof. biology Meml. Center Cancer and Allied Diseases, N.Y.C., 1952-53; asst. prof. pathology Sloan Kettering div. Cornell U., 1953-57, assoc. prof., 1957-58, prof. pathology, 1961—; assoc. attending pathologist N.Y. Hosp., 1961-72, attending pathologist, 1972—; pathologist-in-chief, dir. labs., 1958—; assoc. dir. research Hosp. for Spl. Surgery, N.Y.C., 1958-69, dir. research, 1969—; Mem. research adv. com. NIH, 1962-66; adv. com. Nat. Inst. Environ. Health Sci., 1966-69; com. nomenclature and classification of disease Coll. Am. Pathologists, 1960-64. Author: Analytical Cytology, 1955, 2d edit., 1959, Analytical Pathology, 1957. Served as lt. (j.g.), M.C. USNR, 1944-46. Recipient Kappa Delta award Am. Acad. of Orthopedic Surgeons, 1962. Fellow Royal Coll. Pathologists, Am. Soc. Clin. Pathology; mem. Am. Assn. Pathologists, Am. Assn. Immunologists, Am. Soc. Biol. Chemists, Am. Orthopedic Assn. (hon.). Subspecialties: Pathology (medicine); Immunology (medicine). Current work: Etiology and parthogenesis of rheumatoid arthritis; systemic lupus eryth ematosus and related autoimmune diseases; immunopathology of type C RNAtumor viruses (Retro-viruses). Home: 3 Hardscrabble Circle Armonk NY 10504

MELMON, KENNETH LLOYD, physician, biologist; b. San Francisco, July 20, 1934; s. Abe Irving and Jean (Kahn) M.; m. Elyce Edelman, June 9, 1957; children: Bradley S., Debra W. A.B. in Biology with honors, Stanford U., 1956; M.D., U. Calif. at San Francisco, 1959. Intern, then resident in internal medicine U. Calif. Med. Center, San Francisco, 1959-61; clin. assoc., surgeon USPHS, Nat. Heart and Lung Inst., NIH, 1961-64; chief resident in medicine U. Wash. Med. Center, Seattle, 1964-65; chief div. clin. pharmacology U. Calif. Med. Center, 1965-78; chief dept. medicine Stanford U. Med. Center, 1978—; mem. sr. staff Cardiovascular Research Inst., 1973—; chmn. Joint commn. prescription drug use Senate Subcom. on Health, Inst. Medicine and HEW-Pharm. Mfrs. Assn.; cons. to govt. Author articles, chpts. in books, sects. encys.; Editor: Clinical Pharmacology: Basic Principles in Therapeutics, 2d edit, 1978, Cardiovascular Therapeutics, 1974; assoc. editor: The Pharmacological Basis of Therapeutics (Goodman and Gilman), 1980; editor, 1985. Burroughs Wellcome clin. pharmacology scholar, 1966-71; Guggenheim fellow, 1971; vis. scientist Weizmann Inst., Israel, 1971-72; NIH spl. fellow, 1972. Fellow AAAS; mem. Am. Fedn. Clin. Research (pres. 1973-74), Am. Soc. Clin. Investigation (pres. 1978-79), Assn. Am. Physicians, Western Assn. Physicians (1983-84), Am. Soc. Pharmacology and Exptl. Therapeutics, Inst. Medicine of Nat. Acad. Sci., Am. Physiol. Soc., Calif. Acad. Medicine, Phi Beta Kappa. Jewish. Subspecialty: Pharmacology. Research on cellular response to low molecular weight hormones. Home: 51 Cragmont Way Woodside CA 94062 Office: Stanford U Med Center Dept Medicine Stanford CA 94305

MELNICK, JOSEPH L., educator, virologist; b. Boston, Oct. 9, 1914; s. Samuel and Esther (Melny) M.; m. Matilda Benyesh, 1958; 1 dau., Nancy. A.B., Wesleyan U., 1936; Ph.D. (Univ. scholar), Yale, 1939; D.Sc., Wesleyan U., 1971. Asst. in physiol. chemistry Yale Sch. Medicine, 1937-39; Asst. in physiol. chemistry, Finney-Howell Research Found. fellow, 1939-41, NRC fellow in med. scis., 1941-42, research asst. in preventive medicine with rank of instr., 1942-44, asst. prof., 1944-48, research asso., 1948-49, asso. prof. microbiology, 1949-54, prof. epidemiology, 1954-57; chief virus labs., div. biologics standards NIH, USPHS, 1957-58; prof. virology and epidemiology Baylor Coll. Medicine, Houston, 1958-74, disting. service prof., 1974—, dean grad. studies, 1968—; Mem. com. virus diseases WHO, 1957—; dir. WHO Internat. Center Enteroviruses, 1963-70, mem. cons. group on poliomyelitis vaccine, 1973—; dir. WHO Collaborating Center for Virus Reference and Research, 1974—; mem. com. on live poliovirus vaccines USPHS, 1959-61; virus reference bd. NIH, 1962-70; mem. human cancer virus task force Nat. Cancer Inst., USPHS, 1962-67; nat. adv. cancer council NIH, 1965-69; sec.-gen. Internat. Congresses Virology, 1968-71; chmn. Internat. Conf. Viruses in Water, Mexico City, 1974; mem. exec. bd. Internat. Assn. Microbiol. Socs., chmn. sect. on virology, 1970-75, mem. internat. commn. microbiol ecology, 1972—; mem. research council Am. Cancer Soc., 1971-75; mem. com. on hepatitis Nat. Acad. Sci./NRC, 1972-77; lectr. cons. Chinese Acad. Med. Scis., 1978, 79; mem. adv. com. Comparative Virology Orgn., 1978—. Author: other sci. publs. Textbook of Medical Microbiology; Editor: Monographs in Virology; editor-in-chief: ofcl. jour. virology Intervirology, Internat. Union Microbiol. Socs. Recipient Internat. medal (with Dr. Herids von Magnus of Denmark); award Argentinian Found. Against Infantile Paralysis, for research in immunity to poliomyelitis, 1949; Humanitarian award Jewish Inst. Med. Research, 1964; Modern Medicine Disting. Achievement award, 1965; Eleanor Roosevelt Humanities award, 1965; Indsl. Research-100 award (with Prof. Craig Wallis), 1971, 74; Inventor of Year award Houston Patent Law Assn., 1972; named to Nat. Found.'s Polio Hall Fame, 1958; Gold medal South African Poliomyelitis Research Found., 1979; Maimonides award State of Israel, 1980. Fellow AAAS, Am. Pub. Health Assn., N.Y. Acad. Scis. (Freedman Found. award for research in virology 1973), Am. Acad. Microbiology; mem. Am. Soc. Microbiology, Soc. Exptl. Biol. and Medicine (mem. council 1965-69), Am. Assn. Immunologists, Am. Epidemiol. Soc., Am. Assn. Cancer Research (pres. S.W. sect. 1968), Microbiol. Soc. Israel (hon.), USSR Soc. Microbiologists and Epidemiologists (hon.), Phi Beta Kappa, Sigma Xi. Subspecialties: Virology (biology); Virology (medicine). Current work: Medical virology including a synthetic vaccine for viral hepatitis, possible viral etiologies of atherosclerosis and multiple sclerosis, environmental virology.

MELOUK, HASSAN ALY, plant pathologist; b. Alexandria, Egypt, May 19, 1941; came to U.S., 1965, naturalized, 1975; s. Aly H. and Hanem M. (Kadour) M.; m. Afaf H. Abo-El-Dahab, Dec. 9, 1964; children: Sammy, Sharif. B.S., Alexandria U., 1962; M.S., Oreg. State U., 1967, Ph.D., 1969. Instr., Alexandria (Egypt) U., 1962-64; grad. research asst. Oreg. State U., Corvallis, 1965-69; postdoctoral research assoc. Wash. State U., Prosser, 1969-70; research assoc. Oreg. state U., Corvallis, 1970-76; plant pathologist, prof. dept. plant pathology U.S. Dept. Agr., Agrl. Research Service, Okla. State U., Stillwater, 1976—. Contbr. articles to profl. jours. Den leader Cub Scouts Am., 1972-74. Nat. Mint Research Council research grantee, 1972-76. Mem. Am. Phytopath. Soc., Am. Peanut Research and Edn. Soc. Republican. Subspecialties: Plant pathology; Plant genetics. Current work: Development of peanut germplasm with multiple resistance to diseases. Home: 4802 Country Club Ct Stillwater OK 74074 Office: US Dept Agr Agrl Research Service Dept Plant Pathology Okla State U Stillwater OK 74078

MELSA, JAMES LOUIS, electrical engineer, educator, consultant; b. Omaha, July 6, 1938; s. Louis Fred and Ann (Pelnar) M.; m. Katherine Smith, June 25, 1960; children: Susan, Elisabeth, Peter, Jon, Jennifer, Mark. B.S.E.E., Iowa State U., 1960; M.S.E.E., U. Ariz., 1962, Ph.D., 1965. Assoc. mem. tech. staff Radio Corp. Am., Tucson, 1960-61; instr. elec. engring. U. Ariz., 1961-65, Asst. prof., 1965-67; assoc. prof. info. and control scis. So. Meth. U., 1967-69, prof., 1969-73; prof., chmn. dept. elec. engring. U. Notre Dame, 1973—; cons. to industry; cons. Los Alamos Sci. Lab., 1965—. Author-12 books including: (with J.D. Gibson) Nonparametric Detection with Applications, 1975, (with D.L. Cohn) A Step by Step Introduction to 8080 Microprocessor Systems, 1977, Decision and Estimation Theory, 1978; editor: (with M.K. Sain and J.L. Peczkowski) Alternatives for Linear Multivariable Control, 1978; mem. editorial adv. bd.: Jour. Computers and Elec. Engring, 1972—; assoc. editor: Man and Cybernetics, 1972-79; contbr. numerous articles profl. jours. Fellow IEEE; mem. AIAA, Nat. Engring. Consortium (dir. 1975-79), Am. Soc. Engring. Edn. (Western Electric award Gulf S.W. sect. 1973), Sigma Xi, Tau Beta Pi, Pi Mu Epsilon, Eta Kappa Nu, Phi Kappa Phi. Subspecialty: Electrical engineering. Home: 53222 Martin Ln South Bend IN 46635 Office: Dept Elec Engring Notre Dame Notre Dame IN 46556

MELTON, CHARLES ESTEL, educator; b. Fancy Gap, Va., May 18, 1924; s. Charlie Glenn and Ella (Ayers) M.; m. Una Faye Hull, Dec. 7, 1946; children—Sharon (Mrs. Lawrence Husch), Wayne, Sandra (Mrs. Glenn Allen). B.A., Emory and Henry Coll., 1952, D.Sc., 1967; M.S., Vanderbilt U., 1954; Ph.D., U. Notre Dame, 1964. Physicist Oak Ridge Nat. Lab., 1954-67; prof. chemistry U. Ga., Athens, 1967—, head dept., 1972-77. Author: Principles of Mass Spectrometry and Negative Ions, 1970; Contbr. articles to profl. jours. Served with USNR, 1943-46. Recipient DeFriece medal Emory and Henry Coll., 1959; numerous research grants. Fellow AAAS; mem. Am. Phys. Soc., Am. Chem. Soc., Ga. Acad. Sci. Presbyterian. Subspecialties: Physical chemistry; Geophysics. Current work: Research in the svolution of petroleum, the evolution of the earths atmosphere and oceans, the orgin of the earths magnetic field, and the origin and age of diamonds. Home: Route 2 Box 18 Hull GA 30646 Office: Dept Chemistry Univ Georgia Athens GA 30602

MELTON, RUSSELL PAUL, biochemistry educator; b. London, July 19, 1929; U.S., 1952, naturalized, 1958; s. David Oscar and Frances Clare (Briggs) M.; m. Linda S. Lucci, Sept. 12, 1960; children: Roger, Diana, Leslie. B.S., U. London, 1951; Ph.D., Princeton U., 1956. Research biochemist NIH, Bethesda, Md., 1957-62; research biochemist U. Calif. Berkeley, 1962-66, asst. prof. biochemistry, 1966-70; assoc. prof. Johns Hopkins U., Baltimore, 1970-75, prof. biochemistry, 1976—; cons. NSF, 1979—. Mem. Am. Soc. Biol. Chemists, AAAS, Am. Soc. Microbiology, Genetics Soc. Club: Cosmos (Washington). Subspecialties: Gene actions; Biochemistry (biology). Office: Werik Meml Lab 200 E Lexington St Suite 1600 Baltimore MD 21202

MELTON, WILLIAM GROVER, JR., geologist; b. Oakland, Calif., Jan. 1, 1923; s. William Grover and Vera Athel (Todd) M.; m. Veronica Ann Silverstrand, Aug. 15, 1956; children: Thomas, Douglas, Rebecca, Mary, Phillip. B.A., U. Mont., 1953; M.S., U. Mich., 1969. Geologist U.S. Geol. Survey, Washington, 1953-57; preparator Mus. Paleontology, U. Mich., 1957-66; curator dept. geology U. Mont., Missoula, 1966—; co-chmn. profl. symposia in field. Contbr. articles in field to profl. jours. Mem. Mont. Preservation Review Bd., Helena, 1979—. Served to sgt. U.S. Army, 1940-46. Mem. Soc. Vertebrate Paleontology, Paleontol. Soc., Hellgate Rocks & Minerals Club (N.W. Fedn. scholar 1973), Sigma Xi. Presbyterian. Subspecialties: Geology; Paleoecology. Current work: Found and described first conodont-bearing animals; collection and description of upper Mississippi Age Bear Gulch fauna from central Montana; collection and description of late Wisconsin age Blacktail Cave fauna. Home: 2304 Spring Dr Missoula MT 59801 Office: Dept Geology U Mont Missoula MT 59812

MELTZER, HERBERT YALE, psychiatrist, educator, laboratory director; b. Bklyn., July 29, 1937; s. David and Estelle (Gross) M.; m. Sharon Bittenson, June 12, 1960; children: David, Danielle. B.A., Cornell U., 1958; M.A., Harvard U., 1959; M.D., Yale U., 1963. Diplomate: Am. Bd. Psychiatry and Neurology, Nat. Bd. Med. Examiners. Research asso. Yale U. Sch. Medicine, New Haven, 1959-63; Teaching fellow in psychiatry Harvard U. Med. Sch., Boston, 1964-66; instr. grad. tng. program NIMH, Bethesda, Md., 1967-68; asst. prof. psychiatry U. Chgo., 1968-71, asso. prof., 1971-74, prof., 1974—, dir. lab. biol. psychiatry, 1975—, dir. mental health clin. research center, 1977—; dir. acute psychosis research program Ill. State Psychiat. Inst., 1968-75; cons. NIMH. Contbr. over 250 articles to profl. jours.; editorial bd.: Jour. Clin. Psychopharmacology, Am. Jour. Psychiatry, Psychiatry Research, Biol. Psychiatry, Biochem. Pharmacology. Mem. Am. Coll. Neuropsychopharmacology (pres.-elect, Daniel C. Efron award 1980, exec. com.), Am. Psychiat. Assn., Ill. Psychiat. Soc., Internat. Soc. Psychoneuroendocrinology, AAAS, Soc. Biol. Psychiatry, N.Y. Acad. Scis., Am. Soc. Pharmacology and, Psychiat. Research Soc. Subspecialties: Psychopharmacology; Neuropharmacology. Current work: Biological psychiatry, psychopharmacology, neuroendocrinology, schizophrenia, and depression. Home: 6831 S Euclid Chicago IL 60649 Office: 950 E 59th St Chicago IL 60637

MELTZER, RICHARD STUART, cardiologist, educator; b. N.Y.C., Sept. 6, 1948; s. Ezra and Hilda (Komaroff) M.; m. Colette Haesaerts, Aug. 8, 1971; children: Michelle Ann, Sara Muriel. B.A. Magna cum laude, Harvard U., 1970, M.D., 1974; Ph.D., Erasmus U., Netherlands, 1982. Diplomate: Am. Bd. Internal Medicine; Subcert. in cardiovascular disease. Cardiologist, asst. prof. Thoraxcenter, Rotterdam, Netherlands, 1979-82; vis. assoc. prof. Heart Inst., Sheba Med. Ctr., Tel Aviv, Israel, 1982; asst. prof., cardiologist Mt. Sinai Med. Ctr., N.Y.C., 1983—. Editor: Contrast Echocardiography, 1982; contbr. 80 articles to profl. jours. Fellow ACP, Am. Coll. Cardiology. Jewish. Subspecialty: Cardiology. Current work: Cardiac diagnosis using innovative ultrasound techniques: contrast, Doppler, etc. Office: Cardiology Div Mt Sinai Med Ctr One Gustave Levy Pl New York NY 10029

MELVILLE, RICHARD DEVERN SAMUELS, JR., engineering executive; b. Los Angeles, Nov. 10, 1939; s. Richard Devern Samuels and Dorothy Irene (Dorchester) M.; m. Esther Jean Melville, Sept. 1, 1962; children: Richard Devern Samuels, Donald Scott. B.S. in Elec. Engring, U. So. Calif. Elec. Engring, 1960; M.S. in Physics, U.S. Naval Postgrad. Sch., 1967, Calif. Inst. Tech., 1971, Ph.D., 1975. Commd. ensign U.S. Navy, 1960; advanced through grades to lt. comdr., 1970, served as naval aviator, fighter pilot; resigned, 1970; sr. researcher R & D Assoc., Marina Del Rey, Calif., 1975-81; mgr advanced systems dept TRW, Redondo Beach, Calif., 1981-82; co-founder, dir. advanced tech. programs EOS Technologies Inc., Santa Monica, Calif., 1982—. Contrbr articles to profl. jours. Mem. Optical Soc. Am., Sigma Xi. Subspecialties: Laser technology; Optical image processing. Current work: Corporate director advanced technology programs, principally intelligence and surveillance systems, military systems.

MELVOLD, ROGER WAYNE, microbiology educator, researcher; b. Henning, Minn., Mar. 21, 1946; s. Sam Reder and Palma (Ronning) M. B.S., Moorhead State U., 1968; Ph.D., U. Kans., 1973. Assoc. Harvard U. Med. Sch., Boston, 1972-76, prin. research assoc., 1976-79; asst. prof. dept. microbiology-immunology Northwestern U. Med. Sch., Chgo., 1979—. NIH research grantee, 1979—. Mem. Am. Assn. Immunologists, Transplantation Soc., Sigma Xi. Democrat. Club: Morale Com. Subspecialties: Immunology (agriculture); Animal genetics. Current work: Identifying mutations of genes affecting transplantation in the mouse and their subsequent use in experimental analyses of the genetic components of the immune system. Home: 5858 N Sheridan Rd Chicago IL 60660 Office: Dept Microbiology-Immunology Northwestern Univ Med Sch 303 E Chicago Ave Chicago IL 60611

MENAKER, LEWIS, dental educator, administrator; b. N.Y.C., Apr. 15, 1942; s. David and Sophie (Hochberg) M. D.M.D., Tufts U., 1965; Sc.D., M.I.T., 1971. Prof. dentistry, chmn. dept. oral biology U. Ala. Sch. Dentistry, Birmingham, 1981—; assoc. dean acad. affairs, 1981—, sr. scientist Inst. Dental Research, U. Ala., Birmingham; dir. Office Applied Research, U. Ala.; cons. Warner-Lambert Co., Morris Plains, N.J., State Bd. Dental Examiners Ala., Montgomery. Editor: Biologic Basis of Wound Healing, 1975, Biologic Basis of Dental Caries, 1980; mem. editorial bd.: Jour. Dental Research, 1982—; assoc. editor: Ala. Jour. Med. Scis, 1975—. NIH fellow, 1967-71; recipient Clin. Investigator award NIAMDD, 1972-75; Fedn. Dentaire Internat. Preventive Dentistry award, 1977. Mem. Am. Dental Assn. (Preventive Dentistry award 1975), Am. Assn. Dental Research, Internat. Assn. Dental Research, Am. Assn. Dental Schs., Ala. Dental Assn. Republican. Jewish. Subspecialties: Oral biology; Preventive dentistry. Current work: Animal and clinical studies on the effect of nutrition and other agents on the prevention of dental caries and periodontal disease. Office: University of Alabama School of Dentistry University Station Birmingham AL 35294

MENALDI, JOSÉ LUIS, mathematics educator, researcher; b. Cañada de Gómez, Argentina, July 26, 1952; came to U.S., 1979; s. Miguel Adalberto and Dominga (Bonetto) M.; m. Mariá Cristina Ametrano, May 9, 1981. Lic. in Math, U. Rosario, Argentina, 1973; D.Sc., U. Paris-Dauphine, 1980. Asst. instr. U. Rosario, 1973-75; asst. U. Paris-Dauphine, 1976-77, asst. prof., 1978-79, 80-81, U. Ky., 1979-80; asst. prof. math. Wayne State U., Detroit, 1981—; stagière Inst. Research in Computer Sci. and Automation, Rocquencourt, France, 1975-76, researcher, 1977-78. Contbr. articles to profl. jours. U.S. Army Research grantee, 1983-86. Mem. Union Matemática Argentina, Am. Math. Soc., Soc. Indsl. and Applied Math. Roman Catholic. Subspecialties: Applied mathematics; Probability. Current work: Applied mathematics, stochastic control systems theory, calculus of variation and optimal control, optimization, partial differential equations, probability theory and stochastic processes. Office: Wayne State U Dept Math Detroit MI 48202

MENARD, HENRY WILLIAM, geologist; b. Fresno, Calif., Dec. 10, 1920; s. Henry William and Blanche (Hodges) M.; m. Gifford Merrill, Sept. 21, 1946; children—Andrew O., Elizabeth M., Dorothy M. B.S., Calif. Inst. Tech., 1942, M.S., 1947; Ph.D., Harvard U., 1949; D.Sc. (hon.), Old Dominion U., 1980. Geologist Amerada Petroleum Corp., 1947; from assoc. to supervisory oceanographer Navy Electronics Lab., 1949-55; assoc. prof. marine geology Inst. Marine Resources, U. Calif., 1955-61, prof., 1961—, acting dir., 1967-68; prof. geology Scripps Inst. Oceanography, 1955—; dir. U.S. Geol. Survey, 1978-81; Mem. panel underwater swimmers Nat. Acad. Scis.-NRC, 1955, Pacific sci. bd., 1956-62, panel on ocean resources, 1958; mem. com. sci. and pub. policy; tech. asst. U.S. Office Sci. and Tech., 1965-66; oceanographic expdns., Mid-Pacific, 1949, Capricorn, 1952-53, Chinook, 1956, Downwind, 1957-58, Nova, 1967, Ngendie, 1983; Dir. Geol. Diving Cons., Inc., 1953-58. Served from ensign to lt. comdr. USNR, 1942-46. Decorated Bronze Star.; Guggenheim Meml. Found. fellow, 1962-63; Overseas fellow Churchill Coll., 1970-71; Woods Hole Oceanographic Instn. fellow, 1973. Fellow Geol. Soc. Am., Am. Geophys. Union, AAAS; mem. Nat. Acad. Scis., Am. Acad. Arts and Scis., Am. Assn. Petroleum Geologists, Royal Astron. Soc., Sigma Xi. Subspecialty: Geology. Office: Scripps Inst Oceanography U Calif San Diego Mail Code A-020 La Jolla CA 92093

MENDEL, VERNE EDWARD, animal physiology educator, researcher; b. Lewistown, Mont., Apr. 28, 1923; s. William Claude and Alvina Wilhelmina (Stief) M.; m. Beatrice Helen Dolan, Feb. 6, 1946; m. Velda Fern Haynes, Mar. 24, 1973; children: Frank C., Glen W., Todd O. B.S. in Agr, U. Idaho, 1955, M.S. in Nutrition, 1958; Ph.D. in Physiology, U. Calif., Davis, 1960. Asst. prof. dept. animal sci. U. Alta. (Can.), Edmonton, 1960-63; asst. animal scientist, dept. animal sci. U. Calif., El Centro, 1963-67; asst. prof. dept. animal sci U. Calif.,-Davis, 1967-69, assoc. prof., 1969-74, prof. dept. animal physiology, 1973—, chmn. dept., 1973-78. Contbr. articles to sci. jours. Served with U.S. Army, 1943-46. Mem. Endocrine Soc., Am. Physiol. Soc., Soc. for Neurosci., Fedn. Am. Socs. for Exptl. Biology. Democrat. Presbyterian. Subspecialties: Animal physiology; Endocrinology. Current work: Neural and endocrine control of food intake; researcher food intake control. Office: Dept Animal Physiology Univ Calif Davis CA 95616

MENDELOFF, ALBERT IRWIN, physician, educator; b. Charleston, W.Va., Jan. 29, 1918; s. Morris Israel and Esther (Cohen) M.; m. Natalie Lavenstein, Dec. 19, 1943; children: Henry, John, Katherine. A.B., Princeton U., 1938; M.D., Harvard U., 1942, M.P.H., 1944. Fellow in nutrition Rockefeller Found., 1943-44; nutrition cons. UNRRA mission to Greece, 1944-46; asst. prof. medicine and preventive medicine Washington U. Med. Sch., 1949-54, asso. prof., 1955; gastroenterology cons. Barnes Hosp., St. Louis, 1952-55; asso. prof. medicine Johns Hopkins Med. Sch., 1955-70, prof., 1970—; physician-in-chief Sinai Hosp., Balt., 1955-80, chief research medicine, 1980-83; Sr. surgeon Res. USPHS. Editor-in-chief: Am. Jour. Clin. Nutrition, 1981—. Mem. Assn. Am. Physicians, Am. Soc. Clin. Investigation, Central Soc. Clin. Research, Am. Fedn. Clin. Research, Am. Gastroent. Assn., Phi Beta Kappa, Alpha Omega Alpha. Subspecialty: Nutrition (medicine). Home and Office: 2109 Northcliff Dr Baltimore MD 21209

MENDELS, JOSEPH, psychiatrist; b. Capetown, South Africa, Oct. 29, 1937; came to U.S., 1964, naturalized, 1975; s. Max and Lily (Turecki) M.; m. Ora Kark, Jan. 17, 1960; children: Gilla Avril, Charles Alan, David Ralph. M.B., Ch.B., U. Capetown, 1960; M.D., U. Witwatersrand, 1965. Intern U. Witwatersrand, 1961; resident in psychiatry U. Witwatersrand and U. N.C., 1962-67; chief depression research program, prof. psychiatry and pharmacology U. Pa. and VA Hosp., Phila., 1973-80; med. dir., v.p. Fairmount Inst., Phila., 1980-81; med. dir. Therapeutics Inc., 1981—, Phila. Med. Inst., 1983—; prof. psychiatry, human behavior and pharmacology Thomas Jefferson Med. Coll., Phila., 1982—; ad hoc cons. NIMH, NIH; dir. Spectrum Publs.; lectr. univs. and hosps. Author, editor: Concepts of Depression, 1971, Biological Psychiatry, 1973, Psychobiology of Affective Disorders, 1981; contbr. articles to med. jours. Grantee NIMH, VA. Fellow Inst. for Study Human Issues; mem. Am. Psychiat. Assn. (Lester N. Hofheimer award 1976), Psychiat. Research Soc., Am. Coll. Neuropsychopharmacology, Congressium Internationale Neuropsychopharmacologia. Subspecialties: Psychopharmacology; Pharmacokinetics. Current work: Clinical pharmacology; psychopharmacology; biology and treatment of depression and mania. Office: 1015 Chestnut St Philadelphia PA 19107

MENDELSOHN, LAWRENCE BARRY, educator; b. N.Y.C., Apr. 19, 1934; s. Abraham and Roslyn M.; m. Francine D. Sperling, June 28, 1958; children: Alan, Edward, Robyn. B.A., Bklyn. Coll., 1955; M.A., Columbia U., 1958; Ph.D., N.Y.U., 1965. Physicist TRG, Syosset, N.Y., 1958-62; asst. prof. physics Cooper Union, N.Y.C., 1962-65; asso. prof. Poly. Inst. N.Y., Bklyn., 1965-73; prof. physics Bklyn. Coll., 1973—; cons. Sandia Labs., Engring. Index. Contbr. articles to profl. jours. Mem. Am. Phys. Soc., Sigma Xi. Democrat. Subspecialties: Atomic and molecular physics; Electronic materials. Current work: X-ray and electron scattering from atoms and solids. Home: 73-15 193d St Flushing NY 11366 Office: Physics Dept Brooklyn Coll Brooklyn NY 11210

MENDELSOHN, NATHAN SAUL, mathematician; b. Bklyn, Apr. 14, 1917; s. Samuel and Sylvia (Kirschenbaum) M.; m. Helen Brontman, Oct. 26, 1940; children—Eric, Alan. B.A., U. Toronto, Ont., Can., 1939, M.A., 1940, Ph.D., 1942. Teaching fellow U. Toronto, 1939-42; lectr. Queen's U., 1945-46; asst. prof. math. U. Man., Can., Winnipeg, 1947, now prof., head dept. math.; participant I nternat. Math. Union, Saltzobaden, Sweden, 1962, Dubna, USSR, 1966, Menton, France, 1970. Contbr. numerous articles on geometry, group theory and combinational math. to profl. jours. Fellow Royal Soc. Can. (Can. del., Henry Marshall Tory medal 1979, Disting. prof. 1981); mem. Computing and Data processing Soc. Can. (dir. 1960-65). Subspecialties: Algorithms; Artificial intelligence. Current work: Computer algorithms for interactive proof of mathematical theorems. Home: 364 Enniskillen Winnipeg MB Canada R2V 0J3 Office: U Man 343 Machray Hall Winnipeg MB Canada R3T 2N2

MENDELSOHN, RICHARD, chemistry educator; b. Montreal, Que., Can., July 1, 1946; came to U.S., 1976; s. Israel Charles and Celia (Groper) M.; m. Nancy Rzemieniak, Apr. 25, 1971; 1 dau., Naomi. B.Sc., McGill U., Montreal, 1967; Ph.d., MIT, 1972. Research assoc. King's Coll., London, 1972-73; research assoc. NRC Can Ottawa, 1973-76; mem. faculty dept. chemistry Rutgers U., Newark, 1976—, now prof. chemistry. NSF grantee, 1979-83; NIH grantee, 1982-85. Mem. Am. Chem. Soc., Biophys. Soc. Subspecialties: Biophysical chemistry; Membrane biology. Current work: Molecular vibrational spectroscopy; biological membrane structure. Office: Dept Chemistry Rutgers U 73 Warren St Newark NJ 07102 Home: 505 Hillside Terr South Orange NJ 07079

MENDELSON, NEIL HARLAND, geneticist, microbiologist, educator; b. N.Y.C., Nov. 15, 1937; s. Michael and Rose (Kutner) M.; m. Joan Rintel, July 30, 1959; children: Debora Cybelle, Marie Dianna. B.S., Cornell U., N.Y., 1959; Ph.D., Ind. U., 1964. NSF postdoctoral fellow Med. Research Council, London, 1965-66; asst. prof. biology U. Md., Catonsville, 1967-69; asso. prof. micro/med. tech. U. Ariz., Tucson, 1969-74, prof., 1974-78, prof. cellular and developmental biology, 1978—, head dept., 1979-83; vis. scientist Inst. Pasteur, Paris, 1976-77, U. Lausanne, Switzerland, 1979. Contbr. articles to sci. jours. Bd. dirs. Tucson Jr. Strings, 1976-79; mem. So. Ariz. Symphony, treas., 1981-83, pres., 1983—. Served to capt., Chem. Corps U.S. Army, 1963-65. NIH grantee, 1971-82; NSF grantee, 1968-71, 82—. Mem. Am. Soc., Microbiolgy, AAAS, Genetics Soc. Am. Jewish. Subspecialties: Genetics and genetic engineering (biology); Microbiology. Current work: Genetic regulation of cell growth and division; helical macrofibers; helix clock theory; cell surface and morphology; minicell molecular biology; stress-strain deformations in cell regulation. Home: 7031 Katchina Ct Tucson AZ 85715 Office: U Ariz Tucson AZ 85721

MENDELSON, THEA, biologist, educator, researcher; b. Durham, N.C., May 5, 1935; d. Carl Williams and Mable Dorothy (Gaiser) Borgmann; m. Martin Mendelson, Mar. 29, 1958 (div.); children: Kathryn, Christopher. B.A., Swarthmore Coll., 1957; postgrad., Cornell U., 1957-58; Ph.D., SUNY-Stony Brook, 1978. Muscular Dystrophy fellow Cornell U., Ithaca, N.Y., 1977-80; asst. prof. biology Wells Coll., Aurora, N.Y., 1980—; lectr. Occidental Coll., Los Angeles, 1958-60. Contbr. articles to profl. jours. Recipient Nat. Research Service award NIH, 1980. Mem. AAAS, Soc. Neurosci., N.Y. Acad. Scis., NOW, Sierra Club, Sigma Xi. Democrat. Subspecialties: Neurobiology; Regeneration. Current work: Neurotransmitter uptake at crayfish neuro-muscular junction; role of glial cells.

MENDENHALL, GEORGE DAVID, chemistry educator, researcher; b. Iowa City, Iowa, Feb. 12, 1945; s. George Emery and Eathel (Tidrick) M.; m. Yvonne Astrid Hendricks, Feb. 16, 1972; children: Catherine, Stuart. B.S., U. Mich., 1966; Ph.D., Harvard U., 1970. With NRC Can., Ottawa, Ont., 1970-72, SRI Internat., Menlo Park, Calif., 1972-73; staff scientist Battelle-Columbus, Ohio, 1973-80; assoc. prof. chemistry Mich. Tech. U., Houghton, 1980—. Mem. Am. Chem. Soc. Subspecialties: Organic chemistry; Photochemistry. Current work: Chemiluminescence, oxidation phenomena, ozone chemistry, polymer stability, free radicals. Home: Mich Tech U Dept Chemistry and Chem Engring Houghton MI 49931

MENENDEZ, MANUEL GASPAR, physicist, educator; b. N.Y.C., June 15, 1935; s. Edilio and Rita (Garciá) M.; m. Sandra Guerra, June 22, 1937; children: Mark A., Myra A., Natalie S. B.Ch.E., U. Fla., 1958, Ph.D., 1963. Postdoctoral fellow Oak Ridge Nat. Lab., 1963-65; atomic physicist Nat. Bur. Standards, Washington, 1965-66; staff scientist Martin-Marietta, Orlando, Fla., 1966-69; adj. prof. Rollins Coll., 1967-69; prof. physics U. Ga.; cons. in field. Contbr. articles to profl. jours. Served to 1st lt. USAR, 1958-63. Mem. Am. Phys. Soc. Roman Catholic. Subspecialty: Atomic and molecular physics. Current work: The study of ionization mechanisms in electron and ion-atom collisions. Office: Dept Physics and Astronomy U Ga Athens GA 30602

MENGELING, WILLIAM LLOYD, veterinary virologist; b. Elgin, Ill., Apr. 1, 1933; s. William Paul and Blanche Joyce (Wormwood) M.; m. Barbara Ann Kethcert, Aug. 23, 1958; children: Michelle A., Michael W. Student, Wis. State Coll.-Platteville, 1951-53; B.S., Kans. State U., 1958, D.V.M., 1960; M.S., Iowa State U., 1966, Ph.D., 1969. Practice vet. medicine, specializing in small animal practice, Albuquerque, 1960-61; research veterinarian Nat. Animal Disease Ctr., Agrl. Research Service, Dept. Agr. Ames, Iowa, 1961-69, research leader respiratory diseases of swine, 1969-76; chief Virological Research Lab., 1976—; adj. prof. microbiology Coll. Vet. Medicine, Iowa State U.; cons. prodn. viral vaccines of vet. importance. Contbr. numerous articles to profl. jours.; bd. sci. reviewers: Am. Jour. Vet. Research, 1980—; co-editor: Diseases of Swine, 1981. Served to sgt. U.S. Army, 1953-55. Mem. Am. Coll. Vet. Microbiologists, AVMA, Am. Soc. Virology, Conf. Research Workers in Animal Diseases. Lodge: Kiwanis. Subspecialties: Virology (veterinary medicine); Genetics and genetic engineering. Current work: Virology, especially virus-induced diseases of reproduction; research administration; research on pathogenesis and molecular virology.

MENINO, ALFRED RODRIGUES, JR., reproductive physiology educator; b. Hilo, Hawaii, Sept. 17, 1954; s. Alfred R. and Gertrude V. (Dias) M.; m. Rebecca Suzanne Whelchel, July 7, 1979; 1 dau., Holly Marie Ann. B.A. in Biology, U. Hawaii, 1976; M.S. in Animal Sci, Wash. State U., 1978, Ph.D., 1981. Teaching asst. in animal sci., research asst. in animal sci. Wash. State U., 1977-81; asst. prof. animal sci. U. Hawaii-Hilo, 1981—; cons. in field. Contbr. articles on reproductive physiology to profl. jours. Mem. Sci. Research Soc., Am. Soc. Animal Sci., Internat. Embryo Transfer Soc., Soc. for Study of Reprodn., AAAS, Phi Kappa Phi, Sigma Xi. Roman Catholic. Subspecialties: Animal physiology; Reproductive biology. Current work: Mammalian embryology with emphasis on the domestic species. Office: 1400 Kapiolani St Hilo HI 96720

MENLOVE, HOWARD OLSEN, technical administrator, researcher; b. Mayfield, Utah, Oct. 23, 1936; s. Roy and Lucille (Olsen) M.; m. Frances Lee, Jne 4, 1958; children: Stephen, Lynelle, Spencer, Lauren. B.S., U. Calif.-Berkeley, 1959; M.S., U. Mich., 1961; Ph.D., Stanford U., 1966. Assoc. engr. Boeing Co., Seattle, 1956-59; head teaching fellow U. Mich., Ann Arbor, 1959-61; research asst. Inst. Sci. and Tech., Ann Arbor, 1961-62; scientist Lockheed Missiles & Space Corp., Palo Alto, Calif., 1962-66; vis. scientist Kernforschungszentrum, Karlsruhe, W.Ger., 1966-67; project mgr. Los Alamos Nat. Lab.-U. Calif., 1967—; cons. IAEA, 1971-80. Recpiient Eckhart prize Stanford U., 1967; Radiation in Industry award Am. Nuclear Soc., 1981; Disting. Performance award Los Alamos Nat. Lab., 1982. Mem. Am. Phys. Soc., Am. Nuclear Soc., Inst. Nuclear Materials Mgmt., Tau Beta Pi. Subspecialties: Nuclear engineering; Nuclear fission. Current work: Nondestructive assay research and development for domestic and international safeguards. Patentee subthreshold neutron interrogator; spent fuel assay device. Home: 2880 Arizona St Los Alamos NM 87544 Office: PO Box 1663 Los Alamos NM 87545

MENNINGER, RICHARD PRICE, physiology educator, researcher; b. Columbus, Ohio, Mar. 18, 1937; s. Robert Ingersol and Miriam Louise (Koch) M.; m. Nancy Carol Agee, June 13, 1964; children: Michael A., Laura Anne, Anne Elizabeth. Student, Kenyon Coll., 1955-58; B.S., Ohio State U., 1959; Ph.D., U. Ky., 1971. Research asst. Ind. U. Cardiovascular Pulmonary Lab., Wright-Patterson AFB, Ohio, 1962-67, U. Ky. Physiology and Biophysics Lab., Lexington, 1967-71; asst. prof. physiology U. South Fla., Tampa, 1971-76, assoc. prof., 1976-82; prof. physiology Mercer U. Sch. Medicine, Macon, Ga., 1982—. Contbr. articles on physiology to profl. jours. Served to 1st lt. Air NG, 1961-67. Fla. Heart Assn. grantee, 1972-75; NIH grantee, 1974-79, 80-84. Mem. Am. Physiol. Soc. Mem. Ch. of Christ. Subspecialties: Physiology (medicine); Neuroendocrinology. Current work: The integration of information from cardiac, arterial and osmotic receptors in the control of hypothalamic neurons which release vasopressin using single neuron recordings and hormone assay. Home: 109 Tennyson Trail Macon GA 31210 Office: Mercer Univ Sch Medicine 1550 College St Macon GA 31207

MENNINGER, WILLIAM WALTER, psychiatrist; b. Topeka, Oct. 23, 1931; s. William Claire and Catharine Louisa (Wright) M.; m. Constance Arnold Libbey, June 15, 1953; children: Frederick Prince, John Alexander, Eliza Wright, Marian Stuart, William Libbey, David Henry. A.B., Stanford U., 1953; M.D., Cornell U., 1957. Diplomate: Am. Bd. Psychiatry and Neurology, Am. Bd. Forensic Psychiatry. Intern Harvard Med. Service, Boston City Hosp., 1957-58; resident in psychiatry Menninger Sch. Psychiatry, 1958-61; chief med. officer, psychiatrist Fed. Reformatory, El Reno, Okla., 1961-63; asso. psychiatrist Peace Corps, 1963-64; staff psychiatrist Menninger Found., Topeka, 1965—, coordinator for devel., 1967-69, dir. law and psychiatry, 1981—; clin. supr. Topeka State Hosp., 1969-70, sect. dir., 1970-72, asst. supt., clin., dir. residency tng, 1972-81; adj. prof. Washburn U., Wichita State U.; instr. Topeka Inst. for Psychoanalysis; mem. adv. bd. Nat. Inst. Corrections, 1975—, chmn., 1980—; cons. U.S. Bur. Prisons; mem. Fed. Prison Facilities Planning Council, 1970-73. Syndicated columnist: In-Sights, 1975-83; Author: Happiness Without Sex and Other Things Too Good to Miss, 1976, Caution: Living May Be Hazardous, 1978, Behavioral Science and the Secret Service, 1981; Editor: Psychiatry Digest, 1971-74; mem. bd. editors, 1974-78; Contbr. chpts. to books, articles to profl. jours. Mem. nat. health and safety com. Boy Scouts Am., 1970—, chmn., 1980—, mem. nat. exec. bd., 1980—; mem. Kans. Gov.'s Adv. Commn. on Mental Health, Mental Retardation and Community Mental Health Services, 1983—; bd. dirs. Nat. Com. for Prevention Child Abuse, 1975—; mem. nat. adv. health council HEW, 1967-71; mem. Nat. Commn. Causes and Prevention Violence, 1968-69, Kans. Gov.'s Penal Planning Council, 1970. Served with USPHS, 1959-64. Fellow ACP, Am. Psychiat. Assn., Am. Coll. Psychiatrists; mem. AMA, Group for Advancement of Psychiatry (chmn. com. mental health services 1974-77), Inst. Medicine Nat. Acad. Scis., Am. Psychoanalytic Assn., Am. Acad. Psychiatry and Law, AAAS. Subspecialty: Psychiatry. Current work: Preventive psychiatry, appllications in corrections, law and psychiatry, public education. Office: Menninger Found Box 829 Topeka KS 66601

MENON, MADHAVAN KRISHNA, pharmacologist, researcher; b. Trivandrum, India, Dec. 10, 1936; came to U.S., 1969, naturalized, 1979; s. Krishnapillay Madhavan Nayar and Mookambika Rajamma M.; m. Nirmala Krishna Menon, Aug. 27, 1966; children: Murali Madhav, Anupama. B.S., U. Coll., Trivandrum, 1954; B.Pharm., Madras (India) Med. Coll., 1959; M.S., Sawai Man Singh Med. Coll., Jaipur, India, 1964, Ph.D., 1969. Research fellow, research asst. Sawai Man Singh Med. Coll., 1960-64, asst. research officer, 1964-66; asst. prof. U. Saugar, India, 1966-69; postdoctoral research fellow VA Med. Center, Sepulveda, Calif., 1969-72, pharmacologist, 1974-79, chief, 1979—; research pharmacologist Clarke Inst. Psychiatry, Toronto, Ont., Can., 1972-73; pharmacologist dept. pharmacology Sch. Medicine, U. Calif., San Francisco, 1973-74; asso. research psychopharmacologist dept. psychiatry and biobehavioral scis. Neuropsychiat. Inst., UCLA, 1979—. Contbr. articles to sci. jours. Indian Council Med. Research research fellow, 1960-63; Found.'s Fund for Research in Psychiatry fellow, 1969-72; Found. for Vol. Control of Psychophysiol. States fellow, 1972-73. Mem. Am. Soc. Pharmacology and Exptl. Therapeutics, Soc. Neurosci., Soc. Biol. Psychiatry, AAAS, Western Pharmacology Soc., Brit. Brain Research Assn. (hon.), European Brain and Behavior Soc. (hon.). Subspecialties: Psychopharmacology; Neuropharmacology. Current work: Research on etiological and therapeutic aspects of schizophrenia and movement disorders such as Parkinson's disease, tardive dyskinesias, myoclonus, Huntington's diaease and Tourette's Syndrome. Research on pharmacological, neurochemical and behavioral aspects of drugs of abuse, temperature regulation, electromyography and drug-receptor interactions. Patentee antiparkinsonism drug. Office: 16111 Plummer St Sepulveda CA 91343

MENSE, ALLAN TATE, research scientist; b. Kansas City, Mo., Nov. 29, 1945; s. Martin Conrad and Nancy (Tate) M.; m. Ramona Carol Stelford, Aug. 26, 1983; children: by previous marriage: Melanie Georgia, Eileen Madelaine. B.S., U. Ariz., 1968, M.S., 1970; Ph.D., U. Wis., 1977. Research scientist, fusion energy div. Oak Ridge Nat. Lab., 1976-79; sci. cons., sci. and tech. com. U.S. Ho. of Reps., Washington, 1979-81; sr. scientist, fusion sci. dept. McDonnell Douglas Astronautics Co., St. Louis, 1981—; adj. prof. physics U. Mo., St. Louis, 1982—. Contbr. articles to profl. jours. Served to capt. USAF, 1975. Mem. Am. Phys. Soc., Am. Nuclear Soc., Am. Vacuum Soc., IEEE (sr.), Sigma Xi, Theta Tau. Subspecialties: Nuclear fusion; Plasma engineering. Current work: Supervise and perform state-of-the-art modeling of fusion reactors, their physics and engring. design requirements. Home: 6030A Washington Ave Saint Louis MO 63112 Office: McDonnell Douglas Astronautics Co Dept E-240 PO Box 516 Saint Louis MO 63166

MENSH, IVAN NORMAN, psychology educator; b. Washington; s. Shea Jacob and Rose Clayman M.; m.; 1 dau., Frances Levitas. A.B., George Washington U., 1940, A..M., 1942; Ph.D., Northwestern U., 1948. Diplomate: Am. Bd. Profl. Psychology. Prof., head med. psychology dept. psychiatry Washington U. Sch. Medicine, St. Louis, 1948-58; prof., head div. med. psychology UCLA Sch. Medicine, Los Angeles, 1958—; cons. VA, Los Angeles, 1958—; dir. Am. Bd. Profl. Psychology, Washington, 1954-64. Author: Clinical Psychology: Science and Profession, 1966, Professional Obligations and Approaches to the Aged, 1974; contbr. articles in field to profl. jours. Served to capt. USNR, 1943—. Recipient Silver Psi award Calif. State Psychol. Assn. Mem. Am. Psychol. Assn., AAUP, Nat. Register Health Service Providers in Psychology (mem. council 1964—). Subspecialties: Behavioral psychology; Gerontology. Current work: Stress events and responses to life experiences in the elderly. Office: UCLA Sch Medicine 760 Westwood Plaza Los Angeles CA 90024

MENZIES, ROBERT ALLEN, biochemist, geneticist, biol. oceanographer, educator; b. San Francisco, Nov. 13, 1935; s. Paul Edward and Rachel (Genden) M.; m. Esther Diaz Alvarez, June 22, 1958; children: Edward Aaron, Roland Asa, Sarah Angela. B.S., U. Fla., 1960, M.S. (Woods Hole Marine Biol. Lab. scholar), 1962; Ph.D., Cornell U., 1966. Research chemist VA Hosp., Balt., 1965-67; asst. prof. biochemistry La. State U. Med. Center, New Orleans, 1967-73; asst. prof. biochemistry and oceanography, asso. prof., then prof. Nova U., Ft. Lauderdale, Fla., 1973—; cons. in field. Contbr. articles on biochemistry and oceanography to profl. jours. Served with U.S. Army, 1953-56. NIH grantee, 1971-76; NOAA grantee, 1976-80; NSF grantee, 1980-83. Mem. Am. Soc. Biol Chemists, AAAS, N.Y. Acad. Scis., Am. Fisheries Soc., Gulf and Caribbean Fisheries Inst., Am. Chem. Soc., Am. Soc. Cell Biology, Biophys. Soc. Jewish. Subspecialties: Biochemistry (biology); Oceanography. Current work: Biochemistry, genetics, evolution, biofouling and mariculture.

MERCER, JAMES WAYNE, geological corporation executive, educator, consultant; b. Panama City, Fla., Dec. 23, 1947; s. Wayne James and Hattie Margaret (Haverstick) M.; m. Barbara Joyce Crowley, Sept. 5, 1969 (div. 1980); m. Maria Alice Wieckowski, June 20, 1980; 1 child, Bobby Jan Gac. A.S., Gulf Coast Jr. Coll., 1967; B.S. summa cum laude, Fla. State U., 1969; M.S., U. Ill., 1971, Ph.D., 1973. Cert. profl. geologist. Hydrologist U.S. Geol. Survey, Reston, Va., 1971-79; pres. GeoTrans, Inc., Reston, 1979—; assoc. professorial lectr. George Washington U., 1979—. Author: Ground-Water Modeling, 1981; contbr. chpts. to books in field; editor: Role of the Unsaturated Zone in Radioactive and Hazardous Waste Disposal, 1983. Chevron sr. scholar Fla. State U.; NDEA Title IV fellow U. Ill. Fellow Geol. Soc. Am. (editorial bd. Geology 1979-82); mem. Soc. Petroleum Engrs., Am. Geophys. Union (unsaturated zone com. 1978—), ASCE, Nat. Water Well Assn. (editorial bd. Ground Water 1980—), Phi Beta Kappa. Subspecialties: Ground water hydrology; Hydrogeology. Current work: Numerical simulation, ground-water flow, solute and heat transport, multiphase flow in porous media, hazardous waste disposal, radioactive waste disposal, sea water intrusion, geothermal reservoir analysis. Office: GeoTrans Inc PO Box 2550 Reston VA 22090 Home: 2025 Winged Foot Ct Reston VA 22091

MERCER, KERMIT RAY, biophysics educator; b. Brockport, N.Y., June 1, 1933; s. Harold R. and Elma H. (Case) M.; m. Janet L. Hollinger, Feb. 28, 1953; children: Deborah L., Susan R. B.S., SUNY-Brockport, 1970. 2Electronics asst. Delco div. Gen. Motors Corp., Rochester, N.Y., 1951-53; research assoc. Gen. Dynamics, Rochester, 1957-70; assoc. tchr. Monroe County BOCES, Spencerport, N.Y., 1970-72; assoc. in biophysics U. Rochester, 1972—; owner Electro-Sci., Brockport, N.Y., 1963—. Chmn. Clarkson (N.Y.) Bd. Appeals, 1963—. Served with USAF, 1953-57. Mem. IEEE, U.S. Power Squadron. Club: Oak Orchard Yacht. Subspecialties: Electron Spin resonance; Biophysics (physics). Current work: Radiation damage to DNA constituents using ESR techniques involving magnetic fields, microwaves and cryogenics. Home: 7816 Ridge Rd Brockport NY 14420 Office: Univ Rochester Dept Biophysics Rochester NY 14642

MERCER, L. PRESTON, biochemistry educator, researcher; b. Ft. Worth, Jan. 16, 1941; s. Leonard P. and Margie Mae (Miller) M.; m. Diane Cottingham, Feb. 26, 1962; children: Cindy, Tim, Megan. B.S., U. Tex., 1968; Ph.D., La. State U., 1971. Instr. U. South Ala., Mobile, 1973-74, asst. prof., 1974-77; asst. prof. biochemistry Oral Roberts U., Tulsa, 1977-80, assoc. prof., 1980—, chmn. dept., 1981—; speaker Christian Broadcasting Network; lectr. on sci. in Bible, chs. in, Okla. Deacon Cottage Hill Baptist Ch., Mobile. NSF trainee, 1970; NIH fellow, 1971. Mem. Am. Chem. Soc., Am. Inst. Nutrition, Soc. Nutrition Edn., Nutrition Today Soc., AAAS. Subspecialties: Biochemistry (medicine); Nutrition (medicine). Current work: Nutritional biochemistry, mathematical models of nutritional responces, appetite control. Home: 6925 E 92d St S Tulsa OK 74133 Office: Oral Robert- U 7777 S Lewis Tulsa OK 74171

MERCHANT, HOWARD CARL, engineering consulting company executive, educator; b. Mt. Vernon, Wash., Jan. 9, 1935; s. William Hollis and Marie Louise (Allenbach) M.; m. Corolyn L. McMinimee, June 11, 1960; children: Mark Whitney, Martha Marie. B.S. summa cum laude, U. Wash.-Seattle, 1956; S.M., MIT, 1957; Ph.D., Calif. Inst. Tech., 1960. Asst. prof. mech. engring. U. Wash., 1961-63; group leader Sandia labs., Livermore, Calif., 1963-65; head vulnerability div. Physics Internat., San Leandro, Calif., 1965-67; prof. mech. engring. U. Wash., 1967-82, affiliate prof., 1982—; pres. MerEnCo, Inc., Bellevue, Wash., 1981—; adj. prof. geophysics U. Wash., 1974-82; cons. in field. Editor, author: Productive Applications of Vibrations, 1982; contbr. articles to profl. jours. Precinct committeeman Republican Party, Bothell, 1972-76, del. state conv., 1974. NSF fellow, 1956-59; Woodrow Wilson fellow, 1960; Battelle Lab. fellow, 1973-74; recipient Charles Bassett III award Instrument Soc. Am., 1975. Mem. ASME (chmn. western Wash. sect. 1981-82, chmn. awards com. Region VIII 1983—), Earthquake Engring. Research Inst., Acoustical Soc. Am., Am. Soc. Naval Architects and Marine Engrs., Marine Tech. Soc., Nat. Soc. Profl. Engrs., Enological Soc. Pacific Northwest. Club: Mountaineers. Subspecialties: Mechanical engineering; Theoretical and applied mechanics. Current work: Consulting in vibrations, earthquake engineering, acoustics, project management and design; research in mechanical vibration, earthquake engineering. Office: 16334 84th Ave NE Bothell WA 98011

MERCHANT, MYLON EUGENE, engineer, physicist; b. Springfield, Mass., May 6, 1913; s. Mylon Dickinson and Rebecca Chase (Currier) M.; m. Helen Silver Bennett, Aug. 4, 1937; children—Mylon David, Leslie Ann Merchant Alexander, Frances Sue Merchant Jacobson. B.S. magna cum laude, U. Vt., 1936, D.Sc. (hon.), 1973, U. Cin., 1941, U. Salford, Eng., 1980. Research physicist Cin. Milacron, Inc., 1940-48, sr. research physicist, 1948-51, asst. dir. research, 1951-57, dir. phys. research, 1957-63, dir. sci. research, 1963-69, dir. research planning, 1969-81, prin. scientist, mfg. research, 1981-83; dir. advanced mfg. research Metcut Research Assocs., Inc., 1983—; adj. prof. mech. engring. U. Cin., 1964-69; vis. prof. mech. engring. U. Salford, Eng., 1973—. Contbr. articles to profl. jours. Bd. dirs. Dan Beard council Boy Scouts Am., 1967-80, pres.'s council, 1980—. Recipient Georg Schlesinger prize City of Berlin, 1980; Otto Benedikt prize Hungarian Acad. Scis., 1981. Fellow Am. Soc. Lubrication Engrs. (pres. 1952-53), Am. Soc. Metals, Ohio Acad. Sci.; mem. Nat. Acad. Engring., Soc. Mfg. Engrs. (hon.; pres. 1976-77), ASME (hon.), Internat. Instn. Prodn. Engring. Research (hon.; pres. 1968-69), Engring. Soc. Cin. (pres. 1961-62), Fedn. Materials Socs. (pres. 1974), Phi Beta Kappa, Sigma Xi, Tau Beta Pi. Current work: Research on computer integrated manufacturing (computer automation, optimization and integration of total system of manufacturing). Research on systems approach to mfg. Home: 3709 Center St Cincinnati OH 45227 Office: 4701 Marburg Ave Cincinnati OH 45209

MERCIER, PAUL, oral and maxillofacial surgeon, researcher; b. Montreal, Que., Can., Mar. 3, 1936; s. Oscar and Jeanne (Bruneau) M.; m. Andree Fernet, Aug. 27, 1960; children: Bruno, Brigitte. B.A., U. Montreal, 1956, D.D.S., 1961; diploma, U. Pa., 1964. Diplomate: Am. Bd. Oral and Maxillofacial Surgeons. Asst. prof. oral surgery Faculty Dental Medicine, U. Montreal, 1964-69, assoc. prof., 1969-73; research assoc. St.-Mary's Hosp., Montreal, 1974—; examiner Royal Coll. Dentists, 1969—; Nat. Bd. Dental Examiners, Can., 1970-73; dir. Clinic for Atrophy of Maxillae, 1974—. Editorial bd.: Internat. Jour. Oral Surgery, 1980. Fellow Royal Coll. Dentists Can.; mem. Internat. Assn. Dental Research, Can. Assn. oral and Maxillofacial Surgeons (pres. 1973), Que. Assn. Oral and Maxillofacial Surgeons, Am. Assn. Oral and Maxillofacial Surgeons. Subspecialty: Oral and maxillofacial surgery. Current work: Primary activity centered around the rehabilitation of severe masticatory deficits due to atrophy of jaws. Office: St-Mary's Hosp 3830 Ave Lacombe Montreal Canada H3T 1M5

MEREDITH, DAVID BRUCE, engineering technology educator; b. Dover, Ohio, July 11, 1950; s. C. Curtis and Dorothy Emma (Weaver) M.; m. Linda Marie Singer, May 30, 1978; children: Mark William, Scott David. B.S.M.E., Ohio State U., 1972; M.S. in Mech. Engring, Colo. State U., 1978. Tech. service engr. in fired equipment and environ. control systems Procter & Gamble, Cin., 1972-76; asst. prof. gen. engring., chmn. solar heating and cooling tech. program Pa. State U., Uniontown, 1979—. Eastman Kodak fellow, 1976-78. Mem. Am. Solar Energy Soc., Internat. Solar Energy Soc., ASHRAE, Am. Soc. Engring. Edn. Methodist. Subspecialties: Solar energy; Energy conservation. Current work: Home air infiltration, solar cooling methods, solar design methods. Office: PO Box 519 Uniontown PA 15401

MEREDITH, ORSELL MONTGOMERY, research administrator; b. Jamestown, N.Y., Oct. 19, 1923; s. Orsell Montgomery and Bernardine Elva (Goggin) M.; m. Martha Linnea Helbon, Jan. 29, 1949; 1 son, Michael Wayne. B.S., U. Chgo., 1948; M.S., U. So. Calif., 1951, Ph.D., 1953; M.S., Am. U., 1975. Asst research pharmacologist, asst. prof. biophysics and nuclear medicine UCLA, 1953-61; research scientist Lockheed Palo Alto Research Labs., Calif., 1961-66; ops. research analyst/radiation biologist Naval Radiol. Def. Lab., 1966-69, Naval Ordnance Lab., 1969-75; exec. sec. grants rev. Nat. Cancer Inst., Bethesda, Md., 1976—. Contbr. articles to profl. publs. Served to lt. (j.g.) USNR, 1942-46. Mem. Radiation Research Soc., Am. Soc. Pharmacology and Exptl. Therapeutics, N.Y. Acad. Scis., Soc. Nuclear Medicine, Health Physics Soc. Subspecialties: Biophysics (physics); Operations research (engineering). Current work: Research administration. Home: PO Box 135 Rockville MD 20901 Offic: 5333 Westbard Ave Room 822 Bethesda MD 20205

MEREDITH, RUBY FRANCES, research biologist; b. Sedalia, Mo., Feb. 6, 1948; d. Russell R. and Eunice M. (Curry) M. B.A., U. Mo., 1969; M.A., Ind. U., 1971, Ph.D., 1974. Asso. instr. Ind. U., 1972, research asst., 1973; asst. prof. Baylor U., 1974; postdoctoral fellow Allegheny Gen. Hosp., Pitts., 1976-78; asst. scientist Cancer Research Labs., 1978—; also tchr., cons. Editor: (with Okunewick) Graft-versus-Leukemia in Man and Animal Models, 1981; also articles and monographs. Recipient Harold C. Bold award Am. Inst. Biol. Scis., 1974; Leukemia Soc. Am. fellow, 1977; award for outstanding research Health Research and Services Found. United Way of Pitts., 1978; Landacre Day research award Ohio State U., 1982. Mem. Genetics Soc. Am., Radiation Research Soc., Internat. Soc. Exptl. Hematology, Internat. Assn. Comparative Research on Leukemia and Related Diseases. Subspecialties: Marrow transplant; Transplantation. Current work: Murine model studies of immunobiology of hematoporitic tissue transplatation and ablation of marrow in preparation for marrow transplantation. Office: Cancer Research Labs Allegheny Gen Hosp 320 E North Ave Pittsburgh PA 15212

MERENDA, PETER FRANCIS, psychology and statistics educator, researcher, consultant; b. Everett, Mass., July 18, 1922; s. Frank P. and Sarah (Lino) M.; m. Rose Cafasso, Aug. 31, 1946; children: Anne, Rosemary, Pamela. B.S., Tufts U., 1947, M.A., 1948; C.A.S., Harvard U., 1951; Ph.D., U. Wis.-Madison, 1957. Cert. psychologist, R.I. Dir. guidance Southbridge Pub. Schs., Mass., 1949-50; research psychologist, dir. research U.S. Naval Examining Ctr., Great Lake, Ill., 1950-57; exec. assoc., research dir. Walte V. Clarke Assos., Inc., Providence, 1957-60; asst. prof. psychology U. R.I., 1960-62, assoc. prof., 1963-65, prof., 1965-68, prof. psychology and stats., 1968—, assoc. dean, 1965-68. Author: (with Lindeman) Educational Measurement, 1978, (with Lindeman and Gold) Introduction to Bivariate and Multivariate Analysis, 1980. Chmn., sec. R.I. Bd. Examiners in Psychology, 1978-82. Fulbright-Hays scholar, 1967-68, 74-75; NATO grantee, 1973-74; Pacific Cultural Found. grantee, 1980-81. Mem. Internat. Assn. Applied Psychology (pres. div. psychol. assessment 1982-86), New Eng. Psychol. Assn. (pres. 1982-83), R.I. Psychol. Assn. (pres. 1983—), Internat. Council Psychologists (pres. 1980-81, treas. 1977-79). Club: United Italian-American (dir. 1976—). Subspecialties: Measurement and evaluation; Statistics. Current work: Identification of talent in developing countries, identification of children: with learning problems, public image of international leaders. Home: 258 Negansett Ave Warwick RI 02888 Office: U Rhode Island Kingston RI 02881

MERENSTEIN, GERALD B., physician; b. Pitts., Feb. 14, 1941; s. Morris and Sarah (Shrinsky) M.; m. Barnetta Maryn, Aug. 21, 1960; children: Scott, Stacy, Ray. B.S., U. Pitts., 1962, M.D., 1966. Diplomate: Am. Bd. Pediatrics, Nat. Bd. Med. Examiners. Intern Fitzsimons Army Med. Center, Aurora, Colo., 1966-67, resident in pediatrics, 1967-69, chief newborn service, 1971-79, asst. chief dept. pediatrics, 1973-79, chief dept. pediatrics, 1979—; fellow in neonatal-perinatal medicine Children's Hosp., San Francisco, 1969-71; clin. prof. pediatrics Uniformed Services U. Health Scis., Bethesda, Md., 1982—, U. Colo. Health Sci. Center, Denver, 1983—. Editor: Handbook of Neonatal Intensive Care, 1983; contbr. chpts. to books, articles to profl. jours. Chmn. med. adv. com. Metro Denver March of Dimes, 1976—; chmn. Colo. Perinatal Care Council, 1976-79; mem. Ad Hoc Com. on Nursery Standards, State of Colo., 1975. Named Outstanding Man of Yr. Denver Jaycees, 1974; recipient Apgar award Denver March of Dimnes, 1979. Fellow Am. Acad. Pediatrics (chmn. dist perinatal sect. 1982—, fellow, mem. com. on fetus and newborn); mem. Western Soc. for Pediatric Research, Am. Fedn. for Clin. Research, Alpha Omega Alpha. Subspecialties: Pediatrics; Neonatology. Current work: Neonatal, maternal transport; polychythemia; transcutaneous monitoring; neonatal asphyxia. Home: 2472 Macon Way Aurora CO 80014 Office: Dept Pediatrics Fitzsimons Army Med Center Aurora CO 80045

MERGLER, HARRY WINSTON, educator; b. Chillicothe, Ohio, June 1, 1924; s. Harry Franklin and Letitia (Walburn) M.; m. Irmgard Erna Steudel, June 22, 1948; children—Myra A. L., Marcia B. E., Harry F. B.S., M.I.T., 1948; M.S., Case Inst. Tech., 1950, Ph.D., 1956. Aero. research scientist NACA, 1948-56; mem. faculty Case Inst. Tech., 1957—, prof. engring., 1962—, Leonard Case prof. engring., 1973—; dir. Digital Systems Lab., 1959—; vis. scientist, USSR, 1958; vis. prof. Norwegian Tech. U., 1962; cons. to industry, 1957—; editor Control Engring. mag., 1956—; pres. Digital/Gen. Corp., 1968-72; cons. Exploratory Research div. NSF. Author: Digital Systems Engineering, 1961, also articles, chpts. in books. Served with AUS, 1942-45. Recipient Gold medal for sci. achievement Case Inst. Tech., 1980. Fellow IEEE (pres. Indsl. Electronic Soc. 1977-79, recipient Lamme medal 1978); mem. Cleve. Engring. Soc., N.Y. Acad. Scis., Nat. Acad. Engring., Sigma Xi, Tau Beta Pi, Theta Tau, Pi Delta Epsilon, Zeta Psi, Blue Key. Subspecialties: Computer engineering; Robotics. Current work: Robotic vision, in-process ganging. Home: 1525 Queen Anne's Gate Westlake OH 44145

MERIGAN, THOMAS CHARLES, JR., physician, med. researcher, educator; b. San Francisco, Jan. 18, 1934; s. Thomas C. and Helen M. (Greeley) M.; m. Joan Mary Freeborn, Oct. 3, 1959; 1 son, Thomas Charles III. B.A. with honors, U. Calif., Berkeley, 1955, M.D., 1958. Diplomate: Am. Bd. Internal Medicine. Intern in medicine 2d and 4th Harvard med. services Boston City Hosp., 1958-59, asst. resident medicine, 1959-60; clin. asso. Nat. Heart Inst., NIH, Bethesda, Md., 1960-62; asso. Lab. Molecular Biology, Nat. Inst. Arthritis and Metabolic Diseases, NIH, 1962-63; practice medicine specializing in internal medicine and infectious diseases, Stanford, Calif., 1963—; asst. prof. medicine Stanford U. Sch. Medicine, 1963-67, asso. prof. medicine, 1967-72, head div. infectious diseases, 1966—, prof. medicine, 1972—, George E. and Lucy Becker prof. medicine, 1980—; dir. Diagnostic Microbiology Lab., Univ. Hosp., 1966-72, Diagnostic Virology Lab., 1969—; hosp. epidemiologist; mem. microbiology research tng. grants com. NIH, 1969-73, virology study sect., 1974-78; cons. antiviral substances program Nat. Inst. Allergy and Infectious Diseases, 1970—; mem. Virology Task Force, 1976-78, bd. sci. counselors, 1980—; mem. U.S. Hepatitis panel U.S. and Japan Cooperative Med. Sci. Program, 1979—; co-chmn. interferon evaluation Group Am. Cancer Soc., 1978-81, mem. adv. com. J.A. Hartford Found., 1980—; mem. Albert Lasker awards jury. Contbr. numerous articles on infectious diseases, virology and immunology to sci. jours.; editor: Antivirals with Clinical Potential, 1976, Antivirals and Virus Diseases of Man, 1979, Regulatory Functions of Interferon, 1980, Interferons, 1982; asso. editor: Virology, 1975-78; co-editor: monograph series Current Topics in Infectious Diseases, 1975—; editorial bd.: Archives Internal Medicine, 1971, Jour. Gen. Virology, 1972-77, Infection and Immunity, 1973—, Intervirology, 1973—, Virology, 1975-78, Proc. Soc. Expt. Biology and Medicine, 1978—, Reviews of Infectious Diseases, 1979—, Jour. Interferon Research, 1980—, Antviral Research, 1980—, Jour. Antimicrobial Chemotherapy, 1981—, Molecular and Cellular Biochemistry, 1982—, AIDS Research, 1983—. Recipient Borden award for Outstanding Research Am. Assn. Med. Colls., 1973; Guggenheim Meml. fellow, 1972. Mem. Assn. Am. Physicians, Western Assn. Physicians, Am. Soc. Microbiology, Am. Soc. Clin. Investigation (council 1977-80), Am. Assn. Immunologists, Am. Fedn. Clin. Research, Western Soc. Clin. Research, Soc. Exptl. Biology and Medicine, Infectious Diseases Soc. Am.; mem. Am. Soc. Virology; Mem. Inst. Medicine; mem. Pan Am. Group for Rapid Viral Diagnosis; Mem. AMA; mem. Internat. Soc. Interferon Research (council 1983—); Mem. Calif. Med. Assn., Santa Clara County Med. Soc., Calif. Acad. Medicine, Royal Soc. Medicine, AAAS, Alpha Omega Alpha. Subspecialties: Virology (medicine); Infectious diseases. Current work: Host response to viral infections (including interferon); antivirals. Home: 148 Goya Rd Portola Valley CA 94025 Office: Div Infectious Diseases Stanford Univ Sch Medicine Stanford CA 94305

MERILAN, CHARLES PRESTON, dairy husbandry scientist; b. Lesterville, Mo., Jan. 14, 1926; s. Peter Samuel and Cleo Sarah (Harper) M.; m. Phyllis Pauline Laughlin, June 12, 1949; children—Michael Preston, Jean Elizabeth. B.S. in Agr, U. Mo., 1948, A.M., 1949, Ph.D., 1952. Mem. faculty U. Mo., Columbia, 1950—, prof. dairy husbandry, 1959—, chmn. dept., 1961-62; asso. dir. Mo. Agrl. Expt. Sta., 1962-63, asso. investigator space sci. research center, 1964-74, exec. sec., dir. grad. studies physiology area, 1969-72, chmn. univ. patent and copyright com., 1963-80. Served with USMC, 1944-45. Decorated Purple Heart. Mem. AAAS, Am. Chem. Soc., Am. Dairy Sci. Assn., Am. Soc. Animal Sci, IEEE (profl. group biomed. electronics), Soc. Cryobiology, Sigma Xi, Alpha Zeta, Gamma Sigma Delta, Phi Beta Pi. Subspecialties: Animal physiology; Biophysics (biology). Current work: Reproductive physiology, cell preservation, bioinstrumentation, biophyscal characterization of cellular response to microenvironmental parameters. Researcher on biol. material preservation. Home: 1509 Bouchelle Ave Columbia MO 65201 Office: Univ Missouri Columbia MO 65211

MERILO, MATI, research engineer, research institute administrator; b. Tallinn, Estonia, Jan. 23, 1944; came to U.S., 1977; s. Arkadi and Hermeline (Dampf) M.; m. Kathleen Lorraine Frail, Aug. 21, 1977; children: Erik Grant, Aleksander Evan. B.Eng., McGill U., Montreal, Que., 1966; M.S., Case Inst. Tech., 1968, Ph.D., 1972. Research engr. Atomic Energy of Can. Ltd., Chalk River, Ont., 1971-77; project mgr. Electric Power Research Inst., Palo Alto, Calif., 1977—. Editor: Thermal-Hydraulics of Nuclear Reactors, 1983; contbr. articles to profl. jours. Mem. ASME, Am. Nuclear Soc. Subspecialties: Mechanical engineering; Fluid mechanics. Current work: Two phase flow and heat transfer, aerosol transport and deposition. Office: Electric Power Research Inst 3412 Hillview Ave Palo Alto CA 94303

MERKEL, CHRISTIAN GOTTFRIED, biochemistry educator; b. East Orange, N.J., Sept. 4, 1943; s. Gottfried F. and Winifred L. (Ruter) M.; m. Carolyn Ash, Sept. 9, 1972; children: Holly Sue, Stephen Christian, Sarah Marie. M.A., U. Calif., Santa Barbara, 1971; B.A., U. Cin., 1969, Ph.D. in Biochemistry, 1976. Research fellow div. biology Calif. Inst. Tech., 1976-79; cons. Jet Propulsion Lab., Pasadena, Calif., 1978-79; teaching asst., research U. Cin., 1971-76; asst. prof. biochemistry Coll. Osteo. Medicine Pacific, Pomona, Calif., 1979—. Contbr. articles to profl. jours. Served with USAF, 1963-66. Damon Runyon-Walter Winchell Cancer Fund grantee, 1976-77; HEW grantee, 1977-79. Mem. Sigma Xi. Subspecialties: Genetics and genetic engineering (medicine); Neurochemistry. Current work: Mechanism and molecular genetics of globoid cell leukodystrophy and other neurological disorders involving ganglioside metabolism. Home: 2741 N Lake St Altadena CA 91001 Office: Coll Osteo Medicine of the Pacific College Plaza Pomona CA 91766

MERKLEY, DAVID FREDERICK, veterinary surgeon, educator; b. Pipestone, Minn., Apr. 23, 1945; s. Robert Merle and Betty Jean (Robinson) M.; m. Donna Jean Sweeney, Aug. 14, 1970; children: Erica Lea, Leah Ann. B.A. in Math, U. S.D., 1967; D.V.M., Iowa State U., 1971; M.S., Mich. State U., 1974. Diplomate: Am. Coll. Vet. Surgeons. Instr. Mich. State U., 1971-75, asst. prof., 1975-79; assoc. prof. vet. surgery Iowa State U., Ames, 1979—. Mem. Am. Coll. Vet. Surgeons, Am. Assn. Vet. Clinician, Am. Animal Hosp. Assn., AVMA. Subspecialties: Surgery (veterinary medicine); Cancer research (veterinary medicine). Current work: Develop treatment modalities for spontaneous tumors in dogs and cats. Home: 2009 Buchanan Dr Ames IA 50010 Office: Department of Veterinary Clinical Science Iowa State University Ames IA 50011

MERLUZZI, VINCENT JAMES, immunologist, educator; b. Waterbury, Conn., Mar. 31, 1949; s. Paul John and Julia (Basile) M. B.A., Northeastern U., 1972; Ph.D. in Microbiology, Boston U., 1977. Cert. Am. Assn. Immunologists. Research asso. Boston U., 1977; postdoctoral fellow Sloan-Kettering Cancer Center, Rye, N.Y., 1977-79, research asso., 1979—; asst. prof. CUNY, 1982—. Contbr. articles to profl. jours. Mem. N.Y. Acad. Scis., Am. Assn. Immunologists, Sigma Xi. Roman Catholic. Subspecialties: Immunobiology and immunology; Cancer research (medicine). Current work: Specific repair of drug-induced immune cellular deficits; effect of immunosuppressive therapy on human CTL precursors; in vivo modification of host immunity after chemotherapy. Home: 1212 Franklin Ave Mamaroneck NY 10543 Office: 145 Boston Post Rd 3115W Rye NY 10580

MERRIAM, DANIEL F(RANCIS), geologist; b. Omaha, Feb. 9, 1927; s. Faye Mills and Amanda Frances (Wood) M.; m. Annie Laura Young, Feb. 12, 1946; children: Beth Ann Merriam Wissman, John Francis, Anita Pauline Merriam Howe, James Daniel, Judith Diane Merriam Palley. B.S., U. Kans., 1949, M.S., 1953, Ph.D., 1961; M.Sc., Leicester U., 1969, D.Sc., 1975. Geologist, Union Oil Co. Calif., Rocky Mountains and W. Tex., 1949-51, summer 1952; asst. instr. U. Kans., 1951-53, instr., 1954, research assoc., 1963-71; geologist Kans. Geol. Survey, 1953-58, div. head basic geology, 1958-63, chief geologic research, 1963-71; Jessie Page Heroy prof. geology Syracuse U., 1971-81, chmn. dept., 1971-80; disting prof. natural scis., chmn. dept. geology Wichita State U, Kans., 1981—; Endowment Assn. Disting. prof. natural sci., chmn. dept. geology Wichita State U., 1981—; vis. research scientist Stanford U., 1963; Fulbright-Hays sr. research fellow, U.K., 1964-65; dir. Am. Geol. Inst.'s Internat. Field Inst. to Japan, 1967; vis. prof. geology Wichita State U., 1968-70; Am. Geol. Inst. vis. geol. scientist, 1969; U.S. del. UNESCO-IUGS Internat. Geol. Correlations Program, Budapest, 1969; mem. U.S. Nat. Com. for IGCP, 1976-79, 81—, chmn., 1976-79; participant project COMPUTE, Dartmouth Coll., 1974; Esso disting. lectr. Earth Scis. Found., U. Sydney, Australia, 1979; vis. prof. Centre d'Informatique Geologique, Ecole des Mines de Paris, 1980; mem. U.S. Nat. Commn. for UNESCO, 1979—. Founder, editor-in-chief: Math. Geology, 1968-76; founder, editor-in-chief: Computers & Geoscis., 1975—; editorial cons.: Geosystems, 1971—; co-editor: Pacific Geology, 1971—; founder, editor: Syracuse U. Geology Contbns., 1973-81; editorial rev. bd.: Colo. Sch. Mines Quar., 1974—; editorial com.: Syracuse U. Press, 1978-81; nat. editor: The Compass (Sigma Gamma Epsilon), 1983—; corr.: Open Earth, 1978—; contbr. articles to sci. jours. Served with USNR, 1945-46. Recipient Erasmus Haworth Grad. award in Geology, U. Kans., 1955. Fellow Geol. Soc. Am., AAAA (chmn. sect. E. 1983-84); mem. Am. Assn. Petroleum Geologists (bus. com. 1956-57, research com. 1964-67, chmn. field trip research and coordination com. 1964-73, assoc. editor 1969-75, Matson Award com. 1970, acad. adv. com. 1973-77, ho. of dels. 1974-76, computer applications in geology com. 1971-81, membership com. 1980—, Mem., chmn. tech. program com. regional meeting 1983); mem. Soc. Econ. Paleontologists and Mineralogists (chmn. research group in computer tech. 1970-75, 82-83, nominating com. 1972-73, publs. com. 1980-83, chmn. 1981-82), Classification Soc. (chmn. membership com. 1968-71, dir. 1968-71, council 1968-72); Mem. AAAS (chmn. sect. nominating com. 1983-84), Geosci. Info. Soc. (program com. 1980-82); mem. Internat. Assn. Math. Geology (council 1968—, sec.-gen. 1972-76, pres. 1976-80, William Christian Krumbein medal 1981, Mem., chmn. Pres.'s Prize com. 1981, others, Wm. Christian Krumbein medal 1981); mem. N.Y. State Geol. Assn. (exec. sec. 1972-77, pres. 1977-78, dir. 1978-82), Rocky Mountain Assn. Geologists (research com. 1966-69), Kans. Acad. Sci. (program chmn. geology sect. 1959), Geologists Assn. (Eng.) (field trip dir. 1965), CODATA (ad hoc com. on publs. 1980), Internat. Union Geol. Scis. (chmn. adv. bd. on publs. 1980—), Kans. Geol. Soc. (mem. coms., dir. 1964), Leicester Geol. Soc. (hon. life), Nat. Assn. Geology Tchrs., Sigma Xi, Sigma Gamma Epsilon (Alpha chpt. pres. 1952-53); Mem. Phi Kappa Phi. Subspecialties: Geology; Fuels. Current work: Developing quantitative techniques for stratigraphic correlation, analyzaing cyclic sediments, and plains-type folding; defining similarity functions for comparing thematic maps; study recent sediments in Florida Bay and their analogies in the ancient rock record, especially in the midcontinent. Office: Dept Geology Wichita State Univ Wichita KS 67208

MERRIAM, ESTHER VIRGINIA, biologist, educator, researcher; b. Pitts., Apr. 9, 1940; d. Harold Powell and Esther Margaret (Westerman) Wills; m. John Roger Merriam, Aug. 30, 1963; children: Andrew John, Jeffrey Allen. B.S., Elizabethtown Coll., 1962; Ph.D. (NSF fellow), U. Wash., 1966. NIH predoctoral fellow U. Wash., Seattle, 1963-66; postdoctoral fellow biology div. Oak Ridge Nat. Lab., 1966-67, Calif. Inst. Tech., 1967-69; acting assoc. in molecular biology UCLA, 1969-71; asst. prof. biology Loyola Marymount U.,

1971-75, assoc. prof., 1975-82, prof., 1982—. Contbr. articles to profl. jours. Recipient Eli Lilly award U. Wash., 1964, Faculty Devel. award NSF, 1980-81. Mem. Genetics Soc., Am. Soc. Microbiology, Am. Women in Sci., Calif. Women in Higher Edn. (state pres. 1979-80), Sigma Xi. Subspecialty: Genetics and genetic engineering (agriculture). Current work: Tetrahymena genetics. Home: 518 E Rustic Rd Santa Monica CA 90402 Office: Dept Biology Loyola Marymount U Loyola Blvd at W 80th Los Angeles CA 90045

MERRIFIELD, ROBERT BRUCE, educator, biochemist; b. Ft. Worth, July 15, 1921; s. George E. and Lorene (Lucas) M.; m. Elizabeth Furlong, June 20, 1949; children—Nancy, James, Betsy, Cathy, Laurie, Sally. B.A., UCLA, 1943, Ph.D., 1949. Chemist Park Research Found., 1943-44; research asst. Med. Sch., UCLA, 1948-49; asst. Rockefeller Inst. for Med. Research, 1949-53, asso., 1953-57; asst. prof. Rockefeller U., 1957-58, asso. prof., 1958-66, prof., 1966—. Asso. editor: Internat. Jour. Peptide and Protein Research; Contbr. aritcles sci. jours. Recipient Lasker award biomed. research, 1969, Gairdner award, 1970, Intra-Sci. award, 1970, Nichols medal, 1973; Alan E. Pierce award Am. Peptide Symposium, 1979. Mem. Am. Chem. Soc. (award creative work synthetic organic chemistry 1972), Nat. Acad. Scis., Am. Soc. Biol. Chemists, Sigma Xi, Phi Lambda Upsilon, Alpha Chi Sigma. Subspecialty: Biochemistry (biology). Developed solid phase peptide synthesis; completed (with B. Gutte) 1st total synthesis of an enzyme, 1969. Office: Rockefeller U New York NY 10021

MERRILL, JEAN ELIZABETH, neuroimmunologist, educator; b. Somerville, Mass., July 15, 1947; d. John Edward and Roberta June (Wilson) M. B.A., Smith Coll., 1969; Ph.D., UCLA, 1979, M.B.A., 1982. Research assoc. Yale U., New Haven, 1971-74; postdoctoral fellow Karolinska Inst., Stockholm, 1979-81; asst. research neurologist UCLA, 1981-82, asst. prof. neurology, 1982—. Contbr. articles to profl. jours. Recipient Swedish Multiple Schlerosis Research award, 1980; Harvey Weaver Neurosci. Jr. Faculty award Nat. Multiple Sclerosis Soc., 1982. Mem. Am. Assn. Immunologists, Sigma Xi. Democrat. Subspecialty: Neuroimmunology. Current work: Immunoregulatory defects in patients with multiple sclerosis; tropic influences on glial hyperplasia. Home: 233 S Barrington Ave Apt 308 West Los Angeles CA 90049 Office: Dept Neurology Reed Neurol Research Center UCLA Sch Medicine Los Angeles CA 90024

MERRILL, JOSEPH MELTON, physician, educator, researcher; b. Andalusia, Ala., Dec. 8, 1923; s. Walter Clement and Mary (McLaney) M.; m. Gudrun Wallgren, Sept. 15, 1962; children: Maria, Caroline. M.D., Harvard U., 1948. Diplomate: Am. Bd. Internal Medicine. Intern Louisville Gen. Hosp., 1948-49; resident in internal medicine Vanderbilt U. Hosp., Nashville, 1950-51; assoc. chief staff-research VA Hosp., Nashville, 1958-64; chief Gen. Clin. Research Ctr., NIH, Bethesda, Md., 1964-67; prof. Baylor Coll. Medicine, Houston, 1967—, assoc. dean, then dean, 1967-70, exec. v.p., 1970-78; dir. First City Med. Center Bank, Houston. Author: Relationships between University and National Health Service, 1977; assoc. editor: Lipid Transport, 1964; co-editor: The Arts in Medicine, 1978. Bd. dirs. Southwest Ctr. Urban Research, Houston, 1968-76, Ctr. for Community Design and Research, Rice U., Houston, 1973-75; chmn. bd. dirs. Tex. Med. Center Library, Houston, 1974-76. Served to capt. USAF, 1951-53. Mem. Am. Phys. Soc., ACP, Am. Heart Assn., Am. Fedn. Clin. Research. Democrat. Episcopalian. Club: Cosmos. Subspecialties: Internal medicine; Cardiology. Current work: Ergonomics in medicine, psychobiology. Home: 2234 Inwood Houston TX 77019 Office: Baylor Coll Medicine 1200 Moursund Ave Houston TX 77025

MERRILL, MARSHALL LEIGH, software engineer; b. Paducah, Ky., Oct. 23, 1955; s. Glenn Herschel and Perla Dudley (Hudson) M. B.A. in Math./Computer Sci., U. S. Ala., 1977; M.S. in Computer Sci., Northwestern U., 1978. Cons. Computer Cons., Mobile, Ala., 1975-77, Telemed Corp., Hoffman Estates, Ill., 1978-80; research and devel. software engr. Teletype Corp., Skokie, Ill., 1980—; instr. computer sci. Northwestern U., Evanston, Ill., 1978-80; freelance programmer. Mem. Assn. Computing Machinery, IEEE Computer Soc. Baptist. Subspecialties: Operating systems; Graphics, image processing, and pattern recognition. Current work: Microcomputer networks and operating systems, computer graphics on microcomputers. Home: 8258 Niles Center Rd 0 Skokie IL 60077 Office: 5555 Touhy Skokie IL 60077

MERRILL, ROBERT DAVID, geology educator; b. Orange, Calif., June 22, 1941; s. Archie Lel and Lydia Child (Field) M.; m. Cynthia Leslie Tucker, July 5, 1966; children: Cyrus Benjamin, Nathaniel Tucker. B.A., U. Calif.-Riverside, 1963; M.S., U. Mass., 1965; Ph.D., U. Tex., 1974. Prof. geology Calif. State U.-Fresno, 1970—. Mem. Fresno County Water Adv. Bd., Fresno, 1975-82. Mem. Geol. Soc. Am. Democrat. Subspecialties: Sedimentology; Geology. Current work: Investigationof ancient depositional environments. Home: 4061 N Wishon St Fresno CA 93704 Office: Dept Geology Calif State U Fresno CA 93740

MERRILL, WILLIAM, JR., plant pathologist, educator; b. Haverhill, N.H., Sept. 5, 1933; s. William and Westa D. (Deno) M.; m. Mary Lou Anderson, July 15, 1961; children: Bjorn, Kurt. B.S. in Forestry magna cum laude, U. N.H., 1958; M.S. in Plant Pathology, U. Minn., 1961, Ph.D., 1963. Instr. U. Minn., Mpls., 1961-64; research staff pathologist Yale U., 1964-65; asst. prof. plant pathology Pa. State U., University Park, 1965-70, assoc. prof., 1970-75, prof., 1975—. Contbr. articles to profl. jours. Served with U.S. Navy, 1951-54. Recipient Christian R. and Mary F. Lindback award for disting. undergrad. teaching Pa. State U., 1976. Mem. Am. Phytopathol. Soc., AAAS, Nat. Assn. Coll. Tchrs. Agr. Subspecialties: Plant pathology; Integrated pest management. Current work: Research on conifer diseases; consultant on biodeterioration in relation to historical preservation. Office: 210 Buckhout Lab University Park PA 16802

MERRIN, CLAUDE EMILE ANDRE, physician, clin. researcher; b. Paris, Apr. 18, 1936; U.S., 1966, naturalized, 1975; s. Bernard and Isabelle (Haimo) M. B.S. in Exptl. Sci, U. Paris, 1955; M.D., U. Buenos Aires, 1962. Cert. Am. Bd. Urology. Intern, ACS fellow Cook County Hosp., Chgo., 1966-67, resident in gen. surgery, 1967-68, in urology, 1968-71; cancer research urologist Roswell Park Meml. Inst., Buffalo, 1971-72, sr. cancer research urologist, 1972-73, acting dir. dialysis unit and transplantation, 1973-74, acting chief dept. urologic oncology, 1973-74, dir. dialysis unit and transplantation program, 1973-79, chief dept. urologic oncology, 1974-79; research assoc. in surgery SUNY, Buffalo, 1971-73; urologic oncologist Swedish Covenant Hosp., Chgo., 1979—; assoc. clin. prof. urology Strich Med. Sch., Loyola U., Chgo., 1979—; cons. European Orgn. Research on Treatment of Cancer, Ill. Cancer Council. Contbr articles to profl. jours. Mem. Am. Assn. Cancer Research, Am. Soc. Clin. Oncology, Am. Urol. Assn. Acad. Surgery, Chgo. Med. Soc., Chgo. Urol. Soc., Ill. State Med. Soc., Ill. State Urol. Soc., N.Y. State Cancer Program Assn., Roswell Park Surg. Soc., Soc. Surg. Oncology. Subspecialties: Chemotherapy; Urology. Current work: Clin. oncological researc; urologic oncology, clin. research. Office: 5145 N California Ave Chicago IL 60625

MERRITT, CHARLES RANDALL, software engineer; b. Pensacola, Fla., Sept. 27, 1950; s. Charles Pitman and Livia M. A.A., Pensacola Jr. Coll., 1971; B.S. magna cum laude, U. West Fla., 1974; M.S.E.E., Auburn U., 1978. Data Base adminstr., Eglin AFB, Fla., 1974-76; software engr. Tex. Instruments, Inc., Austin, 1978-81. Mem. Assn. Computing Machinery, IEEE Computer Soc. Subspecialties: Operating systems; Computer architecture. Current work: Data-flow computers; operating system language design; instruction setarchitecture; memory management hierarchies/policies. Home: 3517 N Hills Dr Apt B 203 Austin TX 78731

MERRITT, DAVID ROY, theoretical astrophysicist; b. Van Nuys, Calif., Nov. 16, 1955; s. Samuel Harvey and Marian Lillian (Pruett) M. B.S., U. Santa Clara, 1977; Ph.D., Princeton U., 1981. Research assoc. Nat. Radio Astronomy Obs., Charlottesville, Va., 1981—. NSF fellow, 1977-81. Mem. Am. Astron. Soc., Phi Beta Kappa. Subspecialty: Theoretical astrophysics. Current work: Dynamics and evolution of clusters of galaxies.

MERRITT, JOSEPH CLAUDE, oilfield repair and manufacturing company executive; b. Colorado City, Tex., Aug. 17, 1941; s. Joseph Ford and Imogene (Sanders) M.; m. Carolyn Perkins Shimer, Aug. 30, 1963; children: Sharon, Joe, Bob. B.S. in Mech. Engring, Tex. A&M U., 1964. Registered profl. engr., Tex. With Cooper-Bessemer Corp., 1964-74, mktg. mgr.-customer services, Mt. Vernon, Ohio, 1969-70, mid-continent area mgr.-customer services, Tulsa, 1970-74; owner, pres. Dixie Iron Works Inc., Alice, Tex., 1974—; dir. First City Bank Alice. Served to 1st lt. C.E. U.S. Army; Vietnam. Trustee Alice Ind. Sch. Dist., 1978—. Mem. ASME, Gas Processers Suppliers Assn. Methodist. Club: Alice Country (pres. 1980-81). Subspecialty: Mechanical engineering. Current work: Specializing inservice to production and process industries of South Texas and Gulf Coast. Home: 1320 Southwood Alice TX 78332 Office: 70 Kentucky Alice TX 78332

MERTENS, LAWRENCE EDWIN, engineer, researcher; b. N.Y.C., Mar. 6, 1929; s. John Henry and Marie (Loehr) M.; m. Margarete Anna Waider, July 8, 1975; children: Oliver Larry, Thomas John. B.S., Columbia U., 1951, M.S., 1952, D.Eng.Sci., 1955. Staff engr. RCA Def. Electronic Products, Camden, N.J., 1952-60; mgr. digital communications RCA Surface Communications, Camden, 1960-62; chief scientist RCA Service Co., Patrick AFB, Fla., 1962-72, mgr. deep look project, 1972-75, mgr. aerostat systems, 1975-81, mgr. tech. analysis, chief scientist, 1981—; adj. prof. Fla. Inst. Tech., Melbourne, 1968—; pres. Photoquatics, Palm Bay, 1967-72. Author: In Water Photography, 1970; inventor: patentee pulse transmitter, 1961; Guest editor: Optical Engring, 1977. Samuel Willard Bridgham fellow Columbia U., 1951; Boese fellow, 1952-53. Mem. IEEE (sr.), N.Y. Acad. Scis., Soc. Motion Picture and TV Engrs. (sr.), Soc. Photo-Optical Instrumentation Engrs., Sigma Xi. Subspecialties: Systems engineering. Current work: Advanced radar, signal processing, optical transmission through sea water, imaging systems (laser), aerostat technology. Home: 798 SW Pebble Beach Ave Palm Bay FL 32905 Office: RCA Missile Test Project PO Box 4308 Bldg 989 MU 645 Patrick AFB FL 32925

MERTENS, THOMAS ROBERT, biology educator; b. Ft. Wayne, Ind., May 22, 1930; s. Herbert F. and Hulda C. (Burg) M.; m. Beatrice Janet Abair, Apr. 1, 1953; children: Julia Ann, David Gerhard. B.S., Ball State U., 1952; M.S., Purdue U., 1954, Ph.D., 1956. Research assoc. dept. genetics U. Wis.-Madison, 1956-57; asst. prof. biology Ball State U., Muncie, Ind., 1957-62, assoc. prof., 1962-66, prof., 1966—, dir. doctoral programs in biology, 1974—. Author: Genetics Laboratory Investigations, 1980; co-author: Human Genetics, 1983; contbr. articles to profl. jours. NSF fellow Stanford U., 1963-64; recipient Service award Nat. Assn. Biology Tchrs., 1981, Disting. Alumnus award Ball State U., 1983. Fellow Ind. Acad. Sci., AAAS; mem. Nat. Assn. Biology Tchrs., Am. Genetic Assn., Genetics Soc. Am., others. Lutheran. Club: Muncie. Subspecialty: Genetics and genetic engineering (biology). Current work: Human genetics education, educational needs assessments relative to human genetics and bioethics. Home: 2506 Johnson Rd Muncie IN 47304 Office: Dept Biology Ball State Univ Muncie IN 47306

MERTZ, DAVID BYRON, biology educator; b. Sandusky, Ohio, July 10, 1934; s. Henry William and Mary (Hettrick) M.; m. Sarah Burnham, Feb. 25, 1961. S.B., U. Chgo., 1960, Ph.D., 1965. NSF postdoctoral fellow U. Calif.-Berkeley, 1965-66; asst. prof. U. Calif.-Santa Barbara, 1966-69; assoc. prof. U. Ill.-Chgo., 1969-74, prof. dept. biol. scis., 1974—. Co-editor: Readings in Ecology and Ecological Genetics, 1970; mem. editorial bds. jours.: Ecology, 1970-71, Evolution, 1979. NSF grantee, 1967-79. Mem. AAAS, Am. Soc. Naturalists, Ecol. Soc. Am., Soc. for Study Evolution, Brit. Ecol. Soc. Democrat. Subspecialties: Ecology; Evolutionary biology. Current work: Studies in experimental demography, interspecies competition, ecological genetics, population theory. Office: Dept Biol Scis Univ Ill-Chicago Chicago IL 60680 Home: 1226 E Madison Park Chicago IL 60615

MERTZ, EDWIN THEODORE, biochemist, educator; b. Missoula, Mont., Dec. 6, 1909; s. Gustav Henry and Louise (Sain) M.; m. Mary Ellen Ruskamp, Oct. 5, 1936; children—Martha Ellen, Edwin T. B.A., U. Mont., 1931, M.S. in Biochemistry, 1933, D.Sc. (hon.), 1979; Ph.D. in Biochemistry, U. Ill., 1935; D.Agr. (hon.), Purdue U., 1977. Research biochemist Armour & Co., Chgo., 1935-37; instr. biochemistry U. Ill., 1937-38; research assoc. pathology U. Iowa, 1938-40; instr. agrl. chemistry U. Mo., 1940-43; research chemist Hercules Powder Co., 1943-46; prof. biochemistry Purdue U., West Lafayette, Ind., 1946-76, emeritus, 1976—; vis. prof. U. Notre Dame, South Bend, Ind., 1976-77; cons. in field. Author: Elementary Biochemistry, 1969. Recipient McCoy award Purdue U., 1967; John Scott award City of Phila., 1967; Hoblitzelle Nat. award Tex. Research Found., 1968; Congressional medal Fed. Land Banks, 1968; Distinguished Service award U. Mont., 1973; Browning award Am. Soc. Agronomy, 1974; Pioneer Chemist award Am. Inst. Chemists, 1976. Mem. AAAS, AAUP, Nat. Acad. Scis., Am. Soc. Biol. Chemists, Am. Inst. Nutrition (Osborne-Mendel award 1972), Am. Chem. Soc. (Spencer award 1970), Am. Assn. Cereal Chemists. Presbyterian. Subspecialties: Genetics and genetic engineering (agriculture); Plant genetics. Current work: Improvement of nutritional value of cereal grains through genetics and genetic engineering. Co-discoverer high lysine corn, 1963. Office: Dept Biochemistry Purdue U Lafayette IN 47907

MERTZ, WALTER, nutritionist; b. Mainz, Germany, May 4, 1923; s. Oskar and Aenne (Gablemann) M.; m. Marianne C. Maret, Aug. 8, 1953. M.D., U. Mainz, 1951. Lic. physician, W. Ger. Intern County Hosp., Hersfeld, W. Ger., 1952-53; with NIH, Bethesda, Md., 1953-61, Walter Reed Army Inst. Research, Washington, 1961-69; staff Human Nutrition Research Ctr., U.S. Dept. Agr., Beltsville, Md., 1969—, dir., 1972—. Served with Germany Army, 1941-42. Recipient Research and Devel. Achievement award Dept. Army, 1969; Superior Service award U.S. Dept. Agr., 1972. Mem. Am. Inst. Nutrition (Osborne and Mendel award 1971, Lederle award 1982), Am. Soc. Clin. Nutrition. Subspecialty: Nutrition (biology). Current work: Trace element nutrition and metabolism. Office: US Dept Agr Bldg 308 BARC-East Beltsville MD 20705

MERZBACHER, EUGEN, physicist, educator; b. Berlin, Germany, Apr. 9, 1921; came to U.S., 1947, naturalized, 1953; s. Siegfried and Lilli (Wilmersdoerffer) M.; m. Ann Townsend Reid, July 11, 1952; children: Celia Irene, Charles Reid, Matthew Allen, Mary Letitia. Licentiate, U. Istanbul, 1943; A.M., Harvard U., 1948, Ph.D., 1950. High sch. tchr., Ankara, Turkey, 1943-47; mem. Inst. Advanced Study, Princeton, N.J., 1950-51; vis. asst. prof. Duke, 1951-52; mem. faculty U. N.C., Chapel Hill, 1952—, prof., 1961—, acting chmn. physics dept., 1965-67, 71-72, Kenan prof. physics, 1969—, chmn. dept., 1977-82; vis. prof. U. Wash., 1967-68. Author: Quantum Mechanics, 2d edit, 1970; also articles.; Editorial bd.: Am. Jour. Physics, 1965-70, Atomic Data and Nuclear Data Tables, 1970—, Phys. Rev., Sect. A, 1981—. NSF Sci. Faculty fellow U. Copenhagen, Denmark, 1959-60; recipient Thomas Jefferson award U. N.C., 1972; Humboldt sr. scientist award U. Frankfurt, Germany, 1976-77. Fellow Am. Phys. Soc. (chmn. Southeastern sect. 1971-72, mem.-at-large council 1972-76, chmn. div. electron and atomic physics 1978-79, mem. exec. com. div. nuclear physics 1983-85); fellow AAAS (sect. del. to council); mem. Am. Assn. Physics Tchrs., AAUP, Sigma Xi. Subspecialty: Theoretical physics. Current work: Research on applications of quantum mechanics to study atoms and nuclei. Research on applications of quantum mechanics to study atoms and nuclei. Home: 1396 Halifax Rd Chapel Hill NC 27514

MESEGUER, JOSE GUAITA, computer scientist; b. Murcia, Spain, Mar. 31, 1950; came to U.S., 1977; s. Francisco Meseguer and Fuensanta Guaita. Licenciatura in Math, U. Zaragoza, Spain, 1972, Ph.D., 1975. Asst. prof. math. U. Zaragoza, 1975-76; research fellow in math. U. Santiago, Spain, 1976-77; vis. scholar in computer sci. UCLA, 1977-78; research assoc. in math. U. Calif.-Berkeley, 1977-80; computer scientist SRI Internat., Menlo Park, Calif., 1980—. Contbr. articles on abstract data types, software specification, computer security and pure math. to profl. jours. March Found. fellow, 1976-77; Spanish Ministry Edn. postdoctoral research fellow, 1977-79; Joint U.S.-Spain Com. for Sci. Cooperation fellow, 1979-81. Mem. Assn. Computing Machinery, European Assn. Theoretical Computer Sci. Subspecialties: Software engineering; Programming languages. Current work: Software specification and verification; logic programming languages; computer security; foundations of computer science. Office: SRI Internat 333 Ravenswood Ave Menlo Park CA 94025

MESELSON, MATTHEW STANLEY, educator, biochemist; b. Denver, May 24, 1930; s. Hymen Avram and Ann (Swedlow) M.; m. Sarah Page; children: Zoe, Amy Valor. Ph.B., U. Chgo., 1951, D.Sc. (hon.), 1975; Ph.D., Calif. Inst. Tech., 1957; Sc.D. (hon.), Oakland Coll., 1964, Columbia, 1971. From research fellow to sr. research fellow Calif. Inst. Tech., 1957-60; asso. prof. biology Harvard U., 1960—, prof. biology, 1964-76, Thomas Dudley Cabot prof. natural scis., 1976—. Recipient prize for molecular biology Nat. Acad. Scis., 1963, Eli Lilly award microbiology and immunology, 1964, Alumni medal U. Chgo., 1971; Pub. Service award Fedn. Am. Scientists, 1972; Lehman award N.Y. Acad. Scis., 1975; Alumni Distinguished Service award Calif. Inst. Tech., 1975; Leo Szilard award Am. Phys. Soc., 1978. Mem. Am. Acad. Arts and Scis., Council Fgn. Relations, Accademia Santa Chiara, Nat. Acad. Scis., Inst. Medicine, Am. Philos. Soc. Subspecialty: Biochemistry (biology). Address: Fairchild Biochem Bldg 7 Divinity Ave Harvard U Cambridge MA 02138

MESHII, MASAHIRO, materials science and engineering educator, consultant, researcher; b. Hyogo, Japan, Oct. 6, 1931; s. Masataro and Kazuyo M.; m. Eiko Kumagai, May 21, 1959; children: Alisa, Erica. B.Eng. in Metallurgy, Osaka U., 1954, M.S., 1956; Ph.D. in Materials Sci, Northwestern U., 1959. Lectr., research assoc. Northwestern U., Evanston, Ill., 1959-60, asst. prof. materials sci. and engring., 1960-64, assoc. prof., 1964-67, prof., 1967—, chmn. dept., 1978-82; vis. scientist Nat. Research Inst. for Metals, Tokyo, 1979-71; vis. prof. Osaka U., 1971; NSF summer faculty participant Argonne Nat. Lab., 1975; now cons.; cons. Briggs & Stratton Corp. Editor: Mechanical Properties of BCC Metals; co-editor: Martensitic Transformation; contbr. over 100 articles to profl. jours. Recipient Tech. Teaching award Northwestern U. Technol. Inst., 1978; Fulbright grantee, 1956; Japan Soc. fellow, 1958-59. Mem. Am. Soc. Metals (Henry Marion Howe gold metal 1968), AIME, Am. Phys. Soc., Japan Inst. Metals (Kosekisho achievement award 1972), Phys. Soc. Japan, Am. Soc. for Engring. Edn., Microbeam Analysis Soc. Club: Deerbrook Park Toastmasters (pres. 1983). Subspecialties: Metallurgy; Materials. Current work: Physical metallurgy, radiation damage, lattice defects, electronmicroscopy and mechanical properties of metals. Home: 3051 Centennial Ln Highland Park IL 60035 Office: Technol Inst Northwestern U 2145 Sheridan Rd Evanston IL 60201

MESLER, RUSSELL BERNARD, chemical engineering educator; b. Kansas City, Mo., Aug. 24, 1927; s. James Elmer and Catherine (Knaack) M.; m. Jenny-Lea McGowan, June 9, 1952; children: Diane, Scott, Douglas, Sandra. B.S., U. Kans., 1949; M.S., U. Mich., 1953, Ph.D., 1955. Engr. Colgate Palmolive Corp., Kansas City, Kans., 1949-51, Oak Ridge Sch. Reactor Tech., 1951-52; project engr. Ford Nuclear Reactor Co., Ann Arbor, Mich., 1952-57; asst. prof. U. Mich., 1955-57; Warren S. Bellows prof. chem. engring. U. Kans., 1957—; cons. Spencer Chem. Co., Kansas City, Mo., 1957-61, Farmland Industries, Lawrence, 1980-82, Berkeley Nuclear Labs., Eng., 1975-76; research participant Savannah River Labs., Aiken, S.C., 1981. Served with USNR, 1945-46. Danforth assoc., 1963; Robert T. Knapp award ASME, 1967. Fellow Am. Inst. Chem. Engrs.; mem. Am. Chem. Soc., Am. Soc. for Engring. Edn., Am. Nuclear Soc. Lutheran. Subspecialties: Chemical engineering; Nuclear engineering. Current work: Nucleate boiling. Home: 1629 Dudley Ct Lawrence KS 66044 Office: U Kans 102 Nuclear Reactor Center Lawrence KS 66045

MESSIHA, FATHY SABRY, biochemical pharmacologist, toxicologist, educator; b. Cairo, Feb. 10, 1936; U.S., 1966, naturalized, 1972. Diploma in Pharmacy, U. Basel, Switzerland, 1963; Ph.D. in Physiol. Biochemistry, U. Bern, Switzerland, 1965; postgrad., Med. Chem. Inst., Sch. Medicine, 1965-66, U. Vt. Med. Sch., 1966-67. Research asso. Biochem. Research Lab., Spring Grove State Hosp., Balt., 1967-68; research scientist dept. biochemistry Md. Psychiat. Research Center, Balt., 1968-71, lab. dir. pharmacology unit, 1971-72; asso. prof. pharmacology and therapeutics Tex. Tech. U. Health Sci. Center Sch. Medicine, Lubbock, 1972-77, asso. prof. psychiatry, 1973-80, prof., 1980—, assoc. prof. pathology, 1977-80, prof., 1980—, dir. div. toxicology, 1980—. Editor 3 books, meeting presentations abstracts; contbr. numerous articles to sci. jours., procs., chapt. to books. Mem. Am. Coll. Toxicology, Am. Soc. Pharmacology and Exptl. Therapeutics, Am. Pharm. Assn., Acad. Pharm. Scis., Soc. for Neurosci., N.Y. Acad. Scis., Collegium Internationale Neuropsychopharmacologicum, Soc. Exptl. Biology and Medicine, Internat. Union Pharmacology (sect. on toxicology), Am. Chem. Soc., Western Pharmacology Soc., AAAS, Research Soc. on Alcoholism.

Copt Orthodox. Subspecialties: Toxicology (medicine); Psychopharmacology. Current work: Toxicology, psychopharmacology, CNS-pharmacology, clinical pharmacology, biological psychiatry, enzymology, chemotherapy, drug metabolism, alcoholism. Office: Dept Pathology Tex Tech U Health Sci Center Sch Medicine Lubbock TX 79430

MESSING, RITA BAILEY, pharmacologist; b. Bklyn., July 7, 1945; d. Max and Kate (Katkin) Zimmerman; m. William Messing, June 20, 1965; 1 son, Charles. B.A. magna cum laude, Bklyn. Coll., 1966; Ph.D., Princeton U., 1970. Asst. prof. Rutgers U., Camden, N.J., 1969-72; research asso. M.I.T., 1973-75; asso. research psychobiologist U. Calif., Irvine, 1976-81; Research fellow Organon Pharms, Oss, The Netherlands, 1980; asst. prof. U. Minn., 1981—. Editor: (with others) Endogenous Peptides and Learning and Memory Processes, 1981. NSF fellow, 1966-69; Med. Found. fellow, 1974-75; Organon fellow, 1980. Mem. AAAS, Am. Soc. Pharmacology and Exptl. Therapeutics, Soc. Neursci. Subspecialties: Neuropharmacology; Neurobiology. Current work: Nueropharmacology and Behavioral Toxicology. Patentee in field. Home: 735 Goodrich Ave St Paul MN 55105 Office: Dept Pharmacology U Minn Minneapolis MN 55455

MESULAM, M. MARSEL, neurologist, medical educator; b. Istanbul, Turkey, Apr. 7, 1945; s. Mose and Fani (Rozanes) M.; m.; 1 dau., Semra. B.A., Harvard U., 1968, M.D., 1972. Diplomate: Am. Bd. Psychiatry and Neurology. Assoc. prof. neurology Harvard U. Med. Sch., 1980—; dir. neuroanatomy and behavioral neurology Beth Israel Hosp., Boston, 1979—. Mem. Internat. Brain Research Orgn., Am. Neurol. Assn., Am. Acad. Neurology, Am. Assn. Anatomists, Histochem. Soc., Soc. Neurosci. Subspecialties: Neurology; Neuroanatomy. Current work: Connections of the brain; behavioral neurology. Home: 15 Desmond Watertown MA 02172 Office: Beth Israel Hosp Boston MA 02215

METCALF, JOHN FRANKLIN, medical educator, researcher; b. Phoenix, Dec. 12, 1944; s. John Allen and Lois McIlwain (Willet) M.; m. Susan Wolschina, Sept. 5, 1974; children: Andrew Allen, Jessica Lynn, Christopher John. B.A., Northwestern U., 1966, M.D., 1969. Diplomate: Am. Bd. Internal Medicine. Intern and jr. resident Cleve. Met. Hosp., 1969-71; resident, fellow Duke U. Med. Ctr., Durham, N.C., 1973-75; postdoctoral fellow U. Calif.-San Diego, 1975-77; asst. prof. dept. medicine Med. U. S.C., Charleston, 1977—; staff physician Univ. Hosp., Charleston, 1977. Served to capt. U.S. Army, 1971-73; Vietnam. USPHS grantee, 1978. Fellow Am. Fedn. Clin. Research; mem. Am. Thoracic Soc., ACP, Am. Coll. Chest Physicians, Phi Beta Kappa, Alpha Omega Alpha. Democrat. Subspecialties: Physiology (biology); Cell biology. Current work: Pathophysiology of gas exchange in pulmonary disease; inflammatory secretion of proteases by alveolar macrophages. Home: 588 Coinbow Dr Mount Pleasant SC 29464 Office: 171 Ashley Ave Charleston SC 29425

METIL, IGNATIUS, corrosion protection exec.; b. Kolpec, Ukraine, Dec. 1, 1919; came to U.S., 1951; s. Andrew and Tatianna (Rynda) M.; m. Erika Metil, Nov. 11, 1952; children: Mary T., Andrew A. M.S. in Chemistry, Canisius Coll., 1961; Dipl.Ing., Inst. Tech., Graz, Austria, 1951, D.Sc., 1965. Cert. corrosion engr., corrosion specialist. Chemist, tech. dir. Nukem Products Corp., Buffalo, 1954-59; lab mgr. Amercoat Corp., Buffalo, 1959-69; tech. dir. Hempel's Marine Paints, Inc., Buffalo, 1963-73; pres. IMCO Labs., Inc., Buffalo, 1973—. Contbr. articles to profl. jours. Trustee Ukrainian Community Ch. Mem. Nat. Assn. Corrosion Engrs., ASTM, Fedn. Socs. for Paint Tech., Am. Inst. Chemists, AAAS, N.Y. Acad. Sci., Washington Paint Tech. Group, Ukrainian Engrs. Soc Am., Shevchenko Sci. Soc. Subspecialties: Chemical engineering. Current work: New corrosion inhibiting pigments and corrosion protective materials. Home: 364 Whitfield Ave Buffalo NY 14220 Office: 1800 Broadway Buffalo NY 14212

METROPOLIS, NICHOLAS C., mathematical physicist, computer scientist; b. Chgo., June 11, 1915; s. Constantine and Katherine (Ganas) M.; m.; children: Katharine, Penelope, Christopher. B.S., U. Chgo., 1936, Ph.D., 1941. Staff Metall. Lab., Chgo., 1942-43; group leader Los Alamos Nat. Lab., 1943-46; asst. prof. physics U. Chgo., 1946-48; group leader Los Alamos Nat. Lab., 1948-57; sr. fellow, 1965—; prof. physics U. Chgo., 1957-65, founding dir. Inst. for Computer Research, 1957-63; adv. com. research NSF, Washington, 1974-75, cons., 1975-80, U. Ill. - Urbana, 1971—, Brookhaven Nat. Lab., L.I., 1979—; Tech. adv. UN, India, 1961; mem. Oppenheimer Meml. Com., 1965—. Author, editor: Surveys in Applied Mathematics, 1976, A History of Computing, 1980; editor: Advances in Applied Mathematics, 1980—. NSF grantee, 1976; NSF sci. exchange award, Russia, 1976; session chmn UNESCO, Paris, 1959. Fellow Am. Phys. Soc.; fellow Am. Acad. and Scis.; mem. Am. Math. Soc., Soc. Indsl. and Applied Math. Subspecialties: Theoretical physics; Foundations of computer science. Current work: Combinatorial theory, foundations of computation, theory of universal functions. Office: Los Alamos Nat Lab PO Box 1663 Los Alamos NM 87545

METSGER, ROBERT WILLIAM, geologist; b. N.Y.C., Apr. 27, 1920; s. William Martin and Mabel (Herbst) M.; m. Sylvia Haff, Mar. 29, 1947 (dec. 1961); children: Mary Gwendolen Metsger Chong, Deborah Anne; m. Barbara Lawrence Holbert, Aug. 2, 1963; 1 son, Robert Lawrence. A.B., Columbia Coll., 1948. Cert. profl. geol. scientist. With N.J. Zinc. Co., Ogdensburg, N.J., 1949—, chief geologist, 1981—. Contbr. numerous geol. articles to profl. publs. Sr. warden Christ Ch., Newton, N.J., 1978—; v.p. Western Dist., Episcopal Diocese of Newark, 1972-76, pres, 1976-80; mem. missions dept., 1972-82. Served to lt. comdr. USNR, 1941-46. Fellow Geol. Soc. Am. (chmn. NE sect. 1981-82); mem. Soc. Econ. Geologists, Am. Inst. Profl. Geologists. Republican. Episcopalian. Lodge: Rotary. Subspecialties: Geology; Ground water hydrology. Current work: Studies of the genesis and subsequent alteration and tectonic disruption of a Precambrian ore deposit in New Jersey; studies of Earth strain, both as a result of mining activity and as a natural phenomenon. Home: 69 Hunters Ln Sparta NJ 07871 Office: NJ Zinc Co Inc Plant St Ogdensburg NJ 07439

METTEE, HOWARD DAWSON, chemistry educator, researcher; b. Boston, Aug. 6, 1939; s. Milton Howard and Marjorie (Muther) M.; m. Linda Arlene Band, Feb. 9 1940; children: Michael Stewart, Nancy, Travis Dawson. B.A. in chemistry, Middlebury Coll., 1961; Ph.D. in phys. chemistry, U. Calgary, Alta., 1964. Fellow NRC, 1964-66; fellow U. Tex., 1966-68; asst. prof. chemistry Youngstown State U., 1968-73, asso. prof., 1973-82, prof., 1982—; vis. scientist U. Calif.-Berkeley, 1979-80. Contbr. articles in field to profl. jours. Research Corp. grantee, 1970; Univ. Research Council grantee, 1972, 74, 82. Mem. Am. Chem. Soc., Royal Soc. Chemistry, Interamerican Photochemical Soc., Sigma Xi. Democrat. Subspecialties: Physical chemistry; Solar energy. Current work: Devel. of materials capable of water splitting photoelectrochemistry, thermodynamic and kinetic behavior of excited molecules, and hydrogen bonded complexes. Home: 2271 1/2 Cordova St Youngstown OH 44504 Office: Department Chemistr Youngstown State Universit Youngstow OH 44555

METTLER, RUBEN FREDERICK, electronics and engineering company executive; b. Shafter, Calif., Feb. 23, 1924; s. Henry Frederick and Lydia M.; m. Donna Jean Smith, May 1, 1955; children: Matthew Frederick, Daniel Frederick. Student, Stanford, 1941-43; B.S. in Elec. Engring, Calif. Inst. Tech., 1944, M.S., 1947, Ph.D. in Elec. and Aero. Engring, 1949. Registered profl. engr., Calif. Asso. div. dir. systems research and devel. Hughes Aircraft Co., 1949-54; spl. cons. to asst. sec. def., 1954-55; asst. gen. mgr. guided missile research div. Ramo-Wooldridge Corp., 1955-58; pres. Space Tech. Labs., Inc., Los Angeles, 1962-65, TRW Systems Group, 1965-68; exec. v.p., dir. TRW Inc. (formerly Thompson Ramo Wooldridge, Inc.), 1965, asst. pres., 1968-69, pres., 1969-77, chmn. bd., chief exec. officer, 1977—; dir. Bank Am. Corp., Merck & Co.; past vice-chmn. Ind. adv. council Dept. Def. Author reports airborne electronic systems. Nat. campaign chmn. United Negro Coll. Fund, 1980; chmn. Pres.' Sci. Policy Task Force, 1969; mem. Pres.' Blue Ribbon Def. Panel, 1969-70, Emergency Com. for Am. Trade; chmn. Nat. Alliance Business, 1978-79; vice chmn. bd. trustees Calif. Inst. Tech.; trustee Com. Economic Devel.; bd. dirs. Nat. Action Council for Minorities in Engring.; trustee Cleve. Clinic Found. Served with USNR, 1942-46. Named 1 of ten Outstanding Young Men of Am. U.S. Jr. C. of C., 1955; So. Calif.'s Engr. of Year, 1964; recipient Meritorious Civilian Service award Dept. Def., 1969, Nat. Human Relations award NCCJ, 1979; Excellence in Mgmt. award Industry Week mag., 1979. Fellow IEEE, AIAA; mem. Sci. Research Soc. Am., Bus. Roundtable (Chmn. 1982—), Conf. Bd. (trustee 1982—), Bus. Council (vice chmn. 1981-82), Nat. Acad. Engring., Sigma Xi, Eta Kappa Nu (named nation's outstanding young elec. engr. 1954), Tau Beta Pi, Theta Xi. Clubs: Cosmos (Washington); Union, 50 (Cleve.). Subspecialties: Aerospace engineering and technology; Electronics, engineering management. Patentee interceptor fire control systems. Home: 23555 Euclid Ave Cleveland OH 44117 Office: 1 Space Park Redondo Beach CA 90278

METZGER, ALBERT EMANUEL, space scientist; b. N.Y.C., Sept. 10, 1928; s. Frederic and Hortense (Abrams) M.; m.; children: Joslyn Vahni, Loren Frederic. Student, Bethany (W.Va.) Coll., 1946; B.A., Cornell U., Ithaca, N.Y., 1949; M.A., Columbia U., 1951, Ph.D., 1958. Chemist Sylvania Electric Products, Inc., Bayside, L.I., 1951-54; scientist, research group supr. Jet Propulsion Lab., Pasadena, Calif., 1960—. Contbr. articles to profl. jours. Recipient Exceptional Sci. Achievement medal NASA, 1975. Fellow AAAS; mem. Am. Phys. Soc., Am. Geophys. Union, Am. Astron. Soc., Phi Lambda Upsilon. Subspecialties: Planetary science; Gamma ray high energy astrophysics. Current work: Spacecraft expt. design, devel., data reduction and analysis in fields of planetary composition, magnetospheric interactions, and high energy astrophysics. Office: 4800 Oak Grove Dr Pasadena CA 91109

METZGER, ROBERT MELVILLE, chemistry educator, cons. researcher; b. Yokohama, Japan, May 7, 1940; s. Ferdinand J. and Gabriella J. (Szigeti) M.; m. Christian D. Csoeke-Poeckh, Sept. 12, 1971; 1 son, Gian-Lorenzo. B.S., UCLA, 1962; Ph.D., Calif. Tech. Inst., 1968. Postdoctoral research assoc. Stanford (Calif.) U., 1969-71, 1972; lectr. Italian, 1970-71; asst. prof. chemistry U. Miss., University, 1971-76, assoc. prof., 1976—; vis. prof. U. Heidelberg, W.Ger., 1969-70, Bordeaux U., France, 1969-70. Contbr. articles on chemistry to profl. jours. Mem. Am. Chem. Soc., Am. Phys. Soc., Am. Crystallographic Assn. Subspecialties: Theoretical chemistry; Solid state chemistry. Current work: Organic donor-acceptor crystals, cohesion and ionicity, organic unimolecular rectifiers, polarizability calculations. Home: 2225 Lee Loop Oxford MS 38655 Office: Dept Chemistry U Miss University MS 38677

METZGER, SIDNEY, electrical engineer; b. N.Y.C., Feb. 1, 1917; s. Julius and Molly (Gottesman) M.; m. Miriam Marsha Lipstein, Dec. 3, 1944; children: David J., Sally Ann Metzger Fasman, Philip C. B.S. in Elec. Engring., NYU, 1937; M.E.E., Poly. Inst. Bklyn., 1950. Chief radio relay br. U.S. Army Signal Corps Labs., Ft. Monmouth, N.J., 1939-45; div. head radio relay div. ITT Labs., Nutley, N.J., 1945-54; mem. tech. staff RCA Labs., Princeton, N.J., 1954-58; mgr. communications engring. div. astro electronics RCA, Princeton, 1958-63; v.p., chief scientist Communications Satellite Corp., Washington, 1963-82; pres. Sidney Metzger & Assocs., Cons. Engrs., 1982—; cons. Arthur S. Flemming Awards, 1975-77, judge, 1980—. Fellow IEEE (Eascon Aerospace Electronics award 1975, award in internat. communications 1976, mem. joint telecommunications adv. com. IEEE 1971-79, chmn. joint com. 1975-77), AIAA; mem. Nat. Acad. Engring., Sigma Xi, Tau Beta Pi. Current work: Consulting in communications satellite systems and communications engineering. Designer early radio relay systems in U.S., Can., Mexico, Belgium; designer mil. radio relay systems; communications engring. for early satellites and earth stas.; patentee in field.

METZGER, WALTER JAMES, physician, educator; b. Pitts., Oct. 30, 1945; s. Walter James and Marion Smith (Vine) M.; m. Carol Louise Hughes, Jan. 17, 1942; children: James Andrew, Joel Robert, Anne Elizabeth. B.A., Stanford U., 1967; M.D., Northwestern U., 1971. Diplomate: Am. Bd. Allergy and Clin. Immunology. Staff allergist David Grant Med. Center, Travis AFB, Calif., 1976-78; asst. prof. internal medicine U. Iowa Med. Sch., Iowa City, 1978—. Served to maj. USAF, 1976-78. Fellow ACP, Am. Acad. Allergy and Clin. Immunology; mem N. Central Allergy Soc. (pres.), Iowa Soc. Allergy (v.p.). Subspecialties: Internal medicine; Allergy. Current work: Allergy, immunotherapy, asthma, late asthmatic responses, mediators of late responses. Office: Dept Internal Medicine U Iowa Iowa City IA 52242

METZL, MARILYN NEWMAN, clinical psychologist, educator; b. N.Y.C., Apr. 12, 1938; d. George and Rose (Shanen) Newman; m. Kurt Metzl, June 25, 1961; children: Jonathan, Jordan, Jamie, Joshua. B.A., Queens Coll., CUNY, 1959; M.A., Hunter Coll., 1969; Ph.D., U. Kans., 1978. Therapist Columbia Presbyn. Med. Ctr., N.Y.C., 1960-63; dir. Dependent Clinic, Ismir, Turkey, 1963-65; therapist Menorah Med. Ctr., Kansas City, Mo., 1966-68; dir. Family Devel. Ctr., Spelman Hosp., Smithville, Mo., 1980—; assoc. prof. Avila Coll., Kansas City, 1978—; dir. Psycholednl. Assocs., Kansas City, 1971—; dir. Family Devel. Ctr. Gardner (Kans.) Hosp., 1982—. Mem. Soc. for Research in Child Devel., Am. Psychol. Assn., Council for Exceptional Children (mem. adv. com. div. children with communicative disorders 1980—), Am. Speech and Hearing Assn., NOW. Democrat. Jewish. Subspecialties: Cognition; Developmental psychology. Current work: Parent-child interaction, parenting styles and child development, adolescent development, learning disabilities. Home: 1000 W. 59th St Kansas City MO 64113

METZLER, DWIGHT FOX, state ofcl., civil engr.; b. Carbondale, Kans., Mar. 25, 1916; s. Ross R. and Grace (Fox) M.; m. Lela Ross, June 20, 1941; children—Linda Diane, Brenda Lee, Marilyn Anne, Martha Jeanne. B.S., U. Kans., 1940, C.E., 1947; S.M., Harvard, 1948. Registered profl. engr., Kans., N.Y., diplomate: Am. Acad. Environ. Engrs. Asst. engr. Kans. Bd. Health, 1940-42, san. engr., 1946-48, chief engr., Topeka, 1948-62; asso. prof. dept. civil engring. U. Kans., 1948-59, prof., 1959-66; exec. sec. Kans. Water Resources Bd., Topeka, 1962-66; dep. commr. N.Y. State Dept. Health, Albany, 1966-70, N.Y. State Dept. Environ. Conservation, 1970-74; sec. Kans. Dept. Health and Environment, Topeka, 1974-79, dir. water supply devel., 1979—; Cons. san. engring. Fed. Pub. Housing Authority, USPHS, 1943-46; housing cons. Chgo.-Cook County Health Survey, 1946; cons. water supply and water pollution control USPHS, 1957-66; adviser Govt. India, 1960; mem. ofcl. exchange to USSR on environ. health research and practice, 1962; adviser WHO, 1964; mem. Water Pollution Bd., Internat. Joint Commn., 1967-74, Assembly of Engring. Nat. Research Council, 1977-80. Contbr. articles to profl. jours. Chmn. Kans. Bible Chair Bd., 1957-66; chmn. com. for new bldg. U. Kans. Sch. Religion. Recipient Distinguished Service award U. Kans., 1970. Fellow Royal Soc. Health Gt. Britain (hon.), Am. Pub. Health Assn. (Centennial award 1972, Sedgewick medal 1981, governing council exec. bd., pres., chmn. action bd.), ASCE (sec. san. engring. div. 1959-61, chmn. 1963); mem. Am. Water Works Assn. (Fuller award 1954, Purification div. award 1958), Water Pollution Control Fedn. (Bedell award 1963), Kans. Pub. Health Assn. (Crumbine award 1965), Kans. Engring. Soc. (Outstanding Engr. award 1978), Nat. Acad. Engring., Sigma Xi, Tau Beta Pi. Subspecialties: Water supply and wastewater treatment; Environmental engineering. Current work: Air quality management work-development of regional water system. Research on rainfall and runoff predictions; health effects of organic chemicals; organics in ground water; advanced methods of water purification. Home: 3219 MacVicar Ave Topeka KS 66611 Office: Kansas Dept of Health and Environment Topeka KS 66620

METZNER, ARTHUR BERTHOLD, educator; b. Can., Apr. 13, 1927; s. Reinhold Berthold and Lillian (Bredis) M.; m. Elisabeth Krieger, May 9, 1948; children—Elisabeth L., Arthur P., Rebecca. B.Sc., U. Alta., 1948; Sc.D., Mass. Inst. Tech., 1951; D.A.Sc. (honoris causa), Katholieke U. te Leuven, Belgium, 1975. With Def. Research Bd., Ottawa, 1947-48, Colgate Palmolive Co., Jersey City, 1951-53; instr. Mass. Inst. Tech., 1950-51, Bklyn. Poly. Inst., 1951-53; mem. faculty U. Del., 1953—, H. Fletcher Brown prof. chem. engring., 1962—, chmn. dept., 1970-77; vis. prof. Stanford, 1967; Lacey lectr. Calif. Inst. Tech., 1968; vis. prof. U. Cambridge, Eng., 1968-69. Recipient Wilmington Sect. award Am. Chem. Soc., 1958; Colburn award Am. Inst. Chem. Engrs., 1958; Walker award, 1970; Lewis award, 1977; Bingham medal Soc. Rheology, 1977; Guggenheim fellow, 1968. Mem. Nat. Acad. Engring., Soc. Rheology, Am. Inst. Chem. Engrs. Subspecialty: Chemical engineering. Home: 713 Greenwood Rd Wilmington DE 19807 Office: U Del Newark DE 19711

MEYER, ANDREW U(LRICH), electrical engineering educator, researcher; b. Berlin, Apr. 21, 1927; U.S., 1949; s. Edmund and Elsbeth (Stelzer) M.; m. Elisabeth Voigts, Dec. 26, 1964; children: Michele C., Lydia N. M.S., Northwestern U., 1958, Ph.D., 1961. Devel. engr. Associated Research, Inc., Chgo., 1950, 53; project engr. Sun Electric Corp., Chgo., 1953-55; assoc. elec. engr. Armour Research Found. of Ill. Inst. Tech., Chgo., 1956-57; mem. tech. staff Bell Labs., Whippany, N.J., 1961-65; assoc. prof. elec. engring. N.J. Inst. Tech., Newark, 1965-68, prof. elec. engring., 1968—; vis. prof. Middle East Tech. U., Ankara, Turkey, 1969-70. Author: (with J.C. Hsu) Modern Control Principles and Applications, 1968. Served with U.S. Army, 1951-52. Mem. IEEE and IEEE Contro. System Soc. (chmn. chpt. 1964-65), IEEE New Tech. and Sci. Activities Com. (sec. 1967-68), Biomed. Systems Com. Am. Automatic Control Council, Internat. Fedn. Automatic Control, Soc. Indsl. and Applied Math., Am. Soc. Engring. Edn., AAUP, Internat. Assn. Math. and Computers in Simulation, AAUP, Sigma Xi, Eta Kappa Nu. Subspecialties: Biomedical engineering; Systems engineering. Current work: System analysis with application fo biomedical engineering problems, including modelling and clinical use of computers. Home: 746 Ridgewood Rd Millburn NJ 07041 Office: Elec Engring Dept NJ Inst Technology 323 High St Newark NJ 07102

MEYER, BONNIE JUNE FRANCIS, psychologist; b. Spokane, Wash., June 27, 1948; d. Robert Loring and Phyllis June (Smith) Francis; m. Richard Schlomer Meyr, Aug. 15, 1970; children: Jennifer June, Christina Francis. A.B., Wash. State U., 1970; M.S., Cornell U., 1971, Ph.D., 1974. Lic. psychologist, Ariz. Faculty research assoc. Cornell U., Ithaca, N.Y., 1973-74; asst. prof. Western Conn. State Coll., Danbury, 1975-76; vis. research psychologist Ednl. Testing Services, Princeton, N.J., 1974-76; asst. prof. Ariz. State U., Tempe, 1976-80, assoc. prof., 1980-83, prof., 1983—. Author: Organization of Prose and Its Effects, 1975; editor: Newsletter for Ednl. Psychologists, 1981 ; contbr. articles to profl. jours. Ariz. State U. faculty research awardee, 1982; NIMH grantee, 1978-81; Nat. Inst. Aging grantee, 1982—. Fellow Am. Psychol. Assn. (editor div. 15 1981—); mem. Am. Ednl. Research Assn. (sec.), Internat. Reading Assn. Subspecialties: Cognition; Developmental psychology. Current work: Learning/memory of prose across adult life span. Office: Ariz State U B325 L Payne Hall Tempe AZ 85287

MEYER, CARL DEAN, JR., mathematics educator; b. Greeley, Colo., Nov. 22, 1942; s. Carl and Louise (Fitzler) M.; m. Bethany Burrell, Jan. 7, 1961; children: Martin Dean, Holly Faythe. B.A., U. No. Colo., 1964; M.S., Colo. State U., 1966, Ph.D., 1968. Computer analyst Rocky Flats AEC, Golden, Colo., 1965-66; vis. prof. computer sci. Stanford (Calif.) U., 1978; prof. math. N.C. State U., Raleigh, 1968—. Author: Generalized Inverses of Linear Transformation, 1979, Creative Computer Programming, 1980; mng. editor: Soc. Indsl. and Applied Math. Jour. on Algebric and Discrete Methods, 1982—. Mem. Am. Math. Soc., Soc. Indsl. and Applied Math., Math. Assn. Am. Subspecialties: Numerical analysis; Mathematical software. Current work: Investigations of the effects of perturbations on the stationary distributions of Markov chains. Home: 704 Merwin Rd Raleigh NC 27606 Office: Math Dept NC State U Raleigh NC 27650

MEYER, DALE R., research scientist; b. Cin., Sept. 15, 1946; s. George Ernest and Thelma Louise N.; m. Pamela De Angelo, May 23, 1970; children: Royce J., Justin T. B.A., U. Cin., 1969; Ph.D., U. Louisville, 1975. NIMH research fellow U. Minn., Mpls., 1975-77; sr research toxicologist pathology and toxicology sect. Norwich-Eaton Pharms., Norwich, N.Y., 1977-79; research assoc. dept. ophthalmology U. Louisville, 1979-82; research asst. prof. dept. ophthalmology U. Tex. Health Sci. Center, 1982—; cons. drug safety evaluation. Contbr. articles to profl. jours. NSF fellow, 1969-73. Mem. Midwestern Psychol. Assn., So. Soc. Philosophy and Psychology, Soc. Neuroscience, AAAS, Nat. Acad. Sci., Assn. Research in Vision and Ophthalmology, U. Cin. Alumni Assn., U. Louisville Alumni Assn., Psi Chi. Subspecialties: Cell biology; Toxicology (medicine). Current work: Cell biology of ocular and neural tissues; in vitro model development for study of ophthalmic and neurotoxicity; clinical toxicology and pharmacology. Office: 5323 Harry Hines Blvd Dallas TX 75235

MEYER, DONALD CHARLES, physiologist, educator; b. Jersey City, Apr. 20, 1946; s. Edward C. and Margaret (Hendess) M.; m. Mary Jane Kittler, Oct. 18, 1948; children: Derek Craig, Douglas Scott. B.A., NYU, 1969; Ph.D., Rutgers U., 1973; postdoctoral trainee, U. Wis., 1973-76. Project assoc. Wis. Primate Center, Madison, 1976-77; asst. prof. physiology and biophysics U. Louisville Med. Sch., 1977—. Commr. City of Kingsley, 1982—. NIH grantee, 1980. Mem. Am. Physiol. Soc., Endocrine Soc., Soc. Neuroscience, Soc. Study Reprodn., Am. Soc. Zoologists. Lutheran. Subspecialty: Neuroendocrinology. Current work: Endocrinology, neuroendocrinology; role of neurotransmitters in neuroendocrinology and chronobiology; mechanism of serotonergic control of ovulation. Office: Dept Physiology Louisville U Med Sch Louisville KY 40292

MEYER, EDMOND GERALD, energy and natural resources educator; b. Albuquerque, Nov. 2, 1919; s. Leopold and Beatrice (Ilfeld) M.; m. Betty F. Knobloch, July 4, 1941; children: Lee Gordon, Terry Gene, David Gary. B.S. in Chemistry, Carnegie Mellon U., 1940, M.S., 1942; Ph.D. (research fellow), U. N.Mex., 1950. Chemist Harbison Walker Refractories Co., 1940-41; instr. Carnegie Mellon U., 1941-42; asst. phys. chemist Bur. Mines, 1942-44; chemist research div. N.Mex. Inst. Mining and Tech., 1946-48; head dept. sci. U. Albuquerque, 1950-52; head dept. chemistry N.Mex. Highlands U., 1952-59, dir., 1957-63, dean, 1961-63, Coll. Arts and Sci., U. Wyo., 1963-75, v.p., 1974-80, prof. energy and natural resources, 1981—; exec. cons. Diamond Shamrock Corp., 1980; pres. Coal Tech. Corp., 1981—; chmn. Am. Nat. Bank, Laramie.; Sci. adviser Gov. Wyo.; cons. Los Alamos Nat. Lab., NSF, HHS, GAO, Diamond Shamrock Corp., Wyo. Bancorp.; contract investigator Research Corp., Dept. Interior, AEC, NIH, NSF, Dept. Energy, Dept. Edn.; Fulbright exchange prof. U. Concepcion, Chile, 1959. Co-author: Chemistry-Survey of Principles, 1963, Legal Rights of Chemists and Engineers, 1977, Industrial R&D by Management, 1982; Contbr. articles to profl. jours. Served with USNR, 1944-46. Recipient Disting. service award Jaycees. Fellow AAAS, Am. Inst. Chemists (dir.); mem. Asso. Western Univs. (chmn. 1973-74), Am. Chem. Soc. (councillor), Chilean Chem. Soc., Biophys. Soc., Council Coll. Arts and Scis. (pres. 1971, sec.-treas. 1972-75, dir. Washington office 1973), C. of C. (pres. 1984), Sigma Xi. Subspecialties: Coal. Current work: Invention, development and commercialization of a novel coal based slurry transport system. Home: 1058 Colina Dr Laramie WY 82070 Office: Arts and Scis Bldg U Wyo Laramie WY 82071

MEYER, FRED LEWIS, ops. research scientist, naval officer; b. Queens, N.Y., Sept. 8, 1943; s. Henry and Pauline F. (Bauer) M.; m. Patricia Lynn McGrath, Oct. 12, 1968; children: Julie Anne, James Bryant. A.B. in Psychology, U. Rochester, 1965; M.S. in Ops. Research, U.S. Naval Postgrad. Sch., 1973; diploma, Nat. Def. U., 1981. Commd. ensign U.S. Navy, 1965, advanced through grades to comdr., 1979; dep. planning officer Naval Supply Ctr., Charleston, S.C., 1973-75; asst. supply officer USS Saratoga, 1975-77; dir. stock point and afloat systems Naval Supply Systems Command Hdqrs., Washington, 1977-80; dir. fleet non-tactical logistics info. systems Chief of Naval Material Hdqrs., Washington, 1980-81; dir. systems devel., internat. logistics programs Aviation Supply Office, Phila., 1981-83, dep. comptroller, 1982—; Mem. supervisory com. Navy Fed. Credit Union, 1980-81. Writer: Supply Corps Newsletter, 1983—. Cubmaster Boy Scouts Am., 1977-81, asst. scoutmaster, 1981-82. Recipient Leadership award Gen. Dynamics Corp., 1965; Citizenship award B'nai B'rith, 1961; decorated Navy Commendation (2). Mem. Ops. Research Soc. Am., World Future Soc., Theta Chi. Republican. Lutheran. Subspecialties: Operations research (mathematics); Information systems (information science). Current work: Operations research of navy logistics and business systems; comptrollership; information systems functional applications; design; acquisition; planning; inventory and financial systems modelling; strategic planning; futuristics. Home: 5314 Stonington Dr Fairfax VA 22032 Office: 700 Robbins Ave Philadelphia PA 19111

MEYER, GEORGE WILBUR, physician; b. Cleve., Apr. 30, 1941; s. George Wilbur and Emily Fuller (Campbell) M.; m. Carolyn Edwards Garrett, Apr. 8, 1967; children: Robert James, Elizabeth Jackson. B.S., MIT, 1962; M.D., Tulane U., 1966. Diplomate: Am. Bd. Internal Medicine, Am. Bd. Gastroenterology. Commd. capt. U.S. Air Force, 1967, advanced through grades to col., 1980; fellow in gastroenterology David Grant USAF Med. Ctr., Travis AFB, Calif., 1974-76, asst. chmn. medicine, Keesler AFB, Miss., 1976-78; asst. prof. medicine Uniformed Services U. Health Scis., Bethesda, MD., 1978-80; assoc. prof., 1980—; chmn. dept. medicine Wright Patterson USAF Med. Ctr.; asso. prof medicine Wright State U. Sch. Medicine, 1980-82; chief of medicine Wilford Hall USAF Med. Ctr., Lackland AFB, Tex., 1982—. Contbr. articles to profl. jours. Trustee, mem. Med. adv. com. United Health Services, Dayton, Ohio, 1980-82. Recipient Outstanding Community Service award, 1981; Meritorious Service medal Dept. Def., 1980; decorated Air Force Commendation medal. Fellow ACP; mem. Am. Soc. for Gastrointestinal Endoscopy, Am. Gastroent. Assn., AAAS, Am. Fedn. for Clin. Research. Subspecialties: Gastroenterology; Internal medicine. Current work: Gastroenterology—esophagus. Home: 2915 Hunters Den San Antonio TX 78230 Office: USAF Med Ctr/SGHM Wilford Hall Lackland AFB TX 78236

MEYER, HARRY M., JR., pediatrician; b. Palestine, Tex., 1928. M.D., U. Ark., 1953. Diplomate: Am. Bd. Pediatrics. Intern Walter Reed Army Hosp., 1953-54; asst. resident pediatrics N.C. Meml. Hosp., Chapel Hill., 1957-59; chief lab. viral immunology, div. biologics standards NIH (transferred to Food and Drug Adminstrn.), 1959-72; dir. Bur. Biologics NIH, 1972-82; dir. Nat. Ctr. for Drugs and Biologics, Rockville, Md., 1982—. Served to capt. M.C. AUS, 1953-57. Subspecialties: Microbiology; Virology (medicine). Current work: Program research and regulation drugs and biologics. A pioneer in devel. German measles vaccine. Address: Bur Biologics FDA 8800 Rockville Pike Bethesda MD 20205

MEYER, HOWARD STUART, chemical engineer; b. Chgo., Dec. 19, 1949; s. Sam A. and Melaine (Seldin) M.; m. Carol Renee Lewis, Sept. 3, 1972; children: Amanda Nicole, Sarah Gabrielle. B.S. in Chem. Engring, U. Ill.-Chgo., (1972); M.E. in Chem. engring, U. Idaho, (1978.). Chem. engr. Bee Chem. Co., Lansing, Ill., 1972-74; group leader Exxon, Idaho Falls, Idaho, 1974-80; project mgr. Gas Research Inst., Chgo., 1980. Mem. Am. Inst. Chem. Engrs., Am. Chem. Soc. (Richard A. Glenn award div. fuel chemistry 1982). Subspecialties: Coal; Gas cleaning systems. Current work: Coal gasification research; synthesis gas cleaning and upgrading to substitute natural gas. Office: 8600 W Bryn Mawr Chicago IL 60631

MEYER, JOHN AUSTIN, chemistry educator, administrator; b. St. Marys, Pa., Sept. 18, 1919; s. Henry George and Josephine K. (Hoehn) M.; m. Marion W. Waterman, Aug. 13, 1955. B.S. in Chemistry, Pa. State U., 1949, M.S. in Micro. Chemistry, 1950; Ph.D. in Poly. Chemistry, SUNY, 1958. Asst. chemist Speer Carbon Co., St. Marys, Pa., 1939-41; analytical chemist Gulf Research and Devel., Pitts., 1950-52; asst. dir. analytical lab. Koppers Research Ctr., Pitts., 1952-53; research asst. SUNY Coll. Environ. Sci. and Forestry., Syracuse, 1954-58; prof. nuclear and radiation chemistry, 1958—; cons. Allied Chem. Co., Syracuse, Ovation Instruments, New Hartford, Conn., PermaGrain Co., Media, Pa., Brookhaven Nat. Labs., Upton, N.Y. Contbr. articles to profl. jours. Bd. dirs. Pa. State Coll. Sci. Alumni; exec. com. SUNY Coll. Environ. Sci. Served with USAAF, 1941-45. Recipient Indsl. Arts award Indsl. Art Tchrs., N.Y., 1966; Chancellor's award SUNY, 1978; Borden Chem. award Forest Products Research, Soc., 1979. Mem. Am. Chem. Soc., Am. Nuclear Soc. (pres. sect. 1980-81), Soc. Plastic Engrs. (pres. CNY sect. 1964-65), Forest Products Research Soc. (sec. 1976). Republican. Subspecialties: Analytical chemistry; Polymer chemistry. Current work: Wood-polymers, neutron activation analysis, wood-preservation. Home: 149 Vincent St Syracuse NY 13210 Office: SUNY Coll Environ Sci and Forestry Syracuse NY 13210

MEYER, JOHN (HANS) FORREST, engineering executive; b. Milw., Feb. 6, 1947; s. Maynard W. and Mary H. (Nagler) M.; m. Pamela Dunbar; children: Ezra Joshua, Sydney Welch, Windsor Dunbar. B.S. in Aero. Engring, U. Mich., 1970. Founder, pres. Wind Works, Inc., Mukwonago, Wis., 1970—; spl. asst. to R. Buckminster Fuller, 1970—, cons. in field. Mem. editorial bd.: Solar Age Mag. Mem. Am. Wind Energy Assn., Internat. Solar Energy Soc. Subspecialties: Wind power; Solar energy. Office: Box 44A Route 3 Mukwonago WI 53149

MEYER, JOHN STIRLING, research administrator, neurologist, educator; b. London, Eng., Feb. 24, 1924; came to U.S., 1940; s. William Charles Bernard and Alice Elizabeth (Stirling) M.; m. Wendy Haskell, June 20, 1947; children: Jane, Anne, Elizabeth, Helen, Margaret. B.S., Trinity Coll., 1944; M.S., McGill U., 1949, M.D., C.M., 1948. Diplomate: Am. Bd. Psychiatry and Neurology. Instr. Yale U., New Haven, 1949-50; instr., research asst. Harvard U., Boston, 1950-57; prof. neurology Wayne State U., Detroit, 1957-69, Baylor Coll. Medicine, Houston, 1969—; chief cerebrovascular research VA Med. Center, Houston, 1977—; mem. Pres.'s Commn. on Heart Disease, Cancer, Stroke, Washington, 1964; mem. nat. adv. com. NIH, Bethesda, Md., 1965-69. Editorial staff: Jour. AMA; Author, editor: Diagnosis - Management Stroke, 1978 (Harold G Wolff award 1977, 78, 79); co-author: Medical Neurology, 1977; editor: Cerebral Vascular Disease, 1979; contbr. articles to sci. publs. Chmn. med. adv. bd. Harris County Parkinsons Soc., Houston, 1977—; chmn. stroke council Am. Heart Assn., Dallas. Served to lt. USN, 1952-54. Recipient award Am. Heart Assn., Detroit, 1967, citation of merit Mich. chpt. Multiple Sclerosis Soc., 1972; Disting. Service award Baylor Coll. Medicine, 1980; Pub. Awareness award Nat. Migrane Found., 1982. Fellow Am. Acad. Neurology; mem. World Fedn. Neurology (sec. 1977), Japanese Neurol. Soc. (hon.), Italian Neurol. Soc. (hon.). Republican. Episcopalian. Club: River Oaks Country. Subspecialties: Gerontology; Neurology. Current work: Cerebral blood flow and metabolism, stroke prevention, aging, migrane. Home: 2940 Chevy Chase Dr Houston TX 77019 Office: Cerebrovascular Research VA Med Center 2002 Holcombe Blvd Houston TX 77211

MEYER, PETER, educator, physicist; b. Berlin, Germany, Jan. 6, 1920; came to U.S., 1952, naturalized, 1962; s. Franz and Frida (Lehmann) M.; m. Luise Schützmeister, July 20, 1946 (dec. 1981); children: Stephan S., Andreas S.; m. Patricia G. Spear, June 14, 1983. Dipl.Ing., Tech. U., Berlin, 1942; Ph.D., U. Goettingen, Germany, 1948. Faculty U. Goettingen, 1946-49; fellow U. Cambridge, Eng., 1949-50; mem. sci. staff Max-Planck Inst. fuer Physik, Goettingen, 1950-52; faculty U. Chgo., 1953—, prof. physics, 1965—; dir. Enrico Fermi Inst., 1978—; Cons. NASA, NSF.; Mem. cosmic ray commn. Internat. Union Pure and Applied Physics, 1966-72; mem. space sci. bd. Nat. Acad. Scis., 1975-78. Fellow Am. Phys. Soc. (chmn. div. cosmic physics 1972-73), AAAS; mem. Am. Astron. Soc., Am. Geophys. Union, Max Planck Inst. fuer Physik und Astrophysik (fgn.), Sigma Xi. Subspecialty: Cosmic ray high energy astrophysics. Office: 933 E 56th St Chicago IL 60637

MEYER, RALPH ROGER, biological sciences educator; b. Milw., Feb. 18, 1940; s. Ralph George and Geneva Lorna (Schmidt) M.; m. Marjorie Kathleen Stark, Sept. 24, 1960 (div. 1971); children: Christine, Gregory, Lauren; m. Diane Carla Rein, Sept. 26, 1974; 1 dau. Jocelyn. B.S., U. Wis.-Milw., 1961; M.S., U. Wis.-Madison, 1963, Ph.D., 1966. Postdoctoral fellow in biochemistry Yale U., 1966-67, SUNY-Stony Brook, 1967-69; asst. prof. biol. sci. U. Cin., 1969-75, assoc. prof., 1975-79, prof., 1979—; vis. prof. biochemistry Stanford U., 1977-78. Contbr. articles to profl. jours. NSF grantee, 1969-71; Am. Cancer Soc. grantee, 1969-73, 78—; NIH grantee, 1975—. Mem. Am. Soc. Biol. Chemists, Am. Soc. Cell Biology, Am. Soc. Microbiology, AAAS, Sigma Xi. Subspecialties: Biochemistry (medicine); Cancer research (medicine). Current work: Regulation and mechanism of DNA synthesis in normal and turmor cells and in E. Coli. Home: 4067 Ridgedale Dr Cincinnati OH 45247 Office: Dept Biol Sci U Cin Cincinnati OH 45221

MEYER, RICHARD ARTHUR, research engineer; b. Springfield, Mass., June 26, 1946; s. Arthur Richard and Norma (Saunders) M.; m. Janice Bennett, Apr. 16, 1947; children: Jennifer Norma, Rebecca Joy. B.S.E.E., Valparaiso U., 1968; M.S. in Applied Physics, Johns Hopkins U., 1971. Sr. engr. Applied Physics Lab. Johns Hopkins U., 1968—, instr. neurosurgery, 1980-83, asst. prof. neurosurgery and biomed. engring., 1983—. Contbr. articles to profl. jours. Bd. dirs. Village of Kings Contrivance, Columbia, Md., 1980-82. Mem. IEEE, Soc. Neurosci., Internat. Assn. Study of Pain, Sigma Xi, Tau Beta Pi. Subspecialties: Biomedical engineering; Neurophysiology. Current work: Neurophysiological mechanisms of pain sensations. Office: Applied Physics Lab Johns Hopkins U Laurel MD 20707

MEYER, WALTER, energy engineer; b. Chgo., Jan. 19, 1932; s. Walter and Ruth (Killoran) M.; m. Jacqueline Miscall, May 8, 1953; children: Kim, Holt, Eric, Leah, Suzannah. B.Chem. Engring., Syracuse (N.Y.) U., 1956, M.Chem. Engring., 1957; postgrad. (NSF Sci. Faculty fellow), M.I.T., 1962, Ph.D., Oreg. State U., 1964. Registered nuclear engr., Calif. Prin. chem. engr. Battelle Meml. Inst., Columbus, Ohio, 1957-58; instr., then asst. prof. Oreg. State U., 1958-64; research engr. Hanford Atomic Labs., Richland, Wash., 1959-60, Lawrence Radiation Lab., Livermore, Calif., 1964; from asst. prof. to prof. nuclear engring. Kans. State U., Manhattan, 1964-72; prof., chmn. nuclear engring. U. Mo., Columbia, 1972-82, Robert Lee Tatum prof. engring., 1974—, co-dir. energy systems and resources program, 1974; co-founder Energy and Public Policy Center, 1981—; 1st holder Niagara Mohawk Energy prof. Syracuse U., 1982—; adj. prof. nuclear engring. U. Mo.-Columbia; dir. summer insts. NSF-AEC, 1969, NSF, 1972; co-dir. summer instr. AEC, 1972, dir. workshop, 1973; dir. (ERDA workshops), 1975-79; mem. Columbia Coal Gasification Task Force, 1977, Gov. Kans. Nuclear Energy Council, 1971-72; cons. to govt. and industry. Author. Mem. Manhattan Human Relations Orgn., 1966-72; active local Boy Scouts Am.; co-chmn. No on 11 Com., 1980. Grantee NSF, 1965-67, 73-75, 77-80, AEC, 1969-71, 73-74, Dept. Def., 1969-72, ERDA, 1975, 77—; NRC, 1977-80. Fellow Am. Nuclear Soc. (chmn. pub. info. com. 1981-84, nat. spl. award 1974, outstanding service award 1980, bd. dirs. 1981—); Mem. Am. Inst. Chem. Engrs. (chmn. nuclear engring. div. 1977-78), Am. Chem. Soc. (touring lectr. 1976-82), Am. Soc. Engring. Edn. (chmn. nuclear div. 1976-77), Am. Wind Soc., Sigma Xi, Tau Beta Pi. Presbyterian. Subspecialties: Chemical engineering; Nuclear engineering. Patentee in field Office: Office of the Dean of Engineering Link Hall Syracuse Univ Syracuse NY

MEYERAND, RUSSELL GILBERT, JR., scientist; b. St. Louis, Dec. 2, 1933; s. Russell Gilbert and Elsa Louise (Gebhardt) M.; m. Mary Grace Guillemin, June 16, 1956; 1 dau., Mary Elizabeth. B.S. in Elec. Engring, M.I.T., 1955, M.S. in Nuclear Engring, 1956, Ph.D. in Plasma Physics, 1959. Cons. to atomic power equipment dept. Gen. Electric Co., Schenectady, 1955-56; research asst. M.I.T., 1957-58; prin. scientist plasma physics United Technologies Research Center, East Hartford, Conn., 1958-64, chief research scientist, 1964-67, dir. research, 1967-79, v.p. tech., 1979—; Mem. vis. com. for sponsored research M.I.T.; mem. adv. com. U. Hartford Sch. Arts and Scis., 1965—; mem. engring. adv. com. Rensselaer Poly. Inst., 1965—; mem. vis. com. Nat. Bur. Standards, 1980—. Contbr. articles to profl. jours., chpts. to books. Bd. dirs. Newington Children's Hosp. Recipient Eli Whitney award Conn. Patent Law Assn., 1979. Fellow AIAA; mem. Nat. Acad. Engring., IEEE, Am. Phys. Soc., Sigma Xi. Subspecialty: Plasma physics. Patentee in field. Office: Silver Ln East Hartford CT 06108

MEYERHOF, WALTER ERNST, physicist, educator; b. Kiel, Germany, Apr. 29, 1922; came to U.S., 1941, naturalized, 1946; s. Otto Fritz and Hedwig (Schallenberg) M.; m. Miriam G. Ruben, Aug. 21, 1947; children: Michael O., David L. Ph.D., U. Pa., 1946, U. Frankfurt, W. Ger., 1980. Asst. prof. U. Ill., Urbana, 1946-49; faculty physics Stanford U., 1949—, prof., 1959—, acting exec. head, 1962-63, chmn. dept. physics, 1970-77. Fellow Am. Phys. Soc. Subspecialty: Atomic and molecular physics. Home: 213 Blackburn Ave Menlo Park CA 94025 Office: Physics Dept Stanford U Stanford CA 94305

MEYERHOFF, ARTHUR AUGUSTUS, consulting companies executive, petroleum geologist; b. Northampton, Mass., Sept. 9, 1928; s. Howard Augustus and Anna Sophia (Theilen) M.; m. Kathryn Eleanor Laskaris, Jan. 2, 1951; children: James Charles, Richard Dietrich, Donna Kathryn. B.A., Yale U., 1947; M.S., Stanford U., 1950, Ph.D., 1952. Geologist U.S. Geol. Survey, Mont., Wyo., 1948-52; geologist Standard Oil Co. Calif., Latin Am., 1952-56, sr. geologist, various fgn. and domestic locations, 1956-65; publs. mgr. Am. Assn. Petroleum Geologists, Tulsa, 1965-74; pres. Meyerhoff & Cox, Inc., Tulsa, 1974—; v.p. for research and devel. Associated Resource Cons., Inc., Tulsa, 1978—; dir. Hagen-Greenbriar Corp., Houston, 1980—; cons. exec. br., Washington, 1976—; guest lectr. U. Belgrade, 1973, USSR Ministry Oil Industry, 1974, USSR Acad. Scis. Inst. Physics of Earth, 1974, USSR Acad. Scis. Far East Inst., 1977, 78; Disting. lectr. Geol. Assn. Can. and U. Calgary, 1974; guest TV shows; vis. scholar Gonville and Caius Coll., Cambridge (Eng.) U., 1978-79. Contbr. numerous articles on oceanography, sea floor spreading, geology, politics of petroleum to profl. publs. Fellow Geol. Soc. Am., AAAS, Geol. Soc. (London); mem. Geol. Soc. Australia, Am. Assn. Petroleum Geologists (cert. petroleum geologist, Matson award 1969), Can. Soc. Petroleum Geologists (hon. lectr. 1975), Am. Geophys. Union, Soc. Econ. Paleontologists and Mineralogists, Paleontol. Assn., Petroleum Exploration Soc. Australia (Distig. Lectr. from Overseas 1982), Assn. Earth Sci. Editors (pres. 1968-69), Rocky Mountain Assn. Geologists, Tulsa Geol. Soc. (hon.), Lafayette (La.) Geol. Soc., Soc. Ind. Profl. Earth Scientists (cert. profl. geologist), Assn. Ind. Profl. Geologists, S.E. Asia Exploration Soc., Geol. Soc. Malaysia, German Geol. Soc., Okla. City Geol. Soc., Geol. Assn. India, Sociedad Mexicana Geologica, Houston Geol. Soc., Assn. Geoscientists for Internat. Devel., Asociación Mexicana de Geólogos Petroleros, Soc. Exploration Geophysicist, Sociedade Brasileira de Geologia. Club: Cosmos (Washington). Subspecialties: Tectonics; Geophysics. Current work: Earth dynamics—most research directed toward explaining origin of earth; petroleum geology consultant, especially to Third World nations which are energy poor. Home: 3123 E 28th St Tulsa OK 74114 Office: Associated Resources Cons PO Box 4602 3336 E 32d St Suite 208 Tulsa OK 74135

MEYERHOFF, JAMES LESTER, psychiatrist; b. Phila., Dec. 12, 1937. B.A., U. Pa., 1962, M.D., 1966. Diplomate: Am. Bd. Psychiatry and Neurology; Nat. Bd. Med. Examiners. Intern Misericordia Hosp., Phila., 1966-67; resident in psychiatry U. Chgo. Hosp., 1967-70; postdoctoral fellow dept. pharmacology and exptl. therapeutics Johns Hopkins U. Sch. Medicine, Balt., 1970-71; research assoc. div. neuropsychiatry Walter Reed Army Inst. Research, Walter Reed Army Med. Ctr., Washington, 1971-72, head neurochemistry sect. microwave research, div. neuropsychiatry, 1972-74, chief dept. neuroendocrinology div. neuropsychiatry, 1974-76, chief neuroendocrinology and neurochemistry br. dept. med. neurosci., div. neuropsychiatry, 1976—; research prof. psychiatry Uniformed Services U. Health Scis., 1978; clin. assoc. prof. psychiatry Georgetown U., Washington, 1977; lectr. and cons. in field. Guest reviewer: in endocrinology Jour. Neurochemistry; contbr. numerous articles on psychiatry to profl. jours. Cons. to Mobile Med. Care; soccer coach to D.C., Arlington, Va. and Columbia Women's Leagues. Served to maj. U.S. Army, 1969-72. Fellow Am. Psychiat. Assn.; mem. Washington Psychiat. Soc., Internat. Soc. Chronobiology. Subspecialties: Psychopharmacology; Neurochemistry. Current work: Neuroendocrinology of stress and depression; neurochemistry of epilepsy and depression. Office: Dept Neuropsychiatry Walter Reed Army Med Center Walter Reed Army Inst Research Washington DC 20012

MEYEROWITZ, ELLIOT MARTIN, biol. researcher; b. Washington, May 22, 1951. A.B., Columbia U., 1973; M.Phil., Yale U., 1975, Ph.D., 1977. Research fellow dept. biochemistry Stanford (Calif.) U., 1977-79; asst. prof. biology Calif. Inst. Tech., Pasadena, 1980—. Sloan Found. fellow, 1981-83. Mem. Genetics Soc. Am., AAAS. Subspecialties: Genetics and genetic engineering (biology); Molecular biology. Current work: Gene regulation and pattern formation in animal and plant development. Office: Div Biology Calif Inst Tech Pasadena CA 91125

MEYERS, LAWRENCE STANLEY, psychology educator, consultant; b. Bklyn., Apr. 6, 1943. B.S., Bklyn. Coll., 1964; Ph.D., Adelphi U., 1968. Vis. asst. prof. U. Tex., Austin, 1968-69; NSF postdoctoral fellow, 1968-69, Purdue U., West Lafayette, Ind., 1969-70, vis. asst. prof., 1969-70; asst. prof. Calif. State U.-Sacramento, 1970-74, assoc. prof., 1974-81, prof., 1981—, co-dir., 1977-81; cons. Research Cons. Corp., Sacramento, 1976-81, Calif. Dept. Justice, 1982—. Author: (with Neal Grossen) Behavioral Research, 1974, 78. Grantee U.S. Office Edn., 1979-82, Calif. Dept. Edn., 1979-81, Calif. State U. Chancellor's Office, 1979-80, 80-81. Mem. Am. Psychol. Assn., Psychonomic Soc., Western Psychol. Assn., Eastern Psychol. Assn. Subspecialties: Cognition. Current work: Applied research, job analysis, test validation, human factors, communication disabilities. Office: Psychology Dept Calif State U Sacramento CA 95819

MEYERS, MORTON ALLEN, physician, educator; b. Troy, N.Y., Oct. 1, 1933; s. David and Jeanne Sarah (Dunn) M.; m. Beatrice Applebaum, June 1, 1963; children—Richard, Amy. M.D., SUNY, Upstate Med. Coll., 1959. Diplomate: Am. Bd. Radiology. Intern Bellevue Hosp., N.Y.C., 1959-60; resident in radiology Columbia-Presbyn. Med. Center, N.Y.C., 1960-63; fellow Am. Cancer Soc., 1961-63; prof. dept. radiology Cornell U. Med. Center, N.Y.C., 1973-78; prof., chmn. dept. radiology SUNY Sch. Medicine, Stony Brook, 1978—; vis. investigator St. Mark's Hosp., London, 1976. Author: Diseases of the Adrenal Glands: Radiologic Diagnosis, 1963, Dynamic Radiology of the Abdomen: Normal and Pathologic Anatomy, 1976, 2d edit., 1982, Iatrogenic Gastrointestinal Complications, 1981; series editor: Radiology of Iatrogenic Disorders, 1981—; Founding editor-in-chief: Gastrointestinal Radiology, 1976—; contbr. chpts. to med. textbooks, articles to med. jours. Served to capt. M.C. U.S. Army,

1963-65. Fellow Am. Coll. Radiology; mem. Radiol. Soc. N. Am., Am. Roentgen Ray Soc., Assn. Univ. Radiologists, N.Y. Roentgen Ray Soc., Am. Gastroenterol. Assn., Soc. Gastrointestinal Radiologists, AAAS, Soc. Uroradiology, N.Y. Acad. Gastroenterology, Phila. Roentgen Soc., Harvey Soc., N.Y. Acad. Scis., Soc. Chmn. Acad. Radiology Depts., L.I. Radiologic Soc., Alpha Omega Alpha. Subspecialty: Diagnostic radiology. Home: 14 Wainscott Ln East Setauket NY 11733 Office: Dept Radiology Sch Medicine Health Scis Center SUNY Stony Brook NY 11794

MEYERS, PAUL, microbiologist, educator; b. Phila., Sept. 23, 1939; s. Joseph and Dorothy (Shapin) M.; m. (div.); children: Jonathan, Karen. B.A., Temple U., 1961; Ph.D., SUNY-Syracuse, 1970. Research assoc. dept. microbiology U. Miami (Fla.) Sch. Medicine, 1970-71, instr., 1971-72, asst. prof., 1972-73; asst. prof., assoc. cons. dept. microbiology Mayo Clinic and Found., Rochester, Minn., 1973-74, asst. prof., 1973-77, assoc. prof., 1977-80, cons., 1974-80; research assoc. prof. microbiology SUNY Upstate Med. Center, Syracuse, 1980—. Contbr. articles to profl. jours. Chmm. conservation com. Fla. Sierra Club, 1972-73. Nat. Cancer Inst. grantee, 1973—. Fellow Am. Acad. Microbiology; mem. Soc. Gen. Microbiology, Am. Assn. Cancer Research, Am. Assn. Immunologists, Am. Soc. Microbiology (pres. North Central br. 1977-79), Transplantation Soc., N.Y. Acad. Scis., Am. Soc. Virology. Subspecialties: Virology (biology); Immunobiology and immunology. Current work: Virology; immunity to oncogenic viruses and viral antigens; transplantation antigens, cell-mediated immunity, tolerance. Home: 3 Fir Tree La Jamesville NY 13078 Office: 766 Irving Ave Syracuse NY 13210

MEYERS, PHILIP ALAN, chemist, educator, consultant; b. Hackensack, N.J., Mar. 3, 1941; s. Harold Grove and Gertrude Myra (Smith) M.; m. Judith Arlene Brown, May 15, 1965; children: Shelley, Suzanne, Christopher. B.S. in Chemistry, Carnegie-Mellon U., Pitts., 1964; Ph.D. in Oceanography, U. R.I., Kingston, 1972. Research chemist Inmont Corp., Nutley, N.J., 1967-68; asst. prof. oceanography U. Mich., Ann Arbor, 1972-77, assoc. prof., 1977-82, prof., 1982—; vis scientist Ind. U., Bloomington, 1979-80; cons. Raytheon Ocean Systems Co., NOAA, UCLA; dir. Gt. Lakes and Marine Waters Center of U. Mich., 1982. Contbr. articles and abstracts to profl. jours. Served to lt. (j.g.) USNR, 1964-67. Recipient Disting. Service award U. Mich. Coll. Engring., 1976; NOAA faculty summer fellow, 1981. Fellow Geol. Soc. Am.; mem. AAAS, Am. Geophys. Union, Am. Soc. Limnology and Oceanography, Internat. Assn. Gt. Lakes Research, Geochem. Soc. Club: Ann Arbor Country. Subspecialties: Organic geochemistry; Oceanography. Current work: Organic geochemistry of natural waters and sediments, diagenesis and degradation of fatty acids and hydrocarbons and sterols, biomarker and paleoenvironmental studies in Deep Sea Drilling Project. Home: 4810 Whitman Circle Ann Arbor MI 48103 Office: 2455 Hayward Ave Ann Arbor MI 48109

MEYERS, RICHARD ANTHONY, elec. engr., educator, author; b. Orange, N.J., Mar. 2, 1945; s. Harry Daniel and Evelyn Ruth (Brickman) M. B.A., Hamilton Coll., 1968; B.S., Renssalear Poly. Inst., 1968; M.S.E.E., Columbia U., 1977. With N.J. Bell Telephone Co., 1968-70; Peace Corps vol., India, 1970-72; head physics dept. Thimphu (Bhutan) Pub. Sch., 1973-74; adj. prof. computer sci. N.Y. Inst. Tech., N.Y.C., 1980-82, asst. prof., staff specialist, 1982. Author 2 spl. English series textbooks on elec. engring. and computer sci. Mem. Union Concerned Scientists, Council for a Livable World, Ctr. for Def. Info., SANE, Tau Beta Pi, Eta Kappa Nu. Theosophist. Subspecialties: Computer science education; Programming languages. Current work: I study languages because they're fun—I have no desire to do research, only to learn and teach. Educator—author of 5 study guides used internally at N.Y. Inst. Tech; the pursuit of excellence in both learning and teaching relevant, non-violent technology. Home: 479 Broad Ave Apt 2G Palisades Park NJ 07650 Office: 1855 Broadway 12th Floor New York NY 10023

MEYERSON, LAURENCE R., neuroscientist; b. Chgo., July 4, 1947; s. Edwin W. and Marian (Rappeport) M.; m. Penelope J. Giles, Feb. 19, 1982. B.S., U. N.Mex., 1970; Ph.D., Chgo. Med. Sch., 1974. Sr. staff scientist VA Med. Ctr., Houston, 1974-78; sr. research biochemist Hoechst Roussel Pharm., Somerville, N.J., 1978-80; adj. assoc. prof. dept. psychology CUNY, 1982—; head neurobiochem. labs. Am. Cyanamid Co., Lederle Labs., Pearl River, N.Y., 1980—; cons. dept. psychiatry VA Med. Ctr., N.Y.C. Contbr. articles to profl. jours., chpts. to books, abstracts to jours. Recipient Cyanamid Research Achievement award, 1983; NIMH grantee, 1974-78; co-investigator Sprague Found. Predoctoral fellow, 1972-74. Mem. Soc. Neuroscis., Am. Soc. Neurochemistry, Am. Soc. Pharmacology and Exptl. Therapeutics, AAAS, Sigma Xi. Jewish. Club: Kayak and Canoe of N.Y. (membership and publicity chmn.). Subspecialty: Neurochemistry. Current work: Mechanisms of depression, anxiety and aging; enzymology of pertinent CNS function, peptide function; characterization of endogenous biochemical modulators. Home: 4 Beechwood Rd Montvale NJ 07645 Office: Dept CNS Research Med Research Div Am Cyanamid Co Pearl River NY 10965

MEYSTEL, ALEXANDER MICHAEL, electrical engineering educator; b. Leningrad, USSR, Feb. 25, 1935; came to U.S., 1978, naturalized, 1983; s. M.L. and C.S. (Cotlair) M.; m. Marina Selitsky, Feb. 26, 1971; 1 child, Misha. M.S.E.E., Poly. Inst., Odessa, USSR, 1957; Ph.D., ENIMS, Moscow, 1965. Project leader Design Office Machines, Odessa, USSR, 1957-63; sr. researcher Exptl. Sci. Research Inst. Metalcutting Machines, Moscow, 1963-65, head lab., Erevan, 1965-69, sr. scientist, Moscow, 1969-73; head dept. Informelecto, 1973-77; sr. staff scientist Gould, Inc., Chgo., 1978-79; research/devel. dir. Hyperloop, Inc., Chgo., 1980-81; assoc. prof. elec. engring. U. Fla., Gainesville, 1980—; cons. in field. Contbr. articles to profl. jours.; author: Automated Positioning Controls, 1970, Engineering Computations and Design of Automated Machines, 1976, Computer Aided Decision Making, 1976. Recipient medals for engring. innovations All-Union Exhbn. of Indsl. Achievements, Moscow, 1962-70. Mem. IEEE, N.Y. Acad. Sci., Soc. Indsl. and Applied Math., Am. Assn. Artificial Intelligence, Sigma Xi. Jewish. Subspecialties: Electrical engineering; Artificial intelligence. Current work: Control of intelligent machines, optimum control of multilink robot manipulators, intelligent mobile systems, others. Patentee in field. Office: Dept Elec Engring U Fla 137 Larsen Hall Gainesville FL 32611 Home: 1421 NW 51st Terr Gainesville FL 32605

MICELI, JOSEPH N(ICOLA), pharmacology educator, toxicologist, laboratory administrator; b. N.Y.C., Jan. 13, 1945; s. Salvatore and Jean (Ippolito) M.; m. dau., Laura. B.S., Pace U., 1966; M.S., U. Detroit, 1977, Ph.D., 1977. Lic. lab. dir., Mich. Assoc. research scientist NYU Med. Ctr., 1971-74; sr. research scientist U. Mich., 1974-75; asst. prof. pediatrics and pharmacology Wayne State U., 1975-79, assoc. prof., 1980—; dir. Clin. Pharmacology and Toxicology Labs., Children's Hosp. Mich., Detroit, 1975—; adj. asst. prof. Pace U., 1972-74; adj. asst. prof. chemistry Post-mortum Tissue Distbn., Mich. Dept. Public Health, 1981-82. Contbr. chpts. to books, articles to profl. publs. Recipient Faculty Merit award Wayne State U., 1976, 77, 78, 82. Mem. Am. Acad. Clin. Toxicology, Am. Soc. Pharmacology and Exptl. Therapeutics, Soc. Pediatric Research, Am. Assn. Clin. Chemistry, N.Y. Acad. Scis., Mich. Acad. Scis. Subspecialties: Pharmacology; Toxicology (medicine). Current work: Clinical pharmacology and toxicology, pediatrics, drug disposition, analytical methodology for drugs, toxicological analysis.

MICHAEL, EUGENE JOSEPH, environmental engineer, consultant; b. Buffalo, Nov. 14, 1947; s. Eugene Benjamin and Margret Mary (Burke) M.; m. Arlene Francis Pollard, Aug. 9, 1980; 1 dau., Jennifer Lee. B.S. in Civil Engring, Clarkson Coll. Tech., 1970. Cert. profl. engr., N.Y. Jr. engr. N.Y. State Dept. Environ. Conservation, Albany, 1970-71, asst. engr., 1971-74, sr. engr., 1974-77; environ. engr. Stone & Webster Engring., Boston, 1977-78, prin. engr., 1978-79, lead environment engr., 1979-80, supr. Radiological planning, 1980—. Mem. Health Physics Soc., Am. Nuclear Soc., Atomic Indsl. Forum. Roman Catholic. Subspecialties: Nuclear engineering; Nuclear physics. Current work: Preparation of radiological emergencies response plans for utilities, states and counties or municipalities; development of evaluation time estimates and development of prompt notification systems for utilities. Office: Stone & Webster Engring Corp 245 Summr St Boston MA 02107

MICHAEL, JACOB GABRIEL, microbiology educator; b. Rimavska Sobota, Czechoslovakia, July 2, 1931; s. Stephan and Mary (Laszlo) M.; m. Ann Deborah Blumrosen, Aug. 29, 1958; children: Naomi, Ruth, David, Abigail. B.A., Hebrew U., Jerusalem, Israel, 1955, M.S., 1956; Ph.D., Rutgers U., 1959. Vis. scientist Nat. Cancer Inst., Bethesda, Md., 1959-61; assoc. dir. research Children's Hosp., Harvard Med. Sch., Boston, 1961-66; assoc. prof. U. Cin., 1966-70, prof. microbiology and immunology, 1970—; vis. prof. Inst. Pasteur, Paris, 1979, Karolinska Inst., Stockholm, 1973. Contbr. articles to profl. jours. Am. Heart Assn. fellow, 1962; Am. Soc. Microbiology fellow, 1964; Med. Found. Boston fellow, 1966. Fellow Am. Acad. Microbiology; mem. Am. Soc. Microbiology, N.Y. Acad. Scis., Am. Assn. Immunologists, AAAS. Jewish. Subspecialties: Immunobiology and immunology; Allergy. Current work: Regulation of immune response; immunology, allergens, autoimmune disease, suppression or enhancement of the response. Office: Dept Microbiology U Cin Coll Medicine Cincinnati OH 45267

MICHAELS, ADLAI ELDON, chemistry educator; b. Alma, Wis., Nov. 22, 1913; s. Frederick William and Olga (Wald) M.; m. Josephine Gertrude Blake, Mar. 16, 1940; children: Lee Frederick, Carol Ann Price; m. Opal M. Carfrey, June 4, 1971. B.S., U. Wis., 1935; Ph.D., Ohio State U., 1940. Instr. U. Tenn., Knoxville, 1940-43; research chemist Esso Research & Engring. Co., Elizabeth, N.J., 1949-59; mem. chemistry faculty Washington and Jefferson Coll., Washington, Pa., 1959—, prof. chemistry, 1967—. Sci adviser U.S. Congress. Recipient Arthur Donahue Meml. award Alpha Phi Omega, 1982. Mem. Am. Chem. Soc. Lodge: Elks. Subspecialties: Physical chemistry; Analytical chemistry. Current work: Corrosion, air pollution control. Patentee in field. Office: Washington and Jefferson Coll Washington PA 15301

MICHAELS, HOWARD BRIAN, med. physicist; b. Toronto, Ont., Can., May 29, 1949; s. Isaiah and Rosalind (Rosenberg) M.; m. Lois S. Kwitman, Mar. 15, 1980. B.A.Sc., U. Toronto, 1971, M.Sc., 1973, Ph.D. in Med. Biophysics, 1976. Registered profl. engr., Ont. Postdoctoral fellow Ont. Cancer Inst., Toronto, 1976; research fellow in radiation medicine Mass. Gen. Hosp., 1976-78; research fellow in radiation therapy Harvard Med. Sch., Boston, 1976-78; asst. radiation biophysicist and asst. prof. radiation therapy, 1978-81; chief physicist, head div. clin. physics Ont. Cancer Found., Toronto-Bayview Clinic, 1981—; asst. prof. depts. med. biophysics and radiology, U. Toronto, 1981—, depts. oncology and radiology, 1981—. Contbr. articles to profl. jours. Nat. Cancer Inst. Can. K.J. Hunter fellow, 1975-76; Radiation Research Soc. awardee, 1974, 76, 78, 79. Mem. Am. Assn. Physicists in Medicine, Can. Assn. Physicists, Radiation Research Soc., Assn. Profl. Engrs. Ont. Current work: Medical radiation physics associated with radiotherapy treatment for cancer; radiobiology and radiation chemistry; radiation sensitization by oxygen and chemical radiosensitizers and radioprotectors of mammalian cells; clinical use of radiosensitizers in radiotherapy; chemical modifiers of ionizing radiation damage. Office: 2075 Bayview Ave Toronto ON Canada M4N 3M5

MICHALITSIANOS, ANDREW GERASIMOS, astrophysicist; b. Alexandria, Egypt, May 22, 1947; came to U.S., 1951; s. Gerasimos Andrew and Maria (Soultanakis) M.; m. Kathryn, Mar. 25, 1950. B.S. in Physics, U. Ariz., 1969; Ph.D. in Astrophysics, U. Cambridge, Eng., 1973. Research fellow Calif. Inst. Tech., Pasadena, 1972-75; research assoc. Swiss Fed. Inst. Tech., Zurich, 1976-77; assoc. NRC/ Nat. Acad. Scis., NASA Goddard Space Flight Center, Greenbelt, Md., 1975-77; staff scientist Lab. for Astronomy and Solar Physics, 1977—, head obs. sect., 1981—. Contbr. articles to profl. jours. Recipient NASA Merit Achievement award, 1980. Fellow Am. Astron. Soc., Royal Astron. Soc.; mem. Internat. Astron. Union. Democrat. Greek Orthodox. Subspecialties: Ultraviolet high energy astrophysics; Optical astronomy. Current work: Ultraviolet astronomy mainly with earth orbiting UV astronomical satellites; ground-based observations with radio telescopes and optical instrumentation.

MICHALSKI, FRANK JOSEPH, lab adminstr., researcher, educator; b. St. Louis, Sept. 28, 1938; s. Frank Joseph and Ann Teresa (Kane) M.; m. Julieta Amy Buera, Aug. 11, 1973. Ph.D. in Biology, U. Notre Dame, 1971. Postdoctoral virus research Wistar Inst. U. Pa., 1971-73, Yale U., 1973-76; dir. diagnostic virology lab. St. Michaels Med. Center, Newark, 1976—; asst. prof. N.J. Med. Sch., Newark, 1976—. Mem. Am. Soc. Microbiology, Pan Am. Group for Rapid Viral Diagnosis, Acad. Medicine of N.J. Democrat. Roman Catholic. Subspecialty: Virology (medicine). Current work: Rapid virus diagnosis, herpes virus research, oncogenic viruses.

MICHEJDA, CHRISTOPHER JAN, chemist, cancer researcher; b. Kielce, Poland, Dec. 19, 1937; came to U.S., 1951, naturalized, 1956; s. Jan and Janina (Trybulski) M.; m. Maria Lacki, Feb. 8, 1964; 1 dau., Monika. B.S. in Chemistry, U. Ill., 1959; Ph.D. (Kodah prize), U. Rochester, N.Y., 1963. Postdoctoral fellow Harvard U., 1963-64; from asst. prof. to prof. chemistry U. Nebr., 1964-78; vis. prof. Swiss Fed. Inst. Tech., 1972-73; program officer NSF, 1974-77; head chemistry carcinogens sect. Frederick (Md.) Cancer Research Facility, Nat. Cancer Inst., 1978—; cons. in field. Author papers in field. Mem. Am. Chem. Soc., Am. Assn. Cancer Research, AAAS. Subspecialties: Organic chemistry; Cancer research (medicine). Current work: Mechanistic organic chemistry, chemical carcinogenesis, bio-organic chemistry, synthetic organic chemistry. Office: NCI Frederick Cancer Research Facility Frederick MD 21701

MICHEL, BERNARD, civil engineering educator; b. Chicoutimi, P.Q., Can., May 31, 1930; s. Joseph-Williams and Jeanne (Tremblay) M.; m. Mariette Boivin, Sept. 9, 1954; children: Marianne, Francois, Luc, Jacques, Charles, Christine. B. Applied Sci., Laval U., 1954; Dr.Engr., Grenoble U., 1962. Profl. engr., P.Q. Research engr. Lasalle Hydraulic Lab., Quebec, 1956-60; head dept. civil engring. Laval U., Quebec, 1960-63, prof. civil engring., 1963—; v.p. Arctec Can. Ltd., Ottawa, 1973-78; cons. Recherches B.C. Michel Inc., Quebec, 1978—. Author: Ice Mechanics, 1978. Recipient Gzowski medal Engring. Inst. Can., 1963. Fellow Royal Soc. Can. (mem. Acad. of Scis.), Engring. Inst. Can., Can. Soc. Civil Engring. (Keefer medal 1977, 81), Internat. Assn. Hydraulic Research (pres. com. on ice problems 1970-76). Subspecialty: Civil engineering. Current work: Research and consultations for the development of northern water resources with specialty in ice engineering. Inventor, patentee hydraulics and off-shore installations. Home: 739 des Vignes Ste-Foy PQ Canada G1V 2Y1 Office: Universite Laval Dep de Genie civil Ste-Foy PQ Canada G1K 7P4

MICHELL, RICHARD EDWARD, nematologist, plant pathologist; b. Bronx, N.Y., Mar. 5, 1942; s. Richard Walter and Alice Martha (Bauer) M.; m. Peggy Ann Slater, Oct. 19, 1973; children: Deana, Bonnie. A.A.S., SUNY-Farmingdale, 1962; B.S.A., U. Ga., 1965, M.S., 1967; Ph.D., U. Ill., 1972. Nematologist, plant pathologist Registration div. Office of Pesticide Programs, EPA, Washington, 1972. Contbr. articles to profl. jours. Active Va. Wildlife Soc. Mem. Soc. Nematologists, Am. Phytopath. Soc. Democrat. Roman Catholic. Club: Bass Anglers Sportsman Soc. Subspecialties: Plant pathology; Plant nematology. Current work: Fungicide and nematicide testing and labeling; fungicides; nematicides; pesticide registration; efficacy testing; registered use patterns. Home: 14557 Eastman St Dale City VA 22193 Office: Environmental Protection Agency 401 M St SW Washington DC 20460

MICHELS, CORRINE ANTHONY, biologist, educator, researcher; b. N.Y.C., Jan. 2, 1943; d. James A. and Clorinda C. (Sicari) Anthony; m. Harold T. Michels, Jan. 24, 1964; children: William J., Catherine L. B.S., Queens Coll., Flushing, N.Y., 1963; M.A., Columbia U., 1965, Ph.D., 1969. Postdoctoral work Columbia U., 1969-70; research asso. Albert Einstein Coll. Medicine, 1970-72; faculty Queens Coll., 1972—, asso. prof., 1979—. Contbr. articles to profl. jours. NIH grantee, 1975—. Mem. Genetics Soc. Am., AAAS, Soc. Microbiology. Subspecialties: Molecular biology; Gene actions. Current work: Catabolite repression in yeast. Office: Dept Biolog Queens Coll Flushing NY 11367

MICHELS, DONALD JOSEPH, research physicist, educator; b. Bklyn., Apr. 17, 1932; s. Joseph George and Alice Elizabeth (Sweeney) M.; m. Jacqueline Mary Grace, June 24, 1961; children: Mary, Joseph, Margaret, John, Anne, Thomas. B.S. in Physics, St. Peter's Coll., 1954, M.S., Fordham U., 1956, Ph.D., Cath. U. Am., 1970. Research physicist Naval Research Lab., Washington, 1955-79, supervisory research physicist, 1979—; lectr. in astronomy U. Md. Mem. Optical Soc. Am., Am. Geophys. Union. Subspecialties: Solar-terrestrial physics; Solar physics. Current work: Experimental observation of outer solar corona; advanced instrumentation for space research; solar wind; advanced detectors; extreme ultraviolet; electro-optics; charge-coupled devices. Office: Naval Research Lab Code 4173 Washington DC 20375

MICHELS, LESTER DAVID, medical device manufacturing company research executive; b. Chgo., Feb. 5, 1948; s. Lester William and Irene (Caulder) M.; m. Jane Elizabeth Wittich, Dec. 20, 1969; 1 son, Zachary. B.S.Ch.E., U. Minn.-Mpls., 1971-75, Ph.D., 1975. Research engr. Dow Chem. Co., Midland, Mich., 1970-71; teaching asst. U. Minn.-Mpls., 1971-75, research specialist, 1975-76; research physiologist Mpls. Med. Research and U. Minn., 1976-82; research mgr. Medtronic, Inc., Mpls., 1982—. Contbr. articles to profl. jours. Named Outstanding Lectr. U. Minn. Dental Sch., 1980; Am. Diabetes Assn. grantee, 1979-81; Juvenile Diabetes Assn. grantee, 1980-81; NIH grantee, 1980-83. Mem. Am. Physiol. Soc., Am. Fedn. Clin. Research, Am. Soc. Nephrology. Subspecialties: Psychophysiology; Biomedical engineering. Current work: Pathophysiological consequences of renal disease; development of biomedical devices. Home: 9381 Amsden Way Eden Prairie MN 55344 Office: Medtronic Inc 6951 Central Ave NE PO Box 1453 Minneapolis MN 55440

MICHENER, CHARLES DUNCAN, educator, entomologist; b. Pasadena, Calif., Sept. 22, 1918; s. Harold and Josephine (Rigden) M.; m. Mary Hastings, Jan. 1, 1941; children: David, Daniel, Barbara, Walter. B.S., U. Calif., Berkeley, 1939, Ph.D., 1941. Tech. asst. U. Calif., Berkeley, 1939-42; asst. curator Am. Mus. Natural History, N.Y.C., 1942-46, asso. curator, 1946-48, research asso., 1949—; asso. prof. U. Kans., 1948-49, prof., chmn. dept. entomology, 1949-61, 72-75, Watkins distinguished prof. entomology, 1959—, acting chmn. dept. systematics, ecology, 1968-69, Watkins distinguished prof. systematics and ecology, 1969—; dir. Snow Entomol. Museum, 1974—, state entomologist, 1949-61; Guggenheim fellow, vis. research prof. U. Paraná, Curitiba, Brazil, 1955-56; Fulbright fellow U. Queensland, Brisbane, Australia, 1958-59; research scholar U. Costa Rica, 1963; Guggenheim fellow, Africa, 1966-67. Author: (with Mary H. Michener) American Social Insects, 1951, (with S.F. Sakagami) Nest Architecture of the Sweat Bees, 1962, The Social Behavior of the Bees, 1974; also articles; editor: Evolution, 1962-64; Am. editor: Insectes Sociaux, Paris, 1954-55, 62—; asso. editor: Ann. Rev. of Ecology and Systematics, 1970—. Served from 1st lt. to capt. San. Corps AUS, 1943-46. Fellow Am. Entomol. Soc., Am. Acad. Arts and Scis., Royal Entomol. Soc. London, AAAS; mem. Nat. Acad. Scis., Linnean Soc. London (corr.), Soc. for Study Evolution (pres. 1967), Soc. Systematic Zoologists (pres. 1969), Am. Soc. Naturalists (pres. 1978), Internat. Union for Study Social Insects (pres. 1977-82), Bee Research Assn., Kans. Entomol. Soc. (pres. 1950), Brazilian Acad. Scis. (corr.), Entomol. Soc. Am. Subspecialties: Sociobiology; Systematics. Current work: Origin and evolution of social behavior in insects; kin recognition; ethology of bees; systematics and ecology of bees; principles of systematics. Home: 1706 W 2d St Lawrence KS 66044

MICHENER, H. ANDREW, behavioral science researcher, sociology educator; b. N.Y.C., Dec. 1, 1940; s. Howard Perry and Rosemary (McCabe) M. B.A., Yale U., 1963; M.A., U. Mich.-Ann Arbor, 1964, Ph.D., 1968. Instr. sociology U. Wis.-Madison, 1968, asst. prof., 1968-72, assoc. prof., 1972-75, prof., 1975—. Contbr. articles tech. jours. NSF grantee, 1975—. Mem. Am. Psychol. Assn., Am. Statis. Assn., Am. Sociol. Assn., Ops. Research Soc. Am., Soc. Exptl. Social Psychology. Subspecialties: Social psychology; Operations research (mathematics). Current work: Conduct laboratory experimental research testing predictive accuracy of solution concepts from mathematical theory of n-person games. Office: University of Wisconsin 1180 Observatory Dr Madison WI 53706

MICHENER, JOHN RUSSELL, research engineer, applied physicist, microstructural analyist; b. Oakland, Calif., Oct. 6, 1951; s. John Harold and Ann (Crabtree) M.; m. Ann Gardner, Dec. 30, 1974; children: Robin Ann, Kristin Lee. B.S. in Physics, U. Md., 1973; M.S. in Materials Sci, U. Rochester, 1979; Ph.D. in Mech. Engring. (Materials), U. Rochester, 1983. Research physicist Eastman Kodak Co., Rochester, N.Y., 1974-79; research asst. U. Rochester, 1979-82; microelectronics research engr. Siemens Co., Princeton, N.J., 1982—; cons. on microstructural characterization. Mem. Am. Phys. Soc., Am. Soc. for Metals, AIME, IEEE, AAAS. Subspecialties: Fracture mechanics; Microelectronics. Current work: Failure analysis, microstructural characterization of electronic devices, processing of compound semiconducton devices, cryptography and computer security, internal stress effects on materials. Patentee in field. Home: 177 Moore St Princeton NJ 08540 Office: Siemens Co 105 College Ave Princeton Forestal Center Princeton NJ 08540

MICHOD, RICHARD EARL, theoretical population biologist; b. Chgo., May 11, 1951; s. Charles Louis and Florence (Wise) M.; m. Carol Elaine Santich, Mar. 20, 1977; 1 dau., Kristin Olivia. B.S., Duke U., 1973; M.A., U. Ga., 1979, Ph.D., 1979. Asst. prof. dept. ecology and evolutionary biology U. Ariz., Tucson, 1979-82, assoc. prof., 1982—. NSF grantee, 1979-82, 82-85. Mem. Am. Soc. Naturalists, Soc. Study Evolution, Genetics Soc. Am., Ecol. Soc. Am. Subspecialties: Evolutionary biology; Sociobiology. Current work: Evolution of social behavior; origin of life. Office: Dept Ecology and Evolutionary Biology University of Arizona Tucson AZ 85721

MICKELSON, CLAUDIA A(NN), immunologist, genetic engineer, researcher; b. Detroit, Mar. 8, 1944; d. Gordon Francis and Virginia R. (Randall) Roberts; m. Michael Jay Mickelson, Sept. 17, 1965; 1 son, David Paul. B.Sc. in Biology, Antioch Coll., 1966, Ph.D., U. Rochester, 1974. Research scientist Bigelow Labs., Booth Bay Harbor, Maine, 1976-78; sr. research fellow U. Glasgow, Scotland, 1978-80, U. Melbourne, Australia, 1981—. Mem. Am. Soc. Immunologists. Subspecialties: Genetics and genetic engineering (biology); Immunobiology and immunology. Office: Research Center Dept Pathology U Melbourne Melbourne Australia

MICKEY, GEORGE HENRY, cytogeneticist; b. Claude, Tex., Jan. 26, 1910; s. Luke Ross and Clara Alice (Pennington) M.; m. Alwilda Editha Davis, Aug. 20, 1932; children: Wilda Rhea, Don Davis. B.A., Baylor U., 1931; M.S., U. Okla., 1934; Ph.D., U. Tex., Austin, 1938. Cert. clin. lab. dir., Conn. Instr. zoology U. Tex., Austin, 1935-38; instr. La. State U., Baton Rouge, 1938-42, asst. prof., 1942-44, asso. prof., 1944-48; research fellow biology Calif. Inst. Tech., Pasadena, 1948; asso. prof. biology Northwestern U., Evanston, Ill., 1949-56; prin. biologist Oak Ridge Nat. Lab., 1953; prof., chmn. dept. zoology La. State U., 1956-59, dean, 1959-60; sr. scientist New Eng. Inst., Ridgefield, Conn., 1960-70; dean (Grad Sch.), 1970-74; clin. asso. and dir. cytogenetics Duke U. Med. Center, Durham, N.C., 1975—; vis. prof. U. Bridgeport, Conn., 1971. Co-author: Manual Studies in General Zoology, 1947; contbr. articles to sci. jours. Guggenheim fellow, 1948; grantee Sigma Xi, Rockefeller Found., AEC, Wallace Genetic Found., NSF, Wood Found., Population Council, Hartford Found., Office Naval Research. Fellow AAAS; mem. Am. Soc. History Edn., Am. Genetics Assn., Am. Naturists, Am. Soc. Human Genetics, Am. Inst. Biol. Sci., Am. Soc. Zoologists, AAUP, Genetics Soc. Am., Assn. Southeastern Biologists (pres. 1948), Sigma Xi, Beta Beta Beta (nat. pres. 1957-60). Democrat. Methodist. Subspecialties: Genome organization; Cytology and histology. Current work: Human chromosome analysis; genetic counseling; hazards of genetic damage by environmental agents. Home: 2404 Perkins Rd Durham NC 27706 Office: Duke U Med Center Box 3062 Durham NC 27710

MICKLEY, HAROLD SOMERS, chemical company executive; b. Seneca Falls, N.Y., Oct. 14, 1918; s. Harold Franklin and Marguerite Gladys (Somers) M.; m. Margaret W. Phillips, Dec. 21, 1941; children: Steven P., Richard S. B.S., Calif. Inst. Tech., 1940, M.S., 1941; Sc.D., MIT, 1946. Process engr. Union Oil Co. Calif., 1941-42; civilian project dir. torpedo research U.S. Navy, World War II; mem. faculty Mass. Inst. Tech., 1946-71, Ford prof. engring., 1961-71; dir. Center Advanced Engring., 1963-71, Stauffer Chem. Co., Westport, Conn., 1967—, v.p., 1971-72, exec. v.p., 1972—, vice chmn. bd., 1981—; dir. Montrose Chem. Corp., SWS Silicones Corp. Co-author: Applied Mathematics in Chemical Engineering, 1957, Recent Advances in Heat and Mass Transfer, 1961; Contbr. articles to profl. jours. Recipient Naval Ordnance Devel. award, 1946; Disting. Alumni award Calif. Inst. Tech., 1973. Fellow AAAS, Am. Inst. Chem. Engrs., Am. Acad. Arts and Scis.; mem. Nat. Acad. Engring., Am. Chem. Soc., Soc. Chem. Industry, N.Y. Acad. Scis., Conn. Acad. Sci. and Engring., Sigma Xi, Tau Beta Pi. Clubs: Country of Darien (Conn.); Pinnacle (N.Y.C.). Subspecialties: Chemical engineering; Immunobiology and immunology. Home: 11 Pequot Trail Westport CT 06880 Office: Stauffer Chem Co Westport CT 06881

MICZEK, KLAUS A., psychologist, researcher, educator; b. Burghausen, Germany, Sept. 28, 1944; came to U.S., 1967; s. Erich and Irene (Wirthl) M.; m. Christiane Baerwaldt, Aug. 8, 1970; 1 son, Nikolai. Ph.D., U. Chgo., 1971. Research asst. U. Chgo., 1968-71; asst. prof. psychology Carnegie-Mellon U., Pitts., 1972-76, assoc. prof., 1976-79; assoc. prof. psychology Tufts U., Medford, Mass., 1979—. Contbr. chpts. to books, articles to profl. jours. Nat. Inst. on Drug Abuse grantee, 1973-76, 76-79, 79-82, 82—; Nat. Inst. on Alcohol Abuse and Alcoholism grantee, 1980-83; W.T. Grant Found. grantee, 1973; Pitts. Found grantee, 1973. Mem. AAAS, Am. Psychol. Assn., Soc. Neurosci., Internat. Soc. Research on Aggression, N.Y. Acad. Sci., Am. Soc. Primatologists, Internat. Primatological Soc., Behavioral Pharmacol. Soc. Subspecialties: Psychobiology; Psychopharmacology. Current work: Aggression research, preclinical psychopharmacology, brain-behavior relation, brain mechanisms of animal social behavior. Office: Dept Psycholog Tufts U Medford MA 02155

MIDDLEDITCH, BRIAN STANLEY, biochemistry educator; b. Bury St. Edmunds, Suffolk, Eng., July 15, 1945; came to U.S., 1971; s. Stanley Stafford and Dorothy (Harker) M.; m. Patricia Rosalind Nair, July 18, 1970; 1 dau., Courtney Lauren. B.Sc., U. London, 1966; M.Sc., U. Essex, 1967; Ph.D., U. Glasgow, 1971. Research asst. U. Glasgow, Scotland, 1967-71; vis. asst. prof. Baylor Coll. Medicine, Houston, 1971-75; asst. prof. U. Houston, 1975-80, assoc. prof., 1980—. Author: Mass Spectrometry of Priority Pollutants, 1981; editor: Practical Mass Spectrometry, 1979, Environmental Effects of Offshore Oil Production, 1981. Grantee Nat. Marine Fisheries Service, 1976-80, Sea Grant Program, 1977-81, NASA, 1980-83. Mem. Am. Chem. Soc., Am. Soc. Mass Spectrometry, World Mariculture Soc. Subspecialties: Analytical chemistry; Environmental engineering. Current work: Biochemical ecology, mariculture. Home: 11403 Sagetown Dr Houston TX 77089 Office: U Houston Houston TX 77004

MIDDLETON, HENRY MOORE, III, internist, gastroenterologist, medical educator; b. Winston-Salem, N.C., Feb. 16, 1943; s. Henry Moore and Helen (Wilson) M.; m. Dorothy Finlator Ingram, Dec. 23, 1966; children: Rachel, Henry, Sarah. A.B., U. N.C., 1965, M.D., 1969. Diplomate: Am. B.d. Internal Medicine, subcert. in gastroenterology. Intern Vanderbilt U. Hosps., Nashville, 1969-70, resident in internal medicine, gastroenterology, 1970-74; instr. Vanderbilt Hosp, Nashville, 1973-74; asst. prof. medicine Med. Coll. Ga., Augusta, 1974-78, assoc. prof., 1978—; asst. chief gastroenterology dept. VA Med. Ctr., Augusta, 1974—, prin. investigator, 1974—. Contbr. sci. articles to profl. publs. Mem. So. Soc. Clin. Investigation, Ga. Gastroent. Soc. (sec.-trea. 1982—), Am. Assn. Study Liver Diseases, Am. Inst. Nutrition, Am. Soc. Clin. Nutrition. Subspecialties: Internal medicine; Gastroenterology. Current work: Academic gasteroenterology with patient care; teaching medical students; housestaff and fellows; and research into the absoption of water-soluable vitamins. Office: Va Med Ctr 111C Augusta GA 30910 Home: 3514 Lost Tree Ct Augusta GA 30907

MIDDLETON, RICHARD BURTON, educator, microbial geneticist; b. Rockford, Ill., Nov. 24 1936; s. Ralph Burton and Dorothy Ellen (Bolen) M.; m. Jacqueline Champagne, Apr. 15 1959 (dec. Nov. 13, 1982); children: Jane Bradley, Peter Dalby. A.B. cum laude, Harvard U., 1958, A.M., 1960, Ph.D., 1963. Research assoc. Brookhaven Nat. Lab., Upton, N.Y., 1962-64; asst. prof. biology Am. U., Beirut, 1964-65; asst. prof. genetics McGill U., Montreal, Que., Can., 1965-71; assoc. prof. microbiology and genetics Meml. U., St. John's, Nfld., 1971-75, prof., 1975-77; prof. microbiology, assoc. dean basic, scis. U. Medicine and Dentistry N.J., Piscataway, 1977—; mem. study sect. short-term tng. grants NIH, 1978, mem. site visit team biomed. research devel. grant, 1980; mem. com. experts Internat. Standards Orgn. Water Quality Am. Nat. Standards Inst., 1978-83. Contbr. articles to profl. jours. Pres. Continental Assn. Funeral and Meml. Socs., 1982-84; treas. Princeton Meml. Assn., 1979-83; program chmn. Isle La Motte (Vt.) Hist. Soc., 1973-83. Research grantee Research Corp., 1966-67, Med. and Nat. Research Councils Can., 1966-76, Food and Drug Directorate Can., 1969-71, NIH, 1979-80, Nat. Found. March of Dimes, 1974, Dept. Sec. of State Can., 1975, Dept. Nat. Health and Welfare Can., 1975-78, Nfld. Dept. Consumer Affairs and Environment, 1975-76, Dept. Consumer and Corporate Affairs Can., 1976-77, NSF, 1983. Mem. Acad. Medicine N.J. (dir. 1982—, chmn. basic scis. sect. 1982), AAAS, AAUP, Am. Inst. Biol. Scis., Am. Soc. Microbiology, Genetics Soc. Am., N.J. Health Scis. Group (chmn. edn. com. 1980—), Theobald Smith Soc. (mem. Waksman award com. 1982-85), N.J. Univ. Research Coordinating Council, N.Y. Acad. Sci., Can. Assn. Univ. Tchrs., Can. Coll. Microbiologists, Genetics Soc. Can., Can. Soc. Cell Biology, Can. Soc. Microbiologists, Harvard Alumni Assn. (dir. 1976-79), Sigma Xi. Clubs: Harvard (Boston); (N.J.) (mem. schs. com. 1981-84). Subspecialties: Genetics and genetic engineering (medicine); Microbiology (medicine). Current work: Exchange of genetic material between species of enteric bacteria. Home: Hill House Isle La Motte VT 05463 Office: Univ Medicine and Dentistry New Jersey PO Box 55 Piscataway NJ 08854

MIDLARSKY, ELIZABETH RUTH, psychology educator, researcher; b. Bklyn., Apr. 29, 1941; d. Abraham Allan and Frances Lucille (Wiener) Steckel; m. Manus Issachar Midlarsky, June 25, 1961; children: Susan Rachel, Miriam Joyce, Michael George. B.A., CUNY, 1961; M.A., Northwestern U., 1966, Ph.D., 1968. Lic. psychologist, Mich., Colo. Lectr. Northwestern U., 1964-66; asst. prof. psychology U. Denver, 1968-73; vis. assoc. prof. psychology U. Colo., Denver, 1975; dir. research and evaluation Malcolm X Mental Health Ctr., Denver, 1974-76; assoc. prof., dir. psychiat. service tng. program Met. State Coll., Denver, 1975-77; assoc. prof. psychology U. Detroit, 1977-83, prof. psychology, 1983—, chmn. dept., 1977-81, dir. Ctr. Study Human Devel. and Aging, 1983—; mem. rev. com. NIMH Criminal and Violent Behavior Commn., Rockville, Md., 1979-82, Nat. Ctr. Prevention and Control of Rape, 1976-79. Editor: Acad. Psychology Bull, 1982—; contbr. chpts. to books and articles to profl. jours. AAUW fellow, 1974-75; U. Denver grantee, 1970-73; AARP Andrus Found. grantee, 1982; Nat. Inst. Aging grantee, 1982-85. Mem. Am. Psychol. Assn., Gerontol. Soc. Am., Mich. Psychol. Assn., Soc. Psychol. Study of Social Issues, Midwestern Psychol. Assn., Sigma Xi, Phi Beta Kappa, Psi Chi. Subspecialties: Social psychology; Developmental psychology. Current work: Systematic investigation of antecedents, nature and outcome of altruism and helping; research regarding interpersonal processes and their development across the life-span. Home: 13158 Nadine St Huntington Woods MI 48070 Office: Dept Psychology U Detroit 4001 W McNichols Rd Detroit MI 48221

MIELCZAREK, EUGENIE VORBURGER, biophysics educator; b. N.Y.C., Apr. 22, 1931; m. Stanley Mielczarek; 2 children. B.S., Queens Coll., 1953; M.S., Cath. U. Am., 1957, Ph.D. in Physics, 1963. Physicist Nat. Bur. Standards, Washington, 1953-57; from research asst. to research assoc. in physics Cath. U. Am., Washington, 1957-62, asst. research prof., 1962-65; prof. physics George Mason U., Fairfax, Va., 1965—; vis. scientist NIH, 1977-78; mem. vis. scientist program Am. Inst. Physics, 1964-71; vis. prof. Hebrew U., Jerusalem, 1981. Mem. Am. Phys. Soc., Biophys. Soc., Am. Assn. Physics Tchrs., Assn. for Women in Sci., Sigma Xi. Subspecialties: Biophysics (physics); Condensed matter physics. Current work: Mossbauer spectroscopy of biological compounds; solid state low temperature physics, semiconductors, Fermi surfaces of metals. Office: George Mason U Dept Physics Fairfax VA 22030

MIELENZ, JONATHAN RICHARD, microbiologist; b. Denver, Mar. 25, 1948; s. Richard Childs and Jessie (Lupton) M.; m. Susan Wyatt, Aug. 10, 1974; children: Rhonda Susan, Benjamin David. A.B., Wittenberg U., 1970; M.S., U. Ill.-Urbana, 1973, Ph.D., 1976. Postdoctoral fellow U. Calif.-Davis, 1976-78; researcher CPC Internat. Moffett Tech. Center, Argo, Ill., 1979-80, team leader, sr. researcher, 1981, sect. leader, 1982-83; sect. head research and devel. G.D. Searle and Co., Skokie, Ill., 1983—. Contbr. articles to profl. jours. Mem. Am. Soc. Microbiology, Soc. Indsl. Microbiology. Subspecialties: Genetics and genetic engineering (biology); Enzyme technology. Current work: Microbial genetics and genetic engineering to improve industrial production microorganisms. Home: 937 Newberry St LaGrange Park IL 60525 Office: GD Searle and Co Skokie IL 60077

MIELENZ, RICHARD CHILDS, consulting engineering executive; b. Burlingame, Calif., Dec. 18, 1913; s. Edward Rudolph and Martha May (Childs) M.; m. Jessie Kithcart Lupton, Dec. 30, 1939; children: Geoffrey Carroll, Susan Barbara, Jonathan Richard, Laurel Ann. AA., Marin Jr. Coll., Kentfield, Calif., 1933; B.A. cum laude, U. Calif.-Berkeley, 1936, Ph.D., 1939. Registered profl. civil engr., Ohio. Head Petrographic Lab. Bur. Reclamation, Denver, 1941-56; dir. research Master Builders Martin Marietta Corp., Cleve., 1956-64, v.p. product devel., 1964-68, v.p. research and devel., 1968-78; pres. Richard C. Mielenz, Inc. (P.E.), Gates Mills, Ohio, 1978—; cons. in petrography. Fellow Mineral. Soc. Am. (life), Geol. Soc. Am., ASTM (Sanford Thompson award 1948, 51, 55, Richart award 1966, award of merit 1976), Am. Concrete Inst. (hon., pres. 1977-78, Wason medal 1948, 59, Kennedy award 1973, Bloem award 1977); mem. Nat. Soc. Profl. Engrs., Phi Beta Kappa, Theta Tau. Republican. Presbyterian. Subspecialties: Materials (engineering); Petrology. Current work: Microscopical analysis of hardened concrete; consultation on concrete failures; evaluation of aggregates for concrete; participation in technical committees. Patentee admixture for concrete, 1964, 68. Home: Route 1 Box 103 Brigham Rd Gates Mills OH 44040 Office: Richard C Mielenz PE Inc Route 1 Box 103 Brigham Rd Gates Mills OH 44041

MIERNYK, WILLIAM HENRY, economist, researcher, educator; b. Durango, Colo., Jan. 4, 1918; s. Andrew Taber and Elizabeth (Sopko) M.; m. Mary Lorraine Davis, Oct. 4, 1940; children: Jan, Judith, Jeanne, James. B.A., U. Colo., 1946; M.A., Harvard U., 1952, Ph.D., 1953. Instr. to prof. econs. Northeastern U., 1952-62; prof. econs., dir. Bur. Econ. Research, U. Colo., Boulder, 1962-65; C.W. Benedum prof. econs. W.Va. U., Morgantown, 1965—; dir. Regional Research Inst. U.Va. U., 1965-83; vis. prof. econs. MIT, 1957-58, Harvard U., 1969-70. Author or contbr.39 books, including: Regional Analysis and Regional Policy, 1982; (with others) Regional Impacts of Rising Energy Prices, 1978, (with John T. Sears) Air Pollution Abatement and Regional Economic Development, 1974; contbr. numerous articles to profl. jours. Served with AUS, 1941-45. Fellow So. Regional Sci. Assn.; mem. Am. Econ. Assn., Regional Sci. Assn., Indsl. Relations Research Assn. Subspecialties: Fuels; Coal. Current work: Regional impacts of energy availability and prices.

MIEYAL, JOHN JOSEPH, biochemistry educator; b. Cleve., Feb. 17, 1944; s. Stanley John and Jennie Ann (Miesowicz) M.; m. Donna Celeste Dobscha, Aug. 20, 1966; children: Thomas Joseph, Paul Anthony, Jennifer Lynn, Angela Marie. B.S., John Carrol U., 1965; Ph.D (USPHS fellow), Case-Western Res. U., 1969. USPHS fellow in biochemistry Brandeis U., 1969-71; asst. prof. pharmacology and biochemitry Northwestern U. Med. Sch., Chgo., 1971-76; assoc. prof. pharmacology Case Western Res. U., Cleve., 1976—; assoc. prof. chemistry, 1981—. Contbr. articles to profl. jours. Mem. parish council St. Ann Catholic Ch. Recipient John S. Diekhoff Teaching award Case-Western Res. U., 1980; Chgo. Heart Assn. grantee, 1972-76; Research Corp. grantee, 1972-75; Nat. Inst. Gen. Med. Scis. grantee, 1971-, Los Alamos Stable Isotope Resource grantee, 1980-83, Am. Heart Assn. of N.E. Ohio grantee, 1977-80; Kidney Found. Ohio grantee, 1981-82. Mem. Am. Soc. Biol. Chemists, Am. Soc. Pharmacology and Exptl. Therapeutics, Am. Chem. Soc., AAAS, AAUP, Sigma Xi. Roman Catholic. Subspecialty: Molecular pharmacology. Current work: Molecular basis for differential reactivity of hemoprotein involved in oxygen transport, drug and xenobiotic metabolism, toxic activation of drugs and environmental agents and selective activity of chemotherapeutic agts. Home: 2245 Lamberton Rd Cleveland Heights OH 44118 Office: Dept Pharmacology Case Western Res U Med Sch Cleveland OH 44106

MIGDALOF, BRUCE HOWARD, xenobiologist; b. Bklyn., July 19, 1941; s. Samuel and Jessica (Koch) M.; m. Joan Phullis Selman, June 25, 1967; children: Barrie Ruth, Amanda, Shari-Lynne, Jonathan. B.A., Cornell U., 1962; M.S., Purdue U., 1965; Ph.D., U. Pitts., 1969. Sr. scientist Sandoz Pharms., East Hanover, N.J., 1969-72; sr. scientist McNeil Labs., Inc., Ft. Washington, Pa., 1972-75, group leader drug disposition, 1975-77; dir. drug metabolism Squibb Inst. Med. Research, New Brunswick, N.J., 1977—. Contbr. chpts. to books, articles to profl. jours. Mem. AAAS, Am. Chem. Soc., Am. Soc. Pharmacology and Exptl. Therapeutics, Am. Pharm. Assn., Acad. Pharm. Sci., N.Y. Acad. Sci., Internat. Soc. Study Xenobiotics. Democrat. Jewish. Subspecialties: Xenobiology; Pharmacokinetics. Current work: Drug metabolism and disposition, pharmacodynamic-pharmacokinetic correlations. Office: PO Box 191 New Brunswick NY 08903 Home: 156 Richardson Rd Robbinsville NJ 08691

MIHALAS, DIMITRI MANUEL, scientist; b. Los Angeles, Mar. 20, 1939; s. Emmanuel Demetrious and Jean (Christo) M.; m. Alice Joelen Covalt, June 15, 1963 (div. Nov. 1974); children—Michael Demetrious, Genevieve Alexandra; m. Barbara Ruth Rickey, May 18, 1975. B.A. with highest honors, UCLA, 1959; M.S., Calif. Inst. Tech., 1960, Ph.D., 1964. Asst. prof. astrophys. scis. Princeton U., 1964-67; asst. prof. physics U. Colo., 1967-68; asso. prof. astronomy and astrophysics U. Chgo., 1968-70, prof., 1970-71; adj. prof. astrogeophysics, also physics and astrophysics U. Colo., 1972-80; sr. scientist High Altitude Obs., Nat. Center Atmospheric Research, Boulder, Colo., 1971-79, 82—; astronomer Sacramento Peak Obs., Sunspot, N.Mex., 1979-82; cons. Los Alamos Nat. Lab, 1981—; vis. prof. dept. astrophysics Oxford (Eng.) U., 1977-78; sr. vis. fellow dept. physics and astronomy Univ. Coll., London, 1978; mem. astronomy adv. panel NSF, 1972-75. Author: Galactic Astronomy, 2d edit, 1981, Stellar Atmospheres, 1970, 2d edit., 1978, Theorie des Atmospheres Stellaires, 1971; asso. editor: Astrophys. Jour, 1970-79, Jour. Computational Physics, 1981—; editorial bd.: Solar Physics, 1981—. NSF fellow, 1959-62; Van Maanen fellow, 1962-63; Eugene Higgins vis. fellow, 1963-64; Alfred P. Sloan Found. Research fellow, 1969-71. Mem. U.S. Nat. Acad. Sci., Internat. Astron. Union (pres. commn. 36 1976-79), Am. Astron. Soc. (Helen B. Warner prize 1974), Astron. Soc. Pacific (dir. 1975-77). Subspecialties: Theoretical astrophysics. Current work: Radiation hydrodynamics; radiation transport and spectral-line formation; stellar pulsation. Home: 106 Timber Lane Boulder CO 80302 Office: High Altitude Obser Nat. Center for Atmosphere Research-Box 3000 Boulder CO 80307

MIHALYI, ELEMER, research chemist; b. Deva, Romania, Jan. 11, 1919; came to U.S., Jan. 12, 1949, naturalized, 1955; s. Elemer and Maria (Illyes) M.; m. Elisabeth Sedony, Mar. 23, 1948; children: Christina L., Julia M. M.D., Francis Joseph U., Kolozsvar, Hungary, 1943—; Ph.D., U. Cambridge, Eng., 1963. Instr. in med. chemistry U. Kolozsvar, 1941-44, Intern, 1942-43; instr. in biochemistry U. Budapest, Hungary, 1946-48; research assoc. Inst. Muscle Research, Woods Hole, Mass., 1949-51; research fellow U. Pa., 1951-55; research chemist NIH, Bethesda, Md., 1955—, researcher in clin. pathology, hematology service, 1977—; coordinator NSF U.S.-Spain Cooperative Project in biochemistry, 1974-76. Author: Application of Proteclytic Enzymes to Protein Structure Studies, 1972, 2d edit. in 2 vols., 1978; research numerous publs. in biochemistry; editor: Thrombosis Research, 1972-78. Fellow Internat. Soc. Hematology; mem. Am. Soc. Biol. Chemists, Internat. Soc. Thrombosis and Hemostasis, Am. Heart Assn. Council on Thrombosis. Republican. Roman Catholic. Subspecialties: Biochemistry (biology); Biophysical chemistry. Current work: Mechanism of blood coagulation; application of proteolytic enzymes to structural studies of proteins; pioneering studies of fragmentation of proteins into their constituent domains by limited proteolysis; physicochemistry of proteins. Home: 10210 Fleming Ave Bethesda MD 20814 Office: NIH Bldg 10 Room 2C-390 Bethesda MD 20205

MIHAS, FAQUIR ULLAH, mathematician, researcher; b. Shadiwal, Pakistan, Dec. 4, 1924; emigrated to Can., 1962; s. Atta and Sakeena Minhas. B.A. in Math, Panjab U., Lahore, Pakistan, 1952, M.A., 1957; M.A.Sc., U. Laval, Que., Can., 1965, D.Sc. in Mech. Engring, 1972; M.Eng. in Elec. Engring, McGill U., Montreal, 1982. Edn. officer Pakistan Air Force, 1957-61; devel. engr. John Inglish Co. Ltd., Toronto, Ont., Can., 1963-64; sr. project engr. Dominion Engring. Works Ltd., Montreal, 1966-74, head transport phenomena br., 1974-80; staff specialist in aerodynamics Canadair Ltd., Montreal, 1980-82, staff specialist in systems, 1982—. Fellow Brit. Interplanetary soc.; mem. Can. Aerospace Inst., ASME, AIAA, Order of Engrs., Que. Subspecialties: Aerospace engineering and technology; Plasma. Current work: Electrical propulsion; electromagnetic confinement of plasma. Home: 4000 Maisonneuve West Apt 2404 Westmount Montreal PQ Canada H3Z 1J9 Office: Canadair Ltd 1800 Laurentain Blvd Montreal PQ Canada H3C 3G9

MIHM, MARTIN CHARLES, JR., dermatopathology administrator, researcher; b. Pitts.; s. Martin Charles and Cecelia Matilda (Hepp) M. A.B., Duquesne U., 1955; M.D., U. Pitts., 1961. Diplomate: Am. Bd. Dermatology, Am. Bd. Pathology, Am. Bd. Dermatopathology. Intern Mt. Sinai Hosp., N.Y.C., 1961-62, resident, 1963-64; resident in dermatology Mass. Gen. Hosp., Boston, 1965-67, resident in pathology, 1969-73; asst. prof. pathology Harvard U., 1972-75, assoc. prof., 1975-79, chief dermatopathology, 1982—, Mass. Gen. Hosp., 1974—; prof. pathology Mass. Gen. Hosp./Harvard Med. Sch., 1980—, pres., 1982—; cons. VA Hosp., Boston, U. B.C., Vancouver. Author: Human Malignant Melanoma, 1979, Pigment Cell, 1981, 1981, Dermatology in General Medicine, 1981, Primer of Dermatopathology, 1983. Served to comdr. USPHS-Coast Guard, 1967-69. Fellow Am. Acad. Dermatology, ACP; mem. New Eng. Pathology Soc., New Eng. Dermatol. Soc., Soc. Investigative Dermatology, Am. Soc. Dermatopathology, Am. Soc. Clin. Oncology, Am. Dermatol. Assn., Alpha Omega Alpha, Pi Gamma Nu. Roman Catholic. Subspecialties: Dermatology; Pathology (medicine). Current work: Malignant melanoma, delayed hypersensitivity studies in man, monoclonal antibody studies of inflammation, allograft rejection, bullous diseases, mechanism and pathogenesis. Office: Dermatopathology Unit Mass Gen Hosp Fruit St Boston MA 02114

MIKAT, EILEEN MARIE, medical research scientist; b. Cleve., May 1, 1930; d. Edward and Elizabeth (Seman) M. A.B., Case Western Res. U., 1952; M.A., Duke U., 1969, Ph.D., 1979. Coordinator, supr. med. tech. sch. Cleve. Met. Gen. Hosp., Case Western Res. U., Cleve., 1952-61; research asst. Duke U., Durham, N.C., 1961-69, research assoc., 1969-79, asst. prof. dept. pathology, 1979—. Mem. Am. Assn. Pathologists, Am. Soc. Clin. Pathologists. Subspecialties: Pathology (medicine); Cardiology. Current work: Cardiovascular and shock research. Office: Duke U Med Ctr Pathology Dept Box 3712 Durham NC 27707

MIKESELL, JAN ERWIN, biology educator; b. Macomb, Ill., Feb. 19, 1943; s. Lewis and Sarah Ruth (Bl) M.; m. Pamela G. Prestwood, May 16, 1945; children: Danielle Marie, Lisa Michelle. B.S., Western Ill. U., Macomb, 1965, M.S., 1966; Ph.D., Ohio State U., 1973. With landscape sect. of phys. plant Ohio State U., Columbus, 1969; assoc. prof. biology Gettysburg (Pa.) Coll., 1973—. Contbr. articles on biology to profl. jours. Served with U.S. Army, 1967-69. Mellon Found. grantee, 1976-78. Mem. Bot. Soc. Am., AAAS. Republican. Subspecialty: Morphology. Current work: Effects of enviroment the vegetative and reproductive development of certain dicotyledonous plants, anomalous plant development in dicots. Office: Dept Biology Gettysburg Coll Gettysburg PA 17325

MIKNIS, FRANCIS PAUL, chemist, researcher; b. DuBois, Pa., Jan. 31, 1940; s. Francis A. and Mary L. (Markievich) M.; m. Carol M. Bennett, Aug. 27, 1960; children: Patricia, Gregory Robert, Christine. B.S., U. Wyo., 1961, Ph.D., 1967. With Philco-Ford Corp., Newport Beach, Calif., 1966-67; research assoc. Laramie (Wyo.) Energy Tech. Ctr., 1967—. Contbr. articles to profl. jours. Mem. Am. Chem. Soc. (chmn. geochemistry div.), Sigma Xi. Democrat. Roman Catholic. Subspecialties: Geochemistry; Oil shale. Current work: Applications of nuclear magnetic resonance to fossil fuels; applying solid state 13C NMR techniques to characterize oil shales and to study oil shale thermal decomposition. Office: PO Box 3395 University Station Laramie WY 82071

MIKSCHE, JEROME PHILLIP, botany educator, researcher, administrator; b. Breckenridge, Minn., June 11, 1930; s. Anton Francis and Clara Gertrude (Braun) M.; m. Betty Jane Logan, May 23, 1953; children: Michael Logan, Elizabeth Clare, James Jerome. B.S., Moorhead State U., 1954; M.S., Miami U., Oxford, Ohio, 1956; Ph.D., Iowa State U., 1959. Postdoctoral assoc. Brookhaven Nat. Lab., Upton, N.Y., 1959-60; asst. botanist, 1960-65; prin. plant cytologist Inst. of Forest Genetics U.S. Dept. Agr. Forest Service, Rhinelander, Wis., 1965-77; prof., head dept. botany N.C. State U., Raleigh, 1977—. Contbr. articles to profl. jours. Chmn. Headwaters dist. Boy Scouts Am., 1965-67; mem. N.C. Bd. Sci. and Tech., 1979-85. Served with USMC, 1948-51. Named Outstanding Alumnus Moorhead State U., 1977. Mem. Am. Soc. for Cell Biology, Am. Soc. Plant Physiology, AAAS, Am. Soc. for Histochemistry, Bot. Soc. Am., Sigma Xi, Gamma Sigma Delta. Democrat. Roman Catholic. Subspecialties: Genome organization; Developmental biology. Current work: DNA genome Changes during development in plants. Home: 3212 Ruffin St Raleigh NC 27607 Office: Botany Dept NC State U Raleigh NC 27650

MIKULLA, VOLKER, aeronautical engineer; b. Dzieditz, Silesia, Germany, Jan. 5, 1944; s. Hans Joachim and Elfriede (von Gruchalla) M.; m. Jennifer Mary Bartlett, Aug. 19, 1972; children: Christian, Claire, Peter. B.Aero. Engring., Rensselaer Poly. Inst., 1967; M.Engr., U. Liverpool, Eng., 1969, Ph.D., 1973. Research engr. helicopter div. Messerschmitt Boelkow-Blohm, Munich, Bavaria, W.Ger., 1975—. Assoc. fellow AIAA; mem. Deutsche Gesellschaft fuer Luft-und Raumfahrt. Roman Catholic. Subspecialties: Aeronautical engineering; Fluid mechanics. Current work: Research and development in helicopter aerodynamics, numerical analysis, computational fluid dynamics, turbulence, wind tunnel techniques. Patentee. Home: Hohenbrunnerstrasse 73 8012 Riemerling BavariaFederal Republic of Germany Office: Messerschmitt Boelkow-Blohm Postfach 8011 40 8000 Munich Federal Republic of Germany 80

MILANI, CYRUS SAEED, pathologist; b. Mashad, Iran, May 14, 1941; came to U.S., 1968, naturalized, 1980; s. Boyouk and Habibeh (Sadaghianai) Sadeghi; m. Afsaneh Khavas, Aug. 25, 1966; children: Natalie B., Natasha B. M.D., Pahlavi U., Shiraz, Iran, 1966. Intern Ellis Hosp., Schenectady, 1968-69; resident Hosp. U. Pa., Phila., 1969-73, assoc. pathologist, 1973-74; pathologist Cantonal U. Hosp., Lausanne, Switzerland, 1974-75; asst. prof. Teheran (Iran) U., 1975-77; dir. Central Diagnostic Lab., Tarzana, Calif., 1977—, Preventive Clinic Am., Encino, 1981—, Valley Cryobank div. Tissue Preservation Inst., 1981—. Bd. dirs. French Am. Sch., Van Nuys, Calif., 1979—, Am. Cancer Soc., Van Nuys, 1981-82, Am. Heart Assn., Studio City, Calif., 1982—. Mem. Am. Assn. Immunologists, Coll. Am. Pathologists, Am. Soc. Clin. Pathologists, Soc. Cryobiology, AAAS. Subspecialties: Pathology (medicine); Cellular engineering. Current work: Tissue banking, autologous cell preservation and use anatomic pathology and clinical pathology. Office: Valley Cryobank Div Tissue Preservation Inst 16542 Ventura Blvd Suite 201 Encino CA 91436

MILBURN, RONALD MCRAE, chemistry educator; b. Wellington, N.Z., May 29, 1928; s. Henry Joseph and Sylvia May (Sandlant) M.; m. Josephine Nicholson Fishel, Apr. 26, 1928; children: Rosemary R., Jeffrey R. B.Sc., Victoria U., Wellington, 1949, M.Sc., 1951; Ph.D., Duke U., 1954. Instr., research assoc. Duke U., 1954; lectr. Victoria U., 1955; instr., research assoc. Duke U., 1956; research assoc. U. Chgo., 1956-57; asst. prof. chemistry Boston U., 1957-63, assoc. prof., 1963-68, prof., 1968—; NIH spl. fellow Oxford U., Eng., 1965-66; vis. fellow Australian Nat. U., 1974-75; vis. scientist U. Basel, Switzerland, 1982-83. Contbr. articles to profl. jours. Fulbright grantee, 1952. Mem. Am. Chem. Soc., AAUP. Subspecialty: Inorganic chemistry. Current work: Influence of metal centers on reactions of coordinated ligands, phosphate chemistry. Office: Dept Chemistry Boston U Boston MA 02215

MILES, JOHN WILDER, educator; b. Cin., Dec. 1, 1920; s. Harold M. and Cleopatra (Morton) M.; m. Herberta Blight, June 19, 1943; children—Patricia Marie, Diana Catherine, Ann Leslie. B.S., Calif. Inst. Tech., 1942; M.S., Aero. Engr., 1944; Ph.D., 1944. With research lab. Gen. Electric Co., Schenectady, 1942; instr. Calif. Inst. Tech., 1943-44; with radiation lab. Mass. Inst. Tech., 1944; with Lockheed Aircraft Co., Burbank, Calif., 1944-45; asst. prof. U. Calif., Los Angeles, 1945-49, asso. prof., 1949-55, prof., 1955-61; prof. applied math. Australian Nat. U., 1962-64; prof. applied mechanics and geophysics U. Calif. at San Diego, 1965—, chmn. applied mechanics and engring. sci., 1968-73, vice chancellor acad. affairs, 1980; cons. Northrop Aircraft Co., 1946-49, N.Am. Aviation Co., 1948-51, U.S. Naval Ordnance Test Sta., 1948-51, Douglas Aircraft Co., 1952-54, Ramo Wooldridge Corp., 1955-60, NASA, 1959-60, Aerospace Corp., 1961-68, Gulf Gen. Atomic Corp., 1965-68, Systems Sci. & Software, 1972-75, Sci. Applications, 1978—, Gas Centrifuge Theory Cons. Group, 1978—. Asso. editor: Jour. Soc. Indsl. and Applied Math, 1958-61, Jour. Fluid Mechanics, 1966—; co-editor: Cambridge Monographs on Mechanics and Applied Math, 1963—, Math. Tools for Engrs. Series, 1967-70; bd. editors: U. Calif. Press, 1955-58, Ann. Rev. of Fluid Mechanics, 1967-73, Fluid Mechanics-Soviet Research, 1972—; reviewer: Applied Mechanics Revs, 1947—, Math. Revs, 1946—. Recipient Fulbright awards U. New, Zealand, 1951, Cambridge U., 1969; vis. lectr. Imperial Coll., London, 1952; Guggenheim fellow, 1958-59, 68-69. Fellow Am. Inst. Aeros. and Astronautics, Am. Acad. Arts and Scis., Am. Acad. Mechs.; mem. Nat. Acad. Scis., AAAS, N.Y. Acad. Scis., Fedn. Am. Scientists, Am. Geophys. Union, Sigma Xi. Subspecialties: Applied mathematics; Geophysics. Current work: Water waves, nonlinear dynamics (strange attractors, etc.). Home: 8448 Paseo del Ocaso La Jolla CA 92037

MILES, RICHARD BRYANT, electrical engineer, applied physicist, educator; b. Washington, July 10, 1943; s. Thomas Kirk and Elizabeth (Bryant) M.; m. Susan McCoy. B.S., Stanford U., 1966, M.S., 1967, Ph.D., 1972. Postdoctoral assoc. Stanford U., 1972; asst. prof. mech. and aerospace engring. Princeton U., 1972-78, assoc. prof., 1978-82, prof., 1982—. Contbr. articles to profl. jours. Fannie and John K. Hertz fellow, 1969-72. Mem. Am. Inst. Physics, IEEE, Sierra Club, Aircraft Owners and Pilots Assn. Subspecialties: Optical engineering; Spectroscopy. Current work: Laser diagnostics, molecular dynamics, optics, laser induced fluorescence, nonlinearoptics, laser ranging, picosecond surface spectroscopy, photoacoustics, molecular relaxation. Patentee in field. Home: 3 Newlin Rd Princeton NJ 08540 Office: Princeton U D-414 Engring Quardrangle Princeton NJ 08544

MILESTONE, WAYNE DONALD, mechanical engineering educator, engineering consultant; b. Darlington, Wis., Apr. 27, 1935; s. Arthur E. and Opal M. (Bardell) M.; m. Mary Virginia Hughes, Dec. 28, 1970; children: Sarah K., David W. B.S. in Mech. Engring, Ohio State U., 1959, M.S., 1960, Ph.D., 1966. Registered profl. engr., Ohio. Engr. Rockwell Corp., Bellfontaine, Ohio, 1956-58, E.I. DuPont Co., Newark, Del., 1959; teaching asst. engring. graphics Ohio State U., Columbus, 1959-60, research assoc. dept. mech. engring., 1960-66; mem. faculty U. Wis., Madison, 1966—, prof. mech. engring., 1982—; Cons. engring. Pratt & Whitney Aircraft, West Palm Beach, Fla., 1979-80; mem. adv. com. amusement rides and devices Dept. of Industry, Labor and Human Relations, State of Wis., Madison, 1977—. Contbr. articles to profl. publs. Mem. spl. com. product liability Wis. Legis. Council, Madison, 1978-81. Recipient Service citation Wis. Legis. Council, 1980; Outstanding Mech. Engring. Prof. award Pi Tau Sigma, 1976, 81; Excellence in Teaching award U. Wis., Madison, 1971. Mem. ASME (Centennial medallion 1980), ASTM, Am. Soc. Engring. Educators, Sigma Xi, Tau Beta Pi, Pi Tau Sigma. Club: Nakoma Golf (Madison). Subspecialty: Mechanical engineering. Current work: Mechanical design; machine design; reliability; failure analysis; metal fatigue; fracture mechanics; fretting. Home: 4147 Iroquois Dr Madison WI 53711 Office: Univ Wis Mech Engring Bldg 1513 University Ave Madison WI 53706

MILETICH, DAVID JOHN, medical researcher, educator; b. Akron, Ohio, Feb. 12, 1937; s. John and Katherin (Tressel) M.; m. Marilyn Ellen Schroeder, Mar. 25, 1971; children: Brenna, Elizabeth, Nicklas. B.A. in Biology, Kent State U., 1963; Ph.D. in Physiology, Georgetown U., 1968. Dir. anesthesia research Michael Reese Hosp. and Med. Ctr., Chgo., 1971—; assoc. prof. anesthesiology U. Chgo., 1971—. Served to lt. USN, 1957-61. Mem. Am. Soc. Anesthesiologists, Stroke Council Am. Heart Assn., Am. Physiol. Soc., Internat. Anesthesia Research Soc., N.Y. Acad. Scis. Subspecialties: Anesthesiology; Physiology (medicine). Current work: Cerebral physiology and drug effects, cerebral blood flow and metabolism anesthetic effects. Office: Michael Reese Hosp and Med Ctr 31st and Lake Shore Dr Chicago Il 60616

MILEY, GEORGE HUNTER, nuclear engineering educator; b. Shreveport, La., Aug. 6, 1933; s. George Hunter and Norma Angeline (Dowling) M.; m. Elizabeth Burroughs, Nov. 22, 1958; children: Susan Elizabeth, Hunter Robert. B.S. in Chem. Engring, Carnegie-Mellon U., 1955; M.S., U. Mich., 1956, Ph.D. in Chem.-Nuclear Engring, 1959. Nuclear engr. Knolls Atomic Power Lab., Gen. Electric Co., Schenectady, 1959-61; mem. faculty U. Ill., Urbana, 1961—, prof., 1967—, chmn. nuclear engring. program, 1975—, dir. Fusion Studies Lab., 1976—; Vis. prof. U. Colo., summer, 1967, Cornell U., 1969-70, Lawrence Livermore Lab., summers 1973-74. Author: Direct Conversion of Nuclear Radiation Energy, 1971, Fusion Energy Conversion, 1976; editor: Jour. Nuclear Tech./Fusion, 1980—; U.S. assoc. editor: Laser and Particle Beams, 1982—. Served with C.E. AUS, 1960. Recipient Western Electric Teaching-Research award, 1977; NATO sr. sci. fellow, 1975-76. Fellow Am. Nuclear Soc. (dir. 1980-83, Disting. Service award 1980), Am. Phys. Soc., mem. Am. Soc. Engring. Edn. (chmn. energy conversion com. 1967-70, pres. U. Ill. chpt. 1973-74, chmn. nuclear div. 1975-76, Outstanding Tchr. award 1973), Sigma Xi. Presbyterian. Lodge: Kiwanis. Subspecialties: Fusion; Nuclear fusion. Current work: Fusion plasma engineering; inertial confinement fusion; nuclear pumped lasers; direct energy conversion. Research on fusion, energy conversion, reactor kinectics. Patentee direct energy conversion devices. Office: 214 Nuclear Engring Lab 103 S Goodwin St U Ill Urbana IL 61801 My professional goal has been to insure that future generations have a plentiful supply of economical, readily available energy such as offered by fusion. Not only should this insure a continued improvement in the standard of living for persons in all nations, but it should help maintain peace which is threatened by the struggle to obtain and control limited natural sources of energy.

MILIC-EMILI, JOSEPH, physiologist, educator; b. Sezana, Yugoslavia, May 27, 1931; emigrated to Can., 1963, naturalized, 1969; s. Joseph and Giovanna (Perhavec) Milic-E.; m. Ann Harding, Sept. 2, 1957; four children. M.D., U. Milan, Italy, 1955. Asst. prof. physiology U. Milan, Italy, 1956-58, U. Liège, Belgium, 1958-60; research fellow in physiology Harvard U., Boston, 1960-63; asst. prof. physiology and exptl. medicine McGill U., Montreal, Que., Can., 1963-65, asso. prof., 1965-69, prof. dept. physiology, 1970—, prof. exptl. medicine, 1970—, chmn. dept. physiology, 1973-78; dir. Meakins-Christie Labs., Montreal, 1979—; vis. cons. aeros. Imperial Coll. Tech., London, Eng., 1969-70; vis. cons. medicine Royal Postgrad. Med. Sch., London, 1969-70; cons. Brookhaven Nat. Lab., 1974, Columbia U., 1977. Editorial bds.: American Jour. of Physiology, 1970-76, Jour. of Applied Physiology, 1970-76, Revue Francaise Des Maladies Respiratoires, 1979, Rivista di Biologia, Am. Rev. Respiratory Disease, 1982; assoc. editor: Can. Jour. Physiology and Pharmacology, 1972-76; asso. editor: Jour. Applied Physiology, 1976-78; contbr. over 200 sci. research articles to profl. publs. Fellow Royal Soc. Can.; mem. Association des Physiologistes de Langue Francaise, Am. Physiol. Soc., Can. Physiol. Soc., Can. Soc. Clin. Investigation, Italian Physiol. Soc., Can. Thoracic Soc., Med. Research Council (grants com., heart and lung), Fleischner Soc., Societe Belge de Pneumologie (hon.). Subspecialties: Physiology (medicine); Pulmonary medicine. Current work: Respiratory mechanics, control of breathing; respiratory diseases. Home: 4394 Circle Rd Montreal PQ H3W 1Y5 Canada Office: Meakins-Christie Labs McGill Univ Lyman Duff Medical Sciences Bldg 3775 University St Montreal PQ H3A 2B4 Canada

MILJKOVIC, MOMCILO, chemistry educator, researcher; b. Belgrad, Yugoslavia, Dec. 12, 1931; s. Adam and Dragoslava (Jankovic) M.; m. Irina Beljajev, Oct. 23, 1960; children: Marko, Marija. B.S. in Chemistry, Faculty of Sci., U. Belgrade, 1959; Ph.D. in chemistry, Eidg. Technische Hochschule, Zurich, Switzerland, 1965; M.S., Hershey Med. Center, Pa. State U.-Hershey, 1968. Research assoc. dept. biochemistry Duke U., Durham, N.C., 1967; asst. prof. Hershey Med. Center, Pa. State U., 1968-75, assoc. prof., 1975—. Contbr. articles on chemistry to profl. jours. Mem. Swiss Chem. Soc., Serbian Chem. Soc. Serbian Orthodox. Subspecialties: Organic chemistry; Synthetic chemistry. Current work: Natural product chemistry, carbohydrate chemistry, use of carbohydrate as chiral synthons for synthesis of non-carbohydrate chiral molecules, synthetic reactions, sialic acid chemistry. Home: 1440 Deerfield Dr Hummelstown PA 17036 Office: 500 University Dr Hershey PA 17033

MILLAR, GORDON HALSTEAD, mech. engr., agrl. machinery mfg. co. exec.; b. Newark, Nov. 28, 1923; s. George Halstead and Dill E. (McMullen) M.; m. Virginia M. Jedryczka, Aug. 24, 1957; children—George B., Kathryn M., Juliet S., John G., James H. B.M.E., U. Detroit, 1949, D.Sc. (hon.), 1977; Ph.D., U. Wis., 1952. Supr. new powerplants Ford Motor Co., 1952-57; engring. mgr. Meriam Instrument Co., Cleve., 1957-59; dir. new products McCulloch Corp., Los Angeles, 1959-63; with Deere & Co., 1963—, v.p. engring., Moline, Ill., 1972—; mem. Fed. Adv. Com. Indsl. Innovation, 1979; chmn. West Central Ill. Ednl. Telecommunications Corp. Contbr. numerous articles to profl. publs. Chmn. Quad Cities chpt. United Way, 1976-77; bd. dirs.; adv. council Bradley U. Coll. Engring. and Tech.; mem. exec. com. Illowa council Boy Scouts Am., 1977-79. Served with U.S. Army, World War II. Decorated Purple Heart; recipient Alumnus of Year award U. Detroit, 1976. Mem. Nat. Acad. Engring., Soc. Automotive Engrs., Engrs. Joint Council, Indsl. Research Inst., Engring. Soc. Detroit, ASME, Am. Soc. Agrl. Engrs., Ill. Soc. Profl. Engrs., Moline C. of C., Aviation Council. Subspecialty: Agricultural engineering. Current work: Responsible for company-wide research and development, management of product engineering and overall administration of engineering activities. Patentee in field. Home: 3705 39th St Ct Moline IL 61265 Office: John Deere Rd Moline IL 61265

MILLARD, RONALD WESLEY, medical educator, researcher; b. Bridgeport, Conn., Sept. 10, 1941. B.S., Tufts U., 1963; Ph.D., Boston U., 1969. Asst. prof. Harvard U., 1974-75; asst. prof. Brown U., 1975-78; assoc. prof. pharmacology U. Cin. Coll. Medicine, 1978—. Contbr. numerous articles to profl. jours. Sr. Fulbright fellow, 1973; Am. Council on Edn. fellow, 1982. Fellow Am. Heart Assn. Council on Circulation; mem. Am. Physiol. Soc., Am. Pharmacol. Soc. Subspecialties: Physiology (medicine); Pharmacology. Current work: Mechanisms of calcium channel blocking drugs on cardiovascular system. Office: Coll Medicine U Cin 231 Bethesda Ave ML575 Cincinnati OH 45267

MILLAY, MICHAEL ALAN, science educator; b. Wright Patterson AFB, Ohio; s. Robert Harry and Peggy Joyce (Greene) M. B.A. in Biology, Wittenberg U., Springfield, Ohio, 1970; M.S., U. Ill., Chgo., 1972, Ph.D. in Botany, 1976. Fellow Ohio State U., 1976-77, lectr. botany, 1978-80; asst. prof. botany (U. Md.), 1980—. Contbr. articles to profl. jours. NSF grantee, 1977-78, 82—. Mem. Bot. Soc. Am., Internat. Orgn. Paleobotany, Internat. Soc. Plant Morphologists, Am. Assn. Plant Taxonomists, Am. Assn. Stratigraphic Palynologists, Palaeontological Assn. Club: Lincoln Continental Owners (Nogales, Ariz.). Subspecialties: Paleobiology; Evolutionary biology. Current work: Cladistics fern evolution, pollen biology evolution, spore ultrastructure; evolution of vascular plants, particularly ferns and seed ferns using cladistic techniques; paleobotany, palynology, sedimentology. Home: 4203 College Heights Dr University Park MD 20782 Office: Botany Department University of Maryland College Park MD 20742

MILLEN, JONATHAN KAYE, computer science researcher; b. Malden, Mass., Aug. 13, 1942; s. Leonard Kaye and Genevieve Bernice (Hoberman) M.; m. Helen Magedson, May 7, 1972; children: Samuel, Asher. A.B., Harvard U., 1963; M.A., Stanford U., 1965; Ph.D., Rensselaer Poly. Inst., 1969. With Mitre Corp., Bedford, Mass., 1969—, mem. dept. staff, 1980—. Mem. Assn. for Computing Machinery, IEEE Computer Soc. Subspecialties: Cryptography and data security; Artificial intelligence. Current work: Operating system security; information flow analysis; program verification; network protocols; expert systems. Home: 661 Main St Concord MA 01742 Office: Mitre Corp (K202) Bedford MA 01730

MILLER, ALAN DOUGLAS, neuroscientist; b. N.Y.C., Apr. 19, 1947; s. Alan S. and Helen H. (Marsh) M.; m. Janet R. Sparrow, July 22, 1978. B.S. in Chemistry, Union Coll., 1969, M.S., 1973; Ph.D. in Physiology, U. Western Ont., London, 1980. Asst. prof. neurophysiology Rockefeller U., N.Y.C., 1982—. Mem. Soc. Neurosci. Subspecialties: Neurophysiology; Gravitational biology. Current work: Motion sickness, vestibular system, sensorimotor integration research. Office: Rockefeller U 1230 York Ave New York NY 10021

MILLER, ALLEN H., physicist, educator; b. Bklyn., June 23, 1932; s. Samuel and Rachel (Miller) M.; m. Ann R. Tierney, July 1975. B.S., Bklyn. Coll., 1953; M.Sc., Rutgers U., 1955, Ph.D., 1960. Research assoc. U. Ill.-Urbana, 1960-61, research asst. prof., 1961-62; asst. prof. physics Syracuse U., 1962-67, assoc. prof., 1967—. Contbr. articles to physics jours. Mem. Am. Phys. Soc., Sigma Xi (pres. Syracuse chpt. 1967-69). Subspecialties: Condensed matter physics; Statistical physics. Current work: Phase transitions, many-particle-problem, work in theoretical physics, effects of impurities on semiconductor properties. Home: 300 Roosevelt Ave Syracuse NY 13210 Office: Dept Physics Syracuse U Syracuse NY 13210

MILLER, ARNOLD REED, research scientist; b. Marion, Ind., Dec. 25, 1944; s. Verling Campbell and Marguerite Jane (Reed) M.; m. Carol Anne Hautamaki, June 3, 1967. A.B., Ind. U., Bloomington, 1969; Ph.D., U. Ill-Urbana, 1973. Lectr., asst. dir. Inst. Aging, U. Wis.-Madison, 1978-79; vis. research scientist dept. genetics and devel. U. Ill-Urbana, 1980-82, research scientist, 1982—, vis. asst. prof. dept. med. info. sci., 1982—. Contbr. articles in field to profl. jours. Nat. Inst. Aging grantee, 1982-83. Mem. IEEE Reliability Soc., Soc. Computer Simulation, Soc. Math. Biology, Gerontol. Soc. Am. Subspecialties: Reliability theory; Gerontology. Current work: Evolution of wearout in complex, self-replicating systems; biological aging; reliability theory; computer simulation. Office: 1408 W University Ave Urbana IL 61801

MILLER, BERNARD, chemistry educator, researcher; b. Monticello, N.Y., Sept. 1, 1930; s. Isidore and Sarah (Mandelbaum) M.; m. Ruth Kussner, Dec. 12, 1965; children: Judith Rosalind, Laura Alison. B.S., CUNY, 1951; M.A., Columbia U., 1953, Ph.D., 1955. Sr. research scientist Am. Cyanamid Co., Princeton, N.J., 1957-67; assoc. prof. chemistry U. Mass., Amherst, 1957-71, prof., 1971—. Author: Organic Chemistry: The Basis of Life, 1980; contbr. articles to profl. jours. Subspecialty: Organic chemistry. Current work: Molecular rearrangements; blocked aromatic molecules; organic systehsis; cool conversion. Patentee in field. Home: 120 Columbia Dr Amherst MA 01002 01003

MILLER, BERNARD PAUL, science executive; b. Chelsea, Mass., Apr. 19, 1929; s. Hyman and Rose E. (Feldman) M.; m. Ruth Louise

Whyl, Dec. 15, 1973; children: Steven, Nina Dana, Robert, Simon, Toby. B.S. in Aero. Engring, Pa. State U., 1950, M.S., 1954. Mgmt. positions with RCA, Princeton, N.J., 1957-73; exec. v. p. ECON, Inc., Princeton, 1973-80, pres., 1980—. Mem. Consolidation Study Commn., Princeton, 1980; committeeman Fifth District, Princeton, 1981-82. Served to capt. USAF, 1950-57. Recipient Pub. Service award NASA, 1965. Assoc. fellow AIAA (mem. internat. activities com.); mem. Nat. Assn. Bus. Economists. Subspecialties: Operations research (engineering); Systems engineering. Current work: Costs benefits and market impacts of advanced technical products and systems. Office: ECON Inc 900 State Rd Princeton NJ 08540

MILLER, C ARDEN, physician, educator; b. Shelby, Ohio, Sept. 19, 1924; s. Harley H. and Mary (Thuma) M.; m. Helen Meihack, June 26, 1948; children—John Lewis, Thomas Meihack, Helen Lewis, Benjamin Lewis. Student, Oberlin Coll., 1942-44; M.D. cum laude, Yale, 1948. Intern, then asst. resident pediatrics Grace-New Haven Community Hosp., 1948-51; faculty U. Kans. Med. Center, 1951-66, dir. childrens rehab. unit, 1957-60, dean, provost, 1965-66, dean Med. Sch., dir., 1960-65; prof. pediatrics and maternal and child health U. N.C., Chapel Hill, 1966—, vice chancellor health scis., 1966-71, chmn. dept. maternal and child health, 1977; cons. UNICEF, HEW, World Bank; chmn. exec. com. Citizens Bd. Inquiry into Health Services for Am., 1968—; cons. United Mine Workers Welfare and Retirement Fund, 1973—; mem. adv. com. on nat. health ins. U.S. Ho. of Reps. Ways and Means Com., 1974—. Mem. editorial bd.: Jour. Med. Edn, 1960-66; Author numerous articles in field. Trustee Appalachian Regional Hosps., 1974, Alan Guttmacher Inst., Planned Parenthood Fedn. Am. Markle scholar in med. scis., 1955-60; Recipient Robert H. Felix Distinguished Service award St. Louis U., 1977. Hon. fellow Royal Soc. Health; mem. Am. Pub. Health Assn. (chmn. action bd. 1972—, pres. 1974-75), Soc. Pediatric Research, Assn. Am. Med. Colls. (v.p. 1965-66), Inst. of Medicine of Nat. Acad. Sci., Sigma Xi, Alpha Omega Alpha, Delta Omega. Subspecialties: Pediatrics; Preventive medicine. Current work: Public policy for improved maternal and child health outcomes. Home: 908 Greenwood Rd Chapel Hill NC 27514

MILLER, CARLOS OAKLEY, plant science educator; b. Jackson, Ohio, Feb. 19, 1923; s. William Flay and Marella (Leach) M. B.S. in Agr, Ohio State U., 1948, M.A., 1949, Ph.D., 1951. Project assoc. U. Wis., Madison, 1951-55, asst. prof., 1955-57, Ind. U., Bloomington, 1957-60, assoc. prof., 1960-63, prof., 1963—. Contbr. articles to profl. jours. Served with AUS, 1943-46. Mem. Bot. Soc. Am., Am. Soc. Plant Physiologists, Japanese Soc. Plant Physiologists, AAAS. Democrat. Subspecialties: Plant growth; Plant physiology (biology). Current work: Biosynthesis, action and roles of cytokinins, plant hormones, tissue culture. Home: 103 N Glenwood Ave W Bloomington IN 47401 Office: Biology Dept Indiana U Bloomington IN 47405

MILLER, CAROL ANN, neuropathologist; b. Cleve., May 3, 1939; d. Edward L. and Edith N. (Stein) M.; m. Martin Feldman, June 6, 1965 (div. 1979); children: Renny, Douglas; m. Seymour Benzer, May 11, 1980. A.B. cum laude, Mt. Holyoke coll., 1961; M.D., Jefferson Med. Coll., 1965. Diplomate: Am. Bd. Pathology. Intern Phila. Gen. Hosp., 1965-66; resident in pathology and neuropathology Washington U. Sch. Medicine, St. Louis, 1966-69; fellow in neuropathology Albert Einstein Coll. Medicine, Bronx, N.Y., 1969-70, fellow in cell biology, 1970-73, asst. prof. neurosci., 1973-77; assoc. prof. pathology U. So. Calif. Sch. Medicine, Los Angeles, 1977—; chief Cajal Lab. Neuropathology Los Angeles Hosp.-U. So. Calif. Med. Ctr., 1982—. Contbr. articles to profl. jours. Grantee NIH, 1979—, Hereditary Disease Found., 1979-80, Muscular Dystrophy Assn., 1979—. Mem. Am. Assn. Neuropathologists, Am. Soc. Microbiologists. Jewish. Subspecialties: Neuropathology; Pathology (medicine). Current work: Neurovirology; cellular and molecular bases of viral persistence in the central nervous system. Home: 2075 Robin Rd San Marino Ca 91108 Office: Dept Pathology U So Calif Sch Medicine 2025 Zonal Ave Los Angeles CA 90033

MILLER, CAROLE ANN, neurosurgeon; b. Kalamazoo, May 7, 1939. A.S., Bay City Jr. Coll., 1959; B.A., Ohio State U., 1962, M.D., 1966. Diplomate: Am. Bd. Neurology. Intern Grad. Hosp. U. Pa., 1966-67; resident Ohio State U. Hosp., 1967-72, instr., 1970-71, asst. prof., 1975, assoc. prof., 1980—; asst. prof. surgery U. Mich., Ann Arbor, 1972-74. Contbr. articles to profl. jours. Ohio Heart Assn. grantee, 1970; NIH grantee, 1973, 75. Mem. ACS, Am. Assn. Neurol. Surgeons, AAUP, Surg. Hist. Soc., Neurosurg. Soc. Am., Am. Assn. Neurol. Surgeons, Ohio Neurosurg. Soc., AMA, Ohio Med. Soc. Subspecialty: Neurosurgery. Current work: Vascular smooth muscle physiology; cerebellar stimulation in cerebral palsy; percutaneous spinal cord stimulation in multiple sclerosis. Office: 410 W 10th Ave University Hospital N911 Columbus OH 43210

MILLER, CHRIS H., microbiology educator, researcher; b. Indpls., Feb. 28, 1942; s. James Aaron and Louise (Hungate) M.; m. Sharon Alyce Dye, June 30, 1963; children: Matthew J., Kristine L., Jacob A., Bennett W. B.A., Butler U., 1964; M.S., U.N.D., 1966, Ph.D., 1969; postgrad., Purdue U., 1964-70. Asst. prof. dept. oral microbiology Sch. Dentistry Ind. U., 1970-75, assoc. prof., 1975-81, prof., 1981—, chmn. dept., 1981—; dir. Central Sterilization, 1973—, Sterilization Monitoring Service, 1980—. Editor: Dental Asepsis Rev, 1980—; contbr. chpts. to books, articles to profl. jours. Pres. Allisonville Little League, Inc., Indpls., 1980, 82. NIH fellow, 1969-70; Nat. Inst. Dental Research grantee, 1977-79; recipient Hon. Alumnus award, Ind. U. Sch. Dentistry, 1976. Fellow Am. Acad. Microbiology; mem. Am. Soc. Microbiology, Am. Assn. Dental Schs. (sec. microbiology sect. 1976, chmn. 1977), Am. Assn. Dental Research (pres. microbiology immunology group 1982, councillor Ind. sect. 1979—). Subspecialties: Microbiology; Oral biology. Current work: Microbial carbohydrate metabolism and microbial pathogenic properties as related to dental caries and periodontal diseases. Office: Indiana University School of Dentistry 1121 W Michigan St Indianapolis IN 46202

MILLER, D. MERRILY, educational cons., therapist; b. Yonkers, N.Y., Mar. 3, 1943; d. Stanley and Pearl (Colin) Dulman; m. Edward Richard Miller, Dec. 24, 1964; children: Logan, Sloan, Dane. A.B. cum laude, Vassar Coll., 1965; M.A., Memphis State U., 1968; Ed.M., Columbia U., 1972, Ed.D., 1974. Tchr. Bd. Edn., Yonkers, N.Y., 1968-72; instr. Fairleigh Dickinson U., Teaneck, N.J., 1972-73; dir. edn. Mend Human Devel. Ctr., N.Y.C., 1973-74; ednl. coordinator The Door, N.Y.C., 1974-75; asst. prof. Fordham U. Grad. Sch. Edn., N.Y.C., 1976-82; ednl. cons. Miller Assocs., Katonah, N.Y., 1981—. Contbr. articles to profl. jours. Columbia U. fellow, 1972-73; Bur. Edn. Handicapped grantee, 1977-80. Mem. Am. Psychol. Assn., Council for Exceptional Children, Assn. Women Administrs. in Westchester, Assn. N.Y. Educators of Emotionally Disturbed. Subspecialties: Learning; Learning. Current work: Narrowing gap between research/theory and ednl. practice. Address: 26 Cherry St Katonah NY 10536

MILLER, DAVID BENNETT, psychologist, educator, researcher; b. Cleve., Sept. 21, 1948; s. Joseph and May (Eglin) M.; m. Linda L. Lach, Oct. 16, 1950. B.A., U. Fla., 1970; M.S., U. Miami, 1972, Ph.D., 1973. Research assoc. Div. Mental Health State of N.C., Raleigh, 1973-77; Alexander von Humboldt fellow U. Bielefeld, W.Ger., 1977-78; research assoc. Div. Mental Health State of N.C., Raleigh, 1978-80; asst. prof. psychology U. Conn., Storrs, 1980-81, assoc. prof., 1982—. Contbr. articles to profl. jours. NSF grantee, 1980, 82. Mem. Am. Psychol. Assn. (program chmn. div. 6, 1982-83), Animal Behavior Soc., Internat. Soc. Developmental Psychobiology, Internat. Soc. Human Ethology, Psychonomic Soc. Subspecialties: Psychobiology; Ethology. Current work: Development of species-typical behavior; parent-young interaction; auditory communication in birds, behavioral embryology. Home: Mansfield Apts#89 Storrs CT 06268 Office: U Conn Dept Psychology Storrs CT 06268

MILLER, DONALD SPENCER, geologist, educator; b. Ventura, Calif., June 12, 1932; s. Spencer Jacob and Marquerite Rachael (Williams) M.; m. Carolyn Margaret Losee, June 12, 1954; children: Sandra Louise, Kenneth Donald, Christopher Spencer. B.A., Occidental Coll., 1954; M.A., Columbia U., 1956, Ph.D., 1960. Asst. prof. Rensselaer Poly. Inst., Troy, N.Y., 1960-64, assoc. prof., 1964-69, prof., 1969—, chmn. dept. geology, 1969-76, 80—; research assoc. geology Columbia U., 1960-63; research fellow geochemistry Calif. Inst. Tech., Pasadena, summer 1963; NSF Sci. Faculty fellow U. Bern, Switzerland, 1966-67; sci. guest prof. Max-Planck Inst. Nuclear Physics, Heidelberg, W. Ger., 1977-78, vis. prof., summer 1979, guest scientist, Aug. 1979, 80, 81, 82; vis. prof. Isotope Geology Lab., U. Berne, summer 1979; participant NATO exchange program Demokritos Inst., Athens, Greece, Sept. 1983. Pres., treas. Troy Rehab. and Improvement, Inc., 1968-74; mem. Troy Zoning Bd. Appeals, 1970—. Fellow Geol. Soc. Am.; mem. Am. Geophys. Union, Geochem. Soc., Nat. Assn. Geol. Tchrs., Sigma Xi, Sigma Pi Sigma. Subspecialty: Geochemistry. Current work: Studies of igneous and sedimentary rocks using quantitative age methods to reveal uplift rates and to determine their thermal history. Home: 2198 Tibbits Ave Troy NY 12180 Office: Dept Geology Rensselaer Poly Inst Troy NY 12181

MILLER, EDWARD JOSEPH, biochemist, educator; b. Akron, Ohio, Oct. 27, 1935; s. Edward Joseph and Juanita Mae (Smith) M.; m. Kathleen Ellen Maurin, Dec. 26, 1964; children: John David, Elizabeth Ann, Steven Edward. B.S., Spring Hill Coll., 1960; Ph.D., U. Rochester, 1964. Fellow ADA, Bethesda, Md., 1964-66; chemist NIH, Bethesda, 1966-71; prof. U. Ala., Birmingham, 1971—; cons. NIH, 1974-79, VA, Washington, 1983—. Editor-in-chief: Collagen Jour; contbr. numerous articles to sci. publs. Recipient Internat. Assn. Dental Research Basic Sci. award, 1971, Carol Nachman Soc. Rheumatology Research award, 1978. Mem. Am. Chem. Soc., Am. Soc. Biol. Chemists, Am. Rheumatism Assn. Roman Catholic. Subspecialty: Biochemistry (biology). Current work: research on the chemical properties, biosynthesis and physiological significance of the genetically-distinct collagens of vertebrate organisms. Home: 4429 Fredericksburg Dr Birmingham AL 35213 Office: U Ala University Station Birmingham AL 35294

MILLER, ELIZABETH CAVERT, oncology educator, research laboratory administrator; b. Mpls., May 2, 1920; d. William Lane and Mary Elizabeth (Mead) Cavert; m. James Alexander Miller, Aug. 30, 1942; children: Linda Ann, Helen Louise. B.S., U. Minn., 1941; M.S., U. Wis., 1943, Ph.D., 1945; D.Sc. (hon.), Med. Coll. Wis., 1982. Instr. dept. oncology U. Wis. Med. Center, 1946-48, asst. prof., 1948-58, assoc. prof., 1958-69, prof., 1969—; assoc. dir. McArdle Lab. for Cancer Research, U. Wis., Madison, 1973—; Wis. Alumni Research Found. prof. oncology U. Wis., 1980—, Van Rensselaer Potter prof. oncology, 1982—. Assoc. editor: Cancer Research, 1957-62; contbr. numerous articles on chem. carcinogenesis and microsomal oxidations to profl. jours. Recipient (with J.A. Miller) Langer-Teplitz award for cancer research Ann Langer Cancer Research Found., 1962, Lucy Wortham James award for cancer research James Ewing Soc., 1965, Bertner award M.D. Anderson Hosp. and Tumor Inst., 1971, Wis. div. award Am. Cancer Soc., 1973, Outstanding Achievement award U. Minn., 1973, Papanicolau award for cancer research Papanicolaou Cancer Research Inst., Miami, 1975; Rosenstiel award for basic med. scis. Brandeis U., 1976; Nat. award Am. Cancer Soc., 1977; Bristol-Myers award in cancer research, 1978; Gairdner Found. Ann. award, 1978; Founders award Chem. Industry Inst. Toxicology, 1978; Prix Griffuel Assn. pour Developpement de Recherche sur Cancer, 1978; 3M Life Sci award Fedn. Am. Socs. Exptl. Biology, 1979; Freedman award N.Y. Acad. Sci., 1979; Mott award Gen. Motors Cancer Research Found., 1980. Fellow Am. Acad. Arts and Scis., Wis. Acad. Scis., Arts and Letters; mem. Nat. Acad. Sci., Am. Assn. for Cancer Research, Am. Soc. Biol. Chemists, Japanese Cancer Soc. (hon.). Subspecialty: Oncology. Home: 5517 Hammersley Rd Madison WI 53711 Office: University of Wisconsin McArdle Lab Madison WI 53706

MILLER, FAYNEESE SHERYL, social psychologist, researcher; b. Danville, Va., Mar. 3, 1955; d. Charles Abraham and Essie May (Stanfield) M. B.A., Hampton Inst., 1977; M.S., Tex. Christian U., 1979; Ph.D., 1981. NIMH postdoctoral applied social psychology fellow Yale U., New Haven, 1981—; cons. Pub. Schs., Ft. Worth, 1979-81; Dept. Def., Washington, 1981, Western Oil Co., Ft. Worth, 1981. Contbr. articles to profl. publs. Univ. fellow Tex. Christian U., 1977-79; NSF predoctoral fellow, 1979-81. Mem. Am. Psychol. Assn., Psi Chi (parliamentarian 1975-77, pres. 1978-80, Service award 1981), Beta Kappa Chi, Sigma Gamma Nu, Alpha Kappa Alpha (Chgo. philacter 1979-80). Subspecialty: Social psychology. Current work: Relationship between symbolic beliefs, instrumental beliefs, attitudes and affect within a public policy context. Home: 75 Canton St Apt A-9 West Haven CT 06516 Office: Dept Psychology Yale Univ Box 11A Yale Station New Haven CT 06520

MILLER, FRANCIS PETER, pharmacologist; b. Bklyn., Sept. 5, 1941; s. Milton Peter and Florence Dorothea (Gattavara) M.; m. Marie Suzanne Gaspar, Dec. 24, 1962; m. Jacquelin Joyce McKee, Feb. 25, 1978; children: Marie Suzanne, Francis Peter, Eric Alan, Kristin Elizabeth. B.S. in Chemistry, Manhattan Coll., Riverdale, N.Y., 1959-63; M.S. in Biochemistry, George Washington U., 1965; Ph.D. in Pharmacology, Ind. U., 1968. Chemist NIH, Bethesda, Md., 1963-65; research asst. Ind. U., Bloomington, 1965-68; sr. pharmacologist Lakeside Labs., Milw., 1968-75, Merrell Dow Pharms., Inc., Cin., 1975—. Contbr. articles to profl. jours. Mem. Soc. Neuroscience, N.Y. Acad. Scis., Sigma Xi. Republican. Roman Catholic. Subspecialties: Pharmacology; Biochemistry (medicine). Current work: Study of the effects of drugs on the brain; development of better drugs for treating abnormal behavior. Home: 336 Broadway Loveland OH 45140 Office: 2110 E Galbraith Rd Cincinnati OH 42515

MILLER, FREEMAN DEVOLD, astronomer, educator; b. Somerville, Mass., Jan. 4, 1909; s. Rasmus Kjeldsberg and Ednah Freeman (Weeks) M.; m. Caroline Marie Dresser, June 27, 1933. S.B., Harvard U., 1930, M.A., 1932, Ph.D., 1934. Dir. Swasey Obs., Denison U., Granville, Ohio, 1934-40; mem. faculty U. Mich., Ann Arbor, 1946—, prof. astronomy, 1956-77, prof. emeritus, 1977—; assoc. dean Rackham Schl. Grad. Studies, 1959-66. Contbr. articles to profl. publs. Served to capt. USN, 1931-69. Mem. Am. Astron. Soc., Internat. Astron. Union. Episcopalian. Subspecialty: Cometary physics. Current work: Study of plasma tails of comets and interaction with interplanetary medium. Home: 1614 Shadford Rd Ann Arbor MI 48104 Office: U Mich 1049 Denison Bldg Ann Arbor MI 48109

MILLER, GLENDON RICHARD, microbiology educator; b. Columbus, Ohio, Oct. 28, 1938; s. Glendon Ivory and Gladys (Reicker) M.; m. Suzy Elizabeth von Achen, June 11, 1966; children: Glendon Kenneth, Erica Elizabeth. B.A., So. Ill. U., 1960, M.A., 1962; Ph.D., U. Mo., 1967. Research microbiologist Colgate Palmolive Co., Piscataway, N.J., 1966-68; asst. prof. microbiology Wichita (Kans.) State U., 1968-72, assoc. prof., 1972—. Mem. Am. Soc. Microbiology, AAAS, Sigma Xi. Republican. Methodist. Subspecialties: Microbiology; Microbiology (medicine). Current work: Microbial antibiotic resistance, Beta-Lactamases. Home: 1505 Timothy St Wichita KS 67212 Office: Dept Biol Scis Wichita State Univ Wichita KS 67208

MILLER, GLENN ALLAN, immunologist, educator; b. Allentown, Pa., Jan. 19, 1948; s. Wilbert Edward and Ruth B. (Folk) M.; m. Kay Sara Oberhaltzer, July 19, 1969; children: Jill Ellen, Kimberly Anne. B.S., Muhlenberg Coll., 1969; Ph.D. in Microbiology, Syracuse U., 1973. Postdoctoral fellow Scripps Clinic, LaJolla, Calif., 1973-76; asst. prof. microbiology Med. Coll. Va., 1976-81; dir. microbiology research A.H. Robins Pharm. Co., Richmond, Va., 1981—; adj. assoc. prof. Med. Coll. Va., 1981; adj. prof. Va. State U., 1979—; mem. faculty Alton Jones Coll. Sci. Center, 1980, Richmond Math. and Sci. Center, 1977-80. Contbr. articles to profl. jours. Am. Cancer Soc. grantee; NIH grantee. Mem. Assn. Immunologists. Subspecialties: Microbiology (medicine); Immunology (medicine). Current work: Macrophage heterogeneity, tumor immunity, direct microbiology and immunological research projects. Office: 1211 Sherwood Ave Richmond VA 23220

MILLER, GLENN EDWARD, astronomer, educator; b. Holyoke, Mass., Oct. 2, 1953; s. Ralph August and Mary Louise (Johnston) M.; m. Cherie Vaughn Miller, June 15, 1975; 1 son, Glenn Edward. B.A., Rollins Coll., 1975; M.A., U. Tex.-Austin, 1978, Ph.D., 1981. Asst. prof. astronomy U.Va., Charlottesville, 1981-83; staff astronomer CSC Space Telescope Sci. Inst. Johns Hopkins U., Balt., 1983—. Contbr. articles in field to profl. jours. Mem. AAAS, Am. Astron. Soc. Subspecialty: Theoretical astrophysics. Current work: Galaxies, stellar evolution. Office: Space Telescope Sci Inst Computer Scis Corp Johns Hopkins U Baltimore MD 21218 Home: 3215 N Charles St Apt 1002 Baltimore MD 21218

MILLER, HARRY F., mechanical engineer; b. Olean, N.Y., Sept. 24, 1950; s. Harry Eastman and Eugenia Rose (Kasperski) M.; m. Faye Ann Fischer, Aug. 27, 1977; 1 dau., Kathryn. B.S.M.E., Northeastern U., 1973; M.B.A., Lehigh U., 1977. Cert. engr. in tng., Mass. Constrn. engr. Pa. Power & Light Co., Allentown, 1973-77; contract engr. Dresser Clark., Olean, 1977-80; head product engr. axial compressors Dresser Clark, Olean, 1980—. Mem. ASME, Nat. Mgmt. Assn. Republican. Roman Catholic. Subspecialty: Mechanical engineering. Current work: Axila compressors, turbomachinery, mechanical and aerodynamic design of axial compressor product line, supervision of subordinates. Home: 116 N 10th St Olean NY 14760 Office: PO Box 560 Olean NY 14760

MILLER, HARVEY PHILIP, engineering educator, consultant; b. N.Y.C., Jan. 18, 1945; s. Samuel Abraham and Lila Dorothy (Discount) M. B.Mech. Engring., Rensselaer Poly. Inst., 1966; M.S. in Mech. Engring, MIT, 1968; M.Phil. in Applied Physics, Columbia U., 1973; Postgrad. in applied physics, Columbia U. Profl. engr., Pa. Research asst. Columbia U., N.Y.C., 1970-72; standards engr. ASME, N.Y.C., 1973-75; research assoc. U. Miami, Coral Gables, Fla., 1975-78; cons. engr. United Engrs., Phila., 1978-82; adj. prof. Drexel U., Phila., 1982—; cons. and Seminar Speaker Energy and Environ. Dynamics, San Juan, P.R., 1981. Contbr. articles to profl. jours. Raytheon Co. fellow, 1966. Mem. Am. Phys. Soc., Am. Nuclear Soc., ASME, Am. Soc. Indsl. and Applied Math., Tau Beta Pi, Pi Tau Sigma. Democrat. Subspecialties: Fluid mechanics; Applied mathematics. Current work: Mathematical modeling of hydrodynamics and heat transfer of environmental flows. Office: Drexel U 32nd & Chestnut Sts Philadelphia PA 19103 Home: 2400 Chestnut St Apt 1407 Philadelphia PA 19103

MILLER, HILLARD CRAIG, physicist; b. Northampton, Pa., Dec. 15, 1932; s. Hillard Alvin and Dorothy (Frantz) M.; m. Ruth Hazel Kingsbury, June 16, 1956; children: Eric, Kent, Curtis, Alice. B.A., Lehigh U., 1954, M.S., 1955; Ph.D., Pa. State U., 1960. Physicist Gen. Electric Co., 1960—, prin. physicist, St. Petersburg, Fla., 1972—; adj. instr. St. Petersburg Jr. Coll., 1974—. Served with C.E. U.S. Army, 1955-57. Mem. AAAS, Am. Phys. Soc., Am Vacuum Soc., IEEE, Royal Astron. Soc. Can. Subspecialty: Plasma physics. Current work: Researcher in elec. discharges and breakdown in gases and vacuums. Home: 616 Ruskin Rd Clearwater FL 33575 Office: General Electric Co PO Box 11508 Saint Petersburg FL 33733

MILLER, IRVING FRANKLIN, educator, univ. dean; b. N.Y.C., Sept. 27, 1934; s. Sol and Gertrude (Rochkind) M.; m. Baila Hannah Milner, Jan. 28, 1962; children—Eugenia Lynne, Jonathan Mark. B.S. in Chem. Engring, N.Y. U., 1955; M.S., Purdue U., 1956; Ph.D., U. Mich., 1960. Research scientist United Aircraft Corp., Hartford, 1959-61; asst. prof. to prof., head chem. engring. Poly. Inst. Bklyn., 1961-72; prof., head bioengring. program U. Ill., Chgo., 1973-79, acting head systems engring. dept., 1978-79, dean, 1979—, asso. vice chancellor for research, 1979—; cons. to industry, also Nat. Acad. Scis., NIH. Editor: Electrochemical Bioscience and Bioengineering, 1973; Contbr. articles profl. jours. Mem. Am. Inst. Chem. Engrs., Am. Chem. Soc., AAAS, Biomed. Engring. Soc., N.Y. Acad Scis. Subspecialties: Biomedical engineering; Artificial organs. Current work: Bioelectrochemistry, membrane transport, artificial organs. Home: 2600 Orrington Ave Evanston IL 60201 Office: Box 4348 Chicago IL 60680

MILLER, JACK W., pharmacologist, educator; b. Knoxville, Tenn., Sept. 26, 1925; s. Joseph and Ruth (Weinberg) M.; m. Barbara Bradshaw, Feb. 2, 1952; 1 son, Grant W. B.A., San Diego State Coll., 1949; Ph.D., U. Calif., Berkeley, 1954. From instr. to assoc. prof. pharmacology U. Wis., Madison, 1954-62; from asso. prof. to prof. U. Minn., Mpl., 1962—, also coordinator basic med. sci. courses. Contbr. articles to profl. jours. Served with AC USN, 1943-45. Recipient Disting. Teaching award Minn. Med. Found., 1965; fellow U. Ill., 1974. Mem. Am. Soc. Exptl. Pharmacology and Therapeutics. Subspecialty: Pharmacology. Current work: Pharmacology and physiology of labor; premature labor; narcotic analgesics; drug receptors. Home: 2111 W Hoyt St Saint Paul MN 55108 Office: U Minn Minneapolis MN 55455

MILLER, JACQUELIN NEVA, environmental oceanographer; b. Carlsbad, Calif., Apr. 27, 1934; d. Edward George and Neva Muriel (Sayre) Kentner; m. Gaylord Riggs Miller, Aug. 15, 1958 (dec. Dec. 1976); children: Anna Marie, Amy Sue, Erik Edward. B.A., U. Calif.-Riverside, 1956; M.S., U. Calif.-LaJolla, 1966. Research asst. U. Calif. Scripps Instn., LaJolla, 1957-64; asst. in research Inst. Geophysics U. Hawaii, Honolulu, 1966-68, asst. researcher, 1969-73, assoc. specialist, 1973—. Mem. Hawaiian Acad. Scis., Marine Tech. Soc. (exec. bd. 1979—). Subspecialties: Oceanography; Resource management. Current work: Deep ocean waste disposal, dredge spoil disposal, sediment transport, alternate energy development, biological and chemical studies on discharge from deep water intake pipes associated with ocean thermal energy conversion research, aquaculture development. Home: 6561 Hawaii Kai Dr Honolulu HI 96825 Office:

Environmental Center Univ Hawaii 2550 Campus Rd Honolulu HI 96822

MILLER, JAMES ALEXANDER, oncologist; b. Dormont, Pa., May 27, 1915; s. John Herman and Emma Anna (Stenger) M.; m. Elizabeth Cavert, Aug. 30, 1942; children: Linda Ann, Helen Louise. B.S. in Chemistry, U. Pitts., 1939; M.S., U. Wis., 1941, Ph.D. in Biochemistry, 1943; D. Sc. (hon.), Med. Coll. Wis., 1982. Finney-Howell fellow in cancer research U. Wis., Madison, 1943-44, instr. oncology, 1944-46, asst. prof., 1946-48, asso. prof., 1948-52, prof., 1952—, Wis. Alumni Research Found. prof. oncology, 1980-82, Van Rensselaer Potter prof. on oncology, 1982—; mem. advisory coms. Nat. Cancer Inst., Am. Cancer Soc., 1950—. Contbr. numerous articles on chemical carcinogenesis and microsomal oxidations to profl. jours. Recipient awards (with E.C. Miller); Langer-Teplitz award Ann Langer Cancer Research Found., 1962; Lucy Wortham James award James Ewing Soc., 1965; G.H.A. Clowes award Am. Assn. Cancer Research, 1969; Bertner award M.D. Anderson Hosp. and Tumor Inst., 1971; Papanicolaou award Papanicolaou Inst. Cancer Research, 1975; Rosenstiel award Brandeis U., 1976; award Am. Cancer Soc., 1977; Bristol-Myers award in cancer research, 1978; Gairdner Found. ann. award, Toronto, 1978; Founders award Chem. Industry Inst. Toxicology, 1978; 3M Life Sci. award Fedn. Am. Socs. Exptl. Biology, 1979; Freedman award N.Y. Acad. Sci., 1979; Mott award Gen. Motors Cancer Research Found., 1980. Fellow Am. Acad. Arts and Scis.; mem. Am. Assn. for Cancer Research, Am. Soc. Biol. Chemists, AAAS, Japanese Cancer Soc. (hon.), Am. Chem. Soc., Soc. Toxicology, Soc. for Exptl. Biology and Medicine, Nat. Acad. Scis. Subspecialty: Oncology. Home: 5517 Hammersley Rd Madison WI 53711 Office: McArdle Laboratory University of Wisconsin Madison WI 53706

MILLER, JAMES EDWARD, computer scientist, educator; b. Lafayette, La., Mar. 21, 1940; s. Edward Gustave and Orpha Marie (DeVilbis) M.; m. Diane Moon, June 6, 1964; children: Deborah Elaine, Michael Edward. B.S., U. Southwestern La., 1961, Ph.D., 1972; M.S., Auburn U., 1964. Systems engr. IBM, Birmingham, Ala., 1965-68; asst. prof. U. West Fla., Pensacola, 1968-70, chmn. systems sci., 1972—; grad. researcher U. Southwestern La., Lafayette, 1970-72; computer systems analyst EPA, Washington, 1979; cons., lectr. in field. Author numerous articles for tech. publs. Mem. Assn. Computing Machinery (editor Computer Sci. Edn. spl. interest group bull. 1982), Data Processing Mgmt. Assn. Democrat. Methodist. Subspecialties: Cryptography and data security; Information systems, storage, and retrieval (computer science). Current work: Computer crime; computer science education. Home: Route 5 Box 91A Cantonment FL 32533 Office: U West Fla Pensacola Fl 32514

MILLER, JAMES EDWARD, research biochemist; b. Spring Grove, Pa., Apr. 28, 1942; s. Clair C. and Dorothy M. (Strausbaugh) M.; m. Sara E. Brake, Aug. 4, 1962; children: Ellen, Sally. Ph.D., U. N.D., 1968. Research investigator G.D. Searle & Co., Skokie, Ill., 1970-78, research scientist, 1978-80, sr. research scientist, 1980—. Contbr. articles to profl. jours. Mem. Am. Chem. Soc., Am. Oil Chemist's Soc., N.Y. Acad. Sci., AAAS, Sigma Xi. Subspecialties: Biochemistry (medicine); Pharmacology. Current work: Molecular basis of diseases of aging using cells in culture and animal models. Office: Searle Research/Development 4901 Searle Pkwy Skokie IL 60077

MILLER, JAMES FREDERICK, geology educator; b. Davenport, Iowa; s. Harry Earnest and Frances Elizabeth (Henry) M.; m. Louise Linnette, June 17, 1967; children: Jason Frederick, Michelle Linnette. B.A., Augustana Coll., Rock Island, Ill., 1965; M.A., U. Wis.-Madison, 1968, Ph.D., 1971. Asst. prof. geology U. Utah, 1970-73; asst. prof. Lawrence U., 1973, U. Pitts., 1974, Southwest Mo. State U., Springfield, 1974-77, assoc, prof., 1977-82, prof., 1982—. Mem. Geol. Soc. Am., Paleontol. Soc., Paleontol. Assn., Australasian Assn. Paleontologists, Internat. Paleontol. Assn., Working Group on the Cambrian-Ordovician Boundary (sec. 1980—), Pander Soc., AAAS. Subspecialties: Paleontology; Taxonomy. Current work: Taxonomy, evolution and biostratigraphy of Cambrian and Ordovician Conodonta, redefinition and international correlation of the Cambrian-Ordovician boundary. Home: 2117 S Wellington St Springfield MO 65807 Office: Dept. Geography and Geology Southwest MO State U Springfield MO 65804

MILLER, JAMES KINCHELOE, animal scientist; b. Elkton, Md., June 16, 1932; s. James Zenus and Miriam (Kincheloe) M.; m. Geraldine Speering Miller, Aug. 6, 1960; children: Allen David, Daniel Speering. B.S. in Gen. Agr, Berry Coll., 1953; M.S., U. Ga., 1959, Ph.D. in Animal Nutrition, 1962. Asst. prof. animal sci. U. Tenn., 1961-67, assoc. prof., 1967-74, 1974-81, assoc. prof. dept. animal sci., Knoxville, 1981—. Contbr. to profl. jours. Served with U.S. Army, 1953-56. U.S. Dept. Agr. grantee, 1979-81. Mem. Am. Inst. Nutrition, Am. Dairy Sci. Assn., Am. Soc. Animal Sci. (Gustav Bohstedt award 1974). Baptist. Subspecialties: Animal nutrition; Animal physiology. Current work: Mineral metabolism in ruminants blood platelet function as related to death in hypomagnesemic ruminants; hormonal changes related to hypomagnesemia in ruminants. Home: Route 2 Box 84 Heiskell TN 37754 Office: Dept Animal Sci U Tenn Knoxville TN 37901

MILLER, JAMES WOODELL, research institute administrator, researcher, writer, consultant; b. Detroit, June 30, 1927; s. Harold Ormond and Mary Nettleton (Kenny) M.; m. Ardeth Jean Stevens, Jan. 6, 1951; children: Stephen W., Jeffery R. B.A., Mich. State U., 1949, M.A., 1950, Ph.D., 1956. Research assoc. Kresge Eye Inst., Detroit, 1952-60; staff engr. Hughes Aircraft Co., Fullerton, Calif., 1960-63; dir. engr. psychology Office Naval Research, Washington, 1963-69; dir. ocean tech. Dept. Interior, Washington, 1969—70; dep. dir. undersea sci. and tech. NOAA, Rockville, Md., 1970-82; assoc. dir. Fla. Inst. Oceanography, St. Petersburg, 1982—; dep. chmn. com. on vision Nat. Acad. Scis., NRC, 1960-64; exchange scientist Bulgarian Acad. Sci., Sofia, 1974; mem. bd. advs. Nat. Assn. Underwater Instrs., Montclair, Calif., 1978—, Deep Ocean Tech., Inc., Oakland, Calif., 1981—. Scoutmaster Capitol Area Council Boy Scouts Am., Falls Church, VA., 1963-66; mgr. Little League Baseball, McLean, Va., 1963-69; pres. Pensacola (Fla.) Chamber Orch., 1954-57. Served with USN, 1945-46. Recipient Arthur S. Flemming award U.S. Jr. C. of C., Washington, 1966, Disting. Civilian Service award USN, 1969; Disting. Alumni award Mich. State U., 1973. Fellow Human Factors Soc.; mem. Undersea Med. Soc., Marine Tech. Soc. Subspecialties: Oceanography; Human factors engineering. Current work: Development, implementation, management of integrated ocean science programs, particulary those involving manned systems, such as diving, undersea habitats, robotics; research management, program development of intercisciplinary oceanographic programs. Office: Fla Inst Oceanography 830 First St S Saint Petersburg FL 33701

MILLER, JOAN ELLEN, veterinarian; b. Darby, Pa., Nov. 24, 1951; d. Ralph A. and Lois (Averill) M.; m. Peter M. Briglia, Jr., May 18, 1974. B.A. cum laude, U. Pa., 1973, V.M.D. summa cum laude, 1978. Chemist E. I. duPont de Nemours & Co., Inc., Wilmington, Del., 1973-74; intern U. Calif., Davis, 1978-79; research assoc. U. Wash., Seattle, 1979-80; resident Mich. State U., East Lansing, 1980—. AAUW fellow, 1977; Am. Coll. Vet. Surgeons grantee, 1981; Biomed. Research Support grantee, 1981; Mich. Vet. Med. Assn. grantee, 1981. Mem. AVMA, Am. Animal Hosp. Assn. Subspecialty: Surgery (veterinary medicine). Current work: Cardiopulmonary diseases. Pulmonary function in canine mitral regurgitation. Office: Mich StateU Vet Clin Ctr A-201 East Lansing MI 48824

MILLER, JOHN JOHNSTON, III, physician; b. San Francisco, Apr. 9, 1934; s. John Johnston and Florence Irene (Ratzell) M.; m. Anne Robeson, May 30, 1958; children: John, Daniel, Andrew, Erich. B.S., Wesleyan U., 1955; M.D., U. Rochester, 1960; Ph.D., U. Melbourne, 1965. Diplomate: Am. Bd. Pediatrics, 1969. Intern, resident in pediatrics U. Calif., San Francisco, 1960-62; dir. rheumatic disease service Children's Hosp. at Stanford, Palo Alto, Calif., 1967—; prof. clin. pediatrics Stanford U., 1979—; mem. council of pediatric rheumatology Am. Rheumatism Assn., 1978-82. Editor: Juvenile Rhematoid Arthritis, 1978; Contbr. articles to profl. jours. Served to comdr. M.C. USNR, 1965-71. Mem. Am. Acad. Pediatrics, Soc. Pediatric Research, Am. Assn. Immunologists. Subspecialties: Pediatrics; Immunology (medicine). Current work: Mechanisms of arthritis in children. Home: 160 Locksonart Rd 8 Sunnyvale CA 94087 Office: 520 Willow Rd Palo Alto CA 94304

MILLER, JOSEPH S., nuclear engineer; b. Sacramento, Calif., May 26, 1948; s. Lloyd E. and Mildred M. (Knight) M.; m. Dixie L. Parham, Aug. 28, 1970; 1 son, Jason. B.S. in Indsl. Engring, U. Ark., 1971, 1972; M.S. in Nuclear Engring, Kans. State U., 1974. Cert. profl. engr., Kans. Instrumentation technician Ark. Power & Light, North Little Rock, 1969-70; engr. NUS Corp., Rockville, Md., 1974-76; sr. engr. EG&G Idaho Inc., Idaho Falls, 1976-78; nuclear engr. Black & Veatch, Kansas City, Mo., 1978-80, project design engr., 1980—; coll. recruiter, 1979—80. Judge Kansas City Sci. Expo, 1980. Mem. ASME, Am. Nuclear Soc. (vice chmn. Mo./Kans. sect. 1983), Phi Kappa Phi, Sigma Xi. Baptist. Lodge: Moose. Subspecialties: Nuclear engineering; Software engineering. Current work: Thermal hydraulics; heat transfer. Office: Black & Veatch 11401 Lamar Overland Park KS 66211

MILLER, KENNETH JOHN, chemist, educator; b. Chgo., Mar. 24, 1939; s. John N. and Elsie M. (Sucic) M.; m. Brunhilde Franziska, June 19, 1964; children: Michael N., John L. B.S., Ill. Inst. Tech., 1960; M.S., Johns Hopkins U., 1964; Ph.D., Iowa State U., 1966. Successively Nat. Acad. Sci.-NRC postdoctoral research assoc., asst. prof., assoc. prof. Rensselaer Poly. Inst., Troy, N.Y., 1967-81, prof. theoretical chemistry, 1981—. Contbr. articles to profl. publs. Faculty adviser local chpt. Pi Kappa Phi, 1980—. Recipient Achievement award Am. Inst. Chemists, 1960; grantee Petroleum Research Found., NIH. Mem. Am. Inst. Physics, Am. Chem. Soc., AAUP (v.p. 1976-77, pres. 1977-78), Sigma Xi, Phi Eta Sigma, Sigma Pi Sigma, Phi Lambda Upsilon. Subspecialties: Biophysical chemistry; Molecular pharmacology. Current work: Theoretical chemistry, conformations of DNA, computer assisted drug design and analysis of carcinogenicity, computer graphics, intercalation of molecules with DNA, analysis of carcinogenicity of polynuclear aromatic hydrocarbons, chemical reactivity of sulfur containing systems and diol epoxides of polynuclear aromatic hydrocarbons. Home: 20 Michael St Troy NY 12180 Office: Dept Chemistry Rensselaer Poly Inst Troy NY 12181

MILLER, LANCE ARNOLD, research center executive; b. El Paso, Tex., Aug. 28, 1938; s. L. A. and Helen Adelaide (Wheeler) M.; m. Eva-Maria Mueckstein, Nov. 25, 1977; children: by previous marriage: Kate, Scott. Student, MIT, 1956-58; B.S., Tufts U., 1960; M.A., U. Wis., 1966, Ph.D., 1968. Research psychologist Army Research Inst., Natick, Mass., 1960-63; mem. research staff IBM Watson Research Ctr., Yorktown Heights, N.Y., 1968-75, research mgr., 1975-82, Sr. research mgr., 1983—. Contbr. articles to profl. jours. Pres. S.E. Mus., Brewster, N.Y., 1977. Office Naval Research grantee, 1973-78. Mem. Assn. for Computing Machinery, Am. Psychol. Assn. (assoc.), Eastern Psychol. Assn., Midwestern Psychol. Assn., Human Factors Soc., AAAS, Assn. for Computational Linguistics, Lang. Soc. Am., Am. Assn. for Artificial Intelligence, Assn. for Lit. and Linguistic Computing, Cognitive Scis. Soc. Subspecialties: Artificial intelligence; Automated language processing. Current work: Computational linguistics and artificial intelligence: development of a computer system which understands natural English text input syntactically for text-critiquing and translation applications. Home: Baptist Church Rd Box 368 Yorktown Heights NY 10598 Office: IBM Watson Research Ctr PO Box 218 Yorktown Heights NY 10598

MILLER, LEE W., editor, plant ecologist; b. Decatur, Ill., July 9, 1930; s. Ben G. and Marguerite (Rosenberg) M.; m. Sylvia D. Bordek, Nov. 29, 1953; children: Benjamin A., Danna Ruth. Student, U. Ill., 1948-52; B.B.A., So. Methodist U., 1953; M.F., Yale U., 1961; Ph.D., Duke U., 1966. Time study engr. Bell & Howell, Lincolnwood, Ill., 1953-54; with Borg-Warner, Decatur, 1954-55; mgr. Ben Miller Jewelers, Decatur, 1955-59; asst. ecologist Brookhaven Nat. Lab., Upton, N.Y., 1961-62; asst. prof. Cornell U., 1966-73, adminstrv. mgr. dept. entomology, 1974-76; vis. prof. SUNY Coll. Environ. Sci. and Forestry, 1978; mng. editor Ecol. Soc. Am., Ithaca, N.Y., 1979—. Author: Psychrometry in Water Relations Research; contbr.: articles to Sci., Health Physics Bot. Gazette, Can. Jour. Botany; bd. editors: Who's Who in Frontier Sci. and Tech; mem. publs. com.: Am. Inst. Biol. Scis. Mem. Ecol. Soc. Am., Council Biology Editors. Club: Adirondack Mountain (pres. Finger Lakes chpt.). Subspecialties: Ecology; Theoretical ecology. Current work: Plant ecology. Office: Ecol Soc Am E139 Corson Hall Cornell U Ithaca NY 14853

MILLER, LYNNE CATHY, biological sciences educator; b. Washington, Dec. 25, 1951; d. Albert and Lorraine Shirley (Sweet) M.; m. Gary Franklin Clark, July 25, 1982. B.S. in Pharmacy, U.R.I., 1974; M.S. in Biol. Sci, U. Tex.-El Paso, 1977; Ph.D. in Biology, N. Mex. State U., 1980. Registered pharmacist, Mass., N.Y., R.I., Pa., N. Mex. Clin. pharmacy intern Misericordia Hosp., Bronx, N.Y., 1974-75; grad. teaching asst. dept. biology U. Tex.-El Paso, 1975-77, N. Mex. State U., Las Cruces, 1977-80, postdoctoral fellow, 1980-81; clin. pharmacist Meml. Gen. Hosp., Las Cruces, 1979; prof. parasitology Bloomsburg (Pa.) U., 1981—, pre-med. advisor, 1981—, pre-pharmacy advisor, 1981—, dir. internships, 1982—, coordinator honors program, 1981—; sponsor USPHS program, 1981. Contbr. articles to profl. jours. Sponsor Ronald McDonald House, Children's Oncology Services, 1981. Faculty research grantee, Pa., 1981—. Fellow Sigma Xi (assoc.); mem. AAAS, Pa. Acad. Sci., Phi Kappa Phi, Beta Beta Beta. Jewish. Clubs: Biology (Bloomsburg State Coll.) (sponsor); Rathcamp Matchcover (Allentown) (advisor 1981—). Subspecialties: Parasitology; Pharmacology. Current work: Research on the protective mechanism or self-cure reaction elicited by the host-immune response which causes elimination of parasitic worm burden; implications on reduction of anthelmintic dosages. Office: Dept Biology and Allied Health Bloomsburg U Bloomsburg PA 17815

MILLER, MATTHEW STEVEN, toxicologist; b. N.Y.C., Nov. 28, 1955; s. Matthew Samuel and Patricia Ann M.; m. Mary Jo Ruegg, May 27, 1978. B.S., State Coll. at Bridgewater, Mass., 1977; M.S., U. Ariz., 1979, Ph.D., 1982. Grad. research asst. dept. pharmacology and toxicology U. Ariz., 1977-82; USPHS postdoctoral fellow Inst. for Neurotoxicology, Albert Einstein Coll. Medicine, Bronx, N.Y., 1982—. Contbr. articles to profl. jours. Recipient various fellowships and grants. Mem. Soc. for Neurosci., Am. Soc. for Pharmacology and Exptl. Therapeutics, Soc. for Toxicology. Subspecialties: Neurochemistry; Toxicology (medicine). Current work: Biochemical mechanisms of neurotoxicity. Neuropeptides, axoplasmic transport, neurotrophic agents, peripheral neuropathy. Office: Albert Einstein Coll Medicine Inst. Neurotoxicology Rose Kennedy Center Bronx NY 10461

MILLER, MICHAEL JAMES, medical educator, research physiologist; b. Iron Mountain, Mich., Oct. 11, 1946; s. Robert J. and Jeanne (Barlement) M. B.S., U. Mich., 1968; Ph.D., Dartmouth Med. Sch., 1974, M.D., 1975. Teaching fellow Dartmouth Med. Sch., Hanover, N.H., 1974-75; clin. fellow in medicine Harvard Med. Sch., Boston, 1975-77, instr. medicine, 1981-83, 83—; instr. Med. U. S.C., Charleston, 1978-80; pulmonary cons. Brockton (Mass.) VA Hosp., 1981-83; v.p. Controls Supply Co., Inc., Kingsford, Mich., 1975—; co-dir. ICU, Brockton VA Med. Ctr. Hosp., 1982—. Contbr. articles to sci. jours. Mem. Republican Nat. Com., 1981-83, Brockton VA Exec. Com., 1982—. State of S.C. grantee, 1980; Parker B. Francis Found. grantee, 1978-80; recipient Alice Ryan award Dartmouth Coll., 1970-73; William Ogden award U. Mich., 1964. Mem. Mass. Med. Soc., Soc. Critical Care Medicine, Am. Thoracic Soc., Am. Fedn. Clin. Research (com. chmn. 1982). Methodist. Club: Newton Yacht. Subspecialties: Pulmonary medicine; Physiology (medicine). Current work: Examining the physiology of the respiratory muscles in both in vitro studies and in patients with pulmonary disease. Inventor digital speed limiter and monitor, 1982. Home: 45 Mathews Dr Wayland MA 01778 Office: 75 Francis St Boston MA 02115

MILLER, MORTON W(ILLIAM), radiation and cell biology researcher, educator, consultant; b. Neptune, N.J., Aug. 4, 1936; s. Elwood E. and Francis E. (Senkel) M.; m. Marylynn M. Brown, July 13, 1968; children: Marcus, Heath, Carl. B.A. cum laude, Drew U., 1958; M.S., U. Chgo., 1960, Ph.D., 1962. NATO postdoctoral fellow Oxford (Eng.) U., 1962-63; postdoctoral fellow Brookhaven Nat. Lab., Upton, N.Y., 1963-65; sci. officer IAEA, Vienna, Austria, 1965-67; asst. prof. radiation biology and biophysics U. Rochester, 1967-75, assoc, prof., 1975—. Numerous publs. in field. Bd. dirs. Chili (N.Y.) Pub. Library, 1981—. Dept. Energy grantee, 1967—; Office Naval Research grantee, 1969-72; NIH grantee, 1976—; Empire State Electric Energy Research Corp. grantee, 1981—. Mem. Radiation Research Soc., Environ, Mutagen Soc., N.Y. Acad. Scis., AAAS, Am. Inst. Biol. Scis., Council Biology Editors, Am. Inst. Ultrasound Medicine, Bioelectromagnetics Soc. Subspecialties: Radiation Biology; Cell biology. Current work: Radiation biology of electromagnetic and ultrasonic fields. Office: U Rochester Box RBB Rochester NY 14642

MILLER, NANCY E(LLEN), psychopathology clinical researcher, research administrator, psychoanalyst, clinical psychologist, neuropsychologist; b. Long Beach, N.Y., Aug. 20, 1947; d. Jerome H. and Katherine (Pearlman) M.; m. Walter A. Romanek, Aug. 25, 1983. B.A., N.Y. U., 1969; M.A., Harvard U., 1970; postgrad. fellowship program in mental health adminstrn, Washington Sch. Psychiatry, 1977-78; Ph.D., U. Chgo., 1978; postgrad. advanced psychotherapy program, Washington Sch. Psychiatry, 1978-81; candidate, Washington Psychoanalytic Inst., 1981—. Lic. psychologist, D.C.; cert. psychologist, Md. Clin. psychologist S.E. Community Mental Health Ctr., Chgo., 1971-77; research assoc. dept. psychiatry U. Chgo. Sch. Medicine, 1972-77; chief clin. research program Ctr. for Studies Mental Health of Aging, NIMH, Rockville, Md., 1977—; exec. sec. aging and mental health initial rev. group NIMH, 1977-79; instr. in clin. psychiatry Georgetown U. Sch. Medicine, 1980—. Editor: (with G.D. Cohen) Clinical Aspects of Alzheimer's Disease and Senile Dementia, 1981, Psychodynamic Research Perspectives on Development, Psychopathology and Treatment in the Elderly, 1983, (with L. Erlenmeyer-Kimling) Longitudinal Predictors of Psychopathology Across the Life Span, 1983, Schizophrenia, Paranoia and Schizophreniform Disorders in Later Life, 1983; spl. edit.: Neurobiology of Aging, winter, 1983; editorial bd.: Jour. Edn. Gerontology, 1976—, Neurobiology of Aging, 1980—, Profl. Psychology, 1980—, Psychoanalytic Psychology, 1982—; contbr. numerous articles to profl. publs. Bd. dirs. Montgomery County Mental Health Assn., 1978-80, Montgomery County Mobile Med. Free Clinic, 1978—. Subspecialties: Neuropsychology; Developmental psychology. Current work: Research on clinical aspects of psychopathology in late adulthood and old age, with special emphasis on late life dementias and major affective disorders, including their etiology and treatment; research administration; clinical assessment and treatment; teaching. Home: 10401 Grosvenor Pl Apt 704 Rockville MD 20852 Office: NIMH 5600 Fishers Ln 11-C-03 Rockville MD 20857

MILLER, NEAL ELGAR, psychologist, emeritus educator; b. Milw., Aug. 3, 1909; s. Irving E. and Lily R. (Fuenfstueck) M.; m. Marion E. Edwards, June 30, 1948; children: York, Sara. B.S., U. Wash., 1931; M.S., Stanford U., 1932; Ph.D., Yale U., 1935; D.Sc., U. Mich., 1965, U. Pa., 1968, St. Lawrence U., 1973, U. Uppsala, Sweden, 1977, LaSalle Coll., 1979. Social sci. research fellow Inst. Psychoanalysis, Vienna, Austria, 1935-36; asst. research psychologist Yale U., 1933-35; instr., asst. prof., research asst. psychol. Inst. Human Relations, 1936-41, assoc. prof., research assoc., 1941-42, 46-50, prof. psychology, 1950-52, James Rowland Angell prof. psychology, 1952-66; fellow Berkeley Coll., 1955—; prof. Rockefeller U., N.Y.C., 1966-81, prof. emeritus, 1981—; expert cons. Am. Inst. Research, 1946-62; spl. cons. com. human resources Research and Devel. Bd., Office Sec. Def., 1951-53; mem. tech. adv. panel Office Asst. Sec. Def., 1954-57; expert cons. Ops. Research Office and Human Resources Research Office, 1951-54. Author: (with J. Dollard et al) Frustration and Aggression, 1939, (with Dollard) Social Learning and Imitation, 1941, Personality and Psychotherapy, 1950, Graphic Communication and the Crisis in Education, 1957, N.E. Miller: Selected Papers, 1971; contbr. chpts. to psychol. handbooks; editor: Psychological Research on Pilot Tng, 1947. Chmn. bd. sci. dirs. Roscoe B. Jackson Meml. Lab., Bar Harbor, Maine, 1962-76; bd. sci. counsellors NIMH, 1957-61; fellowship com. Founds. Fund for Research in Psychiatry, 1956-61; mem. central council Internat. Brain Research Orgn., 1964; v.p. bd. dirs. Foote Sch., 1964-65; chmn. NAS/NRC Com. on Brain Scis., 1969-71; bd. sci. counsellors Nat. Inst. Child Health and Human Devel., 1969-72. Served to maj. U.S. A.C., 1942-46; officer in charge research, Psychol. Research Unit 1, Nashville, 1942-44; dir. psychol. research project Hdqrs. Flying Tng. Command, Randolph Field, Tex., 1944-46. Recipient Warren medal for exptl. psychology, 1954, Newcomb Cleveland prize, 1956; Nat. medal of Sci., 1964; Kenneth Craik Research award U. Cambridge, 1966; Wilbur Cross medal Yale U., 1966; Alumnus Summa Laude Dignitatus U. Wash., 1967; Distinguished Alumnus award Western Wash. State Coll.; Gold medal award Am. Psychol. Found., 1975; Mental Health Assn. research achievement award, 1978. Fellow Am. Acad. Arts and Scis. (council 1979-83), Brit. Psychol. Soc. (hon. fgn.); mem. Am. Philos. Soc., N.Y. Acad. Scis. (hon. life), Spanish Soc. Psychology (hon.), Am. Psychol. Assn. (council reps. 1954-58, pres. exptl. div. 1952-53, pres. 1960-61, pres. div. health psychology 1980-81, Disting. Sci. Contbn. award 1959, award for Disting. Contbns. to Knowledge 1983), Eastern Psychol. Assn. (pres. 1952-53), NRC (div. anthropology and psychology 1950-53, chmn. 1958-60), Nat. Acad. Sci. (chmn. sect. psychology 1965-67, sr. fellow Inst. of Medicine 1983), Soc. Exptl. Psychologists, AAAS, Soc. Neurosci. (pres. 1971-72), Biofeedback Soc. Am. (pres.-elect 1983), Acad. Behavioral Medicine Research (pres. 1978-79), Inst. of Medicine (sr.), Sigma Xi (pres. Rockefeller U. chpt.

1968-69), Phi Beta Kappa. Club: Mory's (New Haven). Subspecialty: Psychophysiology. Office: Rockefeller U New York NY 10021

MILLER, OSCAR LEE, JR., cell biologist; b. Gastonia, N.C., Apr. 12, 1925; s. Oscar Lee and Rose (Evans) M.; m. Mary Rose Smith, Dec. 18, 1948; children: Sharon Lee, Oscar Lee, Ill. B.S., N.C. State Coll., 1948, M.S., 1950; Ph.D., U. Minn., 1960. Comml. farm mgr., Horry County, S.C., 1950-56; mem. research staff biology div. Oak Ridge Nat. Lab., 1961-73; prof. Oak Ridge Grad. Sch. Biomed. Scis., U. Tenn., 1967-73; prof. biology U. Va., 1973—, chmn. dept. biology, 1973-79. Contbr. numerous articles on ultrastructural aspects of genetic activity in prokaryotic and eukaryotic cells to profl. jours. Served with USN, 1943-46. USPHS postdoctoral fellow, 1960-61; NSF grantee, 1973-77; NIH grantee, 1973, 77—. Mem. Am. Soc. Cell Biology, AAAS, Soc. Devel. Biology, Nat. Acad. Sci. Subspecialty: Cell biology. Office: Dept Biology U Va Charlottesville VA 22901

MILLER, PATRICIA LYNN, clinical psychologist, consultant; b. Chgo., Jan. 27, 1938; d. Joseph L. and Gertrude R. (Kontek) Lynn; m. Eric E. Miller, Feb. 27, 1960; children: Kurt D., Nathan C., Peter J. A.B., U. Chgo., 1958; M.S., Ill. Inst. Tech., 1971, Ph.D., 1979. Cert. sch. psychologist, Ill.; lic. clin. psychologist, Ill. Pub. relations dir. Chgo. Area Council Camp Fire Girls, 1958-68, asst. exec. dir., 1966-68; tchr. Valley View Sch. Dist., Romeoville, Ill., 1968-70; sch. psychologist Lockport (Ill.) Spl. Edn. Dist., 1971-80; clin. psychologist in pvt. practice, specializing in diagnostics and treatment of women and children, part-time, 1977-80, full-time, 1980—; sec./treas., dir. HEM, Inc., Joliet, 1980—; instr., cons. dept. psychology Ill. Inst. Tech., Chgo., 1975-77; field supr. Chgo. Sch. Profl. Psychology, 1981-82. Author: forms Health and Social History of the Child, 1981, Psychological History, 1982. Mem. Citizens' Com. for Wider Use of the Schs., Mayor Daley's Youth Commn., Tribune Charities Youth Commn., all Chgo., 1958-68; active Women's Network for E.R.A., 1970s. Grad. fellow State of Ill., 1970. Mem. Am. Psychol Assn., Ill. Psychol. Assn., Internat. Neuropsychol. Soc., Nat. Assn. Sch. Psychologists, Ill. Sch. Psychologists Assn., Rialto Sq. Arts Assn., Sigma Xi. Club: Zonta. Subspecialty: Neuropsychology. Current work: Continuing development of Psy-Dx, electronic version of Halstead Neuropsychological Test Battery with computer integration of related test results for comprehensive diagnosis; co-developer Psy-Dx. Co-developer Neuropsychological Test Battery, 1980. Home: 3510 Bankview Ln Joliet IL 60435 Office: 310 N Hammes Ave Joliet IL 60435

MILLER, PAUL LEROY, JR., mechanical engineering educator; b. Guthrie, Okla., June 27, 1934; m., 1955; 3 children. B.S., Kans. State U., 1957, M.S., 1961; Ph.D. in Mech. Engring, Okla. State U., 1966. From instr. to assoc. prof. Kans. State U., 1957-72, prof. mech. engring., 1972—, head dept. mech. engring., 1975—; cons. in field. Mem. Am. Soc. Engring. Edn.; Fellow ASHRAE; Mem. ASME., Am. Soc. Engring. Edn. Subspecialty: Mechanical engineering. Office: Dept Mech Engrin Coll Engring Kans State U Manhatta KS 66506

MILLER, RAYMOND MICHAEL, microbial ecologist; b. Chgo., July 16, 1945; s. John Michael and Margaret Leona (Courtney) M. B.S., Colo. State U., 1969; M.S., Ill. State U., 1971, Ph.D., 1975. Ext. plant pathologist Colo. State U., Ft. Collins., 1969; postdoctoral fellow Environ. Impact Div., Argonne (Ill.) Nat. Lab., 1975-76; asst. scientist Land Reclamation Program, 1976-80, scientist environ. research div., 1980—. Mem. AAAS, Am. Soc. Microbiology, Mycological Soc. Am., Soil Sci. Soc. Am., Bot. Soc. Am. Roman Catholic. Subspecialties: Ecosystems analysis; Plant pathology. Current work: Role of mycorrhizae in natural and disturbed ecosystems; effects of acid deposition on soil ecosystems. Home: 16W 731 Mockingbird Ln Apt 105 Hinsdale IL 60521 Office: Argonne Nat Lab Argonne IL 60439

MILLER, REGIS BOLDEN, wood anatomist; b. Meyersdale, Pa., Aug. 29, 1943; s. Elam D. and Rita L. (Bolden) M.; m.; children: Kelly, Sean. B.S. in Wood Sci, W.Va. U., 1966; M.S. in Botany, U. Wis., 1968, Ph.D., U. Md., 1973. With U.S. Forest Products Lab., Madison, Wis., 1970—, botanist, 1970-80, project leader, 1980—. Mem. Internat. Assn. Wood Anatomists, Internat. Assn. Plant Taxonomists, Bot. Soc. Am. Roman Catholic. Subspecialties: Taxonomy. Current work: Computer-assisted wood identification; wood anatomy of Juglandaceae, Flacourtiaceae and Leguminosae; temperate and tropical wood identification. Developed chem. spot-test for aluminum and its value for wood identification, 1980, wood identification via computer, 1980, standard list of characters suitable for computerized hardwood identification, 1981. Home: 5021 Old Middleton Rd Apt 14 Madison WI 53705 Office: Forest Products Lab PO Box 5130 Madison WI 53705

MILLER, RENE HARCOURT, aeronautical engineering educator, consultant; b. Tenafly, N.J., May 19, 1916; s. Arthur and Elizabeth (Tobin) M.; m. Maureen E. Michael, Nov. 20, 1973; m. Marcelle Hansotte, July 16, 1948; children: Christal L., John J. B.A., Cambridge (Eng.) U., 1937, M.A., 1954. Registered profl. engr., Mass. Aero. engr. G.L. Martin Co., Balt., 1937-39; chief aerodynamics and devel. McDonnell Aircraft, St. Louis, 1939-44; mem. faculty MIT, Cambridge, 1944—, prof., 1957-62, H.N. Slater prof. flight transp., 1962—, head dept. aeros. and astronautics, 1968-78; dir. Space Systems Lab., 1978—. Contbr. articles to sci. jours. Recipient Meritorious Civilian Service award U.S. Army, 1967, 70; I.B. Laskowitz award N.Y. Acad. Scis., 1976. Mem. Nat. Acad. Engring., AIAA (Sylvanus Albert Reed award 1969), Am. Helicopter Soc. (Klemin award 1968), Royal Aero. Soc. Republican. Club: St. Botolph's (Boston). Subspecialties: Aerospace engineering and technology; Wind power. Current work: Space systems, transportation systems, VTOL aircraft, rotary wing aerodynamics and dynamics including helicopter and wind turbines. Home: 321 Beacon St Boston MA 02115 Office: 77 Massachusetts Ave Bldg 33-411 Cambridge MA 02139

MILLER, RICHARD HANS, computer systems support supervisor, consultant; b. Portsmouth, Va., Aug. 12, 1953; s. R. Bruce and Adele (Mohr) M. B.A., Rice U., 1975, M.E.E., 1976. Systems programmer Rice U., Houston, 1974-76; head systems support Baylor Coll. Medicine, Houston, 1976—; cons. Horning, Johnson & Grove, Oklahoma City, 1980, IMSL, Inc., Houston, 1977-80. Mem. Assn. Computing Machinery, AAAS, Data Processing Mgmt. Assn. Democrat. Methodist. Subspecialties: Distributed systems and networks; Statistics. Current work: Implementation of local area networks, development of resource sharing network and network operating system. Interfacing heterogeneous equipment into a single network. Home: 4100 Greenbriar #331 Houston TX 77098 Office: Baylor Coll Medicine 1200 Moursund Houston TX 77030

MILLER, RICHARD KEITH, civil engineering educator, engineering consultant; b. Fresno, Calif., June 12, 1949; s. Albert Keith and Gloria Mae (Pittman) M.; m. Elizabeth Ann Parrish, July 10, 1971; 1 dau., Katherine. B.S., U. Calif.-Davis, 1971; M.S., M.I.T., 1972; Ph.D., Calif. Inst. Tech., 1976. Asst. prof. U. Calif.-Santa Barbara, 1975-79; assoc. prof. U. So. Calif., Los Angeles, 1979—; cons. Astro Research Corp., Carpinteria, Calif., 1977—, Hughes Aircraft Co., Fullerton, Calif., 1982—. Grantee NSF, 1979-83, NASA, 1982-83. Mem. ASME, ASCE, Am. Acad. Mechanics. Subspecialties: Theoretical and applied mechanics; Aerospace engineering and technology. Current work: Research in nonlinear structural dynamics, wave propagation, interface phenomena, structural mechanics, earthquake engineering, membrane mechanics. Office: U So Calif Dept Civil Engring Los Angeles CA 90089-1113

MILLER, RICHARD KERMIT, medical scientist, educator; b. Scranton, Pa., Oct. 17, 1946; s. Roland Kermit and Vera (Edwards) M.; m. Judith Berens. A.B., Dartmouth Coll., 1968, Ph.D., 1973. Postdoctoral fellow Jefferson Med. Coll., Phila., 1972-74; asst. prof. ob-gyn and pharmacology/toxicology U. Rochester, N.Y., 1974-80, dir. div. research, 1978—, assoc. prof., 1980—. Contbr. articles on pharmacology to profl. jours. Goode Found. grantee, 1976-82; Nat. Cancer Inst. grantee, 1978-84; NIMH grantee, 1979-86; Nat. Insts. Environ. Health Scis. grantee, 1982-86. Mem. Teratology Soc., Am. Soc. Pharmacology and Exptl. Therapeutics, Soc. Gynecol. Investigation, Behavioral Teratology Soc. Subspecialties: Pharmacology; Maternal and fetal medicine. Current work: Teratology/transplacental carcinogenesis, diethylstilbestrol, cadmium, placental function, diazepam, neurochemical/behavior alterations. Office: Dept Ob-Gyn 601 Elmwood Ave Rochester NY 14642

MILLER, RICHARD LEE, research biochemist, enzymologist; b. Glendale, Calif., May 30, 1942; s. Nolan Clyde and Anelia (Lumberg) M.; m. Teresa Ann Journey, Aug. 29, 1964; children: Diane Marie, Lynn Kathryn. B.S., San Jose State Coll., 1964; Ph.D., Ariz. State U., 1968. Sr. biochemist Wellcome Research Labs., Research Triangle Park, N.C., 1969—. Bd. dirs. Burroughs Wellcome Credit Union, Research Triangle Park, 1981—. Mem. Am. Soc. Biol. Chemists, Sigma Xi. Subspecialties: Biochemistry (medicine); Molecular pharmacology. Current work: Study of biochemical pathways (enzymes) of purine metabolism in parasites, bacteria and mammalian systems. Patentee in field. Home: 4100 Picardy Dr Raleigh NC 27612 Office: Wellcome Research Labs 3030 Cornwallis Rd Research Triangle Park NC 27709

MILLER, RICHARD LYNN, microbiologist, researcher; b. Stevens Point, Wis., Sept. 27, 1945; s. Gordon L. and Jean E. (Leary) M.; m. Lisa L., Sept. 15, 1973; children: Analiese, Colin, Autumn. B.S. in Biology, U. Wis., Stevens Point, 1968; Ph.D. in Microbiology, U. Minn., Mpls., 1974. Lab technician in diagnostic virology U.S. Army Med. Lab., 1968-70; teaching asst. in microbiology U. Minn., 1970-74, asst. prof., 1977-80; postdoctoral researcher Pa. State U. Med. Sch., 1975-77; sr. microbiologist, researcher Rike Labs., 3M Co., St. Paul, 1977-79, research specialist, 1979—. Contbr. papers to profl. meetings, jours. Served with U.S. Army, 1968-70. Mem. Am. Soc. Microbiology, AAAS, Henrici Soc. Minn., N.Y. Acad. Sci. Subspecialties: Virology (medicine); Microbiology (medicine). Current work: Antiviral drug research, research and supervision antiviral drug program. Office: 3M Center Bldg 270-2S-06 Saint Paul MN 55144

MILLER, ROBERT ALAN, senior staff scientist; b. Montclair, N.J., Jan. 30, 1943; s. George Ulmer and Florence Lahoma (Fairchild) M.; m. Mary Kathleen Sheridan, Jan. 30 1971; children: Brendan Alexander, Stacey Ann. B.S., U. Ill.-Urbana, 1965, M.S., 1966, Ph.D. (NSF fellow), 1970. Research assoc. Coll. of William and Mary, 1970-72, Rutgers U., 1972-74; theoretician, dir. theory, program dir. reactor design, mem. mgmt. com. Fusion Energy Corp., 1974-77; applications software group leader Princeton Gamma Tech., 1977-81; v.p. Sci. Transfer Assocs., N.Y.C., 1981-82, pres., 1982-83; sr. staff scientist Princeton Gamma-Tech, 1983—; cons. Contbr. articles to profl. jours. Mem. Am. Phys. Soc., IEEE, N.Y. Acad. Sci., U. Ill. Alumni Assn., Sigma Xi, Tau Beta Pi. Subspecialties: Fusion; Software engineering. Current work: X-ray fluorecence material analysis. Home: 22 Evans Dr Cranbury NJ 08512 Office: 1200 State Rd Princeton NJ 08540

MILLER, ROBERT VERNE, biochemistry and biophysics educator, author, consultant; b. Modesto, Calif., Dec. 27, 1945; s. Clyde Elwin and Jessie Marian M.; m. Barbara Ryan, Aug. 17, 1968; children: Katherine Lucile, Michael Brian, Colleen Elizabeth. B.A. in Microbiology, U. Calif., Davis, 1967, M.S., U. Ill., 1969, Ph.D., 1972. Research assoc. U. Calif., Berkeley, 1972-74; asst. prof. microbiology U. Tenn., Knoxville, 1974-78, assoc. prof., 1978-80; assoc. prof. biochemistry Stritch Sch. Medicine, Loyla U., Chgo., 1980—; sci. cons. Museum Sci. and Industry, Chgo.; cons. to drugs cos., sci. publs. Author: books, including A Study Guide to Microbiology, 1981, (with J. R. Meyer and with W. H. Yongue Jr.) A guide to the Study of Biology, 1982; contbr. articles, abstracts on molecular genetics of Pseudomonas aeruginosa and other bacterial species to profl. jours. Recipient Research Career Devel. award NIH, 1978-83; Calif. div. Am. Cancer Soc. Dernham postdoctoral fellow, 1972-74. Mem. Am. Soc. Microbiology, Am. Soc. Biol. Chemists, Genetics Soc. Am., AAAS, Calif. Alpha, Phi Beta Kappa, Sigma Xi. Subspecialties: Molecular biology; Genetics and genetic engineering (biology). Current work: Genetics of gene transfer and recombination on Pseudomonas aeruginosa, its bacteriophages and plasmids; molecular and biochemical genetics of the processes of gene transfer and establishment in procaryotic cells. Office: Dept Biochemistry Loyola U Med Center 2160 S 1st Ave Maywood IL 60153

MILLER, ROWLAND SPENCE, psychology educator; b. San Diego, Dec. 31, 1951; s. Donald Wesley and Charlotte (Williams) M.; m. Gale Annette Lewis, Dec. 18, 1982. B.A., Cornell U., 1973; M.A., U. Fla., 1976, Ph.D., 1978. Asst. prof. psychology Sam Houston State U., Huntsville, Tex., 1978—; NIMH trainee, 1976-78. Ad hoc reviewer: Personality & Social Psychology Bull, 1982—; assoc. editor: Jour. Ednl. Studies, 1981—; contbr. articles to profl. jours. Recipient Edwin Newman award for excellence in research Am. Psychol. Assn., 1980. Mem. Am. Psychol. Assn., Soc. Personality and Social Psychology, Southwestern Psychol. Assn., Phi Kappa Phi. Subspecialty: Social psychology. Current work: Psychological research into nature and processes of interpersonal interaction. Office: Sam Houston State Univ Dept Psychology Huntsville TX 77341

MILLER, SIDNEY ISRAEL, chemist, educator, cons., researcher; b. Saskatoon, Sask., Can., May 22, 1923; s. Max and Esther Herstein (Zuckerman) M.; m. Laura Reznick, Jan. 31, 1950; children: Matthew, Naomi, Joel. B.Sc., U. Man., 1945, M.Sc., 1946; Ph.D., Columbia U., 1951. Lectr. Ill. Inst. Tech., 1951-54, asst. prof., 1954-60, assoc. prof., 1960-64, prof., 1964—; tech. cons., tech. witness, analyst, info. specialist. Contbr. numerous articles on chemistry to profl. jours. Mem. Am. Chem. Soc., AAUP. Subspecialties: Organic chemistry; Analytical chemistry. Current work: Mechanisms, kinetics, acetylenes, heterocycles, coal chemistry. Office: Dept Chemistry Ill Inst Tech Chicago IL 60616

MILLER, STEPHEN DOUGLAS, immunologist, educator; b. Harrisburg, Pa., Jan. 22, 1948; s. Bruce Lloyd and E. Virginia M.; m. Kimberley Kohnlein, June 21, 1969; children: Jennifer R., Elizabeth K. B.S. in Microbiology, Pa. State U., 1969, M.S. in Microbiology (USPHS fellow), 1973, Ph.D. in Microbiology-Immunology, 1975. NIH postdoctoral fellow U. Colo. Med. Ctr., 1975-78, instr. medicine, 1978-80, asst. prof. medicine, 1980-81; asst. prof. microbiology-immunology Northwestern U. Med. Sch., 1981—. Contbr. chpt., articles to profl. publs. Served with U.S. Army, 1970-72. Recipient Young Investigator award NIH, 1978-81; NIH individual postdoctoral fellow, 1977-78; individual research fellow, 1981—. Mem. Am. Soc. Microbiology, Chgo Assn. Immunologists, Am. Assn. Immunologists, Phi Kappa Phi. Democrat. Presbyterian. Subspecialties: Cellular engineering; Immunogenetics. Current work: Research in immunoregulation of T cell mediated cellular immune responses; regulation of delayed-type hypersensitivity responses and T cell proliferative and cytotoxic responses mediated by antigen-specific suppressor T cells. Home: 946 S. Gunderson Ave Oak Park IL 60304 Office: 303 E. Chicago Ave 6-093 Morton Chicago IL 60611

MILLER, STEWART EDWARD, elec. engr.; b. Milw., Sept. 1, 1918; s. Walter C. and Martha L. (Ferguson) M.; m. Helen Jeanette Stroebel, Sept. 27, 1940; children—Jonathan James, Stewart Ferguson, Chris Richard. Student, U. Wis., 1936-39; B.S., Mass. Inst. Tech., 1941, M.S., 1941. With Bell Labs., N.Y.C., 1941-49, Holmdel, N.J., 1949—, dir. guided wave research, 1958—. Recipient Ballantine medal Franklin Inst., 1977. Fellow IEEE (Liebmann award 1972, Baker prize 1975), Optical Soc. Am.; mem. AAAS, Nat. Acad. Engring. Subspecialties: Fiber optics; Electrical engineering. Current work: Optical fiber telecommunication. Patentee in field. Home: 67 Wigwam Rd Locust NJ 07760 Office: Bell Labs Box 400 Holmdel NJ 07733

MILLER, TERRY DEE, research exec.; b. Salt Lake City, Mar. 28, 1944; s. Percy and Muriel (Strange) M.; m. Jolene Rae Schuldt, Sept. 16, 1965; children: Jeffrey, Andrea, Bryan, Michelle, Kathleen. B.S., Utah State U., 1966, Ph.D., 1971. Research assoc. U. Calif.-Berkeley, 1969-71; asst. prof. Ohio State U., Columbus, 1971-73; pres. Agrl. Cons. & Testing Lab., 1971-75, Miller Farm Research, 1976—, Western Seeds, Inc., 1976—; Mem. agrl. bd. Ricks Coll. NDEA fellow, 1966-69. Mem. Am. Phytopath. Soc., Am. Potato Soc., Soc. Sugarbeet Technologist, Sigma Xi. Mormon. Subspecialties: Plant pathology; Integrated pest management. Current work: Disease free seeds, optimum production. Home: PO Box 87 Minidoka ID 83343 Office: Route 4 Box 421 Rupert ID 83350

MILLER, THOMAS ALLEN, surgery educator; b. Harrisburg, Pa., July 7, 1944; s. Joseph E. and Marion R. (Corpman) M.; m. Janet Ruth Walters, Dec. 28, 1968; children: David Allen, William James, Laurie Ann. B.S. cum laude, Wheaton (Ill.) Coll., 1966; M.D., Temple U., 1970. Instr. dept. biology Wheaton(Ill.) Coll., 1966; intern in surgery U. Chgo. Hosps., 1970-71; resident in Surgery U. Mich. Hosps., 1971-75; instr. dept. surgery, postdoctoral research fellow in gastrointestinal hormone physiology U. Tex. Med. Branch, Galveston, 1975-76; instr. dept. surgery & physiology, postdoctoral research fellow in gastrointestinal physiology U. Tex. Med. Sch., Houston, 1976-77, asst. prof. surgery, 1977-79, assoc. prof., 1979—, dir. acad. affairs, resident tng., 1981-82, dir. grad. surg. edn., 1982—. Co-editor: (with S.J. Dudrick) The Management of Difficult Surgical Problems, 1981; contbr. articles to profl. jours. Upjohn Co. grantee, 1977-78; Distilled Spirits Council grantee, 1977-78; U. Tex. grantee, 1978-79; NIH grantee, 1979—. Fellow ACS; Mem. AAAS, Am. Digestive Disease Soc., Am. Fedn. Clin. Research, Am. Gastroent. Assn., AMA, Am. Physiol. Soc., Am. Soc. Parenteral and Enteral Nutrition, Assn. Acad. Surgery (chmn. com on legis. issues 1978-79, com. on issues 1979-81, nominating com. 1983—), Coll. Internat. Chirugiae Digestivae, Harris County Med. Soc. (cancer com. 1980—), Houston Gastroent. Assn., Houston Surg. Soc., N.Y. Acad. Scis., Pancreatic Club Inc., Soc. Internat. de Chirurgie, Soc. Exptl. Biology & Medicine, Soc. Surgery of Alimentary Tract (nominating com. 1982, auditing com. 1983), Soc. Univ. Surgeons (councilman-at-large 1983-86), Splanchnic Circulation Group, Tex. Med. Assn. Republican. Presbyterian. Subspecialties: Physiology (medicine); Surgery. Current work: Extensive research on the mechanisms by which prostaglandins mediate their ability to prevent gastric mucosal injury; teaching medical students and residents daily. Office: Dept Surgery U Tex Med Sch 6431 Fannin Rm 4266 Houston TX 77030 Home: 10618 Shady River Houston TX 77042

MILLER, THOMAS PAUL, electrical engineer, corporation executive; b. Milw., May 29, 1930; s. Paul N. and Christnie B. (Mack) M.; m. Dolores N. Schwartz, May 6, 1949; children. B.S. in Physics; M.S. in Indsl. Engring. Mgr. switching devel. Kellogg Switchboard, Chgo., 1959-62; dir. mktg. ITT-Kellogg, Chgo., 1962-64; v.p. Systems Installation, Inc., 1964-67; dir. long range planning ITT, Memphis; pres. Terra Corp., Jackson, Tenn., 1969—. Mem. West Tenn. Mfrs. Council; mem. adv. council Boy Scouts Am., 1977-80. Mem. IEEE (sr. mem.; switching com. 1962-69), Western Soc. Engrs. (sr. mem.), Air Craft Owners and Pilots Assn. Club: Jackson Golf and Country. Subspecialties: Computer engineering; Software engineering. Current work: Micro computers; control; energy; industrial; commercial; power generator control. Patentee in field. Home: PO Box 2182 Jackson TN 38301 Office: Terra Corp 46 Conalco Dr Jackson TN 38301

MILLER, THOMAS WILLIAM, psychology educator; b. Rochester, N.Y., Feb. 7, 1943; s. William John and Evelyn Ann (Weber) M.; m. Jean M. Alderson, June 17, 1967; children: David, Jeanine. B.S., St. John Fisher Coll., 1965; M.S., U. Scranton, 1967; Ph.D., SUNY-Buffalo, 1971. Diplomate: Am. Bd. Profl. Psychology in Clin. Psychology. Vice pres. Daemen Coll., Buffalo, 1970-75; asst. clin. prof. SUNY-Buffalo, 1975-81; clin. psychologist Va. Med. Ctr., Buffalo, 1975-81, chief psychologist, Lexington, Ky., 1981—; assoc. prof. U. Ky., Lexington, 1981—; cons. Child Devel. Ctr., Buffalo, 1971-75, Gateway, Snyder, N.Y., 1974-77; dir. psychol. services Rosary Hill Coll., Amherst, N.Y., 1970-74; mem. Ky. Bd. Psychology, 1983—. Contbr. articles to profl. jours. Com. chmn. Boy Scouts Am., Buffalo, 1970; post adv. Explorer Scouts, Lexington, 1982; cons. Project CARE, Amherst, 1973. Recipient Disting. Achievement award Psychol. Assn. Western N.Y., 1981. Mem. Am. Psychol. Assn., Psychol. Assn. Western N.Y. (pres.), Ky. Psychol. Assn. (pres. elect), Southeastern Psychol. Assn., Eastern Psychol Assn. Subspecialties: Behavioral psychology; Social psychology. Current work: Psychological aspects of major psychiatric disorders. Home: 3433 Fleetwood Dr Lexington KY 40502 Office: Chief Psychol Service 1168 VA Med Center Leestown Div Lexington KY 40511

MILLER, VIRGINIA MAY, physiologist; b. Pitts., Apr. 29, 1948; d. Otto F. and Ruth C. (Williams) May; m. Wayne L. Miller; 1 son, Matthew. B.S., Slippery Rock State Coll., 1970; Ph.D., U. Mo., 1976. Postdoctoral tng. U. Va., 1976-78, vis. asst. prof., 1978-79; assoc. scientist U. Del., 1979—. Contbr. articles to profl. jours. Research grantee Del. affiliate Am. Heart Assn., 1980—. Mem. Assn. Women in Sci., Am. Physiol. Soc., Soc. Neurosci., Research Soc. N.Am., Sigma Xi. Subspecialties: Comparative physiology; Physiology (medicine). Current work: Cardiovascular responses in animals capable of hibernating; neuro peptide distribution and physiological control systems. Office: SLHS U Del Newark DE 19711

MILLER, WALTER L., molecular biologist, pediatric researcher and educator; b. Alexandria, Va., Feb. 21, 1944; s. Luther S. and Beryl R.

M. S.B., MIT, 1965; M.D., Duke U., 1970. Diplomate: Nat. Bd. Med. Examiners, Am. Bd. Pediatrics. Intern, resident Mass. Gen. Hosp., Boston, 1970-72; staff assoc. NIH, Bethesda, Md., 1972-74; sr. resident U. Calif.-San Francisco, 1974-75, fellow in biochemistry, 1975-77, fellow in pediatric endocrinology, 1977-78, asst. prof. pediatrics, 1978-83, assoc. prof. pediatrics and metabolic research unit, 1983—, staff, 1978—, mem. endocrinology grad. faculty, 1982—. Editor: DNA, 1982—. Del., Democratic Nat. Conv., 1976. Served to lt. comdr. USPHS, 1972-74. Giannini fellow Bank of Am., 1977; NIH grantee, 1978. Fellow Am. Acad. Pediatrics; mem. Endocrine Soc., Soc. Pediatric Research (Ross research award 1982), Lawson Wilkins Pediatric Endocrine Soc., Am. Fedn. Clin. Research, AAAS, Theta Delta Chi. Club: Nuit San Wogga Wogga (winemaster 1980-83). Subspecialties: Genetics and genetic engineering (biology); Pediatrics. Current work: Recombinant DNA/molecular biology/endocrinology. Application of molecular biology to pediatric endocrinology and other areas. Molecular evolution. Treatment of disease via recombinant DNA. Office: Dept Pediatrics U Calif-San Francisco San Francisco CA 94143

MILLER, WILBUR CHARLES, mathematician; b. Westcliffe, Colo., Nov. 8, 1930; s. Martin and Fredia (Kitzman) M.; m. Marilyn Louise Franz, Aug. 1, 1953; 1 son, Robert. B.A., U. Wash., 1962; M.B.S., U. Colo., 1967; Ph.D., Colo. State U., 1975. Research test designer Boeing Co., Seattle, 1958-63; tchr. Douglas County Schs., Castle Rock, Colo., 1963-65, Fremont County Schs., Florence, Colo., 1965-67; mem. faculty U. So. Colo., Pueblo, 1967—, assoc. prof. math, 1976-79, prof., 1979—. Served as sgt. USAF, 1953-57. Mem. Soc. Indsl. and Applied Math, Am. Soc. Animal Scientists, Am. Soc. Agrl. Cons. Subspecialties: Integrated systems modelling and engineering; Applied mathematics. Current work: Mathematical modeling in agricultural small scale hydroelectric power generation. Home: Rt 1 Box 48 Westcliffe CO 81252 Office: Dept Math U So Colo Pueblo CO 81001

MILLER, WILLARD, JR., mathematician; b. Ft. Wayne, Ind., Sept. 17, 1937; s. Willard and Ruth (Kemerly) M.; m. Jane Campbell Scott, June 5, 1965; children—Stephen, Andrea. S.B. in Math, U. Chgo., 1958; Ph.D. in Applied Math, U. Calif., Berkeley, 1963. Vis. mem. Courant Inst. Math. Scis., N.Y. U., 1963-65; mem. faculty U. Minn., 1965—, prof. math., 1972—, head, 1978—. Author: Lie Theory and Special Functions, 1968, Symmetry Groups and Their Applications, 1972, Symmetry and Separation of Variables, 1977; Asso. editor: Jour. Math. Physics, 1973-75, Applicable Analysis, 1978—. Mem. Soc. Indsl. and Applied Math. (mng. editor Jour. Math. Analysis 1975-81), Am. Math. Soc., AAAS, AAUP, Sigma Xi. Subspecialties: Applied mathematics; Group Theory. Current work: Relations between the symmetry groups of partial differential equations and the solutions at these equations that arise through variable separation. Home: 4508 Edmund Blvd Minneapolis MN 55406 Office: Sch Math U Minn Minneapolis MN 55455

MILLER, WILLIAM HUGHES, chemistry educator; b. Kosciusko, Miss., Mar. 16, 1941; s. Weldon Howard and Jewell Irene (Hughes) M.; m. Margarete Ann Westbrook, June 4, 1966; children: Alison Leslie, Emily Sinclaire. B.S., Ga. Inst. Tech., 1963; M.S., Harvard U., 1964, Ph.D., 1967. Jr. fellow Soc. of Fellows, Harvard U., Cambridge, Mass., 1967-69; asst. prof. chemistry U. Calif.-Berkeley, 1969-72, assoc. prof., 1972-74, prof., 1974—. Alfred P. Sloan fellow, 1970-72; Camille and Henry Dreyfus tchr.-scholar, 1973-79; Guggenheim fellow, 1975-76; Overseas fellow Churchill Coll., Cambridge U., 1975-76; Alexander von Humboldt sr. scientist awardee, 1981-82. Mem. N.Y. Acad. Scis., Am. Chem. Soc., Am. Phys. Soc. Club: No. Calif. Squash Rackets. Subspecialties: Physical chemistry; Theoretical chemistry. Current work: Theory of chemical reaction dynamics, theory of atomic and molecular scattering processes, semiclassical theory for chemical and molecular phenomena. Office: Dept Chemistry U Calif Berkeley CA 94720

MILLER, WILLIAM ROBERT, JR., physicist, educator; b. Balt., June 17, 1934; s. William Robert and Ida Louise (Werner) M.; m. Barbara Ann Flammer, June 21, 1958; children: Robin Louise, Patricia Ann, Jill Elizabeth. B.A., Gettysburg Coll., 1956; Ph.D., U. Del., 1961. Sr. physicist Westinghouse Electric Corp., Balt., 1965-67; physicist RCA Corp., Lancaster, Pa., 1967-68; assoc. prof. physics Pa. State U., Middletown, 1969—; instr. York Coll. Pa., 1969-78. Contbr. articles to profl. jours. Mem. Am. Phys. Soc., Am. Assn. Physics Tchrs., Sigma Xi. Lutheran. Subspecialties: Condensed matter physics; Atomic and molecular physics. Current work: Experimental solid state physics (interface plasmons at bimetallic junction). Home: 1029 Preston Rd Lancaster PA 17601 Office: Pa State U Middletown PA 17057

MILLER, WILLIAM W., III, biology educator; b. Starkville, Miss., Oct. 4, 1932; s. William W. and Nettie (Sanders) M.; m. Jane Pierce, July 21, 1957; children: William W., Eva Miller Lockwood. B.S., Miss. State U., 1954, M.S., 1957; Ph.D., Auburn U., 1962. Prof. reproductive physiology Samford U., Birmingham, Ala., 1962-67; prof. physiology Northeast La. U., Monroe, 1967—. Served to 1st lt., inf. U.S. Army, 1954-56. Mem. Soc. Study of Reprodn., La. Acad. Sci., Sigma Xi. Republican. Presbyterian. Subspecialties: Endocrinology; Cancer research (medicine). Current work: Bladder cancer. Home: 43 Jana Dr Monroe LA 71203 Office: Dept Biology NE La State U Monroe LA 71209

MILLERO, FRANK JOSEPH, JR., chemist, educator; b. Greenville, Pa., Mar. 16, 1939; s. Frank Joseph and Jennie Elizabeth (Marks) M.; m. Judith Busang, Oct. 2, 1965; children: Marta, Frank, Anthony. B.S., Ohio State U., 1961; M.S., Carnegie-Mellon U., 1964, Ph.D., 1965. Research scientist ESSO Research & Engring. Co., Linden, N.J., 1965-66; prof. Marine and phys. chemistry U. Miami, 1966—; vis. prof. U. Kiel, Germany, 1975, Inst. Ricerca Sulle Aque, Rome, 1979-80; cons. to govt., pvt. corps., univs.; mem. Ocean Sci. Bd., 1981—, UNESCO Panel for Ocean Standard, 1976—, NSF Ocean Panel, 1973-75, 82; chmn. Gordon Conf. Chem. Oceanography, 1983. Contbr. numerous articles to profl. jours. Coach Little League; advisor sch. camp. Mem. AAAS, Am. Chem. Soc., Am. Geophys. Union, Geochem. Soc., Sigma Xi. Democrat. Roman Catholic. Lodge: KC. Subspecialties: Physical chemistry; Thermodynamics. Current work: Physical chemistry of natural waters, thermodynamics of solutions. Home: 7720 SW 90th Ave Miami FL 33173 Office: Rosenstiel School of Maine and Atmospheric Science University of Miami Miami FL 33149

MILLHORN, DAVID EUGENE, physiologist; b. Chattanooga, June 25, 1945; s. Cecil David and Ola (Doggett) M.; m. Sherry Lynn Long, July 5, 1968; children: Amy Lynn, Emily Katherine, Lauren Paige. B.S., U. Tenn., 1974; Ph.D., Ohio State U., 1978. NIH postdoctoral fellow dept. physiology U. N.C., Chapel Hill, 1978-80, Parker B. Francis fellow, 1980-82, asst. prof. physiology, 1980—; established investigator Am. Heart Assn., 1982-87. Contbr. articles to profl. jours. Am. Heart Assn. grantee-in-aid, 1982-85; NIH grantee, 1982-87. Mem. Am. Physiol. Soc., Soc. Neurosci., AAAS, Am. Heart Assn. Subspecialties: Physiology (medicine); Neurophysiology. Current work: Central neural regulation of respiration and cardiovascular function. Home: 1001 Sedwick W Durham NC 27713 Office: U NC 68 Med Research Wing Dept Physiology Chapel Hill NC 27514

MILLICH, FRANK, chemistry educator; b. N.Y.C., Jan. 31, 1928; s. Frank J. and Frances (Cop) M.; children: Theadocia Fran, Frank Theodore. B.S., CUNY, 1949; M.S., Poly. Inst. N.Y., 1956, Ph.D., 1959; postgrad., Cambridge (Eng.) U., 1958-59, U. Calif.-Berkeley, 1959-60. Asst. prof. chemistry U. Mo.-Kansas City, 1960-64, assoc. prof., 1964-67, prof., 1967—; vis. research scientist Ames Research Center, Moffett Field, Calif., 1964, Jet Propulsion Lab., Calif. Inst. Tech., 1966; cons. in fields of photochemistry, organic chemistry, and polymer sci. Author: (with C.E. Carraher) Interfacial Synthesis, vol. I, II, 1977; Contbr. articles in field to profl. jours. Grantee in field. Mem. Am. Chem. Soc., AAAS, Chem. Soc. London. Subspecialties: Polymer chemistry; Organic chemistry. Current work: Polymer syntheses and properties, interfacial synthesis, artificial biopolymers, polyisocyanides, viscosity theory. Home: 301 E New Santa Fe Trail Kansas City MO 64145 Offic: 5100 Rockhill Rd U Mo-Kansas City Kansas City MO 64110

MILLICHAP, J(OSEPH) GORDON, neurologist, neuropharmacologist, educator; b. Wellington, Eng., Dec. 18, 1918; came to U.S., 1953; s. Joseph Profit and Alice (Flello) M.; m. Mary Irene Fortey, Apr. 25, 1946 (dec. 1969); children: Martin Gordon, Paul Anthony; m. Nancy Melanie Kluczynski, Nov. 7, 1970; children: Gordon Thomas, John Joseph. M.B.,B.S. with honors, St. Bartholomew's Hosp. Med. Coll., U. London, 1946, M.D., 1950. Diplomate: Am. Bd. Neurology, Am. Bd. Pediatrics (child neurology, EEG). Intern St. Bartholomew's Hosp., London, 1946-47; resident Mass. Gen. Hosp., Boston, 1958-60; assoc. prof. U. Utah, Salt Lake City, 1954-55; pediatric neurologist NIH, Bethesda, Md., 1955-56; assoc. prof. Albert Einstein Coll. Medicine, N.Y.C., 1956-58; cons. in pediatric neurology Mayo Clinic, Rochester, Minn., 1960-63; assoc. prof. Mayo Grad. Med. Sch., Rochester, 1960-63; prof. neurology and pediatrics Northwestern U. Med. Sch., Chgo., 1963—; attending neurologist Northwestern Meml. Hosp., Chgo., 1963—; cons. Burroughs Wellcome, N.Y.C., 1956-60, Baxter Labs., Morton Grove, Ill., 1963-69, USPHS, Washington, 1955-70. Author: The Hyperactive Child, 1975, Febrile Convulsions, 1968; editor: Learning Disabilities, 1976, Pediatric Neursology, 1964; contbr. articles to profl. publs. Chmn. med. adv. bd. Epilepsy Found. Am., Chgo. chpt.; United Cerebral Palsy, Learning Disabilities Assn., 1963—. Served as squadron leader, med. specialist RAF, 1948-50. Fellow NIH, USPHS, 1958; recipient Citizenship award Chgo. Met. Area, 1965, DAR award, 1966, Advances in Epilepsy award Midwest Epilepsy Found., 1981. Fellow Am. Acad. Neurology (chmn. pediatric neurology sect. 1956—), Royal Coll. Physicians (London); mem. Am. Neurol. Assn., Am. Soc. Pharmacology and Therapeutics, Am. Soc. Pediatric Research, Am. Epilepsy Soc., Pharm. Soc. Gt. Britain. Episcopalian. Club: Saddle and Cycle (Chgo.). Subspecialties: Neurology; Neuropharmacology. Current work: Pediatric neurology; electroencephalography; drug therapy of epilepsy, learning and behavior disorders. Home: Walden Walden Dr Lake Forest IL 60045 Office: Northwestern Meml Hosp 250 E Superior St Suite 1428 Chicago IL 60611

MILLIKAN, ALLAN GROSVENOR, research scientist; b. Charleston, W.Va., July 31, 1927; s. Robert F. and Laura (Grosvenor) M.; m. Nancy McCombs, Sept. 2, 1951 (dec.); children: Mark, David, Laura, Melanie; m. Elizabeth Goodrich, Aug. 6, 1977. A.B., Oberlin Coll., 1949; M.S., Purdue U., 1951. Scientist Color Tech. div. Eastman Kodak Co., Rochester, N.Y., 1951-57, with research labs., 1957—, sr. research assoc., 1973—; vis. scientist Kitt Peak Nat. Obs., Tucson, 1971-72. Contbr. articles to profl. jours. Mem. Am. Astron. Soc. (past chmn. working group on photographic materials), Internat. Astron. Union, Arabian Horse Registry. Unitarian. Subspecialty: Optical astronomy. Current work: Photographic detectors for ground and space optical astronomy; computer-aided research; photographic emulsions. Home: 7061 Boughton Hill Rd Victor NY 14564 Office: Research Labs B-5 Eastman Kodak Co Rochester NY 14650

MILLIKAN, DANIEL FRANKLIN, JR., plant pathology, educator; b. Lyndon, Ill., May 31, 1918; s. Daniel Franklin and Harriet Adeline (Parmenter) M. B.S., Iowa State U., 1947; Ph.D., U. Mo.-Columbia, 1954. Instr. botany U. Mo.-Columbia, 1950-54, asst. prof. horticulture, 1954-59, assoc. prof., 1959-68, prof. plant pathology, 1968—. Contbr. articles to profl. jours. Served with U.S. Army, 1941-45. Fellow AAAS; mem. Am. Phytopathology Soc., Am. Bot. Soc., Am. Soc. Hort. Sci., Mo. Acad. Sci., Scandinavian Plant Physiology Soc., Japanese Physiol. Soc., Polish Acad. Sci. (fgn.), Sigma Xi. Subspecialties: Plant pathology; Plant virology. Current work: Director of state nursery virus certification program; involved with teaching and research dealing with horticulture and forestry crops. Home: 416 Price Ave Apt 3 Columbia MO 65201 Office: U Mo 108 Waters Hall Columbia MO 65211

MILLING, MARCUS EUGENE, geologist; b. Galveston, Tex., Oct. 8, 1938; s. Robert Richardson and Leonora (Currey) M.; m. Sandra Ann Millin Dunlay, Sept. 11, 1959; children: Marcus Eugene. B.S., Lamar U., 1961; M.S., U. Iowa, 1964, Ph.D., 1968; postgrad., U. Tex., Austin, 1964. Registered profl. geologist. Research geologist Exxon Prodn. Research Co., Houston, 1968-75, research supr., 1975-77; prodn. geologist Exxon Co. USA, Kingsville, Tex., 1977-78, dist. geologist, 1978-80; mgr. geol. research ARCO Oil & Gas Co., Dallas, 1980—; dir. GEOSAT, San Francisco, 1980—. Fellow Geol. Soc. Am.; mem. Am. Assn. Petroleum Geologists (editor 1982—), Soc. Econ. Geologists Paleontologists. Lutheran. Subspecialties: Geology; Sedimentology. Current work: Hydrocarbon potential deepwater clastic reservoirs. Home: 2300 Winding Hollow Rd Plano TX 75075 Office: ARCO Oil and Gas Co PO Box 2819 Dallas TX 75221

MILLS, HARRY LEE, JR., psychologist, consultant; b. Huntsville, Ala., Sept. 17, 1944; s. Harry Lee and Florence (Barnett) M.; m. Joyce Reynolds, May 20, 1972; children: Candice Michelle, Courtney Sullivan. B.S., U. So. Miss., 1966; Ph.D., 1973. Lic. clin. psychologist, Tenn., Ala. Research assoc. U. Miss. Med. Ctr., 1972-73; project dir. Youth Achievement House, Gulfport, Miss., 1973-74; unit leader and psychologist Mobile (Ala.) Mental Health Ctr., 1974-75; coordinator adult services Huntsville (Ala.) Mental Health Ctr., 1975-79; assoc. exec. dir. Dede Wallace Ctr., HES, Inc., Nashville, Tenn., 1979—; adj. assoc. prof. Vanderbilt U., 1979-82; clin. dir. Health Edn. Services, Nashville, 1981-82; bd. dirs. Research Rev., Ala. A&M U., Huntsville, 1977-79; chmn, profl. adv. bd. H.E.L.P. Inc., Huntsville, 1975-79. Author Clinical Productivity, 1981. NDEA fellow, 1968; ACTS grantee NIH, 19 77; HEW grantee, 1976. Mem. Am. Psychol. Assn., Assn. Advancement of Behavior Therapy, Tenn. Psychol. Assn., Tenn. Assn. Behavior Therapy. Democrat. Methodist. Subspecialties: Behavioral psychology; Learning. Current work: Treatment of compulsive disorders and agoraphobia, applications of behavioral technology to organizational problems; development and supervision. Home: 657 Hill Rd Brentwood TN 37027 Office: Dede Wallace Ctr. and HES Inc 700 Craighead Ave Nashville TN 37211

MILLS, JOHN ALEXANDER, physician, educator; b. Montreal, Que., Can., June 5, 1929; s. Edward Sadler and Marion (Baile) M.; m. Nancy Gordon. B.A., McGill U., 1950, M.D., 1954. Diplomate: Am. Bd. Internal Medicine. Intern Montreal Gen. Hosp., 1954-55; resident in medicine Mass. Gen. Hosp., Boston, 1956-58, physician, med. service, 1970—; mem. faculty Harvard Med. Sch., Boston, 1961—, assoc, prof. medicine, 1972—. Fellow Royal Coll. Physicians (Can.); mem. Am. Rheumatism Assn., Am. Assn. Immunologists, Portuguese Soc. Rheumatology (hon.). Subspecialties: Internal medicine; Immunology (medicine). Current work: Clinical research relating to rheumatic diseases and immunotherapy of autoimmune disease.

MILLS, MICHAEL DAVID, med. physicist; b. LaFollette, Tenn., Apr. 7, 1952; s. Charles Eugene and Nancy (Wallace) M. B.S. in Physics, Ga. Inst. Tech., 1974, M.S. in Nuclear Sci, 1975; Ph.D., U. Tex., Houston, 1980. Sr. research asst. M.D. Anderson Hosp. and Tumor Inst., Houston, 1976-80; clin. radiotherapy physicist Cleve. Clinic Found., 1981—; cons. in field. Contbr. articles in field to profl. jours. Mem. Am. Assn. Physicists in Medicine, Health Physics Soc. Subspecialties: Radiology; Biophysics (physics). Current work: Electron beam theory; radiotherapy and medical physics. Home: 6920 Carriage Hill Dr Brecksville OH 44141 Office: 9500 Euclid Ave Radiotherapy Cleveland OH 44106

MILLS, PATRICK LEO, industrial chemical engineering researcher, chemical engineering educator; b. Quantico, Va., Sept. 24, 1952; s. Robert Arthur and Nellie (Clark) M.; m. Pamela Marie Jackson, Mar. 30, 1974; children: Patrick Leo, Paul Joseph, Robert Edwin. B.S. in Chem. Engring, Tri-State U., 1973, M.S., Washington U., 1980, D.Sc., 1980. Lectr. Tri-State U., 1973-74; process design engr. Monsanto Co., St. Louis, 1974-75; instr. Washington U., 1975-80, adj. prof., 1980—; research engr. Gen. Electric Co., Schenectady, 1980-81, Monsanto Co., St. Louis, 1981—. Contbr. articles to profl. jours., chpts. to books. Recipient Gen. Motors Co. Scholarship award, 1971. Mem. Am. Inst. Chem. Engrs., Am. Chem. Soc., Soc. Indsl. and Applied Math., Internat. Assn. Math. Modelling, Sigma Xi. Roman Catholic. Subspecialties: Chemical engineering; Applied mathematics. Current work: Industrial chemical reaction engineering research and development and technology assessment; applied mathematics in chemical engineering; chemical engineering education. Home: 1321 Dautel Ln Creve Coeur MO 63146 Office: Monsanto Co 800 N Lindbergh Blvd Saint Louis MO 63167

MILLS, ROBERT LAURENCE, educator; b. Englewood, N.J., Apr. 15, 1927; s. Frederick Cecil and Dorothy Katherine (Clarke) M.; m. Elise Ackley, July 21, 1948; children—Katherine, Edward, Jonathan, Susan, Dorothy. A.B., Columbia Coll., 1948; B.A., Cambridge (Eng.) U., 1950, M.A., 1954; Ph.D., Columbia, 1955. Research asso. Brookhaven Nat. Lab., Upton, N.Y., 1953-55; mem. Sch. Math. Inst. for Advanced Study, Princeton, 1955-56; asst. prof. physics Ohio State U., Columbus, 1956-59, asso. prof., 1959-62, prof., 1962—. Author: Propagators for Many-Particle Systems, 1969, (with C.N. Yang) Rumford Premium, 1980. Mem. Am. Phys. Soc., AAUP, Fedn. Am. Scientists. Subspecialties: Theoretical physics; Condensed matter physics. Current work: Electronic states and other properties of random materials. Home: 3655 Weston Pl Columbus OH 43214

MILLS, WENDELL HOLMES, JR., research mathematician; b. Detroit, July 31, 1945; s. Wendell Holmes and Athaleah (Muse) M.; m. Gloria Darlene Gonzalez, July 4, 1969. B.S.E., U. Mich, 1968, 1968, Ph.D., 1976; M.S., U. Fla., 1972. Mem. tech. staff Rockwell Internat., Los Angeles, 1968-72; asst. prof. math Pa. State U., University Park, 1976-82; project leader Standard Oil Co., Cleve., 1982—. NSF research grantee, 1981. Mem. Am. Math. Soc., Soc. Indsl. and Applied Math. Republican. Subspecialties: Numerical analysis; Applied mathematics. Current work: Numerical analysis and scientific computing. Home: 3820 Meadow Gateway Broadview Heights OH 44147 Office: Standard Oil Co 4440 Warrensville Center Rd Cleveland OH 44147

MILMAN, HARRY ABRAHAM, toxicologist, pharmacologist; b. Cairo, May 16, 1943; s. David and Ruth (Manitsky) M.; m. Caren Susan Weisberg, Aug. 5, 1968; children: Deborah, Jennifer. B.S. in Pharmacy, Columbia U., 1966; M.S., St. John's U., 1968; Ph.D. in Pharmacology, George WashingtonU., 1978. Lic. pharmacist, N.Y., Md. Lab. asst. St. John's U., Jamaica, N.Y., 1967-68; chief pharmacist USPHS Indian Health Center, White Earth, Minn., 1968-70; research pharmacist Nat. Cancer Inst., Bethesda, Md., 1970-77; toxicologist nat. toxicology program NIH, Bethesda, 1977-80; sr. toxicologist EPA, Washington, 1980—; with USPHS, 1968—, comdr., 1978—. Contbr. articles to sci. jours. Recipient Commendation medal USPHS, 1980. Mem. Am. Soc. Pharmacology and Exptl. Therapeutics, Soc. Toxicology, Internat. Assn. Comparative Research on Leukemia and Related Diseases, Internat. Soc. Oncodevelopmental Biology and Medicine. Democrat. Jewish. Subspecialties: Toxicology (medicine); Cancer research (medicine). Current work: Biological markers of carcinogenicity; chemical carcinogenesis; regulatory toxicology.

MILNER, JOEL SHANDY, psychology educator; b. Smithville, Tex., Nov. 29, 1943; s. Jonas B. and Helen E. M.; m. Renanne Brock, Aug. 28, 1971; children: Ryan, Ashleigh. B.S., U. Houston, 1966; M.S., Okla. State U., 1968, Ph.D., 1970. Lic. psychologist, N.C. Postdoctoral fellow Lafayette Clinic, Detroit, 1970-71; prof. psychology Western Carolina U., Cullowhee, N.C., 1971—; clin. cons. Cherokee Children's Hosp., Cherkee, N.C., 1972—. Contbr. articles to profl. jours. Recipient Chancellor's Disting. Teaching award Western Carolina U., 1976, Paul A. Reid Disting. Service award, 1982; grantee NIMH, 1982—; others. Mem. Southeastern Psychol. Assn., Am. Psychol. Assn., Psychonomic Soc., Sigma Xi, Phi Kappa Phi. Current work: Child abuse research, clinical consultation and private practice, biochemical correlates of behavior. Home: PO Box 1 Webster NC 28788 Office: Dept Psychology Western Carolina U Cullowhee NC 28723

MILNER, JOHN A., nutritionist, educator; b. Pine Bluff, Ark., June 11, 1947; s. Austin P. and N. Lourena (Briggs) M.; m. Mary Frances Picciano, June 19, 1976; children: Kristina, Matthew. B.S., Okla. State U., 1969; Ph.D., Cornell U., 1974. Teaching asst. Cornell U., Ithaca, N.Y., 1969-73, USPHS postdoctoral fellow, 1974; asst. prof. dept. food sci. U. Ill., Urbana, 1975-79, assoc. prof., 1979—; asst. dir. Ill. Agrl. Expt. Sta., 1981—, dir. div. nutritional scis., 1981—. Author articles. Served to 2d lt. N.Y. State N.G., 1969-75. Recipient Young Investigator award Nutritional Found., 1976, Am. Diabetes Assn., 1976. Mem. Am. Inst. Nutrition, Am. Assn. for Cancer Research, Soc. for Nutrition Edn., Inst. Food Technologists, Internat. Assn. Bioinorganic Scientists. Lodge: Rotary. Subspecialties: Nutrition (medicine); Cancer research (medicine). Current work: Relationship of diet to cancer. Control of intermediary metabolism as it applied to protein metabolism. Home: 2505 Bedford St Champaign IL 61820 Office: U Ill 455 Bevier Hall 905 S Goodwin St Urbana IL 61801

MILNOR, WILLIAM ROBERT, medical scientist; b. Wilmington, Del., May 4, 1920; s. William Robert and Virginia (Sterling) M.; m.; children: Katherine Alexander, William Henry. A.B., Princeton U., 1941; M.D., Johns Hopkins U., 1944. Diplomate: Am. Bd. Internal

Medicine. Intern Johns Hopkins Hosp., Balt., 1944-45, resident, 1945-46; practice medicine specializing in cardiology, Balt., 1949-62; instr. medicine Johns Hopkins U. Sch. Medicine, Balt., 1951-54, asst. prof., 1954-60, assoc. prof., 1960-69, assoc. prof. physiology, 1962-69, prof., 1969—; physician Johns Hopkins Hosp., Balt., 1952—; pres. Heart Assn. Md., 1963-64; vis. fellow St. Catherine's Coll., Oxford, Eng., 1968. Author: Hemodynamics, 1982; contbr. articles to profl. jours. USPHS research fellow, 1949-51; Clayton scholar Johns Hopkins U., 1962—. Mem. Am. Physiol. Soc., Am. Fedn. Clin. Research, Am. Heart Assn. (chmn. research com. 1966-67). Clubs: Hamilton St. (Balt.); Princeton (N.Y.C.). Subspecialties: Physiology (medicine); Cardiology. Current work: Cardiovascular physiology, hemodynamics, control of heart and blood vessels, electrocardiography, comparative physiology. Home: 6616 N Charles St Baltimore MD 21204 Office: Johns Hopkins Sch Medicine 725 N Wolfe St Baltimore MD 21205

MILSTEAD, ANDREW HAMMILL, aerospace scientist, pilot, flight instructor; b. Statesville, N.C., Mar. 28, 1934; s. Andrew Dallam and Sarah Lee (Kincaid) M.; m. Jennifer Ann Pickering, Feb. 20, 1971 (div. 1981). B.S., U. N.C., 1956; M.S., Ohio State U., 1957; M.A., UCLA, 1964, Ph.D., 1968. Lic. airline transport pilot; cert. flight instr., real estate broker, Calif. Assoc. engr. Douglas Aircraft Co., Santa Monica, Calif., 1958-59; mem. tech. staff Space Tech. Labs. (now TRW), El Segundo, Calif., 1959-60; staff engr., sect. mgr. Aerospace Corp., El Segundo, 1960-70; sr. staff engr. Hughes Aircraft Co., El Segundo, 1970-71, sect. head, 1971-78, dept. mgr., 1978-81, sr. scientist, 1981—. Served as lt. comdr. USNR, 1975—. Aerospace Corp. doctoral fellow, 1965-67. Mem. AIAA, U.S. Naval Inst., Hughes Mgmt. Club, Naval Res. Assn., Aircraft Owners and Pilots Assn., Sigma Xi, Sigma Pi Sigma, Pi Mu Epsilon. Episcopalian. Subspecialties: Aerospace engineering and technology; Satellite studies. Current work: Mission analysis of earth satellite systems. Home: 721 25th St Santa Monica CA 90402 Office: hughes Aircraft Co PO Box 92919 Los Angeles CA 90009

MILTON, CHARLES, geology-mineralogy researcher; b. N.Y.C., Apr. 25, 1896; m. Leona T. Kohn; children: Daniel, Michael. Ph.D., Johns Hopkins U., 1929. Geologist various oil cos., Angola, Africa, Venezuela, S.Am., 1926-31; chemist, mineralogist U.S. Geol. Survey, Washington, 1931—; research prof. George Washington U., 1965-75, Quondam research prof., 1975—; research assoc. Smithsonian Instn., U. Pa. Mus., 1982—. Contbr. numerous articles, chpts. to profl. publs. Recipient Disting. Service award and Gold medal Dept. Interior, 1967. Fellow Geol. Soc. Am., Mineral Soc. Am., Mineral Soc. Can., Soc. Econ. Geologist. Subspecialties: Mineralogy; Petrology. Current work: Mineralogy of oil shale; archeological materials; mineralogy of Arkansas. Home: Beechbank Rd Forest Glen Silver Spring MD 20910 Office: Dept Geology George Washington U Washington DC 20052

MILUSCHEWA, SIMA, systems engineer. B.S. in Mech. Engring, N.J. Inst. Tech., 1955, M.S., 1955. Design engr. design and dvel. dept. Westinghouse Electric Co., 1951-56; project engr. aero. group research div. Curtiss-Wright Co., Clifton, N.J., 1956-58; space systems specialist advanced systems analysis group Astro-Electronics div. RCA, Princeton, N.J., 1958-68; sr. engr. Grumman Aerospace Corp., Bethpage, L.I., N.Y., 1968—. Recipient Salute to Women in Elec. Living award for space vehicle design work Elec. Women's Round Table, Inc., 1963. Mem. AIAA, Robotics Internat., Soc. Women Engrs. Subspecialties: Solar energy; Robotics. Current work: Satellite Power Systems; high energy lasers and other space systems.

MIN, BYUNG KON, aerospace company scientist, researcher, consultant; b. Chungjoo, Korea, June 2, 1947; came to U.S., 1972; s. Kwan Shik and Kil Soon (Shin) M.; m. Heikyung Chun, July 12, 1947; children: David Kunneghee, James Seghie. B.S., Seoul (Korea) Nat. U., 1969; M.S., Brown U., 1974, Ph.D., 1976. Research assoc., lectr. Cornell U., 1976-79; research scientist Lockheed Palo Alto Research Lab. (Calif.), 1979—; cons. in field. Contbr. articles to profl. publs. Served to 1st lt. Republic of Korea Army, 1969-71. Mem. ASME, Am. Soc. Metals, Ma. Acad. Mechanics, Soc. Advancement Materials and Process Engring., Sigma Xi. Subspecialties: Solid mechanics; Composite materials. Current work: Constitutive modeling, mechanics of deformation and fracture, and experimental mechanics of materials. Home: 948 Clinton Rd Los Altos CA 94022 Office: Lockheed Palo Alto Research Lab 3251 Hanover St Palo Alto CA 94304

MIN, KYUNG-WHAN, pathologist, educator; b. Seoul, Korea, May 5, 1937; came to U.S., 1964; s. Chungi and Inyoung (Lee) M.; m. Young-Jin, May 24, 1939; children: K. Christopher, W. David. M.D., Seoul Nat. U., Korea, 1962. Rotating intern Hanil Hosp., Seoul, 1962-63, resident in internal medicine, 1962-64; resident in pathology Balt. City Hosp., 1964-65, Baylor U., 1965-70, asst. instr., 1965-70; assoc. prof. Choson (Korea) U., 1970-71; asst. prof. Baylor U., 1971-78; pathologist Mercy Hosp. Med. Ctr., Des Moines, 1978—; also clin. assoc. prof. Creighton U., Omaha, 1978—. Fellow ACP, Am. Soc. Clin. Pathologists, Coll. Am. Pathologists; mem. Internat. Acad. Pathologists, Am. Assn. Pathologists. Subspecialties: Pathology (medicine); Cancer research (medicine). Current work: Ultrastructural cytopathological changes in relation to cancer and other allied diseases for understanding and diagnostic application of electronmicroscopy in cancer diagnosis. Home: 5109 Aspen Dr West Des Moines IA 50265 Office: Mercy Hosp Med Ctr 6th and University Sts Des Moines IA 50314

MINCH, MICHAEL JOSEPH, chemistry educator; b. Klamath Falls, Oreg., Apr. 7, 1943; m. Sharon K. Pierson, May 1979; 1 son, Matthew Charles. B.S. in Chemistry with honors, Oreg. State U., 1965, Ph.D., U. Wash., 1970. USPHS postdoctoral fellow U. Calif., 1970-72; asst. prof. Tulane U., 1972-74; asst. prof. chemistry U. Pacific, 1974-79, assoc. prof., 1979—. Weyerhaeuser scholar, 1961-65; NIH predoctoral fellow, 1966-68. Mem. Am. Chem. Soc., AAAS, Internat. Soc. Magnetic Resonance. Subspecialties: Biophysical chemistry; Nuclear magnetic resonance (chemistry). Current work: NMR studies of biomolecules. Office: Dept Chemistry U of Pacific 3601 Pacific Ave Stockton CA 95211

MINDLIN, RAYMOND DAVID, educator, civil engineer; b. N.Y.C., Sept. 17, 1906; s. Henry and Beatrice (Levy) M.; m. Elizabeth Roth, 1940 (dec. 1950); m. Patricia Kaveney, 1953 (dec. 1976). Student, Ethical Culture Sch., N.Y.C., 1918-24; B.A., Columbia U., 1928, B.S., 1931, C.E., 1932, Ph.D., 1936; D.Sc. (hon.), Northwestern U., 1975. Research asst. dept. civil engring. Columbia U., N.Y.C., 1932-38, Bridgham fellow, 1934-35, instr. civil engring., 1938-40, asst. prof., 1940-45, assoc. prof., 1945-47, prof. civil engring., 1947-67, James Kip Finch prof. applied sci., 1967-75, prof. emeritus, 1975—; cons. Bell Telephone Labs. Inc., N.Y.C., 1943-51; cons. physicist Dept. Terrestrial Magnetism, Carnegie Instn. of Washington, 1940-42; cons. Nat. Defense Research Com., 1941-42; sect. T Office Sci. Research and Devel., Applied Physics Lab., Johns Hopkins U., Balt., 1942-45. Mem. adv. bd.: Applied Mech. Reviews; Contbr. articles to tech. and sci. jours. Recipient Illig medal Columbia U., 1932, U.S. Naval Ordnance Devel. award, 1945, Presdl. Medal of Merit, 1946, Class of 1889 Sch. Mines medal Columbia U., 1947, research prize ASCE, 1958, Great Tchr. award Columbia U., 1960, von Karman medal ASCE, 1961, Timoshenko medal ASME, 1964, ASME medal, 1976, C.B. Sawyer Piezoelectric Resonator award, 1967, Egleston medal Columbia U., 1971, Trent-Crede medal Acoustical Soc., Am., 1971, Frocht award Soc. Exptl. Stress Analysis, 1974, Nat. medal of Sci., 1979. Fellow Acoustical Soc. Am., Am. Acad. Arts and Scis.; mem. Soc. Exptl. Stress Analysis (founding mem.; exec. com. 1943-50, v.p. 1946, pres. 1947), ASCE (sec. com. applied mechanics 1940-42, chmn. 1942-45), Eastern Photoelasticity Conf. (exec. com. 1938-41), Am. Phys. Soc., Nat. Acad. Engring., Nat. Acad. Sci., Conn. Acad. Sci. and Engring., U.S. Nat. Com. Theoretical and Applied Mechanics, Internat. Union Theoretical and Applied Mechanics, ASME (hon.), Tau Beta Pi, Sigma Xi, Columbia Varsity "C". Subspecialty: Theoretical and applied mechanics. Home: 89 Deer Hill Dr Ridgefield CT 06877

MINER, BRYANT ALBERT, chemist, educator; b. Moroni, Utah, Aug. 9, 1934; s. Glen Bryant and Caroline (Eyring) M.; m. Janice Mabey, June 8, 1960; children: Melanie, JaNae, David, Helen, Heidi, Jason, Jared. B.A., U. Utah, 1961, Ph.D., 1965. Agrl. researcher Am. Smelting & Refining Co., Murray, Utah, summers 1953, 54; chem. researcher U. Utah, summers 1965, 67; prof. chemistry Weber State Coll., Ogden, Utah, 1964—. Served with U.S. Army, 1957-59. Recipient medal Am. Inst. Chemists, 1961. Mem. Am. Chem. Soc., Phi Beta Kappa, Phi Kappa Phi. Mormon. Subspecialties: Physical chemistry; Thermodynamics. Current work: Thermodynamics, high pressure electrochemistry. Home: 4260 Jefferson Ave Ogden UT 84403 Office: Chemistry Dept 2503 Weber State Coll Ogden UT 84408

MINER, ELLIS DEVERE, JR., astrophysicist; b. Los Angeles, Apr. 16, 1937; s. E. Devere and Myrle (Fletcher) M.; m. Beverly Allen, June 19, 1961; children: Steven, Marjorie, David, Jeffrey, Christine, Rebecca, Laura. B.S., Utah State U., 1961; Ph.D., Brigham Young U., 1965. Missionary Latter Day Saints Ch., Germany, 1957-60; space scientist Jet Propulsion Lab., Pasadena, Calif., 1965—. Contbr. articles to profl. jours. Bishop, Sunland ward Latter Day Saints Ch., 1971-77; active Boy Scouts Am. Served to capt., Signal Corps. U.S. Army, 1956-67. Coop. Grad. fellow NSF, 1962-65; recipient Exceptional Sci. Achievement medal NASA, 1981. Mem. Am. Astron. Soc., Nat. Geography Soc., Sigma Xi, Sigma Pi Sigma. Republican. Subspecialties: Planetary science; Infrared optical astronomy. Current work: Photometry and radiometry of solar system objects. Home: 11335 Sunburst St Lake View Terrace CA 91342 Office: 4800 Oak Grove Dr Pasadena CA 91109

MINER, JOHN BURNHAM, psychology research educator, writer, consultant; b. N.Y.C., July 20, 1926; s. John Lynn and Bess (Burnham) M.; m.; children by previous marriage: Barbara, John, Cynthia, Frances; m. Barbara Allen Williams, June 1, 1979. A.B., Princeton U., 1950, Ph.D., 1955; M.A., Clark U., 1952. Lic. psychologist, Ga. Research assoc. Columbia U., 1956-57; mgr. psychol. services Atlantic Refining Co., Phila., 1957-60; faculty mem. U. Oreg., Eugene, 1960-68; prof., chmn. dept. organizational sci. U. Md., College Park, 1968-73; research prof. Ga. State U., Atlanta, 1973—; pres. Organizational Measurement Systems Press, Atlanta, 1976—; cons. McKinsey & Co., N.Y.C., 1966-69; vis. lectr. U. Pa., Phila., 1959-60; vis. prof. U. Calif.-Berkeley, 1966-67, U. South Fla., Tampa, 1972. Author: many books and monographs including Personnel Psychology, 1969, The Challenge of Managing, 1975, (with Mary Green Miner) Policy Issues in Personnel and Industrial Relations, 1977, (with George A. Steiner) Management Policy and Strategy, 1977 (James A. Hamilton-Hosp. Adminstrs. Book award), 1982 (1979), 2d edit.), (with M.G. Miner) Employee Selection Within the Law, 1978, Theories of Organizational Behavior, 1980, Theories of Organizational Structure and Process, 1982; contbr. numerous articles, papers in field to profl. lit. Served with AUS, 1944-46; ETO. Decorated Bronze Star medal, Combat Infantryman's Badge; named Disting. Prof. Ga. State U., 1974. Fellow Acad. Mgmt. (editor Jour. 1973-75, pres. 1977-78), Am. Psychol. Assn., Soc. for Personality Assessment; mem. Indsl. Relations Research Assn. Republican. Club: Princeton (N.Y.C.). Subspecialty: Organizational psychology. Current work: Organizational motivation, theories of organization, human resource utilization, personnel management, business policy/strategy. Home: 651 Peachtree Battle Ave N W Atlanta GA 30327 Office: Dept Mgntl Ga State U University Plaza Atlanta GA 30303

MINES, ALLAN HOWARD, physiology educator, wine importer; b. N.Y.C., Apr. 11, 1936; s. Lawrence Irving and Lillian (Zinkofsky) M.; m. Lela Mae Hunt, 1959 (div. 1972); m. Susan Fay Edmonds, Aug. 7, 1973; children: Russell David, Wendy Suzanne. B.S., U. Ill.-Urbana, 1961, M.A., 1962; Ph.D., U. Calif.-San Francisco, 1968. Asst. prof. physiology U. Calif.-San Francisco, 1968-74, assoc. prof., 1974-82, prof., 1982—. Author: Respiratory Physiology, 1981. Served with U.S. Army, 1958-60. NIH grantee, 1968-78. Mem. Am. Physiol. Soc. Subspecialty: Physiology (biology). Current work: Main research interests pulmonary physiology, especially as related to control of breathing in normal people and in patients; under conditions of hypoxia and increased levels of CO2. Office: Dept Physiology 3d St and Parnassus Ave San Francisco CA 94143

MINICH, CARL EDWARD, data processing exec.; b. Lansing, Mich., Apr. 14, 1935; s. Carl Francis and Doris Ellen M.; m. Lois Elaine Baer, July 12, 1960; children: Michael Christopher, David, Carla, James, Robert. B.S.E.E., Mich. State U., 1956. Engr. Lockheed Missiles Systems, Palo Alto, Calif., 1960-62; systems mgr., designer Sanders Assocs., Nashua, N.H., 1962-77; product mgr., planning mgr. Harris corp., Dallas, 1977—, mgr. interactive products div. planning, 1982—. Served to capt. USAF, 1957-60. Mem. IEEE, Assn. for Computing Machinery. Club: Dallas Venture. Subspecialty: Computer engineering. Current work: Intelligent terminals, telecommunications. Home: 7625 Woodstone St Dallas TX 75240 Office: PO Box 400010 Dallas TX 75240

MINK, DOUGLAS JOHN, planetary astronomer; b. Lincoln, Nebr., Sept. 3, 1951; s. Robert and Lillian M. B.S., MIT, 1973, M.S., 1974. Mem. research staff Remote Sensing Lab., MIT, 1974-76, dept. earth and planetary sci., 1979—; mem. research staff Cornell Lab. for Planetary Studies, 1976-79. Mem. Am. Astron. Soc. Subspecialties: Planetary science; Optical astronomy. Current work: Studies of planets, satellites, comets and asteroids using stellar occultation and CCD imaging techniques; satellites of Uranus and Neptune; development of computer software for astronomy. Co-discoverer of the rings of Uranus. Office: Mass Inst Tech Room 37-556 Cambridge MA 02139

MINSKY, MARVIN LEE, mathematician; b. N.Y.C., Aug. 9, 1927; s. Henry and Fannie (Reyser) M.; m. Gloria Anna Rudisch, July 30, 1952; children: Margaret, Henry, Juliana. B.A., Harvard U., 1950; Ph.D., Princeton U., 1954. Mem. Harvard Soc. Fellows, 1954-57; with Lincoln Lab., Mass. Inst. Tech., 1957-58, prof. math., 1958-61, prof. elec. engring., 1961—, Donner prof. sci., 1973; dir. artificial intelligence group MAC project, 1958—, dir. artificial intelligence lab., 1970—. Author: Computation, 1967, Semantic Information Processing, 1968, Perceptrons, (with S. Papert), 1968. Served with USNR, 1945-46. Recipient Turing award Assn. for Computing Machinery, 1970. Fellow I.E.E.E., Am. Acad. Arts and Scis., N.Y. Acad. Scis., Nat. Acad. Sci. Subspecialty: Artificial intelligence. Office: 545 Main St Cambridge MA 02139

MINTON, NORMAN ALTON, research nematologist; b. Spring Garden, Ala., Oct. 12, 1924; s. Pearson H. and Clara M. (Savage) M.; m. Doris Lorene Pope, Mar. 20, 1944; 1 dau., Cathy Teresa. Student, Jacksonville (Ala.) State Coll., 1946-48; B.S. in Agrl. Edn, Auburn U., 1950; M.S. in Horticulture, Auburn U., 1951; Ph.D. in Zoology, Auburn U., 1960. Asst. county agrl. agt. Coop. Ext. Service, Abbeville and Geneva, Ala., 1951-53; horticulturist Berry Coll., Mt. Berry, Ga., 1953-55; nematologist U.S. Dept. Agr., Auburn U., 1955-64, U.S. Dept. Agr., Coastal Plain Sta., Tifton, Ga., 1964—. Contbr. articles to profl. jours. Served with U.S. Army, 1944-46. Decorated Bronze star (2); recipient research award Ga. Soybean Assn., 1978. Mem. Soc. Nematologists (co-recipient best econ. paper award 1981), Am. Phytopathological Soc., Helminthological Soc. Washington, Orgn. Tropical Am. Nematologists, Ga. Assn. Plant Pathologists, Sigma Xi, Gamma Sigma Delta. Methodist. Current work: Economic nematode control as influenced by crops, cultural practices and chemicals; nematode-fungus relationship to plant disease; development of nematode resistant varieties. Office: Plant Pathology Dept Coastal Plain Sta Tifton GA 31793

MINTON, SHERMAN ANTHONY, microbiology educator; b. New Albany, Ind., Feb. 24, 1919; s. Sherman and Gertrude (Gurtz) M.; m. Madge Shortridge Rutherford, Oct. 10, 1943; children: Brooks, April, Holly. B.A., Ind. U., 1939, M.D., 1942. Asst. to assoc. prof. Ind. U.-Indpls., 1947-58, assoc. prof. microbiology and immunology, 1963-71, prof., 1971—; vis. prof. Postgrad. Med. Ctr., Karachi, Pakistan, 1958-62; research assoc. Am. Mus. Natural History, 1957—. Author: (with M. R. Minton) Venomous Reptiles, 1969, Giant Reptiles, 1973, Venom Diseases, 1974; editor: Snake Venoms and Envenomation, 1971. Served to lt. USN, 1943-45. USPHS grantee, 1964; NSF grantee, 1980; Wyeth Inc. grantee, 1983; La. State U. fellow, 1957. Fellow Ind. Acad. Sci.; mem. Internat. Soc. Toxicology, Am. Soc. Ichthyologists and Herpetologists, Soc. Study Amphibians and Reptiles, Am. Soc. Tropical Medicine and Hygiene. Subspecialties: Taxonomy; Immunotoxicology. Current work: Venomous animals and clinical aspects of envenomation, serologic taxonomy of snakes, evolution and geographic distribution venomous snakes. Office: Ind U Sch Medicine 1100 W Michigan Ave Indianapolis IN 46223

MINTZ, BEATRICE, biologist; b. N.Y.C., Jan. 24, 1921; d. Samuel and Janie (Stein) M. A.B. magna cum laude, Hunter Coll., N.Y.C., 1941; postgrad., N.Y. U., 1941-42; M.S., U. Iowa, 1944, Ph.D., 1946; D.Sc. (hon.), N.Y. Med. Coll., 1980, Med. Coll. Pa., 1980, Northwestern U., 1982. Instr. to assoc. prof. biol. scis. U. Chgo., 1946-60; assoc. mem. Inst. Cancer Research, Phila., 1960-65; sr. mem., 1965—. Lalor Found. Research fellow, 1954, 55; recipient Bertner Found. award, 1977, Papanicolaou award, 1979, Lewis S. Rosenstiel award, 1980. Fellow AAAS, Am. Acad. Arts and Scis.; hon. fellow Am. Gynecol. and Obstet. Soc.; mem. Genetics Soc. Am., Soc. Study Developmental Biology, Internat. Soc. Developmental Biology, Am. Soc. Zoologists, Am. Inst. Biol. Scis., Nat. Acad. Scis., N.Y. Acad. Sci. (recipient award in biol. and med. scis. 1979), Sigma Xi, Phi Beta Kappa. Office: 7701 Burholme Ave Philadelphia PA 19111

MINTZ, LEIGH WAYNE, geology educator, university administrator; b. Cleve., June 12, 1939; s. William Michael and Laverne (Bulicek) M.; m. Carol Sue Jackson, Aug. 4, 1962; children: Kevin Randall, Susan Carol. B.S., U. Mich., 1961, M.S., 1962; Ph.D., U. Calif.-Berkeley, 1966. Asst. prof. geology Calif. State U.-Hayward, 1965-70, assoc. prof., 1970-75, prof., 1975—, assoc. dean sci., 1971-72, dean instrn., 1972-73, dean undergrad. studies, 1973-79, assoc. v.p. acad. programs, 1979—. Author: Historical Geology, 1972, 3d edit., 1981, Physical Geology, 1982; contbr. articles to profl. jours. and encys. Judge Alameda County Fair, Pleasanton, Calif., 1968-82. Recipient Palmer prize Dept. Paleontology U. Calif.-Berkeley, 1965; Wells scholar Gen. Biol. Supply House, Chgo., 1965-66; NSF grad. fellow, 1961-64. Fellow Geol. Soc. Am.; mem. Paleontol. Soc., Nat. Parks and Conservation Orgn., Sierra Club, Phi Beta Kappa, Sigma Xi, Phi Kappa Phi. Subspecialties: Paleontology; Geology. Current work: Plate tectonics in earth history, particularly Western U.S.A. Home: 5940 Highwood Rd Castro Valley CA 94546 Office: Calif State U Hayward CA 94542

MINTZER, DAVID, educator, university dean and official; b. N.Y.C., May 4, 1926; s. Herman and Anna (Katz) M.; m. Justine Nancy Klein, June 26, 1949; children: Elizabeth Amy, Robert Andrew. B.S. in Physics, Mass. Inst. Tech., 1945, Ph.D., 1949. Asst. prof. physics Brown U., 1949-55; research asso. Yale U., 1955-56, asso. prof., dir. lab. marine physics, 1956-62; prof. mech. engring., astronautical scis. Northwestern U., Evanston, 1962—, prof. astrophysics, 1968—; asso. dean Technol. Inst., 1970-73, acting dean, 1971-72, v.p. for research, dean sci., 1973—; mem. mine adv. com. Nat. Acad. Sci.-NRC, 1963-73. Trustee EDUCOM (interuniv. communications council), 1975—, vice chmn., 1977-78, chmn., 1978-81; trustee Adler Planetarium, 1976—. Fellow Am. Phys. Soc., Acoustical Soc. Am.; mem. ASME, Am. Astron. Soc., Sigma Xi, Tau Beta Pi, Pi Tau Sigma. Subspecialties: Acoustics; Statistical physics. Current work: Research administration. Home: 736 Central St Evanston IL 60201

MINVIELLE, FRANCIS PAUL GEORGES, geneticist, researcher; b. Chartres, France, Sept. 12, 1948; s. Robert and Janine (Vauchelle) M.; m. Francine Esterez, July 17, 1971; 1 dau., Carole. Ingenieur Agronome, E.N.S.A., Toulouse, France, 1971; M.S., U. Calif., Davis, 1973, Ph.D., 1979. Asst. prof. Laval U., Quebec, Que., Can., 1979—; vis. prof. Faculty Agr. Haiti, 1981, 82. Contbr. articles to profl. jours. Mem. Genetics Soc. Am., Am. Soc. Animal Sci., Genetics Soc. Can. Subspecialties: Animal genetics; Quantitative genetics. Current work: Limits to selection, natural versus artifical selection, heterosis, swine breeding. Office: Universite Laval FSAA Pav Comtois Ste-Foy PQ Canada G1K 7P4

MINWADA, JUN, molecular immunologist, cancer researcher, microbiologist; b. Kyoto, Japan, Nov. 5, 1927; s. Masuji and Fusae (Nishimura) M.; m. Kimiko Kawazoe, Jan. 9, 1936; children: George, Mary, Nora. M.D., Kyoto Prefectural U. of Medicine, 1952, Ph.D., 1959. Lic. physician Japanese Med. Bur. Assoc. chief dept. molecular immunology Roswell Park Meml. Inst., N.Y. State Dept. Health, Buffalo, 1975-82; prof. Edward Hines Jr. VA Hosp., Maywood, Ill., 1982—; research prof. assoc. chief cancer research scientist, head cell culture lab. SUNY, Buffalo, 1979-82; subcom. human leukocyte antigens WHO. Contbr. numerous articles to sci jours. Recipient numerous grants and awards including NIH, 1982. Mem. Am. Assn. Cancer Research, Am. Soc. Clin. Oncology. Subspecialties: Oncology; Hematology. Current work: Leukemia and lympoma clinical and theoretical aspects in man. Office: Edward Hines Jr VA Hosp Hines ILL 60141

MIR, GHULAM NABI, pharmacologist, toxicologist, researcher; b. Srinagar, Kashmir, June 26, 1939; came to U.S., 1967; s. Habib Lillah and Sara Jahan (Shah) M.; m. Diane Gayle, Sept. 21, 1968; children: Shama, Kahlil. B.Sc., Kashmir U., 1958, B. Pharm., 1962; M. Pharm., Punjab U., 1965; Ph.D., U. Miss., 1971; M.B.A., Temple U., 1979. Research pharmacologist Inst. History of Medicine and Med. Research, New Delhi, India, 1965-67; research assoc. U. Tenn. Med Center, Memphis, 1971-72; group leader pharmacology William H. Rorer, Inc., Fort Washington, Pa., 1972-74, sect. head pharmacology, 1974-81, dept. head, 1981—. Contbr. articles to profl. jours. Research fellow Council Indsl. and Sci. Research, India, 1963-64; NIH fellow, 1967-70. Mem. Am. Soc. Pharmacology and Exptl. Therapeutics, Acad. Pharm. Scis., Soc. Toxicology, Sigma Xi. Muslim. Clubs: Middle

Atlantic Road Runners, Mid Atlantic AAU. Subspecialties: Pharmacology; Toxicology (medicine). Current work: Physiological and biochemical aspetcs of gastrointestinal, cardiovascular and smooth muscle pharmacology and drug toxicity. Patentee in field. Office: William H Roer Inc 500 Virginia Dr Fort Washington PA 19034 Home: 2802 Red Gate Dr Buckingham PA 18912

MIRABEL, IGOR FELIX, astronomer, educator; b. Montevideo, Uruguay, Oct. 23, 1944; s. Mauricio and Zulema (Miquele) M.; m. Silvia Alicia Revora, May 10, 1944; children: Mariana, Paula, Sofia. Ph.D. in Astronomy, Nat. U. La Plata, Argentina, 1974. Researcher NRC Argentina, Buenos Aires, 1974-76; prof. philosophy Nat. U. Buenos Aires, 1975; postdoctoral research fellow Nuffield Radio Astronomy Labs., Jodrell Bank, Eng., 1976-79; research asso. astronomy U. Md., 1979-81; prof., researcher in astronomy, physics dept. U. P.R., 1981—. Contbr. articles to sci. jours. Internat. Astron. Union grantee, 1976; NSF grantee; OCEC U. P.R. grantee, 1981-82. Mem. Internat. Astron. Union, Am. Astron. Soc., Argentine Astron. Soc. Subspecialty: Radio and microwave astronomy. Current work: Scientific research in galactic and extragalactic astronomy. Office: Dept Physics U PR Box AT Rio Piedras PR 00931

MIRAND, EDWIN ALBERT, cancer scientist; b. Buffalo, July 18, 1926; s. Thomas and Lucy (Papier) M. B.A., U. Buffalo, 1947, M.A., 1949; Ph.D., Syracuse (N.Y.) U., 1951; D.Sc. (hon.), Niagara (N.Y.) U., 1970, D'Youville Coll., Buffalo, 1974. Successively undergrad. asst., grad. asst., instr. U. Buffalo, 1946-48; teaching fellow Syracuse U., 1948-51; instr. Utica (N.Y.) Coll., 1950; mem. staff Roswell Park Meml. Inst., Buffalo, 1951—; head W. Seneca labs., 1961—, asso. inst. dir., head dept. edn., 1967—, dir. cancer research, 1968-73, head dept. viral oncology, 1970-73, head dept. biol. recources, 1973—; research prof. biology Grad. Sch., prof. biochem. pharmacology Sch. Pharmacy, State U. N.Y., Buffalo, 1955—; dean Roswell Park Meml. Inst. grad. div., 1967—; research prof. biology (Grad. Sch.); dean Roswell Park Meml. Inst. grad. div. Niagara U.; mem. human cancer virus task force, clin. cancer edn. com. NIH. Author articles cancer research, endocrinology, hematology, virology.; Editorial bd.: Jour. Surg. Oncology. Mem. U.S. nat. com. Union Internat. Contra Cancer; profl. edn. com. cancer control Nat. Cancer Inst; liaison mem. Pres.'s Nat. Cancer Adv. Bd.; mem. N.Y. State Health Research Council. Recipient Billings Silver medal AMA, 1963; award sci. research mammalian tumor viruses Med. Soc. State N.Y., 1963; citation award in sci. Coll. Arts and Scis., State U. N.Y., Buffalo, 1964. Life mem., fellow N.Y. Acad. Sci.; fellow AAAS; mem. Am. Cancer Soc., Assn. Gnotobiotics (pres. 1968-69, dir. 1975-78), Assn. Am. Cancer Insts. (sec.-treas. 1968—), Am. Assn. Cancer Research, Radiation Research Soc., Am. Soc. Zoologists, Soc. Exptl. Biology and Medicine, Buffalo Acad. Medicine, Animal Care Panel, Internat. Soc. Hematology, Pub. Health Cancer Assn. Am., Internat. Union Against Cancer (chmn. U.S. nat. com. 1979—, sec.-gen. 13th Internat. Cancer Congress), Hematology Soc., Am. Soc. Preventive Oncology, Buffalo Hist. Soc. (life), Buffalo Fine Arts Acad. (life), Sigma Xi. Subspecialties: Cancer research (medicine); Hematology. Current work: Role of viruses in tumors, interferon and role of erythroporetin in regulating erythropoiesis. Home: 925 Delaware Ave Buffalo NY 14209 Office: 666 Elm St Roswell Park Meml Inst Buffalo NY 14263

MIRANDA, FRANK JOSEPH, dental educator, dental practitioner; b. Erie, Pa., June 30, 1946; s. Joseph Francis and Vivian Mary (Lewis) M.; m. Leslie Joan Hintze, July 1, 1967 (div. 1975); m. Joan Ethel Antes, Oct. 7, 1976; children: Cory Michael, Erin Christine. Student, UCLA, 1964-67, D.D.S., 1971; M.Ed., Central State U., Edmond, Okla., 1976, M.B.A., 1979. Assoc. dentist UCLA Sch. Dentistry, 1971-74; pvt. practice dentistry Lynwood (Calif.) Children's Found., 1971-72; asst. prof. operative dentistry Okla. U. Coll. Dentistry, 1974-80, assoc. prof., 1980—; mem. adj. faculty Oscar Rose Jr. Coll., 1980—; cons. odontologist Okla. Chief Med. Examiner, 1975—; pvt. practice dentistry, 1974—. Contbr. articles to profl. jours.; author teaching syllabi, 1980. Mem. Okla. Hispanic Assn. for Higher Edn., Oklahoma City, 1981; panelist Sta.-KTVY Unity Program, Oklahoma City, 1981; bd. dirs. Okla. U. Health Sci. Center Fed. Credit Union, 1982; vice chmn. Affirmative Action Council, Oklahoma City, 1982. Named Outstanding Tchr. Okla. U. Coll. Dentistry graduating srs., 1978, 80, 81; Acad. Gen. Dentistry fellow, 1980; Acad. Dentistry Internat. fellow, 1981; Acad. Internat. Dental Studies fellow, 1982. Mem. Internat. Assn. Dental Research, Am. Assn. Dental Research (pres. chpt. 1980-82), Acad. Operative Dentistry, ADA, Okla. Dental Assn., Am. Assn. Dental Schs. (chmn. exec. com. 1982—). Current work: Dental education; dental materials research; business and psychology of dentistry; pain control; dental practice management. Office: U Okla Coll Dentistry 1001 Stanton L Young Blvd Oklahoma City OK 73190

MIRANDA, MANUEL ROBERT, psychology educator, cons., researcher; b. King City, Calif., Oct. 14, 1939; s. Manuel Robert and Mary Ann (Alouso) M.; m. Cynthia Jean, June 23, 1978. B.S., U. Oreg., 1962; M.S., Calif. State U.-San Jose, 1967; Ph.D., U. Wash., 1971. Lic. clin. psychologist, Calif. Asst. prof. Calif. State U., San Jose, 1970-73; assoc. prof. UCLA, 1973-76, prof., 1978—; assoc. prof. U. Minn. Mpls., 1976-78; researcher in field. Author: Spanish Speaking Mental Health, 1976; editor: The Spanish Speaking Elderly, 1980, Chicano Mental Health Issues, 1982; author: The Human Life Cycle, 1983. UCLA grantee, 1975; Dept. Labor grantee, 1978; NIMH grantee, 1980, 83. Mem. Am. Psychol. Assn., Council of Social Work Educators, Western Psychol. Assn., Nat. Hispanic Psychol. Assn. Democrat. Roman Catholic. Current work: Research specialist in hispanic mental health issues, as well as educator in the area of personality and cultural development. Office: Sch Social Welfare UCLA 238 Dodd Hall Los Angeles CA 90024

MIRCETICH, SRECKO MIRKO, research plant pathologist; b. Skela Kod Obrenovca, Yugoslavia, Sept. 2, 1926; came to U.S., 1957, naturalized, 1960; s. Mirko Srecko and Jelena Rajko (Mircetih) M.; m. Jane Velika, Nov. 16, 1956; children: Jon, Kristofer. Agronomy Engr., U. Sarajevo, 1952; M.S., U. Belgrad, 1954; Ph.D., U. Calif.-Riverside, 1966. Research assoc. dept. plant pathology U. Calif.-Riverside, 1958-66; research plant pathologist Nat. Agrl. Research Ctr., U.S. Dept. Agr., Beltsville, Md., 1966-71, supervisory research plant pathologist, 1971-72; research plant pathologist U.S. Dept. Agr., U. Calif.-Davis, 1973—. Contbr. articles in field to profl. jours. Mem. Internat. Soc. Plant Pathology, Am. Phytopath. Soc. Republican. Serbian Orthodox. Subspecialty: Plant pathology. Current work: Researcher in etiology and control of soil-borne diseases of decidious fruit and nut trees.

MIRENBURG, BARRY LEONARD, psychotherapist, publisher, educator; b. N.Y.C., Feb. 16, 1952; s. Fred and Mildred (Solomon) M. B.S., Mercy Coll., 1971; B.F.A., Cooper Union, 1972; M.F.A., Syracuse U., 1982; M.B.A., N.Y. Inst. Tech., 1983; M.Ed., Columbia U., 1980, M.A., 1980, Ed.D., 1982. Pub. Barlenmir House, N.Y.C., 1972—; prof. psychology N.Y. Inst. Tech., N.Y.C., 1979—; cons. psychotherapist IUC, Inc., N.Y.C., 1980—, also Pub. Schs. Adv. Legis. Adv. Com., N.Y.C., 1979—. Mem. N.Y. Acad. Scis. Subspecialties: Social psychology; Sensory processes. Current work: Creative personality, psychology, awareness and consciousness Gestalt perception and creative behavior. Office: 413 City Island Ave New York NY 10064

MIRKIN, LAZARO DAVID, pediatric pathologist, educator; b. Buenos Aires, Argentina, Feb. 22, 1931; s. Moises and Celia (Wexman) M.; m.; children: Alexander, Laura, Martin. M.D., U. Buenos Aires, 1957. Diplomate: Am. Bd. Pathology. Intern Hosp. de Niños de Buenos Aires, 1960, resident, 1961-65; dir. infirmary Sec. Health Santa Ana, Argentina, 1957-60; sub chief dept. pathology Children's Hosp., Buenos Aires, 1966-70; head dept. pathology Nat. Health Service, Santiago, Chile, 1971-76; assoc. prof. pathology Ind. U., Indpls., 1976—, dir. pediatric pathology, 1978-80; cons. Pan Am. Health Orgn. Fellow Coll. Am. Pathologists; mem. Internat. Acad. Pathology, Pediatric Pathology Club, Am. Assn. Pathology, Internat. Soc. Pediatric Pathology. Subspecialties: Pathology (medicine); Pediatrics. Current work: Electron microscopy of pediatric tumors. Home: 811 Lincolnwood Ln Indianapolis IN 46260 Office: Riley Hospital 702 Barnhill Dr Indianapolis IN 46223

MISCONI, NEBIL YOUSIF, astronomer; b. Baghdad, Iraq, Dec. 8, 1939; came to U.S., 1970; s. Yousif Yacoub and Kolomba Salim (Dawood) M.; m. Irene Theresa Donohue, Aug. 31, 1972; 1 son, Michael Nebeel. B.Sc., Istanbul (Turkey) U., 1965; Ph.D., SUNY-Albany, 1975. Instr. Physics Lab., U. Baghdad, 1966-70; grad. research asst. Dudley Obs. and SUNY-Albany, 1970-75; postdoctoral research assoc. Space Astronomy Lab., SUNY-Albany, 1975-77, research assoc., 1978-80; asst. research scientist U. Fla., 1980-82, assoc. research scientist, 1982—. Research publs. in field. NASA grantee, 1978-79; NSF grantee, 1982-84. Mem. Am. Astron. Soc., Royal Astron. Soc., Internat. Astron. Union (Commns. 21 and 22), Sigma Xi. Subspecialty: Planetary science. Current work: Interplanetary medium; zodiacal light; dynamics of cosmic dust; cometary physics; laser particle dynamics.

MISELIS, RICHARD R., neurobiologist, veterinarian, educator; b. Boston, Mar. 13, 1945; s. Frank J. and Theodora (Tringale) M.; m. Karen L. Archambault, Dec. 23,1965; children: Todd, Kristin. B.S., Tufts U., 1967; V.M.D., U. Pa., 1973, Ph.D., 1973. Asst. prof. Sch. Vet. Medicine, U. Pa., Phila., 1975-81, assoc. prof. animal biology, 1982—. Contbr. articles to profl. jours. Sloan Found. fellow, 1978. Mem. AAAS, Soc. Neurosci., Am. Assn. Anatomists, Sigma Xi. Subspecialties: Neurobiology; Anatomy, neuroanatomy. Current work: Neurobiological basis of ingestive behavior, neuroanatomy, hypothalmus, circumventricular organs, thirst, hunger, obesity, brainstem, vagus nerve, solitary nucleus. Home: 4719 Cedar Ave Philadelphia PA 19143 Office: School Vet Medicine U Pa Philadelphia PA 19104

MISHELL, DANIEL RANDOLPH, JR., physician, educator; b. Newark, May 7, 1931; s. Daniel Randolph and Helen M.; m. Carol Goodrich, Sept. 30, 1961; children: Sandra Rose, Daniel R. III, Tanya Ruth. B.A. with great distinction, Stanford U., 1952, M.D., 1955. Intern Harbor Gen. Hosp., Torrance, Calif., 1955-56; resident in Ob-Gyn UCLA service, 1959-63; asst. prof. dept. Ob-Gyn UCLA Sch. Medicine, 1968-69; assoc. prof. U. So. Calif. Sch. Medicine, 1968-69, prof., 1969—, assoc. chmn. dept., 1972-78, chmn. dept., 1978—; cons. internat. adv. commn. WHO, 1972-79; cons. Population Council, Internat. Com. for Contraceptive Research, N.Y.C.; bd. examiner Am. Bd. Ob-Gyn, 1976—; research fellow Univ. Hosp., Uppsala, Sweden, 1961-62. Co-editor: Reproductive Endocrinology, Infertility and Contraception, 1979; editor-in-chief: Contraception, 1969—. Served as capt. M.C. USAF, 1957-59. WHO, Population Council grantee. Mem. Am. Fertility Soc., Am. Assn. Planned Parenthood Physicians, Am. Coll. Obstetricians and Gynecologists, Am. Fedn. Clin. Research, Endocrine Soc., Internat. Soc. Research in Biology of Reproduction, Los Angeles Ob-Gyn Soc., Pacific Coast Fertility Soc., Pan Am. Med. Assn., Soc. Gynecologic Investigation, Soc. Study of Reproduction, Am. Gynecology Soc., Am. Assn. Obstetricians and Gynecologists. Subspecialty: Obstetrics and gynecology. Home: 637 Via Del Monte Palos Verdes Estates CA 90274 Office: Women's Hosp Los Angeles County-U So Calif Med Center 1240 N Mission Rd Los Angeles CA 90033

MISHRA, SHRI KANT, neurologist; b. Raiya Mirzapur, India, July 17, 1941; s. Jai Gopal and Dil Raji (Pathak) M.; m. Annamma Varughese, Mar. 3, 1968; children: Alok, Arvind. B. Medicine and Surgery, Coll. Med. Scis. B.H.U., Varanasi, India, 1968; M.D., U. Toronto, Ont., Can., 1971. Intern U. Tex. Southwestern Med. Sch., Dallas, 1971-72; resident in neurology U. Miss. Med. Ctr., Jackson, 1980—; clin. dir. Muscular Dystrophy Assn. Contbr. articles to profl. jours. Fellow AAAS, Acad. Neurology; mem. Am. Fedn. Clin. Research, N.Y. Acad. Scis., Sigma Xi. Subspecialties: Health services research; Genetics and genetic engineering. Current work: Membrane abnormalities in myotonic muscular dystrophy. Offic: Dept Neurolog VA Hosp 1500 E Woodrow Wilson St Jackson MS 39216

MISRA, ANAND LAL, medicinal chemist, educator; b. Kanpur, India, June 5, 1928; came to U.S., 1969, naturalized, 1977; s. Chotey Lal and Sahodra Ruttondevi (Shukla) M. B.Sc., Allahabad U., 1946, M.Sc., 1948, Ph.D. (Kanta Prasad research scholar), 1950; F.H.-W.C. (Assam Oil Co.-Burmah Shell scholar), Heriot-Watt Coll., Edinburgh (Scotland) U., 1955. Research asst. Allahabad U., 1951-53; sci. office Nat. Chem. Lab., Poona, India, 1956-57; Smith-Mundt Fulbright research fellow U. Mich., Ann Arbor, 1957-60; research officer Central Drug Research Inst., Lucknow, India, 1960-61; research chemist Ciba Ltd., Basel, Switzerland, 1961-64; head product devel. Ciba of India Ltd., Bombay, 1964-67; research assoc. U. Iowa, Iowa City, 1969-70; research scientist V Research Lab., N.Y. State Div. Substance Abuse Services, Bklyn., 1971—; adj. assoc. prof. dept. psychiatry SUNY Downstate Med. Center, Bklyn., 1980—; participant numerous nat. and internat. sci. confs. Contbr. numerous articles on structure-activity relationships, biopharmaceutics, phytochemistry, pharmacokinetics, biol. disposition and metabolism of drugs subject to human abuse to sci. jours. U.P. Govt. scholar, 1951-53; U.S. Army Med. Research and Devel. Command grantee, 1973-76. Fellow Royal Soc. Chemistry (U.K.), Am. Inst. Chemists; mem. Am. Soc. Pharmacology and Exptl. Therapeutics, Am. Chem. Soc., AAAS, Fedn. Am. Scientists, Assn. for Med. Edn. and Research in Substance Abuse, N.Y. Acad. Scis. Subspecialties: Pharmacology; Medicinal chemistry. Current work: Radioactive tracer studies; biological disposition, metabolism, pharmacokinetics of drugs subject to human abuse; mechanism of development of tolerance, psychic and physiological dependence on psychoactive drugs. Office: Research Lab NY State Div Substance Abuse Services 80 Hanson Pl Brooklyn NY 11217

MISRA, BALABHADRA, chemical engineer, research consultant; b. Berhampur, India, Apr. 21, 1926; came to U.S., 1949; s. Gopal Bhatta and Malati (Dash) M.; m. Nalini Misra, Apr. 14, 1949; children: Bijoy, Binoy. B.Sc. with honors, Ravenshaw Coll., India, 1947; B.S., Indian Inst. Sci., 1949; M.S., Columbia U., 1951, Ph.D., 1957. Registered profl. engr., Calif. Research assoc. Columbia U., N.Y.C., 1956-58; tech. specialist Aerojet-Gen. Corp., Sacramento, 1958-71; chem. engr. Argonne Nat. Lab., Ill., 1971—, project mgr., 1971-77. Contbr. numerous articles to sci. jours. Fundraiser Sacramento Symphony, 1967-70; troop leader Boy Scouts Am., 1966-68. Recipient Gold medal Utkal U., Orissa, India, 1947. Mem. Am. Inst. Chem. Engrs., AIAA, Am. Nuclear Soc. Republican. Hindu. Subspecialties: Nuclear fission; Nuclear fusion. Current work: Component development and testing for LMFBR and fusion reactor; safety analysis of nuclear reactors; tritium extraction from fusion reactor breeding blankets. Home: 73 Finch Ct Naperville IL 60565 Office: 9700 S Cass Ave Argonne IL 60439

MISRA, RAGHUNATH PRASAD, physician, pathology educator; b. Calcutta, West Bengal, India, Feb. 1, 1928; came to U.S., 1964, naturalized, 1971; s. Guru Prasad and Anandi (Devi) M.; m. Therese Rettenmund, Sept. 12, 1963; children: Sima, Joya, Maya, Tara. B.Sc., Calcutta (India) U., 1948, M.B.B.S., 1954; Ph.D., McGill U., 1965. Diplomate: Am. Bd. Pathology. Intern Med. Coll. Hosps., Calcutta, 1953-54, resident, 1954-56, Univ. Hosps., Cleve., 1973-76; instr., asst. prof. dept. medicine U. Louisville Sch. Medicine, 1966-68; asst., assoc. investigator Mt. Sinai Hosp., Cleve., 1968-73; asst. prof. exptl. pathology Case Western Res. U., Cleve., 1971-76; asst. prof. pathology La. State U. Sch. Medicine, Shreveport, 1976-80, assoc. prof., 1980—, chief renal immunopathology, 1976—; dir. Kidney Research Lab., U. Louisville Sch. Medicine, 1966-68, Kidney Lab., Mt. Sinai Hosp., Cleve., 1968-73. Author: An Atlas of Skin Biopsy, 1983. Sec. bd. govs. India House Project, Cleve., 1976. Indian Med. Assn. fellow, 1957-60; Can. Heart Found. fellow, 1960-64; Jean Talliamon Meml. fellow, 1971-73. Fellow Coll. Am. Pathologists, Am. Soc. Clin. Pathologists; mem. Am. Assn. Pathologists, Am. Soc. Nephrology, AMA, Sigma Xi. Unitarian-Universalist. Subspecialties: Nephrology; Pathology (medicine). Current work: Molecular pathobiology of Kidney diseases. Home: 6153 River Rd Shreveport LA 71105 Office: La State Sch Medicine 1501 Kings Hwy Shreveport LA 71130

MISTRY, VITTHALBHAI DAHYABHAI, med. physicist; b. Vesma, India, Sept. 30, 1942; came to U.S., 1965, naturalized, 1976; s. Dahyabhai Jivanji and Jamnaben (Dullabhbhai) M.; m. Padmaben Kalyanji Prajapati, May 18, 1974; children: Vandana, Rohitkumar. B.S. with great distinction, Haile Selassie I U., Addis Ababa, Ethiopia, 1964; Ph.D. in Nuclear Physics, U. Tex.-Austin, 1969. Cert. in therapeutic radiol. physics Am. Bd. Radiology. Asst. prof. Haile Selassie I U., 1964-65; research assoc. U. Tex.-Austin, 1967-69; postdoctoral fellow Tex. Christian U., Ft. Worth, 1969-71; inst. x-ray physics Austin (Tex.) State Hosp., 1974-78; med. physicist Capital Area Radiation and Research Ctr., Austin, 1971—; cons. med. radiation physicist. Mem. Am. Phys. Soc., Am. Assn. Physicists in Medicine, Soc. Tex. Regional Physicists, Am. Soc. Therapeutic Radiology, Sigma Xi, Sigma Pi Sigma. Hindu. Subspecialties: Nuclear physics; Medical physics. Current work: Biological effect of therapeutic doses of ionizing radiation for different treatment modalities in radiation therapy. Office: Capital Area Radiation and Research Center 2600 E Martin Luther King Blvd Austin TX 78702

MITALAS, ROMAS, astronomer, educator; b. Lithuania, Feb. 28, 1933; s. Adolfas and Leonarda (Grigonyte) M.; m. Nancy Louise Callaghan, Sept., 1979. B.A. in Applied Math, U. Toronto, 1957, M.A., 1958; Ph.D. in Theoretical Physics, Cornell U., 1964. Asst. prof. physics U. Western Ont., London, 1964-68, asst. prof. astronomy, 1968-73, asso. prof., 1973—. Mem. Can. Astron. Soc., Am. Astron. Soc., Am. Phys. Soc. Subspecialties: Theoretical astrophysics; Stellar structure and evolution. Current work: Stellar structure and evolution. Home: 61 Meridene Circle E London ON Canada N58 2M1 Office: Dept Astronomy U Western Ont London ON Canada N68 3K7

MITCHELL, CLIFFORD L., laboratory chief, researcher, pharmacology and toxicology educator; b. Ottumwa, Iowa, Dec. 7, 1930; s. Clifford Lurton and Margaret Louise (Fisher) M.; m. Lou-Ann Marilyn Sperry, June 20, 1954; children: Steven, Catherine, Michael, Michelle. B.A., U. Iowa, 1952, B.S., 1954, M.S., 1958, Ph.D., 1959. Asst. prof. dept. pharmacology U. Iowa, Iowa City, 1959-60, asst. prof., 1962-66, assoc. prof., 1966-69, prof., 1969-73; fellow pharmacology, dept. pharmacology Stanford U., Palo Alto, Calif., 1960-62; mgr. CNS and Cardiopulmonary Pharmacology, Riker Labs., St. Paul, 1973-76; lectr. dept. pharmacology U. Minn., Mpls., 1973-76; head behavioral toxicology program Nat. Inst. Environ. Health Scis., Research Triangle Park, N.C., 1976-77, chief lab. behavioral and neurol. toxicology, 1977—; adj. prof. dept. pharmacology U. N.C., Chapel Hill, 1976—. Editor: Nervous System Toxicity, 1982; contbr. articles to profl. publs. Served with U.S. Army, 1954-56. Mem. Am. Soc. Pharmacol. Exptl. Therapeutics, AAAS, Soc. Exptl. Biol. Medicine, Soc. Neurosci., Bioelectromagnetic Soc., Rho Chi. Subspecialties: Toxicology (medicine); Neuropharmacology. Current work: Neurotoxic effects of environmental agents, researcher environmental agents such as heavy metals, pesticides and microwaves. Home: 243 Seminole Dr Chapel Hill NC 27514 Office: NIEHS Box 12233 Research Triangle Park NC 27709

MITCHELL, DEAN LEWIS, physicist, research administrator; b. Montour Falls, N.Y., Apr. 9, 1929; s. Robert B. and Helen B. Mitchell; m. Gertrue E. Marburger, Apr. 23, 1930; children: Dale, Claire, Jay, Elaine. Student, Rochester Inst. Tech., 1945-46; B.S. in Physics, Syracuse U., 1952, Ph.D., 1959. Asst. prof. Utica Coll., N.Y., 1958-61; research physicist Semiconds. Br., Naval Research Lab., Washington, 1961-70, head, 1970-74; mem. staff Div. Materials Research NSF, Washington, 1974—. Served with USN, 1946-48, 52-53. Mem. Am. Phys. Soc., AAAS. Subspecialties: Condensed matter physics; Electronic materials. Current work: Electronic band structure; semiconductors; amorphous materials; scientific administrator national user facilities and materials research laboratory program. Office: Div Materials Research NSF Washington DC 20550

MITCHELL, GEORGE ERNEST, JR., animal scientist, educator; b. Duoro, N.Mex., June 7, 1930; s. George Ernest and Alma Thyrza (Hatley) M.; m. Billie Carolyn McMahan, Mar. 14, 1952; children: Leslie Dianne, Karen Leigh, Cynthia Faye. B.S., U. Mo., 1951, M.S., 1954; Ph.D., U. Ill., 1956. Asst. prof. animal sci. U. Ill., 1956-60; assoc. prof. U. Ky., Lexington, 1960-67, prof., 1967—, dir. grad. studies in animal scis., 1964—, coordinator beef cattle and sheep, 1974—; mem. com. on animal nutrition NRC. Contbr. articles to profl. jours. Served with USAF, 1951-53. Fulbright research scholar, New Zealand, 1973-74. Mem. Am. Soc. Animal Sci. (sec. 1969-70, v.p. 1970-71, pres. So. sect. 1971-72), Am. Dairy Sci. Assn., Am. Inst. Nutrition, AAAS, Am. Inst. Biol. Sci., N.Y. Acad. Sci., Council for Agrl. Sci. and Tech., Sigma Xi, Alpha Zeta, Gamma Sigma Delta, Omicron Delta Kappa. Democrat. Methodist. Subspecialties: Animal nutrition; Nutrition (biology). Current work: Ruminant nutrition, comparative nutrition, vitamin metabolism, starch utilization, digestive physiology. Home: 690 Hill 'n' Dale Lexington KY 40503 Office: 809 Agrl Sci So U Ky Lexington KY 40546

MITCHELL, JAMES CURTIS, psychology educator; b. Youngstown, Ohio, July 30, 1935; s. James C. and Mildred E. (Hohlock) M.; m. Linda R. Paterson, Dec. 17, 1977; stepchildren: Stpehanie, Craig Raborn. B.S., Ohio State U., 1957, M.A., 1959, Ph.D., 1962. Research assoc. U. Miss., Jackson, 1959-60; asst. prof. So. Ill. U., Carbondale, 1962-66; assoc. prof. Kans. State U., Manhattan, 1966-71, prof. psychology, 1971—. Contbr. articles to profl. jours. Mem. Soc. Neursci., Psychonomic Soc., Am. Psychol. Assn., Medwestern Psychol. Assn. Democrat. Subspecialties: Physiological psychology; Neuropsychology. Current work: Research concerns neural mechanisms controlling eating and body weight regulation. Home: 2220 Seaton Ave Manhattan KS 66502 Office: Kans State Univ Dept Psychology Manhattan KS 66506

MITCHELL, JAMES KENNETH, civil engineer, educator; b. Manchester, N.H., Apr. 19, 1930; s. Richard N. and Henrietta (Moench) M.; m. Virginia D. Williams, Nov. 24, 1951; children: Richard A., Laura K., James W., Donald M., David L. B.C.E., Rensselaer Poly. Inst., 1951; M.S., M.I.T., 1953; D.Sc., 1956. Mem. faculty U. Calif., Berkeley, 1958—, prof. civil engring., 1968—, chmn. dept., 1979—; geo tech. cons., 1960—. Author: Fundamentals of Soil Behavior, 1976; contbr. articles to profl. jours. Asst. scoutmaster Boy Scouts Am., 1975-82; mem. Moraga (Calif.) Environ. Rev. Com., 1978-80. Served to 1st lt. AUS, 1956-58. Recipient Exceptional Sci. Achievement medal NASA, 1973. Fellow ASCE (Huber prize 1965, Middlebrooks award 1962, 70, 73, Norman medal 1972); mem. Nat. Acad. Engring., Am. Soc. Engring. Edn. (Western Electric Fund award 1979), Transp. Research Bd. (exec. com. 1983—), Clay Minerals Soc., Sigma Xi, Tau Beta Pi, Chi Epsilon. Subspecialty: Civil engineering. Current work: Ground improvement for engineering works, geotechnical engineering, geotechnical aspects of hazardous waste containment. Office: Dept Civil Engring U Calif Berkeley CA 94720

MITCHELL, JAMES VINCENT, JR., educational psychologist; b. Chgo., Oct. 20, 1925; s. James Vincent and Amanda Bertha (Hansen) M.; m. Margaret Mary Mattern, Jan. 6, 1973; children by previous marriage: Steven James, Keith Vincent. A.B., U. Chgo., 1948, M.A., 1950, Ph.D., 1953. Assoc. prof. ednl. psychology U. Tex., Austin, 1959-62; prof. edn. U. Rochester, N.Y., 1962-67, assoc. dean grad. studies, 1967-74; dean Tchrs. Coll., Ball State U., Muncie, Ind., 1974-77; v.p. acad. affairs Bloomsburg (Pa.) State Coll., 1977-80; dir. Buros Inst. Mental Measurements, U. Nebr., Lincoln, 1980—; cons. Brighton Sch. System, Rochester, 1965. Contbr. articles to profl. jours.; editor: Mental Measurement Yearbooks, 1982—. Organizer Buros - Nebr. Symposium on Measurement & Testing, 1982—. Recipient Student Assn. Award for excellence in teaching U. Tex., Austin, 1962; award for exptl. design TV Bur. Advt., 1961. Fellow Am. Psychol. Assn.; mem. Am. Ednl. Research Assn., Sigma Xi, Phi Kappa Phi, Phi Delta Kappa. Subspecialty: Personality theory and assessment. Current work: Measurement of values; structural dimensions of value systems. Home: 2941 S 74th St Lincoln NE 68506 Office: Buros Inst Mental Measurements Univ Nebr Lincoln NE 68588

MITCHELL, JOHN MURRAY, JR., climatologist; b. N.Y.C., Sept. 17, 1928; s. John Murray and Lanier (Comly) M.; m. Pollyanne Bryant, May 5, 1956; children: John Murray, Brian Harrison, Katherine Comly, Anne Stuart. B.S., Mass. Inst. Tech., 1951, M.S., 1952; Ph.D., Pa. State U., 1960; postgrad., Nat. War Coll., 1970-71. Research meteorologist Weather Bur., Commerce Dept., Suitland, Md., 1955-65; project scientist environ. data service NOAA, Silver Spring, Md., 1965-74, sr. research climatologist, 1974-79, sci. advisor, 1980—; mem. various coms. and panels Nat. Acad. Scis., NRC; vis. lectr., prof. U. Calif., U. Wash. Contbr. articles to books, encys., tech. jours.; editor: Meteorol. Monographs, 1965-73. Mem. Fairfax County Air Pollution Control Bd. Served with USAF, 1952-55. Recipient Silver and Gold medals Commerce Dept. Fellow AAAS, Am. Meteorol. Soc. (2d Half Century award), Washington Acad. Scis.; mem. Am. Geophys. Union, Royal Meteorol. Soc., Sigma Xi. Subspecialty: Climatology. Current work: Causes of present-day climatic changes, and assessment of natural and anthropogenic causes of future changes. Home: 1106 Dogwood Dr McLean VA 22101 Office: NOAA 6010 Executive Blvd Rockville MD 20852

MITCHELL, MALCOLM STUART, physician; b. N.Y.C., May 6, 1937; s. Max E. and Sylvia W. M.; m. June Kan, Aug. 14, 1976; children: Jeffrey Scott, Roderick Keith, Derek James. A.B. magna cum laude, Harvard U., 1957; M.D., Yale U., 1962; postgrad. (Fulbright scholar), U. Oxford, Eng., 1959-60. Diplomate: Am. Bd. Internal Medicine. Instr. to asso. prof. medicine Yale U., 1968-78; prof. medicine and microbiology U. So. Calif., 1978—, chief med. oncology, dir. clin. investigations, 1978—; adv. com. NIH, 1975-82, Am. Cancer Soc., 1975-79, U.S. Pharmacopeia, 1975-80; chmn. adv. com. Nat. Cancer Cytology Center, 1981—. Editor in chief: Yale Jour. Biology and Medicine, 1976-78; asso. editor 6 other jours.; contbr. numerous articles to profl. jours.; editor: Hybridomas in Cancer Diagnosis and Treatment, 1982. Served with USPHS, 1963-65. Leukemia Soc. scholar, 1968-73; NIH awardee, 1974-79. Mem. Soc. Clin. Investigation, Am. Assn. Cancer Research, Am. Assn. Immunologists, Phi Beta Kappa, Sigma Xi. Subspecialties: Oncology; Immunology (medicine). Current work: Immunology, immunotherapy. Office: 2025 Zonal Ave Los Angeles CA 90033

MITCHELL, NANCY BROWN, human factors engineer; b. Chgo., Mar. 7, 1941; d. Edward Berrien and Jeannette (Landes) Brown; m. James Evans, Jr., July 18, 1961 (dec. 1979); children: James Evans, Thomas Edward, Janet Lucille. B.A., Montevallo U., 1967; M.S., U. Ga., 1974, Ph.D., 1976. Tchr. remedial reading Hannah J. Mallory Sch., Goodwater, Ala., 1967-71; research and teaching asst. U. Ga., Athens, 1971-75; spl. asst. So. Research Edn. Bd., Atlanta, 1976-77; psychologist Dept. Youth Services, Birmingham, Ala., 1977; human engr. U.S. Army Infantry Sch., Columbus, Ga., 1977-80; human factors engr. Army Research Inst., Alexandria, Va., 1980—. Bd. dirs. PTA, Sylacauga, Ala., 1965. Mem. Am. Psychol. Assn., Human Factors Soc., Women in Sci. and Engrng., Sigma Xi, Kappa Delta Pi. Subspecialties: Graphics, image processing, and pattern recognition; Human factors engineering. Current work: Extension of surrogate training to travel over open terrain; purpose, simulation, tactical raining; development and evaluation of displays for use in remotely piloted reconnaissance vehicles and tracked vehicles. Home: 5478 Mersea Ct Burke VA 22015 Office: Army Research Inst 5001 Eisenhower Ave Alexandria VA 22333

MITLER, HENRI EMMANUEL, physicist, educator, researcher; b. Paris, France, Oct. 26, 1930; came to U.S., 1940, naturalized, 1946; s. Jack J. and Cecile Milter; m.; children: Julia Ann, David Duncan. B.S., CCNY, 1953; Ph.D., Princeton U., 1960. Jr. scientist Nuclear Devel. Assocs., White Plains, N.Y., 1953; instr. Princeton U., 1958-59; lectr. in physics Brandeis U., 1959-61; staff scientist Smithsonian Astrophys. Obs., Cambridge, Mass., 1961-75; lectr. in astronomy Harvard U., Cambridge, 1965-75, sr. research assoc. div. applied scis., 1975—. Contbr. articles to profl. jours. NSF fellow, 1953; Fulbright scholar, 1956. Mem. AAAS, Am. Astron. Soc., Am. Phys. Soc., Combustion Inst., ACLU, Common Cause, Ams. for Democratic Action, Phi Beta Kappa. Democrat. Subspecialties: Theoretical and applied mechanics; Mathematical modeling. Current work: Developing math. model of how fires develop in enclosures. Office: 40 Oxford St Room 416 Cambridge MA 02138

MITLER, MERRILL MORRIS, psychologist; b. Racine, Wis., Jan. 1, 1945; s. Benjamin and Dorothy (Farrell) M.; m. Karin Helen Yock, Jan., 1967 (div. 1976); m. Elizabeth Ann McClements, Aug. 27, 1976; children: Marc, Morris, Maximillian. B.A., U. Wis.-Madison, 1967; M.A., Mich. State U., 1968, Ph.D., 1970. Lic. psychologist. Postdoctoral fellow Stanford U. Sch. Medicine, 1970-73, research assoc., 1973-76, adminstrv. dir., 1976-78; chief Sleep Disorders Ctr., SUNY-Stony Brook Med. Sch., 1978—; Pres. Wakefulness-Sleep Edn. and Research Found., Port Jefferson, N.Y., 1982—. NIH fellow, 1969, 70; grantee, 1979, 82. Mem. AAAS, Am. Psychol. Assn., Assn. Psychophysiol. Study of Sleep. Republican. Jewish. Subspecialties: Physiological psychology; Neuropharmacology. Current work: Sleep disorders, performance impairment due to sleep disorder; standards of practice for sleep specialists. Home: 121 Old Post Rd Jefferson NY 11777 Office: Dept Psychiatry SUNY Health Scis Center Stony Brook NY 11794

MITRA, JYOTI, medical researcher, educator; b. India, July 10, 1934; U.S., 1966, naturalized, 1980; s. Rabindra Kumar and Tara Rani (Dutta) M.; m. Sibani Chatterjee, June 1, 1974; 1 dau., Runa. B.S., U. Calcutta, India, 1955, MS., 1957; Ph.D., U. Cambridge, Eng., 1962. Lectr. Calcutta U., 1962-63; research fellow Max Planch Inst., Munich, W.Ger., 1963-65, Brain Research Inst., UCLA, 1965-66; research assoc. Ctr. Brain Research, Rochester, N.Y., 1966-68, vis. scientist, 1971-74; fellow Montreal (Que., Can.), Neurol. Inst., 1968-69; research assoc. U. Amsterdam, Netherlands, 1969-71; sr. research assoc. U. Pa., 1974-77; sr. research assoc. dept. medicine, asst. prof. physiology Case Western Res. U., 1977—. Mem. Soc. Neurosci., AAAS, N.Y. Acad. Scis., Sigma Xi. Subspecialties: Neurophysiology; Bioinstrumentation. Current work: Nervous control of breathing-processing of information at the cellular level in brain and spinal cord, both in anesthetized and awake free-moving animals. Office: Univ Hosp Hearn Bldg Room 207 Cleveland OH 44106

MITRA, JYOTIRMAY, biology educator; b. Calcutta, India, Nov. 25, 1921; s. Jnan Chandra and Aruna (Basu) M.; m. Bakul Mitra, Mar. 26, 1970. B.A. with honors, Calcutta U., 1942, M.A., 1944; Ph.D., Cornell U., 1955. Research asst. Bot. Survey of India, Calcutta, 1945-46; lectr. botany Calcutta U., 1948; research fellow Imperial Chem. Industries, Calcutta, 1948-51; research asso. Cornell U., Ithaca, N.Y., 1957-61; cytogeneticist-in-charge Beth Israel Med. Center, N.Y.C., 1961-77; assoc. prof. biology N.Y. U., N.Y.C., 1963-67, prof. biology, 1967—; cons. in field. Contbr. articles to profl. jours. Calcutta U. scholar, 1942-44; Sir R.B. Ghosh Research scholar, 1945; Fulbright and Smith-Mundt research scholar, 1952; NIH grantee, 1975; NSF grantee, 1977-80; Damon Runyon Cancer Research grantee, 1967-69; others. Mem. Genetics Soc. Am., Am. Genetic Assn., AAAS, Sigma Xi, Phi Kappa Phi. Club: Synapsis. Subspecialties: Animal genetics; Plant genetics. Current work: Cytogenetics and cytotaxonomy of Eukaryotic organisms; cancer genetics/cytogenetics. Office: Dept Biology NY Univ New York NY 10003 Home: 4 Washington Square Village 4H New York NY 10012

MITRA, SAIBAL K(UMAR), process and engineering and scientific industry specialist, computer systems specialist; b. Calcutta, W.Bengal, India, Jan. 10, 1937; came to U.S., 1978; s. Rasvihari and Umarani (Bose) M.; m. Abantika Kundu, Nov. 25, 1966; 1 child, Sanchayeeta. B.S. with honors, Calcutta (India) U., 1958, M.S., 1961; M.S., Imperial Coll. Sci. and Tech., London, U.K., 1975. Lectr. Indian Statis. Inst., Calcutta, 1962-66; systems engr. IBM World Trade Corp., Calcutta, 1966-69; ops. research/systems analyst Mgmt. Innovation Assos., London, U.K., 1969-70; lectr., part time cons. City Univ., London, 1970-78; industry specialist IBM Corp., Cranford, N.J., 1978—. Mem. Assn. Computing Machinery, Inst. Mgmt. Sci., Am. Prodn. and Inventory Control Soc. Subspecialties: Information systems (information science); Operations research (mathematics). Current work: Design of information system; application marketing; decision support system; consulting in data processing and operations research. Home: 4 Marc Ct Edison NJ 08820 Office: IBM Corp 20 Commerce Dr Cranford NJ 07016

MITSCHER, LESTER ALLEN, chemist, educator; b. Detroit, Aug. 20, 1931; s. Lester and Mary Athelda (Pounder) M.; m. Betty Jane McRoberts, May 29, 1953; children: Katrina, Kurt, Mark. B.S., Wayne U., 1953, Ph.D., 1958. Research scientist, group leader Lederle Labs., Pearl River, N.Y., 1958-67; prof. Ohio State U., Columbus, 1967-75; Univ. disting. prof., chmn. dept. medicinal chemistry U. Kans., Lawrence, 1975—; inter-research prof. Victorian Coll. Pharmacy, Melbourne, Australia, 1975—; cons. NIH, Abbott Labs., Searle Labs., Adria Labs. Author: (with D. Lednicer) The Organic Chemistry of Drug Synthesis, Vol. 1, 1976, Vol. 2, 1980, Vol. 3, 1984, The Chemistry of the Tetracycline Antibiotics, 1978; contbr. over 140 articles to profl. jours. Recipient Disting. Alumnus award Sch. Pharmacy, Wayne State U., 1980, Research Achievement award Acad. Pharm. Scis., 1980. Mem. Am. Chem. Soc. (former chmn. and councilor medicinal chemistry div.), Chem. Soc. London, Japanese Antibiotics Assn., AAAS, Soc. Heterocyclic Chemistry, Am. Soc. Pharmacognosy. Presbyterian. Club: Elks. Subspecialty: Organic chemistry. Office: Dept Medicinal Chemistry U Kans Lawrence KS 66045

MITSON, HERBERT HENRY, JR., mech. engr.; b. Detroit, May 24, 1927; s. Herbert Henry and Mattie Marie (Brown) M.; m. Jacqueline Jean Marentette, June 16, 1951; children: Bryan, Karen Swope. B.S. in Mech. Engring, Mich. State U., 1951; M.S. in Engring. Tech, Rochester Inst. Tech., 1973. Registered profl. engr., Pa. Mech. engr. radio communications Gen. Dynamics, Rochester, N.Y., 1966-69; design engr. hosp. equipment Am. Sterilizer, Erie, Pa., 1973-75; mech. engr. N.Am. Systems, Cleve., 1976-78; mech. engr. machine design, structural and computerized stress analysis Centerline Design, Cleve., 1978-81; analytical engr. heat transfer, dynamic and stress analysis Gould Ocean Systems, Cleve., 1981—; instr. Sch. Engring., Behrens Coll., Erie, 1974. Served with USN, 1945-46. Mem. ASME. Subspecialties: Mechanical engineering; Solid mechanics. Home: 26 Brune Dr Bedford OH 44146 Office: 18901 Euclid Ave Cleveland OH 44117

MITSUMOTO, HIROSHI, neurologist; b. Sapporo, Japan, Mar. 3, 1944; s. Masaji and Etsu M.; m. Chizuko Suzuki, Sept. 26, 1947; children: Ken Joseph, Jun Michael. B.S., Toho U., 1964, M.D., 1968, D. Med. Scis., 1981. Intern Toho U., Tokyo, 1968-69, asst. in internal medicine, 1969-72; intern Balt. City Hosp., 1972-73; resident in neurology Univ. Hosps., Cleve., 1973-76; fellow in neuropathology Cleve. Clinic, 1976-78, assoc. staff, 1978-79; research fellow Tufts-New Eng. Med. Ctr., Boston, 1979-81; asst. prof. neurology and neuropathology Case Western Res. U., Cleve., 1981-83, Cleve. Clinic, 1983—. Contbr. articles to profl. jours. Muscular Dystrophy Assn. Am. fellow, 1979-81; Amyotrophic Lateral Sclerosis Found. grantee, 1979-81, 82—; NIH grantee, 1982. Mem. Am. Acad. Neurology, Soc. Neuroscience, Am. Assn. Neuropathologists, Japanese Soc. Neurology. Subspecialties: Neurology; Neuropathology. Current work: Amyotrophic lateral sclerosis, murine motor neuron disease, neuromuscular diseases, axonal and neuronal regeneration, perhipheral neuropathies. Home: 3387 Ingleside Shaker Heights OH 44122 Office: Cleve Clinic Cleveland OH 44106

MITTAL, CHANDRA KANT, educator, pharmacologist; b. Aligarh, Uttar Pradesh, India, Aug. 5, 1948; s. Murari Lal and Kanti Devi (Singhal) M.; m. Sarita Dayal, July 6, 1976; children: Richa, Hersh. B.S., U. Lucknow, India, 1967, M.S. in Biochemistry, 1969; Ph.D., All-India Med. Inst., 1975. Research fellow All-India Med. Inst., 1969-72, sr. research fellow, 1972-74; fellow in pharmacology U. Va., Charlottesville, 1974-77, research instr. pharmacology, 1977-80; asst. prof. U. Ill. Coll. Medicine, Peoria, 1980—. Contbr. articles on pharmacology to profl. jours. Am. Heart Assn. grantee, 1981-83. Mem. Am. Soc. for Exptl. Therapeutic Pharmacology. Subspecialties: Biochemistry (medicine); Molecular pharmacology. Current work: Hypertension, drug receptors, cyclic nucleotides, calcium modulations, smooth muscle pharmacology. Office: One Illni Dr Peoria IL 61656

MITTAL, KAMAL KANT, immunogeneticist, researcher, laboratory director; b. Pihani, India, July 1, 1935; s. Vishveshwar and Shanti (Devi) M.; m. Sudha Rani, Jan. 23, 1971; children: Parul Divya, Rahul Vikrum. B.Sc., Agra U., 1954; B.V.Sc. & A.H., Vet. Coll., Mathura, India, 1958; M.S., U. Ill., 1962, Ph.D., 1965. Research fellow U. Ill.-Urbana, 1965-66; research fellow Calif. Inst. Tech., 1966-67; research geneticist UCLA, 1967-69, asst. research geneticist, 1970-72; asst. prof. microbiology Baylor Coll. Medicine, 1969-70; asst. prof. Northwestern U., 1972-75; research microbiologist FDA, Bethesda, Md., 1975—, also dir. histocompatibility testing lab. Contbr. articles to profl. jours. Recipient Bronze medal Agra U., 1958; Outstanding New Citizen award Citizenship Council Met. Chgo., 1975. Mem. Transplantation Soc., Am. Assn. Immunologists, Am. Assn. Clin. Histocompatability Testing, AAAS. Hindu. Subspecialties: Immunogenetics; Transplantation. Current work: Transplantation genetics and immunology. Home: 14620 Rolling Green Way Gaithersburg MD 20878 Office: Bldg 29 Room 232 8800 Rockville Pike Bethesda MD 20205

MITTAL, KASHMIRI LAL, research chemist, lecturer; b. Kilrodh, India, Oct. 15, 1945; came to U.S., 1966; s. Parma N and Bhagwan Dai (Gupta) M.; m. Usha Rani Gupta, Dec. 30, 1970; children: Anita, Rajesh, Nisha, Seema. B.Sc., Panjab U., Chandigarh, India, 1964; M.S. in Chemistry, Indian Inst. Tech., New Delhi, 1966, Ph.D., U. So. Calif., 1970. Teaching asst. dept. chemistry U. So. Calif., 1966-67; lectr. in chemistry Calif. State Coll.-Dominguez Hills, 1969; postdoctoral research assoc. Pa. State U., 1970-71; postdoctoral research fellow Electro-chemistry Lab., U. Pa., 1971-72; postdoctoral research scientist gen. products div. IBM, San Jose, Calif., 1972-74, chemist, Poughkeepsie, N.Y., 1974-77, East Fishkill, N.Y., 1977—; chmn., organizer symposia, confs. in field; speaker at nat., internat. meetings, orgns.; tchr. short courses in colloid, surface and adhesion sci. Papers, presentations in field; editor numerous books in areas of surface, colloid and adhesion sci.; editorial bd.: Jour. Coating Tech, 1979, Solid State Tech, 1979—, Advances in Colloid and Interface Sci, 1979—, Progress in Organic Coatings, 1980—, Adhesives Age, 1980-81; book reviewer for jours. Recipient First Invention Plateau award IBM, 1982; Panjab U. merit scholar, 1962; Panjab State merit scholar, 1962; Govt. India's Nat. Merit scholar, 1962-65; Govt. India's Atomic Energy scholar, 1965-66; Anderson Clayton & Co. fellow, 1967-69. Fellow Am. Inst. Chemists, Indian Chem. Soc.; mem. Am. Chem. Soc. (adv. com. div. organic coatings and plastics chemistry 1980—), Electrochem. Soc., Internat. Soc. Hybrid Microelectyronics, Adhesion Soc., Am. Vaccum Soc., Indian Sci. Congress Assn., Soc. Advancement Electrochem. Sci. and Tech., Soc. Plastics Engrs., Royal Soc. Chemistry (colloid and interface sci. group), Indian Vacuum Soc., Internat. Assn. Colloid and Interface Scientists, Fine Particle Soc., India Chemists and Chem. Engrs. Club (v.p. 1982—), Sigma Xi, Phi Lambda Upsilon. Subspecialties: Surface chemistry; Polymer chemistry. Current work: Research on surface, colloid and polymer chemistry. Office: IBM Hopewell Junction NY 12533

MITTELMAN, PHILLIP SIDNEY, computer service co. exec.; b. N.Y.C., Sept. 28, 1925; s. Joseph F. and Rose (Brooks) M.; m. Myra Schoenfeld, Apr. 10, 1948; children: Vicki, David. B.S., Rensselaer Poly. Inst., 1945; M.A., Harvard U., 1948; Ph.D., Rensselaer Poly. Inst., 1953. Scientist Brookhaven Nat. Lab., Upton, N.Y., 1947; mgr. physics and math. United Nuclear Corp., Elmsford, N.Y., 1953-66; pres., chmn. bd. Math. Applications Group, Inc., Elmsford, 1966—. Contbr. articles to profl. jours. Mem. Am. Phys. Soc., Am. Nuclear Soc., Assn. for Computing Machinery. Subspecialties: Graphics, image processing, and pattern recognition; Nuclear fission. Current work: Supervision of computer applications in applied computer graphics and services. Home: Heritage Hills 110B Somers NY 10589 Office: 3 Westchester Plaza Elmsford NY 10523

MITTLER, SIDNEY, biology educator; b. Detroit, Aug. 2, 1917; s. Max and Ida (Shulman) M.; m. Leonore Broder, Aug. 16, 1942; children: Jeanne, Judith, Michele; m. Judith W. Daskovsky, Aug. 10, 1969. B.S., Wayne U., 1938, M.S., 1939; Ph.D., U. Mich., 1944. Instr. Bowling Green (Ohio) State U., 1945-46; asst. prof. Ill. Inst. Tech., 1946-52; research biologist Armour Research Found., 1952-60; prof. biol. scis. No. Ill. U., DeKalb, 1960—. Contbr. articles to profl. jours. Served with M.C. U.S. Army, 1942-45. Recipient Award of Sci. Merit Morton Salt Co. and Chemistry dept. Armour Research Found., 1955. Fellow AAAS; mem. Radiation Research Soc., Genetics Soc. Am., Environ. Mutagenesis Soc., Am. Genetics Assn. Subspecialties: Genetics and genetic engineering (biology); Radiation biology. Current work: Research on effect of hyperthermia upon radiation induced genetic damage. Office: Dept Biol Sci No Ill Univ DeKalb IL 60115

MITYAGIN, BORIS SAMUEL, mathematician, researcher, math. educator; b. Voronezh, USSR, Aug. 12, 1937; came to U.S., 1978; s. Samuel Yakov and Vera Alexander (Mityagin) Gasul; m. Sophia A. Katz, July 13, 1962; 1 dau., Enia. M.S., Moscow State U., 1958, Sc.D., 1963. Asst. lectr. Voronezh State U., USSR, 1961-62, assoc. prof., 1962-64, prof., 1964-67; head lab. sr. researcher Central Econs. - Math, Inst., Acad. Sci., Moscow, USSR, 1967-78; prof. Moscow State U., 1968-73; vis. prof. Purdue U., 1978-79; prof. math. Ohio State U., 1979—; vis. prof. l'Ecole Polytechnique, Paris, 1978; prof. Inst. Math., Bonn (W.Ger.) U., 1979; Rothschild prof. Tel-Aviv (Israel) U., 1980; prof. Central U. Venezuela, Caracas, 1982. Contbr. numerous articles to profl. jours.; mem. editorial bd.: Functional Analysis and Its Applications Jour, 1976-77, Integral Equations and Operator Theory Jour, 1981—; mem. adv. bd.: Jour. Match Econs, 1973—. NFS grantee, 1979-82. Mem. Moscow Match. Soc., Am. Math. Soc., Soc. Indsl. and Applied Match. Club: Ohio State U. Faculty. Subspecialties: Functional analysis and its applications; Applied mathematics. Current work: Research to develop general methods of linear and nonlinear functional analysis and apply them in differential equations, numerical analysis and models of economic dynamics. Home: 3538 La Rochelle Dr Upper Arlington OH 43221 Office: Ohio State U 231 W 18th Ave Columbus OH 43210

MIURA, ROBERT MITSURU, mathematician, educator, researcher; b. Selma, Calif., Sept. 12, 1938; emigrated to Can., 1975; s. Richard Katsuki and Frances Yoneko (Yukutake) M.; m. Joycie Sayeko Yagura, July 29, 1961; children: Derek Katsuki, Brian Robert. B.S., U. Calif.-Berkeley, 1960, M.S., 1962; M.A., Princeton U., 1964, Ph.D., 1966. Postdoctoral assoc. Princeton Plasma Phys. Lab., 1965-67; assoc. research scientist Courant Inst., N.Y.U., 1967-68, asst. prof., 1968-71; assoc. prof. Vanderbilt U., Nashville, 1971-75; vis. and assoc. prof. U. B.C., Vancouver, 1975-78, prof., 1978—. Editor: Backlund Transformations, 1976, Some Mathematical Questions in Biology-Neurophysiology, 1982, (with others) Nonlinear Phenomena in Physics and Biology, 1981; assoc., editor: Can. Jour. Math, 1982—; mem. adv. bd.: Jour. Math. Biology, 1982. Instr. judo Killarney Community Center and Vancouver Judo Club, 1982-83. Guggenheim fellow, 1980-81; Killam Found. fellow, 1980-81. Mem. Am. Math. Soc. (chmn. life scis. com. 1981—), Soc. Indsl. and Applied Math., Can. Applied Math. Soc. (chmn. program com. 1982-83), Can. Math. Soc., AAAS, N.Y. Acad. Sci., Sigma Xi. Subspecialties: Applied mathematics; Neurophysiology. Current work: Nonlinear wave propagation phenomena, wave phenomena in neurophysiology, spreading cortical depression, nonlinear diffusion problems, differential-difference equations, computations. Office: Dept Mat Univ BC 121-1984

Mathematics Rd Vancouver BC Canada V6T 1Y4 Home: 3205 W 35th Ave Vancouver BC Canada V6N 2M9

MIXSON, WAYNE CLARK, plant pathologist, seed company executive; b. Apopka, Fla., Nov. 29, 1944; s. James Clark and Gladys Marie (Goodman) M.; m. Lurlie Ann Evans, Jan. 29, 1978; children: Onnie Lee, Leeann. A.A., U. Fla., 1965, B.S., 1967, M.S., 1969, Ph.D., 1972. Mgr. Southeastern research and devel. O.M. Scott & Sons, Apopka, 1972—. Mem. Am. Soil Assn., Am. Phys. Soc., Weed Sci. Soc., Am. Crop Sci. Soc., Am. Fla Entomol. Soc., Soc. Nematologists, Soil Sci. Soc. Am., Am. Phytopath. Soc. Subspecialties: Plant pathology; Turf management. Current work: Turfgrass and ornamental research. Office: 1151 E Oak St Drawer T Apopka FL 32703

MIYAKODA, KIKURO, research meteorologist; b. Yonago, Japan, Nov. 7, 1927; came to U.S., 1965, naturalized, 1974; s. Kanae and Kimiyo (Nishimura) M.; m. Toyoko Fukuda, Jan. 8, 1954; 1 child, Noriko. Ph.D., Tokyo U., 1961. Assoc. prof. Tokyo U., 1960-65; research meteorologist Geophys. Fluid Dynamics Lab., NOAA, Princeton, N.J., 1965—, sr. exec. service, supervisory meteorologist, 1979—; vis. prof., 1968—. Contbr. articles to profl. jours. Recipient ESSA Outstanding Sci. Paper award, 1970; gold medal U.S. Dept. Commerce, 1972. Fellow Am. Meteorol. Soc.; mem. Japan Meteorol. Soc. (award 1956, Fujiwara award 1983), Am. Geophys. Union, Sigma Xi. Subspecialty: Meteorology. Current work: Long-range weather forecast, numerical weather prediction, meteorological dynamics.

MIYAMOTO, MICHAEL DWIGHT, pharmcologist, educator; b. Honolulu, Apr. 22, 1945; s. Donald Masanobu and Chisako (Moriwaki) M.; m. Janis Ways, June 16, 1973; children: Julie, Scott. B.A., Northwestern U., 1966, Ph.D., 1971. Teaching asst. Northwestern U., 1968-70; instr. Rutgers U., 1970-72; asst. prof. U. Conn. Health Ctr., Farmington, 1972-78; assoc. prof. pharmacology East Tenn. State U., 1978—. Recipient Teaching award East Tenn. State U., 1980, award Pharm. Mfrs. Assn., 1975, Epilepsy Found. Am., 1976; USPHS grantee, 1975-78, 79-80. Mem. Soc. Neurosci., Am. Soc. Pharmacology and Exptl. Therapeutics. Subspecialties: Neuropharmacology; Neurophysiology. Current work: Neuromuscular transmission. Home: 318 Baron Dr Johnson City TN 37601 Office: East Tennessee State University PO Box 19 810A Johnson City TN 37614

MIYASHIRO, AKIHO, geologist, educator; b. Okayama Prefecture, Japan, Oct. 30, 1920; came to U.S, 1967; s. Tsuneshi and Hideyo (Shigemasa) M. B.S., U. Tokyo, 1943, Ph.D., 1953. Jr. instr. U. Tokyo, 1946-58, assoc. prof., 1958-67; vis. prof. Columbia U., N.Y.C., 1967-70; prof. SUNY, Albany, 1970—. Author: Metamorphism and Metamorphic Belts, 1973, Orogeny, 1982. Recipient Geol. Soc. Japan prize, 1958; P. Fourmarier prize Royal Acad. Belgium, 1982. Fellow Geol. Soc. Am. (Arthur L. Day medal 1977), Mineral. Soc. Am.; hon. mem. Geol. Soc. London, Geol. Sco. France. Subspecialty: Petrology. Current work: Metamorphic geology, metamorphic minerals, igneous petrology. Home: 14 Stonehenge Dr Albany NY 12203 Office: Dept Geol Scis SUNY Albany NY 12222

MIZICKO, JOHN RICHARD, plant pathologist; b. Denver, Oct. 4, 1949; s. John P. and Johanna R. (Carson) M.; m. Marlena M., Apr. 9, 1969; children: John, Christian. B.S. with high distinction, Colo. State U., 1971, M.S., 1973. Research specialist plant pathology U. Minn., 1973-75, asst. extension specialist, 1975-78; plant pathologist Moran Seeds, Inc., Davis, Calif., 1978-82, Agrigenetics Corp., Hollister, Calif., 1982—. Coach Little League baseball; referee, coach Am. Youth Soccer Orgn. Mem. Am. Phytopath. Soc. Episcopalian. Subspecialties: Plant pathology; Plant cell and tissue culture. Current work: Vegetable diseases and resistance, vegetable breeding, disease control, disease resistance, tissue culture. Home: 1621 Sunset Dr Hollister CA 95023 Office: 9870 Fairview Rd Hollister CA 95023

MIZUSHIMA, MASATAKA, physics educator; b. Tokyo, Mar. 30, 1923; U.S., 1952, naturalized, 1962; s. Seizo and Shizuko (Ishikawa) M.; m. Yoneko Tsuboi, Sept. 28, 1955; children: Nanako V., Naomi A., Nori G., Nobuko E., Nieret J. B.Sc., U. Tokyo, 1946, D.Sc., 1951. Research asst. U. Tokyo, 1947-52; research assoc. Duke U., Durham, N.C., 1952-55; asst. prof. U. Colo., Boulder, 1955-58, assoc. prof., 1958-60, prof. physics, 1960—; vis. prof. U. Tokyo, 1962-63, U. Rennes, France, 1964, Romanian Acad., Bucharest, 1969-70, U. Nijmegen, Netherlands, 1972, U. Tokyo, 1972, U. London, 1982, U. Electro-Communication, Japan, 1981, Inst. Molecular Sci., 1982. Author: Quantum Mechanics of Atomic Spectra and Atomic Structure, 1970, Theoretical Physics, 1972, Theory of Rotating Diatomic Molecules, 1975. Fellow U.S. Nat. Acad., 1964, 81, Japanese Assn. Promotion of Sci., 1972, Ministry Edn. Japan, 1981. Mem. Am. Phys. Soc. Subspecialties: Atomic and molecular physics; Infrared spectroscopy. Current work: Theoretical and experimental (microwave through infrared) molecular spectroscopy. Home: 523 Theresa Dr Boulder CO 80303 Office: Dept Physics U Colo Boulder CO 80309

MO, SUCHOON, psychology educator; b. Nagoya, Aichi, Japan, Apr. 19, 1932; came to U.S., 1955, naturalized, 1963; s. Chihyun and Chiha (Shin) M.; m. Judith Carol Oslick, Dec. 27, 1969; children: Sage, Daisy, Clifton; m.; children by previous marriage: Blaise, Bernard. B.S., Ida. State U., 1959; Ph.D., U. Pa., 1968. Asst. prof. U. Detroit, 1967-73; asst. prof. U. So. Colo., Pueblo, Colo., 1973-76, assoc. prof., 1976-80, prof. psychology, 1980—; cons. VA Med. Ctr., Ft. Lyon, Colo., 1978-82. Contbr. articles to profl. jours. Bd. dirs. Pueblo W. Met. Dist., 1982. Served to 1st lt. Korean Army, 1953-55. Mem. Am. Psychol. Assn., Internat. Assn. Study of Time (corr.), N.Y. Acad. Sci., Psychonomic Soc. Republican. Current work: Psychopathology of time; brain hemisphere reversal of time info. Office: Univ So Colo Dept Psychology Pueblo CO 81001 Home: 1158 S Yerba Santa Dr Pueblo West CO 81007

MOCHIZUKI, LESLIE YASUKO, systems engineer; b. Los Angeles, May 16, 1959; d. Bruce Koji and Chieko (Asano) M. A.B. in Math, U. So. Calif., 1979. Mem. tech. staff Rockwell Internat., Anaheim, Calif., 1979—; mem. youth motivation task force, 1981-82, conservation com., 1982-83. Active Wintersburg Presbyn. Ch., Garden Grove, Calif. Mem. Am. Math. Soc., Soc. Indsl. and Applied Math., Soc. Women Engr. (assoc.), Alpha Epsilon Delta (life). Republican. Subspecialties: Mathematical software; Software engineering. Current work: Currently designing and developing operational systems software for the Minuteman missile, designing requirements for code processing systems. Home: 3621 S Parton St Santa Ana CA 92707 Office: Rockwell Internat Anaheim CA 92803

MOCK, DOUGLAS WAYNE, sociobiology educator; b. N.Y.C., July 4, 1947; s. Vern F. and Esther (Gillil) M.; m. Karilyn Sue Cummins, Dec. 21, 1969 (div. Apr. 1982); m. Patricia L. Schwagmeyer, Jan. 8, 1983. B.S., Cornell U., 1969; M.S., U. Minn., 1972, Ph.D., 1976. Postdoctoral fellow Smithsonian Instn., Washington, 1976-77; vis. scientist FitzPatrick Inst. African Ornithology, Cape Town, South Africa, 1977; asst. prof. zoology U. Okla., 1978—. Mem. AAAS, Am. Ornithologists Union, Animal Behavior Soc., Ecol. Soc. Am. Subspecialties: Sociobiology; Evolutionary biology. Current work: Sibling competition, parent-offspring conflict, evolution of coloniality. Office: Dept Zoology U Okla Norman OK 73019

MODAK, MUKUND JANARDAN, biochemist, biomed. research scientist; b. Satara, India, Dec. 12, 1942; came to U.S., 1970, naturalized, 1974; s. Janardan B. and Usha J. M.; m. Shanta M. Modak, Oct. 20, 1969; children: Rajiv, Rohit. B.Sc. (hons.), Poona (India) U., 1963; M. Sc., Bombay (India) U., 1965, Ph.D., 1969. NAITO Found. Overseas fellow U. Nagoya (Japan), 1970; research assoc. Columbia U., N.Y.C., 1971-72, Sloan Kettering Inst. Cancer Research, 1972-75, assoc., 1976-79, assoc. mem., 1979—; assoc. prof. biochemistry, molecular biology and genetics Cornell U. Contbr. articles to sci. jours. Haffkine Inst. of Bombay Diamond Jubilee fellow, 1966-69; Internat. Union Against Cancer vis. fellow, 1975; recipient Research Career Devel. award NIH, 1979. Mem. Am. Soc. Biol. Chemists, Am. Soc. Microbiology, Am. Soc. Virology, Am. Assn. Cancer Research. Subspecialties: Gene actions; Molecular biology. Current work: Mechanisms of DNA synthesis, DNA enzymes, T cell differentiation, tumor virus enzymes. Office: 1275 York Ave New York NY 10021

MODEL, PETER, scientist, researcher, consultant; b. Germany, May 17, 1933; s. Leo and Jane (Ermel) M.; m. Marjorie Russel, June 21, 1981; children: Paul, Sascha. B.A., Stanford U., 1954; Ph.D., Columbia U., 1965. Assoc. prof. Rockefeller U., N.Y.C.; cons. dir. Biotechnica Internat. Inst. for Sci. Info. Contbr. articles to sci. publs. Served to 1st lt. U.S. Army, 1954-56. Mem. Am. Soc. Microbiology. Subspecialties: Molecular biology; Biochemistry (biology). Current work: Protein synthesis, phage genetics, membranes. Office: Rockefeller Univ New York NY 10021

MODRICH, PAUL L., biochemist educator; b. N. Mex. Grad., MIT; Ph.D., Stanford U., 1973. Formerly postdoctoral fellow Harvard U.; former asst. prof. chemistry U. Calif.-Berkeley; now assoc. prof. biochemistry Duke U., Durham, N.C. Recipient Career Devel. award NIH; Dreyfus Tchr.-Scholar award; Pfizer award in enzyme chemistry, 1983. Subspecialty: Biochemistry (biology). Office: Dept Biochemistry Duke U Med Ctr Durham NC 27710

MOE, GORDON KENNETH, physiologist, educator; b. Fairchild, Wis., May 30, 1915; s. Sylvester and Ellen Mae (Hanson) M.; m. Janet Woodruff Foster, Aug. 6, 1938; children—Christopher, Melanie, Jonathan, Bruce, Sally, Eric. Student, Virginia (Minn.) Jr. Coll., 1932-34; B.S., U. Minn., 1937, M.S., 1939, Ph.D., 1940; M.D., Harvard, 1943. Instr. Physiology U. Minn., 1939-40; instr. pharmacology Harvard Med. Sch., 1941-44; asst. prof. pharmacology U. Mich. Med. Sch., 1944-46, asso. prof., 1946-50; prof. physiology, chmn. dept. State U. N.Y. Coll. Medicine, Syracuse, 1950-60; dir. Masonic Med. Research Lab., Utica, N.Y., 1960—; Chmn. physiol. test com. Nat. Bd. Med. Examiners, 1956-57; mem. WHO (vis. scientists), Israel, Iran, 1951; cons. Walter Reed Army Med. Center.; Mem. Nat. Heart and Lung Adv. Council, 1970-74. Recipient travel award Internat. Physiol. Congress Oxford, 1947; Outstanding Achievement award U. Minn., 1958; Merit award Am. Heart Assn., 1968; Am. Physiol. Soc. fellow Western Res. U., 1940-41; USPHS fellow Instituto Nacional de Cardiologia, Mexico City, 1948. Fellow AAAS (chmn. sect. 1958), Am. Coll. Cardiology (hon.); mem. Am. Physiol. Soc., N.Y. Soc. Med. Research, Am. Heart Assn. (chmn. basic sci. council 1962-63, chmn. research com. 1959-60), Mexican Acad. Medicine (hon.), Mexican Soc. Cardiology (hon.), Sigma Xi, Alpha Omega Alpha. Club: Mason (33 deg.). Subspecialties: Physiology (medicine); Psychiatry. Current work: Cardiac electrophysiology. Office: Masonic Med Research Lab Utica NY 13503

MOE, JAMES BURTON, army officer, veterinary pathologist; b. Hayfield, Minn., Oct. 4, 1940; s. James Herald and Clara Clemnetine M.; m. Janice Naomi Nackerud, Nov. 27, 1959; children: Carolyn, Alyson, Jennifer, Bryce. B.S., U. Minn., 1962, D.V.M., 1964; Ph.D., U. Calif.-Davis, 1978. Cert. Am. Coll. Vet. Pathologists, 1973. Gen. practice vet. medicine Dodge Center, Minn., 1964-66; practice vet. medicine specializing in preventive vet. medicine, Ft. Bliss, Tex., 1966-68; commd. 2d lt. U.S. Army, 1966, advanced through grade to lt. col., 1979; resident in pathology, Ft. Detrick, Md., 1969-73, microbiol.-path. researcher, 1973-75; 78-80; with research mgmt. div. Walter Reed Army Inst. Research, Washington, 1980—, dir. div. pathology; cons. pathology, cons. to Surgeon Gen. of Army; chmn. instl. Biosafety Com. Contbr. articles on pathology, microbiology and infectious diseases. Decorated Bronze Star.; Calif. Lung Assn. grantee, 1976-77. Mem. Am. Coll. Vet. Pathologists, AVMA, Internat. Acad. Pathology, Am. Assn. Pathologists. Lutheran. Subspecialties: Pathology (veterinary medicine); Microbiology (medicine). Current work: Experimental pathology, director research division engaged in basic and applied aspects of infectious diseases, toxicology and drug and vaccine development. Office: Walter Reed Army Inst Research Washington DC 20012

MOEHRING, JOAN MARQUART, microbiology educator; b. Orchard Park, N.Y., Sept. 23, 1935; d. Carl B. and Marjorie H. (Pearson) Marquart; m. Thomas J. Moehring, June 6, 1964. B.S., Syracuse U., 1961; M.S., Rutgers U., 1963, Ph.D., 1965. Postdoctoral fellow Stanford U., 1965-68; research assoc. U. Vt., 1968-73, research prof. dept. med. microbiology, 1973—; cons. in field. Contbr. articles to profl. jours. NIH grantee, 1973—. Mem. Am. Soc. Microbiology, Tissue Culture Assn., Phi Beta Kappa, Sigma Xi. Subspecialties: Cell and tissue culture; Microbiology. Current work: Study of action of microbial toxins in mammalian cells, biological research. Office: U Vt Dept Med Microbiology B210 Given Bldg Burlington VT 05405

MOELLER, DADE WILLIAM, environmental engineer, educator; b. Grant, Fla., Feb. 27, 1927; s. Robert A. and Victoria (Bolton) M.; m. Betty Jean Radford, Oct. 7, 1949; children: Garland Radford, Mark Bolton, William Kehne, Matthew Palmer, Elisabeth Anne. B.C.E., Ga. Inst. Tech., 1947, M.S. in Civil Engring, 1948; Ph.D., N.C. State U., 1957. Commd. jr. asst. san. engr. USPHS, 1948, advanced through grades to san. engr. dir., 1961; research engr. Los Alamos Sci. Lab., 1949-52; staff asst. Radiol. Health Program, Washington, 1952-54; research asso. Oak Ridge Nat. Lab., 1956-57; chief radiol. health tng. Taft San. Engring. Center, Cin., 1957-61; officer charge Northeastern Radiol. Health Lab., Winchester, Mass., 1961-66; assoc. dir. Kresge Center Environ. Health, Harvard Sch. Pub. Health, 1966-83, prof. engring. in environmental health, head dept. environmental health scis., 1968-83, dir. Office of Continuing Edn., 1982—; cons. radiol. health Profl. Exam. Service, Am. Pub. Health Assn., 1960-62, WHO, 1965—. Contbr. articles to profl. jours. Chmn. Am. Bd. Health Physics, 1967-70; mem. Nat. Council Radiation Protection and Measurements, 1968—; mem. com. 4 Internat. Commn. on Radiol. Protection, 1978—; chmn. nat. air pollution manpower devel. adv. com. U.S. Environ. Protection Agy., 1972-75; mem. adv. com. reactor safeguards U.S. Nuclear Regulatory Commn., 1973—, vice chmn., 1975, chmn., 1976. Fellow Am. Pub. Health Assn., Am. Nuclear Soc.; mem. Am. Acad. Environ. Engrs. (bd. dirs. 1968-73), Nat. Acad. Engring., Health Physics Soc. (pres. 1971-72), AAAS. Subspecialties: Environmental engineering; Gas cleaning systems. Current work: Natural background radiation; control of airborne radon decay products inside buildings; planning for nuclear emergencies; nuclear air cleaning; safety of nuclear power plants. Home: 27 Wildwood Dr Bedford MA 01730 Office: 677 Huntington Ave Boston MA 02115

MOERTEL, CHARLES GEORGE, physician; b. Milw., Oct. 17, 1927; s. Charles Henry and Alma Helen (Soffel) M.; m. Virginia Claire Sheridan, Mar. 22, 1952; children: Charles Stephen, Christopher Loren, Heather Lynn, David Matthew. B.S., U. Ill., 1946-51, M.D., 1953; M.S., U. Minn., 1958. Intern Los Angeles County Hosp., 1953-54; resident in internal medicine Mayo Found., Rochester, Minn., 1954-57; cons. Mayo Clinic, Rochester, 1957—, chmn. dept. oncology, 1975—; dir. Mayo Comprehensive Cancer Center, 1975—; prof. medicine Mayo Med. Sch., 1972-76, prof. oncology, 1976—; mem. oncologic drugs adv. com. FDA; mem. cancer adv. com. AMA; mem. bd. sci. counselors, div. resources, centers and community activities Nat. Cancer Inst. Author: Multiple Primary Malignant Neoplasms, 1966, Advanced gastrointestinal Cancer, Clinical Management and Chemotherapy, 1969; editorial bd.: Cancer, 1974—, Cancer Medicine, 1978—, Current Problems in Cancer, 1978—, Jour. Soviet Oncology, 1979—, Cancer Research, 1979—, Cancer Treatment Rep, 1980-82, Internat. Jour. Radiation Oncology Biol. Physics, 1981. Served with U.S. Army, 1946-47. Walter Hubert lectr. Brit. Assn. Cancer Research, 1976; Ejnar Perman Meml. lectr. Swedish Surg. Assn., 1978. Mem. Am. Soc. Clin. Oncology (pres. 1979-80), Soc. Clin. Trials (dir.), Am. Assn. Cancer Research, Gastrointestinal Tumor Study Group (co-chmn.), A.C.P., Soc. Surg. Oncology, Am. Gastroenterologic Assn., North Central Cancer Treatment Group (co-chmn.), Sigma Xi. Subspecialties: Cancer research (medicine); Oncology. Research in treatment of gastrointestinal cancer, clin. pharmacology. Home: 1009 Skyline Ln SW Rochester MN 55902 Office: 200 SW 1st St Rochester MN 55905

MOFFAT, ANTHONY FREDERICK JOHN, astronomy researcher and educator; b. Toronto, Ont., Can., Jan. 30, 1943; s. Bryce F. and Margaret E. (Boorman) M.; m. Ruth Ann Huntley, Sept. 10, 1966; children: Bryce A., Lesley A. B.Sc., U. Toronto, 1965, M.Sc., 1966; Dr. rer. nat., Ruhr (W.Ger.) U., 1970, Dr. Habil., 1976. Sci. asst. U. Bonn, W.Ger., 1966-69; sci. asst. Ruhr U., 1970-76; assoc. prof. astronomy U. Montreal, 1977-80, prof., 1981—. Contbr. articles to profl. jours. Recipient Silver medal in physics and math. U. Toronto, 1965; Gold medal Royal Astron. Soc. Can., 1965; Imperial Oil Can. Ltd. fellow, 1966-69; NRC of Can. grantee, 1977—; Alexander von Humboldt research fellow, 1982. Mem. Internat. Astron. Union, Am. Astron. Soc., Can. Astron. Soc., Royal Astron. Soc. Can., Astronomische Gesellschaft. Subspecialty: Optical astronomy. Current work: The nature of massive stars, young star clusters; structure and dynamics of milky way system. Office: Dept Physic U Montreal CP 6128 Montreal PQ Canada H3C 3J7

MOFFAT, JOHN WILLIAM, physics educator; b. Copenhagen, Denmark, May 24, 1932; s. George William and Esther (Winther) M. Ph.D., Trinity Coll., Cambridge (Eng.) U., 1958. Sr. research fellow Imperial Coll., London, Eng., 1957-58; scientist Research Inst. Advanced Studies, Balt., 1958-60, prin. scientist, 1961-64; scientist CERN, Geneva, Switzerland, 1960-61; asso. prof. dept. physics U. Toronto, Ont., Can., 1964-67, prof., 1967—. Contbr. articles to profl. jours. Dept. Sci. and Indsl. Research fellow, 1958-60; NRC Can. grantee, 1965. Fellow Cambridge Philos. Soc. (Eng.). Subspecialties: Relativity and gravitation; Particle physics. Current work: Theory of gravitation, unified gauge theories, early universe cosmology, solar physics and astrophysics. Office: Dept Physics Univ Toronto Toronto ON Canada

MOFFATT, JOHN GILBERT, chemical institute research executive; b. Victoria, B.C., Can., Sept. 19, 1930; came to U.S., 1960; s. William James and Esther Christina (Shankey) M.; m. June Collinson, Aug. 15, 1953; children: Susan, Vicki, Janet, Karen. B.A., U. B.C., 1952, M.S., 1953, Ph.D., 1956. Tech. officer Def. Research Bd., Ottawa, Ont., Can., 1952-53; research assoc. B.C. Research Council, Vancouver, 1956-60; sect. leader Calif. Corp. Biochem. Research, Los Angeles, 1960-61; with Syntex Research, Palo Alto, Calif., 1961—; dir. Inst. Bio-Organic Chemistry, v.p., 1967—. Contbr. tech. papers to publs. Mem. Am. Chem. Soc. (exec. com. organic div. 1973-75, exec. com. carbohydrate div. 1975-78), Chem. Soc. (London), Am. Soc. Biol. Chemists. Subspecialty: Organic chemistry. Current work: Synthetic organic chemistry related to nucleosides, nucleotides, carbohydrates, peptides and glycopeptides; medicinal chemistry. Patentee. Home: 495 S Clark Ave Los Altos CA 94022 Office: Inst Bio-Organic Chemistry Syntex Research 3401 Hillview Palo Alto CA 94304

MOFFETT, THOMAS JOSEPH, educator, astronomer; b. New Orleans, Nov. 15, 1944; s. R.L. and Marguerite J. (Frey) M.; m. Carol G., Aug. 24, 1968; 1 son, Thomas Joseph II. B.S., La. State U., 1966, M.S., 1968; Ph.D., U. Tex., 1973. Research asso. dept. astronomy U. Tex., Austin, 1973-75; asst. prof. physics Purdue U., West Lafayette, Ind., 1975-78, asso. prof., 1978—. Mem. Internat. Astron. Union, Am. Astron. Soc., Royal Astron. Soc., Astron. Soc. of Pacific. Subspecialty: Optical astronomy. Current work: Observational studies of variable stars. Home: 180 Bristol Ct West Lafayette IN 47906 Office: Purdue University Dept Physics West Lafayette IN 47907

MOFFIE, ROBERT WAYNE, psychology educator; b. Los Angeles, June 22, 1950; s. Marvin Louis Moffie and Dorothy Ruth (Morris) Miller. B.A., U. Calif.-Riverside, 1972; M.A., U. Notre Dame, 1976, Ph.D., 1978. Adj. faculty Ind. U.-South Bend, 1978-79; asst. prof. psychology Oglethorpe U., Atlanta, 1979—; cons. Michael Wagner Assos., Calgary, Alta., Can., 1980—, TLY Assos., Atlanta, 1982—, Monument Pictures, 1975—. Contbr. articles to jours. Vol. cons. United Way of St. Joseph County, South Bend, Ind., 1974-79, United Way of Scioto County, Portsmoutn, Ohio, 1980-81. Mem. Am. Psychol. Assn., Southeastern Psychol. Assn., Mental Health Assn. Met. Atlanta, Ind. Psychol. Assn., Midwestern Psychol. Assn. Democrat. Episcopalian. Subspecialties: Clinical pathology; Neuropsychology. Current work: Biological bases of behavior pathology; diagnostic criteria in behavior pathology classification; research on foundation of sexual orientations. Home: 5043 Happy Hollow Rd Atlanta GA 30360 Office: Oglethorpe Univ 4484 Peachtree Rd NE Atlanta GA 30319

MOFFROID, MARY THOMPSON, physical therapist, educator; b. Detroit, May 7, 1940; d. Thomas E. and Adelaide (Wood) Thompson; m. Pierre Edmond Moffroid, May 16, 1964; children: Eric, Gregory, Daniel. B.S., U. Mich., Ann Arbor, 1962; M.S., N.Y.U., 1968, Ph.D., 1980. Staff phys. therapist Inst. Rehab. Medicine, N.Y.U. Med. Ctr., N.Y.C., 1964-70; instr. Downstate Med. Ctr., SUNY, Bklyn., 1970-72; asst. prof. U. Vt., Burlington, 1972-76, assoc. prof. phys. therapy, 1976—. Playwright: Fraternal Twins, 1973. Mem. sch. bd. Warren Elem. Sch., 1980-83. NIH life sci. teaching grantee, 1972; U. Vt. instructional resource grantee, 1980. Mem. Am. Phys. Therapy Assn., Soc. Behavioral Kinesiology, Soc. Neurosci. Democrat. Subspecialties: Physical medicine and rehabilitation. Current work: Effects of training on muscle performance, particularly speed of contraction and speed of limb movement.

MOGENSEN, HANS LLOYD, botany educator; b. Price, Utah, Dec. 16, 1938; s. Hans and Phyllis (Lewis) M.; m. Shirley Ann Helen Bowman, Nov. 28, 1958; children: Deserie, Shellie. B.S., Utah State U., 1961; M.S. in Plant Morphology, Iowa State U., 1965, Ph.D., 1965. Prof. biology No. Ariz. U., Flagstaff, 1965—. Contbr. articles to profl. publs. NSF grantee, 1980-85. Mem. Bo. Soc. Am., Internat. Soc. Plant Morphologist. Subspecialty: Developmental biology. Current work: Research in plant embryology. Home: 1525 W University Heights Dr

N Flagstaff AZ 86001 Office: No Ariz Univ Biology Dept Box 5640 Flagstaff AZ 86011

MOHAN, C., research computer scientist; b. Mayuram, Tamilnadu, India, June 3, 1955; came to U.S., 1977; s. K. and Radha Chandrasekaran. B.Tech. in Chem. Engring, Indian Inst. Tech., Madras, 1977; Ph.D. in Computer Sci, U. Tex.-Austin, 1981. Research and teaching asst. U. Tex.-Austin, 1977-81; research computer scientist IBM Research Lab., San Jose, Calif., 1981—; vis. scientist Inst. National de Recherche en Informatique et en Automatique, Le Chesnay, France, 1979, Hahn-Meitner Inst. for Kernforschung, Berlin, West Germany, 1980. Contbr. articles to profl. jours. Mem. com. India Students Assn. U. Tex.-Austin, 1978-79. Mem. Assn. Computing Machinery, IEEE Computer Soc., Bay Area Tamil Assn., Phi Kappa Phi. Subspecialties: Database systems; Distributed systems and networks. Current work: Deadlock management; concurrency control; commit coordination; distributed programming; failure recovery; network protocols; database design; distributed software design and implementation; reliability. Home: 309 Tradewinds Dr Apt 10 San Jose CA 95123 Office: IBM Research Lab 5600 Cottle Rd San Jose CA 95193

MOHANTY, MIRODE CHANDRA, electrical engineer; b. Soro, India, Jan. 18, 1937; s. Chakra Dhar and Sabitri (Das) M.; m. Sneha Das, May 21, 1964; children: Bapi, Lisa. M.S., Utkal U., Bhubaneswar, India, 1960, U. So. Calif., 1968, U. So. Calif., 1971; Ph.D., 1972. Prof. Utkal U., 1960-66; research scholar U. So. Calif., 1966-73; prof. SUNY-Buffalo, 1973-75; research engr. Systems Control Co., 1975-76; sr. engring. specialist Ford Aerospace, Newport Beach, Calif., 1976-81; mem. tech. staff Aerospace Corp., Los Angeles, 1981—; mem. tech. staff Rockwell Internat., Los Angeles, 1976-79. Contbr. articles to profl. jours. Recipient Excellence in Creativity award Rockwell Internat., 1978. Mem. IEEE (sr. mem. chmn. info. theory div. 1973-75, chmn. nat. aerospace and electronics div. 1979), Internat. Telemetry Conf. (chmn. 1980). Subspecialties: Electrical engineering; Information systems (information science). Current work: Signal processing, as related to radar, sonar, electrooptics, speech, music and computer vision, communications systems. Home: 18362 Springtime Huntington Beach CA 92646 Office: Aerospace Corp PO Box 92957 Los Angeles CA 90009

MOHANTY, SASHI B., virologist, educator; b. India, Sept. 4, 1932; came to U.S., 1960; s. Madhu s. and Narayani (Parida) M.; m. Pranoti Samal, July 3, 1957; children: Nibedita, Bibhu, Nihar, Puspa. B.V.Sc., Bihar U., India, 1956; M.S., U. Md., 1961, Ph.D., 1963. Vet. surgeon, Orissa, India, 1956-60; asst. prof. vet. sci. U. Md., College Park, 1963-69, assoc. prof., 1969-74, prof., 1974—; vis. prof. U. Munich, W.Ger., 1972-73; cons. Flow Lab., Rockville, Md., Indsl. Biologic Lab., Rockville. Author: Veterinary Virology, 1981, Electron Microscopy for Biologist, 1982; contbr. chpts. to books, articles to profl.jours. NIH grantee, 1963-65; U.S. Dept. Agr. grantee, 1977—. Mem. AVMA, Am. Soc. Microbiology, Electron Microscopy Soc. Am., Soc. for Exptl. Biology and Medicine, Sigma Xi. Subspecialties: Virology (veterinary medicine); Internal medicine (veterinary medicine). Current work: Pathogenesis and prevention of bovine respiratory viruses; effect of interferon and antiviral drugs on bovine repsitory viruses; electron microscopy of viruses and cells. Home: 4306 Kenny St Beltsville MD 20705 Office: U Md College Park MD 20742

MOHILNER, DAVID MORRIS, chemist, educator; b. Wichita, Kans., Jan. 3, 1930; s. Harry and Helen Mae (Ringer) M.; m. Patricia Reynolds, June 6, 1958 (dec. 1979). B.S., U. Kans.-Lawrence, 1955, Ph.D., 1961. Postdoctoral research assoc. La. State U., 1961-62; asst. prof. chemistry U. Pitts., 1962-64; postdoctoral research assoc. U. Tex., Austin, 1964-65; asst. prof. chemistry Colo. State U., 1965-67, asso. prof., 1967-71, prof., 1971—. Contbr. articles on phys. chemistry to profl. jours., 1957—. Served to 1st lt. U.S. Army, 1952-54. Mem. Am. Chem. Soc., Electrochem. Soc., Internat. Soc. Electrochemistry. Subspecialties: Physical chemistry; Thermodynamics. Current work: Equilibrium properties of electrical double layer computer-controlled electrocapillary measurements.

MOHIUDDIN, SYED MAQDOOM, cardiologist, educator; b. Hyderabad, India, Nov. 14, 1934; came to U.S., 1961, naturalized, 1976; s. Syed Nizamuddin and Amat-Ul-Butool Mahmoodi; m. Ayesha Sultana Mahmoodi, July 16, 1961; children: Sameena J., Syed R., Kulsoom S. M.B., B.S., Osmania U., 1960; M.S., Creighton U., Omaha, 1967; D.Sc., Laval U., Que., Can., 1970. Diplomate: Am. Bd. Internal Medcine (cardiovascular disease). Intern Altoona (Pa.) Gen. Hosp., 1961-62; resident in cardiology Creighton Meml. Hosp., also St. Joseph Hosp., Omaha, 1963-65, mem. staff, 1965—; prof. adjoint Laval U. Med. Sch., 1970; practice medicine specializing in cardiology, Omaha, 1970—; prof. Creighton U. Med. Sch., 1977—, assoc. dir. div. cardiology, 1983—; cons. Omaha VA Hosp. Research fellow Med. Research Council Can., 1968; grantee Med. Research Council Can., 1970, NIH, 1973. Fellow ACP, Am. Coll. Cardiology; mem. Am. Heart Assn. (fellow council clin. cardiology), Am. Fedn. Clin. Research, Nebr. Heart Assn. (chmn. research com. 1974-76, dir. 1973—), Gt. Plains Heart Com. (Nebr. rep. 1976—, pres. 1977—), Nebr. Cardiovascular Soc. (pres. 1980-81). Democrat. Islam. Subspecialties: Cardiology; Internal medicine. Current work: Cardiovascular drugs, cardiac pacemaker, echo-cardiography, cardiovascular education. Home: 12531 Shamrock Rd Omaha NE 68154 Office: 601 30th St Omaha NE 68131

MOHLA, SURESH, endocrinologist; b. Calcutta, India, May 5, 1943; came to U.S., 1970, naturalized, 1983; s. Rai Sahib Karam Ch and Shakuntla Devi (Kumaria) M.; m. Chitra Mohla Harisingh, Aug. 31, 1969; 1 dau. Anjali. B.Sc. with honors, U. Delhi, 1963, M.Sc., 1965, Ph.D., 1968. Postdoctoral fellow U. Delhi, India, 1968-70; research assoc. U. Chgo., 1970-75, 75-76; asst. prof. oncology Howard U., Washington, 1977-82, assoc. prof., 1982—, dir. hormone receptor support lab., 1976—, program dir. endocrinology, 1979—, chmn. research com., 1983—; Mem. adv. com. on anti-smoking Am. Cancer Soc., Washington, 1980-82; mem. speaker's bur. Cancer Coordinating Council and Cancer Info. Service for Met. Washington, 1979—. Recipient Travel award NFS, 1978; NIH research grantee, 1980-83. Mem. Am. Assn. Cancer Research, Am. Physiol. Soc., Endocrine Soc., Am. Soc. Cell Biology. Hindu. Subspecialties: Cancer research (medicine); Endocrinology. Current work: Hormonal control of growth in normal and neoplastic tissues and mechanism of steroid hormone action, clinical significant of steroid hormone receptors in human breast and prostrate cancers, androgen resulation of nitrosamine metabolism in the kidney. Home: 6440 Franconia Ct Springfield VA 22150 Office: Howard U Cancer Ctr 2041 Georgia Ave NW Washington DC 20060

MOHLER, ORREN CUTHBERT, educator, astronomer; b. Indpls., July 28, 1908; s. Charles Mikesell and Mary Ann (Culp) M.; m. Helen Jean Beal, June 10, 1935; children—Alice Beal (Mrs. William George DeLana), Jane Radcliffe (Mrs. Jeffery Wessels Barry). A.B., Mich. State Normal Coll., 1929; M.A. (State Coll. fellow 1929), U. Mich., 1930; Ph.D. (Lawton fellow 1930-33), U. Mich., 1933; D.Sc. (hon.), Eastern Mich. U., 1956. Instr. astronomy Swarthmore Coll., 1933-40; astromomer Cook Obs., Wynnewood, Pa., 1933-40; observer McMath-Hulbert Obs., U. Mich., 1933; dir. Obs., 1961—, mem. faculty univ., 1940—, prof. astronomy, 1955-78, emeritus prof. astronomy, 1978—; chmn. dept. astronomy, dir. observatories, 1962-71; adj. prof. physics Oakland U., 1980—; dir. Assn. Univs. Research Astronomy, Tucson, 1961-74, chmn. scientific com., 1965-73; Vice pres. U.S. Nat. Com. Internat. Astron. Union, 1962-65; Bd. govs. Cranbrook Inst. Sci., 1960—. Author: Photometric Atlas of the Near Infrared Solar Spectrum, 1950, Table of Solar Spectrum Wave Lengths, 1955; also articles. Recipient USN award, 1946; Fulbright research scholar Inst. d'Astrophysique, Liege, 1960-61. Mem. Am., Astron. Soc., Royal Astron. Soc. Can., Phi Beta Kappa, Sigma Xi. Clubs: Univ., Research (pres. 1966-67), Sci. Research (pres. 1970-71), Ann Arbor Golf and Outing; Torch (Oakland County). Subspecialties: Optical astronomy; Solar physics. Current work: Encyclopedic table of wave lengths in solar spectrum; sunspot cycle; design of large solar telescopes, design and development of detectors of raidiation. Home: 405 Awixa Rd Ann Arbor MI 48104

MOHLER, RONALD RUTT, electrical engineering educator; b. Ephrata, Pa., Apr. 11, 1931; s. David Weal and Elizabeth (Rutt) M.; m. Nancy Alice Strickler, May 6, 1950; children: Curtis, Pamela, Susan, Anita, John, Andrew, Jennifer, Lisa. B.S.M.E., Pa. State U., 1956; M.S.E.E., U. So. Calif., 1958; Ph.D., U. Mich., 1965. Trainee Textile Machine Works, Reading, Pa., 1949-56; staff Hughes Aircraft Co., Culver City, Calif., 1956-58, Los Alamos Nat. Lab., 1958-65; assoc. prof. U. N.Mex., Albuquerque, 1965-69; prof. dept. elec. engring. U. Okla., Norman, 1969-72; prof. aerospace, mech. and nuclear engring., chmn. info. and computing sci. Oreg. State U., Corvallis, 1972—, dept. head, 1972-78; pres., chmn. bd. Pace Tech., Inc., Corvallis, 1982—; cons. Bonneville Power Adminstrn., 1975—, Optimization Software, 1972—, Sandia Nat. Lab., Aerojet Gen., 1965-72, others. Author: Bilinear Control Processes, 1973, Optimal Control of Nuclear Reactors, 1970; editor: Theory and Application of Variable Structure Systems, 1972; editorial staff: Annals of Nuclear Energy, 1977—. NATO sr. scientist, 1978-79; Nat. Acad. Scis. vis. scientist, 1981; NSF grantee, 1966—; Office Naval Research grantee, 1981—. Fellow IEEE (chpt. pres. 1973-74); Mem. Am. Soc. Engring. Edn., Sigma Xi, Tau Beta Pi. Democrat. Subspecialties: Information systems, storage, and retrieval (computer science); Electrical engineering. Current work: Mathematical modeling, computer simulation and analysis in signal processing and control; immune system analysis and control; nonlinear system analysis; automation. Home: 2050 NW Dogwood Dr Corvallis OR 97330 Office: Oreg State Univ Dept Elec and Computer Engring Corvallis OR 97331

MOHNEY, LEONE LAURA, microbiologist; b. Raton, N.Mex., May 29, 1935; d. Curtis Gilliam and Ruth Clara (Jillson) M. B.S., U. Ariz., 1957; M.S., U. Calif., Berkeley, 1961. Research asst. George W. Hooper Found., U. Calif., San Francisco, 1958-60; research assoc. U. Calif. Sch. Pub. Health, Berkeley, 1961-77; research asst. dept. microbiology U. Ariz., Tucson, 1978-81, research asst. Environ. Research Lab., 1983—; research asst. U. Ariz. Health Scis. Center, Tucson, 1981-83. Active LWV, Berkeley, 1963-77. Mem. Am. Soc. for Microbiology, Am. Inst. Biol. Scis., Assn. Women in Sci. Republican. Presbyterian. Subspecialties: Microbiology; Genetics and genetic engineering (biology). Current work: Genetic studies on Bacillus species; use of plasmids as vectors of genes that produce medically insecticidal products, study of microbial diseases of shrimp. Office: Dept Aquaculture Pathology Environmental Research Lab Tucson Internat Airport Tucson AZ 85706

MOHS, FREDERIC EDWARD, educator, surgeon; b. Burlington, Wis., Mar. 1, 1910; s. Frederic Carl and Grace Edith (Tilton) M.; m. Mary Ellen Reynolds, June 18, 1934; children—Frederic Edward, Thomas James, Jane Ann. M.D., U. Wis., 1934. Intern Multnomah County Hosp., Portland, Ore., 1934-35; Bowman Cancer Research Fellow U. Wis. Med. Sch., 1935-38, asso. in cancer research, instr. in surgery, 1939-42, asst. prof. chemosurgery, 1942-48, asso. prof. chemosurgery, 1948-69, clin. prof. surgery, 1969—. Author: Chemosurgery: Microscopically Controlled Surgery for Skin Cancer; Contbr. articles, papers on treatment of cancer by means of chemosurgery which is a microscopically controlled method he developed. Recipient Lila Gruber award for cancer research Am. Acad. Dermatology; Internat. Facial Plastic Surgery award 3d Internat. Symposium on Plastic and Reconstructive Surgery of Head and Neck. Fellow AMA; mem. Am. Assn. Cancer Research, AAAS, Dane County Med. Soc. (pres. 1959- 60), Am. Coll. Chemosurgery (founder, 1st pres.). Subspecialty: Surgery. Home: 3616 Lake Mendota Dr Madison WI 53705

MOIR, RALPH WAYNE, physicist, nuclear engineer; b. Bellingham, Wash., Jan. 21, 1940; s. Francis Leroy and Florence Augusta (Hershey) M.; m. Elizabeth Grace Branstead, June 6, 1963; children: Sara L., Steven H., Christina E. B.S., U. Calif.-Berkeley, 1962; Sc.D., MIT, 1967. Registered engr., Calif. Postdoctoral researcher Ctr. Nuclear Studies, AEC, Fontenay-Aux-Roses, France, 1967-68; group leader Lawrence Livermore Nat. Lab., Calif., 1968—. Contbr. articles to profl. publs. Committeeman Boy Scouts Am., 1980—. AEC fellow, 1962-65. Fellow Am. Phys. Soc.; mem. Am. Nuclear Soc. (exec. com. 1982), Sigma Xi, Delta Chi (house mgr. 1960-62). Republican. Unitarian. Subspecialties: Fusion; Plasma physics. Current work: Research on feasibility of using fusion-produced neutrons to convert thorium into uranium-233. Patentee yin-yang magnet, venetian blind direct converter. Home: 1730 Murdell LN Livermore CA 94550 Office: Lawrence Livermore Nat Lab Livermore CA 94550

MOJAVARIAN, PARVIZ, pharmaceutical chemist, radioimmunossay specialist, educator; b. Meshed, Iran, Jan. 22, 1949; came to U.S., 1974; s. Ali and Sonia M.; m. Susan E. Alexander, Aug. 18, 1980. B.Sc. in Chemistry, Pahlavi U., 1972; M.S. in Pharm. Chemistry, Phila. Coll. Pharmacy and Sci., 1976, Ph.D., 1979. Radioimmunoassay specialist DeJohn Med. Lab., Phila., 1979-80; fellow in clin. pharmacology Thomas Jefferson U., Phila., 1980-82, asst. prof. medicine and pharmacology, 1983—; research assoc. U. Kans., Lawrence, 1982-83. Contbr. research papers to profl. jours. Mem. Am. Chem. Soc., Am. Pharm. Assn., Am. Fedn. Clin. Research, Clin. Ligand Assay Soc., Sigma Xi, Rho Chi. Subspecialties: Analytical chemistry; Clinical chemistry. Current work: Current research activities include: kinetic, stability, and formulation of anti-cancer agents. Home: 207 Carter Rd Princeton NJ 08540 Office: U Kans Dept Pharm Chemistry 3068 Malott Hall Lawrence KS 66045

MOK, MACHTELD CORNELIA, genetics educator; b. De Bilt, Netherlands, Aug. 27, 1947; d. Cornelis A. and Francisca (Lindenborn) Nierstrasz; m. David W.S. Mok; 1 dau., Francisca. B.S., Agrl. U., Wageningen, Netherlands, 1969; M.S., U. Wis.-Madison, 1973, Ph.D., 1975. Asst. prof. Oreg. State U., Corvallis, 1975-80, assoc. prof. genetics, 1980—. Contbr. articles to profl. jours. Mem. Am. Soc. Plant Physiologists, Tissue Culture Assn., Genetics Soc. Am. Subspecialties: Plant genetics; Bioinstrumentation. Office: Dept Hort Oreg State U Corvallis OR 97331

MOKLER, CORWIN MORRIS, pharmacologist, educator; b. Forsyth, Ill., Dec. 10, 1925; s. Morris Michael and Mary Lois (Goodrich) M.; m. Margaret Mary Costello, Aug. 15, 1950; children: David James, Gregory Alan. B.A., Colo. Coll., 1950; M.S., U. Nev., 1952; Ph.D., U. Ill., 1958. Sr. investigator G.D. Searle & Co., Skokie, Ill., 1958-61; asst. prof. U. Fla. Coll. Pharmacy, Gainesville, 1961-67; assoc. prof. U. Ga. Sch. Pharmacy, Athens, 1967—. Contbr. articles to profl. jours. Served with USNR, 1943-46. Mem. Am. Soc. Pharmacology and Exptl. Therapeutics, Am. Heart Assn., Athens Choral Soc., Sigma Xi. Lodge: Elks. Subspecialties: Pharmacology; Physiology (medicine). Current work: Cardiovascular physiology, pharmacology. Office: Sch Pharmacy U Ga Athens GA 30602

MOLDENHAUER, JEANNE ELISABETH, microbiologist, consultant; b. Chgo., July 22, 1951; d. Richard E. and Dorothy C. Gibbons. A.A., Wright Coll., Chgo., 1972; B.A. with honors, U. Mo., St. Louis, 1974, M.S., Loyola U., Chgo., 1977. Teaching asst. Loyola U., 1976-77; developer quality systems for immunological and microbiol. new product line Lab Tek div. Miles Labs., Naperville, Ill., 1979-80; sterilization specialist Travenol Labs., Round Lake, Ill., 1980—. Mem. Am. Soc. Quality Control (cert. quality engr., mem. FDS subdiv.), Am. Soc. Microbiology, Ill. Soc. Microbiology, Tissue Culture Assn., Muscular Dystrophy Assn., South Central Assn. Clin. Microbiology. Mem. Christian Ch.. Club: Navigators (Colorado Springs, Colo.). Subspecialties: Cell and tissue culture; Operating systems. Current work: Chemical serum replacements; muscular viral infectivity; diagnostic virology; computer applications for qualification, validation plant assistance, troubleshooting, develop quality systems, and methods of sterilization, improved cycles, trend analysis. Patentee cell culture media supplement. Office: Travelnol Labs Wilson Rd and Belvidere Round Lake IL 60073

MOLDENKE, ALISON FEERICK, insect toxicology researcher; b. N.Y.C., Oct. 18, 1943; d. Martin Joseph and Kathleen Elizabeth (Strain) Feerick; m. Andrew Ralph Moldenke, July 15, 1967; 1 son, Kelsey Galen. B.A., Wellesley Coll., 1964; M.A., Wesleyan U., 1966; Ph.D., Stanford U., 1973. Vis. asst. prof. Vassar Coll., 1977-78; vis. asst. prof. U. Santa Clara, Calif., 1978-79; research assoc. insect toxicology Oreg. State U., Corvallis, 1980—. Contbr. articles to profl. jours. Gen. Motors Nat. Scholar, 1960-64; NSF fellow, 1968-70. Mem. Genetics Soc. Am., AAAS, Sigma Xi. Subspecialties: Cell biology; Toxicology (agriculture). Current work: Biochemical basis of resistence to insecticides; developmental genetics and biochemistry insect biochemistry; insect toxicology, cytochrome P-450; monooxygenase-dependent detoxication. Home: 1850 NW Arthur Circle Corvallis OR 97330 Office: Department of Entomology Oregon State University Corvallis OR 97331

MOLINE, HAROLD E., research plant pathologist; b. Frederic, Wis., Nov. 13, 1939; s. Thorsten J. F. and Agnes V. (Johnson) M.; m. Bonnie G. Larson, Mar. 6, 1965; children: Jenel M., Christopher A. B.Sc., U. Wis.-River Falls, 1967, M.S., 1969; Ph.D., Iowa State U., 1972. U.S. Dept. Agr. postdoctoral intern No. Grain Insects Research Lab., Brookings, S.D., 1972-73; research plant pathologist Hort. Crops Quality Lab., U.S. Dept. Agr.-Agrl. Research Service, Beltsville, Md., 1974—; adj. prof. Howard U., Washington, 1977—. Contbr. articles to profl. jours. Served with USN, 1958-62. Sigma Xi grantee, 1971. Mem. AAAS, Am. Phytopathological Soc., Electron Microscopy Soc. Am. Democrat. Lutheran. Club: Toastmasters Internat. Subspecialties: Plant pathology; Plant virology. Current work: Ultrastructural response of plant tissues to chilling injury; electron microscopic studies of harvested and stored plant tissues to study response to low temperatures and effects on decay. Office: US Dept Agri-ARS Beltsville MD 20705

MOLL, DON L., ecologist, educator, researcher; b. Peoria, Ill., Oct. 3, 1949; s. Edward and Bessie I. (Kennedy) M.; m. Barbara Kay Rogers, Mar. 17, 1972; children: Jane Elisabeth, Bryan Christopher. B.S., Ill. State U., 1971; M.S., Western Ill. U., 1973; Ph.D., Ill. State U., 1977. Asst. prof. biology Southwest Mo. State U., Springfield, 1977-82, assoc. prof. biology, 1982—; cons. Mo. Dept. Conservation, Jefferson City, 1980-82, U.S. Fish and Wildlife Service, Washington, 1980, World Wildlife Fund, 1983, Internat. Union Conservation Nature and Natural Resources, Gland, Switzerland, 1981—. Contbr. numerous articles to profl. jours. Southwest Mo. State U. grantee, 1982, 83. Mem. Am. Soc. Ichthyologists and Herpetologists, Herpetologists League, Soc. Study Amphibians and Reptiles, Mo. Acad. Sci., Nat. Audubon Soc., Bobby Witcher Soc., Sigma Xi (assoc.). Clubs: Alligator Snapper Soc., Gretna, La. Subspecialties: Ecology; Species interaction. Current work: Paleoecology of marine turtles, ecology of endangered reptiles and amphibians, foraging ecology and interactions in aquatic vertebrate communities. Home: 2455 S Aspen Springfield MO 65807 Office: Dept Biology Southwest Mo State U 901 S National Springfield MO 65804

MOLL, JOHN LEWIS, electronics engineer; b. Wauseon, Ohio, Dec. 21, 1921; s. Samuel Andrew and Esther (Studer) M.; m. Isabel Mary Sieber, Oct. 28, 1944; children: Nicolas Josef, Benjamin Alex, Diana Carolyn. B.Sc., Ohio State U., 1943, Ph.D., 1952; Dr. h.c., Faculty Engring., Katholieke U. Leuven, (Belgium), 1983. Elec. engr. RCA Labs., Lancaster, Pa., 1943-45; mem. tech. staff Bell Telephone Labs., Murray Hill, N.J., 1952-58; mem. faculty Stanford U., 1958-69, prof. elec. engring., 1959-69; tech. dir. optoelectronics Fairchild Camera and Instrument Corp., 1969-74; dir. integrated circuits labs. Hewlett-Packard Labs., Palo Alto, Calif., 1974-80, dir. IC structures research, sr. scientist, 1980—. Author: Physics of Semi Conductors, 1964. Guggenheim fellow, 1964; Recipient Howard N. Potts medal Franklin Inst., 1967, Disting. Alumnus award Coll. Engring., Ohio State U., 1970. Fellow IEEE (Ebers award 1971); mem. Am. Phys. Soc., Nat. Acad. Engring., Sigma Xi, Sigma Pi Sigma. Subspecialty: Integrated circuits. Current work: Integrated circuits structure research. Home: 4111 Old Trace Rd Palo Alto CA 94306 Office: 3500 Deer Creek Rd Palo Alto CA 94304

MOLLER, AAGE RICHARD, physiologist; b. Fiderup, Denmark, Apr. 16, 1932; s. Jens and Kristine Marie (Pedersen) M.; m. Margaretta Bjuro, July 26, 1977; children: Peter, Jan. Cand. med., Karolinska Inst., Stockholm, 1975, Ph.D., 1965. Research assoc., asst. prof., research fellow Swedish Med. Research Council, Karolinska Inst., 1966-77; assoc. prof. otolaryngology U. Gothenburg, Sweden, 1977-78; research prof. otolaryngology and physiology U. Pitts. Sch. Med., 1978—; sr. lectr. Carnegie Mellon U., Pitts., 1980—; dir. research Internat. Ctr. Insect Physiology and Ecology, Nairobi, Kenya, 1970-76. Contbr. article to profl. jours. Served as cpl. Danish Army, 1951-53. Fellow Acoustical Soc. Am.; mem. Swedish Physiol. Assn., Swedish Acoustical Soc., Soc. Occupational and Environ. Health, Soc. Neuroscience, N.Y. Acad. Sci., Assn. Research in Otolaryngology, AAAS. Subspecialties: Neurophysiology; Acoustical engineering. Current work: Coding of sounds in auditory nervous system. Home: 5427 Northumberland St Pittsburgh PA 15217 Office: Eye and Ear Hosp 230 Lothrop St Pittsburgh PA 15213

MOLLICONE, RICHARD ANTHONY, research and development executive; b. Bklyn., Feb. 18, 1935; s. Anthony Francis and Bertha Marie (Filas) M.; m. Joyce Elayne McCoy, Nov. 30, 1958. B.S., U.S. Mil. Acad., 1957; M.S., Rensselaer Poly. Inst., 1965, Ph.D., 1970. Assoc. prof. engring. mechanics and materials sci. U.S. Air Force Acad., Colorado Springs, Colo., 1966-71; commd. office U.S. Air Force, 1957, advanced through grade to lt. col, 1978; chief advanced concepts div. USAF (space div.), El Segundo, Calif., 1977-78; v.p., dir. western research ops. Analytic Decisions, Inc., Hawthorne, Calif., 1978—; cons. in field. Contbr. articles to profl. jours. Decorated Air Force Commendation medal, Bronze Star medal, Air medal with oak leaf cluster. Mem. AIAA, ASME, Nat. Def. Preparedness Assn.,

V.F.W., Nat. Rifle Assn., Order of Daedalians, Sigma Xi. Roman Catholic. Lodges: Lions; K.C. Subspecialties: Aerospace engineering and technology; Satellite studies. Current work: Infrared system design, manufacturing technology, focal plane and signal processing technologies, materials science, radiation and laser hardening. Home: 15931 Redlands St Westminster CA 92683 Office: 5155 Rosecrans Ave 307 Hawthorne CA 90250

MOLTENI, AGOSTINO, pathologist, educator, researcher; b. Como-Lombardy, Italy, Nov. 12, 1933; came to U.S., 1963, naturalized, 1970; s. Enrico and Antonia (Signorini) M.; m. Loredana Brizio-Molteni, Sept. 5, 1963; children: Claudio, Ronald. M.D., U. Milano, Italy, 1957; Ph.D., SUNY, Buffalo, 1970. Cert. Specialy Bd. Internal Medicine, Italy, 1963. Intern Univ. Hosp., Milan, Italy, 1957, resident, 1958-63; fellow SUNY, Buffalo, 1965-69, asst. prof. pathology, 1970-72; assoc. prof. U. Kans., Kansas City, 1972-76; prof. Northwestern U., 1976—; cons. in field. Contbr. chpts. to books and articles in field to profl. jours. Recipient Research Career Devel. award NIH, 1972; co-recipient Albert E. Lasker Public Health award, 1980. Mem. AMA, Am. Assn. Pathologists, Internat. Acad. Pathology, Endocrine Soc., Am. Assn. Clin. Chemistry, Sigma Xi. Subspecialties: Pathology (medicine); Receptors. Current work: Systemic and pulmonary hypertension; clin. and exptl. endocrinology; with particular emphasis on hormone receptors. Office: 303 E Chicago Ave Chicago IL 60611

MOMPARLER, RICHARD LEWIS, pharmacologist; b. N.Y.C., Jan. 6, 1935; s. Fred and Frances (Alejandro) M.; m. Louise Farley, Jan. 22, 1966. B.Sc., Mich. State U., 1957; Ph.D., U. Vt., 1966. Asst. prof. biochemistry McGill U., 1967-72; asso. prof. pharmacology U. So. Calif., 1973-77; research prof. pharmacology U. Montreal, Que., Can., 1978—. Served with USNG, 1958-64. Mem. Internat. Assn. Comparative Research on Leukemia, Am. Assn. Cancer Research, Cell Kinetics Soc. Subspecialties: Cancer research (medicine); Cellular pharmacology. Current work: Cancer chemotherapy, leukemia, new drug. devel. Home: 3175 Cote Ste Catherine 2729 Montreal Canada H3T 1C5

MONACO, ANTHONY PETER, surgeon, educator; b. Phila.; s. Donoto Charles and Rose (Consalvi) M.; m. Mary Louise Oudens, Sept. 1, 1938; children: Anthony Peter, Mark Churchill, Christopher and Lisa (twins). B.A., U. Pa., 1952; M.D., Harvard U., 1956. Diplomate: Am. Bd. Surgery. Resident in surgery Mass. Gen. Hosp., Boston, 1956-63; from instro. to prof. surgery Harvard Med. Sch., 1963—; sci. dir. Cancer Research Inst., New Eng. Deaconess Hosp., Boston, 1980—, also dir. div. organ transplantation.; Bd. dirs. Interhosp. Organ Bank of New Eng., 1981—. Contbr. articles to profl. jours. Recipient Lederle Med. Faculty award, 1967. Mem. ACS, Am. Surg. Assn., Am. Soc. Transplant Surgeons. Club: Harvard of Boston. Subspecialties: Transplant surgery; Transplantation. Current work: Transplantation; immunobiology; cancer immunology; kidney transplantation. Home: 25 Farlow Rd Newton MA 02158 Office: New Eng Deaconess Hosp Boston MA 02155

MONARD, JOYCE ANNE, nuclear engineer; b. Bethlehem, Pa., Nov. 5, 1946; d. Charles and Marguerite (Finglis) M.; m. Mitchell C. Gregory, Feb. 23, 1973 (div. 1983); children: Dru Monard, Cheryl Anne. A.B., Bryn Mawr Coll., 1968; Ph.D., U. Tenn., 1972. Postdoctoral U. Calif.-Berkeley, 1972-75; nuclear engr. Gen. Electric Co., Sn Jose, Calif., 1975-78, sr. engr., 1978, tech. leader, 1979, program mgr.-fuels, 1979-82, sr. program mgr.-nuclear services, 1982-83, product devel. mgr.-nuclear services, 1982-83. NSF trainee, 1968-72; European fellow Bryn Mawr Coll., 1968; Charles S. Hinchman scholar, 1967; Maria L. Eastman Brooke Hall scholar, 1967. Mem. Am. Nuclear Soc., Am. Phys. Soc., Sigma Pi Sigma (chpt. sec. 1969-70). Republican. Episcopalian. Subspecialties: Nuclear engineering; Nuclear physics. Current work: Wide range neutron monitor, channel measurement, advanced nuclear fuel, reload core design and core manangement. Home: 1757 Erinbrook Pl San Jose CA 95131 Office: 175 Curtner Ave M/C 853 San Jose CA 95125

MONASH, ELLIS ALAN, geomathematician; b. Phila., Dec. 31, 1938; s. Max and Helen Selma (Streitfeld) M.; m. Carol Joan Bernard, June 11, 1960; children: Temma J., Adam M., Nathan S. A.B., Temple U., 1961; M.Sc., U. Wis., 1962, U. Colo., 1971. Ops research analyst System Devel. Corp., Colorado Springs, Colo., 1968-70; mathematician U.S. Army Air Def. Commn., Ent AFB, Colo., 1970-75; computer systems analyst U.S. Geol. Survey, Lakewood, Colo., 1975-81; engring. cons. Intercomp Resource Devel., Englewood, Colo., 1981-82; sr. research analyst System Tech., Inc., Paonia, Colo., 1982—. Author, editor: Lumped Parameter Models of Hydrocarbon Reservoirs, 1983. Recipient Gubernatorial scholarship State Pa., 1958. Mem. Soc. Computer Simulation (assoc. editor 1980—), Soc. Petroleum Engrs., Soc. Indsl. and Applied Match., Internat. Assn. Match. Geology. Subspecialties: Numerical analysis; Resource management. Current work: Research to maximize ultimate recovery of hydrocarbon reservoirs while minimizing thermodynamic waste from them. Home: 2080 Newcombe Dr Lakewood CO 80125 Office: System Tech Inc PO Box 459 Paonia CO 81428

MONCE, MICHAEL NOLEN, physicist, educator; b. Honolulu, Apr. 13, 1952; s. Norman Herbert and Margaret Eileen (Whisen) M.; m. Dolores E. Perrino, Oct. 16, 1982. B.A., U. Colo., 1974; M.S., Colo. State U., 1976; Ph.D., U. Ga., 1981. Research asst. cyclotron facility U. Colo., Boulder, 1974; research and teaching asst. Colo. State U., Ft. Collins, 1974-76, U. Ga., Athens, 1977-81, grad. asst., 1980-81; asst. prof. dept. physics and astronomy Conn. Coll., New London, 1981—. Contbr.: articles to Jour. Chem. Physics. Mem. Am. Phys. Soc., Am. Assn. Physics Tchrs., Sigma Pi Sigma. Democrat. Subspecialty: Atomic and molecular physics. Current work: Molecular dissociation by high energy protons, molecular spectroscopy; atomic and molecular physics, experimental molecular dissociation, proton impact on molecules. Office: Dept Physics and Astronomy Conn Coll New London CT 06320

MONDAL, KALYAN, electrical and computer engineer, educator, researcher; b. Calcutta, India, Aug. 17, 1951; came to U.S., 1974; s. Dwijendra Nath and Bijali (Das) M.; m. Chitralekha Mandal, Aug. 5, 1981; 1 dau., Indrani. B.Sc. with honors, U. Calcutta, 1969, B.Tech., 1972; M.Tech., 1974; Ph.D., U. Calif.-Santa Barbara, 1978. Research asst. U. Calif.-Davis, 1975-77; research asst. U. Calif.-Santa Barbara, 1977-78; postgrad. researcher, 1978-79, lectr., 1978-79; asst. prof. Lehigh U., Bethlehem, Pa., 1980-81, adj. asst. prof., 1982—; mem. tech. staff AT&T Bell Telephone Labs., Allentown, Pa., 1982—. Contbr. research papers to profl. jours. Recipient Gold medal U. Calcutta, 1972, 74; Pa. Power & Lights Co. grantee, 1981. Mem. IEEE, Soc. Indsl. and Applied Math., Assn. for Computing Machinery, Sigma Xi, Eta Kappa Nu. Subspecialties: Integrated circuits; Computer engineering. Current work: VLSI system design, digital signal processing, microprocessor-based system design, database systems, and digital image processing. Office: AT&T Bell Telephone Labs 1247 S Cedar Crest Allentow PA 18103

MONDER, CARL, biochemist, laboratory administrator; b. N.Y.C., Aug. 24, 1928; s. Frank and Jennie (Black) M.; m. Theodora Rubenstein, June 24, 1959; children: Benjamin, Eric. B.S., CCNY, 1950; M.S., Cornell U., 1952; Ph.D., U. Wis., 1956. Asst. prof. biochemistry Albert Einstein Coll. Medicine, N.Y.C., 1959-69; assoc. prof. biochemistry Mt. Sinai Sch. Medicine, N.Y.C., 1969-78; assoc. dir. research Research Inst. (Hosp. Joint Diseases), N.Y.C., 1964-81; prof. biochemistry Mt. Sinai Sch. medicine, N.Y.C., 1978-81; sr. scientist Ctr. Biomed. Research (Population Council), N.Y.C., 1981—; cons. Squibb Corp., Princeton, N.J., Can. Research Council, Montreal. Corr. editor: Jour. Steroid Biochemistry, Paris. Quillman fellow, 1951; USPHS fellow, 1956; Career Devel. awardee, 1969. Mem. Am. Soc. Biol. Chemists, AAAS, N.Y. Acad. Scis., Harvey Soc., Endocrine Soc. (chmn. subcom. pub. affairs 1980), Phi Beta Kappa. Subspecialties: Biochemistry (medicine); Endocrinology. Current work: Metabolism and physiological significance of corticosteroids; developmental biology. Home: 180 Garth Rd Scarsdale NY 10583 Office: Population Council 1230 York Ave New York NY 10021

MONDER, HARVEY, neuroendocrinologist, psychopharmacologist; b. Bklyn., Jan. 13, 1939; s. Frank and Jennie (Black) M.; m. Marcia Joyce, Jan. 28, 1961; children: Jeffrey, Alan. B.S., CCNY, 1960; M.A., Bklyn. Coll., 1974; Ph.D. (Univ. fellow), SUNY-Binghamton, 1979. In various personnel and investigative positions with fed. govt., 1962-73; research asst. Bklyn. Coll., 1973-75; research assoc. SUNY-Binghamton, 1975-82, research asst. prof. behavioral physiology, 1982-83; statistician U.S. Govt., 1983—. Contbr. articles to profl. jours. H.F. Guggenheim Found. grantee, 1980-81. Mem. Soc. Neuroscis., Eastern Psychol. Assn., AAAS, N.Y. Acad. Scis., Biometrics Soc. Subspecialties: Developmental biology; Psychobiology. Current work: Role of pituitary-adrenal-gonadal systems in expression of aggression in female mice, endocrine and behavioral changes in animals with changes in population density, ACTH and opioid receptor interactions.

MONIER, LOUIS MARCEL, computer scientist; b. La Seyne, Var, France, Mar. 21, 1956; s. Maurice and Ginette (Bernardini) M.; m. Nadine Solange Berenguier, Aug. 2, 1976. Baccalaureat Math. (hon.), France, 1973; M.S., Orsay U., 1978, Ph.D., 1980. With ENS, France, 1975-78; research scientist INRIA, France, 1978-80, Carnegie-Mellon U., 1980-82; mem. research staff Xerox Palo Alto Research Center, 1983—. Served with French Armed Forces. Subspecialties: Algorithms; Computer architecture. Current work: VLSI architectures, systolic architectures, layout of VLSI circuits, CAD tools, algorithms for arithmetic, cryptography. Patentee in field. Office: Computer Sci Dept Carnegie-Mellon U Doherty 3321 Pittsburgh PA 15213

MONISMITH, CARL LEROY, civil engr., educator; b. Harrisburg, Pa., Oct. 23, 1926; s. Carl Samuel and Camille Frances (Geidt) M. B.S. in Civil Engring, U. Calif., Berkeley, 1950, M.S., 1954. Mem. faculty U. Calif., Berkeley, 1951—, prof. civil engring., 1965—, chmn. dept., 1974-79; cons. to govt. and industry. Author numerous papers, reports in field. Served to 2d lt. AUS, 1945-47. Recipient Rupert Meyers medal U. New South Wales, Australia, 1976; Fulbright scholar, Australia, 1971. Fellow ASCE (pres. San Francisco sect. 1979-80, State of Art award 1978); mem. Nat. Acad. Engring., Assn. Asphalt Paving Technologists (pres. 1968, Walter J. Emmons award 1961, 65), Transp. Research Bd. o3(asso.) (chmn. pavement design sect. 1973-79, K.B. Woods award 1972), ASTM, Am. Soc. Engring. Edn. Subspecialty: Civil engineering. Current work: Pavement design and rehabilitation, including overlay design; asphalt paving technology. Office: 215 McLaughlin Hall U Calif Berkeley CA 94720

MONJAN, ANDREW ARTHUR, immunologist, psychologist; b. N.Y.C., Feb. 9, 1938; s. Victor and Sonia (Sherinian) M.; m. Susan Vollenweider, Aug. 2, 1962; m. Usha Bose, Aug. 14, 1969; children: Matthew V., Vanessa K. B.S., Rensselaer Poly. Inst., 1960; Ph.D., U. Rochester, 1965; M.P.H., Johns Hopkins U., 1970. Postdoctoral fellow Ctr. Brain Research, U. Rochester, N.Y., 1964-66; asst. prof. depts. psychology and physiology U. Western Ont., London, Ont., Can., 1966-69; asst. prof. dept. epidemiology Johns Hopkins U. Sch. Hygiene and Pub. Health, Balt., 1971-75, assoc. prof., 1975-83; program dir. for AIDS epidemiology Nat. Cancer Inst., NIH, Bethesda, Md., 1983—. Contbr. articles to profl. jours. NIH grantee; Med. Research Can. grantee; Nat. Research Council Can. grantee. Mem. Am. Assn. Immunologists, Assn. Research in Vision and Ophthalmology, Am. Assn. Virology, Eastern Psychol. Assn., Teratology Soc, AAAS, Sigma Xi. Democrat. Subspecialties: Infectious diseases; Neurobiology. Current work: Stress and the modulation of immune response; pathogenesis of viral infections of the central nervous and visual systems, immunopathology due to viruses, AIDS epidemiology. Home: 6727 Glenkirk Rd Towson MD 21239 Office: Spl Programs Br Div Cancer Cause and Prevention Nat Cancer Inst NIH Landow Bldg Rom 8016 Bethesdae MD 20205

MONNEY, NEIL THOMAS, ocean engineer, researcher; b. St. Paul, Alta., Can., Jan. 19, 1940; s. Paul Benjamin and Katherine Millie (Carpenter) M. B.S., U.S. Naval Acad., 1962; M.S.C.E., U. Wash., 1965, Ph.D., 1967. Resident officer in charge constrn. N.W. div. Naval Facilities Engring. Command, 1962-63, engring. coordination officer, 1965-67; research and devel. project officer U.S. Naval Material Command, 1967-70; sr. policy analyst for ocean engring. U.S. Dept. Commerce, Washington, 1978-79; dir. ocean engring. U.S. Naval Acad., Annapolis, Md., 1970-81; v.p. Marine Resources Devel. Found., Ft. Lauderdale, Fla., 1981—; engring. cons., 1981—; mem. environ. criteria panel of com. offshore energy tech. NRC, 1978-79; ocean engring. cons. U.S. Congress Office of Tech. Assessment, 1976-79, NRC, 1975-76, U.S. State Dept., 1975, others; mem. com. seafloor engring. Nat. Acad. Engring., 1973-76; mem. U.S. marine facilities panel U.S.-Japan Com. Natural Resources, 1972-80. Editor: Marine Tech. Soc. Jour, 1980-82, Jour. Ocean Sci. and Engring, 1979—. Recipient numerous letters of commendation. Mem. ASME (nat. chmn. ocean engring. div. 1979), Marine Tech. Soc., Sigma Xi. Subspecialties: Ocean engineering; Ocean thermal energy conversion. Current work: Ocean mineral recovery, ocean energy development. Office: Marine Resources Devel Found 13106 Port Everglades Station Fort Lauderdale FL

MONROE, RUSSELL RONALD, psychiatrist; b. Des Moines, June 7, 1920; s. Ronald Russell and Mildred (Schmidt) M.; m. Lillian Constance Brooks, June 23, 1945; children—Constance Ellen Teevan, Nancy Brooks Monroe Amoss, Russell Ronald. B.S., Yale U., 1942, M.D., 1944; cert. in psychoanalysis, Columbia U., 1950. Intern New Haven Hosp., 1944-45; resident in psychiatry Rockland State Hosp., Orangeburg, N.Y., 1947-50; asst. prof. psychiatry Tulane U. Sch. Medicine, New Orleans, 1950-53, asso. prof., 1953-60; prof. psychiatry U. Md. Sch. Medicine, Balt., 1960—, chmn. dept., 1976—; dir. Inst. of Psychiatry and Human Behavior, 1976—; vis. prof. Am. U., Beirut, Lebanon, 1966-67; mem. nat. adv. com. Nat. Inst. Law Enforcement and Criminal Justice, 1977-80. Author: Episodic Behavioral Disorders, 1970, Brain Dysfunction in Aggressive Criminals, 1978. Served to capt. AUS, 1945-47. NIMH fellow, 1966-67; Commonwealth fellow, 1966-67; NIMH grantee. Fellow Am. Psychiat. Assn., Am. Acad. Psychoanalysis; mem. Am. Coll. Psychiatrists. Clubs: Hamilton Street, Southern Yacht. Subspecialty: Psychiatry. Home: 236 W Lafayette Ave Baltimore MD 21217 Office: U Md Sch Medicine 645 W Redwood St Baltimore MD 21201

MONSON, ROBERTA ANN MILLS, internist, medical educator; b. Niagara Falls, N.Y., Apr. 10, 1942; d. Stanley Allen and Elizabeth Kathryn (Harding) Mills; m. Thomas Phillip Monson, Apr. 20, 1968; children: Kristina, Eric. B.S., Allegheny Coll., 1963; M.D., Harvard U., 1967. Diplomate: Am. Bd. Internal Medicine. Intern Cleve. Met. Gen. Hosp., 1967-68; resident in radiology Case Western Res. Univ. Hosp., 1968-69; resident in medicine Cleve. Met. Gen. Hosp., 1969-71; chief resident in medicine VA Hosp.-U.Wis.-Madison, 1971-72, Stetler research fellow and fellow in infectious diseases, 1972-74; asst. prof. medicine U. Wis.-Madison, 1974-76; mem. faculty U. Ark. Coll. Medicine, Little Rock, 1976—, assoc. prof. medicine, 1980—; mem. staff Doctors Hosp., Little Rock, St. Vincent's Hosp.; cons. Riverview Med. Ctr., Little Rock, 1982—. Contbr. sci. articles to profl. publs. Mem. sr. bd. Florence Crittendon Home, Little Rock, 1983. Huidekoper-Harvard scholar, 1962-63; Robert Wood Johnson grantee, 1982. Fellow ACP; mem. Soc. Research and Edn. in Primary Care Internal Medicine, Am. Fedn. Clin. Research. Subspecialties: Internal medicine; Health services research. Current work: Health services research; general internal medicine. Home: 47 Gloucester Dr Little Rock AR 72207 Office: 5300 W Markham St Suite G Little Rock AR 72205

MONTAGNA, WILLIAM, scientist; b. Roccacasale, Italy, July 6, 1913; s. Cherubino and Adele (Giannangelo) M.; m. Martha Helen Fife, Sept. 1, 1939 (div. 1975); children: Eleanor, Margaret, James and John (twins); m. Leona Rebecca Swift, Apr. 19, 1980. A.B., Bethany Coll., 1936, D.Sc., 1960; Ph.D., Cornell U., 1944; D. B.S., Universitá di Sassari, 1964. Instr. Cornell U., 1944-45; asst. prof. L.I. Coll. Medicine, 1945-48; asst., asso. prof. Brown U., 1948-52, prof., 1952-63, L. Herbert Ballou univ. prof. biology, 1960-63; prof., head exptl. biology U. Oreg. Health Scis. Center; dir. Oreg. Regional Primate Research Center, Beaverton, 1963—. Author: The Structure and Function of Skin, 1956, 3d edit., 1974, Comparative Anatomy, 1959, Nonhuman Primates in Biomedical Research, 1976, Science Is Not Enough, 1980; co-author: Man, 1969, 2d edit., 1973; editor: The Biology of Hair Growth, 1958, Advances in Biology of Skin, 20 vols, The Epidermis, 1965, Advances in Primatology, 1970, Reproductive Behavior, 1974. Decorated Order di Cavaliere, 1963, Cavaliere Ufficiale, 1969, Commendator della Repubblica Italiana, 1975; Italy; recipient spl. award Soc. Cosmetic Chemists, 1957; Gold award Am. Acad. Dermatology, 1958; Ann. award Consiglio Nazionale delle Ricerche, 1963; gold medal for meritorious achievement Universitá di Sassari, 1964; Aubrey R. Watzek award Lewis and Clark Coll., 1977; Hans Schwarzkopf Research award German Dermatol. Soc., 1980. Fellow AAAS, N.Y. Acad. Scis., Soc. Gerontology; mem. Am. Assn. Anatomists, Am. Assn. Zoology, Histochem. Soc., Acad. Dermatology and Syphilology, Soc. Investigative Dermatology (pres. 1969, recipient Stephen Rothman award 1972, ann. William Montagna lectr. 1975—), Sigma Xi (Pres. 1960-62). Subspecialty: Dermatology. Research in biology mammalian skin with emphasis on primates. Office: 3181 SW Sam Jackson Rd Dept Dermatology Portland OR 97201 Professional success came to me by way of an uncompromising father whom I despised and admired, and eventually loved. Though nearly illiterate, he was one of the most gifted and idealistic men I have ever known. From him I learned doggedness, diligence, dedication, complete involvement in everything I did, and an appreciation of natural beauty.

MONTAGNE, JOHN, geology educator; b. White Plains, N.Y., Apr. 17, 1920; s. Henry and Ella (Spurgeon) de la M.; m. Phoebe Corthell, Dec. 23, 1942; children: Clifford, Mathew H. B.A., Dartmouth Coll., 1942; M.A., U. Wyo., 1952, Ph.D., 1955. Cert. profl. geologist Am. Inst. Profl. Geologists. Tchr. sci. Jackson-Wilson High Sch., Jackson, Wyo., 1946-48; asst. to dir. admissions Dartmouth Coll., Hanover, N.H., 1948-49; instr. to asst. prof. geology Colo. Sch. Mines, Golden, 1953-57; asst. prof. to prof. geology Mont. State U., Bozeman, 1957—; chmn. adv. commn. Mont. Bur. Mines and Geology, 1975-82; chmn. Internat. Snow Sci. Workshop, Bozeman, 1982; adv. Yellowstone Library and Mus. Assn., 1962—. Served to 1st lt. U.S. Army, 1942-46. Fellow Geol. Soc. Am. (sect. chmn. 1981-82); mem. Am. Assn. Petroleum Geologists, Am. Quaternary Assn. (treas. 1970-76), Internat. Glaciological Soc., Mont. Wilderness Assn., Sigma Xi, Phi Kappa Phi. Lodge: Rotary. Subspecialty: Geology. Current work: Geomorphology, glacial geology, Pleistocene and environ. geology; tertiary stratigraphy and structural geology; snow, its mechanics, control. Office: Dept Earth Scis Mont State Univ Bozeman MT 59717 Home: 17 Hodgman Canyon Bozeman MT 59715

MONTE, JUDITH ANN, environmental scientist; b. Mt. Holly, N.J., June 18, 1947; d. Michael and Mary (Davison) M. A.B., Douglass Coll., 1969; M.S., Rutgers U., 1971; Ph.D., La. State U., 1978. Vis. asst. prof. geography U. S.C., Columbia, 1974-79; dir. MAST-Baruch Inst., 1977-79; vis. asst. prof. geography San Diego State U., 1979-80; assoc. prof. geography U. Md., College Park, 1982-83; sr. earth scientist Greenhorne & O'Mara, Inc., Greenbelt, Md., 1980—. Author: Manual Landsat and Groundwater Exploration, 1982. Mem. Assn. Am. Geographers, Ecol. Soc. Am., Am. Soc. Photogrammetry, Coastal Soc., LWV. Club: Sierra (College Park). Subspecialties: Remote sensing (geoscience); Ground water hydrology. Current work: Resource assessments; wetlands; groundwater exploration; remote sensing; vegetational succession on dredge spoil; human impact on coastal environments, especially petroleum impacts. Home: 9314 Cherry Hill Rd Apt 902 College Park MD 20740 Office: Greenhorne & O'Mara Inc 9001 Edmonston Rd Greenbelt MD 20770

MONTEFUSCO, CHERYL MARIE, surgery educator, lung transplantation coordinator; b. New Kensington, Pa., May 14, 1948; d. Benjamin Bosco and Edith (Costanzo) Bongiovanni; m. Frank Joseph Montefusco, Nov. 14, 1970 (div. 1982). B.S., St. Francis Coll., 1970; Ph.D., Coll. Medicine and Dentistry of N.J., 1975. Research assoc. Squibb Inst. Med. Research, Lawrenceville, N.J., 1970-71; editor dept. radiology Harrison S. Martland Hosp., Newark, 1972-73; instr. Coll. Medicine and Dentistry N.J., Newark, 1973-74; instr. dept. biology Kean Coll. of N.J., Union, 1974; research assoc. vascular surgery, dir. scanning electron microscopy lab. Newark Beth Israel Med. Ctr., 1975-76; tech. sales rep. W.L. Gore & Assocs., Inc., Newark, Del., 1976-77; instr. physiology Albert Einstein Coll. Medicine, N.Y.C., 1977—, asst. prof. surgery, 1977—; coordinator lung transplant research Montefiore Hosp. and Med. Ctr., N.Y.C., 1977—; project leader Exptl. Lung Transplantation, 1980—, clin. coordinator lung transplantation, 1981—; pres. Contemporary Weight Control Systems, Ltd., N.Y.C., 1981—; guest lectr. U. Utrecht, Netherlands, 1976. Contbg. author sci. writings in field. Fellow Internat. Coll. Angiology, Am. Coll. Angiology; mem. Am. Physiol. Soc., N.Y. Acad. Scis., N.Y. Transplantation Soc. Republican. Roman Catholic. Subspecialties: Physiology (medicine); Transplantation. Current work: Lung transplantation, mucociliary transport, exercise, physiology, nutrition in health and disease states. Office: Montefiore Med Ctr 111 E 210th St Bronx NY 10467

MONTELONE, BETH ANN, biologist; b. Albany, N.Y., Oct. 2, 1954; d. Louis James and Mary Elizabeth (Weir) M. B.S., Rensselaer Poly. Inst., 1976; M.S., U. Rochester, 1978, Ph.D., 1982. Teaching asst. U. Rochester, N.Y., 1976-78, research asst., 1981-82; Research assoc. dept. biochemistry U. Miami (Fla.) Sch. Medicine, 1982—. Contbr. articles to profl. jours. NIH trainee, 1978-81. Mem. Genetics Soc. Am. Subspecialties: Genetics and genetic engineering (biology); Molecular biology. Current work: Inducible error-prone repair and recombination in animal cells. Home: 1840 NE 142d St Apt 5D North Miami FL 33181 Office: Dept Biochemistry U Miami Sch Medicine PO Box 016129 Miami FL 33101

MONTGOMERY, DEANE, mathematician; b. Weaver, Minn., Sept. 2, 1909; s. Richard and Florence (Hitchcock) M.; m. Katherine Fulton, July 14, 1933; children—Mary, Richard. A.B., Hamline U., 1929; M.S., U. Iowa, 1930; Ph.D., 1933; Ph.D. hon. doctorate, Hamline U., 1954, Yeshiva U., 1961, Tulane U., 1967, U. Ill., 1977. NRC fellow Harvard, 1933-34, Princeton U., 1934-35, vis. assoc. prof., 1943-45; NRC fellow Inst. for Advanced Study, 1934-35, Guggenheim fellow, 1941-42, mem., 1945-46, permanent mem., 1948-51, prof. math, 1951—; asst. prof., prof. Smith Coll., 1935-41, 42-43; asso. prof. Yale, 1946-48. Mem. Am. Math. Soc. (pres. 1961-62), Nat. Acad. Sci., Math. Assn. Am., Am. Philos. Soc., Am. Acad. Arts and Sics., Internat. Math. Union (pres. 1975-78). Subspecialty: Topology. Current work: Action of compact lie groups on manifolds. Home: 33 Rollingmead Princeton NJ 08540 Office: Inst for Advanced Study Princeton NJ 08540

MONTGOMERY, DOIL DEAN, psychologist; b. Buhl, Idaho, June 20, 1939; s. Doil Manion and Rose Lorene (Smith) M.; m. Elena L. Lancaster, July 10, 1965; children: Matthew, Melody, Kenneth. A.A., Diablo Valley Coll., 1964; B.A., Sonoma State Coll., 1966; postgrad., SUNY-Stony Brook, 1966-68; Ph.D., W.Va. U., 1971. Lic. psychologist, Fla. Prof. psychology Nova U., Ft. Lauderdale, Fla., 1971-82, dir. biofeedback clinic, 1976—; psychologist Ctr. Neurol. Services, Ft. Lauderdale, 1980—, dir. biofeedback ctr., 1980—; cons. Goodwill Broward County, 1979-81. Served with U.S. Army, 1958-62. Broward Heart Assn. grantee, 1972-73. Mem. Am. Psychol. Assn., Biofeedback Soc. Am., Biofeedback Soc. Fla. (pres. 1981-82). Subspecialties: Behavioral psychology; Physiological psychology. Current work: Clinical application of biofeedback techniques. Home: 162 SW 53d Ave Plantation FL 33317 Office: 5601 N Dixie Hwy Fort Lauderdale FL 33334

MONTGOMERY, JOHN ATTERBURY, chemist; b. Greenville, Miss., Mar. 29, 1924; s. Daniel and Ruth (Atterbury) M.; m. Jean Kirkman, July 19, 1947; children: John, Elaine, Kirkman, Adrianne. A.B. in Chemistry cum laude, Vanderbilt U., 1946, M.S. in Organic Chemistry, 1947; Ph.D., U. N.C., 1950. Postdoctoral research U. N.C., Chapel Hill, 1951-52; adj. prof. organic chemistry Birmingham-So. Coll., Ala., 1957, 59, 62; adj. sr. scientist Comprehensive Cancer Center, U. Ala., Birmingham, 1978—; with So. Research Inst., Birmingham, 1952—, dir. organic chemistry research, 1956—, v.p., 1974—; sr. v.p. and dir. Kettering-Meyer Lab., 1981—; mem. various study sects. and adv. panels NIH; mem. bd. sci. cons. Sloan-Kettering Inst. Cancer Research; rep. Assn. Am. Cancer Insts.; mem. pharmacology adv. com. M. D. Anderson Hosp. and Tumor Inst., Houston; mem. adv. com. Internat. Soc. Heterocyclic Chemistry. Editorial bd.: Jour. Heterocyclic Chemistry, 1965—, Jour. Medicinal Chemistry, 1972-76, Cancer Treatment Reports, 1976-79, Jour. Organic Chemistry, 1981—, Nucleosides and Nucleotides, 1982—; editor: The Chemistry of Heterocyclic Compunds: Triazoles, 1981; contbr. numerous articles to sci. jours.; numerous lectures. Recipient Herty medal, 1974; Taito O. Soine Meml. award, 1979; So. Chemist award, 1980; Walter H. Hartung Meml. lectr., 1982; Cain Meml. award, 1982. Mem. Am. Chem. Soc. (councilor 1971—, chmn. Ala. sec. 1975-76, chmn. long range planning com. 1979, chmn. medicinal chemistry div. 1980), Am. Assn. Cancer Research, Internat. Soc. Heterocyclic Chemistry (dir. 1968—, nominating com. 1979), AAAS, NY. Acad. Scis., Am. Soc. Pharmacology and Exptl. Therapeutics, Sigma Xi, Alpha Chi Sigma. Episcopalian. Club: Country (Birmingham). Lodge: Rotary. Subspecialties: Medicinal chemistry; Organic chemistry. Current work: Organic and medicinal chemistry, pharmacology, biochemistry, chemotherapy. Home: 3596 Springhill Rd Birmingham AL 35223 Office: PO Box 3307-A Birmingham AL 35255

MONTGOMERY, PAUL CHARLES, immunologist, educator; b. Phila.; s. James Stewart and Marian Charlotte (Pringle) M.; m. Janet May (Henzel) M.; children: Sharon E., Timothy R., Christopher P. B.S., Dickinson Coll., 1965; Ph.D., U. Pa., 1969. Postdoctoral fellow Nat. Inst. Med. Research, London, 1965-70; asst. prof. U. Pa., Phila., 1970-73, assoc. prof., 1974-81, prof., 1981-83; prof., chmn. dept. immunology and microbiology Wayne State Med. Sch., Detroit, 1983—; Fogarty fellow U. Louvain, Brussels, 1978-79; cons. Smith Kline Diagnostics, Phila., 1976-77, Franklin Inst., 1976-80; adhoc reviewer Nat. Cancer Inst., Nat. Inst. Dental Research, Washington. Editor: Vet. Immunology and Immunopathology; contbr. numerous articles to sci. publs. Treas., pres. Pin Oak Farms Civic Assn., West Chester, 1975-80; treas. East Goshen Home and Sch. Assn., West Chester, 1982-83. Nat. Inst. Dental Research grantee, 1970-83; Nat. Inst. Allergy and Infectious Diseases grantee, 1975-78; Fogarty Internat. Ctr. sr. fellow, 1979-80; Nat. Eye Inst. grantee, 1981-84. Mem. Brit. Soc. Immunology, Am. Assn. Microbiology, Am. Assn. Immunologists, AAAS, Assn. Med. Sch. Microbiology Chmn., Sigma Xi, Alpha Chi Rho. Republican. Presbyterian. Subspecialties: Immunobiology and immunology; Microbiology. Current work: immunity at mucosal surfaces; mechanisms of secretory antibody induction; ocular immune system; the control of antibody diversity. Office: Dept Immunology and Microbiolog Wayne State U School Medicine 540 E Canfield Ave Detroit MI 48201

MONTGOMERY, REX, biochemistry educator; b. Halesowen, Eng., Sept. 4, 1923; came to U.S., 1948, naturalized, 1963; s. Fred and Jane (Holloway) M.; m. Barbara Winifred Price, Aug. 9, 1948; children: Ian, David, Jennifer, Christopher. B.Sc., U. Birmingham, 1943, Ph.D., 1946, D.Sc., 1963. Research assoc. U. Minn., Mpls., 1951-55; asst. prof. U. Iowa, Iowa City, 1955-59, assoc. prof., 1959-64, dir. physician asst. program, 1973-76, prof. biochemistry, 1964—, assoc. dean, 1974—; mem. Govs. Commn., High Tech. Commn. Iowa, 1983-85. Author: The Chemistry of Plant Gums and Mucilages, 1959, Qualitative Problems in the Biochemical Sciences, 1976, Biochemistry - A Case Oriented Approach, 1983. Mem. Am. Soc. Biol. Chemists, Am. Chem. Soc., Chem. Soc. London. Subspecialties: Biochemistry (medicine); Cancer research (medicine). Current work: Biochemistry of glycoproteins and the effect of the carbohydrate groups on the conformation of the molecules, specificity of carbohydrase enzymes, ligand receptor sites and tumor specific antigens, antitumor agents, antigens of Legionnaire's disease. Home: 5 Princeton Ct Iowa City IA 52240 Office: U Iowa Coll Medicine Iowa City IA 52242

MONTGOMERY, ROBERT LEW, psychology educator; b. Grayson, Ky., July 2, 1941; s. Everett DeForest and Ruth Agnes (Glass) M.; m. Sallie Stewart Meier, Sept. 6, 1966 (dec. Feb. 1978); m. Frances Marie Haemmerlie, June 16, 1979; children: Melissa, John. B.A., Bethany (W.Va.) Coll., 1964; M.S., Okla. State U., 1967, Ph.D., 1968. Research asst. Okla. State U., 1964-68; asst. prof. psychology U. Mo.-Rolla, 1968-73, assoc. prof., 1973-78, prof., 1978—, head dept. psychology, 1975-81; vis. prof. U. Fla., 1974-75. Contbr. articles to profl. jours. Recipient Phi Kapp Phi Disting. Service award U. Mo.-Rolla, 1975; Nat. Cambell fellow, 1963-64; NDEA fellow, 1967-68; U. Mo.-Rolla Asst. Prof. grantee, 1969, 71. Mem. Am. Psychol. Assn., Psychonomic Soc., Midwestern Psychol. Assn. (local rep. 1981—), Southwestern Psychol. Assn., Mo. Psychol. Assn. Episcopalian. Club: Oak Meadows (Rolla). Subspecialties: Social psychology; Behavioral psychology. Current work: Experimental social psychology, industrial/organizational psychology, evaluation research, group dynamics, leadership, attitude test construction, social influence, community psychology, psychology of aging, and person perception. Home: Route 4 Box 223 Rolla MO 65401 Office: Dept Psycholog U Mo Rolla MO 65401

MONTROLL, ELLIOTT WATERS, educator, scientist; b. Pitts., May 4, 1916; s. Adolph Baer and Esther (Israel) M.; m. Shirley Abrams, Mar. 7, 1943; children—Wendy, Brenda, Nicholas, Heidi, Mark, John, Toby, Andrew, Kim, Charles. B.S., U. Pitts., 1937, Ph.D., 1940. Sterling research fellow Yale U., 1940-41; research fellow Cornell U., 1941-42; instr. physics Princeton, 1942-43; head math, group Kellex Corp., 1943-46; asst., then asso. prof. U. Pitts, 1946-48; head physics br. Office Naval Research, 1948-50, phys. sci. dir., 1952-54; research prof. U. Md., 1950-59; gen. sci. dir. IBM Corp. Research Center, Yorktown Heights, N.Y., 1960-63, v.p. research Inst. Def. Analysis, Washington, 1963-66; Albert Einstein prof. physics U. Rochester, 1966-81; also dir. Inst. Fundamental Studies; prof. Inst. Phys. Sci. and Tech., U. Md., 1981—; chmn. Commn. on Sociotech. Systems, NRC, Washington, 1980—; Lorentz prof. U. Leiden, Netherlands, 1961, 68. Editor: Jour. Math. Physics, 1960-70. Recipient Lanchester prize Ops. Research Soc., 1959; Guggenheim fellow, 1958; Fulbright fellow, 1959. Mem. Nat. Acad. Scis., Am. Acad. Arts and Scis., Phi Beta Kappa, Sigma Xi. Club: Cosmos (Washington). Subspecialty: Applied mathematics. Home: 4 Laurel Pkwy Chevy Chase MD 20015 Office: Inst Phys Sci and Tech U Md College Park MD 20742

MONTY, RICHARD A., exptl. psychologist; b. Sanford, Maine, Apr. 2, 1935; s. Leo James and Evelyn J. (Delahunt) M.; m. Margaret Ellen Pennisten, May 22, 1965. B.A., Boston U., 1956; M.A., Columbia U., 1957; Ph.D., U. Rochester, 1962. Research assoc. HUMRRO, Monterey, Calif., 1957-59; engrng. psychologist Gen. Electric Co., Phila., 1961-62; research psychologist Cornell Aero. Lab., Buffalo, 1962-65, U.S. Army Human Engrng. Lab., Aberdeen Proving Ground, Md., 1965-79, chief behavioral research dir., 1979—; cons. VA Hosp., Syracuse, N.Y., 1966—, Va. Poly. Inst. and State U., 1975—, VA Hosp., Boston, 1975—. Editor: Eye Movements and Psychological Processes, 1976, Eye Movements and the Higher Psychological Processes, 1978-, Choice and Perceived Control, 1979, Eye Movements: Cognition and Visual Perception, 1981. Served with U.S. Army, 1957-59. Recipient Sustained Performance award U.S. Army, 1972, 76, Meritorious Performance award, 1982. Fellow Am. Psychol. Assn., Soc. Engring. Psychologists (exec. com. 1974-77); mem. Psychonomic Soc., Human Factors Soc., Eastern Psychol. Assn. Clubs: Cosmos (Washington); America First Day Cover Soc. (N.J.) (pres. 1979-80). Subspecialties: Cognition. Current work: Learning and memory, cognition, visual search, eye movements, sensory processes. Home: 3004 Whitefield Rd Churchville MD 21028 Office: Behavioral Research Directorate US Army Human Engring Lab Aberdeen Proving Ground MD 21005

MOODY, ARNOLD RALPH, plant pathologist; b. Augusta, Maine, Oct. 8, 1941; s. Norman O. and Annie (Bradstreet) M.; m. Donna Rich, June 19, 1965; 1 dau., Sara Ann. B.S., U. Maine, 1963; M.S., U. N.H., 1965; Ph.D., U. Calif.-Berkeley, 1971. Asst. research plant pathologist U. Calif.-Berkeley, 1971-72; research plant pathologist Station Federale de Recherches Agronomiques de Changins, Nyon, Switzerland, 1972-74; assoc. prof. plant pathology Va. State U., Petersburg, 1975—. Contbr. articles to profl. jours. Mem. Am. Phytopath. Soc., Sigma Xi. Presbyterian. Subspecialties: Plant pathology; Microbiology. Current work: Vegetable diseases, biol. control, root diseases, soil microbiology, resistance. Home: 301 Charlotte Ave Colonial Heights VA 23834 Office: Virginia State U Box 501 Petersburg VA 23803

MOODY, CHARLES EDWARD, JR., immunologist; b. Portsmouth, N.H., Feb. 4, 1948; s. Charles Edward and Doris Mae (Chapman) M.; m. Alexandra C. Soggiu, Feb. 14,1976 (div. 1982). B.A., Providence Coll., 1970; M.Sc., U. R.I., 1973, Ph.D., 1976. Postdoctoral fellow, div. allergy and immunology Cornell U. Med. Coll., N.Y.C., 1976-79, asst. prof. immunology in medicine, div. geriatrics and gerontology, 1979-82; asst. prof. microbiology U. Maine, Orono, 1982—. Contbr. articles to profl. jours. Nat. Inst. on Aging/NIH grantee, 1977-79, 81—. Mem. Am. Assn. Immunologists, Harvey Soc., N.Y. Acad. Scis., AAAS, Sigma Xi. Democrat. Roman Catholic. Subspecialties: Immunobiology and immunology; Microbiology. Current work: Aging and the immune response, immunology education. Aging. tolerance, autoreactivity, self recognition, syngeneic mixed lumphocyte reaction. Home: 11 Fernald Rd Apt 11 Orono ME 04473 Office: Dept Microbiology U Maine Room 262 Hitchner Hall Orono ME 04469

MOODY, DAVID EDWARD, experimental pathologist; b. Oak Lawn, Ill., Mar. 30, 1950; s. John Edward and Marcia May (Lininger) M.; m. Peggy Ann Harmon, May 20, 1972; 1 dau., Anne Renee. B.A., U. Kans.-Lawrence, 1972; Ph.D., U. Kans.-Kansas City, 1977. Postgrad. research pathologist dept. pathology U. Calif., San Francisco, 1977-80, asst. research pathologist, 1980—, vis. asst. research scientist dept. entomology, Davis, 1982—; cons. Chem. Mfrs. Assn., Washington, 1982. Contbr. articles in field to profl. jours. Patroller Nat. Ski Patrol System, Far West div., 1979—. Giannini fellow Bank of Am./ Giannini Found., U. Calif., San Francisco, 1977-78; Monsanto Found. fellow, 1978-80. Mem. Am. Soc. Cell Biology, Am. Assn. Pathologists, Soc. Toxicology. Subspecialties: Pathology (medicine); Toxicology (medicine). Current work: Studies on the response of cellular components to environmental chemicals and the effects of these changes on the physiological role of the cellular component. Home: 438 Yuba Pl Woodland CA 95695 Office: Dept Pathology U Calif San Francisco CA 94143

MOODY, ELIZABETH ANNE, physicist; b. Portland, Me., Oct. 29, 1948; d. Earl Louis and Margaret Mary (Downing) M. B.A., Simmons Coll., 1970; postgrad., Harvard U., 1973. Data analyst Smithsonian Astrophys. Obs., Cambridge, Mass., 1969; research cons. M.I.T. Instrumentation Lab., Cambridge, 1973-74, Ind. Cons., Boston, 1975-76; research scientist Aerodyne Research, Burlington, Mass., 1977-78, Sci. Applications, Bedford, Mass., 1979-80, U.S. Air Force Geophysics Lab., Bedford, 1981; design/devel. engr. Raytheon Co., Bedford, 1982—; cons. Contbr. articles to profl. jours. Recipient Woman of Future citation Mass. Dept. Edn., 1968; Leadership award Am. Legion, 1970; Community Service commendation United Community Services Mass., 1970. Mem. Am. Astron. Soc., Am. Phys. Soc., Astron. Soc. Pacific, Assn. Women in Sci., ACLU. Mem. Ch. of Larger Fellowship, Unitarian Universalist. Club: Cosmology. Subspecialties: Cosmology; Aerospace engineering and technology. Current work: Dynamical astronomy, photometry, space physics, orbital analysis, satellite tracking, high energy astrophysics. Home: PO Box 546 Beverly Farms MA 01915 Office: MSD MS M26-4 Hartwell Rd Bedford MA 01730

MOODY, FRANK GORDON, surgeon; b. N.H., May 3, 1928; s. Frederick G. and Elsie E. (Wilson) M.; m. Maria Charlotta Stolpe, Aug. 23, 1964; children—Anne E., Frank W., Jane A. B.A., Dartmouth Coll., 1953; M.D., Cornell U., 1956. Diplomate: Am. Bd. Surgery. Intern, then resident in surgery N.Y. Hosp.-Cornell U. Med. Center, 1956-63; research fellow U. Calif. Med. Center, San Francisco, 1963-64, advanced research fellow, 1964-65, asst. prof. surgery, 1965-66; asso. prof. surgery, chief div. gastrointestinal and gen. surgery, asst. prof. physiology and biophysics U. Ala. Med. Center, Birmingham, 1966-69, prof. surgery, 1969-71; prof. surgery, chmn. dept. U. Utah Med. Center, Salt Lake City, 1971—. Mem. editorial bds. profl. jours. Served with AUS, 1946-48. Mem. A.C.S., Am. Fedn. Clin. Research, Am. Gastroenterol. Assn., AMA, Am. Physiol. Soc., Am. Surg. Assn., Biophysics Soc., Soc. Surgery Alimentary Tract, Soc. U. Surgeons. Subspecialties: Surgery; Gastroenterology. Current work: Surgical gastroenterology, stress erosive gastritis, morbid obesity. Office: 50 N Medical Dr Salt Lake City UT 84132

MOODY, TERRY WILLIAM, biochemistry educator, researcher; b. Fresno, Calif., Dec. 2, 1949; s. Robert A. and Marie M.; m. Carol D. Linden, July 9, 1977; 1 dau., Elizabeth. B.Sc., U. Calif.-Berkeley, 1972; Ph.D., Calif. Inst. Tech., 1977. USPHS predoctoral trainee Calif. Inst. Tech., 1972-77; research chemist Nat. Inst. Drug Abuse, NIH, Rockville, Md., 1977-79; HEW postdoctoral fellow NIMH, Bethesda, Md., 1979-80; asst. prof. biochemistry George Washington U. Med. Sch., Washington, 1980-83, assoc. prof., 1983—. Contbr. over 30 articles to sci. jours. USPHS grantee, 1981—. Mem. AAAS, N. Y. Acad. Sci., Am. Soc. Biol. Chemists, Soc. Neurosci. Subspecialties: Neurochemistry; Biochemistry (biology). Current work: Neuron and neuroendocrine peptides and their receptors; research concerning the chemistry and biology of bombesin, a peptide present in certain neurons and cancer cells. Office: George Washington U Med Center Washington DC 20037

MOOLTEN, FREDERICK LONDON, physician, cancer research scientist; b. N.Y.C., Oct. 11, 1932; s. Sylvan Elkan and Isabel Marion (London) M.; m. Norma Nadal, June 26, 1960 (; (remarried)), Jan. 9, 1982; children: David, Marjorie, Judith, Amy. A.B. cum laude, Harvard U., 1953, M.D. magna cum laude, 1963. Intern, resident, research fellow in medicine Mass. Gen. Hosp., Boston, 1963-69; research asso.in microbiology Boston U. Sch. Medicine, 1969-71, asst. prof. microbiology, 1971-74, asso. prof., 1974—, chmn. sci. eval. com. Contbr. articles to profl. jours. Served with U.S. Army, 1956-58. Recipient Cancer research scholar award Mass. div. Am. Cancer Soc., 1971; Nat. Cancer Inst. grantee. Mem. Am. Assn. Cancer Research, Phi Beta Kappa, Alpha Omega Alpha. Subspecialties: Cancer research (medicine); Immunopharmacology. Current work: Immunotherapy and chemotherapy of cancer, immune prevention of cancer; immunotherapy, antibody-toxin conjugates; immunization against carcinogens; genetics of anticancer drug sensitivity. Office: Boston U Sch Medicine 80 E Concord St Boston MA 02118

MOONEY, HAROLD ALFRED, plant ecologist; b. Santa Rosa, Calif., June 1, 1932; s. Harold Walter and Sylvia Anita Stefany; m. Sherry Lynn Gulmon, Aug. 15, 1974; 1 dau., Adria. A.B., U. Calif., Santa Barbara, 1957; M.A., Duke U., 1958, Ph.D., 1960. From instr. to assoc. prof. UCLA, 1960-68; asso. prof. Stanford U., 1968-73, prof. biology, 1975—, Paul S. Achilles prof. environ. biology, 1976—; advisor NRC, Dept. Energy; adviser NSF, Electric Power Research Inst., Ford Found. Author: Mediterranean-type Ecosystems, 1973, Convergent Evolution in Chile and California, 1977, Components of Productivity of Mediterranean Climate Regions, 1981. Served with AUS, 1953-55. Guggenheim fellow, 1974; Nat. Acad. Scis. fellow, 1982; Am. Acad. Arts and Scis. fellow, 1982. Fellow AAAS; mem. Ecol. Soc. Am. (Mercer award 1961), Brit. Ecol. Soc. Subspecialties: Ecology; Photosynthesis. Current work: Carbon balance of plants. Home: 2625 Ramona St Palo Alto CA 94306 Office: Biology Dept Stanford Univ Stanford CA 94305

MOONEY, JAMES DONALD, computer science educator; b. Jersey City, Nov. 29, 1946; s. Donald J. and Anita M. (Degross) M.; m. Joan Arbogast, Apr. 29, 1972; 1 dau., Tara. B.S.E.E., U. Notre Dame, 1968; M.Sc., Ohio State U., 1969, Ph.D., 1977. Systems programmer Dymo Graphic Systems, Wilmington, Mass., 1971-79; asst. prof. computer sci. W.Va. U., Morgantown, 1979-82, assoc. prof., 1982—. Scoutleader, Boy Scouts Am., 1967-75. Mem. IEEE, Assn. Computing Machinery. Roman Catholic. Subspecialties: Computer architecture; Operating systems. Current work: Computer architecture, text processing, operating systems. Home: 632 W Virginia Ave Morgantown WV 26505 Office: Dept Stats and Computer Sci W Va U Morgantown WV 26506

MOONEY, RICHARD T., radiol. physicist, cons.; b. N.Y.C., Jan. 12, 1925; s. Michael J. and Rose G. M.; m. Cecilia G. Powers, Sept. 23, 1950; children: Maureen Rose, Michael Richard. B.S., Pratt Inst., 1944; M.S., N.Y.U., 1952. Diplomate: Am. Bd. Radiology, Am. Bd. Health Physics. Physicist Physics Service, N.Y.C. Dept. Hosps., 1950-60, prin. physicist, 1960-67; dir. physics services N.Y.C. Health and Hosps. Corp., 1967—; asst. prof. clin. radiology Sch. Medicine, SUNY-Stony Brook; cons. radiol. physics. Contbr. articles to profl. publs. Served to ensign USN, 1942-46. Fellow Am. Coll. Radiology; mem. Radiol. Soc. N.Am., Am. Assn. Physicists in Medicine, Am. Physics Soc., N.Y. Acad Scis. Roman Catholic. Subspecialties: Imaging technology; Nuclear medicine. Current work: Research and development in radiology imaging techniques and radiation protection. Inventor in field. Home: 4 Edgemont Circle Scarsdale NY 10583 Office: 82-68 164th St Jamaica New York NY 11432

MOORE, ALAN FREDERIC, cardiovascular pharmacologist; b. Birmingham, Eng., Aug. 27, 1948; s. Frederic Hotchin and Olive May (Ballard) M.; m. Valerie Harvey, Dec. 16, 1972; children: Jessica, Eleanor. B.Sc. (hons.), U. Aston, Birmingham, 1970, Ph.D., 1974. Postdoctoral fellow Cleve. Clinic, 1974-76, project research scientist, 1976-77; asst. prof., pharmacology U. Houston, 1977-79; sr. research scientist Norwich Eaton Pharms., N.Y., 1979-82, unit leader, 1982—. Contbr. articles to profl. jours. Recipient Lower award Cleve. Clinic, 1976. Fellow Am. Heart Assn.; mem. Am. Soc. Pharmacology and Exptl. Therapeutics, N.Y. State Neurosci. Soc., Interam. Soc. Hypertension. Subspecialty: Pharmacology. Current work: Etiology and maintenance of hypertension; development of peripherally and centrally acting drugs which regulate cardiovascular system. Home: 9 Eric St Norwich NY 13815 Office: Norwich Eaton Pharms PO Box 191 Norwich NY 13815

MOORE, ALLEN MURDOCH, biology educator; b. Ithaca, N.Y., Mar. 15, 1940; s. Donald Ward and Ella (Miller) M.; m. Joyce Elaine Shields, Sept. 7, 1969; children: Adrienne Alice, Megan Iris. A.B., Cornell U., 1961; Ph.D., U. Tex., 1968; postgrad., U. N.C., 1968-69. Teaching asst. U. Tex., Austin, 1961-63; asst. prof. dept. biology Western Carolina U., Cullowhee, 1969-74, assoc. prof., 1975—. NSF grantee, 1978-81; U.S. Forest Service grantee, 1981-82. Mem. AAAS, Ecol. Soc. Am., Sigma Xi. Subspecialties: Ecosystems analysis; Ecology. Current work: Modelling moisture and decomposition in forest floor leaf litter; studying fate of acidity in rain, in soils and streams of Great Smoky Mountains National Park. Office: Dept Biology Western Carolina Univ Cullowhee NC 28723

MOORE, BARBARA S. P., ocean engr.; b. Phila., July 4, 1942; d. Philip and Eleanor (Eckman) Schneider; m. John Norton Moore, Dec. 12, 1981. B.S. in Chem. Engring, Drexel U., 1965; M.S. Ocean Engring, Cath. U. Am., 1969. U.S. Peace Corps. vol., 1965-66; chem. engr. U.S. Navy, White Oak, Md., 1966, Naval Oceanographic Office, Washington, 1966-69; ocean engr. NOAA, Washington, 1969—; guest lectr. Georgetown U., 1978-82, Am. U., 1983. Mem. Marine Tech. Soc., Am. Oceanic Orgn. (v.p. 1980-81), Tau Beta Pi. Republican. Episcopalian. Subspecialties: Ocean engineering. Current work: Marine pollution, undersea science, man-in-the-sea, marine environmental science policy, instrumentation development. Home: 602 Pendleton St Alexandria VA 22314 Office: NOAA Office Policy Planning 14th and Constitution Aves Washington DC 20230

MOORE, BENJAMIN L., clinical psychologist; b. Atlanta, Jan. 19, 1940; s. Donald Laverne and Carolyn (Carson) M.; m. Mary Evelyn Ratteree, June 8, 1963; children: Donald Todd, Kevin Carson. B.A., Emory U., 1961, M.Div., 1969; M.S., Fla. State U., 1971, Ph.D., 1973. Registered psychologist, Ill.; cert. sch. psychologist, Ill. Child behavior cons. Regional Rehab. Ctr., Tallahassee, Fla., 1971-72; clin. psychology intern W.Va. U. Med. Ctr., Morgantown, 1972-73; asst. prof. dept. psychology Ill. State U., Normal, 1973-80, assoc. prof., 1980—; vis. assoc. prof. dept. psychology Ill. Wesleyan U., Bloomington, 1980; clin. dir. The Baby Fold, Normal, 1976—; impartial due processing hearing officer Ill. Office Edn., 1976-78; cons. several schs. for behavioral and emotionally disordered children, 1974—. Editor: Junction, 1968-69. Mem. pres's sci. adv. com. Ill. Wesleyan U., 1977-80; advisor Gov.'s Commn. on Children and Adolescent Mental Health and Devel. Disabilities, State of Ill., 1979-80; chmn. Council on Ministries, Calvary United Meth. Ch., Normal, 1979-80. Served to 1st lt. USAF, 1962-66. USPHS fellow, 1970-72; decorated Air Force Commendation medal with oak leaf cluster. Mem. Am. Psychol. Assn., Midwestern Psychol. Assn., Southeastern Psychol. Assn., Ill. Psychol. Assn., Assn. Behavior Analysis, Omicron Delta Kappa. Democrat. Subspecialties: Behavioral psychology; Developmental psychology. Current work: Research on behavior and learning disorders/disabilities in children: Psychological assessment: theory and practice of behavior change and psychotherapy. Office: The Baby Fold 108 E Willow St Normal IL 61761 Home: 14 Briarwood Ave Bloomington IL 61701

MOORE, CARLETON BRYANT, educator; b. N.Y.C., Sept. 1, 1932; s. Eldridge Carleton and Mabel Florence (Drake) M.; m. Jane Elizabeth Strouse, July 25, 1959; children—Barbara Jeanne, Robert Carleton. B.S., Alfred U., 1954, D.Sc. (hon.), 1977; Ph.D., Cal. Inst. Tech., 1960. Asst. prof. geology Wesleyan U., Middletown, Conn., 1959-61; mem. faculty Ariz. State U., Tempe, 1961—; now prof., dir. Center for Meteorite Studies; vis. prof. Stanford U., 1974; Prin. investigator Apollo 11-17; preliminary exam. team Lunar Receiving Lab., Apollo, 12-17. Author: Cosmic Debris, 1969, Meteorites, 1971; author: Principles of Geochemistry, 1982; Editor: Researches on Meteorites, 1961; editor: Jour. Meteoritical Soc; Contbr. articles to profl. jours. Fellow Ariz.-Nev. Acad. Sci. (pres. 1979-80), Meteoritical Soc. (pres. 1966-68), Geol. Soc. Am., Mineral. Soc. Am., AAAS (council 1967-70); mem. Geochem. Soc., Am. Chem. Soc., Sigma Xi. Subspecialties: Geochemistry; Space chemistry. Current work: Analytical geochemistry, meteorites, lunar samples, ceramic materials. Home: 507 E Del Rio Dr Tempe AZ 85282

MOORE, ELLIOTT PAUL, educator; b. Ninnekah, Okla., May 24, 1936; s. Paul Victor and Vivian Lulu (Perry) M.; m. Gail Andrea Moore, May 28, 1960. B.A., U. Chgo., 1956, B.S. in physics, 1957; Ph.D., U. Ariz., 1968. Faculty N. Mex. Inst. Mining & Tech., Socorro, 1968—; now assoc. prof. astrophysics, co-dir. Joint Obs. for Cometary Research. Mem. Am. Astorn. Soc., Am. Soc. Physics, Internat. Astron. Union. Subspecialties: Optical astronomy; Graphics, image processing, and pattern recognition. Current work: Astrophysics of comets; development two-dimensional image photon counting techniques. Address: Dept Physic N Mex Inst Mining and Tec Socorro NM 87801

MOORE, FRANCIS DANIELS, surgeon; b. Evanston, Ill., Aug. 17, 1913; s. Philip W. and Caroline (Seymour) D.; m. Laura Benton Bartlett, June 24, 1935; children: Nancy, Peter, Sarah, Caroline, Francis Daniels. A.B., Harvard, 1935, M.D., 1939; M.Ch. (hon.), Nat. U. Ireland, 1961, LL.D., U. Glasgow, 1965, D.Sc., Suffolk U., 1966, Harvard U., 1982, M.D., U. Gotenborg, Sweden, 1975, U. Edinburgh, 1976, U. Paris, 1976, U. Copenhagen, 1979. Diplomate: Am. Bd. Surgery. Surg. intern, asst. resident and resident Mass. Gen. Hosp., 1939-43; instr. surgery Harvard Med. Sch., 1934-46, asso., 1946-47, asst. prof., 1947-48, Moseley prof. of surgery, 1948-76, Elliot Carr Cutler prof. surgery, 1976-80, Moseley prof. surgery emeritus, 1980—; surgeon-in-chief Peter Bent Brigham Hosp., 1948-76, surgeon, 1976-80, surgeon-in-chief emeritus, 1980—; sr. cons. in surgery Sidney Farber Cancer Inst., 1976—, surg. oncology cons. emeritus, 1980—; Harvey lectr., 1956, cons. to surgeon gen., 1952; cons. to Surgeon Gen. of Navy, 1981—; chmn. surgery study sect. USPHS; chmn. subcom. on metabolism in trauma Office of Surgeon Gen.; bioscis. cons. NASA; chmn. com. on pre and postoperative care A.C.S., v.p., 1969; mem. numerous adv. panels; pres. Mass. Health Data Consortium, Inc., 1981—. Contbr.: surg. and investigative articles including Studies on Surgical Manpower and Care Distribution; Editor: book rev. sect. New Eng. Jour. Medicine, 1980—. Bd. regents Uniformed Services U. Health Scis., 1976-81; trustee New Eng. Conservatory Music, 1977—. Recipient Blakeslee award Mass. Heart Assn., 1965; Harvey Allen award Am. Burn Assn., 1970; Purkinje medal Purkinje Soc., Czechoslovakia, 1971; Bigelow medal Boston Surg. Soc., 1973; Lister medal Royal Coll. Surgeons, Eng., 1978. Mem. Am. Acad. Arts and Scis., Am. Surg. Assn. (pres. 1972, Samuel D. Gross medal 1978), Soc. for Clin. Surgery (pres. 1965), Internat. Surg. Soc., Soc. U. Surgeons (pres. 1958), Soc. Clin. Investigation, Polish Acad. Scis., Nat. Acad. Scis. Subspecialties: Surgery; Gerontology. Current work: Health policy research, health manpower, economics. Home: 66 Heath St Brookline MA 02146 Office: Countway Library 10 Shattuck St Boston MA 02115 also Peter Bent Brigham Hospital Boston MA 02115 It is fashionable to be skeptical; the American media thrive on alarm and pessimism; age is alleged to bring bitterness. My own view is directly contrary to all these trends. As I grow towards retirement I become more and more keen in my enthusiasm about the nature of humanity, and most particularly the remarkable flexibility of the American public and our ability, as a nation, to make mid-course corrections with very, very gentle hands on the controls.

MOORE, GEORGE WILLIAM, geologist; b. Palo Alto, Calif., June 7, 1928; s. George Raymond and Grace Amy (Hauch) M.; m. Ellen Louise James, Nov. 27, 1960; children: Leslie Ann, Geoffrey. B.S., Stanford U., 1950, M.S., 1951; Ph.D., Yale U., 1960. Geologist U.S. Geol. Survey, Menlo Park, Calif., 1951—; geologist in charge La Jolla Marine Geology Lab., Calif., 1966-75; research assoc. Scripps Instn. Oceanography, La Jolla, 1972-75; participant Deep Sea Drilling Project, 1977. Author: Speleology, 1978. Fellow Geol. Soc. Am.; mem. Nat. Speleological Soc. (pres. 1963), Am. Assn. Petroleum Geologists (com. chmn. 1977), AAAS. Subspecialties: Geology; Sea floor spreading. Current work: Mechanism of plate tectonics. Home: 828 La Jennifer Way Palo Alto CA 94306 Office: US Geol Survey 345 Middlefield Rd Menlo Park CA 94025

MOORE, GORDON E., electronics company executive; b. San Francisco, Jan. 3, 1929; s. Walter Harold and Florence Almira (Williamson) M.; m. Betty I. Whittaker, Sept. 9, 1950; children: Kenneth, Steven. B.S. in Chemistry, U. Calif., 1950; Ph.D. in Chemistry and Physics, Calif. Inst. Tech., 1954. Mem. tech. staff Shockley Semicondr. Lab., 1956-57; mgr. engring. Fairchild Camera & Instrument Corp., 1957-59, dir. research and devel., 1959-68; exec. v.p. Intel Corp., Santa Clara, Calif., 1968-75, pres., chief exec. officer, 1975-79, chmn., chief exec. officer, 1979—; dir. Micro Mask Inc., Silver King Ocean Farms., Varian Assocs. Inc., Transamerica Corp. Fellow IEEE; mem. Nat. Acad. Engring., Am. Phys. Soc., Electrochem. Soc. Subspecialties: Microchip technology (engineering); 3emiconductors. Current work: Electrical, microchip technology; semiconductors. Office: 3065 Bowers Ave Santa Clara CA 95051

MOORE, JAMES ROBERT, geol. oceanographer; b. Temple, Tex., May 18, 1925; s. James Robert and Mary Louise (Petty) M. B.S. with honors, U. Houston, 1951; M.A., Harvard U., 1954; Ph.D., U. Wales, 1964. Research geologist Standard Oil, Ohio, 1951-52; sr. scientist Texaco Research, 1956-66; chief marine geologist U.K.-Irish Sea Project, 1962-64; prof. U. Wis., Madison, 1966-77; also dir. marine lab.; prof., dir. Marine Sci. Inst., U. Alaska, 1977-79; prof. marine studies, dir. marine sci. inst. U. Tex., Austin, 1979—. Editor: Marine Mining, 1976—; editor-in-chief: Marine Series, 1978—; Contbr. articles to profl. jours. Served with USNR, 1943-46. Mem. Assn. Marine Mining (exec. sec.), Am. Assn. Petroleum Geologists, Soc. Econ. Mineralogists, Geochem. Soc., Challenger Soc. Oceanography, AAAS, Internat. Assn. Sedimentologists, Sigma Xi. Subspecialty: Geology. Home: PO Box 8178 Univ Tex Station Austin TX 78712 Office: Marine Inst Univ Tex Austin TX 78712

MOORE, JAY W., biologist, educator; b. Madison, Wis., Apr. 20, 1942; s. Millard and Leona (Miller) M.; m. Nancy E. Shimits, June 26, 1965; children: Meredith, Steven. B.S., Cedarville Coll., 1964; M.S., U. Nebr., 1966; Ph.D., U. Mass., 1970. Asst. prof. Eastern Coll., 1970-73, assoc. prof., 1973-78, prof., 1978—, chmn. dept. biology, 1974, 76, 82—, chmn. sci. div., 1978-82. Contbr. articles to profl. jours. Recipient Pres.'s trophy Cedarville Coll. Mem. AAAS, Am. Inst. Biol. Scis., Am. Genetics Assn., Sigma Xi, Sigman Zeta. Subspecialties: Animal genetics; Gene actions. Office: Eastern Coll Saint Davids PA 19087

MOORE, JESSE WILLIAM, NASA ofcl.; b. Newberry, S.C.; s. Haskel Woodrow and Eva Pearl (Bundrick) M.; m. Brenda Kay Polson, June 20, 1964; children—William Randolph, Stephanie Kay. B.S. in Elec. Engring, U. S.C., 1961, M.S., 1964. Engr. Gen. Dynamics Corp., Pomona, Calif., 1964-66; sr. engr. Jet Propulsion Labs., Pasadena, Calif., 1966-75; mgr. sci. and mission design Galileo Project, 1976-78; dir. Spacelab Flight Div., Office of Space Sci., NASA, Washington, 1979—. Recipient Creative Mgmt. award NASA, 1980, Exceptional Service award, 1981, Spacelab Level IV Implementation Group Achievement award, 1981. Mem. AIAA, Tau Beta Pi, Eta Kappa Nu, Pi Mu Epsilon. Lutheran. Club: Redeyed Bassmasters Va. (Vienna). Subspecialty: Astronautics. Home: 10406 Hunt Country Ln Vienna VA 22180 Office: NASA Hdqrs Code SM-8 Washington DC 20546

MOORE, JOHN ALEXANDER, biology educator; b. Charles Town, W.Va., June 27, 1915; s. George Douglas and Louise Hammond (Blume) M.; m. Anna Betty Clark, June 4, 1938; children: Sally, Nancy (dec.). A.B., Columbia, 1936, A.M., 1939, Ph.D., 1940. Tutor biology Bklyn. Coll., 1939-41; instr. biology Queens Coll., 1941-43; asst. prof. zoology Barnard Coll., 1943-47, asso. prof., 1947-50, prof., 1950-68, chmn. dept., 1948-52, 53-54, 60-66; asst. zoology Columbia, 1936-39, chmn. dept., 1949-52, prof., 1954-68; prof. biology U. Calif. at Riverside, 1969—; research asso. Am. Mus. Natural History, 1942—; Fulbright research scholar, Australia, 1952-53; Walker-Ames prof. U. Wash., 1966; mem. Commn. Sci. Edn., 1967-72, chmn., 1970-72. Author: Principles of Zoology, 1957, Heredity and Development, 1963, 2d edit., 1972, A Guide Book to Washington, 1963, Readings in Heredity and Development, 1972; co-author: Biological Science: An Inquiry into Life, 3d edit, 1973, Interaction of Man and The Biosphere, 3d edit, 1979; Editor: Physiology of the Amphibia, 1964, Ideas in Modern Biology, 1965, Ideas in Evolution and Behavior, 1970, Dobzhansky's Genetics of Natural Populations, 1981; editorial bd.: Ecology, 1949-52, Jour. Morphology, 1951-54; mng. editor, 1955-60, Am. Zoologist, 1960-63, 77-80. Guggenheim fellow, 1959. Mem. Marine Biol. Lab., AAAS, Genetics Soc. Am., Am. Soc. Zoologists (pres. 1974), Am. Soc. Naturalists (v.p. 1969, pres. 1972), Soc. Study Evolution (v.p. 1961, pres. 1963), Am. Soc. Ichthyologists and Herpetologists, Am. Acad. Arts and Scis., Nat. Acad. Scis. Club: Cosmos. Subspecialty: Evolutionary biology. Current work: Ecology and chromosomal variations in natural populations of Drosophila, history and politics of creationist movement; dynamics of academic libraries. history of technology. Home: 11522 Tulane Ave Riverside CA 92507

MOORE, JOHN DUAIN, plant science educator; b. Lancaster, Pa., Dec. 11, 1913; s. Willis Duain and Charlotte Blanche (Rote) M.; m. Doris Fretz Blakemore, June 13, 1940; children: Barbara Moore Henderson, John Duain. B.S., Pa. State U., 1939, Ph.D., U. Wis., 1943. Asst. prof. plant pathology U. Wis.-Madison, 1945-49, asso. prof., 1949-54, prof. plant pathology, 1954—, dir. Univ. Exptl. Farms, 1974-79; U.S. Dept. Agr. agt., plant pathologist, Madison, 1956-65, collaborator, 1965-71; collaborator U.S. AID/U. Wis. Project, U. Ife, Ile-Ife, Nigeria, 1968-70, head dept. plant sci., 1968-70, dean Grad. Sch., 1969, dean faculty agr., 1969-70; mem. study team Future Nigerian-U.S. Linkages in Higher Edn., Nigeria, 1977. Editor: (with D.C. Arny and R.N. Schwebke) Phytopathological Classic No. 11 (Investigations of the Brand Fungi and the diseases of plants caused by them with reference to grain and other useful plants), 1969; co-editor: Virus Diseases and Noninfectious Disorders of Stone Fruits in North America, 1976; contbr. articles profl. jours. Mem. Am. Phytopath. Soc. (asso. editor 1949-51), Bot. Soc. Am., AAAS, Am. Inst. Biol. Sci., Wis. Acad. Scis., Arts and Letters, Sigma Xi, Phi Eta Sigma, Phi Kappa Phi, Gamma Sigma Delta. Clubs: Rotary, Professional Men's (Madison). Subspecialty: Plant pathology. Home: 2918 Grandview Blvd Madison WI 53713

MOORE, JOHN GEORGE, JR., medical educator; b. Berkeley, Calif., Sept. 17, 1917; s. John George and Mercedes (Sullivan) M.; m. Mary Louise Laffer, Feb. 6, 1946; children: Barbara Ann, Douglas Terence, Bruce MacDonald, Martha Christine. B.A., U. Calif., Berkeley, 1939, M.D., 1942. Diplomate: Am. Bd. Ob-Gyn (pres. 1974-78, now chmn.). Asst. prof. U. Iowa, 1950-51; asso. prof. UCLA, 1951-65, prof., chmn. dept. ob-gyn, 1968—, Columbia U. Coll. Physicians and Surgeons, N.Y.C., 1965-68. Contbr. articles to profl. jours. Served to maj. M.C. U.S. Army, 1942-46. Decorated Silver Star, Bronze Star, Purple Heart.; NIH grantee U. Copenhagen; Royal Postgrad. Sch. Medicine, London. Mem. Soc. Gynecol. Investigation (pres. 1967), Assn. Profs. Gynecology and Obstetrics (pres. 1975), Western Assn. Gynecol. Oncologists (pres. 1976), ACS, Am. Coll. Obstetricians and Gynecologists, Am. Gynecol. Soc., Pacific Coast Obstet. and Gynecol. Soc., Los Angeles Obstet. and Gynecol. Soc., Soc. Gynecol. Investigation. Club: Malibu Riding and Tennis. Subspecialty: Obstetrics and gynecology. Home: 31804 Seafield Dr Malibu CA 90265

MOORE, JOHN HAYS, chemist, educator; b. Pitts., Nov. 6, 1941; s. John Hays and Mary Eva (Welfer) M.; m. Judy Williams, Aug. 10, 1963; children: John Hays, Victoria Inez. B.S., Carnegie Inst. Tech., 1963; M.A., Johns Hopkins U., 1965, Ph.D., 1967. Research assoc. Johns Hopkins U., Balt., 1967-79; asst. prof. chemistry U. Md., College Park, 1969-73, assoc. prof., 1973-77, prof., 1977—; vis. fellow Joint Inst. for Lab. Astrophysics, U. Colo.-Boulder, 1975-76; program officer NSF, 1980-81. Author: Building Scientific Apparatus, 1982; Contbr. numerous articles to profl. jours. Mem. Am. Phys. Soc., Am. Chem. Soc., Sigma Xi. Subspecialties: Physical chemistry; Atomic and molecular physics. Current work: Electron spectroscopy. Office: Chemistry Dept U Md College Park MD 20742

MOORE, LARRY WALLACE, plant pathology educator; b. Menan, Idaho, Aug. 24, 1937; s. Leland Wallace and Florence Martha (Bowles) M.; m. N. Marjean, Apr. 24, 1941; children: Christie Lee, Michael David, Suzanne, Jeffrey, Brian. B.S., U. Idaho, 1962, M.S., 1964; Ph.D., U. Calif.-Berkeley, 1970. Mem. faculty dept. botany Oreg. State U., Corvallis, 1969—, now assoc. prof. plant pathology. Mem. Am. Phytopath. Soc., AAAS, N.Y. Acad. Scis., Sigma Xi. Republican. Mormon. Subspecialties: Plant pathology; Microbiology. Current work: Microbial ecology, biological control of plant diseases; R-DNA biotechnology, beneficial microbes to enhance plant growth. Home: 1545 NW 13th St Corvallis OR 97330 Office: Dept Plant Pathology Oreg State U Corvallis OR 97331

MOORE, LAURENCE D., plant pathologist, educator; b. Danville, Ill., July 12, 1937; s. Jean W. and Marjorie (Spring) M.; m. Mary Ann Pichon, Aug. 16, 1958; children: Jean Martin, Susan Laura, Steven Dale, Ellen Spring. B.S., U. Ill., 1959; M.S., Pa. State U., 1961, Ph.D., 1965. Asst. prof. plant pathology Va. Poly Inst. and State U., Blacksburg, 1965-70, assoc. prof., 1970—, chmn. interdepartmental grad. curriculum in plant physiology, 1981-83. Contbr. chpts. to books, articles to profl. jours. Mem. exec. bd. Blue Ridge Mountain Council Boy Scouts Am., 1975—, dist. scout chmn., 1976-78, scoutmaster, 1972-76. Recipient Silver Beaver award Boy Scouts Am., 1976. Mem. Am. Phytopath. Soc. (pres. Potomac div. 1982-83), Am. Soc. Plant Physiologist, Va. Acad. Sci. Roman Catholic. Subspecialties: Plant pathology; Plant physiology (agriculture). Current work: Disease physiology, mechanism of resistance, role of steroids, effects of pollution stress. Home: 615 Broce Dr NW Blacksburg VA 24060 Office: VA Poly Inst and State U 401 Price Hall Blacksburg VA 24061

MOORE, LEONARD ORO, chemist; b. Payson, Utah, Sept. 20, 1931; s. Oro Huish and Ethel (Nuttall) M.; m. Elsa Rohrig, June 18, 1954; children: Mylan L., Lisa. B.S., Brigham Young U., 1953; Ph.D., Iowa State U., 1957. Cert. profl. chemist. Research chemist Union Carbide Corp., South Charleston, W. Va., 1958-63; project scientist The Ansul Co., 1963-68, project leader, 1968-72, mgr. research, Weslaco, Tex., 1972-76, mgr. research and devel., Marinette, Wis., 1976-82, mgr. extinguishants and specialty chems. research and devel., 1982—; dir. Rich-Mor, Inc., Springman King Printing. Contbr. articles to profl. jours. Charles Maw scholar, 1952-53; ISU DuPont teaching fellow, 1956-57; European Reasearch Assn. postdoctoral fellow, 1957-58. Mem. Am. Chem. Soc., AAAS, Am. Inst. Chemists, Sigma Xi. Mem. Ch. of Jesus Christ of Latter-day Saints. Lodge: Rotary. Subspecialties: Organic chemistry; Polymer chemistry. Current work: Fire extinguishants, chemical synthesis, free radical chemistry. Patentee in field. Office: 1 Stanton St Marinette WI 54143 Home: RB 192 Riverside Blvd Menominee MI 49858

MOORE, MARTHA MAY, genetic toxicologist; b. Durham, N.C., Feb. 17, 1949; d. E. Leon and Dorothea Virginia (McNeal) M.; m. Terry David O'Brian, Apr. 3, 1982. B.A. summa cum laude, Western Md. Coll., 1971; Ph.D. (NIH fellow), U. N.C., 1980. Research biologist genetic toxicology div. EPA, Research Triangle Park, N.C., 1977—. Contbr. articles to profl. jours. Active NOW. Mem. Environ. Mutagen Soc., Genetics Soc. Am., Genetic Toxicology Assn. Lutheran. Subspecialties: Environmental toxicology; Genetics and genetic engineering (biology). Current work: In vitro mammalian mutagenesis, types of genetic damage detected at Thymidine kinase locus of L5178Y/TK /- mouse lymphoma cells. Office: EPA Research MD 68 Toxicology Triangle Park NC 27711

MOORE, MICHAEL HART, automation engineer, consultant; b. Los Angeles, Aug. 7, 1938; s. Harvey Lee and Emily Ann (Hart) M.; m.; children: Douglas K., Andrew K. B.A., Williams Coll., 1960; M.A., UCLA, 1966; Ph.D., Cornell U., 1971. Assoc. prof. U. Fla., Gainesville, 1970-73; sr. system analyst Vector Research, Inc., Ann Arbor, Mich., 1973-76; prin. engr. Orincon Corp., San Diego, Calif., 1977-80; mng. partner MHM Cons., San Diego, 1980—; sr. automation engr. System Devel. Corp., San Diego, 61980-82; mgr. artificial intelligence project Systems Exploration, Inc., San Diego, 1982—. Served to lt. comdr. USN, 1960-63. N.Y. State Regents fellow, 1968, 69. Mem. Am. Assn. Artificial Intelligence, Assn. Unmanned Vehicles Systems, Ops. Research Soc. Am., Soc. Indsl. and Applied Math. Republican. Subspecialties: Artificial intelligence; Robotics. Current work: Automation via computer-hosted algorithms. Inventor controller algorithms for gen. weapon systems, 1980, controller for robot aircraft, 1982. Office: Sci Applications inc 1200 Prospect St San Diego CA 92037

MOORE, POLLY, applied mathematician, scientific computing manager; b. Oberlin, Ohio, Aug. 1, 1947; d. James Warren and Mary (McGuckin) M.; m. Stuart E. Builder, July 4, 1981. B.A., Oberlin Coll., 1969; M.S., Wesleyan U., 1972; Ph.D., U. Wash., 1979. Applied mathematician Boeing Co., Seattle, 1979-80; sr. research mathematician Merck & Co., Rahway, N.J., 1980-82; mgr. sci. computing Genentech, Inc., South San Francisco, Calif., 1982—; math. cons. Polymath, Inc., Belmont, Calif., 1982—. Named to Top 40 Westinghouse Sci. Talent Search, 1965. Mem. Am. Math. Soc., Soc. Indsl. and Applied Math., Phi Beta Kappa. Democrat. Congregationalist. Subspecialties: Applied mathematics; Mathematical software. Current work: Mathematics and associated software as applied to problems of genetic engineering. Home: 2827 Wemberly Dr Belmont CA 94002 Office: Genentech Inc 460 Point San Bruno Blvd South San Francisco CA 94080

MOORE, RICHARD HARLAN, college dean, biology researcher; b. Houston, Sept. 16, 1945; s. Russell Lewis and Hazel (Harlan) M.; m. Robin Morris, May 14, 1977; 1 son, Merlin Morris. B.A., Vanderbilt U., 1967; M.A., U. Tex.-Austin, 1970, Ph.D., 1973. Staff scientist Environ. Cons., Inc., Dallas, 1973-74; asst. prof. biology Coastal Carolina Coll., Conway, S.C., 1974-78, assoc. prof., 1978—, chmn. div. sci., 1978-79, dean Sch. sci., 1979—. Author: (with H. D. Hoese) Fishes of the Gulf of Mexico, 1977; contbr. articles to profl. jours. Leader Coastal Bend council Boy Scouts Am., Pt. Aransas, Tex., 1970-73; cub scout leader Coastal Carolina council, Pawley's Island, S.C., 1978-82; sec. adv. bd. trustees Waccamaw Sch., Pawley's Island, 1981—; bd. dirs. S.C. Crawfish Festival Assn., Pawleys Island, 1980—, Waccamaw chpt. Nat. Audubon Soc., Myrtle Beach, S.C., 1981—. Mem. Am. Soc. Zoologists, Soc. Systematic Zoology, Am. Soc. Ichthyologists and Herpetologists, Am. Fisheries Soc., Southeastern Fishes Council. Episcopalian. Subspecialties: Marine Zoology; Evolutionary biology. Current work: Physiological ecology and evolutionary biology of aquatic organisms especially fish. Home: PO Box 513 Murrell's Inlet SC 29576 Office: Coastal Carolina College PO Box 1954 Conway SC 29526

MOORE, ROBERT HALDANE, III, pharmacologist; b. Brownsville, Tenn., Dec. 10, 1946; s. Robert Haldance and Janie (Burks) M.; m. Jane Kelly Bishop, June 22, 1968; 1 son, Robert; m. Teresa Lee Teter, Feb. 14, 1980; children: Timothy, Patricia. B.S., U. Tenn., 1969, Ph.D. in Pharmacology, 1973. Registered pharmacist, Tenn., Mo. Research assoc. Lafayette Clinic, Detroit, 1973-75; asst. prof. dept. physiology Kirksville (Mo.) Coll. Osteo. Medicine, 1975-81, asst. prof. dept.

pharmacology, 1981-82, assoc. prof., 1982—. Mem. Am. Soc. Pharmacology and Exptl. Therapeutics, Soc. Neuroscis., Nat. Soc. for Med. Research, Sigma Xi. Subspecialties: Neuropharmacology; Pharmacology. Current work: Endorphins, cardiovascular system, neuropharmacology, drug abuse, effects of endorphins on cardiovascular system. Home: Route 3 Kirksville MO 63501 Office: Kirksville Coll Osteo Medicine 800 W Jefferson Kirksville MO 63501

MOORE, W(ALTER) E(DWARD) C(LADEK), microbiologist, researcher, educator; b. Rahway, N.J., Oct. 12, 1927; s. Rollins W. and Anstes D. C. M.; m. Hester J. Barrus, June 18, 1948; children: David R., Howard M., Jay F. B.S., U. N.H., 1951; M.S., U. Wis.-Madison, 1952, Ph.D., 1954. Asst. prof. anaerobic microbiology Va. Poly. Inst. and State U., 1954-59, assoc. prof., 1965—, Univ. Disting. prof., 1980—, head dept. anaerobic microbiology, 1967—; vice-chmn. jud. com. Internat. Com. Systematic Bacterial, 1978—. Contbr. numerous articles to profl. jours., 1955—; editor lab. manual, 1977; Internat. Jour. Systematic Bacteriology, 1982—. Recipient Research award Va. chpt. Gamma Sigma Delta, Kimble award Conf. Pub. Health Dirs., 1973. Mem. Am. Soc. Microbiology, Internat. Commn. Systematic Bacteriology, AAAS, Internat. Assn. Dental Research. Subspecialties: Microbiology (medicine); Systematics. Current work: Anaerobic bacteriology, systematics, infectious disease, normal flora analytical methods, taxonomy. Home: 616 Fairview Ave Blacksburg VA 24060 Office: Dept Anaerobic Microbiology Va Poly Inst and State U Blacksburg VA 24061

MOORE, WILLIAM E., physicist; b. Madison, Wis., Feb. 23, 1947; s. Wayne and Catherine (Hood) M.; m. Joan J. Jordan, Dec. 28, 1974. B.S., U. Wis.-Madison, 1969; M.S., Calif. Inst. Tech., Ph.D. in Physics, 1975. Asst. prof. U. Wash., 1975-82; sr. research physicist Eastman Kodak Research Labs., Rochester, N.Y., 1982—. Contbr. articles to profl. jours. NIH grantee, 1975-82. Mem. AAAS, Am. Inst. Physics, Am. Assn. Physicists in Medicine, Soc. Photog. Scientists and Engrs. Subspecialties: Imaging technology; Diagnostic radiology. Current work: Devel. new med. imaging systems. Office: Eastman Kodak Co Bldg 81 Rochester NY 14650

MOORE, WILLIAM FRED, plant pathologist; b. Selma, Ala., Jan. 24, 1938; s. Fred Charles and Mary Elizabeth (Rentz) M.; m. Mary Beth Hamblin, Apr. 16, 1966; children: Catherine Estelle, Christina Marie. Student, Livingston State Coll., 1957-59; B.S., U. So. Miss., 1963; M.S., Miss. State U., 1965; postgrad., Ark. State U., 1969; Ph.D., Miss. State U., 1974. Lab. teaching asst. U. So. Miss., 1961-63; grad. asst. Miss. State U., 1967-68, extension plant pathologist, 1965-74, leader extension plant pathology dept., 1974—; extension specialist in plant pathology. Recipient 1st Miss. Coop. award of excellence, 1977, Miss. County Agts. Achievement award, 1974, Faculty Achievement award Miss. State U. Alumni Assn., 1981, Disting. Service award So. Soybean Assn., 1981. Mem. Am. Phytopath. Soc., Soc. Nematologists, Miss. County Agts. Assn., Gamma Sigma Delta, Epsilon Sigma Phi. Roman Catholic. Subspecialty: Plant pathology. Current work: Diseases and their control of soybeans, rice and corn.

MOORE, WILLIAM MARSHALL, physical chemist, educator; b. Lincoln, Nebr., Dec. 25, 1930; s. William H. and Emma S. (Rolloff) M.; m. Patricia A. Roth, Sept. 1, 1958; children: William H., Toby B., Sean T., Sophia J.; m. Susan J. Packer, July 23, 1968. B.A. in Chemistry, Colo. Coll., 1952; Ph.D. in Phys. Chemistry, Iowa State U., 1959. Research assoc. Calif. Inst. Tech., 1958-59; NIH research fellow Cambridge (Eng.), U., 1959-60; prof. chemistry Utah State U., Logan, 1960—. Served to cpl. U.S. Army, 1953-55. Mem. Am. Chem. Soc., AAUP, ACLU. Democrat. Subspecialties: Photochemistry; Kinetics. Current work: Studies on kinetics of molecules suitable for solar energy conversion by photochemistry, surface chemistry of halocarbons on hot metal surfaces. Home: 241 W 1st St N Smithfield UT 84335 Office: Dept Chemistry Utah State University Logan UT 84322

MOORE, WILLIAM WALTER, pollution control co. exec.; b. Plainfield, N.J., Apr. 13, 1924; s. George Thomas and Helen Elizabeth (Cullinan) M.; m. (married), Dec. 2, 1950; children: George T., William D., Alan J., Barbara J. B.S. in Mech. Engring, U. Notre Dame, 1945; postgrad., Advanced Mgmt. Program, Harvard U., 1961. Successively service engr., estimator, sales engr., dist. sales mgr., v.p. sales, v.p., div. mgr., dir. Research-Cottrell, Bound Brook, N.J., 1946-71; pres., chmn. bd., chief exec. officer, dir. Belco Pollution Control Corp., Parsippany, N.J., 1971—; chmn. Belco Mexicana; dir. Foster Wheeler Mexicana; past mem. Nat. Adv. Com. on Air Pollution Control Techniques. Contbr. articles to electrostatic precipitation to profl. lit. Served to lt. USNR, 1942-46. Mem. ASME, Air Pollution Control Assn., Indsl. Gas Cleaning Inst. (past pres., dir.). Republican. Roman Catholic. Subspecialties: Environmental engineering; Gas cleaning systems. Current work: Improved techniques on electrostatic precipitators. Home: 34 Hillcrest Rd Martinsville NJ 08836 Office: Belco Pollution Control Corp 119 Littleton Rd Parsippany NJ 07054

MOORE-EDE, MARTIN CHRISTOPHER, physiologist, researcher; b. London, Nov. 22, 1945; s. Roderick and Margaret Hatton (Riggall) M.-E.; m. Donna Smith, May 27, 1977; 1 son, Andrew. B.Sc. in Physiology, U. London, 1967; M.B., B.S., Guy's Hosp. Med. Sch., U. London, 1970; Ph.D. in Physiology, Harvard U., 1974. Instr. physiology Guy's Hosp. Med. Sch., 1970; intern in medicine Toronto East Gen. Hosp., 1970-71; research fellow in surgery Peter Bent Brigham Hosp., 1971-74; assoc. in surgery and physiology Harvard Med. Sch., 1974-75, asst. prof. physiology, 1975-81, assoc. prof. physiology, 1981—; dir. research lab. investigating the circadian timing system, 1974—; physiologist dept. surgery Brigham and Women's Hosp., 1982—; chmn. commn. on circadian rhythms and sleep physiology Internat. Union Physiol. Scis.; chmn. internat. scientific adv. bd. Center for Design of Indsl. Schedules. Author: The Price of Defense, 1978, The Clocks That Time Us, 1982, Mathematical Models of the Circadian Sleep-Wake Cycle, 1983; Assoc. editor: Am. Jour. Physiology, Regulatory, Integrative and Comparative Physiology; contbr. numerous articles on the circadian timing system to profl. jours. Mem. Boston Study Group, 1974—; mem. Internat. Physicians for Prevention of Nuclear War, 1981—. Frank Knox fellow Harvard Med. Sch., 1971; Sci. Research Council fellow, 1971; Warren fellow Harvard U., 1972; recipient Cate award for scholarship Peter Bent Brigham Hosp., 1973; Andrew W. Mellon faculty award Harvard Med. Sch., 1975; Career Devel. award NIH, 1977-82. Fellow Brit. Interplanetary Soc.; mem. AAAS, Am. Fedn. Clin. Research, Internat. Soc. Chronobiology, Am. Physiol. Soc., Am. Soc. for Photobiology, Aerospace Med. Assn., Fedn. Am. Scientists, Sleep Research Soc., Endocrine Soc., Soc. Neurosci. Subspecialties: Physiology (biology); Space medicine. Current work: Anatomy and physiology of the circadian timing system. Regulation of the biological clocks that control the sleep-wake cycle. Applications of circadian theory to clinical medicine, occupational shift work schedules and aerospace medicine. Office: Dept Physiology and Biophysics Harvard Med Sch 25 Shattuck St Boston MA 02115

MOORHEAD, DEBORAH KAY, systems engineer; b. Davenport, Iowa, Oct. 1, 1952; d. Eugene Warren and Velma Jane (Thorson) M. B.S. in Math, Ariz. State U., 1974; M.S.E.E., U. So. Calif., 1982. Data analyst Integrated Systems Support, Inc., Holloman AFB, N.Mex., 1975; systems analyst System Devl. Corp., Santa Monica, Calif., 1975-79; mem. tech. staff Aerospace Corp., El Segundo, Calif., 1979—. Contbr. articles to profl. jour. First v.p. women's com. Aerospace Corp., 1982-83. Mem. Assn. Computing Machinery (chmn. spl. interest group of software engring. Los Angeles chpt. 1982-83), Assn. Women in Computing (membership v.p. 1982-83), AIAA, Women in Mgmt., Phi Beta Kappa, Phi Kappa Phi. Subspecialties: Software engineering; Distributed systems and networks. Current work: Consultant to other departments and Air Force in distributed software and hardware, computer security, software and hardware interfaces, display terminals. Office: Aerospace Corp 2350 E El Segundo Blvd El Segundo CA 90245

MOORHOUSE, DOUGLAS CECIL, engring. cons. co. exec.; b. Oakland, Calif., Feb. 24, 1926; s. Cecil and Lynda (Roe) M.; m. Donis L. Slinker; children—Scott, Jan. B.S. in Civil Engring, U. Calif., Berkeley, 1950, postgrad., 1961; student, Advanced Mgmt. Program, Harvard U., 1973. Research and resident engr. State of Calif. Div. Hwys., 1950-59; dir. San Diego office Woodward-Clyde & Assos., 1959-62; pres. Woodward-Moorhouse & Assos., 1962-73; pres., chief exec. officer Woodward-Clyde Cons., San Francisco, 1973—; dir. J.E. Higgins Lumber Co. Contbr. articles to profl. jours. Trustee World Coll. West; mem. adv. com., dept. engring. U. Calif. Berkeley. Served with inf. U.S. Army. Mem. Am. Arbitration Assn., ASCE (Wesley W. Horner award 1979), Earthquake Engring. Research Inst., Harvard Soc. Engrs. and Scientists, Soc. Am. Mil. Engrs. Clubs: Harvard Bus. Sch. Assn. No. Calif., Engineers. Subspecialty: Civil engineering. Office: 600 Montgomery St 30th Floor San Francisco CA 94111

MOORMAN, GARY WILLIAMS, plant pathologist, educator; b. Albany, N.Y., Jan. 30, 1949; s. Thomas and Mary (Rosenson) M.; m. Frances Elaine Sulya, June 12, 1971; 1 dau., Sara Marian. B.S., U. Maine, 1971; M.S., U. Vt., 1974; Ph.D., N.C. State U., 1978. Asst. Prof. dept. plant pathology Pa. State U., University Park, 1983—. Mem. Am. Phytopath. Soc., Mycological Soc. Am. Subspecialties: Plant pathology; Microbiology. Current work: Epidemiology of soil borne plant diseases. Office: 211 Buckhout Lab University Park PA 16802

MOOS, HENRY WARREN, physicist, educator, cons.; b. N.Y.C., Mar. 26, 1936; s. Henry H. and Dorothy (Warren) M.; m. Doris Elaine McClure, July 13, 1957; children: Janet, Paul, Daniel, David. Sc.B., Brown U., 1957; M.A., U. Mich., 1959, Ph.D. in Physics, 1962. Research assoc. Stanford U., 1961-63, acting asst. prof. physics, 1963-64; asst. prof. physics Johns Hopkins U., Balt., 1964-68, assoc. prof., 1968-71, prof., 1971—; cons.; mem. govt. coms. NASA, Dept. of Energy; co-investigator Apollo 17 Ultraviolet Spectrometer and Voyager Ultraviolet Spectrometer. Contbr. numerous articles profl. jours. Sloan fellow, 1965-69; vis. fellow Joint Inst. Lab. Astrophysics/Lab. Atmospheric and Space Physics, U. Colo., 1972-73, 81-82. Fellow Am. Phys. Soc.; mem. Am. Astron. Soc. Subspecialties: Plasma physics; Ultraviolet high energy astrophysics. Current work: Ultraviolet spectroscopy of high temperature fusion plasmas, ultraviolet astronomy of the planets. Home: 804 Post Boy Ct Towson MD 21204 Office: Physics Dept Johns Hopkins University Baltimore MD 21218

MOPPER, KENNETH, marine organic chemist, oceanography educator; b. Phila., Jan. 14, 1947; s. Samuel S. and Miriam (Kofsky) M.; m. Pia H. Schalin, Apr. 16, 1977; children: Susanna J., Daniel J. B.A. in Chemistry, Queens Coll., 1968; M.S. in Oceanography, MIT, 1971, Ph.D., 1973. Postdoctoral fellow U. Goteborg, Sweden, 1973-75; guest researcher, 1977-80; staff researcher U. Hamburg, W.Ger., 1975-77; asst. prof. Coll. Marine Studies, U. Del., Lewes, 1980—. Contbr. chpts. to books, articles to profl. jours. Am.-Scandinavian Found. grantee, 1973; Swedish Sci. Research Council grantee, 1977; NIH grantee, 1981; NSF grantee, 1982. Mem. AAAS, Geochem. Soc. (organic chemistry div.), Am. Soc. Limnology and Oceanography, Sigma Xi. Democrat. Jewish. Subspecialties: Organic geochemistry; Oceanography. Current work: Research on biogeochemical processes involving organic compounds in marine environment; development of ultrasensitive shipboard methods for measuring dissolved organics. Home: 53 Sussex Dr Lewes DE 19958 Office: Coll Marine Studies 700 Pilotttown Rd Lewes DE 19958

MORABITO, DAVID DOMINIC, engr.; b. Los Angeles, Jan. 27, 1952; s. Dominic Don and Mary (Matosian) M. B.S. in E.E, U. So. Calif., 1974, M.S., 1976, Engr. in E.E., 1983. Lic. radio telephone operator FCC, 1971. Electronics test technician Hoffman Electronics, El Monte, Calif., 1973; engr. Jet Propulsion Lab., Pasadena, Calif., 1973—. Contbr. articles in field to profl. jours. Mem. IEEE, Am. Astron. Soc., Am. Geophys. Union. Subspecialties: Radio and microwave astronomy; Cosmology. Current work: Astronomy, geodesy; studing extragalactic radio sources using very long baseline interferometry. Office: 4800 Oak Grove Dr Pasadena CA 91103

MORAHAN, PAGE SMITH, microbiology educator; b. Newport News, Va., Jan. 7, 1940; d. Robert B. Smith. B.A. in Chemistry with honors, Agnes Scott Coll., 1961; M.A. in Bilogy, Hunter Coll., 1964; Ph.D. in Microbiology, Marquett U., 1969. Asst. prof. microbiology Med. Coll. Va., Va. Commonwealth U., Richmond, 1970-74, assoc. prof., 1974-81, prof., 1981-82; prof. microbiology Med. Coll. Pa., Phila., 1982—, chmn. dept. microbiology and immunology, 1982—; Mem. cancer research manpower rev. com. Nat. Cancer Inst., 1976-81; course dir. W. Alton Jones Cell. Sci. Ctr., Lake Placid, N.Y., 1980. Contbr. articles to profl. publs. Recipient Research Career Devel. award NIH, 1974-79. Mem. Am. Assn. Immunologists, Reticuloendothelial Soc., Soc. Exptl Biology and Medicine, Am. Soc. Mircobiology, Am. Assn. Cancer Research. Subspecialties: Microbiology (medicine); Infectious diseases. Current work: Microbiology and immunology, immunotherpay, immunotoxicology, macrophages and viruses. Home: 2042 Wallace St Philadelphia PA 19130 Office: Med Coll Pa Dept Microbiology and Immunology 3300 Henry Ave Philadelphia PA 19129

MORAN, THOMAS FRANCIS, chemistry educator; b. Manchester, N.H., Dec. 11, 1936; s. Francis Leo and Mamie Marie M.; m. Joan Elinor Belliveau, June 25, 1960; children: Dorothy, Michael, Linda, Mary. B.A., St. Anselm's Coll., 1958; Ph.D., Notre Dame U., 1962. AEC postdoctoral fellow Brookhaven Nat. Lab., Upton, L.I., N.Y., 1962-64, assoc. scientist, 1964-66; asst. prof. Ga. Inst. Tech., Atlanta, 1966-68, assoc. prof., 1968-72, prof., 1972—. Contbr. numerous articles on chemistry to profl. jours. Recipient Ferst research award, 1970; Danforth assoc., 1971—. Mem. Am. Chem. Soc., Am. Phys. Soc., Am. Soc. for Mass Spectrometry, AAAS, Sigma Xi. Subspecialties: Physical chemistry; Kinetics. Current work: Gaseous ionic collision phenomena, kinetics and mechanisms of gaseous ion-molecule reactions, electron impact phenomena, energy conversion processes in chemical reactions. Office: Chemistry Dept Ga Inst Tech Atlanta GA 30332 Home: 2324 Annapolis Ct NE Atlanta GA 30345

MORAN, WILLIAM RODES, exploration company executive; b. Los Angeles, July 29, 1919; s. Robert Breck and Edna Louise (Venable) M. A.B., Stanford U., 1942, postgrad., 1942-43. Registered profl. geologist, Calif.; cert. petroleum geologist. Mgr., instr. Stanford Geol. Survey, 1942; geologist Union Oil Co. Calif., Los Angeles, 1943-46, Union Oil Paraguay, 1946-50; sr. geologist Union Oil Co. Calif., Nev. and Utah, 1950-51, Cia. Petrolera de Costa Rica, 1951, spl. exploration dept., parent co., 1952-59; acting mgr. Union Oil Devel. Corp., Sydney, Australia, 1959-60, sr. geologist fgn. ops., parent co., 1960-63, v.p., mgr., dir., Los Angeles, 1963—, v.p., dir., Australia, 1968—, v.p. exploration, Los Angeles, 1979—; dir. Que. Columbium Ltd., Montreal, Can., Geosat, Inc., San Francisco. Author: (with T. Fagan) Melba, 1983; series Ency. Victor Recordings, 1982—; contbr. numerous tech. papers, hist. sound recs. to profl. lit. Hon. curator Stanford Archive Recorded Sound, 1978—. Fellow Geol. Soc. Am., Geol. Soc. London, AAAS; life mem. Soc. Econ. Geologists, Am. Assn. Petroleum Geologists (assoc. editor Bull. 1959—, Disting. Founder energy minerals div. Denver 1980). Republican. Presbyterian. Clubs: Chaparral of Los Angeles, Univ. of Los Angeles. Subspecialty: Geology. Current work: Mineral exploration; historical sound recordings. Home: 1335 Olive Ln La Cañada CA 91011 Office: Molycorp Inc subs Union Oil Co of Calif PO Box 54945 Los Angeles CA 90054

MORAVEC, HANS PETER, research scientist; b. Kautzen, Austria, Nov. 30, 1948; came to Can., 1952; came to U.S., 1971. Student in Engring, Loyola Coll., Montreal, Can., 1967; B.S. in Math, Acadia U. N.S., Can., 1969; M.Sc. in Computer Sci, U. Western Ont., 1971; Ph.D., Stanford U., 1980. Research asst. Artificial Intelligence Lab., Stanford (Calif.) U., 1971-80; research scientist Robotics Inst. and dept. computer sci. Carnegie-Mellon U., Pitts., 1980—; cons. Contbr. articles to profl. jours., presentations and filmstrips to profl. confs.; developer pamphlets and programs in computer sci. Mem. AAAS, ACM, Brit. Interplanetary Soc., IEEE, Nat. Space Inst., Space Studies Inst. Subspecialties: Artificial intelligence; Graphics, image processing, and pattern recognition. Current work: Mobile robots, exotic approaches to three dimensional modelling and graphics, multi-branched architectures for super-dextrous robot manipulators, large structures for space travel. Office: Robotics Inst Carnegie-Mellon U Pittsburgh PA 15213

MORDFIN, LEONARD, mechanical engineer, consultant; b. Bklyn., June 23, 1929; s. Samuel and Margaret (Flyer) M.; m. Norma Marcia, Oct. 10, 1954; children: Stephen Jay, Theodore Gary, Robin Ilene. B.M.E., Cooper Union, 1946; M.S., U. Md., 1954, Ph.D., 1966. Engr. Nat. Bur. Standards, Washington, 1950-67, 69-77, dep. program mgr. for nondestructive evaluation, 1977—; phys. sci. adminstr. Office Aerospace Research, Arlington, Va., 1967-69; cons. in field. Contbr. numerous articles to profl. jours.; contbr. to ency.; editor: (with Fong and Dobbyn) Critical Issues in Materials and Mechanical Engineering, 1981. Mem. ASME, Am. Soc. Nondestructive Testing, ASTM, Soc. Exptl. Stress Analysis. Jewish. Subspecialties: Materials (engineering); Theoretical and applied mechanics. Current work: Materials testing, both mechanical and nondestructive; program management for developing new and improved standards for nondestructive evaluation of material, structures. Home: 1609 Billman Ln Silver Spring MD 20902 Office: Nat Bur Standards B312 Physics Bldg Washington DC 20234

MORELL, JONATHAN ALAN, psychology educator; b. Bklyn., July 10, 1946; s. Barry and Dorothy (Fleisher) M.; m. Margaret Elizabeth Pacer, Aug. 17, 1975; children: Daniela Sophia, Aliza Maria. B.A., McGill U., Can., 1968; M.S., Northwestern U., 1972, Ph.D., 1974. Chief of evaluation Nat. Tng. Ctr. on Drug Abuse, Chgo., 1974; research assoc. dept. psychiatry U. Chgo., 1974; assoc. prof. psychology Hahnemann U., Phila., 1975—; cons. to various orgns., 1976—. Author: Program Evaluation in Social Research, 1979; editor-in-chief: Evaluation and Program Planning, 1979—; author articles. Mem. Evaluation Research Soc., Eastern Evaluation Research Soc., Am. Psychol. Assn., Evaluation Network, Soc. Psychologists in Substance Abuse. Subspecialty: Social psychology. Current work: Program evaluation; use of social science for decision making; substance abuse. Home: 511 S 18th St Philadelphia PA 19146 Office: Hahnemann U 112 N Broad St Philadelphia PA 19102

MORELL, PIERRE, biochemist, educator; b. Dominican Republic, Dec. 10, 1941; came to U.S., 1942, naturalized, 1955; s. Anatol G. and Halina (Leczynska) M.; m. Bonnie Jean Brown, Sept. 26, 1965; children: Sharon, David. A.B., Columbia U., 1963; Ph.D., Albert Einstein Coll. Medicine, 1967. Research assoc. U. Mich., Ann Arbor, 1968-69; asst. prof. Albert Einstein Coll. Medicine, Bronx, N.Y., 1969-73; assoc. prof. biochemistry U. N.C., Chapel Hill, 1973-77, prof., 1977—, also dir. Mem. editorial bd.: Brain Research. Mem. Am. Soc. Biol Chemists, Am. Soc. for Neurochemistry, Internat. Brain Research Orgn., Internat. Soc. for Neurochemistry, Soc. for Neurosci., Internat. Soc. for Devel. Neurosci. Subspecialty: Neurochemistry. Current work: Metabolism of myelin, axonal transport, environmental pollutants and brain development. Home: 404 Brookside Dr Chapel Hill NC 27514 Office: U NC 322 BSRC 22OH Chapel Hill NC 27514

MORENO, EDGARD CAMACHO, chemist, researcher; b. Bogota, Colombia, Jan. 16, 1927; came to U.S., 1953; s. Ignacio E. and Maria Luisa (Camacho) M.; m. Helena C. Cassis, Dec. 20, 1953; 1 son, Edgard I. B.S., Nat. U. Colombia, 1948; Ph.D., U. Calif.-Berkeley, 1957. Research chemist TVA, Mussle Shoals, 1957-62; research assoc. Nat. Bur. Standards, Washington, 1962-69; sr. staff mem. Forsyth Dental Ctr., Boston, 1969—. Contbr. numerous articles to profl. jours. Nat. Inst. Dental Research grantee, 1969—; recipient cert. recognition Dept. Commerce, 1969. Fellow AAAS; mem. Am. Chem. Soc., Internat. Assn. Dental Research, Am. Assn. Dental Research, Sigma Xi. Subspecialties: Physical chemistry; Cariology. Current work: Chemical models of dental caries formation; adsorption of proteins and peptides onto apatitic surfaces; crystal growth of calcium phosphates from dilute solutions. Home: 182 Wilson Rd Nahant MA 01908 Office: Forsyth Dental Ctr 140 Fenway Boston MA 02115

MOREST, DONALD KENT, neurobiologist, educator; b. Kansas City, Mo., Oct. 4, 1934; s. F. Stanley and Clara Josephine (Riley) M.; m. Rosemary R., June 13, 1963; children: Lydia R., D. Claude. B.A., U. Chgo., 1955; M.D., Yale U., 1960. Asst. prof. anatomy U. Chgo., 1963-65; asst. prof. Harvard Med. Sch., Boston, 1965-70, assoc. prof., 1970-77; prof. U. Conn., Farmington, 1977—. Mem. Internat. Soc. Developmental Neurosci., Am. Assn. Anatomists, Soc. Neurosci., Assn. Research Otolaryngology. Subspecialties: Neurobiology; Anatomy and embryology. Current work: Teacher, researcher in the structure, function and development of the auditory system. Office: U Conn Health Ct Dept Anatomy Farmington CT 06032

MOREWITZ, HARRY ALAN, engineering consultant, researcher; b. Newport, News, Va., June 2, 1923; s. Jacob Louis and Sallie Florence (Rome) M.; m. Myra Kalkin, June 20, 1948; children: Ralph Steven, Dara Beth Morewitz Pearlman. B.S. in Physics, Coll. William and Mary, 1943, M.A., Columbia U., 1949, Ph.D., NYU, 1953. Project engr. Metavac, Inc., Long Island City, N.Y., 1952-53; supervisory scientist Westinghouse Bettis Labs., Pitts., 1953-59; project mgr. Atomics Internat. div. Rockwell Internat., Canoga Park, Calif., 1959-82; engring. cons., prin. H. M. Assocs., Ltd., Tarzana, Calif., 1982—; lectr. in med. physics UCLA, 1967-71; mem. adv. com. on reactor physics AEC, 1967-73; cons. reactor safety Electric Power REsearch Inst., Palo Alto, Calif., 1949—; mem. com. safety of nuclear installations Group of Experts on Nuclear Aerosols In Reactor Safety, OECD, Paris, 1982—. Served to 1st lt. USMCR, 1943-46. Mem. Am. Nuclear Soc., Am. Phys. Soc., Health Physics Soc., Am. Geophys. Union, Sigma Xi. Democrat. Jewish. Subspecialties: Nuclear fission;

Nuclear engineering. Current work: Investigation of nuclear reactor accident source terms, i.e. the amount and timing of released radioactive material following major reactor accidents. Co-inventor, patentee reactor safety fuse. Office: H M Assocs Ltd 5300 Bothwell Rd Tarzana CA 91356

MOREY, ELSIE D., paleobotanist, laboratory executive; b. Cambridge, Mass., Nov. 25, 1940; d. William Culp and Helen Marie (Hilsman) Darrah; m. Philip R. Morey, June 10, 1967. B.A., W. Va. U., 1963; M.S., So. Ill. U., 1965. Research technician Tex. A&M Research Sta., U.S. Dept. Agr., 1973-74; owner Morey Paleobotany Lab., Lubbock, Tex., 1977-82, Morgantown, W. Va., 1982—; histology technician dept. pathology W.Va. U. Med. Center. Contbr. articles to profl. jours. Mem. Bot. Soc. Am., Paleontology Soc., Sigma Xi. Subspecialties: Morphology; Paleobiology. Current work: Fossil woods, ferns, lepidodendrons. Home and Office: 200 Wagner Rd Morgantown WV 26505

MOREY, PHILIP RICHARD, industrial hygienist, educator; b. Cleve., July 3, 1940; s. Everett F. and Helen M. (Szabo) M.; m. Elsie Louise Darrah, June 10, 1967. B.S. in Biology, U. Dayton, 1962; M.S., Yale U., 1964, Ph.D., 1967. Cert. Am. Bd. Indsl. Hygiene. Lectr. Harvard U., Cambridge, Mass., 1967-70; mem. faculty Tex. Tech U., Lubbock, 1970-82, prof. biology, 1967-82; research indsl. hygienist environ. invesitations br. Nat. Inst. Occupational Safety and Health, Morgantown, W.Va., 1982—; instr. Mining Acad., Beckley, W.Va., 1982—; adj. prof. Wood Sci. Sch. Forestry, W.Va. U., Morgantown, 1982—. Author: How Trees Grow, 1973; contbr. articles to profl. publs. Chmn. United Way Drive for Biol. Scis., Tex. Tech. U., 1980. Recipient Faculty Research award Coll. Arts and Scis., Tex. Tech U., 1980. Mem. Am. Indsl. Hygiene Assn., Bot. Soc. Am. (Albert E. Diamond Fund award 1975), Panhellenic Assn. (Tex. Tech U. Teaching award 1972), Am. Conf Govtl. Indsl. Hygienists, Am. Acad. Indsl. Hygiene, Am. Soc. Agronomy. Republican. Roman Catholic. Subspecialties: Health services research; Environmental engineering. Current work: Industral hygiene aspects of indoor air pollution; quantification of airborne microbes in office ventilation systems. Home: 200 Wagner Rd Morgantown WV 26505 Office: 944 Chestnut Ridge Rd Morgantown WV 26505

MORFOPOULOS, ARIS PAUL, oceanographic services company official; b. Montreal, Que., Can., July 19, 1953; s. Christos Anastasios and Athanasia (Nounopoulos) M.; m. Marilyn Bittman, Sept. 27, 1979. B.Commerce with honors, U. Man. Mgr. oceanographic surveying Can-Dive Services Ltd., North Vancouver, B.C., Can., 1977—. Mem. Marine Tech. Soc. (sec. Vancouver 1979-80, vice chmn. 1980-81, 81-82). Subspecialties: Oceanography; Offshore technology. Current work: Development of manned state-of-the-art underwater vehicles; Arctic search and location techniques through ice.

MORGAN, ALEXANDER PAYNE, industrial mathematician; b. Savannah, Ga., Aug. 12, 1945; s. Dimitri Dejanikus and Lucretia Payne M. B.S., U. Ga., 1967; M.A., Yale U., 1968, Ph.D., 1975. Instr. U. Miami, Coral Gables, 1975-76; systems engr. Med. Coll. Ga., Augusta, 1976-77; sr. systems analyst Savannah River Plant, Aiken, S.C., 1977-78; staff research scientist Gen. Motors Research Labs., Warren, Mich., 1978—. Served with U.S. Army, 1970-72. NSF grad. fellow, 1967-68. Mem. Soc. Indsl. and Applied Math. Subspecialties: Applied mathematics; Mathematical software. Current work: Solving systems of nonlinear algebraic equations; qualitative theory of differential equations. Office: Math Dept Gen Motors Research Labs Warren MI 48090

MORGAN, BRIAN LESLIE GORDON, nutrition educator; b. Gillingham, Kent, Eng., Jan. 6, 1947; came to U.S., 1975; s. Sydney William Gordon and Grace Hanna (Milner) M.; m. Roberta Eddie, Apr. 8, 1973. B.Sc., London U., 1971, M.Sc., 1972, Ph.D., 1975. Post-doctoral fellow Inst. Human Nutrition, Columbia U., 1975-78, assoc., 1978-79, asst. prof., 1979—; cons. John Wiley & Sons, Holt, Rheinehart & Winston, Self and Family Circle mags., New York, 1981—. Author: Diet and Nutrition Program for Your Heart, 1982, Diet and Nutrition Program for Hypertension, 1983, The Lifelong Nutrition Guide, 1983, The High Carbohydrate Weight Loss Program, 1983, Brown Adipose Tissue, 1984. Mem. N.Y. Acad. Sci. (animal research com.), N.Y. Acad. Medicine (nutrition edn. in med./dentral schs. com.), British Nutrition Soc., Am. Inst. Nutrition, Harvey Soc., Am. Assn. Dental Schs., N. Am. Assn. Study of Obesity. Subspecialty: Nutrition (medicine). Current work: Thermogenesis, brown adipose tissue, and its role in obesity, function of gangliosides and glycoproteins in neurotransmission, nutrition and brain development. Office: Inst Human Nutrition Columbia U Coll Physicians & Surgeons 701 W 168th St New York NY 10011

MORGAN, CHARLES HERMANN, JR., psychologist, educator; b. N.Y.C., Mar. 14, 1949; s. Charles Hermann and Miriam (Wagner) M.; m. Ruth-Anne Hein, Aug. 16, 1969; children: Jennifer Susan, Benjamin Charles. B.A., Columbia U., 1971; M.A., New Sch. Social Research, 1974; Ph.D., U. Fla., 1979. Lic. clin. psychologist, Ky. Intern Fairfield Hills Hosp., Newtown, Conn., 1978-79; cons. Bur. Rehab., Thelma, Ky., 1979-82; pvt. practice psychology Morehead (Ky.) Clinic, 1979—; asst. prof. psychology Morehead State U., 1979—; inservice trainer Gateway Dist. Health Dept., Owingsville, 1981. Contbr. articles to profl. jours. Mem. United Campus Ministry, Morehead State U., 1980—, chmn., 1981. Mem. Am. Psychol. Assn., Southeastern Psychol. Assn., Eastern Psychol. Assn., Ky. Psychol. Assn. (co-chmn. ethics com.). Democrat. Presbyterian. Current work: Psychotherapy and hypnotherapy with adults, teaching of clinical psychology, research on stressful life events. Home: 224 Lyons Ave Morehead KY 40351 Office: Morehead State Univ UPO 1336 Morehead KY 40351

MORGAN, DONALD O'QUINN, veterinarian, researcher; b. Star, N.C., Mar. 24, 1934; s. Ernest Addision and Lillie (O'Quinn) M.; m. Sue Brown, Jan. 17; children: Ernest, Margaret. B.S., N.C. State U., 1955, M.S., 1963; D.V.M., U. Ga., 1959; Ph.D., U. Ill., 1967. Instr. N.C. State U., 1962; diagnostician N.C. Dept. Agr., 1963; instr. U. Ill., 1966; vet. med. officer U.S. Dept. Agr., 1967-69; asst. prof. U. Ky., 1969-72, assoc. prof., 1972-74, U. Ill., 1974-76; vet. med. officer U.S. Dept. Agr., Greenport, N.Y., 1976—. Contbr. articles to profl. jours. Mem. AVMA, Am. Soc. Microbiology. Subspecialties: Immunobiology and immunology; Virology (veterinary medicine). Current work: Foot and mouth disease vaccines. Home: PO Box 724 Cutchogue NY 11935 Office: US Dept Agr PO Box 848 Greenport NY 11944

MORGAN, JACK BRANDON, steel co. exec.; b. New Kensington, Pa., Jan. 12, 1922; s. Robert and Mamie Recelia (Thompson) M.; m. Peggy Snyder, Sept. 2, 1950. B.S., U. Pitts., 1948. Research engr. Alcoa Co., New Kensington, Pa., 1948-56; sr. supervising metallurgist Allegheny Ludlum Steel Corp., Brackenridge, Pa., 1956—. Contbr. articles in field to profl. jours. Served with U.S. Army, 1943-45. Fellow Am. Soc. Nondestructive Testing; mem. ASTM, ASME, Am. Soc. Metals. Republican. Subspecialties: Metallurgy; Nondestructive Testing. Current work: Sensor development for process monitoring and control. Patentee in field. Home: 2740 Edith St Lower Burrell PA 15068 Office: Research Ctr Brackenridge PA 15014

MORGAN, LUCIAN LLOYD, aerospace engineer; b. Wichita Falls, Tex., Dec. 22, 1928; s. J. Hugh and Myrtle Irene (Huffman) M.; m. Dorothy Rea Dill, Sept. 14, 1950; children: Larry Rea, Lauri Louann. B.S., Tex. A&M U., 1949; M.S. in Nuclear Engring, So. Methodist U., 1958; student, Stanford U., 1964-65; M.S. in Systems Mgmt, U. So. Calif., 1975. Registered profl. engr., Calif., Tex. Supr. chem. engring. U.S. Gypsum Co., Sweetwater, Tex., 1949-50; asst. chief engring. test Gen. Dynamics, Ft. Worth, 1950-56, project nuclear engr., 1956-60, mgr. advanced propulsion, San Diego, 1960-62; project leader Lockheed Missiles & Space Co., Sunnyvale, Calif., 1962-80, mgr. systems engring., 1980—. Served to capt. AUS, 1952-54. Recipient NASA Pub. Service award, 1981. Mem. AIAA, Air Force Assn. Republican. Mem. Ch. of Christ. Subspecialties: Laser weapons; Aerospace engineering and technology. Current work: Engineering of acquisition, tracking and pointing systems for laser weapons. Home: 2029 Kent Dr Los Altos CA 94022 Office: Lockheed Missiles and Space Co Orgn 59-12 Bldg 593 Sunnydale CA 98802

MORGAN, PAUL WINTHROP, chemist; b. West Chesterfield, N.H., Aug. 30, 1911; s. Herbert and Olive (Lermond) M.; m. Elsie Louise Bridges, Aug. 27, 1939; children—Dennis Lee, Patricia Morgan Harding. B.S. in Chemistry, U. Maine, 1937; Ph.D. in Organic Chemistry, Ohio State U., 1940. Postdoctoral fellow Ohio State U., Columbus, 1940-41; with E.I. duPont de Nemours & Co., Wilmington, Del., 1941-76, research fellow, 1957-73, sr. research fellow, 1973-76; chem. cons., West Chester, Pa., 1976—; cons., lectr. in field; chmn. Gordon Research Conf. on Polymers, 1974. Author: Condensation Polymers, 1965; contbr. articles to profl. jours. Asst. dist. commr. Minquas Trail dist. Chester County council Boy Scouts Am., 1956-61; asst. scoutmaster Chester County council, 1961-76, chmn. troop com., 1976—. Recipient Silver Beaver award Boy Scouts Am., 1967, Swinburne award Plastics and Rubber Inst., London, 1978, Engring. Materials Achievement award Am. Soc. Metals, 1978. Mem. Am. Chem. Soc. (Best Publ. of Yr. award Del. chpt. 1959, 78, nat. Polymer Chem. award 1976, Midgley award Detroit chpt. 1979), Nat. Acad. Engring., Franklin Inst. (Howard N. Potts medal 1976), Mineral. Soc. Pa., Sierra Club, Appalachian Trail Conf., Wilderness Soc., Audubon Soc., Nat. Wildlife Assn., Early Am. Industries Assn., Chester County (Pa.) Hist. Soc., Fiber Soc. (hon.). Subspecialties: Polymer chemistry; Polymers. Current work: Consultant in polymer science and technology; specialist in condensation polymers, high tenacity fibers, liquid crystalline systems. Research on low temperature polycondensation, heat resistant fibers, high-strength, high-modulus fibers. Patentee polymers and fibers. Home and Office: 822 Roslyn Ave West Chester PA 19380

MORGAN, STEPHEN LYLE, software engineer; b. Nashville, Oct. 20, 1949; s. Howard Edwin and Helena Mae (Lawson) M.; m. Renee Ellen Lautzenhiser, June 1, 1974; children: Jonathan, Geoffrey. B.A., Johns Hopkins U., 1971; M.S., Pa. State U., 1974. Grad. asst. Pa. State U., State College, 1971-74; cons., State College, 1974-77; engr. HRB-Singer, Inc., State College, 1977-79, advanced engr., 1979-81, sr. engr., 1981—; mayor's fellowship intern City of Balt., 1971. Mem. Nat. Computer Graphics Assn., Am. Assn. Geographers, Am. Congress on Surveying and Mapping, Am. Soc. Photogrammetry, Assn. Computing Machinery. Methodist. Clubs: Optimist (pres. 1981-82), Pa. Guild Craftsman, Nittany Valley Gem and Mineral. Subspecialties: Graphics, image processing, and pattern recognition; Algorithms. Current work: Automated cartography, digital terrain analysis addressing problems of movement in space and locational analysis. Home: 335 Douglas Dr State College PA 16803 Office: HRB-Singer Inc Dept 123 Box 60 Science Park State College PA 16803

MORGAN, WILLIAM WILSON, astronomer, educator; b. Bethesda, Tenn., Jan. 3, 1906; s. William Thomas and Mary McCorkle (Wilson) M.; m. Helen Montgomery Barrett, June 2, 1928 (dec. 1963); children: Emily Wilson, William Barrett; m. Jean Doyle Eliot, 1966. Student, Washington and Lee U., 1923-26; B.S., U. Chgo., 1927, Ph.D., 1931; D.Honoris Causa, U. Cordoba, Argentina, 1971; D.Sc. (hon.), Yale U., 1978. Instr. Yerkes Obs., U. Chgo., Williams Bay, Wis., 1932-36, asst. prof., 1936-43, asso. prof., 1943-47, prof., 1947-66, Bernard E. and Ellen C. Sunny Distinguished prof. astronomy, 1966-74, prof. emeritus, 1974—, chmn. dept. astronomy, 1960-66; dir. Yerkes and McDonald Observatories, 1960-63; mng. editor Astrophys. Jour., 1947-52; Henry Norris Russell lectr. Am. Astron. Soc., 1961. Author: (with P.C. Keenan, Edith Kellman) An Atlas of Stellar Spectra, 1943, (with H.A. Abt and J.W. Tapscott) Revised MK Spectral Atlas for Stars Earlier than the Sun, 1978; contbr. research articles to profl. publs. Recipient Bruce gold medal Astron. Soc. Pacific, 1958, Henry Draper medal Nat. Acad. Scis., 1980. Mem. Am. Acad. Arts and Scis., Nat. Pontifical acads. scis., Royal Danish Acad. Scis. and Letters, Royal Astron. Soc. (assoc., Herschel medal 1983), Nat. Acad. Scis. Argentina, Soc. Royale des Sciences de Liege. Conglist. Office: Yerkes Observatory Box 258 Williams Bay WI 53191

MORGANSTEIN, STANLEY, engineering psychologist, computer scientist; b. N.Y.C., June 4, 1940; s. Louis and Rose M. B.S., Bklyn. Coll., 1962; M.S., Lehigh U., 1964; Ph.D., U. Mass., 1970. Prof. psychology Claflin U., Orangeburg, S.C., 1970-72; asst. prof. psychology William Patterson Coll., Wayne, N.J., 1972-74; analyst Binary Data Corp., Plainview, N.J., 1974-76; systems engr. Four Phase Corp., N.Y.C., 1976-77; programmer, analyst European Am. Bank, Westbury, L.I., 1977-79; sr. programmer Interactive Data Corp., N.Y.C., 1979-80; engring. psychologist Pacific Missile Test Center, Point Mugu, Calif., 1980—. Contbr. articles to profl. jours. Lehigh U. tuition scholar, 1962; MIT fellow, 1965; U. Mass. fellow, 1967. Mem. Am. Psychol. Assn., AAAS, AIAA, Sci. Adv. Bd. (sec. 1981-82), Sigma Xi. Subspecialties: Engineering psychology; Software engineering. Current work: Research in the human machine interface including computers, simulation and aircraft/missile applications. Home: 6250 Telegraph Rd Ventura CA 93003 Office: Pacific Missile Test Center Point Mugu CA 93042

MORIN, LAWRENCE PORTER, neuroscientist, researcher, educator; b. Portsmouth, N.H., Aug. 11, 1947; s. Lawrence J. and Barbara A. (Porter) M.; m. Margaret Burdick, Sept. 14, 1947; children: Abigail A., Jennifer B. A.B., Brown U., 1969; Ph.D., Rutgers U., 1974. NIMH postdoctoral fellow dept. psychology U. Calif., Berkeley, 1974-76; faculty dept. psychology Dartmouth Coll., 1976-81; research scientist L.I. Research Inst. Health Scis. Center; research asst.prof. psychiatry SUNY, Stony Brook, 1982—. Contbr. articles to profl. jours. Mem. Soc. Neurosci., AAAS, Internat. Soc. Chronobiology. Subspecialties: Psychobiology; Chronobiology. Current work: Reproductive system and biological rhythms. Circadian, Ultradian, rhythms, sex differences, hamsters, hormones, neural control, environmental influences. Home: 34 Old Post Rd Setauket NY 11733 Office: Dept Psychiatry and Behavioral Science HSC 10T SUNY Stony Broo NY 11794

MORITA, TOSHIO, nuclear engineer; b. Kanazawa-shi, Japan, Jan. 7, 1930; came to U.S., 1971; s. Ihei and Tokiko M.; m. Mitsuyo Kawai, Mar. 30, 1955. B.Engring., Tokyo Inst. Tech., 1953, D.Engr., 1963. Tech. officer Ministry of Internat. Trade and Industry, Tokyo, 1953-61; engr., mgr. Mitsubishi Atomic Power Industries, Tokyo, 1961-71; fellow; engr. Westinghouse Electric Corp., Pitts., 1971—. Violinist Shinagawa Symphony Orch., Tokyo, 1960, Wilkinsburg Symphony Orch., Pitts., 1970. Mem. Am. Nuclear Soc., Atomic Energy Soc. Japan. Subspecialties: Nuclear engineering; Numerical analysis. Current work: Research and development for pressurized water reactor plant, mostly reactor design and control area. Inventor reactor design, constant axial offsect control. Home: 1601 Penn Ave Apt 913 Pittsburgh PA 15221 Office: Westinghouse Electric Corp Monroeville Mall Blvd Monroeville PA 15146

MORRA, MICHAEL ANTHONY, airport ofcl., psychologist; b. N.Y.C., Jan. 17, 1934; s. Rocco Anthony and Maria Ann (Buchiarelli) M.; m.; 1 son from previous marriage: Rocco. B.S., N.Y. U, 1957; M.A., Columbia U., 1959; Ph.D., U. Tenn., 1965. Instr. computers Unival Corp., N.Y.C., 1958-59; fellow NASA, Houston, 1967; asst. prof. Purdue U., Indpls., 1966-69; pvt. clin. psychologist, Wichita, 1969-72; asst. mgr. KGI Airport, Kansas City, Mo., 1975—; lectr. U. Mo., Kansas City, 1975—; docent Nelson Art Mus., Kansas City, 1976-82. Host TV series, Sta. KAKE, 1971. Lilly Found. grantee, 1968-69. Mem. Am. Psychol. Assn., Assn. Aviation Psychologists, Am. Assn. Airport Execs. (exec.), Mo. Psychol. Assn. Mem. Unity Ch. Club: Fine Arts Singles. Subspecialties: Airport psychology; Behavioral psychology. Current work: Human factors; transportation systems; clinical applications to community problems; positive value systems and mental health; use of simulators in airfield safety. Home: 5016 Grand Ave Kansas City MO 64112 Office: Kansas City Internat Airport PO Box 20047 Kansas City MO 64195

MORRE, D. JAMES, biochemist; b. Owensville, Mo., Oct. 20, 1935; s. Harvey Henry and Donna Marie (Maurer) M.; m. Dorothy M. Wibberg, Aug. 25, 1956; children: Connie Marie, Jeffrey Thomas, Suzanne Anette. B.S., U. Mo., 1957; M.S., Purdue U., 1959; Ph.D., Calif. Inst. Tech., 1962. Asst. prof. Purdue U., 1963-66, asso. prof., 1966-71, prof., 1971—, dir., 1976—. Subspecialties: Biochemistry (biology); Membrane biology. Current work: Research in membrane biochemistry, membrane biogenesis, glycolipids and tumorigenesis. Office: Purdue Cancer Cente Pharmacy Bldg Purdue Uni West Lafayette IN 47907

MORRIS, CHARLES REGINALD, instrument and control engineer; b. Halifax, N.S., Can., July 29, 1953; came to U.S., 1978; s. George Reginald and Lise Ida (Brunet) M.; m. Pamela Jean Blissett, May 11, 1974. B.S. in Elec. Engring, McMaster U., 1975. Registered profl. engr., Ont. Jr. engr. in trng. Ont. Hydro, Rolphton, Ont., 1975-77, asst. tech. supr., Tiverton, 1978; jr. engr. Kans. Gas and Electric, Wichita, 1978-79, engr., 1979-80, engr. III, 1980-81, sr. engr., 1981-82, lead engr., 1982—. Mem. Am. Nuclear Soc., Assn. Profl. Engrs. Province Ont. Subspecialties: Nuclear engineering; Electronics. Current work: Design review for instrumentation and controls for the Wolf Creek Nuclear Plant. Responsible for all nuclear instrumentation and controls, radiation monitoring equipment and plant security. Home: 1637 Timberline Dr Rose Hill KS 67133 Office: Kans Gas and Electric Co 201 N Market Wichita KS 67201

MORRIS, DON MELVIN, surgery educator, surgeon; b. Longview, Tex., Jan. 4, 1946; s. John Raymond and Martha (Walker) M.; m. Judy Ray Miller, Sept. 7, 1970 (div. 1979); 1 son, Curtis John Walker; m. Katherine Ruth Crowe, June 26, 1982. A.S., Kilgore Jr. Coll., 1966; B.A. in Biology, U. Tex.-Austin, 1968; M.D., U. Tex. Med. Br., 1972. Diplomate: Am. Bd. Surgery. Intern Bexar County Hosp., U. Tex. Med. Sch., San Antonio, 1972-73; resident U. Md. Hosp., Balt., 1973-75, 76-78, Balt. Cancer Research Ctr., 1975-76; instr. in surg. oncology U. Md. Hosp., Balt., 1978-80, asst. prof. surg. oncology, 1980-81; asst. prof. surgery La. State U. Med. Ctr., Shreveport, 1981—, dir. breast cancer detection clinic, 1981—, prin. investigator, 1981—; cons. to hosps. Contbr. articles on breast, head and neck and colon cancer and other topics to profl. jours. Am. Cancer Soc. Jr. clin. faculty fellow, 1978-81. Fellow ACS; mem. AMA, Soc. Surg. Oncology, Am. Soc. Clin. Oncology, Alpha Omega Alpha. Republican. Mem. Christian Ch. (Disciples of Christ). Subspecialties: Surgery; Chemotherapy. Current work: Breast cancer, head and neck cancer, colon cancer, surgical techniques and devices, electrical potentials of cancerous tissues, monolclonal antibodies. Home: 5210 Foxglove Dr Bossier City LA 71112 Office: La State U Med Ctr 1501 Kings Hw PO Box 33932 Shreveport LA 71130

MORRIS, EDWARD KNOX, JR., behavioral science educator, researcher; b. Bryn Mawr, Pa., May 4, 1948; s. Edward Knox and Catherine Ruth (Merryman) M. B.S. in Psychology, Denison U., 1970, M.A., U. Ill., 1974, Ph.D., 1976. Asst. prof. dept. human devel. and psychology (courtesy) U. Kans., Lawrence, 1975-81, assoc. prof., 1981—. Bd. editors: Behaviorism, 1979—, Psychological Record, 1978—, Mexican Jour. Behavior Analysis, 1979—, Behavior Analyst, 1982—; contbr. chpts. to books, articles to profl. publs.; invited speaker convs. Active ACLU, NOW, Nat. Abortion Rights Action League. Mem. Am. Psychol. Assn. (pub. info. officer div. 25 1982—), Am. Soc. Criminology, AAAS, Assn. Behavior Analysis, Assn. Advancement of Behavior Therapy, Nat. Council Crime and Delinquency, Soc. Research on Child Devel., Psi Chi. Club: Lawrence Track. Subspecialties: Behavioral psychology; Developmental psychology. Current work: Experimental analysis of child behavior (schedules of reinforcement, stimulus control); applied behavior analysis of crime and delinquency; history, theories, systems of psychology. Home: 2112 Vermont St Lawrence KS 66044 Office: Dept Human Devel Univ Kans Haworth Hall Lawrence KS 66045

MORRIS, GEORGE VINCENT, chemist, educator; b. Providence, Nov. 18, 1930; s. Patrick J. and Mary A. (McKenna) M.; m. Mary Louise Morris, June 6, 1959; children: Susan E., Jennifer M. B.S., Providence Coll., 1952; M.S., U.R.I., 1957, Ph.D., 1962. Research chemist Eltex Chem. Co., 1957-60; Naval Underwater Systems Ctr., Newport, R.I., 1962-65; with Raytheon Co., Portsmouth, R.I., 1966—; now sr. chemist, prof. chemistry and physics Salve Regina Coll., Newport, 1963—; cons. in field. Mem. Republican City Com., Riverside, R.I., 1970-76. Served with U.S. Army, 1952-54. NSF fellow, 1968-69; grantee, 1968. Mem. Am. Chem. Soc. (chmn. R.I. sec 1982—). Subspecialties: Physical chemistry; Materials. Current work: Kinetics of reactions in solution; thermochemistry; decomposition of inorganic solids, physical properties of composite materials, instrumentation. Patentee in field. Home: 41 Merritt Rd Riverside RI 02915 Office: Salve Regina Coll Newport RI 02840

MORRIS, JAMES GRANT, animal scientist, educator; b. Brisbane, Australia, Aug. 30, 1930; came to U.S., 1969; s. Fredrick Grant and Mary (Wadley) M.; m. Jocelyn, June 20, 1959; children: Elizabeth Anne, Ione Julie, James Peter Grant. B.Agr. Sc., U. Queensland, Brisbane, 1953, B.Agr.Sc. with 1st class honors, 1955, B.Sc., 1958, M.Agr.Sc., 1959; Ph.D., Utah State U., 1961. Dir. Husbandry research Animal Research Inst., Yeerongpilly, Brisbane, 1964-69; assoc. prof. animal sci. and physiol. chemistry U. Calif., Davis, 1969-75, prof., 1975—. Research numerous publs. in field. Recipient award F.A. Brodie, Australian Meat & Livestock Corp., 1970. Mem. Am. Inst. Nutrition, Nutrition Soc. (U.K.), Am. Soc. Animal Sci., Brit. Soc. Animal Prodn., Australian Soc. Animal Prodn. Subspecialties: Animal nutrition; Nutrition (biology). Current work: Comparative animal nutrition. Home: Route 1 Box 2034 Davis CA 95616 Office: U Calif Dept Animal Sci Davis CA 95616

MORRIS, JOHN WILLIAM, industrial engineer; b. Huntington Park, Calif., Aug. 3, 1921; s. William and Anna Cecelia (Burke) M.; m.

Kathleen Mary Hinds, Aug. 19, 1923; children: Shawn M., Muriel A., Margaret L. B.Engring., U. So. Calif., 1951. Engr., project engr. chief product engr. Preco Inc., Los Angeles, 1951-63; chief engr., plant mgr. Metal Improvement Co., Hackensack, N.J., 1963-68; sr. supervising engr. Ameron Co., Los Angeles, 1968-69, dir. equipment engring., 1969-75, plant mgr., Spartanburg, S.C., 1975—; lectr. engring. U. So. Calif., 1962-63. Served with inf. AUS, 1942-46. Mem. Am. Soc. Metals, ASME (dir. machine design div. 1958). Republican. Episcopalian. Club: Piedmont. Lodge: Rotary. Subspecialties: Mechanical engineering; Materials processing. Current work: Mechanical design; menagement of plant producing fiber glass reinforced epoxy pipe. Patentee load control devices for r.r. cars, improvements in shot peening machinery. Home: 414 Overland Dr Spartanburg SC 29302 Office: Ameron Co PO Box 5723 Spartanburg SC 29304

MORRIS, J(OSEPH) ANTHONY, microbiologist, public interest organization official; b. Marboro, Md., Sept. 6, 1918; s. Charles Lafayette and Essie (Stokes) M.; m. Ruth Savoy, Nov. 1, 1942; children: Carol Ann, Marilyn T., Joseph A., Larry A. B.Sc., Cath. U. Am., 1940, M.Sc., 1942, Ph.D., 1947. Asst. scientist Josiah Macy Jr. Found., N.Y.C., 1943-44; virologist Depts. Agr., Interior, Laurel, Md., 1944-47; virologist, chief hepatitis virus research Walter Reed Army Inst. Research, Washington, 1947-56; virologist, asst. chief dept. virus and rickettsial diseases U.S. Army Med. Command, Japan, 1956-59; virologist chief sect. respiratory viruses, div. biologics standards NIH, Bethesda, Md., 1959—; dir. slow latent and temperate virus br. FDA, Bethesda, 1972-76; lectr. dept. microbiology U. Md., College Park, 1977-79; vice chmn., Bell of Atri, Inc., College Park, 1979; cons. Commn. on Influenza, Armed Forces Epidemiologic Bd., 1960—, Nat. inst. Neurol. Diseases and Blindness, 1962—. Mem. Soc. Tropical Medicine and Hygiene, Soc. Am. Microbiologists, Soc. Exptl. Biology and Medicine, Am. Assn., Immunologists, N.Y. Acad. Sci. Subspecialties: Microbiology; Virology (biology). Current work: Cause and prevention of infectious diseases; research on infectious hepatitis, respiratory disease of virus etiology and zoonosis. Discoverer of respiratory sycytial virus. Office: PO Box 40 College Park MD 20740 Home: 23-E Ridge Rd Greenbelt MD 20770

MORRIS, LUCIEN ELLIS, anesthesiologist, educator, researcher; b. Mattoon, Ill., Nov. 30, 1914; s. James Lucien and Pearl (Ellis) M.; m. Ethel Jean Pinder, June 27, 1942; children: James Lucien, Robert Pinder, Sara Jean Morris Hoffman, Donald Charles, Laura Lee Morris Bean. A.B., Oberlin Coll., 1936; M.D., Western Res. U., 1943. Diplomate: Am. Bd. Anesthesiology. Intern Grassland Hosp., Valhalla, N.Y.; resident in anesthesia U. Wis., 1946-48, instr. anesthesia, 1948-49; asst. prof. anesthesia U. Iowa, 1949-51, assoc. prof., 1951-54; prof., head anesthesia U. Wash., Seattle, 1954-60, clin. prof., 1961-68; prof. U. Toronto, Ont., Can., 1967-70; chmn. dept. anesthesia Med. Coll. Ohio, 1970-80, prof. anesthesiology, 1970—; mem. WHO travelling med. faculty, Israel, Iran, 1951; mem. com. on anesthesia NRC, 1956-61; dir. Anesthesia Research Labs., Providence Hosp., Seattle, 1960-68; external examiner Coll. Medicine, U. Lagos, Nigeria, 1977; vis. prof. London Hosp. Med. Coll., 1980-81. Author articles and chpts. on anesthesiology and pharmacology; author monographs. Elder Presbyn. Ch., Mercer Island, Wash. Served to capt. M.C. U.S. Army, 1944-46. Fellow Royal Soc. Medicine, Faculty Anaesthetists Royal Coll. Surgeons; mem. Assn. Univ. Anesthetists, Am. Soc. Anesthesiologists (del. World Fedn. Socs. Anesthesiology 1960-64), Internat. Anesthesia Research Soc., Assn. Anaesthetists Gt. Britain and Ireland, Can. Anesthetists Soc., Am. Soc. Pharmacology, Soc. Exptl. Biology and Medicine, Internat. Assn. Study Pain, Am. Soc. Regional Anesthesia, Anaesthetic Research Soc. (Eng.). Subspecialties: Bioinstrumentation; Anesthesiology. Current work: Improvement of anesthesia equipment; monitors for anesthetic gases. Inventor copper kettle vaporizer, other anesthesia equipment. Home: 3425 Bentley Blvd Toledo OH 43606

MORRIS, MARK ROOT, astronomy educator; b. Aberdeen, Wash., Sept. 2, 1947; s. Donald William and Mary Elizabeth (Root) M.; m. Francoise Alice Queval, Mar. 26, 1976; 1 son, Mathieu Alexandre. B.A. in Physics, U. Calif.-Riverside, 1969, Ph.D., U. Chgo., 1974. Research fellow Owens Valley Radio Obs., Calif. Inst. Tech., Pasadena, 1974-77; asst. prof. astronomy Columbia U., N.Y.C., 1977-82; prof. assoc. Groupe D'Astrophysique Universite Scientifique et Médicale de Grenoble, France, 1981-82, 83; vis. research assoc. UCLA, 1982, adj. assoc. prof. astronomy, 1983—. NSF grantee, 1979, 82-84. Mem. Am. Astron. Soc., Internat. Astron. Union, Internat. Union Radio Sci. Subspecialties: Radio and microwave astronomy; Theoretical astrophysics. Current work: Studies of the interstellar medium of the circumstellar envelopes of evolved stars and of the galactic center. Office: Dept Astronom UCLA Los Angeles CA 90024

MORRIS, MELVIN LEWIS, dentist, educator; b. N.Y.C., Nov. 28, 1914; s. David and Rose (Harris) M.; m. Muriel R. Liebling, Sept. 19, 1943; children: Barry, Stephen, David. B.S., CCNY, 1934; M.A., Columbia U., 1937, D.D.S., 1941. Diplomate: Am. Bd. Periodontology. Dental intern Mt. Sinai Hosp., 1941-42; with periodontia dept. Polyclinic Hosp., 1946-47; chief periodontia clinic Mt. Sinai Hosp., 1947-52; instr. periodontia Columbia U., 1950-53, asst. clin. prof., 1953-56, assoc. clin. prof., 1958-69, clin. prof., 1969—; instr. periodontia 1st Dist. Dental Soc., 1948—; cons. periodontia Franklin D. Roosevelt Hosp., 1953-59, Castle Point Hosp., 1953-56; asst. attending dentist Presby. Hosp., 1974-75, assoc. attending dentist, 1975—; lectr. Contbr. articles profl. jours., chpts. to books. Served to capt. Dental Corps AUS, 1942-46; ETO. Recipient Isidor Hirschfeld Meml. award Northeastern Soc. Periodontists, 1979. Mem. Am. Acad. Periodontology, 1st Dist. Dental Soc., Eastern Dental Soc. (pres. 1955), Am. Soc. Periodontists, Sigma Xi, Omicron Kappa Upsilon. Democrat. Jewish. Subspecialty: Periodontics. Office: 654 Madison Ave New York NY 10021

MORRIS, RALPH WILLIAM, pharmacologist; b. Cleveland Heights, Ohio, July 30, 1928; s. Earl Douglas and Viola Minnie (Mau) M.; m. Virginia Myrtha Lynn, June 4, 1955; children—Christopher Lynn, Kirk Stephen, Timothy Allen and Todd Andrew (twins), Melissa Mary. B.A., Ohio U., Athens, 1950, M.S., 1953; Ph.D., U. Iowa, 1955; postgrad., Seabury-Western Theol. Sem., 1979—. Research fellow in pharmacology, then teaching fellow U. Iowa, 1952-55; asst. prof. dept. pharmacognosy and pharmacology Coll. Pharmacy; prof. pharmacology Med. Center, U. Ill., 1969—, adj. prof. edn., 1976—; vis. scientist San Jose State U., Calif., 1982—; mem. adv. com. 1st aid and safety Midwest chpt. ARC, 1972—; cons. in drug edn. to Dangerous Drug Commn., Ill. Dept. Public Aid, Chgo. and suburban sch. dists. Referee and contbr. articles to profl. and sci. jours., lay mags., radio and TV appearances. Trustee Palatine (Ill.) Public Library, 1967-72, pres., 1969-70; trustee N. Suburban Library System, 1968-72, pres., 1970-72, mem. long-range planning com., 1975-81; mem. Title XX Ill. Citizens' Adv. Council, 1981—. Recipient Golden Apple Teaching award U. Ill. Coll. Pharmacy, 1966; cert. of merit Town of Palatine, 1972. Mem. AAAS, Am. Assn. Coll. Pharmacists, Am. Pharm. Assn., Ill. Pharm. Assn., Internat. Soc. Chronobiology, Am. Soc. Pharmacology and Exptl. Therapeutics, Drug Info. Assn., Am. Library Trustee Assn., Ill. Library Trustee Assn. (v.p. 1970-72, dir. 1969-72), Sigma Xi, Rho Chi, Gamma Alpha. Episcopalian. Subspecialties: Chronobiology; Pharmacology. Current work: Medical applications of biological rythms to health, disease and drug therapy; positive drug education instead of drug abuse scare tactics, positive relationships amongst biological rhythms,drug responses and religious beliefs. Home: 9710 Hillandale Ln Richmond IL 60071 Office: 833 S Wood St Chicago IL 60612

MORRIS, RICHARD JULES, clinical psychology educator, researcher; b. Chgo., Feb. 19, 1942; s. George A. and Pearl (Wohl) Koenig; m. Yvonne Paula Goodman, June 3, 1967; children: Stephanie, Michael, Jacquelyn. B.S., U. Wis.-Madison, 1963; M.A., Roosevelt U., 1965; Ph.D., Ariz. State U., 1970. Cert. psychologist, Ariz. Asst. prof. psychology Syracuse U., 1970-74, assoc. prof., 1974-78; clin. asst. prof. pediatrics SUNY Upstate Med. Ctr., Syracuse, 1974-78; assoc. prof. spl. edn. U. Ariz., 1978-80, prof., 1980—; cons. Ariz. Tng. Program, Tucson, 1978—; adv. editor for edn. Scott, Foresman Co., Chgo., 1981—. Author: Behavior Modification with Children, 1976, (with T. Kratochwill) Treating Fears and Phobias in Children, 1982; editor: Perspectives in Abnormal Behavior, 1974, Practice of Child Therapy, 1983; cons. editor: Rehab. Psychology Jour, 1974—, Mental Retardation Jour, 1979—. Vocat. Rehab. Adminstrn. fellow, 1967; USPHS fellow, 1968-69. Mem. Am. Psychol. Assn. (treas. Div. 22 1978-81), AAAS, Assn. Advancement Behavior Therapy, Am. Assn. Mental Deficiency, AAUP, Sigma Xi. Subspecialties: Behavioral psychology; Clinical psychology. Current work: Mental retardation; social skills training in children; fear reduction methods; child therapy. Home: 6881 N Solaz Segundo Tucson AZ 85718 Office: U Ariz Dept Spl Edn Tucson AZ 85721

MORRIS, ROBERT, computer scientist; b. Boston, July 25, 1932; s. Walter White and Helen Hortense Kelly; m. Anne Burr Farlow, June 9, 1962; children: Meredith, Robert, Benjamin. A.B., Harvard U., 1956, A.M., 1957. Analyst Ops. Research Office, Johns Hopkins U., Balt., 1957-60; with Bell Telephone Labs., Murray Hill, N.J., 1960-81, Whippany, N.J., 1981—; vis. prof. elec. engring. U. Calif., Berkeley, 1966-67; cons. Air Force Studies Bd., Washington, 1982—. NRC. Contbr. several sci. articles to profl. publs. Mem. Am. Math. Soc., Sigma Xi. Subspecialties: Computer architecture; Cryptography and data security. Current work: Architectural design of large parallel processors. Patentee in field. Home: 42 Old Mill Rd Millington NJ 07946 Office: Bell Telephone Labs Room 2C-104 Whippany NJ 07981

MORRIS, ROBERT HOWARD, mechanical and nuclear engineer; b. Roanoke, Va., Sept. 27, 1950; s. Warren L. and Margaret (Cummings) M.; m. Diana Humphreys, Apr. 22, 1978. B.S. in Mech. Engring., U. Hawaii, 1974; M.S. in Nuclear Engring., U. Wis., 1976. Registered profl engr., Tenn. Devel. assoc. III Oak Ridge Nat. Lab., Tenn., 1976-80; sr. engr. Burns and Roe, Inc., Oak Ridge, 1980-83, Nutech Engr., San Jose, Calif., 1983—. Author: Single-Phase Sodium Tests in a 61-Pin Full-Length Simulated LMFBR Fuel Assembly, 1980, Two-Phase Transient Sodium Test-in-a 61-Pin Full-Length Simulated LMFBR Fuel Assembly, 1981. Mem. Am. Nuclear Soc.; mem. ASME (assoc.). Subspecialties: Mechanical engineering; Nuclear engineering. Current work: Quality control engneering during startup of a large, twin unit, pressurized water reactor electric generating station. Home: PO Box 672 Avila Beach CA 93424 Office: PO Box 561 Avila Beach CA 93424

MORRISON, DAVID, research planetary astronomer, educator, writer; b. Danville, Ill, June 26, 1940; s. Donald Harlan and Alice (Douglass) M.; m. Nancy Ellen, June 19, 1966; m. Janet Lee, Aug. 23, 1981. B.A., U. Ill.-Urbana, 1962; M.A., Harvard U., 1964, Ph.D., 1969. Prof. astronomy U. Hawaii, Honolulu, 1969—; vis. prof. U. Ariz., Tucson, 1975-76; asst. dep. dir. planetary div. NASA Hdqrs., Washington, 1976-78, agy. acting dep. assoc. adminstr. for space sci., 1981; cons. NASA, Jet Propulsion Lab., Voyager imaging sci. team; interdisciplinary scientist Project Galileo. Author: Voyages to Saturn, 1982, (with J. Samz) Voyage to Jupiter, 1980; editor: Satellites of Jupiter, 1982, (with D.M. Hunten) The Saturn System, 1978, (with W.C. Wells) Asteroids, An Exploration Assessment, 1978; contbr. tech . articles to profl. publs. Mem. Am. Astron. Soc. (past chmn. div. planetary sci., mem. council 1982-85), Internat. Astron. Union, Am. Geophys. Union, Astron. Soc. of the Pacific (dir. pres. 1982-84). Club: Cosmos (Washington). Subspecialties: Planetary science; Infrared optical astronomy. Current work: Planetary exploration, especially smaller bodies (satellites and asteroids) using groundbased astronomy and direct exploration by space vehicles. Home: 1521 Punahou Apt 801 Honolulu HI 96822 Office: 2680 Woodlawn Honolulu HI 96822

MORRISON, DONALD MICHAEL, operations manager, researcher; b. Los Angeles, Jan. 3, 1931; s. Donald Angus and Beatrice (Rohde) M.; m. Frances Louise Allen, Nov. 21, 1959; children: Michelle, Kenneth, Jack, Benjamin. B.S., Pepperdine U., 1955, M.A., 1970; student, U. So. Calif., 1974—, UCLA, 1968-76; Ph.D., Stanton U., 1980. Cinematographer Am. Diving, Inc., Redondo Beach, Calif., 1956-59; mgr. D&B Water Sports, Santa Ana, Calif., 1959-60; comml. diver Oceanics Unlimited, Torrance, Calif., 1960-71; diving instr. Undersea Specialists div. WTS, Inc., Los Angeles, 1971—, ops. mgr., 1977—; cons. Western Tech. Assn. Los Angeles, 1971—; research affiliate Catalina marine sci., hyperbaric specialist center U. So. Calif., 1971—. Author: An Analysis of Methods Used to Teach Scuba Diving, 1970; producer: films Beyond the Anchor, 1961, The Kelp Forest, 1965. Undercover agt. Torrance (Calif.) Police Dept., 1965-67. Served with USMCR, 1950-53. Named Policeman of Year Torrance Police Dept., 1967; Dept Energy grantee, 1977. Mem. Los Angeles County Underwater Instrs. Assn. (dir. 1977-80), Nat. Assn. Underwater Instrs. Democrat. Mem. Ch. of Christ. Subspecialties: Species interaction; Offshore technology. Current work: Fish population dynamics and photogrammetry. Home: 4827 Narrot St Torrance CA 90503 Office: Undersea Specialists Div WTS Inc 5730 Arbor Vitae St Los Angeles CA 90045

MORRISON, FRANCIS SECREST, physician; b. Chgo., July 29, 1931; s. Clifton B. and Marie B. (LaPierre) M.; m. Dorothy Daniels, Nov. 29, 1957; children—Francis, Thomas, Kenneth. Student, U. Ill., Chgo., 1949-51; B.S. with honors, Miss. State U., 1954; M.D., U. Miss., 1959. Diplomate: Am. Bd. Internal Medicine. Intern Hosp. of U. Pa., Phila., 1959-60, resident in internal medicine, 1960-62; trainee in hematology Blood Research Lab., Tufts-New Eng. Med. Center, Boston, 1962-64, research fellow, 1964-65; vis. investigator St. Mary's Hosp., London, 1966; attending physician, dir. div. hematology and oncology Univ. Hosp., Jackson, Miss., 1969-80, dir. div. hematology, 1980—, dir. blood transfusion service, 1974—; cons. in hematology Miss. Meth. Rehab. Center, Jackson, 1976; asst. prof. medicine U. Miss., Jackson, 1969-70, dir. div. hematology, 1969—, assoc. prof., 1970-76, prof., 1976—, mem. faculty, 1971—; profl. adv. Jackson Community Blood Bank, Inc., 1973-75; dir. regional cancer program, also regional blood program Miss. Regional Med. Program, 1971-75; exec. dir. Miss. Regional Blood Center, 1975-79; mem. adv. bd. Jackson-Hinds Comprehensive Health Center, 1973-78; research cons. Alcorn A. and M. Coll., 1973-74; guest lectr. various health orgns. and TV programs; mem., chmn. hemophilia adv. bd. Miss. Bd. Health, 1974—; chmn. task force on regionalization Am. Blood Commn., 1978-80; mem. Miss. Gov.'s Council on Aging, 1976—. Contbr. numerous articles on hematology and oncology to med. jours. Pres. parish council St. Peter's Cathedral, Jackson, 1972-74; chmn. Natchez-Jackson Diocesan Com. Community Services, 1972-76; bd. dirs. Miss. Opera Assn., 1973-78; pres. bd. St. Joseph High Sch., 1974-75. Served to comdr. M.C. USN. Fellow ACP; mem. Am. Assn. Blood Banks (sci. workshop com. 1975-78), Internat. Soc. Blood Transfusion, Am., Internat. socs. hematology, Jackson Acad. Medicine (pres. 1976), Am. Coll. Nuclear Medicine (alt. del. Miss. 1975), Am. Assn. Cancer Edn. (exec. com. 1978-81), Am. Assn. Cancer Research, N.Y. Acad. Scis., Miss. Acad. Scis., World Fedn. Hemophilia, Internat. Soc. Thrombosis and Haemostasis, Central Med. Soc., So. Med. Assn., Miss. Med Assn. (com. on blood transfusion 1976—), Am. Soc. Nuclear Medicine, Am. Soc. Clin. Oncology, S.W. Oncology Group (prin. investigator), Soc. Cryobiology, Am. Cancer Soc. (dir. 1971—, pres. Miss. div. 1977, chmn. exec. com. 1978), S. Central Assn. Blood Banks (program chmn. 1975, v.p. 1977-79), So. Blood Club (pres. 1977), Council Community Blood Centers (trustee 1975-79), Internat. Platform Assn., Sigma Xi, Phi Kappa Phi, Omicron Delta Kappa. Subspecialties: Hematology; Cancer research (medicine). Current work: Research in acuteleukemia-treatment and supportive care, especially blood and blood cell transfusion support. Home: 1402 Hazel St Jackson MS 39202 Office: U Miss Sch Medicine 2500 N State St Jackson MS 39216

MORRISON, JOHN B., cardiologist, educator; b. White Plains, N.Y., Apr. 6, 1938; s. Fredrick A. and Louise (Feldt) M.; m. Barbara A. Otto, June 15, 1970; children: Donna E., Gregory J. B.S., St. Lawrence U., 1960; M.D., Cornell U., 1964. Fellow in cardiology Cornell U. Med. Coll., N.Y.C., 1969-71, asst. prof. medicine, 1971-76, assoc. prof., 1976—; dir. cardiac care unit North Shore U. Hosp., Manhasset, N.Y., 1976—, co-chief div. cardiology, 1981—. Contbr. articles to profl. jours. Served to capt. M.C. U.S. Army, 1965-67. AHA grantee, 1976-83; Herman Goldman Found. grantee, 1977. Mem. AAAS, Am. Fedn. Clin. Research, Harvey Soc., Am. Heart Assn. (2d v.p. Nassau chpt. 1983), N.Y. Acad. Sci. Republican. Subspecialties: Cardiology; Graphics, image processing, and pattern recognition. Current work: Clinical cardiology research; image processing; cardiac radionuclide and enzyme research. Home: 4 Dooley Ct Dix Hills NY 11746 Office: North Shore U Hosp 300 Community Dr Manhasset NY 11030

MORRISON, MARTIN, biochemistry educator; b. Detroit, Dec. 9, 1921; s. David H. and Rose (Zatkin) M.; m. Hattie Joyce Pollard, June 25, 1947; children: Janet, Roger, Eleanor, Julie. B.S., U. Mich., 1947; Ph.D., Wayne State U., 1952. Spl. instr. Wayne State U., Detroit, 1949-51; instr. and asst. prof. U. Rochester, N.Y., 1952-60; postgrad. Cambridge (Eng.) U., 1961; head sect. research enzymes City of Hope Med. Ctr., Duarte, Calif., 1961-67; chmn. dept. biochemistry St. Jude Hosp., Memphis, 1967—; prof. U. Tenn. Ctr. Health Scis., 1967—; site visitor and ad hoc study sect. participant NIH, 1976—. Editor: Oxidases and Related Redox Systems, 1965, 73, 81; editorial bd.: Archives of Biochemistry and Biophysics, 1973—, Membrane Biochemistry, 1976—. Served to 2d lt. USAAF, 1942-45. NIH grantee, 1960—; USPHS Career Devel. award, 1956-61; Damon Runyan Meml. Fund award, 1971-73. Mem. AAAS, Am. Chem. Soc. (chmn. Memphis sect. 1976), Am. Soc. Biol. Chemistry, Am. Soc. Cell Biology, Am. Soc. Hematology, Am. Soc. Physiology, Biochem. Soc., Biophys. Soc., N.Y. Acad. Scis., Sigma Xi. Subspecialties: Biochemistry (biology); Membrane biology. Current work: Arrangement of proteins in membranes and relation to function, biological role of mammalian peroxidases. Office: St Jude Children's Research Hosp 332 N Lauderdale Memphis TN 38117

MORRISON, MARY DYER, psychotherapist; b. Somerville, Mass., Apr. 5, 1927; d. Leon Lee and Gertrude (Quinlan) Dyer; m. Robert Lloyd Morrison, Aprl. 7, 1945; 1 dau., Leigh. B.A., U. N.C., 1976; M.A., U. Mo., 1977, Ph.D., 1979. Prof. Coast Community Coll. Dist., Costa Mesa, Calif., 1980—; pvt. practice psychotherapy, Huntington Beach, Calif., 1980—; cons., 1980—. U. Mo. grantee, 1979; NIMH fellow. Mem. Am. Psychol. Assn., Phi Delta Kappa. Subspecialties: Developmental psychology; Gerontology. Current work: Psychology of aging. Researcher on effect of women's studies program on occupational aspiration (coll. undergrad. women). Address: 8566 Colusa Cir 903A Huntington Beach CA 92646

MORRISON, PHILIP, physics educator; b. Somerville, N.J., Nov. 7, 1915; m. Phyllis Singer. B.S., Carnegie Inst. Tech., 1936; Ph.D., U. Calif.-Berkeley, 1940. Instr. physics San Francisco State Coll., 1941-42; instr. physics U. Ill.-Urbana, 1941-42; physicist Metall. Lab., U. Chgo., 1943-44; physicist, group leader Los Alamos Sci. Lab., 1944-46; assoc. prof. to prof. physics Cornell U., Ithaca, N.Y., 1946-65; prof. physics MIT, Cambridge, 1965—. Author: Powers of Ten; co-editor: The Search for Extra-terrestrial Intelligence; contbr. articles in physics to profl. jours. Recipient Pregel prize, 1955, Babson prize, 1957; decorated Oersted medal, 1965. Fellow Am. Phys. Soc.; mem. Nat. Acad. Scis., Am. Astron. Soc. Subspecialty: High energy astrophysics. Office: Dept Physics MIT Cambridge MA 02139

MORRISON, ROBERT FLOYD, industrial and organizational research psychologist, consultant, educator; b. Mpls., Oct. 30, 1930; s. Floyd William and Ruth Angeline (Foster) M.; m. Kathryn Helene Olson McHenry, Jan. 20, 1953 (div. 1968); children: Scott Robert, Rebekah Kim, Spencer Kirk, Bennett Todd, Holly Faith; m. Anne Deneiko, Oct. 11, 1975. Student, Grinnell Coll., 1948-49; B.S., Iowa State U., 1952, M.S., 1956; Ph.D., Purdue U., 1961. Registered psychologist, Pa., cert. Ont. Employee relations asst. Mobil Oil Co., Casper, Wyo., 1956-59; mgmt. devel. specialist Mead Corp., Chillicothe, Ohio, 1961-62; personnel devel. specialist Martin Co., Balt., 1962-64; personnel research mgr. Sun Co., Phila., 1964-70; assoc. prof. mgmt. U. Toronto, Ont., Can., 1970-76; supervisory research psychologist Navy Personnel Research and Devel. Ctr., San Diego, 1976—; vis. prof. W.Va. U., 1970; adj. prof. U. So. Calif., 1978—; pres. R.F. Morrison & Assocs., Phila., Toronto, 1966-76. Contbr. chpt. to book, articles to publs. in field. Bd. dirs. Central Wyo. Mental Health Clinic, Casper, 1957-59; active Y Men's Club, Casper, 1958-59. Served with U.S. Army, 1952-54. Mem. Am. Psychol. Assn. (div. chmn. sci. affairs 1974-75, James McKeen Cattell award 1982), Acad. Mgmt. (program chmn. research group 1979-80), Internat. Assn. Applied Psychologists, Summit Group, Sigma Xi, Psi Chi. Republican. Club: University City Racquet (tourney chmn. 1979). Subspecialties: Behavioral psychology; Developmental psychology. Current work: Research on career decisions and planning of adults, especially managers and professionals (currently managing seven-year longitudinal, multiple cohort study); management of scientists and professionals. Home: 6137 Syracuse Way San Diego CA 92122 Office: Navy Personnel Research and Devel Center San Diego CA 92152

MORRISON, RODERICK GORDON, nuclear engineer, electronic production company executive; b. Alhambra, Calif., Apr. 16, 1924; s. Emery Alexander and Christine Katherine (McLean) M.; m. Bonnie Jean Parker, Dec. 28, 1943; children: Robert Emery, Allan Keith, James Ross. A.B., Fresno State U., 1948. Test engr. So. Calif. Edison, Big Creek, 1949-51; design engr. Kern County Land Co., Bakersfield, Calif., 1952-53, Los Alamos Nat. Lab., 1953-57, 60-67; test ops. mgr. EG & G, Las Vegas, Nev., 1957-60; devel. mgr. Phillips Petroleum, Idaho Falls, Idaho, 1967-71, GA Technologies, Inc., San Diego, 1978—. Mem. campaign com. Republican Nat. Conv., Carlsbad, Calif., 1980—. Served with USMC, 1941-46. Mem. Am. Nuclear Soc., IEEE. Republican. Subspecialties: Microelectronics; Nuclear engineering. Current work: Improved microprocessor based measurement systems for nuclear source detector applications using semiconductor detectors; emphasis on miniaturization and low power consumption. Patentee gamma compensated fission detection, 1971, isotopic precipitation gauge, 1974, low level snow precipitation gauge, 1976, flow measuring method and apparatus, 1977. Home: 2022 Cima

Ct Carlsbad CA 92008 Office: GA Technologies Inc PO Box 81608 San Diego CA 92138

MORRISON, ROGER BARRON, geologist, writer; b. Madison, Wis., Mar. 26, 1914; s. Frank Barron and Elise (Bullard) M.; m. Harriet Louise Williams, Apr. 7, 1941; children: John Christopher, Craig Brewster, Peter Hallock. B.A. in Geology, Chemistry, Cornell U., 1933, M.S., 1934, Ph.D., U. Nev.-Reno, 1964. Geologist U.S. Geol. Survey, Denver, 1939-76; vis./adj. prof. dept. geosciences U. Ariz., Tucson, 1976-80; consulting geologist Morrison and Assocs., Golden, Colo., 1978—, mng. ptnr., 1978—; adj. prof. MacKay Sch. Mines, U. Nev.-Reno, Lake Bonneville, 1983—. Author: Quaternary Stratigraphy of eastern Jordan Valley, 1965, (with J.C. Frye) Correlation of Middle and Late Quaternary Successions of Lake Lahontan, Lake Bonneville, 1965; also, U.S. Geol. Survey Profl. Papers; Editor: publs. of Internat. Assn. Quaternary Research; contbr. over 50 articles in field to profl. jours. NASA grantee Landsat and Skylab projects, 1972-75. Fellow Geol. Soc. Am.; mem. AAAS, Am. Soc. Photogrammetry, Soil Sci. Soc. Am., Internat. Soil Sci. Assn., Internat. Assn. Quaternary Research, Am. Quaternary Assn. Subspecialties: Geology; Tectonics. Current work: Environmental geology; quaternary geology and geomorphology; soil stratigraphy and pedology; remote sensing of earth resources; airphoto and image analysis/mapping of geologic terrains, soils, and minerals; quaternary stratigraphy and paleoclimatic history of NW Nev.; pliocene and quarternary stratigraphy and tectonics of Arizona. Office: Morrison and Associates 13150 W 9th Ave Golden CO 80401

MORRISSEY, JOHN FIELDING, gastroenterologist, educator; b. Brookline, Mass., June 16, 1924; s. John Thomas and Florence Beatrice (Fielding) M.; m. Ruth Emily Claus, Jan. 14, 1950; children: Ann Emily, Sara Ruth. A.B., Dartmouth Coll., 1946; M.D., Harvard U., 1949. Diplomate: Am. Bd. Internal Medicine (subcert. in gastroenterology). Resident in internal medicine U. Wis., Madison, 1955-56; staff VA Hosp., Madison, 1956-60; fellow U. Wash., Seattle, 1960-62; faculty U. Wis. Med. Sch., Madision, 1962—, prof., vice chmn. dept. medicine, 1974—. Contbr. articles to profl. jours. Served to capt. U.S. Army, 1943-46, 50-52. Mem. Am. Gastroent. Assn., Am. Soc. Gastrointestinal Endoscopy (Schindler award 1983), ACP, Brit. Soc. Gastroenterology (hon. mem.). Club: Blackhawk Country. Subspecialties: Internal medicine; Gastroenterology. Current work: Fiberoptic endoscopy of gastrointestinal tract; teaching technique, developing and improving instruments and techniques for use. Home: 6018 S Hill Dr Madison WI 53705 Office: 600 Highland Ave Madison WI 53792

MORROW, GARY ROBERT, clinical psychology/research educator; b. Pitts., Mar. 4, 1944; s. J. Robert and Frances (Pertes) M.; m. Joan Marilyn Baumgartner, June 7, 1967; children: Andy, J.J., Jennifer. B.A., U. Notre Dame, 1966, B.S. in Mech. Engring, 1967; M.S., U. R.I., 1973, Ph.D., 1974. Lic. psychologist, N.Y. Intern U. Rochester, N.Y., 1975, post-doctoral fellow, 1975-77, asst. prof., 1977-83, assoc. prof. oncology in psychiatry, psychology, 1983—; ad hoc cons. Nat. Cancer inst., Washington, 1978—; mem. adv. com. Am. Cancer Soc., N.Y.C., 1981—, com. chmn., 1982—. Contbr. sci. articles, abstracts, procs. to profl. jours.; editorial bd.: Jour. Psychosocial Oncology, 1982. Com. mem. Cub Scouts Am., Pittsford, N.Y., 1976; div. dir./coach Pittsford Soccer League, 1978; asst. scoutmaster Boy Scouts Am., Pittsford, 1981. Served to lt. Submarine Service USN, 1967-71. Research grantee United Cancer Council, Rochester, 1979, Nat. Cancer Inst., 1976, 79, 83, Am. Cancer Soc., 1983. Mem. Am. Soc. Clin. Oncology, Am. Psychol. Assn., Acad. Behavioral Medicine Research. Repubican. Presbyterian. Subspecialties: Behavioral psychology; Cancer research (medicine). Current work: Behavioral medicine, especially behavioral techniques for the control of the side effects of chemotherapy treatment for cancer. Home: 11 Sutherland St Pittsford NY 14534 Office: Dept Psychiatry/Psychology U Rochester Med Ctr 300 Crittenden Blvd Rochester NY 14642

MORROW, WALTER EDWIN, JR., elec. engr., univ. lab. adminstr.; b. Springfield, Mass., July 24, 1928; s. Walter Edwin and Mary Elizabeth (Ganley) M.; m. Janice Lila Lombard, Feb. 25, 1951; children—Clifford E., Gregory A., Carolyn F. S.B., M.I.T., 1949, S.M., 1951. Staff Lincoln Lab., M.I.T., Lexington, Mass., 1951-55, group leader, 1956-65; head div. communications M.I.T. Lincoln Lab., Lexington, Mass., 1966-68, asst. dir., 1968-71, asso. dir., 1972-77, dir., 1977—. Contbr. articles to profl. publs. Recipient award for outstanding achievement Pres. M.I.T., 1963, Edwin Howard Armstrong Achievement award IEEE Communications Soc., 1976. Fellow IEEE, Nat. Acad. Engring. Subspecialty: Electronics. Patentee synchronous satellite, electric power plant using electrolytic cell-fuel cell combination. Office: PO Box 73 Lexington MA 02173

MORSE, BERNARD S., hematology educator; b. N.Y.C., Sept. 11, 1934; s. Raymond Alvin and Caroline (Harris) M.; m. Rosemary Natalie D'Elia, June 5, 1960; children: Wayne, Jennifer. B.A., NYU, 1955, M.S., 1956; M.D., Seton Hall U., 1960. Diplomate: Am. Bd. Internal Medicine. Asst. prof. medicine Tufts U., Boston, 1967-69; assoc. prof. medicine U. Medicine and Dentistry-N.J.-N.J. Med. Sch., Newark, 1969-77, prof., 1977—. Served to capt. USAF, 1965-67. NIH grantee, 1972-75; Am. Heart Assn. grantee, 1981-83; Lupus Found. grantee, 1982-83. Mem. Am. Soc. Hematology, Internat. Soc. Hematology, Am. Fedn. Clin. Research. Subspecialties: Hematology; Immunology (medicine). Current work: Development of techniques for measurement of platelet antibodies. Office: U Medicine and Dentistry NJ-NJ Med Sch 100 Bergen St Newark NJ

MORSE, FREDERICK H., government energy administrator; b. N.Y.C., Feb. 20, 1936; s. Samuel and Mollie (Vikodetz) M.; m. Naomi S., June 14, 1959; children: Daniel, Joel. B.S., Rensselaer Poly. Inst., 1957; M.S., MIT, 1959; Ph.D., Stanford U., 1966. Instr. dept. mech. engring. Rensselaer Poly. Inst.; mem. faculty dept. mech. engring. U. Md., 1968-76; chief research and devel. br. Office Solar Applications ERDA, 1975-77; dir. Office Solar Applications, Dept. Energy, after 1977; now dir. Office Solar Heat Techs.; chmn. exec. com. for solar heating and colling coop. programs Internat. Energy Agy. Editorial bd.: Solar Age. Served to 1st lt. U.S. Army, 1957-62. Recipient Spl. Achievement award ERDA, Spl. Act of Service award Dept. Energy. Mem. Internat. Solar Energy Soc. (dir.). Subspecialty: Solar energy. Current work: Administration of National Solar Heat Technologies Program, including research and development of active solar heating and cooling, passive and hybrid solar, and solar thermal energy conversion; program planning; budget oversight. Office: Dept Energy Office Solar Heat Techs 1000 Independence Ave Washington DC 20585

MORSE, MELVIN LAURANCE, geneticist, university administrator; b. Hopkinton, Mass., Feb. 23, 1921; s. Clinton D'Arcy and Vilna Louise (Macines) M.; m. Helvise Gantt Glessner, Jan. 25, 1949; children: Margaret Louise, Laurance Clinton. B.S., U. N.H., 1944; M.S., U. Ky., 1947; Ph.D., U. Wis., 1955. Mem. faculty U. Coll. Med. Sch., 1956—, investigator, 1956—, prof. biochemistry, biophysics and genetics, 1960—, dean, 1977—. Contbr. articles to profl. jours. Mem. AAAS, Genetics Soc. Am., Am. Soc. Human Genetics, Am. Genetics Assn. Subspecialty: Genetics and genetic engineering (medicine). Office: U Colo Med Ctr B122 4200 E 9th Ave Denver CO 80262

MORSE, PETER HODGES, ophthalmologist; b. Chgo., Mar. 1, 1935; s. Emerson Glover and Carol Elizabeth (Rolph) M. A.B., Harvard U., 1957; M.D., U. Chgo., 1963. Diplomate: Am. Bd. Ophthalmology. Intern U. Chgo. Hosp., 1963-64; resident Wilmer Inst. Johns Hopkins Hosp., Balt., 1966-69; fellow, retina service Mass. Eye and Ear Infirmary, Boston, 1969-70; asst. prof. ophthalmology, chief retina service U. Pa., 1971-75, asso. prof., 1975, U. Chgo., 1975-77, sec. dept. ophthalmology, 1976-77, chief retina service, prof., 1979—; prof. La. State U., 1978; chmn. dept. ophthalmology, chief retina service Ochsner Clinic and Found. Hosp., New Orleans, 1977-78; clin. prof. Tulane U., 1978. Author: Vitreoretinal Disease: A Manual for Diagnosis and Treatment, 1978; co-editor: Disorders of the Vitreous, Retina, and Choroid; bd. editors: Perspectives in Ophthalmology, 1976—, Retina, 1980—; contbr. articles to profl. jours. Served with USNR, 1964-66. Mem. AMA, New Orleans Acad. Ophthalmology, A.C.S., La., Miss. Ophthalmol. and Otolaryngol. Soc., Assn. Research Vision and Ophthalmology, Retina Soc., Soc. Heed Fellows, Ophthalmol. Soc. U.K., Pan Am. Assn. Ophthalmology, Orleans Parish, La. med. socs., Oxford Ophthalmol. Congress, All-India Ophthalmol. Soc., Soc. Eye Surgeons, Sigma Xi. Republican. Episcopalian. Subspecialty: Ophthalmology. Current work: Clinical research in diseases which affect vitreous, retina, and uvea. Home: 5801 S Dorchester Ave Chicago IL 60637 Office: Dept Ophthalmology U Chgo 939 E 57th St Chicago IL 60637

MORSE, PHILIP MCCORD, research physicist; b. Shreveport, La., Aug. 6, 1903; s. Allen Craft and Edith (McCord) M.; m. Annabelle Hopkins, Apr. 26, 1929; children: Conrad Philip, Annabella. B.S., Case Sch. Applied Sci., 1926, hon. Sc.D., 1940; A.M., Princeton, 1927, Ph.D., 1929. Salesman, reporter, lecturer, 1922-30; instr. in physics Princeton, 1929-30; Internat. Research fellow, 1930-31; asst. prof. physics Mass. Inst. Tech., 1931-34, asso. prof., 1934-37, prof., 1937-69, emeritus, 1969—, dir. Computation Center, 1956-68, dir. Ops. Research Center, 1958-69, chmn. faculty, 1958-60; dir. Brookhaven Nat. Lab., 1946-48; asso. editor Tech. Rev., 1936-46; Trustee Rand Corp., 1948-52, Analytical Services, Inc., 1962-73, Inst. for Def. Analysis, 1956-61; dir. Control Data Corp.; Mem. of NDRC, Div. 6, asst. dir. Office Field Service OSRD, 1943-45; dir. research Weapons Systems Evaluation Group, Office Sec. of Def., 1949-50; dir. Ops. Research Group, U.S. Navy, 1942-46; mem. Ordnance Research Adv. Com., 1950-56; chmn. NATO Adv. Panel Ops. Research, 1960-65, OECD Panel on Ops. Research, 1963-68. Author: (with E.U. Condon) Quantum Mechanics, 1929, Vibration and Sound, 1936 (rev. edit. 1946), Methods of Operations Research, (with G.E. Kimball), 1950, Methods of Theoretical Physics, (with H. Feshbach), 1953, Queues, Inventories and Maintenance, 1958, Thermal Physics, 1961, (with K.U. Ingard) Theoretical Acoustics, 1967, Library Effectiveness, 1969, In At the Beginnings, a Physicist's Life, 1977; Editorial bd.: Sci., Bull. Atomic Scientists; Editor: Annals of Physics, 1957-77, Notes on Operations Research, 1959, Operations Research for Public Systems, 1967, Analysis of Public Systems, 1972; Contbr. to profl. items, mag. articles and sci. papers. Trustee Am. Inst. of Physics, 1948-50, 1953-57, 73-75, chmn. bd. govs., 1975-80; chmn. bd. govs. Research Soc. Am., 1950-58; chmn. Navajo Coop. Commn., Nat. Acad. Scis., 1980—. ; Decorated Medal for Merit, U.S., 1946; recipient silver medal Operational Research Soc. London, 1965. Fellow Ops. Research Soc. Am. (1st pres. 1952-53, Lanchester prize 1969, G.E. Kimball medal 1974), Am. Acad. Arts and Scis., Am. Phys. Soc. (mem. council 1935-40, 72-75, v.p. 1970-71, pres. 1971-72, chmn. panel on pub. affairs 1975-76), Acoustical Soc. Am. (pres. 1950-51, gold medal 1973); mem. Internat. Fedn. Ops. Research Socs. (exec. sec. 1961-65), Nat. Acad. Scis., Sigma Xi, Tau Beta Pi. Democrat. Presbyterian. Club: Cosmos (Washington). Subspecialties: Acoustics; Operations research (mathematics). Current work: Mathematical methods in physics and operations; use of computers in technical problems. Home: 126 Wildwood St Winchester MA 01890 Office: Mass Inst of Tech Cambridge MA 02139

MORSE, RICHARD STETSON, corporate director, consultant, government official; b. Abington, Mass., Aug. 19, 1911; s. Kenneth Lee and Mary Celia (Skinner) M.; m. Marion Elsa Baitz, Nov. 27, 1935; children: Richard Stetson, Kenneth Paul. S.B., MIT, 1933; D.Sc. (hon.), Clark U., 1961, D.Eng., Bklyn. Poly. Inst., 1960. Mem. sci. staff Eastman Kodak, 1935-40; pres. Nat. Research Corp., Cambridge, Mass., 1940-58; dir. research Asst. Sec. Army for Research and Devel., Washington, 1958-62; sr. lectr. Sloan Sch. Mgmt., MIT, 1962-77; now dir. Boston Five Bank, Dresser Industries Inc., Aerospace Corp, MIT Devel. Found., Tracer Tech. Inc., PMC/Beta Corp.; mem. gen adv. com. ERDA; chmn. Army Sci. Adv. Bd., Civilian Adv. Bd., Air Force Systems Command; mem. Commerce Tech. Adv. Bd., founder Minute Maid Corp., 1948. Mem. corp. Woods Hole (Mass.) Oceanographic Inst., Boston Mus. Sci. Recipient Disting. Civilizn Service medal U.S. Dept. Def., 1960. Mem. Am. Chem. Soc., World Affairs Council, Nat. Acad. Engring. Republican. Clubs: Quissett Yacht (Falmouth, Mass.); MIT Faculty (Cambridge, Mass.); St. Botolph, Merchants/Comml. (Boston); Breaburn Country (Newton, Mass.); St. Andrew's (Delray, Fla.). Subspecialties: Batteries and industrial instruments; Management of new enterprises. Current work: Innovation and new enterprises; corporate director; consultant to government and new technical enterprises. Patentee sound reprodn., vacuum dehydration, instrumentation, color reprodn., and other areas; commercialized lens coating, freeze drying.

MORSE, STEPHEN SCOTT, microbiologist, educator; b. N.Y.C., Nov. 22, 1951; s. Murray H. and Phyllis M.; m. Marsden Williams Gresham, Jan. 31, 1981. Student, Harvard U., 1968, Balliol Coll., Oxford U., 1969; B.S., City Coll., CUNY, 1971; M.S., U. Wis.-Madison, 1974, Ph.D., 1977. Research fellow Med. Coll. Va., Richmond, 1977-80; asst. prof. dept. biochemistry and microbiology. Cook Coll., Rutgers U., New Brunswick, N.J., 1981—, dir. Cell and Tissues Culture Lab., 1981—; investigator, mem. Marine Biol. Lab., Woods Hole, Mass., 1981—. Nat. Cancer Inst. fellow, 1978-80; fellow Rutgers Coll. Mem. Am. Soc. Microbiology, Am. Assn. Pathologists, Biophys. Soc., AAAS, Sigma Xi. Subspecialties: Infectious diseases; Immunology (medicine). Current work: Cell biology: phagocytic cells, macrophages; cell culture; pathogenic microbiology; interferon and related compounds; mechanisms of host resistance to viruses and tumor cells; pathogenic mechanisms and host resistance in infectious diseases. Office: Dept Biol Sci Nelson Biol Labs Rutgers U New Brunswick NJ 08903

MORSE, THEODORE FREDERICK, engineer, educator; b. Bklyn., Feb. 28, 1932; s. Albert and Dorothy (Bowen) M.; m. Edelgard B. Morse, Aug. 1, 1955; children: Karin, Peter. B.A., Duke U., 1953, M.A., 1954; B.Sc., U. Hartford, 1957; M.Sc., Rensselaer Poly. Inst., 1959; Ph.D., Northwestern U., 1961. Research engr. Pratt & Whitney Aircraft, East Hartford, Conn., 1955-59; sr. research engr. A.R.A.P., Inc., Princeton, N.J., 1961-63; asst. prof. Brown U., 1963-65, assoc. prof., 1965-68, prof., 1968—. Contbr. numerous articles on kinetic theory, fluid mechanics, applied laser physics to profl. jours. Fulbright sr. research fellow, 1969-70. Mem. Am. Inst. Physics, Am. Inst. Chem. Engrs. Subspecialty: Fluid mechanics. Current work: Optical fiber manufacture; two phase flow. Patentee gas centrifuge, optical fibers. Home: 22 Benevolent St Providence RI 02912 Office: Div Engring 184 Hope St Box D Providence RI 02912

MORTENSEN, EARL MILLER, chemistry educator; b. Salt Lake City, June 25, 1933; s. Earl Emanuel and Serena Mae (Miller) M.; m. Sharlene Wilcox Mortensen; children: Eric, Brian, Russell, Mark. B.A., U. Utah, 1955, Ph.D., 1959. Research assoc. U. Calif., 1960-62; asst. prof. chemistry U. Mass., 1962-69; assoc. prof. Cleve. State U., 1969—. NSF fellow, 1959-60. Mem. Am. Chem. Soc., Am. Phys. Soc., Sigma Xi. Mormon. Subspecialties: Physical chemistry; Theoretical chemistry. Current work: Determine reactive cross sections for atom-diatom reactions using quantum mechanical scattering theory. Home: 24227 LeBern Dr North Olmsted OH 44070 Office: Cleveland State University Department of Chemistry Cleveland OH 44115

MORTENSEN, RICHARD F., immunologist, educator; b. Green Bay, Wis., July 21, 1945; s. Anthony F. and Arvella M. M.; m. Patricia D. Kosmalski, Apr. 8, 1978. B.S., U. Wis., 1967; M.S., Pa. State U., 1969, Ph.D., 1973. Postdoctoral fellow Rush Med. Sch., Chgo., 1973-75, asst. prof., 1975-76; lab. chief Mich. Cancer Found., Detroit, 1976-80; asst. prof. dept. microbiology Ohio State U., Columbus, 1980-83, assoc. prof., 1983—. Contbr. chpts. to books, articles to profl. jours. Served with U.S. Army, 1971-73. NIH grantee, 1975—; Am. Cancer Soc. grantee, 1976-78, 81—. Mem. Am. Assn. Immunologists, Am. Assn. Cancer Research, Am. Soc. Microbiology, N.Y. Acad. Scis. Subspecialties: Immunobiology and immunology; Infectious diseases. Current work: Acute phase serum proteins and host-resistance; cell-mediated immune reactivity to tumors, prognosis. Office: Dept Microbiology Ohio State U 484 W 12th St Columbus OH 43210

MORTILLARO, NICHOLAS A., physiologist, educator; b. Bklyn., Aug. 14, 1936; s. Nicolo and Pietrina (LaCorte) M.; m. Mildred Mary Detloff, July 30, 1962; children: Philip Michael, Susan Patricia. B.S. in Elec. Engring, Heald Engring. Coll., 1964, M.S., Newark Coll. Engring., 1968; Ph.D., Coll. Medicine and Dentistry N.J., 1972. Project engr. Bendix Corp., Teterboro, N.J., 1964-68; research fellow U. Miss. Sch. Medicine, Jackson, 1972-74; instr. U. Miss., 1973-74; asst. prof. U. South Ala., Mobile, 1974-77, assoc. prof., 1977—; vis. assoc. prof. Tex. A&M U. Sch. Medicine, College Station, 1980. Author: Gastrointestinal Function, 1984; editor/contbr.: Physiology and Pharmacology of the Microcirculation Vols. I and II, 1983. Pres. Italian-Am. Cultural Soc. S. Ala., 1982-83. NIH research grantee, 1974—; Am. Heart Assn. research grantee, 1974-76. Mem. Am. Physiol. Soc., Microcirculatory Soc., Internat. Lymphology Soc., Am. Heart Assn., Splanchnic Circulation Group. Roman Catholic. Subspecialties: Physiology (medicine); Gastroenterology. Current work: Local and nervous control of blood flow; capillary, interstitial and lymphatic fluid and solute exchange dynamics; properties of vascular smooth muscle. Home: 5700 Grandee Ct Mobile AL 36609 Office: U S Ala Dept Physiology College Medicine Mobile AL 36688

MORTIMER, ROBERT GEORGE, chemistry educator, researcher; b. Provo, Aug. 25, 1933; s. William Earl and Margaret Eastmond (Johnson) M.; m. Dorothy Jean Randall, June 23, 1960; children: Julia, Paul, Jeannine, John, David. B.S., Utah State U., 1958, M.S., 1959; Ph.D., Calif. Inst. Tech., 1963. Research chemist U. Calif., San Diego, 1962-64; asst. prof. chemistry Ind. U., 1964-70; asst. prof. Southwestern U. Memphis, 1970-72, assoc. prof., 1972-81, prof., 1981—. Author: Mathematics for Physical Chemistry, 1981. Served to 2d lt. USAR, 1959. Woodrow Wilson fellow, 1958-59; NSF fellow, 1959-62; grantee in field. Mormon. Subspecialties: Physical chemistry; Statistical mechanics. Current work: Statistical mechanics, transport processes, chem. edn. instruction and research in phys. chemistry. Home: 2895 Falkirk Rd Memphis TN 38128 Office: 2000 N Parkway Memphis TN 38112

MORTLOCK, ROBERT PAUL, microbiology educator; b. Bronxville, N.Y., May 12, 1931; s. Donald Robert and Florance Mary (Bellaby) M.; m. Florita Mary Welling, Sept. 11, 1954; children: Florita M., Jeffrey R., Douglas P. B.S., Rensselaer Poly. Inst., 1953; Ph.D., U. Ill., 1958. Asst. prof. microbiology U. Mass., Amherst, 1963-68, assoc. prof., 1968-73, prof., 1973-78, head dept., 1975-78; prof., chmn. dept. microbiology Cornell U., Ithaca, N.Y., 1978—. Fellow Am. Acad. Microbiology; mem. Am. Soc. Microbiology. Subspecialties: Microbiology; Biochemistry (biology). Current work: Research in bacteria mutation in acquiring new metabolis pathways. Office: Dept Microbiology Cornell U Stocking Hall Ithaca NY 14853

MORTON, BRUCE ELDINE, neurochemistry educator, researcher; b. Loma Linda, Calif., May 9, 1938; s. Virgil Weston and Charlotte Laureen (Heath) M.; m. Lois Ramey, June 6, 1960 (div.); children: Sylvia Jean, Daniel Weston; m. Abelyn Kaulapaina Lumtto, May 18, 1975. B.A. in Chemistry, La Sierra Coll., 1960; M.S. in Biochemistry, U. Wis., 1963, Ph.D., 1965. Postdoctoral fellow Inst. Enzyme Research, U. Wis., 1965-66, M.I.T., 1966-67, Harvard U. Med. Sch., 1967-69; asst. prof. biochemistry and biophysics John A. Burns Sch. Medicine, U. Hawaii, 1969-74, assoc. prof., 1974-; dir. Brain Systems Research Lab., 1977—. Research numerous pbls. in field. NIH grantee, 1963-73, 74—. Mem. Soc. Neurosci., Internat. Soc. Research on Aggression, Am. Soc. Biol. Chemists, Soc. Study Reprodn., AAS, Hawaiian Acad. Sci., U.S. Hang Gliding Assn. Subspecialties: Neurochemistry; Neuropsychology. Current work: Receptor and activity mapping of neurotransmitter systems of emotion, consciousness, memory and mental illness. Office: U Hawaii 411 Snyder Hall 2538 The Mall Honolulu HI 96822

MORTON, RANDALL EUGENE, nuclear engineer, aerospace electronics manufacturing company consultant; b. Portland, Oreg., May 4, 1950; s. Eugene Randall and Kathryn Hazel (Myers) M.; m. Lori Kay Turner, Mar. 23, 1979. B.S. in Elec. Engring, U. Wash., Seattle, 1972; M.S. in Nuclear Engring, 1974; Ph.D., 1979. Registered nuclear engr., Wash. Exec. cons. Holloran & Assocs., Bellevue, Wash., 1977-81; corp. cons. AGA Cons., Bellevue, 1981-82; sr. mfg. systems analyst Eldec Corp., Lynnwood, Wash., 1982—. Contbr. articles to profl. jours. Mem. Am. Nuclear Soc. (vice chmn. Puget Sound chpt. 1981-82, chmn. 1982-83), Am. Welding Soc. (cert. welding insp.), Am. Soc. for Quality Control (cert. quality engr.). Subspecialties: Systems engineering; Nuclear engineering. Current work: Management systems for high technology manufacturing company; general engineering and technical communications consulting. Home: 10320 181st St NW Redmond WA 98052 Office: Eldec Corp 16700 13th Ave W Po Box 100 Lynwood WA 98036

MOSATCHE, HARRIET SANDRA, psychology educator, writer, researcher; b. N.Y.C., Apr. 10, 1949; d. Charles Morris and Ruth (Green) Rosenberg; m. Ivan Lawner, Feb. 19, 1983. B.A., Bklyn. Coll., 1970; M.A., Hunter Coll., CUNY, 1972; Ph.D., CUNY, 1977. Lectr. in psychology Hunter Coll., CUNY, 1973-76, instr. psychology, 1976-77; asst. prof. psychology Coll. Mt. St. Vincent, Riverdale, N.Y., 1977—, chmn. psychology dept., 1978—; cons., N.Y.C., 1976—. Author: Searching, 1983; contbr. chpt. to book, articles to jours. in field. Active Spring Creek Assn. Retarded Children, Bklyn., 1973-80. Research grantee Nat. Inst. Aging, 1982—, Coll. Mt. St. Vincent, summer 1981. Mem. Am. Psychol. Assn., Eastern Psychol. Assn., N.Y. Acad. Scis., Sigma Xi, Psi Chi. Subspecialties: Developmental psychology; Social psychology. Current work: Sibling relationships, psychological aspects of religious cults, psychology and law. Office: College of Mt St Vincent Riverdale NY 10471

MOSBACH, ERWIN HEINZ, biochemistry educator, research adminstr.; b. Gelsenkirchen, Germany, Feb. 18, 1920; came to U.S., 1938, naturalized, 1944; s. Leopold and Emma (Heymann) M.; m. Marianne Levi, Mar. 5, 1944; 1 son, Peter Alan. B.A., Columbia U., 1943, M.A., 1948, Ph.D., 1950. Biochemist Oak Ridge Nat. Lab., 1950-51; research assoc. Columbia U., 1951-54; asst. prof. Coll. Physicians and Surgeons, N.Y.C., 1954-60, assoc. prof., 1960-61; chief dept. lipid research Pub. Health Research Inst., N.Y.C., 1961-78; dir. surg. research lab. Beth Israel Med. Ctr., N.Y.C., 1978—; research prof. dept. surgery Mt. Sinai Med. Ctr., N.Y.C., 1978—; cons. Manhattan VA Med. Ctr., N.Y.C., 1973—. Contbr. articles to profl. jours.; editor: Jour. Lipid Research, 1976-78. Served with USN, 1944-46. Recipient Windaus prize Falk Found., 1982; N.Y.C. Health Research Couuncil grantee, 1960. Mem. Am. Soc. Biol. Chemists, Am. Inst. Nutrition, Council Arteriosclerosis Am. Heart Assn., Am. Gastroent. Assn., Am. Assn. Study of Liver Diseases. Subspecialties: Biochemistry (biology); Nutrition (biology). Current work: Biochemistry of bile acids. Inventor controlling microorganisms, 1983. Office: Beth Israel Med Center 10 Perlman Pl New York NY 10024

MOSBERG, LUDWIG, educational psychologist, educator; b. Vienna, Austria, June 2, 1938; came to U.S., 1939, naturalized, 1947; s. Armin and Josephine (Schwartz) M. B.S., U. Mich., 1959; Ph.D., UCLA, 1967. Dept. head S.W. Regional Lab. for Ednl. Research and Devel., Los Alamitos, Calif., 1967-70; asst. prof. ednl. studies U. Del., 1970-72, assoc. prof., 1972—, chmn. dept. ednl. studies, 1974—. Contr. numerous articles to prfl. jours. Served to 1st U.S. Army, 1962-64. Mem. Am. Psychol. Assn., Am. Ednl. Research Assn., Eastern Psychol. Assn. Subspecialties: Cognition; Learning. Current work: Language comprehension, memory. Office: Dept Ednl Studies U Del Newark DE 19711

MOSCATELLI, EZIO ANTHONY, biochemistry educator, researcher; b. N.Y.C., Nov. 17, 1926; s. Peter Robert and Silvia (Gatti) M.; m. Lorraine Brako, June 14, 1952 (div. 1965); 1 son, Peter Edmond. A.B., Columbia U., 1948; M.S., U. Ill.-Urbana, 1949, Ph.D., 1958. Chemist NIH, Bethesda, Md., 1958-59; sr. chemist Merck, Sharp & Dohme, Rahway, N.J., 1959-62; asst. prof. biochemistry U. Tex. S.W. Med. Sch., Dallas, 1962-69; assoc. prof. U. Mo.-Columbia, 1970—. Contbr. articles to profl. jours. Bd. dirs. Maplewood Barn Community Theater, Columbia, 1978—. Served with AUS, 1945-46. Mem. Am. Soc. Biol. Chemists, Am. Soc. Neurochemistry, Am. Oil Chemists Soc., Internat. Soc. Neurochemistry, AAAS. Subspecialties: Biochemistry (biology); Membrane biology. Current work: Biochemistry of lipids in biological membranes in the central nervous system. Home: 2309 W Broadway St #536 Columbia MO 65201 Office: U Mo M121 Medical Science Bldg Columbia MO 65212

MOSCONA, ARON ARTHUR, scientist, educator; b. Israel, July 4, 1922; came to U.S., 1955, naturalized, 1965; s. David DeAstogne-Abarbanel and Lola (Krochmaal) M.; m. Malka Kempinsky, July 6, 1954; 1 dau., Anne. M.Sc., Hebrew U., Jerusalem, 1947, Ph.D., 1950; postgrad. fellow, Strangeways Research Lab., Cambridge, Eng., 1950-52. Vis. investigator Cambridge U., 1950-52; with Rockefeller Inst., N.Y.C., 1955-57; prof. biology U. Chgo., 1958—, Luis Block prof. biol. scis., 1972—; chmn. Com. on Developmental Biology, 1973; vis. prof. Stanford U., 1959, U. Montreal, 1960, U. Palermo, Italy, 1966, Hebrew U., Jerusalem, 1972, Tel-Aviv (Israel) U., 1977, 79, Kyoto U., Japan, 1980. Editor: Current Topics in Developmental Biology, 1965, Mechanisms of Aging and Development, 1971, Cell Differentiation, 1975, Cancer Research, 1977, Devel. Neurosci, 1977, Experimental Cell Research, 1962-77; contbr. articles to profl. jours. Mem. Nat. Acad. Scis., Instituto Lombardo (Milan), N.Y. Acad. Scis., AAAS, Internat. Soc. Devel. Biology (pres. 1977-81), Soc. Devel. Biology, Internat. Soc. Cell Biology, Am. Soc. Zoology, Am. Soc. Anatomy, Sigma Xi. Subspecialty: Developmental biology.

MOSELEY, GERARD FRANKLIN, radio astronomer, educator, univ. ofcl.; b. Harrisonburg, Va., May 15, 1940; s. Claude Franklin and Grace Lucille (Hopkins) M.; m. Judith Grace Cooke, June 12, 1965; children: David Franklin, Elizabeth Grace, Benjamin Gerard. B.S. in Physics (Snyder Acad. scholar, Watts Physics scholar), Randolph-Macon Coll., 1962, M.S., Yale U., 1964, Ph.D. in Radio Astronomy, 1969. Research asst., research assoc., faculty lectr. astronomy dept. U. Tex., Austin, 1966-78, asst. acad. dean, 1971-78; prof. physics U. Oreg., Eugene, 1978; assoc. provost for student affairs, 1978; also student affairs cons. Contbr. articles to sci. jours. Pres. Child and Family Service, Austin, 1977, Eugene Family YMCA, 1982; chmn. U. Oreg. YMCA, 1981. Recipient cert. of appreciation Air Force ROTC, U. Tex., 1977, Exceptional Leadership award Child and Family Service Bd. Dirs., Austin, 1977, Service award YMCA, Eugene, 1980. Mem. Am. Astron. Soc., Internat. Astron. Union (conf. participant Brighton, Eng. 1971, Uppsala, Sweden 1971, Grenoble, France 1976), Am. Conf. Acad. Deans, Council Colls. Arts and Scis. (rep.), Yale Sci. and Engring. Assn., Am. Assn. Collegiate Registrars and Admissions Officers, Pacific Assn. Collegiate Registrars and Admissions Officers, Nat. Assn. Student Personnel Adminstrs. (region V adv. council), N.W. Coll. Personnel Assn., Western Deans, Friars Soc., Phi Beta Kappa, Sigma Xi, Omicron Delta Kappa, Pi Delta Epsilon, Chi Beta Phi, Phi Kappa Phi, Alpha Lambda Delta, Phi Eta Sigma, Alpha Phi Omega (nat. adv. council 1980). Methodist. Lodge: Kiwanis. Subspecialty: Radio and microwave astronomy. Current work: Extragalactic research with precise position determination of radio sources.

MOSELEY, JOHN TRAVIS, research physicist, educator; b. New Orleans, Feb. 26, 1942; s. Fred Baker and Gay (Lord) M.; m. Belva McCall Hudson, Aug. 11, 1962 (div. 1978); children: Melanie Lord, John Mark; m. Susan Diane Callow, Aug. 5, 1979; 1 dau., Stephanie Marie. B.S., Ga. Inst. Tech., 1964, M.S., 1966, Ph.D., 1969. Asst. prof. U. West Fla., Pensacola, 1958-59; physicist SRI Internat., Menlo Park, Calif., 1969-74, sr. physicist, 1974-77, program mgr., 1977-79; assoc. prof. physics U. Oreg., Eugene, 1979-82, prof., 1982—, dir., 1980—. Contbr. articles to profl. jours. NSF fellow, 1964-68. Fellow Am. Phys. Soc.; mem. AAAS, AAUP, Sigma Xi (award for thesis research 1969), Phi Kappa Phi, Tau Beta Pi. Subspecialties: Atomic and molecular physics; Laser-induced chemistry. Current work: Processes involving molecular ions, in particular, photodissociation, spectroscopy, reactions and transport properties. Home: 1925 Dogwood Dr Eugene OR 97405 Office: Dept Physics U Oreg Eugene OR 97403

MOSELEY, MAYNARD FOWLE, JR., botany educator; b. Boston, July 15, 1918; s. Maynard Fowle and Muriel Perry (Chase) M.; m.; children: Margery Chase, Andrew McCaskill. B.S., Mass. State Coll., 1940; M.S., U. Ill., 1942, Ph.D., 1947. Instr. botany Cornell U., Ithaca, N.Y., 1947-49; prof. botany U. Calif. Santa Barbara, 1949. Active Boy Scouts Am. Served with U.S. Army, 1942-47. NSF grantee, 1958-60, 66-68, 81-83. Mem. Bot. Soc. Am., Internat. Assn. Wood Anatomists, Internat. Soc. Plant Morphologists, Linnean Soc. London. Democrat. Episcopalian. Subspecialty: Morphology. Current work: Wood and floral anatomy of higher plants and utilization of such in evolution (phylogeny). Home: 5087-B Rhoads Ave Santa Barbara CA 93111 Office: U Calif Santa Barbara CA 93106

MOSELEY, ROBERT DAVID, JR., radiologist, educator; b. Minden, La., Feb. 29, 1924; s. Robert David and Lettie E. (Looney) M.; m. Janet C. Watson, Mar. 15, 1947; children: Robert David III, Richard Havard, Marianne Lee. M.D., La. State U., 1947. Diplomate: Am. Bd. Radiology. Intern Highland Sanitarium and Clinic, Shreveport, La., 1947-48; asst. resident U. Chgo. Clinics, 1949-50; asst. resident Los Alamos spl. research project, dept. radiology U. Chgo., 1950-51; staff mem. U. Calif.-Los Alamos Sci. Lab., 1951-52, radiologist, asso. chief staff med. center, 1951-52; mem. staff dept. radiology U. Chgo., 1954-71, prof., chmn. dept., 1958-71; prof., asst. chmn., dir. div. diagnostic radiology U. N.Mex. Sch. Medicine, Albuquerque, 1971-78, prof., chmn. dept. radiology, 1978—; chief of staff Bernalillo County Med. Center, 1971-72, 78, 79; Bd. dirs. Nat. Council Radiation Protection, 1973-80; mem. radiation study sect. USPHS, NIH, 1971-75; mem. radiology tng. com. NIH, 1966-70; chmn., 1969-70; U.S. rep. UN Sci. Com. on Effects Atomic Radiation, 1976—; chmn. adv. com. on biology and medicine U.S. AEC, 1967-73; chmn. U.S. delegation Internat. Congresses of Radiology, Rio de Janeiro, 1977, Brussels, 1981; pres. XVI Internat. Congress Radiology, Hawaii, 1985. Pres. bd. dirs. James Picker Found., 1972—. Served as lt. (s.g.) USNR, 1952-54. Fellow Chgo. Roentgen Soc. (past sec.-treas. bd. trustees, pres. 1966-67), Am. Coll. Radiology (pres. 1973-74, Gold medal 1980), Royal Coll. Radiologists (Eng.) (hon.), Swedish Soc. Med. Radiology (hon.); mem. Internat. Soc. Radiology (pres.-elect), N.Mex., Albuquerque, Bernalillo County med. socs., Assn. U. Radiologists (founding mem., past pres., Gold medal 1980), Am. Roentgen Ray Soc., Radiation Research Soc., Acad. Soc. Lund (Sweden) (corr.), Deutsche Röntgengesellshaft (corr.), Radiol. Soc. N.Am., Inter Am. Coll. Radiology, Royal Physiographic Soc. (fgn. mem., Lund, Sweden), N.Mex. Soc. Radiologists, Sigma Xi, Sigma Nu, Phi Chi. Clubs: Univ. (Chgo.); Four Hills Country (Albuquerque). Subspecialties: Diagnostic radiology; Imaging technology. Current work: Imaging technology and evaluation. Home: 635 Running Water Circle SE Albuquerque NM 87123

MOSEMAN, JOHN GUSTAV, plant pathologist; b. Oakland, Nebr., Dec. 7, 1921; s. John Gerhart and Bertha Ann (Hopp) M.; m. Marjorie Jean Bell, May 31, 1948; children: David R., Barbara J. Moseman Smith, Thomas B. B.S., U. Nebr., 1943; M.S., Wash. State U., 1948; Ph.D., Iowa State U., 1950. Research plant pathologist U.S. Dept. Agr., N.C. State U., Raleigh, 1950-54, research plant pathologist, Beltsville, Md., 1954-69, leader barley investigations, 1969-72, chmn. plant genetics and germplasm inst., 1972-81, research plant pathologist, 1981—. Served in 1st lt. USAAF, 1943-46. Mem. Am. Phytopath. Soc., Am. Soc. Agronomy. Subspecialties: Plant pathology; Plant genetics. Current work: Genetics, host-pathogen relationships, germplasm enhancement. Home: 1918 Blackbriar St Silver Spring MD 20903 Office: Beltsville Agr Research Center Beltsville MD 20705

MOSER, HUGO WOLFGANG, physician; b. Switzerland, Oct. 4, 1924; came to U.S., 1940, naturalized, 1943; s. Hugo L. and Maria (Werner) M.; m. Ann Boody, Dec. 28, 1963; children—Tracey, Peter, Karen, Lauren. M.D., Columbia U., 1948; A.M. in Med. Sci, Harvard U., 1956. Intern Columbia-Presbyn. Med. Center, N.Y.C., 1948-50; asst. in medicine Peter Bent Brigham Hosp., Boston, 1950-52; research fellow dept. biol. chemistry Harvard U., 1955-57; asst. resident, resident in neurology Mass. Gen. Hosp., 1957-59, asst. neurologist, 1960-67, asso. neurologist, 1967-69, neurologist, 1969-76; teaching fellow neuropathology Harvard Med. Sch., 1959-60, instr. neurology, 1960-64, asso. in neurology, 1964-67, asst. prof., 1967-69, asso. prof., 1969-72, prof., 1972-76; dir. research and tng. Walter E. Fernald State Sch., 1963-68, asst. supt., 1968-73, acting supt., 1973-74, supt., 1974-76; dir. Center for Research on Mental Retardation and Related Aspects of Human Devel., dir. univ. affiliated facilities for mentally retarded, 1965-74; co-dir. Eunice Kennedy Shriver Center for Mental Retardation, Inc., 1969-74; dir. John F. Kennedy Inst., Balt., 1976—; prof. neurology and pediatrics Johns Hopkins U., 1976—. Author: (with others) Mental Retardation: An Atlas of Diseases with Associated Physical Abnormalities, 1972; Contbr. articles to med. jours. Served with AUS, 1943-44; to capt. U.S. Army, 1952-54. Mem. Am. Acad. Neurology, Am. Assn. Mental Deficiency, Am. Assn. Neuropathologists, Am. Neurol. Assn., Internat. Soc. Neurochemistry, Am. Pediatrics Soc., Sigma Xi, Alpha Omega Alpha. Home: 100 Beechdale Rd Baltimore MD 21210 Office: 707 N Broadway Baltimore MD 21205

MOSER, ROBERT HARLAN, physician; b. Trenton, N.J., June 16, 1923; s. Simon and Helena (Silvers) M.; m. Stella Margo Neeson, June 17, 1948; children—Steven Harlan, Jonathan Evan. Student, Loyola Coll., 1940-42; B.S., Villanova Coll., 1943; M.D., Georgetown U., 1948. Diplomate: Am. Bd. Internal Medicine (recertified 1974). Commd. 2d lt. U.S. Army, 1948, advanced through grades to col., 1966; intern D.C. Gen. Hosp., 1948-49, fellow pulmonary diseases, 1949-50; bn. surgeon, Korea, 1950-51; asst. resident Georgetown U. Hosp., 1951-52, chief resident, 1952-53; chief med. service U.S. Army Hosp., Salzburg, Austria, 1953-55, Wurzburg, Germany, 1955-56; resident cardiology Brooke Gen. Hosp., 1956-57, asst. chief dept. medicine, 1957-59, chief, 1967-68; fellow hematology U. Utah Coll. Medicine, 1959-60; asst. chief U.S. Army Tripler Gen. Hosp., 1960-64; chief William Beaumont Gen. Hosp., 1965-67, Brooke Gen. Hosp., 1967-68, Walter Reed Gen. Hosp., 1968-69; ret., 1969; chief staff Maui Meml. Hosp., 1972-73; asso. prof. medicine Baylor U. Coll. Medicine, 1958-59; clin. prof. medicine Hawaii U. Coll. Medicine, 1969—, Washington U. Coll. Medicine, 1970-77, Abraham Lincoln Sch. Medicine, 1974-75; adj. prof. medicine U. Pa. Sch. Medicine, 1977—; vis. prof. Uniformed Services U. Health Scis., 1979; exec. v.p. A.C.P., 1977—; flight controller Project Mercury, 1959-62; cons., mem. med. evaluation team Project Gemini, 1962-66; cons. Project Apollo, 1967-73, Tripler Gen. Hosp., 1970-77, Walter Reed Army Med. Center, 1974—; mem. cardiovascular and renal adv. com. FDA, 1978—. Contbg. editor: Med. Opinion and Rev, 1966—; mem. editorial bd.: Archives of Internal Medicine, 1967-73, Western Jour. Medicine, 1975—, Chest, 1975—, Med. Times, 1977—, Quality Rev. Bull, 1979; chief div. sci. pubis.; editor: Jour. AMA, 1973-75; Contbr. numerous articles to med., sci. jours., also med. books. Fellow A.C.P., Am. Coll. Cardiology, Am. Clin. and Climatol. Assn.; mem. Am. Med. Writers Assn., Am. Therapeutic Soc., Am. Osler Soc., Inst. Medicine, AMA (adv. panel registry of adverse drug reactions 1960-67, mem. council on drugs 1967-73), Coll., Physicians Phila., Soc. Med. Consultants to Armed Forces, Alpha Sigma Nu, Alpha Omega Alpha. Democrat. Jewish. Subspecialty: Internal medicine. Home: 11 Ashbrooke Dr Voorhees Township NJ 08043 Office: 4200 Pine St Philadelphia PA 19104

MOSES, JOEL, computer scientist; b. Petach Tikvah, Israel, Nov. 25, 1941; came to U.S., 1954, naturalized, 1960; s. Bernhard and Golda (Losner) M.; m. Margaret A. Garvey, Dec. 27, 1970; children: Jesse, David. B.A., Columbia U., 1962, M.A., 1963; Ph.D., M.I.T., 1967. Asst. prof. dept. elec. engring. and computer sci. M.I.T., 1967-71, asso. prof., 1971-76, prof., 1976—, asso. dir. Lab for Computer Sci., 1974-78, asso. head computer sci. and engring., dept. elec. engring. and computer sci., 1978-81, head dept., 1981—. Editor: The Computer Age: A Twenty Year View, 1979. Mem. Assn. Computing Machinery. Subspecialties: Mathematical software; Artificial intelligence. Office: Mass Inst Tech 38-401 Cambridge MA 02139

MOSES, RAY NAPOLEON, JR., physicist; b. Clinton, N.C., Jan. 23, 1936; s. Ray Napoleon and Annie Laurie (Hairr) M.; m. Ruth Kathleen Alsbrooks, June 10, 1962; 1 stepson, Mike Jolly. B.S. in Aero. Engring, Ga. Inst. Tech., 1964; Ph.D., Ohio State U., 1974. Registered profl. engr., S.C., Ala.; 1st class radiotelephone comml. pilot. Engr., Apollo project Boeing Aircraft, Huntsville, Ala., 1964-67, specialist engr., 1981—; engr. Lockheed Aircraft, Huntsville, 1968-69, space systems analyst, 1969-70; asst. prof. physics Furman U., 1974-79; vice-pres. Seige Corp., Greenville, S.C., 1980-81. Lt. Col. USAFR. Mem. AAAS, Nat. Soc. Profl. Engrs., Am. Astron. Soc. Current work: Space war; integration of physical systems. Home: Box 11038 Huntsville AL 35805 Office: Boeing Aerospace Co JA99 Box 1470 Huntsville AL 35807

MOSHER, MELVYN WAYNE, chemistry educator, chemist; b. Palo Alto, Calif., June 10, 1940; s. Kenneth H. and Romona M. (Walker) M.; m. Donna P. Baerg, June 6, 1963; children: Michael David, Craig Andrew, Thomas Charles. B.A., U. Wash., 1962; M.S., U. Idaho, 1964, Ph.D., 1968. Chemist U.S. Bur. Comml. Fisheries, Seattle, 1960-64; teaching asst. U. Idaho, Moscow, 1964-67; univ. fellow U. Alta., Edmonton, 1967-69; asst. prof. Marshall U., Huntington, W.Va., 1969-74, Mo. So. State Coll., Joplin, 1974-79, asst. dir. regional crime lab., assoc. prof. chemistry, 1979—. Contbr. articles to profl. jours. Pres. Greater Ozark Soccer Assn., Joplin, 1977—. Mem. Am. Chem. Soc., Mo. Acad. Sci. (chmn. chemistry 1979-80). Lutheran. Subspecialty: Organic chemistry. Current work: Free radical introduction of functional groups into alkanes, both synthetic and mechanism. Office: Dept Chemistry Mo So State Coll Joplin MO 64801

MOSHMAN, DAVID STEWART, psychology educator; b. Bklyn., May 9, 1951; s. Howard Benjamin and Ruth (Silver) M. B.A., Lehigh U., 1971; M.S., Rutgers U., 1975, Ph.D., 1977. Asst. prof. ednl. psychology U. Nebr., Lincoln, 1977-82, asso. prof., 1982—. Contbr. articles, chpts. to profl. jours., books. Mem. Am. Psychol. Assn., AAUP (chpt. pres.), Soc. Research in Child Devel., Am. Ednl. Research Assn., Jean Piaget Soc., AAAS, Nebr. Civil Liberties Union (dir.), Amnesty Internat. Democrat. Subspecialties: Developmental psychology; Cognition. Current work: Development of logical and scientific reasoning; intellectual development beyond childhood; Piagetian theory; developmental approaches to edn. Office: Edn Psychology Dept Univ Nebr Lincoln NE 68588

MOSHY, RAYMOND JOSEPH, food co. exec.; b. Bklyn., Aug. 12, 1925; s. Jamil and Alice (Waked) M.; m. Mary T. Fazzolare, Aug. 15, 1948; children—Helena, Janice, Raymond, Christopher, Douglas. B.S., St. John's U., 1948; M.S. in Organic Biochemistry, Fordham U., 1950; Ph.D. in Organic Chemistry (fellow), Fordham U., 1953. Cons. Am. Lecithin Corp., L.I. City, N.Y., 1950-52; research chemist Heyden-Newport Chem. Corp., Garfield, N.J., 1952-55; project leader chem. research Gen. Foods Corp., Tarrytown, N.Y., 1955-59; with AMF, Inc., Springdale, Conn., 1959-70, unit mgr. exploratory research, 1959-61, sec. mgr. exploratory research and tobacco research, 1961-63, mgr. food and tobacco lab., 1963-64, mgr. chem. devel. lab., 1964-66, v.p., dir. research div., 1966-70; v.p. research and devel. Hunt-Wesson Foods, Inc., Fullerton, Calif., 1970-75, v.p., group exec., 1975—; mem. sci., engring., and indsl. advisory councils various colls. Contbr. numerous articles in chemistry, food, tobacco, process devel. and engring. to profl. jours. Served with USAF, 1943-45. William-Waterman Fund for Dietary Research grantee, 1949-52. Mem. Food and Drug Law Inst. (gov.), Indsl. Research Inst., Inst. Food Technologists, Am. Chem. Soc., Am. Inst. Chemists, Am. Health Found. (mem. food and nutrition bd.), Food Safety Council, Grocery Mfrs. Am., Nat. Food Processor's Assn. Subspecialties: Organic chemistry; Agricultural Biotechnology Management.. Patentee in field. Home: 1712 Antigua Way Newport Beach CA 92660 Office: 1645 W Valencia Dr Fullerton CA 92634

MOSIER, BENJAMIN, analytical chemist, cons., research co. exec.; b. Corsicanna, Tex., July 15, 1926; s. Phillip and Fannie (Zulauff) M.; m. Doreen N. Zidel, Aug. 22, 1954; children: Marc L., David M., Linda J., Adam J. B.S., Tex. A&M U., 1949, M.S., 1951; Ph.D., U. Ill., 1957. Cert. profl. chemist Nat. Certification Com. in Chemistry and Chem. Engring. Instr. chemistry Kilgore (Tex.) Coll., 1949-50; research scientist Gen. Dynamics, Ft. Worth, 1951-52; NSF fellow U. Ill., Urbana, 1954-57; research scientist Humble Oil & Refining Co., Houston, 1957-60; pres., dir. Inst. Research, Inc., Houston, 1960—; cons. Armour Indsl. Chem. Co., Chgo., 1961—; mem. adv. bd. analytical instrument program Houston Community Coll., 1980, Oilfield Service Corp. Am., Lafayette, La., 1980; assoc. research prof. Baylor U. Coll. Medicine, Houston. Author articles. Served with USAF, 1944-46, 52-54. Recipient numerous awards NASA. Mem. AAAS, Am. Council Ind. Labs., Am. Chem. Soc., Am. Inst. Chemists, Am. Petroleum Inst., Electrochem. Soc., Ill. Acad. Sci., Inst. Food Technologists, Nat. Assn. Corrosion Engrs., N.Y. Acad. Scis., Tex. Acad. Sci., Sigma Xi, Phi Lambda Upsilon. Lodge: Rotary. Subspecialties: Analytical chemistry; Polymer chemistry. Current work: Corrosion, medical instrumentation, microencapsulation, phenolic foam, enhanced oil recovery. Patentee in chem. field.

MOSKOWITZ, MICHAEL ARTHUR, neurologist, educator; b. N.Y.C., May 26, 1942; s. Irving Lawrence and Clara (Dranoff) M.; m. Ann Michael, May 29, 1965; 1 dau., Jenna Rachel. B.A., Johns Hopkins U., 1964; M.A., Tufts U., M.D., 1968. Intern Yale New Haven Hosp., 1968-69, resident in medicine, 1969-71; resident in neurology Brigham-Children's-Beth Israel hosps., Boston, 1971-74; asst. prof. dept. neurology Harvard Med. Sch., Boston, 1975-78, assoc. prof., 1979—; assoc. prof. neurosci. MIT, Boston, 1979-81; asst. prof. neurology div. health sci. and tech. Harvard-MIT, 1979—; research dir. Stroke Lab. Mass. Gen. Hosp. Contbr. articles on neurosci. to profl. jours. Sloane fellow, 1976-78; tchr. investigator awardee Nat. Inst. Neurological Communicative Disorders and Stroke, 1975-80; established investigator awardee Am. Heart Assn., 1980-85. Mem. Am. Neurol. Assn., AAAS, Am. Acad. Neurology, Am. Soc. of Pharmacology and Exptl. Therapeutics, Neurosci. Soc. Subspecialties: Neurology; Neurochemistry. Current work: Blood vessel innervation, receptor pharmacology, prostaglandins and arachidonic acid metabolism.

MOSKOWITZ, PAUL A., physicist; b. Bklyn., Nov. 22, 1945; s. Samuel and Dora Z. (Luntz) M. B.S. in Engring. Sci, N.Y.U., 1966, M.S., 1968, Ph.D., 1971. Registered profl. engr., Colo. NSF fellow U. Grenoble, France, 1971; postdoctoral fellow U. Mainz, W.Ger., 1974; staff Joint Inst. Lab. Astrophysics, Boulder, Colo., 1975-77; mem. research staff IBM Thomas J. Watson Research Ctr, Yorktown Heights, N.Y., 1977—. Contbr. articles to profl. jours. Mem. Am. Phys. Soc. Subspecialties: Laser-induced chemistry; Low temperature physics. Current work: Laser enhanced etching techniques and packaging for superconductive instrumentation. Office: IBM TJ Watson Research Ctr Yorktown Heights NY 10598

MOSMANN, TIMOTHY RICHARD, immunologist; b. Birkenhead, Cheshire, Eng., Mar. 7, 1949; came to U.S., 1982; s. Paul George and Joan Doris (Debbage) M.; m. Penella Eunice Farnham, Dec. 22, 1972. B.Sc., U. Natal, South Africa, 1968, Rhodes U., South Africa, 1969; Ph.D., U. B.C., 1973. Postdoctoral fellow Hosp. for Sick Children, Toronto, Ont., Can., 1973-75; postdoctoral fellow dept. biochemistry Glasgow (Scotland) U., 1975-77; asst. prof. dept. immunology U. Alta. (Can.), Edmonton, 1977-81; prin. scientist DNAX Research Inst., Palo Alto, Calif., 1982—. Contbr. articles to profl. jours. MRC of Can. grantee, 1977-81; Alta. Cancer Hosp. Bd. grantee, 1980-81; Alta. Heritage Fund grantee, 1981. Mem. Can. Soc. Immunology, Am. Assn.

Immunologists. Subspecialties: Immunobiology and immunology; Genetics and genetic engineering (biology). Current work: analysis of antigen-specific molecules of lymphocytes and dection characterization and recombinant DNA cloning of immunoregulatory factors. Home: 69 Lloyden Dr Atherton CA 94025 Office: DNAX Research Inst 1450 Page Mill Rd Palo Alto CA 94304

MOSS, FRANK EDWARD, physics educator; b. Paris, Ill., Feb. 10, 1934; s. Arnold Russel and Mary Christine (Bosie) M.; m. Elaine Spero Koumparakis, Aug. 3, 1962; 1 son, Frank. B.S. in Elec. Engring, U. Va., 1957, M.S. in Nuclear Engring, 1961, Ph.D. in Physics, 1964. Sr. scientist U. Va., Charlottesville, 1966-71; prof. physics U. Mo., St. Louis, 1971—; guest researcher U. Rome, 1966-67; cons., dir. Solar Sumac, Inc., St. Louis, 1981—. Contbr. articles to profl. jours. Served to 2d lt. U.S. Army, 1957-58. U.S. AEC fellow, 1962; NSF fellow, 1965; Brit. Sci. Research Council sr. vis. fellow, 1979; NSF grantee, 1975-80. Mem. Am. Phys. Soc. Subspecialties: Low temperature physics; Statistical physics. Current work: Statistical and low temperature physics. Home: 66 Bellerive Acres Saint Louis MO 63121 Office: U Mo Natural Bridge Rd Saint Louis MO 63121

MOSS, JOHN ELIOT BLAKESLEE, computer scientist; b. Staunton, Va., Jan. 1, 1954; s. William Edwin and Mary Frances (Blakeslee) M.; m. Hannah Allen Abbott, May 29, 1976. S.B.E.E., M.I.T., 1975, S.M.E.E., 1978, E.E., 1978, Ph.D., 1981. Commd. 1st lt. U.S. Army, 1981; computer programmer/systems analyst U.S. Army War Coll., Carlisle Barracks, Pa., 1981—. Co-author: CLU Reference Manual, 1981. NSF fellow, 1975. Mem. Assn. for Computing Machinery, IEEE, Armed Forces Communications-Electronics Assn. Episcopalian. Subspecialties: Distributed systems and networks; Programming languages. Current work: Transaction processing; system architecture; theory of distributed systems; advanced compiler technology. Home: 965 Crains Gap Rd Carlisle PA 17013 Office: USACC Box 483 Carlisle Barracks PA 17013

MOSS, ROBERT LOUIS, physiology educator; b. Bklyn., Aug. 24, 1940; s. Robert C. and Susan (Tuccinardi) M.; m. Rita Mary Walsh, Sept. 22, 1962; children: Michele E., Robert C. B.S., Villanova U., 1962; M.S., Claremont Grad. Sch., 1967, Ph.D., 1969. Research asst. Behavioral Research Lab., Patton State Hosp., 1964-65; NIH postdoctoral fellow U. Bristol, Eng., 1969-71; asst. prof. physiology U. Tex. Health Sci. Ctr., Dallas, 1971-76, assoc. prof., 1976-80, prof., 1980-82, prof. physiology and neurology, 1982—, chmn. physiology grad. program, 1982—; co-dir. endocrinology/human reprodn. course Southwestern Med. Sch., 1977—, co-dir. med. neurobiology course, 1977—. Recipient Research Career Devel. award. NIH., 1976-81; Young Scientist award Am. Psychol. Assn., 1969; Disting. Alumni award Villanova U., 1982; NIH grantee, 1972—; NSF grantee, 1974-76; Tex. Salk Inst. grantee, 1982-83. Mem. AAAS, Am. Physiol. Soc., Endocrine Soc., Internat. Soc. Neuroendocrinology (charter mem.), Soc. for Neurosci. Republican. Roman Catholic. Lodge: K.C. Subspecialties: Neurophysiology. Current work: Neural and biochemical mechanisms involved in hypothalmic control over pituitary functions and reproductive behavior. Office: Dept Physiology Tex Health Sci Ctr 5323 Harry Hines Blvd Dallas TX 75235

MOSSBERG, THOMAS WILLIAM, research physicist, educator; b. Mpls., June 28, 1951; s. William Andrus and Rosemary Elizabeth M.; m. Jane Elizabeth Pai, June 22, 1974; 1 child. A.B. with honors, U. Chgo., 1973; M.A., Columbia U., 1975, Ph.D. in Physics, 1978. Research asst. Columbia U., N.Y.C., 1975-78, research asso., 1978-80, asst. prof., 1980-81; asst. prof. physics Harvard U., Cambridge, Mass., 1981—. Contbr. writings to profl. publs. Mem. Am. Phys. Soc., Optical Soc. Am. Subspecialties: Laser data storage and reproduction; Atomic and molecular physics. Current work: Coherent optical laser phenomena in matter. Developer time-domain frequency selective optical data storage

MOSTELLER, FREDERICK, educator, mathematical statistician; b. Clarksburg, W.Va., Dec. 24, 1916; s. William Roy and Helen (Kelley) M.; m. Virginia Gilroy, May 17, 1941; children: William, Gale. B.S., Carnegie Inst. Tech., (now Carnegie-Mellon U.), 1938, M.S., 1939, D.Sc., 1974; A.M., Princeton U., 1942, Ph.D., 1946; D.Sc., U. Chgo., 1973; D.Social Scis., Yale U., 1981. Instr. math. Princeton U., 1942-44; research assoc. Office Pub. Opinion Research, 1942-44; spl. cons. research br. War Dept., 1942-43; research mathematician Statis. Research Group, Princeton, applied math. panel Nat. Devel. and Research Council, 1944-46; mem. faculty Harvard U., 1946—, prof. math. stats., 1951—, chmn. dept. stats., 1957-69, 75-77, chmn. dept. biostats., 1977-81, chmn. dept. health policy and mgmt., 1981—, Roger I. Lee prof., 1978—; vice chmn. Pres.'s Commn. on Fed. Stats., 1970-71; mem. Nat. Adv. Council Equality of Edn1. Opportunity, 1973-78, Nat. Sci. Bd. Commn. on Pre-coll. Edn. in Math., Sci. and Tech., 1982-83; Fund for Advancement of Edn. fellow, 1954-55; nat. tchr. NBC's Continental Class-room TV course in probability and stats., 1960-61; fellow Center Advanced Study Behavioral Sciences, 1962-63, bd. dirs., 1980—; Guggenheim fellow, 1969-70; Miller research prof. U. Calif. at Berkeley, 1974-75. Co-author: Gauging Public Opinion (editor Hadley Cantril), 1944, Sampling Inspection, 1948, The Pre-election Polls, 1948, 1949; author: Stochastic Models for Learning, 1955, Probability with Statistical Applications, 1961, Inference and Disputed Authorship, The Federalist, 1964, The National Halothane Study, 1969, Statistics: A Guide to the Unknown, 1972, On Equality of Educational Opportunity, 1972, Sturdy Statistics, 1973, Statistics By Example, 1973, Cost, Risks and Benefits of Surgery, 1977, Data Analysis and Regression, 1977, Statistics and Public Policy, 1977, Data for Decisions, 1982, Understanding Robust and Exploratory Data Analysis, 1983, Biostatistics in Clinical Medicine, 1983, Beginning Statistics with Data Analysis, 1983; Author articles in field. Trustee Russell Sage Found.; mem. bd. Nat. Opinion Research Center, 1962-66. Recipient Outstanding Statistician award Chgo. chpt. Am. Statis. Assn., 1971; Myrdal prize Evaluation Research Soc., 1978; Paul F. Lazarsfeld prize Council Applied Social Research, 1979. Fellow AAAS (chmn. sect. U 1973, dir. 1974-78, pres. 1980, chmn. bd. 1981), Inst. Math. Statistics (pres. 1974-75), Am. Statis. Assn. (v.p. 1962-64, pres. 1967), Social Sci. Research Council (chmn. bd. dirs. 1966-68), Math. Social Sci. Bd. (acad. governing bd. 1962-67), Am. Acad. Arts and Scis., Royal Statis. Soc. (hon.); mem. Am. Philos. Soc., Internat. Statis. Inst., Am. Math. Soc., Math. Assn. Am., Psychometric Soc. (pres. 1957-58), Inst. Medicine of Nat. Acad. Scis. (council 1978), Nat. Acad. Scis., Biometric Soc. Subspecialties: Statistics; Gerontology. Current work: Health policy research, evaluation, robust methods. Office: 1 Oxford St Cambridge MA 02138

MOSTOW, GEORGE DANIEL, mathematics educator; b. Boston, July 4, 1923; s. Isaac J. and Ida (Rotman) M.; m. Evelyn Davidoff, Sept. 1, 1947; children: Mark Alan, David Jechiel, Carol Held, Jonathan Carl. B.A., Harvard U., 1943, M.A., 1946, Ph.D., 1948. Instr. math. Princeton U., 1947-48; mem. Inst. Advanced Study, 1947-49, 56-57, 75, trustee, 1982—; asst. prof. Syracuse U., 1949-52; asst. prof. math. Johns Hopkins U., 1952-53, asso. prof., 1954-56, prof., 1957-61; prof. math. Yale U., 1961-66, James E. English prof. math., 1966-81, Henry Ford prof. math., 1981—, chmn., 1971-74; vis. prof. Conselho National des Pesquisas, Instituto de Matematica, Rio de Janiero, Brazil, 1953-54, U. Paris, 1966-67, Hebrew U., Jerusalem, 1967, Tata Inst. Fundamental Research, Bombay, 1970, Institut des Hautes Etudes Scientifiques, Bures-Sur-Yvette, 1966, 71, 75; Chmn. U.S. Nat. Com. for Math., 1971-73, 83-85; chmn. Office Math. Scis., NRC, 1975-78. Asso. editor: Annals of Math, 1957-64, Trans. Am. Math. Soc, 1958-65, Am. Scientist, 1970-82; editor: Am. Jour. Math, 1965-69; asso. editor, 1969-79; author research articles. Fulbright research scholar, Utrecht U., Netherlands, also Guggenheim fellow, 1957-58. Mem. Nat. Acad. Sci. (chmn. sect. math 1982-84), Am. Acad. Arts and Scis., Am. Math. Soc. (v.p. 1979—), Internat. Math. Union (chmn. U.S. del. to gen. assembly Warsaw 1982, mem. exec. com. 1983—), Phi Beta Kappa, Sigma Xi. Subspecialty: Group theory. Home: Beechwood Rd Woodbridge CT 06525 Office: Yale U New Haven CT 06520

MOTICKA, EDWARD J., immunology educator; b. Oak Park, Ill., May 21, 1944; s. Edward and Vivian (Charvet) M.; m. Betty Horvath, June 21, 1969; children: Juli, Gabrielle, Danielle. B.A., Kalamazoo Coll., 1966; Ph.D., U. Ill. Med Ctr., Chgo., 1970. Postdoctoral fellow UCLA, 1970-71; vis. scientist Czechoslovak Acad. Scis., Prague, 1971-72; asst. prof. U. Tex. Health Sci. Ctr., Dallas, 1972-78; assoc. prof. So. Ill. U., Carbondale, 1978-80, Springfield, 1980. Contbr. articles to sci. jours. Mem. Am. Assn. Immunologists, Am. Soc. Zoologists. Subspecialty: Cellular engineering. Current work: Immunoregulatory mechanisms of autoantibody production and autoimmune disease. Office: PO Box 3926 Springfield IL 62708

MOTTINGER, JOHN PHILIP, botany educator, researcher; b. Detroit, Nov. 28, 1938; s. Claude W. and Elizabeth E. M. B.A., Old Wesleyan U., 1961; Ph.D., Ind. U., 1968. Instr. botany U. R.I., Kingston, 1967-68, asst. prof., 1968-75, assoc. prof., 1975—; research assoc. U. Calif., Berkeley, 1980. Served with U.S. Army, 1956-57. Mem. Genetics Soc. Am., AAAS, Sigma Xi. Subspecialties: Genetics and genetic engineering (agriculture); Plant genetics. Current work: Virus induced genetic instability and gene regulation in maize. Home: 44 Wood Ave Narragansett RI 02882 Office: Department of Botany University of Rhode Island Kingston RI 02881

MOTTOLA, HORACIO A., chemistry educator; b. Buenos Aires, Argentina, Mar. 22, 1930; s. Carmelo Juan and Ramona Maria (Saborido) M.; m. Maria D. Mottola, Sept. 5, 1958; children: Adriana, Horacio. Licentiate, U. Buenos Aires, 1957, doctorate, 1962. Asst. prof. chemistry U of Pacific, Stockton, Calif., 1964-66; mem. faculty Okla. State U., Stillwater, 1967—, prof. chemistry, 1975—; cons. in field. Contbr. numerous research articles to profl. publs. Recipient Award of Merit Okla. Acad. Scis., 1981; fellow U. Minn., 1958-60, U. Ariz., 1963-64, 66-67; NSF grantee, 1968—. Mem. Am. Chem. Soc., Sigma Xi (lectr. 1981). Subspecialty: Analytical chemistry. Current work: Automated chemical analysis; chromatographic separations; photochromism of metal chelate species; reaction rate methodology. Office: Okla State Univ Chemistry Dept Stillwater OK 74078

MOTULSKY, ARNO GUNTHER, geneticist, physician, educator; b. Fischhausen, Germany, July 5, 1923; s. Herman and Rena (Sass) Molton; m. Gretel C. Stern, Mar. 22, 1945; children: Judy, Harvey, Arlene. Student, Central YMCA Coll., Chgo., 1941-43, Yale U., 1943-44; B.S., U. Ill., 1945, M.D., 1947. Diplomate: Am. Bd. Internal Medicine. Intern, fellow, resident Michael Reese Hosp., Chgo., 1947-51; Staff mem. charge clin. investigation dept. hematology Army Med. Service Grad. Sch., Walter Reed Army Med. Center, Washington, 1952-53; research asso. internal medicine George Washington U. Sch. Medicine, 1952-53; from instr. to assoc. prof. dept. medicine U. Wash. Sch. Medicine, Seattle, 1953-61, prof. medicine, prof. genetics, 1961—; head div. med. genetics, dir. genetics clinic Univ. Hosp., 1959—, Children's Med. Center, 1966-72; dir. Center for Inherited Diseases, 1972—; attending physician Univ. Hosp., Seattle; cons. physician Children's Orthopedic Hosp. and Med. Center, Seattle; cons. various coms. NRC, NIH, WHO, others; mem. Am. Bd. Med. Genetics, 1979-82, Pres.'s Commn. for Study of Ethical Problems in Medicine and Biomed. and Behavioral Research, 1979-83. Editor: Am. Jour. Human Genetics, 1969-75, Human Genetics, 1969—, Progress in Med. Genetics, 1974—; editorial bd.: DM Disease a Month. Spl. Commonwealth Fund fellow in human genetics Univ. Coll., London, Eng., 1957-58; John and Mary Markle scholar in med. sci., 1957-62; fellow Center Advanced Study in Behavioral Scis., Stanford U., 1976-77. Fellow A.C.P.; mem. Internat. Soc. Hematology, Am. Fedn. Clin. Research, AAAS, Genetics Soc. Am., Western Soc. Clin. Research, Am. Soc. Human Genetics, Am. Soc. Clin. Investigation, Am. Assn. Physicians, Nat. Acad. Scis., Inst. of Medicine, Am. Acad. Arts and Scis. Subspecialty: Genetics and genetic engineering (medicine). Research and publs., including books, in field. Home: 4347 53d Ave NE Seattle WA 98105 Office: U Wash Div Med Genetics Seattle WA 98195

MOTZ, ROBIN OWEN, physician, cons. physicist and pharmacologist; b. Bronx, N.Y., Mar. 9, 1939; s. Lloyd and Minne (Rosenbaum) M.; m. Marcia Linda Motz, Aug. 29, 1959; children: Jeremy, Nicole, Benjamin. A.B. in Physics and Astronomy, Columbia U., 1959, Ph.D., 1965, M.D., 1975. Diplomate: Diplomate Am. Bd. Internal Medicine. Intern Columbia-Presbyn. Med. Ctr., N.Y.C., 1975-76, resident in internal medicine, 1976-78; asst. prof. physics Stevens Inst. Tech., 1965-71; lectr. Columbia U., 1971-78, asst. prof. clin. medicine, 1978—; med. dir. Thomas Ferguson Assocs.; cons. physicist Thexon Corp. Astronomy editor, contbg. author: Columbia Ency, 1975; contbr. numerous articles to profl. jours. Fellow N.Y. Acad. Scis., ACP; mem. AMA, Am. Phys. Soc., N.Y. Acad. Scis., Mensa, Sigma Xi. Subspecialties: Internal medicine; Plasma physics. Current work: Spray drying and ionic recombination, applied physics research for med. cures. Office: 404 Tenafly Rd Tenafly NJ 07670

MOUDGIL, BRIJ MOHAN, material engineering educator; b. Pataudi, Haryana, India, Aug. 4, 1945; s. Deviki Nandan and Bhagwati Devi (Bhardwajs) M.; m. Sheela Moudgil, Oct. 1, 1973; children: Suniti, Sarika. B.S., DAV Coll., Jullundur, India, 1963-65, Indian Inst. Sci., Bangalore, 1968; M.S., Columbia U., 1972, Sc.D., 1981. Research engr. Occidental Research Corp., Irvine, Calif., 1976-80; grad. research assoc. Columbia U., N.Y.C., 1978-81; assoc. prof. dept. material s. sci. and engring. U. Fla., Gainesville, 1981—, dir., 1982—. Contbr. articles to profl. jours. Mem. AIME (vice chmn. fundamentals), Am. Inst. Chem. Engrs., Sigma Xi. Subspecialties: Metallurgical engineering; Surface chemistry. Current work: Mineral beneficiation, applied surface chemistry; fine particle processing, mineral industry waste disposal. Patentee in field. Home: 2101 NW 20th St Gainesville FL 32605 Office: Dept Material Sci and Engring U Fla Gainesville FL 32611

MOUDGIL, VIRINDER KUMAR, biological science educator; b. Ludhiana, Panjab, India; came to U.S., 1973; s. Harbhagwan and Lajwanti (Devi) M.; m. Parviz Gandhi; children: Sapna, Rishi. B.Sc., Govt. Coll., Ludhiana, 1967; M.S., Banaras Hindu U., Varanasi, India, 1969, Ph.D., 1972. Jr. research fellow Banaras Hindu U., 1969-71, asst. research officer, 1971-73, sr. research fellow, 1973; postdoctoral fellow Mayo Clinic, Rochester, Minn., 1973-76; asst. prof. biol. sci. Oakland U., Rochester, Mich., 1976-82, assoc. prof., 1982—; adj. assoc. prof. Wayne State U., 1982. NIH grantee, 1978—. Mem. Am. Physiol. Soc., Am. Chem. Soc., Endocrine Soc., Soc. Biol. Chemists, Biophys. Soc. Subspecialties: Biochemistry (biology); Reproductive biology. Current work: Mechanism of action of steroid hormones. Home: 159 Nesbit Ln Rochester MI 48063 Office: Oakland U Biological Sciences Rochester MI 48063

MOUNT, MARK SAMUEL, plant pathology educator, researcher; b. Crawfordsville, Ind., Nov. 18, 1940; s. Verner Mason and Bertha Eulalia (COle) M.; m. Patricia Johnson, Aug. 17, 1963; children: Bradley Alan, Christopher Clayton. B.S., Ill. Wesleyan U., 1963; M.S., Mich. State U., 1965, Ph.D., 1968. Postdoctorate Cornell U., 1968-69; asst. prof. dept. plant pathology, 1969-75, assoc. prof., 1975-82, prof., 1982, head dept., 1981; cons. in field. Editor: (with George Lacy) Phytopathogenic Prokaryotes, 1982; contbr. articles to profl. jours. Coach Little League Baseball, Amherst, Mass., 1979-82. NSF grantee, 1971; NIH grantee, 1974; U.S. Dept. Def. grantee, 1976. Mem. Am. Phytopath. Soc., Am. Soc. Plant Physiologists, Am. Soc. Microbiologists, Sigma Xi. Republican. Subspecialties: Plant pathology; Genetics and genetic engineering (agriculture). Current work: Genetics and enzyme regulation of plant pathogenic bacteria; department administration, graduate teaching, research. Office: Department Plant Pathology University of Massachusetts Amherst MA 01003

MOUNTCASTLE, VERNON BENJAMIN, JR., neurophysiologist; b. Shelbyville, Ky., July 15, 1918; s. Vernon Mountcastle and Anne-Francis Marguerite (Waugh) M.; m. Nancy Clayton Pierpont, Sept. 6, 1945; children: Vernon Benjamin III, Anne Clayton, George Earle Pierpont. B.S. in Chemistry, Roanoke Coll., Salem, Va., 1938, D.Sc. (hon.), 1968; M.D., Johns Hopkins U., 1942; D.Sc. (hon.), U. Pa., 1976, M.D., U. Zurich, 1983. House officer surgery Johns Hopkins Hosp., 1942-43; mem. faculty Johns Hopkins Sch. Medicine, 1946—, prof. physiology, 1959, dir. dept., 1964-80, Univ. prof. neurosci., 1980—; dir. Neurosci. Research Program, Rockefeller U., 1981—; pres. Neurosci. Research Found., 1981—; spl. univ. lectr. Univ. Coll., London, Eng., 1959; Penfield lectr. Am. U. Beirut, 1971; Sherrington lectr. U. Liverpool, Eng., 1974; Harmon lectr. Am. Assn. Anatomists, 1976; Kershman lectr. Eastern EEG Soc., 1976; Stevenson lectr. U. Western Ont., 1976; Mellon lectr. U. Pitts., 1977; Regents lectr. U. Helsinki, 1977; Hughlings Jackson lectr. McGill U., 1978; Harvey lectr., 1979; Bishop lectr. Washington U., 1979; Hines lectr. Emory U., 1979; Scott lectr. Wayne State U., 1981; vis. prof. Coll. de France, Paris, 1980; spl. research physiology brain. Chmn. physiology study sect., mem. physiology tng. com. NIH, 1958-61; adv. council Nat. Eye Inst., 1971-74; mem. sci. adv. bd. USAF, 1969-71; vis. com. dept. psychology Mass. Inst. Tech., 1966-75; bd. biology and medicine NSF, 1970-73; mem. commn. on neurophysiology Internat. Union Physiol. Sci. Editor-in-chief: Jour. Neurophysiology, 1961-64; assoc. editor: Bull. Johns Hopkins Hosp, 1954-62; editorial bd.: Physiol. Revs, 1957-59, Jour. Neuropharm, 1966-71, Exptl. Brain Research, 1966—; editor, contbr.: Med. Physiology, 12th edit, 1968, 13th edit., 1974, 14th edit., 1980, (with G.M. Edelman) The Mindful Brain, 1978; author articles in field. Served to lt. (s.g.) M.C. USNR, 1943-46. Recipient Lashley prize Am. Philos. Soc., 1974, F.O. Schmitt prize and medal MIT, 1975, Sherrington prize and Gold medal Royal Acad. Medicine, London, 1977, Horowitz prize Columbia U., 1978, Fyssen Internat. prize, Paris, 1983, Lasker award, 1983. Mem. Am. Physiol. Soc., Am. Acad. Arts and Scis., Nat. Acad. Sci. (chmn. sect. physiology 1971-74), Harvey Cushing Soc., Am. Neurol. Assn. (Bennett lectr. 1978, hon. mem.), AAAS, Soc. Neurosci. (pres. 1970-72, Gerard Prize 1980), Am. Philos. Soc. (councillor 1979-82), Nat. Inst. Medicine, Phi Beta Kappa, Sigma Chi, Alpha Omega Alpha, Phi Chi. Club: 14 West Hamilton St. (Balt.). Subspecialties: Neurobiology; Neurophysiology. Current work: Physiology of cerebral cortex and its relation to higher function of brain. Home: 31 Warrenton Road Baltimore MD 21210

MOWERY, DWIGHT FAY, chemist, educator, researcher; b. Moorehead, Minn., May 1, 1915; s. Dwight Fay and Elizabeth King (McGiffert) M.; m. Suzanne Frame Lenderman, June 12, 1943. B.A. in Chemistry, Harvard U., 1937; Ph.D., M.I.T., 1940. Research chemist DuPont Co., Wilmington, Del., 1940-42, Hercules Powder Co., Wilmington, 1942-43; prof. chemistry Elms Coll., 1943-46; prof. Franklin Tech. Inst., 1946-49; instr. Trinity Coll., Hartford, Conn., 1949-53; prof. Ripon Coll., 1953-57, Southeastern Mass. U., 1957-64, Commonwealth prof., 1964—; cons. in field. Contbr. numerous articles on organic chemistry, microanalytical chemistry, chromatography, and chem. edn. to profl. jours. L.F. Verges fellow, 1939; NSF research grantee, 1956. Mem. Am. Chem. Soc., AAUP, Sigma Xi, Sigma Pi Sigma. Republican. Unitarian. Subspecialties: Organic chemistry; Mathematical software. Current work: Carbohydrate chemistry; organic analysis employing classical microchem. methods; gas, liquid and thin layer chromatography; mass, nuclear magnetic reasonance and infrared spectrophotometry; reaction mechanism and rate studies employing computer programming. Office: Southeastern Mass U North Dartmouth MA 02747

MOWRY, ROBERT WILBUR, physician, educator; b. Griffin, Ga., Jan. 10, 1923; s. Roy Burnell and Mary Frances (Swilling) M.; m. Margaret Neilson Black, June 11, 1949; children: Janet Lee, Robert Gordon, Barbara Ann. B.S., Birmingham So. Coll., 1944; M.D., Johns Hopkins U., 1946. Rotating intern U. Ala. Med. Coll., 1946-47, resident pathology, 1947-48; sr. asst. surgeon USPHS-NIH, Bethesda, Md., 1948-52; fellow pathology Boston City Hosp., 1949-50; asst. prof. pathology Washington U., St. Louis, 1952-53, U. Ala. Med. Center, Birmingham, 1953-54, assoc. prof. pathology, 1954-57, prof., 1958—, prof. health services adminstrn., 1976—, dir. Anat. Pathology Lab., 1960-64, dir. grad. programs in pathology, 1964-72; sr. scientist U. Ala. Inst. Dental Research, 1967-72, dir. autopsy services, 1975-79; vis. scholar dept. pathology U. Cambridge, Eng., 1972-73; cons. FDA, 1975-81. Author: (with J.F.A. McManus) Staining Methods: Histologic and Histochemical, 1960; Editorial bd.: Jour. Histochemistry and Cytochemistry, 1960-75, Stain Tech, 1965—; editorial bd.: AMA Archives of Pathology, 1967-76. Served with USPHS, 1948-52. Mem. Am Assn. Pathologists, Histochem. Soc. (councillor 1974-78), Internat. Acad. Pathology, Biol. Stain Commn. (v.p. 1974—, pres. 1976-81, trustee 1966—), Am. Assn. U. Profs. Pathology, AMA, Phi Beta Kappa, Sigma Xi, Delta Sigma Phi, Alpha Kappa Kappa. Presbyn. Subspecialties: Pathology (medicine); Cytology and histology. Current work: Cytochemistry of complex carbonohydrates and certain proteins, methods development and applications in diagnostic histopathology of human diseases. Home: 4165 Sharpsburg Dr Birmingham AL 35213

MOYER, MARY PATRICIA, experimental oncologist, educator; b. Arlington, Mass., Apr. 27, 1951; d. Joseph Francis and Virginia Audrey (Rivers) Sutter; m. Rex Carlton Moyer, Dec. 26, 1974; dau., Amy Jo. B.S., Fla. Atlantic U., 1972, M.S., 1974; Ph.D., U. Tex., 1981. Research assoc. Equine Research Inst., Boca Raton, Fla., 1969-74; research asst. S.W. Fedn. Research and Edn., San Antonio, 1974; research scientist Thorman Cancer Lab., Trinity U., San Antonio, 1974-81; research instr. U. Tex. Health Sci. Center, San Antonio, 1981-82, research asst. prof. depts surgery and microbiology, 1982—. Contbr. articles to profl. jours. Grantee NIH, Am. Cancer Soc.; others. Mem. Am. Soc. Microbiology, AAAS, Tissue Culture Assn., N.Y. Acad. Sci., Assn. Women in Sci., Am. Soc. Cell Biology, Soc. Exptl. Biology and Medicine, Sigma Xi. Subspecialties: Cancer research (medicine); Cell study oncology. Current work: In vitro culture and transformation—gastrointestinal epithelium; virology—SV40; mechanisms of metastasis; chemotherapy; genetic engineering; culture of normal human cells; human tumor cells. Home: Rural Route 20 Box 662 San Antoni TX 78218 Office: 7703 Floyd Curl Dr Room 212E San Antonio TX 78284

MOYER, ROBERT FINDLEY, med. physicist; b. N.Y.C., May 12, 1937; s. James Herbert and Ina Ruth (Findley) M.; m. Cecclia Rita; 3 children. B.S., Pa. State U., 1959, M.S., 1961; Ph.D., UCLA, 1965. Cert. radiation equipment safety officer N.Y. State Dept. Health, 1974. Instr. SUNY Upstate Med. Center, Syracuse, 1965-70, asst. prof., 1970-81; chief physicist Reading (Pa.) Hosp. and Med. Ctr., 1981—; cons., researcher in field. Author: Personalized Instruction in Radiation Therapy, 1978, 79; contbr. articles to profl. publs. Elder Presbyterian ch., 1968—; cubmaster Hiawatha Council Boy Scouts Am., 1968-70. Recipient cert. of merit for exhibit Radiol. Soc. N.Am., 1977. Mem. Am. Assn. Physicists in Medicine, Am. Assn. Med. Dosimetrists (assoc.). Republican. Subspecialties: Radiology; Medical physics. Current work: Medical radiation physics and health physics. Home: 342 RD 2 Mertztown PA 19539 Office: Reading Hosp and Med Ctr Radiology Reading PA 19603

MRAZ, GEORGE JAROSLAV, mech. engr.; b. Chocen, Bohemia, Czechoslovkia, Mar. 23, 1922; s. Jaroslav Stepan and Marie Frantiska (Padourova) M.; m. Milada Barbara Mraz, Jan. 25, 1947; children: Magda Mathisen, Jerrold J. B.Sc., Imperial Coll. Sci. and Tech. London, 1943; Ing., Bohemian Coll. Tech., Prague, 1949. Registered profl. engr., Pa. Plant engr. Ing. J. Mraz Co., Hradek, Czechoslovakia, 1946-49; project engr. (Stavo project), Bratislava, Czechoslovakia, 1949-52; prof. State Engring. Sch., Chomutov, Czechoslovakia, 1952-61; design engr. Skoda Works, Ostrov, Czechoslovakia, 1961-67, M.W. Kellogg Co., N.Y.C., 1967-69; engring. mgr. Nat. Forge Co., Irvine, Pa., 1970. Contbr. articles to profl. jours. Served with RAF, 1943-46. Decorated Czech Medal Bravery. Mem. ASME, Nat. Soc. Proft. Engrs., Am. Soc. Metals, Internat. Assn. Advancement of High Pressure Sci. and Tech. Subspecialty: Mechanical engineering. Current work: High pressure technology; hot and cold isostatic pressing; safety of high pressure systems. Home: 106 Cobham Park Rd Warren PA 16365 Office: 100 Front St Irvine PA 16329

MUAN, ARNULF, university dean, geochemistry educator; b. Lökken Verk, Norway, Apr. 19, 1923; came to U.S., 1952, naturalized, 1962; s. Anders O. and Ingeborg (Engen) M.; m. Hildegard Hoss, Jan. 29, 1960; children: Michael, Ingrid. Diploma in chemistry, Tech. U. Norway, 1948; Ph.D. in Geochemistry, Pa. State U., 1955. Asst. prof. metallurgy Pa. State U., University Park, 1955-57, asso. prof., 1957-62, prof., 1962-66, prof. mineral scis., 1966—, head dept. geochemistry and mineralology, 1966-71, head dept. geoscis., 1971-73, asso. dean research, 1976—. Author: (with E.F. Osborn) Phase Equilibria Among Oxides in Steelmaking, 1965. Sr. NSF fellow, 1962; Fulbright-Hays lectr. USSR, 1973-74. Fellow Am. Ceramic Soc. (Ross Coffin Purdy award 1958, John Jeppson gold medal 1978), Mineral. Soc. Am. (v.p. 1973-74, pres. 1974-75), Geol. Soc. Am.; mem. Am. Inst. Metall. Engrs., A.A.A.S., Geochem. Soc., Norwegian Chem. Soc., Sigma Xi. Subspecialties: High-temperature materials; High temperature chemistry. Current work: Chemistry and characterization of metal-, oxide- and carbide phases at high temperatures under various atmospheric conditions. Home: 400 Toftrees Ave Apt 107 State College PA 16801 Office: 416 Walker Bldg University Park PA 16802

MUCHMORE, HAROLD GORDON, physician, educator; b. Ponca City, Okla., Mar. 8, 1920; s. Clyde E. and Iola R. (Winner) M.; m. Donna M. Stevens, Feb. 23, 1954; children—Bruce, Nancy, Steven, Allan. B.A., Rice U., Houston, 1943; M.D., U. Okla., 1946. Intern U. Okla., 1946-47, resident internal medicine, 1954-56, infectious diseases, 1956-60; mem. faculty U. Okla. Coll. Medicine, 1947-62, 66—, prof. medicine, 1970—, Carl Puckett prof. pulmonary diseases, 1970—, adj. prof. microbiology and immunology, 1970—; mem. faculty U. Minn. Med. Sch., 1962-66. Served with AUS, 1943-45; Served with USAF, 1952-54. Fellow A.C.P., Infectious Diseases Soc. Am.; mem. Am. Thoracic Soc., Internat. Soc. Human and Animal Mycology, Explorers Club. Subspecialties: Internal medicine; Infectious diseases. Current work: Immunity during bio-isolation; fungus diseases in Antarctic. Home: 3005 Robin Ridge Rd Oklahoma City OK 73120 Office: PO Box 26901 Oklahoma City OK 73190

MUCHOVEJ, JAMES JOHN, plant pathology educator; b. Elizabeth, N.J., June 3, 1953; s. Stephen and Shirley May (Bryan) Mucha; m. Rosa Maria Castro, Dec. 15, 1975; children: Sarah Cristina Castro, Stephen James Castro. B.S., Purdue U., 1975, M.S., 1976; Ph.D. candidate, Va. Poly. Inst. and State U., 1980—. Lectr. Faculdade de Ciencias Agrarias Do Para, Belem, Brazil, 1977; lectr. U. Fed. de Vicosa, Brazil, 1978-80, asst. prof., 1980—. Contbr. articles to profl. jours. Mem. Am. Phytopath. Soc., Assn. Applied Biologists, Brit. Mycol. Soc., Mycol. Soc. Am., Brit. Soc. for Plant Pathology (founding), Sociodade Brasileira de Fitopatalogia. Subspecialty: Plant pathology. Current work: Phyloplane mycoflora; plant cuticle interactions. Home: 7918 Corbett Rd Pennsauken NJ 08109 Office: Dept Fitopatologia U Fed de Vicosa 36570 Vicosa MG Brazil

MUDGETT-HUNTER, MEREDITH, immunologist, researcher, educator; b. Springfield, Mass., Jan. 9, 1945; d. John S. and Barbara (Rogers) Mudgett; m. Samuel Hunter, Oct. 29, 1977; children: Nathan, Jerome. A.B., Mt. Holyoke Coll., 1967; Ph.D., U. Ill.-Chgo., 1973. Postdoctoral fellow Rockefeller U., 1973-76, trainee, 1973-76, asst. prof., 1976-77; asst. prof. dept. pathology Harvard U. Med. Sch., Boston, 1977; asst. biochemist Cellular and Molecular Research Lab., Mass. Gen. Hosp., Boston, 1977; cons. in field. Contbr. articles to profl. jours. N.Y. Heart Assn. James N. Jarvie Meml. fellow, 1975-77; established investigator Am. Heart Assn., 1979. Mem. Am. Assn. Immunologists, Harvey Soc., N.Y. Acad. Scis., Sigma Xi. Subspecialties: Immunology (medicine); Immunogenetics. Current work: Research centers around the production of monoclonal antibodies; study of structure-function relationships using monoclonal anti-digoxin antibodies as model system. Office: Cellular and Molecular Research Lab Mass Gen Hosp Boston MA 02114

MUELLER, GEORGE E., corporation executive; b. St. Louis, July 16, 1918; m. Maude Rosenbaum (div.); children: Karen, Jean; m. Darla Hix, 1978. B.S. in Elec. Engring, Mo. Sch. Mines, 1939, M.S., Purdue U., 1940; Ph.D. in Physics, Ohio State U., 1951; hon. degrees, Wayne State U., N.M. State U., U. Mo., 1964, Purdue U., Ohio State U., 1965. Mem. tech. staff Bell Telephone Labs., 1940-46; prof. elec. engring. Ohio State U., 1946-56; cons. electronics Ramo-Wooldridge, Inc., 1955-57; from dir. electronic lab. to v.p. research and devel. Space Tech. Labs., 1958-62; asso. adminstr. for manned space flight NASA, 1963-69; corporate officer, sr. v.p. Gen. Dynamics Corp., N.Y.C., 1969-71; chmn., pres. System Devel. Corp., Santa Monica, Calif., 1971-80, chmn., chief exec. officer, 1981—; sr. v.p. Burroughs Corp., 1981—. Author: (with E.R. Spangler) Communications Satellites. Recipient 3 Distinguished Service medals NASA; Eugen Sanger award; Nat. Medal Sci., 1970; Nat. Transp. award, 1979. Fellow AAAS, IEEE, Am. Inst. Aeros. and Astronautics, Am. Phys. Soc., Am. Astronautical Soc. (Space Flight award), Am. Geophys. Union, Brit. Interplanetary Soc.; mem. Nat. Acad. Engring., N.Y. Acad. Scis. Subspecialty: Electrical engineering. Patentee in field. Home: Santa Barbara CA Office: PO Box 5856 Santa Barbara CA 93108

MUELLER, HERBERT JOSEPH, dental research associate, consultant; b. Milw., Feb. 17, 1941; s. Herbert L. and Ann (Geimeiner) M. B.M.E., Marquette U., 1964; M.S., Northwestern U., 1966, Ph.D., 1969. Research fellow Northwestern U., 1964-69, postdoctoral fellow, 1977-80; asst. prof., chmn. dept. Loyola U., 1968-71; research assoc. ADA, Chgo., 1980. Contbr. articles to profl. jours., chpt. in book. Nat. Inst. Dental Research grantee, 1971. Mem. Am. Soc. Metals, Internat. Assn. Dental Research, Soc. Biomaterials, Sigma Xi, Pi Tau Sigma, Tau Beta Pi. Subspecialties: Dental materials; Biomaterials. Current work: Interactions between biomaterials and saliva in terms of electro-chemistry of metals; binding of ions to protein, absorption of protein to materials. Development of improved dental amalgams with studies related to corrosion-fatigue, creep-environment interactions amalgam-cement interactions and amalgams with new chemistries. Characterization, improvement and development of dental phosphate and ethyl silicate investments. Home: 3533 W Lakefield Dr Milwaukee WI 53215 Office: Am Dental Assn 211 E Chicago Ave Chicago IL 60611

MUELLER, NANCY ALICE SCHNIEDER, biology educator; b. Wooster, Ohio, Mar. 8, 1933; d. Gilbert D. and Winifred (Porter) Schneider; m. Helmut Charles Mueller, Jan. 27, 1959; 1 son, Karl Gilbert. A.B. in Biology, Coll. Wooster, 1955; M.S. in Zoology, U. Wis.-Madison, 1957, Ph.D., 1962. Instr. zoology U. Wis.-Madison, 1966; vis. asst. prof. depts. zoology and poultry sci. N.C. State U., Raleigh, 1968-71; vis. prof. biology N.C. Central U., Durham, 1971-74, assoc. prof., 1974-79, prof. 1979; vis. scientist U. Vienna, 1975. Contbr. articles to sci. jours. Democrat precinct chmn., 1982—. Mem. Am. Soc. Zoologists, Am. Ornithologists Union (bibliography com. 1983), Wison Ornithol. Soc. (resolutions com. 1982-84), Women in Cell Biology, Sigma Xi, Beta Kappa Chi. Subspecialties: Developmental biology; Immunobiology and immunology. Home: Route 5 Box 103 Chapel Hill NC 27514 Office: NC Central U Durham NC 27707

MUELLER, PAUL ALLEN, geology educator, researcher; b. Anniston, Ala., Sept. 9, 1945; s. Raymond John and Hazel Ellen (Jones) M. A.B., Washington U., St. Louis, 1967; M.S., Rice U., 1971, Ph.D., 1971. Research assoc. U. N.C.-Chapel Hill, 1971-73; asst. prof. geology U. Fla., 1973-78, assoc. prof., 1978—. Contbr. articles to profl. jours. Recipient Nininger prize Ariz. Ctr. Meteorite Studies, 1969. Mem. Geochem. Soc., Am. Geophys. Union, Geol. Soc. Am. Subspecialties: Geochemistry; Petrology. Current work: Isotope geology, trace element geochemistry, particularly in relation to origin of earth's crust. Office: Dept Geology U Fla Gainesville FL 32611

MUELLER, SHIRLEY MALONEY, neurologist educator; b. LaCrosse, Wis., May 7, 1942; d. Thomas Leo and Ellen (Ludeking) Maloney; m. Thomas Maythem Mueller, Jan. 11, 1968; 1 dau., Ellen. B.S., Clarke Coll., 1963; M.S., U. Iowa, 1967, M.D., 1971. Intern U. Iowa Hosps., Iowa City, 1971-72, resident in pediatrics, 1972-74, resident in neurology, 1974-77; asst. prof. neurology Ind. U.-Indpls., 1978-82; assoc. prof. neurology Ind. U.-Indpsl., 1983—, asst. prof. physiology, 1981-82, assoc. prof. physiology, 1983—, dir. neurovascular lab., 1981—; site visitor NIH, 1981—; dir. neurology Wishard Meml. Hosp., Indpls., 1982—; mem. arteriosclerosis, hypertension and lipid metabolism adv. com. Nat. Herat, Lung, and Blood Council, 1982-85. Recipient Nat. Research Service award NIH, 1977-78, Young Investigator award, 1981-84. Fellow Am. Heart Assn. Stroke Council; mem. Am. Acad. Neurology, Am. Fedn. Clin. Research, Central Soc. Neurol. Research, Central Soc. Clin. Research, Am. Physiol. Soc., Soc. Pediatric Research. Subspecialties: Neurology; Pediatrics. Current work: Basic and clinical research in diabetes and hypertension. Office: Ind U Sch Medicine 1001 W 10th St Indianapolis IN 46202

MUELLER-HEUBACH, EBERHARD AUGUST, medical educator, obstetrician-gynecologist; b. Berlin, Feb. 24, 1942; U.S., 1968; s. Heinrich G. and Elisabeth (Heubach) Mueller; m. Cornelia R. Uffmann, Feb. 6, 1968; 1 son, Oliver Maximilian. Abitur, Lichtenbergschule, Darmstadt, Ger., 1961; M.D., U. Cologne, W.Ger., 1966. Diplomate: Am. Bd. Ob-Gyn (maternal-fetal medicine). Intern U. Cologne, 1967-68; intern Middlesex Gen. Hosp., New Brunswick, N.J., 1968-69; research fellow Columbia U., N.Y.C., 1969-71; resident and chief resident Sloane Hosp. for Women, N.Y.C., 1971-75; asst. prof. U. Pitts. Sch. Medicine/Magee-Women's Hosp., 1975-81, assoc. prof., 1981—. Reviewer: Am. Jour. Ob-Gyn, 1978—, Obstetrics & Gynecology, 1979—; contbr. chpts. to books, articles to profl. jours. Fellow Am. Coll. Obstetricians and Gynecologists (Hoechst award 1972); mem. Tri-State Perinatal Orgn. (v.p. 1981), Pa. Perinatal Assn. (pres.-elect 1982-84), Soc. Gynecologic Investigation, Soc. Perinatal Obstetricians, Am. Fedn. Clin. Research, Pitts. Ob-Gyn Soc. Subspecialties: Maternal and fetal medicine; Reproductive biology (medicine). Current work: Animal studies in fetal and maternal physiology; diabetes mellitus in pregnancy, high risk obstetrics. Office: Magee-Womens Hosp U Pitts Sch Medicine Pittsburgh PA 15213

MUETTERTIES, EARL LEONARD, chemist, educator; b. Elgin, Ill., June 23, 1927; s. Earl Conrad and Muriel Guinevere (Carpenter) M.; m. JoAnn Mary Wood, Mar. 3, 1956; children: Eric Joseph, Mark Conrad, Gretchen Ann, Maria Christine, Martha Alane, Kurt Andrew. B.S. with highest distinction, Northwestern U., 1949; A.M., Harvard U., 1951, Ph.D., 1952. Research chemist central research dept. E.I. du Pont de Nemours & Co., Wilmington, Del., 1952-57, research supr., 1957-65, asso. dir. research, 1965-73; prof. chemistry Cornell U., 1973-78, U. Calif., Berkeley, 1978—; adj. prof. U. Pa., 1969-73; asso. mem. Monell Chem. Sense Center, 1969-74. Author: (with W.H. Knoth) Polyhedral Boranes, 1969, (with J.P. Jesson) Chemist's Guide, 1969; editor-in-chief: Inorganic Syntheses, vol. X, 1967; editor: Chemistry of Boron and Its Compounds, 1967, Transition Metal Hydrides, vol. I of Chemistry of Hydrogen and Its Compounds, 1971, Boron Hydride Chemistry, 1975. Recipient John C. Bailar medal U. Ill., 1976. Fellow Royal Chem. Soc. (hon.); Mem. Nat. Acad. Scis., Am. Acad. Arts and Scis., Am. Chem. Soc. (Tex. Instruments Co. award in inorganic chemistry 1965, Mallinckrodt, Inc., award for disting. service 1979), Am. Phys. Soc., Chem. Soc. London, AAAS. Subspecialties: Inorganic chemistry; Surface chemistry. Current work: Research is directed to an understanding of chemistry of metal atoms, metal complexes, metal clusters and metal surfaces. Home: 12120 Tartan Way Oakland CA 94619 Office: Dept Chemistry U Calif Berkeley CA 94720

MUIRHEAD, VINCENT URIEL, aerospace engineer; b. Dresden, Kans., Feb. 6, 1919; s. John Hadsell and Lily Irene (McKinney) M.; m. Bobby Jo Thompson, Nov. 5, 1943; children: Rosalind, Jean, Juleigh. B.S., U.S. Naval Acad., 1941, U.S. Naval Postgrad. Sch., 1948; Aero. Engr., Calif. Inst. Tech., 1949; postgrad., U. Ariz., 1962, 64, Okla. State U., 1963. Midshipman U.S. Navy, 1937; commd. ensign, 1941, advanced through grades to comdr., 1951; nav. officer U.S.S. White Plains, 1945-46; comdr. Fleet Aircraft Service Squad, 1951-52; with Bur. Aeros., Ft. Worth, 1953-54; comdr. Helicopter Utility Squadron I, Pacific Fleet, 1955-56; chief staff officer Comdr. Fleet Air, Philippines, 1956-58; exec. officer Naval Air Tng. Center, Memphis, 1958-61; ret., 1961; asst. prof. U. Kans., Lawrence, 1961-63, asso. prof. aerospace engring., 1964-76, prof., chmn. dept., 1976—; cons. Black & Veatch (cons. engrs.), Kansas City, Mo., 1964—. Author: Introduction to Aerospace, 1972, Thunderstorms, Tornadoes and Building Damage, 1975. Decorated Air medal. Mem. Am. Acad. Mechanics, AIAA, Am. Soc. Engring. Edn., N.Y. Acad. Scis., Sigma Gamma Tau. Mem. Ch. of Christ (elder). Subspecialties: Aerospace engineering and technology; Fluid mechanics. Current work: Drag reduction on trucks, improvement of environmental conditions in livestock trailers. Research on aircraft, tornado vortices, shock tubes and waves. Home: 503 Park Hill Terr Lawrence KS 66044

MUKHERJEE, ANIL BARAN, physician, medical researcher, emergency medical consultant; b. Suri, West Bengal, India, Jan. 20, 1942; came to U.S., 1962, naturalized, 1978; s. Shyama Pada and Sabasana M.; m. Diane Colleen Cunningham, May 25, 1956. B.Sc. with honors, U. Calcutta, India, 1962; M.S., U. Utah, 1964, Ph.D., 1966; M.D., SUNY-Buffalo, 1975. Postdoctoral fellow Columbia U., 1966-67; asst. prof. biology Queen's U., Kingston, Ont., Can., 1967-68; asst. research prof. pediatrics and human genetics SUNY-Buffalo, 1968-71; cons. in human genetics Children's Hosp., Buffalo, 1971-75; postgrad. in internal medicine Georgetown U. Med. Div., 1975-76; clin. assoc. NIH, Bethesda, Md., 1976-78, sr. investigator, 1978-80; chief sect. on devel. genetics, human genetics br. Nat. Inst. Child Helath and Human Devel., 1981—; commd. capt., med. dir. USPHS, 1982. Contbr. numerous articles, chpts. to profl. publs. Recipient USPHS Clin. Soc. award, 1981, award for excellence in reprodn. research Brazilian Soc. Human Reprodn., 1981-82. Mem. Am. Soc. Human Genetics, Am. Soc. Cell Biology, AAAS, Am. Soc. Gynecologic Investigation, Sigma Xi. Club: Cosmos (Washington). Subspecialties: Internal medicine; Genetics and genetic engineering (medicine). Current work: Genetics of alcholism, fetal alcohol syndrome, developmental genetics. Office: Nat Inst Child Health and Human Devel NIH Bethesda MD 20205

MULICK, JAMES ANTON, psychologist; b. Passaic, N.J., June 17, 1948; s. Andreus and Anna (Petruchkowich) M.; m. Nancy Elizabeth Witt, Aug. 8, 1970. A.B., Rutgers Coll., 1970; M.A., U. Vt., 1973, Ph.D., 1975. Lic. psychologist, R.I., N.C., Mass. Psychol. services dir. I SIB Program, Murdoch Ctr., Butner, N.C., 1976-77; chief psychologist Shriver Ctr., Waltham, Mass., 1977-78; dir. psychology and tng. Child Devel. Ctr., R.I, Hosp., Providence, 1978; vis. asst. prof. psychology Northeastern U., Boston, 1977-80; clin. asst. prof. pediatrics Brown U., Providence, 1978; adj. assoc. prof. psychology U. R.I., North Kingstown, 1979. Editor: Handbook of Mental Retardation, 1983, Parent Professional Partnership in Developmental Disability Services, 1983; mem. editorial bd.: Applied Research in MR, 1980; Contbr. articles to profl. jours. Mem. R.I. Gov.'s Com. on Mental Retardation, 1979-82. Fellow Behavior Therapy and Research Soc.; mem. Am. Psychol. Assn., Am. Assn. on Mental Deficiency (chmn. Region X 1983). Subspecialties: Behavioral psychology; Mental retardation. Current work: Behavior analysis in mental retardation; learning and behavioral development, social policy analysis relating to children. Home: 3 Harris Ave Johnston RI 02919 Office: Child Devel Ctr RI Hosp 593 Eddy St Providence RI 02902

MULLAN, JOHN FRANCIS, neurosurgeon, educator; b. Ireland, May 17, 1925; came to U.S., 1955, naturalized, 1962; s. John and Mary Catherine M.; m. Vivian C. Dunn, June 3, 1959; children—Joan, John, Brian. M.B., B.Ch., B.A.O., Queens U., Belfast, 1947, D.Sc. (hon.), 1976; postgrad., McGill U., 1953-55. Trained gen. surgery Royal Victoria Hosp., Belfast, No. Ireland, 1947-50, in neurosurgery, 1951-53; trained gen. surgery Guy's Hosp. and Middlesex Hosp., London, Eng., 1950-51, Montreal (Que., Can.) Neurosurg. Inst., 1955; asst. prof. neurosurgery U. Chgo., 1955-61, asso. prof., 1961-63, prof., 1962—, John Harper Seeley prof., 1967—, chmn. neurosurgery, 1967—; dir. Brian Research Inst. Chgo. Contbr. articles in field to profl. publs. and textbooks. Recipient McClintock award, 1961. Fellow Royal Coll. Surgeons (Eng.), A.C.S.; mem. Soc. Neurol. Surgeons, Acad. Neurol. Surgery, Am. Assn. Neurol. Surgeons, Am. Neurosurg. Assn., Central Neurosurg. Soc., Chgo. Neurol. Soc. Roman Catholic. Subspecialties: Neurosurgery; Microsurgery. Current work: Vascular diseases of brain; intractable pain; head injury; cerebral blood flow. Condr. research vascular diseases of the brain, pain, head injury. Home: 5844 Stony Island Ave Chicago IL 60637 Office: 950 E 59th Chicago IL 60637

MULLANE, JOHN FRANCIS, pharmaceutical company executive; b. N.Y.C., Mar. 10, 1937; s. John G. and Rita A. (Hoben) M.; m. Ruth Ann Cecka, Nov. 17, 1962; children: Rosemarie, Michael, Kathleen, Therese, Thomas. M.D., SUNY-Downstate Med. Ctr., Bklyn., 1963, Ph.D., 1968; J.D., Fordham U., 1977. Assoc. dir. clin. research Ayerst Labs. div. Am. Home Products Corp., N.Y.C., 1975, dir. clin. research, 1975-76, v.p. clin. research, 1977, v.p. sci. affairs, 1977-82, sr. v.p. sci. affairs, 1982, exec. v.p. sci. affairs worldwide, 1983—. Contbr. numerous articles to profl. jours. Recipient Achievement of Excellence in Medicine award Upjohn Co., 1970. Fellow Am. Coll. Clin. Pharmacology; mem. Am. Assn. Study Liver Disease, Am. Soc. Nephrology, Am. Fedn. Clin. Research. Roman Catholic. Subspecialty: Pharmacology. Current work: Administer pre-clinical, clinical, and technical departments related to drug discovery and development. Home: 203 Cliff Ave Pelham NY 10803 Office: Ayerst Labs Div Am Home Products Corp 685 3d Ave New York NY 10017

MULLANI, NIZAR ABDUL, medical educator, researcher; b. Daressalaam, Tanzania, Oct. 22, 1942; s. Abdulshamsh Husein and Noorbanu Jiwan-Hirjee M.; m. Linda Kay, June 21, 1975. B.S., Washington U., St. Louis, 1967. Research asst. biomed computer lab. Washington U., 1970-80; research assoc. div. radiation scis. Mallinckrodt Inst. Radiology, Sch. Medicine, 1976-80; asst. prof. medicine, tech. dir. positron diagnostic and research ctr., 1982—; U. Tex. Health Sci. Ctr., Houston, 1980—; asst. prof. gen. instrn. U. Tex. Grad. Sch. Biomed. Scis., Houston, 1981—; Mem. site visit rev. com. NIH and Dept. Energy; reviewer Jour. Nuclear Medicine, IEEE Transactions on Nuclear Sci., Jour. Computer Assisted Tomography. Contbr. articles to profl. jours. Grantee Am. Heart Assn. Mem. IEEE Nuclear Sci. Soc. (sr.), Soc. Nuclear Medicine, AAAS. Moslem. Subspecialty: PET scan. Current work: Design, construction and application of positron emission tomographs to in vivo biochemical studies in man using radioactive tracers. Home: 15214 Beacham Houston TX 77070 Office: 6431 Fannin Houston TX 77025

MULLER, ERNEST HATHAWAY, geology educator, geomorphologist, quaternary geologist; b. Tabriz, Iran, Mar. 4, 1923; s. Hugo Arthur and Laura Barnett (McComb) M.; m. Wanda Custis, Apr. 7, 1951; children: Ruth Anne, David Stewart, Katherine Lee. B.A., Wooster Coll., 1947; M.S., U. Ill., 1949, Ph.D., 1952. Geologist U.S. Geol. Survey, Washington, 1947-54; asst. prof. Cornell U., Ithaca, N.Y., 1954-59; assoc. prof. Syracuse (N.Y.) U., 1959-63, prof. geology, 1963—, interim chmn., 1970-71, 80-81; geologist Am. Geog. Soc. Austral Chile Expdn., San Rafael Glacier, 1959; lectr. AID, Sagar, India, 1966; research assoc. Natural History Mus., Reykjavik, Iceland, 1968-69; seasonal geologist N.Y. State Mus., Albany, 1956-76; vis. prof. Alaska Pacific U., Anchorage, 1979. Author lab. manual, bulletins in field; compiler geologic map, 1977. Com. mem. Onondaga County Environ. Mgmt. Council, Syracuse, 1977—. Served to 1st lt. USAAF, 1943-46. Erskine vis. prof. Canterbury U., Christchurch, N.Z., 1974. Fellow Geol. Soc. Am., AAAS; mem. Am. Quaternary Assn. (councilor 1982—), Internat. Glaciol. Soc., Sigma Xi (pres. Syracuse chpt. 1965-67). Democrat. Presbyterian. Subspecialties: Geology. Current work: Glacial geology; geomorphologist engaged in mapping and interpretation of Quaternary deposits N.Y. State; lithology and genesis of glacial drift of modern and Pleistocene glaciers. Home: 874 Livingston Ave Syracuse NY 13210 Office: Syracuse U 204 Heroy Geology Lab Syracuse NY 13210

MULLER, JOHN PAUL, clin. psychologist; b. N.Y.C., Dec. 31, 1940; s. John and Magdolna (Jeromos) M. A.B., Fordham U., 1964, M.A., 1966; Ph.D. in Clin. Psychology, Harvard U., 1971. Diplomate: in clin.

psychology Am. Bd. Profl. Psychology. Instr. Harvard U., Cambridge, Mass., 1970-71; chmn. dept. human services Sinte Gleska Coll., Rosebud, S.D., 1971-74; coordinator psychol. services Convalescent Hosp. for Children, Cin., 1974-75; sr. researcher, mem. therapy staff Austen Riggs Ctr., Stockbridge, Mass., 1975; dir. research and edn. Niobrara Inst., Crookston, Nebr., 1982. Author: (with William J. Richardson) Lacan and Language: A Reader's Guide to Ecrits, 1982. Mem. Am. Psychol. Assn., N.Y. Acad. Scis. Subspecialties: Cognition; Psychoanalysis. Current work: Language in psychoanalysis; Jacques Locan; ego as set of constraints. Office: Austen Riggs Ctr Main St Stockbridge MA 01262

MÜLLER, MIKLÓS, biochemical parasitologist, educator; b. Budapest, Hungary, Nov. 24, 1930; came to U.S., 1964; s. Pal and Ilona (Marsovszky) M.; m. Noemi Mezei, Apr. 28, 1959 (div. 1969); children: Judith, Daniel; m. Janet S. Keithly, June 24, 1973. M.D., Budapest Med. U., 1955. Asst. prof. Budapest (Hungary) Med. U., 1955-62, assoc. prof., 1962-64; research assoc. Rockefeller U., N.Y.C., 1964-65, asst. prof., 1966-73, assoc. prof. biochem. cytology and parasitology, 1973—; guest investigator Carlsberg Lab., Kópenhavn, Denmark, 1965-66; adj. assoc. prof. Cornell U. Med. Coll., N.Y.C., 1975—; advisor WHO, 1977-82; cons. G.D. Searle & Co., Skokie, Ill., 1977. Editor: jour. Molercular and Biochem. Parasitology, 1980—. Mem. Soc. Protozoologists (pres. 1982-83, S.H. Hutner prize 1977), Am. Soc. Parasitology, Am. Soc. Biol. Chemistry, Am. Soc. Microbiology, Am. Veneral Disease Assn. Subspecialties: Biochemistry (biology); Parasitology. Current work: Biochemistry and cell biology of anaerobic parasitic protozoa, study of hydrogenosomes, study of the mode of action of antiparasitic drugs. Home: 450 E 63rd St New York NY 10021 Office: Rockefeller U 1230 York Ave New York NY 10021

MÜLLER-EBERHARD, HANS JOACHIM, med. research scientist; b. Magdeburg, Ger., May 5, 1927; came to U.S., 1959, naturalized, 1973; s. G. Adolf and Emma (Jenrich) Müller-E.; m. Ursula Fleck, 1953-77; children—Monika, Kristina; m. Pilar Canicio Albacar, 1977. M.D., U. Göttingen, Ger., 1953; M.D.Sc., U. Uppsala, Sweden, 1961; D. Medicinae h.c., Ruhr U., Bochum, W.Ger., 1982. Asst. physician dept. medicine U. Göttingen, 1953-54; asst. Rockefeller Inst., N.Y.C., also asst. physician inst. hosp., 1954-57; fellow Swedish Med. Research Council, 1957-59; asst. prof., then asso. prof., also hosp. physician Rockefeller U., 1959-63; mem. dept. exptl. pathology Scripps Clinic and Research Found., La Jolla, Calif., 1963-74, chmn. dept. molecular immunology, 1974-82, chmn. dept. immunology, 1982—, assoc. dir., 1978—; Cecil H. and Ida M. Green investigator med. research, 1972—; lectr. immunochemistry U. Uppsala, 1961—; adj. prof. pathology U. Calif. Med. Sch., La Jolla, 1968—; Harvey lectr., 1970. Co-editor: Textbook of Immunopathology, 1976, Springer Seminars in Immunopathology, 1978—; author numerous sci. articles; adv. editor: Jour. Exptl. Medicine, 1963—; mem. jour. editorial bds. Recipient Squibb award Infectious Diseases Soc. Am., 1970; T. Duckett Jones Meml. award Helen Hay Whitney Found., 1971; Distinguished Achievement award Modern Medicine mag., 1974; Karl Landsteiner Meml. award Am. Assn. Blood Banks, 1974; ann. Internat. award Gairdner Found., Can., 1974; Mayo H. Soley award Western Soc. Clin. Research, 1975; Emil von Behring prize Philipps U., Marburg, Ger., 1977. Mem. Nat. Acad. Scis., Am. Soc. Exptl. Pathology (Parke, Davis award 1966), Assn. Am. Physicians, Western Assn. Physicians, Am. Soc. Clin. Investigation, Am. Assn. Immunologists, Am. Acad. Allergy, Am. Soc. Biol. Chemists, Sigma Xi. Subspecialties: Biochemistry (biology); Immunology (medicine). Current work: Molecular analysis of human complement system including protein-protein interaction that leads to generation of biological activities such as cell activation and cell killing. Address: Research Inst Scripps Clinic 10666 N Torrey Pines Rd La Jolla CA 92037

MULLETT, CHARLES EDWIN, electronic consulting company executive; b. Buffalo, Apr. 25, 1938; s. Charles Beatty and Sarah Mary (Hill) M.; m. Vivian Firko, June 15, 1968; children: Jennifer Lynn, Kevin Charles. B.S.E.E., U. Ill.-Urbana, 1960, M.S.E.E., 1962. Registered profl. engr., Calif., 1968. Field engr. Tektronix, Inc., Pasadena, Calif., 1962-68; Prin. Mullett Assocs., Los Angeles, 1965-68, 70-77, pres., 1977—; chief engr. Datapulse div. Systron-Donner, Culver City, Calif., 1968-70. Mem. IEEE, Calif. Soc. Profl. Engrs., Nat. Soc. Profl. Engrs., Internat. Power Conversion Soc. Republican. Roman Catholic. Subspecialties: Applied magnetics; Electronics. Current work: Design of power converters; design of electronic measurement and control instruments; design of consumer electronic products. Patentee portable transmitter, optical intrusion alarm system, data collection system using telephone lines. Home: 8235 Gulana Ave Playa Del Ray CA 90291 Office: 5301 Beethoven St Apt 102 Los Angeles CA 90066

MULLIGAN, RICHARD C., biologist, educator; b. Summit, N.J., Sept. 9, 1954; s. James H. and Jeanne G. Milligan. B.S. in Biology, MIT, 1976; Ph.D. in Biochemistry, Stanford U., 1980. Postdoctoral fellow Center for Cancer Research, MIT, Cambridge; asst. prof. molecular biology, research fellow in medicine dept. medicine Harvard U. Med. Sch., 1980-81; research fellow in medicine endocrine unit Mass. Gen. Hosp., Boston, 1980-81. Contbr. articles to profl. jours. Recipient John S. Asinari award for undergrad. research in life scis., 1976; NSF fellow, 1976-79; MacArthur Found. fellow, 1981—. Subspecialties: Molecular biology; Biochemistry (medicine). Office: Center for Cancer Research E17-220 MIT Cambridge MA 02139

MULLIKEN, ROBERT SANDERSON, scientist, educator; b. Newburyport, Mass., June 7, 1896; s. Samuel Parsons and Katherine (Mulliken) M.; m. Mary Helen von Noé, Dec. 24, 1929 (dec. Mar. 1975); children: Lucia Maria (Mrs. John P. Heard), Valerie Noé. B.S., MIT, 1917; Ph.D., U. Chgo., 1921; Sc.D. (hon.), Columbia, 1939, Marquette U., Cambridge U., 1967, Gustavus Adolphus Coll., 1975, Ph.D., Stockholm U., 1960. Research on war gases, Washington, 1917-18; tech. research with N.J. Zinc Co., 1919; research on separation isotopes, 1920-22, researches on molecular spectra and molecular structure, 1923—; Nat. research fellow U. Chgo., 1921-23, Harvard, 1923-25; asst. prof. physics Washington Sq. Coll. (N.Y. U.), 1926-28; asso. prof. physics U. Chgo., 1928-31, prof. physics, 1931-61, Ernest DeWitt Burton Disting. Service prof., 1956-61, Disting. Service prof. physics and chemistry, 1961—; Disting. Research prof. chem. physics Fla. State U., 1965-71; Baker lectr. Cornell U., 1960; vis. prof., Bombay, 1962, Indian Inst. Tech., Kanpur, 1962; Silliman lectr. Yale, 1965; Jan Van Geuns vis. prof. Amsterdam U., 1965; J.S. Guggenheim fellow for European study, 1930, 32, leave of absence, 1942-45; as dir. information div. Plutonium Project at Chgo.; editor Plutonium Project Record in Nat. Nuclear Energy Series.; Fulbright research fellow for research at Oxford, 1952-53; vis. fellow St. John's Coll., Oxford, 1952-53; sci. attaché Am. Embassy, London, 1955. Served with C.W.S. U.S. Army, 1918. Recipient medal U. Liege, 1948; Gilbert N. Lewis medal Calif. sect. Am. Chem. Soc., 1960; Theodore W. Richards medal Northeastern sect., 1960; Peter Debye award Am. Chem. Soc., 1963; J.G. Kirkwood award New Haven sect., 1964; Willard Gibbs medal Chgo. sect., 1965; Nobel prize for chemistry, 1966; Golden Plate award Am. Acad. Achievement, 1983. Fellow Am. Phys. Soc. (chmn. div. chem. physics 1951-52), AAAS, Royal Irish Acad. (hon.), Indian Nat. Acad. Sci. (hon.), London Chem. Soc. (hon.); mem. Nat. Acad. Scis., Am. Philos. Soc., Am. Chem. Soc., Internat. Acad. Quantum Molecular Sci., Am. Acad. Arts and Scis., European Acad. Arts, Scis., and Humanities (titular mem.), Royal Soc. (fgn. mem.), Soc. de Chimie Physique (hon.), Royal Soc. Sci. of Liège (corr.), Chem. Soc. Japan (hon.), Gamma Alpha, Phi Lambda Upsilon. Clubs: Quadrangle; Cosmos (Washington).

MULLIN, MICHAEL MAHLON, biological oceanography educator; b. Galveston, Tex., Nov. 17, 1937; s. Francis J. and Alma (Hill) M.; m. Constance W. Hammond, Dec. 29, 1964; children: Stephen Joseph, Keith Alan, Laura Anne. A.B., Shimer Coll., 1957, Harvard U., 1959, M.A., 1960; postgrad., Scripps Inst. Oceanography, 1961; Ph.D., Harvard U., 1964. NSF postdoctoral fellow Internat. Indian Ocean Expedition, R/V Anton Brunn and Dept. Zoology, U. Auckland, N.Z., 1964; vis. pre-doctoral student Woods Hole Oceanographic Inst., 1958-64; lab. instr. Harvard Coll., Cambridge, Mass., 1962-63; vis. asst. prof. oceanography U. Wash., Seattle, summer, 1966, 68, 70, vis. assoc. prof., 1972, 74, 76, vis. prof., 1978; postgrad. research biologist U. Calif., San Diego, 1964-65, IMR Food Chain Research Group, LaJolla, 1964-65; asst. research biologist, asst. prof. Scripps Inst. Oceanograhy, 1965-71, assoc. research biologist, assoc. prof., 1971-77, research biologist, prof., 1977—. Contbr. articles to profl. jours. Mus. dir. Madrigal Singers, U. Calif., San Diego, 1969-83. Sr. queen's fellow in marine sci., Australia, 1981-82. Mem. Am. Soc. Limnology and Oceanography, Phi Beta Kappa. Subspecialties: Oceanography; Ecology. Current work: Ecology of marine plankton. Office: Inst Marine Resources A-018 Univ Calif San Diego LaJolla CA 92093 Home: 7758 Ludington Pl LaJolla CA 92037

MULVEY, JOHN MICHAEL, engineering management educator and consultant; b. Chgo., Oct. 31, 1946; s. John J. and Mary M. (Steinberg) M.; m. Lauri D. Ervin, Sept. 28, 1982. B.S. in Engring, U. Ill.-Champaign, 1969, M.S. in Computer Sci, 1969, UCLA, 1973, Ph.D., 1975. Mem. tech. staff, mgr. TRW Systems Group, Redondo Beach, Calif., 1969-75; asst. prof. Harvard U., Cambridge, Mass, 1975-78; assoc. prof., dir. engring. mgmt. Princeton (N.J.) U., 1978—; co-founder Analysis Research and Computation Inc., Austin, Tex., 1973; cons. U.S. Congress, fed. govt. agys. Editor: Network Applications, 1981, Evaluating Math Programs, 1982; contbr. articles to profl. jours.; developer computer programs in network optimization. UCLA scholar, 1974; recipient dissertation award Inst. for Research Scis., San Francisco, 1975; NSF grantee, 1978—. Mem. Math. Programming Soc. (exec. com.), Ops. Research Soc., Assn. Computing Machinery, Inst. Mgmt. Scis. Subspecialties: Operations research (engineering); Operations research (mathematics). Current work: Design, test, and apply methods of mathematical programming, especially large-scale problems. Build algorithms for solving combinatorial optimization problems. Office: Sch Engring and Applied Sci Princeton U Olden Ave Princeton NJ 08544

MUMA, JOHN RONALD, educator; b. Allegan, Mich., Jan. 19, 1938; s. Clark Benjamin and Bernice (Roe) M.; m. Diane Laure Bush, Sept. 8, 1973; 1 dau., Taylor Diane. B.S., Central Mich. U., 1960, M.A., 1963; Ph.D., Pa. State U., 1967. Asst. prof. U. Ga., Athens, 1967-68; assoc. prof. U. Ala., Tuscaloosa, 1968-73; vis. prof. Memphis State U., 1973-74; assoc. prof. SUNY-Buffalo, 1974-76; vis. prof. So. Ill. U., Carbondale, 1976-78; prof. Tex. Tech U., Lubbock, 1978—; grants reviewer Office Spl. Edn., 1978-80, March of Dimes, 1978-80. Author: Language Handbook, 1978, Language Primer, 1981, (with D. Muma) Muma Assessment Program, (1979); mem. editorial bd.: Issues in Learning, 1980—. Consumer advocate West Tex. Com. for Services for the Deaf, 1979. Recipient alumni recognition award Central Mich. U., 1978; cert. of appreciation Am. Speech-Lang.-Hearing Assn., 1971, 78, 81. Fellow Am. Speech-Lang.-Hearing Assn.; mem. Am. Psychol. Assn., Soc. for Research in Child Devel. Subspecialties: Developmental psychology; Cognition. Current work: Language acquisition, child devel. Home: 3811 62nd Dr Lubbock TX 79413 Office: Speech and Hearing Sci Tex Tech U Lubbock TX 79409

MUMFORD, DAVID BRYANT, mathematics educator; b. Worth, Sussex, Eng., June 11, 1937; came to U.S., 1940; s. William Bryant and Grace (Schiott) M.; m. Erika Jentsch, June 27, 1959; children: Stephen, Peter, Jeremy, Suchitra. B.A., Harvard U., 1957, Ph.D., 1961. Jr. fellow Harvard U., 1958-61, asso. prof., 1962-66, prof. math., 1966—, chmn. dept. math, 1981—. Author: Geometric Invariant Theory, 1965, Abelian Varieties, 1970, Introduction to Algebraic Geometry, 1976. Recipient Fields medal Internat. Congress Mathematicians, 1974. Fellow Tata Inst. (hon.); mem. Nat. Acad. Scis., Am. Acad. Arts and Scis. Subspecialty: Artificial intelligence. Current work: Pure mathematics (algebraic geometry); artifical intelligence. Home: 26 Gray St Cambridge MA 02138 Office: 1 Oxford St Cambridge MA 02138

MUMMA, ALBERT G., mgmt. cons., ret. naval officer; b. Findlay, Ohio, June 2, 1906; m. Carmen Braley, 1927; children—Albert G., John S., David B. Grad., U.S. Naval Acad., 1926; D.Engring., Newark Coll. Engring., 1970. Commd. ensign USN, 1926; advanced through grades to rear adm., now ret.; engr. on staff comdr. Naval Forces Europe, asst. naval attaché, London, mem., head tech. intelligence div., mem., Europe, World War II; chief bur. ships, coordinator shipbldg. (including nuclear and missile fleets), conversion and repair Dept. Def., 1955-59; v.p., group exec. Worthington Corp., 1959-64, exec. v.p., dir., in charge all domestic operations, 1964-67, pres., chief operating officer, 1967, chmn. bd., 1967-71; chmn. Am. Shipbldg. Commn., 1971-73; former dir. Prudential Ins. Co., Kauffelpesser Co., C.R. Bard Co., Murray Hill, N.J., Howard Savs. Bank. Bd. dirs. Eutectic & Castolin Inst.; trustee Drew U., Madison, N.J., St. Barnabas Med. Center, Livingston, N.J. Recipient Adm. Jerry Land Gold medal. Fellow Soc. Naval Architects and Marine Engrs. (hon., past pres.); mem. Am. Soc. Naval Engrs. (hon., past pres.), Nat. Acad. Scis. (past mem. research council, past chmn. numerous coms.), Nat. Acad. Engring. Clubs: Army and Navy, Army and Navy Country (Washington); N.Y. Yacht (N.Y.C.); Baltusrol Golf (Springfield, N.J.); Mountain Lake (Lake Wales, Fla.); Essex. Subspecialties: Mechanical engineering; Systems engineering. Current work: Marine engineering and naval construction; system and weapons engineering, including nuclear. Home: 66 Minnisink Rd Short Hills NJ 07078 After having spent many years in naval and indsl. research, I learned that the most important phase of that field is to identify with accuracy the areas that block progress. Each such area can then be concentrated upon by attacking the two or three most promising approaches to solution simultaneously, with much greater chance of early success than a sequential attack on each approach and with nearly always a saving in research funds. This principle was used in the Manhattan district for development of the atomic bomb, and for the design and construction of the Polaris submarine and missile. In both cases the time element was cut in half and total expense was also reduced.

MUMMERT, THOMAS ALLEN, mfg. co. exec.; b. Toledo, Ohio, Dec. 24, 1946; s. James Allen and Betty Alice (Thomas) M.; m. Icia Linda Shearer, Dec. 17, 1966; children: Sherry Lynn, Robert Thomas, Michael Allen. Student, Toledo U., 1965-66. Pres. Mummert Electric & Mfg. Co. Inc., Toledo, 1969-70; research engr. Am. Lincoln Corp., Bowling Green, Ohio, 1970-73; test engr. Dura div. Dura Corp., Toledo, 1973-74; head research dept. Jobst Inst. Inc., Toledo, 1974—. Served with U.S. Navy, 1968-69. Mem. Assn. for Advancement of Med. Instrumentation, Nat. Mgmt. Assn., Am. Soc. for Engring. Edn., Am. Soc. for Quality Control, AAAS, Ohio Acad. Sci., N.Y. Acad. Scis., Biol. Engring. Soc. Subspecialties: Biomedical engineering; Bioinstrumentation. Current work: Therapeutic and rehabilitation devices for circulatory blood flow disorders; peripheral blood flow imaging technology. Patentee in field. Home: 1448 Palmetto Ave Toledo OH 43606 Office: 653 Miami St Toledo OH 43694

MUNDA, RINO, physician, educator; b. Rome, Italy, Feb. 2, 1943; came to U.S., 1967; s. Salvador and Marina (Tabusso) M.; m. Margarita Landra, Nov. 11, 1968; children: Sergio, Franco. B.S., U.Nat. Mayor de San Marcos-Peru, 1960; M.D., Facultad De Medicina Cayetano Heredia-Peru, 1966. Cert., Am. Bd. Surgery, 1974. Instr. surgery N.Y. Med. Coll., N.Y.C., 1972-73; asst. prof. research surgery U. Cin. Coll Medicine, 1974-75, asst. prof., 1975-79, assoc. prof., 1979—; cons. in field. Mem. adv. com. Cin. Kidney Found., 1979—; mem. med. review bd. Ohio Valley Renal Network, 1981—. Am. Diabetes Assn. grantee, 1981-82. Mem. Surg. Soc. N.Y. Med. Coll., Assn. Acad. Surgery, Am. Soc. Transplant Surgeons, Transplantation Soc., Am. Coll. Surgeons, Surg. Infection Soc. (hon.), Soc. Peruana de Angiologia (Peru). Subspecialties: Transplantation; Transplant surgery. Current work: Academic surgeon. Home: 315 Lafayette Ave Cincinnati OH 45220 Office: 234 Goodman St Cincinnati OH 65267

MUNGER, ROBERT JOHN, veterinary ophthalmologist, educator; b. Memphis, Aug. 6, 1950; s. Donald McClellan and Dorothy Virginia (Arnold) M.; m. Pamela Lynn Welch, Dec. 30, 1972; 1 dau., Erin Rebecca. D.V.M., Tex. A & M U., 1973. Diplomate: Am. Coll. Vet. Ophthalmology. Intern in large animal medicine and surgery Coll. Vet. Medicine, U. Calif., Davis, 1973-74, resident in large animal medicine, 1974-75, resident in vet. ophthalmology, 1976-78; practice vet. medicine specializing in ophthalmology, Dallas and Ft. Worth, 1978—; asst. prof. clin. vet. ophthalmology Coll. Vet. Medicine, U. Tenn., Knoxville, 1979-82; staff veterinarian, research and devel. Alcon Labs., Ft. Worth, 1983—. Contbr. numerous vet. ophthalmology articles to profl. publs. Served with USAFR, 1972-75. Mem. Am. Coll. Vet. Ophthalmologists, Am. Vet. Med. Assn., Am. Animal Hosp. Assn., Am. Soc. Vet. Ophthalmologists, Internat. Soc. Vet. Ophthalmologists, Tex. Vet. Med. Assn., Dallas County Vet. Med. Assn. Republican. Baptist. Subspecialties: Veterinary ophthalmology; Ophthalmology. Current work: Ocular toxicology; ocular pathology; feline herpes virus; ocular manif, systemic disease. Office: Alcan Labs 6201 S Freeway PO Box 1959 Fort Worth TX 76101

MUNIAPPAN, RANGASWAMY NAICKER, entomologist; b. Coimbatore, Madras, India, June 1, 1941; emigrated to U.S., 1967, naturalized, 1976; s. Rangaswamy and Rangammal Naicker; m. Sheila Naidu, Feb. 25, 1972; 1 dau., Brindha. B.S., U. Madras, 1963, M.S., 1965; Ph.D., Okla. State U., 1969. Asst. lectr. Agrl. Coll., Coimbatore, 1965-67; research asst. Okla. State U., Stillwater, 1968-69, postdoctoral fellow, 1969-70; entomologist Dept. Agr., Mangilao, Guam, 1970-71, chief agr., 1971-75; assoc. prof. U. Guam, Mangilao, 1975-76, assoc. dean, 1976-81; assoc. dir. Agrl. Expt. Sta., 1982—. Mem. Pacific Sci. Assn. (entomology com.), Entomol. Soc. Am., Philippines Entomol. Soc., Hawaiian Entomol. Soc., Phi Sigma Phi. Subspecialty: Biological Control. Current work: Biological control of insects, weeds and snails, tropical agricultural research programs. Home: PO Box 8784 Yona Guam 96914 Office: Agrl Expt Sta U Guam Mangilao Guam 96913

MUNK, WALTER HEINRICH, geophysicist; b. Vienna, Austria, Oct. 19, 1917; s. Hans and Rega (Brunner) M.; m. Judith Horton, June 20, 1953; children—Edith, Kendall. B.S., Calif. Inst. Tech., 1939, M.S., 1940; Ph.D., U. Calif., 1947, U. Bergen, Norway, 1975. Asst. prof. geophysics U. Calif., 1947-49, asso. prof., 1949-54; prof. Inst. Geophysics and Planetary Physics and at the Scripps Inst., 1954—; asso. dir. Inst. Geophysics and Planetary Physics systemwide, 1959-82. Guggenheim fellow Oslo U., 1948, Cambridge, 1955, 62; Josiah Willard Gibbs lectr. Am. Math. Soc., 1970; Sr. Queen's fellow, Australia, 1978; Arthur L. Day medal Geol. Soc. Am., 1965; Sverdrup Gold medal Am. Meteorol. Soc., 1966; Alumni Distinguished Service award Calif. Inst. Tech., 1966; gold medal Royal Astron. Soc., 1968; named Calif. Scientist of Year Calif. Mus. Sci. and Industry, 1969; 1st Maurice Ewing medal Am. Geophys. Union and USN, 1976; Capt. Robert Dexter Conrad award Dept. Navy, 1978. Fellow Am. Meteorol. Soc., AAAS, Acoustical Soc. Am.; mem. Nat. Acad. Scis. (chmn. geophysics sect. 1975-78, Agassiz medal 1976), Am. Philos. Soc., Am. Geological Soc., Am. Acad. Arts and Scis., Deutsche Akademie der Naturforscher Leopoldina, Royal Soc. (fgn.). Subspecialties: Geophysics; Oceanography. Current work: Remote sensing of oceans by sound. Home: 9530 LaJolla Shores Dr LaJolla CA 92037 Office: IGPP Mail Code A025 Univ Calif-San Diego LaJolla CA 92093

MUNN, RAYMOND SHATTUCK, structural and materials engineer; b. Springfield, Mass., Feb. 2, 1947; s. Clarence Robert and Anna Robinson (Shattuck) M.; m. Gail Ellen McGibney, Aug. 16, 1969; children: Ray Derek, Todd Robert. B.S.M.E., Northeastern U., 1969, M.S.M.E., 1970, M.Eng., 1974; postgrad., U. Conn., 1975. Registered profl. engr., Conn. Detail designer Ford Motor Co., Cleve., 1965-66; designer Moore Drop Forging Co., Springfield, 1966-68; engr. No. Research & Engring. Corp., Cambridge, Mass., 1968-69; instr. Northeastern U., Boston, 1969-74; mech. engr. U.S. Naval Underwater Systems Ctr., New London, Conn., 1973; corrosion engr. Atlantech Assocs., Inc., New London, 1979; cons. structural and corrosion engring. Contbr. articles to profl. jours. Mem. Zoning Bd. Appeals, Groton Long Point, Conn., 1980; Bd. dirs., Groton Long Point, Conn., 1983; mem. Community Ctr. Planning Com, Mystic, Conn., 1980-81; active Boy Scouts Am. Served to maj. USAR, 1969. NSF fellow, 1969-72; recipient Superior Achievement award U.S. Navy, 1975, 82, 83. Mem. ASME, Am. Soc. for Metals Nat. Assn. Corrosion Engrs. Subspecialties: Materials (engineering); Corrosion. Current work: Corrosion analysis techniques. Structural analysis, fracture mechanics. Home: 26 Middlefield St Groton Long Point CT 06340 Office: Code 44 Naval Underwater Systems Center New London CT 06320

MUNRO, HAMISH NISBET, biochemist, educator; b. Edinburgh, Scotland, July 3, 1915; came to U.S., 1966, naturalized, 1973; s. Donald and Margaret (Nisbet) M.; m. Edith E. Little, Apr. 5, 1946; children: Joan Bruce, Colin Scott, Andrew Fraser, John Michael. B.Sc., U. Glasgow, Scotland, 1936, M.B., 1939, D.Sc., 1956. Physician, pathologist Victoria Infirmary, Glasgow, 1939-45; lectr. physiology U. Glasgow, 1946-47, sr. lectr., reader biochemistry, 1948-63, prof. biochemistry, 1964-66; prof. physiol. chemistry Mass. Inst. Tech., Cambridge, from 1966; now dir. USDA Human Nutrition Research Center on Aging, Tufts U., Boston. Editor: Mammalian Protein Metabolism, vols. 1-4, 1964-70; Contbr. articles in field of protein metabolism to profl. publs. Recipient Osborn Mendel award Am. Inst. Nutrition, 1968, Borden award, 1978; Bristol-Myers award for disting. achievement in nutrition research, 1980; Rank prize for significant advances in nutrition, 1980. Fellow Royal Soc. Edinburgh; mem. Nat. Acad. Sci., Am. Soc. Biol. Chemists, Am. Inst. Nutrition (pres. 1978-79), Brit. Biochem. Soc. Presbyterian. Subspecialties: Cell and tissue culture; Nutrition (biology). Current work: Bioregulatory studies on optimizing genome function throughout life, notably interaction of nutrition in expression of aging genome. Home: 159 Concord Ave Cambridge MA 02138 Office: Human Nutrition Center 711 Washington St Boston MA 02111

MUNSON, JOHN BACON, neuroscience researcher and educator; b. Clifton Springs, N.Y., Nov. 15, 1932; s. Arthur Reynolds and Charlotte (Bacon) M.; m. Faith Witte, Aug. 15, 1959; children: Susan, Johanna, Kibby. A.B., Union Coll., Schenectady, 1957; Ph.D., U. Rochester, 1965. Instr. to prof. Coll. Medicine, U. Fla., Gainesville, 1966—, prof. dept. neurosci., 1981—; mem. Neurology B Study Sect. NIH, Washington, 1981—. Served with USNR, 1951-53. Research grantee NSF, 1968-71, Nat. Eye Inst., 1974-77, Nat. Inst. Neurol. and Communicative Diseases and Stroke, 1979—. Mem. Soc. for Neurosci., Am. Physiol. Soc., Internat. Congress Physiol. Scis., AAAS, Sigma Xi. Subspecialties: Neurophysiology; Regeneration. Current work: Spinal motor mechanisms; spinal cord injury; regeneration and recovery of function; muscle physiology. Office: Coll Medicine U Fla Box J-244 Gainesville FL 32610

MUNSON, JOHN CHRISTIAN, acoustician; b. Clinton, Iowa, Oct. 9, 1926; s. Arthur J. and Frances (Christian) M.; m. Elaine Hendershot, Sept. 2, 1950; children: John Christian, Holly Elizabeth. B.S., Iowa State Coll., 1949; M.S., U. Md., 1952, Ph.D., 1962; Navy Dept. scholar, MIT, 1956. Electronic scientist Naval Ordnance Lab., Washington, 1949-66; tech. dir. navy portion Practice Nine, Naval Air Systems Command, 1967; supt. acoustics div. Naval Research Lab., 1968—; asst. extension prof. elec. engring. U. Md., 1964-66; mem. Underwater Sound Adv. Group, 1969-75, U.S. Sonar Team, 1971—, Mobile Sonar Tech. Com., 1972—. Author. Mem. exec. bd. D.C. Bapt. Conv., 1973—, chmn. fin. com., 1973; trustee Midwestern Bapt. Theol. Sem., 1970-80, D.C. Bapt. Home, 1976—; sec. D.C. Bapt. Home, 1978, v.p., 1979, treas., 1980-82, pres., 1982—. Served with USNR, 1944-46. Fellow IEEE, Acoustics, Speech and Signal Processing Soc. (adminstrv. com. 1974-76, chmn. underwater acoustics com. 1973-76), Acoustical Soc. Am.; mem. Sigma Xi. Subspecialties: Acoustics; Acoustical engineering. Current work: Developing technology base for future genration sonar systems (signal processing algorithms, systems concepts, modeling of temporal and spatial character of acoustic fields, performance prediction.). Patentee in field. Home: 119 Marine Terr Silver Spring MD 20904 Office: Naval Research Lab Washington DC 20375 I have a positive joy for life, and I am an incurable optimist: my basic attitude is that things will work out for the best—but only if we do our very best. Each of us has a responsibility to grow to our maximum capacity and to be of reasonable service to mankind. The proper balance among family, job, service to God, service to others, and attention to yourself is essential. Whatever you are doing, do it from the right motivation and with enthusiasm.

MUNSTER, ANDREW MICHAEL, surgeon, educator; b. Budapest, Hungary, Dec. 10, 1935; came to U.S., 1965, naturalized, 1969; s. Leopold Steven and Marianne (Barcza) M.; m. Joy O'Sullivan, Dec. 7, 1964; children: Andrea, Tara, Alexandra. M.D., U. Sydney, Australia, 1959. Diplomate: Am. Bd. Surgery. Resident in surgery Harvard Med. Sch., Boston, 1965-68; assoc, prof. Med. U. S.C., Charleston, 1971-76; assoc. prof. surgery Johns Hopkins U., Balt., 1976; dir. Regional Burn Ctr., Balt., 1976. Author: Burn Care for the House Officer, 1981, Surgical Anatomy, 1972; editor: Surgical Immunology, 1976. Pres. Chesapeake Research and Edn. Trust, Balt., 1980, Charleston Symphony Orch., 1973-74, Charleston Tricounty Arts Council, 1974-75; v.p. Chesapeake Physicians, Balt., 1979. Served to 1t. col. U.S. Army, 1968-71. Recipient George Allan prize U. Sydney, 1959. Fellow Royal Coll. Surgeons (Eng.) (Edinburgh), So. Surg. Assn., Am. Assn. Surgery of Trauma; mem. Soc. Univ. Surgeons, Am. Burn Assn., others. Subspecialties: Surgery; Immunology (agriculture). Current work: Burn and trauma care and research. Office: Johns Hopkins U Balt City Hosps 4940 Eastern Ave Baltimore MD 21224

MURATA, TADAO, engineering educator; b. Takayama, Japan, June 26, 1938; came to U.S., 1962; s. Yonosuke and Ryu M.; m. Nellie Shin; children: Patricia Emi, Theresa T. B.S., Tokai U., Tokyo, 1962; M.S., U. Ill., 1964; Ph.D., 1966. Research asst. U. Ill.-Urbana, 1962-66, research assoc., 1966; asst. prof. U. Ill., Chgo., 1966-68, assoc. prof., 1972-78, prof. dept. elec. engring. and computer sci., 1978—; vis. assoc. prof. U. Calif.-Berkeley, 1976-77; guest researcher GMDmbH, Bonn, W.Ger., 1979; vis. scientist CNRS, Toulouse, France, 1981; mem. panel NRC, Nat. Acad. Scis. Contbr. articles to profl. jours. NSF grantee, 1978-81, 81-83, 83—. Mem. IEEE (sr.), Info. Processing Soc. Japan, Assn. Computing Machinery, European Assn. for Theoretical Computer Sci. Subspecialties: Computer engineering; Distributed systems and networks. Current work: Concurrent computer systems; petri nets and related computation models, VLSI systems and architectures, data flow computations. Home: 757 N Geneva Ave Elmhurst IL 60126 Office: Department Electrical Engineering and Computer Science University of Illinois Box 4348 Chicago IL 60680

MURAYAMA, MAKIO, biochemist; b. San Francisco, Aug. 10, 1912; s. Hakuyo and Namiye (Miyasaka) M.; children: Bruce Edward Robertson, Gibbs Soga, Alice Myra. B.A., U. Calif., Berkeley, 1938, M.A., 1940; Ph.D. (NIH fellow), U. Mich., 1953. Research biochemist Children's Hosp. of Mich., Detroit, 1943-48, Harper Hosp., 1950-54; research fellow in chemistry Calif. Inst. Tech., Pasadena, 1954-56; research asso. in biochemistry Grad. Sch. Medicine, U. Pa., Phila., 1956-58; spl. research fellow Nat. Cancer Inst. at Cavendish Lab., Cambridge, Eng., 1958; research biochemist NIH, Bethesda, Md., 1958—. Author: (with Robert M. Nalbandian) Sickle Cell Hemoglobin, 1973. Fellow Am. Inst. Chemists; mem. Am. Chem. Soc., Am. Soc. Biol. Chemists, AAAS, Assn. Clin. Scientists, Internat. Platform Assn., W.African Soc. Pharmacology (hon.), Sigma Xi. Subspecialties: Biochemistry (medicine); Microscopy. Current work: Discovered cause of acute mountain sickness and prevention of it,possible cause of decompression disease and intravascular coagulation in arteriostenosis due to Bernoulli effect, and prevention of these as well. Home: 5010 Benton Ave Bethesda MD 20814 Office: NIH Bldg 6 Room 139 Bethesda MD 20205

MURINO, CLIFFORD JOHN, research institute executive; b. Yonkers, N.Y., Feb. 10, 1929; s. Vincent Joseph and Marie (Fuccillo) M.; m.; children: John Clifford, Carolyn Ruth, Kathryn Marie. B.S., St. Louis U., 1950, M.S., 1954, Ph.D., 1957. Prof. atmospheric sci. St. Louis U., 1954-75, v.p. fin. and research, 1969-75; phys. sci. cons. NSF, Washington, 1967-69; dir. atmospheric tech. div. Nat. Ctr. Atmospheric Research, Boulder, Colo., 1975-80; pres. Desert Research Inst., Reno, 1980—. Trustee Univ. Corp. Atmospheric Research, Nev. Devel. Authority. Served in USAF, 1951-52. Recipient Sustained Superior Performance award NSF, 1969; grantee NOAA, 1964-66, NSF, 1963, NASA, 1962, 64. Mem. Am. Meteorol. Soc. Lodge: Elks. Subspecialties: Meteorology; Remote sensing (atmospheric science). Home: 4790 Warren Way Reno NV 89506 Office: P.O. Box 60220 Reno NV 89506

MURPHY, EDWARD THOMAS, manufacturing company executive; b. Boston, Nov. 20, 1947; s. Edward William and Eleanor Catherine (Brown) M.; m. Marianne Scheid, May 1, 1976; 1 son, Edward Robert. B.S. Calif. Inst. Tech, Pasadena, 1969; M.S. in Nuclear Sci. and Engring, Carnegie-Mellon U., Pitts., 1971. Registered profl. engr., Pa., Md. Engr. Westinghouse Electric Corp., Pitts., 1969-74, fuel project engr., 1974-80, mgr. licensing ops., Bethesda, Md., 1980—. Mem. Am. Nuclear Soc. Subspecialty: Nuclear fission. Current work: Facilitate communications between Westinghouse and Nuclear Regulatory Commission staff. Office: Westinghouse Electric Corp 4901 Fairmont Ave Bethesda MD 22814

MURPHY, EUGENE FRANCIS, consultant, retired government official; b. Syracuse, N.Y., May 31, 1913; s. Eugene Francis and Mary Grace (Thompson) M.; m. Helene M. Murphy, Dec. 31, 1955; children: Anne Fitzpatrick, Thomas E. M.E., Cornell U., 1935; M.M.E., Syracuse U., 1937; Ph.D., Ill. Inst. Tech., 1948. Teaching asst. Syracuse U., 1935-36; engr. Ingersoll-Rand Co., Painted Post, N.Y., 1936-39; instr. Ill. Inst. Tech., 1939-41; from instr. to asst. prof. U. Calif., Berkeley, 1941-48; staff engr. Nat. Acad. Scis., Washington, 1945-48; adv. fellow Mellon Inst., Pitts., 1947-48; with VA, N.Y.C., 1948—; chief research and devel. div. Prosthetic and Sensory Aids Service, 1948-73; dir. Research Center for Prosthetics, 1973-78, Office of Tech. Transfer, 1978-83, sci. advisor, 1983-85; Mem. council Alliance for Engring. in Medicine and Biology, 1969—; mem. adv. com. U. Wis., 1978-82, Case Western Res. U., 1981—, Am. Found. for Blind, 1981. Contbg. author: Human Limbs and their Substitutes, 1954, Orthopaedic Appliances Atlas, vol. 1, 2, Human Factors in Technology, 1963, Biomedical Engineering Systems, 1970, Critical Revs. in Bioengring, 1971, CRC Handbook of Materials, Vol. III, 1975, Atlas of Orthotics, 1976, Therapeutic Medical Devices, Application and Design, 1982; editor: Bull. Prosthetics Research, 1978-82; contbr. articles profl. jours. Recipient Silver medal, Paris, France, 1961; Meritorious Service award VA, 1971; Disting. Career award VA, 1983; citation Outstanding Handicapped Fed. Employee, 1971; Profl. Achievement award Ill. Inst. Technology, 1983; Fulbright lectr. Soc. and Home for Cripples, Denmark, 1957-58. Fellow AAAS, ASME, Internat. Soc. for Prosthetics and Orthotics, Rehab. Engring. Soc. N.Am.; assoc. fellow N.Y. Acad. Medicine; mem. Nat. Acad. Engring., N.Y. Acad. Sci., Acoustical Soc. Am., Optical Soc. Am., Sigma Xi, Tau Beta Pi, Phi Kappa Phi. Subspecialties: Biomedical engineering. Current work: Biomedical engineering, broadly, especially rehabilitation research and development (prosthetics, orthotics, sensory aids, selected implants, biomaterials); writing, editing disseminating information. Home: 511 E 20th St New York NY 10010 Office: 252 7th Ave New York NY 10001

MURPHY, MICHAEL ARTHUR, geology educator; b. Spokane, Wash., Mar. 11, 1925; s. Arthur Powell and Hazel Mary (Snyder) M. B.A., UCLA, 1950; Ph.D., 1954. Registered geologist, Calif. Geologist Shell Oil Co., Bakersfield, Calif., 1954; asst. prof. earth scis. U. Calif.-Riverside, 1954-61, assoc. prof., 1961-67, prof., 1967—. Served with U.S. Army, 1943-46; ETO. Fellow Geol. Soc. Am.; mem. Paleontol. Soc., Soc. Econ. Paleontologists and Mineralogists, Internat. Paleontol. Assn., Sigma Xi. Subspecialties: Paleontology; Geology. Current work: Paleontology, stratigraphy, structural geology of middle Paleozoic rocks of Central Great Basin in United States. Office: Dept Earth Scis U Calif Riverside CA 92521

MURPHY, RICHARD ALAN, physiology educator; b. Twin Falls, Idaho, July 4, 1938; s. Albert M. and S. Elizabeth (McClain) M.; m. Genevieve M. M. Johnson, Dec. 16, 1961; children: Hayley McClain, Wendy Louise Marshall. A.B., Harvard U., 1960; Ph.D., Columbia U., 1964. Postdoctoral fellow Max-Planck Inst., Heidelberg, Germany, 1964-66; research assoc. U. Mich.-Ann Arbor, 1966-68; asst. prof. U. Va., Charlottesville, 1968-72, assoc. prof., 1972-77, prof. physiology, 1977—; cons. study sects. NIH, NSF, Washington, 1965—. Mem. editorial bd.: Am. Jour. Physiology, 1975—, Circulation Research, 1974-77, 82—, Blood Vessels, 1973—; contbr. articles in field to profl. jours. Recipient Career Devel. award NIH, 1971-76; NIH postdoctoral fellow, 1964-66. Mem. Am. Physiol. Soc., Biophys. Soc., Soc. Gen. Physiologists, Circulation Group, AAAS. Subspecialties: Biophysics (biology); Physiology (biology). Current work: Contractile and regulatory proteins of muscle, phosphorylation as cellular regulatory mechanism for contractile proteins, structural and biochemical basis for contraction in mammalian smooth muscle. Home: Turkey Run Box 262-B Keswick VA 22947 Office: U Va Dept Physiology Sch Medicine Box 449 Charlottesville VA 22908

MURRAY, JOHN PATRICK, psychologist; b. Cleve., Sept. 14, 1943; s. John Augustine and Helen Marie (Lynch) M.; m. Ann Coke Dennison, Apr. 17, 1971; 1 son, Jonathan Coke. B.A., John Carroll U., 1965; M.A., Cath. U. Am., 1968, Ph.D., 1970. Lic. psychologist, Nebr., D.C. Research dir. Office Surgeon Gen., Washington, 1969-72; asst. prof. Macquarie U., Sydney, Australia, 1973-77, assoc. prof., 1977-79; vis. assoc. prof. U. Mich., Ann Arbor, 1979-80; dir. youth and family policy Boys Town (Nebr.) Ctr., 1980—; mem. State Foster Care Rev. Bd., Lincoln, 1982—, Advocacy Office for Children, Omaha, 1981—, Nat. Council for Children and TV, Princeton, N.J., 1982—; adj. prof. U. Nebr., 1982—, U. Kans., 1983—. Author: Status Offenders: A Sourcebook, 1983, Television and Youth: 25 Years of Research and Controversy, 1980, Small Screen, Big Business, 1979, Children and Families in Australia, 1979. NIMH fellow, 1972-73. Fellow Am. Psychol. Assn.; mem. Am. Sociol. Assn., AAAS, Soc. for Pediatric Psychology, Soc. for Research in Child Devel., Royal Commonwealth Soc. Subspecialties: Developmental psychology; Social psychology. Current work: Research and social policy formulation in area of children, youth and families; special interest in juvenile justice. Office: Boys Town Ctr Boys Town NE 68010

MURRAY, PETER, metallurgist, manufacturing company executive; b. Rotherham, Yorks, Eng., Mar. 13, 1920; came to U.S., 1967, naturalized, 1974; s. Michael and Ann (Hamstead) M.; m. Frances Josephine Glaisher, Sept. 8, 1947; children: Jane, Paul, Alexander. B.Sc. in Chemistry with honors, Sheffield (Eng.) U., 1941, postgrad., 1946-49; Ph.D. in Metallurgy, Brit. Iron and Steel Research Bursar, Sheffield, 1948. Research chemist Steetley Co., Ltd., Worksop, Notts, Eng., 1941-45; with Atomic Energy Research Establishment, Harwell, Eng., 1949-67, head div. metallurgy, 1960-64, asst. dir., 1964-67; tech. dir., mgr. fuels and materials, advanced reactors div. Westinghouse Electric Corp., Madison, Pa., 1967-74; dir. research Westinghouse Electric Europe (S.A.), Brussels, 1974-75; chief scientist advanced power systems divs. Westinghouse Electric Corp., Madison, Pa., 1975-81, dir. nuclear programs, Washington, 1981—; mem. divisional rev. coms. Argonne Nat. Lab., 1968-73; Mellor Meml. lectr. Inst. Ceramics, 1963. Contbr. numerous articles to profl. jours.; editorial adv. bd.: Jour. Less Common Metals, 1968—. Recipient Holland Meml. Research prize Sheffield U., 1949. Fellow Royal Inst. Chemistry (Newton Chambers Research prize 1954), Inst. Ceramics; mem. Brit. Ceramics Soc. (pres. 1965), Am. Ceramic Soc., Am. Nuclear Soc., Nat. Acad. Engring. Roman Catholic. Subspecialty: Nuclear energy research and development administration. Home: 20308 Canby Ct Gaithersburg MD 20760 Office: Westinghouse Electric Corp 1801 K St NW Washington DC 20006

MURRAY, RAYMOND HAROLD, physician; b. Cambridge, Mass., Aug. 17, 1925; s. Raymond Harold and Grace May (Dorr) M.; m.; children—Maureen, Robert, Michael, Margaret, David, Elizabeth, Catherine, Anne. B.S., U. Notre Dame, 1946; M.D., Harvard U., 1948. Diplomate: Am. Bd. Internal Medicine, also Sub-bd. Cardiovascular Disease. Practice medicine, Grand Rapids, Mich., 1955-62; asst. prof. to prof. medicine Ind. U. Sch. Medicine, 1962-77; prof., chmn. dept. medicine Mich. State U. Coll. Human Medicine, 1977—; chmn. aeromed-bioscis panel Sci. Adv. Bd., USAF, 1977-81. Contbr. numerous articles to profl. publs. Served with USNR, 1942-45; Served with USPHS, 1950-53. Fellow A.C.P.; mem. Am. Heart Assn. (fellow council clin. cardiology), Am. Fedn. Clin. Research, Central Soc. Clin. Research, Am. Physiol. Soc. Subspecialty: Cardiology. Office: B220 Life Scis Bldg Mich State U East Lansing MI 48824

MURRAY, ROBERT FULTON, JR., physician; b. Newburgh, N.Y., Oct. 19, 1931; s. Robert Fulton and Henrietta Frances (Judd) M.; m. Isobel Ann Parks, Aug. 26, 1956; children—Colin Charles, Robert Fulton, III, Suzanne Frances, Dianne Akwe. B.S., Union Coll., Schenectady, 1953; M.D., U. Rochester, N.Y., 1958; M.S., U. Wash., Seattle, 1968. Diplomate: Am. Bd. Internal Medicine. Rotating intern Denver Gen. Hosp., 1958-59; resident in internal medicine U. Colo. Med. Center, 1959-62; staff investigator (service with USPHS) Nat. Inst. Arthritis and Metabolic Diseases, NIH, Bethesda, Md., 1962-65; NIH spl. fellow med. genetics U. Wash., 1965-67; mem. faculty Howard U. Coll. Medicine, Washington, 1967—, prof. pediatrics and medicine, 1974—, grad. prof., 1976, prof. oncology, 1976; mem. nat. adv. gen. med. scis. council NIH, 1971-75; sci. adv. bd. Nat. Sickle Cell Anemia Found.; mem. ethics adv. bd. to sec. HEW, 1978-80; chmn. Washington Mayor's Adv. Com. on Metabolic Disorders, 1980—; mem. Med. Com. Human Rights. Author: Genetic, Metabolic and Developmental Aspects of Mental Retardation, 1972; co-editor: Genetic Counseling: Facts, Values and Norms, 1979; asso. editor: Am. Jour. Clin. Genetics, 1977—; editorial adv. bd.: Ency. Bioethics, 1975-77. Trustee Union Coll., 1972-80. Rotary Found. fellow, 1955-56; research grantee NIH, 1969-75. Fellow A.C.P., AAAS, Inst. Medicine, Inst. Society, Ethics and Life Scis.; mem. Am. Soc. Human Genetics, Genetics Soc. Am., D.C. Med. Soc., Neighbors Inc. D.C., Sigma Xi. Unitarian. Subspecialties: Internal medicine; Genetics and genetic engineering (medicine). Current work: Inherited susceptibility to disease; genetic screening and counseling; sickle cell disease and traits; bioethics. Home: 510 Aspen St NW Washington DC 20012 Office: Coll Medicine Box 75 Howard Univ Washington DC 20059

MURRAY, RODNEY BRENT, pharmacologist, microcomputer cons.; b. Phila., July 12, 1949; s. Albert Rodney and Victoria J. (Rinck) M. B.A. in Biology, Temple U., 1971, Ph.D. in Pharmacology (NIH fellow), 1979. Cardiovascular technologist Lankenau Hosp., Phila., 1971-74; instr. biomed. engring. and sci. dept. Drexel U., Phila., 1977—; NIH fellow dept. pharmacology Temple U. Med. Sch., Phila., 1979-82; pres. Micro Computer Specialists, Elkins Park, Pa., 1977—; pharmacologist Biosearch, Inc., Phila., 1982—; cons. for microcomputer software and hardware. Author: (with R.J. Tallarida) Manual of Pharmacologic Calculations with Computer Programs, 1981. Mem. Soc. for Neurosci., Am. Soc. Pharmacology and Exptl. Therapeutics. Subspecialties: Neuropharmacology; Bioinstrumentation. Current work: Application of computers to neuropharmacology; basic research in neuropharmacology and neurotoxicology; computer modeling, database design and statistical and signal analysis. Office: Biosearch Inc PO Box 8598 Philadelphia PA 19101

MURRAY, ROYCE WILTON, chemistry educator; b. Birmingham, Ala., Jan. 9, 1937; s. Royce Leeroy and Justina Louisa (Herd) M.; m. Mirtha Umana, Dec. 11, 1982; children: Katherine, Stewart, Debra, Melissa, Marion. B.S. in Chemistry, Birmingham So. Coll., 1957; Ph.D., Northwestern U., 1960. Faculty U. N.C., Chapel Hill, 1960—, prof. chemistry, 1969-80, acting dept. chmn., 1970-71, vice chmn., 1972-75, chmn., 1980—, Kenan prof., 1980—; On leave from U. N.C. in chemistry sect. NSF, Washington, 1971-72. Author two books, also articles in analytical chemistry and surface chemistry. Sloan research fellow, 1969-72; Guggenheim fellow, 1980-81. Mem. Am. Chem. Soc., Electrochem. Soc., Western Electroanalytical-Theoretical Soc., Phi Beta Kappa, Sigma Xi, Omicron Delta Kappa, Phi Eta Sigma, Theta Sigma Lambda, Theta Chi Delta, Alpha Tau Omega, Alpha Chi Sigma. Subspecialty: Analytical chemistry. Office: Dept Chemistry U North Carolina Chapel Hill NC 27514

MURRAY, STEPHEN S., astrophysicist, cons.; b. N.Y.C., Aug. 28, 1944; s. Leon and Beatrice E. M.; m. Judith G. Gittlen, Aug. 31, 1965; children: Jeffrey J., Micah N. B.S., Columbia U., 1965; Ph.D., Calif. Inst. Tech., 1971. Staff scientist Am. Sci & Engring., Cambridge, Mass., 1971-72; physicist Smithsonian Astrophys. Obs., Cambridge, 1973-82, astrophysicist, 1982—; cons. Space Telescope Sci. Inst.; lectr. Harvard Coll. Obs. Bd. dirs. Temple B'nai Abraham, Beverly, Mass. Mem. Am. Astron. Soc., Internat. Astron. Union, AAAS, Sigma Xi. Subspecialties: High energy astrophysics; 1-ray high energy astrophysics. Current work: Extragalactic x-ray astronomy, high resolution imaging detectors; development of single photon imaging detectors for x-ray astronomy and low light level optical astronomy; research on x-ray background and extragalactic x-ray sources. Office: 60 Garden St B416 Cambridge MA 02138

MURRIN, LEONARD CHARLES, II, pharmacologist, educator, researcher; b. Iowa City, Oct. 9, 1943; s. Leonard Charles and Huberta Frances (Jones) J.; m. Kathryn Grace McDermott, Aug. 17, 1968; children: Leonard Charles III, Rose Colleen, Clare Rita. B.A., St. John's Coll., Camarillo, Calif., 1965; Ph.D., Yale U., 1975. Postdoctoral fellow Yale U., 1975, Johns Hopkins U., 1975-78; asst. prof. pharmacology U. Nebr. Med. Sch., Omaha, 1978—. Contbr. numerous articles, abstracts to profl. jours. March of Dimes Birth Defects Found. grantee, 1979-84; NSF grantee, 1980-83. Mem. AAAS, Soc. Neurosci., Am. Soc. Pharmacology and Exptl. Therapeutics, Am. Soc. Neurochemistry, Internat. Soc. Neurochemistry. Subspecialties: Neuropharmacology; Neurochemistry. Current work: Developmental neurochemistry, neuropharmacology and neuroanatomy of central nervous system. Home: 5204 N 86th St Omaha NE 68134 Office: Dept Pharmacology U Nebr Med Center Omaha NE 68105

MURTHY, VADIRAJA VENKATESA, biochemist, educator; b. Bombay, India, Mar. 27, 1940; s. Ramanathpur Venkatesa and Saroja Venkatesa M.; m. Jayashree Vadiraj, Sept. 21, 1969; children: Deepti Vadiraj, Seema Vadiraj. B.Sc. with honors, Bombay U., 1959, M.Sc., 1961; Ph.D., U. Md., 1968. Research fellow dept. biochemistry U. Md. Med. Sch., Balt., 1963-68; sci. officer St. John's Med. Coll., Bangalore, India, 1968-70; sr. research biochemist, asst. group leader div. biol. research USV Charm. Corp., Yonkers, N.Y., 1970-71; research assoc. toxicology ctr. U. Iowa Sch. Medicine, Iowa City, 1971-72; vis. scientist Nat. Inst. Environ. Health Scis., Research Triangle Park, N.C., 1972-74; sr. research assoc. dept. pharmacology Emory U., Atlanta, 1974-75; adj. asst. prof. chemistry dept. Atlanta U., 1975-76; prof. biology, co-dir Minority Biomed. Research Support Program, Talladega (Ala.) Coll., 1976—; sci. cons. City of Talladega, 1980—. Contbr. articles to profl. jours. NIH grantee, 1976—. Mem. Am. Assn. Cancer Research, Am. Chem. Soc., N.Y. Acad. Scis., Am. Fedn. Clin. Research, AAAS, Southeastern Cancer Research Assn., Sigma Xi. Democrat. Hindu. Subspecialties: Biochemistry (medicine); Molecular pharmacology. Current work: Regulation and control of aromatic aminoacid metabolism, energy from biomass, active-site directed enzyme inhibitors, mechanism of action of cancer chemotherapeutic agents, regulation of sea-urchin embryonic development. Office: Biology Dep Talladega Coll Talladega AL 35160 Home: 111 Dogwood Circle Talladega AL 35160

MURTHY, VARANASI RAMA, geology educator; b. Visakapatnam, India, July 2, 1933; s. Varanasi Rama and Varanasi (Sodemma) Brahmam; m. Monique Danielle Bois, Aug. 21, 1959; children: V. Aanand, V. Katyayini. B.S., Andhra U., Waltair, India, 1951;

A.I.S.M., Indian Sch. Mines, 1954; M.S., Yale, 1955, Ph.D., 1957. Research fellow Calif. Inst. Tech., 1957-59; asst. research geologist U. Calif., San Diego, 1959-62, asst. prof., 1962-65; asso. prof. U. Minn., Mpls., 1965-69, prof. geology and geophysics, 1969, head Sch. Earth Scis., 1971—, acting dean Inst. Tech., 1983. Research grantee NASA and NSF; Tata Found. fellow, 1954-57. Mem. Am., Indian geophys. unions, Geochem. Soc., AAAS, Sigma Xi. Subspecialties: Geochemistry; Petrology. Current work: Geochemistry and chemical evolution of mantle and crust of earth; origin of igneous rocks. Home: 315 Seymour Pl Minneapolis MN 55414

MURTHY, VISHNUBHAKTA SHRINIVAS, physician, pharmacologist, educator; b. Kanker, India, Jan. 1, 1942; s. V. M. and Veda Vati (Lanka) M.; m. Veda Vani Vishnubhakta, Aug. 11, 1978; children: Renuka, Kishori, Ashck, Vikrum. MBBS, M.G.M. Med. Coll., Indore, India, 1965; Ph.D., U. Man., Winnipeg, Can., 1972. Lectr. pharmacology M.G.M. Med. Coll., Indore, 1968-69; sr. scientist Warner Lambert Co., Ann Arbor, Mich., 1972-74; sr. research investigator Squibb Inst. for Med. Research, Princeton, N.J., 1974-78; asst. prof. dept. medicine, dir. cardiovascular research lab. Mt. Sinai Med. Center, Milw., 1981—. Contbr. articles to profl. jours. Mem. N.Y. Acad. Scis., Am. Soc. Clin. Pharmacology and Therapeutics, Am. Soc. Pharmacology and Exptl. Therapeutics. Hindu. Subspecialties: Internal medicine; Pharmacology. Current work: Research in hypertension and coronary artery disease, clin. pharmacology, patient care. Home: 6655 Jean Nicolet Rd Glendale WI 53217 Office: 950 N. 12th St Milwaukee WI 53233

MUSCHLITZ, EARLE EUGENE, JR., Chemist, educator; b. Palmerton, Pa., Apr. 23, 1921; s. Earle Eugene and Ferne Estelle (Altemose) M.; m. Barbara Pfahler, Sept. 17, 1953; children: Robert Earle, Karl William. B.S., Pa. State U., 1941, M.S., 1942, Ph.D., 1947. Instr. chemistry Cornell U., Ithaca, N.Y., 1947-51; from asst. prof. to assoc. prof. chemistry U. Fla., Gainesville, 1951-58, prof., 1958—, chmn. dept., 1973-77. Author articles. NSF sr. fellow, 1963-64; Alexander von Humboldt sr. U.S. scientist, W.Ger., 1978; vis. fellow Joint Inst. for Lab. Astrophysics, 1968. Fellow Am. Inst. Chemists, Am. Phys. Soc.; mem. Am. Chem. Soc., Am. Soc. for Mass Spectrometry, Sigma Xi, Alpha Chi Sigma. Democrat. Subspecialties: Physical chemistry; Kinetics. Current work: The structure and properties of molecules in electronically excited states. Experimental investigations of inelastic collisions of electronically excited atoms and molecules using crossed molecular beam, mass spectrometric, and optical techniques. Home: 4850 NW 20th Pl Gainesville FL 32605 Office: Dept Chemistry U Fla Gainesville FL 32611

MUSCOPLAT, CHARLES CRAIG, medical products executive, immunologist; b. St. Paul, Aug. 13, 1948; s. Abe and Berdie (Shapiro) M.; m. Susan Jean Gallop, July 2, 1950; children: Amy, Marnie. B.A., U. Minn., 1970, Ph.D., 1975. Exec. v.p. ops. med. products Molecular Genetics, Inc., Minnetonka, Minn., 1981. Contbr. articles to profl. jours. Served with USAR, 1969-73. Mem. Am. Assn. Immunologists. Subspecialty: Immunobiology and immunology. Current work: Immunology of infectious disease. Home: 7800 Winsdale St Golden Valley MN 55427 Office: Molecular Genetics Inc 10320 Bren Rd E Minnetonka MN 55343

MUSE, MARK DANA, psychologist; b. Pasadena, Calif., Mar. 1, 1952; s. Harry Lee and Nelda Hayward (Evans) M.; m. Michele Standish, Sept. 10, 1970 (div. 1975); 1 dau., Dana Michele; m. Gloria Frigola, Aug. 15, 1978. B.S., No. Ariz. U., 1973, M.A., 1978, Ed.D., 1980. Lic. psychologist, Md. Grad. asst. No. Ariz. U., 1978-80, adj. faculty, 1981; psychodiagnostician Inst. Human Devel., Flagstaff, Ariz., 1980-81; cons. psychologist Centro Psicologico, Ibiza, Spain, 1981—; dir. Pain Clinic, Sacred Heart Hosp., Cumberland, Md., 1982—. Author: El Stress y El Relax, 1983. Mem. Am. Psychol. Assn., Phi Kappa Phi, Psi Chi. Democrat. Subspecialty: Cognition. Current work: Effect of cognitive style in patients' coping skills with chronic benign pain syndromes. Home: 937 Bishop Walsh St Cumberland MD 21502 Office: Sacred Heart Hosp 900 Seton Dr Cumberland MD 21502

MUSGRAVE, GARY EUGENE, cardiovascular physiologist, educator, pharmacology consultant; b. Chattanooga, Nov. 26, 1947; s. George Newit and Evelyn Louise (Penny) M.; m. Judith Ann Lisle, Dec. 21, 1968 (div. 1978); 1 son, David Westley; m. Jewell Elizabeth Powell, Dec. 17, 1978. B.S., Auburn U., 1969, 1975, Ph.D., 1979. Research asst. Emory U., Atlanta, 1970-74; asst. prof. U. Ga., Athens, 1978-79; NIH postdoctoral fellow U. Tex. Health Sci. Ctr., San Antonio, 1979-80, asst. instr., 1980-82; research physiologist Audie Murphy VA Med. Ctr., San Antonio, 1980-82; asst. prof. cardiovascular physiology Med. Coll. Va., Richmond, 1982—; project dir. NASA contract, Richmond, 1982—. Contbr. chpt. to book in field. NIH fellow Nat. Inst. Gen. Med. Scis., San Antonio, 1979. Mem. Am. Physiol. Soc., Am. Soc. Pharmacology and Exptl. Therapeutics, Am. Fedn. Clin. Research, Am. Soc. Clin. Pharmacology and Therapeutics, AAAS, Sigma Xi, Eta Kappa Nu, Rho Chi. Democrat. Roman Catholic. Subspecialties: Physiology (medicine); Pharmacology. Current work: Autonomic regulation of sinus nodal function, Baroreflex involvement in hypertension. Office: McGuire VA Med Ctr Cardiovascular Physiology 1201 Broad Rock Blvd Richmond VA 23249

MUSHINSKI, J. FREDERIC, research biochemist; b. Beaver Falls, Pa., May 18, 1938; s. Joseph Albert and Ruth Selina (Roberts) M.; m. Elizabeth Bridges, May 1, 1971. B.A., Yale U., 1959; M.D., Harvard U., 1963. Diplomate: Nat. Bd. Med. Examiners. Intern, fellow dept. medicine Duke U., 1963-65; research assoc. Nat. Cancer Inst., NIH, Bethesda, Md., 1965-69; sr. investigator Lab. Cell Biology, 1970—; vis. scientist Max Planck Inst. Exptl. Medicine, Goettingen, W. Ger., 1969-70. Contbr. numerous articles to profl. jours. Served with USPHS, 1965—. William O. Moseley Jr. travelling fellow Harvard U., 1969. Mem. Am. Soc. Biol. Chemists, Am. Assn. Immunologists, Am. Assn. Cancer Research, AAAS. Subspecialties: Biochemistry (medicine); Immunogenetics. Current work: Recombinant DNA studies of antibody and cancer genes. Office: Bldg 37 Room 2826 NIH Bethesda MD 20205

MUSIEK, FRANK EDWARD, audiologist, clin. researcher; b. Union City, Pa., July 4, 1947; s. Steve F. and Nellie A. (Glod) M.; m. Sheila J. Knuth, Aug. 19, 1972; 1 son, Erik. B.S., Edinboro Coll., 1968; M.A., Kent State U., 1971; Ph.D., Case Western Res. U., 1975. Asso. prof., dir. audiology Dartmouth-Hitchcock Med. Center Faculty Neurosci., Dartmouth Coll. Med. Sch., Hanover, N.H., 1977—. Contbr. articles on audiology to profl. jours. Mem. N.H. Gov.'s Council for Hearing Aids, 1981. Hitchcock Found. grantee, 1972-82; NSF Speech and Hearing Assn. award for fellowship of a good time, 1981. Mem. Soc. for Neurosci., Am. Auditory Soc., Am. Speech Lang. and Hearing Assn., Acoustical Soc. Am., Deafness Research Found. Democrat. Roman Catholic. Subspecialties: Otorhinolaryngology; Neurophysiology. Current work: Central auditory function and electrophysiological audiology hearing, audition, audiology, neurophysiology, central auditory function, ear and brain relationships, audition in splitbrain patients. Office: Two Maynard St Hanover NH 03755

MUSLIN, LAWRENCE RICHARD, consulting engineer; b. Chgo., Mar. 29, 1943; s. Isadore Sam and Marjory (Axelrod) M. B.S., Mich. State U., 1964; M.B.A., Loyola U., Chgo., 1972. Various tech. mgmt. positions Abbott Labs., North Chicago, Ill., 1966-82; v.p. tech. ops. Reg Tech Assocs. Inc., Northbrook, Ill., 1982—; guest lectr. Phila. Coll. Pharmacy. Contbr. in field. Served with USAR, 1966. Mem. Am. Soc. Quality Control (cert. quality engr.), Am. Assn. Contamination Control, Soc. Plastics Engrs. Current work: Bioclean pharaceutical device manufacturing environments consulting, engineering, pharmaceuticals, medical devices, materials, sterlization, clean rooms. Office: 880 W Golf Rd Des Plaines IL 60016

MUSTAFA, SYED JAMAL, pharmacologist, educator, researcher; b. Lucknow, India, July 10, 1946; came to U.S., 1971, naturalized, 1978; s. Syed Mohd and Ahmad Jehan M.; m. Yasmeen Khan, June 11, 1973; children: Zishan, Farhan. B.S., Lucknow U., 1962; M.S. in Biochemistry, 1965, Ph.D., 1969. NIH postdoctoral fellow dept. physiology U. Va. Med. Sch., Charlottesville, 1971-74; asst. prof. dept. pharmacology U. South Ala. Med. Sch., Mobile, 1974-77, asso. prof., 1977-80; asso. prof. dept. pharmacology East Carolina U. Med. Sch., Greenville, N.C., 1980—. Contbr. articles to physiol. and pharm. jours., chpts. to books; referee editor several jours. NIH grantee, 1976—; Ala. Heart Assn. grantee, 1975-79. Mem. Am. Soc. Pharmacology and Exptl. Therapeutics, Am. Physiol. Soc., Soc. Exptl. Biology and Medicine, Internat. Soc. Heart Research, Am. Heart Assn., AAAS, N.Y. Acad. Scis., Islamic Med. Assn., Muslim Student Assn., Sigma Xi. Subspecialties: Pharmacology; Cardiology. Current work: Cellular mechanism(s) of coronary flow regulation (metabolic) in normal and pathophysiological models; research in cardiovascular pharmacology. Office: Dept Pharmacology East Carolina U Med Sch Greenville NC 27834

MUSTARD, JAMES FRASER, university research administrator, physician; b. Toronto, Ont., Can., Oct. 16, 1927; s. Alan Alexander and Jean Ann (Oldham) M.; m. Christine Elizabeth Sifton, 1952; children: Cameron, Anne, James, Duncan, John, Christine. M.D., U. Toronto, 1953; Ph.D., Cambridge (Eng.) U., 1956. Intern Toronto Gen. Hosp.; asst. prof. medicine U. Toronto, 1963; assoc. prof. McMaster U., Hamilton, Ont., Can., 1965, prof. pathology, chmn. dept. pathology, 1972-80, prof. pathology, 1982—, v.p. health scis., 1980-82. Contbr. numerous articles to sci. jours. Recipient Internat. award for med. research Gairdner Found., 1967. Fellow Royal Soc., Royal Coll. Physicians, Can. Inst. Advanced Research (pres.). Subspecialties: Arteriosclerosis; Research administration. Office: Dept Pathology McMaster U Hamilton ON Canada L8N 3Z5

MUTSCHLECNER, JOSEPH PAUL, physicist; b. Ft. Wayne, Ind., July 6, 1930; s. Henry and Aurelia Anna (Neeb) M.; m. Alice Diane Schlenker, June 17, 1955; children: David, Timothy, Celia. A.B., Ind. U., 1952, M.A., 1954; Ph.D., U. Mich., 1963. Physicist U.S. Naval Ordnance Test Sta., China Lake, Calif., 1961-63; research scientist Los Alamos (N.Mex.) Nat. Lab., 1963-67, project leader, 1981—; assoc. prof. astronomy Ind. U.-Bloomington, 1967-81. Contbr. numerous astron. articles to profl. publs. NSF fellow, 1954-55. Mem. Internat. Aston. Union, Am. Astron. Soc., AAAS. Subspecialties: Acoustics; Solar physics. Current work: Infrasound measurement and theory. Office: Los Alamos Nat Lab ESS-5 F665 Los Alamos NM 87545

MUTTER, WALTER EDWARD, physicist; b. N.Y.C., Nov. 13, 1921; s. Frederick Charles and Mary Margaret (Aulback) M.; m. Katherine Jeanne Meier, Aug. 31, 1962; children: Walter P., Paul M. B.S., Poly. Inst. N.Y., 1942; Ph.D., M.I.T., 1949. Tube devel. engr. RCA Corp., Lancaster, Pa., 1942-46; research assoc. in physics M.I.T., Cambridge, 1946-49; with IBM Corp., 1949—, sr. physicist, Hopewell Junction, N.Y., 1960—. Contbr. articles in field to profl. jours. Mem. Am. Phys. Soc., Am. Chem. Soc., IEEE, Electrochem. Soc., AAAS, Sigma Xi. Subspecialties: Microchip technology (engineering); Semiconductors. Current work: Researcher in high performance bipolar integrated circuits. Patentee in field. Office: IBM Corp Bldg 300-48A PO Box 390 Hopewell Junction NY 12533

MYER, CAROLE WENDY, veterinary radiology educator; b. N.Y.C., Mar. 14, 1944; d. John Paul and Lydia Estelle (Quinones) M.; m. Charles Edward Bynner, Oct. 30, 1971. D.V.M., Cornell U., 1967; M.S., Ohio State U., 1973. Diplomate: Am. Coll. Vet. Radiology. Practice vet. medicine, S.I., N.Y., 1967-70; resident in radiology Ohio State U., 1970-73, clin. instr., 1973-74, asst. prof. vet. radiology, 1974-81, assoc. prof., 1981. Contbr. articles to profl. jours. Mem. AVMA, Am. Coll. Vet. Radiology (sec. 1976-78, (80-83), pres. 1979-80). Republican. Methodist. Subspecialty: Veterinary radiology. Current work: Clinical veterinary radiology; veterinary ultrasound. Office: Ohio State U Vet Teaching Hosp 1935 Coffey Rd Columbus OH 43210

MYERS, DONALD ARTHUR, geologist; b. Seattle, May 30, 1921; s. Arthur Raymond and Lydia Albertina (Lofgren) M.; m. Margaret Martin Pettus, Aug. 6, 1955; children: David, John, Alan. Student, U. Utah, 1938-40; B.A., Stanford U., 1943, postgrad., 1946-48; postgrad., Johns Hopkins U., 1948-50. Geologist U.S. Geol. Survey, Denver, 1948. Author: Geology, Late Paleozoic Horseshoe Atoll, Tex., 1956. Served to 1st lt. USAF, 1943-46. N.Mex. Bur. Mines research assoc., 1978. Fellow Geol. Soc. Am.; mem. Paleontol. Soc., Soc. Econ. Paleontologists and Mineralogists, Am. Assn. Petroleum Geologists. Subspecialties: Biostratigraphy; Geology. Current work: Fusulinid Foraininifera, research into specific and infraspecific variations in time as a tool to aid in regional and interregional stratigraphic correlation in Pennsylvanian and early Permian sedimentary rocks. Home: 1240 Cody St Lakewood CO 80215 Office: US Geol Survey Fed Ctr Box 25046 Mail Stop 91B Denver CO 80225

MYERS, ERNEST RAY, psychologist, educator, consultant; s. David and Alma (Harper) M.; m. Carole Elaine Ferguson. M.S.W., Howard U., 1964; postgrad., Am. U., 1973-75; Ph.D., Union Grad. Sch., Yellow Springs, Ohio, 1976. Personnel specialist U.S. Air Force, 1956-60, tactical instr., 1956—; research asst. dept. preventive medicine Howard U., 1960-63; psychiat. social worker, social services unit St. Elizabeth's Hosp., Washington, 1963; psychol. services ctr. Lorton (Va.) Reformatory, 1963-64; cons. Pres.'s Task Force on War Against Poverty, Washington, 1964; with VISTA, 1964-67; nat. project devel. officer, project devel. and field support div. OEO, Washington, 1966-67, nat. program plans and policy devel. officer, program plans and policy br., 1967; nat. neighborhood services program officer and coordinator Washington Inter-Agy. Rev. Com. HUD, 1967-68; asst. dir. Nat. Urban League, Washington, 1968; mgr. program devel. social/instl. systems devel. dept. (internat.) Westinghouse Learning Corp., Bladensburg, Md., 1968-69; dir. coll.-community evaluation office Fed. City Coll., 1969-71, program developer, 1969-71, assoc. prof. psychology and community edn., 1970-71; dir. servicemen's early ednl. counseling program Bur. Higher Edn., Office Edn., HEW, Washington, 1971; assoc. prof. human ecol. systems and services, and counseling and mental health U. D.C., 1972—, dir. community psychology program, 1974-78, dir. mental health studies, 1978-81; assoc. trainer Nat. Tng. Labs., Washington, 1969-71; cons./trainer Met. Ecumenical Tng. Ctr., Washington, 1969-70; cons. Social and Rehab. Services Administrn., HEW, 1968-69; nat. chmn. Delivery of Services Work Group, Nat. Citizens' Planning Com. on Rehab. Services for Disabled and Disadvantaged, 1968-69; pres. Ernest R. Myers and Assocs., 1969—; pres./founder Comprehensive Human Devel. Ctr., Washington, 1978-79. Author: books, monographs, including The Community Psychology Concept: Integrating Theory, Education and Practice in Psychology, Social Work and Public Administration, 1977; 2d edit., 1980, Implications of Race and Culture for Mental Health Service Delivery Systems, 1980; contbr. chpts., articles to profl. publs.; assoc. editor: The Pioneer, VISTA OEO Quar, 1965-66. Bd. dirs. D.C. Assn. Mental Health, 1976—, 1st v.p., 1981-82, mem. mental health edn. com., 1982—; bd. dirs. Orgn. Afro-Am. Vets., 1977—, Wider Opportunities for Women, 1979-82, Community Research, Inc., Washington, 1977—; mem. health brain trust Congressional Black Caucus, U.S. Congress, 1980—. Recipient cert. appreciation Nat Assn. Black Social Workers, 1975; Spl. Opportunity Fellowship award Am. U., 1975; leadership and dedication cert. Alumni Assn. Howard U., 1978; Outstanding Alumnus Sch. Social Work, 1982; cert. appreciation Kiwanis Club, Eastern Br., Washington, 1979; Disting. Service award Nat./D.C. Mental Health Assn., 1982. Mem. D.C. Assn. Black Psychologists (cert. appreciation 1975, 82, chmn. 1975-76), Am. Psychol. Assn., Nat. Assn. Black Psychologists (Nat. Scholarship award 1982), Acad. Cert. Social Workers, Nat. Assn. Social Workers. Subspecialty: Community psychology. Home: 5315 Colorado Ave NW Washington DC 20011 Office: U DC 2565 Georgia Ave NW Washington DC 20009

MYERS, EUGENE DOLAN, air force officer, computer communications analyst; b. Radford, Va., Aug. 30, 1952; s. Eugene Garnett and Helen Elizibeth (Webb) M.; m. Helene Wen-Hsin Young, May 1, 1975. B.S. in Computer Sci, Va. Tech. Inst., 1974, M.S., 1976; M.B.A., U. Okla., 1974. Commd. 2d lt. U.S. Army, 1974; officer U.S. Air Force, 1978, advanced through grades to capt., 1979; computer systems analyst, Offutt AFB, Nebr., 1978-79, computer communications analyst, 1979-82, Ft. Meade, Md., 1982—. Mem. Assn. Computing Machinery, IEEE. Subspecialties: Distributed systems and networks; Cryptography and data security. Current work: Research in computer security, in network and computer architectures. Home: 4811-B Ninninger Ct Fort Meade MD 20755 Office: DOD Fort Meade MD 20755

MYERS, LYLE LESLIE, microbiologist, educator; b. Salem, Oreg., June 11, 1938; s. James Elton and Vesta Elizabeth (Carothers) M.; m. Patricia Ann Walter, Aug. 27, 1960; children: Diane Kay, Linda Ellen, Carolyn Sue. B.S. in Animal Sci, Oreg. State U., 1960; M.S., Mont. State U., 1962; Ph.D. in Agrl. Biochemistry, Purdue U., 1966. Asst. prof. dept. vet. sci. Mont. State U., Bozeman, 1966-71, assoc. prof., 1971-79, prof., 1979—, dir. enteric disease research program, 1980—. Contbr. articles to profl. jours. Mem. Conf. of Research Workers in Animal Diseases, Am. Soc. for Microbiology, Sigma Xi, Phi Lambda Upsilon. Republican. Subspecialties: Microbiology (veterinary medicine); Preventive medicine (veterinary medicine). Current work: Developed an E. coli vaccine now used commercially to prevent scours in calves caused by E. coli; now working to prevent other forms of scours in calves and lambs. Home: 1410 Harper-Puckett Rd Bozeman MT 59715 Office: Dept Vet Sci Mont State U Bozeman MT 59717

MYERS, PHILIP CHERDAK, astrophysicist; b. Elizabeth, N.J., Nov. 18, 1944; s. Reney and Anne (Cherdak) M.; m. Anne Hoffman, July 21, 1972; 1 son, David Hoffman Myers. B.A., Columbia U., 1966; Ph.D., MIT, 1972. Staff researcher MIT, 1972-75, asst. prof. physics, 1975-78, asso. prof. physics, 1978-82; staff Harvard-Smithsonian Ctr. Astrophysics, Cambridge, 1982—; sr. research asso. Nat. Acad. Scis., NRC, Goddard Inst. for Space Studies, N.Y.C., 1979-80, 82-83. Contbr. articles to books, referred jours. Mem. Am. Astron. Soc., Internal. Union Radio Sci. Subspecialties: Radio and microwave astronomy; Biomedical engineering. Current work: Dense star-forming interstellar clouds and application of microwave radiometry to medical diagnosis. Office: Harvard-Smithsonian Ctr Astrophysics D312 Cambridge MA 02138

MYERS, PHILLIP SAMUEL, mechanical engineering educator; b. Webber, Kans., May 8, 1916; s. Earl Rufus and Sarah Katharine (Breon) M.; m. Jean Frances Alford, May 26, 1943; children: Katharine Myers Muirhead, Elizabeth Myers Baird, Phyllis Myers Rathbone, John, Mark. B.S. in Math. and Commerce, McPherson Coll., 1940, Kans. State Coll., 1942; Ph.D. in Engring, U. Wis., 1947. Registered profl. engr., Wis. Instr. Ind. Tech. Coll., Ft. Wayne, summer 1942; instr. U. Wis., Madison, 1942-47, asst. prof., 1947-50, assoc. prof., 1950-55, prof., 1955—, chmn. dept. mech. engring., 1979-83; cons. Diesel Engine Mfrs. Assn., U.S. Army, also various oil and ins. cos.; dir. Nelson Industries, Echlin Mfg. Corp. Contbr. articles to profl. jours. Chmn. Pine Lake com. W. Wis. Conf. Meth. Ch., 1955-60; Mem. Village Bd., Shorewood Hills, 1962-67. Recipient B.S. Reynolds Teaching award, 1964, McPherson Coll. Alumni citation of merit, 1971; Dugald Clerk award, 1971. Fellow ASME (Diesel Gas Power award 1971), Soc. Automotive Engrs. (Colwell award 1966, 79, Horning award 1968, nat. pres. 1969, hon. mem.); mem. Am. Soc. for Engring. Edn., AAAS, Nat. Acad. Engring., Blue Key, Sigma Xi, Phi Kappa Phi, Sigma Tau, Pi Tau Sigma (Gold medal 1949), Tau Beta Pi (Ragnar Onstad Service to Soc. award 1978). Mem. Ch. of the Brethren. Subspecialties: Combustion processes; Fuels. Current work: Teaching; understanding the ragaries of internal combustion engines. Patentee in field. Home: Madison WI

MYERS, WADE HAMPTON, JR., engineering company executive; b. Durham, N.C., Nov. 2, 1941; s. Wade Hampton and Pauline Eugenia (Cross) M.; m. Joyce Yvonne Thornburg, Apr. 17, 1964. B.S. in Physics and Math, U. So. Miss., 1965. Sci. programmer Gen. Electric Co., Syracuse, N.Y., 1965, engr., 1966, systems engr., 1967; sr. systems engr. Honeywell Info. Systems, Phoenix, 1974-76; sr. engr. Courier Tech. Terminal Systems, Phoenix, 1977, Intel Corp., Santa Clara, Calif., 1977; sr. mem. sci. staff BNR, Inc., Mountain View, Calif., 1979; sr. engr. Signuetics Corp., Sunnyvale, Calif., 1978, microsystems group software mgr., 1981-82; propr. Wade Myers Enterprises, 1982—; cons. in field. Mem. Assn. Computing Machinery, IEEE. Subspecialties: Computer architecture; Distributed systems and networks. Current work: Operating systems, distributed processing, microprocessor development tools. Home: PO Box 99323 San Francisco CA 94109 Office: PO Box 880068 San Francisco CA 94188-0068

MYERS, WILLIAM GRAYDON, physician, scientist, educator; b. Toledo, Aug. 7, 1908; s. Leo J. and Anna C. (Johnson) M.; m. Florence R. Lenahan, Dec. 24, 1940. B.A., Ohio State U., 1933, M.Sc., 1937, Ph.D. in Physical Chemistry, 1939, M.D., 1941; D.Sc. (hon.), Bucknell U., 1977. Diplomate: Am. Bd. Nuclear Medicine. Grad. asst. chemistry Ohio State U., 1933-37, Comly asst. med. research, 1939-42; intern Univ. Hosp., 1941-42; research asso. Ohio State U. Research Found., 1942-45; Julius F. Stone fellow med. Ohio State U., 1945-49, research asso. prof., 1949-53, research prof. med. biophysics (depts. medicine, physiology and radiology), 1953—; vis. prof. nuclear medicine U. Calif. at Berkeley, winters 1970—; vis. prof. Cornell Grad. Sch. Med. Scis., 1978—; investigator Meml. Sloan-Kettering Cancer Center, 1978—; cons. nuclear medicine to various labs. Monitor radiol. safety sect. atom bomb tests Operation Crossroads, Bikini, 1946; invited participant numerous nat. and internat. confs. on applications nuclear sci. in nuclear medicine. Author articles profl. jours. Exec. com. Columbus (Ohio) Civil Def., 1950—. Recipient Lucy Wortham James research award James Ewing Soc., 1966. Fellow Royal

Soc. Arts (life), A.C.P.; mem. Am. Chem. Soc., Am. Phys. Soc., Am. Physiol. Soc., Am. Assn. Cancer Research, Soc. Exptl. Biology and Medicine, AAAS, Ohio Acad. Sci., Am. Assn. Physicists in Medicine, Am. Radium Soc., Radiation Research Soc., Soc. Nuclear Medicine (1st Paul C. Aebersold award 1973, historian 1973—, Hevesy Nuclear Medicine Pioneer award 1981), Ohio State Med. Assn., Columbus Acad. Medicine, Phi Lambda Upsilon, Sigma Xi (chpt. pres. 1959-60), Alpha Omega Alpha. Clubs: Torch (Columbus); Explorers (N.Y.C.). Subspecialties: Biophysical chemistry; Nuclear medicine. Current work: Applications of radioactive isotopes in med. diagnosis and therapy; presently serving about 25-35%of time at Memorial Sloan-Kettering Cancer Research Center in N.Y.C. doing research in applying radioisotopes in Dx & Tx. Home: [illegible] Rd Columbus OH 43221 Office: Ohio State Univ Hosp 410 W 10th Ave Columbus OH 43210 Savoring the past enriches the present and presages the future. In FUNdamental research, the zest is in the quest and not in the conquest-but only the first three letters really count! History highlights heuretic. Imagination creates opportunities unlimited. History cements facts together. Laughter lubricates life. "Twinkling atoms bring 'scintillating' people together.

MYERS-BORDINI, LINDA LEE, environ. health services monitor, research botanist; b. Monogahela, Pa., May 21, 1942; d. Walter and Lucille L. (Archabold) Warren; m. Jack L. Myers, Jan. 1960; children: Dana M., Jacqueline L., Kimberly A.; m. Primo Bordini, Apr. 28, 1973. B.S. in Biology, California (Pa.) State Coll, 1976, postgrad., 1977—. Lic. beauty culture operator, visible emissions evaluator. Process analyst Clairton (Pa.) works U.S. Steel Corp., 1978—. Mem. Bot. Soc. Am., Pa. Acad. Sci., Sierra Club, Beta Beta Beta, Chi Gamma PS. Democrat. Methodist. Current work: Study of phylloplane microflora of a native plant in Pennsylvania. Home: 609 Hancock St Monongahela PA 15063 Box 411 Sutersville PA 15083

MYRICK, HENRY NUGENT, waste mgmt. cons.; b. Cisco, Tex., Apr. 30, 1935; s. Elbert Porter and Lila Faye (Hill) M.; m. Mary Frances Ross, June 15, 1968. B.S., Lamar State U., 1957; M.S., Rice U., 1959; Sc.D. in Environ. Sci. and Engring, Washington U., St. Louis, 1962. Fellow, instr. environ. engring. Rice U., Houston, 1962-63; fellow, instr. san. engring., 1963-65; assoc. prof. civil and environ. engring. U. Houston, 1965-74; pres. Process Co., Inc., Houston, 1970—. Pub.: Tex. Solid Waste Mgmt. News, 1972—; contbr. articles to profl. jours. Recipient H. P. Eddy award for noteworthy research Water Pollution Control Assn., 1961. Mem. Air Pollution Control Assn., Water Pollution Control Fedn., Nat. Solid Waste Mgmt. Assn., Govtl. Refuse Collection Disposal Assn., ASCE, Am. Inst. Chem. Engrs., Am. Inst. Chemistry. Presbyterian. Lodge: Rotary. Subspecialties: Water supply and wastewater treatment; Resource management. Office: 5704 Valverde Suite 2 Houston TX 77057

MYSLINSKI, NORBERT RAYMOND, neuropharmacologist; b. Buffalo, Apr. 14, 1947; s. Bernard and Amelia Joan M.; m. Patricia Ann Byrne, June 19, 1970 (dec. 1980). B.S. in Biology, Canisius Coll., Buffalo, 1969; Ph.D. in Pharmacology, U. Ill., 1973. Research assoc. dept. pharmacology and biochemistry Tufts U. Sch. Medicine, Boston, 1973-75; asst. prof. dept. pharmacology Sch. Dentistry U. Md., Balt., 1975-80, asso. prof., 1980—, co-dir. neurosci. program, 1981—; mem. faculty dept. dental aux. Community Coll. Balt., 1980-83; mem. dental com. Md. High Blood Pressure Coordinating Council, 1978-80. Editor: Md. Soc. Med. Research Newsletter, 1977-81; Contbr. articles to profl. jours. USPHS awardee, 1969-73. Mem. Md. Soc. Med. Research (dir. 1978—), Am. Soc. Pharmacology and Exptl. Therapeutics, Brit. Brain Research Assn. (hon. mem.), European Brain and Behavior Soc. (hon. mem.), N.Y. Acad. Scis., Soc. Neurosciences, Am. Assn. Dental Schs., Internat. Assn. Dental Research, Pharm. Mfrs. Assn., Internat. Union Pharmacology, Am. Assn. World Health (mem. U.S. com. for WHO 1979-81), Sigma Xi. Subspecialties: Neuropharmacology; Neurophysiology. Current work: Neurophysiology, brain research, pharmacology, neuroscience education. Home: 108 Rockrimmon Rd Reisterstown MD 21136 Office: 666 W Baltimore St Baltimore MD 21201

NACHMAN, ARJE, mathematics and mechanics researcher; b. Salzburg, Austria, Sept. 12, 1946; came to U.S., 1949, naturalized, 1951; s. Nathan and Miriam N.; m. Ruth Cymber, Aug. 15, 1971. B.S., Washington U., St. Louis, 1968; Ph.D., NYU, 1973. Asst. prof. Tex. A&M U., 1973-79; assoc. prof. Old Dominion U., 1979-81; sr. research scientist S.W. Research Inst., San Antonio, 1981. Contbr. articles to profl. jours. Mem. Soc. Indsl. and Applied Math., Am. Acad. Mechanics. Subspecialties: Applied mathematics; Theoretical and applied mechanics. Current work: Theoretical mechanics, singular perturbations, nonlinear differential equations. Office: SW Research Inst 6220 Culebra San Antonio TX 78284

NADIN, MIHAI, semiotics and computer-aided design educator, consultant; b. Brasov, Romania, Feb. 2, 1938; came to U.S., 1980; s. Lowi and Ana (Catap) Nudelman; m. Elvira Palesey, Sept. 14, 1971; children: Ari, Esther, Elisabeth. M.S., Poly. Inst., Bucharest, Romania, 1960; M.A., U. Bucharest, 1968, Ph.D., 1972; Dr. phil. habil., U. Munich, W.Ger., 1980. Editor Astra, Brasov, 1965-75; assoc. prof. U. Brasov, 1969-75; researcher Institut de Filosofie, Bucharest, 1976-79; prof. U. Munich, 1978-80, U. Essen, W.Ger., 1979-80; prof. semiotics and design R.I. Sch. Design, 1980—; pres., cons. Nadin & Ockerse, Ltd., Providence, 1981—. Author: Semiotics of the Energy Crisis, 1981, Zeichen und Wert, 1981; editor: New Elements of the Semiotics of Communication, 1982, Kodikas/Code, 1980. Recipient Richard Merton award U. Braunschweig, Germany, 1977, I. L. Caragiale award Nat. Acad. Romania, 1978. Mem. Assn. Computing Machinery, Semiotic Soc. Am., Deutsche Gesellschaft fur Semiotik. Subspecialties: Semiotics of computer use and applied artificial intelligence; Information systems (information science). Current work: New computer models based on fuzzy logic and parallel processing procedures; computer-related activities and artificial-intelligence methods as forms of semiotic practice; information provider systems. Office: RI Sch Design 2 College St Providence RI 02903

NADLER, HENRY LOUIS, pediatrician, geneticist; b. N.Y.C., Apr. 15, 1936; s. Herbert and Mary (Kartiganer) N.; m. Benita Weinhard, June 16, 1957; children: Karen, Gary, Debra, Amy. A.B., Colgate U., 1957; M.D., Northwestern U., 1961; M.S., U. Wis., 1965. Diplomate: Am. Bd. Pediatrics, Am. Bd. Med. Genetics. Intern NYU Med. Ctr., 1961-62, sr. resident pediatrics, 1962-63, chief resident, 1963-64; teaching asst. NYU Sch. Medicine, 1962-63, clin. instr., 1963-64, U. Wis. Sch. Medicine, 1964-65; practice medicine specializing in pediatrics, Chgo., 1965—; fellow Children's Meml. Hosp. dept. pediatrics Northwestern U., 1964-65; assoc. in pediatrics Northwestern U. Med. Sch., 1965-66, asst. prof., 1967-68, assoc. prof., 1968-70, prof., 1970-81, chmn. dept. pediatrics, 1970-81, prof., 1971-80; mem. staff Children's Meml. Hosp., 1965-81, head div. genetics, 1969-81, chief of staff, 1970-81; dean, prof. pediatrics Wayne State U. Med. Sch., Detroit, 1981—; mem. vis. staff, div. medicine Northwestern Meml. Hosp., 1972-81; cons. Cook County Hosp., Chgo. Editorial bd.: Comprehensive Therapy, 1973—, Am. Jour. Human Genetics, 1979-83, Pediatrics in Rev, 1980-83, Am. Jour. Diseases of Children, 1983—; contbr. articles to profl. jours. Recipient E. Mead Johnson award for pediatric research, 1973; Irene Heinz Given and John La Porte Given research prof. pediatrics, 1970—. Fellow Am. Acad. Pediatrics; mem. Am. Soc. for Clin. Investigation, Am. Soc. Human Genetics, Am. Pediatric Soc., Soc. for Pediatric Research, Midwest Soc. for Pediatric Research, Pan Am. Med. Assn., Soc. for Exptl. Biology and Medicine, Alpha Omega Alpha. Subspecialty: Pediatrics. Home: 4669 Maura Ln West Bloomfield MI 48033 Office: Wayne State U Med Sch Scott Hall 540 E Canfield Detroit MI 48201

NAEYE, RICHARD L., pathologist, educator; b. Rochester, N.Y., Nov. 27, 1929; s. Peter John and Gertrude Ellen (Lookup) N.; m. Patricia Ann Dahl, June 4, 1955; children: Nancy Ellen, Susan Amy, Robert Peter. A.B., Colgate U., 1951; M.D., Columbia U., 1955. Diplomate: Am. Bd. Pathology. Intern N.Y. Hosp., N.Y.C., 1955-56; resident Columbia-Presbyn. Med. Center, 1956-58, Mary Fletcher Hosp., Burlington, Vt., 1958-60; practice medicine, specializing in pathology, Burlington, 1960-67, Hershey, Pa., 1967—; asst. attending pathologist Mary Fletcher Hosp., 1960-63; asso. prof. U. Vt., 1963-67, prof. pathology, 1967; prof., chmn. dept. pathology M.S. Hershey Med. Center, Pa. State U. Coll. Medicine, 1967—; cons. Armed Forces Inst. Pathology, Am. Heart Assn. Council on Cardiopulmonary Disease, 1972—; mem. NIH study sect. USPHS, 1968-72. Mem. editorial bd.: Human Pathology, 1982—; Contbr. articles to med. jours. Markle scholar in acad. medicine, 1960-65; NIH research grantee, 1963-83. Mem. Am. Soc. Exptl. Pathology, Internat. Acad. Pathology, Am. Assn. Pathologists and Bacteriologists, Am. Soc. Clin. Pathologists, Coll. Am. Pathologists, Pediatric Pathology Club, Pa. Soc. Clin. Pathologists, Pa. Med. Soc. Subspecialty: Pathology (medicine). Current work: Diseases of fetus and neonate; sudden infant death syndrome. Home: 50 Laurel Ridge Rd Hershey PA 17033 Office: Dept Pathology Pa State U Coll Medicine 500 University Dr Hershey PA 17033

NAFTCHI, NOSRAT ERIC, rehabilitation medicine, educator; b. Kermenshah, Iran, Mar. 25, 1929; s. Haroon and Malak N.; m. Joyce Claddette; children: Maxim, Glenn, Nicole. A.B., Syracuse U., 1952; postgrad., Columbia U., 1953-55; fellow, 1963; M.S., N.Y.U., 1956; Ph.D., Rutgers U., 1966. Research asst. in medicine Mt. Sinai Hosp., N.Y.C., 1958-66; research assoc. medicine Circulatory Physiology Lab., 1966-67; assoc. medicine CUNY-Mt. Sinai Sch. Medicine Dept. Medicine, 1967-68; asst. prof. pharmacology and rehab. medicine N.Y.U. Med. Center, N.Y.C., 1968-72, dir. labs. biochem. pharmacology and microcirculation, 1968—; assoc. prof. rehab. medicine, 1972-78, prof. rehab. medicine, 1978—; cons., lectr. in field. Rutgers U. fellow, 1964-66. Mem. AAAS, Am. Congress Rehab. Medicine, Am. Heart Assn. (fellow Council on Circulation), Am. Phys. Soc., Am. Soc. Neurochemistry, Am. Soc. Pharmacology and Exptl. Therapeutics, Nat. Hypertension Assn., European Biochem. Soc., European Microcirculation Soc., Internat. Assn. for Study of Pain, Internat. Soc. Biochem. Pharmacology, Internatsoc. Neurochemistry, Internat. Soc. Paraplegia, Sigma Xi. Subspecialties: Neuropharmacology; Petroleum engineering. Current work: Pharmacological manipulations of the neuroendocrine alterations in spinal cord injured animals. Regeneration of the central nervous system. Home: 389 Forest Ave Teaneck NJ 07666 Office: 400 E 34th St New York NY 10016

NAFTOLIN, FREDERICK, physician, reproductive biologist educator; b. Bronx, N.Y., Apr. 7, 1936; s. Nathan and Jean (Pesacov) N.; m. Phyllis Barbara Kaptowsky, July 14, 1957; children: Michael Eugene, Joshua Joseph. A.A., UCLA, 1957; B.A. with honors, U. Calif., Berkeley, 1958, M.D., 1961; D.Phil., U. Oxford, 1970. Intern King County Hosp., Seattle, 1961-62; resident in ob-gyn UCLA, 1962-66; asst. chief gynecology, endocrine fellow USPHS, Seattle, 1966-68; NIH fellow Oxford (Eng.) U., 1968-70; asst. prof. ob-gyn U. Calif., San Diego Sch. Medicine, 1970-73; asso. prof. ob-gyn Harvard Med. Sch., 1973-75; prof., chmn. ob-gyn dept. McGill Faculty Medicine, Montreal, 1975-78; prof., chmn. dept. ob-gyn Yale Med. Sch., New Haven, Conn., 1978—; vis. prof. U. Geneva, 1982-83. Author: Subcellular Mechanisms in Reproductive Neuroendocrinology, 1976, Abnormal Fetal Growth, 1978, Clinical Neuroendocrinology, 1979, Dilatation of the Uterine Cervix, 1980; 3-vol. series Basic Reproductive Medicine, Vol. I, Basis of Normal Reproduction, Vol. II, Male Reproduction, Vol. III, Metabolism of Steroids by Neuroendocrine Tissues; Mem. editorial bd.: Psychoneuroendocrinology; Contbr. articles to med. jours. Fogarty internat. fellow, 1982; John Simon Guggenheim fellow, 1983. Mem. Soc. Gynecological Investigation, Endocrine Soc., Internat. Soc. Neuroendocrinology, Internat. Soc. Psychoneuroendocrinology, New Haven Ob-Gyn Soc., Can. Fertility Soc., Am. Soc. Andrology. Subspecialty: Reproductive biology (medicine). Office: Dept Ob-Gyn Yale Med Sch 333 Cedar St New Haven CT 06510

NAG, SUBIR K., radiation oncologist; b. Calcutta, India, Dec. 10, 1951; came to U.S., 1977; s. Sunil K. and Bela R. (Dawn) N.; m. Sima Dutta, Mar. 9, 1982. M.B.B.S., All India Inst. Med. Sci., New Delhi, 1975. Diplomate: Am. Bd. Radiology. Resident in radiotherapy Montefiore Hosp., Bronx, N.Y., 1977-79, chief resident, 1978-79; fellow in radiotherapy Meml. Sloan Kettering Cancer Ctr., N.Y.C., 1980, spl. brachytherapy fellow, 1980-81; asst. prof. radiation oncology U. Tenn. Hosp., Memphis, 1981—, chief radiation oncology, 1983—; staff radiation oncologist City of Memphis Hosp., 1981—; cons. St. Judes Childrens Research Hosp., Memphis, 1982—, VA Med. Ctr., 1981—. Contbr. articles to profl. jours. Mem. Am. Soc. Therapeutic Radiologists, Radiol. Soc. N.Am., AMA, Am. Coll. Radiology, Am. Soc. Clin. Oncologist, Radiation Research Soc., Am. Assn. Cancer Edn. Subspecialties: Oncology; Cancer research (medicine). Current work: Enhancement of radiation response by hyperthermia, chemotherapy and radiation sensitisers, clinical brachytherapy. Home: 5111 Darlington Dr Memphis TN 38118 Office: Dept Radiation Oncology U Tenn Health Sci Ctr 800 Madison Ave Memphis TN 38163

NAGAR, ARVIND KUMAR, mechanical engineer, researcher, educator; b. Achheja, Ghaziabad, India, July 4, 1939; came to U.S., 1960, naturalized, 1971; s. Kaley Singh and Risalo (Bhati) N.; m. Sampat Dhabhai, June 5, 1971; children: Anil, Sunil, Jayesh. B.S. in Mech. Engring, Okla. State U., 1969; M.M.E., Midwest Coll. Engring., Lombard, Ill., 1978; postgrad., Ohio State U., 1981—. Registered profl. engr., Ohio, Ill. Assoc. mech. engr. Xerox Corp., Webster, N.Y., 1969-71; mech. engr. All Steel Inc., Aurora, Ill., 1973-79; research scientist fatigue and fractures projects Battelle Meml. Inst. Labs., Columbus, Ohio, 1979-81; lectr., grad. teaching assoc. engring. mechanics dept. Ohio State U., Columbus, 1980—. Served with U.S. Army, 1966-68. Recipient cert. of merit Toastmasters Internat., 1962. Mem. ASME (paper prize 1962), Pi Tau Sigma, Sigma Pi Sigma. Subspecialties: Fracture mechanics; Solid mechanics. Current work: Fracture and crack growth of metals at elevated temperatures, linear/nonlinear elasticity.

NAGASAWA, HERBERT TAUKASA, medicinal chemist, educator; b. Hilo, Hawaii, May 31, 1927; s. Yasuzo and Chie (Maeda) N.; m. Katherine Chizuko Imahiro, June 30, 1951; children: LLoyd Stuart, Scott Glenn. Student, U Hawaii, 1947-48; B.S., Western Res. U., 1950; Ph.D., U. Minn., 1955. Postdoctoral fellow in biochemistry U. Minn., Mpls., 1955-57, asst. prof. pharm. chemistry, 1959, assoc. prof. medicinal chemistry, 1963-72, prof., 1973—; sr. scientist VA Med. Center Lab. Cancer Research, Mpls., 1961-78; prin. scientist Gen. Med. Research Labs., VA Med. Center, Mpls., 1961—. Asso. editor: Jour. Medicinal Chemistry, 1972—; contbr. articles to profl. jours. Served with Mil. Intelligence Corps U.S. Army, 1945-47. Mem. Am. Chem. Soc., Am. Soc. Pharmacological Exptl. therapeutics, Soc. Toxicology, Research Soc. Alcoholism, Am. Assn. Cancer Research, AAAS, N.Y. Acad. Sci., Internat. Soc. Biomed. Research on Alcoholism. Subspecialties: Medicinal chemistry; Organic chemistry. Current work: Design and synthesis of trapping agts. for the detoxication of xenobiotic substances (including ethanol) that are activated to toxic metabolites in vivo and latentiated (prodrug) forms of biologically active substances. Home: 6223 Harriet Ave S Richfield MN 55423 Office: 54th St and 48th Ave S Bldg 31 Minneapolis MN 55417

NAGEL, DAVID JOSEPH, physicist; b. Aurora, Ill., Feb. 20, 1938; s. Edgar Anthony and Loretta (Lies) N.; m. Carol Ann Koch, May 11, 1963; children: Ann, Paul. B.S., U. Notre Dame, 1960; M.S. in Physics, U. Md., 1969; Ph.D. in Materials Engring, U. Md., 1977. Tech. liaison officer Naval Research Lab., 1962-64, physicist, 1964-72, sect. head, 1972-80, br. head condensed matter physics, 1980—; cons. in field. Contbr. articles to profl. jours. Served with USN, 1960-64; served to capt. USNR, 1964—. Mem. Am. Phys. Soc., Optical Soc. Am., Tau Beta Pi. Subspecialty: Condensed matter physics. Current work: X-ray physics with applications in x-ray lithography, fusion energy research and nuclear weapons effects simulation. Home: 3273 Rose Glen Ct Falls Church VA 22042 Office: Naval Research Lab Code 6680 Washington DC 20375

NAGHDI, PAUL MANSOUR, engineering educator; b. Tehran, Iran, Mar. 29, 1924; came to U.S., 1944, naturalized, 1948; s. G. H. and A. (Momtaz) N.; m. Patricia Spear, Sept. 6, 1947 (dec. Mar. 15, 1975); children: Stephen, Suzanne, Sondra. B.S., Cornell U., 1946; M.S., U. Mich., 1948, Ph.D., 1951. From instr. to prof. engring. mechanics U. Mich., 1949-58; prof. engring. sci. U. Calif., Berkeley, 1958—, chmn. div. applied mechanics, 1964-69; Miller prof. Miller Inst. Basic Sci., 1963-64, 71-72; cons. theoretical and applied mechanics, 1953—; Mem. U.S. Nat. Com. on Theoretical and Applied Mechanics, 1972—, chmn., 1979-80. Served with AUS, 1946-47. Recipient Distinguished Faculty award U. Mich., 1956, George Westinghouse award Am. Soc. Engring. Edn., 1962; Guggenheim fellow, 1958. Fellow ASME (chmn. applied mechanics div. 1971-72, Timoshenko medal 1980, hon. mem. 1983), Acoustical Soc. Am., Soc. Engring. Sci. (dir. 1963-70); mem. Soc. Rheology, Sigma Xi. Subspecialties: Theoretical and applied mechanics; Solid mechanics. Current work: Continuun mechanics, theory of shells, plasticity, theories of fluid sheets and fluid jets. Home: 530 Vistamont Ave Berkeley CA 94708

NAGODE, LARRY ALLEN, vet. pathologist, educator; b. New Deal, Mont., Nov. 15, 1938; s. Harry Louis and Ruth Berdella (Bair) N.; m. Dorothy Marie Fenlon, June 6, 1963 (div. 1975); children: Patricia Anne, Russell Allen; m. Carole Lynne Steinmeyer, Apr. 2, 1976. D.V.M., Colo. State U., 1963; M.Sc., Ohio State U., 1965, Ph.D., 1968. Research fellow Ohio State U., Columbus, 1963-65, postdoctoral fellow, 1965-68; Nat. Cancer Inst. spl. fellow U. Pa., Phila., 1968-70; research scientist Pathophysiologische Inst., Med. Coll. Bern, U. Bern, Switzerland, 1973; assoc. prof. vet. pathology Ohio State U., 1973—; cons. Upjohn Drug Co. Contbr. articles to profl. jours. Served with U.S. Army, 1956. U.S. Dept. Agr. grantee, 1974-75, 75-76, 79-80; other grants. Mem. AAAS, Phi Kappa Phi, Phi Zeta, Phi Eta Sigma. Unitarian-Universalist. Subspecialties: Pathology (veterinary medicine); Animal pathology. Current work: Vitamin D metabolism and mechanisms, Enterotoxin mechanisms. Alkaline phosphatase isoenzyme significance. Home: 2837 Helston Rd Columbus OH 43220 Office: 1925 Coffey Rd Columbus OH 43210

NAHM, ALEXANDER HONG, metallurgical engineer; b. Seoul, Korea, Aug. 1, 1942; s. Donwoo and Soohae (Kim) N.; m. Soonze Kim, Nov. 20, 1968; children: Isabel Kyongah, Elizabeth Kyonghyon. B.S., Yonsei U., 1967, M.S., 1969; M.S., U. Cin., 1973, Ph.D., 1975. Research engr. Korea Atomic Energy Research Inst., Seoul, 1967-69; instr. Yonsei U., Seoul, Korea, 1969; research assoc. metall. engring. dept. U. Cin., 1975-76; research engr. Gen. Electric Co., Cin., 1976—. Contbr. numerous sci. articles to profl. publs. Yang Young Found. scholar, 1963-67. Mem. Korean-Am. Assn. (pres. Cin. chpt. 1981), Am. Soc. Metals, ASME, Korean Scientists & Engrs. Assn. Am. Subspecialties: Metallurgical engineering; Mechanical engineering. Current work: Research engineering in the area of material and process technology in aircraft engines, particularly expertise in jet engine disk alloys and jet engine bearing materials. Home: 348 Circlewood Ln Wyoming OH 45215 Office: General Electric Co Aircraft Engine Bus Group Mail Drop M-85 Cincinnati OH 45215

NAHM, MOON HEA, pathologist, immunology researcher, educator; b. Seoul, Korea, Mar. 23, 1948; came to U.S., 1965, naturalized, 1971; s. C.C. and A.J. N. A.B., Washington U., St. Louis, 1970, M.D., 1974. Intern in internal medicine Jewish Hosp., Washington U., St. Louis, 1974-75, resident, 1975-76, resident in lab. medicine, 1976-80, asst. prof. pathology, 1980—. Mem. Am. Immunology Assn., Phi Beta Kappa. Subspecialties: Immunology (medicine); Pathology (medicine). Current work: Study of lymphocyte orgn. Home: 8104 Halifax Clayton MO 63105 Office: Dept Pathology Washington U Box 8818 Saint Louis MO 63110

NAHOUM, HENRY ISAAC, denistry educator, orthodontist; b. N.Y.C., Sept. 9, 1919; s. Isaac Joseph and Rebecca (Hasson) N.; m. Lillian deSoto, Mar. 28, 1948; 1 dau., Bonita. A.B., Bklyn. Coll., 1940; D.D.S., Columbia U., 1943. Cert. orthodontics, 1952., diplomate: Am. Bd. Orthodontics. Research asst. Columbia U., N.Y.C., 1947-56, instr., 1957-60, asst. clin. prof., 1960-63, assoc. clin. prof., 1963-69, assoc. prof., 1969-75, prof. denistry, 1975—; attending dental surgeon Presbyn. Hosp., N.Y.C., 1969—. Bd. dirs. Sephardic Jewish Ctr., N.Y.C., 1965—. Served to capt. AUS, 1944-46; ETO. Named Alumnus of Yr. Columbia Orthodontic Alumni, 1980. Fellow Am. coll. Dentists (chmn. N.Y. sect. 1981), N.Y. Acad. Dentistry; mem. ADA, Am. Assn. Orthodontists, Sigma Xi, Omicron Kappa Upsilon. Subspecialties: Orthodontics; Dental growth and development. Current work: Research on open-bite malocclusion, appliance designs, practice orthodontics. Inventor vacuum, forming dental appliances, 1959. Office: 44A E 73d St New York NY 10021

NAHRWOLD, DAVID LANGE, surgeon, educator; b. St. Louis, Dec. 21, 1935; s. Elmer William and Magdalen Louise (Lange) N.; m. Carolyn Louise Hoffman, June 14, 1958; children: Stephen Michael, Susan Alane, Thomas James, Anne Elizabeth. A.B., Ind. U., 1957, M.D., 1960. Diplomate: Am. Bd. Surgery, Am. Bd. Thoracic Surgery. Intern, then resident in surgery Ind. U. Med. Center, Indpls., 1960-65; postdoctoral scholar in gastrointestinal physiology VA Center, UCLA, 1965; asst. prof. surgery Ind. U. Med. Sch., 1968-70; assoc. prof. Pa. State U. Coll. Medicine, 1970-73, vice chmn. dept. surgery, 1971-82, assoc. provost, dean health affairs, 1981-82; assoc. dean patient care Milton S. Hershey Med. Center, Hershey, Pa., 1978-80, prof. surgery, chief div. gen. surgery, 1974-82; Loyal and Edith Davis prof. surgery, chmn. dept. surgery Northwestern U. Med. Sch., Chgo., 1982—; surgeon-in-chief Northwestern Meml. Hosp., Chgo., 1982—. Author articles, abstracts in field, chpts. in books. Served with M.C. U.S. Army, 1966-68. Fellow A.C.S.; mem. Am. Fedn. Clin. Research, Am. Gastroent. Assn., AMA, Am. Physiol. Soc. (asso.), Am. Surg. Assn., Assn. Acad. Surgery, Central Surg. Assn., Chgo. Med. Soc., Midwest Gut Club, Ill. State Med. Soc., Ill. Surg. Soc., Soc. Clin. Surgery, Soc.

Surgery of Alimentary Tract, Soc. Univ. Surgeons, others. Subspecialty: Surgery. Current work: Surgical educator and administrator. Gastrointestinal surgery and research. Home: 512 Roslyn Rd Kenilworth IL 60043 Office: Dept Surgery Northwestern U Med Sch 303 E Chicago Ave Chicago IL 60611

NAIR, CHANDRA KUNJU, medical educator; b. Trichur, India, May 20, 1944; came to U.S., 1973; s. Kunju and Narayani (Pillai) N.; m. Nigar Sultana, June 19, 1976; children: Nisha, Reshma, Tanya. B.S., Bombay U., 1964, M.D., 1972; M.B.B.S., Armed Forces-India Med. Coll., 1968. Diplomate: , Am. Bd. Internal Medicine & Cardiology. Resident in internal medicine Bombay U. Hosp., 1969-70, registrar in cardiology, 1970-73; intern in internal medicine Med. Coll. of Ohio Hosp., Toledo, 1973-74, chief resident, 1975; cardiology fellow Creighton U. Hosp., Omaha, 1976-78, asst. prof. medicine, 1978—. Contbr. articles in field to profl. jours. Fellow Am. Coll. Cardiology, ACP, Am. Heart Assn., Am. Coll. Chest Physicians; mem. Am. Fedn. Clin. Research, Midwest Clin, Soc., Cardiovascular Soc. Omaha, Met. Omaha Med. Soc., Am. Soc. Internal Medicine, Am. Inst. Ultrasound in Medicine. Subspecialty: Cardiology. Current work: Coronary heart disease; valvular heart disease. Home: 9929 Devonshire Omaha NE 68132 Office: Creighton Cardiac Ctr 601 N 30th St Omaha NE 68132

NAIR, MADHAVAN (GOPAL NAIR), biochemistry educator, consultant, researcher drug development; b. Mavelikara, India, May 15, 1940; s. Madhavan V. and Ponnamma K. Pillai; m. Indira G. (Lakshmi), Dec. 19, 1970; children: Raj, Venu. Ph.D., Fla. State U., 1969. Asst. prof. chemistry and medicine U. Ala.-Birmingham, 1971-63; assoc. prof. biochemistry Coll. Medicine, U. South Ala., Mobile, 1973-78, prof. biochemistry, 1979—. Contbr. articles to sci. jours. Co-founder nat. exec. com. Nat. Assn. Americans of Asian Indian Descent. Am. Cancer Soc. grantee, 1975-82; Nat. Cancer Inst. grantee, 1978-83. Mem. Am. Soc. Biol. Chemists, Am. Soc. Pharmacology and Exptl. Therapeutics, Am. Chem. Soc., Sigma Xi. Subspecialties: Cancer research (medicine); Synthetic chemistry. Current work: Antitumor drug development, metabolism and pharmacology of anticancer drugs, synthetic organic chemistry and pesticide chemistry. Office: Dept Biochemistry Univ South Ala Mobile AL 36688

NAJARIAN, HAIG HACOP, parasitologist, educator; b. Nashua, N.H., Jan. 5, 1925; s. Hagop M. and Antaram (Shamlyian) N.; m. Mary Ada Been, May 26, 1957; children: Andrea, John, Steven. B.S., U. Mass., 1948; M.A., Boston U., 1949; Ph.D., U. Mich., 1953. Asst. prof. biology Northeastern U., Boston, 1953-55; research parasitologist Parke Davis & Co., Detroit, 1955-58; scientist WHO, Baghdad, Iraq, 1958-59; asst. prof. microbiology U. Tex. Med. Br., Galveston, 1960-66; prof. biology U. So. Maine, Portland, 1966—; cons. WHO, summer 1966; USPHS fellow U. Tex.-Galveston, 1959-60; lectr. Wayne State U., Detroit, 1955-57; asst. prof. biology Boston U., 1959-64. Author: Patterns in Medical Parasitology, 1982, Sex Lives of Animals Without Backbones, 1976; contbr. articles to profl. jours. Served with U.S. Army, 1944-46. Fellow AAAS; mem. Royal Soc. Tropical Medicine and Hygiene, Am. Soc. Tropical Medicine, Am. Soc. Parsitologists, Am. Microscopy Soc., Am. Inst. Biol. Sci., Sigma Xi. Democrat. Subspecialties: Parasitology; Microbiology. Home: 173 Pleasant Ave Portland ME 04103 Office: Univ So Maine 96 Falmouth St Portland ME 04103

NAKAI, GEORGE S., physician, educator; b. Los Angeles, Feb. 1, 1930; s. Kanesaburo and Yoshiko (Iwatake) N.; m. Nadine M. (Fukagawa), Feb. 27, 1959; children: Richard Jay, Kenneth Jon. B.S., U. Utah, 1952, M.D., 1956. Diplomate: Am. Bd. Internal Medicine. Asst. prof. medicine U. N.Mex., Albuquerque, 1967-71; asst. chief hematology, oncology VA Hosp., Long Beach, Calif., 1971-80; asso. adj. prof. dept. medicine U. Calif., Irvine, 1972-80; asst. corp. med. dir. Atlantic Richfield Co., Los Angeles, 1980—; clin. prof. dept. community and environ. medicine U. Calif. Coll. Medicine, Irvine, 1980—. Served with U.S. Army, 1948-49. NIH fellow, 1965-67. Mem. Am. Fedn. Clin. Research, AAAS, Am. Soc. Hematology, ACP, Am. Soc. Clin. Oncology, Am. Assn. Cancer Research, Am. Occupational Med. Assn., Am. Acad. Occupational Medicine., Los Angeles County Med. Assn. Democrat. Unitarian-Universalist. Subspecialties: Internal medicine; Hematology. Current work: Occupational medicine, environmental toxicology, cancer surveillance, preventive medicine, health education. Office: 515 S Flower St Los Angeles CA 90071

NAKAMOTO, TETSUO, dentist, nutritionist; b. Kure, Hiroshima, Japan, Dec. 20, 1939; came to U.S., 1964; s. Takamori and Masae (Nakamuta) N.; m. Lynda G. Ward, May 14, 1980. D.D.S., Nihon U., Tokyo, 1964; M.S., U. Mich., 1966, 71, U. N.D., 1968; Ph.D., MIT, 1978. Diplomate: Japanese Nat. Dental Bd. Research asst. U. N.D., 1968-69; asst. prof. physiology La. State U. Med. Ctr., New Orleans, 1978—. NIH fellow, 1969-72, 72-78; U. N.D. fellow, 1967-68. Mem. Internat. Assn. Dental Research, Soc. Exptl. Biology and Medicine, Am. Inst. Nutrition, Am. Physiol. Soc., N.Y. Acad. Sci., Sigma Xi. Roman Catholic. Subspecialties: Dental growth and development; Nutrition (medicine). Current work: Bone and tooth growth and development in neonates. Infant nutrition and physiology. Office: La State U Med Center 1100 Florida Ave New Orleans LA 70019

NAKAMURA, ICHIRO, immunologist, educator; b. N.Y.C., Jan. 2, 1941; m. Mitsuko Tanaka, Jan. 14, 1969; children: Lisa, Emi. M.S., Kyoto (Japan) U., 1966, Ph.D., 1969. Research fellow Zool. Inst., Kyoto, 1969; fellow Weizmann Inst. Sci., Rehovot, Israel, 1969-71, sr. postdoctoral fellow, 1973-75; Sigrid Juselius Found. fellow U. Helsinki, Finland, 1971-73; research asst. prof. dept. pathology Sch. Medicine, SUNY, Buffalo, 1975-79, asst. prof., 1979—. Mem. Am. Assn. Immunologists, Am. Soc. Zoologists, N.Y. Acad. Scis., AAAS, Transplantation Soc., Entomol. Soc. Am., Reticuloendothelial Soc., Lepidopterists Soc., Societas Europaea Lepidopterologie, Lepidopterological Soc. Japan. Subspecialties: Immunobiology and immunology; Systematics. Current work: Cell-mediated cytotoxicity, hybrid resistance, autoreactivity, natural killer cells. Butterfly systematics, biology, biogeography. Home: 41 Sunrise Blvd Williamsville NY 14221 Office: Dept Pathology SUNY Buffalo NY 14214

NAKAMURA, RICHARD KEN, research psychologist; b. N.Y.C., July 11, 1946; s. James I. and Teksuko (Fujii) N.; m. Sandra Lee K Nakamura, Dec. 1, 1973; children: Daniel, Susan. B.A., Earlham Coll., 1968; M.A., N.Y.U., 1973; Ph.D., SUNY, Stony Brook, 1976. Postdoctoral fellow Lab. Neuropsychology, NIMH, Bethesda, Md., 1976-79, sr. staff fellow Lab. Psychology and Psychopathology, 1980—. Mem. AAAS, Soc. Neurosci. Subspecialties: Neuropsychology; Neurophysiology. Current work: Neural basis of sensory processing, attention, and consciousness in monkeys and man. Home: 7413 Cedar Ave Takoma Park MD 20912 Office: NIMH Bldg 9 Room 1N107 Bethesda MD 20205

NAKAMURA, ROBERT MOTOHARU, pathologist, educator researcher; b. Montebello, Calif., June 10, 1927; s. Mosaburo and Haru N.; m. Shigeyo Jane Hayashi, July 29, 1957; children: Mary, Nancy. M.D., Temple U., 1954. Cert. in pathologic anatomy and clin. pathology. Intern Los Angeles County Gen. Hosp., 1954-55; resident Long Beach (Calif.) VA Hosp., 1955-59; chief of clin. labs. St. Joseph Hosp. and Children's Hosp. of Orange County, Orange, Calif., 1968-69, cons. pathologist, 1969—; chief clin. labs. Orange County Med. Ctr., 1969-74, acting dir. pathology services, 1971-72; assoc. prof. pathology U. Calif., Irvine, 1969-71, vice chmn., 1970-71, acting chmn., 1971-72, prof. pathology, 1971-74, adj. prof., 1974-75; consulting pathologist Long Beach VA Hosp., 1970—; clin. mem. dept. molecular immunology Scripps Clinic and Research Found., La Jolla, Calif., 1974-80, chmn. dept. pathology, 1974—; adj. prof. pathology U. Calif.-San Diego, La Jolla, 1975—; cons. pathologist Univ. Hosp., San Diego, 1976—; mem. dept. molecular immunology Scripps Clinic and Research Found., 1980—; pres. Scripps Clinic Med. Group, Inc., 1981—. Author, editor: Immunopathology: Clinical Laboratory Concepts and Methods, 1974. Fellow Coll. Am. Pathologists, Am. Soc. Clin. Pathologists, Assn. Clin. Scientists; mem. Am. Assn. Blood Banks, Calif. Soc. Pathologists, Internat. Acad. Pathology, AAAS, N.Y. Acad.Sci., Am. Soc. Exptl. Pathology, Am. Assn. Immunologists, Am. Assn. Pathologists and Bacteriologists, Soc. Exptl. Biology and Medicine, Assn. Univ. Pathologists, Am. Soc. Microbiology, Am. Assn. Clin. Chemistry, Internat. Soc. Oncodevelopmental Biology and Medicine, Nat. Acad. Clin. Biochemistry, Alpha Omega Alpha. Subspecialty: Pathology (medicine). Office: Dept Pathology Scripps Clinic and Research Found 10666 N Torrey Pines Rd La Jolla CA 92037

NAKANISHI, KOJI, chemistry educator, bioorganic institute executive; b. Hong Kong, May 11, 1925; U.S., 1969; s. Yuzo and Yoshiko (Sakata) N.; m. Yasuko Abe, Oct. 5, 1947; children: Kay, Jun. B.Sc., Nagoya U., 1947, Ph.D., 1954. Asst. prof. Nagoya U., 1955-58; prof. Tokyo Kyoiku U., 1958-63, Tohoku U., Sendai, Japan, 1963-69; Prof. Columbia U., N.Y.C., 1969-80, Centennial prof., 1980—; dir. research Internat. Ctr. Insect Physiology and Ecology, Nairobi, Kenya, 1969-80; dir. Suntory Inst. Bioorganic Research, Osaka, Japan, 1979—. Author: Infra-red Spectroscopy—Practical, 1962, Circular Dichroic Spectroscopy—Exciton Coupling in Organic Stereochemistry, 1983; author/editor: Natural Products Chemistry, vol. I, 1974, vol. II, 1975, vol. III, 1983; contbr. over 360 articles to profl. jours. Recipient Asahi Press Cultural award, 1968, E.E. Smissman medal U. Kans., 1979, H.C. Urey award Lambda Upsilon, Columbia U., 1980. Mem. Japan Chem. Soc. (Pure Chemistry award 1954, Chemistry award 1979), Am. Chem. Soc. (E. Guenther award 1978, Remsen award 1981), Brit. Chem. Soc. (Centenary medal 1979), AAAS. Subspecialty: Organic chemistry. Current work: Isolation, structure determination and biorganic studies of biologically active naturally occurring compounds (bioactive natural products); bioorganic studies of visual pigments and bacteriorhodopsin; applications of spectroscopy in structure determinations at sub-mg levels. Office: Dept Chemistry Columbia U W 116th and Broadway New York NY 10027 Home: 560 Riverside Dr New York NY 10027

NALCIOGLU, ORHAN, educator; b. Istanbul, Feb. 2, 1944; U.S., 1966, naturalized, 1974; s. Mustafa and Meliha N. B.S., Robert Coll., Istanbul, 1966; M.S., Case Western Res. U., 1968; Ph.D., U. Ore., 1970. Postdoctoral fellow dept. physics U. Calif.-Davis, 1970-71; Research assoc. dept. physics U. Rochester, N.Y., 1971-74; Research assocs. dept. physics U. Wis., Madison, 1974-76; sr. physicist EMI Med. Inc., Northbrook, Ill., 1976-77; assoc. prof. depts. radiol. scis., elec. engring. and medicine U. Calif.-Irvine, 1977—; cons. UN, 1980-81. Contbr. articles to profl. jours. Mobil scholar, 1961-66. Mem. Am. Phys. Soc., Am. Assn. Physicists in Medicine, IEEE. Democrat. Subspecialty: Medical physics. Current work: Digital radiography, nuclear magnetic resonance imaging; research in image science with applications in medical imaging. Office: Dept Radiol Sci Univ Calif Irvine CA 92717

NAM, SANG BOO, physicist; b. Kyung-Nam, Korea, Jan. 30, 1936; came to U.S., 1959, naturalized, 1979; s. Sae Hi and Boon Hi (Kim) N.; m. Wonki Kim, June 1, 1969; children: Saewoo, Jean Ok. B.S., Seoul Nat. U., 1958; M.S., U. Ill.-Urbana, 1961, Ph.D., 1966. Vis. prof. Seoul Nat. U., 1970; asst. prof. U. Va., Charlottesville, 1968-71; vis. assoc. prof. Belfer Grad. Sch. Sci. Yeshiva U., N.Y.C., 1971-74; sr. fellow Nat. Acad. Scis.-NRC, 1974-76; research prof. U. Dayton (Ohio), 1976-80; sr. research physicist U. Research Center, Wright State U., Dayton, 1980—. Contbr. articles to profl. jours. Served with Korean Army, 1958-59. Ministry Edn. Korea nat. fellow, 1955-58; U. Ill. fellow, 1959-60. Fellow Am. Phys. Soc.; mem. AAAS, N.Y. Acad. Scis., Sigma Xi. Subspecialties: Condensed matter physics; Theoretical physics. Current work: Low temperature physics, superconductivity, semiconductor physics, phase transition. Home: 7735 Peters Pike Dayton OH 45414

NAMBU, YOICHIRO, physicist; b. Tokyo, Jan. 18, 1921; U.S., 1952; s. Kichire and Kimiko (Kikuchi) N.; m. Chieko Hida, 1945; children: Jun-ichi, Albert Kenji. B.S., U. Tokyo, 1942, Sc.D., 1952. Asst. U. Tokyo, 1945-49; from asst. prof. to prof. Osaka (Japan) City U., 1949-56; research assoc. Enrico Fermi Inst Nuclear Studies, U. Chgo., 1954-56, from assoc. prof. to prof., 1956-71, Disting. Service prof. physics, 1971—, Harry Pratt Judson Disting. Service prof., 1977—; mem. Inst. Advanced Study, 1952-54. Recipient J. Robert Oppenheimer prize, 1976; Nat. medal of Sci., 1983. Mem. Nat. Acad. Sci., Am. Phys. Soc. (Dannie Heineman prize in math. physics 1970), Am. Acad. Arts and Sci. Subspecialty: Theoretical physics. Current work: Nuclear physics. Office: Dept Physics Enrico Fermi Inst Nuclear Studies U Chgo Chicago IL 60637

NAMIAS, JEROME, meteorologist; b. Bridgeport, Conn., Mar. 19, 1910; s. Joseph and Saydie (Jacobs) N.; m. Edith Paipert, Sept. 15, 1938; 1 dau., Judith Ellen. Student, MIT, 1932-34; M.S., U. Mich., 1941; Sc.D. (hon.), U. R.I., 1972. Research asst. Blue Hill Meteorol. Obs., Milton, Mass., 1933-35; research asso. Mass. Inst. Tech., 1936-41, Woods Hole Oceanographic Inst.; mgr. extended forecast br. U.S. Weather Bur., Washington, 1941-64; asso. dir. Nat. Meteorol. Center, 1964-66, chief extended forecast div., 1966-71; vis. scientist N.Y. U., N.Y.C., 1966; research meteorologist Scripps Instn. Oceanography, La Jolla, Calif., 1968—; vis. scholar Rockefeller Study and Conf. Center, Bellagio, Italy, 1977; frequent cons. USAAF, USN. Author: An Introduction to the Study of Air Mass and Isentropic Analysis, 1936, Extended Forecasting by Mean Circulation Methods; monograph, 1947, Thirty-Day Forecasting, 1953; Short Period Climatic Variations, Collected Works of Jerome Namias, 1934-74, 1975-82, 83; also tech. articles to sci. jours.; Editorial bd.: Geofisica Internacional, Mexico. Recipient Meisinger award Am. Meteorol. Soc., 1938; citation for weather forecasts North African invasion Sec. of Navy, 1942; Dept. Commerce Meritorious Service award, 1950; Rockefeller Pub. Service award, 1955; award for extraordinary meteorol. accomplishment Am. Meteorol. Soc., 1955; Sverdrup Gold medal Am. Meteorol. Soc., 1981; Gold medal for distinguished achievement Dept. Commerce, 1965; Rossby fellow Woods Hole Oceanographic Instn., 1972. Fellow Am. Geophys. Union, Washington Acad. Scis., AAAS, Am. Meteorol. Soc. (councilor 1940-42, 50-53, 60-63, 70-73), Explorers Club; mem. Am. Acad. Arts and Sci., Royal Meteorol. Soc. Great Britain., Nat. Acad. Sci. Subspecialties: Synoptic meteorology; Climatology. Current work: Short period climatic fluctuations from months to years and their prediction. Developer of system for extending time range of gen. weather forecasts up to a season. Home: 240 Coast Blvd 2C La Jolla CA 92037 Office: Sverdrup Hall A-024 Scripps Instn Oceanography La Jolla CA 92093 I was stimulated by a few exceptionally capable high school teachers to pursue a career in meteorology. This subject became both a vocation and a most challenging hobby and never has ceases to fascinate me. My guiding philosophy has been that one is rewarded if he contributes fundamental ideas which are reasonable and at least partially verifiable. It is not necessary to be highly aggressive to get ahead.

NANCE, RICHARD DAMIAN, geologist, educator; b. St. Ives, Cornwall, Eng., Oct. 25, 1951; came to U.S., 1980; s. Richard William Morton and Edith Eleanor (Leach) N.; m. Rita Felice Carpenter, Aug. 28, 1982; 1 son, Andre Bernard Carpenter. B.Sc. with honours, U. Leicester, Eng., 1972; Ph.D., U. Cambridge, Eng., 1978. Asst. prof. geology St. Francis Xavier U., Antigonish, N.S., Can., 1976-80; cons. geologist La. State U., Baton Rouge, 1978-81; assoc. prof. geology Ohio U., Athens, 1980—; sr. research geologist Exxon Prodn. Research Co., Houston, 1982—; cons. Radwaste Program, Dept. Energy, 1978-81. Contbr. articles to profl. jours. Natural Environment Research Council grantee, 1972-76; St. Francis Xavier U. Council for Research grantee, 1977-79; Natural Sci. and Engring. Research Council of Can. grantee, 1980; N.S. Geosci. Research Council grantee, 1981; Ohio U. Research Council grantee, 1981-82. Fellow Geol. Assn. Can.; mem. Geol. Soc. Am., Am. Geophys. Union, Sask. Geol. Soc., Ussher Soc. Eng., Royal Geol. Soc. Cornwall. Subspecialties: Tectonics; Sea floor spreading. Current work: Episodicity in plate tectonics and its control of eustatic sea level; structural style and hydrocarbon traps in petroleum provinces; tectonic evolution of the Avalon zone of the Appalachians in southern New Brunswick. Home: 6555 Harbor Town Dr Apt 917 Houston TX 77036 Office: Exxon Production Research Co ST4112 PO Box 2189 Houston TX 77001

NANDI, SATYABRATA, zoology educator, laboratory administrator; b. North Lakhimpur, Assam, India, Dec. 1, 1931; came to U.S., 1954, naturalized, 1962; s. Kunja Bihari and Jyotirmoyee (Sen-Gupta) N.; m. Jean Brandt-Erichsen, July 5, 1957. B.Sc. in Zoology with honors, U. Calcutta, India, 1949, M.Sc., 1951; Ph.D., U. Calif. Berkeley, 1958. Demonstrator Bethune Coll., Calcutta, 1949-52; research asst. biophysics div. Saha Inst. Physics, U. Calcutta, 1952-54; jr. and asst. research endocrinology Cancer Research Lab., U. Calif., Berkeley, 1958-62, asst. prof. dept. zoology, 1962-64, assoc. prof., 1964-68, prof., 1968—, vice chmn. dept. zoology, 1965-66, chmn., 1971-73, research endocrinologist, 1968—, dir., 1974—; vis. scientist Virus Research Inst. Kyoto U., Kyoto, Japan, 1965. Recipient outstanding achievement award from Indian ambassador, 1979, Assn. Indians in Am. award, 1980; USPHS grantee; Guggenheim fellow, 1967-68. Mem. Am. Assn. Cancer Research, Tissue Culture Soc., Endocrine Soc. Subspecialties: Cell biology; Cell and tissue culture. Home: 2119 Los Angeles Ave Berkeley CA 94707 Office: Cancer Research Laboratory 230 Warren Hal University of California Berkeley CA 94720

NANOS, GEORGE PETER, JR., naval officer, physicist; b. Torrington, Conn., Apr. 11, 1945; s. George Peter and Margaret Elizabeth (Kelleher) N.; m. Joanne Louise Knowles, July 5, 1969; 1 son, George P. B.S. in Engring, U.S. Naval Acad., 1967; M.A., Princeton U., Ph.D. in Physics, 1974. Commd. ensign U.S. Navy, 1967, advanced through grades to comdr., 1981; anti-submarine warfare officer USS Glennon, 1967-69; engr. officer USS Forrest Sherman, 1974-76; material officer Destroyer Squadron 10, 1976-78; mgr. tech. devel. Navy Directed Energy Weapons Program, Washington, 1978-82; combat systems officer Norfolk Naval Shipyard, Portsmouth, Va., 1982—; instr. physics No. Va. Community Coll. Trident scholar, 1966-67; Burke scholar, 1969-73. Mem. Am. Phys. Soc., Am. Assn. Physics Tchrs., Am. Astrophys. Soc., U.S. Naval Inst., Am. Soc. Naval Engrs. Episcopalian. Subspecialties: Cosmology; Laser physics. Current work: Astrophysics, cosmology, laser physics; adaptive optics; shipboard combat systems; design, development, acquisition, repair and installation of naval weapons systems. Home: Norfolk Naval Shipyard Quarters L Portsmouth VA 23709 Office: Norfolk Naval Shipyard Code 190 Portsmouth VA 23709

NARAHASHI, TOSHIO, pharmacologist, educator; b. Fukuoka, Japan, Jan. 30, 1927; came to U.S., 1961; s. Asahachi Ishii and Itoko Yamasaki; m. Kyoko, Apr. 20, 1956; children: Keiko, Taro. B.S. in Vet. Medicine (equivalent D.V.M.), U. Tokyo, 1948, Ph.D. in Neurotoxicology (equivalent D.Sc.), 1960. Instr. lab. applied entomology U. Tokyo, 1951-60, 63-65; research asso. dept. physiology U. Chgo., 1961-62; asst. prof. dept. physiology and pharmacology Duke U. Med. Center, Durham, N.C., 1962-63, 65-67, asso. prof., 1967-69, prof., 1969-77, head pharmacology div., 1970-73, vice-chmn. dept., 1973-75; prof., chmn. dept. pharmacology Northwestern U. Med. and Dental Schs., Chgo., 1977—. Editor 4 books; specific field editor: Jour. Pharmacology and Exptl. Therapeutics; mem. editorial bd. 4 profl. jours.; contbr. 218 sci. research articles to profl. jours. Recipient Japanese Soc. Applied Entomology and Zoology award, 1955; various NIH research and tng. grants. Mem. Am. Soc. Pharmacology and Exptl. Therapeutics, Am. Physiology Soc., Soc. Neurosci., Biophys. Soc. (Cole award 1981), Soc. Gen. Physiologists, Internat. Soc. Toxinology, Soc. Toxicology, Entomol. Soc. Am., Am. Chem. Soc. (div. pesticide chemistry), N.Y. Acad. Scis., Sigma Xi. Subspecialties: Neuropharmacology; Neurophysiology. Current work: Mechanism of action of toxins, therapeutic drugs and insecticides on excitable membranes, as studied by electrophysiological methods. Home: 2130 Swainwood Dr Glenview IL 60025 Office: Dept Pharmacology Northwestern U Med Sch 303 E Chicago Ave Chicago IL 60611

NARANJO, JENNINGS NEAL, marketing, management and research consultant; b. Lufkin, Tex., July 1, 1948; s. J. Neal and Stella Frances (Jennings) N.; m. Mary Ann Platz, Aug. 18, 1979. B.A., U. Tex., 1971; M.A., U. So. Calif., 1977, Ph.D., 1978. Mem. faculty psychology dept. Bklyn. Coll., 1971-74; mem. faculty dept. mgmt. U. So. Calif., Los Angeles, 1975-78, dept. mktg., 1978; asst. neuroanatomist McLean's Hosp., Belmont, Mass., 1979-81; fellow dept. anatomy Harvard Med. Sch., 1979, instr., 1980-81; teaching fellow dept. neurology Boston U. Sch. Medicine, 1980-81; dir. mktg. and mgmt., cons. Narjen Internat., Lufkin, Tex., 1981—. Author: Organizational Behavior, 1977; contbr. articles on neurosci. and physiol. psychology to various publs. Mem. sch. bd. St. Cyprian's Episcopal Sch., Lufkin, 1983—. Mem. Am. Psychol. Assn., Am. Mktg. Assn., SAR, Phi Beta Kappa, Sigma Xi, Psi Chi, Alpha Kappa Psi, Omicron Delta Pi. Clubs: Harvard (N.Y.C.); Jonathan, Magic Castle (Los Angeles); Crown Colony Country. Lodges: Rotary (Lufkin); Masons; Shriners. Subspecialty: Neuropsychology. Current work: Neuropsychology, marketing, stress, and aging. Office: Narjen International PO Box 1745 Lufkin TX 75901

NARASIMHACHARI, NEDATHUR, pharmacology educator; b. Nellore, Andhra Pradesh, India, July 14, 1944; s. Gopalachari and Ranganayaki N.; m. Indira Narasimhachari; children: Raghavan, Gita, Asha. BSc., Andhra U., Waltair, India, 1946, M.Sc., 1947; Ph.D., Delhi (India) U., 1950. Lectr. Delhi U., 1949-53; chief organic chemist Antibiotics Research Ctr., Pimpri, Poona, India, 1953-68; Nat. Research Council Can. postdoctoral fellow, Saskatoon, Sask., 1960-62; research scientist Ill. State Dept. Mental Health, Galesburg and Chgo., 1968-78; prof. dept. psychiatry and pharmacology, dir. mass spectrometry facility Med. Coll. Va., Richmond, 1978—. Contbr. articles to profl. jours. Mem. Am. Chem. Soc. (biol. chemistry sect.), Am. Soc. Neurochemistry, Am. Soc. Neurosci., AAAS, Am. Soc. Mass

Spectrometry, Soc. Biol. Psychiatry, Soc. Pharmacology and Exptl. Therapeutics, Am. Soc. Biol. Chemistry. Subspecialties: Neurochemistry; Neuropharmacology. Current work: Studies on drug metabolism with special reference to antidepression and antipsychotic drugs, biochemical and biomedical applications of mass spectrometry; biochemical studies in schizophrenia, depression and early infantile autism. Home: 3160 Mounthill Dr Midlothian VA 23113 Office: Med Coll Va Box 710 MCV Station Richmond VA 23298

NARAYANA, ANAND DEO, mechanical engineer; b. Chapra, India, June 23, 1948; s. Brajendra Deo and Sharda (Bihari) N.; m. Pratima Narayana; children: Manu, Himanshu, Alok. B.S., Patna U., 1968; M.S., U. Cin., 1977, Ph.D., 1979. Lectr. mech. engring. Patna U., India, 1969-71; asst. exec. engr. Ministry Def. India, 1971-75; research asst. U. Cin., 1976-79; sr. mech. component design engr. Garett Turbine Engine Co., Phoenix, 1979—. Indian Govt. Nat. scholar, 1963-68. Subspecialties: Mechanical engineering; Solid mechanics. Current work: Design gas turbine disks and blades; solid mechanics including crack propagation and creep behavior in metals and alloys. Home: 2103 W El Prado Rd Chandler AZ 85224 Office: 111 S 34th St Phoenix AZ 85010

NARAYANAN, C. S., radiation med. physicist, cons.; b. India, Sept. 29, 1944; s. C.R.S. Iyer and T.R. Janaki; m. Kamala, July 12, 1946; children: Anita, Anand. B.S. in Physics, Jain Coll, Madras, India, 1963, Madras Inst. Tech., 1968; M.E. in Radiation Physics, U. Va., 1976; M.S. in Bus. Adminstrn, St. Francis. Coll., Ft. Wayne, Ind., 1982. Sales engr. RCA, India, 1968-71; application engr. Capintec, N.Y., 1972-75; radiation physicist Highland Hosp., Rochester, N.Y., 1976-77, Luth. Hosp., Ft. Wayne, 1977—; mem. State Bd. Health, Radiation Emergency Response Com., Ind.; cons. Local hosps. and clinics. Contbr. sci. paper to profl. confs. Mem. Am. Assn. Physicists in Medicine. Subspecialties: Radiology; Medical radiation physics. Home: 5926 Vance Ave Fort Wayne IN 46815 Office: Lutheran Hosp 3024 Fairfield Ave Fort Wayne IN 46807

NARDELLA, FRANCIS ANTHONY, physician, medical researcher, educator, consultant; b. Clarksburg, W.Va., Aug. 5, 1942; s. Christopher and Rose Mary (Palagiano) N.; m. Mary Ann Gintner, Aug. 8, 1970; children: Anne Frances, Christopher William. A.B., W.Va. U.-Morgantown, 1964, M.D., 1968. Diplomate: Am. Bd. Internal Medicine. Rotating intern Hennepin County Gen. Hosp., Mpls., 1968-69; resident in medicine U. Wash., Seattle, 1972-74, fellow in rheumatology, 1974-77, instr. medicine, 1977-79, asst. prof. medicine, 1979—; attending physician U. Wash. U. Hosp., Seattle, 1977—, Harborview Med. Ctr., 1977—. Served to lt. commdr. USNR, 1969-72. Mem. Am. Rheumatism Assn., Am. Assn. Immunologists, Am. Fedn. Clin. Research. Democrat. Roman Catholic. Subspecialties: Immunology (medicine); Internal medicine. Current work: Role of rheumatoid factors in rheumatoid arthritis, initiation of rheumatoid factor synthesis. Office: Dept Medicine RG-20 U Wash Seattle WA 98195

NARDUCCI, LORENZO M., physicist, educator; b. Italy, May 25, 1942; s. Francesco and Franca (Barsanti) N.; m. Rinamaria, Nov., 1965; children: Francesco, Robert, Laura. Ph.D., U. Milan, Italy, 1964. Asst. prof. U. Milan, 1964-66; asst. prof. Worcester (Mass.) Poly. Inst., 1966-71, assoc. prof., 1971-76; assoc. prof. physics Drexel U., Phila., 1976-79, prof., 1979—, dir. quantum optics group. Contbr. articles to profl. jours. Fellow Optical Soc. Am.; mem. Am. Phys. Soc., Sigma Xi. Republican. Subspecialties: Atomic and molecular physics; Theoretical physics. Current work: Quantum optics, lasers, synergetics, nonlinear spectroscopy. Office: Dept Physics Drexel U Philadelphia PA 19104

NASH, DONALD ROBERT, immunologist, educator; b. Pittsfield, Mass., Nov. 15, 1938; s. Joseph and Bernadette (Valley) N.; m. Mary Campbell, June 23, 1963; 1 son, Brendon. B.A., Am. Internat. Coll., 1961; M.S., Boston Coll., 1963; Ph.D., U. N.C., 1967. Postdoctoral fellow Cath. U. Louvain, Belgium, 1968-69; asst. prof. U. Hawaii, Honolulu, 1969-70; sr. research fellow WHO, Lausanne, Switzerland, 1970-72; research assoc. prof. U. Tex. Health Ctr., Tyler, 1972—; cons. M.D. Anderson Hosp. and Tumor, Inst., Houston, 1973-76; assoc. prof. U. Tex., Tyler, 1979—. Belgian-Am. Found. fellow, 1964, 69; recipient Service Recognition award AMA, 1972. Mem. Am. Assn. Immunologists, Am. Thoracic Soc., AAAS, Internat. Assn. Study Lung Cancer, N.Y. Acad. Scis. Subspecialties: Immunobiology and immunology; Infectious diseases. Current work: Infectious diseases, pulmonary immunology, immunopharmacology. Office: U Tex Health Ctr PO Box 2003 Tyler TX 75710

NASH, DONALD ROBERT, IMMUNOLOGIST/ MICROBIOLOGIST, researcher; b. Pittsfield, Mass., Nov. 15, 1938; s. Joseph and Bernadette (Vallee) N; m. Mary Campbell, Aug. 29, 1936; 1 son, Brendon. B.A., Am. Internat. Coll., Springfield, Mass., 1961; M.S., Boston Coll., 1963; Ph.D., U. N.C., Chapel Hill, 1967; postgrad., U. Louvain, Belgium, 1968-69. Asso. prof. U. Hawaii, Honolulu, 1969-70; WHO fellow, Lausanne, Switzerland, 1970-72; research asso. prof. U. Tex. Health Center, Tyler, 1972—. Recipient Belgian Am. Ednl. Found. award, 1968-69; AMA Service Recognition award, 1970. Mem. Am. Assn. Immunologists, Am. Thoracic Soc., AAAS, Internat. Assn. Study of Lung Cancer, N.Y. Acad. Sci. Subspecialties: Immunology (medicine); Microbiology (medicine). Current work: Lung cancer; immunochemistry; toxicology; infectious diseases. Office: Dept Immunology U Tex Box 2003 Tyler TX 75710

NASH, JOHN CHRISTOPHER, information services firm executive, educator; b. Tunbridge Wells, Kent, Eng., Sept. 4, 1947; emigrated to Can., 1957; s. Harry and Margaret (Moss) N.; m. Mary Margaret Frohn, June 28, 1970. B.Sc. with honors, U. Calgary, Alta., Can., 1968; D.Phil., U. Oxford, Eng., 1972. Economist, statistician Agr. Can., Ottawa, 1973-80; sr. cons. Hickling Ptnrs., Ottawa, 1980-81; pres. Nash Info. Services, Ottawa, 1980—; assoc. prof. U. Ottawa, 1980—. Author: Compact Numerical Methods for Computers, 1979; mag. columnist: Micromathematician (for Interface Age), 1980. Treas. Amnesty Internat. Can., 1978-81. Mem. Assn. Computing Machinery, Am. Statis. Assn., Soc. Indsl. and Applied Math. Subspecialties: Mathematical software; Applied mathematics. Current work: Computational methods for small computers. Home: 1975 Bel Air Dr Ottawa ON Canada K2C OX1 Office: Faculty Adminstrn U Ottawa Ottawa ON Canada K1N 9B5

NASH, JONATHON MICHAEL, engineering administrator, researcher; b. Little Rock, Aug. 10, 1942; s. Bertram B. and Nora B. (Shed) N.; m. Meta W. Smith, Aug. 12, 1972; children: Lillian Kendrick, Caroline Michael. B.S. in Mech. Engring, U. Miss., 1966, M.S. in Engring. Sci, 1970, Ph.D., 1973. Registered profl. engr., Ala., Md., Miss. Jr. engr. IBM Fed. Systems Div., Huntsville, Ala., 1967-68, systems engr., mgr., 1973-78, advisory engr., Gaithersburg, Md., 1978-81, tech. planning mgr., Gaithersburg, 1981—; univ. fellow U. Miss., 1970-73; instr. U. Ala., Huntsville, 1977-78; adj. assoc. prof. mech. engring. dept. U. Miss., 1982—; research engring. res. officer U.S. Army Mobility Equipment Research and Devel. Command, Ft. Belvoir, Va., 1971—. Co--editor: Modeling, Simulation, Testing and Measurements for Solar Energy Systems, 1978; author book sect., invention disclosures. Vice pres. Arts Council of Frederick City and County, Md., 1982, dir., 1981-82; exec. com. Lafayette County Miss. Republicans, 1972-73. Served to 1st lt. U.S. Army, 1968-70. Recipient New Tech. award NASA, 1979, Apollo Achievement award, 1971. Assoc. fellow AIAA; mem. ASME (exec. com. solar energy div. 1981—, cert. appreciation 1982), Soc. Am. Mil. Engrs. (chpt. pres. 1976-77, engring. achievement award 1977, Tudor Medal 1978), Nat. Soc. Profl. Engrs. (chpt. dir. 1978, Ala. Young Engr. of Year 1978), Sigma Xi (chpt. pres. 1977-78), Alpha Tau Omega (pres. Huntsville alumni 1976-77). Subspecialties: Systems engineering; Solar energy. Current work: Systems engineering technology and methodology as applied to data processing systems and energy conversion processes. Home: 300 Rockwell Terr Frederick MD 21701 Office: IBM Corp Fed Systems Div 18100 Frederick Pike Gaithersburg MD 20879

NATELSON, SAMUEL, clinical chemist; b. Bklyn., Feb. 28, 1909; s. Max and Betty Ann N.; m. Ethel D. Nathan, Apr. 4, 1937; children: Stephen, Ethan, Elissa, Nina. B.S., CCNY, 1928; Sc.M., N.Y. U., 1930, Ph.D., 1931. Diplomate: Am. Bd. Clin. Chemistry. Teaching fellow N.Y. U., 1928-31; research chemist N.Y. Testing Labs., 1931-31, Jewish Hosp. of Bklyn., 1931-49; lectr. biochemistry and organic analysis Grad. Sch., Bklyn. Coll., 1939-49; head dept. biochemistry Rockford (Ill.) Meml. Hosp., 1949-58, St. Vincents Hosp., N.Y.C., 1958-59, Roosevelt Hosp., 1959-65; Chmn. dept. biochemistry Michael Reese Hosp., Chgo., 1966-79; adj. prof. Vet. Sch. Medicine, U. Tenn., Knoxville, 1980—. Author textbooks in field.; Contbr. articles to profl. jours. Recipient Chgo. Sect. Sr. award Ill. State Soc. Bioanalysts Sci., 1971. Fellow Nat. Acad. Clin. Biochemistry, Am. Inst. Chemists; mem. Am. Assn. Clin. Chemistry (Ames award 1965, Van Slyke award 1961), Am. Chem. Soc. (50 Yr. mem.), Am. Microchem. Soc., Am. Soc. Applied Spectroscopy, AAAS, Sigma Xi. Subspecialties: Clinical chemistry; Biochemistry (biology). Current work: Epilepsy, intermediate nitrogen metabolism; research in guanidino compounds and their relationship to epilepsy. Patentee in field. Home: 925 Southgate Rd Knoxville TN 37919 Office: Coll Vet Medicine U Tenn Knoxville TN 37901

NATH, RAVINDER, radiological physicist; b. Jullunda, India, Apr. 9, 1942; s. Kedar Nath and Rajrani (Katyal) N.; m. Rashmi Duggal, Nov. 20, 1971; children: Anjali, Sameer. B.Sc., Delhi U., 1963, M.Sc., 1965; Ph.D., Yale U., 1971. Cert. Am. Coll. Radiology. Asst. lectr. Delhi Coll, 1965-67; research staff physicist Yale U., 1971-73, research asso., 1973-75, asst. prof., 1975-79; asso. prof., 1979—. Contbr. articles to profl. jours. Recipient Med. Physics award Am. Assn. Physicist in Medicine, 1975. Mem. Am. Assn. Physicists in Medicine, Am. Phys. Soc., Radiation Research Soc., Health Physics Soc., Am. Soc. Therapeutice Radiology. Subspecialty: Radiological Physics. Current work: Radiol. physics related to radiation therapy of cancer. Office: Yale Med Sch 333 Cedar St New Haven CT 06510

NATHAN, MARSHALL I., physicist; b. Lakewood, N.J., Jan. 22, 1933; s. Benjamin and Ruth (Blumenthal) N.; m. Nancy Jennens, Aug. 1955; children: Eric, Barbara; m. Rosalie Spofford, Apr. 10, 1971; 1 stepson, Joseph Spofford. B.S. in Physics, MIT, 1954; Ph.D. in Applied Physics, Harvard U., 1958. With IBM Corp., 1958—; mem. research staff Research Ctr., Yorktown Heights, N.Y., 1960—. Fellow IEEE (David Sarnoff award 1980), Am. Phys. Soc. Subspecialty: Condensed matter physics. Current work: Semiconductor device physics; high speed device physics and measurements. Home: 106 Stephen Dr Tarrytown NY 10591 Office: IBM Corp Research Center Box 218 Yorktown Heights NY 10598

NATHANS, DANIEL, biologist; b. Wilmington, Del., Oct. 30, 1928; s. Samuel and Sarah (Levitan) N.; m. Joanne E. Gomberg, Mar. 4, 1956; children—Eli, Jeremy, Benjamin. B.S., U. Del., 1950; M.D., Washington U., 1954. Intern Presbyn. Hosp., N.Y.C., 1954-55, resident in medicine, 1957-59; clin. asso. Nat. Cancer Inst., 1955-57; guest investigator Rockefeller U., N.Y.C., 1959-62; prof. microbiology Sch. Medicine, Johns Hopkins, 1962-72, prof., dir. dept. microbiology, 1972—. Recipient Nobel prize in physiology or medicine, 1978. Fellow Am. Acad. Arts and Scis.; mem. Nat. Acad. Scis. Subspecialty: Microbiology. Office: Dept Microbiology Sch Johns Hopkins U 725 N Wolfe St Baltimore MD 21205

NATHANSON, LARRY, physician, educator; b. Boston, Dec. 23, 1928; s. Robert B. and Leah (Rabin) N.; m. Anna, May 17, 1963; children: Andrew, Aran, Nicholas. A.B., Harvard U., 1950; M.D., U. Chgo., 1955. Intern Los Angeles County Hosp., 1955-56; resident Stanford U. Hosp., 1958-60, chief resident, 1962-63; asst. in medicine Harvard U., 1965-66; jr. assoc. medicine Peter Bent Brigham Hosp., 1966-68; asst. physician New Eng. Med. Center Hosp., Boston, 1967-69, chief med. oncology services, 1969-79; prof. Tufts U, 1975-80; dir. oncology hematology div. Nassau Hosp., Mineola, N.Y., 1980—; prof. Sch. Medicine SUNY-Stony Brook, 1980—. Contbr. 100 articles in field to profl. jours. Served to capt. U.S. Army, 1957-59. Nat. Cancer Inst fellow, 1960-62; Straus Meml. lectr., 1973. Fellow ACP; mem. Sigma Xi. Clubs: Harvard of N.Y.C., Seawanhaka Corinthian Yacht. Subspecialties: Oncology; Chemotherapy. Current work: Biology of malignant melanoma, experimental therapeutics of cancer. Office: Nassau Hosp 259 1st St Mineola NY 11501

NATHANSON, MORTON, neurologist, educator; b. N.Y.C., May 31, 1918; s. Nathan and Celia N.; m. Margret Regina Maier, Dec. 16, 1948; children: David, Madlyn, Laura. A.B., U. Mich., 1939; M.D., La. State U., New Orleans, 1943. Diplomate: Am. Bd. Neurology. Resident N.Y.U.-Bellevue, 1946-49, dir. Multiple Sclerosis Research program, 1950-54, instr. neurology, 1950-54, asst. prof., 1954-60, asso. prof., 1960-67; clin. prof. neurology Mt. Sinai Sch. Medicine, N.Y.C., 1967-72, profl. lectr., 1973—; prof. neurology Sch. Medicine SUNY-Stony Brook, 1972—; chief neurology Long Island Jewish-Hillside Med. Center, New Hyde Park, N.Y., 1972—; doctoral faculty CUNY Grad. Sch.; resident in neuropathology Columbia Coll. Physicians and Surgeons, 1947. Contbr. chpts. to books and 55 articles in field to profl. jours. Served with M.C. U.S. Army, 1944-46. Recipient NIH contract award, 1978—. Fellow Am. Acad. Neurology; mem. Am. Neurol. Assn., Assn. Univ. Profs. Neurology, Am. Fedn. Clin. Research, Am. Assn. History Medicine, Soc. Neuroscience, N.Y. Acad. Sci., N.Y. Acad. Medicine. Subspecialties: Neurology; Neurophysiology. Current work: Altered states of consciousness, brain stem function, perception (visual and somatic-sensory), myoclonus of palate and related structures. Home: 1 Pebble Ln Roslyn Heights NY 11577 Office: Division of Neurology Long Island Jewish Hillside Medical Center New Hyde Park NY 11042

NATHENSON, STANLEY GAIL, research scientist, educator; b. Denver, Aug. 1, 1933; s. Abe and Esther (Kurl) N.; m. Susan Lawrence, Oct. 16, 1959; children: Matthew, John. M.D., Washington U., St. Louis, 1959. Postdoctoral fellow USPHS, Washington, 1960-62; Helen Hay Whitney fellow Blond Labs., East Grinstead, Sussex, Eng., 1964-67; asst. prof. dept. microbiology and immunology Einstein Coll. Medicine, Bronx, N.Y., 1967-69, assoc. prof., 1969-73, prof., 1973—. Contbr. articles to sci. jours. Served to asst. surgeon USPHS, 1962-64. USPHS grantee, 1966—. Mem. Am. Assn. Immunologists, AAAS, Phi Beta Kappa. Subspecialties: Immunogenetics; Transplantation. Current work: Research on molecular genetics and structural analysis of the major histocompatibility complex (MHC).

NAUMAN, ROBERT VINCENT, chemistry educator; b. East Stroudsburg, Pa., Dec. 6, 1923; s. Carl Arnold and Bernice Irene (Zacharias) N.; m. Jean Marie Hodgeson, Aug. 29, 1955; children: Andrea Carol, Marcus Alan, Stephen Brian, Suzanne Marie. B.S., Duke U., 1944; Ph.D., U. Calif.-Berkeley, 1947. Research assoc. Cornell U., 1947-52; asst. prof. chemistry U. Ark., 1952-53; asst. prof. La. State U., 1953-56, assoc. prof., 1956-63, prof., 1963—, dir. chemistry grad. studies, 1981—. Contbr. articles in field to profl. jours. Served with USMCR, 1943-44. Fulbright grantee, 1966-67; recipient Standard Oil Ind. Outstanding Tchr. award, 1969. Mem. Am. Chem. Soc., Am. Phys. Soc., AAAS, Interamerican Photochemical Soc., Sigma Xi. Republican. Methodist. Subspecialties: Physical chemistry; Photochemistry. Current work: Spectroscopy; photochemistry; photophysics; effect of geometry and conformation on electronic transitions. Office: Dept Chemistry La State Baton Rouge LA 70803

NAVANGUL, HIMANSHOO VISHNU, Chemistry educator; b. Wai, Maharashtra, Inda., Dec. 22, 1941; s. Vishnu Narayan and Usha (Gokhale) N.; m. Neelarani B. Tilak, Sept. 18, 1964; children: Sangeeta, Bharat. B.Sc., U. Poona, India, 1961, 1962, M.Sc., 1963, Ph.D., 1967. Instr., vis. asst. prof. U. Mo., Kansas City, 1971-75; vis. asst. prof. N.E. Mo. State U., 1975-76; asso. prof. Al Fateh U., Tripoli, Libya, 1976-79; vis. faculty mem. Clemson (S.C.) U., 1979-80; asso. prof., chmn. dept. chemistry N.C. Weskeyan Coll., Rocky Mount, 1980—. Contbr. articles to profl. jours. Swiss Govt. scholar, 1967-69. Mem. Am. Chem. Soc., Am. Inst. Chemists, Sigma Xi. Hindu. Subspecialties: Physical chemistry; Use of computers in education.

NAYAK, DEBI PROSAD, virologist, educator, researcher; b. Eadpore, West Bengal, India, Apr. 1, 1937; came to U.S., 1961, naturalized, 1978; s. Sarat Chandra and Durga Rani (Mandal) N.; m. Abantika, June 18, 1965; children: Prasun, Dipak. B.V.Sc., U. Calcutta, India, 1957; M.S., U. Nebr., 1962, Ph.D., 1964. Instr. Bengal Vet. Coll., Calcutta, India, 1958-61; grad. research asst. dept. vet. medicine U. Nebr., 1961-64; acting asst. prof. UCLA, 1964-66, asst. researcher in virology, 1966-68, asst. prof. virology, 1968-71, assoc. prof., 1971-77, prof., 1977—. Am. Cancer Soc. Calif. Div. sr. Dernahm fellow, 1969-74; Grantee Am. Cancer Soc., 1970-74, Nat. Cancer Soc., 1972-77, Nat. Inst. Allergy and Infectious Disease, NIH, 1975—, NSF, 1975—. Mem. Am. Soc. Microbiology, AAAS, Am. Soc. Virology. Subspecialties: Virology (biology); Molecular biology. Current work: Influenza viruses; defective interfering viruses; nature of viral genes; expression of viral genes; recombinant DNA technology; gene expression; development of viral vaccine using recombinant DNA technology; genesis of defective interfering viruses. Home: 1918 Granville Ave Los Angeles CA 90025

NEARY, JOSEPH THOMAS, biochemist, researcher; b. Carbondale, Pa., Oct. 14, 1943; s. Joseph Francis and Mary Cecelia (McDonough) N.; m. Judith A. Clark, May 6, 1967; children: Robert, Suzanne. B.S., U. Scranton, 1965; Ph.D., U. Pitts., 1969. Postdoctoral fellow U. Ill, Urbana, 1969-71; assoc. in biochemistry and medicine Mass. Gen. Hosp. and Harvard U. Med. Sch., Boston, 1971-78; investigator Marine Biol. Lab., Woods Hole, Mass., 1978—. Contbr. articles to sci. jours. NIH fellow, 1969-71; NSF scholar, 1964, 65. Mem. Am. Chem. Soc., AAAS, Soc. Neurosci. Democrat. Roman Catholic. Subspecialties: Biochemistry (medicine); Neurobiology. Current work: Biochemistry of behavior and learning; excitable membranes and ion channels, neuropeptides; protein phosphorylation. Office: Marine Biol Lab Woods Hole MA 02543

NEBERT, DANIEL WALTER, molecular geneticist, pharmacologist, pediatrician; b. Portland, Oreg., Sept. 26, 1940; s. Walter and Marie (Schick) N.; m. Myrna E. Sisk, Mar. 12, 1960; children: Douglas Daniel, Dietrich Andrew; m. Kathleen Dixon, Aug. 15, 1981. B.A., Conn. Wesleyan U., 1961; M.S., U. Oreg., 1964; M.D., U. Oreg. Sch., 1964. Lic. physician, Calif. Pediatric intern and resident UCLA Hosps., 1964-66; postdoctoral fellow Nat. Cancer Inst., Bethesda, Md., 1966-68; sr. investigator Nat. Inst. Child Health and Human Devel., Bethesda, 1968-71, sect. head, 1971-74, acting chief neonatal and pediatric medicine br., 1974-75, chief lab. developmental pharmacology, 1975—; adj. prof. Uniformed Services U. Health Scis., Bethesda, 1980—; also lectr. Contbr. over 260 articles to sci. jours.; editorial bd.: Molecular Pharmacology, 1972—, Biochem. Pharmacology, 1972-83, Archives of Biochemistry and Biophysics, 1973-76, Archives Internationales de Pharmacodynamie et de Therapie, 1975-81, Jour. Environ. Scis. and Health, 1976-81, Chemico-Biol. Interactions, 1977—, Developmental Pharmacology and Therapeutics, 1979—, Teratogenesis, Carcinogenesis, and Mutagenesis, 1980-83, Anticancer Research, 1981-83. Served as med. officer USPHS, 1966—. Ann. Pfizer lectr., 1978, 79; recipent 1st prize Future Scientists Am., 1956; others. Mem. AAAS, Am. Soc. Pharmacology and Exptl. Therapeutics, Am. Soc. Biol. Chemists, Am. Soc. Clin. Investigation, Am. Soc. Pediatric Research, Sigma Xi. Republican. Unitarian. Subspecialties: Gene actions; Genetics and genetic engineering (medicine). Current work: Molecular genetics; pharmacology; enzyme regulation; chemical carcinogenesis; mutagenesis; teratogenesis; drug toxicity; environmental pollution; pediatric genetics.

NECHAMKIN, HOWARD, chemistry educator; b. Bklyn., Aug. 18, 1918; s. Charles J. and Celia J. (Wiener) N.; m. Donna Rae Polensky, July 29, 1956 (div. June 1979); 1 dau., Emily Jean; m. Murielle Wolff, Dec. 22, 1979; 1 son, David A. Katz. B.A., Bklyn. Coll., 1939; M.S., Poly. Inst. Oklyn., 1949; Ed.D., N.Y.U., 1961. Analytical chemist R. H. Macy & Co., N.Y.C., 1939-42; research chemist Hazeltine Electronics, Little Neck, N.Y., 1942-45; assoc. prof. chemistry Pratt Inst., Bklyn., 1945-61; prof. chemistry Trenton (N.J.) State Coll., 1961—. Author: Organic Chemistry—Theory and Problems, 1978, others. Fellow Am. Inst. Chemists.; Mem. Phi Kappa Phi. Lodge: K.P. Subspecialties: Inorganic chemistry; Organic chemistry. Current work: Preparative procedures for reclamation of waste materials, consumer chemistry, reaction kinetics. Home: 325 Glenn Ave Lawrenceville 08648

NECHAY, BOHDAN ROMAN, pharmacologist, toxicologist; b. Prague, Czechoslovakia, Nov. 26, 1925; came to U.S., 1949, naturalized, 1955; s. Simon M. and Maria (Malewicz) N.; m. Brigitta Singhild Ahlm, July 14, 1961; children: Peter, Nicholas. Student, U. Innsbruck, Austria, 1945; student, Sch. Vet. Medicine, Munich, W. Ger., 1946-47, Cand. Med. Vet., Sch. Vet. Medicine, Giessen, W. Ger., 1947-49; D.V.M., U. Minn., 1953. Poultry pathologist Hilltop Labs., Mpls., 1951-53; practice vet. medicine, Mpls., 1954-56; asst. prof. dept. pharmacology and therapeutics Coll. Medicine, U. Fla., 1961-66; asst. prof. div. pharmacology and urology Duke U. Sch. Medicine, 1966-68; assoc. prof. dept. pharmacology and toxicology U. Tex. Med. Br., Galveston, 1968-78, prof., 1978—; vis. scientist Inst. Med. Pharmacology Biomed. Ctr., Uppsala (Sweden) U., 1972-73. Contbr. articles to profl. jours. NIH fellow, 1960-61; Am. Heart Assn. fellow, 1959-60. Mem. Am. Soc. Pharmacology and Exptl. Therapeutics, Am. Soc. Nephrology, Soc. Toxicology, Southwestern Assn. Toxicologists, So. Salt Water and Kidney Club. Subspecialties: Cellular pharmacology; Toxicology (medicine). Current work: Renal pharmacology; pharmacology and toxicology of kidneys; toxic metals; environmental and toxic factors in cardiovascular disease. Patentee in field. Home: 29 Dansby Dr Galveston TX 77551 Office: Dept Pharmacology and Toxicology U Tex Med Br Galveston TX 77550

NEEDLEMAN, ALAN, mech. engr., educator; b. Phila., Sept. 2, 1944; s. Herman and Hannah (Goodman) N.; m. Wanda Sapolsky, Apr. 12, 1970; children: Deborah, Daniel. B.S., U. Pa., 1966; M.S., Harvard U., 1967, Ph.D., 1971. Instr. applied math. M.I.T., 1970-72, asst. prof., 1972-75; asst. prof. engring. Brown U., Providence, 1975-78, assoc. prof., 1978-81, prof., 1981—. Guggenheim fellow, 1977. Mem. ASME, Am. Acad. Mechanics. Subspecialties: Solid mechanics; Fracture mechanics. Current work: Plasticity of solids, ductile fracture mechanics, buckling of structures, finite element methods. Office: Div Engring Brown U Providence RI 02912

NEEFE, JOHN ROBERT, JR., medical oncologist, tumor immunologist, medical educator; b. Phila., July 17, 1943; s. John Robert and Harriet May (Langeluttig) N.; m. Dana Lynne Iverson, June 29, 1968; children: Heather Lynne, Lauren Aimee. B.A. magna cum laude, Harvard U., 1965; M.D., U. Pa.-Phila., 1969. Diplomate: Am. Bd. Internal Medicine. Intern Balt. City Hosps., 1969-70, resident, 1970-71; investigator Nat. Cancer Inst., Bethesda, Md., 1976-78; asst. prof. medicine Georgetown U., Washington, 1978-82, assoc. prof., 1982—. Served with USPHS, 1972-78. Mem. ACP,., Am. Assn. Immunologists, Transplantation Soc., Am. Soc. Clin. Oncology, Am. Assn. Cancer Research. Republican. Subspecialties: Cancer research (medicine); Cellular engineering. Current work: Cancer therapy with biological response modifiers such as interferon, monoclonal antibodies, and anti-idiotypic antibodies. Office: Georgetown U 3800 Reservoir Rd NW Washington DC 20007

NEEL, JAMES VAN GUNDIA, geneticist, educator; b. Hamilton, Ohio, Mar. 22, 1915; s. Hiram Alexander and Elizabeth (Van Gundia) N.; m. Priscilla Baxter, May 6, 1943; children—Frances, James Van Gundia, Alexander Baxter. A.B., Coll. Wooster, 1935, D.Sc. (hon.), 1959; Ph.D., U. Rochester, 1939, M.D., 1944, D.Sc. (hon.), 1974, Med. Coll. Ohio, 1981. Instr. zoology Dartmouth, 1939-41; fellow zoology NRC, 1941-42; intern, asst. resident medicine Strong Meml. Hosp., 1944-46; asso. geneticist lab. vertebrate biology, asst. prof. internal medicine U. Mich. Med. Sch., 1948-51, geneticist Inst. Human Biology, asso. prof. med. genetics, 1951-56, prof. human genetics, chmn. dept., 1956—, prof. internal medicine, 1957—, Lee R. Dice U. prof. human genetics, 1966—; Galton lectr. U. London, 1955; Cutter lectr. Harvard, 1956; Russel lectr. U. Mich., 1966; cons. USPHS, AEC, NRC, WHO.; Pres. 6th Internat. Congress Human Genetics. Author med. articles; mem. editorial bd.: Blood, 1950-62, Perspectives in Biology and Medicine, 1956—, Human Genetics Abstracts, 1962—, Mutation Research, 1964—. Served to 1st lt. M.C. AUS, 1946-47; acting dir. field studies Atomic Bomb Casualty Commn., 1947-48. Recipient Albert Lasker award, 1960, Allan award Am. Soc. Human Genetics, 1965; Nat. Medal Sci., 1974. Mem. Am. Philos. Assn., Am. Acad. Arts and Scis., Inst. of Medicine, Nat. Acad. Scis. (mem. council 1970-72), Genetics Soc. Am., Am. Soc. Human Genetics (v.p. 1952-53, pres. 1953-54), Am. Fedn. Clin. Research, Am. Soc. Naturalists, Assn. Am. Physicians, Japanese, Brazilian socs. human genetics, Phi Beta Kappa, Sigma Xi, Alpha Omega Alpha. Club: Cosmos. Subspecialties: Genetics and genetic engineering (medicine); Environmental toxicology. Current work: Population monitoring for genetic damage. Home: 2235 Belmont Rd Ann Arbor MI 48104

NEFF, WILLIAM DUWAYNE, research scientist, educator; b. Lomax, Ill., Oct. 27, 1912; s. Lyman M. and Emma (Jacobson) N.; m. Ernestine Anderson, Sept. 1937 (div. 1960); m. Florence Palmer Anderson, Sept. 23, 1961 (dec. May 1978); children—Carol Jean (Mrs. William Fritsch), Peter Lyman. A.B., U. Ill., 1936; Ph.D., U. Rochester, 1940. Research asso. Swarthmore (Pa.) Coll., 1940-42; research scientist Columbia, also U. Calif. divs. war research, New London, Conn., 1942-46; asst. prof., prof. U. Chgo., 1946-61; dir. lab. physiol. psychology Bolt, Beranek & Newman, Cambridge, Mass., 1961-63; research prof. Ind. U., Bloomington, 1963—; dir. Center For Neural Scis., 1964-78; Cons. NSF, NIH, NASA, EPA; staff adviser Nat. Acad. Scis.-NRC Com. on Hearing, Bioacoustics and Biomechanics. Editor: Contributions to Sensory Physiology, 1965; mem. editorial bd.: Handbook of Sensory Physiology; Contbr. articles profl. jours. Fellow Am. Acad. Arts and Scis.; mem. Nat. Acad. Scis., A.A.A.S., Am. Inst. Biol. Scis., Am. Physiol. Soc., Psychonomic Soc., Acoustical Soc. Am., Am. Acad. Arts and Scis., Soc. for Neurosci., Soc. Exptl. Psychology, Am. Otol. Soc., Otosclerosis Study Group, Am. Acad. Neurology, Brit. Royal Soc. Medicine (affiliate), N.Y. Acad. Scis., Internat. Soc. Audiology, Internat. Brain Research Orgn., Sigma Xi. Subspecialties: Neuropsychology; Sensory processes. Home: 3505 Bradley St Bloomington IN 47401

NEGISHI, EI-ICHI, chemistry educator; b. Shinkyo City, Manchuria, China, July 14, 1935; came to U.S., 1960; s. Ryuzaburo and Fusae (Nagashima) N.; m. Sumire Suzuki, June 29, 1960; children: Wakaba Charlotte, Michil Florence. B.S., U. Tokyo, 1958; Ph.D., U. Pa., 1963. Research chemist Teijin, Ltd., Iwakuni, Japan, 1958-60, Tokyo, 1963-66; postdoctoral assoc. Purdue U., West Lafayette, Ind., 1966-72, prof. chemistry, 1979—; asst. prof. chemistry Syracuse (N.Y.) U., 1972-76, assoc. prof., 1976-79. Author: Organometallics in Organic Synthesis, 1980; contbr. articles to profl. jours. Fulbright scholar, 1960; Harrison fellow, 1962. Mem. Am. Chem. Soc. Subspecialties: Organic chemistry; Synthetic chemistry. Current work: Exploratory organometallic chemistry, development of transition metal-catalyzed reactions, natural products synthesis. Home: 4957 Taft Rd West Lafayette IN 47906 Office: Dept Chemistry Purdue U West Lafayette IN 47907

NEI, MASATOSHI, geneticist, educator; b. Miyazaki, Japan, Jan. 2, 1931; came to U.S., 1969; s. Tadashi and Masae (Kawasaki) N.; m. Nobuko Hara, Apr. 25, 1963; children: Keitaro, Maromi. B.S., Miyazaki U., 1953; M.S., Kyoto U., 1955, Ph.D., 1959. Instr. Kyoto U., 1958-62; geneticist Nat. Inst. Radiol Scis., Japan, 1962-65; head Population Genetics Lab., 1965-69; assoc. prof. biology Brown U., Providence, 1969-71, prof. biology, 1971-72; prof. population genetics Center Demographic and Population Genetics, U. Tex.-Houston, 1972—. Author: Molecular Population Genetics and Evolution, 1975, Evolution of Genes and Proteins, 1983, also articles and sci. papers.; Assoc. editor, editorial bds. profl. jours. Rockefeller Found. fellow, 1960-61; research grantee NIH, NSF, 1970—; recipient award Japan Soc. Human Genetics, 1977. Mem. AAAS, Genetics Soc. Am., Am. Soc. Naturalists, Am. Soc. Human Genetics. Subspecialties: Evolutionary biology; Genetics and genetic engineering (biology). Current work: Molecular evolution, population genetics, human genetics. Home: 8918 Sager Dr Houston TX 77096 Office: PO Box 20334 Houston TX 77225

NEIDHARDT, FREDERICK CARL, biology educator; b. Phila., May 12, 1931; s. Adam Fred and Carrie (Fry) N.; m. Elizabeth Robinson, June 9, 1956; children: Richard Frederick, Jane Elizabeth; m. Germaine Chipault, Dec. 3, 1977; 1 son, Marc Frederick. B.A., Kenyon Coll., 1952, D.Sc. (hon.), 1976; Ph.D., Harvard U., 1956. Research fellow Pasteur Inst., Paris, 1956-57; H.C. Ernst research fellow Harvard Med. Sch., 1957-58, instr., then assoc., 1958-61; mem. faculty Purdue U., 1961-70, assoc. prof, then prof., assoc. head dept. biol. scis., 1965-70; chmn., then prof. dept. microbiology and immunology U. Mich. Med. Sch., Ann Arbor, 1970—; Found. for Microbiology lectr. Am. Soc. Microbiology, 1966-67; cons. Dept. Agr., 1964-65; mem. grant study panel NIH, 1965-69; mem. commn. scholars Ill. Bd. Higher Edn., 1973-79; mem. test com. for microbiology Nat. Bd. Med. Examiners, 1975—, chmn., 1979—. Author papers in field.; Mem. editorial bd. profl. jours. Recipient award bacteriology and immunology Eli Lilly and Co., 1966; Alexander von Humboldt Found. award for U.S. sr. scientist, 1979; NSF sr. fellow U. Copenhagen, 1968-69. Mem. Am. Soc. Microbiology (pres. 1981-82), Am. Soc. Biol. Chemists, Am. Inst. Biol. Scis., N.Y. Acad. Sci., Soc. Gen. Physiology, Phi Beta Kappa, Sigma Xi. Subspecialties: Microbiology; Molecular biology. Current work: Regulation of gene expression in bacteria; controls related to growth and macromolecule synthesis; heat shock and other stress responses of bacteria. Office: Dept Microbiology & Immunology U Mich Ann Arbor MI 48109

NEILSON, GEORGE CROYDON, nuclear physicist, educator; b. Vancouver, C., Apr. 4, 1928. B.Sc., U. B.C., 1950, M.Sc., 1952, Ph.D., 1955. Physicist radiation sect. Def. Research Bd., Suffield, Alta., Can., 1955-58, head radiation sect., 1958-59; asst. prof., U. Alta., 1959-61, asso. prof., 1961-66, prof. physics, 1966—; dir. Nuclear Research Centre; vis. scientist nuclear physics sect. AERE, Harwell, Eng., 1966-67, TRIUMF, Vancouver, 1973-74. Contbr. numerous articles to profl. jours. Mem. Canadian Assn. Physicists, Am. Phys. Soc. Subspecialty: Nuclear physics. Current work: Nuclear physics at intermediate energies; physics with neutron, induced reactions. Home: 6403 128th St Edmonton AB Canada T6H 3X4 Office: Nuclear Research Center U Alta Edmonton AB Canada T6G 2N5

NELMS, GEORGE ESTES, animal scientist, educator; b. Brookland, Ark., Feb. 6, 1927; s. Cecil Otto and Monico N.; m. Fairy May (Miller), Apr. 5, 1950; children: Randy, David, Lisa, Nancy. B.S., Ark. State U., 1951; M.S., Oreg. State U., 1954; Ph.D., 1956. Instr. Oreg. State U., 1955-56; asst. prof. U. Ariz., 1956-59; prof. animal breeding U. Wyo., Laramie, 1959—; researcher. Treas. Gem City Service Club, 1965—. Served with USNR, 1945-46. Mem. Am. Soc. Animal Sci., Am. Genetics Assn., N.Y. Acad. Scis., Sigma Xi, Gamma Sigma Delta, Alpha Zeta. Democrat. Presbyterian. Club: Jacoby Park Men's. Lodge: Elks (Laramie). Subspecialties: Animal genetics; Animal physiology. Current work: Maternal ability of various genotypes, fatty acid composition of milk from beef cows, beef cattle genetics.

NELSON, ARNOLD BERNARD, animal science educator; b. Valley Springs, S.D., Aug. 26, 1922; s. Joseph Bernard and Huldah Bernhardina N.; m. Dorothy Millicent Larson, Aug. 3, 1943; children: James, Terry, Thomas, Barbara. B.S. in Animal Sci, S.D., State U., 1943, M.S., 1948, Ph.D., Cornell U., 1950. Asst. prof. to assoc. prof. animal sci. Okla. State U., 1950-63; prof. animal and range scis. N.Mex. State U., Las Cruces, 1963—, head dept., 1971—. Served to 1st lt. AUS, 1943-46. Fellow Am. Soc. Animal Sci., AAAS; mem. Am. Dairy Sci. Assn., Soc. Range Mgmt., Council Agr. Sci. and Tech., Nat. Assn. Coll. Tchrs of Agr., Latin Am. Assn. Animal Prodn. Lutheran. Subspecialty: Animal nutrition. Current work: Range livestock nutrition, animal science administration. Home: 2010 Crescent Dr Las Cruces NM 88005 Office: New Mexico State University Box 3-I Las Cruces NM 88003

NELSON, CARNOT EDWARD, psychology educator; b. Milw., Feb. 20, 1941; s. Max T. and Marion (Roth) N.; m. Alice Katz, Sept. 1, 1963; children: Jeremy H., Seth R. B.S., U. Wis., Madison, 1963; Ph.D., Columbia U., 1966. Asst. prof. sociology Emory U., Atlanta, 1966-67; asst. prof. psychology and research assoc. Johns Hopkins U., Balt., 1966-71; assoc. prof. psychology U. S. Fla., Tampa, 1971-76; prof., 1978—; fellow Nat. Inst. Edn., Washington, 1976-78; cons. U.S. Dept. Agr., 1980-81, Northside Community Mental Health Ctr., Tampa, 1979-82. Editor: Communication Among Scientists, 1972; contbr. articles to profl. jours. Bd. dirs. Congregation Rodeph Saalem, Tampa, 1973-76, 78—, Hillel Sch. of Tampa, 1973-76, 79—, Jewish Community Center, Tampa, 1982—. State of Fla. grantee, 1980; NIMH grantee, 1980-82. Fellow Am. Psychol. Assn.; mem. Am. Ednl. Research Assn., AAAS, Sigma Xi. Subspecialties: Social psychology; Organizational psychology. Current work: Organizational change, knowledge utilization. Home: 727 S Edison Ave Tampa FL 33606 Office: Dept Psychology U South Fla Tampa FL 33620

NELSON, DAVID TORRISON, physics educator; b. Decorah, Iowa, May 16, 1927; s. David Theodore and Esther Caroline (Torrison) N.; m. Betty Jane Rikansrud, Apr. 20, 1957; children: Elise, Andrea, Kathryn, Stephen. A.B., Luther Coll., Decorah, 1949; M.A., U. Rochester, 1955; Ph.D., Iowa State U., 1960. Instr. physics Luther Coll., 1954-60, asst. prof., 1960-63, assoc. prof., 1963-67, prof., 1967—. Mayor City of Decorah, 1978—; mem. Decorah City Council, 1972-73. NSF Sci. Faculty fellow, 1967-68. Mem. Am. Phys. Soc., Am. Assn. Physics Tchrs., Optical Soc. Am., Acoustical Soc. Am., Internat. Solar Energy Soc., Iowa Acad. Sci., Sigma Xi. Subspecialty: Solar energy. Current work: Photovoltaic physics and applications. Home: 215 High St Decorah IA 52101 Office: 602 Leif Erikson Dr Valders 152E Decorah IA 52101

NELSON, DENNIS RAYMOND, biochemist, researcher; b. New Rockford, N.D., Feb. 7, 1936; s. Carl Raymond and Marie Ann (Pfeiffer) N.; m. Janice Barbara Nygaard, June 22, 1961; children: David, Barbara, Laurie. B.S., N.D. State U., 1958, M.S., 1959; Ph.D., U. N.D., 1964. Research chemist USDA-ARS, Fargo, N.D., 1964-71, research leader, 1971—. Mem. Am. Soc. Biol. Chemists, Am. Chem. Soc., AAAS, N.D. Acad. Sci., Sigma Xi. Lutheran. Subspecialty: Integrated pest management. Current work: hormonal control and identification of insect surface lipids responsible for water-proofing, sex attraction and parasite/predator attraction. Home: 17 35th Ave NE Fargo ND 58102 Office: Metabolism and Radiation Research Lab USDA-ARS State University Station Fargo ND 58105

NELSON, DIANA FURST, radiation therapist; b. N.Y.C., Dec. 17, 1943; d. Joseph B. Furst and Helen Tunis Burrows; m. Fredric P. Nelson, Jan. 4, 1970; 1 son, Daniel Wolcott. B.A., Mount Holyoke Coll., 1965; M.D., Downstate Med. Ctr., 1969. Diplomate: Am. Bd. Radiology. Intern. U. Vt. Med. Ctr., Burlington, 1969-70; clin. fellow radiology Yale-New Haven Hosp., 1970-71; resident and clin. fellow radiation therapy Harvard Med. Sch., Boston, 1971-74; research fellow radiation biology Harvard Sch. Pub. Health, 1974-75; asst. prof. radiation oncology in radiology U. Rochester, 1975-78; asst. prof. radiation therapy Hosp. of U. Pa., Phila., 1979—; cons. radiation oncologist Pondville Hosp., Norfolk, Mass., 1974-75; assoc. radiation oncologist Strong Meml. Hosp., Rochester, N.Y., 1975-78; asst. attending radiation oncologist Genesee Hosp., Rochester, 1975-78; assoc. attending radiation oncologist Highland Hosp., Rochester, 1978-79; staff radiation therapist Hosp. of U. Pa., Phila., 1979—; staff Presbyn. U. Pa. Med. Ctr., VA Hosp., Phila., 1979—, Grad. Hosp., 1981—; mem., chmn. brain subcom. Radiation Therapy Oncology Group, 1982—, mem. protocol com., 1982—. Contbr. chpts. to books, articles to sci. jours. Mary Dole fellow, 1974; recipient Bernice MacLain Zoology prize Mount Holyoke Coll., 1963. Mem. Am. Soc. Therapeutic Radiologists, Am. Med. Women's Assn., Inc., Pa. Med. Soc., Phila. County Med. Soc., Am. Soc. Clin. Oncology, Am. Radium Soc. Current work: Treatment of cancer patients with radiation therapy; clinical research on brain tumors and lymphoma. Office: Hosp U Pa 3400 Spruce St Philadelphia PA 19104 Home: 324 Llandrillo Rd Bala Cynwyd PA 19004

NELSON, DONALD FREDERICK, research physicist; b. East Grand Rapids, Mich., July 4, 1930; s. Paul Vine and Florence Dorothea (Atchison) N.; m. Margaret Ellen Fuerstenau, Dec. 18, 1954; children: Elizabeth Ellen, Julia Karen. B.S., U. Mich., 1952, M.S., 1953, Ph.D. in Physics (fellow), 1958. Postdoctoral fellow U. Mich., 1958-59; mem. tech. staff research area Bell Telephone Labs., Murray Hill, N.J., 1959-67, 68—; prof. physics U. So. Calif., 1967-68; vis. lectr. Princeton U., 1976. Author: Electric, Optic, and Acoustic Interactions in Dielectrics, 1979; contbr. numerous articles to profl. jours. Councilman City of Summit, N.J., 1981—; mem. N.J. Gov.'s Sci. Adv. Com., 1982—; del. Union County Community Devel. Revenue Sharing Com., 1977—. Fellow Am. Phys. Soc.; mem. Optical Soc. Am., Acoustical Soc. Am. Subspecialties: Condensed matter physics; Laser research. Current work: Nonlinear optics, nonlinear acoustics, mechanics, crystal physics; theory and experiment of high-field drift velocities of electrons and holes in semiconductors. Patentee in field, including end pumping of lasers.

NELSON, EARL EDWARD, plant pathologist; b. New Richmond, Wis., Jan. 11, 1935; s. Gust E. and Myrtle Ellen (Monson) N.; m. Carol A. King, Sept. 17, 1960; children: Barry, Suzanne. Ph.D., Oreg. State U., 1962. Research forester U.S. Forest Service, Dept., Agr., 1957-60, research plant pathologist, 1960-72, project leader, 1975—; assoc. prof. plant pathology Oreg. State U. Mem. Mycol. Soc. Am., Am. Phytopath. Soc., Sigma Xi. Subspecialty: Plant pathology. Current work: Biological control, Silvicultural control. Office: 3200 Jefferson Way Corvallis OR 97331

NELSON, ERIC CHARLES, physician, medical educator, oncology researcher; b. Grinnell, Iowa, Oct. 22, 1946; s. Harry Alfred and Jean Buchanan (Bates) N.; m. Mary Lynn Bryant, Sept. 5, 1971; children: Mary Katherine, Bryant Buchanan. B.S., The Citadel, 1968; M.D., Wake Forest U., 1972. Diplomate: Am. Bd. Internal Medicine (medical oncology). Intern Wilford Hall USAF Med. Center, San Antonio, 1972-73, resident in internal medicine, fellow in hematology/oncology, 1973-77; chief hematology-oncology USAF Hosp., Keesler, Miss., 1977-79; asst. prof. internal medicine U. S.C., Columbia, 1979—; practice medicine specializing in hematology, oncology, Spartanburg, S.C., 1980—; prin. investigator S.W. Oncology Group, 1977-79, Piedmont Oncology Assn., 1978—. Bd. dirs. Am. Cancer Soc.; v.p. United Way. Served as maj. USAF, 1972-79. Mem. ACP, Am. Soc. Clin. Oncology, Am. Soc. Internal Medicine, AMA. Baptist. Subspecialties: Chemotherapy; Hematology. Current work: Clinical trials cancer chemotherapeutic agents. Office: Spartanburg Hematology-Oncology Assocs 122 Dillon Dr Spartanburg SC 29302

NELSON, GORDON LEON, agricultural engineering educator; b. Chippewa County, Minn., Dec. 28, 1919; s. John Anton and Hilda (Weberg) N.; m. Florence Jeanne Wise, June 7, 1942; children: Gordon Leon, Carol (Mrs. James Earl), Linda (Mrs. Arthur Ochsner), Janet (dec.), David, Barbara. B.Agrl. Engring., U. Minn., 1942; certificate naval engring. design, U.S. Naval Acad. Postgrad. Sch., 1945; M.Sc., Okla. State U., 1951; Ph.D., Iowa State U., 1957. Sr. agrl. engr. Portland Cement Assn., Chgo., 1946-47; asso. prof. to prof. agrl. engring. Okla. State U., 1947-69; prof., chmn. dept. agrl. engring. Ohio State U., also Ohio Agrl. Research and Devel. Center, 1969—; dir. Ohio State U.-Ford Found. project Coll. Agrl. Engring., Punjab (India) Agr. Univ., 1969-72; cons. in field.; Mem. 7 engring. edn. and accreditation ad hoc visitation teams to evaluate agrl. engring. curricula Engrs. Council Profl. Devel. Contbr. articles to profl. jours. Chmn. bd. dirs. Stillwater (Okla.) Municipal Hosp., 1956-60; mem. grad. council Ohio State U., 1970-74; bd. dirs. Council for Agrl. Sci. and Tech., 1975-81. Served to comdr. USNR, 1942-68. NSF Sr. Postdoctoral fellow U. Calif., Berkeley and Davis, 1964, 65-66. Fellow Am. Soc. Agrl. Engrs. (dir. awards, bd. dirs., dir. edn. and research 1979, Metal Bldg. Mfg. award 1960, 8 outstanding Paper awards); mem. Am. Soc. Engring. Edn., Am. Assn. Engring. Socs. (chmn. continuing edn. com.), Sigma Xi, Tau Beta Pi, Sigma Tau, Alpha Epsilon, Phi Kappa Phi, Phi Tau Sigma, Gamma Sigma Delta. Republican. Baptist (chmn. deacons 1971). Subspecialties: Agricultural engineering; Environmental engineering. Current work: Research interest include: cracking mechanics for crusted sorts, dynamic pressures in cylindrical grain storages, applications of similitude theory. Home: 6000 Sedgwick Rd Worthington OH 43085

NELSON, HAROLD STANLEY, physician, educator; b. New Britain, Conn., Jan 17, 1930; s. Harold Stanley and Ebba Arvida (Lawson) N.; m. Sarah Milledge, July 25, 1953; children: Erik, Mark, Stanley. A.B., Harvard U., 1951; M.S., U. Mich., 1969; M.D., Emory U., 1955. Diplomate: Am. Bd. Internal Medicine, Am. Bd. Allergy and Immunology. Commd. 2d lt. M.C. U.S. Army, 1956, advanced through grades to col., 1971; chief dept. medicine, Ft. Rucker, Ala., 1962-64, Bad Cannstatt, W. Ger., 1964-67, chief allergy-immunology service, Aurora, Colo., 1969—; clin. prof. medicine U. Colo. Center Health Scis., Denver, 1982—; cons. allergy-immunology Army Surgeon-Gen. Contbr. articles to profl. jours. Fellow Am. Acad. Allergy, ACP, Am. Coll. Allergists; mem. Am. Thoracic Soc., Am. Assn. Immunologists, Assn. Mil. Allergists. Subspecialties: Allergy; Internal medicine. Current work: Beta adrenergic agonists, subsensitivity, allergy immunotherapy, allergens and cross allergenicity; clinical practice, teaching and research. Home: 4970 S Fulton St Englewood CO 80111 Office: Fitzsimons Army Med Center Aurora CO 80045

NELSON, JAMES ARLY, research scientist in pharmacology; b. Livingston, Tex., Feb. 8, 1943; s. Oscar Curtis and Cecil Lorene (Jacobs) N.; children: Kurtis, Michael, Susana. B.S., U. Houston, 1965, M.S., 1967; Ph.D., U. Tex. Med. Br., 1970. Diplomate: Tex. Bd. Pharmacy. Postdoctoral research assoc. Brown U., Providence, 1970-72; sr. biochemist So. Research Inst., Birmingham, Ala., 1972-76; asst. prof. pharmacology U. Tex. Med. Br., Galveston, 1976-79; assoc. prof. pediatrics (pharmacology), assoc. pharmacologist U. Tex.-M.D. Anderson Hosp. Tumor Inst., Houston, 1979—. Recipient Sigma Xi Research award, 1970. Mem. Am. Soc. Pharmacology and Exptl. Therapeutics, Am. Assn. for Cancer Research, Sigma Xi. Subspecialties: Pharmacology; Biochemistry (medicine). Current work: Chemotherapy; biochemical pharmacology. Office: U Tex-MD Anderson Hosp 6723 Bertner Ave Houston TX 77030

NELSON, JEREMIAH I., neurophysiologist; b. N.Y.C., June 2, 1943; s. Gilbert Isaac and Florence (Laikind) N.; m. Robin Hannay, June 13, 1967; 1 son, Lorrin H. B.A., Swarthmore Coll., 1965; Ph.D., SUNY-Stony Brook, 1971. Asst. prof. SUNY-Stony Brook, 1978-79, N.Y.U. Med. Center, N.Y.C., 1979—. NSF fellow, 1971-72; Fulbright fellow, Australia, 1972-74; grantee NIMH, Spencer Found., Fight for Sight, Dept. Def. Mem. Assn. Research in Vision and Ophthalmology, Soc. Neurosci. Subspecialties: Neurophysiology; Sensory processes. Current work: Neurophysiological basis in brain of higher perceptual processes. Home: 3 Hill Rd Saint James NY 11780 Office: 550 First Ave New York NY 10016

NELSON, JOHN FRANKLIN, oral pathology educator, former military officer; b. Twin Falls, Idaho, Sept. 27, 1934; s. John Harold and Esther Louise (Ratcliffe) N.; m. Josephine Ellen Lillihei, Aug. 17, 1958; children; John Michael; Suzanne Ellen. B.S., U. Min.-Mpls., 1957, D.D.S., 1959; M.Ed., George Washington U., 1971; Cert. oral medicine, U. Pa., Phila., 1968; cert. Oral pathology, Armed Forces Inst. Pathology, Washington, 1971. Diplomate: Am. Acad. Oral

Pathology, Am. Acad. Oral Medicine. Commd. officer Dental Corps., U.S. Army, 1958, advanced through grades to col., 1977; mentor oral pathology residency Dental Corps. U.S. Army, Washington, 1978-79; chief div. pathology U.S. Army Inst. Dental Research, 1978-79; chief dental liaison officer U.S. Army Med Research and Devel. Command, Fort Detrick, Md., 1977-78; ret., 1980; prof. oral pathology U. Iowa, Iowa City, 1979—; mem. oral biology and medicine rev. com. NIH, Washington, 1978-79. Author U.S. Army tng. movie, 1980; contbr. articles, abstracts to profl. jours. County commr. Johnson City Broadband Telecommunications, Iowa City, 1980—. Decorated Legion of Merit (2); Recipient instr. Yr. award U. Iowa Coll. Dentistry, 1980. Fellow Am. Coll. Dentists, Internat. Coll. Dentists, Am. Acad. Oral Pathology; mem. ADA, Internat. Assn. Dental Research, AMA (affiliate), Beta Theta Pi, Psi Omega. Republican. Subspecialties: Oral pathology; Oral medicine. Current work: osteogenesis, epidemiology, dental education, radiology, oral disease, oral tumors. Home: 4 Wendram Bluff Iowa City IA 52240 Office: U Iowa Coll Dentistry Iowa City IA 52242

NELSON, MARGARET CHRISTINA, biologist, biological illustrator; b. Louisville, Nov. 13, 1943; d. Norton and Rose Sarah (Cohen) N.; m. Ronald Raymond Hoy, June 1, 1980; 1 child. B.A., Swarthmore Coll., 1965; M.A., U. Pa., 1968, Ph.D., 1970. Postdoctoral fellow biology Tufts U., 1970-72; asst. prof. psychology Brandeis U., 1972-75; research assoc., instr. neurobiology Harvard U., 1975-79; sr. research assoc. neurobiology Cornell U., 1979-82, acting assoc. prof. neurobiology and behavior, 1982—; mem. faculty neurobiology Marine Biol. Lab., Woods Hole, Mass., 1978, neural systems and behavior, 1979—, winter behavior, 1977-79. Contbr. articles to profl. jours. Nat. Merit scholar, 1961-65; NSF and NIH grantee, 1977—. Mem. AAAS, Animal Behavior Soc., Cambridge Entomol. Club, Am. Soc. Zoology, Soc. Neurosci., Guild Natural Sci. Illustrators, Sigma Xi. Subspecialty: Comparative neurobiology. Current work: Animal communication, neural basis of behavior, neural network, invertebrate neurobiology, neuroethology. Office: Cornell University Mudd Hall Ithaca NY 14853

NELSON, NORTON, environmental medicine educator, toxicologist; b. McClure, Ohio, Feb. 6, 1910; s. William and Bertha C. (Ballmer) N.; m. Rose S. Cohen, Sept. 3, 1936; children: Robert, Margaret, Richard. A.B., Wittenberg U., 1932, D.Sc. (hon.), 1964; Ph.D., U. Cin., 1938. Research asst. Children's Hosp. Research Found., Cin., 1934-38, research asso., 1946-47; biochemist May Inst. Med. Research, Jewish Hosp., Cin., 1938-42; asst. prof. biochemistry U. Cin., 1946-47; asso. prof. indsl. medicine NYU, 1947-53, dir. research Inst. Indsl. Medicine, 1947-53, prof. environ. medicine, 1953—; dir. Inst. Environ. Medicine, NYU Med. Center, 1954-80 (on leave 1966-67), chmn. dept. environ. medicine, 1954-80; acting dir. NYU Valley, 1962-66, dir., 1967-80; provost NYU Heights Center, 1966-67; chief chemistry dept. Armored Med. Biochemists; cons. NSF, 1971-72, 74-75, FDA, 1972-77; chmn. adv. com. on protocols for safety evaluation; mem. com. on environ. physiology NRC, chmn. com. air quality standards in space flight, mem. com. research in life sci., com. on nitrate accumulation, 1970-72; mem. Armed Forces Epidemiol. Bd., 1962-77, cons., 1977—, Commn. on Environ. Health, 1962-65; mem. Mayor's Tech. Adv. Com. on Radiation, Mayor's Sci. and Tech. Adv. Council, 1966-74; mem. exec. com. N.Y.C. Health Research Council, chmn. environ. pollution working group, 1965-72; mem. adv. com. Nat. Inst. Environ. Health Sci., NIH, 1967-71, 74-77, cons., 1972—, chmn. task force on research planning in environ. health sci., 1969-70, chmn. 2d task force, 1975-76; mem. Cancer Cause and Prevention adv. com. NIH, Nat. Cancer Inst., 1971-73, Carcinogenesis Program (Etiology), 1972-73, Clearinghouse on Environ. Carcinogenesis, 1976-80; mem. pesticide adv. com. HEW, 1970; mem. panel on herbicides U.S. Office Sci. and Tech., 1969, com. tech. forecasting behalf environ. health, 1970, chmn. task force on hazardous trace substances, 1970; mem. panel on chems. and health Pres.'s Sci. Adv. Com., 1970-73; mem. hazardous materials adv. com. EPA, 1970-74, cons. sci. adv. bd., 1974-75, chmn. environ. health adv. com., 1975-80; mem. Roster Cons. to Adminstr. ERDA, 1976-78; chmn. conf. on protocols for evaluating chems. in environ. NRC, Nat. Acad. Scis., 1972; mem. White House task force on air pollution, 1969, Environ. Studies Bd., 1974-77; chmn. human ecology commn. Internat. Assn. Ecology; mem. Commn. Natural Resources, 1977-80; mem. expert panel on carcinogenicity Internat. Union Against Cancer; mem. com. motor vehicle emission WHO, 1964-68, mem. com. microchem. pollutants, 1964-68; chmn. Expert Com. Manual Toxicity of Chems., 1975-77; mem. panel U.S.-USSR Joint Commn. for Health Cooperation, 1972—, U.S.-Japan Coop. Med. Sci. Program, 1972-78; mem. subcom. toxicology of metals Permanent Commn. and Internat. Assn. Occupational Health, 1972—; vis. com. dept. nutrition and food scis. MIT, 1971-74; mem. Milbank Meml. Fund Commn. for Study Higher Edn. Pub. Health, 1972-75; mem. energy policy project adv. bd. Ford Found., 1972-74; mem. Hudson Basin Project adv. bd. Rockefeller Found., 1973-75, Pres.'s Biomed. Research Panel, chmn. subcluster on environ. health and toxicology, 1975; chmn. Bd. Toxicology and Environ. Health Hazards, Nat. Acad. Scis., 1977-80, ex-officio mem., 1981; bd. dirs. Found. for Advanced Edn. in Sci. Asso. editor: Jour. Occupational Medicine; cons. editor: Environ. Research; mem. editorial bd.: Archives of Environ. Health, Jour. Toxicology and Environ. Health, MIT Press Series of Toxicology, 1981—. Trustee Tuxedo Meml. Hosp., 1957-75, Indsl. Health Found., 1969-75; sci. adviser Indsl. Health Found., 1976—; mem. vis. com. bd. overseers Harvard U. Sch. Pub. Health, 1973-76; mem. adv. com. on occupational safety and health edn. Resource Center, 1980—; chmn. med. and sci. adv. bd. Will Rogers Meml. Hosp., 1960-70, chmn., 1968-70; mem. research adv. com. Boyce Thompson Inst. Plant Research; adv. council dept. stats. Princeton U., 1973-76; bd. dirs. N.Y. Lung Assn., 1974-77; mem. biomed. and environ. sci. adv. com. Los Alamos Sci. Lab., 1976-80; chmn. panel environ. health NIH Fogarty Center, Am. Coll. Preventive Medicine, 1975; cons. NSF, 1971-72, 74-75; mem. project 4 com. Sci. Com. Problems of Environ., Internat. Council Sci. Unions, NRC, 1974-77; chmn. bd. sci. counselors Nat. Toxicology Program, 1980—; chmn. carcinogenesis adv. panel Office Tech. Assessment, U.S. Congress, 1979-81; chmn. and mem. exec. com. Sci. Group on Methodologies for Safety Evaluation of Chems., 1979—; mem. sci. com. on problems of environ. WHO, 1979—; mem. UN Environ. Programme, 1979—; mem. com. health related effects of marijuana use NRC, 1980—; sr. mem. Inst. Medicine, 1981; mem. health research rev. com. State of N.Y., 1977—; co-chmn. coal techs., health and environ. effects of energy techs. Fed. Interagy. Com., Dept. Energy, Dept. Health and Human Services, EPA, 1979-80; mem. environ. research and devel. subpanel energy research adv. bd. Dept. Energy, 1980—; v.p. John B. Pierce Found., 1983—. Served to lt. col. San. Corps, AUS, 1942-46. Recipient Nat. Health Achievement award Blue Cross and Blue Shield Assns., 1979. Mem. Am. Indsl. Hygiene Assn., Soc. Exptl. Biology and Medicine, Harvey Soc., Air Pollution Control Assn., Ecology Soc. Am., Tarrytown Hist. Soc. (pres. 1958-60), AAAS, Am. Acad. Occupational Medicine (hon.), Am. Chem. Soc., Am. Pub. Health Assn. (Com. occupational health and safety), Am. Soc. Biol. Chemists, Am. Soc. Pharmacology and Exptl. Therapeutics (com. environ. pharmacology), Soc. Occupational and Environ. Health, N.Y. Acad. Scis. (hon., life, Gordon Y. Billard award 1976), Indsl. Hygiene Round Table, Soc. Toxicology (hon.), Sigma Xi. Subspecialty: Environmental medicine. Home: 37 DeVries Ave North Tarrytown NY 10591 Office: NY U Med Center 550 1st Ave New York NY 10016

NELSON, JR. OLIVER EVANS, geneticist, educator; b. Seattle, Aug. 16, 1920; s. Oliver Evans and Mary Isabella (Grant) N.; m. Gerda Kjer Hansen, Mar. 28, 1963. A.B., Colgate U., 1941; M.S., Yale, 1943, Ph.D., 1947. Asst. prof. genetics Purdue U., 1947-49, asso. prof., 1949-54, prof., 1954-69; prof. genetics U. Wis., Madison, 1969—; vis. investigator Biochem. Inst., U. Stockholm and Nat. Forest Research Inst., Stockholm, 1954-55; NSF, sr. postdoctoral fellow Calif. Inst. Tech., Pasadena, 1961-62. Recipient John Scott medal City of Phila., 1967; Hoblitzelle award Tex. Research Found., 1968; Browning award Am. Soc. Agronomy, 1974; Donald F. Jones medal Conn. Argl. Exptl. Sta., 1976. Mem. Nat. Acad. Scis., Am. Acad. Arts and Scis., Genetics Soc. Am., Am. Genetics Assn., Am. Soc. Plant Physiologists, Crop Sci. Soc., AAAS, Sigma Xi. Subspecialties: Plant genetics; Gene actions. Current work: Controlling elements; genes affecting starch synthesis. Home: 4197 Barlow Rd Cross Plains WI 53528 Office: Lab Genetics Univ Wis Madison WI 53706

NELSON, PHILIP EDWIN, food science educator; b. Shelbyville, Ind., Nov. 12, 1934; s. Brainard Russell and Alta Edna (Pitts) N.; m. Sue Bayless, Feb. 27, 1955; children: Jennifer, Philip, Bradley. B.S., Purdue U., 1956, Ph.D. in Food Sci, 1967. Plant mgr. Blue River Packing Co., Morristown, Ind., 1957-60; asst. prof. Purdue U., West Lafayette, Ind., 1967-70, assoc. prof., 1970-74, prof., 1975—, dir., 1975—; cons. to industry. Contbr. over 80 articles to sci. jours.; author: (with D. Tressler) Fruit and Vegetable Juice Processing, 1980. Served in U.S. Army, 1957-59. Recipient Eugene L. Grant award The Engring. Economist, 1979. Fellow Inst. Food Technologist (Indsl. Achievement award 1976), AAAS, Sigma Xi. Lodge: Elks. Subspecialty: Food science and technology. Current work: Processing and preservation of foods; aseptic processing and packaging research. Patentee (9). Office: Purdue U West Lafayette IN 47907

NELSON, PHILLIP GILLARD, neurobilogist; b. Albert Lea, Minn., Dec. 3, 1931; s. Conrad Arthur and Calla Jean (Gillard) N.; m. Karin Becker, Mar. 20, 1955; children: Sarah Elizabeth, Rebecca Judith, Jennifer Becker, Peter Tobias. M.D., U. Chgo., 1956, Ph.D., 1957. Intern Phila. Gen. Hosp., 1957-58; with NIH, Bethesda, Md., 1958—; chief lab. devel. neurobiology NICHD, 1968—. Contbr. articles to profl. jours. Pres. PTA. Served with USPHS, 1958—. Subspecialties: Cell and tissue culture; Developmental biology. Current work: Development of the nervous system. Home: 5524 Charles St Bethesda MD 20814 Office: NIH Bldg 36 Room 2A21 Bethesda MD 20205

NELSON, RANDALL JAY, neurophysiologist; b. Evanston, Ill., Feb. 13, 1953; s. Earl Leroy and Georgene Adele (Stieglitz) N.; m. Leslie Catherine Bearden, May 19, 1979. B.S. magna cum laude, Duke U., 1975; Ph.D. in Anatomy, Vanderbilt U., 1980. NIH postdoctoral fellow U. Calif., San Francisco, 1980; staff fellow lab. neurophysiology NIMH, Bethesda, Md., 1981—. Contbr. articles to sci. jours. Mem. Soc. Neurosci. Subspecialties: Neurophysiology; Regeneration. Current work: Information processing in the somatic sensory and motor cerebral cortex of primates; neurophysiological correlates of behavior. Office: NIH Bethesda MD 20205

NELSON, ROBERT M., astronomer; b. Los Angeles; s. Steve and Margaret (Yeager) N.; m.; children: Tom, Chet. B.S. in Physics, CUNY, 1966; M.A. in Astronomy, Coun. Wesleyan U., 1969; Ph.D. in Earth and Planetary Sci., U. Pitts., 1977. Mem. faculty dept. geology Youngstown (Ohio) State U., 1977-78; NRC resident research assoc. Jet Propulsion Lab., Calif. Inst. Tech., Pasadena, 1978-80, sr. scientist, 1980—; co-chmn. So. Calif. Fedn. Scientists. Co-host: radio show about sci. The Wizards, KPFK-FM.; Contbr. articles to profl. jours. Mem. Internat. Astron. Union, Am. Inst. Physics, Am. Astron. Soc. (div. planetary scis.), AAAS, Am. Geophys. Union, Sigma Xi. Subspecialties: Planetary science; Satellite studies. Current work: Research on the chemical and physical composition of the surfaces of solid surface bodies in the solar system.

NELSON, ROBERT NORTON, chemistry educator; b. Cin., Nov. 1, 1941; s. Norton A. and Rose S. (Cohen) N.; m. Anne Milbouer, Feb. 21, 1965; children: David Norton, Louis Myron. Sc.B. in Chemistry, Brown U., 1963; Ph.D. in Phys. Chemistry, MIT, 1969. Asst. prof. chemistry Ga. So. Coll., 1970-76, 78-82, assoc. prof., 1982—; vis. asst. prof. chemistry Colgate U., 1977-78; vis. lectr. chemistry U. Ga., 1981. Scoutmaster, Boy Scouts Am., 1981—. Mem. Am. Chem. Soc., Am. Phys. Soc., Soc. Applied Spectroscopy, Sigma Xi. Subspecialties: Physical chemistry; Spectroscopy. Current work: Laser spectroscopy-CARS, molecular collisions. Office: Georgia Southern College Chemistry 8064 Statesboro GA 30460 8064

NEMER, MARTIN JOSEPH, molecular biologist, researcher; b. Phila., Nov. 26, 1929; s. David and Anna P. (Greenberg) N.; m. Lucia Wehle, Apr. 6, 1967 (div. 1980); children: Jessica, Daniel. B.A., Kenyon Coll., 1952, D.Sc. (hon.), 1977; M.S., Harvard U., 1955, Ph.D., 1958. Research assoc. Inst. for Cancer Research, Fox Chase, Phila., 1960-63, asst. mem., 1963-67, assoc. mem., 1967-78, mem., 1978—. Assoc. editor jour.: Developmental Genetics, 1980—. Mem. Am. Soc. Biol. Chemistry. Subspecialties: Molecular biology; Developmental biology. Current work: Gene expression in cell differentiation and embryonic development. Office: Inst for Cancer Research Fox Chase Philadelphia PA 19111

NEMEROFF, CHARLES BARNET, psychiatrist, neurobiologist, educator; b. Bronx, N.Y., Sept. 7, 1949; s. Philip Peace and Sarah (Greenberg) N.; m. Melissa Ann Pilkington, May 24, 1980; 1 son, Matthew Pilkington N. B.S., CCNY, 1970; M.S., Northeastern U., 1973; Ph.D., U. N.C., 1976, M.D., 1981. Research assoc. Harvard U. Med. Sch., Boston, 1972-73; postdoctoral fellow U. N.C. Sch. Medicine, 1976-81; sr. research fellow Biol. Sci. Research Ctr., 1976—; intern N.C. Meml. Hosp., Chapel Hill, 1981-82, resident in psychiatry, 1981—; asst. prof. dept. pharmocology and psychiatry Duke U. Med. Ctr., Durham, N.C. Contbr. numerous articles, chpts. to profl. publs. Recipient A.E. Bennett award Soc. Biol. Psychiatry, 1979; 2d prize Roche Lab. Award in Neurosci. and Mead Johnson Research Forum, 1979; Nat. Inst. Neurol. and Communicable Diseases and Stroke fellow, 1977; N.C. Alcoholism Research Authority co-grantee, 1980, 81. Mem. Soc. Neurosci., AMA, AAAS, Am. Psychiat. Assn., Endocrine Soc., Am. Soc. Neurochemistry, Am. Pain Soc., Internat. Soc. Psychoneuroendocrinology. Subspecialties: Psychopharmacology; Neuropharmacology. Current work: Biological basis of major mental disorder; role of neuropeptides in central nervous system function. Office: Box 3859 Duke U Med Ctr Durham NC 27710

NES, WILLIAM ROBERT, biochemistry educator; b. Oxford, Eng., May 16, 1926; s. William H. and Mary (Lineback) N.; m. Estelle J. Shirley, May 16, 1946; children: Shirley Anne Nes Warshaw, William David. B.A., U. Okla.-Norman, 1946; Ph.D., U. Va.-Charlottesville, 1950. Fellow Mayo Clinic, Rochester, Minn., 1950-51, NIH, Bethesda, Md., 1951-58; dir. tng. program Worcester Found., Shrewsbury, Mass., 1958-64; prof. chem. and pharm. chemistry U. Miss., Oxford, 1964-67; W.L. Obold prof. biol. scis. Drexel U., Phila., 1967—; vis. prof. Hahnemann Med. Coll., Phila., 1970—. Author: Biochemistry of Steroids and Other Isopentenoids, 1977, Lipids in Evolution, 1980. Served with USN, 1944-46. Research grantee NIH, 1958—; Travel Grantee Am. Cancer Soc., 1955-56, Anna Fuller Fund, 1955-56; research grantee Am. Cancer Soc., 1960-70. Mem. Am. Chem. Soc., Am. Soc. Biol. Chemistry, Endocrine Soc., Phytochem. Soc., Protozool. Soc. Republican. Episcopalian. Subspecialties: Biochemistry (biology); Organic chemistry. Current work: The relationship between the structure, biosynthesis and function of sterols and triterpenoids is being studied. Home: 10 Tanglewood Circle Rose Valley PA 19086 Office: Dept Biol Scis Drexel U 32d and Chestnut Sts Philadelphia PA 19104

NESBIT, RICHARD ALLISON, scientific instrument company executive; b. Whittier, Calif., Jan. 17, 1935; s. Harry H. and Florence L. (Golden) N.; m. Rose Marie Franklin, Sept. 13, 1957; 1 dau., Laura Lynn. B.S., UCLA, 1958, M.S., 1960, Ph.D., 1963. Mem. tech. staff space tech. lab. Aerospace Corp., El Segundo, Calif., 1958-63; sr. scientist Lear Siegler, Inc., Santa Monica, Calif., 1963-64, mgr. engring. Beckman Instruments, Fullerton, Calif., 1964-66, mgr. research and devel., 1966-75, dir. research, 1975-80, v.p. research and devel., 1980—; mgmt. cons., 1966—; owner Nesbit Control Lab.; lectr. mgmt. sci. Redlands U., 1969; lectr. engring. UCLA, 1962-66. Mem. Soc. for Indsl. and Applied Math., Sigma Xi, Phi Eta Sigma, Tau Beta Pi, Delta Tau Delta. Subspecialty: Scientific instruments research and development management. Home: 1849 Avenida Del Norte Fullerton CA 92633 Office: 2500 Harbour Blvd Fullerton CA 92634

NESBITT, PATRICIA MARIE, environ, and farm cons. co. exec., environ. and agrl. researcher, writer, educator; b. Chgo., Nov. 6, 1948; d. Charles John and Elizabeth Patricia (Boehme) N.; m. Victor Jay Habib, Apr. 22, 1981. B.S., U. Dayton, 1970, M.A., 1973. In various cons., research and writing positions, 1968—; mem. faculty dept. agr. Lord Fairfax Community Coll., 1976—; owner, operator Shenandoah Organic Suppliers, Strasburg, Va., 1978—; pres. Patti Nesbitt & Assos., Strasburg, 1980—; speaker in field; research dir. Earth Apple Farm Strasburg; research adv. Signal Knob Farm, Strasburg. Author: (with C. Stoner) Goodbye to the Flush Toilet, 1977, (with C. Keough) Water Fit to Drink, 1980; contbr. articles to profl. publs. Va. coordinator Earth Day '80, 1980. Mem. Va. Assn. Biol. Farmers (exec. bd. 1977-82, coordinator for programming 1977—, newsletter research editor 1981—), Water Pollution Control Fedn., Rural Va. (dir. 1979-82), Ohio Ecol. Farmers Assn., Carolina Farm Stewardship Assn., Am. Agr. Movement, Am. Water Works Assn. Subspecialties: Integrated systems modelling and engineering; Water supply and wastewater treatment. Current work: Development of sustainable agriculture based on renewable resources; organic or biological farming and gardening; alternate water supply and wastewater treatment for small towns and rural areas. Home and Office: Route 2 Box 77-F Strasburg VA 22657

NETLEY, CHARLES THOMAS, hospital psychologist, educator; b. Brighton, Ont., Can., Apr. 25, 1939; s. Samuel and Frances Jean (Thomas) N.; m., Sept. 2, 1961; children: Rebecca, Rachel, Julie. B.A., Queen's U., Kingston, Ont., 1961, M.A., 1962; Ph.D., U. London, 1966. Lectr. U. London, 1965-66; chief psychologist Hosp. for Sick Children, Toronto, 1966—; asst. prof. behavioral sci. U. Toronto, 1971-76, assoc. prof., 1976-83, prof., 1983—. Mem. Am. Psychol. Assn., Can. Psychol. Assn., Internat. Neuropsychol. Soc., Behavior Genetics Assn. Subspecialties: Developmental psychology; Neuropsychology. Current work: Relations between cognitive development and neural processes. Home: 11 Crestview Rd Toronto ON Canada M5N 1H3 Canada

NETZEL, JAMES PHILLIP, mech. engr.; b. Chgo., Nov. 16, 1940; s. Phillip and Ella (Freislinger) N.; m. Anita Francine Pahlke, June 5, 1965; children: Carol Ann, William James. B.S.M.E., U. Ill., Champaign, 1963. Mech. engr. Crane Packing, Morton Grove, Ill., 1963-67, sr. mech. engr., supr. mech. design group, 1967-79, asst. chief engr. product engring., 1979-81; chief engr. John Crane-Houdaille, Morton Grove, 1981—; instr. in engring. Oakton Community Coll., 1974-79. Contbr. articles on mech. seals and wear-related problems to profl. jours. Mem. ASME, Am. Soc. Lubrication Engrs. Subspecialty: Mechanical engineering. Current work: Friction, wear, and lubrication as applied to mech. seals. Office: 6400 W Oakton St Morton Grove IL 60053

NETZER, CAROL, psychologist; b. N.Y.C., July 2, 1931; d. Jack and Pauline (Potofsky) Risika; m. Dick Netzer, Dec. 30, 1952; children: Jenny, Kate. B.A., U. Wis., 1949; M.A., Boston U., 1950. Psychologist S. Beach Psychiat. Ctr., Bklyn., 1973-83, supervising psychology, 1978-83, cons. in family therapy, 1979—; now pvt. practice psychotherapy. Contbr. articles to profl. jours. Mem. Am. Psychol. Assn., Soc. for Advancement Psychoanalytical Devel. Psychology. Subspecialty: Developmental psychology. Current work: Writer in psychoanalytic development psychology (theory) and in the theory of family therapy. Address: 227 Clinton St Brooklyn NY 11201

NEUFELD, HAROLD ALEX, research biochemist, consultant; b. Paterson, N.J., Mar. 23, 1924; s. Meyer and Helen Sarah (Maiden-Stein) N.; m. Shirley Yvonne Lazerowitz, June 25, 1950; children: Howard Scott, Kenneth Paul, Marjorie Gail, Matthew David. B.S. in Chemistry, Rutgers U., 1949; Ph.D. in Biochemistry, U. Rochester Med. Sch., 1953. Instr. U. Rochester (N.Y.) Med. Sch., 1953-55; investigator Crops Div. U.S. Army, Ft. Detrick, Frederick, Md., 1955-57, 1957-71; prin. investigator Phys. Sci. Div. U.S. Army Med. Research Inst. Infectious Diseases, 1972—; instr. chemistry part-time Frederick Community Coll., 1956-70; grad. lectr. in biochemistry Hood Coll., Frederick, 1974-77. Contb. chpt. to book, article, rev. to publ. Scoutmaster, com. chmn. Nat. Capital Area council Boy Scouts Am., Frederick, 1950—; v.p. Frederick County YMCA, 1977-79; pres. Beth Sholom Synagogue, Frederick, 1960; mem. Frederick City Police Commn., 1981—. Mem. Am. Chem. Soc., Am. Soc. Biol. Chemists, Soc. for Exptl. Biology and Medicine, AAAS. Democrat. Jewish. Subspecialties: Biochemistry (biology); Biochemistry (medicine). Current work: Research in mycotoxins and lipid metabolism. Patentee automatic biol. agt. detector. Home: 117 W 14th St Frederick MD 21701 Office: Physical Scis Div Fort Detrick Frederick MD 21701

NEUFFER, MYRON GERALD, geneticist; b. Preston, Idaho, Mar. 4, 1922; s. Myron and Camille (Cole) N.; m. Margaret McGregor, Mar. 18, 1943; children: David, John, Gregory, Dale, Barbara, Peggy, Linda. B.S., U. Idaho, 1947; M.S. in Genetics, U. Mo., 1949, Ph.D., 1952. Asst. prof. field crops U. Mo.-Columbia, 1951-56, assoc. prof., 1956-66, prof., 1966-67, prof. genetics, chmn. dept., 1967-70, prof. biol. scis., 1970-76, prof. agronomy, 1976—. Author: Mutants of Maize, 1968. Mem. AAAS, Genetic Soc. Am., Am. Genetics Assn., Crop Sci. Soc. Am., Sigma Xi. Mem. Ch. Jesus Christ on Latter-day Saints. Subspecialty: Plant genetics. Current work: Maize genetics and mutation.

NEUGEBAUER, GERRY, astrophysicist; b. Gottingen, Germany, Sept. 3, 1932; came to U.S., 1939; s. Otto E. and Grete (Brück) N.; m. Marcia MacDonald, Aug. 26, 1956; children—Carol, Lee. B.S., Cornell U., 1954; Ph.D., Calif. Inst. Tech., 1960. Mem. faculty Calif. Inst. Tech., Pasadena, 1962—, prof. physics, 1970—; mem. staff Hale Obs., 1970-80; acting dir. Palomar Obs., 1980-81, dir., 1981—. Served with AUS, 1961-63. Fellow Am. Acad. Arts and Scis.; mem. Nat. Acad. Scis. Subspecialties: Infrared optical astronomy; Astronomy from space. Current work: Director Palomar Observatory; principal scientist IRAS (Infrared Astronomonical Satellite. Office: Calif Inst Tech Pasadena CA 91125

NEUHAUSER, DUNCAN VON BRIESEN, community health educator; b. Phila., June 20, 1939; s. Edward B.D. and Gernda (von Briesen) N.; m. Elinor Toaz, Mar. 6, 1965; children: Steven, Ann. B.A., Harvard U., 1961; M.H.A., Mich. U., 1963; M.B.A., U. Chgo., 1966, Ph.D., 1971. Research assoc. Ctr. for Health Adminstrn. Studies, Grad. Sch. Bus., U. Chgo., 1965-70; asst. prof. health services adminstrn. Harvard Sch. Pub. Health, Cambridge, Mass., 1970-74, assoc. prof., 1974-79; cons. in medicine Mass. Gen. Hosp., 1975-80; prof. epidemiology and community health, Keck Found. sr. research scholar Case Western Res. U., Cleve., 1979—; adj. prof. orgnl. behavior Weatherhead Sch. Mgmt., 1979—; assoc. dir. Health Systems Mgmt. Ctr., Case Western Res. U., 1979—; cons. Cleve. Metro. Gen. Hosp., 1981—; cons. in field; lectr. in field; chmn. Great Lakes VA Regional Health Services Research & Devel. Field Program adv. com., 1982-83, coordinating com. Cleve. Clin. Decision Analysis Group, 1982-83; assoc. chmn. Program for Health Systems Mgmt., Harvard Bus. Sch. and Sch. Pub. Health, 1972-79; mem. Ctr. for Analysis of Health Practices, 1975-79; dir. Harvard Sch. Pub. Health/VA Region I Health Services Research & Devel. Unit, 1977-79. Adv. editor: Health Care Mgmt. Rev, 1975—; editorial bd.: New Eng. Jour. Human Services, 1980—; Contbr. articles to profl. jours.; author: (with Florence Wilson) Health Services in the U.S., 2nd edit, 1982, (with A. Kovner) Health Services Mgmt., 2nd edit, 1983, (with Milton Weinstein and others) Clinical Decision Analysis, 1980. First vice chmn. Vis. Nurse Assn. of Greater Cleve., 1982—; bd. dirs. New Eng. Grenfell Assn., 1973—, Internat. Grenfell Assn., 1975—, Braintree Hosp., 1975—, Mass. Eye and Ear Infirmary, 1977-79, Hospice Council of No. Ohio, 1980—. Kellog fellow U. Chgo., 1963-65; Neuhauser lectr. Soc. Pediatric Radiology, 1982. Mem. Assn. Univ. Programs in Health Adminstrn., Am. Pub. Health Assn., Soc. for Clin. Decision Making, Forum of Health Services Adminstrs., Inst. Medicine of Nat. Acad. Scis., Beta Gamma Sigma. Club: St. Botolph. Subspecialties: Epidemiology; Health services research. Current work: Clinical decision analysis, evaluation of costs and quality of medical care, health services research, health services management and organization. Home: 2655 N Park Blvd Cleveland Heights OH 44106 Office: Med Sch 2119 Abington Rd Cleveland OH 44106

NEUMANN, HERSCHEL, physicist, educator; b. San Bernardino, Calif., Feb. 3, 1930; s. Arthur and Dorothy (Greenhood) N.; m. Julia Black, June 15, 1951; children: Paul Alfred, Keith Edward. Student, San Bernardino Valley Coll., 1947-49; B.A. in Physics, U. Calif.-Berkeley, 1951, M.S., U. Oreg.-Eugene, 1959, Ph.D., U. Nebr.-Lincoln, 1965. Theoretical physicist Hanford Labs., Gen. Electric Co, Richland, Wash., 1951-57; instr. physics U. Nebr., 1964-65; asst. prof. physics U. Denver, 1965-71, assoc. prof., 1971—; researcher. Contbr. articles to profl. jours. Mem. Am. Phys. Soc., Am. Assn. Physics Tchrs. Subspecialties: Atomic and molecular physics; Numerical analysis. Current work: Theoretical studies of collisions between H/H and atmospheric gases, theoretical studies of collisions between ions and surfaces. Office: Dept Physics U Denver Denver CO 80208

NEUMEYER, JOHN LEOPOLD, medicinal chemist, educator; b. Munich, Germany, July 19, 1930; came to U.S., 1945, naturalized, 1950; s. Albert and Martha (Stern) N.; m. Evelyn, June 24, 1956; children: Ann, David, Elizabeth. B.S., Columbia U., 1952; Ph.D., U. Wis., 1961. Research chemist Ethicon Inc. div. Johnson and Johnson, Somerville, N.J., 1952-57; research chemist FMC Corp., Princeton, N.J., 1961-63; sr. staff scientist Arthur D. Little Inc., Cambridge, Mass., 1963-69; prof. medicinal chemistry Northeastern U., 1969—, prof. chemistry, 1976—, Disting. Univ. prof., 1980—, dir., 1978—; cons. in field; dir. Research Biochems. Inc. Assoc. editor: Jour. Medicinal Chemistry; Research, numerous publs. in field; contbr. chpts. to books. Served with U.S. Army, 1953-55. Recipient First prize Lundsford Richardson Award, 1961; Am. Found. Pharm. Edn. Gustavus N. Pfeiffer Meml. research fellow, 1975; Hayes-Fulbright sr. fellow, 1975. Fellow AAAS, Acad. Pharm. Scis. (Research Achievement award 1982); mem. Am. Chem. Soc. (chmn. div. med. chemistry 1982), Soc. Neurosci., Sigma Xi, Phi Kappa Phi, Rho Chi. Subspecialties: Medicinal chemistry; Organic chemistry. Current work: Neuropharmacology-receptor interactions of dopamine agonists and antagonists, mechanism of action of CNS active agents, cancer chemotherapy, environmental effects of pesticides. Patentee in field.

NEURATH, ALEXANDER ROBERT, virologist, biochemist; b. Bratislava, Czechoslovakia, May 8, 1933; came to U.S., 1964, naturalized, 1970; s. Ernest and Lenka (Weinberger) N. Dipl. Ing., Inst. Tech., Bratislava, 1957; D.Sc., Inst. Tech. Vienna, Austria, 1968. Research scientist Inst. Biology, Slovak Acad. Scis., Bratislava, 1957-59; head biochem. control dept. Inst. for Sera and Vet. Vaccines (Bioveta), Nitra, Czechoslovakia, 1959-61; research scientist Inst. Virology, Czechoslovak Acad. Scis., Bratislava, 1961-64; vis. scientist Karolinska Inst., Stockholm, 1964; fellow Wistar Inst. Anatomy and Biology, Phila., 1964-65; research scientist Wyeth Labs., Inc., Phila., 1965-72; investigator Lindsley F. Kimball Research Inst., N.Y. Blood Center, N.Y.C., 1972—; cons. in field. Contbr. articles to profl. jours. Nat. Heart, Lung and Blood Inst. grantee, 1976-84; N.Y. State Health Research Council grantee, 1981. Mem. Am. Soc. Microbiology, AAAS, N.Y. Acad. Scis., Soc. Gen. Microbiology, Panam. Group for Rapid Viral Diagnosis, Am. Soc., Virology. Subspecialties: Virology (medicine); Infectious diseases. Current work: Viruses transmitted by blood transfusion and products derived from blood; hepatitis B and nonA, nonB, antiviral vaccines, synthetic peptides as immunogens, diagnostic tests. Patentee in field. Home: 230 E 79th St Apt 5F New York NY 10021 Office: 310 E 67th St New York NY 10021

NEVA, FRANKLIN ALLEN, physician, educator; b. Cloquet, Minn., June 8, 1922; s. Lauri Albin and Anna (Lahti) N.; m. Alice Hanson, July 5, 1947; children—Karen, Kristin, Erik. S.B., U. Minn., 1944, M.D., 1946; A.M. (hon.), Harvard, 1964. Diplomate: Am. Bd. Internal Medicine. Intern Harvard Med. Services, Boston City Hosp., 1946-47, resident, 1949-50; research fellow Harvard Med. Sch., 1950-53; asst. prof. U. Pitts. Med. Sch., 1953-55; mem. faculty Harvard Sch. Pub. Health, 1955-69, John LaPorte Given prof. tropical pub. health, 1964-69; chief Lab. Parasitic Diseases, Inst. Allergy and Infectious Diseases, NIH, 1969—; Mem. commn. parasitic diseases, asso. mem. commn. virus infections Armed Forces Epidemiological Bd., 1963-68; mem. Latin Am. sci. bd. Nat. Acad. Scis.-NRC, 1963-68; bd. sci. counselors Inst. Allergy and Infectious Diseases, NIH, 1966-69. Served to lt. (j.G.) USNR, 1947-49. Mem. Soc. Exptl. Biology and Medicine, Infectious Diseases Soc. Am., Am. Soc. Tropical Medicine and Hygiene (Bailey K. Ashford award 1965), Assn. Am. Physicians. Subspecialties: Infectious diseases; Parasitology. Current work: Research on parasitic disease. Spl. research infectious diseases especially tropical, parasitic and virus infections. Home: 10851 Glen Rd Potomac MD 20854 Office: Institute of Allergy and Infectious Diseases NIH Bethesda MD 20014

NEVILLE, WALTER EDWARD, JR., animal scientist; b. Rabun Gap, Ga., May 5, 1924; s. Walter Edward and Kate Elizabeth (Haulbrook) N.; m. Mary Augusta Clark, June 19, 1948; children: Gail Clark, Harry Walter. B.S.A., U. Ga., 1947; M.S., U. Mo., 1950; Ph.D., U. Wis., 1957. Asst. county agrl. agt., Chattooga County, Ga., 1947-49; asst. prof. dept. animal sci. U. Ga., Tifton, 1950-51, assoc. prof., 1952-76, prof., 1977—. Served with USN, 1945-46. Recipient Sigma Xi research award U. Ga., 1975. Mem. Am. Soc. Animal Sci., AAAS, Council for Agrl. Sci. and Tech., Sigma Xi, Phi Kappa Phi, Gamma Sigma Delta. Presbyterian. Subspecialty: Animal genetics. Current work: Economic Value of crossbreeding beef cattle, energy requirements of beef cattle, performance testing of beef cattle. Home: 2429 Madison Dr Tifton GA 31794 Office: Coastal Plain Experiment Sta Box 748 Tifton GA 31793

NEWBRUN, ERNEST, oral biology educator, periodontist; b. Vienna, Austria, Dec. 1, 1932; came to U.S., 1955, naturalized, 1964; s. Victor N.; m. Eva, June 17, 1956; children: Deborah, Daniel, Karen. B.D.S., U. Sydney, Australia, 1954; M.S., U. Rochester, 1957; D.M.D., U. Ala., 1959; Ph.D., U. Calif.-San Francisco, 1965. Dental House surgeon King's Coll. Hosp., London, 1954; resident dental surgeon Edgware Gen. Hosp., Middlesex, Eng., 1954-55; fellow Eastman Dental Dispensary, Rochester, N.Y., 1955-56, research assoc., 1956-57; dental extern Children's Hosp., Birmingham, Ala., 1958-59; research assoc. U. Ala. Med. Ctr., 1957-59; dental research fellow Nat. Health and Med. Research Council Australia, Inst. Dental Research, Sydney, 1960-61; NIH research tchr. trainee U. Calif.-San Francisco, 1961-63, NIH postdoctoral fellow, 1963-64, assoc. research dentist, assoc. prof. oral biology in residence, 1965-70, prof. oral biology in residence, 1970-72, prof. oral biology, 1972—, chmn. sect. biol. scis., 1972-77, postgrad. periodontology resident, 1981-83; pvt. practice dentistry, Sydney, 1960-61, San Francisco, 1970-72; pedodontist St. Mary's Hosp., San Francisco, 1970-72; mem. dental staff Mt. Zion Hosp., San Francisco, 1970; mem. dental drug products adv. com. FDA, 1974-78; mem. nat. adv. com. on fluoridation ADA, 1978—; vis. scientist Inst. Oral Microbiology, U. Lund Sch. Dentistry, Malmo, Sweden, 1967-68; vis. prof. Caries Research Sta., Dental Inst., U. Zurich, Switzerland, summer 1969; vis. prof. preventive dentistry U. Nijmegen, Netherlands, summer 1971; vis. scientist Inst. Dental Research, Sydney, 1979. Author: Cariology, 1978, 2d edit., 1983; contbr. numerous articles to profl. publs.; editor: Fluorides and Dental Caries, 1975, 2d edit., 1977; referee various jours.; abstractor and translator: Oral Research Abstracts, 1966-74. Mem. Internat. Assn. Dental Research, Am. Inst. Oral Biology, Bay Area Bone and Calcium Study Group, Europen Orgn. for Caries Research, Federation Dentaire Internationale, AAUP, AAAS, Am. Acad. Periodontology, Calif. Acad. Scis., Am. Soc. Microbiology, Sigma Xi, Omicron Kappa Upsilon. Subspecialties: Oral biology; Cariology. Current work: Dental plaque, oral microbiology, caries, periodontal disease, fluoride, preventive dentistry. Home: 1823 8th Ave San Francisco CA 94122 Office: Div Oral Biology U Calif San Francisco CA 94143

NEWBURN, RAY LEON, JR., astronomer; b. Rock Island, Ill, Jan. 9, 1933; s. Ray L. and Gertrude G. (Nelson) N.; m. Virginia Lee Gaskin, May 4, 1968; children: Steven Ray and Kevin Francis (twins). B.S., Calif. Inst. Tech., 1954, M.S., 1955, postgrad., 1955-56. Staff Calif. Inst. Tech., Jet Propulsion Lab., 1956—, leader Internat. Halley Watch and leader Cometary Sci. team, 1980—. Contbr. articles to profl. jours. Recipient NASA Group Achievement award, 1976. Mem. Am. Astron. Soc., Am. Geophys. Union, AAAS, Internat. Astron. Union, Sigma Xi. Subspecialties: Planetary science; Optical astronomy. Current work: Models of cometary nucleus and coma, photometry of comets; observation of Comet Halley from ground and space. Office: 4800 Oak Grove Dr T 1166 Pasadena CA 91109

NEWCOMB, MARTIN EUGENE, chemist, educator; b. Mishawaka, Ind., Nov. 17, 1946; s. Martin Eugene and Yolanda F. (Saliani) N.; m. Jill Ruth Wood, Oct. 7, 1968; 1 dau., Jennifer Ruth. B.A., Wabash Coll., 1968; Ph.D., U. Ill., 1973. Postdoctoral research assoc. UCLA, 1973-75; asst. prof. Tex. A&M U., College Station, 1975-81, assoc. prof. chemistry, 1981—. Author numerous research publs. Camille and Henry Dreyfus scholar, 1980—. Mem. Am. Chem. Soc., Royal Chem. Soc., AAAS, N.Y. Acad. Sci. Subspecialties: Organic chemistry; Synthetic chemistry. Current work: Electron transfer reactions, synthetic methodology, mechanistic and synthetic organic chemistry. Office: Dept Chemistry Tex A&M U College Station TX 77843

NEWELL, ALLEN, educator; b. San Francisco, Mar. 19, 1927; s. Robert R. and Jeannette (LeValley) N.; m. Noel Marie McKenna, Dec. 20, 1947; 1 son, Paul Allen. B.S. in Physics, Stanford, 1949; postgrad. Math., Princeton, 1949-50; Ph.D. in Indsl. Adminstrn, Carnegie Inst. Tech., 1957. Research scientist RAND Corp., Santa Monica, Cal., 1950-61; Univ. prof. Carnegie-Mellon U., Pitts., 1961—; U.A. and Helen Whitaker U. Prof. in Computer Sci., 1976—; cons. Xerox Corp. Author: (with G. Ernst) GPS, A Case Study in Generality, 1969, (with G. Bell) Computer Structures, 1971, (with H.A. Simon) Human Problem Solving, 1972, (with C.G. Bell and J. Grason) Designing Computers and Digital Systems, 1972. Recipient Harry Goode award Am. Fedn. Information Processing Socs., 1971; with H.A. Simon A.M. Turing award Assn. Computing Machinery, 1975. Fellow IEEE, AAAS; mem. Am. Psychol. Assn., Assn. Computing Machinery, Nat. Acad. Sci., Am. Acad. Arts and Sci. Inst. Mgmt. Sci. Subspecialties: Information systems, storage, and retrieval (computer science); Artificial intelligence. Office: Carnegie-Mellon Univ Schenley Park Pittsburgh PA 15213

NEWELL, FRANK WILLIAM, ophthalmologist, educator; b. St. Paul, Jan 14, 1916; s. Frank John and Hilda (Turnquist) N.; m. Marian Glennon, Sept. 12, 1942; children: Frank William, Mary Susan Newell O'Connell, Elizabeth Glennon, David Andrew. M.D., Loyola U., Chgo., 1939; M.Sc., U. Minn., 1942. Diplomate: Am. Bd. Ophthalmology (chmn. bd. 1967-69, cons. 1971-74). Intern Ancker Hosp., St. Paul, 1939-40; teaching fellow U. Minn., 1940-42; research fellow, instr., asso. dept. ophthalmology Northwestern U. Med. Sch., 1946-53; now James and Anna Raymond prof. dept. ophthalmology U. Chgo. Med. Sch.; prof. extraordinario Autonomous U. Barcelona, Spain; Montgomery lectr. Trinity Coll., Dublin, Ireland, 1966; Dunphy lectr. Mass. Eye and Ear Infirmary, 1967; McPherson lectr. U. N.C., 1968; May lectr. N.Y. Acad. Medicine, 1968; Wright lectr. U. Toronto, 1973; Irvine lectr. U. So. Calif., 1971; Lang lectr. Royal Soc. Medicine, London, 1974; O'Brien lectr. Tulane U., 1974; Atkinson lectr. Internat. Soc. Eye Surgeons, 1975; Stein lectr. UCLA, Los Angeles, 1976; Towne lectr. U. Louisville, 1977; Duke-Elder lectr. U. London, 1979; Fralick lectr. U. Mich., 1980; McDonald lectr. Loyola U., Chgo., 1980; sci. counselor Nat. Inst. Neurol. Diseases and Blindness, 1959-62, chmn., 1961-62; mem. nat. eye council NIH, 1972-75; mem. Internat. Council Ophthalmology, 1977—; v.p. XXIV Internat. Congress Ophthalmology.; Bd. dirs. Heed Ophthalmic Found., chmn., 1975—; dir. Ophthalmic Pub. Co., sec.-treas., 1971—. Author: (with D. Syndacker) Refraction, 3d edit, 1971, Ophthalmology: Principles and Concepts, 5th edit, 1982; also articles.; Editor-in-chief: Am. Jour. Ophthalmology; editor-in-chief: Trans Glaucoma Conf., Vols. 1-5, 1955-61; editor: (with D. Syndacker) Amblyopia and Strabismus, 1975, Hereditary Diseases of the Eye, 1980; editorial bd.: Survey Opthalmology. Trustee Loyola U., Chgo., 1977-81. Served from 1st lt. to maj. M.C. AUS, 1943-46. Recipient Alumni Citation award Loyola U., 1962, Stritch medal, 1966; Howe prize medal in ophthalmology AMA, 1968; Outstanding Achievement award U. Minn., 1975; Gold Key U. Chgo., 1981; Lang medal Royal Soc. Medicine, London, 1974; medal honor Soc. Eye Surgeons, 1975; medalla Andrés Bello U. Chile, 1977; medalla de Oro Instituto Barraquer, Barcelona, Spain, 1982. Mem. Chgo. Soc. History Medicine, Soc. Med. Cons. Armed Forces, AAUP, Nat. Soc. Prevention Blindness (dir., chmn. research com. 1961-68, 70-81, v.p. 1970-81, pres. 1981-83, Dunnington medal 1976), AMA (chmn. sect. ophthalmology 1964-65), Am. Acad. Ophthalmology and Otolaryngology (honor award 1953, 1st v.p. 1965, counselor 1969-72, pres. 1975), Inst. Barraguer (pres. 1970—), Assn. U. Profs. Ophthalmology (trustee 1966-69, pres., chmn. 1967-69), Assn. Research Ophthal. (trustee 1963-68, chmn. bd. trustees 1967-68), Pan-Am. Assn. Ophthalmology (dir. 1969—, Gradle lectr. 1972, pres. 1981-83), Am. Ophthal. Soc. (council 1977-81, Howe medal 1979), Chgo. Ophthal. Soc. (pres. 1957-58), Pacific Coast Ophthal. Soc. (hon.), Mont. Ophthal. Soc. (hon.), Central Ill. Ophthal. Soc. (hon.), Oxford (Eng.) Ophthal. Soc. (hon., dep. master 1980), Colombia Ophthal. Soc. (hon., Hellenic Ophthal. Soc., hon.), Academia Ophthalmologica Internationalis (1st v.p. 1976-80, pres. 1980-84), Soc. Exptl. Biology and Medicine, Soc. Franc d'ophth., Sigma Xi, Alpha Omega Alpha. Roman Catholic. Clubs: Literary, Quadrangle (Chgo.). Subspecialties: Ophthalmology; Laser medicine. Current work: Laser effects epithelial proliferation. Home: 4500 N Mozart St Chicago IL 60625 Office: 939 E 57th St Chicago IL 60637

NEWELL, ROBERT TERRY, physicist; b. Whittier, Calif., Nov. 13, 1946; s. John Robert and Nan Marion (Bracken) N.; m. Elizabeth Carol Martin, Aug. 26, 1967; children: Darcy Lia, Jenna Michelle. B.S., N.Mex. Inst. Mining and Tech., 1968, Ph.D., 1981; M.S., U. West Fla., 1970. Asso. engr. Boeing Co., Albuquerque, 1967-68; research asst. N.Mex. Inst. Mining and Tech., 1977-79; jr. research asso. Nat. Radio Astronomy Obs., Socorro, N. Mex., 1979-81, research assoc., 1981-82; sr. scientist Scott Sci. and Tech., ALbuquerque, 1983—. Contbr. articles to profl. jours. Active pastor-parish relations com. Meth. Ch. Served with USN, 1968-76; Served with UNSR Ready Res., 1977—. Decorated Air medals.; Recipient Founders Award N.Mex.Inst. Mining and Tech., 1981. Mem. Am. Astron. Soc., AIAA, Naval Res. Assn. Republican. Subspecialties: Radio and microwave astronomy; Theoretical astrophysics. Current work: Seismic instrumentation and data processing, stellar atmospheres, mass loss and evolution. Home: Rural Route 3 Box 1156 Los Lumas NM 87031 Office: Scott Sci and Tech 2601 Wyoming NE Suite C ALbuquerque NM 87112

NEWHOUSE, PAUL ALFRED, psychiatrist; b. Manchester, Eng., Mar., 1953; s. Vernon L. and Joan (Goldstone) N.; m. Mary Louise Newhouse, Aug. 4, 1942; 1 son, Sean Matthew. B.S., Kans. State U., 1974; M.D., Loyola U., Chgo., 1977. Intern Loyola U. Med. Ctr., Chgo., 1977-78; commd. capt. U.S. Army, 1978, advanced through grades to maj., 1983; resident in psychiatry (Walter Reed Army Med. Ctr.), Washington, 1978-81, research trainee neuropsychiatry div., 1980, div. psychiatrist, Schweinfurt, W.Ger., 1981-82, staff psychiatrist, Heidelberg, W.Ger., 1982—. Contbr. articles to profl. jours. Mem. Soc. Neurosci., Am. Psychiat. Assn., AAAS, Assn. U.S. Army. Roman Catholic. Subspecialties: Psychiatry; Psychopharmacology. Current work: Psychopharmacology; movement disorders and depression. Office: Dept Psychiatry 130th Sta Hosp APO New York NY 09102

NEWKIRK, GORDON ALLEN, JR., astrophysicist; b. Orange, N.J., June 12, 1928; s. Gordon Allen and Mildred (Fleming) N.; m. Nancy Buck, Apr. 11, 1956; children—Sally, Linda, Jennifer. B.A., Harvard U., 1950; M.A., U. Mich., 1952, Ph.D., 1953. Research asst. Observatory, U. Mich., 1950-53; astrophysicist Upper Air Research Observatory, 1953; sr. research staff High Altitude Observatory, Boulder, Colo., 1955—; adj. prof. dept. astrogeophysics U. Colo., 1961-65, adj. prof. physics and astrophysics, 1965-76; dir. High Altitude Obs., 1968-79; cons. NASA, 1964—; mem. geophysics research bd. Nat. Acad. Sci., 1976-80. Contbr. articles on solar-interplanetary physics to profl. jours. Served with U.S. Army, 1953-55. Mem. AAAS, Am. Astron. Soc. (chmn. solar physics div. 1972-73), Internat. Astron. Union (pres. com. 10 1975-79), Research Soc. Am., Commn. V Union Radio Sci. Internat., Am. Geophysical Union, Sigma Xi. Subspecialties: Optical astronomy; Solar physics. Current work: Structure of solar corona and heliosphere, solar cycle modulation of galactic cosmic rays. Home: 3797 Wonderland Hill Ave Boulder CO 80302 Office: High Altitude Observatory PO Box 3000 Boulder CO 80307

NEWLEY, PATRICK FOSTER, electrical engineering educator; b. Des Moines, Iowa, Aug. 4, 1945; s. Frederick Foster and Patricia Rose (Delbert) N.; m. Belinda Marie Thorne, Nov. 10, 1970; children: Elizabeth, Robert, William. B.S., George Washington U., 1966; M.S.E.E., Calif. Inst. Tech., 1968; Ph.D., U. Mich., 1972. Asst. prof. Pa. State U., 1973-76; assoc. prof. Okla. State U. Tech. Inst., 1976-81, prof. elec. engring., 1981—; vis. staff Los Alamos Nat. Lab., 1982; cons. Oak Ridge Nat. Lab. Guggenheim Found. fellow, 1978-80. Mem. IEEE (sr.), Assn. Computing Machinery, AAAS. Republican. Methodist. Club: Masons. Subspecialties: Electrical engineering; Computer-aided design. Office: Werik Bldg 3532 NW 23rd Oklahoma City OK 73107

NEWMAN, DARRELL FRANCIS, nuclear research engineer; b. Fort Knox, Ky., Mar. 22, 1940; s. Charles Carlisle and Lillian Evelyn (Karmann) N.; m. Sue Carol Farley, June 18, 1966. B.S., Kans. State U., 1963; M.S., U. Wash.-Seattle, 1970. Registered profl. engr., Wash. Nuclear engr. Gen. Electric Co., Richland, Wash., 1963; sr. research engr. Battelle-Pacific N.W. Lab., Richland, Wash., 1965-81; tech. cons. U.S. Dept. Energy, Washington, 1981-82; mgr. spent fuel program Battelle-Pacific N.W. Lab., Richland, 1982—; sr. experimenter (High Temperature Lattice Test Reactor), Richland, Wash., 1967-71, Richland, 1972-74, 1971-72. Contbr. articles in field to profl. jours. Served to 1st lt. Ordnance Corps U.S. Army, 1963-65. Recipient Excellence in Sci. and Tech. award Battelle Meml. Inst., 1968; Dirs. award Battelle N.W. Lab., 1969. Mem. Am. Nuclear Soc. (treas. reactor physics div. 1982-83), Nat. Soc. Profl. Engrs. (energy com. chmn. Wash. state 1979-81, Engr. of Yr. 1979), AAAS, N.Y. Acad. Scis., Wash. Soc. Profl. Engrs. (chmn. energy 1980). Republican. Baptist. Subspecialties: Nuclear fission; Nuclear engineering. Current work: Research, develop and demonstrate dry storage technologies as licensable options for management of commercial nuclear spent fuels. Home: 1100 McMurray St Richland WA 99352 Office: Battelle-Pacific NW Lab Battelle Blvd Richland WA 99352

NEWMAN, DAVID EDWARD, research physicist; b. Jan. 4, 1947. B.S., M.S., M.I.T., 1970; Ph.D., Princeton U., 1975. Postdoctoral scholar, asst. research scientist U. Mich., 1974-82; staff scientist Gen. Atomic Co., San Diego, 1982—. Mem. Am. Phys. Soc., AAAS. Subspecialties: Nuclear physics; Plasma physics. Current work: Fusion diagnostic techniques. Office: GA Technologies Inc PO Box 81608 San Diego CA 92138

NEWMAN, JERRY OKEY, agricultural engineer, researcher; b. New Martinsville, W.Va., May 9, 1936; s. James Okey and Hazel Edna (Howell) N.; m. Patricia Grace Devericks, May 25, 1958; children: Tamrah M., Lucetta R., Symantha C., Onika M. B.S.Ag.E., W.Va. U., 1958, M.S.Ag.E., 1960; Ph.D. in Engring, U. Md., 1972. Registered profl. engr., W.Va. Grad. asst. W.Va. U., 1958-60; hosp. engr. VA, Clarksburg, W.Va., 1960-62; research engr. in rural housing U.S. Dept. Agr., Beltsville, Md., 1962-73; in rural housing and solar energy Agrl. Research Service, Clemson, S.C., 1973—; auctioneer; real estate broker. Author numerous Dept. Agr. publs.; contbr. numerous articles to profl. jours. Pres. Pickens County (S.C.) Retarded Citizens Assn., 1975-78. Recipient Superior Performance award VA, 1961; Danforth summer fellow, 1957. Mem. Am. Soc. Agrl. Engrs., Nat. Inst. Bldg. Sci., Internat. Solar Energy Soc., Sigma Xi, Tau Beta Pi. Democrat.

Clubs: Red Necks (Clemson); 4H All Stars. Subspecialties: Agricultural engineering; Solar energy. Current work: Energy conservation, housing, solar heating, house design, earth housing, insulation. Home: 375 Route 1 Central SC 29630 Office: PO Box 792 Clemson SC 92633

NEWMAN, JOHN DENNIS, neuroscientist; b. Newark, Sept. 28, 1940; s. Richard and Elsie (Brown) N.; m. Judith Osler Scaffidi, Aug. 18, 1962; children: Peter, Matthew. B.S., Cornell U., 1962; Ph.D., U. Rochester, 1969. Staff fellow Nat. Inst. Child Health and Human Devel., NIH, Bethesda, Md., 1969-74; research physiologist Lab. Devel. Neurobiology, 1982—, reviewer Sci. Books and Films; vis. scientist Max Planck Inst. Psychiatry, Munich, W.Ger., 1974-75. Contbr. in field. Recipient Alexander von Humboldt Soc. award, 1974. Mem. AAAS, Soc. Neuroscience, Animal Behavior Soc., Am. Soc. Primatologist. Mem. United Ch. of Christ. Subspecialties: Neurobiology; Ethology. Current work: Brain mechanisms of primate communication; neurophysiology neuroethology; bioacoustics; evolution; communication; evolution of species-specific communication mechanisms, genetic programming of behavior. Home: 10910 Stillwater Ave Kensington MD 20895 Office: Bldg T-18 NIHAC Bethesda MD 20205

NEWMAN, MELVIN SPENCER, chemist, educator; b. N.Y.C., Mar. 10, 1908; s. Jacob K. and Mae (Polack) N.; m. Beatrice N. Crystal, June 30, 1933; children—Kiefer, Susan, Beth, Robert. B.S., Yale, 1929, Ph.D., 1932; D.Sc. (hon.), U. New Orleans, 1975, Bowling Green State U., 1978, Ohio State U., 1979. Instr. Ohio State U., 1936-39, Elizabeth Clay Howald scholar, 1939-40, asst. prof., 1940-44, prof. chemistry, 1944—, regents prof., 1966; Fulbright lectr. U. Glasgow, 1957, 67. Asso. editor: Jour. Organic Chemistry and Organic Syntheses. Guggenheim fellow, 1951; Recipient award, 1961, E.W. Morley medal, 1969, Columbus (Ohio) award, 1976, Roger Adams award, 1979; all Am. Chem. Soc.; W.L. Cross medal Yale U., 1970; Joseph Sullivant medal Ohio State U., 1976. Mem. Am., Brit. chem. socs., Nat. Acad. Sci., Sigma Xi. Subspecialties: Organic chemistry; Synthetic chemistry. Current work: Synthesis of new compounds of interest in theoretical chemistry and cancer related. Home: 2239 Onandaga Dr Columbus OH 43221

NEWMAN, MORRIS, mathematics educator; b. N.Y.C., Feb. 25, 1924; s. Isaac and Sarah (Cohen) N.; m. Mary Aileen Lenk, Sept. 18, 1948; children: Sally Ann, Carl Lenk. A.B., NYU, 1945; M.A., Columbia U., 1946; Ph.D., U. Pa., 1952. Mathematician applied math. div. Nat. Bur. Standards, Washington, 1951-63, chief numerical analysis sect., 1963-70, sr. research mathematician, 1970-76; prof. math. U. Calif.-Santa Barbara, 1976—; dir. Inst. Interdisciplinary Applications of Algebra and Combinatorics, 1980—. Author: Matrix Representations of Groups, 1968, Integral Matrices, 1972; contbr. articles to math. jours. Recipient gold metal U.S. Dept. Commerce, 1966. Mem. Am. Math. Soc. (council 1980—), Math. Assn. Am., AAAS, London Math. Soc., Washington Acad. Scis., Sigma Xi. Jewish. Subspecialties: Applied mathematics; Algorithms. Current work: Applications of number theory to computation, matrix computations, exact computation. Home: 1050 Las Alturas Rd Santa Barbara CA 93103 Office: Dept Math U Calif Santa Barbara CA 93106

NEWMAN, ROBERT ALWIN, pharmacologist; b. Winchester, Mass., July 11, 1948; s. Winston Kenneth and Alice Wilder (Palmer) N.; m. Marian Frank, May 22, 1976. B.S., U. R.I., 1970; M.S., U. Conn., 1973, Ph.D. in Pharmacology and Toxicology, 1975. Postdoctoral fellow Med. Coll. Ga., Augusta, 1975-76; research assoc. U. Vt., Burlington, 1976-77, assoc. prof. dept. pharmacology and dept. medicine, 1982—; mem. staff Vt. Regional Cancer Center, 1976—; dir. Biochem. Pharmacology Lab., 1976—. Contbr. articles to profl. jours. Bd. dirs. Am. Cancer Soc., Chittenden County, Vt. Mem. Am. Chem. Soc., N.Y. Acad. Sci., Am. Tissue Culture Assn., Am. Assn. for Cancer Research, Am. Soc. Clin. Oncology, AAAS, Am. Soc. for Pharmacology and Exptl. Therapeutics, Sigma Xi, Rho Chi. Subspecialties: Molecular pharmacology; Toxicology (medicine). Current work: Pharmacology and toxicology of cancer chemotherapeutic compounds; connective tissue biochemistry; pulmonary toxicology. Office: Dept Pharmacology Coll Medicine U Vt Burlington VT 05405

NEWMAN, SIMON LOUIS, immunology educator; b. Jacksonville, Fla., Oct. 31, 1947; s. Melvin and Jeanette Juliet (Marks) N.; m. Diana S. Bayar, Mar. 19, 1972. B.A., Emory U., 1969; M.S., U. Ala., 1971, Ph.D., 1978. Research asst. U. Ala. Med. Sch., Birmingham, 1971-72; asst. pathology U. Fla. Coll. Medicine, Gainesville, 1972-73; postdoctoral fellow Nat. Jewish Hosp., Denver, 1978-81, U. N.C., Chapel Hill, 1981-82, research asst. prof., 1982—. Mem. U.S. Coast Guard Aux., Chapel Hill, 1982—. Sigma Xi grantee-in-aid, 1970; Am. Soc. Clin. Immunology and Allergy Young Investigator fellow, 1979; recipient Investigator award Nat. Arthritis Found., 1982. Mem. AAAS, Am. Soc. Microbiology, Am. Assn. Immunologists, N.Y. Acad. Sci. Jewish. Subspecialties: Infectious diseases; Cell biology (medicine). Current work: Macrophage recognition via specific and non-specific membrane receptors, involvement of the complement system with phagocytic cells. Home: 313 Carol St Carrboro NC 27510 Office: U NC 932 FLOB 231-H Chapel Hill NC 27514

NEWSOM, GERALD HIGLEY, astronomer, educator; b. Albuquerque, Feb. 11, 1939; s. Carroll Vincent and Frances (Higley) N.; m. Ann Bricker, June 17, 1972; children: Christine, Elizabeth. B.A., U. Mich., 1961; Ph.D., Harvard U., 1968. Postdoctoral research asst. Imperial Coll. Sci. and Tech., London, 1968-69; asst. prof. Ohio State U., Columbus, 1969-73, assoc. prof. astronomy, 1973-82, prof. astronomy, 1982—; sr. postdoctoral research asst. Physikalisches Institut, U. Bonn, W.Ger., 1978. Author: (with W.M. Protheroe and E.R. Capriotti) Astronomy, 1976, Exploring the Universe, 1979; also articles. Mem. Internat. Astron. Union, Am. Astron. Soc. Subspecialties: Optical astronomy; Atomic and molecular physics. Current work: Spectrum of SS 433; atomic absorption in ultraviolet wavelengths. Home: 46 W Weisheimer Rd Columbus OH 43214 Office: 174 W 18th Ave Columbus OH 43210

NEWTH, CHRISTOPHER JOHN LESTER, physician, educator; b. Hamilton, New Zealand, Sept. 5, 1941; came to U.S., 1982; s. Walter John and Patricia Eleanor (Osborn) N.; m. Claire Veronica Fraser-Jones, Sept. 16, 1972; children: Annemarie, Lara, Joshua. B.Sc., Canterbury U., Christchurch, New Zealand, 1962; M.N., Ch.B., U. Otago, Dunedin, New Zealand, 1967. Intern Waikato Hosp., Hamilton, New Zealand, 1968-69; resident Hosp. Sick Children, Toronto, Ont., Can., 1970-74, asst. prof., 1976-82; postdoctoral fellow U. Calif. Med. Ctr., San Francisco, 1974-76, assoc. prof. pediatrics and anesthesia, 1982—; dir. pediatric critical care unit Moffitt Hosp., 1982—. Recipient Frederick Tisdall Research Prize Hosp. Sick Children, 1971; Am. Thoracic Soc. fellow, 1974-76; Med. Research Council Can. grantee, 1978-80; Canadian Cystic Fibrosis Found. grantee, 1982; Ont. Heart Found./Ont. Ministry Health grantee, 1980-82. Mem. N.Z. Med. Assn., Am. Thoracic Soc., Royal Coll. Physicians and Surgeons Can., Royal Australasian Coll. Physicians, Am. Physiol. Soc., Soc. Critical Care Medicine. Anglican. Subspecialties: Pediatrics; Critical care. Current work: Clinical pediatric, respiratory and critical care, research upper airway obstruction/lung water/pediatric pharmacology/pulmonary vascular bed. Home: 71 Mountain View Ave Mill Valley CA 94941 Office: U Calif Med Ctr 505 Parnassus St San Francisco CA 94143

NEWTON, ROBERTA ANN, physical therapist, educator; b. Providence, May 8, 1947; d. Robert E. and Georgianna R. (La Croix) N. B.S. in Phys. Therapy, Med. Coll. Va., 1975, Ph.D. in Physicology-Neurophysiology, 1973. Cert. phys. therapist, Va. Asst. prof. dept. phys. therapy Med. Coll. Va., Richmond, 1973-80, asso. prof., 1980—; manuscript reviewer. Author: (with Payton and Hirt) Scientific Bases for Neurophysiological Approaches to Therapeutic Exercies, 1977, (with Van Sant) Therapeutic Exercise Competencies for Entry Level Physical Therapists and Therapeutic Exercise Instructor, 1981; contbr. numerous articles to sci. jours. Med. Coll. Va. grantee, 1973, HEW spl. improvement grantee, 1974-79. Mem. Soc. for Neurosci., AAAS, Am. Phys. Therapy Assn. (grant reviewer), Va. Acad. Sci., Sigma Xi. Methodist. Subspecialty: Neurophysiology. Current work: Neurophysiological bases of therapeutic exercise; physiological bases for high voltage galvanic electrotherapy. Office: Dept Physical Therapy Med Coll Va Box 224 Richmond VA 23298

NEWTON, STEVEN ARTHUR, optical engineer, researcher; b. Teaneck, N.J., Feb. 6, 1954; s. Richard A. and Evelyn (Caruccio) N. B.S. in Physics summa cum laude, U. Mass., 1976; M.S. in Applied Physics, Stanford U., 1978, postgrad., 1978-83. Lab. asst. Hasbrouck Lab., U. Mass., Amherst, 1975-76; research asst. E.L. Ginzton Lab., Stanford U., Palo Alto, Calif., 1976-82, research assoc., 1982-83; summer intern Hewlett-Packard Labs., Palo Alto, 1978, mem. tech. staff, 1978—. Contbr. articles on laser light scattering, hollow cathode metal vapor lasers, single-mode fiber-optic components, fiber-optic gyros, fiber signal processing devices to profl. publs. Mem. Optical Soc. Am., Soc. Photo-Optical Instrumentation Engrs., IEEE, Phi Beta Kappa, Phi Kappa Phi. Subspecialties: Fiber optics; Optical signal processing. Current work: High-speed optoelectronics. Home: 2044 Lyon Ave Belmont CA 94002 Office: 1651 Page Mill Rd Bldg 28C Palo Alto CA 94304

NEY, EDWARD PURDY, physicist; b. Mpls., Oct. 28, 1920; s. Otto Frederick and Jessamine (Purdy) N.; m. June Felsing, June, 1942; children—Judy, John, Arthur, William. B.S., U. Minn., 1942; Ph.D., U. Va., 1947. Research asst., research asso. U. Va., 1940-46; cons. Naval Research Lab., 1943-44; asst. prof. U. Va., 1947; mem. faculty dept. physics and astronomy U. Minn., Mpls., 1947—, prof., 1955-74, U. Minn. regents prof., 1974—, chmn. dept. astronomy, 1974-78. Author: Electromagnetism and Relativity, 1962; contbr. articles to profl. jours. Recipient NASA Exceptional Sci. Achievement medal, 1975. Mem. Am. Phys. Soc., Am. Geophys. Union, Am. Astron. Soc., Nat. Acad. Sci., Am. Acad. Arts and Sci., Internat. Astron. Union, Sigma Xi. Subspecialties: Infrared optical astronomy; Geophysics. Home: 1925 Penn Ave S Minneapolis MN 55405 Office: Sch Physics and Astronomy 116 Church St SE Minneapolis MN 55455

NEZU, ARTHUR MAGUTH, psychologist; b. N.Y.C., Nov. 24, 1952; s. Tetsuo and Mary (Yabutani) N.; m. Christine Maguth, June 12, 1983; stepchildren: Frank, Alice, Linda. B.A., SUNY-Stony Brook, 1974, M.A., 1976, Ph.D., 1979. Lic. psychologist, N.Y., N.J. Clin. psychology intern Norwich Hosp., Conn., 1977-78; clin. asst. prof. dept. community dentistry Fairleigh Dickinson U., Hackensack, N.J., 1981—, dept. psychology, 1982—, coordinator tng. div. psychol. services, asst. dir., 1978—, co-dir Vets. Readjustment Program, 1982—, assoc. dir. Natural Setting Therapeutic Mgmt. Program, 1980—; cons. Pilgrim State Hosp., L.I., 1977; adj. prof. psychology Ramapo Coll. N.J., 1980. Contbr. articles to profl. jours.; editor books. N.Y. State Regents scholar, 1970-74; recipient SUNY-Stony Brook undergrad. psychology award, 1974. Mem. Am. Psychol. Assn., Assn. for Advancement Behavior Therapy, Assn. for Children with Down's Syndrome, Eastern Psychol. Assn., Am. Assn. Mental Deficiency, Soc. Behavioral Medicine, AAAS, Assn. Am. Dental Schs., Phi Beta Kappa. Subspecialties: Behavioral psychology; Cognition. Current work: Cognitive factors in psychopathology; problem-solving; clinical decision-making and judgement; behavioral problems of mentally retarded, depression, behavioral medicine/dentistry. Office: Div Psychol Services Fairleigh Dickinson U 452 Churchell Rd Teaneck NJ 07666

NG, LORENZ KENG YONG, physician, consultant, researcher; b. Singapore, Aug. 6, 1940; came to U.S., 1958, naturalized, 1969; s. Seak Khuan and Poh Hiang (Tan) N.; m. Roberta Melia, Dec. 7, 1981. A.B., Stanford U., 1961; M.D., Columbia U., 1965. Diplomate: Am. Bd. Psychiatry and Neurology, Nat. Bd. Med. Examiners. Chief resident in neurology Hosp. of U. Pa., Phila., 1968-69; spl. research fellow NIMH, 1969-72, spl. asst. to dir. drug abuse, 1972-76; chief Intramural Research Lab., Nat. Inst. on Drug Abuse, Bethesda, Md., 1976-78, chief pain studies program, 1978-81; med. dir. Washington Pain Ctr., 1982—; asst. prof. psychiatry Johns Hopkins U., 1978—; asst. clin. prof. neurology George Washington U., 1978—; med. dir. Consumers United Ins. Co., 1982—. Author: The Population Crisis: Implications and Plans for Action, 1964, Alternatives to Violence, 1968, Pain, Discomfort and Humanitarian Care, 1981, Strategies for Public Health, 1981, New Approaches to Treatment of Chronic Pain. Pres. World Man Fund. Served with USPHS, 1974-81. Recipient Young Investigators award for research in acupuncture Am. Soc. Chinese Medicine, 1975; Commendation medal USPHS, 1981. Mem. Am. Acad. Neurology (S. Weir Mitchell award 1971), Soc. for Neurosci., Soc. Biol. Psychiatry (A.E. Bennett award 1972), World Acad. Art and Sci.; mem. Am. Inst. Stress (founding dir.). Club: Cosmos (Washington). Subspecialties: Neurology; Preventive medicine. Current work: Treatment of pain and stress disorders; acupuncture; behavioral medicine and health enhancement. Office: 2026 R St NW Washington DC 20009

NG, THOMAS K., microbiologist; b. Macau, China, Sept. 30, 1954; s. Ming and Tsui Wah (Lee) N.; m. Theresa Puifun Chow, June 24, 1954. B.S., U. Wis.-Madison, 1975, M.S., 1977, Ph.D., 1981. Enzymologist Solar Energy Research Inst., Dept. of Energy, Golden, Colo., 1980-81; research scientist E.I. du Pont de Nemours & Co., Wilmington, Del., 1981—. Contbr. articles to profl. jours. Mem. Am. Soc. for Microbiology, Am. Chem. Soc., Sigma Xi. Subspecialties: Microbiology; Enzyme technology. Current work: Optimization and process control in fermentation; commodity and specialty chemicals production from biomaterials. Office: E I du Pont de Nemours & Co Wilmington DE 19898

NGAI, KIA LING, physicist; b. Canton, China, May 20, 1940; came to U.S., 1962; s. Hei Ming and Hok Yee (Cheung) N.; m. Linsen Hsia, Dec. 23, 1967; children: Siann, Seagan, Serin. B.Sc., U. Hong Kong, 1962; M.S. in Math, U. So. Calif., 1964; Ph.D. in Physics, U. Chgo., 1969. Research assoc. James Franck Inst., U. Chgo., 1969; staff mem. Lincoln Lab., M.I.T., 1969-71; research physicist Naval Research Lab., Washington, 1971-80, cons., 1980—. Recipient Navy Civilian Superior Service award Dept. Def., 1976. Mem. Am. Phys. Soc., Am. Chem. Soc. Subspecialties: Condensed matter physics; Polymer physics. Current work: Unified model of universal relaxation properties of condensed matter. Home: 6500 Bellamine Ct McLean VA 22101 Office: Naval Research Lab Washington DC 20375

NGHIEM, DAI DAO, surgeon; b. Thai-Nguyen, Vietnam, July 14, 1941; came to U.S., 1975, naturalized, 1983; s. Ho Xuan and Uc Thi (Nguyen) N.; m. Phuong Thi Le, May 20, 1969; children: Giang-Chau, Giang-Huong, Giang-Tien. M.D., Saigon Med. Sch., 1969. Diplomate: Am. Bd. Surgery. Intern Binh Dan Hosp., Saigon, 1965, resident, 1966-69, Pitts. Univ. Hosp., 1973-74, U. Ark., Little Rock, 1975-77, W.Va. U., Charleston, 1977-78; asst. prof. surgery Saigon Med. Sch., 1975; fellow in transplantation Med. Coll. Va., 1978-80; asst. prof. surgery U. Iowa, 1981—. Contbr. articles to profl. jours. Served to 1st lt. Army Republic of Vietnam, 1970-75. NIH grantee, 1981-82. Mem. A.C.S., AMA, Am. Soc. Transplant Surgeons, Assn. Acad. Surgery. Subspecialties: Transplantation; Transplant surgery. Current work: Prevention of ischemic organ injuries by prostacyclin and superoxide dismutase, organ preservation, transplantation biology; role of total lymphoid irradiation and cyclosporine A. in the presensitized recipient; transplantation of kidneys and solid organs. Office: Dept Surgery Univ Hosp Iowa City IA 52242

NGUYEN, CHINH TRUNG, electrical engineer, researcher; b. Hanoi, Vietnam, Nov. 19, 1946; came to Can., 1971; s. Hau Van and Nga Thi (Vu) N.; m. Hallouma Khediri, Feb. 11, 1974; 1 child, Tien Gia. Diploma in Engring, Swiss Fed. Inst. Tech., Lausanne, 1970; M.S.E.E., Laval U., Que., 1972; Ph.D., U. Que., 1982. Research asst. Swiss Fed. Inst. Tech., Lausanne, 1970-71; computer analyst Hydro-Quebec, Varennes, Que., Can., 1972-76; sr. researcher IREQ, 1976—; invited prof. U. Que., Varennes, 1973-74. Contbr. articles to profl. publs. Mem. IEEE (sr., Paper award 1982, W.R.G. Baker award 1983), Order of Engrs. Que., Soc. Indsl. and Applied Math. Subspecialties: Computer engineering; Mathematical software. Current work: Numerical control systems, computer simulation of power systems, parallel processing, computer networking, digital measurement and instrumentation. Patentee frequency digital meter. Office: Institute de Recherche d'Hydro-Quebec Varennes PQ Canada JOL 2P0

NGUYEN, VIETCHAU, research and development company executive; b. Hue City, Vietnam, Oct. 1, 1953; came to U.S., 1971, naturalized, 1981; s. Hai V. and Ha-Lanh (Ton-Nu) N.; m. Amy Catherine Hamlin, June 21, 1980; 1 son, Gabriel. B.S.C.E. with honors, U. Minn., 1974, M.S.C.E., 1976; M.A., Princeton U., 1977, Ph.D., 1979. Research staff St. Anthony Falls Hyd. Lab., U. Minn, Mpls., 1973-75; research lectr. Sch. Applied Sci. & Engring., Princeton (N.J.) U., 1979-82; prin. investigator, head Research Div. MKA Inc., Mpls., 1982—. Contbr. articles to profl. jours. Shell Found. fellow, 1977; NSF fellow, 1978; Wallace Meml. award Princeton Faculty, 1979. Mem. Am. Geophys. Union, Am. Inst. Chem. Engrs., ASCE, Internat. Assn. Math. and Computers in Simulation, Soc. Indsl. and Applied Math., Soc. Petroleum Engrs. of AIME. Subspecialties: Theoretical and applied mechanics; Numerical analysis. Current work: Theoretical and exptl. studies of multiphase thermochem. flow in porous and fractured media. Office: 12800 Industrial Park Blvd Plymouth MN 55441

NIBHANUPUDY, JAGANNADHA RAO, medical physicist, consultant; b. Bapatla, India, June 19, 1940; came to U.S., 1968; s. Narasimha Rao and Tayaramma (Rachapudy) N.; m. Janaki Anuradha Poluri, Aug. 21, 1964; children: Narasimha Rao, Padmavathi. B.Sc., Andhra (India) U., 1959; M.S., Vikram (India) U., 1963; M.S.E., U. Wash., Seattle, 1970. Diplomate: Am. Bd. Radiology. Demonstrator Loyola Coll., Vijayawada, India, 1960-61; sci. officer Bhabha Atomic Research Center, Bombay, India, 1964-68; med. physicist Howard U. Hosp., Washington, 1971—; asst. prof. Howard U. Coll. Medicine, 1978—. Contbr. articles to profl. jours. Sec. Greater Washington Telugu Soc., 1976. Mem. Am. Assn. Physicists in Medicine. Subspecialties: Medical physics; Radiology. Current work: Physics aspects of intra-operative radiotherapy, Brachy therapy, and hyperthermia in cancer treatment; teaching resident physicians, medical students and radiotherapy technology students. Home: 8509 Brae Brooke Dr Lanham MD 20706 Office: Dept Radiotherapy Howard Univ Hosp 2041 Georgia Ave NW Washington DC 20060

NICCOLAI, MARINO JOHN, computer science educator, consultant; b. Pitts., Feb. 9, 1943; s. William and Marzia (Niccolai) N.; m. Frances Ramey, Nov. 13, 1965; children: Anne-Marie, Andrew. B.S. in Math. and Physics, Auburn U., 1963, M.S., U. Del., 1968; Ph.D. in Applied Math, N.C. State U., 1976. System analyst Mitre Corp., Washington, 1969-72; asst. prof. math. and computer sci. U. Louisville, 1976-81; assoc. prof. computer sci. U. South Ala., 1981—; cons. computer sci., Mobile, Ala., 1981—. Served to lt. USN, 1964-69. Mem. Math. Assn. Am., Soc. Indsl. and Applied Math., Am. Math. Assn., Assn. Computing Machines, Ala. Council Computing in Edn., Am. Assn. Engring. Edn. Roman Catholic. Subspecialties: Mathematical software; Software engineering. Current work: Computer simulation and math modelling of engineering/business system—integration of softward and graphics. Office: U South Ala Mobile AL 36688

NICHOLS, BUFORD LEE, JR., medical educator, physician, nutritionist, researcher; b. Ft. Worth, Dec. 21, 1931; m. (married); 3 children. B.A., Baylor U., 1955, M.S., 1959; M.D., Yale U., 1960. Diplomate: Am. Bd. Pediatrics. Instr. in physiology Baylor U. Coll. Medicine, Houston, 1956-57, instr. in physiology and pediatrics, 1964-66, from asst. prof. to assoc. prof. pediatrics, 1966-67, instr. in physiology, 1966-74, assoc. prof. community medicine, chief sect. nutrition and gastroenterology, 1970—; prof. physiology and pediatrics, 1977—; sci. dir. Children's Nutrition Research Ctr.; instr. pediatrics Yale U., 1963-64. Mem. Am. Acad. Pediatrics, Am. Soc. Clin. Nutrition, Am. Coll. Nutrition (pres. 1975-76). Subspecialties: Nutrition (medicine); Pediatrics. Current work: Environmental effects upon growth and development of the infant, especially alterations in body composition and muscle physiology in malnutrition; diarrhea and infectious deseases. Office: Dept Pediatrics Baylor Coll Medicine Tex Med Ctr Houston TX 77030

NICHOLS, PAUL ARTHUR, nuclear engineer; b. Manchester, N.H., Feb. 6, 1956; s. Richard Arthur and Barbara Helen (Deans) N.; m. Gail Renee Lawyer, Feb. 2, 1980; children: Tracy, Kelly. B.Nuclear Engring., Ga. Inst. Tech., 1978. Jr. engr. Cleve. Electric Illuminating Co., North Perry, Ohio, 1978-80, assoc. engr., 1980-82, lead NSSS engr., 1982—. Scoutmaster Greater Cleve. council Boy Scouts Am., 1979-83. Mem. Am. Nuclear Soc., ASME (assoc.). Republican. Fundamental Christian. Subspecialties: Nuclear engineering; Mechanical engineering. Current work: Design and installation engineering at nuclear plant. Office: Perry Nuclear Power Plant 10 Center Rd North Perry OH 44081

NICHOLS, WALTER KIRT, surgeon, educator; b. Flint, Mich., Sept. 3, 1941; s. Walter H. and Hazel G. (Marshal) N.; m. C. Paula Cooper, June 19, 1965 (div. Dec. 1978); children: Krystin Jennifer, Kreg Kirt, Kecia Liane. A.S., Flint Jr. Coll., 1961; M.D., U. Mich., 1966. Diplomate: Am. Bd. Surgery. Intern. St. Joseph Mercy Hosp., Ann Arbor, Mich., 1966-67; resident U. Mo. Hosp., Columbia, 1967-68, 70-73, asst. prof. surgery, 1973-79, assoc. prof., 1979—; asst. chief surgery VA Hosp., Columbia, 1974—, acting chief, 1977-78. Contbr. articles to profl. jours. Served to capt. USAF, 1968-70. Fellow ACS; mem. Assn. Acad. Surgery, Internat. Cardiovascular Surg. Soc., Frederick A. Coller Surg. Soc., Midwestern Vascular Surg. Soc., Assn. Surg. Edn., Mo. State Surg. Soc. Subspecialty: Surgery. Current work: Non-invasive vascular diagnosis: imaging; computer applications to medical care. Office: M580 Univ Med Center Columbia MO 65212

NICHOLS, WARREN WESLEY, research physician; b. Collingswood, N.J., May 16, 1929; s. David W. and Marie A. (Ringheiser) N.; m. June Helms, Aug. 28, 1953; children: Warren Wesley, Sean H., Lisa Karin. B.S., Rutgers U., 1950; M.D., Jefferson Med. Coll., 1954; Fil. Lic., U. Lund, Sweden, 1964, Ph.D., 1966. Diplomate: Am. Bd. Pediatrics. Chief med. staff Camden Mcpl. Hosp. Contagious Diseases, 1959-61; asst. physician Children's Hosp. Phila., 1959-65; asso. Inst. Med. Research, Camden, N.J., 1959-65, mem., 1966—, asst. dir., 1963-81, Emlen Stokes prof. genetics, 1975—, v.p. research, 1981—, also head dept. cytogenetics; adj. prof. human genetics U. Pa.; clin. prof. pediatrics N.J. Coll. Medicine and Dentistry. Contbr. articles to profl. jours and books. Served with USAF, 1957-59. Recipient Career Devel. award NIH, 1963-72, Alumni Achievement award Jefferson Med. Coll., 1980. Mem. AMA, Am. Acad. Pediatrics, Soc. Pediatric Research, Mendelian Soc., Genetics Soc. Am., Am. Soc. Human Genetics, John Morgan Soc., N.Y. Acad. Scis., Am. Soc. Clin. Investigation, Am. Assn. Cancer Research, Gerontol. Soc. Subspecialties: Gene actions; Cancer research (medicine). Current work: Research on spontaneous and induced gene and chromosome mutations and their role in carcinogenesis, aging and hereditary disease. Emphasis on virus induced cellular genetic changes and high risk cancer individuals and families. Office: Inst Med Research Copewood St Camden NJ 08103

NICHOLSON-GUTHRIE, CATHERINE SHIRLEY, geneticist; b. Jackson, Miss., May 26, 1938; d. James Benjamin and Catherine (Steele) Nicholson; m. George Drake Guthrie, Aug. 5, 1961; 1 son, George Drake. B.S., Auburn U., 1957; M.S., Fla. State U., 1960; Ph.D., Ind. U., 1972. Biologist So. Research Inst., Birmingham, Ala., 1957; instr. dept. biology Fla. State U., Tallahassee, 1960; research asst. biology Calif. Inst. Tech., Pasadena, 1960-62; instr. dept. biology Boston State Coll., 1963-64; vis. asst. prof. biology U. Evansville, Ind., 1972-73; instr., researcher dept. med. genetics Ind. U. Sch. Medicine, Indpls., 1974—; profl. staff U.S. Ho. of Reps., Washington, 1981; also cons., seminar speaker, lectr. Contbr. articles to profl. lit. State bd. dirs. Citizens Energy Coalition, Indpls, 1975-76; vol. genetic counselor, Evansville, 1975—. Recipient Ala. Acad. Sci. research award, 1956; NIH tng. grantee, 1967-71; Sarah Berliner fellow AAUW, 1978-79; AAAS mass media sci. fellow, 1979. Mem. Genetics Soc. Am., AAAS, Sigma Xi. Subspecialty: Gene actions. Current work: Genetic control of chlorophyll synthesis; chloroplast function. Home: 700 Drexel Dr Evansville IN 47712 Office: Ind U Sch Medicine 8600 University Blvd Evansville IN 47712

NICKAS, GEORGE DEMOSTHENES, physicist; b. Chgo., Mar. 7, 1942; s. Andrew George and Delia N.; m. Mildred Anne McKee, Mar. 9, 1966; m. Jeanette Chelf, June 8, 1972; 1 son, George Demosthenes. B.S., Ill. Inst. Tech., 1964; M.S., U. Ill., 1967, Ph.D., 1972. Grad teaching and research asst. U. Ill., Urbana, 1965-72; postdoctoral research and teaching fellow U. B.C., Vancouver, 1972-74; writer book reviewer Vancouver (B.C.) Sun newspaper, 1974-78; research dir. McDonald Research Assocs. Ltd., Vancouver, 1978-80; instr. One-To-One Ednl. Services, Vancouver, 1980-81, Capilano Coll., North Vancouver, B.C., 1981-82; asst. prof. physics Trinity U., San Antonio, 1982—. Recipient Delta Star Scholarship award, 1961-64. Mem. Am. Astron. Soc., Capilano Coll. Faculty Assn., Sigma Pi Sigma. Greek Orthodox. Subspecialties: Theoretical astrophysics; Theoretical physics. Current work: The dynamics of asymmetries in galaxies, theoretical astrophysics, theoretical physics. Home: 5907 Hillman St San Antonio TX 78218 Office: Dept Physics Trinity U San Antonio TX

NICKERSON, NORTON HART, biologist, ecologist, educator, consultant; b. Quincy, Mass., Apr. 14, 1926; s. Norton Hart and Mary Almina (Whitney) N.; m. Joan Young, Jan. 30, 1954; children: William, Susan, Jonathan. B.S., U. Mass., Amherst, 1949; M.A., U. Tex., Austin, 1951; Ph.D., Washington U., St. Louis, 1953. Instr. U. Mass., 1953-56; Cornell U., Ithaca, N.Y., 1956-58; asst. prof. botany Washington U., 1958-62, assoc. prof., 1962-63; assoc. prof. biology Tufts U., Medford, Mass., 1963-81, prof. environ. studies, 1981—; morphologist Mo. Bot. Garden, 1958-63; mem. commn. ecology Internat. Union Conservation of Nature; chmn. Mass. Hazardous Waste Facility Site Safety Council, 1980—; mem. Mass. Bd. Environ. Mgmt., 1977-82, Mass. Agrl. Lands Preservation Com., 1978—. Contbr. articles to tech. jours. Mem. Dennis (Mass.) Conservation Commn., 1964-82; health agt., Dennis, 1971-83. Served to cpl. USAF, 1944-45; 2d lt. Res., 1949-55. Mem. Soc. Study of Evolution, AAAS, Bot. Soc. Am., New Eng. Bot. Club (editor Rhodora 1982—), Ecol. Soc. Am., Explorers Club, Sigma Xi. Subspecialties: Ecology; Resource conservation. Current work: Functioning of coastal ecosystems (Mangroves, salt-marshes, barrier beaches), management of park resources. Office: Dept Biology Tufts U Medford MA 02155

NICKLAS, JANICE ANN, immunogeneticist; b. Buffalo, Sept. 1, 1953; d. James P. and Janet D. (Fink) N.; m. David S. Dummit, June 14, 1975. B.S., Calif. Inst. Tech., 1975; M.A., Princeton U., 1977, Ph.D., 1981. Postdoctoral researcher Cancer Research Ctr., Tufts U. Sch. Medicine, 1981-82, Immunobiology Research Ctr. dept. lab. medicine and pathology U. Minn.-Mpls., 1982—. NSF fellow, 1975-78. Mem. Genetics Soc. Am. Subspecialties: Immunogenetics; Genetics and genetic engineering (biology). Current work: Determining the protein and mRNA. differences between T cell types selecting mutants of HLA in cell culture. Office: Immunobiology Research Center U Minn Minneapolis MN 55455

NICKLAS, ROBERT BRUCE, cell biologist; b. Lakewood, Ohio, May 29, 1932; s. Ford Adelbert and Marthabelle (Beckett) N.; m. Sheila Jean Counce, Sept. 24, 1960. B.A., Bowling Green State U., 1954; M.A. (Eugene Higgins fellow), Columbia U., 1956, Ph.D., 1958. Instr. in zoology Yale U., 1958-61, asst. prof. zoology, 1961-64, asso. prof., 1964-65, Duke U., 1965-71, prof., 1971—, chairperson dept., 1983—; mem. NSF Postdoctoral Fellowship Panel, 1969-71, Am. Cancer Soc. Sci. Adv. Com. for Virology and Cell Biology, 1975-78. Contbr. numerous articles to profl. publs.; editorial bd.: Chromosoma, 1966-83, Jour. Exptl. Zoology, 1970-72, Jour Cell Biology, 1980-81. Recipient award for disting. teaching Duke Alumni, 1975; Yale fellow in scis., 1963-64; John Simon Guggenheim fellow, 1972-73; grantee Inst. Gen. Med. Scis. USPHS, 1960—. Fellow AAAS; mem. Am. Soc. Cell Biology (exec. com. 1976-78, council 1975-78), Am. Soc. Naturalists, Genetics Soc. Am., Soc. Gen. Physiologists, Sigma Xi. Subspecialty: Cell biology. Home: 3101 Camelot Ct Durham NC 27705 Office: Dept Zoology Duke U Durham NC 27706

NICKLAS, WILLIAM JOHN, biochemist, research, educator; b. Phila., Jan. 13, 1939; s. Julius and Ida (Perl) N.; m. Ruth Joan Horvay, Aug. 8, 1964; 1 son, John. B.A. in Chemistry (scholar), La Salle Coll., Phila., 1964; Ph.D. in Biochemistry (NASA fellow), Fordham U., 1969. Teaching asst. Fordham U., Bronx, N.Y., 1964-65; NIH postdoctoral fellow Johnson Research Found., U. Pa., Phila., 1968-70; research assoc. Columbia U. Coll. Physicians and Surgeons, N.Y.C., 1970-73; research asst. prof. Mt. Sinai Sch. Medicine, N.Y.C., 1973-77, research assoc. prof., 1977-80; assoc. prof. dept. neurology Univ. Medicine and Denistry N.J.-Rutgers Med. Sch., Piscataway, 1980—, adj. assoc. prof. dept. pharmacology, 1980—, mem. grad. faculty, 1980—, co-dir. Mollie and Jerome Levine Neuroscis. Lab., 1980—. Contbr. numerous articles and abstracts to sci. publs., chpts. to books. Del. to local and county Democratic convs., 1980, 81, 82, active mem., officer local civic assns., 1974—. Served with U.S. Army Res., 1962-68. NIH grantee, 1973-80, 81—. Mem. Soc. for Neurosci., AAAS, Biochem. Soc. (editorial advisor Biochem. Jour. 1981—), Am. Soc. for Neurochemistry (membership com. 1981—), Phi Beta Kappa, Sigma Xi, Phi Lambda Upsilon. Subspecialties: Neurochemistry; Neurobiology. Current work: Research involves role of neuronal-glial interactions in regulation of transmitter amino acid metabolism and function. This involves classic biochemical techniques with newer aspects of cell biology. Office: Dept Neurology UMDNJ-Rutgers Med Sch Piscataway NJ 08854

NICKOL, BRENT BONNER, parasitology educator; b. Agosta, Ohio, June 22, 1940; s. J. Russell and Marjorie Maxine (Powelson) N.; m. Lynn Nickol, Aug. 22, 1964; children: Devin Reed, David Andrew. B.A., Coll. Wooster, 1962; M.S., La. State U., 1963, Ph.D., 1966. Asst. prof. U. Nebr., Lincoln, 1966-71, assoc. prof., 1971-75; dir. Cedar Point Biol. Sta., 1975-79, prof. dept. life sci., 1975—. Contbr. articles to profl. jours.; editor: Host-Parasite Interfaces, 1979; editorial bd.: Am. Soc. Parasitologists, 1981—, Helminthol. Soc. Washington, 1978-80; editor: Transactions, Nebr. Acad. Sci., 1981—. Mem. Am. Soc. Parasitologists, Wildlife Disease Assn., Helminthol. Soc. Washington, La. Acad. Sci., Nebr. Acad. Sci. Subspecialties: Species interaction; Epidemiology. Current work: Study of parasite flow thru communities; factors of transmission, density-dependent regulation of populations, dispersal of acanthocephalan parasites. Office: Sch Life Sci Univ Nebr Lincoln NE 68588

NICKOLOFF, EDWARD L., med. radiation physicist; b. Harrisburg, Pa., Dec. 6, 1942; s. Boris and Pearl Margaretta (Rosenberry) N.; m. Eileen Lanigan, June 28, 1968. Student, Carnegie Inst. Tech., 1961-63; B.S., Lebanon Valley Coll., 1965; M.S., U.N.H., 1968; Sc.D., Johns Hopkins U., 1977. Asst. prof. radiology Johns Hopkins U. Med. Inst., Balt., 1974-81; asso. prof. physics dept. radiology Columbia U. Coll. Physicians and Surgeons, N.Y.C., 1981—; hosp. physicist univ., 1981—. Contbr. articles to profl. jours. Mem. Soc. Nuclear Medicine, Health Physics Soc., Am. Assn. Physicists in Medicine, Am. Coll. Radiology, Radiol. Assn. Med. Physicists. Subspecialties: Imaging technology; Nuclear medicine. Current work: Medical physics in diagnostic radiology and nuclear medicine. Office: Columbia-Presbyn Med Center 622 W 168th St New York NY 10032

NICOLAE, GHEORGHE, research chemist; b. Focsani, Romania, Oct. 2, 1943; came to U.S., 1979; s. Gheorghe Enache and Victoria (Vasiliu) N.; m. Mariana Lorisa Vasiliu, July 9, 1977; children: Ovidiu-Bogdan, Iulia-Cristina. M.S., U. Bucharest, 1966, Ph.D., 1974. Prof. asst. U. Bucharest, 1967-73, Poly. Inst., Bucharest, 1973-79; research chemist Wallace A. Erickson, Chgo., 1980; research mgr. Confi-Dental Co., Chgo., 1981—. Mem. Internat. Assn. Dental Research, Am. Assn. Dental Research; mem. Am. Chem. Soc.; Mem. Assn. Finishing Processes of Soc. Mfg. Engrs. (sr.). Club: Amoco Motor (Chgo.). Subspecialties: Organic chemistry; Synthetic chemistry. Current work: Heterocyclic chemistry, phenoxathiins derivatives, octahydro-xanthylium compounds, octahydroacridinium compounds, pyrylium salts, acrylic monomers, urethanes derivatives. Inventor in field. Home: 3709 W Leland St Chicago IL 60625 Office: Confi-Dental Products Co 1900 N Clybourne Ave Chicago IL 60614

NICOLOSI, STEPHEN LOUIS, research laboratory radiochemical engineering scientist; b. N.Y.C., Feb. 13, 1950; s. Louis and Angelina (Detomasso) N. B.S., SUNY-Stony Brook, 1973, M.S., 1979. Chemistry assoc. Brookhaven Nat. Lab., Upton, N.Y., 1974-80; research scientist Battelle Columbus Labs., Columbus, Ohio, 1980—. Mem. Am. Nuclear Soc., AAAS. Subspecialties: Nuclear fission; Chemical engineering. Current work: Reactor safety research and nuclear waste management research; computer modeling of radionuclide behavior. Home: Apt O 560 W 4th Ave Columbus OH 43201 Office: Battelle Columbus Labs 505 King Ave Columbus OH 43201

NICOLSON, GARTH L., cancer cell biologist, educator; b. Los Angeles, Oct. 1, 1943; s. Garth F. and Joan D. (Lamb) N. B.S. in Chemistry, UCLA, 1965; M.S. in Biophysics, U. Hawaii, 1967; Ph.D. in Cell Biology, U. Calif., San Diego, 1970. Research assoc. Salk Inst., La Jolla, Calif., 1971-74, head dept. cancer biology, 1974-76; prof. cell biology U. Calif., Irvine, 1975-80, prof. physiology, 1977-80; Florence M. Thomas prof. cancer research, head dept. tumor biology U. Tex. Cancer Ctr., M.D. Anderson Hosp. and Tumor Inst., Houston, 1980—; adj. prof. vet. pathology Tex. A&M U., 1981—; adj. prof. pathology U. Tex. Med. Sch., Houston, 1981—. Editor 10 books in field; contbr. numerous articles to profl. joursl; editor: Cell Surface Revs, 1976—, Clin. and Exptl. Metastasis, 1981—, assoc. editor: Cancer Biology Revs, 1979—, Exptl. Cell Research, 1976—, Invasion and Metastasis, 1982—, Jour. Cellular Biochemistry, 1973—, Yearbook of Cancer, 1980—, Cancer Research, 1978—, Cancer Metastasis Revs, 1982—, Biochimica et Biophysica Acta, 1974-79, Gamate Research, 1978—, Molecular Cell Biochemistry, 1982—. Recipient Presdl. award Electron Microscopy Soc. Am., 1971; Upjohn Biology Edn. award, 1976; Ann. award Japan Histochemical Soc., 1976; Guy Lipscomb Meml. award, 1979. Mem. Am. Soc. Cancer Research, Am. Soc. Biol. Chemists, Am. Soc. Cell Biology, Biophysical Soc., N.Y. Acad. Scis. Subspecialties: Cancer research (medicine); Cell biology. Current work: Mechanisms of cancer metastasis, cell surface properties of cancer, development of new procedures to treat metastatic cancer. Office: Dept Tumor Biology U Tex Cancer Ctr MD Anderson Hosp and Tumor Inst Houston TX 77030

NIEDBALSKI, JOSEPH FRANCIS, research horticulturist; b. Dunkirk, N.Y., Nov. 20, 1931; s. Joseph Frank and Dorothy (Sczublewski) N; m. Joan Marie Christy, Oct. 3, 1953; children: Molly Niedbalski Cline, Joseph S., Peter John, Jamie John. Student, Gannon Coll., 1949-50, SUNY-Fredonia, 1950-51; B.S. in Pomology and Entomology, Rochester Inst. Tech., 1961. Cert. Am. Registry Profl. Entomologists. Research horticulturist Gerber Products, 1958-66; tech. extention rep., head, research head, mgr. Upjohn Co., Kalamazoo, from 1966, now research head, plant health research and devel., product support and field devel. Contbr. papers to profl. confs., publs. Served to capt. U.S. Army, 1952-55. Mem. Am. Soc. Hort. Sci., Am. Phytopathol. Soc., Entomol. Soc. Am., Weed Sci. Soc. Am. Democrat. Roman Catholic. Subspecialty: Agricultural chemicals. Current work: Development of plant health products and agricultural chemicals. Home: 1333 Lakeway Ave Kalamazoo MI 49001 Office: Upjohn Co Unit 9700 Bldg 50-1 Kalamazoo MI 49001

NIEDERKORN, IOAN STEFAN, research metallurgist; b. Oradea, Romania, June 14, 1930; s. Ioan Gavril and Adela Terezia Bekics (Dengyel) N.; m. Taisia Liapina, May 17, 1954; children: Emilia, Irena, Mihel. B.S., M.S., Urals Poly, Inst., Sverdlovsk, USSR, 1954; D.Sci., Inst. of Non-Ferrous Metals and Gold, Moscow, 1958. Researcher Mettall. Research Inst., Bucharet, Romania, 1958-59; sr. researcher Research Inst. Chemistry, Bucharest, 1959-66, head research lab., 1960-66; head research div. and labs. Research and Engring. Ctr. for Raioactive Metals, Bucharest, 1966-81; research metallurgist N.Mex. Inst. Mining and Tech., Socorro, 1982-83; dir. research and devel. Ultra Instrument Lab., Dallas, 1983—; postdoctoral fellow French Atomic Energy Commn., Paris, 1971; sr. research fellow, prof. Sci. and Devel. Com. Romania, Bucharest, 1976. Author 7 books in Romanian and Hungarian; Contbr. articles to profl. jours.; invited lectr. numerous confs. Mem. Am. Nuclear Soc., Metall. Soc. of AIME. Club: Scorro Tennis. Subspecialties: Nuclear engineering; Metallurgical engineering. Current work: Research on radioactive decay products removal from uranium mill tailings; Kinetics of catalitic oxidation of sphalerit concentrates. Patentee in field. Office: Ultra Instrument Lab 10031 Monroe Dr Dallas TX 75229

NIEH, BILL, computer scientist; b. Foochow, Fukien, China, July 1, 1947; came to U.S., 1970, naturalized, 1983; s. Yun-Tiao and Pi-Kung (Liu) N.; m. Wendy C. Nieh, Dec. 23, 1972; children: Rose, Audrey. B.S., Chung-Hsing U., 1969; M.S., West Coast U., 1973; Ph.D., UCLA, 1979. Research engr. UCLA, 1974-79; jr. specialist U. Calif., Berkeley, 1977-79; sr. tech. staff Logicon, San Pedro, Calif., 1979—; lectr. Sch. Bus. Adminstrn., Calif. State U., Long Beach, 1981-83. Mem. Ops. Research Soc. Am., Inst. Mgmt. Sci. Subspecialties: Software engineering; Operations research (engineering). Current work: Software reliability studies, computer security and systems engineering of command, control and communication systems; software verification and validation. Office: 255 W 5th St San Pedro CA 90733

NIELSEN, KLAUS H.B., immunology educator; b. Copenhagen, Denmark, Aug. 18, 1945; s. K.B. and and D. (Sorensen) N.; m. Elizabeth Vincent., Sept. 1, 1966; children: Andrew, David. B. Sc. in Agr, U. Guelph, Can., 1969, M.Sc., 1971; Ph.D., U. Glasgow, 1974. Asst. prof. vet. microbiology U. Guelph, 1974-77; research scientist Agr. Can., Ottawa, Ont., 1977-81; assoc. prof. immunology Tex. A&M U. Vet. Coll., 1981—. Author: (with others) Radioimmunoassay of Antibody, 1972; contbr. articles to sci. jours., chpts. to books. Mem. Am. Assn. Immunologists, Brit. Soc. Immunology, Can. Soc. Immunlogy, Sigma Xi, Phi Zeta. Subspecialties: Immunobiology and immunology; Microbiology. Current work: Immunochemistry of immune response to infectious agents, complement, antigenic composition of bacteria. Home: 1204 Dominik Dr College Station TX 77840 Office: Dept of Veterinary Pathology Texas A&M University College Station TX 77843

NIEMAN, TIMOTHY ALAN, analytical chemistry educator; b. Cin., Dec. 31, 1948; s. Everett Orville and Emma W. (Hoffmeier) N.; m. Sandra Louise Toth, Aug. 29, 1970; 1 dau., Hilary Anne. B.S. in Chemistry, Purdue U., 1971; Ph.D., Mich. State U., 1975. Grad. asst. in chemistry Mich. State U., 1971-75; asst. prof. chemistry U. Ill.-Urbana, 1975-81, assoc. prof., 1981—. L. L. Quill fellow Mich. State U., 1971-74; recipient teaching award Chemistry Dept U. Ill., 1981. Mem. Am. Chem. Soc., Soc. Applied Spectroscopy. Subspecialty: Analytical chemistry. Current work: Analytical chemiluminescence; bipolar pulse conductance; ion selective electrodes; coulostatics. Home: 204 E McHenry St Urbana IL 61801 Office: Chemistry Dept ll 1209 W California St Urbana IL 61801

NIER, ALFRED OTTO CARL, physicist; b. St. Paul, May 28, 1911; s. August Carl and Anna J. (Stoll) N.; m. Ruth E. Andersen, June 19, 1937; children—Janet, Keith; m. Ardis L. Hovland, June 21, 1969. B.S., U. Minn., 1931, M.S., 1933, Ph.D., 1936. Nat. Research fellow Harvard, 1936-38; asst. prof. physics U. Minn., 1938-40, asso. prof., 1940-43, prof., 1946—; physicist Kellex Corp., N.Y.C., 1943-45. Mem. AAAS, Am. Phys. Soc., Minn., Nat. acads. scis., Geochem. Soc., Am. Geophys. Union, Am. Philos. Soc., Am. Assn. Physics Tchrs., Geol. Soc. Am., Am. Soc. Mass Spectrometry, Am. Vacuum Soc., Soc. Applied Spectroscopy, Am. Acad. Arts and Scis., Royal Swedish Acad. Scis., Sigma Xi. Subspecialty: Planetary atmospheres. Current work: Mass spectrometry of planetary atmospheres. Research activities include devel. of mass spectrometer and its application to problems in physics, chemistry, geology, medicine, space sci.; first to separate rare isotope of uranium, U-235, 1940; with J.R. Dunning, E.T. Booth and A.V. Grosse of Columbia, demonstrated it was source of atomic energy when uranium is bombarded with slow neutrons. Home: 2001 Aldine St St Paul MN 55113

NIERENBERG, WILLIAM AARON, oceanography educator; b. N.Y.C., Feb. 13, 1919; s. Joseph and Minnie (Drucker) N.; m. Edith Meyerson, Nov. 21, 1941; children—Victoria Jean (Mrs. Tschinkel), Nicolas Clarke Eugene. Aaron Naumberg scholar, U. Paris, 1937-38; B.S., Coll. City N.Y., 1939; M.A., Columbia U., 1942, Ph.D. (NRC predoctoral fellow), 1947. Tutor Coll. City N.Y., 1939-42; sect. leader Manhattan Project, 1942-45; instr. physics Columbia, 1946-48; asst. prof. physics U. Mich., 1948-50; asso. prof. physics U. Calif. at Berkeley, 1950-53, prof., 1954-65; dir. Scripps Instn. Oceanography, 1965—; vice chancellor for marine scis. U. Calif. at San Diego, 1969 ; dir. Hudson Labs., Columbia, 1953-54; asso. prof. U. Paris, 1960-62; asst. sec. gen. NATO for sci. affairs, 1960-62; spl. cons. Exec. Office Pres., 1958-60, White House Office Sci. and Tech. Policy, 1976—. Contbr. papers to profl. jours. E.O. Lawrence lectr. Nat. Acad. Sci., 1958; Miller Found. fellow, 1957-59, Sloan Found. fellow, 1958, Fulbright fellow, 1960-61; mem. U.S. Nat. Commn. UNESCO, 1964-68, Calif. Adv. Com. on Marine and Coastal Resources, 1967-71; adviser-at-large U.S. Dept. State, 1968—; mem. Nat. Sci. Bd., 1972-78, 82—, Nat. Adv. Com. on Oceans and Atmosphere, 1971-77; chmn. Nat. Adv. Com. on Oceans and Atmosphere, 1971-75; mem. sci. and tech. adv. Council Calif. Assembly; mem. adv. council NASA, 1978-83, chmn. adv. council, 1978-82; mem. council Nat. Acad. Scis., 1979—. NATO Sr. Sci. fellow, 1969; Decorated officer Nat. Order of Merit, France; recipient Golden Dolphin award Assn. Artistico Letteraria Internazionale, Disting. Pub. Service medal NASA, 1982, Compass award Marine Tech. Soc., 1975; Procter prize Sigma Xi, 1977. Fellow Am. Phys. Soc. (council, sec. Pacific Coast sect. 1955-64); mem. Am. Acad. Arts and Scis., Nat. Acad. Engring., Nat. Acad. Scis., Am. Philos. Soc., Am. Assn. Naval Architects, Navy League, Fgn. Policy Assn. (mem. nat. council), Sigma Xi (pres. 1981—). Club: Cosmos. Subspecialties: Nuclear physics; Oceanography. Current work: Expertise in a number of disparate fields. Original field was nuclear physics; that is continuing interest. Current interest in ocean science and ocean engineering with a secondary interest in civil aviation. Deeply concerned with defense matters in a variety of ways, including membership on Defense Science Board and chairmanship of Jason group. Home: 9581 La Jolla Farms Rd La Jolla CA 92037 Office: Scripps Instn Oceanography Dir's Office A-010 U Calif San Diego La Jolla CA 92093

NIGAM, RAJENDRA C., systems analyst, computer scientist; b. Ghatampur, India, June 28, 1930; came to U.S., 1957, naturalized, 1968; s. Ram S. and Ram P. N.; m. Vimala Nigam, Mar. 5, 1953; children: Neeraj, Kamana, Alok, Arti. M.Sc. in Math, U. Lucknow, India, 1950; M.S. in Astronomy, U. Ill., 1958. Astronomer Smithsonian Astrophys. Obs., Cambridge, Mass., 1959-63; with IIT Intelcom Inc., Bellcom, Inc., Washington, 1963-67, MTS, Bedford, Mass., 1967-71; sr. fellow East-West Center, Honolulu, 1976; vis. prof. mgmt. B.H.U., Varanasi, India, 1979-80; systems analyst and designer MTS, Computer Scis. Corp., Silver Spring, Md., 1981-83; astronomer M/A-Com Sci. Data Center, Nat. Space Sci. Data Ctr., Greenbelt, Md., 1983—. Author: Happier Living, 1979; contbr. articles to profl. jours. Mem. World Future Soc., Am. Astron. Soc. Hindu. Subspecialties: Database systems; Astronautics. Current work: DBMS and computer software related to information storage and on-line retrieval. Home: 4813 Quimby Ave Beltsville MD 20705 Office: NSSDC Code 601 Goodard Space Flight Ctr Greenbelt MD 20771

NIGAM, VIJAI NANDAN, biologist, cancer researcher; b. Aligarh, India, July 3, 1932; emigrated to Can., 1960, naturalized, 1969; s. Ganga Prasad and Krishna Kumari N.; m. Uma Nigam, Jan. 18, 1956; children: Amar Jyoti, Sonya, Tara, Anil. M.Sc., Lucknow (India) U., 1950; Ph.D., Bombay U., 1956. Asst. scientist dept. agr. biochemistry U. Minn., St. Paul, 1956-57; research assoc. cancer research unit Tufts Med. Sch., Boston, 1957-60; research assoc. Montreal Cancer Inst., 1960-67; assoc. prof. dept. biochemistry U. Montreal, 1967-71; assoc. prof. dept. cell biology U. Sherbrooke, 1971-81, prof. dept. anatomy and cell biology, 1982—; research assoc. Nat. Cancer Inst. Can., 1967-78. Contbr. articles in field to profl. jours. Eleanor Roosevelt Internat. Cancer fellow, 1960-70. Mem. Am. Chem. Soc., AAAS, Am. Soc. Cell Biology, Am. Assn. Cancer Research, Can. Biochemistry Soc., Can. Immunology Soc., N.Y. Acad. Sci. Subspecialties: Cell biology (medicine); Cellular engineering. Current work: Immunotherapy of cancer, transformation, neoplasia by carcinogens. Home: Rural Route 1 Vaughan Rd North Hatley PQ Canada JOB 2CO Office: Dept d' Anatomie et de Biologie Cellulaire Centre Hospitalier Univ Sherbrooke Sherbrooke PQ Canada

NIGG, HERBERT NICHOLAS, environmental scientist, educator; b. Detroit, July 9, 1941; s. Herbert Lee and Emily Ruth (Croxton) N.; m. Kirsten Halkjaer, May 20, 1964; children: Herbert Lee, Karen Margrete. B.S., Mich. State U., 1967; Ph.D., U. Ill., 1972. Postdoctoral fellow NRC-U.S. Dept. Agr., Beltsville, Md., 1972-74; assoc. prof. dept. entomology U. Fla., Lake Alfred, 1974—. Editor-in-chief: Bull. Environ. Contamination and Toxicology, 1983—. Served with U.S. Army, 1961-64. Mem. Soc. Toxicology, Am. Chem. Soc., Entomol. Soc. Am., Sigma Xi. Subspecialties: Toxicology (agriculture); Environmental toxicology. Current work: Environ. behavior of pesticides; exposure of agrl. labor to pesticides. Home: PO Box 562 Lake Alfred FL 33850 Office: U Fla 700 Experiment Station Rd Lake Alfred FL 33850

NIGHTINGALE, ELENA OTTOLENGHI, physician, scientist, educator, genetic counselor; b. Leghorn, Italy, Nov. 1, 1932; came to U.S., 1939, naturalized, 1948; d. Mario L. and Elisa V. (Levi) Ottolenghi; m. Stuart L. Nightingale, July 1, 1965; children: Elizabeth, Marisa. A.B. summa cum laude, Barnard Coll., 1954; postgrad., Columbia U., 1954-56; Ph.D. (fellow), Rockefeller U., 1961; M.D., NYU, 1964. Intern, Georgetown U. Hosp., 1973-74; resident instr. medicine NYU, 1964-65; asst. prof. microbiology Cornell U., 1965-70; asst. prof. Johns Hopkins U., 1970-73; sr. staff officer Nat. Acad. Scis./NRC, 1975-76; clin. instr. pediatrics Georgetown U., 1975-80, clin. asst. prof., 1980-82, adj. prof., 1982—; acting dir. div. health scis. policy Inst. Medicine Nat. Acad. Sci., 1979-80, dir. div. health promotion and disease prevention, 1977-80, sr. program officer, 1979-82, sr. scholar in residence, 1982-83; vis. assoc. prof. social medicine health policy Harvard U., 1980—; genetics cons. Birth Defects Clinic, Georgetown U., 1975—; spl. adv. to pres. Carnegie Corp., 1983—. Contbr. articles to profl. jours. Nat. Acad. Sci./Sloan fellow, 1974-75. Fellow N.Y. Acad. Sci.; mem. The Harvey Soc., AAAS, Genetics Soc. Am., Am. Soc. Microbiology, Am. Soc. Human Genetics, Sigma Xi, Phi Beta Kappa. Subspecialties: Genetics and genetic engineering (biology); Preventive medicine. Current work: Health and Science policy; disease prevention; genetics molecular and clinical; medical education.

NIKLAS, KARL J., botanist, educator; b. N.Y.C., Aug. 23, 1948; s. Karl and Ann (Cingel) N. B.S., CCNY, 1970; M.S., U. Ill., Urbana, 1971, Ph.D., 1974; postdoctoral fellow, U. London, 1974. Assoc. curator N.Y. Bot. Garden, 1974-78; asst. prof Cornell U., Ithaca, N.Y., 1978-81, assoc. prof., 1981—. Woodrow Wilson fellow, 1970-71; Fulbright-Hayes fellow, 1974-75; NSF grantee, 1975—. Mem. Soc. Study of Evolution, Am. Bot. Soc. Subspecialties: Paleobiology; Reproductive biology. Current work: Aerodynamics of wind pollination; plant evolution. Home: 1005 Danby Rd Ithaca NY 14850 Office: Cornell U Ithaca NY 14853

NIKOLAI, ROBERT JOSEPH, univ. adminstr., biomech. engr., cons., educator; b. Rock Island, Ill., Apr. 6, 1937; s. Joseph Lawrence and Martha Marie (Holt) N.; m. Susan Eloise Shannon, June 10, 1961; children: Catherine, Teresa, Margaret, David, Philip. Student, Ill. Benedictine Coll., 1955-57; B.S. in Mech. Engring, U. Ill, Urbana, 1959; M.S. in Engring. Mechanics, U. Ill, Urbana, 1961, Ph.D., 1964. Registered profl. engr., Mo. Asst. prof. engring. and engring. mechanics St. Louis U., 1964-68, asso. prof. engring. and engring. mechanics, 1968-71, asso. prof. orthodontics, 1971-75, prof. biomechanics in orthodontics, 1975—, asst. dean, 1971-72, asso. dean, 1972—; adj. prof. civil and mech. engring. Washington U., St. Louis, 1980—; cons. in field. Research, publs. in field; editorial panel: Jour. Biomechanics, 1982—; manuscript reviewer: Am. Jour. Orthodontics, 1981—. Mem. University City (Mo.) Traffic Commn., 1972-79, chmn, 1975-77. Orthodontic Edn. and Research Found. grantee, 1981. Mem. Am. Acad. Mechanics, ASME, Internat. Assn. Dental Research, Am. Assn. Dental Research, Orthodontic Edn. and Research Found. (hon.), Pi Tau Sigma. Subspecialties: Solid mechanics; Orthodontics. Current work: Structural behavior of orthodontic appliances; graduate teaching and research in orthodontic bioengineering. Home: 7134 Stanford Ave University City MO 63130

NIKORA, ALLEN PETER, software systems designer and analyst; b. Stuttgart, W.Ger., July 17, 1955; s. Eugene S. and Lise L. (Kohler) N. B.S., Calif. Inst. Tech., 1977. With McDonnell-Douglas Astronautics Co., Huntington Beach, Calif., 1977-78; software systems designer, analyst Jet Propulsion Lab., Pasadena, Calif., 1978—. Mem. Assn. Computing Machinery, AIAA (public policy com. Los Angeles sect.). Subspecialties: Distributed systems and networks; Software engineering. Current work: Design tests to validate the command and data susbystem flight software for Galileo Jupiter Orbiter. Office: 4800 Oak Grove Dr Pasadena CA 91103

NILAVER, GAJANAN, neurologist, immunocytochemist, educator; b. Bangalore, India, Aug. 30, 1946; came to U.S., 1971, naturalized, 1973; s. Shanker and Nalini (Katre) N. M.B., B.S., U. Madras, 1968. Resident in neurology St. Vincent's Hosp., N.Y.C., 1972-75; NIH Nat. Research Service postdoctoral fellow in neuroendocrinology Columbia U., N.Y.C., 1975-79, research assoc., asst. neurologist, 1979-81, asst. prof. neurology, 1982—. Contbr. articles to sci. jours. NIH grantee, 1979—. Mem. AMA, Soc. Neurosci., Am. Acad. Neurology, AAAS, N.Y. Soc. Electron Microscopists, N.Y. Acad. Scis. Subspecialties: Neuroendocrinology; Immunocytochemistry. Current work: Neuroendocrinology, immunocytochemistry, immunology. Home: 151 Prospect Ave Apt 6E1 Hanckensack NJ 07601 Office: Room 309 Black Bldg Columbia U 630 W 168th S New York NY 10032

NIPPO, MURN MARCUS, vet. scientist, educator; b. N.Y.C., Feb. 8, 1944. B.S. in Agrl. Sci, U. Maine, 1965, M.S. In Animal Sci, 1968; Ph.D. in Animal Nutrition, U. R.I., 1976. Cert. Am. Registry Cert. Animal Scientists. Assoc. prof. animal and vet. sci. U. R.I., Kingston, 1972—; researcher. Recipient Teaching Excellence award U. R.I., 1979. Mem. Am. Soc. Animal Sci., Am. Assn. Lab. Animal Sci., R.I. Nutrition Council. Subspecialty: Animal nutrition. Current work: Nutrition and behavior in laboratory animals. Office: Dept of Animal Sci URI Kingston RI 02881

NIRENBERG, LOUIS, mathematician, educator; b. Hamilton, Ont., Can., Feb. 28, 1925; came to U.S., 1945, naturalized, 1954; s. Zuzie and Bina (Katz) N.; m. Susan Blank, Jan. 25, 1948; children: Marc, Lisa. B.Sc., McGill U., Montreal, 1945; M.S., N.Y. U., 1947, Ph.D., 1949. Mem. faculty N.Y. U., 1949—, prof. math., 1957—, dir. Courant Inst., 1970-72; visitor Inst. Advanced Study, 1958. Author research articles. Recipient Crafoord prize Royal Swedish Acad., 1982; NRC fellow, 1951-52; Sloan Found. fellow, 1958-60; Guggenheim Found. fellow, 1966-67, 75-76; Fulbright fellow, 1965. Mem. Nat. Acad. Scis., Am. Acad. Arts and Scis., Am. Math. Soc. (v.p. 1976-78, M. Bôcher prize 1959). Home: 221 W 82d St New York NY 10024 Office: Courant Inst 251 Mercer St New York NY 10012

NIRENBERG, MARSHALL WARREN, biochemist; b. N.Y.C., Apr. 10, 1927; s. Harry Edward and Minerva (Bykowsky) N.; m. Perola Zaltzman, July 14, 1961. B.S. in Zoology, U. Fla., 1948, M.S., 1952; Ph.D. in Biochemistry, U. Mich., 1957. Postdoctoral fellow Am. Cancer Soc. at NIH, 1957-59; postdoctoral fellow USPHS at NIH, 1959-60; mem. staff NIH, 1960—; research biochemist, chief lab. biochem. genetics Nat. Heart, Lung and Blood Inst., 1962—. Recipient Molecular Biology award Nat. Acad. Scis., 1962, award in biol. scis. Washington Acad. Scis., 1962, medal HEW, 1964, Modern Medicine award, 1963, Harrison Howe award Am. Chem. Soc., 1964, Nat. Medal Sci. Pres. Johnson, 1965, Hildebrand award Am. Chem. Soc., 1966, Research Corp. award, 1966, A.C.P. award, 1967, Gairdner Found. award merit, Can., 1967, Prix Charles Leopold Meyer French Acad. Scis., 1967, Franklin medal Franklin Inst., 1968, Albert Lasker Med. Research award, 1968, Priestly award, 1968; co-recipient Louisa Gross Horowitz prize Columbia, 1968, Nobel prize in medicine and physiology, 1968. Fellow AAAS, N.Y. Acad. Scis.; mem. Am. Soc. Biol. Chemists, Am. Chem. Soc. (Paul Lewis award enzyme chemistry 1964), Am. Acad. Arts and Scis., Biophys. Soc., Nat. Acad. Scis., Washington Acad. Scis., Soc. for Study Devel. and Growth, Harvey Soc. (hon.), Leopoldina Deutsche Akademie der Naturforscher, Pontificial Acad. Scis. Subspecialty: Biochemistry (biology). Spl. research mechanism protein synthesis, genetic code, nucleic acids, regulatory mechanisms in synthesis macromolecules, neurobiology. Office: Nat Heart Inst NIH Lab Biochem Genetics Bethesda MD 20014

NISBET, JOHN STIRLING, electrical engineering educator; b. Darval, Scotland, Dec. 10, 1927; s. Robert George Jackson and Kathleen Agnes (Young) N.; m. J. Valerie Payne, Jan. 10, 1953; children: Robert John, Alexander Stevens. B.S. with honors, London U., 1950; M.S., Pa. State U., 1957, Ph.D., 1960. Trainee Nash & Thompson Ltd., Surbiton, Eng., 1944-51; engr. Decca Radar Ltd., Surbiton, 1951-53, Can. Westinghouse, Hamilton, Ont., 1953-55; research assoc. Pa. State U., University Park, 1955-60, mem. faculty, 1960—, prof. elec. engring., 1966—, dir. ionosphere research lab., 1971—; chmn. bd. trustees Upper Atmosphere Research Corp., 1981—. Contbr. articles to sci. jours. NSF fellow, Brussels, 1965; Fulbright-Hays lectr., Kharkov, USSR, 1979; fellow NRC/Nat. Acad. Scis., Goddard Space Flight Ctr., NASA, 1980. Mem. IEEE (sr.), Am. Geophys. Union, Internat. Sci. Radio Union, Sigma Xi, Phi Kappa Phi. Unitarian. Subspecialties: Aeronomy; Planetary atmospheres. Current work: Aeronomy, ionospheric current systems, planetary atmospheres and ionospheres. Home: 618 Glenn Rd State College PA 16801 Office: Pa State U University Park PA 16802

NISFELDT, MICHAEL LEE, immunologist, educator; b. Davenport, Iowa, June 15, 1950; s. Melvin Lawrence and Wanda Irene (Dee) M.; m. Mary Alice Griffin, Aug. 4, 1973; 1 son, Andrew Michael. B.S., U. Ill., 1972; Ph.D., U. Iowa, 1977. Research asst. dept. microbiology U. Iowa, Iowa City, 1973-77; staff fellow Lab. Molecular Genetics, Nat. Inst. Child Health and Devel.; NIH, Bethesda, Md., 1977-81; asst. prof. microbiology U. Mo., Columbia, 1981—. Contbr. articles to profl. jours. Mem. Am. Soc. Microbiology, N.Y. Acad. Scis., Am. Assn. Immunologists, AAAS. Subspecialties: Cellular engineering; Infectious diseases. Current work: Immunomodulation by bacterial products, T-lymphocyte differentiation. Office: U Mo-Columbia M264 Med Scis Bldg MO 65212

NISONOFF, ALFRED, biochemist, educator; b. N.Y.C., Jan. 26, 1923; s. Hyman and Lillian (Klein) N.; m. Sarah Weiseman, July 17, 1946; children: Donald Michael, Linda Ann. B.S. (State scholar), Rutgers U., 1942; M.A., Johns Hopkins U., 1948, Ph.D. (AEC fellow), 1951. Postdoctoral fellow Johns Hopkins Med. Sch., 1951-52; research chemist U.S. Rubber Co., Naugatuck, Conn., 1952-54; sr. cancer research scientist Roswell Park Meml. Inst., Buffalo, 1954-57, assoc. cancer research scientist, 1957-60; assoc. prof. microbiology U. Ill., Urbana, 1960-62, prof., 1962-66; prof. microbiology U. Ill. Coll. Medicine, Chgo., 1966-69, head dept. biol. chemistry, 1969-75; prof. biology Rosenstiel Research Center, Brandeis U., Waltham, Mass., 1975—; Mem. grant rev. bds. allergy and immunology study sect. NIH, 1965-67, 71-74, Nat. Multiple Sclerosis Soc., 1972-75, 77-80; invited lectr. 100th Anniversary of Birth of Dr. Jules Bordet, Brussels, 1970. Author: The Antibody Molecule, 1975, Introduction to Molecular Immunology, 1982; Editorial bd.: Jour. Immunology, 1962-67, 69-74; sect. editor, 1971-74; sr. editor, 1975-79; editorial bd.: Immunochemistry, 1964-70, Bacteriological Revs, 1968-70, Jour. Exptl. Medicine, 1974-78, Critical Reviews in Immunology, 1980—; contbr. articles to profl. jours. Served to lt. (j.g.) USNR, 1943-46. Recipient Research Career award NIH, 1962-69; Pasteur Inst. medal, 1970. Fellow AAAS; Mem. Am. Assn. Immunologists, Am. Soc. Biol. Chemists, Belgian Royal Acad. Medicine (fgn. corr.), Phi Beta Kappa, Phi Lambda Upsilon. Subspecialty: Immunobiology and immunology. Current work: Studying genetic control of antibodies of defined specificity; also regulation of immune response. Home: 16 Winter St Waltham MA 02154 Office: Dept Biology Rosenstiel Research Center Brandeis Univ Waltham MA 02254

NITECKI, MATTHEW HENRY, museum curator; b. Sosnowiec, Silesia, Poland, Apr. 30, 1925; came to U.S., 1950; s. Henry W. and Antonina (Janicz) N.; m. Doris June Vinton, Dec. 29, 1964; children: Mark Raymond, David Allan. Ph.D., U. Chgo., 1968. Curator U. Chgo. Mus., 1955—, Field Mus., Chgo., 1965—. Co-author: Orientation of Receptaculitids, 1982; editor: Biochemical Aspects of Evolutionary Biology, 1982, Biotic Crises, 1981, Mazon Creek Fossils, 1979; contbr. articles to profl. jours. Served with U.S. Army, 1942-47. NSF grantee, 1963, 68-69, 77-79, 80-82. Fellow Geol. Soc. Am.; mem. Paleontol. Soc., Soc. Econ. Paleontologists and Mineralogists (treas. 1971-73), Am. Assn. Petroleum Geologists. Subspecialties: Paleobiology; Evolutionary biology. Current work: Evolution, morpholoy and systematics of Paleozoic problematic organisms, particularly algae; models of scientific theories. Office: Field Mus Natural History Roosevelt Rd and Lake Shore Dr Chicago IL 60605 Home: 5649 S Blackstone Chicago IL 60637

NITOWSKY, HAROLD MARTIN, physician, educator; b. Bklyn., Feb. 12, 1925; s. Max and Fannie (Gershowitz) N.; m. Myra Heller, Nov. 28, 1954; children—Fran Ellen, Daniel Howard. A.B., N.Y. U., 1944, M.D., 1947; M.S., U. Colo., 1952. Intern Mt. Sinai Hosp., N.Y.C., 1947-48; resident pediatrics U. Colo. Med. Center, 1948-50; USPHS postdoctoral fellow U. Colo., 1950-51; staff Sinai Hosp., Balt., 1953-67, dir. pediatric research, 1960-67; faculty Johns Hopkins Sch. Medicine, 1953-67, asso. prof. pediatrics, 1962-67; prof. pediatrics and genetics Albert Einstein Coll. Medicine, 1967—; cons. Nat. Inst. Child Health and Human Devel., 1966—; Sr. surgeon USPHS, 1951-53. Contbr. articles on nutrition, metabolism, genetics to profl. jours. Mem. Am. Pediatric Soc., Soc. Pediatric Research. Subspecialties: Genetics and genetic engineering (medicine); Pediatrics. Home: 25 Devonshire Rd New Rochelle NY 10804

NITZ, DAVID EDWIN, physics educator, researcher; b. Dallas, Feb. 19. 1951; s. Ralph A. and Thelma I. (Hanson) N.; m. Debra D. Ralstin, June 25, 1977. B.A., St. Olaf Coll., Northfield, Minn., 1973; M.A., Rice U., 1976, PH.D., 1978. Research assoc. Joint Inst. Lab. Astrophysics, Boulder, Colo., 1978-79; instr. physics St. Olaf Coll., 1979-80, asst. prof., 1980—. Contbr. articles in field to profl. jours. Research Corp./Northwest Area Found. grantee, 1981-82. Mem. Am. Phys. Soc.; mem. Am. Assn. Physics Tchrs. Lutheran. Subspecialty: Atomic and molecular physics. Current work: Interaction of radiation with atoms and molecules, molecular srucucture; molecules structure; undergraduate physics teaching; research in hyperfine structrue of polar diatomic molecules. Office: Dept Physics Saint Olaf Coll Northfield MN 55057

NIXON, SCOTT WEST, marine ecologist, educator, cons.; b. Phila., Aug. 24, 1943; s. Robert S. and Elizabeth (Nixon) West; m. Pendleton Hall, Aug. 28, 1965. B.A., U. Del., 1965; Ph.D., U. N.C., 1970. Research assoc. U. R.I., Kingston, 1969-70, asst. prof. oceanography, 1970-75, assoc. prof., 1975-79, prof., 1979—; dir. Coastal Ecology Environ. Cons. Contbr. articles on oceanography to profl. jours. NSF grantee; EPA grantee. Mem. Am. Soc. Limnology and Oceanography, Estuerine Research Fedn. Subspecialties: Ecology; Ecosystems analysis. Current work: Coastal Marine ecology, wetlands, nutrient dynamics, productivity, fisheries, environmental management. Office: Dept Oceanography U RI Kingston RI 02881

NOBE, KEN, chemical engineering educator; b. Berkeley, Calif., Aug. 26, 1925; s. Sidney and Kiyo (Uyeyama) N.; m. Mary Tagami, Aug. 31, 1957; children: Steven Andrew, Keven Gibbs, Brian Kelvin. B.S., U. Calif., Berkeley, 1951; Ph.D., UCLA, 1956. Jr. chem. engr. Air Reduction Co., Murray Hill, N.J., 1951-52; asst. prof. chem. engring. UCLA, 1957-62, asso. prof., 1962-68, prof., 1968—, chmn. dept. chem. engring., 1978—. Div. editor: Jour. Electrochem. Soc, 1967—, Electrochimica Acta, 1977—. Served with U.S. Army, 1944-46. Mem. Electrochem. Soc., Am. Chem. Soc., Nat. Assn. Corrosion Engrs., Internat. Soc. Electrochemistry, Sigma Xi. Subspecialties: Chemical engineering; Corrosion. Current work: Electrode kinetics; photelectrochemistry and applied catalysis. Office: UCLA 405 Hilgard Ave Los Angeles CA 90024

NOBEL, PARK S., biology educator, research plant physiological ecologist; b. Chgo., Nov. 4, 1938; s. James Dodman and Ruth Eleanor (Uetz) N.; m. Eiko Mizuguchi, Feb. 7, 1965; children: Catherine, Elizabeth. B. Engring. Physics, Cornell U., 1961; M.S., Calif. Inst. Tech., 1963; Ph.D., U. Calif., Berkeley, 1965. NSF postdoctoral fellow U. Tokyo, 1965-66, King's Coll., U. London, 1966-67; asst. prof. molecular biology UCLA, 1967-71, assoc. prof. biology, 1971-75, prof., 1975—; cons. in field. Contbr. numerous articles on biophysical plant physiology and ecology to profl. jours.; author 3 books on plant physiology; editor 4 books on plant ecology. Guggenheim fellow Australian Nat. U., Canberra, 1973-74. Mem. Ecol. Soc. Am., Am. Soc. Plant Physiologists, Bot. Soc. Am. Subspecialties: Plant physiology (biology); Ecology. Current work: Plant physiological ecology, especially desert plants; research, writing and teaching in plant physiology and ecology. Office: Dept Biology U Calif Los Angeles CA 90024

NOBLE, CHARLES CARMIN, civil engr.; b. Syracuse, N.Y., May 18, 1916; s. Anthony and Julia (Samar) N.; m. Edith Margaret Lane, Sept. 12, 1942; children—Jeanne Lane (Mrs. Richard Davey, Jr.), Robert Charles, Barbara Lynn (Mrs. James P. O'Looney), Carol Anne, Stephen Gary. B.S., U.S. Mil. Acad., 1940; M.S. in Civil Engring, Mass. Inst. Tech., 1948; M.A. in Internat. Affairs, George Washington U., 1964; grad., Army War Coll., 1957, Nat. War Coll., 1964. Registered profl. engr., D.C., La., Mass., Nev., Nebr., Tex., N.Y., Va., Miss. Commd. 2d lt., C.E. U.S. Army, 1940, advanced through grades to maj. gen., 1969; comdr. engr. combat bn., Europe, World War II, exec. officer Manhattan Dist., Oak Ridge, 1946-47; mem. AEC staff, 1947, Army Gen. staff, 1948-51; planner Supreme Hdqrs., Allied Powers Europe, 1951-54; dep. dist. engr., N.Y., 1954-56, comdr. engr. combat group, Ft. Benning, Ga., 1957-58, dist. engr., Louisville, 1958-60; dir. Atlas and Minuteman ICBM Constrn. Program, 1960-63; theater engr. UNC/8th U.S. Army, Korea, 1964-66; dir. constrn. Office Sec. Def., 1966-67; dir. civil works Office Chief Engrs., U.S. Army, 1967-69; chief engr. U.S. Army Europe, 1969-70, U.S. Army Vietnam; comdg. gen. U.S. Army Engr. Command, Vietnam, 1970-71; div. engr., Lower Mississippi River Valley; pres. Mississippi River Commn., 1971-74; mgr. spl. projects Chas. T. Main Engrs., Inc., Boston, 1974-75, v.p., 1975-77; group v.p., 1977-78, exec. v.p., 1978-81; pres., chief operating officer C.T. Main Corp., 1981—, also dir.; engring. agt. Atlantic-Pacific Interoceanic Canal Study Presdl. Commn.; Dept. Def. rep. Mexico-U.S. Joint Com. Mut. Disaster Assistance; army rep. Water Resources Council; mem. indsl. com. Am. Power Conf.; govt. rep. governing bd. U.S. Power Squadrons; army liaison mem. Fed. Adv. Council Regional Econ. Devel., 1967-70; mem. Pres.'s Task Force Bridge Safety, 1968; pres. U.S. Army Coastal Engring. Research Bd.; def. mem. com. Multiple Uses of the Coastal Zone, 1969-70; chmn., fed. mem. Red River Compact Commn. Decorated D.S.M. with 2 oak leaf clusters, Legion of Merit with 2 oak leaf clusters, Bronze Star, Army Commendation medal. Fellow ASCE, Am. Cons. Engring. Council, Soc. Am. Mil. Engrs. (Wheeler medal 1962); mem. Internat. Assn. Navigational Congresses (chmn. Am. sect. 1969), U.S. Com. Large Dams, Nat. Acad. Engring. U.S., Nat. Soc. Profl. Engrs., Newcomen Soc., Sigma Xi. Subspecialty: Civil engineering. Home: Gloucester Apt 3-G 770 Boylston St Boston MA 02199 Office: CT Main Corp Prudential Center Boston MA 02199

NOBLE, NANCY LEE, biochemist, University dean, researcher; b. Chattanooga, Mar. 1, 1922; d. Samuel Edward and Agnes Lee (Caulkins) N. B.S. in Chemistry, U. Chattanooga, 1943; M.S. in Biochemistry (Univ. fellow), Emory U., 1949; Ph.D. (USPHS predoctoral research fellow), Emory U., 1953. Research asst. Organic Research Lab., Chattanooga Medicine Co., 1943-48; lab. asst. dept. biochemistry Emory U. Sch. Medicine, 1949-50; research asst. prof. dept. biochemistry U. Miami Sch. Medicine, 1953-57, asst. prof. biochemistry, 1957-63, assoc. prof. biochemistry and medicine, 1963—, assoc. dean for faculty affairs, 1981—; dir. Biochemistry Research Lab., Miami Heart Inst., 1953-56; investigator Labs. for Cardiovascular Research, Howard Hughes Med. Inst., Miami, Fla., 1956-70; bd. dirs., mem. com. sch. health and sci. careers Miami affiliate Am. Heart Assn., bd. dirs., mem. com. sch. health and sci. careers Fla. affiliate. Contbr. numerous articles abstracts, chpts. to profl. publs. Mem. Episcopal Cursillo Movement, Emory U. Grad. Council, 1980—. Co-recipient award for basic research relevant to problems of aging Ciba Found., 1957, 58, Outstanding Achievement in Research recognizfion Miami Profl. Women's Club, 1963, recognition for meritorious service in fight against heart disease Fla. Heart Assn., 1967, 69, 71, Miami Heart Assn., 1968, 70; Spl. Service award Fla. affiliate Am. Heart Assn., 1980; recognition for 15 yrs. of dedicated service, 1981. Fellow Gerontol. Soc., Am. Inst. Chemists (cert.profl. chemist); mem. Am. Chem. Soc., AAAS, Am. Soc. Study

Arteriosclerosis, N.Y. Acad. Scis., Soc. Exptl. Biology and Medicine (councilor Southeastern sect. 1974-79), Fla. Acad. Scis., Am. Heart Assn. Council on Basic Sci., Biochem. Soc., AAUP, Assn. Women in Sci., Am. Heart Assn. Council on Thrombosis, Am. Physiol. Soc., Emory Alumni Assn. (treas. Miami chpt. 1963-64, 79-81, v.p. chpt. 1964-65, pres. chpt. 1965-66), U. Tenn. Alumni Assn. (pres. S.E. Fla. chpt. 1976-78), Women's Panhellenic, Mortar Bd. (v.p. chpt. 1977-79), Sigma Xi (treas. chpt. 1974-78), Chi Omega. Subspecialties: Biochemistry (medicine); Animal physiology. Current work: Connective tissue, cardiovascular disease.

NOBLE, ROBERT CUTLER, medical educator; b. Raleigh, N.C., Feb. 19, 1938; s. Roy Fred and Mamie (Cutler) N.; m. Audrey Joyce Melman, May 8, 1965; children: Lisa Meredith, Amy Brooke. B.A., U. N.C., 1960; M.D., Duke U., 1964. Intern Duke U. Hosp., Durham, N.C., 1964-65; resident in internal medicine, 1965-66; surgeon USPHS (streptococcus lab.), Atlanta, 1966-68; sr. resident dept. medicine Stanford U., Palo Alto, Calif., 1968-69, fellow infectious disease, 1969-71; prof. medicine U. Ky., Lexington, 1971—, hosp. epidemiologist, 1974-78, 80-83; cons. Lexington VA Hosp., 1972—. Author: Sexually Transmitted Diseases, 1982; editorial bd., 1979-85, Medical Aspects of Human Sexuality, 1981—. Served with USPHS, 1966-68. NIH fellow, 1969-71. Fellow ACP, Infectious Diseases Soc. Am.; mem. Am. Veneral Diseases Assn., Central Soc. Clin. Research, Am. Soc. Microbiology, Phi Beta Kappa. Subspecialties: Infectious diseases; Microbiology (medicine). Home: 3408 Bellefonte Dr Lexington KY 40502 Office: Div Infectious Diseases Dept Medicine U Ky Coll Medicine Lexington KY 40536

NOCENTI, MERO RAYMOND, physiologist, editor; b. Masontown, Pa., Sept. 7, 1928; s. Silvio and Josephine (Pisani) N.; m. Louise I. Norante, Feb. 5, 1955; children: Mary, Ann, David. B.A., W.Va. U.-Morgantown, 1951, M.S., 1952; Ph.D., Rutgers U., 1955. Waksman-Merck postdoctoral fellow Rutgers U., New Brunswick, N.J., 1955-56; prof. physiology Columbia U., N.Y.C., 1956—. Recipient Tchr. of Year award Columbia Med. Students, 1973, Columbia Dental Students, 1978. Mem. AAAS, Am. Physiol. Soc., Soc. Exptl. Biology and Medicine (exec. sec. 1974—, editor proceedings 1974—, Council Biology Editors), Sigma Xi. Roman Catholic. Subspecialties: Physiology (medicine); Endocrinology. Current work: Hormonal control of electrolyte and water balances. Office: Columbia U 630 W 168th St New York NY 10032

NODA, HIROHARU, neurophysiologist, educator, researcher; b. Okayama, Japan, Apr. 29, 1936; s. Itaro and Kimie (Goda) N.; m. Ritsuko Noda, Mar. 20, 1961; children: Hiroko, Seiichi, Naomi, Rie. M.D., Osaka U., 1961, Ph.D. in Neurophysiology, 1966. Lic. physician, Japan. Asst. research anatomist UCLA Sch. Medicine, 1966-69, asso. research anatomist, 1971-75, adj. prof. physiology and anatomy, 1976-81; prof. Sch. Optometry, Ind. U., Bloomington, 1981—; asst. prof. Max-Planck-Inst., Munich, W.Ger., 1969-71. Editorial bd.: Exptl. Neurology; contbr. articles to sci. jours. NIH grantee, 1973—. Mem. Internat. Brain Research Orgn., Soc. Neurosci., AAAS, Assn. Resarch in Vision and Ophthalmology. Subspecialties: Neurophysiology; Comparative physiology. Current work: Vision physiology; oculomotor physiology; sensori-motor integration; cerebellar control of eye movements; information processing in the higher nervous system; neuronal mechanisms of smooth eye movement controls. Home: 2226 Maxwell Ln Bloomington IN 47401 Office: 800 E Atwater Bloomington IN 47405

NOGGLE, JOSEPH HENRY, research chemist, writer, consultant; b. Harrisburg, Pa., Mar. 19, 1936; s. David J. H. and Mabel (Atticks) N.; m. Carol Innis, Sept. 12, 1960; children: Jennifer, Pamela. B.S., Juniata Coll., Huntingdon, Pa., 1960; M.S., Harvard U., 1963, Ph.D., 1965. Asst. prof. chemistry U. Wis.-Madison, 1965-71; assoc. prof. U. Del., Newark, 1971-77, prof., 1977—. Author: The Nuclear Overhauser Effect, 1971. Served with U.S. Army, 1954-57. Mem. Am. Chem. Soc., AAUP. Subspecialties: Nuclear magnetic resonance (chemistry); Physical chemistry. Current work: Characterization of polymers by NMR; writing a text book of physical chemistry. Home: 804 Cambridge Dr Newark DE 19711 Office: University of Delaware Chemistry Dept Newark DE 19711

NOLAND, PAUL ROBERT, animal scientist, cons.; b. Chillicothe, Ill., Sept. 28, 1924; s. Paul Ennis and Ruby Norma (Couch) N.; m. Eunice Claire Hazenfield, Dec. 27, 1947; children: Steven, Stewart, Robert, Brian. Student, Blackburn Coll., 1942-43, U. Detroit, 1943-44; B.S., U. Ill., 1946, M.S., 1948; PH.D., Cornell U., 1951. Asst. prof. animal sci. U. Ark., Fayetteville, 1951-55, asso. prof., 1955-60, prof., 1960—; scientist (U. Ark. Mission to Panama), 1955-57; cons. to Latin Am. Contbr.: articles to Jour. Animal Sci, Animal Feed Sci.and Tech, Ark. Agrl. Expt. Sta. bulls; editorial bd.: Jour. Animal Sci, 1979-82. Mem. Fayetteville City Bd. Dirs., 1972-82, mayor, 1982. Served with inf. U.S. Army, 1943-46. Decorated Bronze Star; recipient Alpha Zeta Teaching award U. Ark., 1955, 58, 67; named Outstanding Tchr. U. Ark. Alumni Assn., 1963. Mem. Am. Soc. Animal Sci. (cert. animal scientist), Federated Soc. Biology and Medicine. Methodist. Lodge: Lions. Subspecialty: Animal nutrition. Current work: Amino acids, swine, energy, mgmt., utilization waste products. Office: Dept Animal Sci U Ark Fayetteville AR 72701

NOLTIMIER, HALLAN COSTELLO, geology educator; b. Los Angeles, Mar. 19, 1937; s. Frederick Henry and Corinne Marie (Scheel) N.; m. Judith Walford Summers, July 14, 1961; children: Mark Andrew, Romy Ann. B.S., Calif. Inst. Tech., 1958; Ph.D. in Geophysics, U. Newcastle upon Tyne, Eng., 1965. Lectr. U. Newcastle upon Tyne, 1966-68; asst. prof. geology U. Okla., Norman, 1968-71, assoc. prof., 1971-72; assoc. prof. geology Ohio State U., Columbus, 1972-78, prof., 1978—; cons. in field. Contbr. chpts. to books. Fellow Royal Astron. Soc.; mem. Am. Geophys. Union, Geol. Soc. Am., AAAS, Ohio Acad. Sci., Sigma Xi (sec. Okla. 1969-71). Democrat. Subspecialties: Geophysics; Low temperature physics. Current work: Research in paleomagnetism of fossiliferous sediments using cryogenic magnetometry. Office: Dept Geology and Mineralogy Ohio State U Columbus OH 43210

NOMURA, ABRAHAM MICHAEL YOZABURO, physician; b. Honolulu, July 23, 1939; s. George Giichi and Kaneyo N.; m. Susan Hitomi, Feb. 20, 1972; children: Liane Miyoko, Ryan Yoichi. B.S., John Carroll U., 1962; M.D., Loyola U., Chgo., 1966; M.P.H., Johns Hopkins U., 1972, Dr.P.H., 1974. Diplomate: Nat. Bd. Med. Examiners, Am. Bd. Preventive Medicine. Intern Michael Reese Med. Ctr., Chgo., 1966-67; resident in preventive medicine Johns Hopkins Hosp., Balt., 1971-74; asst. dir. Japan-Hawaii cancer study Kuakini Med. Ctr., Honolulu, 1974-76, dir., 1976—; epidemiologist Cancer Center Hawaii, U. Hawaii, Honolulu, 1974—; assoc. prof. Sch. Pub. Health, 1976-82, prof., 1982—; cons. NIH, 1983—. Assoc. editor: Am. Jour Epidemiology, 1982—; contbr. articles on research in gastrointestinal and breast cancer in profl. jours. Served to lt. USN, 1967-69. Nat. Cancer Inst. grantee, 1976—. Mem. Hawaii Med. Assn. Subspecialties: Epidemiology; Cancer research (medicine). Current work: Activities devoted to identifying environmental causes of common cancers. Home: 486 Luakini St Honolulu HI 96817 Office: 347 N Kuakini St Honolulu HI 96817

NOMURA, YASUMASA, engineering educator; b. Kumamoto, Japan, Sept. 25, 1921; s. Masaki and Aya N.; m. Kimi Higashi, Oct. 24, 1949; children: Kazuo, Etsuko. B.Eng., Tokyo Inst. Tech., 1945, D.Eng., 1962. Tchr. Yatsushiro (Japan) Jr. High Sch., 1946-47, Kumamoto (Japan) High Sch., 1948-52; asst. prof. Coll. of Coast Guard, Kure, Japan, 1952-57, Nat. Defence Acad., Yokosuka, Japan, 1957-62, prof. dept. aero. sci., 1962—, head aeor. sci. dept., 1971-73, 81—. Mem. AIAA, Japan Soc. Aero. and Space Sci. Subspecialties: Aeronautical engineering; Aerospace engineering and technology. Current work: Research of aerodynamics, especially three-dimensional boundary layer problems. Home: Kotsubo 1-Chome 24-43 Zushi Kanagawa PrefectureJapan 239 Office: National Defence Academy Hashirimizu 1-Chome Yokosuka KanagawaJapan 239

NOODEN, LARRY D., educator, botanist, consultant; b. Oak Park, Ill., June 10, 1936; s. Larry and Ruth (Backstrom) N.; m. Sarah M. Nooden, Dec. 27, 1963; children: Lars D., Linnea S. B.S. with highest distinction, U. Ill., 1958; M.S., U. Wis., 1959; Ph.D., Harvard U., 1963. From asst. prof. to prof. botany U. Mich., Ann Arbor, 1965—; vis. research fellow Australian Nat. U., 1981, Fulbright sr. research fellow, 1983; vis. scientist Boyce Thompson Inst., Ithaca, N.Y., 1979; NIH spl. research fellow Calif. Inst. Tech., 1971-72; NIH postdoctoral fellow U. Edinburgh, Scotland, 1964-65; cons. Contbr. articles to sci. jours. Scoutmaster Boy Scouts Am., 1977-83. Wis. Alumni Research Found. fellow, 1958-59; NIH fellow, 1960-63, 64-65, 71-72. Mem. AAAS, Am. Soc. Plant Physiology, Am. Inst. Biol. Scis., Japan Soc. Plant Physiology, Crop Sci. Soc. Am., Agronomy Soc. Am., Plant Growth Regulator Soc. Am., Scandinavian Soc. Plant Physiology, Bot. Soc. Am., Phi Beta Kappa, Phi Kappa Phi. Lutheran. Subspecialties: Plant physiology (biology); Plant growth. Current work: Physiology of senescence and seed growth in soybean; soybean maturation, senescence and seed development, yield limitations, cytokinin translocation and metabolism. Office: U Mich Ann Arbor MI 48109

NOONAN, THOMAS WYATT, educator; b. Glendale, Calif., July 16, 1933; s. Gustave Vincent and Florence (Seely) N.; m. Annabel Smith, Aug. 24, 1966; children: Rachel, John, Matthew. B.S., Calif. Inst. Tech., 1955, Ph.D., 1961. Vis. asst. prof. physics dept. U. N.C., Chapel Hill, 1962-65, asst. prof., 1965-68; assoc. prof. physics Brockport State Coll., 1968-71, prof., 1971—. Author: Relativity and Cosmology, 1968. Mem. Am. Astron. Sco., Royal Astron. Soc., Internat. Astron. Union, Astron. Soc. Pacific. Subspecialties: Relativity and gravitation; Cosmology. Current work: Astrophysics, gravitation, cosmology research, teaching. Office: Physics Dept State Coll Brockport NY 14420

NOOR, AHMED KHAIRY, engineering educator; b. Cairo, Aug. 11, 1938; U.S., 1971, naturalized, 1976; s. Mohamed Sayed and Fatma Mohamed (El-Zeini) N.; m. Zakia Mahmoud Taha, Aug. 18, 1966; 1 son, Mohamed. B.S. with honors, Cairo U., 1958; M.S., U. Ill.-Urbana, 1961, Ph.D., 1963. Asst. prof. aeros. and astronautics Stanford U., 1963-64; sr. lectr. structural mechanics Cairo U., 1964-67; vis. sr. lectr. structural mechanics U. Baghdad, Iraq, 1967-68; sr. lectr. structural mechanics U. New South Wales, Australia, 1968-71; prof. engring. and applied sci. George Washington U., NASA Langley Research Ctr., Hampton, Va., 1971—; mem. coms. computational mechanics and large space systems Nat. Acad. Engring., 1978—. Co-editor: Trends in Computerized Structural Analysis and Synthesis, 1978, Computational Methods in Nonlinear Structural and Solid Mechanics, 1980, Advances and Trends in Structural and Solid Mechanics, 1982; contbr. articles in field to profl. jours. Mem. AIAA, ASME, ASCE, Am. Acad. Mechanics, Sigma Xi. Subspecialties: Solid mechanics; Aerospace engineering and technology. Current work: Computational mechanics, large space structures, fibrous composite structures. Home: 31 Towler Dr Hampton VA 23666 Office: MS-246 GWU-NASA Langley Research Ctr Hampton VA 23665

NOORDERGRAAF, ABRAHAM, biophysics educator; b. Utrecht, Netherlands, Aug. 7, 1929; s. Leendert and Johanna (Kool) N.; m. Geertruida Alida Van Nee, Sept. 6, 1956; children: Annemiek,(Mrs. James A. Young), Gerrit Jan, Jeske Inette, Alexander Abraham. B.Sc., U. Utrecht, 1953, M.S., 1955, Ph.D., 1956; M.A. (hon.), U. Pa., 1971. Teaching asst. U. Utrecht, 1949-50, asst. dept. physics, 1951-53, research asst. dept. med. physics, 1953-55, research fellow dept. med. physics, 1956-58, sr. research fellow dept. med. physics, 1959-65; tchr. math. and physics Vereniging Nijverheidsonderwijs, Utrecht, 1951; research asst. U. Amsterdam, Netherlands, 1952; vis. fellow dept. therapeutic research U. Pa., Phila., 1957-58; asso. prof. biomed. engring. Moore Sch. Elec. Engring., 1964-70, acting head electromed. div., 1968-69, prof. biomed. engring., 1970—; prof. physiology Sch. Vet. Medicine, 1976 ; asso. dir. biomed. engring. tng. program Moore Sch. Elec. Engring., 1971-76, asso. dir. sch., 1972-74, chmn. grad. group in biomed. electronic engring., 1973-75, chmn. dept. bioengring., 1973-76, chmn. grad. group bioengring., 1975-76, dir. systems and integrative biology tng. program, 1979—; vis. prof. biomed. engring. U. Miami, 1970-79, Erasmus U. Med. Sch., Rotterdam, Netherlands, 1970-71, Tech. U., Delft, 1970-71, prof. Dutch culture Sch. Arts and Scis., 1983—; Mem. spl. study sect. NIH, 1966-68; cons. sci. affairs div. NATO, 1973—; participant numerous internat. confs. in field. Author: (with I. Starr) Ballistocardiography in Cardiovascular Research, 1967, Circulatory System Dynamics, 1978; contbg. author: Biological Engineering, 1969; Editor: (with G.N. Jager and N. Westerhof) Circulatory Analog Computers, 1963, (with G.H. Pollack) Ballistocardiography and Cardiac Performance, 1967, (with E. Kresch) The Venous System: Characteristics and Function, 1969, (with J. Baan and J. Raines) Cardiovascular System Dynamics, 1978, Biophysics and Bioengring. Series, 1976—; contbr. numerous articles to profl. jours.; Referee: Biophys. Jour., 1968—, Physics in Medicine and Biology, 1969—, Bull. Math. Biophysics, 1972—, Circulation Research, 1973—; mem. editorial adv. bd.: Jour. Biomechanics, 1969—; assoc. editor: Bull. Math. Biology, 1973—. Vice pres. Haverford Friends Sch. PTA, 1968-70. Recipient Herman C. Burger award, 1978. Fellow IEEE (mem. adminstrv. com. engring. in medicine and biology group 1967-70, mem. edn. com. group biomed. engring. 1968-70, sec. Phila. chpt. 1974-75, mem. regional council profl. group engring. in medicine and biology 1974-77), N.Y. Acad. Scis., AAAS, Explorers Club, Coll. Physicians Phila., Am. Coll. Cardiology; mem. Nederlandse Natuurkundige Vereniging, Ballistocardiograph Research Soc. U.S.A. (sec.-treas. 1965-67, pres. 1968-70), Biophys. Soc. (charter), European Soc. for Noninvasive Cardiovascular Research (co-founder 1960, sec.-treas. 1960-61, mem. com. on nomenclature 1960-61, officer 1961-62), Cardiovascular System Dynamics Soc. (co-founder 1976, pres. 1976-80), Franklin Inst., John Morgan Soc., Biomed. Engring. Soc. (founding mem., chmn. membership com. 1978-79, dir. 1972-75), Am. Heart Assn., Instrument Soc. Am. (sr. mem.), Soc. Math. Biology (charter mem.), Am. Physiol. Soc., Microcirculatory Soc., Am. Assn. Med. Systems and Informatics, Pa. Acad. Sci., Sigma Xi. Presbyterian. Subspecialties: Biophysics (physics); Cardiology. Current work: Cardiovoscular system dynamics. Home: 620 Haydock Ln Haverford PA 19041 Office: 460 Moore Bldg D2 Univ Pa Philadelphia PA 19104

NORCROSS, DAVID WARREN, research physicist; b. Cin., July 18, 1941; s. Gerald Warren and Alice Elizabeth N.; m. Mary Josephine Boydrias, Aug. 26, 1967; children: Joshua David, Sarah Elizabeth. A.B., Harvard U., 1963; M.Sc., U. Ill., 1965; Ph.D., Univ. Coll. London, 1970. Research physicist Sperry Rand Research Center, Sudbury, Mass., 1965-67; research assoc. Joint Inst. Lab. Astrophysics, Boulder, Colo., 1970-74; physicist Quantum Physics div. Nat. Bur. Standards, Boulder, 1974—; lectr. U. Colo., 1974—; vis. staff Los Alamos Nat. Lab., 1977—; cons. Lawrence Livermore Lab., 1978—; assoc. Harvard Coll. Obs., 1980-81; chmn. Joint Inst. Lab. Astrophysics, 1983—. Contbr. articles to profl. jours. Recipient Sustained Superior Performance award Nat. Bur. Standards, 1979, Bronze medal Nat. Bur. Standards, 1982. Fellow Am. Phys. Soc.; mem. Fedn. Am. Scientists. Subspecialty: Atomic and molecular physics. Current work: Interaction of radiation, structure and collisions of atoms and molecules. Office: U Colo Box 440 Boulder CO 80309

NORDEN, CARROLL RAYMOND, zoology educator; b. Escanaba, Mich., May 20, 1922; s. Emil and Naomi (Carroll) N. B.A., No. Mich. U., 1948; M.S., U. Mich., 1951, Ph.D., 1958. Fish biologist Mich. Dept. Conservation, Ann Arbor, 1953-56, U.S. Fish & Wildlife Service, 1957-58; asst. to assoc. prof. biology U. Southwestern La., Lafayette, 1957-63, U. Wis., Milw., 1963-71, prof. zoology, 1971—; sr. scientist Ctr. for Great Lakes Studies, Milw., 1975—. Served with USAAF, 1942-46. Mem. Am. Fisheries Soc., Ichthyologists and Herpetologists Soc., Am. Soc. Zoologists, Soc. of Limnology and Oceanography. Subspecialties: Ecology; Systematics. Current work: Ichthyology, fishery biology. Home: 5328 Thornapple Ln Cedarburg WI 53012 Office: U Wis Dept Zoology PO Box 413 Milwaukee WI 53201

NORDGAARD, JOHN THOMAS, agrl. cons.; b. Bottineau, N.D., Apr. 24, 1952; s. Vern L. and Helen J. N. B.S., N.D. State U., 1975, M.S., 1976. Dir. research Agvise, Inc., Northwood, N.D., 1976-79, chief operating officer, v.p., Benson, Minn., 1979—; also dir.; dir. Swift Pro-Ag, Benson, 1980—. Mem. Am. Soc. Phytopathology, Entomology Soc. Am., Weed Sci. Soc. Am, Nat. Alliance Ind. Crop Cons., Minn. Ind. Crop Cons. Assn. (pres. 1984). Subspecialties: Integrated pest management; Plant pathology. Current work: Plant health and commercial production, research includes soil amendments, pesticides, fertilizers and varietal interactions. Home: 707 11th St. S Benson MN 56215 Office: Box 187 Benson MN 56215

NORDGREN, RONALD PAUL, research engineer; b. Munising, Mich., Apr. 3, 1936; s. Paul A. and Martha B. N.; m. Joan McAfee, Sept. 12, 1959; children: Sonia, Paul. B.S.E., U. Mich., 1957, M.S.E., 1958; Ph.D., U. Calif.-Berkeley, 1962. Research asst. U. Calif.-Berkeley Inst. Engring. Research, 1959-62; with exploration and prodn. research Shell Devel. Co., Houston, 1963—, on spl. assignment, The Hague, Netherlands, 1970-71, sr. staff research engr., 1974-80, research assoc., 1980—; lectr. mech. engring. Rice U., 1965-68, U. Houston, 1980. Contbr. articles to profl. jours.; assoc. editor: Jour. Applied Mechanics, 1972-76, 81—. Fellow ASME; mem. Soc. Indsl. and Applied Math., Sigma Xi, Tau Beta Pi. Subspecialties: Solid mechanics; Petroleum engineering. Current work: Applied mechanics-mechanics of solids, offshore engineering, petroleum engineering, Arctic engineering. Patentee in field. Home: 14935 Broadgreen Dr Houston TX 77079 Office: Shell Devel Co PO Box 481 Houston TX 77001

NORDLANDER, J(OHN) ERIC, chemist, educator; b. Schenectady, July 3, 1934; s. Birger William and Emily Margaret (Brown) N.; m. Ruth Hallett, Oct. 9, 1965; children: Theodore Hallett, Elizabeth Kathryn. A.B. in Chemistry with honors, Cornell U., 1956, Ph.D., Calif. Inst. Tech., 1961. Instr. in chemistry Western Res. U. (became Case Western Res. U. 1967), 1961-62, asst. prof. chemistry, 1962-67, asso. prof., 1967-75, prof., 1975—. Research publs. in field. Ruling elder Fairmount Presbyterian Ch., Cleveland Heights, Ohio. Mem. Am. Chem. Soc., Chem. Soc. (London). Subspecialties: Organic chemistry; Synthetic chemistry. Current work: Mechanisms of organic reactions; carbocations; organic systhesis. Home: 2184 Delaware Dr Cleveland Heights OH 44106 Office: Dept Chemistry Case Western Res U Cleveland OH 44106

NORDLIE, ROBERT CONRAD, biochemistry educator, researcher; b. Willmar, Minn., June 11, 1930; s. Peder Conrad and Myrtle (Spindler) N.; m. Sally Ann Christianson, Aug. 23, 1959; children: Margaret, Melissa, John. Student, Gustavus Adolphus Coll., 1948-49; B.S., St. Cloud State Coll., 1952; M.S., U. N.D.-Grand Forks, 1957, Ph.D., 1960. Postdoctoral research assoc. U. Wis.-Madison, 1960-62; Hill research prof. U. N.D., Grand Forks, 1962-74, Chester Friz disting. prof., 1974-76, prof. biochemistry, 1976—, chmn. dept., 1983—; cons. enzymology Oak Ridge Associated Univs., 1961-80. Editorial bd.: Biochem. Biophys. Acta, 1965—; contbr. articles to profl. jours. Served to cpl. U.S. Army, 1953-55. Recipient Golden Apple award U. N.D., 1968; Hon. Prof. San Marcos U., Lima, Peru, 1982; Disting. Alumnus St. Cloud State U., 1982; Edgar Dale award for Outstanding Teaching, 1983. Mem. Am. Soc. Biol. Chemists, Am. Inst. Nutrition, Am. Chem. Soc. (sec. local sect. 1970, organizer nat. symposium 1974), Sigma Xi (award 1969). Democrat. Lutheran. Lodge: Elks. Subspecialties: Biochemistry (medicine); Nutrition (medicine). Current work: The enzymes of carbohydrate metabolism in mammals, blood glucose homeostatis, glucose-6-phosphate metabolism, glucose-6-phosphatase, mechanisms of hepatic glucose phosphorylation, non-hormonal and hormonal control of blood glucose. Office: Dept Biochemistry U ND University Sta Grand Forks ND 58202 Home: 162 Columbia Ct Grand Forks ND 58201

NOREM, JAMES H., physicist. B.S., U. Chgo., 1964; Ph.D., Rutgers U., 1969. With Rutgers U., 1969-71, 74-75, U. Liverpool, Eng., 1971-74; physicist Argonne (Ill.) Nat. Lab., 1975—. Subspecialties: Accelerator physics; Fusion. Office: Argonne Nat Lab Bldg 360 Argonne IL 60439

NORIEGA, BRIAN KEITH, medical physicist; b. Miami, Fla., Sept. 14, 1955; s. Rogelio and Mary Scott (Dorman) N.; m. Penny Proctor, Aug. 19, 1978. B.S., U. Fla., 1977, M.S., 1978. Radiol. physics cons. Radiation Services, Plant City, Fla., 1978-81; chief med. physicist Mobile (Ala.) Infirmary, 1981-83; cons. Mem. Am. Cancer Soc. Speakers' Bur., 1981—. Mem. Am. Assn. Physicists in Medicine, Health Physics Soc. Democrat. Methodist. Subspecialty: Radiology. Current work: Electron arc therapy, radiation oncology; medical physics; radiological physics. Office: 3001 W Buffalo Ave Tampa FL 33607

NORIN, ALLEN J., microbiology educator; b. Chgo., July 30, 1944; m. Harriett Wolf; children: Lary, Andy. Student, U. Ill., Chgo., 1962-63, Wright Jr. Coll., Chgo., 1963-65; B.S. in Biology, Roosevelt U., Chgo., 1967; M.S. in Microbiology, U. Houston, 1970, Ph.D., 1972. USPHS postdoctoral trainee fellow, dept. microbiology U. Chgo., 1972-75, research assoc., 1973-75; cellular immunologist Montefiore Hosp.-USPHS Program Project on Lung Transplantation, N.Y.C., 1975—, project leader prevention of lung allograft rejection, 1976—, assoc. dir. project, 1977—; asst. prof. microbiology, immunology and surgery Albert Einstein Coll. Medicine, Bronx, N.Y., 1976—; cons. Nat. Heart, Lung and Blood Inst., NSF. USPHS fellow, 1972-75; Ill. State scholar, 1966-67; recipient Henry L. Moses prize, 1983. Mem. Am. Soc. Immunologists, Am. Soc. Microbiology, AAAS. Subspecialties: Immunobiology and immunology; Cell and tissue culture. Current work: Transplantation, diagnostic immunology, immunotoxicology, Carcinogenesis. Office: 111 E 210th St New York NY 10467

NORMAN, DONALD ARTHUR, psychology educator; b. N.Y.C., Dec. 25, 1935; s. Noah and Miriam (Friedman) N.; m. Julie Norman; children: Cynthia, Michael, Eric. B.S.E.E., MIT, 1957; M.S.E.E., U. Pa., 1959, Ph.D. in Math. Psychology, 1962. Asst. instr. to instr. U. Pa., 1957-59; NSF postdoctoral fellow Harvard Ctr. Cognitive Studies, 1962-64; lectr., research fellow dept. psychology and Ctr. Cognitive Studies, Harvard U., 1963-66; assoc. prof. to prof. psychology U. Calif.-San Diego, 1966—, dir. Inst. Cognitive Sci., 1981—, chmn. dept., 1974-78; fellow Ctr. Advanced Studies Behavorial Scis., Stanford, Calif., 1973-74; vis. prof. U. Hawaii, 1977; acting dir. Ctr. Human Info. Processing, U. Calif.-San Diego, 1971-72, 78, 82; cons. to industry; adv. com. machine intelligence and robotics NASA, 1977-79; mem. faculty NATO Advanced Studies Inst. Author: Memory and Attention: An Introduction to Human Information Processing, 1969, Human Information Processing, 1972, (with Rumelhart and LNR Research Group) Explorations in Cognition, 1975, Learning and Memory, 1982; editor: Models of Human Memory, 1970, Perspectives on Cognitive Science, 1981, Jour. Cognitive Sci, 1981—; editorial bd.: Cognitive Sci. Series, 1979—, Cognition and Brain Theory, 1980—, Cognitive Psychology, 1969, Jour. Cognitive Sci., 1977-81, Jour. Exptl. Psychology: Human Learning and Memory, 1977-82; contbr. articles profl jours. Fellow AAAS, Am. Psychol. Assn.; mem. Am. Assn. Artificial Intelligence, Assn. Computing Machinery, Cognitive Sci. Soc. (chmn. bd.), Fedn. Behavioral, Psychol. and Cognitive Scis., Human Factors Soc. (dir. San Diego chpt. 1982-83), Human Factors in Computer Systems, Psychonomics Soc., San Diego Human Factors Soc., Soc. Exptl. Psychologists, Soc. Lab. Users of Computers in Psychology (steering com. 1974-79), Sigma Xi. Subspecialty: Cognition. Current work: Cognitive science, human-machine interface, human memory and attention. Office: University of California Institute for Cognitive Science C 015 University of California La Jolla CA 92093

NORMAN, DOUGLAS JAMES, physician, researcher; b. Seattle, Mar. 6, 1946; s. James Carroll and Pearl Barbara (Ondrasek) N.; m. Jennie Sage, Nov. 23, 1977; 1 son, Alex. B.A., Stanford U., 1968; M.D., U. Wash., 1972. Diplomate: Am. Bd. Internal Medicine and Nephrology. Fellow in nephrology Harvard Med. Sch., Boston, 1976-79; dir. Lab. of Immunogenetics, Oreg. Health Scis. U., Portland, 1979—, dir. end stage renal disease program, 1979—, dir. hemodialysis unit, 1982—. Mem. Internat. Soc. Nephrology, Am. Soc. Nephrology, Am. Assn. Histocompatibility Testing, Am. Soc. Transplant Physicians. Subspecialties: Immunogenetics; Nephrology. Current work: Transplantation immunology. Home: 721 NW Albemarle Terr Portland OR 97210 Office: Oreg Health Scis U 3181 SW Sam Jackson Park Rd Portland OR 97210

NORMAN, JACK C., chemist, educator; b. Taunton, Mass., June 16, 1938; s. Chesley and Marjory (Owen) N.; m. Carol Bee Owens, Jan. 12, 1964; children: Jeffrey Jay, Jon Robin. B.S., U. N.H., 1960; Ph.D. in Phys. Chemistry, U. Wis., 1965. Postdoctoral instr. U. Wash., Seattle, 1965-66; asst. prof. U. Ky., Lexington, 1966-68; assoc. prof. U. Wis., Green Bay, 1968—. Contbr. articles to profl. jours. Wis. Pub. Service Corp. grantee, 1981-82; Exxon Ednl. Found. grantee, 1978; others. Mem. Am. Chem. Soc., Am. Phys. Soc., AAAS. Subspecialties: Nuclear chemistry; Solar energy. Current work: Energy conservation and solar energy. Office: U Wis Green Bay WI 54302

NORMAN, PHILIP SIDNEY, educator, physician; b. Pittsburg, Kans., Aug. 4, 1924; s. P. Sidney and Mildred Ada (Lawyer) N.; m. Marion Birmingham, Apr. 15, 1955; children: Margaret Reynolds, Meredith Andrew, Helen Elizabeth. A.B., Kans. State Coll., 1947; M.D., Washington U., St. Louis, 1951. Diplomate: Am. Bd. Internal Medicine and subplty. in allergy. Intern Barnes Hosp., St. Louis, 1951-52; asst. resident Vanderbilt U. Hosp., 1952-54; USPHS fellow Rockefeller Inst., N.Y.C., 1954-56; mem. faculty Johns Hopkins U., 1959—, prof. medicine, 1975—, head div. clin. immunology, 1970—. Contbr. articles to profl. jours. Served with USAF, 1943-46. Mem. Am. Fedn. Clin. Research, Am. Acad. Allergy, Am. Soc. Clin. Investigation, N.Y. Acad. Scis., Soc. Exptl. Biology and Medicine, Am. Clin. and Climatological Soc., Assn. Am. Physicians. Episcopalian. Subspecialties: Allergy; Immunopharmacology. Current work: Clinical trials in allergy and lung diseases. Home: 13500 Manor Rd Baldwin MD 21013 Office: Good Samaritan Hosp 5601 Loch Raven Blvd Baltimore MD 21239

NORMAN, RICHARD DAVIESS, dentist, educator; b. Franklin, Ind., Feb. 7, 1927; s. William Byron and Edith May (Grubb) N.; m. Joan May Roler, July 15, 1951; children: Beverly Joan and Elizabeth Jane. A.B., Franklin Coll. Ind., 1950; D.D.S., Ind. U., 1958, M.S.D., 1964. Analytical chemist Eli Lilly & Co., Indpls., 1950-54; research asst. to prof. Ind. U., Indpls., 1955-76; dir. dental clin. research Johnson & Johnson Dental Products Co., Heightstown, N.J., 1976-79; cons., 1979—; clin. prof. Fairleigh-Dickinson U., Teaneck, N.J., 1979-80; prof. chmn. dept. restorative dentistry So. Ill. U.-Alton, 1980—. Author: (with others) Materials for the Practicing Dentist, 1969. Mem. pres. Greenwood (Ind.) Community Sch. Bd., 1965-75. Served with AUS, 1945-47; ETO. Recipient Alumni citation Franklin Coll., 1967. Fellow Am. Coll. Dentists (sec.-treas. Ind. sect. 1969-75); mem. Am. Dental Assn., Am. Assn. for Dental Research (councillor St. Louis sect.), Internat. Assn. Dental Research, Dental Materials Group (pres. 1979-80), Sigma Xi, Omicron Kappa Upsilon, Sigma Phi Alpha. Republican. Presbyterian. Lodge: Masons. Current work: Wear of materials, metals in dentistry, fluoride uptake, cements, ceramic fused to metals, physical properties of dental materials, analytical chemistry. Home: 5 Monterey Pl Alton IL 62002 Office: Sch Dental Medicine So Ill Univ 2800 College Ave Alton IL 62002

NORMANN, SIGURD JOHNS, pathologist; b. Cin., Oct. 24, 1935; s. Theodore Frederick and Bird (Krause) N.; m.; children: Jennifer, Elizabeth. M.D., U. Wash., 1960, Ph.D., 1966. Intern U. Calif., San Francisco, 1960-61; resident in pathology U. Wash., 1960-66; prof. pathology U. Fla., 1968—. Contbr. articles to profl. jours. Served to capt. U.S. Army, 1966-68. Mem. Reticuloendothelial Soc., Am. Assn. Pathologists. Subspecialty: Pathology (medicine). Office: U Florida Box J-275 JHMHC Gainesville FL 32610

NORMENT, HILLYER GAVIN, research scientist, meteorologist, cons.; b. Washington, Jan. 13, 1928; s. Hillyer Gavin and Mary Thomas (Quisenberry) N.; m. Reva Lucille Shepherd, Mar. 28, 1953 (dec.); children: Eric Stuart, Jeffrey Leland, Philip Evan; m. Jean Eleanor Porter, July 21, 1973. B.S. in Chemistry, U. Md., 1951, Ph.D. in Phys. Chemistry, 1956. Group leader Callery Chem. Co., Pa., 1956-59; phys. chemist Naval Research Lab., Washington, 1959-62; ops. analyst Research Triangle Inst., Durham, N.C., 1962-63; sr. scientist Tech. Ops., Inc., Burlington, Mass., 1963-67, ARCON Corp., 1967-71; prin. scientist Mt. Auburn Research Assn., Newton, Mass., 1971-75; chief scientist propr. Atmospheric Sci. Assn., Bedford, Mass., 1975—; cons. Contbr. articles to profl. jours. Mem. Am. Meterol. Soc., Royal Meteorol. Soc. London, Am. Geophys. Union, Air Pollution Control Assn., AAAS, Am. Chem. Soc., Am. Crystallographic Soc., Sigma Xi. Subspecialties: Meteorology; Fluid mechanics. Current work: Nuclear fallout modeling, aircraft icing, air pollution transport, aerosol physics boundary layer physics. Home: 186 Peter Spring Rd Concord MA 01742 Office: 363 Great Rd PO Box 307 Bedford MA 01730

NORRIS, DOUGLAS MONROE, research engr.; b. Boston, Apr. 14, 1927; s. Douglas Monroe and Ruth (Blum) N.; m. Barbara Meythaler, June 22, 1951; children: Lisa Ann, Jane Leslie, Douglas Monroe III. B.S. in M.E, Tufts Coll., 1951, M.Ed., 1955; Ph.D. in Applied Mechanics, Mich. State U., 1962. Registered profl. engr., N.H. Instr. engring. Jackson (Mich.) Community Coll., 1955-62; assoc. prof. applied mechanics U. N.H., Durham, 1962-69; sect. leader, engring. mechanics Lawrence Livermore (Calif.) Nat. Lab., 1969-79; tech. mgr. structural mechanics program Electric Power Research Inst., Palo Alto, Calif., 1979—. Contbr. articles to profl. jours. Served to sgt. US. Army, 1945-46. NSF fellow, 1961-62. Mem. ASME, ASTM. Subspecialties: Theoretical and applied mechanics; Fracture mechanics. Current work: Fracture of ductile materials. Office: 3412 Hillview Ave Palo Alto CA 94303

NORTHUP, SHARON JOAN, toxicologist, researcher; b. Union, Mo., Dec. 26, 1942; d. Arthur Benneth and Elizabeth Gladys (Hickman) Carlson; m. William Carlton Northup, June 24, 1970; children: Richard Carlton, Karen Frances. B.S., Washington U., St. Louis, 1965; M.S., U. Mo.-Columbia, 1968, Ph.D., 1971. Diplomate: Am. Bd. Toxicology. Research assoc. Cancer Research Center, Columbia, Mo., 1971-72; instr. U. Mo., Columbia, 1973; research chemist Vets. Hosp., Columbia, 1974-76; mgr. pharmacology Travenol Labs., Morton Grove, Ill., 1976-77, asst. dir. pharmacology, 1977-78, assoc. dir. toxicology, 1978—; subcom. chmn. Pharm. Mgrs. Assn., Washington, 1980-82; mem. biocompatibility task force Health Industries Mfrs. Assn., Washington, 1980—; adv. Pan Am. Health Orgn., Sao Paulo, Brazil, 1980; cons. Chem. Mfrs. Assn., Washington, 1982-83. Trustee Bethel Baptist Ch., Columbia, Mo., 1976. Campus Chest scholar, 1961-62; Sarah Gentry Elston award, 1962-63; NSF scholar, 1967-70; Young Investigator's award Am. Cancer Soc., 1971. Mem. Am. Soc. Pharmacology and Exptl. Therapeutics, Soc. Toxicology, Am. Coll. Toxicology, Am. Chem. Soc., Soc. Biomaterials, Iota Sigma Pi (nominating com. 1983). Democrat. Baptist. Club: Student Wives (Columbia, Mo.) (pres. 1973-74). Subspecialties: Toxicology (medicine); Biomaterials. Current work: Toxicology and biochemistry using in vitro models, hazard assessment, mathematical modeling, alternatives to animal testing, immunotoxicology, mutagenicity, carcinogenicity. Home: 24 Williamsburg Terrace Evanston IL 60203 Office: Travenol Labs Inc 6301 Lincoln Ave Morton Grove IL 60053

NORTON, DON CARLOS, plant nematologist; b. Toledo, Ohio, May 22, 1922; s. Gideon and Eva (Russell) N.; m. Joanne Parker, Aug. 6, 1952; children: Shirley, Glenn, Wayne, Carol. B.S., U. Toledo, 1947; M.S., Ohio State U., 1949, Ph.D., 1950. From asst. prof. to assoc. prof. Tex. A&M U., 1951-59; assoc. prof., then prof. Iowa State U., 1959—. Author: Ecology of Plant-Parasitc Nematodes, 1978; contbr. articles to profl. jours. Served with AUS, 1943-46. Named Disting. Iowa Scientist Iowa Acad. Sci., 1982. Mem. Soc. Nematologists (best paper award 1978), Iowa Acad. Sci., Am. Phytopath. Soc., Sigma Xi, Phi Kappa Phi. Subspecialties: Plant pathology; Ecology. Current work: Ecology and economics of plant-parasitcnematodes. Home: 2305 Broadmoor Ames IA 50010 Office: Iowa State U 321 Bessey Hall Ames IA 50011

NORTON, HORACE WAKEMAN, III, statistics educator; b. Lansing, Mich., Jan. 17, 1914; s. Horace Wakeman and Mabel Mann (Reeves) N.; m. Anne Wallace, Dec. 20, 1937; m. Winifred McCue, Apr. 7, 1963; children: Karl, David, Philip, Paul; m.; 1 stepson, James Alleman. B.Sc. in Chem. Engring, U. Wis., 1935; M.S. in Math, Iowa State U., 1937; Ph.D. in Math. Stats, U. London, Eng., 1940. Research assoc. physics U. Chgo., 1940-42; agt. Agrl. Mktg. Service, U.S. Dept. Agr., 1940-42; sr. meteorologist U.S. Weather Bur., Washington, 1942-47; sr. statistician AEC, Oak Ridge, Tenn., 1947-50; prof. statls. design and analysis dept. animal sci. U. Ill., 1950—; cons. in field. Mem. Champaign County, Ill.) Crime Prevention Council. Mem. AM. Soc. Animal Sci., Blometric Soc., Inst. Math. Stats., Am. Statis. Assn., AAAS, Am. Genetic Assn., AAUP. Subspecialties: Statistics. Home: 2120 Cureton Dr Urbana IL 61801 Office: 327 Mumford Hall 1301 W Gregory Dr Urbana IL 61801

NORTON, JULIA ANNE, statistics educator; b. Birmingham, Ala., Oct. 19, 1948; d. Robert Louis and Betty Sue (Hawkins) N.; m. John Dewey Lovell, Dec. 25, 1978; children: Tanzy Lovell, Robert Norton. B.S. in Math, MIT, 1970; M.A. in Stats, Harvard U., 1974, Ph.D., 1977. Assoc. prof. stats. Calif. State U., Hayward, 1974—; ednl. cons. U. South Pacific, Suva, Fiji Islands, 1980-81. Author: Communication in the Eighties, 1981, Applied Time Series, 1982; reviewer, John Wiley Co., N.Y.C., 1982, Dellen Pub., San Francisco. Bookkeeper Univs. Nat. Anti-War Fund, Cambridge, 1970-73; cons. Hayward (Calif.) Area Recreational Dist., 1974, St. Christophers Episcopal Ch., San Lorenzo, Calif., 1982-83; treas. Starr King Unitarian Ch., Hayward, 1977-79. NSF trainee MIT, 1970. Mem. Am. Statis. Assn., Inst. Math. Stats., Internat. Time Series Assn. Subspecialty: Statistics. Current work: Teaching statistics and data analysis; applied time series; statistics in psychology. Home: 27152 Grand View Ave Hayward CA 94542 Office: Calif State Dept Stats 25700 Carlos Bee Blvd Hayward CA 94542

NORTON, WILLIAM NICHOLSON, JR., biology educator; b. Nashville, May 6, 1945; s. William and Marge Ellen (Steven) N.; m. Betty Jo Jones, Jan. 20, 1968; children: Jennifer, Gregory, Brian. B.S., Troy State U., 1968; M.S., Miss. State U., 1969; Ph.D., Tex. A&M U., 1975. Instr. Troy (Ala.) State U., 1969-71; postdoctoral fellow Baylor Coll. Medicine, Houston, 1975-77; assoc. prof. biology Southeastern La. U., Hammond, 1977—; dir. Research Ctr. for Coll. of Sci. and Tech., 1982—. Contbr. articles to profl. jours. Fulbright research award Council Internat. Exchange of Scholars, Portugal, 1983; U.S. Air Force Office Sci. Research grantee, 1982; NSF grantee, 1978. Mem. La. Soc. Electron Microscopy, Electron Microscopy Soc. Am., AAAS, Sigma Xi, Phi Sigma. Democrat. Methodist. Subspecialties: Microscopy; Cytology and histology. Current work: Study of ultrastructural and cytochemical effects of aromatic hydrocarbons on specific animal model systems; study effects of riboflavin deficiency on internal organs of rats. Home: 23 Cherry St Hammond LA 70402 Office: Southeastern La U PO Box 335 Hammond LA 70402

NORTON, WILLIAM THOMPSON, neuroscience educator, consultant, editor; b. Damariscotta, Maine, Jan. 27, 1929; s. Carroll P. and Josephine G. (Eales) N.; m. Lila Mazur, Sept. 13, 1957; children: Hamish, Adam. A.B., Bowdoin Coll., 1950; M.A., Princeton U., 1952, Ph.D., 1954. Research chemist duPont Co., 1953-57; instr. biochemistry in medicine Albert Einstein Coll. Medicine, N.Y.C., 1957-59, asst. prof. biochemistry in neurology, 1959-64, assoc. prof. neurochemistry, 1965-71, prof. neurochemistry, 1971—; cons. Cerebral Palsy Found., 1966-69, Nat. Multiple Sclerosis Soc., 1969-72, 73-75, 78-80, NIH, 1971-75, 79—. Chief editor: Jour. Neurochemistry, 1982—; editorial bd.: Jour. Lipid Research, 1969-72, Brain Research, 1973—, Devel. Neurosci, 1978—, Jour. Neurochemistry, 1972-81; contbr. numerous articles to sci. jours., chpts. to books. USPHS grantee, 1958—; Nat. Multiple Sclerosis Soc., 1969—; USPHS sr. postdoctoral fellow, 1957-60, 67-68. Mem. Am. Soc. Biol. Chemists, Am. Soc. Neurochemistry, Internat. Soc. Neurochemistry, AAAS, Soc. Neurosci., Sigma Xi. Subspecialties: Neurochemistry; Neurobiology. Current work: Biochemistry of nervous system, research in myelin, demyelination, oligodendroglia, cytoskeletal proteins, tissue culture of isolated brain cells. Office: 1300 Morris Park Ave Bronx NY 10461

NOSANOW, LEWIS HAROLD, univ. adminstr.; b. Phila., July 9, 1931; s. Harry and Pearl (Schware) N.; m. Barbara J. Shissler, Oct. 19, 1973; children—Maria Nicole, James Harry. B.A., U. Pa., 1954; Ph.D., U. Chgo., 1958. NSF postdoctoral fellow, 1968-69; asst. prof. dept. physics U. Minn., Mpls., 1962-64, asso. prof., 1964-68, prof., 1968-73; chmn. dept. physics U. Fla., Gainesville, 1973-74; head condensed matter scis. sect. NSF, Washington, 1974-80; asso. provost, prof. physics U. Chgo., 1981—. Guggenheim fellow, 1967-68. Club: Cosmos. Subspecialties: Materials research administration; Condensed matter physics. Current work: Macroscopic quantum systems, quantum crystals, spin-polarized atomic hydrogen. Home: 5333 Hyde Park Blvd Chicago IL 60615

NOSIL, JOSIP, medical physicist; b. Zagreb, Yugoslavia, June 16, 1944; s. Ivan and Durda (Matijevic) N.; m. Helena Bucar, May 2, 1969; children: Patrick, Cynthia. B. Sc., U. Zagreb, 1968, M.Sc. in Atomic and Nuclear Physics, 1971, D.Sci. in Physics, 1977. Cert. hosp. physicist IAEA. Physicist R. Bošković Inst., Zagreb, 1968-75; med. physicist dept. nuclear medicine and oncology Dr. M. Stojanovic Hosp., Zagreb, 1975-78; vis. fellow Royal Post-Grad. Med. Sch., Hammersmith Hosp., Lond, 1974-75; cons. physicist Foothills Hosp., Calgary, Alta., Can., 1978—. Contbr. articles to profl. jours. Mem. Am. Assn. Physicists in Medicine, Soc. Nuclear Medicine, Can. Assn. Physicist. Mormon. Subspecialties: Medical physics. Current work: Mathematical organ modelling, nuclear cardiology, medical imaging technology, instrumentation. Home: 120 Whitefield Close NE Calgary AB Canada T1Y 4X7 Office: 1403 29th St NW Calgary AB Canada T2N 2T9

NOVACK, GARY DEAN, pharmacologist; b. Oakland, Calif., Nov. 21, 1953; s. Robert Lloyd and Dorothy Louise (Scheibner) N.; m. Dona Ann, Aug. 28, 1977; 1 dau., Rebecca Adrienne. A.B. in Biology with honors, Kresge Coll., U. Calif., Santa Cruz, 1973; Ph.D. in Pharmacology and Environ. Toxicology (Anthony fellow), U. Calif., Davis, 1977. NIH postdoctoral scholar UCLA Mental Retardation Research Ctr., 1977-79; neuropharmacologist Merrell Dow Research Center, Cin., 1979-82; clin. coordinator Allergon Pharm., Irvine, Calif., 1982—. Contbr. articles to profl. jours. Recipient Cum Laude award Assn. Ind. Secondary Schs., 1970. Mem. Soc. Neurosci., Assn. Psychophysiol. Study Sleep, Western Pharmacology Soc. Subspecialty: Neuropharmacology. Office: Allergan Pharmaceuticals 2525 Dupont Dr Irvine CA 92713

NOVAK, JOHN ALLEN, zoology educator; b. Bedford, Ohio, Aug. 17, 1943; s. John George and Mildred Bertha (Olbreys) N.; m. Rose Mary Miller, Sept. 11, 1966; children: Pamela Jane, Emily Melissa. B.Ed., Kent State U., 1965, M.A., 1967; Ph.D., Wash. State U., 1972. Asst. prof. Colgate U., 1972-81, assoc. prof. zoology, 1982—. Free-lance photographer Animals, Animals-Earth Scenes, N.Y.C., 1981—, photographs appeared cover photos mags., books, advt.; contbr. articles to profl. jours. Sloan grantee, 1974; Carter-Wallce Fellowship awardee, 1975; Mellon Fund grantee, 1976-77; Colgate Research grantee, 1976, 79, 80. Mem. Entomol. Soc. Am., Wash. State Entomol. Soc., Sigma Xi. Episcopalian (vestryman, 1974-76, 78-80). Subspecialties: Systematics; Behaviorism. Current work: Systematics, behavior, biogeography and evolution of certain Diptera belonging to families Tephritidae and Syrphidae. Home: Preston Hill Rd Hamilton NY 13346 Office: Dept Biology Colgate U Hamilton NY 13346

NOVICK, DAVID MILES, physician, educator; b. N.Y.C., June 30, 1948; s. Bernard and Rose Elizabeth (Ritterman) N.; m. Jane Gillman Phillips, Apr. 11, 1976; children: Batya Leslie, Elana Marian. B.A., Duke U., 1970; M.D., SUNY-Downstate Med. Ctr., Bklyn., 1974. Diplomate: Am. Bd. Internal Medicine. Intern in medicine Beth Israel Med. Ctr., N.Y.C., 1974-75, resident in medicine, 1975-77; staff physician Morris J. Bernstein Inst., 1977-79, asst. chief medicine, 1979—; asst. prof. clin. medicine Mt. Sinai Sch. Medicine, N.Y.C., 1981—. Contbr. articles to med. jours. Trustee Tifereth Israel Town and Village Synagogue, N.Y.C., 1982. Named Intern of Yr. Dept. Medicine Beth Israel Med. Ctr., 1975. Mem. ACP, N.Y. County Med. Soc., Med. Soc. State of N.Y., Am. Assn. for Study of Liver Diseases, Phi Lambda Upsilon. Democrat. Jewish. Subspecialty: Internal medicine. Current work: Research, teaching, and patient care in alcoholism, drug addiction, and liver diseases. Office: Beth Israel Med Ctr 10 Nathan D Perlman Pl New York NY 10003

NOVICK, MELVIN ROBERT, statistian, educator, consultant; b. Chgo., Sept. 21, 1932; m. Naomi Spark, Sept. 15, 1957; children: Laura Renee, Raymond Lloyd. B.A., Roosevelt U., Chgo., 1957, M.A. and B.S. in Math. and Psychology, 1959; Ph.D. in Math. Stats, U. N.C.-Chapel Hill, 1963. Research asst. Assn. Am. Med. Colls., 1956-59; research asst. U. N.C.-Chapel Hill, 1959-62, research assoc., 1962-63; assoc. research statistician Ednl. Testing Service, Princeton, N.J., 1963-64, research statistician, 1964-68, sr. research statistician, 1968-70; dir. psychometric research Am. Coll. Testing Program, Iowa City, 1970-74; prof. measurement and stats. U. Iowa, Iowa City, 1970—; lectr. Univ. Coll. Wales, 1967-68; hon. lectr. Univ. Coll. London, 1967-73; vis. prof. edn. U. Pa., 1970; cons. U. London, 1968, Office of Edn., 1970, Am. Assn. Dental Schs., 1973; Nat. Inst. Edn., 1974, govt. agys. and corps.; participant profl. confs. Author: (with others) Tables for Bayesian Statisticians, 1974; author: Statistical Theories of Mental Test Scores, 1968, Bayesian Statistics, 1974, (with P.H. Jackson) Statistical Methods for Educational and Psychological Research, 1974; contbr. chpts. to books, articles to profl. jours. Served in USAF, 1952-55; ETO. Grantee Office Naval Research, U.S. Office Edn., NSF, NIH, Carnegie Corp., others. Fellow Am. Psychol. Assn. (pres. div. 5 1980-81), Am. Statis. Assn.; mem. Inst. Am. Ednl. Research Assn., Math. Stats., Nat. Council Measurement in Edn. (dir. 1980-82), Psychometric Soc. (trustee 1977-81, pres. 1979-80), Iowa Ednl. Research and Evaluation Assn. Subspecialties: Psychometrics; Statistics. Current work: Behavioral statistics and computer-assisted data analysis. Home: 306 Mullin Ave Iowa City IA 52240 Office: U Iowa 356 Lindquist Center Iowa City IA 52242

NOVICK, RICHARD PAUL, molecular biologist, research institute administrator, educator; b. N.Y.C., Aug. 10, 1932; s. Samuel John and Mollie (Forster) N.; m. Barbara Jane Zabin, June 1, 1958; children: Lynn, Dorothy. B.A., Yale U., 1954; M.D. with honors, NYU, 1959. Intern Yale Med. Ctr., New Haven, 1959-60; resident Vanderbilt U. Med. Ctr., Nashville, 1962-63; postdoctoral fellow Nat. Inst. Med. Research, London, 1960-62, Rockefeller U., N.Y.C., 1963-65; assoc. Pub. Health Research Inst., N.Y.C., 1968-71, assoc. mem., 1971-75, mem., chmn. dept. plasmid biology, 1976—, dir., 1980—; research prof. dept. microbiology NYU Med. Sch.; mem. Nat. Recombinant DNA Adv. Com., 1978—. Founder, editor-in-chief: Plasmid; contbr. articles to profl.jours. Grantee NIH, 1967—, NSF, 1972—, Am. Cancer Soc., 1978—. Mem. Am. Soc. Microbiology, Am. Genetics Soc., Harvey Soc. Subspecialties: Molecular biology; Genetics and genetic engineering (biology). Current work: Plasmids, transposons, antibiotic resistance, molecular cloning, gene expression, replication of DNA. Home: 10 W 86th St Apt 11A New York NY 10024 Office: Pub Health Research Inst 455 First Ave New York NY 10016

NOVICK, WILLIAM JOSEPH, JR., pharmacologist; b. Revoloc, Pa., Dec. 14, 1931; s. William J. and Rugh (Jones) N.; m. Joanne C. Imgrund, Sept. 17, 1955; children: William, Susan, Carol, Mary. B.S., St. Francis Coll., 1953; Ph.D., Duke U., 1961. Sect. head pharmacology Smith Kline & French, Phila., 1955-58, 61-67; mngr. pharmacology William H. Rorer, Fort. Washington, Pa., 1967-70; dir. biol. scis. Hoechst Roussel Pharms Inc., Somerville, N.J., 1970-83, sr. dir. internat. product devel., 1983—. Friends of St. Francis scholar, 1949-53; Walter Karr fellow, 1958-61. Mem. Am. Soc. Pharmacology and Exptl. Therapeutics, Am. Soc. Microbiology, N.Y. Acad. Sci. Republican. Roman Catholic. Subspecialties: Pharmacology; Microbiology (medicine). Office: Hoechst-Roussel Pharms Inc Route 202-206N Somerville NJ 08876

NOVIKOFF, ALEX BENJAMIN, cell biologist; b. Russia, Feb. 28, 1913; came to U.S., 1913, naturalized, 1918; s. Jack and Anna (Tretyakoff) N.; m. Phyllis M. Iaciofano, Dec. 1, 1968; children—Kenneth, Lawrence. B.S., Columbia U., 1931, M.A., 1933, Ph.D., 1938; D.Sc., U. Vt., 1983. Fellow biology dept. Bklyn. Coll., 1931-35, instr., 1936-47, asst. prof. biology, 1947-48; Am. Cancer Soc. research fellow U. Wis. Med. Sch., 1946-47; prof. exptl. pathology, asso. prof. biochemistry U. Vt. Coll. Medicine, 1948-53; sr. investigator Waldemar Med. Research Found., Port Washington, N.Y., 1954-55; assoc. research prof. pathology Albert Einstein Coll. Medicine, 1955-57, prof. pathology, 1958—. Author: (with Eric Holtzman) Cells and Organelles, 1970, 2d edit., 1976, 3d edit., 1983; author books on sci. for children, films on enzymes and carbohydrate metabolism for, U.S. Army; contbr. 270 sci. articles to profl. publs. Recipient Distinguished Service award Columbia U., 1960, E.B. Wilson award Am. Soc. Cell Biology, 1982; Nat. Cancer Inst. research career award, 1962—; Mem. pathology study sect. NIH, 1972-76. Mem. Nat. Acad. Scis., Japanese Soc. Cell Biology (hon.), Societe Francaise de Microscopie Electronique (hon.), Japanese Histochem. Soc. (hon.). Subspecialties: Cell biology (medicine); Pathology (medicine). Office: Albert Einstein Coll Medicine Dept Pathology 1300 Morris Park Ave Bronx NY 10461

NOVOTNY, DONALD WAYNE, electrical engineering educator; b. Chgo., Dec. 15, 1934; s. Adolph and Margaret N.; m. Louise J. Eenigenburg, June 26, 1954; children: Donna Jo, Cynthia Jean. B.E.E., Ill. Inst. Tech., 1956, M.S., 1957; Ph.D., U. Wis., 1961. Registered profl. engr., Wis. Instr. Ill. Inst. Tech., 1957-58; mem. faculty U. Wis., Madison, 1958—, prof. elec. engring., 1969—, chmn. dept. elec. and computer engring., 1976—; vis. prof. Mont. State U., 1966, Eindhoven Tech. U., Netherlands, 1974; Fulbright lectr. Tech. U., Ghent, Belgium, 1981; dir. Wis. Electric Machines and Power Electronics Consortium, 1981—; cons. to industry. Author: Introductory Electromechanics, 1965; also research papers; assoc. editor: Electric Machines and Electromechanics, 1976—. Fellow Gen. Electric Co., 1956; Ford Found., 1960; recipient Kiekhofer Teaching award U. Wis., 1964; Outstanding Paper award Engring. Inst. Can., 1966; grantee numerous industries and govt. agencies. Mem. IEEE, Am. Soc. Engring. Edn., Sigma Xi, Tau Beta Pi, Eta Kappa Nu. Congregationalist. Subspecialty: Electrical engineering. Home: 1421 E Skyline Dr Madison WI 53705 Office: Dept Elec and Computer Engring 1415 Johnson Dr U Wis Madison WI 53706

NOVOTNY, JIRI, research biochemist; b. Kladno, Czechoslovakia, Dec. 15, 1943; came to U.S., 1979; s. Jaroslav and Eva (Foustkova) N.; m. Jarmila Novotna, Feb. 28, 1966; 1 dau., Paula. R.N.Dr., Charles U., 1970; Ph.D., Czechoslovak Acad. Sci., 1970. Scientist, Inst. of Organic Chemistry and Biochemistry Czechoslovak Acad. Sci., 1965-76, sr. scientist, 1976-79; vis. scientist, dept. molecular biophysics U. Oxford, Eng., 1979; research assoc. Harvard Med. Sch.-Mass. Gen. Hosp., 1980-81; asst. prof. biochemistry Harvard Med. Sch., 1981—. Recipient prize Czechoslovak Acad. Sci., 1979. Mem. Am. Assn. Immunologists; mem. Am. Chem. Soc.; Mem. N.Y. Acad. Sci. Club: Gull Point Yacht (Quincy, Mass.). Subspecialties: Genetics and genetic engineering (biology); Molecular biology. Current work: Study of protein structure and evolution, in particular antibody binding site by genetic engineering and computer modeling. Home: 2 Howe St Quincy MA 02169 Office: Fruit St Boston MA 02114

NOWAK, THOMAS L., biochemist, enzymologist; b. Niagara Falls, N.Y., Nov. 25, 1942; s. Sigmund C. and Jane (Marzec) N.; m. Marlene R. Nowak, June 3, 1967; children: Stephen T., Natasha J., Thomas Stanislaus. B.S., Case Inst. Tech., 1964; Ph.D., U. Kans., 1969. Postdoctoral fellow Inst. Cancer Research, Phila., 1969-71; research assoc., 1971-72; asst. prof. U. Notre Dame, Ind., 1972-78, assoc. prof., 1978—. Contbr. articles to profl. jours. Research Career Devel. awardee NIH, 1976—; vis. scientist dept. phys. chemistry U. Groningen, Netherlands, 1981-82. Mem. Am. Chem. Soc., Am. Soc. Biol. Chemists, Internat. Soc. Magnetic Resonance, Biophysics Soc., AAAS, Alpha Chi Sigma. Roman Catholic. Subspecialties: Biophysical chemistry; Biochemistry (biology). Current work: Enzyme catalysts; regulation of metabolism, molecular basis of regulation; cations in biochemical systems, biochemical applications of NMR. Home: 1709 E Wayne St South Bend IN 46615 Office: Dept Chemistry Univ Notre Dame Notre Dame IN 46556

NOWAK, WELVILLE BERENSON, mechanical engineering; b. Hartford, Conn., Oct. 6, 1921; s. Abraham and Ann (Segal) N.; m. Ruth Goldberg, Mar. 27, 1950; children: Ann Leslie, Michael David. B.S., M.I.T., 1942, Ph.D., 1949. Staff mem. Radiation Lab. M.I.T., Cambridge, 1942-45, staff mem. metall. project, 1949-52; physicist Microwave Assocs., Boston, 1952-54; project mgr., dir. electronic research Nuclear Metals Co., Concord, Mass., 1954-62; prof. mech. engring. Northeastern U., Boston, 1962—, Donald W. Smith prof. mech. engring., 1980—, chmn. dept., 1975-81; cons. in field. Contbr. articles to profl. jours. Mem. Am. Phys. Soc., Am. Soc. for Metals, Am. Vacuum Soc., ASME, Am. Soc. for Engring. Edn. Subspecialties: Electronic materials; Corrosion. Current work: Electronic materials/processing (especially thin films); wear-resistant films; corrosion-resistant metal films; solar cells; teaching materials science. Patentee in field. Home: 17 Furbush Ave West Newton MA 02165 Office: 360 Huntington Ave Boston MA 02115

NOWELL, PETER CAREY, educator, pathologist; b. Phila., Feb. 8, 1928; s. Foster and Margaret (Matlack) N.; m. Helen Worst, Sept. 9, 1950; children—Sharon, Timothy, Karen, Kristin, Michael. B.A., Wesleyan U., Middletown, Conn., 1948; M.D., U. Pa., 1952. Intern Phila. Gen. Hosp., 1952-53; resident pathology Presbyn. Hosp., Phila., 1953-54; med.-teaching, research specializing in exptl. pathology, Phila., 1956—; from instr. to prof. pathology Sch. Medicine U. Pa., 1956—, chmn. dept. pathology, 1967-73; dir. (Cancer Center), 1973-75. Served to lt., M.C. USNR, 1954-56. Recipient Research Career award USPHS, 1964-67, Parke-Davis award, 1965, Lindback Distinguished Teaching award, 1967. Mem. Am. Soc. Exptl. Pathology (pres. 1970), Am. Assn. Cancer Research, Radiation Research Soc., Nat. Acad. Sci., Transplantation Soc., Am. Assn. Immunologists. Subspecialties: Pathology (medicine); Immunology (medicine). Current work: Tumor cytogenetics; lymphocyte biology. Home: 345 Mt Alverno Rd Media PA 19063 Office: Dept Pathology Sch Medicine Univ of Pennsylvania Philadelphia PA 19104

NOWLIN, WORTH D., JR., oceanographer, educator; b. Smithville, Tex., Oct. 1, 1935; s. Worth Dabney and Celeste (Lynch) N.; m. Toni Joanetta Powers, June 5, 1959; children: Keith E., Julie E. Student, So Meth. U., 1956, Georg-August U., Gottingen, Germany, 1956-57; B.A. in Math, Tex. A&M U., 1958, M.S., 1960, Ph.D. in Oceanography (NDEA grad. fellow), 1966. Oceanographer Ocean and Tech. Group, Office Naval Research, 1967-68, program dir. phys. oceanography, 1968-69; dep. head Office for Internat. Decade of Ocean Exploration, NSF, Washington, 1971; prof. oceanography Tex. A&M U., 1974—, dir. Sea Grant Program, 1976-78, head oceanography dept., 1976-79; panel on oceanography com. polar research Nat. Acad. Scis., NRC, 1972-75; co-chmn., exec.com. Internat. So. Ocean Studies, 1974-80; bd. govs. Joint Oceanographic Instn., Inc., 1976-79; ocean scis. bd. Nat. Acad. Scis/NRC, 1978-81; mem. ocean climate research strategies com. Nat. Acad. Scis./NRC, 1983-85. Assoc. editor for oceanography: Revs. of Geophysics and Space Physics, 1972-75, Jour. Phys. Oceanography, 1969-79, Antarctic Research Series of Am. Geophys. Union, 1974-78; contbr. articles profl. jours. Recipient Faculty Disting. Achievement award in research Tex. A&M U., 1979. Fellow Am. Geophys. Union (pres. oceanographic sect. 1978-80); mem. Sigma Xi (Research award 1979), Phi Eta Sigma. Subspecialty: Oceanography. Current work: Meso and large-scale ocean circulation and property distributions, ocean current measurements, ocean dynamics, Antarctic Circumpolar Current, Carribbean Sea, Gulf of Mexico. Office: Dept Oceanography Tex A&M U College Station TX 77843

NOYCE, ROBERT NORTON, manufacturing company executive; b. Burlington, Iowa, Dec. 12, 1927; s. Ralph B. and Harriet (Norton) N.; m. Ann S. Bowers, Nov. 27, 1975; children: William B., Pendred, Priscilla, Margaret. B.A., Grinnell Coll., 1949; Ph.D., Mass. Inst. Tech., 1953. Research engr. Philco Corp., Phila., 1953-56; research engr. Shockley Semicondr. Lab., Mountain View, Calif., 1956-57; founder, dir. research Fairchild Semicondr., Mountain View, 1957-59, v.p., gen. mgr., 1959-65; group v.p. Fairchild Camera & Instrument, Mountain View, 1965-68; founder, pres. Intel Corp., Santa Clara, Calif., 1968—, chmn., 1968-75; vice chmn., 1979—; dir. Rolm Corp., Santa Clara, Calif., Diasonics Inc., Milpitas, Calif. Trustee Grinnell Coll., 1962—; regent U. Calif., 1982—. Recipient Stuart Ballentine award Franklin Inst., 1967, Harry Goode award AFIPS, 1978; Nat. Medal of Sci. Pres. of U.S., 1979; I.E.E. Faraday medal, 1979; Harold Pender award U. Pa., 1980. Fellow IEEE (Cledo Brunetti award 1978, medal of honor 1978); mem. Nat. Acad. Engring. Subspecialty: 3emiconductors. Patentee in field. Home: Los Altos CAOffice: 3200 Lakeside Dr Santa Clara CA 95051

NOYES, RICHARD MACY, physical chemist, educator; b. Champaign, Ill., Apr. 6, 1919; s. William Albert and Katharine Haworth (Macy) N.; m. Winninette Arnold, July 12, 1946 (dec. Mar. 1972); m. Patricia Jean Harris, Jan. 26, 1973. A.B. summa cum laude, Harvard U., 1939; Ph.D., Calif. Inst. Tech., 1942. Research asso. rocket propellants Calif. Inst. Tech., 1942-46; mem. faculty Columbia U., 1946-58, asso. prof., 1954-58; Guggenheim fellow, vis. prof. U. Leeds, Eng., 1955-56; prof. chemistry U. Oreg., 1958—, head dept., 1963-68, 75-78. Editorial adv. com.: Chem. Revs, 1967-69; editorial adv. com.: Jour. Phys. Chemistry, 1973-80; assoc. editor, 1980-82, Internat. Jour. Chem. Kinetics, 1972—; Contbr. to profl. jours. Fulbright fellow; Victoria U., Wellington, New Zealand, 1964; NSF sr. postdoctoral fellow Max Planck Inst. fûr Physikalische Chemie, Göttingen, Germany, 1965; sr. Am. scientist awardee Alexander von Humboldt Found., 1978-79. Fellow Am. Phys. Soc.; mem. Nat. Acad. Scis., Am. Chem. Soc. (chmn. div. phys. chemistry 1961-62, exec. com. div. 1960-75, mem. council 1960-75, chmn. Oreg. sect. 1967-68, com. on nominations and elections 1962-68, com. on publs. 1969-72), Chem. Soc. (London), Wilderness Soc., ACLU, Soc. Social Responsibility in Sci., Phi Beta Kappa, Sigma Xi. Club: Sierra (past chmn. Atlantic and Pacific N.W. chpts., N.W. regional v.p. 1973-74). Subspecialty: Kinetics. Current work: Mechanisms of oscillating chemical reactions; unstable states of chemical systems. Research mechanisms chem. reactions, developing gen. theories, intrepretation phys. properties chemicals. Home: 2014 Elk Dr Eugene OR 97403 When I was young, I wanted to be an "explorer." I am fortunate to have a job in which I can make discoveries as exciting as those of the explorers who first sailed uncharted seas. Then I can try to convey the excitement to another generation. As an avocation, I try to influence government policies toward our least developed lands. It is a gratifying mix of satisfying curiousity and serving society.

NUNAN, CRAIG SPENCER, scientist; b. Medford, Oreg., Dec. 22, 1918; s. Charles J. and Mignon I. (Thompson) N.; m. Analee P. Freeman, Sept. 11, 1948; children: Lori, Shary, Kevin, Scott. Student engring. sci., Stanford U., 1968-69; B.S., U. Calif.-, Berkeley, 1940, M.S., 1949. Electronics engr. Bur. Ships, U.S. Navy, 1940-46; project engr. Lawrence Radiation Lab., U. Calif., 1946-53; dir. research Chromatic TV, 1953-55; gen. mgr. radiation div. Varian Assocs., Palo Alto, Calif., 1955-68, sr. scientist, 1968—. Subspecialties: Biomedical engineering; Biomedical engineering. Current work: Invention of equipment for field of diagnostic and therapeutic radiology and industrial x-ray. Patentee fields of color TV, electron linear accelerators, radiotherapy equipment, x-ray CT scanners, orthopedic x-ray equipment. Home: 26665 St Francis Rd Los Altos Hills CA 94022 Office: Varian 611 Hansen Way Palo Alto CA 94303

NUNN, WALTER MELROSE, JR., elec. engr., educator; b. New Orleans, Sept. 16, 1925; s. Walter Melrose and Leah Agnes (Hennessey) N.; m. Hortense Rosa Hillery, Aug. 14, 1949. B.S.E.E., Tulane U., 1950; M.S.E.E., Okla. State U., 1952; Ph.D. in Elec. Engring, U. Mich., 1961; M.S. in Physics, U. Ill., 1969. Registered profl. engr., Fla. Research engr. Hughes Aircraft Co., Los Angeles, 1954-56; research assoc. U. Mich., Ann Arbor, 1956-60; asst. prof. elec. engring. U. Minn. - Mpls., 1960-63; prof. elect. engring. Tulane U., New Orleans, 1963-67; prof. elec. engring. Fla. Inst. Tech., Melbourne, 1969—; cons. U.S. Army Missile Command, NASA, Harris Corp., U.S. Dept. Navy, Nat. Acad. Sci., Mpls. Honeywell Co. Contbr. articles to profl. jours. Served with USMC, 1943-46. Mem. Am. Phys. Soc., Tau Beta Pi, Sigma Pi Sigma. Subspecialties: Electrical engineering; Electronics. Current work: Electromagnetic radiation, antennas, microwave measurements and techniques, electromagnetic radiation from highly-ionized plasmas, electron and atomic physics, lasers and quantum electronics. Office: Dept Elec Engring Fla Inst Tech Melbourne FL 32901

NUSS, DONALD LEE, research microbiologist; b. Murfreesboro, Tenn., May 15, 1947; s. Horace William and Virginia Ruth (Hord) N.; m. Christa L. Mitchell, Dec. 29, 1965; 1 dau., Dana Christine. B.S., Edinboro State Coll., 1969; Ph.D., U. N.H., 1973. Postdoctoral fellow Roche Inst. Molecular Biology, Nutley, N.J., 1973-75, research assoc., 1975-76; research scientist N.Y. Dept. Health, Albany, 1976—. Contbr. articles in field to profl. jours. Mem. Am. Soc. Microbiology, Am. Soc. Virology. Lutheran. Subspecialties: Plant virology; Animal virology. Current work: Researcher in molecular biology of virus transmission by insect vectors. Home: 55 Nathaniel Blvd Delmar NY 12054 Office: NY Dept Health Div of Labs and Research Tower Bldg Empire State Plaza Albany NY 12201

NUSSENZWEIG, VICTOR, immunologist, educator; b. Sao Paulo, Brazil, Nov. 2, 1928; s. Michel and Regina (Kupferbloom) N.; m. Ruth Sonntag, May 13, 1954; children: Michel, Sonia, Andre. M.D., U. Sao Paulo, 1953, Ph.D., 1957. Asst. prof. U. Sao Paulo, 1956-58; research assoc. Pasteur Inst., Paris, 1958-65; asst. prof. pathology N.Y. U., 1965-67, assoc. prof., 1967-71, prof., 1971—. Research numerous publs on parasitology and immunolgy. Grantee NIH, WHO, Rockefeller Found. Mem. Am. Assn. Immunologists, Harvey Soc. Subspecialties: Immunology (medicine); Parasitology. Current work: Research on biochemistry and physiology of complement system and on immunity to malaria. Office: 550 First Ave Room 599 New York NY 10016

NUTTER, GENE DOUGLAS, engineer, physicist, university engineering administrator, consultant; b. Columbus, Tex., June 9, 1929; s. William F. and Susie M. (Baker) N.; m. Mary Ann Souder, June 9, 1956. B.Sc. in Physics, U. Nebr., 1951, M.S., 1956. Physicist Nat. Bur. Standards, 1952-54; engring. supr. Atomics Internat. div. Rockwell Internat., Canoga Park, Calif., 1956-67; asst. dir. Instrumentation System Ctr., U. Wis.-Madison, 1967—; cons. radiation thermometry, precision radiometry. Contbr. articles to profl. publs.; editor: (with N. E. Huston) Management Systems for Laboratory Instrument Services, 1980. Mem. Instrument Soc. Am., ASTM, Optical Soc. Am., IEEE, Am. Inst. Physics. Democrat. Unitarian. Lodge: Masons. Subspecialties: Optical engineering; Metrology. Current work: Detailed analysis of radiation thermometry and its organization into a formal engineering discipline; development improved methods. Home: 6117 Old Middleton Rd Madison WI 53705 Office: Instrumentation Systems Center 1500 Johnson Dr Madison WI 53706

NUTTING, EHARD FORREST, pharmaceutical company research manager, researcher, endocrinologist; b. Milw., Oct. 4, 1929; s. Charles William and Mary M. (Beck) N.; m. Shirley Irene Richards, Nov. 7, 1951; children: Ron D., Bryan R., Bruce L. Tamara S., Amy L., Eric F. B.S., Utah State U., 1948-51; M.S., U. Wis., 1954-56, Ph.D., 1959-60. Research fellow dept. genetics U. Wis., 1954-56, dept. zoology, 1958-60; research biologist Endocrine Labs., Madison, Wis., 1958-60; with D. G. Searle & Co., 1960—; dir. dept. biol. research Searle Labs., Skokie, Ill., 1973-75, sr. scientist pioneering research dept., 1975-80, dir. cellular and endocrine diseases sect., dept. biol. research, 1980—; report reviewer Office Tech., Congress U.S., 1981; mem. research proposal rev. panels Nat. Inst. Child Health and Diseases, 1972, 73, 77, 78, 80,83; research proposal reviewer NSF, 1979, 80, 82,83; participating scientist spl. program research in human reprdon., human reprodn. unit WHO, 1972-78; guest lectr. Rice U., 1963, U. Sherbrooke, Quebec, Que., Can., 1970; lectr. postdoctoral workshops, dept. ob-gyn and physiology Wayne State U., 1971-72. Author chpts. and articles on contraception, reproduction, physiology, pharmacology and endocrinology. Served to capt. U.S. Army, 1951-54. Mem. Endocrine Soc., Soc. Exptl. Biology and Medicine, Soc. Study Fertility, Soc. Study Reprodn., Am. Soc. Pharmacology and Exptl. Therapeutics, Fedn. Am. Socs. Exptl. Biology, AAAS. Roman Catholic. Subspecialties: Reproductive biology; Endocrinology. Office: 4901 Searle Pkwy Skokie IL 60077

NUTTLI, OTTO WILLIAM, geophysicist; b. St. Louis, Dec. 11, 1926; s. Otto Peter and Marie Bertha (Wehinger) N. B.S., St. Louis U., 1948, M.S., 1950, Ph.D., 1953. Instr. geophysics St. Louis U., 1952-56, asst. prof., 1956-59, asso. prof., 1959-62, prof., 1962—; vis. research scientist U. Mich., summer 1962, U. Calif., Berkeley, summers 1964, 67; mem. adv. panel NSF and C.E.; cons. to fed. govt. and industry; mem. U.S. Nat. Com. on Geology, 1975-79, Com. on Seismology, 1976-79. Fellow Am. Geophys. Union (asso. editor Jour. Geophys. Research 1978-80), Royal Astron. Soc.; mem. Seismol. Soc. Am. (pres. 1976-77, editor Bull 1971-75), Soc. Exploration Geophysicists, Earthquake Engring. Research Inst. Subspecialty: Geophysics. Current work: Seismicity and earthquake hazard in Eastern U.S. earthquake source physics. Home: 5422 Finkman St Saint Louis MO 63109 Office: 221 N Grand Blvd Saint Louis MO 63103

NUZZO, SALVATORE JOSEPH, electronics co. exec.; b. Norwalk, Conn., Aug. 6, 1931; s. Rocco and Angelina (Ranzulli) N.; m. Lucille Cocco, Oct. 3, 1953; children—James, David, Thomas, Dana. B.S. in Elec. Engring, Yale U., 1953; M.S. in Bus, Columbia U., 1974. With Hazeltine Corp., Greenlawn, N.Y., 1953—, v.p. govt. products and mktg., 1969-73, v.p. govt. products div., 1973-74, sr. v.p. ops., 1974-76, exec. v.p., chief operating officer, 1976, pres., chief operating officer, 1977—, chief exec. officer, 1980—. Mem. Armed Forces Communications and Electronics Assn., Am. Def. Preparedness Assn., Assn. U.S. Army, Electronic Industries Assn., Nat. Security Indsl. Assn. (trustee), Am. Mgmt. Assn., NAM, Yale Sci. and Engring. Assn. (pres.), Columbia Sch. Bus. Alumni Assn. (bd. dirs.). Subspecialty: Electronics company management. Home: 94 Holst Dr W Huntington NY 11743 Office: Hazeltine Corp Commack NY 11725

NWANGWU, PETER UCHENNA, clinical pharmacologist; b. Umuahia, Imo, Nigeria, June 13, 1949; came to U.S., 1972; s. Sidney Nwokeke and Phoebe (Akueke) N.; m. Patience Okaro, June 3, 1978; children: David, Daniel, Joy. B.A. in Chemistry, U. Nebr., 1974, M.S., 1976, Pharm.D., 1979, Ph.D. in Med. Scis, 1979. Registered pharmacist, Nebr. Proctor of physics U. Nebr., Lincoln, 1973-74, instr. biology, 1974; instr. chemistry SUNY-Syracuse, 1975; asst. prof. pharmacology and toxicology Sch. Pharmacy, Fla. A&M U., Tallahassee, 1979-81, dir. clin. research, 1979-81; assoc. prof. pharmacology and toxicology St. John's U., Jamaica, N.Y., 1981-83; mem. Pharm.D. program adv. bd.; clin. research monitor Ayerst Labs., N.Y.C., 1983—; internat. rep. Fine Future Co., Nigeria, 1976-77; v.p. ops. Finecoop, Nigeria, 1977-79; mem. vis. faculty Oak Ridge Associated Univs., Oak Ridge Nat. Labs. Author: Concepts and Strategies in New Drug Development, 1983, (with V.A. Skoutakis) Clinical Toxicology of Non Drugs, 1983; editorial adv. bd., Clin. Toxicology Cons.; program dir. NBC-TV series on new drug devel., 1977-81; Nat. Symposium on Concepts and Strategies in Clin. Research, N.Y.C., 1982; contbr. book reviews, abstracts, manuscripts and articles to med. jours. Bd. dirs. Calvary Baptist Ch., Omaha. NIH grantee, 1981—. Fellow Am. Coll. Clin. Pharmacology, Am. Soc. Cons. Pharmacists, Am. Coll. Tropical Medicine; mem. Nat. Pharm. Alliance (tech. and drug regulatory com.), Am. Biog. Inst. (nat. bd. advs.), Am. Fedn. Clin. Research, Am. Pharm. Assn., Acad. Pharm. Scis., Am. Assn. Colls. Pharmacy, AAAS, N.Y. Acad. Scis., Smithsonian Instn., Drug. Info. Assn., Jaycees (charter), Fla. Pharm. Assn. (continuing edn. provider), Internat. Platform Assn., Union African Socs. Pharmacology, Phi Eta Sigma, Rho Chi, Sigma Xi. Subspecialties: Pharmacology; Toxicology (medicine). Current work: Clinical trials of new drugs; clinical toxicology; preventive medicine. Developed rapid in vivo technique for screening antiarrhythmic agents in mice, 1977; rapid technique for identification of time of myocardial infarction employing Tc-99 pyrophosphate. Home: 157 Cocoanut St Brentwood NY 11717 Office: Ayerst Labs 685 3d Ave New York NY 11717

NYBERG, DENNIS, biologist; b. Oklahoma City, Feb. 11, 1944; s. Roy Carl and Veryl (Thornstrom) N.; m. Nancy Holmes, May 6, 1967; children: Carl, Ralph, Gustaf, Roy Darwin. B.S., MIT, 1965; M.S., U. Ill., 1969, Ph.D., 1971; NSF fellow, U. Sussex, Brighton, Eng., 1971-72. Research assoc. Ind. U., Bloomington, 1972-73; asst. prof. U. Ill., Chgo., 1973-77, assoc. prof., 1977—. Contbr. articles to profl. jours. U. Ill. fellow, 1970-71; NSF fellow, 1967-70; NSF grantee, 1976-78; NIH grantee, 1979-82. Mem. AAAS, Genetics Soc. Am., Am. Genetic Assn., Soc. Study Evolution, Soc. Protozoologists. Subspecialty: Evolutionary biology. Current work: Research in genetics of adaptation, evolution of sex, aging, recombination. Home: 323 S Humphrey St Oak Park IL 60302 Office: Dept Biol Sci U Il Chicago IL 60680

NYQUIST, GERALD WARREN, mechanical engineer; b. Detroit, Dec. 28, 1940; s. Paul Gustave and Lucille Phyllis (Reiter) N. B.S. in Civil Engring, Lawrence Inst. Tech., 1963; M.S. in Engring. Mechancis, Wayne State U., 1967; Ph.D., Mich. State U., 1970. Registered profl. engr., Mich. Product test engr. Ford Motor Co., Dearborn, Mich., 1963-65; research assoc. Biomechanics Research Ctr., Wayne State U., 1965-72; sr. research engr. Gen. Motors Research Lab., Gen. Motors Corp., 1972-76, sect. supr. environ. activities staff, 1976-80, staff analysis engr. engring. staff, 1980-82; pres. Gerald W. Nyquist, Inc., East Detroit, Mich., 1982—; research assoc. Bioengring. Ctr. Wayne State U., 1983—; cons. biomechanics and vehicle crash safety. Contbr. articles to tech. jours. Mem. Soc. Automotive Engrs., ASME, Sigma Xi. Subspecialties: Biomedical engineering; Theoretical and applied mechanics. Current work: Human impact tolerance, mechanical simulation of the human, automotive crash safety. Address: 19059 Holbrook St East Detroit MI 48021

NYQUIST, WYMAN ELLSWORTH, biometry educator, researcher; b. Scobey, Mont., June 13, 1928; s. Rudolph Ephraim and Alyce Maria (Nordberg) N.; m. Ruth Malene, June 15, 1952; children: Elaine Annette, Craig Evan. B.S., Mont. State U., 1950; Ph.D., U. Calif., Davis, 1953. Instr. U. Calif., Davis, 1953-57, asst. prof., 1958-63; assoc. prof. Purdue U., West Lafayette, Ind., 1963-68, prof., 1968—. Served to 1st lt. USAF, 1954-56. NIH spl. research fellow, 1969-70. Mem. Am. Soc. Agronomy, Biometrics Soc., Council of Agr., Sci. and Tech., Crop Sci. Soc. Am., Genetics Soc. Am., Sigma Xi. Republican. Subspecialties: Statistics; Plant genetics. Current work: Research in biometry, experimental design, statistical genetics, quantitative genetics. Home: 1600 Ravinia Rd West Lafayette IN 47906 Office: Dept Agronomy Purdue Univ West Lafayette IN 47907

NYSSEN, GERARD ALLAN, chemist, educator, researcher, translator; b. Hattiesburg, Miss., Nov. 9, 1942; s. Howard C. and Margreth Elizabeth (Faul) N.; m. Mary Jane, Aug. 21, 1965; children: John Mark, David, Amy. A.B., Olivet Nazarene Coll., 1965; Ph.D., Purdue U., 1970. Asst. prof. chemistry Trevecca Nazarene Coll., 1970-72, asso. prof., 1972-76, prof., 1976—; vis. prof. Belmont Coll., 1970—; postdoctoral researcher Okla. State U., summer 1971, Oak Ridge Nat. Lab., summer 1974, Vanderbilt U., 1976-77, and summer 1978; translator Lang. Services, Nashville and Knoxville, Tenn., 1979—; mem faculty Gov.'s Sch. of S.C. at Coll. Charleston, summer 1981; tchr. consumer electricity, energy alts. Contbr. articles to profl. jours. Mem. PTA, Glenview Elem. Sch., Nashville; blood donor, CPR instr., lang. bank vol. Nashville sect. ARC. Mem. Am. Chem. Soc. (sec.-treas. Nashville sect. 1978-80). Subspecialties: Inorganic chemistry; Analytical chemistry. Current work: Therapeutic chelating agents. Home: 1902 Elanor Dr Nashville TN 37217 Office: 333 Murfreesboro Rd Nashville TN 37203

OATLEY, DAVID HERBERT, health physicist, health physics consultant; b. St. John's, Mich., June 10, 1953; s. Herbert Lewis and Esther (Ferguson) O.; m. Eileen Doyle, Nov. 27, 1981. B.S., Central Mich. U., 1975; cert. secondary teaching, Mich. State U., 1976; M.S., U. Mich., 1980. Pharm. rep. Upjohn Co., Kalamazoo, Mich., 1977-78; research asst. U. Mich., 1978-80; radiol. engr. Portland Gen. Electric, Rainier, Oreg., 1979; health physicist Wash. Publ. Power Supply System, Elma, 1980-83, Pacific Gas and Electric Co., San Francisco, 1983—; pvt. practice health physics cons., Olympia, Wash., 1982—, Pleasant hill, Calif., 1983—. Contbr. articles to profl. jours. Mem. Am. Nuclear Soc., Health Physics Soc. Congregationalist. Subspecialties: Nuclear fission; Physiology (biology). Current work: Radiation dosimetry, power reactor health physics and emergency preparedness for power reactors. Home: 9 Janin Pl Pleasant Hill CA 94523 Office: Pacific Gas and Electric Co 77 Beale St San Francisco CA 94106

OBENDORF, RALPH LOUIS, agronomist, educator; b. Milan, Ind., July 11, 1938; s. Louis Eugene and Miriam Clara (Stegemoller) O.; m. Sharon Kay Randel, Mar. 11, 1967; children: Michael Bradley, Kevin Andrew. B.S. in Agr, Purdue U., West Lafayette, Ind., 1960; M.S. in Agrnomy, U. Calif.-Davis, 1962; Ph.D. in Plant Physiology, U. Calif.-Davis, 1966. Asst. prof. crop sci. Cornell U., Ithaca, N.Y., 1966-71, assoc. prof., 1971-77, prof., 1977—; vis. scientist Inst. Cancer Research, Fox Chase, Phila., 1971-72; vis. plant physiologist Plant Growth Lab., U. Calif., Davis., 1983. Mem. Am. Soc. Plant Physiologists, Crop Sci. Soc. Am., Am. Soc. Agronomy, AAAS, Am. Assn. Cereal Chemists. Subspecialties: Plant cell and tissue culture; Plant physiology (agriculture). Current work: In vitro soybean seed and pod cultures; plant tissue/organ culture; seed biology; seed formation; seed germination; seed physiology; plant physiology. Home: 24 Dart Dr Ithaca NY 14850 Office: 619 Bradfield Hall Ithaca NY 14953

OBERDORSTER, GUNTER, toxicologist, educator, researcher; b. Cologne, Germany, Feb. 27, 1939; came to U.S., 1979; s. Ewald and Liesel (Selbach) O.; m. Ingeborg Gerda Karden, Mar. 23, 1968; children: Jan, Eva, Uta. D.V.M., U. Giessen, W.Ger., 1964, Dr. med. vet., 1966. Cert. in pharmacology and toxicology. Researcher Lab. Pharmacology, Cologne, 1965-67; asst. prof. U. Cologne, 1968-71; sci. staff mem. Fraunhofer Gesellschaft, Schmallenberg, W.Ger., 1971-79; vis. assoc. prof. inhalation toxicology U. Rochester, 1979-81, assoc. prof., 1981—; mem. Contact Group on Heavy Metals EC, Brussels, 1977-79. Recipient Joseph V. Fraunhofer prize Fraunhofer Gesellschaft, 1982; NIH grantee, 1982. Fellow Am.Coll. Vet. Toxicologists; mem. German Physiol. Soc., Assn. Aerosol Research, Am. Thoracic Soc., Soc. Toxicology., ctl Conf. Govt. Indsl. Nygienists. Subspecialties: Environmental toxicology; Toxicology (medicine). Current work: Effects of oxidants and N02 on respiratory system; measurement of lung clearance and epithelial permeability of airways; toxicity and carcinogenicity of heavy metals after inhalation; therapeutic use of chelating agents. Home: 121 Southern Pkwy Rochester NY 14618 Office: U Rochester Elmwood Ave Rochester NY 14642

OBERFIELD, RICHARD ALAN, physician; b. N.Y.C., July 29, 1932; s. George B. and Frances (Hurwitz) O.; m. (widower); children: Elizabeth Ann, Alice Amy. B.A., Alfred (N.Y.) U., 1953; M.D., N.Y. Med. Coll., 1957. Diplomate: Am. Bd. Internal Medicine. Intern Greenwich (Conn.) Hosp., 1957-58; sr. asst. surgeon USPHS, Detroit, 1958-60; tng. fellow N.Y.U. Med. Ctr., 1960-61; resident in medicine Dartmouth Med. Sch. Affiliated Hosp., Hanover, N.H., 1961-62, fellow hematology and cancer chemotherapy, 1963-65; staff physician Lahey Clinic Found., Boston, 1965—, head sect. med. oncology, Burlington, Mass., 1969—; clin. instr. Harvard U. Med. Sch., 1972—. Author numerous papers, reports in field. Mead Johnson postgrad. scholar ACP, 1962-63. Mem. AMA (cert. merit 1966, hon. mention sci. exhibit 1973), Mass. Med. Soc., Am. Assn. Cancer Research, Am. Fedn. Clin. Research, Am. Soc. Clin. Oncology, Eastern Coop. Oncology Group. Subspecialties: Internal medicine; Chemotherapy. Current work: Advanced cancer related to cancer management and clinical investigational program employing chemotherapeutic agents. Office: Lahey Clinic Found 41 Mall Rd Burlington MA 01805

OBERLEY, TERRY DE WAYNE, physician, pathologist; b. Effingham, Ill., Jan. 23, 1946; s. James Donald and Ruby Eloise (Moore) O.; m. Edith Marjorie Toole, June 19, 1968; children: Mathew James, Alexander John. B.A., Northwestern U., 1968, Ph.D., 1973, M.D., 1974. Diplomate: Am. Bd. Pathology. Intern pathology U. Wis., Madison, 1974-75, USPHS pathology trainee, 1975-77, resident pathology, 1977-78, asst. prof. pathology, 1977-82, assoc. prof. pathology, 1983—; chief electron microscopy William S. Middleton VA Hosp., Madison, 1983—. Author: Understanding Your New Life with Dialysis, 1983; Contbr. articles to profl. jours. NIH postdoctoral fellow, 1975; research grantee, 1975; Am Cancer Soc. resident fellow, 1977; March of Dimes Basil O'Connor grantee, 1978. Mem. Am. Assn. Pathologists, Am. Soc. Nephrology, Am. Soc. Microbiology, AMA, Nat. Kidney Found. Wis. (chmn. research com. 1983). Democrat. Unitarian. Subspecialties: Cell and tissue culture; Tissue culture. Current work: Research on the regulation of cell growth, with emphasis on studies of regulation of growth of normal and cancerous kidney cells, studies of the role of oxygen metabolites in cell growth. Home: 42 S Meadow Ln Madison WI 53706 Office: Dept Pathology William S Middleton VA Hosp 2500 Overlook Terr Madison WI 53705

OBLAD, ALEXANDER GOLDEN, chemist, chemical engineer, educator; b. Salt Lake City, Nov. 26, 1909; s. Alexander H. and Louie May (Brewster) O.; m. Bessie Elizabeth Baker, Feb. 23, 1933; children: Alex Edward, Elizabeth (Mrs. D. Sonne), Virginia (Mrs. M. Christensen), John R.B., Hayward B., Jean Rio B.(Mrs. S. Calder). B.A. in Chemistry, U. Utah, 1933, M.A. in Phys. Chemistry, 1934, D.Sc. (hon.), 1980; Ph.D. in Phys. Chemistry, Purdue U., 1937; D.Sc. (hon.), Purdue U., 1959. Research chemist Standard Oil Co., Whiting, Ind., 1937-42, Magnolia Petroleum Co., Dallas, 1942-43, sect. leader, 1943-46, chief chem. research, 1946; head indsl. research Tex. Research Found., Dallas, 1947; dir. chem. research Houdry Process Corp., Marcus Hook, Pa., 1947-52, asso. mgr. research and devel., 1952-55, mgr., 1955, v.p., dir., 1955-57; v.p. research and devel. M.W. Kellogg Co., N.Y.C., 1957-66, v.p. research and engring. devel., 1966-69; v.p. IRECO Chems., Salt Lake City, 1969-70; prof. metallurgy and fuels engring. U. Utah, 1969-75, distinguished prof. metallurgy and fuels engring., 1975—, prof. chemistry, 1975—, asso. dean, 1970-72, acting dean, 1972-75; Dir. Ireco Chem. Co., 1974—; Mem. Sec. of Interior's Saline Water Conversion Adv. Com., Office of Saline Water, Washington, 1959-61; mem. fossil energy research working group Dept. Energy, 1980-82. Mem. editorial bd.: Catalysis Revs., Fuel Processing Tech, 1976; mem. internat. adv. bd.: Ency. of Chem. Processing and Design, 1973—; Contbr. numerous articles on phys. chemistry, catalysis, petroleum chemistry and chem. engring. to tech. and sci. jours. Mem. alumni research council Purdue Research Found., Lafayette, Ind., 1960-63; chmn. sustaining membership campaign Orange Mountain council Boy Scouts Am., 1966-67; mem. adv. council Brigham Young U., Provo, Utah, 1967—; Bd. dirs. Internat. Congress on Catalysis, 1956-65, Energy Inst. Recipient First Purdue Chemist's award, 1959, Distinguished Alumni award U. Utah, 1962. Mem. Am. Chem. Soc. (mng. editor publs. div. petroleum chemistry 1857-69, sec. treas. div. petroleum chemistry 1952-54, E.V. Murphree award 1969), Am. Inst. Chem. Engrs., AAAS, Calif. Catalysis Soc., Rocky Mountain Fuel Soc., Nat. Acad. Engring., Am. Inst. Chemists (Chem. pioneer award 1972), Sigma Xi, Phi Lambda Upsilon, Sigma Pi Sigma, Phi Kappa Phi, Tau Beta Pi. Mem. Ch. of Jesus Christ of Latter-Day Saints. Club: Rotarian. Subspecialties: Chemical engineering; Physical chemistry. Patentee in field. Home: 1415 Roxbury Rd Salt Lake City UT 84108 Office: 302 Browning Mineral Science Bldg Univ of Utah Salt Lake City UT

O'BRIEN, KENNETH STANLEY, mfg. co. ofcl.; b. Boston, Aug. 2, 1942; s. Stanley and Alice (Carroll) O'B.; m. Annette Marie, Jan. 16, 1966; children: Michelle, Neal, Evan. B.S.M.E., Northeastern U., 1965, M.S. in Engring. Mgmt, 1972. Mech. devel. engr. Itek Corp., Lexington, Mass., 1965-68; mech. engr. Instron Corp., Canton, Mass., 1968-72; mgr. research and devel. A. W. Chesterton Co., Stoneham, Mass., 1972-76, corp. quality assurance mgr., 1976—. Mem. ASME, Am. Soc. Quality Control, Am. Mgmt. Assn., Pi Tau Sigma. Subspecialties: Quality assurance; Mechanical engineering. Current work: Application of computer tech. to quality assurance; devel. of quality assurance systems to multi-div., multi-nat. orgn. Office: A W Chesterton Co Middlesex Indsl Park Route 93 Stoneham MA 02180

O'BRIEN, MORROUGH PARKER, educator; b. Hammond, Ind., Sept. 21, 1902; s. Morrough and Lulu (Parker) O'B.; m. Roberta Libbey, May 16, 1931 (div.); children—Sheila, Morrough; m. Mary Wallner Kremers, 1963. B.S., Mass. Inst. Tech., 1925; D.Sc., Northwestern U., 1959; Dr. Engring., Purdue U., 1961; LL.D., U. Cal., 1968. Registered profl. engr., N.Y., Calif. Engr. Hudson River Regulating Dist., 1925-27; research asst. Purdue U., 1925-27; Freeman scholar Am. Soc. C.E., 1927-28; asst. Royal Coll. Engring., Stockholm, Sweden, 1927-28; successively asst. prof., asso. prof., prof. U. Calif., 1928-59, prof. emeritus, 1959—, chmn. dept. mech. engring., 1937-43, dean, 1943-59, dean emeritus, 1959—; adj. prof. coastal engring. U. Fla., 1983—; Dir. research and engring. Air Reduction, Inc., 1947-49; cons. engr. aerospace and def. group Gen. Electric Co., 1949—; Mem. U.S. Coastal Engring. Research Bd., Def. Sci. Bd., 1961-65, Army Sci. Adv. Panel, 1955-74; mem. nat. sci. bd. NSF, 1958-60. Author: Applied Fluid Mechanics, 1937. Mem. ASCE (hon.), ASME (hon.), Am. Soc. Engring. Edn. (Lamme medal), Am. Shore and Beach Preservation Assn. (pres.), Nat. Acad. Engring., Sigma Xi, Tau Beta Pi, Delta Tau Delta. Clubs: Bohemian (San Francisco); Cosmos (Washington); Athenian-Nile (Oakland, Calif.); Faculty (Berkeley); Sleepy Hollow Country (Scarborough-on-Hudson, N.Y.). Subspecialties: Fluid mechanics. Current work: Processes resulting from the interaction of an oceanic environment with sandy coasts. Home: PO Box 265 Cuernavaca Mexico

O'BRIEN, RICHARD LEE, physician, educator; b. Shenandoah, Iowa, Aug. 30, 1934; s. Thomas Lee and Grace Ellen (Sims) O'B.; m. Joan Frances Gurney, June 29, 1957; children: Sheila, Kathleen, Michael, Patrick. M.S., Creighton U., 1958, M.D., 1960; postgrad., U. Wis., Madison, 1962-64. Resident 1st med. div. Bellevue Hosp., N.Y.C., 1960; asst. prof. pathology U. So. Calif., 1966-70, assoc. prof., 1970-76, 1976-82, dep. dir., 1975-80, dir. research and edn., 1980-81, dir., 1981-82; dean Creighton U. Sch. Medicine, 1982—; vis. prof. U. Geneva, 1973-74. Contbr. numerous articles to profl. jours. Served to capt. U.S. Army, 1964-66. Mem. Am. Cancer Soc. (dir.), ACS, AAAS, Am. Assn. Cancer Edn., Am. Assn. Cancer Insts. (bd. dirs.), Alpha Omega Alpha. Subspecialties: Cell biology (medicine); Cancer research (medicine). Current work: Mechanisms of lymphocyte activation.

O'CALLAGHAN, JAMES PATRICK, research pharmacologist; b. West Palm Beach, Fla., Mar. 6, 1949; s. James Patrick and Paula Ann (Reinholtz) O'C. B.S. in Biology, Purdue U., 1971; Ph.D. in Pharmacology, Emory U., 1975. NIH postdoctoral fellow N.Y. State Div. Substance Abuse Services, Bklyn., 1975-78; pharmacology research assoc. Nat. Inst. Gen. Med. Scis., NIH, Bethesda, Md., 1978-80; research pharmacologist EPA, Research Triangle Park, N.C., 1980—. Contbr. chpts. to books, articles to profl. jours. NIH predoctoral trainee, 1971-75. Mem. AAAS, Am. Soc. Pharmacology and Exptl. Therapeutics, N.Y. Acad. Sci. Subspecialties: Neuropharmacology; Toxicology (medicine). Current work: Nervous-system-specific proteins as biochemical indicators of neurotoxicity. Office: EPA Neurotoxicology Div (MD-74B) Research Triangle Park NC 27711

OCHILLO, RICHARD FREDERICK, pharmacologist, toxicologist, educator, consultant, researcher; b. Yala, Kenya, Oct. 10, 1942; came to U.S., 1973; s. Yusuf O. and Klaris O. (Ragot) O.; m. Yvonne Hyacinth Richards, Aug. 30, 1969; children: Odera, Owino (dec.), Oremo, Fracella, Okelo. B.Sc., U. Victoria, B.C., Can., 1969; M.Sc., Dalhousie U., 1971; Ph.D., Vanderbilt U., 1977. Lectr. Siriba Coll., Maseno, Kenya, 1969, U. Nairobi, Kenya, 1971-73; asst. prof. pharmacology and toxicology Xavier U., New Orleans, 1977-80, assoc. prof., 1981—, chmn. div. basic pharm. scis., 1983—. Contbr. articles to profl. jours. NIH grantee, 1978—; Edward Schlieder grantee, 1978-82; NSF grantee, 1981-82. Mem. Am. Soc. Pharmacology and Exptl. Therapeutics, AAAS, S.E. Pharmacology Soc., Internat. Soc. Study Xenobiotics, Acad. Pharm. Scis., Am. Assn. Colls. Pharmacy, N.Y. Acad. Scis., Gerontol. Sco. Am., Soc. Toxicology, Am. Pharm. Assn. Subspecialties: Pharmacology; Toxicology (medicine). Current work: Pharmacodynamics of cholinergic and cardiovascular agents with special interests on muscarine and its analogs; toxicology of natural toxins particularly African arrow poisons. Home: 6 Cocodrie Ct Kenner LA 70062 Office: 7325 Palmetto St New Orleans LA 70125

OCHOA, SEVERO, biochemist; b. Luarca, Spain, Sept. 24, 1905; came to U.S., 1940, naturalized, 1956; s. Severo and Carmen (Albornoz) O.; m. Carmen G. Cobian, July 8, 1931. A.B., Malaga (Spain) Coll., 1921; M.D., U. Madrid, Spain, 1929; D.Sc., Washington U., U. Brazil, 1957, U. Guadalajara, Mexico, 1959, Wesleyan U., U. Oxford, Eng., U. Salamanca, Spain, 1961, Gustavus Adolphus Coll., 1963, U. Pa., 1964, Brandeis U., 1965, U. Granada, Spain, U. Oviedo, Spain, 1967, U. Perugia, Italy, 1968; Dr. Med. Sci. (hon.), U. Santo Tomas, Manila, Philippines, 1963, U. Mich., U. Buenos Aires, 1968, U. Tucuman, Argentina, 1968; L.H.D., Yeshiva Univ., 1966; LL.D., U. Glasgow, Scotland, 1959. Lectr. physiology U. Madrid Med. Sch., 1931-35; head physiol. div. Inst. for Med. Research, 1935-36; guest research asst. in physiology Kaiser-Wilhelm Inst. for Med. Research, Heidelberg, Germany, 1936-37; Ray Lankester investigator Marine Biol. Lab., Plymouth, Eng., 1937; demonstrator Nuffield research asst. biochemistry Oxford (Eng.) U. Med. Sch., 1938-41; instr., research asso. pharmacology Washington U. Sch. of Medicine, St. Louis, 1941-42; research asso. medicine N.Y. U. Sch. Medicine, 1942-45, asst. prof. biochemistry, 1945-46, prof. pharmacology, chmn. dept., 1946-54, prof., chmn. dept. biochemistry, 1954—; distinguished mem. Roche Inst. Molecular Biology. Author publs. on biochem. of muscles, glycolysis in heart and brain, transphosphorylations in yeast fermentation, pyruvic acid oxidation in brain and role of vitamin B1; RNA and Protein biosynthesis; genetic code. Decorated Order Rising Sun, Japan; recipient (with Arthur Kornberg) 1959 Nobel prize in medicine, Albert Gallatin medal N.Y. U., 1970, Nat. Medal of Sci., 1980. Fellow N.Y. Acad. Scis., N.Y. Acad. Medicine, Am. Acad. of Arts and Sci., A.A.A.S.; mem. Nat. Acad. Sci., Am. Philos. Soc., Soc. for Exptl. Biology and Medicine, Soc. of Biol. Chemists (pres. 1958, editor jour. 1950-60), Internat. Union Biochemistry (pres. 1961-67), Biochem. Soc. (Eng.), Harvey Soc. (pres. 1953-54), Alpha Omega Alpha (hon.); fgn. mem. German Acad. Nat. Scis., Royal Spanish, USSR, Polish, Pullian, Italian, Argentinian, Barcelona (Spain), Brazilian acads. sci., Royal Soc. (Eng.), Pontifical Acad. Sci., G.D.R. Acad. Scis., Argentinian Nat. Acad. Medicine. Subspecialty: Molecular biology. Home: 530 E 72d St New York NY 10021 Office: Roche Inst Molecular Biology Nutley NJ 07110

OCKERMAN, HERBERT WOOD, agricultural educator; b. Chaplin, Ky, Jan. 16, 1932; s. Herbert Newton and Addie Mae (Simpson) O.; m. Frances Ockerman. B.S., U. Ky., 1954, M.S., 1958; Ph.D., N.C. State U., 1962. Prof. meat sci. area dept. animal sci. Ohio State U., 1961—; cons. in field. Author: Source Book for Food Scientists, 1978; contbr. to books and profl. jours. Served to capt. USAF, 1955-58. Recipient Prof. award Pingtung U., Taiwan, 1978; commendation for internat. work in agr. Ohio Ho. of Reps., 1977; Badge of Merit for service to agr. Polish Govt., 1977; Plaque Argentine Nat. Met Bd., 1981, U. Cordoba, Spain, 1982. Mem. Inst. Food Technology, Am. Meat Sci. Assn., Am. Soc. Animal Sci., Polish Vet. Soc. (hon.), Sigma Xi, Phi Tau Sigma, Gamma Sigma Delta. Subspecialties: Food science and technology; Statistics. Current work: Teaching and research in the meat and food areas, both domestically and internationally. Office: 2029 Fyffe Rd Columbus OH 43210

OCKERMAN, MRS. H. L. *See also* **VERNADAKIS, ANTONIA**

O'CLOCK, GEORGE DANIEL, JR., elec. engr., cons., educator, researcher; b. Chgo., Sept. 25, 1939; s. George Daniel and Estella June (Taylor) O'C.; m. Priscilla Marie, June 14, 1969; children: Michael Anthony, Kathleen Marie. B.S.E.E., S.D. Sch. Mines and Tech., 1962, M.S.E.E., 1967; Ph.D., 1979; M.B.A. in Fin, UCLA, 1977. Registered profl. engr., Calif., Minn. Sr. engr. RCA Advanced Tech. Labs., Van Nuys, Calif., 1970-73; research engr. Northrup Corp., Hawthorne, Calif., 1973-74; mem. tech. staff Rockwell Internat., Anaheim, Calif., 1974-76, mgr. research and engring., 1977-80; staff scientist Perkin-Elmer, Eden Prairie, Minn., 1979-82; prof. physics and electronics engring. tech. Mankato State U., 1982—; v.p. and gen. mgr. RCD & A Cons., Cypress, Calif., 1975—. Contbr. numerous articles to profl. jours. Served to 1st C.E. U.S. Army, 1962-64. Mem. IEEE, Soc. Photo-Optical Instrumentation Engrs., Sigma Xi, Eta Kappa Nu. Lodge: K.C. Subspecialties: Microelectronics; Semiconductors. Current work: Communications systems, semiconductor materials and devices, electronics, communications, microwaves, optics, photovoltaics, spectroscopy. Patentee in field. Office: Dept Physics and Electronics Engring Tech Mankato State U Mankato MN 56001

O'CONNELL, JOHN PAUL, chemical engineering educator; b. Morristown, N.J., Sept. 19, 1938; s. Hugh Mellen and Helen Mae (Evans) O'C.; m. Verna Lee Hovey, June 20, 1959; children: Michael Andrew, Kathleen Elizabeth, Daniel Hovey. A.B., Pomona Coll., 1961; B.S., MIT, 1961, M.S., 1962; Ph.D., U. Calif.-Berkeley, 1967. Asst. prof. chem. engring. U. Fla., Gainesville, 1966-68, assoc. prof., 1969-74, prof., 1974—, chmn. dept. chem. engring., 1981—; vis. prof. chem. engring. UCLA, 1973, Stanford (Calif.) U., 1973-74; chmn. 3d Internat. Conf. on Fluid Properties, Callaway Gardens, Ga., 1983; cons. in field. Co-author: Computer Calculations for VLE, 1967, Computer Calculations for VLE and LLE, 1981. NSF grad. fellow, 1962-66; Danforth Found. assoc., 1968—. Fellow AAAS; Mem. Am. Inst. Chem. Engring., Am. Soc. Engring. Edn. (named Outstanding Tchr. 1982). Democrat. Methodist. Subspecialties: Chemical engineering; Thermodynamics. Current work: Application of molecular theory to correlating and predicting the properties of fluids, fluid mixtures and adsorption of nonelectrolyte, electrolyte and surfactant systems; teaching of chemical engineering thermodynamics. Home: 715 NW 40th Terr Gainesville FL 32607 Office: U Fla Dept Chem Engring Gainesville FL 32611

O'CONNELL, ROBERT FRANCIS, physicist, educator; b. Athlone, Ireland, Apr. 22, 1933; came to U.S., 1958, naturalized, 1969; s. William and Catherine (O'Reilly) O'C.; m. Josephine Mary Buckley, Aug. 3, 1963; children: Adrienne, Fiona, Eimear. B.Sc., Nat. U. Ireland, 1953; D.Sc., Nat. U. Reland, 1975; Ph.D., U. Notre Dame, 1962. Scholar Inst. Advanced Studies, Dublin, 1962-64; asst. prof., then assoc. prof. La. State U., Baton Rouge, 1964-69, prof. physics, 1969—. Contbr. articles on theoretical atomic, condensed-matter gravitation to profl. publs. Recipient Sir J.J. Larmor prize in physics, 1953; named Disting. Research Master La. State U., 1975. Fellow Am. Phys. Soc.; mem. Internat. Astron. Union, Am. Astron. Soc., Gen. Relativity and Gravitation Soc. Subspecialty: Condensed matter physics. Current work: Two-dimensional systems; high Rydberg states; statistical mechanics. Home: 522 Bancroft Way Baton Rouge LA 70808 Office: Dept Physics La State U Baton Rouge LA 70803

O'CONNOR, G(EORGE) RICHARD, ophthalmologist; b. Cin., Oct. 8, 1928; s. George Leo and Sylvia Johanna (Voss) O'C. A.B., Harvard U., 1950; M.D., Columbia U., 1954. Resident in ophthalmology Columbia-Presbyn. Med. Center, N.Y.C., 1957-60; research fellow Inst. Biochemistry, U. Uppsala, Sweden, 1960-61, State Serum Inst., Copenhagen, 1961-62; asst. prof. ophthalmology U. Calif., San Francisco, 1962-68, prof., 1972—; dir. Francis I. Proctor Found. for Research in Ophthalmology, 1970—; mem. Nat. Adv. Eye Council NIH, 1974-78. Author: (with G. Smolin) Ocular Immunology, 1981; asso. editor: Am. Jour. Ophthalmology, 1976-81. Served with USPHS, 1955-57. Recipient Janeway prize Coll. of Physicians and Surgeons, Columbia U., 1954; NIH grantee, 1962-81. Mem. Am. Bd. Ophthalmology (examiner), Assn. for Research in Vision and Ophthalmology (trustee 1979—), AMA, Am. Ophthal. Soc., Calif. Med. Assn., Frederic C. Cordes Eye Soc., Pan Am. Ophthal. Assn. Republican. Presbyterian. Club: Faculty. Subspecialties: Ophthalmology; Infectious diseases. Home: 22 Wray Ave Sausalito CA 94965 Office: 95 Kirkham St San Francisco CA 94122

ODELL, ANDREW PAUL, astronomer; b. Galesburg, Ill., May 6, 1949; s. Athol F. and Mary Louise (Lagomarcino) O. B.A. in Physics and Astronomy, U. Iowa, 1970; Ph.D. in Astronomy, U. Wis., 1974. Asst. prof. U. No. Iowa, Cedar Falls, 1974-79; reaearch assoc. Lunar and Planetary Lab., Steward Obs., U. Ariz., Tucson, 1979-81; asst. prof. physics and astronomy No. Ariz. U., Flagstaff, 1981—; researcher radiation transfer in planetary atmospheres, stellar evolution, pulsating stars. Mem. Am. Astron. Soc., AAAS., Internat. Astron. Union. Subspecialties: Theoretical astrophysics; Optical astronomy. Current work: Physics and astronomy education, pulsating star research. Office: Dept Physics No Ariz U Flagstaff AZ 86011

ODELL, WILLIAM DOUGLAS, physician, educator; b. Oakland, Calif., June 11, 1929; s. Ernest A. and Emma L. (Mayer) O.; m. Margaret F. Reilly, Aug. 19, 1950; children—Michael, Timothy, John D., Debbie, Charles. A.B., U. Calif., Berkeley, 1952; M.S., also M.D. in Physiology, U. Chgo., 1956; Ph.D. in Biochemistry and Physiology, George Washington U., 1965. Intern, resident, chief resident in medicine U. Wash., 1956-60, postdoctoral fellow in endocrinology and metabolsim, 1956-58; sr. investigator Nat. Cancer Inst., Bethesda, Md., 1960-65; chief endocrine service NIH, 1965-66; chief endocrinology Harbor-UCLA Med. Center, Torrance, Calif., 1966-72, chmn. dept. medicine, 1972-79; vis. prof. medicine Auckland Sch. Medicine, New Zealand, 1979-80; prof. medicine and physiology, chmn. dept. medicine U. Utah Sch. Medicine, Salt Lake City, 1980—. Mem. editorial bds. med. jours.; Author 4 books in field; contbr. numerous articles to med. jours. Served with USPHS, 1960-66. Recipient Distinguished Service award U. Chgo., 1973, Pharmacie award for outstanding contbns. to clin. chemistry, 1977, also research awards. Mem. Am. Soc. Clin. Investigation, Am. Physiol. Soc., Assn. Am. Physicians, Endocrine Soc. (V.p.), Soc. Study of Reproduction (dir.), Pacific Coast Fertility Soc. (pres.), Western Assn. Physicians (dir.), Western Soc. Clin. Research, Soc. Pediatric Research, Alpha Omega Alpha. Subspecialties: Internal medicine; Endocrinology. Current work: Physiology and pathophysiology of protein hormones. Office: U of Utah Med Center 30 Medical Dr Salt Lake City UT 84132

ODEN, JOHN TINSLEY, engineering mechanics educator, consultant; b. Alexandria, La., Dec. 25, 1936; s. John James and Sara Elizabeth (Lyles) O.; m. Barbara Clare Smith, Mar. 19, 1965; children: John Walker, Elizabeth Lee. B.S., La. State U., 1959; M.S., Okla. State U., 1960, Ph.D., 1962. Registered profl. engr., Tex. Teaching asst. La. State U., 1959; asst. prof. Okla. State U., 1961-63; sr. structures engr. Gen. Dynamics, Ft. Worth, 1963-64; prof. engring. mechanics U. Ala.-Huntsville, 1964-73, head dept., 1969-71; prof. aerospace engring. and engring. mechanics U. Tex.-Austin, 1973—, Carol and Henry Groppe prof. engring., 1979—; prof. Coppe Universidade Federal, Rio de Janeiro, Brazil, summer 1974; dir. Tex. Inst. Computational Mechanics, Austin, 1974—. Author: numerous books, including Finite Elements of Nonlinear Continua, 1972, Russian transl., 1976, Japanese transl., 1980, Chinese transl., 1983, Mechanics of Elastic Structures, 1967, Finite Elements, 6 vols, 1980-83, Applied Functional Analysis, 1978, Variational Methods in Theoretical Mechanics, 1976; author numerous tech. reports and articles on applied math. and mechanics; editor: (with others) numerous books, including Finite Elements in Fluids, vol. I, 1975, vol. II, 1975, vol. III, 1978, Computational Methods in Nonlinear Mechanics, 1980; assoc. editor for: U.S.A., Computer Methods in Applied Mechanics and Engring, 1978—; assoc. editor: Internat. Jour. Engring. Sci, 1976—, Jour. Nonlinear Analysis, 1980—; editorial bd.: various jours., including Soc. Indsl. and Applied Math. Jour. Sci. Computing and Stats, 1979—, Jour. Numerical Functional Analysis and Optimization, 1979—, Internat. Jour. Numerical Methods in Engring, 1975—. Recipient Outstanding Service award ASCE, 1968, Walter Huber Research prize, 1973; Engring. Found. Disting. Service award U. Tex.-Austin, 1975, 76, 78; SECTAM Research award Southeastern Conf. on Theoretical and Applied Mechanics, 1978; Disting. Grad. Teaching award U. Tex., 1979. Mem. Soc. Engring. Sci. (chmn. com. computational methods 1976-79, v.p. 1977-78, pres. 1979), Am. Acad. Mechanics (dir. S.W. region 1976-79), Soc. Natural Philosophy, Soc. Indsl. and Applied Math. (vis. lectr. 1979, 80). Current work: Nonlinear partial differential equations, nonlinear elasticity, continuum mechanics, finite element methods, numerical analysis. Office: Dept Aerospace Engring and Engring Mechanics U Tex WRW Bldg 305 Austin TX 78712

ODER, FREDERIC CARL EMIL, aerospace co. exec.; b. Los Angeles, Oct. 23, 1919; s. Emil and Katherine Ellis (Pierce) O.; m. Dorothy Gene Brumfield, July 2, 1941; children—Frederic E., Barbara Oder Debes, Richard W. B.S., Calif. Inst. Tech., 1940, M.S., 1941; Ph.D., UCLA, 1952. Commd. 2d lt. U.S. Army Air Force, 1941; advanced through grades to col. U.S. Air Force, 1960; ret., 1960; asst. dir. and program mgr. for research and engring. Apparatus and Optical div. Eastman Kodak Co., Rochester, N.Y., 1960-66; with Lockheed Missiles & Space Co., Sunnyvale, Calif., 1966—, v.p., asst. gen. mgr. div. space systems, 1972-73, v.p., gen. mgr. div. space systems, 1973—; mem. Def. Intelligence Agy. Sci. Adv. Com., 1972-76, asso. mem., 1976-78; mem. Air Force Studies Bd., NRC Assembly Engring., 1975-79, Def. Sci. Bd. Summer Study, 1975, Rev. Panel, 1979. Contbr. articles to profl. publs. Decorated Legion of Merit. Fellow AIAA; mem. Nat. Acad. Engring., Sigma Xi. Episcopalian. Clubs: Masons (Los Altos, Calif.); Cosmos, Commonwealth. Subspecialty: Aerospace engineering and technology. Home: 400 San Domingo Way Los Altos CA 94022 Office: 1111 Lockheed Way Sunnyvale CA 94086

ODISHAW, HUGH, university dean; b. N. Battleford, Can., Oct. 13, 1916; came to U.S., 1922, naturalized, 1941; s. Abraham and Miriam (Davajan) O.; m. Marian Lee Scates, 1958. A.B., Northwestern U., 1939, M.A., 1941; student, Princeton, 1939-40; B.S., Ill. Inst. Tech., 1944; Sc.D., Carleton Coll., 1958. Instr. Ill. Inst. Tech., 1941-44; with Westinghouse Electric Corp., 1944-45, 45-46; staff OSRD, 1945; asst. to dir. Nat. Bur. Standards, 1946-54; exec. dir. U.S. nat. com. IGY, 1954-65; dir. IGY World Data Center A., 1954-72; exec. sec., div. phys. scis. Nat. Acad. Scis., 1966-72; dean Coll. Earth Scis., U. Ariz., Tucson, 1972—. Author, editor: (with L. V. Berkner) Science in Space, 1961; Editor: (with E.U. Condon) The Handbook of Physics, 1958; Author sci articles. Fellow Am. Geophys. Union; mem. Am. Phys. Soc., AAAS, Phi Beta Kappa. Presbyn. Subspecialties: Geophysics; Earth sciences education. Home: 1531 Entrada Sexta Tucson AZ 85718 Office: U Ariz Tucson AZ 85721

ODOM, JAMES VERNON, exptl. psychologist; b. Laurinburg, N.C., Aug. 26, 1948; s. James Calvin and Elizabeth Edna (Norton) O. Student, U. Montpellier, France, 1968-69, Goethe Institut, Radolfzell, W.Ger., summer 1969; A.B. in Psychology (Whitaker scholar 1966, Guttman scholar 1968, Dane scholar 1969), Davidson Coll, 1970, M.A., U. N.C., Greensboro, 1974, Ph.D., 1978. Adminstrv. asst., coordinator counselling services Consult Ed., Inc., Greensboro, 1970-71; research asst. U. N.C., Greensboro, 1971-75, instr., 1975-76, Guilford Coll., 1975; fellow Ctr. Creative Leadership, Greensboro, summer 1975, research asst., 1975-77; instr. N.C. Agrl. and Tech. State U., fall 1977; research assoc. Case Western Res. U., 1978-79; Nat. Research Service research fellow U. Calif., Berkeley, 1979-80, U. Fla., 1980-81, asst. research scientist, 1981-82; asst. prof. W. Va. U., Morgantown, 1982—; vis. fellow U. Calif., San Francisco, 1979-80. Contbr. chpt., articles to profl. publs.; author profl. papers. NIH grantee, 1981. Mem. Acad. Mgmt., AAAS, Am. Psychol. Assn., Assn. Research in Vision and Ophthalmology, N.Y. Acad. Scis., Soc. Neurosci., Soc. Research in Child Devel., Internat. Soc. Clin. Electrophysiology in Vision, Optical Soc. Am., Sigma Xi. Democrat. Subspecialties: Psychophysiology; Sensory processes. Current work: Research on use of electrophysiological and psychophysiological methods to investigate normal and abnormal development of human visual functions, especially binocularity. Office: Dept Ophthalmology W Va Med Ctr Morgantown WA 25606 Home: 416 Grant Ave Morgantown WV 26505

O'DONNELL, MICHAEL JAMES, computer science educator; b. Spartanburg, S.C., Apr. 4, 1952; s. William Joseph and Linnie Lucille (Hynds) O'D.; m. Julie Ann Nerini, Feb. 6, 1982. B.S., Purdue U., 1972; Ph.D., Cornell U., 1976. Asst. prof. Purdue U., West Lafayette, Ind., 1976-81, assoc. prof. dept. computer sci., 1981—; research assoc. U. Toronto, Ont., Can., 1976-77. Subspecialties: Foundations of computer science; Programming languages. Current work: Interapplications of computing and logic; nonprocedural programming languages; equational interpreter; programming logic; type theory. Home: 421 N Chauncey Ave West Lafayette IN 47906 Office: Department of Computer Sciences Purdue University West Lafayette IN 47907

O'DONOGHUE, JOHN (LIPOMI), pathologist, toxicologist, neuroscientist; b. Lowell, Mass., Apr. 12, 1947; s. James Gregory and Sarafina Frances (Lipomi) O'D.; m. Sandra G. Piekos, June 24, 1967; children: Shawn Michael, Bevin Ruth. Student, U. Mass.-Amherst, 1964-66; V.M.D., U. Pa., 1970, Ph.D., 1979. Cert. Am. Bd. Toxicology. Pathologist Eastman Kodak Co., Rochester, N.Y., 1974-79, pathology group leader, 1979—; asst. prof lab. animal medicine U. Rochester, 1975—, asst. prof. toxicology, 1982—. Contbr. chpts. to books, articles to profl. publs. Leader Otetiana council Boy Scouts Am. Student research fellow, 1967; USPHS fellow, 1970; recipient Elks scholarship award, 1964, Phi Zeta award, 1970, Sigma Xi award, 1970. Mem. AVMA, AAAS, Am. Assn. Neuropathologists, Electron Microscopy Soc. Am. Roman Catholic. Subspecialties: Pathology (medicine); Toxicology (medicine). Current work: Research on neurotoxicology and neuropathology of industrial chemicals. Home: 3915 Clover St Honeoye Falls NY 14472 Office: Eastman Kodak Co B320 Kodak Park Rochester NY 14650

O'DONOHUE, THOMAS LEO, pharmacologist, biochemist; b. Jersey City, Jan. 10, 1954; s. Thomas J. and Frances O'D.; m. Gail Harndelmann, Aug. 19, 1979. B.S., Bucknell U., 1976; Ph.D., Howard U., 1980. Staff fellow Nat. Inst. Gen. Med. Scis., Bethesda, Md., 1980-82; asst. prof. adj., depts. medicine and biochemistry Howard U. Med. Sch., 1982—; chief, neuroendocrinology unit Nat. Inst. Neurol. Communicative Diseases and Stroke, NIH, Bethesda, 1982—. Contbr. articles to profl. jours. NIMH fellow, 1977; NIH research asssoc., 1980; Internat. Soc. Neuroendocrinology fgn. travel awardee, 1980. Mem. Soc. Neuroscience, Found. Advanced Edn. in Scis., AAAS. Subspecialties: Neuropharmacology, Neurochemistry. Current work: Chemistry of intercellular communication. Home: 25 Eastmoor Dr Silver Spring MD 20901 Office: NIH Bldg 10 Rm 50106 Bethesda MD 20205

O'DUFFY, JOHN DESMOND, internist, rheumatologist; b. Monaghan, Ireland, May 8, 1934; came to U.S., 1959; s. Frank and Elizabeth (Moore) O'D.; m. Margaret Teresa Slattery, Jan. 10, 1959; children: Siobhan, Anne, Gavan, Brendan. M.B., B.Ch., Nat. U. Ireland, 1958. Diplomate: Am. Bd. Internal Medicine. Intern St. Joseph's Hosp., Parkersburg, W.Va., 1959-60; resident Cleve. Clinic, 1960-63, mem. staff, 1963-68, Anderson Clinic, Arlington, Va., 1968-69; cons. Mayo Clinic, Rochester, Minn., 1969—, assoc. prof., 1976-82, prof., 1982—. Contbr. numerous articles to med. jours. ACP scholar, 1972. Fellow ACP; mem. Am. Rheumatism Assn., Central Soc. Clin. Research. Roman Catholic. Club: Rochester Tennis (vice chmn. 1972-75). Subspecialty: Internal medicine. Current work: Behcets disease, rheumatoid arthritis, pseudogout. Home: 2323 Merrihills Dr Rochester MN 55901 Office: Mayo Clinic 1st SW Rochester MN 55901

OELTMANN, THOMAS NAPIER, biochemist, educator; b. Covington, Ky., Dec. 28, 1941; s. Jerome Francis and Anna Mae (Napier) O.; m. Margaret Ann Turner, Aug. 26, 1965; children: Tim, John Ethan, Andrew Napier. Ph.D., U. Ga., 1967. Research asst. prof. U. Pitts., 1974-76; sr. research assoc. U. Iowa, Iowa City, 1976-79; asst. prof. medicine, research asst., prof. biochemistry Vanderbilt U. Med. Ctr., Nashville, 1979—. Contbr. articles to profl. jours. Served to capt., Chem, Corps U.S. Army, 1967-69. NIH grantee. Mem. AAAS, Am. Assn. for Cancer Research, Complex Carbohydrate Soc., Am. Chem. Soc., Am. Soc. Biol. Chemists, Am. Assn. Immunologists, N.Y. Acad. Scis., Cancer Center Task Force on Immunology. Roman Catholic. Subspecialties: Biochemistry (biology); Molecular biology. Current work: Biochemistry of hybrid proteins and their applications to cell biology, tumor biology, and cancer research. Home: 158 Cottonwood Dr Franklin TN 37064 Office: Dept Medicine Div Oncolog Vanderbilt U Med Center Nashville TN 37232

OERTEL, GOETZ KUNO HEINRICH, physicist; b. Stuhm, Germany, Aug. 24, 1934; came to U.S., 1957, naturalized, 1968; s. Egon F.K. and Margarete (Wittek) O.; m. Brigitte Beckmann, June 17, 1960; children: Ines M.H., Carsten K.R. Vordiplom, U. Kiel, 1956; Ph.D. in Physics, U. Md., 1964. With NASA, Langley AFB, 1963-68; dir. Office of Nuclear Def. Waste and Byproducts, Dept. of Energy, Washington, 1977—; Mem. profl. adv. com. engring. Pa. State U., 1979—. Contbr. articles to profl. jours. Fulbright fellow, 1957; Fed. Exec. Devel. Program grantee, 1974; recipient NASA Exceptional Service medal, 1972, others. Mem. Am. Phys. Soc., Am. Astron. Soc., German Sch. Soc. (bd. dirs. 1972-81), Sigma Xi. Club: Potamac Tennis. Current work: Plasma spectroscopy, nuclear waste, astronomy. Patentee in field. Home: 9609 Windcroft Way Potomac MD 20854

OESTERREICHER, HANS KARL, chemist, educator; b. Innsbruck, Austria, May 16, 1939; came to U.S., 1965, naturalized, 1973; s. Johann and Johanna (Rihak) O.; m. (married), 1969; 1 son, George. Ph.D., U. Vienna, 1965. Postdoctoral U. Pitts., 1965-66; Brookhaven Nat. Lab., Upton, N.Y., 1966-67; instr. Cornell U., 1967-68; asst. prof. Oreg. Grad. Center, Portland, 1968-73; mem. faculty U. Calif., San Diego, 1973—, prof., 1982—. Contbr. numerous articles to profl. jours. NSF grantee, 1972-83; NASA grantee, 1975-79; Dept. Energy grantee, 1969-73. Subspecialty: Solid state chemistry. Current work: Magnetism and hydrides of intermetallics. Address: Dept Chemistry U Calif San Diego Box 109 La Jolla CA 92093

OESTERWINTER, CLAUS, research astronomer; b. Hamburg, Germany, Jan. 18, 1928; s. Franz and Katharina (Boediger) O.; m. Ursula (Eckelmann); children: Angelica, Cary. M.S., Yale U., 1964, Ph.D., 1965. Research astronomer Naval Surface Weapons Ctr., Dahlgren, Va., 1959—. Mem. Am. Astron. Soc., Internat. Astron. Union. Subspecialties: Celestial mechanics; Astronautics. Current work: Orbital motion of planets, natural and artificial satellites. Home: PO Box 431 Dahlgren VA 22448 Office: Naval Surface Weapons Ctr Code K10 Dahlgren VA 22448

OETTGEN, HERBERT F., physician; b. Cologne, Germany, Nov. 22, 1923; came to U.S., 1958; s. Peter and Minna (Kaul) O.; m. Gertrud Hesberg, Feb. 16, 1957; children: Hans Christoph, Joerg Peter, Anne Barbara. M.D., U. Cologne, 1951. Intern U. Cologne, 1951, resident, 1952-54, 55-58, U. Marburg, Germany, 1954-55; mem. Sloan-Kettering Inst., 1958—; attending physician Meml. Hosp., N.Y.C., 1967—; prof. medicine and biology Cornell U., N.Y.C., 1966—. Contbr. articles to profl. jours. Served with German Navy, 1942-45. Recipient Wilhelm Warner award for cancer research, 1970; Lisec-Artz award for cancer research, 1982; NIH grantee; Am. Cancer Soc. grantee; other grants. Mem. Am. Assn. Cancer Research, Am. Soc. Clin. Oncology, Am. Soc. Hematology, Am. Assn. Immunologists, ACP, Harvey Soc., Internat. Soc. Hematology, Am. Fedn. Clin. Research. Presbyterian. Subspecialties: Oncology; Cancer research (medicine). Current work: Cancer research, cancer immunology, cancer therapy, cancer medicine. Research in biological approaches to cancer therapy—vaccines, monoclonal antibodies, lymphokines, differentiating proteins. Home: 48 Overlook Dr New Canaan CT 06840 Office: 1275 York Ave New York NY 10021

OETTINGER, ANTHONY GERVIN, educator, mathematician; b. Nuremberg, Germany, Mar. 29, 1929; came to U.S., 1941, naturalized, 1947; s. Albert and Marguerite (Bing) O.; m. Marilyn Tanner, June 20, 1954; children: Douglas, Marjorie. A.B., Harvard U., 1951, Ph.D., 1954; Henry fellow, U. Cambridge, Eng., 1951-52. Mem. faculty Harvard, 1955—, asso. prof. applied math., 1960-63, prof. linguistics, 1963-75, Gordon McKay prof. applied math., 1963—, chmn. program on info. resources policy, 1972—, mem. faculty of govt., 1973—, prof. info. resources policy, 1975—; mem. command control communications and intelligence bd. Dept. Navy, 1978-83; mem. sci. adv. group Def. Communications Agy., 1979—; cons. Arthur D. Little, Inc., 1956-80, Office Sci. and Tech., Exec. Office of Pres., 1960-73, Bellcomm, Inc., 1963-68, Systems Devel. Corp., 1965-68, Nat. Security Council, Exec. Office of Pres., 1975-81, Pres.'s Fgn. Intelligence Adv. Bd., 1981—; chmn. Computer Sci. and Engring. Bd., Nat. Acad. Scis., 1968-73; commr. Mass. Community Antenna TV Commn., 1972-79, chmn., 1975-79; mem. research adv. bd. Com. for Econ. Devel., 1975-79. Author: A Study for the Design of an Automatic Dictionary, 1954, Automatic Language Translation: Lexical and Technical Aspects, 1960, Run Computer Run: The Mythology of Educational Innovation, 1969, High and Low Politics: Information Resources for the 80s, 1977; Editor: Proc. of a Symposium on Digital Computers and Their Applications, 1962. Fellow Am. Acad. Arts and Scis., AAAS, IEEE; mem. Assn. Computing Machinery (mem. council 1961-68, chmn. com. U.S. Govt. relations 1964-66, editor computational linguistics sect. Communications 1964-66, pres. 1966-68), Soc. Indsl. and Applied Math. (mem. council 1963-67), Council on Fgn. Relations, Phi Beta Kappa, Sigma Xi. Clubs: Cosmos (Washington); Harvard (N.Y.C.). Subspecialties: Automated language processing; Information resources policy. Home: 65 Elizabeth Rd Belmont MA 02178 Office: 33 Oxford St Cambridge MA 02138

OETTINGER, PETER ERNEST, aeronautical engr., research scientist; b. Antwerp, Belgium, Apr. 14, 1937; came to U.S., 1943, naturalized, 1947; s. Hale Noa and Ruth Minka (Katzenstein) O.; m. Marlys Sandra (Klug), Jan. 28, 1967; children: Philip, Marc, Ariane, Andrea. B.S. in Agrl. Engring, Cornell U., 1959; M.S. in Aeros, Calif. Inst. Tech., 1960; Ph.D. in Aeros. and Astros, Stanford U., 1966. Research engr. Applied Physics Lab., Johns Hopkins U., 1959, Lockheed Missiles and Space Co., 1960-63; research assoc., lectr. physics and engring. Joint Inst. Lab. Astrophysics, U. Colo., 1966-68; mem. tech. staff Aerospace Corp., 1968-71; sr. scientist Centre de Recherches en Physique des Plasmas, Lausanne, Switzerland, 1971-72; mgr. research and devel. direct energy conversion dept., sr. scientist Thermo Electron Corp., Waltham, Mass., 1978—. Contbr. articles to profl. jours. Served to 1st lt. U.S. Army, 1960-61. Fellow AIAA (asso.); mem. Am. Phys. Soc., Optical Soc. Am., Soc. Applied Spectroscopy, ASME, AAAS. Subspecialties: Spectroscopy; Plasma physics. Current work: Laser-induced electron beam generator by photo-and thermionic-emission processes; thermionic energy conversion. Patentee optical devices with liquid coating, thermionic laser, laser generated high electron density source, high current density photoelectron generator; inventor 3-mirror laser, laser surface densitometer, particle thermionic converter, controllable thermal gradient furnace. Home: 4 Phlox Ln Acton MA 01720 Office: 85 1st Ave Waltham MA 02154

O'FARRELL, PATRICK HENRY, biochemistry educator; b. Halifax, N.S., Can., July 2, 1949; came to U.S., 1969; s. John Ross Reid and Fleurette (Dufour) O'F; m. Patricia Claire Zambryski, July, 1969 (div. 1978). B.S., McGill U., 1969; Ph.D., U. Colo., 1974. Postdoctoral fellow U. Calif.-San Francisco, 1974-79, asst. prof., 1979—. Mem. Soc. Biol. Chemists, Genetics Soc. Am. Subspecialties: Developmental biology; Gene actions. Current work: Investigation of molevular nature of genes which act to define the pathways of development in the fruit fly Drosophila melanogaster. Patentee dynamic equilibrium electrophoresis. Home: 165 Joose San Francisco CA 94131 Office: Dept Biochemistry and Biophysics Univ Calif San Francisco CA 94143

O'FARRELL, TIMOTHY JAMES, clinical psychologist, educator; b. Lancaster, Ohio, Apr. 22, 1946; s. Robert James and Helen Loretta (Tooill) O'F.; m. Jayne Sara Tlamage, May 19, 1973; 1 son, Colin. B.A., U. Notre Dame, 1968; M.A., Boston U., 1969, Ph.D., 1975. Diplomate: in clin. psychology Am. Bd. Profl. Psychology, Internat. Acad. Profl. Counseling & Psychotherapy. Mem. psychol. staff Boston City Hosp., 1974-75; mem. psychol. staff VA Med. Center, Brockton, Mass., 1975-77, chief alcoholism clinic, 1978—, chief psychol. studies lab., 1981—; adj. assoc. prof. Boston State Coll., 1974-77; instr. psychology Harvard Med. Sch., Cambridge, Mass., 1977-82, asst. prof.

psychology, 1982—. Contbr. chpts. to books and articles in field to profl. jours. Boston U. research grantee, 1974; VA med. research grantee, 1978, 81; Ayerest Labs. research grantee, 1981. Fellow Mass. Psychol. Assn., Behavior Therapy & Research Soc.; mem. Am. Psychol. Assn., Assn. Advancement Behavior Therapy, Eastern Psychol. Assn., Am. Assn. Sex Educators, Counselors & Therapists. Subspecialties: Behavioral psychology. Current work: Development and evaluation of marital therapies for alcoholics; research on characteristics of marriages and families of alcoholics; alcoholism treatment outcome evaluation. Home: 260 High St Duxbury MA 02332 Office: VA Med Ctr 151D 940 Belmont St Brockton MA 02401

OGAWA, JOSEPH MINORU, plant pathology, educator; b. Sanger, Calif., Apr. 24, 1925; s. Joseph Shosaku and Naomi (Yamaka) O.; m. Margie Hiroko Kawasaki, Nov. 7, 1954; children: Julie M., Martin K., Jo Ann. B.S., U. Calif.-Davis, 1950, Ph.D., 1954. Mem. faculty U. Calif.-Davis, 1953—, prof. plant pathology, 1968—. Author: (with E.E. Wilson) Fungal Bacterial and Certain Nonparasitic Diseases of Fruit and Nut Crops in California, 1979; contbr. over 100 articles to profl. jours. Served with C.I.C. U.S. Army, 1945-46. Recipient plaques Calif. Freezer's Assn., San Joaquin Cherry Growers and Industries Found., 1972. Mem. Am. Phytopathol. Soc., Calif. Aggie Alumni Assn (treas. 1953-68), Sigma Xi. Democrat. Methodist. Club: Commonwealth (San Francisco). Subspecialties: Plant pathology; Integrated pest management. Current work: Plant pathology, epidemiology and disease control, research on fruit and nut disease, postharvest diseases, fungicides, also tomato and hop diseases. Home: 806 Linden Ln Davis CA 95616 Office: Dept Plant Pathology U Calif Davis CA 95616

OGILVIE, RICHARD IAN, clinical pharmacologist; b. Sudbury, Ont., Can., Oct. 9, 1936; s. Patrick Ian and Gena Hilda (Olson) O.; m. Ernestine Tahedl, Oct. 9, 1965; children—Degen Elisabeth, Lars Ian. M.D., U. Toronto, 1960. Intern Toronto (Ont.) Gen. Hosp., 1960-61; resident Montreal Gen. and Univ. Alta. hosps., 1962-66; fellow in clin. pharmacology McGill U., Montreal, 1966-68, asst. prof. medicine, pharmacology and therapeutics, 1968-73, asso. prof., 1973-78, prof., 1978—, chmn. dept. pharmacology and therapeutics, 1978-83; clin. pharmacologist Montreal Gen. Hosp., 1968—, dir. div. clin. pharmacology, 1976-83; prof. medicine and pharmacology U. Toronto, 1983—; dir. div. cardiology and clin. pharmacology Toronto Western Hosp., 1983—. Grantee Med. Research Council Can. Quebec Heart Found., Can. Kidney Found., J.C. Edwards Found., Can. Found. Advancement Therapeutics, Conseil de la recherche en santé du Québec. Fellow Royal Coll. Physicians of Can., A.C.P.; mem. Can. Soc. Clin. Investigation (council 1977-80), Can. Hypertension Soc. (dir. 1979—), Can. Found. Advancement Clin. Pharmacology (dir. 1978—), Canadian Soc. for Clin. Pharmacology (pres. 1979-82), Internat. Union Pharmacology (council mem. clin. pharmacology sect. 1981—), Que. Heart Found. (pres. med. adv. com. 1976-81), Pharm. Soc. Can., Can. Cardiovascular Soc., Am. Soc. Pharmacology and Exptl. Therapeutics, Am. Soc. Clin. Pharm., Am. Fedn. Clin. Research. Subspecialty: Pharmacology. Current work: Clinical pharmacology of drugs affecting Cardiovascular system. Home: 79 Collard Dr Rural Route 1 King City ON Canada L0G 1K0 Office: 399 Bathurst St Toronto ON Canada M5T 2S8

OH, JANG, microbiologist; b. Seoul, Korea, Jan. 15, 1927; came to U.S., 1953, naturalized, 1971; s. Ki-yang and Moo-duk (Lee) O.; m. Won-Y Hyun, June 12, 1955; 1 son, Dennis. M.D., Yonsei U., Seoul, 1948; Ph.D., U. Wash.-Seattle, 1960. Intern Severance Union Med. Coll. Hosp., Seoul; resident in pathology Hamot Hosp., Erie, Pa.; pathologist Carle Clinic Hosp., Urbana, Ill., 1956-57; research instr. U. Wash., Seattle, 1957-61; asst. prof. U. B.C., Vancouver, 1961-66; research microbiologist U. Calif.-San Francisco, 1966—; cons. NIH, USPHS, Washington, 1979-83. Contbr. articles to profl. jours. Elder Korean United Presbyterian Ch., San Francisco, 1977—. Recipient Lederle Med. Faculty award, 1963-66; NIH research grantee, 1973—; Fight for Sight research grantee, 1973; Nat. Soc. Prevention Blindness grantee, 1972. Mem. Am. Soc. Microbiology, Am. Assn. Pathologists, Soc. Exptl. Biology and Medicine, Assn. Research in Vision and Ophthalmology (program com. 1982—). Subspecialties: Ophthalmology; Virology (medicine). Current work: Pathogenesis of virua infection of humans, chemotherapy of virus infection, prevention of virus infection, eye infection of herpes simplex virus and its treatment and presentation with antiviral agents and immunization. Office: U Calif 3d and Parnassus Sts San Francisco CA 94143

OH, JUNG HEE, transplantation immunologist, nephrologist, educator; b. Hamhung, Korea, Aug. 16, 1925; came to U.S., 1978; s. Yae Mook and Jee Sun (Kang) O.; m. Soon Ok Kim, Dec. 19, 1959; children: Helen, Frederick, Kenneth, Christopher. M.D., Seoul (Korea) Nat. U., 1951; Ph.D., McGill U., Montreal, Que., Can., 1969. Intern Mt. Sinai Hosp., Mpls., 1956-57; resident in medicine and nephrology Highland Park Hosp., Receiving Hosp., Detroit, 1957-63; research fellow in nephrology and transplantation Royal Victoria Hosp., Montreal, Que., Can., 1963-69; asst. prof. exptl. medicine and surgery McGill U., 1969-78; assoc. prof. medicine Emory U., 1978—; dir. Transplantation Lab. Grady Meml. Hosp., Atlanta. Served with M.C. Republic of Korea Army, 1950-56. Mem. Transplantation Soc., Am. Assn. Clin. Histocompatibility Testing, Am. Soc. Nephrology, Am. Soc. Transplant Physicians. Subspecialties: Transplantation; Nephrology. Current work: Histocompatibility testing; induction of immunological tolerance. Office: 80 Butler St SE Atlanta GA 30303

OH, RICHARD YOUNG, computer scientist; b. Kwangju, Korea, Mar. 21, 1940; came to U.S., 1967; s. Y.G. and Y.S. (Kook) O.; m. Julie J. Lee, Mar. 7, 1970; children: Viviane, Karen, Benjamin. M.S., Ohio State U., 1977. Programmer U. Calif., Irvine, 1972-74; Systemed Corp., Dayton, Ohio, 1972-74; sr. systems programmer Accuray Corp., Columbus, Ohio, 1974-79; mem. tech. staff Bell Telephone Labs., Columbus, 1979—. Mem. Assn. Computing Machinery. Subspecialties: Distributed systems and networks; Computer architecture. Current work: Reliable and fault-tolerant distributed systems, computer networks, local area networks, distributed operating systems, data communications. Home: 806 Loch Lomond Ln Worthington OH 43085 Office: 6200 E Broad St Columbus OH 43213

OHATA, CARL ANDREWS, physiologist; b. Pearl City, Hawaii, Sept. 22, 1947; s. Robert Osamu and Alice Hatsuyo (Osaki) O.; m. Sue Sachiko, Jan. 16, 1971; children: Heather Hatsuko, Brent Ryan. B.A., U. Hawaii, 1969, M.S., 1972; Ph.D., U. Alaska, 1976. Research asst. U. Hawaii, Honolulu, 1970-72, U. Alaska, Fairbanks, 1972-76; research asso. U. Tex. Med. Br., Galveston, 1976-77; NIH postdoctoral fellow U. Okla. Health Scis. Center, Oklahoma City, 1977-79; Commd. capt. U.S. Army, 1979; research physiologist U.S. Army Research Inst. Environ. Medicine, Natick, Mass., 1979—. Contbr. articles on neural control of circulation, temperature regulation and physiology of marine mammals to sci. jours. Okla. affiliate Am. Heart Assn. grantee, 1978-79. Mem. Am. Physiol. Soc., Soc. for Neurosci., AAAS, Sigma Xi. Congregationalist. Subspecialties: Physiology (biology); Neurology. Current work: Neural control of circulation, neuroanatomy of central autonomic regulation, temperature regulation, physiology of marine mammals. Home: 59 Pine St Medfield MA 02052 Office: US Army Research Inst Environ Medicine Natick MA 01760

OHKI, KENNETH, plant physiologist; b. Livingston, Calif., June 13, 1922; s. Zenjiro and Yaye (Watanabe) O.; m. Kiyoki N., Apr. 15, 1945; children: Suzanne S. Hammond, Stephen K. B.S., U. Calif., Berkeley, 1949, M.S., 1951, Ph.D., 1963. Supr., specialist Internat. Minerals and Chem. Corp., Libertyville, Ill., 1964-70; asst. prof. agronomy U. Ga., Experiment, 1971-73, asso. prof., 1973-79, prof., 1979—. Served with U.S. Army, 1945-46. Mem. Am. Soc. Plant Physiology, Am. Soc. Agronomy, Crop Sci. Soc. Am., Soil Sci. Soc. Am., Gamma Sigma Delta. Subspecialties: Plant physiology (agriculture); Plant growth. Current work: Mineral nutrition related to plant growth, development and physiology, particularly with microelements; critical deficiency and toxicity levels related to plant growth and physiological processes. Home: 206 Larcom Ln Griffin GA 30223 Office: Dept Agronomy U Ga Experiment GA 30212

OHNUMA, TAKAO, oncologist, researcher; b. Sendai, Japan, May 16, 1932; came to U.S., 1958; s. Hachiji Inomata and Teruko (Ohnuma) O.; m. Yoko Nishikawa, July 24, 1967; children: Nancy, Mary, Kenneth. M.D., Tohoku U., Sendai, 1957; Ph.D. in Biochemistry, U. London, 1965. Cancer research clinician Roswell Park Meml. Inst., Buffalo, 1968-73; research asst. prof. medicine SUNY, Buffalo, 1970-73; assoc. prof. dept. neoplastic diseases Mt. Sinai Sch. Medicine, N.Y.C., 1973-79; assoc. attending staff Mt. Sinai Hosp., N.Y.C., 1973-79; prof. neoplastic diseases Mt. Sinai Sch. Medicine, 1980—; attending physician Mt. Sinai Hosp., 1980—. USPHS research grantee; recipient Nat. Cancer Inst. contracts; research grantee United Leukemia Fund, Inc. Mem. Am. Chem. Soc., Am. Assn. Cancer Research, Am. Soc. Clin. Oncology, Cancer and Leukemia Group B. Subspecialties: Cancer research (medicine); Chemotherapy. Current work: Experimental chemotherapy, clinical pharmacology, cellular pharmacology. Home: 11 Paxford Ln Scarsdale NY 10583 Office: Mount Sinai Sch of Medicine Dept of Neoplastic Diseases One Gustave Levy Pl New York NY 10029

OHTA, ALAN TAKASHI, geneticist; b. Honolulu, Feb. 4, 1947; s. Kenneth Hiroshi and Umeyo Ann (Kanai) O.; m. Gayle Sanae Tabusa, July 3, 1971; children: Leigh, Aaron Takami. B.A., U. Hawaii, 1970, 5th yr. teaching diploma, 1972, Ph.D., 1977. Entomologist Dept. Primary Industries, Brisbane, Queensland, Australia, 1977-79; asst. researcher entomology U. Hawaii, 1980—. Served with Air NG, 1967-75. Mem. Soc. Study Evolution, Genetics Soc. Am., Am. Soc. Naturalists, Am. Genetic Assn., AAAS, Hawaii Acad. Sci., Sigma Xi. Subspecialties: Evolutionary biology; Gene actions. Current work: Researcher in genetic basis of ovipositional behavior in Hawaii Drosophila. Office: Dept Entomology U Hawaii Honolulu HI 96822

OILER, LARRY WAYNE, educator, cons.; b. Bethany, Mo., Feb. 11, 1941; s. Harold Jay and Paula Rogene (Ballantyne) O.; m. Lila Mae White; children: Michael Scott, Kristin Renae, Bradley James. B.A., Graceland Coll., 1963; M.A., Drake U., 1967; Ph.D., Purdue U., 1974. Asst. prof. dept. biology Purdue U., 1967-71; asst. prof. biology Wabash Coll., Crawfordsville, Ind., 1974-78; sci. curriculum coordinator Villisca (Iowa) High Sch., 1978—; cons. sci. curriculum. Pres. Reorganized Latter Day Saints Social Services Corp., 1980—; Ordained priest, elder Reorganized Ch. of Jesus Christ of Latter Day Saints. Mem. Am. Soc. Microbiology. Republican. Club: Page County Amateur Radio. Subspecialties: Gene actions; Molecular biology. Current work: Autoassembly of bacterial flagellins. Mil. amateur radio service operator. Home: PO Box 203 Stanton IA 51573 Office: 3rd St & 4th Ave Villisca IA 50864

OJALVO, MORRIS S(OLOMON), engineering educator and administrator; b. N.Y.C., July 6, 1923; s. Salomon Y. and Bessie (Barlia) O.; m. Neysa Gold, June 6, 1949; children: Steven I., Philip K., Beth J. B.M.E., Cooper Union, 1944; M.M.E., U. Del., 1949; Ph.D., Purdue U., 1962. Registered profl. engr., Md., 1953. Research asst. Pa. State U., 1946-47; instr. physics U. Ill.-Chgo., 1947-48; instr. and research asst. U. Del., Newark, 1948-50, assoc. prof. mech. engring., 1955-56; asst. prof., then assoc. prof. mech. engring. U. Md., College Park, 1950-55; instr. and research asst. Purdue U., Ind., 1956-60; assoc. prof., then prof. engring. and applied sci. George Washington U., Washington, 1960-66, adj. faculty, 1966—; program dir. NSF, Washington, 1965—; expert UNESCO, Mexico City, 1967-69. Author: Quieting: A Practical Guide to Noise Control, 1976. Served with USN, 1944-46; PTO. Recipient medal Cooper Union, 1944; Hausner award Fine Particle Soc., 1980; NSF fellow, 1961; NSF grantee, 1964, 65. Fellow ASME; mem. Soc. Automotive Engrs., Am. Soc. Engring. Edn., Fine Particle Soc., Am. Assn. Aerosol Research. Subspecialties: Mechanical engineering; Chemical engineering. Current work: Thermo dynamics, heat transfer, fluid mechanics, thermal power, particulate and multiphase processing, acoustics and noise control. Home: 8502 Barron St Takoma Park MD 20912 Office: 1800 G St NW Washington DC 20550

OKA, TAKAMI, chemist; b. Tokyo, Jan. 1, 1940; U.S., 1964; s. Kiyomi and Fumi O.; m. Keiko, Mar. 7, 1969; children: Hidemi, Tomomi. Ph.D., Stanford Med. Sch., 1969. Sr. investigator Nat. Inst. Arthritis, Diabetes and Digestive and Kidney Diseases, NIH, Bethesda, Md., 1974—. Mem. Am. Soc. Biol. Chemists, Am. Soc. Cell Biology. Subspecialties: Biochemistry (medicine); Cell biology (medicine). Current work: Mechanism of action of hormones on cell growth and differentiation. Home: 7219 Barnett Rd Bethesda MD 20817 Office: NIAMDD NIH Bldg 10 Room 9B15 Bethesda MD 20205

OKA, TAKESHI, physicist; b. Tokyo, June 10, 1932; Can., 1963, naturalized, 1973; s. Shumpei and Chiyoko O.; m. Keiko Nukui, Oct. 24, 1960; children: Ritsuko, Noriko, Kentaro, Yujiro. B.Sc., U. Tokyo, 1955, Ph.D., 1960. Research asso. U. Tokyo, 1960-63; fellow NRC Can., Ottawa, Ont., 1963-65, asst., 1965-68, asso., 1968-71, sr. research physicist, 1971—; prof. U. Chgo., 1981—. Mem. editorial bd.: Chem. Physics, 1972—, Jour. Molecular Spectroscopy, 1973—; Jour. Chem. Physics, 1975-77. Recipient Steacie prize, 1972; Earle K. Plyler prize, 1981. Fellow Royal Soc. Can., Am. Phys. Soc.; mem. Can. Assn. Physicists, Am. Astron. Soc. Subspecialties: Spectroscopy; Atomic and molecular physics. Current work: Laser spectroscopy; double resonance; molecular ions; astrophysics; interstellar molecules; energy transfer. Home: 1463 E Park Pl Chicago IL 60637 Office: Dept Chemistry and Dept Astronomy and Astrophysics U Chgo Chicago IL 60637

OKADA, ROBERT DEAN, physician, educator; b. Seattle, Sept. 18, 1947; s. Yoshitaka Robert and Jane Reiko (Sugawara) O.; m. Carolyn Rose Casella, May 19, 1973; 1 son, David. B.A., U. Wash., 1969; M.D., U. Pa., 1973. Diplomate: Am. Bd. Internal Medicine. Intern U.Ariz. Health Scis. Center, Tucson, 1973-74, resident, 1974-76, cardiology fellow, 1976-78; clin. and research fellow in cardiology Mass. Gen. Hosp., Boston, 1978-79, sr. staff cardiology lab., 1979—, asst. in medicine, 1982—, cons. nuclear medicine, 1981—; instr. medicine Harvard U. Med. Sch., Boston, 1979-81, asst. prof. medicine, 1981—. Author: Noninvasive Cardiac Imaging, 1982; contbr. chpts. to books and articles in field to profl. jours. Fellow Am. Coll. Cardiology, Am. Heart Assn. (established investigator 1982—), ACP, Am. Coll. Chest Physicians; mem. Am. Fedn. Clin. Research. Subspecialties: Cardiology; Internal medicine. Current work: Cardiac nuclear imaging; nuclear magnetic resonance. Home: 62 Oxford St Winchester MA 01890 Office: Cardiac Catherization Unit Mass Gen Hosp Fruit St Boston MA 02114

O'KEEFE, EDWARD JOHN, psychologist; b. Yonkers, N.Y., Apr. 15, 1937; s. Edward M. and Florence (King) O'K.; m. Marilyn J. Keller, June 3, 1961; children: Kevin, Kenneth, Christopher. Student, Iona Coll., 1959; Ph.D., Fordham U., 1970. Lic. psychologist, N.Y. Lectr. Fairfield (Conn.) U., 1960-61; dir. clin. services Cardinal Hayes Home for Children, Millbrook, N.Y., 1961—; pvt. practice psychology, Poughkeepsie, 1980—; prof. Marist Coll., Poughkeepsie, 1961—; workshop leader IBM, 1972—. Contbr. articles to profl. jours. Named Outstanding Educator of Am. Marist Coll., 1975. Mem. Mid-Hudson Psychol. Assn. (pres. 1970), Am. Psychol. Assn., Assn. Advancement of Behavior Therapy, N.Y. Soc. Clin. Psychologists, AAUP, Assn. Behavior Analysis. Roman Catholic. Subspecialties: Behavioral psychology; Multimodal psychology. Current work: Research on using multimodal psychology to assess and treat psychol. disorders; multimodal treatment for mental retardation. Address: 12 Priscilla Ln Poughkeepsie NY 12603

O'KELLEY, JOSEPH CHARLES, educator; b. Unadilla, Ga., May 9, 1922; s. Thomas Landrum and Maude (Hall) O'K.; m. Hallie LaVonne Held, June 16, 1951; children—Susan Elaine, Celia Anne, Thomas Burnell, Ellen Catherine. A.B., U. N.C., 1943, M.A., 1948; Ph.D., Iowa State U., 1950. Mem. faculty dept. biology U. Ala., University, 1951—, asso. prof., 1958-61, prof., 1961—, chmn. dept., 1969-74, research prof., 1977—. Served with USAAF, 1943-46. Decorated Air medal.; NSF postdoctoral fellow U. Wis., 1953-54; NIH spl. fellow Johns Hopkins, 1965-66. Fellow AAAS; mem. Am. Inst. Biol. Scis., Phi Beta Kappa. Subspecialties: Sedimentology; Precambrian geology. Current work: History of science. Home: 2910 17th St E Tuscaloosa AL 35404 Office: PO Box 1927 University AL 35486

OKEN, MERTIN M., hematologist, oncologist, researcher; b. Pen Yan, N.Y., Dec. 24, 1939; s. Benjamin and Ellan (Schwartz) O.; m. Naomi Keber, June 25, 1968; children: Jonathan, Dara, Juliette. B.S., U. Mich., 1961; M.D., Duke U., 1965. Med. intern U. Rochester, N.Y., 1965-66, med. resident, 1966-67, NYU Bellevue Hosp., N.Y.C., 1967-68; hematoloty trainee U. Rochester, 1968-70; chief hematology/oncology William Beaumont Gen. Hosp., El Paso, 1970-72; asst. prof. U. Minn., 1972-81, assoc. prof., 1982—; prin. investigator U. Minn., 1973—; dir. cancer day treatment ctr. VA Med. Ctr., Mpls., 1981—. Served with U.S. Army, 1970-72. Mem. Am. Soc. Clin. Oncology, ACP, Am. Soc. Hematology, Eastern Coop. Onclogy Group (exec. com.). Subspecialties: Oncology; Hematology. Current work: Cancer treatment studies, pathphysiolosy of hematologic malignancies, pyrogen production and effects. Office: VA Med Ctr Hematology Sect 54th St and 48th Ave S Minneapolis MN 55417

OKRENT, DAVID, educator; b. Passaic, N.J., Apr. 19, 1922; s. Abram and Gussie (Pearlman) O.; m. Rita Gilda Holtzman, Feb. 1, 1948; children—Neil, Nina, Jocelyne. M.E., Stevens Inst. Tech., 1943; M.A., Harvard, 1948, Ph.D. in Physics, 1951. Mech. engr. NACA, Cleve., 1943-46; sr. physicist Argonne (Ill.) Nat. Lab., 1951-71; regents lectr. U. Calif. at Los Angeles, 1968, prof. engring., 1971—; vis. prof. U. Wash., Seattle, 1963, U. Ariz., Tucson, 1970-71; Isaac Taylor chair Technion, 1977-78. Author: Fast Reactor Cross Sections, 1960, Computing Methods in Reactor Physics, 1968, Reactivity Coefficients in Large Fast Power Reactors, 1970, Nuclear Reactor Safety, 1981; contbr. articles to profl. jours. Mem. adv. com. on reactor safeguards AEC, 1963—, also chmn., 1966; sci. sec. to sec. gen. of Geneva Conf., 1958; mem. U.S. del. to all Geneva Atoms for Peace Confs. Guggenheim fellow, 1961-62, 77-78; recipient Distinguished Appointment award Argonne Univs. Assn., 1970. Fellow Am. Phys. Soc., Am. Nuclear Soc. (Tommy Thompson award 1980), Nat. Acad. Engring. Subspecialties: Nuclear fission; Nuclear engineering. Current work: Nuclear reactor safety and technology; societal risks. Home: 439 Veteran Ave Los Angeles CA 90024

OKUN, DANIEL ALEXANDER, educator; b. N.Y.C., June 19, 1917; s. William Howard and Leah (Seligman) O.; m. Elizabeth Griffin, Jan. 14, 1946; children—Michael Griffin, Tema Jon. B.S., Cooper Union, 1937; M.S., Calif. Inst. Tech., 1938; D.Sc., Harvard, 1948. Diplomate: Am. Acad. Environ. Engrs. (pres. 1969-70); registered profl. engr., N.C., N.Y. With USPHS, 1940-42; teaching fellow Harvard, 1946-48; with Malcolm Pirnie (cons. environ. engrs.), N.Y.C., 1948-52; asso. prof. dept. environ. scis. and engring. U. N.C. at Chapel Hill, 1952-55, prof., 1955-73, Kenan prof., 1973—, head dept. environ. scis. and engring., 1955-73; dir. Water Resources Research Inst., 1965—, chmn. faculty, 1970-73; Dir. Wapora, Inc.; Vis. prof. Technol. U. Delft, 1960-61, Univ. Coll., London, 1966-67, 73-75, Tianjin U., 1981; editor environ. scis. series Acad. Press, 1968-75; cons. to industry, cons. engrs., govtl. agys. World Bank, WHO. Author: (with Gordon M. Fair and John C. Geyer) Water and Wastewater Engineering, 2 vols, 1966, 68, Elements of Water Supply and Wastewater Disposal, 1971, (with George Ponghis) Community Wastewater Collection and Disposal, 1975, Regionalization of Water Management—A Revolution in England and Wales, 1977; Editor: (with M.B. Pescod) Water Supply and Wastewater Disposal in Developing Countries, 1971; Contbr. to publs. in field. Adv. bd. Ackland Meml. Art Mus., 1973-78; bd. dirs. Warren Regional Planning Corp., 1971-77, Inter-Ch. Council Housing Corp., 1975—, N.C. Water Quality Council, 1975-77; adv. com. for med. research Pan Am. Health Orgn., 1976-79; chmn. Washington Met. Area Water Supply Study Com., 1976-80, Nat. Acad Scis.-NRC; bd. sci. and tech. for internat. devel. NRC, 1978—, vice chmn. environ. studies bd., 1980—. Served from lt. to maj. AUS, 1942-46. Recipient Harrison Prescott Eddy medal for research Water Pollution Control Fedn., 1950, Catedratico Honorario, Univ. Nacional de Ingenieria, Lima, Peru, 1957, Gordon Maskew Fair award Am. Acad. Environ. Engrs., 1973, Thomas Jefferson award U. N.C. at Chapel Hill, 1973, Gordon Y. Billard award N.Y. Acad. Scis., 1975; Gordon Maskew Fair medal Water Pollution Control Fedn., 1978; NSF fellow, 1960-61; Fed. Water Pollution Control Adminstrn. fellow, 1966-67; Fulbright-Hayes lectr., 1973-74. Fellow ASCE (chmn. environ. engring. div. 1967-68, Simon W. Freese award 1977); corp. fellow Inst. Water Pollution Control (Eng.); mem. Nat. Acad. Engring., Inst. Medicine, AAUP (chpt. pres. 1963-64), Water Pollution Control Fedn. (hon. mem.; chmn. research com. 1961-66, dir.-at-large 1969-72), Sigma Xi (pres. chpt. 1968-69). Subspecialty: Water supply and wastewater treatment. Current work: Water quality management; water reuse; regionalization of water management. Home: Linden Rd Route 7 Chapel Hill NC 27514

OLAH, GEORGE ANDREW, educator, chemist; b. Budapest, Hungary, May 22, 1927; came to U.S., 1964, naturalized, 1970; s. Julius and Magda (rasznai) O.; m. Judith Agnes Lengyel, July 9, 1949; children: George John, Ronald Peter. Ph.D., Tech. U. Budapest, 1949. Mem. faculty Tech. U. Budapest, 1949-54; asso. dir. Central Chem. Research Inst., Hungarian Acad. Scis., 1954-56; research scientist Dow Chem. Can. Ltd., 1957-64, Dow Chem. Co., Framingham, Mass., 1964-65; prof. chemistry Case-Western Res. U., Cleve., 1965-69, C.F. Mabery prof. research, 1969-77; Disting. prof. chemistry, dir. Hydrocarbon Research Inst., U. So. Calif., Los Angeles, 1977—; vis. prof. chemistry Ohio State U., 1963, U. Heidelberg, Germany, 1965, U. Colo., 1969, Swiss Fed. Inst. Tech., 1972, U. Munich, 1973, U. London, 1973-79, L. Pasteur U., Strasbourg, 1974, U. Paris, 1981; hon. vis. lectr. U. London, 1981; cons. to industry. Author: Friedel-Crafts

Reactions, Vols. I-IV, 1963-64, (with P. Schleyer) Carbonium Ions, Vols. I-V, 1969-76, Friedel-Crafts Chemistry, 1973, Carbocations and Electrophilic Reactions, 1973, Halonium Ions, 1975; also chpts. in books, numerous papers in field. Recipient Leo Hendrik Baekeland award N.J. sect. Am. Chem. Soc., 1966, Morley medal Cleve. sect., 1970; Alexander von Humboldt sr. U.S. scientist award, 1979; Guggenheim fellow, 1972. Fellow Chem. Inst. Can., AAAS; mem. Nat. Acad. Scis., Italian Nat. Acad. Scis., Am. Chem. Soc. (award petroleum chemistry 1964, award synthetic organic chemistry 1979), German Chem. Soc., Brit. Chem. Soc. (Centenary lectr. 1978), Swiss Chem. Soc., Sigma Xi. Subspecialty: Organic chemistry. Current work: Mechanistic and synthetic organic chemistry; hydrocarbon chemistry; reactive intermediates; biological alkylating agents. Patentee in field. Home: 2252 Gloaming Way Beverly Hills CA 90210 Office: Univ So Calif Los Angeles CA 90007 America still is offering a new home and nearly unlimited possibilities to the newcomer who is willing to work hard for it. It is also where the "main action" in science and technology remains.

OLAH, STEPHEN, mechanical engineer; b. Medgyesbodzas, Hungary, July 15, 1928; s. Gregory and Rozalia (Forgo) O.; m. Adel Csentery, Sept. 3, 1954; children: Gabriella, Martin, Gregory, Stephanie; m. Edith Elfriede, Apr. 21, 1973. Mech. Engr., Poly. Metalworking Inst. Tech., Budapest, Hungary, 1952. Forging tooling designer Ironworks of Gyor/Hungary, 1953-56; design engr. Wells Industries, North Hollywood, Calif., 1957-65; sr. research and devel. engr. Ultra Violet Products, Inc., San Gabriel, Calif., 1965-73; sr. design engr. Markhon Industries, Wabash, Ind.; mech. engring. mgr. Applied Solar Energy, Inc., Industry, Calif., 1975—. Author: UV in Graphic Arts, 1970, Automated Concentrator Cell Module Assembly, 1980, Terrestrial Solar Modules, 1981, Point Focus Concentrators, 1982; patentee in field. Mem. ASME, Soc. Plastic Engrs. Roman Catholic. Subspecialties: Solar energy; Mechanical engineering. Current work: Photovoltaic solar energy application and development, design and development of photovoltaic devices, design development of equipment used in manufacturing these devices. Home: 145 N Sunnyside St Sierra Madre CA 91024 Office: 15251 E Don Julian Rd City of Industry CA 91749

OLAJOS, EUGENE JULIUS, toxicologist; b. Steyr, Austria, July 13, 1945; came to U.S., 1950, naturalized, 1956; s. Eugene and Margaret (Huber) O.; m. Helen M. Bourgoin, Nov. 25, 1970; children—Elizabeth, Stephanie, Andrew. B.A., Wayne State U., 1969, M.S., 1972; Ph.D., U. Mich., 1976. Nat. Research Service fellow MIT, Cambridge, 1975-76, Albany (N.Y.) Med. Coll., 1976-77, research asst. prof. toxicology, 1977-80; asst. prof. chemistry Old Dominion U., Norfolk, Va., 1980-82; phys. scientist Chem. Systems Lab., U.S. Army, Aberdeen Proving Ground, Md., 1982—. Contbr. articles on toxicology to profl. jours., also chpts. to books. Mem. Am. Chem. Soc., Internat. Soc. Study Xenobiotics, Soc. Toxicology, Soc. for Neurosci. Subspecialties: Toxicology (medicine); Neurochemistry. Current work: Researcher in biochemical mechanisms of neurotoxicity, enzymatic markers as correlates of neuropathology and toxicant-induced neurotransmitter alterations. Home: 605 Webb Rd Newark DE 19711 Office: Chemical Systems Laboratory US Army Aberdeen Proving Ground MD 21010

OLD, LLOYD JOHN, cancer biologist; b. San Francisco, Sept. 23, 1933; s. John H. and Edna A. (Marks) O. B.A., U. Calif. at Berkeley, 1955, M.D. at, 1958. Research fellow Sloan-Kettering Inst. Cancer Research, 1958-59, research asso., 1959-60, asso., 1960-64, asso. mem., 1964-67, mem., 1967—, chief div. immunology, 1967-72; research asso. biology Sloan-Kettering div. Grad. Sch. Med. Scis., Cornell U. Med. Coll., 1960-62, asst. prof. biology, 1962-66, asso. prof. biology, 1966-69, prof. biology, 1969—; acting asso. dir. research planning Sloan-Kettering Inst. Cancer Research, 1972, v.p., asso. dir., 1973-76, v.p., asso. dir. for sci. devel., 1976—; asso. dir. for research Meml. Sloan-Kettering Cancer Center and Meml. Hosp., 1973—; cons. Nat. Cancer Inst., 1967-70, mem. developmental research working group, 1969, spl. virus cancer program, 1970, mem. virus cancer program adv. com., 1975-78, mem. bd. sci. counselors div. cancer cause and prevention, 1978-81; asso. med. dir. N.Y. Cancer Research Inst., Inc., 1970, med. dir., 1971—. Adv. editor: Jour. Exptl. Medicine, 1971-75, Progress in Surface and Membrane Science, 1972-74; asso. editor: Virology, 1972-74.; Editorial adv. bd.: Cancer Research, 1967-70, Cancer, 1968—, Recent Results in Cancer Research, 1972—; Contbr. articles to profl. jours. Mem. med. and sci. adv. bd. Leukemia Soc. Am., Inc., 1970-73, trustee, 1970-73; mem. sci. adv. com. Ludwig Inst. Cancer Research, 1974—; mem. sci. adv. bd. Jane Coffin Childs Meml. Fund for Med. Research, 1970-75; mem. research council Public Health Research Inst. City N.Y., 1977-80, bd. dirs., 1979-81. Recipient Roche award, 1957, Alfred P. Sloan award cancer research, 1962, Lucy Wortham James award James Ewing Soc., 1970, Louis Gross award, 1972, Cancer Research Inst. award for immunology research, 1975, Rabbi Shai Shacknai Meml. award, 1976; Research Recognition award Noble Found., 1978; G.H.A. Clowes Meml. lectr., 1980; Robert Roesler de Villiers award, 1981. Mem. N.Y. Acad. Scis., Reticuloendothelial Soc., Soc. Exptl. Biology and Medicine, Harvey Soc., Am. Acad. Arts and Scis., Am. Assn. Cancer Research (bd. dirs. 1980-83), AAAS, Am. Assn. Immunologists, Inst. Medicine of Nat. Acad. Scis., Phi Beta Kappa, Sigma Xi, Alpha Omega Alpha. Subspecialty: Cancer research (medicine). Office: Sloan-Kettering Inst Cancer Research 410 E 68th St New York NY 10021

OLDFIELD, JAMES EDMUND, animal nutritionist, educator; b. Victoria, C., Can., Aug. 30, 1921; s. Henry C. and Doris O. O.; m. Mildred E. Atkinson, Sept. 4, 1942. B.S., U.B.C., 1941, M.S., 1949; Ph.D., Oreg. State U., 1951. Instr. U. B.C., Can., 1947-49; grad. research asst. Oreg. State U., Corvallis, 1949-51, instr. dept. animal sci., 1952-57, assoc. prof., 1958-63, prof., 1964—, head dept. animal sci., 1967—. Contbr. articles to profl. jours. Mem. Corvallis Sch. Bd., 1965-73, chmn., 1972. Served to capt. Can. Army, 1943-46. Decorate Milit. Cross; Fulbright research scholar Massey U., New Zealand, 1974. Mem. Agrl. Inst. Can., Am. Chem. Soc., Am. Soc. Animal Sci. (Morrison award 1972, hon. fellow 1978), Am. Inst. Nutrition, Am. Council on Sci. and Health, Council for Agrl. Sci. and Tech. Club: Kiwanis (Corvallis). Subspecialties: Animal nutrition; Biochemistry (biology). Current work: Animal nutrition, specializing in metabolic effects on mineral elements; selenium teaching, higher education, research, administration. Home: 1325 NW 15th St Corvallis OR 97330 Office: Oregon State University Department of Animal Science Corvallis OR 97331

OLDHAM, ROBERT KENNETH, physician, med. oncologist, immunologist; b. Pocatello, Ida., Sept. 16, 1941; s. John Harold and Leona G. (Gaines) O.; m. Anne Catherine Moon, Jan. 27, 1963; children: Robert K., Andrew G., J. Clark, Daniel H., John B. M.D., U. Mo., 1968. Diplomate: Am. Bd. Internal Medicine, Med. Oncology. Intern Vanderbilt U. Med. Ctr., Nashville, 1968-69, resident, 1969-70; fellow Nat. Cancer Inst., Bethesda, Md., 1970-72; sr. investigator (Lab. Cell Biology), 1973-74, 1974-75; dir. div. oncology, assoc. dir. Cancer Research Ctr. and assoc. prof. medicine Vanderbilt U., Nashville, 1975-80; asso. dir. Biol. Response Modifiers program Div. Cancer Treatment, Nat. Cancer Inst., Frederick, Md., 1980—. Editor-in-chief: Jour. Biol. Response Modifiers; mem. editorial bd.: Cancer Immunology and Immunotherapy; contbr. articles to profl. jours. Fellow ACP; mem. So. Soc. Clin. Investigation, Am. Soc. Clin. Oncology, Am. Fedn. Clin. Research, Am. Assn. Cancer Research, Am. Assn. Immunologists, Internat. Soc. Exptl. Hematology, Alpha Omega Alpha. Subspecialties: Cancer research (medicine); Cellular engineering. Office: Nat Cancer Inst Frederick Cancer Research Facility DCT Bldg 567 Frederick MD 21701

OLDS, DURWARD, animal scientist, educator; b. Conneaut, Ohio, Apr. 12, 1921; s. Benjamin Harrison and Sada Hannah (Raudabaugh) O.; m. Gertrude M. Walk, Sept. 21, 1947; children: Beverly, John. D.V.M., Ohio State U., 1943; M.S., U. Ill., 1954, Ph.D. (AVMA research fellow), 1956. Lic. in vet. medicine, Ky. Artificial insemination technician Clark County (Wis.) Breeders Coop., 1944-46; researcher, prof. animal sci. U. Ky., Lexington, 1946—. Contbr. articles to profl. jours. Mem. Am. Dairy Sci. Assn., AVMA, Am. Soc. Animal Sci., Soc. Study of Reprodn. Lutheran. Subspecialties: Animal breeding and embryo transplants; Animal physiology. Current work: Artificial insemination and infertility of cattle, estrous behavior, management of breeding animals and embryo culture. Holder copyright time of insemination dial for cattle; performed world's first non-surg. embryo transfer in cattle, 1964. Home: 1605 Elizabeth St Lexington KY 40503 Office: Dept Animal Science U Ky Lexington KY 40546

OLDSHUE, JAMES Y., manufacturing company executive; b. Chgo., Apr. 18, 1925; s. James and Louise (Young) O.; m. Betty Ann Wiersema, June 14, 1947; children: Paul, Richard, Robert. B.S. in Chem. Engring, Ill. Inst. Tech., 1947, M.S., 1949, Ph.D. in Chem. Engring, 1951. Registered engr., N.Y. With Mixing Equipment Co., Rochester, N.Y., 1950—, dir. research, 1960-63, tech. dir., 1963-70, v.p. mixing tech., 1970—. Contbr. chpts., articles to books, jours. Chmn. budget com. Internat. div. YMCA; bd. dirs. Rochester YMCA. Served with AUS, 1945-47. Recipient 1st Disting. Service award NE YMCA Internat. Com., 1979; named Rochester Engr. of Yr., 1980. Fellow Am. Inst. Chem. Engrs. (pres. 1979, treas. 1983, Eminent Engr.); mem. Am. Assn. Engring. Socs., Am. Chem. Soc., Internat. Platform Assn., Nat. Acad. Engring., Sigma Xi. Mem. Reformed Ch. in Am. Subspecialties: Chemical engineering; Fluid mechanics. Current work: Fluid mixing, scale-up technology. Home: 141 Tyringham Rd Rochester NY 14617 Office: 135 Mt Read Blvd Rochester NY 14611

O'LEARY, DENNIS PATRICK, neurophysiologist, educator; b. Fargo, N.D., Dec. 24, 1939; s. Henry J. and Margaret L. (Smith) O'L; m. Ann M. Merker, Aug. 29, 1964; children: Brenda, Kathleen. B.S., U. Chgo., 1962; Ph.D., U. Iowa, 1969. Asst. prof. UCLA, 1969-74; faculty U. Pitts., 1974—, now asso. prof. otolaryngology and physiology; cons. in field. Mem. Internat. Brain Research Orgn., Soc. Neurosci. Subspecialties: Neurophysiology; Biophysics (biology). Current work: Neurophysiology of inner ear; vestibular system, oculomotor sytem, balance system disorders. Office: Eye & Ear Hosp 230 Lothrop St Pittsburgh PA 15213

O'LEARY, GERARD PAUL, JR., biology educator; b. Bridgeport, Conn., Oct. 16, 1940; s. Gerard Paul and Ethal (Shortell) O'L.; m. Janice E. Murray, Dec. 19, 1981. B.S., Mount St. Mary's Coll., 1962; M.S., N. Mex. State U., 1964; Ph.D., U. N.H., 1967. Mem. faculty Providence Coll., 1969–-, prof. biology, 1982–-; Head Eastern Colls. Sci. Conf., Inc., Providence, 1981. Can. Research Council fellow McGill U., 1967-69. Mem. Am. Soc. Microbiologists, N.Am. Apiotherapy Soc., AAAS, Warren Meml. Arthritis Found. (research council 1981). Roman Catholic. Subspecialties: Biochemistry (medicine); Cell biology (medicine). Current work: Arthritis; oncology; effects of heavy metals; muscle biochemistry; infection agents detection by isoelectric focusing; plasma emission spectroscopy. Office: Providence Coll Biology Dept 101 Hicke Research Bldg Providence RI 02918

OLESEN, DOUGLAS EUGENE, research institute executive; b. Tonasket, Wash., Jan. 12, 1939; s. Magnus and Esther Rae (Myers) O.; m. Michaele Ann Engdahl, Nov. 18, 1964; children: Douglas Eugene, Stephen Christian. B.S., U. Wash., 1962, M.S., 1963, postgrad., 1965-67, Ph.D., 1972. Research engr. space research div. Boeing Aircraft Co., Seattle, 1963-64; with Battelle Meml. Inst., Pacific NW Labs., Richland, Wash., 1967—, mgr. water resources systems sect., water and land resources dept., 1970-71, mgr. dept., 1971-75, dep. dir. research labs., 1975, dir. research, 1975-79, v.p. inst., dir. NW div., 1979—; affiliate asso. prof. U. Wash., 1973—; mem. study adv. com. Joint Center for Grad. Study, 1979—. Bd. dirs. United Way of Benton and Franklin Counties, 1978—, chmn. 1983 campaign; Bd. dirs. Tri-City Nuclear Indsl. Council, 1979—; bd. dirs., mem. exec. com. United Way of Wash., 1980—, v.p., 1981-83, pres., 1983; gov.'s appointee Energy Fair '83 Commn., 1980—; mem. Kadlec Hosp. Med. Center Found., 1980—; mem. adv. bd. Sta. KCTS/9 Pub. TV, 1983—; mem. Fed. Interagy. Task Force on Acid Precipitation, 1980—; chmn. Mid-Columbia U.S. Savs. Bond campaign, 1981. Mem. Water Pollution Control Fedn., Pacific NW Pollution Control Assn. (pres. 1977). Subspecialty: Water supply and wastewater treatment. Patentee process and system for treating waste water. Office: NW Div Lab Battelle Meml Inst Battelle Blvd Richland WA 99352

OLESON, NORMAN LEE, physicist, educator; b. Detroit, Aug. 19, 1912; s. Christian Gad and Mathilde Lorenzo (Halversen) O.; m. Gabrielle Dorothy Sauve, June 18, 1939; children: Karen A., Norman Lee, Richard P. B.S., U. Mich., 1935, M.S., 1937, Ph.D., 1940. Instr. U.S. Coast Guard Acad., 1940-46; physicist Gen. Electric Co., Cleve., 1946-48; assoc. prof. physics Naval Postgrad. Sch., 1948, prof., 1952-69; chmn. dept. physics U. South Fla., 1969-78, prof., 1978—; cons. physics Lawrence Livermore Lab., 1959-69; vis. prof. physics Queen's U., Belfast, 1955-56; vis. prof. nuclear engring. M.I.T., 1967-68. Contbr. articles to sci. jours. Served with USCGR, 1940-46. Fellow Am. Phys. Soc.; mem. AAAS, Sigma Xi, Sigma Pi Sigma. Subspecialty: Plasma physics. Current work: Investigation of turbulent plasmas in magnetic fields. Home: 11003 Carrollwood Dr Tampa FL 33618 Office: Dept Physics U South Fla Tampa FL 33620

OLHOEFT, GARY ROY, research geophysicist, photographer; b. Akron, Ohio, Feb. 15, 1949; s. Roy Carl and Helen Francis (Landefeld) O.; m. Jean Kane, Sept. 2, 1972. B.S.E.E., M.I.T., 1971, M.S.E.E., 1972; Ph.D., U. Toronto, 1975. Registered profl. engr., Ont. With Kennecott Copper Corp., Lexington, Mass., 1970-72, NASA Manned Spacecraft Ctr., Houston, 1971, M.I.T., Cambridge, 1971-72, Lockheed Electronics Co., Houston, 1972-73, U. Toronto, Ont., Can., 1973-75; research geophysicist U.S. Geol. Survey, Denver, 1975—, chief, 1981—; adj. prof. U. So. Calif., U. Colo.; advisor NASA, EPA, Bur. Mines, U.S. Dept. Energy, Bur. Reclamation, Dept. Justice. Editorial bd.: Internat. Jour. Thermophysics; contbr. articles to profl. publs. in field. Vol. reader Recs. for the Blind; bd. dirs. Kipling Klub Townhouses, Amberwick Townhomes. Recipient Lunar Sci. Group achievement award, 1975. Mem. Am. Geophys. Union, ASTM (chmn. D-18.06), AAAS, Am. Astron. Soc. (div. planetary sci.), Soc. Exploration Geophysics (presentation award 1972), Soc. Photographic Scientists and Engrs., Soc. Profl. Well Log Analysts, Royal Soc. Chemists, IEEE. Subspecialties: Geophysics; Electrochemistry. Current work: Physical and chemical properties and processes of earth materials petrophysics, electrochemistry, induced polarization, complex resistivity, dielectrics, electrical properties, radar, geophysics, geotechnics, geothermal, minerals, permafrost, lunar and planetary materials. Office: PO Box 2505 DFC MS 964 Denver CO 80225

OLINS, DONALD EDWARD, biomedical science educator, researcher; b. N.Y.C., Jan. 11, 1937; s. Bernard David and Sylvia (Elfman) O.; m. Ada Mariam Levy, July 2, 1961; children: Joshua Daniel, Barak Levi. B.A. in Biology, U. Rochester, 1958; postgrad., Albert Einstein Coll. Medicine, N.Y.C., 1958-60; Ph.D. in Biochemistry, Rockefeller Inst., 1964. Postdoctoral fellow Dartmouth Med. Sch., Hanover, N.H., 1964-67; asst. prof. U. Tenn., Oak Ridge Grad. Sch. Biomed. Sci., 1967-69, assoc. prof., 1969-76, prof., 1976—; vis. scientist dept. biophysics King's Coll., London, 1970-71; Humboldt award scientist German Cancer Research Ctr., Heidelberg, 1979-80; mem. research council U. Tenn.-Knoxville, 1982-83; Searle Lectr. Brit. Soc. Cell Biology, Glasgow, Scotland, 1975. Editorial bd.: Molecular and Cellular Biochemistry, 1982—. Helen Hay Whitney Found. postdoctoral fellow, 1965-67; recipient Career Devel. award NIH, 1968-73, Sr. U.S. Scientist award Alexander Von Humboldt Found., 1979-80. Fellow AAAS; mem. Biophys. Soc., Am. Soc. Biol. Chemists. Democrat. Jewish. Subspecialties: Cell biology; Graphics, image processing, and pattern recognition. Current work: Eukaryotic chromosome structure; combined biochemical, biophysical and ultrastructural methods. Development of 3-D EM techniques. Home: 901 West Outer Dr Oak Ridge TN 37830 Office: Univ Tenn Oak Ridge Grad Sch Biomed Sci PO Box Y Biology Div Oak Ridge Nat Lab Oak Ridge TN 37830

OLIVE, LINDSAY SHEPHERD, botanist, researcher; b. Florence, S.C., Apr. 30, 1917; s. Lindsay Shepherd and Sada (Williamson) O.; m. Anna Jean Grant, Aug. 28, 1942. A.B., U. N.C., 1938, M.A., 1940, Ph.D., 1942. Instr. U. N.C., Chapel Hill, 1942-44, prof., 1968-69, univ. disting. prof., 1969-82, univ. disting. prof. emeritus, 1982—; mycologist Dept. Agr., 1944-45; asst. prof. U. Ga., 1945-46; assoc. prof. La. State U., 1946-49, 1949-57; prof. Columbia U., 1957-67. Author: The Mycetozoans, 1975; contbr. 162 articles to profl. jours.; Ency. Americana; chpts. to books. Guggenheim fellow, 1956. Fellow AAAS; mem. Brit. Mycol. Soc. (hon.), Mycol. Soc. Am. (pres. 1966, Disting. Mycol. award 1982), Nat. Acad. Sci., Am. Bot. Soc., Soc. Protozoology, Nature Conservancy (hon.). Subspecialties: Taxonomy; Morphology. Current work: Fungi and mycetozoans. Home: PO Box 391 Highlands NC 28741 Office: Highlands Biological Station Highlands NC 28741

OLIVER, BERNARD MORE, electrical engineer, technical consultant; b. Soquel, Calif., May 27, 1916; s. William H. and Margaret E. (More) O.; m. Priscilla June Newton, June 22, 1946; children: Karen, Gretchen, Eric. A.B. in Elec. Engring., Stanford U., 1935; M.S., Calif. Inst. Tech., 1936, Ph.D., 1940. Mem. tech. staff Bell Telephone Labs., N.Y.C., 1940-52; dir. R&D, Hewlett-Packard Co., Palo Alto, Calif., 1952-57, v.p. R&D, 1957-81, tech. adv. to pres., 1981—; lectr. Stanford U., 1957-60; cons. Army Sci. Adv. Com., from 1966; mem. sci. and tech. adv. com. Calif. State Assembly, 1970-76. Contbr. articles on electronic tech. and instrumentation to profl. jours. Mem. Pres.'s Commn. on the Patent System, 1966, State Senate Panel for the Bay Area Rapid Transit System, 1973; trustee Palo Alto Unified Sch. Dist., 1961-71. Recipient Disting. Alumni award Calif. Inst. Tech., 1972. Fellow IEEE (Lamme medal 1977); mem. Am. Astron. Soc., Nat. Acad. Sci., Nat. Acad. Engring., AAAS. Republican. Club: Palo Alto. Subspecialty: Radio and microwave astronomy. Current work: Adjunct professor of astronomy. Patentee electronic circuits and devices. Home: 13310 La Paloma Rd Los Altos Hills CA 94022 Office: 1501 Page Mill Rd Palo Alto CA 94304

OLIVER, DOUGLAS LAMAR, neuroanatomy researcher; b. Cin., Mar. 11, 1949; s. John Brewster and Margery (Weber) O.; m. Francis Elizabeth Carmichael, July 14, 1972. B.A., Emory U., 1971; Ph.D., Duke U., 1977. Trainee in neursci./psychology Duke U., 1971-75, predoctoral fellow, 1975-77; postdoctoral fellow Harvard Med. Sch. and U. Conn. Health Center, 1977-80; instr. U. Conn. Health Center, 1980-83, asst. prof., 1984—. Contbr. articles to profl. jours. USPHS-NIH grantee, 1982; U. Conn. Research Found. grantee, 1982. Mem. AAAS, Am. Assn. Anatomists, Soc. Neurosci., Sigma Xi. Democrat. Subspecialties: Neurobiology; Morphology. Current work: Morphology of the mammalian auditory system, synaptic organization. Office: Dept Anatomy U Conn Health Center Farmington CT 06032

OLIVER, JACK (WALLACE), veterinary pharmacology educator; b. Ellettsville, Ind., Jan. 6, 1938; s. Levi Webster and Helen Marie (Stultz) O.; m. Martha Lou (Young); children: Jason B., Kevin M. B.S. in Agr, Purdue U., 1960, M.S., 1963, D.V.M., 1966, Ph.D. in Vet. Physiology, 1969. Asst. prof. vet. pharmacology Purdue U., West Lafayette, Ind., 1969-70, Tex. A&M U., College Station, 1970-71, Ohio State U., Columbus, 1972-75; assoc. prof. vet. pharmacology U. Tenn., Knoxville, 1975-82, prof., 1982—, Lindsay Young Prof. vet. medicine, 1980, lab. dir. Fellow Am. Acad. Vet. Pharmacology and Therapeutics; mem. Am. Soc. Vet. Physiology and Pharmacology, Assn. Am. Vet. Med. Colls., Conf. Research Workers in Animal Disease, AVMA, Tenn. Vet. Med. Assn., Phi Zeta. Subspecialty: Veterinary pharmacology. Current work: Thyroid physiology relationship to aberrant metabolism. Office: Dept Environmental Practice Coll Vet Medicine Univ Tenn Knoxville TN 37916

OLIVER, JOHN PARKER, astonomer; b. New Rochelle, N.Y., Nov. 24, 1939; s. James Parker and Evelyn Smithen (Grant) O.; m. Barbara McKenna, Nov. 2, 1963; children: Jennifer, Keith, Rebecca. B.S. in Physics, Rensselaer Poly. Inst., 1962; Ph.D. in Astronomy, UCLA, 1974. Asst. prof. astronomy U. Fla., 1970-77, assoc. prof. astronomy, 1977—, dir. Mem. Internat. Astron. Union. Subspecialties: Optical astronomy; Astronomical instruments. Current work: Observation of binary stars; lunar occultation photometry; computer based instruments. Office: U Fla 211 SSRB Astronomy Gainesville FL 32611

OLIVER, KELLY HOYET, JR., biology educator, aquatic biology consultant; b. Rosboro, Ark., June 22, 1923; s. Kelly H. and Willye B. (Long) O.; m. Carley Dickey, Dec. 26, 1947; children: Carol A., Mark L., Rush H. B.S., So. Meth. U., 1952, M.S., 1953; Ph.D., Okla. State U., 1963. Assoc. prof. biology Ark. State U., Jonesboro, 1962-64; chief of biolab Vitro Services, Eglin AFB, Fla., 1964-68; prof. biology Henderson State U., Arkadelphia, Ark., 1968—; pres. Kocomoro, Inc., Arkadelphia, 1973—. Chmn. com. Boy Scouts Am., Ft. Walton Beach, Fla., 1964. Served with U.S. Army, 1943-46; ETO. Mem. Ark. Acad. Sci., Sigma Xi, Phi Delta Kappa. Democrat. Methodist. Club: Pullen's (Arkadelphia) (v.p. 1980-82). Lodges: Masons (De Queen, Ark.); Lions. Subspecialties: Ecology; Environmental toxicology. Current work: Effects of pesticides on fauna; bioassay of residues on fishes and selected invertebrates. Home: 204 N 27th St Arkadelphia AR 71923 Office: Henderson State U Henderson St Arkadelphia AR 71923

OLIVER, LARRY RAY, indsl. engr.; b. Springfield, Mo., Oct. 7, 1941; s. Ray Thomas and Audrey Arlene O.; m. Judith Ann Pyle, Dec. 2, 1961; children: Elise, Elaine. B.S. in Mech. Engring, So. Meth. U., 1964; M.S. in Physics, Brown U., 1967; Ph.D. in Indsl. Engring, U. Ark., Fayetteville, 1972. Registered profl. engr., Calif. Math. analyst Dayco, Springfield, Mo., 1967-77, mgr. advanced engring., 1977—. Contbr. articles to profl. jours. Mem. ASME (chmn. subcom. on belts and chains, chmn. nat. conf. power transmission belt drives), Am. Inst. Indsl. Engrs. Subspecialties: Industrial engineering; Mechanical engineering. Current work: Continuously variable automotive transmissions, belt drive analysis, reliability, quality control, computer graphics. Patentee asymetric variable speed drive, torque-sensing pulley and drive system. Home: 3461 S Pinehurst Ct Springfield MO 65807 Office: 2601 W Battlefield Rd Springfield MO 65808

OLIVO, MARGARET ANDERSON, educator; b. Omaha, June 17, 1941; d. Clarence Lloyd and Anita Emma (Kruse) Anderson; m. Richard Francis Olivo, Sept. 4, 1971. B.A., Augustana Coll., 1963; Ph.D., Stanford U., 1967. NIH fellow Harvard U., Cambridge, Mass., 1968-70; research assoc., neurobiology lab. U.P.R., San Juan, 1970-71; asst. prof. Bennington (Vt.) Coll., 1972; asst. prof. biol. sci. Smith Coll., Northampton, Mass., 1973-79, assoc. prof., 1979—. NSF fellow, 1963-67. Mem. Am. Soc. Zoologists, Soc. for Neurosci., Biophys. Soc., Soc. Gen. Physiologists. Subspecialties: Neurobiology; Physiology (biology). Current work: Physiology of excitable cells. Office: Smith Coll Dept Biol Sci Northampton MA 01063

OLKIN, INGRAM, statistics educator; b. Waterbury, Conn., July 23, 1924; s. Julius and Caroline (Bander) O.; m. Anita Mankin, May 1, 1945; children: Vivian L., Rhoda J., Julia A. B.S., CCNY, 1946; M.S., Columbia U., 1949; Ph.D., U. N.C., 1951. Faculty Mich. State U., East Lansing, 1951-60, U. Minn., Mpls., 1960-61; faculty Stanford (Calif.) U., 1961—, now prof. stats. and edn.; vis. prof. U. Chgo., 1955-56; overseas fellow Churchill Coll., Cambridge, 1967-68; with Ednl. Testing Service, 1971-72; vis. prof. Imperial Coll. London, 1976-77. Author: (with L. Gleser and C. Derman) A Guide to Probability Theory and Application, 1973, (with J.D. Gibbons and M. Sobel) Selecting and Ordering Populations, 1977, (with A.W. Marshall) Inequalities: Theory of Majorization and Its Applications, (with L. Gleser & C. Derman) Probability Models and Applications, 1980. Served to 1st lt. USAAF, 1943-46. Mem. Inst. Math. Stats., Am. Statis. Assn., Royal Statis. Soc., Internat. Statis. Inst., Math. Assn. Am. Psychometric Soc., Am. Ednl. Research Assn. Jewish. Subspecialty: Statistics. Current work: Multivariate analysis; probability models. Office: Stanford Univ Sequoia Hall Stanford CA 94305

OLLENDICK, THOMAS HUBERT, psychology educator, consultant; b. Tilden, Nebr., Aug. 6, 1945; s. Alfred Bernard and Ivy Mae (Buettgenbach) O.; m. Mary Catherine Haley, July 29, 1967; children: Laurie Kristine, Kathleen Marie. B.A., Loras Coll., 1967; Ph.D., Purdue, U., 1971. Asst. prof. Ind. State U., Terre Haute, 1972-76, assoc. prof., 1976-79; vis. assoc. prof. U. Pitts. Med. Sch., 1978-79; assoc. prof. Va. Poly. Inst. and State U., Blacksburg, Va., 1980-82, prof., 1983—; cons. Rockville (Ind.) Tng. Ctr., 1972-79; vis. lectr. U. Pitts. Med. Sch., 1980-82, prof., 1983—. Author: Clinical Behavior Therapy with Children, 1981; co-author, editor: Handbook of Child Psychopathology, 1982; co-editor: Child Behavioral Assessment, 1984; sr. editor: Guilford series on Child Psychopathology, 1982—. NIMH predoctoral fellow Purdue U., 1967-69; Yale U. research fellow, 1969; Devereux Found. postdoctoral fellow, 1971. Mem. Am. Psychol. Assn., Assn. Advancement Behavior Therapy, Midwestern Psychol. Assn., N.Y. Acad. Sci., Southeastern Psychol. Assn., Sigma Xi. Subspecialties: Behavioral psychology; Learning. Current work: Clinical behavior therapy with children: developmental psychopathology, social dysfunction with children: anxiety disorders, and maladaptive behaviors of the retarded and emotionally disturbed. Home: 1417 Crestview Dr Blacksburg VA 24060 Office: Va Poly Inst and State U Dept Psychology Blacksburg VA 24060

OLSEN, GEORGE DUANE, physician, clinical pharmacologist, medical educator; b. DeKalb, Ill., Jan. 5, 1940; s. George Meyer and Ida Mae (Hahn) O.; m. Deborah Morgan, June 24, 1965; children: Derrick Meyer, Sonja Julia. A.B., Dartmouth Coll., 1962, B.M.S., 1964; M.D., Harvard U., 1966. Intern U. Hosps., Cleve., 1966-67; NIH fellow dept pharmacology Sch. Medicine Oreg. Health Scis. U., Portland, 1969-70, instr. medicine, 1972—, asst. prof. pharmacology, 1970-78, assoc. prof., 1978—; Fogarty Sr. Internat. fellow Nuffield Inst. Med. Research, U. Oxford, Eng., 1981-82. Served with USPHS, 1967-69. NATO sr. scientist fellow, 1981-82. Mem. Am. Soc. Clin. Pharmacology and Therapeutics (med. edn. com. 1978-81), Am. Soc. Pharmacology and Exptl. Therapeutics, Perinatal Research Soc., Internat. Narcotics Research Conf. Subspecialties: Pharmacology; Developmental biology. Current work: Developmental pharmacology; respiratory pharmacology; control of breathing; drug abuse; obstetrical pharmacology; narcotic analgesics; pharmacology of central nervous system depressants and stimulants. Office: 3181 SW Sam Jackson Park Rd Portland OR 97201

OLSEN, JOHN STUART, biology educator, researcher, lectr.; b. Chgo., Mar. 10, 1950; s. John S. and Lucille S. (Altman) O.; m. Julie Ann Suero, Aug. 19. 1972; children: Kimberly Ann, Kevin John. B.S., U. Ill., Chgo., 1971, M.S., 1973; Ph.D., U. Tex.-Austin, 1977. Instr. U. Tex.-Austin, 1976-77; asst. prof. Southwestern U. at Memphis, 1977—, curator, dir.; instr. Memphis Acad. Art, 1977—; vis. asst. prof. botany U. Tex.-Austin, summers 1978-82, Mountain Lake Biol. Sta., U. Va., summer 1981. Contbr. articles on biology to profl. jours. Mem. AAAS, Am. Inst. Biol. Scis., Am. Soc. Plant Taxonomists, Bot. Soc. Am. (recipient Ralph Alston award 1977), New Eng. Bot. Club, Internat. Assn. Plant Taxonomists, Sigma Xi. Subspecialties: Systematics; Evolutionary biology. Current work: Systematics and evolution within the flowering plant family compositae; phytochemistry, cytotaxonomy, systematics. Office: 2000 N Parkway Memphis TN 38112

OLSEN, KENNETH HARRY, manufacturing company executive; b. Bridgeport, Conn., Feb. 20, 1926; s. Oswald and Svea (Nordling) O.; m. Eeva-Liisa Aulikki Valve, Dec. 12, 1950. B.S. in Elec. Engring, MIT, 1950, M.S., 1952. Elec. engr. Lincoln Lab., MIT, 1950-57; founder, 1957; since pres. Digital Equipment Corp., Maynard, Mass.; dir. Polaroid Corp., Shawmut Assn., Ford Motor Co. Mem. Pres.'s Sci. Adv. Com., 1971-73; Mem. corp., v.p. Joslin Diabetes Found., Wentworth Inst., Boston, MIT, Cambridge; trustee Gordon Coll., Wenham, Mass. Served with USNR, 1944-46. Named Young Elec. Engr. of Year Eta Kappa Nu, 1960. Subspecialty: Electrical engineering. Patentee magnetic devices. Home: Weston Rd Lincoln MA 01773 Office: Digital Equipment Corp Maynard MA 01754

OLSEN, NORMAN HARRY, dentist, educator; b. Omaha, June 9, 1926; s. Carl and Johanna (Jorgensen) O.; m. Donna Roberts, Sept. 14, 1947; children—Karl Raymond, Heidi Roberts, Holly Roberts. Student, U. Idaho, 1945-46; D.D.S., Creighton U., 1951; M.S.D., Northwestern U., 1953. Diplomate: Am. Bd. Pedodontics (bd. examiners). Montgomery Ward fellow Northwestern U., 1951-53, prof. pedodontics, 1954—, chmn. dept., 1954-72, dean, 1972—; prof., chmn. dept. pedodontics U. Kansas City Sch. Dentistry, 1953-54; staff Meml. Mercy Hosp., Kansas City, 1953-54, Northwestern Meml. Hosp., 1974—; head dental service Children's Meml. Hosp., Chgo., 1954—; courtesy staff Evanston Hosp., St. Francis Hosp., Evanston; mem. Cleft Palate Inst., Northwestern U. and Childrens Meml. Hosp.; cons. Throat Culturing Research Program, Stickney Twp. Health Dept.; chmn. Am. Bd. Pedodontic Examiners, 1973-74; Commr. Joint Commn. on Accreditation Dental Labs. Co-editor: 4th edit. Current Therapy in Dentistry, 1970; editor pedodontic sect., 5th edit., 1974; Contbr.: chpt. Dental Clinics of N.Am, 1961; articles profl. jours. Sec. faculty com. intercollegiate athletics Northwestern U., 1962-68; chmn. profl. div. Evanston United Fund, 1967-68; Pres. First Ward Non-Partisan Civic Assn., Evanston; former trustee Roycemorce Sch., Evanston; mem. exec. com. Dist. 65 Sch. Bd.; bd. dirs. Infant Welfare Soc., Chgo. Served with USNR, 1944-46. Recipient Service award Northwestern U., 1966; Merit award Am. Soc. Dentistry for Children, 1951; Alumni Merit award Creighton U., 1975. Fellow Am. Coll. Dentistry (vice chmn. Ill. sect. 1979-80, regent 1979—), Internat. Coll. Dentists, Chgo. Inst. Medicine; mem. ADA (pedodontic adv. com. Joint Commn. Accreditation, past sect. chmn.), Ill. Dental Soc. (past program chmn.), Chgo. Dental Soc. (past program chmn.), Am. Acad. Pedodontics (pres. 1973-74), Am. Soc. Dentistry for Children (past pres. Ill., award of Excellence 1974, Man of Yr. award Ill. unit), Internat. Assn. Dental Research, Am. Assn. Cleft Palate Rehab., Am. Acad. Gold Foil Operators, Odontographic Soc. Chgo. (bd. govs., past program chmn.), Ill. Pedodontics (charter), Am. Dental Soc. of Europe (hon.) GV Black Soc. Northwestern U. (sec.), Northwestern U. Dental Alumni Assn. (past pres.), Am. Assn. Dental Schs. (past sect. chmn.), Sigma Alpha Epsilon, Delta Sigma Delta, Omicron Kappa Upsilon. Clubs: Evanston (dir.), Rotary (pres. Evanston chpt. 1980-81), John Evans (Northwestern U.) (dir.). Subspecialties: Pediatric dentistry; Dental education/admininstration. Home: 221 Winnetka Ave Winnetka IL 60093 Office: 311 E Chicago Ave Chicago IL 60611

OLSEN, RALPH A., chemistry educator; b. Moroni, Utah, Jan. 30, 1925; s. John L. and Ethel E. (Oman) O.; m. Vanona Fisher, Aug. 30, 1949; children: Ravona Beverly, Karen, Loren, Paulette, Miriam, Amy. B.S., Brigham Young U., 1949; M.S., Cornell U., 1951, Ph.D., 1953. Soil scientist U.S. Dept. Agr., Beltsville, Md., 1953-56; prof. chemistry Mont. State U., Bozeman, 1956—. Served with U.S. Army, 1944-46. Postdoctoral fellow Mineral Nutrition Pioneering Research Lab., Inst. Biol. Chemistry, U. Copenhagen, Plant Stress Lab., Beltsville. Mem. Soil Sci. Soc. Am., Sigma Xi. Subspecialties: Soil chemistry; Plant growth. Current work: Chemistry of soil as it relates to plant nutrition, chemical modification of the rhizosphere by plant roots. Home: 1905 Willow Way Bozeman MT 59715 Office: Chemistry Dept Mont State U Bozeman MT 59715

OLSEN, RICHARD GEORGE, virologist, research cons.; b. Independence, Mo., June 25, 1937; s. Benjamin B. and Ruth N. (Myrtle) O.; m. Judith M. Olsen, June 22, 1957; children: Cynthia G., David G., Susan B., John D. B.A., U. Mo., Kansas City, 1959; M.S., U. Atlanta, 1963; Ph.D., SUNY, Buffalo, 1969. With Roswell Park Meml. Inst., Buffalo, 1967-69; prof. virology Ohio State U., Columbus, 1969—; cons. Vet. Biol. Corp. Author: Immunology and Immunopathology, 1979, Feline Leukemia, 1981; contbr. numerous articles to profl. jours. NIH predoctoral fellow, 1967-68; N.Y. State Dept. Health fellow, 1968-69. Mem. Am. Soc. Microbiology, Internat. Assn. Corp. Research on Leukemia, Tissue Culture Assn., Am. Assn. Cancer Research, Am. Assn. Tissue Banks. Subspecialties: Immunobiology and immunology; Virology (veterinary medicine). Current work: Immunology of virus diseases in animals. Patentee method of recovering cell antigen, preparation of feline leukemia vaccine.

OLSEN, WARD ALAN, gastroenterologist, researcher, educator; b. Holmen, Wis., Sept. 13, 1934; s. Orville Cornelius and Sophia (Vosseteig) O.; m. Margaret Mary Grant, June 24, 1961; children: Eric, Edward, Julia. B.S., U. Wis.-Madison, 1956, M.D., 1959. Intern Boston City Hosp., 1959-60, resident in internal medicine, 1961-62, 64-65; research fellow Boston U., 1965-67; research assoc. Thorndike Lab., Boston City Hosp., 1967-68; chief gastroenterology Madison VA Hosp., 1968—; instr. in medicine Harvard U., 1967-68; asst. prof. medicine U. Wis.-Madison, 1968-72, assoc. prof., 1972-78, prof., 1978—. Contbr. numerous articles, abstracts to profl. publs. Served to capt. U.S Army, 1962-64. Ger. NIH grantee, 1968—; VA grantee, 1968—. Mem. Am. Soc. Clin. Investigation, Am. Gastroent. Assn., Cenral Soc. Clin. Research, Phi Beta Kappa, Alpha Omega Alpha. Subspecialties: Internal medicine; Gastroenterology. Current work: Physiology, biochemistry, morphology of small intestine in health and disease. Office: VA Hosp 2500 Overlook Terr Madison WI 53705 Home: 173 N Prospect Ave Madison WI 53705

OLSON, DAVID LEROY, metallurgical engineer educator; b. Oakland, Calif., Mar. 17, 1942; s. Roy T. and Adelia H. (Gustafson) O.; m. Judith Ellen Perrine, Sept. 15, 1963; children: Katherine, Eric, Ivan, Anna. B.S., Wash. State U., 1965; Ph.D., Cornell U., 1969. Registered profl. engr., Colo. Mem. tech. staff Tex. Instruments, Dallas, 1969-70; research assoc. Ohio State U., Columbus, 1970-72; asst. to prof. Colo. Sch. Mines, Golden, 1972—, prof. metall. engring. and dir., 1981—; vis. research scientist Norwegian Inst. Tech., Trondheim, 1979. Contbr. articles to profl. jours. Ohio State U. postdoctoral fellow, 1971-72; recipient ASM Bradley Stoughton award, 1976; Am. Welding Soc. Comfort A. Adams award, 1978; Am. Nuclear Soc./MSTD Lit. award, 1978; AMOCO Found. teaching award, 1982. Mem. ASM, Am. Welding Soc., Am. Ceramic Soc., AIME, Nat. Assn. Corrosion Engrs., ASTM, Am. Foundrymen's Soc., Am. Phys. Soc., Sigma Xi, Theta Tau, Tau Beta Pi, Alpha Sigma Mu. Subspecialties: Materials processing; Metallurgy. Current work: Welding metallurgy and engring., corrosion. Home: 13943 W 20th Pl Golden CO 80401 Office: Ctr Welding Research Colo Sch Mines Golden CO 80421

OLSON, DONALD RICHARD, mechanical engineering educator; b. Sargent, Nebr., Dec. 26, 1917; s. Harry T. and Gyneth E. (Wittemyer) O.; m. Nancy Walker Benton, June 17, 1944; children: Walter H., Sally, Timothy W. B.S., Oreg. State U., 1942; M.Engring., Yale U., 1944, D.Engring., 1951. Profl. engr., Conn. Asst. prof., asso. prof. mech. engring. Yale U., New Haven, 1951-62; prof. mech. engring. Pa. State U., University Park, 1962—; head underwater power plants Applied Research Lab., 1962-72, head dept. mech. engring., 1972-83; mem. engring. accreditation commn., 1979—. Contbr. tech. papers in field to publs. Mem. ASME, Soc. Automotive Engrs. (dir. 1968-71), Sigma Xi. Subspecialties: Mechanical engineering; Systems engineering. Current work: Underwater powerplants; high energy reaction systems; simulation of power systems. Home: 621 Glenn Rd State College PA 16801

OLSON, EDWARD COOPER, astronomer; b. Worcester, Mass., June 7, 1930; s. Nils and Marion (Cooper) O.; m. Margaret Jean Edmundson, May 31, 1959; children: Eric, Jeffrey. Ph.D., Ind. U., 1960. Faculty Smith Coll., 1960-64, Rensselaer Poly. Inst., Troy, N.Y., 1965-66; faculty U. Ill., Urbana, 1966—, now astronomer. Contbr. articles to profl. jours. Served with U.S. Army, 1953-55. NSF grantee, 1962—. Mem. Am. Astron. Soc., Internat. Astron. Union, Astron. Soc. Pacific, Internat. Amateur and Profl. Photoelectric Photometry. Subspecialties: Optical astronomy; Close binary stars. Current work: Evolution of close binary systems. Office: Dept Astronomy U Ill 1011 W Springfield Urbana IL 61801

OLSON, EVERETT CLAIRE, biology educator; b. Waupaca, Wis., Nov. 6, 1910; s. Claire Myron and Aimee (Hicks) O.; m. Lila Richardson, July 15, 1939; children Claire, George, Mary Ellen. B.S., U. Chgo., 1932, M.S., 1933, Ph.D., 1935. Instr. to prof. geology U. Chgo., 1935-69; prof. biology UCLA, 1969—, dept. chmn., 1970-72. Author: Evolution of Life, 1965, (with Agnes Whitmarsh) Foreign Maps, 1944, (with Robert Miller) Morphological Integration, 1958, Vertebrate Paleozoology, 1971, (with Jane Robinson) Concepts of Evolution, 1975; editor: jour. Geology, 1960-66, Evolution, 1952-58. NSF grantee. Mem. Soc. Vertebrate Paleontology, Soc. Study Evolution, Am. Soc. Zoologists, Geol. Soc. Am., Nat. Acad. Sci., Phi Beta Kappa, Sigma Xi. Subspecialty: Evolutionary biology. Current work: Chronotaunal evolution; vertebrates. Home: 13760 Bayliss Rd Los Angeles CA 90049 Office: 405 Hilgarde St Los Angeles CA 90024

OLSON, GORDON LEE, physicist; b. Ortonville, Minn., Oct. 1, 1951; s. Gilbert E. and Zella Mae (Fetterly) O.; m. Linda Fae Madsen, Apr. 20, 1974; 1 son, Brian. B.Phys., U. Minn., 1973; M.S., U. Wis., 1975, Ph.D., 1977. Sci. collaborator Astrophys. Inst., Free U. Brussels, Belgium, 1977-79; research assoc. Joint Inst. for Lab. Astrophysics, U. Colo., Boulder, 1979-81; scientific staff Los Alamos Nat. Lab., 1981—. Contbr. articles to profl. jours. Mem. Am. Astron. Soc., AAAS. Subspecialties: Ultraviolet high energy astrophysics; Theoretical astrophysics. Current work: Stellar winds in hot stars and radiative transfer. Home: 470 Aragon Ave White Rock NM 87544 Office: Los Alamos Nat Lab PO Box 1663 Los Alamos NM 87545

OLSON, HARRY F., research dir.; b. Mt. Pleasant, Iowa, Dec. 28, 1902; s. Frans O. and Nelly (Benson) O.; m. Lorene E. Johnson, June 11, 1935. B.E., U. Iowa, 1924, M.S., 1925, Ph.D., 1928, E.E., 1932; D.Sc., Iowa Wesleyan U., 1959. Acoustical research Radio Corp. Am., Princeton, 1928—; dir. Acoustical Lab., RCA Labs., 1945—, also staff v.p. acoustical and electromech. research, 1966—; lectr. acoustical engring., Columbia, 1939-42. Author: Dynamical Analogies, 1943, Elements of Acoustical Engring, 1947, Musical Engineering, 1952, Acoustical Engineering, 1958, Music, Physics and Engineering, 1966, Modern Sound Reproduction, 1972; Contbr. articles to sci. jours. Recipient John H. Potts medal Audio Engring. Soc., 1949; Warner medal Soc. Motion Picture and Television Engrs.; Scott award Engrs. Club, Phila., 51956; John Ericsson medal Am. Soc. Swedish Engrs.; Mervin Kelly medal and Lamme medal IEEE, 1967, 69. Fellow IEEE (chmn. profl. group audio 1957-58), Acoustical Soc. Am. (pres. 1952, First Silver medal 1974, Gold medal 1981), Am. Phys. Soc., Soc. Motion Picture and TV Engrs., Audio Engring. Soc. (pres. 1960); mem. of Nat. Acad. Scis., Sigma Xi, Tau Beta Pi. Subspecialty: Acoustical engineering. Pioneered in research and devel. of various types of directional microphones; patentee in field Office: RCA Laboratories Princeton NJ 08540

OLSON, JAMES ALLEN, biochemistry and biophysics educator and administrator, researcher; b. Mpls., Oct. 10, 1924; m. Giovanna F. Del Nero, 1953; children: Daniel, Lisa, Eric. Student, Gustavus Adolphus Coll., 1941-44, B.S., 1946; Ph.D., Harvard U., 1952. Postdoctoral assoc. Internat. Center Clin. Microbiology, Rome, 1952-54, Harvard U., Cambridge, Mass., 1954-56; asst. prof. to prof. U. Fla., Gainesville, 1956-66; field staff mem., prof. Rockefeller Found., Bangkok, Thailand, 1966-74, Salvador, Brazil, 1974-75; prof. dept. biochemistry and biophysics, dept. chmn. Iowa State U., Ames, 1975—; cons. NIH, various times 1976—, NSF, 1962-64, U.S. AID, 1978-81, Govt. Indonesia, 1980-82. Served to lt. USN, 1944-46. Fellow AAAS; mem. Am. Soc. Biol. Chemists, Am. Inst. Nutrition, Am. Chem. Soc. (alt. councilor 1971-73), Soc. Exptl. Biology and Medicine (councilor 1963-64), Sigma Xi. Subspecialties: Biochemistry (medicine); Nutrition (medicine). Current work: Metabolism, nutrition and function of vitamin A; relation of nutrition to cancer. Office: Dept Biochemistry and Biophysics Iowa State Univ Ames IA 50011

OLSON, ROBERT ALLEN, computer scientist; b. Houston, Nov. 15, 1949; s. Robert Jerome and Joan Millicent (Seline) O.; m. Tatiana Marie Granoff, Oct. 29, 1977; 1 son: Aaron Edward. B.S., Stanford U., 1973, M.S., 1973. With gen. system div. Hewlett Packard Co., Cupertino, Calif., 1973-80; mgr. software devel. Elxsi, San Jose, Calif., 1980—. Mem. Assn. Computing Machinery. Subspecialties: Computer architecture; Operating systems. Current work: High performance operating systems. Office: Elxsi 2334 Lundy Pl San Jose CA 95131

OLSON, ROBERT EUGENE, physician, biochemist, educator; b. Minn., Jan. 23, 1919; s. Ralph William and Minnie (Holtin) O.; m. Catherine Silvoso, Oct. 21, 1944; children: Barbara Lynn, Robert E., Mark Alan, Mary Ellen, Carol Louise. A.B., Gustavus Adolphus Coll., 1938; Ph.D., St. Louis U., 1944; M.D., Harvard, 1951, Chiang Mai U., Thailand, 1983. Diplomate: Nat. Bd. Med. Examiners, Am. Bd. Nutrition (pres. 1962-63). Postgrad. research asst. biochemistry St. Louis U. Sch. Medicine, 1938-43, asst. biochemistry, 1943-44, Alice A. Doisy prof. biochemistry, chmn. dept. biochemistry, 1965—, asso. prof. medicine, 1966-72, prof. medicine, 1972—; vis. prof. (sabbatical) dept. biochemistry U. Freiburg, Breisgau, West Germany, 1970-71; also Hoffman-La Roche Co., Basel, Switzerland, 1970-71; instr. biochemistry and nutrition Harvard Sch. Pub. Health, 1946-47; research fellow Nutrition Found., 1947-49, Am. Heart Assn., 1949-51, established investigator, 1951-52; house officer Peter Bent Brigham Hosp., Boston, 1951-52; prof., head dept. biochemistry and nutrition Grad. Sch. Pub. Health U. Pitts.; lectr. medicine Sch. Medicine, 1952-65; mem. panel malnutrition Japan-U.S. Med. Scis. Program, 1965-69; dir. Nutrition Clinic, Falk Clinic, 1953-65; mem. sr. staff Presbyn. Hosp., dir. metabolic unit, 1960-65; mem. staff St. Louis U. Hosp., 1965-81; prof. biochemistry, prof. medicine, assoc. dean acad. affairs U. Pitts. Sch. Medicine, 1982—; cons. Mercy Hosp., U. Pitts. Med. Center; assoc. in medicine St. Margaret's Meml. Hosp., Pitts., dir. metabolic unit, 1954-60; cons. div. research grants USPHS, 1954-69, 73—; dir. Anemia and Malnutrition Center, Chiang Mai, Thailand, 1967-77; vis. scholar dept. biochemistry Oxford (Eng.) U., 1961-62; vis. prof. dept. biochemistry U. Freiburg, West Germany, 1970-71; Mem. food and nutrition bd. Nat. Research Council, 1977—. Asso. editor: Nutrition Reviews, 1954-56; editor, 1978—; asso. editor: Am. Jour. Medicine, 1956-65, Circulation Research, 1956-67, Am. Heart Jour, 1958-65, Am. Jour. Clin. Nutrition, 1960-66, Methods in Med. Research, 1963-70, Biochem. Medicine, 1967, Molecular and Cellular Cardiology, 1969, Ann. Rev. Nutrition, 1980—; co-editor: Vitamins and Hormones, 1975—. Bd. dirs. Nat. Nutrition Consortium, 1977—. Served as lt. (j.g.) USNR, 1944-46. Recipient Fulbright award, 1961-62; Guggenheim Found. award, 1961-62, 70-71; McCollum award, 1965; Joseph Goldberger award, 1974; Atwater Meml. lectr., 1978; Geiger Meml. lectr., 1979. Fellow Am. Pub. Health Assn. (chmn. food and nutrition sect. 1960-61), A.C.P., Assn. Am. Physicians; mem. Am. Assn. Cancer Research, Am. Heart Assn., AMA (mem. council food and nutrition 1959-67, vice chmn. 1962-67), Royal Soc. Health (London), N.Y. Acad. Scis., Am. Fedn. Clin. Research, Am. Soc. Clin. Investigation, Boyleston Med. Soc., Am. Chem. Soc. (pres. biochemistry group Pitts. sect. 1960-61), Am. Soc. Biol. Chemists, Am. Inst. Nutrition (pres. 1981), AAAS (sec. med. scis. N sect. 1965-67), Soc. Exptl. Biology and Medicine, Am. Soc. Clin. Nutrition (pres. 1961-62, McCollum award 1965), Assn. Med. Sch. Depts. Biochemistry (pres. 1979—), Pa., St. Louis, Allegheny County med. socs., Am. Soc. Study Liver Diseases, Phi Beta Kappa, Sigma Xi, Phi Lambda Upsilon, Alpha Omega Alpha, Alpha Sigma Nu. Clubs: University (Pitts.); Sewickley Heights Golf (Sewickley, Pa.); Cosmos (Washington). Subspecialty: Nutrition (medicine). Home: 906 Amberson Ave Pittsburgh PA 15232 Office: 3550 Terrace St Pittsburgh PA 15261

OLSON, WILMA KING, chemistry educator; b. Phila., Dec. 1, 1945; d. Seth D. and Wilma C. (Herdle) King; m. Gary L. Olson, June 28, 1969; 1 son, Andrew J. B.S., U. Del., 1967; Ph.D., Stanford U., 1971. Postdoctoral fellow Stanford U., 1970-71, Columbia U., N.Y.C., 1971-72; asst. prof. Rutgers U., New Brunswick, N.J., 1972-76, assoc. prof. chemistry, 1976-79, prof., 1979—. Contbr. articles in field to profl. jours. Recipient Research Career Devel. award USPHS, 1975-81; Damon Runyon Found. fellow, 1971-72; A. P. Sloan Found. fellow,

1975-77; Guggenheim fellow, 1978-79. Mem. AAAS, Am. Chem. Soc., Biophys. Soc., N.Y. Acad. Scis. Subspecialties: Biophysical chemistry; Polymer chemistry. Current work: Relationship of chemical architecture to conformation and properties of biol. macromolecules, with emphasis on nucleic acids. Office: Rutgers U New Brunswick NJ 08903

OLSSON, CARL ALFRED, urologist; b. Boston, Nov. 29, 1938; s. Charles Rudolph and Ruth Marion (Bostrom) O.; m. Mary DeVore, Nov. 4, 1962; children: Ingrid, Leif Eric. Grad., Bowdoin Coll., 1959; M.D., Boston U., 1963. Diplomate: Am. Bd. Urology. Asst. prof. urology Boston U. Sch. Medicine, 1971-72, assoc. prof., 1972-74, prof., chmn. dept., 1974-80; dir. urology dept. Boston City Hosp., 1974-77; chief urology dept. Boston VA Med. Center, 1971-75; urologist-in-chief Univ. Hosp., Boston, 1971-80; John K. Lattimer prof., chmn. dept. urology Coll. Phys. and Surgs., Columbia U., N.Y.C., 1980—; dir. Squier Urol. Clinic, urology service Presbyn. Hosp., N.Y.C.; lectr. surgery Tufts U. Sch. Medicine. Boston Interhosp. Organ Bank, 1976-79; mem. working cadre Nat. Prostate Cancer Project, Nat. Cancer Inst., 1979—. Editorial bd.: Jour. Microsurgery and Prostate; asst. editor: Jour. Urology; Contbr. chpts. to books, articles to med. jours. Fellow A.C.S.; mem. Am. Urol. Assn. (coordinator continuing med. edn. New Eng. sect. 1977-80, del. research com., Gold Cystoscope award 1979, Grayson-Carroll award 1971, 73), Boston Surg. Soc. (exec. com. 1976—), Am. Assn. Clin. Urologists, Am. Assn. Genitourinary Surgeons, Clin. Soc. Genitouninary Surgeons, Am. Fertility Soc., AMA, Assn. Acad. Surgery, Am. Soc. Artificial Internal Organs, Am. Soc. Transplant Surgeons, Assn. Med. Colls., Transplant Soc., Societe Internationale d'Urologie, Internat. Urodynamics Soc., Mass. Med. Soc., Soc. Govt. Urologists, New Eng. Handicapped Sportsmen's Assn. (exec. com. 1977—), Soc. Univ. Urologists, Alpha Omega Alpha. Episcopalian. Clubs: U.S. Yacht Racing Union, Yacht Racing Union L.I. Sound, Cottage Park Yacht., Larchmont Yacht. Subspecialties: Urology; Oncology. Current work: Cancer research. Home: 18 Elm St Larchmont NY 10538 Office: 630 W 168th St New York NY 10032

OLSSON, RICHARD KEITH, geology educator, researcher, consultant; b. Newark, Mar. 23, 1931; s. Harold and Martha (Klenke) O.; m. Nancy Hunter, Apr. 27, 1957; children: Richard, Robert, Jeffrey. B.S., Rutgers U., 1953, M.S. in Geology, 1954; M.A., Princeton U., 1956, Ph.D. in Geology, 1958. Instr. geology Rutgers U., New Brunswick, N.J., 1957-60, asst. prof., 1960-63, assoc. prof., 1963-72, prof., 1972—, chmn. dept., 1977—; cons. Humble Oil Co., Md. Geol. Survey, ERT. Rutgers U. fellow, 1970-71; grantee Petroleum Research Fund-Am. Chem. Soc., Union Oil Calif. Found. Fellow Geol. Soc. Am.; mem. Am. Assn. Petroleum Geologists, Soc. Econ. Paleontologists and Mineralogists, Paleontol. Soc., Petroleum Exploration Soc. N.Y. (pres. 1979-80). Subspecialties: Paleoecology; Biostratigraphy. Current work: Biostratigraphy and paleoecology of planktonic and benthic foraminifer in the Mesozoic and Cenozoic eras of geologic time. Home: 115 Dodds Ln Princeton NJ 08540 Office: Rutgers U New Brunswick NJ 08903

OLSTAD, WALTER BALLARD, aerospace industry executive; b. Greenport, N.Y., Oct. 24, 1932; s. Martin H. and Helen D. (Oskison) O.; m. Helen H. Thompson, Apr. 13, 1957; children: Lauren K., David W., Linda J. B.S., Brown U., 1954; M.S., Va. Poly. Inst., 1958; Ph.D., Harvard U., 1967. With NASA (and predecessor), 1954—, head advanced entry analysis br., 1970-75, chief space systems div., 1975-79, asso. adminstr. for aeros. and space tech., Washington, 1979-81, assoc. adminstr. for mgmt., 1981-83; v.p. govt. requirements Lockheed-Calif. Co., Washington, 1983—; asso. professorial lectr. George Washington U.; vis. prof. mech. engring. and mechanics Old Dominion U., 1978, 79. Author articles, reports in field. Mem. Elmer A. Sperry bd. of award. Recipient NASA Disting. Service medal, 1981, NASA Outstanding Leadership medal, 1983, Presdl. award for meritorious service, 1981. Mem. AIAA (asso. fellow, recipient thermophysics award 1981). Subspecialties: Aeronautical engineering; Applied mathematics. Current work: Concepts and technical requirement for advanced tactical and strategic aircraft. Home: 908 Danton Ln Alexandria VA 22308 Office: Lockheed-Calif Co 1825 I St NW Suite 1100 Washington DC 20006

OLTON, DAVID STUART, neuropsychologist; b. Montclaire, N.J., Jan. 15, 1943; s. Robert Matthew and Minnie (Tiepel) O. B.A., Haverford Coll., 1964; M.A., U. Mich., 1968, Ph.D., 1969. Mem. faculty dept. psychology Johns Hopkins U., Balt., 1969—, prof., 1979—, mem. faculty dept. psychiatry and behavioral scis., 1971—. Author: Biofeedback Clinical Applications in Behavioral Medicine, 1980. Treas. Jacksonville Vol. Fire Co., 1972—. Subspecialties: Neuropsychology; Physiological psychology. Current work: The neurochemical and neuroanatomical mechanisms of memory. Home: 3704 Stansbury Mill Rd Phoenix MD 21131 Office: Dept Psychology Johns Hopkins U Baltimore MD 21218

OMAN, CHARLES MCMASTER, scientist, educator, aeronautical engineer, consultant; b. Bklyn., Feb. 22, 1944; s. William Morse and Janet Lyle (McMaster) O.; m. Cherryl L. Huested, Oct. 26, 1974; children: Katherine, Peter. B.S.E. in Aerospace and Mech. Scis., Princeton U., 1966; S.M., MIT, 1968, Ph.D. in Instrumentation, Guidance and Control, 1972. Asst. prof. dept. aeros. and astronautics MIT, Cambridge, 1972-76, assoc. prof., 1976-78, prin. research scientist, 1979-80, sr. research engr., 1981, asso. dir. Man Vehicle Lab., 1982; cons. NASA, U.S. Air Force, Nat. Acad. Scis.; lectr. to govt. agys., univs., coll., hosps., confs., 1977—; Herman von Helmholtz assoc. prof. Harvard U.-MIT Program in Health Scis. and Tech., 1977-78. Contbr. articles to sci. jours. Grass fellow Marine Biol. Lab., Woods Hole, 1974. Mem. IEEE (sr.), Soc. Neurosci., Barany Soc., Aerospace Med. Assn., Assn. Research in Otology. Subspecialties: Aeronautical engineering; Biomedical engineering. Current work: Research on vestibular system, spatial orientation, motion sickness; investigator on space shuttle, Spacelab Missions 1, 4 and D-1 (vestibular experiments). Home: 5 Highland Terr Winchester MA 01890 Office: MIT Man Vehicle Lab Room 37-219 Cambridge MA 02139

OMAYE, STANLEY TERUO, biochem. nutritionist, toxicologist, pharmacologist; b. Detroit, Jan. 25, 1945; s. Shigeru and Teruye (Kanagaki) O. B.A. in Chemistry, Sacramento State Coll., 1968; M.S. in Pharmacology, U. Pacific, 1972; Ph.D. in Biochem. Nutrition, U. Calif., 1975. Postdoctoral fellow Calif. Primate Center, U. Calif., Davis, 1975-76; research chemist Letterman Army Inst. Research, San Francisco, 1976080, cons. div. cutaneous hazards and toxicology group, 1980—; research nutritionist Western Regional Research Center, U.S. Dept. Agr., Berkeley, Calif., 1980—, project leader, 1980—; adj. prof. dept. pharmacology U. of Pacific, 1976—. Contbr. articles on nutrition, pharmacology and toxicology to profl. jours. Mem. Am. Inst. Nutrition, Am. Soc. Pharmacology and Exptl. Therapeutics, Soc. Toxicology, Am. Coll. Toxicology, Sigma Xi, Rho Chi. Subspecialties: Nutrition (medicine); Pharmacology. Current work: Nutrient and xenobiotic interactions, nutrient and nutrient interactions. Office: Western Regional Research Center 800 Buchanan St Berkeley CA 94710

OMENN, GILBERT STANLEY, physician, university dean; b. Chester, Pa., Aug. 30, 1941; s. Leonard and Leah (Miller) O.; m.; children by previous marriage: Rachel Andrea, Jason Montgomery. A.B., Princeton U., 1961; M.D., Harvard U., 1965; Ph.D. in Genetics, U. Wash., 1972. Intern Mass. Gen. Hosp., Boston, 1965-66, asst. resident in medicine, 1966-67; research asso NIH, Bethesda, Md., 1967-69; fellow U. Wash., 1969-71, asst. prof. medicine, 1971-74, asso. prof., 1974-79, prof., 1979—, prof., chmn. environ. health, 1981—, dean Sch. Pub. Health and Community Medicine, 1982—; asso. dir. Office Sci. and Tech. Policy, The White House, 1977-80; asso. dir. human resources Office Mgmt. and Budget, 1980-81; vis. sr. fellow Woodrow Wilson Sch. Public and Internat. Affairs, Princeton U., 1981—; sci. and public policy fellow Brookings Instn., Washington, 1981-82; cons. govt. agys., Howard Hughes Med. Inst., Corps., Cable News Network; investigator, 1976-77. Editor: (with others) Genetics, Environment and Behavior: Implications for Educational Policy, 1972; editorial bd.: Am. Jour. Med. Genetics, Ann. Rev. Pub. Health, Jour. Neurogenetics; Contbr. articles on human biochem. genetics, prenatal diagnosis of inherited disorders, susceptibility to environ. agts., clin. medicine and health policy to profl. publs. Mem. President's Council on Spinal Cord Injury, Nat. Heart, Lung and Blood Adv. Council.; Mem. Wash. State Gov.'s Commn. on Social and Health Services; mem. Nat. Cancer Adv. Bd.; mem. adv. com. Woodrow Wilson Sch., Princeton U.; Trustee Pacific Sci. Center, Seattle Symphony Orch., Seattle Youth Symphony Orch., Santa Fe Chamber Music Festival; chmn. rules com. Democratic Conv., King County, Wash., 1972. Served with USPHS, 1967-69. Recipient Research Career Devel. award USPHS, 1972; White House fellow, 1973-74. Fellow ACP, Hastings Ctr.; mem. White House Fellows Assn., Am. Soc. Neurochemistry, Am. Soc. Human Genetics, Behavior Genetics Assn., Western Soc. Clin. Research, Soc. Study Social Biology (mem. council), Inst. Medicine of Nat. Acad. Sci. Jewish. Subspecialty: Genetics and genetic engineering (medicine). Current work: Human genetics; ecogenetics; cancer chemoprevention. Office: Dean Sch Pub Health U Wash SC-30 Seattle WA 98195

OMMAYA, AYUB K., neurosurgeon, neuroscientist, educator; b. Mianchanun, Pakistan, Apr. 14, 1930; s. Nadir Khan and Ida (Counil) O.; m.; children: David, Alex, Shana. M.D., U. Punjab, 1953; M.A., U. Oxford, 1960; diploma in Psychol. Medicine, U. London, 1956. Diplomate: Am. Bd. Neurosurgery. Intern Mayo Hosp., Lahore, Pakistan, 1953-54; resident in neurosurgery Radcliffe Infirmary, Oxford, Eng., 1957-61; vis. scientist NIH, Bethesda, Md., 1961-63, neurosurgeon Neurol. Inst., 1963-69, chief applied research in neurosurgery, 1969-80; chief med. adviser Nat. Hwy. Traffic Safety Adminstrn., Dept. Transp., Washington, 1980—; clin. prof. neurosurgery George Washington U., Washington, 1980—; cons. in biomechanics and neural trauma mechanisms. Contbr. over 150 sci. articles to profl. publs. Rhodes scholar U. Oxford, 1954-60; J.W.K. Univ. prize, 1956; Hunterian prof. Royal Coll. Surgeons Eng., 1968; 1st Lewin Meml. lectr. U. Cambridge, Eng., 1983. Fellow ACS, Royal Coll. Surgeons Eng.; mem. Assn. Pakistani Physicians (pres. 1979-82), Assn. Brit. Neurosurgeons, Am. Assn. Neurol. Surgeons, ASME, Internat. Brain Research Orgn., Soc. Neurosci. Subspecialties: Neurosurgery; Regeneration. Current work: Neurological trauma mechanisms; mechanisms of neurological reintegration and regeneration after trauma; mechanisms of consciousness and the mind brain problem. Patentee inflatable neck collar for neck injury prevention. Home: 8901 Burning Tree Rd Bethesda MD 20817 Office: 5530 Wisconsin Ave Chevy Chase MD 20015

O'NEILL, EUGENE FRANCIS, communications engr.; b. N.Y.C., July 2, 1918; s. John J. and Agnes (Willmeyer) O'N.; m. Kathryn M. Walls, Oct. 24, 1942; children—Kathryn Anne, Kevin, Jane A., Andrew Thomas. B.S. in Elec. Engring, Columbia, 1940, M.S., 1941; D.Sc. (hon.), Bates Coll., D.Engring., Politecnico di Milano, D.Sc., St. John's U., N.Y.C. With Bell Telephone Labs., Holmdel, N.J., 1941—, engaged in radar devel., 1941-45, coaxial and submarine cable and microwave radio relay, 1945-56, headed devel. of speech interpolation terminals which doubled capacity submarine telephone cables, 1956-60; dir. Telstar satellite projects, 1960-66, exec. dir. network projects, 1966-81. Pulitzer prize; scholar Columbia, 1936-40. Fellow IEEE; mem. Nat. Acad. Engring., Sigma Xi, Tau Beta Pi. Subspecialty: Communication engineering. Home: 17 Dellwood Ct Middletown NJ 07748 Office: Bell Telephone Labs Holmdel NJ 07733

O'NEILL, RUSSELL RICHARD, engineering educator; b. Chgo., June 6, 1916; s. Dennis Alysious and Florence Agnes (Mathurin) O'N.; m. Margaret Bock, Dec. 15, 1939; children: Richard A., John R.; m. Sallie Boyd, June 30, 1967. B.S. in Mech. Engring, U. Calif. at Berkeley, M.S., 1940; Ph.D., U. Calif. at Los Angeles, 1956. Registered profl. engr., Calif. Design engr. Dowell, Inc., Midland, Mich., 1940-41; design engr. Dow Chem. Co., Midland, 1941-44, Airesearch Mfg. Co., Los Angeles, 1944-46; lectr. engring. UCLA, 1946-56, prof. engring., 1956, asst. dean engring., 1956-61, asso. dean., 1961-73, acting dean, 1965-66, 74-83; staff engr. Nat. Acad. Sci.-NRC, 1954; dir. Data Design Labs., 1977—; Mem. engring. task force Space Era Edn. Study Fla. Bd. Control, 1963; mem. regional Export Expansion Council Dept. Commerce, 1960-66, Los Angeles Mayor's Space Adv. Com., 1964-69; mem. Maritime Transp. Research Bd., 1974-81; bd. adv. Naval Postgrad. Sch., 1976—. Trustee West Coast U., 1981—. Mem. Nat. Acad. Engring., Am. Soc. Engring. Edn., Soc. Naval Architects and Marine Engrs., Triangle, Sigma Xi, Tau Beta Pi. Subspecialties: Systems engineering; Species interaction. Home: 15430 Longbow Dr Sherman Oaks CA 91403 Office: 405 Hilgard Ave Los Angeles CA 90024

ONORATO, HOWARD LOUIS, nuclear engineer, consultant; b. Sea Isle City, N.J., Apr. 24, 1947; s. Frank Thomas and Kathryn (Everingham) O.; m. Nancy Carolyn Denty, Jan. 12, 1974; children: Rodney, Amy. Student, Southeastern Community Coll., Whiteville, N.C., 1977-79, Miss. Coll., 1980, Glouchester Community Coll., 1981-82. Cons. engr. Energy Inc., Idaho Falls, Idaho, 1981—; asst. engr. Pratt & Whitney Aircraft, West Palm Beach, Fla., 1971-75; reactor operator Carolina P & L, Southport, N.C., 1977-79; cons. engr. Quadrex, Campbell, Calif., 1979-81, Energy, Inc., Idaho Falls, 1981—. Served with USN, 1965-71, 75-77. Mem. Am. Nuclear Soc. Subspecialties: Nuclear engineering; Human factors engineering. Current work: Provide operational support to run Salem nuclear support, update of procedures and equipment considering human factors. Home: 154 Edward Dr Swedesboro NJ 08085 Office: Energy Inc PO Box 736 Idaho Falls ID 83402

OPGRANDE, JOHN DONALD, medical educator; b. Dickinson, N.D., Mar. 8, 1941; s. J. Lionel and Esther Caroline (Malkewick) O.; m. Carolyn Thea Holm, June 20, 1964; children: Heidi, Kristen, Julie, John. B.A., Concordia Coll.-Minn., 1963; B.S., U. N.D., 1963-65; M.D., U. Kans., 1967; Master's, U. Minn., 1979. Intern St. Luke's Hosp., Saginaw, Mich., 1967-68; resident Mayo Clinic, 1970-73, fellow, 1973-74; adj. prof. engring. N.D. State U., Fargo, 1976—, chmn. dept. orthopedic surgery, 1980—; hand surgeon Dakota Clinic Ltd., Fargo, 1974—, bd. dirs., 1980-82, Dakota Med. Found., Fargo, 1981—. Contbr. articles in field to profl. jours. Mem. Lake Agassiz Arts Council, Fargo, 1979; mem. Arthritis Found., 1975—. Served to Capt. U.S. Army, 1968-70. Regional Med. Program grantee, 1977; Pfizer Pharm. grantee, 1983. Fellow Am. Acad. Orthopedic Surgeons (com. mem. 1979—, councilor 1983); mem. U.S. Army Assn. Flight Surgeons, AMA (del. 1980—), Orthopedic Research Soc., Am. Soc. Surgery of the Hand (com. mem. 1981—). Lutheran. Subspecialties: Orthopedics; Microsurgery. Current work: Biomechanical engineering; study of static and dynamic forces affecting function of the hand; pharmacology; investigation of a new drug to treat hand infections. Home: 2485 Lilac Ln Fargo ND 58102 Office: U N D Dept Orthopedics Dakota Clinic Ltd PO Box 6001 1702 S University Fargo ND 58108-60001

OPIE, WILLIAM ROBERT, metallurgical engineer; b. Butte, Mont., Apr. 3, 1920; s. Ellison Stuart and Myrtle (Williams) O.; m. Constance E. Kickuth, Oct. 14, 1944; children: Lyle Margaret, Guy William. B.S., Mont. Sch. Mines, 1942, M.E. (hon.), 1965; Sc.D. (hon.), MIT, 1949; student, Advanced Mgmt. Program, Harvard U., 1967; Sc.D. (hon.), Mont. Coll. Mineral Sci. and Tech., 1980. Foundry metallurgist Wright Aero Corp., Paterson, N.J., 1942-45; research asso. MIT, Cambridge, 1946-48; research metallurgist Am. Smelting and Refining, Perth Amboy, N.J., 1948-50; research supr. Nat. Lead Co., Sayreville, N.J., 1950-60; pres. Amax Base Metals Research & Devel., Inc., Carteret, N.J., 1960—. Contbr. articles to profl. jours. Served with U.S. Navy, 1945-46. Fellow AIME, Am. Soc. Metals; mem. Can. Inst. Mining and Metallurgy, Nat. Acad. Engring., Metals Soc. (London). Club: Mining (N.Y.C.). Subspecialty: Metallurgy. Patentee in field. Home: 119 Crawfords Corner Rd Holmdel NJ 07733 Office: 400 Middlesex Ave Carteret NJ 07008

OPITZ, JOHN MARIUS, medical scientist; b. Hamburg, Germany, Aug. 15, 1935; came to U.S., 1950, naturalized, 1957; s. Friedrich and Erica Maria (Quadt) O.; children: Elisabeth, Gabriella, John, Chrisanthi, Felix. B.A., State U. Iowa, 1956, M.D., 1959; D.Sc. h.c., Mont. State U., 1983. Diplomate: Am. Bd. Pediatrics, Am. Bd. Med. Genetics. Intern State U. Iowa Hosp., 1959-60, resident in pediatrics, 1960-61; resident and chief resident in pediatrics U. Wis. Hosp., Madison, 1961-62; fellow in pediatrics and med. genetics U. Wis., 1962-64, asst. prof. med. genetics/pediatrics, 1964-69, assoc. prof., 1969-72, prof., 1972-79; dir. Wis. Clin. Genetics Ctr., 1974-79, Shodair Mont. Regional Genetics Program, Helena, 1979—; clin. prof. med. genetics and pediatrics U. Wash., Seattle, 1979—; adj. prof. medicine, biology, history and philosophy of sci., vet. research and vet. sci. Mont. State U., Bozeman, 1979—; dir. Shodair-Mont. Regional Genetic Services Program, Helena, 1979—; chmn. dept. med. genetics Shodair Children's Hosp., 1983—; adj. prof. med. genetics and pediatrics U. Wis.-Madison, 1982—; editor-in-chief, founder Am. Jour. Med. Genetics, 1977—; mng. editor European Jour. Pediatrics, 1977—. Mem. Am. Soc. Human Genetics, Am. Pediatric Soc., Soc. Pediatric Research, Am. Bd. Med. Genetics, Birth Defects Clin. Genetic Soc., Am. Inst. Biol. Scis., Am. Soc. Zoologists, AAAS, Teratology Soc., Genetic Soc. Am., European Soc. Human Genetics, Soc. Study Social Biology, Am. Acad. Pediatric, German Pediatrics Soc., Western Soc. for Pediatrics Research, Sigma Xi. Democrat. Roman Catholic. Subspecialties: Developmental biology; Genetics and genetic engineering (medicine). Current work: Human developmental genetics. Home: 1806 Livingston St Helena MT 59601 Office: Shodair Children's Hospital PO Box 5539 840 Helena Ave Helena MT 59604

OPPENHEIM, ANTONI KAZIMIERZ, mech. engr.; b. Warsaw, Poland, Aug. 11, 1915; came to U.S., 1948, naturalized, 1954; s. Tadeusz and Zuzanna (Zuckerwar) O.; m. Lavinia Stephens, July 18, 1945; 1 dau., Terry Ann. Dipl. Ing., Warsaw Inst. Tech., London, 1943; D.I.C., Ph.D. in Engring, City and Guilds Coll., Imperial Coll. Sci. and Tech., U. London, 1945, D.Sc., 1976; Dr. h.c., U. Poitiers, France, 1981. Registered profl. engr., Calif. Research asst. City and Guilds Coll., 1942-48, lectr., 1946-48; asst. prof. mech. engring. Stanford U., 1948-50; faculty U. Calif. at Berkeley, 1950—, prof. mech. engring., 1958—, Miller prof., 1961-62; vis. prof. Sorbonne, Paris, 1960-61, U. Poitiers, France, 1973, 80; staff cons. Shell Devel. Co., 1952-60. Editor-in-chief: Acta Astronautica, 1974-79; contbr. articles to profl. jours., also monographs. Chmn. Heat Transfer and Fluid Mechanics Inst., 1958; IAA Com. on Gasdynamics of Explosions, 1968—; organizer Internat. Colloquia on Gas Dynamics of Explosions and Reactive Systems, 1967, 69, 71, 73, 75, 77, 79, 81; mem. NASA; adv. com. fluid mechanics, 1963-69. Numa Manson medal, 1981. Fellow Am. Rocket Soc. (nat. dir. 1959-62, founder, pres. No. Calif. sect. 1957), ASME (chmn. profl. conf. 1958, program chmn. San Francisco 1959-60), Am. Inst. Aeros. and Astronautics (Pendray award 1966); elected mem. Internat. Acad. Astronautics, Nat. Acad. Engring.; mem. Inst. Mech. Engrs. (Water Arbitration prize 1948), Am. Phys. Soc., Am. Chem. Soc., Am. Soc. Engring. Edn., Sigma Xi, Psi Tau Sigma, Tau Beta Psi. Subspecialty: Mechanical engineering. Spl. research compressible fluid flow, gas turbines and internal combustion engines, heat transfer, combustion, detonation and blast waves. Home: 54 Norwood Ave Berkeley CA 94707

ORAN, DAVID ROBERT, computer scientist; b. Phila., Jan. 11, 1949; s. Max and Minerva (Smith) O.; m. Silvia Marina Arrom, Dec. 30, 1972; 1 dau., Christina Alexandra. B.A., Haverford Coll., 1970. Sr. programmer NASA/Bartol Research Found., Swarthmore, Pa., 1969-71; sr. systems programmer Bank of Calif. (N.A.), San Francisco, 1971-74; ind. cons. Networks and Communications Engring., Mexico City, 1974-76; network architect Digital Equipment Corp., Tewksbury, Mass., 1976—; sr. instr. integrated computer systems, 1980. Mem. Assn. Computing Machinery, IEEE Computer Soc. Subspecialties: Distributed systems and networks; Operating systems. Current work: Internetwork gateways, network operating systems, network architecture and design of distributed algorithms and protocols, distributed operating systems, local area network design.

ORCHIN, MILTON, educator, chemist; b. Barnesboro, Pa., June 4, 1914; s. Morris and Mary (Rivkin) O.; m. Ruth Wilner, June 4, 1941; children: Morton Lewis, Michael David. B.A., Ohio State U., 1936, M.A., 1937, Ph.D., 1939. Chemist FDA, Cin., Chgo., 1939-42; chief organic chemistry sect. U.S. Bur. Mines, 1943-53; asso. prof. chemistry U. Cin., 1953-56, prof., head dept., 1956-62, prof. chemistry, dir. basic sci. lab., 1962-70; dir. Hoke S. Greene Lab. for Catalysis, 1970—, disting. service prof., 1981—. Co-author 7 books.; Contbr. numerous articles research publs. Bd. govs. Ben Gurion U. of Negev, Israel, 1974-80. Recipient E. J. Houdry award in applied catalysis N. Am. Catalysis Soc., 1983; Guggenheim fellow, 1947-48. Mem. AAAS (chmn. chemistry sect. 1963), Am. Chem. Soc. (chmn. Cin. sect. 1962, Eminent Chemist award Cin. sect. 1957, E.V. Murphree award in indsl. and engring. chemistry 1979, Morley medal Cleve. sect. 1980), Phi Beta Kappa, Sigma Xi. Subspecialty: Catalysis chemistry. Current work: Homogeneous catalytic reactions with transition metal complexes. Patentee in field. Home: 3763 Middleton Ave Cincinnati OH 45220

ORDER, STANLEY ELIAS, radiation oncologist, educator; b. Vienna, Austria, Nov. 1, 1934; s. Albert A. and Bertha O.; m. Mary Singer, 1958; children: Jeffrey, Paul, Leanne. B.S., Albright Coll., Reading, Pa., 1956; M.D., Tufts U., 1961, ; Sc.D., Albright Coll., 1978, Elizabeth Coll., 1980. Diplomate: Am. Bd. Therapeutic Radiology. Instr. radiation therapy Harvard U., 1969, asst. prof., 1970-73, assoc. prof., 1973-75; dir. radiation oncology, prof. oncology and radiation scis. Johns Hopkins U. Hosp., 1975—; prof. div. allied health Essex Community Coll., Baltimore County, 1980—; cons. staff Ch. Hosp. Corp., Balt. Contbr. articles to profl. jours. Served to maj. U.S. Army, 1963-65. Named Willard and Lillian Hackerman prof., 1981; Pfahler orator Phila. Roentgen. Soc., 1980. Mem. Am. Soc. Therapeutic Radiologists, Am. Soc. Clin. Oncology, Am. Assn. Cancer Research, Md. Radiologic Soc., Am Acad. Sci. Subspecialties: Radiology;

Oncology. Current work: Isotopic I-131 Antiferritin IgG in exptl. and clin. hepatoma.

O'REILLY, JAMES EMIL, chemistry educator; b. Cleve., Jan. 14, 1945; s. James Emmit and Nella (Del Col) O'R.; m. Carol Anne Doehner, Aug. 10, 1968; children: J. Kevin, M. Shannon. B.S., U. Notre Dame, 1967; Ph.D., U. Mich., 1971. Postdoctoral assoc. U. Ill., Urbana, 1971-73; asst. prof. U. Ky., Lexington, 1973-79; vis. scientist FDA, Washington, 1980-81; assoc. prof. chemistry U. Ky., Lexington, 1979—. Editor: Instrumental Analysis, 1978; contbr. articles to profl. jours. Referee Central Ky. Soccer Ofcls. Assn., Lexington, 1980—, Lexington Youth Soccer Assn., 1978—. Grantee Petroleum Research Fund-Am. Chem. Soc., 1973, Ross Labs., 1977; NSF doctoral fellow, 1972; NDEA fellow, 1967-70. Mem. Am. Chem. Soc. (sec.-treas. sect. 1981—), Ky. Acad. Sci., Sigma Xi, Phi Kappa Phi. Roman Catholic. Subspecialty: Analytical chemistry. Current work: Analytical methods development, atomic absorption/emission spectrometry, gas and liquid chromatography, electrochemistry, ion-selective electrodes. Home: 369 Henry Clay Blvd Lexington KY 40502 Office: Dept Chemistry U Ky Lexington KY 40506

OREL, ANN ELIZABETH, physicist; b. Lake Charles, La., Oct. 26, 1955; d. Bernard Anthony and Bernice Josephine (Toplikar) O. B.S. cum laude, Calif. Inst. Tech., 1977; Ph.D., U. Calif.-Berkeley, 1980. Staff scientist Lawrence Livermore Lab., Calif., 1980—. Contbr. articless to profl. jours. Mem. Am. Phys. Soc., Sigma Xi, Iota Sigma Pi (A.L. Hoffman award 1980). Democrat. Roman Catholic. Club: Vaqueros del Mar. Subspecialty: Atomic and molecular physics. Current work: Low energy atomic and molecular physics; chemical physics. Office: Lawrence Livermore Nat Lab PO Box 808 L-486 Livermore CA 94550

ORGEL, LESLIE ELEAZER, chemist; b. London, Jan. 12, 1927; U.S., 1964; s. Simon and Deborah (Gnivish) O.; m. Hassia Alice Levinson, July 30, 1950; children—Vivienne, Richard, Robert. B.A., Oxford (Eng.) U., 1948, D.Phil., 1951. Reader univ. chemistry lab. Cambridge (Eng.) U., 1963-64; fellow Peterhouse Coll., 1957-64; sr. fellow Salk Inst. Biol. Studies, La Jolla, Calif., 1964—; adj. prof. U. Calif., San Diego, 1964—. Author: An Introduction to Transition-Metal Chemistry, Ligand-Field Theory, 1960, The Origins of Life: Molecules and Natural Selection, 1973; co-author: The Origins of Life on the Earth, 1974. Recipient Harrison prize Chem. Soc. London, 1957. Fellow Royal Soc. Subspecialty: Prebiotic chemistry. Home: 6102 Terryhill Dr La Jolla CA 92037 Office: Salk Inst PO Box 85800 San Diego CA 92138

ORKAND, RICHARD KENNETH, physiologist, educator; b. N.Y.C., Apr. 23, 1936; s. Sidney and Anna (Jaffee) O.; m. Paula Marie Makinen, Dec. 28, 1960; 1 son, Adam Reino. B.S., Columbia U., 1956; Ph.D., U. Utah, 1961; M.S. (hon.), U. Pa., 1975. Hon. research asst. Univ. Coll. London, 1961-64; USPHS spl. fellow Harvard U. Med. Sch., Boston, 1964-66; asst. prof. physiology U. Utah, Salt Lake City, 1966-68; from assoc. prof. to prof. biology UCLA, 1968-74; vis. prof. U. Geneva, 1981-82; Humboldt sr. U.S. scientist U. Heidelberg, Ger., 1982-83; prof. physiology U. Pa., Phila., 1974—. Co-author: Introduction to Nervous Systems, 1977. Pres. Larsage Neighbors, 1978-79; mem. Democratic Ward Com., 1979-81. USPHS fellow, 1956-61, 61-64, 64-68; grantee, 1968-82. Mem. Am. Physiol. Soc., Soc. Neurosci., Am. Soc. Pharmacology and Exptl. Therapeutics, Internat. Brain Research Orgn., AAAS, Physiol. Soc. (assoc.), Greater Phila. Com. Med. Pharm. Scis. Subspecialties: Neurophysiology; Neuropharmacology. Current work: Physiology of neuroglia and neuron-glia interactions, physiology and pharmacology of synaptic transmission. Office: 4001 Spruce St Philadelphia PA 19104

ORKIN, STUART H., pediatrician, molecular biologist, educator; b. N.Y.C., Apr. 23, 1946; s. Lazarus A. and Sylvia (Holl) O.; m. Roslyn W., Aug. 15, 1970; 1 dau., Jane. B.S., MIT, 1967; M.D., Harvard U., 1972. Research assoc. NIH, 1973-75; intern Children's Hosp., Boston, 1972-73, resident, 1975-76, fellow in hematology, 1976-78; asst. prof. pediatrics Harvard U., 1978-81, assoc. prof., 1982—. Research numerous publs. in field. Served with USPHS, 1973-75. Recipient Basil O'Connor award March of Dimes, 1978-80, Career Devel. award NIH, 1979—. Mem. Am. Soc. Hematology, Am. Soc. Clin. Investigation, Soc. Pediatric Research. Subspecialties: Genetics and genetic engineering (medicine); Hematology. Current work: Molecular genetics of human diseases: hemoglobin disorders, prenatal diagnosis of genetic disease. Office: 300 Longwood Ave Boston MA 02115

ORLAND, FRANK JAY, oral microbiology educator, historiographer; b. Little Falls, N.Y., Jan. 23, 1917; s. Michael and Rose (Dorner) O.; m. Phyllis Therese Mrazek, May 8, 1943; children: Frank R., Carl, June, Ralph. A.B., U. Chgo., 1937, S.M., 1945, Ph.D., 1949; B.S., U. Ill.-Chgo., 1939, D.D.S., 1941. Diplomate: Am. Bd. Microbiology. Intern U. Chgo. Hosps. and Clinics, 1941-42; clin. fellow Zoller Meml. Dental Clinic, 1942-49, from instr. to prof. dental surgery and microbiology, 1949-58, dir., 1954-66; prof. Fishbein Ctr. for Study of History of Sci. and Medicine, 1976—, prof. oral microbiology and dental surgery, 1958—; mem. commn. on research Fedn. Dentaire Internationale, 1976-81. Author: The First Fifty Year History of the International Association for Dental Research, 1973; editor: Jour. Dental Research, 1958-69; editor, contbr.: Microbiology in Clinical Dentistry, 1982; chief advisor: Medical History of Dentistry, 1971-74; editor: Centennial Brochure, Loyola U. Sch. Dentistry, 1983. Chmn. Candidates Forum, Village Election, Forest Park, 1979, 83; pres. Hist. Soc. Forest Park, 1976—; chmn. exec. com. Forest Park Centenary Celebration, 1981-85; chmn. hist. com. Centennial Loyola U. Sch. Dentistry, 1982-83; chmn. heritage com. Bicentennial Commn., Forest Park, 1975-76; mem. adv. humanities council Triton Coll. Recipient numerous award including Hon. Alumnus award Loyola U. Sch. Dentistry, 1983. Fellow Inst. Medicine (chmn. com. pub. info.), AAAS (chmn. sect. Chgo.); mem. Internat. Assn. for Dental Research (pres. 1971-72), Am. Acad. History of Dentistry (pres. 1976-77, Hayden-Harris award 1980), Soc. Med. History (pres. Chgo. 1980-82, mem. council), Ill. State Dental Soc. (chmn. com. on history 1977—), Sigma Xi, Gamma Alpha (pres. Chgo. chpt. 1975—). Club: Chgo. Lit. Subspecialties: Microbiology; Oral biology. Current work: Historical writings of research in field of oral microbiology and dental sciences, conducting research seminars, providing lectures and demonstrations for graduate students in oral biology at different universities. Home: 519 Jackson Blvd Forest Park IL 60130 Office: U Chgo 950 E 59th St Chicago IL 60637

ORLANS, FLORA BARBARA, physiologist, health science administrator; b. Birmingham, Eng., Jan. 14, 1928; d. Christopher and Flora Christine (Brookes) Hughes; m. Herbert C. Morton, June 19, 1982; children: by previous marriage; Andrew, Nicholas. B.Sc., Birmingham U., (Eng.), 1949; M. Sc., London U., 1954, Ph.D., 1956. Sr. staff scientist George Washington U., Med. Ctr., Washington, 1973-74; exec. sec. Nat. Heart, Lung, and Blood Adv. Council, NIH, Bethesda, Md., 1977-79; health scientist adminstr. Nat. Heart, Lung, and Blood Inst., 1979—; pres. Scientists Ctr. for Animal Welfare, Washington, 1978—. Author: Animal Care: From Protozoa to Small Mammals, 1977; editor: (with W. Jean Dodds) Scientific Perspectives on Animal Welfare, 1982; contbr. articles to profl. jours. Mem. Am. Soc. Pharmacology and Exptl. Therapeutics. Democrat. Subspecialty: Psychophysiology. Current work: Coronary heart disease; animal welfare. Home: 7106 Laverock Ln Bethesda Md 20817 Office: NHLBI Fed Bldg 310 NIH Bethesda MD 20205

ORLIK-RÜCKEMANN, KAZIMIERZ JERZY, aeronautical engineer; b. Warsaw, Poland, May 20, 1925; emigrated to Can., 1955; m. Krystyna Lgocka; children: George, Andrew. M.S., Royal Inst. Tech., Stockholm, Sweden, 1947, Tech. Lic., 1968, Tech.D., 1970. Profl. engr., Ont. Aero. scientist Royal Inst. Tech., Stockholm, 1947-51; sr. aero. scientist Aero. Research Inst., Stockholm, 1951-55; research officer NRC Can., Ottawa, Ont., 1955-58, head high speed aero. lab., 1958-62, head unsteady aero. lab., 1963—; chmn. fluid dynamics panel, adv. group aero. research and devel. NATO, 1979-81; docent Royal Inst. Tech., Stockholm, 1971—; pres. Supersonic Tunnel Assn., 1963-64; cons. orgns. Can., U.S., Sweden. Contbr. writings to profl. publs. Fellow Can. Aeros. and Space Inst. (assoc. editor transactions 1969-75), AIAA. Subspecialties: Aeronautical engineering; Fluid mechanics. Current work: Stability of aircraft, dynamic wind tunnel experiments, flight mechanics, applied aerodynamics. Office: NRC Montreal Rd Ottawa ON Canada K1A 0R6

ORLOFF, JACK, physician, research physiologist; b. Bklyn., Dec. 22, 1921; s. Samuel and Rebecca (Kaplan) O.; m. Martha Vaughan, Aug. 4, 1951; children: Jonathan Michael, David Geoffrey, Gregory Joshua; 1 child by previous marriage, Lee Frances. Student, Columbia U., 1937-38, Harvard U., 1938-40; M.D., N.Y. U., 1943. Intern Mt. Sinai Hosp., N.Y.C., 1944; resident medicine Montefiore Hosp., N.Y.C., 1944-46; research fellow Yale Med. Sch., 1948-50; mem. sr. research staff Nat. Heart Inst., Bethesda, Md., 1950-57, dep. chief lab. kidney and electrolyte metabolism, 1957-62, chief lab., 1962-75; dir. intramural research Nat. Heart, Lung and Blood Inst., 1974—; professorial lectr. physiology, med. Georgetown U. Med. Sch., 1962—; disting. alumni lectr. N.Y. U. Sch. Medicine, 1966; Del.-at- large exec. council Fedn. Am. Sci., 1962-64, exec. com., treas., 1963-64. Author: Essays in Metabolism, 1957, Metabolic Disturbances in Clinical Medicine, 1958, The Metabolic Basis of Inherited Disease, 1972, Heart, Kidney and Electrolytes, 1962, Diseases of the Kidney, 1971, Cellular Function of Metabolic Transport, 1964, Hormones and the Kidney, 1964, Nobel Symposium on Prostaglandins, 1967, Physioli of Diuretic Agents, 1966, Ocytocin, Vasopressin and their Structural Analogues, 1964; Editorial bd.: Am. Jour. Physiology; sect. editor, 1964-68, Jour. Applied Physiology, 1964-68, Renal Physiology sect. 8, Handbook of Physiology, 1973; asso. editor: Kidney Internat; cons. editor: Life Scis, 1973-78. Trustee Greenacres Sch., Rockville, Md., 1960-64, v.p. bd. trustees, 1962-63; Mem. sci. adv. bd. Nat. Kidney Found., 1962-71; mem. Inst. Medicine Nat. Acad. Scis. Served to capt. M.C. AUS, 1946-48. Recipient Homer Smith award N.Y. Heart Assn., 1973, Meritorious Service award USPHS, 1974, Distinguished Alumni Achievement award in basic sci. N.Y. U. Sch. Medicine, 1976; Distinguished Service medal HEW, 1977. Mem. Am. So. Clin. Investigation, Am. Physiol. Soc., Fedn. Am. Scientists (vice-chmn. 1963-64), Assn. Am. Physicians, Am. Soc. Nephrology (sec.-treas. 1970-72, pres. 1973-74), Sigma Xi, Alpha Omega Alpha. Club: Cosmos. Subspecialties: Internal medicine; Nephrology. Current work: Director intramural research. Home: 11608 W Hill Dr Rockville MD 20852 Office: Nat Heart Lung and Blood Inst Bethesda MD 20205

ORLOFF, MARSHALL J., surgeon, educator; b. Chgo., Oct. 12, 1927; s. Leo and Minnie Lea (Schnitzer) O.; m. Ann Stuart, May 7, 1953; children: Mark, Bruce, Eric, Susan, Lisa, Karen. B.S., U. Ill., 1949; M.S. in Pharmacology, 1951; M.D., 1951; D.M.Sc. (hon.), Chung Ang U., Korea, 1974. Diplomate: Am. Bd. Surgery, Am. Bd. Thoracic Surgery. Asst. instr. pharmacology U. Ill., 1949-51; dir. research in convulsive disorders Manteno State Hosp., 1949-51; assoc. epilepsy clinic U. Ill., 1950-51; asst. instr. surgery U. Pa., 1952-53, 55-57; instr. surgery, 1957-58; instr. U. Colo., 1958-59, assoc. prof. surgery, 1959-61, lectr. pharmacology, 1960-61; asst. chief surgery Denver Gen. Hosp., 1958-60, Denver VA Hosp., 1960-61; head gastrointestinal surgery Colo. Gen. Hosp., 1960-61; Markle scholar in med. sci. U. Colo., 1959-61; Markle scholar in acad. medicine UCLA Sch. Medicine, 1961-64, prof. surgery, 1961-67; chief surgery Harbor Gen. Hosp., 1961-67, dir. surg. research lab., 1961-67; cons. in surgery Wadsworth VA Hosp., 1962-67; prof., dept. surgery U. Calif.-San Diego, 1965—, chmn. dept. surgery, 1967-81; chief surg. services U. Calif. Med. Ctr., 1967-81; cons. in surgery VA Hosp., San Diego, 1971-78; mem. surgery tng. com. NIH, 1966-68, mem. surgery study sect., 1968-72, chmn. ad hoc study sects. on trauma research ctr. grants, 1970-73. Author numerous books in field; editor-in-chief: World Jour. Surgery, 1976-81; assoc. editor: Surgery, 1971—; mem. editorial bd.: Jour. Surg. Oncology, 1969—, Surgery in the 70s, 1973—, Current Surg. Techniques, 1973—; mem. editorial adv. bd.: Indian Jour. Cancer, 1977—; chmn. editorial bd.: Societe Internationale de Chirugie, 1974-81; contbr. numerous chpts. to books, articles to profl. jours. Bd. visitors U. San Diego Sch. Law, 1976-82. Served with M.C. U.S. Army, 1953-55. Recipient Outstanding Alumnus award U. Ill., 1978, Kaiser award for excellence in teaching U. Calif.-San Diego, 1979, 82; established in his honor Marshall J. Orloff Soc. Fellow ACP (pres. San Diego chpt. 1973-74, numerous coms.); mem. Am. Surg. Assn., Soc. Univ. Surgeons (pres. 1971-72), Am. Fedn. Clin. Research, Calif. Med. Assn. (chmn. sect. surgery 1971-72), Soc. Clin. Surgery, Pacific Coast Surg. Assn. (chmn. com. sci. sessions 1971-72), Soc. Gen. Surgeons San Diego (council 1970-73), Internat. Microsurgery Soc. (pres. 1972), Internat. Microsurg. Soc. (pres. 1981-83), Sigma Xi, Alpha Omega Alpha, numerous others profl. assns. Subspecialties: Surgery; Transplant surgery. Current work: Interests are in the fields of liver physiology and disease, portal hypertension, alcoholism, gastrointestinal physiology and disease, biliary disease, diabetes, organ transplantation and transplantation immunology, cancer, neuropharmacology, endocrinology, metabolism, microsurgery, medical education and health care. Home: 8861 Kilbourn Dr La Jolla CA 92037 Office: Dept Surgery U Calif San Diego 225 Dickinson St San Diego CA 92103

ORME-JOHNSON, WILLIAM H., chemistry educator; b. Phoenix, Apr. 23, 1938. B.Sc., U. Tex., Austin, 1959, Ph.D. in Chemistry, 1964. Fellow Biochem. Inst., U. Tex., 1964-65; fellow Inst. Enzyme Research, U. Wis., 1965-67, asst. prof. enzyme chemistry, 1967-70, from asst. prof. to prof. biochemistry, 1970-79; prof. chemistry MIT, Cambridge, 1979—. Fellow AAAS; mem. Am. Soc. Biol. Chemists, Am. Chem. Soc., Am. Soc. Microbiologists. Subspecialty: Biochemistry (biology). Office: Dept Chemistry Mass Inst Tech Cambridge MA 02139

ORMES, JONATHAN F., physicist, researcher; b. Colorado Springs, Colo., July 18, 1939; s. Robert Manly and Suzanne (Viertel) O.; m. Karen Lee Minnick; children: Laurie Kylee; m. Janet Carolyn Dahl, Sept. 12, 1964; children: Marina Elizabeth, Nicholas Stuart. B.S. in Physics, Stanford U., 1961, M.S., U. Minn., 1966, Ph.D., 1967. Teaching asst. U. Minn., 1961-62, research asst., 1962-67; postdoctoral research assoc. Nat. Acad. Sci., 1967-69; with NASA/Goddard Space Flight Center, 1969—, head high energy cosmic ray studies group, 1976—, head, 1982—; prin. investigator High Energy Astrophysics Obs., NASA, 1971, 1979. Contbr. articles to profl. jours. Tchr. re-eval. counseling. Mem. Am. Geophys. Union, Am. Assn. Physics Tchrs., AAAS, Am. Phys. Soc., Am Astron. Soc. Unitarian. Subspecialties: High energy astrophysics; Cosmic ray high energy astrophysics. Current work: Develop hypotheses and design and implement expts. to test them concerning sites and mechanisms for origin, acceleration and propagation of cosmic ray nuclei in the galaxy. Home: 4519 Greenwood Rd Beltsville MD 20705 Office: NASA Goddard Space Flight Center Code 661 Greenbelt MD 20771

ORNSTEIN, DONALD SAMUEL, mathematician; b. N.Y.C., July 30, 1934; s. Harry and Rose (Wisner) O.; m. Shari Richman, Dec. 20, 1964; children—David, Kara, Ethan. Student, Swarthmore Coll., 1952; Ph.D., U. Chgo., 1956. Fellow Inst. for Advanced Studies, Princeton, N.J., 1955-57; faculty U. Wis., Madison, 1957-58, 58-59, Stanford (Calif.) U., 1959—, prof. math., 1966—; faculty Hebrew U., Jerusalem, 1975-76. Author: Ergodic Theory Randomness and Dynamical Systems, 1974. Recipient Bocher prize Am. Math. Soc., 1974. Mem. Nat. Acad. Sci. Jewish. Office: Dept Math Stanford U Stanford CA 94305

ORO, JUAN, biochemist, educator; b. Lerida, Spain, Oct. 26, 1923; came to U.S., 1952; s. Juan and Maria Florensa (Rue) Oro-V.; m. Francisca Forteza, May 19, 1948; children: Maria Elena, Juan, Jaime, David. Licenciate in Chem. Scis, U. Barcelona, Spain, 1947; Ph.D., Baylor U., 1956; Dr. honoris causa, U. Granada, Spain, 1972. Mem. faculty U. Houston, 1955—, asst. prof., 1956-58, asso. prof., 1958-63, prof., 1963—, chmn. dept. biophysical scis., 1967-69; research chemist Lawrence Radiation Lab., U. Cal. at Berkeley, summer 1962; Catedratico Universidad Autonoma de Barcelona, 1971; prin. investigator NASA Lunar Sample Analysis, 1967—; molecular analysis team mem. NASA Unmanned Mars Landing, 1970—; dir. Instituto de Biofisica y Neurobiologia, Barcelona, Spain, 1977; Founding mem., v.p. Spain and Tex. Soc., Inc., Houston, 1970—; mem. Parliament of Catalunya (Spain), 1980-81; pres. Sci. and Technol. Council Catalunya, 1980-81; pres. Sci. Council Inst. Fundamental Biology, Universidad Autonoma de Barcelona, 1981; mem. space sci. bd. NRC-Nat. Acad. Scis. U.S., 1980-83; mem. Jurado Premio Principe de Asturias de Investigacion Cientifica y Tecnica, Fundacion Principado de Asturias. Editor: Biogenesis sect. Jour. Molecular Evolution, 1972, (with others) Prebiotic and Biochemical Evolution, 1971, Viral Replication and Cancer, 1973, Cosmochemical Evolution and the Origins of Life, 1974, Reflections on Biochemistry, 1976, Avances de la Bioquimica, 1977, The Viking Mission and the Question of Life on Mars, 1979, Virus and Cancer, 1971, Energy Sources and Development, 1978, Catalunya Agricola, 1978, Planetes Comparats, 1980; contbr. articles to profl. publs. Recipient medal and hon. councilor Consejo Superior de Investigaciones Cientificas, 1969; Gran Cruz de la Orden Civil de Alfonso X el Sabio, Madrid, 1974; Life Scientist award NASA, 1974-75; Gold medal City of Lerida, 1976; medal Narcis Monturiol al Merit Cientific y Tecnologic Catalunya, Spain, 1982; Gran Cruz de la Orden del Merito Aeronautico, Madrid, 1983; named Important Scientist of Year, Barcelona, 1970, Madrid, 1977. Mem. Am. Soc. Biol. Chemists, Am. Chem. Soc., Nat. Assn. Chemists Spain, Spanish Soc. Biochemists, Royal Acad. Arts and Scis. of Barcelona, Inst. Hispanic Culture (founding mem.), Internat. Soc. for Study Origin of Life (exec. council 1974), Asociacion de Amigos de Gaspar de Portola (pres. 1981—), Spanish Soc. Biochemistry (hon.), Sigma Xi. Subspecialties: Biochemistry (biology); Organic chemistry. Current work: Synthesis of biochemical compounds under plausible primitive earth conditions; planetary exploration; studies on possibility of extra-terrestrial life. Home: 11306 Endicott Ln Houston TX 77035 Office: Dept Biochem and Biophysical Scis SRI U Houston Houston TX 77004

OROSS, JOHN WILLIAM, botanist; b. Toledo, Ohio, Feb. 28, 1949; s. Stephen and Bernita Luella (McGill) O. B.S. Ed., Miami U., Oxford, Ohio, 1971, M.S., 1977; Ph.D., U. Calif., Riverside, 1983. Postdoctoral research assoc. dept. botany U. Calif., Davis, 1983—. Contbr. articles to profl. jours. Mem. Am. Bot. Soc., Am. Soc. Plant Physiologists, Sigma Xi. Democrat. Subspecialties: Cell biology; Plant physiology (biology). Current work: Structure-function relationships in secretory structures. Salt glands, transport, ultrastructure, electron microprobe analysis, cytochemistry. Office: Dept Botany U Calif Davis CA 92616

ORR, J. RICHIE, physicist; b. Waukon, Iowa, Jan. 16, 1933; s. Lester Duncan and Carolyn (Dayton) O.; m. Suzanne Alice, June 20, 1961. B.S. in Physics with honors, U. Iowa, 1955; Ph.D., U. Wash., 1965. Research engr. Boeing Aircraft Co., Seattle, 1956-60; research asst. U. Wash., 1960-65; research asst. prof. U. Ill., 1966-70; with Fermilab, Batavia, Ill., 1972—; now head accelerator div. and project mgr. Energy Saver. Contbr. articles to sci. jours. Mem. Am. Phys. Soc., Phi Beta Kappa, Sigma Xi. Subspecialty: Particle physics. Current work: Weak interaction physics; electromagnetic interactions.

ORSZAG, STEVEN ALAN, applied mathematician; b. N.Y.C., Feb. 27, 1943; s. Joseph and Rose (Siegel) O.; m. Reba Karp, June 21, 1964; children—J. Michael, Peter Richard, Jonathan Marc. B.S., M.I.T., 1962; postgrad. (Henry fellow), St. John's Coll., Cambridge (Eng.) U., 1962-63; Ph.D., Princeton U., 1966. Mem. Inst. Advanced Study, Princeton, N.J., 1966-67; asst. prof. math. M.I.T., 1967-70, asso. prof., 1970-75, prof. applied math., 1975—; pres. CHI Inc., 1975—; cons. in field. Author: Studies in Applied Mathematics, 1976, Numerical Analysis of Spectral Methods, 1977, Advanced Mathematical Methods for Scientists and Engineers, 1978; editor: Springer Series in Computational Physics, 1977—, Springer Lecture Notes in Engineering, 1981—; editorial bd.: Jour. Computational Physics, 1977—, Studies in Applied Math, 1975—, Ency. Math, 1975—, Computers & Mathematics, 1975—; research numerous publs. in field. A.P. Sloan Found. fellow, 1970-74. Mem. Am. Inst. Physics, AIAA, Soc. Indsl. and Applied Math. Subspecialties: Fluid mechanics; Numerical analysis. Office: Dept Math MIT Cambridge MA 02139

ORTHWEIN, WILLIAM COE, mech. engr.; b. Toledo, Ohio, Jan. 27, 1924; s. William Edward and Millie Minerva (Coe) O.; m. Helen Virginia Poindexter, Feb; children—Karla Frances, Adele Diana, Maria Theresa. B.S., M.I.T., 1946; M.S., U. Mich., 1957, Ph.D., 1959. Registered profl. engr., Ill., Ind., Ky. Aerophysicist Gen. Dynamics Co., Ft. Worth, 1951-52; research asso. U. Mich., 1952-59; adv. engr. IBM Corp., Owego, N.Y., 1959-61; dir. computer centers U. Okla., Norman, 1961-63; research scientist Ames Lab., NASA, Moffett Field, Calif., 1963-65; mem. faculty So. Ill. U., Carbondale, 1965—, prof. engring., 1967—; cons. in field. Author papers, revs. in field. Pres. Jackson County (Ill.) Taxpayers Assn., 1976. Served with AUS, 1943-46. Mem. ASME (Outstanding Service award 1972), Soc. Exptl. Stress Analysis, Tensor Soc., Soc. Mining Engrs., Am. Gear Mfrs. Assn., Am. Acad. Mechanics, Ill. Soc. Profl. Engrs. (chmn. salary and employment com. 1974, chmn. ad hoc com. continuing edn. 1975), Nat. Rifle Assn., Aircraft Owners and Pilots Assn., Sigma Xi. Mormon. Subspecialties: Mechanical engineering; Computer-aided design. Current work: Machine design, theoretical stress analysis, computer-aided engineering and computer-aided product performance testing. Home: PO Box 3332 Carbondale IL 62901 Office: So Ill Univ Carbondale IL 62901 Success in engineering is, I believe, contingent upon one's ability to see the world as it really is, to quickly gain insight enough to detect fundamental parameters that determine behavior of the system in question, to conduct a straightforward check of one's analysis, and to simply synthesize a means of modifying and/or controlling the parameters to obtain the desired results. These ingredients apply to both physical mechanisms and to human organizations—only the means of implementation differ.

ORTIZ-SUAREZ, HUMBERTO JOSE, neurosurgeon; b. Santurce, P.R., Oct. 29, 1941; s. Humberto Ortiz and Antonia Suarez; m. Conchita Z. Ortiz, Dec. 21, 1964; children: Humberto, Elena. B.S., U. P.R., 1961, M.D., 1965; Ph.D., U. Minn., 1974. Intern Univ. Hosp. P.R. Med. Ctr., 1965-66; resident U. Tex., San Antonio, 1968-69; resident in neurosurgery U. Minn. Hosps., Mpls., 1969-74; asst. prof. neurosurgery U. P.R. Med. Sch., San Juan, 1974-77, assoc. prof., 1977—; assoc. attending in neurol. surgery Univ. Dist. Hosp., 1977—, San Juan City Hosp., 1977—; attending in neurol. surgery VA Hosp., San Juan, 1982—. Contbr. articles to profl. jours. Bd. dirs. P.R. Soc. Epilepsy, 1975-78; mem. exec. com. Univ. Hosp., 1978-80. Served to capt. M.C. AUS, 1966-68; Vietnam. Decorated Air medal, Combat Med. badge, Bronze Star medal. Mem. AMA (Physicians Recognition award 1972, 76), P.R. Assn. Neurol. Surgeons (sec. 1980-82), Am. Assn. Neurol. Surgeons, P.R. Med. Assn., Soc. Neurosci., World Fedn. Neurol. Surgeons, Caribbean Assn. Neurol. Surgeons, ACS, Congress Neurol. Surgeons, Neurosurg. Soc. Am., Alpha Omega Alpha. Roman Catholic. Subspecialties: Neurosurgery; Enzyme technology. Current work: Central nervous system trauma, vascular diseases of the nervous system, computerized tomography. Office: Neurological Surgery Sect Med Scis Campus GPO Box 5067 San Juan PR 00936

ORTON, COLIN GEORGE, med. physicist; b. London, June 4, 1938; U.S., 1966, naturalized, 1981; s. Frederick G. and Audrey V. (Sewell) O.; m. Barbara Orton, July 25, 1964; children: Nigel, Susanne, Philip. B.Sc. with honors in physics, Bristol U., 1959; M.Sc., London U., 1961, Ph.D., 1965. Instr. St. Bart's Hosp. Med. Coll., London, 1961-66; chief physicist N.Y.U. Med. Center, 1966-75; R.I. Hosp., 1975-81; assoc. prof. Brown U., Providence, 1975-81; chief physicist Harper-Grace Hosps., Detroit, 1981—; prof. radiation oncology/radiology Wayne State U., Detroit, 1981—. Contbr. articles to profl. jours.; editor: Bull. Am. Assn. Physicists in Medicine, 1971-73, Radiological Physics Examination Review Book, Vol. 1, 1971, Vol II, 1978, Practical Aspects of Electron Beam Treatment Planning, 1978, Progress in Medical Radiation Physics, Vol. I, 1982. Inst. Physics London fellow, 1976—. Mem. Am. Assn. Physicists in Medicine (pres. 1981), Am. Inst. Physics, Brit. Inst. Radiology, Inst. Physics London, Health Physics Soc. Subspecialties: Radiology; Medical physics. Current work: Research in radiotherpay physics, radiation biology, med. physics, radiation dosimetry.

ORWOLL, ROBERT ARVID, chemist, educator; b. Mpls., Aug. 28, 1940; s. Arvid L. and Agnes (Christiansen) O.; m. Betty M., Feb. 24, 1972; children: Katherine S., Karen E. B.A., St. Olaf Coll., 1962; Ph.D., Stanford U., 1966. Research instr. Dartmouth Coll, Hanover, N.H., 1966-68; research fellow U. Conn., Storrs, 1968-69; asst. prof. chemistry Coll. William and Mary, Williamsburg, 1969-72, assoc. prof., 1972-82, prof., 1982—. Mem. Am. Chem. Soc., AAAS, Va. Acad. Sci. Methodist. Subspecialties: Physical chemistry; Polymer chemistry. Current work: Liquid crystal thermodynamics, polymer solution thermodynamics. Office: Dept Chemistry William and Mary Coll Williamsburg VA 23185

ORZACK, MARESSA HECHT, psychology educator, researcher, clin. behavior therapist; b. Boston; d. Selig and Cecilia (Huchschman) Hecht; m. Louis H. Orzack, Sept. 4, 1947; children: Deborah, Steven, Elizabeth. B.A., U. Rochester, 1945, M.A., 1947; Ph.D., Columbia U., 1951. Lic. psychologist, Mass. Research psychologist Central Wis. Colony, Madison, 1959-61; research assoc. Boston U. Med. Sch., 1961-67, asst. to assoc. prof., 1967-79, adj. assoc. prof., 1979—; dir. drug abuse liability project McLean Hosp., Harvard U. Med. Sch., Boston, 1980—; clk., dir. Pub. Responsibility in Medicine Research, Boston, 1976—; mem. instnl. rev. bd. Boston U. Med. Sch., 1977-80, Edith Nourse Rogers VA Authority, Bedford, Mass., 1982—. Contbr. numerous articles to profl. jours. Fellow Mass. Psychol. Assn.; mem. Am. Coll. Neuropsychopharmacology, Am. Psychol. Assn., LWV (dir. 1967-71). Subspecialties: Psychopharmacology; Neuropsychology. Current work: Study of mood changes in recreational drug users to assess potential abuse liability of psychotrpic drugs. Office: McLean Hosp 115 Mill St Belmont MA 02178

OSBORN, ELBURT FRANKLIN, research scientist; b. Winnebago County, Ill., Aug. 13, 1911; s. William Franklin and Anna (Sherman) O.; m. Jean Thomson, Aug. 12, 1939; children: James Franklin, Ian Charles. B.A., DePauw U., 1932; M.S., Northwestern U., 1934; Ph.D., Calif. Inst. Tech., 1938. Teaching fellow geology Northwestern U., 1932-34, instr. geology, 1937; teaching fellow Calif. Inst. Tech., 1934-37; geologist Que-on-Gold Mines, Ltd., Que., 1936, Val d'Or, 1938; petrologist geophys. lab. Carnegie Instn., Washington, 1938-42; phys. chemist div. I Nat. Def. Research Com., 1942-45; research chemist Eastman Kodak Co., Rochester, N.Y., 1945-46; prof. geochemistry Pa. State U., 1946-70, chmn. div. earth sci., 1946-52, asso. dean coll., 1952-53, dean, 1953-59, v.p. research, 1959-70, prof. emeritus geochemistry, v.p. research emeritus, 1971—; dir. U.S. Bur. Mines, 1970-73; distinguished prof. Carnegie Instn. Washington, 1973-77, emeritus, 1977—; sr. postdoctoral fellow Cambridge U., 1958; Com. adviser geophysics br. Office Naval Research, 1947-50; adviser panel mineral products div. Nat. Bur. Standards, 1958-62, chmn., 1958-59, adviser panel metallurgy div., 1958-64; adviser to com. basic research U.S. Army Research Office in Ceramics, 1960-63; mem. earth sci. panel NSF, 1953-55, div. com. math., phys. and engring. scis., 1955-59, chmn., 1957-58, adv. panel phys. scis. facilities, 1960-64; exec. com. earth scis. div. Nat. Acad. Sci., 1969-72; mem. materials adv. com. Office Tech. Assessment, U.S. Congress, 1975-80; chmn. adv. com. on mining and mineral resources research U.S. Dept. Interior, 1978-83; mem. geoscis. adv. panel Los Alamos Sci. Lab., Dept. Energy, U. Calif., Berkeley, 1975-78. Mem. bd. Univ. Corp. Atmospheric Research, Pa. Health Research Inst., Geisinger Med. Center. Recipient Tech. Meeting award Am. Iron and Steel Inst., 1954. Fellow Geol. Soc. Am. (council 1959-62), Mineral. Soc. Am. (pres. 1960-61, asso. editor Am. Mineralogist 1953-55, Roebling medal 1972), Geochem. Soc. (pres. 1967-68), A.A.A.S. (council), Am. Ceramic Soc. (pres. 1964, chmn. basic scis. div. 1951-52, tech. adv. com. Nat. Bur. Standards 1954-58, chmn. 1957-58, chmn. publ. com. 1960-63, 70-71, Jeppson award 1973, Bleininger award 1976), Am. Geophys. Union; mem. Am. Chem. Soc., Geol. Soc. Washington, Am. Inst. Mining, Metall. and Petroleum Engrs. (Hardinge award 1974), Soc. Econ. Geologists (v.p. 1965, pres. 1972-73), Am. Geol. Inst. (dir. 1956-59), Nat. Acad. Engring. (mem. engring. aspects environ., quality com.), NRC (chmn. mineral sci. and tech. com. 1966-69, mem. materials adv. bd. 1965-69, chmn. bd. mineral resources 1974-77, mem. commn. natural resources 1974-77, geophysics research bd. 1974-77, chmn. workshop on continental drilling for sci. purposes 1978, chmn. com. geol. aspects of indsl. waste disposal 1980-82), Internat. Union Geol. Sci. (nat. com. 1961-64, 73-77, chmn. 1975-77), Can. Ceramic Soc. (hon.), Internat. Assn. Volcanologists, Nat. Assn. State Univs. and Land Grant Colls. (chmn. com. mineral resources 1969, mem. water resources com. 1965-71, council research policy adminstrn. 1965-71), Phi Beta Kappa, Sigma Xi (nat. exec. com. 1961-64, com. nat. lectureships 1969-71), Phi Lambda Upsilon, Delta Tau Delta, Phi Kappa Phi, Keramos. Subspecialties: Geochemistry; Ceramics. Current work: Laborotory research on silicate systems at high temperatures and pressures; especially applicable to origins of volcanic images. Home: 330 E Irvin Ave State College PA 16801 Office: Deike Bldg Pa State U University Park PA 16802

OSBORN, JEFFREY LYNN, physiologist, researcher; b. Angola, Ind., Jan. 27, 1952; s. Richard Fern and Mary Ann (ahckett) O.; m. Debra Kay McLauchlin, June 14, 1975; children: Barrett Jeffrey, Melissa Claire. B.A., Amherst Coll., 1974; M.S., Mich State U., 1976, Ph.D., 1979. Postdoctoral fellow U. Iowa, Iowa City, 1979-81; asst. prof. Med. Coll. Wis., Milw., 1981—. NIH grantee, 1979; Kidney Found. Iowa fellow, 1979. Mem. Am. Physiol. Soc., Am. Soc. Nephrology, Internat. Soc. Nephrology, Am. Fedn. Soc. Clin. Research, Soc. Exptl. Biology and Medicine. Subspecialties: Physiology (medicine); Nephrology. Current work: Neural control of renal function, neural and endocrine regulation of cardiovascular function, reflex control of renal sympathetic nerve activity. Office: Med Coll Wis 8701 Watertown Plank Rd Milwaukee WI 53226

OSBORN, JOHN EDWARD, mathematics educator; b. Onamia, Minn., July 12, 1936; s. John Max and Edna Mary (Sehlin) O.; m. Janice Marian Matson, Aug. 22, 1959; children: Nancy, Kevin, Michael. B.S., U. Minn.-Mpls., 1958, M.S., 1963, Ph.D., 1965. Asst. prof. math. U. Md., 1965-69, assoc. prof., 1969-75, prof., 1975—, chmn. dept., 1982—. Contbr. articles to profl. jours. Mem. Am. Math. Soc., Soc. Indsl. and Applied Math. Subspecialty: Applied mathematics. Current work: Numerical analysis, partial differential equations. Home: 9612 Wire Ave Silver Spring MD 20901 Office: Dept Math U Md College Park MD 20742

OSBORN, MARY JANE MERTEN, biochemist; b. Colorado Springs, Colo., Sept. 24, 1927; d. Arthur John and Vivien Naomi (Morgan) Merten; m. Ralph Kenneth Osborn, Oct. 26, 1950. B.A., U. Calif., Berkeley, 1948; Ph.D., U. Wash., 1958. Postdoctoral fellow, dept. microbiology N.Y. U. Sch. Medicine, N.Y.C., 1959-61, instr., 1961-62, asst. prof., 1962-63; asst. prof. dept. molecular biology Albert Einstein Coll. Medicine, Bronx, N.Y., 1963-66, asso. prof., 1966-68; prof. dept. microbiology U. Conn. Health Center, Farmington, 1968—, dept. head, 1980—; mem. bd. sci. counselors Nat. Heart, Lung and Blood Inst., 1975-79; mem. Nat. Sci. Bd., 1980—. Asso. editor: Jour. Biol. Chemistry, 1978-80; contbr. articles in field of biochemistry and molecular biology to profl. jours. NIH fellow, 1959-61; NIH grantee, 1962—; NSF grantee, 1965-68; Am. Heart Assn. grantee, 1968-71. Fellow Am. Acad. Arts and Scis., Nat. Acad. Scis.; mem. Am. Chem. Soc. (chmn. div. biol. chemistry 1975-76), Am. Soc. Biol. Chemists (pres. 1981-82), Am. Soc. Microbiologists. Democrat. Subspecialty: Biochemistry (biology). Office: Dept Microbiology U Conn Health Center Farmington CT 06032

OSBORN, WAYNE HENRY, astronomer, physicist, educator; b. Los Angeles, Oct. 8, 1942; s. Wyman Henry and Margaret Lois (Nash) O.; m. Marie-Therese Zehnder, Aug. 7, 1971; children: Nathalie, Christine, Wayne Henry. A.B., U. Calif.-Berkeley, 1964; postgrad., U. Md., 1964-65; M.A., Wesleyan U., Middletown, Conn., 1966; Ph.D., Yale U., 1971. With Universidad de Los Andes, Merida, Venezuela, 1971-76, Venezuelan Consejo de Investigaciones Cientificas y Tecnicas, 1972-76; assoc. prof. dept. physics Central Mich. U., Mt. Pleasant, 1976—, chmn. dept., 1979—. Contbr. articles to profl. jours. Fellow Royal Astron. Soc.; mem. Internat. Astron. Union., Astron. Soc. Pacific (life), Am. Astron. Soc., Sigma Pi Sigma. Subspecialty: Optical astronomy. Current work: Variable stars, globular clusters, occultations. Office: Dept Physics Central Mich U 316 Brooks Hall Mount Pleasant MI 48859

OSBORNE, ADAM, computer company executive; b. Bangkok, Thailand, Feb. 6, 1939; m. Cynthia Osborne (div.); children: Ian, Paul, Alexandra; m. Barbara Ann Burdick, Dec. 4, 1982. B.S., U. Birmingham, Eng., 1961; Ph.D. in Chemistry, U. Del. With M.W. Kellogg Co., 1961-64, Shell Devel. Corp., 1967-70; prin. Osborne & Assocs., 1970-81; founder, pres. Osborne Computer Corp., Hayward, Calif., 1981-83, chmn. bd., 1983—. Author computer guides and handbooks. Subspecialty: Computer engineering. Office: 26538 Danti Ct Hayward CA 94545

O'SHAUGHNESSY, JAMES COLIN, environmental engineering educator; b. Cambridge, Mass., June 1, 1943; s. James Nicholas and Dorothy Ann (Dube) O'S.; m. Patricia L. Van Den Berghe, Sept. 21, 1963; children: Christine A., James P. B.S., U. N.H., 1965; M.S., Pa. State U., 1970, Ph.D, 1973. Registered profl. engr., Mass., N.H., Pa. Engr. Calif. Div. Water Resources, Los Angeles, 1965-66; engr. N.H. Dept. Hwys., Concord, 1966; instr. U.S. Army Engring. Sch., Ft. Belvoir, Va., 1966-68; research asst. Pa. State U., 1968-72; assoc. prof. dept. civil engring. Northeastern U., Boston, 1972—; dir. Environ. Engring. Labs., 1975—. Contbr. articles to profl. jours. Mem. ASCE, Water Pollution Control Fedn., Internat. Assn. Water Pollution Research, Boston Soc. Civil Engrs., Sigma Xi. Subspecialties: Environmental engineering; Water supply and wastewater treatment. Current work: Industrial waste treatment, toxic materials treatment and impact on water quality. Home: 25 Sweetwater Ave Bedford MA 01730 Office: Dept Civil Engring Northeastern U Boston MA 02115

O'SHEA, DONALD CHARLES, physicist, educator, optical cons., author; b. Akron, Ohio, Nov. 14, 1932; s. Donald Joseph and Sarah (Walsh) O'S.; m. Helen Spustek, Oct. 20, 1962; children: Kathleen, Sean, Sheila, Patrick. B.S. in Physics, U. Akron, 1960, M.S., Ohio State U., 1963, Ph.D., Johns Hopkins U., 1968. Research fellow McKay Labs., Harvard U., 1968-70; asst. prof. physics Ga. Inst. Tech., Atlanta, 1970-75, assoc. prof., 1975—; Prin. lectr. laser system design U. Wis. Extension, 1979—. Author: (with Callen and Rhodes) Introduction to Lasers and Their Applications, 1976, Elements of Modern Optical Design, 1983. Mem. Am. Phys. Soc., Optical Soc. Am., Am. Assn. Physics Tchrs., Sigma Xi. Subspecialties: Spectroscopy; Optical design. Current work: Laser light scattering spectroscopy; optics education; optical design. Office: Sch Physics Ga Inst Tech Atlanta GA 30332

OSHEROFF, DOUGLAS DEAN, physicist; b. Aberdeen, Wash., Aug. 1, 1945; s. William and Bessie Anne (Ondov) O.; m. Phyllis S.K. Liu, Aug. 14, 1970. B.S. in Physics, Calif. Inst. Tech., 1967; M.S., Cornell U., 1969, Ph.D., 1973. Mem. tech. staff Bell Labs., Murray Hill, N.J., 1972-82, head solid state and low temperature physics research dept., 1982—. Bd. dirs. Plainfield Sci. Edn. Coalition, 1982—. John D. & Catherin T. MacArthur prize fellow, 1981; recipient Simon Meml. Prize Brit. Inst. Physics, 1976, Oliver E. Buckley Solid State Physics prize, 1981. Fellow Am. Acad. Arts and Scis., Am. Phys. Soc. Office: 600 Mountain Ave Murray Hill NJ 07974

OSMER, PATRICK STEWART, astronomer; b. Jamestown, N.Y., Dec. 17, 1943; m., 1973; 2 children. B.S., Case Inst. Tech., 1965; Ph.D. in Astronomy, Calif. Inst. Tech., 1970. Research assoc. in astronomy Cerro Tololo Inter-Am. Obs., La Serena, Chile, 1969-70, asst. astronomer, 1970-73, assoc. astronomer, 1973-76, astronomer, 1977—, dir., 1981—. Mem. Am. Astron. Soc., Am. Astron. Soc. Pacific, Internat. Astron. Union. Subspecialty: Astrophysics. Office: Casilla 603 La Serena Chile

OSTAZESKI, STANLEY ANTHONY, plant pathologist; b. Superior, Wis., Apr. 17, 1926; s. Stanley Anthony and Annette Marie (Majeski) O.; m. Marjorie Gae Cross, Aug. 12, 1950; children: Stanley Anthony III, Theresa M., Paul A., Edward W., Lisa K. B.S., Wis. State U., Superior, 1952; M.S., U. Ill., 1955, Ph.D., 1957. Research plant pathologist U. S. Dept. Agr., various locations, 1957-62, Beltsville, Md., 1962—; cons. FAO, Argentina. Contbr. articles to profl. jours. Served with U.S. Navy, 1944-47. Mem. Am. Phytopath. Soc., Mycological Soc. Am. Christian. Subspecialties: Plant pathology; Microbiology. Current work: Breeding for resistance to diseases of forage legumes.

O'STEEN, WENDALL KEITH, anatomy educator; b. Meigs, Ga., July 3, 1928; s. Wellna Hubert and Lillian (Powell) O'S.; m. Sandra Kraeer, July 30, 1983; children by previous marriage: Lisa Diane, Kerry Keith. B.A., Emory U., 1948, M.S., 1950; Ph.D., Duke U., 1958. From instr. to asst. prof. Emory U., Atlanta, 1948-51, prof. anatomy, 1967-77, dir. grad. studies, 1967-77; from instr. to prof. anatomy U. Tex. Med. Br., Galveston, 1958-67; prof. anatomy Bowman Gray Sch. Medicine, Wake Forest U., Winston-Salem, N.C., 1977—, chmn. dept. anatomy, 1977—. Contbr. articles to profl. publs. Served to lt. col. U.S. Army, 1953-76. Recipient Emory Williams Disting. Tchr. award Emory U., 1974; NIH grantee, 1959-67. Mem. AAUP, Am. Assn. Anatomists, Soc. Neurosci., Endocrine Soc., AAAS. Republican. Methodist. Club: Res Officers. Subspecialties: Anatomy and embryology; Neurobiology. Current work: Retinal photodamage; endocrinology research (neuroscience). Home: 5020 Mountain View Rd Winston-Salem NC 27104 Office: Bowman Gray Sch Medicine Wake Forest Univ Anatomy Dept Winston-Salem NC 27103

OSTENDORF, DAVID WILLIAM, civil engineering educator, researcher; b. Stamford, Conn., Feb. 6, 1950; s. Bernard and Dorothy (Pennie) O.; m. Judith Rae Ireland, Dec. 9, 1978; 1 son, Raymond. B.S.E., U. Mich., 1972; S.M., MIT, 1978, Sc.D., 1980. Hydraulic engr. Stone & Webster, Boston, 1973-75; research assoc. MIT, 1977-80; asst. prof. civil engring. U. Mass., 1980—; cons. Applied Tech. Ctr., 1982—. Mem. ASCE (assoc.), Am. Geophys. Union. Democrat. Roman Catholic. Subspecialties: Oceanography; Fluid mechanics. Current work: Modeling of transport processes in coastal and estuarine environments. Office: Dept Civil Engring U Mass Amherst MA 01003

OSTENSON, NED ALLEN, oceanographer; b. Fargo, N.D., June 22, 1930; s. Nels A. and Estella T. (Temple) O.; m. Grace E. Laudon, June 29, 1963. B.S in Geology, U. Wis., 1952, M.S., 1953, Ph.D., 1962; postgrad., Johns Hopkins Sch. Advanced Inernat. Studies, 1975. With Geophys. Services, Inc., Houma, La., 1953-56, Arctic Inst. N.Am., Washington, 1956-59; geophysicist Office Naval Research, Washington, 1966-76; oceanographer NOAA, Washington, 1977—; dir. Nat. Sea Grant Coll. Program, 1977—. Contbr. articles to sci. jours., chpts. to books. Served to 1st lt. Signal Corps U.S. Army, 1953-55. Mountain in Antarctica named for him; recipient Superior Accomplishment award USN, 1965; Meritorious Service citation Nat. Acad. Scis., 1959. Mem. Marine Tech. Soc. (v.p.), Acad. Polit. Sci., Am. Polar Soc. (pres.), Am. Geophys. Union; fellow Geol. Soc. Am., Explorers Club. Methodist. Club: Cosmos (Washington). Subspecialties: Geophysics; Oceanography. Current work: Polar geophysics; oceanographic and atmospheric research administration. Home: 2871 Audubon Terr NW Washington DC 20008 Office: 6010 Executive Blvd Rockville MD 20852

OSTER, MARTIN WILLIAM, oncologist; b. N.Y.C., Apr. 9, 1947; s. Joseph A. and Bella (Lanzberg) O.; m. Karen Anita Oster, May 18, 1975; children: Bonnie Felice, Michelle Rae, Nancy Meredith. B.A., Columbia U., 1967, M.D., 1971. Diplomate: Am. Bd. Internal Medicine and Med. Oncology. Intern, resident in medicine Mass. Gen. Hosp., Boston, 1971-73; oncology fellow Nat. Cancer Inst., Bethesda, Md., 1973-76; asst. prof. clin. medicine Columbia U., 1976—; attending physician Columbia-Presbyn. Med. Center, 1976—. Contbr. articles med. jours. Served with USPHS, 1973-76. Recipient Asher Green award Columbia Coll., 1967, Mosby Med. award, 1970; jr. faculty clin. fellow Am. Cancer Soc., 1976-79; recipient Physician Recognition awards AMA, 1976-78, 78-81. Fellow ACP; mem. Am. Soc. Clin. Oncology, Am. Assn. Cancer Research, N.Y. Cancer Soc., Phi Beta Kappa, Alpha Omega Alpha. Subspecialties: Oncology; Chemotherapy. Current work: Mgmt. and treatment of cancer patients, med. oncologist, chemotherapist, med. sch. asst. prof., clin. oncology research. Home: 5 Birch Grove Dr Armonk NY 10504 Office: Columbia Presbyterian Medical Center 161 Fort Washington Ave New York NY 10032

OSTER, ZVI HERMAN, nuclear medicine physician; b. Burdwjeni, Rumania, Mar. 2, 1932; s. A. Edmond and Shifra (Sperber) O.; m., July 7, 1951; children: Michal, Ady, Shai. M.D., Hebrew U., 1962. Resident in medicine Hadassah U. Hosp., Jerusalem, 1962-67; assoc. dean Hebrew U. Med. Sch., Jerusalem, 1967-73; resident, fellow Johns Hopkins Instn., Balt., 1973-75; asst. prof. Johns Hopkins U., 1975-76, NYU, N.Y.C., 1976-77; dir. nuclear medicine Sheba Hosp., Israel, 1977-79; assoc. prof. SUNY-Stony Brook, 1979—, co-chief nuclear medicine, 1979—. Served to capt. M.C. Israeli Army, 1973. Mem. Soc. Nuclear Medicine, Assn. Univ. Radiologist, Radiol. Soc. N.Am. Subspecialty: Nuclear medicine. Current work: Development and evaluation of new radiopharmaceuticals; tumor imaging. Home: 16 Stratton Ln Stony Brook NY 11790 Office: Health Scis Center SUNY Stony Brook NY 11794

OSTERBERG, ARNOLD CURTIS, pharmacologist; b. Rochester, Minn., Sept. 14, 1921; s. Arnold Erwin and Ann Elizabeth (Curtis) O.; m. Barbara Jane Towey, Apr. 6, 1946; children: Ann, Jane, Eric, John. B.S., U. Iowa, 1942; Ph.D., U. Minn., 1953. Lab. asst. U. Minn., Mpls., 1948-53; prin. research scientist Lederle Labs., Pearl River, N.Y., 1953—. Contbr. articles on pharmacology to profl. jours. Served to lt. j.g. USN, 1942-46. Mem. Am. Soc. for Pharmacology and Exptl. Therapeutics, Internat. Narcotics Research Club. Roman Catholic. Subspecialties: Neuropharmacology; Pharmacology. Current work: Analgesics, antidepressants, anxiolytics, antipsychotics. Home: 179 Crooked Hill Rd Pearl River NY 10965 Office: Lederle Labs Pearl River NY 10965

OSTERBROCK, DONALD EDWARD, astronomy educator; s. William Carl and Elsie (Wettlin) O.; m. Irene L. Hansen, Sept. 19. 1952; children: Carol Ann, William Carl, Laura Jane. Ph.B., U. Chgo., 1948, B.S., 1948, M.S., 1949, Ph.D., 1952. Mem. faculty Princeton, 1952-53, Calif. Inst. Tech., 1953-58; faculty U. Wis.-Madison, 1958-73, prof. astronomy, 1961-73, chmn. dept. astronomy, 1966-67, 69-72; prof. astronomy U. Calif., Santa Cruz, 1972—; dir. Lick Obs., 1972-81; mem. staff Mt. Wilson Obs., Palomar Obs., 1953-58; vis. prof. U. Chgo., 1963-64, Ohio State U., 1980; Hill Family vis. prof. U. Minn., 1977-78. Author: Astrophysics of Gaseous Nebulae, 1974; Editor: (with C.R. O'Dell) Planetary Nebulae, 1968; Letters editor: Astrophys. Jour, 1971-73. Served with USAAF, 1942-46. Recipient profl. achievement award U. Chgo. Alumni Assn., 1982; Guggenheim fellow Inst. Advanced Studies, Princeton, N.J., 1960-61, 82-83; NSF Sr. Postdoctoral Research fellow U. Coll., London, 1968-69. Mem. Nat. Acad. Scis. (chmn. astronomy sect. 1971-74, sec. class math. and phys. sci. 1980-83), Am. Acad. Arts and Scis., Internat. Astron. Union (pres. commn. 34 1967-70), Royal Astron. Soc. (asso.), Am. Astron. Soc. (councilor 1970-73, v.p. 1975-77), Astron. Soc. Pacific. (chmn. history com. 1982—). Congregationalist. Subspecialties: Optical astronomy; High energy astrophysics. Current work: Observational astronomy, astrophysics; astronomical spectroscopy. Home: 120 Woodside Ave Santa Cruz CA 95060

OSTERCAMP, DARYL LEE, chemistry educator, researcher; b. Garner, Iowa, Feb. 5, 1932; s. Otto Sam and Ruth Marie (Lee) O.; m. Janet Louise Freed, June 21, 1958; children: Stephen, Daniel. A.A., Waldorf Coll., 1951; B.A., St. Olaf Coll., 1953; M.S., U. Wis.-Madison, 1954; Ph.D., U. Minn.-Mpls., 1959. Instr. Luther Coll., Decorah, Iowa, 1954-56; research assoc. Pa. State U., University Park, 1959-60; Fulbright asst. prof. U. Mosul, Iraq, 1964-65; prof. U. Petroleum, Dhahran, Saudia Arabia, 1977-80; prof. chemistry Concordia Coll., Moorhead, Minn., 1960—. Referee: Arabian Jour. Sci, Dhahran, Saudi Arabia, 1977-80; contbr. articles in field to profl. jours. Precinct chmn. Republican party, Moorhead, Minn., 1962; deacon Our Savior's Lutheran Ch., Moorhead, 1962-65, Christ the King Luth. Ch., Moorhead, 1981-83. NSF predoctoral fellow, 1958; NSF sci. faculty fellow U. East Anglia, 1969; F.G. Cottrell grantee Research Corp., 1962-64; named Outstanding Tchr. Mondamin Soc., 1968. Mem. Am. Chem. Soc. (sec. local sect. 1964-65), Phi Lambda Upsilon, Gamma Alpha. Lutheran. Subspecialties: Organic chemistry; Nuclear magnetic resonance (chemistry). Current work: Chemistry teacher of undergraduates, research in organic chemistry, synthesis and spectroscopic properties of vinylogous systems. Home: 1415 22d Ave S Moorhead MN 56560 Office: Concordia Coll 920 S 8th St Moorhead MN 56560

OSTER-GRANITE, MARY LOU, anatomist, developmental neurogeneticist; b. Hobart, Okla., Oct. 24, 1947; d. Raymond Glen and Eleanor Frances (Ulichny) Oster; m. David Samuel Granite, July 1, 1970; 1 son, Stephen Jacob. B.A. in Biology, U. Rochester, 1969; Ph.D. in Anatomy, Johns Hopkins U., 1974. Postdoctoral fellow dept. neurology Johns Hopkins U. Sch. Medicine, Balt., 1975-76; asst. prof. dept. anatomy U. Md. Sch. Medicine, Balt., 1976-82, research asso. prof. anatomy and pediatrics, 1982—. Contbr. articles to profl. jours. Recepient Disting. Young Scientist of Yr. award Md. Acad. Sci., 1982; Alfred Sloan Found. fellow, 1977-81. Mem. Electron Microscopic Soc. Am., N.Y. Soc. Electron Microscopy, AAAS, Soc. Neuroscience, Assn. Women in Sci., Am. Assn. Anatomists, Johns Hopkins Med. and Surg. Assn., U. Md. Alumni Assn., N.Y. Acad. Scis., Am. Genetics Assn. Subspecialties: Anatomy and embryology; Developmental neurogenetics. Current work: Utilize murine chimeras (allophenics) to determine cell lineage in normal and chromosomally aneuploid neuraxis during development by immunocyto chemical methods. Home: 6620 Auburn Ave Lanham MD 20706 Office: Dept of Anatomy University Maryland School Medicine 655 W Baltimore St Baltimore MD 21201

OSTERHELD, R(OBERT) KEITH, chemistry educator; b. Bklyn., Apr. 19, 1925; s. Albert Henry and Hilda Pearl (Heatlie) O.; m. Jean Drake Evans, June 28, 1952; children: Robert Keith, Albert Laighton, James Evans, Thomas Heatlie. B.S. in Chemistry, Poly. Inst. Bklyn., 1945; Ph.D. in Inorganic Chemistry, U. Ill.-Urbana, 1950. Instr. Cornell U., Ithaca, N.Y., 1950-54; asst. prof. chemistry U. Mont., Missoula, 1954-58, assoc. prof., 1958-65, prof., 1965—, chmn. chemistry dept., 1973—. Mem. Florence (Mont.) Sch. Bd., 1969-75, chmn., 1972-73, 74-75; bd. dirs. Mont. Sch. Bd. Assn., Helena, 1973-75; treas. Florence-Carlton Community Ch., 1965—. Served with USAAF, 1945-47. Mem. Am. Chem. Soc., N.Am. Thermal Analysis Soc., AAUP, Sigma Xi. Subspecialties: Inorganic chemistry; High temperature chemistry. Current work: Kinetics and mechanisms of thermal decomposition of solids, particularly by interfacial mechanisms; hydrolysis of condensed phosphates. Home: NW 4610 Larry Creek Loop Florence MT 59833 Office: Univ Mont Chemistry Dept Missoula MT 59812

OSTHOFF, ROBERT C., elec. engr., elec. mfg. co. exec.; b. Buffalo, June 7, 1924; s. Eugene Charles and Anna Marie (Schultz) O.; m. Jane A. Halsted, Sept. 12, 1953; children: William J., Susan M., John D. B.S., U. Buffalo, 1948, M.S., 1949; Ph.D., Harvard U., 1951. Research assoc. research lab. Gen. Electric Co., Schenectady, 1951-57, mgr. advance engring. lamp metals and components dept., Cleve., 1957-60; mgr. electrochemistry br. Mar-Tech. Resources, Pittsfield, Mass., 1969-81, mgr. energy storage and conversion programs advanced energy programs dept., Ballston Spa, N.Y., 1981. Contbr. articles to profl. jours. Served with AUS, 1943-46. Mem. Am. Chem. Soc., IEEE, Nat. Elec. Mfrs. (coms. dielectric fluids, polychlorinate biphenyls). Subspecialties: Physical chemistry; Materials. Current work: Carbonate fuel cells, beta batteries, pressurized fluidized bed technology, electrochemistry, materials, fuels, coal purification, ionic conductors. Patentee in field. Home: 1179 Bellemead Ct Schenectady NY 12309 Office: Route 3 Plains Rd Ballston Spa NY 12020

OSTRACH, SIMON, engineering educator; b. Providence, Dec. 26, 1923; s. Samuel and Bella (Sackman) O.; m. Gloria Selma Ostrov, Dec. 31, 1944 (div. Jan. 1973); children: Stefan Alan, Louis Hayman, Naomi Ruth, David Jonathan, Judith Cele; m. Margaret E. Stern, Oct. 29, 1975. B.S. in Mech. Engring, U. R.I., 1944, M.E., 1949; Sc.M., Brown U., 1949, Ph.D. in Applied Math, 1950. Research scientist NACA, 1944-47; research assoc. Brown U., 1947-50; chief fluid physics br. Lewis Research Center NASA, 1950-60; prof. engring., head div. fluid, thermal and aerospace scis. Case Western Res. U., Cleve., 1960-70, Wilbert J. Austin Distinguished prof. engring., 1970—; Distinguished vis. prof. City Coll. CUNY, 1966-67; Lady Davis fellow, vis. prof. Technion-Israel Inst. Tech., 1983-84; cons. to industry, 1960—; Mem. research adv. com. fluid mechanics NASA, 1963-68. Contbr. papers to profl. lit. Recipient Conf. award for best paper Nat. Heat Transfer Conf., 1963; Richards Meml. award Pi Tau Sigma, 1964. Fellow Am. Acad. Mechanics, AIAA, ASME (Heat Transfer Meml. award 1975, Freeman scholar 1982, Max Jacob meml. award 1983); mem. Nat. Acad. Engring., Soc. Natural Philosophy, Sigma Xi (nat. lectr. 1978-79). Subspecialties: Fluid mechanics; Applied mathematics. Current work: Natural convection; transport phenomena in industrial processes; physicochenmical fluid dynamics. Home: 28176 Belcourt Rd Pepper Pike OH 44124 Office: Case Western Res U Cleveland OH 44106

OSTRIKER, JEREMIAH PAUL, astrophysicist, educator; b. N.Y.C., Apr. 13, 1937; s. Martin and Jeanne (Sumpf) O.; m. Alicia Suskin, Dec. 1, 1958; children—Rebecca, Eve, Gabriel. A.B., Harvard, 1959; Ph.D. (NSF fellow), U. Chgo., 1964; postgrad., U. Cambridge, Eng., 1964-65. Research assoc., lectr. astrophysics Princeton U. (N.J.), 1965-66, asst. prof., 1966-68, assoc. prof., 1968-71, prof., 1971—; chmn. dept. astronomy, dir. obs. Princeton (N.J.) U., 1979—; Charles A. Young prof. astronomy Princeton U. (N.J.), 1982—. Editorial bd., trustee Princeton U. Press; Contbr. articles to profl. jours. Alfred P. Sloan Found. fellow, 1970-72. Mem. Am. Astron. Soc. (councilor 1978-80, Warner prize 1972, Russell prize 1980), Internat. Astron. Union, Nat. Acad. Scis., Am. Acad. Arts and Scis. Subspecialties: Theoretical astrophysics; High energy astrophysics. Current work: Origin and evolution of galaxies. Home: 33 Philip Dr Princeton NJ 08540 Office: Princeton Univ Obs Peyton Hall Princeton NJ 08544

OSTROW, DAVID G., psychopharmacologist, research and psychiatry consultant; b. Bklyn., Apr. 28, 1947; s. Bernard R. and Miriam (Travaini) O.; m. Alice Pinsley, June 14, 1970. Diplomate: Am. Bd. Psychiatry and Neurology. Psychopharmacology research fellow Ill. Psychiat. Inst., Chgo., 1976; resident in psychiatry Michael Reese Hosp., Chgo., 1976-79; coordinator biol. psychiatry programs VA Lakeside Med. Ctr., Chgo., 1979-82, dir., 1982—; dir. biol. psychiatry programs Northwestern Meml. Hosp., 1983—; dir. affective disorder evaluation unit; dir. research Howard Brown Meml. Clinic; clin. cons. in psychopharmacology, sexually transmitted diseases; spl. asst. to Chgo. Commr. Health, 1983—; chmn. Chgo AIDS Task Force, 1982—; organizer, chmn. Task Force Vaccination Strategies for Sexually Transmitted Hepatitis B, 1982—. Editor: (with T.A. Sandholzer and Y. Felman) Sexually Transmitted Diseases in Homosexually Active Men: A Guide to Diagnosis and Treatment, 1983; Contbr. articles and abstracts to profl. jours. Recipient Nat. Sci. award Bausch & Lomb, 1965; Med. Alumni prize for research U. Chgo., 1976; Med. Scientist Tng. Program fellow, 1969-75; Am. Psychiat. Assn. Falk fellow, 1977-79; Career Devel. awardee VA Lakeside Med. Ctr., 1979-82. Mem. Am. Psychiat. Assn., Ill. Psychiat. Soc. (research award 1980), Soc. Biol. Psychiatry, Am. Pub. Health Assn., Am. VD Assn., AAAS, N.Y. Acad. Sci., Sigma Xi, Alpha Omega Alpha. Jewish. Subspecialties: Neuropharmacology; Epidemiology. Current work: Research in use of biological measurements in diagnosis and treatment of major mental disorders; mechanism of action of lithium; epidemiology of major mental disorders and sexually transmitted diseases; behavioral-biological interactions. Office: 333 E Huron St Room 1537 Chicago IL 60611

OTERO, RAYMOND B., microbiologist, educator; b. Rochester, N.Y., May 8, 1938; s. Charles I. and Emma G. (Gonzalez) O.; m. (div.); children: Raymond B., Bryan C. B.S., U. Dayton, 1960; M.S., U. Rochester, 1963; Ph.D., U. Md., 1968. Postdoctoral fellow U. Ky., Lexington, 1975-76; quality control biologist Lederle Lab., Pearl River, N.Y., 1963-65; prof. microbiology Eastern Ky. U., Richmond, 1968—; cons. St. Joseph Hosp., Lexington, 1969—; lectr. in field. Author 2 manuals, numerous articles. Recipient Excellence in Teaching award Eastern Ky. U., 1980. Mem. Am. Soc. Microbiology, Ky. Acad. Sci., N.Y. Acad. Sci., South Central Assn. Clin. Microbiologists, Assn. Practitioners in Infection Control. Roman Catholic. Club: Blue Grass Sportsman (Lexington). Subspecialty: Microbiology. Current work: Clinical microbiology as it involves patient care.

OTHMER, DONALD FREDERICK, chemical engineer, educator; b. Omaha, May 11, 1904; s. Frederick George and Fredericka Darling (Snyder) O.; m. Mildred Jane Topp, Nov. 18, 1950. Student, Ill. Inst. Tech., Chgo., 1921-23; B.S., U. Nebr., 1924, D.Eng. (hon.), 1962; M.S., U. Mich., 1925, Ph.D., 1927; D.Eng. (hon.), Poly. Inst. N.Y., 1977, N.J. Inst. Tech., 1978. Registered profl. engr., N.Y., N.J., Ohio, Pa. Devel. engr. Eastman Kodak Co. and Tenn. Eastman Corp., 1927-31; prof. Poly. Inst. N.Y., Bklyn., 1933, Distinguished prof., 1961—, sec. grad. faculty, 1948-58; head dept. chem. engring., 1937-61; Hon. prof. U. Conception, Chile, 1951; cons. chem. engr., dir. various engring. and mfg. corps., licensor of process patents, 1931—, to numerous cos., govtl. depts., U.S., Can., Mexico, Cuba, P.R., Central and S.Am., Norway, Sweden, Finland, Denmark, Germany, France, Eng., Belgium, Switzerland, Italy, Spain, South Africa, India, Burma, Yugoslavia, Korea, Japan, Taiwan, Peoples Republic of China, P.I., Dominican Republic, United Arab Emirates, Poland, People's Republic of China; in field of chem. engring., cons. UN, UNIDO, WHO, Dept. Energy, Office Saline Water of U.S. Dept. Interior, Chem. Corps. and Ordnance Dept. U.S. Army, USN, WPB, Dept. State, HEW, Nat. Materials Advisory Bd., NRC sci. advisory bd., U.S. Army Munitions Command; mem. Panel Energy Advisers to Congress, other depts. U.S., fgn. govts.; Sr. gas officer Bklyn. Citizens Def. Corps.; Lectr. sci. and engring. fields; lectr. Swiss univs. for Am. Swiss Found. Sci. Relations, 1950; lecture tour Chem. Inst. Can., 1944-52, Am. Chem. Soc., Shri RAM Inst., India, 1980; plenary lectr. Peoples Republic of China, hon. del. Engring. Congresses, Japan, 1983, plenary lectr., hon. del., Germany, Greece, Mexico, Czechoslovakia, Yugoslavia, Poland, P.R., France, Can., Argentina, India, Turkey, Spain, Rumania, Kuwait, Iran, Iraq, Algeria, China, United Arab Emirates; lectr. U.S. Army War Coll., 1964, Canadian Royal Mil. Coll., 1981. Contbr. over 350 articles on chem. engring., chem mfg., thermodynamics to tech. jours.; Co-founder, Co-editor: Kirk-Othmer Ency. Chem. Tech., 17 vols, 1947-60, 24 vols., 2d edit., 1963-71, 25 vols., 3d edit., 1976-83, Spanish edit., 1960-66; editor: Fluidization, 1956; Co-author: Fluidization and Fluid Particle Systems, 1960; Mem. adv. bd.: Perry's Chem. Engr.'s Handbook; tech. editor: UN Report, Technology of Water Desalination, 1964. Bd. regents L.I. Coll. Hosp.; bd. dirs. numerous ednl. and philanthropic instns. Recipient Golden Jubilee award, 1975, Profl. Achievement award, 1978, also; named to Hall of Fame, 1981; (all Ill. Inst. Tech.). Fellow AAAS, Am. Inst. Cons. Engrs., Am. Inst. Chemists (Honor Scroll 1970, Chem. Pioneer award 1977), ASME (chmn. chem. processes div. 1948-49), N.Y. Acad. Scis. (hon. life fellow, chmn. engring. sect. 1972-73), Instn. Chem. Engrs. (London) (hon. life), Am. Inst. Chem. Engrs. (Tyler award 1958, chmn. N.Y. sect. 1944, dir. 1956-59); mem. Am. Chem. Soc. (council 1945-47, E.V. Murphree-Exxon award 1978, hon. life mem.), Soc. Chem. Industry (Perkin medal 1978), Am. Soc. Engring. Edn. (Barber Coleman award 1958), Engrs. Joint Council (dir. 1957-59), Societe de Chimie Industrielle (pres. 1973-74), Chemurgic Council (dir.), Japan Soc. Chem. Engrs., Assn. Cons. Chemists and Chem. Engrs. (award of Merit 1975), Newcomen Soc., Am. Arbitration Assn. (panel mem. or sole arbitrator numerous cases), Sigma Xi (citation disting. research 1983), Tau Beta Pi, Phi Lambda Upsilon, Iota Alpha, Alpha Chi Sigma, Lambda Chi Alpha; hon. life mem. Deutsche Gesellschaft für Cheme. Apparatewesen. Clubs: Norwegian, Chemists (pres. 1974-75), Rembrandt (Bklyn.)). Subspecialties: Chemical engineering; Fuels and sources. Current work: Production methods and fuel uses of methanol; waste water treatment systems. Designer chem. plants and processes for numerous corps., U.S., fgn. countries. Holder over 135 U.S. and fgn. patents on methods, processes and engring. equipment in mfg. of pharms., sugar, salt, acetic acid, acetylene, fuel-methanol, synthetic rubber, petro-chems., pigments, zinc, aluminum, titanium, also wood pulping, refrigeration, solar and other energy conversion, water desalination, sewage treatment, peat utilization, coal desulferization, methanol production, pipeline heating, etc. Home: 140 Columbia Heights Brooklyn NY 11201 Office: 333 Jay St Brooklyn NY 11201 The reward to the educator lies in his pride in his students' accomplishments. The richness of that reward is the satisfaction in knowing that the frontiers of knowledge have been extended.

OTT, EDGAR ALTON, animal nutritionist, educator; b. Ft. Wayne, Ind., Jan. 10, 1938; s. Earl S. and Luella Fay (Keister) O.; m. Judith Carlene Smith, June 9, 1963; children: Gregory Mark, Ronda Fay. B.S., Purdue U., 1959, M.S., 1963, Ph.D., 1965. Research assoc., mgr. product devel. for horses and sheep Ralston Purina Co., St. Louis, 1965-70; assoc. prof. animal nutrition U. Fla., Gainesville, 1970-79, prof., 1979—, prof.-in-charge, 1970—. Contbr. articles to profl. jours. Leader 4-H Club, 1974-82. Served as 1st lt. U.S. Army, 1959-61. U.S. Brewers Assn. grantee, 1977-78; Fla. Canners Assn. grantee, 1974-75. Mem. AAAS, Am. Soc. Animal Sci., Equine Nutrition and Physiology Soc., Sigma Xi. Republican. Baptist. Subspecialties: Animal nutrition; Animal physiology. Current work: Equine nutrition and physiology. Office: U Fla 210G Animal Sci Bldg Gainesville FL 32611

OTT, WALTER RICHARD, ceramic engineer, university dean; b. Bklyn., Jan. 20, 1943; s. Harold V. and Mary E. (Butler) O.; m. Jeannette Winter, June 27, 1964; children: Regina, Christina, Walter. B.S. in Ceramic Engring, Va. Poly. Inst., 1965, M.S., U. Ill., 1967; Ph.D., Rutgers U., 1969. Registered profl. engr., Pa. Proce-s engr. Corhart Refractories, Buckhannon, W.Va., 1965-66; staff research engr. Champion Spark Plug Co., Detroit, 1969-70; assoc. prof. and asst. dean dept. ceramic engring. Rutgers U., New Brunswick, N.J., 1970-80; dean N.Y.S. State Coll. Ceramics, Alfred (N.Y.) U., 1980—; cons. to industry. Recipient Ralph Teetor award Soc. Automotive Engrs., 1973. Mem. Nat. Inst. Ceramic Engrs. (Profl. achievement award 1975), Am. Ceramic Soc., Am. Soc. Engring. Edn., Ceramic Assn. N.Y., Soc. Glass Tech., Keramos. Subspecialties: Ceramic engineering; Materials (engineering). Current work: Phosphate glasses, microwave ferrites, reaction kinetics and thermal analysis. Patentee in field. Office: NY State Coll Ceramics Alfred NY 14802

OTTENBRITE, RAPHAEL MARTIN, chemistry educator; b. Claybank, Sask, Can., Sept. 20, 1936; s. Joseph G. and Teresa O.; m. Nancy Louise Schrot; children: Shelley Ann, Carol Louis, Raphael Martin. B.Sc., Assumption U., 1958, M.Sc., 1962; Ph.D., U. Windsor, 1967. Tchr. Western Ont. Inst. Tech., 1960-65; research assoc. U. Fla., 1966; with Va. Commonwealth U., 1967; prof. chemistry, coordinator grad. studies for (Coll. Humanities and Scis.), 1981; guest prof. Nagasaki (Japan) U., 1983. Contbr. numerous articles in field to profl. jours. Mem. Am. Chem. Soc., Sigma Xi. Subspecialties: Polymer chemistry; Biophysical chemistry. Current work: Synthesis and characterization of electrolytic polymers with emphasis, on polyanions with antitumor activity. Home: 2781 Brigstock Rd Midlothian VA 23113 Office: Department Chemistry Virginia Commonwealth University Richmond VA 23284

OTTENSMEYER, DAVID JOSEPH, physician, health facility exec.; b. Nashville, Jan. 29, 1930; s. Raymond Stalney and Glenda Jessie (Helpingsteine) O.; m. Mary Jean Langley. B.A., Wis. State U., Superior, 1951; M.D., U. Wis., Madison, 1959. Diplomate: Am. Bd. Neurol. Surgery. Intern Univ. Hosps., Madison, Wis., 1959-60, resident in neurosurgery, 1960-64; faculty U. Wis., 1964-65; neurol. surgeon Marshfield (Wis.) Clinic, 1965-76, pres., chief exec. officer, 1972-75, Lovelace Med. Found., Albuquerque, 1976—; clin. prof. surgery U. N.Mex., 1976—. Contbr. articles to profl. jours. Trustee Albuquerque Acad., 1977—. Served to 1st lt. USAF, 1951-55; col. M.C. U.S. Army Res. Fellow A.C.S., Am. Coll. Physician Execs. (regent 1980—); mem. Am. Group Practice Assn. (dir. 1978-81, polit. action com. 1977-81), AMA, Am. Assn. Neurol. Surgeons, Congress Neurol. Surgeons. Republican. Episcopalian. Club: Rotary. Subspecialty: Health care administration.. Home: 1405 Quimera Trail SE Albuquerque NM 87123 Office: Lovelace Med Found 5400 Gibson Blvd SE Albuquerque NM 87108

OVE, PETER, educator, biologist; b. Dellstedt, Germany, Aug. 31, 1930; came to U.S., 1957, naturalized, 1963; s. Paul and Thekla (Brockmann) O.; m. Maria Siert, July 20, 1956; children—Norman, Roger, Torsten. B.S., U. Pitts., 1963, Ph.D., 1967. Research technician Connaught Med. Research, Toronto, Ont., Can., 1953-57; research asst. dept. microbiology U. Pitts., 1957-63; asst. prof. U. Durham Med. Sch., 1967-69; asst. prof. dept. anatomy and cell biology U. Pitts. Med. Sch., 1969-73, asso. prof., 1973-78, prof., 1978—; Mem. study sect. NIH, 1976-79. Recipient NIH research grants, 1970-73, 74-75, 76-79. Mem. Am. Assn. Cancer Research, Am. Assn. Cell Biology. Subspecialties: Cell biology (medicine); Cancer research (medicine). Current work: Control of growth of normal and malignan cells; hormonal factors and regulation of DNA sylnthesis; mechanisms of hepatocarcinogenesis. Home: 6225 Heberton St Verona PA 15147 Office: U Pitts Sch Medicine Pittsburgh PA 15213

OVED, YOEL, aeronautical engineering educator; b. Basra, Iraq, May 13, 1942; came to U.S., 1981; s. David and Nitza (Salomon) O.; m. Anny Hana Metoudi, Mar. 26, 1973; children: Gal, Iris. B.S., Technion, Israel Inst. Tech., 1968, M.Sc., 1973, D.Sc., 1976. Sr. research scientist Armament Devel. Authority, Haifa, Israel, 1968-77; research assoc. U. Victoria, B.C., Can., 1977-79, lectr. math., 1980-81; assoc. prof. aeronautical engring. Embry-Riddle Aeronautical U., Prescott, Ariz., 1981—. Contbr. articles to profl. jours. Mem. Israel Phys. Soc., Canadian Assn. Physicists, AIAA. Subspecialties: Fluid mechanics; Aeronautical engineering. Current work: Teaching aerodynamics, fluid-mechanics, calculus, differential equations, research in aerodynamics and gasdynamics. Office: Embry-Riddle Aeronautical U 3200 N Willow Creek Rd Prescott AZ 86301

OVENS, WILLIAM GEORGE, mechanical engineer, educator, consultant; b. Paterson, N.J., July 18, 1939; s. William George and Dora Jane (Mingle) O.; m. Jill Janet Whiton, Aug. 24, 1963; children: Bevan Jane, Janine Elise. B.S.E. in Mech. Engring, U. Mich., 1964; M.S., U. Conn., 1969, Ph.D., 1971. Registered engr.-in-tng., N.Y. State, 1976. Engr. Pratt & Whitney Aircraft, East Hartford, Conn., 1964-65; research specialist U. Conn., 1966-71; lectr. Papua New Guinea U. Tech., Lae, 1972-75; asst. prof. mech. and indsl. engring., area coordinator mfg. engring. Clarkson Coll. Techl., 1976-81; asso. prof. mech. engring. Rose-Hulman Inst. Techl., 1981; cons. Martin Marietta Aerospace Co., Tri Industries. Author profl. publs. Recipient Ralph R. Teetor award Soc. Automotive Engrs., 1976, Teaching Excellence award Pi Tau Sigma, 1977, Tau Delta Kappa, 1980, Tau Beta Pi, 1980; named Outstanding Adv. Clarkson Coll. Tech., 1980. Mem. U. Mich. Alumni Assn., ASME (chmn. materials div. 1982-83), Am. Soc. Metals. Subspecialties: Mechanical engineering; Materials processing. Current work: Effect of processing on service life of advanced aerospace castings; research, development and teaching in mechanical properites of aerospace metals affected by various manufacturing processes; design of custom manufacturing machinery. Home: 4693 Woodshire Dr Terre Haute IN 47803 Office: 5500 Wabash Ave Terre Haute IN 47803

OVERBERGER, CHARLES GILBERT, chemistry educator; b. Barnesboro, Pa., Oct. 12, 1920; s. Charles Edward and Beatrice (MacAnulty) O.; m. Elizabeth Chase; children by previous marriage: Erica Marie, Carla Lee, Charles, Ellen Ann. B.S., Pa. State U., 1941; Ph.D., U. Ill., 1944; D.Sc. (hon.), Holy Cross Coll., 1966, Long Island U., 1968. Research assoc. U. Ill., 1944-46; post-doctoral research fellow Mass. Inst. Tech., 1946-47; asst. prof. organic chemistry Poly. Inst. Bklyn., 1947-50, assoc. prof., 1950-51; prof., assoc. dir. Polymer Inst., 1951-55, head dept. chemistry, 1955-64, acting v.p. research, 1963-64, dean sci., 1964-67; dir. Polymer Research Inst., 1964-67; chmn. dept. chemistry U. Mich., Ann Arbor, 1967-72, v.p. research, 1972-82, dir. Macromolecular Research Center, 1968—; Chem. cons. indsl. cos.; chmn. U.S. Nat. Com. IUPAC, Nat. Acad. Sci., NRC, 1972-77; pres. macromolecular div. IUPAC, 1975-77; chmn. exec. com. council research policy and grad. edn. Nat. Assn. State Univs. and Land-Grant Colls., 1977. Author: (with J.P. Anselme and J.G. Lombardino) Organic Compounds with Nitrogen-Nitrogen Bonds, 1966; Editor: Macromolecular Syntheses; Past mem. adv. bd.: Polymer News; Contbr. articles to profl. jours. Chmn. bd. trustees Gordon Research Confs., 1966. Recipient Witco award in polymer chemistry Am. Chem. Soc., 1968, Charles Lathrop Parsons award, 1978; Disting. Alumni award Pa. State U., 1975; Internat. award Soc. Plastics Engrs., 1979; How and N. Potts medal Franklin Inst., 1983; named Mich. Scientist of Yr. Impression 5 Mus., 1983. Fellow Am. Inst. Chemists; mem. Am. Chem. Soc. (dir., nat. councilor, pres. 1967, chmn. com. on chem. and pub. affairs 1971-77), AAAS, Soc. Chem. Industry, Chem. Soc. Gt. Britain, Brazilian Acad. Scis., Sigma Xi, Phi Lambda Upsilon, Alpha Chi Sigma, Phi Kappa Pi, Phi Eta Sigma, Sigma Pi. Subspecialties: Polymer chemistry; Organic chemistry. Home: 436 Huntington Pl Ann Arbor MI 48104

OVERBYE, KAREN MARIE, molecular geneticist; b. Bklyn., Apr. 16, 1951; d. John and Catherine Grace (Chase) O.; m. Thomas Richard Fulton, Jan. 19, 1974; 1 son, Erik John. A.B., Hunter Coll., CUNY, 1974; M.S., NYU, 1978, Ph.D., 1982. Adj. lectr. Hunter Coll., CUNY, 1974, 80; lab. instr. NYU Med. Ctr., 1978; instr. MIT, 71982, NIH postdoctoral fellow, 1982—. Contbr. articles to profl. jours. NIH grantee. Mem. Genetics Soc. Am., Am. Soc. for Microbiology, AAAS. Subspecialties: Gene actions; Genetics and genetic engineering (biology). Current work: Control of gene expression; mechanism of promoter function. Promoters, gene fusions, regulatory sequences, gene expression, site-specific mutagenesis. Home: 48 Sterling St Somerville MA 02144 Office: 77 Massachusetts Ave 56-623 Cambridge MA 02139

OVERCASH, MICHAEL RAY, chem. engring. educator; b. Kannapolis, N.C., July 17, 1944; s. Ray L. and Ruth C. O.; m. Mary Y.; 1 dau.: Rachael T.Y. B.S., N.C. State U., 1966; M.S., U. N.S.W., Australia, 1967; Ph.D., U. Minn., 1972. Asst. prof. chem. engring. dept., biol. and agrl. engring. dept. N.C. State U., Raleigh, 1972-76, assoc. prof., 1976-80, prof., undergrad. adminstr., dir. biotech. research programs in engring., 1981—; cons. in field. Fulbright scholar, 1966; U.S. Dept. Agr. grantee, 1972-75; U.S. Dept. Interior grantee, 1972-80; U.S. EPA grantee, 1974-83. Subspecialties: Terrestrial systems; Chemical engineering. Home: 2908 Chipmunk Ln Raleigh NC 27607 Office: NC State Univ PO Box 5035 Raleigh NC 27650 Home: 2908 Chipmunk Ln Raleigh NC 27607

OVERHAUSER, ALBERT WARNER, physicist; b. San Diego, Aug. 17, 1925; s. Clarence Albert and Gertrude Irene (Pehrson) O.; m. Margaret Mary Casey, Aug. 25, 1951; children—Teresa, Catherine, Joan, Paul, John, David, Susan, Steven. A.B., U. Calif. at Berkeley, 1948, Ph.D., 1951; D.Sc. (hon.), U. Chgo., 1979. Research asso. U. Ill., 1951-53; asst. prof. physics Cornell U., 1953-56, asso. prof., 1956-58; supr. solid state physics Ford Motor Co., Dearborn, Mich., 1958-62, mgr. theoret. scis., 1962-69, asst. dir. phys. scis., 1969-72, dir. phys. scis., 1972-73; prof. physics Purdue U., West Lafayette, Ind., 1973-74, Stuart prof. physics, 1974—. Served with USNR, 1944-46. Recipient Oliver E. Buckley Solid State Physics prize Am. Phys. Soc., 1975, Alexander von Humboldt sr. U.S. scientist award, 1979. Fellow Am. Phys. Soc., Am. Acad. Arts and Scis.; mem. Nat. Acad. Scis. Subspecialties: Theoretical physics; Solid state physics. Current work: Theory of metals. Home: 236 Pawnee Dr West Lafayette IN 47906 Office: Dept Physics Purdue U West Lafayette IN 47907

OVERMAN, AMEGDA JACK, nematologist; b. Tampa, Fla., May 17, 1920; d. Nicholas George and Eloise Urquhart (Smith) Jack; m. Richard Douglas Overman, July 5, 1953. B.S., U. Tampa, 1942; M.S., U. Fla., 1951. Asst. soil chemistry U. Fla., Bradenton, 1951-56, asst. soil microbiologist, 1956-68, assoc. nematologist, 1968-73, nematologist, 1973—. Vice-chmn. Planning Commn., City of Bradenton, Fla., 1975—. Mem. Soc. Nematologists, European Soc. Nematologists, Orgn. Tropical Am. Nematologists, Fla. State Hort. Soc., Soil and Crop. Sci. Soc. Fla. Eastern Orthodox. Subspecialties: Plant pathology; Integrated pest management. Current work: Biology and control of plant nematodes, vegetables, ornamentals, agronomic crops, soil fumigants, crop rotation, nematicides. Office: 5007 60th St E Bradenton FL 34203

OVERMIER, JAMES BRUCE, psychology educator; b. Queens, N.Y., Aug. 2, 1938; s. James and Annette (Carleton) Wheelwright; m. Judith Ann Smith, Aug. 19, 1962; 1 dau., Larisa Nicole. A.B., Kenyon Coll., 1960; M.A., Bowling Green State U., 1962, U. Pa., 1964, Ph.D., 1965. Lic. cons. psychologist, Minn. Asst. prof. psychology U. Minn., Mpls., 1965-68, assoc. prof., 1968-71, prof., 1971—; exec. officer Ctr. for Research in Human Learning, Mpls., 1973-78, 81—; cons. Learning Strategies Corp., Mpls., 1981—; adv. panel psychobiology NSF, 1976-79. Editor: Learning & Motivation, 1973-76; assoc. editor: Physiol. Psychology, 1982—; regional editor: Behavioral Brain Research, 1980—; author: (with others) Animal Learning: Survey and Analysis, 1979. Fulbright-Hays fellow, Yugoslavia, 1980; Sci. Exchange fellow Nat. Acad. Sci., Poland, 1972; NSF sr. postdoctoral fellow, 1971; USPHS predoctoral fellow, 1963-65. Fellow Am. Psychol. Assn.; mem. Psychonomic Soc. (governing bd. 1983—, sec.-treas. 1981-83), AAUP, Am. Acad. Behavioral Medicine, Delta Kappa Epsilon. Subspecialties: Learning; Psychobiology. Current work: Mechanism of learning, motivation, behavior and the CNS substrates; emphasis on avoidance, punishment, stress. Home: 1965 E River Ter Minneapolis MN 55414 Office: Center for Research in Human Learning 75 E River Rd Minneapolis MN 55455

OVERTON, DONALD A., psychopharmacologist; b. Troy, N.Y., Dec. 9, 1935; s. Ernest C. and Evangeline S. O. Student, Shimer Coll., Mt. Carroll, Ill., 1952-54; B.S.E.E., Rensselaer Poly. Inst., 1958; M.A., Harvard U., 1961; Ph.D. in Psychology, McGill U., Montreal, 1962. Postdoctoral fellow in brain research U. Rochester, N.Y., 1962-64; postdoctoral fellow in psychiatry and physiology Albert Einstein Coll. Medicine, Bronx, N.Y., 1964-66; asst. prof. psychiatry Temple U. Sch. Medicine, 1966-69, asso. prof., 1969-77, prof. psychiatry, 1977—. Contbr. articles to profl. jours. NSF fellow, 1962-64; NIMH fellow, 1964-65. Fellow Am. Psychol. Assn., AAAS; mem. Psychonomic Soc., Behavioral Pharmacology Soc., Eastern Psychol. Assn., Am. Coll. Neuropsychopharmacology, Neurosci. Soc., Soc. Stimulus Properties of Drugs, Soc. Biol. Psychiatry, Sigma Xi. Subspecialties: Learning; Neuropharmacology. Current work: Psychopharmacology with primary emphasis on state dependent learning produced by drugs and on stimulus properties of drugs; electrophysiological correlates of mental illness. Office: Dept Psychiatry Temple University School Medicine 3401 N Broad St Philadelphia PA 19140

OVSHINSKY, STANFORD ROBERT, engring. co. exec.; b. Akron, Ohio, Nov. 24, 1922; s. Benjamin and Bertha T. (Munitz) O.; m. Iris L. Miroy, Nov. 24, 1959; children—Benjamin, Harvey, Dale, Robin Dibner, Steven Dibner. Student public schs., Akron; D.Sc. (hon.), Lawrence Inst. Tech., 1980, D.Engring., Bowling Green State U., 1981. Pres. Stanford Roberts Mfg. Co., Akron, 1946-50; mgr. centre drive dept. New Britain Machine Co., Conn., 1950-52; dir. research Hupp Corp., Detroit, 1952-55; pres. Gen. Automation, Inc., Detroit, 1955-58, Ovitron Corp., 1958-59; pres., chmn. bd. Energy Conversion Devices, Inc., Troy, Mich., 1960-78, pres., chief exec. officer, chief scientist, 1978—; adj. prof. engring. scis. Coll. Engring., Wayne State U. Contbr. articles on physics of amorphous materials, neurophysiology and neuropsychiatry to profl. jours. Recipient Diesel Gold medal German Inventors Assn., 1968. Mem. Am. Phys. Soc., IEEE, Soc. Automotive Engrs., N.Y. Acad. Scis., Detroit Physiol. Soc., Electrochem. Soc., Engring. Soc. Detroit. Subspecialty: Solid state chemistry. Office: 1675 W Maple Rd Troy MI 48084

OWCZAREK, JERZY ANTONI, mechanical engineering educator, researcher; b. Piotrkow, Lodz, Poland, Nov. 2, 1926; came to U.S., 1954, naturalized, 1959; s. Bronislaw and Waclawa (Grzembo) O.; m. Elisa M. Basoli, Dec. 19, 1959; children: Deborah, Diana, Maria-Ines, Jennifer, David. Dipl. Ing., Polish Univ. Coll., London, 1950; Ph.D., U. London, 1954. Asst. lectr. Battersea Poly. Inst., London, 1950-53; lectr. Queen Mary Coll., U. London, 1953-54; devel. engr. Gen. Electric Co., Schenectady, 1955-60; assoc. prof. mech. engring. Lehigh U., 1960-64, prof., 1964—; cons. Transam. Delaval Inc., Trenton, N.J., 1967—, Bell Labs., Allentown, Pa., 1974—. Author: Fundamentals of Gas Dynamics, 1964, Introduction to Fluid Mechanics, 1968. Served with Polish Underground Army, 1944-45. Mem. ASME, AIAA, Tau Beta Pi, Pi Tau Sigma, Sigma Xi. Roman Catholic. Subspecialties: Fluid mechanics; Aeronautical engineering. Current work: Turbomachinery fluid mechanics, turbine and compressor blade vibration; encapsulation of electronic circuits; thick film screen printing process. Patentee turbo machine blading. Home: 2345 Overlook Dr Bethlehem PA 18017 Office: Lehigh U Dept Mechanical Engring and Mechanics Bethlehem PA 18015

OWEN, CHARLES SCOTT, biochemist, educator; b. Springfield, Mo., July 4, 1942; s. Frank Charles and Helen B. (Scott) O.; m. Mary L. Kraybill, 1967 (; (widowed 1969)), 1967; m. Judith A. Hunter, Dec. 15, 1973; children: Patricia, Scott. B.S., U. Rochester, 1964; Ph.D., U. Pa., 1969. Postdoctoral researcher Washington U., St. Louis, 1969-71, U. Calif., Santa Barbara, 1971-72, U. Pa., 1972-77, asst. prof., 1979-82; assoc. prof. biochemistry Jefferson Med. Coll., Phila., 1982—. NIH Research Career Devel. awardee, 1977-82. Mem. Am. Assn. Immunologists, Biophys. Soc. Subspecialties: Biophysics (biology); Cellular engineering. Current work: magnetic cell sorting methods, applications in immunology; artificial biomembrane vesicles (Liposomes), applications to immunology. Office: Jefferson Med College Dept Biochemistry 1020 Locust St Philadelphia PA 19107

OWEN, DANIEL LEE, management consultant, educator; b. Flint, Mich., Mar. 12, 1947; s. Ben and Angeline (Tomczak) O.; m. Patricia Anne Weir, May 2, 1969. B.S., U. Mich., 1969; M.S., Carnegie-Mellon U., 1972; Ph.D., Stanford U., 1979. Registered profl. engr., Calif. Sr. engr. Westinghouse, Pitts., 1970-74; decision analyst SRI Internat., Menlo Park, CAlif., 1976-80; cons. Strategic Decision Group, Menlo Park, 1981—; cons. prof. Stanford U., 1980-82. Mem. Am. Nuclear Soc., Sigma Xi. Subspecialties: Probability. Current work: Structuring complex decision problems; designing hazardous products; management of research and development. Home: 189 Arbuelo Way Los Altos CA 94022 Office: Strategic Decision Group 3000 Sand Hill Rd Bldg 3 Suite 150 Menlo Park CA 94025

OWEN, DAVID GRAY, pediatric dentistry educator, researcher; b. Lowville, N.Y., Aug. 9, 1935; s. Allen Fulton and Ruth (Gray) O.; m. Patricia Ruth MacLeod, July 11, 1964; children: Sandra Elenora, Linda MacLeod. A.B., Syracuse U., 1960; D.D.S., McGill U., 1964; A.M., U. Chgo., 1969; certificate pedodontics, U. Ill., Chgo., 1970. Amanuensis and research asst. Royal Dental Coll., Copenhagen, 1964-66; instr., asst. prof. U. Chgo. Hosp. and Clinic, 1967-72; chief pediatric dentistry Wyler Children's Hosp., Chgo., 1970-72; asst. prof. pediatric dentistry U. Md. Dental Sch., Balt., 1972-75; assoc. prof. pediatric dentistry Balt. Coll. Dental Surgery, U. Md., 1975—; mem. Internat. Biol. Program, Chgo., 1970-72; cons. Kernan's cleft palate team, Balt., 1975—, John F. Kennedy Inst. Hosp., 1972—, Mercy Hosp., 1980—. Editor: McGill Dental Rev, 1964. Scoutmaster Boy Scouts Am., Troop 175, Balt., 1975-77; ruling elder Govans Presbyterian Ch., Balt., 1976-82; bd. mem. Cedarcroft Community Orgn., Balt., 1981-82. Served with U.S. Army, 1956-58. Zoller fellow U. Chgo., 1966-69. Mem. Internat. Assn. Dental Research, Am. Assn. Dental Schs., Md. Soc. Dentistry for Children (pres. 1981-82). Republican. Presbyterian. Subspecialties: Dental growth and development; Environmental toxicology. Current work: Control mechanisms -craniofacial growth and development; and image enhancement determination of heavy metal dentin incorporation. Home: 402 Hollen Rd Baltimore MD 21212 Office: Balt Coll Dental Surgery U Md 666 Baltimore S Baltimore MD 21201

OWEN, GEORGE MURDOCK, pharmaceuticals company medical director, educator; b. Alliance, Ohio, Feb. 16, 1930; s. Roger N. and Doris (Murdock) O.; m. Barbara L. Davis, Aug. 30, 1952 (div. 1976); children: David, Suzanna, John; m. Anita L. Vangarelli, June 29, 1976; 1 son, Gregory. A.B., Hiram Coll., 1951; M.D., U. Cin., 1955. Diplomate: Am. Bd. Pediatrics; med. lic. Ohio, N.Mex., Mich., N.Y. Intern U. Iowa Hosps., Iowa City, 1955-56, resident, 1956-58; clin. assoc. NIH, Bethesda, Md., 1958-60; research fellow U. Iowa, Iowa City, 1960-62, asst. prof., 1962-66; prof. Ohio State U., Columbus, 1966-73; prof., dir. U. N.Mex., Albuquerque, 1973-77, U. Mich.-Ann Arbor, 1977-81; med. dir. Bristol-Myers Co., N.Y.C., 1981—; clin. prof. Cornell Med. Center, N.Y.C., 1982—; attending physician N.Y. Hosp.; mem. nutrition study sect. NIH, 1972-76. Contbg. editor: Manual Clin. Nutrition, 1979—; mem. editorial bd.: Am. Jour. Clin. Nutrition, 1966-77; contbr. writings to profl. publs.; editor monographs. Served as sr. asst. surgeon USPHS, 1958-60. USPHS tng. grantee, 1978; postdoctoral fellow NIH/USPHS, 1960; research grantee in maternal and child health USPHS, 1966, 73. Fellow Am. Acad. Pediatrics; mem. Am. Soc. Clin. Nutrition, Soc. Pediatric Research, Am. Pediatric Soc. Subspecialties: Pediatrics; Nutrition (medicine). Current work: Nutritional research and development. Home: 251 Wilton Rd Westport CT 06880 Office: Bristol-Myers Co Internat Div 345 Park Ave New York NY 10154

OWEN, RAY DAVID, biology educator; b. Genesee, Wis., Oct. 30, 1915; s. Dave and Ida (Hoeft) O.; m. June J. Weissenberg, June 24, 1939; 1 son, David G. B.S., Carroll Coll., Wis., 1937, Sc.D., 1962; Ph.D., U. Wis., 1941, Sc.D., 1979; Sc.D., U. of Pacific, 1965. Asst. prof. genetics, zoology U. Wis., 1944-47; Gosney fellow Calif. Inst. Tech., Pasadena, 1946-47, asso. prof. div. biology, 1947-53, prof. biology, 1953—, also chmn., v.p. for student affairs, dean of students, 1975-80; research participant Oak Ridge Nat. Lab., 1957-58; Cons. Oak Ridge Inst. Nuclear Studies; mem. Pres.'s Cancer Panel. Author: (with A.M. Srb) General Genetics, 1952, 2d edit. (with A.M. Srb, R. Edgar), 1965; Contbr. articles to sci. jours. Mem. Genetics Soc. Am. (pres.), Am. Assn. Immunologists, Am. Soc. Human Genetics, Western Soc. Naturalists, Am. Soc. Zoologists, Am. Genetics Assn., Nat. Acad. Scis., Am. Acad. Arts and Scis., Am. Acad. Polit. and Social Sci., Sigma Xi. Subspecialty: Genetics and genetic engineering (biology). Home: 1583 Rose Villa St Pasadena CA 91106

OWEN, ROBERT BARRY, research physicist; b. Chgo., Oct. 16, 1943; s. Jack Saunders and Dorothy Orleen (Riley) O.; m. Lyn Irvin, Oct. 31, 1970; children: Catherine Anne, Ruth Riley. B.S. in Physics, Va. Poly. Inst., 1966, Ph.D., 1972. Student trainee NASA Marshall Space Flight Center, Huntsville, Ala., 1962-66, aerospace technologist, 1966-72, research physicist, 1972—; grad. teaching asst. Va. Poly. Inst., Blacksburg, 1966-69; chief scientist Optical Research Inst.; cons. in optics. Contbr. articles to profl. jours. Recipient sect. prize IEEE, 1978; Service citation Soc. Photo-Optical Instrumentation Engrs., 1980; cert. of recognition NASA, 1979, 80, 81; others. Mem. Optical Soc. Am, Sigma Xi, Sigma Pi Sigma. Republican. Unitarian. Subspecialties: Optical Measurement Systems; Holography. Current work: Original and novel application of advanced optical measurement techniques to space science, materials science, fluids, atmospheric science, oceanography, and medicine. Home: 703 Clinton Ave E Huntsville AL 35801 Office: Marshall Space Flight Center ES 74 AL 35812

OWEN, S. JOHN T., electrical engineering educator and administrator; b. Ironbridge, Eng.; s. Albert Edward and Isabel (Jasper) O.; m. Rosemary Molly, Aug. 6, 1960; children: Matthew, Rebekah, Harriet. B.Sc., U. Nottingham, Eng., Ph.D., 1961. Lectr. U. Nottingham, 1961-68; vis. prof. U. Ala., 1968-69; reader in phys. electronics U. Nottingham, 1970-77; prof. electrophysics Oreg. State U., Corvallis, 1977—, head dept. elec. and computer engring., 1978—. Author: (with J.E. Parton) Applied Electromagnetics, 1974; contbr. articles to profl. jours. Recipient Lord Rutherford award Inst. Electronic and Radio Engrs., U.K., 1971. Mem. IEEE, Instn. Elec. Engrs. (U.K.). Anglican. Club: Beaver. Subspecialties: 3emiconductors; Electronics. Current work: Semiconductor materials and devices, semiconductor heterojunctions, ternary and quaternary materials. Home: 7835 NW Mitchel Dr Corvallis OR 97330 Office: Dept Elec and Computer Engring Oreg State U Corvallis OR 97330

OWEN, TOBIAS CHANT, astrophysicist, educator; b. Oshkosh, Wis., Mar 20, 1936; s. George Colville and Monica (Volkert) O.; m. Linda Lewis, Sept. 2, 1960; children: Jonathan, David. B.A., U. Chgo., 1955, B.S., 1959, M.A., 1960; postgrad., Goethe U., Frankfurt, Germany, 1956-57; Ph.D., U. Ariz., 1965. Asst. physicist to sci. advisor IIT Research Inst., Chgo., 1965-69; vis. assoc. prof. planetary sci. Calif Inst. Tech., Pasadena, 1970; faculty SUNY-Stony Brook, 1970—, prof. astronomy, 1972—; Vernadsky Inst lectr., USSR, 1982; cons. NASA, NAS, NRC. Contbr. articles to profl. jours.; author: (with Donald Goldsmith) The Search for Life in the Universe, 1980. Recipient NASA Medal for Exceptional Sci. Achievement, 1977, Group Achievement awards, 1971, 81; medal Aero Club of Cairo, 1977; AAAS Group award of Newcomb Cleve. Prize, 1977; U. Chgo. Alumni Profl. Achievement award, 1983; NASA grantee, 1966—. Mem. AAAS, Am. Geophys. Union, Am. Astron. Soc., Astron. Soc. Pacific, Internat. Assn. Geochem. and Cosmochemistry, Internat. Astron. Union, Internat. Soc. Study Origin of Life, Royal Astron. Soc., N.Y. Acad. Sci., Com. for Space Research. Subspecialties: Planetary science; Planetary atmospheres. Current work: Composition, structure, origin of planetary atmospheres, comets, icy satellites. Office: ESS SUNY Stony Brook NY 11794

OWEN, WALTER SHEPHERD, educator; b. Liverpool, Eng., Mar. 13, 1920; s. Walter L. and Dorothea (Lunt) O. B.Engring., U. Liverpool, 1940, M.Engring., 1942, Ph.D., 1950, D.Eng., 1972. Metallurgist English Electric Co., 1940-46; mem. research staff MIT, 1951-57; prof. metallurgy U. Liverpool, 1957-66; prof., dir. materials sci. and engring. Cornell U., 1966-70; dean Tech. Inst., Northwestern U., 1970-73; prof. materials sci. and engring. MIT, 1973—; Cons. to industry. Author research papers. Commonwealth Fund fellow, 1951. Mem. Am. Inst. Metall. Engrs., Nat. Acad. Engring., Am. Soc. Metals, Instn. Metallurgists, Inst. Metals, Iron and Steel Inst., Sigma Xi. Subspecialty: Metallurgy. Office: Dept Materials Sci and Engring Mass Inst Tech Cambridge MA 02139

OWENS, CHARLES WESLEY, chemistry educator; b. Billings, Okla., Oct. 27, 1935; s. Fred and Mary (Metheney) O.; m. Barbara Jeanne, Dec. 18, 1955; children: Charles, Wesley, Michael, Janet. B.S., Colo. Coll., 1957; Ph.D., U. Kans., 1963. With Phillips Chem. Co., Borger, Tex., 1957-58; mem. faculty U. N.H., Durham, 1963—, now prof. chemistry. Contbr. articles to sci. jours. Served to 1st lt. U.S. Army, 1957-58. Mem. Am. Chem. Soc., Northeastern Assn. Chemistry Tchrs., Phi Beta Kappa, Sigma Xi. Subspecialties: Physical chemistry; Plant pathology. Current work: Chemistry of wood. Office: U NH Durham NH 03824

OWENS, FREDRIC NEWELL, animal science educator; b. Baldwin, Wis., Sept. 1, 1941; s. Fred and Stella (Jorstad) Ownes. Student, Wis. State U.-River Falls, 1959-61; B.S., U. Minn., 1964, Ph.D., 1968. Asst. prof. animal sci. U. Ill., 1968-74; assoc. prof. Okla. State U., 1974-78, prof., 1978—. Contbr. numerous articles to profl. jours.; sect. editor: Jour. Animal Sci., 1975-78. Served with Army NG, 1959-61. Recipient Young Researcher award Am. Soc. Animal Soc., 1974; Tyler award Okla. State U., 1980. Lutheran. Subspecialty: Animal nutrition. Current work: Rumnant nutrition. Office: Animal Sci Bldg Stillwater OK 74078 Home: 121 W 35th St Stillwater OK 74074

OWENS, JAMES CARL, physicist; b. Saginaw, Mich., May 24, 1937; s. James Samuel and Marion Alice (Borgerding) O.; m. Emily Ann Warren, June 24, 1961; children: Lauran Elizabeth, Matthew Warren. A.B. in Physics summa cum laude, Oberlin Coll., 1959; A.M. Harvard U, 1960; Ph.D. in Physics, 1965. Physicist, project leader U.S. Nat. Bur. Standards, Boulder, Colo., 1964-65, Environ. Sci. Services Adminstrn., Boulder, 1965-69; research assoc. physics div. Research Labs., Eastman Kodak Co., Rochester, N.Y., 1969-80, 83—; lab. head. Research labs., Eastman Kodak Co., Rochester, N.Y., 1969-71, 76-79; research assoc., mem. sr. staff color photography div. Research Labs., Eastman Kodak Co., 1980-83. Contbr. articles to profl. publs. NSF fellow Harvard U., 1959-64. Mem. Optical Soc. Am. (pres. Rochester chpt. 1981-82), Am. Phys. Soc., IEEE, Soc. Photog. Scientists and Engrs. Democrat. Unitarian. Club: Torch (Rochester). Subspecialties: Graphics, image processing, and pattern recognition; Laser data storage and reproduction. Current work: Computer and optical image processing; laser and electro-optical systems; photographic system design. Patentee in field. Home: 3 Woods End Pittsford NY 14534 Office: Eastman Kodak Co Research Labs Bldg 59 Rochester NY 14650

OZER, HARVEY LEON, biochemist, educator; b. Boston, July 6, 1938; s. Samuel L. and Naomi (Smith) O.; m. Joy Hochstadt, Feb. 3, 1960; children: Juliane Hochstadt-Ozer. B.A., Harvard Coll., 1960; M.D., Stanford U., 1965. Staff fellow Nat. Cancer Inst., NIH, Bethesda, Md., 1969-72; sr. scientist Worcester Found. Exptl. Biology, Shrewsbury, Mass., 1972-77; prof. biochemistry/biology Hunter Coll., CUNY, N.Y.C., 1977—; cons., mem. virology study sect. NIH, 1978-82. Contbr. research publs. to profl. lit. Served with USPHS, 1966-68. NIH research grantee, 1972—. Mem. Am. Soc. Genetics, Am Soc. Microbiology, Am. Soc. Virology (charter), N.Y. Acad. Scis., AAAS, Sigma Xi. Subspecialties: Genetics and genetic engineering (biology); Virology (biology). Current work: Gene transfer, virus-cell interaction, DNA tumor viruses, cell mutants, DNA replication. Home: 300 Central Park W Apt 14G New York NY 10024 Office: 695 Park Ave New York NY 10021

PACEY, GILBERT ELLERY, chemistry educator, consultant; b. Peoria, Ill., May 2, 1952; s. Gilbert Joseph and Grace Ellory (Baker) P.; m. Vicki Ellen Miller; 1 son, Ian Joseph. B.S., Bradley U., 1974; Ph.D., Loyola U., Chgo., 1979. Asst. prof. Loyola U., Chgo., 1978-79, Miami U., Oxford, Ohio, 1979—; cons. Abbott Labs., North Chicago, Ill., 1980—, Tecator Inc., Hoganass, Sweden, 1982—. Contbr. articles to profl. jours. Recipient Young Faculty Research award Soc. Analytical Chemists Pitts., 1982; research award Abbott Labs., 1982-84. Mem. Am. Chem. Soc., Soc. Applied Spectroscopy, Sigma Xi. Presbyterian. Subspecialties: Analytical chemistry; Organic chemistry. Current work: Development of new analytical reagents, and research into flow injection analysis; the reagent group under investigation are Chromogenic and flourogenic crown ethers for clinical analysis. Patentee: Triflouremethyl Substituted Chromogenic Crown Ether and Methods of Using, 1982. Home: 114 Beechpoint Oxford OH 45506 Office: Dept Chemistry Miami U Oxford OH 45056

PACHECO-MALDONADO, ANGEL MANUEL, psychologist; b. San Juan, P.R., Aug. 10, 1946; s. Angel Manuel and Catalina (Pacheco) Maldonado; m. Blanca G. Silvestrini, July 25, 1968; children: Angel Jaime, Javier Francisco. B.A., U. P.R., 1968; Ph.D., SUNY-Albany, 1972; postdoctoral student, Harvard U., 1972-73. Dir. P.R. Research Inst., Rio Piedras, 1975-76; asst. prof. U. P.R., Rio

Piedras, 1975-78, assoc. prof., 1978—; fellow Clark U., Worcester, Mass., 1980—; research fellow in edn. Harvard U., Cambridge, Mass., 1980—; adj. prof., cons. Cayey Sch. Medicine, San Juan, 1980-81; sci. adv. Drug Abuse Control Dept., San Juan, 1976-77; research cons. Govt. Venezuela, Caracas, 1976; tng. cons. Hilton Internat., San Juan, 1978. NRC sr. postdoctoral fellow, 1980; NIMH fellow, 1981-83; Nat. Inst. Drug Abuse grantee, 1978-80; Social Sci. Research Council fellow, 1973. Mem. Am. Psychol. Assn., InterAm. Soc. Psychology (nat. rep. 1981—), Phi Delta Kappa. Subspecialties: Developmental psychology; Social psychology. Current work: Critical person in environment transitions, such as migration; parental values and goals for self development across cultures. Office: Sch Edn Harvard U Cambridge MA 02138 Home: PO Box 22275 University Sta Rio Piedras PA 00931

PACHMAN, LAUREN MERLE, physician, pediatrician; b. Durham, N.C., Mar. 16, 1937; d. Daniel James and Vivian Allison (Futter) P.; m. Mark A. Satterthwaite; children: Emily Ann, Theodore Daniel. B.A., Wellesley Coll., 1957; M.D., U. Chgo., 1961. Diplomate: Am. Bd. Pediatrics, Am. Bd. Allergy and Clin. Immunology. Intern Phila. Gen. Hosp., 1961-62; resident Babies Hosp., N.Y.C., 1962-64; asst. prof. Hosp. of Rockefeller U., 1964-66; asst. dept. pediatrics Columbia U., 1964-66; instr. dept. pediatrics LaRabida-U. Chgo., 1966-69, asst. prof., 1969-71; assoc. prof. pediatrics Northwestern U., 1971-78, prof., 1978—; head div. immunology and rheumatology Children's Meml. Hosp., 1971—. Contbr. articles to profl. jours. Nat. Found. fellow, 1958; NIH fellow, 1964-66; William K. Ishmael vis. prof. and lectr., Oklahoma City, 1983. Fellow Am. Acad. Pediatrics; mem. Soc. Pediatric Research, Am. Assn. Immunologists, Am. Rheumatism Assn., AAAS. Subspecialties: Pediatrics; Rheumatology. Current work: Pediatric rheumatic diseases. Home: 1521 Brummel St Evanston IL 60202 Office: 2300 Children's Plaza Chicago IL 60614

PACHUT, JOSEPH FRANCIS, JR., geology educator; b. Little Falls, N.Y., Mar. 6, 1950; s. Joseph F. and Marion E. (Steinberg) P.; m. Elizabeth A. Battisti, June 3, 1972; children: Jennifer, Melissa, Geoffrey. B.A., SUNY-Oneonta, 1972; Ph.D., Mich. State U., 1977. Grad. asst. Mich. State U., East Lansing, 1972-77, instr., 1975, 77, Albion (Mich.) Coll., 1975; assoc. prof. geology Ind. U.-Purdue U., Indpls., 1977—; speaker at invited seminars, 1977—. Contbr. articles in field to profl. jours. N.Y. State Regents scholar, 1968-72; NDEA fellow, 1975; recipient Sigma Xi research award, 1976; NSF grantee, 1982—. Mem. Paleontol. Soc., AAAS, Internat. Bryozoology Assn., Soc. Systematic Zoology, Sigma Xi. Democrat. Roman Catholic. Subspecialties: Paleobiology; Paleoecology. Current work: Influence of environmental conditions (inferred) on the genetic adaptive strategies and evolution of fossil bryozoans; biometrics, evolution and paleoecology of Paleozoic bryozoans. Home: 63 Trails End Greenwood IN 46142 Office: Dept Geology Ind U.-Purdue U 425 Agnes St Indianapolis IN 46202

PACKARD, DAVID, manufacturing company executive, electrical engineer, former deputy secretary of defense; b. Pueblo, Colo., Sept. 7, 1912; s. Sperry Sidney and Ella Lorna (Graber) P.; m. Lucile Salter, Apr. 8, 1938; children: David Woodley, Nancy Ann Packard Burnett, Susan Packard Orr, Julie Elizabeth. A.B., Stanford U., 1934, E.E., 1939; LL.D. (hon.), U. Calif., Santa Cruz, 1966, Catholic U., 1970, Pepperdine U., 1972, D.Sc., Colorado Coll., 1964, Litt.D., So. Colo. State Coll., 1973, D.Eng., U. Notre Dame, 1974. With vacuum tube engrng. dept. Gen. Electric Co., Schenectady, 1936-38; co-founder, partner Hewlett-Packard Co., Palo Alto, Calif., 1939-46, pres., 1947-64, chmn. bd., 1964-68, 72—, chief exec. officer, 1964-68; dep. sec. defense, Washington, 1969-71; dir. Caterpillar Tractor Co., 1972—, Standard Oil Co. of Calif., 1972—, The Boeing Co., 1978—. Mem. Palo Alto Bd. Edn., 1947-56; mem. President's Commn. Personnel Interchange, 1972-74, Trilateral Commn., 1973-81; chmn. bd. regents Uniformed Services U. of Health Scis., 1974-82; mem. U.S.-USSR Trade and Econ. Council, 1974-82; bd. dirs. Santa Clara (Calif.) County Mfrs. Group, 1978—; mem. bd. overseers Hoover Instn., 1972—; bd. dirs. Nat. Merit Scholarship Corp., 1963-69, Found. for Study of Presidential and Congressional Terms, 1978—, Alliance to Save Energy, 1977—, Atlantic Council, 1972—; vice chmn. Atlantic Council, 1972-80; bd. dirs. Am. Enterprise Inst. for Public Policy Research, 1978—; trustee Herbert Hoover Found., 1974—, Colo. Coll., 1966-69, U.S. Churchill Found., 1965-69, Stanford U., 1954-69; pres. bd. trustees Stanford U., 1958-60. Decorated Grand Cross of Merit Fed. Republic of Germany, 1972; recipient numerous awards including Medal of Honor Electronic Industries Assn., 1974, Silver Helmet Defense award AMVETS, 1973, Washington award Western Soc. Engrs., 1975, Hoover medal ASME, 1975, Gold Medal award Nat. Football Found. and Hall of Fame, 1975, Good Scout award Boy Scouts Am., 1975, Vermilye medal Franklin Inst., 1976, Internat. Achievement award World Trade Club of San Francisco, 1976, Merit award Am. Consulting Engrs. Council Fellows, 1977, Achievement in Life award Ency. Britannica, 1977, Engrng. Award of Distinction San Jose State U., 1980, Thomas D. White Nat. Def. award U.S. Air Force Acad., 1981, Disting. Info. Scis. award Data Processing Mgmt. Assn., 1981, Sylvanus Thayer award U.S. Mil. Acad., 1982, Environ. Leadership award Natural Resources Def. Council, 1983. Fellow IEEE (Founders medal 1973); mem. Nat. Acad. Engrng. (Founders award 1979), Instrument Soc. Am. (hon. mem.), Calif. C. of C. (dir. 1962-78), Am. Mgmt. Assn. (dir. 1956-59), Wilson Council, The Bus. Roundtable Advt. Council, Bus. Council, Am. Ordnance Assn. (Crozier Gold medal 1970), Sigma Xi, Phi Beta Kappa, Tau Beta Pi, Alpha Delta Phi (named Disting. Alumnus of Yr. 1970). Clubs: Bohemian, Commonwealth, Pacific Union, Engrs. (San Francisco); Exec. (Chgo.); The Links (N.Y.C.); Alfalfa, Capitol Hill (Washington); California (Los Angeles). Subspecialty: Electrical engineering. Office: 1501 Page Mill Rd Palo Alto CA 94304

PACKER, KATHERINE HELEN, librarian, educational administrator; b. Toronto, Ont., Can., Mar. 20, 1918; d. Cleve Alexander and Rosa Ruel (Dibblee) Smith; m. William A. Packer, Sept. 27, 1941; 1 dau., Marianne Katherine. B.A., U. Toronto, 1941; A.M.L.S., U. Mich., 1953; Ph.D., U. Md., 1975. Cataloguer William L Clements Library, U. Mich., 1953-55, U. Man. (Can.) Library, Winnipeg, 1956-59; cataloguer U. Toronto Library, 1959-63; asst. prof. Faculty Library Sci., 1967-75, asso. prof., 1975-78, prof., dean, 1979—; head cataloguer York U. Library, Toronto, 1963-64; chief librarian Ont. Coll. Edn., Toronto, 1964-67; Can. Council Library Schs. rep. to Adv. Bd. on Sci. and Tech. Info., NRC Can., 1976-78. Author: Early American School Books, 1954. Recipient Disting. Alumnus award U. Mich., 1981. Mem. Am. Assn. Library Schs., ALA, Am. Soc. Info. Sci., Can. Library Assn. (Howard Phalin award 1972), Internat. Fedn. Library Assns., Phi Kappa Phi. Subspecialties: Information systems (information science); Information systems, storage, and retrieval (computer science). Current work: Research interests; social impact of videotex, electronic mail systems. Home: 53 Gormley Ave Toronto ON M4V 1Y9 Canada Office: Faculty Library and Info Sci U Toronto 140 Saint George St Toronto ON M5S 1A1 Canada

PACKMAN, ALLAN B., research engineer; b. Detroit, May 21, 1939; s. Victor William and Rose (Lander) P.; m. Judith Francis Berman, Aug. 26, 1961; children: Debra, Cheryl. B.S., U. Mich., 1961; M.S., Rensselaer Poly. Inst., 1964. Research engr. Pratt & Whitney Aircraft, East Hartford, Conn., 1961—; adj. prof. U. Hartford, West Hartford, Conn., 1979—. Mem. AIAA (aero-acoustics tech. com. 1980—). Republican. Subspecialties: Acoustics; Acoustical engineering. Current work: Direct jet noise reduction program. Inventor aircraft engine noise reduction devices, 1979, 80, 81. Home: 58 Pilgrim Rd West Hartford CT 06117 Office: Pratt and Whitney Aircraft 400 Main St East Hartford CT 06108

PADBERG, DANIEL IVAN, university dean, consultant; b. Summersville, Mo., Nov. 9, 1931; s. Christopher E. and Ruth (Bagley) P.; m. Mildred Frances True, Aug. 5, 1956; children: Susan E., Jean E., Carol N. B.S., U. Mo., 1953, M.S., 1955; Ph.D., U. Calif.-Berkeley, 1961. Asst. prof. Ohio State U., Columbus, 1961-64; project leader Nat. Commn. Food Mktg., Washington, 1965-66; from assoc. prof. to prof. Cornell U., Ithaca, N.Y., 1966-75; prof., head dept. agrl. econs. U. Ill.-Urbana, 1975-81; dean Coll. Food and Natural Resources, U. Mass., Amherst, 1981—; cons. Author: Economics of Food Retailing, 1968, Today's Food Broker, 1971, also articles. Served to lt. (j.g.) USN, 1955-58. Simon fellow, Manchester, Eng., 1972-73. Mem. Am. Agrl. Econs. Assn. (Disting. Research award 1975, 77), Northeastern Agrl. Econs. Council. Episcopalian. Club: Rotary (Amherst, Mass.). Subspecialties: Agricultural economics; Nutrition (biology). Current work: Interaction between large food manufacturers and consumers involving new product development, advertising, nutrition, information processing, consumer behavior. Home: 148 Aubinwood Rd Amherst MA 01002 Office: U Mass Amherst MA 01003

PADGETT, WILLIAM JOWAYNE, statistics educator, researcher; b. Walhalla, S.C., May 15, 1943; s. J.J. and Edith (Abercrombie) P.; m. Faye S. Swayngham, July 3, 1965; children: Carla, Scott. B.S. in Math, Clemson U., 1966, M.S., 1968; Ph.D. in Stats, Va. Poly. Inst. and State U., 1971. Asst. prof. stats. U. S.C., 1971-73, assoc. prof., 1973-77, prof., 1977—. Author: (with C.P. Tsokos) Random Integral Equations, 1974. Mem. Inst. Math. Stats., Am. Stat. Assn., Soc. Indsl. and Applied Math. Subspecialties: Statistics; Probability. Office: Dept Math and Stats U SC Columbia SC 29208

PADILLA, GERALDINE V(ALDES), nursing researcher; b. Manila, Feb. 28, 1940; d. Jose M. and Ruby (Baugh) Valdes; m. Gilbert J. Padilla, June 30, 1966; children: Mark J., Mathew L. B.A., Assumption Coll, Manila, 1962; M.A., Ateneo U., Manila, 1965; Ph.D., UCLA, 1970. Dir. nursing research City of Hope Nat. Med. Ctr., Duarte, Calif., 1970—; lectr. nursing Calif. State U.-Los Angeles, 1981—; cons. in field. Author: Interacting with Dying Patients, 1975. Dept. HHS grantee, 1980-84; Nat. Cancer Inst. grantee, 1983-86. Mem. Am. Psychol. Assn., Western Soc. Research in Nursing. Subspecialties: Behavioral psychology; Social psychology. Current work: Study of effective patient education programs; methods of reducing patient distress associated with medical treatments and procedures; and quality assurance in nursing care. Office: City of Hope Nat Med Center 1500 E Duarte Rd Duarte CA 91010

PAECH, CHRISTIAN GEROLF, chemistry and plant science educator, researcher; b. Leipzig, Saxony, Germany, July 10, 1942; came to U.S., 1975; s. Karl and Birgit (Fischer) P.; m. Sigrid Bläsig, May 15, 1970; children: Reikja Ailien, Nikolas Kolja, Maju Loriann, Suenia Yvonne. Diploma in Chemistry, U. Freiburg, W.Ger., 1970, Dr. rer. nat., 1975. Research assoc. dept. chemistry U. Freiburg, 1975; dept. biochemistry Mich. State U., East Lansing, 1975-78; asst. research biochemist U. Calif.-San Francisco, 1978-81; asst. prof. chemistry S.D. State U., Brookings, 1981-82, asst. prof. chemistry and plant sci., 1982—. Grantee Am. Soybean Assn., 1981, U.S. Dept. Agr., 1982. Mem. Am. Soc. Biol. Chemists, Am. Soc. Plant Physiologists, AAAS, Gesellschaft Deutscher Chemiker, Sigma Xi. Subspecialties: Biochemistry (biology); Photosynthesis. Current work: Agricultural research; enzymology and physiology of photosynthesis; soybean flower and pod setting. Home: 214 Lincoln Ln S Brookings SD 57006 Office: SD State Univ Station Biochemistry Box 2170 Brookings SD 57007

PAFFENBARGER, GEORGE CORBLY, health foundation executive; b. McArthur, Ohio, Nov. 3, 1902; s. Andrew Wolf and Ida Priscilla (Seal) P.; m. Rachel Ada Appleman, July 9, 1925; children: George Corbly, Gretchen Minners, Anne. D.D.S., Ohio State U., 1924, D.Sc., 1944; D.Sc., Nihon U., Tokyo, 1961, N.J. Coll. Medicine and Dentistry, 1975, Georgetown U., 1978. Externe Palama Dental Clinic, Honolulu, 1925-27; instr. Ohio State U., Columbus, 1928-29; research assoc. Nat. Bur. Standards, Washington, 1929-42, 46-68; sr. research assoc. ADA Health Found. Research Unit, Nat. Bur. Standards, Washington, 1946—; pres. William J. Gies Found. Advancement of Dentistry, N.Y.C., 1966—; praelector faculty medicine St. Andrews (Scotland) U., 1959. Author: (with W. Souder) Physical Properties of Dental Materials, 1942; editor: (with S. Pearlman) Frontiers of Dental Science, 1962. Founder, pres. Seneca Creek Watershed Assn., Montgomery County, Md., 1954; tchr., stated clk., sec. bd. trustees Darnestown Presbyterian Ch., Germantown, Md., 1946-79. Served with USN, 1942-46. Recipient Wilmer Souder award Dental Materials Group, Internat. Assn. Dental Research, 1958; Internat. Miller prize Fedn. Dentaire Internat., 1972; Hollenback Meml. prize Acad. Restorative Dentistry, 1976; Disting. Service award ADA, 1982. Fellow Am. Coll. Dentists (sec., pres. Washington sect. 1930-38, Gies award 1935), N.Y. Acad. Dentistry; mem. Japan Dental Assn., Deutsche Gesellschaft fur Zahn Mund und Kieferheilkunde, Fedn. Dentaire Francaise. Lodge: Masons. Current work: Physical and chemical characterization of dental materials, foundation executive. Home: 407 Russell Ave Apt 804 Gaithersburg MD 20877 Office: ADA Health Found Research Unit Nat Bur Standards Washington DC 20234

PAGALA, MURALI KRISHNA, muscle physiologist, researcher; b. Sri Kalahasti, Andhra, Chittoor, India, Oct. 2, 1942; came to U.S., 1970, naturalized, 1979; s. Lakshmaiah and Radhamma (Bhimavaram) P.; m. Vijaya Bhimavaram, Dec. 12, 1969; 1 child, Sobhana. Ph.D., Sri Venkateswara U., Tirupati, Andhra, India, 1969. Postdoctoral fellow Inst. Muscle Disease, N.Y.C., 1970-73, asst. mem., 1974; assoc. research scientist N.Y. U., N.Y.C., 1974-75; electrophysiologist Maimonides Med. Ctr., Bklyn., 1975—; vis. research scientist U. Saarland, Homburg, W.Ger., 1981, 82. Muscular Dystrophy Assn. postdoctoral fellow, 1970-73; Maimonides Research Fund grantee, 1983. Mem. Am. Physiol. Soc., Cell and Gen. Physiology Assn., N.Y. Acad. Scis. Democrat. Hindu. Subspecialties: Neurophysiology. Current work: Physiological studies on in vitro electromyography, neuromuscular transmission, neuromuscular fatigue, myasthenia gravis, human muscle electrophysiology, computer aided data and signal processing. Office: Maimonides Med Center 4802 10th Ave Brooklyn NY 11219

PAGANELLI, CHARLES VICTOR, physiology educator; b. N.Y.C., Feb. 13, 1929; s. Charles V. and Mary (Spalla) P.; m. Barbara Harriet Slauson, Sept. 18, 1954; childrn: William, Kathryn, Peter, Robert, John. B.A., Hamilton Coll., 1950; M.A., Harvard U., 1952, Ph.D., 1957. From instr. to prof. physiology SUNY-Buffalo, 1958—. Recipient Elliott Coues award Am. Ornithol. Union, 1981. Mem. Am. Physiol. Soc., Undersea Med. Soc., AAAS. Roman Catholic. Subspecialty: Physiology (medicine). Current work: Diffusion and transport of substances across biological membranes; gas exchange in avian eggs. Office: SUNY 120 Sherman Hall Buffalo NY 14214

PAGANO, JOSEPH STEPHEN, physician; b. Rochester, N.Y., Dec. 29, 1931; s. Angelo and Marian (Vinci) P.; m. Nancy Louise Reynolds, June 8, 1957; children: Stephen and Christopher. A.B. with honors, U. Rochester, 1953; M.D., Yale U., 1957. Intern medicine Mass. Meml. Hosp., Boston, 1957-58; research asso. Wistar Inst., Phila., 1958-60; asst. resident Peter Bent Brigham Hosp., Boston, 1960-61; research asso. Karolinska Inst., Stockholm, Sweden, 1961-62; instr. pediatrics Children's Hosp., Phila., 1959-60; asso. medicine U. Pa., Phila., 1962-65; asst prof. medicine and bacteriology and immunology U. N.C. Sch. Medicine, Chapel Hill, 1965-68, co-dir. div. infectious diseases, 1969-72, asso. prof. medicine and bacteriology and immunology, 1973—, dir., 1974—; vis. prof. Swiss Inst. Cancer Research, Lausanne, 1970-71; dir. virology lab. N.C. Meml. Hosp., U. N.C. Sch. Medicine, Chapel Hill, 1969-74, attending physician medicine and infectious diseases, 1965—; dir. Nat. Research Service Award Tng. Grant, Cancer Research Center, 1975—; cons. Nat. Inst. Neurol. Diseases and Stroke, 1974—; ad hoc mem. Cancer Research Ctr. Rev. Com., 1975—, Cancer Spl. Program Adv. Com., 1975—. Mem. bd. assoc. editors: Cancer Research, 1976-80, Jour. Immunology, 1977-80; Contbr. articles to profl. jours. Bd. dirs. N.C. div. Am. Cancer Soc., 1980—. Served with USPHS, 1958-60. Recipient Research Career award USPHS, 1968-73, Sinsheimer Found award, 1966-71; Nat. Found. fellow, 1961-62. Mem. Infectious Diseases Soc. Am. (chmn. immunization com.), Am. Soc. Microbiology (chmn. div. DNA viruses 1982-83), Am. Fedn. Clin. Research, Elisah Mitchell Sci. Soc., Am. Assn. Immunologists, Soc. Gen. Microbiology, Am. Soc. Clin. Investigation, Am. Assn. Cancer Research, Am. Assn. Comparative Research Leukemia and Related Diseases, Assn. Am. Cancer Insts., So. Soc. Clin. Investigation, Sigam Xi. Subspecialties: Virology (medicine); Internal medicine. Current work: Cancer and virus infections, tumor virology and molecular biology of herpes viruses, especially Epstein-Barr biruses. Home: 114 Laurel Hill Rd Chapel Hill NC 27514 Office: U NC Cancer Research Center Box 3 Swing Bldg Chapel Hill NC 27514

PAGE, CHARLES HENRY, neurophysiology educator; b. East Orange, N.J., Aug. 28, 1941; s. Henry Harrison and Ruth Claire (Hancock) P.; m. Carol Ann Laundy, Aug. 17, 1963; children: Christopher, Frederick. B.S., Allegheny Coll., 1963; M.S., U. Ill., Ph.D., 1969. NIH postdoctoral fellow Stanford U., 1968-70; asst. prof. dept. zoology Ohio U., 1970-74; asst. prof. physiology Rutgers U., New Brunswick, 1974-76, assoc. prof., 1976—, also dir. physiology grad. program. Grass Found. fellow, 1971. Mem. Am. Physiol. Soc., Soc. Neurosci., Soc. Exptl. Biology, Am. Soc. Zoologists, AAAS, Sigma Xi. Subspecialties: Neurophysiology; Comparative neurobiology. Current work: Neural control of movement in crustaceans, cellular and comparative neurobiology. Office: Nelson Biol Lab Rutgers U Piscataway NJ 08854

PAGE, DON NELSON, theoretical physicist, educator, researcher; b. Bethel, Alaska, Dec. 31, 1948; s. Nelson Monroe and Zena Elizabeth (Payne) P. A.B., William Jewell Coll., 1971; M.S., Calif. Inst. Tech., 1976, Ph.D., 1976; M.A. (hon.), U. Cambridge, 1978. Research assoc. Calif. Inst. Tech., 1976; research asst. dept. applied math. and theoretical physics Cambridge (Eng.) U., 1976-77, 78-79; asst. prof. physics Pa. State U.-University Park, 1979—; vis. research fellow U. Tex.-Austin, 1982-83. Contbr. articles and research papers to profl. jours., conf. proceedings, encyclopedia. Served to lt. jr. grade USNR, 1971-76. NSF fellow, 1971-74; Woodrow Wilson fellow, 1971-72; Robert A. Millikan fellow, 1971-72; Danforth fellow, 1971-76; NATO fellow, 1977-78; Darwin Coll. fellow, 1977-79; Alfred P. Sloan fellow, 1982-84. Mem. Internat. Astronomical Union, Am. Phys. Soc. Baptist. Subspecialties: Relativity and gravitation; General relativity. Current work: Research in quantum radiation from black holes, gravitational instantons, predictability and time symmetry in quantum gravity, tharrow of time, the future of the universe, foundations of quantum mechanics, supergravity. Office: Pa State U 104 Davey Lab University Park PA 16802

PAGE, IRVINE HEINLY, physician; b. Indpls., Jan. 7, 1901; s. Lafayette and Marian (Heinly) P.; m. Beatrice Allen, Oct. 28, 1930; children: Christopher, Nicholas. B.A. in Chemistry, Cornell U., 1921, M.D., 1926; LL.D., John Carroll U., 1956; D.Sc. (hon.), Union U., 1957, Boston U., 1957, Ohio State U., 1960, U. Brazil, 1961, Cleve. State U., 1970, M.D., U. Siena, Italy, 1965, Med. Coll. Ohio, 1973, Ind. U., 1975, Rockefeller U., 1977. Intern Presbyn. Hosp., N.Y.C., 1926-28; head chem. div. Kaiser Wilhelm Inst., Munich, Germany, 1928-31; assoc. mem. Hosp. of Rockefeller Inst. for Med. Research, 1931-37; dir. Lab. for Clin. Research, Indpls. City Hosp., 1937-44; dir. research div. Cleve. Clinic Found., 1945-66, sr. cons. research div., 1966-68, cons. emeritus, 1968—; Past mem. nat. adv. heart council, USPHS; chmn. gov. bd. Methods in Medical Research; mem. subcom. shock Nat. Acad. Scis.-NRC; mem. adv. com. Whitaker Found. Author: Chemistry of the Brain, 1937, Hypertension, 1943, Arterial Hypertension-Its Diagnosis and Treatment, 1945, Experimental Renal Hypertension, 1948, Strokes, 1961, Serotonin, 1968, (with J. W. McCubbin) Renal Hypertension, 1968, Speaking to the Doctor, 1972, (with F. M. Bumpus) Anglotensin, 1974; also articles.; Editorial bd. various profl. jours.; editor-in-chief: Modern Medicine. Trustee Whitehead Inst. Med. Research. Recipient Lasker award Am. Heart Assn., 1958; alumni award of distinction Cornell U. Med. Coll., 1961; John Phillips Meml. award A.C.P., 1962; Gairdner award, 1963; Distinguished Service award AMA, 1964; Med. Communications award Am. Med. Writers Assn., 1965; Achievement award AMA, 1966; Oscar B. Hunter Meml. award, 1966; Passano Found. award, 1967; Sheen award AMA, 1968; Heart of the Year award Am. Heart Assn., 1969; Stouffer prize, 1970; Gifted Tchr. award Am. Coll. Cardiology, 1970. Mem. A.C.P. (master), Am. Soc. Pharmacology and Exptl. Therapeutics (hon.), Central Soc. Clin. Research, Am. Heart Assn. (past pres., founder and mem. council on high blood pressure research), AMA (chmn. sci. adv. com., past chmn. sect. on exptl. med.), Am. Soc. Biol. Chem., Nat. Acad. Scis., Am. Acad. Arts and Scis., Inst. Medicine (founder, founding mem.), Soc. Exptl. Biology and Medicine, Am. Physiol. Soc., Am. Chem. Soc., AAAS (v.p.), Am. Acad. Arts and Scis., Am. Soc. Study Arteriosclerosis (founding mem.), German Med. Soc. (corr.), Swedish Royal Acad. Sci. (fgn.), Sigma Xi. Subspecialty: Cardiology. Current work: Consultant, author. Home: Box 516 Hyannis Port MA 02647 Office: Cleve Clinic 9500 Euclid Ave Cleveland OH 44106

PAGE, JOHN BOYD, physics educator; b. Columbus, Ohio, Sept. 4, 1938; s. John B. and Heley (Young) P.; m. Norma Kay Christensen, July 28, 1966; children: Rebecca, Elizabeth. B.S., U. Utah, 1960, Ph.D., 1966. Research assoc. Institut fur Theoretische Physik, U. Frankfurt, W.Ger., 1966-68, Cornell U., Ithaca, N.Y., 1968-69; asst. prof. physics Ariz. State U., Tempe, 1969-75, assoc. prof., 1975-80, prof., 1980—; vis. researcher Max Planck Inst. Solid State Physics, 1975-76; vis. prof. Tech U. Munich, summer 1971. NSF grantee, 1972—. Mem. Am. Phys. Soc., Am. Assn Physics Tchrs., Sigma Xi, Phi Beta Kappa. Subspecialties: Condensed matter physics; Theoretical physics. Current work: Resonance raman scattering; phonon properties of perfect and imperfect crystals; electron-phonon interaction in light scattering. Home: 12434 S 71st St Tempe AZ 85287 Office: Dept Physics Ariz State Univ Tempe AZ 85287

PAGE, JOHN GARDNER, pharmacologist, toxicologist, researcher; b. Milw., Sept. 14, 1940; s. Raymond Gardner and Leone Bertha

(Churchill) P.; m. Joyce Ann Krueger, July 7, 1962; children: Teresa Ann, Kimberly Christine. B.S. in Pharmacy, U. Wis., 1964, M.S. in Pharmacology/Biochemistry, 1966, Ph.D., 1967. Diplomate: Am. Bd. Toxicology. Staff fellow NIH, Bethesda, Md., 1967-69; sr. toxicologist Eli Lilly Co., Greenfield, Ind., 1969-77; dir. toxicology/pathology Rhone Poulenc, Inc., Ashland, Ohio, 1977-79; dir. toxicology Toxigenics, Decatur, Ill., 1979-81; dir. pharmacology/toxicology White Sands Research Center, Alamogordo, N.Mex., 1981-83; sr. research adv. Battelle Meml. Inst., Columbus, Ohio, 1983—; adj. prof. environ. toxicology U. Ill., Urbana, 1980—. Contbr. articles to sci. jours. Active Hancock County unit Am. Cancer Soc. Recipient Rennebohm Teaching award U. Wis., 1964. Fellow Fedn. Am. Socs. Exptl. Biology; mem. AAAS, Am. Soc. Pharmacology and Exptl. Therapeutics, Soc. Toxicology, Am. Coll. Toxicology, Internat. Soc. Study Xenobiotics, Internat. Soc. Biochem. Pharmacology, Sigma Xi, Rho Chi. Subspecialties: Pharmacology; Toxicology (medicine). Current work: Genetic and environmental factors in drug metabolism; pharmacology; toxicology; drug metabolism; pharmacognetics; pharmacokinetics. Home: 1269 Castleton Rd N Columbus OH 43220 Office: 505 King St Columbus OH 43201

PAGEL, DEBORAH JOANNE, health physicist; b. Chgo., Apr. 25, 1955; d. Raymond Frank and Amelia Emelda (Suchecki) Heppeler; m. Richard Arlin Pagel, Aug. 4, 1979. B.S. in Biology, Elmhurst Coll., 1981. Health physics sr. technician Argonne Nat. Lab., Ill., 1977-81; health physicist Commonwealth Edison, Dresden Sta., Morris, Ill., 1981—. Contbr. articles to profl. jours. Mem. Am. Nuclear Soc., Health Physics Soc., Assn. Women in Sci. Republican. Roman Catholic. Subspecialties: Nuclear fission; Nuclear engineering. Current work: Radiation protection methods in the design and operation of a boiling water reactor. Home: 4035 Washington St Westmont IL 60559 Office: Commonwealth Edison Dresden Nuclear Power Station RR 1 Morris IL 60450

PAGELS, HEINZ RUDOLF, institute director, physicist, educator; b. N.Y.C., Feb. 19, 1939; m. Elaine L. Hiesey, June 7, 1969; children: Mark William. B.S. in Physics magna cum laude, Princeton U., 1960; Ph.D., Stanford U., 1965. Research assoc. Stanford Linear Accelerator Ctr., 1965, U. N.C., Chapel Hill, 1965-66; research assoc. Rockefeller U., N.Y.C., 1966-67, asst. prof., 1967-69, assoc. prof., 1969-82, adj. prof., 1982—; exec. dir., chief exec. officer N.Y. Acad. Scis., 1982—; vis. scientist/prof. Inst. Theoretical Physics, Serpukov, USSR, Oxford (Eng.) U., Cambridge (Eng.) U., CERN, Geneva, Fermi Nat. Accelerator Lab., Batavia, Ill., Los Alamos Sci. Lab., Lawrence Radiation Lab., Berkeley, Calif., Inst. Theoretical Physics, SUNY-Stony Brook; participant sci. confs., speaker; cons. Brookhaven Nat. Lab., 1967-74, White House Spl. Task Force to Sec. HEW, 1972, Los Alamos Sci. Lab., 1975, Inst. Advanced Study, Princeton, N.J., 1974-79, Dept. Energy, 1979. Author: The Cosmic Code: Quantum Physics as the Language of Nature, 1982, over 70 sci. articles. Trustee Aspen Ctr. for Physics, 1972-78, treas., 1973-79, sec., 1972, mem. exec. com., 1977; trustee Internat. League Human Rights, 1978; mem. Com. Concerned Scientists. U.S. Steel Found. Sci. Writing award Am. Inst. Physics, 1982. Mem. N.Y. Acad. Scis. (bd. govs. 1976-82, v.p. 1976-80, pres.-elect 1980, pres., chief exec. officer 1981, chmn. Albert Einstein Lecture com. 1976, chmn. phys. scis. sect. 1973-76, chmn. human rights com. 1976-78, chmn. Einstein Centennial Com. 1976-79, chmn. affiliates com. 1979), Am. Phys. Soc., AAAS. Subspecialties: Particle physics; Cosmology. Current work: Relativistic quantum field theory; cosmology and the early universe. Office: 2 E 63d St New York NY 10021

PAHWA, ASHOK, computer scientist; b. Ranikhet, India, Sept. 11, 1953; came to U.S., 1977; s. Jagdish Mitra and Sarla P.; m. Heman, Dec. 2, 1956. B.E. with honors, Birla Inst. Tech., Pilani, India, 1975; P.G. diploma, U. Roorkee, India, 1976; M.S., U. Kans., 1978. Specialist performance evaluation Gen. Electric Credit Co., Stamford, Conn., 1978-80; mem. tech. staff Bell Labs., Naperville, Ill., 1980—. Mem. Assn. Computing Machinery, IEEE, Computer Measurement Group. Subspecialties: Database systems; Computer architecture. Current work: Computer database management system; computer performance evaluation.

PAI, SHIH I., research educator; b. Tatung, Anhwei, China, Sept. 30, 1913; came to U.S., 1947, naturalized, 1955; s. Hsi Chuan and Swe Lin (Cha) P.; m. Yu Feng Chi, Feb. 4, 1934 (dec. Feb. 1958); children: Stephen Ming, Sue Pai Yang, Robert Yang, Lou Lung; m. Alice Jen-lan Wang, July 2, 1960. B.S. in Elec. Engring, Nat. Central U., Nanking, China, 1935; M.S. in Aero. Engring, MIT, 1938; Ph.D. in Aero. Engring. and Math, Calif. Inst. Tech., Pasadena, 1940; Dr. Techn. (hon.), Tech. U. Vienna, Austria, 1968. Prof. aero. engring. Nat. Central U., Chunking, China, 1940-47; vis. prof. aero. engring. Cornell U., Ithaca, N.Y., 1947-49; research prof. Inst. Fluid Dynamics and Applied Math., U. Md., College Park, 1949-76, Inst. for Phys. Sci. and Tech., 1976-83, prof. emeritus, 1983—; vis. prof. Tokyo U., 1966-67, U. Karlsruhe, W.Ger., 1980-81, U. Paris, 1981; hon. prof. Northwestern Poly. U., Xian, China, 1980—. Author: Magnetodynamics, 1962, Radiation Gasdynamics, 1966, Two-Phase Flows, 1977, Modern Fluid Mechanics, 1981. Served to capt. Chinese Air Force, 1937-40. Guggenheim fellow, 1957; Academia Sinica (China) fellow, 1962; NSF sr. postdoctoral fellow, 1966; recipient Sr. U.S. Scientist award Alexander von Humboldt Found., W.Ger., 1980. Fellow AIAA (assoc.); mem. Internat. Acad. Astronautics (corr.), Am. Phys. Soc., Sigma Xi (Achievement award 1975). Republican. Episcopalian. Subspecialties: Aeronautical engineering; Aerospace engineering and technology. Current work: Jet flows, coherent structure of turbulent flow, magnetogasdynamics, radiation gasdynamics, two-phase flows, lunar ash flow. Home: 4301 Sarasota Pl Beltsville MD 20705 Office: Inst for Phys Sci and Tech U Md College Park MD 20742

PAINE, THOMAS OTTEN, high technology consultant; b. Berkeley, Calif., Nov. 9, 1921; s. George Thomas and Ada Louise (Otten) P.; m. Barbara Helen Taunton Pearse, Oct. 1, 1946; children: Marguerite Ada, George Thomas, Judith Janet, Frank Taunton. A.B. in Engring, Brown U., 1942; M.S. in Phys. Metallurgy, Stanford, 1947; Ph.D., Stanford U., 1949. Research asso. Stanford, 1947-49; with Gen. Electric Co., 1949-68, 70-76, GE Research Lab., Schenectady, mgr. center advanced studies, Santa Barbara, Calif., 1963-68, v.p., group exec. power generation, 1970-73, sr. v.p. tech. planning and devel., 1973-76; pres. Northrop Corp., Los Angeles, 1976-82; chmn. Thomas Paine Assocs., Los Angeles, 1982—; dep. adminstr., then adminstr. NASA, 1968-70; dir. Eastern Air Lines. Contbr. articles to tech. publs. Trustee Occidental Coll., Brown U., Asian Inst. Tech., Bangkok. Served to lt. USNR, World War II. Decorated Submarine Combat insignia with stars, USN Commendation medal; grand ufficiale della Ordine al Merito, Italy; recipient Distinguished Service medal NASA, 1970, Washington award Western Soc. Engrs., 1972; John Fritz medal United Engring. Socs., 1976; Faraday medal Inst. Elec. Engrs., London, 1976. Fellow AIAA; mem. Nat. Acad. Engring., N.Y. Acad. Scis., Am. Phys. Soc., IEEE, U.S. Naval Inst., Sigma Xi. Clubs: Explorers, Lotos, Sky (N.Y.C.); Cosmos, Army and Navy (Washington); Calif. (Los Angeles). Subspecialty: Aerospace engineering and technology. Current work: High technology. Co-inventor lodex R magnets. Home: 765 Bonhill Rd Los Angeles CA 90049 Office: Thomas Paine Assocs 10880 Wilshire Blvd 2011 Los Angeles CA 90024

PAINTER, GENEVIEVE, clinical psychologist, consultant, author; b. Chgo., Sept. 5, 1919; d. Max and Amelia (Swartz) Berkowitz; m. David R. Schneidman, Aug., 1941 (div. 1956); children: Bruce, Terry; m. John Paul Painter, July 20, 1960 (dec. Dec. 1972). B.S., U. Ill.-Urbana, 1963, M.S., 1964, Ed.D., 1967; cert. psychotherapy, Alfred Adler Inst., Chgo., 1969. Music tchr. Lyon and Healy Sch. Music, Chgo., 1950-64; asst. prof. edn. U. Ill.-Urbana, 1967-68; pvt. practice psychology, Champaign, Ill., 1968-73; founder, dir. Family Edn. Ctrs., Champaign, 1968-73; instr. radio Hawaii Pacific Coll., Honolulu, 1975—; instr. U. Hawaii, 1983; dir. Assocs. for Human Devel., Honolulu, 1973—; cons. schs. and agys., Honolulu, 1973—; Chmn. profl. standards com. Family Edn. Ctrs. Hawaii, Honolulu, 1974—; tchr., cons. Kapiolani/Children's Med. Ctr., Honolulu, 1982. Author: Teach Your Baby, 1971, The Practical Parent, 1975. Chmn. career women United Jewish Appeal, Honolulu, 1980-82. U.S. Office Edn. fellow, 1964; U.S. Office Vocat. Rehab. fellow, 1963. Mem. Am. Psychol. Assn., N.Am. Soc. Adlerian Psychology (del. assembly 1965—), Soc. Advancement Psychology, Hawaii Psychol. Assn. Subspecialties: Behavioral psychology; Developmental psychology. Current work: Infant intellectual developmental acceleration, new psychological techniques for changing debilitating fixed beliefs of individuals, new psychological techniques for adolescents. Home: 2333 Kapiolani Blvd Apt 3214 Honolulu HI 96826 Office: Assocs for Human Devel 750 Amana St Honolulu HI 96814

PAINTER, JEFFREY FARRAR, mathematician; b. Munich, W.Ger., Mar. 3, 1951; s. Paul Hawley and Nancy (Farrar) P. B.S., Harvey Mudd Coll., 1973; M.A., U. Wis.-Madison, 1975, M.S., 1977, Ph.D., 1979. Mathematician Lawrence Livermore Lab., Calif., 1979—. Contbr. articles to profl. jours. Mem. Soc. Indsl. and Applied Math. Subspecialties: Applied mathematics; Numerical analysis. Office: Lawrence Livermore Lab PO Box 808 Livermore CA 94550

PAINTER, LINDA ROBINSON, physics educator; b. Lexington, Ky., May 4, 1940; d. J. Kenneth and Juanita Marie (Crosier) Robinson; m. Roy Allen Painter, May 6, 1967; children: Holly Suzanne, Brent Allen. B.S., U. Louisville, 1962; M.S., U. Tenn., 1963, Ph.D., 1968. Asst. prof. physics U. Tenn., Knoxville, 1968-75, assoc. prof., 1975-82, prof., 1982—; cons. Health Phys. Div., Oak Rige Nat. Lab., 1967-77, adj. research and devel. participant, 1977—. Grantee AEC, ERDA, Dept. Energy. Fellow Am. Phys. Soc.; mem. Health Physics Soc., Radiation Research Soc. (assoc. editor 1983—). Subspecialties: Condensed matter physics; Atomic and molecular physics. Current work: Developing systems to allow optical, photoionization and photoemission measurements on liquids; optical properties of polymers; calculating mean free paths for organic insulators. Home: 7708 Devonshire Dr Knoxville TN 37919 Office: U Tenn Circle Dr - Physics Bldg Knoxville TN 37996

PAINTER, RICHARD GRANT, research scientist; b. Jacksonville, Fla., Dec. 10, 1945; s. George Latimer and Sarah Mae (Lancaster) P.; m. Rosemary Sodek, Dec. 23, 1967; children: Robert Aaron, Thomas Scott, Michele Marie. B.S., U. Fla., 1967; Ph.D. in Biochemistry, Duke U., 1971. Postdoctoral fellow U. Calif., San Diego, 1971-74; staff scientist Syntex Research, Palo Alto, Calif., 1974-77; asst. mem. Scripps Clinic and Research Found., La Jolla, 1977—; cons. Nat. Inst. Heart and Lung Diseases. Contbr. articles to profl. jours. NASA fellow, 1967-70; NIH fellow, 1971-73; Am. Cancer Soc. sr. fellow, 1977; research career devel. award NIH, 1978-83. Mem. Am. Soc. Cell Biology, AM. Assn. Pathology, Am. Soc. Immunologists, AAAS. Democrat. Episcopalian. Subspecialties: Cell biology; Membrane biology. Current work: Membrane structure-function relationships in leukocytes and platelets.

PAIS, ABRAHAM, educator, physicist; b. Amsterdam, Holland, May 19, 1918; s. Jesaja and Kaatje (van Kleeff) P.; m. Lila Atwill, Dec. 15, 1956 (div. 1962); 1 son, Joshua; m. Sara Ector Via, May 29, 1976; 1 stepson, Daniel. B.Sc., U. Amsterdam, 1938; M.Sc., U. Utrecht, 1940, Ph.D., 1941. Via. Research fellow Inst. Theoretical Physics, Copenhagen, Denmark, 1946; prof. Inst. Advanced Study, Princeton, N.J., 1950-63; prof. physics Rockefeller U., N.Y.C., 1963-81, Detler W. Bronk prof., 1981—; Balfour prof. Weizmann Inst., Israel, 1977. Author: Subtle is the Lord (Am. Book award 1983). Recipient J.R. Oppenheimer Meml. prize, 1979; Guggenheim fellow, 1960. Fellow Am. Phys. Soc.; mem. Royal Acad. Scis. Holland (corr.), Am. Acad. Arts and Scis., Am. Philos. Soc., Nat. Acad. Scis., Council on Fgn. Relations. Club: Knickerbocker. Subspecialties: Particle physics; Theoretical physics. Current work: Quantum field theory, relativity, history of science. Home: 450 E 63d St New York NY 10021 Office: Rockefeller Univ New York NY 10021

PAISNER, JEFFREY ALAN, physicist; b. Bronx, N.Y., Aug. 28, 1948; s. David and Janet (Manius) P.; m. Deborah Leah Freiberg, June 15, 1975. B.S., Cooper Union, 1969; M.S., Stanford U., 1970, Ph.D., 1974. Laser physicist in exptl. research group-laser fusion program Lawrence Livermore Nat. Lab., 1974-75, laser spectroscopist, non-linear optician in exptl. physics group-laser isotope separation program, 1975-81, group leader process sci. advanced isotope separation program, 1981—; cons. to laser cos. Contbr. articles to profl. jours. Fellow Optical Soc. Am.; Mem. Am. Phys. Soc. Subspecialties: Spectroscopy; Laser-induced chemistry. Current work: Multi-step laser spectroscopy of atomic uranium; tunable laser sources and laser isotope separation; science and design of atomic vapor laser isotope separation process. Patentee in laser isotope separation. Home: 712 SilverLake Dr Danville CA 94583 Office: Lawrence Livermore Nat Lab PO Box 808 L-459 Livermore CA 94550

PAJAK, THOMAS FRANCIS, clinical biostatistician; b. Chgo., Dec. 15, 1942; s. Francis Martin and Charlotte Ann (Janiga) P.; m. Karin Bergerson, Aug. 1, 1969; children: Christine Marie, Catherine Ann. B.A., Divine Sem. Coll., 1965; M.A., DePaul U., 1967; M.S., SUNY, Buffalo, 1970, Ph.D., 1972. Asst. prof., dir Oncology Stats. Center, Bowman Gray Sch. Medicine, Winston-Salem, N.C., 1972-74; biostatistician, dir. Studies Analysis Center, Western Cancer Study Group, Los Angeles, 1974-75; asso. biostatistician Cancer and Leukemia Group B., Scarsdale, N.Y., 1976-81; group biostatistician Radiation Therapy Oncology Group, Phila., 1982; also chief statisitician Am. Coll. Radiology, Phila., 1982; asst. prof. U. So. Calif., 1974-75; asso. prof. Mt. Sinai Sch. Medicine, N.Y.C., 1976-81, Thomas Jefferson U. Hosp., Phila., 1982; cons in field. Contbr. numerous articles to profl. jours. Mem. Am. Statis. Assn., Am. Soc. Clin. Oncology, Am. Assn. Cancer Research. Democrat. Roman Catholic. Subspecialties: Statistics; Cancer research (medicine). Current work: Evaluation of treatment strategies, particular in cancer. Office: Am Coll Radiology 925 Chestnut St Philadelphia PA 19107

PAK, CHARLES Y.C., research physician, educator; b. Seoul, Korea, Nov. 27, 1935; s. You Pyung and Wechun (Kim) P.; m. Jane Ellen Riechers, June 15, 1963; children: Laura, Gregory, Marjorie. Student, Shimer Coll., 1953-55; B.S. with honors, U. Chgo., 1957, M.D., 1961. Intern U. Chgo. Clinics, 1962-63, sr. investigator, 1963-68; chief sect. mineral metabolism NIH, Bethesda, Md., 1968-72; asso. prof. medicine Southwestern Med. Sch., U. Tex., Dallas, 1972-75, prof., 1975—, program dir. gen. clin. research center, 1974—. Contbr. 230 articles to sci. jours.; author: Calcium Urolithiasis, 1978. Served to lt. comdr. USPHS, 1963-67. Mem. Endocrine Soc., Assn. Am. Physicians, Am. Soc. Clin. Investigation, Am. Soc. Nephrology. Subspecialties: Endocrinology; Nephrology. Current work: Cause and management of renal stones, metabolic bone diseases and disorders of parathyroid function and vitamin D metabolism. Home: 7107 Churchill Way Dallas TX 75230 Office: 5323 Harry Hines Blvd Dallas TX 75235

PAKE, GEORGE EDWARD, research executive, physicist; b. Jeffersonville, Ohio, Apr. 1, 1924; s. Edward Howe and Mary Mabel (Fry) P.; m. Marjorie Elizabeth Semon, May 31, 1947; children—Warren E., Catherine E., Stephen G., Bruce E. B.S., M.S., Carnegie Inst. Tech., 1945; Ph.D., Harvard U., 1948. Physicist Westinghouse Research Labs., 1945-46; mem. faculty Washington U., St. Louis, 1948-56, 62-70, prof. physics, provost, 1962-69, exec. vice chancellor, 1965-69, Edward Mallinckrodt prof. physics, 1969-70; v.p. Xerox Corp.; mgr. Xerox Palo Alto (Calif.) Research Center, 1970-78, v.p. corp. research, 1978-83, group v.p., 1983—; prof. physics Stanford U., 1956-62. Author: (with E. Feenberg) Quantum Theory of Angular Momentum, 1953, Paramagnetic Resonance, 1962, (with T. Estle) The Physical Principles of Electron Paramagnetic Resonance, 1973. Mem. gov. bd. Am. Inst. Physics, 1957-59; bd. dirs. St. Louis Research Council, 1964-70; mem. physics adv. panel NSF, 1958-60, 63- 66; chmn. physics survey com. Nat. Acad. Sci.-NRC, 1964-66; Mem. St. Louis County Bus. and Indl. Devel. Commn., 1963-66; chmn. bd. Regional Indsl. Devel. Corp., St. Louis, 1966-67, St. Louis Research Council, 1967-70; mem. President's Sci. Adv. Com., 1965-69; Bd. dirs. St. Louis Country Day Sch., 1964-70, Central Inst. for Deaf, 1965-70; trustee Washington U., 1970—, Danforth Found., 1971—. Fellow Am. Phys. Soc. (pres. 1977); mem. Am. Assn. Physics Tchrs., AAUP, AAAS, Am. Acad. Arts and Scis., Nat. Acad. Sci., Sigma Xi, Tau Beta Pi. Club: University (Palo Alto). Subspecialty: Research management. Home: 10 Arastradero Rd Portola Valley CA 94025 Office: Xerox Palo Alto Research Center Palo Alto CA

PALADE, GEORGE EMIL, cell biologist, educator; b. Jassy, Romania, Nov. 19, 1912; came to U.S., 1946, naturalized, 1952; s. Emil and Constanta (Cantemir) P.; m. Irina Malaxa, June 12, 1941 (dec. 1969); children—Georgia Teodora, Philip Theodore; m. Marilyn G. Farquhar, 1970. Bachelor, Hasdeu Lyceum, Buzau, Romania; M.D., U. Bucharest, Romania. Instr., asst. prof., then asso. prof. anatomy Sch. Medicine, U. Bucharest, 1935-45; vis. investigator, asst. asso., prof. cell biology Rockefeller U., 1946-73; now prof. cell biology Yale; correlated biochem. and morphological analysis cell structures. Author sci. papers. Recipient Albert Lasker Basic Research award, 1966, Gairdner Spl. award, 1967, Horwitz prize, 1970, Nobel prize, 1974. Fellow Am. Acad. Arts and Scis.; mem. Nat. Acad. Sci., Pontifical Acad. Sci. Subspecialty: Cell biology. Office: Cell Biology Dept Yale U Sch Medicine New Haven CT 06510

PALADINO, FRANK VINCENT, biology educator; b. Bklyn., Apr. 13, 1952; s. Frank Anthony and Grace Ann (Abbene) P.; m. Jacqueline Ann Corral, Dec. 27, 1974 (div. Jan. 1980); children: Michael F., Gregory F. B.A., SUNY-Plattsburh, 1974; M.A., SUNY-Buffalo, 1976; Ph.D. in Physiology, Wash. State U., 1979. Research asst. Wash. State U., Pullman, 1976-79; asst. prof. zoology Miami U., Oxford, Ohio, 1979-82; asst. prof. biology Purdue U., Ft. Wayne, Ind., 1982—; asst. dir. Crooked Lake Biol. Field Sta., Ft. Wayne, 1982—. Contbr. articles to profl. jours. Grantee NIH, Am. Soc. Zoologists, Purdue Research Found., others. Mem. Am. Physiol. Soc., Am. Soc. Zoologists, Am. Ornithologists Union, Cooper Ornithol. Soc., Sigma Xi. Roman Catholic. Lodge: K.C. Subspecialties: Comparative physiology; Environmental toxicology. Current work: Comparative vertebrate environmental physiology. Aquatic fish toxicology, thermoregulation and bioenergetics. Office: Dept Biology Ind-Purdue U Fort Wayne IN 46805 Home: 1534 J Reed Rd Fort Wayne IN 46815

PALAFOX, ANASTACIO LAIDA, animal scientist, nutritionist, educator, researcher; b. Philippines, Apr. 24, 1914; came to U.S., 1931, naturalized, 1943; s. Emeterio Salvacion and Geronima Baisa (Laida) P.; m. Jesusa AVecilla, Mar. 16, 1950; children: Brian A., Neal A., Riley A., Edgar A. B.S., Wash. State U., 1940, M.S., 1941; Ph.D., Mich. State U., 1970. Asst. instr. U. Hawaii, 1946-47, instr., 1947-57, asst. prof. animal nutrition, 1957-70, assoc. prof., 1970-78, prof., 1978-80, prof. emeritus, 1980—; prof. animal scis. U. Guam, Mangilao, 1981—; cons. in field. Research numerous publs. in field. Chmn. ednl. com. United Filipino Council of Hawaii. Served with USN, 1942-46. Named Father of Yr. Filipino Community, Honolulu, 1978, Outstanding Filipino Overseas Pres. Marcos, Philippines, 1982; Research Council grantee; E. I. Dupont research grantee, 1948-51; Dept. Agr. grantee, 1982—. Mem. Poultry Sci. Assn., Worlds Poultry Sci. Assn., AAUP, AAAS, Inst Biol Scis., Sigma Xi, Gamma Sigma Delta. Roman Catholic. Club: Toastmasters (Honolulu) (pres. club 1968-70). Subspecialties: Animal nutrition; Animal physiology. Current work: Nutrition, biochemistry, protein energy, metabolism, physiology, interrelationships, amino acids, carbohydrates energy, enzymes, mgmt. systems, growth, hormones. Home: 5276 Kimokeo St Honolulu HI 96821 Office: Coll Agr U Guam Mangilao Guam 96913

PALATINO, RICHARD DUANE, pharmacology researcher; b. Clearwater, Fla., Apr. 10, 1937; s. Michael Fors and Edna Suzanne (Carlton) P.; m. Roberta Dorne Casey, May 30, 1959; 1 son, Michael. B.S., U. N.C., 1958, M.S., 1960; Ph.D., SUNY-Syracuse, 1963. Postdoctoral fellow dept. pharmacology Upstate Med Ctr., Syracuse, 1963-64; asst. prof. dept. pharmacology Upstate Med. Ctr., Syracuse, 1964-68; sect. head Sterling-Winthrop Research Inst., Rensselaer, N.Y., 1969-72; assoc. prof. pharmacology Case Western Res. U., Cleve., 1972-77, prof., 1977—; cons. Eli Lilly and Co., Upjohn Co. Contbr. articles to profl. jours. Bd. dirs. Am. Heart Assn. Fellow Am. Soc. Pharmacology and Exptl. Therapeutics; mem. Internat. Soc. Heart Research, AAAS, N.Y. Acad. Scis. Subspecialties: Pharmacology; Cardiology. Office: Werik Lab The Superior Bldg Rm 111 Cleveland OH 44114

PALAY, SANFORD LOUIS, educator, physician; b. Cleve., Sept. 23, 1918; s. Harry and Lena (Sugarman) P.; m. Victoria Chan Curtis, 1970; children: Victoria Li-Mei, Rebecca Li-Ming. A.B., Oberlin Coll., 1940; M.D. (Hoover prize scholar 1943), Western Res. U., 1943. Teaching fellow medicine, research asso. anatomy Western Res. U., Cleve., 1945-46; NRC fellow med. scis. Rockefeller Inst., 1948, vis. investigator, 1953; from instr. anatomy to asso. prof. anatomy Yale U., 1949-56; chief sect. neurocytology, lab. neuroanatomical scis. Nat. Inst. Neurol. Diseases and Blindness, NIH, Washington, 1956-61, chief lab. neuroanatomical scis., 1960-61; Bullard prof. neuroanatomy Harvard, Boston, 1961—; Linnean Soc. lectr., London, 1959; vis. investigator Middlesex Hosp. (Bland-Sutton Inst.), London, Eng., 1961; Phillips lectr. Haverford Coll., 1959; Ramsay Henderson Trust lectr. U. Edinburgh, Scotland, 1962; Disting. Scientist lectr. Tulane U. Sch. Medicine, 1969, 75; vis. prof. U. Wash., 1969; Rogowski Meml. lectr. Yale, 1973; Disting. lectr. biol. structure U. Miami, 1974; Disting. Scientist lectr. U. Ark., 1977; other Disting. lectureships; vis. prof. U. Osaka, Japan, 1978, Nat. U. Singapore, 1983; Mem. fellowship bd. NIH, 1958-61, cell biology study sect., 1959-65, adv. com. high voltage electron microscope resources, 1973-80, mem. rev. com. behavioral and neurol. scis. fellowships, 1979—; chmn. Gordon Research Conf. Cell Structure and Metabolism, 1960; asso. Neuroscis. Research Program, 1962-67, cons. asso., 1975—; mem. anat. scis. tng. com. Nat. Inst. Gen. Med. Scis., 1968-72; mem. sci. adv. com. Oreg.

Regional Primate Research Center, 1971-76. Author: The Fine Structure of the Nervous System, 1970, 2d edit., 1976, Cerebellar Cortex, Cytology and Organization, 1973; Editor: Frontiers of Cytology, 1958, The Cerebellum, New Vistas, 1981; Mem.: sci. council Progress in Neuropharmacology and Jour. Neuropharmacology, 1961-66; editorial bd.: Exptl. Neurology, 1959-76, Jour. Cell Biology, 1962-67, Brain Research, 1965-71, Jour. Comparative Neurology, 1966—, Jour. Ultrastructure Research, 1966—, Jour. of Neurocytology, 1972—, Exptl. Brain Research, 1965-76, Neurosci, 1975—, Zeitschrift fur Anatomie and Entwicklungsgeschichte, 1968; co-mng. editor, 1978—; editor-in-chief: Jour. Comparative Neurology, 1981—; mem. adv. bd. editors: Jour. Neuropathology and Exptl. Neurology, 1963—, Internat. Jour. Neurosci., 1969-71, Tissue and Cell, 1969—; Contbr. articles to profl. jours. Served to capt. M.C. AUS, 1946-47. Recipient 50 Best Books of 1974 award Internat. Book Fair, Frankfurt, Germany, Best Book in Profl. Readership award Am. Med. Writers Assn., 1975; Guggenheim fellow, 1971-72; Fogarty scholar-in-residence NIH, Bethesda, 1980-81. Fellow Am. Acad. Arts and Scis.; mem. Nat. Acad. Scis., Am. Assn. Anatomists (chmn. nominating com. 1964, mem. exec. com. 1970-74, anat. nomenclature com. 1975-78, pres. 1980-81), Histochem. Soc., Electron Microscope Soc. Am., AAAS, Am. Soc. Cell Biology (program com. 1975), Internat. Soc. Cell Biology, Soc. for Neurosci., Washington Soc. Electron Microscopy (organizing com., sec.-treas. 1956-58), Soc. Francaise de Microscopie Electronique (hon.), Royal Microscopical Soc. (hon.), Golgi Soc. (hon.), Phi Beta Kappa, Sigma Xi, Alpha Omega Alpha. Club: Cajal (pres. 1973-74). Subspecialties: Anatomy and embryology. Current work: Fine structure of nerve cells and tissues, cytochemical localization of heuroactive substances in nerve cells of cerebellar cortex. Home: 78 Temple Rd Concord MA 01742 Office: Dept Anatomy Harvard U Med Sch Boston MA 02115

PALISIN, HELEN, developmental psychologist, researcher, consultant; b. Lakewood, Ohio, Jan. 9, 1934; d. John George and Theresa (Morgosh) P. R.N., St. Elizabeth Sch. Nursing, 1955; B.S., Western Res. U., 1962; M.N., U. Wash., 1968, Ph.D., 1977. Asst. chief nurse No. Ohio chpt. ARC, Cleve., 1958-65, chief nurse, Detroit, 1966; asst. prof., coordinator Creighton U., Omaha, 1971-74; postdoctoral research fellow U. Wash.-Seattle, 1977-80, research cons., 1980-82, research assoc., 1983—; research cons. State of Wash., Olympia, 1981—; vis. lectr. U. Wash., 1980. USPHS fellow, 1977-80; ARC scholar, 1960; recipient Disting. Tchr. award Creighton U., 1972. Mem. Soc. Research Child Devel., Am. Psychol. Assn. Roman Catholic. Subspecialty: Developmental psychology. Current work: Social emotional development, temperament, mental health, program evaluation. Office: U Wash DMRC So Bldg WJ-10 Seattle WA 98195

PALL, MARTIN LAWRENCE, biochemical geneticist, researcher; b. Montreal, Que., Can., Jan. 20, 1942; came to U.S., 1947, naturalized, 1979; s. Gordon and Eleanor (Dresdner) P.; m. Linda Blackwelder, May 30, 1970; 1 son, Zachary Aaron. B.A., Johns Hopkins U., 1962; Ph.D., Calif. Inst. Tech., 1968. Asst. prof. Reed Coll., 1967-72; asst. prof. genetics and biochemistry Wash. State U., Pullman, 1972-75, assoc. prof., 1975-83, prof. genetics, cell biology and biochemistry, 1983—. Mem. Am. Soc. Biol. Chemists, Genetics Soc. Am. Subspecialties: Biochemistry (biology); Gene actions. Current work: Cellular regulatory mechanisms; cyclic AMP and regulation in microorganisms; gene amplification and carcinogenesis. Home: S E 560 Jackson Pullman WA 99163 Office: Dept Molecular Genetics Wash State U Pullman WA 99164

PALLADINO, JOSEPH JAMES, psychology educator; b. Bronx, Aug. 16, 1950; s. Gennaro and Anna Marie (Sabatello) P.; m. Marie Dolores Assante, Aug. 12, 1972; children: Karin, Sharin. B.S., Fordham U., 1972, M.A., 1974, Ph.D., 1982. Asst. prof. psychology Brescia Coll., Owensboro, Ky., 1977-78; asst. prof. psychology St. Francis Coll., Loretto, Pa., 1978-81, Ind. State U., Evansville, 1981—; cons. CBS Coll. Pub. Co., N.Y.C., 1980—; stress mgmt. cons. Ind. State U., 1981—. Mem. Am. Psychol. Assn., Southeastern Psychol. Assn., Council Undergrad Psychology Depts., Midwestern Psychol. Assn. Subspecialties: Biofeedback; Cardiology. Current work: Biofeedback in stress management; personality correlates of coronary-prone behavior. Home: 1400 N Boehne Camp Rd Evansville IN 47712 Office: Dept Psychology Ind State U Evansville IN 47712

PALLADINO, NUNZIO JOSEPH, nuclear engineer; b. Allentown, Pa., Nov. 10, 1916; s. Joseph and Angelina (Trentalange) P.; m. Virginia Marchetto, June 16, 1945; children: Linda Susan, Lisa Anne, Cynthia Madaline. B.S., Lehigh U., 1938, M.S., 1939, D.Eng. (hon.), 1964. Registered profl. engr., Pa. Engr. Westinghouse Electric Co., Phila., 1939-42; nuclear reactor designer, Oak Ridge Nat. Lab., 1946-48; staff asst. to div. mgr. Argonne Nat. Lab., Lemont, Ill., 1948-50; mgr. PWR reactor design subdiv. Westinghouse Electric Corp., Pitts., 1950-59; head nuclear engrng. dept. Pa. State U., University Park, 1959-66, dean, 1966-81; chmn. Nuclear Regulatory Commn., Washington, 1981—; Past mem. Pa. Gov.'s Sci. Advisory Com., Gov.'s Energy Council; past mem. Pa.'s Commn. to Investigate TMI. Contbr. tech. articles to profl. jours. Served to capt. AUS, 1942-45. Recipient Order of Merit Westinghouse Electric Corp. Fellow ASME (Prime Movers award), Am. Nuclear Soc. (past pres.); mem. Am. Soc. Engrng. Edn., Nat. Soc. Profl. Engrs., Argonne Univs. Assn. (interim pres., past dir.), Nat. Acad. Engrs. Roman Catholic. Club: Rotary. Subspecialty: Nuclear engineering. Office: 1717 H St NW Washington DC 20555 Do it right the first time

PALLARDY, STEPHEN GERARD, plant science educator; b. St. Louis, Mar. 19, 1951; s. Cedric Charles and June Josephine (Van Leer) P.; m. Judy Siebert, Dec. 26, 1981. B.S., U. Ill., 1973, M.S., 1975; Ph.D., U. Wis.-Madison, 1978. Asst. prof. plant sci. Kans. State U.-Manhattan, 1979-80; asst. prof. plant sci. U. Mo.-Columbia, 1980—. Author: Water Relations of Closely Related Woody Plants, 1981; assoc. editor: Forest Sci, 1983—. Mem. Am. Soc. Plant Physiologists, Ecol. Soc. Am. Subspecialty: Plant physiology (biology). Current work: Stress physiology, plant water relations, physiological plant ecology. Home: Rural Route 1 Ashland MO 65010 Office: Sch Forestry U Mo Columbia MO 65211

PALLMANN, ALBERT JOSEF, earth and atmospheric scientist, sci. cons., educator; b. Wiesbaden, Ger., Dec. 12, 1926; came to U.S., 1963, naturalized, 1978; s. Joseph Ferdin and Helene Maria (Schueller) P.; m. Margot Maria Pallmann, June 26, 1958; 1 son, Thomas R.S. Doctorand, U. Cologne, W. Ger, 1953, Dr.rer.nat., 1958. Research assoc. Geophys. Inst., U. Cologne, 1954-58; chief meterologist Nat. Weather Analysis Center, El Salvador, 1959-62; indsl. cons. and prof. meterology Escuela Nacional de Marina and Academia Nacional de la Fuerza Aerea, El Salvador, 1959-62; asst. prof. meterology St. Louis U., 1962-65, assoc. prof., 1965-69, prof., 1969—; indsl. cons., El Salvador, 1959-62; sci. councilor Univ. Corp. for Atmospheric Research, Boulder, Colo., 1964-67, St. Louis U. sci. rep., 1967-80; cons. space div. McDonnell-Douglas Corp., 1966-67. Author: Basic Concepts of Radiative Transfer, 1983; contbr. articles sci. jours. Mem. St. Louis Council World Affairs, Inc., 1981—. Research grantee NASA, NSF, Nat. Environ. Satellite Service, Dept. Energy. Mem. Am. Meteorol. Soc., Am. Geophys. Union, AAAS, Sigma Xi. Subspecialties: Meteorologic instrumentation; Planetary atmospheres. Current work: Radiative heating/cooling in Martian ground-atmosphere system, meteorol. radiometry, solar and terrestrial, satellite- and air-borne, ground-based, events of climatic extremes. Office: PO Box 8099 Laclede Station St Louis MO 63156

PALLOTTA, BARRY S, biophysicist; b. Bklyn., Jan. 18, 1951; s. Ralph and Harriet (Glicksberg) P.; m. Patricia A. Jones, July 21, 1979. B.S. in Physics, SUNY-Stony Brook, 1973; Ph.D. in Physiology and Biophysics, U. Vt., 1978. Grad. teaching asst. U. Vt., 1973-78; postdoctoral fellow U. Miami Sch. Medicine, 1978-81, sr. research assoc., 1981—. Contbr. articles to profl. jours. Recipient Nat. Research Service award NIH, 1978-81. Mem. Soc. Neurosci., Biophys. Soc., Soc. Gen. Physiologists. Subspecialties: Biophysics (biology); Neurobiology. Current work: Kinetics of drug-receptor interactions; recor from single ion channels and stat. descriptions of single ion channel kinetics. Office: Dept Physiology and Biophysics R-430 U Miami Sch Medicine Miami FL 33101

PALMER, EDWARD LEO, social psychologist, educator, consultant; b. Hagerstown, Md., Aug. 11, 1938; s. Ralph Leon and Eva Irene (Brandenburg) P.; m. Ruth-Ann Pugh, June 2, 1962; children: Edward Lee, Jennifer Lynn. B.A., Gettysburg Coll., 1960; B.D., Lutheran Theol. Sem., 1964; M.S., Ohio U., 1967, Ph.D., 1970. Asst. prof. Western Md. Coll., Westminster, 1968-70; asst. prof. Davidson (N.C.) Coll., 1970-77, assoc. prof. dept. psychology, 1977—; guest researcher Harvard U., Cambridge, Mass., 1977; cons. Council on Children, Media and Merchandising, Washington, 1978-79; NSF project grant reviewer, Washington, 1981—. Editorial cons.: Jour. of Broadcasting, Ohio State U., Columbus, 1980—; invited contbr.: Wiley Ency. Psychology, 1982; co-editor: Children and the Faces of Television, 1981; task force book Telecommunications 2000, 1982; contbr. articles to jours. Sec. North Mecklenburg Child Devel. Assn., Davidson, 1972-74. N.C. gov. research grantee, 1971; Cronhardt scholar Luth. Theol. Sem., 1964. Mem. Am. Psychol. Assn., Internat. Communication Assn., Soc. Southeastern Social Psychologists, So. Assn. Pub. Opinion Research, Southeastern Psychol. Assn., Phi Beta Kappa. Subspecialty: Social psychology. Current work: The effects of television advertising and programming upon young children. Home: Route 1 Box 1792 Davidson NC 28036 Office: Dept Psychology Davidson Coll Davidson NC 28036

PALMER, GENE CHARLES, neuropharmacologist, research director, educator; b. Elmhurst, Ill., Nov. 4, 1938; s. Thomas Harold and Charolette Carolyn (Ritzler) P.; m. Shelby Jo Herston, Sept. 14, 1938; 1 dau., Pamela Jo. B.S. in Biology, Tenn. Tech. U., 1960; Ph.D. in Anatomy, Vanderbilt U., 1968; postgrad. in pharmacology, Vanderbilt U., 1968-70. Asst. prof., asso. prof. depts. anatomy and pharmacology U. N.Mex. Sch. Medicine, Albuquerque, 1970-76; asso. prof. dept. pharmacology U. South Ala. Coll. Medicine, Mobile, 1976-78, prof., 1978-82; dir. research Frist-Massey Neurol. Inst., Vanderbilt U. Sch. Medicine, Nashville, 1983; prof. dept. anatomy First-Massey Neurol. Inst., Vanderbilt U. Sch. Medicine, Nashville, 1983; legal cons. Sterling Winthrop Drug Co., 1979; sci. reviewer Robert J. Brady Co., Bowie, Md., 1981—. Author: Neuropharmacology of Cyclic Nucleotides, 1979, Neuropharmacology of Central Nervous System and Behavioral Disorders, 1981; mem. editorial bd.: Jour. PharmacologyMethods; contbr. over 100 articles and revs. to sci. publs. Served to capt. USMC, 1960-63. Recipient 1st Dan Burrow's Meml. award N.Mex. Heart Assn., 1971; NIMH grantee, 1974-77; NSF grantee, 1977-79, 80-82; Am. Epilepsy Found. grantee, 1977-78, 80-91; Frist-Massey Neurol. Inst. grantee, 1981-82. Mem. Am. Soc. Neurochemistry, Soc. Neurosci., Am. Soc. Pharmacology and Exptl. Therapeutics, Western Pharmacology Soc., Microcirculatory Soc. Independent. Roman Catholic. Subspecialties: Neuropharmacology; Neurochemistry. Current work: Using drugs as tools, my interests are to determine the roles and functions of cellular receptor enzymes to disorders of the nervous system and hopefully develop new or alternate methods of treatment for stroke, epilepsy, schizophrenia, mania and depression. Home: Route 3 High Point Ridge Rd Franklin TN 37064 Office: Frist-Massey Neurol Inst 356 24th Ave N Suite 104 Nashville TN 37203

PALMER, MELVILLE LOUIS, agricultural engineering educator; b. Dobbinton, Ont., Can., Aug. 30, 1924; s. Louis Grange and Laura Lavina (Peacock) P.; m. Shirley Adams, Aug. 2, 1952; children: Laura, Melanie, Bradley. B.Sc., U. Toronto, 1950; M.Sc., Ohio State U., 1955; doctoral candidate, U. Wis.-Madison, 1969-70. Asst. dean of men Ont. Agrl. Coll., Can., Guelph, 1950-52; asst. mgr. farm machinery United Coops., Toronto, Ont., 1952-53; research asst. Ohio Agrl. Expt. Sta., Wooster, 1953-55; extension agrl. engr. (water mgmt.) Ohio State U., Columbus, 1955—, prof., 1970—. Contbr. articles to profl. jours. Served with RCAF, 1943-46. Fellow Am. Soc. Agrl. Engrs. (Hancor award 1979), Soil Conservation Soc. Am.; mem. AAAS, Nat. Water Well Assn., Water Mgmt. Assn. Ohio. Subspecialties: Agricultural engineering; Resource conservation. Current work: Extension education in farm drainage, rural water supply, irrigation, rural sanitation, resource conservation, erosion control.

PALMER, SUSHMA MAHYERA, biochemist/nutritionist, researcher; b. Sirhind, India, Jan. 13, 1944; d. Jagdish Chandra and Santosh (Mohindra) Mahyera; m. Robie M.H. Palmer, July 14, 1941. B.S., U. Delhi, 1963, M.S., 1965; D.Sc. in Biochemistry, Belgrade U., 1977. Registered dietitian, Am. Dietetic Assn., 1968. Asst. lectr. advanced nutrition Lady Irwin Coll., U. Delhi, 1965-66; therapeutic dietitian Washington Hsp. Ctr., 1966-68; research nutritionist George Hyman Research Found., Washington, 1968; postgrad. fellow 2d Moscow Med. Inst., 1969-71; dir. div. nutrition, asst. prof. pediatrics Georgetown U. Sch. Medicine, Washington, 1971-75; project dir. Nat. Acad. Sci., 1978-82; adj. asst. prof. pediatrics Georgetown U. Sch. Medicine, Washington, 1978-82; exec. dir. Food and Nutrition Bd., Nat. Acad. Scis., 1982—; mem. program adv. com. Nat. Inst. Dental Research, NIH. Author textbook, articles; contbr. chpts. to books. Nat. Cancer Inst. grantee. Mem. Am. Inst. Nutrition, Am. Dietetic Assn., Soc. Nutrition Edn., AAAS, N.Y. Acad. Scis. Subspecialties: Biochemistry (biology); Nutrition (biology). Current work: Nutrition and immune response, cancer; initiating and directing projects on food, nutrition and health. Office: 2101 Constitution Ave NW Rm 344B Washington DC 20418

PALMER, THOMAS YEALY, environmental consultant; b. Kiowa, Kans., Sept. 11, 1924; s. Thomas Joseph and Maude May (Alexander) P.; m. Jeraldine May Palmer, Aug. 15, 1955; children: Kevin Alexander, Keith Elliot. B.S., St. Louis U., 1950; M.S., U. Wash., 1961. Forecaster. Northwest Airlines, Seattle, 1951-54; staff mem. Sandia Corp., Albuquerque, 1955-59; project leader A.F. Geophysics Lab., Bedford, Mass., 1961-62; research specialist Boeing Co., Seattle, 1962-66; project leader U.S. Forest Service, Riverside, Calif., 1966-77; pres. Southwest Environ. Tech. Lab., Fallbrook, Calif., 1977—; cons. Phys. Dynamics Co., LaJolla, Calif., 1977—. Contbr. articles to profl. jours. Bd. dirs. Fallbrook Farm Bur., Calif., 1981—; v.p. Ridge Drive Rd Assn., Fallbrook, 1979-81. Served with U.S. Army, 1943-45. Fellow AIAA (assoc.); mem. Am. Meteorol. Soc., Am. Geophys. Union, AAAS, Combustion Inst. Subspecialties: Meteorologic instrumentation; Combustion processes. Current work: Nuclear weapons effects; observation instrumentation development and computer modelling. Patentee method for measuring water vapor and oxygen, stable mist generation device, environmentally resistant anemometer, wind component anemometer. Home: 823 Ridge Dr Fallbrook CA 92028

PALMITER, RICHARD DEFOREST, biochemistry educator; b. Poughkeepsie, N.Y., Apr. 5, 1942; s. DeForest Harold and Viola Louise (Antholt) P.; m. Kathleen Lockwood (div.); m. Lynne T. Smith (div.); 1 dau., Carolyn Reeve. Student, U. Alaska, 1960-61; A.B., Duke U., 1964; Ph.D., Stanford U., 1968. Postdoctoral fellow Stanford (Calif.) U., 1968-71, Searle Research Labs., High Wycombe, Eng., 1971-73, Harvard U., Cambridge, Mass., 1973-74; asst. prof. dept. biochemistry U. Wash., Seattle, 1974-78, assoc. prof., 1978-81, prof., 1981—. Contbr. articles to biochem. jours. Recipient George Thorn award Howard Hughes Med. Inst., 1982. Subspecialties: Gene actions; Developmental biology. Current work: Regulation of metallothionein genes; developmental expression of genes introduced into animals (mice). Office: Department of Biochemistry University of Washington Seattle WA 98195

PALMQUIST, JOHN CHARLES, geology educator; b. Omaha, Sept. 6, 1934; s. Wilbur Nathaniel and Amy Irene (Anderson) P.; m. Carol Jayne Anderson, Aug. 25, 1956; children: David, Kirsten, Douglas. A.B., Augustana Coll., 1956; M.S., U. Iowa, 1958, Ph.D., 1961. Exploration geologist Chevron Oil Co., New Orleans, 1961-62; asst. prof. geology Monmouth (Ill.) Coll., 1962-68; assoc. prof. geology Lawrence U., Appleton, Wis., 1968—; assoc. Samuel T. Pees & Assocs., Meadville, Pa., 1980—. Fellow Geol. Soc. Am.; mem. Am. Assn. Petroleum Geologists, Am. Soc. Photogrammetry. Republican. Lutheran. Subspecialties: Tectonics; Remote sensing (atmospheric science). Current work: Precambrian basement structure and its relationship to later deformation; role of remote sensing in exploration. Home: 8 Arborea Ln Appleton WI 54915 Office: Dept Geology Lawrence U Box 599 Appleton WI 54912

PAMATMAT, MARIO MACALALAG, marine biologist, researcher, educator; b. Santa Cruz, Laguna, Philippines, Oct. 17, 1928; came to U.S., 1958; s. Dominador Bonifacio and Alejandra Puhawan (Macalalag) P.; m. Nellie Bangonan Smith, July 7, 1961; children: Timothy Carl, Phillip James. Cert. in fish culture, Philippine Inst. Fish Tech., 1950; B.S. in Fisheries Mgmt, Auburn U., 1958, M.S., 1960; Ph.D. in Oceanography, U. Wash., 1966. Fish warden Philippines Bur. Fisheries, Manila, 1952-54, asst. fisheries biologist, 1954-57; research assoc. U. Wash., Seattle, 1966-73; assoc. prof. Auburn (Ala.) U., 1973-78; vis. prof. U. Kiel, W.Ger., 1978; adj. prof. San Francisco State U., 1979—. Contbr. articles to sci. jours. Grantee Dept. Energy, EPA, NSF, NOAA. Mem. Am. Soc. Limnology and Oceanography, Marine Biol. Assn. U.K., AAAS, Pacific Estuarine Research Soc., Sigma Xi. Episcopalian. Subspecialties: Ecology; Oceanography. Current work: Research on benthic ecological energetics, direct measurements of heat flow in relation to metabolic activity of individual organisms as well as benthic communities, application of direct calorimetry to environmental pollution problems. Home: 109 Robinhood Dr San Rafael CA 94901 Office: 3150 Paradise Dr Tiburon CA 94920

PAN, CHAI-FU, chemistry educator; b. Loshon, Szechwan, China, Sept. 8, 1936; came to U.S., 1960, naturalized, 1977; s. I-chen and Shih-Liang (Shih) P.; m. Chia-Yao Shih, Aug. 18, 1962; children: Lawrence Shou-pung, Mariette Shou-jung. B.S., Nat. Taiwan U., 1956; Ph.D., U. Kans., 1966. Assoc. prof. chemistry Ala. State U., Montgomery, 1966-71, prof., 1971—. Contbr. articles to profl. jours. Fellow Am. Inst. Chemists; mem. Am. Chem. Soc., Phi Lambda Upsilon. Subspecialties: Physical chemistry; Thermodynamics. Current work: Thermodynamics of electrolyte solutions. Home: 2420 Wentworth Dr Montgomery AL 36106 Office: Ala State U 915 S Jackson St Montgomery AL 36195

PAN, VICTOR YAKOVLEVICH, computer science educator, researcher, consultant; b. Moscow, USSR, Sept. 8, 1939; came to U.S., 1977; s. Yakov S. and Raissa-Rievka (Kogan) P.; m. Lidia R. Perelman, Sept. 8, 1972. M.S. in Math, Moscow State U., 1961, Ph.D. equivalent in Physics and Math, 1964. Jr., sr. researcher Inst. Electronic Control Machines, Moscow, USSR, 1964-69; sr. researcher Inst. Econs., Acad. of Sci., Moscow, 1969-76; vis. scientist IBM Research Ctr., Yorktown Heights, N.Y., 1977-79, 80; vis. mem. Inst. for Advanced Studies, Princeton, N.J., 1980-81; vis. prof. computer sci. dept. Stanford (Calif.) U., 1981; prof. computer sci. SUNY-Albany, 1979—; cons. Gen.Electric, Schenectady, 1980—; referee profl. jours. Contbr. articles to jours. in field. Univ. awardee SUNY, 1980; NSF grantee, 1980-82, 82—; grantee Inst. Advanced Study, Princeton, 1980-81. Mem. Assn. Computing Machinery, Am. Math. Soc., Soc. Indsl. and Applied Math., Moscow Math. Soc. Subspecialties: Algorithms; Numerical analysis. Current work: Evaluation of Polynomials, bilinear forms, matrix multiplication, ranks of tensors, lower bounds on arithmetic complexity of computation, bit-operation complexity of arithmetic problems, linear and millilinear algebra (computational aspects). Home: 146 Lancaster St Albany NY 12210 Office: SUNY Albany 1400 Washington Ave Albany NY 12222

PAN, YUH KANG, chemistry educator, researcher; b. Guangzhou, China, Feb. 14, 1937; s. Shu Y. and Lie C. (Lian) P.; m. Su C. Wang, Jan. 3, 1938; children: Irene, Elsie. B.Sc., Nat. Taiwan U., 1959; Ph.D., Mich. State U., 1966. Postdoctoral fellow U. So. Calif., 1966, Harvard U., 1967; prof. chemistry Boston Coll., 1967—; vis. prof. Stuttgart (Ger.) U., 1974, Max-Planck Inst., Mulheim, Ger., 1975; hon. prof. Academia Sinica Jilin U., China; hon. prof. academia sinica Lanzhou U., China. Author: General Chemistry, 1956, Acids and Bases, 1959; editor: Jour. Molecular Sci, 1981—; contbr. book revs. and articles to profl. jours. Exec. dir. Nat. Assn. Chinese Ams. Mem. Am. Phys. Soc., Sci. and Edn. Soc. Subspecialty: Theoretical chemistry. Current work: Quantum chemistry, theortical chemistry. Office: 140 Commonwealth Ave Newton MA 02167

PANAGIDES, JOHN, pharmacologist, immunologist, educator; b. N.Y.C., Aug. 15, 1944; s. Chris John and Sophie (Marmar) P.; m. Kathleen A. Heimann, July 9, 1967; children: Christopher, Melissa, Adrienne. B.S., CCNY, 1966; M.S. (NDEA Title IV fellow), U. N.C., 1968; Ph.D., SUNY, Buffalo, 1972. Research assoc. in biochem. cytology Rockefeller U., N.Y.C., 1972-73; sr. scientist in inflammation and immunology Lederle Labs., Pearl River, N.Y., 1973-82, clin. research assoc., 1982—; adj. asst. prof. Pace U., 1978. Contbr. numerous articles on inflammation and immunology to sci. jours. Mem. Am. Soc. Pharmacology and Exptl. Therapeutics, AAAS, N.Y. Acad. Scis. Republican. Greek Orthodox. Subspecialties: Immunopharmacology; Immunobiology and immunology. Current work: Regulation of the immune response; immunopharmacology; vaccine development. Office: Lederle Labs Pearl River NY 10965

PANDEY, JANARDAN PRASAD, geneticist, educator; b. Gonda, U.P., India, May 4, 1946; s. Ram Chandra and Suryakala (Dubey) P. M.S., U. Wis., 1970, Ph.D., 1972. Postdoctoral fellow U.Wis., Madison, 1972-73, U. Conn., Storrs, 1973-74, U. Calif., San Francisco, 1974-75; instr. Med. U. S.C., Charleston, 1976-77, asst. prof. immunology, 1978-83, assoc. prof. immunology, 1983—. Contbr. articles to sci. jours. Mem. Genetics Soc. Am., Am Assn. Immunologists. Subspecialties: Gene actions; Immunogenetics. Current work: Genetics of human immunoglobulin allotypes and their role in immune response and diseases. Home: 5 Lauden St Isle of Palms SC 29451 Office: Med U SC 171 Ashley Ave Charleston SC 29425

PANDEY, SUDHAKAR, nuclear engineer; b. Howrah, India, Nov. 27, 1947; came to U.S., 1974; s. Bhaskar Rao and Vimala (Kashikar) P.; m. Pramila Mahajan, June 28, 1974; children: Suchitra, Sanjeev. B.S. in Electronics, Calcutta U., 1969, M.S., 1970; M.S. in Nuclear Engring, Pa. State U., 1976, Ph.D., 1978. Engr. trainee Bhaba Atomic Research Ctr., Bombay, India, 1970-71; sci. officer Reactor Research Ctr., Kalpakkam, India, 1971-74; dir. engring. Nuclear Research Corp., Warminster, Pa., 1978-80; prin. engr. Franklin Research Ctr., Phila., 1980—. Mem. Am. Nuclear Soc., IEEE (affiliate). Subspecialty: Nuclear engineering. Current work: Safety and licensing of nuclear power plants; special interest in radiation protection, radiation monitoring instrumentation, ALARA, off-site dose calculation and consequence analysis. Home: 3009 Ashley CT Bensalem PA 19020 Office: Franklin Research Ctr 20th and Race Sts Philadelphia PA 19103

PANDOLFO, JOSEPH PETER, meteorologist, oceanographer, researcher, consultant, educator; b. N.Y.C., Sept. 26, 1930; s. Paolo and Lucia (Ganci) P.; m. Dolores Gertrude Maple, Sept. 6, 1952; children: Lucia Pandolfo Featherstone, Paul, Philip, Joseph, John, Robert. B.S., Fordham U., 1951; M.S., N.Y. U., 1956, Ph.D., 1961. Instr. to asst. prof. meteorology and oceanography N.Y.U., 1957-62; research scientist Travelers Research Center, Inc., 1962-69; research fellow Center for Environ. and Man, Inc., Hartford, Conn., 1969-71, v.p., 1971-76, pres., 1976—; dir., 1977—; lectr Hartford Grad. Center of Rensallear Poly. Inst.; cons. NOAA, NSF. Contbr. numerous articles to profl. jours. Served to 1st lt. USAF, 1951-55. Mem. AAAS, Am. Meteorol. Soc., Conn. Acad. Sci. and Engring. Roman Catholic. Subspecialties: Meteorology; Oceanography. Current work: Dynamic oceanography, dynamics and physics of planetary boundary layers, computer simulation, atmospheric and oceanic systems, air-land-sea interactions, solar energy. Office: 275 Windsor St 5 th Floor Hartford CT 06120

PANDYA, KRISHNAKANT HARIPRASAD, pharmacologist, educator, researcher; b. Mahemadavad, India, Oct. 19, 1935; came to U.S., 1970, naturalized, 1977; s. Hariprasad Jivanlal and Savitaben Chunilal P.; m. Ramaben Shankerlal Vyes, Sept 10, 1954; children: Prashant, Darshak. B.S. in Pharmacy, Gujarat U., 1958, M.S., 1961, Ph.D., 1968. Demonstrator L. M. Coll. Pharmacy, Gujarat U., 1958-61, tutor, 1961-63, jr. lectr., 1963-69; instr. Kirksville (Mo.) Coll. Osteo. Medicine, 1971-72, asst. prof., 1972-77, assoc. prof., 1977—. Contbr. articles on pharmacology to profl. jours. Am. Osteo Assn. grantee, 1972,79,83; Warner Found. grantee, 1976; NIH grantee, 1978; Am. Lung Assn./Am. Thoracic Soc. grantee, 1979-83. Mem. Am. Soc. for Pharmacology and Exptl. Therapeutics. Subspecialties: Pharmacology; Molecular pharmacology. Current work: Postnatal developmental changes in autonomic receptor function. Home: 1413 E Cottage Ln Kirksville MO 63501 Office: Kirksville Coll Osteo Medicine 800 W Jefferson Kirksville MO 65301

PANEM, SANDRA, virologist; b. Bklyn., June 25, 1946. B.S. in Biochemistry, U. Chgo., 1966, Ph.D. in Microbiology, 1970. Postdoctoral fellow Hosp. St. Louis, 1970-71; instr. to asst. prof. pathology U. Chgo., 1971-76, asst. prof. pathology and virology, 1975—. Contbr. articles to profl. jours. Sci. Policy fellow Brookings Instn., 1982-83; Kellogg nat. fellow, 1981-83; Leukemia Soc. Am. scholar, 1978-83; spl. fellow, 1975-77. Mem. Am. Assn. Cancer Research, Am. Soc. Virology, Am. Soc. Microbiology, Am. Assn. Women in Sci., Soc. Gen. Microbiology, AAAS. Subspecialties: Virology (biology); Cancer research (medicine). Current work: Interferon and autoimmunity, primate retroviruses, molecular biology.

PANG, JONG-SHI, management/administration educator; b. Saigon, Vietnam, Aug. 18, 1953; came to U.S., 1973; s. Tien-Ohn P.; m. Cindy S. C. Pang, Dec. 31, 1977; 1 son, Michael M. B.S., Nat. Taiwan U., 1973; M.S., Stanford U., 1975, Ph.D., 1976. Research asst. Stanford (Calif.) U., 1974-76; asst. scientist U. Wis.-Madison, 1976-77; asst. prof. Carnegie-Mellon U., Pitts., 1977-80, assoc. prof., 1980-81, U. Tex.-Dallas, Richardson, 1981—; prin. investigator NSF, 1980-83. Contbr. articles to profl. jours. Mem. Math. Programming Soc., Ops. Research Soc. Am., Soc. Indsl. and Applied Math. Subspecialties: Operations research (engineering); Numerical analysis. Current work: Research interests, mathematical programming, linear algebra and its applications. Office: Sch Mgmt and Adminstrn Box 688 Richardson TX 75080

PANG, KIM-CHING SANDY, pharmacologist, pharmacokineticist, educator, researcher; b. Hong Kong, Jan. 4, 1947; came to U.S., 1971; s. Yin Loon and Shiu Ching (Leung) P. B.S. in Pharmacy, U. Toronto, 1971, Ph.D. in Pharm. Chemistry, U. Calif., San Francisco, 1976. Vis. fellow NIH, 1976-78; asst. prof. pharmaceutics U. Houston, 1978-82; adj. asst. prof. Baylor Coll. Medicine, Houston, 1978-82; assoc. prof. Faculty Pharmacy, U. Toronto, Ont., Can., 1982; vis. prof. U. Groningen, Netherlands, 1980; ad hoc cons. NIH, 1978—. Mem.: editorial rev. bd. Durg Metabolism Disposition; contbr. articles to profl. jours. Nat. Inst. Gen. Med. Scis. grantee, 1979-82; Fungo grantee, Netherlands, 1980; Nat. Inst. Diabetic, Digestive and Kidney Disorders grantee, 1982; NATO grantee, 1982—. Mem. AAAS, Am. Soc. Pharmacology and Exptl. Therapeutics. Subspecialties: Pharmacokinetics; Pharmacology. Current work: Metabolite kinetics and clearance concepts; enzyme localization in organs with respect to drug and metabolite elimination and kinetics. Home: 7515 Teal Run Houston TX 77071 Office: 1441 Moursund St Houston TX 77030

PANGBURN, MICHAEL KENT, biochemist; b. Washington, Sept. 7, 1946; m. Kerry L.W. Pangburn, July 1, 1978; 1 son: Todd Owen. B.S., UCLA, 1969; Ph.D., U. Wash, 1974. Postdoctoral research fellow Scripps Clinic and Research Found., La Jolla, Calif., 1976-79, asst. mem., 1979—. Served to capt. USAF, 1974-76. Mem. AAAS, Am. Assn. Immunologists. Subspecialties: Biochemistry (medicine); Immunology (medicine). Current work: Proteins and enzymes of the complement system in human blood which combat disease. Office: 10666 N Torrey Pines Rd La Jolla CA 92037

PANITZ, JOHN ANDREW, physicist; b. Flushing, N.Y., May 31, 1944; s. William and Louise P.; m. Janda Kirk, Sept. 21, 1968. B.S., Pa. State U., 1965, M.S., 1966, Ph.D., 1969. Postdoctoral fellow Ionosphere Research Lab. Pa. State U., 1969-70; mem. tech. staff Sandia Nat. Labs., Albuquerque, 1970—. Contbr. numerous articles to profl. jours. Mem. Am. Phys. Soc., Am. Chem. Soc., AAAS, N.Y. Acad. Sci., Pi Eta Sigma, Sigma Pi Sigma. Subspecialties: Condensed matter physics; Biophysics (physics). Current work: Interface between materials and biological sciences; investigation of electrical breakdown phenomena, studies of the vacuum-solid interface using high field analytical techniques and the development of a novel field-ion tomographic microscope to image unstained organic and biological molecules. Patentee imaging atom-probe field ion microscope. Office: Division 1134 Sandia National Laboratory Albuquerque NM 87185

PANKOW, JAMES FREDERICK, chemistry educator, researcher; b. Mexico City, Jan. 15, 1951; U.S., 1954; s. Bernard John and Irma Lillian (Cordts) P.; m. Kathy S. M. Kim, Mar. 18, 1979. B.S., SUNY-Binghamton, 1973; M.S., Calif. Inst. Tech., 1976, Ph.D., 1978. Instr. Oreg. Grad. Ctr., Beaverton, 1978-80, asst. prof., 1980-82, assoc. prof. chemistry, 1982—. Grantee EPA, 1980, NOAA, 1980, U.S. Geol. Survey, 1982. Mem. Am. Chem. Soc. Subspecialties: Analytical chemistry; Environmental engineering. Current work: Trace organics in ground water - determination and modelling; trace organics in the atmosphere; fates of organic pollutants in the environment. Home: 527 S Emerald Loop Cornelius OR 97113 Office: Oreg Grad Ctr 19600 NW Walker Rd Beaverton OR 97006

PANOFSKY, WOLFGANK KURT HERMANN, physics educator; b. Berlin, Apr. 24, 1919; s. Erwin and Dorothea (Mosse) P.; m. Adele Irene DuMond, July 24, 1942; children: Richard, Margaret, Edward, Carol, Steven. B.A., Princeton U., 1938; Ph.D., Calif. Inst. Tech., 1942; hon. degrees, Case Inst. Tech., 1963, U. Sask., 1964, Columbia U., 1977. Dir. Office of Sci. Research and Devel. Calif. Inst. Tech., 1942-43; cons. Manhattan Project, Los Alamos, N.Mex., 1943-45; physicist U. Calif., Berkeley, 1945-46, asst. prof. to assoc. prof. physics, 1946-51; prof. physics Stanford U., 1951—, dir. high energy physics lab., 1953-61; dir. Stanford Linear Accelerator Ctr., 1961—; cons. arms control and disarmament, high energy physics. Author: (with M. Phillips) Classical Electricity and Magnetism, 1955, 2d edit., 1962. Recipient Ernest Orlando Lawrence Meml. award, 1961; Disting. Alumni award Calif. Inst. Tech., 1966; Calif. Scientist of Yr. award, 1967; Nat. Medal of Sci., 1969; Franklin Inst. award, 1970; Public Service award Fedn. Am. Scientists, 1973; Enrico Fermi award, 1979; Leo Szilard award Am. Phys. Soc., 1982. Mem. Nat. Acad. Scis., Am. Acad. Scis., Am. Phys. Soc., Phi Beta Kappa, Sigma Xi. Subspecialty: Particle physics. Current work: High energy elementary particle physics and arms control and disarmament. Home: 25671 Chapin Ave Los Altos Hills CA 94022 Office: PO Box 4349 Stanford CA 94305

PANTON, RONALD LEE, mechanical engineering educator, consultant; b. Neodesha, Kans., Feb. 14, 1933; s. Charles Wilson and Catheryne (McDowell) P.; m. Ruth Elaine Gulbrondsen, May 6, 1960; children: William, Theodore, Henry. A.B., Wichita State U., 1956, B.M.E., 1956; M.M.E., U. Wis.-Madison, 1962; Ph.D., U. Calif.-Berkeley, 1966. Registered profl. engr., Tex. Engr. N.Am. Aviation, Los Angeles, 1956-58; prof. mech. engring. Okla. State U., 1966-71, U. Tex., Austin, 1971—. Contbr. articles to profl. jours. Served to 1st lt. USAF, 1956-60. Mem. ASME, AIAA, AAUP, Sigma Xi. Subspecialties: Fluid mechanics; Aeronautical engineering. Current work: Research interests in experimental and theoretical aspects of incompressible flows. Home: 5901 Overlook Dr Austin TX 78731 Office: Mech Engring Dept U Tex Austin TX 78712

PANUSH, RICHARD SHELDON, physician; b. Detroit, Nov. 9, 1942; s. Bernard and Anne (Hecker) P.; m. Rena Joffe, May 20, 1965; children: Rachel Beth, Jonathan Boris, David Matthew. B.A., U. Mich., 1965, M.D., 1967. Diplomate: Am. Bd. Internal Medicine; subplty. cert. in rheumatology. Intern Duke U. Hosps., Durham, N.C., 1967-68, resident, 1968-69; instr. U. Colo., 1972-73; asst. prof. U. Fla., 1973-77, chief div. clin. immunology, 1976—, assoc. prof., 1977—. Contbr. numerous articles to profl. jours.; editorial cons. numerous jours. Served to maj. M.C. USAR, 1971-73. Fellow ACP; mem. Am. Rheumatism Assn., Am. Fedn. Clin. Research, So. Soc. Clin. Investigation, Fla. Soc. Rheumatism (past pres.), Am. Assn. Immunologists, Am. Acad. Allergy, Phi Beta Kappa, Alpha Omega Alpha. Subspecialties: Immunology (medicine); Internal medicine. Current work: Immunologic aspects of rheumatoid arthritis.

PAPAGIANNIS, MICHAEL D., astronomer; b. Athens, Greece, Sept. 3, 1932; s. Demetrios M. and Themitsa D. (Lucas) P.; m. Mary Hutton, May 29, 1961; children—Dimitrios, Christina. M.S., Nat. Poly. U. Athens, 1955, U. Va., 1960; Ph.D. in Physics and Astronomy, Harvard U., 1964. Lectr. astronomy Harvard U., 1964-66; asst. prof. astronomy Boston U., 1965-68, asso. prof., 1968-70, prof., 1970—, chmn. dept. astronomy, 1969-82; assoc. Harvard Obs., 1966-80; mem. exec. com. of corp. operating Haystack Radio Obs. Author: Space Physics and Space Astronomy, 1972; editor: 8th Texas Symposium on Relativistic Astrophysics, 1978, Strategies for the Search for Life in the Universe, 1980; contbr. articles to profl. publs. Trustee Hellenic Coll. and Holy Cross Sch. Theology, 1970—. Served with Greek Army, 1957-58. Fellow AAAS; mem. Am. Astron. Soc., Am. Geophys. Union, Internat. Astron. Union (Com. Sl-Search for Extraterrestrial Life 1982—), Internat. Union Radio Sci., Helicon (pres. 1968-69, 75-76), Acad. of Athens (corr.). Greek Orthodox. Clubs: Order Ahepa (sec. Ednl. Found. 1978-80, pres. Lexington Minuteman chpt. 1979-80, cert. of merit. Subspecialties: Radio and microwave astronomy; Space colonization. Current work: Active in new field of bioastronomy, which is the search for extraterrestrial life using astronomical techniques. President of new commission of International Astronomical Union on this subject. Home: 37 Coolidge Ave Lexington MA 02173 Office: Dept Astronomy Boston U Boston MA 02215 The people most likely to succeed are those who know exactly where they want to go, are fully aware of all the difficulties involved, and are still eager to devote all of their energy to the realization of this goal.

PAPAYANNOPOULOU, THALIA, hematologist, researcher, consultant, educator; b. Korinth, Greece, Apr. 28, 1937; d. Vassilios and Maria (Adrianou) P.; m. George Stamatoyannopoulos, Nov. 19, 1964; children: John Alkinoos, Vassilis-Alexis. M.D., U. Athens, Greece, 1961, D. Med. Scis., 1964. Postdoctoral fellow in pathology and hematology U. Wash., then research asst. prof., research assoc. prof., to 1980, assoc. prof. medicine, div. hematology, 1980—; cons. in field. Contbr. numerous articles on regulation of erythropoiesis and hemoglobin expression to profl. jours. Mem. Am. Fedn. Clin. Research, Am. Soc. Human Genetics, N.Y. acad. Scis., Am. Soc. Hematology, Am. Soc. Clin. Investigation. Subspecialties: Cell biology (medicine); Hematology. Current work: Research on regulation of globin expression in development and differentiation. Office: U Wash Seattle WA 98195

PAPE, HARRY RUDOLPH, JR., dental educator; b. Chgo., Mar. 31, 1937; s. Harry Rudolph and Vina Pearl (Derry) P.; m. Diana June Seagert, June 24, 1961; children: Nancy, Dale, Jerry, Yvonne. M.S., U. Mich., 1975, D.D.S., 1961. Pvt. practice dentistry, Quincy, Mich., 1963-71; instr. U. Mich., Ann Arbor, 1971-75, asst. prof., 1975-78, assoc. prof., 1978—; cons. Blissfield Migrant Workers Program, 1981—. Sec. Ridgeway Twp. Bd. Appeals; pres C. of C.; leader 4-H Garden Club. Served as lt. USN, 1961-63. Recipient Outstanding Instr. award Jr. Class Sch. Dentistry, U. Mich., 1974, 75, 76, 78, 82. Mem. Am. Assn. Dental Schs. Republican. Lutheran. Clubs: Clark Lake Yacht (Jackson, Mich.); Western Square Dancing (Adrain, Mich.). Subspecialties: Cariology; Preventive dentistry. Current work: Caries and periodontal prevention in areas of flourides, nutrition, diet analysis, plaque control, patient education, TV projection of phase microscopy and bacteriology. Home: 7241 Ridge Rd Ridgeway MI 49275 Office: U Mich Sch Dentistry Ann Arbor MI 48109

PAPP, KIM ALEXANDER, astronomy and astrophysics educator; b. Calgary, Alta., Can., Mar. 21, 1953; s. Alexander and Alice Doreen (Meding) P.; m. Joanne Enis Pianosi, July 11, 1981. B.Sc., U. Calgary, 1974, postgrad., 1982—; M.Sc., York U., Can., 1976, Ph.D., 1980, U. Chgo., 1980-81. Cons. Algas Engring., Calgary, 1974; asst. prof. U. Waterloo, Ont., Can., 1981—. Contbr. articles to profl. jours. Natural Sci. and Engring. Research Council grantee, 1982. Mem. Am. Phys. Soc., Am. Astron. Soc., Can. Astron. Soc., Can. Med. Assn., Alta. Med. Assn. Subspecialties: Theoretical astrophysics; Optical astronomy. Current work: Galaxy formation, galaxy evolution, galactic dynamics, stellar dynamics, planetary nebulae formation, supernovae production. Offic: MIR Health Scis Cente U Calgary 3330 Hospital Dr NW Calgary AB Canada T2N 4N1

PAPPALARDO, ROMANO G., physicist; b. Genova, Italy, Nov. 23, 1932; s. Francesco V. and Maria G. (Canepa) P.; m. Sheena C. MacColl; children: Anna, Elena. Ph.D. in Physics, U. Pavia, Italy, 1955. Asst. prof. U. Pavia, 1955-56; postgrad. fellow Bristol (Eng.) U., 1956-57; research asst. Pitts. U., 1957-59; vis. scholar Bell Telephone Labs., Murray Hill, N.J., 1959-61; research scientist Cyanamid European Inst., Geneva, 1961-66; research assoc. Argonne Nat. Lab., 1966-68; mem. tech. staff, research scientist GTE Labs., Waltham, Mass., 1968. Contbr. articles to profl. jours. Fulbright scholar, 1957. Mem. Electrochemical Soc. Subspecialties: Condensed matter physics; Spectroscopy. Current work: Synthesis and spectroscopy of inorganic luminescent materials for lighting industry. Patentee in field. Office: 40 Sylvan Rd Waltham MA 02254

PAPPAS, CAROL LYNN, internal medicine educator, neurophysiologist; b. Burlington, Vt., Apr. 21, 1949; d. William F. and Carolyne R. (Pearce) P. B.A. magna cum laude, Hobart and William Smith Coll., 1971; Ph.D., SUNY Upstate Med. Ctr., 1980, M.D., 1981. Postdoctoral fellow dept. physiology SUNY Upstate Med. Ctr., 1980-81; internal medicine intern U. South Fla., 1981-82, fellow in neurology, 1982—. Contbr. articles to profl. jours. NIH predoctoral fellow, 1971-75. Mem. Soc. Neurosci., A.C.P., Phi Beta Kappa. Subspecialties: Neurology; Neurophysiology. Current work: Evoked responses, cortical mapping, motor control, biofeedback, neuromuscular retraining. Office: Dept. Internal Medicine VA Hosp 13000 N 30th St Tampa FL 33612

PAPPAS, PETER WILLIAM, zoology educator; b. Pasadena, Dec. 9, 1944; s. William and Rosalie (Ashton) P.; m. Carolyn Ann Clague, Jan. 9, 1945; children: Allyson, Nicholas. B.A., Humboldt State Coll, 1966, M.A., 1968; Ph.D., U. Okla., 1971. Instr. biology Coll. Redwoods, Eureka, Calif., 1968; postdoctoral fellow Rice U., Houston, 1971-73, research assoc., 1973; asst. prof. Ohio State U., Columbus, 1973-77, assoc. prof., 1977-82, prof. zoology, 1982—. Editor: (with others) Cellular Interactions in Symbiosis and Parasitism, 1980, (with C. Arme) Biology of the Eucestoda, 2 vols, 1983; contbr. articles to profl. jours. NIH fellow, 1971-73; recipient Outstanding Research award Sigma Xi, 1975. Mem. Am. Soc. Parasitologists, Am. Soc. Tropical Medicine and Hygiene, Ann. Medwestern Conf. Parasitologists, Southwestern Assn. Parasitologists, Ohio Acad. Sci. Subspecialties: Parasitology; Cell biology. Current work: Biochemistry and physiology of parasites; membrane biogenesis and turnover in parasitic helminths. Home: 1709 Andover Rd Upper Alrington OH 43212 Office: Ohio State Univ Zoology Dept 1735 Neil Ave Columbus OH 43210

PAPPAS, SOCRATES PETER, chemist, educator, cons.; b. Hartford, Conn., Feb. 3, 1936; s. Peter and Valetina (Apostolidou) P.; m. Betty C. Thompson, Sept. 30, 1961; children: Niki Thais, Paul Alexander. B.A in Chemistry, Dartmouth Coll., 1958, Ph.D., U. Wis., 1962. Postdoctoral research asso. U. Wis., Madison, 1962-63; postdoctoral research asso. Brandeis U., 1963-64; asst. prof. Emory U., Atlanta, 1964-68, N.D. State U., Fargo, 1968-73, prof., 1973—; research asso. DeSoto, Inc., 1973; vis. prof. U. Stuttgart, W. Ger., 1977; cons. coatings sci. Contbr. numerous articles on coatings sci. to profl. jours.; editor: UV Curing: Sci. and Tech, 1978; co-editor: Photodegradation and Photostabilization of Coatings, 1981. NATO grantee, 1981-82; Dept. Interior grantee, 1978-80; Recipient Roon award, First Prize Fedn. Socs. of Coatings Tech., 1974,75,76,80; recipient DAAD award Ger. Acad. Exchange Service, 1977. Mem. Am. Chem. Soc., Inter-AM Photochem. Soc. Subspecialties: Organic chemistry; Photochemistry. Current work: Photoinitiators, thermally activated latent catalysts and rectants, photostabilization of coatings. Office: Polymers and Coatings Dept N D State U Fargo ND 58105

PAPPENHEIMER, JOHN RICHARD, educator, physiologist; b. N.Y.C., Oct. 25, 1915; s. Alwin Max and Beatrice (Leo) P.; m. Helena Gilder Palmer, Sept. 2, 1949; children—Glenn Alwin, William DeKay, Rosamond Gilder, Frank Richard. B.S., Harvard, 1936; Ph.D., Cambridge (Eng.) U., 1940. Demonstrator pharmacology Univ. Coll., London, Eng., 1939-40; research fellow physiology Columbia Med. Sch., 1940-41, instr., 1941-42; fellow biophysics Johnson Found. Med. Research, U. Pa., 1942-45; asso. physiology Harvard Med. Sch., 1946-49, asst. prof., 1949-53, vis. prof., 1953-69, Higginson prof. physiology, 1969—; career investigator Am. Heart Assn., 1953—; Overseas fellow Churchill Coll., Cambridge, Eng., 1971-72; Eastman prof. Oxford U., 1975-76. Fellow Am. Acad. Arts and Scis.; mem. Am. Physiol. Soc. (pres. 1964-65), Nat. Acad. Sci., Soc. Gen. Physiology, Brit. Physiol. Soc. (hon.), Harvey Soc., Phi Beta Kappa (hon.). Subspecialty: Physiology (medicine). Home: 15 Fayerweather St Cambridge MA 02138

PAQUE, RONALD EDWARD, immunologist, educator; b. Green Bay, Wis., Apr. 29, 1938; s. E. A. and Lucille R. (Aerts) P. B.S. with honors, Wis. State U., Oshkosh, 1960; M.S., U. Wis.-Madison, 1963; Ph.D. (scholar), U. Ariz., Tucson, 1966. Assoc. dir. immunology div. Microbiol. Assocs., Bethseda, Md., 1966-68; USPHS fellow U. Ill. Med Ctr., Chgo., 1968-70, asst. prof., 1970-74; assoc. prof. U. Tex. Health Sci. Ctr., San Antonio, 1974—; cons. NIH, NSF, Nat. Cancer Inst.; pres. Outdoor Cons. Pub., San Antonio, 1981—; participant Internat. Symposium on RNA in Devel. and Differentiation, Peking, 1980. Contbr. articles, revs. to sci. jours. and books. Served to cpl. USMCR, 1956-62. Mem. Am. Assn. Immunologists, Am. Soc. Microbiology, N.Y. Acad. Scis., AAAS, Am. Assn. Cancer Research, North San Antonio C. of C. Republican. Club: Los Amigos Ski. Subspecialties: Immunobiology and immunology; Cellular engineering. Current work: Nucleic acid programming of lymphoid cells; immunological cellular engineering; immunology of viral-induced myocarditis; immunology of infectious diseases, lymphokine technology; tumor immunotherapy. Office: 7703 Floyd Curl Dr San Antonio TX 78284

PARCE, DONALD LEWIS, computer scientist; b. San Francisco, Sept. 11, 1946; s. Lewis Howe and Diane (Ballard) P.; m. Joan Michele Padget, June 17, 1967; 1 dau., Lisa Michele. A.B., U. Calif.-Berkeley, 1968. Computer scientist U.S. Navy Electronics Lab., San Diego, 1968-74; mem. tech. staff Data Gen. Corp., Westborough, Mass., 1974-76, co-mgr. products, Research Triangle Park, N.C., 1976-77; co-founder, v.p. research and devel. Bus. Application Systems, Raleigh, N.C., 1977-81; dir. software devel. SCI Systems, Inc., Research Triangle Park, 1981-83; cons., 1983—; co-founder Interactive Devel. Corp., San Diego, 1971-72. Mem. IEEE, Assn. Computing Machinery, Cousteau Soc., Sierra Club, Audubon Soc. Baptist. Subspecialties: Computer architecture; Operating systems. Current work: Software development tools; machine independent system software. Home: Route 4 Box 412 Raleigh NC 27606

PARDEE, ARTHUR BECK, educator, biochemist; b. Chgo., July 13, 1921; s. Charles A. and Elizabeth B. (Beck) P.; m. Ruth Sager; children by previous marriage: Michael, Richard, Thomas, Elizabeth. B.S., U. Calif. at Berkeley, 1942; M.S., Calif. Inst. Tech., 1943, Ph.D., 1947. Merck postdoctoral fellow U. Wis., 1947-49; mem. faculty U. Calif. at Berkeley, 1949-61, asso. prof., 1957-61; NSF fellow Pasteur

Inst., 1957-58; prof. biology, chmn. dept. biochem. scis. Princeton, 1961-67, prof. biochemistry, 1961-75, Donner prof. sci., 1966; prof. Sidney Farber Cancer Inst. and pharmacology dept. Harvard Med. Sch., Boston, 1975—; Mem. research adv. council Am. Cancer Soc., 1967-71. Co-author: Experiments in Biochemical Research Techniques, 1957; Editor: Biochemica et Biophysica Acta, 1962-68, 74—. Trustee Cold Spring Harbor Lab. Quantitative Biology, 1963-69. Recipient Young Biochemists travel award NSF, 1952, Krebs Medal Fedn. European Biochem. Socs., 1973, Rosenstiel award Brandeis U., 1975, 3M award Fedn. Am. Socs., Exptl. Biology, 1980. Mem. Nat. Acad. Sci. (editorial bd. proc. 1971-73, com. on scis. and pub. policy 1973-76), Am. Chem. Soc. (Paul Lewis award 1960), Am. Soc. Biol. Chemists (treas. 1964-70, pres. 1980-81), Am. Assn. Cancer Research (dir. 1983—), Am. Acad. Arts and Scis., Am. Soc. Microbiologists, Japanese Biochem. Soc., Phi Beta Kappa, Sigma Xi. Subspecialties: Biochemistry (biology); Cell biology. Current work: Cancer research on growth control basic mechanisms, chemotherapy and carcinogenesis research. Home: 30 Codman Rd Brookline MA 02146

PARDUE, MARY LOU, biology educator; b. Lexington, Ky., Sept. 15, 1933; d. Louis A. and Mary A. (Marshall) P. B.S., Coll. William and Mary, 1955; M.S., U. Tenn.-Knoxville, 1959; Ph.D., Yale U., 1970. Am. Cancer Soc. postdoctoral fellow U. Edinburgh, 1970-72; assoc. prof. biology MIT, 1972-80, prof., 1980—; organizer Molecular Cytogenetics Course Cold Spring Harbor Lab., 1971-78, organizer Drosphila Course, 1979, 80; sci. adv. panel Wistar Inst., 1976; mem. cellular and molecular basis of disease rev. com. NIH, 1980. Editorial bd.: Genetics, 1973, Jour. Cell Biology, 1973-76, Cell, 1973-77, Chromosoma, 1976, Cell Biology Internat. Reports, 1982, Molecular and Cellular Biology, 1983; contbr. articles to sci. jours. Recipient Ester Langer award for cancer research, 1977. Fellow AAAS; mem. Genetics Soc. Am. (pres. 1982-83), Nat. Acad. Scis., Am. Soc. Cell Biology. Subspecialties: Cell biology; Molecular biology. Office: Biology 16 717 MIT Cambridge MA 02139

PARESCE, FRANCESCO, astrophysicist; b. London, Apr. 11, 1940; U.S., 1949, naturalized, 1973; s. Gabriele and Degna (Marconi) P.; m. Didatta Malvezzi Campeggi, Mar. 5, 1966; children: Donata, LLaria, Camilla. Baccalaureat en philosophie, U. Paris, 1958; Laurea in Fisica, U. Rome, 1964; Ph. D. in astronomy, U. Calif., Berkeley, 1972. Research fellow Centro Di Astrofisica Spazzale, Frascati, Italy, 1964-67; research asst. dept. astronomy U. Calif., Berkeley, 1967-72, asst. research astronomer, 1972-80, assoc. research astronomer, 1980—. Contbr. articles in field to profl. jours. Accademia Nazionale Dei Lincei, Rome fellow, 1976; NATO fellow, 1982; Fulbright scholar, 1967; recipient NASA tech. achievement award, 1972, group achievement award, 1976, cert. recognition, 1978. Mem. Internat. Astron. Union, Am. Astron. Soc., Am. Geophysical Union. Subspecialties: Ultraviolet high energy astrophysics; Aeronomy. Current work: Design and devel. of lab. and flight instrumentation for soft x-ray and uv astronomy and aeronomy; devel. of theoretical phys. models for the interpretation of exptl. data. Patentee in field. Home: 597 San Luis Rd Berkeley CA 94707 Office: Space Sciences Laboratory University of California Berkeley CA 94720

PARIS, STEVEN MARK, software engineer, educator; b. Boston, May 26, 1956; s. Julius Louis and Frances (Keleishik) P. B.S., Rensselaer Poly. Inst., 1978; M.S., Boston U., 1980, postgrad., 1980—. Sr. software engr. Prime Computer Inc., Framingham, Mass., 1978-82; sr. analyst Computervision Corp., Bedford, Mass., 1982—; instr. Babson Coll. Recipient Boston Sci. Fair 1st prize, 1973, 74; State of Mass. Sci. Fair 3d prize, 1973; 2d prize, 1974; nat. merit scholarship letter of Commendation. Mem. Assn. for Computing Machinery, IEEE, Boston Computer Soc., Planetary Soc. Jewish. Subspecialties: Software engineering; Operating systems. Current work: Research in massively parallel processing and functional programming. Home: 27 Colwell Ave Brighton MA 02135 Office: 201 Burlington Rd Bedford MA 01730

PARISI, ALFRED FRANCIS, cardiologist, researcher; b. N.Y.C., Dec. 11, 1937; s. Alfred Avino and Lucretia (Cecere) P.; m. Joan Teresa Severino, June 18, 1960; children: Mark, Christianne, John. A.B., Georgetown U., 1959; M.A., Harvard U., 1969; M.D., Cornell U., 1963. Diplomate: Am. Bd. Internal Medicine. Intern Peter B. Brigham Hosp., Boston, 1963-64, resident, 1964-66; NIH fellow Harvard U. Med. Sch., Boston, 1966-69; staff cardiologist Wilford Hall Air Force Med. Center, San Antonio, 1969-71; chief cardiology VA Hosp., Balt., 1971-74, VA Med. Ctr., West Roxbury, Mass., 1974—; asst. prof. U. Md., Balt., 1971-73, assoc. prof., 1973-74; asst. prof. Harvard U. Med. Sch., Boston, 1974-77, assoc. prof., 1977—. Author: Non-Invasive Approaches to Cardiovascular Diagnosis, 1979; author and editor: Non-Invasive Cardiac Imaging, 1983; editor: Jour. Cardiovascular Medicine, 1980—. Served to maj. USAF, 1969-71. Recipient Communication award Assn. Advancement of Med. Instrumentation, 1981. Fellow ACP, Am. Coll. Cardiology, Council Clin. Cardiology-Am. Heart Assn.; mem. Am. Soc. Echocardiography, Am. Fedn. Clin. Research. Roman Catholic. Subspecialties: Cardiology; Internal medicine. Current work: Quantative approaches to non-invasive cardiovascular diagnosis, primarily with two-dimensional echocardiography. Office: VA Med Center 1400 VFW Pkwy West Roxbury MA 02132

PARISI, JOSEPH THOMAS, microbiologist, educator; b. Chgo., Apr. 28, 1934; s. Joseph A. and Josephine P.; m. Elaine Ann Smithbower, June 8, 1963; 1 dau., Melissa. B.S., Loyola U. Chgo., 1956; M.S., Ohio State U., 1958, Ph.D., 1962. With Duquesne U., Pitts., 1962-65; with U. Mo. Sch. Medicine, Columbia, 1965—, prof. microbiology, 1975. Contbr. articles to sci. jours. Mem. Am. Soc. Microbiology, Sigma Xi. Subspecialties: Microbiology (medicine); Genetics and genetic engineering (medicine). Current work: Genetics of antibiotic resistance, plasmid biology, epidemiology. Home: 313 Defoe Dr Columbia MO 65201 Office: Dept Microbiology U Mo Sch Medicine Columbia MO 65212

PARIZA, MICHAEL WILLARD, food microbiologist, educator; b. Waukesha, Wis., Mar. 10, 1943; s. Willard and Dorothy Violet (Miller) P. B.S. in Bacteriology, U. Wis., 1967; M.S in Microbiology, Kans. State U., 1969, Ph.D., 1973. Postdoctoral trainee McArdle Lab. Cancer Research, U. Wis., 1973-76, asst. prof. dept. food microbiology and toxicology, 1976-81, asso. prof., 1981—; chmn. dept., 1982—. Contbr. articles to profl. jours. Served with AUS, 1969-71. NIH grantee, 1979, 81. Mem. Am. Assn. Cancer Research, Am. Soc. Microbiology, Inst. Food Tech., AAAS, Sigma Xi. Subspecialties: Cancer research (medicine); Cell and tissue culture. Current work: Chemical carcinogenesis in relation to diet, nutrition and food safety. Office: 1925 Willow Dr Madison WI 53706

PARK, CHAN HYUNG, cell biologist, physician; b. Seoul, Korea, Aug. 16, 1936; s. Chung Suh and Yoon Sook Yuh; m. Mary Hyungrok Kim, Apr. 16, 1966; 1 son, Christopher Myungwoo. M.D., Seoul Nat. U., 1962, M.S., 1964; Ph.D., U. Toronto, Ont., Can., 1972. Diplomate: Am. Bd. Internal Medicine with subplty. of med. oncology. Asst. prof. U. Kans. Med. Center, 1974-80, assoc. prof., 1980—. Transl. novel from German to Korean; contbr. articles to biomed. and sci. jours. Recipient research career devel. award USPHS, NIH, 1979—. Mem. ACP (fellow), Am. Assn. Cancer Research, Am. Soc. Clin. Oncology, Am. Soc. Hematology, Internat. Soc. Exptl. Hematology, Cell Kinetics Soc., Am. Fedn. Clin. Research. Subspecialties: Cell and tissue culture; Oncology. Current work: Clinical application of in vitro chemotherapy human tumor cell culture. Home: 9137 Grandview Dr Overland Park KS 66212 Office: 39th and Rainbow Blvd Kansas City KS 66103

PARK, CHUL, aerospace engineer; b. Taegu, Korea, June 8, 1934; came to U.S., 1964; s. Hyung-jin and Ha-Woon (Ryang) P.; m. Chyon Sue Sohn, Sept. 29, 1962; children: Sora, Pora, Marie. B.S., Seoul (Korea) Nat. U., 1957, M.S., 1960; Ph.D., Imperial Coll. Sci. and Tech., U. London, 1964. Research assoc. NASA Ames Research Ctr., Moffett Field, Calif., 1964-67, aerospace engr., 1967—; vis. engr. M.I.T., 1971-72. Served to 1st lt. Korean Air Force, 1958-61. Mem. AIAA. Subspecialty: Aerospace engineering and technology. Current work: High temperature gas physics and fluid mechanics related to hypersonic flight and space travel. Home: 21194 Bank Mill Rd Saratoga CA 95070 Office: Ames Research Center NASA Bldg 229-4 Moffett Field CA 94035

PARK, CHUNG SURU, dairy nutrition educator; b. Seoul, Korea, Jan. 14, 1942; s. Myung S. and Shin D. (Lee) P.; m. Myung S. Lee, Sept. 6, 1972; children: Austin M., Derrick E. B.S., Seoul Nat. U., 1964; M.S., U. Ga., 1972; Ph.D., Va. Poly. Inst., 1975. Research asst. dairy sci. dept. Va. Poly. Inst., Blacksburg, 1975-77; research assoc. animal sci. dept. Purdue U., West Lafayette, Ind., 1977-78; asst. prof. dept. animal sci. N.D. State U., Fargo, 1978-82, assoc. prof., 1982—. Contbr. articles to profl. jours. Mem. Am. Dairy Sci. Assn., Am. Soc. Animal Sci. Subspecialties: Animal nutrition; Cell and tissue culture. Current work: Dairy nutrition, protein and energy interaction, cholesterol metabolism, amino acid transport in mammary cell culture. Home: 330 21st St N Fargo ND 48102 Office: ND State U 169 Hultz Hall Fargo ND 58105

PARK, DONG HWA, neurobiologist, educator; b. Seoul, Mar. 3, 1937; U.S., 1963, naturalized, 1976; s. Chi Ho and Ok Nam (Shin) P.; m. Min Jung, Sept. 9, 1967; children: Henry, Bernard. B.S. in Chemistry, Seoul Nat. U., 1961; M.S. in Biochemistry, Brigham Young U., 1968, Ph.D., 1970. Nat. Vitamin Found. postdoctoral fellow Columbia U.-St. Luke's Hosp. Center, N.Y.C., 1970-72; asst. scientist N.Y.U., N.Y.C., 1972-75; instr. neurobiology Cornell U. Med. Coll., N.Y.C., 1975-78, asst. prof., 1978—. Contbr. articles to sci. jours. Mem. Am. Chem. Soc., Am. Soc. for Neurochemistry, Soc. for Neurosci. Baptist. Subspecialties: Neurochemistry; Neurobiology. Current work: Neurotransmitter synthesizing enzymes, purification, characterization, production of antibodies to the above enzymes and immunochemical studies. Home: 50-21 59th Pl Woodside NY 11377 Office: 411 E 69th St New York NY 10021

PARK, JOHN THORNTON, physicist, educator; b. Phillipsburg, N.J., Jan. 3, 1935; s. Dawson J. and Margaret M. (Thornton) P.; m. Dorcas Marshall, June 1, 1956; children: Janet Ernst, Karen. B.A., Nebr. Wesleyan U., 1956; Ph.D., U. Nebr., 1963. Teaching asst. U. Nebr., 1956-58, research asst., 1958-62, research assoc., 1962-63; NSF postdoctoral fellow Univ. Coll., London, 1963-64; asst. prof. U. Mo.-Rolla, 1964-68, assoc. prof., 1968-71, prof. physics, 1971—, chmn. dept., 1977—; vis. assoc. prof. NYU, 1970-71; mem. NRC panel, 1975, 76; mem. gen. com. Internat. Conf. Physics of Acad. and Atomic Physics, 1975-79, Small Accelerator Conf., 1978—; honors lectr. Mid-Am. State Univ. Assn., 1973-77; mem. panel NSF-NATO postdoctoral fellowship, 1977, mem. program com. div. electron and atomic physics, 1974-75, mem. nominating com., 1978. Contbr. papers to profl. lit. Recipient Outstanding Tchr. award U. Mo.-Rolla, 1970; Young Alumni Service award Nebr. Wesleyan U., 1968; NSF research grantee, 1966—. Fellow Am. Phys. Soc. (div. electron and atomic physics); mem. Am. Assn. Physics Tchrs., Am. Vacuum Soc. Methodists. Lodge: Rotary. Subspecialty: Atomic and molecular physics. Current work: Measurements of ion-atom collision cross sections with emphasis on proton-atomic hydrogen collisions. Office: Physics Dept U Mo Rolla MO 65401

PARK, MYUNG KUN, pediatric cardiologist, researcher, educator; b. Suhung, Korea, Sept. 30, 1934; s. Jung Jin nd and Sonnyu (Lee) P.; m. Issun Kim, Oct. 15, 1942; children: Douglas, Christepher, Warren. Student, Seoul Nat. U., 1954-56; M.D., 1960. Diplomate: Am. Bd. Pediatrics. Instr. pediatrics U. Wash. Sch. Medicine, Seattle, 1966-68; asst. prof. U. Kans. Med. Sch., Kansas City, 1973-76; asso. prof. pediatrics and pharmacology U. Tex. Med. Sch., San Antonio, 1976-83, prof. pediatrics, 1983; dir. pediatric cardiology Med. Center Hosp., San Antonio, 1977—. Author: How to Read Pediatric ECGs, 1981, Pediatric Cardiology for Practitioners, 1983; also articles. NIH grantee, 1971-73. Mem. Am. Acad. Pediatrics, Am. Coll. Cardiology, Am. Soc. Pharmacology and Exptl. Therapeutics, N.Y. Acad. Sci., Am. Heart Assn. Prsbyterian. Subspecialties: Pediatrics; Pharmacology. Current work: Developmental pharmacology; digitalis pharmacology; cardiovascular physiology. Home: 3318 Buckhaven Dr San Antonio TX 78230 Office: 7703 Floyd Dr San Antonio TX 78284

PARK, OK-CHOON, psychologist; b. Keumsan, Choongnam, Korea, Dec. 30, 1944; came to U.S., 1974; s. Dal-moon and Eon-yon (Yook) P.; m. Young-soon Yang, Sept. 10, 1974; children—Michael H., Christine J. M.A., U. Minn., 1976, Ph.D., 1978. Prin. researcher Control Data Corp., Mpls., 1979—; asst. prof. SUNY-Albany, 1981—. Contbr. numerous articles to profl. jours. Research fellow Health Service Research Center UMinn., 1977-78; SUNY Research found. fellow, Albany, 1981. Mem. Am. Psychol. Assn., Am. Ednl. Research Assn., Assn. Devel. Computer-based Instructional Systems, Assn. Ednl. Communication and Technology, Phi Kappa Phi. Subspecialties: Learning; Artificial intelligence. Current work: Combination of human cognitive info. processes and computers information processes into an artificial intelligence system; development computer-based instructional system. Home: 3098 Evelyn St Roseville MN 55113 Office: Control Data Corp 511 11th Ave S Minneapolis MN 55415

PARK, U. YOUNG, nuclear engineer, mechanical engineer; b. Seoul, Korea, Oct. 12, 1940; s. M.W. amd and D.C. (Chang) P.; m. Linda L. Rugh; children: Tara, Thomas. B.S., Seoul Nat. U., 1963; M.S., U. Cin., 1970. Registered profl. engr., Calif. Power engr. State of Ohio, Columbus, 1975-78; nuclear engr. Battelle, Columbus, 1978-81; nuclear system engr. Bechtel Power Corp., San Francisco, 1981—. Contbr. articles to profl. jours. Mem. Am. Nuclear Soc. Subspecialties: Nuclear engineering; Mechanical engineering. Current work: Design of nuclear power plant. Home: 2850 Madigan Ct Concord CA 94518

PARKE, ROSS DUKE, psychology educator; b. Huntsville, Ont., Can., Dec. 17, 1938; came to U.S., 1965; s. Horace and Dora (Henderson) P. B.A., U. Toronto, 1962, M.A., 1963; Ph.D., U. Waterloo, Ont., 1965. From asst. prof. to prof. U. Wis.-Madison, 1965-71; sr. scientist Fels Research Inst., Yellow Springs, Ohio, 1971-75; prof. psychology U. Ill., Champaign, 1975—; cons. NSF, 1976-78, Nat. Inst. Child Health and Human Devel., 1979-83. Editor: Recent Trends in Social Learning Theory, 1972; author: Fathers, 1981; co-author: Child Psychology: A Contemporary Perspective, 1975, 79; contbr. articles to sci. jours. NSF grantee, 1968-70, 71-75; March of Dimes grantee, 1980-84; others. Fellow Am. Psychol. Assn. (exec. com. div. developmental psychology 1980-83), Internat. Soc. Study of Behavioral Devel. (exec. com. 1979-83); mem. Soc. Research in Child Devel. (governing council 1983—). Subspecialty: Developmental psychology. Current work: Social development of infants; father's role in child development. Office: U Ill Dept Psychology Champaign IL 61820

PARKER, ALICE CLINE, electrical engineering educator, consultant; b. Birmingham, Ala., Apr. 10, 1948; d. Joseph Kalman and Elizabeth (Wenk) Cline; m. Donald Joseph Bebel, Aug. 9, 1980. B.S.E.E., N.C. State U., Raleigh, 1970; M.S.E.E., Stanford U., 1971; Ph.D., N.C. State U., 1975. Asst. prof. Carnegie-Mellon U., Pitts., 1975-80; asst. prof. elec. engring. U. So. Calif., Los Angeles, 1980-83, assoc. prof., 1983—; cons. Aerospace Corp., 1980-82, Xerox Corp., 1981. NSF fellow, 1970-71. Mem. IEEE, Assn. Computing Machinery, Sigma Xi. Subspecialties: Computer-aided design; Computer engineering. Current work: Automatic synthesis of integrated circuits from high-level behavioral specifications; includes optimization of cost-speed, testability requirements. Office: Elec Engring Systems Dep U So Cali Los Angeles CA 90089-0781

PARKER, CHARLES WARD, physician, researcher, educator; b. St. Louis, Mar. 23, 1930; s. William B. and Florence (Mershon) P.; m. Mary L. Langston, June 13, 1953; children: Keith, Charles, Kathy, Christy, Sandy. M.D., Washington U., St. Louis, 1953. Intern Barnes Hosp., St. Louis, 1953-54, resident, 1956-58; USPHS research fellow, 1961-62; head div. immunology Washington U., St. Louis, 1962—, prof. medicine, 1971—, prof. microbiology and immunology, 1975—, dir., 1977—. Served with USNR, 1954-56. Recipient Hixon award, Bausch & Lomb award, Mosby award; NIH grantee, 1962-72. Mem. Central Soc. Clin. Research, Am. Acad. Allergy, Am. Assn. Immunologists, Collegium Internationale Allergologicum, Am. Acad. Allergy (fellow), Sigma Xi, Alpha Omega Alpha, Phi Eta Sigma. Subspecialties: Allergy; Immunology (medicine).

PARKER, DONALD EDWARD, psychology educator, researcher, consultant; b. Chgo., Apr. 6, 1936; s. Kenneth Calwell and Florence (Wilson) P.; m. Mary Lynn Goodrich; children: Katherine, Susan, Geoffrey, Rebecca. B.A., DePauw U., 1958; Ph.D., Princeton U., 1961. Postdoctoral fellow Auditory Research Lab., Princeton U., N.J., 1961-62; lectr. in psychology Miami U., Oxford, Ohio, 1962-65, asst. prof., 1966-69, assoc. prof., 1969-72, prof. psychology, 1973—, chair, 1977-80; postdoctoral fellow Max Planck Institut fur Verhaltensphysiologie, Seewiesen, W. Ger., 1965-66; vis. asst. prof. U. Victoria, B.C., 1969; vis. scientist Johnson Space Ctr., NASA, 1983. Contbr. articles profl. jours. Served to capt. USAFR, 1962-65. Grantee USAF, NASA; NIH fellow; recipient award Spacelab I, 1977. Mem. Barany Soc., Assn. Research in Otolaryngology, Acoustical Soc. Am., Psychonomics Soc., Midwestern Psychol. Assn., Sigma Xi. Subspecialties: Sensory processes; Gravitational biology. Current work: Adaptation to weightlessness during space flight. Home: 400 E Chestnut St Oxford OH 45056 Office: Pychology Dep Miami Univ Oxford OH 45056

PARKER, EARL RANDALL, educator; b. Denver, Nov. 22, 1912; s. Sam and Rebecca Rose (Presley) P.; m. Mary Mildred Larkin, June 2, 1935; children—Robert Earl (dec.), Margaret Mary, William John. Met.E., Colo. Sch. Mines, 1935. Research metallurgist Gen. Electric Research Lab., 1935-44; research metallurgist U. Calif. at Berkeley, 1944-45, asso. prof. metallurgy, 1945-49, prof., 1949—, chmn. div. mineral tech., 1953-57, dir., 1958-64; Campbell Meml. lectr., 1957, Robert S. Williams lectr., 1957; Mem. Nat. Materials Bd., 1971. Author book, tech. papers in field. Recipient Distinguished Citizens award, Denver, 1958, Vincent Bendix gold medal Am. Soc. Engring. Edn., 1969; Guggenheim fellow, 1960; named Calif. Scientist of Yr., 1970. Fellow Am. Inst. Mining, Metall. and Petroleum Engrs. (Mathewson gold medal 1956), Am. Phys. Soc.; mem. Am. Soc. Metals (pres. 1968, Sauveur Achievement award 1964, Gold medal 1972), Nat. Acad. Engring. (U.S. Pres.'s medal of sci. award 1980), Am. Soc. Engring. Edn. Subspecialty: Metallurgy. Home: 15629 Kavin Ln Monte Sereno CA 95030

PARKER, EDWIN BURKE, business executive; b. Berwyn, Alta., Can., Jan. 19, 1932; m. Frances G. Spigal, 1976; children: David Kendall, Karen Liane. B.A., U. B.C., Can., 1954; M.A., Stanford U., 1958, Ph.D. in Mass Communications, 1960. Staff reporter Vancouver (B.C.) Sun, 1954-55; info. officer U. B.C., Vancouver, 1955-57; research asst. Inst. Communications Research, Stanford U., 1957-60, asst. prof., 1962-63, assoc. prof., 1963-71, prof., 1971-79; v.p. Equatorial Communications Co., 1979—; asst. prof. communications U. Ill.-Urbana, 1960-62. Ctr. for Advanced Studies in Behavioral Sci. fellow, 1969-70. Current work: Social effects of communication technology. Office: Equatorial Communications Co 300 Ferguson Dr Mountain View CA 94043

PARKER, EUGENE NEWMAN, physics educator; b. Houghton, Mich., June 10, 1927; s. Glenn H. and Helen D. (MacNair) P.; m. Niesje M. Parker, Nov. 24, 1954; children: Joyce M., Eric Glenn. B.S. in Physics, Mich. State U., 1948, Ph.D., Calif. Inst. Tech., 1951; D.Sc. (hon.), Mich. State U., 1975. Instr. dept. math. U. Utah, 1951-53, asst. prof. dept. physics, 1953-55; with U. Chgo., 1955—, prof. physics, 1962—, prof astronomy and astrophysics, 1967—. Author: Interplanetary Dynamical Processes, 1963, Cosmical Magnetic Fields, 1979. Recipient Space Sci. award AIAA, 1962; Sydney Chapman medal Royal Astron. Soc., 1979; Disting. Alumnus award Calif. Inst. Tech., 1980. Mem. Nat. Acad. Sci. (Henryk Arctowski medal 1969), Am. Phys. Soc., Am. Astron. Soc. (Henry Norris Russel lecture 1969, George Ellery. Hale award 1978), Am. Geophys. Union (recipient John Fleming award 1968). Subspecialties: Theoretical astrophysics; Plasma physics. Current work: Theoretical researches into the nature and causes of the activity of the sun and other stars. Office: Laboratory for Astrophysics and Space Research 933 E 56th St Chicago Il 60637

PARKER, FRANK S., biochemistry educator, researcher; b. Boston, Jan. 25, 1921; s. Louis J. and Jennie D. P.; m. Gladys Baker, Sept. 1, 1946; children: Judith Ann, George Edward (dec.). B.S. in Chemistry, Tufts U., 1942, M.S., 1944, Ph.D., Johns Hopkins U., 1950. Asst. prof. Bryn Mawr Coll., 1950-54; assoc. prof. SUNY Downstate Med. Ctr., 1960-63, N.Y. Med. Coll., N.Y.C., 1963-70, prof. biochemistry, Valhalla, 1970—. Author: Applications of Infrared, Raman, and Resonance Raman Spectroscopy in Biochemistry, 1983; editorial bd.: Applied Spectroscopy, 1968-76, Can. Jour. Spectroscopy, 1973—. Served with USN, 1944-46. Recipient Career Scientist award N.Y.C. Health Research Council, 1963. Mem. Am. Soc. Biol. Chemists, Am. Chem. Soc., Biophys. Soc., Soc. Applied Spectroscopy, Coblentz Soc., AAUP, Harvey Soc., Sigma Xi. Subspecialties: Biophysical chemistry; Biochemistry (biology). Current work: Conformational analysis; infrared and raman studies. Office: Dept Biochemistry NY Med Coll Basic Sci Bldg Valhalla NY 10595

PARKER, FRANK WAYNE, mech. engr., state police officer; b. Lordsburg, N. Mex., Feb. 7, 1945; s. Paul M. and Viola Catherine (Nol) P. B.S.M.E., N. Mex. State U., 1971. Lic. pvt. investigator, Ariz. Research engr. Autotronics, El Paso, Tex., 1971; police officer Ariz. Dept. Public Safety, 1972—; officer II, accident engr., Glendale, 1974—, accident reconstructionist, 1977—; prin. P & R Reconstrns. (cons. engring. and accident engring.), Glendale, 1978—; tchr. advanced investigative techniques. Author: Nat. Transp. Safety Bd. publs, 1975. Served with USAF, 1963-67; ETO. Mem. Soc. Automotive Engrs., ASME (asso.). Subspecialties: Mechanical engineering; Theoretical and applied mechanics. Current work: Sci.

devel. of field of accident investigation/reconstrn. Home and Office: 5747 W Missouri Apt 31 Glendale AZ 85301

PARKER, GARALD GORDON, geologist; b. Leona, Oreg., July 2, 1905; s. Arta William and Bertha Mona (Smith) P.; m. Bernadette Elizabeth Spalding, Oct. 29, 1955 (div. 1955); children: Robert A., Elizabeth A., Carole L., Deborah E., Lisa E.; m. Martha Rosetta Harriger, July 3, 1925 (dec. Nov. 1954); children: Avonne A., Garald Gordon. A.B. magna cum laude, Central Wash. State Coll., Ellensburg, 1935, M.S., U. Wash., Seattle, 1946. Cert. profl. geologist. Geologist, hydrologist U.S. Geol. Survey, Miami, Richland, Wash., Washington, Phila., Denver, Albany, N.Y., 1939-69; chief hydrologist and sr. scientist S.W. Fla. Water Mgmt. Dist., Brooksville, Fla., 1969-75; pres Garald G. Parker Sr. & Assocs., Tampa, 1978—; sr. hydrologist P.E. Lamoreaux & Assocs., Tampa, 1975-76; sr. scientist Geraghty & Miller, Tampa, 1976-78; pres., owner Garald G. Parker Sr. & Assocs., Tampa, 1978—; expert examiner geology U.S. CSC, 1950-59; mem. Water Resources Task Force Nat. Water Assessment, Washington, 1966-67; pres. N.Y. State Capitol dist. Fed. Exec. Council, 1967-68; teaching fellow U. Wash., Seattle, 1938-40; affiliate prof. hydrogeology Am. U., Washington, 1955-56; vis. lectr. Columbia U., N.Y.C., 1957-58; affiliate prof. hydrology Colo. State U., Ft. Collins., 1962-63; vis. prof. dept. agrl. engring. U. Fla., 1975-81. Author: Water Resources of Southeast Florida, 1955, Effect of Pleistocene Epoch on Geology and Water Resources of Florida, 1945 (Gold medal Fla. Acad. Scis.), Water Resources, Delaware River Basin, 1959 (Outstanding Achievement award Fla. Acad. Scis. Survey); assoc. editor: Am. Water Resources Assn. Jour, 1966-70 (pres.'s award 1970); contbr. over 100 articles to profl. jours. Bd. dirs. Fla. Canal Authority, Tallahassee, 1978-79, Chem. and Environ. Services Ctr., U. South Fla. Coll. Natural Scis., Tampa, 1980—; mem. Tampa Environ. Com., 1975-77. Recipient Outstanding Achievement awards S.W. Fla. Water Mgmt. Dist., 1975, Fla. Audubon Soc., 1975, S.E. Geol. Soc., 1976. Fellow Geol. Soc. Am., Am. Geophys. Union, Am. Water Resources Assn. (dir. Rocky Mountain dist. 1964-66, editorial bd. 1965-80, dir. 1965-68, pres. 1968-69); mem. Am. Inst. Profl. Geologists (pres. Fla. sect. 1967-68), Fla. Engring. Soc., Washington Geol. Soc. (sec., councilor 1953-54). Democrat. Club: Cosmos (Washington). Lodges: Kiwanis; Masons. Subspecialties: Hydrogeology; Resource management. Current work: Geologic, water resources and environmental management consulting. Home: 3414 Reynoldswood Dr Tampa FL 33618 Office: PO Box 270089 Tampa FL 33688

PARKER, HARRY WILLIAM, chem. engr. and cons.; b. Tulia, Tex., June 4, 1923; s. A. D. and Effie Mae (Sorrenson) P.; m. Phyllia Ann Spidy, July 4, 1954; children: Andrew David, Kevin Philip. B.S., Tex. Tech. U., Lubbock, 1954; M.S., Northwestern U., 1954, Ph.D., 1956. Registered professional engineer Tex., Okla. Research group leader Phillips Petroleum Co., Bartlesville, Okla., 1956-70; prof. chem. engring. Tex. Tech. U., Lubbock, 1970-79, 81—; engr. in residence Engring. Socs. Commn. on Energy, Washington, 1979-81. Mem. Am. Inst. Chem. Engrs., Soc. Petroleum Engrs. Republican. Presbyterian. Subspecialties: Biomass (energy science and technology); Comparative energy costs. Current work: Relative energy costs, fuels from biomass. Patents, publs. in energy field. Address: Chem Engring Dept Tex Tech U Lubbock TX 79409

PARKER, JERRY CALVIN, psychologist, researcher; b. Jacksonville, N.C., July 31, 1947; s. Calvin C. and L. Estelle; ; s. Calvin C. and Sanders (Parker); m. Alice Jane Prosch, Aug. 21, 1971; children: Aaron, Adam. B.A., Otterbein Coll., 1969; M.A., Xavier U., 1976; Ph.D., U. Mo.-Columbia, 1976. Cert. psychologist, Mo., W.Va. Staff psychologist Ohio Dept. Corrections, Columbus, Ohio, 1969-72; staff psychologist VA Med. Ctr., Martinsburg, W.Va., 1976-78, Truman Meml. Vets. Hosp., Columbia, Mo., 1978-80, asst. chief psychology, 1980-82, acting chief psychology, 1982—; clin. asst. prof. psychology U. Mo. Sch. Medicine, 1978—, mem., 1981—; assoc. tng. dir. Mid-Mo. Psychology Consortium, 1980-82, tng. dir., 1982—. Vol. Am. Cancer Soc., 1977; mem. Neo-Fight, Columbia, Mo., 1982. VA Med. Research Program grantee, 1979, 82; Ohio Dept. Mental Health and Mental Retardation Tng. grantee, 1972. Mem. Am. Psychol. Assn., Internat. Neuropsychol. Soc., Nat. Acad. Neuropsychologists, Arthritis Health Professions Assn., Lambda Gamma Epsilon. Democrat. Subspecialties: Pain management; Neuropsychology. Current work: Health psychology, pain, stress management, behavioral medicine; neuropsychology, cerebrovascular disease effects on cognition, outcomes following carotid endarterectomy. Home: 3240 Brampton Ct Columbia MO 65201 Office: Harry S Truman Meml Vets Hosp 800 Stadium Blvd Columbia MO 65201

PARKER, JOHN CLARENCE, research company executive, microbiologist; b. Washington, Sept. 13, 1935; s. Marion W. and Katherine L. (Hagan) P.; m. Mary Ann Baker, June 10, 1957 (div. 1973); children: John C., Robert C.; m. Norma Leona Justmann, Jan. 9, 1975. B.S. in Zoology, U. Md., 1957, M.S. in Parasitology, 1961, Ph.D., 1965. Project dir. Microbiol. Assocs., Bethesda, Md., 1961-78, dir. clin. diagnostic lab., 1973—, v.p. ops., 1976-79, pres., 1979—; cons. various sci. research orgns. and programs. Contbr. chpts. to research books; assoc. editor: Am. Assn. for Lab. Animal Sci, 1976—. Mem. AAAS, Am. Assn. Immunologists, Am. Assn. for Animal Sci. (Charles A. Griffin award 1979), Am. Soc. for Microbiology, Pan Am. Group for Rapid Viral Diagnosis, Soc. for Exptl. Biology and Medicine, Tissue Culture Assn. Republican. Episcopalian. Subspecialties: Microbiology (medicine); Infectious diseases. Current work: Virology: diseases of research animals useful as models of human diseases; biotechnology research for diagnosis of human diseases. Home: 5023 Mussetter Rd Ijamsville MD 21754 Office: Microbiol Assocs 5221 River Rd Bethesda MD 20816

PARKER, LAWRENCE NEIL, physician, endocrinologist; b. N.Y.C., Nov. 8, 1943; s. Norman Samuel and Lee (Shapiro) P.; m. Joyce Hope Steinfeld, July 18, 1970; children: Jill Monica, Gregory Robert. A.B., Columbia U., 1964; M.D., Stanford U., 1969. Diplomate: Am. Bd. Internal Medicine. Asst. chief endocrinology VA Med. Ctr., Long Beach, Calif., 1977—; asst. prof. medicine U. Calif.-Irvine Med. Sch., 1977-82, assoc. prof., 1982—. Served to lt. comdr. USPHS, 1970-72. VA grantee, 1978—. Fellow ACP.; Mem. Endocrine Soc., Am. Fedn. for Clin. Research, Am. Diabetes Assn. Subspecialty: Endocrinology. Current work: Control of the androgens of the adrenal gland. Office: Endocrinology Div VA Med Ctr 5901 E 7th St Long Beach CA 90822

PARKER, LINDA ALICE, psychology educator, researcher; b. Los Angeles, Sept. 3, 1948; emigrated to Can., 1974; d. James Everett and Evelyn Geneva (Lukkasson) P.; m. Ernest Ira Shockley, Aug. 24, 1980; 1 son, Loren Everett. B.A. in Psychology, Calif. State U.-Long Beach, 1971, M.A., 1974; Ph.D., Meml. U. Nfld., St. John's, 1979. Neurophysiol. research asst. Long Beach VA Hosp., 1973-74; research asst. Meml. U., 1974-77; lectr. U. N.B., St. John, 1977-79, asst. prof., 1979-83, assoc. prof., 1983—. Contbr. articles to profl. jours. Research on Drug Abuse scholar Non Med. Drugs Directorate in Can., 1975; U. N.B. research awards, 1979-83; Excellence in Teaching award, 1982; Nat. Scis. and Engring. Research Council grantee, 1980-83. Mem. Am. Psychol. Assn., Can. Psychol. Assn., Eastern Psychol. Assn. Subspecialty: Learning. Current work: Investigation of behavioral conditioned responses elicited by drug-paried flavor stimuli and environmental stimuli. Office: Div Social Science UNB PO Box 5050 Saint John NB Canada E2L 4L5

PARKER, LLOYD ROBINSON, JR., chemistry educator, researcher; b. Rome, Ga., Sept. 24, 1950; s. Lloyd R. and Virginia L. (Holl) P. B.A., Berry Coll., 1972; M.S., Emory U., 1974; Ph.D., U. Houston, 1978. Asst. prof. chemistry dept. Vassar Coll., Poughkeepsie, N.Y., 1978—. Mem. Am. Chem. Soc., Sigma Xi. Republican. Subspecialty: Analytical chemistry. Current work: Atomic absorption spectroscopy; optimization; computer interfacing; statistics and data analysis. Home: Vassar Coll Box 501 Poughkeepsie NY 12601 Office: Chemistry Dept Vassar Coll Poughkeepsie NY 12601

PARKER, MICHAEL ANDREW, materials scientist; b. Chgo., Oct. 1, 1949; s. Andrew Stanley and Elizabeth Rose P.; m. Mary Ellen Donahue, June 19, 1982. B.S., Ill. Inst. Tech., 1974, M.S., 1978. Staff engr., electronic materials scientist IBM, Rochester, Minn., 1978—; temporary assignment materials sci. dept. Stanford (Calif.) U. Ill. State scholar; Ill. Inst. Tech. Grad. fellow; Argonne (Ill.) U. Assn. summer research fellow, 1976; IBM grad. research fellow, 1982. Mem. Am. Phys. Soc., IEEE, Microbeam Analysis Soc., AIME, Electron Microscopy Soc. Am. Subspecialties: Electronic materials; Microchip technology (materials science). Current work: Microanalysis of submicron structures by means of various particle beams, electrons, ions and photons by such techniques as STEM, SEM, ISS, SIMS, XPS, EDX, WDX, Auger, etc. Conversely, microfabrication of submicron structures using particle beams for electronics and microchips. Home: 42858 Roberts Ave Fremont CA 94538 Office: IBM Dept 579 Bldg 004-1 Rochester MN 55901 Materials Sci Dept Peterson Bldg Stanford U Stanford CA

PARKER, NORMAN FRANCIS, electronics company executive; b. Fremont, Nebr., May 14, 1923; s. Frank Huddleston and Rose Johanna (Launer) P.; m. Carol Hope Watt, June 12, 1949; children: Leslie Ann, Kerry Irene, Sandra Jean, Noel Louise. Student, W.Va. U., 1943-44; B.S. in Elec. Engring. (George Westinghouse scholar), Carnegie-Mellon U., 1947, M.S., 1947, D.Sc. in Engring. (Buhl fellow), 1948. Engr. Westinghouse Electric Corp., East Pittsburgh, Pa., summer 1941, Bloomfield, N.J., summer 1942, U. Calif. Radiation Lab., Oak Ridge, 1944-45; with Autonetics div. N.Am. Aviation, Inc., Anaheim, Calif., 1948-67, asst. chief engr., 1956-59, v.p., gen. mgr., 1959-62, exec. v.p., 1962-66, pres. autonetics div., 1966-67, v.p., 1966-67; exec. v.p., dir. Bendix Corp., 1967-68; pres., dir. Varian Assos., Inc., 1968-81, chief exec. officer, 1972-81; dir. U.S. Leasing Internat., San Francisco., Internat. Game Tech., Reno. Life trustee Carnegie-Mellon U.; mem. bus. adv. com. Grad. Sch. Indsl. Adminstrn. Served with AUS, 1943-46. Fellow IEEE, AIAA; mem. Nat. Acad. Engring., Sigma Xi, Tau Beta Pi, Eta Kappa Nu, Phi Kappa Phi, Pi Mu Epsilon. Subspecialty: Systems engineering. Patentee curve follower and gyroscopes. Office: 611 Hansen Way Palo Alto CA 94303

PARKER, RICHARD C., chemistry educator; b. Coleman, Tex., July 17, 1939; s. E.C. and Bernice L. (Trapp) P.; m. Marilyn Ankeney; children: Melissa, Christopher. Ph.D. in Phys. Chemistry, U. Wash., 1966. Apollo research engr. N.Am. Aviation, Downey, Calif., 1962; asst. prof. chemistry N.J. Inst. Tech., 1966-75, assoc. prof., 1975-81, prof., 1981—. Environ. commr., Oceanport, N.J., 1981. Mem. Am. Chem. Soc., Am. Phys. Soc., AAAS, Sigma Xi, Omega Xi Epsilon. Subspecialties: Physical chemistry; Kinetics. Current work: Biomass conversion, effect of sound on reaction rates. Office: NJ Inst Tech 323 High St Newark NJ 07757

PARKER, ROBERT B., pharmacologist; b. Springfield, Mass., Apr. 4, 1944; s. Richard and Irene (Rumpal) P.; m. Julie R. Rinehart, Aug. 10, 1968; 1 dau., Sarah L.; m. Robin L. Boyle, Nov. 14, 1980. B.S., Bates Coll., 1966; Ph.D., Harvard U., 1971. Research pharmacologist Warner Lambert/Parke-Davis, Ann Arbor, Mich., 1972-77, sr. scientist, 1977-79; sr. prin. investigator Pennwalt Corp., Rochester, N.Y., 1979—. Co-chmn. fund dr. United Way, 1976-77. Recipient award Am. Inst. Chemists, 1966; Andelot fellow, 1966; NRC postdoctoral research asso., 1971-72. Mem. Am. Heart Assn., Am. Lung Assn., Am. Soc. Pharmacology and Exptl. Therapeutics. Mem. United Ch. Christ. Subspecialties: Pharmacology; Information systems, storage, and retrieval (computer science). Current work: Cardiovascular pharmacology; respiratory pharmacology; obesity; geropharmacology; acquisition, storage, retrieval and analysis of data. Home: 193 Mason Rd Fairport NY 14450 Office: 755 Jefferson Rd Rochester NY 14623

PARKER, ROBERT HALLETT, ecological consultant; b. Springfield, Mass., Feb. 14, 1922; m., 1945; 3 children. B.Sc., U. N.Mex., 1948, M.Sc., 1949; Mag. Sci. and Doctorand, Copenhagen U., 1962. Asst. in biology U. N.Mex., 1948-49; asst. in zoology Duke U., 1949-50; marine biologist State Game and Fish Commn., Tex., 1950-51; geophys. trainee Phillips Petroleum Co., 1951; research biologist Scripps Inst., U. Calif., 1951-58, jr. research ecologist, 1958-63; resident ecologist Systems Ecology Program, Marine Biol. Lab., Woods Hole, Mass., 1963-66; assoc. prof. biology and geology Tex. Christian U., 1966-70; pres., chmn. bd. Coastal Ecosystems Mgmt., Inc., Ft. Worth, 1970—; cons. Am. Mus. Natural History, 1957, Standard Oil Co. N.J., 1956-58, Pneumodyn Corp., 1960-61; research scientist Tex. Christian Research Found. Mem. Soc. Systematic Zoology, Ecol. Soc. Am., Am. Soc. Limnology and Oceanography, Am. Assn. Petroleum Geology (assoc., award 1956). Subspecialty: Ecosystems analysis. Office: Coastal Ecosystems Management Inc 3600 Hulen St Fort Worth TX 76106

PARKER, WILLIAM SKINKER, biology educator, ecology researcher; b. St. Louis, Aug. 28, 1942; s. Norman Cecil and Betty (Skinker) P.; m. Therese Butler, July 22, 1966 (div.); 1 son, Vincent; m. Elisabeth Wells, Jan. 2, 1979. B.A., Wabash Coll., 1964; M.S., Ariz. State U., 1966; Ph.D., U. Utah, 1974. Asst. prof. Miss. U. for Women, Columbus, 1974-77, assoc. prof., 1977-83, prof., 1983—. Author: Comparative Ecology of Two Colubrid Snakes, 1980; editor: Jour. and Newsletter Herpetological Rev, 1975-76; contbr. numerous articles to profl. jours. Bd. dirs. Contact Teleministries, Columbus, 1982—, vol. worker, 1977-82. Mem. AAAS, Ecol. Soc. Am., Am. Soc. Ichtyologists and Herpetologists, Am. Soc. Naturalists, Sigma Xi. Subspecialties: Ecology; Population biology. Current work: Ecology of reptiles and amphibians; demography and disperal of pond turtles. Home: 1219 Mapleview Dr West Point MS 39773 Office: Miss U for Women Box W-100 Columbus MS 39701

PARKOLA, WALTER R(OBERT), design engineer; b. Bridgeport, Conn., Dec. 24, 1954. B.S.M.E., U. Bridgeport, 1978. Mfg. engr. IBM, Lexington, Ky., 1978-79; staff devel. engr. Exxon Office Systems, Brookfield, Conn., 1979-81; devel. engr. Wentworth Labs., Brookfield, 1981-83; design engr. Perkin-Elmer Corp., Danbury, Conn., 1983—. Mem. ASME, Am. Soc. Metals, Soc. Mfg. Engrs. Subspecialty: Ink jet printing. Current work: Microelectronics test probe devel. Inventor ink jet printing tech. Home: Box C230 RD 2 Sandy Hook CT 06482 Office: 100 Wooster Heights Rd Danbury CT 06810

PARR, ALBERT CLARENCE, research physicist; b. Tooele, Utah, June 22, 1942; s. Trafton C. and Esther L. (Schuldheisz) P.; m. Ruth E. Pieplow, June 27, 1965; children: Robin, Trafton. B.S. in Math, Oreg. State U., 1964; M.S. in Physics, U. Chgo., 1965, Ph.D., 1971. Research assoc. U. Chgo., 1970; asst. prof. U. Ala.-Tuscaloosa, 1971-76, assoc. prof., 1976-80; research physicist far ultraviolet physics sect. Nat. Bur. Standards, Washington, 1980—. Contbr. articles to sci jours. Active Nat. Capital Area council Boy Scouts Am., 1979—. Served with USNR, 1960-66. Recipient cert. of recognition Nat. Bur. Standards, 1979, 1981. Mem. Optical Soc. Am., Am. Phys. Soc. Lodge: Elks. Subspecialty: Atomic and molecular physics. Current work: Photoelectron spectroscopy, synchrotron radiation.

PARR, GARY RAYMOND, biochemist, educator; b. Janesville, Wis., Nov. 22, 1942; s. Harold Lawrence and Thelma Irene (Quade) P.; m. Susan Dale Resneck, Sept. 27, 1968 (div. Feb. 1977); 1 dau., Alexandra Lynn. B.S., U. So. Calif.-Los Angeles, 1965; Ph.D., U. Wis.-Madison, 1973. Postdoctoral assoc. Cornell U., Ithaca, N.Y., 1973-75; research assoc. Ithaca Coll., 1975-76; staff fellow NIH, Bethesda, Md., 1976-78, sr. staff fellow, 1978-81; asst. prof. Clemson (S.C.) Univ., 1982—. Served to 1st lt. U.S Army, 1966-68; Vietnam. Decorated Bronze Star. Mem. Am. Soc. Biol. Chemists, Am. Chem. Soc., N.Y. Acad. Scis., AAAS, Sigma Xi. Subspecialties: Biophysical chemistry; Kinetics. Current work: Mechanism of protein folding, fast kinetic techniques, differential scanning calorimetry. Office: Clemson U Dept Biochemistry 132 Long Hall Clemson SC 29631

PARR, JAMES FLOYD, JR., soil microbiologist; b. Seattle, Feb. 20, 1929; s. James Floyd and Clara Georgian (Kestner) P.; m. Carol June Cunningham, Aug. 29, 1964; children: Lauren Melissa, James Floyd. B.S. in Agr, Wash. State U., 1952; M.S. in Soil Microbiology, Purdue U., 1957, Ph.D., 1961. Irrigation extension agt. Agrl. Extension Service, Wash. State U., Pullman, 1953-54; county extension agt. Agrl. Extension Service, Mont. State U., Bozeman, 1954-55; chemist Calif. Dept. Water Resources, Sacramento, 1957-58; instr. agronomy dept. Purdue U., West Lafeytte, Ind., 1958-61; research assoc. dept. botany U. Mich., Ann Arbor, 1961-63; research chemist, microbiologist TVA, Muscle Shoals, Ala., 1963-67; research leader Agrl. Research Service U.S. Dept. Agr., Baton Rouge, 1967-75, chief biol. waste mgmt. and organic resources Lab., Beltsville, Md., 1975—; cons. to developing countries on waste mgmt. and recycling of organic wastes. Contbr. numerous articles on soil microbiology to profl. jours. Served with USN, 1946-48. Recipient Dept. Agr. Superior Service award, 1977, Cert. of Merit, 1981. Fellow Am. Soc. Agronomy, Soil Sci. Soc. Am.; mem. Sigma Xi, Alpha Zeta, Phi Kappa Phi. Subspecialties: Resource conservation; Soil chemistry. Current work: Conduct research to develop methods for handling, processing, composting and utilizing, municipal, and industrial wastes on agricultural lands. Patentee in field. Office: Biol Waste Mgmt Lab Agrl Research Ctr US Dept Agr Beltsville MD 20705

PARR, ROBERT GHORMLEY, chemistry educator; b. Chgo., Sept. 22, 1921; s. Leland Wilbur and Grace (Ghormley) P.; m. Jane Bolstad, May 28, 1944; children: Steven Robert, Jeanne Karen, Carol Jane. A.B. magna cum laude with high honors in Chemistry, Brown U., 1942; Ph.D. in Phys. Chemistry, U. Minn., 1947. Asst. prof. chemistry U. Minn., 1947-48; mem. faculty Carnegie Inst. Tech., 1948-62, prof. chemistry, 1957-62, Johns Hopkins U., 1962-74, chmn. dept., 1969-72; William R. Kenan, Jr. prof. theoretical chemistry U. N.C., Chapel Hill, 1974—; Vis. prof. chemistry, mem. Center Advanced Study U. Ill., 1962; distinguished vis. prof. State U. N.Y. at Buffalo, also Pa. State U., 1967; vis. prof. Japan Soc. Promotion Sci., 1968, 79, U. Haifa, 1977, Free U. Berlin, 1977; Firth prof. U. Sheffield, 1976; Chmn. com. postdoctoral fellowships in chemistry Nat. Acad. Sci.-NRC, 1961-63; chmn. panel theoretical chemistry Westheimer com. survey chemistry Nat. Acad. Sci., 1964; mem. council Gordon Research Conf., 1974-76; mem. Commn. on Human Resources, NRC, 1979-82. Author: Quantum Theory of Molecular Electronic Structure, 1963, also numerous articles.; Asso. editor: Jour. Chem. Physics, 1956-58, Chem. Revs, 1961-63, Jour. Phys. Chemistry, 1963-67, 77-79, Am. Chem. Soc. Monographs, 1966-71, Theoretica Chimica Acta, 1966-69; bd. editors: Jour. Am. Chem. Soc, 1969-77; adv. editorial bd.: Internat. Jour. Quantum Chemistry, 1967—, Chem. Physics Letters, 1967-79. Recipient Outstanding Achievement award U. Minn., 1968, N.C. Disting. Chemist award, 1982; fellow U. Chgo., 1949; research asso., 1957; Fulbright scholar U. Cambridge, Eng., 1953-54; Guggenheim fellow, 1953-54; NSF sr. postdoctoral fellow U. Oxford (Eng.) and Commonwealth Sci. and Indsl. Research Orgn., Melbourne, Australia, 1967-68; Sloan fellow, 1956-60. Fellow Am. Phys. Soc. (chmn. div. chem. physics 1963-64), AAAS; mem. Am. Chem. Soc. (chmn. div. phys. chemistry 1978), AAUP, Am. Acad. Arts and Sci., Nat. Acad. Scis., Internat. Acad. Quantum Molecular Sci. (v.p. 1973-79, hon. pres. 1979—), Phi Beta Kappa, Sigma Xi, Phi Lambda Upsilon, Pi Mu Epsilon. Subspecialty: Theoretical chemistry. Home: 701 Kenmore Rd Chapel Hill NC 27514 Office: Dept Chemistry U North Carolina Chapel Hill NC 27514

PARRISH, ALAN DESCHWEINITZ, astronomer, researcher; b. Albany, N.Y., Nov. 17, 1942; s. William Judson and Mary (deSchweinitz) P. B.E.E., Yale U., 1965; Ph.D., Cornell U., 1971. Research assoc. Arecibo Obs., P.R., 1971-72; research engr. Nat. Radio Astronomy Obs., Green Bank, W.Va., 1972-74; mem. sponsored research staff MIT, 1974-78; research asso. SUNY-Stony Brook, 1978-81, sr. research assoc., 1981—; sr. scientist Millitech Inc., Amherst, Mass., 1982—. Contbr. articles to profl. jours. Mem. Internat. Sci. Radio Union, Internat. Astron. Union. Subspecialties: Remote sensing (atmospheric science); Radio and microwave astronomy. Current work: Application of radio astronomy techniques to remote sensing of stratospheric minor constituents. Office: Millitech Corp Amherst Fields Research Park Amherst MA 01002

PARRISH, JOHN WESLEY, JR., biologist, educator; b. Dennison, Ohio, Mar. 5, 1941; s. John Wesley and Dorothy Irene (Dickinson) P.; m. Paula Schmanke, July 9, 1966; children: Corinne Danelle, Wesley Allen. B.S., Denison U., 1963; M.A., Bowling Green State U., 1970, Ph.D. (Univ. fellow), 1974. Tchr. sci. Northside Jr. High Sch., Norfolk, Va., 1967; vis. instr. dept. biology Kenyon Coll., 1973-74; NIH postdoctoral fellow dept. zoology U. Tex., Austin, 1974-76; assoc. prof. biology Emporia State U., 1976—. Vice pres. Kanza Audubon Soc., Emporia, Kans., 1978, pres., 1979-80, bd. dirs., 1981-83. Served to lt. USNR, 1964-67. Josselyn Van Tyne Fund grantee, 1968; Emporia State U. Faculty Research Com. grantee, 1977-79, 79-80, 81-82, 83-84. Mem. Am. Ornithologist's Union (student award 1971), Am. Soc. Andrology, Am. Soc. Zoologists, Cooper Ornithol. Soc., Kans. Acad. Sci., Kans. Ornithol. Soc., Soc. Study Reprodn. Democrat. Methodist. Subspecialties: Comparative physiology; Reproductive biology. Current work: Photoreception, photoperiodism, biochronometry, pineal gland function, placental function, testicular function. Office: Emporia State U Div Biol Sci Emporia KS 66801

PARROTT, ROBERT HAROLD, physician, educator; b. Jackson Heights, N.Y., Dec. 29, 1923; s. Harold Leslie and Ruth Mabel (Hargrove) P.; m., June 2, 1951; children: Timothy, Maureen, Daniel, Theresa, Christopher, Edward. Student, Fordham U.; M.D., Georgetown U., 1949. Intern Hosp. of St. Raphael, New Haven, 1949-50; resident Children's Hosp. of D.C., 1950-52; staff pediatrician, chief Pediatric Unit, Lab. of Clin. Investigation, Nat. Inst. Allergy and Infectious Diseases, NIH, Bethesda, Md., 1952-56; physician-in-chief, dir. Research Found., Children's Hosp. of D.C., Washington, 1956-62, dir., 1962—; prof. child health and devel. George Washington U. Mem. Am. Acad. Pediatrics, Am. Pediatric Soc., Am. Acad. Med. Dirs., Am. Coll. Physician Execs., D.C. Med. Soc., Infectious Diseases Soc. Subspecialty: Virology (medicine). Home: 9852 Singleton Dr

Bethesda MD 20817 Office: 111 Michigan Ave NW Washington DC 20010

PARRY, JOHN O., health physicist, researcher; b. Oceanside, Calif., Apr. 16, 1952; s. John O. and Helen A. (Phetteplace) P.; m. Valorie D. Steller, June 9, 1979; children: Jennifer Lynn, Jason Charles. B.S. in Applied Physics, Mich. Technol. U., 1974. Health physicist, unit 1 Dresden-Commonwealth Edison, Morris, Ill., 1974-76, lead health physicist, 1976-78, radiation protection/chem. supr., 1978-80; corp. health physicist Wash. Public Power System, Richland, 1980-82, sr. health physicist, 1982-83, prin. health physicist, 1983—; public speaker on radiation protection, 1974-82. Served to capt. C.E. U.S. Army, 1974. Mem. Health Physics Soc. (chmn. public info. 1982-83, sec. 1983-84, nat. com. state and Fed. legislation 1983-86); mem. Am. Nuclear Soc. (assoc.). Club: Am. Bridge (Kennewick, Wash.). Subspecialties: Nuclear fission; Nuclear engineering. Current work: Radiation protection in nuclear power plants; radiation effects on materials. Office: Wash Public Power Supply System 3000 George Washington Way Richland WA 99352

PARSHALL, GEORGE WILLIAM, chemistry researcher; b. Hackensack, Minn., Sept. 19, 1929; s. George C. and Frances (Virnig) P.; m. Naomi Bess Simpson, Oct. 9, 1954; children: William E., Jonathan L., David C. B.S. in Chemistry, U. Minn., 1951; Ph.D. in Organic Chemistry, U. Ill.-Urbana, 1954. With central research dept. E.I. duPont de Nemours & Co., Wilmington, Del., 1954—, research supr., 1965-79, dir. chem. sci., 1979—. Author: Inorganic Syntheses, vol. 15, 1974, Homogeneous Catalysis, 1980; contbr. over 80 articles to sci. jours. Mem. Am. Chem. Soc. (award 1983). Episcopalian. Subspecialty: Catalysis chemistry. Current work: Homogeneous catalysis of organic reactions; direction of programs in organic chemistry, photoimaging and electronics. Patentee in field (18). Office: DuPont Exptl Sta Wilmington DE 19898

PARSLEY, RONALD LEE, educator; b. Madison, Wis., July 14, 1937; s. Palo and Gertrude (Heidel) P.; m. Nancy Joan Johnson, Aug. 18, 1962 (div. 1973); children: Rodney Alexander, Andrew Arthur. A.B., UCLA, 1960; M.S., U. Cin., 1963; Ph.D., 1969. Asst. prof. Tulane U., New Orleans, 1966-71, assoc. prof., 1971-79, prof., 1979—; Colo. river boatman Hatch River Expdns., Vernal, Utah, 1976—. Contbr. articles to profl. jours. Nat. Acad. Scis., exchange scientist, Prague, Czechoslovakia, 1981. Mem. Paleontol. Soc., Paleontol. Research Inst., Geol. Soc. Am., AAAS. Club: Sierra Club. Subspecialties: Paleobiology; Paleoecology. Current work: Paleobiology of primitive echinodermata; their systematics (including modes of evolution), functional morphology and paleoecology. Home: 1314 Audubon St New Orleans LA 70118 Office: Dept Geology Tulane Univ New Orleans LA 70118

PARSLY, LEWIS FULLER, nuclear and chemical engineer; b. Phila., Jan. 27, 1918; s. Lewis Fuller and Hester Carroll (Anderson) P.; m. Marjorie Norwood Pickens, Aug. 11, 1951; children: Thomas, Anne, Mary, James. B.S. in Chem. Engring, U. Pa., 1940, M.S., 1947, Ph.D., 1948. Jr. engr. Am. Chem. Paint Co., Ambler, Pa., 1940, Electric Storage Battery Co., Phila., 1940-41, Day & Zimmermann, Inc., 1948-51; devel. engr. nuclear div. Union Carbide, Oak Ridge, 1951—. Served to capt. U.S. Army, 1941-46; PTO. Mem. Am. Inst. Chem. Engrs. (sect. chmn.), Am. Chem. Soc., Am. Nuclear Soc., Nat. Soc. Profl. Engrs. Subspecialty: Nuclear engineering. Current work: Decay heat removal systems of nuclear power plants. Home: 108 Hutchinson Pl Oak Ridge TN 37830 Office: Nuclear Div Union Carbide Corp PO Box Y Oak Ridge TN 37830

PARSONS, JAMES EUGENE, biology educator; b. Lima, Ohio, Nov. 10, 1939; s. Virgil Lorain and Dorothy Mae (Sandy) P. B.S., Ohio State U., 1961, M.S., 1963, Ph.D., 1977. Registered microbiologist, specialist microbiologist Am. Acad. Microbiology; clin. lab. specialist Nat. Cert. Agy. for Med. Lab. Personnel. Clin. microbiologist Akron Gen. Med. Ctr., Akron, Ohio, 1977-81; adj. prof. biology U. Akron, 1979-81, adj. prof. Inst. Biomed. Engring. Research, 1979-81; asst. prof. clin. microbiology Northeastern Ohio U. Coll. Medicine, Rootstown, Ohio, 1978-81; asst. prof. biology Bloomsburg (Pa.) U., 1982—; chmn. bd. Animalcule, Ltd., Akron, 1978—; cons. Medina (Ohio) Community Hosp., 1978—. Bd. dirs. Canton (Ohio) Coll., 1978-81. Recipient Tchr of Yr. award Akron Gen. Med. Technologists, 1979. Mem. Am. Soc. Microbiology, Electron Microscopy Soc. Am., Med. Mycology Soc. Am., N.Y. Acad. Scis., Ohio Acad. Sci., South Central Assn. Clin. Microbiology, Sigma Xi, Phi Kappa Phi. Subspecialties: Microbiology (medicine); Epidemiology. Current work: Rapid identification of mycobacteria and pathogenic fungi; automated analysis in the clinical microbiology laboratory; clinical parasitology. Home: 3018 Olney Rd Kalamazoo MI 49007 Office: Bloomsburg State College 102 Harline Science Center Bloomsburg PA 17815

PARSONS, MICHAEL LOEWEN, chemist, educator, cons.; b. Oklahoma City, Apr. 20, 1940; s. Orville Loewen and Eva (O'Neill) P.; m. Virginia E. Thomas, Mar. 23, 1972; children: Steven M., David R., Guinn E., L. Erin. B.A., Kans. State Coll., 1962, M.S., 1963; Ph. D., U. Fla., 1966. Research chemist Phillips Petroleum Co., 1966-67; asst. prof. dept. chemistry Ariz. State U., Tempe, 1967-72, assoc. prof., 1972-78, prof., 1978—, acting assoc. chmn., 1981-82; cons. Motorola Inc., Kerley Chems., Goodyear Aerospace, Beckman Instruments, Engrs. Testing Lab., SRP, Sperry Flight Systems, Maurice B. Berenter Co., various law firms. Contbr. numerous articles to sci. publs., also books. Recipient William F. Meggers award Soc. Applied Spectroscopy, 1967, 75. Mem. Am. Chem. Soc. (Chemist of Yr. award Ariz. sect. 1981), Soc. for Applied Spectroscopy, Optical Soc. Am., Can. Spectroscopy Soc., Sigma Xi, Alpha Chi Sigma. Subspecialties: Analytical chemistry; Artificial intelligence. Current work: Atomic spectroscopy (atomic emission techniques, atomic absorption techniques, atomic fluourescence techniques); trace organics in environmental samples; pattern recognition (geosamples, air particulate samples, radioactive wastes); infrared spectroscopy data. Home: 8402 E Mangum and 84th St Mesa AZ 85207 Office: Dept Chemistry Ariz State U Tempe AZ 88287

PARSONS, ROBERT WILTON, plastic surgery educator; b. Los Angeles, Dec. 25, 1929; s. Harold Hunt and Ellen Grace (Henderson) P.; m. Elise Hampton, Dec. 12, 1953; children: Ann, Keith, Kimberly, Susan. Ph.B., U. Chgo., 1949; M.D., Washington U., St. Louis, 1954. Diplomate: Am. Bd. Surgery, Am. Bd. Plastic Surgery. Intern Salt Lake County Gen. Hosp., 1954-55; resident in surgery Fitzsimons Army Med. Ctr., 1956-59, Brooke Army Med. Ctr., 1960-61, Barnes Hosp., St. Louis, 1961-62; asst. chief plastic surgery Walter Reed Army Med. Ctr., Washington, 1962-66; chief plastic surgery Landstuhl Army Med. Ctr., Germany, 1966-69, Walter Reed Army Med. Ctr., 1969-77; assoc. prof. plastic surgery Ind. U., Indpls., 1977-79; prof. plastic surgery U. Chgo., 1979—; cons. to Surgeon Gen., U.S. Army, 1969-77. Served to col. U.S. Army, 1955-77. Decorated Legion of Merit, Army Commendation Medal. Fellow ACS, Am. Assn. Plastic Surgeons; mem. Am. Soc. Plastic and Reconstructive Surgeons, Am. Soc. Maxillofacial Surgeons, Am. Soc. Aesthetic Plastic Surgery. Presbyterian. Subspecialty: Surgery. Current work: Graduate surgical education, biocompatibility of implanted silicones, cleft lip and palate and related deformities. Office: U Chicago 950 E 59th St Chicago Ill 60637

PARTHASARATHY, MANDAYAM VEERAMBHUDI, plant biologist, educator; b. Bangalore, India; m. Rama. B.S., U. Madras, India, 1953; M.S., U. Poona, India, 1957; Ph.D., Cornell U., 1966; Montgomery fellow, Technische Hochschule, Zurich, Switzerland, 1968-69. Asst. prof. plant biology Cornell U., Ithaca, N.Y., 1969-70, assoc. prof., 1977-80, assoc. prof. sect. plant biology, div. biol. scis., 1980—, also dir. electron microscopy facility. NSF grantee, 1976—. Mem. Bot. Soc. Am., Microbeam Analysis Soc., Electron Microscopy Soc. Subspecialties: Cell biology; Developmental biology. Current work: Structure and function of the food conducting tissues in plant (phloem), mechanism of streaming in higher plants. Office: Sect Plant Biology 204 Plant Sci Cornell U Ithaca NY 14853

PARTHASARATHY, SRISAILAM V., radiological physicist; b. Ellore, India, Nov. 22, 1937; came to U.S., 1966; s. S. and Lakshmamma Varadacharya; m. Kousalya, Mar. 18, 1971; children: Anand, Aruna. M.S., U. Mo.-Kansas City, 1970. Diplomate: Am. Bd. Radiology. Sci. officer Bhabha Atomic Research Ctr., Bombay, India, 1957-67; radiation biophysicist Research Hosp. Med. Ctr., Kansas City, Mo., 1970-74; radiol. physicist West Coast Cancer Found., San Francisco, 1975—, now sr. radiol. physicist. Mem. Am. Assn. Physicists in Medicine (sec.-treas. Bay area chpt. 1982—), Health Physics Soc., Am. Coll. Radiology. Subspecialties: Radiation therapy; Radiological physics. Current work: Radiation therapy physics; radiation therapy machine calibration; treatment planning and dosimetry. Home: 243 Appian Ct Martinez CA 94553 Office: West Coast Cancer Fedn 50 Francisco St San Francisco CA 94133

PARTRIDGE, BRIAN LLOYD, researcher, biologist, educator; b. Berea, Ohio, May 16, 1953; s. Bruce James and Mary Janice (Smith) P.; m. Nancy Jane Meredith, May 11, 1974; children: Brett James, Justin Ross. Student, Stanford U., 1970-71; B.Sc. with honors (Govt. B.C. scholar, C.A. Banks Meml. fellow), U. B.C., 1974; D.Phil (NRC Can. fellow, NSF grad. fellow), Oxford (Eng.) U., 1978. NSF postdoctoral fellow Scripps Instn. Oceanography, 1978-79; lectr. U. Calif., San Diego, 1979; asst. prof. biology U. Miami, Coral Gables, Fla., 1979—; vis. research fellow Scripps Instn. Oceanography, 1980; cons. NSF, NIH. Reviewer numerous jours.; contbr. articles profl. jours. Grantee NIMH, 1982-83. Mem. Soc. Neurosci., Internat. Soc. Neuroethology (charter), Animal Behavior Soc., Assn. Study of Animal Behaviour, Sigma Xi. Subspecialties: Comparative neurobiology; Ethology. Current work: Neuroethology, spatial sensing by teleost lateral lines, researcher in neuroethology. Home: 6221 SW 61st St South Miami FL 33143 Office: Dept of Biology University of Miami Coral Gables FL 33124

PARTRIDGE, L. DONALD, neurophysiologist; b. Phila., May 10, 1945; s. Lloyd D. and Jean M. (Rutledge) P.; m. Ann Clark, Dec. 29, 1965 (div. Dec. 82); children: Erika, Daniella. B.S., M.I.T., 1967; Ph.D., U. Wash., 1973. Asst. prof. U. N.Mex., Albuquerque, 1976—; research fellow U. Bristol, Eng., 1973-74, U. Wash., Friday Harbor, 1974-76. Served to 1st lt. U.S. Army, 1967-69. NIH research grantee, 1980-83; NSF research grantee, 1980-83. Mem. Soc. Neurosci., Biophys. Soc. Democrat. Anglican. Subspecialties: Neurophysiology; Physiology (medicine). Current work: Electrical properties of nerve cell membranes. Home: 311 Bryn Mawr SE Albuquerque NM 87106 Office: Dept Physiology U NM Albuquerque NM 87131

PARTRIDGE, ROBERT BRUCE, astronomy educator, university dean; b. Honolulu, May 16, 1940; s. Robert B. and Laura Lea (Johnson) P.; m. Jane C. Widseth, Aug. 28, 1976; 1 son, John. A.B., Princeton U., 1962; D.Phil., Oxford U., 1965. Inst. to asst. prof. Princeton (N.J.) U., 1965-70; asso. prof. to prof. astronomy Haverford (Pa.) Coll., 1970-82, dean, 1982—. Contbr. articles to profl. jours. Rhodes scholar, 1962; A.P. Sloan fellow, 1971-75; Fulbright fellow, 1979. Mem. Am. Phys. Soc., Internat. Astron. Union, Am. Astron. Soc., Sigma Xi. Subspecialties: Cosmology; Radio and microwave astronomy. Current work: Cosmology, microwave background radiation, galaxy formation, radio astronomy. Home: 628 Overhill Rd Ardmore PA 19003 Office: Haveford Coll Haverford PA 19041

PARTYKA, ROBERT EDWARD, horticulturist, plant pathologist; b. Wakefield, R.I., Sept. 5, 1930; s. John L. and Mary (Wilk) P.; m. Victoria L. Sanders, June 29, 1957; children: Robert J., Betsy J., Christopher A. B.S., U. R.I., Kingston, 1952; Ph.D., Cornell U., 1957. Extension plant pathologist Ohio State U., Columbus, 1957-74; plant pathologist ChemLawn Corp., Columbus, 1974-79, horticulturist, 1979—. Author: (with others) Woody Ornamentals, 1980. Mem. Phytopathology Soc., Nematology Soc., Am. Inst. Biol. Scis., Sigma Xi. Subspecialties: Plant pathology; Plant physiology (agriculture). Current work: Direct horticulture program for tree and shrub division of ChemLawn. Home: 3405 Kirkham Rd Columbus OH 43221 Office: ChemLawn Corp 8275 N High St Columbus OH 43085

PARZEN, EMANUEL, statistical scientist; b. N.Y.C., Apr. 21, 1929; s. Samuel and Sarah (Getzel) P.; m. Carol Tenowitz, July 12, 1959; children: Sara Leah, Michael Isaac. A.B. in Math, Harvard U., 1949; M.A., U. Calif. at Berkeley, 1951, Ph.D., 1953. Research scientist Columbia, 1953-56, asst. prof. math. statistics, 1955-56; faculty Stanford, 1956-70, asso. prof. statistics, 1959-64, prof., 1964-70; prof. statistics State U. N.Y. at Buffalo, 1970-73, prof. statis. sci., 1973-78; distinguished prof. statistics Tex. A and M U., College Station, 1978—; Guest prof. Imperial Coll., London, Eng., 1961-62; vis. prof. Mass. Inst. Tech., 1964-65, Harvard U., 1976. Author: Stochastic Processes, 1962, Modern Probability Theory and its Applications, 1960, Time Series Analysis Papers, 1967, also articles. Fellow Internat. Statis. Inst., Am. Statis. Assn., AAAS, Royal Statis. Soc., Inst. Math. Statistics; mem. Am. Math. Soc., Soc. Indsl. and Applied Math., Math. Assn. Am., Assn. Computing Machinery, Bernoulli Soc., Biometric Soc., Econometric Soc., N.Y. Acad. Scis., Phi Beta Kappa, Sigma Xi. Subspecialty: Statistics. Research in extended limit theorems probability theory to uniform convergence in a parameter. Introduced reproducing kernel Hilbert space formulation founds. time series analysis; introduced Parzen window for statis. spectral analysis; developed approach to empirical time series analysis; developer theory autoregressive/maximum entropy spectral analysis, also density-quantile estimation approach to nonparametric statis. data modeling. Office: Inst Statistics Tex A & M U College Station TX 77843

PASCHALL, LEE MCQUERTER, telecommunications company executive; b. Sterling, Colo., Jan. 21, 1922; s. Lee M. and Agnes W. P.; m. Bonnie E. Edwards, Oct. 24, 1942; children: Patricia, Stephen, David. A.B., U. Ala., 1957; M.A., George Washington U., 1964. Served with U.S. Army, 1940-46; commd. capt. U.S. Air Force, 1947, advanced through grades to lt. gen., 1974; ret., 1978, cons. telecommunications, Springfield, Va., 1978-81; pres., chief exec. officer Am. Satellite Co., Rockville, Md., 1981—; dir. Radiation Systems, Inc., Gen. Data Communications Industries Inc. Decorated Def. Disting. Service medal, Air Force Disting. Service medal, Legion of Merit with oak leaf cluster.; Named EASCON Man of Yr. IEEE, 1979. Mem. armed Forces Communications Electronics Assn., Air Force Assn., Phi Beta Kappa. Subspecialty: Aerospace engineering and technology. Current work: Digital telecommunications services provided to government and businesses via satellite transmission. Office: 1801 Research Blvd Rockville MD 20850

PASELK, RICHARD ALAN, biochemistry educator; b. Inglewood, Calif., July 20, 1945; s. Robert Arthur and Doris Mae (Miller) P.; m. Gail Annette Gulliver, Mar. 18, 1967; children: Laura Ann, Deborah Aileen. B.S., Calif. State U.-Los Angeles, 1968; Ph.D., U. So. Calif., 1975. Research technician U. So. Calif. Sch. Medicine, Los Angeles, 1968-69; lectr. Calif. State U.-Long Beach, 1974-76; asst. prof. biochemistry Humboldt State U., Arcata, 1976-82, assoc. prof., 1982—. Contbr. articles to profl. jours. Docent Clarke Meml. Mus., Eureka, Calif., 1980—. Subspecialties: Biochemistry (biology); Clinical chemistry. Current work: Analytical biochemistry; origins and evolution of life. Home: 3201 Zelia Ct Arcata CA 95521 Office: Dept Chemistry Humboldt State U Arcata CA 95521

PASHLEY, DAVID HENRY, physiology and biology educator, researcher; b. Seattle, Apr. 24, 1939; s. Harry B. and Frances L. (Henry) P.; m. Edna L. O'Sullivan, Aug. 4, 1962; children: Steven James, Thomas Matthew. B.S., U. Portland, 1960; D.M.D., U. Oreg., 1964; Ph.D., U. Rochester, 1970. Asst. prof. oral biology Med. Coll. Ga., 1970-73, assoc. prof., 1973-78, prof., 1978—; cons. U.S. Army, Ft. Gordon, Ga., 1981—. Nat. Inst. for Dental Research grantee, 1973-81, 78-83, 82-84. Mem. ADA, AAAS, Am. Physiol. Soc., Internat. Assn. Dental Research, Sigma Xi. Republican. Presbyterian. Subspecialties: Oral biology; Physiology (biology). Current work: Fluoride metabolism, especially soft tissue distribution, renal handling, effects of fluoride on organ and tissue function; dentin permeability, dentin pain, dentin desensitization. Inventor use of oxalate as a dentin desensitizer, 1977. Office: Med Coll Ga Augusta GA 30912

PASK, JOSEPH ADAM, emeritus ceramic engineering educator; b. Chgo., Feb. 14, 1913; s. Adam Poskoczem and Catherine (Ramanauskas) P.; m. Margaret J. Gault, June 11, 1938; children: Thomas Joseph, Kathryn Edyth. B.S., U. Ill., 1934, Ph.D., 1941; M.S., U. Wash., 1935. Ceramic engr. Willamina Clay Products Co., Oreg., 1935-36; teaching asst. ceramic engring. U. Ill., 1938, instr., 1938-41; asst. ceramic engr. electrotech. lab. U.S. Bur. Mines, 1941; assoc. ceramic engr. N.W. Exptl. Sta., 1942-43; asst. prof. ceramic engring., head dept. Coll. of Mines, U. Wash., Seattle, 1941-43; research ceramist, lamp div. Westinghouse Electric Corp., N.J., 1943-46, research engr. ceramic sect., 1946-48; assoc. prof. ceramic engring., head ceramic group div. materials sci. and engring. U. Calif. at Berkeley, 1948-53, founder program ceramic engring. and sci., 1948, prof., 1953-80, prof. emeritus, 1980—, vice chmn. div., 1956-57, chmn. dept., 1957-61, asso. dean grad. student affairs, 1969-80; research ceramist materials and molecular research div. Lawrence Berkeley Lab.; John Dorn Meml. lectr. Northwestern U., 1977; Mem. clay mineral com. NRC; mem. materials adv. bd., chmn. ad hoc com. ceramic processing, adv. commn. metallurgy div. U.S. Bur. Standards; chmn. NSF study objective criteria in ceramic engring. edn., U.S.-China Seminar on Basic Sci. of Ceramics, Shanghai, 1983. Recipient John F. Bergerson Meml. Service award Ceramic Engring. div. U. Wash., Seattle, 1969; gold medal for research and devel. French Soc. for Research and Devel., 1979; Berkeley citation U. Calif., 1980; Alumni honor award for disting. service in engring. U. Ill. Coll. Engring., 1982; Outstanding Achievement in Edn. award Com. of Confucius, 1982. Fellow Am. Ceramic Soc. (hon. life mem., v.p. 1953-54, pres. ednl. council 1954-55, trustee 1959-62, chmn. electronics div. 1959-60, John Jeppson award 1967, Ross Coffin Purdy award 1979), AAAS, Mineral. Soc. Am.; mem. Nat. Inst. Ceramic Engrs., Nat. Acad. Engring., Am. Soc. Metals, Brit. Ceramic Soc., Am. Soc. Engring. Edn. (chmn. materials com. 1961-63), Clay Minerals Soc., Keramos, Sigma Xi (Tau Beta Pi), Alpha Sigma Mu. Subspecialties: Ceramics; Materials processing. Current work: Bonding and joining of ceramics/metals at high temperatures; phase equililibria in the Al_2O_3-SiO_2 systems. Home: 994 Euclid Ave Berkeley CA 94708 Office: Dept Material Sci and Mineral Engring U Calif Berkeley CA 94720 Whatever success I have had can be attributed to hard work generated by a desire for success and recognition by my peers. This attitude was generated by my mother who came to this country, a land of opportunities, as an immigrant from Lithuania at the age of 16 without any knowledge of English. There was no question in her mind—and consequently mine—that I would get an education and be successful.

PASQUA, PIETRO, F(ERNANDO), nuclear engineering educator, consultant; b. Englewood, Colo., May 30, 1922; s. Frank Daniel and Josephine (Liveratti) P.; m. Madonna Murphy, Feb. 22, 1945; children: Linda, Randall, Michael, Karen. B.S., U. Colo., 1944; M.S., Northwestern U., 1947, Ph.D., 1952. Registered engr., Tenn. Instr. U. Colo., Boulder, 1944-45, Northwestern U., Evanston, Ill., 1947-52; assoc. prof. mech. engring. U. Tenn.-Knoxville, 1952-57, acting head mech. engring. dept., 1957-58, prof., head nuclear engring. dept., 1958—; cons. Union Carbide Nuclear Div., Oak Ridge, 1954-70, Sci. Applications, Inc., 1975—, Tech. for energy Corp., Knoxville, 1976—. Mem. Alcoa (Tenn.) City Bd. Edn., 1974-82. Recipient Coll. Brooks award Coll. Engring. Faculty, 1976. Fellow Am. Nuclear Soc., Am. Soc. Engring. Edn. Subspecialties: Nuclear fission; Nuclear engineering. Current work: Heat transfer, fluid flow, thermodynamics, administration nuclear fission research. Home: 319 West Hunt Rd Alcoa TN 37701 Office: Univ Tenn Knoxville TN 37996

PASS, BOBBY CLIFTON, entomologist, educator; b. Cleveland, Ala., Nov. 4, 1931; s. Rufus Clifton and Alma A. (Payne) P.; m. Ann Rutherford, Aug. 17, 1953; 1 son, Kevin Clifton. B.S., Auburn U., 1952, M.S., 1960; Ph.D., Clemson U., 1962. Farmer, Cleveland, 1952-53; in sales and bookkeeping U.S. Pipe & Foundry Co., Birmingham, Ala., 1955-57; research asst. Auburn (Ala.) U., 1958-60, Clemson (S.C.) U., 1960-62; asst. prof. dept. entomology U. Ky., Lexington, 1962-67, assoc. prof., 1967-71, prof., 1972—, chmn. dept., 1969—. Contbr. chpts. to books, articles to profl. jours. Served to cpl. U.S. Army, 1953-55. Mem. Entomol. Soc. Am. (governing bd. 1982—, pres. North Central br. 1976-77), Am. Registry Profl. Entomologists (pres. 1978-79), Entomol. Soc. Can., AAAS, Sigma Xi, Gamma Sigma Delta. Democrat. Methodist. Subspecialty: Integrated pest management. Current work: Integrated pest management in forage ecosystems; especially alfalfa. Crop director CIPM-Alfalfa project. Home: 3234 Pepperhill Rd Lexington KY 40502 Office: Dept Entomolog U Ky S225 Agrl Sci Center N Lexington KY 40546

PASSERO, MICHAEL ANTHONY, medical educator; b. Newark, Mar. 28, 1945; s. Anthony John and Connie (Blasi) P.; m. Mary Ann Cecere, June 14, 1969; children: Michael A., Christopher J. B.A., Dartmouth Med. Coll., 1966, B. Med. Sci., 1967; M.D., Harvard U., 1969. Diplomate: Am. Bd. Internal Medicine. Mem. med. house staff Duke U. Med. Ctr., Durham, N.C., 1969-71, fellow pulmonary and environ. tng. program, 1971-74; asst. prof. medicine Brown U. Program in Medicine, Providence, 1976—; assoc. chief medicine, head allergy Roger Williams Gen. Hosp., Providence, 1982—; dir. ICU, 1976; head occupational medicine program Brown U., 1982. Served to maj. U.S. Army, 1974-76. Fellow ACP, Am. Coll. Chest Physicians; mem. Am. Fedn. Clin. Research, Soc. Complex Carbohydrates, Am. Thoracic Soc. Subspecialties: Pulmonary medicine; Toxicology (medicine). Current work: Pulmonary toxicology, effects of oxidant substances and solvents on pulmonary metabolism. Home: 10 Briarfield Rd Barrington RI 02806 Office: Roger Williams Gen Hosp 825 Chalkstone Ave Providence RI 02908

PASSMAN, RICHARD HARRIS, psychology educator; b. Providence, Sept. 22, 1945; s. Carl and Ruth Ilse (Oberlander) P.; m.

Jane Aline (Horvitz); June 22, 1969; m.; children: Elana Michele, Joshua Lee. A.B., Brown U., 1967; M.A., Temple U., 1970; Ph.D., U. Ala., 1972. Dir. psychol. services Head Start, Pawtucket, R.I., 1968, psychology cons., staff psychol. instr., 1969, dir. psychol. services, East Providence, 1969; instr. staff psychologist U. Ala., Tuscaloosa, 1971; intern clin. psychology U. Minn. Hosp., Mpls., 1971-72; asst. prof. psychology U. Wis.-Milw., 1972-78, assoc. prof., 1978—. Contbr. articles to profl. jours. Bd. dirs. Milw. County Mental Health Assn., 1975-82. USPHS fellow, 1969-71; NIMH fellow, 1971-72. Mem. Am. Psychol. Assn., Midwestern Psychol. Assn., Soc. for Research Child Devel. Subspecialties: Developmental psychology; Clinical psychology. Current work: Children's attachment bond to mothers, fathers, nonsocial objects; factors related to parental administration of discipline; child abuse. Home: 310 E Calumet Rd Fox Point WI 53217 Office: Dept Psychology U Wis-Milw TEM 115 3415 N Downer Ave Milwaukee WI 53201

PASTORELLE, PETER JOHN, nuclear diagnostic laboratories executive; b. White Plains, N.Y., Jan. 23, 1933; s. Dominic John and Marguerite (Xavier) P.; m. Maria Rita Del Campo, Oct. 10, 1970. B.S., Fordham U., 1955; student, Juilliard Sch. Music, 1955-56; M.A., N.Y. U., 1961; B.D., Maryknoll (N.Y.) Sem., 1963. Producer/dir. UN-TV, N.Y.C., 1965-70; pres., founder Nuclear Diagnostic Labs., Peekskill, N.Y., 1965—; Peter Pastorelle Prodns., Inc., Mount Kisco, N.Y., 1970—; panelist Nat. Acad. TV Arts and Scis., N.Y.C., 1981-83; lectr. Am. Chem. Soc., Health Physics Soc. Writer, producer, dir., music score: films Fung Seui, 1970 (N.Y. Internat. Film Festival Gold award), Face of Hunger, 1976 (N.Y. State Dept. Commerce award), Water, 1978 (N.Y. Internat. Film Festival Silver award), Samoa - Culture in Crisis, 1982 (CINE Gold medal). Served with AUS, 1956-58. Mem. Health Physics Soc., Am. Nuclear Soc., Am. Pub. Health Assn., Soc. Motion Picture and TV Engrs., Am. Fedn. Musicians, Radiation Research Soc. Republican. Roman Catholic. Subspecialties: Nuclear engineering; Nuclear medicine. Current work: Radioactive waste disposal, radioactivity services and supplies. Home: RFD 3 Armonk Rd Mount Kisco NY 10549 Office: Nuclear Diagnostic Labs Inc PO Box 791 Peekskill NY 10549

PATE, JOHN RAY, mech. engr.; b. Memphis, July 17, 1942; s. Stoy L. and Edith F. (Sawyer) P.; m. Sandra Snell, Dec. 16, 1962; children: Phillip, Eric. B.S., David Lipscomb Coll., 1965; B.Engring., Vanderbilt U., Nashville, 1966. Engr. Trane Co., La Crosse, Wis., 1966-73; pres., chief exec. officer Environ. Enterprises, Little Rock, 1973—; instr. U. Ark. Sch. Engring., Little Rock, 1976-79; Mem. adv. bd. U. Ark. Sch. Engring., 1979—. Mem. Christian Businessmen's Com., Little Rock, 1980—. Mem. ASHRAE (Lincoln Boullion Nat. award 1973, Ernest Pettit State award 1974, pres. Ark. chpt. 1971-72, sec. nat. tech. com. on testing and balancing 1975-80), Ark. Council of Engring. and Related Socs. (chmn. 1979, dir. 1980—), Joint Engring. Soc. (founder), Met. Little Rock C. of C. (mem. indsl. devel. com. 1975-80), Ark. Fedn. Water & Air Users, Assn. Energy Engrs. Club: Little Rock Engrs. Subspecialties: Biomass (energy science and technology); Solar energy. Current work: Fuel alcohol via biomass conversion.

PATEL, BHIKHUBHAI L., mechanical engineer, air pollution control engineer; b. India, Oct. 12, 1942; s. Lallubhai D. and Benaben L. P.; m. Jashuben B., May 30, 1967; children—Sapna B., Sema B. B.E. in Mech. Engring, U. Karnatak, India, 1967; M.S. in Metall. Engring, S.D. Sch. Mines and Tech., 1969. Registered profl. engr., N.J., Pa. Design engr. Utah Copper div. Kennecott Copper Corp., Salt Lake City; project engr., design engr. U.S. Filter Corp., Summit, N.J., until 1974; sr. mech. engr. Research-Cottrell Inc., Sommerville, N.J., 1974—, lead proposal engr., proposal mgr.; cons. and designer air pollution control equipment and systems. Recipient award for tech. paper Power Engring. Jour., 1979. Mem. ASME, Air Pollution Control Assn. Subspecialties: Mechanical engineering; Metallurgical engineering. Home: RD 2 Box 321 East Stroudsburg PA 18301 Office: PO Box 1500 Sommerville NJ 08876

PATEL, CHANDRA KUMAR NARANBHAI, research company executive; b. Baramati, India, July 2, 1938; came to U.S., 1958, naturalized, 1970; s. Naranbhai Chaturbhai and Maniben P.; m. Shela Dixit, Aug. 20, 1961; children: Neela, Meena. B.Engring., Poona U., 1958; M.S., Stanford U., 1959, Ph.D., 1961. Mem. tech. staff Bell Telephone Labs., Murray Hill, N.J., 1961—, head infrared physics and electronics research dept., 1967-70, dir. electronics research dept., 1970-76, dir. phys. research lab., 1976-81, exec. dir. research physics div., 1981—; trustee Aerospace Corp., Los Angeles, 1979—. Contbr. articles to tech. jours. Recipient Adolph Lomb medal Optical Soc. Am., 1966; Ballantine medal Franklin Inst., 1968; Coblentz award Am. Chem. Soc., 1974; Honor award Assn. Indians in Am., 1975; Zworykin award Nat. Acad. Engring., 1976; Lamme medal IEEE, 1976; Founders prize Tex. Instruments Found., 1978; award N.Y. sect. Soc. Applied Spectroscopy, 1982. Fellow Am. Acad. Arts and Scis., Am. Phys. Soc., IEEE, AAAS, Optical Soc. Am. (Townes medal 1982), Indian Nat. Sci. Acad. (fgn.); mem. Nat. Acad. Scis., Nat. Acad. Engring., Gynecologic Laser Surgery Soc. (hon.). Subspecialties: Spectroscopy; Atomic and molecular physics. Current work: Atomic and molecular physics. Home: 5 Manor Hill Rd Summit NJ 07901 Office: Bell Telephone Labs 600 Mountain Ave Murray Hill NJ 07974

PATEL, DHANOOPRASAD GORDHANBHAI, pharmacologist, research investigator; b. Baroda, India, Mar. 17, 1936; came to U.S., 1968, naturalized, 1977; s. Grodhanbhai Tulshibhai and Kashiben (Gordhanbhai) P.; m. Niranjana Petlikar, July 12, 1967; 1 son, Ashesh. B. Pharm., Gujarat U., 1957; M.Sc. Pharm., U. Baroda, 1963, Ph.D. in Pharmacology, 1967; M.A. in Health Facility Mgmt, Webster Coll., St. Louis, 1976. Research fellow Council Sci. and Indsl. Research, Gov't. of India, Baroda, 1963-67, sr. sci. officer, 1967-68; research asso. Ohio State U., Columbus, 1968-71; NIH trainee in diabetes N.Y. Med. Coll., N.Y.C., 1971-73; research pharmacologist VA Med. Center, North Chicago, Ill., 1973-79; research investigator Barbara Kopp Research Center, Auburn, N.Y., 1979—. Author research papers and abstracts, also chpts. in sci. books. Founding mem., bd. dirs. Paul K. Kennedy Child Care Center, VA med. Center, North Chicago, 1975-78; bd. dirs. Cayuga (N.Y.) County Sub-Area Health Council, 1981—. Mem. Am. Diabetes Asssn., AAAS, Am. Soc. Exptl. Pharmacology and Therapeutics, Endocrine Soc., Soc. Exptl. Biology and Medicine, N.Y. Acad. Scis., Rho Chi. Subspecialties: Pharmacology; Endocrinology. Current work: Diabetes and its complications; effects of ethanol on carbohydrate metabolism. Office: 100 Thornton Ave Auburn NY 13021

PATEL, KANTILAL CHATURBHAI, chemistry educator, researcher; b. Gutal, Gujarat, India, May 12, 1934; came to U.S., 1969, naturalized, 1975; s. Chaturbhai Lallubhai and Kashiben P.; m. Shantaben K. Patel, 1953; children: Kiron, Dilip, Pankaj. Ph.D. U. Surrey, Guilford, Eng., 1967; Assoc. prof. dept. chemistry Bklyn. Coll., 1976—. Pres. Bochasanwasi Swaminarayan Sanstha, N.Y.C., 1973—. Subspecialties: Inorganic chemistry; Synthetic chemistry. Current work: Spectral and magnetic properties of coordination compounds. Office: Dept Chemistr Bklyn Coll Bedford Ave and Ave H Brooklyn NY 11210

PATEL, RAMESH CHANDRA, chemistry educator; b. Fiji Islands, Sept. 22, 1940; came to U.S., naturalized, 1973; s. Ambalal P. and Taraben A. P.; m. Rasila R. Patel, May 15, 1965; children: Rohini, Bharat. B.S., Tech. U. Braunschweig, 1962, M.S., 1965; Ph.D., Boston U., 1969. Research assoc. U. Md., 1969-71; vis. scientist Humboldt fellow Max Planck Inst. Biophys. Chemistry, 1971-72; vis. asst. prof. U. Okla., Norman, 1972-73; asst. prof. Clarkson Coll. Tech., Potsdam, N.Y., 1974-81, assoc. prof. chemistry, 1981—. Contbr. articles in field to profl. jours. Petroleum Research Fund–Am. Chem. Soc., grantee, 1974-76; Research Corp. grantee, 1974-76; Clorox Co. grantee, 1976; NIH grantee, 1978-80. Mem. Am. Chem. Soc., Am. Phys. Soc., Biophys. Soc., AAUP, Inst. Colloid Surface Sci., Sigma Xi. Subspecialties: Biophysical chemistry; Inorganic chemistry. Current work: Fast reaction kinetic studies of biologically important reactions. High temperature aqueous solution chemistry, biomolecule interaction with inorganic colloidal particles. Home: 80 Elm St Potsdam NY 13676 Office: Chemistry Dept Clarkson Coll Tech Potsdam NY 13676

PATEL, VITHAL A., mathematics educator, researcher; b. Sertha, Gujarat, India, Dec. 26, 1936; s. Ambala H. and Laxmiben A. P.; m. Latika V., May 15, 1960; children: Mihir, Hemakshi. B.Sc., Gujarat U., Ahmedabad, India, 1957; M.Sc., Sardar Patel U., Vallabh Vidyanagar, India, 1959; Ph.D., U. Calif.-Berkeley, 1970. Lectr. M.G. Sci. Inst., Ahmedabad, 1963-65; teaching asst. U. Calif., Berkeley, 1965-67, research asst., 1967-69; asst. prof. math. Humboldt State U., 1969-73, prof., 1973—. Mem. Soc. Engring. Sci., Soc. Indsl. and Applied Math. Subspecialties: Applied mathematics; Fluid mechanics. Current work: Stability of the flow around a circular cylinder. Home: 495 Lynn St Arcata CA 95521 Office: Humboldt State U Arcata CA 95521

PATENAUDE, ANDREA FARKAS, clinical psychologist; b. Newark, Aug. 28, 1946; d. Joseph and Sophie (Kahan) Farkas; m. Leonard Patenaude, Oct. 7, 1978. B.A., U. Chgo., 1967; M.A., Mich. State U., 1969, Ph.D., 1973. Diplomate: Am. Bd. Profl. Psychology. Staff psychologist dept. psychiatry Children's Hosp., Boston, 1970-74, psychology tng. coordinator, devel. evaluation clinic, 1975-83, psychologist bone marrow transplant team, 1976—; staff psychologist Human Resource Inst., Brookline, Mass., 1974-75; acting supr. psychology program Sidney Farber Cancer Ctr., Boston, 1983—; instr. in psychology dept. psychiatry Harvard Med. Sch., Boston, 1973-74, 75—. NIMH predoctoral research grantee, 1971-73; tng. fellow in clin. psychology, 1969-70, 67-68. Fellow Mass. Psychol. Assn.; mem. Am. Psychol. Assn. Subspecialties: The psychological aspects of critical care medicine; Marrow transplant. Current work: Coping with life-threatening illness. Psychological management during bone marrow transplantation; survival guilt. Office: Sidney Farber Cancer Ctr 44 Binney St Boston MA 02115

PATERSON, PHILIP Y., physician, educator; b. Mpls., Feb. 6, 1925; s. Donald Gildersleeve and Margaret (Young) P.; m. Virginia Lee Bray, Mar. 22, 1947; children: Anne, Peter, Benjamin. B.S., U. Minn., 1946, M.B., 1947, M.D., 1948. Intern Mpls. Gen. Hosp., 1948-49; research fellow div. infectious diseases Tulane U. Sch. Medicine, 1949-50, instr. in medicine Am. Heart Assn. research fellow, 1950-51; Am. Heart Assn. research fellow dept. microbiology U. Va. Sch. Medicine, 1953, asst. resident and co-resident in medicine univ. hosp., 1953-55; asst. prof. microbiology, instr. in medicine, established investigator Am. Heart Assn., 1955-57; med. officer Lab. Immunology, Nat. Inst. Allergy and Infectious Diseases, NIH, Bethesda, Md., 1957-60; vis. asst. prof. microbiology N.Y. U., 1957-60, assoc. prof. medicine, 1960-65; assoc. prof. medicine, dir. Samuel J. Sackett Research Labs.; and chief sect. infectious diseases, dept. medicine Northwestern U., 1965-66; Samuel J. Sackett prof. medicine and chief infectious diseases, hypersensitivity sect., dept. medicine Northwestern U.-McGaw Med. Center, 1966-72, Samuel J. Sackett prof. medicine and microbiology, chief infectious diseases-hypersensitivity sect., 1972-75; prof. microbiology-immunology, chmn. dept. microbiology-immunology Northwestern U. Med. and Dental Schs., 1975—, Guy and Anne Youmans prof. microbiology/immunology, 1982—; prof. neurobiology and physiology Northwestern U., Evanston, Ill., 1983—; Christine Larsen lectr. Sch. Medicine, Albuquerque, 1973; Disting. lectr. Ann. Meeting Assn. Am. Physicians and Western Sect. Am. Fedn. Clin. Research, Carmel, Calif., 1974; Grace Faillace Meml. lectr. Leo Goodwin Inst. Cancer Research, Nova U., 1978; Ernest Witebsky Meml. lectr. 6th Internat. Convocation on Immunology, Niagara Falls, N.Y., 1978; Joseph E. Smadel Meml. lectr. Ann. Meeting, Infectious Diseases Soc. Am., Atlanta, 1978; mem. adv. com. on drug devel. Nat. Multiple Sclerosis Soc., 1973—; mem. research study com. Am. Heart Assn., 1978-81; mem. internat. med. adv. bd. Internat. Fedn. Multiple Sclerosis Socs., 1973—; participant Mid-West Immunol. Conf., sec.-treas., 1974-76, chmn., 1977-78; mem. adv. panel Am. Bd. Med. Lab. Immunology, 1978—. Editor and contbg. author: The Biological and Clinical Basis of Infectious Diseases, 1975, 2d edit., 1980, 3d edit., 1984, contbr. over 200 articles, chpts. and revs. on neuroimmunology, host-parasite interactions and infectious disease to books and profl. jours.; mem. editorial bd.: Procs. Soc. Exptl. Biology and Medicine, 1968-73, 1977—, Cellular Immunology, 1969—, Infection and Immunity, 1970-81, Clin. and Exptl. Immunology, 1970—, Clin. Immunology and Immunopathology, 1978—, Jour. Clin. and Lab. Immunology, 1979—; assoc. editor: Jour. Infectious Diseases, 1979-83. Served to capt., M.C. U.S. Army, 1951-53. Mem. Am. Fedn. Clin. Research, AAAS, Am. Soc. Microbiology, Am. Assn. Immunologists, Harvey Soc., Am. Soc. Clin. Investigation, Infectious Diseases Soc. Am. (Gold medal 1978), Central Soc. Clin. Research, Am. Rheumatism Assn., Assn. Am. Physicians, Am. Clin. and Climatol. Assn., Sigma Xi, Alpha Omega Alpha. Subspecialties: Infectious diseases; Neuroimmunology. Current work: Neuroimmunology. Home: 1025 Chestnut St Wilmette IL 60091 Office: 303 E Chicago Ave Chicago IL 60611

PATI, JOGESH CHANDRA, physicist; b. Barpada, Orissa, India; s. Kailash Chandra and Indumar P.; m. Geeta Dash, Oct. 8, 1961; children—Sangeeta, Yagyensh, Sibani, Devesh. Richard C. B.Sc. with honors, Ravenshaw Coll., Cuttack, India, 1955; M.Sc., Delhi (India) U., 1957; Ph.D., U. Md., 1960. Tolman fellow Calif. Inst. Tech., 1960-62; mem. staff Inst. Advanced Study, Princeton, N.J., 1962-63; from asst. prof. to asso. prof. U. Md., 1963-72, prof. physics, 1972—. NSF grantee, 1963-81; John Simon Guggenheim fellow, 1979-80. Mem. Am. Phys. Soc., Washington Acad. Scis. (Disting. Scientist award 1973). Subspecialties: Particle physics; Theoretical physics. Current work: Unification of particles and their forces. Composite models of quarles and leptons. Home: 8604 Saffron Dr Lanham MD 20801 Office: Dept Physics U Md College Park MD 20742

PATO, MARTIN LEON, molecular biologist; b. Manhattan, N.Y., July 19, 1940; s. Harry and Florence (Reznickoff) P. B.S., M.I.T., 1962; Ph.D., U. Calif., Berkeley, 1968. Postdoctoral fellow U. Copenhagen, 1968-70; mem. staff dept. molecular and cellular biology Nat. Jewish Hosp. and Research Center, Denver, 1971—; asso. prof. dept. microbiology U. Colo. Sch. Medicine, Denver, 1974—. Contbr. articles to profl. jours. Am. Cancer Soc. fellow, 1968-70; NIH Research grantee, 1972-75, 77-80, 81—. Mem. Am. Soc. Microbiology, AAAS. Subspecialties: Molecular biology; Virology (biology). Current work: Research on the biology of transposition of genetic elements, especially bacteriophage MU. Home: 2007 Race St Denver CO 80205 Office: 3800 E Colfax Ave Denver CO 80206

PATRICK, GEORGE WALTER, anatomy educator; b. Kansas City, Mo., Oct. 31, 1951; s. George Andrew and Eunice Marie (Brown) P.; m. Charlene M., Aug. 10, 1973. B.S., Rockhurst Coll., Kansas City, Mo., 1973; Ph.D., Ind. State U., 1980. Research assoc. dept. anatomy W.Va. U. Sch. Medicine, Morgantown, 1979-81; asst. prof. anatomy Ind. U. Sch. Medicine, Ft. Wayne, 1981—. Contbr. articles to profl. jours. Mem. Soc. Neurosci. Subspecialties: Neurobiology; Anatomy and embryology. Current work: The effects of various toxicological (pb, Al) and pharmacological (THC) substances on developing nervous system (in vivo and in vitro) using anatomical, membrane physiological and behavioral approaches. Home: 4437 Stellhorn Rd Fort Wayne IN 46815 Office: Ind U Sch Medicine 2101 Coliseum Blvd E Fort Wayne IN 46805

PATRICK, MICHAEL HEATH, biology educator; b. Chgo., Mar. 31, 1936; s. William Washington and Dorothy Francis (Heath) P.; m. Catherine Anne Russell, Nov. 23, 1956 (div. 1975); children: Kelly, Kevin Michael; m. Beverly Jane Crane, Oct. 18, 1976; 1 son, Jeremy William. B.A., U. Calif.-Santa Barbara, 1958; Ph.D., U. Chgo., 1964. Research assoc. Johns Hopkins U., Balt., 1964-66; research scientist Southwestern Center Advanced Studies, Richardson, Tex., 1966-67, asst. prof., 1967-69; Asst. prof. U. Tex.-Dallas, Richardson, 1969-72, assoc. prof., 1972-80, prof. dept. biology, 1980—. Coordinator: book Photochemistry Photobiology Nucleic Acids, 1974. NIH research awardee, 1970, 74. Mem. AAAS, Biophys. Soc., Radiation Research Soc., Fedn. Am. Scientists, Sigma Xi. Democrat. Unitarian. Subspecialties: Biophysics (biology); Molecular biology. Current work: Structure and function of nucleic acids. Home: 6024 Worth Dallas TX 75214 Office: Univ Tex Dallas PO Box 688 Dallas TX 75080

PATRICK, ROBERT L., neurosciences educator; b. Phila., May 25, 1945; s. Edward and Rosalee (Mersky) P.; m. Saundra L. Brown, June 15, 1969; children: Melissa, Jennifer. B.A. cum laude in Chemistry, Temple U., 1967; Ph.D. (USPHS fellow) in Biochemistry, Duke U., 1972; postdoctoral (USPHS fellow) in neuropharmacology, Stanford U., 1971-73. Research assoc. psychiatry Stanford U., Calif., 1973-77; asst. prof. medi. scis., neurosci. sect., div. biology and medicine Brown U., Providence, 1977-82, assoc. prof., 1982—. Alfred P. Sloan Found. fellow, 1977-81. Mem. AAAS, Soc. Neurosci., Am. Soc. Pharmacology and Exptl. Therapeutics. Subspecialties: Neuropharmacology; Neurochemistry. Current work: Regulation of neurotransmitter synthesis and release in the central nervous system. Office: Brown U Box G Providence RI 02912

PATSCH, JOSEF RUDOLF, med. educator, researcher; b. Vienna, Austria, Jan. 13, 1942; came to U.S., 1975; s. Josef Ferdin and Lydia (Wolf) P.; m. Brenda Sue Wittneben, Dec. 16, 1978; 1 dau., Katherin Michelle. M.D., U. Innsbruck, Austria, 1966. Diplomate: Am. Bd. Internal Medicine. Research fellow U. Vienna, 1967-68; resident in medicine U. Innsbruck, 1968-71, asst. prof., 1971-75; vis. scientist Baylor Coll., Houston, 1975-77, asst. prof., 1977-81, assoc. prof., 1981—; prin. investigator NIH, 1979-83, 1982-87. Contbr. articles to med. jours. Recipient Irvine H. Page atherosclerosis research prize Am. Heart Assn., 1982. Fellow Am. Fedn. Clin. Research, Atherosclerosis Council. Roman Catholic. Subspecialties: Biochemistry (medicine); Internal medicine. Current work: Atherosclerosis, coronary heart disease, blood lipids, cholesterol, serum lipoproteins, preventive medicine, physical exercise. Home: 4814 O'Meara Houston TX 77035 Office: Baylor Coll Medicine Tex Med Ctr Houston TX 77030

PATSCH, PETER *See also* **PATSCH, WOLFGANG**

PATSCH, WOLFGANG (PETER PATSCH), physician, researcher; b. Wels, Austria, Aug. 10, 1946; came to U.S., 1978; s. Josef and Lydia (Wolf) P.; m. Eva Maria Haiden, Sept. 12, 1948. M.D., U. Innsbruck, Austria, 1971; cert. in Biochemistry, U. Uppsala, Sweden, 1978. Cert. internal medicine, Austria. Resident Med. Sch., Innsbruck, 1971-74, asst. prof., 1975-78; vis. asst. prof. Washington U., St. Louis, 1978-80, asst. prof., 1980-82; asst. prof. medicine and preventive medicine Baylor Coll., 1982—, dir. clin. lab. atherosclerosis, 1982—. Contbr. articles to profl. jours. Recipient Hoechst-Stiftung award Hoechst, Austria, 1975, 76; Price award Arztekammer, Innsbruck, 1979. Mem. Am. Fedn. Clin. Research, Am. Soc. Clin. Nutrition, European Lipoprotein Club, Atherosclerosis Council, European Soc. Clin. Investigation. Roman Catholic. Subspecialties: Internal medicine; Biochemistry (medicine). Current work: Atherosclerosis, metabolism, lipoprotein, cell biology. Office: Dept Medicin Baylor College Mail Station A601 6565 Fannin St Houston TX 77030

PATT, YEHUDA Z., physician, oncologist, educator; b. Tel-Aviv, Oct. 30, 1941; s. Abraham Hanoch and Shoshana R. (Witorz) P.; m. Nurit Hale Hale, Apr. 15, 1970; children: Iddo, Hanoch, Avinoam, Suzanne. M.D., Hadassah Med. Sch., Hebrew U., Jerusalem, 1967. Intern, Jaffa Govt. Hosp., Tel-Aviv, 1966-67; resident Tel-Hashomer Hosp., Ramat Gan, Israel, 1970-75; mem. faculty U. Tex. System Cancer Center-M.D. Anderson, Hosp. and Tumor Inst., Houston, 1977—, asst. prof., 1977-81, assoc. prof. medicine, 1981—, chief, 1981—. Contbr. numerous sci. articles to profl. publs. Vice pres. Congregation United Orthodox Synagogue, Houston, 1981-82. Served to capt. Israel Def. Forces, 1967-70. Nat. Cancer Inst. Grantee, 1978-81. Mem. AMA, Harris County Med. Soc., Am. Soc. Clin. Oncology, AAAS. Subspecialties: Cancer research (medicine); Chemotherapy. Current work: Regional therapy of cancers confined to the liver or peritoneal cavity. Reconstitution of the immune response in Acquired Immunodeficiency Syndrome (AIDS). Office: 6723 Bertner St Houston TX 77030

PATTEE, PETER ARTHUR, microbiology educator; b. Bklyn., Nov. 15, 1932; s. Arthur Lincoln and Barbara (Strout) P.; m. Kathleen Blevins, Sept. 6, 1958; children: Paul, Karen, Diane, Scott. B.S., U. Maine, 1955; M.S., Ohio State U., 1957, Ph.D., 1961. Asst. prof. dept. microbiology Iowa State U., Ames, 1961-65, assoc. prof., 1965-69, prof., 1969—. Contbr. articles to profl. jours. Served with U.S. Army, 1957-58. Mem. Am. Soc. Microbiology. Subspecialty: Microbiology. Current work: Genetic analysis/characterization of Staphyloccocus aureus; genetic exchange transfer mechanisms in bacteria, especially Gram positive bacteria. Home: 2614 Northwood Dr Ames IA 50010 Office: Iowa State Univ Dept Microbiology 205 Sci I Ames IA 50011

PATTEN, ETHEL DOUDINE, hematologist, educator; b. Staten Island, N.Y., Feb. 21, 1942; d. Serge and Ethel Campbell (Bruno) Doudine; m. Bernard M. Patten, June 27, 1964; children: Allegra, Craig Dustin. A.B., Barnard Coll., 1963; M.D., N.J. Coll. Medicine, 1967. Diplomate: Am. Bd. Internal Medicine (subcert. in hematology); cert. in blood banking, 1974. Intern USPHS Hosp., Staten Island, 1967-68, resident in internal medicine, 1968-70; fellow in hematology NYU Coll. Medicine, N.Y.C., 1970-71, NIH, Bethesda, Md., 1971-73; acting med. dir. Washington Regional ARC, 1973; asst. prof. medicine, dir. Blood Bank, U. Tex. Med. Br., Galveston, 1974—. Author numerous sci. papers. Served to maj. USPHS, 1967-72. Recipient N.J. Coll. Medicine Merck Manual award, 1967; Moody Found. grantee, 1975. Fellow ACP; mem. S. Central Assn. Blood Banks (pres. 1980-81). Subspecialties: Internal medicine; Hematology. Current work: Cryopreservation of platelets; immuno-hematology; plasma exchange. Home: 1019 Baronridge Dr Seabrook TX 77586 Office: U Tex Med Branch 8th and Mechanic Galveston TX 77550

PATTERSON, DAVID, geneticist, biologist; b. Medford, Mass., Aug. 24, 1944; s. David and Mildred (Hughes) P.; m. Norma Jean Riggs, June 3, 1967; children: Matthew, Jennifer. B.S., MIT, 1966; Ph.D., Brandeis U., 1971. With Eleanor Roosevelt Inst. Cancer Research, Denver, 1971—, sr. fellow, 1974-78, asst. prof. dept. biophysics/genetics, 1974-78, assoc. prof., 1978-83, assoc. dir., 1978—; assoc. prof. dept. medicine U. Colo. Health Scis. Ctr., Denver, 1982—, prof. biochemistry/biophysics/genetics, 1983—; bd. dirs. Therapeutic Day Care Ctr., Denver, 1981-84. Editor: Stomatic Cell and Molecular Genetics, 1983, Trisomy 21, 1983. Mem. biotech. com. Colo. Commn. Higher Edn., Denver, 1983. Grantee Nat. Inst. Aging, 1975-86, March of Dimes, 1975-83, Kroc Found., 1982-84, Nat. Inst. Childhood Diseases, 1979-85. Subspecialties: Gene actions; Cancer research (medicine). Current work: Gene regulation in mammalian cells related to aging, development, genetic disease, and cancer using molecular methods such as cloning of human genes, gene transfer, recombinant DNA, and nucleotide biochemistry. Home: 625 Albion St Denver CO 80220 Office: Eleanor Roosevelt Inst Cancer Research 4200 E 9th Ave B129 Denver CO 80262

PATTERSON, DAVID ANDREW, computer science educator; b. Evergreen Park, Ill., Nov. 16, 1947; s. David Dwight and Lucie Jeanette; ; s. David Dwight and Ekstrom P.; m. Linda Ann Crandall, Sept. 4, 1967; children: David, Michael. A.B. in Math, UCLA, 1969, M.A. in Computer Sci, 1970, Ph.D., 1976. Mem. tech. staff Hughes Aircraft Co., Culver City, Calif., 1972-76; research engr. Digital Equipment Co., Tewksbury, Mass., fall 1979; mem. faculty U. Calif., Berkeley, 1977—, assoc. prof. computer sci., 1981—; cons. Xerox Palo Alto (Calif.) Research Ctr., 1982. Contbr. numerous sci. articles to profl. publs. Recipient Disting. Teaching award U. Calif. Acad. Senate, Berkeley, 1982; Disting. Lectr. award Stanford U., 1981; Hughes fellow, 1974-76. Mem. IEEE, Assn. Computing Machinery. Subspecialty: Computer architecture. Current work: Build cost effective computer systems using very large scale integrated circuits. Office: Univ Calif 573 Evans Hall Computer Science Dept Berkeley CA 94720

PATTERSON, DWIGHT ROBERT, nuclear engineer; b. Detroit, Oct. 4, 1924; s. Ulysses Lee and Edith Mae P.; m. Reva Dell Sutton, May 1, 1943; 1 son, Mark B. B.S.M.E., U. Tenn., 1948. Registered profl. engr., Tenn., Ill. Mech. engr. TVA, Knoxville, Tenn., 1949-54, nuclear devel. engr., Argonne, Ill., 1954-59, chief mech. engr., Knoxville, 1973-79, chief nuclear engr., 1979-80, asst. to mgr., 1980—. Contbr. chpts. to books. Recipient Outstanding Engring. Alumnus award U. Tenn., 1980. Mem. ASME, Am. Nuclear Soc. (exec. com. power div. 1980), Tau Beta Pi. Methodist. Subspecialties: Mechanical engineering; Nuclear engineering. Current work: Nuclear safety in design and construction of nuclear power plants. Home: 4405 Deerfield Dr Knoxville TN 37921 Office: TVA 400 W Summit Hill Dr Knoxville TN 37902

PATTERSON, GARY DAVID, polymer physicist, researcher; b. Honolulu, July 31, 1946; s. Eugene Edgar and Alice Hariette (Titus) P.; m. Susan Ilene Stone, Sept. 1, 1967; 1 son, Kenneth Isaac. B.S., Harvey Mudd Coll., 1968; Ph.D., Stanford U., 1972. Mem. tech. staff Bell Labs., Murray Hill, N.J., 1972—. Contbr. numerous articles to profl. jours. Recipient Initiatives in Research award Nat. Acad. Scis., 1981. Fellow Am. Phys. Soc., Royal Soc. Chemistry; mem. Am. Chem. Soc., Soc. Rheology. Subspecialties: Polymer physics; Physical chemistry. Current work: Light scattering spectroscopy of polymers, dynamics of polymers, the glass transition. Home: 197 Stirling Rd Watchung NJ 07060 Office: Bell Labs 600 Mountain Ave Murray Hill NJ 07974

PATTERSON, LOYD THOMAS, microbiologist; b. Grove Oak, Ala., Feb. 23, 1930; s. Oz Vernon and Ludie (Strange) P.; m. Dorothy Jean Dupree, May 4, 1947; children: Karen Lynne, Alan Loyd. B.S., Auburn U., 1959; M.S., 1960, Ph.D., 1963. Postdoctoral fellow U. Tex. Med. Br., Galveston, 1963-65; asst. prof. microbiology U. Ark., Fayetteville, 1965-68, assoc. prof., 1968-73, prof., 1973—. Contbr. articles to profl. jours. Served to sgt. U.S. Army, 1952-55; Korea. Mem. Am. Soc. Microbiology, AAAS, Reticuloendothelial Soc., Poultry Sci. Assn., Sigma Xi. Baptist. Subspecialties: Immunology (medicine); Animal virology. Current work: Regulation of immune response; nonspecific resistance mechanisms; oncogenic and immunosuppressive viruses. Office: Dept Animal Scis U Ark Fayetteville AR 72701

PATTERSON, MANFORD K., JR., research inst. adminstr.; b. Muskogee, Okla., Aug. 20, 1926; s. Manford K. and Sara Lou (Patten) P.; m. Nancy Beverly Wilson, Aug. 21, 1953; 1 dau., Shelly Lynne. B.S., U. Okla., 1953, M.S., 1954; Ph.D., Vanderbilt U., 1962. Research chemist Noble Found, Ardmore, Okla., 1951-53, 54-57, sr. research chemist, 1961-66, sect. head, 1966—, sr. scientist, 1980—, v.p., dir. biomedical div., 1973—; pres. S.W. Am. Assn. Cancer Research, 1974. Co-editor: Tissue Culture: Methods and Applications, 1973; editor-in-chief: In Vitro, 1978; contbr. articles to profl. jours. Bd. trustees Crosstimbers Hospice, Inc., 1982—; bd. dirs. Chickasaw Hist. Soc., 1982—; mem. Carter County Bd. Health, 1976—. Served with USN, 1944-46. Mem. Am. Soc. Biol. Chemistry, Tissue Culture Assn. (treas. 1972-76, exec. bd., council 1976-78, council 1982—), AAAS, Am. Chem. Soc., Soc. Exptl. Biol. Medicine, N.Y. Acad. Scis., Council Biol. Editors. Lodge: Kiwanis. Subspecialties: Biochemistry (biology); Cell and tissue culture. Current work: Cell metabolism and growth control. Home: 2215 Hickory Ardmore OK 73401 Office: Noble Found Route 1 Ardmore OK 73401

PATTERSON, MICHAEL MILTON, biomed. researcher, educator, univ. ofcl., cons.; b. Muscatine, Iowa, Mar. 17, 1942; s. Harvey Milton and Vivian Doris Ann (Bridgeman) P.; m. Janice Pauline Ficke, June 11, 1966; children: Michael Shane, Shad Milton. Student, Grinnell Coll., 1964, Ind. U., 1964-66; Ph.D., U. Iowa, 1969. NIH postdoctoral fellow U. Calif., Irvine, 1969-71; assoc. prof. Kirksville (Mo.) Coll. Osteo. Medicine, 1971-77; prof. psychology Coll. Osteo. Medicine, Ohio U., Athens, 1977—, prof. osteo. medicine, 1977—, dir. research affairs, 1977—; Louisa Burns Meml. lectr., 1980, cons. to research and ednl. groups; lectr. in field; grant reviewer. Cons. editor sci. jours.; contbr. numerous articles to sci. jours., chpts. to books. Asst. scoutmaster, round table commr. Boy Scouts Am. NIH fellow, 1966-71; Am. Osteo. Assn. grantee, 1973—; NIH grantee, 1975—. Mem. Am. Psychol. Assn., Psychonomic Soc., Midwest Psychol. Assn., Sec. for Neurosci., Sigma Xi. Republican. Mem. Ch. of Christ Disciples. Club: Tuesday. Subspecialties: Neuropsychology; Neurophysiology. Current work: Spinal cord function including the effects of experience on spinal reflex patterns; brain-behavior relationships including alterations in brain activity during learning; effects of manipulative therapy on disease process. Home: 88 Wonder Hills Dr Athens OH 45701 Office: Coll Osteo Medicine Ohio U Athens OH 45701

PATTERSON, ROBERT LOGAN, systems analyst; b. Pitts., Mar. 12, 1940; s. Walter Glenn and June (Logan) P.; m. Christine Borgmann, Feb. 6, 1965 (div. 1973); 1 dau., Caitlin Williams. B.A., U. Vt., 1962; M.A., San Jose State U., 1965; M.Ed., Boston State Coll., 1979; cert., Northeastern U., 1982; postgrad., Lesley Coll. Grad. Sch., 1982—. Info. officer IBM U.K., London, 1966-68; systems analyst Info. Dynamics, Reading, Mass., 1968-70; automation librarian Boston Theol. Inst., Cambridge, Mass., 1971-73; systems analyst U. Mass., Boston, 1973—; psychology intern various mental health ctrs., 1978—; Active tng. program Boston Inst. Psychotherapists, 1981—, Cambridge Hosp. Psychotherapy Ctr., 1982—. Mem. Am. Psychol. Assn., Friends of Boston Psychoanalytic Soc. and Inst., Mass. Psychol. Assn., Mass. Tchrs. Assn., Northeastern Soc. Group Psychotherapy, SAR, Kappa Delta Pi. Psi Chi. Subspecialties: Psychotherapy; Library systems automation. Current work: Research in the implications of the rapprochement crisis (M.S. Mahler) as a theory for the understanding separation experiences in adults. Home: 14 Dudley St Cambridge MA 02140 Office: U Mass Harbor Campus Boston MA 02125

PATTERSON, ROSALYN VICTORIA MITCHELL, cell biology educator; b. Madison, Ga., Mar. 25, 1939; d. Walter Melvin and Hazeltine (Jones) Mitchell; m. Joseph William Patterson, June 1, 1961; children: Hazelyn Mammette, Joseph William, Rosman Victor. B.A., Spelman Coll., 1958; M.S., Atlanta U., 1960; Ph.D., Emory U., 1967. Instr. biology Spelman Coll., Atlanta, 1960-66, asst. prof., 1966-68, assoc. prof., 1968-69, prof., 1969-70; postdoctoral research fellow Ga. Inst. Tech., Atlanta, 1969-70; staff specialist to commr. and cons. Bur. Reclamation, U.S. Dept. Interior, Washington, 1970-71; coordinator nat. environ. edn. devel. program Nat. Park Service, 1971-72; NIH postdoctoral fellow div. biologics standards Bur. Biologics, FDA, Bethesda, Md., 1972-73; assoc. prof. biology Ga. State U., Atlanta, 1974-76; assoc. prof., then prof. Atlanta U., 1977—, chairperson dept. biology, 1980—. Mem. adv. bd. Mays Acad., Atlanta, 1980—. Mem. AAAS, Am. Soc. Cell Biology, Soc. Developmental Biology. Baptist. Subspecialties: Cell and tissue culture; Genome organization. Current work: The role of free radical activity in the alteration of chromosome structure and mechanisms by which antioxidants (i.e. vitamin E) may reduce the free radical effects. Home: 109 Burre Ln SW Atlanta GA 30331 Office: Atlanta U Dept Biology 223 Chestnut St Atlanta GA 30314

PATTERSON, ROY, physician, educator; b. Ironwood, Mich., Apr. 26, 1926; s. Donald I. and Helmi (Lantta) P. M.D., U. Mich., 1953. Diplomate: Am. Bd. Internal Medicine, Am. Bd. Allergy and Immunology. Intern U. Mich. Hosp., Ann Arbor, 1953-54, med. asst. research, 1954-55, med. resident, 1955-57, instr. dept. medicine, 1957-59; attending physician VA Research Hosp., Chgo., Northwestern Meml. Hosp.; mem. faculty Northwestern U. Med. Sch., Chgo., 1959—, prof. medicine, 1964—, Bazley prof. allergy-immunology, 1967—, chmn. dept. medicine, 1973—. Editor: Jour. Allergy and Clin. Immunology, 1973-78. Served with USNR, 1944-46. Fellow Am. Acad. Allergy (pres. 1976), A.C.P.; mem. Central Soc. for Clin. Research (pres. 1978-79). Subspecialty: Allergy. Current work: Allergy; immunologic lung disease; immunology; occupational immunologic lung disease; asthma models of asthma and allergy. Address: Dept Medicine Northwestern Univ Med Sch 303 Chicago Ave Chicago IL 60611

PATTERSON, TERENCE EDWARD, psychophysiological research scientist; b. Belfast, No. Ireland, Mar. 8, 1947; came to U.S., 1978, naturalized, 1982; s. Edward and Alice Elizabeth (Walsh) P.; m. Nancy Jean Crawford, 1983. B.A. with honors, Queens U., Belfast, 1969; Ph.D., Reading (Eng.) U., 1976. Clin. unit. dir. Ont. Hosp., Whitby, Ont., Can., 1969-71; postdoctoral fellow York (Eng.) U., 1974-77; pres. Patterson, Donnelly & Assocs., Toronto, Ont., 1977-78; prin. investigator dept. research Menninger Found., Topeka, 1978—; mem. faculty Menninger Sch. Psychiatry; cons. in field. Contbr. articles to profl. jours.; editorial cons. to jours. NIMH grantee, 1981. Mem. Soc. Psychophysiol. Research, AAAS, European Neuroscis. Assn. Subspecialties: Neurophysiology; Cognition. Current work: Three-dimensional representation of brain electrical activity during complex information processing in schizophrenics and normal human subjects. Rifle designer/builder. Office: Menninger Found Box 829 Topeka KS 66601

PATTI, ROBERT DALE, electronic engr.; b. San Diego, Dec. 14, 1940; s. Angelo John and Frances Marie (Miosi) P.; m. Christine Cummings, June 17, 1967; children: Elaine T., Janel C., Robert Dale. B.S. in E.E, Calif. Poly. State U., 1963. Registered profl. engr., Calif. Reliability project engr. Gen. Precision, Inc., San Marcos, Calif., 1963-66; sr. reliability engr. Gen. Dynamics Corp., San Diego, 1966-68; supr. quality engring. Westinghouse Electric Corp., Mansfield, Ohio, 1968-76; mgr corporate reliability and quality engring. Baxter Travenol Labs., Inc., Round Lake, Ill., 1976-78; dir. quality assurance G.D. Searle Co., Skokie, Ill., 1978-82; dir. product assurance Medtronic, Inc., Mpls., 1982—. Contbr. articles to profl. jours. Mem. Am. Soc. Quality Control, Instrument Soc. Am., Am. Assn. Advancement Med. Instrumentation. Republican. Roman Catholic. Subspecialties: Electronics; Bioinstrumentation. Current work: Assurance scis., reliability and quality engring. Office: PO Box 1453 Minneapolis MN 55440 Home: 17020 32d Ave N Plymouth MN 55447

PATTILLO, ROLAND A., physician, scientist; b. DeQuincy, La., June 12, 1933; s. James T. and Rhena F. P.; m. Marva Parks, June 20, 1959; children: Catherine, Michael, Patrick, Sheri, Mary E. B.S., Xavier U., 1955; M.D., St. Louis U., 1959. Diplomate: Am. Bd. Obstetrics and Gynecology. Intern, resident Milwaukee County Gen. Hosp., 1959-63; resident Boston Lying-in Hosp., Harvard U., 1963-64; fellow in reproductive oncology Johns Hopkins U., 1965-67; dir. Cancer Research and Reproductive Biology Labs., vice chmn. dept. ob-gyn Med. Coll. Wis., Milw., 1980—. Contbr. articles to profl. jours. Bd. dirs. Milw div. Am. Cancer Soc., Jr. Acad. Medicine, Milw. Grantee NIH, Am. Cancer Soc., DeRance Found., Cudahy Found., Stackner Found., McBeath Found., Steimke Found. Mem. AAAS, Am. Coll. Obstetricians and Gynecologists, Soc. Gynecologic Oncologists, Soc. Gynecologic Investigations. Roman Catholic. Subspecialties: Gynecological oncology; Reproductive biology (medicine). Current work: Gynecologic oncology, reproductive biology, endocrinology, immunology and chemotherapy.

PATTON, DENNIS DAVID, radiologist, medical educator, researcher; b. Oakland, Calif., Aug. 4, 1930; s. Owen and Norma (Barnes) Gotzian P.; m. Pamela Hughes, Feb. 14, 1965; children: James Patrick, William Christopher. A.B. in Physics, U. Calif.-Berkeley, 1953; M.D., UCLA, 1959; Dolmetscher cert., U. Heidelberg, W.Ger., 1951. Diplomate: Am. Bd. Radiology, Am. Bd. Nuclear Medicine. Intern Wadsworth VA Hosp., Los Angeles, 1959-60; resident in radiology U. Calif.-Irvine, 1965-68, asst. prof. radiology, 1968-70; prof. radiology Vanderbilt U., Nashville, 1970-75, U. Ariz., Tucson, 1975—; mem. residency rev. com. AMA, 1980—. Author: The Abnormal Brain Scan, 1975, Current Status of Liver Imaging, 1982; editor: (with Sol Nudelman) Imaging for Medicine, 1980. Fellow Am. Coll. Radiology, Am. Coll. Nuclear Medicine; mem. Soc. Nuclear Medicine (trustee 1976—), Am. Coll. Nuclear Physicians (bd. regents 1980—). Republican. Methodist. Subspecialties: Nuclear medicine; Nuclear medicine. Current work: Development and evaluation of novel imaging systems; measurement of cerebral blood flow, theory and application of medical decision analysis. Home: 6502 Pontatoc Rd Tucson AZ 85718 Office: Dept Radiology and Nuclear Medicine U Ariz Health Scis Ctr Tucson AZ 85724

PATTON, PETER CLYDE, computer center administrator; b. Wichita, Kans., June 11, 1935; s. Claude and Beryl Inez (Jones) Barney) P.; m. Naomi Julia Lawson, Aug. 21, 1957; children: Peter, Claudia, Theresa, Richard, Phillip. A.B. in Engring. and Applied Physics, Harvard U., 1957; M.A. in Math, Kans. U., 1959; Dr. in Aerospace Engring, Stuttgart (W. Ger.) U., 1966. Sci. cons. Sperry Internat., Lausanne, Switzerland, 1961-67; mgr. system design Sperry Univac, Roseville, Minn., 1967-69; gen. mgr. Analysts Internat., Mpls., 1969-71; dir. U. Minn. Computer Center, Mpls., 1971—. Author: Data Structures & Computer Architecture, 1976; author, editor: Computing in the Humanities, 1981, Computer System Requirement, 1982. Fellow Inst. Math. & Its Applications; mem. IEEE (sr.), IEEE Computer Group, Assn. Computing Machinery. Subspecialties: Computer architecture; Software engineering. Current work: Software automation, design and implementation of application generator software, knowledge based systems. Home: 3471 Churchill St Paul MN 55112 Office: U Minn 208 Union St SE Minneapolis MN 55455

PAUDLER, WILLIAM WOLFGANG, chemistry educator, university dean; b. Varnsdorf, Czechoslovakia, Feb. 11, 1932; s. Fred and Johanna (Drabovsky) P.; m. Renee Eva West, Dec. 21, 1932; children: Gary, Leslie, David. B.S., U. Ill., 1954; Ph.D., Ind. U., 1958. Research chemist Proctor & Gamble Co., 1958-59; postdoctoral research assoc. Princeton U., 1960-61; asst. prof. chemistry Ohio U., Athens, 1961-65, assoc. prof., 1965-67, prof., 1967-70, disting. prof. chemistry, 1970-73; chmn., prof. U. Ala.-Tuscaloosa, 1973-81; dean Coll. Liberal Arts and Scis., prof. chemistry Portland (Oreg.) State U., 1981—. Author: Nuclear Magnetic Resonance, 1971, 74, (with George Newkome) Contemporary Heterocyclic Chemistry, 1982; contbr. articles to profl. jours. Fellow Am. Inst. Chemists; mem. Internat. Soc. Heterocyclic Chemists, Am. Chem. Soc. Jewish. Subspecialties: Organic chemistry; Nuclear magnetic resonance (chemistry). Current work: Heterocyclic chemistry, natural product, nuclear magnetic resonance, coal and wood chemistry. Home: 19521 Hidden Springs West Linn OR 97068 Office: Portland State U Portland OR 97207

PAUKER, STEPHEN GARY, internist, educator; b. N.Y.C., Nov. 21, 1942; s. Carl Jacob and Helen (Yurdin) P.; m. Susan Perlmutter, Sept. 2, 1967; children: Sheridan Joanna, Scott Gregory. A.B., Harvard U., 1964, M.D., 1968. Diplomate: Am. Bd. Internal Medicine. Intern Boston City Hosp., 1968-69; resident in internal medicine Mass. Gen. Hosp., Boston, 1969-70; chief div. clin. decision making New Eng. Med. Ctr., Boston, 1972—; assoc. prof. medicine Tufts U., Boston, 1972—; mem. health care tech. study sect. NIH, Bethesda, Md., 1980—; mem. bd. sci. counsellors Nat. Library of Medicine, Bethesda, 1981—. Contbr. over 50 sci. articles to profl. publs. Research Career Devel. awardee NIH, 1977-82. Fellow ACP, Am. Coll. Cardiology, Am. Heart Assn.; mem. Soc. Clin. Decision Making (trustee 1979-83). Subspecialties: Cardiology; Health services research. Current work: Directs the only medical division of clinical decision making in this country devoted to the application of computers and formal decision theory to the care of individual patients and to the development of policies and controversial health care issues. Home: 48 Concord St Newton MA 02162 Office: New England Med Center 171 Harrison Ave Boston MA 02111

PAUL, ARA GARO, univ. dean; b. New Castle, Pa., Mar. 1, 1929; s. John Hagop and Mary (Injejikian) P.; m. Shirley Elaine Waterman, Dec. 21, 1962; children—John Bartlett, Richard Goyan. B.S. in Pharmacy, Idaho State U., 1950; M.S., U. Conn., 1953, Ph.D. in Pharmacognosy, 1956. Cons. plant physiology Argonne (Ill.) Nat. Lab., 1955; asst. prof. pharmacognosy Butler U., Indpls., 1956-57; mem. faculty U. Mich., Ann Arbor, 1957—, prof. pharmacognosy, 1969—, dean, 1975—; vis. prof. microbiology Tokyo U., 1965-66; mem. vis. chemistry faculty U. Calif., Berkeley, 1972-73. Contbr. articles to profl. jours. Del. U.S. Pharmacopeial Conv., 1980. G. Pfeiffer Meml. fellow Am. Found. Pharm. Edn., 1965-66; fellow Eli Lilly Found., 1951-53, Am. Found. Pharm. Edn., 1954-56, NIH, 1972-73; recipient Outstanding Tchr. award Coll. Pharmacy, U. Mich., 1969, Outstanding Alumnus award Idaho State U., 1976. Fellow AAAS; mem. Am., Mich. pharm. assns., Am. Soc. Pharmacognosy, Acad. Pharm. Scis., Am. Assn. Colls. Pharmacy, Washtenaw County Pharm. Soc., Am. Soc. Hosp. Pharmacists. Subspecialties: Pharmacognosy; Medicinal chemistry. Current work: Biosynthesis of Alkalords; phytochemistry of fungi; phytochemistry of cacti. Home: 1415 Brooklyn Ave Ann Arbor MI 48104 Office: Coll Pharmacy Univ Mich Ann Arbor MI 48109

PAUL, DEREK A. L., physicist; b. Brussels, Belgium, Oct. 1, 1929; s. Russell and Kathleen M. G. (Wrathall) P.; m. Normelia Isabel Falls; m. Hanna Louisa Emily Little; children: Antonia, Julian, Lara, John Mark. Student, Marlborough Coll., Wiltshire, Eng., 1943-47; B.A., Clare Coll., Cambridge (U.K.), 1950; Ph.D., Queen's U., Kingston, Ont., Canada, 1958. With B.U.S.M. Co., Leicester, Eng., 1950-53; lectr. Royal Mil. Coll. Can., 1953-58, asst. prof., 1958-61, asso. prof., 1961-63; asso. prof. physics U. Toronto, 1964-76, prof. physics, 1976—; research dir. Sci. for Peace, 1982—; participant Pugwash, 1976—. Contbr. articles to profl. jours. Mem. Am. Phys. Soc., Can. Assn. Physicists, IEEE, UN Assn., Can. Inst. Internat. Affairs, Can. Peace Research and Edn. Assn. New Democratic Party. Subspecialties: Atomic and molecular physics; Nuclear physics. Current work: Fundamental questions; nuclear physics; position scattering and positionium; positronium scattering. Office: Physics Dept U Toronto ON Canada M5S 1A7

PAUL, DONALD ROSS, chemical engineer, educator; b. Yeatesville, N.C., Mar. 20, 1939; s. Edgar R. and Mary E. (Cox) P.; m. Sally Annette Cochran, Mar. 28, 1964; children: Mark Allen, Ann Elizabeth. B.S., N.C. State Coll., 1961; M.S., U. Wis., 1963, Ph.D., 1965. Research chem. engr. E.I. DuPont de Nemours & Co., Richmond, Va., 1960-61; instr. chem. engring. dept. U. Wis., Madison, 1963-65; research chem. engr. Chemstrand Research Center, Durham, N.C., 1965-67; asst. prof. chem. engring. U. Tex., Austin, 1967-70, assoc. prof., 1970-73, prof., 1973—, T. Brockett Hudson prof., 1978—, chmn. dept. chem. engring., 1977—, dir. Center for Polymer Research, 1981—; cons. in field. Author: (with F.W. Harris) Controlled Release Polymeric Formulations, 1976, (with S. Newman) Polymer Blends, 2 vols, 1978; Mem. editorial adv. bds.: Jour. Membrane Sci, 1976—, Polymer Engring. and Sci, 1975—, Jour. Applied Polymer Sci, 1979—; contbr. articles to profl. jours. Recipient award Engring. News Record, 1975, Ednl. Service award Plastics Inst. Am., 1975, awards U. Tex. Student Engring. Council, 1972, 75, 76; award for engring. teaching Gen. Dynamics Corp., 1977; Joe J. King Profl. Engring. Achievement award, 1981. Mem. Am. Chem. Soc. (Doolittle award 1973), Am. Inst. Chem. Engrs., Soc. Plastics Engrs. (Outstanding Achievement in Research award 1982), Fiber Soc., Phi Eta Sigma, Tau Beta Pi, Phi Kappa Phi, Sigma Xi. Subspecialties: Chemical engineering; Polymer engineering. Current work: Basic research in polymer blends (or alloys); transport behavior of polymers; polymeric membranes for separation processes and drug delivery. Home: 7104 Spurlock Dr Austin TX 78731 Office: Dept Chemical Engineering U Texas Austin TX 78712

PAUL, EDWARD, chemistry educator; b. Newark, July 16, 1944; s. Albert and Sylvia (Wiederhorn) P.; m. Janis Brewer, June 14, 1970; children: Katherine, Jean. B.A., Brandeis U., 1966; Ph.D., U. Oreg., 1970. Researcher U. Calif.-Santa Cruz, 1970-71, NIH, Bethesda, Md., 1971-72; dean Stockton State Coll., Pomona, N.J., 1979-82, assoc. prof. dept. chemistry, 1972—; ISEP awards reviewer NSF, Washington, 1978; reviewer Benjamin Cummings, Reading, Mass.,

1979—. Mem. Port Republic Planning Bd., 1981—, chmn., 1983—. NDEA fellow, Brussels, 1966-69; NATO summer program fellow, Sweden/Norway, 1969. Mem. N.J. Sci. Adv. Panel. Subspecialties: Statistical mechanics; Thermodynamics. Current work: Phase equilibria, computers, education, mathematical modeling, statistical thermodynamics. Office: Stockton State Coll Dept Chemistry Pomona NJ 08240 Home: 216 Old New York Rd Port Republic NJ 08241

PAUL, JOHN FRANCIS, physical oceanographer; b. Massillon, Ohio, July 6, 1947; s. John Francis and Irene Rose (Jenei) P.; m. Maureen Skelley, May 3, 1975; 1 son, Justin Morgan. B.S. in Engring. with high honors, Case Western Res. U., 1969, M.S., 1971, Ph.D., 1974. Teaching asst. dept. fluid and thermal scis. Case Western Res. U., Cleve., 1969-73, research assoc., 1973-75, sr. research assoc., 1975-77; environ. scientist EPA, Grosse Ile, Mich., 1977-81, research environ. scientist, Narragansett, R.I., 1981—; cons. and lectr. in field. Contbr. tech. reports and articles to profl. jours. NDEA fellow Case Western Res. U., 1969-72. Mem. N.E. Area Remote Sensing System Assn. (dir. 1982—), Am. Geophys. Union, Assn. for Computing Machinery, Internat. Assn. Gt. Lakes Research, AIAA, AAAS. Roman Catholic. Subspecialties: Oceanography; Fluid mechanics. Current work: Research interests include the development and application of quantitative assessment methodologies for predicting the impacts of waste disposal in the marine environment. Home: 1 Enterprise Terr Kingston RI 02881 Office: EPA S Ferry Rd Narragansett RI 02882

PAUL, ROBERT WILLIAM, JR., biology educator, researcher; b. Jersey City, Nov. 27, 1946; s. Robert William and Grace (Dorbandt) P.; m. Susan Lane Geeson, June 21, 1972; 1 dau., Lindsey Anne. B.A., Westminster Coll., Fulton, Mo., 1969; M.S., St. Louis U., 1974; Ph.D., Va. Poly. Inst. and State U., 1977. Instr. St. Mary's Coll., Md., 1977-79; asst. prof. biology, 1979—, dir., 1978-80. Contbr. articles to sci. jours. Adviser St. Mary's County Pub. Schs., 1979—. Served to 1st lt. U.S. Army, 1969-71; Okinawa. Rockefeller Found. grantee, 1973; Mellon fellow, 1981-82. Mem. N.Am. Benthol. Soc., Ecol. Soc. Am., AAAS, Sigma Xi. Presbyterian. Subspecialty: Environmental toxicology. Current work: Ecology of standing and flowing water; decomposition processes in aquatic ecosystems; aquatic pollution; ecology of estuarine organisms; salt marsh ecology; structure and function of aquatic systems with an interest in the effects of pollutants. Office: Saint Mary's College Saint Mary's City MD 20686

PAUL, THOMAS DANIEL, biophysical sciences educator; b. Butte, Mont., June 10, 1948; s. Thomas Anthony and Helen (O'Brien) P.; m. Carolyn Hicks, Dec. 20, 1976; children: Thomas Richard, Jennifer Ann. A.B., Carroll Coll., Helena, Mont., 1970; M.S., Ind. U., 1975, Ph.D., 1977. Diplomate: Am. Bd. Med. Genetics. Asst. instr. dept. pediatrics SUNY-Buffalo, 1977-78, asst. prof., 1978-80, asst. prof. dept. biophys. scis., 1980—; biochem. genetics cons. J.N. Adams Devel. Ctr., Perrysburg, N.Y., 1978—. Contbr. chpts. to books, articles to profl. jours. Fellow Internat. Soc. for Twin Studies; mem. Am. Soc. Human Genetics, Soc. for Inherited Metabolic Diseases, Am. Soc. Mass Spectroscopy. Subspecialties: Genetics and genetic engineering (medicine); Biochemistry (medicine). Current work: Study of metabolic disease by using stable isotopes, and gas chromatography mass spectrometry; In vivo metabolic studies. Home: 150 Huntingdon Ave Buffalo NY 14214 Office: Dept Biophys Sci SUNY-Buffalo 132 Cary Hall Buffalo NY 14214

PAUL, WILLIAM ERWIN, immunologist; b. Bklyn., June 12, 1936; s. Jack and Sylvia (Gleicher) P.; m. Marilyn Heller, Dec. 25, 1958; children: Jonathan M., Matthew E. A.B., Bklyn. Coll., 1956; M.D., SUNY, Downstate Med. Ctr., 1960. Intern, asst. resident in medicine Univ. Hosp., Boston, 1960-62; clin. assoc. Nat. Cancer Inst., Bethesda, Md., 1962-64; fellow in pathology NYU Sch. Medicine, 1964-66, trainee to instr. in medicine, 1966-68; sr. investigator lab. immunology Nat. Inst. Allergy and Infectious Diseases, Bethesda, 1968-70, chief lab. immunology, 1970—; commd. med. officer USPHS, 1975—; mem. sci. rev. bd. Howard Hughes Med. Inst.; sci. adv. bd. Jane Coffin Childs Fund for Med. Research; bd. sci. visitors Okla. Med. Research Found.; mem. sci. adv. com. New Eng. Regional Primate Ctr. Contbr. numerous articles on immunology to profl. jours.; editor Ann. Rev. of Immunology, 1982—. Served with USPHS, 1962-64. Recipient Founders' prize Tex. Instruments Found., 1979; G. Burroughs Mider Lectureship award NIH, 1982. Mem. Am. Assn. Immunologists (councillor), Am. Soc. for Clin. Investigation (Councillor 1976-79, pres. 1980-81), Nat. Acad. Scis., Assn Am. Physicians. Subspecialties: Immunobiology and immunology; Immunogenetics. Current work: Mechanisms of lymphocyte activation, antigen regocnition, immune response gene function, and cellular interactions. Office: Bldg 10 Room 11N31 NIH Bethesda MD 20205

PAULE, MARVIN R., biochemistry educator; b. San Antonio, Sept. 3, 1943; s. Melvyn Leslie and Naomi (Hicks) P.; m. Linda Adele Giovannetti, Aug. 30, 1964; 1 child, Lauren. B.S., Calif. State U.-Los Angeles, 1966; Ph.D., U. Calif.-Davis, 1970. Jane Coffin Childs Meml. Fund fellow U. Calif. Health Sci. Ctr., San Francisco, 1971-72; mem. faculty Colo. State U., Ft. Collins, 1977—, prof. biochemistry, 1981—; sr. internat. fellow Labaratoire Genetique Moleculaire des Eucaryotes, Strasbourg, France, 1980-81. Contbr. sci. articles to profl. publs. Mem. com. Colo. Commn. Higher Edn., Denver, 1982-83. NDEA fellow, 1966; Fogarty Sr. Internat. fellow, 1980. Mem. Am. Soc. Biol. Chemists, AAAS, Phi Kappa Phi. Subspecialties: Genetics and genetic engineering (biology); Genetics and genetic engineering (agriculture). Current work: Molecular mechanisms of selective gene transcription in eukaryotes; eukaryotes gene structure; genetic engineering of crop plants. Office: Biochemistry Dept Colo State U Fort Collins CO 80523

PAULING, LINUS CARL, educator; b. Portland, Oreg., Feb. 28, 1901; s. Herman Henry William and Lucy Isabelle (Darling) P.; m. Ava Helen Miller, June 17, 1923; children—Linus Carl, Peter Jeffress, Linda Helen, Edward Crellin. B.S., Oreg. State Coll., Corvallis, 1922, Sc.D. (hon.), 1933; Ph.D., Calif. Inst. Tech., 1925; Sc.D. (hon.), U. Chgo., 1941, Princeton, 1946, U. Cambridge, U. London, Yale, 1947, Oxford, 1948, Bklyn. Poly. Inst., 1955, Humboldt U., 1959, U. Melbourne, 1964, U. Delhi, Adelphi U., 1967, Marquette U. Sch. Medicine, 1969; L.H.D., Tampa, 1950; U.J.D., U. N.B., 1950; LL.D., Reed Coll., 1959; Dr. h.c., Jagiellonian U., Montpellier (France), 1964; D.F.A., Chouinard Art Inst., 1958; also others. Teaching fellow Calif. Inst. Tech., 1922-25, research fellow, 1925-27, asst. prof., 1927-29, asso. prof., 1929-31, prof. chem., 1931-64, chmn. div. chem. and chem. engring., dir., 1936-58, mem. exec. com., bd. trustees, 1945-48; research prof. (Center for Study Dem. Instns.), 1963-67; prof. chemistry U. Calif. at San Diego, 1967-69, Stanford, 1969-74; pres. Linus Pauling Inst. Sci. and Medicine, 1973-75, 78—, research prof., 1973—; George Eastman prof. Oxford U., 1948; lectr. chemistry several univs. Author several books, 1930—, including, Cancer and Vitamin C, 1979; Contbr.: articles to profl. jours. Cancer and Vitamin C. Fellow Balliol Coll., 1948, NRC, 1925-26, John S. Guggenheim Meml. Found., 1926-27; Numerous awards in field of chemistry, including; U.S. Presdl. Medal for Merit, 1948; Nobel prize in chemistry, 1954; Nobel Peace prize, 1962; Internat. Lenin Peace prize, 1972; U.S. Nat. Medal of Sci., 1974; Fermat medal; Paul Sabatier medal; Pasteur medal; medal with laurel wreath of Internat. Grotius Found., 1957; Lomonosov medal, 1978; U.S. Nat. Acad. Sci. medal in Chem. Scis., 1979, Priestley medal, 1984. Hon., corr., fgn. mem. numerous assns. and orgns. Subspecialties: Physical chemistry; Biophysical chemistry. Home: Salmon Creek Big Sur CA 93920 Office: Linus Pauling Inst Sci and Medicine 440 Page Mill Rd Palo Alto CA 94306

PAULK, CHARLES JASPER, JR., electrical engineer; b. Pensacola, Fla., Dec. 21, 1950; s. Charles Jasper and Ethel Mai (Harris) P.; m. Marilynn Jane Overton, Aug. 12, 1972; children: Charles Christopher, Monika Michelle. B.S., SUNY-Albany, 1980; A.S.E.E., Internat. Corr. Sch., 1981; B.A., N.C. Wesleyan Coll., 1982. Engr. Carolina Power & Light Co., New Hill, N.C., 1980-83, sr. engr., 1983—. Served with USN, 1971-80. Mem. Am. Nuclear Soc., IEEE. Baptist. Subspecialties: Electrical engineering; Nuclear physics. Current work: Start-up and test engineer-electrical, includes reactor control. Home: 8513 Holly Springs Rd Apex NC 27502 Office: Carolina Power & Light Co PO Box 165 New Hill NC 27562

PAULS, JOHN FREDERICK, biostatistician; b. Washington, Iowa, Dec. 7, 1928; s. George A. and Hannah Esther (Jones) P.; m. Margaret M. Delashmutt, June 10, 1950; children: Debra J. O'Leary, Cathy Sue Aafjes, Douglas J. B.S., Iowa State U., 1952, M.S., 1954. Sr. Statistician Smith Kline Corp., Phila., 1956-58, group leader applied math. services, 1958-64, asst. sect. head stats., 1964-67, assoc. dir., 1967-75; asst. dir statis. services Carter-Wallace, Inc., Cranbury, N.J., 1975-82, dir. statis. services, 1982—; adj. assoc. prof. biometrics dept. Temple U., Phila., 1974-75. Chmn. Towamencin Mcpl. Swimming Club; mem. steering and operating com. Towamencin Twp., Montgomery County, Pa., 1967-75; treas. N. Pa. Sch. Dist. Authority, Lansdale, 1975-76. Served with USN, 1946-48. NSF fellow, 1952-53. Mem. Am. Statis. Assn., Biometrics Soc., Am. Soc. Quality Control, Soc. for Clin. Trials, Sigma Xi. Republican. Methodist. Subspecialty: Statistics. Current work: Experimental design, statistical analysis, design and analysis clinical trials.

PAULSEN, MARVIN RUSSELL, agricultural engineering educator; b. Minden, Nebr., July 17, 1946; s. LeRoy Harold and Mildred L. (Madsen) P.; m. Karen Anne Petricek, June 19, 1970; children: David J., Sarah B. B.S., U. Nebr., 1969, M.S., 1972; Ph.D., Okla. State U., 1975. Registered profl. engr., Ill. Engr. Chevrolet div. Gen. Motors Corp., Warren, Mich., 1969; research asst. dept. agrl. engring. U. Nebr., Lincoln, 1971-72, Okla. State U., Stillwater, 1972-75; research assoc. dept. agrl. engring. U. Ill., Urbana, 1975-77, asst. prof., 1977-81, assoc. prof., 1981—; cons. Ill.-Iowa Moisture Meter Task Force, 1980-82. Contbr. articles to profl. jours. Active YMCA Indian Guides, 1981—. Served with U.S. Army, 1969-71. Mem. Am. Soc. Agrl. Engrs. (vice chmn. materials handling com. 1982—, award for excellence on pub. work 1980), Am. Assn. Cereal Chemists, Sigma Xi, Gamma Sigma Delta, Alpha Epsilon. Republican. Lutheran. Subspecialty: Agricultural engineering. Current work: Determining grain quality of corn and soybeans in export shipment and from combine harvesting; determining corn varietal effects of breakage susceptibility; testing the calibration of electronic moisture meters.

PAULSON, BOYD COLTON, JR., civil engineer, educator, researcher; b. Providence, R.I., Mar. 1, 1946; s. Boyd Colton and Barbara Jean (McKinstry) P.; m. Jane Margaret, Feb. 12, 1970; children: Jeffrey Boyd, Laura Jane. B.S., Stanford U., 1967, M.S., 1969, Ph.D., 1971. Asst. prof. civil engring. U. Ill.-Urbana, 1972-73; assoc. prof. civil engring. Stanford U., after 1974, now prof., assoc. chmn. dept. civil engring.; mem. adv. com. civil and environ. engring. div. NSF. Author: (with Donald S. Barrie) Professional Construction Management, 1978; Contbr. articles to profl. jours. Recipient Walter L. Huber Civil Engring. Research prize ASCE, 1980; Alexander von Humboldt Found. fellow, W. Ger., 1983. Mem. Assn. Computing Machinery, ASCE, Am. Soc. Engring. Edn., Project Mgmt. Inst., Sigma Xi, Tau Beta Pi. Subspecialties: Civil engineering; Construction engineering. Current work: Computer applications in construction engineering. Microcomputers; interactive graphics; simulation; automated real-time data acquisition and process control; robotics. Office: Dept Civil Engring Stanford Univ Stanford CA 94305

PAULY, JOHN EDWARD, anatomist; b. Elgin, Ill., Sept. 17, 1927; s. Edward John and Gladys (Myhre) P.; m. Margaret Mary Oberle, Sept. 3, 1949; children: Stephen John, Susan Elizabeth, Kathleen Ann, Mark Edward. B.S., Northwestern U., 1950; M.S., Loyola U., Chgo., 1952, Ph.D., 1955. Grad. asst. gross anatomy Stritch Sch. Medicine, Loyola U., Chgo., 1953-54; research asst. anatomy Chgo. Med. Sch., 1952-54, research instr., 1954-55, instr. in gross anatomy, 1955-57, asso. gross anatomy, 1957-59, asst. prof. anatomy, 1959-63, asst. to pres., 1960-62; asso. prof. anatomy Tulane U. Sch. Medicine, 1963-67; prof., head dept. anatomy U. Ark. for Med. Scis., Little Rock, 1967—, prof., head dept. physiology and biophysics, 1978-80; tech. adviser Ency. Brit. Films, 1956, mem. safety and occupational health study sect. Nat. Inst. Occupational Safety and Health, Center for Disease Control, 1975-79. Author: (with Hans Elias) Human Microanatomy, 1960, (with Elias and E. Robert Burns) Histology and Human Microanatomy, 1978; editor: (with Lawrence E. Scheving and Franz Halberg) Chronobiology, 1974, (with Heinz von Mayersbach and Lawrence E. Scheving) Biological Rhythms in Structure and Function, 1981, Am. Jour. Anatomy, 1980—; co-mng. editor: Advances in Anatomy, Embryology and Cell Biology, 1980—; adv. editorial bd.: Internat. Jour. Chronobiology; contbr. articles to profl. jours. Served with USNR, 1945-47. Recipient merit certificates AMA, 1953, 59; Bronze award Ill. Med. Soc., 1959; Lederle Med. Faculty award, 1966. Fellow AAAS; mem. Am. Assn. Anatomists (sec.-treas. 1972-80, pres. 1982-83), So. Soc. Anatomists (pres. 1971-72), Assn. Anatomy Chmn. (sec.-treas. 1969-71), Am. Soc. Cancer Research, Am. Physiol. Soc., Assn. Am. Med. Colls., Internat. Soc. Chronobiology, Pan-Am. Assn. Anatomy, Internat. Soc. Electrophysiol. Kinesiology, Internat. Soc. Steriology, Consejo Nacional de Profesores de Ciencias Morphológicas (hon.), Sigma Xi, Sigma Alpha Epsilon. Roman Catholic. Subspecialties: Anatomy and embryology; Chronobiology. Current work: Chronopharmacology; chronotoxicology; chronochemotherapy of neoplasms and effect of various peptides on DNA synthesis. Home: 11 Hearthside Dr Little Rock AR 72207

PAUSTIAN, HAROLD HERMAN, nuclear engr.; b. Davenport, Iowa, July 13, 1951; s. Herbert Herman and Ruby Elanor (Poppe) P.; m. Mary Theresa Scott, Mar. 29, 1980. B.S. in Engring. Sci, Iowa State U., 1973, M.S. in Nuclear Engring, 1975. Registered profl. engr., Calif. Engr. Gen. Electric Co., San Jose, Calif., 1975-81, sr. engr., tech. leader, 1981—. Nat. Merit scholar, 1969. Mem. Am. Nuclear Soc. (assoc.). Republican. Lutheran. Subspecialties: Nuclear engineering; Nuclear fission. Current work: Design of nuclear fuel for commercial fission power plants, with an emphasis on improving fuel cycle economics.

PAVELIC, ZLATKO PAUL, pathologist, researcher; b. Slavonski Brod, Yugoslavia, Aug. 14, 1943; came to U.S., 1975, naturalized, 1982; s. Mirko and Zlata (Godic) P.; m. Ljiljana Duic Pavelic, May 7, 1947. M.D., U. Zagreb, Croatia, Yugoslavia, 1969, M.S., 1971, Dr.S., 1974. Intern U. Zabreg Hosp., 1969-71, Gen. Hosp., Pakrac, Yugoslavia, 1969-71; resident in pathologic anatomy U. Zagreb, 1971-74, asst. prof. pathologic anatomy, 1970-75; assoc. prof., 1975; vis. scientist Roswell Park Meml. Inst., Buffalo, 1975-77, assoc. cancer research scientist, 1977—; clin. asst. prof. pathology SUNY, Buffalo, 1978-80, assoc. prof., 1980—, asst. research prof. dept. pharmacology, 1978—. Contbr. numerous articles in field to profl. jours. and books. Grantee in field. Mem. Med. Assn. Croatia, Yugoslavia Assn. Immunology, Yugoslavia Assn. Pathology, Am. Assn. Cancer Research, Am. Soc. Clin. Oncology, Am. Assn. Pathology, N.Y. Acad. Sci., AAAS, Sigma Xi. Subspecialties: Pathology (medicine); Pharmacology. Current work: Preclinical toxicological pathology; cloning and drug sensitivity of human solid tumors in soft agar. Home: 58 Hiler Ave Kenmore NY 14217 Office: 666 Elm St Buffalo NY 14263

PAVKOVIC, STEPHEN FRANK, chemistry educator; b. Highland Park, Mich., Oct. 29, 1932; s. Steve and Barbara P.; m. Julia Daniels, June 24, 1961; children: Karen, Kristin, Steven, David. B.S., Wayne State U., 1955, M.S., 1961; Ph.D., Ohio Stae U., 1964. Asst. prof. Hillsdale (Mich.) Coll., 1959-61; postdoctoral fellow Kettering Research Lab., Yellow Springs, Ohio, 1964-65; prof. chemistry Loyola U., Chgo., 1965—. Contbr. articles, reviewer: Acta Crystallographica, 1977—. Mem. Am. Chem. Soc., Am. Crystallographic Soc. Subspecialties: Chemical education; X-ray crystallography. Current work: Preparation of coordination compounds with sterically hindered ligands, and the determination of molecular structures via X-ray methods. Office: Dept Chemistry Loyola U Chgo 6525 N Sheridan Rd Chicago IL 60626

PAVLETIC, MICHAEL MARK, veterinary surgeon, educator; b. Waukegan, Ill., Oct. 8, 1950; s. Merle R. and Geraldine I. (Hibbard) P.; m. Adria C. Pavletic, Sept. 10, 1977. B.S., U. Ill., 1972, D.V.M., 1974. Diplomate: A. Coll. Vet. Surgeons. Intern Angell Meml. Animal Hosp., Boston, 1974-75, resident in companion animal surgery, 1975-77; asst. prof. compansion animal surgery Sch. Vet. Medicine, La. State U., 1977-82; asst. prof. Tufts U., Boston, 1982—; lectr. in field. Contbr. articles to profl. jours. Mem. AVMA, Am. Coll. Vet. Surgeons, Am. Animal Hosp. Assn., Assn. Am. Vet. Med. Colls., Phi Zeta, Phi Kappa Phi. Episcopalian. Subspecialty: Surgery (veterinary medicine). Current work: Companion animal plastic and reconstructive surgery; companion animal general surgery; skin flaps and free graft research. Office: Dept Small Animal Surgery Tufts University School of Veterinary Medicine Boston MA 02130

PAVLIDIS, GEORGE THEOPHILOU, psychiatry educator, researcher; b. Thessaloniki, Greece, Apr. 16, 1949; came to arrived Eng., 1981; s. Theophilos and Parthena (Katseas) P. B.A. with honors, Aristotelean U., Thessaloniki, Greece, 1972; Ph.D., U. Manchester, Eng., 1980. Lectr. U. Salford, Eng., 1972-75; tutor U. Manchester, Eng., 1973-80, research fellow, research asst. prof., 1974-78, research fellow, research assoc. prof., 1978-80, sr. research fellow, research prof., 1981—; vis. assoc. prof. psychiatry and pediatrics UMDNJ-Rutgers Med. Sch., Piscataway, N.J., 1981—; co-dir. Reading Disability Research Inst., 1982. Contbr. articles to profl. jours.; Editor: Dyslexia Research and Its Applications to Education, 1981. Fellow Internat. Acad. Research in Learning Disabilities; mem. Brain Research Assn., Exptl. Psychology Soc., Am. Psychol. Assn., British Psychol. Soc. Subspecialties: Neuropsychology; Developmental psychology. Current work: The relationship between eye movements and evoked potentials to: dyslexia, learning disabilities, schizophrenia, hyperactivity, alcoholism, higher cognitive processes and intelligence. Patentee biodigitizer. Office: UMDNJ-Rutgers Med Sch Dept Psychiatry Box 10 Piscataway NJ 08854

PAWLOWICZ, EDMUND FRANK, geophysicist; b. Toledo, Apr. 2, 1941; s. Edmund Edward and Genevieve Rose (Kurdziel) P.; m. Melinda Jo Lieser, Oct. 2, 1982; m. Barbara Paulette Gruszczynski, June 20, 1964 (div. Nov. 1980); children: Karen Ann, Sharon Jean, Mark Edward. B.E.E., Ohio State U., 1964, M.Sc., 1965, Ph.D., 1969. Registered profl. engr., Ohio. Elec. engr. Toledo Scale Corp., summers 1963, 64, Army Electronics Command, Ft. Monmouth, N.J., summer 1966; research geophysicist Naval Civil Engring. Lab., Port Hueneme, Calif., 1968-70; assoc. prof. Bowling Green (Ohio) State U., 1970-81; staff geophysicist Amoco Prodn. Co., Denver, 1981—. Served to lt. USN, 1968-70. Mem. Soc. Exploration Geophysicists, European Assn. Exploration Geophysicists, Can. Soc. Exploration Geophysicists, Am. Geophys. Union, Soc. Profl. Well-Log Analysts, Tau Beta Pi, Sigma Gamma Epsilon, Eta Kappa Nu., Mensa., Intertel. Club: Polish (Denver). Subspecialties: Geophysics; Tectonics. Current work: Borehole geophysics (vertical-seismic profiling, velocity logging); potential fields (gravity, magnetics, electrical methods); rock mechanics (wave velocity studies). Office: Amoco Prodn Co 1670 Broadway Denver CO 80202

PAXSON, CHARLES L., JR., neonatologist, educator; b. Columbus, Kans., Jan. 4, 1946; s. Charles L. Sr. and Freida (Derfelt) P.; m. Cathleen Joy Carlson, Oct. 13, 1974; children: Katherine Joy, Benjamin Charles. B.A., Pitts. Coll., 1968; M.D., U. Kans., 1971. Diplomate: Am. Bd. Pediatrics. Asst. prof. pediatrics U. Nebr., Omaha, 1976-79, asst. prof. physiology, 1976-79, dir. neontology research, 1977-79; dir. neonatology Central Plains Clinic, Sioux Falls, S.D., 1979-81; dir. newborn service St. Elizabeth Hosp., Youngstown, Ohio, 1981—; asst. prof. pediatrics Northeastern Ohio Univs. Coll. Medicine, Rootstown, 1982—. Editor: Van Leeuwen's New Medicine, 1979; contbr. articles to profl. jours. Am. Heart Assn. grantee, 1976-79. Fellow Am. Fedn. Clin. Research. Mem. Assemblies of God Ch. Subspecialty: Neonatology. Current work: Primary interest is in blood and volume expansion of hypotensive infants, and techniques to rapidly assess circulating blood volume. Home: 209 Edna Poland Village OH 44514 Office: St Elizabeth Hosp Med Ctr 1044 Belmont Ave Youngstown OH 44501

PAXTON, HAROLD WILLIAM, steel company executive; b. Yorkshire, Eng., Feb. 6, 1927; came to U.S., 1953, naturalized, 1961; s. John Wilfrid and Hilda Annie (Vasey) P.; m. Ann Dorothy Davies, May 13, 1953; children: Jane Elizabeth, Sally Patricia, Anthony Charles, Nigel John. B.Sc. with 1st class honours, U. Man., 1947, M.Sc., 1948; Ph.D., U. Birmingham, Eng., 1952. Univ. fellow U. Birmingham, 1950-53; mem. faculty Carnegie-Mellon U., 1953-74, prof. metall. engring., 1962-74, Firth Sterling prof. metall. research, 1958-63; head dept. metallurgy and materials sci., dir. Metals Research Lab., 1966-71, dir. research, 1973-74; adj. sr. fellow Mellon Inst., 1965-67; dir. div. materials research NSF, 1971-73; v.p. research U.S. Steel Corp., 1974—; vis. prof. Mass. Inst. Tech., 1970; Campbell meml. lectr., 1978, cons. to industry, 1953-74; Chmn. Internat. Conf. High Velocity Deformation, 1960, Internat. Conf. Fracture, 1962. Author: (with E. C. Bain), 3d edit.) Alloying Elements in Steel, 1966, also numerous articles. NSF sr. postdoctoral fellow Imperial Coll., London, Eng., 1962-63. Fellow Am. Soc. Metals (Bradley Stoughton Young Tchrs. award 1960), Metall. Soc. of Am. Inst. M.E. (pres. 1976-77, v.p. inst. 1977-78), AAAS; mem. AIME (pres. 1982—), Am. Iron and Steel Inst., Nat. Acad. Engring. Subspecialties: Alloys; Metallurgy. Home: 115 Eton Dr Pittsburgh PA 15215 Office: Room 1365 600 Grant St Pittsburgh PA 15230

PAXTON, HUGH CAMPBELL, nuclear physicist; b. Los Angeles, Apr. 29, 1909; s. Charles Hugh and Grace Beatrice (Dorn) P.; m. Jean Nellis Thomson, May 28, 1937; children: Susan Jean Paxton Gomez, Alan Hugh. A.B., UCLA, 1930; Ph.D., U. Calif.-Berkeley, 1937. Physicist Coll. of France, Paris, 1937-38; instr. Columbia U., N.Y.C., 1938-42, sr. physicist, 1942-44, Union Carbide Corp., Oak Ridge, 1944-45; staff mem. Sharples Corp., Phila., 1945-48, Los Alamos Nat. Lab., 1948-76; cons. Lawrence Livermore (Calif.) Nat. Lab., 1980—; part-time mem. Atomic Safety and Licensing Bd. Panel, U.S. NRC,

Washington, 1963—. Author: (with J.R. Dunning) Matter, Energy and Radiation, 1941. Fellow Am. Phys. Soc., Am. Nuclear Soc. (award Trinity sect. 1966, Contbn. Recognition award 1972, editorial adv. bd. Jour. Nuclear Sci. & Engring. 1963—), Am. Contract Bridge League. Democrat. Subspecialties: Nuclear fission; Nuclear engineering. Current work: Critical experiments, nuclear criticality safety, reactor safety. Patentee tube forming device, sylphon-sealed pump, shrinkless refractory synthesis. 87544

PAXTON, JACK DUNMIRE, plant pathology educator; b. Oakland, Calif., Feb. 17, 1936; s. Glenn Ernest and Frances Willa (Dunmire) P.; m. Sarah Elizabeth Clough, Sept. 4, 1960; children: Anne, Paul. B.S., U. Calif.-Berkeley, 1958; Ph.D., U. Calif.-Davis, 1964. Research asst. U. Calif.-Davis, 1958-64; research assoc. U. Ill., Urbana, 1964-65, asst. prof., 1965-70, assoc. prof. plant pathology, 1970—, chmn. plant physiology faculty, 1976-78. Contbr. articles, chpts. to jours., books. Fulbright fellow, 1972; NATO grantee, 1970, 75, 80; U.S. Dept. Agr. grantee. Mem. Am. Chem. Soc., Am. Phytopath. Soc., Am. Soc. Plant Physiologists, Phytochem. Soc. N.Am. Subspecialties: Plant pathology; Cell and tissue culture. Current work: Investigate plant disease resistance especially role of phytoalexins and their elicitors, role of lectins in cell-cell recognition and interaction. Home: 2603 Brownfield Rd Urbana IL 61801 Office: Dept Plant Pathology U Ill 1102 S Goodwin St Urbana IL 61801

PAXTON, KENNETH BRADLEY, elec. engr.; b. Norwich, N.Y., Dec. 31, 1938; s. Kenneth W. and Eva P. (Thompson) P.; m. Joyce A. Passero, May 5, 1962; children: Kenneth W., Holly A. B.S.E.E., Rensselaer Poly. Inst., 1960; M.S., U. Rochester, 1966, Ph.D.E.E., 1971. Registered profl. engr. N.Y. Engr. Eastman Kodak Co., Rochester, N.Y., 1960-73, supr. devel. copy products, 1973-81, asst. to mgr. consumer products, 1981-82, mgr. advanced devel., 1982—. Mem. Math. Assn. Am., Soc. Indsl. and Applied Math., Soc. Photog. Sci. and Engring. Subspecialties: Electrical engineering; Applied mathematics. Current work: Research and development of imaging products. Home: 814 Lithuanica Ln Webster NY 14580 Office: Eastman Kodak Co 901 Elmgrove Rd Rochester NY 14650

PAYNE, FRED RAY, engineering educator; b. Mayfield, Ky., Jan. 26, 1931; s. Joe L. and Bonnie B. (Vincent) P.; m. Marilyn Maassen, Oct. 12, 1957; children: John, Kevin, Joel. B.S. in Physics, U. Ky., Lexington, 1952; M.S. in Aero. Engring, Pa. State U., 1964, Ph.D., 1966. Registered profl. engr., Tex. Commd. 2d lt. U.S. Air Force, 1952, advanced through grades to maj., 1966; ret., 1966; asst. prof. aero. engring. Pa. State U., State College, 1966-68; design specialist Gen. Dynamics, Fort Worth, 1968-69; prof. aerospace engring. U. Tex.-Arlington, 1969—; NASA grantee Ames Research Center, Moffett AFB, Calif., 1975-80; cons. Office Naval Research, U.S. Navy, State College, 1963-68, NSF, 1966-68, Bell Aerospace, Houston, 1974. Youth dir. Ft. Worth Chess Club, 1972-77; organizer Community Center, Ft. Worth, 1972. Decorated Air Force Commendation medal. Mem. N.Y. Acad. Sci., Am. Phys. Soc., Am. Geophys. Union, Soc. Natural Philosophy, Am. Acad. Mechanics, Soc. Indl. and Applied Math. Subspecialty: Aeronautical engineering. Current work: Boundary layer and turbulence computations and experiments, integral methods. Inventor airborne hot-wire system on powered aircraft, 1964, new integral method, 1979. Office: Dept Aerospace Engring Univ Texas Arlington TX 76019

PAYNE, GERALD LEW, nuclear physics educator; b. Columbus, Ohio, Mar. 11, 1938; s. Harry Joseph Payne and Lucy Loretta (Frabott) Casa; m.; children: Tracy, Lucy, Karen. B.S., Ohio State U., 1961, M.S., 1961; Ph.D., U. Calif.-San Diego, 1967. Research assoc. U. Md., College Park, 1967-69; faculty U. Iowa, Iowa City, 1969—, prof. physics, 1980—; cons. Los Alamos Nat. Lab., 1976—, Lockheed Research Lab., Palo Alto, Calif., 1978—. Contbr. articles to profl. jours. Served to lt. (j.g.) USNR, 1961-64. NSF fellow, 1965. Mem. Am. Phys. Soc., Am. Assn. Physics Tchrs., AAAS. Democrat. Subspecialties: Nuclear physics; Plasma physics. Current work: Nuclear structure and scattering, three-body dynamics; non-linear wave effects in plasmas. Home: 1834 Hafor Dr Iowa City IA 52240 Office: Dept Physics U Iowa Iowa City IA 52242

PAYNE, PHILIP WARREN, chemistry educator; b. New Castle, Ind., Feb. 26, 1950; s. Robert Hedges and Esther Lois (Lamb) P.; m. Caroline Lelear, May 27, 1978. B.A., Pomona Coll., Claremont, Calif., 1971; Ph.D., Princeton U., 1976. Research assoc. U. N.C., Chapel Hill, 1976-77; asst. prof. chemistry U. Hawaii at Manoa, 1977—. Contbr. articles to profl. jours. Mem. Am. Chem. Soc. (grantee 1980-82), Am. Phys. Soc. Subspecialties: Theoretical chemistry; Atomic and molecular physics. Current work: Methodological development and applications of ab initio electronic structure theory to molecular and solid state science: reduced density matrices, electron correlation, quantum dynamics. Home: 1822 Punahou St Honolulu HI 96822 Office: U Hawaii at Manoa Dept Chemistry 2545 The Mall Honolulu HI 96822

PAYNE, RICHARD EARL, oceanographer, researcher; b. Holyoke, Mass., Apr. 2, 1936; s. Lester and Marjorie (Stead) P.; m. Sheila Hammond Tulk, Aug. 23, 1958; children: Heather, Stephanie. B.S., Bowdoin Coll., 1958; M.S., U. Md., 1961; Ph.D., U. R.I., 1971. Physicist Nat. Bur. Standards, Washington, 1961-63; research asst. Woods Hole (Mass.) Oceanographic Instn., 1963-71, postdoctoral investigator, 1971-72, research assoc., 1973—; vis. research fellow U. Southampton, Eng., 1972-73. Mem. Am. Meteorol. Soc., Am. Geophys. Union. Subspecialties: Meteorologic instrumentation; Oceanography. Current work: Development and routine deployment of ocean buoy-mounted meteorological sensors and data collection and transmission systems. Office: Woods Hole Oceanographic Instn Woods Hole MA 02543

PAYNE, ROBERT BRYAN, psychology educator; b. Mt. Sterling, Ill., Oct. 23, 1914; s. George Merrimon and Iva Tipton (Bryan) P.; m. Ruth Helen Sorensen, Aug. 4, 1941; 1 dau.: Cynthia Anne. A.B., Maryville Coll., 1936; M.A., Ind. U., 1938, Ph.D., 1942. Instr. psychology Ind. U., 1940-41; commd. 2d lt. U.S. Army Air Force, 1942, advanced through grades to col., 1962-68; research psychologist, various locations, 1941-49; head dept. psychology U.S. Air Force Sch. Aviation Medicine, Randolph AFB, Tex., 1949-57; dir. bioscis. Arctic Aero-med. Lab., Ladd AFB, Alaska, 1957-59; dir. med. research Air Force Sch. Aerospace Medicine, Brooks AFB, Tex., 1961-64; dir. fgn. tech. Aerospace Med. Div., 1964-68; ret., 1968; prof. psychology U. Ga., 1968—; cons. psychology Surgeon Gen., U.S. Air Force, 1964-68, CIA, Washington, 1964-68, U.S. Air Force Fgn. Tech. Div., Wright-Patterson AFB, Ohio, 1972-75. Contbr. numerous articles to profl. jours.; Assoc. editor: Jour. Motor Behavior, 1972—. Fellow Am. Psychol. Assn.; mem. Psychonomic Soc., So. Soc. Philosophy and Psychology, Sigma Xi. Subspecialty: Learning. Current work: Research psychomotor learning and performance; chairman doctoral program in experimental psychology. Home: 280 Lullwater Rd Athens GA 30606 Office: Dept Psychology U Ga Athens GA 30602

PAYNE, ROBERT WALTER, psychologist, educator; b. Calgary, Alta., Can., Nov. 5, 1925; s. Reginald William and Nora Winnifred (Cowdery) P.; m. Helen June Mayer, Dec. 4, 1948 (div. 1972); children: Raymond William, Barbara Joan, Margaret June; m. Josephine Mary Riley, Mar. 26, 1977 (div. 1982); children: George Reginald Alexander, Robin Charles. B.A. with honors, U. Alta., Edmonton, 1949; Ph.D., U. London, 1954. Cons. psychologist West Park Hosp., Epsom, Eng., 1950-52; lectr. in psychology Inst. Psychiatry, London, 1952-59; assoc. prof. to prof. Queen's U., Kingston, Ont., 1949-65; prof., chmn. dept. behavioral sci. Temple U. Med. Sch., Phila., 1965-78; sr. med. research scientist Eastern Pa. Psychiat. Inst., Phila., 1965-78; prof. psychology, dean Faculty of Human and Social Devel., U. Victoria, B.C., 1978—. Contbr. chpts. to books, articles to profl. jours. Served wth Can. Army, 1943-45. Recipient Stratton Research award. Am. Psychopath. Assn., 1964. Fellow Am. Psychol. Assn., Brit. Psychol. Assn., Can. Psychol. Assn.; mem. Am. Psychopathol. Assn. Subspecialty: Experimental psychopathology. Current work: Thought disorder in psychosis. Home: 949 Pattullo Pl Victoria BC Canada V8S 5H6 Office: Univ Victoria PO Box 1700 Victoria BC Canada V8W 2Y2

PAYNE, ROBERT WILLIAM, metallurgist, quality assurance executive; b. Indpls., Mar. 29, 1922; s. Francis William and Mary Louise (Hughel) P.; m. Eileen Ruth McCann; m. Maryan June Browning; children: Gregory Robert, Sheila Eileen, Linda Suzan. B.S. in Metallurgy, U. Calif.-Berkeley, 1947. Metallurgist and specifications examiner Columbia-Geneva div. U.S. Steel Corp., San Francisco, 1947-50, distbn. asst. wire and wire products, 1951-64; chief metallurgist, dir. quality assurance Western Forge & Flange Co., Santa Clara, Calif., 1964—; cons. in field. Served with USCG, 1942-43. Mem. ASME, Am. Soc. Metals, ASTM. Democrat. Mormon. Subspecialties: Metallurgical engineering; Materials (engineering). Current work: Physical ferrous/non ferrous metallurgy; forgings, quality assurance. Inventor in field. Home: 5222 Colgate Ave Santa Clara CA 95051 Office: PO Box 327 Santa Clara CA 95052

PAZDERNIK, THOMAS LOWELL, pharmacologist; b. Detroit Lakes, Minn., Jan. 3, 1943; s. Alvin Joseph and Irene Helen (Kersting) P.; m. Betty Catherine Platt, June 20, 1967; children: Lisa Ann, Nancy Lea. Student, Coll. St. Thomas, St. Paul, 1961-63; B.S. in Pharmacy, U. Minn., 1967; Ph.D. in Medicinal Chemistry, U. Kans., Lawrence, 1971. Postdoctoral fellow U. Kans. Med. Center, Kansas City, 1971-73, asst. prof., 1973-77, asso. prof. pharmacology, 1977—; vis. research scientist U. Helsinki, Finland, 1976—. Contbr. articles to profl. jours. NIH fellow, 1979-71, 71-73; research career devel. awardee, 1975-80. Mem. Am. Chem. Soc., Am. Soc. Pharmacology and Exptl. Therapeutics, Tissue Culture Assn., Sigma Xi, Rho Chi, Phi Lambda Upsilon. Roman Catholic. Subspecialties: Immunotoxicology; Neuropharmacology. Current work: Effects of drugs and chemicals on the immune system, autoradiographic localization of neurotransmitters in the brain. Office: U Kans Med Center 39th and Rainbow Sts Kansas City KS 66103

PAZDUR, RICHARD, oncologist, educator; b. Hammond, Ind., May 6, 1952; s. John Joseph and Joan Patricia (Hudzik) P.; m. Mary Patricia Bagby, Mar. 12, 1982. B.A., Northwestern U., 1973; M.D., Loyola U., 1976. Diplomate: Nat. Bd. Med. Examiners, Am. Bd. Internal Medicine. Intern Loyola U. Hosp., Maywood, Ill., 1976-77, resident in internal medicine, 1977-79; fellow med. oncology Rush Presbyn. St. Luke's Hosp., Chgo., 1979-81, U. Chgo. Med. Ctr., 1981-82; asst. prof. oncology Wayne State U., Detroit, 1982—. Am. Cancer Soc. clin. fellow, 1979. Mem. ACP, Am. Soc. Clin. Oncology, Am. Fedn. Clin. Research. Roman Catholic. Subspecialties: Chemotherapy; Oncology. Current work: Cancer chemotherapy, investional clinical drug investigation. Home: 1431 Washington Blvd Apt 2308 Detroit MI 48226 Office: Dept Oncology Wayne State U 3990 John R 5 Hudson St Detroit MI 48201

PAZOLES, CHRISTOPHER JAMES, neurobiologist; b. Chgo., Jan. 17, 1950; s. Lewis J. and Maria J. (Katsaros) P.; m. Pamela Peri, July 2, 1972. B.A., Oberlin Coll., 1971; Ph.D., U. Notre Dame, 1975. Staff fellow Nat. Inst. Child Health and Human Devel., NIH, Bethesda, Md., 1975-76, Nat. Inst. Arthritis, Metabolism and Digestive Diseases, 1976-78, sr. staff fellow, 1978-81; project leader Pfizer Central Research, Groton, Conn., 1981—. Mem. Soc. Neurosci., AAAS; mem. Am. Pain Soc., Internat. Assn. for Study of Pain. Eastern Orthodox. Subspecialties: Neurobiology; Neuropharmacology. Current work: Modulation of neuropeptides; mechanisms of pain and analgesia. Office: Pfizer Central Research Groton CT 06340

PAZZAGLINI, MARIO PETER, psychologist; b. Endicott, N.Y., Mar. 9, 1940; s. Mario and Dina (Albertini) P. B.A., SUNY-Binghamton, 1961; M.A., U. Del., 1965, Ph.D., 1969. Lic. clin. psychologist, Del. Staff psychologist, co-dir. adolescent program Del. State Hosp., New Castle, 1968-77; psychologist-adminstr. Bur. Alcoholism and Drug Abuse, New Castle, 1970—; pvt. practice psychology, Newark, Del., 1973—; clin. instr. Jefferson Med. Sch., Phila., 1972—; clin. psychologist St. Francis Hosp., Wilmington, Del., 1977—; cons. Dept. Pub. Instrn., Dover, Del., 1972—, Dept. Health and Social Services, New Castle, 1972—, ct. system, Wilmington-Dover, 1972—, Dept. Corrections, 1972—. Co-author psychol. inventory, 1980; painter, illustrator, 1965—. Cons. community based free drug clinics, N.Y.C., 1968-75; cons. community based free drug clinics State of Del., 1968-74, community based emergency hot lines, 1968—, coms. for services to children and youth, 1971—. HEW grantee, 1971-72. Mem. AAAS, Am. Psychol. Assn., N.Y. Acad. Scis., N.Y. Gestalt Inst., Del. Psychol. Assn., Sigma Xi. Democrat. Roman Catholic. Subspecialties: Behavioral psychology; Clinical psychology. Current work: Research in the field of imagery—crosscultural studies, historical-ancient studies, current therapeutic uses. Office: 523 Capitol Trail Newark DE 19711

PE, MAUNG HLA, educator; b. Mandalay, Burma, Nov. 14, 1920; s. U Soe Min and Daw Tin; m. Ma Tin Hlyne, Apr. 16, 1957. B.S., Rangood U., 1946; M.S., E.E., Lehigh U., 1951; Ph.D., 1957; M.S. in Applied Math, N.Y. Poly Inst., 1952; M.Nuclear Engring., N.Y. U., 1957. Lectr. CUNY, 1957-61; asst. prof. physics Manhattan Coll., Bronx, N.Y., 1961-64; asso. prof. physics, 1964—; faculty fellow for space physics Columbia U., 1965; NATO fellow for advanced study Inst. U. of Keele, Staffordshire, Eng., 1966; vis. prof. Center for Research and Advanced Study, Mexico City, 1969. Contbr.: sects. to Acad. Am. Ency, 1981; article to profl. jours. Mem. N.Y. Acad. Scis., AAAS, Am. Assn. Physicists in Medicine. Subspecialties: Plasma physics; Medical physics. Current work: Radiation biophysics edn., physics for radiation therapy, nuclear medicine and ultrasonics. Office: Dept Physics Manhattan Coll Bronx NY 10471

PEACEMAN, DONALD WILLIAM, chemical engineer; b. Miami, June 1, 1926; s. Morris and Ida B. (Rabinowitz) P.; m. Ruth C. Klein, Jan. 27, 1952; children: Caren Cowan, Alan. B.Ch.E., CCNY, 1947; Sc.D., M.I.T., 1952. Research engr. Exxon Prodn. Research Co., Houston, 1951—. Author: Fundamentals of Numerical Reservoir Simulation, 1977. Mem. Soc. Petroleum Engrs. of AIME (Robert Earll McConnell award 1979), ACM, Soc. Indsl. and Applied Math., Am. Inst. Chem. Engrs. Subspecialties: Petroleum engineering; Numerical analysis. Current work: Numerical reservoir simulation, numerical mathematics, computer programming. Home: 4907 Glenmeadow St Houston TX 77096 Office: Exxon Prodn Research Co PO Box 2189 Houston TX 77001

PEACH, MICHAEL JOE, pharmacologist; b. Morgantown, W.Va., Aug. 22, 1940; s. Joseph George and Juanita Olivia (White) P.; m. JoAnne Hutchinson, Dec. 23, 1965; 1 dau., Elizabeth Anne. Ph.D., 1967. Instr. pharmacology W.Va. U., 1967; fellow research div. Cleve. Clinic, 1967-68; asst. prof. pharmacology U. Va., 1968-71, assoc. prof., 1971-76, prof., 1976—; cons. drug cos. NIH. Mem. Soc. Neurosci., Am. Soc. Pharmacology and Exptl. Therapeutics, Soc. Exptl. Biology and Medicine, Council High Blood Pressure Research, Am. Heart Assn., Sigma Xi. Dem. Episcopalian. Subspecialties: Cellular pharmacology; Receptors. Current work: Mechanism of action of the renin-angiotensin system; cardiovascular regulation; vasoactive polypeptides. Office: Dept Pharmacology Box 448 U Va Med Center Charlottesville VA 22908

PEACH, ROBERT WESTLY, consulting engineer; b. Balt., May 5, 1921; s. Eggelston W. and Ola (English) P.; m. Jean M. McClure, July 11, 1953; children: John R., Helen E. B.S.E., U. Mich.-Ann Arbor, 1944, M.S.E., 1994. Registered profl. engr., Conn., Md. Project engr. Reaction Motors, Inc., Dover, N.J., 1950-52; supr. Electric Boat Div., Groton, Conn., 1952-62; asst. chief naval architect Md. Ship & Drydock Co., Balt., 1962-67; mgr. mech. engring. Ocean Research and Engring. Ctr., Westinghouse Elec. Corp., Annapolis, Md., 1967-73; owner R. W. Peach Engring. Assocs., Arnold, Md., 1973—. Fellow Royal Instn. Naval Architects; mem. Soc. Naval Architects and Marine Engrs., Am. Soc. Naval Engrs., Marine Tech. Soc., N.E. Coast Ins. Engrs. and Shipbuilders. Clubs: Sailing of Chesapeake (race com. chmn. 1975), Windjammers.). Subspecialties: Mechanical engineering; Marine engineering. Current work: Engineering mechanics, problems in the marine field covering surface ships, submarine and deep submersibles. Home: 888 Pine Trail Arnold MD 21012 Office: R W Peach Engring Assocs 888 Pine Trail Arnold MD 21012

PEACHEY, LEE DEBORDE, biology educator; b. Rochester, N.Y., Apr. 14, 1932; s. Clarence Henry and Eunice (DeBorde) P.; m. Helen Pauline Fuchs, June 7, 1958; children—Michael Stephen, Sarah Elizabeth, Anne Palmer. B.S., Lehigh U., 1953; postgrad., U. Rochester, 1953-56; Ph.D. (Leitz fellow), Rockefeller U., 1959; M.A. (hon.), U. Pa., 1971. Research asso. Rockefeller U., 1959-60; asst. prof. zoology Columbia U., 1960-63, asso. prof., 1963-65; asso. prof. biochemistry and biophysics U. Pa., Phila., 1965-70, prof. biology, 1970—; adj. prof. molecular, cellular and developmental biology U. Colo., 1969—; mem. molecular biology study sect. NIH, 1969-73; Mem. Mayor's Sci. and Tech. Adv. Council, Phila., 1972—. Editor: Third and Fourth Conferences on Cellular Dynamics, N.Y. Acad. Scis., 1967, First and Second Confs. on Cellular Dynamics, 1968, Am. Physiol. Soc. Handbook on Skeletal Muscle, 1983; mem. editorial bd.: Tissue and Cell, 1969—, Jour. Cell Biology, 1970-73, Pitman Series in Cellular and Development Biology, 1977—, Jour. Electron Microscopy Technique, 1982—, Advances in Optical and Electron Microscopy, 1983—; Contbr. articles to sci. jours. Trustee Keith R. Porter Endowment for Cell Biology, Narberth, Pa., 1981—. Guggenheim and Fulbright-Hays fellow, 1967-68; Overseas fellow Churchill Coll., Cambridge, Eng., 1967-68; NSF grantee, 1960-72; NIH grantee, 1973—; Muscular Dystrophy Assn. Am., Inc. grantee, 1973—; Fogarty sr. internat. fellow, 1979-80; hon. research fellow Univ. Coll., London, 1979-80. Fellow AAAS; Mem. Electron Microscope Soc. Am. (council 1975-78, pres. 1982), Am. Soc. Cell Biology (program chmn. 1965, council 1966-69), Biophys. Soc. (program chmn. 1976, council 1976-80, exec. com. 1976-82, pres. 1981-82), Internat. Union Pure and Applied Biophysics (council 1978—, chmn. commn. on cell and membrane biophysics 1981—), Physiol. Soc. (Eng.).; mem. Internat. Soc. Stereology (internat. stereology software com. 1982—), Soc. Gen. Physiologists. Subspecialties: Biophysics (biology); Cell biology. Current work: Cell physiology of the mechanism and control of muscle contraction; electron microscopy; quantitative and three-dimensional analysis of cell structure. Home: 606 Old Gulph Rd Narberth PA 19072 Office: Dept Biology/G7 U Pa Philadelphia PA 19104

PEALE, STANTON JERROLD, educator; b. Indpls., Jan. 23, 1937; s. Robert Frederick and Edith (Murphy) P.; m. Priscilla Laing Cobb, June 1960; children: Robert Edwin, Douglas Andrew. B.S., Purdue U., 1959; M.S., Cornell U., 1962, Ph.D., 1965. Research engr. Gen. Electric Co., Cin., 1959; research asso. Cornell U., 1964-65; asst. prof. astronomy UCLA, 1965-68; asst. prof. physics U. Calif.-Santa Barbara, 1968-70, asso. prof., 1970-76, prof., 1976—; cons. in field. Contbr. numerous articles to profl. jours. NASA grantee, 1969—; recipient Newcomb Cleveland prize AAAS, 1980, James Craig Watson award Nat. Acad. Scis., 1982. Fellow AAAS; mem. Am. Astron. Soc., Am. Geophys. Union, Internat. Astron. Union. Subspecialty: Planetary science. Current work: Dynamical histories of solar system bodies. Home: 6233 Cumberland Dr Goleta CA 93117 Office: Dept Physics U Calif Santa Barbara CA 93106

PEARCE, MALCOLM BULKELEY, JR., mech. engr.; b. West Haven, Conn., Nov. 12, 1929; s. Malcolm Bulkeley and Heneritta Louise (Bowyer) P.; m. Helen Christine Korn, Oct. 13, 1951; children: Malcolm Bulkeley III, April, Holly. B.S. in M.E, Ind. Inst. Tech., 1950; M.E., UCLA, 1953; M.D., U. N.H., 1968. Registered Engr., N.Y., Conn. Flight test NAA, Los Angeles, 1951-54; power plant engr. Pratt & Whitney Aircraft, East Hartford, Conn., 1954-57; aerospace engr. AVCO, Stratford, Conn., 1957-61; nuclear engr. Canel Pratt & Whitney Aircraft, Middletown, Conn., 1961-63; fuel controls CECO, West Hartford, Conn., 1963-66; spl. weapons E-B, Simsbury, Conn., 1966-68; dir. mgr. IBC Loctite, Newington, Conn., 1968-72; tech. market mgr. Rawl, N.Y.C., 1972-78; mgr. material engring. United Tech., Sikrosky, Conn., 1978—; mgr. spl. advance projects materials group Sikorsky Aircraft, Stratford, Conn., 1981—; cons. consumer preports. Dir. civil def., Durham, Conn., 1954—, dir. police services, 1970-82; mem. Republican Town Com., 1976—. Served with AUS, 1945-47. Mem. ASME, Soc. Automotive Engrs., Am. Nat. Standards Inst., Expansion Anchor Mfrs. Inst., Structural and Materials Process Engring. Assn. Republican. Mem. United Ch. (deacon, trustee). Lodges: Masons; Shriners. Subspecialties: Composite materials; Aerospace engineering and technology. Current work: Aerospace investigation advanced projects management. 0H. Patentee in field. Home: 242 Bayberry Hill Durham CT 06422 Office: Sikorsky Aircraft N Main St Stratford CT 06602

PEARCE, WILLIAM RICHARD, nuclear engineering consultant; b. Port Arthur, Ont., Can., Feb. 28, 1929; s. Carl Richard Pearce and Margaret (Calder) Atkins; m. Piera Cegan, Aug. 2, 1958; 1 son: William Richard. B.S. in Mech. Engring, Mich. Technol. U., 1951; student, Oak Ridge Sch. Reactor Tech., 1952. Registered profl. engr., Md. Nuclear engr. Bendix Aviation Corp., Detroit, 1952-55, Monsanto Chem. Co., Clayton, Mo., 1955; project mgr. Internuclear Co., Clayton, 1955-62, Allis-Chalmers Corp. Nuclear Div., Bethesda, Md., 1963-69; ind. nuclear cons., Bethesda, 1970—. Patentee flux-trap research nuclear reactor. Mem. Am. Nuclear Soc., Atomic Indsl. Forum. Subspecialty: Nuclear engineering. Current work: Assistance to electric utilities, engineering firms, national laboratory and others in design, safety analysis and operation of nuclear power plants. Address: 6848 Glenbrook Rd Bethesda MD 20814

PEARLMUTTER, A. FRANCES, biochemistry educator; b. Chelsea, Mass., Oct. 28, 1940; d. Albert and Esther Leah (Wool) Goldberg; m. Fishel A. Pearlmutter, Sept. 5, 1960; children: Barak Avrum, Nili Naomi. B.S., Tulane U., 1962; M.A., Case Western Res. U., 1967, Ph.D., 1969. Instr. Med. Coll. Ohio, Toledo, 1969-73, asst. prof., 1973-80, assoc. prof. biochemistry, 1980—. Pres. Toledo Bd. Jewish Edn.,

1980-82. Mem. Am. Chem. Soc., AAAS, Endocrine Soc., Am. Soc. Biol. Chemists, Sigma Xi. Subspecialties: Neurochemistry; Biochemistry (medicine). Current work: Role of peptide hormones. Home: 2727 Meadowwood Toledo OH 43606 Office: Med Coll Ohio CS 10008 Toledo OH 43699

PEARLSON, GODFREY DAVID, physician, neuroscience researcher; b. Sunderland, Eng., Jan. 30, 1950; s. Elias and Blanche (Book) P.; m. Judith Gail Sirote, May 23, 1982. M.B.B.S., Newcastle on Tyne Med. Sch., Eng., 1974; M.A. in Philosophy, Columbia U., 1976. Diplomate: Am. Bd. Psychiatry and Neurology. Intern Newcastle Hosp., 1974-75; resident Johns Hopkins Hosp., Balt., 1975-77; asst. prof. psychiatry and behavioral sci. Johns Hopkins U., Balt., 1976—. Recipient Surgery prize Newcastle on Tyne Med. Sch., 1974, Psychiatry prize, 1974; Annual Prize award Md. Psychiat. Assn. Mem. AAAS, Soc. for Neurosci., Am. Psychiat. Assn., Eastern Psychol. Assn. Subspecialties: Psychiatry; Neuropsychology. Current work: Image processing applied to CNS radiology; treatment of chronic pain. Office: Johns Hopkins Hosp Meyer 279 Baltimore MD 21120

PEARLSTEIN, SOL, nuclear physicist; b. N.Y.C., Feb. 21, 1930; s. David and Tillie P.; m. Susan Fuchs, June 3, 1951; children: Teri, Debra, Judith. B.S., Poly. Inst. Bklyn., 1951; M.S., N.Y. U., 1952; Ph.D., Rensselaer Poly. Inst., 1964. Nuclear engr. Knolls Atomic Power Lab., Gen. Electric Co., Schenectady, 1952-64; physicist Brookhaven Nat. Lab., Upton, N.Y., 1964—; dir. nat. nuclear data ctr. Fellow Am. Nuclear Soc. Subspecialties: Nuclear physics; Information systems, storage, and retrieval (computer science). Current work: Reactor physics, nuclear data evaluation, computer science. Office: Bldg 197D Brookhaven Nat Lab Upton NY 11973

PEARSON, ERWIN GALE, veterinary medicine educator; b. McMinnville, Oreg., July 19, 1932; s. Emil Walberg and Clarice Francis (Hurner) P.; m. Evelyn Kohler, Sept. 4, 1954; m. Marianne Mackay, Jan. 8, 1982; children: Julie, Deborah, Timothy, Cheri. B.S., Oreg. State U, 1954; D.V.M., Cornell U., 1958; M.S., Oreg. State U., 1979. Diplomate: Am. Coll. Vet. Internal Medicine 1979. Gen. practice vet. medicine, Oreg., 1960-79; asst. prof. vet. medicine Cornell U., 1979-82; assoc. prof. medicine Sch. Vet. Medicine, Oreg. State U., 1982—. Contbr. to profl. jours. Served with U.S. Army, 1958-60. Mem. AVMA, Am. Assn. Equine Practitioners, Am. Assn. Bovine Practioners. Subspecialty: Internal medicine (veterinary medicine). Current work: Liver diseases, neonatal diarrhea. Home: 2145 NW 14th St Corvallis OR 97330 Office: School Veterinary Medicine Oregon State University Corvallis OR 97331

PEARSON, GERALD LEONDUS, electronics engineer; b. Salem, Oreg., Mar. 31, 1905; s. David Shafer and Sarah (Allen) P.; m. Mildred Oneta Cannoy, June 30, 1929; children: Ray Leon, Carol Ann. A.B., Willamette U., 1926; Sc.D. (hon.), 1956; A.M. in Physics, Stanford, 1929. Mem. tech. staff Bell Telephone Labs., 1929-57, head applied physics solids dept., 1957-60; prof. elec. engring. Stanford, 1960-70, emeritus, 1970—; dir. Center Materials Research, 1965-66, Solid-State Electronics Lab., 1970-80, emeritus, 1980—; vis. prof. U. Tokyo, 1966; mem. ad hoc com. materials and processes for electron devices NRC, 1971-72; mem. com. on recommendations for U.S. Army basic sci. research, 1978-81. Author. Recipient John Scott award City Phila., 1956; Gold Plate award Am. Acad. Achievement, 1963; Marian Smoluchowski medal Polish Phys. Soc., 1975; Solid State Sci. and Tech. award, medal and prize Electrochem. Soc., 1981; Alfred Krupp von Bohlen und Holbach prize for energy research, 1981; Gallium Arsenide and Related Compounds Symposium award, 1981; Heinrich Welker medallion Siemens AG, 1983. Fellow Am. Phys. Soc., Am. Inst. Elec. and Electronics Engrs.; life mem. Franklin Inst. (John Price Wetherill medal 1963); mem. Nat. Acad. Scis., Nat. Acad. Engring., Telephone Pioneers Am., Sigma Xi. Subspecialties: Semiconductors; Condensed matter physics. Current work: Semiconductor devices including solar cells, transistors and rectifiers. Patentee in field. Home: 501 Portola Rd Portola Valley CA 94025 Office: Stanford Electronics Labs Stanford CA 94305

PEARSON, JAMES ELDON, nephrologist; b. Newton, Mass., Oct. 8, 1926; s. James Elmer and Mary Ann (Kelley) P.; m. Gertraud Linser, July 11, 1953; children: Mary, Alfred, Richard, Kathryn, James. B.S. in Biology, Loyola U., New Orleans, 1956, M.S. in Biochemistry, 1958. Asst. in angiology Touro Research Inst., New Orleans, 1964-68; research assoc. in ecology Pacific Biomed. Research, U. Hawaii, 1971-72; research assoc. in pharmacology La. State U. Med. Ctr., New Orleans, 1968-71, research assoc. in nephrology, 1972-74, 76—, Med. Ctr., U. Miss., 1974-76. Served to sgt. USMC, 1943-46. Mem. Am. Soc. Pharmacology and Exptl. Therapeutics, Am. Soc. Nephrology, Internat. Soc. Nephrology, Am. Soc. Artificial Internal Organs. Subspecialties: Nephrology; Toxicology (medicine). Current work: Nephrology, as it concerns the artificial kidney and drug effects on the kidney. Home: 6540 Fleur de Lis New Orleans LA 70124 Office: 1542 Tulane Ave New Orleans LA 70112

PEARSON, JEROME, aerospace engineer, consultant, researcher; b. Texarkana, Ark., Apr. 19, 1938; s. Elbert Sharp and Angela Lucille (Bond) P.; m. Jeannie Thompson, Aug. 22, 1965; children: Edmund B., Marcus A., Laura E. B.S. in Engring, Washington U., St. Louis, 1961; M.S. in Geology, Wright State U., 1977. Aerospace technologist NASA Langley Research Ctr., Hampton, Va., 1962-66; aerospace engr. NASA Ames Research Ctr., Mountain View, Calif., 1966-71, Air Force Wright Aero. Labs., Wright-Patterson AFB, Ohio, 1971—; tech. cons. to motion picture industry Disney EPCOT Ctr.; lectr. in field. Contbr. articles to profl. jours. Soccer coach Little League, Dayton, Ohio, 1976—. Served with USMCR, 1961-65. Recipient Apollo Achievement award NASA, 1969; Sci. Achievement award Air Force Systems Command, 1975; Eminent Engr. award Tau Beta Pi, 1979. Fellow Brit. Interplanetary Soc., AIAA (assoc.), Am. Astronautical Soc. (sr.), Soc. Automotive Engrs., Scientists' Inst. for Public Info., Mensa. Republican. Methodist. Subspecialties: Aerospace engineering and technology; Astronautics. Current work: Managing contracted and in-house research on vibration control of high-energy lasers and dynamics and control of large-space structures; independent space research. Conceived orbital tower, anchored lunar satellite, rotary rocket; inventor satellite sail; patentee zero-G massmeter; authored proposals to use rotating structures for spacecraft launching; developed Apollo lunar staff sensor, force-feedback control system. Office: Air Force Wright Aero Labs AFWAL/FIB Wright-Patterson AFB OH 45433

PEARSON, LORENTZ C., biology educator; b. American Fork, Utah, Jan. 28, 1924; s. Clarence N. and Johanna (Peterson) P.; m. Ingvi Linnea Lindblad, Mar. 17, 1952; children: Suzanne Pearson Benson, Beth Pearson Cazier, Solveig Pearson Nelson, Kjerstin, Ellen, Laura. B.S., Utah State U., 1952; M.S., U. Utah, 1952; Ph.D., U. Minn., 1958. Missionary Mormon Ch., Sweden, 1948-50; tchr. sci. Swedish Secondary Schs., 1955-56; tchr. agronomy and biology Ricks Coll., Rexburg, Idaho, 1952—; Mem. Madison County Weed Control Bd., 1960—. Author: Principles of Agronomy, 1967; contbr. articles to profl. jours. Bd. dirs. Madison County 4-H, 1964—; chmn. Democratic Central Com., Rexburg, 1974-75; active Boy Scouts Am., 1950—. Served with Signal Corps U.S. Army, 1942-46. NSF grantee and fellow, 1960-62, 63-64, 66, 70-73. Mem. AAAS, Bot. Soc. Am., Ecol. Soc. Am., Am. Bryological and Lichenological Soc., Idaho Acad. Sci., Sigma Xi. Subspecialties: Ecosystems analysis; Plant genetics. Current work: Effects of air pollution on lichen physiology; effects of air pollution on cell membranes; microclimatic studies; desert ecosystems. Patterns of evolution in plants. Home: 366 S 3d E Rexburg ID 83440 Office: Dept Biology PSB 106 Ricks Coll Rexburg ID 83440

PEARSON, PHILLIP THEODORE, veterinarian, university dean; b. Ames, Iowa, Nov. 21, 1932; s. Theodore B. and Hazel C. (Christianson) P.; m. Mary Jane Barlow, Aug. 28, 1954; children: Jane Catherine, Bryan Theodore, Todd Wallace, Julie Ann. D.V.M., Iowa State U., 1956, Ph.D., 1962. Intern Angell Meml. Animal Hosp., Boston, 1956-57; instr. Coll. Vet. Medicine Iowa State U., 1957-59, asst. prof., 1959-63, assoc. prof., 1963-64, prof. vet. clin. scis. and bio-med. engring., 1965-72, dean, 1972—, dir. Vet. Med. Inst.; prof. Sch. Vet. Medicine, U. Mo., 1964-65. Bd. dirs. Iowa Sate Meml. Union, 1970—; v.p. Iowa State Meml. Union, 1975; chmn. council of deans Assn. Am. Vet. Medicine Colls., 1978-79. Recipient Riser award, 1956; distinguished Tchr. award Norden Labs., 1962; Gaines award Gen. Foods Corp., 1966; Outstanding Tchr. award Iowa State U., 1968; Faculty citation, 1974. Mem. Am., Iowa vet. med. assns., Am. Animal Hosp. Assn., Am. Assn. Vet. Clinicians, Am. Coll. Vet. Surgeons (bd. regents 1972, pres. 1977), Sigma Xi, Phi Kappa Phi, Phi Zeta, Alpha Zeta, Gamma Sigma Delta. Lodge: Kiwanis (dir., pres. Ames 1966—). Subspecialty: Surgery (veterinary medicine). Current work: Research administration. Home: 1610 Maxwell Ave Ames IA 50010

PEARSON, RALPH GOTTFRID, educator; b. Chgo., Jan. 12, 1919; s. Gottfrid and Kerstin (Larson) P.; m. Lenore Olivia Johnson, June 15, 1941 (dec. June 1982); children—John Ralph, Barry Lee, Christie Ann. B.S., Lewis Inst., 1940; Ph.D., Northwestern U., 1943. Faculty Northwestern U., 1946-76, prof. chemistry, 1957-76, U. Calif., Santa Barbara, 1976—; Cons. to industry and govt., 1951—. Co-author 5 books. Served to 1st lt. USAAF, 1944-46. Guggenheim fellow, 1951. Mem. Am. Chem. Soc. (Midwest award 1966, Inorganic Chemistry award 1969), Nat. Acad. Sci., Phi Beta Kappa, Sigma Xi, Phi Lambda Upsilon (hon.). Lutheran. Subspecialties: Inorganic chemistry; Theoretical chemistry. Current work: Mechanisms of chemical reactions; nature of chemical bond. Originator prin. of hard and soft acids and bases.

PEARSON, SAMUEL DIBBLE, III, civil engineer; b. Augusta, Ga., Mar. 28, 1944; s. Samuel Dibble and Adelaide (DeBeaugrine) P.; m. Patricia Ann Johnson, Dec. 26, 1967 (dec. Dec. 1974); 1 son, Pierce DeBeaugrine; m. Deborah McGahee, Aug. 30, 1975; 1 dau., Elizabeth Allison. B.S., U. Ga., 1966; B.C.E., Auburn U., 1973. Registered profl. engr., Ga., S.C. Project engr. Babcock & Wilcox, Inc., Augusta, Ga., 1973-78; project mgr. Kimberly-Clark Corp., Waynesboro, Ga., 1978-80; engring. div. mgr. Chem-Nuclear Systems, Inc., Barnwell, S.C., 1980—. Served to capt. USAF, 1968-72. Mem. ASCE, Am. Nuclear Soc., Tau Beta Pi, Auburn U. Alumni Assn. (dir. 1976-78). Republican. Baptist. Subspecialties: Civil engineering; Nuclear engineering. Current work: Design and development of state of the art low level nuclear waste disposal sites, design and development of state of the art process equipment to handle low level nuclear waste generated by operating nuclear power plants. Home: 4038 Indian Hills Dr Augusta GA 30906 Office: Chem-Nuclear Systems Inc Osborn Rd Snelling SC 29812

PEARSON, WILLIAM DEAN, biology educator; b. Moline, Ill., Dec. 6, 1941; s. Paul C. and Virginia (Conlon) P.; m. B. Juanelle Lozano, July 19, 1977; children: Leslie G., Eric C. B.S., Iowa State U., 1964; M.S., Utah State U., 1967, Ph.D., 1970. Mem. faculty, asst. prof. N. Tex. State U., Denton, 1970-75; assoc. research prof. dept. biology U. Louisville, 1975—; pres. Lake Mgmt. Analysts, Louisville, 1978—. Contbr. articles to profl. jours. Mem. AAAS, Am. Fisheries Soc., Am. Ecol. Soc., N.Am. Benthological Soc., Am. Soc. Ichthyologists and Herpetologists. Subspecialties: Ecology; Population biology. Current work: Ichthyology, fisheries management, population dynamics of fishes, larval fishes and early life history of fishes. Home: 4513 Shennandoah Dr Louisville KY 40222 Office: U Louisville Water Resources Lab Louisville KY 40292 Home: 4513 Shennandoah Dr Louisville KY 40222

PEASE, DAVID NATHANIEL, environmental consulting firm executive; b. Concord, N.H., June 16, 1952; s. Nathaniel John and Helen Mabel (Edmark) P.; m. Donna-Marie Desautels, May 15, 1977. B.A. in Marine Biology, U. N.H., 1974; postgrad., U. B.C., Vancouver, 1974; M.S., U. N.H., 1983. Technician U. N.H., Durham, 1972-74; staff scientist Normandeau Assocs., Inc., Bedford, N.H., 1974-81; project mgr. offshore services Peck Environ. Lab., Inc., Hampton Falls, N.H., 1981—, exec. v.p., 1983—; also dir.; pres. Resource Analysts, Inc., 1983—. Mem. Estuarine Research Fedn., Marine Tech. Soc., Phi Beta Kappa, Phi Kappa Phi, Phi Sigma. Subspecialties: Resource management; Analytical chemistry. Current work: Management of environmental laboratory with specializations in organic contamination of soils, groundwater, and tissues. Manage projects related to environmental concerns in offshore petroleum exploration and development. Home: PO Box 212 Hampton NH 03842 Office: Resource Analysts Inc 1 Lafayette Rd Hampton Falls NH 03844

PEASE, ROBERT LOUIS, physics educator, researcher; b. Fitchburg, Mass., July 13, 1925; s. Daniel and Emma Marguerite (Wippert) P. A.B., Miami U., Oxford, Ohio, 1943; Ph.D., MIT, 1950. Prof. physics SUNY-New Paltz, 1967—. Served to lt. (j.g.) USNR, 1944-46; PTO. Mem. Am. Phys. Soc., IEEE, Ops. Research Soc. Am. Subspecialties: Theoretical physics; Electrical engineering. Current work: Nonlinear quantum field theory, microwave design. Office: Dept Physics SUNY - New Paltz NY 12561

PEATMAN, JOHN GRAY, consulting firm executive, psychologist; b. Centerville, Iowa, Mar. 16, 1904; s. Clarence Albert and Binnie Oriel (Gray) P.; m. Lillie Burling; children: Alice Peatman Dettmers, John, William; m. Madeline Martin; 1 dau., Mary Peatman Fitzpatrick. B.A., Columbia U., 1927, M.A., 1928, Ph.D., 1931. Pres. Research Cons., Inc., Norwalk, 1950—; also dir.; ast. prof. psychometrics CCNY, 1929-42, assoc. prof., 1942-48, prof., 1948-70, Emeritus prof., 1970—; pres. Office of Research, Inc., N.Y.C., 1945-58, also dir.; cons. in field. Author: Descriptive and Sampling Statistics, 1947, Introduction to Applied Statistics, 1963; editor: (with Eugene Hartley) Festschrift to Gardner Murphy, 1960. Fellow Am. Psychol. Assn. (chmn. policy and planning bd. 1950-51); mem. Psychonomic Soc., Phi Beta Kappa, Sigma Xi. Congregationalist. Club: University (N.Y.C.). Subspecialties: Applied mathematics; Statistics. Current work: Research in applied mathematics and its applications. Home: Stonewood Comstock Hill Norwalk CT 06852 Office: Research Cons Inc 83 East Ave Norwalk CT 06851

PECK, DALLAS LYNN, geologist; b. Cheney, Wash., Mar. 28, 1929; s. Lynn Averill and Mary Hazel (Carlyle) P.; m. Tevis Sue Lewis, Mar. 28, 1951; children—Ann, Stephen, Gerritt. B.S., Calif. Inst. Tech., 1951, M.S., 1953; Ph.D., Harvard U., 1960. With U.S. Geol. Survey, 1954—, asst. chief geologist, office of geochemistry and geophysics, Washington, 1967-72, geologist, geologic div., 1972-77, chief geologist, 1977-81, dir., 1981—; mem. Lunar Sample Rev. Bd., 1970-71; chmn. earth scis. adv. com. NSF, 1970-72; vis. com. dept. geol. scis. Harvard U., 1972-78. Recipient Meritorious Service award Dept. Interior, 1971, Disting. Service award, 1979. Mem. Geol. Soc. Am., Soc. Econ. Geologists, Am. Geophys. Union (pres. sect. volcancology, geochemistry and petrology 1976-78). Subspecialty: Geology. Home: 2524 Heathcliff Ln Reston VA 22091 Office: US Geol Survey Reston VA 22092

PECK, RALPH BRAZELTON, engineering educator; b. Winnipeg, Man., Can., June 23, 1912; s. Orwin K. and Ethel Indie (Huyck) P.; m. Marjorie Elizabeth Truby, June 14, 1937; children: Nancy Jeanne Peck Young, James Leroy. C.E., Rensselaer Poly. Inst., 1934, D.C.E., 1937, D.Eng., 1974; postdoctoral, Harvard U., 1938. Structural detailer Am. Bridge Co., Ambridge, Pa., 1937; asst. subway engr. soil investigations City of Chgo., 1939-43; chief engr. testing Holabird & Root, Scioto Ordnance Plant, Marion, Ohio, 1943; research asst. prof. soil mechanics U. Ill., Urbana, 1943, research prof. found. engring., 1948-57, prof. found. engring., 1957-74, emeritus, 1974—; cons. in field. Author: (with K. Terzaghi) Soil Mechanics in Engineering Practice, 1948, 2d edit., 1967, (with W.E. Hanson, T.H. Thornburn) Foundation Engineering, 1953, 2d edit., 1973; contbr. articles to profl. jours. Recipient Dept. Army Disting. Civilian Service award, 1973; Nat. medal of Sci., 1974; Washington award, 1976. Fellow Geol. Soc. Am., Am. Cons. Engrs. Council; mem. ASCE (Norman medal 1944, Wellington prize 1965, Terzaghi award 1969), Nat. Soc. Profl. Engrs. (award 1972), Internat. Soc. Soil Mechanics and Found. Engring. (past pres.), Am. Ry. Engring. Assn., Nat. Acad. Engring., Sigma Xi, Tau Beta Pi, Chi Epsilon, Phi Kappa Phi. Subspecialty: Civil engineering. Current work: Dams, tunnels. Home: 1101 Warm Sands Dr SE Albuquerque NM 87123

PECK, ROBERT F(RANKLIN), psychology researcher, educator, consultant; b. Buffalo, Sept. 22, 1919; s. Charles R. and Jessie M. (Kelley) P.; m.; children: Joanne, Brian. B.Sc. in Edn, SUNY-Buffalo, 1941; M.Sc., SUNY-Albany, 1942; Ph.D., U. Chgo., 1951; LL.D. (hon.), Brock U., St. Catherines, Ont., Can., 1974. Instr. U. Chgo., 1947-50, research assoc., 1950-54; assoc. prof. ednl. psychology U. Tex.-Austin, 1954-59, prof., 1959—, dir., 1958—, 1965-77, Mgmt. Devel. Services, Austin, 1957—. Author: Psychology of Character Development, 1960; author: 5 vol. research report Coping and Achievement: A Cross National Study, 1972-81; test Styles of Coping, 1981; ednl. system Personalized Teacher Education, 1982. NIMH grantee, 1958-63, 62-65; Office Edn., Nat. Inst. Edn. grantee, 1965-72; Nat. Inst. Edn. grantee, 1973-82. Fellow Am. Psychol. Assn.; mem. Internat. Soc. for Study Behavioral Devel., Soc. Cross-Cultural Psychology, Internat. Assn. Applied Psychology. Club: Headliners (Austin). Subspecialties: Developmental psychology; Management psychology. Current work: Cross-cultural, longitudinal research on coping skills and motives that explain human competence; research on coping and addiction; training systems to promote career effectiveness. Home: 3304 Glen Rose Dr Austin TX 78731 Office: U Tex EDA 3.203 Austin TX 78712

PECORA, ROBERT, chemistry educator, researcher; b. Bklyn., Aug. 6, 1938; s. Alfonso E. and Helen (Buscavage) P. A.B., Columbia U., 1959, A.M., 1960, Ph.D., 1962. Asst. prof. chemistry Stanford U., 1964-70, assoc. prof., 1971-78, prof., 1978—; vis. prof. U. Manchester, Eng., 1971, U. Nice, France, 1978. Author: Dynamic Light Scattering, 1976; contbr. numerous articles to profl. jours. Nat. Acad. Scis. postdoctoral fellow, Brussels, 1963. Fellow Am. Phys. Soc.; mem. Am. Chem. Soc., AAAS. Subspecialties: Physical chemistry; Polymer chemistry. Current work: Physical chemistry of liquids and polymers; fluctuation spectroscopy; dynamic light scattering. Office: Stanford U Dept Chemistry Stanford CA 94305

PEDDICORD, KENNETH LEE, nuclear science educator, researcher; b. Ottawa, Ill., Apr. 5, 1943; s. Kenneth Charles and Elizabeth (Hughes) P.; m. Patricia Ann Cullen, Aug. 2, 1969; children: Joseph, Clare. B.S. in Mech. Engring, U. Notre Dame, 1965; M.S., U. Ill., 1967, Ph.D., 1972. Research asst. U. Ill., Urbana, 1966-71; research nuclear engr. Swiss Fed. Inst. Reactor Research, Wuerenlingen, Switzerland, 1972-75; asst. prof. nuclear engring. Oreg. State U., Corvallis, 1975-78, assoc. prof. nuclear engring., 1978-82; vis. scientist Joint Research Ctr., Ispra, Italy, 1981-82; prof. nuclear engring. Tex. A&M U., College Station, 1983—; cons. EG&G Idaho Ins., Idaho Falls, 1979, Portland Gen. Electric, 1980; organizer and dir. Nuclear Tech. Workshop, 1976-81. Contbr. articles to profl. jours. Danforth Found. Assoc., 1973; recipient Oreg. State U. outstanding teaching award, 1980, faculty devel. award, 1973. Mem. Am. Nuclear Soc., Am. Soc. Engring. Edn., ASME, Pi Tau Sigma, Alpha Nu Sigma. Roman Catholic. Subspecialties: Nuclear engineering; Nuclear fission. Current work: Research and teaching in advanced nuclear fuels and reactor thermal hydraulics. Office: Dept Nuclear Engring TEX A&M U College Station TX 77843

PEDERSEN, LEE GRANT, chemistry educator; b. Oklahoma City, June 15, 1938; s. M.L. and N. (Shinn) P.; m. Barbara L. Barrett, Feb. 15, 1964; children: Kurt, Lars. B.Ch., U. Tulsa, 1958-61; Ph.D., U. Ark., 1965. Research assoc. dept. chemistry Columbia U., N.Y.C., 1965-66, Harvard U., Cambridge, Mass., 1966-67; prof. chemistry U. N.C.-Chapel Hill, 1967—. Co-author: Problems in Quantum Chemistry and Physics, 1974, Passing Freshman Chemistry, 1981. NSF fellow, 1965; NIH fellow, 1966; recipient Tanner teaching award U. N.C., 1970. Mem. Am. Phys. Soc., Sigma Xi (chpt. sec. 1977-79). Subspecialties: Theoretical chemistry; Physical chemistry. Current work: Application of quantum mechanics to chemical systems of interest to life scientists. Home: 308 Severin St Chapel Hill NC 27514 Office: Venable Hall U NC Dept Chemistry Chapel Hill NC 27514

PEDERSEN, NIELS CHRISTIAN, veterinary medicine educator, immunologist; b. Altadena, Calif., Oct. 28, 1943; s. Christian G. and Evelyn A. (Lund) P.; m. Geri Hutchings, Sept. 29, 1964; children: Stephanie, Holly, Collin, Megan. B.S., U. Calif.-Davis, 1965, D.V.M., 1967; Ph.D. with high commendation, Australian Nat. U., 1972. Intern small animal surgery, medicine Colo. State U., 1967-68; faculty U. Calif-Davis Sch. Vet. Medicine, 1972—, now prof. small animal medicine. Contbr. articles to profl. jours.; patentee in field. Recipient 1st award for sci. Cat Fanciers Assn., 1977, Outstanding Research award Purina Co., 1982. Mem. Am. Assn. Vet. Immunologists, Am. Soc. Viology, Acad. Vet. Allergists. Mormon. Subspecialties: Cancer research (veterinary medicine); Animal virology. Current work: Infectious and immunologic diseases of dogs and cats.

PEEK, JAMES MACK, physicist; b. Unionville, Mo., Sept. 5, 1933; s. Everett R. and Ramah I. (McNabb) P.; m. Marilyn May, Sept. 23, 1962; children: Cassandra L., Corinne Lee. B.S. in Edn, Western Ill. State Coll, Macomb, 1955; M.Sc., Ohio State U., 1958, Ph.D., 1962. Mem. tech. staff Sandia Nat. Labs., Albuquerque, 1962—. Author: Dissociation in Heavy Particle Collisions, 1972; contbr. articles in field to profl. jours. Trustee Nat. Kidney Found. N.Mex., 1973—; mem. exec. council, 1976-82, treas, 1977-79. Kennan Found. fellow, 1958-59; vis. fellow Joint Inst. Lab. Astrophys. U. Colo., 1971-72. Fellow Am. Phys. Soc.; mem. Am. Math. Soc., Am. Chem. Soc. Subspecialties: Atomic and molecular physics; Theoretical physics. Current work: Application of atomic and molecular theory to the modeling of material in extreme high temperature environments. Home: 724 Carlisle St SE Albuquerque NM 87106 Office: Division 1231 Sandia National Laboratories Albuquerque NM 87185

PEEK, LEON ASHLEY, psychology educator; b. DeLand, Fla., Mar. 31, 1945; s. Cecil McIntosh and Margaret Virginia (Taylor) P.; m. Roberta Harper Brent, Nov. 27, 1968; children: Jacob, Ashley, Margaret. B.S. in Psychology, Va. Commonwealth U., 1970, M.S. in Clin. Psychology, 1973, Ph.D. in Psychology, 1976. Lic. psychologist, Tex. Research assoc. Med. Coll. Va., Richmond, 1973-74; researcher Kirk, Buttler & Assocs., Richmond, 1973-75; assoc. prof. psychology N. Tex. State U., Denton, 1974—; cons. U.S. Dept. Agr., Denton, 1978, Escuela Desarrollo Infantil, Guadalajara, Mex., 1978, Med. U. S.C., Charleston, 1979. Cons. editor: Jour. Forensic Psychology, 1979-82; contbr. numerous articles to profl. publs. Bd. dirs. Selwyn Sch., Denton, 1978; NIH grantee, 1971; Tex. Dept. Hwys. grantee, 1977-79; Wacker Found. grantee, 1978; NSF fellow, 1973-76. Mem. Am. Psychol. Assn., Tex. Pediatric Soc. (environ. hazards com. 1978-80), Psi Chi. Democrat. Episcopalian. Subspecialties: Psychopharmacology; Cancer research (medicine). Current work: Psychological treatments in medical disorders, biofeedback, measurement in medical research. Home: 1121 Egan St Denton TX 76201 Office: N Tex State Univ Psychology Dept Denton TX 76203

PEELLE, HOWARD ARTHUR, computer science educator; b. Schenectady, Feb. 9, 1944; s. Robert B. and Gertrude (Maginniss) P.; m. Carolyn Curtiss, Sept. 9, 1967; children: Juliet, Jessica, Caleb, Mariah. B.S., Swarthmore Coll., 1965; Ed.D., U. Mass., 1971. Tchr. math. Oakwood Sch., Poughkeepsie, N.Y., 1965-67; researcher Sci. Research Assocs., Inc., Chgo., 1967-68; faculty mem. U. Mass., Amherst, 1971—, prof. edn. and computer sci., 1971—; cons. Computer Camps Internat., Vernon, Conn., 1981—; seminar leader Aetna Co., Hartford, Conn., 1981—. Author: APL - An Introduction, 1978, Learn to Solve Rubik's Cube, 1981, Instructional Applications of Computers, 1983. Recipient Heuristic Lab. award HEW, 1975; Tex. Instruments, Inc. grantee, 1983. Mem. Assn. Computing Machinery (Nat. Lectr. award 1975-76). Subspecialties: Learning; Programming languages. Current work: Instructional Applications of computers, APL. Office: Sch Edn U Mass Amherst MA 01003

PEERENBOOM, JAMES PETER, energy systems engineer; b. Appleton, Wis., Nov. 17, 1950; s. Robert Vincent and Rosalie (Diny) P.; m. Jane Roberts, Jan. 8. 1977; 1 dau., Katherine Ann. B.S. in Nuclear Engring, U. Wis.-Madison, 1973, M.S., 1974, Ph.D. in Land Resources, 1981. Research assoc. Oak Ridge Nat. Lab., 1974-76; cons. Resource Mgmt. Assocs., Madison, 1980-81; research asst. U. Wis.-Madison, 1977-81; energy systems engr. Argonne Nat. Lab., Ill., 1981—; fellow East-West Environ. and Policy Inst., Honolulu, 1980; vis. scientist Brookhaven Nat. Lab., Upton, N.Y., 1977, Internat. Inst. for Applied Systems Analysis, Laxenburg, Austria, 1977. Contbr. articles to profl. jours. Mem. Am. Nuclear Soc., Soc. for Risk Analysis, Tau Beta Pi. Subspecialties: Fuels and sources; Systems engineering. Current work: Utility systems analysis and applications of decision analysis techniques to energy/environmental decision problems. Home: 6091 Elm St Lisle IL 60532 Office: Argonne Nat Lab EES Bldg 362 9700 S Cass Ave Argonne IL 60439

PEES, SAMUEL THOMAS, exploration geologist, consultant; b. Meadville, Pa., Nov. 16, 1926; s. H. Chester and Dorothy Marie (Cook) P. B.S., Allegheny Coll., 1950; M.S., Syracuse U., 1959; postgrad., Colo. Coll., 1949, Tulsa U., 1968. Cert. petroleum geologist; cert. profl. geol. scientist. Exploration geologist Tex. Petroleum Co., Venezuela and Peru, 1953-61; fgn. exploration rep. Skelly Oil Co., S.Am. and Indonesia, 1962-68; exploration geologist Exploration Consortium, South Pacific, 1969-71; exploration cons. Patrick J. Delaney & Assocs., Rio de Janerio, Brazil, 1973-76; exploration project leader James A. Lewis Engrs., Tex. and, Argentina, 1976-78; sr. exploration geologist Samuel T. Pees & Assocs., Meadville, 1978—; sr. assoc. geologist Geoinformacao, Rio de Janeiro, 1976—; guest lectr. Allegheny Coll., 1971-76; cons. explorationist Pluspetrol S.A., Buenos Aires, Argentina, 1978; founder, donor S.T. Pees Caribbean Research Fund (Paleontol. Research Inst.), Ithaca, N.Y., 1977—. Contbr. articles to profl. jours. Collector Samuel T. Pees Art Collection, Allegheny Coll.; artists' sponsor Sanggar Ligar Sari, Bandung, Indonesia, 1969—; patron Meadville Arts Council, 1977. Served in U.S. Army, 1945-46; PTO, Korea. Fellow Geol. Soc. Am.; mem. Am. Assn. Petroleum Geologists, Am. Inst. Profl. Geologists, Pakontol. Research Am. Inst. (life), Am. Soc. Photogrammetry, AAAS, Malaysian Geol. Soc., Am. Legion, Sigma Alpha Epsilon, 749th Ry. Bn. Vets. Orgn., Crawford County Hist. Soc. Republican. Episcopalian. Subspecialties: Geology; Remote sensing (geoscience). Current work: Application of remote sensed morphotectonics in oil, gas and mineral exploration; basin mapping from satellite MSS imagery, return-beam vidicon and radar; determination of fractured zones via remote sensing; sub-surface mapping utilizing well data and remote sensed information, Appalachia Plateau (U.S.), Argentina and Brazil. Home: 889 Porter St Meadville PA 16335 Office: Masonic Bldg #224 Meadville PA 16335

PEGG, ANTHONY EDWARD, physiology educator; b. Matlock, Derbyshire, Eng., April 13, 1942; came to U.S., 1975; s. Edward William and Rose Lilian (Hollingbery) P.; m. Elizabeth Rosemary Heffer, Aug. 21, 1965; children: James, Jonathan. B.A., Cambridge Eng. (Eng.) U., 1963, M.A., 1965, Ph.D., 1966. Postdoctoral fellow Johns Hopkins U., Balt., 1966-69; sr. lectr. Middlesex Hosp., U. London, 1967-76; asso. prof. physiology Hershey (Pa.) Med. Center, Pa. State U. Coll. Medicine, 1975-77, prof. physiology, 1977—. Michael Sobell Fellow, 1969-75; Am. Heart Assn. grantee, 1976-81; others. Mem. John Hopkins U. Soc. Scholars, Am. Assn. for Cancer Research, Biochem. Soc., Am. Soc. Physiologists, Am. Soc. Biochemists. Subspecialties: Biochemistry (medicine); Cancer research (medicine). Current work: Biochemistry and function of polyamines, mechanism of cancer induction by nitrosamines. Office: Dept Physiology Hershey Med Center Hershey PA 17033

PEHLKE, ROBERT DONALD, materials and metallurgical engineering educator; b. Ferndale, Mich., Feb. 11, 1933; s. Robert William and Florence Jenny (McLaren) P.; m. Julie Anne Kehoe, June 2, 1956; children: Robert Donald, Elizabeth Anne, David Richard. B.S. in Engring, U. Mich., 1955; S.M., Mass. Inst. Tech., 1958, Sc.D., 1960; postgrad., Tech. Inst., Aachen, Ger., 1956-57. Registered profl. engr., Mich. Mem. faculty U. Mich., 1960—, prof. materials and metall. engring., 1968—, chmn. dept., 1973—; cons. to metall. industries. Author: Unit Processes of Extractive Metallurgy, 1973; Editor, contbr. numerous articles to profl. jours. Pres. Ann Arbor Amateur Hockey Assn., 1977-79. NSF fellow, 1955-56; Fulbright fellow, 1956-57. Fellow Am. Soc. Metals (mem. tech. divs. bd. 1982—, sec. metals acad. com. 1977); fellow AIME (Howe Meml. lectr. 1980); mem. Am. Inst. Metall. Engrs. (Gold medal sci. award extractive metallurgy div. 1976), Iron and Steel Soc. of AIME (Disting. life mem.; chmn. process tech. div. 1976-77, dir. 1976-79), Metall. Soc. of AIME, Germany, London, Japan socs. iron and steel, Am. Foundrymen's Soc., Am. Soc. Engring. Edn., N.Y. Acad. Sci., Nat. Soc. Profl. Engrs., Sigma Xi, Tau Beta Pi, Alpha Sigma Mu (pres. 1977-78). Subspecialties: Metallurgy; Materials. Current work: Metallurgical engineering, iron and steel making, extractive metallurgy, computer applications. Home: 9 Regent Dr Ann Arbor MI 48104 Office: 3062 Dow Bldg North Campus Univ Mich Ann Arbor MI 48109

PEI, RICHARD YUSIEN, engineer; b. Soochow, China, July 24, 1927; came to U.S., 1955; s. Tseziang and Shihju (Char) P.; m. Paula C. Cheng, Dec. 3, 1951; children: Gabriel, Raphael, Michael. B.S., U. l'Aurore-China, 1947, M.S., 1948; Ph.D., Rensselaer Polytech. Inst., 1964. Supr. Gen. Electric, Schenectady, 1956-62; mem. staff Neptune Research Lab., Wallingford, Conn., 1962-64, Bellcomm Inc., Washington, 1964-67, Inst. Def. Analyses, Arlington, Va., 1967-69; sr. engr. TRW Systems Group, McLean, Va., 1969-71, Rand Corp., Washington, 1971—; adj. faculty Cath. U. Am., Washington, 1964-67. Fellow AAAS; mem. ASME. Subspecialties: Oil shale; Operations research (mathematics). Current work: Operations research; energy science and technology. Home: 1474 Waggaman Circle McLean VA 22101 Office: Rand Corp 2100 M St NW Washington DC 20037

PEKALA, RONALD JAMES, clinical psychologist, researcher; b. Windber, Pa., July 15, 1952; s. John Stanley and Susan Elizabeth (Marcinko) P.; m. Debra Kay Rager, Feb. 14, 1976; 1 son, Michael James. B.S., Pa. State U., 1975; M.A., Mich. State U., 1978, Ph.D., 1981. Claims rep. Social Security Administrn., Johnstown, Pa., 1975-76; psychol. trainee Battle Creek (Mich.) VA Med. Ctr., 1978-79, Chillicothe (Ohio) VA Med. Ctr., 1979-80; psychology intern Pitts. VA Consortium, 1980-81; staff psychologist Coatesville (Pa.) VA Med. Ctr., 1981—; asst. prof. dept. psychiatry Jefferson Med. Coll., Phila., 1982—; co-dir. Biofeedback Clinic, Coatesville VA Med. Ctr., 1982—. Recipient Dotterer award Dept. Philosophy, Pa. State U., 1973; NIMH fellow, 1976, 77. Mem. Am. Psychol. Assn., Pa. Psychol. Assn. (assoc.), Am. Soc. Clin. Hypnosis, Assn. Transpersonal Psychology, Am. Soc. Psychical Research, Phi Beta Kappa, Phi Lambda Phi. Democrat. Roman Catholic. Subspecialties: Behavioral psychology; Cognition. Current work: Biofeedback, hypnosis, psychophenomenological approaches to consciousness, individual and group psychotherapy, cognition. Home: 182 Blackberry Ln Malvern PA 19355 Office: Coatesville VA Med Ctr Coatesville PA 19320

PEKARY, ALBERT EUGENE, research chemist, medical educator; b. Santa Monica, Calif., Aug. 13, 1943; s. Raymond Henry and Thelma (Yeager) P.; m. Jean Nicolas, Mar. 29, 1969; children: Dianne Nicolas, Leslie Ann. B.S., Stanford U., 1965; Ph.D., U. Calif.-Berkeley, 1970. Population Council fellow Calif. Inst. Tech., Pasadena, 1970, research assoc., 1970-72; postdoctoral fellow U. So. Calif. Sch. Medicine, Los Angeles, 1972-73; supervisory chemist VA Wadsworth Med Ctr., Los Angeles, 1973—; adj. asst. prof. dept. medicine UCLA, 1976-82, adj. assoc. prof., 1982—; computer cons. Harbor Gen. Hosp., Los Angeles, 1981. State of Calif. scholar, 1961; NIH biophysics tng. grantee, 1966; recipient performance award VA Wadsworth Med. Ctr., 1979. Mem. AAAS, Am. Fedn. Clin. Research, Endocrine Soc., Pacific Coast Fertility Soc. Democrat. Presbyterian. Subspecialties: Neuroendocrinology; Reproductive biology. Current work: Biosynthesis, structure and function of thyrotropin-releasing hormone (TRH) and TRH-like peptides in extrahypothalamic tissues of mammals especially the reproductive system. Office: VA Wadsworth Med Ctr Bldg 114 Rm 200 kWilshire and Sawtelle Blvds Los Angeles CA 90230

PELC, ROBERT EDWARD, psychologist, consultant; b. Flint, Mich., July 14, 1946; s. Jim Emil and Mary (Maxa) P.; m. Sharon Walsh, Sept. 12, 1980; children: Sean, Colum, Liam. B.S., Central Mich. U., 1968; M.A., U. Denver, 1973, Ph.D., 1975. Lic. psychologist, Colo. Program dir. Forensic Service Ctr., Denver, 1975-77; dir. Drug Abuse Treatment Program, Denver, 1977-79; clin. dir. Profl. Drug Treatment Program, Denver, 1979-80; psychologist, Denver, 1980—; mem. Colo. Bd. Psychol. Examiners, 1978-81. Author: Process of Clinical Assessment, 1982; contbr. articles to profl. jours.; columnist: Denver Bus. World. Bd. dirs. Wilderness Experience Program, 1972-75. Served with U.S. Army, 1968-70. NIMH fellow, 1971. Mem. Am. Psychol. Assn., Adminstrv. Mgmt. Soc., Am. Soc. Profl. Cons. Republican. Presbyterian. Subspecialties: Behavioral psychology; Applied psychology. Current work: Work and personal stress, business psychology, forensic psychology. Home: 475 Kearney St Denver CO 80220 Office: 1900 Wazee St Suite 260 Denver CO 80202

PELLERIN, CHARLES JAMES, JR., astrophysicist, research scientist; b. Shreveport, La., Dec. 11, 1944; s. Charles James and Joy Jefferson (Stewart) P.; m. Frances Kern, June 15, 1968; children: Julie, Charles. B.S. in Physics, Drexel U., 1967, M.S., Catholic U. Am., 1970, Ph.D., 1974. Research scientist Goddard Space Flight Ctr., Greenbelt, Md., 1970-74; staff scientist high energy astrophysics NASA Hdqrs., Washington, 1975, chief shuttle payloads analysis, 1976-79, dep. dir. spacelab flight div. astrophysics div., 1979-82, dir. astrophysics div., 1982—. Contbr. articles to profl. jours. Fellow Washington Acad. Scis.; mem. Am. Phys. Soc. Subspecialty: Astrophysics. Current work: Astronomy, solar physics, high energy astrophysics; director astrophysics programs. Patentee two axis fluxgate magnetometer. Office: 600 Independence Ave SW Washington DC 20546

PELLETIER, CHARLES A., engineer; b. New Britain, Conn., Jan. 10, 1932; s. Philip Hughes and Rose Alvina (Carlson) P.; m. Linda Jane Holland, Feb. 25, 1961; children: Philip, Julie. B.C.E., Rensselaer Poly. Inst., 1956; M.S. in Radiation Biology, U. Rochester, 1957; Ph.D. in Environ. Health, U. Mich., 1966. Diplomate: Am. Bd. Health Physics. Radiation control engr. Bethlehem Steel Co., Pa., 1958-60; asst. prof. U. Mich., Ann Arbor, 1960-66; br. chief U.S. AEC, Idaho Falls, Idaho, 1967-71, NRC, Bethesda, Md., 1971-73; v.p., dept. mgr. Sci. Applications Inc., Rockville, Md., 1973—. Contbr. articles to profl. jours.; patentee in field. Served with U.S. Army, 1953-55. Mem. Health Physics Soc., Am. Nuclear Soc., Soc. Risk Analysis, Sigma Xi. Republican. Subspecialties: Nuclear fission. Current work: Product and methods development, manufacturing and services for the electric power industry. Radioactive effluent measurement and control, waste management and condenser leak detection. Home: 5803 Greenlawn Dr Bethesda MD 20814 Office: Science Applications Inc 3 Choke Cherry Rd Rockville MD 20850

PELLETIER, KENNETH R., clinical psychologist, educator; b. Nashua, N.H., Apr. 27, 1946; s. Roger N. and Lucy B. (Leonetti) P.; m. Elizabeth Anne Berryhill, Oct. 26, 1980. B.A., U. Calif.-Berkeley, 1969, Ph.D., 1974; postgrad., C.G. Jung Inst., Zurich, Switzerland, 1970. Lectr. psychiatry U. Calif. Sch. Medicine, San Francisco, 1974-78, research psychologist, 1974-78, asst. clin. prof. psychiatry, 1978—, asst. clin. prof. internal medicine, 1980—; asst. prof. pub. health U. Calif.-Berkeley, 1980—. Author: Mind as Healer, Mind as Slayer, 1977, Toward a Science of Consciousness, 1978, Holistic Medicine: From Stress to Optimum Health, 1979, Longevity: Fulfilling Our Biological Potential, 1981, Healthy People in Unhealthy Places, 1983; editorial bd.: Am. Jour. Clin. Biofeedback, 1978—; assoc. editor: Med. Self Care, 1979—. Bd. dirs. Calif. Health and Med. Found., 1976—, Golden Gate Nat. Recreation Area, 1976-79; mem. Gov.'s Council on Wellness and Phys. Fitness, 1980—; bd. advisors Inst. Advancement of Health, N.Y.C. and San Francisco, 1981—. Woodrow Wilson fellow, 1969; USPHS fellow, 1972-74. Mem. Am. Psychol. Assn., Biofeedback Soc. Am., AAAS, Acad. Psychosomatic Medicine, Am. Holistic Med. Asn. (founding mem., editorial bd. 1981—), Phi Beta Kappa. Subspecialties: Psychophysiology; Neuroimmunology. Current work: Behavioral and holistic medicine; psychosomatic medicine; psychoneuroimmunology; neurophysiology; brain and consciousness; human longevity; optimum health. Office: Dept Internal Medicine Univ Calif Sch Medicine 400 Parnassus Ave A402 Francisco CA 94143

PELLETIER, LAWRENCE LEE, JR., physician, educator; b. Bangor, Maine, Dec. 26, 1942; s. Lawrence Lee and Louise (Collins) P.; m. Mary Beacon Rowland, June 6, 1968; children: Christina Louise, Anne Elizabeth, Caron Elise. B.A., Bowdoin Coll., 1964; M.D., Columbia U., 1968. Diplomate: Am. Bd. Internal Medicine. Resident in internal medicine U. Kans. Med. Ctr. (Kansas Ctr.), 1968-71, 73; fellow in infectious diseases U. Wash. Sch. Medicine, Seattle, 1973-75; asst. prof. medicine U. N.D. Sch. Medicine, Fargo, 1975-78, assoc. prof., 1978-79; assoc. prof. med. U. Wash. Sch. Medicine, Seattle, 1979-82; profl. medicine U. Kans. Sch. Medicine, Wichita, 1982—; chief med. service American Lake VA Med. Ctr., Tacoma, 1979-82, Wichita VA Med. Ctr., 1982—. Contbr. chpts. to textbooks, articles to profl. jours. Fellow ACP, Infectious Disease Soc. Am.; mem. Am. Venereal Disease Assn., N.Y. Acad. Scis., Assn. VA Chiefs of Medicine, Soc. Research and Edn. in Primary Care Internal Medicine, Am. Soc. Microbiology, Am. Fedn. Clin. Research, Soc. Exptl. Biology and Medicine, Sigma Xi, Phi Beta Kappa. Presbyterian. Subspecialties: Infectious diseases; Internal medicine. Current work: Infection in the aged. Infectious endocarditis. Strongyloidiasis. Home: 14509 Willowbend Circle Wichita KS 67230 Office: Med Service Wichita VA Med Center 5500 E Kellogg Wichita KS 67230 Office: Dept Internal Medicine U Kans Med Sch 1001 N Minneapolis Wichita KS 67241

PELLETT, HAROLD MELVIN, horticulture educator; b. Atlantic, Iowa, Feb. 17, 1938; s. Melvin Ambrose and Elizabeth (Dallinger) P.; m. Shelby Jean Anderegg, Feb. 28, 1960; children: Lori, Darwin, Loren, Linda, Deanne, Cindy. B.S., Iowa State U., 1960, M.S., 1961, Ph.D., 1964. Instr., grad. asst. Iowa State U., Ames, 1960-64; asst. prof. U. Nebr., Lincoln, 1964-66; from asst. prof. to prof. U. Minn., St. Paul, 1966—. Mem. Bd. Edn. Westonka Dist. 277 Schs., Mound, Minn., 1981—; mem. cable commn., City of Mound, 1982—, Minnetonka Cable Commn., 1982—. Served to capt. U.S. Army, 1960; with USAR, 1960-68. Mem. Internat. Plant Propagators, Am. Soc. for Hort. Sci. Subspecialties: Plant genetics; Plant physiology (agriculture). Current work: Breeding, culture and physiology of landscape plants. Home: 1450 Game Farm Rd N Mound MN 55364 Office: U Minn Landscape Arboretum 3675 Arboretum Dr Chanhassen MN 55317

PELLICANE, PATRICK JOSEPH, wood science educator; b. Elmhurst, N.Y., Jan. 24, 1951; s. Anthony James and Theresa Veronica (Adams) P.; m. Nancy Jean Fagan, Dec. 7, 1975; 1 dau., Amanda. B.S., CCNY, 1972; M.A., St. John's U., 1974; M.S., Colo. State U., 1978, Ph.D., 1980. Statis. analyst Automobile Ins. Plans Service Office, N.Y.C., 1974-76; instr. Colo. State U., Ft. Collins, 1979-80, asst. prof. wood sci., 1981—. Rotary Found. fellow, 1983. Mem. Forest Products Research Soc. (internat. chmn. tech. coms.), Soc. Wood Sci. and Tech., ASCE. Republican. Roman Catholic. Lodge: Rotary. Subspecialties: Materials (engineering); Statistics. Current work: Statistical prediction of material properties for use in probabilistic design of structures. Home: 3927 Windom St Fort Collins CO 80526 Office: Dept Forestry Colo State U Fort Collins CO 80526

PELLMAR, TERRY C., neurobiologist; b. Bklyn., Nov. 4, 1951; s. Ruben and Frances (Freilich) P.; m. Howard L. Leikin, Jan. 4, 1981. B.Sc., Brown U., 1973; Ph.D. in Physiology and Pharmacology, Duke U., 1977. Postdoctoral fellow dept. neurobiology Armed Forces Radiobiology Research Inst., Bethesda, Md., 1977-80, research physiologist dept. physiology, 1982—. Contbr. numerous articles to profl. jours. Postdoctoral fellow Lab. Preclin. Studies, Nat. Inst. Alcoholism and Alcohol Abuse, Rockville, Md., 1980-82. Mem. Neurosci. Soc., N.Y. Acad. Sci., Sigma Xi. Subspecialties: Neurobiology; Neurophysiology. Current work: Actions of neurotransmitters; synaptic mechanisms. Office: Dept Physiology AFRRI Bethesda MD 20814

PELTIER, EUGENE JOSEPH, civil engineer, former naval officer, bus. exec.; b. Concordia, Kans., Mar. 29, 1910; s. Frederick and Emma Helen (Brasseau) P.; m. Lena Evelyn Gennette, June 28, 1932; children: Marion Joyce, Eugene Joseph, Carole Josephine, Kenneth Noel, Judith Ann. B.S. in Civil Engring., Kans. State U., 1933, LL.D., 1961. Registered profl. engr., Mo., N.Y., Kans., Fla., Va., Calif. Commd. lt. (j.g.) U.S. Navy, 1936, advanced through grades to rear adm., 1957; asst. public works officer, Great Lakes, Ill., 1940-42, sr. asst. supt. civil engr., Boston, 1942-44, officer in charge 137th Constrn. Bn., Okinawa, 1945, officer in charge 54th Constrn. Regt., 1945, officer various public works assignments, Pensacola, Fla., 1945-46, Memphis, 1946-49, Jacksonville, Fla., 1949-51, dist. public works officer, 1951-53, asst. chief maintenance and materials Bur. Docks, Washington, 1953-56, comdg. officer, Pt. Hueneme, 1956-57, chief Bur. Yards and Docks, Navy Dept., Washington, 1957-62, chief of civil engrs., 1957-62, ret., 1962; instrumentman, resident engr. Kans. Hwy. Commn., Norton, Topeka, Chanute, 1934-40; v.p. Sverdrup & Parcel & Assos., Inc., St. Louis, 1962-64; sr. v.p. Sverdrup & Parcel & Assos., Inc., 1964-66, exec. v.p., 1966-67, pres., dir., 1967-75, chief exec. officer, 1972-75, ptnr., 1966-75, cons., 1975-82; pres., dir. Sverdrup & Parcel & Assos., N.Y., Inc., 1967-75; dir. Sverdrup & Parcel Internat., Inc., 1967-75; v.p., dir. ARO, Inc., Tullahoma, Tenn., 1966-75; cons. EPA, 1976-80; dir. Merc. Trust Co., St. Louis, 1971-81. Mem. emeritus Civic Progress, Inc.; bd. dirs. YMCA, St. Louis, 1972-76. Decorated Legion of Merit; recipient citation Am. Inst. Steel Constrn., 1973. Mem. ASCE (hon.), Am. Public Works Assn. (1 of Top Ten Public Works Men of Year 1960), Soc. Mil. Engrs. (pres. 1960-61), Am. Concrete Inst., Am. Road and Transp. Builders Assn. (pres. 1972-73), Nat. Soc. Profl. Engrs., Mo. Soc. Profl. Engrs., Public Works Hist. Soc. (pres. 1977-78, trustee 1975—), Nat. Acad. Engring., Cons. Engrs. Council (award of Merit 1962), Sigma Tau, Phi Kappa Phi. Clubs: Army-Navy Country (Washington); Old Warson Country (St. Louis). Subspecialty: Civil engineering. Current work: Engineering consultant, primarily on construction project. 800 N 12th St Saint Louis MO 63101

PELZER, CHARLES FRANCIS, geneticist, educator; b. Detroit, June 5, 1935; s. Francis J. and Edna Dorothy (Ladach) P.; m. Veronica A. Killeen, July 7, 1972; 1 dau., Mary Elizabeth. B.S., U. Detroit, 1957; Ph.D., U. Mich., 1964. Kettering Found. fellow Wabash Coll., Crawfordsville, Ind., 1965-66; instr. U. Detroit, 1966-68; asst. prof., then asso. prof., then prof. biology Saginaw Valley State Coll. University Center, Mich., 1969—; research asso. Mich. State U., East Lansing, 1976-77; research fellow Henry Ford Hosp., Detroit, 1982. Contbr. articles to profl. jours. Vice pres. Saginaw Valley Retinitis Pigmentosa Found., 1978-80. Recipient Alumni award Saginaw Valley State Coll., 1971; grantee NIH, 1961-64, Kellog Found., 1961, Saginaw Valley State Coll. Found., 1976-82. Mem. AAAS, Am. Soc. Human Genetics, N.Y. Acad. Scis., others. Subspecialties: Genetics and genetic engineering (medicine); Gene actions. Current work: Biochemical genetics, especially in humans and mammals. Gene action in health and disease. Analysis of genetic variant proteins by electrophoresis and isoelectric focusing. Office: Biology Dept Saginaw Valley State Coll University Center MI 48710

PENCE, DENNIS DALE, mathematician; b. Huntington, Ind., Feb. 28, 1948; s. Dale M. and Marjorie Ann (Williams) P.; m. Rebecca Sue McLaughlin, June 13, 1970; 1 son, Adam Gregory. B.S., Purdue U., 1970, M.S., 1972, Ph.D., 1976. Instr. Purdue U., West Lafayette, Ind., 1977; vis. asst. prof. Tex. A&M U., College Station, 1977—; asst.

scientist Math. Research Ctr., U. Wis., Madison, 1977-78; asst. prof. U. Vt., Burlington, 1978—. Mem. MAth. Assn. Am., Am. Math. Soc., Soc. Indsl. and Applied Math, Sigma Xi, Phi Kappa Phi. Mem. United Ch. Christ. Club: Burlington Oratorio Soc. (treas. 1980-82). Subspecialties: Applied mathematics; Numerical analysis. Current work: Approximation theory, especially spline interpolation and approximation; optimization; mathematics of space flight. Home: 44 Greenfield Rd Essex Junction VT 05452 Office: Dept Mat Univ Vt Burlington VT 05405

PENDLETON, J. DAVID, electronic engineer, research consultant; b. Stephensville, Tex., Jan. 12, 1945; s. Vern Odell and Verba Christine (Jones) P. B.S. in Mech. Engring, U. Tex.-Arlington, 1968; M.S. in Physics, Old Dominion U., Norfolk, Va., 1975, Ph.D., U. Tenn., Knoxville, 1982. Grad. teaching asst. Old Dominion U., 1973-75; grad. research asst. U. Tenn., Tullahoma, 1976-82; electronic engr. Naval Electronic Systems Command, Washington, 1982—. Served with U.S. Army, 1968-70; Korea. Mem. Optical Soc. Am., Soc. Indsl. and Applied Math., Am. Assn. Physics Tchrs. Subspecialties: Particle sizing instrument simulation; Electromagnetic wave scattering theory. Current work: Developing simulation algorithms and codes for instruments which measure the size and shape of atmospheric and biological particles by laser light scattering. Home: 250 S Van Dorn N-419 Alexandria VA 22304 Office: Naval Electronics Systems Command ELEX 621 NAVELEX Washington DC 20363

PENG, ANDREW C(HUNG-YEN), horticulture educator, researcher, food technologist; b. Peiping, Hepei, China, Feb. 14, 1924; came to U.S., 1958; s. Ming C. and Pao Y. (Lu) P.; m. Louise Moe, Feb. 27, 1964; children: Laura C., Cora C. B.S. with honors, Wash. State U., 1961; M.S. in Food Sci, Mich. State U., 1962, Ph.D., 1965. Project leader Swift & Co. Research and Devel. Ctr., Chgo., 1965-67; asst. prof. Ohio State U., Columbus, 1968-72, assoc. prof., 1972-78, prof., 1978—. Mem. Inst. Food Technologists, Am. Oil Chemists Soc. (MacGee award 1965), Am. Assn. Cereal Chemists. Subspecialty: Food science and technology. Current work: Technology of vegetable fats and oils, plant proteins, thermal processing of canned foods; research on fruit and vegatable lipids and utilization of vegetable proteins. Patentee milk substitute, soy-cheese whey curd. Office: Ohio State Univ 2001 Fyffe Ct Columbus OH 43210

PENG, YING-SHIN CHRISTINE, entomology educator; b. Harbin, China, Dec. 26, 1945; came to U.S., 1970, naturalized, 1981; d. Fang and Wan-Chen (Tsai) Liu; m. Raymond Peng, Sept. 7, 1974; children: Phyllis, Leslie. B.Sc., Nat. Taiwan U., Taipei, 1968; Ph.D., U. Man., Winnipeg, 1973. Research assoc. Fisheries Research Inst., Taipei, 1969-70; research asst. U. Man., 1970-73; asst. prof. entomology U. Calif.-Davis, 1974-81, assoc. prof., 1981—. NRC Can. fellow, 1973-74. Mem. Internat. Bee Research Assn., Entomol. Soc. Am., Entomol. Soc. Can., Internat. Assn. Social Insect. Subspecialties: Animal physiology; Plant cell and tissue culture. Current work: Honey bee nutrition, digestion and absorption, reproductive physiology BGE semen storage, artificial insemination of queen bees. Office: Dept Entomology U Calif Davis CA 95616

PENHOS, JUAN CARLOS JACOBO, physiology and biophysics educator; b. Buenos Aires, Argentina, Feb. 12, 1918; came to U.S., 1964, naturalized, 1970; s. Salomon and Sara (Bochi) P.; m. Elena Nusbaum, Oct. 28, 1944; 1 son, Juan C. B.A., Nat. U., Buenos Aires, 1935, M.D., 1942, Endocrinologist, 1952, Physiologist, 1955. Intern Instituto Modelo, Hosp. Rawson, Buenos Aires, 1940-41, resident, 1942-46; chief research endocrinologist Inst. Biol. Medicine Exptl., Buenos Aires, 1952-64; assoc. prof. medicine N.Y. Med. Coll., N.Y.C., 1964-67, George Washington U., Washington, 1968-80; prof. physiology and biophysics Georgetown U. Med. Ctr., Washington, 1970—; diabetes research cons. VA Med. Ctr., Washington, 1970—. Argentine Assn. Advancement Sci. fellow, 1953-54; Lederle Internat. fellow Cyanamid Co., 1960-62; Career Scientist award Nat. Council Sci. and Tech. Research, 1962-63. Mem. Am. Diabetes Assn., Endocrine Soc., Am. Physiol. Soc., Soc. Exptl. Biology and Medicine (pres. D.C. chpt. 1975), N.Y. Acad. Scis., Sigma Xi. Subspecialties: Endocrinology; Physiology (medicine). Current work: Hormones, carbohydrates, lipid and protein metabolism, experimental diabetes, studies on prostaglandins and steroids. Home: 5402 Surrey St Chevy Chase MD 20815 Office: Georgetown U Med Center Dept Physiology and Biophysics Washington DC 20007

PENICK, ELIZABETH CARNEL, clinical psychologist, educator, researcher; b. New Orleans, July 17, 1934; d. Rawley Martin and Marie (Sells) P. B.A., Sophie Newcomb Coll., 1957; postgrad., U. London, 1958; M.Sc., Tulane U., 1960; Ph.D., Washington U., St. Louis, 1975. Cert. psychologist, Kans. Research asst. Washington U., 1960-62, intern in clin. psychology, 1962-63, instr., 1966-74; dir. psychology lab. St. Louis Children's Hosp., 1964-68; pvt. practice psychology, cons. to local schs., St. Louis, 1968-69; clin. psychologist III, coordinator Community Psychol. Services, Malcolm Bliss Mental Health Ctr., St. Louis, 1969-74; research project dir. dept. psychiatry U. Ky.-Lexington, 1974-77; asst. prof. dept. psychiatry U. Ky. Med. Ctr., 1976-77; research project dir. Research Service, VA Hosp., Kansas City, Mo., 1977-78, coordinator Alcohol Dependency Treatment Unit, 1978-80; assoc. prof. psychiatry Kans. U. Med. Ctr., Kansas City, 1977-82, prof., 1982—, dir. div. psychology, 1980—; assoc. Gerontology Ctr. U. Kans., Kansas City, 1981—; mem. adv. bd. Home Study Project, Central Inst. for Deaf, St. Louis, 1969-74; mem. profl. adv. bd. Judevine Found. for Autistic Children, St. Louis, 1970-74; mem. adv. bd. Head Start, Health Services, Human Devel. Corp., St. Louis, 1972-74; invited mem. Lt. Gov.'s Task Force on Mental Health Problems in Times of Disaster, 1973-74; mem. Kansas City Nat. Council on Alcoholism, 1978—. Contbr. articles to profl. jours. Recipient prize in psychology Chi Omega, 1957; NIMH research fellow and research trainee Washington U. Sch. Medicine, 1963-64; Spl. Office Drug and Alcohol Programs grantee, 1974-77; Nat. Inst. Drug Abuse grantee, 1978-82; VA merit rev. research grantee, 1981-85. Mem. Am. Psychol. Assn., Evaluation Research Soc., Kans. Psychol. Assn., N.Y. Acad. Sci., Soc. Behavioral Medicine, Sigma Xi. Subspecialty: Behavioral psychology. Current work: Diagnosis, alcoholism, treatment evaluation. Home: 4725 Black Swan Dr Shawnee KS 66216 Office: Dept Psychiatry Kans U Med Center 39th and Rainbow Blvd Kansas City KS 66103

PENISTON, EUGENE G., clinical psychologist; b. Osceola, Iowa, June 23, 1931; s. James Milton and Delia B. (Young) P.; m. Helen M. Kerr, Oct. 16, 1959; children: Denise R., Eugene Lyle. A.B., Central State U., Wilberforce, Ohio, 1953; M.A., S.D. State U., 1962; Ed.D., Okla. State U., 1972. Lic. psychologist. Chief sch. psychologist Clarance (N.Y.) public schs., 1965-69; chief psychol. service and tng. center Va. State Hosp., Petersburg, 1972-75; assoc. prof. Va. State U., Petersburg, 1975-76; mental health cons. USPHS, Roosevelt, Utah, 1976-79, cons. clin. psychologist, Redfield, S.D., 1979-81; clin. psychologist VA Med. Center, Ft. Lyon, Colo., 1981—. Contbr. articles to profl. jours. Served to lt. U.S. Army, 1953-55. Mem. Am. Psychol. Assn., Am. Ednl. Research Assn., Phi Delta Kappa. Democrat. Lutheran. Subspecialties: Behavioral psychology; Neuropsychology. Current work: Behavior therapy; memory; cerebral dysfunction; betaendorphins (neuropsychology); mental retardation (habilitation) deinstitutionalization; alcoholism, biofeedback training and relaxation training. Home: 1919 Cimarron Ave La Junta CO 81050 Office: VA Med Center Fort Lyon CO 81038

PENNEBAKER, JAMES WHITING, psychologist, researcher; b. Midland, Tex., Mar. 2, 1950; s. William F. and Elizabeth (Whiting) P.; m. Ruth Burney, Dec. 30, 1972; 1 dau., Catherine Teal. Student, U. Ariz., 1968-70; B.A., Eckerd Coll., 1970-72; Ph.D., U. Tex-Austin, 1977. Asst. prof. dept. psychology U. Va., Charlottesville, 1977-83; adj. asst. prof. Int. Clin. Psychology, U. Va., 1980-83; assoc. prof. So. Meth. U., Dallas, 1983—; cons. FBI, Quantico, Va., 1982—. Author: The Psychology of Physical Symptoms, 1982; editor: Mass Psychogenic Illness, 1982. NIH grantee, 1981-84. Mem. Soc. Psychophysiol. Research, Am. Psychol. Assn. Subspecialties: Social psychology; Health psychology. Current work: Research and teaching in medical psychology, primary work deals with how individuals perceive physical symptoms and sensations relative to physiological activity, interests also include stress, psychosomatics, emotions and deception. Office: Dept Psychology So Meth U Dallas TX 75275

PENNER, STANFORD SOLOMON, educator; b. Unna, Germany, July 5, 1921; came to U.S., 1936, naturalized, 1943; s. Heinrich and Regina (Saal) P.; m. Beverly Preston, Dec. 28, 1942; children: Merrilynn Jean, Robert Clark. B.S., Union Coll., 1942; M.S., U. Wis., 1943, Ph.D., 1946; Dr. rer. nat. (hon.), Technische Hochschule Aachen, W. Ger., 1981. Research asso. Allegany Ballistics Lab., Cumberland, Md., 1944-45; research scientist Standard Oil Devel. Co., Esso Labs., Linden, N.J., 1946; sr. research engr. (Jet Propulsion Lab.), Pasadena, Calif., 1947-50; mem. faculty Calif. Inst. Tech., 1950-63, prof. div. engring., jet propulsion, 1957-63; dir. research engring. support div. Inst. Def. Analyses, Washington, 1962-64; prof. engring. physics, chmn. dept. aerospace and mech. engring. U. Calif. at San Diego, 1964-68, vice chancellor for acad. affairs, 1968-69, dir., 1968-71, 1973—; U.S. mem. adv. group aero. research and devel. NATO, 1952-68, chmn. combustion and propulsion panel, 1958-60; mem. adv. com. engring. scis. USAF-Office Sci. Research, 1961-65; mem. subcom. combustion NACA, 1954-58; research adv. com. air-breathing engines NASA, 1962-64; mem. coms. on gas dynamics and edn. Internat. Acad. Astronautics, 1969-80; mem. coms. NRC; cons. to govt., univs. and industry, 1953—; chmn. NRC/U.S. com. Internat. Inst. Applied Systems Analysis, 1978-81; nat. Sigma Xi lectr., 1977-79. Author: Chemical Reactions in Flow Systems, 1955, Chemistry Problems in Jet Propulsion, 1957, Quantitative Molecular Spectroscopy and Gas Emissivities, 1959, Chemical Rocket Propulsion and Combustion Research, 1962, Thermodynamics, 1968, Radiation and Reentry, 1968; sr. author: Energy, Vol. I (Demands, Resources, Impact, Technology and Policy), 1974, 81, Energy, Vol. II (Non-nuclear Energy Technologies), 1975, 77, 83, Energy, Vol. III (Nuclear Energy and Energy Policies), 1976; Editor: Chemistry of Propellants, 1960, Advanced Propulsion Techniques, 1961; Asso. editor: Jour. Chem. Physics, 1953-56; editor: Jour. Quantitative Spectroscopy and Radiative Transfer, 1960—, Jour. Missile Def. Research, 1963-67, Energy, The internat. Jour, 1975—. Recipient spl. awards People-to-People Program, NATO, pub. service award U. Calif. San Diego; N. Manson medal Internat. Colloquia on Gasdynamics of Explosions and Reactive Systems.; Internat. Columbus award Internat. Inst. Communications, Genoa, Italy; Guggenheim fellow, 1971-72. Fellow Am. Phys. Soc., Optical Soc. Am., AAAS, N.Y. Acad. Scis., AIAA (dir. 1964-66, past chmn. com., G. Edward Pendray award 1975, Thermophysics award 1983, Energy Systems award 1983), Am. Acad. Arts and Scis.; mem. Nat. Acad. Engring., Internat. Acad. Astronautics, Am. Chem. Soc., Combustion Inst., Sigma Xi. Subspecialties: Fuels and sources; Combustion processes. Current work: Aerospace engineering, synthetic fuels, propellants and combustion, laser spectroscopy,atomic and molecular physics. Home: 5912 Ave Chamnez La Jolla CA 92037 Office: U Calif San Diego CA 92093

PENNEY, DAVID GEORGE, physiology researcher and educator; b. Detroit, Jan. 11, 1940; s. George Donald and Gertrude Ellen (Goodhew) P.; m. Margaret White McMath, June 29, 1963 (div. 1975); children: Loren David, Morgan Donald; m. Susan Vera Carter, Aug. 13, 1978; 1 dau., Elizabeth Caroline Carter-Penney. B.S. in Biology, Wayne State U., 1963; M.S. in Zoology, UCLA, 1966, Ph.D., 1969. Asst. prof. U. Ill.-Chgo., 1969-75, assoc. prof., 1975-77; assoc. prof. dept. physiology Wayne State U. Sch. Medicine, Detroit, 1977—. Contbr. chpts. to books, articles to profl. jours. Mem. Am. Physiol. Soc., Internat. Heart Research Soc. Subspecialty: Physiology (medicine). Current work: Response of the cardiovascular system to carbon monoxide; processes of cardiomegaly; cellular, hematologic, hemodynamic, morphologic. Home: 1403 Sunset Blvd Royal Oak MI 48201 Office: Wayne State Univ Sch Medicine 540 E Canfield Detroit MI 48067

PENNINGTON, WAYNE D(AVID), geology educator, researcher; b. Rochester, Minn., Dec. 19, 1950; s. Stanley Robert and Shirley (Heckart) P.; m. Laura Leitermann, Dec. 23, 1978; 1 son, Matthew. A.B., Princeton U., 1972; M.S., Cornell U., 1976; Ph.D., U. Wis.-Madison, 1979. Asst. prof. geol. scis. U. Tex.-Austin, 1979—. Mem. Am. Geophys. Union, Seismol. Soc. Am., Geol. Soc. Am., Soc. Exploration Geophysicists, Sigma Xi. Subspecialties: Geophysics; Tectonics. Current work: Seismicity, tectonics, induced earthquakes. Office: Dept Geol Scis U Tex Austin TX 78712

PENNISTON, JOHN THOMAS, biochemistry educator, researcher; b. St. Louis, Sept. 10, 1935; s. Alonzo Schofield and Esther Rosella (Thomas) Pennsiton; m. Joyce Carolyn Kendall, June 11, 1960; children: Sarah Constance, Mary Grace. A.B., Harvard U., 1957, A.M., 1959, Ph.D., 1962. Vis. asst. prof. chemistry Pomona Coll., Claremont, Calif., 1962-63; research fellow Enzyme Inst., U. Wis.-Madison, 1964-66, asst. prof., 1966-71; assoc. prof. chemistry U. N.C.-Chapel Hill, 1971-76; assoc. prof. biochemistry Mayo Clinic, Rochester, Minn., 1976-79, prof. biochemistry, 1979—; mem. phys. biochemistry study sect. NIH, 1982-86. Contbr. numerous articles to prof. jours. NIH grantee, 1980-85, 1978-81, 1974-83. Mem. Am. Soc. Biol. Chemists, Am. Chem. Soc., Fedn. Am. Scientists, AAAS. Democrat. Presbyterian. Club: Harvard of Minn. Subspecialties: Biochemistry (biology); Membrane biology. Current work: Biological membranes and biochemical aspects of transport, red blood cells, Ca2 as an intracellular messenger of hormones, plasma membrane Ca2 pumps. Home: 835 10-1/2 St SW Rochester MN 55902 Office: Mayo Clinic 200 First St SW Rochester MN 55905

PENSE, ALAN WIGGINS, metallurgical engineer; b. Sharon, Conn., Feb. 3, 1934; s. Arthur Wilton and May Beatrice (Wiggins) P.; m. Muriel Drews Taylor, June 28, 1958; children—Daniel Alan, Steven Taylor, Christine Muriel. B.Metall. Engring., Cornell U., 1957; M.S., Lehigh U., 1959, Ph.D., 1962. Research asst. Lehigh U., Bethlehem, Pa., 1957-59, instr., 1960-62, asst. prof., 1962-65, asso. prof., 1965-71, prof., 1971—, chmn. dept. metallurgy and materials engring., 1977-83; cons. adv. com. on reactor safeguards NRC, 1965—. Author: (with R.M. Brick and R.B. Gordon) Structure and Properties of Engineering Materials, 4th edit, 1978; also articles. Recipient Robinson award Lehigh U., 1965, Stabler award, 1972; Danforth fellow, 1974—. Mem. Am. Soc. Metals, Internat. Inst. Welding, Am. Welding Soc. (William Spraragan award 1963, Adams membership award 1966, Jennings award 1970, William Hobart medal 1982, Adams lectr. 1980, William Hobart medal 1982), ASTM, Am. Soc. Engring. Edn. (Western Elec. award 1968). Republican. Evang. Congregationalist (pres. bd. trustees Evang. Sch. Theology). Subspecialties: Alloys; Nuclear fission. Current work: Metallurgy and mechanical properties of high strength steels; fracture mechanics; cryogenic alloys; welding of metals. Home: 2227 West Blvd Bethlehem PA 18017 Office: Lehigh U 5 Bethlehem PA 18015 Achievement of significant goals in our life must be balanced by the quality of that life itself, for what we are is as important as what we do.

PENTECOST, JOSEPH LUTHER, ceramic engineer; b. Winder, Ga., Apr. 23, 1930; s. Joseph Edwin and Vera (Sims) P.; m. Joanne Jackson, July 30, 1949 (dec.); children: David Edwin, Stephen Joseph, Richard Jackson, Brian Alan, Madge Annette; m. Maxene L. Meyer, Jan. 3, 1969. B.Ceramic Engring., Ga. Inst. Tech., 1951; M.S., U. Ill., 1954, Ph.D. (Owens Corning fellow), 1956. Sr. engr., later dir. research Melpar Inc., Falls Church, Va., 1956-59, 62-68; chief research engr. Aeronca Mfg. Corp., Middletown, Ohio, 1959-61; research engr., supr. W.R. Grace Co., Clarksville, Md., 1968-72; prof., dir. Sch. Ceramic Engring., Ga. Inst. Tech., Atlanta, 1972—. Contbr. articles to profl. jours., chpts. to books. Served with USMC, 1951-53. Recipient PACE award, 1967. Mem. Nat. Inst. Ceramic Engrs. (pres. 1976-77), Am. Ceramic Soc. (trustee 1980-82, v.p. 1982-83), Am. Soc. Metals, Am. Radio Relay League. Republican. Presbyterian. Subspecialty: Ceramic engineering. Patentee in field. Home: 3605 Clubwood Trail Marietta GA 30067 Office: Sch Ceramic Engring Ga Inst Tech Atlanta GA 30332

PENTO, J. THOMAS, pharmacologist, educator, cons.; b. Masontown, Pa., Sept. 1, 1943; s. Joseph and Pearl (Marchando) P.; m. Maureen Mae Daley, July 13, 1946; children: Kelly Lynn, Christopher Thomas. Student, Waynesburg Coll., 1961-62; B.A., W.Va. U., 1965, M.S., 1967; Ph.D., U. Mo., Columbia, 1970. Postdoctoral fellow Maimonides Med. Center, N.Y.C., 1970-71; from asst. prof. to prof. and sect. chief Coll. Pharmacy, U. Okla., Oklahoma City, 1971—; also cons. Contbr. articles to proft. jours. NDEA fellow, 1967-69; NSF grantee, 1974-78. Mem. Endocrine Soc., Am. Soc. Pharmacology and Exptl. Therapeutics, Am. Soc. Bone and Mineral Research, AAAS, Gideon Soc. Methodist. Subspecialties: Pharmacology; Endocrinology. Current work: Endocrine pharmacology; calcium metabolism; hormone secretion; antiestrogen mechanisms. Home: 2217 Forister Ct Norman OK 73069 Office: 1110 N Stonewall Ave Oklahoma City OK 73190

PENZIAS, ARNO ALLAN, astrophysicist, electronics company executive; b. Munich, Germany, Apr. 26, 1933; came to U.S., 1940, naturalized, 1946; s. Karl and Justine (Eisenreich) P.; m. Anne Pearl Barras, Nov. 25, 1954; children: David Simon, Mindy Gail, Laurie Ruth. B.S., CCNY, 1954; M.A., Columbia U., 1958, Ph.D., 1962; Dr. honoris causa, Observatoire de Paris, 1976; Sc.D. (hon.), Rutgers U., 1979, others. Mem. tech. staff Bell Labs., Holmdel, N.J., 1961-72, head radiophysics research dept., 1972-76, dir. radio research lab., 1976-79, exec. dir. research, communications scis. div., 1979-81, v.p. research, 1981—; lectr. Princeton U., 1967-72, vis. prof., 1972—; research asso. Harvard Coll. Obs., Cambridge, Mass., 1968-80; adj. prof. SUNY, Stony Brook, 1974—; Edison lectr. U.S. Naval Research Lab., 1979; Kompfner lectr. Stanford U., 1979; Gamow lectr. U. Colo., 1980; mem. astronomy adv. panel NSF, 1978-79; trustee Trenton State Coll., 1977-79; mem. vis. com. Calif. Inst. Tech., 1977-79, Max Planck Inst., 1978—; mem. Fachbeirat, 1978-80, chmn., 1981-83. Mem. editorial bd.: Ann. Rev. Astronomy and Astrophysics, 1974-78; asso. editor: Astrophys. Jour, 1978-82; contbr. numerous articles to profl. jours. Served to lt. Signal Corps U.S. Army, 1954-56. Recipient Henry Draper medal Nat. Acad. Sci., 1977; Herschel medal Royal Astronom. Soc., 1977; Nobel prize in physics, 1978. Mem. Nat. Acad. Scis., Am. Astron. Soc., Am. Acad. Arts and Scis., Am. Phys. Soc., Internat. Astron. Union, Com. Concerned Scientists (vice chmn.). Republican. Jewish. Subspecialties: Cosmology; Planetary science. Current work: Recent work includes study of chemical molecules in outer space, with particular emphasis on how elements in these molecules are formed, as well as studies of structure of our galaxy, the Milky Way. Research on cosmology, interstellar molecules, astrophysics, communication techniques. Office: Murray Hill NJ 07924

PENZIEN, JOSEPH, educator; b. Philip, S.D., Nov. 27, 1924; s. John Chris and Ella (Stebbins) P.; m. Jeanne Ellen Hunson, Apr. 29, 1950; children—Robert Joseph, Karen Estelle, Donna Marie, Charlene May. Student, Coll. Idaho, 1942-43; B.S., U. Wash., 1945; Sc.D., Mass. Inst. Tech., 1950. Mem. staff Sandia Corp., 1950-51; sr. structures engr. Consol. Vultee Aircraft Corp., Fort Worth, 1951-53; asst. prof. U. Calif. at Berkeley, 1953-57, asso. prof., 1957-62, prof. structual engring., 1962—; dir. Earthquake Engring. Research Center, 1968-73, 77—; Cons. engring. firms; chief tech. adv. Internat. Inst. of Seismology and Earthquake Engring., Tokyo, Japan, 1964-65. NATO Sr. Sci. fellow., 1969. Fellow Am. Acad. Mechanics; mem. ASCE (Walter Huber Research award), Am. Concrete Inst., Structural Engrs. Assn. Calif., Seismol. Soc. Am., Nat. Acad. Engring. Subspecialty: Civil engineering. Current work: Research in earthquake engineering with emphasis on seismic performance of civil engineering structures. Home: 800 Solana Dr Lafayette CA 94549 Office: Davis Hall Univ Calif Berkeley CA 94720

PEPELKO, WILLIAM EDWARD, toxicologist; b. Youngstown, Ohio, June 6, 1934; s. Edward and Mary Mihoci (Campbell) P.; m. Barbara Carol Bollaert, Aug. 25, 1962; children: Anne, Daniel, Douglas, Katherine. B.S., Pa. State U., 1958; M.S., Mich. State U., 1958-60; Ph.D., U. Calif.-Davis, 1964. Physiologist USAF, San Antonio, 1964-74, U.S. EPA, Cin., 1974-78, toxicologist, 1978—. Editor: Health Effects of Diesel Engine Emission, 1980. Mem. Am. Physiol. Soc., Soc. Exptl. Biology and Medicine, Am. Coll. Toxicology, Sigma Xi, Am. Contract Bridge League. Democrat. Roman Catholic. Subspecialties: Toxicology (medicine); Animal physiology. Current work: Responsible for the preparation of health risk assessment documents for environmental pollutants. Home: 10386 Lochcrest Dr Cincinnati OH 45231 Office: US EPA 26 W Saint Clair St Cincinnati OH 45268

PEPINE, CARL JOHN, physician, educator; b. Pitts., June 8, 1941; s. Charles John and Elizabeth (Hovan) P.; m. Lynn Divers, Aug. 3, 1963; children: Mary Lynn, Anne, Elizabeth. B.S., U. Pitts., 1962; M.D., N.J. Coll. Medicine, 1966. Intern Allegheny Gen. Hosp., U. Pitts., 1966-67; resident in internal medicine Jefferson Med. Coll. Hosp., Phila., 1967-68, fellow in physiology and cardiovascular disease, 1969-71; asst. prof. medicine Jefferson Med. Coll., Phila., 1971-74, U. Fla., Gainesville, 1972-75, assoc. prof., 1975-79, prof., 1979—, dir. cardiovascular research, 1974—, assoc. dir. div. cardiovascular medicine, 1982—; dir. cardiology catheterization lab. Shands Hosp., U. Fla., Gainesville, 1974—; dir., chief cardiology VA Regional Med. Ctr., Gainesville, 1974—. Mem. editorial bd.: Circulation; mem. editorial bd. Am. Jour. Cardiology, Jour. Am. Coll. Cardiology, Cardiac Catherization, Cardiovascular Diagnosis; contbr. articles to profl. jours.; developer catheters to measure blood flow and heart circulation. Served to comdr. USN, 1968-74. VA grantee, 1975—; Am. Heart Assn. grantee, 1977—. Fellow Am. Coll. Cardiology, Am. Heart Assn. (council on clin. cardiology and on circulation), Am. Fedn. Clin. Research, Soc. Cardiac Angiography, Am. Soc. Clin. Investigation; mem. N.Y. Acad. Scis., AAAS, Phi Kappa Alpha. Democrat. Roman Catholic. Subspecialties: Cardiology; Laser medicine. Current work:

Dynamics of the coronary circulation; effects of coronary artery spasm and stenosis; effects of lasers on blood vessels. Office: U Fla 1400 SW Archer Rd Box J-277 Gainesville FL 32610

PEPLINSKI, DANIEL RAYMOND, physical chemist; b. Chgo., Sept. 23, 1951; s. Alexander J. and Florence B. P. B.S., Loyola U., Chgo., 1973; M.S., Ill. Inst. Tech., 1977. Research asst. ADA, Chgo., summer 1978; mem. tech. staff The Aerospace Corp., El Segundo, Calif., 1978—. Contbr. articles to profl. jours. Mem. Am. Chem. Soc., Am. Phys. Soc. Republican. Roman Catholic. Subspecialties: Atomic and molecular physics; Kinetics. Current work: Molecular beam research, spacecraft contamination, gas surface studies, space environment simulation. Home: 1864 W 260th St Lomita CA 90717 Office: PO Box 9295 M2/271 Los Angeles CA 90009

PEPPAS, NIKOLAOS ATHANASSIOU, chemical engineering educator, consultant; b. Athens, Greece, Aug. 25, 1948; came to U.S., 1971; s. Athanassios Nikolaou and Alice Petrou (Roussopoulou) P. Diploma Chem. Engring., Nat. Tech. U. Athens, Greece, 1971; Sc.D. in Chem. Engring, MIT, 1973. Profl. engr., Greece.; Research assoc. chem. engring. MIT, Cambridge, 1975-76; asst. prof. chem. engring. Purdue U., West Lafayette, Ind., 1976-78, assoc. prof., 1978-82, prof. chem. engring., 1982—; vis. prof. faculty scis. U. Geneva, 1972-83; vis. prof. chem. engring. Calif. Inst. Tech., Pasadena, 1983; cons. four polymer and pharm. cos., 1976—. Editor: Jour. Biomaterials, 1982—; cons. editor: Jour. Polymer News, 1979—; editorial bd.: Jour. Applied Polymer Sci, 1976—; Author: five books including Biomaterials: Interfacial Phenomena and Applications, 1982; contbr. articles to profl. publs. Served to 2d lt. Greek Army, 1973-75. Recipient Zyma Found. award, Nyon, Switzerland, 1982; A.A. Potter award Purdue U., 1978; R.N. Shreve award, 1978, 80, 82. Mem. Am. Inst. Chem. Engrs. (Best Counselor award 1982), N.Y. Acad. Scis., Am. Chem. Soc. (chmn. program com. indsl. chemistry div. 1982), Am. Phys. Soc., Am. Soc. Artificial Internal Organs, Internat. Soc. Artificial Organs, Soc. Biomaterials, AAAS, Soc. Rheology, Soc. Engring. Sci., Soc. Controller Release, Am. Soc. Engring. Edn. (Western Electric Fund award 1980), European Soc. Membrane Soc., Soc. Plastics Engrs., Sigma Xi. Greek Orthodox. Subspecialties: Polymer engineering; Chemical engineering. Current work: Diffusion in polymers, macromolecular structure, biomaterials, biomedical engineering, transport phenomena, membrand science. Patentee. Home: 2501 Soldiers Home Rd West Lafayette IN 47906 Office: Sch Chem Engring Purdue Univ West Lafayette IN 47907

PERCICH, JAMES ANGELO, plant pathologist educator; b. Detroit, Sept. 14, 1944; s. Peter James and Virginia Constance (Ceriotti) P.; m. April Lynn Narcisse, June 12, 1982. B.Sc., Mich. State U., 1965; M.S., 1970; Ph.D., 1974. Postdoctoral research fellow U. Wis., Madison, 1974-76; research fellow dept. plant pathology U. Minn., St. Paul, 1976-77, asst. prof., 1977—. Mem. AAAS, Am. Agronomy Soc., Am. Phytopath. Soc., Can. Phytopath. Soc. Subspecialty: Plant pathology. Current work: Diseases of wild rice and sugarbeet; chemical control of plant diseases; integrated pest management. Office: U Minn Dept Plant Pathology 304 Stakman Hall St Paul MN 55108

PERCY, ALAN KENNETH, pediatric neurology educator; b. Watertown, N.Y., May 12, 1938; s. Kenneth Edwin and Carolyn Emily (Wood) P.; m. Linda Karen Andersen, Aug. 26, 1961 (div. May 1979); children: Stephen Eric, Karen Elizabeth, Christopher Alan; m. Phyllis Diane Wilhite, June 16, 1979. A.B., Harvard U., 1960; M.D., Stanford U., 1965. Diplomate: Am. Bd. Pediatrics, Am. Bd. Psychiatry and Neurology. Intern Stanford U., 1965-66, resident in pediatrics, 1968-69; fellow in child neurology Johns Hopkins U., 1969-72; physician, mem. faculty UCLA, 1972-79; mem. faculty Baylor U., 1979—, assoc. prof. pediatric neurology, 1979—; assoc. dean for research Drew Med. Sch., 1977-79; cons. neurologist Tex. Children's Hosp., Houston, 1979—. Contbr. articles to profl. jours., chpts. in books. Served to lt. comdr. USPHS, 1966-68. NIH research grantee, 1973, 75. Mem. Soc. Pediatric Research, Am. Soc. Neurochemistry, Am. Acad. Neurology, Am. Neurol. Assn., Soc. Neurosci., Child Neurology Soc. Subspecialties: Pediatrics; Neurochemistry. Current work: Metabolism of complex lipids in neural tissue; regulation of neural lipids; role of lipids in neural function. Home: 13515 Barryknoll Ln Houston TX 77079 Office: Baylor Coll of Medicine 1200 Moursund Ave Houston TX 77030

PERCY, JOHN REES, astronomer, educator; b. Windsor, Eng., July 10, 1941; s. George Francis and Christine (Holland) P.; m. Maire Ede Robertson, June 16, 1962; children: Carol Elaine. B.Sc., U. Toronto, 1962, M.A., 1963, Ph.D., 1968. Tchr. Bloor Collegiate Sch., Toronto, 1964-65; mem. faculty U. Toronto, 1967—, assoc. prof. astronomy, 1973-78, prof., 1978—. Contbr. articles to profl. jours. Leverhulme fellow Cambridge (Eng.) U., 1972-73; Recipient Royal Jubilee medal Govt. of Can., 1977. Mem. Internat. Astronom. Union. Subspecialty: Optical astronomy. Current work: Structure, evolution, and stability of stars. Office: U Toronto 60 St George St Canada M5S 1A7

PERDUE, PHILIP TAW, research health physicist; b. Salem, Va., June 11, 1917; s. Peter Taw and Willie Mae (Doss) P.; m. Ione C. Sisson, Nov. 15, 1937 (div. 1953); children: Julia, Patricia Ann; m. Stella Wilson, Sept. 20, 1959; 1 dau.: Karen Rae. Student, Roanoke Coll., Salem, 1935-37. Registered radiation protection technologist; radio telephone lic. FCC. Health physicist Oak Ridge Nat. Lab., 1954—. Contbr. articles to profl. jours. Bd. dirs. Oak Ridge Community Playhouse, 1964-66, 73-75. Mem. Health Physics Soc. (treas. East Tenn. chpt. 1982-84, Disting. Service award 1980), Am. Nuclear Soc. (Appreciation cert. 1982), N.Y. Acad. Scis., AAAS. Unitarian. Subspecialties: Geochemistry; Environmental engineering. Current work: Quantitative dtermination of radioisotopes in homes and soil, particularly daughters of the radon isotopes. Patentee determine radon in air, fast-neuron solid-state dosemeter. Home: 103 Oak Ln Oak Ridge TN 37830 Office: Oak Ridge Nat Lab PO Box X Oak Ridge TN 37830

PEREIRA, MICHAEL ALAN, pharmacologist, toxicologist, cancer researcher; b. Bronx, N.Y., May 3, 1944; s. Frank F. and Eleanor (Minowitz) P.; m. Bette Ann Weinstein, Sept. 17, 1967; children: Arlene, Steve. B.S. in Microbiology, Ohio State U., 1967, Ph.D. in Pharmacology, 1971. Postdoctoral fellow NIH, Bethesda, Md., 1971-73; research asso. N.Y. Blood Center, N.Y.C., 1973-74; research scientist NYU Med. Ctr., Tuxedo, N.Y., 1974-78; br. chief, pharmacologist EPA, Cin., 1978—; adj. asst. prof. Ohio State U. Coll. Medicine, Columbus, 1979—. Contbr. articles to profl. jours. Damon Runyon fellow, 1971-73. Mem. Soc. Toxicology, Am. Assn. for Cancer Research, Environ. Mutagen Soc., Am. Coll. Toxicology. Jewish. Subspecialties: Toxicology (medicine); Cancer research (medicine). Current work: Research in cause and prevention of cancer; mechanism of chemical carcinogenesis; tumor promotion; chemical mutagenesis; environ. carcinogenesis and toxicology; biochem. toxicology. Home: 12165 Brookston Dr Cincinnati OH 45240 Office: EPA 26 W St Clair St Cincinnati OH 45268

PEREJDA, ANDREA JEANNE, biochemistry researcher, medical educator; b. East Lansing, Mich., Mar. 15, 1952; d. Andrew D. and Alice Jeanne (French) P. B.A., Kalamazoo Coll., 1974; Ph.D., Washington U., St. Louis, 1979. Postdoctoral fellow Washington U., 1980-81; advanced research fellow Harbor-UCLA Med. Ctr., Torrance, Calif., 1981-83, asst. research prof. medicine, 1983—. Contbr. articles to profl. jours. Recipient Grad. Research award Central States Electron Microscopy Soc., 1980; AAUW scholar, 1970-74. Mem. AAAS, Am. Diabetes Assn., Am. Fedn. Clin. Research. Subspecialties: Biochemistry (biology); Dermatology. Current work: Collagen and elastin biochemistry, glycosaminoglycan production by human skin fibroblasts in tissue culture, diabetes mellitus. Office: Harbor UCLA Med Ctr Box 9 1000 W Carson St Torrance CA 90509

PEREL, JAMES MAURICE, pharmacologist, educator, cons.; b. Argentina, Mar. 30, 1933; came to U.S., 1947, naturalized, 1954; s. Adolph and Della (Silverberg) P.; m. Audrey Feldman, Apr. 10, 1972; children: Allan B., Alissa A., Stephen M. B.S., CCNY, 1956; M.S., N.Y.U., 1961, Ph.D. (NSF fellow), 1964. Nuclear chemist N.Y. Naval Shipyard, 1956-58; asso. research scientist N.Y.U. Med. Sch., 1963-67; asst. prof. medicine and chemistry Emory U., Atlanta, 1967-70; asst. prof. psychiatry Coll. Physicians and Surgeons, Columbia U., 1970-76, asso. prof. clin. pharmacology and psychiatry, 1976-80; chief psychiat. research N.Y. State Psychiat. Inst., 1976-80; prof. psychiatry and pharmacology U. Pitts. Sch. Medicine, 1980—; dir. clin. pharmacology program Western Psychiat. Inst. and Clinic, 1979—; cons. in clin. pharmacology VA Med. Center, Pitts., Found. for Research in Mania and Depression, N.Y.C., FDA, N.Y. State Dept. Mental Health and Mental Retardation, Obstetric Anesthesiology Found., N.Y.C. Contbr. chpts. to books, articles to profl. jours. Recipient Founder's Day award N.Y.U., 1964; USPHS fellow, 1963-65; grantee, 1967, 70, 73, 76, 79-80, 80—. Fellow Am. Inst. Chemists; mem. Am. Soc. Pharmacology and Exptl. Therapeutics, Am. Fedn. Clin. Research, Soc. Biol. Psychiatry, Am. Soc. Clin. Pharmacology and Therapeutics, N.Y. Acad Scis., Harvey Soc. Subspecialties: Pharmacology; Pharmacokinetics. Current work: Mechanisms of drug actions in humans, psychopharmacology of antidepressants and neuroleptics, drug metabolism and pharmacokinetics. Office: 3811 O'Hara St Pittsburgh PA 15213

PERELLE, IRA B., psychologist, researcher; b. Mt. Vernon, N.Y., Sept. 16, 1925; s. Joseph Y. and Lillian (Schaffer) P.; m. Diane A. Granville, July 18, 1982. B.S., Fordham U., 1969; E.E., RCA Inst.; M.S., Fordham U., 1970, Ph.D., 1972. Pres. Westlab, Inc., Yonkers, N.Y., 1949-60; exec. dir. Interlink Ltd., Bronxville, N.Y., 1960-72; prof. psychology Mercy Coll., 1972—; cons. Boces-Westchester, Port Chester, N.Y., 1976—, Mt. Vernon Pub. Schs., 1977—, Reader's Digest, Pleasantville, N.Y., 1979-81, G.T.E., Stanford, Conn., 1975-79. Contbr. articles to profl. jours. Mem. Am. Psychol. Assn., Acoust. Soc. Am., Am. Ednl. Research Assn., Am. Soc. Naturalists, Am. Genetic Soc., AAAS, Am. Inst. Physics, Animal Behavior Soc., Audio Engring. Soc., IEEE, N.Y. Acad. Sci., Brain and Behavioral Scis. Subspecialties: Behavioral psychology; Ethology. Current work: Laterality - investigation of behavior and etiology of differential use of body (hands) and brain in humans and other animals. Discovered psychol. perceptual attention phenomenon; Perelle phenomenon; laterality; genetics. Home: East Branch Rd Box 297 RD 3 Patterson NY 12563 Office: Mercy Coll 555 Broadway Dobbs Ferry NY 10522

PERELSON, ALAN STUART, theoretical biologist, researcher; b. Bklyn., Apr. 11, 1947; s. Morris L. and Ruth D. (Ferst) P.; m. Janet A. Gerard, June 23, 1968; children: Elissa Danielle, Andrew Jacob. B.S. in Life Scis, M.I.T., 1967, B.S.E.E., 1967; Ph.D. in Biophysics, U. Calif.-Berkeley, 1972. Acting asst. prof. div. med. physics U. Calif.-Berkeley, 1973; research assoc., NIH postdoctoral fellow dept. chem. engring. and material scis. U. Minn.-Mpls., 1973-74; mem. staff theoretical div. Los Alamos Nat. Lab., 1974—; asst. prof. med. scis. Brown U., 1978-80. Editor: Theoretical Immunology, 1978, Cell Surface Dynamics: Concepts and Models, 1983; assoc. editor: Bull. Math. Biology, 1977—; adv. editor: Jour. Mth. Biology, 1977—; editorial bd.: Lecture Notes in Biomath, 1978—, Math. Biosci, 1981—; book rev. editor, 1982—. Mem. Am. Assn. Immunologists, Soc. Indsl. and Applied Math., IEEE, Soc. Math. Biology (dir. 1982—). Current work: Development of theoretical and mathematical models of biological phenomena, particularly in immunology and cell biology. Office: Theoretical Div Los Alamos Nat Lab Los Alamos NM 87545

PERERA, THOMAS BIDDLE, psychology educator; b. N.Y.C., Nov. 20, 1938; s. Lionel C. and Dorothy (Bittel) P.; m. Gretchen G. Gifford, Aug. 28, 1960; children: Daniel G., Thomas B. A.B., Columbia U., 1961, M.A., 1963, Ph.D., 1968. Lic. psychologist, N.Y. Asst. prof. psychology Barnard Coll., N.Y.C., 1966-74; assoc. prof. psychology Montclair State Coll., Upper Montclair, N.J., 1974-80, prof., 1980—; vis. prof. Barnard Coll., 1975—; sr. research scientist N.Y. State Psychiat. Inst., 1964-71; cons. Omni Systems Assocs., West Caldwell, N.J., 1978—, Columbia U., 1960—. Contbr. articles to profl. jours. Mem. Am. Psychol. Assn., Eastern Psychol. Assn., AAAS, Biofeedback Soc. Am., Southeastern Psychol. Assn. Subspecialties: Physiological psychology; Behavioral psychology. Current work: Extensive research in innovative computer applications in psychology, physiological psychology with emphasis on biofeedback as applied to human anxiety. Home: 11 Squire Hill Rd North Caldwell NJ 07006 Office: Dept Psycholog Montclair State Coll Valley and Normal Aves Upper Montclair NJ 0704

PEREZ, FRANCISCO IGNACIO, clinical psychologist, behavioral researcher; b. Havana, Cuba, May 21, 1947; came to U.S., 1960, naturalized, 1969; s. Francisco Jose and Maria F. (Villa) P.; m. Georgina M. Montero, Aug. 21, 1971; children: Francisco A., Teresa M. B.A., U. Fla.-Gainesville, 1969, M.A., 1970, Ph.D., 1972. Lic. psychologist, Tex. Pvt. practice clin. psychology, Houston, 1972—; asst. prof. psychology U. Houston, 1975—, dir. clin. edn. lab., 1972-75; clin. asst. prof. neurology and physical medicine Baylor Coll. Medicine, 1980—, dir. Neuropsychology Lab., 1975-80. Contbr. in field. Bd. dirs. Houston Youth Community Council. NIH grantee, 1975-80, 78-80; Marion Labs. grantee, 1978-80. Mem. Am. Psychol. Assn., Internat. Neuropsychology Soc., Tex. Psychol. Assn., Biofeedback Soc. Am. Roman Catholic. Subspecialties: Behavioral psychology; Neuropsychology. Current work: Behavioral research in dementia and the elderly; biofeedback applications; computerized cognitive retaining in the brain injured. Home: 3407 Fawn Creek Dr Kingswood TX 77339 Office: PGO and Associates PC 6560 Fannin St Suite 1224 Houston TX 77030

PEREZ-PERAZA, JORGE A., astrogeophysicist; b. Mexico D.F., Mexico, July 23, 1944; s. Carlos E. and Esther (Peraza-Nunez) Perez-Rivero; m. Silvia C. Robles-Municha, Apr. 11, 1970; children: Jorge D. Perez-Robles, Esther C. Perez-Robles; m. Carmen Rivera-Ramirez, Mar. 21, 1981; 1 son, David D. Perez-Rivera. B.S., Instituto Politecnico Nacional, Mexico, 1968; M.Sc., U. Paris, 1970, Ph.D., 1972. Prof. physics Instituto Politecnico Nacional, 1966-68; asst. cosmic ray physics courses Laboratoire de Physique Cosmique, Verriere le Buisson, France, 1971-72; titular researcher on solar physics and cosmic rays physics Instituto Astronomy, Universidad Nacional Autonoma de Mex., 1973-81, titular researcher, 1981—; guest investigator Tata Inst. Fundamental Research, Bombay, India also; CEA, Saclay, France, 1980. Contbr. articles to profl. jours. Served with Mexican Army, 1965. Instituto Nacional de la Investigacion Cientifica grantee, 1968-69; Consejo Nacional de Ciencia y Tecnologia grantee, 1969-73, 80-81. Mem. Am. Geophys. Union, Internat. Astron. Union, Am. Astron. Soc., Astron. Soc. Pacific, Am. Inst. Physics, Indian Astron. Soc., Indian Phys. Soc., Instituto Panamericano de Georgrafia e Historia. Subspecialties: Cosmic ray high energy astrophysics; Solar physics. Current work: Origin of cosmic rays, solar flares, solar cosmic rays, energy spectra and chemical composition of cosmic rays, acceleration in astrophysics, energy losses of fast ions in plasmas and atomic media. Office: Instituto de Geofisica UNAM 04510-CU Coyoacan Mexico 20 DF Mexico

PERHAC, RALPH MATTHEW, research institute executive; b. Bklyn., July 29, 1928; s. George and Irene (Harpas) P.; m. Constance Main, Feb. 4, 1950 (dec. 1973); children: Ralph M., Janet I. A.B., Columbia U., 1949; A.M., Cornell U., 1952; Ph.D., U. Mich., 1961. Sr. research geochemist Exxon Research, Houston, 1960-67; prof. geochemistry U. Tenn., Knoxville, 1967-74; program mgr. NSF, Washington, 1974-76; dir. environ. assessment dept. Elec. Power Research Inst., Palo Alto, Calif., 1976—; lab. research cons. Oak Ridge Nat. Lab., 1969-74, lunar investigator, 1972-74. Fellow Geol. Soc. Am.; mem. Internat. Assn. Geochem. Cosmochemistry (editor 1970-72), Soc. Environ. Geochemistry and Health, Am. Assn. Petroleum Geologists, Geochem. Soc. Subspecialties: Geochemistry; Environmental assessment. Current work: Mineral chemistry, environmental science. Home: 675 Sharon Park Dr Menlo Park CA 94025 Office: Electric Power Research Inst Box 10412 Palo Alto CA 94303

PERHACH, JAMES LAWRENCE, JR., pharm. co. exec., educator, cons.; b. Pitts., Oct. 26, 1943; s. James Lawrence and Elizabeth Louise (Hoffman) P.; m. Judith Irene Selter, April 15, 1967; children: Laura Anne, Amy Elizabeth. B.S., U. Dayton, Ohio, 1966; M.S., U. Pitts., 1969, Ph.D. (NIH predoctoral fellow), 1971. Sr. Scientist dept. pharmacology Mead Johnson Research Center, 1971-74, sr. investigator dept. biol. research, 1974-76, sr. research assoc. dept. biol. research, 1976-77, sr. research assoc. dept. pathology and toxicology, 1977-78, prin. research assoc. dept. pathology and toxicology, 1978-80; dir. pharmacology Wallace Biol. Research, Wallance Labs. div. Carter-Wallace, Inc., Cranbury, N.J., 1980, exec. dir. biol. research, 1980—; adj. prof. toxicology Phila. Coll. Pharmacy and Sci., 1981—; assoc. faculty Evansville Center Med. Edn., Ind. U., 1973-1980, mem. addictions med. edn. program, 1972-78; lectr. grad. physiology U. Evansville, 1973-79; grad. teaching asst. U. Pitts., 1967-69, instr. pharmacology and exptl. therapeutics, 1968-69; mem. coordinating group 1969 meeting Am. Soc. Pharmacology and Exptl. Therapeutics; mem. substance abuse com. Tri-State Area Health Planning Council, Evansville, 1972-75. Contbr. articles and abstracts to profl. jours. Mem. Am. Soc. Pharmacology and Exptl. Therapeutics, Soc. Neurosci., Soc. Exptl. Biology and Medicine, N.Y. Acad. Scis., Physiol. Soc. Phila., Internat. Soc. Immunopharmacology, European Soc. Toxicology, Am. Coll. Toxicology, Internat. Union Pharmacology, Behavioral Teratology Soc., Genetic Toxicology Assn., AAAS, Sigma Xi. Republican. Roman Catholic. Subspecialties: Pharmacology; Toxicology (medicine). Current work: New drug discovery, elucidation of mechanism of action and safety evaluation of new therapeutic agents, CNS control of blood pressure and hypertension, behavioral pharmacology and teratology. Home: 13 Hopkins Dr Lawrenceville NJ 08648 Office: PO Box 1 Cranbury NJ 08512

PERKINS, COURTLAND DAVIS, aeronautical engineer, educator; b. Phila., Dec. 27, 1912; s. Harry Norman and Emily Cramp (Taylor) P.; m. Jean Elizabeth Enfield, Sept. 27, 1941 (dec. Oct. 1980); children: William Enfield, Anne Taylor. B.S., Swarthmore Coll., 1935, D.Eng. (hon.), 1977; M.S., Mass. Inst. Tech., 1941; D.Eng. (hon.), Lehigh U., 1977, Rensselear Poly. Inst., 1977. Br. engr. Am. Radiator Co., Phila., 1935-39; prof. aero. engring. Princeton U., 1945—, chmn. dept., 1951-63, chmn. dept. aerospace and mech. scis., sch. engring. and applied sci., 1963-74, asso. dean Sch. Engring., 1974-75; dir. Fairchild-Industries Corp., 1962-75, Am. Airlines, 1967-75; Keuffel & Esser Corp., 1969-75; trustee Mitre Corp., C. Stark Draper Labs.; Aero engr. USAF, Air Materiel Command, Dayton, Ohio, 1941-45; chief scientist USAF, 1956-57; asst. sec. Research and Devel., 1960-61; mem. sci. adv. bd. USAF, chmn., 1969-73, 77-78; mem. def. sci. bd. Def. Dept., 1969-73, 77-78; chmn. adv. group for aero. research and devel. NATO, 1963-67; chmn. space systems com. NASA, 1973-77. Author: (with Robert E. Hage) Airplane Performance Stability and Control, 1949. Decorated Legion of Honor, France). Fellow AIAA (hon., pres. 1964), Am. Acad. Arts and Scis., Royal Aero. Soc.; mem. Nat. Acad. Engring. (pres. 1975—), Nat. Acad. Sci. (mem. space sci. bd. 1964-70), Sigma Xi, Tau Beta Pi. Subspecialty: Aeronautical engineering.

PERKINS, DAVID D(EXTER), geneticist; b. Watertown, N.Y., May 2, 1919; s. Dexter M. and Loretta F. (Gardiner) P.; m. Dorothy L. Newmeyer, Aug. 1, 1952; 1 dau., Susan J. A.B. in Biology, U. Rochester, 1941; Ph.D. in Zoology, Columbia U., 1949. Mem. faculty Stanford U., 1949—, prof. biology, 1961—; research fellow U. Glasgow, Scotland, 1954-55, Columbia U., 1962-63, Australian Nat. U., Canberra, 1968-69; vis. scholar univs. Wash., Hawaii and Calif., San Diego, 1975-76; mem. India-U.S. Exchange Scientists Program, 1974; mem. genetics tng. com. USPHS, 1961-65. Editor Genetics, 1963-67. Served with USAAF, 1943-45. Recipient Research Career award USPHS, 1964—. Mem. Internat. Genetics Fedn. (exec. bd. 1978—), Genetics Soc. Am. (pres. 1977), Nat. Acad. Scis. Subspecialty: Genetics and genetic engineering (biology). Office: Dept Biol Scis Stanford U Stanford CA 94305

PERKINS, GLENN RICHARD, nuclear engineering cons.; b. Portsmouth, N.H., Nov. 3, 1947; s. Milo Orman and Rosamond (Thorner) P.; m. Linda Soley Swalm, May 23, 1970; 1 child, Dain Eyrikur. A.S., Wentworth Inst., 1969; B.S., Northeastern U., 1972. Asst. test engr. Newport News Shipbldg. and Drydock Co., Va., 1972-73; devel. engr. Combustion Engring. Co., Windsor, Conn., 1973-78, prin. devel. engr., 1978-79, supr. inservice inspection group, 1979-81; pres. NDE Engring. Cons., Storrs, Conn., 1981—; cons. NUSCO, Hartford, Conn., 1982, Yankee Atomic Electric Co., Framingham, Mass., 1981-82, Fla. Power & Light Co., Miami, 1981-82, NES, Danbury, Conn., 1981. Cubmaster Boy Scouts Am., 1982. Mem. Am. Nuclear Soc., Am. Soc. NonDestructive Testing. Subspecialties: Non-destructive testing; Data management. Current work: Development and implementation of Non Destructive Examination programs to meet NRC and industry codes and standards for commercial nuclear power plants. Home: 241 Woodland Rd Storrs CT 06268 Office: NDE Engring Cons PO Box 535 Storrs CT 06268

PERKINS, HENRY FRANK, agronomy educator; b. Georgetown, Ga., June 19, 1921; s. Frank Lee and Ada (Jones) P.; m. Laura McClain, Apr. 18, 1945; children: Douglas W., Donald L. B.S., U. Ga., 1945, M.S., 1951; Ph.D., Rutgers U., 1954. Soil scientist U.S. Dept. Agr., Elbertson, Ga., 1945-47; soil analyst U. Ga., Athens, 1947-51, asst. prof. agronomy, 1951-58, assoc. prof., 1958-64, prof., 1964—. Recipient D. W. Brooks Disting. Teaching award U. Ga., 1981. Fellow Am. Soc. Agronomy, Soil Sci. Soc. Am.; mem. Internat. Soil Sci. Soc., Soil Conservation Soc. Am., So. Assn. Agrl. Scientists. United Methodist. Subspecialty: Soil chemistry. Current work: Research on characterization and utilization of sub-tropical soils. Home: 540 Westview Dr Athens GA 30606 Office: U Ga Agronomy Dept Room 3111 Miller Bldg Athens GA 30602

PERKINS, PORTER J., JR., aerospace engineer; b. Beverly, Mass., May 3, 1922; s. Porter J. and Sue (Thistle) P.; m. Hedda Nordestgaard, May 17, 1952; children: Diane Sue, Kimberly Anne. B.S. in Mech. Engring, Northeastern U., Boston, 1944. Aero. engr. NACA, Cleveland, Ohio, 1944-58; aerospace engr. NASA, Cleveland, 1958-80; aero cons. ANALEX Corp., Dayton, Ohio, 1980—. Contbr. articles to profl. jours.; patentee light weight cryogenic insulation. Sr. mem. , AIAA. Subspecialties: Aeronautical engineering; Aerospace engineering and technology. Current work: aircraft icing clouds, and ice protection systems for general aviation and helicopters. Home: 2743 Lakeview Ave Rocky River OH 44116 Office: NASA 21000 Brookpark Rd Cleveland OH 44135

PERKINS, RICHARD SCOTT, chemist, educator; b. Hammond, La., June 21, 1940; s. Gerald and Louise Signe (Young) P.; m. Bonnie Ann Bench, May 15, 1964; children: Laura, Sara. B.S., La. State U., 1962; Ph.D., U. Utah, 1966. Postdoctoral fellow U. Ottawa, Ont., Can., 1966-68, U. Utah, Salt Lake City, 1968-69; asst. prof. chemistry U. Southwestern La., Lafayette, 1969-74, assoc. prof., 1974-80, prof., 1980—. Bd. dirs. Lafayette Natural History Mus. Recipient disting. Teaching award U. Southwestn La. Found., 1979. Mem. Am. Chem. Soc. Subspecialties: Physical chemistry; Electrochemistry. Current work: Luminescence in electrochemical systems. Office: Box 44370 Lafayette LA 70504

PERKINS, ROGER ALLAN, research scientist; b. Milw., May 28, 1926; s. Sumner Verney and Lucy Norma (Grimm) P.; m. Jean Ellen Geerin, Sept. 12, 1953; children: Daniel, Robert, Jane. B.S. in Metall. Engring, Purdue U., 1949, M.S., 1951. Registered profl. engr., Calif. Research metallurgist Union Carbide Metals Co., Niagara Falls, N.Y., 1951-59; sect. head Aeroject Gen. Corp., Sacramento, 1959-60; with Lockheed Palo Alto (Calif.) Research Lab., 1960—; sr. mem., 1965—. Contbr. over 30 tech. articles to profl. publs. Scoutmaster Boy Scouts of Am., Los Altos, Calif., 1966-73. Served with USNR, 1944-46. Fellow Am. Soc. Metals (chmn. Santa Clara Valley chpt. 1972-73). Subspecialties: High-temperature materials; Clad metals and coating technology. Current work: Materials for rocket propulsion and energy conversion; research on high temperature corrosion and protective coatings. Holder 5 U.S. patents. Home: 1142 Lincoln Dr Mountain View CA 94040 Office: 3251 Hanover St Palo Alto CA 94304

PERKINS, STERRETT THEODORE, physicist; b. Oakland, Calif., July 25, 1932; s. Frank Bernard P.; m. Carol Louise Russell, 1969 (div. 1981); children: Charles, Pamela, Jill, Terri, Lisa, Sheila. B.S., U. Calif.-Berkeley, 1956, M.S., 1957, Ph.D., 1965. Mech. engr. Lawrence Berkeley Nat. Lab., 1956-57; sr. nuclear engr. Aerojet Gen. Nucleonics, San Ramon, Calif., 1957-59, prin. nuclear engr., 1960-65, cons., 1959-60; nuclear engr. Lawrence Livermore (Calif.) Nat. Lab., 1959-60, sr. physicist, 1965—. Contbr. numerous articles to profl. jours. Mem. Am. Nuclear Soc., Am. Phys. Soc., Livermore Heritage Guild (dir. 1975-77), Sigma Xi, Tau Beta Pi, Pi Tau Sigma. Subspecialties: Nuclear physics; Plasma physics. Current work: Particle transport in high temperature plasmas, including theoretical development, cross section evaluation, and simulation with large scale computer programs. Office: Lawrence Livermore Nat Lab PO Box 808 Livermore CA 94550

PERKOWITZ, SIDNEY, physicist, educator; b. Bklyn., May 1, 1930; s. Morris and Sylvia (Gray) P.; m. Sandra F. Price, Jan. 14, 1967; 1 son, Michael. B.S. in Physics, Poly. Inst. Bklyn., 1960, M.S., U. Pa., 1962, Ph.D. in Solid State Physics, 1967. Solid state physicist Gen. Telephone & Electronics Lab., Bayside, N.Y., 1966-69; asst. prof. physics Emory U., Atlanta, 1969-74, asso. prof., 1974-79, prof., 1979—, chmn. dept. physics, 1980—; tech. and ednl. cons. Contbr. articles to tech. publs., chpts. to books; editor 2 conf. proc.; mem. editorial bd. 2 jours. Research grantee Alfred P. Sloan Found., Research Corp., NSF, AEC, Office Naval Research, Dept. Energy, NIH. Mem. Am. Phys. Soc., AAAS, Optical Soc. Am., Phi Beta Kappa. Subspecialties: Condensed matter physics; Low temperature physics. Current work: Optical properties of condensed matter (superconductors, semiconductors); far infrared spectroscopy. Office: Physics Dept Emory U Atlanta GA 30322

PERL, MARTIN LEWIS, physicist, educator; b. N.Y.C., June 24, 1927; s. Oscar and Fay (Rosenthal) P.; m. Teri Hoch, June 19, 1948; children: Jed, Anne, Matthew, Joseph. B.Chem. Engring., Poly. Inst. Bklyn., 1948; Ph.D., Columbia U., 1955. Chem. engr. Gen. Electric Co., 1948-50; asst. prof. physics U. Mich., 1955-58, asso. prof., 1958-63; prof. Stanford, 1963—. Author: High Energy Hadron Physics, 1975; contbr. articles on high energy physics and on relation of sci. to soc. to profl. jours. Served with U.S. Mcht. Marine, 1944-45; Served with AUS, 1945-46. Recipient Wolf prize in physics, 1982. Fellow Am. Phys. Soc.; mem. Nat. Acad. Scis., AAAS. Subspecialty: Particle physics. Current work: Experimental particle physics. Home: 525 Lincoln Ave Palo Alto CA 94301 Office: Stanford Linear Accelerator Center Stanford U Stanford CA 94305

PERLIS, ALAN J., educator, computer scientist; b. Pitts., Apr. 1, 1922; s. Louis Phillip and Zelda Anne (Gilfond) P.; m. Sydelle Gordon, Oct. 28, 1951; children—Mark Lawrence, Robert Gordon, Andrea Lynn. B.S. in Chemistry, Carnegie Inst. Tech., 1943; postgrad., Calif. Inst. Tech., 1946-47; M.S., Mass. Inst. Tech., 1950, Ph.D. in Math, 1950; D.Sc. (hon.), Davis and Elkins Coll., 1968, Purdue U., 1973, Waterloo U., 1974, Sacred Heart U., 1979. Asst. prof. math., dir. computer center Purdue U., 1952-56; mem. faculty Carnegie Inst. Tech., 1956-71, prof. math., dir. computer center, 1960-71, head dept. math., 1961-64, head dept. computer sci., 1965-71; Eugene Higgins prof. computer sci. Yale U., 1971—, chmn. dept., 1976—; Gordon and Betty Moore vis. prof. engring. Calif. Inst. Tech., 1977; mem. NSF computer com. Nat. Joint Computer Com., 1954-56, Gov. Pa. Council Sci. and Tech., 1963—; com. computers research NIH, 1963—; mem. computer sci. and engring. research bd. Nat. Acad. Scis., 1968—; mem. Assembly of Engring., NRC, 1978—. Served to 1st lt. USAAF, 1942-45; ETO. Mem. Assn. Computing Machinery (pres. 1962-64, editor-in-chief jour. Communications 1958-62), Soc. Indsl. and Applied Math., Am. Math. Soc., Math. Assn., Am. Acad. Arts. and Scis., Conn. Acad. Sci. Engring., Nat. Acad. Engring. Am. Subspecialties: Programming languages; Software engineering. Home: 19 Tumblebrook Rd Woodbridge CT 06525 Office: Yale Univ New Haven CT 06520

PERLMAN, DANIEL, psychologist, educator; b. Kingston, N.Y., July 4, 1942; s. Paul and Rosalene (Preston) P.; m. Elizabeth Isitt; 1 son, Anton. A.B., Bard Coll., 1964; M.A., Claremont Grad. Sch., 1969, Ph.D., 1971. Mem. faculty U. Man., Winnipeg, 1970-83, prof. psychology, 1981-83, assoc. dept. head, 1977-78; prof. family sci., dir. family studies grad. program U. B.C., Vancouver, 1983—; grant reviewer NSF, 1977. Editor: Can. Psychology, 1979-82; editorial bd., 1983—, Jour. Social Issues, 1983—, Applied Social Psychology Ann, 1983—, Jour. Sex Research; co-editor: Loneliness: A Sourcebook of Current Theory, Research and Therapy, 1982, Social Psychology, 1983. N.Y. State Regents scholar, 1960-63; John Bard scholar, 1963; NDEA fellow, 1963-64; Ford Found. teaching intern, 1969-70; USPHS grantee, 1970-71, 79-80; UCLA sabattical fellow, 1976-77; Can. Council grantee, 1974-76; leave fellow, 1977; Man. Mental Health Research Found. grantee, 1977-78; Social Scis. and Humanities Research Council Can. grantee, 1981-82; recipient Clifford J. Robson Disting. Psychologists in Man. award, 1982; Brit. Council exchange visitor, 1983. Fellow Can. Psychol. Assn.; mem. Am. Psychol. Assn., Nat. Council Family Relations, Soc. Psychol. Studey of Social Issues (exec. council 1982—), Man. Psychol. Soc. (treas. 1974-76, 82-83), Soc. Advancement of Social Psychology, Soc. Exptl. Social Psychology. Subspecialty: Social psychology. Current work: Interpersonal relations. Office: U BC 2075 Wesbrook Mall Vancouver BC Canada V6T 1W5

PERLMUTTER, BARRY, geology educator, researcher; b. N.Y.C., Jan. 7, 1945; s. Leon and Jessie (Monder) P.; m. Rebecca Ann Robertson, Mar. 20, 1971. B.S., Bklyn. Coll., CUNY, 1966; M.A., Ind. U.-Bloomington, 1968; Ph.D., U. Iowa, 1971. Teaching asst. Ind. U., 1966, U. Iowa, 1968; asst. prof. geosci. and geography Jersey City State Coll., 1972-82, assoc. prof., 1982—. N.Y. State Regents scholar, 1962; Kans. Geol. Survey grantee, 1970. Mem. Paleontol. Soc., Mus. Natural History, Smithsonian Instn. Jewish. Subspecialties: Paleontology; Paleoecology. Current work: Biostratigraphic descriptions of paleozoic conodonts of Pennsylvanian and Permian age. Office: Jersey City State Coll Geosci and Geography Dept 2039 Kennedy Blvd Jersey City NJ 07305

PERLMUTTER, DANIEL D., educator; b. Bklyn., May 24, 1931; s. Samuel and Fannie (Kristal) P.; m. Felice Davidson, Oct. 23, 1954; children—Shira, Saul, Tova. B.S. in Chem. Engring, N.Y. U., 1952; D.Eng., Yale, 1956. Prof. U. Pa., Phila., 1965—. Author: Introduction to Chemical Process Control, 1965, Stability of Chemical Reactors, 1972. Guggenheim fellow, 1964; Fulbright fellow, 1968, 72. Mem. Am. Inst. Chem. Engrs., AAAS. Subspecialty: Chemical engineering. Office: 311 Towne U Pa Philadelphia PA 19119

PERLOVSKY, LEONID ISAACOVICH, oil company researcher; b. Odessa, Ukrain, USSR, Nov. 11, 1948; came to U.S., 1978; s. Isaac E. and Riva B. (Bormashenko) P.; m. Olga J. Belayeva, Jan. 3, 1974; children: Ilya, Boris. M.S. with honors, Novosibirsk U., 1971; Ph.D., Joint Inst. Nuclear Research, Dubna, 1975. Asst. prof. Siberia Civil Engring. Inst., Novosibirsk, U.S.S.R., 1975-77; assoc. prof., 1977-78; asst. prof. NYU, N.Y.C., 1979-80; sr. research physicist Exxon Prodn. Research, Houston, 1980-81, research specialist, 1981—; cons. Sibera Agrl. Inst., Novosibirsk, USSR, 1975-78, Software Devel., Inc., N.Y.C., 1979-80. Contbr. articles in field to profl. jours. Mem. Soc. Exploration Geologists, Soc. Profl. Well Log Analysts, Soc. Indsl. and Applied Math., Am. Statis. Assn. Subspecialties: Geophysics; Applied mathematics. Current work: Seismic data processing, well log data processing, deconvolution, stacking, pattern recognition. Home: 8837 Dunlap Houston TX 77074 Office: Exxon Prodn Research PO Box 2189 Houston TX 77001

PERLOW, MARK JACOB, neurologist, neurobiologist; b. Chgo., Feb. 26, 1942; s. Samuel and Bertha (Shapiro) P. M.D., Northwestern U., 1967. Intern Cook County Hosp., Chgo., 1967-68; resident, fellow Albert Einstein Coll. Medicine, Bronx, N.Y., 1968-72; staff physician NIH, Bethesda, Md., 1974-80; assoc. prof. neurology Mt. Sinai Sch. Medicine, N.Y.C., 1980-81; practice medicine specializing in neuroloy, Chgo., 1981—; chief neurology service West Side VA Hosp., Chgo., 1981—; assoc. prof. neurology U. Ill. Med. Sch. Contbr. articles on neurology to profl. jours.; prin. author paper demonstrating for first time brain tissue transplanted between Mammals can function appropriately in host. Served as maj. U.S. Army, 1972-74. Recipient George Cotzias prize Am. Parkinson's Disease Found., 1980. Mem. Am. Acad. Neurology, Endocrine Soc. Subspecialties: Neurobiology; Chronobiology. Current work: Nervous tissue transplantation, growth and regeneration, biological rhythms, neuroendocrinology.

PERMUTT, MARSHALL ALAN, physician, medical educator; b. Birmingham, Ala., Nov. 20, 1939; s. James L. and Marguerite P.; m. Madelaine Robin; children: Joelle, Robin; m. Sally Barker; 1 child, Alexander Barker. B.A., Johns Hopkins U., 1961; M.D., Washington U., 1965. Intern Yale-New Haven Med. Ctr., 1965-66; resident in medicine U. Wash., Seattle, 1966-68; asst. prof. medicine Washington U. Med. Sch., St. Louis, 1973-78, assoc. prof., 1978—; external advisor Diabetes Edn. and Research Ctr., Baylor Coll. Sch. Medicine, 1982-83. Mem. editorial bd.: Endocrinology, 1977—, Diabetes, 1979—. Mem. Am. Diabetes Assn. (research com.), Endocrine Soc., Am. Fedn. Clin. Research, Internat. Diabetes Fedn., Am. Soc. Clin. Investigation. Subspecialties: Endocrinology; Genetics and genetic engineering (medicine). Current work: Studies of regulation of insulin biosynthesis in experimental laboratory animals and man, studies of the etiology and pathogenesis of diabetes mellitus. Home: 6341 Washington St Saint Louis MO 63130 Office: Washington U Sch Medicine 660 S Euclid Ave Saint Louis MO 63130

PERNICK, BENJAMIN, research scientist; b. N.Y.C., June 9, 1931; s. Abraham and Rose (Federbush) P.; m. Nancy R. Horn, June 10, 1956; children: Jonathan, Gary, Alice. B.S., CCNY, 1954; M.S., Stevens Inst. Tech., 1958, Ph.D., 1965. Sr. staff scientist Grumman Aerospace Corp., Bethpage, N.Y., 1965—; adj. assoc. prof. physics CUNY, 1964—. Contbr. articles to profl. jours.; patentee. Mem. Optical Soc. Am., Am. Phys. Soc., Sigma Pi Sigma. Subspecialties: Laser research; Optical signal processing. Current work: Research and development in laser technology and applications, signal processing, guided waves, electro-optics, acousto-optics, photoconductivity. Home: 2771 E 64th St Brooklyn NY 11234 Office: Research Dept Grumman Aerospace Bethpage NY 11714

PERO, JANICE GAY, molecular biologist; b. Lowell, Mass., June 11, 1943; d. Henry Leland and Helen Elizabeth (Bradway) P.; m. Richard Marc Losick, Aug. 8, 1970; children: Eric, Vicki. A.B., Oberlin Coll., 1965; Ph.D., Harvard U., 1971. Research fellow in biology Harvard U., 1971-75, tutor biochem. scis., 1973-77, asst. prof. biology, 1975-78, assoc. prof., 1978-83; sr. research cons. Biotechnica Internat., Inc., Cambridge, Mass., 1981-82, sr. scientist and program dir., 1982—; sr. research cons. Biotechnica Internat., Inc. Contbr. articles to profl. jours. NSF fellow, 1966-70; Am. Cancer Soc. fellow, 1970-72. Mem. Am. Soc. Microbiology, Am. Soc. Virology, Phi Beta Kappa. Subspecialties: Molecular biology; Genetics and genetic engineering (biology). Current work: RNA polymerase, sigma factors, promoters and the regulation of phage SP01 gene expression; gene expression and genetic engineering in Bacillus subtilis. Home: 4 Russell Rd Lexington MA 02173 Office: 85 Bolton St Cambridge MA 02140

PERRAULT, JACQUES, microbiology educator, researcher; b. Montreal, Que., Can., June 25, 1944; came to U.S., 1964; s. Jean-Paul and Irene (Girard) P.; m. Marcella McClure, Mar. 1, 1974. B.Sc., McGill U., 1964; Ph.D., U. Calif., 1972. Postdoctoral fellow McGill U., 1972; acting instr., postdoctoral fellow U. Calif., Irvine, 1972-74, asst. research biologist, San Diego, 1976-77; asst. prof. microbiology and immunology Washington U., St. Louis, 1977—. Contbr. articles to profl. jours. Recipient NIH Research Career Devel. award, 1980—. Mem. AAAS, Am. Soc. Microbiology, Soc. Gen. Microbiology (London). Subspecialties: Virology (medicine); Molecular biology. Current work: Defective interfering virus particles, molecular mechanisms of action and role in long term virus persistence. Office: Washington U Sch Medicine PO Box 8093 Saint Louis MO 63110

PERRENOD, STEPHEN CHARLES, reservoir engr., astrophysicist, software engr.; b. Natchez, Miss., Jan. 31, 1951; s. Charles Albert and Sylvia (Stevens) P. B.S. in Physics, M.I.T., 1972; M.A. in Astrophysics, Harvard U., 1973, Ph.D., 1977. Teaching fellow Harvard U., 1973-77; research assoc. dept. physics U. Ill.-Urbana, 1977-78, Kitt Peak Nat. Obs., Tucson, 1978-80; systems analyst Sohio Petroleum Co., San Francisco, 1980-82, reservoir engr., 1982—. Research, publs, in astrophysics. NASA research grantee Einstein X-Ray Obs., 1978-79. Mem. Soc. Petroleum Engrs. Libertarian. Club: Harvard (San Francisco). Subspecialties: Petroleum engineering; Software engineering. Current work: Reservoir simulation, vector computing, expert systems, fluid dynamics, interactive graphics, oil lease analysis. Office: 100 Pine St San Francisco CA 94111

PERRILL, STEPHEN ARTHUR, biology educator; b. Dayton, Ohio, Mar. 13, 1941; s. George Hayward and Jane (Reynolds) P.; m. Helen Nichols, Jan. 21, 1968; children: Jennifer, Stephanie. B.A., Ohio Wesleyan U., 1963; M.S., So. Conn. State Coll., 1967; Ph.D., N.C State U., 1973. Tchr. sci. Grove Sch., Madison, Conn., 1964-66, Savannah (Ga.) Day Sch., 1968-70; mem. faculty biology Butler U., Indpls., 1973—, assoc. prof., 1979—. Mem. AAAS, Am. Soc. Ichthyologists and Herpetologists, Herpetologists League, Soc. Study of Amphibians and Reptiles. Subspecialty: Ethology. Current work: Behavioral ecology of anuran amphibians. Home: 257 W 46 St Indianapolis In 46208 Office: Butler Univ Dept Zoology Indianapolis IN 46208

PERRIN, CHARLES LEE, chemistry educator; b. Pitts., July 22, 1938; s. Samuel R. and Ethel (Katz) P.; m. Marilyn Heller, June 14, 1964; children: David M., Edward J. A.B. summa cum laude, Harvard U., 1959; Ph.D., 1963. Asst. prof. chemistry U. Calif.-San Diego, La Jolla, 1964-71, assoc. prof., 1971-80, prof., 1980—. Author: Mathematics for Chemists, 1970, (with P. Zuman) Organic Polarography, 1969; contbr. articles to profl. jours. NSF fellow, 1963; NSF grantee, 1974—. Mem. Am. Chem. Soc., AAAS, Sigma Xi, Phi Beta Kappa. Subspecialties: Organic chemistry; Kinetics. Current work: Mechanisms of organic reactions; applications of nuclear magnetic resonsance to chemical kinetics. Patentee in field. Home: 8844 Robinhood Ln La Jolla CA 92037 Office: Dept Chemistry U Calif PO Box 109 La Jolla CA 92093

PERRIN, EDWARD BURTON, statistician, biostatis. research adminstr.; b. Greensboro, Vt., Sept. 19, 1931; s. Justus N. and Dorothy E. (Willey) P.; m. Carol Anne Hendricks, Aug. 18, 1956; children—Jenifer, Scott. B.A., Middlebury Coll., 1953; postgrad. (Fulbright scholar) in stats, Edinburgh (Scotland) U., 1953-54; M.A. in Math. Stats, Columbia U., 1956; Ph.D., Stanford U., 1960. Asst. prof. dept. biostats. U. Pitts., 1959-62; asst. prof. dept. preventive medicine U. Wash., Seattle, 1962-65, asso. prof., 1965-69, prof., 1969-70, chmn. dept. biostats., 1970-72, prof. dept. health services, 1975—; clin. prof. dept. community medicine and internat. health Sch. Medicine, Georgetown U., Washington, 1972-75; dep. dir. Nat. Center for Health Stats., HEW, 1972-73, dir., 1973-75; research scientist Health Care Study Center, Battelle Human Affairs Research Centers, Seattle, 1975-76, dir., 1976-78, Health and Population Study Cent er, Battelle Human Affairs Research Centers, Seattle, 1978—; sr. cons. biostats. Wash./Alaska regional med. programs, 1967 -72; biometrician VA Co-op Study on Treatment of Esophageal Varices, 1961-73; mem. panel on health services research NRC, 1981. Contbr. articles on biostats., health services and population studies to profl. publs.; mem. editorial bd.: Jour. Family Practice, 1978—. Mem. tech. bd. Milbank Meml. Fund, 1974-76. Recipient Outstanding Service citation HEW, 1975. Fellow AAAS, Am. Public Health Assn. (Spiegelman Health Stats. award 1970, program devel. bd. 1971, chmn. stats. sect. 1978-80), Am. Statis. Assn. (mem. adv. com. to div. statis. policy 1975-77); mem. Inst. Medicine of Nat. Acad. Sci., Population Assn. Am., Biometrics Soc. (pres. N.Am. region 1971), Internat. Union for Sci. Study of Population, Inst. Math. Stats., Internat. Epidemiologic Assn., Sigma Xi, Phi Beta Kappa. Subspecialty: Statistics. Home: 4900 NE 39th St Seattle WA 98105 Office: 4000 NE 41st St Seattle WA 98105

PERRIN, EUGENE VICTOR DEBS, physician, educator, environmental scientist; b. Detroit, Mar. 7, 1927; s. Emanuel Anatol and Frances Theresa (Levin) Paperno; m. Jane Carol Schutter, Mar. 31, 1956; children: Daniel Claude, Miriam Shoshanah, Adam Yehudah, Joshua Lindsay. A.B., Wayne U., 1948; student, Yale U., 1945-46, U. Chgo., 1948-49; M.D., U. Mich., 1953; D.D. (hon.), Universal Life Ch., Modesto, Calif., 1981-82. Lic. in medicine, Mich., Ohio. Intern Sinai Hosp., Detroit, 1953-54; gen. practice medicine, Detroit, 1954-55; resident Children's Hosp., Boston, 1955-56, Lying-in and Women's Free Hosp., 1956-57, Lahey Clinic Hosps., 1957-58, Babies Hosp., N.Y.C., 1958-59; mem. faculty U. Cin., 1959-66; physician Children's Hosp. and Cin. Gen. Hosp., 1959-66; mem. faculty, physician Univ. Hosps. Case Western Res. U., Cleve., 1966-74; prof. pathology, assoc. in pediatrics, ob-gyn anatomy Wayne State U., 1974—; dir. anat. pathology Children's Hosp. Mich., Detroit, 1974—, attending physician, 1974—; assoc. physician Hutzel, Sinai and Receiving hosps., Detroit, 1974—; chmn. adv. bd. Cystic Fibrosis Found., Cleve., 1970-72; cons. com. aquatic ecosystems objectives com. Internat. Joint Commn., Windsor, Ont., Can., 1979—; chmn. adv. bd. March of Dimes Met. Detroit, 1982—. Editor: Pathbiology of Development, 1973, Moral Problems in Medicine, 1975; contbr. chpts., articles, book revs. and poems. Vice chmn. Fellowship House, Cin., 1961-65; bd. dirs. East Mich. Environ. Action Council, Detroit, 1975—; chmn bd. Mich. chpt. Sierra Club, 1979-81; founder, chmn. Physicians for Social Responsibility, Detroit, 1981—; bd. dirs. Ctr. for Peace and Conflict Studies, Detroit, 1980—. Served with AUS, 1945-46. Mem. Pediatric Pathology Club (pres. 1971-72), Teratology Soc. (pres. 1977-78), Am. Assn. Pathologists, Am. Soc. Cell Biology, Tissue Culture Assn., Am. Coll. Toxicology, Am. Acad. clin. Toxicology, Soc. Occupational and Environ. Health, Internat. Acad. Pathology. Clubs: Faculty, Emma Goldman, Mazel (Detroit). Subspecialties: Environmental toxicology; Pathology (medicine). Current work: Ontology of receptors and receptor recycling as affected by environmental pollutants; environmental induction of malformations; transplacental carcinogenesis. Office: Children's Hosp Wayne State U 3901 Beaubien St Detroit MI 48201 Home: 26318 Dundee St Huntington Woods MI 48070

PERRIN, JAMES STUART, scientific corporation executive; b. Superior, Wis., June 19, 1936; s. Stuart Henry and Esther Catherine P.; m. Brenda Madge Mendenhall. B.S. in Metall. Engring, MIT, 1958, M.S., U. Ill., 1961; Ph.D. in Materials Sci, Stanford U., 1966. Registered profl. engr., Ohio, Calif. Sr. research leader Batelle Columbus (Ohio) Labs., 1966-81; pres. Fracture Control Corp., Goleta, Calif., 1981—. Editor: Proc. Symposium on Effects of Radiation on Materials, 1982. Fellow ASTM (award 1980, sec. com. E-10 1978-80, vice chmn. 1980); Mem. Am. Soc. Metals, AIME, Am. Nuclear Soc., ASME. Subspecialties: Metallurgical engineering; Nuclear engineering. Current work: Effects of nuclear radiation on structural materials in nuclear power reactors, especially in relation to mechanical properties. Home: PO Box 6098 Santa Barbara CA 93111 Office: Fracture Control Corp 340 G S Kellogg Ave Goleta CA 93117

PERRINE, RICHARD LEROY, environmental engineering educator; b. Mountain View, Calif., May 15, 1924; s. George Alexander and Marie (Axelson) P.; m. Barbara Jean Gale, Apr. 12, 1945; children: Cynthia Gale, Jeffrey Richard. A.B., San Jose State Coll., 1949; M.S., Stanford U., 1950, Ph.D. in Chemistry, 1953. Research chemist Calif. Research Corp., La Habra, 1953-59; asso. prof. U. Calif. at Los

Angeles, 1959-63, prof. engring. and applied sci., 1963—, chmn. environ. sci. and engring., 1971-82; cons. environ. sci. and engring., energy resources, flow in porous media; mem. Los Angeles County Energy Commn., 1973-81; mem. adv. council South Coast Air Quality Mgmt. Dist., 1977—; mem. air conservation com. Los Angeles County Lung Assn., 1970—. Served with AUS, 1943-46. Recipient Outstanding Engr. Merit award in environ. engring. Inst. Advancement Engring., 1975. Mem. Am. Chem. Soc., Soc. Petroleum Engrs., Am. Inst. Chem. Engrs., Can. Inst. Mining and Metallurgy, Nat. Assn. Environ. Edn., Nat. Assn. Environ. Profls., Air Pollution Control Assn., Am. Water Resources Assn., AAAS, N.Y. Acad. Scis., Western Regional Sci. Assn., Soc. Environ. Toxicology and Chemistry, Calif. Tomorrow, Sierra Club, Wilderness Soc., Audubon Soc., Sigma Xi, Tau Beta Pi, Phi Lambda Upsilon. Subspecialty: Environmental engineering. Home: 22611 Kittridge St Canoga Park CA 91307 Office: Engring I Room 2066 U Calif Los Angeles CA 90024

PERRONE, RONALD DAVID, physician, researcher; b. Newark, Mar. 24, 1949; s. John and Ethel P. B.S. magna cum laude, Pa. State U., 1971; M.D., Hahnemann Med. Coll., 1975. Lic. N.Y., Mass., Ga.; cert. internal medicine. Intern, resident in internal medicine Grady Meml. Hosp., Atlanta, 1975-77, 78; clin. fellow nephrology Boston U. Med. Sch., 1979-80, research fellow nephrology, 1980-82; asst. profl. medicine U. Rochester, N.Y., 1982—. Recipient Lange Med. award Lange Pub., 1975. Mem. ACP, Am. Soc. Nephrology, Am. Fedn. Clin. Research, Alpha Omega Alpha. Subspecialties: Nephrology; Membrane biology. Current work: Hormonal modification of epithelial transport. Office: U Rochester Med Ctr Renal Sect Rochester NY 14642

PERRY, CHARLES LEWIS, educator; b. Culver, Ind., Dec. 9, 1933; s. Leujay Joseph and Bertha Elizabeth (Lebold) P.; m.; 1 son, Bradford Jay. B.A., Ind. U., 1955; Ph.D., U. Calif., Berkeley, 1965. Research assoc. Kitt Peak Nat. Obs., Tucson, 1963-65; research fellow Mount Stromlo Obs., Canberra, Australia, 1965-66; asst. prof. La. State U., Baton Rouge, 1966-71, assoc. prof., 1971-79, prof. dept. physics and astronomy, 1979—. Contbr. articles in field to profl. jours. Mem. Internat. Astron. Union, Am. Astron. Soc., Astron. Soc. Pacific. Subspecialty: Optical astronomy. Current work: Multicolor photometry with applications to the study of dust component of interstellar medium and galactic clusters. Home: 442 Raintree Rd Baton Rouge LA 70810 Office: Dept Physic La State U Baton Rouge LA 70803

PERRY, DAVID ANTHONY, natural resources researcher, educator; b. Kansas City, Kans., Sept. 19, 1933; s. Everett Cecil and Maxine (Sharkey) P.; m. Peggy Louise Cayton, Nov. 17, 1962 (dec. Nov. 1981); children: Kyna, David Anthony; m. Carol Ivy Rosenblum, Aug. 17, 1982. B.S. in Forestry, U. Fla., 1961, M.S., 1966; M.S. in Physics, Mont. State U., 1971; Ph.D. in Ecology, Mont. State U., 1974. Range ecologist Mont. Dept. Natural Resources, Miles City, 1974-75; research forester U.S. Dept. Agr. Forest Service, Bozeman, Mont., 1975-77; asst. prof. dept. forest sci. Oreg. State U., Corvallis, 1977-82, assoc. prof., 1982—. Co-editor: Nitrogen Fixing in Temperate Forests, 1979; contbr. articles to profl. jours. Served with USMC, 1956-57. Mem. ACLU, Amnesty Internat. Democrat. Subspecialties: Ecology; Ecosystems analysis. Current work: Biology of managed ecosystems. Home: 5633 NW Fairoaks corvallis OR 97330 Office: Dept Forest Sci Oreg State U Corvallis OR 97331

PERRY, DAVID JOHN, physician, army officer; b. Eugene, Oreg., June 6, 1948; s. Raymond Lee and Andree Eugenie (Torralba) P. A.B., Princeton U., 1970; M.D., U. Pa., 1979. Cert. Am. Bd. Internal Medicine. Commd. capt. U.S. Army, 1975, advanced through grades to maj., 1979; intern U. Ariz., 1974-75; gen. med. officer Europe, Schweinfurt, W.Ger., 1975-77; resident Walter Reed Army Med. Ctr., Washington, 1977-79, hematology oncology fellow, 1979-82, staff physician, 1982—. Mem. ACP, Am. Soc. Clin. Oncology, Am. Soc. Hematology, Am. Fedn. Clin. Research, Assn. Mil. Surgeons U.S. Subspecialties: Chemotherapy; Hematology. Current work: Clinical research in solid tumor chemotherapy in particular head and neck cancer. Office: Hematology Oncology Service Walter Reed Army Medical Center Washington DC 20307

PERRY, ERIK DAVID, mechanical and structural engineer; b. Dallas, Oct. 17, 1952; s. John Wilson and Rotraud Anne-Ilse (Mezger) P.; m. Isabelle Marie Diana, Aug. 17, 1974; 1 son, Daniel Allen. B.S. in M.E, Cornell U., 1974; M.S.E., U. Mich., 1975; cert. in mgmt, Mercer County Community Coll., Hamilton Square, N.J., 1982. Teaching asst. U. Mich., 1974-75; engr. Lockheed Missiles & Space, Sunnyvale, Calif., 1975-76; sect. head Plasma Physics Lab. Princeton U., 1977—. Mem. Am. Nuclear Soc., ASME, AIAA. Subspecialties: Fusion; Mechanical engineering. Current work: Mechanical and structural design of tokamak type fusion reactors. Home: 545 Flock Rd Hamilton Square NJ 08690 Office: Plasma Physics Lab Princeton U PO Box 451 Princeton NJ 08544

PERRY, KATHARINE BROWNE, horticultural science educator; b. Washington, Mar. 19, 1952; d. Arthur Vincent and Mary Elizabeth (King) Browne; m. Steven Gerard Perry, Oct. 2, 1976; children: Matthew Thomas, James Arthur. B.S., Pa. State U., 1974, M.S., 1976, Ph.D., 1979. Research asst. dept. horticulture Pa. State U., State College, 1979-80; extension specialist N.C. State U., Raleigh, 1980-81, asst. prof. dept. horticultural sci., 1982—. Mem. Am. Meteorol. Soc., Am. Soc. Hort. Sci., Nat. Weather Assn., Sigma Xi, Alpha Zeta, Gamma Sigma Delta. Home: 1226 Kilmory Dr Cary NC 27511 Office: Dept Hort Sci NC State U 123 Kilgore Hall Raleigh NC 27607

PERRY, MARY JANE, oceanographer; b. N.Y.C., Mar. 30, 1948; d. William Joseph and Alice Margaret (Tierney) P.; m. Peter Alfred Jamars, Mar. 6, 1976. B.A., Coll. New Rochelle, 1969; Ph.D., U. Calif.-San Diego, 1974. Research fellow, research asst. U. Calif.-San Diego, 1969-74; research instr. Washington U. Med. Sch., St. Louis, 1974-76; research asst. prof. U. Wash., Seattle, 1976-80, research assoc. prof., 1982—; assoc. program dir. NSF, Washington, 1980-82. NSF research grantee, 1977-83. Mem. AAAS, Am. Geophys. Union, Phycol. Soc., Am., Soc. Limnology and Oceanography (meetings program coordinator), Am. Inst. Biol. Sci., Am. Women in Sci, Ecol. Soc. Am. Subspecialty: Biological oceanography. Current work: Biological oceanography, primary production, phytoplankton physiology and ecology, remote sensing, marine microbiology. Office: Sch Oceanography WBIO U Wash Seattle WA 98195

PERRY, MICHAEL CLINTON, physician, educator, researcher; b. Wyandotte, Mich., Jan. 27, 1945; s. Clarence Clinton and Hilda Grace (Wigginton) P.; m. Nancy Ann Kaluzny, June 22, 1968; children: Rebecca Carolyn, Katherine Grace. B.A., Wayne State U., 1966, M.D., 1970; M.S., U. Minn., 1975. Diplomate: Am. Bd. Internal Medicine, Hematology, Oncology. Intern Mayo Grad. Sch. Medicine, Rochester, Minn., 1970-71, resident in internal medicine, 1971-72, fellow hematology, 1972-74, fellow oncology, 1974-75, instr. dept. medicine, 1974-75; asst. prof. dept. medicine U. Mo., Columbia, 1975-80, assoc. prof. medicine, 1980—, dir. div. hematology-oncology, 1982—; cons. Ellis Fischel State Cancer Hosp., Harry S. Truman VA Hosp. Recipient Young Internist of Yr. award Am. Soc. Internal Medicine, 1981. Fellow ACP; mem. Mo. Soc. Internal Medicine (pres.-elect), Am. Soc. Clin. Oncology, Am. Soc. Hematology, Am. Soc. Internal Medicine, AMA, Am. Fedn. Clin. Research. Subspecialties: Chemotherapy; Cancer research (medicine). Current work: Clinical research-chemotherapy of cancer, teaching, patient care. Home: 1112 Pheasant Run Columbia MO 65201 Office: N408 Dept Medicine U Mo Health Scis Ctr One Hospital Dr Columbia MO 65212

PERRY, MILDRED ELIZABETH, analytical chemist, environ. researcher; b. Clarksville, Tenn., June 18, 1943; d. Elwyn Woodrow and Bessie (Robertson) Bateman; m. Don Merrill Perry, Aug. 12, 1966 (div. Mar. 1970). B.S., Austin Peay State Coll., 1965, M.S., 1969; Ph.D. in Analytical Chemistry, U. Tenn., 1983. Tchr. Met. Nashville Bd. Edn., 1966-69, Montgomery County Bd. Edn., Clarksville, 1969-77, teaching asst. dept. chemistry U. Tenn., Knoxville, 1977-81, instr., summer 1981, research asst., 1981—. Treas. Montgomery County Beautiful Com., Clarksville, 1973-75. Fed. Soil and Conservation Service grantee, 1976. Fellow Tenn. Acad. Sci. (named Disting. Tchr. 1975); me. Am. Chem. Soc., Soc. Applied Spectroscopy, Optical Soc. Am., Kappa Delta Pi, Beta Beta Beta. Subspecialties: Analytical chemistry; Spectroscopy. Current work: Characterization of coal-derived materials and shale oil by laser-induced fluorescence of matrix-isolated samples for polycyclic aromatic hydrocarbon content. Office: Dept Chemistry U Tenn W Cumberland Ave Knoxville TN 37996

PERRY, NELSON ALLEN, radiological physicist, cons.; b. Louisville, Mar. 26, 1937; s. Leslie I. and Sue H. (Harris) P.; m. Sarita S. Cornn, July 1, 1938; children: Melody S., Kimberly D. A.A., Campbellsville Jr. Coll., 1956; B.A., U. Louisville, 1962; M.S., U. Okla., 1966. Cert. hazard control exec., health care safety profl. Radiation safety officer Michael Reese Hosp., Chgo., 1966-67; instr., radiol. physicist St. Francis Hosp., Beech Grove, Ind., 1968-70; asst. prof., radiation safety officer U. Ind. Med. Center, Indpls., 1970-73; assoc. prof., radiation cons. Perry Radiol. Cons., Inc., Indpls., 1973-76; asst. prof., radiation safety officer U. South Ala., Mobile, 1976—. Contbr. articles to profl. jours. Active CD, Indpls., 1969-76 Active CD, Mobile, 1978—. USPHS trainee, 1965-66; Named hon. Ky. Col. Mem. Health Physics Soc., Am. Assn. Physicists in Medicine, Jaycees (state dir. 1968-73, also v.p., conv. club pres.). Subspecialties: Nuclear engineering; Radiology. Current work: Radiation safety; radiological instrument calibration and shielding design; teaching radiation physics and radiation biology. Home: 1150 Byronell Dr Mobile AL 36609 Office: 370 Cancer Center Mobile AL 36688

PERRY, WILLIAM JAMES, IV, geologist; b. Staunton, Va., Dec. 20, 1935; s. William James and Virginia (Wilbur) P.; m. Diane Alford, Jan. 31, 1959; children: Virginia Ellen, William J., Joel Huston. A.B., Johns Hopkins U., 1958; M.S., U. Mich., 1960; M.Phila., Yale U., 1968, Ph.D., 1971. Geologist United Fuel Gas Co., Charleston, W.Va., 1960-64; teaching asst., asso. Yale U., New Haven, 1964-68; area geologist U.S. Geol. Survey, Metairie, La., 1968-71, geologist, Arlington, Va., 1971-74, Washington, 1971-74, Reston, Va., 1974-80, Lakewood, Colo., 1980—. Contbr. articles to profl. jours. Vol. No. Va. Hotline, 1978-80. Sigma Xi summer research grantee, 1966; recipient U.S. Geol. Survey unit citation, 1977. Fellow Geol. Soc. Am.; mem. Am. Geophys. Union, Am. Assn. Petroleum Geologists (assoc. editor 1980-81), Rocky Mountain Assn. Geologists, Sigma Xi. Club: Tudor and Stuart. Subspecialties: Geology; Tectonics. Current work: Determination of structural geometry, sequence and style of deformation, burial history and petroleum geology of Mont. Thrust Belt. Home: 2695 Vassar Dr Boulder CO 80303 Office: U S Geol Survey MS 940 Federal Center Box 25046 Denver CO 80225

PERRYMAN, JAMES HARVEY, neuromuscular researcher, biological science educator; b. Kansas City, Mo., Aug. 18, 1918; s. Alfred Fey and Christine Snowden (Moore) P.; m. Lucile Cooney, July 3, 1948. A.B., Stanford U., 1940; M.A., U. Calif.-Berkeley, 1942; Ph.D., U. Calif.-San Francisco, 1953. Mem. in biomechanics U. Calif. Med. Sch.-San Francisco, 1949-51, research physiologist, 1954-56; asst. prof. U. San Francisco, 1954-56; assoc. prof. NYU, 1956-70; prof. biol. sci. U. Medicine and Dentistry N.J., Newark, 1970—; research prof. Queens Coll., 1966-70; hon. lectr. Collogium, 1970. Contbr. articles on ocular system, nervous system, muscles to profl. publs.; editor, contbg. author: Oral Physiology, 1978. Served in U.S. Army, 1942-46; Philippines. NIH-USPHS grantee, 1957-67; NRC-VA grantee, 1949-56; Coll. Medicine and Denistry N.J. Found. grantee, 1973-75; State N.J. Research grantee, 1971-73. Fellow Internat. Orgn. Psychophysiology (hon.), AAAS; mem. Am. Assn. Anatomists, Am. Assn. Physiologists, Internat. Orgn. For Study Paine. Subspecialties: Neurophysiology; Biofeedback. Current work: Muscle coordination and automic nervous system function; sensory and muscular feedback controlling coordinated muscular action of head and neck. Home: 330 3d Ave New York NY 10010 Office: U Medicine and Dentistry NJ 100 Bergen St Newark NJ 07103

PERSEK, STEPHEN CHARLES, management science educator, nonlinear differential equations researcher; b. Long Island City, N.Y., May 4, 1945; s. Stephen George and Zora Jane (Duzbaba) P. B.S. in Applied Math, MIT, 1966, M.S., Courant Inst., 1968, Ph.D., 1976. Instr. in bus. stats. N.Y. Inst. Tech., 1968-76; asst. prof. math. Marist Coll., 1977-79; assoc. prof. mgmt. sci. St. John's U., Jamaica, N.Y., 1979—; staff cons. GSP Cons., N.Y.C., 1981—; stats. cons. Purchasing Mgmt. Assn. N.Y., N.Y.C., 1981—. Contbr. articles to math. jours. Mem. Soc. Indsl. and Applied Math., Am. Math. Soc., Math. Assn. Am., Inst. Mgmt. Sci. Roman Catholic. Subspecialties: Applied mathematics; Operations research (mathematics). Current work: Applied mathematics: Nonlinear ODE's, nonlinear PDE's, stability theory, uniform approximations, bifurcation theory, nonlinear oscillations and wave motions. Home: 160 Banbury Rd Mineola NY 11501 Office: CBA-Mgmt St John's U Grand Central and Utopia Pkwys Jamaica NY 11439

PERSON, DONALD AMES, pediatrician, rheumatologist; b. Fargo, N.D., July 17, 1938; s. Ingwald Haldor and Elma Wilhelmina (Karlstrom) P.; m. Blanche Durand, Apr. 28, 1962; children: Donald Ames, David Wesley. Student, Gustavus Adolphus Coll., 1956-58, U. Minn., 1958-59; B.S., U. N.D., 1961; M.D., U. Minn., 1963. Intern Mpls.-Hennepin County Gen. Hosp., 1963-64; resident neurol. surgery Mayo Clinic and Mayo Grad. Sch. Medicine, Rochester, Minn., 1967, fellow in microbiology, 1968-70; research asso. Baylor Coll. Medicine, Houston, 1971, Arthritis Found. fellow, 1972-74, mem. faculty, 1971—, asst. prof. pediatrics, 1980—, resident in pediatrics, 1978-80; asst. attending pediatrics Harris County Hosp. Dist., 1980—; rheumatologist Tex. Children's Hosp., 1980—, attending pediatrician, 1982—; cons. Kelsey Seybold Clinic, 1980—. Contbr. articles to profl. jours. Served with AUS, 1964-66. Arthritis Found. sr. investigator, 1975-77. Mem. Am. Acad. Pediatrics, AAAS, Am. Fedn. Clin. Research, AMA, Am. Rheumatism Assn., Am. Soc. Microbiology, Soc. Pediatric Research, Am. Soc. Tropical Medicine and Hygiene, Arthritis Found. (dir., med. adv. bd.), Assn. Mil. Surgeons U.S., Harris County Med. Soc., Houston Acad. Medicine, Houston Pediatric Soc., Internat. Orgn. Mycoplasmologists, N.Y. Acad. Sci., N.D. Acad. Sci., Soc. Exptl. Biology and Medicine, So. Soc. Pediatric Research, S.W. Sci. Forum, Tex. Med. Assn., Tex. Pediatric Soc., Tex. Rheumatism Assn., Tissue Culture Assn., U.S. Fedn. Culture Collections. Subspecialties: Pediatrics; Virology (biology). Office: Texas Children's Hosp Dept Rheumatology Box 20269 Houston TX 77030

PERSON, WILLIS BAGLEY, chemistry educator; b. Salem, Oreg., Apr. 23, 1928; s. Carl Waldo and Grace Cassity (Bagley) P.; m. Loretta Fay Ferren, Dec. 26, 1949; children: Jean Kathryn, George David. B.S., Williamette U., 1947; M.S., Oreg. State Coll.-Corvallis, 1949; Ph.D., U. Calif.-Berkeley, 1953. Research fellow U. Minn., 1952-54; inst. Harvard U., 1954-55; asst. prof. U. Iowa, 1955-61, assoc. prof., 1961-66; NSF postdoctoral fellow, vis. assoc. prof. U. Chgo., 1965-66; prof. chemistry U. Fla., Gainesville, 1966—; vis. staff mem. Los Alamos Nat. Lab., 1975—; UNESCO cons. State U. Campinas, Brazil, 1980. Author: (with R. S. Mulliken) Molecular Complexes, 1969; editor: (with G. Zerbi) Vibrational Intensities in Infrared and Raman Spectroscopy, 1982; contbr. over 130 articles in field to profl. jours. Guggenheim fellow U. Chgo., 1960-61; Chem. Soc. sr. postdoctoral fellow, 1978. Mem. Am. Chem. Soc., Optical Soc. Am., Chem. Soc. (London), AAAS, Coblentz Soc. Subspecialties: Physical chemistry; Spectroscopy. Current work: Molecular spectroscopy, particularly vibrational intensities and laser spectroscopy, with applications to molecular structure, photochemistry and biophysics. Office: Dept Chemistry U Fla Gainesville FL 32611

PERTSCHUK, LOUIS PHILIP, pathologist; b. London, July 4, 1925; s. Isaac M. and Rose P.; m. Jeanette Smith, Aug. 20, 1953; children: Eric, Shawn, Brandy. A.B., NYU, 1946; D.O., Phila. Coll. Osteo. Medicine, 1950. Diplomate: Am. Bd. Pathology. Instr. Downstate Med. Ctr., SUNY-Bklyn., 1974-75, asst. prof., 1975-79, assoc. prof., 1979—; cons. Corning (N.Y.) Glass Works, 1982—, Zeus Sci. Co., 1982—, Abbott Labs., 1982—. Editor: Morphology of Steroid Hormone Binding in Human Tumors, 1984. Served with U.S. Army, 1943-46. NCI/NIH grantee, 1979, 82. Fellow Coll. Am. Pathologists, Am. Soc. Clin. Pathologists; mem. Am. Assn. Pathologists, AAAS, Internat. Acad. Pathology, N.Y. Acad. Sci. Subspecialty: Pathology (medicine). Current work: Identification of steroid hormone binding sites in human neoplasms by histochemical and immunohistological techniques. Home: 33-01 255 St Little Neck NY 11363 Office: Box 25 Downstate Med Center 450 Clarkson Ave Brooklyn NY 11203

PESANTI, EDWARD LOUIS, physician; b. Deer Lodge, Mont., June 23, 1944; s. Edward and Agnes (Hogue) P.; m. Madeline Gonsky, June 13, 1969; children: Deborah, Heather, Stephen. B.A., Carroll Coll., Helena, Mont., 1965; M.D., U. Chgo., 1969. Resident Barnes Hosp., St. Louis, 1969-71; med. officer USPHS, Window Rock, Ariz., 1971-73; fellow Stanford Hosp., 1973-75; asst. prof. U. Iowa, Iowa City, 1975-82; assoc. prof. medicine U. Conn., VA Med. Ctr., Newington, 1982—. Served to lt. comdr. USPHS, 1971-73. Fellow ACP, Infectious Diseases Soc. Am. Subspecialties: Parasitology; Infectious diseases. Current work: Biology and Immunology of Pneurmocystis carinii; pulmonary defenses against microorganisms. Home: 242 Burnham Rd Avon CT 06001 Office: VA Med Center 555 Willard St Newington CT 06111

PESCH, PETER, astronomer; b. Zurich, Switzerland, June 29, 1934; s. Roland Henry and Hildegard May (Stoecklin) P.; m. Donna Marie Lehmann, July 25, 1975; children—Marina, Ada Ines. B.S. in Physics, U. Chgo., 1955, M.S., 1956, Ph.D. in Astronomy, 1960. Research asso. Case Western Res. U., 1960-61, mem. faculty, 1961—, prof. astronomy, 1974—; chmn. dept., dir. Warner and Swasey Obs., 1975—; on leave with Kitt Peak Nat. Obs., Tucson, 1970-71; program dir. astron. instrumentation and devel. program NSF, summer 1972. Mem. Am. Astron. Soc., Internat. Astron. Union. Subspecialty: Optical astronomy. Current work: Studies of galactic structure and cosmology by means of objective prism surveys with schmidt-type telescope. Home: 2361 Euclid Heights Blvd Cleveland Heights OH 44106 Office: 1975 Taylor Rd East Cleveland OH 44112

PESCH, WILLIAM ALLAN, systems engineer, photographic company executive; b. Buffalo, Jan. 18, 1948; s. William Adam and Anna Martha (Hinz) P.; m. Cheryle Lynn Ryer, Oct. 3, 1969; children: Pamela Ann, William Allan, Jennifer Lynn. B.S.I.E., SUNY-Buffalo, 1970; M.S.I.E., Purdue U., 1971, postgrad., 1972. Instr. Purdue U., West Lafayette, Ind., 1970-72; engr. Eastman Kodak Co., Rochester, N.Y., 1972-74, mgr. domestic shipping, 1974-76, mgr. adminstrn., 1976-78, sysaems supr., 1978-81, mgr. tech., planning, and research, 1981—; mgmt. cons. Western N.Y. Hosp. Assn., Buffalo, 1970; computer cons. Gates Pub. Library, Rochester, 1982-83; Chpt. reviewer Handbook of Industrial Engineering, 1983. Committeeman Monroe County Dem. Com., Rochester, 1975-80; budget dir. Washington Irving Sch. PTA, Rochester, 1981. Named Student Engr. of Yr. N.Y. State Soc. Profl. Engrs., Erie County, 1970; Engring. Student of Yr. Sch. Engring., SUNY-Buffalo, 1970. Mem. Inst. Mgmt. Sci., Ops. Research Soc., Am. Inst. Indsl. Engrs. Subspecialties: Operations research (engineering); Information systems, storage, and retrieval (computer science). Current work: Working to provide operations research-based dicision support tools for company management; automating information and control systems and office automation systems. Home: 23 Pine Knoll Dr Rochester NY 14624 Office: Eastman Kodak Co/Distbn Div 343 State St B-205 Rochester NY 14650

PESCHMANN, KRISTIAN RALF, physicist; b. Hamburg, W.Ger., Dec. 11, 1940; s. Erich and Kathe Lucy (Bliesath) P.; m.; children: Kristin, Konrad. Diplom Physicist, U. Kiel, 1966, Dr. rer. nat., 1968. Asst. prof. U. Kiel; scientist Philips Research Lab., Aachen, W.Ger., 1969-80; asso. prof. dept. radiology U. Calif., San Francisco, 1980—; dir. advanced devel. Imatron Assoc., 1981. Mem. Soc. Photo-optical Instrumentation Engrs., Am. Assn. Physicists in Medicine. Club: Acad. Rowing (Kiel). Subspecialties: Imaging technology; Atomic and molecular physics. Current work: Computerized radiologic imaging devices; silverless radiography; advancement of fast X-ray detection tech. Home: 762 Madrid St San Francisco CA 94112 Office: 460 Carlton Ct South San Francisco CA 94080

PESKE, PATRIC O'CONNELL, diagnostic psychologist; b. Akron, Ohio, Sept. 21, 1942; s. Robert Wilhelm Peske and Eileen Michele (Doherty) Bordeaus; m. Nancy Pokorosky, Nov. 18, 1966; 1 son, Arthur Zygmunt Aleksandor. B.A., U. Akron, 1968, M.A., 1972; postgrad., Case Western Res. U., 1968, Temple U., 1970-72, U. Chgo., 1975—. Lic. psychologist, Ohio, Mich. Sr. research asst. E. N. Hay Assoc., Inc., Phila., 1970-72; intern in child study psychology Wadsworth (Ohio) Edn. System, 1972-73; psychologist Genesee Intermediate Edn. Office, Flint, Mich., 1973—; sr. analyst Pro-Media Cons., Flint, 1973—; diagnostic cons., 1972—; instr. Mott Community Coll., 1978—; speaker at numerous nat. internat. congresses and orgns.; U.S. del. Internat. Rorschach Congress, Bern, Switzerland, 1978. Research numerous publs. in field. Public service speaker, radio, TV, Flint, 1979—; mem. Republican Nat. Com., 1963—. Recipient award Nat. Assn. Sch. Psychology, 1973, 74, 75, 76, 77, 78. Mem. Am. Psychol. Assn., Brit. Soc. Projective Psychology, Internat. Soc. Study Art and Psychopathology, Internat. Rorschach Congress, Soc. Personality Assessment, Nat. Research Soc. Subspecialties: Developmental psychology; Learning. Current work: Mnemonic and semiotic learning and development in young children and adults and their influences on unconscious perceptual decision-making in presence of ambiguous visual stimuli. Office: Care Nat Rorschach Soc PO Box 7149 Flint MI 48507

PETER, GEORGES, physician; b. Cambridge, Mass., June 23, 1938; s. J. Georges and Helen Fairchild (Mann) P.; m. Carolyn Lisle

McClintock, Oct. 3, 1964; children: Allison Fairchild, Marc Phillips. A.B., Harvard U., 1959, M.D., 1964; B.Med.Sci., Dartmouth Med. Sch., 1962, Brown U., 1979. Diplomate: Am. Bd. Pediatrics. Postgrad. tng. in pediatrics and medicine Strong Meml. Hosp., Rochester, N.Y., 1964-66; clin. assoc. NIH, Bethesda, Md., 1966-68; resident in pediatrics, fellow in immunology and invectious deseases Children's Hosp. Med. Center, Boston, 1968-72; research and teaching fellow Harvard Med. Sch., Boston, 1968-72; asst. prof. pediatrics Brown U., Providence, 1972-78, assoc. prof., 1978—; physician Roger Williams Gen. Hosp., Providence, 1972-76; dir. div. infectious diseases, dept. pediatrics and medicine R.I. Hosp., Providence, 1976—. Contbr. 60 articles to profl. jours. Mem. Harvard-Radcliffe Schs. Com. R.I., 1976—, vice chmn., 1982—. Served with USPHS, 1966-68. NIH spl. fellow, 1966-68; FDA grantee, 1975-77; Charles H. Hood Found. grantee, 1978-79. Fellow Am. Acad. Pediatrics, Infectious Disease Soc. Am.; mem. Am. Fedn. Clin. Research, Am. Soc. Microbiology, Soc. Pediatric Research. Democrat. Clubs: Annisquam Yacht (Gloucester, Mass.); Harvard of R.I., Univ. (Providence). Subspecialties: Pediatrics; Infectious diseases. Current work: Epidemiology and prevention of invasive Haemophilus influenza disease; role of serum antibody to the core glycolipid of Enterobactericeae; in vitro and clinical studies of antibodies. Office: RI Hosp 593 Eddy St Providence RI 02902

PETER, THOMAS DOUGLAS, geneticist, educator, researcher; b. Washington, Mar. 27, 1947; s. Joseph and Helen Barbara (Szalanczy) P.; m. Rosann Alexander Farber, July 20, 1973; 1 dau., Laura Elizabeth. Sc.B., Brown U., 1969; Ph.D. in genetics, U. Wash.-Seattle, 1973. Postdoctoral fellow Nat. Inst. Med. Research, London, 1973-75; M.I.T., 1975-77; asst. prof. dept. microbiology U. Chgo., 1977-82, assoc. prof., 1982—. Recipient Research Career Devel. award Nat. Inst. Aging, 1980—. Subspecialties: Genome organization; Genetics and genetic engineering (biology). Current work: I am interested in the organization and genetic behavior of the yeast gernone. Home: 5618 S Kimbark Ave Chicago IL 60637 Office: Dept Microbiolog U Chicago 920 E 58th St Chicago IL 60637

PETERLE, TONY JOHN, zoologist, educator; b. Cleve., July 7, 1925; s. Anton and Anna (Katic) P.; m. Thelma Josephine Coleman, July 30, 1949; children—Ann Faulkner, Tony Scott. B.S., Utah State U., 1949; M.S., U. Mich., 1950, Ph.D. (Univ. scholar), 1954; Fulbright scholar, U. Aberdeen, Scotland, 1954-55; postgrad., Oak Ridge Inst. Nuclear Studies, 1961. With Niederhauser Lumber Co., 1947-49, Macfarland Tree Service, 1949-51; research biologist Mich. Dept. Conservation, 1951-54; asst. dir. Rose Lake Expt. Sta., 1955-59; leader Ohio Coop. Wildlife Research unit U.S. Fish and Wildlife Service, Dept. Interior, 1959-63; asso. prof., then prof. zoology Ohio State U., Columbus, 1959—, chmn. faculty population and environmental biology, 1968-69, chmn. dept. zoology, 1969-81, dir. program in environ. biology, 1970-71; co-organizer XIII Internat. Congress Game Biology, chmn. internat. affairs com., mem. com. ecotoxicology, 1979-80; mem. com. rev. EPA pesticide decision making Nat. Acad. Scis.-NRC; mem. vis. scientists program Am. Inst. Biol. Scis.-ERDA, 1971-77; mem. com. pesticides Nat. Acad. Scis., com. on emerging trends in agr. and effects on fish and wildlife; mem. ecology com. of sci. adv. council EPA, 1979—; mem. research units coordinating com. Ohio Coop. Wildlife and Fisheries, 1963—. Editor: Jour. of Wildlife Management, 1969-70, 84-85. Served with AUS, 1943-46. Fellow AAAS, Am. Inst. Biol. Scis., Ohio Acad. Sci.; mem. Wildlife Disease Assn., Wildlife Soc. (regional rep. 1962-67, v.p. 1968, pres. 1972), Ecol. Soc., INTECOL-NSF panel U.S.-Japan Program, Xi Sigma Pi, Phi Kappa Phi. Subspecialties: Environmental toxicology; Ecology. Current work: Professor of zoology, research on transport, fate, effects of toxic substances in environment; editor-in-chief Journal of Wildlife Management, 1984-85. Home: RFD 3 Delaware OH 43015 Office: Dept Zoology Ohio State U 1735 Neil Ave Columbus OH 43210

PETERS, ALEXANDER ROBERT, mechanical engineering educator; b. Pender, Nebr., Nov. 17, 1936; s. Alexander and Maude (Murphy) P.; m. Jane Catherine Spence, Feb. 14, 1959; children: Catherine Marie, Jane Louise. B.S.M.E., U. Nebr., 1959, M.S., 1963; Ph.D., Okla. State U., 1967. Registered profl. engr., Nebr. Engr. aerodynamics space and info. systems div. N.Am. Aviation, Downey, Calif., 1963-64; instr. mech. engring. U. Nebr., Lincoln, 1966-67, asst. prof., 1967-70, assoc. prof., 1970-75, prof., chmn. dept., 1975—; engring. specialist Ford Motor Co., Dearborn, Mich., 1970, summers 1971, 73. Served to 1st lt. U.S.M.C., 1959-62. NASA trainee, 1964-66; NSF and NIH grantee, 1967-77. Mem. ASME (Chmn. region VII dept. heads 1982-84, 1982), Am. Soc. Engring. Edn., AIAA, Soc. Automotive Engrs. (Ralph R. Teetor ednl. award 1969), Fine Particle Soc., Accreditation Bd. Engring. and Tech. (bd. dirs. 1983-86), Pi Tau Sigma (v.p. 1975-80, pres. 1980—). Republican. Presbyterian. Clubs: Lincoln University, Knolls Country. Lodge: Elks. Subspecialties: Mechanical engineering; Fluid mechanics. Current work: Fluidized bed heat transfer, slurry flow, pneumatic conveying. Home: 3235 S 29th St Lincoln NE 68502 Office: U Nebr-Lincoln 255NEC Lincoln NE 68588

PETERS, CHARLES FREDERICK, systems engr., astronomer; b. Columbus, Ohio, Oct. 11, 1935; s. Samuel W. and Mary K. (Cooper) P.; m. Martha J. Burkhart, Nov. 21, 1959; children: Charles, Kathleen, Kevin. B.A., Vanderbilt U., 1957, M.A., 1962; M.S., Yale U., 1966, Ph.D., 1968. Mathematician U.S. Naval Weapons Lab., Dahlgren, Va., 1958-70; mem. tech. staff Jet Propulsion Lab., Calif., Inst. Tech., Pasadena, 1970-81; sr. systems engr. Valley Forge Space div. Gen. Electric Co., Phila., 1981—; mem. NASA Pioneer 10 Nav. Team, 1972, Voyager Satellite Ephemeris Devel. Team, 1980, Voyager Nav. Team, 1980. Contbr. articles on satellites and celestial mechanics to profl. jours. Mem. Am. Astron. Soc. Roman Catholic. Subspecialties: Astronautics; Satellite studies. Current work: Celestial mechanics; satellites; civilian and mil. space contracts at General Electric. Home: 1017 Derwydd Ln Berwyn PA 19312 Office: GE Space Div Box 8555 Philadelphia PA 19101

PETERS, CHARLES WILLIAM, corp. ofcl., nuclear physicist; b. Pierceton, Ind., Dec. 9, 1927; s. Charles Frederick and Zelda May (Line) P.; m. Katharine Louise Schuman, May 29, 1953; children: Susan K., Bruce Merkle, Leslie Sanaie, Philip Merkle, William Merkle; m. Patricia Ann Miles, Jan 2, 1981. B.A., Ind. U., 1950; postgrad., U. Md., 1952-58. Supervisory research physicist Naval Research Lab., 1950-71; physicist EPA, 1971-76; mgr. advanced systems Consol. Controls Corp., Springfield, Va., 1976—. Contbr. numerous articles to profl. jours. Served to cpl. AUS, 1945-47. Mem. IEEE, Am. Phys. Soc., AAAS. Subspecialties: Nuclear physics; Operations research (mathematics). Current work: Radiation measurement systems, radiation detectors, robotic applications, applications of inelastic-gamma ray spectrography, fast-neutron radiography. Home: 12303 Mulberry Ct Woodbridge VA 22192 Office: PO Box 726 Springfield VA 22150

PETERS, GERALD JOSEPH, physicist; b. Balt., Sept. 29, 1941; s. Harold Raymond and Mary Elizabeth P.; m. Helena Catherine Madrzykowski, Aug. 25, 1967; children: Gregory, Tamara. B.S., Loyola Coll., Balt., 1963; M.S., U. Toledo, 1965, postgrad., 1966-68; postgrad., U. Md., 1969-74. Electronics engr., physicist U.S. Naval Ordnance Lab., Silver Spring, Md., 1965-66; research physicist U.S. Naval Surface Weapons Ctr., Silver Spring, 1968-80; physicist U.S. Dept. Energy, Washington, 1980—; teaching asst. U. Toledo, 1963-65, grad. trainee, 1966-68. Loyola Coll. scholar, 1959-63. Mem. IEEE, AAAS, Am. Nuclear Soc. Subspecialties: Nuclear physics; Particle physics. Current work: Management in basic research fields of high energy physics and nuclear physics with special interest in particle accelerator and instrumentation technologies. Office: Office High Energy and Nuclear Physics ER 20.1 US Dept Energy Washington DC 20545

PETERS, LAWRENCE HARVEY, management educator; b. St. Louis, Oct. 19, 1945; s. Albert and Anna Sarah (Schmidt) P.; m. Jennifer Rosemary Ruhl, Jan. 28, 1945; 1 dau., Andrea Jean. A.B., Washington U., 1968; M.S., So. Ill. U., 1969; Ph.D., Purdue U., 1975. Asst. prof. mgmt. U. Tex.-Dallas, 1975-80, assoc. prof. mgmt., 1980-81, So. Ill. U., Carbondale, 1981—; cons. Tex. Instruments, 1979, ERA Realtors, 1980. Contbr. articles to profl. jours. Served with U.S. Army, 1969-72. AMOCO Teaching award, 1981; Organized Research awards U. Tex. at Dallas, 1976-81. Mem. Am. Psychol. Assn., Acad. Mgmt., Midwest Psychol. Assn., S.W. Acad. Mgmt. (program coordinator), So. Acad. Mgmt. Current work: Individual effectiveness at work: Situational factors which influence person's performance and affective responses; race/sex bias. Home: 3107 W Kent Dr Carbondale IL 62901 Office: Dept Adminstrv Scis So Ill Univ Carbondale IL 62901

PETERS, MARVIN ARTHUR, pharmacologist, educator; b. Saginaw, Mich., June 23, 1933; s. Walter Albert and Evelyn C. (Maier) P.; m. Beverly Jean Clevenger, June 23, 1957; children: Bruce, Sharon, Coleen. B.S., Ferris State Coll.; M.S., Loma Linda U.; Ph.D., U. Iowa. Pharmacist Hinsdale (Ill.) Hosp., 1957-63; mem. faculty Loma Linda (Calif.) U., 1969—, now prof. physiology and pharmacology; researcher Nat. Inst. Environ. Health Sci., 1976-77; vis. prof. Universidad de Montemorelos, Mex. Contbr. numerous sci. articles and abstracts to profl. publs. Mem. AAAS, Western Pharmacology Soc., Sigma Xi. Republican. Subspecialties: Developmental biology; Pharmacology. Current work: Transplacental toxicology and perinatal pharmacology; transplacental effects of drugs of abuse. Office: Loma Linda Univ Pharmacology Dept Loma Linda CA 92350

PETERS, MAX STONE, chemical engineer, educator; b. Delaware, Ohio, Aug. 23, 1920; s. Charles Clinton and Dixie Mae (Stone) P.; m. Laurnell Louise Stephens, June 29, 1947; children—Margaret Dixie, M. Stephen. B.S. in Chem Engring, Pa. State U., 1942, M.S., 1947, Ph.D. (Shell Oil Co. grad. fellow 1949-51), 1951. Registered profl. engr., Pa., Colo. Prodn. supr. Hercules Powder Co., 1942-44; research asst. Pa. State U., 1946-47; tech. plant supr. George I. Treyz Chem. Co., 1947-49; mem. faculty U. Ill., 1951-62, prof. chem. engring., 1957-62, head dept., 1958-62; dean engring. U. Colo., 1962-78, prof. chem. engring., 1978—, chmn. dept., 1981—; Mem. adv. com. engring. div. NSF, 1962-66; chmn. Pres.'s Nat. Medal Sci. Com., 1969—, Colo. Environ. Commn., 1970-72. Author: Elementary Chemical Engineering, 1954, Plant Design and Economics for Chemical Engineers, rev. edit, 1980; Cons. editor: McGraw-Hill series chem. engring, 1960—. Served with AUS, 1944-46. Recipient Merit award Am. Assn. Cost Engrs., 1969; Distinguished Alumnus award Pa. State U., 1974, U. Colo., 1971; Phillips Lecture award Okla. State U., 1980. Mem. Nat. Acad. Engring., Am. Inst. Chem. Engrs. (dir. 1961-64, pres. 1968, Founders award 1974, Lewis award 1979), Am. Soc. Engring. Edn. (chmn. chem. engring. div. 1962, sec. engring. coll. adminstrn. council 1965-67, George Westinghouse award 1959, Lamme award 1973), Am. Chem. Soc. (adv. bd. jour. 1956-59), Am. Assn. Cost Engring., Sigma Xi, Alpha Chi Sigma, Phi Eta Sigma, Phi Lambda Upsilon, Tau Beta Pi, Sigma Tau. Subspecialties: Chemical engineering; Biomass (energy science and technology). Current work: Biomass processes for alchol production and economics of processes. Spl. research biomass, kinetics, mechanisms. Home: 1965 Vassar Circle Boulder CO 80303

PETERS, PAUL STANLEY, JR., linguistics educator; b. Houston, Dec. 7, 1941; s. Paul Stanley and Mary Beth P.; m.; children by previous marriage: Craig, Rebecca, Daniel. B.S. in Math, MIT, 1963. Prof. linguistics U. Tex.-Austin, 1966-83, chmn. dept., 1977-81, co-dir., 1978-83; prof. linguistics Stanford U., Calif., 1983—, assoc. dir., 1983—. Guggenheim fellow, 1973-74. Mem. Linguistic Soc. Am. Subspecialties: Cognition; Automated language processing. Current work: Formal syntax, model-theoretic semantics. Office: Linguistics Dept Stanford U Stanford CA 94305

PETERS, TERENCE MALCOLM, medical physicist; b. Dunedin, N.Z., Jan. 5, 1948; s. Alfred Malcolm and Lylah Campbell (Munro) P.; m. Leigh Sandrey, Jan. 17, 1970; children: Emma, Timothy, Carla. B.E. with 1st class honors, U. Canterbury, Christchurch, N.Z., 1970, Ph.D., 1973. Physicist, bioengr. dept. med. physics Christchurch Hosp., 1973-78; med. physicist Montreal (Que. Can.) Neurol. Inst., 1978—; asst. prof. depts. neurology and neurosurgery, radiology, med. physics McGill U., Montreal, 1980—. Mem. Australasian Coll. Phys. Scientists in Medicine, Assn. Physiciens et Ingenieurs Biomedicaux des Centres Hospitaliers du Que., Am. Assn. Physicists in Medicine, IEEE. Mem. United Ch. Can. Club: Mt. Royal Operatic Soc. (Montreal). Subspecialties: Imaging technology; CAT scan. Current work: Devel. image processing techniques in radiology. Home: 421 Greenoch Ave Town of Mount Royal PQ Canada H3P 2H3 Office: 3801 University St Montreal PQ Canada H3A 2B4

PETERS, THOMAS GUY, surgical educator, consultant, researcher; b. Cin., Oct. 3, 1945; s. Robert Lewis and Martha (Renter) P.; m. Dorothy Jean Geers, Apr. 15, 1977; children: Elizabeth Jan, Andrew Thomas. A.V., Miami (Ohio) U., 1966; M.D., U. Cin., 1970. Diplomate: Am. Bd. Surgery. Intern Milwaukee County Gen. Hosp., 1970-71; resident in surgery Med. Coll. Wis., Milw., 1970-77; fellow in clin. transplantation U Colo., Denver, 1977-78; asst. prof. surgery U. Tenn., Memphis, 1978—; pres. med. staff U. Tenn. Ctr., Memphis, 1982. Contbr. articles and abstracts to profl. jours., chpts. in books. Pres. West Tenn. Kidney Found., 1981-82; mem. affiliate relations com. Nat. Kidney Found., 1982. Served to maj. USAR, 1972-74. U. Cin. research fellow, 1967; Wis. Surg. Soc. awardee, 1976, 77; named Outstanding Tchr. U. Tenn. Nat. Alumni Assn., 1980. Fellow ACS; mem. Assn. Acad. Surgery, Am. Soc. Transplant Surgeons, Transplantation Soc., AMA. Subspecialties: Surgery; Transplant surgery. Current work: Kidney and liver transplantation surgery, multiorgan transplantation surgery, nutritional support in severely ill. Home: 1555 Vinton Ave Memphis TN 38104 Office: Dept Surgery U Tenn 956 Court Ave Memphis TN 38163

PETERS, WILLIAM CALLIER, mining consultant; b. Oxford, Ohio, July 12, 1920; s. Thomas Richard and Isabelle (Callier) P.; m. ruth Benseler, Sept. 6, 1942; 1 son, Stephen. B.A., Miami U., Oxford, Ohio, 1942; M.S., U. Colo., 1948, Ph.D., 1957. Registered profl. engr., Ariz. Mine geologist N.J. Zinc Co., Gilman, Colo., 1948-49; exploration geologist FMC Corp., Denver, 1950-59; chief geologist Utah Copper div. (Kennecott Copper Corp.), Salt Lake City, 1959-64; prof. mining and geol. engring. U. Ariz., Tucson, 1964-82; mining cons., Tucson, 1982—. Author: Exploration and Mining Geology, 1978; editor: Mining and Ecology in the Arid Environment, 1970. Served to capt. U.S. Army, 1942-46; ETO. Fellow Geol. Soc. Am.; mem. Soc. Econ. Geologistss, Instn. Mining and Metallurgy, Can. Inst. Mining and Metallurgy, Mining and Metall. Soc. Am. Club: Mining of S.W. (Tucson, Ariz.). Subspecialties: Geology. Current work: Mineral exploration and development, mining geology. Home: 5702 E 7th St Tucson AZ 85711

PETERSDORF, ROBERT GEORGE, internist, medical educator, university chancellor and dean; b. Berlin, Feb. 14, 1926; s. Hans H. and Sonja P.; m. Patricia Horton Qua, June 2, 1951; children: Stephen Hans, John Eric. B.A. cum laude, Brown U., 1948; M.D., Yale U., 1952; D.Sc. (hon.), Albany Med. Coll., 1979, Bethel Coll. Pa., 1982, A.M., Harvard U., 1980. Diplomate: Am. Bd. Internal Medicine, 1959, past chmn. Intern Yale U., 1952-53, resident, 1953-54, chief resident, 1957-58; researcher Johns Hopkins U., 1955-57, asst. prof. medicine, 1958-60; assoc. prof. U. Wash., 1960-62, prof., 1962-79; chmn. dept. medicine, 1964-79; prof. Harvard U., 1969-81; pres. Brigham Young Women's Hosp., 1979-81; prof. U. Calif.-San Diego, 1981—, now vice chancellor for health scis., dean; dir. Am. Hosp. Supply Corp. Contbr. numerous articles to profl. publs.; editor numerous books. Served with USAAF, 1944-46. Recipient Lederl Med. Faculty award, Wiggers award Albany Med. Coll., 1979, Keen Disting. Alumni award Bonn U., 1980. Mem. ACP (pres. 1975-76, Alfred E. Stengler Meml. award 1980), Am. Soc. Clin. Investigation (counselor 1969-72), Assn. Am. Physicians (pres. 1977), Western Assn. Physicians, Western Soc. Clin. Investigation (pres. 1971-72), Am. Acad. Arts and Scis., Inst. Medicine. Club: Cosmos (Washington). Subspecialties: Internal medicine; Infectious diseases. Current work: Pathogenesis of bacterial infections; biomedical administration; clinical medicine and teaching.

PETERSEN, JOHN DAVID, chemistry educator; b. Los Angeles, Nov. 21, 1947; s. John S. and Betsy J. (Trausch) P.; m. Carol D. Birck, Aug. 23, 1949; children: Melissa, Andrew. B.S. in Chemistry, Calif. State U., 1970; Ph.D., U. Calif.-Santa Barbara, 1975. Asst. prof. chemistry Kans. State U., Manhattan, 1975-80; asst. prof. chemistry Clemson (S.C.) U., 1980-81, assoc. prof., 1981—, assoc. dean, 1983—. Contbg. author: Concepts of Inorganic Photochemistry, 1965. Research grantee Dept. Energy, 1980—; Petroleum Research Bd., 1975-78, 82—, Research Corp., 1976-79. Mem. Am. Chem. Soc. (alt. Councillor local sect. 1978-80), Interam. Photochem. Soc., Phi Lambda Upsilon. Club: North Central Kans. Ski. Subspecialties: Photochemistry; Inorganic chemistry. Current work: Photochemistry and electron transfer reactions of transition metal complese; basic research and applications for solar energy conversion. Home: 324 Woodland Way Clemson SC 29631 Office: Clemson Univ Dept Chemistry Clemson SC 29631

PETERSEN, LAWRENCE JOHN, electron microscopist; b. Clinton, Iowa, Mar. 26, 1927; s. Holger Hedegaard and Irma Viola (Heldt) P.; m. Merna Arlene, Sept. 26, 1948; children: Carol Ann Petersen McCarthy, Nancy Sue Petersen McClish. Student, Denver U., 1951-52; B.Sc., Colo. State U., 1955, M.S., 1958; postgrad., U. Calif.-Davis., 1958-62. Plant pathologist Experiment Sta., Colo. State U., Fort Collins, 1955-58; prin. lab. technician U. Calif.-Davis, 1958-65, staff research asso. dept. plant pathology, 1965—. Contbr. articles in field to profl. jours. Served with USN, 1945-46; 1st lt. U.S. Army, 1955-58. Research award Colo. Wyo. Acad. Sci., 1958. Mem. Am. Phytopath. Soc., Sigma Xi, Xi Sigma Pi, Phi Kappa Phi. Subspecialty: Virology (biology). Current work: Researcher in plant virology, scanning and transmission electron microscopy.

PETERSEN, NANCY SUE, molecular biology educator; b. Paris, Tex., Mar. 11, 1943; s. Joseph Eugene and Elizabeth (Rindt) Dennehy; m. Robert A. Petersen, Aug. 28, 1965. B.S., Harvey Mudd Coll., Calif., 1965; M.A., Brandeis U., 1968; Ph.D., U. Calif-Irvine, 1972. Postdoctoral fellow dept. microbiology UCLA, 1972-74; postdoctoral fellow div. biology City of Hope Med. Ctr., Duarte, Calif., 1974-76; sr. research fellow div. biology Calif. Inst. Tech., 1977-83; asst. prof. dept. biochemistry U. Wyo., Laramie, 1983—. Contr. articles to profl. jours. Mem. AAAS, Am. Soc. Cell Biology, Genetics Soc. Am. Subspecialties: Gene actions; Developmental biology. Current work: Gene expression during differenliation effects of heat shock on gene expression heat induced developmental defects. Office: Dept Biochemistry U Wyo Laramie WY 82071

PETERSEN, ROBERT V., pharmacy educator; b. S. Jordan, Utah, Apr. 21, 1926; s. Edgar Ray and Martha (Smith) P.; m. Betty Jayne Bigler, Oct. 23, 1950; children: Robyn, Kent Earl, Susan, Marilyn. B.S., U. Utah, 1950; Ph.D., U. Minn., 1955. Teaching asst. U. Minn., 1951-54; asst. prof. pharm. chemistry Oreg. State Coll., 1955-57; asst. prof. pharmacy U. Utah, 1957-61, asso. prof., 1961-67, chmn. dept. applied pharm. scis., 1965-78, chmn. dept. pharms., 1978-82, prof., 1967—; exec. com. Am. Assn. Colls. Pharmacy, 1968-74, v.p., 1971-72, pres., 1972-73. Contbr. articles sci., scholastic jours. Served with AUS, 1944-46. Mem. Am. Chem. Soc., Am., Utah pharm. assns., Acad. Pharm. Sci., Am. Inst. History Pharmacy, Sigma Xi, Rho Chi (exec. council 1970-72, 74-76), Phi Lambda Upsilon. Current work: Nonaqeous emulsions, plastics composition and toxicology, biodegradable polymers as drug delivery devices; cosmetics. Research field nonaqueous emulsions, plastics composition and toxicology, biodegradable polymers as drug delivery devices, skin permeation of cosmetics. Home: 4639 Meadow Rd Murray UT 84107 Office: Coll Pharmacy U Utah Salt Lake City UT 84112

PETERSON, BRADLEY MICHAEL, astronomer; b. Mpls., Nov. 26, 1951; s. Harry C. and Dona M. (Erickson) P.; m. Janet Rae, Oct. 19, 1978; 1 son, Evan. B. Physics, U. Minn., 1974; Ph.D., U. Ariz., 1978. Research asso. U. Minn., 1979; postdoctoral fellow Ohio State U., 1979-80, asst. prof. astronomy, 1980—. Contbr. numerous articles to sci. jours. NSF grantee, 1981-83. Mem. Am. Astron. Soc., Astron. Soc. Pacific, Internat. Astron. Union. Roman Catholic. Subspecialty: Optical astronomy. Current work: Spectra of Seyfert galaxies and quasars. Office: Dept Astronomy Ohio State U 174 W 18th Ave Columbus OH 43210

PETERSON, CHARLES JOHN, astronomer, consultant; b. Seattle, Oct. 13, 1945; s. Charles John and Edna Jerome (Bishop) P.; m. Virginia Erlene Selvage, June 10, 1967. B.S., U. Wash., Seattle, 1967; M.A., U. Calif., Berkeley, 1968, Ph.D., 1974. Postdoctoral fellow dept. terrestrial magnetism Carnegie Instn., Washington, 1974-76; research assoc Cerro Tololo Inter-Am. Obs., LaSerena, Chile, 1976-78; asst. prof. dept. physics U. Mo., Columbia, 1978-83, assoc. prof. dept. physics, 1983—. Contbr. articles to profl. jours. Served with USMC, 1967-69. Mem. Internat. Astron. Union, Am. Astron. Soc., Astron. Soc. of the Pacific, N.Y. Acad. Scis., Nat. Audubon Soc., Phi Beta Kappa, Sigma Xi. Subspecialty: Optical astronomy. Current work: Structure and dynamics of stellar systems; motion of Milky Way galaxy; photometry of galaxies; archaeoastronomy and history of astronomy. Office: Univ Mo 223 Physics Columbia MO 65211

PETERSON, CURT MORRIS, botanist, educator; b. Fargo, N.D., Jan. 16, 1942; s. Justus Kenneth and Elizabeth Jane (Morris) P.; m. Judity Diane Pollock, June 13, 1964; children: Craig Michael, Jill Elizabeth, John Eric. B.S. in Biology, Moorhead State U., 1966; Ph.D. in Biology (NDEA fellow), U. Oreg., 1970. Asst. prof. botany Auburn U., 1971-76, assoc. prof., 1976—; plant physiologist agrl. Research Service, Dept. Agr., Pendleton, Oreg., 1979-80. Author: Plant Biology Laboratory Manual 2d edit, 1976. Mem. Auburn Planning Commn., 1978-79; mem. troop com. Chatahoochee council Boy Scouts Am., 1979, 80-81, chmn. troop com. Blue Mountain council, 1979-80. Mem. Agronomy Soc. Am., Am. Soc. Plant Physiologists, Ala. Acad. Sci.,

Tot. Soc. Am., Crop Sci. Soc. Am., Sigma Zi, Kappa Delta Pi, Gamma Sigma Delta, Phi Kappi Phi. Subspecialties: Developmental biology; Plant physiology (biology). Current work: Vegetative and reproductivie devel. of soybean, Glycine max. Home: 727 McKinley Ave Auburn AL 36830 Office: Dept Botany Plant Pathology and Microbiology Auburn U Auburn AL 36849

PETERSON, DARWIN WILSON, clinic administrator, researcher; b. Redmond, Utah, Mar. 23, 1938; s. Gleave E. and Velda Jane (Aiken) P.; m. Patricia Ann White, Feb. 13, 1959; children: Wayne, Kathi, Gail, Cheryl, Gary. B.S., U. Nev.-Reno, 1966, M.S., 1967; Ph.D., U. Ala.-Birmingham, 1972, postgrad., 1972-73. Instr. U. Nev.,-Reno, 1967-68; asst. prof. Bowman Gray Sch. Medicine, Winston-Salem, N.C., 1973-82; dir. med. info. Dialysis Clinics, Inc., Nashville, 1982—. Mem. Am. Physiol. Soc. (assoc.), Sigma Xi. Subspecialties: Information systems, storage, and retrieval (computer science); Physiology (medicine). Current work: Computerized medical information system - design and implementation. Home: 432 Cages Bend Rd Gallatin TN 37066 Office: Dialysis Clinics Inc 1600 Hayes St Suite 300 Nashville TN 37203

PETERSON, DAVID MAURICE, plant physiologist; b. Woodward, Okla., July 3, 1940; s. Maurice Llewellyn and Katherine Ann (Jones) P.; m. Margaret Ingegerd Sundberg, June 19, 1965; children: Mark David, Elise Marie. B.S., U. Calif., Davis, 1962; M.S., U. Ill., 1964; Ph.D., Harvard U., 1968. Research biologist Allied Chem. Corp., Morristown, N.J., 1970-71; plant physiologist Agrl. Research Service, U.S. Dept. Agr., Madison, Wis., 1971—. Contbr. articles to profl. jours. Served to capt. U.S. Army, 1968-70. U.S. Dept. Agr. grantee, 1981. Mem. Am. Soc. Plant Physiologists, Am. Soc. Agronomy, Crop Sci. Soc. Am. (assoc. editor 1975-78), Am. Assn. CerealChemists, AAAS. Methodist. Subspecialty: Plant physiology (agriculture). Current work: Synthesis and characterization of cereal seed storage proteins. Seed development. Long distance transport in cereal plants. Home: 6328 Piping Rock Rd Madison WI 53711 Office: Dept Agronomy U Wis 1575 Linden Dr Madison WI 53706

PETERSON, DEAN FREEMAN, JR., civil engineering educator; b. Delta, Utah, June 3, 1913; s. Dean Freeman and Lucille (Crookston) P.; m. Bessie Virginia Carter, Sept. 16, 1938; children: Linda Peterson Thorne, Doris Peterson Rusch, Dean Freeman III, Lucille Peterson Randall, Nicholas R. B.S., Utah State U., 1934, D.Sc. (hon.), 1978; M.S. in Civil Engring, Rensselaer Poly. Inst., 1935; Dr. Civil Engring. (Russell Sage fellow 1937-39), Rensselaer Poly. Inst., 1939; D.Sc. (hon.), Mahatma Phule U., India, 1981. Jr. engr. Utah WPA, 1935-36; road engr. Indian Service, 1936-37; jr. hydraulic engr. U.S. Geol. Survey, 1937-39; instr. U. Wash., 1939-40; project engr. Upper Potomac River Commn., 1940-41; progress engr. Sanderson & Porter, Pine Bluff, Ark., 1941-44; asso. prof., prof. civil engring. Utah State U., 1946-49, prof., dean engring. Coll. Agrl. and Applied Sci., 1957-73, v.p. for research, 1973-76, prof. civil engring., 1982—; chief soil and water div. AID, Washington, 1976-78; dir. Office of Agr., 1978-79; agrl. research and irrigation advisor AID, New Delhi, India, 1979-82; prof., head civil engring. Colo. State U., 1949-57; cons. engr., 1946—; tech. asst., office sci. and tech. Exec. Office Pres., 1965-66; dir. Office of Water for Peace, 1968-69; mem. advisory com. on soil, water and fertilizer research Dept. Agr., 1952-59; chmn. panel weather modification NSF, 1963-65; irrigation cons. Near East-S.Asia AID, 1959-64, 67—; sr. U.S. del. 3d-5th Near East-S.Asia irrigation seminars, Lahore, Pakistan and; Ankara, Turkey, 1960, 62, New Delhi, 1964; cons. water resources Office Sci. and Tech., Exec. Office of Pres., 1963, 64; chmn. com. water resources research Fed. Council Sci. and Tech., 1965-66; chief Agrl. Rev. Team, Afghanistan, 1967; chmn. Utah Advisory Council on Sci. and Tech., 1973-75. Author publs. on irrigation and drainage, hydraulic engring. Mem. Cache Valley council Boy Scouts Am. Served to lt. USNR, 1944-46; capt. Res. Fellow Am. Geophys. Union, AAAS (mem. and chmn. com. arid lands 1970-78); mem. ASCE (hon.; v.p. 1972-74, mem. exec. com. irrigation and drainage div. 1963-66, vice chmn. 1964, chmn. exec. com. 1966, Royce J. Tipton award 1968, Julian Hinds award 1980), Univs. Council on Water Resources (dir., chmn. 1965), Am. Water Resources Assn. (Icko Iben award 1979), Am. Soc. Engring. Edn. (chmn. civil engring. div. 1956-57, mem. commn. on grad. edn. 1956-58), AAUP, Nat. Acad. Engring., Am. Acad. Arts and Scis., Sigma Xi, Phi Kappa Phi, Chi Epsilon, Sigma Tau. Mormon. Club: Cosmos (Washington). Subspecialties: Agricultural engineering; Civil engineering. Current work: Planning design and operation of irrigation systems. Home: 765 E 8th N Logan UT 84321 Office: UMC 41 Utah State U Logan UT 84321

PETERSON, DOUGLAS EDWARD, hospital dentist, researcher, educator; b. Omaha, Nov. 19, 1947; s. DonaldLynne and Miriam Rosalie (Goodsell) P.; m. Leslie Angell, May 26, 1979; 1 son, Matthew Craig. B.A., Colgate U., 1969; D.M.D., U. Pa., 1972, Ph.D., 1976. Instr. in pathology U. Pa. Dental Sch., 1973-75; vis. scientist Frederick (Md.) Cancer Research Ctr., 1975-76; asst. prof. oral diagnosis U. Md. Dental Sch., 1976-81, assoc. prof., 1981—; asst. prof. oncology Med. Sch., 1982—; cons. in field; dir. dental program U. Md. Cancer Ctr. (formerly Balt. Cancer Research Ctr.), 1978—. Contbr. articles to profl. jours.; editor: Oral Complications of Cancer Chemotherapy, 1983. Instr. biology tutoria program Phial. High Sch. System, 1970-72, Am. Heart Assn., Balt., 1978—, Md. High Blood Pressure Council, Balt., 1980-82. NIH fellow, 1970, 71; grantee, 1978-81. Mem. Internat. Assn. Dental Research, Orgn. Tchrs. Oral Diagnosis, Am. Fedn. Clin. Research, Assn. Gnotobiotics, Omicron Kappa Upsilon. Lutheran. Subspecialties: Chemotherapy. Current work: Oral complications of cancer therapy, including chemotherapy and radiation therapy; microbial shifts in myelosuppressed patients; role of oral intervention in reducing or preventing acute oral complications occurring during cancer therapy. Office: Dept Oral Diagnosis U Md 666 W Baltimore St Baltimore MD 21201

PETERSON, ELMOR LEE, mathematician, educator; b. McKeesport, Pa., Dec. 6, 1938; s. William James and Emma Elizabeth (Scott) P.; m. Sharon Louise Walker, Aug. 1957 (div.); 1 dau., Lisa Ann; m. Miriam Drake Mears, Dec. 23, 1966; 1 son, David Scott. B.S. in Physics, Carnegie-Mellon U., 1960, M.S. in Math, 1961, Ph.D., 1964. Sr. mathematician Westinghouse R&D, Churchill Boro, Pa., 1963-66; asst. prof. math. U. Mich., Ann Arbor, 1967-69; vis. assoc.prof. Math. Research Ctr. U. Wis.-Madison, 1968-69; assoc. prof. math. and mgmt. sci. Northwestern U., Evanston, Ill., 1969-73, prof. math. and mgmt. sci., 1973-79; vis. prof. ops. research. Stanford U., Palo Alto, Calif., 1976-77; prof. math. ops. research., N.C. State U., Raleigh, 1979—; cons. Westinghouse Telecomputer, Pitts., 1971-73, McGraw Edison, 1972, Montgomery Ward, Chgo., 1977. Co-author: Geometric Programming, 1967; author numerous sci. papers. Socony Mobil fellow, 1962-63; Mobil Found. grantee, 1967-69; Urban Systems grantee Northwestern U., 1971-73; Air Force Office Sci. Research grantee, 1973-78; Mellon scholarship; U.S. Steel scholar, 1956-60. Mem. Soc. Indsl. and Applied Math., Math. Assn. Am., Am. Math. Soc., Ops. Research Soc. Am., Math. Programming Soc. Subspecialties: Applied mathematics; Operations research (mathematics). Current work: Optimization and equilibration, especially geometric programming and its interfaces with economics, engineering, operations research, and physical sciences. Home: 3717 Williamsborough Ct Raleigh NC 27609 Office: NC State U Hillsboro St Raleigh NC 27650

PETERSON, ERNEST A., acoustic physiologist, educator; b. N.Y.C., June 16, 1931; s. Ernest A. and Mildred G. P.; m. Catherine D. Peterson, Sept. 28, 1977; children: Leslie, Jason. B.S. summa cum laude, Rutgers U., 1959; Ph.D., Princeton U., 1962. Research asst., research assoc. psychopharmacology lab. Rutgers U., New Brunswick, N.J., 1958-62, acting dir. psychopharmacology lab., 1962-63; postdoctoral fellow in sensory psychology Princeton U., 1962-64; dir. Evoked Auditory Response Lab., U. Miami (Fla.) Sch. Medicine, 1974-78; asst. prof. dept. otolaryngology, div. auditory research U. Miami Sch. Medicine, 1964-68, assoc. prof., 1968—, chief div. auditory research, 1965—, mem. faculty, 1969—, assoc. prof. dept. psychology, 1969—; pres. Noise Analysis Cons., Inc., 1970—. Contbr. articles to profl. jours. EPA grantee, 1977-82. Mem. AAAS, AAUP, Am. Psychol. Assn., Acoustical Soc. Am., N.Y. Acad. Scis., Soc. Neurosci., Nat. Assn. Noise Com. Ofcls. (assoc.), Am. Auditory Soc., Am. Assn. Lab. Animal Sci., Sigma Xi, Phi Beta Kappa. Subspecialty: Otorhinolaryngology. Office: Division Auditory Research D-71 U Miami PO Box 016960 Miami FL 33101

PETERSON, FRANK LYNN, geology educator, consultant; b. Klamath Falls, Oreg., May 8, 1941; s. Burton H. and Elizabeth M. (Ritsch) P.; m. Barbara Ann Bennett, July 1, 1967. B.A., Cornell U., 1963; M.S., Stanford U., 1965, Ph.D., 1967. Research asst. Stanford U., Calif., 1963-67; ground-water geologist U.S. Geol. Survey, Menlo Park, Calif., 1965-67; prof. geology U. Hawaii, Honolulu, 1967—; Hydrology cons.; research geologist Hawaii Water Research Ctr., 1967—. Author: Volcanoes in the Sea, 1983; contbr. numerous articles to profl. jours., chpts. to books, monographs. Numerous research grants, 1967-82. Fellow Geol. Soc. Am.; mem. Am. Geophys. Union, Nat. Water Well Assn., Am. Water Resorces Assn., Assn. Engring. Geologists, Hawaiin Acad. Sci., Sigma Xi. Subspecialties: Hydrogeology; Geology. Current work: Teach and conduct research at the university level in: general geology, geology of the Hawaiin Islands, groundwater and hydrology, engineering geology, and geologic hazards and environmental geology. Home: 1341 Laukahi St Honolulu HI 96821 Office: Dept Geology and Geophysics U Hawaii Honolulu HI 96822

PETERSON, GERALD LEONARD, psychology educator; b. Neenah, Wis., Aug. 22, 1945; s. Gerald F. and Mary (Letwon) P. B.S., U. Wis.-Oshkosh, 1968; M.A., U. Mo.-Kansas City, 1972; Ph.D., Kans. State U., 1975. Lic. psychologist, Pa. Asst. prof. Duquesne U., Pitts., 1975-81; assoc. prof. psychology Saginaw State Coll., University Center, Mich., 1981—. Served with U.S. Army, 1968-70; Germany. Mem. Am. Psychol. Assn., Midwestern Psychol. Assn., AAAS, N.Y. Acad. Scis., Soc. Advancement Social Psychology, Psi Chi, Phi Kappa Phi. Roman Catholic. Subspecialty: Social psychology. Current work: Communication in groups, group processes and social cognition; examining persuasive techniques in small groups, organizational settings. Office: Dept Psychology Saginaw Valley State Coll 2250 Pierce Rd University Center MI 48710

PETERSON, JAMES ROBERT, engineering psychologist; b. St. Paul, Minn., Apr. 16, 1932; s. Palmer Elliot and Helen Evelyn (Carlson) P.; m. Marianna J. Stockvig, June 26, 1954; 1 dau., Anne Christine. B.A. in Psychology cum laude, U. Minn., 1954, M.A. in Exptl. Psychology, 1958; Ph.D. in Engring. Psychology, U. Mich., 1965. Engring. aide Honeywell, Inc., Mpls., 1958-59, devel. engr., 1961-65, sr. devel. engr., 1965-67, staff engr., Clearwater, Fla., 1967—; sponsor student insect flight expt. on space shuttle, Mar. 1982. Contbr. numerous articles and papers to profl. jours. Recipient Founder's medal Tampa Bay sect. AIAA, 1982. Fellow AIAA (assoc.); mem. Human Factors Soc., Am. Psychol. Assn., Soc. Engring. Psychologists. Lodge: Masons. Subspecialties: Aerospace engineering and technology; Human factors engineering. Current work: Analysis, design and development of man-machine systems; development of manned spaceflight systems. Home: 3303 San Gabriel St Clearwater FL 33519 Office: Space and Strategic Avionics Div Honeywell Inc 13350 US Hwy 19 Clearwater FL 33546

PETERSON, PETER ANDREW, educator; b. Bristol, Conn., Mar. 17, 1925; m. (married); children: Sara, Susan. B.S., Tufts U., 1947; Ph.D., U. Ill., Champaign-Urbana, 1953. Geneticist U. Calif., Riverside, 1953-56; prof. agronomy Iowa State U., 1956—. Editor: Maydica, 1974—. Served with USN, 1942-45. Subspecialties: Plant genetics; Genetics and genetic engineering (biology). Current work: Genetics. Office: Department of Agronomy Iowa State University Ames IA 50010

PETERSON, ROY JEROME, physicist, educator; b. Everett, Wash., Oct. 18, 1939; s. Roy Adolph and Evelyn Katherine (Anderson) P.; m. Faith Alloway, Oct. 22, 1971; children: James, Michael, Stephen, Marcus. B.S., U. Wash., 1961, Ph.D., 1966. Instr. Princeton U., 1966-68; research assoc. Yale U., 1968-70; mem. faculty dept. physics U. Colo., Boulder, 1970—, now prof.; program dir. NSF, 1978-79; fellow Niels Bohr Inst., Copenhagen, 1982-83. Contbr. articles to profl. jours. Mem. Am. Phys. Soc. Republican. Subspecialty: Nuclear physics. Current work: Nuclear structure and reactions as revealed by experiments with beams of light ions on pi mesons. Office: U Colo Nuclear Physics Lab Campus Box 446 Boulder CO 80309

PETERSON, RUDOLPH NICHOLAS, pharmacologist, educator; b. N.Y.C., June 6, 1932; s. Peter and Christina Mary (Kavanagh) Pantalakis. B.S., St. John's U., 1957; M.A., Bklyn. Coll., 1962; Ph.D., U. Fla., 1965. Mem. faculty dept. pharmacology N.Y. Med. Coll., Valhalla, 1966-76, assoc. prof., 1973-76; prof. pharmacology dept. physiology-pharmacology Sch. Medicine, So. Ill. U., Carbondale, 1976—. Contbr. chpts. to books, articles to profl. jours. Served with USAF, 1951-55. Mem. Am. Soc. Pharmacology and Exptl. Therapeutics, Am. Soc. Cell Biology. Subspecialties: Gynecological oncology; Biochemistry (medicine). Current work: Cell surface of mammalian gametes. Office: Sch Medicine So Ill Univ Carbondale IL 62901

PETERSON, THOMAS MARK, orthodontist; b. Norfolk, Va., Mar. 3, 1954; s. Thomas W. and Margaret (Hogard) P. B.S., Va. Poly. Inst. and State U., 1975; D.D.S., Med. Coll. Va., 1979, ortho-certificate, 1981. Asst. prof. orthodontics Med. Coll. Va., Richmond, 1981—; cons. Craniofacial Deformities Clinic, Richmond, 1981-83. Med. Coll. Va. Orthodontic Found. grantee, 1982-82; So. Soc. Orthodontics grantee, 1981-82. Mem. Internat. Assn. Dental Researchers, Am. Dental Assn., Am. Assn. Orthodontics, Va. Dental Assn., Med. Coll. Va. Orthodontic Found., Am. Assn. Dental Schs., So. Soc. Orthodontics. Subspecialties: Orthodontics; Dental growth and development. Current work: Electromyography, craniofacial morphology, orthograthic surgery, dento-facial orthopedics. Home: 1505 Largo St T-2 Richmond VA 23229 Office: Med Coll Va 566 Med Coll Va Sta Richmond VA 23226

PETERSON, VERN LEROY, research scientist; b. Gothenburg, Nebr., Nov. 8, 1934; s. Elmer Robert and Vera Theresa (Maline) P.; m. Roberta Vye Elsey, June 3, 1961; children: Susan, Stephen, Scott. B.S. in Engring. Physics, U. Colo., 1956; M.A. in Astronomy, Ind. U., 1961; Ph.D. in Astrophysics, Ind. U., 1963. Research physicist-atmosphere NOAA, Boulder, Colo., 1963-69; assoc. prof. physics Utah State U., Logan, 1969-74; prof. adj. in meteorology U. Sao Paulo, Brazil, 1974-77; sr. scientist Research Analysis and Devel., Colorado Springs, Colo., 1977-78; pres., chief scientist Centennial Sciences, Inc., Colorado Springs, 1978-80; mgr. geophys. systems dept. Ocean Data Systems, Inc., Monterey, Calif., 1980—. Author articles. Served with U.S. Army, 1957-59. Named Outstanding Prof. Physics, Utah State U., 1973. Mem. Am. Geophys. Union, Am. Meteorol. Soc., Am. Astron. Soc. Subspecialties: Meteorology; Satellite studies. Current work: Atmospheric radar, satellite cloud imagery, ocean surface currents, data buoy studies. Home: 22610 Murietta Rd Salinas CA 93908 Office: 2400 Garden Rd Monterey CA 93940

PETRI, WILLIAM HENRY, III, oral and maxillofacial surgeon, researcher, consultant; b. Hopkinsville, Ky., Nov. 8, 1938; s. William Henry and Edna Alice (Walker) P. B.A., U. Louisville, 1961, D.M.D., 1965. Diplomate: Am. Bd. Oral and Maxillofacial Surgery. Intern U. Louisville and VA Hosp., Louisville, 1965-66; commd. engisn U.S. Navy, 1966, advanced through grades to capt., 1981; resident Naval Regional Med. Ctr., Portsmouth, Va., 1972-75, staff oral and maxillofacial surgeon, 1975-76, 80-81, head oral and maxillofacial surgery, Subic Bay, Philippines, 1976-77, Naval Regioal Dental Ctr., Little Creek, Va., 1977-79; head dental dept. USS Nimitz, Norfolk, Va., 1979-80; prin. investigator Naval Med. Research Inst., Bethesda, Md., 1981—; cons. in field Nat. Naval Med. Ctr., Bethesda, 1981—; cons. instrumentation Walter Lorenz Inc., Jacksonville, Fla., 1981—; lectr., temporomandibular joint diseases Nat. Naval Grad. Dental Sch., Bethesda, 1981—. Author: One Year, 1970; contbr. monogram, articles to profl. jours. Pres. Merrifields Civic League, Portsmouth, 1976, Hatton Point Civic Assn., Portsmouth, 1977; mem. administrv. bd. Centenary Methodist Ch., Portsmouth, 1979; tchr. adult Sunday Sch., 1977-80. Fellow Am. Assn. Oral and Maxillofacial Surgeons; mem. ADA, Tidewater Soc. Oral Surgeons (pres. 1980), Psi Chi. Republican. Subspecialties: Oral and maxillofacial surgery; Physiology (medicine). Current work: Wound healing oral and maxillofacial injuries, bone physiology; researcher in skin wound closures, microvascular anastamosis of vascularized, freeze-dried, free bone allografts. Osteogenesis in anti-biotic supplemented allograft. Inventor surg. instruments. Office; US Naval Med Research Inst Mail Stop 18 Bethesda MD 20814

PETRI, WILLIAM HUGH, develop mental geneticist, apiculturist, consultant; b. San Francisco, Dec. 30, 1944; s. Marino and Aida Clair P.; m. Arlene Ruth Wyman, July 30, 1976; children: Jonah Wyman, Sonya Ruth,. B.A., U. Calif.-Berkeley, 1966, Ph.D. in Genetics, 1972. Research fellow Harvard U., 1972-76; asst. prof. biology Boston Coll., 1976-81, assoc. prof., 1981—. Mem. Genetics Soc. Am., Soc. Develop mental Biology. Subspecialties: Gene actions; Developmental biology. Current work: Teaching and research in developmental genetics. Office: Dept Biology Boston College Chestnut Hill MA 02167

PETRIE, ROY HOWARD, reproductive medicine educator, physician; b. Bardwell, Ky., Nov. 9, 1940; s. Randolph Hazel and Glodine (Brown) P. B.S., Western Ky. U., 1961; M.D., Vanderbilt U., 1965. Diplomate: Am. Bd. Ob-Gyn. Intern U. Rochester (N.Y.) Hosp., 1965-66; resident Columbia-Presbyn. Med. Ctr., N.Y.C., 1966-70; asst. prof. dept. ob-gyn Coll. Physicians and Surgeons, Columbia U., N.Y.C., 1972-79, assoc. prof., 1979—. Author books in field of fetal monitoring; contbr. articles to profl. jours. Mem. vestry Christ Ch. Riverdale, Bronx, N.Y., 1982. Served to lt. comdr. USNR, 1970-72. Fellow ACS, N.Y. Obstet. Soc., Am. Coll. Ob-Gyn; mem. Soc. Perinatal obstetricians (sec.-treas. 1981-82), Soc. Gynecologic Investigation. Subspecialties: Maternal and fetal medicine; Perinatal diagnosis and therapy. Current work: Fetal pharmacology, fetal heart rate, acid-base surveillance of the intra and ante partum fetus. Office: Dept Ob-Gyn Coll Physicians and Surgeons Columbia U 630 W 168th St New York NY 10032

PETRIE, WILLIAM MARSHALL, physician, researcher; b. Louisville, Oct. 19, 1946; s. Garner McReynolds and Claire (Samuels) P.; m. Patricia Roberts, Aug. 3, 1968; children: Christopher William, Ellen McLaine, Shelley Marshall. B.A., Vanderbilt U., 1968, M.D., 1972. Diplomate: Am. Bd. Psychiatry and Neurology, Nat. Bd. Med. Examiners. Resident in psychiatry Duke Med. Ctr., Durham., N.C., 1972-75; lectr. in pharmacology Howard U., Washington, 1976; clin. instr. psychiatry Georgetown U., Washington, 1976-77; clin. assoc. prof. psychiatry Vanderbilt U., Nashville, 1977—; chief geriatric research service Tenn. Neuropsychiat. Inst., Nashville, 1977-82; clin. asst. prof. psychiatry Meharry Med. Coll., Nashville, 1981—; research psychiatrist psychopharmacology br. div. extramural research programs NIMH, Rockville, Md., 1975-77, mem. life course review com., 1982—; cons. Social Action Group in Aging, Nashville, 1981—; asst. examiner Am. Bd. Psychiatry and Neurology, 1981. Contbr. articles to profl. jours. Bd. advisors Middle Tenn. Long-Term Care Gerontology Ctr., Nashville, 1981—. Served to lt. comdr. USPHS. Mem. AMA, Am. Psychiat. Assn., Nashville Acad. Medicine. Methodist. Subspecialties: Psychopharmacology; Gerontology. Current work: Psychopharmacology, geriatric psychopharmacology, geriatric psychiatry. Office: 356 24th Ave N Suite 405 Nashville TN 37203 Home: 5680 Cloverland Dr Brentwood TN 37027

PETROFSKY, JERROLD SCOTT, psyiology and engineering educator; b. St. Louis, May 5, 1948; m., 1974; 1 child. A.B., Washington U., St. Louis, 1970; Ph.D. in Physiology, St. Louis U., 1974. Electron technician dept. physiology St. Louis U., 1968-69, fellow, 1974-76, asst. prof. physiology, 1976—; asst. prof. engring. Wright State U., 1978—. Mem. Am. Physiol. Soc., IEEE, Biomed. Engring. Soc., Am. Coll. Sports Medicine, Sigma Xi. Subspecialties: Biomedical engineering; Physiology (biology). Office: Dept Physiology Med Sch Saint Louis U Saint Louis MO 63104

PETRONE, ROCCO A., aerospace manufacturing executive; b. Amsterdam, N.Y., Mar. 31, 1926; s. Anthony and Theresa (DeLuca) P.; m. Ruth Holley, Oct. 29, 1955; children—Teresa, Nancy, Kathryn, Michael. B.S., U.S. Mil. Acad., 1946; degree Mech. Engring, Mass. Inst. Tech., 1952; D.Sc. (hon.), Rollins Coll., 1969. Devel. officer Redstone Missile Devel., Huntsville, Ala., 1952-55; mem. army gen. staff Dept. Army, Washington, 1956-60; mgr. Apollo program Kennedy Space Center, 1960-66, dir. launch ops., 1966-69; Apollo program dir. NASA, Washington, 1969-73; dir. Marshall Space Flight Center, Huntsville, Ala., 1973-74; asso. adminstr. NASA, Washington, 1974-75; pres., chief exec. officer Nat. Center for Resource Recovery, Washington, 1975-81; exec. v.p. Space Transp. and Systems Group, Rockwell Internat., Downey, Calif., 1981-82, pres., 1982—. Decorated D.S.M. with 2 clusters NASA; Commendatore Ordine al Merito, Italy). Fellow Am. Inst. Aeros. and Astronautics; mem. Nat. Acad. Engring., Sigma Xi. Subspecialty: Aerospace engineering and technology. Home: 1329 Granvia Altamira Palos Verdes Estates CA 90274

PETSCHEK, ALBERT GEORGE, physicist; b. Prague, Czechoslovakia, Jan. 31, 1928; s. Hans and Eva (Epler) P.; m. Marilyn Adiene Poth, June 25, 1949; children: Evelyn A., Rolfe G., Elaine L., Mark A. B.S., M.I.T., 1947; M.S., U. Mich., 1948; Ph.D., U. Rochester, 1952. Jr. research physicist Carter Oil Co., Tulsa, 1948-49; staff mem., group leader Los Alamos (N.Mex.) Sci. Lab., 1952-66; prof. physics N.Mex. Inst. Mining and Tech., Socorro, 1966-68, 71—; sr. research scientist Systems Sci. and Software, La Jolla, Calif., 1968-71; also dir.; fellow Los Alamos Nat. Lab., 1980—; cons. Los Alamos Sci. Lab., Sandia Corp., Systems Sci. and Software, Mission Research

Corp.; dir. Interhealth, Inc., 1970-71. Contbr. sci. papers to profl. publs. Mem. Am. Phys. Soc., Am. Astron. Soc., AAAS. Subspecialties: Applied Physics; Theoretical astrophysics. Current work: Inertial confinement fusion, thunderstorm modeling supernovae, quasars, gamma bursts. Home: 122 Piedra Loop Los Alamos NM 87544 Office: MS 434 Los Alamos Nat Lab Los Alamos NM 87545 Dept Physics New Mexico Tech Socorro NM 87801

PETSKO, GREGORY ANTHONY, chemistry educator; b. Washington, Aug. 7, 1948; s. John and Mary Virginia (Santoro) P.; m. Carol Bannister Chamberlain, July 3, 1971 (div. 1931). A.V., Princeton U., 1970; D.Phil., Oxford U., Eng., 1973. Instr. Wayne State U., Detroit, 1973-76, asst. prof., 1976-79; assoc. prof. chemistry MIT, Cambridge, 1979—; cons. Genentech, Inc., South San Francisco, Calif., 1982—. Rhodes scholar, 1970; Danforth fellow, 1971; Sloan fellow, 1979. Mem. Am. Chem. Soc., AAAS, Am. Crystallographic Assn. (Sidhu award 1981), Biophys. Soc., N.Y. Acad. Sci. Roman Catholic. Club: Campur (Princeton). Subspecialties: Biochemistry (biology); Biophysical chemistry. Current work: Mechanism of action of enzymes; protein structure and its relation to function; dynamics of proteins; biochem. signalling systems, especially bacterial chemotaxis; antibody-antigen interactions; site-specific mutagenesis. Home: 8 Jason Rd MA 02178 Office: MIT Rom 18 025 Cambridge MA 02139

PETTENGILL, GORDON H(EMENWAY), physicist; b. Providence, Feb. 10, 1926; s. Rodney Gordon and Frances (Hemenway) P.; m. Pamela Anne Wolfenden, Oct. 28, 1967; children—Mark Robert, Rebecca Jane. B.S., M.I.T., 1948; Ph.D., U. Calif., Berkeley, 1955. Staff mem. Lincoln Lab., M.I.T., Lexington, 1954-63, 65-68; asso. dir. Arecibo (P.R.) Obs., 1963-65, dir., 1968-71; prof. planetary physics, dept. earth and planetary scis. M.I.T., Cambridge, 1971—. Served with inf., Signal Corps AUS, 1944-46. Decorated Combat Inf. badge. Mem. Am. Phys. Soc., Am. Astron. Soc., AAAS, Internat. Astron. Union, Internat. Radio Sci. Union, Nat. Acad. Sci., Am. Acad. Arts and Sci., Nat. Audubon Soc., Mass. Audubon Soc. Subspecialties: Planetary science; Remote sensing (geoscience). Current work: Study of planetary surfaces using radio/radar techniques. Pioneer several techniques in radar astronomy for describing properties of planets and satellites; discovered 59-day rotational period of planet Mercury. Office: Room 54-620 Mass Inst Tech Cambridge MA 02139

PETTERS, RICHARD ALAN, ocean engineer, marine consultant; b. Fredericksburg, Va., Sept. 14, 1951; s. James Richard and Grace Esco P. B.S., Fla. Atlantic U., 1974; M.S., U. Hawaii, 1976. Registered profl. engr., Wash. Technician Vector Cable Co., Houston, 1972, McClelland Engrs., 1973; grad. asst. Hawaiian Inst. Geophysics, Honolulu, 1975; ocean engr. Internat. Nickle Co., Bellevue, Wash., 1976-78; v.p., engr. Sound Ocean Systems, Seattle, 1978—. Mem. Marine Tech. Soc. Subspecialties: Ocean engineering; Offshore technology. Current work: Underwater mining for manganese nodules, silver and gold tailing recovery and offshore placier deposits, remotely operated submersibles, towed vehicles for deep ocean side scan surveys, deck handling equipment, and offshore operations. Office: Sound Ocean Systems Inc PO Box 31699 Seattle WA 98103

PETTEY, DIX HAYES, mathematics educator; b. Salt Lake City, Mar. 16, 1941; s. Leo Melvin and Kathleen (Hayes) P. B.S., U. Utah, 1965, Ph.D., 1968. Asst. prof. U. Mo., Columbia, 1968-74, assoc. prof., 1974-83, prof. math., 1983—. Contbr. articles to profl. jours. Mem. Am. Math. Soc., Math. Assn. Am., Soc. Indsl. and Applied Math. Mormon. Subspecialties: Operations research (mathematics). Current work: Reliability of communications networks; scheduling problems; P-minimal and P-closed topological spaces. Office: MO 208 Math Science Bldg Columbia MO 65211

PETTINATI, HELEN MARIE, psychologist; b. Washington, Mar. 31, 1951; d. John Charles and Mary Josephine (Bearly) Rimkus; m. Joseph Vincent Pettinati, June 23, 1973; 1 dau., Lisa Marie. B.S., Drexel U., 1973; Ph.D., Med. Coll. Pa., 1979. Research asst. Med. Coll. Pa., Phila., 1974, E. Pa. Psychiat. Inst., 1975; research asst/assoc. Inst. of Pa. Hosp., Phila., 1975-79; coordinator research activities Carrier Found., Belle Mead, N.J., 1979-81, asst. dir. research, 1981—; adj. asst. prof. Rutgers Med. Sch., 1981—; referencing asst. Internat. Jour. Clin. & Exptl. Hypnosis, 1975-80. Contbr. articles to profl. jours.; guest interviewee various radio broadcasts. Med. Coll. Pa. fellow, 1974-75; recipient J. Peterson Ryder award for Women Drexel U., 1973. Mem. AAAS, Am. Psychol. Assn., Am. Soc. Clin. Hypnosis, Internat. Soc. Hypnosis (co-chmn. film program 1976), Soc. Clin. and Exptl. Hypnosis (co chmn. sci. program 1983 84), Eastern Psychol. Assn., Gerontol. Soc. N.J. Roman Catholic. Subspecialties: Psychiatry; Cognition. Current work: Clinical psychiatric research in etiology of and current treatment for problems in aging, alcoholism, anorexia nervosa, bulimia, depression, cognition and memory disturbance; investigation of nature of hypnosis and its usefulness in medicine, especially psychiatry. Home: 568 Fernwood Ln Fairless Hills PA 19030 Office: Carrier Found Research Div Belle Mead NJ 08502

PETTINGER, WILLIAM A., physician, researcher, educator; b. Cumberland, Iowa, May 26, 1932; s. Adolph P. and Virginia E. (Lauhoff) P.; m. Margaret C. Carney, Aug. 12, 1961; children: Maria, Tom, Elise, Will. B.S. in Math, Creighton U., 1954, M.S. in Physiology, 1957, M.D., 1959. Intern, asst. resident in medicine Jersey City Hosp., 1959-61; clin. investigator exptl. therapeutics Nat. Heart and Lung Inst., NIH, Bethesda, Md., 1961-63; sr. resident in medicine Yale-Grace New Haven Hosp., New Haven, 1963-64; postdoctoral fellow, asst. prof. pharmacology and internal medicine Vanderbilt U., Nashville, 1964-67; dir. cardiovascular renal research Hoffman-LaRoche, Nutley, N.J., 1967-71; asso. prof. medicine N.J. Coll. Medicine, 1967-71; assoc. prof. pharmacology and internal medicine U. Tex. Health Sci. Ctr., Dallas, 1971-74, prof., 1974—, dir. clin. pharmacology, 1971—; cons. pharm. industries; William N. Creasy vis. prof. in clin. pharmacology, 1977, 81; mem. pharmacology test com. Nat. Bd. Med. Examiners. Contbr. numerous articles to sci. jours.; editor-in-chief: Jour. Cardiovascular Pharmacology, 1978-80; mem. editorial bd.: Annals of Internal Medicine; mem. editorial adv. bd.: Jour. Pharmacology and Exptl. Therapeutics. Burroughs-Wellcome scholar awardee in clin. pharmacology, 1974-79; recipient Rawls-Palmer award Am. Soc. Clin. Pharmacology and Exptl. Therapeutics, 1982. Fellow ACP, Am. Coll. Cardiology; mem. So. Soc. Clin. Investigation, Am. Soc. Clin. Investigation, Soc. Exptl. Biology and Medicine, Am. Soc. Clin. Pharmacology and Therapeutics (dir.), Am. Heart Assn. (med. adv. bd.). Subspecialties: Internal medicine; Pharmacology. Current work: Hypertension, antihypertensive drugs, cardiovascular and renal regulation, alpha adrenergic receptors. Home: 3548 University Blvd Dallas TX 75205 Office: 5323 Harry Hines Blvd Dallas TX 75235

PETTIT, FLORA HUNTER, biochemist, educator; b. Bellvue, Pa., Sept. 17, 1928; d. Thomas P. and Nancy Ellen (Pryor) Hunter; m. Rowland Pettit, July 21, 1959 (dec. Dec. 10, 1981); children: George Hunter, Nancy Selina. B.A., U. Houston, 1952; Ph.D., U. Tex., 1961. Research assoc. U. Tex., Austin, 1961-63, lectr. in nutrition, 1966-71; research scientist assoc. Clayton Found., 1963-74, research scientist, 1974—. Mem. AAAS, Fedn. Am. Soc. Exptl. Biology, Sigma Xi. Subspecialties: Biochemistry (biology); Enzyme technology. Current work: purification, characterization and regulation of the alpha-Keto acid dehydrogenase complexes from mammalian tissue. Home: 2900 Wade Ave Austin TX 78703 Office: Clayton Found Biochemical Research U Tex Austin TX 78712

PETTIT, FREDERICK SIDNEY, engineering educator, consultant; b. Wilkes-Barre, Pa., Mar. 10, 1930; s. Edwin H. and Edith M. (Barnecut) P.; m. Lou Jean M. Corso, Aug. 30, 1958; children: Frederick N., Theodore E., John C., Charles A. B.Eng., Yale U., 1952, M.Eng., 1960, Dr.Eng., 1962. Postdoctoral fellow Max Planck Inst., Gottingen, W. Ger., 1962-63; sr. staff scientist Pratt & Whitney Aircraft, East Hartford, Conn., 1963-79; chmn., prof. metall. and materials dept. U. Pitts., 1979—. Served with USMC, 1954-57. Mem. Am. Soc. for Metals, Metall. Soc., Electrochem. Soc., Sigma Xi. Roman Catholic. Subspecialties: High-temperature materials; Clad metals and coating technology. Current work: Materials high temperature, degradation of materials and use of coatings for protection. Home: 201 Ennerdale Ln Pittsburgh PA 15237 Office: U Pitts 848 BEH Pittsburgh PA 15261

PETTIT, GEORGE ROBERT, chemist, educator; b. Long Branch, N.J., June 8, 1929. B.S. in Chemistry, Wash. State U., 1952; M.S., Wayne State U., 1954, Ph.D., 1956. Teaching asst. Wash. State U., 1950-52, lecture demonstrator, 1952; research chemist E.I. dupPont de Nemours & Co., 1953; grad. teaching asst. Wayne State U., 1952-53, research fellow, 1954-56; sr. research chemist Norton Norwich Co., 1956-57; asst. prof. chemistry U. Maine, 1957-61, assoc. prof., 1961-65, prof., 1965; prof. chemistry Ariz. State U., 1965—, chmn. organic div., 1966-68, dir., 1974-75, 1975—, Dalton prof. chemistry, 1982—; vis. prof. Stanford U., 1965; mem. cancer treatment adv. bd. div. cancer treatment Nat. Cancer Inst., 1971-74, adv. div., 1965-76; vis. prof. So. African Univs., 1978; lectr. in field. Mem. editorial bd.: Synthetic Communications, 1970—; contbr. numerous articles to profl. jours. Fellow Am. Inst. Chemists; mem. Am. Chem. Soc., Chem. Soc. London, Pharmacognosy Soc., Am. Assn. Cancer Research, Sigma Xi, Phi Lambda Upsilon. Subspecialties: Organic chemistry; Cancer research (medicine). Office: Cancer Research Inst Dept Chemistry Arizona State University Tempe AZ 85281

PETTUS, DAVID, zoology educator; b. Goliad, Tex., Nov. 28, 1925; s. Thomas Lott and Burneice (Lockhart) P.; m. Shirley Rose McKinley, June 11, 1947; children: Alana, Lynette, Thomas, Bligh, Virginia. B.a., Ariz. State U., 1951, M.A., 1952; Ph.D., U.Tex.-Austin, 1956. Mem. faculty dept. zoology Colo. State U., Ft. Collins, 1956—, prof., 1968—; cons. Colo. Div. Wildlife, 1974—. Contbr. articles to profl. jours. Served with USN, 1943-46; PTO. Recipient Krause award Ariz. State U., 1951. Mem. AAAS, Genetics Soc. Am., Soc. Study Amphibians and Reptiles, Soc. Ichthologists and Herpetologists, Soc. Study Evolution, Sigma Xi. Democrat. Presbyterian. Subspecialties: Evolutionary biology; Population biology. Current work: Mechanisms of adaptation in amphibians and reptiles. Home: 1317 Stover Fort Collins CO 80524 Office: Dept Zoology Colo State U Fort Collins CO 80521

PETTY, HOWARD RAYMOND, biological sciences educator, researcher; b. Toledo, Aug. 1, 1954; s. Dale Eugene and Patricia Ann (Purvis) P.; m. Leslie Ellen Isler, June 8, 1980; 1 son, Aaron Raymond. B.S., Manchester Coll., 1976; Ph.D., Harvard U., 1979. Fellow Stanford (Calif.) U., 1979-81; asst. prof. dept. biol. scis. Wayne State U., Detroit, 1981—; cons. Oak Ridge Nat. Lab., 1976—. Contbr.: articles to profl. jours. including Jour. Ultrastructure Research. Damon Runyon-Walter Winchell Cancer Fund fellow, 1979-81; grantee NIH, 1982—, NSF, 1982—, Research Corp., 1982—. Mem. Am. Assn. Immunologists, Am. Soc. Cell Biologists, Reticuloendothelial Soc., Biophys. Soc., Am. Soc. Microbiology. Democrat. Subspecialties: Membrane biology; Biophysics (biology). Current work: Membrane biochemistry and biophysics of immune cell-cell recognition, with special emphasis on macrophages. Office: Dept Biological Sciences Wayne State Univ Detroit MI 48202 Home: 18224 Middlebelt Ave Livonia MI 48152

PETURSSON, SIGURDUR RAGNAR, physician, educator; b. Sida, Iceland, May 22, 1948; came to U.S., 1962, naturalized, 1970; s. Esra and Asta (Einarsdottir) P. A.B., Cornell U., 1970; M.D., U. Rochester, 1974. Diplomate: Am. Bd. Internal Medicine, Am. Bd. Med. Oncology, Am. Bd. Hematology. Intern Presbyn. U. Hosp., Pitts., 1974-75, resident in internal medicine, 1975-77; fellow in hematology and oncology U. Pitts. Sch. Medicine, 1977-80, asst. prof. medicine, 1980—. Contbr. chpts. to book. Leukemia soc. Am. fellow, 1981-83; United Way grantee, 1977-81. Mem. ACP, Am. Fedn. Clin. Research, Am. Soc. Clin. Oncology, Internat. Soc. Exptl. Hematology. Lutheran. Subspecialties: Hematology; Oncology. Current work: In vitro cultures of megakaryocytes; graft vs. host disease in marrow transplantation. Office: U Pitts 922 Scaife Hall Pittsburgh PA 15261

PEUSNER, KENNA DALE, anatomist, educator; b. Lowell, Mass., Aug. 19, 1946; d. Samuel Gilbert and Gladys Gertrude (Levy) Kaplan; m. Leonardo Peusner, Oct. 16, 1970 (div. Nov. 15, 1974). B.S., Simmons Coll., 1968; Ph.D. in Anatomy (NIH fellow), Harvard U., Boston, 1974. Postdoctoral research fellow Harvard U. Med. Sch., Boston, 1974-75; instr. in anatomy Jefferson Med. Coll., 1975-77, asst. prof. anatomy, 1977-81; asst. prof. George Washington U. Med. Sch., 1981-83, assoc. prof., 1983—. Recipient Christian R. and Mary F. Lindback award Jefferson Med. Coll., 1981; Nat. Inst. Neurol. and Communicative Diseases and Strokes grantee, 1979—. Mem. AAAS, Am. Assn. Anatomists, AAUP, Cajal Club, Neurosci. Soc., N.Y. Acad. Scis., Washington Soc. Electron Microscopy. Club: George Washington U. Subspecialties: Neurobiology; Cytology and histology. Current work: Development of nervous system, in particular the formation of synaptic endings and synaptic organization; neuroembryology, synaptogenesis, development of vestibular system, role of sensory receptor on development of central sensory systems.

PÉWÉ, TROY LEWIS, geologist, educator; b. Rock Island, Ill., June 28, 1918; s. Richard E. and Olga (Pomrank) P.; m. Mary Jean Hill, Dec. 21, 1944; children—David Lee, Richard Hill, Elizabeth Anne. A.B. in Geology, Augustana Coll., 1940; M.S., State U. Iowa, 1942; Ph.D., Stanford, 1952. Head dept. geology Augustana Coll., 1942-46; civilian instr. USAAC, 1943-44; instr. geomorphology Stanford, 1946; geologist Alaskan br. U.S. Geol. Survey, 1946—; chief glacial geologist U.S. Nat. Com. Internat. Geophys. Year, Antarctica, 1958; prof. geology, head dept. U. Alaska, 1958-65; prof. geology Ariz. State U., 1965—, chmn. dept., 1965-76; dir. Mus. Geology, 1976—; lectr. in field, 1942—; Mem. organizing com. 1st Internat. Permafrost Conf. Nat. Acad. Sci., 1962-63, chmn. U.S. planning com. 2d Internat. Permafrost Conf., 1972-74, chmn. U.S. del. 3d Internat. Permafrost Conf., 1978, chmn. U.S. organizing com. 4th Internat. Permafrost Conf., 1979-83; com. to study Good Friday Alaska Earthquake Nat. Acad. Scis., 1964-70, mem. glaciological com. polar research bd. Good Friday Alaska Earthquake, 1971-73, chmn. permafrost com., mem. polar research bd. Good Friday Alaska Earthquake, 1975-81; organizing chmn. Internat. Assn. Quarternary Research Symposium and Internat. Field Trip Alaska, 1965; mem. Internat. Commn. Periglacial Morophology, 1964-71, 80—; mem. polar research bd. NRC, 1975-78, late Cenozoic study group, sci. com. Antarctic research, 1977-80. Contbr. numerous papers to profl. lit. Recipient Antarctic Service medal, 1966; Outstanding Achievement award Augustana Coll., 1969. Fellow AAAS (pres. Alaska div. 1956, com. on arid lands 1972-79), Geol. Soc. Am. (editorial bd. 1975-82, chmn. cordilleran sect. 1979-80, chmn. geomorphology div. 1981-82), Arctic Inst. N.Am. (bd. govs. 1969-74, exec. bd. 1972-73), Iowa Acad. Sci., Ariz. Acad. Sci.; mem. Assn. Geology Tchrs., Glaciological Soc., N.Z. Antarctic Soc., Am. Soc. Engring. Geologists, Am. Quaternary Assn. (pres. 1984—), Internat. Geog. Union. Subspecialty: Geology. Current work: Arctic and desert environmental geology; geology for land use planning. Home: 538 E Fairmont Dr Tempe AZ 85282

PEZZUTO, JOHN MICHAEL, biochemist, cons., researcher, educator; b. Hammonton, N.J., Aug. 29, 1950; s. Michael Louis and Elizabeth (Brown) P.; m. Arlene Angela, May 2, 1969; 1 dau., Jennifer Anne. A.B., Rutgers U., 1973; Ph.D., Coll. Medicine and Dentistry N.J., 1977. NIH fellow M.I.T., 1977-79; instr. chemistry, NIH fellow U. Va., 1979-80; asst. prof. U. Ill.-Chgo., 1980—. Mem. Am. Chem. Soc., AAAS, Am. Soc. Pharmacognosy, Am. Assn. Cancer Research. Subspecialties: Biochemistry (medicine); Cancer research (medicine). Current work: Mechanisms of mutagenesis and carcinogenesis, inhibition of tumorigenesis, biospecific evaluation of cancer chemotherapeutic agts. Home: 97 S Park Blvd Glen Ellyn IL 60137 Office: Coll Pharmacy U Ill Med Center Chicago IL 60612

PFAFF, DONALD W., brain research scientist, educator; b. Rochester, N.Y., Dec. 9, 1939; s. Norman Joseph and Eleanor Blakeslee (Wells) P.; m. Stephanie Strickland, Mar. 1, 1963; children: Robin, Alexander, Douglas. A.B. magna cum laude, Harvard U., 1961; Ph.D., M.I.T., 1965. Research assoc. M.I.T., 1965-66; postdoctoral fellow Rockefeller U., 1966-69, asst. prof., 1969-71, asso. prof., 1971-77, prof. neurobiology and behavior, 1977—; pharm. cons. Author: Steroid Hormones and Brain Function, 1974, Estrogens and Brain Function, 1980, Physiological Mechanisms of Motivation, 1982; contbr. articles to profl. jours. NIH fellow, 1963-65, 66-68. Mem. Am. Physiol. Soc., Am. Assn. Anatomy, Endocrine Soc., Soc. Neurosci., Internat. Brain Reasearch Orgn., Hastings Inst. Soc., Ethics and Life Scis. Subspecialties: Neurobiology; Neurophysiology. Current work: Effects of hormones on brain and behavior; nerve cell mechanisms for emotional behavior; effects of steroid hormones on hypothalamic nerve cells. Home: 5 Edgewood Rd Scarsdale NY 10583 Office: Rockefeller University 1230 York Ave New York NY 10021

PFAFFMANN, CARL, educator; b. Bklyn., May 27, 1913; s. Charles and Anna (Haaker) P.; m. Hortense Louise Brooks, Dec. 26, 1939; children—Ellen Anne, Charles Brooks (dec.), William Sage. Ph.B., Brown U., 1933, M.A., 1935; B.A., Oxford (Eng.) U., 1937; Ph.D. (Rhodes scholar), Cambridge (Eng.) U., 1939; D.Sc., Brown U., 1965, Bucknell U., 1966, Yale, 1972. Research asso. Johnson Found., U. Pa., 1939-40; instr. psychology Brown U., 1940-42, asst. prof., 1945-48, asso. prof., 1949-51, prof., 1951-65, Florence Pirce Grant U. prof., 1960; vis. prof. Yale, 1958-59, Harvard, 1962-63; Nat. Sigma Xi lectr., 1963; v.p. Rockefeller U., 1965-78, prof., 1965-80, Vincent and Brooke Astor prof., 1980—; chmn. div. behavorial scis. NRC, 1962-64; chmn. 17th Internat. Congress Psychology, 1963; mem. exec. com. Internat. Union Psychol. Scis., 1966-72; chmn. sect. exptl. psychology and animal behavior Internat. Union Biol. Scis., 1969-76; chmn. com. olfaction and taste Internat. Union Physiol. Scis., 1971-77. Bd. fellows Brown U., 1968—. Served from lt. (j.g.) to comdr. USNR, 1942-45. Recipient Kenneth Craik Research award St. John's Coll., Cambridge U., 1968; Guggenheim fellow, 1960-61. Fellow Am. Psychol. Assn. (Distinguished Sci. Contbn. award 1963, pres. div. exptl. psychology 1956-57), AAAS, Am. Acad. Arts and Scis.; mem. Am. Philos. Soc., Nat. Acad. Sci., Soc. Exptl. Psychologists (Howard Crosby Warren medal research psychology 1960), Am. Physiol. Soc., Eastern Psychol. Assn. (pres. 1958-59). Subspecialties: Physiological psychology; Sensory processes. Current work: Chemical senses; taste and smell; physiology; central nervous function; behavioral processes. Home: One Gracie Terr New York City NY 10028 also Pond Meadow Rd Killingworth CT 06417

PFAHLER, PAUL LEIGHTON, geneticist, educator; b. Essex County, Ont., Can., Nov. 3, 1930; s. Charles and Vivian (Gawley) P.; m. Mary Lucile Neely, Aug. 25, 1967; children: Diane, Beth. B.S., U. Mich., 1952; M.S., Mich. State U., 1954; Ph.D., Purdue U., 1957. Research assoc. Stanford (Calif.) U., 1955; postdoctoral fellow Purdue U., West Lafayette, Ind., 1957-58; asst. prof. U. Fla., Gainesville, 1958-65, assoc. prof., 1965-71, prof., 1971—; vis. scientist Czechoslovak Acad. Sci., 1973, 83; vis. prof. Nijmegen U., The Netherlands, 1977. Contbr. articles to profl. jours. AEC grantee, 1963-67; U. Fla. Faculty Devel. awardee, 1969, 77. Mem. Genetic Soc. Am., Am. Genetics Assn., Crop Sci. Soc. Am., AAAS. Presbyterian. Subspecialties: Plant genetics; Population biology. Current work: Population and quantitative genetics with maize, oats, rye. Home: 2609 NW 12th Ave Gainesville FL 32605 Office: Univ Fla Dept Agronomy 304 Newell Hall Gainesville FL 32611

PFANN, WILLIAM GARDNER, scientist; b. N.Y.C., Oct. 25, 1917; s. John G. and Anna F. (Liedtke) P.; m. Mary L. Gronewold, May 31, 1946; children—Jean L., Susan D., Anna V. B.Chem.Engring., Cooper Union, N.Y.C., 1940. Lab. asst. Bell Telephone Labs., Murray Hill, N.J., 1935-43, staff scientist, 1943-59, dept. head research, 1959-74, cons., 1975—; vis. scientist dept. metallurgy Cambridge (Eng.) U., 1962. Author: Zone Melting, 2d edit, 1966. Recipient Clamer medal Franklin Inst., 1957; Moisson medal Sch. Chemistry U. Paris, 1962; 1st award for Creative Invention Am. Chem. Soc., 1968; award for Excellence Carborundum Co., 1972; award in Solid State Sci. and Tech. Electrochem. Soc., 1973; Profl. Progress award Am. Inst. Chem. Engrs., 1960. Fellow AIME (Mathewson medal 1955), Am. Soc. Metals (Sauveur achievement award 1958); mem. Nat. Acad. Scis., Am. Phys. Soc. (award for New Materials 1976). Subspecialty: Metallurgy. Patentee in field. Office: Bell Labs Murray Hill NJ 07974

PFEFFER, JANICE MARIE, cardiovascular physiologist, researcher, educator; b. Rockford, Ill., Oct. 31, 1943; d. William and Mary Frances (Olszewski) Sikorski; m. Marc Alan Pfeffer, Aug. 22, 1970; 1 dau., Kathryn Marie. B.A., Rockford (Ill.) Coll., 1969; Ph.D., U. Okla., 1977. Biologist VA Hosp., Oklahoma City, 1969-76; Nat. Heart Lung Blood Inst. fellow Harvard Med. Sch., Boston, 1977-80, instr. medicine, 1980-81, asst. prof. medicine, 1981—. Contbr. articles to profl. jours. Recipient Young Investigator Research award NIH, 1980-83. Mem. Am. Physiol. Soc., Am. Heart Assn., Internat. Soc. Hypertension, Internat. Soc. Heart Research, Am. Fedn. Clin. Research. Subspecialties: Physiology (medicine); Physiology (biology). Current work: The performance of the heart in various cardiovascular disorders, such as hypertension and myocardial infarction, and the response of the heart to therapy. Office: Brigham and Women's Hosp 75 Francis St Boston MA 02115

PFEFFER, LAWRENCE MARC, cell biologist, researcher; b. Bronx, N.Y., Nov. 28, 1951; s. Paul and Bess (Wilkens) P.; m. Susan Ritterstein Pfeffer, Sept. 19, 1976. B.S., SUNY-Albany, 1972; Ph.D., Cornell U., 1977. Grad. fellow Sloan Kettering Inst., 1972-77; postdoctoral fellow Rockefeller U., 1977-80, research assoc., 1980-81, asst. prof. virology, 1981—. Contbr. articles to profl. jours. Recipient Jr. Faculty Research award Am. Cancer Soc., 1982. Mem. N.Y. Acad. Sci., Am. Soc. Microbiology, Am. Soc. Virology, Cell Cycle Soc., Sigma Xi. Democrat. Jewish. Subspecialties: Cell and tissue culture; Virology (biology). Current work: Animal virus host cell interactions, interferon; biologic regulatory molecules, interferon, membrane biology. Office: 1230 York Ave New York NY 10021

PFEFFER, RICHARD LAWRENCE, educator; b. Bklyn., Nov. 26, 1930; s. Lester Robert and Anna (Newman) P.; m. Roslyn Ziegler, Aug. 30, 1953; children—Bruce, Lloyd, Scott, Glenn. B.S. cum laude, CCNY, 1952; M.S., Mass. Inst. Tech., 1954, Ph.D., 1957. Research asst. Mass. Inst. Tech., 1952-55, guest lectr., 1956; atmospheric physicist Air Force Cambridge Research Center, Boston, 1955-59; sr. scientist Columbia U., 1959-61, lectr., 1961-62, asst. prof. geophysics, 1962-64; asso. prof. meteorology Fla. State U., Tallahassee, 1964-67; prof., dir. Geophys. Fluid Dynamics Inst., 1967—; cons. NASA, 1961-64, N.W. Ayer & Son, Inc., 1962, Ednl. Testing Service, Princeton, N.J., 1963, Voice of Am., 1963, Grolier, Inc., 1963, Naval Research Labs., 1971-76; Mem. Internat. Commn. for Dynamical Meteorology, 1972-76. Editor: Dynamics of Climate, 1960; Contbr. articles to profl. jours. Bd. dirs. B'nai B'rith Anti-Defamation League; chmn. religious concern and social action com. Temple Israel, Tallahassee, 1971-72. Mem. Am. Meteorol. Soc. (program chmn. ann. meeting 1963), Am. Geophys. Union, N.Y. Acad. Scis. (chmn. planetary scis. sect. 1961-63), Sigma Xi, Chi Epsilon Pi, Sigma Alpha. Subspecialty: Meteorology. Current work: Global atmospheric circulation; hurricane formation; atmospheric dynamics, laboratory and computer modeling. Home: 926 Waverly Rd Tallahassee FL 32312

PFEFFERKORN, ELMER ROY, microbiologist, educator; b. Manitowoc, Wis., Dec. 13, 1931; s. Elmer R. and Mollie (Meyer) P.; m. Lorraine M. Cassidy, Apr. 4, 1964. B.A., Lawrence College., 1954, Oxford (Eng.) U., 1956, M.A., 1959; Ph.D., Harvard Med. Sch., Boston, 1960. Instr. Harvard Med. Sch., 1960-64, asst. prof., 1964-67; assoc. prof. microbiology Dartmouth Med. Sch., 1967-70, prof., 1970—, chmn. dept. microbiology, 1980—; mem. virology study sect. NIH, 1970-74, tropical medicine parasitology study sect., 1981—. NIH grantee, 1960—. Fellow AAAS; mem. Am. Soc. Biol. Chemists, Am. Soc. Tropical Medicine and Hygiene, Am. Soc. Parasitologists. Subspecialties: Parasitology; Biochemistry (medicine). Current work: Biochemistry and genetics of intracellular protozoan parasites. Home: 24 Rope Ferry Rd Hanover NH 03755 Office: Dept Microbiology Dartmouth Med Sch Hanover NH 03756 Home: 24 Rope Ferry Rd Hanover NH 03755

PFEIFFER, HEINZ GERHARD, chemist; b. Pforzheim, Germany, Mar. 31, 1920; came to U.S., 1926, naturalized, 1935; s. Gottlieb F. and Helene Berta (Banz) P.; m. Alma Roorda, Mar. 21, 1948; children: Karen Wolk, James. B.A., Drew U., 1941; M.A., Syracuse U., 1944; Ph.D., Calif. Inst. Tech., 1949. Radio technician USN, 1944-46; research assoc. Gen. Electric Research & Devel. Ctr., Schenectady, 1948, mgr. dielectric studies, 1954, liaison scientist to electric utility group, 1964, dir. ednl. tech. planning for the, 1968, mgr. ednl. tech. br., 1969; dir. Ctr. for Study of Sci. and Soc., SUNY, 1971; mem. and mgr. tech. energy assessment Pa. Power & Light Co., Allentown, Pa., 1972—; dir. Pa. Sci. & Engring. Found., 1974-80; adj. prof. Lehigh U., 1975; chmn. breeder and nuclear fuel subcom. IEEE, 1982; chmn. magnetohydrodynamics task force Edison Electric Inst., 1980-82. Co-author: X-Ray Absorption and Emission in Analytical Chemistry, 1960, Advances in X-Ray Analysis, Vol. II, 1968, X-Rays, Electrons, and Analytical Chemistry, 1972; contbr. chpts. to books, articles to profl. jours. Mem. Niskayuna Sch. Bd., 1962-72, pres., 1968-70. Served with USN, 1944-46. Recipient Bausch & Lomb medal, 1937; Disting. Alumnus award Drew U., 1963. Mem. Am. Chem. Soc., IEEE, Am. Inst. Chemists, N.Y. Acad. Sci., AAAS. Unitarian. Subspecialties: Fuels; Resource management. Current work: Direct electric utility research efforts in new energy sources, conservation, use of reject heat. Home: Barnes Ln Allentown PA 18101 Office: 2 N 9th St Allentown PA 18101

PFENNINGER, KARL HANS, cell biologist; b. Stafa, Switzerland, Dec. 17, 1944; came to U.S., 1971; s. Hans R. and Delie M. (Zahn) P.; m. Marie-France Maylié, July 12, 1974; children: Jan Patrick, Alexandra Christina. M.D., U. Zurich, Switzerland, 1971. Predoctoral researcher Brain Research Inst., U. Zurich, 1966-71; postdoctoral researcher dept. anatomy Washington U., St. Louis, 1971-73, sect. cell biology Yale U., 1973-76; assoc. prof. Columbia U., 1976-81, prof. anatomy and cell biology, 1982—. Contbr numerous articles to profl. jours. Served to 1st lt. Swiss Army, 1965-71. Recipient C. J. Herrick award in comparative neurology Am. Assn. Anatomists, 1977; I. T. Hirschl Career Scientist award, 1977. Mem. AAAS, Am. Soc. Cell Biology, Soc. Neurosci., N.Y. Acad. Sci., Harvey Soc. Subspecialties: Cell biology; Neurobiology. Current work: Cellular and developmental neurobiology; molecular mechanisms of neuronal sprouting and synaptogenesis. Office: Dept Anatomy and Cell Biology Coll Physicians and Surgeons Columbia U 630 W 168th St New York NY 10032

PFISTER, PHILIP CARL, mechanical engineer, educator, consultant; b. Bklyn., Apr. 12, 1925; s. Philip and Katie (Brauchler) P.; m. Elizabeth R. Dawson, June 6, 1959; 1 son, David A. B.S.M.E., CCNY, 1947; M.S.M.E., Columbia U., 1949; Ph.D., Ill. Inst. Tech., 1962. Registered profl. engr., Utah, N.D. Instr. CCNY, 1947-50, W.Va. U., 1950-52; asst. prof. U. Utah, 1952-55; instr. Ill. Inst. Tech., 1955-62, asst. prof., 1962-64, 1966-67; prof. Kabul (Afghanistan) U., 1964-66; prof. mech. engring. N.D. State U., 1967—, chmn. dept. mech. engring., 1967-70; cons. in field. Ford Found. grantee, 1961-62. Mem. ASME, Am. Soc. Engring. Edn., AAUP (pres. chpt. 1975-76), ACLU, Scientists for Nuclear Freeze. Democrat. Methodist. Subspecialties: Mechanical engineering; Wind power. Current work: Wind power. Home: 30 Meadowlark Ln Fargo ND 58102 Office: Dept Mech Engring ND State U Fargo ND 58105

PFLEDERER, FRED RAYMOND, mechanical engineer; b. Elgin, Ill., Mar. 30, 1928; s. Fred and Emma (Schmidgall) P.; m. Mary Jane Bauernfeind, May 24, 1927; children: Janet Lee, Gail Jean, Raymond Fred. B.S. in Mech. Engring, Northwestern U., 1952. Registered profl. engr., Wis., 1957. Devel. engr. A. O. Smith Corp., Milw., 1952-64, supr. new product devel., 1970-75, project mgr., research and devel., 1978—; instr. in engring. Marquette U., 1956-57; supr. performance standards and design A. O. Smith Inland, Little Rock, 1965-70; mgr. research and devel. Stearn Sailing Systems, Sturgeon Bay, Wis., 1975-78. Contbr. articles to profl. publs. Deacon, elder Wauwatosa (Wis.) Presbyterian Ch. Served with U.S. Army, 1946-48. Mem. ASME. Subspecialties: Mechanical engineering; Composite materials. Current work: Product development, including design and fabrication, product development of filament wound composite materials primarily for commercial applications. U.S., Can. patentee in field. Home: 4349 N Raymir Pl Wauwatosa WI 53222 Office: 3533 N 27 St Milwaukee WI 53216

PFORZHEIMER, HARRY, JR., oil co. exec.; b. Manila, Philippines, Nov. 19, 1915; s. Harry and Mary Ann (Horan) P.; m. Jean L. Barnard, June 2, 1945; children: Harry, Thomas. B.S. in Chem. Engring, Purdue U., 1938; postgrad., Case Western Res. U., 1939-41, in law, George Washington U., 1942-43. With Standard Oil Co. Ohio, Cleve., 1938-80, v.p. oil shale and tar sands, 1971-80; pres., chmn. bd., chief exec. officer Paraho Devel. Corp., Grand Junction, Colo., 1980-82, sr. mgmt. advisor, dir., 1982—; dir. IntraWest Bank of Grand Junction, Colo. Contbr. articles to profl. jours. Wayne Aspinall Found.; mem. adv. bd. St. Mary's Hosp.; mem. Petroleum Adminstrn. for War in Washington, 1942-45. Recipient Disting. Student award Purdue U., 1938. Mem. Am. Inst. Chem. Engrs., Am. Mining Congress, Colo. Mining Assn., others. Republican. Roman Catholic. Clubs: Denver Petroleum, Army and Navy, Bookcliff Country. Subspecialty: Oil shale. Current work: Management of shale oil company. Office: 101 S 3d St Suite 300 Grand Junction CO 81501 Home: 2700 G Rd Apt 1C Grand Junction CO 81501

PFRANG, EDWARD OSCAR, association executive; b. New Haven, Aug. 9, 1929; s. Luitpold and Anna P.; m. Jacquelyn Marcia Montefalco, June 7, 1958; children: Lori Ann, Leslie Jean, Philip Edward. B.S., U. Conn., 1951; M.E., Yale U., 1952; Ph.D., U. Ill., 1961. Registered profl. engr., Md., N.Y., Calif. Sect. chief structures sect., bldg. research div. Nat. Bur. Standards, Washington, 1966-83, mgr. housing tech. program, 1970-73, chief structures materials safety div., 1973-83; exec. dir. ASCE, Washington, 1983—. Contbr. articles to profl. jours. Served with USNR, 1953-56. Fellow ASCE, Am. Concrete Inst. (pres.); mem. Earthquake Engring. Inst., Sigma Xi, Tau Beta Phi, Chi Epsilon, Sigma Tau. Subspecialty: Civil engineering. Office: Nat Bur Standards B368 226 Washington DC 20234

PHAN, SEM HIN, pathology educator, researcher; b. Jakarta, Indonesia, Sept. 15, 1949; came to U.S., 1967, naturalized, 1980; s. Yoek Sioe and Hong Tek (Hauw) P.; m. Katherine A. Assimos, June 19, 1976; children: Nicholas N., Louis N. B.Sc., Ind. U., Bloomington, 1971, Ph.D., 1975, M.D., 1976. Diplomate: Am. Bd. Pathology. Intern U. Conn. Health Ctr., Farmington, 1976-78, research fellow, 1978-80, resident anat. pathology, 1976-78; asst. prof. pathology U. Mich. Med. Sch., Ann Arbor, 1980—; research assoc. VA Med. Ctr., Ann Arbor, 1980—. NIH research grantee, 1982. Mem. Am. Assn. Pathologists, AAAS, Am. Thoracic Soc., Mich. Soc. Pathologists, N.Y. Acad. Scis., Phi Beta Kappa. Roman Catholic. Subspecialties: Pathology (medicine); Biochemistry (medicine). Current work: Biochemical mechanisms in tissue injury and repair, as well as membrane biochemistry and biophysics of neutrophil function. Office: Dept Pathology M045 U Mich Med Sch 1335 E Catherine St Ann Abror MI 48109

PHARRISS, BRUCE BAILEY, pharm. co. exec.; b. Springfield, Mo., Dec. 12, 1937; s. McNatt and Mary Louise (Bailey) P.; m. Joyce Ann Onions, 1959; 1 dau., Lotti Louise. B.S., M.A., Ph.D., U. Mo. Research Scientist Upjohn Co., 1966-70. Research dir. Alza Corp., 1970-79; v.p. sci. affairs Collagen Corp., Palo Alto, Calif., 1980—. Mem. Endocrine Soc., N.Y. Acad. Scis., World Population Soc., Soc. Biomaterial (Founding). Subspecialties: Physiology (biology); Reproductive biology. Office: 2455 Faber Pl Palo Alto CA 94303

PHELAN, ROBERT JOSEPH, JR., physicist; b. Pasadena, Calif., Oct. 18, 1933; s. Robert Joseph and Rachael Susan (Garrison) P.; m. Mary Elizabeth Doran, June 19, 1960; children: Robert Joseph III, Rebecca E., William E. B.S. in Physics, Calif. Inst. Tech., 1958; Ph.D., U. Colo., 1962. Staff physicist M.I.T. Lincoln Lab., Lexington, Mass., 1962-69; physicist Nat. Bur. Standards, Boulder, Colo., 1969—. Contbr. articles to profl. jours. Served with U.S. Army, 1955-56. Mem. Am. Phys. Soc., IEEE, Optical Soc. Am. Subspecialties: Semiconductors; Fiber optics. Current work: Electro-optic device research—the development of unique optica. Patentee in field. Home: 1780 Ithaca Dr Boulder CO 80303 Office: 325 Broadway Boulder CO 80303

PHELPS, HARRIETTE LONGACRE, biology educator; b. Exeter, N.H., Mar. 15, 1936; d. Andrew and Marian Chandler (Sykes) Longacre; m. John Bedford Phelps, Aug. 30, 1958 (div. 1972); children: David, Elizabeth. B.A., Carleton Coll., 1956; M.S., Mount Holyoke Coll., 1958; Ph.D., Ohio State U., 1965. Staff fellow NIH, Bethesda, Md., 1965-69; asst. prof. biology dept. Fed. City Coll., Washington, 1968-72, assoc. prof., 1972-78; prof. biology U. D.C., Washington, 1978—. NIH fellow, 1977-78; Dept. Energy fellow, 1982; recipient Research Grants NASA, 1974-82, NIH, 1977-80, Water Resources Ctr., 1980-81, Title III, 1975-77. Mem. Atlantic Estuarine Research Fedn., Am. Chem. Soc., Am. Soc. Limnology and Oceanography, AAAS, Geol. Soc., Am., Sigma Xi. Democrat. Congregationalist. Subspecialty: Environmental toxicology. Current work: Biogeochemistry of heavy metals in estuaries with special reference to shellfish, including uptake, accumulation, and effects on reproduction and behavior. Home: 7822 Hanover Pkwy Apt 303 Greenbelt MD 20770 Office: Biology Dept U DC 1321 H St N Washington DC 20005

PHELPS, JAMES PARKHURST, nuclear engineering educator; b. Medford, Mass., Feb. 13, 1924; s. Edward Parkhurst and Ethel Alice (Cottle) P. B.S., U. Maine, 1951; Ph.D., Mich. State U.-East Lansing, 1956. Sr. engr. Martin Co., Balt., 1956-57; scientist Brookhaven Nat. Lab., Upton, N.Y., 1957-70; prof. in charge of reactor Lowell (Mass.) Tech. Inst., 1970-75; prof. nuclear engring. U. Lowell, 1975—, chmn. nuclear engring. dept., 1975-82. Contbr. numerous nuclear engring. articles to profl. publs. Nuclear Regulatory Commn. fellow Mich. State U., 1955-56. Mem. Am. Phys. Soc., Am. Nuclear Soc., Sigma Xi. Republican. Subspecialties: Nuclear engineering; Nuclear fission. Current work: Reactor physics and reactor operations; interaction between technology and human values; public educator of technology. Home: 64 Melendy Rd Hudson NH 03051 Office: U Lowell 1 University Ave Lowell MA 11854

PHEMISTER, ROBERT DAVID, university president; b. Framingham, Mass., July 15, 1936; s. Robert Irving and Georgia Nora (Savignac) P.; m. Ann Christine Lyon, June 14, 1960; children: Katherine, David, Susan. D.V.M., Cornell U., 1960; Ph.D., Colo. State U., Ft. Collins, 1967. Diplomate: Am. Coll. Vet. Pathologists. Research assoc. U. Calif., Davis, 1960-61; staff scientist Armed Forces Inst. Pathology, Washington, 1962-64; sect. leader to dir. collaborative radiol. health lab. Colo. State U., 1964-77; mem. faculty Coll. Vet. Medicine and Biomed. Scis., 1968—, prof. vet. pathology, 1973—, assoc. dean, 1976-77, assoc. dir. expt. sta., 1977—, dean, 1977-82, acad. v.p., 1982, pres. (interim), 1983—; vis. research pathologist U. Calif., Davis, 1974-75; cons. Miss. State U., 1977-81. Author papers in field. Mem. AAAS, AVMA, Assn. Am. Vet. Med. Colls., Internat. Acad. Pathology, Radiation Research Soc., Colo. Vet Med. Assn., Sigma Xi, Phi Zeta, Phi Kappa Phi. Subspecialties: Pathology (veterinary medicine); Reproductive biology. Current work: Pathology of development; pathology of low-level irradiation; reproductive physiology of the dog; canine renal pathology. Home: 1909 Pawnee Dr Fort Collins CO 80525 Office: Coll Vet Medicine and Biomed Scis Colo State Univ Fort Collins CO 80523

PHILBRICK, DAVID ALAN, researcher and state official; b. Ross, Calif., Apr. 20, 1948; s. Shirley Seavey, Jr. and Emily King (Browning) P.; m. Cathey Lee McPhaden, July 15, 1972; children: Kenneth Alan, John Lee. A.B. magna cum laude, Brown U., 1970; Ph.D. in Biophysics, U. Calif.-Berkeley, 1976. Researcher Donner Lab., Berkeley, Calif., 1972-76; environ. specialist Oreg. Dept. Energy., Salem, 1976-78; adminstr. renewable resources div. Oreg. Dept. Energy, 1978-83; program leader Oreg. Energy Extension Service, 1983—. Mem. Internat. Solar Energy Soc. Subspecialties: Solar energy; Geothermal power. Current work: Energy policy development and evaluation of energy technologies. Office: Oreg State Extension Service Oreg State U Corvallis OR 97331

PHILIP, A G DAVIS, astronomer, educator; b. N.Y.C., Jan. 9, 1929; s. Van Ness and Lilian (Davis) P.; m. Kristina Drobavicius, Apr. 25, 1964; 1 dau., Kristina Elizabeth. B.S., Union Coll., Schenectady, 1951; M.S., N.Mex. State U., 1959; Ph.D., Case Inst. Tech., 1964. Tchr. Brooks Sch., 1954-59; instr. Case Inst. Tech., 1962-64; asst. prof. astronomy U. N.Mex., 1964-66, SUNY, Albany, 1966-67, assoc. prof., 1967-76; prof. astronomy Union College., 1976—; astronomer Dudley Obs., 1967-81, Van Vleck Obs., 1982—; Frank L. Fullam chair astronomy, 1980-81; vis. prof. Yale U., 1972, 73, vis. fellow, 1976; vis. prof. La. State U., 1973, 76, Acad. Scis. Lithuania, 1973, 76, 77, Stellar Data Center, Strasbourg, France, 1978, 79, 80, 82; Harlow Shapley lectr. Am. Astron. Soc., 1973-81; dir., sec.-treas. N.Y. Astron. Corp.; 1st U.S. observer Soviet Union's 6-meter reflector telescope, 1979 -82; exhibited 2d Ann. Photog. Regional, Albany, 1980; pres., treas. L Davis Press, Inc. Exec. com. Arts. and Scis. Council, 1975-76. Author: (with M. Cullen and R.E. White) UBV Color—Magnitude Diagrams of Galactic Globular Clusters; editor: (with D.S. Hayes) Galactic Structure in the Directon of the Galactic Polar Caps, (with M.F. McCarthy) Spectral Classification of the Future, (with M.F. McCarthy and Coyne) In Memory of Henry Norris Russel, (with D. DeVorkin) The HR Diagram: The One-Hundredth Anniversary of Henry Norris Russel, (with D.S. Hayes) IAU Colloquium No. 68, Astrophysical Parameters for Globular Clusters, Dudley Obs. Reports, 1977-81, ASNY News Letter, 1976—; contbr. articles profl. jours., chpts. in books. Served with AUS, 1951-53. Fellow Royal Astron. Soc., AAAS; mem. Am. Phys. Soc., Am. Astron. Soc., Internat. Astron. Union (sec. commn galactic structure 1973-76, v.p. commn. radial velocity 1979-82, pres. 1982—, chmn working group spectroscope and photometric data 1982—, mem. working group standard stars), N.Y. Acad. Scis., Astron. Soc. Pacific, Astron. Soc. N.Y. (sec.-treas.), Sigma Xi. Subspecialty: Plant growth. Home: 1125 Oxford Pl Schenectady NY 12380 Office: 69 Union Ave Schenectady NY 12308

PHILIPS, GERALD JOHN, mech. engr., researcher; b. Bklyn., Feb. 20, 1945; s. William and Eugenia (Halasa) P.; m. Dolores Grace, Mar. 30, 1964; children: Jeffrey John, Jean Marie; m. Sarah Kaplan, Mar. 14, 1981. B.E. in Mech. Engring, Pratt Inst., 1968. Research mech. engr. David Taylor Naval Ship Research and Devel. Ctr., Annapolis, Md., 1968—. Mem. Annapolis Jaycees, 1974-79. Mem. ASME (assoc). Subspecialties: Mechanical engineering; Fiber optics. Current work: Tribology; Bearings; vibrations; machine condition monitoring; design and devlopement of machinery condition monitoring systems; research to improve bearing service life. Patentee rolling contact bearing design, fiber optic machinery performance monitor. Office: David Taylor Naval Ship Research and Devel Center Code 2832 Annapolis MD 21402

PHILLIPS, BARRIE MAURICE, toxicologist; b. Highland Park, Mich., Feb. 18, 1934; s. Clyde Maurice and Meribah Catherine (McClung) P.; m. Beverly Jean Kingsley, Sept. 8, 1956; children: Gail M., Elizabeth L., Barrie M., Jennifer K. B.S., Ferris State Coll., 1957; M.S., Purdue U., 1960, Ph.D., 1962. Dir. toxicology Miles Labs., Inc., Elkhart, Ind., 1962-77; v.p. Ind. Bio-Test Labs., Inc., Northbrook, Ill., 1977-78, pres., 1978-79; owner, mgr. Tox-Con Assocs., Elkhart, Ind., 1979-80; assoc. dir. Bioassay div. Internat. Research and Devel. Corp., Mattawan, Mich., 1980-81, dir. adminstrv. services, 1981—. Contbr. articles to profl. jours. Mem. Soc. Toxicology, Am. Soc. Pharmacology and Exptl. Therapeutics, Sigma Xi, Rho Chi, Phi Lambda Upsilon. Republican. Subspecialties: Toxicology (medicine); Pharmacology. Current work: Evaluation of the safety of drugs, foods, food additives, industrial and agricultural chemicals. Home: 58418 Hilly Ln Elkhart IN 46517 Office: Internat Research and Devel Corp 500 N Main St Mattawan MI 49071

PHILLIPS, CHRISTINE ELAINE, research scientist; b. Detroit, Sept. 4, 1949; d. Gordan F. and Normal R. (Wendt) P.; m. John A. Wilson, Dec. 30, 1977. B.A., U. Mich., 1972; M.S., Wayne State U., 1974; Ph.D., U. Oreg., 1978. Postdoctoral fellow in zoology Cambridge (Eng.) U., 1978-79; research assoc. in biology U. Va., Charlottesville, 1979-81, research assoc. dept. neurosurgery, 1981—; Bd. dirs. Stonehenge Homeowner's Assn. Contbr. articles to profl. publs. Grantee NIH, 1974-79, 81-82, Sigma Xi, 1977. Mem. Soc. Neurosci., Soc. Exptl. Biology, Am. Zool. Soc. Democrat. Subspecialties: Neurobiology; Neurophysiology. Current work: Comparative neuroanatomy/physiology; structure and function of synapses, neuronal circuitry and local non-spiking interneurons. Home: 13 Cheshire Ct Charlottesville VA 22901 Office: U Va Sch Medicine Dept Neurosurgery Box 420 Charlottesville VA 22908

PHILLIPS, GREGORY CONRAD, plant geneticist; b. Covington, Ky., Sept. 11, 1954; s. Ronald Edward and Mary Susan (Conrad) P.; m. Ella Louise Dedman, July 18, 1975; 1 son: Isaac Clarke. B.A., U. Ky., 1975, Ph.D., 1981. Research asst. U. Ky., Lexington, 1976-81; asst. prof. horticulture N.Mex. State U., Las Cruces, 1981—. Contrb. articles to sci. jours. Recipient Oswald award U. Ky., 1974; N.Mex. Water Resources Research Inst. grantee, 1982-83. Mem. Crop Sci. Soc. Am., Am. Soc. Agronomy, Am. Soc. Hort. Sci., Internat. Assn. Plant Tissue Culture, Tissue Culture Assn. Subspecialties: Plant genetics; Plant cell and tissue culture. Current work: Development of tissue and cell culture for plant improvement; somatic cell genetics, plants, tissue and cell culture, cell selection, cloning, plant regeneration, plant breeding, agriculture. Office: N Mex State U Box 3530 Las Cruces NM 88003

PHILLIPS, JAMES ALFRED, research scientist; b. Johannesburg, South Africa, May 17, 1919; s. Ray Edmund and Dora Amanda (Larson) P.; m. Marilyn Virginia Hopkins, Aug. 14, 1948; children: Raymond Bruce, Margaret Ann, Kathryn Amanda. B.S., Carlton Coll., Northfield, Minn., 1942; M.S., U. Ill., Urbana, 1943, Ph.D., 1949. Mem. staff Tenn. Eastman Corp., Oak Ridge, 1944-45; research asst. U. Ill., 1942-49; mem. staff Los Alamos Nat. Lab., 1949-75, 82, fellow, 1982; head physics sect. IAEA, Vienna, Austria, 1975-79; vis. scientist Culham Plasma Physics Lab., 1964-65. Contbr. articles to profl. jours. Fellow Am. Phys. Soc. (sec. div. plasma physics 1960-64, chmn. div. 1967, regional sec. 1970-75). Republican. Congregationalist. Subspecialties: Nuclear fusion; Plasma. Current work: Controlled fusion research-magnet confinent of high temperature plasmas, physics and engineering. Patentee in field. Home: 48 Loma Del Escolar Los Alamos NM 87544 Office: Los Alamos Nat Lab Los Alamos NM 87544

PHILLIPS, JAMES CHARLES, educator, physicist; b. New Orleans, Mar. 9, 1933; s. William D. and Juanita (Hahn) P.; m. Erika Muehl, Oct. 24, 1976. B.A., U. Chgo., 1952, B.S., 1953, M.S., 1955, Ph.D.,

1956. Mem. tech. staff Bell Labs., 1956-58; NSF fellow U. Calif. at Berkeley, 1958-59, Cambridge (Eng.) U., 1959-60; faculty U. Chgo., 1960-68, prof. physics, 1965-68; mem. tech. staff Bell Labs., 1968—. Sloan fellow, 1962-66; Guggenheim fellow, 1967. Fellow Am. Phys. Soc. (Buckley prize 1972); mem. Nat. Acad. Scis. Subspecialties: Condensed matter physics; Solid state chemistry. Home: 204 Springfield Ave Summit NJ 07901

PHILLIPS, JAMES WOODWARD, engineering educator, consultant, researcher; b. Washington, Mar. 8, 1943; s. Clinton Woodward and Ethel Catherine (Lynch) P.; m. Constance Ann Rodrigue, Sept. 6, 1965; children: Andrew W., Matthew J., Thomas J., Michael L. B.M.E., Cath. U. Am., 1964; Sc.M. in Engring, Brown U., 1966, Ph.D., 1969. Registered profl. engr., Ill. Research assoc. Brown U., Providence, 1969; asst. prof. U. Ill., Urbana, 1969-74, assoc. prof., 1974-81, prof. theoretical and applied mechanics, 1981—; cons.; vis. assoc. prof. U. Md., 1978-79; comml. tester. Editor: Mechanics, 1975-78; contbr. articles to engring. jours. Univ. fellow Brown U., 1964-65; summer faculty fellow U. Ill., 1970, 71; research initiation grantee NSF, 1973-74. Mem. Soc. Exptl. Stress Analysis, Am. Acad. Mechanics. Roman Catholic. Subspecialties: Solid mechanics; Mechanical engineering. Current work: Stress waves in solids; waterhammer waves; wire rope analysis; theoretical and experimental analysis of wire rope; commercial testing of wire rope, oil-well casing and other components. Home: 2302 Aspen Dr Champaign IL 61821 Office: 216 Talbot Laboratory 104 S Wright St Urbana IL 61801

PHILLIPS, JOHN HOWELL, JR., microbiologist, consultant; b. Fresno, Calif., Dec. 19, 1925; s. John Howell and Daisy Isabel (Trott) P.; m.; children: John Howell III, Melissa Louise, Charles Henry. A.B., U. Calif.-Berkeley, 1949, M.A., 1954, Ph.D., 1955. Asst. prof. U. Calif.-Berkeley, 1957-63; assoc. prof. Stanford U., Calif., 1963-80; prof. Hopkins Marine Sta., Pacific Grove, Calif., 1966, dir. emeritus, 1980, 1965-72; cons. Monterey (Calif.) Abalone Farms, 1980—, NSF, Washington, 1970; adviser pesticide adv. com. Dept. Agr., Sacramento, 1969—, 1972-73. Contbr. articles to profl. jours. Mem. Coordinating Community Council for Higher Edn. Calif., 1968; bd. dirs. Del Monte Forest Property Owners Assn., Pebble Beach, Calif., 1970-75, Pacific Grove Mus. Natural History Assn., Pacific Grove, 1971—; mem. com. Monterey Inst. Fgn. Studies Scholarship, Monterey, Calif., 1968. Fellow Am. Cancer Soc., 1955-56; Waksman Merck fellow Rutgers U., 1956-57. Mem. Am. Assn. Immunologists, AAAS, N.Y. Acad. Scis., Sigma Xi. Democrat. Episcopalian. Club: Sierra (Monterey, Calif.) (exec. com. 1972-74). Subspecialties: Microbiology. Current work: Aquaculture and molluscan diseases. Home: 2834 Treasure Rd Pebble Beach CA 93953 Office: Monterey Abalone Farms 300 Cannery Row Monterey CA 93940

PHILLIPS, JOHN HUNTER, cardiologist, medical educator; b. Huston, Nov. 2, 1930; s. John Hunter and Ida (Sholars) P.; m. Mary Jo Holland, Dec. 18, 1954; children: Cynthia K., J. Hunter, Virginia M. B.S., Tulane U., 1952, M.D., 1955. Diplomate: Am. Bd. Internal Medicine, Am. Bd. Cardiovascular Disease. Intern Charity Hosp., New Orleans, 1955-56; asst. instr. medicine dept. Tulane Med. Sch., 1956-57, instr. medicine, 1957-62, asst. prof. medicine, 1962-66, assoc. prof. medicine, 1966-70, prof. medicine, 1970—, chief cardiology sect., 1974—; chief Tulane cardiology div. New Orleans Charity Hosp., 1974—; chief cardiology sect. VA Hosp., New Orleans, 1966-74; cons. cardiology East Jefferson Hosp., Metairie, La., 1971—. Cons. editor: Cardiology Monograph series, Addison-Wesley Pub. Co., 1978—; mem. editorial bd.: Primary Cardiology Jour, 1978—, Biomaterials, Medical Devices and Artificial Organs Jour, 1971—; asst. editor: Am. Heart Jour, 1959-73. Mem. adv. bd. St. Anna's Asylum, New Orleans, 1967—; Bd. dirs. La. Heart Assn., 1978-82. Served to lt. comdr. USNR, 1963-65. Recipient AMA Physician Recognition award, 1981—; Querens-Rives-Shore award in cardiology, 1955; Angiology Research Found. Honor Achievement award, 1968. Fellow Am. Coll. Cardiology (gov. for La. 1976-79), Am. Coll. Chest Physicians, ACP, Council on Clin. Cardiology, Am. Heart Assn. Subspecialties: Cardiology; Internal medicine. Home: 108 Duplessis St Metairie LA 70005 Office: Cardiology Sect Tulane Med Sch 1430 Tulane Ave New Orleans LA 70112

PHILLIPS, LAWRENCE S(TONE), medical clinician-investigator, educator; b. Washington, Sept. 16, 1941; s. Hiram S. and Ruth (Kusner) P.; m. Carol Minette Newman, June 5, 1971; children: Elizabeth, David. B.A. in Chemistry with honors, Swarthmore Coll., 1963; M.D., Harvard U., 1967. Diplomate: Am. Bd. Internal Medicine, (Endocrinology). Intern Presbyterian-St. Luke's Hosp., Chgo., 1967-68, resident, 1968-69; mem. lab. program USPHS, Center Disease Control, Atlanta, 1969-71; staff assoc. Grady Meml. Hosp., Atlanta, 1969-71; instr. in medicine Emory U., 1970-71; fellow in metabolism Washington U., St. Louis, 1971-74; asst. physician Barnes Hosp., St. Louis, 1972-74; asst. prof. medicine Northwestern U., 1974-79, assoc. prof., 1979—, mem., 1974—; adj. attending physician Northwestern Meml. Hosp., Chgo., 1974-75, assoc. attending physician, 1975-81, attending physician, 1981—. Served to lt. comdr. USPHS, 1969-71. Recipient Elliot P. Joslin Research and Devel. award Am. Diabetes Assn. Ill., 1977; Chgo. Diabetes Assn. grantee, 1975-77; Am. Diabetes Assn. grantee, 1977-78; NIH grantee, 1978-86. Mem. Endocrine Soc., Am. Diabetes Assn., Am. Fedn. Clin. Research, Chgo. Endocrine Club (pres. 1978-79), Central Soc. Clin. Research, AAAS, Am. Soc. Clin. Investigation. Subspecialties: Endocrinology; Nutrition (biology). Current work: Interaction of insulin and nutrition inccontrol of somatomedin production and growth. Office: Northwestern U Med Sch 303 E Chicago Ave Chicago IL 60611

PHILLIPS, MICHAEL IAN, physiologist, educator; b. London, Eng., July 30, 1938. B.Sc., U. Exeter, U.K., 1962; M.Sc., U. Birmingham, U.K., 1965, Ph.D., 1967. Postdoctoral fellow Brain Research Labs., U. Mich., 1967-80, vis. asst. prof., 1968-69; research fellow Calif. Inst. Tech., 1969-70; asst., then asso. prof. physiology U. Iowa, 1970-76, acting chmn. dept., 1976-77, prof., 1977; prof., chmn. dept. physiology U. Fla., Gainesville, 1980—; vis. scientist U. Zurich, 1975, NIH, 1979-80; Humboldt scholar U. Heidelberg, 1976-77; mem. NIH Exptl. Cardiovascular Research Rev. Bd., 1980-85, NSF Neurobiology Rev. Bd., 1977-80. Author numerous articles in field.; Editor: Brain Unit Activity During Behavior, 1973, IBRO News, 1982—, (with others) The Renin Angiotens in System in the Brain, 1982. Named Tchr of Yr. U. Iowa, 1976, 79; recipient NIMH research career devel. award, 1974-80; grantee NIH, NSF. Mem. Am Physiol. Soc., Neurosci. Soc., Internat. Brain Research Orgn. Subspecialties: Neuroendocrinology; Neurophysiology. Current work: Neuroendocrinology: peptides, brain, hypertension. Office: U Fla Dept Physiology Box J-274 Gainesville FL 32610

PHILLIPS, RAYMOND BRUCE, botanist, educator; b. Urbana, Ill., May 31, 1949; s. James Alfred and Marilyn Virginia (Hopkins) P. B.A., Pomona Coll., 1973; Ph.D. (NSF grantee, Inst. Profl. Edn., scholar), U. Calif., Berkeley, 1982. Asst. prof. dept. botany and microbiology U. Okla., Norman, 1980—. Fellow Linnean Soc. London; mem. Am. Soc Plant Taxonomist, Internat. Assn. Plant Taxonomy, Bot. Soc. Am., Soc. Systematic Zoology, Classification Soc., AAAS, Sigma Xi. Subspecialties: Systematics; Taxonomy. Current work: Plant systematics: application of computer-assisted methods in biological classification, especially cladistic (phylogenetic) approaches; classificatory bases in evolutionary history; evolutionary relationships of Parnassia, Saxifragaceae, Scrophulariaceae; information storage and retrieval for herbarium specimens; flora of Okla. Home: 1915 Oakhollow Dr Norman OK 730071 Office: Dept Botany and Microbiology U Okla Norman OK 73019

PHILLIPS, RICHARD ARLAN, research adminstr.; b. Detroit, May 30, 1933; s. William George and Sylvia Ann (Meldrum) P.; m. Eleanor Louise Dorn, Jan. 19, 1956; children: Gail Eleaine, Dawn Ellen. B.S., U. Mich., 1955, M.S., 1962, Ph.D., 1965. Reservoir engr. Exxon, Venezuela, 1955-60; asst. prof. U. Minn., Mpls., 1966-71; dir. research and devel. Foster Grant Corp., Leominster, Mass., 1971—. Co-author: Contemporary Optics for Scientists and Engineers, 1976; contbr. articles to profl. jours. Mem. Optical Soc. Am. Unitarian. Subspecialty: Optics research. Patentee in field. Home: 3 Betsy Ross Circle Acton MA 01720 Office: Foster Grand Corp 289 N Main St Leominster MA 01453

PHILLIPS, ROBERT BOONE, radio astronomer, researcher; b. Syracuse, N.Y., Sept. 18, 1952; s. Harold Boone and Marilyn Bernadette (Bordeau) P. B.S., Syracuse U., 1973; M.S., U. Iowa, 1975, Ph.D., 1979. Postdoctoral research fellow Brandeis U., Waltham, Mass., 1979-80; asst. prof. physics and astronomy U. Kans., Lawrence, 1980-81; mem. research staff Haystack Observatory, Westford, Mass., 1981—. Contbr. articles in field to profl. jours. Mem. Am. Astron. Soc., Phi Kappa Phi. Subspecialties: Radio and microwave astronomy; High energy astrophysics. Current work: Radio astronomer in high resolution mapping of cores of galaxies and quasars with very long baseline interferometry. Office: Haystack Observatory Westford MA 01886

PHILLIPS, THOMAS L., corp. exec.; b. May 2, 1924. B.S.E.E., Va. Poly. Inst., 1947, M.S.E.E., 1948; hon. doctorates, Stonehill Coll., 1968, Northeastern U., 1968, Lowell Technol. Inst., 1970, Gordon Coll., 1970, Boston Coll., 1974, Babson Coll., 1981. With Raytheon Co., Lexington, Mass., 1948—, exec. v.p., 1961-64, pres., chief operating officer, 1964-68, chief exec. officer, 1968—, chmn. bd., 1975—, dir., 1962—; dir. John Hancock Life Ins. Co., State St. Investment Corp. Trustee Gordon Coll., Northeastern U.; mem. corp. Joslin Diabetes Found., Mus. Sci., Boston. Recipient Meritorious Pub. Service award for work in Sparrow III missile system, U.S. Navy, 1958. Mem. Nat. Acad. Engring., SRI Internat. Council, Bus. Council, Bus. Roundtable. Clubs: Pilgrims U.S., Algonquin, Comml. (Boston); Weston Golf. Current work: Engineering company management. Office: Raytheon Co 141 Spring St Lexington MA 02173

PHILLIPS, WILLIAM EVANS, mechanical engineer; b. Pascagoula, Miss., May 25, 1944; s. Jack Oliver and Emma Josephine (Evans) P.; m. Sherril Ann Poche, Sept. 28, 1968; 1 dau., Julie Anne. B.S., Miss. State U., 1966; M.B.A., U. Houston, 1973. Registered profl. engr., Wis. Prodn. technician Union Carbide Corp., Taft, La., 1966-68; project group asst., Houston, 1968-70; vessel engr. M.W. Kellog Co., Houston, 1970-73, Foster Wheeler Energy Corp., 1973-77, lead vessel engr., 1977-81, sr. lead vessel engr., 1981—; cons. high temperature tech. Author corp. design manuals. Mem. ASME. Democrat. Methodist. Subspecialties: Mechanical engineering; High-temperature materials. Current work: High temperature and erosion resistant refractory and other materials, high temperature solids transfer systems. Inventor valves, vessels, insulating and erosion resistant material systems, high temperature fluid solids systems apparatus. Office: PO Box 22395 Houston TX 77227

PHILLIPS, WILLIAM REVELL, geology educator, researcher; b. Salt Lake City, Jan. 9, 1929; s. William L. and Della (Weight) P.; m. Harriet Vail LaRue, June 21, 1950; children: Lee Revell, Lyle Vail, Lane William, Kathryn Ann. B.S., U. Utah, 1950, M.S., 1951, Ph.D., 1954. Research mineralogist Kennecott Copper Corp., Salt Lake City, 1954-56; asst. prof. geology La. Poly. Inst., 1956-57; asst. prof. Brigham Young U., 1957-60, assoc. prof., 1960-66, prof., chmn. dept., 1966-75, prof., 1975—; supervisory park naturalist Nat. Park Service, Yellowstone Nat. Park, Wyo., summers, 1956-66; Fulbright prof. U. Sind, Hyderabad, Pakistan, 1963-64, Middle-East Tech. U., Ankara, Turkey, 1966-67; vis. prof. U. Waterloo, Ont., Can., 1971-72; vis. research prof. Hacetepe U., Ankara, Turkey, 1975-76. Author: Mineral Optics, 1971; co-author: Optical Mineralogy, 1981. Fellow Geol. Soc. Am.; mem. Mineral. Soc. Am., Sigma Xi, Phi Kappa Phi. Mormon. Subspecialties: Mineralogy; Petrology. Current work: Archeological geology; application of geology to the history and archeological excavations of countries of the Middle East, e.g., Egypt, Turkey. Home: 1839 N 1500 East St Provo UT 84604 Office: Brigham Young U Provo UT 84602

PHILLIPS, WINFRED MARSHALL, university administrator, mechanical engineering educator; b. Richmond, Va., Oct. 7, 1940; s. Claude Marshall and Gladys Marian (Barden) P.; m. Lynda Bartlett, July 14, 1960; children: Stephen, Sean. B.S.M.E., Va. Poly. Inst., 1963; M.A.E., U. Va., 1966, D.Sc., 1968. Asst. prof. aerospace engring. Pa. State U., 1968-74, assoc. prof., 1974-76, prof., 1976-80, assoc. dean, 1979-80; prof., head Sch. Mech. Engring., Purdue U., 1980—; vis. prof. Institut de Pathologie Cellulaire, U. Paris, 1976-77; mem. Ind. Boiler and Pressure Vessel Bd., Indpls., 1981—; mem. com. quality of engring. edn. Nat. Assn. State Univ. and Land-Grant Colls., 1982. Contbr. numerous articles, abstracts on biofluid dynamics, cardiovascular devices, fluid mechanics, rheology to profl. jours.; assoc. editor: Jour. Biomech. Engring, 1977—. Recipient Outstanding Young Faculty Mem. award Am. Soc. Engring. Edn. and Dow Chem. Co., 1971, Research Career Devel. award USPHS, 1975-80; named Eminent Engr. Tau Bet Pi, 1981; NSF trainee, 1965-67. Assoc. fellow AIAA; fellow AAAS; mem. Am. Soc. Engring. Edn., am. Soc. Artificial Internal Organs (trustee 1982—, councilman-at-large 1982-83), ASME. Lodge: Rotary. Subspecialties: Mechanical engineering; Fluid mechanics. Current work: Fluid mechanics, cardiovascular fluid dynamics, biofluid dynamics, biorheology, gas dynamics. Office: Sch Mech Engring 211 Mech Engring Bldg Purdue U West Lafayette IN 47907 Home: 1806 Bent Tree Trail West Lafayette IN 47906

PHILLIPS-JONES, LINDA, research scientist; b. South Bend, Ind., Mar. 3, 1943; d. Robert Milton and Priscilla (Tancy) Phillips; m. G. Brian Jones, Feb. 16, 1980; stepchildren: Laurie, Tracy. B.A., U. Nev., 1964; A.M., Stanford U., 1965; Ph.D., UCLA, 1977. Lic. psychol. asst., Calif. Cons. mgmt. and career devel., Los Angeles, 1972-77, Los Gatos, Calif., 1979—; research scientist Am. Insts. for Research, Palo Alto, Calif., 1979—; counselor Coalition of Counseling Centers Staffed by Christians, Inc., San Jose and Los Altos, Calif., 1980—. Author: Mentors and Proteges, 1982. Bd. dirs. Indochinese Resettlement and Cultural Center, San Jose, 1980—. Recipient Labor medal Republic of Vietnam, 1970; U.S. Dept. Edn. fellow, 1972-76. Mem. Am. Psychol. Assn., Am. Personnel and Guidance Assn., Nat. Vocat. Guidance Assn., Assn. Religious and Values Issues in Counseling. Baptist. Subspecialty: Psychological counseling. Current work: Mentoring; stress management; personal individual and group counseling; adaptation of refugees to American culture. Office: Am Insts for Research PO Box 1113 Palo Alto CA 94302

PHILLIS, JOHN WHITFIELD, physiologist; b. Port of Spain, Trinidad, April 1, 1936; came to U.S., 1982; s. Ernest and Sarah Anne (Glover) P.; m. Beverly Shane Wright, Jan. 24, 1969; children: David, Simon, Susan. B.V.Sc., Sydney (Australia) U., 1958, D.V.Sc., 1976; Ph.D., Australian Nat. U., Canberra, 1961; D.Sc., Monash U., Melbourne, 1970. Cert. veterinarian. Sr. lectr. Monash U., 1963-69; vis. prof. Ind. U., 1969; prof. and assoc. dean U. Man., Winnipeg, 1970-73; prof., chmn. U. Sask., Saskatoon, 1973-81, Wayne State U., Detroit, 1982—; mem. scholarship, grants coms. Med. Research Council, Ottawa, Can., 1973-79. Author: Pharmacology of Synapses, 1970; editor: Veterinary Physiology, 1976, Physiology and Pharmacology of Adenosine Derivatives, 1983. Mem. sci. bd. Muscular Dystonia Found., Beverley Hills, Calif., 1980—; mem. sci. adv. panel World Soc. for Protection of Animals, 1982—. Mem. Can. Physiol. Soc. (pres. 1978-79), Brit. Pharmacol. Soc., Am. Physiol. Soc., The Physiol. Soc., Soc. Neuroscience. Anglican. Subspecialties: Neuropharmacology; Neurophysiology. Current work: Identification of synaptic transmitters in the central nervous system; development of psychoactive agents. Home: 25501 Circle Dr Southfield MI 48075 Office: Dept Physiol Wayne State U 540 E Canfield Ave Detroit MI 48201

PHILPOT, VAN BUREN, JR., research pathologist; b. Houston, Miss., Mar. 3, 1923; s. Van Buren and Lois (Atkinson) P.; m. Gene Philpot, Aug. 1, 1946 (div. Sept. 1972); children: Marjorie Gene, Eloise, James George, Van Buren, Rachel. M.D., Tulane U., 1950. Diplomate: Am. Bd. Pathology. Resident in pathology U. Wis.-Madison, 1954-57; asst. prof. U. Tenn., Memphis, 1957-59; pathologist Cary Meml. Hosp., Caribou, Maine, 1959-61; fellow in pathology Boston Area Hosp., 1961-64; dir. Philpot Meml. Lab., Houston, Miss., 1964—. Contbr. articles to sci. jours. Active Boy Scouts Am.; pres. N.E. Miss. Hist. Soc., 1968. Served to capt. M.C. U.S. Army, 1950-54; Korea. Decorated Bronze Star. Fellow Coll. Am. Pathologists, Am. Soc. Clin. Pathology; mem. AMA, Miss. Assn. Pathologists (pres. 1973). Republican. Roman Catholic. Subspecialty: Pathology (medicine). Home: PO Box 312 Houston MS 38851 Office: 132 Umberson Rd Houston MS 38851

PHINNEY, RALPH E(DWARD), research and development engineer; b. Cleve., Mar. 3, 1928; s. Myron and Marie (Foreman) P.; m. Deborah Laurie Cushing, Sept. 9, 1961; children: Susan, Mike. B.S. in Math, U. Mich., 1950, M.S. in Aeros, 1951, Ph.D., 1954. Research asst. U. Mich., Ann Arbor, 1951-55; postgrad. Johns Hopkins U., Balt., 1955-57; research engr. Martin Marietta, Balt., 1959-66, Research Inst. Advanced Studies, 1966-68; aerospace engr. Naval Surface Weapons Center, White Oak, Md., 1968—. Contbr. articles to profl. jours. Served with U.S. Army, 1957-59. Mem. AAAS, AIAA. Subspecialties: Fluid mechanics; Aerospace engineering and technology. Current work: Boundary layer flow; jet breakup; aerodynamic test evaluation. Home: 220 W Lanvale St Baltimore MD 21217 Office: Naval Surface Weapons Center New Hampshire Ave White Oak MD 20910

PHIPPS, PATRICK MICHAEL, plant pathologist; b. New Martinsville, W.Va., Oct. 19, 1945; s. Glenn Isaac and Bertha Emily (Huey) P.; m. Janet Lynn Miller, Nov. 20, 1967; 1 son, James Patrick. B.S., Fairmont State Coll., 1970; M.S., Va. Poly Inst., Blacksburg, 1972; Ph.D., W.Va. U., 1974. Vis. asst. prof. N.C. State U., Raleigh, 1974-78; assoc. prof. Va. Polytech. Inst., Tidewater Research and Continuing Edn. Ctr., Suffolk, 1978—. Served with U.S. Army, 1966-68. Mem. Am. Phytopath. Soc., Mycological Soc. Am., Am. Peanut Research and Edn. Soc. Subspecialties: Plant pathology; Microbiology. Current work: Improving strategies for chemical and biological control of plant disease in field crops. Office: Tidewater Ctr Suffolk VA 23437

PHOEL, WILLIAM C., oceanographer; b. Jamaica, N.Y., Nov. 17, 1939; s. William C. and Margaret (Fischlein) P.; m. Dolores M. Schulze, Sept. 18, 1965. B.S. in Biology, C.W. Post Coll., 1963; M.S. in Marine Sci, L.I. U., 1970; postgrad., Va. Inst. Marine Sci., 1975-76, 81—. Agr. technician L.I. Oyster Farms, Oyster Bay, N.Y., 1968-69; comml. diver Internat. U/W Contractors, City Island, N.Y., 1969; asst. research scientist NYU Med. Ctr., 1969-70; oceanographer, diving officer Sandy Hook lab. U.S. Dept. Commerce, Highlands, N.J., 1971—; diving officer Nat. Marine Fisheries Service, 1983—. Served to lt. USN, 1963-67. Mem. Am. Soc. Ichtyologists and Herpetologists, Am. Soc. Limnology and Oceanography, Marine Tech. Soc., Undersea Med. Soc., U.S. Naval Inst. Subspecialties: Oceanography; Ecology. Current work: Oxygen consumption and nutrient flux of seabed and of individual benthic and dermersal animals from stressed and unstressed environments. Home: 35 Freeman Ct Toms River NJ 08753 Office: US Dept Commerce NOAA NMFS Sandy Hook Lab Highlands NJ 07732

PIANKA, ERIC RODGER, population biologist; b. Hilt, Calif., Jan. 23, 1939; s. Walter Henry and Virginia Lincoln (High) P.; m. Helen Louise Dunlap, Dec. 20, 1965 (div. Dec. 1980); children—Karen Elizabeth, Gretchen Anna. B.A., Carleton Coll., 1960; Ph.D. (NIH fellow), U. Wash., 1965. NIH postdoctoral fellow Princeton U., 1965-68, U. Western Australia, Nedlands, 1966-67; asst. prof. zoology U. Tex., Austin, 1968-72, asso. prof., 1972-77, prof., 1977—; vis. prof. U. Kans., 1978, U. P.R., 1981. Author: Evolutionary Ecology, 3d edit, 1983, also, Japanese, Spanish, Polish and Russian transl; co-editor: Lizard Ecology: Studies of a Model Orgranism, 1983; mng. editor: The Am. Naturalist, 1971-74; mem. editorial bd., 1975-77, BioSci, 1975-80; contbr. articles to profl. publs. Guggenheim fellow, 1978-79; NSF grantee, 1966-82; Nat. Geog. Soc. grantee, 1975-79. Fellow AAAS; mem. Am. Soc. Naturalists, Ecol. Soc. Am., Am. Soc. Ichthyologists and Herpetologists, Soc. for Study Evolution, Herpetologists League, Western Australian Naturalists. Subspecialties: Population biology; Evolutionary biology. Current work: Natural history and ecology of desert lizards; life historical phenomena; community structure and resource partitioning. Research on ecology and diversity of desert lizards. Office: Dept Zoology U Tex Austin TX 78712 Differential reproductive success is all pervasive, but awfully short sighted. The disparity between what humans *could be* versus what we in fact have achieved is simply pitiful. For the first time in the history of life on Earth, a product of natural selection has looked back at itself and said "Ah ha, I see you!" Yet we are totally unable to put this wisdom to use to save ourselves, let alone the other creatures sharing this poor beleaguered planet. A wild rattlesnake has as much "right to life" as you or I. Watch as *Homo* the sap blindly outreproduces himself right to extinction.

PIATAK, MICHAEL, JR., genetic engineer; b. Carbondale, Pa., Feb. 10, 1950; s. Michael and Anna (Mondak) P.; m. Lisa Rosanne Langlois, Nov. 7, 1981. B.S., Pa. State U., 1971; M.Ph., Yale U., 1973, Ph.D., 1978. NIH postdoctoral fellow Yale U., 1978-80, Leopold Schepp Found. fellow, 1980-81, research asst., 1981-82; scientist Cetus Corp., Berkeley, Calif., 1982—. Contbr. chpts., articles to profl. publs. Mem. Am. Soc. Microbiology. Eastern Rite Catholic. Subspecialties: Genetics and genetic engineering (biology); Molecular biology. Current work: Genetic engineering, gene expression in foreign host, species hybrid molecules, animal virology, toxins, immunobiology. Office: 1400 53d St Emeryville CA 94608

PICAZO, ESTEBAN DAVID, environmental engineer; b. Siloam Springs, Ark., Mar. 19, 1947; s. Mardoqueo Evaristo and Rachael (Overstreet) P.; m. Brenda Grace Markley, Jan. 27, 1973; children: Steven David, Andrew Jon. B.A., Asbury Coll., 1969; M.S., Morehead (Ky.) State U., 1978. Radiation control safety officer Nuclear Engring. Co., Morehead, 1977-79; cons. health physics Dames & Moore, White

Plains, N.Y., 1979-81, in-house cons., 1982—. Served with USN, 1969-75. Mem. Health Physics Soc., Ky. Acad. Sci., Am. Nuclear Soc. Republican. Methodist. Subspecialties: Environmental engineering; Radioactive waste management. Current work: Providing environmental surveillance for nuclear fuel reprocessing waste solidification project. Home: 135 Newman St Springville NY 14141 Office: Dames & Moore 20 Haarlem Ave White Plains NY 10603

PICKARD, DAVID KENNETH, statistical educator; b. Summerside, Can., July 8, 1945; came to U.S., 1977; s. George Campbell and Molly Patricia (Joyce) P.; m. Dale E. Dixon, May 24, 1968; children: Damon Lee, Darcy Christine. B.S., Mount Allison U., Can., 1967; M.S., Stanford U., 1970; Ph.D., Australian Nat. U., 1977. Lectr. dept. math. Mount Allison U., Sackville, Can., 1967-68, 71-73; math. master Jaima (Sierra Leone) Secondary Sch., 1968-69; research scholar Australian Nat. U., Canberra, 1974-77; asst. prof. dept. stats. Harvard U., Cambridge, Mass., 1977-80, assoc. prof., 1980—; research assoc. Nat. Sci. and Engring Research Council Can., 1967, 68, 71, 74, 78-81. Contbr. articles to profl. jours. NSF grantee, 1981—. Mem. Am. Statis. Assn., Math. Assn. Am., Am. Math. Soc., Inst. Math. Stats. Subspecialties: Probability; Chemical engineering. Current work: Stochastic sedimentation; random packing; statistical inference for Markov fields; spatial processes; feline leukemia; biomedical modelling; pharmacokinetics and anesthetic uptake. Office: Dept Statistics Harvard U 1 Oxford St Cambridge MA 02138

PICKERING, RANARD JACKSON, hydrologist; b. Goshen, Ind., Mar. 24, 1929; s. Carlyle Whitehead and Bertha May (Ranard) P.; m. Joyce Elizabeth Holt, Sept. 10, 1950; children: Pamela Elizabeth, Nancy Jeanne, Michael Jackson. B.A. in Geology, Ind. U., 1951, M.A., 1952; Ph.D., Stanford U., 1961. Geologist Ind. Geol. Survey, Bloomington, 1950-52, N.J. Zinc Co., Franklin, 1952-55; raw minerals engr. Columbia Iron Mining Co., U.S. Steel, Provo, Utah, 1955-58; geologist U.S. Geol. Survey, Oak Ridge, 1961-65, hydrologist, Columbus, Ohio, 1965-67, assoc dist. chief, 1967-70, asst. chief quality water br., Washington, 1970-72, chief quality water br., Reston, Va., 1972—. NSF fellow, 1959, 61; recipient Dept. Interior Meritorious Service award, 1981. Fellow Geol. Soc. Am.; mem. Geochem. Soc., AAAS, Am. Water Resources Assn., Soc. Environ. Geochemistry and Health, Phi Beta Kappa, Sigma Xi. Subspecialties: Geochemistry; Hydrology. Current work: Composition of atmospheric deposition (acid rain) in the United States, geochemistry of weathering processes, fate of radionuclides in a fluvial environment, factors affecting water quality. Office: US Geol Survey 12201 Sunrise Valley Dr MS 412 Reston VA 22092

PICKERING, WILLIAM HAYWARD, physics educator, scientist; b. Wellington, N.Z., Dec. 24, 1910; s. Albert William and Elizabeth (Hayward) P.; m. Muriel Bowler, Dec. 30, 1932; children—William B., Anne E. B.S., Calif. Inst. Tech., 1932, M.S., 1933, Ph.D. in Physics, 1936; hon. degrees, Clark U., 1966, Occidental Coll., 1966, U. Bologna, 1974. Mem. Cosmic Ray Expdn. to, India, 1939, Mexico, 1941; faculty Calif. Inst. Tech., 1940—, prof. elec. engring., 1946—, dir. jet propulsion lab., 1954-76; Mem. sci. adv. bd. USAF, 1945-48; chmn. panel on test range instrumentation (Research and Devel. Bd.), 1948-49; mem. U.S. nat. com. tech. panel Earth Satellite Program, 1955-60; mem. Army Sci. Adv. Panel, 1960-64. Decorated Order of Merit, Italy; knight comdr. Order Brit. Empire; recipient James Wyld Meml. award Am. Rocket Soc., 1957; Columbus medal Genoa, 1964; Prix Galabert for Astronautics; Goddard award Nat. Space Club, 1965; NASA Distinguished Service medal, 1965; Army Distinguished Civilian Service award, 1959; Spirit of St. Louis medal, 1965; Crozier medal Am. Ordnance Assn., 1965; Man of Year award Indsl. Research Inst., 1968; Interprofl. Coop. award Soc. Mfg. Engrs., 1970; Marconi medal Marconi Found., 1974; Nat. Medal of Sci., 1976; Fahrney medal Franklin Inst., 1976; award of merit Am. Cons. Engrs. Council, 1976. Fellow Am. Inst. Aeros. and Astronautics (pres. 1963, Louis W. Hill Transp. award 1968), Am. Acad. Arts and Sci., IEEE (Edison medal 1972); mem. AAAS, Nat. Acad. Scis., Am. Geophys. Union, Internat. Astronautical Fedn. (pres. 1965-66), Royal Soc. N.Z., Nat. Acad. Engring., AAUP. Subspecialty: Aerospace engineering and technology. Current work: Unmanned spacecraft systems. Home: 292 St Katherine Dr Flintridge CA 91011

PICKERLING, HOWARD WILLIAM, metallurgy educator; b. Cleve., Dec. 15, 1935; s. Howard William and Marian Amelia (Vittes) P.; m. Judith Ann Burch, Apr. 20, 1963; children: John, Kim Anne, Scott, Carolyn. B.S. in Metallurgy Engring, U. Cin., 1958, M.S., 1959; Ph.D., Ohio State U., 1961. Sr. scientist U.S. Steel Corp., Monroeville, Pa., 1962-72; prof. metallurgy Pa. State U., University Park, 1972—; vis. scientist Max Planck Inst. for Phys. Chemistry, Gottingen, W. Ger., 1964-65; vis. research prof. physics U. Tokyo, Japan, 1982. Editor: Jour. Corrosion Science, 1975—; contbr. articles to profl. jours. Grantee NSF, Dept. of Energy, others; Shell fellow; Nat. Steel fellow; Bethlehem Steel fellow. Mem. AIME, Sigma Xi, Tau Beta Pi. Subspecialties: Alloys; Corrosion. Current work: Thin film analysis by atom probe film and electrochemical studies of metals; educator and research director in the metal-environmental reaction area.

PICKETT, JACKSON BRITTAIN ELBRIDGE, neurologist, researcher; b. San Antonio, Apr. 30, 1943; s. Jackson B. E. and Mary Ruth (Brittain) P. B.A., Occidenta Coll., 1964; M.D., Yale U., 1968. Diplomate: Am. Bd. Psychiatry and Neurology. Intern Grady Meml. Hosp., Atlanta, 1968-69; resident U. Calif.-San Francisco, 1969-72; asst. prof. neurology U. Calif., San Francisco, 1975-81; assoc. prof. Med. U. S.C., 1981—; staff neurologist VA Med. Ctr., Charleston, S.C., 1981—. Served to maj. USAF, 1971-74. Mem. Am. Acad. Neurology. Subspecialty: Neurology. Office: Neurology Service VA Med Ctr Bldg 378 Charleston SC 29403

PICKETT-HEAPS, JEREMY DAVID, research biologist, educator; b. Bombay, India, June 3, 1940; came to U.S., 1970; s. Harold Arthur and Edna (Azua) P.-H.; m. Charmain Daphne Scott, Jan. 18, 1964; children: David, Rebecca; m. Julianne Francis Jack, July 14, 1977; 1 son, Christopher. B.A., Clare Coll., Cambridge, Eng., 1962, Ph.D., 1965. Research fellow Australian Nat. E., Canberra, 1965-70, fellow, 1968-70; asst. prof. then assoc. prof. dept. molecular, cellular and devel. biology U. Colo., Boudler, 1970-78, prof., 1978—. Recipient Darbaker Prize Am. Bot. Soc., 1974; Colo. Teaching Recognition award, 1975; Faculty Teaching award, 1979. Mem. Am. Soc. Cell Biology, Phycological Soc. Am., Am. Micros. Soc. Subspecialty: Cell biology. Current work: Ultrastructure and function in cell division and photogenesis, particulary in green algae and diatoms. Home: 107 Timberlane St Boulder CO 80302 Office: U Colo MCD Biology Campus 347 Boulder CO 80309

PICKOFF, ARTHUR STEVEN, pediatric cardiology educator; b. N.Y.C., Dec. 24, 1948; s. Bernard Benjamin and Katie (Seigal) P.; m. Carrell Phillips, Aug. 29, 1982. B.A., Queens Coll., 1970; M.D., Albert Einstein Coll. Medicine, 1975. Diplomate: Am. Bd. Pediatrics (subcert. in cardiology). Asst. prof. pediatric cardiology U. Miami, Fla., 1977—. Contbr. chpt. in book. Nat. Heart, Lung and Blood Inst. grantee, 1981-86; recipient award Miami chpt. Am. Heart Assn., 1981-83. Fellow Am. Acad. Pediatrics, Am. Coll. Cardiology. Democrat. Jewish. Subspecialties: Pediatrics; Cardiology. Current work: Manner in which neonates, infants and children respond to cardiovascular drugs used in treating cardiac arrhythmias. Home: 980 NW North River Dr Miami FL 33136 Office: U Miami PO Box 016820 Miami FL 33101

PIECZYNSKI, WILLIAM JOHN, toxicologist; b. Chgo., Feb. 27, 1945; s. Edward Anthony and Virginia (Maloblocki) P.; m. Margaret Jean Carey, June 21, 1969; children: Jeanne, Andrew, Edward. B.S. in Biochemistry, Ill. Benedictine Coll., 1969. Med. technologist VA West Side Hosp., Chgo., 1968-69; research supr. Biolabs Inc., Northbrook, Ill., 1969-76; sr. toxicologist Baxter Labs., Morton Grove, Ill., 1976-79; regional tech. rep. and tech. supr. Lab-Tek div. Miles Lab., Naperville, Ill., 1979—. Served with U.S. N.G., 1967-73. Mem. AAAS, Am. Soc. Microbiology, Am. Chem. Soc., Tissue Culture Assn., Ill. Soc. Microbiology, Midwest Tissue Culture Assn. Subspecialties: Tissue culture; Cell study oncology. Current work: Effect of chemical carcinogen in cell culture. Home: 26 W 587 Grand Ave Wheaton IL 60187

PIENKOWSKI, ROBERT L(OUIS), entomologist, educator; b. Cleve., Aug. 22, 1932; s. Oliver E. and Elaine I. (Groh) P.; m. Joan K. Moore, Sept. 6, 1958; children: Sara, Nathan, Samuel. B.S., Ohio State U., 1954; M.S., U. Wis.-Madison, 1958, Ph.D., 1961. Asst. prof. entomology Va. Poly. Inst. and State U., 1961-67, assoc. prof., 1967-71, prof., 1971—. Served to 1st lt. U.S. Army, 1954-56. Mem. Entomol. Soc. Am., Entomol. Soc. Can., Brit. Ecol. Soc., Internat. Orgn. Biol. Control, Va. Acad. Sci. Presbyterian. Subspecialties: Integrated pest management; Ecology. Current work: Ecology of insects in forage and pasture crops. Office: Va Poly Inst and State U Dept Entomology Blacksburg VA 24061

PIEPER, GEORGE F(RANCIS), aerospace center administrator; b. Boston, Jan. 1, 1926; s. George Francis and Katherine Gertrude (Cross) P.; m. Barbara Ferguson, Dec. 27, 1950; children: Pamela, Lynell Pieper Smillie. B.A., Williams Coll., 1946; M.S. in Engring, Cornell U., 1949; Ph.D., Yale U., 1952. Staff mem. Radiation Lab. M.I.T., 1944-45; instr. physics Yale U., 1952-55, asst. prof., 1955-60; head exptl. satellites project Applied Physics Lab. Johns Hopkins U., 1960-64; dep. asst. dir. advanced research NASA Goddard Space Flight Ctr., Greenbelt, Md., 1964-65, dir. scis., 1965-83, asst. ctr. dir. for policy planning and devel., 1983—. Contbr. in field. Recipient NASA medal for exceptional sci. achievement, 1969, medal for outstanding leadership, 1977. Fellow Am. Phys. Soc.; mem. Am. Astron. Soc., Am. Geophys. Union, AAAS. Subspecialties: Satellite studies; Nuclear physics. Current work: Senior staff in overall direction of Goddard Space Flight Center. Office: NASA Goddard Space Flight Center Code 100 Greenbelt MD 20771

PIEPER, REX DELANE, range scientist, educator, researcher; b. Idaho Falls, Idaho, Jan. 19, 1934; s. Gustave Henry and Maude Ethyl (Beam) P.; m. Susan Jeanne Twyeffort, June 11, 1965; children: Julie, Loren, Tracy. B.S., U. Idaho, Moscow, 1956; M.S., Utah State U., 1958; Ph.D., U. Calif., Berkeley, 1963. Asst. prof. range sci. N.Mex. State U., 1963-69, asso. prof., 1969-75, prof., 1975—; cons. to various firms. Contbr. numerous articles on range mgmt., plant ecology, range nutrition to profl. jours. Recipient Disting. Teaching award N. Mex. State U. Coll. Agr. and Home Econs., 1971, Disting. Research award, 1979. Fellow AAAS; mem. Soc. Range Mgmt., Am. Inst. Biol. Scis., So. Assn. Naturalists, Sigma Xi. Republican. Methodist. Subspecialties: Ecology; Ecosystems analysis. Current work: Plant ecology, ecosystem analysis, range nutrition. Home: 4825 Senita Rd Las Cruces MN 88001 Office: Dept Animal and Range Scis N Mex State U Las Cruces MN 8800

PIERCE, BENJAMIN ALLEN, biology educator; b. Birmingham, Ala., Nov. 15, 1953; s. John Rush and Frances Amanda (Allen) P. B.S., So. Meth. U., 1976; Ph.D., U. Colo., 1980. Asst. prof. zoology Conn. Coll., New London, 1980—. Contbr. numerous articles in field to profl. jours. David Flyr Meml. scholar, 1974; U. Colo. fellow, 1979-80. Mem. Genetics Soc. Am., Soc. for Study Evolution, Am. Soc. Icthyologists and Herpetologists, AAAS, Soc. for Study Amphibians and Reptiles. Subspecialties: Evolutionary biology; Genetics and genetic engineering (biology). Current work: Evolutionary biology of amphibians.

PIERCE, CARL WILLIAM, immunologist, pathologist; b. Buffalo, Oct. 2, 1939; s. William Wright and Dorothea (Fuerch) P.; m. Judith Anne Kapp, Dec. 1, 1973. A.B., Colgate U., 1960; M.D., U. Chgo., 1966, Ph.D., 1966. Intern U. Colo. Hosps., Denver, 1966-67; asst. in pathology U. Colo., Denver, 1966-67; research assoc. NIH, Bethesda, Md., 1967-70; asst. prof. pathology Harvard Med. Sch., Boston, 1970-73, assoc. prof., 1973-76; pathologist-in-chief Jewish Hosp., St. Louis, 1976—; prof. pathology and microbiology-immunology Washington U., St. Louis, 1976; mem. immunobiology study sect. NIH, Bethesda, Md., 1976-80. Contbr. articles to profl. jours. Served with USPHS, 1967-70. Mem. Am. Assn. Immunologists, Am. Assn. Pathologists (Parke Davis award 1979), AAAS, Assn. U. Pathologists, N.Y. Acad. Scis. Subspecialties: Immunobiology and immunology; Immunogenetics. Current work: Study regulatory mechanisms in immune responses to antigens controlled by immune response genes. Home: 701 Dominion Dr St Louis MO 63131 Office: Jewish Hosp St Louis 216 S Kings highway St Louis MO 63110

PIERCE, DANIEL THORNTON, research physicist; b. Los Angeles, July 16, 1940; s. Daniel Gordon and Celia Francis (Thornton) P.; m. (div.); children: Jed, Maia. B.S. in Physics, Stanford U., 1962, Ph.D. in Applied Physics, 1970; M.A., Conn. Wesleyan U., 1966. Research assoc. Stanford Electronics Lab., 1970-71; research physicist Swiss Fed. Inst. Tech., 1971-75; research physicist radiation physics div. Nat. Bur. Standards, Washington, 1975—, physicist radiation physics div., 1977—. Contbr. chpts. to books, articles to profl. jours. Recipient Dept. Commerce Silver medal, 1978; Washington Acad. Sci. Achievement award, 1981. Mem. Am. Phys. Soc., Am. Vacuum Soc., Swiss Phys. Soc. Subspecialty: Condensed matter physics. Current work: Experimental surface physics, spin polarized electron scattering and emission, surface magnetism. Home: 1353 Carlsbad Dr Gaithersburg MD 20879 Office: Nat Bur Standards Washington DC 20234

PIERCE, JOHN GRISSIM, biochemistry educator; b. San Jose, Calif., May 9, 1920; s. James Lester and Elise (Furst) P.; m. Elizabeth Copeland, Dec. 18, 1949; children—Sarah Thurston, John Pieronnet, Anne Moore, James Lester. A.A., San Jose State Coll., 1939; B.A., Stanford U., 1941, M.A., 1942, Ph.D., 1944. Instr. to asst. prof. Cornell U. Med. Coll., 1948-52; from asst. prof. to prof. biochemistry UCLA, 1953—, vice chmn. dept. biol. chemistry, 1963-79, chmn. dept. biol. chemistry, 1979—; Eli Lilly lectr. Endocrine Soc., 1971; chmn. biochemistry and endocrinology sect. NIH, 1981-83. Editorial bd.: Jour. Biol. Chemistry, 1969-74, 78-80. Served from pvt. to lt. (j.g.) USNR, 1944-46. Fellow Nutrition Found., 1942-44, Am. Chem. Soc., 1946-47, Arthritis and Rheumatism Found., 1952-53, Guggenheim Found., 1960-61, 75-76. Mem. Am. Chem. Soc., Am. Soc. Biol. Chemists, Harvey Soc., A.A.A.S., Phi Beta Kappa. Subspecialty: Biochemistry (biology). Research in chemistry, biology of pituitary hormones. Home: PO Box 95 Cambria CA 93428 Office: U Calif Center Health Sci Los Angeles CA 90024

PIERCE, NATHANIEL FIELD, immunologist, educator; b. Rudyard, Mich., July 27, 1934; s. Warren David Pierce and Mabel Lamson Field; m. Diane June Baxter, Mar. 26, 1966; children: Shanti Elizabeth, Christopher Baxter, Matthew Warren. M.D., U. Mich., 1958. Diplomate: Am. Bd. Internal Medicine. Intern, resident in internal medicine Los Angeles County Gen. Hosp., 1958-60, 61-63; instr. medicine U. So. Calif., Los Angeles, 1964-65, Johns Hopkins U., Balt., 1966-68, asst. prof. medicine, 1968-72, assoc. prof. medicine, 1972-79, prof. medicine, 1979—; vis. fellow St. Cross Coll., Oxford U., 1973-74; chmn. U.S. cholera panel, NIH, Sci. Working Group Bacterial Infections, WHO; cons. Immunology/Therapy Diarrhoeal Diseases; mem. Health and Biomed. R&D Adv. Com., Nat. Acad. Scis. Contbr. numerous articles to sci. publs. NIH Research Career Devel. awardee, 1971-76; NIH grantee, 1977-83. Fellow ACP, Infectious Diseases Soc. Am.; mem. Am. Fedn. Clin. Research, Am. Soc. Clin. Investigation, Alpha Omega Alpha. Democrat. Episcopalian. Subspecialties: Infectious diseases; Internal medicine. Current work: research on mucosal immunology, especially as related to development of vaccines for mucosal infections; pathophysiology and treatment of intestinal infections. Home: 13823 Manor Glen Rd Baldwin MD 21013 Office: Dept Medicine Baltimore City Hosps Baltimore MD 21224

PIERCE, SAM, aerospace scientist; b. Los Angeles, July 4, 1943; s. William J. and Josephine P.; m. (div.); children: Robert M., Steven M. B.S., UCLA, 1965, M.S., 1967, Ph.D., 1970. Cert. tchr., Calif. Instr. UCLA, Loyola U.-Los Angeles, El Camino Coll., 1965-70; sr. cons. Rand Corp., N.Am. Sci. Ctr., Systems Devel. Corp., Santa Monica, Calif., 1965-70; asst. prof. Calif. State U.-Fullerton, 1970-75; sr. staff Aerospace Corp., El Segundo, Calif., 1975-78; sr. staff cognizant engr. Jet Propulsion Lab., Pasadena, Calif., 1978-81; sr. scientist lab. staff Hughes Aircraft, El Segundo, 1981—; chmn. computer sci. com., dir. computer sci. internship and grad. programs Calif. State U., 1973-75. Author tech. reports and articles in field. Grantee NASA, NSF, 1971-75, UCLA, 1969-70; Calif. State scholar, 1961; NDEA fellow, 1965-68; UCLA faculty fellow, 1975. Mem. N.Y. Acad. Scis., Soc. Indsl. and Applied Math, Assn. Computing Machinery, Numerical Analysis Interest Group (chmn. 1980-82), Am. Astron. Soc., Am. Inst. Physics, AIAA, AAAS, 1st Mars Landing Soc., Mensa, Sigma Xi, Pi Mu Epsilon. Subspecialties: Numerical analysis. Current work: Numerical astrodynamics, satellite systems design; mission planning; numerical and statistical modeling; computer software; operations research; error analysis.

PIERCE, WILLIAM GAMEWELL, geologist; b. Gettysburg, S.D., Sept. 24, 1904; s. Lee I. and Mary Clyde (Gamewell) P.; m. May Bell Henry, Oct. 6, 1930; children: William Henry, Kenneth Lee, Diane May. A.B., U. S.D., 1927; M.A., Princeton U., 1929, Ph.D., 1931. From jr. geologist to research geologist U.S. Geol. Survey, Washington, 1929-54, Menlo Park, Calif., since 1955. Contbr. articles to profl. jours. Mem. Pub. Works Com., Los Altos Hills, Calif., 1976—. Recipient Disting. Service award U.S. Dept. Interior, 1965; NSF grantee, 1962-63, 76. Fellow Geol. Soc. Am.; mem. Am. Assn. Petroleum Geologists (rep. D.C. 1948), Geol. Soc. Washington (sec. 1951), Am. Geol. Inst. Democrat. Episcopalian. Subspecialties: Geology; Tectonics. Current work: Structural geology of the northern Rocky Mountain region. Home: 14380 Manuella Rd Los Altos Hills CA 94022 Office: US Geol Survey 345 Middlefield Rd Menlo Park CA 94025

PIERCE, WILLIAM SCHULER, cardiac surgeon; b. Wilkes Barre, Pa., Jan. 12, 1937; s. William Harold and Doris Louis (Schuler) P.; m. Peggy Jayne Stone, June 12, 1965; children: William Stone, Jonathan Drew. B.S., Lehigh U., 1958; M.D., U. Pa., 1962. Resident in surgery Hosp. U. of Pa., 1962-70; asst. prof. M.S. Hershey Med. Ctr., Pa. State Coll. Medicine, Hershey, 1970-73, assoc. prof., 1973-77, prof. surgery, 1977—. Contbr. numerous articles on cardiac surgery to profl. jours. Served with USPHS, 1965-67. Fellow ACS; mem. AMA, Pa. Med. Soc., Am. Soc. Artificial Internal Organs, Internat. Cardiovascular Soc., Soc. Vascular Surgery, Soc. Univ. Surgeons, Am. Surg. Assn., Soc. Clin. Surgery. Subspecialties: Cardiac surgery; Artificial organs. Current work: Mechanical ventricular assistance, artifical heart, pneumatic blood pump development, implantable, motor driven blood pumps, long-term circulatory support in animals, polyurethanes for flexible blood contacting. surfaces. Inventor cardiac valve, blood pump. Office: Dept Surgery MS Hershey Med Ctr Hershey PA 17033

PIERCEY, MONTFORD FREDERIC, pharmacologist; b. Merident, Conn., July 25, 1942; s. Montford Thomas and Adeline Katherine (Goodsell) P.; m. Renay S. Schwartz, June 20, 1965; children: Eric, Victor. A.B., Boston U., 1966, M.A., 1967; Ph.D., Albert Einstein Coll. Medicine, Yeshiva U., 1972. Research fellow Albert Einstein Coll. Medicine, Yeshiva U., 1972-74; research scientist Upjohn Co., Kalamazoo, 1978-82, sr. Scientist, 1982—. Contbr. articles to sci. jours. Treas. Oakwood Little League, 1979; mem. promotional com. Downtown Kalamazoo Assn., 1981—. Mem. AAAS, Soc. for Neurosci., Am. Soc. Pharmacology and Exptl. Therapeutics, Internat. Research Cofn., Am. Pain Soc., Winter Neuropeptide Conf., Am. Mgmt. Assn. Jewish. Subspecialties: Pharmacology; Neurophysiology. Current work: Discovery of novel therapeutic agents through basic research; pain and analgesia, electrophysiology, neuropharmacology, neuropeptides, gastrointestinal pharmacology, antidiarrheal agents, structure-activity relations, single neuron studies. Home: 3720 Blackberry Ln Kalamazoo MI 49008 Office: Upjohn Co Kalamazoo MI 49001

PIERRE, DONALD ARTHUR, electrical engineering educator; b. Bloomington, Wis., July 2, 1936; s. Joseph J. and Odile M. (LeGrave) P.; m. Mary Louise Pierre, Nov. 21, 1959; children: Michael, Louise, John. B.S.E.E. with honors, U. Ill., 1958; M.S.E.E., U. So. Calif., 1960; Ph.D., U. Wis., 1962. Registered profl. engr., Mont. Part-time instr. U. So. Calif., Los Angeles, 1959-60; mem. tech. staff Hughes Aircraft Co., Los Angeles, 1958-62; asst. prof. elec. engring. Mont. State U., Bozeman, 1962-65, assoc. prof. elec. engring., mem. staff, 1965-69, prof. elec. engring., group leader systems group, 1969-79, head elec. engring. and computer sci. dept., 1979—. Author: Optimization Theory with Applications, 1969, Mathematical Programming Via Augmented Lagrangians, 1975; contbr. articles to profl. jours.; mem. editorial adv. bd.: Jour. Computers and Electrical Engring, 1971—. Francis Rogers Bacon fellow, 1958-59; FIER-GE fellow, 1959-60; recipient Wiley award for meritorious research Mont. State U., 1982. Mem. Am. Soc. Engring. Edn., IEEE, IEEE Control System Soc. (chmn. edn. com.), Instrument Soc. Am., Sigma Xi, Tau Beta Pi, Eta Kappa Nu, Pi Mu Epsilon. Roman Catholic. Subspecialty: Electrical engineering. Current work: Control systems, computer control, robotics. Home: 6343 Aajker Circle Rd Bozeman MT 59715 Office: Elec Engring and Computer Sci Dept Mont State U Bozeman MT 59717

PIERSON, RICHARD NORRIS, JR., physician, educator; b. N.Y.C., Sept. 22, 1929; s. Richard Norris and Dorothy (Stewart) P.; m. Anne Bingham, July 10, 1954 (div. 1974); children: Richard, Olivia, Alexandra, Cordelia; m. Alice Roberts, Aug. 26, 1974. B.A., Princeton U., 1951; M.D., Columbia U., 1955. Diplomate: Am. Bd. Internal Medicine, Am. Bd. Nuclear Medicine. Intern St. Luke's Hosp., N.Y.C., 1955-56, resident, 1958-61; dir. nuclear medicine div. St. Luke's-Roosevelt Hosp. Ctr., N.Y.C., 1965—; dir. medicine Hackensack (N.J.) Hosp., 1973-74; prof. clin. medicine Columbia U., N.Y.C., 1980—; cons. CAPINTEC, Montvale, N.J., 1979-82. Editor:

Quantitative Nuclear Cardiography, 1975; contbr. med. articles to profl. jours. Pres. Englewood (N.J.) Bd. Health, 1966-73. Served as lt. USN, 1956-58. Fellow ACP, N.Y. Acad. Medicine (chmn. bioengring. com. 1980); mem. N.Y. County Health Services Rev. Orgn. (chmn. 1980-82), N.Y. Country Med. Soc. (pres. 1977-78), N.Y. State Med. Soc. (chmn. med. sch. relations 1980--), AMA (del. Chgp. 1978--), Am. Med. Peer Rev. Assn. (vice speaker 1981--), Am. Bur. Med. Advancement in China (pres. N.Y.C. chpt. 1980--). Episcopalian. Club: Century (N.Y.C.). Subspecialties: Nuclear medicine; Biomedical engineering. Current work: Nuclear cardiology, body composition research. Home: 94 Beech Rd Englewood NJ 07631 Office: Nuclear Medicine Div Saint Luke's Hosp Ctr Amsterdam Ave at 114th St New York NY 10025

PIET, STEVEN JAMES, nuclear engineer; b. Greenville, N.C., Mar. 1, 1956; s. James V. and Diane (Wallace) P. B.S., MIT, 1979, M.S. in Nuclear Engring, 1979, Sc.D., 1982. Various summer positions Savannah River Lab., Aiken, S.C., 1977-80, Sandia Nat. Lab., Albuquerque, 1977-80, Argonne Nat. Lab., Ill., 1977-80; U.S. Dept. Energy magnetic fusion energy tech. fellow MIT, Cambridge, 1981-82; sr. engr. EG&G Idaho, Inc., Idaho Falls, 1982—. Vice-pres. MIT Technology Community Assn., 1974-82; mem. MIT Mass. Voice of Energy, 1980-82, chmn., 1977-78. Recipient Compton award MIT, 1978, Stewart award, 1977. Mem. Am. Nuclear Soc., Sigma Xi, Tau Beta Pi, Alpha Nu Sigma. Republican. Roman Catholic. Subspecialties: Nuclear fusion; Nuclear engineering. Current work: Research and development in national fusion reactor safety research program: fusion probabilistic risk assessment, involvement with the blanket comparison and selection study, and program management. Home: 1487 Vega Circle #7 Idaho Falls ID 83402 Office: PO Box 1625 Idaho Falls ID 83415

PIETREWICZ, ALEXANDRA THERESA, psychology educator, researcher; b. Worcester, Mass., Aug. 13, 1949; d. Edward Stanley and Alexandra Elizabeth (Rydzewski) P.; m. John Eugene Mauldin, July 11, 1981. B.S., U. Mass., 1972, M.S., 1975, Ph.D. in Psychology, 1977. Vis. lectr. U. Mass., Amherst, 1976-77; asst. prof. psychology Emory U., Atlanta, 1977—. Co-editor: Issues in the Ecological Study of Learning; contbr. articles to profl. jours. NSF grantee, 1977-79; Emory U. grantee, 1977-78, 79-81, 82-83. Mem. Am. Psychol. Assn. (div. comparative and physiol. psychology). Subspecialties: Psychobiology; Ethology. Current work: Ecological determinants of behavior; learning; social mechanisms of population regulation. Office: Dept Psychology Emory U Atlanta GA 30322

PIETRI, CHARLES EDWARD, government official; b. N.Y.C., July 6, 1930; s. Charles Palmer and Marie Rosemond (Cropper) P.; m. Jeanne Muir Gaub, May 25, 1951; children: Randolph, Dianna, Richard. B.A., N.Y.U., 1951; postgrad., Rutgers U., 1972-74. Chemist E. I. duPont de Nemours & Co., Oak Ridge, 1951-53, Aiken, S.C., 1953-56; research chemist Curtiss-Wright Corp., Quehanna, Pa., 1956-58; chief plutonium chemistry U.S. AEC, New Brunswick, N.J., 1958-72; chief analytical chemistry (U.S. Dept. Energy), New Brunswick, 1972-76, asst. dir., Argonne, Ill., 1976—; dir. Planter Assocs., Chgo., 1980—. Dist. chmn. Boy Scouts Am., New Brunswick, 1960-75, council commr., LaGrange, Ill., 1976-82, council pres., 1982—. Recipient Silver Beaver award Boy Scouts Am., 1974. Fellow Am. Inst. Chemists; mem. Royal Soc. Chemistry, Am. Chem. Soc., Inst. Nuclear Materials Mgmt., Sigma Xi. Episcopalian. Subspecialties: Analytical chemistry; Nuclear fission. Current work: Analytical chemistry of plutonium and uranium, reference materials, automated instrumentation, computerized systems, separations. Patentee in field. Home: 8253 Lakeside Dr Downers Grove IL 60516 Office: US Dept Energy 9800 S Cass Ave Argonne IL 60439

PIETTE, LAWRENCE HECTOR, educator, biophysicist; b. Chgo., Jan. 4, 1932; s. Gerald John and Lillian (Bumgardner) P.; m. Mary Irene Harris, Aug. 15, 1957; children—Jeffrey, Martin. B.S., Northwestern U., 1953, M.S., 1954; Ph.D., Stanford, 1957. Mgr. research biochemistry and biophysics Varian Assos., 1956-65; prof. biophysics U. Hawaii, 1965—, chmn. dept., 1968—, dir., 1970—; exec. dir. Cancer Center Hawaii, 1974; Chmn. cancer adv. com., regional med. program, research com. Hawaii div. Am. Cancer Soc. Contbr. articles to profl. jours.; Asso. editor: Jour. Organic Magnetic Resonance. Mem. Am. Chem. Soc., Biophys. Soc., A.A.U.P. Subspecialties: Biophysics (biology); Cancer research (medicine). Current work: Mechanism of carcinogenesis, electron spin resonance of free radicals. Home: 4954 Kolohala Ave Honolulu HI 96816

PIGFORD, THOMAS HARRINGTON, nuclear engineering educator; b. Meridian, Miss., Apr. 21, 1922; s. Lamar and Zula Vivian (Harrington) P.; m. Catherine Kennedy Cathey, Dec. 31, 1948; children: Cynthia Pigford Naylor, Julie Pigford Earnest. B.S. in Chem. Engring, Ga. Inst. Tech., 1943, S.M., M.I.T., 1948, Sc.D., 1952. Asst. prof. chem. engring., dir. Sch. Engring. Practice, M.I.T., 1950-52, asst. prof. nuclear and chem. engring., 1952-55, asso. prof., 1955-57; head engring., dir. nuclear reactor projects and asst. dir. research lab. Gen. Atomic Co., La Jolla, Calif., 1957-59; prof. nuclear engring., chmn. dept. nuclear engring. U. Calif., Berkeley, 1959—; mem. Pres.'s Commn. on Accident at 3-Mile Island, 1979; mem. bd. radioactive waste mgmt. NRC; chmn. waste isolation systems panel, mem. waste isolation pilot plant panel; chmn. adv. council Inst. Nuclear Power Op.; cons. in field. Author: (with Manson Benedict) Nuclear Chemical Engineering, 1958, 2d edit., 1981; contbr. numerous articles to profl. jours. Served with USNR, 1944-46. Recipient John Wesley Powell award U.S. Geol. Survey, 1981; named Outstanding Young Man of Greater Boston Boston Jaycees, 1955; E. I. DuPont DeNemours research fellow, 1948-50; Japan Soc. for Promotion Sci. fellow, 1974-75; NSF grantee, 1960-75; EPA grantee, 1973-78; Dept. Energy grantee, 1979—; Ford Found. grantee, 1974-75; Electric Power Research Inst. grantee, 1974-75. Fellow Am. Nuclear Soc. (Arthur H. Compton award 1971); mem. Nat. Acad. Engring., Am. Chem. Soc., Am. Inst. Chem. Engrs. (Robert E. Wilson award 1980), Atomic Indsl. Forum (dir.), AIME, Sigma Xi, Phi Kappa Phi, Tau Beta Pi. Subspecialty: Nuclear engineering. Patentee in field. Home: 1 Garden Dr Kensington CA 94708 Office: Dept Nuclear Engring U Calif Berkeley CA 94720

PIGOTT, JOHN DOWLING, geology educator; b. Gorman, Tex., Feb. 2, 1951; s. Edwin and Emma Jane (Poe) P.; m. Patricia Karen Bettis, June 17, 1978. B.A. in Zoology, U. Tex.-Austin, 1974, B.S. in Geology, 1974, M.A., 1977; Ph.D., Northwestern U., 1981. Geologist Amoco Internat. Oil Co., Chgo., 1978-79; sr. petroleum geologist Amoco Prodn. Co., Houston, 1979-81; asst. prof. geology U. Okla., Norman, 1981—. Penrose research grantee Geol. Soc. Am., 1977; Mobil scholar, 1979; U. Okla. jr. faculty research fellow, 1983. Mem. Am. Assn., Petroleum Geologists, Geol. Soc. Am., Soc. Exploration Geophysicists, Sigma Xi. Democrat. Subspecialties: Geochemistry; Sedimentology. Current work: Investigation of secular trends in the geochemistry of the earth's surface environment; interactions between tectonics and sedimentary geochemistry. Office: Univ Okla Sch Geology and Geophysics 830 Van Vleet Oval Norman OK 73019 Home: 1511 Barwick St Norman OK 73069

PIHL, ROBERT OLANDER, psychology educator; b. Milw., Feb. 2, 1939; emigrated to Can., 1966; s. Howard Oscar and Evelyn Lenore (Olander) P.; m. Joyce Blanche Swiderski, June 20, 1960 (div. Aug. 1973); children: Christen, Eric; m. Sandra Olga Woticky, Oct. 8, 1975. B.A., Lawrence U., 1961; M.A., Ariz. State U., 1965, Ph.D., 1966. Clin. intern Barrow Neurol. Inst., Phoenix, 1964-66, psychologist, convulsive disorder unit, 1964-65; faculty assoc. Ariz. State U., Tempe, 1965-66; dir. dept. psychology Lakeshore Gen. Hosp., Pointe Claire, Que., Can., 1967—; cons. Constance Lethbridge Rehab. Ctr., Montreal, 1968-72; dir. clin. psychology program McGill U., Montreal, 1969-75, prof. psychology, 1979—; pres. Pihl-Ervin Inc., Montreal, 1981—. Mem. AAAS, Am. Psychol. Assn., Can. Psychol. Assn., N.Y. Acad. Scs. Subspecialties: Behavioral psychology; Psychopharmacology. Current work: Drugs, mineral elements, and psychological factors and aggression; the identification and treatment of learning disabilities; the response to alcohol. Home: 225 Bedbrook St Montreal West PQ Canada H4X 1S2 Office: Dept Psychology McGill U 1205 Dr Penfield Ave Montreal PQ Canada H3A 1B1

PIHLSTROM, BRUCE LEE, dental educator, researcher; b. Mpls., July 4, 1943; s. Earl W. and Elsie R. (Knutson) P.; m. Carol A. Minelli, July 29, 1967; children: Chris, Daniel. B.S., U. Minn.-Mpls., 1965, D.D.S., 1967; M.S., U. Mich.-Ann Arbor, 1969. Diplomate: Am. Bd. Periodontology. Instr. periodontics Washington U., St. Louis, 1969-71; clin. asst. prof. U. Minn., 1971-72, asst. prof., 1972-78, assoc. prof., 1978—; pvt. practice periodontist, Golden Valley, Minn., 1975—; cons. in field. Leader Indian head council Boy Scouts Am., St. Paul, 1982. Served to capt. USAF, 1969-71. NIH grantee, 1980—. Mem. Am. Acad. Periodontology (chmn. edn. com.), Am. Dental Assn. (cons. commn. dental accreditation 1979—). Lutheran. Subspecialty: Periodontics. Current work: Current research interests involve clinical trials testing various methods of periodontal therapy and microbial etiology of periodontal disease. Office: University of Minnesota School of Dentistry 515 Delaware St SE Minneapolis MN 55455

PIKE, CHARLES P., physicist; b. Winthrop, Mass., Feb. 21, 1941. B.S., Boston Coll., 1963; M.S., Northeastern U., 1969. Physicist USAF Geophysics Lab, Hanscom AFB, 1967—. Coeditor: Space Systems and Their Interactions with the Earth's Space Environment. Recipient Scientific Achievement award USAF Geophysics Lab, 1971, 73, 74. Mem. AIAA (past chmn. space sci. and astronomy com., mem. nat. tech. com.). Subspecialties: Satellite studies; Geophysics. Current work: Magnetospheric physics, aurora, spacecraft charging. Home: 69 Locksley Rd Lynnfield MA 01940 Office: USAF Geophysics Lab Hanscom AFB MA 01730

PIKE, RALPH WEBSTER, university research administrator; b. Tampa, Fla., Nov. 10, 1935; s. Ralph Webster and Macey (Adams) P.; m. Patricia Jennings, Aug. 23, 1958; children: Brian, Charlene. B.Ch.E., Ga. Inst. Tech., 1957, Ph.D., 1962. Registered profl. engr., La., Tex. Research engr. Exxon Research & Devel., Baytown, Tex., 1962-64; asst. prof. to prof. chem. engring. La. State U., Baton Rouge, 1964—, now also asst. vice chancellor research; cons. Exxon Corp., La Dept. Natural Resources, Wright-Patterson Air Devel. Command, OAS, others. Author: Formulation and Optimization of Mathematical Models, 1970, Optimizacion en Ingenieria, 1982. NASA grantee, 1966-72; NSF grantee, 1966-69; U.S. Dept. Energy grantee, 1978-81; U.S. Dept. Interior grantee La. Mineral Inst., 1979-82. Mem. Am. Inst. Chem. Engrs., Am. Chem. Soc., AAAS, Am. Soc. Engring. Edn., AIAA. Methodist. Subspecialties: Chemical engineering. Current work: Computer-aided process design with microcomputers. Home: 6053 Hibiscus Dr Baton Rouge LA 70803 Office: La State U 118 David Boyd Hall Baton Rouge LA 70803

PIKUNAS, JUSTIN, psychology educator; b. Alytus, Lithuania, Jan. 7, 1920; came to U.S., 1950; s. Baltrus and Anele (Radzius) P.; m. Regina Liesunaitis, Aug. 8, 1953; children: Justas, Kristina, Ramona. Ph.D., U. Munich, W.Ger., 1946, 1949. Lic. psychologist, Mich. Instr. U. Detroit, 1951-52, asst. prof., 1952-56, assoc. prof., 1956-61, prof., 1961—, chmn. dept. psychology, 1974-78, prof. and dir., 1970—. Author: Human Development: An Emergent Science, 1976, Manual for the Pikunas Graphoscopic Scale, 1982. Recipient Outstanding Service award U. Detroit, 1981. Mem. Am. Psychol. Assn., Nat. Council Grad. Depts. Psychology, ATEITIS Fedn. (pres. 1967-73), Lithuanian Cath. Fedn. Am. (pres. 1978-80). Subspecialties: Developmental psychology; Clinical psychology. Current work: Child management, the self-concept, Rorschach and drawing tests. Home: 8761 W Outer Dr Detroit MI 48219 Office: Dept Psychology U Detroit Detroit MI 48221

PILBEAM, DAVID ROGER, paleontologist, educator; b. Brighton, Eng., Nov. 21, 1940. B.A., Cambridge (Eng.) U., 1962; Ph.D. in Biology, Geology and Anthropology, Yale U., 1967. Demonstrator in phys. anthropology Cambridge U., 1965-68; from asst. prof. to assoc. prof. Yale U., New Haven, 1968-74, prof. anthropology, geology and geophysics, chmn. dept. anthropology, 1974-81; prof. anthropology Harvard U., Cambridge, Mass., 1981—. Author: The Evolution of Man, 1970, The Ascent of Man, 1972. Wenner-Gren Found. fellow, 1968. Mem. Am. Anthrop. Assn., Am. Assn. Phys. Anthropology, Brit. Soc. Study Human Biology. Subspecialty: Paleontology. Current work: paleontology. Discoverer of Sivapithecus indicus, Pakistan, 1980. Office: Dept Anthropology Peabody Museum Harvard U 11 Divinity Ave Cambridge MA 02138

PILGER, REX HERBERT, JR., geophysics educator; b. North Platte, Nebr., June 8, 1948; s. Rex Herbert and Eva Louise (Hastings) P.; m. Rita Mary Daly, Sept. 8, 1973; children: Elizabeth, Paul, Andrew. B.S., U. Nebr., 1970, M.S., 1972; Ph.D., U. So. Calif., 1976. Asst. prof. La. State U., Baton Rouge, 1976-79; research geophysicist Naval Ocean Research Devel. Activity Nat. Space Tech. Lab., 1979-82; assoc. prof. geophysics La. State U., Baton Rouge, 1979—. Editor: Procs. of Conf. on Origin of Gulf of Mexico, 1980. Mem. Am. Geophys. Union, Geol. Soc. Am., Soc. Exploration Geophysicists (assoc.). Roman Catholic. Subspecialties: Geophysics; Tectonics. Current work: Inverse modeling of gravity magnetics, plate tectonic kinematics; intraplate stress, tectonics of sedimentary basins and mountain belts. Home: 9869 Kinglet Dr Baton Rouge LA 70809 Office: Dept Geology La State U Baton Rouge LA 70803

PILGERAM, LAURENCE OSCAR, research biochemist; b. Great Falls, Mont., June 23, 1924; s. John R. and Bertha R. (Phillips) P.; m. Cynthia Ann Moore, Feb. 11, 1951; children: Karl E., Kurt J. A.A., U. Calif.-Berkeley, 1948, B.A., 1949, Ph.D., 1953. Postdoctoral fellow U. Calif.-Berkeley, 1953-54; instr. physiology U. Ill. Sch. Medicine, Chgo., 1954-55; asst. prof. biochemistry Stanford (Calif.) U. Sch. Medicine, 1955-57; dir. arteriosclerosis research lab. U. Minn. Sch. Medicine, Mpls., 1957-71; dir. coagulation lab. Baylor U. Coll. Medicine, Houston, 1971-75; dir. Cell Culture Lab., Santa Barbara, Calif., 1975—; cons. NIH, 1973-75. Contbr. over 70 articles to research jours.; guest editor: Nat. Dairy Council Jour, 1974, Biomedicine, 1982. Recipient award CIBA, London, 1959; Karl Thomas award, 1973; grantee NIH, Am. Heart Assn. Fellow Am. Heart Assn. Council on Stroke and Council on Thrombosis; mem. Am. Soc. Biol. Chemists (nat. corr. office pub. affairs 1969-75). Subspecialties: Biochemistry (medicine); Cell biology (medicine). Current work: Research on thrombosis and cell biology. Patentee detecting and measuring intravascular blood clotting. Office: PO Box 1583 Goleta Sta Santa Barbara CA 93116

PILLAY, SIVASANKARA K.K., scientist, educator; b. Puliyoor, Kerala, India, Jan. 28, 1935; came to U.S., 1960; s. Raman T. N. and Janaki (Amma) P.; m. Revathi Krishnamurthy, Mar. 22, 1964; 1 son, Gautam. B.S. with honors, U. Mysore, Bangalore, India, 1955, M.S., 1956; Ph.D., Pa. State U., 1965. Lectr. chemistry U. Mysore, Chitradurga, 1956-60; teaching asst. Pa. State U., University Park, 1960-65, assoc. prof., 1971-81; resident research assoc. Argonne (Ill.) Nat. Lab., 1965-66; sr. research scientist Western Nuclear Research ctr., Buffalo, 1966-71; staff scientist Los Alamos Nat. Lab., 1981—; con. Brookhaven Nat. Lab., Upton, N.Y., 1976-80, Pa. State Police, Harrisburg, 1971-75, U.S. Dept. Energy, Washington, 1977-80, Radiation Mgmt. Corp., Phila., 1978-81. Author: Laboratory Experiments in Applied Nuclear and Radiochemistry, 1977, Nuclear Technology Laboratory Experiments, 1979; contbr. articles on nuclear tech. to profl. jours. Warden Silver Jubilee Orphanage, Chitradurga, India, 1958-60; sec. Univ. Scout Troop, Chitradurga, 1957-60; troop treas. Boy Scouts Am., State College, Pa., 1975-76. Govt. of India scholar, 1954-56; Kopper Chem. fellow, 1962-63; ERDA fellow, 1976. Fellow Am. Inst. Chemists; mem. Am. Nuclear Soc. (asst. tech. program chmn. 1980), Am. Chem. Soc., AAAS, N.Y. Acad. Scis., Inst. Nuclear Materials Mgmt., Health Physics Soc., Can. Nuclear Soc., Sigma Xi. Hindu. Subspecialties: Nuclear and Radiochemistry; Inorganic chemistry. Current work: Applications of nuclear technologies, industrial nuclear process chemistry, radioactive waste management, nuclear material safeguards, chemistry of reactor coolants and fuel cycles. Office: Los Alamos Nat Lab Mailstop E541 Los Alamos NM 87545 Home: 369 Cheryl Ave Los Alamos NM 87544

PILSON, MICHAEL EDWARD QUINTON, oceanographer, educator; b. Ottawa, Ont., Can., Oct. 25, 1933; s. Edward Charles and Frances A. (Ferguson) P.; m. Joan E. Johnstone, July 4, 1977; children: Diana, John. B.S., Bishops U., 1954; M.S., McGill U., 1959; Ph.D., U. Calif.-San Diego, 1964. Chemist Windsor Mills (Que.)Paper Co., 1954-55; research asst. McGill U., U. Calif.-San Diego, 1956-63; asst. research biologist San Diego Zool. Soc., 1963-66; asst. to assoc. prof. U.R.I., Narragansett, 1966-78; prof. oceanography, 1978—; dir. Marine Ecosystems Research Lab., 1976—; cons. in field. Author numerous articles in profl. publs. Scripps Inst. Oceanography Sverdrup fellow, 1959. Mem. AAAS, Am. Geophys. Union, Am. Soc. Limnology and Oceanography, Am. Soc. Mammalogists, Sigma Xi. Club: Saunderstown Yacht (treas.). Subspecialties: Marine ecosystems; Environmental toxicology. Current work: Experimental analysis of marine ecosystems. Structure and function of marine mesocosms; fates and effects of chemicals. Office: URI Grad Sch Oceanography Narragansett RI 02882 Home: Box 27 Saunderstown RI 02874

PILTCH, MARTIN STANLEY, physicist, cons.; b. Bklyn., Aug. 11, 1939; s. Nathan and Sylvia (Karon) P.; m. Susan E. Quilter, Nov. 10, 1969; children: Sarah C., Emily M. A.B., Columbia Coll., N.Y.C., 1960; M.S., Poly. Inst. N.Y., 1968, Ph.D., 1971; M.B.A., U. N.Mex., 1980. Staff physicist TRG Control Data Corp., Melville, N.Y., 1961-66; NSF fellow Poly. Inst. N.Y., Bklyn., 1966-70; fellow Max-Planck Inst. Fur Biophysikalische Chemie, Gottingen, W. Ger., 1970-72; staff mem. Los Alamos (N.Mex.) Nat. Lab., 1972—, dep. group leader applied photochemistry div., 1974—; sr. fellow Max-Planck Inst. Fur Quantenoptik, Garching bei Munchen, W.Ger., 1981. Contbr. articles to profl. jours. Mem. Am. Phys. Soc., IEEE. Subspecialties: Laser photochemistry; Spectroscopy. Current work: Research and development of lasers for photochemistry, laser physics, spectroscopy, photochemistry, solid state physics, gas discharge physics. Patentee in field.

PIMENTEL, DAVID, entomology educator; b. Fresno, Calif., May 24, 1925; s. Frank Freitas and Marion (Silva) P.; m. Marcia Hutchins; children: Christina Pimentel Piper, Susan Pimentel Duran, Mark David. B.S., U. Mass., 1948; Ph.D., Cornell U., 1951. Scientist USPHS, Savannah, Ga., 1951-55; asst. prof. entomology Cornell U., 1955-61, assoc. prof., 1961-63, prof., 1963—; mem. Energy Research Adv. Bd., Washington; mem. environ. com. Pa. Power & Light, Allentown. Author: Food, Energy, and Society, 1979; editor: Pest Control: Cultural and Environmental Aspects, 1980, Handbook of Energy Utilization in Agriculture, 1980, Handbook of Pest Management in Agriculture, 1981. Trustee, village forester Cayuga Heights (N.Y.) City Council. Served to capt. USPHS, 1951-54. Mem. Ecol. Soc. Am., Entomol. Soc. Am., AAAS, Am. Inst. Biol. Scis., Am. Soc. Naturalists. Subspecialties: Ecology; Biomass (agriculture). Current work: Ecology, land and water resources; energy use in food system; biomass energy. Home: 147 N Sunset Dr Ithaca NY 14850 Office: Cornell U Comstock Hall 50-A Ithaca NY 14853

PIMENTEL, GEORGE CLAUDE, chemist; b. Rolinda, Calif., May 2, 1922; s. Emile J. and Lorraine Alice (Reid) P.; m.; children: Anne Christine, Tess Loren, Janice Amy. A.B., UCLA, 1943; Ph.D. in Chemistry, U. Calif. - Berkeley, 1949. From instr. to asso. prof. chemistry U. Calif., Berkeley, 1949-59, prof., 1959—; dep. dir. Nat. Sci. Found., Washington, 1977-80; dir. Lab. Chem. Biodynamics, U. Calif., Berkeley, 1981—; participant U.S.-Japan Eminent Scientists Exchange Program, 1973-74. Editor: Chemistry—An Experimental Science, 1963; co-author: Understanding Chemistry, 1971, Introductory Quantitative Chemistry, 1956; editor: Chem. Study, 1960; contbr. papers to profl. jours. Served with USNR, 1944-46. Recipient Campus Teaching award U. Calif., Berkeley, 1968, Coll. Chemistry Teaching award Mfg. Chemists Assn., 1971, Joseph Priestley Meml. award Dickinson Coll., 1972, Spectroscopy Soc. Pitts., 1974, Alexander von Humboldt Sr. Scientist award, 1974, Pauling medal, 1982, Wolf prize, 1982, Debye award, 1983, Madison Marshall award, 1983; Guggenheim fellow, 1955. Fellow Am. Acad. Arts and Sci.; mem. Nat. Acad. Scis., Am. Chem. Soc. (Precision Sci. award 1959, award Calif. sect. 1957), Am. Phys. Soc. (Earle K. Plyer prize 1979, Lippencott medal 1980), Optical Soc. Am., Phi Beta Kappa, Sigma Xi, Phi Eta Sigma, Phi Lambda Epsilon, Alpha Chi Sigma. Subspecialty: Biophysical chemistry. Home: 754 Coventry Rd Kensington CA 94707 Office: Lab Chem Biodynamics U Calif Berkeley CA

PINAPAKA, MURTHY V.L.N., physiology educator; b. India, Jan. 10, 1934; s. Rao V. and Seetharamamma V. P.; m. Stayavati Turlapati, May 7, 1961; children: Vijara, Sekhar. B.V.Sc., India, 1956; M.S., Tex. A&M U., 1966, Ph.D., 1968. Vet. surgeon, India, 1957-64; prof. biology Fayetteville (N.C.) State U., 1968—, coordinator med. and allied health sci., 1978—. Fulbright grantee, 1964-68. Mem. Am. Inst. Biol. Scis., AAAS, N.C. Acad. Sci., Sigma Xi, Phi Kappa Phi, Gamma Sigma Delta. Hindu. Subspecialties: Animal physiology; Nutrition (medicine). Current work: Retardation of atherosclerosis induced by dietary factors. Office: Dept Med & Allied Health Fayetteville State U Fayetteville NC 28301

PINCUS, IRWIN J., medical educator, retired physician; b. Braddock, Pa., Dec. 2, 1912; s. Jacob M. and Ethel (Broida) P.; m. Lena Magaziner, Sept. 24, 1939; children: David F., Robert F., Carol Fiacco. B.A., U. Pa., 1934, M.Sc., 1949, D.Sc. in Medicine, 1951; M.D., Jefferson Med. Coll., Phila., 1937. Intern Los Angeles County Gen. Hosp., 1937-39; resident in internal medicine and gastroenterology U. Pa., Phila., 1939-41; Ross V. Patterson fellow Jefferson Med. Coll., Phila., 1941-42, assoc. prof. physiology, 1946-56; assoc. prof. medicine and physiology Woman's Med. Coll., Phila., 1956-60; asst. clin. prof. medicine U. So. Calif., Los Angeles, 1963-70, clin. prof. medicine, 1970—. Served to maj. U.S. Army, 1942-46. Fellow ACP; mem. AMA, Am. Physiol. Soc. Jewish. Subspecialties: Physiology (biology); Gastroenterology. Current work: Gastrointestinal physiology; endocrinology; gastroenterology. Home: 610 N Roxbury Dr Beverly Hills CA 90210

PINCUS, JACK HOWARD, research institute executive; b. N.Y.C., Jan. 4, 1939; s. Sam and Sylvia (Baron) P.; m. Gladys Bernstein, Feb. 20, 1966. B.A., N.Y. U., 1959; Ph.D., Columbia U., 1966. Research fellow in medicine Mass. Gen. Hosp., Boston, 1966-69; sr. staff fellow NIH, Bethesda, Md., 1969-73; research chemist VA Lakeside Hosp., Chgo., 1973-75; assoc. dir. Biomed. Research Lab., SRI Internat., Menlo Park, Calif., 1975-81; dir. research and devl. Becton Dickinson Immunodiagnostics, Orangeburg, N.Y., 1981-82; exec. dir. Pub. Health Research Inst., N.Y.C., 1982—; NIH spl. fellow, 1969-70. Contbr. numerous articles to tech. jours. Fellow Am. Inst. Chemists; mem. Am. Chem. Soc., AAAS, Am. Assn. Immunologists. Subspecialties: Biochemistry (biology); Genetics and genetic engineering (medicine). Current work: Application of advanced biological processes to the production of commercial products; technology management; research management. Home: 172 Treetop Circle Nanuet NY 10954 Office: Pub Health Research Inst 455 1st Ave New York NY 10016

PINDELL, MERLE HERBERT, pharmacologist, lab. co. exec.; b. Bloomington, Ill., Jan. 22, 1926; s. Ira Francis and Bertha Elnora (Schutt) P.; m. F. Arlene Kimler, Aug. 5, 1945; children: Terry, Pamela, James. B.S., U. Ill., 1948, M.S. in Physiology, 1953; Ph.D. in Pharmacology, Med. Coll. Va., 1956. Pharmacologist Miles Labs., Elkhart, Ind., 1949-53; instr. dept. pharmacology U. Colo. Med. Sch., 1956-57; dir. pharm. research Bristol Labs., Syracuse, N.Y., 1957-63, dir. biol. research, 1963-67, dir. research, 1967-70; pres. Panlabs, Inc., Sunapee, N.H., 1970—; cons. to pharm. cos. abroad; vice chmn. Gordon Research Conf. on Med. Chemistry, 1969, chmn., 1970. Contbr. articles to sci. jours. Pres. Home and Sch. Assn., Fayetteville, N.Y., 1959-63. Mem. Am. Soc. Pharmacology and Exptl. Therapeutics, AAAS, Am. Chem. Soc., European Soc. for Study Drug Toxicology, Inflammation Research Assn., N.Y. Acad. Scis. Clubs: Lake Sunapee Country, Lake Sunapee Yacht. Subspecialties: Pharmacology; Physiology (medicine). Current work: Direction and administration of Panlabs, a scientific service organization engaged in pharmacological and fermentation reserach as a worldwide service to the pharmaceutical and chemical industries. Home: 1 Fishers Bay PO Box P Sunapee NH 03782 Office: PO Box P Sunapee NH 03782

PINDERA, JERZY TADEUSZ, mechanical engineer; b. Czchow, Poland, Dec. 4, 1914; emigrated to Can., 1965, naturalized, 1975; s. Jan Stanislaw and Natalia Lucia (Knapik) P.; m. Aleksandra-Anna Szal, Oct. 29, 1949; children: Marek Jerzy, Maciej Zenon. B.S. (equivalent) in Mech. Engring, Tech. U., Warsaw, 1936, M.S. in Aero. Engring, 1947; D.Tech.Scis., Polish Acad. Scis., 1959; D.Habil. in Applied Mechanics, Tech. U., Cracow, 1962. Registered profl. engr., Ont. Asst. Lot Polish Airlines, Warsaw, 1947; head lab. Aero. Inst., Warsaw, 1947-52, Inst. Metallography, 1952-54; dep. prof., head lab. Polish Acad. Scis., 1954-59; head lab. Bldg. Research Inst., Warsaw, 1959-62; vis. prof. mechanics Mich. State U., E. Lansing, 1963-65; prof. mechanics U. Waterloo, Ont., Can., 1965—; pres. J.T. Pindera & Sons Engring. Services, Inc., Waterloo; chmn. Internat. Symposium Exptl. Mechanics at univ., 1972, 10th Can. Fracture Conf., 1983; vis. prof. in, France and W. Ger., cons. in field. Author tech. books, also articles, chpts. in books.; Editorial adv. bd.: Mechanics Research Communications, 1974—. Served with Polish Army, 1939. Mem. Canadian Soc. Mech. Engring., Gesellschaft angewandte Mathematik und Mechanik, N.Y. Acad. Scis., Soc. Engring. Sci., Soc. Exptl. Stress Analysis (N.M. Frocht award 1978), ASME, Soc. Française des Mécaniciens, Assn. Profl. Engrs. Ont. Subspecialties: Theoretical and applied mechanics; Fracture mechanics. Current work: Experimental mechanics; theory of modelling, observations and measurements, reliabilitu of analytical and experimental evaluations, viscoelasticity of polymers; flow of energy in deformed bodies; photoelasticity; fracture mechanics; composite structures. Home: 310 Grant Crescent Waterloo ON N2K 2A2 Canada Office: Dept Civil Engring 200 University Ave Waterloo ON N2L 3G1 Canada It is true that "Nothing is more practical than a theory," provided however, that the assumptions on which the theory is founded are well understood. But, indeed, nothing can be more disastrous than a theory when applied to a real problem outside of the practical limits of the assumptions made simply because of an homonymous identity with the problem under consideration.

PINDOK, MARIE THERESA, physiologist, educator; b. Chgo., Jan. 20, 1941; d. Michael and Laura Pearl (Piaskowy) P. B.S., Loyola U., Chgo., 1963; Ph.D., Chgo. Med. Sch., 1978. Research asst. Loyola U. Sch. Medicine, Maywood, Ill., 1963-70; research assoc. U. Health Scis., Chgo. Med. Sch., North Chicago, Ill., 1978; instr. physiology, 1978-83, asst. prof., 1983—. Mem. Am. Physiol. Soc., AAAS, N.Y. Acad. Scis., Sigma Xi. Subspecialty: Physiology (medicine). Current work: Relationship between cardiovascular hemodynamics and cardiac metabolism, neurotransmitters and cyclic nucleotides during environmentally or pharmacologically induced stress. Office: U Health Scis-Chgo Med Sch 3333 Green Bay Rd North Chicago IL 60064

PINE, CHARLES JOSEPH, clinical psychologist, health services administrator; b. Excelsior Springs, Mo., July 13, 1951; s. Charles Edison and LaVern (Upton) P.; m. Mary Day, Dec. 30, 1979; 1 son, Charles Andrew. B.A., U. Redlands, 1973; M.A., Calif. State U.-Los Angeles, 1975; Ph.D., U. Wash., 1979. Lic. psychologist, Calif. Teaching asst. U. Wash., 1976-78; intern in psychology VA Outpatient Clinic, Los Angeles, 1978-79; asst. prof. Okla. State U., 1979-80; postdoctoral scholar UCLA, 1980-81; asst. prof. Wash. State U., 1981-82; clin. psychologist, dir. behavioral health services Riverside/San Bernardino County (Calif.) Indian Health Inc., 1982—. Editorial cons.: White Cloud Jour, 1982—. Recipient stipend Inst. Indian Studies, U. Wash., 1975-76; UCLA Inst. Am. Cultures grantee, 1981-82. Mem. Am. Psychol. Assn. (mem. minority edn. tng. task force 1982), Nat. Indian Cousnelros Assn., Soc. Indian Psychologists (pres. 1981-83), Western Psychol. Assn., Internat. Council Psychologists, Sigma Alpha Epsilon. Republican. Baptist. Club: Calif. Paso-Fino Assn. (Saugus). Subspecialties: Health psychology; Clinical psychology. Current work: Studying psychological factors related to such health problems as obesity, hypertension and diabetes from a cross-cultural perspective, especially as they relate to American Indians. Home: 365 W Grove Rialto CA 92376 Office: Riverside/San Bernardino County Indian Health Inc 11555-1/2 Petrero Rd Banning CA 92220

PINEDA, MAURICIO HERNAN, reproductive physiologist; b. Santiago, Chile, Oct. 17, 1930; came to U.S., 1970, naturalized, 1982; s. Teofilo Pineda-Garcia and Bertila Pinto-Bouvret; m. Rosa A. Gomez, July 26, 1956; children: Anamaria, George H., Monserrat. D.V.M., U. Chile, 1955; M.S., Colo. State U., 1965, Ph.D., 1968. Prof. Coll. Vet. Medicine, Austral U. Chile, Valdivia, 1958-63, prof., head animal reprodn. lab., 1968-70; postdoctoral trainee U. Wis., Madison, 1970-72; postdoctoral fellow Colo. State U., Ft. Collins, 1972-74, research assoc., 1972-78; assoc. prof. physiology, dept. physiology and pharmacology Coll. Vet. Medicine, Iowa State U., Ames, 1979—. Contbr. chpts. to books, articles to profl. jours. Recipient Best Student award U. Chile Coll. Vet. Medicine, 1954; Rockefeller Found. scholar, 1963; Morris Animal Found. fellow, 1974. Mem. Chilean Vet. Med. Assn., Soc. for Study of Reprodn., Soc. for Study of Fertility (Eng.), Sigma Xi, Beta Beta Beta, Phi Kappa Phi. Subspecialties: Reproductive biology; Animal contraception. Current work: Development of contraceptives to control pet population; research on male and female gametes as related to the development of methods of contraception. Office: Dept Physiology and Pharmacology Vet Med Iowa State U Ames IA 50011

PINERO, GERALD JOSEPH, anatomy educator, researcher; b. New Orleans, Feb. 26, 1943; s. Jules Sebastian and Margaret Catherine (Matthews) P.; m. Ann Cousar, Dec. 27, 1969; children: Brian Joseph, Daniel Thomas. B.S., La. State U.-New Orleans, 1964; Ph.D., La. State U.-Baton Rouge, 1970. Research asst. La. State U.-Baton Rouge, 1970-71; asst. prof. U. Tex.-Houston, 1971-75, assoc. prof., 1975-80, prof. anatomy, 1980—; mem. (faculty Grad. Sch. Biol. Sci.), 1981-82. Mem. Am. Assn. Anatomists, Internat. Assn. Dental Research, Tissue Culture Soc., AAAS. Subspecialties: Anatomy and embryology; Cell and tissue culture. Current work: Function of fibroblasts, particularly in relationship to oral disease. Home: 8302 Burning Hills St Houston TX 77071 Office: University of Texas PO Box 20068 Houston TX 77225

PINES, ARIEL LEON, psychologist, educator; b. Bulawayo, South Rhodesia, Dec. 29, 1947; s. Michael Y. and Neima (Ratner) P.; m. Mary Lynne Mormann, Sept. 2, 1977. B.Sc., Hebrew U., Jerusalem, 1973, M.Sc., 1974, Dip.Ed., 1974; Ph.D., Cornell U., 1977. Instr. IDF, Israel, 1968-70; research asst. Hebrew U., 1971-74, Cornell U., Ithaca, N.Y., 1974-77; asst. prof. U. Maine - Farmington, 1977-79, assoc. prof., 1979—; vis. prof. Monash U., Melbourne, Australia, 1981, Hebrew U., Rehovot, Israel, 1978. Bd. editors: Jour. Contemporary Ednl. Psychology, 1981—; Author monographs; contbr. articles to profl. jours. Mem. Am. Psychol. Assn., Can. Psychol. Assn., Nat. Sci. Tchrs. Assn., Am. Ednl. Research Assn., Internat. Congress for Individualized Instrn., Nat. Assn. Biology Tchrs., Northeastern Research Assn., Maine Assn. for Human Genetics (chmn. ethics com. 1980), No. New Eng. Philos. Assn. (v.p. 1982—), Maine Sci. Tchrs. Assn., Environ. Action, Cousteau Soc., World Wildlife Soc., Common Cause, Handgun Control Assn., Union of Concerned Scientists, others. Club: BAL (Farmington). Subspecialties: Learning; Animal genetics. Current work: Learning and the structure of memory. How do young children and adults acquire scientific concepts? How can we evaluate or assess what a person knows? Curriculum development and instructional planning within an epistemological-psychological framework. Office: Univ Maine - Farmington 23 Mantour Library 41 High St Farmington ME 04938 Home: PO Box 487 Temple ME 04984

PINES, DAVID, physicist; b. Kansas City, Mo., June 8, 1924; s. Sidney and Edith (Adelman) P.; m. Aronelle Siegerman, June 15, 1948; children: Catherine Deirdre, Jonathan David. A.B., U. Calif. at Berkeley, 1944; M.A., Princeton, 1948, Ph.D., 1950. Instr. U. Pa., 1950-52; research asst. prof. U. Ill. at Urbana, 1952-55; asst. prof. Princeton U., 1955-58; mem. Inst. Advanced Study, Princeton, 1958-59; prof. physics and elec. engring. U. Ill., 1959—, dir., 1967-70, asso., 1972-73, prof., 1978—; prof. asso. Faculty des Scis., U. Paris, France, 1962-63; vis. prof. NORDITA, Copenhagen, 1970; Lorentz prof. U. Leiden, Netherlands, 1971; Fritz London Meml. lectr. Duke, 1972; Guilio Racah Meml. lectr. Hebrew U., Jerusalem, 1974; Marchon lectr. Newcastle-Upon-Tyne U., 1976; Sherman Fairchild distinguished scholar Calif. Inst. Tech., 1977-78; exchange prof. U. Paris, 1978; Eugene Feeberg Meml. lectr., Washington, 1982; Disting. lectr. U. Rochester, 1983; dir. W.A. Benjamin, Inc., N.Y.C., 1963-71; mem. physics survey com. Nat. Acad. Scis., 1963-65; v.p. Aspen Center for Physics, 1968-76, chmn. exec. com. bd. trustees, 1976-77; mem. council for biology and human affairs Salk Inst., 1969-73; co-chmn. Joint Soviet-U.S. Symposia on Theory of Condensed Matter, 1968, 70, 71, 73, 74, 76; mem. NRC Commn. Internat. Relations, 1974-79, Com. Scholarly Communication with People's Republic of China, 1974-78; mem. bd. advisors Inst. Theoretical Physics, U. Calif., Santa Barbara, 1979-81; mem. theoretical div. adv. com. Los Alamos Nat. Lab., 1977—, chmn., 1980-81; Chmn. Nat. Acad. Scis.-NRC Bd. Internat. Sci. Exchange, 1974-77; chmn. Working Group on Physics US-USSR Joint Commn. on Sci. and Tech. Coop., 1974—, NRC Commn. US-USSR Coop. Physics, 1978—; mem. space sci. bd. NRC, 1978-81; trustee Assn. Univ. Research Astronomy, 1975-78, 79—, exec. com., 1981-82. Author: The Many Body Problem, 1961, Elementary Excitations in Solids, 1963, (with P. Nozières) Theory of Quantum Liquids, 1966; Editor: Frontiers of Physics, 1961—, Lecture Notes and Supplements in Physics, 1962—, Revs. of Modern Physics, 1973—; Mem. editorial adv. bd.: Jour. of Non-Metals. Recipient Frieman prize in comdensed matter physics, 1983; NSF sr. postdoctoral fellow, 1957-58; Guggenheim fellow, 1962-63, 70. Fellow AAAS, Am. Phys. Soc., Am. Astron. Soc.; mem. Nat. Acad. Scis., Am. Acad. Arts and Scis., Internat. Astron. Union, Phi Beta Kappa, Sigma Xi. Club: Cosmos. Subspecialty: Theoretical physics. Home: 403 W Michigan St Urbana IL 61801

PINKER, STEVEN ARTHUR, experimental psychologist; b. Montreal, Que., Can., Sept. 18, 1954; came to U.S., 1976, naturalized, 1980; s. Harry and Roslyn (Weisenfeld) P.; m. Nancy L Etcoff, June 15, 1980. B.A., McGill U., 1976; Ph.D., Harvard U., 1979. Postdoctoral fellow M.I.T., 1979-80, asst. prof. psychology, 1982—; asst. prof. Harvard U., 1980-81, Stanford U., 1981-82; mem. adv. bd. Cognitive Sci. Series M.I.T. Press, 1980—; reviewer grant proposals NSF, Washington, 1980—. Contbr. chpts. to books and articles to profl. jours. NRC Can. fellow, 1976-79; Frank Knox Found. fellow, 1976-78; NSF grantee, 1980-83, 82—; Ctr. for Advanced Study in Behavioral Scis. fellow-elect, 1981—. Mem. Am. Psychol. Assn., Psychonomic Soc., Cognitive Sci. Soc. Jewish. Subspecialties: Cognition; Developmental psycholinguistics. Current work: Formulating and testing theories of how humans learn their first language, how they mentally represent 3-D space and how they comprehend graphs and charts. Office: Dept Psychology MIT Cambridge MA 02139 Home: 51 Cottage Farm Rd Brookline MA 02146

PINNAS, JACOB LOUIS, physician, allergist-immunologist; b. Newark, Jan. 31, 1940; s. David and Sally (Goldberg) P.; m. Ellen Susan Reiff, Jan. 20, 1942; children: Michelle, Deborah, Laura, Shira. M.D., M.S. in Pathology, U. Chgo., 1965. Resident Upstate Med. Center, Syracuse, N.Y., 1965-66, 68-70; med. epidemiologist Center Disease Control, Atlanta, 1966-68; research and clin. fellow in allergy and immunology Scripps Clinic and Research Found., La Jolla, Calif., 1970-73; assoc. prof. internal medicine, attending staff U. Ariz., 1973—; attending staff, dir. allergy service Univ. Hosp., 1973—, VA Hosp., Tucson, 1973—. Contbr. articles to profl. jours. Served with USPHS, 1966-68. Robert Wood Johnson Found. scholar, 1961-65. Fellow Am. Acad. Allergy, ACP; mem. Am. Assn. Immunologists, Western Soc. Allergy and Immunology (pres. 1984). Subspecialties: Internal medicine; Immunology (medicine). Current work: Hypersentivity reactions, antiallergic drugs, asthma. Office: 1501 N Campbell Tucson AZ 85724

PINNEY, FRANK BATCHELDER, jet engine co. exec.; b. Montpelier, Vt., May 12, 1933; s. Perry Batchelder and Eva Jane (Littlefield) P.; m. Alice Marie Garrity, Nov. 26, 1960; children: Joyce Marie, Anne Marie, Thomas Batchelder. B.S.M.E., U. N.H., 1955; M. Engring. Sci. in Mettal. Engring, Renssaeler Poly. U., 1973. Engr. Bethlehem Steel, Sparrows Point, Md., 1955-56, Quincy, Mass., 1958-63, Hamilton Standard, Windsor Locks, Conn., 1963-64; with Pratt & Whitney Aircraft, East Hartford, Conn., 1968—, supr., 1973-81, gen. supr. exptl. process planning, 1981—. Bd. dirs. Metacomet Homes, Inc., East Granby, Conn., 1973—; pres. Parish Council, St. Bernard's Ch., Tariffville, Conn., 1975-78, East Granby (Conn.) Soccer Club, 1978-81. Served with U.S. Army, 1956-58. Mem. ASME, Hartford Engrs. Club. Republican. Subspecialties: Materials processing; Computer-aided manufacturing. Current work: Mgmt. activities in computer-aided mfg. and computer process planning, particularly related to computer graphics. Patentee, inventor fiber reinforced composites, 1968-70. Home: 59 Wynding Hills Rd East Granby CT 06026

PIORE, EMANUEL RUBEN, physicist; b. Wilno, Russia, July 19, 1908; came to U.S., 1917, naturalized, 1924; s. Ruben and Olga (Gegusin) P.; m. E. Nora Kahn, Aug. 26, 1931; children—Michael Joseph, Margot Deborah, Jane Ann. A.B., U. Wis., 1930, Ph.D., 1935, D.Sc. (hon.), 1966, Union U., 1962. Asst. instr. U. Wis., 1930-35; research physicist RCA, 1935-38; engr. in charge TV lab. CBS, 1938-42; head spl. weapons group, bur. ships U.S. Navy, 1942-44; head electronics br. Office Naval Research, 1946-47, dir. phys. sci., 1947-48, dep. for natural sci., 1949-51, chief sci., 1951-55; v.p., dir. Avco Mfg. Corp., 1955-56; dir. research IBM Corp., 1956-61, v.p. research and engring., 1961-63, v.p., group exec., 1963-65, v.p., chief scientist, 1965—; also dir.; physicist research lab. electronics Mass. Inst. Tech., 1948-49; Dir. Sci. Research Assos., Inc., Health Advancement, Inc., Paul Revere Investors, Guardian Mut. Fund. Mem. Pres.'s Sci. Adv. Com., 1959-62; mem. Nat. Sci. Bd., 1961—; bd. dirs. N.Y. State Found. for Sci.; past bd. dirs. NSF; chmn. vis. com. Nat. Bur. Standards; chmn. bd. Hall of Science, N.Y.C.; mem. corp. Woods Hole Oceanographic Instn.; mem. exec. com. Resources for Future; bd. dirs. Stark Draper Lab., Nat. Info. Bur.; mem. vis. com. to elec. engring. dept. Mass. Inst. Tech., 1956-57; vis. com. Harvard Coll., 1958-70; trustee Sloan-Kettering Inst. Cancer Research; mem. N.Y.C. Bd. Higher Edn., 1976—. Served to lt. comdr. USNR, 1944-46. Recipient Indsl. Research Inst. award, 1967; Distinguished Civilian medal Dept. Navy; Kaplun award Hebrew U., 1975. Fellow AAAS, Royal Soc. Arts (London, Eng.), Am. Phys. Soc., IEEE, Am. Acad. Arts Scis.; mem. Sci. Research Soc. Am., Sci. Research Assn. (dir.), Nat. Acad. Sci., Nat. Acad. Engring., Am. Inst. Physics (dir.), Am. Philos. Soc., Sigma Xi. Clubs: University (N.Y.C.); Cosmos (Washington). Subspecialties: Atomic and molecular physics; Superconductors. Home: 115 Central Park W New York City NY 10023 Office: Armonk NY 10504

PIPES, ROBERT BYRON, university administrator, mechanical and aerospace engineer, educator, consultant; b. Shreveport, La., Aug. 14, 1941; s. Walter H. and Mattye Mae (Wilson) P.; m. Ruth Ellen, June 27, 1964; children: Christopher Franz, Mark Robert. B.S., La. Poly. Inst., 1964, M.S., 1965; M.S.E., Princeton U., 1969; Ph.D., U. Tex., 1972. Sr. structures engr. Gen. Dynamics Corp., Ft. Worth, 1969-72; asst. prof. mech. engring. Drexel U., 1972-74; assoc. prof. mech. and aerospace engring. U. Del., 1974-80, prof., 1980—, dir. Ctr. Composite Materials, 1978—; cons. in field; mem. nat. materials adv. bd. NRC. Author: Experimental Mechanics of Fiber-Reinforced Composite Materials, 1982; contbr. numerous articles to profl. jours. Active Boy Scouts Am. Mem. ASME, Soc. Exptl. Stress Analysis, ASTM, Tau Beta Pi, Pi Tau Sigma, Omicron Delta Kappa. Methodist. Subspecialty: Composite materials. Current work: Composite materials. Home: 100 Mason Dr Newark DE 19711 Office: Center Composite Materials U Del Newar DE 19716

PIPPEN, RICHARD WAYNE, biology and botany educator, systematic botany research; b. Villa Grove, Ill., Sept. 8, 1935; s. William Scott and Martha Jane (Bailey) P.; m. Sally Virginia Hitchings, Jan. 27, 1961; 1 son, Jeffrey Scott. B.S. in Edn, Eastern Ill. U., 1957; M.A., U. Mich., 1959, Ph.D., 1963. Asst. prof. Western Mich. U., Kalamazoo, 1963-68, assoc. prof., 1968-78, prof., 1978—, chmn. biology, 1977—; dir. tropical studies program Assn. Univ. Internat. Edn., 1973-77. Contbr. articles to profl. jours. Recipient outstanding tchr. award Alumni Assn. Western Mich. U., 1970, 82. Mem. Mich. Bot. Club (pres. 1975-79), Am. Soc. Plant Taxonomists, Bot. Soc. Am., Eco. Soc. Am., Sigma Xi. Subspecialties: Systematics; Reproductive biology. Current work: Reproductive biology and systematics of Aureolaria (Scrophulariaceae), Costus (Zingiberaceae) and Cacalioid genera of Asteraceae. Home: 1136 Berkshire Kalamazoo MI 49007 Office: Western Mich U Biology Dept Kalamazoo MI 49007

PIRAINO, FRANK FRANCIS, clin. virologist; b. LaSalle, Ill., Dec. 24, 1925; s. Frank and Marie (Buttitta) P.; m. Alberta Joan, July 27, 1936; 4 children. B.S., U. Wis., 1954, M.S., 1955, Ph.D., 1958. Research microbiologist Victo Chem. Co., Ashland, Ohio, 1958-60; research virologist Regional Poultry Lab. U.S. Dept. Agr., 1960-62; chief virologist Dept. Health, City of Milw., 1962-70; clin. virologist St. Joseph's Hosp., Milw., 1970-82, dir., 1982—. Served with U.S. Army, 1944-46. Am. Cancer Soc. grantee, 1962-63; NIH grantee, 1966-69. Mem. Am. Soc. Microbiology, AAAS. Subspecialties: Animal virology; Immunology (agriculture). Home: 2178 W 71st St Milwaukee WI 58213

PIROFSKY, BERNARD, physican, educator; b. N.Y.C., Mar. 27, 1926; s. Hyman and Yetta (Herman) P.; m. Elaine Friedwald, June 19, 1953; children: Daniel Niles, Tandy Ellen, Jillann Yetta. A.B., NYU, 1946, M.D., 1950. Intern, Bellevue Hosp., N.Y.C., 1950-51, resident medicine, 1951-52, 54-55; research fellow U. Oreg., 1955-56; dir. Pacific N.W. Regional Blood Center, 1956-58; instr. medicine U. Oreg., 1956-59, asst. prof., 1959-63, asso. prof., 1963-66, prof., 1966—, prof. microbiology, 1973—, head div. immunology, allergy and rheumatology, 1965—; cons. immunohematologist Portland VA Hosp., Pacific N.W. Regional Blood Center; mem. med. adv. bd. Leukemia Soc., 1965-69; vis. prof. Nat. Inst. Nutrition, Mexico, 1966-67, Med. Research Council, South Africa, 1977; vis. scientist Nat. Acad. Sci., Japan, 1978-79. Author: Autoimmunization and the Autoimmune Hemolytic Anemias, 1969, Blood Banking Principles, 1973; contbr. articles to profl. jours. Served as capt. USAF, 1952-54. Commonwealth Fund fellow, 1966-67; recipient Gov.'s N.W. Sci. award, 1968; Emily Cooley award Am. Assn. Blood Banks, 1972. Fellow ACP; mem. Internat. Soc. Hematology, Internat. Soc. Blood Transfusions, Am. Soc. Hematology, Mexican Soc. Hematology (hon. mem.), Transplantation Soc., Am. Assn. Immunologists. Subspecialties: Immunology (medicine); Hematology. Research on immunology and hematology. Home: 10370 SW Ridgeview Ln Portland OR 97219

PIRRUNG, MICHAEL CRAIG, chemistry educator; b. Cin., July 31, 1955; s. Joey Matthew and Grace (Fielman) P.; m. Jane Elizabeth, Sept. 12, 1982. B.A., U. Tex.-Austin, 1976; Ph.D., U. Calif.-Berkeley, 1980. Postdoctoral assoc. Columbia U., 1980-81; asst. prof. chemistry Stanford U., 1981—. Author: Total Synthesis of Sesquiterpenes, 1983. NSF fellow, 1980-81; recipient Research award Dreyfus Found., 1981. Mem. Am. Chem. Soc. Subspecialty: Organic chemistry. Current work: Organic synthesis; biosysthesis of plant growth hormones. Office: Stanford Dept Chemistry Stanfor CA 94305

PIRTLE, EUGENE CLAUDE, microbiologist; b. Wichita Falls, Tex., Nov. 17, 1921; s. Claude Howell and Jettie Alma P.; m. Julianne Pirtle, Nov. 29, 1944; children: Edith, Victor, Margaret, David, James. B.S., U. Colo., 1948; M.S., U. Iowa, 1950, Ph.D., 1952. Asst. prof. microbiology U. S.D., 1952-56, assoc. prof., 1956-60; dir. Virus Lab. Dept. Health, State of Hawaii, 1960-61; microbiologist Agrl. Research Service, Nat. Animal Disease Ctr., Ames, Iowa, 1961—. Contbr. numerous articles to profl. jours. Served with U.S. Army, 1942-44. Recipient Cert. of Appreciation U.S. Dept. Agr., 1978. Mem. Tissue Culture Assn., Sigma Xi, Phi Zeta. Subspecialties: Microbiology (medicine); Animal virology. Current work: Pathogenesis of viral diseases; DNA fingerprinting. Office: PO Box 70 Ames IA 50010

PISKO, EDWARD JOHN, rheumatologist, immunologist, medical educator; b. Phila., July 11, 1945; s. John and Rita Delores (Cipparone) P.; m. Wendy Packer, July 26, 1973; 1 dau., Cheryl R.M. A.B., U. Pa., 1967; M.D., U. Pitts., 1971. Diplomate: Am. Bd. Internal Medicine (subcert. in rheumatology). Intern Thomas Jefferson U. Hosp., Phila., 1971-72; resident in internal medicine U. Minn. Hosp., Mpls., 1972-74; internist Permanente Med. Group, Denver, 1974-76; rheumatology fellow Bowman Gray Sch. Medicine, Wake Forest U., Winston-Salem, N.C., 1976-78, instr., 1977-78, asst. prof. medicine, 1978—. Contbr. articles to med. jours., chpts. to books. Mem. avd. bd. N.C. chpt. Arthritis Found., Durham, 1978—. Fellow ACP; mem. Am. Rheumatism Assn., Am. Fedn. Clin. Research, Am. Soc. Clin. Pharmacology and Therapeutics, Internat. Soc. Immunopharmacology. Subspecialties: Immunopharmacology; Pharmacology. Current work: Current investigations include studying the cellular immune abnormalities that areiimportant in the pathogenesis of rheumatoid arthritis and evaluating newer drugs, including immunopharmacological agents, for the treatment of a variety of arthritic disorders. Home: 719 Roslyn Rd Winston-Salem NC 27104 Office: Rheumatology Sect Dept Medicine Bowman Gray Sch Medicine 300 S Hawthorne Rd Winston-Salem NC 27103

PISTER, KARL STARK, university dean; b. Stockton, Calif., June 27, 1925; s. Edwin LeRoy and Mary Kimball (Smith) P.; m. Rita Olsen, Nov. 18, 1950; children: Francis, Therese, Anita, Jacinta, Claire, Kristofer. B.S. with honors, U. Calif., Berkeley, 1945, M.S., 1948; Ph.D., U. Ill., Urbana, 1952. Instr. theoretical and applied mechanics U. Ill., 1949-52; mem. faculty U. Calif., Berkeley, 1952—, prof. engrng. scis., 1962—; dean Coll. Engrng., 1981—; Richard Merton guest prof. U. Stuttgart, W. Ger., 1978; cons. to govt. and industry. Author research papers in field; assoc. editor: Computer Methods in Applied Mechanics and Engring, 1972, Jour. Optimization Theory and Applications, 1982; editorial bd.: Res Mechanica, 1978. Served with USNR, World War II. Recipient Wason Research medal Am. Concrete Inst., 1960; Fulbright scholar, Ireland, 1965, W. Ger., 1973. Fellow Am. Acad. Mechanics; mem. Nat. Acad. Engrng., ASCE, ASME, Earthquake Engrng. Research Inst., Soc. Engrng. Sci. Subspecialties: Civil engineering; Theoretical and applied mechanics. Current work: Mechanics of solids and structures, particularly finite deformation problems with inelastic material behavior; constitutive modeling of inelastic materials. Computer-aided design of structures, with emphasis on optimazation-based design of dynamically loaded structures and applications to earth-quake resistence design. Home: 828 Solana Dr Lafayette CA 94549 Office: Coll Engring Univ Calif Berkeley CA 94720

PISU (PISUNYER), F. XAVIER, medical educator, researcher; b. Barcelona, Spain, Dec. 3, 1933; came to U.S., 1941; s. James and Mercedes (Diaz) Pi-Sunyer; m. Penelope Wheeler, June 24, 1961; children: Andrea M., Olivia A., Joanna L. B.A., Oberlin Coll., 1955; M.D., Columbia U., 1959; M.P.H., Harvard U., 1963. Intern and resident St. Luke's Hosp. Ctr., N.Y.C., 1959-60; resident in medicine St. Bartholomew's Hosp., London, 1961-62; fellow in medicine Thorndike Lab., Harvard Med. Sch., Boston, 1964-65; from instr. to asst. prof. medicine Columbia U., N.Y.C., 1965-76, assoc. prof. clin. medicine, 1976-78, assoc. prof. medicine, 1978—; mem. staff Inst. Human Nutrition, 1965—; from asst. attending physician to attending St. Luke's Hosp., N.Y.C., 1965—; dir. div. endocrinology St. Luke's-Roosevelt Hosp. Ctr., N.Y.C., 1977—; councilor Am. Bd. Nutrition, 1983—; mem. nutrition study sect. NIH, 1983—. Pres. bd. dirs. N.Y. affiliate Am. Diabetes Assn., 1970—, mem. nat. com. research, 1982—; chmn. com. diet N.Y. Heart Assn., 1977-79; Mem. alumni bd. Oberlin Coll., 1970-73, 82—. N.Y. Heart Assn. fellow, 1966-68; Fogarty Internat. fellow, 1979-80; vis. scientist U.K. Med. Research council, London, 1979-80. Mem. AAAS, Am. Inst. Nutrition, Am. Diabetes Assn., Am. Soc. Clin. Nutrition, N.Y. Diabetes Assn., Endocrine Soc., Am. Fedn. Clin. Research, Soc. Exptl. Biology and Medicine, Am. Bd. Nutrition, Harvey Soc., N.Y. Acad. Medicine, Physicians for Social Responsibility, Amnesty Internat., Sierra Club. Subspecialties: Endocrinology; Nutrition (medicine). Current work: Diabetes mellitus; obesity; carbohydrate and lipid metabolism; food intake regulation; pancreatic islet transplantation; adipocyte metabolism; trace metal effects on carbohydrate and lipid metabolism. Home: 305 Riverside Dr New York NY 10025 Office: St Luke's-Roosevelt Hosp Ctr 114th St and Amsterdam Ave New York NY 10025

PITCHER, GEORGIA ANN, psychology educator, psychologist; b. Indpls., Feb. 22, 1927; d. Arling Edgar and Lyda Lucille (Doty) P.; m. Donald Aubrey Baker, Aug. 21, 1948 (div. Apr. 1968); children: Catherine Lucille Baker Black, Martha Ann Baker, Susan Jane Baker Oakley, Daniel Pitcher Baker. B.S., Butler U., 1948, M.S., 1951; Ph.D., Purdue U., 1969. Lic. Psychologist, Nat. Register of Health Service Providers in Psychology. Asst. prof. Butler U., 1964-68; asst. prof. Purdue U.-West Lafayette, Ind., 1969-74; dir. psychol. services St. Elizabeth Hosp., Lafayette, Ind., 1974-81; instr. Butler U., 1981—; pvt. practice psychology, Indpls., 1980—; psychologist Cary Home Diagnostic Center, Lafayette, Ind., 1980-81; cons. Lafayette Head Start, 1976-81; cons. psychologist Comprehensive Devel. Centers, Monticello, Ind., 1975-76; evaluator Pre-Sch. Project Title III, Mishawaka, Ind., 1971-72. Contbr. articles to profl. jours. Mem. Nat. Govtl. Activities Com., United Cerebral Palsy Assns. Inc., N.Y.C., 1976-77; v.p. United Cerebral Palsy Ind., Indpls., 1976—; mem. Protective Services Task Force Com., State of Ind., 1976-77. Mem. Am. Psychol. Assn., Am. Ednl. Research Assn., Nat. Acad. Neuropsychology, Com. for Found. of Internat. Orgn. Psychophysiology. Democrat. Subspecialties: Developmental psychology; Neuropsychology. Current work: Development and traumatic causes of learning disorders, particularly brain behavior relationships; cognitive and affective. Home: 3725 E Thompson Rd Indianapolis IN 46227

PITCHUMONI, CAPECOMORIN S., medical educator, gastroenterologist; b. Madura, India, Jan. 20, 1938; came to U.S., 1967; s. Sankar Harihar and Jaya Lekshmi P.; m. Prema Iyer, Nov. 11, 1964; children: Sheila, Shoba, Suresh. M.B.B.S., Med. Coll., Trivandrum, Kerala, India, 1960, M.D., 1965. Intern Med. Coll. Hosp., Trivandrum, Kerala, India, 1960-61; resident Norwalk (Conn.) Hosp., 1967-68; asst. prof. N.Y. Med. Coll., Valhalla, 1972-75, assoc. prof., 1975-80, prof. clin. medicine, 1980—; dir. div. gastroenterology Misericordia Hosp. Med. Ctr., Bronx, N.Y., 1980—. Co-author: Progress in Gastroenterology, Vol. III, 1978, Hand Book of Infectious Diseases, 1982. Trustee Hindu Temple N. Am., 1978. Recipient Hechst Omprakash award Indian Soc. Gastroenterology, 1976. Fellow Royal Coll. Physicians and Surgeons Can., ACP, Am. Coll. Gastroenterology; mem. Am. Gastroenterological Assn. Subspecialties: Gastroenterology; Nutrition (medicine). Current work: Nutritional disorders of the pancreas. Home: 178 Fairmount Ave Glenrock NJ 07452 Office: Misericordia Hosp Med Ctr 600 E 233d St Bronx NY 10466

PITCOCK, JAMES ALLISON, pathologist, cardiovascular researcher; b. Little Rock, Ark., Sept. 13, 1929; s. Radford Bolling and Anne (Whitelaw) P.; m. Cynthia DeHaven, June 18, 1954; children: Allison, James. B.S., MIT, 1951; M.D., Washington U., St. Louis, 1955. Cert. Am. Bd. Pathology. Intern Vanderbilt U. Hosp., Nashville, 1955-56; resident in pathology Barnes Hosp., St. Louis, 1956-59, 61-62, assoc. pathologist, 1962, St. Vincent's Infirmary, Little Rock, 1963, Baptist Meml. Hosp., Memphis, 1964-75, assoc. dir. labs., 1975—; instr. pathology Washington U., 1961-62; clin. asst. prof. U. Ark., 1963, U. Tenn., Memphis, 1965-74, clin. assoc. prof., 1974-81, clin. prof., 1981—. Served to capt. M.C. USAF, 1959-61. Fellow Am. Soc. Clin. Pathologists; mem. Council High Blood Pressure Research, Internat. Soc. Hypertension, Arthur Prudy Stout Soc. Surg. Pathology, Internat. Acad. Pathology, Am. Assn. Pathologist, Sigma Xi, Alpha Omega Alpha. Episcopalian. Subspecialty: Pathology (medicine). Current work: High blood pressure research; surgical pathology. Home: 4828 Briarcliff St Memphis TN 38117 Office: Baptist Memorial Hospital 899 Madison Ave Memphis TN 38146

PITELKA, LOUIS FRANK, biology educator, researcher; b. Berkeley, Calif., Mar. 28, 1947; s. Frank Alois and Dorothy Gretchen (Riggs) P.; m. Sandra Lea Sanders, Sept. 20, 1969; children: Erik Loren, Jessica Kristine. B.S., U. Calif., Davis, 1969; Ph.D. in Biol. Scis., Stanford U., 1974. Asst. prof. biology Bates Coll., Lewiston, Maine, 1974-81, assoc. prof., 1981—, chmn. biology dept., 1982—; program dir. population biology and physiol. ecology; program NSF, Washington, 1983—. Contbr. articles on biol. scis. to profl. jours. NSF grantee, 1980-82, 1982-84; Research Corp. grantee, 1979-81. Mem. Ecol. Soc. Am., Bot. Soc. Am. (chmn. ecol. sect. 1982-83), Am. Inst. Biol. Scis., Internat. Soc. Plant Population Biologists, Torrey Bot. Club. Subspecialties: Ecology; Population biology. Current work: Plant ecology, plant population biology, population biology of forest understory herbs, including demography, life history patterns and physiological ecology. Home: 5613 Herberts Crossing Dr Burke VA 22015 Office: NSF Population Biology Program Washington DC 20550

PITKIN, ROY MACBETH, medical educator; b. Anthon, Iowa, May 24, 1934; s. Roy and Pauline (McBeath) P.; m. Marcia Alice Jenkins, Aug. 17, 1957; children: Barbara, Robert, Kathryn, William. B.A. with highest distinction, U. Iowa, 1956, M.D., 1959. Intern King County Hosp., Seattle, 1959-60; resident in ob-gyn U. Iowa Hosp., Iowa City, 1960-63; asst. prof. ob-gyn U. Ill., Chgo., 1965-68; assoc. prof. U. Iowa, Iowa City, 1968-72, prof., 1972—, head dept., 1977—; examiner Am. Bd. Ob-Gyn, 1974—, dir., 1980—; mem. Council on Resident Edn. in Ob-Gyn, 1979—; mem. human embryology and devel. study sect. NIH, 1980—. Editor: Yearbook of Ob-Gyn, 1975—; editorial bd.: Perinatal Press, 1976—, Ob-Gyn, 1977-80; editor: Clinical Obstectrics and Gynecology, 1978. Served to lt. comdr., M.C. USNR, 1963-65. Nat. Inst. Arthritis, Metabolism and Digestive Diseases grantee, 1973-77. Mem. Am. Coll. Obstetricians and Gynecologists (residency rev. com. 1982—), AMA (Joseph B. Goldberger award 1982), Central Assn. Obstetricians and Gynecologists (award 1968), Soc. Gynecologic Investigation, Soc. Perinatal Obstetricians (pres. 1978-79), Am. Fedn. Clin. Research, Soc. Exptl. Biology and Medicine, Perinatal Research Soc., Sigma Xi, Phi Eta Sigma, Omicron Delta Kappa, Alpha Omega Alpha. Presbyterian. Subspecialties: Obstetrics and gynecology; Maternal and fetal medicine. Current work: Nutrition in pregnancy and lactation; calcium metabolism during pregnancy and perinatal period. Home: 517 Templin Rd Iowa City IA 52240 Office: U Iowa Hosps Iowa City IA 52242

PITKOW, HOWARD SPENCER, physiologist, researcher; b. Phila., May 21, 1941; s. Jack and Molly (Zelinger) P.; m. Ellice Spector, Feb. 9, 1969; children: Seth, Stephanie. B.A., U. Pa., Phila., 1962, M.S., 1963; Ph.D., Rutgers U., 1971. Instr. histology and embroyology Pa. Coll. Podiatric Medicine, Phila., 1965-66, asst. prof. anatomy, 1967-68, asst. prof. physiology and biology, 1969-71, assoc. prof. physiology sci., 1972-77, chmn., 1974—, prof., 1978—; guest lectr. Phila. Community Coll., 1972-76; lectr. Bucks County Community Coll., 1978-81; assoc. prof. Hahnemann Med. Sch., Phila., 1980—; adj. prof. Drexel U., Phila., 1980; research prof. Temple Health Sci. Ctr., Phila., 1981—. Contbr. articles to profl. jours.; reviewer: others. Am. Podiatry Assn. fellow, 1966, 68; Internat. Union Physiol. Sci. travel grantee, 1977; Argonne Nat. Lab. travel grantee, 1978, 79. Mem. Am. Physiol. Soc., Fedn. Am. Socs. Exptl. Biology, Endocrine Soc., Nat. Assn. Biology Tchrs., Soc. Exptl. Biology and Medicine, Pa. Acad. Scis. Jewish. Subspecialties: Reproductive biology; Environmental toxicology. Current work: Effects of carcinogens on pregnancy, lactation and neonatal development, environmental toxicologic and teratologic effects of carcinogens on reproduction, role of various hormone mimicking amino acids of mulifying the deleterious effects of carcinogens on lactational physiology. Home: 3245 Ethan Allen Ct Bensalem PA 19020 Office: Pa Coll Podiatric Medicine 8th at Race Sts Philadelphia PA 19107

PITOT, HENRY CLEMENT, physician, educator; b. N.Y.C., May 12, 1930; s. Henry Clement and Bertha (Lowe) P.; m. Julie S. Schutten, July 29, 1954; children: Bertha, Anita, Jeanne, Cathy, Henry, Michelle, Lisa, Patrice. B.S. in Chemistry, Va. Mil. Inst., 1951; M.D., Tulane U., 1955, Ph.D. in Biochemistry, 1959. Instr. pathology Tulane U. Med. Sch., 1955-59; postdoctoral fellow McArdle Lab., U. Wis.-Madison, 1959-60; mem. faculty U. Wis. Med. Sch.-Madison, 1960—, prof. pathology and oncology, 1966—, chmn. dept. pathology, 1968-71; acting dean Sch. Medicine, 1971-73; dir. McArdle Lab., 1973—. Mem. Nat. Cancer Adv. Bd., 1976-82, chmn., 1979-82. Recipient Borden undergrad. research award, 1955; Lederle Faculty award, 1962; Career Devel. award, 1965; Parke-Davis award in exptl. pathology, 1968. Fellow N.Y. Acad. Scis.; mem. AAAS, Am. Soc. Cell. Biology, Am. Assn. Cancer Research, Am. Soc. Biol. Chemists, Am. Chem. Soc., Am. Assn. Pathologists (pres. 1976-77), Soc. Exptl. Biology and Medicine, Soc. Developmental Biology, Soc. Surg. Oncology (Lucy J. Wortham award 1981), Tissue Culture Assn., Am. Soc. Preventive Oncology. Roman Catholic. Subspecialties: Oncology; Pathology (medicine). Home: 1812 Van Hise Ave Madison WI 53705 Where and who we are today is the result of those whom we have met and known and loved until now.

PITT, BERTRAM, cardiologist; b. Kew Gardens, N.Y., Apr. 27, 1932; s. David and Shirley (Blum) P.; m. Elaine Liberstein, Aug. 10, 1962; children: Geoffrey S., Jessica M., Jillian A. B.A., Cornell U., 1953; M.D., U. Basel, Switzerland, 1959. Diplomate: Am. Bd. Internal Medicine. Intern Beth Israel Hosp., N.Y.C., 1959-60, resident in medicine, Boston, 1960-63; fellow in cardiology Johns Hopkins Hosp., 1966-67; instr. to prof. medicine Johns Hopkins U., 1967-77; prof. medicine, dir. div. cardiology U. Mich., 1977—; pres. Cardiovascular Research Cons., Inc. Served to capt. U.S. Army, 1963-65. Mem. Am. Physiol. Soc., Am. Heart Assn., Am. Soc. Clin. Investigation, Am. Assn. Physicians, Am. Coll. Cardiology. Jewish. Subspecialty: Cardiology. Current work: Nuclear cardiology; central nervous system control of coronary circulation; myocardial infarction and ischemia; role of prostaglandins in ischemic heart disease. Home: 24 E Ridgeway St Ann Arbor MI 48104 Office: University Hospital Ann Arbor MI 48109

PITT, LOREN DALLAS, mathematics educator, researcher; b. Chewelah, Wash., Sept. 18, 1939; s. Dallas Richard and Gertrude (Kruger) P.; m. Kay Francis Salyer, 1961; children: Dallas Eric, Kristin Elaine. B.S. in Math, U. Idaho, 1961, M.A., Cath. U. Am., 1964, Princeton U., 1966, Ph.D., 1967. Research assoc. Rockefeller U., 1967-68, asst. prof., 1968-70; asst. prof. math. U. Va., 1970-72, assoc. prof., 1972-77, prof., chmn. dept., 1977—. Editorial bd.: Jour. Multivariate Analysis, 1978—; contbr. articles to profl. jours. NSF grantee, 1974-83. Mem. Am. Math. Soc., Math. Assn. Am., Inst. Math. Stats., Soc. Indsl. and Applied Math. Subspecialties: Probability; Statistics. Current work: Prediction theory for random fields and processes, correlation inequalities, sample function properties for random fields. Home: 2412 Angus Rd Charlottesville VA 22901 Office: Dept Mathematics University of Virginia Charlottesville VA 22903

PITTA, PATRICIA JOYCE, psychologist; b. N.Y.C., July 3, 1947; d. John Joseph and Mildred (Glosa) P.; m. Eric E. Kirk, June 26, 1976; children: Eric Jon, Kevin. B.A., Queens Coll., 1968; M.S., Hunter Coll., 1972; Ph.D., Fordham U., 1975. Intern Roosevelt Hosp., N.Y.C., 1973-74; staff psychologist Bellevue Hosp., N.Y.C., 1974-77; cons. psychologist St. Johns Episcopal Hosp., Smithtown, N.Y., 1977-78; clin. instr. N.Y. U. Med. Ctr., N.Y.C., 1975—; cons. psychologist N. Shore Univ. Hosp., Manhasset, 1978—; pvt. practice psychology, Manhasset, 1971—. Contbr. articles to profl. jours. Mem. Am. Psychol. Assn., Nassau County Psychol. Assn. Subspecialties: Behavioral psychology; Cognition. Current work: Exploring and trying to define emotional components in obesity in children/adults with goal of developing appropriate program for control.

PITTENDRIGH, COLIN STEPHENSON, biologist; b. Whitley Bay, Eng., Oct. 13, 1918; came to U.S., 1945, naturalized, 1950; s. Alexander and Florence Hemy (Stephenson) P.; m. Margaret Dorothy Eitelbach, May 1, 1943; children—Robin Ann, Colin Stephenson. B.Sc. with 1st class honors (Lord Kitchener nat. meml. scholar), U. Durham, Eng., 1940; A.I.C.T.A., Imperial Coll., Trinidad, B.W.I., 1942; Ph.D. (Univ. fellow), Columbia, 1947. Biologist internat. health div. Rockefeller Found., 1942-45; adviser Brommellad-malaria Brazilian govt., 1945; asst. prof. biology Princeton, 1947-50, asso. prof., 1950-57, prof., 1957-69, Class of 1877 prof. zoology, 1963-69; dean Grad. Sch., 1965-69; profl. biology Stanford, 1969—, Bing prof. human biology, 1970-76; Miller prof. biology, dir. Hopkins Marine Sta., Stanford U., 1976—; vis. prof. Rockefeller Inst., 1962; Phillipps lectr. Haverford Coll., 1956; Timothy Hopkins lectr. Stanford, 1957; Phi Beta Kappa nat. lectr., 1976-77, Sigma Xi nat. lectr., 1977-78, 78-79; Adviser Brit. colonial office on Bromeliad-malaria, 1958; mem. Nat. Acad. Scis.-NRC Com. oceanography. Author: (with George Gaylord Simpson, L. H. Tiffany) Life, 1957; Contbr. to books, profl. jours. Past v.p., trustee Rocky Mountain Biol. Lab. Guggenheim fellow, 1959. Fellow Am. Acad. Arts and Scis.; mem. Am. Philos. Soc., Am. Soc. Zoologists, Am. Soc. Naturalists (pres. 1968), Nat. Acad. Sci., AAAS (v.p. for zoology 1968). Subspecialty: Biological research management. Home: 26240 Jeanette Rd Salinas CA 93908

PITTER, RICHARD LEON, atmospheric sciences researcher, atmospheric physics educator; b. Whittier, Calif., Apr. 4, 1947; s. Jack F. and Eva L. (Ketchum) P.; m. Keiko Murata, Nov. 21, 1971; children: Gregory M., Jacqueline K. A.B., UCLA, 1969, M.S., 1970, C.Phil., 1972, Ph.D., 1973. Instr., asst. prof. Oreg. Grad. Ctr., Beaverton, 1973-77; asst. proof. U. Md., College Park, 1977-81; assoc. research prof. Desert Research Inst., Reno, Nev., 1981—; mem. tech. staff, cons. Computer Scis. Corp., Silver Spring, Md., 1978-79; mem. tech. staff The Mitre Corp., McLean, Va., 1978-81. NSF trainee UCLA, 1969-73; UCLA regent scholar, 1965-69. Mem. Am. Meteorol. Soc., Am. Geophys. Union, Royal Meteorol. Soc., Meteorol. Soc. Japan, Air Pollution Control Assn., AAAS. Subspecialties: Meteorology; Atmospheric chemistry. Current work: Cloud physics, aerosol chemistry and dynamics, air pollution meteorology and chemistry. Office: Desert Research Inst PO Box 60220 Reno NV 89506

PITTILLO, J(ACK) DANIEL, biology educator, researcher, consultant; b. Hendersonville, N.C., Oct. 25, 1938; s. Louis and Mattie (Hill) P.; m. Jean Farr, Aug. 28, 1966; children: Heather Ann, Shane Keven. A.B., Berea Coll., 1961; M.S., U. Ky., 1963; Ph.D., U. Ga., 1966. Teaching and research asst. U. Ky., 1961-63, U. Ga., 1963-66; asst. prof. biology Western Carolina U., 1966-71, assoc. prof., 1971-77, prof., 1977—; researcher Nat. Park Service, 1975-77; cons. N.C. State Planning, 1972-74, Nat. Olivine Co., Asheville, N.C., 1980; Prin. investigator U. Ga. NSF, 1982-83. Author: Natural Landmarks Southern Blue Ridge Province, 1976, Important Plant Habitats of the Blue Ridge Parkway, 1977. Pres. N.C. Bartram Trail Soc., 1980—. Mem. Assn. Southeastern Biologists, Ecol. Soc. Am., N.C. Acad. Sci. (dir. 1975-79), Bot. Soc. Am., So. Appalachian Bot. Club (pres. 1977-78), Sigma Xi. Current work: Vegatational history of the Southern Appalachians and China, vascular plant life forms of Coweeta Hydrologic Laboratory basin. Office: Dept Biology Western Carolina U Cullowhee NC 28723

PITTS, CHARLES WILLIAM, entomologist, educator; b. Corinth, Miss., Dec. 11, 1933; s. Charles William and Annie Mae (O'Shields) P.; m. Karol, Nov. 24, 1956; children: Timothy, Kimberly, Kelly, Kerrie. B.S., Miss. State U., 1960; M.S., Kans. State U., 1962, Ph.D., 1965. Asst. prof. entomology Kans. State U., 1965-68, assoc. prof., 1968-75, prof., 1975-78; prof. entomology Pa. State U.-University Park, 1978—, now head entomology dept. Grantee in field. Mem. Entomol. Soc. Am., Kans. Entomol. Soc., Pa. Entomol. Soc. Republican. Methodist. Clubs: Kiwanis, Toastmasters. Subspecialties: Ethology; Physiology (biology). Current work: Behavior, sensory physiology, scanning and transmission electron microscope. Office: 106 Patterson Bldg University Park PA 16802

PITTS, NATHANIEL GILBERT, neurophysiologist; b. Macon, Ga., Apr. 2, 1947; s. Raymond J. and Kathleen Lenora (Cook) P.; m. Carol Ann Shaw, June 19, 1971. B.A., Whittier Coll., 1969; Ph.D., U. Calif., Davis, 1974. Sci. fellow Rockefeller U., 1974-77; asst. program dir. in neurobiology NSF, Washington, 1977-79, assoc. program dir. in neurobiology, 1979-83, program dir. in integrative neural systems, 1983—. Contbr.: numerous articles to Exptl. Brain Research. Recipient Sustained Superior Performance award NSF, 1979; Porter Devel. Found. fellow, 1972-74; NIH postdoctoral fellow, 1975-77. Mem. AAAS, Soc. Neurosci. Soc., European Neurosci. Soc., N.Y. Acad. Scis., Internat. Brain Research Orgn. Subspecialties: Neurophysiology; Neurobiology. Current work: Neurophysiology of motor and sensory systems. Office: 1800 G St NW Washington DC 20550

PITTS, THOMAS GRIFFIN, nuclear engineer; b. Clinton, S.C., Aug. 19, 1935; s. Pascal Mark and Esther (Holland) P.; m. Sue Randolph Rucker, Oct. 14, 1961; children: Suzanne H., Thomas G. B.S., Presbyterian Coll., Clinton, 1957. Mathematician atomic energy div. Babock & Wilcox, Lynchburg, Va., 1957-58, engr., physicist, 1959-63, engr. research and devel. div., 1963-73, sr. engr., 1973—; Bd. dirs. Nuclear Power Fed. Credit Union, Lynchburg, 1982—. Com. chmn. Blue Ridge Mountains council Boy Scouts Am., Lynchburg, 1981—.

Mem. Am. Nuclear Soc., IEEE. Methodist. Subspecialties: Mathematical software; Nuclear physics. Current work: Computer applications on non destructive methods; computer operating systems, nuclear reactor instrumentation. Home: 1804 Parkland Dr Lynchburg VA 24505 Office: Babocks & Wilcox Lynchburg Research Center PO Box 239 Lynchburg VA 24505

PITZER, KENNETH S., chemist, educator; b. Pomona, Calif., Jan. 6, 1914; s. Russell K. and Flora (Sanborn) P.; m. Jean Mosher, July 1935; children—Ann, Russell, John. B.S., Calif. Inst. Tech., 1935; Ph.D., U. Calif., 1937; D.Sc., Wesleyan U., 1962; LL.D., U. Calif. at Berkeley, 1963, Mills Coll., 1969. Instr. chemistry U. Calif., 1937-39, asst. prof., 1939-42, assoc. prof., 1942-45, prof., 1945-61, asst. dean letters and sci., 1947-48, dean coll. chemistry, 1951-60; pres., prof. chemistry Rice U., Houston, 1961-68, Stanford, Calif., 1968-70; prof. chemistry U. Calif. at Berkeley, 1971—; tech. dir. Md. Research Lab. for OSRD, 1943-44; dir. research U.S. AEC, 1949-51, mem. gen. adv. com., 1958-65, chmn., 1960-62; Centenary lectr. Chem. Soc. Gt. Britain, 1978; mem. adv. bd. U.S. Naval Ordnance Test Sta., 1956-59, chmn., 1958-59; mem. commn. chem. thermo-dynamics Internat. Union Pure and Applied Chemistry, 1953-61; dir. Owens-Ill., Inc.; mem. Pres.'s Sci. Adv. Com., 1965-68. Author: (with others) Selected Values of Properties of Hydrocarbons, 1947, Quantum Chemistry, 1953, (with L. Brewer) Thermodynamics, rev, 1961; Editor: Prentice-Hall Chemistry series, 1955-61; Contbr. articles to profl. jours. Trustee Pitzer Coll., 1966—; Mem. program com. for phys. scis. Sloan Found., 1955-60. Guggenheim fellow, 1951; Precision Sci. Co. award in petroleum chemistry, 1950; 1 of 10 Outstanding Young Men U.S. Jr. C. of C., 1950; Clayton prize Instn. Mech. Engrs., London, 1958; Priestley Memorial award Dickinson Coll., 1963; Priestley medal Am. Chem. Soc., 1969; Nat. medal for sci., 1975. Fellow Am. Nuclear Soc., Am. Inst. Chemists (Gold medal award 1976), Am. Acad. Arts and Scis., Am. Phys. Soc.; mem. Am. Chem. Soc. (award pure chemistry 1943, Gilbert N. Lewis medal 1965, Williard Gibbs medal 1976), Faraday Soc., AAAS, Nat. Acad. Scis. (councilor 1964-67, 73-76), Am. Philos. Soc., Chem. Soc. (London), Am. Council Edn. Clubs: Chemists (hon.), Bohemian; Cosmos (Washington). Subspecialties: Physical chemistry; Theoretical chemistry. Home: 12 Eagle Hill Berkeley CA 94707 Office: Coll Chemistry: U Calif Berkeley CA 94720

PITZER, RUSSELL MOSHER, chemistry educator; b. Berkeley, Calif., May 10, 1938; s. Kenneth Sanborn and Jean Elizabeth (Mosher) P.; m. Martha Ann Seares, Sept. 3, 1959; children: Susan M., Kenneth R., David S. B.S., Calif. Inst. Tech., 1959; A.M., Harvard U., 1961, Ph.D., 1963. Research fellow in physics M.I.T., 1963; instr., asst. prof. chemistry Calif. Inst. Tech., 1963-68; assoc. prof. Ohio State U., Columbus, 1968-79, prof. chemistry, 1979—; cons. Lawrence Berkeley Lab., Lawrence Livermore Lab., 1978—. Contbr. articles in field to profl. jours. Mem. Am. Chem. Soc., Am. Phys. Soc., AAUP. Subspecialty: Theoretical chemistry. Current work: Quantum chemical calculations of molecular structure and spectra,computational chemistry, chemical instruction. Home: 1308 Castleton Rd N Columbus OH 43220 Office: 140 W 18th Ave Columbus OH 43210

PIVIK, RUDOLPH TERRY, psychologist, educator, researcher; b. Rock Springs, Wyo., Jan. 11, 1943; emigrated to Can., 1975; s. Rudolph Lewis and Flossie Mae (Sturgeon) P.; m. Margaret McGarvey, Sept. 8, 1962; children: Jane, Lisa, Suzanne, Christopher. Student, So. Methodist U., 1961-62, Western Wyo. Jr. Coll., 1962; B.A., U. Wyo., 1965; M.A. in Exptl. Psychology, U. Wyo., 1966; Ph.D. in Physiol. Psychology, Stanford U., 1970. Biosci. trainee Stanford U., 1966-70; NIMH research fellow in psychiatry Harvard U. Med. Sch., 1970-72; asst. prof., research assoc. depts. psychiatry and psychology U. Chgo., 1972-75; asst. prof. psychiatry U. Ottawa, Ont., Can., 1975-79, asst. prof. psychology, 1976-79, assoc. prof. psychiatry and psychology, 1979—; research assoc. psychology Carleton U., Ottawa, 1978—; dir. Clin. Psychophysiology Lab., Ottawa Gen. Hosp., 1975—. Contbr. chpts., articles, numerous abstracts to profl. publs.; mem. editorial bd.: Psychiat. Jour, U. Ottawa, 1976—. Ont. mental health research scholar, 1977-82; Ont Mental health research assoc., 1982-86; U. Ottawa Rector's Fund grantee, 1976-78; Med. Research Council Can. grantee, 1977-81; Sunnybrook Med. Centre and Sunnybrook Hosp.-U. Toronto Centre Funds grantee, 1977-78; Physicians' Services Incorporate Found. grantee, 1978-79; U. Ottawa Faculty Social Sci. grantee, 1979, 81-83; Ont. Mental Health Found. grantee, 1981-82, 83-85; U. Ottawa Dean's Funds grantee, 1981-82; Hosp. for Sick Children Found. grantee, 1982-84. Fellow Internat. Orgn. Psychophysiology; mem. Sleep Research Soc., Soc. Neurosci., Ottawa Neuroscis Soc. (treas. 1981-82, co-organizer ann. symposium for 1982), Can. Psychiat. Assn., Cans. for Health Research, Can. Coll. Neuropsychopharmacology (interim sec. 1979-80), Group-Without-A-Name Internat. Psychiat. Research Soc., Can. Psychol. Assn., Phi Beta Kappa, Sigma Xi. Democrat. Roman Catholic. Subspecialties: Psychophysiology; Neurophysiology. Current work: Biological psychiatry; psychophysiology and neurophysiology of sleep; neurophysiology and psychophysiology of mental illness and sleep. Office: Ottawa Gen Hosp 501 Smyth Rd Ottawa ON Canada K1H 8L6

PIZZICA, PHILIP ANDREW, nuclear engr.; b. Chgo., Sept. 22, 1945; s. Viro Manuel and Irene Mathilda (Peterson) P.; m. Hazel D. Burns, Apr. 5, 1975. B.S., U. Ill., Chgo., 1968, M.A., 1973. Nuclear Engr. Argonne (Ill.) Nat. Lab., 1967-71, nuclear engr., 1974—; systems analyst United Aircraft Research Lab., East Hartford, Conn., 1973-74. Mem. Am. Nuclear Soc. Subspecialties: Nuclear engineering; Nuclear fission. Current work: Creation and use of mathematical models to analyze accidents in fast breeder reactors, especially computer models. Office: Argonne Nat Lab 9700 S Cass Argonne IL 60439

PIZZO, SALVATORE VINCENT, pathology researcher, physician; b. Phila., June 22, 1944; s. George J. and Aida R. (Alcaro) P.; m. Carol A (Kurkowski), Dec. 28, 1968; children: Steven, David, Susan. B.S., St. Joseph's Coll., 1966; Ph.D., Duke U., 1972, M.D., 1973. Diplomate: Am. Bd. Pathology. Asst. prof. Duke U., Durham, N.C., 1976-80, assoc. prof. dept. pathology and biochemistry, 1980—; intern Duke U. Med. Ctr., 1973-74, resident, 1974-76; cons. Huntleigh Group, London, 1980—, Kaki Co., Stockholm, 1981—, Mitsubishi Co., Tokyo, 1981—, Dade Corp., Miami, 1980—. Contbr. articles to profl. jours. NIH med. scientist tng. program fellow, 1966-73; named Best Basic Sci. Tchr. Duke Med. Sch., 1980. Mem. Am. Chem. Soc., AAAS, Am. Heart Assn., Am. Assn. Pathologists, Am. Soc. Biol. Chemists, Am. Soc. Coagulationists, Sigma Xi, Alpha Sigma Nu, Phi Beta Kappa, Alpha Omega Alpha. Subspecialties: Biochemistry (medicine); Pathology (medicine). Current work: Study of blood coagulation and clinical diseases, host tumor interactions, regulation of proteases. Office: Duke U Med Center Box 3712 Durham NC 27710

PLAKOSH, PAUL, JR., psychologist, clinic administrator, researcher; b. Pitts., May 17, 1949; s. Paul and Leonora (Durso) P. Student, Case Western Res. U., 1967-70; B.S. summa cum laude, U. Pitts., 1973; M.A., U. Iowa, 1976; Ph.D., Palo Alto Sch. Profl. Psychology, 1979. Research psychologist U. Calif.-SanFrancisco Med. Ctr.-Langley Porter Inst., 1978-81; exec. dir. Franklin Clinic, San Francisco, 1980—. Contbr. articles to profl. jours. Mem. Internat. Neuropsychol. Soc., Am. Psychol. Assn., AAAS. Subspecialties: Behavioral psychology; Neuropsychology. Home: 291 Broderick St San Francisco CA 94117 Office: Franklin Clinic 2509 Bush St San Francisco CA 94115

PLASSE, TERRY FREEMAN, physician, researcher; b. Bklyn., Mar. 13, 1947; s. Herman and Sherley (Puner) P.; m. Barbara May Friedman, Feb. 1, 1970; children: Amitai, Ori, Eitan. A.B., Brandeis U., 1969; M.D., Washington U., St. Louis, 1973. Diplomate: Am. Bd. Internal Medicine. Resident in internal medicine Beth Israel Med Ctr., N.Y.C., 1973-76; fellow in neoplastic diseases Mt. Sinai Med. Center, N.Y.C., 1978-80, research fellow, 1980-81; pres. BioCell Tech. Corp., N.Y.C., 1981-83; med. dir. Interferon Scis. Inc., New Brunswick, N.J., 1983—; clin. instr. Mt. Sinai Med. Center, 1982—. Author articles on infusion pump tech., chemotherapy, cell culture. Served as surgeon, lt. comdr. USPHS, 1976-78. Mem AMA, Am. Soc. Clin. Oncology, ACP, Assn. Advancement of Med. Instrumentation. Subspecialty: Cell study oncology. Current work: Development of automated techniques for cell culture and testing of anti-cancer drugs. Office: Interferon Scis Inc 783 Jersey Ave New Brunswick NJ 08901

PLASTOCK, ROY A., mathematics educator; b. N.Y.C., Nov. 17, 1945; s. Irving M. and Agnes (Fertig) P.; m. Helen D. Sommer, Nov, 6, 1971 (div. 1980); 1 son, Adam. B.S., Bklyn. Coll., 1966; Ph.D., Yeshiva U., 1972. Research fellow U. Sussex, Eng., 1972-73; asst. prof. Cooper Union, N.Y.C., 1973-75; assoc. prof. math. N.J. Inst. Tech., Newark, 1975—. Research grantee Sci. Research Council Eng., 1973; recipient merit award N.J. Inst. Tech., 1981. Mem. Am Math. Soc., Soc. Indsl. and Applied Math. Jewish. Subspecialties: Applied mathematics; Artificial intelligence. Current work: Artificial intelligence, software engineering, graphics. Office: NJ Inst Tech 323 High St Newark NJ 07102

PLATE, JANET MARGARET-DIETERLE, scientist; b. Minot, N.D., Nov. 27, 1943; d. David and Bertha (Hoffer) Dieterle; m. Charles Alfred Plate, June 12, 1964; children: Jason, Stacey, Aileen. B.A., Jamestown Coll., 1964; Ph.D., Duke U., 1970. Am. Cancer Soc. postdoctoral fellow Mass. Gen. Hosp., Boston, 1970-72; research assoc. Harvard U./Mass. Gen. Hosp., Boston, 1972-78; asst. prof. Harvard Sch. Pub. Health, Boston, 1977-78; assoc. prof. Rush Med. Coll., Chgo., 1978—; sci. reviewer immunobiology study sect. NIH, Bethesda, Md., 1979-83, Ill. div. Am. Cancer Soc., Chgo., 1979—. Ruling elder Community Presbyterian Ch., Clarendon Hills, Ill., 1980-82. Nat. Cancer Inst./NIH grantee, 1975, 78—. Mem. Am. Assn. Immunologists, Am. Assn. Clin. Histocompatibility Testing, Transplantation Soc., Chgo. Assn. Immunologists (chmn. 1980-81). Presbyterian. Subspecialties: Immunobiology and immunology; Cancer research (medicine). Current work: Cellular immunobiologist, immunogeneticist, immune response to cancer specific antigens. Office: Rush Presbyn St Lukes Med Center 1753 W Congress Pkwy Chicago IL 60612

PLATSOUCAS, CHRIS DIMITRIOS, immunology educator and researcher; b. Athens, Greece, Apr. 17, 1951; came to U.S., 1973; s. Dimitrios and Maria (Tsonidis) P. B.S., U. Patras, Greece, 1973; postgrad., Purdue U., 1974-75; Ph.D., MIT, 1978. Research asst. MIT, 1975-78; research fellow Meml. Sloan-Kettering Cancer Ctr., N.Y.C., 1978-79, research assoc., 1979-80, asst. mem., 1980—, asst. prof., 1981—, head lab. biol. response modifiers, 1981—; spl. rev. study sect. NIH, Bethesda, Md., 1982. Contbr. articles to profl. jours. Nat. Research Service awardee NIH, 1978-79; Am. Cancer Soc. grantee, 1980—; NIH grantee, 1982—; NSF grantee, 1982—. Mem. Am. Assn. Immunologists, Am. Assn. Cancer Research, Am. Soc. Hematology, Biophys. Soc., N.Y. Acad. Scis., Sigma Xi. Subspecialty: Immunobiology and immunology. Current work: Human immunology, cytotoxic T cells, monoclonal antibodies, lymphoproliferative disorders, T-T cell hybrids. Office: Memorial Sloan Kettering Cancer Center 1275 York Ave New York NY 10021

PLATT, ALLISON MICHAEL, nuclear technologist; b. Schenectady, Jan. 4, 1922; m. Marinel Dean., Sept. 2, 1945; children: Terri Linda, Victoria Gail. B.S., Carnegie Inst. Tech., 1943; M.S., Tex. A&M U., 1950. Engr., sr. engr. Gen. Electric Co., Richland, Wash., 1951-58; mgr. chem. devel. and waste calcination dept. Battelle Pacific N.W. Lab., Richland, 1958-64, mgr. chem. tech. dept., 1965-72, mgr. nuclear tech., 1972—. Editor jour.: Radioactive Waste Mgmt, 1980—; contbr. to ency. Fellow Am. Nuclear Soc. (chmn. nuclear fuelcycle div. 1974-75, recipient spl. award for outstanding achievements in nuclear waste isolation tech., 1982), Tau Beta Pi. Subspecialties: Nuclear engineering; Nuclear engineering. Current work: Nuclear fuel cycle activities. Home: 2401 Alexander St Richland WA 99352 Office: Battelle Pacific NW Lab PO Box 999 Richland WA 99352

PLATZER, EDWARD GEORGE, parasitologist; b. Vancouver, C., Can., Oct. 3, 1938; came to U.S., 1965; s. George and Diane Christine (Bailey) P.; m. Anne Colville Cooper, June 30, 1962; 1 dau., Linda. Ph.D. in Zoology, U. Mass., 1968. Asst. prof. dept. nematology U. Calif.-Riverside, 1971-76, assoc. prof., 1977-82, prof., 1982—. Contbr. articles to profl. jours. Mem. Am. Soc. Parasitologists, Soc. Nematologists, So. Calif. Parasitologists (sec.-treas. 1979—), Sigma Xi. Subspecialties: Parasitology; Physiology (biology). Current work: Host-parasite interactions; nutrition and physiology of parasites; biological control of insects. Office: Dept Nematology U Calif Riverside CA 92521

PLAUT, ANDREW GEORGE, physician, researcher; b. Leipzig, Ger., Feb. 19, 1937; s. Otto L. and Johanna (Lowenstein) P.; m. Linda Fields, June 23, 1965; children: Julie, John. B.S., Ohio State U., 1958; M.D., Tufts U., 1962. Diplomate: Am. Bd. Internal Medicine. Intern Bellevue Hosp. Cornell U. and Meml. Hosp. for Cancer, N.Y.C., 1962-63, resident, 1963-64, fellow, 1964-65; asst. prof. SUNY, Buffalo, 1968-73; physician, prof. medicine Tufts U.-New Eng. Med. Ctr. Hosp., 1973—; cons. NIH. Contbr. articles to profl. jours. Served to capt. U.S. Army, 1966-68. Mem. Am. Soc. Clin. Investigation, Am. Soc. Microbiology, Infectious Disease Soc. Am., Am. Gastroent. Assn., ACLU, Union Concerned Scientists. Subspecialties: Infectious diseases; Gastroenterology. Current work: Immunity of bacterial infection, subversion of immunity by bacteria, enzymology. Office: 171 Harrison Ave Boston MA 02111

PLAUT, SIMON MICHAEL, psychology educator; b. N.Y.C., July 8, 1941; s. Isidore Irving and Rae (Katz) P.; m. Judith Moore, Aug. 3, 1968. B.A., Adelphi U., 1965; Ph.D., U. Rochester, 1969. Lic. psychologist, Md. Research scientist Galesburg (Ill.) State Research Hosp., 1968-73; assoc. prof., asst. dean U. Md. Sch. Medicine, Balt., 1973—; mem. State Bd. Examiners of Psychologists, Md., 1982—; councilor Am. Psychosomatic Soc., 1982—. Contbr. articles to profl. jours. Mem. Internat. Soc. Developmental Psychobiology, Animal Behavior Soc., Am. Psychosomatic Soc., AAAS, Am. Psychol. Assn. Democrat. Jewish. Subspecialties: Psychobiology; Developmental psychology. Current work: Psychosomatic medicine, animal behavior and development, research methodology, student administration, human sexuality. Office: Univ. Md Sch Medicine 645 W Redwood St Baltimore MD 21201

PLAUT, THOMAS FRANZ ALFRED, research psychologist; b. N.Y.C., Dec. 29, 1925; s. Alfred and Margaret (Blumenfeld) P.; m. Evelyn Z. McPurhoff, Dec. 26, 1950 (div. Sept. 1976); children: Melanie, Tony, Jeffrey, Daphne, Iris, Roger; m. Bonnie A. Cox, Nov. 27, 1976; stepchildren: Carole, Susan. B.A., Swarthmore Coll., 1949; Ph.D., Harvard U., 1955, M.P.H., 1956. Diplomate: Md. Bd. Psychologists. Research assoc. Stanford U., 1962-67; asst. chief Alcoholism ctr. NIMH, HEW, Bethesda, Md., 1969-71; dir. div. manpower and tng. programs NIMH, Bethesda, 1969-71, assoc. dir., program coordinator, Rockville, Md., 1971-72, counselor to dr., 1972-74, dep. dir., 1974-79, dir., 1979-80, assoc. dir. div. mental health services, 1980-82, research psychologist, research div., 1982—; lectr., sr. assoc. Johns Hopkins Sch. Health, Balt., 1967-71, 81—. Co-author: Personality in Communal Society, 1956, Alcohol Problems, 1967; author: The Treatment of Alcoholism, 1967; mem. editorial bd.: Evaluation Rev, 1979-82; assoc. editor: Jour. Health and Social Behavior, 1968-71; contbr. articles to profl. jours. Bd. dirs. Bannockburn Community Club, Bethesda, 1979-82. Served with U.S. Army, 1944-46. Recipient Sec.'s Spl. Citation HEW, 1971. Fellow Am. Psychol. Assn. (Cert. of Recognition 1982); mem. Am. Publ. Health Assn. (life), Am. Sociol.Assn., Phi Beta Kappa, Delta Omega. Subspecialties: Clinical psychology; Social psychology. Current work: Behavioral medicine, impact of lifestyles on health; application of knowledge regarding health and behavior. Home: 6410 W Halbert Rd Bethesda MD 20817 Office: NIMH 5600 Fishers Ln Rockville MD 20857

PLAVEC, MIREK JOSEF, astronomer; b. Sedlcany, Bohemia, Czechoslovakia, Oct. 7, 1925; came to U.S., 1970, naturalized, 1981; s. Antonin and Anezka (Beranova) P.; m. Zdenka Bazikova, Mar. 15, 1930; children: Helena Kirkpatrick, Jirka G. B.Sc., Charles U., Prague, 1949, Ph.D., 1955, D.Sc., 1968. Prin. sci. officer Astron. Inst. Ondrejov, Czechoslovak Acad. Scis., 1965-69; prof. astronomy UCLA, 1970—, chmn. dept., 1975-77. Author: Komety a Meteory, 1957, Clovek a Hvezdy, 1960, (with Popper and Ulrich) Close Binary Stars: Observations and Interpretation, 1980. Mem. Internat. Astron. Union (pres. commn. 42 1970-73), Am. Astron. Soc., Astron. Soc. Pacific. Subspecialties: Ultraviolet high energy astrophysics; Theoretical astrophysics. Current work: Studies of interacting binary stars; their evolution, structure, spectra in ultraviolet and optical regions; observations and theoretical modeling. Home: 767 Jacon Way Pacific Palisades CA 90272 Office: UCLA 8907 MS Bldg Los Angeles CA 90024

PLAZEK, DONALD JOHN, polymer engineering educator; b. Milw., Jan. 12, 1931; s. Stanley and Marian (Parker) P.; m. Patricia Lenore Filkins, Oct. 29, 1955; children: Mary, Joseph, Caroline, Daniel, John, David, Anne. B.S., U. Wis., 1953, Ph.D., 1957. Postdoctoral research asst. U. Wis.-Madison, 1957-58; research fellow Mellon Inst., Pitts., 1958-67; assoc. prof. polymer engrng. U. Pitts., 1967-74, prof., 1975—; Sr. vis. research fellow U. Glasgow, Scotland, 1976-77. Fellow Am. Phys. Soc.; mem. Am. Chem. Soc., Soc. Rheology. Democrat. Roman Catholic. Subspecialties: Polymer engineering; Physical chemistry. Current work: Physical properties of polymers; structure-property correlations; viscoelastic behavior; instrument design; properties of organic glasses. Home: 34 N Harrison Ave Pittsburgh PA 15202 Office: Dept Metall and Materials Engring U Pitts 848 Benedum Hall Pittsburgh PA 15261

PLEBUCH, RICHARD KARL, aerospace manager, nuclear engineer; b. Longview, Wash., Sept. 7, 1935; s. Donald Wiley and Bernadine Perkins (Chitty) P.; m. Barbara Joyce Jacobson, Sept. 22, 1956; children: Donald Andrew, Sharyl Lynn, Ronald Paul, Karolyn Joyce. Sc. D. in Nuclear Engring, MIT, 1963, M.S., U. Mich., 1958; B.S. in Chem. Engring, U. Wash., 1957. Engr. Boeing Aerospace Co., Seattle, 1957; mem. tech. staff Space Tech. Labs., Redondo Beach, Calif., 1963-65; sect. head Systems Group, TRW, Redondo Beach, 1965-68; dept. mgr. Def. Systems Group, 1968—; lectr. N.Y. Acad. Scis., 1967-69, UCLA-Westwood, 1968. Contbr. articles on manned Mars landing mission to profl. jours., 1965-68, on nuclear rocket propulsion, 1967-69. Asst. scoutmaster Los Angeles Area council Boy Scouts Am., 1969-72, scoutmaster, 1972-82; camping chmn. South Bay dist., 1977-80. Served to 1st lt. Chem. corps U.S. Army, 1958-60. Recipient award of merit Boy Scouts Am., 1979; Silver Beaver award, 1980. Fellow AIAA (assoc.); mem. Am. Nuclear Soc., Sigma Alpha Epsilon. Republican. Episcopalian. Lodge: Lions. Subspecialties: Nuclear engineering; Systems engineering. Current work: Space radiation environments, radiation shielding, nuclear radiation environments and effects, missile and space system design. Patentee nuclear reactor application, 1965. Home: 28563 Blythewood Dr Rancho Palos Verdes CA 90274 Office: TRW Def Systems Group One Space Park Redondo Beach Ca 90278

PLECK, JOSEPH HEALY, social science researcher, educator writer; b. Evanston, Ill., July 14, 1946; s. Joseph Harold and Katherine (Healy) P.; m. Elizabeth H. Hafkin, June 14, 1968. B.A., Harvard U., 1968, M.A., 1971, Ph.D., 1973. Asst. research scientist Inst. for Social Research, U. Mich., Ann Arbor, 1974-77; assoc. dir. Ctr. for Family, U. Mass.-Amherst, 1977-78; program dir. Ctr. for Research on Women, Wellesley (Mass.) Coll., 1978—. Author: The Myth of Masculinity, 1981; editor: Men and Masculinity, 1974, The American Man, 1980. NIMH grantee, 1976-77, 79-81; Adminstrn. for Children, Youth and Families grantee, 1979-80; NSF grantee, 1979-81. Fellow Am. Psychol. Assn.; mem. Am. Sociol. Assn. (exec. com. sect. on sex and gender 1978-80), Nat. Council on Family Relations. Subspecialties: Social psychology; Behavioral psychology. Current work: Male roles; work and family roles. Office: Center for Research on Women Wellesley Coll Wellesley MA 02159

PLESHKEWYCH, ALEXANDER, biology educator, cancer research consultant; b. Lviv, Ukraine, USSR, Nov. 3, 1939; came to U.S., 1949; s. Omelan and Emilia (Romanowsky) P.; m. Johanna A. Diakiw, Sept. 6, 1966; children: Mark, Sonya, Alexander. B.S. in Zoology, Kans. State U., 1963; M.S. in Biology, Emporia (Kans.) State U., 1964, Ph.D., Wayne State U., 1970. Lab. asst. Emporia State U., 1963-64; instr. Wayne State U., Detroit, 1967-68, U. Mich., Dearborn, 1969; prof. biology Daemen Coll., Amherst, N.Y., 1970—; prof. Attica (N.Y.) Consortium, 1976-83; research cons. Roswell Park Meml. Inst., Buffalo, 1979-83. Contbr. articles in field to sci. Publs. Exec. mem. Ukrainian Congress Com., Buffalo, 1974-83. NSF predoctoral research grantee, 1963-69; named outstanding faculty Daemen Coll., 1977. Mem. Ukrainian Med. Assn. Subspecialties: Cell biology; Cancer research (medicine). Current work: Research in cancer drugs and their effects on cell structures. Home: 296 Seabrook Williamsville NY 14221 Office: Daemen Coll 4380 Main St Amherst NY 14226

PLESKOT, LARRY KENNETH, planetary scientist, systems programmer, computer systems designer, investment cons.; b. East St. Louis, Ill., Oct. 1, 1949; s. Kenneth Joseph and Phyllis Earlene (Wilhoyte) P. Researcher Research Lab. Electronics, M.I.T., 1970-71; teaching asst. dept. geophysics and space physics UCLA, 1975-76, research asst. dept. planetary and space sci. and dept. geophysics and space physics, 1973-78, research assoc. dept. earth and space sci., 1978-80; scientist Jet Propulsion Lab., Pasadena, Calif., 1979—; v.p., ops. mgr. Comtramark, Westwood, Calif., 1981—; cons. securities, remote sensing, efficiency, environ. engring. Contbr. numerous articles to profl. jours. Mem. Am. Geophys. Union, Am. Astron. Soc., AAAS, Am. Inst. Physics. Mem. Christian Ch. (Disciples of Christ). Subspecialties: Planetary science; Remote sensing (atmospheric science). Current work: Lab. and planetary reflectance spectroscopy, spectrogoniometry, infra-red radiometry, mineral exploration via remote sensing, minicomputer systems design and operation, fin.

markets modeling. Home: 1380 Midvale Ave Apt 405 Los Angeles CA 90024 Office: 1516 Westwood Blvd Suite 203 Los Angeles CA 90024

PLETCHER, RICHARD HAROLD, mechanical engineering educator, researcher; b. Elkhart, Ind., May 21, 1935; s. Raymond Harold and Annabelle (Aur) P.; m. Carol Jean Robbins, June 9, 1957; children: Douglas, Laura, Cynthia. B.S., Purdue U., 1957; M.S., Cornell U., 1962, Ph.D., 1966. Diplomate: Registered profl. engr., Iowa. Instr. Cornell U., Ithaca, N.Y, 1961-64; sr. research engr. United Aircraft Research Lab., East Hartford, Conn., 1965-67; asst. prof. Iowa State U., Ames, 1967-70, assoc. prof., 1970-76, prof. mech. engring., 1976—; cons. Caterpillar Tractor, Peoria, Ill., 1971-74, Arnold Research Orgn., Arnold AFB, Tenn., 1977, Gen. Electric Co., Phila., 1980-82. Author: (with D. Anderson, J. Tannehill) Computational Fluid Mechanics and Heat Transfer, 1984; tech. editor: Jour. Heat Transfer, 1981—; mem. editorial adv. bd.: Numerical Heat Transfer, 1980—; contbr. articles in field to profl. jours. Com. mem. Boy Scouts Am., Ames, Iowa, 1975; bd. deacons Collegiate Presbyn. Ch., Ames, 1975. Served to lt. (j.g.) USN, 1957-60. Grantee NSF, NASA, Army Research Office. Mem. ASME (com. chmn. 1980-83), AIAA, Am. Soc. Engring. Edn., AAUP, Sigma Xi. Club: Osborn Research (Ames, Iowa). Subspecialties: Mechanical engineering; Fluid mechanics. Current work: Research in heat transfer and fluid mechanics, application of computational methods to predict complex lamina and turbulent flows, turbulence modeling. Home: 411 Oliver Ave Ames IA 50010 Office: Dept Mech Engring Iowa State U Ames IA 50011

PLOCKE, DONALD JOSEPH, biologist, educator; b. Ansonia, Conn., May 5, 1929; s. Joseph Alvin and Stella Rose (Loda) P. B.S. in Physics (Charles H. Pine scholar), Yale U., 1950; M.A. in Philosophy, Boston Coll., 1956; Ph.D. in Biophysics (NIH fellow 1957-59, NSF fellow 1959-61), M.I.T., 1961. Lic. Theol., Weston Coll., 1965. Joined S.J., Roman Catholic Ch., 1950, ordained priest, 1964; asst. prof. biology Boston Coll., 1966-71, assoc. prof., 1971—, chmn. dept. biology, 1971-80; Consultor New Eng. Province S.J., 1976-81. Trustee Cheverus High Sch., Portland, Maine, 1978-84. Mem. AAAS, Am. Soc. Plant Physiologists, Biophys. Soc. (charter), Sigma Xi. Subspecialties: Biophysics (biology); Molecular biology. Current work: Role of metal ions in enzyme and nucleic acid function. Home: 246 Beacon St Chestnut Hill MA 02167 Office: Dept Biology Boston Col Chestnut Hill MA 02167

PLONSEY, ROBERT, electrical and biomedical engineer, educator; b. N.Y.C., July 17, 1924; s. Louis B. and Betty (Vinograd) P.; m. Vivian V. Vucker, Oct. 1, 1948; 1 son, Daniel. B.E.E., Cooper Union, 1943; M.S. in Elec. Engring, NYU, 1948; Ph.D., U. Calif. at Berkeley, 1955; postgrad., Med. Sch., Case Western Res. U., 1969-71. Registered profl. engr., Ohio. Asst. prof. elec. engring. U. Calif. at Berkeley, 1955-57; asst. prof. elec. engring. Case Inst. Tech., Cleve., 1957-60, assoc. prof., 1960-66, prof., 1966-68, dir. bioengring. group, 1962-68; prof. biomed. engring. Sch. Engring. and Sch. of Medicine, Case Western Res. U., 1968-83, chmn. dept., 1976-80; vis. prof. biomed. engring. Duke U., 1980-81, prof. biomed. engring., 1983—; Mem. biomed. fellowships rev. com. NIH, 1966-70; mem. tng. com. Engr. in Medicine and Biology, 1972-73, cons., 1974—, NSF, 1973; mem. com. on electrocardiography Am. Heart Assn., 1976-82. Author: (with R. Collin) Principles and Applications of Electromagnetic Fields, 1961, Bioelectric Phenomena, 1969; with J. Liebman and P. Gillette Pediatric Electrocardiography, 1982; with T. Pilkington Engineering Contributions to Biophysical Electrocardiography, 1982; Mem. editorial bd.: Trans. IEEE Biomed. Engring, 1965-70; asso. editor, 1977-79; editorial bd.: T.I.T. Jour, 1971-81, Electrocardiology Jour, 1974—; proc. editor: Engring. in Medicine and Biology, 17th Ann. Conf, 1964; co-editor: (with R. Collin) Biomed. Engring. Series. Vice pres. Your Schs., Cleveland Heights, Ohio, 1968-69, 73-75; provisional trustee Am. Bd. Clin. Engrs., 1973-74, pres., 1975, trustee, 1976—. Served with AUS, 1944-46. NIH sr. postdoctoral award, 1980-81. Fellow IEEE (chmn. Cleve. chpt. group on biomed. electronics 1962-63, chmn. publs. com. group on engring. in medicine and biology 1968-70, v.p. adminstrv. com. 1970-72, pres. 1973-74, chmn. fellows com. 1977-79, William S. Morlock award 1979); mem. Alliance for Engring. in Medicine and Biology (treas. 1976-78), Biophys. Soc., Biomed. Engring. Soc. (dir. 1975-78, 79—, pres. 1981-82), AAUP, Am. Physiol. Soc., Am. Soc. Engring. Edn. (dir. biomed. engring. div. 1978-83, chmn. 1982-83), AAAS. Subspecialties: Biomedical engineering; Cardiac electrophysiology. Current work: Biomedical engineering educator; research on bioelectricity including cardiac electrophysiology and electrocardiography and mathematical modelling. Office: Duke U Durham NC 27707 External recognition of success is not nearly so important as the inner awareness of coming to full grips with life, to be fully involved, bending all strengths to fulfill one's goals and philosophies. And of all involvements, those with people are most meaningful (to be aware of and share the feelings of colleagues, students, friends, and family—and to enrich these relationships)—and for me most difficult.

PLOTKA, EDWARD DENNIS, endocrinologist; b. Utica, N.Y., Oct. 10, 1938; s. Maxim Jay and Marian Vivian (LaPoten) P.; m. Marie Christina Fischer, July 2, 1966. B.S., Delaware Valley Coll., 1960; M.S., Oreg. State U.-Corvallis, 1963; Ph.D., Purdue U., 1966. Research asst. Oreg. State U., Corvallis, 1961-63; research asst. Purdue U., Lafayette, Ind., 1963-66, asst. prof., 1966-67; asst. research prof. U. Ga., Athens, 1967-69; sr. scientist Marshfield (Wis.) Med. Found., 1969—. Pres. Wildwood Park Zool. Soc., Marshfield, Wis., 1982-83; bd. dirs. North Wood County Humane Soc., Marshfield, 1980-83. Purdue Research Found. fellow, 1963-66. Mem. Soc. Study Fertility, Soc. Exptl. Biology and Medicine, Soc. Study Reprodn., Am. Physiol. Soc., Endocrine Soc. Lodge: Elks. Subspecialties: Neuroendocrinology; Reproductive biology. Current work: Pineal and ovarian function, reproductive hormone receptors, circannual rhythms. Office: Marshfield Med Found 510 N St Joseph Ave Marshfield WI 54449

PLOTKIN, ALLEN, aerospace engineering educator; b. N.Y.C., May 4, 1942; s. Oscar and Claire (Chasick) P.; m. Selena Berman, Dec. 18, 1966; 1 dau., Samantha Rose. B.S., Columbia U., 1963, M.S., 1964; Ph.D., Stanford U., 1968. Asst. prof. U. Md., College Park, 1968-72, assoc. prof., 1972-77, prof. dept, aerospace engring., 1977—; vis. assoc. in engring. sci. Calif. Inst. Tech., Pasadena, 1975-76; cons. Naval Surface Weapons Ctr., White Oak, Md., 1980—. Research grantee NASA, 1970-73, 80—, NSF, 1975-77; recipient 1981 award in engring. scis. Washington Acad. Scis., 1982. Assoc. fellow AIAA (young engr. sci. award Nat. Capitol Sect. 1976); mem. ASME, Soc. Naval Architects and Marine Engrs. Subspecialties: Aeronautical engineering; Fluid mechanics. Current work: Fluid mechanics, aerodynamics. Home: 806 Malta Ln Silver Spring MD 20901 Office: Dept Aerospace Engring Univ Md College Park MD 20742

PLOTKIN, GARY ROBERT, internist, infectious disease specialist; b. Jersey City, June 2, 1944; s. Charles and Rose (Wexler) P. B.S., B.Ph., Rutgers U., 1967; M.D., N.J. Coll. Medicine, Newark, 1972. Intern Kings County Hosp.-Downstate Med. Ctr., Bklyn., 1972-73, resident in internal medicine, 1973-75; fellow in infectious disease U. Fla.-Shands Teaching Hosp., Gainesville, 1974-76, Bellevue Hosp.-NYU, N.Y.C., 1976-77; internist, infectious disease specialist, assoc. Geisinger Med. Ctr., Danville, Pa., 1977-81; internist, infectious disease specialist Trover Clinic, Madisonville, Ky., 1981-82; internist, cons. Danville Internal Medicine, Inc., 1982—. Contbr. articles to profl. jours., chpts. in book. Fellow ACP; mem. Am. Soc. Microbiology, Am. Thoracic Soc., N.Y. Acad. Scis., Sigma Xi (assoc.). Jewish. Subspecialties: Internal medicine; Infectious diseases. Home: PO Box 377 Danville VA 24541 Office: Danville Internal Medicine 115 S Main St Danville VA 24541

PLOTKIN, KENNETH JAY, research engineer; b. N.Y.C., Apr. 13, 1945; s. Moe and Sarah (Kravitz) P.; m. Barbara Bernice Zesk, Aug. 19, 1972; 1 dau., Sarah. B.S., Poly. Inst. Bklyn., 1965; M.Aero. Engring., Cornell U., 1966, Ph.D., 1971. Research asst. Poly. Inst. Bklyn., 1964-65; asst, scientist AVCO Everett (Mass.) Research Lab., 1966; research assoc. Cornell U., Ithaca, N.Y., 1965-70; sr. research scientist Aerodyne Research, Inc., Burlington, Mass., 1970-71; sr. staff research specialist Wyle Labs., Arlington, Va., 1972—. Contbr. papers to profl. publs. and confs. Regents scholar N.Y., 1961-65; NASA trainee, 1965-68. Mem. AIAA, Acoustical Soc. Am., Soc. Automotive Engrs. Subspecialties: Acoustics; Aeronautical engineering. Current work: Acoustics, transportation noise, structural vibration, fluid dynamics, unsteady flows, turbulence. Home: 9511 Liberty Tree Ln Vienna VA 22180 Office: Wyle Labs 2361 Jefferson Davis Hwy Suite PL 404 Arlington VA 22202

PLOTNICK, GARY DAVID, cardiology educator, researcher; b. Balt., Nov. 11, 1941; s. Alvin Bernard and Evelyn Ruth (Altschul) P.; m. Leslie Karol Parker, Feb. 11, 1967; children: Michael Aaron, Daniel Brian. B.A., Johns Hopkins U., 1962; M.D., U. Md., 1966. Intern U. Md. Hosp., Balt., 1966-67, jr. asst. resident, 1967-68, sr. asst. resident, 1970-71; chief med. resident U. Md. and VA Hosp., Balt., 1971-72; cardiology research fellow Johns Hopkins Hosp., Balt., 1972-74; asst. prof. medicine U. Md. Sch. Medicine, Balt., 1974-78, assoc. prof., 1978—, asst. dean student affairs, 1975—; instr. medicine John Hopkins Sch. Medicine, Balt., 1974-79, asst. prof., 1979—; dir. cardiac graphics lab. Balt. VA Med. Ctr., 1974—, dir. non-invasive lab., 1974—, assoc. chief staff for research (acting), 1981-82, dir. cardiolgy edn., 1983—. Served to lt. commdr. USN, 1968-70; Vietnam. Recipient Humanitarian award U. Md. Sch. Medicine, 1982. Fellow Council Clin. Cardiology (Am. Heart Assn.), Am. Coll. Cardiology, ACP; mem. Am. Fedn. Clin. Research, Am. Soc. Echocardiography, Alpha Omega Alpha. Democrat. Jewish. Subspecialty: Cardiology. Current work: Diagnosis and treatment of clinical subgroups of ischemic heart disease, particularly unstable angina, evaluation of cardiac medications such as newer beta blockers and calcium blockers. Home: 7918 Winterset Rd Baltimore MD 21208 Office: Balt VA Med Center 3900 Loch Raven Blvd Baltimore MD 21218

PLOTSKY, PAUL MITCHELL, biomed. researcher and educator; b. Kansas City, Mo., Feb. 1, 1952; s. Herbert and Frances (Kern) P.; m. Andrea Gayle, Jan. 10, 1953; children: Melissa Michelle, Alyson Rose. B.A., U. Kans., 1974; Ph.D., Emory U., 1980. Teaching fellow Brown U., Providence, 1980-82; research assoc. R.I. Hosp., Providence, 1980-82; asst. prof. Salk Inst., San Diego, 1982—; cons. biomed., instrument and pharm. mfrs. Contbr. articles to sci. jours. Mellon Found. grantee, 1982—. Mem. AAAS, Endocrine Soc., Soc. Neurosci. Subspecialties: Neuroendocrinology; Neurophysiology. Current work: Neural control of hypohysiotrophic factors regulating anterior pituitary gland hormone secretions. Office: PO Box 85800 San Diego CA 92138

PLOTZ, PAUL HUNTER, physician; b. N.Y.C, Oct. 19, 1937; s. Milton B. and Helen D. (Ratnoff) P.; m. Judith A. Plotz, Sept. 1, 1963; children: John, David. A.B., Harvard U., 1958, M.D., 1963. Intern and resident in medicine Beth Israel Hosp., Boston, 1963-65; clin. assoc. NIH, Bethesda, Md., 1965-68; Helen Hay Whitney Found./Nat. Inst. Med. Research fellow, London, 1968-70; sr. investigator arthritis and rheumatism br. Nat. Inst. Arthritis, Diabetes, Digestive and Kidney Diseases, NIH, Bethesda, 1970—; clin. prof. medicine U.S. Univ. Health Scis., Bethesda, 1980—. Recipient prize French Soc. Rheumatology, 1981. Mem. Am. Soc. Clin. Investigation, ACP, Am. Assn. Immunology, Arthritis and Rheumatism Assn. Subspecialties: Immunology (medicine); Internal medicine. Home: 3221 Livingston St NW Washington DC 20015 Office: 10/9N244 NIH Bethesd MD 20205

PLUNKETT, ROBERT, mech. engr., educator; b. N.Y.C., Mar. 15, 1919; s. Charles Robert and Helen Rebecca (Edwards) P.; m. Helen Catharine Bair, May 11, 1946; children—Christopher Robert, Brian Charles, Margaret Louise. B.C.E., Mass. Inst. Tech., 1939, Sc.D. in Mech. Engring, 1948; Docteur (hon.), U. Nantes, France, 1966. Research asst. in elec. engring. Mass. Inst. Tech., Cambridge, 1939-41, instr. mech. engring., 1941, asst. prof., 1946-48; asst. prof. mech. engring. Rice U., Houston, 1948-51; cons. engr. acoustics and mechanics Gen. Electric Co., Schenectady, 1951-60; prof. applied mechanics U. Minn., Mpls., 1960—; cons. engr. Hughes Tool Co., Hamilton Standard Co., Ford Co., USN, U.S. Army, USAF, Honeywell Corp., Control Data Corp., Worthington Corp., Westinghouse Co. Served to maj. C.E. AUS, 1942-46. Fulbright scholar, Nantes, 1964, Technion, Israel, 1971. Fellow ASME, AAAS, Acoustical Soc. Am., Am. Inst. Aeros. and Astronautics (asso.); mem. Nat. Acad. Engring. Subspecialties: Materials; Theoretical and applied mechanics. Current work: Visocelostic behavior and damage in composite materials, wave propogation in solids, solid-fluid interaction. Home: 3122 W Owasso Blvd Saint Paul MN 55112 Office: 107 Akerman Hall U Minn 110 Union St SE Minneapolis MN 55455

PLUNKETT, ROBERT DALE, oceanographer, consultant; b. Raub, Ind., Jan. 28, 1925; s. George Peter and Minnie (Watson) P.; m. Barbara Ann Kubly, Apr. 3, 1949; children: Mele Cathleen, Paula Ann. M.S., Scripps Instn. UCLA, 1952; B.S., Duke U., 1946. Commd. U.S. Navy, 1943, advanced through grades to comdr., 1962; ret., 1964; mgr. studies dept. Bendix Marine Advisers, La Jolla, Calif., 1964-69; oceanog. cons., San Diego, 1969-70; mgr. oceanography Dillingham Corp., San Diego, 1970-71; v.p. Woodward Clyde Cons., San Diego, 1971—; assoc., marine adv. com. San Diego Mesa Coll., 1971—. Mem. Clairemont Planning Com., San Diego, 1968-72. Fellow Marine Tech. Soc.; mem. Naval Inst., Ret. Officers Assn. Democrat. Club: Convair Rockhound (bd.control 1978-80). Subspecialties: Oceanography; Environmental engineering. Current work: Russian oceanography; ship characteristics; ocean instrumentation; environmental impact analysis; coastal processes; gemology; gem faceting; naval history. Office: Woodward Clyde Cons 3489 Kurst St San Diego CA 92110

PLYMATE, STEPHEN REX, endocrinologist, army officer, educator; b. Omaha, Aug. 7, 1943; s. Oliver and Lenore Marie (Monahan) P.; m. Lisa Catherine Goldiamono, Dec. 2, 1978; children: Stephanie, Duncan, Sarah, Corinne. M.S., M.D., U. Nebr., 1968. Diplomate: Am. Bd. Internal Medicine, 1972. Commd. lt. col. U.S. Army, 1968; sr. fellow endocrinology Madigan Army Med. Ctr., Tacoma, 1971-73; chief endocrine clinic William Beaumont Army Med. Ctr., El Paso, Tex., 1973-75, Fitzsimons Army Med. Ctr., Denver, 1975-76; active staff Lovelace Bataan Med. Ctr., Albuquerque, 1976-78; asst. dir. endocrinology U. N. Mex., Albuquerque, 1976-78; chief clin. studies service Madigan Army Med. Ctr., Tacoma, 1978—; assoc. dir. endocrine fellowship program, clin. asst. prof. medicine U. Wash. Sch. Medicine, Seattle; asst. prof. medicine U. N. Mex., Albuquerque, 1973-77. Reviewer: Jour. AMA; contbr. numerous articles to sci. publs. Adv. Am. Youth Diabetes Assn., 1978, Blue Cross-Blue Shield Med. Adv. com. mem., 1976-77, Am. Heart Assn. grant review com., 1976-78, Am. Heart Assn. grant review com., Albuquerque. Recipient U. Nebr. Dr. David Richardson Ob-Gyn Senior Student award, 1968. Mem. Endocrine Soc., Pacific Coast Fertility Soc., Am. Fertility Soc., ACP (assoc.), Am. Fedn. Clin. Research, Alpha Delta Omicron, Alpha Omega Alpha. Subspecialties: Endocrinology; Internal medicine. Current work: research in reproductive endocrinology; regulation of sex hormone binding globulin and testicular steroidogenesis. Office: Dept Clinical Investigation Madigan Army Med Center Box 99 Tacoma WA 98431 Home: 2820 N Warner Tacoma WA 98407

POCCIA, DOMINIC LOUIS, biology educator; b. Utica, N.Y., Aug. 8, 1945; s. Louis Joseph and Frances (Surace) P.; m. Alison Gordon, June 12, 1971 (div. 1980); 1 son, Joseph Dominic. B.S., Union Coll., 1967; M.A., Harvard U., 1968, Ph.D., 1971. Postdoctoral fellow Harvard U., Cambridge, Mass., 1971-72; asst. prof. Wellesley (Mass.) Coll., 1971-72; postdoctoral fellow U. Calif.-Berkeley, 1972-74; asst. prof. SUNY-Stony Brook, 1974-78; assoc. prof. dept. biology Amherst (Mass.) Coll., 1978—. NIH grantee, 1975—; postdoctoral fellow, 1972-74; predoctoral fellow, 1967-71; NSF grantee, 1982. Mem. Am. Soc. Cell Biology, Soc. Developmental Biology, AAAS, N.Y. Acad. Sci. Subspecialties: Developmental biology; Molecular biology. Current work: Activation of gene expression, chromation structure, fertilization, centrioles. Home: 83 Woodside Ave Amherst MA 01002 Office: Dept Biology Amherst Coll Amherst MA 01002

POCHAN, JOHN MICHAEL, research scientist, educator; b. Kittanning, Pa., Apr. 8, 1942; s. John and Pauline Carol (Trop) P.; m. Darlyn Dawn Faulkner; children: Lisa, Darrin, Sommyr, Shawna. B.Chem. Engring., Rensselaer Poly. Inst., 1964; M.S., U. Ill.-Urbana, 1967, Ph.D., 1969. Sr. scientist Xerox Corp., Rochester, N.Y., 1969-81; sr. research assoc. Kodak Research, Rochester, 1981—; adj. faculty mem. U. Rochester, 1981—. Contbr. articles to profl. jours. Bd. dirs. Webster (N.Y.) Child Care Ctr., 1980—. Mem. Am. Chem. Soc., Am. Phys. Soc., Soc. Plastic Engrs. (dir. polymer div. 1982—). Democrat. Roman Catholic. Subspecialties: Polymers; Polymer engineering. Current work: Polymer physics, structure property relationships, conducting polymers, dielectric relaxation, adhesives; liquid crystals, structure property relationships. Patentee in field. Home: 1085 Willits Rd Ontario NY 14519 Office: Eastman Kodak Bldg 81 Kodak Park Rochester NY 14650

PODOWSKI, MICHAEL ZBIGNIEW, nuclear engineering educator, researcher; b. Warsaw, Poland, May 15, 1940; came to U.S., 1979; s. Roman Damazy and Halina Eugenia (Paprocka) P.; m. Irene Gryszko, Mar. 26, 1966; children: Raphael M., Martin L. M.S. in Nuclear Engring, Warsaw Tech. U., 1965, Ph.D. in Nuclear Engring, 1972, Habilitation in Nuclear Engring, 1975; M.S. in Math, U. Warsaw, 1970. Teaching asst. to assoc. prof. Warsaw Tech. U., 1965-79; vis. assoc. prof. nuclear engring. Oreg. State U., 1979-80; vis. assoc. prof. Rensselaer Poly. Inst., Troy, N.Y., 1980-82, assoc. prof., 1982—; cons. Westinghouse Electric Corp., Yankee Atomic Electric Corp. Author: Nuclear Radiation Detection, 1973, Reactor Thermal-Hydraulics, 1977, Nuclear Reactor Safety, 1979; contbr. numerous articles to profl. jours. Mem. Am. Nuclear Soc., Soc. Engring. Sci. Subspecialties: Nuclear engineering; Applied mathematics. Current work: Reactor systems modeling under transient and accident conditions, degraded core thermal-hydraulics, stability analysis for multidimensional BWR models, stability methods for nonlinear reactor systems. Office: Dept Nuclear Engineerin Rensselaer Poly Inst Tibbits Ave Troy NY 12181

PODUSLO, SHIRLEY ELLEN, biochemistry researcher and educator; b. Richeyville, Pa., Dec. 24, 1941; d. Joseph and Helen (Kondor) P. B.S. in Anatomy, Ohio State U., 1963; Ph.D. in Biochemistry, Johns Hopkins U., 1980. Research asst. dept. neurology Albert Einstein Coll. Medicine, 1964-73; asst. dept. neurology Johns Hopkins U., 1973-76, asst. prof. neurochemistry dept., 1976—. Contbr. articles to profl. jours. NIH grantee, 1980, 83; Multiple Sclerosis Soc. grantee, 1978, 80, 83. Mem. Am. Soc. Neurochemistry, Internat. Soc. Neurochemistry, Soc. Neurosci., Am. Soc. Cell Biology. Subspecialties: Neurochemistry; Cell biology (medicine). Current work: Characterization of oligodendroglia, neurons, gliomas, as to their metabolism, plasma membranes, glycoproteins. Office: 600 N Wolfe St Meyer 6 119 Baltimore MD 21205

POE, ROBERT HILLEARY, pulmonary disease educator, physician; b. Cin., Apr. 29, 1934; s. Hilleary W. and Louis (Stark) P.; m. Sonja M. Carlton, July 23, 1963; children: Michael, Jennifer, Mark. B.S., U. Cin., 1955, M.D., 1959. Diplomate: Nat. Bd. Med. Examiners. Intern Walter Reed Gen. Hosp., Washington, 1959-60; resident U. Cin. Med. Ctr., 1963-66; asst. med. dir. Union Central Life Ins. Co., Cin., 1966-67; asst. chief pulmonary sect. VA Hosp., Cin., 1967-72; practice medicine specializing in pulmonary diseases, Covington, Ky., 1972-74; chief pulmonary unit Highland Hosp., Rochester, N.Y., 1974—; asst. prof. medicine U. Cin., 1967-72; assoc. prof. medicine U. Rochester, 1974—; cons. VA Hosp., Canandaigua, N.Y., 1977—. Editor, author: Problems in Pulmonary Medicine for the Primary Physician, 1982. Scoutmaster Boy Scouts Am., Pittsford, N.Y., 1982—. Served to lt. col. U.S. Army, 1968-69; Vietnam; to col. USAR, 1977-81. Decorated Legion of Merit. Fellow ACP, Am. Coll. Chest Physicians, Rochester Acad. Medicine; mem. Am. Thoracic Soc., AMA. Republican. Presbyterian. Subspecialties: Pulmonary medicine; Internal medicine. Current work: Airway disease and diagnostic procedures. Office: Highland Hosp 1000 South Ave Rochester NY 14620

POETTMANN, FRED HEINZ, oil co. exec.; b. Germany, Dec. 20, 1919; s. Fritz and Kate (Hussen) P.; m. Anna Bell Hall, May 29, 1952; children—Susan Trudy, Phillip Mark. B.S., Case Western Res. U., 1942; M.S., U. Mich., 1944, Sc.D., 1946; grad., Advanced Mgmt. Program, Harvard U., 1966. Registered profl. engr., Colo., Okla. Research chemist Lubrizol Corp., Wickliffe, Ohio, 1942-43; mgr. production research Phillips Petroleum Co., Bartlesville, Okla., 1946-55, asso. research dir., 1955-72; mgr. comml. devel. Marathon Oil Co., Littleton, Colo., 1972—. Contbr. articles to numerous publs.; Co-author, editor 5 books in field. Chmn. S. Suburban Met. Recreation and Park Dist., 1966-71; chmn. Littleton Press Council, 1967-71; bd. dirs. Hancock Recreation Center, Findlay, Ohio, 1973-77. Mem. Nat. Acad. Engring., Soc. Petroleum Engrs., Am. Inst. Chem. Engring., Am. Chem. Soc., Am. Petroleum Inst., Sigma Xi, Tau Beta Pi, Alpha Chi Sigma, Phi Kappa Phi. Republican. Subspecialties: Chemical engineering; Petroleum engineering. Current work: Research in enhanced oil recovery, natural gas engineering, hydocarbon phase behavior and oil production operations. Holder 45 patents. Home: 47 Eagle Dr Littleton CO 80123 Office: PO Box 269 Littleton CO 80160

POGO, A. OSCAR, cell biologist; b. Buenos Aires, Argentina, Aug. 28, 1927; came to U.S., 1959, naturalized, 1976; s. Bernard and Sara (Braverman) Pogo, A.; m. Beatriz G. T. Pogo, Jan. 13, 1956; children: Gustave, Gabriela. B.S., M. Moreno, Buenos Aires, 1945; M.D., U. Buenos Aires, 1953, D.M.Sci., 1959. Research assoc. faculty med. sci. U. Buenos Aires, Argentina, 1955-59; asst. prof. faculty med. sci. U. Cordoba, Argentina, 1961-64; research assoc. Rockefeller U., N.Y.C., 1964-66, asst. prof., 1966-67; investigator, head Lab. Cell Biology N.Y. Blood Ctr., N.Y.C., 1967-74, sr. investigator, head, 1974—. Guggenheim Found. fellow, 1959-61; recipient Career Devel. award NIH, 1966, grantee, 1967—; NSF grantee, 1976—. Mem. Am. Soc. Biol. Chemistry, Am. Chem. Soc., AAAS, Am. Soc. Cell Biology.

Subspecialties: Biochemistry (biology); Molecular biology. Current work: Biochemistry of the cell nucleus, nonhistone proteins and nuclear architecture, interaction of nucleosomes with nuclear matrix, transcription and processing of primary transcripts. Home: 237 Nyac Ave Pelham NY 10803 Office: L F Kimball Research Inst NY Blood Center 310 E 67th St New York NY 10021

POHL, HERBERT ACKLAND, chemist, educator; b. Lisbon, Portugal, Feb. 17, 1916; came to U.S., 1916; s. Lucien Charles and Emily May (Williams) P.; m. Eleanor Kathleen Rich, Aug. 23, 1941; children—Douglas, Patricia (Mrs. Robert Langdon), Elaine (Mrs. Gene Roy Oltmans), Charles, William. A.B. magna cum laude, Duke, 1936, Ph.D., 1939. Rockefeller fellow in anatomy Johns Hopkins Med. Sch., 1939; Carnegie Instn. of Washington fellow, 1940; NDRC fellow chem. engring. Johns Hopkins, 1941-42; sr. chemist Naval Research Lab., 1942-45; research assoc. E.I. duPont de Nemours & Co., 1945-57; sr. research assoc., lectr. plastics dept. Sch. Engring., Princeton, 1957-62; vis. prof. materials sci. Poly. Inst. Bklyn., 1962-63; vis. prof. quantum chemistry group Uppsala U., Sweden, 1963-64; prof. physics Okla. State U., Stillwater, 1964—; regional dir. Okla. Cancer Research Lab., 1980-81; pres., dir. Pohl Cancer Research Lab. Inc., 1982—; vis. prof. U. Calif., Riverside, 1970, Cavendish Lab., U. Cambridge, Eng., 1971; NATO sr. sci. fellow, 1971; Vice pres. Sci-Tech. Corp., N.Y.C. Author: Semiconduction in Molecular Solids, 1960, Quantum Mechanics, 1967, (with W.F. Pickard) Dielectrophoretic and Electrophoretic Deposition, 1969, How to Tell Atoms from People, 1970, Dielectrophoresis-The Behavior of Matter in Nonuniform Electric Fields, 1978; Editor: Jour. Biol. Physics, 1977—; editorial bd.: Jour. Cell Biophysics; Contbr. articles to profl. jours. Pres. Princeton chpt. United World Federalists, 1959-61; v.p. Stillwater Unitarian Ch., 1978-79. Recipient Rensselaer award in sci., 1933, Soc. Plastics award, 1959; A. & K. Wallanberg fellow, 1963-64. Fellow Am. Inst. Chemists, Explorers Club; mem. Am. Phys. Soc., AAAS, Am. Assn. Physics Tchrs., AAUP, Am. Chem. Soc., Okla. Acad. Sci. (past chmn.), Philos. Soc. Subspecialties: Biophysics (biology); Polymer chemistry. Current work: Biogical dielectrophoresis; cellular spin resonance; organic semiconducting polymers. Patentee in field.

POHL, JENS GERHARD, architect, educator; b. Wetzlar, W.Ger., Sept. 18, 1940; s. Ernst Richard and Hildegard Wilhelmine (Gorschlueter) P.; m. Barbara Moyra Penrose, June 30, 1962; children: Sonya Karen, Kym Jason. B.Arch., U. Melbourne, Australia, 1965; M.Bldg. Sci., U. Sydney, Australia, 1967, Ph.D. in Archtl. Sci, 1970. Registered architect, Victoria and N.S.W., Australia. Architect Victoria Dept. Pub. Works., 1965-68; sr. lectr. Sch. Bldg., U. New South Wales, Sydney, 1969-72; ptnr. Archtl. Design and Research Group, Sydney, 1969-72; prof. architecture Calif. Poly. State U.-San Luis Obispo, 1973—; pres. Educol. Inc., San Luis Obispo, 1981—. Author 8 books on computer applications, bldg. lighting, acoustics, bldg. sci.; contbr. articles to profl. jours. Life bd. govs. Sports Union, U. Sydney. Recipient Recognition award NASA, 1977. Fellow Royal Australian Inst. Architects; mem. Am. Inst. Constructors, Australian Inst. Bldg. Subspecialties: Solar energy; Software engineering. Current work: Solar energy in architecture and agriculture, microcomputer applications in architecture and construction, lightweight systems in building construction. Patentee lightweight concentrating solar collectors (2). Home: 650 Highlands Hill Rd Nipomo CA 93444 Office: Calif Poly State U San Luis Obispo CA 93407

POHLMAN, EDWARD WENDELL, psychology educator, researcher; b. Chuchokee, Malyan, Punjab, Pakistan, Jan. 30, 1933; came to U.S., 1946; s. Edward Wendell and Edna Mabel (Kennedy) e7P.); m. Sharon Freitas, Sept. 1979; children: Douglas, Sharon. B.A., Loma Linda U., 1953; M.A., Andrews U., 1956, Ohio State U., 1958, Ph.D., 1960. Asst. prof. U. Pacific, Stockton, Calif., 1961-64, assoc. prof., 1964-69, prof., 1969—, dir. birth planning research ctr., 1971—; vis. prof. Ctr. Family Planning Inst., New Delhi, 1967-69, U Calif.-Berkeley, summer 1972; cons. WHO, Geneva, 1969, 75, and others. Author: The Psychology of Birth Planning, 1969, How to Kill Population, 1971, The God of Planet 607, 1972, Population: A Clash of Prophets, 1973; contbr. numerous articles to profl. jours. Planned Parenthood/World Population grantee, 1962-66; Fulbright-Hays grantee, 1967; Carolina Population Ctr. grantee, 1967-79; NIMH grantee, 1971-72; NIH grantee, 1972-75; recipient outstanding and unusual service to sci. and profession psychology award Am. Psychol. Assn., 1977. Fellow Am. Psychol. Assn. Subspecialty: Population psychology. Current work: Psychological effects of male and female sterilizing operations (voluntary); children's attitudes toward their own future children. Home: 2618 Canyon Creek Dr Stockton CA 95207 Office: Dept Edn and Counseling Psychology U Pacific Stockton CA 95211

POHORECKY, LARISSA A., neuropharmacology educator; b. Cholm, Poland, Jan. 16, 1942; d. Roman and Maria P.; m. Adrian Dolinsky, Oct. 29, 1972. B.S., U. Ill., 1963; Ph.D., U. Chgo., 1967. Postdoctoral trainee MIT, 1967-71; asst. prof. Rockefeller U., N.Y.C., 1971-79; assoc. prof. neuropharmacology Rutgers U., New Brunswick, N.J., 1979—; cons. Nat. Inst. Alcoholism and Alcohol Abuse, VA. Contbr. numerous articles to profl. jours. NIH fellow. Mem. AAAS, Am. Soc. Pharmacology and Exptl. Therapeutics, Nat. Research Soc. Alcoholism, Sigma Xi. Subspecialties: Neuropharmacology; Neurochemistry. Current work: Mechanisms involved in development of alcohol tolerance and physical dependence; interaction of stress and alcohol. Office: Center Alcohol Studies Rutgers U Piscataway NJ 08854

POINAR, GEORGE ORLO, JR., biologist, educator; b. Spokane, Wash., Apr. 25, 1936; s. George Orlo and Helen Louise (Ladd) P.; m.; children: Hendrik, Maya, Gregory. B.S., Cornell U., 1958, M.S., 1960, Ph.D., 1962. Research asst. Cornell U., Ithaca Coll., 1958-62; postdoctoral fellow U. Calif.-Riverside, Hilo, Hawaii, 1962-63, Rothamsted Exptl. Sta., Eng., 1963-64; invertebrate pathologist U. Calif.-Berkeley, 1964—, also mem. faculty. Author: The Natural History of Nematodes, 1983, Nematodes for the Biological Control of Insects, 1979, Entomogenous Nematodes, 1975, Diagonostic Manual for Identification of Insect Diseases, 1978; mem. editorial bds. profl. jours. Fulbright scholar, 1962-63; NIH fellow, 1963-64; Nat. Acad. Scis. award, 1969. Mem. Soc. Invertebrate Pathology, Nematologists Socs. (U.S. and Europe). Subspecialties: Parasitology; Integrated pest management. Current work: Entomogenous nematodes, amber. Office: 336 Hilgard Hall Dept Entomology U Calif Berkeley CA 94720

POIRIER, JOHN ANTHONY, physics educator; b. Lewistown, Mont., May 15, 1932; s. Anton Alexis and Evelyn (Shannon) P.; m. Margaret Jule Griffin, Jan. 3, 1976; children: Michael, Steven, Gregory, Maureen, Laine. B.S., Notre Dame U., 1954; M.S., Stanford U., 1956, Ph.D., 1959. Staff physicist Lawrence Berkeley Lab., 1959-63; lectr. U. Calif., Berkeley, 1959-63; NSF fellow CERN, 1963-64; assoc. prof. physics U. Notre Dame, Ind., 1964-69, prof., 1969—; cons. NASA, 1962; mem. elem. particle physics sect. NSF, 1977-78. Contbr. articles to profl. jours. Mem. Am. Phys. Soc., Am. Inst. Physics, Sigma Xi. Subspecialty: Particle physics. Current work: Experimental work in elementary particle physics. Office: Dept Physics Notre Dame IN 46556

POLACEK, LAURIE ANN, zoologist; b. Milw., Nov. 19, 1955; d. Joseph Stephen and Ann Apalona (Dornak) P.; m. Glenn James Warren, Mar. 14, 1981. B.S. in Zoology, U. Wis., Eau Claire, 1977, M.S., 1981. Project assoc. Bone Marrow Cryopreservation Lab., Blood Ctr. of Southeastern Wis., Milw., 1982—. Author articles. Mem. Genetics Soc. Am., Am. Soc. Human Genetics, AAAS. Subspecialties: Gene actions; Human genetics. Current work: Cryopreservation, bone marrow, human. Home: 1725 E Park Pl Apt 6 Milwaukee WI 53211

POLAND, ALAN PAUL, research scientist, pharmacology, oncologist, toxicologist; b. Balt., June 5, 1940; s. Henry and Mary (Katz) P.p.; m. Helen Dwyer, Feb. 14, 1941; 1 dau., Carolyn Mary. Student, Rensselaer Poly. Inst., 958-61; M.D., M.S. in Pharmacology, U. Rochester, 1966. Intern Bellevue Hosp., N.Y.C., 1966-67; with USPHS, Center for Disease Control, Atlanta, 1967-69, postdoctoral fellow Rockefeller U., N.Y.C., 1969-71; asst. prof. pharmacology U. Rochester (N.Y.) Sch. Medicine, 1971-77; asst. prof. oncology U. Wis., Madison, 1977-79, asst. prof., 1979—. Recipient award USPHS, 1975; Burroughs Wellcome scholar, 1981-83. Mem. Am. Soc. Pharmacology and Exptl. Therapeutics (John Jacob Abel award 1976), Am. Assn. Cancer Research, Soc. Toxicology. Jewish. Subspecialties: Molecular pharmacology; Cancer research (medicine). Current work: Mechanism of action of halogenated aromatic hydrocarbons. Home: 2918 Nottingham Way Madison WI 53713 Office: U Wis Madison WI 53706

POLAND, ARTHUR IRA, astrophysicist; b. Asbury Park, N.J., Mar. 30, 1943; s. Harris David and Dorothy (Epstein) P.; m. Helen Mantell, June 14, 1964. B.S., U. Mass., 1964; Ph.D., Ind. U., 1969. Scientist Nat. Ctr. for Atmospheric Research, 1969-80; astrophysicist NASA/Goddard Space Flight Ctr., Greenbelt, Md., 1980—. Recipient tech. advancement award Nat. Ctr. for Atmospheric Research, 1974. Mem. Am. Astron. Soc., Internat. Astron. Union, Sigma Xi. Subspecialties: Optical astronomy; Solar physics. Current work: Solar chromosphere, corona, ultraviolet spectroscopy computer modeling, image processing, data handling. Home: PO Box 1107 Warrenton VA 22186 Office: NASA/Goddard Code 682 Greenbelt MD 20771

POLANYI, JOHN CHARLES, chemist, educator; b. Jan. 23, 1929; m. Anne Ferrar Davidson, 1958; 2 children. B.Sc., Manchester (Eng.) U., 1949, M.Sc., 1950, Ph.D., 1952, D.Sc., 1964; D.Sc. (hon.), U. Waterloo, 1970, Meml. U., 1976, McMaster U., 1977, Carleton U., 1981, Harvard U., 1982, LL.D., Trent U., 1977, Dalhousie U., 1983. Mem. faculty dept. chemistry U. Toronto, Ont., Can., 1956—, prof., 1962—, Univ. prof., 1974—; William D. Harkins lectr. U. Chgo., 1970; Reilly lectr. U. Notre Dame, 1970; Purves lectr. McGill U., 1971; F.J. Toole lectr. U. N.B., 1974; Philips lectr. Haverford Coll., 1974; Kistiakowsky lectr. Harvard U., 1975; Camille and Henry Dreyfus lectr. U. Kans., 1975; J.W.T. Spinks lectr. U. Sask., Can., 1976; Laird lectr. U. Western Ont., 1976; Disting. lectr. Simon Fraser U., 1977; Gucker lectr. Ind. U., 1977; Jacob Bronowski meml. lectr. U. Toronto, 1978; Hutchinson lectr. U. Rochester, N.Y., 1979; Priestley lectr. Pa. State U., 1980; Barré lectr. U. Montreal, 1982; Chute lectr. Dalhousie U., 1983; Redman lectr. McMaster U., 1983. Co-editor: (with F.G. Griffiths) The Dangers of Nuclear War, 1979; contbr. articles to jours., mags., newspapers; producer: film Concepts in Reaction Dynamics, 1970. Decorated officer Order of Can., companion Order of Can.; recipient Marlow medal Faraday Soc., 1962; Centenary medal Chem. Soc. Gt. Brit., 1965; with N. Bartlett Steacie prize, 1965; Noranda award Chem. Inst. Can., 1967; award Brit. Chem. Soc., 1971; medal Chem. Inst. Can., 1976; Henry Marshall Tory medal Royal Soc. Can., 1977; Sloan Found. fellow, 1959-63; Guggenheim fellow, 1979-80. Mem. Nat. Acad. Scis. U.S. (fgn.), Am. Acad. Arts and Sci. (hon. fgn.). Subspecialty: Physical chemistry. Home: A6 3 Rosedale Rd Toronto ON Canada M4W 2P1 Office: U Toronto Dept Chemistry 80 St George St Toronto ON Canada M5S 1A1

POLATNICK, JEROME, research biochemist; b. N.Y.C., Oct. 4, 1922; s. Jack and Gussie (Seiden) P.; m. Selma Amster, Aug. 21, 1948; children: Lois, Judith, Barbara. Research chemist Schenley Research Inst., Ind. and N.Y., 1943-47; research chemist N.Y. Bot. Gardens, N.Y.C., 1948-50; biochemist Columbia U., 1950-54; prin. investigator Manhattan Eye and Ear Hosp., N.Y.C., 1954-57; research chemist Plum Island Animal Disease Center, U.S. Dept. Agr., Greenport, N.Y., 1957-80, 1980—, acting lab. chief, 1980. Contbr. numerous articles to profl. jours. Mem. biochem. and biophys. investigations unit receiving Presdl. citation, 1965. Mem. Am. Soc. Microbiology, Am. Chem. Soc., N.Y. Acad. Sci., Sigma Xi. Subspecialties: Molecular biology; Virology (biology). Current work: Viral protein and ribonucleic acid synthesis; viral-induced enzymes; intracellular transport of viral components. Patentee in field. Home: 1310 Crittens Ln Southold NY 11971 Office: PO Box 848 Greenport NY 11944

POLAVARAPU, PRASAD LEELA, chemistry educator, researcher; b. Gudlavalleru, Andhra, India, May 21, 1952; came to U.S., 1977; s. Rao Rajeswara and Navaratnamma (Chaparala) P.; m. Bharathi Krishna, May 23, 1981. M.Sc., Birla Inst. Tech., Pilani, Rajastan, India, 1972; Ph.D., Indian Inst. Tech., Madras, 1976. Mem. faculty dept. chemistry Vanderbilt U., Nashville. NIH grantee, 1982—. Mem. Am. Chem. Soc., AAAS., Sigma Xi. Subspecialties: Physical chemistry; Infrared spectroscopy. Current work: Research on infrared circular dichroism and raman optical activity. Office: Dept Chemistry Vanderbilt U Nashville TN 37235

POLAY, JANET SKINNER, mfg. co. exec.; b. Newark, Apr. 4, 1945; d. Lester Albert and Ruth (Jacobsen) Skinner; m. Michael S. Polay, Jan. 8, 1977; children: John Bryce, Robert Michael, Andrew Evan. R.N., B.S., Wagner Coll., 1967. Clin. coordinator for clin. research Columbia U., 1972-77; clin. researcher Organon, Inc., West Orange, N.J., 1977-82; mgr. product devel. market research The West Co., Phoenixville, Pa., 1982—; cons. Procter & Gamble, Cin., 1972-77. Contbr. articles to profl. jours. Served with USNR, 1967-69. Mem. AAAS, N.Y. Acad. Scis., Am. Fedn. Clin. Research, Am. Soc. Bone and Mineral Research, Am. Soc. Parenteral and Enteral Nutrition. Republican. Episcopalian. Subspecialties: Biomaterials. Current work: Product development for medical device and pharmaceutical components. Home: PO Box 295 Chester Springs PA 19425 Office: The West Co W Bridge St Phoenixville PA 19460

POLCYN, STANLEY JOSEPH, robotics company executive; b. Chgo., July 16, 1930; s. Stanley Frank and Lorrane (Wojcachowski) P.; m. Carole J., Sept. 30, 1955; children: Mark, David, Carla, Linda, James, Suzanne. B.S. in Indsl. Engring, Northwestern U., M.B.A. Various engring. and mktg. positions Otis Elevator Co.; mgr. sales Roller Bearing Co. Am., West Trenton, N.J.; mgr. mktg. Dresser Crane & Hoist, Muskegon, Mich.; now sr. v.p. Unimation, Inc., Danbury, Conn.; instr. Loyola U., Chgo. Active United Way; mem. Newtown Sch. Bd. Served with Air Corps USN, 1950-52. Mem. Robot Inst. Am. (pres.), Robotics Internat. (dir.), Soc. Mfg. Engrs. (mem. mfg. mgmt. council). Roman Catholic. Subspecialty: Robotics. Current work: Robotics technology and its use. Office: Shelter Rock Lane Danbury CT 06810

POLIN, DONALD, animal scientist, educator; b. Arlington, Mass., Dec. 7, 1925; s. Ralph and Bessie (Dickerman) P.; m. Ruth B. Meyer, Mar. 15, 1954; children: Barbara Dawn, Diane Lynn, Richard Bennett. B.S., U.S. Maritime Acad., 1946, Rutgers U., 1951, Ph.D., 1955. Research fellow Merck Inst. Therapeutic Research, Merck Sharpe & Dohme, Rahway, N.J., 1955-67; units leader Eaton Labs., Norwick (N.Y.) Pharm. Co., 1967-69; prof. dept. animal sci. Mich. State U., East Lansing, 1969—; mem. subcom. Nat. Acad. Sci.-NRC, 1979—. Contbr. writings to booklets, jours. in field. Judge U.S. Figure Skating Assn. Served to lt. USN, 1946-64. Recipient various research grants govt. and industry. Mem. Am. Inst. Nutrition, Poultry Sci. Assn., Soc. Toxicology, Soc. Exptl. Biology and Medicine, N.Y. Acad. Scis., World's Poultry Sci. Assn., Sigma Xi, Alpha Zeta. Jewish. Club: Lansing Skating (v.p. 1982-83). Subspecialties: Animal nutrition; Toxicology (agriculture). Current work: Research bioenergetics of farm animals; nutrition versus toxicology of farm animals; regulation of growth and carcass composition, obesity and regulation of food intake. Office: Dept Animal Sci Mich State Univ East Lansing MI 40024

POLING, ALAN DALE, psychology and behavioral pharmacology educator; b. Grafton, W.Va., Oct. 24, 1950; s. Howard Taft and Ruth Jean (Cool) P. B.A., Alderson-Broaddus Coll., 1972; M.A., W.Va. U., 1974; Ph.D., U. Minn., 1977. Research assoc. U.S.c., Columbia, 1977-78; asst. prof. Western Mich. U., Kalamazoo, 1978—. Co-author, editor: Drugs and Mental Retardation, 1982, Applied Psychopharmacology, 1982, Control of Human Behavior, 1982; contbr. articles in field to profl. jours. Mem. Am. Psychol. Assn., Behavioral Pharmacology Soc., Psychonomic Soc., Assn. Behavior Analysis. Subspecialties: Behavioral psychology; Behavioral pharmacology. Current work: Behavioral pharmacology, species-typical behavior, mental retardation, conditioning and learning. Home: 2722 Mount Olivet St Kalamazoo MI 49004 Office: Dept Psychology Western Mich U Kalamazoo MI 49008

POLITIS, DEMETRIOS JOHN, plant pathologist, researcher, consultant; b. Athens, Greece, Oct. 10, 1935; came to U.S., 1969; s. John Demetrios and Violetta (Koutsoyannis). B.S., Coll. Agr., U. Athens, 1960; M.S., McGill U., 1965; Ph.D. (dissertation fellow), U. Ky., 1974. Research asst. Benaki Research Inst., Kifissia, Greece, 1960-63; research assoc. McGill U., Montreal, Que., Can., 1967-69; research specialist U. Ky., Lexington, 1969-72, postdoctoral fellow, 1974-75; research microbiologist U. Mo.-Columbia, 1975-79; team leader Dynamac Corp., Rockville, Md., 1979-82; project assoc. Pa. State U., Frederick, Md., 1982—; cons. in agrl. scis., 1982—. Contbr. articles to sci. jours. Calif. Dept. Food and Agr. grantee, 1981-84. Mem. Am. Phytopath. Soc., Hellenic Phytopath. Soc. Subspecialties: Plant pathology; Cell biology. Current work: Biological control of noxious weeds using exotic (imported) pathogens such as rust fungi from Eurasia. Home: 18638 Grosbeak Terr Gaithersburg MD 20879 Office: Bldg 1301 Fort Detrick Frederick MD 21701

POLITIS, MICHAEL JAMES, neuroscientist; b. Hoboken, N.J., Apr. 12, 1951; s. Andrew J. and Helen K. P. Ph.D. in Physiology, Coll. Medicine and Dentistry N.J., 1978. Postdoctoral research fellow in neuropathology Albert Einstein Coll Medicine, Bronx, N.Y., 1978-80, instr. dept. neurosci, 1980—; Mem. Physicians for Social Responsibility. Contbr. articles to sci. jours. Mem. Soc. Neurosci., Am. Soc. Neurochemistry. Subspecialties: Regeneration; Neuropathology. Current work: Investigation into intracellular interactions in traumatized nerve tissue to find surgical/chemical means of restoring function after injury. Office: 1410 Pelham Pkwy S Bronx NY 10461

POLIVY, JANET, psychology educator, clinician; b. N.Y.C., Feb. 9. 1951; Can., 1976; d. Calvin and Bernice (Malat) P.; m. C. Peter Herman, Aug. 3, 1975; children: Lisa Cesia, Eric Murray. B.A. magna cum laude, Tufts U., 1971; M.A. (USPHS fellow), Northwestern U., 1973, Ph.D., 1974. Lic. Ont. Bd. Examiners in Psychology. Asst. prof. psychology Loyola U., Chgo., 1974-76; vis. asst. prof. dept. psychology U. Toronto, 1976-77, assoc. prof., 1977—, assoc. prof. dept. psychiatry, 1980—; research assoc. clin. investigation unit Clarke Inst. Psychiatry, 1977-79, psychosomatic medicine unit and psychotherapy unit, 1979—; scientist Addiction Research Found. Ont., 1978—; research assoc. Toronto Gen. Hosp., 1982—. Mem. editorial bd.: Jour. Personality, 1979-81; assoc. editor, 1981—; cons. editor: Jour. Abnormal Psychology; contbr. reviews and articles to profl. jours. Clarke Inst. Psychiatry Research Fund grantee, 1977-78, 78-80, 81-82; Ont. Mental Health Found. grantee, 1977-78; Social Sci. and Humanities Research Council grantee, 1979-80, 83-85; Natural Scis. and Engring. Research Council grantee, 1979-82, 82—. Mem. Am. Psychol. Assn., Midwestern Psychol. Assn., Assn. Advancement of Behavior Therapy (chairperson nominations and elections com. 1982-83), Soc. Psychotherapy Research. Subspecialty: Personality/abnormal psychology. Current work: Research and psychotherapy in eating and eating disorders; research on emotion. Office: U. Toront Dept Psychology Erindale Campus Mississauga ON Canada L5L 1C6

POLLACK, BARY WILLIAM, software development company executive, consultant; b. San Francisco, Aug. 10, 1944; s. Seymour and Evelyn H. (Honnell) P.; m. Kimberly Ann Armstrong., Dec. 18, 1971; children: Craig Euen, Scott Eric. S.B., M.I.T., 1965; M.S., Stanford U., 1966, Ph.D., 1975. Asst. prof. computer sci U. Calif.-Berkeley, 1972-75, U. B.C. (Can.), Vancouver, 1975-80; dir. software devel. Datapoint Corp., Berkeley, 1980-82; pres. Vulcan systems, El Cerrito, Calif., 1982—. Author: (with J. Friedman) Model of Transformational Grammar, 1970; editor: Compiler Techniques, 1972; editor-in-chief quar.: Computer Graphics, 1977—. Mem. Assn. Computing Machinery, IEEE, Sigma Xi. Club: Nat. Ski Patrol (patrol leader). Subspecialties: Software engineering; Graphics, image processing, and pattern recognition. Current work: Compiler design, language design, graphics systems design and development. Office: Vulcan Systems 1329 Scott St El Cerrito CA 94530

POLLACK, JAMES BARNEY, space scientist; b. N.Y.C., July 9, 1938; s. Michael and Jeanne (Joseph) P. A.B., Princeton U., 1960; M.A., U. Calif.-Berkeley, 1962; Ph.D., Harvard U., 1965. Research physicist Smithsonian Astrophys. Obs., Cambridge, Mass., 1965-68; sr. research assoc. Cornell U., Ithaca, N.Y., 1968-70; research scientist, chief scientist of ACE program NASA Ames Research Ctr., Moffett Field, Calif., 1970—, mem. space sci. adv. com., 1978-82; editorial bd. ICARUS, 1979-82. Contbr. articles to profl. jours. Recipient NASA medal for exceptional sci. achievement, 1976, 79; Space Sci. award AIAA, 1978. Fellow AAAS, Am. Geophys. Union; mem. Am. Astron. Soc. Subspecialties: Planetary science; Climatology. Current work: Origin and evolution of planets, history of earth's climate, other planets climates. Address: MS 245-3 NASA Ames Research Ctr Moffett Field CA 94035

POLLACK, RALPH MARTIN, chemistry educator, researcher; b. Boston, May 25, 1943; s. Hyman and Elsie (Margolin) P.; m. Ellen Joan Banner, Dec. 20, 1964; children: Robert, Lauren. Sc.B., Brown U., 1965; Ph.D., U. Calif.-Berkeley, 1968. Asst. prof. U. Md. Baltimore County, Catonsville, 1970-75, assoc. prof., 1975-80, prof. chemistry, 1980—; assoc. prof. U. Montpellier, France, 1978-79. Mem. Am. Chem. Soc., AAAS, Sigma Xi. Subspecialties: Organic chemistry; Biochemistry (biology). Current work: Enzyme mechanisms, organic reaction mechanisms. Office: Chemistry Dept U Md Baltimore County 5401 Wilkens Ave Catonsville MD 21228

POLLACK, RICHARD STUART, hospital department administrator; b. N.Y.C., Apr. 20, 1946; s. David and Ruth (Vrouble) P.; m. Joanne M. Cobb, Sept. 24, 1972; children: Deborah, Aaron. B.A., L.I. U., 1968, M.S., 1970. Cert. technologist Nuclear Medicine Tech. Cert. Bd.

Tech. dir. depts. nuclear medicine, radiation therapy and ultrasound JFK Med. Center, Edison, N.J., 1970—; edni. dir. Sch. Nuclear Medicine Tech., 1973—; adj. instr. radiologic tech. Middlesex County Coll., Edison, 1971—; cons. physicist Somerset West Essex Gen. and South Bergen hosps., 1974—; chmn. hosp.-wide coms. on tech. assessment and data processing.; also cons. pvt. corps.; cons. N.J. Commn. on Radiation Protection, vice chmn. N.J. adv. com. on nuclear medicine. Author manuals and tng. programs. Active Jewish Community Ctr. of No. Middlesex County. Recipient commendation Waldemar Cancer Research Found.; USPHS grantee. Mem. AAAS, Soc. Nuclear Medicine (computer council, pres. Greater N.Y. chpt. technologists sect. 1974—, chmn. program com. nat. technologist sect. 1975—), Am. Assn. Physicists in Medicine, Am. Registry Radiologic Technologists, Am. Inst. Ultrasound in Medicine, Am. Hosp. Radiology Adminstrs. Jewish. Subspecialties: Nuclear medicine; CAT scan. Current work: Nuclear magnetic resonance, information systems, radiology, imaging technology. Home: 19 Fox Hill Rd Edison NJ 08820 Office: Nuclear Medicine Dept JFK Med Ctr Edison NJ 08820

POLLACK, ROBERT ELLIOT, college dean, biology educator, researcher; b. Bklyn., Sept. 2, 1940; s. Hy Ephraim and Molly P.; m. Amy L. Steinberg, Dec. 12, 1961; 1 dau., Marya E. B.A., Columbia U., 1961; Ph.D., Brandeis U., 1966. Fellow in pathology NYU Med. Ctr., 1966-68, instr., 1968-69, adj. asst. prof., 1969-70; scientist Cold Spring Harbor (N.Y.) Lab., 1969-74; prof. microbiology SUNY-Stony Brook, 1975-78; prof. biol. scis. Columbia U., N.Y.C., 1978—; dean Columbia Coll., 1982—; vis. prof. pharmacology Albert Einstein Coll. Medicine, 1977—; McGregory lectr. dept. chemistry Colgate U., 1979; mem. adv. bd. U.S.-Israel Binational Sci. Found., 1976—; mem. governing bd. Scientists's Inst.; mem. Damon Runyon Fund Sci. Adv. Bd., 1979-83; mem. genetic biology panel NSF, 1982-83; mem. sci. adv. bd. Nat. Alzheimer's Disease Found., 1982—; cons. in field. Contbr. numerous articles to profl. jours.; assoc. editor: Jour. Virology, 1976-79, Jour. Cell Biology, 1977-80, Virology, 1978-81, Molecular and Cellular Biology, 1980-85; cons. editor: Environ, 1978—; mng. editor: BBA Revs. on Cancer, 1980—; editorial adv. bd.: Cells and Their Interactions, 1980—. NIH spl. fellow Weizmann Inst., Rehovot, Israel, 1970; NIH research career devel. awardee, 1974-75; Nat. Cancer Inst. grantee, 1968—. Mem. AAAS, Am. Soc. Cell Biology, Am. Soc. Microbiology, Fedn. Am. Scientists, N.Y. Acad. Scis. Subspecialties: Cell and tissue culture; Virology (biology). Office: Columbia Coll 208 Hamilton New York NY 10027

POLLACK, ROBERT LEON, biochemistry and nutrition educator, consultant; b. Phila., Apr. 29, 1926; s. Louis A. and Mary D. P.; m. Lydia Aureli, June 22, 1952; children: Janine, Linda. B.Sc. in Chemistry, Phila. Coll. Pharmacy and Sci., 1948, 1949, M.Sc., 1950; Ph.D., U. Tenn., 1954. Sr. research scientist U.S. Dept. Agr., Wyndmoor, Pa., 1954-62; chmn. dept. biochemistry and nutrition Temple U. Sch. Dentistry, Phila., 1962—, dir. Health Ctr., 1975—. Contbr. articles to profl. jours. Served with Hosp. Corps, USN, 1944-46. Fellow Am. Coll. Nutrition; mem. AAAS, Am. Chem. Soc., Internat. Assn. Dental Research, Sigma Xi. Subspecialties: Biochemistry (biology); Nutrition (biology). Current work: Biochemistry of normal and pathologic dental tissues, human nutrition. Office: Temple U Sch Dentistry 3223 N Broad St Philadelphia PA 19140

POLLACK, STEPHEN LEWIS, psychologist, educator; b. Morristown, N.J., Sept. 1, 1948; s. E. Lewis and Margaret (Williams) P.; m. Sarah A. Green, Oct. 24, 1981; children: Russell, Justin. B.A., U. Calif.-Santa Cruz, 1970; M.A., No. Ill. U., 1974; Ph.D., U. Houston, 1978. Intern Baylor Coll. Medicine, Houston, 1974-75; postdoctoral fellow U. Rochester (N.Y.) Sch. Medicine, 1977-79; clin. instr. Houston Child Guidance Ctr., Houston, 1979—; sr. faculty/pvt. pratice Houston-Galveston Family Inst., Houston, 1982—; cons. Tex. Head Injury Found., 1982—; clin. asst. prof. psychology Tex. Research Inst., for Mental Scis., Houston, 1979—; clin. asst. prof. psychiatry Baylor Coll. Medicine, 1979; exec. com. Houston-Galveston Family Therapy Consortium, 1982—. Contbr. articles to profl. jours. Hogg Found. grantee, 1981-82. Mem. Am. Psychol. Assn., Tex. Psychol. Assn., Houston Psychol. Assn. Subspecialties: Clinical psychology; Systems theory. Current work: Psychotherapy and application of systemic models to human behavioral problems. Office: 1020 Holcombe Blvd Suite 1200 Houston TX 77030

POLLACK, SYLVIA BYRNE, tumor immunologist, educator; b. Ithaca, N.Y., Oct. 18, 1940; d. Raymond Tandy and Elsie Frances (Snell) Byrne; m.; children: Seth Benjamin, Ethan David. B.A., Syracuse U., 1962; Ph.D., U. Pa., 1967. Instr. dept. anatomy Woman's Med. Coll. Pa., 1967-68; research assoc. microbiology U. Wash., 1968-73, research asst. prof. microbiology and immunology, 1973-77, research assoc. prof., 1977-81, research assoc. prof. biol. structure, 1981—; asst. mem. Fred Hutchinson Cancer Research Center, 1975-79, assoc. mem., 1979-81; mem. exptl. immunology study sect. NIH, 1983—. Assoc. editor: The Journal of Immunology, 1977-79, the Journal of the Reticuloendothelial Society, 1982—; contbr. articles in field. NIH fellow, 1967-67; Nat. Cancer Inst. grantee, 1973—; Am. Cancer Soc. grantee, 1973-79. Mem. Am. Assn. Immunologists, Soc. Developmental Biology, Am. Assn. Cancer Research, Reticuloerdothelial Soc., Phi Beta Kappa. Subspecialties: Immunobiology and immunology; Cancer research (medicine). Current work: Tumor immunology, natural killer cells, antibody-dependent cellular cytotoxicity, cell-cell interactions, immune regulation, immunogenetics. Home: 8327 S Franklin Rd Clinton WA 98236 Office: Biological Structure SM-20 University of Washington Seattle WA 98195

POLLACK, WILLIAM, research found. exec.; b. London, Eng., Feb. 26, 1926; s. David and Rose (Weis) P.; m. Alison Elizabeth Calder, Dec. 4, 1954; children—Malcolm Trevor, David Calder. B.S., Imperial Coll. Sci. and Tech., 1948; M.S., St. George's Hosp. Med. Sch., London U., 1950; Ph.D., Rutgers U., 1964. V.p., dir. research, diagnostics, 1969; v.p., dir. Ortho Research Inst. Med. Scis., 1975—; asso. clin. prof. pathology Coll. Phys. and Surg., Columbia, 1969—; asso. prof. Rutgers U. Med. Sch. Contbr. articles immunology, immunochemistry to sci. jours. Served with Brit. Royal Navy, 1943-46. Recipient John Scott award Bd. Dirs. City Trusts, City of Phila., 1976; Joseph Bolivar DeLee Humanitarian award, Chgo., 1979; Albert Lasker Clin. Med. Research award, 1980. Fellow N.Y. Acad. Scis., N.Y. Acad. Medicine; mem. A.A.A.S., Am. Soc. Clin. Pathologists, Am. Assn. Blood Banks (Karl Landsteiner award 1969), Brit., Am. socs. immunology, Sigma Xi. Subspecialty: Immunology (medicine). Co-developer of vaccine to prevent Rh disease of babies. Home: Sunset Rd Belle Mead NJ 08509 Office: Ortho Diagnostic Research Raritan NJ 08869

POLLAK, MICHAEL, physics educator; b. Ostrava, Moravia, Czechoslovakia, Sept. 1, 1926; came to U.S., 1956, naturalized, 1969; s. Leo Alois and Elsa (Winterstein) P.; m. Rosemarie Yanni, June 13, 1964; children: Michelle, Tania. B.Sc. in Elec. Engring, Israel Inst. Tech., Haifa, 1953; M.S. in Physics, U. Pitts., 1956, Ph.D., 1959. Jr. scientist Westinghouse Research, Pitts., 1956-59, sr. scientist, 1961-66; mem. tech. staff Bell Labs., 1960-61; mem. faculty U. Calif.-Riverside, 1966—, prof. physics, 1969—; vis. prof. physics Israel Inst. Tech., Haifa, 1970, 77, UCLA, 1975, U. Cambridge, 1977; cons. Jet Propulsion Labs., Pasadena, Calif., 1982—. Co-editor: Conduction in Low Mobility Materials, 1971, Electron Interactions in Disordered Systems, 1983; contbr. articles to profl. publs. Mem. Amnesty Internat., Irvine, Calif., 1982—. Served with IsraeliNavy, 1949-51. Sci. Research Council of U.K. Sr. fellow, 1972, 77. Mem. AAAS. Subspecialties: Condensed matter physics; Statistical physics. Current work: Transport and many-body problems in disordered solids; impurity conduction; amorphous semiconductors. Home: 3051 Mountainview Dr Laguna Beach CA 92651 Office: Physics Dept U. Calif Riverside CA 92521

POLLARD, GERALD TILMAN, research pharmacologist; b. Dunn, N.C. Ph.D., N.C. State U., 1981. Research pharmacologist Wellcome Research Labs., Research Triangle Park, N.C., 1974—. Mem. AAAS; mem. N.Y. Acad. Scis.; Mem. Am. Psychol. Assn., Soc. for Stimulus Properties Drugs, Soc. for Neurosci.; mem. Behavioral Pharmacology Soc.; Mem. Southeastern Pharmacology Soc., So. Soc. for Philosophy and Psychology, MLA. Subspecialties: Behavioral psychology; Psychopharmacology. Current work: Researcher in animal models of psychopathology, animal behavioral methods for identification of psychoactive drugs. Office: Wellcome Research Labs Dept Pharmacology 3030 Cornwallis Rd Research Triangle Park NC 27709

POLLARD, HARVEY BRUCE, biochemist; b. San Antonio, May 26, 1943. B.A., Rice U., 1964; M.D., U. Chgo., 1969, Ph.D., 1973. Research assoc. Nat. Inst. Arthritis, Digestive Diseases, Kidney and Diabetes, NIH, Bethesda, Md., 1969-71, sr. investigator, 1975-79, sect. chief, 1979-81, lab. chief, 1981—; sr. investigator Nat. Inst. Child Health and Devel., 1971-75. Recipient Commendation medal USPHS, 1982. Subspecialty: Cell and tissue culture. Current work: Biochemistry and cell biology of hormone secretion. Office: Nat Inst Arthritis Digestive Diseases Kidney and Diabetes NIH Lab Cell Biology and Genetics Bldg 4 Room 310 Bethesda MD 20205

POLLARD, MORRIS, microbiologist, educator; b. Hartford, Conn., May 24, 1916; s. Harry and Sarah (Hoffman) P.; m. Mildred Klein, Dec. 29, 1938; children: Harvey, Carol, Jonathan. D.V.M., Ohio State U., 1938; M.S., Va. Poly. Inst., 1939; Ph.D. (Nat. Found. Infantile Paralysis fellow), U. Calif.-Berkeley, 1950; D.Sc. (hon.) Miami U., Ohio, 1981. Mem. staff Animal Disease Sta., Nat. Agrl. Research Center, Beltsville, Md., 1939-42; asst. prof. preventive medicine Med. br. U. Tex., Galveston, 1946-48, asso. prof., 1948-50, prof., 1950-61; prof. biology U. Notre Dame, Ind., 1961-66, prof., chmn. microbiology, 1966-81, prof. emeritus, 1981—; dir. (Lobund Lab.), 1961—; vis. prof. Fed. U. Rio de Janeiro, Brazil, 1977; mem. tng. grant com. NIH, 1965-70; mem. adv. bd. (Inst. Lab. Animal Resources), 1965-68, mem. adv. com. microbiology, 1966-68, chmn., 1968-70, mem. sci. adv. com., 1966-70; cons. U. Tex., M.D. Anderson Hosp. and Tumor Inst., 1958-66; mem. project rev. com. United Cancer Council, 1966-70, 74—; mem. colon cancer com. Nat. Cancer Inst., 1972-76, chmn. tumor immunology com., 1976-79; mem. com. cancer cause and prevention NIH, 1979-81; program rev. com. Argonne Nat. Lab, 1979—, chmn., 1983—; lectr. Found. Microbiology, 1978. Editor: Perspectives in Virology Vol. I to XI, 1959-80; contbr. articles to profl. jours. Served from 1st lt. to lt. col. Vet. Corps, AUS, 1942-46. McLaughlin Faculty fellow Cambridge U., 1956; Raine Found. prof. U. Western Australia, 1975; vis. scientist Chinese Acad. Med. Scis., 1979, 81; hon. prof. Chinese Acad. Med. Scis., 1982; Disting. Alumnus Ohio State U., 1979; decorated Army Commendation medal. Mem. Am. Acad. Microbiology (charter), Brazilian Acad. Scis., Soc. Exptl. Biology and Medicine, Am. Soc. Microbiology, Am. Assn. Pathologists, Am. Soc. Lab. Animal Sci., Assn. Gnotobiotics (pres.), Internat. Commn. Lab. Animal Sci., AAAS, Sigma Xi. Subspecialty: Microbiology (medicine). Home: 3540 Hanover Ct South Bend IN 46614 Office: Lobund Lab Notre Dame IN 46556

POLLNOW, GILBERT FREDERICK, physical chemistry educator, researcher, consultant; b. Oshkosh, Wis., Jan. 17, 1925; s. Arthur Ewald and Alma Clara (Sonnenberg) P.; m. Geneva I. Yancy, Oct. 6, 1948; m. Katherine L. Kaye, Apr. 7, 1973; children: Nicole Denise, Steven John. B.S., U. Wis.-Oshkosh, 1950; M.S., U. Iowa, 1951, Ph.D. in Phys. Chemistry, 1954. Research scientist Dow-Corning Corp., Midland, Mich., 1954-58; research scientist Allis-Chalmers Mfg. Co., Milw., 1958-61; prof. phys. chemistry U. Wis., Oshkosh, 1961—; cons. in field. Contbr. chpts. to books and articles to profl. jours. Served with USN, 1942-46. Office Naval Research fellow, 1952-54; U. Wis.-Oshkosh grantee, 1975, 81; Curriculum Devel. grantee, 1976-80; NSF grantee, 1976-78; Digital Equipment Corp. grantee, 1982. Mem. Am. Chem. Soc., Midwest Sociol. Soc., AAAS, Sigma Xi. Subspecialties: Polymer physics; Algorithms. Current work: Application of microcomputers in the laboratory for teaching and research including simulation, experimental design and process control; specific current research includes automated viscometric studies of epoxy and other thermo-setting polymeric resins and composites as a function of temperature and composition. Patentee in field. Home: 103 W S Park Ave Oshkosh WI 54901 Office: Dept Chemistry U of Wisconsin Oshkosh WI 54901

POLLNOW, ROBERT EDWARD, oncologist, internist; b. Rockford, Ill., Jan. 30, 1946; s. Frank Charles and Dorothy Elizabeth (Jeanguenat) Mattingly; m. Karla Ann Sladky, Apr. 24, 1971; children: Kayla Lynne, William Robert, Jonathan Edward. B.S., U. Ill., 1968; M.D., U. Ill.-Chgo., 1972. Diplomate: Am. Bd. Internal Medicine. Intern Baylor Affiliated Hosps., Houston, 1972-73, resident in medicine, 1973-75, chief med. resident, 1975-76; fellow in oncology M.D. Anderson Hosp., Houston, 1976-78; staff oncologist Naval Regional Med. Ctr., Portsmouth, Va., 1979-81; oncologist, internist The Med. Group, Memphis, 1981—; instr. medicine Baylor Coll. Medicine, 1973-76, U. Tex. Health Sci. Ctr., Houston, 1977-78; asst. prof. medicine Eastern Va. Med. Sch., Norfolk, 1978-81; instr. medicine U. Tenn., Memphis, 1981—; conf. presenter. Bd. dirs. Am. Cancer Soc., 1981—. Served to lt. comdr. USNR, 1978-81. Recipient letter of appreciation Naval Regional Med. Ctr., 1981. Mem. AMA (Physicians Recognition award 1977), Tenn. Med. Soc., Memphis Acad. Medicine, Am. Soc. Clin. Oncology. Methodist. Lodge: Kiwanis. Subspecialties: Chemotherapy; Internal medicine. Current work: Clinical medical education, natural history and investigative treatment of hematologic and solid tumor malignancies, infections in the compromised host. Office: The Medical Group 1734 Madison Ave Memphis TN 38104

POLLOCK, GERALD ARTHUR, toxicologist; b. St. Louis, Jan. 12, 1950; s. Glenn E. and Paula (Waldes) P.; m. Rebecca Ann Parker, Nov. 22, 1975; children: Rachel Christine, Nathan Jeremy, Sarah Elizabeth. B.S., U. Calif., Davis, 1972, Ph.D., 1977. Research assoc. environ. toxicology U. Calif., Davis, 1973-77; asst. prof. Wash.-Oreg.-Idaho regional program vet. medicine U. Idaho, Moscow, 1977-82; research supr. Diamond Shamrock Corp., Painesville, Ohio, 1982—. Contbr. articles to profl. jours. Mem. Soc. Toxicology, AAAS, Am. Chem. Soc., Am. Sci. Affiliation, Sigma Xi. Democrat. Presbyterian. Subspecialties: Toxicology (agriculture); Toxicology (medicine). Current work: General toxicology, metabolism, disposition, elimination of pesticides and mycotoxins, analysis of residues and mutagenicity assay. Home: 7825 Skyline View Dr Mentor OH 44060 Office: Safety Assessmen Diamond Shamrock Corp 11211 Spear Rd Painesvill OH 44077

POLLOCK, JERRY JOSEPH, oral biology educator; b. Toronto, Ont., Can., Sept. 10, 1941; came to U.S., 1969; s. Jack and Anne London P.; m.; children: Melanie, Seth, Sean. B.Sc., U. Toronto, 1963, M.Sc., 1966; Ph.D., Weizmann Inst. Sci., Rehovot, Israel, 1969. Postdoctoral fellow N.Y.U., N.Y.C., 1969-71, asst. prof., 1971-73; assoc. prof. SUNY-Stony Brook, 1973-82, prof. oral biology, 1982—; cons. Abbott Labs., Chgo., 1981-82; vis. scientist U. Liege, Belgium, 1972-73; cons. NIH, NSF, Washington, 1975—. Mem. editorial bd.: Jour. Dental Research, 1978—. Recipient Pharmacy Poulenc award Poulenc Co., 1964; NRC Can. predoctoral scholar, 1966-68; Med. Research Council Can. postdoctoral fellow, 1969-71. Mem. Canadian Pharm. Co., Internat. Assn. Dental Research, Am. Soc. Microbiology. Subspecialties: Oral biology; Biochemistry (medicine). Current work: Host defense mechanisms, lysozyme, antibacterial mechanisms, cationic proteins, cell surface studies. Office: Sch Dental Medicine DeptOral Biology and Pathology SUNY Stony Brook NY 11794

POLONIS, DOUGLAS HUGH, engineering educator; b. North Vancouver, B.C., Can., Sept. 2, 1928; came to U.S., 1955, naturalized, 1963; s. William and Ada (Burrows) P.; m. Vera Christine Brown, Jan. 30, 1953; children: Steven Philip, Malcolm Eric, Douglas Hugh, Christine Virginia. B.A.Sc., U. B.C., 1951, Ph.D., 1955; M.A.Sc., U. Toronto, 1953. Metall. engr. Steel Co. Can., Hamilton, Ont., 1951-52; mem. faculty U. Wash., Seattle, 1955—, prof. metall. engring., 1962—, chmn. dept. mining, metall. and ceramic engring., 1969-71, 73-82; metall. cons., 1955—. Contbr. articles to profl. jours. Mem. Am. Soc. Metals, AIME, Am. Soc. Engring. Edn., Tau Beta Pi, Alpha Sigma Mu, Sigma Phi Delta. Mem. Christian Ref. Ch. Subspecialties: Materials (engineering); Metallurgy. Current work: Structure and properties of alloys, mechanical behavior of materials phase transformations in solids microstructure-property relationships. Home: 19227 46th Ave NE Seattle WA 98155

POLONSKY, IVAN PAUL, computer scientist; b. Bklyn., Aug. 23, 1929; s. Murray and Sadie (Futoran) P.; m. Maxine Minoff, Aug. 13, 1953; children: Jonathan, Linda, Amy. B.A., NYU, 1949, M.S., 1953, Ph.D., 1957. Instr. Queens Coll., 1953-60; mem. tech. staff Bell Telephone Labs., Holmdel, N.J., 1960—; adj. prof. Rutgers U., 1972—. Author: (with others) The Snobol 4 Programming Language, 1971; contbr.: chpt. to Handbook of Mathematical Functions, 1964. Pres. Red Bank (N.J.) Bd. Edn., 1967-72, Red Bank Regional Bd. Edn., 1978—; vice chmn. Monmouth Adult Edn. Commn., Eatontown, N.J., 1967—. Mem. ACM. Subspecialties: Programming languages; Distributed systems and networks. Current work: Archiving of distributed databases. Office: Bell Telephone Labs Crawfords Corner Rd Holmdel NJ 07733 Home: 151 Harding Rd Red Bank NJ 07701

POLTROCK, STEVEN EDWARD, psychologist, educator; b. Seattle, Sept. 6, 1946; s. Robert W. and Jeanne (Coulson) P. B.S. in Engring, Calif. Inst. Tech., 1968; M.A., UCLA, 1970; Ph.D., U. Wash., 1977. Mem. tech. staff TRW Systems, Redondo Beach, Calif., 1967-70; programmer Honeywell, Inc., West Covina, Calif., 1970-72; research asst. U. Wash., Seattle, 1972-77, research assoc., 1977; asst. prof. U. Denver, 1977—; mem. tech. staff Bell Labs., Murray Hill, N.J., 1982—. Nat. Inst. Child Health & Human Devel. grantee, 1979; NIMH grantee, 1978. Mem. Am. Psychol. Assn., Rocky Mountain Psychol. Assn., Psychonomic Soc. Subspecialties: Cognition; Perception. Current work: Study of basic perceptual and cognitive processes and the manner in which these processes are combined. Home: 1040 Oakland Ave Plainfield NJ 07060 Office: 2D 444 Bell Labs 600 Mountain Ave Murray Hill NJ 07974

POMERANTZ, JAMES R(OBERT), psychologist, university dean; b. N.Y.C., Aug. 21, 1946; s. Mihiel Charles and Elizabeth; ; s. Mihiel Charles and Solheim P.; m. Sandra E Jablonski, Mar. 1, 1969; children: Andrew Emil, William James. B.A., U. Mich., 1968; Ph.D., Yale U., 1974. Asst. prof. psychology Johns Hopkins U., Balt., 1974-77; assoc. prof. SUNY, Buffalo, 1977—, assoc. dean, 1983—. Editor: Perceptual Organization, 1981. Mem. Am. Psychol. Assn., Eastern Psychol. Assn., Psychonomic Soc., Sigma Xi. Club: Johns Hopkins (dir. 1975-77). Subspecialties: Cognition; Sensory processes. Current work: Visual pattern recognition and form perception in humans; visual motion perception; attention; memory; texture perception. Home: 83 Wickham Dr Williamsville NY 14221 Office: Dept Psychology SUNY 4230 Ridge Lea Rd Buffalo NY 14226

POMERANTZ, MARC ABRAHAM, general surgeon; b. Phila., Dec. 27, 1937; s. Jacob and Mollie (Orloff) P.; m. Rhoda Solomon, Aug. 14, 1958; children: Lauren Eve, Susan Leigh. B.A., U. Pa., 1958; M.D., Temple U., 1963. Diplomate: Am. Bd. Surgery. Intern Presbyn.-St. Luke's Hosp., Chgo, 1963-64, resident in surgery, 1964-66, 68-70; asst. in surgery U. Ill.-Chgo., 1964-66; asst. attending staff Rush-Presbyn.-St. Luke's Med. Ctr., Chgo., 1972—; med. dir. employees health service, 1973-80; instr. gen. surgery Rush Med. Coll., Chgo., 1972-73, asst. prof., 1973-80; practice medicine specializing in surgery, Chgo., 1970—; chmn. utilization com. St. Francis Xavier Hosp., 1972-76, chmn, credential com., 1973-76. Contbr.: chpt. in Surgical Intensive Care, 1981; author med. films. Sci. advisor Ill. Agt. Orange Commn., Springfield, 1982. Served to lt. USNR, 1966-68. Fellow ACS; mem. Chgo. Med. Soc., Ill. Med Soc., AAUP, Am. Geriatric Soc. Subspecialty: Surgery. Current work: Obesity surgery. Home: 1315 Sutton Pl Chicago IL 60610 Office: 1725 W Harrison St Suite 322 Chicago IL 60612

POMERANTZ, MARTIN, chemistry educator; b. N.Y.C., May 3, 1939; s. Harry and Pauline (Sietz) P.; m. Maxine Miller, June 4, 1961; children: Lee Allan, Wendy Jane, Heidi Lauren. B.S., CCNY, 1959; M.S., Yale U., 1961, Ph.D. (NSF, Woodrow Wilson, hon. Sterling and Leeds and Northrup fellow), 1964. NSF postdoctoral fellow U. Wis., 1963-64; asst. prof. Case Western Res. U., Cleve., 1964-69; assoc. prof. Belfer Grad. Sch. Sci., Yeshiva U., N.Y.C., 1969-74, prof., 1974-76, acting chmn. dept., 1971-72, chmn. chemistry dept., 1973-76; vis. assoc. prof. U. Wis., 1972; vis. prof. Columbia U., N.Y.C., part-time, 1970-75; Alfred P. Sloan Found. fellow, 1971-76; vis. prof. Ben-Gurion U. Negev, Israel, summer, 1981; prof. chemistry U. Tex.-Arlington, 1976—; cons. in field. Contbr. numerous articles on chemistry to profl. jours. Robert A. Welch Found. grantee, 1977—; Petroleum Research Fund grantee, 1964-71; NSF grantee, 1967-69, 80-84. Mem. Am. Chem. Soc., Royal Soc. Chemistry, Phi Beta Kappa, Sigma Xi. Subspecialty: Organic chemistry. Current work: Organic chemical reaction mechanisms, synthesis and study of phosphorus-nitrogen compounds, reactive intermediates, strained molecules. Home: 5521 Williamstown Dallas TX 75230 Office: Dept Chemistry U Tex at Arlington PO Box 19065 Arlington TX 76019

POMERANZ, YESHAJAHU, cereal chemist, food analyst; b. Tlumacz, Poland, Nov. 28, 1922; came to U.S., 1959, naturalized, 1967; s. David and Rysia (Bildner) P.; m. Ada Waisberg, Oct. 27, 1948; children: Shlomo, David. B.Sc., Israeli Inst. Tech., 1945; Ph.D., Kans. State U., 1962. Dir. Central Food Testing Lab., Govt. Israel, Haifa, 1948-59; research chemist U.S. Dept. Agr., Manhattan, Kans., 1962-69, dir., 1973—; dir. Barley and Malt Lab., prof. U. Wis., Madison, 1969-73; dir. short courses on cereal sci. and tech. and vis. prof. throughout world, 1975—. Author: (with Meloan) Food Analysis, 1971, Food Analysis-Experimental, 1976, (with Shellenberger) Bread, 1971; editor, co-author 11 books; contbr. over

350 articles to sci. and tech. publs. Recipient Disting. Service award U.S. Dept. Agr., 1983; M.P. Neuman medal, W. Ger., 1978; Von Humboldt award sr. U.S. scientist W. Ger., 1981. Fellow Inst. Food Technologists, AAAS; mem. Internat. Assn. Cereal Chemists (exec. com.), Am. Assn. Cereal Chemists (sci. editor Advances in Cereal Sci. and Tech., T.B. Osborne medal 1980, W.F. Geddes medal 1982), Am. Chem. Soc., Assn. Ofcl. Analytical Chemistry (H. Wiley award 1980), Nat. Assn. Chem. Engring. Mex. (hon.), Gamma Sigma Delta. Subspecialties: Food science and technology; Biochemistry (biology). Current work: Director of research on the relation between the structure, chemical composition, and end use properties of cereal grains, and on the biochemistry of breadmaking. Patentee in field. Home: 1952 Bluestem Terrace Manhattan KS 66502 Office: 1515 College Ave Manhattan KS 66502

POMEROY, LAWRENCE RICHARDS, zoology educator; b. Sayre, Pa., June 2, 1925; s. Rupert Cole and Doris Atkinson (Richards) P.; m. Janet Klerk, Apr. 6, 1952; children: Cheryl S., Russel J. B.S., U. Mich., 1947, M.S., 1948; Ph.D., Rutgers U., 1951. Assoc. prof. U. Ga., Athens, 1958-66, prof., 1966-79, Alumni Found. disting. prof. zoology, 1979—; assoc. program dir. NSF, Washington, 1966-67, mem. ocean scis. panel, 1981—; mem. com. engring. deep-sea drilling NRC, Washington, 1978-82; mem. vis. com Argonne Nat. Lab., Ill., 1976-78. Editor: Cycles of Essential Elements, 1974, Ecology of a Salt Marsh, 1981; editorial com.: Ann. Rev. Ecology and Systematics, 1972-77; contbr. articles to profl. jours. Fellow AAAS; mem. Am. Soc. Limnology and Oceanography (pres. 1983-84), Ecol. Soc. Am. (council 1967-69). Subspecialties: Oceanography; Ecology. Current work: The ocean's food web, especially microbial aspects, origin and fate of particulate and dissolved organic material in ocean water. Office: Dept Zoology U Ga Athens GA 30602

POMILLA, FRANK R., physics educator; b. Bklyn., Oct. 1, 1926; s. Anthony and Carmela (Policano) P.; m. Clara R. Mercurio, June 13, 1953; children: Anthony, Mary, Frank, Paul. B.S., Franklin Coll., 1948; M.S., Fordham U., 1949, Ph.D., 1963. Instr. St. John's U., Queens, N.Y., 1949-51, asst. prof., 1951-59, assoc. prof., 1959-64; research scientist Grumman Aerospace Corp., Bethpage, N.Y., 1964-67; prof. physics York Coll., CUNY, Jamaica, 1967—, Grad. Faculty, CUNY, N.Y.C., 1970—. Author: (with N. Tralli) Atomic Theory, 1969; contbr. articles in field to profl. jours. Trustee Elmont (N.Y.) Union Free Sch. Dist., 1968—, Sewanhaka Central High Sch. Dist., Elmont, 1978—. NSF sci. faculty fellow, 1960-62; NSF grantee York Coll., 1969—; U.S. Dept. Edn. grantee, 1982—; Alfred P. Sloan Found. grantee, 1975. Mem. Am. Phys. Soc., Am. Assn. Physics Tchrs., IEEE, AAUP, Sigma Xi. Subspecialties: Atomic and molecular physics; Planetary atmospheres. Current work: Research and development of materials for science education using latest technologies. Home: 51 Goshen St Elmont NY 11003 Office: York Coll CUNY Jamaica NY 11451

POMPER, SEYMOUR, microbiologist, fermentation technologist; b. N.Y.C., May 8, 1925; s. Israel and Ruth (Graubard) P.; m. Judith Irene Grossman, Oct. 3, 1953; children: Joseph, Laurie, Roni, William. B.S., Cornell U., 1945; M.S., Yale U., 1948, Ph.D., 1949. Biologist Oak Ridge Nat. Lab., 1949-52; chemist Fleischmann Labs., Stamford, Conn., 1953-56, asst. sect. head, 1956-59, sect. head, 1969-70, dept. dir., 1970-71, div. research dir., Peekskill, N.Y., 1971-75, research dir., Stamford, 1975-77; dept. dir. Nabisco Brands, Wilton, Conn., 1977—. Served with AUS, 1945-46. Standard Brands fellow, 1946-49; Nat. Cancer Inst. fellow, 1952-53. Mem. Am. Chem. Soc., AAAS, Genetics Soc., Am. Assn. Cereal Chemists, Am. Soc. Microbiology. Subspecialties: Microbiology; Enzyme technology.

POND, WILSON GIDEON, educator; b. Mpls., Feb. 16, 1930; s. Frank Wilson and Gladys Emma (Miller) P.; m.; 1 son: Kevin Roy. B.S., U. Minn., 1952; M.S., Okla. State U., 1954, Ph.D., 1957. Cert. animal scientist, CAS, 1977. Asst. prof. Cornell U., Ithaca, N.Y., 1957-60, assoc. prof., 1969-70, prof., 1970-78; supervisory animal scientist and research leader nutrition U.S. Meat Animal Research Ctr., Clay Center, Nebr., 1978—; prof. U. Nebr., 1978—. Contbr. articles to profl. jours. Served with U.S. Army, 1954-56. Recipient Am. Feed Mfrs. award for nutrition research Am. Soc. Animal Sci., 1969; Moorman Travel fellow, 1974; Gustav Bohstedt award Am. Soc. Animal Sci., 1979. Fellow AAAS; mem. Am. Soc. Animal Sci., Am. Inst. Nutrition, Soc. Exptl. Biology and Medicine, N.Y. Acad. Sci., Sigma Xi, Alpha Zeta. Republican. Presbyterian. Subspecialties: Animal nutrition; Nutrition (medicine). Current work: Maternal nutrition and progeny development, protein-energy malnutrition, mineral requirements and interrelationships, obesity. Office: US Meat Animal Research Ctr PO Box Clay Center NE 68933 Home: 1203 N Kansas Ave Hastings NE 68901

PONDER, HERMAN, geologist; b. Light, Ark., Jan. 31, 1928; s. Herman Cook and Sylvia Adell (Cameron) P.; m. Barbara Elaine Sando, May 10, 1947; children: Teresa Elaine, David Mark. B.A., U. Mo., 1955, Ph.D., 1959. Registered profl. engr., Colo. Research engr. A.P. Green Refractories Co., Mexico, Mo., 1959-61, lab. mgr., 1961-63; project engr., then mgr. mining div. Colo. Sch. Mines Research Inst., Golden, 1963-67, dir. research, 1967-70; pres. Research Inst. Colo. Sch. Mines, prof. geology, 1970—; dir. La. Land and Exploration Co.; chmn. bd. Colo. Nat. Bank, Golden. Served with USN, 1946-47. Mem. AIME, Rocky Mountain Assn. Geologists, Colo. Mining Assn. Subspecialties: Geology; Materials processing. Current work: Mineral occurrences and mineral recovery processes. Home: 1919 Mt Zion Dr Golden CO 80401 Office: PO Box 112 Golden CO 80401

PONG, RAYMOND S., urological surgeon, cancer researcher; b. Hong Kong, Nov. 21, 1937; s. David C. and Saufong (Wong) P. A.B., Dartmouth Coll., 1960; M.S., M.I.T., 1966, Ph.D., 1970; M.D., Case Western Res. U., 1975. Diplomate: Am. Bd. Med. Examiners, Am. Bd. Urology. Research asso. M.I.T., 1970-71; research fellow Case Western Res. U., 1971-73, surg. resident, 1975-77, fellow, 1976-77; urology resident Mass. Gen. Hosp., 1977-80, chief resident, 1980; fellow in surgery Harvard Med. Sch., 1979-80; sr. surgeon urology City of Hope Med. Center, Duarte, Calif., 1980—; asst. clin. prof. U. So. Calif., 1980—; asso. clin. prof. U. Calif., Irvine, 1982—. Contbr. articles to profl. jours. NIH grantee, 1968, 71. Mem. N.Y. Acad. Sci., Am. Soc. Cell Biology, AAAS, Los Angeles Urol. Soc., Sigma Xi. Club: Pasadena Athletic. Subspecialties: Cancer research (medicine); Urology. Current work: Surgical oncology and carcinogenesis. Office: 11411 Brookshire Ave # 402 Downey CA 90241

PONNAMPERUMA, CYRIL ANDREW, chemist, educator; b. Galle, Sri Lanka, Oct. 16, 1923; came to U.S., 1959, naturalized, 1967; s. Andrew and Grace (Siriwardene) P.; m. Valli Pal, Mar. 19, 1955; 1 dau., Roshini. B.A., U. Madras, 1948; B.Sc., U. London, 1959; Ph.D., U. Calif., Berkeley, 1962. Research assoc. Lawrence Radiation Lab., U. Calif., 1960-62; research scientist, then chief chem. evolution br. Ames Research Center, NASA, 1962-70; prof. chemistry U. Md., College Park, 1971—; dir. Lab. Chem. Evolution, 1971—. Author books, articles in field; editor-in-chief: Origins of Life, 1973-83. Recipient A.I. Oparin Gold medal Internat. Soc. for Study Origin of Life, 1980. Fellow Royal Inst. Chemistry, Indian Nat. Sci. Acad. (fgn. fellow); mem. Am. Chem. Soc. (chmn. internat. activities com. 1981), Astron. Assn., Am. Soc. Biol. Chemists, Explorers Club. Club: Cosmos (Washington). Subspecialties: Biochemistry (biology); Space chemistry. Current work: Chemical studies on the origin of life—search for extraterrestrial life. Office: Lab Chem Evolution U Md College Park MD 20742

PONS, MARCEL WILLIAM, virologist, research inst. adminstr.; b. Bronz, N.Y., Oct. 9, 1932; s. William and Marie Francis (Van Remoortel) P.; m. Joyce Marie, July 7, 1933; children: Lisa, Melissa. B.A., N.Y.U., 1954; M.S., U. Mich., 1956, Ph.D., 1958. USPHS postdoctoral fellow Harvard U., 1958-60; asst. Public Health Research Inst. City N.Y., 1960-66, assoc., 1966-73; assoc. mem., dir. Lab. Molecular Virology, Christ Hosp. Inst. MED. Research, Cin., 1977—, asso. dir. inst., 1981—; assoc. prof. microbiology U. Cin., 1977—, prof. infectious diseases, 1977—; mem. Internat. Com. on Taxonomy Viruses. Contbr. articles, chpts. to profl. jours., textbooks; assoc. editor: Virology. Pres. Ryerson Sch. PTA, Wayne, N.J., 1970-72. USPHS-NIH grantee, 1968, 71, 76, 78, 79, 81. Mem. Am. Soc. Microbiology, Am. Soc. Virologists, AAAS, Sigma Xi. Subspecialties: Virology (biology); Molecular biology. Current work: Biochemistry of virus replication; regulation of RNA synthesis by respiratory viruses. Home: 727 Woodfield Dr Cincinnati OH 45231 Office: 2141 Auburn Ave Cincinnati OH 45219

POOL, PETER EDWARD, research cardiologist, educator; b. Chgo., Dec. 30, 1936; s. Edward Augustus and Ruth (Hebert) P. B.A. in Philosophy, Yale U., 1958; M.D., NYU, 1962. Diplomate: Am. Bd. Internal Medicine. Research assoc. in pharmacology Nat. Heart Inst., Bethesda, Md., 1965-68; asst. prof. medicine (cardiology) U. Calif.-San Diego, 1968-71, assoc. clin. prof., 1971—; pres. North County Cardiovascular Med. Group, Inc., Encinitas, Calif., 1971—; cons. cardiovascular pharmacology. Editorial bd.: Circulation Research, 1969-74; contbr. over 70 articles to sci. and med. jours. Bd. dirs. Calif. affiliate Am. Heart Assn., 1975—, pres., 1980-81; pres. San Diego County Heart Assn., 1978-79. Served as surgeon USPHS, 1965-68. Recipient Founders Day award NYU, 1962; Research Career Devel. award USPHS, 1969; Pfizer fellow, Montreal, 1968; Silver Service medallion Calif. affiliate Am. Heart Assn., 1981. Fellow Am. Coll. Cardiology, Am. Coll. Chest Physicians, ACP, Am. Heart Assn. Council Clin. Cardiology; mem. Am. Soc. Pharmacology and Exptl. Therapeutics, Am. Physiol. Soc., Am. Fedn. Clin. Research, Cardiac Muscle Soc., Western Soc. Clin. Research, Alpha Omega Alpha. Subspecialties: Cardiology; Pharmacology. Current work: Cardiology and cardiovascular pharmacology. Office: 1087 Devonshire Dr # 300 Encinitas CA 92024

POOLE, GEORGE DOUGLAS, mathematics and computer science educator; b. Miami, Nov. 30, 1942; s. Pope Adex and Ora (Noblitt) P.; m. Lois Ann Chesbro, Aug. 31, 1963; children: Aaron, Audra, Ryan, Austin. B.S.E., Emporia (Kans.) State U., 1964; M.S., Colo. State U., Ft. Collins, 1966; Ph.D., Tex Tech U., Lubbock, 1971. Instr. Washburn U., Topeka, Kans., 1966-67; instr. Tex. Tech U., Lubbock, 1967-68; asst. prof. Emporia (Kans.) State U., 1971-80, prof., chmn. dept. math., 1980—; cons. Sauder Industries, Emporia; computer sci. cons. Emporia Sch. System. Author: Algebraic Development of Real Numbers, 1975, Basic Arithmetic Skills, 1978. Mem. Math. Assn. Am., Am. Math. Soc., Soc. Indsl. and Applied Math., Kans. Acad. Scis., Nat, Council Tchrs. Math. Subspecialties: Numerical analysis; Programming languages. Current work: Numerical linear algebra, matrix theory. Home: 1414 Patrick St Emporia KS 66801 Office: Dept Mat Emporia State U 1200 Commercial St Emporia KS 66801

POPE, ALAN THOMAS, aerospace researcher; b. Madison, Tenn., Sept.27, 1943; s. Roger and Grace (Thomas) P.; m. Sara Ellen Webb; 2 children. B.S. in Elec. Engring, Tenn. Technol. U., Cookeville, 1965, M.S., U. Tenn.-Knoxville, 1969; Ph.D. in Clin. Psychology, U. Fla., 1975. Lic. psychologist. Clin. research psychologist VA Hosp., Montrose, N.Y., 1974-80; aerospace technologist NASA Langley Research Ctr., Hampton, Va., 1980—. USPHS fellow, 1969-73. Mem. Biofeedback Soc. Am., Phi Beta Kappa, Tau Beta Pi, Phi Kappa Phi. Subspecialties: Biofeedback; Cognition. Current work: Research and development of psychophysiological methods for addressing man-machine interface issues in aviation. Office: NASA Langley Research Ctr MS 152E Hampton VA 23665

POPOV, DAN, clinical psychologist; b. Butte, Mont., Sept. 27, 1945; s. Frederick Michael and Patricia (Tauson) P.; m. Carol Lark, Mar. 6, 1973 (div. 1980); m. Linda Kavelin, July 20, 1981. B.S., U.S. Mil. Acad., 1968; M.A., U. Colo.-Boulder, 1971, Ph.D., 1973. Lic. psychologist. Commd. 2d lt. U.S. Army, 1968, advanced through grades to capt., 1970; chief behavior sci. cons. service (Letterman Army Med. Ctr.), San Francisco, 1974-76, psychologist, Ft. Ord, Calif., 1976-78, ret., 1978; v.p. research and devel. Systems Effectiveness Assn., Monterey, Calif., 1977—, Info. Access Systems, Boulder, Colo., 1981—; developer Mgmt. Support Tech., Inc., Denver, 1981—, dir., 1981—; v.p. research and devel. Advanced Info. Design, Portland, Oreg., 1982—, chmn. bd., 1982—; exec. dir. LRI Assocs., Inc., Boulder, 1980-82. Co-author: Person, Place, World, 1982. Mem. Am. Psychol. Assn., Am. Soc. Tng. and Devel., AAAS, Mid-Coast Psychol Assn. (pres. 1976-77), Monterey Peninsula Psychol. Assn. (pres. 1977-78). Subspecialties: Organizational/Clinical Psychology; Artificial intelligence. Current work: Application of artificial intelligence on microcomputers to problems in management; organizational, and behavioral sciences using a descriptive psychological approach. Office: Mgmt Support Tech Inc 455 Sherman St Denver CO 80203

POPOV, EGOR PAUL, emeritus civil engineering educator; b. Kiev, Russia, Feb. 19, 1913; s. Paul T. and Zoe (Derabin) P.; m. Irene Zofia Jozefowski, Feb. 18, 1939; children—Katherine, Alexander. B.S., U. Calif., 1933; M.S., Mass. Inst. Tech., 1934; Ph.D., Stanford, 1946. Registered civil, structural and mech. engr., Calif. Structural engr., bldg. designer, Los Angeles, 1935-39; asst. prodn. engr. Southwestern Portland Cement Co., Los Angeles, 1939-42; machine designer Goodyear Tire & Rubber Co., Los Angeles, 1942-43; design engr. Aerojet Corp., Calif., 1943-45; asst. prof. civil engring. U. Calif. at Berkeley, 1946-48, assoc. prof., 1948-53, prof., 1953-83, prof. emeritus, 1983—, chmn. structural engring. and structural mechanics div., dir. structural engring. lab., 1956-60; Miller research prof. Miller Inst. Basic Research in Sci., 1968-69. Author: Mechanics of Materials, 1952, 2d edit., 1976, Introduction to Mechanics of Solids, 1968; Contbr. articles profl. jours. Recipient Disting. Tchr. award U. Calif.-Berkeley, 1976-77, Berkeley citation U. Calif.-Berkeley, 1983. Mem. Am. Soc. Metals, Internat. Assn. Shell Structures (v.p., mem. exec. council), ASCE (Ernst E. Howard award 1976, J. James R. Croes medal 1979, 82, Nathan M. Newmark medal 1981), Soc. Exptl. Stress Analysis (Hetenyi award 1967), Am. Soc. Engring. Edn. (Western Electric Fund award 1976-77, Disting. Educator award 1979), AAAS, Am. Concrete Inst., Nat. Acad. Engring., Soc. Engring. Sci., Internat. Assn. Bridge and Structural Engring., Am. Inst. Steel Constrn. (adv. com. specifications), Sigma Xi, Chi Epsilon, Tau Beta Pi. Subspecialties: Civil engineering; Solid mechanics. Current work: Seismic behavior of steel and reinforced concrete strctures. Hysteretic material behavior. Home: 2600 Virginia St Berkeley CA 94709

POPOVIC, VOJIN, physiologist, eductor; b. Belgrade, Yugoslavia, Sept. 18, 1922; s. Pavle and Bojana (Pavlovic) P.; m. Pava Jovanovic, Oct. 2, 1948; 1 son, Ray. B.Sc., U. Belgrade, 1948, M.Sc., 1949, D.Sc., 1951. From instr. to assoc. prof. U. Belgrade, 1949-56; vis. prof. Nat. Research Ctr., Paris, 1956-57, U. Rochester, N.Y., 1957-58; vis. scientist Nat. Research Council, Ottawa, Ont., Can., 1958-60; assoc. prof. U. Houston, 1960; assoc. prof. physiology Emory U. Sch. Medicine, Atlanta, 1961-65, prof., 1965—. Author: Hypothermia in Biology and in Medicine; contr. chpts. to books, articles to profl. jours. Grantee Am. Cancer Soc., NIH, Office of Naval Research, NASA, Aerospace Med. Sch. Mem. Am. Physiol. Soc., Internat. Soc. for Cryobiology, Am. Heart Assn., Aerospace Med. Assn., others. Greek Orthodox. Subspecialties: Space medicine; Physiology (medicine). Current work: Weightlessness and cardiovascular system; environmental effects and circulation, hypothermia.

POPOVICH, ROBERT PETER, biomedical engineering educator, consultant; b. Sheboygan, Wis., Jan. 9, 1939; s. Anton John and Ida (Selke) P.; m. Diane Evelyn Sohre; children: Kathleen, Steven, Scott, Robert; m. Lou Ellen Addison. B.S. in Chem. Engring, U. Wis., 1963; M.S., U. Wash., 1968, Ph.D. in Chem. Engring, 1970. Registered profl. engr., Tex. Fellow U. Wash., 1965-72; asst. prof. chem. and biomed. engring. U. Tex., Austin, 1972-74, assoc. prof., 1974-80, prof., 1980—; cons.; dir. Moncrief-Popovich Research Inst. Served with U.S. Army, 1957-58. Recipient Dialysis Pioneering award Nat. Kidney Found. Mem. Am. Soc. Artificial Internal Organs, Internat. Soc. Peritoneal Dialysis, Am. Soc. Nephrology. Subspecialties: Artificial organs; Biomedical engineering. Current work: Artificial kidney, particularly peritoneal dialysis; biomedical instrumentation; plasmapheresis. Patentee in field. Office: University of Texas ENS 617 Dept Chemical Engineering Austin TX 78704

POPP, DIANA MARRIOTT, biologist; b. Pontiac, Mich., Sept. 4, 1933; s. Ralph Arthur and Stella Marie (Kruszewski) Marriott; m. Raymond Arthur Popp, Sept. 4, 1954; children: Raymond, Carolyn, David, Stevan. B.S., U. Mich., 1955. Med. technologist U. Mich., 1955-56; lab. asst. Operations Research Nat. Lab., 1960-65, research assoc., 1965—. Mem. Am. Soc. Clin. Pathologists, Transplantation Soc., Am. Assn. Immunologists, Southeastern Immunology Conf., Sigma Xi. Subspecialties: Genetic Toxicology; Immunogenetics. Current work: Utilize genetic and immunologic methodology to study natural and environmentally induced alterations that affect the life cycle of mammals. Home: 9901 Emory Rd Knoxville TN 37921 Office: Biology Div Oak Ridge Nat Lab Oak Ridge TN 37830 Office: Department of Chemistry University MO Kansas City MO 64110

POPP, FRANK DONALD, chemistry educator; b. N.Y.C., Dec. 25, 1932; s. Frank B. and Julia (Brown) P.; m. Barbara L. Popp, June 11, 1955; m. Jane C. Popp, Apr. 30, 1977; children: Bruce D., James W. B.S. in Chemistry, Rensselaer Poly. Inst., 1950-54; Ph.D., U. Kans., 1957. Asst. prof. U. Miami, 1959-62; asst. prof. Clarkson Coll. Tech., Potsdam, N.Y., 1962-64; assoc. prof., 1964-66, prof., 1966-76; prof., chmn. dept. chemistry U. Mo., Kansas City, 1976—. Contbr. to books and articles in field. USPHS fellow, 1958-59, 64-65. Fellow Am. Inst. Chemistry (cert. profl. chemist), AAAS, N.Y. Acad. Sci., Am. Assn. Cancer Research; mem. Am. Chem. Soc., Royal Soc. Chemistry, Mo. Acad. Sci., Internat. Soc. Heterocyclic Chemistry. Subspecialties: Organic chemistry; Medicinal chemistry. Current work: Synthesis and reactions of nitrgoen heterocyclic compounds and medicinal agents

POPPER, DAVID HENRY, fgn. service officer; b. N.Y.C., Oct. 3, 1912; s. Morris and Lilliam (Greenbaum) P.; m. Florence C. Maisel, Mar. 8, 1936; children—Carol, Lewis, Katherine, Virginia. A.B., Harvard, 1932, A.M., 1934. Research asso. Fgn. Policy Assn., Inc., N.Y.C., 1934-40, asso. editor publs., 1941-42; specialist internat. orgn. affairs Dept. of State, 1945—, asst. chief div. internat. orgn. affairs, 1948-49; officer in charge UN Gen. Assembly affairs, 1949-51; dept. dir. Office UN Polit. and Security Affairs, 1951-54, dir., 1954-55; Nat. War Coll., Washington, 1955-56; dep. U.S. rep. for internat. orgns., Geneva, 1956-59; adviser U.S. delegation to UN Gen. Assembly, 1946-53, prin. exec. officer, 1949-50; dep. U.S. rep. Conf. on Discontinuance Nuclear Weapons Tests, Geneva, 1959-61; sr. adv. disarmament affairs U.S. Mission to UN, 1961-62; dir. Office Atlantic Polit. and Mil. Affairs, Dept. State, 1962-65, dep. asst. sec. state for internat. organizational affairs, 1965-69; U.S. ambassador to, Cyprus, 1969-73, to, Chile, 1974-77, asst. sec. state for internat. orgn. affairs, 1973-74; dep. for Panama Canal Treaty Affairs, Dept. of State; with rank of ambassador, 1977-78; spl. rep. of sec. of state for Panama Treaty Affairs, 1978-80; adj. prof. Georgetown U. Sch. Fgn. Service, 1981—. Author: The Puzzle of Palestine, 1938; Contbr. profl. publs. Mem. U.S. delegations to NATO Ministerial Meetings, 1962-65; pres. Am. Friends of Cyprus. Served as capt. C.W.S., M.I. Service AUS, 1942-45. Mem. Latin Am. Studies Assn., Am. Soc. Internat. Law, Washington Inst. Fgn. Affairs, Acad. Polit. Sci., Am. Acad. Polit. and Social Sci. Club: Cosmos. Subspecialties: Pathology (medicine); Gastroenterology. Home: 6116 33d St NW Washington DC 20015

POPPER, STEVEN HERBERT, structural engineer, project engineer, consultant; b. N.Y.C., July 25, 1942; s. William and Ann (Steiner) P.; m. Barbara K. Krouner, Nov. 28, 1965; children: Jeffrey, Nancy, Lee, Sarah. B.S. in Civil Engring, Rensselaer Poly. Inst., 1964, M.S., 1965; postgrad., Babson Coll., 1978—. Engr. Stone & Webster Engring. Corp., Boston, 1965-70, structural engr., 1971-74, sr. structural engr., 1975—. Contbr. article to profl. jour. Mem. Am. Nuclear Soc. (standards com. Dry Storage-Spent Fuel 1981), ASCE (nuclear fuel cycle com.), Am. Concrete Inst. Subspecialties: Fuels and sources; Civil engineering. Current work: Storage concepts for high level radioactive wastes. Home: 31 Wilshire Pk Needham MA 02192 Office: Stone & Webster Engring Corp 245 Summer St Boston MA 02107

PORCH, WILLIAM MORGAN, atmospheric physicist; b. Athens, Ohio, Nov. 8, 1944; s. Virgil and Nellie (Evans) P.; m. Laura Virginia Lee, Nov. 20, 1978. Ph.D. (HEW spl. fellow), U. Wash., 1971. Research optical engr. Boeing Co., 1966-67; research asst. U. Wash., 1967-68, trainee, 1968-69, spl. fellow, 1969-71, research assoc., 1971-72; physicist Lawrence Livermore (Calif.) Nat. Lab., 1972—; lectr. U. Calif., Davis, 1975. Contbr. articles on atmospheric aerosol, climate, remote sensing, air pollution, wind energy, and complex terrain meteorology to profl. jours. Mem. Am. Meteorol. Soc., Optical Soc. Am., Am. Geophys. Union, Am. Assn. Aerosol Research. Subspecialties: Meteorologic instrumentation; Atmospheric optics. Current work: Remote sensing of air pollutants and regional meteorology. Patentee in field.

PORESKY, ROBERT HAROLD, child development educator; b. Allentown, Pa., Sept. 10, 1941; s. Milton N. and Betty (Chernoff) P.; m. Barbara Keebaugh, Sept. 2, 1964; children: Pamela, Laura. A.B., Cornell U., 1963, M.A., 1967, Ph.D., 1969. Cert. psychologist, Kans. Research assoc. Cornell U., Ithaca, N.Y., 1968-72; asst. prof. child development Kans. State U., Manhattan, 1972-77, acting dept. head, 1979-80, assoc. prof., 1977—. Author: Caring for Children Today, 1979; contbr. articles to profl. jours. Bd. dirs. Blue Valley Nursery Sch., Manhattan, 1973—; pres. Kans. Council for Children and Youth. Mem. Soc. for Research in Child Devel., Am. Psychol. Assn., Nat. Assn. for Edn. Young Children, Kans. Assn. for Edn. Young Children (pres.), Am. Home Econs. Assn. Subspecialty: Developmental psychology. Current work: Impact of proximate environment on children's development; development of rural children and families; infant development as affected by child care in home and alternative settings. Home: 3016 Claflin Rd Manhattan KS 66502 Office: Kans State Univ 312 Justin Hall Manhattan KS 66506

PORGES, STEPHEN WILLIAM, psychology educator, researcher; b. New Brunswick, N.J., Jan. 6, 1945; s. Frederick A. and Anna (Berkow) P.; m. Carol Sue Carter, June 12, 1970; 2 sons, Eric S. Carter, Seth Colin. Student, Rutgers U., 1962-64; B.A., Drew U., 1966; M.A., Mich. State U., 1963, Ph.D., 1970. Asst. prof. W. Va. U., 1970-72; asst. prof. psychology U. Ill.-Champaign, 1972-75, assoc. prof., 1975-82, prof., 1982—. Editor: book Psychophysiology, 1976; assoc. editor: jour. Psychophysiology; Contbr. numerous articles to profl. jours. Recipient Research Scientist Devel. award NIMH, 1975-80, 81-86; NIMH grantee, 1975—; Nat. Inst. Edn. grantee, 1973-73; NIH grantee, 1979-81,82—. Mem. Am. Psychol. Assn., Psychonomic Soc., Soc. Child Devel., Perinatal Research Soc., Soc. Psychophysiol. Research (sec.-treas. 1975-78, dir. 1978-79). Subspecialties: Psychobiology; Developmental psychology. Current work: Development of methods to decode spontaneous physiological activity to provide information regarding the condition of the central nervous system in humans. Office: U Ill 603 E Daniel St Champaign IL 61820 Home: 1407 Grandview Champaign IL 61820

PORIES, WALTER J., surgeon; b. Munich, Ger., Jan. 18, 1930; came to U.S., 1940, naturalized, 1945; s. Theodore F. and Frances (Lowen) P.; m. Mary Ann Rose, June 4, 1977; children: Susan, Mary Jane, Carolyn, Kathy, Liza, Michael. B.A., Conn. Wesleyan U., 1972; M.D. with honors, U. Rochester, 1955. Intern in surgery and obstetrics Strong Meml. hosp., U. Rochester, N.Y., 1955-56; fellow in head and neck cancer Center du Cancer, U. Nancy, France, 1956-58; grad. research fellow AEC, U. Rochester, 1958-59; resident in gen. and thoracic surgery, 1958-62; with East Carolina U., 1977—, prof., chmn. dept. surgery, 1977—; chief surgery Pitt County Meml. Hosp., Greenville, N.C., 1977—. Mem. Acad. Medicine Cleve. (dir.), AAAS, Am. Assn. Cancer Edn., Am. Assn. Cancer Research, Am. Burn Assn., Am. Coll. Cardiology, Am. Coll. Nutrition, Am. Coll. Surgeons, Am. Fedn. Clin. Research, Am. Nuclear Soc., Am. Soc. Clin. Oncology, Am. Surg. Assn., Am. Trauma Soc., Assn. Acad. Surgery, Assn. Mil. Surgeons of U.S., Central Surg. Assn., Cleve. Surg. Soc., Cleve. Vascular Soc., Internat. Assn. Bioinorganic Scientists, Internat. Cardiovascular Soc., Ohio State Med. Assn., Pan-Am. Med. Assn., Royal Soc. Health, Soc. U. Surgeons, Soc. Vascular Surgery, Soc. Surg. Oncology, So. Assn. Vascular Surgery, Soc. Surg. Chmn., Soc. Environ. Geochemistry and Health (pres.-elect), Surg. Biology Club. Subspecialties: Surgery; Biochemistry (medicine). Office: East Carolina U Sch Medicine Greenville NC 27834

PORTER, ALAN LESLIE, industrial engineering educator; b. Jersey City, June 22, 1945; s. Leslie Frank and Alice Mae (Kaufman) P.; m. Claudia Loy Ferrey, June 14, 1968; children: Brett, Douglas, Lynn. B.S., Calif. Inst. Tech., 1967; Ph.D., UCLA, 1972. Research asst. prof. U. Wash., Seattle, 1972-74; asst. prof. indsl. engring. Ga. Inst. Tech., Atlanta, 1975-78, assoc. prof., 1979—; cons. U.S. Dept. Commerce, Washington, 1980-81, State of Md., Balt., 1979-80. Author: (with others) Guidebook for Technology Assessment and Impact Analysis, 1980; editor: (with Thomas Kuehn) Science, Technology and National Policy, 1981; co-editor-in-chief: Impact Assessment Bull, 1981—. Grantee NSF, 1974-83, U.S. Dept. Transp., 1977-79, Fund for Improvement Post Secondary Edn., 1976-78, Hazen Found., 1973-74. Mem. Internat. Assn. Impact Assessment (sec. 1981-84), IEEE, Systems, Man and Cybernetics Soc. (chmn. tech. forecasting 1981-83), Am. Soc. Engring. Edn. (chmn. engring. and pub. policy div. 1982-83). Subspecialties: Systems engineering; Technology forecastingand assessment. Current work: Technology forecasting and assessment, quasi-experimental evaluation methodology, interdisciplinary research processes, future of work. Home: 110 Lake Top Ct Roswell GA 30076 Office: Ga Inst Tech Atlanta GA 30332

PORTER, CHASTAIN KENDALL, dentistry educator, consultant, researcher; b. Kansas City, Mo., Oct. 6, 1932; s. Chastain Gant and Mattie (Lantz) P.; m. Dawna Gayle Morgan, July 12, 1958 (; widowed), July 12, 1958; children: Lisa, Christopher; m. m. Linda Jeanne Woolworth, Aug. 3, 1963; children: Keith, Laura, Kevin. B.A., U. Kans. City, Mo., 1954, D.D.S., 1957; M.S., Ohio State U., Columbus, 1959. Instr. oral pathology U. Kans. City, 1959-62, assoc. prof., 1962-63; assoc. prof. dentistry U. Mo.-Kansas City, 1963-69, assoc. prof., 1969-77, prof., 1977—; cons. VA, Leavenworth, Kans., 1966—. Contbr. numerous articles to profl. jours. Mem. Am. Assn. Dental Research, Internat. Assn. Dental Research, Omicron Kappa, Upsilon. Subspecialties: Oral pathology; Dental growth and development. Current work: Research in oral histology and pathology; clinical research in dental caries and periodontics. Office: U Mo Sch Dentistry 650 E 25th St Kansas City MO 64108

PORTER, FREDERICK CHARLES, aerospace company executive, consultant; b. New Orleans, Jan. 10, 1937; s. Frederick Charles and Jane (Currens) Fitz-Randolph; m. Gayle Mae Johnson, June 7, 1959; children: Steve, Linda. B.S. in M.E, U. Colo., 1959; mgmt. cert., Alexander Hamilton Inst., N.Y.C., 1965. Registered profl. engr., Calif. Research asst. U. Colo., 1957-59; structures engr. Rockwell Internat., Los Angeles, 1959-60; sr. project engr. Gen. Dynamics/Convair, San Diego, Calif., 1960—; Contbr. in field. Mem. nat. adv. bd. Am. Security Council; mem. Republican Nat. Com. Fellow AIAA (assoc. treas. San Diego sect. 1970-71), Nat. Mgmt. Assn. Republican. Club: Hacienda Recreation Assn. (pres. 1974-76). Subspecialty: Aerospace engineering and technology. Current work: Develop new aerospace programs in fields of advanced missile weapon systems, space transportation and advanced space applications. Home: 1950 WoodGlen Way El Cajon CA 92020 Office: General Dynamics Convair PO Box 85357 San Diego CA 92138

PORTER, JOHN ROBERT, JR., consulting company executive; b. Oklahoma City, Feb. 27, 1935; s. John Robert and Margaret Florence (Nicholson) P.; m. Amelie Wallace, June 2, 1962; children: Jennifer, Amy. A.B., Dartmouth Coll., 1957; M.S., Okla. U., 1964. Sci. project officer govt. agy., Washington, 1962-66; chief earth resources program NASA, Washington, 1966-69; pres. Earth Satellite Corp. (EarthSat), cons. specializing in remote sensing applied to oil and mineral exploration, crop forecasting and image enhancement software, Chevy Chase, Md., 1969—; dir. Stauffer Communications, Inc., 1962—; mem. Remote Sensing Adv. Com. Dept. Commerce.; mem. Space Applications Com. NRC. Contbr. articles on clay mineralogy, remote sensing to profl. jours. Trustee Washington Gallery Modern Art, 1964-65. Served to 1st lt. Signal Corps U.S. Army, 1960-62. Mem. Am. Assn. Petroleum Geologists (cert. petroleum geologist). Clubs: University, Chevy Chase (Washington). Subspecialties: Remote sensing (geoscience); Satellite studies. Office: 7222 47th St Chevy Chase MD 20815

PORTER, KEITH ROBERTS, cytologist, educator; b. Yarmouth, N.S., Can., June 11, 1912; s. Aaron C. and Josephine (Roberts) P.; m. Elizabeth Lingley, June 16, 1938. B.S., Acadia U., 1934, D.Sc. (hon.), 1964; M.A., Harvard U., 1935, Ph.D., 1938; LL.D. (hon.), Queen's U., 1966, D.Sc., Med. Coll. Ohio, 1975, Rockefeller U., 1976, U. Toronto, 1978, U. Pierre et Marie Curie de Paris, 1979, Med. U. S.C., 1982, U. Colo., 1982, U. Western Ont., 1983. NRC fellow Princeton, 1938-39; research asst. Rockefeller Inst., 1939-45, asso., 1945-50, asso. mem., 1950-56, mem., prof., 1956-61; prof. biology Harvard, 1961-70, chmn. biology dept., 1965-67; prof., chmn. molecular, cellular and developmental biology U. Colo., Boulder, 1968-74; Cons. Armed Forces Inst. Pathology, 1957-70; mem. adv. panel developmental biology NSF, 1960-66; Dir. Marine Biol. Lab., Woods Hole, Mass., 1975-76. Mng. editor: (now Jour. Cell Biology) Jour. Biophys. and Biochem. Cytology, 1955-64; co-editor: Protoplasma, 1964—; cons. editor: McGraw-Hill Ency. Sci. and Tech, 1960-76. Recipient (with George Palade) Warren Triennial award, 1962; (with Palade); Passano award, 1964; Gairdner Found. Ann. award, 1964; with Palade and Albert Claude Louisa Gross Horwitz prize, 1970; Waterford Biomed. award Scripps Clinic and Research Found., 1979; Guggenheim fellow, 1967-68; (with Palade); Dickson prize, 1971; (with Claude, F.S. Sjostrand); Paul Ehlich-Ludwig Darmstaedter prize, 1971; Robert L. Stearns award U. Colo., 1973; Nat. Medal Sci., 1977; Waterford Bio-Med. award Scripps Clinic and Research Found., 1979. Mem. Am. Soc. Zoologists, Am. Assn. Anatomists (Henry Gray award 1981), Harvey Soc., Electron Microscope Soc. Am. (pres. 1962-63, award 1975), Tissue Culture Assn. (1st pres. 1946, pres. 1978-80), Am. Soc. Cell Biology (1st pres. 1961, pres. 1977-78), Am. Acad. Arts and Scis., Internat. Inst. Embryology, Internat. Soc. for Cell Biology, Am. Assn. Pathologists and Bacteriologists, Nat. Acad. Scis. (adv. com. on USSR and Eastern Europe 1962-65), Sigma Xi. Subspecialties: Cell biology; Electron miroscopy. Current work: The cytoplasmic matrix and its involvement in selective transport and the determination of cell form and function. Office: Molecular Cellular and Developmental Biology Dept U Colo Campus Box 347 Boulder CO 80309

PORTER, ROGER JOHN, neurologist, pharmacologist; b. Pitts., Apr. 4, 1942; s. John K. and Margaret P. P.; m. Candace Leland, Feb. 17, 1968; children: David L., Stacey K. B.S., Eckerd Coll., 1964; M.D., Duke U., 1968. Diplomate: Am. Bd. Psychiatry and Neurology. Intern U. Calif., San Diego, 1968-69; commd. officer USPHS, 1969-71, 74—, med. dir., 1979—; resident in neurology U. Calif., San Francisco, 1971-73, chief resident, 1973-74; chief epilepsy br. Nat. Inst. Neurol. and Communicative Disorders and Stroke, NIH, Bethesda, Md., 1979—; clin. prof. neurology, adj. asso. prof. pharmacology Uniformed Services U. Health Scis., Bethesda, 1979—; cons.-lectr. Nat. Naval Med. Center, Bethesda, 1978—. Contbr. numerous articles on epilepsy research and antiepileptic drug pharmacology to sci. jours. Recipient Commendation medal USPHS, 1977; MacArthur Disting. Alumnus award Eckerd Coll., 1977. Mem. Am. Acad. Neurology, Am. Neurol. Assn., Am. Epilepsy Soc., Am. EEG Soc., Soc. Neurosci., Am. Soc. Clin. Pharmacology and Therapeutics. Subspecialties: Neurology; Pharmacology. Current work: Clinical epilepsy research. Office: Fed Bldg Room 114 NIH Bethesda MD 20205

PORTER, STEPHEN CUMMINGS, geological sciences educator; b. Santa Barbara, Calif., Apr. 18, 1934; s. Lawrence J. and Frances C. (Cummings) Seger P.; m. Anne M. Higgins, Apr. 2, 1959; children: John, Maria, Susannah. B.S., Yale U., 1955, M.S., 1958, Ph.D., 1962. Asst. prof. geol. scis. U. Wash., Seattle, 1962-66, assoc, . prof., 1966-69, prof., 1969—; dir. Quaternary Research Ctr., 1981—; v.p. Cambria Corp., Seattle, 1981—. Editor: Late pleistocene Environments of the United States, 1982; editor: Quaternary Research, 1977—; assoc. editor: Radiocarbon; editorial bd.: Quartenary Scis. Revs.; bd. earth scis.: Nat. Acad. Scis./NRC, 1983-85. Served to lt. USNR, 1955-57. Recipient Benjamin Silliman prize Yale U., 1962; Fulbright fellow, 1973-74; vis. fellow Cambridge U., 1980-81. Fellow Arctic Inst. N.Am. (gov. 1972-77), Geol. Soc. Am.; mem. AAAS, Internat. Glaciological Soc. Club: Am. Alpine (N.Y.). Subspecialty: Geology. Current work: Quaternary glaciation of alpine regions, especially Alps, Himalayas, Andes, North American Cordillera, Hawaii; volcanology and tephrochronology; quaternary stratigraphy and chronology. Home: 18034 15th Ave NW Seattle WA 98177 Office: Quaternary Research Ctr U Wash Seattle WA 98195

PORTER, WARREN PAUL, zoology educator; b. Madison, Wis., Jan. 26, 1939; s. Robert Howard and Lorayn Margaret (Huybrecht) P.; m.; children: Lisa, Paul. B.S., U. Wis.-Madison, 1961; M.A., UCLA, 1963, Ph.D., 1966. NASA predoctoral fellow UCLA, 1964-66; NIH postdoctoral research assoc. Washington U., St. Louis, 1966-1968; asst. prof. zoology U. Wis.-Madison, 1968-71, assoc. prof., 1971-74, prof., 1974—; cons. Electric Power Research Inst., A.O. Smith Corp., Fed. Wildlife Disease Control Ctr. Recipient award for outstanding undergrad. teaching AMOCO, 1969; Romnes Faculty fellow, 1977; Guggenheim fellow, 1979. Mem. AAAS, Ecol. Soc. Am., Am. Soc. Ichthyologists and Herpetologists, Am. Inst. Biol. Scis., Am. Soc. Zoologists. Subspecialties: Biophysical ecology; Physiological ecology. Current work: Climate, disease, toxicant interactions and effects on animal maintenance growth and reproduction. Office: Dept Zoology U. Wis 1117 W Johnson St Madison WI 53706

PORTER, WILLIAM HUDSON, biochemist educator; b. Wilson, N.C., Mar. 12, 1940; s. Frank L. and Mildred (McCollum) P.; m. Faye Hardee, Mar. 23, 1963; children: Tracy Michele, Mark William. B.S., The Citadel, 1962; M.S., Med. U. S.C., 1966; Ph.D., Vanderbilt U., 1970. Diplomate: Am. Bd. Clin. Chemistry. Postdoctoral fellow Fla. State U., Tallahassee, 1970-72; research asst. prof. U. Tenn. Meml. Research Ctr., Knoxville, 1972-74; asst. prof. lab. medicine Med. U. S.C., Charleston, 1974-78; assoc. prof. pathology and assoc. dir. clin. chemistry U. Ky. Med. Ctr., Lexington, 1978—, co-dir. postdoctoral tng. program in clin. chemistry, 1976—. Contbg. author: Clinical Guide to Laboratory Tests, 1983; contbr. articles to profl. jours. NIH predoctoral fellow, 1966-70; postdoctoral fellow, 1970-72. Fellow Nat. Acad. Clin. Biochemistry; mem. Am. Soc. Biol. Chemists, Am. Assn. Clin. Chemistry (chmn). Subspecialties: Clinical chemistry; Toxicology (medicine). Current work: Analytical and clinical toxicology; clinical enzymology; analytical methods in clinical chemistry. Home: 356 Atwood Dr Lexington KY 40503 Office: Dept Pathology Univ Ky Med Center 800 Rose St Lexington KY 40536

PORTERFIELD, WILLIAM WENDELL, educator; b. Winchester, Va., Aug. 24, 1936; s. Donald Kennedy and Adelyn (Miller) P.; m. Dorothy Elizabeth Dail, Aug. 24, 1957; children—Allan Kennedy, Douglas Hunter. B.S., U. N.C., 1957, Ph.D., 1962; M.S., Calif. Inst. Tech., 1960. Sr. research chemist Hercules, Inc., Cumberland, Md., 1962-64; asst. prof. chemistry Hampden-Sydney (Va.) Coll., 1964-65, assoc. prof., 1965-68, prof. chemistry, 1968—, chmn. natural sci. div., 1973-77, chmn. dept. chemistry, 1982—. Author: Concepts of Chemistry, 1972, Inorganic Chemistry, 1984; Contbr. articles to profl. jours. Mem. Am. Chem. Soc., Royal Chem. Soc. (London, Eng.), Phi Beta Kappa. Subspecialty: Inorganic chemistry. Current work: Synthesis, spectroscopic and electrochemical properties of room-temperature molten salts. Home: Box 697 Hampden-Sydney VA 23943

POSCHEL, BRUNO PAUL HENRY, neuropsychopharmacologist; b. Bklyn., June 6, 1929; s. Alfred B. and Elvira (Liukkonen) P.; m. Rose Marie Holmes, Mar. 17, 1956; children: Curtis P., Jane E. B.S., Roosevelt U., 1951; Ph.D., U. Ill., 1956. NIMH postdoctoral fellow U. Ill., 1956-57; asst. prof. Wayne State U., Detroit, 1957-59; research scientist Parke-Davis & Co., Ann Arbor, Mich., 1959-82; sr. research fellow Warner-Lambert Co., Ann Arbor, 1982—. Patentee in field; contbr. articles to profl. jours. Fellow Am. Psychol. Assn.; mem. Am. Soc. Pharmacology, Am. Coll. Neuropsychopharmacology. Subspecialties: Neuropharmacology; Neuropsychology. Current work: Discovery of drugs to enhance cognition, learning, memory, and application to disorders such as Alzheimer's disease. Office: Warner-Lambert Co 2800 Plymouth Rd Ann Arbor MI 48105

POSKANZER, ARTHUR M., nuclear chemist; b. N.Y.C., June 28, 1931; s. Samuel I. and Adele (Kerman) P.; m. Lucille Block, June 12, 1954; children; Deborah, Jeffrey, Harold. A.B., Harvard Coll., 1953; M.A., Columbia U., 1954; Ph.D., MIT, 1957. Chemist Brookhaven Nat. Lab., Upton, N.Y., 1957-66; sr. scientist Lawrence Berkeley (Calif.) Lab., 1966—; chmn. nuclear chemistry Godron Conf., 1970; mem. panel on future of nuclear sci. Nat. Acad. Sci., 1976. Numerous isotopes of light elements and contbr. to study of relativistic nuclear collisions. Guggenheim fellow, 1970-71; NATO Sr. fellow in sci., 1975-76; CERN, sci. assoc., 1979-80; recipient nuclear chemistry award Am. Chem. Soc., 1980. Fellow Am. Phys. Soc.; mem. Am. Chem. Soc. (chmn. div. nuclear chemistry and tech. 1977). Subspecialties: Nuclear physics; Nuclear chemistry. Current work: Research on high energy nuclear collisions, and nuclei far from stability. Discoverer of. Office: Lawrence Berkeley Lab Berkeley CA 94720

POSKY, LEON, biochemist, nutritionist; b. N.Y.C., Aug. 2, 1933; s. Myer and Beckie (Prosk) P.; m. Heather Dee Dublin, Apr. 29, 1973; children: Melissa Shelley, Rebecca Evelyn. B.S., Bklyn. Coll., 1954; M.S., Rugers U., 1955, Ph.D., 1958. Postdoctoral fellow dept. pharmacology Washington U. Sch. Medicine, St. Louis, 1960-62; research assoc. biochemistry Albert Einstein Coll. Medicine, Bronx, N.Y., 1963-65; sect. chief div. nutrition FDA, Washington, 1965-79, program mgr. nutrition, 1974-79, dep. chief exptl. nutrition br., 1979—; cons. U. Colo. Med. Ctr., Denver, 1959; adj. prof. biochemistry George Washington U. Med. Sch., Washington, 1980—. Contbr. numerous articles to med jours. Recipient Commendable Service award FDA, 1980; award of Merit, 1982. Fellow Am. Inst. Chemists, AAAS; mem. N.Y. Acad. Scis., Soc. for Exptl. Biology and Medicine (Program chmn. and pres.-elect D.C. sect. 1974-75, awards com. 1967-68, 1973-74, councilor 1976—), Am. Coll. Nutrition, Am. Chem. Soc., European Assn.for Study of Diabetes, Am. Inst. Nutrition, Sigma Xi. Subspecialties: Animal nutrition; Biochemistry (biology). Current work: Nutrition research, dietary protein, carbohydrate metabolism, food additives, obesity. Home: 1004 Challedon Rd Great Falls VA 22066 Office: Expt Nutrition Br FDA Washington DC 20204

POSNER, BARRY ZANE, management scientist and consultant, educator; b. Hollywood, Calif., Mar. 11, 1949; s. Henry and Delores Ann; ; s. Henry and Ginsberg P.; m. Jackie Schmidt, July 23, 1972; 1 dau., Amanda Delores. B.A., U. Calif.-Santa Barbara, 1970; M.A., Ohio State U., 1972; Ph.D., U. Mass., 1976. Asst. prof. mgmt. U. Santa Clara, 1976-82, assoc. prof., 1982—; mgmt. cons., 1974—; founder, chief exec. officer Spectrum Assocs., Santa Clara, 1976—. Author: (with W.H. Schmidt) Manager's Values and Expectations, 1982; contbr. numerous articles to profl. jours. Recipient Pres.'s Disting. Faculty award U. Santa Clara, 1981. Mem. Acad. Mgmt., Am. Inst. Decision Scis., Orgn. Behavior Teaching Soc. (dir. 1982—). Democrat. Jewish. Club: Faculty (Santa Clara) (dir. 1982—). Subspecialties: Social psychology. Current work: Managerial value systems; socialization processes; recruitment and selection; experiential learning; organization development; high performing systems. Home: 2330 Forbes Ave Santa Clara CA 95050 Office: Sch Bus Adminstrn Dept Mgmt U Santa Clara Santa Clara CA 95053

POSSO, MANUEL, physician, pathologist; b. Colombia; s. Napoleon and Anna P.; m. Sylvia Buenaventura; 4 children. M.D., Universidad del Valle, Columbia, 1959. Diplomate: Am. Bd. Pathology, Am. Bd. Family Practice. Chief med. examiner Tompkins County, N.Y., 1977—; assoc. pathologist Tompkins Community Hosp., 1968—; adj. prof. Cornell U., 1980—. Contbr. articles to profl. jours. Bd. dirs. Tompkins County Cancer Soc. Fellow Am. Coll. Pathologists; mem. Am. Acad. Family Physicians. Roman Catholic. Subspecialties: Pathology (medicine); Family practice. Current work: General pathology.

POST, DOUGLASS EDMUND, JR., physicist; b. Gulfport, Miss., Mar. 16, 1945; s. Douglass Edmund and Olive Brown (Shaw) P.; m. Susan Elizabeth Yocky; children: Alison Elizabeth, Alan Douglas. B.S. in Physics, Southwestern U. at Memphis, 1967; M.S. in Physics (Woodrow Wilson fellow, 1967, NSF fellow, 1967), Stanford U., 1968; Ph.D. in Physics (Fannie and John Hertz Found. fellow, 1972-75), Stanford U., 1975. Physicist Lawrence Livermore (Calif.) Lab., 1968-71; physicist Plasma Physics Lab., Princeton (N.J.) U., 1975—, head Tokamak Modeling Br., 1980—; cons. Lawrence Livermore Lab., 1972—. Contbr. articles to profl. publs. and confs. Active N.J. Conservation Found., Zero Population Growth, Sierra Club. Mem. Am. Phys. Soc., Am. Nuclear Soc. (exec. bd. fusion sect.), Am. Vacuum Soc., Phi Beta Kappa, Omicron Delta Kappa. Presbyterian. Subspecialties: Fusion; Numerical analysis. Current work: Computational modeling of fusion experiments, application of atomic and molecular processes to fusion research. devel. of fusion diagnostics. Inventor: cheap high vacuum feedthrough, Alpha-Particle Diagnostic, light atom neutral heating beam, He030 ICRF Diagnostic. Home: 11 Buttonwood Ct Belle Meade NJ 08502 Office: Princeton Univ Plasma Physics Lab PO Box 451 Princeton NJ 08544

POST, MADISON JOHN, physicist; b. Detroit, Oct. 4, 1946; s. Madison and Clara Agnus (Krause) P.; m. Donna Darlene Horky, Oct. 25, 1953. B.S. in Elec. Engring, U. Ill., 1969; M.S. in Astro-geophysics, U. Colo., 1975, U. Ariz., 1983. Asst. sta. sci. leader ESSA, Byrd Sta., Antarctica, 1969-70; engr. NOAA, Boulder, 1970-75, physicist, 1975—, Wave Propagation Lab., 1975—. Contbr. numerous articles in optics and atmospheric research to profl. jours. Mem. Optical Soc. Am., Sigma Xi. Clubs: Colo. Mountain (Boulder); Porsche (Denver). Subspecialties: Remote sensing (atmospheric science); Optical engineering. Current work: Design, analysis of coherent Doppler lidar for atmospheric science; remote wind sensing by coherent laser radar. Office: 325 Broadway R E WP2 Boulder CO 80303

POST, ROBERT MORTON, psychiatrist, researcher; b. New Haven, Sept. 16, 1942; s. William and Esther P.; m. Susan Lee Wolf, Dec. 23, 1966; children: Laura A., David F. B.A. cum laude in Psychology, Yale U., 1964; M.D., U. Pa., 1968. Mixed intern Albert Einstein Sch. Medicine, Jacobi Hosp., Bronx, N.Y., 1968-69; resident in psychiatry Mass. Gen. Hosp., Boston, 1969-70; clin. assoc. psychiatry Clin. Sci. Lab., NIMH, Bethesda, Md., 1970-72, research fellow in psychiatry, 1972-73, chief clin. research unit sect. on psychobiology, 1973-77, chief sect. on psychobiology, 1977-81, acting chief biol. psychiatry br., 1981-83, chief biol. psychiatry br., 1983—; lectr; resident in psychiatry George Washington U., 1973-75; lectr. Assoc. editor: Psychiatry Research; mem. editorial bd.: Biol. Psychiatry; contbr. numerous articles to profl. jours. Recipient Kenneth E. Appel prize U. Pa., 1968; A. E. Bennett Neuropsychiat. Research Found. award Soc. Biol. Psychiatry, 1973; Adminstrs. award HEW, 1978. Mem. AAAS, Am. Coll. Neuropsychopharmacology, Am. Psychiat. Assn. (research prize 1983), Soc. Biol. Psychiatry, Psychiat. Research Soc., Soc. Neurosci., Collegium Internationale Neuro-Psychopharmacologicum, Alpha Omega Alpha. Subspecialties: Psychopharmacology; Neuropharmacology. Current work: Manic-depressive illness, panic anxiety disorders, schizophrenia; behavioral, biochemical and electrophysiological alterations during ill and well states; investigation with clinical treatment probes of neurotransmitter and peptide systems; regional approaches to localization of function; implications of efficacy of anticonvulsant carbamazepine in treatment of affective illness; dissection of biological and behavioral interactions in animal models. Home: 3502 Turner Ln Chevy Chase MD 20015 Office: Bldg 10 Room 3N212 NIMH 9000 Rockville Pike Bethesda MD 20205

POSTE, GEORGE HENRY, experimental pathologist, pharmaceuticals company executive; b. Polegate, Eng., Apr. 30, 1944; came to U.S., 1972; s. John Henry and Kathleen Betty (Brook) P.; m. Mary E., Mar. 9, 1968; 1 dau., Ellie. D.V.M., U. Bristol, Eng., 1966; Ph.D., 1969. Faculty U. London, 1969-72, Roswell Pk. Meml. Inst., SUNY-Buffalo, 1972-80; v.p., dir. research Smith Kline Corp., Phila., 1980-83, v.p., dir. research and devel., 1983—; research prof. dept. pathology U. Pa., Phila., 1980—; mem. NIH study sect. Mem. Royal Coll. Pathologists, London. Subspecialties: Cancer research (medicine); Cell biology (medicine). Current work: Cancer metastasis, cell recognition, macrophage function. Address: 1500 Spring Garden St Philadelphia PA 19101

POSTLETHWAIT, JOHN HARVEY, biology educator, molecular genetics researcher; b. Kittery, Maine, July 15, 1944; s. Samuel Noel and Sara Madeline (Cover) P.; m. Juanita Ann, Mar. 22, 1964; children: Holly Ann, Heather Lynne. B.A. in Biology, Purdue U., 1966; Ph.D. in Devel. Genetics, Case Western Res. U., 1970. Postdoctoral researcher Harvard U., Cambridge, Mass., 1970-71; asst. prof. biology U. Oreg.-Eugene, 1971-76, asso. prof., 1976-81, prof., 1981—; vis. research scientist Institut für Molekularbiologie, Austrian Acad. Scis., Salzburg, 1977-78, CNRS, Faculte de Médécin, Strasbourg, France, 1982-83; program dir. grad. tng. in genetics NIH; textbook editor Eirik Borve Inc.; editorial bd. Wilhelm Roux Archives. Contbr. writings to profl. publs. in field. Woodrow Wilson fellow, 1966-67; Career Devel. awardee UPHS, 1974-79; Fulbright fellow, 1977-78; recipient Ersted Disting. Teaching award U. Oreg., 1981; U.S.-France Endl. Exchange Program awardee, 1982-83. Mem. Genetics Soc. Am., Soc. Developmental Biology, AAAS. Subspecialties: Gene actions; Reproductive biology. Current work: Drosophila development and molecular genetics; vitellogenin; insect endocrinology. Home: 2356 Floral Hill Dr Eugene OR 97403 Office: Dept Biology U Oreg Eugene OT 97403

POSTLETHWAIT, SAMUEL NOEL, biology educator, researcher, consultant; b. willeysville, W.Va., Apr. 16, 1918; s. Albert Franklin and Marietta (Mason) Postlehwait; m. Sara Cover, Mar. 22, 1942; children: John Harvey, Robert Neil. A.B. in Biology, Fairmont State Coll., 1940; M.S. in Botany, W.Va. State U., 1947; Ph.D. in Plant Anatomy, State U. Iowa, 1949; Dr. Pedagogy (hon.), Doan Coll., 1982. Tchr., W.Va. Pub. Schs., 1940-41; instr. botany and biology State U. Iowa, Iowa City, 1948-49; asst. prof. biology Purdue U., West Lafayette, Ind., 1949-56, assoc. prof. biology, 1956-63, prof., 1963—; adv. bd. Community Coll. Forum. Author: (with J. Novak, H.T. Murray) An Integrated Experience Approach to Learning - With Emphasis on Independent Study, 1969, 72; author: Exploring Teaching Alternatives, 1979; editor: (with J. Novak, H. T. Murray) Minicourses in Biology, 1978; contbr. articles on biol. scis. to profl. jours. NSF fellow Manchester U., 1957-58; Fulbright grantee Macquarie U., Australia, 1968; recipient Purdue Student Govt. Best Tchr. award, 1965, Best Tchr. award Sigma Delta Chi, 1965; Ind. Acad. Sci. Speaker of Yr. award, 1978; Internat. Congress Individualized Instrn., Postlethwait award, 1980. Fellow Ind. Acad. Sci., AAAS; mem. Bot. Soc. Am., Nat. Assn. Biology Tchrs., Internat. Soc. Plant Morphology, Am. Inst. Biol. Scis., Internat. Soc. Sterology, Am. Genetic Assn., NEA, Ind. Sci. Tchrs. Assn., Internat. Platform Assn., Nat. Assn. Resources in Sci. Teaching, Torrey Bot. Club, Sigma Xi, Omicron Delta Kappa, Kappa Delta Kappa, Phi Eta Sigma. Subspecialties: Plant growth; Developmental biology. Current work: Plant morphology and science education, audio-tutorial instruction, instructional design, developmental morphology of maize, gene control of form, three-dimension of structure and its interpretation. Home: 3180 Soldiers Home Rd West Lafayette IN 47906 Office: 221 Chemistry Bldg Purdue West Lafayette IN 47907

POSTMA, HERMAN, research instn. exec.; b. Wilmington, N.C., Mar. 29, 1933; s. Gilbert and Sophie Hadrian (Verzaal) P.; m. Patricia Dunigan, Nov. 25, 1960; children: Peter, Pamela. B.S. summa cum laude, Duke U., 1955; M.S., Harvard U., 1957, Ph.D., 1959. Registered profl. engr., Calif. Summer staff Oak Ridge Nat. Lab., 1954-57, physicist thermonuclear div., 1959-62, co-leader, 1962-66, asst. dir. thermonuclear div., 1966, asso. dir. div., 1967, dir. div., 1967-73, dir. nat. lab., 1974—; vis. scientist FOM-Instituut voor Plasma-Fysica, The Netherlands, 1963; cons. Lab. Laser Energetics, U. Rochester.; mem. energy research adv. bd. spl. panel Dept. Energy. Editorial bd.: Nuclear Fusion, 1968-74. Bd. dirs. The Nucleus; chmn. bd. trustees Hosp. of Meth. Ch.; mem. adv. bd. Coll. Bus. Adminstrn., U. Tenn., 1976—, Energy Inst., State of N.C.; bd. dirs., exec. com. Tenn. Tech. Found., 1982—. Fellow Am. Phys. Soc. (exec. com. div. plasma physics), AAAS, Am. Nuclear Soc. (dir.); mem. C. of C. (exec. com., treas. 1981-83), Phi Beta Kappa, Beta Gamma Sigma, Sigma Pi Sigma, Omicron Delta Kappa, Sigma Xi, Pi Mu Epsilon, Phi Eta Sigma. Subspecialties: Plasma physics; Atomic and molecular physics. Current work: Technical management of research and development, particularly large interdisciplinary teams. Home: 104 Berea Rd Oak Ridge TN 37830 Office: Oak Ridge Nat Lab PO Box X Oak Ridge TN 37830

POSTON, HUGH ARTHUR, research physiologist; b. Canton, N.C., May 5, 1929; s. William Basil and Edna Elizabeth (Pless) P.; m. Birdie Lou Mugge, June 11, 1954; children: Mark, Martha, Phyllis. B.S., Berea Coll., 1951; M.S., N.C. State U., 1953, Ph.D., 1961. Cert. fisheries scientist, N.Y. Mgr. dairy research sta. N.C. State U., Raleigh, 1956-58, research asst., 1958-61; research physiologist U.S. Fish and Wildlife Service, Cortland, N.Y., 1961—; interagy. cons. EPA, Narragansett, R.I., 1975-78; subcom. mem. Nat. Acad. Scis./NRC, 1978-81; adj. assoc. prof. N.Y. State Coll. Vet. Medicine, Ithaca, 1977—. Mem. Environ. Mgmt. Council, Cortland; scouting coordinator Troop 94 Boy Scouts Am. Recipient Cert. Appreciation EPA, 1977. Fellow Am. Inst. Fishery Research Biologists; mem. Am. Fisheries Soc., Am. Inst. Nutrition, World Maricultural Soc., Sigma Xi, Gamma Sigma Delta. Club: Y's Men (pres.). Subspecialties: Nutrition (biology); Physiology (biology). Current work: investigations in nutrition and physiology of fish; provide technical advice and information to users about fish nutrition. Office: Tunison Lab Fish Nutrition 28 Gracie Rd Cortland NY 13045 Home: 2644 Sugar Bush Ln Dryden NY 13053

POSTON, JOHN MICHAEL, research biochemist; b. Kalispell, Mont., Oct. 16, 1935; s. Howard Joseph and Mabel Lenore (Iverson) P.; m. Annette Marlane Dapp, July 15, 1967; children: Janice Marie, Susanne Marlene, Todd Russell. B.S., Mont. State Coll., 1958; M.S., U. N.D., 1960, Ph.D., 1970. Research fellow U. N.D., Grand Forks, 1960-61; chemist NIH, Bethesda, Md., 1961-69; research biochemist Nat. Heart, Lung, Blood Inst., 1970—. NIH fellow, 1969. Mem. Am. Chem. Soc., Am. Soc. Biol. Chemists, Am. Soc. Microbiology, Sigma Xi. Lutheran. Subspecialties: Biochemistry (medicine); Microbiology. Current work: Study of metabolism of branched-chain amino acids, especially with regard to involvement of cobalamins (B0120) in man, animals, and bacteria: interest in in-born errors of metabolism that relate to these studies. Home: 29 Orchard Way S Rockville MD 20854 Office: Biochemistry Lab Nat Heart Lung Blood Inst NIH Bethesda MD 20205

POTASH, LOUIS, virologist; b. Boston, Sept. 14, 1924; s. Morris and Ethel (Gerstein) P.; m. Carole Ina Appleberg, Dec. 25, 1951; children: Moira Elayne, Stuart Alan, Neil Bradley. Student, Northeastern U., 1942-43; B.A., Boston U., 1947; M.Sc. in Hygiene, Harvard U., 1951; Ph.D. in Microbiology, Tulane U., 1958. Lab. technician Harvard Med. Sch., 1947-48, Sch. Pub. Health, 1948-49; bacteriologist Chem. Warfare Labs., Ft. Detrick, Md., 1951-52; research assoc. microbiology Tulane U., 1952-53, grad. asst., 1953-54, research assoc. in epidemiology, 1954-58, instr., 1958-59; research assoc. Merck Inst. Therapeutic Research, West Point, Pa., 1959-64; sr. scientist, program mgr., supr. Flow Labs., Inc., McLean, Va., 1964—. Served with U.S. Army, 1943-44. USPHS fellow, 1956-58. Mem. Am. Soc. Microbiology, N.Y. Acad. Scis., Union Concerned Scientists, Fedn. Am. Scientists, Sigma Xi. Democrat. Jewish. Subspecialties: Virology (biology); Virology (medicine). Current work: Virology, respiratory disease agents, influenza, parainfluenza and respiratory syncytial viruses; non-bacterial gastroenteritis agents; rotavirus, Norwalk agent. Home: 5337 Pooks Hill Rd Bethesda MD 20814 Office: 7655 Old Springhouse Rd McLean VA 22102

POTCHEN, EDWARD JAMES, radiologist, educator; b. Queens County, N.Y., Dec. 2, 1932; s. Joseph Anton and Eleanore (Joyce) P.; m. Geraldine J. Jeplawy, Sept. 1, 1956; children: Michelle M., Kathleen A., Michael J., Joseph E. B.S., Mich. State U., 1954; M.D., Wayne State U., 1958; postgrad. (fellow), Harvard U., 1964-66; M.S. in Mgmt. (Sloan fellow), MIT, 1972. Intern Butterworth Hosp., Grand Rapids, Mich., 1958-59; pvt. practice gen. medicine, Grand Rapids, Mich., 1959-61; resident in radiology Peter Bent Brigham Hosp., Boston, 1961-64, chief resident radiologist, 1964; dir. div. nuclear medicine Harvard Med. Sch. dept. radiology, 1965; assoc. radiologist Barnes Hosp., St. Louis, 1966; dir. nuclear medicine div. Edward Mallinckrodt Inst. Radiology, Washington U. Sch. Medicine, St. Louis, 1967; chief diagnostic radiology Edward Mallinckrodt Inst., 1971-73; prof. radiology, dean mgmt. resources Johns Hopkins U. Sch. Medicine, 1973-75; prof. radiology and health systems mgmt. Mich. State U., East Lansing, 1975—, chmn. dept. radiology, 1973—; prof. radiology Washington U. Sch. Medicine, 1969-73; cons. Nat. Heart Inst., Nat. Inst. Gen. Med. Scis., div. biology and medicine U.S. AEC; mem. liaison com. on med. edn. Assn. Am. Med. Colls./AMA, 1980—, chmn. liaison com. on med. edn., 1982—; mem. Med. Radiation Adv. Bd. FDA, 1982—; examiner Am. Bd. Radiology. Editor: Endocrine Radiology, 1966, Frontiers of Pulmonary Radiology, 1969, Diagnostic Radiology, 1971, Fundamentals of Tracer Method, 1972, Neuro-Nuclear Medicine, 1972, Current Concepts in Radiology, 1972, 3d edit., 1977; also contbr. to jours. Recipient John J. Larkin award for basic med. research, 1963; James Picker Found. fellow in acad. radiology Nat. Acad. Scis.-NRC, 1965-66; Sloan fellow Mass. Inst. Tech., 1972-73. Mem. AMA (mem. ho. of dels., sect. on nuclear medicine 1976—, council on med. edn. 1979), Am. Fedn. Clin. Research, Central Soc. Clin. Research, Am. Soc. Clin. Investigation, Soc. Nuclear Med. (trustee, nat. pres. 1975-76), Am. Physiol. Soc., Am. Radium Soc., Assn. of Univ. Radiologists, Liaison Com. on Med. Edn., Alpha Omega Alpha. Home: 4810 Arapaho Dr Okemos MI 48864 Office: Dept Radiology Mich State U B220 Clinic Center East Lansing MI 48824

POTENZA, JOSEPH ANTHONY, chemistry educator; b. N.Y.C., Nov. 13, 1941; s. Nicholas J. and Stella (Sabonis) P.; m. Janet Martha Koch Potenza, June 13, 1964; children: Marc Nicholas, Michael Robb. B.S. in Chemistry, Poly. Inst. Bklyn., 1962, Ph.D., Harvard U., 1967. Asst. prof. chemistry Rutgers U., 1968-72, assoc. prof., 1972-77, prof., 1977-81, prof. II, 1981—; dir. Sch. Chemistry, 1977-80, Grad. Program in Chemistry, 1980—. Contbr. articles to profl. jours. Served to capt. U.S. Army, 1966-68. Alfred P. Sloan fellow, 1971-73; recipient Outstanding Tchr. award Rutgers Coll., 1973-74, Alexander von Humboldt Sr. U.S. Scientist award, 1974-75. Mem. Am. Chem. Soc., Am. Phys. Soc. Democrat. Subspecialties: Physical chemistry; X-ray crystallography. Current work: Molecular structure, molecular dynamics of liquids. Home: 10 Lawrence Ave Highland Park NJ 08904 Office: Department Chemistry Rutgers University New Brunswick NJ 08903

POTTASH, A.L.C., physician psychopharmacologist, researcher; b. Phila., Nov. 30, 1948; s. R.R. and M.E. P. B.S., Trinity Coll., Hartford, Conn., 1970; M.D., Yale U., 1974. Intern Baystate Med. Ctr., Springfield, Mass., 1974-75; resident, fellow Yale U. Sch. Medicine, 1975-78, lectr. dept. psychiatry, 1978-81, cons., 1981—; clin. dir. Whiting Forensic Inst., Portland, Conn., 1976-78; med. dir. Psychiat. Diagnostic Labs., Summit, N.J., 1978—, Regent Hosp., Summit, 1981—, PANJ, Fair Oaks Hosp., Summit, N.J., 1978—; psychiatrist-in-chief Falkirk Hosp., Central Valley, N.Y., 1982—; co-dir. 800-Cocaine, toll-free helpline; numerous appearances radio and TV. Contbr. numerous articles to profl. jours. Fellow Nat. Acad. Clin. Biochemistry, Am. Coll. Clin. Pharmacology; mem. Soc. Neurosci., Am. Psychiat. Assn., AMA, Am. Acad. Clin. Psychiatrists, N.J. Med. Soc., Phi Beta Kappa. Clubs: N.Y. Athletic; Canoe Brook Country, Beacon Hill (Summit). Subspecialties: Psychopharmacology; Neuropharmacology. Current work: Psychopharmacology; neuroendocrinology and depression; drug abuse; clinical laboratory testing in psychiatry.

POTTER, FRANK WALTER, JR., biology educator, museum adminstr., researcher; b. Worcester, Mass., June 27, 1942; s. Frank Walter and Josephine (Hunter) P.; m. Leslie Alden, Nov. 28, 1968; children: John, Kathryn. B.S., Pa. State U., 1965, M.S., 1970; Ph.D. (NSF grantee,), Ind. U., 1975. Research asso. Ind. U., Bloomington, 1975; asst. prof. biology Ft. Hays State U., Hays, Kans., 1976—; curator of paleobotany Sternberg Meml. Mus. Contbr. articles on biology to profl. jours. Dir. Hays Recreation Commn. Soccer Program. Served ith U.S. Army, 1968-69. NSF grantee, 1980-82. Mem. Bot. Soc. Am., Ecol. Soc. Am., Am. Assn. Stratigraphic Polynologists, AAAS, Internat. Assn. Angiosperm Paleobotany. Subspecialties: Paleoecology; Paleobotany. Current work: Plant fossils of Kansas with emphasis on creataceous and permian material. Office: Dept Biology Fort Hays State U Hays KS 67601

POTTER, JOHN FRANCIS, physician, educator; b. N.Y.C., July 26, 1925; s. John Albert and Isabelle Cecelia (Sullivan) P.; m. Tanya Agnes Kristof, Nov. 19, 1955; children: Tanya Jean, Miriam Isabelle, John Mark. Student, Holy Cross Coll.; M.D., Georgetown U., 1949; hon. degree, Universidad Peruana. Diplomate: Am. Bd. Surgery. Instr. to prof. surgery Georgetown U., Washington, 1957-83; now dir. Vincent T. Lombardi Cancer Research Center. Contbr. articles to profl. jours. Served with USN, 1943-45, 51-53. Mem. Peruvian Cancer Soc. (hon.). Subspecialty: Cancer research (medicine). Office: Vincent T Lombardi Cancer Research Center Georgetown U 3800 Reservoir Rd NW Washington DC 20007

POTTER, MILES BUTTLES, sci. cons.; b. Old Forge, Pa., June 5, 1909; s. Lewis M and Ruth (Buttles) P.; m. Helen Rose Bocci, Dec. 31, 1937 (div.); children: Patsy Carol, Barbara Ann, Donna Helen, Miles Milton. Student, Gettysburg (Pa.) Coll., 1927-29; B.S., Bloomsburg (Pa.) State Coll., 1933; postgrad., U. Del.; D.Sc. (hon.), U. London, 1972. Registered profl. engr., N.J., Fla., Maine, Pa., N.Y. Mem. faculty The Citadel, 1943-44, U. North Ga., 1944-45, Va. Mil. Inst., 1945-47, Washington and Lee U., 1946-47, Villanova U., 1947-57, assoc. dir. research, 1950-56; pres. Harris, Henry & Potter, Inc., Buckingham, Pa., 1956-72; with Mcpl. Environ. Assocs., Inc., Spring House, Pa., 1972—; chmn. exec. com. MDD Inc., 1980—; dir. East Coast Chem. Disposal Inc., Pa.; treas. Pa. Cons. Engrs. Council, 1960-69; sci. cons. Mem. PTA, 1947-55, 74-78. Served with Va. Militia, 1945-47. Republican. Methodist. Subspecialties: Civil engineering; Water supply and wastewater treatment. Current work: Providing a transition of scientific (engineering) information to an engineering basis in order to enable technical planning and engineering design. Office: 908 Bethlehem Pike Spring House PA 19477

POTTER, NEIL HARRISON, chemist; b. Lancaster, Pa., Oct. 14, 1938; s. Forrest Dale and Laura (Lewis) P.; m. Mary Walter, Dec. 15, 1962; children: Amy, Andy, Abby, Benjamin. B.S., Franklin and Marshall Coll., 1960; M.S., Middlebury Coll., 1962; Ph.D., Pa. State U., 1966. Mem. faculty Susquehanna U., Selingsgrove, Pa., 1966—, assoc. prof., 1973-81, prof. chemistry, 1981—. Subspecialties: Organic chemistry; Biochemistry (biology). Current work: Oxidation of organic compounds by Permanganate, dichromate, and bromine water. Office: Dept Chemistry Susquehanna U Selingsgrove PA 17870

POTTER, PAUL EDWIN, geology educator; b. Springfield, Ohio, Aug. 30, 1925; s. Edwin F. and Mabel (Yanser) P. M.S., U. Chgo., 1950; Ph.D., 1952; M.S., U. Ill., Urbana, 1959. Assoc. geologist Ill. Geol. Survey, Urbana, 1952-61; Guggenheim fellow Johns Hopkins U., Balt., 1961-62; sr. NSF fellow U. Ill., Urbana, 1959-60; prof. geology Ind. U., Bloomington, 1963-71, U. Cin., 1971—; vis. prof. U. Sao Paul, Brazil, 1974; dir. Orbit Gas Co., Owensboro, Ky., 1974—. Asst. editor: Ency. Britannica, Chgo., 1972-83; Author: Paleocurents and Basin Analysis, 1963, Atlas and Glossary of Sedimentary Structures, 1964, Petrography of Fossils, 1971, Sand and Sandstone, 1972, Sedimentology of Shale, 1980. Served with U.S. Army, 1944-46. Fellow Soc. Grad. Fellows U. Cin.; mem. Johns Hopkins Soc. Fellows, Socidad Geologia Brazilera, Socidad Geologia Mexico, Ky. Geol. Soc., Ind.-Ky. Geol. Soc. Republican. Subspecialties: Sedimentology; Fuels. Home: 1911 Bigelow St Cincinnati OH 45219 Office: Dept Geolog U Cin Mail Location 1 Cincinnati OH 45221

POTTER, ROSARIO H. YAP, dental educator, dental geneticist researcher; b. Manila, Philippines, Aug. 21, 1928; came to U.S., 1959, naturalized, 1970; d. Thomas Eng-Chong and Sun-Tee (Ho) Yap; m. Norman E. Potter, July 17, 1964; 1 son, Brent E. D.M.D. summa cum laude, U. East Coll. Dentistry, Manila, 1952; M.S. in Pedodontics, U. Oreg.-Portland, 1963, Ind. U.-Indpls., 1966. Clin. fellow, research asst. Eastman Dental Dispensary, Rochester, N.Y., 1959-61; research asst. U. Oreg. Dental Sch., Portland, 1961-63; research assoc. med. genetics Ind. U., Indpls., 1963-67, asst. prof. dentistry, 1967-71, assoc. prof., 1971-80, prof., 1980—, cons. to grad. theses research, 1967—; guest lectr. U. East, 1975. Contbr. chpts. to books and articles to profl. jours. Eastman Dental Dispensary fellow, 1959; USPHS trainee, 1966; grantee in field. Mem. Am. Soc. Human Genetics, Internat. Assn. Dental Research, Am. Assn. Dental Research, Craniofacial Biology Group, Behavioral Sci. Group, Sigma Xi. Subspecialties: Dental growth and development; Behaviorism. Current work: The genetics of dentofacial growth and development research; research on dental students professional school performance; computerization of dental education and dental clinical management. Office: Ind U Sch Dentistry 1121 W Michigan St Indianapolis IN 46202

POTTS, JOHN THOMAS, JR., medical administrator, biochemical/biomedical researcher; b. Phila., Jan. 19, 1932; s. John Thomas and Florence (McGann) P.; m. Susanne Kuttner, June 17, 1961; children: Martha, John, Stephan. B.A. maxima cum laude, La Salle Coll., Phila., 1953; M.D., U. Pa., 1957. Diplomate: Am. Bd. Internal Medicine, 1964. Intern in medicine Mass. Gen. Hosp., Boston, 1957-58, asst. resident, 1958-59; resident (clin. assoc.) Nat. Heart Inst., NIH, Bethesda, Md., 1959-60, chief resident, 1960-61, research fellow lab. cellular physiology and metabolism, 1960-63, mem. sr. research staff, lab. metabolism, 1963-66, head sect. polypeptide hormones, lab. molecular diseases, 1966-68; chief endocrine unit Mass. Gen. Hosp., 1968-81, chief gen. med. services, 1981—; dir. Genentech, Inc.; asst. prof. medicine Harvard Med. Sch., Boston, 1968-69, assoc. prof., 1969-75, prof., 1975-81, Jackson prof. clin. medicine, 1981—; vis. lectr. MIT, 1969-78; vis. prof. U. So. Calif., Los Angeles, 1976, Washington U. Sch. Medicine, St. Louis, 1979; Disting. lectr. Royal Coll. Physicians Ireland, Dublin, 1977; Upjohn vis. prof. U. Miami Sch. Medicine, 1978; prof. medicine Harvard-MIT Div. Health Scis. and Tech., 1978—; Solomon A. Berson Meml. lectr. 6th Internat. Congress Endocrinology, Melbourne, Australia, 1980. Editorial bd.: Endocrinology, 1968-71, Clin. Endocrinology, 1971-74, Endocrine Research Communications, 1974—, Calcified Tissue Internat, 1979—; contbr. numerous articles to sci. and med. jours. Recipient Ernst Oppenheimer award Endocrine Soc., 1968; prize Andre Lichwitz, 1968. Mem. Can. Soc. Endocrinology (hon.), Endocrine Soc. (nominating com. 1975-76, future programs com. 1980—), Am. Soc. Biol. Chemists, Am. Soc. Clin. Investigation, Am. Fedn. Clin. Research, Internat. Confs. Calcium Regulating Hormones (dir. and exec. planning com. 1971-79, pres. 1980—), Assn. Am. Physicians, Peripatetic Club, Aesculapian Club, N.Y. Acad. Scis., Med. Exchange Club, Am. Soc. Bone and Mineral Research (council 1978), AAAS, Interurban Clin. Club, ACP, Assn. Am. Med. Colls., Assn. Profs. Medicine, Am. Acad. Arts and Scis., Mass. Med. Soc., Alpha Omega Alpha. Roman Catholic. Subspecialties: Endocrinology; Biochemistry (medicine). Current work: Endocrinology: calcium and bone metabolism, parathyroid hormone, calcitonin, Vitamin D; protein and polypeptide isolation and sequence analysis; peptide synthesis. Office: Mass Gen Hosp Boston MA 02114

POTTS, RICHARD ALLEN, chemistry educator; b. Massillon, Ohio, Jan. 2, 1940; s. Henry M. and Lillian (Lehman) P.; m. Carolyn A. Stuber, Mar. 26, 1965; 1 son, Alan. A.B., Hiram Coll., 1962; Ph.D., Northwestern U., 1966. Postdoctoral assoc. Iowa State U., Ames, 1965-66; asst. prof. U. Mich., Dearborn, 1966-69, assoc. prof., 1969-73, prof. chemistry, 1973—. Mem. AAAS, Am. Chem. Soc. (chmn. local sect. 1980). Subspecialty: Inorganic chemistry. Current work: Coordination chemistry of gold and mercury in lower oxidation states. Office: U Mich 4901 Evergreen Rd Dearborn MI 48128 Home: 9551 Stark Rd Livonia MI 48150

POUND, ROBERT VIVIAN, physicist; b. Ridgeway, Ont., Can., May 16, 1919; came to U.S., 1923, naturalized, 1932; s. Vivian Ellsworth and Gertrude C. (Prout) P.; m. Betty Yde Andersen, June 20, 1941; 1 son, John Andrew. B.A., U. Buffalo, 1941; A.M. (hon.), Harvard U., 1950. Research physicist Submarine Signal Co., 1941-42; staff mem. Radiation Lab. Mass. Inst. Tech., 1942-46; jr. fellow Soc. Fellows, Harvard U., 1945-48; asst. prof. physics Harvard U., 1948-50, asso. prof., 1950-56, prof., 1956-68, Mallinckrodt prof. physics, 1968—, chmn. dept. physics, 1968-72, dir. Physics Labs., 1975-83; Fulbright research scholar Oxford, 1951; Fulbright lectr., Paris, 1958; vis. prof. Coll. de France, 1973; vis. fellow Joint Inst. Lab. Astrophysics, U. Colo., 1979-80; vis. research fellow Merton Coll., Oxford U., 1980. Author, editor: Microwave Mixers, 1948; Contbr. articles to profl. jours. Trustee Asso. Univs., Inc., 1976—. Recipient B. J. Thompson Meml. award Inst. Radio Engrs., 1948; John Simon Guggenheim fellow, 1957-58, 1972-73; Eddington medal Royal Astron. Soc., 1965. Fellow Am. Phys. Soc., Am. Acad. Arts and Scis.; mem. Nat. Acad. Scis., Soc. Franc. de Physique (membre du conseil 1958-61), Acad. des Scis. (fgn. asso.) (France), Phi Beta Kappa, Sigma Xi.

POUNDER, ELTON ROY, physicist, educator; b. Montreal, Que., Can., Jan. 10, 1916; s. Roy M. and Norval (McLeese) P.; m. Marion Crane Wry, Feb. 15, 1941; children—David Crane, Norval Gillian. B.Sc., McGill U., Montreal, 1934, Ph.D., 1937. Field engr. Bell Telephone Co. Can., Ottawa, Ont., Montreal, 1937-39; faculty McGill U., 1945—, prof. physics, 1959—. Author: (with J.S. Marshall and R.W. Stewart) Physics, 2d edit, 1967, The Physics of Ice, 1965, also numerous articles. Fellow Royal Soc. Can.; mem. Canadian Assn. Physicists (past pres.), Am. Phys. Soc., Am. Geophys. Union. Subspecialties: Geophysics; Oceanography. Current work: Analysis of oceanographic data collected in the Arctic. Research on constrn. cyclotron, neutron prodn., doppler radar used mid-Can. line, geophysics of sea ice. Office: Physics Dept McGill U 3600 University St Montreal PQ H3A 2T8 Canada

POUR, PARVIZ M., pathologist; b. Tehran, Iran, Jan. 4, 1933; came to U.S., 1971, naturalized, 1979; s. Timour and Nasimeh (Omid) P.; m. Adelheid Guldimann, July 25, 1951; children: Schahrzad, Schahrameh, Farid. M.D., Duesseldorf (Ger.) Med. Sch., 1963. Diplomate: Am. Bd. Anatomic Pathology. Asst. prof. Hannover (Ger.) Med. Sch., 1968-71; assoc. prof. U. Nebr. Med. Center, Omaha, 1971-72, prof., 1972—; cons. in toxicology. Contbr. articles to profl. jours. Nat. Cancer Inst./NIH grantee, 1972—. Mem. Am. Assn. for Cancer Research, Soc. Toxicologic Pathologists, Am. Pancreatic Assn., Internat. Assn. on Prevention and Detection of Cancer. Moslem. Subspecialties: Pathology (medicine); Cancer research (medicine). Current work: Pancreas cancer: prevention, diagnosis, therapy. Home: 9727 Spring St Omaha NE 68124 Office: 42d and Dewey Omaha NE 68105

POWE, RALPH ELWARD, university administrator; b. Tylertown, Miss., July 27, 1944; s. Roy Elward and Virginia Alyne (Bradley) P.; m. Sharon Eve Sandifer, May 20, 1962; children: Deborah, Ryan, Melanie. B.S.M.E., Miss. State U., 1967, M.S.M.E., 1968; Ph.D. in Mech. Engring, Mont. State U., 1970. Student trainee NASA, 1962-65; research asst., lab. instr. Miss. State U., 1968, instr. in mech. engring., 1968, assoc. prof. mech. engring., 1974-78, prof., 1978-79, assoc. dean engring., dir. engring. and indsl. research sta., 1979-80, assoc. v.p. for research, 1980—; research asst. Mont. State U., 1968-69, teaching asst., 1969-70, asst. prof., 1970-74; cons. energy conservation programs, coal fired power plants, torsional vibration, accident analysis. Contbr. numerous articles to profl. jours., procs. Mem. ASME (Miss. sect. Outstanding Service award 1980), Soc. Automotive Engrs. (Ralph E. Teetor award 1971), Am. Soc. Engring. Edn., Internat. Centre Heat and Mass Transfer, Miss. Acad. Scis., Miss. Engring. Soc., Blue Key, Sigma Xi (Miss. State U. chpt. Research award 1976), Tau Beta Pi (Outstanding Engring. Faculty Mem. 1978), Pi Tau Sigma, Phi Kappa Phi, Omicron Delta Kappa, Kappa Mu Epsilon. Subspecialties: Mechanical engineering; Fluid mechanics. Current work: Energy systems, magnetohydrodynamics, heat transfer, energy conservation. Home: 110 Pinewood Dr Starkville MS 39759 Office: Drawer G Mississippi State U Mississippi State MS 39762

POWELL, CEDRIC JOHN, physicist; b. Subiaco, Western Australia, Australia, Feb. 19, 1935; s. Harley Robert and Hilda King (Drabble) P.; m. Marie Allison Gardner, Jan. 4, 1958; children: Andrew J., Bruce M., David F. B.Sc. with honors, U. Western Australia, 1956, Ph.D. in Physics, 1962. Research physicist Imperial Coll., London, 1960-62; physicist Nat. Bur. Standards, Washington, 1962-78, chief surface sci. div., 1978—; chmn. Gordon Research Conf. on Electron Spectroscopy, 1982; trustee, 1982—; gen. chmn. Phys. Electronics Conf., 1980-84. Contbr. articles to profl. jours.; mem. editorial bd.: Surface and Interface Analysis, Applications of Surface Sci. Recipient Bronze medal U.S. Dept. Commerce, 1980, Silver medal U.S. Dept. Commerce, 1983. Fellow Am. Phys. Soc., Brit. Inst. Physics; mem. Am. Chem. Soc., Am. Vacuum Soc. (chmn. surface sci. div. 1983-84), ASTM (chmn. com. E-42 on surface analysis 1980-83), Australian Inst. Physics. Subspecialties: Condensed matter physics; Surface chemistry. Current work: Surface science, surface analysis, electron spectroscopy, small-particle science. Office: Chemistry B-248 Nat Bur Standards Washington DC 20234

POWELL, DON WATSON, medical educator, research physiologist; b. Gasden, Ala., Aug. 29, 1938; s. Gordon C. and Ruth (Bennett) P.; m. Jo Ann Beason, June 4, 1960; children: Mary Paige, Drew Watson, Shawnne Margaret. B.S. with honors, Auburn U., 1960; M.D. with highest honors, Med. Coll. Ala., 1963; spl. fellow in physiology, Yale U., 1969-71. Diplomate: Am. Bd. Internal Medicine. Intern, resident Peter Bent Brigham Hosp., Boston, 1963-65; resident in internal medicine Yale U. Sch. Medicine, New Haven, 1968-69; asst. prof. medicine U. N.C., Chapel Hill, 1971-74, assoc. prof., 1974-78, prof., 1978—, chief div. digestive disease and nutrition, 1977—; cons. WHO, 1980—, Burroughs-Wellcome Co., Research Triangle Park, N.C., 1982-83, Hoffman-LaRoche, Nutley, N.J., 1983—; chmn. merit rev. bd. gastroenterology VA, Washington, 1979-80. Contbr. chpts. to books, articles in medicine, physiology to profl. jours. Served to capt. U.S. Army, 1965-68. NIH grantee, 1971, 1973, 1975; Nat. Found. Ileitis and Colitis grantee, 1982. Mem. Am. Soc. Clin. Investigation, Am. Gastroent. Assn., Am. Physiol. Soc., Am. Fedn. Clin. Research, Biophys. Soc. Democrat. Subspecialties: Gastroenterology; Physiology (medicine). Current work: Mechanisms of water and electrolyte transport. Home: 102 Porter Pl Chapel Hill NC 27514 Office: U North Carolina 326 Burnett-Womack Bldg Chapel Hill NC 27514

POWELL, DONALD A., psychologist, researcher; b. Spartanburg, S.C., Oct. 29, 1938; s. Russell and Mignon (Cox) P.; m. Palmyra, July 5, 1960. B.S., U. S.C., 1960; M.A., Fla. State U., 1962, Ph.D., 1967. Postdoctoral fellow U. Miami, Fla., 1967-69; research psychologist VA Hosp., Columbia, S.C., 1969—; asst. prof., then prof. U. S.C., 1969—. Contbr. articles to profl. jours. Recipient award VA, 1972, 73, 77; USPHS fellow, 1967-69. Mem. Am. Psychol. Assn., Soc. Neurosci., Psychonomic Soc., Soc. Psychophysiol. Research. Democrat. Subspecialties: Psychobiology; Learning. Current work: Brain-behavior relationships. Home: 619 S Ott Rd Columbia SC 29205 Office: VA Med Center Columbia SC 29201

POWELL, EDWARD GORDON, research physicist; b. Washington, Apr. 5, 1946; s. Edward Carter and Hilda Rae (Blanchard) P.; m. Lona Noel Carlson, Oct. 1, 1977; children: Mary, Soren, Piri, Lynis. B.S., U. Md., 1968, M.S., 1972, postgrad. in astrophysics, 1972-73. Research physicist U.S. Naval Ordnance Sta., Indian Head, Md., 1968-73, U.S. Naval Ordnance Lab., White Oak, Md., 1973-74, U.S. Surface Weapons Center, White Oak, 1973-83, Mitre Corp., McLean, Md., 1983—. Contbr. articles to profl. jours. Mem. Am. Astron. Soc., Assn Computational Linguistics., Am. Assn. Artificial Intelligence. Subspecialty: Artificial intelligence. Current work: Artificial intelligence for robotics, robotic vision. Patentee in field. Office: Code RW-27 Mitre Corp 1820 Dolley Madison Blvd McLean VA 22102

POWELL, GARY LEE, biochemical educator; b. Fullerton, Calif., Jan. 24, 1941; s. Lee D. and Dorothy (Dunphy) P.; m. Constance Marie Andrews, June 26, 1965; children: Gregory Lee, Andrew Scott. B.S., UCLA, 1962; Ph.D., Purdue., 1967. Postdoctoral fellow Washington U., St. Louis, 1967-69; asst. prof. biochemistry Clemson (S.C.) U., 1969-73, assoc. prof., 1973-79, prof., 1979—; vis. prof. chemistry U. Oreg., Eugene, 1975-76. Cubmaster Boy Scouts Am., Clemson, 1981—; rev. bd. Boy Scout Troop 235, 1982—. Mem. Am. Chem. Soc., Am. Soc. Biol. Chemists, S.C. Acad. Sci., Am. Heart Assn., S.C. Heart Assn. (chmn. research rev. and allocations), Sigma Xi (chpt. pres.). Subspecialties: Biochemistry (biology); Membrane biology. Current work: Regulation of lipid metabolism; specificity of lipid-protein interactions in biological membranes. Offic: Dept Biochemistr Clemson Clemson SC 29631 Home: 405 Skyview Dr Clemson SC 29631

POWELL, JAMES CHARLES, educator, consultant; b. Edmonton, Alta., Can., Nov. 18, 1931; s. Edgar Ernest Charles and Hazel Fern (Alcorn) P.; m. Frances Adelaide Cochrane, May 7, 1954. B.Ed., U. Alta., 1956, B.A., 1957, Ph.D., 1970; M.Ed., U. Toronto, 1962; diploma in bus. adminstrn, U. Western Ont., 1973. Tchr. various locations, 1957-65; lectr. edn. U. Alta., Edmonton, 1965-69; tchr. Roman Cath. Schs., Edmonton, 1969-70; sch. psychologist County of Lac Ste. Anne, Sangudo, Alta., 1970-71; asst. prof. edn. U. Sask., Saskatoon, 1971-72; assoc. prof. edn. U. Windsor, Ont., 1973—; cons. Ball, Powell Agy., Windsor, 1980—, T.D. Wearne & Assocs., 1976-80. Author: The Exploratory Loop, 1981, Achievement Information from Wrong Answers, 1970. IBM fellow, 1979. Mem. AAAS, Am. Ednl. Research Assn., Am. Psychol. Assn., Nat. Council Measurement in Edn., Psychometric Soc. Subspecialties: Cognition; Information systems (information science). Current work: Exploring instructional and measurement implications that current achievement measurement practices are invalid when applied to cognition higher the retrieval. Home: 408 Moy Ave Windsor ON Canada N9A 2N4 Office: Faculty of Edn U Windsor 600 3d Concession Rd Windsor ON Canada N9E 1A5

POWELL, JAY RAYMOND, educator, researcher; b. Waukegan, Ill., Apr. 22, 1936; s. Raymond Jean and Josephine R. (Penca) P.; m. Nancy Susan Scheetz, June 20, 1969; 1 son, Tyler Graham. B.S., U. Ill.-Urbana, 1962; M.A., So. Ill. U., 1968, Ph.D., 1970. Med. research scientist Behavior Research Lab., Anna, Ill., 1966-70; coordinator grad. psychology program U. Vera Cruz, Xalapa, Mexico, 1970-72; asst. prof. spl. edn. California U. of Pa., 1972, assoc. prof., 1972-74, prof., 1974—. Contbr. articles to psychology jours. Served to sgt. USMC, 1953-56; Japan. NIMH grantee, 1982. Mem. Eastern Psychol. Assn., Am. Psychol. Assn., Assn. Behavior Analysis, Assn. Advancement Behavior Therapy. Democrat. Subspecialty: Behavioral psychology. Current work: Quantification of behavioral phenomena. Home: RD 1 Box 142 Marianna PA 15345 Office: California U of Pa California Pa 15419

POWELL, JOHN EDWARD, geologist, hydrologist; b. Galesville, Wis., Dec. 2, 1918; s. John Justin and Gertrude (Vickerman) P.; m. Margaret Jane Owens, Apr. 2, 1943; children: Diane Kay, John Everett. Ph.B., Marquette U., 1940; B.S.C.E., U. N.D., 1950, 1951; postgrad., U. Wis., 1940-41. Registered profl. engr., N.D., S.D. Geologist U.S. Geol. Survey, Grand Forks, N.D., 1946-51, hydraulic engr., 1951-58, dist. engr., Huron, S.D., 1958-66, dist. chief, 1966-78; cons. hydrologist/geologist, Huron, 1978—. Contbr. articles to profl. jours. Mem. Mcpl. Water Bd., Huron, 1979—. Served to lt. col. USAF, 1941-46; ETO. Fellow Geol. Soc. Am.; mem. Nat. Soc. Profl. Engrs., Nat. Water Well Assn., S.D. Soc. Profl. Engrs., Res. Officers Assn. (life), Am. Legion, VFW. Roman Catholic. Lodges: Masons; Shriners. Subspecialties: Ground water hydrology; Hydrogeology. Current work: Design of monitoring systems for ground water pollution; availability of ground water in vicinity of igneous intrusives. Home: 1831 McClellan Dr Huron SD 57350

POWELL, JOHN EDWIN, business educator; b. Reading, Pa., June 21, 1936; s. Vernon S. and Theresa (Speece) P.; m. Kay Jean Herbold, June 8, 1958; children: Vern, Rhonda. B.S., Morningside Coll., 1958; M.A., U. S.D., Vermillion, 1962; M.B.A., Ind. U., 1971, D.B.A., 1972. Instr. math. dept. U. S.D., Vermillion, 1962-64, asst. prof., dir., 1964-68; assoc. instr. Ind. U., Bloomington, 1970-72; asst. prof. dept. math. Sch. Bus., U. S.D., Vermillion, 1972-73, assoc. prof., 1973-77, prof. quantitative methods, 1977—; cons. Bomgaars Stores, Sioux City, Iowa, 1975-76, C.P.A. firms, Sioux Falls, S.D., 1977—, S.D. Cement Plant, Rapid City, 1978, S.D. Bd. Regents, Pierre, 1982—. Author: Review of Mathematical Concepts, 1975. Mem. Bd. Regents Computer Task Force, Pierre, S.D., 1982. NSF grantee, 1961; USAF research grantee, 1981. Mem. Assn. Computing Machinery, Inst. Mgmt. Sci., Am. Inst. Decision Scis., Sigma Xi, Beta Gamma Sigma. Republican. Methodist. Club: Eagles (Vermillion, S.D.). Subspecialties: Information systems, storage, and retrieval (computer science); Operations research (mathematics). Current work: Integrating micro-computers into the business school curriculum, auditing computer-maintained information, applications of quantitative procedures to business problems. Home: 514 Poplar St Vermillion SD 57069 Office: Sch Bus U SD Vermillion SD 57069

POWELL, LOUISA ROSE, child guidance center executive; b. Highland Park, Mich., Oct. 10, 1942; d. Albert and Mildred Lorraine (Bos) Feldman; m. Philip Melancthon Powell, Dec. 29, 1962; children: David Edward, Aaron Matthew, Robert Allen. Student, Grinnell Coll., 1960-62; B.S., Roosevelt U., 1966; M.A., U. Chgo., 1969, Ph.D., 1973. Lic. sch. psychologist, Calif.; lic. psychologist, health service provider, Tex. Lectr. Conn. State Coll., New Haven, 1975-76; instr. S.W. Tex. State U., San Marcos, 1978-79; psychologist Austin (Tex.) Child Guidance and Evaluation Ctr., 1979-80, 81-82, dep. dir., 1982—; psychologist San Rafael (Calif.) City Schs., 1980-81; pvt. practice psychology, Austin, 1979-80, 81—. Asst. den mother pack 54 Cub Scouts Am., Austin, 1977, pack chmn., 1977-78; active Doss Elem. Sch. PTO, 1977—; mem. Austin Community Orch., 1983—. Research grantee Tex. Devel. Disabilities Group, 1980, 82. Mem. Am. Psychol. Assn., Southwestern Psychol. Assn., Am. Orthopsychiat. Assn., AAAS, Austin Assn. for Edn. Young Children (hon. affiliate), Mensa (proctor 1979-80). Democrat. Subspecialties: Social psychology; Developmental psychology. Current work: Counseling with families in stress with high risk premature infants, with hyperactive children and with children with other behavioral or emotional difficulties; formulating research to decrease children's behavioral and emotional difficulties and to increase parenting skills and to investigate family correlates of behaviorally and emotionally disturbed children; developmental, emotional and psychoeducational assessment of children. Home: 3910 Edgerock Dr Austin TX 78731 Office: Austin Child Guidance and Evaluation Ctr 612 W 6th St Austin TX 78701

POWELL, RICHARD CONGER, physics educator, educational administrator; b. Lincoln, Nebr., Dec. 20, 1939; s. William Charles and Allis (Conger) P.; m. Gwendolyn Cline, June 24, 1962; children: Douglas, David. B.S., U.S. Naval Acad., 1962; M.S., Ariz. State U., 1964, Ph.D., 1967. Research physicist Air Force Cambridge Research Labs., Bedford, Mass., 1964-68, Sandia Nat. Lab., Albuquerque, 1968-71; physics prof. Okla. State U., Stillwater, 1971—, assoc. dean, dir. research, 1981—; program chmn. Internat. Conf. on Lasers and Applications, 1982; mem. organizing coms. various internat. confs. Co-author: Phonons and Resonances in Solids, 1976; editor: Lasers, 1982; mem. editorial bd.: Jour. Photoacoustics, 1981—; contbr. articles to profl. publs. Served to capt. USAF, 1962-68. Recipient Sci. Achievement award USAF, 1967; hon. lectr. Mid-Am. State Univs. Assn., 1979-80. Fellow Am. Phys. Soc.; mem. Optical Soc. Am., Sigma Xi. Subspecialties: Spectroscopy; Condensed matter physics. Current work: Laser spectroscopy; optical properties of solids; energy transfer and exciton dynamics; nonlinear optical processes; development of new laser materials. Home: 1324 N Washington Stillwater OK 74074 Office: Physics Dept Okla State U Stillwater OK 74078

POWERS, DANIEL DUFFY, chemistry educator; b. Wichita, Kans., Jan. 23, 1935; s. Earl M. and Mary V. (Miller) P.; m. Doris E. Bosworth, Aug. 12, 1960; children: Earl Edward, Sarah Elizabeth. B.S., U. Wichita, 1956, M.S., 1958; Ph.D., U. Kans., 1966. Mem. faculty Sterling (Kans.) Coll., 1963—, assoc. prof., 1967-70, prof. and chmn. dept. chemistry, 1970—. Recipient Outstanding Teaching award Sterling Coll., 1980, Disting. Service medal, 1981. Mem. Am. Chem. Soc. Presbyterian. Subspecialties: Organic chemistry; Biochemistry (biology). Current work: Chemical education; protein chemistry; chemicals from plant sources. Office: Sterling College Sterling KS 67579

POWERS, DAVID LEUSCH, mathematics educator, researcher; b. Abington, Pa., Feb. 17, 1939; s. Everett S. and Alma (Leusch) P.; m. Lya Marquez, Aug. 6, 1966; children: Christopher, Evan, Emily. B.S., Carnegie Inst. Tech., 1960, M.S., 1961; Ph.D., U. Pitts., 1965. Prof. math. Universidad Técnica F. Santa María, Valparaiso, Chile, 1965-67; assoc. prof. math. Clarkson Coll. Tech., Potsdam, N.Y., 1967—; Fulbright lectr. Dept. State, Chile, 1971. Author: Boundary Value Problems, 1972, (with Coxeter and Frucht) Zero-symmetric Graphs, 1981. Served to 1st lt. C.E. U.S. Army, 1961-62. Mellon fellow, 1964-65; OAS research fellow, Chile, 1980-81. Mem. Soc. Indsl. and Applied Math., Sigma Xi. Current work: Relations of graphs to groups and matrices. Office: Clarkson Coll Tech Potsdam NY 13676

POWERS, DENNIS ALPHA, scientist; b. Dearborn, Mich., May 4, 1938; s. Earl Wilson and Virginia (Williams) P.; m. Dianne Williamson, 1963; children: Kathy, Wendy, Julie. B.A., Ottawa U., 1963; Ph.D., U. Kans.-Lawrence, 1970. AEC postdoctoral fellow Argonne (Ill.) Nat. Lab., 1970-71; NSF postdoctoral fellow SUNY-Stony Brook, also Woods Hole Marine Biol. Lab., 1972; asst prof. biology Johns Hopkins U., 1972-78, assoc. prof., 1978-82, prof., 1982—; cons. UN, FAO, Rome, Italy, Bendix Corp., Columbia, Md., Tex. Instruments, N.Y.C., Cousteau Soc., S. Am. Contbr. numerous articles to profl. jours. Served with USMCR, 1957-59. Grantee NSF NIH, Nat. Geographic Soc., NOAA. Mem. AAAS, Md. Acad. Sci., N.Y. Acad. Sci., Fedn. Biol. Chemists, Nat. Geog. Soc., Sigma Xi. Subspecialties: Evolutionary biology; Biochemistry (biology). Current work: Molecular evolution of fishes; biotechnology and genetic engineering of marine organisms; biochemistry of metallothioneins, molecular ecology of fish hemoglobins. Office: The Johns Hopkins U Dept Biology 3400 N Charles St Baltimore MD 21218

POWERS, EDWARD JOSEPH, JR., electrical engineer; b. Winchester, Mass., Nov. 29, 1935; s. Edward Joseph and Mary Elaine (Hennessey) P.; m. Barbara Ann Burns, June 27, 1959; children: Marianne, Joseph Edward, Andrew John, William Julian. B.S., Tufts U., 1957; M.S., M.I.T., 1959; Ph.D., Stanford U., 1965. Engr., scientist Lockheed Missiles & Space Co., Sunnyvale, Calif., 1959-65; mem. faculty U. Tex., Austin, 1965—, prof. elec. engring., 1973—, chmn. dept., 1981—, dir., 1977—; cons. in field. Editor: IEEE Trans. on Plasma Sci, 1979—. Fellow IEEE; Mem. Internat. Sci. Radio Union, Am. Phys. Soc., Sigma Xi, Tau Beta Pi, Eta Kappa Nu. Subspecialty: Electrical engineering. Home: 8703 Mountainwood Circle Austin TX 78759 Office: Dept Elec Engring U Texas Austin TX 78712

POWERS, EDWARD LAWRENCE, biologist; b. Columbia, S.C., Dec. 30, 1915; s. Edward Lawrence and Emilie (Devereux) P.; m. Mary Eleanor Fogarty, Dec. 27, 1939; children: Mary Eugenia (Mrs. Russell Anderson), Emilie Devereux (Mrs. L. Dacunto), Judith Ann (Mrs. Russell Cykoski), Catherine (Mrs. Joseph Schourek), Patricia (Mrs. Bruce Swanney), Christina (Mrs. Thomas Barnett), Barbara Clare. Student, U. Notre Dame, 1933-34; B.S., Coll. of Charleston, S.C., 1938; Litt.D., 1974; Ph.D., Johns Hopkins, 1941. Teaching asst. Coll. Charleston, 1936-38; asst. zoology Johns Hopkins, 1938-41; vis. lectr. genetics Fordham U., 1941, vis. lectr. biology, 1943; instr., then asst. prof. U. Notre Dame, 1941-45; asso. biologist Argonne Nat. Lab., 1946-49, sr. biologist, 1949-65, asso. dir. div. biol. and med. research, 1950-59; prof. zoology, dir. Lab. Radiation Biology, U. Tex., Austin, 1965—; dir. Center for Fast Kinetics Research, U. Tex., Austin, 1965—. Editor Oxygen and Oxy-radicals in Chemistry and Biology, 1982; contbr. articles to profl. jours. Mem. Park Forest (Ill.) Planning and Zoning Commn., 1949-52, chmn., 1950-52; mem. Park Forest Zoning Bd. Appeals, 1951-52, Park Forest Bldg. Code Bd. Appeals, 1951-52; Bd. dirs. Park Forest Family Counseling Services, 1951-52. Tomlinson fellow Coll. Charleston, 1937; Guggenheim fellow, 1958; Douglas Lea Meml. lectr. U. Leeds, Eng., 1961; vis. prof. Christie Hosp., Holt Radium Inst., Manchester, Eng., 1972. Fellow A.A.A.S.; mem. Am. Soc. Human Genetics, Am. Soc. Microbiologists, Am. Soc. Naturalists, Am. Soc. Zoologists, Brit. Assn. Radiation Research, Genetic Assn. Am., Genetics Soc., Radiation Research Soc. (sec.-treas. 1958-64, pres. 1964-65), Phi Beta Kappa, Sigma Xi. Subspecialties: Cell biology; Radiation biology. Current work: Basic physical and chemical mechanisms in radiation-induced changes in cells. Spl. research expt. and theory radiation damage biol. systems. Home: 2504 Inwood Pl Austin TX 78703

POWERS, J. BRADLEY, research scientist; b. Framingham, Mass., Dec. 16, 1937; s. Robert Francis and Sarah Ellen (Ford) P.; m. Maureen Kennedy, Aug. 20, 1966; 1 son, Dana Prescott Kennedy. A.B., Harvard U., 1959; M.A., Brown U., 1961; Ph.D., U. Calif., Berkeley, 1970. Asst. research scientist U. Mich., 1974-77; sr. fellow U. Wash., 1978-79; research assoc. prof. psychology Vanderbilt U., Nashville, 1980—. Contbr. articles to profl. jours. Served with USN, 1961-65. Mem. Soc. Neurosci., AAAS, Animal Behavior Soc., Sigma Xi. Subspecialty: Physiological psychology. Current work: Role of hormones in behavioral development and regulation of adult social behavior. Home: 617 Clarkdun Ct Brentwood TN 37027 Office: Nickolas Hobbs Lab Human Devel John F. Kennedy Center Vanderbilt University Nashville TN 37240

POWERS, JAMES MATTHEW, neuropathologist, physician; b. Cleve., Sept. 15, 1943; s. Alfred Patrick and Margaret Anne (Gunther) P.; m.; children: Kristin, Scott. B.S., Manhattan Coll., 1965; M.D., Med. U. S.C., 1969. Diplomate: Am. Bd. Pathology. Intern Med. U. S.C., Charleston, 1969-70, resident, 1970-71, Albert Einstein Med. Ctr., N.Y.C., 1971-73; asst. prof. pathology Med. U. S.C., Charleston, 1973-76; dir. EM Lab., 1973-76, assoc. prof. pathology, 1976-79, prof., 1979—; mem. merit rev. bd. neurobiology VA, Washington, 1981-84; ad hoc cons. NIH, Bethesda, Md., 1980—. Contbr. articles to profl. jours. Mem. Internat. Acad. Pathology, Am. Assn. Neuropathologists (Moore award 1975, 76, 77, 80, asst. sec.-treas. 1978, 80-84), AAAS, Am. Assn. Pathologists. Roman Catholic. Subspecialty: Pathology (medicine). Current work: Adreno-leukodystrophy, Alzheimer's disease. Office: Dept Pathology Med U SC 171 Ashley Ave Charleston SC 29425

POWERS, MAUREEN KENNEDY, educator, researcher; b. San Francisco, Mar. 12, 1946; d. Kenneth Wayne and Edith (Crofts) Kennedy; m. John Bradley Powers, Aug. 20, 1966; 1 son, Dana Prescott Kennedy. A.B. in Sociology, U. Calif., Berkeley, 1968; Ph.D. in Psychology, U. Mich., 1977. Fellow dept. ophthalmology U. Mich., 1976-77; fellow dept. psychology U. Wash., 1978-79; asst. prof. psychology Vanderbilt U., 1980-83, assoc. prof., 1983—. Contbr.

articles to profl. jours. Vanderbilt U. Research Council grantee, 1980-81; Nat. Eye Inst. NIH grantee, 1980—, 82-83; NSF grantee, 1982—; NIH Research Career Devel. award, 1983—. Mem. AAAS, Optical Soc. Am., Soc. Neurosci., Assn. Research in Vision and Ophthalmology, Sigma Xi. Subspecialties: Sensory processes; Neurobiology. Current work: Research on visual systems structure and function teaching and research in university environment. Home: 617 Clarkdun Ct Brentwood TN 37027 Office: Dept Psychology 134 Wesley Hall Vanderbilt U Nashville TN 37240

POWERS, THOMAS ALLEN, nuclear medicine service administrator, physician; b. Louisville, Mar. 30, 1947; s. Nathaniel Thomas and Helen Elizabeth (Kleinsteuber) P.; m. Sandra Newbill, Apr. 21, 1974. B.S., Duke U., 1969; M.D., Vanderbilt U., 1973. Diplomate: Am. Bd. Internal Medicine, Am. Bd. Nuclear Medicine. Intern Vanderbilt Hosp., Nashville, 1973-74, resident, 1977-80; instr. radiology Vanderbilt U., 1980-81; asst. prof., 1981—; chief nuclear med. service VA Med. Ctr., Nashville, 1980—. Author: Exercises in Diagnostic Radiology-Nuclear Radiology, 1983. Served to lt. USNR, 1974-77. Mem. ACP, Soc. Nuclear Medicine, Am. Inst. Ultrasound in Medicine. Subspecialties: Nuclear medicine; Internal medicine. Current work: Application of radiopharmaceuticals to evaluation of differential renal function and renal blood flow measurement. Home: 4409 Honeywood Dr Nashville TN 37205 Office: Nuclear Medicine Service VA Med Ctr 1310 24th Ave S Nashville TN 37203

POWIS, GARTH, pharmacologist, researcher; b. West Bromwich, Staffordshire, Eng., June 12, 1946; came to U.S., 1977; m. Pauline Hill; children: Adela, Zoe. B.Sc. with 1st class honors, Birmingham (Eng.) U., 1967; D.Phil., Merton Coll., Oxford (Eng.) U., 1970. Lectr. Glasgow (Scotland) U., 1970-77; cons. Mayo Clinic, Rochester, Minn., 1977—. Mem. Am. Assn. Cancer Research, Am. Pharmacological Soc., Am. Soc. Clin. Pharmacology, Brit. Physiol. Soc. Subspecialties: Pharmacology; Cancer research (medicine). Current work: Metabolism, distribution, mechanism of action of anticancer drugs. Home: 403 15th Ave SW Rochester MN 55901 Office: Mayo Clinic Rochester MN 55905

POYDOCK, MARY EYMARD, cancer researcher, educator, nun; b. Sykesville, Pa., Dec. 3, 1910; d. John Andrew and Anna Mary (Dryna) P. A.B., Mercyhurst Coll., 1943; M.A., U. Pitts., 1946; Ph.D., St. Thomas Inst., 1965. Joined Sisters of Mercy, 1932; tchr. elem. sch., Erie, Pa., 1935-41, tchr. high sch., 1941-47; faculty biology St. Justin Sch., 1960-75; now dir. cancer research Mercyhurst Coll., Erie, Pa. Contbr. articles to profl. jours; author lab. guides. Mem. Am. Cancer Soc. (sec. bd. dirs., chmn. cancer prevention study II), Pa. Acad. Sci., AAAS, Tri Beta. Club: Zonta. Subspecialties: Cancer research (medicine); Nutrition (medicine). Current work: Cancer research using Vitamins C and B 12. Address: 501 E 38th St Erie PA 16546

POZUELO, JOSE, psychiatrist; b. Fuentesclaras, Spain, Mar. 19, 1933; s. Leonardo and Felisa (Utanda) P.; m. Lola Claros Halcon, Sept. 20, 1936; children: Leopoldo, Fatima, Marcarena. Bachiller, Instituto Afonso VIII, 1952; M.D., U. Madrid, 1959, Dr. Sci., 1960. Sec. Coll. Antonio de Nebrija, U. Madrid, 1960; resident in internal medicine U. Madrid, 1959-62; resident in psychiatry Mayo Clinic, Rochester, Minn., 1966-68, staff psychiatrist, 1969-73, Cleve. Clinic Found., 1974—. Contbr. articles to profl. jours. Mem. AMA, Am. Psychiat. Assn., Internat. Council on Alcohol and Addictions, Internat. Narcotic Research Conf. Roman Catholic. Subspecialties: Psychopharmacology; Neurochemistry. Current work: Research in substance abuse and depression. Patentee treatment of narcotic addiction, alcoholism and schizophrenia. Office: Cleveland Clinic Found 9500 Euclid Ave Cleveland OH 44106

PRABHU, VENKATRAY G., pharmacologist, educator, researcher; b. Shirali, India, Mar. 15, 1930; came to U.S., naturalized, 1979; s. Govind N. and Radhabai G. P.; m. Nalini V., May 19, 1957; children: Nirmala, Satish. Ph.D., Loyola U., 1962. Sr. research pharmacologist Arnar Stone Lab., Inc., Mt. Prospect, Ill., 1962-63; assoc. dir., dept. head Sarabhai Chems. Research Inst., Ahmedabad, India, 1963-67; instr. Chgo. Coll. Osteo. Medicine, 1967-68, asst. prof., 1968-71, assoc. prof., acting chmn., 1971-75, prof., chmn., 1975—; researcher in pathophysiology of neuromuscular system. Contbr. articles on pharmacology to profl. jours. Mem. Am. Soc. Pharmacology and Exptl. Therapeutics, AAAS. Subspecialties: Pharmacology; Neuropharmacology. Office: Chicago Coll Osteo Medicine Dept Pharmacology 1122 E 53d St Chicago IL 60615

PRABHU, VILAS ANANDRAO, medicinal chemistry educator, researcher; b. Bombay, India, Oct. 11, 1948; came to U.S., 1971; s. Anandrao Baburao and Lalita A. (Kini) P.; m. Sneha Vilas, Aug. 15, 1975; children: Shilpa, Ajay. B.S., U. Bombay, 1970; M.S., Idaho State U., 1973; Ph.D., U. Tex., 1977. Teaching asst. Idaho State U., Pocatello, 1971-73, U. Tex., Austin, 1973-77; asst. prof. Wash. State U., Pullman, 1977-80; asst. prof. medicinal chemistry Sch. Pharmacy, Southwestern Okla. State U., Weatherford, 1980—. Participant: revision U.S. Pharmacopeia XIX, 1975; contbr. articles to profl. jour. Mem. Am. Pharm. Assn., Acad. Pharm. Scis., Am. Assn. Colls. Pharmacy, Okla. Acad. Scis., Sigma Xi, Rho Chi, Phi Kappa Phi, Phi Delta Chi (sponsor 1982—). Hindu. Subspecialty: Medicinal chemistry. Current work: Synthesis and structive-activity relationship of potentially active medicinal agents; synthesis of novel psychotropic agents and adrenergic blockers. Home: 1304 Lee St Weatherford OK 73096 Office: Sch Pharmacy Southwestern Okla State U Weatherford OK 73096

PRADHAN, ANIL KUMAR, educator; b. Fategarh, India, July 26, 1951; emigrated to Can., 1968, U.S., 1983; s. Mahesh Chandra and Sarojini (Saxena) P.; m. Indira, Aug. 20, 1975; 1 dau., Alka. B.Sc., U. Windsor, 1972, M.Sc., 1973; Ph.D., U. London, 1977. Research fellow Univ. Coll. London, 1974-78; research asso. Joint Inst. Lab. Astrophysics, Boulder, Colo., 1978-80, mem., 1983—. Contbr. articles in field to profl. jours. Natural Scis. and Engring. Research Council fellow dept. physics U. Windsor, Ont., Can., 1980-83. Mem. Am. Phys. Soc. Subspecialties: Atomic and molecular physics; 1-ray high energy astrophysics. Current work: Atomic processes in lab. and astrophys. plasmas; teaching. Office: Joint Inst Lab Astrophysics U Colo Boulder CO 80309

PRADHAN, SACHIN N., pharmacologist, educator; b. Midnapur, West Bengal, India, Apr. 3, 1919; s. Bhut Nath and Ahalya (Maity) P.; m. Sikta Ghosh, Dec. 30, 1977. B.Sc., Ripon Coll., Calcutta (India) U., 1939; M.B.B.S., Med. Coll., Calcutta U., 1945; D.T.M., Sch. Tropical Medicine, Calcutta, 1948; Ph.D. in Pharmacology, George Washington U., 1959. Jr., sr. house physician Med. Coll. Hosp., Calcutta, 1945-46; May and Baker fellow dept. pharmacology Sch. Tropical Medicine, Calcutta, India, 1950-51; jr. sci. officer, div. pharmacology Central Drug Research Inst., Lucknow, India, 1951-53; vis. scientist lab. chem. pharmacology Nat. Cancer Inst.-NIH, Bethesda, Md., 1953-55; asst. prof. pharmacology Coll. Medicine Howard U., Washington, 1955-60, assoc. prof., 1960-65, prof., 1965—, grad. prof. pharmacology, 1976—, dist. prof. pharmacology, 1983—; head div. pharmacology Central Drug Research Inst., Lucknow, 1962-64. Author: (with B. Mukerji) Progress in Physiology, 1963; editor: Drug Abuse: Clinical and Basic Aspects, 1977, Fifty Years of Science in India; contbr. articles to profl. jours. Recipient Disting. research cert. award Howard U. Grad. Sch. Arts and Scis., 1976; Outstanding Faculty research award Howard U. Coll. Medicine, 1977; Outstanding research award in health affairs Howard U., 1983. Fellow AAAS, Am. Coll. Neuropsychopharmacology; mem. Am. Soc. Pharmacology and Exptl. Therapeutics, Am. Assn. Cancer Research, Soc. Exptl. Biology and Medicine, N.Y. Acad. Scis., Soc. Neurosci., Behavioral Pharmacology Soc. and Behavioral Terotology Soc., Internat. Soc. Research on Agression. Subspecialties: Neuropharmacology; Psychopharmacology. Office: Dept Pharmacology Howard U Coll Medicine Washington DC 20059

PRAKASH, LOUISE, molecular biologist, educator; b. Lyon, France, Apr. 11, 1943; came to U.S., 1949, naturalized, 1957; d. Aaron and Helene (Palgan) Burlant; m. Satya Prakash, Dec. 4, 1965; children: Ulka, Ravi. B.A., Bryn Mawr Coll., 1963; M.A. in Microbiology, Washington U., St. Louis, 1965, Ph.D., U. Chgo., 1970. Asst. prof. dept. radiation biology and biophysics U. Rochester (N.Y.) Sch. Medicine, 1972-78, asso. prof., 1978—. Contbr. articles to profl. jours. NIH grantee, 1972—; Nat. Inst. Environ Health Scis. - grantee, 1980—. Mem. Biophys. Soc., Am. Soc. for Microbiology, Environ. Mutagens Soc., Genetics Soc. Am. Subspecialties: Molecular biology; Genetics and genetic engineering (biology). Current work: Molecular mechanisms of DNA repair and mutagenesis in Saccharomyces cerevisiae (yeast); alkylation mutagenesis. Office: Dept Radiation Biology and Biophysics U Rochester Sch Medicine Rochester NY 14642

PRAKASH, SATYA, geneticist; b. Pilkhuwa, U.P. India, July 8, 1938; s. Suraj and Atar (Kai) Kali; m. Louise Prakash, Dec. 4, 1965; children: Ulka, Ravi. B.Sc., Meerut Coll., 1956; B.V.Sc. and A.H., Vet. Coll., Mhow, India, 1960; M. Vet. Sci., I.V.R.I. Izatnagar, India, 1962; Ph.D., Washington U., St. Louis, 1966. Reserach asso. biology U. Chgo., 1968-69; asst. prof. biology U. Rochester, 1969-74, assoc. prof., 1974-80, prof., 1980—. NSF grantee, 1970-72; NIH grantee, 1972-82; Nat. Cancer Inst. grantee, 1982—. Mem. Genetics Soc. Am., AAAS. Hindu. Subspecialties: Genetics and genetic engineering (biology); Gene actions. Current work: Research in molecular mechanisms of DNA repair in Saccharomyces cerevisiae. Office: Dept Biology U Rochester NY 14627

PRASAD, ANANDA SHIVA, medical educator; b. Buxar, Bihar, India, Jan. 1, 1928; came to U.S., 1952, naturalized, 1968; s. Radha Krishna and Mahesha (Kaur) Lall; m. Aryabala Ray, Jan. 6, 1952; children: Rita, Sheila, Ashok, Audrey. B.Sc. with honors, Patna (India) Sci. Coll., 1946, M.B. B.S., 1951; Ph.D., U. Minn., 1957. Diplomate: Am. Bd. Nutrition. Intern Patna Med. Coll. Hosp., 1951-52; resident St. Paul's Hosp., Dallas, 1952-53, U. Minn., 1953-56, VA Hosp., Mpls., 1956; instr. dept. medicine Univ. Hosp., U. Minn., Mpls., 1957-58; vis. asso. prof. medicine Shiraz Med. Faculty, Nemazee Hosp., Shiraz, Iran, 1960; asst. prof. medicine and nutrition Vanderbilt U., 1961-63; mem. faculty, dir. div. hematology dept. medicine Wayne State U., Detroit, 1963—, asso. prof., 1964-68, prof., 1968—; mem. staff Harper-Grace Hosp., VA Hosp., Allen Park, Mich.; Mem. trace elements subcom. Food and Nutrition Bd., NRC-Nat. Acad. Scis., 1965-68; chmn. trace elements com. Internat. Union Nutritional Scis.; reviewer several profl. jours. Author: Zinc Metabolism, 1966, Trace Elements in Human Health and Disease, 1976, Trace Elements and Iron in Human Metabolism, 1978, Zinc in Human Nutrition, 1979; editor: Clinical, Biochemical and Nutritional Aspects of Trace Elements, Am. Jour. Hematology; co-editor: Zinc Metabolism, Current Aspects in Health and Disease, 1977, Clinical Applications of Recent Advances in Zinc Metabolism; mem. editorial bd.: Nutrition Research; contbr. articles to profl. jours. Trustee Detroit Internat. Inst., Detroit Gen. Hosp. Research Corp., 1969-72. Pfizer scholar, 1955-56; Recipient research recognition award Wayne State U., 1964, award Am. Coll. Nutrition, 1976. Fellow ACP, AAAS, Internat. Soc. Hematology; mem. Am. Soc. Clin. Nutrition (awards com. 1969-70), Am. Fedn. Clin. Research (pres. Mich. 1969-70), Am. Inst. Nutrition (trace elements panel), Am. Physiol. Soc., Am. Soc. Clin. Investigation, Am. Soc. Hematology, Assn. Am. Physicians, Central Soc. Clin. Research, Soc. Exptl. Biology and Medicine (councillor Mich. 1967-71), Wayne County Med. Soc., AMA (Goldberger award 1975), Internat. Soc. Internal Medicine, Sigma Xi. Club: Cosmos (Washington). Subspecialty: Nutrition (medicine). Home: 4710 Cove Rd Orchard Lake MI 48033 Office: 540 E Canfield Ave Detroit MI 48201

PRASAD, BIRENDRA, research scientist; b. Bihar, India, June 30, 1949; came to U.S., 1973, naturalized, 1977; s. B. P. Gupta and Srimati Ramrati Devi. B.S., Bihar Coll. Engring., 1969; M.S., Indian Inst. Tech., 1971; D.Eng., Stanford U., 1975, Ph.D. in Mech. and Aerospace Engring, 1977. Sr. research engr. Ill. Inst. Tech., Chgo., 1976-78; prin. research scientist Ford Motor Co. Sci. Research Lab., Taylor, Mich., 1980—. Guest editor: spl. issue Internat. Jour. Vehicle Design, 1983; contbr. numerous articles to profl. jours. Mem. AIAA (edn. chmn. Mich. sect.), ASCE, ASME, Soc. Automotive Engrs., Soc. Mfg. Engrs., Am. Acad. Mechanics. Subspecialties: Mechanical engineering; Theoretical and applied mechanics. Current work: Novel concepts for constraint treatments and approximations in large-scale structural optimization. Address: 5421 S Piccadilly Circle West Bloomfield MI 48033

PRASAD, SATISH CHANDRA, radiol. physicist; b. Chapra, Bihar, India, Apr. 1, 1944; s. Shib Chandra and Sitapati Devi P.; m. Jayshri Prasad, July 1, 1954; children: Monica, Anita. B.S., Patna (India) U., 1963; M.S. in Physics, U. Mass., 1968, Ph.D., 1972; M.S. in Radiol. Physics, U. Colo. Med. Ctr., 1976. Cert. in diagnostic and therapeutic radiol. physics Am. Bd. Radiology, 1978. Research asst. physics U. Mass., Amherst, 1967-72; research assoc. physics U. Rochester, N.Y., 1972-74; research asst. radiation physics U. Colo. Med. Ctr., Denver, 1974-76; asst. prof. radiation physics Washington U. Med. Ctr., St. Louis, 1976-81, Upstate Med. Ctr., SUNY, Syracuse, 1981—. Contbr. articles to profl. jours. Mem. Am. Assn. Physicists in Medicine, Am. Phys. Soc., AAAS, Am. Coll. Radiology, Phi Kappa Phi. Current work: Radiation therapy and diagnostic radiological physics; computed tomography, radiation dose to tissues from x-rays, gamma rays and electron beams. Office: 750 E. Adams St Syracuse NY 13210

PRATT, ARNOLD W., govt. ofcl., physician; b. Binghamton, N.Y., Nov. 24, 1920; s. Donald Patrick and Agnes Kate (Smith) P.; m. Mary Durfee, June 17, 1945; children: Mary Pratt Grant, Susan Broomfield, Janet Pratt Oliver. Sc.D., Hobart Coll., 1973; M.D., U. Rochester, 1946. Mem. house staff N.Y. Hosp. (Cornell U. Med. Center), N.Y.C., 1946-47; research asso. Cornell U. Med. Sch., 1947-48; head sect. energy metabolism lab. physiology Nat. Cancer Inst., NIH, Bethesda, Md., 1948-66, dir. div. computer research and tech., 1966—. Editorial bd.: Computers and Biomed. Research, 1967—, Methods of Info. in Medicine, 1970—, Computer Programs in Biomedicine, 1972—. Recipient Superior Service award HEW, 1968, Meritorious Exec. award Sr. Exec. Service, 1980. Subspecialties: Information systems, storage, and retrieval (computer science); Computers in medicine.. Home: 6 North Dr Bethesda MD 20014 Office: 9000 Rockville Pike Bldg 12A Room 3033 Bethesda MD 20814

PRATT, CHARLES WALTER, microbiology educator; b. Kansas City, Mo., Feb. 5, 1944; s. Ivan and Elizabeth (Hartberg) P.; m. Marilyn Hunten, Dec. 27, 1968. B.A., Oreg. State U., Corvallis, 1966; Ph.D., U. Wash., Seattle, 1971. Postdoctoral fellow U. Oreg., Eugene, 1971-73; postdoctoral assoc., instr. MIT, Cambridge, Mass., 1973-79; asst. prof. microbiology U. Ill., Urbana, 1979—, asst. head microbiology, 1983—. Mem. Am. Soc. Microbiology, Genetics Soc. Am., Am. Genetics Assn., AAAS. Subspecialties: Microbiology; Molecular biology. Current work: Regulation of the synthesis of enzymes involved in energy metabolism in methanogenic bacteria. Home: 605 Eliot Dr Urbana IL 61801 Office: Microbiology Dept U Ill 131 Burrill Hall 407 S Goodwin Ave Urbana IL 61801

PRATT, DAVID WIXON, chem. physicist, educator; b. Providence, R.I., Sept. 14, 1937; s. Norman Twombly and Barbara Frances (Fisher) P.; m. Jo Ann Ferrell, Aug. 19, 1961; children: Susan Rounds, Jonathan Wixon. A.B., Princeton U., 1959; Ph.D., U. Calif., Berkeley, 1967. Research asst. U. Calif., Berkeley, 1962-67; asst. prof. chemistry U. Pitts., 1968-73, assoc. prof., 1973-78, prof., 1978—. Contbr. articles to profl. jours. Served to lt. (j.g.) USN, 1959-62. NIH fellow, 1967-68; Fulbright fellow, 1979; recipient Merck Faculty Devel. award, 1969. Mem. AAAS, Am. Chem. Soc., Am. Phys. Soc. Subspecialties: Atomic and molecular physics; Physical chemistry. Current work: Laser spectroscopy, magnetic resonance, dynamics and structure of molecular excited states. Office: Department of Chemistry University of Pittsburgh Pittsburgh PA 15260

PRATT, GEORGE L., agrl. engr., educator; b. Fargo, N.D., Jan. 31, 1926; s. Robert W. and Anne S. (Mangach) P.; m. Patricia Jones, Nov. 23, 1955; children: Thomas, Nancy Pratt Coash. B.S. in Agr, N.D. State U., 1950; M.S. in Agrl. Engring, Kans. State U., 1951; Ph.D., Okla. State U., 1967. Registered profl. engr., N.D. Sales Clay Equipment Corp., Cedar Falls, Iowa, 1952-53; instr. dept. agrl. engring. N.D. State U., Fargo, 1951-52, asst. prof., 1953-57, assoc. prof., 1957-69, prof., 1967—, chmn. dept., 1976—. Contbr. articles to profl. jours. Served with USMC, 1944-46. Mem. Am. Soc. Agrl. Engrs. (Metal Bldgs. Mfrs. award 1978), Sigma Xi., Phi Kappa Phi. Congregationalist. Lodges: Kiwanis; Elks; Masons. Subspecialties: Agricultural engineering; Biomass (agriculture). Current work: Energy demonstration farm involving solar collectors, methane digesters, and minimum tillage crop production. Processing and use of vegetable oil as fuel for diesel engines. Home: 2519 Willow Rd Fargo ND 58102 Office: Dept Agrl Engring ND State U Fargo ND 58105

PRATT, LAWRENCE R., chemistry educator; b. Flint, Mich., Apr. 12, 1950; s. Cecil L. and Alice D. (Dodge) P.; m. Kathleen Ladd, Sept. 6, 1971; children: Jane, Julia. B.S., Mich. State U., 1972; M.S., U. Ill., 1974, Ph.D., 1977. Postdoctoral fellow in chemistry Harvard U., 1977-78; asst. prof. chemistry U. Calif.-Berkeley, 1979—. Mem. Am. Phys. Soc., AAAS, Am. Chem. Soc., Phi Beta Kappa. Subspecialties: Theoretical chemistry; Statistical mechanics. Current work: Theoretical chemical physics; statistical mechanics. Office: Dept Chemistr U Calif Berkeley CA 94720

PRATT, LEE HERBERT, botany educator; b. Oakland, Calif., Dec. 3, 1942; s. Richard W. and Sammy J. (Adams) P.; m. Martha C. Hendrickson, June 14, 1963; children: Robert W., Eric N. B.A., Stanford U., 1963, M.A., 1964; Ph.D., Oreg. State U., 1967. Postdoctoral fellow U. Calif.-San Diego, La Jolla, 1967-68; asst. prof. Vanderbilt U., Nashville, 1969-73, assoc. prof., 1973-79; research assoc. U. Freiburg, W. Ger., 1975; prof. botany U. Ga., Athens, 1979—. Assoc. editor: Photochemistry and Photobiology, 1978—; mem. editorial bd.: Plant Physiology, 1979—, Plant, Cell and Environment, 1979—. NSF predoctoral fellow, 1963, 64; NSF postdoctoral fellow, 1967; NIH postdoctoral fellow, 1968. Mem. Am. Soc. Plant Physiologists (Charles Albert Shull award 1981), Am. Soc. Photobiology (councillor 1977-80), AAAS, Bot. Soc. Am. Subspecialties: Plant physiology (biology); Immunocytochemistry. Current work: Phytochrome, immunochemistry, spectrophotometry, biochemistry, with emphasis on investigations oriented toward phytochrome function at molecular and cellular levels, recent work emphasizes development and use of custom-built spectrophotometers. Office: Botany Dept U Ga Athens GA 30602

PRATT, PHILIP CHASE, pathologist, educator; b. Livermore Falls, Maine, Oct. 19, 1920; s. Harold Sewell and Cora Johnson (Chase) P.; m. Helen Clarke Deitz, Feb. 4, 1945; children: William Clarke (dec.), Charles Chase (dec.). A.B., Bowdoin Coll., 1941; M.D., Johns Hopkins U., 1944. Diplomate: Am. Bd. Pathology. Intern in pathology Johns Hopkins Hosp., 1944-45, asst. resident in pathology, 1945-46; pathologist Saranac Lab., Saranac Lake, N.Y., 1946-52, asst. dir., 1952-55; instr. Ohio State U., 1955-57, asst. prof. pathology, 1957-62, asso. prof., 1962-66, Duke U. Med. Center, 1966-71, prof., 1971—. Contbr. numerous articles to profl. publs. Fellow Am. Coll. Chest Physicians; mem. Am. Thoracic Soc., Am. Soc. Exptl. Pathology, Am. Assn. Pathologists and Bacteriologists, Internat. Acad. Pathology, Royal Soc. Health. Unitarian. Subspecialties: Pulmonary pathology; Occupational and environmental lung disease. Current work: Use of pulmonary morphometry and post mortem epidemiology to study interrelationships of personal, occupational, therapeutic and ambient inhalation exposures in human lung disease. Office: Duke Med Center Davison Bldg Durham NC 27710 The innovative idea is the essential commodity of the academic life. Origins of such ideas are varied but for me usually begin with realization that an existing concept does not adequately explain observed phenomena. When direct reasoning does not produce a new, better concept, the problem is put aside. Weeks later a return to the question often promptly reveals a logical new solution which has arisen without conscious effort. Of course, this must then be subjected to investigation to be either confirmed or refuted.

PRAUSNITZ, JOHN MICHAEL, chemical engineer, educator; b. Berlin, Germany, Jan. 7, 1928; came to U.S., 1937, naturalized, 1944; s. Paul Georg and Susi (Loewenthal) P.; m. Susan Bergmann, June 10, 1956; children: Stephanie, Mark Robert. B.Chem. Engring., Cornell U., 1950; M.S., U. Rochester, 1951; Ph.D., Princeton, 1955. Mem. faculty U. Berkeley, 1955—, prof. chem. engring., 1963—; cons. to cryogenic, polymer, petroleum and petrochem. industries. Author: (with others) Computer Calculations for Multicomponent Vapor-Liquid Equilibria, 1967, (with P.L. Chueh) Computer Calculations for High-Pressure Vapor-Liquid Equilibria, 1968, Molecular Thermodynamics of Fluid-Phase Equilibria, 1969, (with others) Regular and Related Solutions, 1970, Properties of Gases and Liquids, 1977, Computer Calculations for Multicomponent Vapor-Liquid and Liquid-Liquid Equilibria, 1980; contbr. to profl. jours. Guggenheim fellow, 1962, 73; Miller research prof., 1966, 78; Recipient Alexander V. Humboldt Sr. Scientist award, 1976. Mem. Am. Inst. Chem. Engrs. (Colburn award 1962, Walker award 1967), Am. Chem. Soc. (E.V. Murphree award 1979), Nat. Acad. Engring., Nat. Acad. Scis. Subspecialties: Chemical engineering; Physical chemistry. Current work: Thermodynamic properties of fluid mixtures for chemical process design. Home: 52 the Crescent Berkeley CA 94708

PRAVDA, MILTON FRANK, nuclear engineering company executive; b. N.Y.C., Dec. 25, 1923; s. Frank and Nellie (Preplata) P. B.S. in Elec. Engring, Newark Coll. Engring., 1947. Project engr. Gen. Electric Co., Phila., 1947-50, resident engr., Schenectady, 1950-54, mgr. reactor design, 1954-60; mgr. reactor systems Martin Marietta Corp., Balt., 1960-64, dir. engring. and research, 1964-68; chmn. bd., pres. Dynatherm Corp., Cockeysville, Md., 1968—; chmn. bd. Bossalina Machine Co., Inc., Cockeysville, 1968—. Editor: Heat Pipe Design Handbook, 1972; Contbr. articles to profl. jours. Mem. IEEE (sr.), Am. Nuclear Soc. (sr.), AAAS, Tau Beta Pi. Subspecialties: Nuclear engineering; Electrical engineering. Current work: Research on a very thin films generated on tapered surfaces rotating at high rotational velocities. This is the basis of a new absorption heat pump for use on automobiles and in industry to save energy. Patentee in fields of nuclear reactors, isotopic generators, heat pipes, heat exchangers, and absorption systems. Home: 7708 Greenview Terr Towson MD 21204 Office: Dynatherm Corp One Industry Ln Cockeysville MD 21030

PRAVDO, STEVEN H., astrophysicist, researcher aerospace; b. N.Y.C., Dec. 9, 1951; s. William and Elaine Phyllis (Cohen) P. B.S., Haverford Coll., 1972; M.S., U. Md., College Park, 1974, Ph.D. in Physics, 1976. Research assoc. Goddard Space Flight Ctr., Greenbelt, Md., 1976-79; sr. research fellow Calif. Inst. Tech., Pasadena, 1979-81; mem. tech. staff (Jet Propulsion Lab.), Pasadena, 1981—. Contbr.: articles to jours. including Nature. Recipient various NASA research grants. Mem. Am. Phys. Soc., Am. Astron. Soc., AAAS, Phi Beta Kappa. Club: JPL Chess. Subspecialties: 1-ray high energy astrophysics; Optical astronomy. Current work: X-ray sources and identifications, pulsars, supernova remnants, X-ray stars, Herbig-Haro objects. Office: 238-420 JPL 4800 Oak Grove Dr Pasadena CA 91109

PREECE, JOHN EARL, hort. physiologist, educator; b. Woodsville, N.H., Mar. 4, 1952; s. Daniel Platt and Jean Margaret (Page) P.; m. Barbara Joan Grondin, June 2, 1973; 1 dau., Ellen Platt. B.S., U. N.H., 1974; M.S., U. Minn., 1977, Ph.D., 1980. Instr. U. Minn., St. Paul, 1978-79; asst. prof. plant and soil sci So. Ill. U., Carbondale, 1980—. Author lab. manual. McIntyre-Stennis grantee, 1981; U.S. Dept. Agr. Forest Service grantee, 1982. Mem. Am. Soc. Hort. Sci. (chmn.-elect plant propagation working group 1982-83), Council for Agrl. Sci. and Tech., Internat. Plant Propagators Soc., Am. Soc. Plant Physiologists, Pi Alpha Xi, Sigma Xi, Gamma Sigma Delta. Subspecialties: Plant cell and tissue culture; Hydroponics. Current work: In vitro micropropagation; in vitro selection for environmental stress including heavy metals; plant growth regulation; hydroponics and allelochemicals. Home: 112 Parrish Ln Carbondale IL 62901 Office: So Ill U Dept Plant and Soil Sci Carbondale IL 62901

PREISS, IVOR LOUIS, educator; b. N.Y.C., Mar. 24, 1933; s. Louis H. and Carolyn P.; m. Jane M. Rose, Jan. 28, 1956; children: Susan Lai, Sharon, Sandra, Bradley, Michelle; m. Lorraine M. Dixson, June 7, 1970. B.S., Rensselaer Poly. Inst., 1955; M.S., U. Ark., 1957; Ph.D., 1960. Research fellow U. Ark., 1957-60; research assoc. physics Yale U., 1960-66; asst. dir. Heavy Ion Lab., Yale U., 1961-66, lectr. chemistry, 1961-66; prof. chemistry Rensselaer Poly. Inst., Troy, N.Y., 1966—; adj. prof. physics SUNY-Albany, 1967—; bd. dirs. Found Analytical Research in Art, 1979-82. Contbr. articles to profl. jours. Bd. dirs. N.Y. State PTA, 1971-78. Mem. Am. Chem. Soc., Am. Phys. Soc., Soc. Applied Spectroscopy, Zeta Psi. Roman Catholic. Subspecialties: Analytical chemistry; Nuclear physics. Current work: Analysis of trace elements, nuclear scattering, radioisotope utilization, production and application, radiation detection, x-ray fluorescence. Home: 6 County View Rd Latham NY 12110 Office: Cogswell Lab Rensselaer Poly Inst Troy NY 12110

PREISS, JACK, biochemistry educator; b. Bklyn., June 2, 1932; s. Erool and Gilda (Friedman) P.; m. Judith Weil, June 11, 1959; children: Jennifer Ellen, Jeremy Oscar, Jessica Michelle. B.S., CCNY, 1953; Ph.D., Duke U., 1957. Postdoctoral fellow Washington U., St. Louis, 1958-59, Stanford Sch. Medicine, Palo Alto, Calif., 1959-60; scientist NIH, Bethesda, Md., 1960-62; asst. prof. U. Calif.-Davis, 1962-65, assoc. prof., 1965-68, prof. biochemistry, 1968—; cons. Zoecon Corp., Palo Alto, 1982-83. Editor: Carbohydrates: Structure and Function, 1980; editorial bd.: Jour. Biol. Chemistry, 1971-76, 78—, Archives of Biochemistry and Biophysics, 1969—. Guggenheim Meml. fellow, 1969-70; USPHS sr. postdoctoral fellow, 1971. Mem. Am. Chem. Soc. (Charles Pfizer award 1971), AAAS, Biochem. Soc., Fedn. Am. Soc. Exptl. Biologists, Am. Soc. Protozoologists, Am. Soc. Microbiologists, Am. Soc. Plant Physiologists, Soc. Complex Carbohydrates, N.Y. Acad. Scis., Sigma Xi. Subspecialties: Biochemistry (biology); Molecular biology. Current work: Genetic and allosteric regulation of the bacterial glycogen and plant starch biosynthetic enzymes, protein chemistry, structure, function of catalytic and effector sites of the glycogen biosynthetic enzymes, regulation of starch degradation, cloning of the E. coli Glycogen biosynthetic structural genes, DNA sequencing. Home: 912 Colby Dr Davis CA 95616 Office: Dept Biochemistry/Biophysics U Calif Davis CA 95616

PRELAS, MARK ANTHONY, nuclear engineer, educator; b. Pueblo, Colo., July 7, 1953; s. George B. and Katheryn (Beck) P.; m. Rosemary Roberts, May 21, 1979. B.S., Colo. State U.-Ft. Collins, 1975; M.S., U. Ill.-Urbana, 1976, Ph.D., 1979. Asst. prof. nuclear engring. U. Mo.-Columbia, 1979—, dir. Mo. Mirror Project, 1982—, research scientist, 1980—; v.p. Nuclear Pumped Laser Corp., Kingston, N.J., 1980—. Contbr. chpts. to books, articles to profl. publs. NSF grantee, 1980, 82, 83; McDonnell Douglas Corp. grantee, 1981; Gas Research Inst. fellow, 1981. Mem. AAAS, Am. Nuclear Soc., Am. Phys. Soc., IEEE, Sigma Xi. Subspecialties: Nuclear engineering; Plasma physics. Current work: Superconducting magnet design and operation; charged particle transport in various media; plasma chemistry; laser design; gaseous electronics; energy conversion; plasma engineering. Home: 1904 Love Joy Ln Columbia MO 65202 Office: Mo Engring Dept Columbia MO 65211

PREM, KONALD ARTHUR, physician, educator; b. St. Cloud, Minn., Nov. 6, 1920; s. Joseph E. and Theresa M. (Willing) P.; m. Phyllis Edelbrock, June 14, 1947; children: Mary Kristen, Stephanie, Timothy. B.S., U. Minn., 1947; M.B., 1950; M.D., 1951. Diplomate: Am. Bd. Ob-Gyn (with spl. competence in gynecologic oncology). Intern Mpls. gen. Hosp., 1950-51; fellow dept. obstetrics and gynecology U. Minn., Mpls., 1951-54, instr., 1955-58, asst. prof., 1958-60, asso. prof., 1960-69, prof., 1969—, dir. div. gynecologic oncology, 1969—, head dept. obstetrics and gynecology, 1976—. Served to capt. USAR, 1941-46; brig. gen. M.C. USAR (Ret.). Decorated Legion of Merit. Mem. Am. Coll. Ob-Gyn, Am. Gynec. and Obstet. Soc., Central Assn. Ob-Gyn, Hennepin County Med. Soc., Soc. Pelvic Surgeons, Minn. Ob-Gyn Soc., Soc. Gynecologic Oncologists, Soc. Gynecologic Surgery, Minn. Acad. Medicine, Am. Radium Soc., Gynecologic Urology Soc., Mpls. Surg. Soc., Soc. Med. Cons. to Armed Forces. Roman Catholic. Clubs: U. Minn. Alumni, Decathlon. Subspecialty: Gynecological oncology. Home: 4806 Sunnyside Rd Edina MN 55424 Office: PO Box 395 Mayo Bldg 420 Delaware St SE Minneapolis MN 55455

PRENDERGAST, JOCELYN, neuroscientist; b. Boston, Nov. 8, 1942; d. John Clemons and Barbara (Miller) P.; m. Louis James Misantone, May 13, 1972; children: Nina, Elizabeth. B.S., U. Vt., Burlington, 1964; Ph.D., SUNY-Syracuse, 1975. Cert. Tchr., Pa., Mass., Del. Tchr. sci. Eastern Jr. High Sch., Greenwich, Conn., 1964-68, White Plains, (N.Y.) High Sch., 1970-71; asst. prof. dept. anatomy Med. Coll. Pa., 1979-82. Contbr. articles to profl. jours. Sarah A. Luse fellow, 1968-69; NIH fellow, 1971-75, 78-79; Paralyzed Veterans Am. grantee, 1977-79; NIH grantee, 1979-82; AOA/NOF grantee, 1982—. Mem. Am. Assn. Anatomists, Electron Microscopy Soc. Am., Am. Physiol. Soc., Soc. Neurosci., Sigma Xi. Episcopalian. Subspecialties: Neurobiology; Anatomy and embryology. Current work: The morpiological and behavioral effects of spinal damage. Office: Westtown Sch Westtown PA 19395

PRENTKY, ROBERT ALAN, psychology educator; b. L.I.C., Aug. 21, 1947; s. Peter Isaac and Janet (Weinberger) P. Ph.D., Northwestern U., 1975. Dir. Research Mass. Treatment Ctr., Bridgewater, 1980—; research assoc. Brandeis U., Waltham, Mass., 1981—; research asst. prof. Boston U., 1981—; vis. asst. prof. Wheaton Coll. Author: Creativity and Psychopathology, 1980; editor: Biological Aspects of Normal Personality, 1979. NIMH postdoctoral fellow U. Mass., 1975-77, U. York, Eng., 1977-78, U. Rochester Med. Ctr., 1978-80. Mem. Union of Concerned Scientists, Physicians for Social Responsibility, Am. Psychol. Assn., Soc. for Psychophysiol. Research, N.Y. Acad. Sci., Soc. for Life History Research in Psychopathology, AAAS. Subspecialties: Psychobiology; Neurochemistry. Current work: Bio-social antecedents of human sexual and nonsexual aggression; psychopathy; episodic dyscontrol; biogenic aspects of normal personality; biological aspects of creativity. Home: 35 Howland Ln Hingham MA 02043 Office: Mass Treatment Ctr PO Box 554 Bridgewater MA 02324

PRESANT, CARY A., med. oncologist; b. Buffalo, Dec. 16, 1942; s. Allen N. and R. (Reeta) (Coplon) P.; m. Sheila Lassman, June 11, 1966; children: Seth, Sean, Jaron, Jaclyn. M.D., SUNY, Buffalo, 1966. Diplomate: Am. Bd. Internal Medicine, 1970. Staff assoc. Nat. Cancer Inst., 1967-69; asst. prof. medicine and radiology Washington U., St. Louis, 1973-79; dir. med. oncology City of Hope Med. Ctr., Duarte, Calif., 1979-82; med. oncologist Wilshire Oncology Med. Group, West Covina, Calif., 1982—; chmn. melanoma-sarcoma com. S.E. Cancer Study Group, 1975—; prof. medicine U. So. Calif.; cons. Scott Labs. and Oncology Labs., Providence; bd. sci. advs. Vestar Research Corp., Pasadena, Calif. Contbr. articles to profl. jours., chapts. to books. Served with USPHS, 1967-69. Roswell Park Meml. Inst. fellow, 1963; USPHS grantee, 1972-82; Louis Sklarow Meml. lectr. Maimonides Med. Soc., Buffalo, 1982. Fellow ACP; mem. Am. Soc. Hematology, Am. Soc. Clin. Oncology, Am. Assn. Cancer Research, Phi Beta Kappa, Alpha Omega Alpha. Jewish. Subspecialties: Chemotherapy; Cancer research (medicine). Current work: Leukemia and solid tumor treatment; diagnosis and treatment with liposomes. Office: 933 S Sunset Suite 301 West Covina CA 91790

PRESS, FRANK, educator, geophysicist; b. Bklyn., Dec. 4, 1924; s. Solomon and Dora (Steinholz) P.; m. Billie Kallick, June 9, 1946; children—William Henry, Paula Evelyn. B.S., CCNY, 1944, LL.D. (hon.), 1972; M.A., Columbia U., 1946, Ph.D., 1949; D.Sc. (hon.), Notre Dame U., 1973, U. R.I., U. Ariz., Rutgers U., City U. N.Y., 1979. Research asso. Columbia, 1946-49, instr. geology, 1949-51, asst. prof. geology, 1951-52, asso. prof., 1952-55; prof. geophysics Cal. Inst. Tech., 1955-65, dir. seismol. lab., 1957-65; prof. geophysics, chmn. dept. earth and planetary scis. Mass. Inst. Tech., 1965-77; sci. advisor to Pres., dir. Office Sci. and Tech. Policy, Washington, 1977-80; Inst. prof. M.I.T., 1981; pres. Nat. Acad. Scis., 1981—; cons. USAF, 1958, U.S. Geol. Survey, 1957, USN, 1957—, Office Spl. Asst. for Sci. and Tech., 1961-64, Office Sci. and Tech., 1964—, ACDA, 1961-68, AID, 1961-64; dir. United Electro Dynamics; expert UNESCO Tech. Assistance Adminstrn., 1953; seismology and glaciology panel U.S. nat. com. IGY, 1955; chmn. U.S. del. Nuclear Test Ban Conf., 1960; mem. Pres.'s Sci. Adv. Com. and Internat. Geophysics Com., 1961-64, Lunar and Planetary Mission Bd., 1965-70; chmn. adv. bd. Nat. Center Earthquake Research, 1966—; mem. Nat. Sci. Bd., 1970—. Author: (with M. Ewing, W.S. Jardetzky) Propagation of Elastic Waves in Layered Media, 1956, (with R. Siever) Earth, 1974; Co-editor: Physics and Chemistry of the Earth, 1957—. Recipient Columbia medal for excellence, 1960, pub. service award U.S. Dept. Interior, 1972, gold medal Royal Astron. Soc., 1972, pub. service medal NASA, 1973. Mem. Am. Acad. Arts and Scis., Geol. Soc. Am. (councilor), Am. Geophys. Union (pres. 1973), Soc. Exploration Geophysicists, Seismol. Soc. Am. (pres. 1963), AAUP, Nat. Acad. Scis. (councilor), Phi Beta Kappa. Subspecialty: Geophysics. Office: Nat Acad Scis 2101 Constitution Ave Washington DC 20418

PRESS, NEWTOL, zoology educator, consultant; b. N.Y.C., Nov. 27, 1930; s. Morris and Sylvia (Finkelstein) P.; m. Ethel Francine Greenfield, Dec. 23, 1951; children: Betty, David, Michael. B.A., NYU, 1951; M.S., U. Iowa, 1955, Ph.D., 1956. Research asst. U. Iowa, Iowa City, 1952-56; instr. U. Wis.-Milw., 1956-58, asst. prof., 1958-61, assoc. prof., 1961-77, prof. zoology, 1977—; pres. Found. for Advancement Ethical and Moral Values in Biology and Medicine. Pres. Maple Dale-Indian Hill Sch. Bd., Fox Point, Wis., 1975-77, mem., 1972-79. Mem. AAUP (pres. Milw. chpt. 1975-76), Phi Kappa Phi. Jewish. Subspecialties: Cell biology; Genetics and genetic engineering (biology). Current work: Aging as a social disease, bioethical limitations on development. Home: 1055 W Ravine Ln Milwaukee WI 53217 Office: Dept Zoology U Wis PO Box 413 Milwaukee WI 53201

PRESS, WILLIAM HENRY, astrophysicist; b. N.Y.C., May 23, 1948; s. Frank and Billie (Kallick) P.; m. Margaret Ann Lauritsen, Sept. 9, 1969 (div. 1982); 1 dau., Sara Linda. A.B., Harvard Coll., 1969; M.S., Calif. Inst. Tech., 1971, Ph.D., 1972. Asst. prof. theoretical physics Calif. Inst. Tech., 1973-74; asst. prof. physics Princeton (N.J.) U., 1974-76; prof. astronomy and physics Harvard U., Cambridge, Mass., 1976—, chmn. dept. astronomy, 1982—; mem. adv. coms. and panels NSF, NASA, Nat. Acad. Scis., NRC; vis. mem. Inst. Advanced Study, spring 1983. Contbr. articles to profl. jours. Sloan Found. research fellow, 1974-78. Mem. Am. Astron. Soc. (Helen B. Warner Prize 1981), Am. Phys. Soc., Internat. Astron. Union, Internat. Soc. Relativity and Gravitation. Subspecialties: Cosmology; Theoretical astrophysics. Current work: Galaxy and cluster of galaxy formation, astrophysical fluid dynamics; computational methods; relativistic astrophysics. Office: 60 Garden St Cambridge MA 02138

PRESTON, GEORGE W., III, astronomer; b. Los Angeles, 1930; m. (married); 4 children. B.S., Yale, 1952; Ph.D., U. Calif., Berkeley, 1959. Carnegie fellow Mt. Wilson Obs., Carnegie Instn. Washington, 1959-60; asst. astronomer Lick Obs., U. Calif., Santa Cruz, 1960-64, asso. astronomer, 1964-68; staff Mt. Wilson and Las Campanas Obs. (formerly Hale Obs.), 1969-80; asst. dir. Mt. Wilson Obs., 1975-80; acting dir. Mt. Wilson and Las Campanas Obs., 1981—; Robert J. Trumpler lectr. Astron. Soc. Pacific, 1963. Mem. Am. Astron Soc. (Helen B Warner prize 1965), Internat. Astron. Union, Nat. Acad. Scis. Subspecialty: Observatory administration. Address: Mount Wilson and Las Campanas Obs Carnegie Instn Washington 813 Santa Barbara St Pasadena CA 91101

PRESTON, JOAN MURIEL, psychology educator, researcher; b. St. Thomas, Ont., Can., July 28, 1939; d. George Arthur and Vera Victoria (Atkinson) P.; m.; children: Eric Peter John, Stephen George. B.A. with honors, U. Western Ont., London, 1964, M.A., 1965, Ph.D., 1967; teaching cert., London Tchrs. Coll., 1958. Tchr. Newmarket (Ont.) Sch. Bd., 1958-59, North York (Ont.) Bd., 1959-61; asst. prof. psychology U. Toronto, 1967-71; asst. prof. Brock U., St. Catharines, Ont., 1971-73, assoc. prof., 1973—, dir. child studies program, 1981-82; chmn. Lincoln Childhood Edn. Com., St. Catharines, 1979—. Contbr. articles to profl. jours. Sec. Niagara Peninsual Refugee Support Group, St. Catharines, 1980. Can. Council fellow, 1976-77; Social Scis. and Humanities Research Council grantee, 1978-80, 82; Nat. Scis. and Engring. Council grantee, 1978-81. Mem. Can. Psychol. Assn., Am. Psychol. Assn., Soc. Research in Child Devel., Lake Ont. Visual Establishment. Subspecialties: Cognition; Developmental psychology. Current work: Microcomputers and problem solving; psychology of television cognitive processing of pictures and words. Office: Brock U Merrittville Hwy Saint Catharines ON Canada L2S 3A1

PRESTWOOD, ANNIE KATHERINE, veterinary parasitologist, educator; b. Lenoir, N.C., July 4, 1935; d. John Howard and Emily Beatrice (Adkins) P. Student, U. N.C., Greensboro, 1954-56; B.S., N.C. State U., Raleigh, 1958; D.V.M., U. Ga., M.S., Ph.D., 1968. Research assoc. Southeastern Coop. Wildlife Disease Study, U. Ga., Athens, 1966-76; assoc. prof. dept. parasitology Coll. Vet. Medicine, U. Ga., Athens, 1976-81, prof., 1981—. Contbr. articles to profl. jours. Nat. Inst. Heart, Lung and Blood Disorders grantee, 1979—; U.S. Dept. Agr. grantee, 1981—. Mem. Am. Soc. Parasitologists, Am. Vet. Med. Assn., Can. Soc. Zoologists, Wildlife Disease Assn., Wildlife Soc. Subspecialties: Parasitology; Pathology (veterinary medicine). Current work: Pathophysiology of parasitic diseases; coagulation defects, pulmonary thromboembolism. Office: Dept Parasitology Coll Veterinary Medicine U Ga Athens GA 30607

PRETLOW, THOMAS GARRETT, pathologist, educator; b. Warrenton, Va., Dec. 11, 1939; s. William Ribble and May (Tiffany) P.; m. Theresa Pace, June 29, 1963; children: James Michael, Joseph Peter, David Mark. A.B., Oberlin Coll., 1960; M.D., U. Rochester, 1965. Diplomate: Nat. Bd. Med. Examiners. Intern in medicine U. Wis. Hosp., Madison, 1965-66, resident in pathology, 1966-67; research assoc. Nat. Cancer Inst., Bethesda, Md., 1967-69; asst. prof. pathology Rutgers Med. Sch., Piscataway, N.J., 1969-70; assoc. prof. pathology U. Ala., Birmingham, 1971-73, assoc. prof. biochemistry, 1976-82, prof. pathology, 1974—, prof. biochemistry, 1982—; cons. NIH, 1976-80. Editorial bd.: Cell Biophysics, 1979—; editor: Cell Separation: Methods and Selected Applications, 3 vols, 1982, 83, 84; contbr. articles to profl. jours. Exec. bd. Birmingham Council Boy Scouts Am., 1979—; chmn. Birmingham Catholic Com. on Scouting, 1979-82, chmn. diocesan com., 1981—. Served to lt. comdr. USPHS, 1967-69. Recipient Career Devel. award Nat. Cancer Inst., 1973-78; Nat. Cancer Inst. grantee, 1969—; Am. Cancer Soc. grantee, 1971—. Mem. Am. Assn. Pathologists, Am. Assn. Immunologists, Am. Soc. Biol. Chemists, Internat. Acad. Pathology, Am. Soc. Clin. Oncology, Am. Assn. Cancer Research. Roman Catholic. Club: Serra (pres. 1982-83). Subspecialties: Pathology (medicine); Cancer research (medicine). Current work: Neoplastic diseases, cell separation, immunology, free-flow electrophoresis, monoclonal antibody technology. Home: 160 Peachtree Rd Birmingham AL 35213 Office: Dept Pathology Univ Ala University Station Birmingham AL 35294

PREWITT, CHARLES THOMPSON, geoscience educator; b. Lexington, Ky., Mar 3, 1933; s. John Burton and Margaret Porter (Thompson) P.; m. Gretchen Hansen, Jan. 31, 1958; 1 son, Daniel Hansen. S.B., MIT, 1955, S.M., 1960, Ph.D., 1962. Research scientist E.I. duPont deNemours & Co., Wilmington, Del., 1962-69; mem. faculty dept. earth-space scis. SUNY-Stony Brook, 1969—, prof., 1971—; vis. prof. H.C. Orsted Inst., Copenhagen, 1982, Ariz. State U., 1983. Editor: Physics and Chemistry of Minerals, 1976—; contbr. articles to profl. jours. Japan Soc. Promotion of Sci. fellow, 1983. Mem. Mineral. Soc. Am. (v.p. 1982-83). Subspecialties: Mineralogy; Geochemistry. Current work: Crystal chemistry of inorganic materials, application of synchrotron x-radiation experiments to problems in geochemistry and mineralogy. Home: 12 Intervale Rd Setaukel NY 11733 Office: Dept Earth and Space Scis SUNY Stony Brook NY 11794

PREWITT-DIAZ, JOSEPH ORLANDO, teacher educator, school psychologist; b. San Juan, P.R., Nov. 23, 1943; s. Joe Crawford and Leonor (Diaz) Prewitt-D.; m. Maria Dolores Rodriguez, Nov. 3, 1969; children: Joseph Orlando, Maria J., Ana J., Victoria J. B.A., U. P.R., 1967, B.Ed., 1970, M.Ed., 1972, Ph.D., 1979. Tchr. head tchr., prin. Dept. Instrn., Cayey, P.R., 1968-70; instr. U. P.R., Cayey, 1970-72; counseling specialist Sch. Allied Health Professions, U. Conn., Storrs, 1973-74; project dir. Hartford (Conn.) Public Schs., 1974-75; vice prin. Buckeley High Sch., Hartford, Conn., 1975-79; asst. prof. edn. and sch. psychology Pa. State U., University Park, 1979—. Contbr. articles to profl. jours. Commr. Human Relations Commn., Hartford, 1976-79; trustee Community Coll. System, 1976-79 trustee Community Coll. System, Hartford; bd. dirs. World Edn. Fellowship, Storrs, 1977-79; chmn. Pa. Assn. Bilingual Edn., 1979—. World Edn. fellow, 1976; Gestalt Inst. Cleve. fellow, 1982; Kellogg fellow, 1982—. Mem. Am. Psychol. Assn., Nat. Assn. Bilingual/Bicultural Edn., Am. Ednl. Research Assn. Lodge: Masons. Current work: Cross cultural research methodology; personality test translation and validation. Home: 119 Cedar Ln State College PA 16801 Office: Sch Edn Pa State U University Park PA 16801

PREZELIN, BARBARA BERNTSEN, marine biologist, educator; b. Portland, Oreg., Apr. 13, 1948; d. Walter Stanley and Doreen (Mugford) Fitzpatrick) Berntsen; m. Louis M. Prezelin, Jan. 22, 1972; 1 dau., Christine. B.S., U. Oreg.-Eugene, 1966-70; Ph.D., Scripps Instn. Oceanography, 1975. Postdoctoral researcher U. Calif.-Santa Barbara, 1975-77, asst. prof., 1977-83, assoc. prof. dept. biol. scis., 1983—. Assoc. editor: Phycological Soc. Am. Jour. Phycotology, 1982; mem. editorial bd., 1981-83. NSF grantee, 1982-84, 77-82; U.S. Dept. Agr. grantee, 1979-81. Mem. Am. Soc. Limnology (sec.-treas. western sect.), Am. Soc. Photobiology. Democrat. Roman Catholic. Subspecialties: Photosynthesis; Ecology. Current work: algal physiological ecologist, specializing in defining biochemical mechanisms regulating photosynthesis and primary productivity in the oceans. Office: Marine Sci Inst and Dept Biol Scis U Calif Santa Barbara CA 93106

PRIBOR, HUGO CASIMER, physician; b. Detroit, June 12, 1928; s. Benjamin Harrison and Wanda Frances (Mioskowski) P.; m. Judith Eleanor Smith, Dec. 22, 1955; children: Jeffrey D., Elizabeth F., Kathryn A. B.S., St. Mary's Coll., 1949; M.S., St. Louis U., 1951, Ph.D., 1954, M.D., 1955. Diplomate: Am. Bd. Pathology. Intern Providence Hosp., Detroit, 1955-56; resident pathologic anatomy and clin. pathology NIH, Bethesda, Md., 1956-59; field investigator gastric cytology research project Nat. Cancer Inst., Bowman-Gray Sch. Medicine, Winston-Salem, N.C., 1959-60; assoc. pathologist, dir. clin. lab. Bon Secours Hosp., Grosse Pointe, Mich., 1960-63; pathologist, dir. labs. Samaritan Hosp. Assn., East Side Gen. Hosp., Detroit, 1963-64, Anderson Meml. Hosp., Mt. Clemens, Mich., 1963-64; cons. pathologist Middlesex County Med. Examiners Office, New Brunswick, N.J., 1964-73; dir. dept. labs., chief pathologist, sr. attending physician Perth Amboy (N.J.) Gen. Hosp., 1964-73; chmn., chief exec. officer Ctr. Lab. Medicine, Inc., Metuchen, N.J., 1968-77; v.p. med. affairs Damon Corp., Med. Services Group, 1977-78; exec. med. dir. MDS Health Group, Inc., Red Bank, N.J., 1978-80; med. dir. Internat. Clin. Labs., Inc., Nashville; physician Assoc. Pathologists (P.C.), Nashville, 1981—; research assoc. dept. pathology St. Louis U. Sch. Medicine, 1954-55; instr. pathology Bowman-Gray Sch. Medicine, Winston-Salem, N.C., 1959-60; asst. prof. chemistry U. Detroit, 1961-64; instr. pathology Wayne State U. Sch. Medicine, Detroit, 1962-64; clin. assoc. prof. dept. pathology Rutgers Med. Sch.,

Rutgers, The State U., New Brunswick, N.J., 1966-68; cons. Health Facilities Planning and Constrn. Service, USPHS, HEW, Rockville, Md., 1970-71; prof. biomed. engring. Coll. Engring., Rutgers, The State U., New Brunswick, N.J., 1971-75, 80—; chmn. bd. trustees St. Mary's Coll., Winona, Minn., 1972-74, chmn. fin. com., 1971-72; clin. prof. pathology Vanderbilt U. Sch. Medicine, Nashville, 1981—. Author: (with G. Morrell and G. H. Scherr) Drug Monitoring and Pharmacokinetic Data, 1980; contbr. articles in field to profl. jours. Fellow Am. Soc. Clin. Pathologists (Silver award 1968); mem. AMA (physicians recognition award 1969, 74), Am. Assn. Exptl. Pathology, Coll. Am. Pathologists (chmn. subcom. 1974-78), Internat. Acad. Pathology, Pan Am. Med. Assn. (life), Assn. Advancement Med. Instrumentation, N.J. State Med. Soc., Acad. Medicine N.J. (chmn. clin. pathology sect. 1965-67), N.J. Soc. Pathologists (exec. com. 1965-67), Sigma Xi. Republican. Roman Catholic. Subspecialties: Pathology (medicine); Information systems, storage, and retrieval (computer science). Current work: Application of computer technology to the practice of medicine, laboratory automation, and laboratory management. Home: 200 Olive Br Rd Nashville TN 37205

PRIBUSH, ROBERT ALLEN, chemistry educator, consultant; b. Elizabeth, N.J., Sept. 27, 1946; s. Adolph J. and Anne J. (Saraka) P.; m. Eileen C. Paige, July 27, 1968 (div. Feb. 1981); 1 son, David A.; m. Bonita L. Colvin, Mar. 11, 1981; children: 1 son, Robert D. Student, Union Coll., Cranford, N.J., 1964-66; B.S., U. Del., 1968; Ph.D., U. Mass., 1972. Fellow U. So. Calif., Los Angeles, 1972-74; assoc. prof. chemistry Butler U., Indpls., 1974—; cons. Wolf Tech. Services, Indpls., 1976—. Contbr. articles to profl. jours. Mem. Lawrence Twp. Sch. Screening Com., Indpls., 1979-82; adv. Tau Kappa Epsilon, Butler U., 1979—, chmn. univ. student affairs com., 1981—; mem. Twp. Sch. Test Com., 1981-82; chmn. Regional Sci. Fair, Central Ind., 1982. Mem. Am. Chem. Soc. (nat. com. chmn. 1979-81, local chpt. chmn. 1980-81), AAAS, Phi Kappa Phi. Republican. Lutheran. Subspecialties: Inorganic chemistry; Photochemistry. Current work: Photoinduced catalysis by transition metal compounds. Home: 6025 Thornwood Ct Indianapolis IN 46250 Office: Dept Chemistry Butler U Indianapolis IN 46208

PRICE, ALAN ROGER, biochemistry educator, university executive; b. Pontiac., Mich., Jan. 15, 1942; s. Ralph Eugene and Helen Grace (Van Atta) P.; m. Katherine Jean Ralph, July 14, 1962; children: Anita Marie, Michael Ned, Mark Alan, Audra Katherine. B.S. in Chemistry, Fla. State U., 1964; Ph.D. in Biochemistry, U. Minn., 1968; post-grad., Mich. State U., 1968-69. Asst. prof. biol. chemistry Med. Sch. U. Mich., 1970-75, asst. dean research and devel., 1979-81, assoc. prof., 1975—, asst. v.p. research, 1980—; researcher Univ. Inst. Biol. Chemistry, Copenhagen, Denmark, 1976-77. Contbr. articles in field to profl. jours. Am. Council on Edn. fellow, 1980; U.S. Dept. Energy grantee, 1970-82; recipient Rumsey Meml. research award Fla. State U., 1968; Disting. Service award U. Mich., 1973. Mem. AAAS, Am. Soc. Microbiology, Am. Soc. Biol. Chemists, Sigma Xi. Subspecialties: Biochemistry (biology); Virology (biology). Current work: Enzymology of nucleotide and nucleic acid biosynthesis; biochemistry of viral infections, biosynthesis and function of unusual nucleosides in deoxyribonucleic acids. Home: 1450 Covington St Ann Arbor MI 48103 Office: U Mich 4070 Administration Bldg Ann Arbor MI 48109

PRICE, ALVIN AUDIS, veterinarian, coll. adminstr.; b. Lingleville, Tex., Oct. 8, 1917; s. Samuel C. and Lillie M. (Bays) P.; m. Helen Bachschmid, Mar. 9, 1948; children: Robert Alan, Mona Annette Price Agree. A.S., Tarleton State U., 1938; B.S., Tex. A&M U., 1940, D.V.M., 1949, M.S., 1956. Dairy prodn. mgr., Lockhart, Tex., 1940-42; instr. vet. medicine Tex. A&M U., College Station, 1949-52, asst. prof., 1952-57, dean vet. medicine, 1957-73, dir. biomed. sci., 1973—; Mem. Nat. Health Resources Adv. Com.; mem. adv. com. Selective Service System. Contbr. articles to profl. and popular mags. Served to lt. col. U.S. Army, 1942-46. Decorated Bronze Star; recipient Faculty Disting. Achievement award Tex. A&M U., 1956, Faculty Recognition award, 1970, Disting. Alumnus award, 1979. Mem. AVMA, Tex. Vet. Med. Assn. (Recognition award 1970), Brazos Valley Vet. Med. Assn. Democrat. Baptist. Lodge: Lions. Subspecialties: Physiology (medicine); Embryo transplants. Home: 1203 Walton Dr College Station TX 77840 Office: Biomedical Sci Tex A&M U College Station TX 77843

PRICE, BOBBY EARL, civil engr., educator, cons.; b. Henderson, Tex., Nov. 21, 1937; s. Earl and Mary Maurine (Grigsby) P.; m. Patsy Ruth Patrick, Mar. 8, 1958; children: Barry Earl, Kami Kay. B.S. in Civil Engring, Arlington State Coll., (now U. Tex.-Arlington), 1962, M.S., Okla. State U., 1963, Ph.D., U. Tex.-Austin, 1967. Registered profl. engr., La. Civil engr. City of Dallas, 1960-62; grad. asst. Okla. State U., 1962-64; civil engr. City of Austin, Tex., 1964; USPHS trainee U. Tex.-Austin, 1964-66, research engr., 1966-67; assoc. prof. civil engring., div. engring. research La. Tech. U., 1967-76, prof., 1976—, dir., 1967—, dir. engring. grad. studies, 1978—. Contbr. articles to profl. jours., procs.; editor, dir. confs., Sch. Engring., La. Poly. Inst., 1969, La. Tech. U., 1970. Pres. Cedar Creek Sch. Bd. Dirs., Ruston, La., 1980-83. Recipient Ednl. Achievement award La. Tech. Engring. Found., 1969, 74, 76, 77. Mem. ASCE (La. Tech. U. Student chpt. Outstanding Tchr. award 1970-71, 72-73, 76-77, 81-82, Faculty Adv. award 1977, 78), Am. Soc. Engring. Edn., Am. Water Works Assn., La. Engring. Soc., Tau Beta Pi (Excellence in Teaching award 1972-73, 74-75), Chi Epsilon, Omicron Delta Kappa, Epsilon Nu Gamma. Republican. Baptist. Subspecialties: Civil engineering. Current work: Channel and pipe hydraulics; computer applications to hydraulics. Home: 1610 Hodges Rd Ruston LA 71270 Office: PO Box 10348 T S Ruston LA 71272

PRICE, CLIFTON WILLIAM, physics educator; b. Burlington, Vt., Nov. 24, 1945; s. Clifton W. and Blanche Y. (Lanctot) P.; m. Judith K. Breidenback, June 5, 1977. B.S., U. Vt., 1967; Ph.D., Brown U., 1971. Asst. prof. physics Millersville (Pa.) State Coll., 1971-75, assoc. prof., 1975—. Bd. dirs. North Mus. Assocs., Lancaster, Pa., 1980—. Mem. Am. Assn. Physics Tchrs. (sec. central Pa. sect. 1978-79, v.p. 1979-80, pres. 1980-81), Phi Beta Kappa, Sigma Xi, Phi Kappa Phi. Subspecialty: Theoretical astrophysics. Current work: White dwarf stars - model atmospheres; photometry. Home: 305 E Charlotte St Millersville PA 17551 Office: Roddy Sci Ctr Millersville State Coll Millersville PA 17551

PRICE, DEREK DE SOLLA, educator, science historian; b. London, Eng., Jan. 22, 1922; s. Philip and Fanny Marie (de Solla) P.; m. Ellen Hjorth, Oct. 30, 1947; children—Linda Marie, Jeffrey Phillip, Mark de Solla. B.S., U. London, 1942, Ph.D. in Physics, 1946, Cambridge (Eng.) U., 1954; M.A., Yale, 1960. Lab. asst., research asst., lectr. war research S.W. Essex Tech. Coll. U. London, 1938-46; Commonwealth Fund fellow math. physics Princeton, 1946-47; lectr. applied math. U. Malaya, Singapore, 1947-50; cons. history physics, astronomy Smithsonian Inst., Washington, 1957; Donaldson fellow Inst. Advanced Study, Princeton, 1958-59; prof., Avalon prof. history sci. Yale U., 1960—, chmn. dept. history sci. and medicine, 1960-64, 74-77; cons. NSF.; Chmn. UNESCO Working Group Sci. Policy, 1967-70; pres. Internat. Commn. for Science Policy Studies, 1971-75. Author: The Equatorie of the Planetis, 1955, (with Needham and Wang) Heavenly Clockwork, 1959, Science since Babylon, 1961, enlarged edit., 1975, Little Science, Big Science, 1963, Gears From the Greeks, 1974; Editor: (with I. Spiegel-Rösing) Science, Technology and Society—A Cross-Disciplinary Perspective, 1977. Recipient Leonardo da Vinci medal Soc. History Tech., 1976, J.D. Bernal award Soc. Social Studies of Sci., 1983; hon. research assoc. Smithsonian Instn.; Guggenheim fellow, 1969. Mem. Internat. Acad. History Sci., Soc. History Tech., History Sci. Soc., Soc. Social Studies Sci. (founding council mem.), Royal Swedish Acad. Sci. (fgn.), Sigma Xi. Current work: History of science and technology science and technology policy studies; sciento–metrics. Home: 50 Trumbull St New Haven CT 06510 Office: 2036 Yale Station New Haven CT 06520

PRICE, GARY GLEN, childhood education educator; b. Tulsa, June 29, 1945; s. Glen Angelo Price and Elizabeth Sue (Bowles) Bratton; m. Lanette Yuksein Lee, July 12, 1970; children: Joel Frederic Wei-ming, Alana Yu-lan. B.A. in Philosophy, Northwestern U., 1968; M.Ed. in Urban Edn, Boston Coll., 1968-69, U. Lowell, 1972; Ph.D. in Edn, Stanford U., 1976. Elem. tchr. Lowell (Mass.) pub. schs., 1969-72; research asst. Stanford (Calif.) U., 1972-76; program evaluator State of Calif., San Francisco, 1974-75; asst. prof. U. Wis., Madison, 1976-82, assoc. prof. early childhood edn., 1982—; cons. Wis. Ednl. TV Network, 1981-82. Bd. dirs. Community Coordinator Child Care of Dane County, Madison, 1981-83. Spencer fellow Nat. Acad. Edn., 1981. Mem. Am. Ednl. Research Assn., Am. Psychol. Assn., Am. Statis. Assn., Merrill-Palmer Soc., Psychometric Soc., Soc. for Research in Child Devel., AAAS. Democrat. Subspecialties: Developmental psychology. Current work: Environmental influences during early childhood on the formation of intellectual abilities; cognitive and neuropsychological elucidation of psychometric abilities. Office: U Wisconsin 225 N Mills St Madison WI 53706 Home: 6409 Keelson Dr Madison WI 53705

PRICE, HENRY LOCHER, anesthesiologist, educator, researcher; b. Phila., Oct. 21, 1922; s. Henry Locher and Sara Millechamps (Anderson) P.; m. Mary Lowe Buckley, Dec. 12, 1953; children: Susan Garrison, Kathryn Anderson. A.B., Swarthmore Coll., 1945; M.D., U. Pa., 1946. Diplomate: Am. Bd. Anesthesiology. Intern Presbyn. Hosp., Phila., 1946-47, resident in anesthesia, 1947-48, Hosp. of U. Pa., Phila., 1950-51; NRC fellow in physiology Harvard Med. Sch., Boston, 1952-53; spl. NIH fellow in physiology Cardiovascular Research Inst., U. Calif., San Francisco; instr. anesthesiology Hosp. of U. Pa., 1950-51, research assoc., 1951-53, asst. prof. dept. anesthesia, 1953-56, Wellcome assoc. prof., 1956-60, prof., research dir., 1960-70; prof. anesthesia, research dir. Health Sci. Center, Temple U. Hosp., Phila., 1970-71; prof., chmn. dept. anesthesia Hahnemann Med. Coll. and Hosp., Phila., 1971-78, prof., 1978—; sr. examiner Am. Bd. Anesthesiology, 1963—; ad hoc cons. NIH, 1967—, mem. clin. research tng. com., 1963-65; mem. com. on anesthesia NRC, 1961-68; mem. anesthesiology tng. com. Nat. Inst. Gen. Med. Scis., NIH, 1965-67. Contbr. articles to profl. jours.; mem. editorial bd.: Circulation Research, 1964-68, Active Humane Soc, 1965—; chmn. admissions com., 1971-72. Served to capt. M.C. U.S. Army, 1948-50. USPHS grantee, 1954-81. Mem. Am. Soc. for Clin. Investigation, Am. Physiol. Soc., Soc. for Pharmacology and Exptl. Therapeutics, Soc. Anesthesiologists, Assn. Univ. Anesthetists, Acad. Anesthesia Chairmen, Humane Soc., Sigma Xi, Alpha Omega Alpha. Clubs: Martin's Dam Swimming and Tennis (Wayne, Pa.); Phila. Skating (Ardmore, Pa.). Subspecialties: Anesthesiology; Pharmacology. Current work: Effects of anesthetics on the circulation and its regulation. Home: 510 Lynmere Rd Bryn Mawr Pa 19010 Office: Hahnemann U Dept Anesthesia Mail Stop 310 230 N Broad St Philadelphia PA 19102

PRICE, JAMES GORDON, physician; b. Brush, Colo., June 20, 1926; s. John Hoover and Rachel Laurette (Dodds) P.; m. Janet Alice McSween, June 19, 1949; children: James Gordon II, Richard Christian, Mary Laurette, Janet Lynn. B.A., U. Colo., 1948, M.D., 1951. Diplomate: Charter diplomate Am. Bd. Family Practice (dir., pres. 1979). Intern Denver Gen. Hosp., 1951-52; practice medicine specializing in family medicine, Brush, 1952-78; prof. family practice U. Kans. Med. Center, 1978—, chmn. dept., 1982—; Mem. Inst. Medicine, Nat. Acad. Scis., 1973—. Editorial bd.: Med. World News, 1969-79; editor: Am. Acad. Family Practice Home Study Self Assessment Program, 1978-83; author: nationally syndicated column Your Family Physician, 1973—. Trustee Family Health Found. Am., 1970-82. Served with USNR, 1943-46. Charter fellow Am. Acad. Family Physicians (pres. 1973-74); mem. Phi Beta Kappa, Alpha Omega Alpha. Club: Mason. Subspecialty: Family practice. Home: 12736 St Andrew Dr Kansas City MO 64145 Office: U Kans Med Center Coll Health Scis and Hosp Rainbow Blvd at 39th St Kansas City KS 66103

PRICE, PAUL BUFORD, physicist, educator; b. Memphis, Nov. 8, 1932; s. Paul Buford and Eva (Dupuy) P.; m. JoAnn Margaret Baum, June 28, 1958; children—Paul Buford III, Heather Alynn, Pamela Margaret, Alison Gaynor. B.S. summa cum laude, Davidson Coll., 1954, D.Sc., 1973; M.S., U. Va., 1956, Ph.D., 1958. Fulbright scholar U. (Eng.) Bristol, 1958-59; NSF postdoctoral fellow Cambridge (Eng.) U., 1959-60; physicist Gen. Elec. Research & Devel. Center, Schenectady, 1960-69; vis. prof. Tata Inst. Fundamental Research, Bombay, India, 1965-66; adj. prof. physics Rensselaer Poly. Inst., 1967-68; prof. physics U. Calif. at Berkeley, 1969—, dir., 1979—; dir. Terradex Corp., Walnut Creek, Calif.; cons. for NASA (on Lunar Sample Analysis Planning Team); mem. space sci. bd. Nat. Acad. Scis. Author: (with others) Nuclear Tracks in Solids; Contbr. articles to profl. jours. Recipient Distinguished Service award Am. Nuclear Soc., 1964, Indsl. Research awards, 1964, 65, E.O. Lawrence Meml. award AEC, 1971, medal exceptional sci. achievement NASA, 1973; John Simon Guggenheim fellow, 1976-77. Fellow Am. Phys. Soc., Am. Geophys. Union; mem. Nat. Acad. Scis. (chmn. geophysics sect.), Am. Astron. Soc. Subspecialties: Cosmic ray high energy astrophysics; Nuclear physics. Current work: Cosmic ray origin, relativistic nucleus-nucleus collisions, particle physics. Research on space and astrophysics, nuclear physics, particularly devel. solid state track detectors and their applications to geophysics, fossil and nuclear physics problems. Patentee in field; discovery of fossil particle tracks and fission track method of dating; discovery of ultra-heavy cosmic rays.

PRICE, R(ICHARD) MARCUS, physics and astronomy educator, researcher; b. Colorado Springs, Colo., Jan. 18, 1940; s. George Marcus and Letha Belle Gladys (Smith) P.; m. Elaine Haley, Sept. 13, 1968; 2 children. B.S., Colo. State U., 1961; Ph.D., Australian Nat. U., 1966. Cert. sr. exec. service Office Personnel Mgmt., U.S. Govt, 1979. Asst. prof. physics M.I.T., 1968-74, assoc. prof., 1974-75; spectrum mgr., head astronomy research sect. NSF, Washington, 1975-79; prof., chmn. dept. physics and astronomy U. N.Mex., Albuquerque, 1979—; cons. in field. Contbr. articles to profl. jours. Committeeman Monmouth council Boy Scouts Am. Served to capt., Signal Corps U.S. Army, 1965-67; mem. USNR, 1961-71. Fulbright fellow, Australia, 1961-62; M.I.T. Research Lab. Electronics fellow, 1971; NSF grantee. Mem. Internat. Astron. Union, Am. Astron. Soc., Optical Soc. Am., Am. Assn. Phys. Tchrs. Subspecialties: Radio and microwave astronomy; Infrared optical astronomy. Current work: Studies of nuclear regions of spiral galaxies, methods of science education, use of the radio spectrum for scientific research. Office: Department of Physics and Astronomy University of New Mexico Albuquerque NM 87131

PRICE, STEVEN, physiology researcher, educator; b. N.Y.C., May 13, 1937; s. Robert Julius and Mary (Kern) P.; m. Bexcetta Jo-Marie Arnett, May 31, 1958; children: David Andrew, Lisa Michelle. A.B., Adelphi Coll., 1958; M.A., Princeton U., 1960, Ph.D., 1961. NIH postdoctoral fellow Fla. State U., 1961-63; research group leader Monsanto Research Corp., Everett, Mass., 1963-66; asst. prof. physiology and biophysics Va. Commonwealth U., 1966-68, assoc. prof., 1968-75, prof., 1975—; cons. in field., 1966-76. Contbr. numerous articles to profl. jours. Pres. Old Gun-Robious Civic Assn., Midlothian, Va., 1981-83. Recipient Career Devel. award NIH, 1972-76; NSF fellow, 1958; NIH fellow, 1960. Mem. European Chemoreception Research Orgn., Assn. Chemoreception Scis., Am. Physiol. Soc., Soc. Gen. Physiologists, AAAS, Sigma Xi. Subspecialties: Physiology (biology); Neurobiology. Current work: Research on receptor mechanisms in taste and smell; teaching in all areas of physiology. Home: 3901 Victoria Ln Midlothian VA 23113 Office: Dept Physiology and Biophysics Med Coll Va Richmond Va 23298

PRICHARD, JOHN B., research scientist, educator; b. Buffalo, July 31, 1943; s. Willis Brown and Betty Eugenia (Richardson) P.; m. Vivian Leigh Dees, June 18, 1967; 1 dau.; Meredyth Leigh. B.A., Oberlin Coll., 1965; Ph.D., Harvard U., 1970. NSF postdoctoral fellow Upstate Med. Center, Syracuse, N.Y., 1970-71; NIH postdoctoral fellow Mt. Desert Island Biol. Lab., Salsbury Cove, Maine, 1971-72; asst. prof. physiology Med. U. S.C., Charleston, 1972-76, assoc. prof., 1976; research physiologist Nat. Inst. Environ. Health Scis., Research Triangle Park, N.C., 1976-79; head Marine Lab., St. Augustine, Fla., 1979—; adj. assoc. prof. pharmacology U. Fla. Sch. Medicine, Gainesville, 1979. Contbr. articles on physiology and pharmacology to profl. jours. Recipient Sonoco Corp. award, 1976. Mem. Am. Physiol. Soc., Am. Soc. Pharmacology and Exptl. Therapeutics, Am. Zool. Soc., Mt. Desert Island Biol. Lab. Methodist. Subspecialties: Physiology (medicine); Pharmacology. Current work: Role of the kidney in regulation of the excretion of foreign chemicals, mechanisms of anion transport by the kidney, membrane effects of xenobiotics. Office: C V Whitney Lab FRD 1 PO Box 121 St Augustine FL 32084

PRIEDHORSKY, WILLIAM CHARLES, physicist; b. Tacoma, Wash., Oct. 17, 1952; s. Frantisek and Mary Lucille (Ponton) P.; m. Barbara Royer, Dec. 21, 1975; children: Anna Beth, Reid Royer. B.A. in Physics, Whitman Coll., 1973; Ph.D., Calif. Inst. Tech., 1977. Staff mem. fusion expts. and diagnostics group Los Alamos Nat. Lab., 1978-80, mem. fast transient plasma group, 1980-81, mem. space astronomy and astrophysics group, 1981—. Contbr. articles to sci. jours. Mem. Am. Phys. Soc., Am. Astron. Soc. Subspecialties: 1-ray high energy astrophysics; Fusion. Current work: High energy astrophysics instrumentation, analysis and modeling, pulsed x-ray diagnostics. Home: 2887 Woodland Rd Los Alamos NM 87544 Office: Los Alamos National Laboratory MS D436 Los Alamos NM 87545

PRIEN, SAMUEL DAVID, electron microscopist, plant physiologist; b. Amarillo, Tex., May 30, 1956; s. Lester Joseph and Joyce Ann (Chesney) P.; m. Cynthia Kay, Dec. 28, 1982. B.S. in Botany, Tex. Tech. U., 1978, M.S., 1980. Student asst. Tex. Tech. U., 1977-79, research asst., 1979, teaching asst., 1979-80, electron microscope technician II, 1980-82, coordinator, 1982—. Contbr. papers to profl. meetings. Mem. AAAS, Sigma Xi (assoc.). Subspecialties: Microscopy; Plant physiology (biology). Current work: Structure-function relationships in biol. organisms; biol. ultra-structural exam, using the techniques of electron microscopy, freeze fracture, and micro-elemental analysis. Home: 2210 20th Lubbock TX 79411 Office: Dept Anatomy Tex Tech U Health Scis Center Lubbock TX 79430

PRIETO, ROBERT, planning and development engineer; b. N.Y.C, Apr. 22, 1954; s. Marcelino and Betty (Perez) P. B.S. in Nuclear Engring, NYU, 1976, M.S., Poly. Inst. N.Y., 1977. Nuclear engr. Gibbs & Hill, Inc., N.Y.C., 1976-78, project engr., 1978-82, planning and devel. engr., 1982-83, dir. sales, 1983—; mem. assoc. staff Poly. Inst. N.Y., 1976; mem. Atomic Indsl. Forum, Washington, 1981—. Recipient Founders Day award NYU, 1976. Mem. Am. Nuclear Soc., AAAS, N.Y. Acad. Scis., Smithsonian Instn. Republican. Roman Catholic. Subspecialties: Nuclear engineering; Solar energy. Current work: Advanced energy technologies; nuclear waste management; cogeneration. Office: Gibbs & Hill Inc 11 Penn Plaza New York NY 10001

PRIGGE, EDWARD CHRISTIAN, JR., reuminant nutrition researcher, animal sci. educator; b. Bklyn.; s. Edward Christian and Annette (Melamendorf) P.; m. Domenica Prigge, Aug. 6, 1966; children: Christopher, Jonathan, Michelle. B.S., Delaware Valley Coll., 1964; M.S., U. Tenn., 1966; Ph.D., U. Maine, 1973. Fellow Okla. State U., 1973-75; farm advisor Coop. Extension U. Calif., 1975-77; asst. prof. dept. animal sci. W.Va. U., 1977—. Editorial bd.: Jour. Animal Sci, 1982—; contbr. articles to profl. jours. Mem. Am. Soc. Animal Sci. Lutheran. Subspecialties: Animal nutrition; Biochemistry (biology). Current work: Carbohydrate and protein and fatty acid metabolism in ruminants; examining methods to increase production for ruminants by altering intermediate metabolism of carbohydrate fatty acids and protein. Home: Route 7 Cedarhurst Morgantown WV 26505 Office: Division of Animal and Veterinary Science W Va U Morgantown WV 26506

PRIGOGINE, ILYA, physics educator; b. Moscow, Jan. 25, 1917; s. Roman and Julie (Wichmann) P.; m. Marina Prokopowicz, Feb. 25, 1961; children: Yves, Pascal. Ph.D., Free U. Brussels, 1942; hon. degrees, U. Newcastle (Eng.), U. Poitiers (France), U. Chgo., U. Bordeaux (France), numerous others. Prof. U. Brussels, 1947—; dir. Internat. Insts. Physics and Chemistry, Solvay, Belgium, 1962—; dir. Ilya Prigogine center statis. mechanics and thermodynamics U. Tex., Austin, 1967—. Author: (with R. Defay) Traite de Thermodynamique, conformement aux methodes de, Gibbs et de De Donder, 1944, 50, Etude Thermodynamique des Phenomenes Irreversibles, 1947, Introduction to Thermodynamics of Irreversible Processes, 1962, (with A. Bellemans, V. Mathot) The Molecular Theory of Solutions, 1957, Statistical Mechanics of Irreversible Processes, 1962, (with others) Non Equilibrium Thermodynamics, Variational Techniques and Stability, 1966, (with R. Herman) Kinetic Theory of Vehicular Traffic, 1971, (with R. Glansdorff) Thermodynamic Theory of Structure, Stability and Fluctuations, 1971, (with G. Nicolis) Self-Organization in Nonequilibrium Systems, 1977, From Being to Becoming-Time and Complexity in Physical Sciences, 1979, Order Out of Chaos, 1979, La Nouvelle Alliance, Les Métamorphoses de la Science, 1979. Recipient Prix Francqui, 1955; Prix Solvay, 1965; Nobel prize in chemistry, 1977; Honda prize, 1983; medal Assn. Advancement of Sci., France, 1975; Rumford gold medal Royal Soc. London, 1976; Descartes medal U. Paris, 1979. Mem. Royal Acad. Belgium, Am. Acad. Sci., Royal Soc. Scis. Uppsala (Sweden), Fgn. Assn. Nat. Acad. Scis. U.S.A., Soc. Royale des Scis. Liege Belgium (corr.), Acad. Gottingen Ger., Deutscher Akademie der Naturforscher Leopoldine (medaille Cothenius 1975), Osterreichische Akademie der Wissenschaften (corr.), Chem. Soc. Poland (hon.), others. Subspecialties: Statistical mechanics; Thermodynamics. Address: 67 ave Fond'Roy 1180 Brussels Belgium also Ilya Prigogine Ctr Statis Mechanics U Tex Austin TX 78712

PRINCE, ALFRED MAYER, research virologist; b. Berlin, Germany, Dec. 16, 1928; s. Hugo M. and Madeleine (Salm) P.; m. Noriko, Apr. 25, 1962; children: Lisa, Boku. A.B., Yale U., 1949; M.A., Columbia U., 1951; M.D., Western Res. U., 1955. Diplomate: Am. Bd. Pathology. Intern Grace-New Haven Hosp., 1955-56, resident in pathology, 1956-59; assoc. mem. Wistar Inst., Phila., 1962-63; asst. prof. Med. Sch., Yale U., New Haven, 1963-64; with N.Y. Blood Ctr., N.Y.C., 1965—, sr. investigator, 1965—. Contbr. over 200 articles to profl. jours. Served with AUS, 1959-62. Recipient Landsteiner award Am. Assn. Blood Banks, 1979. Mem. Am. Soc. Immunology, Sigma Xi. Subspecialties: Virology (medicine); Infectious diseases. Current work: Hepatitis viruses, chronic liver diseases, vaccines. Office: 310 E 67th St New York NY 10021

PRINCE, HELEN DODSON, astronomy educator, solar cons.; b. Balt., Dec. 31, 1905; d. Henry Clay and Helen Falls (Walter) Dodson; m. Edmond Lafayette Prince (dec.), ; Oct. 24, 1956. A.B., Goucher Coll., 1927, Sc.D. (hon.), 1952; M.A., U. Mich., 1932, Ph.D., 1933. Asst. prof. Wellesley Coll., 1933-45; mem. staff radiation lab. M.I.T., 1943-45; prof. astronomy Goucher Coll., 1945-50; mem. faculty U. Mich., 1947-76, prof. astronomy, 1957-76, prof. emeritus, 1976—; assoc. dir. McMath-Hulbert Obs., 1962-76; solar cons. Applied Physics Lab. Johns Hopkins U., 1979—. Contbr. articles to profl. jours. Recipient Disting. Faculty Achievement award U. Mich., 1975. Mem. Am. Astron. Soc. (Annie Jump Cannon prize 1954), Internat. Astron. Union, Am. Geophys. Union, Phi Beta Kappa. Episcopalian. Subspecialties: Optical astronomy; Solar physics. Current work: Study of solar activity, especially solar flares with consideration of geophysical effects. Home: 4800 Fillmore Ave Apt 820 Alexandria VA 22311

PRIOR, JOHN THOMPSON, pathologist, educator; b. Syracuse, N.Y., Oct. 8, 1917; s. Thomas William and Pauline (Thompson) P.; m. Elizabeth Titus Troy, July 24, 1948; children: Anne, Polly, John Thompson, Thomas, Jeffrey, Timothy. B.S., U. Vt., 1939, M.D., 1943. Diplomate: Am. Bd. Pathology. Intern Morrisania City Hosp., N.Y.C.; resident Syracuse U. Med. Coll.; lab. dir. Crouse Irving Hosp., Syracuse, 1950-71, Community Gen. Hosp., 1971—; prof. pathology Syracuse Med. Ctr., 1972—; mem. exec com. bd. dirs. ARC, Syracuse, 1966—; mem. N.Y. State Hosp. Rev. and Planning Com., 1980-82. Served to maj., M.C. U.S. Army, 1944-46; Served to maj., M.C. Res., 1962-77. Decorated Legion of Merit, Bronze Star medal, Silver Star. Mem. AMA, Am. Assn. Clin. Pathologists, N.Y. Med. Soc. (del. 1976—), Am. Assn. Pathology and Bacteriology, N.Y. Assn. Pub. Health Labs. (pres. 1959-60), Onondaga County Med. Soc. (pres. 1973-74). Republican. Roman Catholic. Current work: Neoplasms and arteriosclerotic research. Address: 100 Lansdowne Rd Dewitt NY 13214

PRISCO, JOHN JOSEPH, research scientist; b. Jersey City, Jan. 6, 1956; s. Salvatore and Lillian Maria (Deputato) P.; m. Jeanne Marie Harkins, July 19, 1980. B.S., Columbia U., 1978; M.S., M.I.T., 1982. Mem. tech. staff GTE Labs., Waltham, Mass., 1980—. Contbr. articles to profl. jours. Mem. IEEE, Optical Soc. Am., Sigma Xi, Tau Beta Pi, Eta Kappa Nu. Subspecialties: Nonlinear optics; Fiber optics. Current work: Adaptive optical communications by nonlinear optical techniques; fiber optics research; integrated optical switching research. Office: 40 Sylvan Rd Waltham MA 02254

PRITCHARD, DONALD WILLIAM, oceanographer; b. Santa Ana, Calif., Oct. 20, 1922; s. Charles Lorenzo and Madeleine (Sievers) P.; m. Thelma Lydia Amling, Apr. 25, 1943; children—Marian Lydia, Jo Anne, Suzanne Louise, Donald William, Albert Charles. B.A., UCLA, 1942; M.S. Scripps Instn. Oceanography, La Jolla, 1948, Ph.D., 1951. Research asst. Scripps Instn. Oceanography, 1946-47; oceanographer USN Electronics Lab., 1947-49; asoc. dir. Chesapeake Bay Inst., Johns Hopkins, Balt., 1949-51, dir., 1951-74, chief scientist, 1974-79, prof., 1958-79, chmn. dept. oceanography, 1951-68; asso. dir. for research, prof. Marine Scis. Research Center, SUNY at Stony Brook, 1979—; cons. C.E., U.S. Army, USPHS, AEC, Internat. AEC, NSF, Adv. Panel Earth Scis.; mem. adv. bd. to Sec. Natural Resources, State of Md. Bd. editors: Jour. Marine Research, 1953-70, Bull. Bingham oceanographic Collection, 1960-70, Geophys. Monograph Bd., Am. Geophys. Union, 1959-70, Johns Hopkins Oceanographic Studies, 1962—; Author articles in field. Served from 2d lt. to capt. USAAF, 1943-46. Fellow Am. Geophys. Union (past pres. oceanography sect.); mem. Am. Soc. Limnology and Oceanography (past v.p.), Am. Meteorol. Soc., AAAS, Nat. Acad. Scis. (past mem. com. oceanography, past chmn. panel radioactivity in marine environment), Sigma Xi (past chpt. pres.). Subspecialty: Oceanography. Office: MSRC SUNY Stony Brook NY 11794

PRITCHARD, HAYDEN NELSON, biologist; b. Bangor, Pa., Feb. 13, 1933; s. Hayden Benjamin and Marion (Flory) P.; m. Mary Carolyn Sollie, Aug. 12, 1962; children: Paul Hayden, David Nelson. A.B., Princeton U., 1955; M.S., Lehigh U., 1960, Ph.D., 1963. Research asst. Walter Reed Army Inst. Research, Washington, 1957-58; predoctoral fellow cytology U. Fla. Med. Sch., Gainesville, 1962-63; asst. prof. biology, 1963-64, Lehigh U., Bethlehem, Pa., 1964-70, assoc. prof., 1970—. Author: Biology of Nonvascular Plants, 1983; contbr. articles in field to profl. jours. Mem. Bangor (Pa.) Area Bd. Edn., 1975—, pres., 1978-82; trustee Easton (Pa.) Hosp., 1977-78. Served with U.S. Army, 1956-58. Recipient Bernard A. Briody, Jr. award Lehigh U., 1979. Mem. Bot. Soc. Am., Audubon Soc., Pa. Acad. Sci., Sigma Xi. Democrat. Club: Exchange (Bangor, Pa.). Subspecialties: Cell biology; Plant growth. Current work: Taxonomy, cell biology, cytochemistry of plant cells and plant growth and development. Home: South High St Bangor PA 18013 Office: Lehigh U Bethlehem PA 18015

PRITCHARD, WILBUR LOUIS, engineering executive; b. N.Y.C., May 31, 1923; s. Harmon and Jessie H. (Roth) P.; m. Kathleen Hunton Moss, Apr. 24, 1949; children: Hugh Arthur, Sarah Margaret, Ruth Wells. B.E.E., CCNY, 1943; postgrad., MIT, 1948-52. Registered profl. engr., Mass., Md. Antenna engr. Raytheon Co., Waltham, Mass., 1946-49, dept. mgr., 1949-59, dir. engring., surface radar and nav. ops., 1959-60, dir. engring.-Europe, 1960-62, dir. range and space support systems, 1962; group dir. communications satellite systems Aerospace Corp., El Segundo, Calif., 1962-67; dir. COMSAT Labs., Washington, 1967-73; v.p. COMSAT Corp., Washington, 1967-73; pres. Fairchild Space & Electronics Co., Germantown, Md., 1973-74, Satellite Systems Engring., Inc., Bethesda, Md., 1974—, Direct Broadcast Satellite Corp., 1981—; dir. Montgomery County Corp. Technol. Tng., Rockville, Md. Author: (with James Harford) China Space Report, 1980; contbr. articles to profl. jours., chpts. to books. Recipient Air Force Systems Command award, 1967; medal CCNY, 1969. Fellow IEEE, AIAA (Aerospace Communications award 1972); mem. Eta Kappa Nu. Club: Cosmos (Washington). Subspecialty: Satellite studies. Current work: Satellite systems engineering; direct-to-home TV satellite systems. Patentee microwave applications and radar (12). Home: 9201 Laurel Oak Dr Bethesda MD 20817 Office: 7315 Wisconsin Ave Bethesda MD 20814

PRIVETTE, CHARLES VICTOR, JR., extension agricultural engineer; b. Hartsville, S.C., Oct. 29, 1941; s. Charles Victor and Ila Ruth (Kelly) P.; m. Esther Arlene Bauknight, June 1, 1968; children: Margaret Ruth, Charles Victor. B.S., Clemson Coll., 1963, M.S., 1968. Registered profl. engr., S.C. Constrn. engr. C.E. U.S. Army, Anchorage, Alaska, 1965-66; extension engr. Clemson U., 1966-67, engr. research asst., 1967-68, ext. engr., 1968—; ednl. adviser S.C. Irrigation Soc., Clemson, S.C., 1978—; water adviser S.C. Farm Bur., Columbia, 1982—. Served to lt. AUS, 1963-65. Mem. Am. Soc. Agrl. Engrs. (treas. S.C. chpt. 1970-75), S.C. Irrigation Soc., N.C. Irrigation Soc. Methodist (chmn. fin. com. 1978—, chmn. adminstrv. bd. 1983—). Subspecialty: Agricultural engineering. Current work: Educational work irrigation-drainage design; water supply development and treatment; chemical application through irrigation. Home: Route 1 Box 574 Central SC 29630 Office: Agrl Engring Dept Clemson U McAdams Hall Room 216 Clemson SC 29631

PRIVITERA, PHILIP JOSEPH, pharmacologist; b. Bklyn., Aug. 29, 1938; s. Joseph P. and Catherine Elvira (Palma) P.; m. Ann Theresa Carraturo, July 22, 1961; children: Laura, Susan, Thomas. B.S. in Pharmacy, St. John's U., 1959; Ph.D., Albany Med. Coll. of Union U., 1 966. Postdoctoral fellow U. Cin., 1966-78, instr., 1968-69, asst. prof., 1969-72, asst. prof. exptl. medicine, 1972; asso. prof. Med. U. S.C., Charleston, 1972-79, prof., 1979—, dir. course med. pharmacology for sophomore med. students, 1978—. Mem. Am. Soc. Pharmacology and Exptl. Therapeutics, N.Y. Acad. Scis., Sigma Xi, Rho Chi. Roman Catholic. Subspecialty: Pharmacology. Current work: Central nervous system control of blood pressure and renin secretion; pharmacology renin-angiotensin system. Office: 171 Ashley Ave Charleston SC 29425

PROAKIS, ANTHONY GEORGE, pharmacologist; b. Chios, Greece, May 26, 1940; came to U.S., 1946, naturalized, 1953; s. George John and Fotine (Parikakis) P.; m. Nancy Lee Pietranton, Feb. 23, 1963; children: Lisa, Andrea, Steven. B.S. in Pharmacy, W.Va. U., 1962; Ph.D. in Pharmacology (NIH fellow), Purdue U., 1972. Registered pharmacist, W.Va., Va., Calif. Pharmacist Medco Pharmacies Inc., Los Angeles, 1962-63, mgr., 1963-66, v.p., 1966-67; sr. research biologist A.H. Robins Research Labs., Richmond, Va., 1972-77, research asso., 1977-80, mgr. cardiovascular pharmacology, 1980—. Contbr. articles to profl. jours. Mem. Soc. Exptl. Biology and Medicine, Am. Soc. Pharmacology and Exptl. Therapeutics, Sigma Xi, Phi Kappa Phi, Rho Chi. Club: AHEPA. Subspecialties: Pharmacology; Cardiology. Current work: Mechanism of action of antihypertensive, antiarrhythmic and antianginal agts.; direct research in discovery and devel. of drugs for therapeutic utility in cardiovascular diseases. Patentee method of increasing coronary blood flow in mammals. Home: 1811 Carbon Hill Dr Midlothian VA 23113 Office: 1211 Sherwood Ave Richmond VA 23220

PROBSTEIN, RONALD FILMORE, engineering educator; b. N.Y.C., Mar. 11, 1928; s. Sidney and Sally (Rosenstein) P.; m. Irene Weindling, July 30, 1950; 1 son, Sidney. B.M.E., N.Y. U., 1948; M.S.E., Princeton U., 1950, A.M., 1951, Ph.D., 1952; A.M. (hon.), Brown U., 1957. Research asst. physics N.Y. U., 1946-48, instr. engring. mechanics, 1947-48; research asst. dept. aero. engring. Princeton U., 1948-52, research assoc., 1952-53, asst. prof., 1953-54; asst. prof. divs. engring., applied math. Brown U., 1954-55, assoc. prof., 1955-59, prof., 1959-62; prof. mech. engring. M.I.T., 1962—; disting. prof. engring. U. Utah, 1973; sr. partner Water Purification Assos., Cambridge, 1974-82; chmn. bd. Water Gen. Corp., Cambridge, 1982-83. Author: Hypersonic Flow Theory, 1959, Hypersonic Flow, Inviscid Flows, 1966, Water in Synthetic Fuel Production, 1978, Synthetic Fuels, 1982; editor: Introduction to Hypersonic Flow, 1961, Physics of Shock Waves, 1966; contbr. articles to profl. jours. Guggenheim fellow, 1960-61. Fellow Am. Acad. Arts and Scis. (councilor 1975-79), Am. Phys. Soc., AIAA, AAAS; mem. Internat. Acad. Astronautics, Nat. Acad. Engring., ASME (Freeman award 1971), Am. Inst. Chem. Engrs., Water Pollution Control Fedn. Democrat. Jewish. Subspecialties: Fluid mechanics; Water supply and wastewater treatment. Current work: Fluid mechanics and its application to water treatment technologies. Patentee in field. Home: 5 Seaver St Brookline MA 02146 Office: 77 Massachusetts Ave Cambridge MA 02139

PROBSTFIELD, JEFFREY LYNN, medical educator; b. Fargo, N.D., June 27, 1941; s. George Berg and Alda Gail (Abbott) P.; m. Margaret Helen Belgum, Dec. 28, 1965; children: Erik, Kathryn, Cindy, Dawn, Sahnnon, Laura. B.A. in Chemistry, Pacific Luth. U., 1963; M.D., U. Wash., 1967; postgrad., U. Minn., 1972-76. Med. fellow U. Minn., Mpls., 1968-71, instr. internal medicine, 1972-75, instr. pharmacology, 1975-76, asst. prof. medicine, 1976-78, Baylor Coll. Medicine, Houston, 1978—; trial dir. Baylor-Meth. Lipid Research Clinic, Houston, 1978—. Contbr. articles to profl. jours. Regent Pacific Luth. U., Tacoma, 1981—. Served to lt. comdr. USN, 1967 . Fellow ACP; Mem. Am. Heart Assn., Soc. Clin. Trials, Am. Fedn. Clin. Research, Tex. State Med. Soc., Harris County Med. Soc.; mem. Alumni Assn. Pacific Luth. U. (pres. 1983-84). Lutheran. Subspecialties: Internal medicine; Pharmacology. Current work: Clinical trials. Home: 5206 Loch Lomond Houston TN 77096 Office: Baylor Coll Medicine 6535 Fannin St MS-A601 Houston TX 88030

PROCTOR, CHARLES LAFAYETTE, II, mechanical engineering educator; b. Crawfordsville, Ind., Nov. 21, 1954; s. Charles Lafayette and Marjorie E. (Purdue) P.; m. Dixie Lee Huffer, May 22, 1976. B.S.M.E., Purdue U., 1976, M.S.M.E., 1979, Ph.D., 1981. Asst. prof. and dir. Combustion Lab., U. Fla., Gainesville, 1981—; cons. Gainesville Regional Utilities, 1982—, MUVA, Gainesville, 1982—. Mem. Combustion Inst., Am. Soc. Engring. Edn., ASME, AAAS. Subspecialties: Combustion processes; Fluid mechanics. Current work: Combustion research investigating the interaction of fluid mechanics and chemistry using advanced optical diagnostic techniques; emphasis on soot formation and turbulence. Home: 6051 NW 19th Pl Gainesville FL 32605 Office: Univ Fla Dept Mech Engring Gainesville FL 32611

PROCTOR, KENNETH GORDON, physiologist, microcirculationist; b. Los Angeles, Aug. 6, 1952; s. Peter Edward and Helen Phyllis (Mularski) P.; m. Marcia Culbertson, Dec. 26, 1976; children: Andrea Michelle, Matthew Scott. B.A., Calif. State U.-Fullerton, 1974; Ph.D., Ind. Sch. Medidine, 1979. Research asst. dept. pediatrics U. Calif.-Irvine Sch. Medicine, 1974-76; grad. asst. dept. physiology Ind. U. Sch. Medicine, Indpls., 1976-79; postdoctoral fellow dept. physiology U. Va. Sch. Medicine, Charlottesville, 1980-82; asst. prof. dept. physiology U. Tenn. Coll. Medicine, Memphis, 1982—. Contbr. articles to profl. publs., papers to profl. confs. NIH Research Service awardee, 1980-82. Mem. Am. Physiol. Soc., Microcirculatory Soc. Episcopalian. Subspecialties: Physiology (medicine); Pharmacology. Current work: Regulation of blood flow by tissues. Home: 4923 Greenway St Memphis TX 38117 Office: Dept Physiology U Tenn Coll Medicine 894 Union Ave Memphis TN 38163

PROCUNIER, JAMES DOUGLAS, biology educator; b. Demerrara River, Guyana, Dec. 7, 1944; came to U.S., 1979; s. James Albert and Kay Rose (Brody) P.; m.; children: Mark, Michael. M.Sc., U. B.C., 1968; Ph.D., U. Calgary, 1973. Fellow Inst. for Cancer Research, Phila., 1973-76, U. B.C., Vancouver, Can., 1976-79; asst. prof. dept. biology Rice U., Houston, 1979—. NRC scholar, 1971-73; fellow, 1973-75; NIH research grantee, 1980-83. Mem. Genetics Soc. Am. Subspecialties: Gene actions; Molecular biology. Current work: Genetics, cloning, Drosophila melanogaster. Office: Dept Biology Rice U Houston TX 77001

PROEBSTING, EDWARD LOUIS, agrl. researcher; b. Woodland, Calif., Mar. 2, 1926; s. Edward Louis and Dorothy (Critzer) P.; m. Patricia Connolly, Aug. 3, 1925; children: William Martin, Patricia Louise, Thomas Alan. B.S. in Pomology, U. Calif., Davis, 1948; Ph.D. in Horticulture, Mich. State U., 1951. Horticulturist Irrigated Agr. Research and Extension Ctr., Wash. State U., Prosser, 1951—. Contbr. numerous articles to profl. jours. Served to lt. USNR, 1943-46, 52-54. Fellow Am. Soc. Hort. Sci. (Gourley award 1955, Woodbury award 1958, Nat. Food Processors award 1979), Japan Soc. Promotion of Sci. Methodist. Subspecialties: Pomology; Plant physiology (agriculture). Current work: Supercooling and ice nucleation in low temperature resistance of Prunus flower buds and fruit. Fruit tree response to controlled plant water deficits. Home: 1929 Miller Ave Prosser WA 99350 Office: Irrigated Agr Research and Extension Center PO Box 30 Prosser WA 99350

PROENZA, LUIS MARIANO, neurophysiologist, educator; b. Mexico City, Dec. 22, 1944; U.S., 1956, naturalized, 1977; s. Luis Proenza Abreu and Sara Maria Gonzalez de P. B.A., Emory U., 1965; M.A. (USPHS fellow), Ohio State U., 1966; Ph.D., U. Minn., 1971. Research asst. dept. psychology U. Conn., 1966-67; asst. prof. psychology U. of Ams., Mexico City, 1967-68; rsearch asst. Neuropsychology Lab., U. Minn. Med. Sch., 1968-69, research fellow, 1969-71; asst. prof. dept. psychology U. Ga., Athens, 1971-78, asso. prof. dept. zoology, 1978—, coordinator faculty neurobiology div. biol. scis., 1975—; study dir. com. on vision Nat. Acad. Scis.-NRC, 1977-79; mem., 1979-82; participant 13th Latin Am. Congress Physiol. Scis., Mexico City, 1977. Editor: (with Enoch and Jampolski) Clinical Applications of Visual Psychophysics, 1981; contbr. articles and abstracts to sci. jours. Recipient Richard M. Griffith Meml. award Soc. for Philosophy and Psychology, 1976; NIH research grantee, 1972—; research career devel. award, 1976—. Mem. AAAS, Assn. for Research in Vision and Ophthalmology, Soc. for Neurosci., Signa Xi. Subspecialties: Neurophysiology; Neurobiology. Current work: Electrophysiology of retinal cells in vision; with emphasis on potassium flux and ionic dependence of glial (Muller) cells; relationship of ionic changes and glial responses to components of the electroretinogram. Office: Dept Zoology U Ga Athens GA 30602 Home: 355 Rambling Rd Athens GA 30606

PROFANT, RICHARD THOMAS, JR., physicist; b. Jefferson, Ohio, Apr. 23, 1954; s. Richard Thomas and Frances Marie (Martin) P.; m. Linda Sue Stuart, Oct. 30, 1982; children: Stuart Alan Turner, Shona Ann Turner. B.S. in Physics, Kent State U., 1976; M.S. in Nuclear Engring, Ohio State U., Columbus, 1978. Grad. research assoc. Ohio State U., Columbus, 1976-80, sr. research assoc., 1980-81; mgr. data systems group Transp. Research Ctr., East Liberty, Ohio, 1981-82; project engr. U.S. Dept. Energy, Aiken, S.C., 1982—. Mem. Instrument Soc. Am. (sec. standards com. 1980—), Am. Nuclear Soc., IEEE. Subspecialties: Nuclear physics; Nuclear engineering. Current work: Working with contractors to develop special isotope separation techniques for various isotopes of plutonium. Special isotope separation (SIS) technology is basic research using laser isotope separation techniques. Home: 114 Heather Way Aiken SC 29801 Office: Dept Energy Savannah River PO Box A Aiken SC 29801

PROFIO, A(MEDEUS) EDWARD, nuclear engineer; b. New Castle, Pa., Apr. 18, 1931; m., 1954; 3 children. S.B., MIT, 1953; Ph.D. in Nuclear Engring, 1963. Scientist Westinghouse Bettis Atomic Power Lab., 1953-55; research assoc. nuclear engring. MIT, 1957-63, asst. prof., 1963-64; mem. research and development staff Gen. Atomic Div., Gen. Dynamics Corp., 1964-69; assoc. prof. nuclear engring. U. Calif., Santa Barbara, 1969-74, prof., 1974—. Mem. Am. Nuclear Soc. Subspecialties: Nuclear engineering; Bioinstrumentation. Office: Dept Chem and Nuclear Engring Univ Calif-Santa Barbara Santa Barbara CA 93106

PROMISEL, NATHAN E., materials scientist, metallurgical engineer; b. Malden, Mass., June 20, 1908; s. Solomon and Lyna (Samwick) P.; m. Evelyn Sarah Davidoff, May, 17, 1931; children: David Mark, Larry Jay. B.S., M.I.T., 1929, M.S., 1930; postgrad., Yale U., 1932-33; D.Engring. (hon.), Mich. Tech. U., 1978. Asst. dir. lab. Internat. Silver Co. Meriden, Conn., 1930-40; chief materials scientist and engr. Navy Dept., Washington, 1940-66; exec. dir. nat. materials adv. bd. Nat. Acad. Scis., Washington, 1966-74; cons. on materials and policy, internationally, Washington, 1974—; mem., chmn. NATO Aerospace Panel, 1959-71; U.S. rep. (materials) OECD, 1967-70; U.S. chmn. U.S./USSR Sci. Exchange Program (materials), 1973-77; hon. guest USSR Acad. Scis.; permanent hon. pres. Internat. Conf. on Materials Behavior; dir. Value Engring. Co.; mem. Nat. Materials Adv. Bd.; adv. com. Oak Ridge Nat. Lab., Lehigh U., U. Pa., U.S. Navy Dept. Labs, U.S. Congress Office Tech. Assessment. Contbr. 65 articles to profl. publs.; contbr., editor: Advances in Materials Research, 1963, Science and Technology of Refractory Metals, 1964, Science, Technology and Application of Titanium, 1970; other books. Named Nat. Capitol Engr. of Yr. Council Engring. and Archtl. Socs., 1974; recipient Outstanding Accomplishment awards Navy Dept., 1955-64; annual hon. lectr. Electrochem. Soc., 1970. Fellow Am. Soc. Metals (pres. 1972, hon. mem., Carnegie lectr. 1959, Burgess award 1961, lectr. 1967), Brit. Instn. Metallurgists, AIME (hon. mem.); mem. Nat. Acad. Engring., Fedn. Materials Socs. (pres. 1972-73, Decennial award), Soc. Automotive Engrs. (chmn. aerospace materials div. 1959-74), Brit. Metals Soc., Soc. Advanced Materials and Process Engring., ASTM (hon.; ann. disting. lectr. 1964), Alpha Sigma Mu (hon.). Subspecialties: Materials (engineering); Composite materials. Current work: Consultant internationally and to US government and industry on transfer of technology, advanced processes for productivity, high temperature and high strengtn materials, national policy on research and resources. Inventor in electroplating, 1930-40; metall. devels., 1941-66. Home and Office: 12519 Davan Dr Silver Spring MD 20904 Ten key words and phrases for a professional career: identified goals, long range vision, can-do attitude, integrity, objectivity, understanding and tolerance, faith and trust, professionalism, dedication and perseverance, sense of humor.

PROSSER, FRANCIS W(ARE), physics educator; b. Wichita, Kans., June 30, 1927; s. Francis Ware and Harriet (Osborne) P.; m. Nancy Lou Baugh, May 31, 1952; children: David, Rebecca, Martha. B.S., U. Kans., 1950, M.S., 1954, Ph.D., 1955. Research assoc. Rice U., Houston, 1955-57; asst. prof. U. Kans., Lawrence, 1957-62, assoc. prof., 1962-67; prof., 1967—; sr. research assoc. Aerospace Research Labs., Dayton, Ohio, 1969-70; vis. scientist Argonne (Ill.) Nat. Labs., 1975-76, 77, 78. Served with USNR, 1945-46. Fellow Am. Phys. Soc.; mem. Am. Assn. Physics Tchrs., Kans. Acad. Sci., N.Y. Acad. Scis., Sigma Xi. Subspecialty: Nuclear physics. Current work: Heavy ion nuclear reactions; nuclear spectroscopy; nuclear structure. Office: U Kans Lawrence KS 66045

PROTHRO, JOHNNIE WATTS, nutrition educator; b. Atlanta, Feb. 26, 1922; d. John Devine and Theresa Louise (Young) Hines; m. Alden Faulkner Watts, Sept. 21, 1949 (div. 1953); 1 dau., Darielle Louise Watts Jones; m. Charles E. Prothro, Jr., Sept. 4, 1964. B.S., Spelman Coll., 1941; M.S., Columbia U., 1946; Ph.D., U. Chgo., 1952. Cert. Specialist in Human Nutrition. Am. Bd. Nutrition. Nutrition advisor Communicable Disease Ctr., Atlanta, 1972-73; prof. Tuskegee Inst., Ala., 1973-75, 79-80, Emory U., Atlanta, 1975-79, Ga. State U., 1980—. Spl. fellow NIH, UCLA, 1958-59; OEEC fellow NSF,

London, summer 1961; recipient Borden award Am. Home Econs. Assn., 1950-51. Mem. Am. Inst. Nutrition, Am. Dietetic Assn., Soc. Nutrition Edn., Nutrition Today Soc. Democrat. Subspecialty: Nutrition (biology). Current work: Nutrient requirements of the elderly. Home: 919 Falcon Dr SW Altanta GA 30311 Office: Ga State U University Plaza Atlanta GA 30311

PROTOPAPAS, DIMITRIOS, electrical engineer; b. Apiranthos, Naxos, Greece, Dec. 13, 1939; came to U.S., 1970; s. Antonios and Maria Petrou (Bardanis) P.; m. Helen Michalopoulos, June 1, 1967; children: Alexander, Anthony. B.Sc., U. Athens, 1962, Dipl. Radio-Elec. Engring., 1963; M.A.Sc., U. Toronto, 1968; Ph.D. in Elec. Engring, Poly. Inst. N.Y., 1980. Design engr. Honeywell Corp., Scarboro, Ont., Can., 1969-70; sr. design engr. Gould Inc., Cleve., 1970-72; sr. project engr. Cambridge Memories, Inc., Bedford, Mass., 1972-75; mgr. Technicon Corp., Tarrytown, N.Y., 1976-78; prin. mem. tech. staff ITT Advanced Tech. Ctr., Shelton, Conn., 1978—; adj. assoc. prof. Poly. Inst. N.Y., 1980—. Author: Microcomputer Hardware Design, 1984; contbr. articles to tech. jours. Mem. IEEE, Assn. Computing Machinery, Assn. Profl. Engrs. Subspecialties: Computer architecture; Computer engineering. Current work: Design of microprocessor-based systems; computer architecture; multiprocessing systems, modeling and analysis of computer systems; computer communication networks. Home: 65 Cloverdale Ave Huntington CT 06484 Office: 1 Research Dr Shelton CT 06484

PROUDFIT, HERBERT KERR, III, pharmacologist, researcher, educator; b. Kansas City, Kans., Mar. 6, 1940; s. Herbert Kerr and Constance Catherine (Rice) P. B.A., U. Kans.-Lawrence, 1964; Ph.D., U. Kans.-Kansas City, 1971. Instr., U. Ill. Coll. Medicine, Chgo., 1972-73, asst. prof. pharmacology, 1974-81, assoc. prof., 1981—. Contbr. chpt., numerous articles and abstracts to profl. publs. USHPS grantee, 1975, 78, 80, 82. Mem. AAAS, Am. Pain Soc., Soc. Pharmacology and Exptl. Therapeutics, Internat. Assn. Study Pain, Soc. Neurosci. Subspecialty: Neuropharmacology. Current work: Research central nervous system pathways involved in pain perception; neuronal systems on which morphine and other opiates act to modulate pain perception. Office: Dept Pharmacolog U Ill-Chgo 901 S Wolcott Chicago IL 60680

PROUDFOOT, RUTH ELLEN, research psychologist, educator; b. N.Y.C., Aug. 9, 1940; d. Benjamin and Sylvia (Klein) Levine; m. Harold Galper, Dec. 29, 1961 (div. 1963); m. Wayne Proudfoot, Jan. 16, 1972; 1 son, Nicholas. A.B., Radcliffe Coll., 1961; Ph.D., NYU, 1969. Cert. psychologist, N.Y. Intern psychology Postgrad. Ctr. for Mental Health, N.Y.C., 1968-69; asst. prof. psychology Ferkauf Grad. Sch., Yeshiva U., 1969-71, CCNY, CUNY, 1971-74, assoc. prof., 1975—. Contbr. articles to profl. jours., 1967—. Danforth Found. assoc., 1976. Mem. Am. Psychol. Assn., Eastern Psychol. Assn., Sigma Xi. Subspecialties: Cognition; Neuropsychology. Current work: Research on functional cerebral asymmetries for cognitive processing. Office: Dept Psychology CCNY CUNY 138 St at Convent Ave New York NY 10031

PROUGH, RUSSELL ALLEN, biochemist, educator, researcher; b. Twin Falls, Idaho, Nov. 5, 1943; s. Elza LeRoy and Beulah Elsie (Huddleston) P.; m. Betty Marie Ehlers, Dec. 26, 1965; children: Jennifer Sally, Kimberly Marie. B.S., Coll. of Idaho, 1965; Ph.D., Oreg. State U., 1969. Postdoctoral fellow VA Hosp., Kansas City, Mo., 1969-72; instr. biochemistry Southwestern Med. Sch., Dallas, 1972-73; asst. prof. biochemistry U. Tex. Health Sci. Center, Dallas, 1973-77, asso. prof., 1977-82, prof., 1982—. Recipient Career Devel. award USPHS, 1976-81. Mem. Am. Assn. Cancer Research, Am. Soc. Biol. Chemistry, Am. Soc. Pharmacology and Exptl. Therapeutics, Sigma Xi. Luthern. Subspecialties: Biochemistry (biology); Cancer research (medicine). Current work: Drug metabolism, azo, azoxy, hydrazine toxicity and carcinogenesis. Office: Dept Biochemistry U Tex Health Sci Center 5323 Harry Hines Blvd Dallas TX 75235

PROUT, GEORGE RUSSELL, JR., med. educator; b. Boston, July 23, 1924; s. George Russell and Marion (Snow) P.; m. Loa Katherine Wheatley, Oct. 17, 1950; children—George Russell III, Elizabeth Louise. Student, Union Coll., 1943; M.D., Albany Med. Coll., 1947; M.A. (hon.), Harvard. Resident N.Y. Hosp., 1952-56; asst. attending Meml. Center for Cancer and Allied Disease, N.Y.C., 1956-57; asst. clinician surgery James Ewing Hosp., N.Y.C., 1956-57; asso. prof., chmn. div. urology U. Miami, 1957-60; prof., chmn. div. urology Med. Coll. Va., 1960-69; chief urol. service Mass. Gen. Hosp., Boston; also prof. surgery Harvard Med. Sch., 1969—; Chmn. Adjuvants in Surg. Treatment of Bladder Cancer; mem. advisory task force to Nat. Cancer Inst., 1968—; chmn. Nat. Bladder Coop. Group, 1973—. Bd. editors of: Contemporary Medicine. Served with USNR, 1950-52. Fellow A.C.S., Acad. Medicine Toronto (corr.); mem. Am., Canadian urol. assns., Am. Cancer Soc., Soc. Pelvic Surgeons, Soc. Surg. Oncology, Soc. Univ. Urologists, AAUP, AMA, Dallas So. Clin. Soc. (hon.), Am. Assn. Genitourinary Surgeons, Soc. Pediatric Urology, Alpha Omega Alpha. Subspecialties: Surgery; Urology. Current work: Urological oncology. Home: 174 Puritan Rd Swampscott MA 01907 Office: Fruit St Boston MA 02114

PROVINE, ROBERT RAYMOND, psychologist, educator; b. Tulsa, May 11, 1943; s. Robert William and Thelma Fern (Morgan) P.; m. Helene Vona; children: Kimberly, Robert. B.S., Okla. State U., 1965; Ph.D., Washington U., St. Louis, 1971. Research asso., research asst. prof. Washington U., St. Louis, 1971-74; asst. prof. psychology U. Md.-Baltimore County, Catonsville, 1974-76, assoc. prof., 1976-83-, prof., 1983—. Subspecialties: Psychobiology; Neurobiology. Current work: Behavioral neuroembryology, neuroethology, comparative neuroscience. Office: Dept Psychology U Md Balt County Catonsville MD 21228

PRUITT, BASIL ARTHUR, surgeon, army officer; b. Nyack, N.Y., Aug. 21, 1930; s. Basil Arthur and Myrtle Flo (Knowles) P.; m. Mary Sessions Gibson, Sept. 4, 1954; children: Scott Knowles, Laura Sessions, Jeffrey Hamilton. A.B., Harvard U., 1952, postgrad., 1952-53; M.D., Tufts U., 1957. Diplomate: Am. Bd. Surgery. Intern Boston City Hosp., 1957-58, resident in surgery, 1958-59, 61-62; commd. capt., M.C. U.S. Army, 1959, advanced through grades to col., 1972; resident (Brooke Gen. Hosp.), Ft. Sam Houston, Tex., 1962-64, chief clin. div., Ft. Sam Houston, 1965-67, chief profl. services, Vietnam, 1967-68; comdr., dir. U.S. Army Inst. Surg. Research, Brooke Army Med. Center, Ft. Sam Houston, 1968—; clin. prof. surgery U. Tex. Med. Sch., San Antonio, 1975—; Uniformed Sers. U. Health Scis., Bethesda, Md., 1978—; cons. U. Tex. Cancer Center, 1974—; mem. surgery, anaesthesiology and trauma study sect. NIH, 1978-82. Author med. books; contbr. chpts. to textbooks, articles to profl. jours.; asso. editor: Jour. Trauma; mem. editorial bd.: Archives Surgery, Consultations in Surgery. Decorated Bronze Star. Fellow A.C.S. (gov. 1973-79, pre- and postoperative care com. 1973-75, com. on trauma 1974—, internat. relations com. 1983—); mem. Am. Burn Assn. (pres. 1975-76), Internat. Soc. Burn Injuries (co-chmn. disaster planning com.), Smoke Burn and Fire Assn. (adv. council), Am. Trauma Soc. (dir., pres. Tex. div. 1974-75), Soc. Univ. Surgeons, Am. Surg. Assn. (2d v.p. 1980-81), Tex. Surg. Assn., Western Surg. Assn., So. Surg. Assn., Halsted Soc., Am. Assn. Surgery of Trauma (recorder 1976-80, pres. 1982-83), Surg. Biol. Club, Internat. Soc. Surgeons, Assn. Acad. Surgery, Surg. Infection Soc. (recorder 1980—), Internat. Surg. Group. Subspecialty: Surgery. Current work: Burn and trauma care and research in shock, infection, alteration of host resistance, metabolic response to injury, surgical nutrition, and the pathophysiology of injury. Home: 402 Tidecrest Dr San Antonio TX 78239 Office: US Army Inst Surgical Research Fort Sam Houston TX 78234

PRUSS, THADDEUS PAUL, research administrator, pharmacologist; b. Balt., Jan. 13, 1934; s. Victor William and Helen (Rutkowski) P.; m. Caorle Davis, Nov. 2, 1964; children: Thad, Keith. B.S. in Pharmacy, U. Md., 1956, M.S. in Pharmacology, 1959, Ph.D., Georgetown U., 1962. Research asst. dept. pharmacology Georgetown U., Washington, 1960-61; sect. head cardiopharmacology Cutter Labs., Berkeley, Calif., 1962-66, Mc Neil Labs., Ft. Washington, Pa., 1966-68, acting dir., 1968-69, dir. pharm. research, 1969-80; div. dir. biol. research and devel. Revlon Health Care Group, Tuckahoe, N.Y., 1980—. Contbr. articles to profl. jours. Coach Little League Baseball and Football, Pa., 1971-78; organizer Neighborhood Watch, Dresher, Pa., 1970-75. Mem. AAAS, Am. Assn. Lab. Animal Sci., Am. Soc. Pharmacology and Exptl. Therapeutics, Fedn. Am. Socs. Exptl. Biology, Gastro-intestinal Pharmacology Group, Heart Assn. Southeastern Pa., Inflammation Research Assn., Internat. Soc. Biochem. Pharmacology, N.Y. Acad. Scis., Physiol. Soc. Phila. (v.p. 1976-77, pres. 1977-78), Soc. Exptl. Biology and Medicine. Democrat. Roman Catholic. Subspecialties: Pharmacology; Health services research. Current work: Responsibility for the biological involvement in drug discovery and development. Home: 14 Greenridge Dr Chappaqua NY 10514 Office: Revlon Health Care Group 1 Scarsdale Rd Tuckahoe NY 10707

PRUTKIN, LAWRENCE, anatomist, educator; b. N.Y.C., Dec. 6, 1935; s. Sol and Beverly (Sherwin) P.; m. Beverly Richman, Nov. 28, 1970; children: Brad, Jordan. B.S., Bklyn. Coll., 1957; M.A., N.Y.U., 1960, Ph.D., 1964. Asst. prof. dept. anatomy N.Y.U. Sch. Medicine, 1965-70, assoc. prof. dept. cell biology, 1970—. Contbr. articles to profl. jours. NIH grantee; Hoffmann-LaRoche grantee. Fellow N.Y. Acad. Scis., Am. Assn. Anatomists, AAAS. Subspecialties: Anatomy and embryology; Cell biology (medicine). Current work: Cancer research as related to dermatology. Office: New York U Med Center 550 1st Ave New York NY 10016

PRYSTOWSKY, HARRY, physician, educator; b. Charleston, S.C., May 18, 1925; s. Moses Manning and Raye (Karesh) P.; m. Rhalda Betsy Bressler, Mar. 8, 1951; children—Michael Wayne, Ray Ellen, Jay Bressler. B.S., The Citadel, 1944, D.Sc., 1974; M.D., Med. Coll. S.C., 1948, L.H.D., 1975. Diplomate: Am. Bd. Obstetrics and Gynecology (dir., asso. examiner). Intern Johns Hopkins Hosp. and Med. Sch., 1948-49, resident, 1950-51, 53-55, instr., 1955-56, asst. prof., 1956-58; research fellow U. Cin. Sch. Medicine, 1949; research fellow physiology Yale Med. Sch., 1955-56; prof., chmn. obstetrics and gynecology U. Fla. Coll. Medicine, 1958-73; provost Milton S. Hershey Med. Center, Pa. State U., 1973—; dean Coll. Medicine, 1973—; Cons. surgeon general USAF. Contbr. articles to med. jours. Served as capt., M.C. U.S. Army, 1951-53. Named 1 of 10 outstanding young men U.S. Jr. C. of C. Fellow A.C.S. (gov.); mem. AMA, Council of Deans, Assn. Am. Med. Colls., Assn. Acad. Health Centers, AAAS, A.M.A., Soc. Gynecol. Investigation, Am. Coll. Obstetrics and Gynecology, Am. Assn. Obstetricians and Gynecologists, Am. Fedn. Clin. Research, Pa. Med. Soc., Am. Gynecol. Soc., S. Atlantic Assn. Obstetricians and Gynecologists, N.Y. Acad. Scis., Assn. Profs. Gynecology and Obstetrics (pres.), U. Fla. Alumni Assn. (hon.), Alpha Omega Alpha. Subspecialties: Medical administration; Obstetrics and gynecology. Home: 1141 Cocoa Ave Hershey PA 17033

PSAROS, GEORGE EMANUEL, project engineer; b. Weirton, W.Va., Oct. 24, 1939; s. Emanuel and Catherine (Kladakis) P.; m. Adamantia E. Karamiha, Nov. 25, 1977; children: Emanuel, Steven. B.S. in Aerospace Engring, W.Va. U.-Morgantown, 1961; M.S., U. Wash.-Seattle, 1965; postgrad., U. Mich.-Ann Arbor, 1965-71, Eastern Mich. U., Ypsilanti, 1968-73. Registered profl. engr., Mich. Assoc. research engr. Boeing Co., Seattle., 1961-65; sr. engr. Bendix Corp., Ann Arbor, 1965-73; sr. project engr. Chrysler Corp., Detroit, 1973-82, Gen. Dynamics, Center Line, Mich., 1982—. Mem. AIAA (sec. 1971-72, vice chmn. 1972-73, chmn. 1973-75), Engring. Soc. Detroit, Nat. Soc. Profl. Engrs., Mich. Soc. Profl. Engrs. (vice chmn. 1973-75). Greek Orthodox. Subspecialties: Aerospace engineering and technology; Mechanical engineering. Current work: Spacecraft thermal control; environmental control; engine and transmission design (Cooling); engine air filtration (turbine and Diesel) Air flow through screen, grills and ducts. Office: General Dynamics 25999 Lawrence St Center Line MI 48015 Home: 14370 Four Lakes Sterling Heights MI 48078

PSUTY, NORBERT PHILLIP, geomorphologist, university administrator; b. Hamtramck, Mich., June 13, 1937; s. Phillip and Jessie (Proszykowski) P.; m. Sylvia Helen Zurinsky, June 13, 1959; children: Eric, Scott, Ross. B.S., Wayne State U., 1959; M.S., Miami U., Oxford, Ohio, 1960; Ph.D., La. State U., 1966. Instr. depts. geography and geology U. Miami, 1964-65; asst. prof. dept. geography U. Wis.-Madison, 1965-69; assoc. prof. depts. geography and geology Rutgers U., New Brunswick, N.J., 1969-73, prof. depts. geography and geology, 1973—, dir., 1972-76, 1976—; cons. Nat. Park Service, State of N.J.; vis. scientist U. Liverpool, Eng., 1983. Mem. East Brunswick Twp. (N.J.) Water Policy Adv. Bd., 1981—. Grantee Nat. Park Service, 1975, 76, 77, 81, 82, NSF, 1970, State of N.J., 1975, 77, 78, 81, 82, Minerals Mgmt. Service, 1982. Mem. AAAS, Assn. Am. Geographers, Coastal Soc. (pres. 1980), N.J. Acad. Sci. (pres. 1981-82). Subspecialties: Resource management; Sedimentology. Current work: Coastal geomorphologic process-response studies, coastal natural resource management, public policy in the coastal zone, coastal ecology.

PTASHNE, MARK STEPHEN, molecular biologist; b. Chgo., June 5, 1940; s. Fred and Mildred P. B.A. in Chemistry, Reed Coll., Portland, Oreg., 1961; Ph.D. in Molecular Biology, Harvard U., 1968. Lectr. dept. biochemistry and molecular biology Harvard U., 1968-71, prof., 1971—, chmn. dept. biochemistry and molecular biology, 1980-83; Harvey lectr., 1975. Contbr. articles profl. jours. Jr. fellow Harvard Soc. Fellows, 1965-68; recipient Ledlie award Harvard U., 1968; Guggenheim fellow, 1973-74, Eli Lilly award in biol. chemistry, 1975; NATO sr. scientist awardee, 1977-78, le Prix Charles-Leopold Mayer, 1977, award in molecular biology U.S. Steel Found., 1979. Fellow Am. Acad. Arts and Scis., N.Y. Acad. Scis.; mem. Nat. Acad. Scis., AAAS. Subspecialties: Molecular biology; Biochemistry (biology).

PUCHTLER, HOLDE, pathology educator; b. Kleinlosnitz, Germany, Jan. 1, 1920; came to U.S., 1955; d. Gottfried and Gunda (Thoma) P. Candidate in Medicine, U. Würzburg, Germany, 1944; M.D., U. Köln, Germany, 1949, Dr. med., 1951. Research assoc. U. Köln, 1949-51, resident in pathology, 1951-55; research fellow Damon Runyon Found., McGill U., Montreal, Que., Can., 1955-58; research assoc. to assoc. research prof. Med. Coll. Ga., Augusta, 1959-68, prof. pathology, 1968—. Mem. editorial bd.: Histochemistry, 1977—, Jour. Histotech, 1982—; contbr. articles to sci. jours. Honoree Symposium on Connective Tissues in Arterial and Pulmonary Diseases, 1980. Fellow Royal Microscopical Soc., Am. Inst. Chemists; mem. Gesellschaft für Histochemie, Histochem. Soc., Am. Assn. Anatomists, Anatomishe Gesellschaft, Am. Soc. for Cell Biology, Internat. Acad. Pathologists, Am. Assn. Pathologists, Am. Soc. Zoologists, Biol. Stain Commn., So. Soc. Anatomists, N.Y Acad. Scis. Subspecialties: Histochemistry; Pathology (medicine). Current work: Histochemistry of collagen, elastin, myosin, amyloid. Development of new technics for light, polarization, visible and infrared fluorescence microscopy based on chemistry of textile and leather dyeing, theoretical chemistry and x-ray diffraction data. Office: Dept Pathology Med Coll Georgia Augusta GA 30912

PUCHYR, PETER JOSEPH, petroleum engineer; b. Lac La Biche, Alta., Can., Nov. 12, 1944; s. John and Mary (Lech) P.; m. Dianne J. Mutual, May 30, 1970; children: David M., Jason P. B.Sc. with honors, U. Alta., 1965; M.Sc., Queen's U., Kingston, Ont., Can., 1967. Registered profl. engr., Alta. Programmer, analyst Imperial Oil Ltd., Edmonton, Alta., 1968-72, Aramco, Dhahran, Saudi Arabia, 1972-74; petroleum engr. Imperial Oil Ltd., Calgary, Alta., 1974-78; gen. mgr. Intercomp Ltd., Calgary, 1978-83; ptnr. Simtech Ltd., 1983—. Mem. Soc. Petroleum Engrs. Subspecialties: Petroleum engineering; Software engineering. Current work: Petroleum reservoir simulation and software development. Home: 743 Lake Lucerne Dr SE Calgary AB Canada T2J 3H6 Office: Intercomp Ltd 603 7th Ave SW Calgary AB Canada T2P 2T5

PUCK, THEODORE THOMAS, educator, geneticist, biophysicist; b. Chgo., Sept. 24, 1916; s. Joseph and Bessie (Shapiro) P.; m. Mary Hill, Apr. 17, 1946; children: Stirling, Jennifer, Laurel. B.S., U. Chgo., 1937, Ph.D., 1940. Mem. commn. airborne infections Office Surgeon Gen., Army Epidemiol. Bd., 1944-46; asst. prof. depts. medicine and biochemistry U. Chgo., 1945-47; sr. fellow Am. Cancer Soc., Calif. Inst. Tech., Pasadena, 1947-48; prof. biophysics U. Colo. Med. Sch., 1948—, chmn. dept., 1948-67; dir. Eleanor Roosevelt Inst. Cancer Research, 1962—; Disting. research prof. Am. Cancer Soc., 1966—; nat. lectr. Sigma Xi, 1975-76. Author: The Mammalian Cell as a Microorganism: Genetic and Biochemical Studies in Vitro, 1972. Mem. Commn. on Physicians for the Future. Recipient Albert Lasker award, 1958; Borden award med. research, 1959; Louisa Gross Horwitz prize, 1973; Gordon Wilson medal Am. Clin. and Climatol. Assn., 1977; award Environ. Mutagen Soc., 1981. Fellow Am. Acad. Arts and Scis.; mem. Am. Chem. Soc., Soc. Exptl. Biology and Medicine, AAAS (Phi Beta Kappa award and lectr. 1983), Am. Assn. Immunologists, Radiation Research Soc., Biophys. Soc., Genetics Soc. Am., Nat. Acad. Sci., Inst. Medicine, Phi Beta Kappa, Sigma Xi. Subspecialties: Genetics and genetic engineering (biology); Cancer research (medicine). Current work: The genetic-biochemical nature of cancer; human gene mapping; reversal of cancer; somatic cell genetics and its application to medicine. Address: 10 S Albion St Denver CO 80222 Our age is threatened by distorted emphasis on power, material wealth, and competitiveness, and by an explosive increase in population which exceeds our traditional regulative capacities. But it also holds promise for new and profound understanding of ourselves-of our basic human biological intellectual and emotional needs. There is room for hope.

PUCKETT, ALLEN EMERSON, aeronautical engineer; b. Springfield, Ohio, July 25, 1919; s. Roswell C. and Catherine C. (Morrill) P.; m. Betty J. Howlett; children—Allen W., Nancy L., Susan E.; m. Marilyn I. McFarland; children—Margaret A., James R. B.S., Harvard, 1939, M.S., 1941; Ph.D., Calif. Inst. Tech., 1949. Lectr. aeros., chief wind tunnel sect. Jet Propulsion Lab., Calif. Inst. Tech., 1945-49; tech. cons. U.S. Army Ordnance, Aberdeen Proving Ground, Md., 1945-60; mem. sci. adv. com. Ballistic Research Labs., 1958-65; with Hughes Aircraft Co., Culver City, Calif., 1949—, exec. v.p., 1965-77, pres., 1977-78, chmn. bd., chief exec. officer, 1978—; dir. Am. Mut. Fund, Lone Star Industries; mem. steering group OASD adv. panel on aeros.; cons. Pres.'s Sci. Adv. Com.; chmn. research adv. com. control, guidance and navigation NASA, 1959-64; vice chmn. Def. Sci. Bd., 1962-66; mem. Army Sci. Adv. Panel, 1965-69, NASA tech. and research adv. com., 1968-72, space program adv. council, 1974-78; Wilbur and Orville Meml. lectr. Royal Aero. Soc., London, 1981. Author: (with Hans W. Liepmann) Introduction to Aerodynamics of a Compressible Fluid, 1947; editor: (with Simon Ramo) Guided Missile Engineering, 1959; contbr. tech. papers on high-speed aerodynamics. Trustee U. So. Calif. Recipient Lawrence Sperry award Inst. Aero. Scis., 1949, Lloyd V. Berkner award Am. Astronautical Soc., 1974; named Calif. Mfr. of Yr., 1980. Fellow AIAA (pres. 1972); mem. Aerospace Industries Assn. (chmn. 1979), Los Angeles World Affairs Council (pres.), Nat. Acad. Scis., Nat. Acad. Engring., A.A.A.S., Sigma Xi, Phi Beta Kappa. Subspecialty: Aeronautical engineering. Office: Hughes Aircraft Co PO Box 1042 El Segundo CA 90245

PUENTE, ANTONIO ENRIQUE, psychologist, educator; b. Havana, Cuba, Feb. 14, 1952; s. Antonio A. and Silvia (Llanso) P.; m. Linda Newman, June 11, 1977; 1 dau., Krista L. B.A., U. Fla., 1973; Ph.D., U. Ga., 1978. Lic. psychologist, N.C. Asst. prof. neuroanatomy St. George's (Granada) U., 1978-79; clin. psychologist Northeast Fla. State Hosp., Maclenny, 1979-81; asst. prof. psychology U. N.C., Wilmington, 1981—; cons. Author, editor, translator various sci. books and articles. Fulbright scholar to Argentina, 1982. Mem. Am. Psychol. Assn., Southeastern Psychol. Assn., Soc. Neurosci., Soc. Psychophysiol. Research, Nat. Acad. Neuropsychologists, N.C. Psychol. Assn., Sociedad Interamericana de Psicologia, So. Soc. Philosophy and Psychology, Psi Chi, Phi Theta Kappa. Roman Catholic. Subspecialties: Neuropsychology; Neurophysiology. Current work: Assessment of brain damage and schizophrenia using neuropsychological and psychophysiological techniques. Translation and dissemination of research into Spanish. Behavioral medicine. Home: 103 Parmele Blvd Wrightsville Beach NC 28480 Office: Dept Psychology U NC Wilmington NC 28406

PUEPPKE, STEVEN G., plant pathologist; b. Fargo, N.D., Aug. 1, 1950; s. Glenn Howard and Letha Pauline (Mitchell) P.; m. Jill Ann Winter, Aug. 19, 1978. B.S., Mich. State U., 1971; Ph.D., Cornell U., 1975. Research asso. Kettering Research Lab., Yellow Springs, Ohio, 1975-76; asst. prof. biology U. Mo.-St. Louis, 1976-69; asst. prof. plant pathology U. Fla., Gainesville, 1979-83, assoc. prof., 1983—. Mem. Am. Soc. Microbiology, Am. Soc. Plant Physiologists, Am. Phytopathol. Soc. Subspecialties: Plant pathology; Nitrogen fixation. Current work: Agrobacterium, Phizobium, lectins. Office: Dept Plant Pathology U Fla Gainesville FL 32611

PUGLISI, SUSAN GRACE, biologist, educator, researcher; b. Bridgeport, Conn., Feb. 15, 1952; d. John Benjamin and Raffaela Susan (DeFelice) P. B.A., U. Bridgeport, 1973, M.S., 1975; Ph.D., Clark U., 1979. Cert. tchr. grades 7-12. NIH postdoctoral fellow U. Mass. Med. Sch., 1979-81; research assoc. U. Conn., 1980-81; substitute asst. prof. Baruch Coll., CUNY, 1981-83. Contbr. articles to profl. jours. Mem. AAAS, Am. Soc. Zoologists, Am. Micro. Soc., Soc. Neurosci, Nat. Assn. Biology Tchrs., N.Y. Acad. Scis., Conn. Audubon Soc. Subspecialties: Comparative neurobiology; Morphology. Current work: Cytoarchitecture of vertebrate nervous system, auditory system, histology; science education. Home: 190 McKinley Ave Bridgeport CT 06606

PUHVEL, SIRJE MADLI, dermatology researcher, educator; b. Tallinn, Estonia, July 26, 1939; came to U.S., 1960; d. George Jüri and Tina (Toodo) Hansen; m. Jaan Puhvel, June 4, 1960; children: Peter Jaan, Andres Jaak, Markus Jüri. B.A., UCLA, 1961, Ph.D., 1963. Postdoctoral fellow UCLA Med. Sch., 1963-64, asst. research immunologist, 1965-72, assoc. research immunologist, 1972-73, adj. assoc. prof., 1973-78, adj. prof., 1978—. Grantee NIH, 1975—. Mem. Soc. Investigative Dermatology (councilor western region 1978-80, chmn. western region 1980-81), Am. Fedn. Clin. Research, Am. Soc. Microbiology, Western Soc. Clin. Research, Sigma Xi. Subspecialties: Dermatology; Microbiology (medicine). Current work: Pathogenesis of acne vulgaris and chloracne; microbiology of skin; keratinization of sebaceous follicle epithelial cells. Home: 15739 High Knoll Rd Encino CA 91436 Office: UCLA Med Sch Los Angeles CA 90024

PUKKILA, PATRICIA JEAN, biologist; b. Woodbury, N.J., Sept. 28, 1948; d. Arnold Oliver and Virginia Ruth (Pickering) P.; m. Gordon Worley, Sept. 14, 1974; 1 son, Read Pukkila-Worley. B.S. with honors, U Wis., 1970; Ph.D. (Woodrow Wilson fellow), Yale U., 1975. NIH postdoctoral fellow Nat. Inst. Med. Research, London, 1975-77; Leukemia Soc. postdoctoral fellow Harvard U., Cambridge, Mass., 1977-79; asst. prof. biology U. N.C., Chapel Hill, 1979—. Contbr. articles to profl. jours. USPHS grantee, 1980—. Mem. Am. Soc. Cell Biology, Genetics Soc. Am., Phi Beta Kappa. Subspecialties: Genetics and genetic engineering (biology); Molecular biology. Current work: Researcher in meiosis, mechanisms of chromosome pairing and genetic recombination. Home: 307 Birch Circle Chaple Hill NC 27514 Office: Dept Biology Univ NC Chapel Hill NC 27514

PUN, PATTLE PAK TOE, microbiologist, educator; b. Hong Kong, Sept. 30, 1946; s. Sak-Chi and Ngan-Chu (Kwan) P.; m. Gwen Yam Qun, Aug. 22, 1970; children: Patrick Hank, Benjamin Tim. B.S. in Chemistry with high honors and distinction, San Diego State U., 1969; M.A. in Biology, SUNY-Buffalo, 1972, Ph.D., 1974. Resident assoc. div. biol. and med. research Argonne Nat. Lab., Ill., 1974-76; vis. microbiologist U. Ill. Med. Ctr., Chgo., 1980-81, 1981; prof. biology Wheaton Ill. Coll., 1973—. Author: Evolution: Nature and Scripture in Conflict?, 1982; contbr. articles to profl. jours. Research Corp. grantee, 1975, 76, 79, 80, 82, 83. Mem. Am. Soc. Microbiology, N.Y. Acad. Scis., Am. Sci. Affiliation. Mem. Chinese Bible Ch. Subspecialties: Microbiology; Genetics and genetic engineering (biology). Current work: Cloning of the conditional asporogenous and rifampicin resistant gene of bacillus subtilis and its genetic and biochemical analysis. Office: Dept Biology Wheaton Coll Wheaton IL 60187

PUNNETT, HOPE SUZANNE, geneticist; b. N.Y.C., Jan. 24, 1927; d. Irving and Mathilde (Fromkes) Handler; m. Thomas Roosevelt Punnett, June 15, 1950; children: Laura, Susan, Jill. B.A. summa cum laude, Smith Coll., 1948; M.S., Yale U., 1949, Ph.D., 1955; postgrad., U. Ill., 1950-54, Johns Hopkins U., 1961. Research asst. U. Ill., 1950-52, NSF fellow in genetics, 1952-53; asst. lectr. biology dept. U. Rocchester, 1959-62; research instr. dept. ob-gyn U. Rochester, 1962-63; assoc. in research pediatrics (genetics) Temple U. Med. Sch., 1963-66, research asst. prof., 1966-69, research assoc. prof., 1969-74, prof. pediatrics, 1974—; vis. scientist dept. human genetics and biometry Univ. Coll., London, Eng., 1969; chief genetics sect. St. Christopher's Hosp. for Children, 1963-77, 82—, dir. genetics lab., 1963—; assoc. biosci. staff in pediatrics Albert Einstein Med. Center, 1979—. Editorial bd.: Am. Jour. Human Genetics, 1975-78; Contbr. numerous articles, papers and chpts. in books to sci. lit. NIH grantee, 1963-68, 69-72, 76-82, 79—; Cystic Fibrosis Found. grantee, 1979-82. Mem. Bot. Soc. Am., Genetics Soc. Am., Am. Soc. Human Genetics (Sec. 1977-80), Am. Genetic Assn., AAAS, Soc. Pediatric Research, Coll. Physicians of Phila., Sigma Xi, Phi Beta Kappa. Subspecialties: Genetics and genetic engineering (medicine); Tissue culture. Current work: Human genetics and cytogenetics; research in genetic diseases such as cystic fibrosis and alpha 1-antitrypsin deficiency; tissue culture; prenatal diagnosis. Office: St Christopher's Hosp for Children Philadelphia PA 19133

PURBRICK, ROBERT LAMBURN, physics educator; b. Salem, Oreg., Dec. 5, 1919; s. Harold and Violet (Barker) P.; m. Ursula Myrtha Glaeser, May 15, 1921; children: Sally Ann, Edith Ingrid, Carol Jean. B.A., Willamette U., 1942; M.A., U. Wis.-Madison, 1944, Ph.D., 1947. Grad. asst. U. Wis.-Madison, 1942-43, Alumni Research fellow, 1946-47; staff Manhattan Project, U. Chgo., 1944-46; asst. prof. physics Willamette U., 1947-49, assoc. prof., 1954—. Contbr. articles to profl. jours. Mem. Am. Phys. Soc., Am. Assn. Physics Tchrs., Optical Soc. Am, AAAS, IEEE, Sigma Xi. Democrat. Methodist. Subspecialties: Atomic and molecular physics; Nuclear physics. Current work: Atomic and molecular physics. Home: 255 W Vista Ave S Salem OR 97302 Office: Willamette U Salem OR 97301

PURCELL, EDWARD MILLS, physicist; b. Taylorville, Ill., Aug. 30, 1912; s. Edward A. and Mary Elizabeth (Mills) P.; m. Beth C. Busser, Jan. 22, 1937; children—Dennis W., Frank B. B.S. in Elec. Engring, Purdue U., 1933, D. Engring. (hon.), 1953; Internat. Exchange student, Technische Hochschule, Karlsruhe, Germany, 1933-34; A.M., Harvard U., 1935, Ph.D., 1938. Instr. physics Harvard U., 1938-40, asso. prof., 1946-49, prof. physics, 1949-58, Donner prof. sci., 1958-60, Gerhard Gade Univ. prof., 1960-80, emeritus, 1980—; sr. fellow Soc. of Fellows, 1949-71; group leader Radiation Lab., Mass. Inst. Tech., 1941-45. Contbg. author: Radiation Lab. series, 1949, Berkeley Physics Course, 1965; contbr. sci. papers on nuclear magnetism, radio astronomy, astrophysics, biophysics. Mem. Pres.'s Sci. Advisory Com., 1957-60, 62-65. Co-winner Nobel prize in Physics, 1952; recipient Oersted medal Am. Assn. Physics Tchrs., 1968; Nat. Medal of Sci., 1980. Mem. Am. Philos. Soc., Nat. Acad. Sci., Phys. Soc., Am. Acad. Arts and Scis. Subspecialties: Theoretical astrophysics; Biophysics (biology).

PURCELL, KEITH FREDERICK, educator; b. St. Louis, Sept. 12, 1939; s. Kenneth Howard and Marjorie Vivian (Phelps) P.; m. Susan Caroline Hartzog, Nov. 18, 1941; children: Kristan Lee, Karen Jean. A.B., Central Col., Fayette, Mo., 1961; Ph.D., U. Ill., 1965. Asst. prof. Wake Forest U., Winston-Salem, N..C., 1965-67; asst. prof. Kans. State U., Manhattan, 1967-70, assoc. prof., 1970-78, prof. chemistry, 1978—. Author: (with J. C. Kotz) Inorganic Chemistry, 1977, An Introduction to Inorganic Chemistry, 1980. Recipient Outstanding Chemistry Teaching award Student Affiliate, Am. Chem. Soc., 1976; named pProfesseur Associe, France, 1980-81. Mem. Am. Chem. Soc., AAAS, Sigma Xi, Phi Lambda Upsilon. Subspecialty: Inorganic chemistry. Current work: Teaching, text writing, research on transition metal complexes, molecular orbital calculations, mossbauer spectroscopy, intersystem crossing, electron transfer reactions, mixed valence compounds. Home: 214 Drake Dr Manhattan KS 66502 Office: Willard Hall Kans State U Manhattan KS 66506

PURCELL, ROBERT HARRY, medical virologist, researcher; b. Keokuk, Iowa, Dec. 19, 1935; s. Edward Harold and Elsie Thelma (Melzl) P.; m. Carol Joan Moody, June 11, 1961; children: David, John. A.S. in Chemistry, Eastern Okla. A&M Jr. Coll., 1955, B.A., Okla. State U., 1957; M.S. in Biochemistry, Baylor U., 1960; M.D., Duke U., 1962. Student fellow Baylor U., 1959-60, Duke U., 1960-62; intern in pediatrics Duke U. Hosp., 1962-63; commd. USPHS, 1963; officer Epidemic Intelligence Service, Communicable Disease Ctr., Atlanta, 1963-65; assigned to Vaccine Devel. br. Nat. Inst. Allergy and Infectious Diseases, 1963-65; sr. surgeon USPHS Lab. Infectious Diseases, 1965-69, med. officer, Bethesda, Md., 1969-72, med dir., 1972-74; head sect. hepatitis virus Lab. Infectious Diseases, Nat. Inst. Allergy and Infectious Diseases, Bethesda, 1974—; speaker in field; vis. prof. Mich. State U. Sch. Medicine, 1979; vis. prof., Noble Wiley Jones lectr. U. Oreg., Med. Sch., 1980; Howard B. Tickton Meml. lectr. George Washington U. Med. Sch., 1980. Contbr. numerous chpts., articles to profl. publs. Recipient Superior Service award USPHS, 1972, Meritorious Service medal, 1974, Gorgas medal, 1977, Disting. Service medal, 1978; Disting. Alumni award Duke U. Sch. Medicine, 1978; Eppinger prize 5th Internat. FALK Symposium on Virus and Liver, Basel, Switzerland, 1979. Fellow Washington Acad. Scis.; mem. Am. Epidemiol, Soc., Am. Soc. Microbiology, AAAS, Soc. Epidemiol. Research, Infectious Disease Soc. Am. (Squibb award 1980), Am. Soc. Clin. Investigation, N.Y. Acad. Scis., Assn. Am. Physicians, Am. Coll. Epidemiology. Subspecialties: Virology (medicine); Epidemiology. Current work: Viral hepatitis.

PURCHASE, HARVEY GRAHAM, research veterinarian, administrator; b. Zambia, Africa, Aug. 8, 1936; came to U.S., 1961, naturalized, 1966; s. Harvey Spurgeon and Vera Margaret (Cooper) P.; m. Nancy Ruth Schneider, July 6, 1963; children: Deborah Ruth, Kenneth Graham. B.Sc., U. Witwatersrand, 1955; B.V.Sc., U. Pretoria, 1959; M.R.C.V.S., U. London, 1961; M.S., Mich. State U., 1965, Ph.D., 1970. Gen. practice vet. medicine, South Africa and Eng., 1960-61; research veterinarian U.S. Dept. Agr., East Lansing, Mich., 1961-74, staff scientist poultry, Beltsville, Md., 1974-78, chief livestock and vet. scis. staff, 1978-82, nat. program leader for bioregulation, 1982—; assoc. prof. Mich. State U., 1972-74. Mem. editorial bd.: Avian Diseases, 1974—, Poultry Sci, 1978—; contbr. numerous articles in field. Leader Nat. Capital Area Council Boy Scouts Am., 1979—. Recipient Sir Arnold Theiler medal, 1959, Arthur S. Fleming award, 1971, Tom Newman Meml. award of Gt. Britain, 1973. Mem. AVMA, Am. Assn. Avian Pathologists, Brit. Vet. Assn., U.S. Animal Health Assn., D.C. Vet. Med. Assn., Poultry Sci. Assn. Soc. Virology, Am. Soc. Microbiology. Presbyterian. Subspecialties: Cancer research (veterinary medicine); Virology (veterinary medicine). Current work: Major research was on avian tumor viruses, work interests are administration of research and research on problems of animal agriculture. Patentee vaccine for immunizing poultry against marek's disease. Home: 8905 Eastbourne Ln Laurel MD 20708 Office: US Department Agriculture BARC West Beltsville MD 20705

PURDOM, PAUL WALTON, JR., computer scientist, educator; b. Atlanta, Apr. 5, 1940; s. Paul Walton and Bettie (Miller) P.; m. Donna Lee Armstong, Aug. 16, 1965; children: Barbara, Linda, Paul. B.S., Calif. Inst. Tech., 1961, M.S., 1962, Ph.D., 1966. Asst. prof. computer sci U. Wis.-Madison, 1965-70, assoc. prof., 1970-71; mem. tech. staff Bell Labs., Naperville, Ill., 1970-71; assoc. prof. computer sci. Ind. U., Bloomington, 1971-82, prof., 1983—, chmn., 1978-82. Mem. Assn. Computing Machinery, Soc. Indsl. and Applied Math. Methodist. Subspecialties: Theoretical computer science; Compiler design. Current work: Analysis of search algorithms, game playing, and compiler design. Home: 2212 Belhaven St Bloomington IN 47401 Office: Ind U Bloomington IN 47405

PURDY, JAMES AARON, medical physicist; b. Tyler, Tex., July 16, 1941; s. Walter Bethel and Florence (Hardy) P.; m. Marilyn Janette, Jan. 29, 1965; children: Katherine Janette, Laura Elizabeth. B.S., Lamar U., 1967; M.A., U. Tex., 1969, Ph.D., 1971. Teaching asst. dept. physics U. Tex., Austin, 1967-68, asst. research scientist, 1969-71; research asst. dept. med. physics M.D. Anderson Hosp. & Tumor Inst., Houston, 1968-69, fellow in med. physics, 1972-73; instr. radiation physics Washington U., St. Louis, 1973-76, asst. prof., 1976-79, asso. prof., 1979-83, prof., 1983—, head, physics sect., 1976—. Contbr. articles in field to profl. jours. Served with USMC, 1961-64. Mem. Am. Assn. Physicists Medicine (pres. elect. 1984), Am. Soc. Therapeutic Radiologists, Am. Coll. Radiology, Health Physics Soc., Radiation Research Soc., N.Y. Acad. Scis., Sigma Pi Sigma, Phi Kappa Phi. Methodist. Subspecialties: Radiology; Biophysics (physics). Current work: Med. physicist in high energy x-ray and electron beam dosimetry, computer, treatment planning, med. linear accelerator devel. and hyperthermia.

PURDY, RALPH EARL, pharmacologist, educator; b. Stockton, Calif., Mar. 15, 1941; s. Earl and Mary Jane (Gardner) P.; m. Leslie Noble, Aug. 23, 1969; children: Christopher Hugh, George Colin. A.A., Modesto Jr. Coll., 1961; B.S., U. Pacific, 1965, M.S. in Physiology and Pharmacology, 1967; Ph.D., UCLA, 1973. Fellow dept. pharmacology UCLA, 1969-74; asst. prof. U. Calif.-Irvine, 1974-81, assoc. prof., 1981—; chmn. United Ministries in Higher Edn. Com., 1978—, dir. seminars dept. pharmacology, 1981-82; cons. and lectr. in field. Reviewer: Rapid Communications for Behavioral and Neural Biology, 1979—, Life Scis, 1980—; contbr. articles to profl. jours. Pheiffer Found. fellow, 1965-66; Calif. State fellow, 1969-70; USPHS grantee, 1969-73; Am. Heart Assn. grantee, 1975-78, 80-83; NIH grantee, 1981—. Mem. Western Pharmacology Soc., Soc. Pharmacology and Exptl. Therapeutics, Phi Kappa Phi, Rho Chi. Democrat. Presbyterian. Subspecialties: Pharmacology; Biochemistry (medicine). Current work: Pharmacological and biochemical characterization of vascular alpha adrenoceptors; blood vessels, alpha adrenoceptors, radioligand binding, affinity labeling. Office: Dept Pharmacology Coll of Medicine U Calif Irvine CA 92717

PURDY, WILLIAM CROSSLEY, educator, chemist; b. Bklyn., Sept. 14, 1930; s. John Earl and Virginia (Clark) P.; m. Myrna Mae Moman, June 17, 1953; children—Robert Bruce (dec.), Richard Scott, Lisa Patrice, Diana Lori. B.A., Amherst Coll., 1951; Ph.D., MIT, 1955. Instr. U. Conn., Storrs, 1955-58; faculty U. Md., College Park, 1958-76, prof. chemistry, 1964-76, head div. analytical chemistry, 1968-76; prof. chemistry, asso. in medicine McGill U., 1976—; vis. prof. Institut für Ernahrungswissenschaft, Justus Liebig-Universitat, Giessen, Germany, 1965-66; Nat. lectr. Am. Assn. Clin. Chemistry, 1971; Fisher Sci. Lecture award Chem. Inst. Can., 1982; cons. Surg. Gen., U.S. Army, 1959-75; sci. adviser Balt. dist. FDA. Author: Electro-analytical Methods in Biochemistry, 1965, also numerous articles.; Bd. editors: Clin. Chemistry, 1971-80, Anal. Letters, 1979—, Anal. Chim. Acta, 1979—, Clin. Biochemistry, 1983—; adv. bd. editors: Analytical Chemistry, 1971-73. Fellow Nat. Acad. Clin. Biochemistry, Royal Soc. Chemistry (London); mem. Am. Chem. Soc., Am. Assn. Clin. Chemistry, Chem. Inst. Can., Can. Soc. Clin. Chemists, Sigma Xi. Subspecialties: Analytical chemistry; Clinical chemistry. Research on application of modern analytical methods to biochem. and clin. systems, separation sci., trace metal analysis. Home: 1321 Sherbrooke St W C-40 Montreal PQ H3G 1J4 Canada Office: McGill U 801 Sherbrooke St W Montreal PQ H3A 2K6 Canada

PURISCH, ARNOLD DAVID, neuropsychologist; b. Bklyn., Feb. 20, 1951; s. Sigmund L. and Tess (Lazar) P.; m. Ellen C. Maslow, Aug. 10, 1975. B.A., U. Md., 1973; M.A., Fairleigh Dickinson U., 1975; Ph.D., U. S.D., 1978. Lic. psychologist, N.J. Neuropsychologist J.F.K. Med. Ctr., Edison, N.J., 1978-80, VA Med. Ctr., East Orange, N.J., 1980-83, Long Beach, Calif., 1983—; asst. prof. N.J. Coll. Medicine, Newark, 1980-83; dir. Neuropsychol. Assocs. of Calif., Long Beach, 1982—. Author: Item Interpretation of the Luria-Nebraska Neuropsychological Battery, 1982, The LNNB Manual, 1980; mem. editorial bd.: Clin. Neuropsychology, 1980—; contbr. articles to profl. jours. Mem. Am. Psychol. Assn., Assn. for Advancement of Psychology, Internat. Neuropsychol. Soc., N.Y. Neuropsychology Group. Subspecialty: Neuropsychology. Current work: Relationship of behavior, emotion, and cognition to the brain; neuropsychology of schizophrenia; recovery of function after brain damage; development of assessment techniques to measure the effects of brain damage, especially the Luria-Nebraska Neuropsychological Battery. Office: Psychology Service VA Med Ctr 5901 E 7th St Long Beach CA 90822

PURPURA, DOMINICK PAUL, educator, med. scientist; b. N.Y.C., Apr. 2, 1927; s. John R. and Rose (Ruffino) P.; m. Florence M. Williams, May 31, 1948; children—Craig, Kent, Keith, Allyson. A.B., Columbia U., 1949; M.D., Harvard U., 1953. Intern Columbia-Presbyn. Med. Center, N.Y.C., 1953-54; resident Neurol. Inst., N.Y.C., 1954-55; asso. prof. neurosurgery, neurology and anatomy Columbia Coll. Physicians and Surgeons, 1958-66; prof., chmn. dept. anatomy Albert Einstein Coll. Medicine, Bronx, N.Y., 1967-74, chmn. dept. neurosci., 1974-81; dean, asso. v.p. med. affairs Stanford (Calif.) U., 1981—; sci. dir. Rose F. Kennedy Center for Research in Mental Retardation, N.Y.C., 1969-72, dir., 1972—; NIH adv. com. on epilepsies USPHS; councillor Soc. Neuroscis. Chief editor: Brain Research. Served with USAAF, 1945-47. Mem. Am. Physiol. Soc., Am. Neurol. Assn., Am. Epilepsy Assn., AAAS. Subspecialties: Neurobiology; Medical education and research adminstration. Research on structure and function of mature and immature brain; origin of brain waves; developmental neurobiology; mechanisms of epilepsy. Office: Stanford U Sch Medicine Stanford CA 94305

PURSER, PAUL EMIL, systems engineer, consultant; b. Amite, La., Dec. 9, 1918; s. Brittain Birdsong and Ethel Elizabeth (Hungate) P.; m. Carlotta Mary King, Aug. 5, 1939; children: Mary Elizabeth Beeman, Margaret King. B.Sc. in Aero. Engring, La. State U., 1939; postgrad., U. Va. Extension, 1942-44. Registered profl. engr., La., Tex. Research engr. Nat. Adv. Com. for Aeros., Hampton, Va., 1939-58; research engr. NASA, Hampton, 1958-62; spl. asst. to dir. Manned Spacecraft Ctr., Houston, 1962-70; cons. Paul E. Purser, P.E., Houston, Humble, Tex., 1970—; sr. systems engr. GURC, Bellaire, Tex., 1974—; cons. Nat. Acad. Engring.-Marine Bd., Washington, 1972-82, Stanford U. Sch. Medicine, Palo Alto, Calif., 1971-76; spl. asst. to pres. U. Houston, 1968-69. Sr. editor: Manned Spacecraft: Engineering Design and Operation, 1964. Bd. dirs. ARC, Houston chpt., 1965-71; dir. Houston Mus. Natural Sci., 1965—. Recipient Cert. of Commendation NASA-Manned Spacecraft Ctr., 1965, Group Achievement awards, 1964, 67. Assoc. fellow AIAA (mem. local council 1962-67); mem. Marine Tech. Soc. (mem. local council 1976-79, meritorious service award and cert. of appreciation 1978), Am. Soc. Engring. Edn., Systems Safety Soc., Soc. Petroleum Engrs., C. of C., mem. sci. and tech. com. (1970-77, cert. of appreciation 1974), Tau Beta Pi, Omicron Delta Kappa, Sigma Gamma Tau. Republican. Presbyterian. Subspecialties: Systems engineering; Aerospace engineering and technology. Current work: Application of systems engineering and aerospace technology to cardiovascular medicine, petroleum and geothermal energy, offshore and marine engineering, diving, marine salvage, explosive ordnance disposal, oil spill prevention. Home: 8950 Shoreview Ln Humble TX 77346 Office: GURC 5909 W Loop S Bellaire TX 77401

PURTILO, DAVID THEODORE, pathology educator, researcher; b. Duluth, Minn., Apr. 13, 1939; m. Ruth Bryant, Aug. 12, 1961. B.A., U. Minn., 1961; M.S., U. N.D., 1963; M.D., Northwestern U., 1967. Diplomate: Am. Bd. Pathology. Intern Boston City Hosp., 1967-68; resident in pathology Brigham and Women's Hosp., Boston, 1968-69, U. Minn. Hosps., Mpls., 1969-71; clin. pathologist Project Hope, Colombia, 1971; research pathologist St. Vincent Hosp., Worcester, Mass., 1973-74; pathologist U. Mass., 1973-81; prof., chmn. dept. pathology U. Nebr.-Omaha, 1981—; teaching faculty hematology Am. Soc. Clin. Pathology; adv. bd. Internat. Soc. Study of Environ. and Cancer, 1981—; ad hoc coms. Am. Cancer Soc., 1979—; mem. ad hoc resources cancer therapy com. Nat. Cancer Inst., Bethesda, Md., 1982—. Author: Survey of Human Diseases, 1978; guest editor: Cancer Research, 1981. Served to maj. M.C. AUS, 1971-73. Wellcome Research grantee, 1979; NIH Fogarty fellow, 1980; Am. Cancer Soc.-Roosevelt Cancer fellow, 1979; Fgn. fellow Smith, Kline and French, Africa, 1966. Mem. Am. Soc. Clin. Pathologists, Internat. Acad. Pathologists, Reticuloendothelial Soc., New Eng. Soc. Pathologists, Mass. Soc. Pathologists, Am. Soc. Microbiology, Am. Fedn. Clin. Research, Am. Assn. Immunologists, Pediatric Pathology Club, AAAS, Coll. Am. Pathologists, AMA, Internat. Assn. Comparative Research on Leukemia and Related Diseases, Am. Assn. Pathologists, Omaha Med. Soc., Sigma Xi. Subspecialties: Immunogenetics; Cancer research (medicine). Current work: Investigating role of immune deficiency and susceptibility to viruses in lymphomagenesis and leukemogenesis. Home: 10606 Canyon Rd Omaha NE 68112 Office: Pathology and Lab Medicine U Nebr Med Ctr 42d St and Dewey Ave Omaha NE 68105

PUSKAS, ELEK, parachute company executive; b. Kassa, Hungary, Nov. 29, 1942; came to U.S., 1961; s. Elek and Olga (Derfinyak) P.; m. Lone Lee Zimmerman, Dec. 30, 1969 (dec. 1977); m. Holly Ann LeClair, Aug. 25, 1982. Auto mechanic Keystone Motors, Berwyn, Pa., 1961-64; engring. technician S. Snyder Enterprises Inc., Cherry Hill, N.J., 1964-69; dir. ops. Para-Flite, Inc., Pennsauken, N.J., 1969-77, pres., 1977—; pres. Ripcord Para Ctr., Inc., Medford, N.J., 1967-76. Contbr. articles to publs. Mem. Parachute Equipment Industry Assn. (chmn. 1981—), AIAA (sr., tech. com.), Am. Mgmt. Assn., U.S. Parachute Assn., AAAS. Democrat. Subspecialties: Aeronautical engineering; Aerospace engineering and technology. Current work: Lifting aerodynamic decelerators. Patentee. Home: Conestoga Ln Mount Holly NJ 08060 Office: Para-Flite Inc 5800 Magnolia Ave Pennsauken NJ 08109

PUTMAN, STEPHEN HOWARD, city and regional planning educator; b. Newark, Feb. 23, 1941; s. Robert S. and Mildred Rose (Pearl) P.; m. Mary Tobias Whiting, Apr. 17, 1962; 1 dau., Rachel C. B.S.E.E, Carnegie Inst. Tech., 1962; M.B.A., U. Pitts., 1963, Ph.D., 1968. Systems analyst Consad Research Corp., Pitts., 1963-69; assoc. prof. dept. city and regional planning U. Pa., Phila., 1969—; prin. S.H. Putman Assocs., Phila., 1980—. Author: Empirical Model of Regional Growth, 1975, Urban Residential Models, 1979, Integrated Urban Models, 1983. Mem. Transp. Research Bd., Regional Sci. Assn. Current work: Computer simulation of transportation, land use, environmental interactions theory development, application and policy analysis of urban and regional systems. Home: 7030 Wissahickon Ave Philadelphia PA 19119 Office: Dept City-Regional Planning U Pa Philadelphia PA 19104

PUTNAM, PAUL A(DIN), govt. agrl. adminstr.; b. Springfield, Vt., July 12, 1930; s. Horace A. and Beatrice Nellie (Baldwin) P.; m. Elsie Mae Ramseyer, June 12, 1956; children—Pamela Ann, Penelope Jayne, Adin Tyler II, Paula Anna. B.S. (Danforth fellow 1951, Borden fellow 1951), U. Vt., 1952; M.S., Wash. State U., 1954; Ph.D. in Animal Nutrition, Cornell U., 1957. With Dept. Agr., Agrl. Research Service, NE Region Beltsville (Md.) Agrl. Research Center, 1957—, chief, 1968-72, asst. dir., 1972-80, dir., 1980—. Chmn. adminstr. bd. Methodist Ch., Silver Spring, Md., 1971-79, mem. council ministries, 1980; chmn. Knollwood Council, Citizens Assn., Adelphi, Md., 1973-74; mem. Prince George's County (Md.) Council Gen. Plan Amendment Citizens' Adv. Com., 1976-78; mem. transp./met. com. Prince George's County C. of C., 1977, hon. bd. dirs., 1978—. Recipient Kidder medal U. Vt., 1952, Outstanding Performance award Dept. Agr., 1964, 80; Purina research fellow, 1954-56. Mem. Am.

Dairy Sci. Assn., Am. Soc. Animal Sci. (pres. North Atlantic sect. 1970, chmn. various coms. 1969-76), Orgn. Profl. Employees of U.S. Dept. Agr. (pres. Beltsville chpt. 1972-74), Council Agrl. Sci. and Tech. Republican. Subspecialties: Animal nutrition; Agricultural research adminstration. Current work: Agricultural research administration. Research, numerous publs. in field, primarily in ruminant nutrition, salivary secretion, animal feeding behavior, 1954-68. Home: 10532 Edgemont Dr Adelphi MD 20783 Office: Dept Agr AR NER Beltsville Agrl Research Center Beltsville MD 20705

PUTT, MARK STUART, dental research chemist, consultant; b. Ft. Wayne, Ind., Sept. 27, 1950; s. Robert Jerome and Thelma (Keith) P.; m. Rhonda Lee Stackhouse, June 22, 1974; 1 son, Karson S. B.S. in Chemistry, Purdue U., 1972; M.S.D., Ind. U. Sch. Dentistry-Indpls., 1979. Research assoc. Preventive Dentistry Research Inst., Ind. U. Sch. Dentistry, Ft. Wayne, 1972—. Recipient cert. for speaking Ind. Dental Hygienists' Assn., 1982. Mem. Internat. Assn. Dental Research, Am. Chem. Soc., Am. Assn. Dental Research. Subspecialties: Preventive dentistry; Cariology. Current work: Basic and applied research into identification and evaluation of therapeutic systems for prevention or inhibition of dental caries and periodontal disease. Patentee tooth-testing system and machine, anticariogenic maloaluminate complexes, dentifrice preparations comprising aluminum, alkaline oral compositions. Home: 4034 Westland Dr Fort Wayne IN 46815 Office: Preventive Dentistry Research Inst Ind U 2101 Coliseum Blvd E Fort Wayne IN 46805

PYE, EARL LOUIS, university dean; b. Merino, Colo., Aug. 18, 1926; s. Guy William and Gladys (Cooper) P.; m. Shirley M. Adicks, Mar. 10, 1949 (div.); children: Deborah, Deanna, Douglas, Cynthia, Mark, Michael. A.B., Chico State U., 1958; M.S., U. Calif.-Davis, 1961; Ph.D., La. State U., 1966. Registered profl. engr., Calif. Pres. Corrosion Control Specialists, San Dimas, Calif., 1976—, Alpha Research Co., San Dimas, 1976—; dean grad. studies and research, prof. chemistry Calif. State Poly. U., Pomona, 1961—. Served with USNR, 1944-46. Recipient Pomona City Civic award, 1972. Mem. Am. Chem. Soc., Nat. Assn. Corrosion Engrs., Sigma Xi, Phi Kappa Phi. Republican. Lodge: Elks. Subspecialties: Corrosion; Physical chemistry. Current work: Instantaneous and instrumental methods of corrosion rate measurements. Patentee in field. Office: Calif Poly U Pomona CA 91768

PYERITZ, REED EDWIN, genetics researcher, educator; b. Pitts., Nov. 2, 1947; s. Paul L. and Ida M. (Meier) P.; m. Jane E. Tumpson, May 28, 1972; 1 dau., Allyson. S.B., U. Del., 1968; A.M. and Ph.D., Harvard U., 1972, M.D., 1975. Diplomate: Am. Bd. Internal Medicine, Am. Bd. Med. Genetics. Intern and resident Peter Bent Brigham Hosp., Boston, 1975-77; resident Johns Hopkins U. Med. Sch., Balt., 1977-78, asst. prof. medicine and pediatrics, 1978-83, assoc. prof. medicine and pediatrics, 1983—; mem. Md. Com. on Hereditary Disorders, 1983—; cons. Md. State Athletic Commn., 1980—. Editor: genetics clinics Johns Hopkins Med. Jour, 1978-83; contbr. articles to sci. jours. NIH fellow, 1978-80; Am. Heart Assn. fellow, 1978-79. Fellow ACP, Am. Bd. Med. Genetics; mem. Am. Heart Assn., AAAS. Club: Balt. Road runners. Subspecialties: Genetics and genetic engineering (medicine); Internal medicine. Current work: Genetic nosology; application of molecular biologic techniques to the diagnosis and management of hereditary disorders; basic defects among the heritable disorders of connective tissue; medical ethics as applied to genetic engineering; computer data-based systems as applied to genetics. Office: Johns Hopkins U Sch Medicine Baltimore MD 21205

PYKETT, IAN LEWIS, physicist; b. Nottingham, Eng., Jan. 11, 1953; came to U.S., 1980; s. William L. and Margaret (Grieves) P.; m. Gwen A. Brown, July 16, 1977. B.Sc. with honors in Physics, London U., 1974; M.Sc. in Radiobiology, Birmingham U., 1975; Ph.D. in Physics, Nottingham U., 1978. Exptl. officer dept. physics U. Nottingham, Eng., 1975-76, postdoctoral research fellow, 1978-80; prin. physicist Technicare Corp., Solon, Ohio, 1980-83; research fellow in physics Mass. Gen. Hosp., Boston, 1980-82, dir. NMR research, 1983; pres. Advanced NMR Systems, Inc., Woburn, Mass., 1983—. Contbr. articles to profl. jours. Mem. Hosp. Physicists Assn. U.K., Am. Assn. Physicists in Medicine, Soc. Magnetic Resonance Imaging, Internat. Soc. Magnetic Resonance, Inst. Physics (U.K.), Soc. Magnetic Resonance in Medicine. Methodist. Subspecialties: Nuclear magnetic resonance (biotechnology); Bioinstrumentation. Current work: Development and clinical applications of nuclear magnetic resonance imaging technology.

PYLE, JAMES LAWRENCE, organic chemist; b. Wilmington, Ohio, Aug. 5, 1938; s. Lawrence D. and Mary Hisle (Gowin) P.; m. Elisabeth Anne Walter, Aug. 6, 1960; children: Jennifer, Laura, Stephen, Edward, James Scott. B.S., Ohio U., 1960; Ph.D., Brown U., 1967. Research assoc. U. Calif.-Davis, 1964-66; mem. faculty dept. chemistry Miami U., Oxford, Ohio, 1966-83, dir. research, 1975-83; rep. Ohio Inter-Univ. Energy Research Council, Columbus, 1977-83; prof. Ball State U., Muncie, Ind., 1983—. Author: Chemistry and the Technological Backlash, 1974; contbr. articles to profl. jours. Chmn. sect. on campus ministry and higher edn. for fin. and property W. Ohio Conf., United Methodist Ch., Columbus, 1977-83. NSF grantee, 1980; Exxon Edn. Found. grantee, 1982. Mem. Am. Chem. Soc., AAAS, Nat. Council Univ. Research Adminstrs., Ohio Acad. Scis., N.Y. Acad. Scis., Phi Beta Kappa, Omicron Delta Kappa. Methodist. Subspecialty: Organic chemistry. Current work: Organic chemistry, synthesis, toxic and hazardous substances, energy, chemical education, research adminstration, higher education. Home: 4301 W University Ave Muncie IN 47304 Office: Ball State U Research Office Muncie IN 47306

QUADE, DANA, biostatistics educator; b. Cardston, Alta., Can., Jan. 11, 1935; s. Edward Schaumberg and Sylvia Pauline (Anthony) Q.; m. Erna Adriana Goetz, Aug. 18, 1962; children: Jonathan, Christopher, Ingrid. B.A., UCLA, 1955; Ph.D., U. N.C., 1960. Research assoc. dept. stats. U. N.C., Chapel Hill., 1960, faculty mem., 1962—, prof. biostats., 1970—, assoc. dean, 1981—; sr. asst. scientist USPHS, Ctr. Disease Control, Atlanta, 1960-61, Nat. Inst. Neurol. Diseases and Blindness, Bethesda, Md., 1961-62. Recipient Research Career Devel. award Nat. Inst. Gen. Med. Scis., NIH, 1968-73. Fellow Am. Statis. Assn.; mem. Biometric Soc., Inst. Math. Stats. Lutheran. Subspecialty: Statistics. Current work: Nonparametric statistics. Home: 2119 Markham Dr Chapel Hill NC 27514 Office: Dept Biostats U NC Sch Pub Health Chapel Hill NC 27514

QUAN, STUART FUN, medical educator, pulmonary physiologist; b. San Francisco, May 16, 1949; s. Stuart Fun and Mabel (Wing) Q.; m. Diana Lee, Dec. 18, 1971; children: Jason Stuart, Jeremy Ryan-Stuart. A.B., U. Calif.-Berkeley, 1970; M.D., U. Calif.-San Francisco, 1974. Diplomate: Am. Bd. Internal Medicine (pulmonary subsplty.). Intern U. Wis.-Madison, 1974-75, resident in internal medicine, 1975-77; fellow in critical care medicine U. Calif.-San Francisco, 1977-78, fellow in emergency medicine, 1978-79; fellow in pulmonary medicine U. Ariz., Tucson, 1979-80, instr. medicine, 1980-81, asst. prof., 1981—. Co-author: Respiratory Disorders: A Pathophysiologic Approach, 1983. Ariz. Thoracic Soc. grantee, 1981. Mem. Am. Thoracic Soc., Soc. Critical Care Medicine, Am. Fedn. Clin. Research, Ariz. Thoracic Soc. Club: La Mariposa (Tucson). Subspecialties: Critical care; Pulmonary medicine. Current work: Physiologic effects of chemical mediators on lung function, especially prostaglandins; effects of high frequency ventilation. Home: 5712 E Waverly Tucson AZ 85712 Office: 1501 N Campbell Tucson AZ 85724

QUARLES, JOHN MONROE, microbiologist, educator, researcher; b. Chattanooga, May 24, 1942; s. John Monroe and Helen Marie (Taylor) Q.; m. Joan Engelhardt, Aug. 26, 1971; 1 son, Bryan Stephen. Ph.D., Mich. State U., 1973. Research microbiologist Center for Disease Control, Atlanta, 1965-66; postdoctoral researcher (Oak Ridge Nat. Labs.), 1973-76; asst. prof. med. microbiology Coll. Medicine, Tex. A&M U., 1976-82, assoc. prof., 1982—. Contbr. articles to sci. jours. Served to lt. USN, 1966-69. Mem. Am. Soc. Microbiology, Tissue Culture Assn., Am. Soc. Virology, Sigma xi. Subspecialties: Virology (medicine); Microbiology (medicine). Current work: Research in virology, antiviral chems. and vaccines for influenza, finluenza virus, antiviral chems., vaccines, matrix perfusion, systems, cell culture, interferon. Home: 2012 Langford College Station TX 77840 Office: Dept Med Microbiology Coll Medicine Tex A&M U College Station TX 77843

QUARTERMAIN, DAVID, experimental psychologist, physiologist; b. N.Z., Dec. 15, 1929; came to U.S., 1964; s. Raynal Parkyn and Christina Jesse (McKenzie) Q.; m. Mary Diana Patterson, June 29, 1957; children: Linda Jane, Michael David. B.A., U. N.Z., Christchurch, 1955, M.A. with 1st class honors in Psychology, 1957, Ph.D., 1962. Jr. lectr. Victoria U. of Wellington, 1957-59; lectr. to sr. lectr. U. Auckland, 1960-63; research fellow Yale U., New Haven, 1964-66; asst. prof. Rockefeller U., N.Y.C., 1967-70; assoc. prof. to prof. neurology N.Y.U. Sch. Medicine, 1970-76, prof. neurology and physiology, 1976—, dir. div. behavioral neurology, 1977—. Contbr. numerous articles on neurosci. and physiol. psychology to profl. jours. Mem. Soc. Neurosci., Am. Psychol. Assn. Subspecialties: Learning; Physiological psychology. Home: 72 Quinn Rd Briarcliff Manor NY 10510 Office: 341 E 25th St New York NY 10010

QUATE, CALVIN FORREST, engineering educator; b. Baker, Nev., Dec. 7, 1923; s. Graham Shepard and Margie (Lake) Q.; m. Dorothy Marshall, June 28, 1945; children: Robin, Claudia, Holly, Rhodalee. B.S. in Elec. Engring, U. Utah, 1944; Ph.D., Stanford U., 1950. Mem. tech. staff Bell Telephone Labs., Murray Hill, N.J., 1949-58; dir. research Sandia Corp., Albuquerque, 1959-60, v.p. research, 1960-61; prof. dept. applied physics and elec. engring. Stanford (Calif.) U., 1961—, chmn. applied physics, 1969-72, 78-81; assoc. dean Sch. Humanities and Scis., 1972-74, 82-83. Served as lt. (j.g.) USNR, 1944-46. Fellow IEEE, Am. Acad. Arts and Scis., Acoustical Soc.; mem. Nat. Acad. Engring., Nat. Acad. Scis., Am. Phys. Soc., Sigma Xi, Tau Beta Pi. Subspecialty: Electrical engineering. Office: Dept Applied Physics Stanford Stanford CA 94305

QUAY, JOHN FERGUSON, pharmaceutical company research scientist; b. Galion, Ohio, Feb. 29, 1932; s. Paul Blair and Annabelle (Ferguson) Q.; m. Jean Marie Stebbins, June 14, 1957; children: Nana Marie, Katherine Ann, Paul Dale. B.Sc., Ohio State U., 1957, M.Sc., 1958; Ph.D., Ind. U., 1968. Phys. chemist Lilly Research Labs., Indpls., 1959-65, sr. phys. chemist, 1968-76, research scientist, 1976—; instr. chemistry Ind. Central U., Indpls., 1971-73. Served with U.S. Army, 1953-55. Mem. Am. Chem. Soc., Biophys. Soc., Am. Soc. Microbiology, Am. Physiol. Soc., Sigma Xi. Republican. Episcopalian. Subspecialties: Pharmacokinetics; Pharmacology. Current work: Evaluation of absorption, distribution and excretion of new antibiotic and antiviral chemotherapeutic agents in dogs. Home: 8950 Carriage Ln Indianapolis IN 46256 Office: Lilly Research Labs 731 S Alabama St Indianapolis IN 46285

QUAY, STEVEN CARL, pathology and biochemistry educator, physician; b. Coldwater, Mich., Nov. 7, 1950; s. LaGene Madsen and Roberta Jean (Yarrington) Q. B.A., Western Mich. U., 1971; M.S., U. Mich., 1974, Ph.D., 1975, M.D., 1977. Diplomate: Am. Bd. Pathology. Research assoc. MIT, Cambridge, 1976-79; instr. pathology Harvard U., 1979-80; staff physician Stanford (Calif.) U. Med. Ctr., 1981—, asst. prof. pathology, 1980—. Recipient Dean Research award U. Mich. Med. Sch., 1977, Disting. Alumnus award Western Mich. U., 1981. Mem. Am. Soc. for Microbiology, Internat. Acad. Pathology, Biophys. Soc., Am. Assn. Pathology. Subspecialties: Biochemistry (medicine); Pathology (medicine). Current work: Membrane structure-function relationships. Home: 4401 Fair Oaks Ave Menlo Park CA 94025 Office: Dept Pathology Stanford U Sch Medicine Stanford CA 94305

QUEBBEMANN, ALOYSIUS JOHN, pharmacology educator; b. Chgo., Jan. 19, 1933. B.S., U. Alaska, 1960; Ph.D., SUNY at Buffalo, 1968. Asso. prof. pharmacology U. Minn., 1975—. Served with U.S. Army, 1951-54. Mem. Am. Soc. Pharmacology and Exptl. Therapeutics, Am. Soc. Nephrology, Internat. Soc. Nephrology. Subspecialty: Pharmacology. Current work: Renal pharmacology and toxicology. Home: 1564 Fulman St Saint Paul MN 55108 Office: Med Sch U Minn 3-260 Millard Hall Minneapolis MN 55455

QUEK, SWEE-MENG, elec. engr.; b. Republic of Singapore, Dec. 15, 1955; came to U.S., 1972; s. Soo-Tong and Soo-Kiang Q.; m. Yee-Chin Lee, Jan 7, 1981; 1 son: Joo-Kwang Justin. B.S.E.E., Stanford U., 1977; M.S.E.E., 1977. Teaching asst. Stanford U., 1975-76; design engr. Data Gen. Corp., Westboro, Mass., 1977-78; sr. mem. tech. staff, project leader Four Phase Systems, Cupertino, Calif., 1978-81; engring. mgr. Storage Tech. Corp., Santa Clara, Calif., 1981—. Mem. IEEE, Assn. for Computing Machinery, SIGARCH, SIGMICRO. Subspecialties: Computer engineering; Computer architecture. Current work: High speed processors and algorithms. Engineering management and technical design of main-frame computers using VLSI CMOS. Home: 2874 Grafton Way San Jose CA 95148 Office: 3450 Central Expressway Santa Clara CA 95051

QUENEAU, PAUL ETIENNE, engineer, educator; b. Phila., Mar. 20, 1911; s. Augustin L. and Jean (Blaisdell) Q.; m. Joan Osgood Hodges, May 20, 1939; children: Paul Blaisdell, Josephine Downs (Mrs. George Stanley Patrick). B.A., Columbia U., 1931, B.Sc., 1932, E.M., 1933; postgrad. (Evans fellow), Cambridge (Eng.) U., 1934; D.Sc., Delft (Netherlands) U. Tech., 1971. With Internat. Nickel Co., 1934-69, dir. research, 1940-41, 46-48, v.p., 1958-69, tech. asst. to pres., 1960-66, asst. to chmn., 1967-69; vis. scientist Delft U. Tech., 1970-71; prof. engring. Dartmouth Coll., 1971—; cons. engr., 1972—; vis. prof. U. Minn., 1974-75; geographer Perry River Arctic Expdn., 1949; chmn. U.S. Navy Arctic Research Adv. Com., 1957; gov. Arctic Inst. N.Am., 1957-62; mem. engring. council Columbia U., 1965-70; mem. vis. com. MIT, 1967-70. Author: (with Hanson) Geography, Birds and Mammals of the Perry River Region, 1956, Cobalt and the Nickeliferous Limonites, 1971; Editor: Extractive Metallurgy of Copper, Nickel and Cobalt, 1961, (with Anderson) Pyrometallurgical Processes in Nonferrous Metallurgy, 1967, The Winning of Nickel, 1967; Contbr. articles to profl. jours. Bd. dirs. Engring. Found., 1966-76, chmn., 1973-75. Served with U.S. Army, World War II; ETO; col. C.E. (ret.). Decorated Bronze Star; Recipient Egleston medal Columbia U., 1965. Fellow Metall. Soc. of AIME (dir. 1964, 68-71, pres. 1969, Extractive Metallurgy Lecture award 1977); mem. AIME (Douglas Gold medal 1968, v.p. 1970, dir. 1968-71), Nat. Soc. Profl. Engrs., Nat. Acad. Engring., Can. Inst. Mining and Metallurgy, Inst. Mining and Metallurgy U.K. (overseas mem. council 1970-80, Gold medal 1980), Australasian Inst. Mining and Metallurgy, Sigma Xi, Tau Beta Pi. Subspecialties: Metallurgical engineering; Metallurgy. Current work: Research and development in the extraction of metals from their ores, for improved productivity and environmental conservation. Patentee in field. Home: Cornish NH 03746 Office: Thayer Sch Engring Dartmouth Coll Hanover NH 03755

QUIGLEY, MICHAEL J., prosthetics service company executive; b. Chgo., Sept. 4, 1947; s. Daniel Lawrence and Kathleen (Waddick) Q.; m. Nancy Ann, Dec. 30, 1974; children: Kelly Ann, Michael Richard. B.S., N.Y.U.; postgrad. cert., Northwestern U. Cert. prosthetist-orthotist Am. Bd. for Certification in Orthotics and Prosthetics. Staff prosthetist Nat. Acad. Sci., 1974-76; dir. patient engring. and orthotics and prosthetics edn. U. So. Calif., Los Angeles, 1976-79; pres. Oakbrook Prosthetics & Orthotics, Ltd., Oakbrook Terrace, Ill.; mem. faculty Prosthetic-Orthotic Center, Northwestern U. Editor: Jour. Orthotics and Prosthetics. Macomber Fund grantee, 1976. Mem. Am. Acad. Orthotists and Prosthetists (past pres.). Lodge: Rotary. Subspecialties: Prosthetics; Biomedical engineering. Current work: Modular plastic artificial limbs for children; myoelectric systems; prosthetics research and development. Patentee knee pad system. Home: 1303 N Jackson Ave River Forest IL 60305

QUILLIGAN, JAMES JOSEPH, JR., pediatrics educator, medical laboratory director; b. Phila., Oct. 18, 1912; m. Estelle I. Quilligan, Mar. 1, 1944; children: Kathleen, Maureen, Laura, Susan. B.A., Ohio State U., 1936; M.D., U. Cin., 1940. Intern U. Pitts. Med. Ctr., 1940-41; resident in pediatrics Univ. Hosp., Ann Arbor, Mich., 1941-43; mem. faculty U. Mich., Ann Arbor, 1946-50, asst. prof. pediatrics and epidemiology, 1950; assoc. prof. pediatrics U. Tex., Dallas, 1951-54, Loma Linda (Calif.) U., 1954-58, research prof. pediatrics, 1958—, dir. Virus Lab., 1954—; mem. sci. adv. bd. Inst. Cancer and Blood Research. NIH grantee, 1962-82. Mem. AMA (cons. council on drugs 1962—). Subspecialties: Immunology (agriculture); Virology (biology). Home: 2921 San Jacinto San Clemente CA 92672 Office: Pediatrics Dept Loma Linda U Virus Lab Loma Linda CA 92350

QUIMBY, FRED WILLIAM, pathologist, educator, consultant; b. Providence, Sept. 19, 1945; s. Edward Harold and Isabelle Bella (Barber) Q.; m. Cynthia Claire Connelly, Aug. 21, 1965; children: Kelly Ann, Cynthia Jane. V.M.D., U. Pa., 1970; Ph.D., 1974. Accredited U.S. Dept. Agr., 1970, diplomate: Am. Coll. Lab. Animal Medicine, 1980. Fellow Tufts-New Eng. Med. Center, Boston, 1974-75, research assoc. in hematology, instr. surgery, 1975-76; asst. prof. pathology and surgery Tufts Med. Sch., Boston, 1976-79; assoc. prof. pathology Cornell U. Med. Coll., N.Y.C., 1979—, N.Y. State Coll. Vet. Medicine, Ithaca, N.Y., 1979—; dir. Ctr. for Research Animal Resources, Cornell U., Ithaca, 1979—; cons. Harvard U., 1977-79, Sidney Farber Cancer Inst., Boston, 1978-79, St. Elizabeth's Hosp., 1975-79. Contbr. chpts. to books, articles to profl. jours. Mem. AVMA, Am. Assn. Lab. Animal Sci. (Bernard F. Trum award New Eng. br. 1979), Am. Soc. Primatology, N.Y. Acad. Scis. Subspecialties: Animal pathology; Immunology (agriculture). Current work: Genetic control of the immunologic abnormalities associated with autoimmune disease. Home: 700 Warren Rd 19-1F Ithaca NY 14850 Office: NY State Coll Veterinary Medicine 221 VRT Ithaca NY 14853

QUINLAN, DENNIS CHARLES, research cell biologist, biology educator; b. Detroit, Jan. 29, 1943; s. Charles John and Lucy Novella (McKenzie) Q.; m. Kathleen Daly, July 19, 1969; children: Aaron, Emily. B.S., Wayne State U., 1967, M.S., 1968; Ph.D., U. Rochester, 1973. Research assoc. Worcester Found. Exptl. Biology, Shrewsbury, Mass., 1972-76; asst. prof. W.Va. U., Morgantown, 1976-80, assoc. prof. biology, 1980—. NIH research grantee, 1978, 82; Am. Cancer Soc. grantee, 1980. Mem. Am. Soc. Biol. Chemists, AAAS. Subspecialties: Cell and tissue culture; Biochemistry (biology). Current work: Biochemical and structural effects on cultured cells by hormones, growth factors; role of hormones in regulating amino acid transport in rat liver paranchymal cells; membrane-associated protein kinases. Office: Dept Biology W Va U Brooks Hall Morgantown WV 26506

QUINN, ART JAY, vet. clinician, educator, vet. ophthalmic cons.; b. Bennington, Kans., Aug. 2, 1936; s. Arthur Jess and Edith May (Reigle) Q.; m. Judith Marie Feight, June 6, 1964; m. Helen Marie Evans, June 3, 1967. B.S., Kans. State U., 1959, D.V.M., 1961. Diplomate: Am. Coll. Vet. Ophthalmologists. Gen. practice vet. medicine, Albuquerque, 1961-72, 64-67, 69-75; field rep. Am. Animal Hosp. Assn., South Bend, Ind., 1968-69; assoc. prof. dept. medicine and surgery Coll. Vet. Medicine, Okla. State U., Stillwater, 1975—; adj. asst. prof. medicine dept. ophthalmology Coll. Medicine-McGee Eye Inst., Oklahoma City. Contbr. articles to profl. jours. Served to capt. U.S. Army, 1962-66. Recipient Kans. Vet. Med. Assn. Small Animal Proficiency award, 1961; Upjohn award, 1961; Sarkey Found. grantee, 1981. Mem. AVMA, Am. Animal Hosp. Assn., Am. Coll. Vet. Ophthalmologists, Am. Soc. Vet. Ophthalmology, Internat. Soc. Vet Ophthalmology, Am. Assn. Vet. Clinicians, North Central Okla. Vet. Med. Assn. Democrat. Current work: Comparative ophthalmology, veterinary clinician and educator with specialty in veterinary ophthalmology. Home: 1820 August Stillwater OK 74074 Office: Coll Ver Medicine Okla State U Stillwater OK 74078

QUINN, HELEN RHODA, physicist; b. Melbourne, Australia, May 19, 1943; m., 1966; 2 children. B.S., Stanford U., 1963, M.S., 1964, Ph.D. in Physics, 1967. Research assoc. in physics Stanford (Calif.) Linear Accelerator, 1967-68, vis. scientist, 1977-78, research assoc., 1978-79, mem. staff, 1979—; guest scientist Deutsches Elektronen Synchroton, Hamburg, Germany, 1968-70; research fellow Harvard U., 1971-72, asst. prof., 1972-76, assoc. prof. physics, 1976-77; vis. assoc. prof. physics Stanford U., 1976-78. Mem. Am. Phys. Soc. Subspecialty: Theoretical physics. Office: Stanford Linear Accelerator Ctr PO Box 4349 Stanford CA 94305

QUINN, JAMES ALLEN, research plant pathologist; b. Gary, Ind., Mar. 29, 1954; s. Gerald N. and Helen L. (Sparks) Q.; m. Mildred A. Creager, Sept. 11, 1976; 1 son: John N. B.S. cum laude, Ohio U., 1975, M.S., 1978; Ph.D., Ohio State U., 1980. Research plant pathologist Rohm and Haas Co., Springhouse, Pa., 1980—. Mem. Council of Southside Orgns., Columbus, Ohio, 1979-80; trustee Livingston Park Neighborhood, 1979-80. Mem. Am. Phytopath. Soc., Mycol. Soc. Am., Can. Phytopath. Soc., AAAS. Subspecialty: Plant pathology. Current work: Fungicides, bactericides, structure-activity relationships, mathematical modelling. Home: 216 Lower Valley Rd North Wales PA 19454 Office: Research Lab Rohm & Haas Co Springfield PA 19477

QUINN, JAMES AMOS, JR., biology educator, ecologist, researcher; b. Chickasha, Okla., Aug. 12, 1939; s. James Amos and Esther Ann (Roth) Q. B.S., Panhandle State U., 1961; M.S., Colo. State U., 1963, Ph.D. (NSF fellow), 1966. Asst. prof. botany Rutgers U., New Brunswick, N.J., 1966-71, assoc. prof., 1971-77, prof. biol. scis., 1977—, dir. grad.program botany and plant physiology, 1983-86, assoc. chmn. for personnel dept. biol. scis., 1981-82, mem. exec. council Grad. Sch., 1983-86; sec. Rutgers Coll. Biol. Scientists, 1969-71, 1973-75; instr. N.J. Conservation Commn., 1970; research fellow U. New Eng., N.S.W., Australia, 1981; cons. Ont. (Can.) Council Grad. Studies, 1982. Contbr. articles on biology and ecology to profl.

jours.; editorial bd.: Am. Jour. Botany, 1980-82; editoral bd.: Bull. Torrey Bot. Club, 1983—. NSF grantee, 1968-70; Rockefeller grantee, 1970; Rutgers Research Council faculty fellow, Australia, 1972-73. Mem. Am. Forage and Grassland Council, Am. Inst. Biol. Scis., Bot. Soc. Am. (vice-chmn. 1977, chmn. 1978, ecol. sect.), Ecol. Soc. Am., N.J. Acad. Sci. (council 1972-76, treas. and exec. com. 1976-79, 1979-81), Soc. Range Mgmt., Torrey Bot. Club (council 1970-82, pres. 1982-83), Sigma Xi. Subspecialties: Ecology; Population biology. Current work: Adaptive differences among populations of a species along environmental gradients, evolution of pollution-tolerant populations, reproductive biology of amphicarpic species, life-histories and sex ratios in populations of dioecious species. Office: Dept Biol Scis Rutgers U Piscataway NJ 08854

QUINN, JAMES GERARD, oceanography educator; b. Providence, Oct. 28, 1938; s. Walter E. and Mary Z. (Victory) Q.; m. Vivian J. Dion, Nov 20, 1965; children: Kevin, Jennifer, Brian. B.S., Providence Coll., 1960; M.S., U. R.I., 1964; Ph.D., U. Conn., 1967. Faculty Grad. Sch. Oceanography, U. R.I., Kingston, 1968—, prof., 1978—. Mem. Am. Chem. Soc., AAAS, Am. Geophys. Union, Geochem. Soc., Sigma Xi. Subspecialties: Organic geochemistry; Oceanography. Current work: Organic chemistry of seawater and marine sediments. Address: Grad Sch Oceanography Univ RI Kingston RI 02881

QUINN, JOHN ALBERT, chemical engineering educator, consultant; b. Springfield, Ill., Sept. 3, 1932; s. Edward Joseph and Marie Regina (Von de Bur) Q.; m. Frances W. Daly, June 22, 1957; children: Sarah D., Rebecca V., John E. B.S., U. Ill., 1954; Ph.D., Princeton U., 1959. Asst. prof. dept. chem. engring. U. Ill., Urbana, 1958-64, assoc. prof., 1964-66, prof., 1966-70; prof. dept. chem. engring. U. Pa., Phila., 1971—, chmn. dept., 1980—, Robert D. Bent prof., 1978—; adj. prof. U. Technologie de Compiegne, 1974—; 6th Mason lectr. Stanford U., 1981; OAS lectr., Argentina, 1981. Editorial bd.: Jour. Membrane Sci, 1976—, Rev. Chem. Engring, 1982—; contbr. articles to profl. jours.; patentee in field. Recipient S. Reid Warren award U. Pa., 1974, award Alpha Chi Sigma, 1978; NSF sr. postdoctoral fellow, 1965. Mem. Am. Inst. Chem. Engrs. (Allan P. Colburn award 1966), Am. Chem. Soc., AAAs, Internatl. Soc. on Oxygen Transfer to Tissue. Subspecialties: Chemical engineering; Enzyme technology. Current work: Synthetic membranes and the application of membranes in chemical processes, bioseparations, biochemical engineering. Office: U Pa 311A Towne Bldg D3 Philadelphia PA 19104

QUINTON, PAUL MARQUIS, physiologist, educator; b. Houston, Sept. 17, 1944; s. Curtis Lincoln and Mercedes Genale (Danley) Q.; m. Bonnie Sue Casey, Aug. 5, 1967; 1 son, Marquis Casey. B.A., U. Tex., 1967; Ph.D., Rice U., 1971. Research physiologist UCLA Med. Sch., 1973-75, asst. prof. medicine/physiology, 1975-79, assoc. prof. physiology, 1981—; asst. prof. biomedicine U. Calif.-Riverside, 1979-81, assoc. prof. biomedicine, 1981—; vis. prof. physiology Harvard Med. Sch., Boston, 1978-79; mem. adv. bd. Calif. Sec. Health, 1979-82; mem. med. adv. council, chmn. cystic fibrosis com. Nat. Cystic Fibrosis Found. Editor: Fluid and Electrolyte Abnormalities in Exocrine Glands in Cystic Fibrosis, 1982. Nat. Cystic Fibrosis Found. research scholar, 1973; recipient NIH Research Career Devel. award, 1977. Mem. Am. Physiol. Soc., Am. Cell Biology Soc., Microbeam Analysis Soc. Democrat. Subspecialty: Physiology (medicine). Current work: Physiology of fluid and electrolyte transport research; abnormalities of fluid movements in cystic fibrosis; physiology of exocrine glands, with emphasis on eccrine sweat glands. Home: 5173 Colina Riverside CA 92507 Office: Biomedical Scis U Calif Riverside CA 92521

QUIRION, REMI, pharmacologist; b. Lac-Drolet, Que., Can., Jan. 9, 1955; s. Joseph and Fernande (Lessard) Q.; m. Pierrette Gaudreau, July 26, 1980; 1 son: Sylvain. B.Sc., U. Sherbrooke, 1976, M.Sc., 1977, Ph.D., 1980. Med. Research Council of Can. fellow NIMH-NIH, Bethesda, Md., 1980-82, pharmacologist, 1982—; vis. lectr. U. Calgary, Can., 1981. Conseil de la Recherche Ensante du Quebec scholar, 1978-80. Mem. AAAS, Am. Soc. Neurosci., Internat. Narcotics Conf. Subspecialties: Neuropharmacology; Pharmacology. Current work: Neuropeptides pharmacology, brain receptors, autoradiography, bioassay; drug abuse. Office: Neurosci Branch NIMH NIH Bethesda MD 20205 Home: Apt 101 13214 Twinbrook Pkwy Rockville MD 20851

QUIRK, RODERIC PAUL, polymer research chemist, educator; b. Detroit, Mar. 26, 1941; s. Raymond Paul and Margaret Katherine (MacBeth) Q.; m. Donna Jeanne Duncan, Aug. 20, 1962; children: Scott Duncan, Brian MacBeth, Marion Maxwell. B.S. cum laude in Chemistry, Rensselaer Poly. Inst., 1963; M.S. in Chemistry, U. Ill.-Champaign-Urbana, 1965, Ph.D., 1967. Research assoc. U. Pitts., 1967-69; asst. prof. organic chemistry U. Ark., Fayetteville, 1969-74, assoc. prof., 1974-79; research scientist Mich. Molecular Inst., Midland, 1979-80, sr. scientist, 1980—; cons. in field; adj. assoc. prof. macromolecular sci. Case Western Res. U., Cleve., 1979—; adj. assoc. prof. chemistry Central Mich. U., Mount Pleasant, 1980—. Contbr. numerous articles on chemistry to profl. publs. NSF grantee, 1981-82; Petroleum Research Fund grantee, 1969-73, 74-76. Mem. Am. Chem. Soc., Sigma Xi, Phi Lambda Upsilon. Unitarian. Subspecialties: Polymer chemistry; Organic chemistry. Current work: Organic polymer synthesis; organometallic polymerization catalysts; mechanism of anionic polymerization. Office: 1910 W Saint Andrews St Midland MI 48640

QUIRKE, TERENCE THOMAS, JR., geologist, systems analyst; b. Mpls., Aug. 18, 1929; s. Terence Thomas and Anne Laura (McIlraith) Q.; m. Ruth Carter, Jan. 18, 1958; 1 dau. Grace Anne. B.S., U. Ill.-Urbana, 1951; M.S., U. Minn., 1953, Ph.D., 1958. Research geologist Internat. Nickel Co. Can. Ltd., Thompson, Man., 1960-69, asst. mgr., 1969-71, mgr., 1971-75; dist. geologist Am. Copper and Nickel Co., Milw., 1975-79, supervisory sr. staff geologist, Denver, 1979. Author: (with others) International Geological Congress Field Guide Book, 1972. Fellow Geol. Soc. Am., Geol. Assn. Can.; mem. Soc. Econ. Geologists, Can. Inst. Mining and Metallurgy, AIME, Mensa, Intertel, Sigma Xi. Episcopalian. Subspecialty: Information systems, storage, and retrieval (computer science). Current work: Information systems to promote and support active exploration for base and precious metals in United States. Office: Am Copper & Nickel Co 1726 Cole Blvd Suite 110 Golden CO 80401 Home: 2310 Juniper Ct Golden CO 80401

RAAB, HARRY FREDERICK, JR., nuclear power administrator; b. Johnstown, Pa., May 9, 1926; s. Harry Frederick and Marjorie (Stiff) R.; m. Phebe Ann Duerr, June 16, 1951; children: Constance Diane, Harry Frederick III, Cynthia Ann Raab Morgenthaler. B.S. in Elec. Engring, M.I.T., 1951, M.S., 1951. Reactor control engr. Bettis Atomic Power Lab, West Mifflin, Pa., 1951-54, mgr. surface physics, 1955-61, mgr. light water breeder physics, 1962-71; chief physicist Navy Nuclear Power Directorate, Washington, 1972—. Lay reader Episcopal Ch., 1957—; stewardship chmn. Episc. Ch. of Good Shepherd, Burke, Va., 1979-82, sr. warden, Burke, Va., 1983. Served with USN, 1944-46. Fellow Am. Nuclear Soc.; mem. Sigma Xi, Tau Beta Pi, Eta Kappa Nu. Republican. Subspecialties: Nuclear fission; Nuclear physics. Patentee (5) on water breeders, 1965, 75. Home: 8202 Ector Ct Annandale VA 22003 Office: Naval Sea Systems Command 08 Washington DC 20360

RAAF, JOHN HART, surgeon, cancer researcher; b. Portland, Oreg., Aug. 10, 1941; s. John E. and Lorene (Rardin) R.; m. Heather N. Neilson, June 15, 1965; children: Jennifer, John, Sabrina. A.B. magna cum laude, Harvard U., 1963, M.D. cum laude, 1970; D.Phil. (Rhodes scholar), Oxford (Eng.) U., 1966. Diplomate: Am. Bd. Surgery. Surg. intern, resident Mass. Gen. Hosp., 1970-73, 75-77; research fellow in immunology Sloan Kettering Inst., 1973-75; asso. prof. surgery Cornell U. Med. Coll., 1981—; faculty asso. M.D. Anderson Hosp., Houston, 1977-78, asst. prof., 1978-79. Contbr. articles to profl. jours.; Editorial bd.: Clin. Bul., 1978-82; editor-in-chief: Your Patient and Cancer, 1981—. Jr. faculty clin. fellow Am. Cancer Soc., 1980—. Mem. Am. Assn. Med. Systems and Informatics, Am. Assn. Cancer Research, AAAS, AMA, Assn. Acad. Surgery, Biochem. Soc. (London), Am. Soc. Clin. Oncology, Soc. Surg. Oncology (publs. com.). Subspecialties: Surgery; Oncology. Current work: Cancer biology, tumor immunology, vascular access for chemotherapy. Home: 47 Woodland Park Dr Tenafly NJ 07670 Office: 1275 York Ave New York NY 10021

RAAFAT, FERAIDOON (FRED RAAFAT), engineering educator; b. Tehran, Iran, Jan. 5, 1953; came to U.S., 1967; s. Manucher and Parvaneh R.; m. Hamideh Oloomi, Aug. 7, 1977; 1 child, Guissu-Naz. B.S., Phillips U., Enid, Okla., 1974, Okla. State U., 1975, M.I.E., 1977, Ph.D., 1982. Registered profl. engr., Okla. Jr. engr. Williams Bros. Engring. Co., Tulsa, Okla., 1976; grad. teaching asst. Okla. State U., 1977-81; asst. prof. Wichita (Kans.) State U., 1981—. Mem. Ops. Research Soc. Am., Inst. Mgmt. Sci., Am. Inst. Decision Sci., Am. Inst. Indsl. Engring., Am. Prodn. and Inventory Control Soc. (dir. Wichita chpt. 1981—). Subspecialties: Industrial engineering; Operations research (engineering). Current work: Production/operations management; inventory theory, simulation. Office: Wichita State U Wichita KS 67208

RAAFAT, FRED *See also* **RAAFAT, FERAIDOON**

RAAM, SHANTHI, cancer researcher, immunochemist, consultant; b. Madras, India, Nov. 26, 1941; came to U.S., 1964, naturalized, 1976. B.S., Madras U., 1960, M.S., 1962; Ph.D., U. Ga., Athens, 1973. Research asst. in parasitology U. Tenn., 1964-65; teaching asst. dept. microbiology Med. Coll. Ga., Augusta, 1968-70, U. Ga., 1970-73; postdoctoral fellow Cancer Research Center, Tufts U., Boston, 1973-75; chief oncology lab. Tufts U. Med. Cancer unit Lemuel Shattuck Hosp., Boston, 1975-77, dir. oncology lab., 1977—; research assoc. dept. medicine Tufts U.; cons. NIH, New Eng. Nuclear, Leary Labs. Contbr. articles to profl. jours. Mem. LWV. Grantee Am. Cancer Soc., N.Y., 1979-83. Subspecialties: Oncology; Receptors. Current work: The significance of steroid hormone receptors in cancers of the human breast, uterus and the brain (immunology of estrogen receptors of human breast carcinoma). Office: Tufts University Medical Center Unit Lemuel Shattuck Hosp 170 Morton St Boston MA 02130

RABI, ISIDOR ISAAC, emeritus physics educator; b. Austria, July 29, 1898; came to U.S. in infancy; s. David and Jennie (Teig) R.; m. Helen Newark, 1926; children: Nancy Elizabeth, Margaret Joella. B.Chem., Cornell U., 1919; Ph.D., Columbia U., 1927, Sc.D., 1968; grad. study in, Munich, Copenhagen, Hamburg, Leipzig, Zurich, 1927-29; D. Sc. (hon.), Princeton, 1947, Harvard, 1955, Williams Coll., 1958, U. Birmingham, 1960, Clark U., 1962, Adelphi Coll., 1962, Technion, 1963, Franklin Marshall Coll., 1964, Brandeis U., 1965, U. Coimbra, Portugal, 1966, Hebrew U., Jerusalem, 1972, Coll. City N.Y., 1977, Bates Coll., 1977, L.I. U., 1977; L.H.D., Hebrew Union Coll., Cin., 1958, Oklahoma City U., 1960; LL.D., Dropsie Coll., 1956; D.H.L., Yeshiva U., 1964; Litt. D., Jewish Theol. Sem., 1966. Tutor in physics Coll. City N.Y., 1924-27; lectr. Columbia U., 1929-30, asst. prof., 1930-35, asso. prof., 1935-37, prof., 1937-64, Higgins prof., 1950-64, U. prof., 1964-67, U. prof. emeritus, 1967—, also exec. officer, dept. physics, 1945-49; Karl Taylor Compton vis. prof. physics, 1968-71; lectr. U. Mich., summer 1936; Stanford U., summer 1938; cons. sci. adv. com. Ballistic Research Lab., Aberdeen, 1939-65; mem. sci. bd. Itek Corp.; Staff mem. and asso. dir. Radiation Lab., M.I.T., 1940-45; mem. gen. adv. com. AEC, 1946—, chmn., 1952-56, cons., 1956—; chmn. sci. adv. Com. ODM, 1953-57; cons. Dept. State, 1958—; mem. Naval Research Adv. Com., 1952—; cons., sci. adv. com., v.p. Internat. Conf. on Peaceful Uses Atomic Energy, Geneva, 1955, 58, 64, v.p. UN Conf. on Peaceful Uses Atomic Energy, 1971—; U.S. rep. adv. com. to sec. gen. UN, 1955—; mem. President's Sci. Adv. Com., 1957-68, chmn., 1957; cons. Los Alamos Sci. Lab., 1943-45, 56—; mem. NATO Sci. Com., 1958—; cons. Research and Devel. Bd., 1946-49; U.S. del. UNESCO Conf., Florence, Italy, 1950; mem. U.S. Nat. Commn. UNESCO, 1950-53, 58, UN Sci. Com., 1954—, IAEA Sci. Com., 1958-72; vis. prof. Rockefeller U. (formerly Inst.), 1957-79; Shreve fellow Princeton, 1961-62; Karl Taylor Compton lectr. Mass. Inst. Tech., 1962; gen. adv. com. ACDA, 1962—. Author: My Life and Times as a Physicist, 1960, Science: The Center of Culture, 1970; Asso. editor: Physical Review, 1935-38, 1941-44; Contbr. to sci. jours. in field. Trustee Assoc. Univs., Inc., 1946—, pres., 1961-62, chmn. bd., 1962-63; bd. govs. Weizmann Inst. Sci., Rehovoth, Israel, 1949—; trustee Mt. Sinai Hosp., 1960—. Served in S.A.T.C., 1918. Decorated Officer French Legion of Honor, 1956; comdr., 1968; Barnard fellow, 1927-28; Internat. Ednl. Bd. fellow, 1928-29; Ernest Kempton Adams fellow, 1935; Sigma Xi Semicentennial prize for physical scis., 1936; Henrietta Szold award, 1956; 1,000 prize from AAAS, for study of radio frequency spectra of atoms and molecules, 1939; Elliot Cresson medal of Franklin Inst., 1942; Nobel Prize in Physics, 1944; Barnard medal, 1960; U.S. Medal for Merit, 1948; King's Medal, British, 1948; Comdr., Order So. Cross, Brazil, 1952; Priestley Meml. award Dickenson Coll., 1964; Niels Bohr Internat. Gold Medal, 1967; co-recipient Atoms for Peace award, 1967; Tribute of Appreciation State Dept., 1978; Pupin Gold medal Columbia U. Sch. Engring., 1981. Fellow Am. Phys. Soc. (pres. 1950-51); mem. Council on Fgn. Relations, Am. Philos. Soc., Japan Acad. Sci. (fgn. mem.), Nat. Acad. Scis., N.Y. Acad. Scis., Sigma Xi. Clubs: Cosmos (Washington); Faculty (Columbia U.); Athenaeum (London). Subspecialty: Nuclear physics. Home: New York NY 10027

RABIN, MICHAEL O., computer science and mathematics educator; b. Breslau, Ger., Sept. 1, 1931; s. Israel A. and Else (Hess) R.; m. Ruth Scherzer, May 31, 1954; children: Tal, Sharon. M.S. in Math, Hebrew U. of Jerusalem, 1953, Ph.D., Princeton U., 1956. Instr. Princeton U., 1956-57, mem., Princeton, N.J., 1957-58; vis. prof. U. Calif.-Berkeley, 1962, MIT, 1963, 72-78; U. Paris, 1965, Yale U., 1967, N.Y.U., 1970; Albert Einstein prof. Hebrew Univ. of Jerusalem, 1965—; prof. Harvard U., 1981—; cons. computer industry. Recipient Rothschild prize in Math. Rothschild Found., 1974; A. M. Turing Award in Computer Sci., 1976; Harvey prize Sci. and Tech. Technion Soc., 1980. Subspecialties: Algorithms; Cryptography and data security. Current work: Theory of algorithms, randomizing algorithms, computer security.

RABINER, LAWRENCE RICHARD, electrical engineer; b. Bklyn., Sept. 28, 1943; s. Nathan Marcus and Gloria Hannah (Bodinger) R.; m. Suzanne Login, June 23, 1968; children: Sheri, Wendi, Joni. B.S., M.I.T., 1964, M.S., 1964, Ph.D., 1967. With Bell Labs., Murray Hill, N.J., 1962—, supr. human-machine voice communication group, 1970—; cons. U.S. govt. Author: Theory and Application of Digital Signal Processing, 1975, Digital Processing of Speech Signals, 1979, Multirate Digital Signal Processing, 1983; contbr. articles to profl. jours. Mem. IEEE (award for paper 1971, Piori award 1979), Acoustical Soc. Am. (Biennial award 1974). Jewish. Subspecialty: Computer engineering. Current work: Speech recognition by machine. Patentee. Home: 58 Sherbrook Dr Berkeley Heights NJ 07922 Office: Bell Laboratories Room 2D533 Murray Hill NJ 07974

RABINOW, JACOB, electrical engineer; b. Kharkov, Russia, Jan. 8, 1910; came to U.S., 1921, naturalized, 1930; s. Aaron and Helen (Fleisher) Rabinovich; m. Gladys Lieder, Sept. 26, 1943; children: Jean Ellen, Clare Lynn. B.S. in Elec. Engring, Coll. City N.Y., 1933, E.E., 1934; D.H.L. (hon.), Towson State U., 1983. Radio serviceman, N.Y.C., 1934-38; mech. engr. Nat. Bur. Standards, Washington, 1938-54; pres. Rabinow Engring. Co., Washington, 1954-64; v.p. Control Data Corp., Washington, 1964-72; research engr. Nat. Bur. Standards, 1972—; Regent's lectr. U. Calif., Berkeley, 1972; lectr., cons. in field. Author. Recipient Pres.'s Certificate of Merit, 1948; certificate appreciation War Dept., 1949; Exceptional Service award Dept. Commerce, 1949; Edward Longstreth medal Franklin Inst., 1959; Jeffersom medal N.J. Patent Law Assn., 1973; named Scientist of Yr. Indsl. R&D mag., 1980. Fellow IEEE (Harry Diamond award 1977), AAAS; mem. Nat. Acad. Engring., Philos. Soc. Washington, Audio Engring. Soc., Sigma Xi. Club: Cosmos (Washington). Subspecialties: Electrical engineering; Mechanical engineering. Current work: Study of invention;patents; innovation and productivity; research and develeopment in robotics; sound reproduction; post office automation. Patentee in field. Home: 6920 Selkirk Dr Bethesada MD 20817 Office: Nat Bur Standards Washington DC 20234 I believe that inventions enrich both the material wealth and the cultural life of a nation and the world. Being a product of original thought, they are an art form and should be supported as such.

RABINOWITZ, JESSE CHARLES, biochemist, educator; b. N.Y.C., Apr. 28, 1925; s. Julius and Frances (Pincus) R. B.S., Poly. Inst. Bklyn., 1945; Ph.D., U. Wis., Madison, 1949. Chemist NIH, Bethesda, Md., 1953-57; assoc. prof. biochemistry U. Calif., Berkeley, 1957-63, prof., 1963—, chmn. dept. biochemistry, 1978-83. Guggenheim fellow, 1977-78. Mem. Am. Soc. Biol. Chemists, Am. Soc. Microbiology, Am. Chem. Soc., Nat. Acad. Scis. Subspecialty: Biochemistry (biology). Home: 321 Vassar Ave Berkeley CA 94708 Office: Dept Biochemistry U Calif Berkeley CA 94720

RABJOHN, NORMAN, educator; b. Rochester, N.Y., May 1, 1915; s. Alfred A. and Elizabeth Mary (Hooper) R.; m. Dora Isabella Taylor, Sept. 9, 1943; 1 son, James Norman. B.S., U. Rochester, 1937; M.S., U. Ill., 1939, Ph.D., 1942. Chemist Eastman Kodak Co., Rochester, N.Y., 1937-38; instr. chemistry U. Ill., Urbana, 1942-44; research chemist Goodyear Tire & Rubber, Akron, Ohio, 1944-48; prof. chemistry U. Mo., Columbia, 1948—; dir. ABC Labs., Columbia, 1970—, Organic Syntheses Inc., Detroit, 1950-81; cons. Gulf Research and Devel., Kansas City, Mo., 1952-69, McNeil Labs., Phila., 1949-59. Editor: Organic Syntheses, Coll. Vol. 4, 1963; contbr. articles to profl. jours. Fellow AAAS; mem. Am. Chem. Soc., Phi Beta Kappa, Phi Lambda Upsilon, Omicron Delta Kappa. Republican. Subspecialties: Organic chemistry; Medicinal chemistry. Current work: Structure-property relationship of branched alkanes; physiologically active organic molecules. Home: 100 E Ridgeley Rd Columbia MO 65201 Office: Dept Chemistry Univ Mo Columbia MO 65211

RABKIN, MITCHELL THORNTON, physician, hospital administrator, educator; b. Boston, Nov. 27, 1930; s. Morris Aaron and Esther (Quint) R.; m. Adrienne M. Najarian, June 24, 1956; children: Julia Margaret, David Gregory. A.B. magna cum laude, Harvard U., 1951, M.D. cum laude, 1955; D.Sc. (hon.), Brandeis U., 1983. Intern Mass. Gen. Hosp., Boston, 1955-56, resident in internal medicine, 1956-57, 59-60, chief resident, 1962, mem. staff, 1963-72, bd. consultation, 1972-80, hon. physician, 1981—; clin. fellow NIH, Bethesda, Md., 1957-59; gen. dir. Beth Israel Hosp., Boston, 1966-80, pres., 1980—; asst. prof. medicine Harvard U., 1969-70, assoc. prof., 1971-83, prof., 1983—; vis. lectr. Harvard Sch. Public Health, 1977—; pres., dir. Commonwealth Health Care Corp. Served with USPHS, 1957-59. Fellow A.C.P.; mem. AAAS, Am. Fedn. Clin. Research, Mass. Med. Soc., Soc. Med. Adminstrs., Hastings Inst. Soc., Ethics and Life Scis., Assn. Am. Med. Colls. (past chmn. Council Teaching Hosps.), Conf. Boston Teaching Hosps. (past chmn.), Inst. Medicine, Nat. Acad. Scis. Jewish. Clubs: Century Assn. (N.Y.C.); Harvard (Boston). Subspecialty: Medical administration. Office: Beth Israel Hospital Boston MA 02215

RABSON, ALAN SAUL, physician, educator; b. N.Y.C., July 1, 1926; s. Abraham and Florence (Shulman) R.; m. Ruth L. Kirschstein, June 11, 1950; 1 son, Arnold B. B.A., U. Rochester, N.Y., 1948; M.D., State U. N.Y. Downstate, 1950. Intern Mass. Meml. Hosp., Boston, 1951-52; resident in pathology N.Y. U. Hosp., 1952-54, USPHS Hosp., New Orleans, 1954-55; pathologist Nat. Cancer Inst., Bethesda, Md., 1955—; prof. pathology Georgetown U. Med. Sch., 1974—, Uniformed Services U. Health Scis., 1978—; professorial lectr. pathology George Washington U., 1978—. Asso. editor: Am. Jour. Pathology; Contbr. articles to med. jours. Mem. Am. Assn. Pathologists, Phi Beta Kappa, Sigma Xi, Alpha Omega Alpha. Subspecialties: Pathology (medicine); Virology (medicine). Current work: Experimental pathology; viral oncology. Address: Room 3A03 Bldg 31 NIH Bethesda MD 20205

RABUSSAY, DIETMAR PAUL, biochemist; b. Wolfsberg, Karnten, Austria, Aug. 9, 1941; came to U.S., 1972; s. Paul Joseph and Emmy (Morianz) R.; m. Liesel Rosenkranz, Oct. 27, 1966; children: Christoph, Michael. M.Sc., Techn. U., Graz, Austria, 1967; Ph.D., U. Munich, 1971. Asst. research prof. U. Calif.-San Diego, 1975-79; vis. scientist Max Planck Inst. Biochemistry, Munich, 1978-79; assoc. prof. Fla. State U., Tallahassee, 1979-81; sect. head Bethesda Research Labs., Gaithersburg, Md., 1981-82, lab. dir., 1982, dir. research/devel., 1982—; mem. NIH Study sect. for Microbial Genetics and Physiology, Bethesda, 1982—. Contbr. articles to profl. jours. Austrian Govt. fellow, 1962-67; Deutscher Akademisher Austausch Dienst fellow, 1967-70; U. Calif. grantee, 1975-77; NIH grantee, 1980; others. Mem. Am. Soc. Biol. Chemistry, Am. Soc. Microbiology, AAAS, Gesellschaft Deutscher Chemiker, Gesellschaft fur Biologische Chemie, European Molecular Biology Orgn. Roman Catholic. Subspecialties: Molecular biology; Genetics and genetic engineering (biology). Current work: Enzymology and nucleic acid chemistry as related to the manipulation of DNA and gene expression; protein purification; novel approaches to DNA manipulation. Patentee in field. Home: 9 Irish Ct Gaithersburg MD 20878 Office: Bethesda Research Labs 8717 Grovemont Circle Gaithersburg MD 20877

RACE, GEORGE JUSTICE, physician, educator; b. Everman, Tex., Mar. 2, 1926; s. Claude Ernest and Lila Eunice (Bunch) R.; m. Annette Isabelle Rinker, Dec. 21, 1946; children: George William Daryl, Jonathan Clark, Mark Christopher, Jennifer Anne (dec.), Elizabeth Margaret Rinker. M.D., U. Tex., Southwestern Med. Sch., 1947; M.S. in Pub. Health, U. N.C., 1953; Ph.D. in Ultrastructural Anatomy and Microbiology, Baylor U., 1969. Intern Duke Hosp., 1947-48, asst. resident pathology, 1951-53; intern Boston City Hosp., 1948-49; asst. pathologist Peter Bent Brigham Hosp., Boston, 1953-54; pathologist St. Anthony's Hosp., St. Petersburg, Fla., 1954-55; staff pathologist Children's Med. Center, Dallas, 1955-59; dir. labs. Baylor U. Med. Center, Dallas, 1959—, chief dept. pathology, 1959—, vice chmn. exec. com. med. bd., 1970—; cons. pathologist VA Hosp., Dallas, 1955—; adj. prof. anthropology and biology So. Meth. U., Dallas, 1969; instr. pathology Duke, 1951-53, Harvard Med. Sch., 1953-54; asst. prof. pathology U. Tex. Southwestern Med. Sch., 1955-58, clin. asso. prof., 1958-64, clin. prof., 1964-72, prof., 1973—, dir. Cancer Center, 1973-76, asso. dean for continuing edn., 1973—; pathologist-in-chief Baylor U. Med. Center, 1959—; prof. microbiology Baylor Coll. Dentistry, 1962-68, prof. pathology, 1964-68, prof., chmn. dept. pathology, 1969-73; dean A. Webb Roberts Continuing Edn. Center, 1973—; spl. adv. on human and animal diseases Gov. Tex., 1979—. Editor: Laboratory Medicine (4 vols.), 1973, 10th edit., 1983; Contbr. articles to profl. jours., chpts. to textbooks. Pres., Tex. div. Am. Cancer Soc., 1970, chmn. Gov.'s Task Force on Higher Edn., 1981. Served with AUS, 1944-46; from 1st lt. to maj. USAF, 1948-51. Decorated Air medal. Fellow Coll. Am. Pathologists, Am. Soc. Clin. Pathologists, A.A.A.S., Am. Coll. Legal Medicine; mem. A.M.A. (chmn. multiple discipline research forum 1969), Am. Assn. Pathologists, Internat. Acad. Pathology, Am. Assn. Med. Colls., Am. Assn. Cancer Research, Am. Assn. Phys. Anthropologists, Sigma Xi. Subspecialties: Pathology (medicine); Parasitology. Current work: Immunopathology due to animal parasites; Hypertension and adrenal diseases; surgical pathology and oncology. Home: 3429 Beverly Dr Dallas TX 75205

RACKER, EFRAIM, biochemist; b. Neu Sandez, Poland, June 28, 1913; came to U.S., 1941, naturalized, 1947; m. Franziska Weiss, Aug. 25, 1945; 1 dau., Ann Racker Costello. M.D., U. Vienna, Austria, 1938; D.S. (hon.), U. Chgo., 1968. Research asst. in biochemistry Cardiff (South Wales) Mental Hosp., 1938-40; research asso. dept. physiology U. Minn., 1941-42; intern, then resident Harlem Hosp., N.Y.C., 1942-44; asst. prof. microbiology N.Y. U. Sch. Medicine, 1944-52; asso. prof. biochemistry Yale U., 1952-54; chief div. nutrition and physiology Public Health Research Inst. City N.Y., 1954-66; Albert Einstein prof. biochemistry Cornell U., 1966—; mem. biochemistry study sect., div. research grants NIH, 1957-70; chmn. bd. sci. counselors Cancer Inst., 1975-76; 73d Christian A. Herter lectr., 1975, 10th Sir Frederick Gowland Hopkins Meml. lectr., 1975, Walker-Ames lectr., 1977. Author: Mechanism in Bioenergetics, 1965, A New Look at Mechanisms in Bioenergetics, 1976; editor: Membranes of Mitochondria and Chloroplasts, 1970, Energy Transducing Mechanisms, 1975; editorial bd.: Jour. Biol. Chemistry, 1959-72. Recipient Warren Triennial prize Mass. Gen. Hosp., 1974, Nat. Medal Sci., 1976. Mem. Am. Soc. Biol. Chemists, Harvey Soc., Brit. Biochem. Soc., Nat. Acad. Scis., Am. Acad. Arts and Scis., AAAS. Subspecialty: Biochemistry (biology). Home: 305 Brookfield Rd Ithaca NY 14850 Office: Sect Biochemistry Wing Hall Cornell U Ithaca NY 14853

RADFORD, ALBERT ERNEST, botany educator, plant taxonomist; b. Augusta, Ga., Jan. 25, 1918; s. Albert Furman and Eloise Harriet (Moseley) R.; m. Laurie Marguerite Stewart, Oct. 10, 1941; children: David Eugene, John Stewart, Linda Katherine Radford Vinson. B.S., Furman U., 1939; Ph.D. in Botany, U. N.C.-Chapel Hill, 1948. Lab. asst. Furman U., 1938-39; teaching asst. botany U. N.C., Chapel Hill, 1946-49, instr., 1949-53, asst. prof., 1953-58, asso. prof., 1958—, prof., 1960—, now dir. herbarium, sr. prof. Author: Manual of the Vascular Flora of the Carolinas, 1968, Vascular Plant Systematics, 1974, Potential Ecological Natural Landmarks, 1975, Natural Heritage, 1981. Chmn. bd. Highlands Biol. Sta.; trustee N.C. Nature Conservancy. Served to capt. U.S. Army, 1941-46. Decorated Croix de Guerre avec Palme; Order of Leopold, Belgium; Croix de Guerre avec Etoile d'Argent, France; recipient Tanner award U. N.C., 1956; Conservation award-Oak Leaf award Nature Conservancy, 1978. Mem. N.C. Acad. Scis., Assn. Southeastern Biologists (Meritorious Teaching award 1978), Am. Soc. Plant Taxonomists, Internat. Assn. Plant Taxonomists, AAAS. Republican. Presbyterian. Current work: Plant systematics. Home: 111 Purefoy Rd Chapel Hill NC 27514 Office: U N C 402 Coker Hall Chapel Hill NC 27514

RADIN, DAVID N., geneticist, researcher; b. Boston, July 15, 1936; s. Aaron H. and Irene (Lieberman) R. B.A. in Biochemistry, U. Calif., Berkeley, 1970, Ph.D. in Genetics, 1976. Research fellow in biology Calif. Inst. Tech., Pasadena, 1976-78; research asso. Los Angeles State and County Arboretum, Araadia, Calif., 1978-82; research specialist in plant sci. U. Calif.-Riverside, 1979-80; research geneticist U. Calif.-Irvine, 1980—. Contbr. articles to profl. jours. Mem. Genetics Soc. Am., Am. Bot. Soc., Crop Sci. Soc., Plant Molecular Biology Assn. Subspecialties: Gene actions; Plant cell and tissue culture. Current work: Plant developmental genetics; plant tissue culture and somatic cell genetics; biochemistry of plants; crop improvement through genetic manipulation; molecular and cellular regulation of fruit ripening, plasticl differentiation, corotenoid synthesis and function. Office: Dept Developmental and cell Biology U Calif Irvine CA 92717

RADONOVICH, LEWIS JOSEPH, chemistry educator; b. Curtisville, Pa., July 2, 1944; s. Lewis L. and Ann (Marino) R.; m. Barbara Ann Radonovich, July 16, 1966; children: Lewis Joseph, David. A.B., Thiel Coll., 1966; Ph.D., Wayne State U., 1970. Research assoc. Cornell U., 1970-73; asst. prof. chemistry U. N.D., 1973-77, assoc. prof., 1977—. Contbr. articles to profl. jours. NDEA fellow, 1966-69; Petroleum Research fellow, 1970; Research Corp. grantee, 1977-81. Mem. Am. Chem. Soc., Am. Crystallographic Assn., Sigma Xi. Subspecialties: X-ray crystallography; Inorganic chemistry. Current work: Structure, bonding and chemistry of -arena complexes of the transition metals. Home: 1522 Cherry St Grand Forks ND 58201 Office: Dept Chemistry U ND Grand Forks ND 58202

RADOSKI, HENRY ROBERT, physicist, govt. phys. sci. adminstr.; b. Jersey City, Aug. 18, 1936; s. Henry Thomas and Stephanie Agatha (Gasior) Radozycki; m. Elizabeth Ann Patton, June 27, 1959; children: Raymond Regis, Henry Zachary, Derek Peter. B.S. in Physics, Holy Cross Coll., 1958, Ph.D., M.I.T., 1963. Assoc. prof. physics Boston Coll., Weston, Mass., 1963-68; research physicist Air Force Cambridge (Mass.) Research Labs., 1968-76; program mgr. Air Force Office Sci. Research, Washington, 1976—. Contbr. articles to profl. jours.; Co-editor: Physics of the Magnetosphere, 1968. Mem. Am. Phys. Soc., Am. Geophys. Union, Am. Astron. Soc., Sigma Xi. Subspecialties: Plasma physics; Satellite studies. Current work: Plans, develops and manages a program in astronomy and astrophysics, including solar, interplanetary, magnetospheric, ionospheric physics and aeronomy relevant to USAF. Office: Air Force Office Sci Research/NP Bldg 410 Bolling AFB Washington DC 20332

RADZIEMSKI, LEON JOSEPH, physicist; b. Worcester, Mass., June 18, 1937; s. Leon Joseph and Josephine Elizabeth (Janczukowicz) R.; m.; children: Michael Leon, Timothy Joseph. B.S., Coll. Holy Cross, Worcester, Mass., 1958; M.S., Purdue U., 1961, Ph.D., 1964. Staff physicist Los Alamos Nat. Lab., 1967-83; head dept. physics N. Mex. State U, 1983—; vis. scientist Laboratoire Aime Cotton, Orsay, Frances, 1974-75; vis. assoc. prof. dept. nuclear engring. U. Fla., Gainesville, 1978-79. Contbr. articles to profl. jours. Served with USAF, 1964-67. Mem. Am. Phys. Soc., Optical Soc. Am., Soc. Applied Spectroscopy, Laser Inst. Am. Subspecialties: Atomic and molecular physics; Spectroscopy. Current work: Applications of lasers and spectroscopy, laser spectrochemistry, laser-induced breakdown spectroscopy, plasma spectroscopy. Patentee in field. Home: 4709 Falcon Dr Las Cruces NM 88001 Office: Dept Physics Box 3D N Mex State U Las Cruces NM 88003

RAFALKO, EDWARD DENNIS, computer scientist, researcher; b. Natrona Heights, Pa., July 31, 1949; s. Edward Francis and Helen Marie (Mazinski) R.; m. Anna Mae Gibbons, May 17, 1974. B.S.E., U. Mich., 1971, M.S.E. in Computer Info. and Control, 1973. Sci. programmer, project engr., sr. project engr. Gen. Motors Corp., Pontiac, Mich., 1973-80; mem. tech. staff Bendix Advanced Tech. Ctr., Columbia, Md., 1980—. Mem. Assn. for Computing Machinery, IEEE, IEEE Computer Soc. Republican. Subspecialties: Computer architecture; Computervision. Current work: Special purpose, computer architectures. Distributive processing, fault tolerance, real-time control, functional simulation, modelling. Office: 9140 Old Annapolis Rd Columbia MD 21045

RAGAN, DONAL MACKENZIE, geology educator; b. Los Angeles, Oct. 4, 1929; s. Rex. and Helena (Mackenzie) R.; m. Janne Stewart, Sept. 7, 1952; children: Paul W., Anneliese. B.A., Occidental Coll., 1951; M.S., U. So. Calif., 1954; Ph.D., U. Wash., 1961; D.I.C., Imperial Coll., London, 1969. Registered geologist, Ariz., Calif. Asst. prof. U. Alaska, Fairbanks, 1960-65, assoc. prof., 1965-67; assoc. prof. geology Ariz. State U., Tempe, 1967-70, prof., 1970—. Author: Structural Geology, 2d edit, 1973; contbr. articles to profl. jours. Served with U.S. Army, 1953-55. Fulbright fellow, 1959-60; NSF sci. faculty fellow, 1966-67. Fellow Geol. Soc. Am., Geol. Soc. London; mem. Am. Geophys. Union, Internat. Soc. for Rock Mechanics, Sigma Xi. Subspecialties: Tectonics; Geophysics. Current work: Special interest in physical processes of rock deformation. Home: 1887 E Concorda Dr Tempe AZ 85282 Office: Dept Geology Ariz State U Tempe AZ 85287

RAGAVAN, VANAJA VIJAYA, internist; b. Madras, India, Mar. 7, 1949; came to U.S., 1970, naturalized, 1975; d. M. D. Vijaya and Vimala (Vasudevan) R.; m. Robert Jim Berrier, June 2, 1979; 1 son, Justin Vikram. A.B., Harvard U., 1972; M.D., N.Y. U., 1976. Diplomate: Am. Bd. Internal Medicine. Intern Kings County Hosp., Bklyn., 1976-77; resident in medicine, 1977-79; fellow in endocrinology Columbia U.-Presby N. Med. Centers, N.Y.C., 1976-81, U. Pa. Sch. Medicine, Phila., 1981-82; staff physician Coatesville (Pa.) VA Med. Center, 1982—; mem. staff dept. medicine Wilmington (Del.) Med. Center, 1982—. USPHS tng. grantee, 1979-81, 81-82. Mem. Am. Fedn. Clin. Research. Subspecialty: Neuroendocrinology. Current work: Clinical and basic investigation in neuroendocrinology and reproductive endocrinology; medical practice limited to general endocrinology and metabolism. Home: 51 Chapel Hill Rd Media PA 19063 Office: Coatesville VA Med Center Coatesville PA 19320 Heritage Profl Plaza Suite 8 Wilmington DE 19808

RAGHEB, MAGDI, nuclear engineering researcher and educator; b. Nov. 25, 1946; m. Barbara Rose Wesolek, Feb. 16, 1980. M.Sc., U. Wis., 1974, Ph.D., 1978. Research asst. dept. nuclear engring. U. Wis.-Madison, 1973-78, post-doctoral project assoc. dept. nuclear engring., 1978-79, cons., 1979-80; asst. prof. dept. nuclear engring. U. Ill., Urbana-Champaign, 1979—; vis. research scientist Brookhaven Nat. Lab., Upton, N.Y., 1975, cons., 1981—; research assoc. Oak Ridge Nat. Lab., 1978. Contbr. articles, reports, revs. to profl. jours. Mem. Am. Nuclear Soc. (program com. 1979—), AAAS, AAUP, N.Y. Acad. Scis., Sigma Xi. Subspecialties: Nuclear engineering; Nuclear fusion. Current work: Advanced energy systems; fusion, fission reactor design; advanced fusion fuel cycles; transport theory; monte carlo theory; neutronics and shielding design of fusion reactors; numerical computations in energy research. Home: 401 Edgebrook Dr Champaign IL 61802 Office: U Ill Nuclear Engring Lab 103 S Goodwin Ave Champaign-Urbana 61801

RAGHUNATHAN, RENGACHARI, biochemist, educator, researcher; b. Amoor, India, Mar. 22, 1943; s. Rengachari and Sakunthala R.; m. Kamala Raghunathan, Dec. 1, 1972; children: Anand, Adithya. B.S. in Agr., Mysore U., 1962; M.S. in Microbiology, Annamalai U., 1965; Ph.D. in Biochemistry, U. Bombay, 1974; D.I.M. in Mgmt., U. Bombay, 1973; H.M.B. in Homeopathic Medicine, U. Bangalore, 1960, D.H.M.S., U. Madras, 1963. Research asst. Tb Chemotherapy Centre, Madras, India, 1965-68; jr. sci. officer Bhabha Atomic Research Centre, Bombay, India, 1968-72, sr. sci. officer, biochemistry, 1972-75; instr. ob-gyn Meharry Med. Coll., Nashville, 1976-78, asst. prof. ob-gyn, 1978—. Contbr. articles to nat., internat. profl. jours. and popular papers and mags. Bd. dirs. March of Dimes, Nashville, 1976-78; trustee Hindu Cultural Ctr. Tenn., Nashville, 1981—. Fellow Phytopath. Soc.; mem. Internat. Soc. Preventive Oncology, Microbiol. Soc., Soc. Biol. Chemists. Subspecialties: Plant pathology; Developmental biology. Current work: Plant disease resistance mechanism; prenatal diagnosis of inborn errors; carbohydrate metabolism and regulation; enzymology, proteins. Home: 822 Kendall Dr Nashville TN 37209 Office: Dept Ob Gy Meharry Med Coll Nashville TN 37208

RAGLAND, WILLIAM LAUMAN, III, veterinarian, educator, research co. exec.; b. Richmond, Va., Aug. 24, 1934; s. William Lauman and Alma Josephine (Tatum) R.; m. Lois Camilla Witcher, July 16, 1961; children: Karen Renee, Alexander Shelton, Amy Elizabeth. B.S., Coll. of William and Mary, 1956; D.V.M., U. Ga., Athens, 1960; Ph.D., Washington State U., 1966. Research asst. Tulane U., New Orleans, 1960-61, instr., 1961-62; NIH fellow dept. vet. pathology Wash. State U., Pullman, 1962-66; NIH Spl. Research fellow dept. oncology and pathology U. Wis.-Madison, 1966-68, asst. prof. pathology, 1968-70; assoc. prof. avian medicine dept. U. Ga., Athens, 1970-76, prof. biochem. pathology/immunology, 1976—; pres. Ragland Research, Inc., Athens, 1980—. Contbr. numerous sci. articles to profl. publs. Mem. AAAS, Internat. Acad. Pathology, Biochem Soc., Soc. Toxicology, Am. Assn. Cancer Research, Am. Assn. Avian Pathologists, Am. Assn. Pathologists. Subspecialties: Immunobiology and immunology; Pathology (veterinary medicine). Current work: The development of precocious immune competence in chickens; enzyme immunoassays for animal disease diagnosis and tissue concentrations of drugs; development of biologics and pharmaceuticals for veterinary use. Developer: Flourescent Gel Scanner, 1978. Home: 155 Frontier Rd Athens GA 30605 Office: 953 College Station Rd Athens GA 30605

RAGOTZKIE, ROBERT AUSTIN, oceanography educator; b. Albany, N.Y., Sept. 13, 1924; s. Robert W. and Edith N. (Van Wormer) R.; m. Elizabeth M. Post, Aug. 1, 1925; children: Peter D., Kim E., Susan J. B.S. in Biology, Rutgers U., 1948, M.S. in Sanitation, 1950; Ph.D. in Zoology-Meteorology, U. Wis.-Madison, 1953. Asst. prof., then assoc. prof. biology, dir. Marine Inst., U. Ga., Athens, 1954-59; from asst. prof. to prof. meteorology U. Wis.-Madison, 1959—; dir. Sea Grant Inst., 1968—. Author 1 book, numerous sci. articles. Served to 2d lt. USAF, 1943-45. Decorated Air medal. Fellow AAAS; mem. Am. Meteorol. Soc., Am. Geophys. Union, Am. Soc. Limnology and Oceanography, Internat. Soc. Limnology, Internat. Assn. Great Lakes Research (past pres.). Subspecialties: Oceanography; Aquatic ecology. Current work: Great Lakes limnology, heat budgets of lakes. Home: 2334 Tanager Trail Madison WI 53711 Office: 1800 University Ave Madison WI 53705

RAHMAN, FAZLUR, med. oncologist, hematologist; b. Pora Bari, Bangladesh, Jan. 12, 1945; came to U.S., 1969; s. Anwar Ali and Hasina (Khatun) Mollah; m. Jahanara Akhtar, Jan. 26, 1969; children: Prince Rahman, Gulshan Ara, Yasmin J., Happy Jennifer. M.D., Dacca Med. Coll., 1968. Diplomate: Am. Bd. Med. Oncology, Hematology and Internal Medicine. Intern St. John's Hosp., Yonkers, N.Y., 1969; resident Queen's Hosp. Ctr., L.I. Jewish Hosp., N.Y.C., 1970-71, Baylor Coll. Medicine, Houston, 1971-73, fellow, 1973-75; oncologist/hematologist West Tex. Med. Assocs., San Angelo, Tex., 1975—; chief oncology service Angelo Community Hosp., 1975—, chief staff, 1980, chmn. cancer com., 1978—. Contbr. articles to med. jours. Dist. med. dir. Am. Cancer Soc., San Angelo, 1980—; mem. selection com. Roy Moon Disting. Lectureship in Sci., Angelo State U., San Angelo, 1982—. Dacca Med. Coll. Merit Scholar, 1962-68, Daulatpur Coll., 1960-62. Fellow ACP; mem. Am. Fedn. Clin. Research, Am. Soc. Clin. Oncology, Am. Soc. Hematology. Subspecialties: Chemotherapy; Hematology. Current work: Care of cancer patients, studies on coping with cancer and cancer quackery, medicine and society. Home: 3105 Ridgecrest Ln San Angelo TX 76904 Office: Dept Oncology and Hematology West Tex Med Assocs 3555 Knickerbocker Rd San Angelo TX 76904

RAHN, HERMANN, physiologist; b. East Lansing, Mich., July 5, 1912; s. Otto and Bell S. (Farrand) R.; m. Katharine F. Wilson, Aug. 29, 1939; children—Robert F., Katharine B. A.B., Cornell U., 1933; student, U. Kiel, 1933-34; Ph.D., U. Rochester, 1938, D.Sc. (hon.), 1973; Docteur Honoris Causa, U. Paris, 1964; LL.D. (hon.), Yonsei U., Korea, 1965; Titulo de Profesor Honorario U. Peruana Cayetano Heredia, Peru, 1980; Dr. Medicinae h.c., U. Bern, Switzerland, 1981. NRC fellow Harvard, 1938-39; instr. physiology U. Wyo., 1939-41; asst. physiology Sch. Medicine, U. Rochester, 1941-42, instr., 1942-46, asst. prof., 1946-50, asso. prof., vice chmn. dept., 1950-56; Lawrence D. Bell prof. physiology, chmn. dept. Sch. Medicine, U. Buffalo (now State U. N.Y. at Buffalo), 1956-73, distinguished prof. physiology, 1973—; vis. prof. Med. Faculty San Marcos U., Lima, Peru, 1955, Dartmouth Med. Sch., 1962, Lab. de Physiologie Respiratoire, CNRS, Strasbourg, France, 1971, Max-Planck-Inst. für experimentelle Medizin, Göttingen, W.Ger., 1977; Mem. adv. com. biol. sci. Air Force Office Sci. Research and Devel., 1958-64; physiol. study sect. NIH, 1958-62; mem. working committee space sci. bd. Nat. Acad. Sci.-NRC, 1962-65; mem. gen. med. research program project com. NIH, 1964-67; mem. research career award com. Nat. Inst. Gen. Med. Scis., 1968-72; cardiopulmonary adv. com. Nat. Heart Inst., 1968-71; mem. nat. adv. bd. R/V Alpha Helix, 1968-71; chmn. com. on underwater physiology and medicine NRC, 1972-74. Author: (with W.O. Fenn) A Graphical Analysis of the Respiratory Gas Exchange, 1955, (with others) Blood Gases: Hemoglobin, Base Excess and Maldistribution, 1973; Editorial bd.: Am. Jour. Physiology, Jour. Applied Physiology, 1953-62; sect. editor for respiration, 1962; bd. publ. trustees, 1959-62; co-editor: (with W.O. Fenn) Handbook of Physiology-Respiration, Vols. I, II, 1964-65; editor: Physiology of Breath-Hold Diving and the Ama of Japan, 1965. Recipient Sr. U.S. Scientist award Alexander von Humboldt Found., 1976. Mem. Am. Inst. Biol. Sci. (adv. com. physiol. 1957-64, adv. panel 1967-71), Am. Physiol. Soc. (council 1960-65, pres. 1963-64), Nat. Acad. Sci., Inst. Medicine, Nat. Acad. Sci., Harvey Soc. (hon.), Soc. Exptl. Biology and Medicine, Internat. Union Physiol. Scis. (council 1965-74, U.S. nat. com. 1966-74, v.p. 1971-74, exec. com. 1971-74), Am. Soc. Zoologists, Am. Acad. Arts and Sci., Sigma Xi. Subspecialties: Animal physiology; Comparative physiology. Current work: Problems of gas exchange. Research in pulmonary physiology, gas exchange, environmental physiology. Home: 75 Windsor Ave Buffalo NY 14209

RAHN, JOAN ELMA, sci. writer; b. Cleve., Feb. 5, 1929; d. George William and Elsie Edna (Thiele) R. B.S., Case Western Res. U., 1950; A.M., Columbia U., 1952, Ph.D., 1956. Asst. prof. biology Thiel Coll., Greenville, Pa., 1956-58, assoc. prof., 1958-59; instr. Ohio State U., Columbus, 1959-60, Internat. Sch. of Am., 1959-60; asst. prof. Lake Forest (Ill.) Coll., 1961-67; freelance sci. writer, 1967—. Author: Seeing What Plants Do, 1972 (an Outstanding Sci. Book for Children, Nat. Sci. Tchrs. Assn./Children's Book Council); author: How Plants Travel (an Outstanding Sci. Book for Children), 1973 (Disting. Service award Soc. of Midland Authors), Biology: The Science of Life, 2d edit, 1974, Grocery Store Botany, 1974 (an Outstanding Sci. Book for Children), More About What Plants Do, 1975, How Plants Are Pollinated, 1975 (an Outstanding Sci. Book for Children), The Metric System, 1976, Alfalfa, Beans & Clover, 1976, Grocery Store Zoology, 1977, Nature in the City: Plants, 1977, Seven ways to Collect Plants, 1978 (an Outstanding Sci. Book for Children), Watch It Grow, Watch It Change, 1978, Traps & Lures in the Living World, 1980 (an Outstanding Sci. Book for Children), Eyes & Seeing, 1981, Plants Up Close (an Outstanding Sci. Book for Children), 1981 (honorable mention N.Y. Acad. Scis. 11th Ann. Children's Sci. Book award), Plants that Changed History, 1982; sci. adv. staff: Experiments in Science, 1967, Looking at the Microscopic World, 1967, Plant Growth Experiments with the LaPine Botanarium, 1969, Science Experimnts for Elementary Schools, 2d edit, 1971. Mem. AAAS, Am. Inst. Biol. Scis., Bot. Soc. Am., Sigma Xi. Lutheran. Subspecialty: General biology. Office: 1656 Hickory St Highland Park IL 60035

RAIKOW, RADMILA BORUVKA, medical researcher; b. Prague, Czechoslovakia, Mar. 20, 1939; d. Svatopluk and Paula (Erba) Boruvka; m. Robert Jay Raikow, June 10, 1966; children: David F., Steven B. B.A., N.Y.U., 1960; M.A., CUNY, 1965; Ph.D., U. Calif., 1970. Fellow dept. biochemistry U. Pitts., 1971-72, lectr. dept. biology, 1974-75; acting asst. prof. dept. biology Chatham Coll. Pitts., 1973-74; fellow, research assoc. Allegheny Gen. Hosp., Pitts., 1975-80; assoc. scientist Allegheny-Singer Research Inst., Pitts., 1980—. Contbr. articles to profl. jours. NIH fellow, 1975-78. Mem. Am. Assn. Cancer Research, Soc. Exptl. Biology and Medicine, AAAS, Sigma Xi. Subspecialties: Immunobiology and immunology; Cell biology. Current work: Immunobiology role of immunity in cancer control pre-clinical research. Home: 1229 Winterton St Pittsburgh PA 15206 Office: 320 S North Ave Pittsburgh PA 15206

RAIMONDI, ANTHONY JOHN, medical educator, neurosurgeon; b. Chgo., July 16, 1928; m., 1954; 3 children. B.A., B.S., U. Ill., 1950; M.D., U. Rome, 1954. Instr. in neurosurgery U. Chgo., 1961-62, clin. asst. prof., 1964-66, clin. assoc. prof., 1966-67; instr. Northwestern U., 1962-64, prof. neurosurgery, chmn. div., 1969—, prof. anatomy, 1974—; assoc. prof. neurol. surgery U. Ill., 1967-69; attending neurosurgeon Children's Meml. Hosp., 1962-63, chmn. div. neurol. surgery, 1969—; prof. Cook County Grad. Sch. Medicine, 1963-70; chmn. dept. Cook County Hosp., 1963-70; chmn. neurosurgery Vet. Res. Hosp., Chgo., 1972—; attending physician in neurosurgery Surg. Service, VA Hosp.; cons., lectr. Gt. Lakes Naval Hosp.; chmn. med. adv. com. Am. Spina Bifida Assn., Epilepsy Fund Am., Assn. Brain Tumor Research. Mem. ACS, Am. Assn. Neurol. Surgery, Am. Assn. Neuropathology, Internat. Soc. Pediatric Neurosurgery (sec.), Am. Assn. Surgeons of Trauma. Subspecialty: Neurosurgery. Address: 330 E Chicago Ave Chicago IL 60611

RAIN, ROBERT L., pharmaceuticals co. exec.; b. St. Louis, July 21, 1933; s. Sidney G. and Golda M. (Gibson) R.; m. Mary Louise Mabis, May 19, 1933; children: Steven B., Susan D. B.S. in Chem. Engring, Purdue U., 1955; postgrad., U. Houston, 1958-61. Quality control mgr. Marbon Chems. Co., Washington, W.Va., 1966-67; quality control mgr. Travenol Labs., Cleveland, Miss., 1969-70, plant mgr., 1970-72, dir. mfg., Deerfield, Ill., 1972-75, dir; quality assurance, 1976-81; v.p. Baxter Travenol Labs., Deerfield, 1981—. Served to lt. (j.g.) USN, 1955-58. Mem. Am. Soc. Quality Control, Pharm. Mfrs. Assn., Pi Kappa Alpha. Presbyterian. Subspecialties: Biomedical engineering; Pharmaceutical quality control. Current work: Quality Assurance, quality control. Office: 1 Baxter Pkwy Deerfield IL 60015

RAINBOW, THOMAS CHARLES, neurochemist, educator; b. Trenton, N.J., Jan. 27, 1954; s. Thomas Francis and Doris Marie (DeCarlo) R.; m. Marsha Louise Kness, Apr. 25, 1981. B.A., U. Pa., 1975, Ph.D., 1979. NIMH trainee U. Pa., Phila., 1975-79; NIMH fellow Rockefeller U., N.Y.C., 1979-81, asst. prof. neurochemistry, 1981-82; asst. prof. pharmacology U. Pa., 1983—. Contbr. articles in field to profl. jours. Alfred P. Sloan Found. fellow, 1982—; Ester A. and Joseph Klingenstein fellow, 1983—. Mem. AAAS, Soc. for Neurosci., Sigma Xi. Subspecialties: Neurochemistry; Neuropsychology. Current work: Neurochemical studies of plasticity and behavior in mammals; neurochemical autoradiography. Home: 44 Colonial Lake Dr Lawrenceville NJ 08648 Office: Pa Med Sch Philadelphia PA 19104

RAINNIE, WILLIAM OGG, JR., ocean engineer; b. Ashtabula, Ohio, Apr. 27, 1924; s. William Ogg and Beryl Naomi (McBride) R.; m. Sara Ann Cross, July 25, 1925; children: William Ward, Michael Curtis. B.S., U.S. Naval Acad., 1946. Commd. ensign U.S. Navy, 1946, advanced through grades to lt., 1952, resigned, 1954; project engr. Fairbanks Morse & Co., Beloit, Wis., 1954-57, tech. rep., Washington, 1957-60; tech. asst. Nat. Acad. Sci., Washington, 1960-61; ocean engr. Woods Hole (Mass.) Oceanographic Inst., 1961-77; chief engring. div. Data Buoy Office NOAA, NASA Space Tech. Labs. station, Miss., 1977—; Program mgr. deep research vehicle Alvin, 1964. Vestryman Trinity Episcopal Ch., Pass Christian, Miss., 1977—. Recipient civilian service award USN, 1966, 73. Mem. Am. Soc. Naval Engrs., Marine Tech. Soc. Clubs: Pass Christian Yacht (commodore 1982, dir. 1979—. Lodge: Rotary (sec. Falmouth, Mass. 1973). Subspecialty: Ocean engineering. Current work: Engineering activity to develop, test and operate environmental monitoring systems for marine uses. Home: 104 Hursey Ave Pass Christian MS 39571 Office: NOAA Data Buoy Ctr NSTL Station MS 39529

RAIZADA, MOHAN KISHORE, endocrinologist, educator; b. Fatehpur, India, Oct. 21, 1948; came to U.S., 1973; s. Maharaj Bahadur and Bittan Kuvar; m. Laura M. Carlson, Nov. 20, 1979; children: Kristen, Keely. B.S., U. Lucknow, India, 1966, M.S., 1968; Ph.D., U. Kanpur, India, 1972. Postdoctoral fellow Med. Coll. Wis., Milw., 1973-74; postdoctoral assoc. Lady Davis Inst., Montreal, Que., Can., 1974-76; assoc. U. Iowa, Iowa City, 1976-78, asst. prof., 1979-80; assoc. prof. U. Fla., Gainesville, 1981—. Nat. Acad. Sci. Young Scientist model, India, 1974. Mem. Am. Physiol. Soc., Endocrine Soc., Tissue Culture Assn., Am. Soc. Cellular Biology. Subspecialties: Endocrinology; Cell biology (medicine). Current work: mechanism of development of diabetes mellitus; central regulation of hypertension. Home: 6509 33d St Gainesville FL 32606 Office: U Fla JHMHC Box J274 Gainesville FL 32610

RAJ, HARKISAN DUNICHAND, microbiology educator; b. Sehawan, Sindh, Pakistan, Jan. 1, 1926; came to U.S., 1956, naturalized, 1969; s. Dunichand Chetanram and Moar Nihchaldas (Chandanani) Tejwani; m. Anita Devika Hemraj Jhurani, Sept. 1, 1956; children: Robin Calvin, Arnaz Ken. B.S. with honors, U. Bombay, India, 1947; M.S., U. Poona, India, 1952, Ph.D., 1955. Bacteriologist Pub. Health Dept., Bombay and Poona, 1948-56; postdoctoral fellow Tex. A&M U., 1956-57; instr. Oreg. State U., 1957-58; asst. prof. U. Wash., Seattle, 1959-62; prof. Calif. State U.-Long Beach, 1962—; cons. pvt. corps. Contbr. articles to profl. jours. and books. Grantee NIH, USPHS, 1960—. Mem. Am. Soc. Microbiology, So. Calif. Soc. Electron Microscopists, Sigma Xi. Subspecialties: Microbiology (medicine); Molecular biology. Current work: Bacterial metabolism, physiology, taxonomy and ultra structures. Address: 16251 Gentry Ln Huntington Beach CA 92647 Office: Dept Microbiology Calif State U 1250 Bellflower Blvd Long Beach CA 90840

RAJA, SRINIVASA N., anesthesiologist, researcher; b. Madras, India, Nov. 5, 1950; s. Venkataraman and Sarada S.; m. Geetha Rajam, Feb. 7, 1979. M.B.B.S., Patna U., India, 1974. Diplomate: Am. Bd. Anesthesiology. Resident in anesthesiology U. Wash., Seattle, 1977-79; research fellow U. Va., 1979-81; asst. prof. Johns Hopkins U. Sch. Medicine, Balt., 1981—. Contbr. articles to profl. jours. Recipient award Nat. Inst. Gen. Med. Scis., 1979-81. Mem. Am. Soc. Aneshtesiologists, Internat. Assn. Study Pain, Soc. Neurosci., Am. Soc. Regional Anesthesiology, Internat. Anesthesia Research Soc. Subspecialties: Neuropharmacology; Neurophysiology. Current work: Mechanism of action of drugs of abuse; interactions of drugs of abuse and anesthetics; psychophysical and neurophysiological aspects of peripheral neural mechanism of pain; effects of anesthetics on response properties of primary afferents. Office: John Hopkins Hosp 600 N Wolfe St Meyer 8-134 Baltimore MD 21205

RAJAMANI, P.N., pharm. co. exec.; b. Bombay, India, June 7, 1943; came to U.S., 1964; s. Puthucode Sethu and Rajam P. (Swamy) Narayanswamy; m. Sheila Aiyer, Nov. 17, 1968; children: Prakash, Ashok. B.Pharm., U. Nagpur, India, 1964; M.S. in Pharmacy, Phila. Coll. Pharmacy and Sci., 1966. Cert. quality engr. Am. Soc. for Quality Control. Research assoc. Squibb Corp., New Brunswick, N.J., 1968-76; quality control mgr. C.I.S. Radiopharms., Inc., Bedford, Mass., 1976-77; quality control supr. Union Carbide Corp., Wallingford, Conn., 1977-78; prin. engr. Baxter-Travenol Labs., Inc., Deerfield, Ill., 1978—. Contbr. articles to profl. jours. Advisor Jr. Achievement, 1980. Mem. Am. Soc. Quality Control (sr., NEI Sect. bd. dirs. 1982-83, dir. Edn. and Tng. Inst. 1982-83), Indian Hosp. Pharm. Assn. (life). Current work: Reliability and quality engineering and fitness for use determination for medical devices and drugs. Vendor quality assurance in areas of pharmaceuticals and devices. Patentee in field. Office: One Baxter Pkwy Deerfield IL 60015

RAJAN, KANNAN RAMALINGAM, physician, researcher; b. Bangalore, India, July 16, 1931; came to U.S., 1963, naturalized, 1981; s. Ramalingam and Amarthammal (Amartham) R.; m. Kalanidhi Doraisawmy, Oct. 15, 1960. B.Sc., St. Joseph's Coll., Bangalore, 1951; M.B., B.S., Mysore Med. Coll., 1958. Resident med. officer Misericordia Gen. Hosp., Winnipeg, Man., Can., 1969-72; staff physician Vets. Hosp., Batavia, N.Y., 1972-73; practice medicine St. Joseph's Hosp., Toronto, Ont., Can., 1973-74; staff physician VA Med. Ctr., Long Beach, Calif., 1974—. Contbr. articles to profl. jours. Recipient cert. Am. Vets. World War II, Korea and Vietnam, 1976, Jewish War Vets. U.S.A., 1976, 78, 79; named Physician of Year Jewish War Vets. U.S.A., 1982. Fellow Royal Coll. Physicians and Surgeons Can., Am. Coll. Gastroenterology; mem. ACP, Am. Fedn. Clin. Research, AMA, N.Y. Acad. Scis. Democrat. Hindu. Subspecialties: Gastroenterology; Internal medicine. Current work: Colonic function in spinal cord injury patients. Home: 14021 Montgomery Dr Westminster CA 92683 Office: VA Med Center 5901 E 7th St Long Beach CA 90822

RAJCHMAN, JAN ALEKSANDER, electronic researcher, consultant; b. London, Aug. 10, 1911; U.S., 1935; s. Ludwik W. and Maria C. (Bojanczyk) R.; m. Ruth T. Teitrick, June 30, 1944; children: Alice Hammond, John. Diploma in Elec. Engring, Swiss Fed. Inst. tech., Zurich, 1935, Dr. Tech. Scis., 1938. Mem. tech. staff RCA Labs., Princeton, N.J., 1936-59; assoc. dir. System Lab., RCA Corp., 1959-61, dir., 1961-71, staff v.p. info. scis., 1971-76; cons. electronics, computer engineering, Princeton. Contbr. numerous articles to profl. publs. Recipient cert. recognition NASA, 1975; Harold Pender award U. Pa., 1977. Fellow IEEE (Liebman award 1960, Edison medal 1974); mem. Am. Phys. Soc., Nat. Acad. engring., Assn. Computing Machinery, AAAS, Franklin Inst. (Lerry medal 1977), N.Y. Acad. Scis., Soc. Info. Display, Am. Optical Soc., Sigma Xi. Subspecialties: Electronics; Civil engineering. Current work: Electronc devices, optical devices, display devices, VLSI, non-impact printers. Patentee in field.

RAJU, NAMBOORI BHASKARA, biologist, researcher; b. Pothumarru, Andhra Pradesh, India, Jan. 1, 1943; came to U.S., 1974; s. Venkatrama and Suramma (Chintalapati) R.; m. Swarajya Rudraraju, Aug. 20, 1950; children: Geeta, Suja, Meena. M.Sc. in Agr, Banaras Hindu U., 1967; Ph.D., U. Guelph, 1972. Research asso. in biol. scis. Stanford U., 1974—. Contbr. numerous articles to biol. jours. Mem. Mycol. Soc. Am. Subspecialties: Plant genetics; Microbiology. Current work: Cytogenetics of fungi, especially Neurospora and the mushroom fungus Coprinus. Office: Biol Scis Dept Stanford U Stanford CA 94305

RAKES, JERRY MAX, animal scientist, educator, farmer, researcher; b. Bentonville, Ark., Dec. 7, 1932; s. Sidney B. and Maruine M. (Harral) R.; m. Betty Jo, Aug. 27, 1949; children: Jerry Randal, Michael Jo, Melissa Ann. B.S., U. Ark., 1955, M.S., 1956; Ph.D, Iowa State U., 1958. Grad. asst. U. Ark., 1955-56, asst. prof. animal sci., 1958-64, asso. prof., 1964-68, prof., 1968—, supr., 1969—; grad. asst. Iowa State U., 1956, research asso., 1957-58; in charge several research projects. Contbr. numerous articles to profl. publs. Served to cpl. F.A. U.S. Army, 1948-50. Recipient award Nat. Assn. Artificial Breeders, 1965. Mem. Am. Dairy Sci. Assn., Am. Soc. Animal Sci., Sigma Xi, Gamma Sigma Delta, Alpha Zeta. Baptist. Club: Band Boosters. Lodge: Masons (Bentonville). Subspecialties: Animal genetics; Animal breeding and embryo transplants. Current work: Genetics research. Home: Route 2 Bentonville AR 72712 Office: U Ark 104 Animal Sci Bldg Fayetteville AR 72701

RAKIC, PASKO, neuroscientist, educator; b. Ruma, Yugoslavia, May 15, 1933; came to U.S., 1969, naturalized, 1974; s. Toma and Julijana R.; m. Ljiliana Lekic, Feb. 20, 1969 (div. 1973); m. Patricia Shoer Goldman, Jan. 19, 1978. M.D., Belgrade U., 1959, Sc.D., 1969. Asst. prof. path. physiology Belgrade U., 1960-61; asst. prof. neuropathology Harvard Med. Sch., Boston, 1969-72, assoc. prof., 1972-77; prof. neurosci. Yale U., New Haven, 1977—, chmn. sect. neuroanatomy, 1977—. Author 3 books and numerous articles. Mem. Am. Assn. Anatomists, Am. Assn. Nuropathology, Assn. Research in Mental and Nervous Disease, Internat. Brain Research Orgn., Soc. Neurosci. Club: Cajal. Subspecialties: Comparative neurobiology; Developmental biology. Current work: Developmental neurobiology. Office: Yale Med Sch 333 Cedar St New Haven CT 06510

RAKOFF, VIVIAN MORRIS, psychiatrist, educator; b. Capetown, South Africa, Apr. 28, 1928; m., 1959; 3 children. B.A., U. Capetown, 1947, M.A., 1949; M.B., B.S., U. London, 1957; D. Psych., McGill U., Montreal, Que., Can., 1963. Psychologist Tavistock Clinic, 1950-511; house officer in surgery St. Charles Hosp., 1957; house officer in medicine Victoria Hosp., 1958; registrar Groote Schuur Hosp., 1958-61; resident in psychiatry McGill U., 1961-63; assoc. dir. research Jewish Gen. Hosp., 1963-67, asst. prof. and dir. research, 1967-68; from assoc. prof. to prof. psychiatry U. Toronto, Ont., Can., 1968-74, dir. postgrad. edn., 1968-71, prof. psychiat. edn., 1974—, prof. psychiatry, chmn. dept. psychiatry, 1980; now also psychiatrist in chief dept. psychiatry; dir. and psychiatrist in chief Clarke Inst. Psychiatry. Fellow Royal Coll. Physicians (Can.); mem. Am. Psychiat. Assn., Can. Psychiat. Assn., Am. Coll. Psychiatrists. Subspecialty: Psychiatry. Address: 250 College St Toronto ON Canada M5T 1R8

RALL, DAVID PLATT, pharmacologist, educator; b. Aurora, Ill., Aug. 3, 1926; s. Edward Everett and Nell (Platt) R.; m. Edith Levy, July 17, 1954; children: Jonathan D., Catharyn E. B.S., North Central Coll., Naperville, Ill., 1946; M.S., Northwestern U., 1948, M.D., Ph.D., 1951. Intern Bellevue Hosp., N.Y.C., 1952-53; commd. officer USPHS, 1953—, asst. surgeon gen., 1971—; sr. investigator Lab. Chem. Pharmacology, Nat. Cancer Inst., NIH, Bethesda, Md., 1953-55, Clin. Pharmacology and Exptl. Therapeutics Service, 1956-58, head service, 1958-63, chief, 1963-69; assoc. sci. dir. for exptl. therapeutics Nat. Cancer Inst., 1966-71; dir. Nat. Inst. Environ. Health Scis., 1971—, dir. Nat. Toxicology Program, 1978—; adj. prof. pharmacology U. N.C., Chapel Hill, 1972—. Mem. AAAS, Am. Assn. Cancer Research, Am. Coll. Preventive Medicine, Am. Soc. Clin. Investigation, Am. Soc. Pharmacology and Exptl. Therapeutics, Inst. Medicine, Soc. Occupational and Environ. Health, Soc. Toxicology. Subspecialty: Pharmacology. Office: Nat Inst of Environ Health Scis NIH Research Triangle Park NC 27709

RALL, JOSEPH EDWARD, physician; b. Naperville, Ill., Feb. 3, 1920; s. Edward Everett and Nell (Platt) R.; m. Caroline Domm, Sept. 28, 1944 (dec. Apr. 1976); children—Priscilla, Edward Christian; m. Nancy Lamontagne, Apr. 15, 1978. B.A., North Central Coll., 1940; M.S., Northwestern U., 1944, M.D., 1945; Ph.D., U. Minn., 1952; D.Sc. (hon.), N. Central Coll., 1966; Dr.h.c., Faculty of Medicine, Free U. Brussels, Belgium, 1975. Asso. mem. Sloan Kettering Inst., N.Y.C., 1950-55; chief clin. endocrinology br. Nat. Inst. Arthritis, Metabolism and Digestive Diseases, NIH, 1955-62, dir. intramural research, 1962-83; dep. dir. intramural research NIH, 1983—; Mem. NRC, 1960-65. Author numerous articles, chpts. in books on thyroid gland and radiation. Served to capt. M.C. AUS, 1946-48. Recipient Van Meter prize Am. Goiter Assn., 1950, Fleming award, 1959, Outstanding Achievement award Mayo Clinic and U. Minn., 1964; Disting. Service award Am. Thyroid Assn., 1967, HEW, 1968; named Outstanding Alumnus N. Central Coll., 1966. Mem. AAAS, Am. Soc. Clin. Investigation, Am. Phys. Soc., Endocrine Soc., Assn. Am. Physicians, Societe de Biologie (France), Royal Acad. Medicine (Brussels), Nat. Acad. Scis. Subspecialties: Biochemistry (medicine); Endocrinology. Current work: Endocrinology and biochemistry, including work on mechanism of action of thyroid hormones and effects of radiation on thyroid gland. Home: 3947 Baltimore St Kensington MD 20795 Office: National Institutes Health Bldg 1 Room 122 Bethesda MD 20205

RALL, LOUIS B(AKER), mathematics researcher, educator; b. Kansas City, Mo., Aug. 1, 1930; m. Mary Frances Landram, Mar. 1, 1952; children: Denise, Alyse. B.S., U. Puget Sound, 1949; M.S., Oreg. State U., 1954; Ph.D., 1956. Assoc. prof. Lamar U., 1957-60; prof. Va. Poly. and State U., 1960-62; prof. math. U. Wis.-Madison, 1962—; vis. prof. U. Innsbruck, Austria, 1970, Oxford (Eng.) U., 1972-73, U. Copenhagen/Tech. U. Denmark, 1980. Author: Computational Solution of Nonlinear Operator Equations, 1969, 2d edit., 1979, Automatic Differentiation, 1981; editor: Error in Digital Computation, vols. 1 and 2, 1965, Nonlinear Functional Analysis, 1971. Served to cpl. U.S. Army, 1951-53; Korea. Fellow Inst. for Math. and Its Applications; mem. Am. Math. Soc., Math. Assn. Am., Soc. Indsl. and Applied Math., Circolo Mathematico di Palermo. Democrat. Unitarian. Club: Univ. (Madison). Subspecialties: Numerical analysis; Mathematical software. Current work: Interval analysis, automatic differentiation, high speed algorithms for optimization, solution of systems of equations, integral equations, ordinary and partial differential equations, automatic error estimation, accurate numerical computation. Office: Math Research Ctr U Wis 610 Walnut St Madison WI 53706

RALL, WILFRID, neuroscientist; b. Los Angeles, Aug. 29, 1922; s. Udo and Doris (Keiser) R.; m.; children from previous marriage: Sara E., Madelyn W. B.S. summa cum laude, Yale U., 1943; M.S., U. Chgo., 1948; Ph.D., U. N.Z., 1953. Jr. physicist Manhattan Project, U. Chgo., 1943-46; biophysics fellow U. Chgo., Woods Hole, Mass., 1946-48; lectr., sr. lectr. physiology biophysics U. Otago, Dunedin, N.Z., 1949-56; head biophysics div. Naval Med. Research Inst., Bethesda, Md., 1956-57; biophysicist, math. research br. Nat. Inst. Arthritis and Metabolic Diseases, Bethesda, 1957-67, sr. research physicist, 1967—; mem. Nat. Acad Sci./NRC Com. on Brain Scis., 1968-73. Contbr. articles to profl. jours. Rockefeller Found. fellow, 1954-55. Mem. Soc. Neuroscience (nat. council 1970-72, chpt. pres. 1981-82), Internat. Brain Research Orgn. (central council 1968-73, U.S. nat. com. 1972-76), Biophys. Soc., Am. Physiol. Soc., Physiol. Soc. U.K., AAAS, Wilderness Soc. Subspecialties: Neurophysiology; Biophysics (physics). Current work: Theory, mathematical models, related to experimental neuroanatomy and neurophysiology; dendritic branching; synaptic function and integration; computation of intracellular and extracellular potential distributions during neuronal activity. Office: NIH Bldg 31 Rm 4B-54 Bethesda MD 20205

RALSTON, DOUGLAS EDMUND, biochemist, educator; b. Cherokee, Iowa, July 9, 1932; s. Edmund G. and Grace A. R.; m. Jane Schroeder, Dec. 5, 1953; children. B.S., Wayne (Nebr.) State Coll., 1953, M.S., 1957; M.S., State U. S.D., 1959; Ph.D., U. Minn., 1967. Asst. prof. chemistry Mankato State U., 1962-68, assoc. prof., 1968—, chmn. dept. chemistry, 1982—. Served in U.S. Army, 1953-55. Subspecialties: Biochemistry (biology); Membrane biology. Office: Mankato State U PO Box 40 Mankato MN 56001

RAM, MADHIRA DASARADHI, surgical educator, surgical researcher; b. Visakhapatnam, Andhra, India, Apr. 6, 1935; came to U.S., 1969, naturalized, 1975; s. Subba Rao and Lakshmi (Madduri) Madhira; m. Noreen Mary Gearon, Sept. 16, 1967; children: Ravi, Ian, Chandra, Colin. B.S., Andhra U., 1952, M.B.B.S., 1957, M.S., 1961; Ph.D., Case Western Res. U., 1975. Lectr. in surgery Andhra U., 1961-65; registrar Royal Postgrad. Med. Sch., London, 1965-69; asst. dir. surgery Mt. Sinai Hosp., Cleve., 1969-71; dir. surgery Huron Road Hosp., Cleve., 1971-77; assoc. prof. surgery U. Ky. Med. Ctr., 1977-80, prof. surgery, 1980—, chief gen. surgery, 1979-80; chief surgery VA Hosp., Lexington, Ky., 1977—. Author: Surgery, 1977, 81; editor: Self-Assessment of Current Knowledge in Surgery, 1981. Fellow ACS, Royal Coll. Surgeons Eng., Royal Coll. Surgeons of Edinburgh, Royal Coll. Surgeons Can., Royal Soc. Medicine; mem. Lexington Surg. Soc. (pres. 1982). Subspecialties: Surgery; Gastroenterology. Current work: Cancer research, endocrinology; biliary pancreatic surgery. Home: 943 Edgewater Dr Lexington KY 40502 Office: U Ky Med Ctr 800 Rose St Lexington KY 40536

RAMAKER, DAVID ELLIS, chemistry educator, consultant; b. Sheboygan, Wis., Aug. 11, 1943; s. Allen J. and Alma (Ver Gowe) R.; m. Beverly Ann Back, Sept. 3, 1966; children: Julie, Jacqueline, Jan, Jason. B.S., U. Wis.-Milw., 1965; M.S., U. Iowa, 1968, Ph.D., 1971. Teaching assoc. U. Minn., Mpls., 1967-68; postdoctoral fellow Sandia Nat. Lab., Albuquerque, 1970-72; research assoc., assoc. instr. U. Utah, Salt Lake City, 1972-74; vis. asst. prof. Calvin Coll., Grand Rapids, Mich., 1974-75; asst. prof. George Washington U., Washington, 1975-78, assoc. prof., 1979-83, prof., 1983—; research chemist Naval Research Lab., Washington, 1976—; expert, cons. Nat. Bur. Standards, Washington, 1982—. Recipient Spl. Achievement awards Naval Research Lab., 1979, 81. Mem. Am. Chem. Soc., Am. Vacuum Soc. Mem. Christian Reformed Ch. Subspecialties: Surface chemistry; Theoretical chemistry. Current work: Auger and photoelectron spectroscopy, electron/photon stimulated desorption, many-body effects in Auger lineshapes, mechanisms for stimulated desorption. Office: Dept Chemistry George Washington U Washington DC 20052 Home: 6943 Essex Ave Springfield VA 22150

RAMAKUMAR, RAMACHANDRA GUPTA, electrical engineer, educator, consultant; b. Coimbatore, Tamil Nadu, India, Oct. 17, 1936; came to U.S., 1967; s. Gopalakrishna Ramachandra and Saraswathi Bai (Swamy) Gupta; m. Tallam Gokuladevi, June 13, 1963; children: Sanjay, Malini. B.E., U. Madras, 1956; M.Tech., Indian Inst. Tech., Kharagpur, 1957; Ph.D. Tech. Coop. Mission scholar, Cornell U., 1962. Registered profl. engr., Okla. Asst. lectr. elec. engring. Coimbatore (India) Inst. Tech., 1957-59, lectr., 1959-62, asst. prof. elec. engring., 1962-67; vis. assoc. prof. Okla. State U., 1967-70, asso. prof., 1970-76, prof., 1976—; cons. in field. Contbr. numerous articles, chpts. to profl. jours., conf. procs., books. Recipient Outstanding Engring. Achievement award Okla. Soc. Profl. Engrs., 1972; NSF internat. travel grantee, 1978. Mem. IEEE (sr.), Power Engring. Soc., Internat. Solar Energy Soc., Am. Solar Energy Soc. Hindu. Subspecialties: Electrical engineering; Solar energy. Current work: Integrated development and utilization of renewable energy sources (solar, wind, hydro, and biomass); electrical power engineering problems arising with penetration of alternate generation sources. Patentee energy conversion, energy storage. Home: 2623 N Husband St Stillwater OK 74075 Office: Oklahoma State U 202 Engring St Stillwater OK 74078

RAMAMOORTHY, CHITTOOR V., computer science educator, consultant; b. Burma, May 5, 1926; s. Chittoor V. Naidu and Lakshmikanthamma; m. Daulat; children: Vijay, Sonia, Maya. M.S. in Mech. Engring, U. Calif.-Berkeley; A.M., Harvard U., Ph.D. With electronic data processing div. Honeywell Co., Waltham, Mass., 1956-71, sr. staff scientist, to 1971; prof. elec. engring. and computer sci. U. Tex.-Austin, 1967-72, U. Calif.-Berkeley, 1972—; cons. in field. Contbr. articles to profl. jours. Fellow IEEE; mem. IEEE Computer Soc. (v.p.). Subspecialties: Computer architecture; Software engineering. Current work: Software engineering and computer architecture. Patentee in field.

RAMANAN, SUNDARAM VENKATA, physician, hematologist, oncologist; b. Calicut, Kerala, India, June 21, 1933; came to U.S., 1969, naturalized, 1975; s. Tarakad Appadoraier and Seethalakshmi (Iyer) S.; m. Chitraleka Rajagopalan, May 2, 1963; children: Kumar, Radhika. G.C.E., U. London, 1954; M.D., U. Madras, India, 1960; M.S., W.Va. U., 1974. Med. officer State Health Service, Sri Lanka, 1960-62; resident and fellow Brit. Nat. Health Service, London and Manchester, Eng., 1963-69; mem. faculty W.Va. U., 1969-76; assoc. prof. medicine and pharmacology, 1973-76; chief hematology/oncology Mt. Sinai Hosp., Hartford, Conn., 1976—; assoc. prof. depts. medicine and lab. medicine U. Conn. Sch. Medicine, 1976—; clin. prof. biology U. Hartford, 1978—; cons. hematology VA Hosp., Newington, Conn., 1976—. Author: Case Studies in Hematology, 1972, 2d edit., 1984. Named Clinician of Yr., W.Va. U. Sch. Medicine, 1973; recipient Outstanding Tchr. award W.Va. U., 1973; Physician

Educator award Mt. Sinai Hosp., Hartford, Conn., 1981. Fellow ACP; fellow Royal Soc. Medicine (London); Fellow Royal Coll. Physicians (Edinburgh), Internat. Soc. Hematology; mem. Am. Soc. Hematology, Am. Fedn. Clin. Research. Hindu. Subspecialties: Hematology; Cancer research (medicine). Current work: Investigator cancer and leukemia; mem. Eastern Cooperative Oncology Group. Office: Mt Sinai Hosp 500 Blue Hills Ave Hartford CT 06112

RAMANATHAN, VEERABHADRAN, meteorologist; b. Madras, India, Nov. 24, 1947; S. Veerabhadran and Janaki R.; m. Girija Ramanathan. B.S.M.E., Annamalai U., 1965; M.S.M.E., Indian Inst. Sci., 1970; Ph.D. in Atmospheric Sci, SUNY-Stony Brook, 1973. Engr. Shri-Ram Refrigeration Industries, 1965-67; Nat. Acad. Scis.-NRC postdoctoral research assocs. Langley Research Ctr., NASA, Hampton, Va., 1974-75; vis. scientist George Washington U.-NASA Langley Research Ctr., 1975-76; vis. scientist, staff scientist Climate Sensitivity Group Nat. Ctr. Atmospheric Research, Boulder, Colo., 1976-82, sr. scientist, 1982—, mem., 1979-84, leader, 1980—; invited expert climate rev. panel Nat. Acad. Scis., 1981-82; co-chmn. working group Internat. Radiation Commn., 1981-83, mem., 1981-83; steering com. on changes affecting habitability of earth NASA. Assoc. editor: Jour. Atmos. Scis, 1979-82, Jour. Applied Meteorology, 1982—; contbr. articles to profl. jours. Recipient Spl. Achievement award NASA, 1975, Incentive award NASA-Langley Research Ctr., 1979, Outstanding Publ. award Nat. Ctr. Atmospheric Research, 1981. Mem. Am. Meteorol. Soc., Am. Geophys. Union, AAAS. Subspecialties: Climatology; Meteorology. Current work: Climate sensitivity; cloud-climate interactions. Office: National Center Atmospheric Research Box 3000 Boulder CO 80307

RAMANUJA, JAYALAKSHMI KRISHNADESIKACHAR, nuclear engineer, health physicist; b. Mysore, India, Dec. 1, 1945; came to U.S., 1971, naturalized, 1979; d. Krishnadesikachar and Rukminamma Vidwam; m. Teralandur K. Ramanuja, Jan. 18, 1971; children: Srinivasan, Rekha. B.Sc., U. Mysore, 1964, M.Sc. in Physics, 1966; M.S. in Radiol. Health, U. Mich., 1976; Ph.D., Indian Inst. Sci., Bangalore, 1971. Postdoctoral research fellow U. Notre Dame, Ind., 1972-73; mgr. nuclear studies Environ. Research Group, Ann Arbor, Mich., 1976-80; sr. engr. Bechtel Power Corp., Ann Arbor, 1980—. Mem. Am. Nuclear Soc., Health Physics Soc. Subspecialties: Nuclear physics; Nuclear engineering. Current work: Advances in radiological health in nuclear power plant and hospitals, and in related industry. Home: 3476 Gettysburg Rd Ann Arbor MI 48105 Office: Bechtel Power Corp 777 E Eisenhower Pkwy Ann Arbor MI 48106

RAMANUJA, TERALANDUR KRISHNASWAMY, structural engineer; b. Mysore, Mysore, India, June 23, 1941; came to U.S., 1967, naturalized, 1979; s. Teralandur and Padmammal Krishnaswamy; m. Jayalakshmi Ramanuja, Jan. 18, 1971; children: Srinivasan, Rekha. B.S. in Civil Engring, U. Mysore, 1962, M.S., U. Notre Dame, 1969. Registered profl. engr., Ill., Ind., Mich. Design engr. Mil. Engring. Service, Bangalore, India, 1962-67; structural engr. Clyde E. Williams, Inc., South Bend, Ind., 1969-73; head structural dept. Ayres, Lewis, Norris & May, Ann Arbor, Mich., 1973-76; sr. project mgr. Johnson & Anderson Inc., Pontiac, Mich., after 1980; now engring. supr. Bechtel Power Corp., Ann Arbor, Mich. Mem. ASCE, Am. Concrete Inst., Chi Epsilon. Subspecialty: Civil engineering. Current work: Structural and foundation design of nuclear and fossil power plants, industrial, petrochemical plants, water and waste treatment facilities. Office: Bechtel Power Corporation 777 E Eisenhower Pkwy Ann Arbor MI 48106

RAMENOFSKY, SAMUEL DAVID, educator, consultant; b. LaSalle, Ill., Aug. 21, 1944; s. Abraham I. and Charlotte (Whitebook) R.; m. Cynthia Ann Wagner, Mar. 21, 1970; children: Carrie, Gregory, David. B.S. in Math, Iowa State U., 1966; Ph.D. in Econs, U. Okla., 1972. Asst. prof. U. Mo., Kansas City, 1970-72; asst. prof. Loyola U., Chgo., 1972-77, assoc. prof., chmn., 1978—. Contbr. articles to profl. jours. Coach Wilmette (Ill.) Park Dist., 1980-82. Gen. Electric fellow, 1975; NSF grantee, 1973. Mem. Am. Econ. Assn., Am. Inst. Decision Sci. Subspecialties: Statistics; Applied mathematics. Current work: Applied statistical and computer applications in economics and business. Office: Loyola U 820 N Michigan Ave Chicago IL 60611

RAMIG, ROBERT ERNEST, research soil scientist; b. McGrew, Nebr., June 22, 1922; s. Carl James and Lydia (Hardt) R.; m. Lois F. Franklin, Nov. 28, 1943; children: Robert Franklin, Mary Kathleen, John Carl. Sc.B., U. Nebr., 1943, Ph.D., 1960; M.Sc., Wash. State U., 1948. Asst. agronomist U. Nebr., North Platte, 1948-51, coop. agt., 1951-57; soil scientist U.S. Dept Agr., North Platte, 1957-61, Pendleton, Oreg., 1961-71, dir., 1971-81, research soil scientist, 1981—. City Councilman, Pendleton, Oreg, 1973—, chmn. mayor's downtown study group. Served with USNR, 1943-46. Sears, Roebuck and Co. fellow, 1939-40; recipient cert. merit U.S. Dept. Agr., 1960; First Citizen award Pendleton C. of C., 1980. Mem. AAAS, Am. Soc. Agronomy, Soil Sci. Soc. Am., Western Soil Sci. Soc., Soil Conservation Soc. Am., Sigma Xi. Democrat. Lutheran. Lodge: Rotary. Subspecialty: Resource conservation. Current work: Soil and water conservation using balanced fertility to give maximum production per unit water used. Home: 1208 NW Johns Ave Pendleton OR 97801 Office: USDA PO Box 370 Pendleton OR 97801

RAMIREZ, FAUSTO, chemistry educator, researcher; b. Zulueta, Cuba, June 15, 1923; s. Arturo and Noeli (Benet) R.; m. Joan Schwartz, May 13, 1949; children: Melissa, Colin. B.S., U. Mich., 1946, M.S., 1947, Ph.D., 1949. Instr. Columbia U., N.Y.C., 1950-53, asst. prof., 1953-58; assoc. Ill. Inst. Tech., Chgo., 1958-59; prof. chemistry SUNY-Stony Brook, 1959—; cons. Amoco Chem. Co., Naperville, Ill., 1955—, Am. Cyanamic Co., Bound Brook, N.J., 1955-75; vis. prof., fellow Alexander von Humboldt Found., Germany, 1973-74. Contbr. chpts. to books, articles to jours. Recipient Silver Medal of Paris Nat. Ctr. Sci. Research, 1969; Guggenheim fellow, 1967-68. Mem. Am. Chem. Soc., N.Y. Acad. Scis. (A. Cressy Morrison award 1968). Subspecialties: Organic chemistry; Synthetic chemistry. Current work: Synthesis of phospholipids, structure of biomembranes, biophosphorus compounds: synthesis and function; mechanisms of enzymatic phosphoryl transfer. Patentee in field. Office: Dept Chemistry SUNY Stony Brook NY 11794

RAMM, DIETOLF, computer scientist, educator, researcher; b. Berlin, June 17, 1942; U.S., 1948, naturalized, 1957; s. Wolfgang Julius and Dora Christiana (Ordnung) R.; m. Mary Kathlyn Brigmon, June 25, 1966; children: Karl, Lenore. Student, U. Munich, W.Ger., 1961-62; B.A., Cornell U, 1964; Ph.D., Duke U., 1969. Physicist U.S. Army, 1961-64; computer scientist Duke Ctr. Study of Aging, Durham, N.C., 1969-82; assoc. dept. computer sci. Duke U., 1972-82, research assoc. prof. computer sci., dir. undergrad. studies in computer sci., 1982—. Contributed numerous articles in low-temperature physics, psychiatry, gerontology and computer sci. to profl. jours. Mem. Assn. Computing Machinery, Am. Phys. Soc. Subspecialties: Medical applications; Microelectronics. Current work: Application of computers to medical problems; data analysis; use of micro-computers in computer science education. Office: 205 North Bldg Duke U Durham NC 27706

RAMMING, KENNETH PAUL, surgery educator; b. Ft. Wayne, Ind, Apr. 17, 1939; s. Leonard Christian and Ruth Dorothy (Becker) R.; m. Mary Ann Koeheman, Aug. 24, 1963; children: Peter, Paul, James. B.A. in English and Chemistry, Valparaiso U., 1961; M.D., Duke U., 1965. Diplomate: Am. Bd. Surgery. Researcher div. immunology Duke U. Sch. Medicine, 1963-64; intern in surgery Duke U. Med. Center, 1965-66, jr. asst. resident in surgery, 1966-67, sr. asst. resident in gen. and thoracic surgery, 1970-73, chief resident in gen. and thoracic surgery, 1973-74; clin. asso. surgery br. Nat. Cancer Inst., NIH, Bethesda, Md., 1967-70; asst. prof. surgery UCLA Med. Center, 1974-78, assoc. prof., 1978-82, prof., 1982—; chief thoracic surg. sect. surg. service VA Hosp., Sepulveda, Calif., 1975—; mem. surgery com. gastrointestinal tumor study group, immunology com., exec. com., lung cancer study group Nat. Cancer Inst., 1979; co-dir. project USPHS, 1981—. Contbr. numerous chpts., articles, abstracts to profl. publs.; author numerous profl. papers. Recipient Golden Scalpel award dept. surgery UCLA, 1980; USPHS grantee, 1975-79, 76-81, 79-82; VA research dept. grantee, 1978-81. Mem. Transplantation Soc., Am. Assn. Cancer Research, Assn. Acad. Surgery, Soc. Univ. Surgeons, Am. Soc. Clin. Oncology, Los Angeles County Med. Assn., Los Angeles Surg. Soc., Internat. Assn. Study Lung Cancer, ACS, Pan-Pacific Surg. Assn., Pacific Coast Surg. Assn. Lutheran. Subspecialties: Surgery; Oncology. Current work: Surgery and immunotherapy of gastrointestinal and thoracic cancer; surg. oncology; thoracic surgery; hyperthermia. Office: UCLA John Wayne Clinic 10083 La Conte Los Angeles CA 90024

RAMO, SIMON, engring. exec.; b. Salt Lake City, May 7, 1913; s. Benjamin and Clara (Trestman) R.; m. Virginia Smith, July 25, 1937; children—James Brian, Alan Martin. B.S., U. Utah, 1933, D.Sc. (hon.), 1961; Ph.D., Calif. Inst. Tech., 1936; D.Eng. (hon.), Case Inst. Tech., 1960, U. Mich., 1966, Poly. Inst. N.Y., 1971, D.Sc., Union Coll., 1963, Worcester Poly. Inst., 1968, U. Akron, 1969, Cleve. State U., 1976, LL.D., Carnegie-Mellon U., 1970, U. So. Calif., 1972. With Gen. Electric Co., 1936-46; v.p. ops. Hughes Aircraft Co., 1946-53; with Ramo-Woolridge Corp., 1953-58; sci. dir. U.S. intercontinental guided missile program, 1954-58; dir. TRW Inc., 1954—, vice chmn. bd., 1961-78, chmn. exec. com., 1969-78; chmn. bd. TRW-Fujitsu Co., 1980—; pres. The Bunker-Ramo Corp., 1964-66; vis. prof. mgmt. sci. Calif. Inst. Tech., 1978—; Regents lectr. UCLA, 1981-82, U. Calif. at Santa Cruz, 1978-79; chmn. Center for Study Am. Experience, U. So. Calif., 1978-80; Faculty fellow John F. Kennedy Sch. Govt., Harvard U., 1980—; dir. Union Bank, Times Mirror Co.; Mem. White House Energy Research and Devel. Adv. Council, 1973-75; mem. adv. com. on sci. and fgn. affairs U.S. State Dept., 1973-75; chmn. Pres.'s Com. on Sci. and Tech., 1976-77; mem. adv. council to Sec. Commerce, 1976-77; co-chmn. Transitition Task Force on Sci. and Tech. for Pres.-elect Reagan; mem. roster consultants to adminstr. ERDA, 1976-77; bd. advisors for sci. and tech. Republic of China, 1981—. Author sci., engring. and mgmt. books. Bd. dirs. Los Angeles World Affairs Council, Music Center Found., Los Angeles, Los Angeles Philharm. Assn.; trustee Calif. Inst. Tech., Nat. Symphony Orch. Assn.; trustee emeritus Calif. State Univs.; bd. visitors UCLA Sch. Medicine, 1980—. Recipient award IAS, 1956, Am. Inst. Elec. Engrs., 1959, Arnold Air Soc., 1960; Am. Acad. Achievement award, 1964; award Am. Iron and Steel Inst., 1968; Distinguished Service medal Armed Forces Communication and Electronics Assn., 1970; medal of achievement WEMA, 1970; awards U. So. Calif., 1971, 79; Kayan medal Columbia U., 1972; award Am. Cons. Engrs. Council, 1974; medal Franklin Inst., 1978; award Harvard Bus. Sch. Assn., 1979, Nat. Medal Sci., 1979; others. Fellow IEEE (Electronic Achievement award 1953, Golden Omega award 1975, Founders medal 1980), Am. Acad. Arts and Scis.; mem. Nat. Acad. Engring. (founder, council mem., Bueche award, 1983), Nat. Acad. Scis., Am. Philos. Soc., Eta Kappa Nu (eminent mem. award 1966). Subspecialty: Engineering management. Office: One Space Park Redondo Beach CA 90278

RAMOND, PIERRE M., physics educator; b. Neuilly sur Seine, France, Jan. 31, 1943; m. s. Lillian T. Cymbala, June 10, 1967; children: Tanya, Lisa, Jennifer. B.S.E.E., N.J. Inst. Tech., 1965; Ph.D., Syracuse U., 1969. Research assoc. Fermilab, Batavia, Ill., 1969-71; instr. Yale U., New Haven, 1971-72, asst. prof. physics, 1972-75; Millikan fellow Calif. Inst. Tech., Pasadena, 1976-79; prof. physics U. Fla., Gainesville, 1980—; trustee Aspen Ctr. for physics. Author: Field Theory: A Modern Primer, 1980. Subspecialties: Particle physics; Cosmology. Current work: Fundamental interactions. Home: 2502 NW 27th Terr Gainesville FL 32605 Office: Dept Physics U Fla Gainesville FL 32611

RAMOS-GABATIN, ANGELITA, internist; b. Manila, Philippines, Aug. 27, 1941; came to U.S., 1965, naturalized, 1975; d. Silvestre Fermin and Paz (San Agustin) Ramos; m. Joe Q. Gabatin, Sept. 14, 1968; children: Anthony, Jo Elizabeth. A.A., U. St. Thomas, 1959, M.D., 1964. Diplomate: Am. Bd. Internal Medicine. Intern Bristol (Conn.) Hosp., 1965-66; resident Hartford (Conn.) Hosp., 1966-68, Charleston (W.Va.) Meml. Hosp., 1968-69; staff internist VA Ctr., Wichita, Kans., 1970-75; internist Wichita Clinic, 1975; commd. maj. U.S. Air Force, 1975; advanced through grades to lt. col.; gen. internist Kadena AFB, Okinawa, Japan, 1976-78; fellow U.S. Air Force Med. Ctr., Wilford Hall, San Antonio, 1978-80; chief endocrine service U.S. Air Force Med. Ctr., Keesler AFB, Miss., 1980—; vis. clin. asst. prof. U. South Ala., 1981—, Tulane U., 1982—, VA Ctr., New Orleans, 1983. Contbr. articles to profl. jours. Arthritis Found. grantee, 1973; Physician's Recognition award AMA, 1980. Fellow ACP; mem. Filipino-Am. Assn. Gulf Coast, Endocrine Soc., Am. Fedn. Clin. Research, Soc. Air Force Physicians, Aerospace Med. Assn., U.S. Air Force Flight Surgeons Soc. Roman Catholic. Subspecialties: Endocrinology; Cytology and histology. Current work: Thyroid hormones; fine needle aspirations of thyroidal and nonthyroidal neck masses - branchial cysts and parathyroid cysts; calcitonin and 1 deg. HPT; MEN syndromes; pituitary tumors; lingual and ectopic thyroids. Home: 6013 Corban Pl Ocean Springs MS 39564 Office: Dept Medicine Endocrinology USAF Med Ctr Keesler AFB Biloxi MS 39534

RAMSAY, WILLIAM CHARLES, energy policy researcher; b. Jamaica, N.Y., Nov. 6, 1930; s. Claude Barnett and Myrtle Marie (Scott) R.; m.; children: Alice, John, Carol, David. B.A. in English Lit, U. Colo., 1952; M.A. in Physics, UCLA, 1957, Ph.D., 1962. NSF postdoctoral fellow, research assoc. U. Calif., San Diego, 1962-64, asst. prof. physics, Santa Barbara, 1964-67; tech. mgr., sr. staff scientist Systems Assoc., Inc., Long Beach, Calif., 1967-72; sr. environ. economist Directorate Regulatory Standards, AEC, Bethesda, Md., 1972-75; tech. adviser to commr. U.S. Nuclear Regulatory Commn., Washington, 1975-76; sr. fellow AID project mgr. Center for Energy Policy Research, Resources for Future, Washington, 1976-83; sr. fellow, dir. bioenergy project Ctr. Strategic and Internat. Studies, Georgetown U., Washington, 1983—; tchr., cons.; participant NATO Summer Insts., Edinburgh, Scotland, 1960, Bergen, Norway, 1962, Fed. Exec. Inst., 1976. Author: (with Claude Anderson) Managing the Environment: An Economic Primer, 1972. Buenos Aires Conv. scholar, Tegucigalpa, Honduras, 1952-53. Mem. Am. Phys. Soc., Am. Astron. Soc., Internat. Assn. Energy Economists. Subspecialties: Energy policy research; Biomass (energy science and technology). Current work: Energy economic and environmental problems, especially in solving policy problems for developing areas. Office: 1800 K St NW Washington DC 20006

RAMSEY, LAWRENCE WILLIAM, astrophysicist, educator; b. Louisville, Mar. 14, 1945; s. Cleve Murray and Mildred Ann (Schultheis) R.; m. Mary Ellen Gessling, Apr. 12, 1970. B.A., U. Mo., 1968; M.S., Kans. State U., 1972; Ph.D., Ind. U., 1976. Electronic systems engr. McDonnell-Douglas Corp., St. Louis, 1966-70; research asst. Kitt Peak Nat. Obs., Tucson, 1972-73; assoc. prof. Pa. State U., State College, 1976—. Contbr. articles to profl. jours. Mem. Am. Astron. Soc., Optical Soc. Am., Astron. Soc. Pacific, Internat. Astron. Union. Subspecialty: Astronomical instrumentation. Current work: Spectroscopy of cool active stars and red giants, the solarstellar connection, astronomical instrumentation. Office: 525 Davey Lab University Park PA 16802

RAMSEY, NORMAN, physicist; b. Washington, Aug. 27, 1915; s. Norman F. and Minna (Bauer) R.; m. Elinor Jameson, June 3, 1940; children: Margaret, Patricia, Janet, Winifred. A.B., Columbia U., 1935; B.A., Cambridge (Eng.) U., 1937, M.A., 1941, D.Sc., 1954; Ph.D., Columbia U., 1940; M.A. (hon.), Harvard U., 1947, D.Sc., Case Western Res. U., 1968, Middlebury Coll., 1969, Oxford (Eng.) U., 1973. Kellett fellow Columbia U., 1935-37, Tyndall fellow, 1938-39; Carnegie fellow Carnegie Inst. Washington, 1939-40; asso. U. Ill., 1940-42; asst. prof. Columbia U., 1942-46; asso. MIT Radiation Lab., 1940-43; cons. Nat. Def. Research Com., 1940-45; expert cons. sec. of war, 1942-45; group leader, asso. div. head Los Alamos Lab., 1943-45; asso. prof. Columbia U., 1945-47; head physics dept. Brookhaven Nat. Lab. of AEC, 1946-47; asso. prof. physics Harvard U., 1947-50, prof. physics, 1950-66, Higgins prof. physics, 1966—; sr. fellow Harvard Soc. of Fellows, 1970—; Eastman prof. Oxford U., 1973-74; Luce prof. cosmology Mt. Holyoke Coll., 1982-83; prof. U. Va., 1983-84; dir. Harvard Nuclear Lab., 1948-50, 52-53, Varlan Assos., 1963-66; mem. Air Forces Sci. Adv. Com., 1947-54; sci. adviser NATO, 1958-59; mem. Dept. Def. Panel Atomic Energy; exec. com. Cambridge Electron Accelerator and gen. adv. com. AEC. Author: Nuclear Moments and Statistics, 1953, Nuclear Two Body Problems, 1953, Molecular Beams, 1956, Quick Calculus, 1965; contbr.: articles Phys. Rev.; other sci. jours. on nuclear physics, molecular beam experiments, radar, nuclear magnetic moments, radiofrequency spectroscopy, masers, nucleon scattering. Trustee Asso. Univs., Inc., Brookhaven Nat. Lab., Carnegie Endowment Internat. Peace, 1962—, Rockefeller U., 1977—; pres. Univs. Research Assos., Inc., 1966-72, 73-81, pres. emeritus, 1981—. Recipient Presdl. Order of Merit for radar devel. work, 1947, E.O. Lawrence award AEC, 1960; Columbia award for excellence in sci., 1980; Guggenheim fellow Oxford U., 1954-55. Fellow Am. Acad. Sci., Am. Phys. Soc. (council 1956-60, pres. 1978-79, Davisson-Germer prize 1974); mem. N.Y., Nat. acads. sci., Am. Philos. Assn., AAAS (chmn. physics sect. 1977), Am. Inst. Physics (chmn. bd. govs. 1980—), Phi Beta Kappa (senator 1979—, v.p. 1982—), Sigma Xi. Subspecialties: Atomic and molecular physics; Particle physics. Current work: Experiments on time reversal symmetry and parity; molecular beams,neutron beams. Home: 55 Scott Rd Belmont MA 02178 Office: Lyman Lab Harvard Univ Cambridge MA 02138

RAMSEY, ROBERT BRUCE, mktg. and tech. oncology specialist; b. Moline, Ill., Jan. 4, 1944; s. Ralph Samuel and Florence Isabelle (Adams) R.; m. Penny Tina Germain, June 10, 1967; children: Anne M., Sarah E. A.B. in Chemistry, Augustana Coll., Rock Island, Ill., 1966; Ph.D. in Biochemistry, St. Louis U., 1971. Postdoctoral fellow Inst. Neurology, U. London, 1971-73; asst. prof. neurology St. Louis U. Sch. Medicine, 1972-79, assoc. prof., 1979-80, assoc. clin. prof., 1980—; product mgr. Lancer div. Sherwood Med., St. Louis, 1980—. Contbr. articles to profl. jours. Mem. Am. Soc. Biol. Chemists, Am. Chem. Soc., Biochem. Soc. (U.K.), Am. Assn. Clin. Chemists. Republican. Presbyterian. Subspecialty: Immunobiology and immunology. Home: 415 N Price Olivette MO 63132 Office: 1831 Olive St Louis MO 63103

RANALLI, DENNIS NICHOLAS, pedodontist, educator, consultant; b. Pitts., Dec. 18, 1946; s. Nick A. and Rose Marie (Cavalieri) R.; m. Linda M. Schlemmer, June 17, 1972; children: Dennis N., Michael A., Christine L. B.S., U. Pitts., 1968, M.D.S., 1982; D.D.S., Temple U., 1972. Gen. practice residency Lancaster Cleft Palate Clinic, 1972-74; staff dentist Lancaster (Pa.) Cleft Palate Clinic, 1973-74; dentist H.W. Zwicker & Assocs., Pitts., 1974-81; clin. instr. in pedodontics U. Pitts., 1974-76, asst. prof. pedodontics, 1977—, dir. pedodontic cleft palate program, 1974—, dir. grad. pedodontic clinic, 1978—; cleft palate team pedodontist, 1981—, asst. chmn. dept. grad pedodontic div., 1982—; mem. staff Children's Hosp., Pitts., 1981—; cons. in field. Contbr. articles to profl. jours.; editorial cons., abstractor: Cleft Palate Jour, 1982—. Mem. Parish Council, mem. edn. com. St. Alphonsus Ch., Wexford, Pa., 1981-85. Recipient Lactona award Temple U., 1972. Mem. ADA, Pa. Dental Assn. (del. 1981), Odontological Soc. Western Pa. (chmn. and mem. coms. 1978—), Western Pa. Soc. Dentistry for Children (pres. 1982-83), Am. Cleft Palate Assn. (editorial cons. 1982—), Am. Cleft Palate Ednl. Found., Sigma Xi, Omicron Kappa Upsilon. Current work: Pedodontics, growth and development; congenital anomalies—dental, cleft palate, syndromes; pediatric oral pathology. Home: 8596 Harvest Manor Dr Pittsburgh PA 15237 Office: U Pitts Sch Dental Medicine 333 Salk Hall Pittsburgh PA 15261

RAND, STEPHEN COLBY, physicist; b. Seattle, Nov. 20, 1949; s. Charles Gordon and Margaret (Colby) R.; m. Paula Dian Fraser, Sept. 6, 1976; 1 son, Spencer Fraser. B.Sc., McMaster U., 1972; M.Sc., U. Toronto, Ont., Can., 1974, Ph.D., 1978. World Trade postdoctoral fellow IBM Research, San Jose, Calif., 1978-80; research assoc. dept. physics Stanford (Calif.) U., 1980-82; scientist Hughes Research, Malibu, Calif., 1982—; referee Phys. Rev. Letters and Optics Letters. Mem. Optical Soc. Am., Am. Phys. Soc. Subspecialties: Spectroscopy. Current work: Research in nonlinear pair interactions in solids, nonlinear optics in fibers and applications of femtosecond optical pulses. Office: 3011 Malibu Canyon Rd Malibu CA 90265

RANDALL, DAVID CLARK, medical educator, research physiologist; b. St. Louis, Apr. 23, 1945; s. Walter Clark and Gwendolyn Ruth (Niebvel) R.; m. Pamela Kaye Reynolds, June 14, 1968; children: Christopher Clark, Matthew Faubion. B.A., Taylor U., 1967; Ph.D., U. Wash., 1971. Asst. prof. Johns Hopkins U., Balt., 1972-75; asst. prof. U. Ky. Coll. Medicine, Lexington, 1975-78, assoc. prof. dept. physiology and biophysics, 1978—. Contbr. article to profl. jour. Mem. Pavlovian Soc. (pres. 1983). Subspecialties: Physiology (medicine); Psychophysiology. Current work: Autonomic neural control of cardiovascular function in intact, unanesthetized animals; bio-behavioral bases of cardiovascular disease; brainstem control of heart rate. Office: Univ Ky Coll Medicine Lexington KY 40536-0084

RANDALL, LINDA LEA, molecular biologist; b. Montclair, N.J., Aug. 7, 1946; d. Lowell Neal and Helen (Watts) R.; m. Gerald Lee Hazelbauer, Aug. 29, 1970. B.S., Colo. State U., 1968; Ph.D., U. Wis., 1971. Postdoctoral fellow Pasteur Inst., Paris, 1971-73; lectr. molecular biology U. Uppsala, Sweden, 1973-75, asst. prof., 1975-80; assoc. prof. biochemistry Wash. State U., Pullman, 1981-83, prof., 1983—. Co-editor: Virus Receptors, Part I, 1980; mem. editorial bd.: Jour. Bacteriology. Swedish Natural Sci. Research Council grantee, 1975-80; NIH grantee, 1980—; WHO grantee, 1981-83; NATO grantee, 1982. Mem. Am. Soc. Microbiology, AAAS, Am. Soc. Biol. Chemists. Subspecialties: Molecular biology; Membrane biology. Current work: Mechanisms of export of protein through biological membranes. Office: Biochemistry/Biophysics Program Wash State U Pullman WA 99164

RANDALL, ROGER ELLIS, educator; b. Browntown, La., Feb. 2, 1925; s. Wilson and Calvie (Haughton) R.; m. Mildred Hamm, Aug. 25, 1950; children: Rogers Ellis, Della Carol, Carolyn Jean. B.A., Dillard U., 1950; M.A., U. Mich., 1951; M.A.T., Miami U., Ohio, 1961; Ph.D., Ohio State U., 1974. Asst. prof. and acting chmn. physics and math. dept. So. U., Baton Rouge, 1951-57; tchr., chmn. sci. dept. Roosevelt High Sch., Gary, Ind., 1958-74; asso. prof. and chmn. dept. sci. and math. Calumet Coll., Hammond, Ind., 1971—. Author: Experiments Inorganic Chemistry, 1977. Served with U.S. Army, 1942-46. Named Chemistry Tchr. of the Yr. Midwest Chem. Ent., 1972; Outstanding Layman United Meth. Men, 1963; Outstanding Coll. Tchrs. Inland-Ryerson Found., 1980. Fellow Am. Inst. Chemists; mem. Ind. Acad. Sci., Nat. Tchrs. Assn., AAUP (treas.), NAACP, Alpha Phi Alpha. Democrat. Methodist. Subspecialty: Inorganic chemistry. Current work: Science education and high energy technology. Home: 2395 W 20th Pl Gary IN 46404 Office: Calumet Coll 2400 New York Ave Whiting IN 46394

RANDALL, RUSSEL R., well log specialist; b. Tulsa, Mar. 9, 1948; s. Russel R. and Shirley L. (Light) R.; m. Bonnie Jean Carpenter, May 1970; children: Brent E., Regina Kay. B.A., Kans. State Tchrs. Coll., 1970; M.S., Kans. State U., 1972, Ph.D., 1975. With Dresser Atlas Industries Inc., Houston, 1975—, sr. project physicist, 1977-81, mgr. pulsed neutron devices, 1981—. Mem. Am. Phys. Soc., Soc. Profl. Well Log Analysis, Soc. Petroleum Engrs. Republican. Subspecialties: Nuclear physics; Atomic and molecular physics. Current work: Research and development of sealed Deuterium-Tritium neutron sources for pulsed neutron oil well logging; research and development of nuclear oil well logging systems. Patentee in field. Office: PO Box 1407 Houston TX 77001

RANDERATH, KURT, pharmacology educator, researcher; b. Dusseldorf, Ger., Aug. 2, 1929; came to U.S., 1963, naturalized, 1972; s. Edmund M. and Mathilde A. (Sachs) R.; m. Erika Randerath, Dec. 19, 1962. M. D., Heidelberg U., 1955; M.S. in Chemistry, 1959. Intern Suttgart (Ger.) City Hosp.; asst. prof. pharmacology Harvard U., 1968-71; assoc. prof. Baylor Coll., 1971-74; prof., 1974—. Author: Thin-Layer Chromatoraphy, 1966; editorial bd.: Jour. Chromatographic Sci, 1969; contbr. in field. Recipient NIH Career Devel. award, 1968; Am. Cancer Soc. Faculty Research award Am. Chem. Soc., 1971. Mem. Am. Soc. Bio . Chemists, Am. Assn. Cancer Research, AAAS. Subspecialties: Cancer research (medicine); Molecular pharmacology. Current work: Actions of anticancer drugs and carcinogens at the molecular level; nucleic acid structure; development of ultrasensitive analytical methods. Office: Depart Pharmacology Baylor U College Medicine Houston TX 77030

RANDOLPH, JAMES EUGENE, aerospace mission designer; b. Los Angeles, Jan. 19, 1940; s. Wallace L. and Katherine L. R.; m. Marilyn Miller, May 19, 1968; 1 son, John James. B.S., Calif. State U.-Los Angeles, 1964; M.S., U. So. Calif., Los Angeles, 1967. Systems engr. advanced s/c studies to Venus and Mars Jet Propulsion Lab., Pasadena, Calif., 1968-70; mission engr. Mariner, Voyager, Shuttle Radar, 1970-77; sci. integration team chief Voyager, 1978, Starprobe study mgr., 1977—, advanced mission engrng. group supr., mission design sect., 1980—. Fellow AIAA (assoc.); mem. Am. Astron. Soc. Subspecialties: Aerospace engineering and technology; Astronautics. Current work: Advanced spacecraft mission and system design management, aerospace engineering and technology, astronautics, systems engineering. Office: 4800 Oak Grove Dr 156-220 Pasadena CA 91109

RANHOTRA, GURBACHAN SINGH, nutritionist, researcher; b. India, Aug. 8, 1935; came to U.S., 1960; s. Moti S. and Surjit K. R.; m. Tejinder K. Suri, May 27, 1960; children: Gurdeep, Anita. Ph.D., U. Minn., 1964. Assoc. prof. Punjab Agrl. U., Ludhiana, India, 1965-67; dir. nutrition Am. Inst. Baking, Manhattan, Kans., 1968—; adj. prof. Kans. State U., Manhattan, 1977—. Assoc. editor: Cereal Chemistry, 1979—; contbr. research articles to profl. publs. U.S. Dept. Agr. research grantee, 1981, 82. Mem. Am. Inst. Nuitrition, Inst. Food Technologists, Am. Assn. Cereal Chemists (chmn. nutrition div. St. Paul 1981-82, chmn. Manhattan sect. 1980—). Democrat. Sikh. Subspecialties: Nutrition (medicine); Food science and technology. Current work: Basic and applied research in minerals and lipids area. Home: 1308 Givens Manhattan KS 66502 Office: Am Inst Baking 1213 Bakers Way Manhattan KS 66502

RANKIN, JOANNA MARIE, astronomy educator; b. Denver, Mar. 10, 1942; d. Robert McCurdy and Julia Bernice (Pelsor) R. B.S., So. Meth. U., 1965; M.S., Tulane U., 1966; Ph.D., U. Iowa, 1970. Research assoc. U. Iowa, Iowa City, 1970-74; vis. scientist Arecibo Obs., P.R., 1970-74, acting head computer dept., 1975; vis. scientist Radiophysics Div., CSIRO, Sydney, Australia, 1972; asst. prof. Dept. Astronomy, Cornell U., Ithaca, N.Y., 1974-78, sr. research assoc. dept. history, 1978-80; assoc. prof. physics U. Vt., Burlington, 1980—. Contbr. articles to profl. jours. Am. Philos. Soc. grantee, 1972; NSF grantee, 1973, 74, 78, 80. Mem. Am. Astron. Soc., Internat. Astron. Union, Internat. Radio Sci. Union. Subspecialty: Radio and microwave astronomy. Current work: Radio astronomy of pulsars and interstellar medium, physical science educator, researcher. Address: A405 Cook Bldg Dept Physics U Vt Burlington VT 05405

RANNEY, J. WARREN, research forester; b. Rockville Centre, N.Y., June 9, 1944; s. Vere J. and Edna Louise (Reyer) R. B.S. in Forestry, N.C. State U., 1969, M.Landscape Arch., 1971; Ph.D. in Ecology, U. Tenn., 1978. Registered landscape architect. Research asst. N.C. State U., Raleigh, 1969-71; coordinator/planner/landscape architect U.S. Dept. Agr., Forest Service, Gainesville, Ga., 1971-75; research fellow U. Tenn., Knoxville, 1975-78, sr. research assoc., 1978-79; research assoc., field research program mgr. Oak Ridge Nat. Lab., 1979—; U.S. rep. to Internat. Energy Agy., Forest Growth and Prodn., 1979—. Sr. author: Forest Island Dynamics of Man-Dominated Landscapes, Ecol. Studies, vol. 14, 1981. Recipient Cert. of Merit U.S. Dept. Agr.-Forest Service, 1974. Mem. Soc. Am. Foresters, Forest Products Research Soc. Subspecialties: Biomass (agriculture); Resource management. Current work: Intensive forest management, forest dynamics, environmental effects of forest management, program management in short-rotation intensive culture of woody biomass for energy and other uses. Office: Environ Sci Div Bldg 1505 Oak Ridge Nat Lab PO Box X Oak Ridge TN 37830

RANSOM, BRUCE ROBERT, medical scientist, neurologist, educator; b. Santa Fe, Aug. 5, 1945; s. H. Robert and Alberta J. (Hoenig) R.; m.; children: Rebecca Kay, Christopher Bruce. B.A., U. Minn., 1967; M.D., Washington U., St. Louis, 1972, Ph.D. in Neurophysiology, 1972. Intern Washington U. Med. Ctr., St. Louis, 1972-73; research assoc. NIH, Bethesda, Md., 1973-76; resident in neurology Stanford (Calif.) Med. Ctr., 1976-79, assoc. prof. neurology, 1979—. Contbr. writings to books and jours. in field. Served to lt. comdr. USPHS, 1973-76. NIH Research Career Devel. awardee, 1980—; research grantee, 1981—; program project grantee, 1981—. Mem. Neurosci. Soc., Am. Acad. Neurology, Am. Neurol. Assn., Am. Soc. Neurol. Investigation. Roman Catholic. Subspecialties: Neurology; Neurophysiology. Current work: Cellular mechanisms of anticonvulsants. Cellular physiology of mammalian central nervous system. Home: 13278 Paramount Dr Saratoga CA 95070 Office: Stanford Hosp Stanford U Med Ctr Stanford CA 94305

RANSON, JOHN HUGH CHARLES, surgeon; b. Bangalore, India, Oct. 28, 1938; came to U.S., 1949, naturalized, 1970; s. Charles Wesley and Jessie Grace (Gibb) R.; m. Patricia Vignolo, Nov. 20, 1982. B.A. with honors, Oxford U., 1960, M.A., 1963, B.M., B.Ch., 1963. Diplomate: Am. Bd. Surgery. Intern Tindel Gen. Hosp./Aylesbury Bucks/Churchill Hosp., 1964-65; surg. resident Bellevue Hosp., NYU Med. Ctr., N.Y.C., 1965-69; instr. surgery NYU Med. Ctr., N.Y.C., 1969-71, asst. prof. surgery, 1971-74, assoc. prof., 1974-79, prof., 1979—; assoc. dir. surg. services NYU Hosp., 1979—. James IV Surg. Traveler, 1979; recipient Rousing-Tscherning medal Danish Surg. Soc., 1982. Fellow Am. Surg. Assn., Soc. Univ. Surgeons, So. Surg. Assn., ACS; mem. Royal Soc. Medicine (London), Am. Gastroenterol. Assn. Subspecialty: Surgery. Current work: Research, teaching and practice in area of pancreatic, hepatic and biliary surgery. Office: NYU Med Ctr 550 1st Ave New York NY 11215

RAO, ADISESHAPPA RAMACHANDRA, civil engineering educator; b. Kanakapura, Karnataka, India, Oct. 23, 1939; came to U.S., 1962; s. N. Adieseshappa and Jayalakshmi (Gopaliah) R.; m. Mamatha Shiama Rao, Aug. 26, 1971; children: Malini B., Karthik A., Siddhartha S. B.E., U. Mysore, India, 1960; M.S.C.E., U. Minn., 1964; Ph.D., U. Ill., 1968. Asst. prof. civil engrng. Purdue U., West Lafayette, Ind., 1968-73, assoc. prof., 1973-80, prof., 1980—; cons. Atty. Gen.'s Office, State of Ind., 1981-82. Author: (with R. L. Kashyap) Dynamic Stochastic Models From Empirical Data, 1976. Sr. research fellow NRAC, Govt. N.Z., 1977-78. Mem. ASCE, Am. Geophys. Union, AAAS. Subspecialties: Surface water hydrology; Civil engineering. Current work: Hydraulic engineering, sedimentation, hydrology, modeling mathematical models. Home: 108 Pawnee Dr West Lafayette IN 47906 Office: Dept Civil Engrng Purdue U West Lafayette IN 47907

RAO, DABEREU CHANDRASEKHARA, educator, administrator; b. Santabommali, India, Apr. 6, 1946; came to U.S., 1972; s. Rama and Venkataratnam R.; m. Sarada Patnaik, July 31, 1974; children: Ravi, RLakshmi. B.Stat., Indian Statis. Inst., 1967, M.Stat., 1968, Ph.D., 1971. Postdoctoral fellow dept. problems and statis. U. Sheffield, Eng., 1971-72; asst. geneticist pop. genetics lab. U. Hawaii, Honolulu, 1972-78, assoc. geneticist pop. genetics lab., 1978-80; assoc. prof., dir. biostatis., assoc. prof. psychiatry and genetics, adj. prof. math. Wash. U., St. Louis, 1980-82, prof., dir. biostatistics, prof. psychiatry and genetics, adj. prof. math., 1982—. Author: A Source Book for Linkage in Man, 1979, Methods in Genetic Epidemiology, 1983. NIMH grantee, 1981; Nat. Inst. Gen. Med. Scis. grantee, 1981. Mem. Am. Soc. Human Genetics, Am. Assn. Phys. Anthropologists, AAAS, Biometric Soc., Soc. Epidemiological Research. Subspecialties: Genome organization; Genetics and genetic engineering (medicine). Current work: Familial transmission of diseases and risk factors; statistical methods in human genetics; genetic epidemiology; genetics of common diseases. Home: 6316 Pershing St Louis MO Office: Div Biostatistic Dept Preventive Medicine Wash U Sch Medicine Box 8067 4566 Scott Ave St Loui M 63110

RAO, DANDAMUDI VISHNUVARDHANA, nuclear physicist, educator; b. Maredumaka, India, Apr. 5, 1944; came to U.S., 1968, naturalized, 1979; s. Veeraraghaviah and Sarojini D. (Koneru) R.; m. Sujata L. Rao, Feb. 27, 1967; children: Saroja, Neeraja. M.S., U. Mass., 1970, Ph.D., 1972. Instr. radiology Albert Einstein Coll. Medicine, Bronx, N.Y., 1972-76; asst. prof. radiology U. Medicine and Dentistry, Newark, 1975-78, assoc. prof. radiology, 1978—, dir. health physics, 1974-78; tech. expert IAEA. Contbr. articles to profl. jours.; Author: Physics of Nuclear Medicine, 1977. Am. Cancer Soc. grantee, 1975-77; Biomed. Research grantee, 1977-78; Nat. Cancer Inst., NIH grantee, 1982-84. Mem. Am. Assn. Physicists in Medicine (program dir. summer sch. 1983), Soc. Nuclear Medicine. Subspecialties: Nuclear medicine; Imaging technology. Current work: In vivo study of radiation effects in Spermatoponial cells from low energy electrons emitted by nuclear medicine radiopharmaceuticals. Patentee radioactive erbium complexes. Office: U Medicine and Dentistry of NJ 100 Bergen St Newark NJ 07103

RAO, GOPAL SUBBA, pharmacologist, researcher; b. Mangalore, India, Aug. 12, 1938; came to U. S., 1961, naturalized, 1980; s. Subba Gopal and Sharada Bai (Bhat) R.; m. Harsha Purushottam Udeshi, May 29, 1972; 1 son, Raveen. B.Sc., Madras U., India, 1958; M.S., Howard U., 1965; Ph.D., U. Mich., 1969. Chemist Pub. Health Inst., Bangalore, India, 1958-61; research asst. Howard U. Coll. Pharmacy, Washington, 1961-65, instr., 1962-65; research asst. U. Mich. Coll. Pharmacy, Ann Arbor, 1965-69; internat. fellow Nat. Heart and Lung Inst., NIH, Bethesda, Md., 1969-72, spl. fellow, 1972-74, NIH grantee, 1974 82; dir., chief research scientist div. biochemistry Research Inst., Am. Dental Assn. Health Found., Chgo., 1978—, head pharmacology lab., 1974—; research adv. grad. students Northwestern U. Med. Dental Schs. Contbr. numerous articles on pharmacology to profl. jours.; abstractor: Dental Abstracts, 1975—. Am. Fund for Dental Health grantee, 1978-80. Mem. Assn. Scientists of Indian Origin (councillor 1980-82, chmn. membership com. 1982), Am. Soc. Pharmacology and Exptl. Therapeutics, Am. Coll. Toxicology, Soc. Toxicology, AAAS, Am. Chem. Soc., Am. Pharm. Assn., Acad. Pharm. Sci., Internat. Assn. Dental Research, Am. Assn. Dental Research, Am. Soc. Pharmacognosy, Sigma Xi (lectr. U. Miss. Med. Center 1982), Rho Chi. Subspecialties: Pharmacology; Oral biology. Current work: Biochemical etiology of periodontal diseases, development of new diagnositc methods and novel drugs and procedures, useful in the treatment of oral diseases, salivary nitrite and carcinogenic nitrosamine formation, occupational hazards in dental practice. Office: ADAHE Research Inst 211 E Chicago Ave Chicago IL 60611

RAO, GUNDU HIRISAVE RAMA RAO, pathology educator, researcher; b. Tunkur, Karnataka, India, Apr. 17, 1938; came to U.S., 1965; s. Rama H. V. and Annapoorna T.S. R.; m. Yashoda T.R. Rao, June 11, 1965; children: Anupama T.G., Prashanth T. G. B.S., U. Mysore, India, 1957, U. Poona, India, 1958, M.S., 1959; Ph.D., Kans. State U., 1968. Research fellow Council Sci. and Indsl. Research, Central Food Tech. Research Inst., Mysore, India, 1960-65; research asst. Kans. State U., Manhattan, 1965-68; postdoctoral fellow Tex. A&M U., College Station, 1968-72, U. Minn., Mpls., 1968-72, scientist dept. pediatrics, 1972-75, asst. prof. lab. med. pathology, 1975-81, assoc. prof. lab. med. pathology, 1981—. Contbr. writings to publs. in field. Bd. dirs. India Club., Mpls., 1982; founding mem. Sch. Indian Langs. and Culture, Mpls., 1978—; life mem. Friends of Vellore Club (Christian Med. Coll.), India, 1982. Fellow Nat. Acad. Clin. Biochemists; mem. Am. Assn. Pathologists, Internat. Soc. on Thrombosis and Haemostasis, Am. Assn. Clin. Chemists, Nat. Council on Thrombosis, Am. Heart Assn. Hindu. Subspecialties: Molecular pharmacology; Cell biology (medicine). Current work: Biochemical mechanisms involved in Thrombosis and Haemostasis. Role of platelets and prostglandins in cancer biology. Home: 1989 Warbler Ln Saint Paul MN 55119 Office: Mayo Hosps Lab Med Pathology U Minn P B 198 Minneapolis MN 55455

RAO, KALIPATNAPU NARASIMHA, biochemistry educator; b. Narasapur, A.P., India; came to U.S., 1971, naturalized, 1981; s. Sambasivarao and Raghavamma (Gaidiraju) K.; m. Rama Rao, Aug. 15, 1965; children: Padmavati, Babu, Uma. B.S., Bombay U., 1958; M.S., Nagpur U., 1960; Ph.D., Indian Agrl. Research Ins., 1965. Research officer Ministry of Health, Govt. India, 1964-71; asst. prof. U. Pitts., 1971—. Nat. Cancer Inst. grantee, 1980. Mem. Am. Inst. Nutrition, Am. Pancreatic Assn., AAAS. Hindu. Subspecialties: Biochemistry (medicine); Cancer research (medicine). Current work: Biochemical pathology of pancreas, experimental nutrition, lipid metabolism. Home: 1298 Lakemont Dr Pittsburgh PA 15243 Office: Pathology Dept Sch Medicine U Pitts Pittsburgh PA 15260 Home: 1298 Lakemont Dr Pittsburgh PA 15243

RAO, PENMARAJU VENUGOPALA, physicist, educator; b. Tirupatipuram, India, Sept. 1, 1932; came to U.S., 1959; s. Pemmaraju Subbarao and Pemmaraju Satyavati; m. P.S. Lakshmi, Aug. 15, 1938; children: Nalini, Saleena. B.Sc. with honors, Andhra U., India, 1953, M.S., 1954; Ph.D., U. Oreg., 1964. Demonstrator Andra U., 1955-57, lectr., 1958-59; teaching fellow U. Oreg., 1959-64, research assoc., 1964-65, Ga. Inst. Tech., 1965-67; assoc. prof. physics Emory U., 1967—. Editor: Telngu Bhasha Patrika, Atlanta, 1970-76. Pres. India Am. Cultural Assn., 1975. Subspecialties: Atomic and molecular physics; Nuclear physics. Current work: Research in atomic, molecular, and nuclear physics. Home: 1550 Diamond Head Circle Decatur GA 30033 Office: Dept Physic Emory U Atlanta GA 30322

RAO, RAVINDRA P., radiation physicist; b. India; S. Purushotham U. and Indra P. R.; m. Sudha, May 30, 1975; 1 child, Harish. B.Sc., U. Bombay, India, M.Sc., 1969; M.E., U. Va., 1977. Mgr. Bombay Synthetics Pvt. Ltd., 1970-76; physicist C.V. Meml. Hosp., Johnstown, Pa., 1976-82, Bapt. Med. Ctr., Jacksonville, Fla., 1982—. Mem. Am. Assn. Physicists in Medicine. Hindu. Subspecialties: Radiology; Biophysics (physics). Home: 5800 University Blvd W Apt 317 Jacksonville FL 32216 Office: 800 Prudential Dr Jacksonville FL 32207

RAO, SAMBASIVA M., pathologist, educator; b. Chiluvur, India, Oct. 19, 1942; d. Bhushaiah and Tulasamma Chaudury; m. Nagasiromani, May 18, 1968; children: Sanjoy, Sunila. M.D., Guntur Med. Coll., 1972. Diplomate: Am. Bd. Pathology, 1978. Asst. prof. pathology Northwestern U., 1977-81, asso. prof., 1981—. Contbr. articles to profl. jours. Mem. Am. Assn. Cancer Research, Am. Assn. Pathologists, Internat. Acad. Pathology, AAAS. Subspecialty: Pathology (medicine). Current work: Carcinogenesis. Home: 1536 Walters Northbrook IL 60062 Office: Dept Pathology Northwestern U 303 E Chicago Ave Chicago IL 60611

RAO, SURENDAR PURUSHOTHAY, nuclear engineer; b. Bombay, India, Jan. 5, 1946; s. Purushotham and Indira R.; m. Alka Surendar, Jan. 23, 1976; children—Avinash, Ansali. B.Sc., U. Bombay, 1968, M.Sc., 1974; M.E., U. Va., 1976. Vice prin. Teaching Inst., Bombay, 1970-73; research asst. U. Va., Charlottesville, 1974-76; nuclear engr. Combustion Engring. Inc., Windsor, Conn., 1976-80; lead engr. EDS Nuclear Inc., Melville, N.Y., 1980; tech. specialist Birchwood, Warrington, Eng., 1981—. Mem. Am. Nuclear Soc. Subspecialties: Nuclear fission; Nuclear engineering. Current work: Nuclear safety kinetics, nuclear systems safety, probabilistic risk assesments for the U.K. pressurized water reactor, systems interactions for U.K. pressurized water reactor. Office: EDS Nuclear Inc Genesis Centre Garretfield Birchwood Warrington England WA3 7BH

RAO, SURYANARAYANA KOPPAL, toxicologist; b. Hyderabad, India, Feb. 20, 1939; s. Hanumanth K. and Rukmini (Bai) R.; m. Batnam K. Saripalli, Nov. 16, 1967; children: Anil, Padmaja. D.V.M., Osmania U., Hyderabad, 1961; Ph.D., Magadh U., 1968. Diplomate: Am. Bd. Toxicology. Toxciologist G. D. Searle, Chgo., 1971-77, Dow Chem. Co., Midland, Mich., 1977—. Recipient Gold medal Osmania and Magadh U. Mem. Soc. Toxicology, Teratology Soc., India Assn. (pres.). Subspecialties: Toxicology (medicine); Teratology. Current work: Safety eval. chems., drugs and pesticides; risk assessment to humans. Home: 504 Harper Ln Midland MI 48640

RAO, VENKATESWARA KOPPANADHAM, nephrologist, clinical investigator; b. Varigunta Padu, India, Sept. 8, 1946; came to U.S., 1969; d. Narasimhaiah and Venkatakkamma (Sunkara) Koppanadham; m. Lata K. Rao, Aug. 9, 1970; children: Ganguly, Sreedhar. M.B.B.S., Andhra U. Guntur Med. Coll., 1968. Diplomate: Am. Bd. Internal Medicine. Instr. medicine U. Minn., Mpls., 1974-76, asst. prof., 1976—; med. dir. renal transplantation Hennepin County Med. Ctr., 1976—. Fellow ACP; mem. Internat. Congress Nephrology, Transplantation Soc., Am. Soc. Nephrology, Am. Fedn. Clin. Research. Subspecialties: Internal medicine; Nephrology. Current work: Clinical research in kidney transplantation. Office: Hennepin County Medical Ctr 701 Park Ave Minneapolis MN 55415

RAPAPORT, FELIX THEODOSIUS, surgeon, educator; b. Munich, Ger., Sept. 27, 1929; s. Max W. and Adelaide (Rathaus) R.; m. Margaret Birsner, Dec. 14, 1969; children: Max, Benjamin, Simon, Michael. A.B., N.Y. U., 1951, M.D., 1954. Diplomate: Am. Bd. Surgery, 1963. Intern Mt. Sinai Hosp., N.Y.C., 1955-56; resident, chief resident N.Y. U. Surg. Services, 1958-62, USPHS postdoctoral fellow in pathology, 1956; trainee in allergy and infectious diseases N.Y. U., 1958-61; head, transplantation and immunology div. N.Y. U. Surg. Services, 1965-77; dir. research Inst. Reconstrn. and Plastic Surgery, N.Y. U., 1965-77, assoc. prof. surgery, 1965-70, prof., 1970-77; prof., dep. chmn. dept. surgery, prof. pathology, dir. transplantation service SUNY, Stony Brook, 1977—, attending, 1980—; cons. VA Hosp., N.Y.C., 1963-77, Northport, N.Y., 1977—. Editor-in-chief: Transplantation Proc, 1968—; assoc. editor: Am. Jour. Kidney Diseases, 1981—, Am. Jour. Craniofacial Genetics and Developmental Biology, 1980—; contbr. over 300 articles to profl. jours.; author/editor 9 books on transplantation. Served to lt. comdr. M.C. USNR, 1956-58. Decorated comdr. Order Sci. Merit, chevalier Ordre National du Merite, France, 1970; recipient Gold medal Societe d'Encouragement au Bien, 1979; grand croix Ordre des Palmes Academiques, 1981. Mem. Soc. Univ. Surgeons, N.Y. Surg. Soc., Am. Surg. Assn., ACS, Am. Assn. Immunologists, Soc. Exptl. Biology and Medicine, Harvey Soc., Am. Assn. Transplant Surgeons, Am. Assn. Clin. Histocompatibility Testing, Internat. Soc. Exptl. Hematology, Transplantation Soc. (founding sec., v.p., pres.), Alpha Omega Alpha. Democrat. Jewish. Subspecialties: Transplant surgery; Transplantation. Current work: Induction of permanent tolerance to major transplantable organs in man; research concerned with effects of total body irradiation and bone marrow transplantation in the production of host unresponsiveness to tissue allografts. Office: Dept Surgery Health Scis Ctr SUNY Stony Brook Stony Brook NY 11794

RAPAPORT, JACOBO, physics educator; b. Santiago, Chile, Nov. 30, 1930; came to U.S., 1965; s. Adolfo and Sarah (Turk) R.; m. Irma Butensky, June 7, 1958; children: Adolfo, Sonia. E.E., U. Chile, 1956; M.Sc. in Physics, U. Fla., 1958, Ph.D., MIT, 1963. Prof. physics U. Chile, Santiago, 1956-65; asst. prof. physics MIT, Cambridge, Mass., 1965-69; researcher Oak Ridge Nat. Lab., 1968-69; mem. faculty dept. physics Ohio U., 1969—, prof., 1973—, disting. prof., 1981—. Fellow Am. Phys. Soc. Subspecialty: Nuclear physics. Current work: Experimental low and intermediate nuclear physics with emphasis on neutron and in charge-exchange reactions. Office: Dept Physics Ohio U Athens OH 45701

RAPER, CARLENE ALLEN, biologist; b. Plattsburgh, N.Y., Jan. 9, 1925; d. Benjamin I. and Cornelia Nichols (Hagar) Allen; m. John Robert Raper, Aug. 9, 1949; children: Jonathan Arthur, Linda

Carlene. B.S., U. Chgo., 1946, M.S., 1948; Ph.D., Harvard U., 1977. Research scientist Argonne Nat. Labs., Ill., 1947-52; research scientist in biology Harvard U., 1961-74, lectr., research asso., 1974-77; postdoctoral fellow developmental plant biology Rijksuniversiteit Groningen, Haren, Netherlands, 1977; vis. lectr. Bridgewater State Coll., 1978; asst. prof. biol. scis. Wellesley (Mass.) Coll., 1978—; co-chmn. Grodon Conf. Fungal Metabolism, 1982-84; cooperating investigator U.S. Israel Binat. Sci. Found., 1974-77. Asso. editor: Jour. Exptl. Mycology, 1979—; contbr. articles to prof. jours. Mem. Masterworks Chorale, 1955—, exec. bd., 1958-60, 65-67; discussion leader LWV, 1957-59; co-founder citizens Com. Lexington Public Schs., 1957, exec. council, 1957-60; co-founder Fund for Urban Negro Devel., 1966-69. Campbell Soup Co. grantee, 1974-75, Maria Moors Cabot Found. grantee, 1975-77; Netherlands Orgn. Advancement Pure Research grantee, 1977; William and Flora Hewlett Found. grantee Research Corp., 1979-81; HEW grantee, 1979; NSF Opportunity Award, 1982. Mem. Genetics Soc. Am., Mycol. Soc. Am., N.Y. Acad. Scis., Women in Sci. and Engring., Sigma Xi. Subspecialties: Developmental biology; Gene actions. Current work: Research in genetic control of devel. and differentiation in higher fungi, especially Agaricus and Schizophyllum. Home: 2 Constitution Rd Lexington MA 02173 Office: Dept Biol Scis Wellesley Coll Wellesley MA 02181

RAPOPORT, STANLEY I., physician, government research administrator; b. N.Y.C., Nov. 24, 1932; m., 1961; 2 children. A.B., Princeton U., 1954; M.D., Harvard U., 1959. Intern in medicine Bellevue Hosp., N.Y.C., 1959-60; research scientist in neurophysiology NIMH, 1960-62, research scientist in physiology, 1964-78; chief lab. neurosci. Gerontology Research Center, Nat. Inst. on Aging, Balt., 1978—; professorial lectr. Georgetown U. Sch. Medicine, 1971—. NSF fellow in biophysics Physiol. Inst. Uppsala, Sweden, 1962-64. Mem. AAAS, Biophys. Soc., Soc. Neurosci., Soc. Gen. Physiology, Am. Physiol. Soc. Subspecialty: Gerontology. Address: 3010 44th Pl NW Washington DC 20016

RAPP, DONALD, research scientist, author; b. Bklyn., Sept. 27, 1934; s. Jacob and Irene (Levenson) R.; m. Zolita Sue Sverdlove, May 30, 1956; children: Erica, Melissa. B.Chem. Engring., Cooper Union, 1955; M.S., Princeton U., 1956; Ph.D., U. Calif.-Berkeley, 1960. Staff scientist Lockheed Co., Palo Alto, Calif., 1959-65; assoc. prof. Poly. Inst. Bklyn., 1965-69; prof. U. Tex., Dallas, 1969-79; div. technologist Jet Propulsion Lab., Calif. Tech., Pasadena, 1979—, sr. research scientist, 1980—. Author: Quantum Mechanics, 1971, Statistical Mechanics, 1972, Solar Energy, 1981; also articles. Fellow Am. Phys. Soc. Subspecialties: Atomic and molecular physics; Solar energy. Current work: Energy systems, concurrent processing computers, micromechanical devices. Home: 1445 Indiana Ave South Pasadena CA 91030 Office: Jet Propulsion Lab Calif Inst Tech MS 157-316 4800 Oak Grove Dr Pasadena CA 91009

RAPP, ULF FUEDIGER, cancer scientist, researcher; b. Wernigerode/Harz, Germany, Dec. 22, 1943; came to U.S., 1970, naturalized, 1981; s. Albert C. and Martha J. (Weindel) R.; m. Marieluise Gertrud, Dec. 21, 1970. Teaching asst. in anatomy U. Freiburg, 1965-66, in biochemistry, 1967-69, intern univ. clinics, 1970; postdoctoral fellow McArdle Lab. Cancer Research, Madison, Wis., 1970-75; vis. scientist Nat. Cancer Inst. NIH, Frederick, Md., 1975—, chief viral pathology sect., 1980—. Contbr. articles to profl. jours. Subspecialties: Cell study oncology; Virology (biology). Current work: Tumor genes; retroviruses; chem. transformed cells; devel. biology; membrane receptors. Home: 5226 39th St NW Washington DC 20015 Office: Nat Cancer Inst Bldg 560 Room 21-77 Frederick MD 21701

RAPPORT, ROBERT, diagnostic radiologist; b. Havana, Cuba, Sept. 23, 1953; came to U.S., 1961; s. Morris and Suzy (Pearl) R. B.S., U. Fla., 1974; M.D., U. Miami, 1978. Diplomate: Nat. Bd. Med. Examiners. Clin. assoc. U. So. Fla. Coll. Medicine, Tampa, 1978-82, chief resident in diagnostic radiology, 1981-82. Mem. Radiol. Soc. N.Am., Hillsborough County Med. Assn. Jewish. Subspecialty: Diagnostic radiology. Current work: Ophthalmoplegia due to spontaneous thrombosis in a patient with bilateral cavernous carotid aneurysms. Home: 3132 W Lambright Ave Unit 906 Tampa FL 33614 Office: Robert E Scherzer CT Body Scanner 7001 N Dale Mabry Hwy Tampa FL 33614

RAPSEY, LAURIE ADELE, software engr.; b. Toronto, Ont., Can., July 28, 1953; d. William Woodside and Ruby Adelaide (Barnett) R.; m. Barrie John Ashworth, Sept. 7, 1979. B.Sc. (Univ. scholar), U. Toronto, 1976; M.Math. in Computer Sci, U. Waterloo, 1979. Programmer analyst Imperial Oil, Toronto, 1976-77; software engr. Central Dynamics, Ltd., Montreal, Que., Can., 1979-81, software mgr., Ottawa, Ont., Can., 1981—. Mem. Assn. Computing Machinery. Subspecialties: Operating systems; Real time work. Current work: Realtime operating systems work and applications. Office: 1775 Courtwood Crescent Ottawa ON Canada K2C 3J2

RASCHKE, CURT ROBERT, electronic device development scientist, electronic materials researcher; b. N.Y.C., Sept. 11, 1944; s. Charles Frederick and Grace Evelyn (van Nostr) R.; m. Susan Michele Hourigan, Oct. 14, 1972; 1 dau., Kimberly Darcy. B.A., Whitman Coll., Walla Walla, Wash., 1966; M.S., Cornell U., 1968, Ph.D., 1971. Sr. physicist A.M. Internat., Warrensville, Ohio, 1972-74; sr. chemist Union Carbide Corp., Bound Brook, N.J., 1974-78; mem. tech. staff Xerox Corp., Dallas, 1978-82, Tex. Instruments, 1982—. Referee: Jour. Applied Physics, 1976—. Mem. Am. Phys. Soc., Sigma Xi, Phi Beta Kappa. Republican. Episcopalian. Clubs: Cornell Tex., Brookhaven Country (Dallas). Subspecialties: Microelectronics; Electronic materials. Current work: Electroacoustic devices; research and engineering into materials development and processing for microwave surface acoustic wave, ink jet printing and polymer electret devices. Patentee in field. Home: 10140 Bettywood Ln Dallas TX 75243 Office: Texas Instruments MS 255 13500 N Central Espressway Dallas TX 75222

RASKA, KAREL FRANTISEK, JR., pathologist, virologist, educator; b. Prague, Czechoslovakia, Mary 26, 1939; came to U.S., 1968, naturalized, 1980; s. Karel and Helena (Heller) R.; m. Jana Raskova, Feb. 16, 1960; children: Karel, Francis. M.D., Charles U., Prague, 1962; Ph.D. in Biochemistry, Czechoslovak Acad. Sci., 1965. Diplomate: Fedn. State Med. Bd. U.S.A., Am. Bd. Pathology. Resident in pathology Coll. Medicine and Dentisty N.J. - Raritan Valley Hosp., Green Brook, 1973-76; fellow Yale U. Sch. Medicine, New Haven, 1965-66; research assoc. Waksman Inst. Microbiology Rutgers U., 1966-67, mem. grad. faculty, 1969—; scientist Inst. Organic Chemistry and Biochemistry Czechoslovak Acad. Sci., Prague, 1967-68; asst. prof. dept. microbiology Coll. Medicine and Dentistry N.J.-Rutgers Med. Sch., 1968-71, assoc. prof. microbiology, 1971-78, assoc. prof. pathology, 1973-76, prof. pathology, 1976—, prof. microbiology, 1978—; pathologist Coll. Medicine and Dentistry N.J.-Raritan Valley Hosp., 1976-81; sr. attending pathologist Middlesex Gen. Hosp., New Brunswick, N.J., 1978—, chief immunopathology, 1982—. Contbr. in field. Commonwealth Fund fellow, 1965; Damon Runyon-Walter Winchel Cancer Fund grantee; Nat. Cancer Inst. grantee; Nat. Inst. Arthritis, Metabolic and Kidney Diseases and Diabetes grantee. Mem. Am. Assn. Cancer Research, Am. Assn. Immunologists, Am. Soc. Cell Biology, Am. Soc. Microbiology, Am. Soc. Virology, Internat. Acad. Pathology, N.J. Soc. Pathologists, Soc. Exptl. Biology and Medicine, Am. Assn. Pathologists, Pluto Soc. Subspecialties: Pathology (medicine); Virology (medicine). Office: PO Box 101 Piscataway NJ 08854

RASLEAR, THOMAS GREGORY, research psychologist; b. N.Y.C., Nov. 25, 1947; s. John William and Catherine (Turchin) R.; m. Lois T. Keck, Aug. 7, 1971. B.S. with honors, CCNY, 1969; Sc.M., Brown U., 1972, Ph.D., 1974. Vis. asst. prof. Boston U., 1974-75; asst. prof. Wilkes Coll., Wilkes-Barre, Pa., 1975-79; research psychologist Walter Reed Inst. Research, Washington, 1979—; vis. asst. prof. (research) Brown U., Providence, 1976. Contbr. numerous articles to profl. jours. N.Y. State Regents scholar, 1965-69; USPHS research fellow, 1970-72. Mem. Acoustical Soc. Am., Am. Psychol. Assn., Eastern Psychol. Assn., Phi Beta Kappa, Sigma Xi. Subspecialties: Psychophysics; Behavioral psychology. Current work: Use behavior analytic methods to study sensory and perceptual processes in animals, including auditory perception and time perception. Office: Dept Med Neuroscis Walter Reed Army Inst Research Washington DC 20012

RASMUSSEN, NORMAN CARL, nuclear engineer; b. Harrisburg, Pa., Nov. 12, 1927; s. Frederick and Faith (Elliott) R.; m. Thalia Tichenor, Aug. 23, 1952; children: Neil, Arlene. B.A., Gettysburg (Pa.) Coll., 1950, Sc.D. (hon.), 1968; Ph.D., MIT, 1956; Dr.h.c., Cath. U. Leuven, Belgium, 1980. Mem. faculty MIT, 1958—, prof. nuclear engring., 1965—, McAfee prof. engring., 1983—, head dept., 1975-81; McAfee prof. engring. M.I.T., 1983—; dir. reactor safety study AEC, 1972-75; mem. Def. Sci. Bd., 1974-78, sr. cons., 1978—; mem. Nat. Sci. Bd., 1982—; mem. Nat Sci Bd. NSF, 1982—; mem. Presdl. Adv. Group on Contbns. Tech. to Econs. Strength, 1975; dir. Bedford Engring. Corp.; adv. com. components tech. div. Argonne Univs. Assn., 1977-80; chmn. sci. rev. com. Idaho Nat. Engring. Lab., 1977—; trustee N.E. Utilities, 1977—. Co-author: Modern Physics for Engineers, 1966; contbr. articles to profl. jours. Chmn. Lincoln-Sudbury Sch. Com., 1971-72; bd. dirs. Atomic Indsl. Forum, 1980—; mem. Utility Sci. Review Council for Nuclear Safety Analysis Center, Electric Power Research Inst., 1979—. Served with USNR, 1945-46. Recipient Disting. Achievement award Health Physics Soc., 1976; Disting. Service award NRC, 1976; Disting. Alumni award Gettysburg Coll., 1983. Fellow AAAS, Am. Nuclear Soc. (spl. award nuclear power reactor safety 1976, Theos J. Thompson award nuclear reactor safety div. 1981, dir. 1976-80), Am. Acad. Arts and Scis.; mem. Inst. Nuclear Materials Mgmt., Nat. Acad. Engring., Nat. Acad. Sci., Soc. Risk Analysis, Health Physics Soc. Subspecialty: Nuclear fission. Current work: Nuclear power plant probabalistic risk assessment and nuclear safety. Home: 80 Winsor Rd Sudbury MA 01776 Office: 77 Massachusetts Ave Cambridge MA 02139

RASOR, NED SHAURER, engring. physicist, corp. exec.; b. Dayton, Ohio, Jan. 2, 1927; s. Floyd Olen and Anna Belle (Shaurer) R.; m. Genevieve Mercia Eads, Oct. 7, 1926; children: Julia Suzanne, Dina Lynn. B.S. in Physics, Ohio State U., 1948, M.S., U. Ill., 1951; Ph.D. in Physics (Westinghouse fellow), Case Inst. Tech., 1954. Research engr. Atomics Internat. div. N.Am. Aviation, Los Angeles, 1949-55, supr., group leader, 1956-60, dept. dir., 1961-62; v.p. research Thermo Electron Corp., Waltham, Mass., 1963-65; cons. physicist, Dayton, Ohio, 1966-70; pres. Rasor Assocs., Inc., Sunnyvale, Calif., 1971—; lectr. on thermionic energy conversion MIT, George Washington U., UCLA, U. Ariz., NATO-AGARD. Contbr. articles to profl. jours., chpts. to books. Served with AUS, 1944-46. Assoc. fellow AIAA; mem. Am. Phys. Soc., Am. Inst. Physics, IEEE (sr.). Democrat. Subspecialties: Engineering physics; Biomedical engineering. Current work: Thermionic energy conversion, space nuclear power, surface physics, gaseous electronics, cardiovascular devices, solid fuel gasification. Patentee energy conversion and biomed. devices. Home: 15601 Montebello Rd Cupertino CA 95014 Office: 253 Humboldt Ct Sunnyvale CA 94086

RASSIGA, ANNE LOUISE, hematology and oncology educator; b. Oceanside, N.Y., June 19, 1942; d. William August and Edna (Chickray) R.; m. George Bernard Pidot, Jr., Sept. 5, 1962 (div. 1972); m. Charles H. Pimlott, Jr., July 5, 1974; children: Andrew William, Christopher Thomas. A.B., Bryn Mawr Coll., 1962; M.D., Harvard U., 1966. Diplomate: Am. Bd. Hematology, Am. Bd. Oncology, Am. Bd. Internal Medicine. Intern Mary Hitchcock Meml. Hosp., Hanover, N.H., 1966-67; resident in internal medicine Dartmouth Affiliated Hosps., Hanover, 1967-69; assoc. dir. div. hematology and oncology St. Luke's Hosp., Cleve., 1979—; asst. prof. medicine Case Western Res. U., 1972—; asst. chief hematology and oncology Cleve. VA Med. Ctr., 1972-79; instr. Dartmouth Coll., 1971-72; med. adv. com. Am. Cancer Soc., Cleve., 1981—. Alumnae dist. counselor Bryn Mawr Coll., 1983—, alumnae recruiting coordinator No. Ohio, 1977-83; patroller central div. Nat. Ski Patrol System, Inc., 1974—. NIH spl. postdoctoral research fellow in hematology, 1969-71. Fellow ACP; mem. Am. Women's Med. Assn., Hospice Council No. Ohio, Am. Fedn. Clin. Research, Am. Soc. Hematology, Am. Soc. Clin. Oncology (clin. practice com. 1982—), Acad. Medicine Cleve. (pres. elect.), Women's Med. Soc. Cleve. (sec.-treas. 1981-83). Club: Cleve. Skating. Subspecialties: Hematology; Oncology. Current work: Clinical and therapeutic trials of cancer treatment; undergraduate and postgraduate medical education. Home: 2025 Chestnut Hills Dr Cleveland Heights OH 44106 Office: Saint Luke's Hospital 11311 Shaker Blvd Cleveland OH 44106

RASSIN, DAVID KEITH, biochemist; b. Liverpool, Eng., Dec. 1, 1942; came to U.S., 1953; s. Meyer and Ella Rosetta Laura (House) R.; m. Glennda McConnell, Feb. 5, 1965; children: Meya Glynne, Keith David. A.B., Columbia U., 1965; Ph.D., Mt. Sinai Grad. Sch., CUNY, 1974. Research asst. Columbia U., 1966-67; asst. research scientist N.Y. State Inst. Basic Research Mental Retardation, 1967-70, research scientist, 1970-77, asso. research scientist, 1977-80; asso. prof. pediatrics, human biol. chemistry and genetics U. Tex. Med. Br., Galveston, 1980—. Contbr. articles to profl. jours. Bd. dirs. East End Hist. Dist. Assn., Galveston, 1981-82. Mem. Soc. Pediatric Research, Internat. Soc. Neurochemistry, Am. Soc. Clin. Nutrition, Am. Soc. Neurochemistry, Am. Inst. Nutrition, Am. Soc. Pharmacology and Exptl. Therapeutics, Am. Soc. Neursci. Subspecialties: Neurochemistry; Nutrition (medicine). Current work: Amino acid metabolism, especially sulfur containing compounds in neurochemistry, nutrition and inherited metabolic diseases. Home: 1318 Sealy Galveston TX 77550 Office: Dept Pediatrics U Tex Med Br Galveston TX 77550

RATAJCZAK, HELEN VOSSKUHLER, immunologist, educator, researcher; b. Tucson, Apr. 9, 1938; d. Max P. and Marion H. (Messer) Vosskuhler; m. Edward F. Ratajczak, June 1, 1959 (div. 1968); children: Lorraine, Eric, Peter, Eileen. B.S., U. Ariz., 1959, M.S., 1970, Ph.D., 1976. Asst. research scientist U. Iowa Coll. medicine, Iowa City, 1976-78; instr. immunology U. Pitts., 1978-80, research assoc., 1980-81; asst. prof. Loyola U., Maywood, Ill., 1981—. Am. Thoracic Soc. fellow, 1974-76; NIH fellow, 1978; Loyola U. grantee, 1981. Mem. Am. Thoracic Soc., Assn. Research in Vision and Ophthalmology, Am. Assn. Immunologists, Sigma Xi, Phi Lambda Upsilon. Republican. Roman Catholic. Subspecialties: Immunology (medicine); Cancer research (medicine). Current work: Cell mediated immunity of hypersensitivity diseases of the lung, eye; tumor immunology; chemotaxis. Office: Department of Pathology Stritch School Medicine Loyola University 2160 S 1st Ave Maywood IL 60153

RATCHESON, ROBERT ALLAN, neurosurgeon; b. Chgo., Aug. 24, 1940; s. Maurice and Kate (Davidow) R.; m. Peggy Steiner; children: Alexey, Rachael. B.S., Northwestern U., 1962, M.D., 1965. Intern Johns Hopkins Hosp., Balt., 1965-66, resident in surgery, 1966-67; resident in neurol. surgery Barnes Hosp., St. Louis, 1969-72; practice medicine specializing in neurosurgery, 1973—; fellow in neurol. surgery Wash. U. Sch. Medicine, 1971, instr. neurol. surgery, 1971-73; vis. scientist Brain Research Lab. Univ. Hosp., Lund, Sweden, 1972-73; asst. prof. neurol. surgery Wash. U. Sch. Medicine, 1973-77, assoc. prof. neurol. surgery, 1977-81; Harvey Huntington Brown Jr. prof. neurol. surgery, dir. div. neurol. surgery Case Western Res. U., 1981—; sr. asst. surgeon USPHS, NIH, 1967-69. William P. Van Wagenen fellow, 1972. Mem. Am. Assn. Neurol. Surgeons, Congress of Neurol. Surgeons (pres. elect 1983-84), ACP, Stroke Council Am. Heart Assn., AMA, Research Soc. Neurol. Surgeons, Scandinavian Neurosurg. Soc., Soc. Neurosci., Internat. Soc. Cerebral Blood Flow and Metabolism, Acad. Medicine Cleve., Ohio Med. Assn. Subspecialty: Neurochemistry. Current work: Cerebral metabolism, cerebral ischemia; neurovascular surgery. Home: 2871 Attleboro Rd Shaker Heights OH 44120 Office: 2074 Abington Rd Cleveland OH 44106

RATH, ROBERT MICHAEL, mech. engr., mfg. co. exec.; b. Hamilton, Ohio, Sept. 20, 1952; s. Robert Frederick and Mildred Rose (Burger) R.; m. Wendy Michele Brower, Dec. 2, 1977; children: Robert F., II, Erik C. M.E., Gen. Motors Inst., 1975. Process engr. Gen. Motors Assembly, Norwood, Ohio, 1970-75; sales engr. B.K. Sweeney Tool Co., Denver, 1975-77; warranty coordinator Regional Transp. Dist., Denver, 1977-79; gen. mgr. Distbrs. Remanufacturing Ctr., Salt Lake City, 1979—. Mem. Soc. Automotive Engrs., ASME, Am. Mgmt. Assn., Am. Prodn. and Inventory Control Soc., Alpha Tau Omega. Roman Catholic. Clubs: New Yorker, Salt Lake. Subspecialties: Mechanical engineering; Systems engineering. Current work: Methods for remanufacturing diesel engines. Home: 2196 Raintree Circle Sandy UT 84092 Office: 2494 Directors Row Salt Lake City UT 84104

RATHJENS, GEORGE WILLIAM, scientist, educator; b. Fairbanks, Alaska, June 28, 1925; s. George William and Jennie (Hansen) R.; m. Lucy van Buttingha Wichers, Apr. 5, 1950; children: Jacqueline, Leslie, Peter. B.S., Yale U., 1946; Ph.D., U. Calif., Berkeley, 1951. Instr. chemistry Columbia U., 1950-53; staff weapons systems evaluation group Dept. Def., 1953-58; research fellow Harvard U., 1958-59; staff spl. asst. to Pres. U.S. for sci. and tech., 1959-60; chief scientist Advanced Research Projects Agy., Dept. Def., 1961, dep. dir., 1961-62; dep. asst. dir. U.S. ACDA, 1962-64, spl. asst. to dir., 1964-65; dir. weapons systems evaluation div. Inst. Def. Analyses, 1965-68; prof. dept. polit. sci. MIT, 1968—. Fellow Am. Acad. Arts and Scis.; mem. AAAS, Council for a Livable World (chmn.), Fedn. Am. Scientists, Council Fgn. Relations, Inst. Strategic Studies, Sigma Xi. Current work: Arms control and defense policy; energy and environmental policy; science and public policy; technology assessment. Office: Mass Inst Tech Cambridge MA 02139

RATHMANN, GEORGE BLATZ, engineering company executive; b. Milw., Dec. 25, 1927; s. Louis and Edna Lorle (Blatz) R.; m. Frances Joy Anderson, June 24, 1950; children—James, Margaret, Laura, Sally, Richard. B.S. with honors, Northwestern U., 1948; M.S., Princeton U., 1950, Ph.D. in Phys. Chemistry, 1952. With 3M Co., St. Paul, 1951-72; various positions including research chemist, research dir., group tech. dir., mgr. X-ray products; with Litton Med. Systems, Des Plaines, Ill., 1972-75; v.p. research and devel. diagnostics div. Abbott Labs., North Chicago, Ill., 1975-80; pres., chief exec. officer Applied Molecular Genetics, Inc., Newbury Park, Calif., 1980—; chmn., dir. Assn. Life, Milw., 1948-73. Mem. Am. Chem. Soc., AAAS, Phi Beta Kappa, Sigma Xi. Subspecialty: Genetics and genetic engineering (biology). Home: 49 Hawthorne Ln Barrington IL 60010 Office: 1900 Oak Terrace Ln Thousand Oaks CA 91320

RATLIFF, FLOYD, scientist, educator; b. La Junta, Colo., May 1, 1919; s. Charles Frederick and Alice (Hubbard) R.; m. Orma Vernon Priddy, June 10, 1942; 1 dau., Merry Alice. B.A.; magna cum laude, Colo. Coll., 1947, D.Sc. honoris causa, 1975; M.Sc., Brown U., 1949, Ph.D., 1950; NRC postdoctoral fellow, Johns Hopkins, 1950-51. Head Lab. Biophysics, 1974—; Instr., then asst. prof. Harvard, 1951-54; asso. Rockefeller Inst., 1954-58; mem. faculty Rockefeller U., 1958—, prof. biophysics and physiol. psychology, 1966—; pres. Harry Frank Guggenheim Found., 1983—; cons. to govt., 1957—. Author: Mach Bands: Quantitative Studies on Neural Networks in the Retina, 1965, also articles.; Editor: Studies on Excitation and Inhibition in the Retina, 1974; editorial bd.: Jour. Gen. Physiology, 1969—. Served to 1st lt. AUS, 1941-45; ETO. Decorated Bronze Star; recipient Howard Crosby Warren medal Soc. Exptl. Psychologists, 1966; Edgar D. Tillyer medal Optical Soc. Am., 1976; medal for disting. service Brown U., 1980. Fellow Am. Acad. Arts and Scis.; mem. Nat. Acad. Scis., Am. Inst. Physics, A.A.A.S., Am. Psychol. Assn., Manhattan Philos. Soc., Internat. Brain Research Orgn., Am. Philos. Soc., China Inst. Am., Oriental Ceramic Soc. (London), Oriental Ceramic Soc. (Hong Kong), Asia Soc., Japan Soc., Phi Beta Kappa, Sigma Xi. Subspecialties: Neurophysiology; Physiological psychology. Home: 500 E 63d St New York NY 10021 Office: Rockefeller U 1230 York Ave New York NY 10021 Office: Harry Frank Guggenheim Found Woolworth Bldg 233 Broadway New York NY 10021

RATLIFF, THOMAS A., JR., quality control engr.; b. Phila., May 5, 1919; s. Thomas A. and Edna Dorothea (Overman) R.; m. Lucy Graydon, Aug. 15, 1942; children: Deborah G. Ratliff Miller, Anne O. Ratliff Schuck. Student, U. Cin., 1936-40. Registered profl. engr., Calif. Quality engr. Am. Standard, Cin., 1953-56; quality engr., chief insp. Gruen Watch Co., Cin., 1956-60; sales mgr. Leyman Corp., Cin., 1960-62; partner Badgett & Smith, Inc., Cin., 1962-66; purchasing mgr. Access Corp., Cin., 1966-70; partner R&R Assocs., Quality Control Cons., Cin., 1970—; tchr. lab. quality control courses Nat. Inst. Occupational Safety and Health. Served to col. U.S. Army, 1941-46. Mem. Am. Soc. Quality Control (sr.). Republican. Quaker. Club: Glendale (Ohio) Lyceum. Subspecialty: Quality control. Current work: Laboratory quality control; quality control engineering. Office: PO Box 46181 Cincinnati OH 45246

RATNOFF, OSCAR DAVIS, physician, educator; b. N.Y.C., Aug. 23, 1916; s. Hyman L. and Ethel (Davis) R.; m. Marian Foreman, Mar. 31, 1945; children: William Davis, Martha. A.B., Columbia U., 1936, M.D., 1939; LL.D. (hon.), U. Aberdeen, 1981. Intern Johns Hopkins Hosp., Balt., 1939-40; Austin fellow in physiology Harvard Med. Sch., Boston, 1940-41; asst. resident Montefiore Hosp., N.Y.C., 1942; resident Goldwater Meml. Hosp., N.Y.C., 1942-43; asst. in medicine Columbia Coll. Physicians and Surgeons, N.Y.C., 1942-46; fellow in medicine Johns Hopkins, 1946-48, instr. medicine, 1948-50, instr. bacteriology, 1949-50; asst. prof. medicine Case Western Res. U., Cleve., 1950-56, asso. prof., 1956-61, prof., 1961—; asst. physician (Univ. Hosps.), Cleve., 1952-56, asso. physician, 1956-67, physician, 1967—. Author: Bleeding Syndromes, 1960; Mem. editorial bd.: Jour. Lab. Clin. Medicine, 1956-62, Circulation, 1961-65, Blood, 1963-69, 78—, Am. Jour. Physiology, 1966-72, Jour. Applied Physiology, 1966-72, Jour. Lipid Research, 1967-69, Jour. Clin. Investigation, 1969-71, Circulation Research, 1970-75, Annals Internal Medicine, 1973-76, Perspectives in Biology and Medicine, 1974—, Thrombosis Research, 1981—, Jour. Urology, 1981—; Contbr. articles to med. jours. Career investigator Am. Heart Assn., 1960—. Served to maj. M.C., 1943-46; Ind. Recipient Henry Moses award Montefiore Hosp., 1949; Disting. Achievement award Modern Medicine, 1967; James F. Mitchell award, 1971; Murray Thelin award Nat. Hemophilia Found., 1971; H.P. Smith award Am. Soc. Clin. Pathology, 1975; Joseph Mather Smith prize Columbia Coll. Physicians and Surgeons, 1976. Mem. A.C.P. (John Phillips award 1974, master 1983), A.M.A., Am. Fedn. Clin. Research (emeritus), Nat. Acad. Scis., Soc. Scholars of Johns Hopkins U., Am. Soc. Clin. Investigation (emeritus), Central Soc. Clin. Research, Assn. Am. Physicians, Am. Soc. Hematology (Dameshek award 1972), Internat. Soc. Hematology, Internat. Soc. Thrombosis (Grant award 1981), Am. Physiol. Soc., Am. Soc. Biol. Chemists, Royal Coll. Physicians and Surgeons Glasgow (hon.), Sigma Xi, Alpha Omega Alpha. Subspecialty: Hematology. Current work: Hemostasis and thrombosis. Home: 2916 Sedgewick Rd Shaker Heights OH 44120 Office: University Hospitals of Cleve Cleveland OH 44106

RATTNER, BARNETT ALVIN, research physiologist, biology educator; b. Washington, Oct. 4, 1950; s. Sydney and Lucille (Shpvitz) R.; m. Francine Rachelle Koplin, Jan. 14, 1978. B.S., U. Md., 1972, M.S., 1974, Ph.D., 1977. Instr. zoology U. Md., College Park, 1972-77; research physiologist Naval Med. Research Inst., Bethesda, Md., 1977-78, Patuxent Wildlife Research Ctr., Laurel, Md., 1978—. Contbr. articles to profl. publs. Active Wheaton-Md. Vol. Rescue Squad, 1971-74; mem. Voting Dist. 18 Scholarship Com., Silver Spring, Md., 1975-78. NRC postdoctoral fellow, 1977. Mem. Am. Physiol. Soc., Am. Soc. Zoologists, Soc. Exptl. Biology and Medicine, Soc. for Study of Reproduction, Soc. Environ. Toxicology and Chemistry, Sigma Xi, Phi Sigma, Phi Kappa Phi. Subspecialties: Environmental toxicology; Comparative physiology. Current work: Effects of petroleum hydrocarbons and organophosphorus insecticides on physiological function in wildlife; effects of hyperbaria and hypobaria in physiological function. Home: 2177 Branchwood Ct Gambrills MD 21054 Office: Patuxent Wildlife Research Ctr US Fish and Wildlife Service Laurel MD 20708

RAUP, DAVID MALCOLM, paleontology educator; b. Boston, Apr. 24, 1933; s. Hugh Miller and Lucy (Gibson) R.; m. Susan Creer Shepard, Aug. 25, 1956; 1 son, Mitchell D. B.S., U. Chgo., 1953; M.A., Harvard U., 1955, Ph.D., 1957. Instr. Calif. Inst. Tech., 1956-57; faculty Johns Hopkins, 1957-65, asso. prof., 1963-65; mem. faculty U. Rochester, 1965-78, prof. geology, 1966-78, chmn. dept. geol. scis., 1968-71, dir., 1977-78; curator geology, chmn. dept. geology Field Mus. Natural History, Chgo., 1978—, dean of sci., 1980-82; prof. geophys. sci. U. Chgo., 1980—, chmn. dept., 1982—; geologist U.S. Geol. Survey, part-time, 1959-77; vis. prof. U. Tubingen, Germany, 1965, 72. Author: (with S. Stanley) Principles of Paleontology, 1971, 78; editor: (with B. Kummel) Handbook of Paleontological Techniques, 1965; Contbr. articles to profl. jours. Recipient Best Paper award Jour. Paleontology, 1966; Schuchert award Paleontol. Soc., 1973; Calif. Research Corp. grantee, 1955-56; Am. Assn. Petroleum Geologists grantee, 1957; Am. Philos. Soc. grantee, 1957; NSF grantee, 1960-66, 75—; Chem. Soc. grantee, 1965-71; NASA grantee, 1983—. Fellow Geol. Soc. Am.; mem. Nat. Acad. Sci., Paleontol. Soc. (pres. 1976-77), Am. Soc. Naturalists (v.p. 1983), Soc. Econ. Paleontology and Mineralogy, Soc. for Systematic Zoology, Soc. for Study Evolution, AAAS, Sigma Xi. Subspecialty: Paleontology. Home: 5801 S Dorchester Ave Chicago IL 60637 Office: Dept Geophys Scis U Chgo Chicago IL 60637

RAUSEN, AARON REUBEN, pediatric oncologist-hematologist; b. Jersey City, June 30, 1930; s. David and Ruth (Schwartz) R.; m. Emalou Watkins Rausen, Apr. 7, 1968; children: David Jacob, Susan Dinah, Elisabeth Ann. Student, Dartmouth Coll., 1947-50; M.D., SUNY Downstate Med. Center, N.Y.C., 1954. Cert. Am. Bd. Pediatrics (subbd. pediatric hematology-oncology). Intern. in pediatrics Bellevue Hosp., N.Y.C., 1954-55; resident in pediatrics, 1955-56; chief resident in pediatrics Mt. Sinai Hosp., N.Y.C., 1958-59; fellow in pediatric hematology Children's Hosp. Med. Center, Boston, 1959-61; chief pediatrics Greenpoint Hosp., Bklyn., 1962-64, City Hosp. Center, Elmhurst, N.Y., 1964-73; dir. pediatrics Beth Israel Hosp., N.Y.C., 1973-81; prof. pediatrics Mt. Sinai Sch. Medicine, N.Y.C., 1972-81, N.Y.U. Sch. Medicine, 1980—, chief pediatric oncology, 1981—; adj. prof. Rockefeller U., 1980—; chief sect. pediatric hematology Lenox Hill Hosp., N.Y.C., 1981—. Contbr. chpts. to books and articles to profl. jours. Served to capt., M.C. U.S. Army, 1956-58. Mem. Am. Pediatrics Soc., Am. Soc. Hematology, Am. Assn. Cancer Research, Am. Soc. Clin. Oncology, Am. Acad. Pediatrics, N.Y. Acad. Medicine, Phi Beta Kappa, Alpha Omega Alpha. Club: Dartmouth (N.Y.C.). Subspecialties: Oncology; Hematology. Current work: Pediatric oncology-hematology, childhood leukemia, Hodgkin's disease, cell differentiation, edn. and clin. care. Office: 530 1st Ave Suite 6A New York NY 10016

RAVEN, PETER HAMILTON, bot. garden exec.; b. Shanghai, China, June 13, 1936; s. Walter Francis and Isabelle Marion (Breen) R.; m. Tamra Engelhorn, Nov. 29, 1968; children—Alice Catherine, Elizabeth Marie, Francis Clark. A.B., U. Calif., Berkeley, 1957; Ph.D., U. Los Angeles, 1960. Taxonomist Rancho Santa Ana Bot. Garden, Claremont, Calif., 1961-62; asst. prof., asso. prof. biol. scis. Stanford, 1962-71; dir. Mo. Bot. Garden, Engelmann prof. biology Washington U., St. Louis, 1971—; mem. Nat. Mus. Services Bd., 1977-82. Author: Biology of Plants, 1971, 3d edit., 1981, Principles of Tzeltal Folk Taxonomy, 1974, others.; Contbr. articles to profl. jours. Mem. Bd. Commrs., Tower Grove Park, 1971—; Vice-pres. XIII Internat. Bot. Congress, Sydney, 1981. Recipient Distinguished Service award Am. Inst. Biol. Scis., 1981. Fellow Am. Acad. Arts and Scis., Calif. Acad. Scis., AAAS; mem. Nat. Acad. Scis., Bot. Soc. Am. (past pres.), Soc. for Study Evolution (past pres.), Am. Soc. Plant Taxonomists (past pres.), Am. Assn. Bot. Gardens and Arboreta, Assn. Systematics Collections (pres.), Orgn. Tropical Studies (dir.), Japan Am. Soc. (past pres. St. Louis), Phi Beta Kappa, Sigma Xi. Clubs: University, Noonday (St. Louis). Subspecialties: Evolutionary biology; Systematics. Current work: Evolution and systematics of higher plants; biogeography of the southern hemisphere. Home: 2361 Tower Grove Ave Saint Louis MO 63110 Office: PO Box 299 Saint Louis MO 63166

RAW, CECIL JOHN GOUGH, chemistry educator; b. Ixopo, Natal, South Africa, Oct. 20, 1929; s. Cecil H. and Beryl Natalie (Gough) R.; m. Gillian Carole Galt, Jan. 7, 1956; children: Jeremy, Timothy, Matthew, Rebecca. B.S. with honors, U. Natal, 1951, M.S., 1952, Ph.D., 1956. Lectr., sr. lectr. in chemistry U. Natal, 1954-59, African Explosives and Chem. Industries Research fellow, 1957-59; research fellow U Minn., Mpls., 1959-60; vis. prof. molecular physics U. Md., 1962; asst. prof. phys. chemistry St. Louis U., 1960-62, assoc. prof., 1962-66, prof., 1966—. Contbr. articles to profl. jours. Mem. Am. Chem. Soc., AAUP, Sigma Xi. Subspecialties: Physical chemistry; Kinetics. Current work: Oscillatory chemical reactions, molecular reaction dynamics, thermodynamics of systems far from equilibrium, microcomputers in physical chemistry. Office: Dept Chemistry St Louis U Saint Louis MO 63103

RAWAT, ARUN KUMAR, biological chemist, researcher, educator, consultant; b. Agra, U.P., India, Sept. 19, 1945; came to U.S., 1969, naturalized, 1973; s. Pyre L. and Shyama L. Rawar; m. Anu R. Sharma, July 26, 1976; children: Atul, Angeli. B.S., U.Lucknow, 1962, M.S., 1964; D.Sc., U. Copenhagen, 1969. Asst. prof. SUNY-Bklyn., 1971-72; assoc. prof. Med. Coll. Ohio, Toledo, 1973-78; prof. U. Toledo, 1979—; exec. dir. Midwest Inst. for Treatment and Study of Alcoholism, Toledo, 1973—. Contbr. articles on biol. chemistry to profl. jours. Mem. Am. Soc. Neurochemistry, Fedn. Am. Soc. Biol. Chemists, Am. Coll. Toxicologists. Subspecialties: Biochemistry (medicine); Neuropharmacology. Current work: Neurobehavioral toxicology, brain development, alcohol metabolism, fetal alcohol syndrome enzymology. Office: 3433 Upton Ave Toledo OH 43613

RAWLINSON, STEPHEN JOHN, computer design engineer; b. Portland, June 11, 1945; s. Alfred John and Margaret Frances (Schwartz) R.; m. Linda Carranza, Dec. 17, 1972. B.S., MIT, 1967; M.S., U. Santa Clara, 1981. Systems analyst Alcoa, Vancouver, Wash., 1968-69; programmer NCR Corp., San Diego, 1972-77; computer design engr. Amdahl Corp., Sunnyvale, Calif., 1977—. Treas., bd. dirs. Sunset Park of Sunnyvale Homeowner's Assn., 1978—. Served with USNR, 1970-72. Mem. Assn. Computing Machinery. Republican. Subspecialties: Microprogramming; Computer engineering. Home: 125-62 Connemara Way Sunnyvale CA 94087 Office: PO Box 470 Sunnyvale CA 94086

RAWLS, HENRY RALPH, biomaterials science educator, consultant; b. Chattahoochee, Fla., Nov. 19, 1935; s. Ralph Brian and Mildred Elizabeth (Evans) R.; m. Judith Anne Brock, Aug. 1, 1964 (div. 1972); m. Andrea Noel Geoffray, June 11, 1978; 1 son, Randell Granier. B.S., La. State U., 1957; Ph.D., Fla. State U., 1964. Research chemist Unilever Research Lab., Vlaardingen, Holland, 1964-67; research mgr. Gulf South Research Inst., New Orleans, 1968-76; assoc. prof. La. State U. Sch. Dentistry, New Orleans, 1976—; vis. scientist Forsyth Dental Ctr., Boston, 1978—; adj. prof. research and devel. program Biomed. Engring. Tulane U., New Orleans, 1977-81; mem. peer rev. panel State of La.; cons. Gulf South Research Inst., 1976—, NIH, 1977-78. Contbr. sci. articles to profl. publs. Served to lt. USNR, 1957-60. Fogerty sr. fellow NIH, Netherlands, 1980; Nat. Inst. Dental Research career awardee, 1977-82; research grantee, 1977, 80, 82, 83; FDA grantee, 1973. Mem. Am. Chem. Soc. (auditor La. sect. 1983—), Internat. Assn. Dental Research (treas. New Orleans sect. 1982—), Am. Assn. Dental Sch., Soc. for Biomaterials, ORCA-European Orgn. for Caries Research. Democrat. Methodist. Club: Bayou Flying (sec. 1981—). Subspecialties: Biomaterials; Cariology. Current work: Physical chemistry and materials science investigations of dental caries and development of polymeric materials for biological applications. Office: La State U Sch Dentistry 1100 Florida Ave New Orleans LA 70119

RAWSON, JAMES RULON YOUNG, botanist, educator; b. Boston, July 28, 1943; s. Rulon Wells and Jane (Young) R.; m. Judy MacDonald, Aug. 15, 1970; children: Donald, David. B.S., Cornell U., 1965; Ph.D., Northwestern U., 1969. Postdoctoral fellow U. Chgo., 1969-72; asst. prof. U. Ga., Athens, 1972-78, assoc. prof., 1978—. Mem. editorial bd.: Plant Physiology. NIH fellow, 1969-71; NSF research grantee, 1973—; U.S. Dept. Agr. research grantee, 1980—. Mem. Am. Soc. Biol. Chemists, Plant Physiology Soc. Subspecialties: Genetics and genetic engineering (biology); Photosynthesis. Current work: Plant molecular biology; genetics of photosynthesis. Home: 360 Sandstone Dr Athens GA 30605 Office: Botany Dept U Ga Athens GA 30602

RAY, PRASANTA KUMAR, tumor immunologist, researcher; b. Calcutta, W. Bengal, India, Sept. 29, 1941; came to U.S., 1969; s. Benode Behari and Labanya Prava (Das) R.; m. Khana Basu, Aug. 7, 1969; children: Partha, Amartya. B.S. with honors, U. Calcutta, 1962, M.S., 1964, Ph.D., 1968, D.Sc., 1974. Research specialist, instr. dept. surgery U. Minn., 1969-73; sr. sci. officer, head cancer immunobiology Bhabba Atomic Research, Bombay, India, 1973-76; dir. Chittaranjan Nat. Cancer Research Ctr., Calcutta, 1976-77; research dir. Bengal Immunity Research Ctr., Calcutta, 1977-78; dir. Alma Dea Morani Lab. Medical Coll. Pa., Phila., 1978—; Mem. editorial bd. Jour. Plasma Therapy, Boston, 1981—; chief com. immunology and nutrition Internat. Federation Human Health, 1980—. Editor: Immunobiology of Transplantation, Cancer and Pregnancy, 1983. Recipient cancer research award Indian Council Medical Research, 1977; fellow Indian Coll. Allergy and Applied Immunology, 1974-78; grantee cancer research W.W. Smith Charitable Trust, 1980-83, R.J. Reynolds Industries, Inc., 1981-83; grantee biomedical research NIH, 1980. Mem. Indian Immunology Soc., Am. Assn. Immunologists, N.Y. Acad. Scis., Am. Assn. Apheresis. Subspecialties: Cancer research (medicine); Immunobiology and immunology. Current work: Teaching, guiding and directing research on tumor immunology and immunotherapy and chemo-immunotherapy. Office: Alma Dea Morani Lab Surg Immunobiology Medical Coll Pa 3300 Henry Ave Philadelphia PA 19129

RAY, RADHARAMAN, biochemist; b. Calcutta, West Bengal, India, Jan. 21, 1943; came to U.S., 1971, naturalized, 1981; s. Satish Chandra and Prafulla Bala R.; m. Prabhati Bandyopadhyay, July 29, 1971; children: Ranjan, Pulak. B.S. with honors in Chemistry, Calcutta. U., 1963, M.S. in Biochemistry, 1965, Ph.D., 1970. Research assoc. Howard U., Washington, 1971-72, asst. prof., 1972-75; guest worker Lab. Biochem. Genetics, Nat. Heart, Lung, and Blood Inst., NIH, Bethesda, Md., 1975-78, staff fellow, 1978-81; research chemist U.S. Army Med. Research Inst. Chem. Def., Edgewood, Md., 1981—; lectr. in field. Govt. India Univ. Grants Commn. fellow, 1967; recipient Nat. Research Service award Nat. Inst. Gen. Med. Scis., NIH, 1975. Mem. Soc. Neurosci., Brit. Brain Research Assn., European Brain and Behavior Soc. Subspecialties: Neurobiology; Neurochemistry. Current work: Ion channels - composition and function; neurotransmitter secretion; synapse formation; membrane receptors; drug action; developmental molecular biology. Office: US Army Med Research Inst of Chem Def Edgewood Aberdeen Proving Ground MD 21010

RAY, RICHARD HALLETT, neurophysiologist; b. Charlotte, N.C., May 31, 1951; s. George Irving and Katherine Knight (Hallett) R.; m. Elizabeth Cadieu, Aug. 6, 1971; children: Richard Hallett, Scott Howell. Student, Bucknell U., 1969-70; B.A. in psychology, U. N.C., Charlotte, 1972; Ph.D. in Physiology, Med. Coll. Ga., 1980. Grad. teaching asst. dept. physiology Med. Coll. Ga., 1974-80; Ahmanson fellow Brain Research Inst., dept. anatomy UCLA, 1980-82; asst. prof. physiology East Carolina U., Greenville, N.C., 1982—. Contbr. in field. Recipient Travel award Soc. Exptl. Biology and Medicine, 1977. Mem. Soc. Neuroscience, Am. Physiol. Soc., AAAS, Sigma Xi. Democrat. Presbyterian. Subspecialties: Neurophysiology; Sensory processes. Current work: Coding in sensory systemsof vertebrates; cerebral cortical devel. and plasticity; physiology of sensory and motor systems; functional orgn. of cerebral cortex. Home: 230 Chippendale Dr Greenville NC 27834 Office: Dept Physiolog Sch Medicine East Carolina Greenville NC 27834

RAY, RICHARD SCHELL, veterinarian, educator; b. Antwerp, Ohio, May 21, 1928; s. Alton D. and Dorothy Fransis (Schell) R.; m. Diane Maxine Foster, June 12, 1954; children: Kathleen F., David A., Elizabeth A. B.A., Ohio State U., 1950, D.V.M., 1955, M.S., 1958, Ph.D., 1963. Diplomate: Ohio Bed. Bet. Examiners; accredited U.S. Dept. Agr. Practice vet. medicine, Toledo, 1955. Instr. vet. clin. scis. and Grad. Sch. Ohio State U., Columbus, 1955-63, asst. prof., 1963-67, assoc. prof., 1967-73, prof., 1973—, teaching team leader, 1969-74, 77—; cons. forensic pharmacology and metabolic disease; dir. Pre- and Post-Race Drug Detection Lab., Ohio State U., 1969—. Contbr. articles to profl. jours. Grantee Harness Racing Inst., 1965-68, N.Y. Racing Asn., 1965-68, Jockey Club, 1965-68, U.S. Trotting Assn., 1969, 71, Horseman's Benevolent Protective Assn., 1971, Nat. Assn. State Racing Commrs., 1976, Snyder Mfg. Co., 1970, USPHS, 1967-71, HEW, NIH, 1973, Ohio Thoroughbred Fund, 1967—. Fellow Am. Coll. Vet. Pharmacology and Therapeutics; mem. Am. Soc. Vet. Physiologists and Pharmacologists, World Assn. Vet. Physiologists, Pharmacologists and Biochemists, Assn. Ofcl. Racing Chemists, Assn. Drug Detection Labs., Am. Chem. Soc. (div. medicinal chemistry), Am. Assn. Vet Clinicians, Am. Assn. Equine Practitioners, Phi Zeta, Omega Tau Sigma, Alpha Sigma Phi. Republican. Methodist. Lodge: Masons. Subspecialties: Pharmacology; Biochemistry (medicine). Current work: Development of new methods for the detection and identification o. Home: 2752 Folkstone Rd Columbus OH 43220 Office: 1935 Coffey Rd Columbus OH 43210

RAY, TUSHAR KANTI, membrane biologist, biochemist, educator; b. Calcutta, West Bengal, India, Oct. 31, 1939; came to U.S., 1967; s. Kamalakshya and Dhumabati R.; m. Mukta Mala Ghose, July 14, 1966; children: Amit, Asim. B.Sc. with honors, U. Calcutta, 1960, M.Sc., 1962, Ph.D., 1966. Research fellow U. Pitts., 1967-68; research assoc. U. Rochester, 1968-70; asst. research physiologist U. Calif., Berkeley, 1970-75; sr. research scientist U. Tex., Houston, 1975-78; assoc. prof. membrane biology SUNY-Upstate Med. Ctr., Syracuse, 1978—, dir. surg. research labs., 1978—. NIH grantee, 1977; NIH award, 1977. Mem. Am. Physiol. Soc., Am. Soc. Biol. Chemists, Biophys. Soc., N.Y. Acad. Sci., AAAS. Democrat. Hindu. Subspecialty: Membrane biology. Current work: Studying the molecular mechanism and control of gastric acid secretion and trying to use the knowledge for anti-ulcer drug development. Home: 302 Greenwood Rd Dewitt NY 13214 Office: SUNY Upstate Med Ctr 750 E Adams St Syracuse NY 13210

RAY, USHARANJAN, microbiologist, researcher; b. Sylhet, Bengal, India, Oct. 9, 1943; s. Pramatha Nath and Charubala (Saha) R.; m. Nupur, Apr. 30, 1949; 1 child, Neelanjan. M.S., Calcutta U., 1965; Ph.D., U. Ill., 1970. Guest worker Roche Inst. Molecular Biology, Nutley, N.J., 1972-74; research assoc. Albert Einstein Coll. Medicine, Bronx, N.Y., 1974-75, Harvard Med. Sch., Boston, 1975-79; vis. assoc. NIH, Bethesda, Md., 1979—. Contbr. articles to research publs. Mem. Am. Soc. Microbiology. Subspecialties: Microbiology (medicine); Virology (medicine). Current work: Microbial genetics, virology, immunology. Office: Nat Inst Health Bldg 30 Room 121 Bethesda MD 20205

RAY, WILLIAM EDWARD, nuclear engineer; b. Altoona, Pa., Sept. 30, 1928; s. Blair P. and Mary E. (Fowler) R.; m. Jocelyn Wilson, July 5, 1952 (div. 1972); children: Nora, Jonathan, Jennifer; m. Rita Horton, Dec. 29, 1973. B.S., Pa. State U., 1950, M.S., 1952. Engr., Hanford ops. Gen. Electric Co., Richland, Wash., 1952-55; cognizant engr. Knolls Atomic Power, Schenectady, 1955-60; v.p. research and devel. Dresser Products, Inc., Great Barrington, Mass., 1960-63; project mgr. atomic fuels Westinghouse Corp., Cheswick, Pa., 1963-65, mgr. advanced materials-atomic power, Pitts., 1965-68, mgr. materials and processes-advancd reactors div., Madison, Pa., 1968—. Mem. Indiana Twp. (Pa.) Zoning Bd., 1971. Mem. Am. Soc. Metals, Am. Nuclear Soc. Republican. Presbyterian. Lodge: Masons. Subspecialties: High-temperature materials; Materials processing. Current work: Development and qualification of high temperature heat transfer materials and components; cladding, coating and surface treatment of special-purpose materials; manufacturing process development. Inventor cadmium-rare earth borates; metal dispersions. Home: RD 1 Box 200 Latrobe PA 15650 Office: PO Box 158 Madison PA 15663

RAYBURN, CAROLE A(NN), clinical psychologist, consultant, researcher; b. Washington, Feb. 14, 1938; d. Carl Frederick and Mary Helen (Milkie) Miller; m. Ronald Allen Rayburn (dec. Apr. 1970). B.A., Am. U., 1961; M.A., George Washington U., 1965; Ph.D., Cath. U. Am., 1969; M. Div., Andrews U. Seventh-day Adventist Theol. Sem., 1980. Lic. psychologist, D.C., Md., Mich. Clin. psychologist Spring Grove State Hosp., Catonsville, Md., 1966-68; clin. psychologist Instl. Care Services, Laurel, Md., 1970-78; pvt. practice clin. and cons. psychology, Washington, Silver Spring, Md., 1971—, pvt. practice research psychology, Silver Spring, 1979—; cons. psychologist Julia Brown Montessori Schs., Laurel, 1972-75, 82—; pvt. practice forensic psychology, Washington, Silver Spring, Md., 1973—. Dist. coordinator sch. bd. candidate, Olney, Md., 1982; del. NOW conv., Md., 1982; mem. religious com. NAACP Montgomery County, Md., 1982-83. Recipient Sustained Superior Service award D.C. Dept. Human Resources, 1971, 72; Service award Council Advancement Psychol. Professions and Scis., 1975. Fellow Am. Orthopsychiat. Assn. (coordinator com. to study women 1977), Md. Psychol. Assn. (editor 1975-76, rep.-at-large 1976-78, cert. recognition 1978); mem. Am. Psychol. Assn. (pres. sect. on clin. psychology of women 1983), Balt. Assn. Cons. Psychologists (pres. 1976-78), D.C. Psychol. Assn. (cert. recognition 1976), Internat. Council Psychologists, AAUW (interbr. rep. to women's fair 1982-83), Psi Chi. Democrat. Adventist. Current work: Basic research in effect of religious society on women, especially women in religion; counseling these women and others in field of religion. Office: 1200 Morningside Dr Silver Spring MD 20904

RAYFORD, PHILLIP LEON, physiology educator; b. Roanoke, Va., July 25, 1927; s. Roosevelt Theodore and Eva Elizabeth (Robinson) R.; m. researcher; m. Geraldine Kimber. B.S., A & T U., Greensboro, N.C., 1949; M.D., U. Md., 1970, Ph.D., 1973. Various positions as research biologist NIH, 1955-73; assoc. prof. biochemistry U. Tex. Med. Br., Galveston, 1973-76, prof., 1977-80, dir. surg. biochem. lab., 1973-80, asst. dean medicine, 1978-80; prof. physiology U. Ark.-Little Rock Coll. Medicine, 1980—, chmn. dept. physiology, 1980. Served with U.S. Army, 1946-47; Philippines. NIH fellow, 1970; NIH Nat. Cancer Inst. grantee, 1981-86. Mem. Assn. Chairmen Depts. Physiology, AAUP, N.Y. Acad. Scis., Am. Physiol. Soc., Nat. Minority Health Affairs Assn. (charter), Omega Psi Phi. Democrat. Presbyterian. Lodge: Masons. Subspecialties: Physiology (biology); Gastroenterology. Current work: Endocrinology of gastrointestinal tract. Office: U Ark Coll Medicine 4301 W Markham Little Rock AR 72205 Home: 3302 Montrose Little Rock AR 72212

RAYMOND, BETH, psychologist, educator; b. N.Y.C., Apr. 12, 1943; d. Abraham and Edna (Meyer) Goldstein; m. Steven Raymond, Aug. 31, 1963; Children: Andrew, Jonathan. B.S., Jackson Coll. of Tufts U., 1963; Ph.D., NYU, 1968. Lic. psychologist, N.Y. Assoc. prof psychology Hofstra U., Hempstead, N.Y., 1968—; clin. psychologist in pvt. practice, Gt. Neck, N.Y., 1977—. Contbr. articles to profl. jours. Mem. Am. Psychol. Assn., Nassau County Psychol. Assn., Phi Beta Kappa. Subspecialty: Clinical psychology. Current work: Marriage; sex differences, therapy outcome. Home: 102 Clover Dr Great Neck NY 11021 Office: Hofstra U Hempstead NY 11550

RAYUDU, GARIMELLA V. S., nuclear med. scientist; b. Gangalakurru, India, Oct. 1, 1936; came to U.S., 1961, naturalized, 1967; S. Somayajulu V.P. and Lakshmi (Vadali) Garimella; m. Vijayalakshmi Rayudu, May 9, 1965; children: Lalitha, Devi, Sridevi, Kumar. B.Sc., Andhra U., India, 1956, M.Sc., 1957; Ph.D., McGill U.,

Montreal, 1961. Cert. Am. Bd. Radiology, Am. Bd. Sci. in Nuclear Medicine, Radiation Safety Physicist Atomic Energy Estate, Bombay. Teaching asst. McGill U., 1958-61; research assoc. Carnegie Inst. Tech., 1961-65; sr. research assoc. U. Toronto, 1965-67; asst. prof. Loyola U., New Orleans, 1967-68; assoc. scientist, asst. prof. Rush-Prosbyn.-St. Luke's Med. Center, Chgo., 1968-72; sr. scientist, assoc. prof., central radiopharmacy dir., guest scientist Argonne (Ill.) Nat. Lab., 1972—; sr. scientist, assoc. prof. Rush U. Med. Center, 1972—; cons. nuclear med. sci., radiopharm. sci. Aurthor: Advances in Radiopharmacology, 1981, Radiotracers for Medical Applications, Vol. I, 1982, Vol. II, 1982; contbr. numerous articles to sci. jours. NRC of Can. fellow, 1960-61. Mem. Am. Chem. Soc., Soc. Nuclear Medicine (Silver medal for sci. exhbns. 1972, Bronze medal 1975), Am. Assn. Physicists in Medicine, AAAS, Royal Soc. Chemistry, Radiopharm. Sci. Council, Sigma Xi. Hindu. Subspecialties: Nuclear medicine; Nuclear and radiochemistry. Current work: High energy nuclear reactions, neutron activation analysis and high resolution gamma spectrometry, nuclear and cosmochemistry, radiopharmaceuticals for tumor, brain, bone, marrow imaging, trace elements in pathological specimens. Office: 1753 W Congress Pkwy Chicago IL 60612

RE, GERALD JAMES, dental educator; b. Oak Park, Ill., July 22, 1943; s. Donald Charles and Ruth (Fishback) Swenson; m. Judy Carol Colvin, Jan. 9, 1965; children: Craig Allen, Christine Angela. B.S., U. Ill., 1966; D.M.D., U. Ky., 1972. Pvt. practice dentistry, Berea, Ky., 1973-77; restorative instr. U. Ky., Lexington, 1973-77; assoc. prof. operative dentistry U. Tex., San Antonio, 1977—. Mem. Am. Assn. Dental Research, Am. Assn. Dental Schs. Current work: Causes and prevention of vertical tooth fracture, clinical caries detection and removal, methods which facilitate performance of operative dental restorations. Office: Restorative Dentistry Dept 7703 Floyd Curl Dr San Antonio TX 78284

REA, DAVID KENERSON, research scientist and educator; b. Pitts., June 2, 1942; s. Cleveland D. and Margaret K. R.; m., Feb. 11, 1967; children: Gregory, Margaret. Asst. prof. Oreg. State U., Corvallis, 1974-75; asst. prof. U. Mich., Ann Arbor, 1975-80, assoc. prof. oceanography, 1980—; cons. Contbr. geol. and geophys. articles to profl. jours. NSF grantee, 1975—. Mem. Am. Geophys. Union, Geol. Soc. Am., Soc. Econ. Paleontology and Mineralogy, AAAS. Subspecialties: Oceanography; Sedimentology. Current work: Marine and lacustrine sediments, paleoclimatology, ocean basin tectonics. Address: Dept Atmospheric and Oceanic Sci U Mich Ann Arbor MI 48109

READ, GEORGE WESLEY, pharmacologist, educator; b. Los Angeles, June 24, 1934; s. Earl George and Gertrude Louise (Mason) R.; m. Dorothy Davis, Aug. 22, 1954 (div. Nov. 1981); children: Gregory Cecil, Lani Louise (dec.), Bonnie Alice. A.A., Menlo Coll., 1957; B.A. in Biology, Stanford U., 1959; M.S. in Physiology (Rosenberg scholar), Stanford U., 1962; Ph.D. in Pharmacology, U. Hawaii, 1969. Cert. tchr., Calif. Instr. U. Hawaii, Honolulu, 1963-64, asst. prof., 1968-74, assoc. prof., 1974—; vis. scientist U. Tex., 1975, U. Wash., 1975, NIH, 1978. Contbr. articles to profl. jours., chpts. to books. Mem. Am. Soc. Pharmacology and Exptl. Therapeutics, Western Pharmacology Soc. Subspecialties: Molecular pharmacology; Receptors. Current work: Research on histamine secretion, drug-receptor interactions. Office: Pharmacology Dept John A Burns Sch Medicine U Hawaii Honolulu HI 96822

READING, JOHN FRANK, physicist, educator; b. West Bromwich, Eng., Oct. 19, 1939; s. Frank Leslie and Winnifred Gertrude (Mason) R.; m. Elizabeth Anne Reading, Aug. 24, 1963; children: Daniel, Emma, Louise, Patience. B.A., Christ Church, Oxford (Eng.) U., 1960; M.A., Oxon, 1963; Ph.D., Birmingham (Eng.) U., 1964. Instr. physics MIT, 1964-66; sr. research assoc. U. Wash., Seattle, 1966-68; assoc. prof. physics Northeastern U., Boston, 1969-71; assoc. prof. Tex. A&M U., 1971-80, prof., 1980—; cons. Oak Ridge Nat. Lab. Contbr. articles in field to profl. jours. Hoff fellow, 1968-69. Mem. Am. Phys. Soc. Subspecialties: Atomic and molecular physics; Nuclear physics. Current work: Scattering theory in atomic, nuclear and solid state physics; ab initio calculator of ion-atom cross sections for fusion. Home: 1223 Merry Oaks College Station TX 77840 Office: Dept Physics Texas A&M U College Station TX 77843

READNOUR, JERRY MICHAEL, chemistry educator; b. Muncie, Ind., Oct. 19, 1940; s. Garell Ralph and Rhea Madell (LaMotte) R.; m. Lillith June Calvert, June 2, 1962; children: Robin Shane, Scott Michael. B.S., Ball State Tchrs. Coll., 1962; Ph.D., Purdue U., 1969. Asst. prof. chemistry S.E. Mo. State U., 1968-74, assoc. prof., 1974-79, prof., 1979—. Mem. Am. Chem. Soc., Mo. Acad. Sci. Mem. Chs. of Christ. Subspecialty: Thermodynamics. Current work: Thermodynamic properties of aqueous solutions; chemical education. Office: Dept Chemistry SE Mo State U Cape Girardeau MO 63701

REAM, LLOYD WALTER, JR., molecular biologist, educator; b. Chester, Pa., Mar. 20, 1953; s. Lloyd Walter and Mary Elizabeth (Alexander) R.; m. Nancy Jane Smith, May 17, 1975. B.A., Vanderbilt U., 1975; Ph.D., U. Calif.-Berkeley, 1980. Sr. research fellow dept. microbiology U. Wash., Seattle, 1980-83; asst. prof. dept. biology Ind. U., 1983—. Contbr. articles to profl. jours. Am. Cancer Soc. fellow, 1981-83. Mem. Genetics Soc. Am., Am. Soc. for Microbiology, AAAS, Sierra Club, Appalachian Trail Conf., Seattle Sailing Assn. Presbyterian. Subspecialties: Molecular biology; Genetics and genetic engineering (biology). Current work: Molecular mechanism of crown gall tormorigenesis in plants; genetic manipulation of higher plants; control of gene expression; regulation of plant cell growth; bacterial-plant interactions. Home: 6566 Lampkins Ridge Rd Bloomington IN 47401 Office: Dept Biology Ind U Bloomington IN 47405

REAMAN, GREGORY HAROLD, pediatric oncology educator, researcher; b. Barberton, Ohio, Sept. 9, 1947; s. Harold J. and Margaret (D'Alfonso) R.; m. Susan J. Pristo, Sept. 7, 1974; 1 dau., Emily M. B.S., U. Detroit, 1969; M.D., Loyola U., Chgo., 1973. Diplomate: Am. Bd. Pediatrics. Intern Loyola U. Med. Ctr., Chgo., 1973; resident in pediatrics Montreal Children's Hosp., 1974-76; clin. assoc. Nat. Cancer Inst., NIH, Bethesda, Md., 1976-78, investigator, 1978-79; asst. prof. pediatrics George Washington U., 1979-82, assoc. prof., 1982—; attending physician Children's Hosp. Nat. Med. Ctr., Washington, 1979—; reviewer merit rev. bd. VA Med. Research. Served to lt. comdr. USPHS, 1976-79. Spl. fellow Leukemia Soc. Am., 1980-82; Am. Cancer Soc. fellow, 1980-83. Mem. Am. Fedn. Clin. Research, Am. Assn. Cancer Research, Am. Soc. Clin. Oncology, Am. Soc. Pediatric Hematology and Oncology. Roman Catholic. Subspecialties: Oncology; Immunology (medicine). Current work: Immunology of pediatric lymphoreticular neoplasms; treatment of acute leukemia; purine metabolism of lymphoid cells. Home: 3807 Underwood St Chevy Chase MD 20815 Office: Dept Hematology-Oncology Childrens Hosp Nat Med Ctr 111 Michigan Ave NW Washignton DC 20010

REASENBERG, ROBERT DAVID, research physicist; b. N.Y.C., Apr. 27, 1942; s. Julian Robert and Lillian May (Gerber) R.; m. Wendy Schoenbach, June 20, 1965; 1 dau., Suzanne. B.S. in Physics, Poly. Inst. Bklyn., 1963, Ph.D., Brown U., 1970. Research assoc. dept. earth and planetary scis. M.I.T., 1969-71, sponsored research staff, 1971-79, prin. research scientist, 1980-82; physicist Smithsonian Astrophys. Obs., Cambridge, 1982—; lectr. Internat. Sch. Cosmology and Gravitation, Ettore Majorana Centre, Erice. Contbr. articles to profl. jours. Recipient Newcomb Cleveland award, 1976. Mem. AAAS, Am. Astron. Soc., Am. Geophys. Union, Am. Phys. Soc., Internat. Astron. Union, Internat. Soc. Gen Relativity and Gravitation, Sigma Xi. Subspecialties: Relativity and gravitation; Planetary science. Current work: Gravity research, experimental relativity, solar-system dynamics, planetology. Home: 16 Garfield St Lexington MA 02173 Office: 60 Garden St Cambridge MA 02138

REAZIN, GEORGE HARVEY, JR., chemist; b. Chgo., Feb. 3, 1928; s. George Harvey and Katherine Maria (Cole) R.; m. Ruth, June 3, 1950; children: Diane Reazin Clinton, David, Elizabeth. B.S., Northwestern U., 1949; M.S., U. Mich., 1951, Ph.D. (Univ. fellow), 1954. Research assoc. Brookhaven Nat. Lab., Upton, N.Y., 1954-56; with Joseph E. Seagram & Sons, Louisville, 1956—, head chemistry sect., 1971-81, mgr. chemistry research and services, 1982—; J. Guymon Meml. lectr. Am. Soc. Enologists, 1981. Contbr. articles to profl. jours. Mem. Bot. Soc. Am., Am. Chem. Soc., Am. Soc. Plant Physiology, Sigma Xi. Episcopalian. Subspecialties: Plant physiology (biology); Biochemistry (biology). Current work: Chemistry of flavors and biochemistry of their formation; research into ways to measure organoleptic impact of different flavors. Home: 1544 Cliftwood Dr Clarksville IN 47130 Office: PO Box 240 Louisville KY 40201

REBACH, STEVE, marine ecologist, biology educator; b. N.Y.C., Nov. 15, 1942; s. Arthur and Rose Esther (Glick) R.; m. Judith Ann Osborn, Jan 21, 1968; children: Benjamin John, Ari Michael. B.S., CCNY, 1963; Ph.D., U. R.I., Kingston, 1970. Grad. teaching asst. U. Md., College Park, 1963-65; grad. research asst. U. R.I., Kingston, 1965-69, instr., 1969-70; asst. prof. biology St. Mary's Coll. Md., St. Mary's City, 1970-72, U. Md.-Eastern Shore, Princess Anne, 1972-81, assoc. prof., 1981—, dir. grad. program in marine scis., 1981—; cons. Aquaculture, 1972—; Mem. Delmarva Sci. and Engrng. Assn., Salisbury, Md., 1972—. Editor: Studies in Adaptation, 1983; contbr. articles to profl. jours. Grantee U.S. Dept. Agr., 1973, 76, 78, Coastal Zone Mgmt., Md., 1981, Mid-Atlantic Fisheries Devel. Found., 1982. Mem. Animal Behavior Soc. (parliamentarian 1970-82), Atlantic Estuarine Research Soc., The Crustacean Soc., AAAS, Am. Inst. Biol. Scis. Clubs: Computer, Coin (Salisbury, Md.). Subspecialties: Ethology; Behavioral ecology. Current work: Orientation and migration of marine organisms, especially crustacea, biological rhythms in marine organisms, especially crustacea, aquaculture of marine crustacea. Home: 104 Southwood Terr Salisbury MD 21801 Office: Dept Biology U Md Eastern Shore Princess Anne MD 21853

REBEC, GEORGE VINCENT, neuroscientist, cons.; b. Harrisburg, Pa., Apr. 6, 1949; s. George Martin and Nadine (Bosko) R. A.B., Villanova U., 1971; M.A., U. Colo., 1974, Ph.D., 1975. Postdoctoral research fellow dept. psychiatry U. Calif., San Diego, 1975-77; asst. prof. psychology Ind. U., Bloomington, 1977-81, asso. prof., 1981—; cons. NIMH. Contbr. articles to profl. jours. NIMH fellow, 1975-77; recipient Eli Lilly award Ind. U., 1979-80; NSF grantee, 1979-82; Nat. Inst. Drug Abuse grantee, 1979—. Mem. AAAS, Soc. Neurosci. (chpt. chmn.). Subspecialties: Neuropharmacology; Neurochemistry. Current work: Mechanisms of action of psychotropic drugs, including hallucinogens, neuroleptics, and stimulants; drug-induced changes in neuronal activity and electrochemical signals are recorded simultaneously. Office: Dept Psychology Ind U Bloomington IN 47405

REBEIZ, CONSTANTIN ANIS, plant physiology educator; b. Beirut, July 11, 1936; U.S., 1969, naturalized, 1975; s. Anis C. and Valentine A. (Choueyri) R.; m. Carole Louise Conness, Aug. 18, 1962; children: Paul A., Natalie, Mark J. B.S., Am. U., Beirut, 1959; M.S., U. Calif. - Davis, 1960, Ph.D., 1965. Dir. dept. biol. scis. Agrl. Research Inst., Beirut, 1965-69; research assoc. biology U. Calif. - Davis, 1969-71; assoc. prof. plant physiology U. Ill., Urbana-Champaign, 1972-76, prof., 1976—. Contbr. articles to sci. publs. plant physiology and biochemistry. Fellow Explorers Club; mem. Am. Soc. Plant Physiologists, Comite Internat. de Photobiologie, Am. Soc. Photobiology, AAAS, Lebanese Assn. Advancement Scis. (exec. com. 1967-69), Sigma Xi. Greek Orthodox. Subspecialties: Nitrogen fixation; Biochemistry (biology). Current work: Bioengineering of photosynthetic reactors of photosynthetic membranes; chloraphyll chemistry and biochemistry. Research on pathway of chlorophyll biosynthesis, chloroplast devel., bioengrng. of photosynthetic reactors; pioneered biosynthesis of chlorophyll in vitro; duplication of greening process of plants in test tube, demonstration of operation of multibranched chlorophyll biosynthetic pathway in nature. Discovered several novel chlorophyll chem. species in plants. Home: 301 W Pennsylvania Ave Urbana IL 61801 Office: Vegetable Crops Bldg U Ill Urbana-Champaign IL 61801 Meaningful scientific discoveries are those that help humans achieve a better understanding of themselves, of their environment or of the universe at large, as well as those that contribute to the betterment of the human spiritual, psychological and physical condition.

REBOIS, RAYMOND VICTOR, plant pathologist, nematologist, researcher; b. San Francisco, Sept. 14, 1924; s. Leon Victor and Ida R.; m. Ann Richardson, Mar. 28, 1948; children: Robert Victor, Suzanne. Ph.D., Auburn U., 1970. Nematologist Dept. Agr., Orlando, Fla., 1956-65, Auburn, Ala., 1965-70; supervisory research plant pathologist Dept. Agr. Beltsville Agrl. Research Ctr., Beltsville, Md., 1970—, also chief lab.; cons. in field. Contbr. numerous articles on plant nematode pathology, resistance, host-parasite interactions and nematode control to profl. jours. Served to comdr. USN, 1943-45. Mem. Helminthological Soc. Washington, Soc. Nematologists, Am. Phyto-pathol. Soc., AAAS. Subspecialties: Plant pathology; Animal physiology. Current work: Biochemistry of host plants and nematodes; physiology of plant parasitic nematodes; molecular basis for plant resistance. Home: 2701 Muskogee St Adelphi MD 20783 Office: Dept Agr Beltsville Agrl Research Center W Bldg 011A Beltsville MD 20705

RECH, RICHARD HOWARD, pharmacology educator; b. Irvington, N.J., Mar. 20, 1928; s. Walter Edward and Francis (Sanders) R.; m. Barbara Jane Rech, Oct. 4, 1952; children: Sharon Rech Graham, Michelle Rech Barrett, Charles E.L. B. S., Rutgers Coll. Pharmacy, 1952; M.S., U. Mich., 1955; Ph.D., 1959. Postdoctoral USPHS fellow U. Utah, 1959-61; instr., then asst. prof., then assoc. prof. dept. pharmacology and toxicology Dartmouth Med. Sch., Hanover, N.H., 1961-71; vis. scholar Mario Negri Research Inst., Milan, Italy, 1978-79; prof. pharmacology and toxicology Mich. State U., East Lansing, 1971—. Contbr. articles to profl. jours.; field editor: Pharmacol. Biochem. Behavior, 1974—. Instr. CPR ARC, 1973—; mem. Nat. Ski Patrol, 1969—; lectr. drug abuse to lay public. Served with AUS., 1946-48. USPHS fellow, 1957-59; USPHS sr. internat. fellow, 1978-79; NIH grantee, 1962—; Smith Kline & French grantee, 1964; Lederle grantee, 1973; Dow Corning grantee, 1974; Adria grantee, 1977; 3M grantee, 1981. Fellow Am. Coll. Neuropsychopharmacology; mem. Am. Soc. Pharmacology and Exptl. Therapeutics, AAAS, Soc. Neurosci., Collegium Internat. Neuro Psycho Pharmacology. Episcopalian. Subspecialties: Psychopharmacology; Environmental toxicology. Current work: Effects of Psychoactive agents on brain neurotransmitters; mechanis of action of hallucinogenic drugs, anxiolytics, neurobehavioral toxicology and teratology of organophosphates. Home: 1474 Mercer Dr Okemos MI 48864 Office: Dept Pharmacology and Toxicology Mich State U East Lsnsing MI 48824

RECHCIGL, MILOSLAV, JR., govt. ofcl.; b. Mlada Boleslav, Czechoslovakia, July 30, 1930; came to U.S., 1950, naturalized, 1955; s. Miloslav and Marie (Rajtrova) R.; m. Eva Marie Edwards, Aug. 29, 1953; children—John Edward, Karen Marie. B.S., Cornell U., 1954, M. Nutrition Sci., 1955, Ph.D., 1958. Teaching asst. Cornell U., 1953-57, grad. research asst., 1957-58, research assoc., 1958; USPHS research fellow Nat. Cancer Inst., 1958-60, chemist enzymes and metabolism sect., 1960-61, research biochemist, tumor host relations sect., 1962-64, sr. investigator, 1964-68; grants assoc. program NIH, 1968-69; spl. asst. for nutrition and health to dir. Regional Med. Programs Service, Health Services and Mental Health Adminstrn., HEW, 1969-70; exec. sec. nutrition program adv. com. Health Services and Mental Health Adminstrn., 1969-70; nutrition adviser AID, Dept. State, Washington, 1970-71, chief, 1970-73, exec. sec. research and instl. grants council, 1970-74; exec. sec. AID research adv. com., 1971-82; AID rep. USC/FAR com., 1972-82; asst. dir. Office Research and Instl. Grants, 1973-74, acting dir., 1974-75, dir. interregional research staff, 1975-78, devel. studies program, 1978, chief research and methodology div., 1979—; del. White House Conf. on Food, Nutrition and Health, 1969; cons. Office Sec. Agr., 1969-70, Dept. Treasury, 1973, Office Technol. Assessment, 1977—, FDA, 1979—; del. Agrl. Research Policy Adv. Com. Conf. on Research to Meet U.S. and World Food Needs, 1975; Organizer, mem. council Montrose Civic Assn., Rockville, Md. Author: (with Z. Hruban) Microbodies and Related Particles: Morphology, Biochemistry and Physiology, 1969, also Russian edit., 1972, Enzyme Synthesis and Degradation in Mammalian Systems, 1971, Food, Nutrition and Health: A Multidisciplinary Treatise Addressed to the Major Nutrition Problems from a World Wide Perspective, 1973, Man, Food and Nutrition: Strategies and Technological Measures for Alleviating the World Food Problem, 1973, World Food Problem: A Selective Bibliography of Reviews, 1975, Carbohydrates, Lipids, and Accessory Growth, 1977-80; 1976, Nutrient Elements and Toxicants, 1977, Nutritional Requirements, 1977, Diets, Culture Media and Food Suplements, 4 vols, 1977-78, Nutritional Disorders, 3 vols, 1978, Nitrogen, Electrolytes, Water and Energy Metabolism, 1979, Nutrition and the World Food Problem, 1979, (with Eva Rechcigl) Biographical Directory of the Members of the Czechoslovak Society of Arts and Sciences in America, 1972, 78, The Czechoslovak Contribution to World Culture, 1964, Czechoslovakia Past and Present, 1968, Educators with Czechoslovak Roots: A U.S. and Canadian Faculty Roster, 1980, others.; Co-editor: Internat. Jour. of Cycle Research, 1969-74, Jour. of Applied Nutrition, 1970—; series editor: Comparative Animal Nutrition, 1976—; editor-in-chief: Handbook Series in Nutrition and Food, 1977—; mem. editorial bd.: Nutrition Reports Internat, 1977-80; translator, abstractor: Chemical Abstracts, 1959—; Contbr. sci. articles to profl. jours. Chmn. cult com. Trustees for Czech chapel in Honor St. John Neumann Nat. Shrine, Washington, 1983; active PTA, Boy Scouts Am., Rockville Alliance for Better Environment, Program Com. Earth Day. Recipeint cert Achievement, adminstr. AID; hon. cert. AID Research Com., 1982; Nat. Acad. Scis. grantee, 1962; USPHS grantee, 1968. Fellow AAAS, Am. Inst. Chemists (councilor 1972-74, program chmn. 1974 ann. meeting, mem. program com. 1980 meeting), Internat. Coll. Applied Nutrition, Intercontinental Biog. Assn.. Washington Acad. Scis. (del. 1972—); mem. Am. Inst. Nutrition (com. Western Hemisphere Nutrition Congress 1971, 74, program com. 1979—), Am. Soc. Biol. Chemists, Am. Chem. Soc. (joint bd.-council com. on internat. activities 1975-76), D.C. Inst. Chemists (pres. 1972-74, councilor 1974-80), Am. Inst. Biol. Scis., Soc. for Exptl. Biology and Medicine, Am. Soc. Animal Sci., Internat., Am. socs. cell biology, Soc. for Developmental Biology, Am. Assn. for Cancer Research, Soc. for Biol. Rhythm, Am. Pub. Health Assn., N.Y. Acad. Scis., Chem. Soc. Washington (symposium com. 1970, 71), Internat. Coll. Applied Nutrition, Internat. Soc. for Research on Civilization Diseases and Vital Substances, Soc. for Geochemistry and Health, Soc. for Internat. Devel., Internat. Platform Assn., Am. Assn. for Advancement Slavic Studies, Czechoslovak Soc. Arts and Scis. in Am. (hon., dir.-at-large 1962—, dir. publs. 1962-68, 70-74, v.p. 1968-74, pres. 1974-78, pres. collegium 1978—), History of Sci. Soc., Soc. Research Adminstrs., Sigma Xi, Phi Kappa Phi, Delta Tau Kappa (hon.). Clubs: Cosmos, Cornell (Washington). Current work: Manages innovative research relating to international development in biotechnology, biomass conversion technology, biolgical control of disease vectors, chemistry for food production, engineering technology, marine and earth sciences; genetic resources. Home: 1703 Mark Ln Rockville MD 20852 Office: Research and Methodology Div AID Dept State Washington DC 20523

RECHNITZER, ANDREAS BUCHWALD, technology advisor, science advisor; b. Escondido, Calif., Nov. 30, 1924; s. Ferdinand Martin and Dagmar (Buchwald) R.; m. Martha Jean Mitchell, Aug. 18, 1946; children: David Franklin, Andrea Jeanne, Martin Allan, Michael Jon. B.S., Mich. State U., 1947; M.A., UCLA, 1951; Ph.D., U. Calif.-San Diego, 1956. Coordinator Naval Electronics Lab., San Diego, 1956-61; chief scientist deep submergence N.Am. Aviation, Anaheim, Calif., 1961-70; tech. advisor Office Chief Naval Ops., Washington, 1970—; pres. Sci. Diving Cons., San Diego, Cedam Internat., Dallas, 1968-77; chmn. bd. Gyrotor Inc., San Diego, 1960-61. Exec. producer: films Treasure of Scorpion Reef, 1968 (gold medal), Five Fathoms to a Lost Ship's Grave, 1968 (gold medal). Served to capt. USNR. Recipient gold medal Chgo. Geography Soc., 1960, Richard Hopper Day Phila. Acad. Sci, 1960, Underwater Photog. Soc. Am. award, 1961. Fellow Marine Tech. Soc. (sect. pres.); mem. Am. Oceanic Orgn., Explorers Club, Nat. Geog. Soc. (hon. life). Club: Cosmos. Subspecialties: Ocean engineering; Robotics. Current work: Advanced concepts in ocean engineering and deep ocean technology. Scientist in charge record dive 35,800 feet, 1960 (disting. civilian service award). Home: 6368 Dockser Terr Falls Church VA 22041 Office: Office Chief Naval Ops Dept Navy Washington DC 20350

RECHTIN, EBERHARDT, aerospace systems co. exec.; b. East Orange, N.J., Jan. 16, 1926; s. Eberhardt Carl and Ida H. (Pfarrer) R.; m. Dorothy Diane Denebrink, June 10, 1951; children—Andrea C., Nina, Julie Anne, Erica, Mark. B.S., Calif. Inst. Tech., 1946, Ph.D. cum laude, 1950. Asst. dir., dir. Deep Space Network, Calif. Inst. Tech. Jet Propulsion Lab., 1949-67; dir. Advanced Research Projects Agy., Dept. Def., 1967-70, prin. dep. dir. def. research and engring., 1970-71, asst. sec. def. for telecommunications, 1972-73; chief engr. Hewlett-Packard Co., Palo Alto, Calif., 1973-77; pres. Aerospace Corp., El Segundo, Calif., 1977—. Served to lt. USNR, 1943-56. Recipient major awards NASA, Dept. Def. Fellow Am. Inst. Aeros. and Astronautics (major awards), IEEE (major awards); mem. Nat. Acad. Engring., Tau Beta Pi. Subspecialty: Electrical engineering. Home: 1665 Cataluna Pl Palos Verdes Estates CA 90274 Office: 2350 E El Segundo Blvd El Segundo CA 90245

RECHTSCHAFFEN, ALLAN, psychology educator, sleep researcher; b. N.Y.C., Dec. 8, 1927; s. Philip and Sylvia (Yeager) R.; m. Karen Ann Wold, Mar. 16, 1980; children: Laura, Katherine, Amy. B.S.S., CCNY, 1949, M.A., 1951; Ph.D., Northwestern U., 1956. Staff psychology Fergus Falls (Minn.) State Hosp., 1951-53; research psychologist VA, Chgo., 1956-57; prof. psychology U. Chgo., 1957—, dir. sleep research lab. Contbr. articles to profl. jours. NIMH research scientist awardee, 1967—. Mem. Sleep Research Soc. (pres. 1979-82). Subspecialty: Physiological psychology. Current work: Sleep and dreams; research on function of sleep. Home: 5800 S Harper Ave Chicago IL 60637 Office: University of Chicago 5743 S Drexel Ave Chicago IL 60637

RECKARD, CRAIG REGINALD, surgeon, researcher; b. Phila., Dec. 21, 1940; s. C. John and Mary Eser (Craig) R.; m.; children: Jonathan Craig, Justin Michael. B.S., Ursinus Coll., 1962; M.D., U. Pa., 1967. Diplomate: Am. Bd. Surgery. Intern U. Pa., Phila., 1967-68, resident in surgery, 1968-74, transplantation fellow, 1970-72, asst. instr., 1968-73, instr., 1973-74; staff surgeon Walter Reed Army Med. Ctr., Washington, 1974-76; asst. prof. surgery U. Chgo., 1976-80, assoc. prof. surgery, 1980—; attending surgeon Little Co. Mary Hosp., Evergreen Park, Ill., 1980—. Mem. ACS, Assn. Acad. Surgery, Am. Soc. Tranplant Surgeons, Transplantation Soc., Soc. Univ. Surgeons. Subspecialties: Surgery; Transplant surgery. Current work: Pancreatic transplantation. Home: 5466 S Cornell Chicago IL 60615 Office: U Chgo 950 E 59th S Box 77 Chicago IL 60637

RECKNAGEL, RICHARD OTTO, physiology educator; b. Springfield, Mo., Jan. 11, 1916; s. Emil and Ella (Moench) R.; m. Maesine G. Recknagel, June 19, 1943; children—Frank Otto, Judith Susan. B.S., Wayne U., 1939; Ph.D., U. Pa., 1949. Asst. prof. physiology Sch. Medicine, Case Western Res. U., 1956-60, asso. prof., 1960-65, prof., 1965, dir. dept. physiology, 1978—. Served to 1st lt. USAAF, World War II. Recipient Kaiser-Permanente Teaching award, 1977; NIH grantee, 57-83. Subspecialty: Physiology (medicine). Office: Case Western Res U Sch Medicine Dept Physiology Cleveland OH 44106

RECORDS, RAYMOND EDWIN, ophthalmology educator, physician; b. Ft. Morgan, Colo., May 30, 1930; s. George Harvey and Sara Barbara (Louden) R.; m.; 1 dau., Lisa Rae. B.S., U. Denver, 1956; M.D., St. Louis, 1961. Diplomate: Am. Bd. Ophthalmology. Intern St. Louis U. Hosp. Group, 1961-62; resident in ophthalmology U. Colo. Med. Ctr., 1962-65; asst. prof. surgery U. Colo., Denver, 1965-70; prof. ophthalmology U. Nebr. Coll. Medicine, Omaha, 1970—, chmn. dept. ophthalmology, 1970—; cons. VA, Omaha, 1970—. Author: Physiology of Human Eye, 1979, Biomedical Foundations of Ophthalmology, 1982. Fellow Am. Acad. Ophthalmology; mem. Nebr. Med. Assn., Omaha Ophthal. Soc. (pres. 1980-81), Nebr. Acad. Ophthalmology. Subspecialty: Ophthalmology. Current work: Bacterial endophthalmitis, ocular physiology. Home: 9916 Devonshire Dr Omaha NE 68114 Office: U Nebr Med Ctr 42d St and Dewey Ave Omaha NE 68105

REDDY, BANDARU S., nutritional oncologist; b. Nellore, India, Dec. 30, 1932; s. Venkata Subba and Ramamma (Pulim) Bandaru; m. Subhashini Reddy Bandaru, Dec. 22, 1962; children: Sudhakar, Sadalakshmi, Srikanth. D.V.M., U. Madras, 1955; M.S., U. N.H., 1960; Ph.D., Mich. State U., 1963. Teaching asst., research asst. U. N.H., 1958-60; research asst. Mich. State U., 1960-63; postdoctoral research assoc. U. Notre Dame, 1963-65; asst. research prof. Lobund Labs., 1965-68, assoc. research prof., 1968-71; research prof. dept. microbiology N.Y. Med. Coll., Valhalla, 1976—; mem., assoc. chief div. nutrition Am. Health Found., Naylor Dana Inst. for Disease Prevention, Valhalla, 1971—; mem. adv. bd. Lobund Labs., U. Notre Dame, 1980-83; mem. reactor panel of diet, nutrition and cancer project Nat. Acad. Scis., 1982. Contbr. articles to profl. jours. Pres. Telugu Lit. and Cultural Assn. N.Y., 1975; v.p. Telugu Assn. N.Am., 1979, Nat. Assn. Ams. of Asian Indian Descent, 1982. NIH grantee. Mem. Assn. Gnotobiotics (pres. 1978-79), Am. Inst. Nutrition, Am. Assn. Pathologists, Soc. Exptl. Biology and Medicine, Am. Assn. Cancer Research, Soc. Toxicology, Internat. Soc. Study of Xenobiotics. Hindu. Subspecialties: Cancer research (medicine); Nutrition (medicine). Current work: Diet, nutrition and cancer; inhibitors and promotors of carcinogenesis; mechanism of carcinogenesis, nutritional toxicology. Office: Am Health Found Dana Rd Valhalla NY 10595

REDDY, CHADA SUDERSHAN, veterinary toxicologist; b. Hyderabad, India, Nov. 5, 1949; came to U.S., 1974; s. Chada Ramchandra and Chada Lalitha (Devi) R.; m. Chaya Dharmapuram, May 18, 1972; 1 son, Chada Ramchander. Ph.D., U. Miss., 1980. Asst. vet. surgeon Govt. Andhra Pradesh, India, 1972-73; grad. research asst. Ala. A&M U., Normal, 1974-76, U. Miss. Med. Ctr., Jackson, 1976-78, NIH predoctoral fellow, 1978-80; asst. prof. preventive medicine Ohio State U., Columbus, 1980—. Contbr. articles to profl. jours. Recipient Nat. Research Service award NIH, 1978-80, New Investigator Research Award NIH, 1983—. Mem. AVMA, Ohio Pub. Health Vet. Assn., Assn. Am. Vet. Med. Colls. Hindu. Subspecialties: Toxicology (medicine); Preventive medicine (veterinary medicine). Current work: Mechanisms of teratogeneis. Toxicology and safety evaluation of naturally occurring and environmental contaminant chemicals in man and animals. Office: 1900 Coffey Rd Columbus OH 43210

REDDY, CHINTHAMANI CHANNA, biochemistry educator; b. Anantapur, India, Sept. 8, 1947; came to U.S., 1975, naturalized, 1978; s. Chinthamani Venkata Rami Reddy and Chinthamani (Obulamma) R.; m. Chinthamani Santha Reddy, Aug. 27, 1972; children; Chinthamani Deepika, Chinthamani Radhika. B.Sc., Regional Coll. Edn., India, 1968, B. Ed., 1969; M.Sc., Mysore U., 1971; Ph.D., Indian Inst. Sci., 1975. Postdoctoral fellow dept. chemistry Pa. State U., University Park, 1975-77, sr. research technologist, 1977-79, research assoc. Ctr. for Air Environ. Studies, 1979-81, asst. prof. vet. sci. dept., 1981—. Recipient Gold medal Mysore U., 1971; Research Career Devel. award NIH, 1983—; Nat. Sci. Talent Research scholar, 1966-75. Mem. Am. Soc. Biol. Chemists, N.Y. Acad. Scis., AAAS. Subspecialties: Biochemistry (biology); Environmental toxicology. Current work: Lipid peroxidation; biochemical interaction of vitamin E and selenium; prostaglandins; prostacyclins; thromboxanes and leukotrienes; pulmonary toxicity; glutathione S-transferases. Home: 229 Canterbury Dr State College PA 16803 Office: 226 Fenske Lab University Park PA 76802

REDDY, JANARDAN KATANGOORY, educator, researcher; b. Andhra Pradesh, India, Oct. 7, 1938; s. Raghava K. and Shanta M. R.; m. Kamalesh K., May 19, 1962; children: Namrata, Neerad. M.B.B.S., Osmania U., Hyderabad, India, 1961; M.D., All India Inst. Meld. Scis., New Delhi, 1965. Cert. Am. Bd. Pathology. Intern Osmania Gen. Hosp., Hyderabad, 1961-62; resident U. Kans. Med. Center, Kansas City, 1966-68; resident fellow U. Kans., Kansas City, 1966-70, asst. prof. pathology, 1970-73, assoc. prof., 1973-76, prof., 1976; prof. pathology Northwestern U. Med. Sch., Chgo., 1976—. Contbr. numerous articles in field to profl. jours. Chmn. awards com. Asian Indians in N.Am. Conv., 1982. NIH grantee, 1972—; Nat. Cancer Inst. grantee, 1972—. Mem. Am. Soc. Cell Biology, Am. Assn. Pathologists, Am. Assn. Cancer Research, Biochem. Soc. U.K., N.Y. Acad. Scis., Histochem. Soc., Internat. Acad. Pathology, Assn. Scientists of Indian Origin in Am. (pres.-elect 1982). Subspecialties: Cancer research (medicine); Cell biology (medicine). Current work: Biology of the cytoplasmic organelle peroxisome; implications of peroxisome proliferation, mechanism of peroxisome proliferation associated carcinogenesis; pancreatic carcinogenesis and differentiation of neoplastic pancreatic acinar cells. Office: 303 E Chicago Ave Chicago IL 60611

REDDY, MOHAN M., immunologist; b. Chittoor, India, Apr. 25, 1942; s. Annaiah M. and Lakshmi R.; m. Sukumari, Feb. 4, 1972; children: Sridhar, Rekha. M.Sc., IARI, New Delhi, Italy, 1966; Ph.D., U. Chgo., 1969. Asst. prof. medicine U. Rochester (N.Y.), 1976-79, assoc. in medicine, 1973-76; specialist in immunology Am. Soc. Clin. Pathology, 1982; dir. allergy and clin. immunology lab. R.A. Cooke Inst. Allergy, St. Luke's-Roosevelt Hosp., N.Y.C., 1979—. Contbr. articles to profl. jours. Mem. Am. Soc. Clin. Pathologists (assoc.), Am. Assn. Immunologists, Am. Acad. Allergy and Immunology, N.Y. Allergy Soc., Soc. Exptl. Biology and Medicine, AAAS, Sigma Xi. Subspecialties: Cellular engineering; Infectious diseases. Current work: Acquired immunodeficiency syndrome and immunotherapy.

REDDY, NEELUPALLI BOJJI, biochemist; b. Panchalingala, Andhra Pradesh, India, July 1, 1943; came to U.S., 1972; s. Vema and (Govidmaa) R.; m. Prabhavathi Reddy, June 25, 1966; 1 child, Venkat N. B.Sc., Osmania Coll., 1964; M.Sc., S.V. U., India, 1966, Ph.D., 1971. Postdoctoral fellow NIH, Bethesda, Md., 1972-75, research assoc., 1975-81; assoc. prof. research U. So. Calif., Los Angeles, 1981—. Vice pres. Telugu Assn. of Greater Washington Area, 1979. Muscular Dystrophy Assn. grantee, 1981-82, 83—. Mem. Am. Physiol Soc., N.Y. Acad. Scis. Subspecialties: Membrane biology; Biochemistry (medicine). Current work: Biochemical analysis of muscle membranes in normal and diseased states; nerve-muscle interaction and their contributions to pathology of skeletal muscle in neuromuscular diseases. Home: 1113 Thompson St Glendale CA 91201 Office: U So Calif Neuromuscular Ctr 637 S Lucas Ave Los Angeles CA 90017

REDEI, GYORGY PAL, geneticist, educator; b. Vienna, Austria, June 14, 1921; came to U.S., 1956, naturalized, 1962; s. Kalman and Margit (Wallenstein) Re.; m. Magdolna M. Nagy, Oct. 21, 1953; 1 dau., Maria Charlotte. B.S., Agrartudomanyi Egyetem, Budapest, 1949; C.S., Magyar Tudomanyos Akademia Genetikai Intezete, Budapest, 1955. Early sci. career in, Budapest, 1948-56; faculty U. Mo., Columbia, 1957—, prof. genetics, 1969—. Author 3 textbooks, also numerous research papers.; Editor: Stadler Genetics Symposia, 14 vols. Research grantee various govt. agys. Mem. Genetics Soc. Am. Roman Catholic. Subspecialties: Genetics and genetic engineering (biology); Gene actions. Current work: Biochemical and molecular genetics, mutation, history of genetics. Home: 3005 Woodbine St Columbia MO 65201 Office: U Mo 117 Curtis Hall Columbia MO 65211

REDFEARN, PAUL LESLIE, JR., biology educator; b. Sanford, Fla., Oct. 5, 1926; s. Paul Leslie and Carolyn (Spencer) R.; m. Alice R. Whitten, June 17, 1949; children: Paul, James. B.S., Fla. So. Coll., 1948; M.S., U. Tenn., 1949; Ph.D., Fla. State U., 1957. Interim instr. U. Fla., Gainesville, 1949-50; assoc. prof. biology S.W. Mo. State Coll., Springfield, 1957-62, prof., 1962—. Author: How to Know the Mosses, 1979. Mem. city council, Springfield, 1973-79, mayor, 1979-81. Served to 1st lt. U.S. Army, 1951-54. Fellow AAAS. Democrat. Methodist. Subspecialties: Systematics; Ecology. Current work: Geography of plants, bryophytes of North America. Home: 655 McCann St Springfield MO 65804 Office: Dept Biology SW Mo State U Springfield MO 65804

REDISH, EDWARD FREDERICK, physicist, educator; b. N.Y.C., Apr. 1, 1942; s. Jules and Sylvia (Coslow) R.; m. Janice Copen, June 18, 1967; children: Aaron D., Deborah M. A.B. magna cum laude, Princeton U., 1963; Ph.D., M.I.T., 1968. Postdoctoral fellow Ctr. for Theoretical Physics, U. Md., College Park, 1968-70, asst. prof., 1970-74, assoc. prof., 1974-79, prof. physics, 1979—, chmn. dept., 1982—; vis. fgn. collaborator C.E.N., Saclay, France, 1973-74; resident sr. research assoc. Goddard Space Flight Center, Greenbelt, Md., 1977-78. Recipient Kusaka Meml. prize in physics, 1963; Inst. Sci. medal Central Research Inst. Physics, Budapest, 1979; hon. Woodrow Wilson fellow; NSF predoctoral fellow. Fellow Am. Phys. Soc.; Mem. AAAS. Subspecialties: Nuclear physics; Theoretical physics. Current work: Research in theoretical nuclear physics and many particle scattering theory. Office: Dept Physics and Astronomy U Md College Park MD 20742

REDMOND, ROBERT FRANCIS, nuclear engineering educator; b. Indpls., July 15, 1927; s. John Felix and Marguerite Catherine (Breinig) R.; m. Mary Catherine Cangany, Oct. 18, 1952; children: Catherine, Robert, Kevin, Thomas, John. B.S. in Chem. Engring, Purdue U., 1950; M.S. in Math, U. Tenn., 1955; Ph.D. in Physics, Ohio State U., 1961. Engr. Oak Ridge Nat. Lab., 1950-53, scientist, adviser-cons. Battelle Meml. Inst., Columbus, Ohio, 1953-70; prof. nuclear engring. Ohio State U., Columbus, 1970—, asso. dean Coll. Engring., dir. Engring. Experiment Sta., 1977—. Contbr. articles to profl. jours. V.p. Argonne Univs. Assn., 1976-77, trustee, 1972-80; mem. Ohio Power Siting Commn., 1978-82. Served with AUS, 1945-46. Mem. Am. Nuclear Soc. (chmn. Southwestern Ohio sect.), AAAS, Am. Soc. Engring. Edn., Sigma Xi, Tau Beta Pi. Subspecialties: Nuclear engineering; Nuclear fission. Current work: University research administration. Home: 3112 Brandon Rd Columbus OH 43221

REED, ALBERT PAUL, molecular biologist; b. St. Marys, Pa., July 4, 1954; s. Albert Paul and Veronica Cecilia (DeLullo) R.; m. Cynthia Lou Reed, July 1, 1978; 1 son, Brandon Charles. B.S., U. Pitts., 1976; M.S., Clarion State Coll., 1981. Research assoc. Genex Corp., Gaithersburg, Md., 1981—. Mem. Am. Soc. Microbiology, AAAS. Republican. Roman Catholic. Subspecialties: Genetics and genetic engineering (biology); Molecular biology. Current work: DNA sequencing, cloning into bacteriophage. Home: McCullough Ln apt 303 Gaithersburg MD 20877 Office: Genex Corp 16020 Industrial Dr Gaithersburg MD 20877

REED, CHARLES EMMETT, physician, educator; b. Boulder, Colo., Mar. 13, 1922; s. Charles E. and Helene (Hadady) R.; m. Janice Tullock, July 19, 1962; children—Jocelyn, David, Marian, James, Walter, Barbara, Lawrence. Student, Harvard, 1943; M.D., Columbia, 1945. Diplomate: Am. Bd. Internal Medicine, Am. Bd. Allergy and Immunology (co-chmn. 1972-75). Intern Colo. Gen. Hosp., Denver, 1945-46; resident medicine Roosevelt Hosp., N.Y.C., 1948-51; pvt. practice Corvallis (Oreg.) Clinic, 1951-61; mem. faculty U. Wis. Med. Sch., 1961—, prof. medicine, 1967-78, Mayo Med. Sch., Rochester, Minn., 1978—; cons. Mayo Clinic. Co-editor: Allergy Principles and Practice, 1978; mem. editorial bd. jour.: Allergy; editor: Jour. Allergy and Clin. Immunology, 1977—; Contbr. articles to profl. jours. Served with AUS, 1946-48. Mem. Am. Acad. Allergy (pres. 1973-74), A.C.P. Subspecialty: Anesthesiology. Current work: Allergic lung diseases. Home: 1026 Plummer Circle Rochester MN 55901

REED, IRVING STOY, electrical engineering and computer science educator, consultant; b. Seattle, Nov. 12, 1923; s. Irving McKenny and Eleanor Doris (Stoy) R.; m.; children by previous marraige: Ann Lenore, Irving L., Thomas E., Mark N.; m. Bernice Margaret Schmidt, Dec. 11, 1965; 1 son, Henry George. Student, U. Alaska, 1940-42; B.S., Calif. Inst. Tech., 1944, Ph.D., 1949; postgrad., Boston Coll. Sch. Law, Newton, Mass., 1957-58. Research engr. Northrop Aircraft, Hawthorne, Calif., 1949-50; dir. research Computer Research Corp., Torrance, Calif., 1950-51; assoc. group leader MIT Lincoln Lab., Lexington, Mass., 1951-60; sr. staff mem. Rand Corp., Santa Monica, Calif., 1960-63, cons., 1963—; assoc. prof. elec. engring. U. So. Calif., 1963-64, prof. elec. engring. and computer sci., 1964—, Charles Lee Powell prof. computer engring., 1978—; research affiliate Jet Propulsion Lab., Pasadena, Calif., 1967—; dir. Tech. Service Corp., Santa Monica, Adaptive Sensors Inc.; cons. Mitre Corp., Bedford, Mass. Author: Theory and Design of Digital Machines, 1962; contbr. numerous articles to profl. jours. Served with USN, 1944-46. Fellow IEEE; mem. Am. Math. Soc., Math. Assn. Am., Nat. Acad. Engring., Sigma Xi. Republican. Subspecialties: Electrical engineering; Computer engineering. Current work: Have recently published papers on adaptive array antennas, signal processing, biomedical engineering, computer circuits and adaptive antennas. Patentee in field. Office: U So Calif 510 Powell Hall Los Angeles CA 90007

REED, JAMES LESLIE, chemistry educator; b. Chgo., May 6, 1947; s. James S. and Stella K. R.; m. Dorothy Jackson, Sept. 2 1972; children: James C., Stephanie I. B.S., U. Ill.-Chgo., 1969; M.S., Northwestern U., 1972, Ph.D., 1974. Asst. prof. to assoc. prof. chemistry Atlanta U., 1973—. Contbr. articles to sci. jours. Urban homesteader Soc. St. Vincent DePaul, Roman Catholic Ch. Mem. Am. Chem. Soc., Ga. Acad. Scis. Subspecialties: Inorganic chemistry; Photochemistry. Current work: Chemistry and photochemistry of metal complexes; energy converting systems. Office: 223 Chestnut St SW Atlanta GA 30314

REED, LESTER JAMES, educator, biochemist; b. New Orleans, Jan. 3, 1925; s. John T. and Sophie (Pastor) R.; m. Janet Louise Gruschow, Aug. 7, 1948; children—Pamela, Sharon, Richard, Robert. B.S., Tulane U., 1943; D.Sc. (hon.), 1977; Ph.D., U. Ill., 1946. Research asst. NDRC, Urbana, Ill., 1944-46; research asso. biochemistry Cornell U. Med. Coll., 1946-48; faculty U. Tex., Austin, 1948—, prof. chemistry, 1958—; research scientist Clayton Found. Biochem. Inst., 1949—, asso. dir., 1962-63, dir., 1963—. Contbr. articles profl. jours. Mem. Nat. Acad. Scis., U.S., Am. Acad. Arts and Scis., Am. Soc. Biol. Chemists, Am. Chem. Soc. (Eli Lilly & Co. award in biol. chemistry 1958), Phi Beta Kappa, Sigma Xi. Subspecialties: Biochemistry (biology); Molecular biology. Current work: Structure, function and regulation of multienzyme complexes; enzyme chemistry. Home: 3502 Balcones Dr Austin TX 78731

REED, WILLIAM PATRICK, immunology educator; b. Denver, Feb. 28, 1933; s. William Patrick and Martha Marie (Walhood) R.; m. Ellen Ann Hill, Aug. 25, 1957; children: Martha, Steven, Richard. A.B., Harvard U., 1955, M.D., 1959. Diplomate: Am. Bd. Internal Medicine. Intern George Washington U., 1959-60; resident in medicine D.C. Gen. Hosp., 1960-61, Madigan Gen. Hosp., Tacoma, 1963-65; chief of medicine U.S. Army Hosp., Zama, Japan, 1965-68; prof. medicine U. N.Mex., Albuquerque, 1978—; assoc. chief of staff for research VA Med. Ctr., Albuquerque, 1974—. Contbr. articles to sci. jours.; contbr. to: Clinical Neurosciences, 1983. Served with U.S. Army, 1961-63. Research assoc. VA, 1969; clin. investigator, 1971. Fellow ACP, Infectious Disease Soc.; mem. Am. Fedn. Clin. Research, Western Soc. Clin. Research, Am. Soc. Microbiology. Subspecialties: Infectious diseases; Internal medicine. Current work: Host defences to infection. Office: VA Hosp 2100 Ridgecrest Dr SE Albuquerque NM 87108

REED, WILLIAM PIPER, JR., surgeon, educator; b. Melrose, Mass., May 24, 1942; s. William Piper and Gertrude Harriet (Irons) R.; m. Martine Francoise Billet, Oct.16, 1963; children: Antoinette Elsa, Christopher Llewellyn. A.B., Harvard U., 1964, M.D., 1968; diploma in head and neck cancer, U. Paris, 1977. Diplomate: Am. Bd. Surgery. Surg. intern Stanford U., Palo Alto, Calif., 1968-69, fellow in transplantation, 1969-70, resident in surgery, 1972-76; fellow in cancer surgery Inst. Gustave-Roussu, Villejuif, France, 1976-77; asst. prof. surgery U. Md., Balt., 1978—, gen. surgeon, 1978-81; cancer surgeon U. Md. Hosp., Balt., 1981-83, dir. surg. intensive care unit, 1979-81. Contbr. articles to profl. jours. Chmn. profl. edn. Am. Cancer Soc., Towson, Md., 1981-83. Served to capt. U.S. Army, 1970-72; Vietnam. NIH fellow, 1969-70; Am. Cancer Soc. fellow, 1972-73; NIH trainee, 1973-75. Fellow ACS; mem. Assn. for Acad. Surgery, Soc. Surg. Oncology, Am. Soc. Clin. Oncology, Soc. Critical Care Medicine, Am. Soc. Transplant Surgeons. Club: Harvard of Md. (Balt.). Subspecialties: Surgery; Artificial organs. Current work: Implantable vascular access devices for chemotherapy. Microwave scalpel for use in spleen and liver resection. Office: Oncology Program Dept Surgery U Md Hosp 22 S Greene St Baltimore MD 21201

REEDER, JOHN R(AYMOND), botanist, researcher; b. Grand Ledge, Mich., July 29, 1914; s. Raymond and Hazel Blanche (Ingersoll) R.; m. Charlotte Goodding, Aug. 15, 1941. B.S., Mich. State U., 1939; M.S., Northwestern U., 1941; M.A., Harvard U., 1946, Ph.D., 1947. Instr. Yale U., 1947-51, asst. prof., 1951-57, assoc. prof., 1957-68, curator herbarium, 1947-68; prof. botany, curator Rocky Mountain Herbarium, U. Wyo., 1968-76; vis. scholar U. Ariz., Tucson, 1976—. Contbr. numerous articles to profl. jours., 1943—; editor-in-chief: Brittonia, 1966-70. Served to sgt. U.S. Army, 1942-45; PTO. Fellow Linnean Soc. London; mem. Am. Soc. Naturalists, Bot. Soc. Am. (chmn. systematics sect. 1956, 74-75), Internat. Assn. Plant Taxonomy, Am. Soc. Plant Taxonomists, Sigma Xi (pres. Yale chpt. 1958-59). Subspecialties: Evolutionary biology; Taxonomy. Current work: Plant systematics and evolution, especially Gramineae. Office: Herbarium 113Agricultural Sciences U Ariz Tucson AZ 85721

REEL, JERRY ROYCE, research director, educator; b. Washington, Ind., May 4, 1938; s. Royce Howard and Anna Belle (Valin) R.; m. Joan Kay Wedberg, Aug. 14, 1965; 1 dau., Justine Jeanette. B.A., Ind. State U., 1960; M.S., U. Ill., 1963, Ph.D., 1966. Diplomate: Am. Bd. Toxicology. Am. Cancer Soc. postdoctoral fellow Oak Ridge Nat. Lab., 1966-68; sect. dir. encodrinology Warner-Lambert/Parke-Davis Research Labs., Warner Lambert Pharm. Co., Ann Arbor, Mich., 1968-78; dir. life scis. and toxicology div. Research Triangle Inst., Research Triangle Park, N.C., 1978—; adj. assoc. prof. Wayne State U., Detroit, 1973—; Mem. adv. bd. Kildaire Homeowners Assn., Cary, N.C., 1979-82. Editor: Hypothalamic Hormones, 1973, Steroid Hormones and Cancer, 1975. Rev. panel mem. WHO, 1979; grant reviewer NSF, 1975, 78; mem. contract rev. panel NIH, 1979. Mem. Endocrine Soc., Am. Physiol. Soc., Am. Chem. Soc., Soc. Toxicology. Subspecialties: Reproductive biology; Endocrinology. Current work: Research in reproductive endocrinology/toxicology; computer applications in research and development/biotechnology. Home: 200 Rosecommon Ln Cary NC 27511 Office: Research Triangle Inst PO Box 12194 Research Triangle Park NC 27709

REEMTSMA, KEITH, physician; b. Madera, Calif., Dec. 5, 1925; m.; children—Lance Brewster, Dirk Van Horn. B.S., Idaho State Coll.; M.D., U. Pa., 1949; Med.Sci.D., Columbia U., 1958. Asst. surgery Columbia Coll. Phys. and Surg., 1957; faculty Tulane U. Sch. Medicine, 1957-66, prof. surgery, 1966; prof., head dept. surgery U. Utah Coll. Medicine, Salt Lake City, 1966-71; chmn. dept. surgery Valentine Mott prof. surgery; Johnson and Johnson distinguished prof. surgery Columbia U., N.Y.C., 1971—. Contbr. articles to profl. jours. Mem. Soc. Clin. Surgery, Am. Surg. Assn., Soc. U. Surgeons, A.C.S., So. Soc. for Clin. Research, Am. Fedn. for Clin. Research, Am. Assn. for Thoracic Surgery, Soc. for Vascular Surgery, Internat. Cardiovascular Soc., So. Thoracic Surg. Assn., New Orleans Surg. Soc., Surg. Assn. La., AMA, La. State, Orleans Parish med. socs., Alpha

Omega Alpha. Subspecialty: Surgery. Office: Dept Surgery Columbia U 622 W 168th St New York City NY 10032

REES, EARL DOUGLAS, medical educator; b. Cleve., May 1, 1928; s. Earl Henry and Mildred G. (Bloom) R.; m. Mary Alice Klingensmith, Aug. 26, 1950; children: Douglas C., David F., Katherine M., Jennifer A.; m. Margaret L. Jordan, July 27, 1973. A.B. cum laude (scholar), Harvard U., 1950; M.D., Yale U., 1954. Mem. biophysics sect., chief biochemistry sect. Armed Forces Inst. Pathology, Washington, 1956-58; instr. U. Chgo., 1958-60; mem. faculty U. Ky. Med. Center, Lexington, 1960—, dir. clin. research lab., 1966—, dir. lipid clinic, 1971—, prof. medicine and pharmacology, 1970—, pres. univ. senate, chmn. univ. senate council U. Ky., 1983-84; cons. VA Hosp. Lexington, Inst. Mining and Minerals, Lexington. Author articles on chem. carcinogenesis; hormone dependency of cancers; clin. chemistry and serum proteins; therapy of serum lipid disorders. Pres. Unitarian Ch. of Lexington, 1966-77. Served to capt., M.C. U.S. Army, 1956-58. Nat. Found. fellow, 1954-55; U. Ky. Research Found awardee, 1971; Blackford Meml. lectr., 1980. Mem. AAUP (chpt. pres. 1979-80), Am. Physiol. Soc., Am. Soc. Pharmacology and Exptl. Therapeutics, Central Soc. Clin. Research, Soc. Exptl. Biology and Medicine, AAAS, Am. Assn. Clin. Chemists, Am. Assn. Cancer Research, Endocrine Soc., Sigma Xi. Democrat. Subspecialties: Endocrinology; Cancer research (medicine). Current work: hormone dependent cancers; chemical carcinogens; cancer and chromosomes; clinical lipoprotein disorders. Home: Route 3 Nicholasville KY 40356 Office: U Ky Lexington KY 40536

REESE, BRUCE ALAN, engineering development center executive; b. Provo, Utah, Aug. 3, 1923; s. David J. and Bessie (Johnson) R.; m. Barbara Taylor, June 20, 1945; children: Bruce Taylor, Michael David, Pamela Kay. Student, Brigham Young U., 1941-43; B.S. in Mech. Engring, U. N.Mex., 1944; Ph.D., Purdue U., 1953. Mem. faculty Purdue U., Lafayette, Ind., 1946-79, asso. prof., 1955-57, prof. mech. engring., 1957-73, head, 1973-79; chief scientist USAF, Arnold Engring. Devel. Center, Tullahoma, Tenn., 1979—; dep. dir. Nike Zeus Research and Devel.; also tech. dir. Nike X Project Office, U.S. Army, 1961-63; dir. Jet Propulsion Center, Purdue U., 1965-73; Cons. aerospace govt. agys., industry; mem. sci. adv. bd. USAF, 1969-76, mem. adv. group fgn. tech. div., 1970-74, adv. group air systems div., 1973-76; mem. or cons. to sci. adv. panel U.S. Army, 1967-78, chmn., 1976-77, chmn. tank and automotive command sci. adv. group, 1970-74, mem. missile command adv. com., 1972-76. Active Boy Scouts Am., 1959-67. Served to ensign USNR, 1943-46. Mem. Am. Inst. Aeros. and Astronautics (chmn. central Ind. sect. 1964-65), ASME (chmn. aircraft gas turbine 1965-68), Sigma Xi, Phi Kappa Phi, Pi Tau Sigma, Tau Beta Phi. Mem. Ch. Jesus Christ of Latter Day Saints. Subspecialty: Aerospace engineering and technology. Research on heat transfer, gas dynamics, hybrid fueled rockets, laser diffusers, slurry fuel combustion. Home: 211 Lake Circle Dr Tullahoma TN 37388 Office: Arnold Engring Devel Center Arnold AF Station TN 37389

REESE, LYMON CLIFTON, civil engineering educator; b. Murfreesboro, Ark., Apr. 27, 1917; s. Samuel Wesley and Nancy Elizabeth (Daniels) R.; m. Eva Lee Jett, May 28, 1948; children: Sally Reese Melant, John, Nancy. Rodman. B.S., U. Tex. at Austin, 1949, M.S., 1950; Ph.D. (Gen. Edn. Bd. Calif. fellow, NSF fellow), U. Calif. at Berkeley, 1955. Diplomate: Registered profl. engr., Tex. Internat. Boundary Commn., San Benito, Tex., 1939-41; layout engr. E.I. DuPont Co. de Nemours & Co., Inc., Pryor, Okla., Childersburg, Ala., 1941-42; surveyor U.S. Naval Constrn. Bns., Aleutian Islands, Okinawa, 1942-45; field engr. Asso. Contractors & Engrs., Houston, 1945; draftsman Phillips Petroleum Co., Austin, 1946-48; research engr. U. Tex., Austin, 1948-50; asst. prof. civil engring. Miss. State Coll., State College, 1950-51, 53-55; asst. prof. U. Tex., Austin, 1955-57, asso. prof., 1957-64, prof., 1964—, chmn. dept., 1965-72, asso. dean engring. for program planning, 1972-79, Nasser I. Al-Rashid Chair, 1981—; cons. Shell Oil Co., Shell Devel. Co., 1955-65, Dames & Moore, 1970—, McClelland Engrs., Houston, 1970—, Assn. Drilled Shaft Contractors, 1971—, Fla. Dept. Transp., 1975—, others; Taylor prof. engring. Tex. U., 1972—. Contbr. articles to profl. jours. Served with USNR, 1942-45. Recipient Thomas Middlebrooks award ASCE, 1958; Joe J. King Profl. Engring. Achievement award, 1977. Fellow ASCE (Karl Terzaghi lectr. 1976, Terzaghi award 1983); mem. Nat. Acad. Engring., Am. Soc. Engring. Edn., Nat. Soc. Profl. Engrs., ASTM, Sigma Xi, Tau Beta Pi, Chi Epsilon, Phi Kappa Phi. Baptist (deacon). Subspecialty: Civil engineering. Home: 11512 Tin Cup Dr 109 Austin TX 78750 Office: Dept Civil Engring U Tex Austin TX 78712

REESE, THOMAS SARGENT, neurobiologist; b. Cleve., May 20, 1935; s. Thomas S. and Jane (Andrews) R.; m. Barbara Hall, Sept. 2, 1976; children: Andrea Cooney, Devin Andrews. B.S., Harvard U., 1957; M.D., Columbia U., 1962. Research asst. Psycho-Acoustic Lab. Harvard U., 1957-58, research fellow in anatomy, 1965-66; intern Boston City Hosp., 1962-63; research asso. NIH, Bethesda, Md., 1963-65, research med. officer Lab. Neruopathology and Neuroanat. Scis., 1966-70, head sect. functional neuroanatomy, 1970-83; instr. neurobiology course Marine Biol. Lab., Woods Hole, Mass., 1975-80, instr.-in-chief, 1980—, chief Lab. Neurobiology, 1983—. Contbr. chpts., numerous articles to profl. publs.; mem. editorial bd.: Jour. Neurocytology, 1976-78, Jour. Cell Biology, 1976-79, Anat. Record, 1977—, Neurosci, 1978—, Cryoletters, 1979—, Jour. Neurosci, 1980—, Jour. Cellular and Molecular Neurobiology, 1980—. Served with USPHS, 1963-65. Recipient Superior Service award USPHS, 1978; Mathilde Solowey award NIH, 1979; W. Alden Spencer award Columbia U., 1980. Mem. Am. Assn. Anatomists (C. Judson Herrick award 1970), Am. Soc. Cell Biology, Soc. Neurosci., Internat. Brain Research Orgn., Cajal Club, Biophys. Soc. Subspecialty: Neurobiology. Current work: Cellular neurobiology: cell and membrane structure, synapses, sensory systems, blood-brain barrier. Office: Lab Neurobiology Marine Biol Lab Woods Hole MA 02543

REEVES, FONTAINE BRENT, botany educator; b. Eufaula, Ala., May 16, 1939; s. Fontaine and Marie Isabel (Carroll) R.; m. Margaret Elizabeth McBrearty, Aug. 22, 1964; children: Margaret Anne, Michael John. B.S., Tulane U., 1961, M.S., 1963; Ph.D., U. Ill., 1966. Asst. prof. dept. botany-plant pathology Colo. State U., Ft. Collins, 1966-71, assoc. prof., 1971-79, prof., 1979—. Contbr. articles to profl. jours. NSF grantee, 1974-76; NIH grantee, 1974-77; U.S. Dept. Agr. grantee, 1980-82; U.S. Dept. Energy grantee, 1975—. Mem. Mycol. Soc. Am., Bot. Soc. Am., Colo.-Wyo. Acad. Sic., AAAS, Sigma Xi, Phi Sigma. Democrat. Subspecialties: Microbiology; Ecology. Current work: Mycorrhizal relationships of plants in disturbed habitats. Home: 1931 Sandalwood Ln Fort Collins CO 80526 Office: Botany and Plant Pathology Dept Colo State Univ Fort Collins CO 80523

REEVES, ROBERT DONALD, nutritionist, educator; b. Lubbock, Tex., Jan. 14, 1942; s. Samuel Winston and Annie (Hamilton) R.; m. Sue Sloan, June 29, 1967; children: Alan Robert, Sherman Winston. B.A., Tex. Tech. U., 1964, M.S., 1965; Ph.D., Iowa State U., 1971. Research assoc. Iowa Agrl. Exptl. Sta., Ames, 1965-71; instr. depts. medicine and biochemistry U. Ark. Sch. Med. Sci., Little Rock, 1972-74, asst. prof. medicine and biochemistry, 1974-77; dir. renal metabolic program dept. med. nephrology services VA Hosp., Little Rock, 1972-77; assoc. prof. nutrition dept. foods and nutrition Kans. State U., Manhattan, 1977—. U.S. Dept. Agr. grantee, 1979-83; Am. Diabetes Assn. grantee, 1980-81. Mem. Am. Inst. Nutrition, Am. Fedn. Clin. Research, Am. Dietetics Assn., Sigma Xi, Gamma Sigma Delta. Republican. Baptist. Subspecialties: Nutrition (biology); Nutrition (medicine). Current work: Nutrition and its effect on metabolic regulation. Home: 3208 Gary Manhattan KS 66506 Office: Dept Foods and Nutrition Kans State U Manhattan KS 66505

REGENSTEIN, JOE MAC, food scientist, educator; b. Bklyn., Sept. 22, 1943; s. Alfred B. and Fannie (Kleinmeyer) R.; m. Carrie E. Forsheit, June 12, 1966; children: Elliot, Scott. B.A., Cornell U., 1965, M.S., 1966; Ph.D. in Biophysics, Brandeis U., 1970; Postdoctoral fellow Children's Cancer Research Center, Boston, 1973-74, Rosenstiel Basic Med. Research Ctr., 1973-74; asst. prof. food sci. Cornell U., 1964-80, assoc. prof., 1980—; vis. scientist Torry Research Sta., Aberdeen, Scotland, 1980-81. Contbr. articles to profl. jours. Grantee in field, 1975—. Mem. Am. Chem. Soc., Am. Council Sci. and Health, Inst. Food Technologists, Poultry Sci. Assn., Am. Meat Sci. Assn., Atlantic Fisheries Technologists Soc. Subspecialty: Food science and technology. Current work: Fish and poultry, new product devel., frozen and fresh storage, Kosher foods. Office: Cornell U 112 Rice Hall Ithaca NY 14853 Home: 301 Muriel St Ithaca NY 14850

REGEV, ODED, astrophysicist; b. Walbrzych, Poland, Oct. 25, 1946; s. Abraham and Ida (Erdmann) Gruber; m. Sarah Remez, June 15, 1980. B.Sc., Hebrew U., Jerusalem, 1968; M.Sc., Tel-Aviv (Israel) U., 1975, Ph.D., 1980. Asst. Tel-Aviv U., 1976-80; research assoc. dept. physics U. Fla., Gainesville, 1980—. Contbr. articles to profl. jours. Served to maj. Israel Def. Forces, 1969-73. Mem. Am. Astron. Soc. Subspecialty: Theoretical astrophysics. Current work: Theory of star formation, supernovae. Theory of stellar pulsation, stellar evolution. Office: U Fla Dept Physics 215 Williamson Hall Gainesville FL 32601

REGGIA, JAMES ALLEN, computer science educator; b. Takoma Park, Md., Oct. 31, 1949; s. Frank and Betty Jo (Patterson) R.; m. Carol G. Garlet, July 31, 1981. M.D., U. Md., 1975, Ph.D. in Computer Sci, 1981. Asst. prof. computer sci. U. Md., Balt., 1982—. Mem. AAAS, Assn. Computing Machinery. Subspecialties: Artificial intelligence; Neurology. Current work: Artificial intelligence, expert systems, decision making, natural language processing, neurolinguistics. Office: Dept Computer Sci University of Maryland College Park MD 20742

REHKUGLER, GERALD EDWIN, agrl engr., educator; b. Lyons, N.Y., Apr. 11, 1935; s. Charles J. and Minnie S. (Tange) R. Student, Syracuse U., 1952-53; B.S., Cornell U., M.S., 1958; Ph.D., Iowa State U., 1966. Registered profl. engr., N.Y. Asst. prof., then assoc. prof. Cornell U., Ithaca, N.Y., 1958-77, prof. agrl. engring., 1977—; cons. agrl. and food machinery design and devel. Contbr. articles to profl. publs. Mem. bd. edn. Dryden (N.Y.) Central Sch., 1977—, pres., 1981-82. NSF sci. faculty fellow, 1964-66. Mem. Inst. Food Technologists, Am. Soc. Agrl. Engrs. (paper awards 1965, 75, 77, 79), Sigma Xi, Phi Kappa Phi, Gamma Sigma Delta. Republican. Lodge: Masons. Subspecialty: Agricultural engineering. Current work: Reasearch and teaching in design of agricultural and food processing equipment developing systems for detection of flaws in agricultural products; investigations of solar refrigeration systems for food processing and examination of dynamics of agricultural tractor motion. Office: Cornell U 228 Riley Robb-Hall Ithaca NY 14853

REHM, LYNN PAUL, psychology educator; b. Chgo., May 20, 1941; s. Stanley F. and Bernice (Stiebler) R.; m. Susan H. Higginbotham, Feb. 28, 1964; children: Elizabeth S., Sarah A. B.A., U. So. Calif., 1963; M.A., U. Wis.-Madison, 1966, Ph.D., 1970. Lic. psychologist, Tex. Asst. prof. Neuropsychiat. Inst., UCLA, 1968-70; assoc. prof. psychology U. Pitts., 1970-79; prof. psychology U. Houston, 1979—. Editor: Behavior Therapy for Depression, 1981. Fellow Behavior Therapy and Research Soc.; mem. AAAS, Am. Psychol. Assn., Assn. for Advancement of Behavior Therapy, Phi Beta Kappa. Subspecialties: Behavioral psychology; Clinical psychology. Current work: Behavior therapy and analysis of depression and of self-management. Home: 7906 Burning Hills Dr Houston TX 77071 Office: Dept Psychology U Houston Houston TX 77004

REICH, HANS JURGEN, chemistry educator; b. Danzig, Germany, May 6, 1943; came to arrived Can., 1950; s. Oswald Daniel and Martha (Adam) R.; m. Ieva Lazdins, June 29, 1970. B.Sc., U. Alta., Can., 1964; Ph.D., UCLA, 1968. Postdoctoral fellow Calif. Inst. Tech., Pasadena, 1968-69; asst. prof. chemistry U. Wis.-Madison, 1970-76, assoc. prof., 1976-79, prof., 1979—; vis. prof. Philipps-Universität, Marburg, W.Ger., 1978-79; cons. Dow Chem., Midland, Mich., 1977-81, Arco Chem., Phila., 1979-80, Marion Labs., Kansas City, Mo., 1982—. Woodrow Wilson fellow, 1964-65; A.P. Sloane fellow, 1975-79; Can. NRC fellow, 1968-70. Mem. Am. Chem. Soc., Chem. Soc. (London), Sigma Xi. Subspecialties: Organic chemistry; Synthetic chemistry. Current work: Mechanistic, stereochemical and synthetic studies of organo selenium, silicon and tin compounds. Office: Dept Chemistry U Wis 1101 University Ave Madison WI 53706

REICH, IEVA L., research chemist, chemistry educator; b. Riga, Latvia, June 30, 1942; came to U.S., 1949, naturalized, 1956; d. Arvids Osvalds and Ilze (Kundzins) Lazdins; m. Hans Jurgen Reich, June 28, 1969. B.S. in Chemistry, U. Wash., 1964; Ph.D. in Organic Chemistry, UCLA, 1969. Project assoc. dept. chemistry U. Wis., Madison, 1970-82, lectr., 1976, 81, asst. scientist, 1982—. NIH grantee, 1982—. Mem. Am. Chem. Soc. Subspecialty: Organic chemistry. Current work: Lithium/metalloid exchange reactions, synthesis of polychlorinated biphenyl arene oxides and their reactivity. Home: 514 Edward St Madison WI 53711 Office: Dept Chemistry U Wis Madison WI 53706

REICHEN, JUERG, medical educator; b. Aarau, Aargau, Switzerland, Jan. 23, 1946; came to U.S., 1976; s. Hans A. and Susi K. (Aeberhard) R.; m. Suzi Graden, May 29, 1970; children: Hansjakob, Annemarie, Katharina. B.A., B.S., Staedt. Gymnasium Burgdorf, 1964; M.D., U. Berne, Switzerland, 1971. Fellow in pharmacology Hoffmann-LaRoche, Basel, Switzerland, 1972; fellow in clin. pharmacology U. Berne (dept. clin. pharmacology), 1973-76; guest scientist NIH, Bethesda, Md., 1976-78; resident VA Med. Ctr., Georgetown U., Washington, 1978-79; fellow in gastroenterology U. Colo. Hosp., Denver, 1979-80, asst. prof. medicine, 1980—. Swiss Nat. Fedn. Sci. Research fellow, 1976; Pharm. Mfrs. Assn. Found. grantee, 1981; NIH grantee, 1981. Mem. Swiss Med. Soc., Am. Fedn. Clin. Research, Western Soc. Clin. Investigation, Am. Gastroent. Assn., Am. Assn. Study Liver Disease. Subspecialties: Gastroenterology; Pharmacology. Current work: Physiology of bile secretion and pathophysiology of cholestasis, quantitation of para and transcellular fluid movement, microcirculation of the liver, quantitation of liver function, pharmacological manipulation of portal hypertension. Home: 13104 E Exposition Ave Aurora CO 80012 Office: U Colo Med Sch 4200 E 9th Ave PO Box B-158 Denver CO 80262

REICHENBECHER, VERNON EDGAR, JR., biochemist; b. Meyersdale, Pa., Mar. 29, 1948; s. Vernon and Violet Eugenia (Hetz) R.; m. Linda Lee Jernigan, June 26, 1976; 1 dau., Jennifer Lee. B.S., W.Va. U., 1970; Ph.D., Duke U., 1976. Postdoctoral fellow Baylor Coll. Medicine, Houston, 1976-79, research assoc., 1980-81; asst. prof. Marshall U. Sch. Medicine, Huntington, W.Va., 1981—. Contbr. articles to profl. jours. Potomac Edison Co. scholar, 1966-70; Hamilton Watch award, 1970; Am. Cancer Soc. fellow, 1977-78. Mem. Genetics Soc. Am., Tissue Culture Assn., Sigma Xi, Appalachian Fiddlers Assn. Subspecialties: Biochemistry (biology); Molecular biology. Current work: The mechanism of protein synthesis in mammalian cells; the study of the structure and function of mammalian ribosomes by means of monoclonal antibodies and mutant cells. Home: 9 Lillian Ct Barboursville WV 25504 Office: Dept Biochemistry Marshall Univ Sch Medicine Huntington WV 25701

REICHGOTT, MICHAEL J., physician, educator; b. Newark, July 20, 1940; s. Leo and Gertrude (Millman) R.; m. Lynn H. Haar, Dec. 22, 1962; children: Jay, Seth, Douglas. A.B., Gettysburg Coll., 1961; M.D., Albert Einstein Coll. Medicine, 1965; Ph.D. in Pharmacology, U. Calif., San Francisco, 1972. Diplomate: Am. Bd. Internal Medicine. Resident in internal medicine U. Calif., San Francisco, 1965-67, fellow in clin. pharmacology, 1969-72; asst. prof. medicine and pharmacology U. Pa., 1972-81, assoc. prof., 1981—, dir. clin. partice dept. medicine, 1978-80; assoc. chief of staff for ambulatory care VA Hosp., Phila., 1980-81, chief sect. gen. medicine, 1981—; chmn. South Eastern Pa. High Blood Pressure Control Program, 1980—; cons. Teaching Nursing Home Program, Robert Wood Johnson Found., 1981—. Contbr. articles, revs. to profl. lit. Vice pres. Beth David Reform Congregation, Phila.; cubmaster, also mem. troop com. Boy Scouts Am. Served to maj. M.C. U.S. Army. Decorated Army Commendation medal. Fellow Phila. Coll. Physicians, ACP; mem. Am. Fedn. Clin. Research, Soc. Research and Edn. in Primary Care Internal Medicine, Am. Soc. Pharmacology and Exptl. Therapeutics, Alpha Omega Alpha. Democrat. Jewish. Subspecialties: Health services research; Pharmacology. Current work: Drug evaluation, non-physician health care provision, compliance program director for general medicine activities; clinical research, trainee education. Office: Phila VA Hosp 39th and Woodland Ave Philadelphia PA 19104

REICHLE, FREDERICK ADOLPH, surgeon, educator; b. Neshaminy, Pa., Apr. 20, 1935; s. Albert and Ernestine R. B.A. summa cum laude, Temple U., 1957, M.D., 1961, M.S. in Biochemistry, 1961, 1966. Diplomate: Am. Bd. Surgery. Intern Abington Meml. Hosp., 1962; resident Temple U. Hosp., Phila., 1966, surgeon, 1966—; practice medicine specializing in surgery, Phila., 1966—; assoc. attending surgeon Epis. Hosp., St. Mary's Hosp., St. Christopher's Hosp. for Children, Phoenixville Hosp.; cons. VA Hosp., Wilkes Barre, Pa., Germantown Dispensary and Hosp.; chmn. dept. surgery Presbyn.-U. Pa. Med. Center, 1980—; prof. surgery U. Pa., 1980—. Contbr. articles to profl. jours. Recipient Surg. Residents Research Paper award Phila. Acad. Surgery, 1964, 66, Gross Essay prize, 1976; Am. Heart Assn. grantee, 1973. Fellow A.C.S., Coll. Physicians Phila.; mem. Am. Surg. Assn., Soc. Univ. Surgeons, AMA, Pa. Med. Soc., Assn. Acad. Surgery, N.Y. Acad. Sci., AAAS, Am. Fedn. Clin. Research, Nat. Assn. Professions, Am. Gastroent. Assn., Am. Assn. Cancer Research, Am. Heart Assn., Phila. Acad. Surgery, Heart Assn. Southeastern Pa., Internat. Soc. Thombosis and Haemostasis, Nat. Kidney Found., Soc. for Surgery Alimentary Tract, Soc. Vascular Surgery, Colleguim Internationale Chirurgie Digestivae, Am. Soc. Pharmacology and Exptl. Therapeutics, Am. Inst. Ultrasound in Medicine, Am. Physiol. Soc., Soc. Internationale de Chirurgie, Am. Soc. Abdominal Surgeons, Surg. Hist. Soc., Am. Aging Assn., Am. Geriatrics Soc., Gerontol. Soc., Am. Diabetes Assn., Surg. Biology Club, Omega Alpha, Sigma Xi, Phi Rho Sigma. Subspecialties: Surgery; Biochemistry (medicine). Current work: Lipid metabolism, blood vessel research, portal hypertension. Home: 771 Easton Rd Warrington PA 18976 Office: 51 N 39th St Philadelphia PA 19104

REICHMANN, MANFRED ELIEZER, microbiologist, researcher, educator; b. Trencin, Czechoslovakia, Apr. 16, 1925; came to U.S., 1964, naturalized, 1972; s. Moritz and Amalia (Lowy) R.; m. Irene Christine Oakley, Aug. 23, 1957; children: Dianne Brink, David, Lindsay. M.Sc. in Physics and Chemistry, Hebrew U., Jerusalem, 1949, Ph.D. in Chemistry, 1951. USPHS postdoctoral fellow Harvard U., 1951-53; research officer Can. Dept. Agr., 1955-64; prof. biochemistry U. B.C., Vancouver, 1955-64; prof. botany U. Ill., Urbana, 1964-71, prof. microbiology, 1971—; assoc. mem. Center for Advanced Studies, U. Ill., 1977-78. Contbr. revs., numerous articles to profl. jours. Served to lt. Sci. Br. Israeli Army, 1948-49. Am. Cancer Soc. scholar, 1977-78; NRC Can. grantee, 1953-55; USPHS grantee, 1965—; NSF grantee, 1970-74; Am. Cancer Soc. grantee, 1972-74. Mem. Am. Soc. Microbiology, Am. Soc. Virology, Am. Soc. Biol. Chemists, AAAS. Subspecialties: Virology (biology); Molecular biology. Current work: Mechanism of interference by defective interfering particles of rhabdoviruses, inhibition of protein, DNA and RNA synthesis of host cell by lytic viruses, viral macromolecular synthesis in doubly infected cells. Home: 205 Pell Circle Urbana IL 61801 Office: U Ill Dept Microbiology 131 Burrill Hall Urbana IL 61801

REID, ALLEN FRANCIS, biophysicist, educator, research and devel. consultant; b. Deer River, Minn., July 31, 1917; s. Allen Roy and Rose Cordelia; ; s. Allen Roy and Seidel R.; m. Dorothy Mary Cullen, May 31, 1943; children: Sally Anne, David Mark. B.Chemistry, U. Minn., 1940; A.M., Columbia U., 1942, Ph.D., 1943; M.D., U. Tex., Dallas, 1959. Clin. chemist Nat. Registry Clin. Chemists, 1968; Dir. radioactivity labs. Columbia U., N.Y.C., 1942-45; indsl. cons., Phila. and N.Y.C., 1945-47; prof., chmn. dept. biophysics and phys. chemistry Grad. Research Inst., Baylor U., Dallas, 1947-50; assoc. prof. Southwestern Med. Sch., U. Tex., Dallas, 1947-51, prof., 1951-60, chmn. dept. biophysics, 1947-60; prof., chmn dept. biology U. Dallas, 1960-68, head sci. div., 1961-68; clin. prof. pathology Downstate Med. Ctr., SUNY, N.Y.C., 1968-74; dir. clin. biochemistry Bklyn.-Cumberland Med. Ctr., 1968-74; dir. pathology Cumberland Hosp., Bklyn., 1970-74; prof., chmn. dept. biology SUNY, Geneseo, 1974—; ptnr. Halff & Reid, Dallas, 1961—; profl. cons., 1947—. Contbr. articles to profl. jours. Bd. dirs. Am. Cancer Soc., Livingston County, N.Y., 1977—; mem. profl. adv. com. Livingston County Dept. Health, 1980—. Fellow AAAS, Am. Inst. Chemists; mem. Am. Chem. Soc., Am. Phys. Soc., Am. Physiol. Soc., AMA, N.Y. Acad. Sci., Am. Assn. for Cancer Research, Sigma Xi, Phi Lambda Upsilon. Republican. Roman Catholic. Subspecialties: Ocean thermal energy conversion; Chemical engineering. Current work: Methodology of improving thermal energy conversion efficiency by increasing the temperature of heat supplied to a turbine working fluid above the warm sea water temperature. Patentee in field. Office: 4736 Reservoir Rd Geneseo NY 14454

REID, DONALD HOUSE, aerospace physiologist; b. Phillipsburg, Pa., May 31, 1935; s. Roger Delbert and Erma Marie (House) R.; m. Mary Alice Rush, Aug. 15, 1964; children: Douglas Charles, Joan Elizabeth. B.S., Cornell U., 1958; M.S., S.D. State U., 1960; Ph.D., U. So. Calif., 1968. Physiol. tng. officer Marine Corps Air Sta., El Toro, Calif., 1962-65; research aerospace physiologist Naval Aerospace Med. Inst., Pensacola, Fla., 1967-69; head biomed. sci. Naval Aerospace Recovery Facility, El Centro, Calif., 1969-73; head crew systems br. Pacific Missile Test Center, Point Mugu, Calif., 1973-78; dep. dir. biol. sci. div. Office Naval Research, Arlington, Va., 1978-80; mgr. sci. support Gen. Electric Co./Matsco, Arlington, 1980—. Served to comdr. USN, 1976-80. Recipient Meritorious Service Medal Dept. Def., 1980. Fellow Aerospace Med. Assn.; mem. Am. Physiol. Soc., Aerospace Physiol. Soc. (pres. 1976-77, Paul Bert award 1975), Space Medicine Br., Life Scis. and Biomed. Engring. Soc. Republican.

Methodist. Club: Potomac Pedalers Touring (Washington). Lodge: Masons. Subspecialties: Physiology (biology); Gravitational biology. Current work: Aerospace and exercise physiology. Home: 8103 West Point Dr Springfield VA 22153 Office: Gen Electric/Matsco 1755 Jeff Davis Hwy Arlington VA 22202

REID, JOHN REYNOLDS, geology educator, researcher; b. Melrose, Mass., Jan. 4, 1933; s. John Reynolds and Elva Margarite (Taylor) R.; m. Barbara Ann Pulsford, June 30, 1956; children: Valerie Ann, William Craig, Karen Louise, Linda Elizabeth. B.S. in Geology, Tufts U., 1955, M.S., U. Mich., 1957, Ph.D., 1961. Asst. prof. geology U. N.D., Grand Forks, 1961-65, assoc. prof., 1965-71, prof., 1971—, assoc. dean., 1967-75; assoc. acting dir. N.D. Regional Assessment Program, Bismarck, N.D., 1975-78; glaciologist Arctic Inst. N.Am., Camp Michigan, Antarctica, 1958-59; geologist Martin River Glacier Research Expdn., Alaska, 1962-63, prin. investigator, 1965, 66,68, U.S. Army Cold Regions Research and Engring. Lab., Orwell Lake, Minn., 1980—. Contbr. articles to profl. publs. Recipient Outstanding Tchr.award U. N.D., 1967; William Herbert Hobbs fellow U. Mich., 1957; Olmstead fellow Tufts U., 1954. Fellow Geol. Soc. Am.; mem. Am. Quaternary Assn., AAAS, N.D. Acad. Sci. (pres. 1972-73), Sigma Xi (pres. U. N.D. 1967-68). Baptist. Subspecialties: Geology; Environmental geology. Current work: Process geomorphology (slope erosion processes); environmental geology, glacial geology, glaciology, quaternary paleoclimatology. Home: 420 25th Ave S Grand Forks ND 58201 Office: U ND Grand Forks ND 58202

REID, ROBERT CLARK, chemical engineering educator; b. Denver, June 11, 1924; s. Frank B. and Florence (Seerley) R.; m. Anna Marie Murphy, Aug. 26, 1950; children: Donald M., Ann Christine. Student, Colo. Sch. Mines, 1946-48; B.S., Purdue U., 1950, M.S., 1951; Sc.D., M.I.T., 1954. Chevron prof. chem. engring. M.I.T., Cambridge, 1954—; Olaf A. Hougen prof. chem. engring. U. Wis., 1980-81; cons. A.D. Little, Inc., 1956—, Nestle Corp., 1978—. Author: (with T.K. Sherwood, J.M. Prausnitz) Properties of Gases and Liquids, 1966, 3d edit., 1976, (with M. Ohara) Modeling Crystal Growth Rates from Solution, 1973, (with M. Modell) Thermodynamics and Its Applications, 1974, 2d edit., 1983; Contbr. articles to profl. jours. Recipient Warren K. Lewis award, 1976; Chem. Engring. award Am. Soc. Engring. Edn., 1977; research fellow Harvard U., 1963-64. Mem. Am. Inst. Chem. Engrs. (Ann. lectr. 1967, council 1969-71, editor jour. 1970-76), Nat. Acad. Engring., Blue Key, Sigma Alpha Epislon, Tau Beta Pi. Subspecialty: Chemical engineering. Home: 22 Burroughs Rd Lexington MA 02173 Office: 66-540 Mass Inst Tech Cambridge MA 02139

REID, ROBERT LELON, mech. engr., educator, researcher, cons.; b. Detroit, May 20, 1942; s. Lelon and Verna Beulah (Custer) R.; m. Judy Nestell, July 21, 1962; children: Robert, Bonnie, Matthew. B.S.E. in Chem. Engring, U. Mich., 1963; M.S.E., So. Methodist U., 1966, Ph.D., 1969. Registered profl. engr., Tenn., Tex. Research engr. ARCO Prodn. Research, Dallas, 1964-65; staff engr. Linde div. Union Carbide, Tonawanda, N.Y., 1966-68; asst. prof. mech. engring. U. Tenn., Knoxville, 1969-75, assoc. prof., then prof., 1977-82; then asst. dir. Energy, Environment, and Resources Ctr., 1979-82; asso. prof. Cleve. State U., 1975-77; prof. mech. and indsl. engring., chmn. dept. mech. and indsl. engring. U. Tex., El Paso, 1982—; cons. Oak Ridge Nat. Lab. Contbr. numerous articles on heat transfer, solar energy, energy conservation to profl. jours.; assoc. editor: Jour. Solar Energy Engring, 1981—. Mem. ASME (Centennial medallion 1980, vice chmn. div. solar energy 1983—), ASHRAE, Am. Solar Energy Soc., Am. Soc. Engring. Edn. Lutheran. Subspecialties: Mechanical engineering; Chemical engineering. Current work: Solar energy, energy conservation. Office: Dept Mech and Indsl Engring U Tex El Paso TX 79968

REID, ROBERT WILLIAM, nuclear engineer; b. Portland, Oreg., Apr. 16, 1924; s. Ralph and Marjorie Grace (Turner) R.; m. Lucretia Lorene Agee, Mar. 4, 1924; children: Karen, Judy, Kimberly, Jeffrey, Georgene, Susan. B.S. in Chem. Engring, Oreg. State U., 1950. Mgr. process tech. Gen. Electric Co., Richland, Wash., 1950-65; mgr. process tech. Douglas United Nuclear Co., Richland, 1965-71; br. chief U.S. NRC, Bethesda, Md., 1971-81; sr. engr. Portland Gen. Electric, Oreg., 1981—. Council chmn. PTA, Kennewick, Wash., 1964; dist. chmn. Boy Scouts Am., Kennewick, 1969. Served to lt. (j.g.) USNR, 1942-45. Mem. Am. Nuclear Soc., Am. Inst. Chem. Engrs. (sect. chmn. 1960). Republican. Subspecialty: Nuclear engineering. Home: 6450 Curl Rd Tillamook OR 97141 Office: Portland Gen Electric Co 121 S W Salmon St Portland OR 97201

REIDENBERG, MARCUS MILTON, clinical pharmacologist, educator; b. Phila. M.D., Temple U., 1958. Diplomate: Am. Bd. Internal Medicine. Intern Community Gen. Hosp., Reading, Pa., 1958-59; resident Temple U. Hosp., Phila., 1962-65; practice medicine specializing in pharmacology; mem. staff N.Y. Hosp.; vis. phys. Rockefeller U. Hosp.; mem. faculty Cornell U. Med. Coll., N.Y.C., 1975—, prof. pharmacology and medicine, 1980—, acting assoc. dean, 1981-82. Served with USN, 1960-62. Subspecialties: Pharmacology; Internal medicine. Current work: Research and teaching in clinical pharmacology, care of patients with problems of drug therapy. Office: Cornell University Medical College Dept Pharmacology 1300 York Ave New York NY 10021

REIDER, BRUCE, surgeon, educator; b. N.Y.C., Feb. 9, 1949; s. Edward and Blanche (Goodman) R. A.B., Yale U., 1971; M.D., Harvard U., 1975. Diplomate: Am. Bd. Orthopaedic Surgery. Intern, vis. clin. fellow dept. surgery Columbia U. Hosp., N.Y.C., 1975-76; resident, fellow in surgery Cornell U. Hosp., N.Y.C., 1976-80, resident, instr. surgery, 1978-80; fellow sports medicine U. Wis., Madison, 1980-81; asst. prof. surgery/orthopedics Pritzker Sch. Medicine, U. Chgo., 1981—, dir. sports medicine, 1981—, head team physician, 1981—. Contbr. articles to profl. jours. Yale U. Alumni Fund rep., 1971—. Yale Nat. scholar, 1967-71. Mem. Am. Coll. Sports Medicine, Am. Orthopedic Soc. for Sports Medicine, Phi Beta Kappa. Club: Quadrangle. Subspecialties: Orthopedics. Current work: Athletic injuries of knee and shoulder, biomechanics of knee ligament injuries, disorders of the patella.

REIF, ARNOLD EUGENE, cancer researcher; b. Vienna, Austria, July 15, 1924; came to U.S., 1947, naturalized, 1956; s. Henry and Margaret (Gestetner) R.; m. Jane C. Chess; m. Katherine E. Hume, July 7, 1979; children: Betrand Paul, John Henry, Joseph Peter. B.A., Cambridge (Eng.) U., 1945, M.A., 1949; B.Sc., London U., 1946; M.S., Carnegie-Mellon U., 1949, D.Sc., 1950; grad. course in basic physics and med. applications of radioisotopes, New Eng. Roentgen Ray Soc. and M.I.T., 1962. Postdoctoral fellow McArdle Lab. Cancer Research, U. Wis. Med. Sch., 1950-53, research assoc. dept. physiol. chemistry, 1953; research assoc. dept. biochemistry Lovelace Found. Med. Edn. and Research, Albuquerque, 1953-57; asst. prof. surgery Tufts U. Sch. Medicine, 1957-69, assoc. prof., 1969-75, lectr. in surgery, 1975—; Am. Cancer Soc. faculty research assoc. Sch. Medicine, 1967-68; research pathologist Mallory Inst. Pathology, Boston City Hosp., 1973—; chief Exptl. Cancer Immunotherapy Lab., 1979—; research prof. pathology Boston U. Sch. Medicine, 1975—; exec. v.p. Vols. for Health Awareness, 1970—; vice chmn. Boston Cancer Research Assn., 1976-77; vis. prof. U. Conn., summer 1979. Contbr. numerous articles, abstracts, letters to profl. jours.; editor: Immunity and Cancer in Man: an Introduction, 1975. Mem. Am. Assn. Immunologists, Transplantation Soc., Am. Assn. Cancer Research (session chmn. immunology and genetics 1971, 72, 74, 77), Brit. Soc. Immunology, N.Y. Acad. Scis., Health Physics Soc. (dir. New Eng. chpt. 1976-77), AAAS, Tissue Culture Assn., Sigma xi. Subspecialties: Cancer research (medicine); Immunology (medicine). Current work: Immunotherapy of cancer in mouse model systems. Office: Boston City Hosp Boston MA 02118

REIFF, PATRICIA HOFER, physicist; b. Oklahoma City, Mar. 14, 1950; d. William Henry and Maxine (Hoffer) R.; m. John Fincher Moore, Jan. 2, 1971 (div. 1974); m. Thomas Westfall Hill, July 4, 1976. B.S., Okla. State U., 1971; M.S., Rice U., 1974, Ph.D., 1975. Nat. Acad. Scis./NRC research assoc. Marshall Space Flight Ctr., Huntsville, Ala., 1975-76; adj. asst. prof. Rice U., Houston, 1976-78, asst. prof., 1978-81, assoc. research scientist space physics and astronomy, 1981—, asst. chmn. dept., 1979—, adj. assoc. prof., 1983—; mem. com. solar terrestrial research Nat. Acad. Scis., 1979—, chmn. panel post IMS data analysis, 1981—; v.p. Wind Power Systems, Inc., Houston, 1981—. Contbr. articles to profl. jours. Pres. Citizens Environ. Coalition, 1980—; mem. Houston Galveston Area Council Air Quality, 1980—. Nat. Acad. Scis. resident research assoc., 1975; NRC faculty assoc., San Antonio, 1979. Mem. Am. Geophys. Union, Internat. Union Geodesy and Geophysics (del. 1975, 81), AAAS, Sigma Xi, Sigma Pi Sigma, Phi Kappa Phi. Clubs: Magellan Soc. (navigator 1979—), Galveston Yacht.). Subspecialties: Satellite studies; Planetary science. Current work: Theoretical and observational studies of the solar wind interaction with the magnetosphere and atmosphere; co investigator on dynamics Explorer spacecraft; wind electric generation. Office: Rice U Dept Space Physics and Astronomy PO Box 1892 Houston TX 77251 Home: 4214 Southwestern St Houston TX 77005

REIFFEL, LEONARD, sci. cons., physicist; b. Chgo., Sept. 30, 1927; s. Carl and Sophie (Miller) R.; m. Judith Eve Blumenthal, 1952 (div. 1962); children—Evan Carl, David Lee; m. Nancy L. Jeffers, 1971. B.Sc., Ill. Inst. Tech., 1947, M.Sc., 1948, Ph.D., 1953. Physicist Perkin-Elmer Corp., Conn., 1948; engring. physicist U. Chgo. Inst. Nuclear Studies, 1948-49; with Ill. Inst. Tech. Research Inst., Chgo., 1949-65, dir. physics research, 1956-63, v.p., 1963-65; cons. to Apollo program NASA Hdqrs., 1965-70, cons., 1970—; tech. dir. Manned Space Flight Expts. Bd., 1966-68; chmn. bd. Instructional Dynamics, Inc., 1966-81, INTERAND Corp., 1969—, TELESTRATOR Industries, Inc., 1970-73; sci. editor WBBM-CBS radio, Chgo.; sci. cons./commentator WBBM-TV, 1971-72; host Backyard Safari, 1971-73; sci. feature broadcaster WEEI-CBS radio, Boston, 1965-75; syndicated newspaper columnist World Book Ency. Sci. Service, Inc. (later Universal Sci. News, Inc.), 1966-72, Los Angeles Times Syndicate, 1972-76; sci. cons. CBS Network, 1967-71; Cons. Korean Govt. on establishment atomic energy research program; mem. adv. com. isotope and radiation devel. AEC; com. research reactors Nat. Acad. Scis., 1958-64; cons. U.S. Army, 1976—. Author numerous sci. papers; novel The Contaminant, 1979. Bd. dirs. Student Competitions on Relevant Engring. Named Outstanding Young Man of Year Chgo. Jr. C. of C., 1954, 61; recipient Merit award Chgo. Tech. Socs., 1968; Peabody award for radio edn., 1968; IR-100 award for inventing Telestrator, 1970; award for coverage space events Aviation Writers Assn., 1971; IR-100 award for invention underwater diver communications system, 1972; also for invention Audiografix, 1973; Distinguished Alumni Achievement award Ill. Inst. Tech., 1974. Fellow Am. Phys. Soc.; mem. AAAS, Sigma Xi, Tau Beta Pi, Eta Kappa Nu. Subspecialties: Information systems (information science); Graphics, image processing, and pattern recognition. Current work: Viedo graphic teleconferencing systems; real-time animation systems and video effects systems. Responsible for world's 1st indsl. nuclear reactor, 1956. Home: 602 W Deming Pl Chicago IL 60614 Office: 666 N Lake Shore Dr Chicago IL 60611

REILLY, CHRISTOPHER ALOYSIUS, JR., research scientist, program adminstr.; b. Tucson, Ariz., Aug. 8, 1942; s. Christopher Aloysius and Bess McCord (Callaway) R.; m. Georgia Kay Cole, Apr. 27, 1968; children: Colleen Marie, Megan Kathleen, Erin Maureen. Student, Mt. Carmel Coll., Niagara Falls, Ont., 1960-61; B.S. in Biology, Loyola U., Los Angeles, 1964; M.S. in Microbiology, U. Ariz., 1966, Ph.D., 1968. Postdoctoral appointment Argonne (Ill.) Nat. Lab., 1968, microbiologist, 1973—, group leaser carcinogenesis and chem. toxicology, 1979—, dep. program mgr. synfuels environ. research program, 1979—; McMullan Meml. lectr. Monmouth (Ill.) Coll., 1979. Contbr. articles in field to profl. jours. NIH grantee, 1966-67, 67-68. Mem. Am. Soc. Microbiology, Am. Assn. Cancer Research, AAAS, Soc. Exptl. Biology and Medicine, Internat. Assn. Comparative Research on Leukemia and Related Diseases, Sigma Xi. Subspecialties: Environmental toxicology; Cancer research (medicine). Current work: Toxicological effects of chemical agents related to energy production; specifically this includes applied and basic aspects of tumor induction mechanisms and gen. mammalian toxicology. Office: 9700 S Cass Ave Argonne IL 60439

REILLY, JOSEPH FRANCIS, pharmacologist; b. Waucoma, Iowa, May 14, 1915; s. Joseph Francis and Grace Mary (Lynch) R.; m. Joan Marilyn Cowie, Oct. 16, 1948; children: Joseph Francis, Joan Elizabeth, John Cowie, Elizabeth, Andrew Owen. Student, Elmhurst Coll., 1933-35; B.A., U. Ill., 1937; M.A., Harvard U., 1939; Ph.D., U. Chgo., 1947; postgrad., Cornell U. Med. Sch., 1948-49. Chemist chem. research dept. Armour & Co., Chgo., 1939-43; research asso. OSRD antimalarial program U. Chgo., 1943-46; pharmacologist Army Chem. Center, Edgewood Arsenal, Md., 1947-48; research fellow Cornell U. Med. Coll., 1948-49; instr. pharmacology Cornell U. Med. Center, 1949-53, asst. prof., 1953-54, asst. and assoc prof. pharmacology in psychiatry, 1954-62; chief pharmacodynamics sect. FDA, Washington, 1963-70, chief drug bioanalysis br., 1971—, acting dep. dir. div. drug biology, 1978-79; cons. council pharmacology and chemistry AMA, 1956, council drugs, 1965; mem. basic sci. council Am. Heart Assn., 1959. Contbr. tp profl. jours. articles on antimalarial drugs, cardiac drugs, blood levels of neurohumoral agents in psychiat. patients, antidepressant drugs, toxicity of insecticides, biol. testing of hormones. Recipient cert. NRC-OSRD, 1946. Mem. Am. Soc. Pharmacology and Exptl. Therapeutics, Soc. Toxicology, Soc. Exptl. Biology and Medicine (sec.-treas. N.Y. sect. 1955-60), Harvey Soc. N.Y., Sigma Xi, Gamma Alpha, Tau Kappa Epsilon. Roman Catholic. Subspecialties: Pharmacology; Toxicology (medicine). Current work: Biological testing of drugs and research on toxic effects of drugs and chemicals. Home: 9623 Alta Vista Terr Bethesda MD 20814 Office: 200 C St. SW HFD 412 Washington DC 20204

REILLY, JOSEPH GARRETT, molecular biologist; b. Washington, Nov. 2, 1950; s. John Garrett and Allyn L. R. B.S., St. Francis Coll., 1972; M.S., U. Md., 1976, Ph.D., 1978. Postdoctoral fellow Scripps Clinic and Research Found., 1978-81; asst. research scientist City of Hope Research Inst., 1981—. Contbr. articles to profl. jours. Am. Cancer Soc. fellow, 1979-80. Mem. Genetics Soc. Am., AAAS, N.Y. Acad. Medicine. Subspecialties: Genetics and genetic engineering (medicine); Molecular biology. Current work: Gene regulation and expression; recombinant DNA research; gene cloning. Office: Dept Molecular Biology City of Hope Research Inst Duarte CA 91010

REILLY, KEVIN DENIS, biostatistics and biomathematics educator; b. Omaha, Sept. 12, 1937; s. Brian A. and Dorothy (Dare) R.; m. JoAnn Grace Caniglia, Feb. 14, 1961; children: Martin L., Eileen M., Shannon D. B.S., Creighton U., 1959; M.S., U. Nebr., 1962; Ph.D., U. Chgo., 1966. Reporter, writer Southwest News Herald, Chgo., 1964-66; teaching asst. U. Nebr., Lincoln, 1959-62; sr. lectr. U. So. Calif., 1969-70; lectr. Sch. Engring. UCLA, 1969-70, research scientist, 1966-70; assoc. prof. biostats. and biomath. computer sci. dept. U. Ala.-Birmingham, 1970—, acting chmn. info. sci. dept., 1971-72; cons. U. Ala., 1970—, U. Mo.-St. Louis, 1970, Thiokol Corp., Huntsville, Ala., 1981. Rev. editor: Jour. Edn. Data Processing, 1974-79. Woodrow Wilson Found. fellow, 1960. Mem. Assn. Computing Machinery, Am. Assn. Artificial Intelligence, Am. Assn. Computational Linguistics, Simulation Councils Inc., Soc. Math. Biology. Subspecialties: Artificial intelligence; Software engineering. Current work: Teaching and research in computer information sciences, subareas of artificial intelligence, simulation and modeling ; software systems analysis and design. Office: Computer Sci Dept U Ala 901 S 15th St UC 4 Room 242 Birmingham AL 35226

REILLY, MARGARET ANNE, biochem. pharmacologist, educator; b. Port Chester, N.Y., Mar. 21, 1937; d. Thomas and Margaret Drake (Byrnes) R. B.A., Coll. of New Rochelle, 1959; M.S., N.Y. Med. Coll., 1978, Ph.D., 1981. Sr. lab. technician Sloan-Kettering Inst., Rye, N.Y., 1959-65; research scientist and assoc. N.Y. State Dept. Mental Hygiene, Rockland Research Inst., Orangeburg, N.Y., 1966—; adj. asst. prof. pharmacology Coll. of New Rochelle Sch. Nursing and N.Y. Med. Coll. Contbr. articles on biochem. pharmacology to profl. jours. Recipient Cath. Youth Orgn. Parish Vol. award N.Y. Archdiocese, 1976. Mem. Am. Soc. for Pharmacology and Exptl. Therapeutics, Histamine Research Soc. (sec.-treas.), N.Y. Acad. Scis., Council of Research Scientists, N.Y. State Office of Mental Health, Assn. of Women in Sci., Women in Neurosci., Am. Irish Assn. of Westchester, Coll. of New Rochelle Alumnae Assn. (dir., sec.). Subspecialties: Pharmacology; Neuropharmacology. Current work: Histamine, benzodiazepines, central nervous system neurotransmitters, mechanism of action of antidepressants. Office: Rockland Research Inst Orangeburg NY 10962

REILLY, PETER JOHN, chem. engr., educator, cons.; b. Newark, Dec. 26, 1938; s. Edward Thomas and Anita (Galdieri) R.; m. Rae Messer, July 3, 1976; children: Diane, Karen. A.B. in Chemistry, Princeton U., 1960; Ph.D. in Chem. Engring, U. Pa., Phila., 1964. Research engr. E. I. Du Pont de Nemours & Co., Deepwater, N.J., 1964-68; asst. prof. chem. engring. U. Nebr., Lincoln, 1968-74; assoc. prof. chem. engring. Iowa State U., Ames, 1974-79, prof., 1979—. Mem. Am. Chem. Soc., Am. Inst. Chem. Engrs., AAUP, Sigma Xi, Phi Kappa Phi. Subspecialties: Chemical engineering; Enzyme technology. Current work: Kinetics of soluble and immobilized enzymes, enzymatic hydrolysis of polysaccharides, utilization of agrl. residues. Home: 1807 Wilson Ave Ames IA 50010 Office: Dept of Chem Engring Iowa State U Ames IA 50011

REINER, ANTON JOHN, neuroscientist, researcher; b. Feffernitz, Austria, Oct. 29, 1950; s. John and Magdalena (Praschinger) R.; m. Marcia Gail Honig, July 25, 1982. B.S. in Biology, St. Joseph's U., Phila., 1972; M.S., Bryn Mawr Coll., 1975, Ph.D., 1977. Postdoctoral fellow neuroanatomy dept. psychiatry and behavioral scis. SUNY-Stony Brook, 1977-80; research asst. prof. dept. neurobiology and behavior, 1980-82; asst. research scientist dept. anatomy and cell biology U. Mich.-Ann Arbor, 1982—; researcher neurobiology. Contbr. articles profl. jours., chpts. in books. USPHS postdoctoral fellow, 1977-79; NIH research grantee, 1980-83; ADAMHA grantee, 1980. Mem. Soc. Neurosi., Am. Assn. Anatomists, AAAS. Subspecialties: Neurobiology; Immunocytochemistry. Current work: Evolution of brain, histochemistry and connectivity of amniote brain, LM and EM localization of neuropeptides within amniote nervous system. Office: University of Michigan Ann Arbor MI 48109

REINERT, CHARLES PETER, solar co. exec., physicist, educator; b. Tracy, Minn., May 23, 1939; s. Ervin Vernon and Lucille Margaret (Peterson) R.; m. Judith Ann Geegh, June 30, 1962; children: Peter, Jennifer (dec.); m. Caryl Ann Keith, Mar 21, 1981. B. Physics, U. Minn., 1961, M.S., 1963, Ph.D. in Physics and Math, 1969. Asst. prof. S.W. State U., Marshall, Minn., 1969-72, assoc. prof., 1973—; v.p. Winona Research, Inc., 1972—, Solarein, Inc., 1977-82, pres., dir., 1982—; solar cons.; tchr. solar heating. Methodist. Subspecialty: Solar energy. Current work: High performance, low cost, site-built solar collecting devices, research, development, and marketing efforts to establish concept nationally. Patentee in field.

REINES, FREDERICK, physicist, educator; b. Paterson, N.J., Mar. 16, 1918; s. Israel and Gussie (Cohen) R.; m. Sylvia Samuels, Aug. 30, 1940; children—Robert G., Alisa K. M.E., Stevens Inst. Tech., 1939, M.S., 1941; Ph.D., N.Y. U., 1944; D.Sc. (hon.), U. Witwatersrand, 1966. Mem. staff Los Alamos Sci. Lab., 1944-59, group leader, 1945-59; dir. (AEC expts. on Eniwetok Atoll), 1951; prof. physics, head dept. Case Inst. Tech., 1959-66; prof. physics U. Calif., Irvine, 1966—, also dean phys. scis., 1966-74. Contbr. numerous articles to profl. jours.; Contbg. author: Effects of Atomic Weapons, 1950, Methods of Experimental Physics, 1961. Mem. Cleve. Symphony Chorus, 1959-62. Recipient J. Robert Oppenheimer meml. prize, 1981; Guggenheim fellow, 1958-59; Sloan fellow, 1959-63. Fellow Am. Phys. Soc., AAAS; mem. Am. Assn. Physics Tchrs., Argonne U. Assn. (trustee 1965-66), Am. Acad. Arts and Scis., Nat. Acad. Sci., Phi Beta Kappa, Sigma Xi, Tau Beta Pi. Subspecialty: Particle physics. Current work: Stability of proton; neutrino physics. Co-discoverer elementary nuclear particle, free antineutrino, 1956. Office: U Calif at Irvine Irvine CA 92717

REINHARDT, JUERGEN, geologist; b. Eutingen, Baden, W.Ger., Oct. 27, 1946; s. Eugene and Herta Anna (Kaelber) R.; m. Judith Grace Twiggar, June 29, 1968; children: Kirstan G., Stefan M. A.B., Brown U., 1968; Ph.D., Johns Hopkins U., 1973. Geologist U.S. Geol. Survey, Reston, Va., 1975—; Md. Geol. Survey, Balt., 1973-75; instr. Johns Hopkins U., Balt., 1973-75. Contbr. articles to profl. jours.; assoc. editor: bull. Geol. Soc. Am, 1983—. Chmn. com. Boy Scouts Am., 1982—. Recipient Outstanding Performance award Dept. Interior, 1980. Fellow Geol. Soc. Am.; mem. Soc. Econ. Paleontologists and Mineralogists, Internat. Assn. Sedimentologists, Am. Assn. Petroleum Geologists, Geol. Soc. Washington. Clubs: Optimist, Loudoun Tennis. Subspecialties: Sedimentology; Tectonics. Current work: Geologic evolution of the eastern Gulf Coastal Plain both from a sedimentological and a tectonic perspective. Home: 158 Edwards Ferry Rd Leesburg VA 22075 Office: US Geol Survey 12201 Sunrise Valley Dr Reston VA 22092

REINKE, LESTER ALLEN, pharmacologist, educator, researcher; b. Davenport, Nebr., Sept. 29, 1946; s. Herman Dick and Alma Ida (Grosshans) R.; m. Carol Sue Paulsen, Sept. 1, 1968; children: Jonathan Paul, Lisa Sue. B.S. in Pharmacy, U. Nebr., 1969, M.S. in Medicinal Chemistry, 1975, Ph.D., 1977. Registered pharmacist, Nebr. Research asst. prof. U. N.C. Sch. Medicine, Chapel Hill, 1980; asst. prof. pharmacology U. Okla. Health Scis. Center, Oklahoma City, 1980-86. Contbr. articles to sci. jours. Served with U.S. Army, 1969-73. Nat. Cancer Inst. fellow, 1977-80; research award, 1980-83. Mem. Am. Soc. Pharmacology and Exptl. Therapeutics, Internat. Soc. for Study Xenobiotics, Sigma Xi, Rho Chi. Lutheran. Subspecialty:

Pharmacology. Current work: Drug metabolism; alcohol; chemical carcinogenesis; intermediary metabolism. Home: 1302 NW 21st St Oklahoma City OK 73106 Office: Pharmacology Dept U Okla PO Box 26901 Oklahoma City OK 73190

REINSTEIN, LAWRENCE ELLIOT, med. physicist, educator; b. Bklyn.; s. Herman and Hilda (Rubinstein) R.; m. P. Gila Steinlisht, May 28, 1967; children: Ezra, David, Gabriel. B.S., Bklyn. Coll., 1966; M.S., Yale U., 1968; Ph.D., Boston U., 1975. Sr. physicist R.I. Hosp., Providence, 1975—; assoc. prof. Brown U., Providence, 1981—. Contbr. articles to profl. jours. Mem. Am. Assn. Physicists in Medicine, Am. Soc. Therapeutic Radiologists. Jewish. Subspecialties: Radiology; Medical physics. Current work: Research in radiation dosimetry, radiotherapy. Address: Dept Radiation Oncology RI Hosp Providence RI 02902

REIS, ARTHUR HENRY, JR., university administrator, chemistry educator; b. Chgo., Nov. 6, 1946; s. Arthur H. and Ardell (Tholotowsky) R.; m. Karen Wessell, Aug. 22, 1970; children: Sally Wessell, Rodger Henry. B.A., Cornell Coll., Mt. Vernon, Iowa, 1968; M.A., Harvard U., 1969, Ph.D., 1969-72. Orbital analyst U.S. Air Force, 1972-74; staff chemist Argonne Nat. Lab., 1974-79; administr. chemistry dept. Brandeis U., 1979-82, assoc. prof. chemistry, 1980—, dir. sci. resources and planning, 1982—. Contbg. author: Neutron Scattering, 1976, Molecular Metals, 1979, Solid State Chemistry, 1980, Extended Linear Chain Compounds, Vol. 1, 1982. Served to 1st lt. USAF, 1970-82. Mem. Am. Chem. Soc. (auditor 1982-83), Am. Crystallographic Assn., Council Chem. Research, N.Y. Acad. Scis., Phi Beta Kappa. Subspecialties: Solid state chemistry; Inorganic chemistry. Current work: Synthesis and characterization of inorganic and organic one-dimensional materials which have anisotropic properties. Home: 436 Weston Rd Wellesley MA 02181 Office: Brandeis U Dean of Faculty Office Waltham MA 02254

REIS, DONALD JEFFERY, neurologist, neurobiologist, educator; b. N.Y.C., Sept. 9, 1931; s. Samuel H. and Alice (Kiesler) R. A.B., Cornell U., 1953, M.D., 1956. Intern N.Y. Hosp., N.Y.C., 1956; resident in neurology Boston City Hosp.-Harvard Med. Sch., 1957-59; Fulbright fellow, United Cerebral Palsy Found. fellow, London and Stockholm, 1959-60; research asso. NIMH, Bethesda, Md., 1960-62; spl. fellow NIH, Nobel Neurophysiology Inst., Stockholm, 1962-63; asst. prof. neurology Cornell U. Med. Sch., N.Y.C., 1963-67, asso. prof. neurology and psychiatry, 1967-71, prof., 1971—; Mem. U.S.-Soviet Exchange Program; adv. councils NIH; bd. sci. advisers Merck, Sharpe and Dohm; cons. biomed. cos. Contbr. articles to profl. jours.; mem. editorial bd. various profl. jours. Mem. Am. Physiol. Soc., Am. Neurol. Assn., Am. Pharmacol. Soc., Am. Assn. Physicians, Telluride Assn., Am. Soc. Clin. Investigation, Phi Beta Kappa, Sigma Xi, Alpha Omega Alpha. Subspecialties: Neurochemistry; Neurophysiology. Current work: Neurochemical, and molecular biological substrates of emotional behavior, brain and autonomic nervous system, brain control of the circulation. Home: 190 E 72d St New York NY 10021 Office: 1300 York Ave New York NY 10021

REISER, MORTON FRANCIS, educator, psychiatrist; b. Cin., Aug. 22, 1919; s. Sigmund and Mary (Roth) R.; m. Lynn B. Whisnant, Dec. 19, 1976; children: David E., Barbara, Linda. B.S., U. Cin., 1940, M.D., 1943; grad., N.Y. Psychoanalytic Inst., 1960. Diplomate: Am. Bd. Psychiatry and Neurology. Intern King's County Hosp., Bklyn., 1944; resident Cin. Gen. Hosp., 1944-49; practice medicine, specializing in psychiatry, Cin., 1947-52, Washington, 1954-55, N.Y.C., 1955-69; mem. faculty Cin. Gen. Hosp., also U. Cin. Coll. Medicine, 1949-52, Washington Sch. Psychiatry, 1953-55; faculty Albert Einstein Coll. Medicine, Yeshiva U., N.Y.C., 1955-69, prof. psychiatry, 1958-69, dir. research dept. psychiatry, 1958-65; chief div. psychiatry Montefiore Hosp. and Med. Center, N.Y.C., 1965-69; chmn. dept. psychiatry Yale Med. Sch., 1969—, prof., 1969-78, Charles B.G. Murphy prof., 1978—; cons. Walter Reed Army Inst. Research, 1957-58, High Point Hosp., Port Chester, N.Y., 1957-69; com. WHO, 1963; mem. profl. adv. com. Jerusalem Mental Health Center, 1972—; mem. clin. program projects rev. com. NIMH, 1970—, chmn., 1973-74; also lectr.; mem. Josiah Macy, Jr. Found. Commn. on Present Condition and Future Acad. Psychiatry, 1977. Author: (with H. Leigh) The Patient: Biological, Psychological, and Social Dimensions of Medical Practice, 1980; editor: American Handbook of Psychiatry, vol. IV, 1975; editor-in-chief: Psychosomatic Medicine, 1962-72; editorial bd.: AMA Archive of Gen. Psychiatry, 1961-71, (with H. Leigh) Psychiatry Medicine and Primary Care, 1978; contbr. articles to profl. jours. and books. Recipient Stella Fels Hoffheimer Meml. prize U. Cin. Coll. Medicine, 1943. Fellow Am. Coll. Psychiatrists, Am. Psychiat. Assn.; mem. Am. Soc. Clin. Investigation, Am. Psychosomatic Soc. (pres. 1960-61), Am. Fedn. Clin. Research, Am. Assn. Chairmen Depts. Psychiatry (exec. com. 1971—, pres. 1975-76), Acad. Behavioral Medicine Research (exec. council 1978), Am. Psychoanalytic Assn. (pres.-elect 1980-82, pres. 1982-84), Internat. Psycho-Analytical Assn., Assn. Psychophysiol. Study of Sleep, Internat. Coll. Psychosomatic Medicine (pres. 1975), Psychiat. Research Soc., A. Graeme Mitchell Undergrad. Pediatric Soc., World Psychiat. Assn. (organizing com. sect. psychosomatic medicine 1967), Sigma Xi, Phi Eta Sigma, Pi Kappa Epsilon, Alpha Omega Alpha. Subspecialties: Psychopharmacology; Psychobiology. Current work: Neurobiology and psychoanalysis; neurobiology of major psychiatric disorders. Home: 99 Blake Rd Hamden CT 06517 Office: 25 Park St New Haven CT 06519

REISKIN, ALLAN BURT, radiologist, educator; b. N.Y.C., Apr. 24, 1936; s. Bernard B. and Bernice B. (Berezin) R.; m. Joan Barenkopf, Nov. 21, 1962; children: Julie Anne, Edward David. B.A., CCNY, 1963; D.D.S., U. Pa., 1963; D.Phil., Oxford U., 1966. Diplomate: Am. Bd. Maxillofacial Radiology, 1982. Brit.-Am. exchange fellow Am. Cancer Soc.; fellow Churchill Hosp., Oxford, Eng., 1963-66; assoc. biologist Argonne (Ill.) Nat. Lab., 1966-70; prof. radiology Sch. Dental Medicine U. Conn., Farmington, 1970—; cons. Newington Children's Hosp., Conn. State Police, Council on Materials and Devices of ADA; police surgeon State of Conn.; cons. Conn. Dept. Health Services, Conn. State Legislature. USPHS fellow, 1960-61. Mem. Radiol. Soc. Conn., Radiol. Soc. N.Am., Brit. Inst. Radiology, Radiation Research Soc., Am. Assn. Cancer Research, ADA, AAAS, Sigma Xi. Jewish. Subspecialties: Radiology; Imaging technology. Current work: Radiation biology, carinogenesis, imaging technology, radiological diagnosis. Office: U Conn Health Ctr Farmington CT 06032

REISMAN, SCOTT, psychologist; b. N.Y.C., Oct. 9, 1951; s. Emanuel and Myra (Deutch) R.; m. Susan A. Levy, Apr. 3, 1982. B.G.S., Ohio U., Athens, 1974; M.S. in Counseling, Nova U., Ft. Lauderdale, 1978, Nova U., Ft. Lauderdale, 1981. Mental health technician Dade County (Fla.), Miami, 1974-77; program evaluator Nova Clinics, Inc., Ft. Lauderdale, 1979-82; adj. faculty Nova Coll., 1981. Mem. Am. Psychol. Assn., Assn. for Advancement of Behavior Therapy, Biofeedback Soc. Am., Fla. Psychol. Assn. Subspecialty: Behavioral psychology. Current work: Research on physiological and psychological effects of several methods of stress management training for professionals. Office: Nova U 3301 College Ave Fort Lauderdale FL 33314

REISS, CAROL SHOSHKES, cellular immunologist; b. Boston, Mar. 14, 1950; d. Milton and Lila (Topal) Shoshkes; m. David Simon Reiss, June 5, 1977. A.B., Bryn Mawr Coll., 1972; M.S., Sarah Lawrence Coll., 1973; Ph.D. in Microbiology, CUNY, 1978. Postdoctoral fellow Harvard U. Med. Sch., Boston, 1978-81; instr. pathology and pediatric oncology, dir. animal facility Sidney Farber Cancer Inst., Boston, 1981—. Mem. Assn. Women in Sci. (chpt. dir.), AAAS, N.Y. Acad. Scis., Am.Assn. Immunologists. Subspecialties: Immunobiology and immunology; Virology (biology). Current work: Regulation and specificity of immune interactions; cytolytic T cells; modulation and regulation of induction of helper/suppressor activity; subcellular/subviral antigens. Office: 44 Binney St Boston MA 02115

REISS, ERROL, microbial immunochemist; b. N.Y.C., Jan. 16, 1942; s. Jack and Claire (Litman) R.; m. Cheryl Linda Aaronson, Jan. 20, 1968; children: Brendan K., Merryl D. B.Sc. (N.Y. State Regents scholar), CCNY, 1963; Ph.D., Rutgers U.-New Brunswick, N.J., 1972. Bacteriologist VA Hosp., Washington, 1966-67; postdoctoral fellow NIH, Bethesda, Md., 1972-74; research microbiologist Centers for Disease Control, Atlanta, 1974—, head Immunochemistry Lab. div. mycotic diseases, 1980—; adj. prof. U. N.C.; lectr. Morehouse Coll. Medicine; adj. prof. Ga. State U. Contbr. articles to profl. jours.; mem. editorial bd.: Jour. Clin. Microbiology, 1982—. Served to 1st lt. U.S. Army, 1963-65. Recipient Service award USPHS, 1980. Mem. Am. Chem. Soc., Am. Assn. Immunologists; mem. Am. Acad. Microbiology; Mem. Am. Soc. Microbiology (cert. recognition 1982). Subspecialties: Microbiology; Immunocytochemistry. Current work: Molecular immunology of microbial infections, especially mycotic infections; immunochemistry; carbohydrate biochemistry; enzymology; cellular immunology; preceptor for graduate students. Home: 1369 Holly Ln Atlanta GA 30329 Office: 1600 Clifton Rd NE 5-B18 Atlanta GA 30333

REISS, HOWARD, chemistry educator; b. N.Y.C., Apr. 5, 1922; s. Isidor and Jean (Goldstein) R.; m. Phyllis Kohn, July 25, 1945; children: Gloria, Steven. A.B. in Chemistry, N.Y.U., 1943, Ph.D., Columbia, 1949. With Manhattan Project, 1944-46; instr., then asst. prof. chemistry Boston U., 1949-51; with Central Research Lab., Celanese Corp. Am., 1951-52, Edgar C. Bain Lab. Fundamental Research, U.S. Steel Corp., 1957, Bell Telephone Labs., 1952-60; asso. dir., then dir. research div. Atomics Internat., div. N.Am. Aviation, Inc., 1960-62; dir. N.Am. Aviation Sci. Center, 1962-67, v.p. co., 1963-67; v.p. research aerospace systems group N.Am. Rockwell Corp., 1967-68; vis. lectr. chemistry U. Calif. at Berkeley, summer 1957; vis. prof. chemistry U. Calif. at Los Angeles, 1961, 62, prof., 1968—; vis. prof. U. So. Calif., 1964, 67; Cons. to chem.-physics program Air Force Cambridge Research Cambridge Research Labs., 1950-52; chmn., editor proc. Internat. Conf. Nucleation and Interfacial Phenomena, Boston; mem. Air Force Office Sci. Research Physics and chemistry Research Evaluation Groups, 1966—, Oak Ridge Nat. Lab. Reactor Chemistry Adv. Com., 1966-68; adv. com. math. and phys. scis. NSF, 1970-72, ARPA Materials Research Council, 1968—. Author articles; editor in field.; Editor: Progress in Solid State Chemistry, 1962-71, Jour. Statis. Physics, 1968-75, Jour. Colloid Interface Sci; editorial adv. bd.: Chemistry; Author: The Methods of Thermodynamics, 1965. Guggenheim Meml. fellow, 1978. Fellow Am. Phys. Soc. (exec. com. div. chem. physics 1966-69); mem. Am. Chem. Soc. (chmn. phys. chemistry sect. N.J. sect. 1957, Richard C. Tolman medal 1973, prize in Colloid and surface chemistry 1980), Nat. Acad. Sci., Sci. Research Soc. Am., AAUP, Phi Beta Kappa, Sigma Xi, Phi Lambda Upsilon. Subspecialties: Physical chemistry; Analytical chemistry. Current work: Statistical mechanics of cooperative system; kinetics of gas phase polymerization; nucleation; membrane science. Office: Dept Chemistry U Calif Los Angeles CA 90024 Many styles can lead to success. My style is however characterized by persistence, even in the face of adversity, and by consideration, as much as possible within the bounds of constructive reality, of the welfare of other human beings.

REISS, HOWARD R., physicist, educator; b. N.Y.C., July 29, 1929; s. Edward and Fannie (Metz) R.; m.; children: Stephanie Jane, John Eden. B.Ae.E., Poly. Inst. N.Y., 1950, M.Ae.E., 1951; Ph.D. in Physics, U. Md., 1958. Research fellow Poly. Inst. N.Y., 1950-51; physicist David Taylor Model Basin, Carderock, Md., 1951-55; physicist, chief nuclear physics div. Naval Ordnance Lab., White Oak, Md., 1955-69; vis. scientist Institute di Fisica dell Universita di Torino, Italy, 1963-64; prof. physics Am. U., Washington, 1969—; research prof. Ariz. Research Labs., U. Ariz., 1981—; cons. Standard Oil Co., Ind., 1980-81. Contbr. articles to profl. jours. Served with U.S. Navy, 1951. Recipient Outstanding Student award Inst. Aero. Scis., 1949; Disting. Civilian Service award U.S. Navy, 1965. Fellow Am. Phys. Soc.; mem. AAAS. Subspecialties: Nuclear physics; Energy science. Current work: Electromagnetic field effects in nuclear physics; novel energy sources. Patentee in field. Office: Ariz Research Labs U Ariz Tucson AZ 85721

REISS, ROBERT, mech. engr., researcher, cons., educator; b. Bklyn., Mar. 12, 1942; s. Ben A. and Margaret (Schmick) R.; m. Sandra Ellen, June 18, 1966; children: Sharon, Daniel. B.S. in Applied Math, Brown U., 1963; Ph.D. in Solid Mechanics, Ill. Inst. Tech., 1967. Asst. project engr. Pratt and Whitney Aricraft, East Hartford, Conn., 1967-70; asst. prof. engring. mechanics U. Mo., Rolla, 1970-73; asst. prof. mechanics U. Iowa, 1973-77; prof. mech. engring., assoc. chmn. dept. mech. engring. Howard U., 1977—; cons. in field. NASA-Langley grantee, 1980; NSF grantee, 1980-82. Mem. ASME, Am. Acad. Mechanics, Soc. Women Engrs., Sigma Xi. Subspecialties: Solid mechanics; Theoretical and applied mechanics. Current work: Theory of structural optimization; formulation and application of variational principles in applied mechanics; mechanics of composite materials. Office: Dept Mech Engring Howard U Washington DC 20059

REISSIG, JOSE LUIS, molecular biology educator; b. Buenos Aires, Argentina, June 21, 1926; came to U.S., 1946; s. Luis and Herminia (Portatadino) R.; m. (div.); children: Celia, Peter, Nora. B.S., U. Mich., 1948; Ph.D., Calif. Inst. Tech., 1952. Researcher Instituto de Investigaciones Bioquimicas, Buenos Aires, 1952-55; researcher Inst. Animal Genetics, Edinburgh, Scotland, 1955-59, U. Genetics Inst., Copenhagen, 1959-60, Pasteur Inst., Paris, 1960-61; prof. molecular biology U. Buenos Aires, 1961-67, C.W. Post Coll., L.I.U., 1967—. Contbr. numerous articles to sci. jours.; Editor: Microbial Interaction, 1977. Recipient L.I.U. trustee award for scholarly achievement, 1978. Subspecialties: Molecular biology; Microbiology. Current work: Fungal morphogenesis. Office: C W Post Coll Room 246 LS Greenvale NY 11548

REISSNER, MAX ERICH, applied mathematician; b. Aachen, Ger., Jan. 5, 1913; came to U.S., 1936, naturalized, 1945; s. Hans and Josephine R.; m. Johanna Siegel, Apr. 19, 1938; children: John E., Eva. M. Dipl. Ing., Technische Hochschule, Berlin, 1935, Dr. Ing., 1936; Ph.D., MIT, 1938; D.Eng. (hon.), U. Hanover, Ger., 1964. Mem. faculty MIT, Cambridge, 1936-69, prof. math., 1949-66, prof. applied math., 1966-69; prof. applied mechanics U. Calif.-San Diego, 1970-78, prof. emeritus, 1978—; vis. prof. U. Mich., Ann Arbor., 1949; aero. research scientist NACA, Langley Field, 1948, 51; cons.; mem. Addison-Wesley Pub. Co., 1949-60. Editor: Jour. Maths. and Physics, 1945-67; assoc. editor: Quar. Applied Maths, 1946—; cons. editor: Math. and Mechanics; contbr. chpts. to books and articles to profl. jours. Recipient Clemens Herschel award Boston Soc. Civil Engrs., 1956; Theodore von Karman medal ASCE, 1964; Timoshenko medal ASME, 1973; Guggenheim fellow, 1962-63; NSF fellow, 1968-69. Fellow Am. Acad. Arts and Scis. (council 1957-61), Am. Acad. Mechanics (pres. 1975-76), ASME (exec. com. applied mechanics div. 1958-63), AIAA; mem. Am. Math. Soc., Nat. Acad. Engring. Subspecialties: Applied mathematics; Theoretical and applied mechanics. Office: U Calif San Diego La Jolla CA 92093 Home: 422 San Lucas Dr Solana Beach CA 92075

REITAN, DANIEL KINSETH, electrical computer engineering educator; b. Duluth, Minn., Aug. 13, 1921; s. Conrad Ulfred and Joy Elizabeth R.; m. Marian Anne Stemme, July 18, 1946; children: Debra Leah, Danielle Karen. B.S.E.E., N.D. State U., 1946; M.S.E.E., U. Wis., 1949, Ph.D., 1952. Registered profl. engr., Wis. Control engr. Gen. Electric Co., Schenectady, N.Y., 1946-48; transmission line engr. Gen. Telephone Co., Madison, Wis., 1949-50; mem. faculty Coll. Engring. U. Wis., Madison, 1952—, prof. elec. and computer engring., 1962—, dir. power systems simulation lab., 1968—, also dir. wind power research; cons. in field. Contbr. articles to profl. jours. Served with U.S. Army, World War II. Recipient Outstanding Tchr. award Polygon Engring. Council., Gov.'s citation for service to State of Wis. Mem. IEEE, IEEE Power Engring., Computer, Control, Indsl. Applications, and Edn. Socs., Conf. Internat. des Grands Reseaux Electriques a Haute Tension, Am. Soc. Engring. Edn., Wis. Acad. Scis., Am. Wind Energy Assn., Sigma Xi, Tau Delta Pi, Tau Beta Pi, Eta Kappa Nu, Kappa Eta Kappa. Lutheran. Subspecialties: Wind power; Energy Management. Current work: Compouter solutions of EHV-AC/DC electric power network; analysis of co-generation and other small power producers interfaces with large electric utility grids. Patentee in field. Office: Elec and Computer Engring Dept 1425 Johnson Dr Madison WI 53706 I believe that in one's career professionalism and perseverance are key factors in success. In one's personal life, the family should be the center, but not the circumference, about which all activities revolve.

REITAN, RALPH MELDAHL, psychologist; b. Beresford, S.D., Aug. 29, 1922; s. John O. and Anna (Meldahl) R.; m. Ann Kirsch, Feb. 15, 1952 (div. 1978); children: Ellen, Jon, Ann, Richard, Erik. B.A., Central YMCA Coll., Chgo., 1944; Ph.D., U. Chgo., 1950. Asst. prof. to prof. psychology Ind. U. Med. Ctr., Indpls., 1951-70; prof. U. Wash., Seattle, 1970-77; prof. psychology U. Ariz., Tucson, 1977—; cons. NASA, Washington, 1964-65; NIH, 1959-70. Author: Clinical Neuropsychology, 1974; contbr. articles to profl. jours. Served with AUS, 1943. Recipient Barrows award Ind. Psychol. Assn., 1968; Research award Ariz. Psychol. Assn., 1982. Fellow Am. Psychol. Assn.; mem. Am. Neurol. Assn., Am. Acad. Neurology. Subspecialties: Neuropsychology; Physiological psychology. Current work: Human brain-behavior relationships. Home: 1338 E Edison St Tucson AZ 85719 Office: Dept Psychology U Ariz Tucson AZ 85721

REITEMEIER, RICHARD JOSEPH, physician; b. Pueblo, Colo., Jan. 2, 1923; s. Paul John and Ethel Regina (McCarthy) R.; m. Patricia Claire Mulligan, July 21, 1951; children: Mary Louise, Paul, Joseph, Susan, Robert, Patrick, Daniel. A.B., U. Denver, 1944; M.D., U. Colo., 1946; M.S. in Internal Medicine, U. Minn., 1954. Diplomate: Am. Bd. Internal Medicine (gov. 1971-79, chmn. 1978-79, rep. to Federated Council Internal Medicine 1977-80, 83—, accreditation council grad. med. edn. 1979—, chmn. 1982-83), Am. Bd. Gastroenterology. Intern Corwin Hosp., Pueblo, 1946-47; resident Henry Ford Hosp., Detroit, 1949-50, Mayo Found., Rochester, Minn., 1950-53; cons. internal medicine and gastroenterology Mayo Clinic, Rochester, 1954—, chmn. dept. internal medicine, 19.67-74, prof., 1971—, bd. govs., 1970-74. Author: (with C. G. Moertel) Advanced Gastrointestinal Cancer, Clinical Management and Chemotherapy, 1969; contbr. numerous articles to med. jours. Trustee Mayo Found., 1970-74, St. Mary's Hosp., Rochester, 1976-82. Served with U.S. Army, 1947-49. Recipient Alumni award U. Colo. Sch. Medicine. Fellow A.C.P. (regent 1979-82, gov. for Minn. 1975-79, pres. 1983-84), Am. Gastroenterol. Assn., AMA, Am. Clin. and Climatol. Assn., Am. Fedn. Clin. Research, Am. Soc. Clin. Oncology, Council Med. Splty. Socs., Inst. Medicine, Am. Assn. Cancer Research, Am. Assn. Study Liver Disease, Alpha Omega Alpha. Republican. Roman Catholic. Subspecialties: Gastroenterology; Chemotherapy. Current work: Medical education; healthcare delivery. Home: 707 12th Ave SW Rochester MN 55901 Office: 200 1st Ave SW Rochester MN 55901

REITER, LAWRENCE W., research pharmacologist; b. Kansas City, Mo., Apr. 21; s. Ted J. and Rosemary L. (Comiskey) R.; m. Margurite L. Scheier, June 11, 1966; children: Jeremy Nash, David Allen. B.S., Rockhurst Coll., Kansas City, Mo., 1965; Ph.D., U. Kans. Med. Center, 1970. Postdoctoral fellow dept. environ. toxicology U. Calif., Davis, 1972-73; research pharmacologist EPA, Research Triangle Park, N.C., 1973-79, neurotoxicology program coordinator, 1978-79, dir. neurotoxicology div., 1980—; adj. asso. prof. N.C. State U., 1982—; adj. asst. prof. U. N.C., 1977—. Contbr. articles to profl. jours. NIH fellow; recipient. Spl. Achievement award EPA, 1978, Bronze medal, 1979. Mem. Soc. Toxicology, N.C. Sec. Toxicology, Soc. Neuroscience, Behavioral Teratology Soc. Subspecialty: Neurotoxicology. Current work: Study of specific agent (toxicants and drugs) on function of nervous system, particular emphasis on behavior. Office: EPA/NTD (MD-74B)/HERL Research Triangle Park NC 27711

REITER, RUSSEL JOSEPH, medical educator, researcher; b. St. Cloud, Minn., Sept. 22, 1936; s. Bernard William and Bernice Philomena (Friedman) R.; m.; children: David Russel, Michael Kendrick. B.S., St. John's U., 1959; M.S., Wake Forest U., 1961, Ph.D., 1964. Asst. prof. U. Rochester, N.Y., 1966-69, assoc. prof., 1969-71, U. Tex. Health Sci. Ctr., San Antonio, 1971-73, prof. dept. anatomy, 1973—. Editor-in-chief: Pineal Research Revs; contbr. articles in field to profl. jours. Vice-pres. Central Park Lions Little League, San Antonio, 1974-79, coach, 1973-79; football coach YMCA, San Antonio, 1977-79. Served to capt., M.C. U.S. Army, 1964-66. Recipient Career Devel. award USPHS, 1969-71, 72-74. Mem. AAAS, Am. Soc. Zoologists, Am. Assn. Anatomists, Am. Physiol. Soc., Endocrine Soc., Soc. Study Reprodn., Internat. Soc. Neuroendocrinology, Soc. Neurosci., Soc. Exptl. Biology and Medicine, Am. Aging Assn., Internat. Brain Research Orgn., Sigma Xi. Republican. Roman Catholic. Subspecialties: Neuroendocrinology; Reproductive endocrinology. Current work: Research in reproductive physiology; teaching of neuroscience. Home: 148 Garrapata St San Antonio TX 78232 Office: Dept Anatomy U Tex Health Sci Ctr San Antonio TX 78284

REITH, MAARTEN EDUARD ANTON, neurochemist; b. Utrecht, Netherlands, Dec. 29, 1946; came to U.S, 1978; s. Johannas Franciscus and Catherina (Poelmann) R.; m. Paula van den Brom, Mar. 8, 1968; 1 son, Michiel; m. Irma Vara, Apr. 26, 1980; 1 dau., Catherina. B.S., State U. Utrecht, 1968, M.S., 1971, Ph.D., 1975. Research fellow Center for Neurochemistry, Strasbourg, France, 1971; research scientist Lab. for Pharmacology, State U. Utrecht, 1971-74; sr. research scientist Inst. Molecular Biology, 1974-78, Ctr. for Neurochemistry, N.Y.C., 1978—. Contbr. articles on neurochemistry to profl. jours. N.Y. State Health Research Council grantee, 1981-82. Mem. Am. Soc. for Neurochemistry, Am. Soc. Biol. Chemists, Internat. Soc. for Neurochemistry. Subspecialties: Neurochemistry; Neuropharmacology. Current work: My research interest focuses on the role of proteins and peptides in the brain; measurement of rates of synthesis and degradation of brain proteins, and study of membrane proteins as receptors for neurotransmitters or recognition sites for pharmaca. Home: 11 Balint Dr Apt 542 Yonkers NY 10710 Office: Ctr for Neurochemistry Wards Island New York NY 10035

REITSEMA, HAROLD JAMES, astronomer, aerospace engr.; b. Kalamazoo, Mich., Jan. 19, 1948; s. Harold and Bernicejean (Hoogsteen) R.; m. Mary Jo Gunnink, Aug. 6, 1970; children: Ellen Celeste, Laurie Jean. B.A., Calvin Coll., 1972; Ph.D., N.Mex. State U., 1977. Research asst. U. Denver, 1973, Kitt Peak Nat. Obs., Tucson, Ariz., 1974, N.Mex. State U., Las Cruces, 1974-77; research assoc. U. Ariz., Tucson, 1977-80, sr. research assoc., 1980-82; sr. mem. tech. staff Ball Aerospace, Boulder, Colo., 1982—; vis. scientist U. Ariz., 1981—; cons. Royal Obs., Edinburgh, Scotland, 1979-81. Contbr. articles in field to internat. jours. Phys. Sci. Lab. fellow N.Mex. State U., 1974. Mem. Am. Astron. Soc., Soc. Photo-optical Instrumentation Engrs., AAAS, Astron. Soc. Pacific, Sigma Xi, Phi Kappa Phi, Beta Theta Pi. Subspecialties: Planetary science; Aerospace engineering and technology. Current work: Design of solid state device imaging systems, solar systems studies, CCD detectors, infrared detectors, spacecraft sci. instrumentation, digital image processing. Office: Ball Aerospace PO Box 1062 Boulder CO 80306

REITZ, BRUCE ARNOLD, cardiac surgeon, educator; b. Seattle, Sept. 14, 1944; s. Arnold and Ruth (Stillings) R.; m. Nan Norton, Oct. 3, 1070; children: Megan, Jay. B.S., Stanford U., 1966; M.D., Yale U., 1970. Diplomate: Am. Bd. Surgery, 1980, Am. Bd. Thoracic Surgery, 1981. Intern Johns Hopkins U., Balt., 1970-71; resident Stanford (Calif.) Med. Ctr., 1971-72, 74-77; asst. prof. cardiovascular surgery Stanford U. Sch. Medicine, 1977-81, assoc. prof., 1982; prof. surgery Johns Hopkins U. Sch. Medicine, Balt., 1982—; also cardiac surgeon-in-charge Univ. Hosp. Served with USPHS, 1972-74. Mem. AMA, Soc. Univ. Surgeons, Samson Thoracic Soc., Transplantation Soc., Assn. Clin. Cardiac Surgeons. Subspecialty: Cardiac surgery. Current work: Heart and Heart-lung transplantation; developing techniques for combined heart and lung transplants. Office: 600 N Wolfe St Baltimore MD 21205

REITZ, RICHARD ELMER, physician, laboratory administrator; b. Buffalo, Sept. 18, 1938; s. Elmer Valentine and Edna Anna (Guenther) R.; m. Gail Ida Pounds, Aug. 20, 1960; children: Richard Allen, Mark David. B.S., Heidelberg Coll., 1960; M.D., SUNY-Buffalo, 1964. Intern Hartford (Conn.) Hosp., 1964-65, resident in medicine, 1966-67; asst. resident in medicine Yale U., 1965-66; research fellow in medicine Harvard Med. Sch., Mass. Gen. Hosp., Boston, 1967-69; vis. research assoc. NIH, Bethesda, Md., 1967-68; dir. Endocrine Metabolic Center, Oakland, Calif., 1973—; asst. prof. medicine U. Calif.-San Francisco, 1971-76; assoc. clin. prof. medicine U. Calif.-Davis, 1976—; chief endocrinology Providence Hosp., Oakland, Calif., 1972—. Contbr. articles to profl. jours., chpt. to book. Mem. Scholarship Com., Bank of Am., San Francisco, 1983. Served to lt. comdr. USNR, 1969-71. Mem. Endocrine Soc., Am. Soc. Bone and Mineral Research, Am. Fedn. Clin. Research, Am. Fertility Soc., Am. Soc. Internal Medicine, Am. Fedn. Clin. Research. Democrat. Lodge: Rotary. Subspecialties: Critical care; Receptors. Current work: Cytoreceptor assay for 1, 25 dihydroxy Vitamin D.; parathyroid hormone and calcitonin radioimmunoassay; role of calcium regulating hormones in metabolic bone disease of renal failure. Home: 867 Stonehaven Dr Walnut Creek CA 94598 Office: Endocrine Metabolic Center 3100 Summit St Oakland CA 94623

REITZ, RONALD CHARLES, biochemist, educator; b. Dallas, Feb. 27, 1939; s. Percy A. and Hazel A. (Thomison) R.; m. Jeanne M. Geiger, Jan. 23, 1965; children: Erica Anne, Pieter Brett. B.S., Tex. A&M U, 1961; Ph.D., Tulane U., 1966. NIH postdoctoral fellow U. Mich., Ann Arbor, 1966-69; vis. scientist Unilever Ltd., Welwyn, Hertsfordshire, Eng., 1968; asst. prof. dept. biochemistry U. N.C., Chapel Hill, 1969-75; assoc. prof. U. Nev., Reno, 1979-80, prof., 1980—. Contbr. articles on biochemistry to profl. jours. Coach soccer YMCA, 1982. Mem. Am. Soc. Biol. Chemists, Am. Soc. Pharmacology and Exptl. Therapeutics, Research Soc. on Alcoholism, AAAS. Methodist. Lodge: Elks. Subspecialties: Biochemistry (medicine); Neuropharmacology. Current work: Biochemistry of alcohol tolerance and dependence. Home: 3237 Susilen Dr Reno NV 89509 Office: Dept Biochemistry U Nev Reno NV 89557

REKERS, ROBERT GEORGE, chemistry educator; b. Rochester, N.Y., Feb. 1, 1920; s. Gerret and Jessie Emma (French) R.; m. Shirley Kathryn Gilbert, Aug. 24, 1951; children: William E., Martha N. B.Sc., U. Rochester, 1942; Ph.D., U. Colo., 1951. Spectroscopist U.S. Naval Ordnance Test Sta., China Lake, 1951-55; faculty dept. chemistry Tex. Tech U., Lubbock, 1955—, asst. chmn., 1970-76, assoc. prof., 1969—. Fellow AAAS, Am. Inst. Chemists; mem. Am. Chem. Soc., Coblentz Soc. (charter). Subspecialty: Analytical chemistry. Home: 3028 67th St Lubbock TX 79413 Office: Dept Chemistry Tex Tech U Lubbock TX 79409

RELF, PAULA DIANE, educator, cons.; b. Dallas, Dec. 3, 1944; d. Robert Eugene and Irma Beryl (Christian) R.; m. Murray Wayne Helfey, Dec. 28, 1964; m. David Aaron Angle, May 17, 1980. B.S., Tex. Tech. U., 1967; M.S. (NSF fellow), U. Md., 1972, Ph.D., 1976. Dir. prodn. and sales Melwood Hort. Tng. Center, Upper Marlboro, Md., 1973-74; exec. sec. Nat. Council Therapy and Rehab. through Horticulture, 1974-75; asst. prof., extension specialist dept. horticulture Va. Poly. Inst. and State U., Blacksburg, 1976—, assoc. prof., 1982—. Contbr. articles on horticulture therapy to profl. jours. Recipient award Am. Soc. Agrl. Engrs., 1981. Mem. Nat. Council Therapy and Rehab. Through Horticulture (award 1979, cert. 1981), Garden Writers Assn. Am., Internat. Soc. Hort. Sci., Am. Soc. Hort. Sci., Phi Alpha Xi. Subspecialty: Plant growth. Current work: Use of horticulture for therapy/rehabilitation, horticultural therapy, rehabilitation through horticulture. Office: Dept Horticulture Va Poly Inst and State Univ Blacksburg VA 24061

RELMAN, ARNOLD SEYMOUR, physician, educator; b. N.Y.C., June 17, 1923; s. Simon and Rose (Mallach) R.; m. Harriet Morse Vitkin, June 26, 1953; children: David Arnold, John Peter, Margaret Rose. A.B., Cornell U., 1943; M.D., Columbia U., 1946; M.A. (hon.), U. Pa., Sc.D., Med. Coll. Wis., Sc. D., Union U., D.M.Sc., Brown U., D.L.H., SUNY. Diplomate: Am. Bd. Internal Medicine. House officer New Haven Hosp., Yale, 1946-49; NRC fellow Evans Meml., Mass. Meml. hosps., 1949-50; practice medicine, specializing in internal medicine, Boston, 1950-68, Phila., 1968-77; asst. prof., prof. medicine Boston U. Sch. Medicine, 1950-68; dir. Boston U. Med. Services, Boston City Hosp., 1967-68; prof. medicine, chmn. dept. medicine U. Pa.; chief med. services Hosp. of U. Pa., 1968-77; editor New Eng. Jour. Medicine, Boston, 1977—; sr. physician Brigham and Women's Hosp., Boston, 1977—; prof. medicine Harvard Med. Sch., 1977—; Cons. NIH, USPHS. Editor: Jour. Clin. Investigation, 1962-67, (with F.J. Ingelfinger and M. Finland) Controversy in Internal Medicine, Vol. 1, 1966, Vol. 2, 1974; Contbr. articles profl. jours. Recipient Columbia Alumni Gold medal, 1980. Fellow Am. Acad. Arts and Scis., A.C.P. (master); mem. Assn. Am. Physicians (council, pres. 1983-84), Am. Physiol. Soc., AMA, Mass. Med. Soc., Inst. Medicine of Nat. Acad. Scis. (council 1979-82), Am. Soc. Clin. Investigation (past pres.), Am. Fedn. Clin. Research (past pres.), Phi Beta Kappa, Alpha Omega Alpha. Subspecialty: Internal medicine. Office: New Eng Jour Medicine 10 Shattuck St Boston MA 02115

REMBERT, DAVID HOPKINS, JR., biologist, educator; b. Columbia, SC., Jan. 14, 1937; s. David Hopkins and Mary Aldrich (Wyman) R.; m. Margaret Rose Rainey, Apr. 23, 1960; children: Rainey, Augusta, Llewellyn, David. B.S., U. S.C., 1959, M.S. Biology, 1964; Ph.D., U. Ky., 1967. Asst. prof. biology U. S.C., Columbia, 1967-72, assoc. prof., 1972-81, prof., 1981—; asst. dean Coll. Sci. and Math., 1972-76, acting dean, 1975. Mem. Central Midlands Planning Council, 1973-77, Hist. Columbia Found., 1980—. Served to lt. USAF, 1961-62; with S.C. Air N.G., 1957-61. Belser fellow, 1959; Haggin fellow, 1966. Mem. Bot. Soc. Am., Assn. Southeastern Biologists, Internat. Soc. Plant Morphology, Soc. Bibliography of Natural History, Appalachian Bot. Club, So. Garden Hist. Soc., S.C. Acad. Sci., Sigma Xi. Episcopalian. Subspecialties: Morphology; Botanical history. Current work: Ovule development in legumes with special reference to phylogeny; floristics of Southeastern U.S.; botanical history. Office: U SC Dept Biology Columbia SC 29208

REMMEL, RONALD SYLVESTER, physiology educator; b. West Bend, Wis., July 18, 1943; s. Elroy Nicholas and Bernice Amanda (Degner) R.; m. Effie Lau, July 8, 1972; 1 dau., Charlotte. B.S. with honors, Calif. Inst. Tech., 1965; Ph.D., Princeton U., 1971; postgrad., U. Calif.-Berkeley, 1972-74, Johns Hopkins U., 1974-75. Asst. prof. physiology and biophysics U. Ark. for Med. Scis., Little Rock, 1975-81, assoc. prof., 1981—, mem. research council, 1975-78, med. neurosci. course exec. com., 1980—. Contbr. articles on physiology and biophysics to profl. jours.; editor: Med. Neurosci. Course Manual, 1980—. NSF fellow, 1965-69; Charles Grosvenor Osgood fellow, 1966-67; NIH fellow, 1972-75; grantee, 1975-82. Mem. Soc. for Neurosci., Am. Physiol. Soc., Assn. for Research in Vision and Ophthalmology, Ark. Acad. Sci., Caltech Alumni Assn., Princeton Alumni Assn., Tau Beta Pi. Subspecialties: Neurophysiology; Biomedical engineering. Current work: Neurons located in the brainstem of the cat which generate eye movements. Inventor pulse-width modulated high-fidelity amplifier, analog computer of average transients. Home: 2024 Vancouver Dr Little Rock AR 72204 Office: Dept Physiology and Biophysics U Ark for Med Sci 4301 W Markham St Little Rock AR 72205

REMO, JOHN LUCIEN, physicist; b. Bklyn., Dec. 13, 1941; s. John G. and Mary (DiVitis) R.; m. Claudia J. Kyser, Aug. 28, 1977; children: John Christopher, Allison Mary. B.S., Manhattan Coll., 1963; M.S. in Earth and Space Sci, SUNY-Stony Brook, 1971, Poly. Inst. N.Y., 1973, Ph.D., 1979. Research assoc. Inst. for Space Studies, NASA, N.Y.C., 1967-69; Amanuensis Copenhagen (Denmark) U. Obs., 1969-70; faculty Wash. U., Pullman, 1973-75, Hofstra U., Hempstead, N.Y., 1975—, adj. assoc. prof. physics and geology, 1979—; adj. prof. engring. SUNY-Stony Brook, 1983—; adj. prof., bd. advisers energy mgmt. N.Y. Inst. Tech., 1983; research assoc. Poly. Inst. N.Y. Farmingdale, 1978-81; pres., chief scientist ERG Cons., St. James, N.Y., 1981—. Contbr. articles in geophysics, laser optics and solar energy to profl. jours. Mem. Am. Phys. Soc., Am. Geophys. Union, Optical Soc. Am., IEEE, Meteoritical Soc., Met. Solar Energy Soc., Sigma Xi, Sigma Pi Sigma. Subspecialties: Lasers/resonator theory; Solar energy. Current work: Research in math. theory of laser resonators, solar energy systems, planetary geophysics and computer based expert systems. Home: Brackenwood Path Head of the Harbor NY 11780

REMPEL, WILLIAM EWERT, animal science educator; b. Lowe Farm, Man., Can., July 6, 1921; came to U.S., 1948, naturalized, 1955; s. Peter K. and Margaretha (Ewert) R.; m. Leola I. Seip, Dec. 23, 1948; children: R. Barrie, Bonnie Gail. B.S., U. Man., 1944, M.Sc., 1946; Ph.D., U. Minn., St. Paul, 1952. Lectr. U. Man., 1946-47; agr. rep. Man. Dept. Agr., Swan River, 1947-48; research asst. U. Minn., 1948-49, research fellow, 1949-50, instr., 1950-52, asst. prof., 1952-54, assoc. prof., 1954-64, prof. animal sci., 1964—; tchr. genetics Summer Inst., U. Wis., River Falls, 1962, 65; dir. Genetics Center, U. Minn., 1965-68; tchr. animal breeding Beijing Agr. U., 1981. Mem. Am. Soc. Animal Sci., Genetics Soc. Am., Am. Genetics Soc., NOW, Amnesty Internat., War Resisters League. Democrat. Subspecialties: animal breeding; Animal genetics. Current work: Swine and sheep breeding, teaching principles of animal breeding; research in swine breeding related to porcine stress system and efficiency of lean tissue growth. Home: 1424 Belmont Ln W Saint Paul MN 55113 Office: 125 Peters Hall U Minn Saint Paul MN 55108

RENARD, KENNETH GEORGE, hydraulic engineer; b. Sturgeon Bay, Wis., May 5, 1934; s. Harry H. and Margaret (Buechner) R., m. Virginia R. Heibel, Sept. 6, 1956; children: Kenlynn, Craig, Andrew. B.S., U. Wis., 1957, M.S., 1959; Ph.D. in Civil Engring, U. Ariz., 1972. Registered profl. engr., Ariz. With Agrl. Research Service, Dept. Agr., 1957—; resident engr. Walnut Gulch Expt. Watershed, Tombstone, Ariz., 1959-64; hydraulic engr. Southwest Watershed Research Ctr., Tucson, 1964-71; dir. S.W. Watershed Research Ctr., 1972—. Contbr. to publs. in field. Recipient Outstanding Performance award Dept. Agr., 1969. Mem. ASCE (past pres. Ariz. sect.), Am. Soc. Agrl. Engrs., Soil Conservation Soc. Am., Am. Geophys. Union, Sigma Xi. Subspecialties: Hydrology; Resource management. Current work: Modeling of natural resources, especially semiarid water resources, erosion and sedimentation.

RENARDY, MICHAEL, mathematician; b. Stugart, Germany, Apr. 9, 1955; s. Heinz and Eva-Maria (George) R.; m. Yurik Yamamuro, Apr. 9, 1981. Dipl. Math, U. Stuttgart, 1977, Dipl. Phys., 1978, Dr. rer. nat., 1980. Research assoc. U. Stuttgart, 1978-80; postdoctoral fellow Deutsche Forschungsgemeinshaft U. Wis.-Madison, 1980-81, asst. prof., 1982—. Fellow U. Minn., Mpls., 1981-82. Mem. Soc. Indsl. and Applied Math, Soc. Rheology. Roman Catholic. Current work: Nonlinear partial differential equations, viscoelastic liquids, bifurcation theory. Home: 4801 Sheboygan Ave Madison WI 53705 Office: Math Research Ctr U Wis Madison WI 53705

RENARDY, YURIKO, mathematician; b. Sapporo, Hokkaido, Japan, Jan. 15, 1955; came to U.S., 1980; d. Sadayuki and Akiko (Maeda) Yamamuro; m. Michael Renardy, Apr. 9, 1981. B.Sc., Australian Nat. U., 1976; Ph.D., U. Western Australia, 1980. Research assoc. Math. Research Ctr., U. Wis.-Madison, 1980-81, 82—; lectr. U. Minn., Mpls., 1981-82. Mem. Soc. Indsl. and Applied Math. Subspecialty: Applied mathematics. Current work: Fluid mechanics, computational methods. Office: Math Research Ctr 610 Walnut St Madison WI 53705

RENEKER, DARRELL HYSON, physicist; b. Birmingham, Iowa, Dec. 5, 1929; s. Glenn B. and Edith (Syfert) R.; m. Joan Eddleman, July 10, 1953; children: Douglas Alan, Elizabeth Ann. B.S., Iowa State U., 1951; M.S., U. Chgo., 1955, Ph.D., 1959. Mem. tech. staff Bell Telephone Labs., Murray Hill, N.J., 1951-53; NSF fellow U. Chgo., 1953-59; physicist E.I. duPont de Nemours Co., Wilmington, Del., 1959-69; sect. chief polymers div. Nat. Bur. Standards, Washington, 1969-75, dep. chief polymers div., 1975-80, dep. dir., 1980—; vis. prof. U. Mass.-Amherst, 1982. Contbr. articles in field to profl. publs. Recipient Dept. Commerce Silver medal, 1978. Fellow Am. Phys. Soc. (exec. com. div. high polymer physics 1981—); mem. Soc. Plastic Engrs., Electron Microscopy Am., ASTM. Subspecialties: Polymer physics; Condensed matter physics. Current work: Polymer morphology; structure, motion, energy and effects of defects in crystalline polymers. Office: National Bur Standards Materials B308 Washington DC 20234

RENNELS, MARSHALL LEIGH, neuroanatomist; b. Marshall, Mo., Sept. 2, 1939; s. Ivory Paul and Alfrieda (Schuetz) R.; m. Margaret Baker, Dec. 28, 1971. B.S., Eastern Ill. U., 1961; M.A., U. Tex.-Galveston, 1964, Ph.D., 1966. Asst. prof. anatomy and neurology U. Md. Med. Sch., Balt., 1966-71, assoc. prof., 1971-79, prof. anatomy, assoc. prof. neurology, 1979—. Contbr. articles to profl. jours. Mem. Am. Assn. Anatomists, Soc. Neuroscience, Internat. Soc. Cerebral Blood Flow and Metabolism, AAAS, Cajal Club, Sigma Xi. Subspecialties: Neurobiology; Morphology. Current work: Brain capillaries; central innervation of cerebral vasculature; cerebral extracellular space; tracer studies of nerve cells. Office: U Md Med Sch 655 W Baltimore St Baltimore MD 21201

RENNER, WENDEL DEAN, med. physicist; b. Indpls., Aug. 28, 1948; s. Donald Wayne and Thelma Lydia (Slaybaugh) R.; m. Constance Witter Beaman, May 15, 1976; children: Elizabeth Witter, Samuel Oak. B.S. in Physics, U. Cin., 1970, M.S. in Radiol. Sci, 1973, Purdue U., 1980. Cert. therapeutic radiol. physics Am. Bd. Radiology, 1977. Asst. prof. radiology W.Va. U. Med. Center, Morgantown, 1973-75; radiation physicist Community Hosp. of Indpls., 1975—. Contbr. writings to profl. publs. in field. Served with U.S. Army, 1971-73. Recipient hon. mention, Talbert Abrams award Am. Soc. Photogrammetry, 1977. Mem. Am. Assn. Physicists in Medicine (exhibit award 1977). Presbyterian. Subspecialties: Medical physics; Imaging technology. Current work: Radiation therapy, imaging, improvement in technology for delivering radiation therapy treatments.

RENNERT, JOSEPH, chemist, educator, cons., researcher; b. Mannheim, Germany, July 26, 1919; came to U.S., 1936, naturalized, 1944; s. Peisach and Sara (Rubner) R.; m. Helen P., Feb. 14, 1947; children: Karen Stephanie, Michael Paul. B.S., CCNY, 1948; M.S., Syracuse U., 1951, Ph.D., 1953. Research chemist Gen. Aniline & Films, Johnson City, N.Y., 1952-54, Charles Bruning Co., Teterboro, N.J., 1954-55; project engr. Balco Research Labs., Newark, 1956-62; sr. research scientist Inst. Math. Scis., N.Y. U., 1956-62; asst. prof. chemistry CCNY, 1962-67, assoc. prof., 1968-72, prof., 1972—; cons. in field. Contbr. articles, chpt. to profl. publs. Served with USN, 1943-46. Mem. Am. Chem. Soc., N.Y. Acad. Scis., Am. Soc. Photobiology, Inter-Am. Photochem. Soc., AAUP, Sigma Xi. Subspecialties: Physical chemistry; Photochemistry. Current work: Imaging processes, photochem. cyclo-additions and scissions. Patentee data storage with organic memory, methods and apparatus for rec. info., apparatus and procedures using internal polarization. Home: 525 Fordham Pl Paramus NJ 07652 Office: CCNY Convent Ave at 138th New York NY 10031

RENNIE, IAN DRUMMOND, nephrologist; b. Leeds, Yorkshire, Eng., Jan. 31, 1936; came to U.S., 1967; s. John King and Isabel Brownlee (Wiese) R.; m. Silvia Gabriella Nussio, July 3, 1958; children: Caroline, Nicholas. M.B., B.Chir., U. Cambridge, Eng., 1960, M.A., M.D., 1969. Assoc. prof. medicine Rush-Presbyn.-St. Luke's Med. Sch., Chgo., 1967-77; assoc. prof. medicine Harvard Med. Sch., 1977-81; prof. medicine Rush U., Chgo., 1982—; chmn. medicine West Suburban Hosp., Oak Park, Ill., 1981—; cons. Pan Am. Health Orgn., 1975—, Rand Corp., 1981—. Dep. editor: New Eng. Jour. Medicine, 1977-81; contbg. editor: Jour. AMA, 1983—; contbr. articles to profl. jours. Recipient Gold medal in medicine Guy's Hosp., London, 1958, Gold medal in ophthalmology, 1959. Fellow Royal Coll. Physicians, ACP; mem. AMA, Council Biology Editors, AAAS. Club: Alpine (London). Subspecialty: Physiology (medicine). Current work: High altitude physiology; societal and economic aspects of high technology treatment. Home: 7900 Greenfield St River Forest IL 60305 Office: West Suburban Hosp 518 N Austin Blvd Oak Park IL 60302

RENTZEPIS, PETER M., chemist; b. Kalamata, Greece, Dec. 11, 1934; s. Michael T. and Levki G. R.; m. Alma Elizabeth Keenan, Dec. 30, 1960; children—Michael John, John Peter. B.S., Dennison U.; M.S., Syracuse U.; Ph.D., Cambridge U., Eng. Mem. tech. staff Research Labs. Gen. Electric Co., Schenectady; mem. tech. staff Bell Labs., Murray Hill, N.J.; later head phys. and inorganic chemistry research, adj. vis. prof. Yale U., U. Tel-Aviv, Mass. Inst. Tech. Asso. editor: Jour. Lasers and Chemistry; mem. editorial bd.: Biophys. Jour; Contbr. articles to profl. jours. Recipient I. Langmuir prize in chem. physics, 1973, Scientist of Yr. award, 1977, A. Crosby Morrison award in natural scis., 1978, ISCO award Am. Chem. Soc., 1979. Fellow N.Y. Acad. Scis., Am. Phys. Soc.; mem. Nat. Acad. Scis. Subspecialty: Physical chemistry. Patentee in field. Home: 1682 Valley Rd Millington NJ 07946 Office: Bell Labs 600 Mountain Ave Murray Hill NJ 07974

REPENNING, CHARLES ALBERT, mammalian paleontologist, researcher; b. Oak Park, Ill., Aug. 4, 1922; s. Albert Ellsworth and Estelle Lorraine (Vallincourt) R.; m. Derryberry, Sept. 17, 1939 (div.); children: Jean, John, Patricia, William. B.S., U. N.Mex., 1949; M.A., U. Calif.-Berkeley, 1964. With U.S. Geol. Survey, Reston, Va., 1949—60, mammalian paleontologist, Menlo Park, Calif., 1960—; research assoc. Smithsonian Instn., Washington, 1970—. Contbr. over 80 sci. articles to profl. publs. Served with AUS, 1942-45. Decorated Bronze Star, Purple Heart; recipient Meritorious Service Honor award U.S. Dept. Interior, 1982, Superior Performance award U.S. Geol. Survey, 1981. Mem. Am. Soc. Mammalogists, Soc. Systematic Zoology, Paleontology Soc., Soc. Vertebrate Paleontologists, Am. Quaternary Assn. Republican. Subspecialties: Paleobiology; Chronobiology. Current work: Mammalian biochronology; evolution; biogeography; Neogene; biochronology of rodents, primarily cricetid. Office: US Geol Survey Mail Stop 15 345 Middlefield Rd Menlo Park CA 94025

REPINE, JOHN EDWARD, physician, scientist, educator; b. Rock Island, Ill., Dec. 26, 1944; s. Joseph Charles and Dorothy Gertrude (Diedrich) R.; children: Mike and Tom (twins), Alex. B.S., U. Wis.-Madison, 1967; M.D., U. Minn., 1971. Intern U. Minn., Mpls., 1971-72; resident and fellow, 1972-75; assoc. prof. medicine and pediatrics U. Colo., 1979—; asst. dir. Webb-Waring Lung Inst., 1979—. Contbr. articles to profl. jours. Established investigator Am. Heart Assn., 1977-82; recipient Borden prize; Basil O'Connor award. Mem. Am. Heart Assn., Am. Assn. Immunologists, Reticuloendothelial Soc., Am. Lung Assn. Subspecialties: Internal medicine; Neuroimmunology. Current work: Role of Phagocytes and O-2- radicals in lung defense and lung injury. Home: 3440 S Columbine Circle Englewood CO 80110 Office: 4200 E 9th Ave Denver CO 80262

RESCH, JOSEPH ANTHONY, neurologist; b. Milw., Apr. 29, 1914; s. Frank and Elizabeth (Zetsch) R.; m. Rose Catherine Ritz, May 25, 1939; children—Rose, Frank, Catherine. Student, Milw. State Tchrs. Coll., 1931-34; B.S., U. Wis., Madison, 1936, M.D., 1938. Intern St. Francisco Hosp., LaCrosse, Wis., 1938-39; gen. practice medicine, Holmen, Wis., 1939-40; med. fellow in neurology U. Minn., 1946-48, clin. instr. neurology, 1948-51, clin. asst. prof., 1951-55, clin. asso. prof., 1955-62, asso. prof., 1962-65, prof., 1965—, head dept. neurology, 1976-82, asst. v.p. health sci., 1970-79, prof. lab. medicine and pathology, 1979—; Practice medicine specializing in neurology,

Mpls., 1948-62. Contbr. articles and abstracts to profl. jours., chpts. in books. Served to lt. col. M.C. U.S. Army, 1940-46; col. Med. Res., 1946-53. Mem. Hennepin County Med. Soc., Minn. Med. Assn., AMA, Minn. Soc. Neurol. Scis., Central Assn. Electroencephalographers, Am. Acad. Neurology, Am. Neurol. Assn., Am. Assn. Neuropathologists, Am. EEG Soc., Am. Heart Assn., Am. Epilepsy Soc., Soc. Epidemiol. Soc., Assn. Am. Med. Colls. Subspecialties: Neurology; Neuroimmunology. Current work: Neuroimmunology, Alzheimer's disease. Home: 200 N Mississippi River Blvd Saint Paul MN 55104 Office: University of Minnesota Medical School 420 SE Delaware St Minneapolis MN 55455

RESCIGNO, THOMAS NICOLA, physicist; b. N.Y.C., s. Joseph and Leona R. (Lewellyn) R.; m. Claudia Ascione, Mar. 14, 1970. B.A., Columbia Coll., 1969; M.A., Harvard U., 1971, Ph.D. in Chem. Physics, 1973. Research fellow Calif. Inst. Tech., Pasadena, 1973-75; staff physicist Lawrence Livermore Lab., Livermore, Calif., 1975-79, group leader, 1979-81, sect. head chemistry dept., 1982—. Editor: Electron-Molecule and Photon-Molecule Collisions, 1979; Contbr. articles to profl. jours. Recipient medal Am. Inst. Chemists, 1969; teaching fellow award Harvard U., 1971; NSF predoctoral fellow, 1969-72. Mem. Am. Phys. Soc., Phi Beta Kappa, Phi Lambda Upsilon. Club: Vaqueros del Mar (Livermore, Calif.). Subspecialty: Atomic and molecular physics. Current work: Low energy theoretical physics and chemistry. Office: Lawrence Livermore Lab L-487 Livermore CA 94550

RESHOTKO, ELI, aerospace engineer, educator; b. N.Y.C., Nov. 18, 1930; s. Max and Sarah (Kalisky) R.; m. Adina Venit, June 7, 1953; children: Deborah, Naomi, Miriam Ruth. B.S., Cooper Union, 1950; M.S., Cornell U., 1951; Ph.D., Calif. Inst. Tech., 1960. Aero. research engr. NASA-Lewis Flight Propulsion Lab., Cleve., 1951-56, head fluid mechanics sect., 1956-57; head high temperature plasma sect. NASA-Lewis Research Center, 1960-61, chief plasma physics br., 1961-64; asso. prof. engring. Case Western Res. U., Cleve., 1964-66, prof. engring., 1966—, chmn. dept. fluid thermal and aerospace scis., 1970-76, chmn. dept. mech. and aerospace engring., 1976-79; Susman vis. prof. dept. aero. engring. Technion-Israel Inst. Tech., Haifa, Israel, 1969-70; cons. United Technologies Research Ctr., Gould Corp., United Research Corp., Scott-Fetzer Co., Dynamics Tech. Inc., Arvin/Calspan Inc., Rockwell Internat.; Mem. adv. com. fluid dynamics NASA, 1961-64; mem. aeros. adv. com. NASA, 1980—; chmn. U.S. Boundary Layer Transition Study Group, NASA/USAF, 1970—; U.S. mem. fluid dynamics panel AGARDINATO, 1981—; chmn. steering com. Symposium on Engring. Aspects Magneto-hydro-dynamics, 1966. Contbr. articles to tech. jours. Chmn. bd. govs. Cleve. Coll. Jewish Studies. Guggenheim fellow Calif. Inst. Tech., 1957-59. Fellow Am. Phys. Soc., AAAS, Am. Acad. Mechanics, AIAA (Fluid and Plasma Dynamics award) (1980), ASME; mem. AAUP, Sigma Xi, Tau Beta Pi, Pi Tau Sigma. Subspecialties: Fluid mechanics; Aeronautical engineering. Current work: Fluid mechanics; propulsion, power generation; studies in boundary layer stability and transition as related to drag reduction. Office: Case Western Res Univ University Circle Cleveland OH 44106

RESLER, STEVEN CHARLES, environmental analyst, field researcher; b. Ft. Worth, Jan. 20, 1953; s. Louis C. and Augusta R. (Salers) R. Cert., SUNY-Cornell Coop. Extension Scope Marine Sci. Inst., 1976; cert., Adelphi U. Inst. Suburban Studies, 1977, Va. Inst. Marine Sci., 1977; postgrad., SUNY-Stonybrook, Empire State Coll., 1979. Environ. analyst Town of Smithtown (N.Y.) Conservation Commn., 1973-81; founder, pres. Sub-Sea Diving Systems, Inc., Smithtown, 1973—; co-investigator Octopus dofleini research team Ministry Fisheries and Oceans, Vancouver Island, B.C., Can., 1981-81; research coordinator octopus joubini research team Brandeis U., St. Joseph Bay, Fla., 1979-80; diving officer in charge SUNY Lab. Undersea Lab., Smithtown Bay, L.I., N.Y., 1977—; exec. dir. Marine Research Found., Inc., St. James, N.Y., 1981—; mem. estuarine sanctuaries steering com. U.S. Dept. Commerce, NOAA, Washington, Hauppauge, N.Y., 1979-81; mem. N.Y. State Shellfish Adv. Com., Stony Brook, 1978—; mem. regional marine resources council Dredging Adv. Commn., Hauppauge, N.Y., 1975-77; adv. ocean engring. program SUNY-Stony Brook, 1976—. Author: Nissequogwe River - A Screening Study, 1978. Mem. Suffolk County Solar Energy Commn., Hauppauge, N.Y., 1978; mem. L.I. Assn. Town Environ. Ofcls., 1979. Recipient 1st place awards N.Y. State Assn. Conservation Commn., 1978, 79, 80; Ford Found. grantee, 1973. Mem. Marine Tech. Soc., Internat. Oceanographic Found., Earthwatch, N.Y. Acad. Scis., N.Y. State Divers Assn. Club: Grumman Scuba (Bethpage, N.Y.). Subspecialties: Resource management; Marine environmental sciences. Current work: Marine field research and scientific diving procedures, current interests in sonic tracking of marine vertebrates and invertebrates (octopuses) and behavior of same. Home: 37 Baylor Dr Smithtown NY 11787 Office: Marine Research Found 115 Long Beach Rd St James NY 11780

RESNEKOV, LEON, medical educator; b. Cape Town, South Africa, Mar. 20, 1978; s. Charles and Alice (Mitchell) R.; m. Carmella, July 28, 1955; children: Orna, Charles Dean. M.B.Ch. B., U. Cape Town, 1952, M.D., 1965. Intern Evoote Schuur Hosp., U. Cape Town, 1952-53; resident in medicine Kings Coll. Hosp., London, 1954-57, regsitrar in cardiology, 1957-61; sr. registrar Nat. Heart Hosp., Inst. Cardiology, London, 1961-67; assoc. prof. medicine (cardiology) U. Chgo. Med. Sch., 1967-71, prof., 1971-81, joint dir. sect. cardiology, 1971-81, Frederick H. Rawson prof., 1981—; Viscount St Cyres lectr. Inst. Cardiology, London, 1979; Welker vis. prof. medicine U. Kans., 1981; Harry and Ann Borun vis. prof. cardiology UCLA, 1981; Elson Meml. lectr., St. Louis, 1982; John J. Sampson vis. prof. cardiology U. Calif-San Francisco, 1983. Contbr. numerous articles to sci./med. jours. Recipient Heart of Yr. award Chgo. Heart Assn., 1981. Fellow Royal Soc. Medicine; mem. Brit. Cardiac Soc., Am. Coll. Cardiology, Chgo. Heart Assn. (pres. 1978-79), Am. Coll. Chest Physicians (pres. 1979-80), Chgo. Soc. Internal Medicine, Chgo. Med. Soc., Am. Fedn. Clin. Research, Central Soc. Clin. Research, Soc. Med. History, Assn. Univ. Cardiologists, Alpha Omega Alpha. Subspecialty: Cardiology. Current work: Cardiovascular physiology; clinical cardiology; radioisotopes in cardiac diagnosis. Office: U Chgo Med Center 950 E 59th St Chicago IL 60637

RESNICK, MARTIN I., physician; b. Bklyn., Jan. 12, 1943; s. Daniel and Bertha (Becker) R.; m. Vicki Ann Klein, July 4, 1965; children: Andrew Howard, Jeffrey Scott. B.A., Alfred (N.Y.) U., 1964; M.D., Bowman Gray Sch. Medicine, Wake Forest U., 1969; M.S., Northwestern U., 1973. Diplomate: Am. Bd. Urology. Mem. faculty Bowman Gray Sch. Medicine, Wake Forest (N.C.) U., 1975-81; prof., chmn. div. urology Case Western Res. U. Sch. Medicine, Cleve., 1981—. Mem. ACS, Soc. Univ. Surgeons, Am. Urol. Soc., Alpha Omega Alpha. Subspecialty: Urology. Current work: Urinary stone disease. Home: 36 Lyman Circle Shaker Heights OH 44122 Office: 2065 Adelbert Rd Cleveland OH 44106

RESNICK, OSCAR, neuroscientist; b. Bayonne, N.J., Apr. 27, 1924; s. Samuel and Rebecca (Rubinstein) R.; m. Janice Zelda Ravitz, July 13, 1949; children: Sandra, Scott. A.B., Clark U., 1944; M.A., Harvard U., 1945; Ph.D., Boston U., 1955. Research fellow U. Iowa Med. Sch., 1945-46; instr. St. Petersburg Jr. Coll., 1946-47; research fellow U. Kans., 1947-49; instr. U. Minn., 1949-50; editorial asst. Biol. Abstracts, U. Pa., 1950-51; scientist Nat. Drug Co., Phila., 1951-53, Worcester Found. Exptl. Biology, Shrewsbury, Mass., 1953—; now sr. scientist; lectr. Boston U., 1961—, Clark U., 1965—; dir. research Worcester County Rehab. and Detention Ctr., West Boylston, Mass., 1965-76; cons. Medfield (Mass.) State Hosp., 1958-68, Norwich (Conn.) State Hosp., 1964-68; mem. mental retardation research com. NIH, 1975-78. Contbr. articles to profl. jours. NIH grantee, 1957—. Fellow Am. Coll. Neuopsychopharmacology; mem. AAAS, Soc. Biol. Psychiatry, Am. Psychopatho. Soc., N.Y. Acad. Sci., Soc. Neurosci., Sigma Xi. Subspecialties: Neurochemistry; Nutrition (medicine). Current work: Prenatal nutrition in role of developing central nervous system; transgenerational effects. Home: 5 Meadow Ln Worcester MA 01602 Office: 222 Maple Ave Shrewsbury MA 01545

RESTAINO, ALFRED JOSEPH, research director; b. Bklyn., Feb. 18, 1931; s. Clement and Celía (Orlando) R.; m. Raffaela B. Sessa, June 27, 1954; children: Stephen, Alfred, Peter, Mario, Lisa. B.S. in Chemistry magna cum laude, St. Francisco Coll., 1952; M.S. in Phys. and Polymer Chemistry, Poly. Inst. Bklyn., 1954, Ph.D., 1956. Unit supr. Martin Aircraft Corp., 1956-58; supr. radiation research ICI Americas (formerly Atlas Chem. Co.), Wilmington, Del., 1958-63, mgr. radiation research, 1963-68, mgr. radiation and plymer research, 1968-70, asst. dur. dept. chem. research, 1970-75, dir. dept. corp. research, 1975—; mem. faculty U. Del.; mem. tech. exchange with USSR, 1972; sci. advisor to gov. State of Del.; mem. adv. council U.S. AEC. Author: Encyclopedia of Materials Science and Engineering; contbr. articles to profl. jours. Mem. Soc. Chem. Industry, Am. Chem. Soc., N.Y. Acad. Scis., Assn. Research Dirs., Indsl. Research Inst., Sigma Xi. Subspecialties: Polymer chemistry; Materials (engineering). Current work: Specialty chemicals; high performance plymers; research administration. Patentee in fields of organic and polymer chemistry. Office: ICI Americas Wilmington DE 19897

RESWICK, JAMES BIGELOW, rehab. engr., educator; b. Ellwood City, Pa., Apr. 16, 1922; s. Maurice and Katherine (Parker) R.; m. Irmtraud Orthlies Hoelzerkopf, Dec. 27, 1973; children—(by previous marriage) James Bigelow, David Parker (dec.), Pamela Patchin. S.B. in Mech. Engring, Mass. Inst. Tech., 1943, S.M., 1948, Sc.D., 1952; D.Eng. (hon.), Rose Poly. Inst., 1968. Asst. prof., then asso. prof., head machine design and graphics div. Mass. Inst. Tech., 1948-59; Leonard Case prof. engring., dir. Engring. Design Center, Case Western Res. U., 1959-70; dir. Rehab. Engring. Center, Rancho Los Amigos Hosp.; prof. biomed. engring. and orthopaedics U. So. Calif., also dir. of research dept. orthopaedics, 1970-80; now asso. dir. tech. Nat. Inst. Handicapped Research, U.S. Dept. Edn.; engring. cons. on automatic control, product devel., automation and bio-med. engring. Mem. com. prosthetics research and devel. Nat. Acad. Scis., 1962—; chmn. design and devel. com.; mem. bd. rev. Army Research and Devel. Office, 1965—; mem. applied physiology and biomed. engring. study sect. NIH, 1972—. Author: (with C.K. Taft) Introduction to Dynamic Systems, 1967; also articles.; Editor: (with F.T. Hambrecht) Functional Electrical Stimulation, 1977; series on engring. design, 1963—. Chmn. Mayor's Commn. for Urban Transp., Cleve., 1969. Served to lt. (j.g.) USNR, 1943-46; PTO. NSF sr. postdoctoral fellow Imperial Coll., London, Eng., 1957; Recipient Product Engring. Master Designer award, 1969; Isabelle and Leonard H. Goldenson award United Cerebral Palsy Assn., 1973. Fellow IEEE; mem. ASME (honor award for best paper 1956, sr. mem.), Am. Soc. Engring. Edn., Instrument Soc. Am., Biomed. Engring. Soc. (sr. mem., pres. 1973, dir.), Am. Acad. Orthopedic Surgeons (asso.), Inst. Medicine of Nat. Acad. Scis., Nat. Acad. Engring., Internat. Soc. Orthotics and Prosthetics, Orthopaedics Research Soc., Rehab. Engring. Soc. N.Am. (founding pres.), Sigma XI. Subspecialties: Biomedical engineering; Mechanical engineering. Current work: Technology for handicapped persons; functional electrical stimulations; design, development, evaluation production and distribution of technology for the handicapped. Patentee in field. Home: 1003 Dead Run Dr McLean VA 22101 Office: NIHR/OSERS US Dept Edn Switzer Bldg 3511 Washington DC 20202

RETTENMEYER, CARL WILLIAM, biology educator, museum director; b. Meriden, Conn., Feb. 10, 1931; s. Frederick William and Gertrude Emma (Mielke) R.; m. Marian Elizabeth Wolf, June 26, 1954; children: Susan, Ronald. B.A., Swarthmore Coll., 1953; Ph.D., U. Kans., 1962. Asst. prof. Kans. State U., Manhattan, 1960-65, assoc. prof., 1965-71; prof. biology U. Conn., Storrs, 1971—, head systematic and evolutionary biology, 1980—, exec. officer bilo. scis. group, 1983—, dir. Mus. Natural History, 1982—; bd. dirs. Orgn. Tropical Studies, Inc., Durham, N.C., 1971—. NSF grantee, 1952-79. Fellow AAAS; mem. Am. Inst. Biol. Scis., Ecol. Soc. Am., Entomol. Soc. Am. Democrat. Subspecialties: Behavioral ecology; Ethology. Current work: Behavior and ecology of ants and social wasps, especially tropical species of army ants and polybiine social wasps, alarm-defense behavior of wasps. Home: 21 Holly Dr Storrs CT 06268 Office: U Conn Biol Scis U-43 Storrs CT 06268

RETTNER, CHARLES THOMAS, research chemist; b. London, Dec. 8, 1953; U.S., 1979; s. Charles William and Bella (Anastasio) R.; m. Elizabeth Jane Wakelin, June 26, 1976; 1 dau., Emily Jane. B.S., Birmingham (Eng.) U., 1975, Ph.D., 1978. Postdoctoral asst. MIT, Cambridge, 1979, Stanford U., 1980, research assoc. dept. chemistry, 1981—. Contbr. articles on chemistry to profl. jours. Hills Meml. fellow, 1977-79. Mem. AAAS, Planetary Soc. Subspecialties: Laser photochemistry; Laser-induced chemistry. Current work: Studies of the dynamics of elementary chemical reactions using lasers both to prepare reagents and to probe reaction products. Office: Dept Chemistry Stanford CA 94305

RETZ, KONRAD CHARLES, neuropharmacologist; b. Oelwein, Iowa, Feb. 19, 1952; s. Donald A. and Erana E. (Anderson) R. B.A. cum laude in Chemistry, Augustana Coll., 1974; Ph.D. in Pharmacology, U. Iowa, 1979. Research postdoctoral fellow dept. pharmacology and exptl. therapeutics Sch. Medicine, Johns Hopkins U., Balt., 1979-81, postdoctoral fellow dept. neurosci., 1981-82; research scientist II L. I. Research Inst., SUNY-Stony Brook, 1982—. Contbr. articles to profl. jours. Recipient various fellowships. Mem. AAAS, Am. Chem. Soc., Soc. for Neurosci., N.Y. Acad. Sci., Alpha Chi Sigma. Republican. Lutheran. Subspecialties: Neuropharmacology; Neurochemistry. Current work: Mechanisms of analgesia; mode of action of excitatory amino acid neurotransmitters; regulation of energy metabolism in the central nervous system; calcium involvement in neurotransmission. Office: LI Research Inst SUNY Health Sci Ctr 10T Stony Brook NY 11794

REUTHER, JAMES JOSEPH, fuel science educator; b. Passaic, N.J., Jan. 29, 1950; s. Frederick William and Virginia (McBride) R.; m. Theresa Mary, Aug. 27, 1972; children: Adam James, Laura Ann. B.A., SUNY-Oneonta, 1973; M.A., SUNY-Binghamton, 1976; Ph.D., Pa. State U., 1979. Asso. prof. fuel sci. Pa. State U., University Park; also dir. Fuels and Combustion Lab., 1978—. Contbr. articles to profl. jours. Mem. Am. Chem. Soc., Am. Phys. Soc., Sigma Xi, Phi Kappa Phi, Phi Lamda Upsilon. Subspecialty: Combustion processes. Current work: Teaching combustion of coal and coal derived alternative fuels.

REVEAL, JAMES LAURITZ, botany educator, researcher; b. Reno, Mar. 29, 1941; s. Jack L. and Arlene H. R.; m. Caroline G. Powell, Jan. 18, 1961; children: Mark L., Kimberly S., Darin S.; m. C. Rose Broome, Jan. 3, 1978. B.S., Utah State U., 1963; M.S., 1965; Ph.D., Brigham Young U., 1969. Asst. prof. botany U. Md., College Park, 1969-74, assoc. prof., 1974-81, prof., 1981—; research assoc. Nat. Mus. Natural History, Smithsonian Instn., Washington, 1970—. Contbr. articles on botany to profl. jours. Mem. AAAS, Am. Inst. Biol. Scis., Am. Soc. Plant Taxonomists, Bot. Soc. Am., Bot. Soc. Washington (pres. 1982), Calif. Bot. Soc., Internat. Assn. Plant Taxonomy, New Eng. Bot. Club, So. Appalachian Bot. Club, Internat. Congress Systematic and Evolutionary Biology (sec.-gen.), Sigma Xi. Subspecialties: Taxonomy; Botanical history. Current work: Systematics of Polygonaceae, subfamily Eriogonoideae, history of botanical explorations in North America, phytogeography of North American plants. Office: Dept Botany U Md College Park MD 20742

REVESZ, GEORGE, radiologic physicist, educator, consultant; b. Budapest, Hungary, July 29, 1923; s. Nicholas and Elizabeth (Wallerstein) R.; m. Gabrielle Sophia Stern, Dec. 23, 1948; children: Julie Ann, Barbara Eva. M.S., Swiss Fed. Inst. Tech., Zurich, 1948; Ph.D., U. Pa., 1964. Staff engr. Gen. Electric Co. of Eng., 1949-54; sr. engr., tech. dir. Robertshaw Controls Co., 1954-61; research sect. mgr., mgr. instrumentation Philco-Ford Co., 1961-68; assoc. prof. Sch. Medicine Temple U., Phila., 1966-76, prof. radiology, 1976—; cons. v.p. Info/Consult; dir. Accu-Sort Systems Co. Contbr. articles to profl. publs. Recipient Bowen award Brit. Inst. Physics, 1954. Mem. Assn. Univ. Radiologists (Stauffer award 1982), Am. Assn. Physicists in Medicine, Optical Soc. Am., Am. Thermographic Soc., Sigma Xi. Subspecialties: Imaging technology; Optical image processing. Current work: Optical and computer processing of radiologic images; medical decision analysis.

REVUSKY, SAM H., psychology educator; b. Tel Aviv, Israel, Aug. 2, 1933; emigrated to U.S., 1935; emigrated to Can., 1971; s. Abraham and Hannah (Starkstein) R.; m. Bow Tong Lett, Dec. 24, 1963; 1 son, Jonathan. A.B., Columbia U., 1955; Ph.D., Ind. U., 1961. Research psychologist U.S. VA Hosp., Northampton, Mass., 1961-64, U.S. Army Med. Research Lab., Fort Knox, Ky., 1964-68; assoc. prof. No. Ill. U., DeKalb, 1968-71; research prof. U. Sussex Falmer, Eng., 1976-77; prof. Meml. U., St. John's, Nfld., Can., 1971—. Contbr. chpts. to books. Grantee NIH, 1968-71, Can. Cancer Inst., 1982—, Can. NRC, 1971—, Drug Council Can., 1974-76. Fellow Am. Psychol. Assn., Can. Psychol. Assn. Subspecialties: Learning; Psychobiology. Current work: Antisickness responses to the management of cancer patients; lithium aversion therapy for alcoholism. Office: Dept Psychology Meml U St Johns NF Canada A1C 5S7 Home: 102 Newtown Rd St John's NF Canada A1B 3A9

REYES, PHILIP, biochemist, educator; b. Tulare, Calif., Sept. 5, 1936; s. Genaro R. and Carmen (Chavez) R.; m. Valerie Jean Grant, Jan. 27, 1962; children: Jeffrey, David, Timothy. B.S., U. Calif., Davis, 1958, M.S., 1959, Ph.D., 1963. Postdoctoral fellow McArdle Lab., U. Wis., Madison, 1963-65, Scripps Clinic and Research Found., La Jolla, Calif., 1965-67; research assoc. Children's Cancer Research Found., Boston, 1967-70; asst. prof. biochemistry U. N.Mex. Sch. Medicine, 1970-75, assoc. prof. biochemistry, 1975-83, prof. biochemistry, 1983—. Leukemia Soc. Am. scholar, 1970-75. Mem. 1Am. Soc. Biol. Chemists, Am. Assn. Cancer Research, Am. Soc. Pharmacology and Exptl. Therapeutics. Subspecialties: Biochemistry (medicine); Cancer research (medicine). Current work: Purine and pyrimidine metabolism; biochemical mode of action of antitumor drugs; biochemistry of malarial parasites. Office: Dept Biochemistry U NMex Sch Medicine Albuquerque NM 87131

REYNAFARJE, BALTAZAR DAVILA, physiology and biochemistry educator, researcher; b. Chachapoyas, Amazonas, Peru, Sept. 21, 1925; s. Baltazar and Rosa Victoria (Davila) R.; m. Victoria Salome, Mar. 11, 1956; children: Jaime Baltazar, Patricia Orfelia, Lourdes Victoria, Alberto, Mariela Asunta. B.S. in Medicine, Universidad San Marcos, Peru, 1953, M.D., 1953, Ph.D., 1971. Instr. Faculty of Medicine, U. San Marcos, 1953-55, head dept. enzymology, 1957-65, assoc. prof., 1965-71, prof., 1971-75; research scientist Johns Hopkins U., Balt., 1975-79, research assoc., 1979—. Mem. AAAS, Am. Soc. Biol. Chemists, Peruvian Chem. Soc. (pres. biochem. br. 1974-75), Sociedad Peruana de Ciencias Fisologicas (founding), Sociedad Latino Americana de Ciencias Fisologicas, Sociedad Peruana de Bioquimica, Sociedad Peruana de Patologia, Sigma Xi. Catholic. Club: Regatas (Lima). Subspecialties: Cell biology; Molecular biology. Current work: Mechanisms of energy transformation in mitochondria and its efficiency in organisms exposed to hypoxia and/or high altitudes. Home: 9 Haymarket Ct Baltimore MD 21236 Office: Johns Hopkins U 725 N Wolfe St Baltimore MD 21205

REYNARD, KENNARD ANTHONY, materials scientist; b. Phila., Jan. 13, 1939; s. Albert Hinkle and Rita Eileen (Brennan) R.; m. (married); 1 son, Daniel Anthony. B.S. in Chemistry, St. Louis U., 1960, M.S., 1964, Ph.D., 1967. Polymers group leader Horizons, Inc., Cleve., 1969-73, head chemistry dept., 1973-74, mgr. contract research and devel., 1974-76; tech. mgr. SWS Silicones Corp., Adrian, Mich., 1976-79, tech. dir., 1979—. Publs. in field. Served to capt. USAF, 1966-69. Recipient Indsl. Research Mag. award, 1971. Mem. Am. Chem. Soc., Am. Wine Soc. Roman Catholic. Subspecialties: Polymer chemistry; Ceramics. Current work: Silicones, phosphazenes, elastomers, coatings, emulsions, room temperature vulcanizing materials, extreme service materials, passivation of electronics. Patents. Office: SWS Silicones Corp Sutton Rd Adrian MI 49221

REYNOLDS, ARDEN FAINE, JR., neurosurgeon, neurophysiologist, neurosurgery educator; b. Woodbury, N.J., Mar. 20, 1944; s. Arden F. and Charlotte (Janeka) R.; m. Mary J. Hicks, June 8, 1965; children: Joshua, Seth. A.A., San Bernardino Valley Coll., 1963; B.A., Loma Linda U., 1965, M.D., 1969. Intern in surgery U. Minn., 1969-70; clin. assoc. Nat. Cancer Inst., Balt., 1970-72; resident in neurosurgery U. Wash., Seattle, 1972-76; asst. prof. surgery U. N.Mex., 1976-77, U. Ariz., Tucson, 1977-82; neurosurgeon, neurophysiologist dept. neurosurgery U. Okla., Oklahoma City, 1982—, assoc. prof. surgery, 1983—. Served with USPHS, 1970-72. Recipient Resident Research award Western Neurosurg. Soc., 1976. Mem. Am. Assn. Neurol. Surgeons, Congress Neurol. Surgeons, Neurosci. Soc., Rocky Mountain Neurosurg. Soc. (program dir. 1983). Democrat. Subspecialties: Neurosurgery; Neurophysiology. Current work: Surgical therapy of intractable epilepsy; spinal cord stimulation for pain and peripheral vascular disease. Home: 824 New Bond Crescent Edmond OK 73034 Office: Dept Neurosurgery U Okla PO Box 26901 Oklahoma City OK 73190

REYNOLDS, DAVID BURKMAN, engineering educator; b. Arlington, Va., July 5, 1949; s. Robert Theodore and Rosemary (Burkman) R.; m. Joyce Ann Barkalow, May 6, 1978; children: Mark David, Allison Jean. B.S.M.E., U. Va., 1971, Mech. Engr., 1972, Ph.D., 1978. Research asst. U. Va., Charlottesville, 1975-77; asst. prof. Wright State U., Dayton, Ohio, 1980-81, asst. prof. dept. engring., 1981—; research fellow Thoracic Diseases Research Unit, Mayo Clinic, Rochester, Minn., 1978-80. Mem. Biomed. Engring. Soc., Am. Physiol. Soc., Sigma Xi, Tau Beta Pi. Roman Catholic. Club: Lake Monticello Golf (Palmyra, Va.). Subspecialties: Biomedical engineering; Biomedical engineering. Current work: Biomechanics, pulmonary/exercise physiology, mathematical modeling of physiological systems, computer simulation of physiological systems. Home: 225 E Dorothy

Ln Kettering OH 45419 Office: Wright State U 371 Fawcett Hall Dayton OH 45435

REYNOLDS, G(ARTH) FREDRIC, chemist, educator, researcher; b. Springfield, Ohio, Sept. 6, 1929; s. William John and Stella Marie (Grauer) R.; m. Gail Anna Youngs, Nov. 5, 1960; children: Jenny Lynn, Mark Fredric. B.A., Wittenberg U., 1953; Ph.D. in Phys. Chemistry, U. Cin., 1959. Fulbright research fellow U. Paris, 1957; mem. tech. staff Bell Telephone Labs., Murray Hill, N.J., 1959-61; asst. prof. chemistry Colo. Sch. Mines, 1961-65; assoc. prof. Marshall U., 1965-68; prof. Mich. Technol. U., 1968—. Contbr. chpts., articles on phys. chemistry, phys.-organic chemistry to profl. pubis. Served in USAF, 1947-50. Recipient Research Participation for Coll. Tchrs. award NSF, 1964; USAF faculty research fellow, USAF Academy, Colo., 1982-84. Mem. Am. Chem. Soc., Sigma Xi, Phi Lambda Upsilon. Subspecialties: Physical chemistry; Nuclear magnetic resonance (chemistry). Current work: Nuclear magnetic resonance, structure and bonding of organic molecules, thermodynamics, kinetics and mechanism of reaction. Office: Dept Chemistry and Chem Engrin Mich Technol U Houghton MI 49931

REYNOLDS, GEORGE OWEN, physicist, educator; b. Haverhill, Mass., July 5, 1937; s. William I. and Marjorie E. (Walker) R.; m. Anne Learnard, June 17, 1961; children: William, Gordon, Andrew. B.S., U. N.H., 1959, M.S. in Physics, 1961. Cons. A.S. Thomas Inc., Westwood, Mass., 1961-63; dir. applied scis. Tech. Ops. Inc., Burlington, Mass., 1963-75; dir. optical scis. Aerodyne Research Inc., Bedford, Mass., 1975-76; sr. staff scientist Arthur D. Little Inc., Cambridge, Mass., 1976-81; sr. staff engr. Honeywell Electro Optics Ops., Lexington, Mass., 1981—; lectr. Northeastern U., Boston, 1966—. Co-author: Theory and Applications of Holography, 1967; contbr. to: Handbook of Optical Holography, 1979; contbr. articles to profl. jours. Fellow Optical Soc. Am., Photo-Optical Instrumentation Engrs. (dir. 1982-83, sec. 1984—); mem. IEEE, Am. Phys. Soc., New Eng. Optical Soc. (exec. council 1975—, rep. to Mass. Engring. Council 1979-81, dir. 1981—), Sigma Xi. Subspecialties: Holography; Optical image processing. Current work: Holography, image processing, phase conjugation, atmospheric effects on light propagation, imaging systems, Fourier optics. Patentee in field. Home: 21 Dwhinda Rd Waban MA 02168 Office: 2 Forbes Rd Lexington MA 02173

REYNOLDS, JOHN HAMILTON, physicist, educator; b. Cambridge, Mass., Apr. 3, 1923; s. Horace Mason and Catharine (Coffeen) R.; m. Ann Burchard Arnold, July 19, 1975; children from previous marriages: Amy, Horace Marshall, Brian Marshall, Karen Leigh, Petra Catharine. A.B., Harvard U., 1943; M.S., U. Chgo., 1948, Ph.D., 1950. Research asst. Electroacoustic Lab., Harvard, 1941-43; asso. physicist Argonne Nat. Lab., 1950; physicist U. Calif. at Berkeley, 1950—, prof. physics, 1961—, chmn. dept. physics, 1984—, faculty research lectr., 1974. Contbr. articles to profl. jours. Served to lt. USNR, 1943-46. Recipient Wetherill medal Franklin Inst., 1965, Golden Plate award Am. Acad. Achievement, 1968, Exceptional Sci. Achievement award NASA, 1973, Guggenheim fellow U. Bristol, Eng., 1956-57; NSF fellow U. São Paulo, Brazil, 1963-64; Fulbright-Hays research grantee U. Coimbra, Portugal, 1971-72; U.S.-Australia Coop. Sci. Program awardee U. Western Australia, 1978-79. Mem. Nat. Acad. Scis. (J. Lawrence Smith medal 1967), Am. Phys. Soc., Am. Geophys. Union, Geochem. Soc., Meteoritical Soc. (Leonard medal 1973), AAAS, AAUP, Phi Beta Kappa. Democrat. Clubs: Faculty (Berkeley); Cal Sailing. Subspecialties: Planetary science; Geochemistry. Current work: Mass spectrometry of noble gases to read a record these trace gases provide in both terrestrial and extra-terrestrial samples. Office: Dept Physics Univ of California Berkeley CA 94720

REYNOLDS, JOHN SPENCER, architect, educator, researcher; b. Muncie, Ind., Jan. 31, 1938; s. Eugene S. and Lura Mae (Schield) R.; m.; children: Nathan, Vaughan, Hannah. B.Arch., U. Ill., 1962; M.Arch., M.I.T., 1967. Registered architect, Mass., Oreg. Structural designer Scholer & Fuller Architects, Tucson, 1962-63; archl. designer Shepley Bulfinch Richardson & Abbott, Boston, 1964-67; prof. architecture U. Oreg., Eugene, 1967—; cons. Author: (with McGuinness and Stein) Mechanical and Electrical Equipment for Buildings, 6th edit, 1980, (with Brown and Ubbelohde) Inside Out, 1982. Bd. dirs. The Solar Lobby, 1978-81; mem. Eugene Water and Electric Bd., 1973-76. Recipient Ersted award for disting. teaching U. Oreg., 1976. Mem. Internat. Solar Energy Assn., Oreg. Solar Energy Assn. Democrat. Subspecialty: Solar energy. Current work: Passively cooled and solar heated buildings, utilizing daylighting.

REYNOLDS, LARRY OWEN, electrical enginee, biomedical engineer; b. Norfolk, VA., Dec. 11, 1940; s. Herman Jewell and Kathryn (Key) R.; m. Laurel Lee Cutts, June 6, 1966 (div. 1971). B.S.E.E., U. Wash., 1969, M.S.E.E., 1970, Ph.D., 1975. NIH sr. fellow ctr. for Bioengring. U. Wash., Seattle, 1975-76, research asst. prof., 1981-82, research assoc. prof., 1982—, research assoc. dept. elec. engring., 1976-79, research asst. prof. dept. nuclear engring., 1979-82, research assoc. prof., 1982—; cons. scientist Puget Sound Blood Ctr., 1976-77, Math. Scis. N.W., Bellevue, Wash., 1978—, Physio Control, Redmond, Wash., 1982—, Abbott Labs., Chgo., 1982—. Contbr. articles to profl. jours. NSF grantee, 1978; NIH grantee, 1980. Mem. Optical Soc. Am., Soc. for Biomaterials, Soc. for Indsl. and Apllied Math., AAAS, Sigma Xi, Tau Beta Pi. Subspecialties: Applied magnetics; Biomedical engineering. Current work: Theoretical and experimental investigations of inverse and direct electromagnetic wave propagation and radiative transport techniques for characterizing dense scattering media. Inventor blood plasma optical pH measurement system, particle size analyser for multiple scattering media. Office: U Washington Seattle WA 98195

REYNOLDS, RAY THOMAS, research scientist; b. Lexington, Ky., Sept. 2, 1933; s. Oscar Ray and Margaret Louise (Gudgel) R.; m. Yolanda Maria de la Luz Gallegos, Oct. 15, 1962; children: Mark Andrew, Daniel Alan. B.S. in Chemistry, U. Ky., 1954, M.S. in Physics, 1960. Research scientist Am. Geog. Soc., Thule, Greenland, 1960-61, U. Calif. (Los Alamos Nat. Lab.), 1961; research scientist Ames Research Center, NASA, Moffett Field, Calif., 1062-69, 78—, chief theoretical studies br., 1969-78. Contbr. articles, reports, abstracts in field to profl. publs. Served with USAF, 1955-57. Recipient Exceptional Sci. Achievement award NASA, 1980. Mem. Am. Astron. Soc. (div. planetary sci.), Am. Geophys. Union, Meteoritical Sec., AAAS (Newcome-Cleveland prize 1979), AIAA. Subspecialties: Planetary science; Planetology. Current work: Theoretical studies of planetary formation, structure and evolution; basic theoretical research, planetary and satellite structure and thermal history, observational, search for undiscovered planetary bodies in near vicinity of solar system, mission support, planning for spacecraft missions. Home: 1650 Shasta Ave San Jose CA 95128 Office: N245-3 Theoretical and Planetary Studies Br Ames Research Center Moffett Field CA 94035

REYNOLDS, ROBERT DAVID, research chemist, nutrition educator; b. Mansfield, Ohio, June 25, 1943; s. Harold David and Ruth Ilene (Kern) R.; m. Rebecca Lynn Black, Sept. 5, 1964; children: Sara Lynn, Daniel David. B.S., Ohio State U., 1965; Ph.D., U. Wis., 1971. Postdoctoral fellow Biochemisches Inst., Freiburg, W. Ger., 1971-72; asst. mem. Fred Hutchinson Cancer Research Ctr., Seattle, 1972-73; research assoc. dept. biochemistry U. Wis.-Madison, 1973-75; adj. asst. prof. dept. food, nutrition, and instn. adminstrn. U. Md., College Park, 1980-82, adj. assoc. prof., 1982—; research chemist U.S. Dept. Agr., Beltsville, Md., 1975—. Editor: Methods in Vitamin B-6 Nutrition, 1981. Cancer research fellow Damon Runyon Fund for Cancer Research, 1971; recipient cert. of merit U.S. Dept. Agr., 1982. Mem. Am. Inst. Nutrition, Am. Assn. Cancer Research, AAAS. Lutheran. Subspecialties: Nutrition (medicine); Cancer research (medicine). Current work: Nutritional requirements of pregnant and lactating women and their infants; nutritional influences on cancer prevention. Home: 2413 Blooming Way Gambrills MD 21054 Office: USD A Bldg 307 Rm 217 Beltsville MD 20705

REYNOLDS, ROBERT DONALD, pharmacologist; b. Butler, Pa., Dec. 11, 1944; s. Frank Edward and Betty Jean (Orner) R.; m. Mary C. Gruenwald, Aug. 17, 1974; children: Ty Douglas, Bret David, Alysia Marie, Scott Edward, Steven Joseph. B.A., Clarion State Coll., 1970; Ph.D., U. Cin., 1974. Research asso. dept. pharmacology Med. Coll. Pa., 1974-76; research investigator Am. Critical Care, McGaw Park, Ill., 1976-78, sr. research investigator, 1978-81, group leader, 1981-83, sr. research fellow, 1983—; adj. asst. prof. Med. Coll. Wis., Milw., 1982—. Served with USMC, 1966-68. Am. Heart Assn. fellow, 1975-76. Mem. Am. Soc. Pharmacology and Exptl. Therapeutics, Am. Heart Assn., Phila. Physiol. Soc. Subspecialties: Pharmacology; Physiology (medicine). Current work: Cardiovascular, physiology and pharmacology, arrhythmia, myocardial infarction, antiarrhythmic drugs; beta-adrenergic blockers. Home: 971 Dunbar Mundelein IL 60060 Office: Am Critical Care McGaw Park IL 60085

REYNOLDS, WILLIAM CRAIG, mechanical engineer; b. Berkeley, Calif., Mar. 16, 1933; s. Merrill and Patricia Pope (Galt) R.; m. Janice Erma, Sept. 18, 1953; children—Russell, Peter, Margery. B.S. in Mech. Engring, Stanford U., 1954, M.S., 1955, Ph.D., 1957. Faculty mech. engring. Stanford U., 1957—, chmn. dept. mech. engring., 1972—, chmn., 1974-81. Author: books, including Energy Thermodynamics, 2d edit, 1976; contbr. numerous articles to profl. jours. NSF sr. scientist fellow, Eng., 1964. Fellow ASME; fellow Am. Phys. Soc.; mem. AAUP, AIAA, Nat. Acad. Engring., Sigma Xi, Tau Beta Pi. Subspecialties: Mechanical engineering; Fluid mechanics. Research in fluid mechanics and applied thermodynamics. Office: Stanford U Dept Mech Engring Stanford CA 94305

REZ, PETER, materials scientist, researcher, cons.; b. London, May 11, 1952. B.A. (Hon. scholar, 1973), Churchill Coll., Cambridge U., 1973; Ph.D. (sr. scholar, 1974), St. Catherine's Coll., Oxford U., 1975. Sci. Research Council research fellow dept. metallurgy Oxford U., 1976-77; researcher dept. materials sci. U. Calif.-Berkeley, 1977-78, 81—; devel. microanalysis equipment Kevex Corp., Foster City, Calif., 1978-81. Mem. U.K. Inst. Physics, Am. Phys. Soc., Electron Microscopy Soc. Am., Microbeam Analysis Soc. Subspecialty: Electron interaction with materials. Current work: Fundamentals of electron interaction with materials applied to electron microscopy, analysis and electron beam microfabrication techniques. Office: VA Microscopes Ltd Charlwoods Rd East Grinstead Sussex England RH19 2JQ

REZVANI, IRAJ, physician, educator; b. Tehran, Iran, Apr. 1, 1939; s. Ebrahim and Maryam (Jarrahi) R.; m. Dorothy A. Wipper, July 3, 1971; children: Andrew Reza, Geoffrey Abraham. M.D., Tehran U., 1961. Diplomate: Am. Bd. Pediatrics. Rotating intern Albert Einstein Med. Ctr., Phila., 1966-67; resident in pediatrics N.J. Coll. Medicine, Newark, 1967-68, U. Louisville, 1968-69; fellow in pediatric endocrinology St. Christopher's Hosp. for Children, Phila., 1969-72, asst. dir. Clin. Research Ctr., 1972-82, dir. Endocrine Lab., attending physician, 1972—; asst. prof. Temple U., Phila., 1972-78, assoc. prof., 1978—. Fellow Am. Acad. Pediatrics; mem. Endocrine Soc., Soc. Pediatric Research, Pediatric Endocrine Soc., Am. Fedn. Clin. Research. Subspecialties: Pediatrics; Endocrinology. Current work: pituitary hormones (HGH, TSH, ACTH); thyroid disorders; adrenal steroids; congenital adrenal hyperplasia; amino acid disorders. Office: St Christopher's Hosp for Children 2600 N Lawrence St Philadelphia PA 19133 Home: 203 Kent Rd Wyncote PA 19095

RHEE, HEE MIN, pharmacologist, educator; b. Choongnam, Korea, June 5, 1941; came to U.S., 1967, naturalized, 1977; s. Chang G. and Oak Y. (Cong) R.; m. Seungae Hong, July 4, 1948; 1 son, Horace H. Ph.D. in Pharmacology, Ohio State U. Sch. Medicine, 1973; postdoctoral study, U. Wis. Med. Sch., 1973-76. Registered pharmacist. Research assoc. Ohio State U. Med. Sch., Columbus, 1968-73; Wis. heart fellow U. Wis. Med. Sch., Madison, 1973-74, NIH fellow, 1975-76; asst. prof. pharmacology Med. Coll. Ohio, Toledo, 1976-80; assoc. prof. pharmacology Oral Roberts U. Med. Sch., Tulsa, 1980—. Contbr. articles to profl. publs. Served to lt. Korean Army, 1963-65. Recipient Individual research service award NIH, 1974-76, Young Investigator award, 1977-80; recipient Internat. Conf. award, 1982. Mem. AAAS, Am. Heart Assn., Am. Soc. Clin. Pharmacologists and Therapeutics, Internat. Soc. Cardiovascular Research, N.Y. Acad. Sci. Methodist. Subspecialties: Cellular pharmacology; Cardiology. Current work: General research on cardiovascular system in health and diseases, that is, congestive heart failure, myocardial ischemic diseases and cardiac arrhythmias, their prevention and treatments. Molecular and cellular characterization of heart diseases, hypertension and diabetes by the study of ion transport, Na , K -ATPase, Ca - ATPase, and interaction with drug receptors. Home: 7021 E 79th St Tulsa OK 74133 Office: Dept Pharmacology Oral Roberts U Med Sch Tulsa OK 74171

RHEINBOLDT, WERNER CARL, educator; b. Berlin, Sept. 18, 1927; U.S., 1956, naturalized, 1963; s. Karl Leo and Gertrud Anna (Hartwig) R.; m. Cornelie J. Hogewind, 1959; children—Bernd Michael, Matthew Cornelius. Dipl. Math, U. Heidelberg, Ger., 1952; Dr.rer.nat., U. Freiburg, Ger., 1955. Mathematician Computer Lab., Nat. Bur. Standards, Washington, 1957-59; dir. Computer Center, asst. prof. math. Syracuse (N.Y.) U., 1959-62; dir. Computer Sci. Center, U. Md., 1962-65, prof. math. and computer sci., 1965-78, dir. interdisciplinary applied math. program, 1974-78; A.W. Mellon prof. math. U. Pitts., 1978—; vis. prof. Gesellschaft für Mathematik und Datenverarbeitung, Bonn, W. Ger., 1969; adv. panel computer sci. NSF, 1972-75; adv. com. Army Research Office, 1974-78; chmn. applied math. com. NRC, 1979—; cons. editor Acad. Press, N.Y.C., 1967—; cons. in field. Author: (with J. Ortega) Iterative Solution of Nonlinear Equations in Several Variables, 1970, Methods of Solving Systems of Nonlinear Equations, 1974; also articles. Served with German Army, 1943-45. Grantee NSF, 1960-62, 65—, NASA, 1963-73, Office Naval Research, 1972—. Mem. Soc. Indsl. and Applied Math. (editor 1964—, v.p. publs. 1976, pres. 1977-78, council 1979-80, trustee 1982—), Math. Assn. Am., Am. Math. Soc., AAAS. Lutheran. Subspecialties: Numerical analysis; Applied mathematics. Current work: Numerical problems for nonlinear problems; finite element methods; adaptive strategies for numerical computations; numerical approaches to problems in continuum mechanics; numerical data structures. Office: Dept Math and Statistics Univ Pitts Pittsburgh PA 15260

RHIM, JOHNG SIK, physician, researcher; b. Korea, July 24, 1930; U.S., 1958, naturalized, 1968; s. Hac Woon and Moo Duc (Choi) R.; m. Mary Margaret Lytle, Aug. 25, 1962; children: Jonathan, Christopher, Peter, Andrew, Michael, Kathleen. M.D., Seoul (korea) Nat. U., 1957. Intern Seoul Nat. U. Hosp., 1957-58; research fellow Children's Hosp. Research Found., Cin., 1958-60, Baylor U., 1961; Grad. Sch. Public Health, U. Pitts., 1962; research assoc. La. State U. Sch. Medicine, New Orleans, 1962-64; vis. scientist Nat. Inst. Arthritis and Infectious Diseases, NIH, Bethesda, Md., 1964-76; project dir. cancer research Microbiol. Assocs. Inc., Bethesda, 1976-78; sr. research scientist Nat. Cancer Inst., NIH, Bethesda, 1978—. Research, publs. on cancer and viral diseases. Bd. dirs. Winchester Sch., Silver Spring, Md. Mem. AAAS, Am. Assn. Cancer Research, Am. Assn. Immunologists, AMA, Am. Soc. Microbiologts, Soc. Exptl. Biology and Medicine, N.Y. Acad. Sci., Internat. Assn. Comparative Leukemia Research, Internat. Soc. Preventive Oncology. Democrat. Subspecialties: Cancer research (medicine); Virology (medicine). Current work: Viral carcinogenesis, chem. carcinogenesis, factors regulating cellular transformation, mechanism of carcinogenesis, identification and rescue of human sarcoma genes. Home: 8309 Mebdy Ct Bethesda MD 20817 Office: 9000 Rockville Pike Bethesda MD 20205

RHOADS, ALLEN ROY, biochemistry educator, researcher; b. Reading, Pa., Dec. 19, 1941; s. and Ethel M. Odum; m. Marcia L. Szczepanek, June 28, 1969. B.S., Kutztown State Coll., 1966; Ph.D., U. Md.-College Park, 1970. Teaching asst. U. Md.-College Park, 1966-70; postdoctoral fellow Howard U., Washington, 1971-72, asst. prof., 1972-77, assoc. prof. biochemistry, 1977—. Mem. Patuxent Conservation League, Fulton, Md., 1982. NIH grantee, 1980-83; Am. Heart Assn. grantee, 1982; Gillette Research Inst. fellow, 1969. Mem. Am. Chem. Soc., AAAS, Soc. Exptl. Biology and Medicine, N.Y. Acad. Scis., Sigma Xi. Subspecialty: Biochemistry (medicine). Current work: The role of cyclic nucleotides and calcium in cellular regulation. Office: Howard U Coll Medicine 520 W St NW Washington DC 20059

RHODES, DONALD FREDERICK, research physicist, educator; b. Johnstown, Pa., July 1, 1932; s. Frederick D. and Irene M. (Ankney) R.; m. Patricia J. Beaumariage, Dec. 22, 1956. B.S., U. Pitts., 1954, M.Litt., 1956; Ph.D., Pacific Western U., 1982. Instr. physics U. Pitts., 1954-55; engr. Westinghouse Electric Pitts., 1956-57; research physicist Gulf Research & Devel., Pitts., 1958—; educator aviation tech., Pitts., 1975—. Recipient IR 100 award Indsl. Research, 1968. Mem. Am. Nuclear Soc., Health Physics Soc. Club: Aero of Pitts. Subspecialties: Nuclear physics; Electronics. Current work: Research in nuclear applications, instrumentation and ultrasonics. Patentee nuclear instrumentation. Home: 439 Trestle Rd Pittsburgh PA 15239 Office: Gulf Research & Devel Co PO Drawer 2038 Pittsburgh PA 15230

RHODES, JAMES BENJAMIN, medical educator, researcher, physician; b. Kansas City, Mo., July 22, 1928; s. B. B. and Helen Elizabeth (Davis) R.; m. Betty Louise Alfini, Mar. 17, 1960; children: Benjamin, Joan, Stephen. A.B., U. Kans.-Lawrence, 1954; M.D., U. Kans.-Kansas City, 1958. Intern U. Chgo., 1958-59, resident in medicine, 1959-62, fellow in gastroenterology, 1962-64, research assoc. in biochemistry, 1964-66; asst. prof. medicine and physiology U. Kans.-Kansas City Med. Ctr., 1966-70, assoc. prof. medicine and physiology, 1970-79, prof. medicine, 1979—; cons. Kansas City VA Hosp., 1966—. Contbr. articles to med. jours. Served with USN, 1946-49. Recipient Teaching award, med. students U. Kans.-Kansas City, 1968, U. Kans. Alumni Assn., 1972; Spl. Recognition award U. Kans.-Kansas City Dept. Medicine, 1976. Fellow ACP; mem. Am. Physiol. Soc., Am. Fedn. Clin. Research, Am. Soc. Gastrointestinal Endoscopy, Am. Gastroent. Assn., Alpha Omega Alpha. Subspecialties: Gastroenterology; Physiology (biology). Current work: clinician; teacher; researcher; digestion and absorption; pancreas and small bowel. Home: 4701 Mullen Rd Shawnee Mission KS 66216 Office: U Kans Med Ctr 39th and Rainbow Blvd Kansas City KS 66103

RHODES, MITCHELL LEE, medical educator and researcher; b. Chgo., Feb. 12, 1940; s. Nathan and Dorothy (Weinstein) R.; m. Rhoda Ellen Krichilsky, Aug. 4, 1963; children: Steven, Dana, Jeffrey. B.S., U. Ill-Urbana, 1961; M.D., U. Ill.-Chgo., 1965. Diplomate: Am. Bd. Internal Medicine (pulmonary medicine). Successively intern, resident, fellow U. Chgo., 1965-70; clin. asst. prof. U. Calif.-San Francisco, 1970-72; assoc. prof. U. Iowa, Iowa City, 1972-76; prof. medicine Ind. U., Indpls., 1976—; chmn. edn. com. Sch. Medicine, 1981—, pres.-elect faculty, 1983-84. Author, editor: Chronic Obstructive Lung Disease, 1978; author: Manual of Pulmonary Procedures, 1980; contbg. author: Internal Medicine, 1983. Served to lt. comdr. USPHS, 1970-72. Pulmonary acad. awardee NIH-Nat. Heart-Lung Inst., 1973. Fellow ACP; mem. Am. Coll. Chest Physicians (Cecil Lehman Mayer research award 1974, bd. govs. Ind. 1979-85), Am. Thoracic Soc., Ind. Thoracic Soc. (v.p. 1982-84), Am. Fedn. Clin. Research. Subspecialties: Pulmonary medicine; Internal medicine. Current work: Study of oxidant injury on lung, endoscopic diagnostic techniques in pulmonary disease; application of computers to medical education. Office: Ind U Sch Medicine C508 University Hosp 926 W Michigan St Indianapolis IN 46223

RHODES, WILLIAM CLIFFORD, chemist; b. Birmingham, Ala., Aug. 8, 1932; s. John Clifford and Catherine Louise (Maddux) R.; m. Barbara Wagner, June 29, 1957; children: Thomas Allen, Lia Susanne, Benjamin Edward. A.B., Samford U., 1954; Ph.D., Johns Hopkins U., 1958. Am. Cancer Soc. postdoctoral fellow Fla. State U., Tallahassee, 1958-59, 60-61, asst.prof., 1959-60, 61-65, assoc. prof., 1965-70, prof. chemistry, 1970—, exec. dir. Inst. Molecular Biophysics, 1975-79, dir., 1979-80. NSF sr. postdoctoral fellow, 1964-65; John A. Southern Commemorative lectr. Furman U., 1980; recipient Career Devel. award NIH, 1965-70; named Alumnus of Year Samford U., 1981. Mem. Sigma Xi. Democrat. Presbyterian. Subspecialties: Theoretical chemistry; Biophysics (physics). Current work: Quantum theory of selective excitation, relaxation and energy channeling processes in molecular systems, laser excitation dynamics, electronic properties of biol. polymers. Home: 1506 Belleau Wood Dr Tallahassee FL 32312 Office: Inst Molecular Biophysic Fla State Tallahassee FL 32306

RHODIN, THOR NATHANIEL, applied physicist, educator, industrial consultant; b. Buenos Aires, Argentina, Dec. 9, 1920; came to derivative citizen; s. Thor N. and Pearl R. R.; m. Elspeth Lindsay, Sept. 21, 1949; children: Robert, Ann, Lindsay, Jeffrey. B.S., Haverford Coll., 1942; A.M., Princeton U., 1945, Ph.D., 1946. Research asst. Manhattan Project, Princeton, N.J., 1944-46; research assoc. Inst. for Study Metals; instr. chemistry U. Chgo., 1946-51; research asso. Engring. Research Lab., E.I. Du Pont de Nemours & Co., Inc., Wilmington, Del., 1951-58; asso. prof. applied and engring. physics Cornell U., Ithaca, N.Y., 1958-65, prof., 1965—. Editor, contbr. 6 books.; editorial advisor: Progress in Surface Sci.; contbr. over 200 articles to sci. jours. Mem. Am. Phys. Soc., Am. Vacuum Soc., AAUP. Quaker. Subspecialties: Condensed matter physics; Surface chemistry. Current work: Physics and chemistry of surfaces and solid interfaces; solid state chemical physics; solid state semiconductor surfaces and interfaces; synchrontron radiation; electron spectroscopy. Solid state physics, physical chemistry. Home: 222 Miller St Ithaca NY 14850 Office: 217 Clark Hal Cornell Ithaca NY 14853

RHODINE, CHARLES NORMAN, electrical, biological engineering educator; b. Denver, Jan. 10, 1931; s. Carl Edward and Ora Loraine (Chivington) R.; m. Barbara Irene Sanborn, Aug. 16, 1957; children: Brent Norman, Craig William, Susan Lynn. B.S. in Elec. Engring, U.

Wyo., 1957, M.S., 1959; Ph.D., Purdue U., 1973. Registered profl. engr., Wyo. Asst. prof. elec. engring. U. Wyo., Laramie, 1959-64, assoc. prof., 1964-79, prof., 1979—; cons. cos., Laramie, 1959—; bd. dirs. Rocky Mountain bioengring symposium U.S. Air Force Acad., 1972—. Served in USN, 1940-42; Korea. NSF grantee, 1967, 1969. Mem. IEEE, Sigma Xi, Eta Kappa Nu, Sigma Tau. Republican. Lutheran. Subspecialties: Biomedical engineering; Computer-aided design. Current work: Analysis of biological systems using statistical communications theory; delivery of computer assisted instruction using microcomputers. Office: Dept Elec Engring U Wyo Box 3295 Univ Station WY 82071 Home: 1403 Reynolds St Laramie WY 82070

RIBIERO, LAIR G(ERALDO) T(HEODORO), clinical research director, educator; b. Juiz De Fora, Minas, Brazil, July 6, 1945; came to U.S., 1976; s. Francisco and Ruth (Reis) R.; m. Edna May Ottoni Porto, Jan. 1, 1968 (div. 1978); children: Frederico, Claudia; m. Mary Miller, May 22, 1979; 1 dau., Christine. B.S., Fundação Machado Sobrinho, 1967; M.D., Juiz De Fora Med. Sch., 1972. Teaching asst. in anatomy Med. Sch. of Fed. U., Juiz de Fora, 1969-71, teaching asst. in cardiology, 1971-72; resident in cardiology Pontificia Universidade Catolica do Rio de Janeiro, Brazil, 1973, instr. cardiology, 1974; cardiologist Cantral Army Hosp., Rio de Jeneiro, 1974; asst. prof. cardiology Med. Sch., Barbacena, Brazil, 1975-76; research fellow in medicine Peter Bent Prigham Hosp. and Harvard Med. Sch., Boston, 1976-78; fellow in cardiology Meth. Hosp.-Baylor Coll. Medicine, Houston, 1978-80; asst. dir. Deborah Cardiovascular Research Inst., Browns Mills, N.J., 1980-82; dir. clin. research-domestic Merck Sharp & Dohme Research Labs., West Point, Pa., 1982—; adj. asst. prof. physiology Thomas Jefferson Coll. Medicine, Phila., 1981—. Author: Coronary Spasm, 1983; co-author: Myocardial Ischemia, 1978, Platelets and Prostaglandins, 1981; contbr. articles to profl. jours. Served with Brazilian Army, 1964. Recipient 1st place in cardiology postgrad. tng. Cath. U., 1973. Fellow Am. Coll. Cardiology; mem. Brazilian Cardiology Soc., Am. Fedn. Clin. Research, N.Y. Acad. Scis. Subspecialties: Cardiology; Pharmacology. Current work: Protection of ischemic mycardium and reduction of myocardial infarct size; prostaglandins; sudden death. Home: 14 Robbins Way Vincentown NJ 08088 Office: Merck Sharp & Dohme Research Labs West Point PA 19486

RICE, BERNARD FRANCIS, endocrinologist, educator; b. St. Augustine, Fla., June 15, 1931; s. Harry and Rose (Nussbaum) R.; m. Joan Feinsilver, July 11, 1954; children: Matthew, Jonathan, Adam. A.B., Temple U., 1952; M.D., 1956; M.S. in Medicine, Mayo Found., Rochester, Minn., 1960. Intern Temple U., Phila., 1956-57; NIH postdoctoral research fellow U. Miami Sch. Medicine, 1962-64; head endocrine research A. Ochsner Med. Found., New Orleans, 1964-76; instr. to assoc. prof. Tulane U. Sch. Medicine, 1964-76; vis. physician Charity Hosp., New Orleans, 1964-76; prof. medicine U. Mo. Sch. Medicine, Kansas City, 1976—. Grantee Nat. Cancer Inst., NIH, 1964-76. Fellow ACP; mem. Endocrine Soc., So. Soc. Clin Investigation, Am. Fedn. Clin Research, Soc. Study Reprodn. Jewish. Subspecialty: Endocrinology. Current work: Reproductive endocrinology; metabolism; calcium metabolism; oncology. Home: 8531 Juniper Ln Prairie Village KS 66207 Office: 8901 W 74th St Shawnee Mission KS 66204

RICE, DONALD LESTER, geochemistry educator, researcher; b. Chamblee, Ga., Sept. 10, 1949; s. Lester C. and Clara M. (Warbington) R. B.S., Ga. Inst. Tech., 1970, M.S., 1974, Ph.D., 1979. Research asst. Skidaway Inst. Oceanography, Savannah, Ga., 1977-79; asst. prof. marine sci. U. S.C., Conway, 1979-80; asst. prof. geology sci. SUNY-Binghamton, 1980—; editorial cons. Merrill Pub. Co., Columbus, Ohio, 1981—. Served with U.S. Army, 1970-72. Recipient award for best dissertation in sci. Sigma Xi, 1979. Mem. Geochem. Soc., Am. Soc. Limnology and Oceanography, AAAS, Estuarine Research Fedn. Episcopalian. Subspecialties: Geochemistry; Oceanography. Current work: Nutrient and trace element biogeochemistry in marine environments. Home: 15 Grand Blvd Apt 2 Binghamton NY 13905 Office: Dept Geol Scis SUNY Binghamton NY 13901

RICE, JAMES ROBERT, engring. scientist, geophysicist; b. Frederick, Md., Dec. 3, 1940; s. Donald Blessing and Mary Celia (Santangelo) R.; m. Renata Dmowska, Feb. 28, 1981; children by previous marriage—Douglas, Jonathan. B.S., Lehigh U., 1962, Sc.M., 1963, Ph.D., 1964. Postdoctoral fellow Brown U., Providence, 1964-65, asst. prof. engring., 1965-68, asso. prof., 1968-70, prof., 1970-81, Ballou prof. theoretical and applied mechanics, 1973-81; McKay prof. engring. sci. and geophysics Harvard U., Cambridge, Mass., 1981—. Recipient awards for sci. publs. ASME, ASTM, U.S. Nat. Com. Rock Mechanics. Fellow ASME, AAAS; mem. Nat. Acad. Engring., Nat. Acad. Sci., ASCE, Am. Geophys. Union, Soc. Rheology. Subspecialties: Solid mechanics; Geophysics. Research contbns. to solid mechanics, materials. Sci. and geophysics. Office: Harvard U Cambridge MA 02138

RICE, JAMES THOMAS, wood technology educator, researcher; b. Birmingham, Ala., Feb. 7, 1933; s. Virgil Pedro and Eva (Childs) R.; m. Fleta Elaine Denney, Jan. 1, 1954 (dec. 1974); children: Lynn, Ann, James; m. Lucy Yancey West, Mar. 30, 1975; 1 son, John. B.S. in Forestry, Auburn U., 1954; M.S. in Wood Tech, N.C. State U., 1960, Ph.D., 1964. Instr. N.C. State U.-Raleigh, 1959-64, asst. prof., 1964-65; assoc. prof. U. Ga., Athens, 1965-69, 70—; mgr. research and devel. lab. Ga. Pacific Corp., Atlanta, 1969-70; tech. coordinator Adhesive and Sealant Council, Arlington, Va., 1966-69, 71-78; cons. Ashland Chem. Co., Columbus, Ohio, 1978—. Contbr. articles to prof. jours. Served to 1st lt. U.S. Army, 1954-56. Mem. Forest Products Research Soc. (sect. chmn. 1969, Wood award 1953), ASTM (vice chmn. Com. D-14 on Adhesives 1982-83), Soc. Wood Sci. and Tech. Subspecialties: Materials; Polymers. Current work: Wood adhesives, wood gluing, glued wood products, plywood, particleboard, bonded composites. Patentee in novel adhesives. Office: Sch Forest Resources U Ga Athens GA 30602

RICE, KENNER CRALLE, chemist; b. Rocky Mount, Va., May 14, 1940; s. Kenner Cralle and Annie Grace (Early) R. B.S., Va. Mil. Inst., 1961; Ph.D., Ga. Inst. Tech., 1966. Sr. scientist Ciba-Geigy Pharm. Corp., Summit, N.J., 1969-71; sr. staff fellow NIH, Bethesda, Md., 1972-76, research chemist, 1977—. Contbr. articles to profl. jours. Served to capt. U.S. Army, 1966-68. Recipient cert. of achievement Walter Reed Army Inst. Research, 1968; Tech. Excellence award 9th ann. World Fair for Tech. Exchange, 1981; Sato Meml. Internat. award Found. for Advanced Edn. in the Scis. and Japanese Pharm. Soc., 1983. Mem. Am. Chem. Soc. Methodist. Subspecialties: Organic chemistry; Synthetic chemistry. Current work: Organic and medicinal chemistry; the chemistry and mechanism of action of drugs which affect the central nervous system. Patentee in field.

RICE, MICHAEL JOHN, theoretical physicist, cons.; b. Cowes, Isle of Wight, Eng., Dec. 25, 1940; came to U.S., 1968; s. Thomas John and Elizabeth Emma (Keeping) R.; m. Annegret Thekla Richter, Sept. 17, 1965; children: Juliet, Jeremy, Jennifer. B.S. with spl. honors, Queen Mary Coll., U. London, 1961, Ph.D., 1965. Research fellow Imperial Coll., U. London, 1965-68; mem. sci. staff Gen. Electric Research and Devel. Ctr., Schenectady, 1968-71; theoretical physicist Brown Boveri Research Ctr., Baden, Switzerland, 1971-74; sr. scientist Xerox Webster Research Ctr., Webster, N.Y., 1974-79, prin. scientist, 1979—; cons. govt. research labs.; lectr. in field. Contbr. numerous articles on physics to profl. jours. Nordita prof. of physics U. Copenhagen, 1979-80. Fellow Am. Phys. Soc. Subspecialties: Condensed matter physics; Theoretical physics. Current work: Frontier problems in theoretical condensed matter physics, conducting polymers, molecular metals and superconductors, fractionally charged particles. Office: 800 Phillips Rd Bldg 114 Webster NY 14580

RICE, STEPHEN LANDON, mech. engr., educator; b. Oakland, Calif., Nov. 23, 1941; s. Landon Frederick and Elda Genevieve (Hunt) R.; m. Penny Baum, Dec. 29, 1965; children: Andrew Landon, Katherine Grace. B.S., U. Calif., Berkeley, 1964, M. Eng., 1969, Ph.D., 1972. Registered profl. engr., Conn. Design engr. U. Calif. Lawrence Berkeley Lab., 1982-83; asst. prof. mech. engring. U. Conn., Storrs, 1972-77, asso. prof., 1977-82, prof. mech. engring. and elec. engring. and computer sci., 1982—; prof. engring. U. Central Fla., Orlando, 1983—, chmn. dept. mech. engring. and aerospace scis., 1983—; dir. automation Robotics and Mfg. Lab. Contbr. articles to profl. jours. Recipient Teetor Edn. award Soc. Automotive Engrs., 1975; Fulbright-Hays sr. research awardee, 1978-79. Mem. ASME, Am. Soc. for Engring. Edn. (Outstanding Young Faculty award New Eng. sect. 1975), Am. Soc. Lubrication Engrs., Soc. Mfg. Engrs. Subspecialties: Wear of Materials; Robotics. Current work: Fundamental research in wear of materials, including dental restoratives; applied research in robotics, computer-aided-design and manufacturing (CAD/CAM); devel. of computer-based educational materials; ednl. evaluation. Office: Dept Mech Engring and Aerospace Scis Central Fla Orlando FL 32816

RICE, STUART ALAN, chemist, educator; b. N.Y.C., Jan. 6, 1932; s. Harry L. and Helen (Rayfield) R.; m. Marian Ruth Coopersmith, June 1, 1952; children—Barbara, Janet. B.S., Bklyn. Coll., 1952; M.A., Harvard, 1954, Ph.D., 1955. Jr. fellow Harvard, 1955-57; faculty U. Chgo., 1957—, prof. chemistry, 1960-69, Louis Block prof. phys. scis., 1969—, chmn. dept. chemistry, 1971-76, Frank P. Hixon disting. service prof., 1977—, dean phys. scis. div., 1981, dir., 1962-68; Mem. Nat. Sci. Bd., 1980. Author: Polyelectrolyte Solutions, 1961, Statistical Mechanics of Simple Liquids, 1965, Physical Chemistry, 1980; bd. dirs.: also numerous articles. Bull. Atomic Scientists. Guggenheim fellow, 1960-61; Falk-Plautt lectr. Columbia, 1964; Riley lectr. Notre Dame U., 1964; NSF sr. postdoctoral fellow, 1965-66; USPHS spl. postdoctoral fellow U. Copenhagen, 1970-71; Univ. lectr. chemistry U. Western Ont., 1970; Seaver lectr. U. So. Calif., 1972; Noyes lectr. U. Tex., Austin, 1975; Foster lectr. SUNY, Buffalo, 1976; Frank T. Gucker lectr. Ind. U., 1976; Fairchild lectr. Calif. Inst. Tech., 1979. Mem. Am. Chem. Soc. (award Pure Chemistry 1963, Leo Hendrik Backeland award 1971), Nat., Am. acads. scis., Am. Phys. Soc., AAAS, Faraday Soc. (Marlowe medal 1963), N.Y. Acad. Scis. (A. Cressy Morrison prize 1955), Danish Acad. Sci. and Letters (fgn.). Subspecialties: Physical chemistry; Theoretical chemistry. Current work: Theory of liquids; photophysical and photochemical processes; statistical mechanics; energy transfer; molecular dynamics.

RICH, ALEXANDER, molecular biologist; b. Hartford, Conn., Nov. 15, 1924; s. Max and Bella (Shub) R.; m. Jane Erving King, July 5, 1952; children—Benjamin, Josiah, Rebecca, Jessica. A.B. magna cum laude in Biochem. Scis, Harvard U., 1947, M.D. cum laude, 1949. Research fellow Gates and Crellin Labs., Calif. Inst. Tech., Pasadena, 1949-54; chief sect. phys. chemistry NIMH, Bethesda, Md., 1954-58; vis. scientist Cavendish Lab., Cambridge (Eng.) U., 1955-56; asso. prof. biophysics Mass. Inst. Tech., Cambridge, 1958-61, prof. biophysics, 1961—, William Thompson Sedgwick prof. biophysics, 1974—; Fairchild Disting. scholar Calif. Inst. Tech., Pasadena, 1976; mem. com. career devel. awards NIH, 1964-67, mem. postdoctoral fellowship bd., 1955-58; mem. com. exobiology space sci. bd. NAS, 1964-65; mem. U.S. nat. com. Internat. Orgn. Pure Applied Biophysics, 1965-67; mem. vis. com. dept. biology Weizmann Inst. Sci., 1965-66; mem. life scis. com. NASA, 1970-75, mem. lunar planetary missions bd., 1968-70; mem. biology team Viking Mars Mission, 1969—; mem. corp. Marine Biol. Lab., Woods Hole, Mass., 1965-77; mem. sci. rev. com. Howard Hughes Med. Inst., Miami, Fla., 1978—; mem. vis. com. biology div. Oak Ridge Nat. Lab., 1972-76; mem. sci. adv. bd. Stanford Synchrotron Radiation Project, 1976—, Mass. Gen. Hosp., Boston, 1978—; mem. U.S. Nat. Sci. Bd., 1976—; mem. bd. govs. Weizmann Inst. Sci., 1976—; mem. research com. Med. Found., Boston, 1976-80; mem. U.S.-USSR Joint Commn. on Sci. and Tech., Dept. State, Washington, 1977—; sr. cons. Office of Sci. and Tech. Policy, Exec. Office of Pres., Washington, 1977-81; mem. council Pugwash Confs. on Sci. and World Affairs, Geneva, 1977—; chmn. basic research com. Nat. Sci. Bd., Washington, 1978—; mem. U.S. Nat. Com. for Internat. Union for Pure and Applied Biophysics, Nat. Acad. Scis., 1979—; mem. nominating com. Nat. Acad. Scis., 1980. Editor: (with Norman Davidson) Structural Chemistry and Molecular Biology, 1968; editorial bd.: Biophys. Jour, 1961-63, Currents Modern Biology, 1966-72, Science, 1963-69, Analytical Biochemistry, 1969—, Bio-Systems, 1973-78, Molecular Biology Reports, 1974—, Procs. Nat. Acad. Sci, 1973—, Jour. Molecular and Applied Genetics, 1980—, Recombinant DNA, 1981—; editorial advisory bd.: Jour. Molecular Biology, 1959-66, Accounts of Chemical Research, 1980—; contbr. articles to profl. jours. Served with USN, 1943-46. Recipient Skylab Achievement award NASA, 1974, Theodore von Karmin award Viking Mars Mission, 1976, Presidential award N.Y. Acad. Scis., 1977; James R. Killian faculty achievement award M.I.T., 1980; Jabotinsky medal Jabotinsky Found., 1980; NRC fellow, 1949-51; Guggenheim Found. fellow, 1963; mem. Pontifical Acad. Scis. The Vatican, 1978. Fellow Am. Acad. Arts and Scis., AAAS; mem. Nat. Acad. Scis., Am. Chem. Soc., Biophys. Soc. (council 1960-69), Am. Soc. Biol. Chemists, Am. Crystallographic Soc., Internat. Soc. for Sty of Origin of Life, Phi Beta Kappa. Subspecialty: Biophysics (physics). Office: Dept Biology Mass Inst Tech Cambridge MA 02139

RICH, ARTHUR, physics educator; b. N.Y.C., Aug. 30, 1937; s. Paul and Augusta (Zelman) R.; m. Paula Shwam, Aug. 25, 1959; children: Donna, Marjorie, Daniel. B.S., Bklyn. Coll., 1959; M.A., Columbia U., 1961; Ph.D., U. Mich., 1965. Asst. prof. physics U. Mich., 1966-70, asso. prof., 1970-75, prof., 1975—; vis. prof. Inst. Fundamental Astrophysics, U. Paris, 1973; guest lectr. Scuola Normale Superior, Pisa, Italy, 1975; mem. program com. 7th Internat. Conf. Atomic Physics, M.I.T., 1980; mem. internat. adv. com. 2d Internat. Conf. on Precision Measurement and Fundamental Constants, Nat. Bur. Standards, 1981; vis. prof. Institut des Hautes Etudes Scientifiques, Paris and Univ. Coll., London, 1981. Contbr. articles to profl. jours. Nat. Bur. Standards grantee, 1971-74. Fellow Am. Phys. Soc.; mem. Mich. Soc. Fellows, Sigma Xi. Subspecialty: Atomic and molecular physics. Current work: Experimental research on fundamental aspects of atomic and nuclear physics and biophysics; measurements of the properties of positronium; use of polarized positrons to investigate chiral molecules; precision measurement of positron polarization in beta decay with application to fundamental weak interaction effects. Office: Dept Physics U Mich Ann Arbor MI 48109

RICH, AVERY EDMUND, plant pathology educator; b. Charleston, Maine, Apr. 9, 1915; s. Nathan Harold and Myrtle (Schermerhorn) R.; m. Erma Littlefield, June 15, 1938; children: Alice Ann Fowler, Donna Rich Moody. B.S., U. Maine, 1937, M.S., 1939; Ph.D., Wash. State U., 1950. Asst. agronomist R.I. State Coll., Kingston, 1943-47; instr. plant pathology Wash. State U., Pullman, 1947-50, asst. prof., 1950-51; assoc. prof. plant pathology U. N.H., Durham, 1951-56, prof., 1956-82, assoc. dean, 1972-82, prof. plant pathology emeritus, 1982—. Author: Potato Diseases, 1983. Mem. Am. Phytopathol. Soc. Subspecialty: Plant pathology. Current work: Plant pathology; potato diseases. Office: Dept Plant Pathology U NH Nesmith Hall Durham NH 03824

RICH, CLAYTON, university administrator; b. N.Y.C., May 21, 1924; s. Clayton Eugene and Leonore (Elliot) R.; m. Mary Bell Hodgkinson, Dec. 19, 1953 (div. May 2, 1974); 1 son, Clayton Greig.; m. Carolyn Sue Miller, Apr. 8, 1982. Grad., Putney Sch., 1942; student, Swarthmore Coll., 1942-44; M.D., Cornell U., 1948. Diplomate Am. Bd. Internal Medicine. Intern Albany (N.Y.) Hosp., 1948-49, asst. resident, 1950-51; research asst. Cornell U. Med. Coll., 1949-50; asst. Rockefeller U., 1953-58, asst. prof., 1958-60; asst. prof. medicine U. Wash. Sch. Medicine, 1960-62, assoc. prof., 1962-67, prof., 1967-71, assoc. dean, 1968-71; chief radioisotope service VA Hosp., Seattle, 1960-70, assoc. chief staff, 1962-71, chief staff, 1968-70; v.p. med. affairs, dean Sch. Medicine; prof. medicine Stanford U., 1971-79, Carl and Elizabeth Naumann prof., 1977-79; chief staff Stanford U. Hosp., 1971-77, chief exec. officer, 1977-79; sr. scholar Inst. Medicine, Nat. Acad. Sci., Washington, 1979-80; Mem. gen. medicine B study sect. NIH, 1969-73, chmn., 1972-73; mem. spl. med. adv. group VA, 1977-81; provost U. Okla., Oklahoma City, 1980—, v.p. for health scis., 1983—, also exec. dean, prof., 1980-83. Editorial bd.: Calcified Tissue Research, 1966-72, Clin. Orthopedics, 1967-72, Jour. Clin. Endocrinology and Metabolism, 1971-72; Contbr. numerous articles to med. jours. Bd. dirs. Children's Hosp. at Stanford, Stanford U. Hosp., 1974-79; chmn. Gordon Research Conf. Chemistry, Physiology and Structure of Bones and Teeth, 1967; bd. dirs., exec. Com. Okla. Med. Research Inst. Served to lt. USNR, 1951-53. Fellow ACP; mem. Assn. Am. Physicians, Western Assn. Physicians, Am. Soc. Mineral and Bone Research (adv. bd. 1977-80), Am. Soc. Clin. Investigation, Assn. Am. Med. Colls. (exec. council 1975-79), Inst. of Medicine, Western Soc. Clin. Research (v.p. 1967-68), Endocrine Soc., Sigma Xi, Alpha Omega Alpha. Subspecialties: Endocrinology; Health services research. Home: 115 Lake Aluma Dr Oklahoma City OK 73121 Office: Provost Office U Okla Oklahoma City OK 73190

RICH, JIMMY RAY, nematologist; b. Collins, Ga., Oct. 29, 1950; s. Bernease and Carolyn Murtle (Guten) R. Student, Abraham Baldwin Agrl. Coll., 1968-70; B.S.A., U. Ga., 1972, M.S., 1973; Ph.D., U. Calif., Riverside, 1976. Asst. prof. nematology U. Fla., Live Oak, 1976-81, assoc. prof., 1981—; also acting dir. Agrl. Research Center; chmn. Tobacco Disease Council, 1978, Fla. Nematology Forum, 1980. Assoc. editor: Nemtropica, 1977-82; author articles. Bd. dirs. Fla. Kiwanis Found., 1982-84. Mem. Soc. Nematologists, Am. Phytopathol. Soc., Orgn. Tropical Am. Nematologists, Sigma Xi, Phi Kappa Phi, Gamma Sigma Delta, Omicron Delta Kappa, Alpha Zeta, Blue Key. Clubs: Suwannee Ind. Football League, Kiwanis. Subspecialties: Plant pathology; Nematology. Current work: Nematode management systems and nematode-plant host-parasite physiology. Office: ARC PO Box 1210 Live Oak FL 32060

RICH, JOHN CHARLES, astrophysicist, air force officer; b. Wichita, Kans., Oct. 12, 1937; s. Hubert E. and Alma Lorene (Sadler) R.; m. Kathleen Susan Hall, June 14, 1963; children: Susan, John N., Daniel. A.B. in Applied Physics, Harvard U., 1959, A.M., 1960, Ph.D. in Astrophysics, 1967. Commd. 2d lt. U.S. Air Force, 1960, advanced through grades to col., 1975; physicist (Air Force Spl. Weapons Center), 1960-63, physicist, br. chief, Alexandria, Va., 1966-70, div. chief, Kirtland AFB, N.Mex., 1970-75, comdr., 1977-78, dir. advanced radiation tech., 1978—. Contbr. articles to profl. jours. Decorated Legion of Merit. Mem. Am. Astron. Soc., AIAA, Am. Geophys. Union, Air Force Assn., Sigma Xi. Episcopalian. Club: Harvard of N.Mex. Subspecialties: High-energy lasers; Astrophysics. Current work: Development of high energy lasers; atomic and molecular physics and astrophysics. Home: 701 Sagebrush Trail SE Albuquerque NM 87123

RICHARD, JOSEPH, physician; b. Boston, Aug. 28, 1932; s. Samuel Richard and Bertha (Grossman) R.; m. Naomi Pearl Noble, Mar. 17, 1957; children: Mark, Jonathan. B.S. in Math, U. Chgo., 1953, M.D. with honors, 1957. Attending physician Community Hosp., Peekskill, N.Y., 1964—. Served with USAF, 1959-61. Fellow ACP; mem. Am. Soc. Clin. Oncology, Eastern Coop. Oncology Group. Democrat. Subspecialties: Chemotherapy; Hematology. Current work: Chemotherapy clinic trials. Home: 2217 Parker Lane Yorktown Heights NY 10598 Office: Jefferson Valley Med Bldg 3505 Hill Blvd Yorktown Heights NY 10598

RICHARDS, FREDERIC MIDDLEBROOK, educator, biochemist; b. N.Y.C., Aug. 19, 1925; s. George Huntington and Marianna (Middlebrook) R.; m. Sarah Wheatland, June 6, 1959. S.B., Mass. Inst. Tech., 1948; Ph.D., Harvard U., 1952. Research fellow phys. chemistry Harvard U., 1952-53; NRC postdoctoral fellow Carlsberg Lab., Denmark, 1954; NSF fellow Cambridge (Eng.) U., 1955; faculty Yale, New Haven, 1955—, prof. biochemistry, 1963—, chmn. dept. molecular biology and biophysics, 1963-67, chmn. dept. molecular biophysics and biochemistry, 1969-73. Mem. editorial bd.: Jour. Biol. Chemistry, 1963-69, Advances in Protein Chemistry, 1963—, Jour. Molecular Biology, 1973-76. Dir. Jano Coffin Childs Meml. Fund for Med. Research, 1976—; mem. corp. Woods Hole Oceanographic Instn., 1977—. Served with AUS, 1944-46. Recipient Pfizer-Paul Lewis award in enzyme chemistry, 1965, Kai Linderstrøm-Lang prize in protein chemistry, 1978; Guggenheim fellow, 1967-68. Mem. Am. Chem. Soc., Am. Soc. Biol. Chemists (pres. 1979-80), Biophys. Soc. (pres. 1972), Am. Crystallographic Assn., Nat. Acad. Scis., Am. Acad. Arts and Scis. Subspecialty: Biophysical chemistry. Home: 69 Andrews Rd Guilford CT 06437 Office: PO Box 6666 260 Whitney Ave New Haven CT 06511

RICHARDS, FREDERICK, II, medical educator; b. Charleston, S.C., Aug. 28, 1938; s. Gustave Patrick and Lizetta Allene (Wagener) R.; m. Anne Irvine Walters, June 2, 1962; children: Frederick Jr., Laura Anne, Charles Patrick. B.S., Davidson Coll., 1960; M.D., Med. Coll. S.C., 1964. Intern Columbia (S.C.) Hosp., Richland County, 1964-65; resident in internal medicine N.C. Bapt. Hosp., Winston-Salem, 1965-66, 68-69; fellow hematology/oncology Bowman Gray Sch. Medicine, Winston-Salem, 1969-71, instr. medicine, 1971-73, asst. prof., 1973-77, assoc. prof., 1977—. Contbr. chpts. to books, articles to profl. jours. Served as capt. USAF, 1966-68. Mem. ACP, N.Y. Acad. Scis., Am. Soc. Clin. Oncology, Am. Soc. Hematology, Am. Soc. Preventive Oncology, Gynecologic Group, Forsyth County Med. Soc. (adv. cancer com., exec. com.), N.C. Med. Soc., Am. Cancer Soc. (bd. dirs., exec. com., pub. edn. com., speaker's bur.), AMA, Cancer and Leukemia Group, Nat. Surgical Adjuvant Project for Breast and Bowel Cancer. Subspecialties: Oncology; Hematology. Current work: Clinical research in oncology. Home: 413 Springdale Ave Winston-Salem NC 27104 Office: Bowman Gray Sch Medicine 300 S Hawthorne Rd Winston-Salem NC 27103

RICHARDS, GARY PAUL, microbiologist, researcher, lab. adminstr.; b. Springfield, Mass., June 4, 1950; s. Norman Raymond and Irene Elizabeth (Clarkin) R.; m. Brenda Grace, Aug. 18, 1973; children: David Raymond, Kevin Vincent. B.A. in Microbiology cum laude, U. N.H., 1973. Dir. quality control Rockland Shrimp Corp., Maine, 1973-75; food insp. Nat. Marine Fisheries Service, Gloucester, Mass., 1975-

77, microbiologist, College Park, Md., 1977-78; research microbiologist Charleston (S.C.) Lab., 1978—; dir. Bio Research & Testing Lab., Charleston, 1979—. Contbr. numerous articles to sci. jours. Recipient Spl. Act award Nat. Marine Fisheries Service, 1978, Incentive award, 1980, Outstanding Achievement award, 1981. Mem. Am. Soc. Microbiology, Assn. Ofcl. Analytical Chemists, Phi Sigma. Roman Catholic. Subspecialties: Microbiology; Virology (biology). Current work: Devel. methodologies for detecting enteric viruses in shellfish and for assay of mutagenic and carcinogenic compounds. Home: 1042 Arborwood Dr Charleston SC 29412 Office: PO Box 12607 Charleston SC 29412

RICHARDS, HOWELL ALAN (SKIP RICHARDS), computer programming adminstr.; b. Omaha, May 7, 1950; s. Yale and Ida (Epstein) R.; m. Karen L. Jones, July 11, 1976; children: Lisa, Sarah. S.B. in Computer Sci. and Engring., MIT, 1975. With Data Gen. Corp., Westboro, Mass., 1975—, project leader, 1977-79, group leader, 1979-80, sect. mgr., 1980-81, dept. mgr., 1981—; mem. X3J1 com. for programming lang. PL/I, Am. Nat. Standards Inst. Recipient Outstanding Devel. award Data Gen. Corp., 1981. Mem. Assn. Computing Machinery. Subspecialties: Programming languages; Software engineering. Current work: Development of electronic office information and management systems. Office: 4440 Computer Dr Suite E111 Westboro MA 01580

RICHARDS, PAUL GRANSTON, geological science educator; b. Cirencester, Eng., Mar. 31, 1943; s. A.G.G. and K.M. (Harding) R.; m. Jody Porterfield, June 1, 1968; children: Mark, Jessica, Gillian. B.A. in Math, U. Cambridge, Eng., 1965; M.S., Calif. Inst. Tech., 1966, Ph.D., 1970. Asst. research geophysicist U. Calif.-San Diego, 1970-71; asst. prof. geology Columbia U., N.Y.C., 1971-76, assoc. prof., 1976-79, prof., 1979—, chmn. dept. geol. scis., 1980—; mem. com. seismology Nat. Acad. Scis./NRC. Author: (with K. Aki) Quantitative Seismology: Theory and Methods, 2 vols, 1980. Ch. organist, mem. choir, active peace movement. Sloan fellow, 1973-74; Guggenheim fellow, 1977-78; MacArthur Prize fellow, 1981—. Mem. Am. Geophys. Union (James B. MacElwane award 1977), Royal Astron. Soc., Seismol. Soc. Am., Soc. Exploration Geophysicists. Episcopalian. Subspecialty: Seismology. Current work: Interpretation of seismograms, seismic waves: their generation, propagation, scattering and attenuation. Home: 240 N Broadway Nyack NY 10960 Office: Lamont-Doherty Geol Obs Palisades NY 10964

RICHARDS, PAUL LINFORD, educator, physicist; b. Ithaca, N.Y., June 4, 1934; s. Lorenzo A. and Zilla (Linford) R.; m. Audrey Jarratt, Aug. 24, 1965; children—Elizabeth Anne, Mary Ann. A.B., Harvard, 1956; Ph.D., U. Calif. at Berkeley, 1960. Research fellow Cambridge U., 1959-60; mem. tech. staff Bell Telephone Labs., 1960-66; prof. physics U. Calif. at Berkeley, 1966—; Cons., U.S. govt. agys., bus. corps. Contbr. articles to tech. jours. Guggenheim Found. fellow Cambridge (Eng.) U., 1973-74; recipient Calif. Scientist of Yr. award, 1981; Alexander von Humboldt sr. scientist award, Stuttgart, Germany, 1982. Fellow Am. Phys. Soc.; mem. Phi Beta Kappa, Sigma Xi. Subspecialty: Low temperature physics. Home: 900 Euclid Ave Berkeley CA 94708

RICHARDS, ROGER THOMAS, acoustical scientist; b. Akron, Ohio, June 19, 1942; s. Clyde Irvin and Thelma Jo (Whitaker) R. B.S., Westminster Coll., 1964; M.S., Ohio U., 1968; Ph.D., Pa. State U.-State College, 1980. Research asst. Ohio U., Athens, 1966-68; engr. Gen. Dynamics, Rochester, N.Y., 1968-71; grad. asst. Pa. State U., State College, 1974-80; staff assoc. Applied Research Lab., State College, 1977-80; mem. tech. staff Rockwell Internat., Groton, Conn., 1980—; dir. U.S. Othello Assn., Falls Church, Va., 1979—. Mem. Pa. State U. Alumni Exec. Bd., State College, 1973-74, mem. grad. faculty council, 1974-76, edn. policy com. bd. trustees, 1973-74. NASA fellow, 1971-74. Mem. Acoustical Soc. Am., AIAA, Nat. Speleological Soc. (vice-chmn. Nittany Grotto 1977-78, life), Am. Cryptogram Assn. (editorial bd. 1982—), Sigma Xi, Sigma Pi Sigma. Subspecialties: Acoustics; Cryptography and data security. Current work: Structural vibration and underwater sound propagation. Home: Edgemere Manor RD 1 Stonington CT 06378 Office: Rockwell Internat 1028 Poquonnock Rd PO Box L Groton CT 06340

RICHARDS, SKIP *See also* **RICHARDS, HOWELL ALAN**

RICHARDSON, CARL REED, animal scientist, educator, cons.; b. Monticello, Ky., Dec. 20, 1947; s. Ervin Rossevelt and Cretia Marie (Dodson) R.; m. Nora Jean Fletcher, Dec. 18, 1971; 1 son, Kevin Reed. B.S., U. Ky., 1971, M.S., 1973; Ph.D., U. Ill., 1976. Research assoc. U. Ill., Urbana, 1976; asst. prof. animal sci. Tex. Tech. U., Lubbock, 1976-81, assoc. prof., 1981—; cons. beef cattle nutrition. Contbr. numerous articles to profl. jours. Served as lt. U.S. Army, 1971-76. Named Outstanding Researcher in Coll. AGrl. Scis. Tex. Tech. U., 1982; grantee in field of beef cattle nutrition research. Mem. Am. Inst. Nutrition, Am. Soc. Animal Sci., Am. Dairy Sci. Assn., Plains Nutrition Council, Sigma Xi, Alpha Zeta, Gamma Sigma Delta. Republican. Baptist. Subspecialty: Animal nutrition. Current work: Improvement in grain and roughage feedstuffs through chemical, biological and mechanical processing. Home: 8010 Raleigh Lubbock TX 79424 Office: Dept Animal Sc Tex Tech U Lubbock TX 79409

RICHARDSON, CAROL LYNN, microbiologist, biochemist; b. Little Rock, Oct. 18, 1948; d. Gerald H. Parnell and Violet Jane Fields; m. Graham T. Richardson, Aug. 1970. B.S., Purdue U., 1969, Ph.D., 1976; M.A., Cornell U., 1970. Clin. microbiologist Riley Meml. Hosp., Meridian, Miss., 1970-72; asst. supr. microbiology Med. Center Hosps., Norfolk, Va., 1972-73; microbiologist Purdue U. Cancer Center, West Lafayette, Ind., 1976; scientist sr., scientist, prin. scientist Meloy Labs., Springfield, Va., 1977-81; dir. mktg. Gibco Labs., Chagrin Falls, Ohio, 1981—. Contbr. articles to profl. jours. Treas. Scholarship Com. for Grad. Women in Sci. Recipient award for support of cancer research Phi Beta Psi, 1972. Mem. Am. Assn. for Cancer Research, N.Y. Acad. Scis., AAAS, Sigma Delta Epsilon, Iota Sigma Pi. Subspecialties: Biochemistry (medicine); Microbiology (medicine). Current work: Microbial media development for biotechnology applications, manager of tissue culture and microbial products for research and clinical laboratories. Patentee in field. Office: Gibco Div Pine and Mogul Sts Chagrin Falls OH 44022

RICHARDSON, CATHERINE KESSLER, research geochemist, educator; b. Washington, June 18, 1949; d. Richard Calvin and Mary Ellenore (Munroe) Kessler; m. Steven McAfee Richardson, May 22, 1971; children: Meredith Nichols, Colin Peter. A.B., Boston U., 1971; M.A., Harvard U., 1972, Ph.D., 1977. Research asst. Harvard U., Cambridge, Mass., 1972-73, teaching fellow, 1973-75; lectr. Boston Coll., Chestnut Hill, Mass., 1976-77; collaborating asst. prof. geochemistry Iowa State U., Ames, 1977-79, adj. asst. prof., 1979—. Contbr. articles to profl. jours. NSF grantee, 1979, 81, 82. Mem. Geochem. Soc., Mineral. Soc. Am. Clubs: Octagon Art Center, Figure Skating (Ames). Subspecialty: Geochemistry. Current work: Solubility and complexing of ore metals; fluid inclusion studies; the genesis of metallic ore deposits; the use of high temperature spectroscopy to determine the identity of metal-complexes; mineral-fluid alternation and replacement reactions. Office: Dept Earth Scis Iowa State U 253 Science I Ames IA 50011

RICHARDSON, CLARENCE ROBERT, research administrator; b. Lovelock, Nev., Jan. 10, 1931; s. James Harold and Mary Lorraine (Ostberg) R.; m. Donna Ruth Ames, Sept. 6, 1955; children: Karen Ann, Jeffrey C. B.S., U. Nev.-Reno, 1957; Ph.D., Johns Hopkins U., 1963. Jr. physicist Naval Ordnance Test Sta., Calif., 1957; physicist Applied Physics Lab. Johns Hopkins U., 1958-59; from asst. physicist to assoc. physicist Brookhaven Nat. Lab., 1963-67; physicist high energy physics Div. Research AEC, 1967-73, physicist nuclear sci. program, 1973-75, dept. mgr. solar inst. project office, 1975-76, dir. program planning div., 1979, physics sci. planning specialist, program planning div. basic energy sci., 1976-79; program mgr. medium energy nuclear physics Dept. Energy, Washington, 1979—; vis. physicist European Orgn. Nuclear Research, Switzerland, 1970-71. Contbr. articles in field to profl. jours. Served with USAF, 1948-52. Recipient Superior Performance award AEC, 1973, Dept. Energy, 1981. Mem. Am. Phys. Soc. Subspecialties: Nuclear physics; Particle physics. Current work: Administrator of basic nuclear research program, program management, budget defense, contract technical management, program planning. Discoverer of Eta Meson, 1962, Omega Hyperon, 1964. Home: 20513 Topridge Dr Boyds MD 20841 Office: Nuclear Physics Div ER-23 Dept Energy Washington DC 20545

RICHARDSON, GEORGE CAMPBELL, plastics engineer, educator; b. Rahway, N.J., Apr. 9, 1942; s. George and Annie Maxwell (Clel) R. B.S., Rutgers U., 1964, M.S., 1966, Ph.D., 1971. Plastics engr. Amerace Esna Corp., Union, N.J., 1970-71; asst. research prof. Rutgers U., New Brunswick, N.J., 1972-77, asst. prof., materials sci., 1977—. Mem. ASTM, Soc. Plastics Engrs., N.Y. Acad. Scis., Am. Phys. Soc. Subspecialties: Composite materials; Polymers. Current work: Mechanical properties of reinforced plastics, both thermoset and thermosplastics, using glass, graphite and aramid as reinforcement. Office: Rutgers U PO Box 909 Piscataway NJ 08854

RICHARDSON, HERBERT HEATH, mechanical engineer; b. Lynn, Mass., Sept. 24, 1930; s. Walter Blake and Isabel Emily (Heath) R.; m. Barbara Ellsworth, Oct. 6, 1973. S.B., S.M. with honors, Mass. Inst. Tech., 1955; Sc.D., 1958. Registered profl. engr., Mass. Research asst., research engr. Dynamic Analysis and Control Lab., Mass. Inst. Tech., 1953-57; instr. Dynamic Analysis and Control Lab., MIT, 1957-58, mem. faculty, 1958—, prof. mech. engring., 1968—, head dept., 1974-83; dean engring. MIT, 1983—; with Ballistics Research Lab., Aberdeen Proving Ground, 1958; chief scientist Dept. Transp., 1970-72; sr. cons. Foster-Miller Assos. Author: Introduction to System Dynamics, 1971; also articles. Served as officer U.S. Army, 1968. Recipient medal Am. Ordnance Assn., 1953, Gold medal Pi Tau Sigma, 1963; Meritorious Service award and medal Dept. Transp., 1972. Fellow ASME (Moody award fluids engring. div. 1970); mem. N.Y. Acad. Sci., AAAS, ASME, Nat. Acad. Engring., Am. Soc. Engring. Edn., Advanced Transit Assn., Soc. Automotive Engrs., Sigma Xi, Tau Beta Pi. Subspecialties: Mechanical engineering; Systems engineering. Current work: Systems and control; transportation; manufacturing. Home: 4 Latisquama Rd Southboro MA 01772 Office: Room 3-173 Mass Inst Tech Cambridge MA 02139

RICHARDSON, JAMES WYMAN, chemistry educator; b. Sioux Falls, S.D., Aug. 8, 1930; s. Lewis Gerhard and Maurine Katherine (Withey) R.; m. Eileen Mae Johnson, Dec. 24, 1952; children: Janilyn, James, Barbara, Gregory. B.S., S.D. Sch. Mines and Tech., 1952; Ph.D., Iowa State U., 1956. Research assoc. U. Chgo., 1956-57; with Purdue U., 1957—, prof. chemistry, 1973—; vis. scientist Philips Research Labs., Eindhoven, Netherlands, 1967-68. Contbr. articles to profl. jours. NSF grantee. Mem. Am. Chem. Soc., Am. Phys. Soc. Subspecialties: Theoretical chemistry; Atomic and molecular physics. Current work: Quantum-mechanical theory of bonding, spectra, magnetic and electrical properties of transition-metal compounds. Office: Dept Chemistry Purdue U West Lafayette IN 47907

RICHARDSON, JOHN PAUL, biochemist, educator; b. Pittsfield, Mass., June 27, 1938; s. Henry Martyn and Mary Estella (Larrick) R.; m. Lislott Rosemarie Richardson, Sept. 16, 1966; children: John Paul Jr., Alexandra, Gabriella. A.B., Amherst Coll., 1960; Ph.D., Harvard U., 1966. NIH postdoctoral fellow Inst. de Biologie Physico-chimique, Paris, 1965-67; Am. Cancer Soc. postdoctoral fellow Inst. de Biologie Moleculaire, Geneva, 1967-69; research assoc. genetics U. Wash., Seattle, 1969-70; asst. prof. Ind. U., Bloomington, 1970-74, assoc. prof., 1974-78, prof., 1978—. Violinist, Bloomington Symphony Orch.; also bd. dirs. NIH Research Career Devel. awardee, 1972-77. Mem. Am. Soc. Biol. Chemists, Am. Soc. Microbiology, AAAS. Democrat. Subspecialties: Biochemistry (biology); Gene actions. Current work: Biosynthesis of RNA in bacteria and higher organisms; enzymology of RNA polymerases and transcription termination factors. Office: Dept Chemistr Ind U Bloomington IN 47405

RICHARDSON, JOSEPH HILL, physician; b. Rensselaer, Ind., June 16, 1928; s. William Clark and Vera (Hill) R.; m. Joan Grace Meininger, July 7, 1950; children: Lois N., Ellen M., James K. B.S. in Medicine, Northwestern U., 1950, M.D., 1953. Diplomate: Am. Bd. Internal Medicine. Intern U.S. Naval Hosp., Great Lakes, Ill., 1953-54; resident, fellow in internal medicine Cleve. Clinic, 1956-59; med. staff Marion (Ind.) Gen. Hosp., 1957-67, Parkview Hosp., Ft. Wayne, Ind., 1967—; med. cons. Marion Gen. Hosp., 1977—. Contbr. articles to profl. jours. Served with USN, 1953-56. Fellow ACP, AAAS; mem. Am. Assn. Clin. Research, Am. Assn. Med. Writers. Subspecialty: Internal medicine. Home: 8726 Fortuna Way Fort Wayne IN 46815 Office: 3010 E State Blvd Fort Wayne IN 46805

RICHARDSON, KEITH ERWIN, biochemistry educator, cons.; b. Tucson, Apr. 22, 1928; s. Edmund Arthur and Ivie (Romney) R.; m. Dorthene Beck, June 5, 1952; children: DeeAnn, Jay Scott, Kerri Lee, Steven Grant, Mark Allen, Stacy. B.S., Brigham Young U., 1952, M.S., 1955; Ph.D., Purdue U., (1958.). Grad. asst. Purdue U., West Lafayette, Ind., 1955-58; postdoctoral fellow Mich. State U., East Lansing, 1958-60; prof. Ohio State U., Columbus, 1960—; cons. Ohio State Dept. Health, Columbus, 1979—. Contbr. articles on biochemistry to profl. jours. Served with U.S. Army, 1952-54. Mem. Am. Soc. Biol. Chemists, Am. Soc. Plant Physiologists, Am. Chem. Soc., Am. Inst. Nutrition. Republican. Mormon. Subspecialties: Biochemistry (medicine); Clinical chemistry. Current work: Biochemical research in kidney stone formation, oxalate metabolism, ethylene glycol toxicity and primary hyperozaluria. Office: Ohio State U 1645 Neil Ave Columbus OH 43210 Home: 3133 Edgefield Rd Columbus OH 43221

RICHARDSON, RICK LEE, mechanical engineer; b. Los Angeles, Dec. 24, 1955; s. Alex and Kathleen R. B.S.M.E., U. So. Calif., 1978. Project mgr. Horgren Steel Fabricators, Downey, Calif., 1975-77; cons. mech. engring., 1977-78; sr. mech. engr. B.J. Hughes, Long Beach, Calif., 1978—. Mem. ASME. Club: Artesta (Calif.) Racquetball. Subspecialties: Mechanical engineering; Materials. Current work: Mobile oil field equipment design advancements; advanced pump design utilizing state-or-art materials. Office: 6505 E Paramount Blvd Long Beach CA 90805

RICHART, FRANK EDWIN, JR., civil engr., educator; b. Urbana, Ill., Dec. 6, 1918; s. Frank Erwin and Fern (Johnson) R.; m. Elizabeth Goldthorp, Feb. 21, 1945; children—John Douglas, Betsy, Willard Clark. B.S., U. Ill., 1940, M.S., 1946, Ph.D., 1948; postgrad, U. Mich., 1940-41; D.Sc., U. Fla., 1972. Research asst., then asso. dept. civil engrng. U. Ill., 1946-48; asst. prof. mech. engring. Harvard, 1948-52; asso. prof. dept. civil engring. U. Fla., 1952-54, prof., 1954-62; prof. civil engring U. Mich., 1962-77, W. J. Emmons Distinguished prof., 1977—, chmn. dept., 1962-69; cons. Moran, Proctor, Mueser, Rutledge, N.Y.C., summers 1953-55, 57, also Office Engrs., U.S. Army, NASA. Contbr. tech. papers to profl. jours. Served to lt. comdr. USNR, 1941-46. Fellow ASCE (T.A. Middlebrooks award 1956, 59, 60, 67, Wellington prize 1963, Terzaghi lectr. 1974, Terzaghi award 1980); mem. Nat. Acad. Engring., Earthquake Engring. Research Inst. Subspecialties: Civil engineering. Current work: Soil dynamics, vibrations of soils and foundations, dynamic soil behavior, machine vibrations. Home: 2210 Hill St Ann Arbor MI 48104

RICHMAN, JACK WILLIAM, physicist; b. Gillingham, England, Feb. 20, 1941; s. Henry George and Ellen Rebecca (Williams) R.; m. Sylvia Helen Brennish; children: Oliver Page, Melanie Blair. B.Sc., U. Wales, 1963; M.Sc., U. Man., Can., 1968, Ph.D., 1972; diploma, Von Karman Inst., Brussels, 1969. Physicist Bristol Aerospace Co., Winnipeg, Man., 1966-68; cons. engr. Dilworth Secord, Toronto, Ont., Can., 1972-74; group leader, Bechtel Corp., Toronto, 1974-77; design engr. specialist Ont. Hydro, Toronto, 1977-78, supr. design engr., 1978-79; nuclear equipment and processes engr., 1979-81, tech. mgr. fusion fuel tech. project, 1981—. Contbr. articles to profl. jours. Served as officer RCAF, 1964-67. NRC scholar, 1966-74; postdoctoral fellow, 1972-74. Mem. Am. Nuclear Soc., Can. Nuclear Soc., Canadian Nuclear Assn., Assn. Profl. Engrs. Ont. (chmn. 1980-82), Royal Can. Mil. Inst. Subspecialties: Nuclear fission; Nuclear fusion. Current work: Fusion energy and fusion fuels. Office: Canadian Fusion Fuel Technology Project 620 University Ave Toronto ON Canada M5G 1X6

RICHMOND, BARRY JAY, research neurologist; b. Altus, Okla., July 13, 1943; s. Julius and Rhee (Chidekel) R.; m. Dorothy Anne Essman, Feb. 4, 1967; children: Jay Benjamin, Nathaniel Jacob, Ian Abraham. B.A., Harvard U., 1965; M.D., Case Western Res. U., 1971. Diplomate: Am. Acad. Pediatrics, 1976, Am. Acad. Neurology, 1978. Intern and resident in pediatrics Univ. Hosps. Cleve., 1971-73; resident in neurology Peter Bent Bridham Hosp., Children's Hosp. Med. Center and Beth Israel Hosp., all Boston, 1973-76; sr. staff fellow Nat. Inst. Mental Health and Nat. Eye Inst., Bethesda, Md., 1976-79; research neurologist lab. neuropsychology NIMH, Bethesda, 1979—; clin. asst. prof. pediatrics Georgetown U. Med. Sch., Washington, 1980—. Mem. Soc. Neurosci., Am. Acad. Neurology. Subspecialties: Neurology; Neurophysiology. Current work: Neurology; neurophysiology (visual); psychophysics (visual); behavorial psychology; visual function in man and primates. Home: 5306 Elsmere Ave Bethesda MD 20814 Office: NIMH Bldg 9 Rm IN-107 Bethesda MD 20205

RICHMOND, JONAS E., biochemist, educator; b. July 17, 1929; s. Benhamin F. and Bessie (Collins) R.; m. Mattie Lee Humes, Aug. 2, 1957; children: Perry Keith, Gigi Renato. M.S., U. Rochester, 1950, Ph.D., 1953. Vis. scientist Oxford (Eng.) U., 1955-57; instr. U. Rochester (N.Y.), 1957; assoc. Harvard U., Boston, 1957-63; biochemist U. Calif.-Berkeley, 1963—; cons. NIH, Bethesda, Md., 1956—, mem. study sect., 1964-78, tng. com., 1971—, research com., 1973-80. Contbr. numerous articles on biochemistry to profl. jours. Pres. Alameda County Heart Assn., Oakland, Calif., 1975; bd. dirs. Burlingame, Calif., 1968—, v.p., 1975-78; trustee Pacific Sch. Religion, Berkeley, 1966—. Recipient Meritorious Service awards Am. Heart Assn., 1975, NIH, 1969, 1972, 80. Mem. Am. Soc. Biol. Chemists (chmn. equal opportunity 1975-80, disting. service award 1980), Am. Physiol. Soc., Biophys. Soc. (chmn. equal opportunity 1975—, Am. Soc. Protozoology.). Congregationalist. Subspecialty: Biophysical chemistry. Current work: Chemistry of growth, differentiation and the role of cell, cell interactions and membrane activities. Office: Dept Nutritional Sci U Calif 219 Morgan Hall Berkeley CA 94720

RICHMOND, JULIUS BENJAMIN, physician, educator; b. Chgo., Sept. 26, 1916; s. Jacob and Anna (Dayno) R.; m. Rhee Chidekel, June 3, 1937; children: Barry J., Charles Allen, Dale Keith (dec.). B.S., U. Ill., 1937, M.S., M.D., 1939; D.Sc. (hon.), Ind. U., 1978, Rush-Presbyn.-St. Luke Med. Center, 1978, U. Ill., 1979, Georgetown U., 1980, D.Med. Sci., Med. Coll. Pa., 1980; D. Public Service, Nat. Coll. Edn., Evanston, Ill., 1980. Intern Cook County Hosp., Chgo., 1939-41, resident, 1941-42, 46, Municipal Contagious Disease Hosp., Chgo., 1941; mem. faculty U. Ill. Med. Sch., Chgo., 1946-52, prof. pediatrics, 1950-53, dir. Inst. Juvenile Research, 1952-53; prof., chmn. dept. pediatrics Coll. Medicine, State U. N.Y. at Syracuse, 1953-65, dean med. faculty, chmn. dept. pediatrics, 1965-70; prof. child psychiatry and human devel., prof., chmn. dept. preventive and social medicine Harvard Med. Sch., 1971-77, prof. health policy, 1981—; also faculty Harvard Sch. Public Health; psychiatrist-in-chief Children's Hosp. Med. Center, Boston, 1971-77, adv. on child health policy, 1981—; dir. Judge Baker Guidance Center, Boston, 1971-77; asst. sec. health and surgeon gen. HHS, 1977-81; mem. Pres.'s Commn. on Mental Health, 1977. Author: Pediatric Diagnosis, 1962, Currents in American Medicine, 1969. Nat. dir. Project Head Start; dir. Office Health Affairs OEO, 1965-66. Served as flight surgeon USAAF, 1942-46. Recipient Agnes Bruce Greig Sch. award, 1966; C. Anderson Aldrich award Am. Acad. Pediatrics, 1966; Myrdal Prize, 1977; ann. award Sect. on Community Pediatrics, 1977; outstanding contbn. award Section Community Pediatrics, 1978; Parents Mag. award, 1966; Disting. Service award Office Econ. Opportunity, 1967; Martha May Eliot award Am. Public Health Assn., 1970; Family Health Mag. award, 1977; award for disting. sci. contbn. Soc. for Research in Child Devel., 1979; Dolly Madison award Inst. on Clin. Infants Programs, 1979; Public Health Disting. Service award HEW, 1980. Fellow Am. Orthopsychiat. Assn.; distinguished fellow Am. Psychiat. Assn.; hon. mem. Am. Acad. Child Psychiatry; asso. mem. New Eng. Council Child Psychiatry; mem. Inst. Medicine of Nat. Acad. Scis., AMA, Am. Pediatric Soc., Am. Acad. Pediatrics, Soc. Pediatric Research, Am. Psychosomatic Soc., Am. Public Health Assn., Sigma Xi, Alpha Omega Alpha, Phi Eta Sigma. Subspecialties: Pediatrics; Psychiatry. Current work: Child development, maternal and child health. Office: 398B Brookline Ave Suite 12 Boston MA 02215

RICHMOND, ROLLIN CHARLES, geneticist, biology educator; b. Nairobi, Kenya, May 31, 1944; came to U.S., 1950, naturalized, 1962; s. Charles and Harriet W. (Truesdell) R.; m. Ann W. Willbern; children: Kathryn, John, Camilla, James. B.A. in Zoology, San Diego State U., 1966; Ph.D. in Genetics, Rockefeller U., 1971. Asst. prof. biology Ind. U., Bloomington, 1970-74, assoc. prof., 1975-81, prof., 1981—, chmn. dept. biology, 1982—. Editorial bd.: Genetica, Jour. Heredity; assoc. editor: Evolution; Contbr. articles to profl. jours. Mem. Soc. Study of Evolution, Am. Soc. Naturalists, Genetics Soc. Am., Am. Genetics Assn., AAAS, Behavior Genetics Assn., Cosmos Club, Sigma Xi, Phi Kappa Phi. Club: Cosmos (Washington). Subspecialties: Evolutionary biology; Gerontology. Current work: Molecular evolution of enzyme systems, genetic bases of aging in Drosophila. Office: Dept Biology Ind U Bloomington IN 47405

RICHSTONE, DOUGLAS ORANGE, astronomer; b. Alexandria, Va., Sept. 20, 1949; s. Morris and Jeanette (Orange) R.; m. Marilyn May Mantei, May 13, 1949. B.S., Calif. Inst. Tech., 1971; Ph.D.,

Princeton U., 1975. Research fellow Calif. Inst. Tech., 1975-77; asst. prof. physics U. Pitts., 1977-80; vis. asst. prof. U. Mich., 1980-82, asst. prof., 1982-83, assoc. prof., 1983—. Mem. Am Astron. Soc., Royal Astron Soc. Subspecialties: Theoretical astrophysics; Cosmology. Current work: Dynamics of galaxies and clusters of galaxies; stellar dynamics; formation and evoluation of galaxies and clusters of galaxies. Office: Dept Astronomy U Mich Ann Arbor MI 48109

RICHTER, BURTON, physicist, educator; b. N.Y.C., Mar. 22, 1931; s. Abraham and Fanny (Pollack) R.; m. Laurose Becker, July 1, 1960; children: Elizabeth, Matthew. B.S., MIT, 1952, Ph.D., 1956. Research asso. Stanford U., 1956-60, asst. prof. physics, 1960-63, asso. prof., 1963-67, prof., 1967—, Paul Pigott prof. phys. sci., 1980—, tech. dir. Linear Accelerator Ctr.; cons. NSF, Dept. Energy. Contbr. articles to profl. publs. Recipient E.O. Lawrence medal ERDA, 1975; Nobel prize in physics, 1976. Fellow Am. Phys. Soc., N.Y. Acad. Sci.; mem. Nat. Acad. Sci. Research elementary particle physics. Subspecialty: Particle physics. Office: Stanford U Stanford CA 94305

RICHTER, GEORGE BROWNELL, technical educator; b. St. Cloud, Minn., Feb. 21, 1927; s. Charles Herman and Mazie Katherine (Brownell) R.; m. Patricia Grain, Mar. 31, 1951; children: Mary Baker, Elisabeth Gibson, Mark Richter, Melodie Hart, Joseph. B.Phys., U. Minn., 1950, M.A., 1968, Ph.D., 1972; B.A., St. Johns U., 1951. Registered profl. engr., Minn. Lectr. graphics St. Thomas Coll., St. Paul, 1966—; tech. div. mgr. St. Paul Tech. Vocat. Inst., 1959—. Served with USNR, 1945-46. Mem. Soc. Mfg. Engrs., John Henry Newman Hon. Soc. Democrat. Roman Catholic. Subspecialties: Mechanical engineering; Statistics. Current work: Maintaining technician programs at current technology levels to meet industrial demands. Home: 2016 Merriam Ln Saint Paul MN 55104 Office: Saint Paul Tech Vocat Inst 235 Marshall Ave Saint Paul MN 55102

RICHTER, JUDITH ANNE, pharmacologist, educator; b. Wilmington, Del., Mar. 4, 1942; d. Henry John and Dorothy Madeline (Schroeder) R. B.A., U. Colo., 1964; Ph.D., Stanford U., 1969. Postdoctoral fellow Cambridge (Eng.) U., 1969-70, Inst. Psychiatry, U. London, 1970-71; asst. prof. pharmacology Ind. U. Sch. Medicine, Indpls., 1971-78, assoc. prof. pharmacology and neurobiology, 1978—; mem. biomed. research rev. com. Nat. Inst. Drug Abuse, 1981—. Mem. editorial bd.: Jour. Neurochemistry; contbr. numerous articles to sci. jours. Mem. Am. Soc. Pharmacology and Exptl. Therapeutics, Am. Soc. Neurochemistry, Internat. Soc. Neurochemistry, Soc. Neurosci., AAAS, Sigma Xi. Subspecialty: Neuropharmacology. Current work: Effects of barbiturates on cholinergic and other neurons; scientific research, teaching. Office: Inst Psychiatric Research Ind U Med Sch 791 Union Dr Indianapolis IN 46223

RICK, CHARLES MADEIRA, JR., geneticist, educator; b. Reading, Pa., Apr. 30, 1915; s. Charles Madeira and Miriam Charlotte (Yeager) R.; m. Martha Elizabeth Overholts, Sept. 3, 1938; children: Susan Charlotte Rick Baldi, John Winfield. B.S., Pa. State U., 1937; A.M., Harvard U., 1939, Ph.D., 1940. Asst. plant breeder W. Atlee Burpee Co., Lompoc, Calif., 1936, 37; instr., jr. geneticist U. Calif., Davis, 1940-44, asst. prof., asst. geneticist, 1944-49, asso. prof., asso. geneticist, 1949-55, prof., geneticist, 1955—; chmn. coordinating com. Tomato Genetics Coop., 1950-82; mem. genetics study sect. NIH, 1958-62; mem. Galapagos Internat. Sci. Project, 1964; mem. genetic biology panel NSF, 1971-72; mem. nat. plant genetics resources bd. Dept. Agr., 1975-82; Carnegie vis. prof. U. Hawaii, 1963; vis. prof. Universidade São Paulo, Brazil, 1965, U. Rosario, Argentina, 1980; vis. scientist U. P.R., 1968; centennial lectr. Ont. Agr. Coll. U. Guelph, Ont., Can., 1974. Contbr. articles in field to books and sci. jours. Fellow Calif. Acad. Sci., AAAS (Campbell award 1959); mem. Nat. Acad. Scis., Bot. Soc. Am. (Merit award 1976), Am. Soc. Hort. Sci. (M.A. Blake award 1974, Vaughan Research award 1946), Am. Genetics Assn. (Frank N. Meyer medal 1982). Subspecialties: Plant genetics; Evolutionary biology. Current work: Genetics and cytogentics of the tomato (Lycopersicon) species; crossability; F1 features; controlled introgression; genetic distance based on allozyme context. Office: U Calif Davis CA 95616

RICKBORN, BRUCE F., chemistry educator; b. New Brunswick, N.J., Feb. 23, 1935; s. Frederick R. and Myrtle M. (Woitscheck) R.; m. Ida R. Vorse, Aug. 13,1955; 1 son, Steven. B.A., U. Calif.-Riverside, 1956; Ph.D., UCLA, 1960. Asst. prof. chemistry U. Calif.-Berkeley, 1960-62; asst. prof. chemistry U. Calif.-Santa Barbara, 1962-66, assoc. prof., 1966-70, prof., 1971—, dean, 1973-78. Co-author: Electrophilic Substitution of Organomercurials, 1968; contbr. articles to profl. jours. Mem. Am. Chem. Soc., AAAS, Sigma Xi. Subspecialty: Organic chemistry. Current work: Synthetic methods and mechanistic studies. Home: 4661 La Espada Dr Santa Barbara CA 93111 Office: Dept Chemistry U Calif Santa Barbara CA 93106

RICKERT, DOUGLAS EDWARD, toxicologist; b. Sioux City, Iowa, Jan. 27, 1946; s. Waldo and Genevieve (Hahn) R.; m. Terrie Sue Baker, Jne 27, 1981. B.S., U. Iowa, 1968, M.S., 1972, Ph.D. in Pharmacology, 1974. Diplomate: Am. Bd. Toxicology. Asst. prof. pharmacology Mich. State U., East Lansing, 1974-77; analytical biochemist Chem. Industry Inst. Toxicology, Research Triangle Park, N.C., 1977—. Contbr. articles to profl. jours. Served in U.S. Army, 169-71. Mem. Soc. Toxicology, Am. Soc. Pharmacology and Exptl. Therapeutics, Am. Soc. Mass Spectrometry, AAAS. Subspecialties: Toxicology (medicine); Pharmacology. Current work: Correlation of mammalian response to toxic agents with hepatic and extrahepatic biotransformation; mechanisms of chemically-induced carcinogenesis; foreign compound absorption, distribution, metabolism, excretion. Office: PO Box 12137 Research Traingle Park NC 27709

RICKS, STEPHEN ANDREW, mechanical engineer; b. Lafayette, Ind., June 25, 1947; s. Michael Theodore and Veronica Katherine (Jordan) R.; m. Diana Mucker, Apr. 15, 1972; children: Douglas Robert, Mary Ellen. B.S.M.E., Rose-Hulman Inst., 1969; M.S.E. in Transp. Engring, George Washington U., 1977. Registered profl engr., Va., 1975, Ind., 1978. Sales engr. Industry and Def. Group, Westinghouse Electric Corp., Chgo., 1969-73; project engr. for facility design DeLeuw, Cather & Co., Washington, 1975-77; asst. project mgr. for HUD emergency flood mapping program, 1973-75; sr. sales engr. Rostone div. Allen-Bradley Co., Lafayette, Ind., 1977-83; asst. city engr. City of Lafayette, 1983—; cons. engr., 1983—; speaker in field. Contbr. articles to profl. jours. Bb. dirs. Tippecanoe Arts Fedn., Lafayette; mem. Tippecanoe County Area Plan Commn., Lafayette, 1979—; pres. bd. dirs. Lafayette Symphony, Inc., 1981. Democrat. Roman Catholic. Subspecialties: Mechanical engineering; Composite materials. Current work: Municipal and consulting engineering. Home: 2540 Lafayette Dr Lafayette IN 47905 Office: 20 N 6th St Lafayette IN 47901

RIDDLE, DONALD LEE, geneticist; b. Vancouver, Wash., July 26, 1945; s. Joseph Gerald and Marjory Helen (Shelley) R.; m. Beverly Dianne Riddle, July 5, 1969; children: Brian Patrick, David Joseph. B.S., U. Calif.-Davis, 1968; Ph.D., Berkeley, 1971. Research assoc. U. Calif.-Santa Barbara, 1971-72; Jane Coffin Childs Meml. Fund fellow MRC Lab. Molecular Biology, Cambridge, Eng., 1973-75; asst. prof. biol. scis. U. Mo.-Columbia, 1975-81, assoc. prof., 1981—; dir. Caenorhabditis Genetics Ctr., 1979—. Contbr. articles to profl. jours. NIH grantee, 1977—; career devel. awardee, 1981—. Mem. Genetics Soc. Am., Soc. Devel. Biology, Soc. Nematologists. Subspecialties: Gene actions; Developmental biology. Current work: Research in developmental genetics, neurobiology. Office: Dept Biol Scis U Mo Columbia MO 65211

RIDER, PAUL EDWARD, educator; b. Des Moines, Nov. 22, 1940; s. John Wesley and Thelma Marie (Hulett) R.; m. Carole Catherine Teresavich, Dec. 28, 1963; children: Paul Edward, Susan, Kathleen. B.A., Drake U., 1962; M.S., Iowa State U., 1964; Ph.D., Kans. State U., 1969. Instr. chemistry Drake U., 1964-66; vis. prof. Coe Coll., Cedar Rapids, Iowa, 1969; asst. prof. U. No. Iowa, Cedar Falls, 1969-73, assoc. prof., 1973-79, prof., 1979—; chem. cons. Shell Devel. Co., Houston, 1975—. AEC summer fellow Iowa State U., 1965; NSF summer fellow Kans. State U., 1966, Shell Devel. Co., 1974; NSF grad. fellow, NASA grad. fellow Kans. State U., 1967-69. Mem. Am. Chem. Soc., Am. Fedn. Musicians, AAAS, Iowa Acad. Sci., Sigma Xi, Kappa Mu Epsilon, Phi Kappa Phi, Delta Phi Alpha, Phi Lambda Upsilon. Subspecialties: Physical chemistry; Theoretical chemistry. Current work: Thermodynamic studies of hydrogen bonding, theoretical aspects of chem. bonding. Home: 1321 Catherine St Cedar Falls IA 50613 Office: Chemistry Dept U No Iowa Cedar Falls IA 5061

RIDGE, DOUGLAS POLL, chemist; b. Portland, Oreg., Nov. 9, 1944; s. Alfred and Ruth Marie (Poll) R.; m. Julie Brown, Mar. 23, 1971; children: Emily, Andrew, Timothy, Claron, Clark. A.B., Harvard Coll., 1968; Ph.D., Calif. Inst. Tech., 1972. Asst. prof. chemistry U. Del., Newark, 1972-78, assoc. prof., 1978—. Contbr. articles to profl. jours. NSF grantee, 1977—; Petroleum Research Fund grantee, 1973-80. Mem. Am. Chem. Soc., Am. Phys. Soc. Mem. Ch. Jesus Christ Latter-day Saints. Subspecialties: Physical chemistry; Kinetics. Current work: Ion molecule reactions; ion cyclotron resonance spectroscopy; gas phase chemistry of transition metal ions; dynamics of ion molecule collisions. Office: Dept Chemistry U Del Newark DE 19711

RIECKEN, HENRY WILLIAM, psychologist; b. Bklyn., Nov. 11, 1917; s. Henry William and Lilian Antoinette (Nieber) R.; m. Frances Ruth Manson, Aug. 7, 1955; children—Mary Susan, Gilson, Anne. A.B., Harvard, 1939, Ph.D., 1950; M.A., U. Conn., 1941. Social sci. analyst Dept. Agr., 1941-46; teaching fellow Harvard, 1947-49, lectr. social psychology, research asso. clin. psychology, 1949-54; asso. prof., then prof. sociology and sr. mem. Lab. Research Social Relations, U. Minn., 1954-58; program dir. social sci. research NSF, 1958-59; head Office Social Sci., 1959-60, asst. dir. social scis., 1960-64, asso. dir. sci. edn., 1964-66; v.p. Social Sci. Research Council, 1966-69; pres., 1969-71; prof. behavioral scis. U. Pa., 1972—; now on leave as Sr. program adviser Nat. Library Medicine; fellow Center for Advanced Study in Behavioral Scis., Stanford, Calif., 1971-72; Paterson Meml. lectr. U. Minn., 1970; Jensen lectr. Duke U., 1973; charter mem. inst. medicine Nat. Acad. Scis.; adv. com. to dir. NIH, 1966-70, chmn. internat. centers com., 1970-73; pres. Am. Psychol. Found., 1971-73; vice chmn., chmn. com. on nat. needs for biomed. and behavioral research personnel NRC, 1975-80; mem. commn. on sociotech. systems Nat. Acad. Scis., 1976-79. Author: The Volunteer Work Camp, 1952, When Prophecy Fails, 1956, Social Experimentation, 1974, Experimental Testing of Public Policy, 1976, also articles.; Contbr. profl. jours. Bd. dirs. Found. for Child Devel. (formerly Assn. Aid Crippled Children), N.Y.; chmn. bd. trustees Bur. Social Sci. Research. Served with USAAC, 1943-45. Recipient Harold M. Hildreth award Am. Psychol. Assn., 1971. Fellow Am. Psychol. Assn., Am. Acad. Arts and Scis.; mem. Am. Sociol. Assn., Am. Assn. Pub. Opinion Research, Sociol. Research Assn. (pres. 1966). Clubs: Harvard (N.Y.C.); Cosmos (Washington). Office: Nat Library Medicine Bethesda MD

RIEDEL, RICHARD ANTHONY, orthodontist, educator; b. Milw., Feb. 26, 1922; s. Otto Louis and Julia (Kub) R.; m. Marie Emma (Myers), Apr. 7, 1945; children: Richard, Thomas, Corinne, Carol. D.D.S., Marquette U., 1945; M.S.D., Northwestern U., 1948. Diplomate: Am. Bd. Orthodontics. Instr. orthodontics Northwestern U., Evanston, Ill., 1948-49, U. Wash., Seattle, 1949-55, asst. prof., 1954-59, assoc. prof. to prof., 1959-83, prof. emeritus, 1983—, acting dean, 1980-81, chmn. orthodontics dept., 1965-75, assoc. dean for acad. affairs and admissions, 1978-80; pvt. practice orthodontics, Seattle, 1949—. Author: Vistas in Orthodontics, 1966. Charter mem. Seattle Big Bros., 1950. Served to lt. (j.g.) USN, 1945-47; PTO. Recipient Alfred Ketchum award for research in orthodontics Am. Assn. Orthodontists, 1983. Mem. AAAS, Am. Bd. Orthodontics (pres. 1978-79), E.H. Angle Soc. Orthodontics (pres. 1971-72), Sigma Xi, Omicron Kappa Upsilon. Roman Catholic. Lodge: Sons of Norway. Subspecialty: Orthodontics. Current work: Long-term changes in human dentition; evaluation of post-treated dentitions in humans.

RIEGLER, GEROLD ERNST, research scientist, project manager; b. Enns, Austria, June 8, 1956; s. Ernst Ignaz and Gertrude Rosina (Roitinger) R. Diploma, Tech. U. Vienna, 1979; Ph.D., Tech. U. Graz, 1982. Engr. FAG, Schweinfurt, F.R. Germany, 1975-76; scientist Patent Agy., Vienna, 1979; research scientist, project mgr. wind energy Research Ctr. Graz, Austria, 1979—. Author: Windturbines for Tethered Systems, 1983. Mem. AIAA. Roman Catholic. Subspecialties: Wind power; Fluid mechanics. Current work: Investigations of innovative wind energy concepts like tethered wind systems; principles of the energy extraction from a free stream, wind tunnel tests. Patentee a wind power plant, 1982. Home: Raiffeisenstrasse 50b Graz Austria A-8010 Office: Research Ctr Graz Inffeldgasse 12 Graz Austria A-8010

RIEHL, ROBERT MICHAEL, physiologist, educator; b. Borger, Tex., Aug. 8, 1951; s. Glenn Herald Riehl and Mary June (Denney) Balacki; m. Mary Elizabeth Wolf, Dec. 30, 1978; 1 dau., Frances Elizabeth. B.S., West Tex. State U., 1973, M.S., 1975; Ph.D., U. Tex.-San Antonio, 1980. Lab. instr. dept. biology West Tex. State U., Canyon, 1971-73, grad. teaching asst., 1973-75; grad. teaching asst. dept. physiology U. Tex. Health Sci. Ctr., San Antonio, 1975-80, research assoc. dept. ob/gyn, 1980-81; postdoctoral assoc. Mayo Found., Rochester, Minn., 1981—. NIH postdoctoral trainee, 1981-83. Student mem. Am. Physiol. Soc., Soc. for Study of Reprodn., Sigma Nu. Subspecialties: Receptors; Reproductive biology. Current work: Mechanisms of action of hormones and related biologically active molecules. Home: 624 7th St SW Rochester MN 55901-3757 Office: Dept Cell Biology Mayo Found Rochester MN 55905

RIELY, CAROLINE ARMISTEAD, internist; b. Washington, Feb. 1, 1944; d. John William and Jean Roy (Jones) R. A.B., Mt. Holyoke Coll., 1966; M.D., Columbia U., 1970. Diplomate: Am. Bd. Internal Medicine, 1973. Intern/resident Presbyn. Hosp., N.Y.C., 1970-73; fellow in liver disease Yale U. Sch. Medicine, New Haven, 1973-75, asst. prof. medicine, 1975-80, assoc. prof. internal medicine and pediatrics, 1978—; cons. Backus Hosp., Norwich, Conn., Hosp. St. Raphael, New Haven. ACP teaching and research scholar, 1975-78. Fellow ACP (v.p. Conn. chpt.); mem. Am. Fedn. Clin. Research. Subspecialties: Internal medicine; Pediatrics. Current work: Pediatric liver diseases; cholestatic liver diseases; the liver in pregnancy, imaging techniques in the liver. Home: 355 Willow St New Haven CT 06511 Office: Yale U School Medicine 333 Cedar St New Haven CT 06510

RIFKIN, BARRY RICHARD, experimental pathologist, dental educator; b. Trenton, Mar. 30, 1940; s. Samuel and Ida (Rosenthal) R.; m.; children: Avery, Carl. B.S., Ohio State U., 1961; M.S., U. Ill.-Champaign, 1964; D.D.S., Temple U., 1968; Ph.D., U. Rochester, 1974. NIH trainee U. Rochester, 1968-73, asst. prof. pathology, 1973-80; assoc. pathologist Strong Meml. Hosp., Rochester, N.Y., 1974-80; assoc. prof. dept. oral medicine chmn. dept. oral medicine NYU, 1980—. Contbr. numerous articles on bone resorption and pathogenesis of periodontal disease to profl. jours., 1976—. NIH grantee, 1976-80. Mem. Internat. Assn. Dental Research, AAAS, Internat. Acad. Pathology, Am. Assn. Dental Schs., Am. Soc. Bone and Mineral Research, Sigma Xi. Democrat. Jewish. Subspecialties: Oral biology; Pathology (medicine). Current work: Pathogenesis of bone loss in periodontal disease; mechanisms of bone resorption origin, structure and function of osteoclast. Office: NYU Dental Ctr 421 First Ave New York NY 10010

RIFKIND, ARLEEN B., pharmacologist, educator; b. N.Y.C., June 29, 1938; d. Michael C. and Regina (Gottlieb) Brenner; m. Robert S. Rifkind; children: Amy, Nina. B.A., Bryn Mawr Coll., 1960; M.D., NYU, 1964. Diplomate: Nat. Bd. Med. Examiners. Intern, then resident NYU-Bellevue Hosp., N.Y.C., 1964-65; clin. assoc. NIH, Bethesda, Md., 1965-68; research asso. Rockefeller U., N.Y.C., 1968-71, adj. asst. prof., 1971-74; asst. prof. pediatrics Cornell U. Med. Coll., N.Y.C., 1971-75, asst. prof. medicine, 1971-83, asso. prof., 1983—, asst. prof. pharmacology, 1973-78, asso. prof. pharmacology, 1978-83, prof., 1983—; asst. resident physician Rockefeller U. Hosp., 1968-71; physician outpatient dept. N.Y. Hosp., 1968-71, asst. attending physician, 1971-77, clin. affiliate in medicine, 1977—. Contbr. numerous articles and chpts. to sci. lit. Mem. environ. health scis. revue com. Nat. Inst. Environ. Health Scis.; mem. ad hoc study sects. NIH. Recipient Andrew W. Mellon Tchr.-Scientist award Cornell U. Med. Coll., 1976-78; Nat. Inst. Child Health and Human Devel. staff fellow, 1965-68; USPHS spl. fellow, 1968-70, 71-72. Mem. Am. Soc. Clin. Investigation, Am. Soc. Pharmacology and Exptl. Therapeutics, Am. Soc. Clin. Pharmacology, Endocrine Soc., AAAS, N.Y. Acad. Scis. Subspecialties: Pharmacology; Toxicology (medicine). Current work: Research in heme synthesis and P-450 function, mechanisms of polyhalogenated hydrocarbon toxicity and developmental pharmacology and toxicology. Office: Cornell U Med Coll 1300 York Ave New York NY 10021

RIFKIND, RICHARD ALLEN, physician; b. N.Y.C., Oct. 26, 1930; s. Simon H. and Adele (Singer) R.; m. Carole Lewis, June 24, 1956; children—Barbara, Nancy. B.S., Yale U., 1951; M.D., Columbia U., 1955. Diplomate: Am. Bd. Internal Medicine. Intern. Presbyn. Hosp., N.Y.C., 1955-56, resident, 1957-61, dir. hematology, 1972-81; asst. prof. dept. medicine Columbia U., 1963-67, asso. prof., 1967-70, prof., 1970-81, dir. comprehensive, 1970-81, chmn. dept. genetics, 1980-81; dir. Grad. Sch. Meml. Sloan-Kettering Cancer Center, N.Y.C., 1981—. Contbr. articles in field. Served to capt. M.C. USAF, 1957-59. Mem. Am. Soc. Clin. Investigation, Am. Assn. Physicians, Am. Soc. Hematology. Democrat. Jewish. Subspecialties: Genetics and genetic engineering (medicine); Hematology. Office: 1275 York Ave New York NY 10021

RIGG, CARL WILSON, computer system architect, co. administrator; b. Penderyn, Wales, Apr. 24, 1947; s. Alfred Carroll and Catherine Coronwen (Price) R.; m. Pamela Michele Gross, June 14, 1979. B.S. in Physics, U. Birmingham, Eng., 1968. Software mgr. Bracknell & Stevenage, 1968-80; mgr. Synertek, Santa Clara, Calif., 1980-81; Software devel. mgr. Summit subs. Exxon, Palo Alto, Calif., 1981-82; UNIX project mgr. Onyx, San Jose, Calif., 1982—. Mem. Assn. Computing Machinery. Subspecialties: Computer architecture; Operating systems. Current work: Computer systems development; administrator operating system development and computer system design. Office: 25 E Tremble Rd San Jose CA 95131

RIGGIN, CHARLES HENRY, JR., molecular biologist, researcher; b. Little Rock., Aug. 1, 1941; s. Charles Henry and Mary Ellen (McDonald) R.; m. Mary Dale Chenault, Aug. 27, 1966. M.S., Memphis State U., 1970; Ph.D., Vanderbilt U., 1975. Research assoc. St. Louis U., 1975-77; vis. lectr. So. Ill. U.; fellow Mayo Clinic, Rochester, Minn., 1978-79; research assoc. oncology Johns Hopkins U., Balt., 1979—83; virologist Becton, Dickinson & Co., Balt., 1983—. Contbr. articles in field to proftl. jours. Mem. Am. Soc. Microbiology, AAAS. Subspecialties: Molecular biology; Genetics and genetic engineering (medicine). Current work: Virology, molecular biology of interferon gene expression, interferon, gene expression, DNA cloning. Office: Dept Oncology Johns Hopkins 600 N Wolfe St Baltimore MD 21205

RIGGS, BYRON LAWRENCE, JR., physician; b. Hot Springs, Ark., Mar. 24, 1931; s. Byron Lawrence and Elizabeth Ann (Patching) R.; m. Janet Templeton Brewer, June 24, 1955; children—Byron Kent, Ann Templeton. B.S., U. Ark., 1953, 1955, M.D., 1955; M.S. in Medicine, U. Minn., 1962. Diplomate: Am. Bd. Internal Medicine. Intern Letterman Army Hosp., San Francisco, 1958-59; resident in internal medicine Mayo Grad. Sch. Medicine Hosp., Rochester, Minn., 1958-61; asst. to staff Mayo Clinic, 1961; mem. staff internal medicine and metabolism Mayo Clinic and Found., 1962—; mem. faculty U. Minn. Med. Sch., 1962—, asso. prof., 1970—; prof. medicine Mayo Med. Sch., 1974—; chmn. div. endocrinology and metabolism Mayo Clinic and Med. Sch., 1974—; mem. gen. medicine B study sect. NIH, 1979-82. Contbr. articles to med. jours. Served with M.C. AUS, 1956-58. Recipient Mayo Found. postgrad. travel award, 1961; Kappa Delta award Am. Acad. Orthopedic Surgery, 1972; traveling fellow Royal Soc. Medicine, 1973. Fellow A.C.P.; mem. Am. Diabetes Assn., AMA, Am. Soc. Clin. Investigation, Endocrine Soc., Am. Fedn. Clin. Research (councillor Midwest sect. 1979-81), Am. Soc. Bone and Mineral Research Soc., Central Soc. Clin. Research (councillor), AAAS. Subspecialty: Endocrinology. Current work: Calcium and bone metabolism; osteoporosis. Home: 432 SW 10th Ave Rochester MN 55901 Office: 200 SW 1st St Rochester MN 55902

RIGGS, JACK EDWARD, neurology educator; b. Toledo, Oct. 6, 1949; s. Paul Henson and Bertha Jean (Terry) R.; m. Christine Marie Gunther, Sept. 18, 1978; 1 dau., Allison Jean. B.A., U. Toledo, 1972, B.S., 1972; M.D., U. Rochester, 1976. Diplomate: Am. Bd. Internal Medicine. Intern, resident in medicine Ohio State U. Hosps., Columbus, 1976-78; resident in neurology Strong Meml. Hosp., Rochester, N.Y., 1978-81; asst. prof. neurology W.Va. U., 1981—; mem. med. adv. bd. Myasthenia Gravis Found. Contbr. articles to profl. jours. Served to lt. comdr. M.C. USNR, 1980—. Mem. Am. Acad. Neurology. Subspecialty: Neurology. Current work: Neuromuscular disease, primarily metabolic myopathies. Office: Dept Neurology W Va U Med Ctr Morgantown WV 26505

RIGGS, KARL A., JR., geologic cons., educator; b. Thomasville, Ga., Aug. 12, 1929; s. Karl. A. and Marjorie Elizabeth (Urquhart) R.; m. Patricia Ann Hartrick, June 28, 1952; children: George, Kathryn Ann Riggs Keen, Linda Kay. B.S. with honors in Geology, Mich. State U., 1951; M.S., 1952; Ph.D. in Geology, Iowa State U., 1956. Cert. profl. geologist., Am. Inst. Profl. Geologists. Research assoc. Iowa State U., 1953-56, instr., 1952-56; sr. research technologist Mobil Research and Devel. Lab., Dallas, 1956-59; asst. prof. Western Mich. U., Kalamazoo, 1966-68; mem. faculty dept. geology Miss. State U., 1968—, assoc. prof., 1973—. Author: Principles of Rock Classification, 1975; contbr. articles to profl. jours. Fellow Geol. Soc. Am.; mem. Am. Assn. Petroleum Geologists, Mineral. Soc. Am., Assn. Engring.,

Geologists. Republican. Methodist. Subspecialty: Geology. Current work: Petroleum, mining and engineering geology. Home: 109 Grandridge Dr Starkville MS 39759 Office: Box KR Mississippi State MS 39762

RIGGS, LORRIN ANDREWS, psychologist; b. Harput, Turkey, June 11, 1912; s. Ernest Wilson and Alice (Shepard) R.; m. Doris Robinson, 1937; 2 children. Ed., Dartmouth Coll., Clark U. NRC fellow biol. scis. U. Pa., 1936-37; instr. U. Vt., 1937-38, 39-41; faculty Brown U., 1938-39, 41—, asst. prof. psychology, assoc. prof., to 1951, prof., 1951—, L. Herbert Ballou Found. prof., 1960-68, Edgar J. Marston U. prof., 1968-77, prof. emeritus, 1977—. Contbr. articles to profl. jours. Guggenheim fellow, 1971-72; recipient Charles F. Prentice award Am. Acad. Optometry, 1973, Disting. Sci. Contbn. award Am. Psychol. Assn., 1974, Kenneth Craik award Cambridge U., 1979. Mem. Am. Psychol. Assn. (div. pres. 1962-63), Eastern Psychol. Assn. (pres. 1975-76), AAAS, Optical Soc. Am. (recipient Edgar D. Tillyer award 1969), Nat. Acad. Scis., Am. Physiol. Soc., Internat. Brain Research Orgn. Soc. Neurosci., Internat. Soc. Clin. Electrophysiology of Vision, Soc. Exptl. Psychologists (recipient Howard Crosby Warren medal 1957), Am. Acad. Arts and Scis., Assn. for Research in Vision and Ophthalmology (pres. 1977, recipient Jonas S. Friedenwald award 1966). Subspecialties: Sensory processes; Psychophysics. Current work: Research in human vision. Office: Hunter Lab Psycholog Brown U Providenc RI 02912

RIGGS, ROBERT DALE, plant pathology educator, nematologist; b. Pocahontas, Ark., June 15, 1932; s. Rosa MacDowell and Grace (Million) R.; m. Jennie Lee Willis, June 6, 1954; children: Rebecca Dawn, Deborah Lee, Robert Dale, James Michael. B.S.A., U. Ark., 1954, M.S., 1956; Ph.D., N.C. State U., 1958. Asst. prof. dept. plant pathology U. Ark., Fayetteville, 1958-62, assoc. prof., 1962-68, prof., 1968—. Editor: Nematology in the Southern Region of the United States, 1982. Mem. Soc. Nematologists, Gamma Sigma Delta. Baptist. Subspecialty: Plant pathology. Current work: Biology and control of nematodes on crop plants in Arkansas, especially soybean cyst nematode on soybean. Home: 1840 Woolsey Ave Fayetteville AR 72701 Office: Dept Plant Pathology U Ark PS 217 Fayetteville AR 72701

RIGTERINK, PAUL VERNON, astronomer, cons.; b. Summit, N.J., May 14, 1942; s. Merle D. and Eleanor (Karch) R.; m. Ellen K. Saftlas, June 6, 1969; children: Amanda, Evan. B.A. in Math, Carleton Coll., 1964; Ph.D. in Astronomy, U. Pa., 1971. Vis. asst. prof. U. South Fla., 1971-72; successively task leader, sect. mgr., dept. mgr, staff Computer Sci. Corp., 1972-82, sr. staff, 1982—; also cons. Contbr. articles to profl. jours. Mem. Am. Astron. Soc., Am. Geophys. Union. Subspecialties: Satellite studies; Numerical analysis. Current work: Develop computer programs for analysis of satellite data: specialties include remote sensing, altitude determination and control and astronomy. Home: 9208 Orchard Brook Dr Potomac MD 20854 Office: Computer Sci Corp 8728 Colesville Rd Silver Spring MD 20910

RIKER, DONALD KAY, biomed. scientist, cons.; b. N.Y.C., Oct. 22, 1945; s. Walter F. and Virginia H. (Jaeger) R.; m. Leigh Bartley, Oct. 30, 1965; children: Scott B., Hal S. Student, Hamilton Coll., 1963-65; B.A., U. Kans., 1969; postgrad., Rockefeller U., 1968-70; Ph.D., Cornell U., 1977. Postdoctoral fellow in pharmacology Yale U. Sch. Medicine, New Haven, 1976-79; research staff scientist, 1979-80, research assoc., 1980-82, research affiliate, 1982—; research health scientist West Haven (Conn.) VA, 1980-81; sr. research investigator biomed. research Vicks Research Center, Richardson-Vicks, Inc., Shelton, Conn., 1982—; asst. to med. adviser Nat. Football League, N.Y.C., 1980—; reviewer Irma T. Hirschl Trust, N.Y.C., 1981—; Scientist U.S. Antarctic Research Program, NSF, 1968. Contbr. articles, abstracts to profl. lit. Recipient U.S. Antarctic Service medal U.S. Congress, 1969; Frank M. Chapman award Am. Mus. Natural History, 1971, 72; NIH nat. research service fellow, 1977-78; grantee Dystonia Med. Research Found., 1979-81. Mem. AAAS, Soc. Neurosci., Internat. Soc. Neurochemistry. Club: Yale (N.Y.C.). Subspecialties: Neuropharmacology; Neurochemistry. Current work: Neuropharmacology and neurochemistry of central nervous system; brain diseases; neurotransmitters; evaluation and clinical trial of psychoactive drugs. Home: 160 Haverford St Hamden CT 06517 Office: One Far Mill Crossing Shelton CT 06484

RIKER, JOSEPH THADDEUS, III, environmental consultant; b. Klamath Falls, Oreg., Oct. 24, 1940; s. Joseph Thaddeus Jr. and Joyce Lucille (Packard) R.; m. Barbara Lulay, June 25, 1960 (div.); children: Michelle L., Nicole M.; m. Joan Faith Marsh, Jan. 11, 1974; 1 dau., Justina D. B.S., Oreg. State U., 1962, M.S., 1964, Ph.D., Purdue U., 1966. Asst. prof. animal sci. U. N.H., Durham, 1966-69, Colo. State U., Ft. Collins, 1969-71; assoc. prof. environ. tech. Oreg. Inst. Tech., Klamath Falls, 1972-82; project dir. Klamath Cons. Service, Inc., Klamath Falls, 1978—; mem. Klamath County Rep. Central Com., 1973—. Bd. dirs. Westminster Found., Presbytery of the Cascades, 1978—. Mem. Am. Inst. Biol. Scis., Am. Soc. Animal Sci., Nat. Assn. Biology Tchrs., Phi Sigma. Republican. Presbyterian. Lodge: Elks. Subspecialties: Ecosystems analysis; Resource management. Current work: Water quality evaluation surface and groundwaters, environmental assessments, water resource management. Home: 5127 Hwy 39 Klamath Falls OR 97601 Office: 5127 Hwy 39 Suite A Klamath Falls OR 97601

RILEY, DAVID JOSEPH, pulmonary physiologist, educator; b. N.Y.C., Sept. 6, 1942; s. Edwin Glover and Gertrude (Pfanner) R.; m. Katherine Moran, June 9, 1969; children: Meredith Ann, Gavin Douglas. A.B., Johns Hopkins U., 1964; M.D., U. Md. Sch. Medicine, 1968. Diplomate: Am. Bd. Internal Medicine (subspecialty Pulmonary Diseases). Intern, jr. asst. resident Balt. City Hosps., 1968-70; fellow in pulmonary disease U. Pa., Phila., 1970-72; sr. asst. resident Johns Hopkins Hosp., Balt., 1972-73; asst. prof. medicine Rutgers Med. Sch., New Brunswick, N.J., 1973-77, assoc. prof. medicine, 1977—, adj. assoc. prof. physiology and biophysics, 1980—, assoc. mem. grad. faculty, 1981—. Contbr. numerous articles to sci. publs. Served to capt. USNG, 1966-73. Recipient NIH Pulmonary Acad. award, 1979. Fellow ACP (Alfred Stengel scholar London 1975), Am. Coll. Chest Physicians; mem. N.J. Thoracic Soc. (v.p. 1982), Am. Thoracic Soc. (program chmn. Eastern sect. 1982), Fedn. Clin. Research, Am. Physiol. Soc. Episcopalian. Clubs: Johns Hopkins (Balt.); Rutgers Faculty (New Brunswick). Subspecialties: Pulmonary medicine; Physiology (medicine). Current work: Experimental lung injury; lung corrective tissue; oxident injury; pulmonary physiology. Home: 282 Easton Ave New Brunswick NJ 08901 Office: Dept Medicine CN 19 Rutgers Med School New Brunswick NJ 08903

RILEY, JOHN THOMAS, chemist, educator; b. Bardstown, Ky., Apr. 2, 1942; s. John N. and Mary F. (Jury) R.; m. Rita Carroll, Dec. 23, 1963; children: Sheila, John Paul. B.S., Western Ky. U., 1964; Ph.D., U. Ky., 1968. Asst. prof. Western Ky. U., Bowling Green, 1968-76, assoc. prof., 1976-81, prof. chemistry, 1981—, also dir.; cons. to local industry. Contbr. articles to profl. jours. Bd. dirs. Eastland Park Pool, 1974-80, pres., 1979, v.p., 1978. Grantee Inst. Mining and Minerals Research, 1979-81. Mem. Am. Chem. Soc., Ky. Acad. Scis., Sigma Xi. Democrat. Roman Catholic. Subspecialties: Coal Chemistry; Analytical chemistry. Current work: Development of new methods of analysis of components in coal and coal-derived products; anodic stripping Voltammetric analysis of trace materials. Home: 1511 Woodhurst Dr Bowling Green KY 42101 Office: Dept Chemistry Western Ky U Bowling Green KY 42101

RILEY, MICHAEL VERITY, biochemistry educator; b. Bradford, Yorkshire, Eng., Dec. 27, 1933; came to U.S., 1967; s. Maurice William and Muriel (Verity) R.; m. Stephanie C. Mueller, June 29, 1963; children: Christopher, Robert, Kirsten, Stuart. B.A., Cambridge (Eng.) U., 1955, M.A., 1960; Ph.D., Liverpool (Eng.) U., 1961. Postdoctoral fellow Johns Hopkins Sch. Medicine, Balt., 1961-62; sr. lectr. Inst. Ophthalmology, London U., 1962-69; assoc. prof. biochemistry Oakland U., Rochester, Mich., 1969-78, prof., 1978—; vis. prof. Washington U. Sch. Medicine, St. Louis, 1967-68, Welsh Nat. Sch. Medicine, Cardiff, Wales, 1975; grants reviewer Nat. Eye Inst., NIH, 1971-74; research cons. Cilco, Inc., Lcsage, W.Va., 1982-83. Contbrg. author: The Biochemistry of the Eye, 1970, Cell Biology of the Eye, 1982. Co-founder Rochester Youth Soccer League, 1979. Served to 2d lt. Brit. Army, 1955-57. NIH grantee, 1970-75; Nat. Eye Inst. grantee, 1969—. Mem. N.Y. Acad. Scis., Assn. for Research in Vision and Ophthalmology (sect. chmn. 1978), Sigma Xi (chpt. pres. 1983). Subspecialties: Biochemistry (medicine); Ophthalmology. Current work: Biochemistry and physiology of the eye; mechanisms of ion transport in the cornea and the energetic processes that control transparency; formation of aqueous humor; metabolism of the ciliary body and retina; free radical damage in the eye. Office: Inst Biol Sci Oakland U Rochester MI 48063

RILEY, MONICA, biochemist; b. New Orleans, Oct. 4, 1926; d. Chauncey and Maude R.; m. Vincent J. Lusby, Jan. 1, 1949 (div.); children: Adam, Christine, Katherine. A.B., Smith Coll., 1947; Ph.D., U. Calif.-Berkeley, 1960. Asst. prof. bacteriology U. Calif.-Davis, 1960-66; asso. prof. biochemistry SUNY-Stony Brook, 1966-75, prof., 1975—. Contbr. articles to profl. jours. Grantee NSF, NIH, Am. Cancer Soc. Mem. Am. Soc. Biol. Chemists, Am. Soc. Microbiologists, Genetics Soc. Am. Subspecialties: Genetics and genetic engineering (biology); Evolutionary biology. Current work: Researcher in molecular mechanisms of evolution of DNA and genome organization in bacteria. Office: Dept Biochemistry SUNY Stony Brook NY 11794

RIMEL, REBECCA WEBSTER, neurosurgery educator, researcher; b. Charlottesville, Va., Apr. 10, 1951; d. John Malangthon IV and Gladys Yvonne (Winebarger) R. B.S. in Nursing, U. Va., 1973; M.B.A., James Madison U., 1982. Head nurse U. Va., Charlottesville, 1973-74, coordinator med. out-patient facilities, 1974-75, nurse practitioner, 1975-77, instr. neurosurgery, 1977-80, asst. prof. neurosurgery, 1981—; dir. NIH Head Injury Study, Charlottesville, 1979—; prin. investigator Robert Wood Johnson Found., Princeton, N.J., 1981—, Pew Meml. Trust, Phila., 1982—. Co-author: Clinical Neurosurgery, 1982; contbr. articles to profl. jours. Bd. Dirs. Nat. Head Injury Found., Boston, 1982—; sec. Neurol. Reserach Inst., Culpeper, Va., 1981—; bd. dirs. Airport Commn., Charlottesville, 1983—. Kellogg Found. fellow, 1982-85. Mem. Am. Assn. Automotive Med. (membership com. 1983), Am. Assn. Neurol. Nurses, Am. Nurses Assn., Am. Pub. Health Assn., Va. State Nurses Assn. Methodist. Subspecialties: Neurology. Current work: Clinical research in neurosurgery, specifically in head injury; restoring social competence after injury; minor head injury and athletic head injuries. Home: 1703 Galloway Dr Charlottesville VA 22901 Office: Dept Neurosurgery U Va Box 180 Charlottesville VA 22908

RINE, DAVID C., computer science educator; b. Bloomington, Ill., Nov. 16, 1941. B.S., Ill. State U., 1963; Ph.D., U. Iowa, 1970. Prof. computer sci. Western Ill. U., Macomb, 1977—; chief examiner Ednl. Testing Service, Princeton, N.J., 1982—. Co-author 6 books in computer sci.; contbr. articles to profl. jours. Active Boy Scouts Am., 1973-82. Recipient cert. achievement Johns Hopkins U., 1981. Mem. IEEE Computer Soc. (sr.; past officer, chmn. ednl. activities 1977-82, 3 honor roll awards, spl. award 1977). Lutheran. Club: YMCA. Subspecialties: Information systems, storage, and retrieval (computer science); Programming languages. Current work: Bridging the gap between computer practice in business and educational studies by students; database systems, personal computing. Office: Computer Sci Div 447 Stipes Hall Western Ill U Macomb IL 61455

RINEHART, JOHN SARGENT, physicist, cons.; b. Kirksville, Mo., Feb. 8, 1915; s. Rupert Lloyd and Gertrude Jane (Upright) R.; m. Marion Sladky, Aug. 10, 1940; children: Margot, Eric. B.S., N.E. Mo. State Tchrs. Coll., 1934, A.B. in Physics, 1935, M.S., Calif. Inst. Tech., 1937, Ph.D., State U. Iowa, 1940. Tech. aide Nat. Def. Research Com., 1942-46; sr. physicist N.Mex. Sch. Mines, 1946-49; research scientist Naval Ordnance Test Sta., China Lake, Calif., 1950-53; asst. dir. Smithsonian Astro-phys. Obs., Cambridge, Mass., 1955-58; prof. mining engring. Golo. Sch. Mines, Golden, 1958-64; dir. research and devel. Coast and Geodetic Survey, Washington, 1964-66; sr. research fellow NOAA, Boulder, Colo., 1966-73; cons. in field. Author 3 books on explosives and their effects, 2 books on geysers and geothermal energy; contbr. articles to profl. jours. Mem. AAAS, Am. Phys. Soc., Am. Geophys. Union, Explorers Club, Sigma Xi. Club: Cosmos (Washington). Subspecialties: Geothermal power; Solid mechanics. Current work: Mechanical effects of explosives; geothermal and related geophysical phenomena. Home: PO Box 392 Santa Fe NM 87051

RINES, HOWARD WAYNE, geneticist; b. Portland, Ind., Feb. 19, 1942; s. Ray G. and N. Norene (Redford) R.; m. Donna R. Olsen, Aug. 22, 1965; children: David, Deborah, Kenneth. B.S. in Agrl. Sci, Purdue U., 1964, M.S. in Genetics, 1965; Ph.D., Yale U., 1969. Asst. prof. dept. botany U. Ga., Athens, 1971-76; research geneticist Agrl. Research Service, U.S. Dept. Agr., 1976—; adj. assoc. prof. dept. agronomy and plant genetics U. Minn., St. Paul, 1976—. Served to capt. U.S. Army, 1969-71. Nat. Merit scholar, 1960-64; Louis G. Ware fellow, 1964-67; NIH fellow, 1967-69. Mem. Am. Soc. Agronomy, Genetics Soc. Am., Internat. Soc. Plant Tissue Culture. Subspecialties: Plant genetics; Plant cell and tissue culture. Current work: Genetics of oats, anther culture, tissue culture, plant breeding, mutagenesis. Home: 4292 Nancy Pl Saint Paul MN 55112 Office: Dept Agronomy Univ of Minn Saint Paul MN 55108

RINKER, GEORGE ALBERT, JR., theoretical physicist, computer scientist; b. Lubbock, Tex., Feb. 7, 1945; s. George Albert and Georgia Jean R.; m. Helen Carol Rinker, Aug. 26, 1967 (div.); children: Thomas Kern, Kenneth Lewis, Daniel Wilson. B.A., Franklin Coll. of Ind., 1967; M.A., U. Calif., Irvine, 1970, Ph.D., 1971. Postdoctoral fellow Los Alamos Sci. Lab., 1971-73, staff mem., 1973-76, 77—; vis. scientist Kernforschungsanlage Julich, W.Ger., 1976-77; vis. lectr. U. Fribourg, Switzerland, 1977; cons. computer problems in environ. sci. Contbr. articles sci. jours. Grassroots organizer Alaska Coalition and other environ. groups. Mem. Am. Phys. Soc. Club: Los Alamos Mountaineers. Subspecialties: Atomic and molecular physics; Condensed matter physics. Current work: Exotic atoms, atomic processes in dense plasmas, studies of elec. and thermal conduction in stellar interiors. Office: MS B212 Los Alamos National Laboratory Los Alamos NM 87544

RIPPIE, EDWARD GRANT, pharmacy educator; b. Beloit, Wis., May 29, 1931; s. Edward George and Esther Audella (Stevens) R.; m. Dorothy Ruth Tegtmeyer, Sept. 24, 1955; 1 son, E. Glenn. B.S., U. Wis., 1953, M.S., 1956, Ph.D., 1959. Asst. prof. pharm. tech. Coll. Pharmacy, U. Minn., Mpls., 1959-62, assoc. prof. pharmaceutics, 1962-66, prof. pharmaceutics, 1966—, head. pharmaceutics dept., 1966-74. Mem. editorial adv. bd.: Jour. Pharm. Scis, Washington, 1978—; contbr. numerous articles to sci. publs. Served with U.S. Army, 1956-58. Fellow Acad. Pharm. Scis., Am. Pharm. Assn. (Ebert prize 1982); mem. Am. Chem. Soc., AAAS. Current work: Particulate solids mass transport within powder beds; thermodynamics of protein binding; hydrodynamics of drug dissolution; the study of pharmaceutical tablet internal structure via stress/strain viscoelastic analysis. Home: 2 N Mallard Rd North Oaks Saint Paul MN 55110 Office: Univ Minn Coll Pharmacy Minneapolis MN 55455

RIS, HANS, zoologist, educator; b. Bern, Switzerland, June 15, 1914; came to U.S., 1938, naturalized, 1945; s. August and Martha (Egger) R.; m. Hania Wislicka, Dec. 26, 1947 (div. 1971); children: Christopher Robert, Annette Margo; m. Theron Caldwell, July 14, 1980. Diploma high sch. teaching, U. Bern, 1936; Ph.D., Columbia, 1942. Lectr. zoology Columbia U., 1942; Seessel fellow in zoology Yale U., 1942; instr. zoology Johns Hopkins U., 1942-44; asst. Rockefeller Inst., N.Y.C., 1944-46, asso., 1946-49; asso. prof. zoology U. Wis., Madison, 1949-53, prof., 1953—. Mem. Am. Acad. Arts Scis., Nat. Acad. Scis., AAAS. Subspecialty: Cell biology. Current work: Cell biology. Researcher mechanisms of nuclear div., chromosome structure, cell ultrastructure, electron microscopy. Office: Zoology Research U Wis Madison WI 53706

RISCALLA, LOUISE BEVERLY, clinical psychologist, researcher; b. Paterson, N.J., Mar. 26, 1932; d. Houston Wheeler and Mary Louise (Cook) Mead; m. Albert Riscalla, July 19, 1958. B.A., Upsala Coll., East Orange, N.J., 1954; M.A., St. Louis U., 1956; Ph.D., Union Grad. Sch., Cin., 1978. Lic. psychologist, N.J.; cert. rehab. counselor, N.J. Rehab. counselor N.J. Rehab. Commn., Newark, 1956-58; psychiat. social worker Bergen Pines County Hosp., Paramus, N.J., 1958-65; clin. psychologist N.J. Diagnostic Ctr., Menlo Park, N.J., 1965-75, Woodbridge Emergency Reception and Child Diagnostic Ctr., Avenel, N.J., 1975-77, Greystone Park (N.J.) Psychiat. Hosp., 1977—. Mem. Am. Psychol. Assn., Nat. Rehab. Assn., Nat. Acad. Neuropsychology, Am. Acad. Behavioral Medicine (diplomate), Nat. Rehab. Counseling Assn. Episcopalian. Subspecialties: Neuropsychology; Psychobiology. Current work: Electromagnetic field theory, how psychological factors control the immune system, relationship of prostaglandin deficiency to mental illness, neurological and medical aspects of psychiatric illness. Home: 8 Lahiere Ave Edison NJ 08817

RISEBERG, LESLIE ALLEN, physicist, research and devel. co. exec.; b. Malden, Mass., July 23, 1943; s. George and Mollie (Linsky) R.; m. Marilyn Jane Oxman, Jan. 5, 1943; children: Jocelyn, Andrew. A.B. cum laude in Physics, Harvard U., 1964; Ph.D. in Physics (NASA fellow, Gilman fellow), Johns Hopkins U., 1968. Mem. tech. staff Bell Labs., Murray Hill, N.J., 1968, Tex. Instruments, Dallas, 1970-72; guest lectr. Hebrew U., Jerusalem, Israel, 1968-69; mem. tech. staff GTE Labs., Waltham, Mass., 1972-75, research mgr., 1975-81, dir., 1981—; Cons. Research div. Raytheon Co., 1969-70, Lawrence Libermore Lab., 1974. Contbr. articles to profl. jours. Mem. Am. Phys. Soc., Optical Soc. Am., IEEE, Electrochem. Soc., Sigma Xi. Subspecialties: Solid state physics; Electrical engineering. Current work: Solid state physics, optics, lasers, plasma physics, materials science, solid state devices, management of research and development. Patentee in field. Home: 47 Cedar Creek Rd Sudbury MA 01776 Office: GTE Labs Inc 40 Sylvan Rd Waltham MA 02254 Home: 47 Cedar Creek Rd Sudbury MA 01776

RISEN, WILLIAM MAURICE, JR., chemist; b. St. Louis, July 22, 1940; s. William Maurice and Frances Hazel R.; m. Katherine M. Daniel, June 20, 1964; children—Mary Elizabeth, Ann Katherine. Sc.B., Georgetown U., 1962; Ph.D., Purdue U., 1967. Asst. prof. chemistry Brown U., Providence, 1967-72, asso. prof., 1972-75, prof., 1975—, chmn. dept., 1972-80; cons. NSF. Contbr. articles to profl. jours. Mem. Am. Chem. Soc., Am. Phys. Soc. Subspecialties: Solid state chemistry; Materials. Current work: Spectroscopic and chemical studies of inorganic glasses; charge-transfer materials; ionic polymers; organometallic compounds. Office: Dept Chemistry Brown U Providence RI 02912

RISER, MARY ELIZABETH, geneticist, cell biology educator; b. Richland, Wash., Aug. 1, 1945; s. Manning Walker and Mary Virginia (Dillard) R.; m. Robert D. Colligan, Sept. 1, 1978. B.S., Tulane U., 1967; M.S., U. Tex.-Houston, 1970, Ph.D., 1973. Postdoctoral fellow Baylor Coll. Medicine, Houston, 1974-77, asst. prof. dept. cell biology, 1977—. Contbr. articles in field to profl. jours. Named Outstanding Trainee M.D. Anderson Hosp. and Tumor Inst., 1982. Mem. Am. Soc. Cell Biology, Tissue Culture Assn. (mem. council 1979-81, pres. Tex. br. 1980-81), Tex. Genetics Soc., Sigma Xi. Methodist. Subspecialties: Genetics and genetic engineering (biology); Cell and tissue culture. Current work: Genetic research into the controls of endocrine system using somatic cell hybrids and recombinant DNA technologies. Home: 5623 Dumfries St Houston TX 77096 Office: Baylor Coll Medicine 1200 Moursund St Houston TX 77030

RISLOVE, DAVID JOEL, educator; b. Rushford, Minn., Nov. 16, 1940; s. Elmer S. and Beatrice H. (Otis) R.; m. Susan M. Schacht, June 15, 1963; children: Kaye E., Lori J. B.A., Winona State U., 1962; Ph.D., N.D. State U., 1968. Chemist Mayo Clinic, Rochester, Minn., 1962; research and teaching asst. Iowa State U., Ames, 1963; Am. Petroleum Inst. research asst. N.D. State U., Fargo, 1964-68; prof. chemistry Winona (Minn.) State U., 1968—. Recipient Merit awards Winona State U., 1977-81. Mem. Am. Chem. Soc., No. Intercollegiate Conf. Episcopalian. Club: Barbershoppers. Lodge: Elks (Winona). Subspecialties: Organic chemistry; Synthetic chemistry. Current work: Synthesis and characterization of pyrroles, porphyrians and metalloporphyrins and polyporphyrins; kinetics of acid-base reactions in hydrocarbon solvents. Home: 745 47th Ave Winona MN 55987 Office: Pasteur Hall Winona State U Winona MN 55987

RISO, RONALD RAYMOND, neurophysiologist; b. Teaneck, N.J., Dec. 6, 1945; s. R. Richard and Beatrice (Winzenreid) R. B.S. in Elec. Engring, Cornell U., 1968; Ph.D. in Neurosci, U. Rochester, 1979. Research assoc. prof. dept. biomed. engring. Case Western Res. U., Cleve., 1979—; with Peace Corps, Brazil, 1968-71. Contbr. articles to profl. jours. Recipient John R. Bartlett Meml. award for excellence in med. research; NIMH trainee, 1972-75; USPHS grantee, 1975-77. Mem. Soc. for Neurosci., Rehab. Engring. Soc. N.Am. Roman Catholic. Subspecialties: Biomedical engineering; Neurophysiology. Current work: Neuroprosthetic devices research and development; mechanisms of tactile and proprioceptive sensation; sensory physiology research; experimental techniques for motor control in cerebral palsy. Office: Rehab Engring Ctr Cleve Met Gen Hosp 3395 Scranton Rd Cleveland OH 44109

RISSING, JOHN PETER, physicisn, medical educator; b. Ft. Wayne, Ind., Apr. 16, 1943; s. Walter J. and Dorothy E. (Eckrich) R.; m. Kathleen Louise Bloom, Apr. 16, 1972; children: Brian Gregory, Erika Rene, David Creighton. A.B. in Philosophy, Ind. U., 1965; M.D., St. Louis U., 1969. Diplomate: Am. Bd. Internal Medicine (subcert. in infectious diseases). Postgrad. tng. Ind. U. Med. Ctr., Indpls., 1969-74; asst. prof. medicine Med. Coll. Ga., Augusta, 1976-80, assoc. prof.

medicine, 1980—, chief infectious disease, 1976—, VA Med. Ctr. Ga., Augusta, 1980—, assoc. chief of staff for research, 1981—, also mem. clin. exec. bd. Author: Handbook of Alcoholism, 1982; contbr. articles to med. jours. Served to maj. U.S. Army, 1974-76. Recipient Excellence in Teaching award Med. Coll. Ga., 1977; NEH grantee, 1982. Fellow ACP; mem. Infectious Disease Soc. Am., Surg. Infection Soc., Am. Fedn. Clin. Research, So. Soc. Clin. Investigation, Alpha Omega Alpha. Roman Catholic. Subspecialties: Internal medicine; Infectious diseases. Current work: Antigen and antibody detection of Bacteroides fragilis infections; therapeutic efficacy of antimicrobials in abdominal infections; tissue cage animal models of monomicrobial and polymicrobial infections. Home: 16 Woodside Circle Evans GA 30809 Office: VA Med Ctr Downtown Div Augusta GA 30910

RITCHIE, JOSEPH MURDOCH, educator, pharmacologist, researcher scientist; b. Aberdeen, Scotland, June 10, 1925; came to U.S., 1956; s. Alexander Farquharson and Agnes Jane (Bremner) R.; m. Brenda Rachel Bigland, July 28, 1951; children: Alasdair John, Jocelyn Anne. B.S., Aberdeen U., 1944, London U., 1949, Ph.D., 1952, D.Sc., 1960; M.A. (hon.), Yale U., 1968. Research physicist Radar in Telecommunications Research Establishment, Malvern, Eng., 1944-46; research student biophysics dept. Univ. Coll. London, 1946-49, jr. lectr. physiology, 1949-51; mem. staff Nat. Inst. Med. Research, 1951-55; vis. asst. prof. pharmacology Albert Einstein Coll. Medicine, N.Y.C., 1956-57, assoc. prof. pharmacology, 1958-63, prof., 1963-68; overseas fellow Churchill Coll., Cambridge, Eng., 1964-65; prof., chmn. dept. pharmacology Yale U. Sch. Medicine, New Haven, 1968-74, dir. div. biol. scis., 1975-78, Eugene Higgins prof. pharmacology, 1968—. Contbr. articles to profl. jours. Fellow Univ. Coll. London, 1979. Fellow Royal Soc.; mem. Physiol. Soc., Brit. Pharmacol. Soc., Brit. Biophys. Soc., Can. Physiol. Soc., Am. Soc. Pharmacology and Exptl. Therapeutics, Am. Physiol. Soc. Club: Yale (N.Y.C.). Subspecialties: Neuropharmacology; Biophysics (biology). Current work: Molecular basis of conduction in nerve especially in nonmyelinated and demyelinated fibers. Home: 47 Deepwood Dr Hamden CT 06517 Office: Dept Pharmacology Yale U Med Sch 333 Cedar St New Haven CT 06510

RITCHIE, WALLACE PARKS, JR., surgeon, clinician, educator; b. St. Paul, Nov. 4, 1935; s. Wallace Parks and Alice (Otis) R.; m. Barbara C. Jewell, Aug. 20, 1960; children: Stephanie, David, Jessica. B.A., Yale U., 1957; M.D., Johns Hopkins U., 1961; Ph.D., U. Minn., 1971. Diplomate: Am. Bd. Surgery. Intern Yale-New Haven Med. Ctr., 1961-62; resident in surgery U. Minn., Mpls., 1969, instr., 1969-70; asst. prof. surgery U. Va., 1973-75, assoc. prof., 1975-76, prof., 1976-83; prof., chmn. dept. surgery Temple U. Sch. Medicine, Phila., 1983—. Mem.: editorial bd. Surgery, 1981—, Jour. Surg. Research, 1980—, Surg. Gastroenterology, 1981—; contbr. numerous articles to profl. jours., chpts. in books. Mem. Nat. Trust Hist. Preservation, Albermarle County Hist. Soc., Friends of Ash Lawn Adv. Com. Served to lt. col. U.S. Army, 1970-73. Decorated Army Commendation Medal; established investigator Am. Heart Assn.; USPHS research grantee, 1974. Fellow ACS; mem. Am. Gastroent. Assn., Assn. Acad. Surgery (pres. 1976-77), Soc. Univ. Surgeons (Am. Assn. Med. Colls. rep. 1979-82), Am. Surg. Assn. Subspecialties: Surgery; Gastroenterology. Current work: Gastrointestinal physiology. Office: Dept Surgery U Va Sch Medicine Box 181 Charlottesville VA 22908

RITT, PAUL EDWARD, electronics company executive; b. Balt., Mar. 3, 1928; s. Paul Edward and Mary (Knight) R.; m. Dorothy Ann Wintz, Dec. 30, 1950; children: Paul Edward, Peter M., John W., James T., Mary Carol, Matthew J. B.S. in Chemistry, Loyola U., Balt., 1950, M.S., 1952, Ph.D., Georgetown U., 1954. Research asso. Harris Research Lab., Washington, 1950-52; aerospace research chemist Melpar, Inc., Falls Church, Va., 1952-60, research dir., 1960-62, v.p. research, 1962-65, v.p. research and engring., 1965-67; v.p., gen. mgr. Tng. Corp. Am., 1965-67; pres. applied sci. div., applied tech. div. Litton Industries, Bethesda, Md., 1967-68; v.p., dir. research GTE Labs., Waltham, Mass., 1968—; instr. U. Va., 1956-58, Am. U., 1959—. Contbr. articles to profl. jours. Mem. dean's adv. council U. Mass. Sch. Engring.; mgmt. bd. advs. Worcester Poly. Inst.; adv. council Stanford U.; mem. Mass. High Tech. Council. Fellow Am. Inst. Chemists; mem. Am. Phys. Soc., Royal Soc. Chemistry, AAAS, IEEE, Am. Inst. Physics, Electrochem. Soc., Am. Vacuum Soc., Am. Ceramic Soc., Am. Chem. Soc., Washington Acad. Sci., N.Y. Acad. Sci., Sigma Xi. Subspecialties: Materials; Information systems (information science). Current work: Direction of research in materials; electronic devices, telecommunication systems and services. Patentee in field. Home: 36 Sylvan Ln Weston MA 02193 Office: 40 Sylvan Rd Waltham MA 02154

RITTER, ALFRED, aerospace research and development company executive; b. Bklyn., Mar. 15, 1923; m. Joyce Rimer, June 15, 1947; children: Michael Glenn, Erica Anne, Theodore William. B.S. in Aero. Engring, Ga. Inst. Tech., 1943, M.S., 1947; Ph.D., Cornell U., 1951. Research engr. Office of Naval Research, Washington, 1951-54; supr. aerophysics Ill. Inst. Tech., Chgo., 1954-58, instr., 1956-58; pres. Therm Advanced Research, Inc., Ithaca, N.Y., 1958-68; head aero. research dept. Calspan Corp., Buffalo, 1968-80; dir. tech. Calspan Field Services, Inc., Arnold Engring. Devel. Ctr. Div., Arnold Air Force Station, Tenn., 1981—; mem. adv. council U. Tenn. Space Inst., Tullahoma, Tenn., 1982—. Mem. Bd. Edn., Ithaca, N.Y., 1965-66. Served to 2d lt. AC U.S. Army, 1943-46; PTO. Mem. AIAA (chmn. Niagara Frontier sect. 1973, 74, v.p. tech. activities 1981—), AAAS, N.Y. Acad. Scis., Sigma Xi. Subspecialties: Aerospace engineering and technology; Fluid mechanics. Current work: Direct a comprehensive technology program which supports aerospace flight dynamics testing at Arnold Engineering Development Center. Home: 1409 Country Club Dr Tullahoma TN 37388 Office: Calspan Field Services Inc Arnold Engring Devel Ctr Division Mail Stop 430 Arnold Air Force Station TN 37389

RITTER, CARL ALAN, pharmacologist, educator; b. Confluence, Pa., Jan. 23, 1932; s. John M. and Louise (Frantz) R.; m. Jeanette Lois Smart, Jan. 18, 1936; 1 son, Alan B. Student, Gettysburg Coll., 1950-52, Syracuse U., 1952-55; Ph.D., SUNY, 1964. Postdoctoral fellow, Johnson Research Found. U. Pa., 1963-66, asst. prof., Sch. Vet. Medicine, 1966-71, head, Lab. Pharmacology, 1968-71, asso. prof., 1972—. Contbr. numerous articles to profl. jours. Mem. AAAS, Am. Soc. Microbiology, Am. Soc. Pharmacology and Exptl. Therapeutics, N.Y. Acad. Scis., Am. Coll. Vet. Toxicologists, Am. Coll. Vet. Pharmacology and Therapeutics, Phi Zeta. Subspecialties: Membrane biology; Cellular pharmacology. Current work: Altering permeability of cells to drugs and metabolites using vesicles made of normal phosholipds. Office: 3800 Spruce St Philadelphia PA 19174

RITTER, GERHARD X., mathematics educator, educator; b. Bochum, Westphalia, Ger., Oct. 27, 1936; came to U.S., 1955, naturalized, 1966; s. Karl Friedrich and Luzie (Golla) R.; m. Cheri Ann Reiche, June 1, 1963; children: Andrea Ann, Erika Renee. B.A., U. Wis., 1966, Ph.D., 1971. Instr. U. Wis. Ext. Ctr., Madison, 1967-70; prof. dept. math. U. Fla., Gainesville, 1971—, joint appointments dept. math, computer sci. and, 1980-82; research cons. NASA, 1979, Ctr. Info. Research, 1979—, co-dir., 1980-81; research geophysicist Texaco, Inc., Bellaire, Tex., 1981. Contbr. articles to profl. jours.; referee various tech. jours. Served with USAF, 1958-62. NASA fellow, 1979; USAF fellow, 1982; NSF grantee, 1982; Air Force Office Scientific Research grantee, 1983. Mem. Am. Math. Soc., Math. Assn. Am., IEEE, Soc. Indsl. and Applied Math. Democrat. Roman Catholic. Subspecialties: Applied mathematics; Graphics, image processing, and pattern recognition. Current work: Parallel image processing and pattern recognition techniques as applied to guided weapons systems. Home: 4107 NW 33 Pl Gainesville FL 32601 Office: Dept Math Fla Gainesville FL 32611

RITTER, ROGERS CHARLES, physics educator; b. Pleasanton, Nebr., Oct. 27, 1929; s. Fredrick Julian and Bertha Marie (Kareitzer) R.; m. Marlene Marilyn Hill, Aug. 27, 1950 (div. Nov. 1977); children: James Robert, William Charles, Michael George; m. Diane Irene Swope, Jan. 8, 1983. B.Sc., U. Nebr., 1952; Ph.D., U. Tenn., 1961. Instrument engr. Oak Ridge Gaseous Diffusion Plant, 1952-61; prof. physics U. Va., Charlottesville, 1961—; cons., Organon, Netherlands, 1973-76. Treas., bd. dir. Ephphata Village Corp., Charlottesville. Mem. Am. Phys. Soc. Subspecialties: Relativity and gravitation; Biophysics (physics). Current work: Precision rotations-test of general relativity and gravitation; fundamental constants. Patentee urological apparatus and method. Home: 117 Chestnut Ridge Rd Charlottesville VA 22901 Office: U Va Dept Physics Charlottesville VA 22901

RITTS, ROY ELLOT, JR., physician; b. St. Petersburg, Fla., Jan. 16, 1929; s. Roy Ellot and Dorothy (Bliss) R.; m. Hilda Joan Stump, June 19, 1953 (div. 1978); children—Leslie Sue, Graham Douglas, Ian Christopher; m. Kristin Gunderson, 1979. A.B., George Washington U., 1948, M.D., 1951. Diplomate: Am. Bd. Microbiology. Intern D.C. Gen. Hosp., 1951; fellow infectious diseases George Washington U., 1952; resident medicine George Washington U. Hosp., 1953; asso. prof. microbiology Georgetown U. Sch. Medicine, 1958-61, chmn. dept., 1959-64, professorial lectr. medicine, 1960, prof. microbiology and tropical medicine, 1961-64; dir. Inst. Biomed. Research, A.M.A., 1964-68, asst. dir. div. sci. activities, 1964-66, dir. med. research, 1966-68; professorial lectr. microbiology U. Chgo., 1964-68; chmn. dept. microbiology Mayo Clinic, 1968-79, head microbiology research, 1979—; prof. microbiology and oncology Mayo Grad. Sch. Medicine, Mayo Med. Sch., U. Minn.; asst. medicine Peter Bent Brigham Hosp., 1954; research fellow medicine Harvard, 1954; vis. investigator, research asso. Rockefeller Inst., 1955-58; cons. clin. pathology NIH; cons. immunology Nat. Naval Med. Center, Bethesda, Md.; lectr. microbiology Naval Dental Sch., 1958-64; mem. Am. Bd. Med. Microbiology, 1965-67; chmn. sci. adv. com. FDA, HEW, 1970-71, chmn. nat. adv. com. on diagnostic products, 1972-74, mem. commr.'s steering com., 1974-75, nat. food and drug adv. com., 1976-78; mem. nat. working party on therapy of lung cancer Nat. Cancer Inst., 1972-75, mem. carcinogenesis rev. com. A, 1974-79, mem. gastrointestinal tumor study group, 1975-80, immunology com. radiation therapy oncology group, 1974-79; chmn. immunology standards com. WHO/Internat. Union Immunology Socs., 1973—; cons. cancer immunology VA Central Office, 1974—; mem. Am. Bd. Med. Lab. Immunology, 1976—; mem. clin. immunology com. Nat. Com. Clin. Lab. Standards, 1980—; immunology com. Internat. Study Group for Cardiac Transplantation, 1981—. Contbr. articles on tumor transplantation and cellular immunology, immunotherapy to med. and sci. publs. Served from ensign to lt. (j.g.), M.C. USNR, 1948-56. Life Ins. Fund for Med. Research fellow, 1954-57. Fellow Royal Soc. Tropical Medicine, Am. Acad. Microbiology, Royal Soc. Health, A.C.P., Am. Coll. Chest Physicians, Assn. Clin. Scientists, Infectious Disease Soc. Am.; mem. Reticuloendothelial Soc., SAR, Am. Assn. Pathologists, Am. Fedn. Clin. Research, Am. Soc. Clin. Oncology, Am. Soc. Exptl. Pathology, Am. Soc. Microbiology, Am. Rheumatism Assn., Soc. Gen. Microbiology (Eng.), Soc. Exptl. Biology and Medicine, Am. Assn. Immunologists, Am. Soc. Cancer Research, Internat. Assn. Study of Lung Cancer (dir. 1974-76, sec. 1976-79), Am. Soc. Surg. Oncology, Brit. Soc. Immunologists, Sigma Xi, Sigma Chi, Phi Chi, Alpha Chi Sigma, Alpha Omega Alpha. Subspecialties: Cellular engineering; Cancer research (medicine). Current work: Human cell fusion; immunotherapy; tumor immunology; immune deficiency states; standardization. Office: Microbiology Research Mayo Clinic Rochester MN 55901

RIVERA, EZEQUIEL RAMIREZ, electron microscopist, biology educator; b. Alpine, Tex., Oct. 17, 1942; s. Ezequiel Gutierrez and German (Ramirez) R.; m. Dorkmai Koonwong, July 12, 1970; 1 dau., Angela Malee. B.S., Sul Ross State Coll., 1964; M.S., Purdue U., 1966; Ph.D., U. Tex., 1971-73. Mem. research staff, botany and plant pathology Purdue U.-West Lafayette, Ind., 1966-67; instr. botany U. Tex.-Austin, 1972; vis. asst. prof. biology U. Notre Dame (Ind.), 1973-74; asst. prof. biol. sci. U. Lowell, Mass., 1974-79, assoc. prof., 1980—; cons. electron microscopy local hosps., also dept. tropical medicine Mahidol U., Bangkok. Contbr. articles to profl. jours. Served to capt., Med. Service Corps. U.S. Army, 1967-70. U. Lowell grantee, 1979, 80; NSF grantee, 1981; WHO grantee, 1982. Mem. Bot. Soc. Am., New Eng. Soc. Electron Microscopy (dir. biol. scis. 1982—), Electron Microscopy Soc. Am., Am. Soc. Plant Physiologists, Sigma Xi, Gamma Sigma Epsilon, Alpha Chi Sigma. Subspecialties: Cell biology; Parasitology. Current work: Use of new electron microscopic and x-ray techniques for study of plant cells, development of molluscs, and reproduction and development of parasitic flukes. Office: Dept Biol Sci U Lowell Lowell MA 01854

RIVIER, CATHERINE LAURE, research endocrinologist; b. Lausanne, Switzerland, June 21, 1943; came to U.S., 1969; d. Raymond E. and Marcelle G. (Lambert) Gafner; m. Jean E.F. Rivier, Dec. 30, 1967; children: Lauraine, Cedric. Licence es Sciences, U. Lausanne, 1968, Ph.D., 1972. Research asst. Inst. Microbiology, Lausanne, 1968-69; grad. research fellow Baylor U. Coll. Medicine, Houston, 1969-70; postdoctoral fellow The Salk Inst., La Jolla, Calif., 1972-74, sr. research assoc., 1974-79, asst. research prof., 1979—. Mem. Endocrine Soc., Am. Physiol. Soc., Soc. Neurosci., Soc. Study of Reprodn., Research Soc. on Alcoholism, AAAS. Subspecialties: Neuroendocrinology; Reproductive biology. Current work: Elucidation of biological role of brain peptides regulating pituitary function. Home: 9674 Blackgold Rd La Jolla CA 92037 Office: 10010 N Torrey Pines Rd La Jolla CA 92037

RIVIER, NICOLAS YVES, physics educator, researcher; b. Lausanne, Vaud, Switzerland, Aug. 5, 1941; m. Lynn A. McElroy; children: Andre, Catherine. Dipl. Physics, U. Lausanne, 1964; Ph.D., Cambridge (Eng.) U., 1968. Asst. prof. UCLA, 1968-69; lectr. U. Calif.-Riverside, 1969-70; lectr. physics Imperial Coll., U. London, 1970—. Contbr. numerous articles to profl. jours. Mem. Am. Phys. Soc., Inst. Physics, European Phys. Soc. Subspecialties: Low temperature physics; Statistical physics. Current work: Glasses, structure and topology of disordered materials; gauge aspects of condensed matter. Home: 15 Coalecroft Rd London England SW15 6LW Office: Dept Physics Imperial Coll Prince Consort Rd London England SW7 2BZ

RIVIERE, GEORGE ROBERT, dental immunologist; b. Decatur, Ill., Feb. 26, 1943; s. Robert Frank and Mary Frances (Cowan) R.; m. Holliston Lee Brown, Aug. 14, 1971; children: Michael Andrew, Kathryn Holliston. B.A., Drake U., 1966; B.S.D., U. Ill., 1966, D.D.S., 1968, M.S., 1970; Ph.D., UCLA, 1973. Practice pediatric dentistry, Chgo., 1968-70; USPHS postdoctoral trainee UCLA Schs (Dentistry and Medicine), 1970-73, asst. prof., 1975-76, assoc. prof., 1977-82, prof., 1982—. Contbr. articles to profl. jours. Served to lt. comdr. USN, 1973-75. NIH awardee, 1976-81. Mem. AAAS, Am. Assn. Immunologists, Am. Assn. Dental Research, Internat. Assn. Dental Research. Subspecialties: Immunology (medicine); Pediatric dentistry. Current work: Tooth transplantation and dental caries immunology research. Office: UCLA Dental Research Inst Los Angeles CA 90024

RIVKIN, MAXCY CALVIN, indsl. research mgr.; b. Columbia, S.C., Mar. 31, 1937; s. Lewis Stanley and Jennie (Winter) R.; m. Judith Frances Hirschman, June 7, 1959; children: Victor Jay, Jan Winter. B.S. in Mech. Engring, U. S.C., 1959. Tech. service engr. Kraft div. Westvaco Corp., 1963, group leader tech. service, 1963-70; group leader process systems Covington Research Center, 1970-78; dir. research Laurel Research Center, Md., 1978—; mem. indsl. steering com. Dept. Energy/Nat. Bur. Standards. Served to lt. (j.g.) USN, 1959-62. Mem. AAAS, TAPPI, Instrument Soc. Am., N.Y. Acad. Sci., Soc. Rheology, History of Sci. Soc., Soc. History of Tech., Am. Mgmt. Assn. Subspecialties: Materials processing; Applied mathematics. Current work: Process dynamics, control and implementation of computer-based process control systems in the pulp and paper industry; directing research in paper product and process devel. Office: Westvaco Corp 11101 Johns Hopkins Rd Laurel MD 20707

RIZACK, MARTIN ARTHUR, biochemist, pharmacologist, physician, educator; b. N.Y.C., Nov. 19, 1926; s. Pincus and Lillie (Finerman) R.; m. Lea van Leeuwen, Mar. 26, 1964; children: Jonathan, Lillie, Joshua, Michele, Tina. A.B., Columbia U., 1946, M.D., 1950; Ph.D., Rockefeller U., 1960. Diplomate: Am. Bd. Internal Medicine. Intern, then asst. resident Bellevue Hosp.-Cornell U., 1950-51; asst., then chief resident St. Luke's Hosp., N.Y.C., 1953-56; fellow Rockefeller U., N.Y.C., 1957-60, asst. prof., 1960-65, asso. prof. biochemistry and pharmacology, also physician, 1965—; head Lab. Cellular Biochemistry and Pharmacology, 1967—. Contbr. articles to biochem. and pharmacol. jours.; Cons. editor: Med. Letter on Drugs and Therapeutics. Served to capt. M.C. USNR, 1951-53. Fellow ACP; mem. Am. Soc. Biol. Chemists, Am. Soc. Pharmacology and Exptl. Therapeutics. Subspecialties: Biochemistry (biology); Cellular pharmacology. Current work: Biochemistry and pharmacology of peptide hormones and their cellular mechanisms of action. Office: Rockefeller U 1230 York Ave New York City NY 10021

ROALES, ROBERT R., biologist, educator; b. N.Y.C., July 17, 1944; s. John and Gertrude (Buxo) R.; m. Francoise Andree Galland, June 6, 1945; 1 dau., Nicole. B.S., Iona Coll, New Rochelle, N.Y., 1966; M.S., NYU, 1969, Ph.D., 1973. Adj. asst. prof. biology NYU, N.Y.C., 1972-74, asst. research scientist, 1973-74; adj. asst. prof. biology LaGuardia Coll., N.Y.C., 1973-74; mem. faculty Ind. U.-Kokomo, 1974—, assoc. prof. anatomy and physiology, 1979—. Contbr. biol. articles to profl. publs. Recipient Amicus award Ind. U., 1976, 78. Mem. AAAS, Am. Fisheries Soc., Am. Inst. Biol. Scis., Am. Soc. Ichthyologists and Herpetologists, N.Y. Acad. Sci., Nat. Assn. Advisors for Health Profession, Sigma Xi. Subspecialties: Environmental toxicology; Physiology (biology). Current work: Effects of environmental contaminants on fish physiology; effects of growth factors on fish physiology. Office: Ind Univ 2300 S Washington St Kokomo IN 46902

ROANE, PHILIP RANSOM, JR., virologist; b. Balt., Nov. 20, 1927; s. Philip Ransom and Mattie (Brown) R.; m. Vernice Reed, Roane, Aug. 1, 1981. B.Sc., Morgan State Coll., 1952; Sc.M., Johns Hopkins U., 1960; Ph.D., U. Md., 1965-70. Cert. dir. clin. labs., Md., 1973. Virologist, dir. quality control Microbiol. Assocs., Inc., Bethesda, Md., 1964-72; asst. prof. virology Howard U., 1972-77, assoc. prof., 1977—; cons. HEM Research Inc., Rockville; mem. virology study sect. NIH, 1976-80; mem. rev. com. viral and rickettsial diseases U.S. Army, 1979-82. Contbr. articles in field. Served with USAAF, 1946-47. NIH grantee, 1973-74, 72-80; recipient Kaiser-Permanente award for Excellence in Teaching, 1979; Med. Student Council award for Inspirational Leadership, 1982; arad. Student Council Merit Award, 1982. Mem. Am. Assn. Immunologists, Am. Soc. Microbiology, N.Y. Acad. Sci. Subspecialties: Virology (biology); Virology (medicine). Current work: Molecular virology, viral immunology; current research involves studies of the molecular interaction beteen herpes simplex virus and hepatitis B virus. Home: 7503 Harpers Dr Fort Washington MD 20744 Office: Howard U Coll Medicine 520 W St NW Washington DC 20059

ROBB, WALTER LEE, manufacturing company executive; b. Harrisburg, Pa., Apr. 25, 1928; s. George A. and Ruth (Scantlin) R.; m. Anne Gruver, Feb. 27, 1954; children: Richard, Steven, Lindsey. B.S., Pa. State U., 1948; M.S., U. Ill., 1950, Ph.D., 1951. With Gen. Electric Co., 1951—; mgr. research/devel. Silicone Products Dept., Waterford, N.Y., 1966-68; venture mgr. Med. Devel. Ops., Schenectady, 1968-71; v.p., div. gen. mgr. Med. Systems Div., Milw., 1973—; industry rep. Med. Systems Ops. OTA Health Care Adv. Council. Recipient IR-100 award. Mem. Nat. Acad. Engring., Health Industry Mfg. Assn. (exec. com., bd. dirs.). Subspecialty: Chemical engineering. Patentee in field of membranes and gas separation. Home: 3665 Mary Cliff Ln Brookfield WI 53005 Office: PO Box 414 Milwaukee WI 53201

ROBBINS, FREDERICK CHAPMAN, physician; b. Auburn, Ala., Aug. 25, 1916; s. William J. and Christine (Chapman) R.; m. Alice Havemeyer Northrop, June 19, 1948; children—Alice, Louise. A.B., U. Mo., 1936, B.S., 1938; M.D., Harvard, 1940; D.Sc. (hon.), John Carroll U., 1955, U. Mo., 1958, D.Sci., U. N.C., 1979, Tufts U., 1983, Med. Coll. Ohio, 1983; LL.D., U. N.Mex., 1968. Diplomate: Am. Bd. Pediatrics. Sr. fellow virus disease NRC, 1948-50; staff research div. infectious diseases Children's Hosp., Boston, 1948-50, assoc. physician, assoc. dir. isolation service, asso. research div. infectious diseases, 1950-52; instr., assoc. in pediatrics Harvard Med. Sch., 1950-52; dir. dept. pediatrics and contagious diseases Cleve. Met. Gen. Hosp., 1952-66; prof. pediatrics Case-Western Res. U., 1952-80, dean, 1966-80, dean emeritus, 1980—; pres. Inst. Medicine, Nat. Acad. Scis., 1980—; vis. scientist Donner Lab., U. Calif., 1963-64. Served as maj. AUS, 1942-46; chief virus and rickettsial disease sect. 15th Med. Gen. Lab.; investigations infectious hepatitis, typhus fever and Q fever. Decorated Bronze Star, 1945; recipient 1st Mead Johnson prize application tissue culture methods to study of viral infections, 1953; co-recipient Nobel prize in physiology and medicine, 1954; Med. Mut. Honor Award for, 1969; Ohio Gov.'s award, 1971. Mem. Am. Epidemiol. Soc., Am. Acad. Arts and Scis., Am. Soc. Clin. Investigation (emeritus mem.), Am. Acad. Pediatrics, Soc. Pediatric Research (pres. 1961-62, emeritus mem.), Am. Assn. Immunologists, Am. Pediatric Soc., Am. Philos. Soc., Phi Beta Kappa, Sigma Xi, Phi Gamma Delta. Subspecialty: Pediatrics. Home: 7021 Oak Forest Ln Bethesda MD 20817 Office: 2101 Constitution Ave NW Washington DC 20418

ROBBINS, HERBERT ELLIS, mathematics educator; b. New Castle, Pa., Jan. 12, 1915; s. Mark Louis and Celia (Klebansky) R.; m. Mary Dimock, 1943 (div. 1955); children: Mary Susannah, Marcia (Mrs. Weston T. Borden); m. Carol Hallett, 1966; children—Mark Hallett, David Herbert, Emily Carol. A.B., Harvard, 1935, Ph.D., 1938; Sc.D., Purdue U., 1974. Instr. math. N.Y. U., 1939-42; asso. prof. math. statistics U. N.C., 1946-50, prof., 1950-52; prof. math. statistics Columbia, 1953-66, 68—; prof. math. U. Mich., Ann Arbor, 1966-68; mem. Inst. Advanced Study, 1938-39, 52-53. Author: (with R. Courant) What is Mathematics?, 1941, (with Y.S. Chow and D.

Siegmund) Great Expectations: The Theory of Optimal Stopping, 1971, (with J. Van Ryzin) Introduction to Statistics, 1975; Contbr. articles to profl. jours. Served with USNR, 1942-46. Guggenheim fellow, 1952, 75. Mem. Inst. Math. Statistics (pres. 1966), Internat. Statis. Inst., Nat. Acad. Scis., Am. Acad. Arts and Scis. Subspecialties: Probability; Statistics. Current work: Theory of sequental experimentation; empirical Bayes methods; stochastic approximation, applications of statistics in the law. Office: Dept of Statistics Columbia U New York NY 10027

ROBBINS, KELLY ROY, animal nutrition scientist, educator, researcher; b. Watertown, N.Y., May 23, 1953; s. Carlton James and Rosemary Elaine (Kelly) R.; m. Regina Ellen Baker, June 7, 1975; children: Reigan Jacoba, Kelly Roy. A.S., SUNY-Canton, 1972; B.S., Cornell U., 1975; M.S., U. Ill., Champaign, 1977, Ph.D., 1979. Research asst. U. Ill., 1975-79; asst. prof. animal sci. U. Tenn., 1979-82, assoc. prof., 1982—; cons. to animal feeds industry. Contbr. articles, abstracts to profl. jours. Recipient Biomed. Research award U. Ill., 1977, Faculty Devel. award U. Tenn., 1982; Dept. Agr. Sci. Edn. Administrn.-Coop. Research grantee, 1981. Mem. Inst. Food Technologists, Poultry Sci. Assn., Am. Inst. Nutrition, N.Y. Acad. Scis. Republican. Methodist. Subspecialties: Animal nutrition; Food science and technology. Current work: Nutrition, research, edn. Home: 4604 Silverhill Knoxville TN 37921 Office: U Tenn 208 Brehm Hall Knoxville TN 37901

ROBBINS, LEONARD G(ILBERT), geneticist; b. Bklyn., Aug. 10, 1945; s. William Henry and Bertha (Elkind) R.; m. Ellen Swanson; 1 son, Daniel Chaim. B.A., CUNY, 1965; Ph.D., U. Wash., 1970. Postdoctoral trainee U. Tex., Austin, 1970-72; vis. scientist U. Wash., Seattle, summer 1976; asst. prof. Mich. State U., East Lansing, 1972-77, assoc. prof. dept. zoology, 1977-83, prof. dept. zoology, 1983—; vis. scientist U. Calif., San Diego, summer 1978; NIH sr. internat. fellow Centro de Biologia Molecular Universidad Autonoma de Madrid, 1980-81. Contbr. articles to sci. jours. NSF grantee, 1975-79, 79-81, 82—. Mem. Fedn. Am. Scientists, AAAS, Genetics Soc. Am. Jewish. Club: Metropolitan Flying (Lansing). Subspecialties: Genome organization; Genetics and genetic engineering (medicine). Current work: Organization of chromosomes with particular attention to analysis of ribosomal gene clusters and recombination among them. Role of maternal gene information during development and the transition to zygotic gene activity. Office: Mich State U East Lansing MI 48915

ROBBINS, MARION LERON, plant breeder, educator; b. Inman, S.C., Aug. 18, 1941; s. Jack Dennis and Tena (Champion) R.; m. Margaret Eleanor Wilson, Aug. 25, 1965; children: Jack, Rona, Jeff, Kyle. B.S., Clemson U., 1964; M.S., La. State U., 1966; Ph.D., U. Md., 1968. Asst. prof. horticulture Iowa State U., Ames, 1968-72; assoc. prof. horticulture Clemson U., Charleston, S.C., 1972-79, prof., 1979-82; acting resident dir. Coastal Expt. Sta., Charleston, 1983—. Columnist: contbr. editor Am. Vegetable Grower, 1978—. Mem. Am. Soc. Hort. Sci. (pres. So. region 1982-83, J.C. Miller award 1967, pres. assn. coll. brs. 1963-64, L.M. Ware award 1963, dir. 1982-83), So. Assn. Agrl. Scientists (dir. 1982-83, J.B. Edmond award 1963), Agrl. Soc. S.C. (dir. 1983). Republican. Presbyterian. Clubs: Exchange (Charleston) (pres. 1976-77, dir. 1977-78. Subspecialties: Genetics and genetic engineering (agriculture); Plant genetics. Current work: Development of improved plant varieties through breeding and genetic engineering. Development of integrated production systems for crops. Administering agricultural research. Developer plant varieties. Home: 308 Cessna Ave Charleston SC 29407 Office: Clemson U Coastal Expt Sta 2865 Savannah Hwy Charleston SC 29407

ROBBINS, PHILLIPS WESLEY, biochemistry educator; b. Barre, Mass., Aug. 10, 1930; m.; children. A.B., DePauw U., 1952; Ph.D., U. Ill., 1955. Research assoc. Mass. Gen. Hosp., Boston, 1955-57; asst. prof. Rockefeller Inst., 1957-59; asst. prof. biochemistry MIT, 1959-62, assoc. prof., 1962-65, prof., 1965—; Am. Cancer Soc. prof. biochemistry Ctr. for Cancer Research, 1978—. Contbr. chpts., numerous articles to profl. publs. Recipient Eli Lilly award, 1966. Mem. Nat. Acad. Scis. Subspecialties: Biochemistry (biology); Cell and tissue culture. Home: 32 Temple St Boston MA 02114 Office: Center Cancer Research E17-23 MIT Cambridge MA 02139

ROBBINS, ROBERT JOHN, educator; b. Niles, Mich., May 10, 1944; s. Robert A. and A. Marjorie (McKinney) R. A.B., Stanford U., 1966; B.S., Mich. State U., 1974, M.S., 1975, Ph.D., 1977. Asst. prof. dept. zoology Mich. State U., East Lansing, 1976-82, assoc. prof., 1982—. Author: Genetics: A Brief Introduction, 1982, The Sweep of Life, 1982. Mem. Philosophy of Sci. Assn., Am. Inst. Biol. Sci., Ecol. Soc. Am., AAAS. Subspecialties: Ethology; Genome organization. Current work: Animal behavior, genetics. Home: 1015 Morgan St Lansing MI 48912 Office: Dept Zoology Mich State U East Lansing MI 48824

ROBBINS, ROBERT RAYMOND, botany educator; b. Des Moines, May 28, 1946; s. Harold Raymond and Marjorie Eunice (Brown) R.; m. Frances Bernice Robbins, Aug. 30, 1968 (div. Sept. 1972). B.S., Iowa State U., 1968; M.S., U. Ill., 1973, Ph.D., 1977. Undergrad. teaching asst. Iowa State U., Ames, 1966-68; grad. teaching asst. U. Ill., Urbana, 1971-75; instr. Ctr. Electron Microscopy, 1973, 75, lectr. botany, 1976, postdoctoral research assoc., 1977; asst. prof. botany U. Wis.-Milw., 1977—. Contbr. articles to profl. jours. Vol., Sta. WUWM. Served with U.S. Army, 1969-71. Mem. Bot. Soc. Am., Am. Inst. Biol. Scis., AAAS, Am. Bryological and Lichenological Soc., Nature Conservancy, Am. Fern Soc., ACLU, Bot. Club Wis., Wis. Acad. Scis., Arts, and Letters, Sigma Xi. Subspecialties: Cell biology; Reproductive biology. Current work: Research in cell biology and cell development of various reproductive structures of plants, including pollen, flagellated sperms of lower land plants, and mechanisms of fertilization in plants. Office: Dept Botany U Wis Milwaukee WI 53201

ROBECK, GORDON G., environmental engineer; b. Denver, Feb. 3, 1923; s. Martin J. and Anna (Selstad) R.; m. Ephrosinia Yaremko, Feb. 3, 1951; children: John, Paul, Mark. B.S.C.E., U. Wis., 1944; S.M., M.I.T., 1950. Commd. officer, san. engr. USPHS, various locations, 1944-74; dir. drinking water research div. U.S. EPA, Cin., 1974—. Contbr. articles on public health aspects of treating drinking water, nat. drinking water standards to profl. jours., chpts. to books. Recipient Gold medal for exceptional service EPA, 1978. Fellow ASCE (research prize 1965); mem. Am. Water Works Assn. (hon. life, outstanding service award 1979, research award 1970, numerous publ. awards), Internat. Water Supply Assn., Water Pollution Control Fedn., Nat. Acad. Engring. Subspecialties: Water supply and wastewater treatment; Civil engineering. Current work: Scientific and technical basis for national drinking water standards. Home: 7104 Bestview Terr Cincinnati OH 45230 Office: 26 W Saint Clair St Cincinnati OH 45268

ROBERDS, RICHARD MACK, nuclear engr.; b. Lawrence, Kans., June 22, 1934; s. Wesley M. and Dorothy (McBroom) R.; m. Marian Marchena, June 29, 1936; children: Michael R., Catherine M., Wendy M. Student, UCLA, 1952-54; A.B., Kans. U., 1956, M.A., 1963; Ph.D., Air Force Inst. Tech., 1975. Commd. 2d lt. U.S. Air Force, 1956, advanced through grades to col., 1978; dep. chief tech. div. (Air Force Weapons Lab.), Kirtland AFB, N.Mex., 1973-77, chief reconnaissance and weapon delivery div., Wright Patterson AFB, Ohio, 1977-80, ret., 1980; head engring. tech. dept. Coll. Engring. Clemson (S.C.) U., 1980—. Decorated Legion of Merit, D.F.C. (2), Air medal (8). Mem. Am. Nuclear Soc., Am. Phys. Soc., Am. Soc. Engring. Edn., Air Force Assn. Baptist. Club: Sertoma. Subspecialties: Nuclear engineering; Nuclear physics. Current work: Management consultant engineering technology. Home: 108 Brookwood Ln Clemson SC 29631 Office: Clemson U 126 Freeman Hall Clemson SC 29631

ROBERT, ANDRE, meteorologist; b. N.Y.C., Apr. 28, 1929; emigrated to Can., 1937, naturalized, 1967; s. Mathias and Irene (Grindler) R.; m. Marguerite Mercier, Aug. 15, 1953; children: Claire, Lise. B.A.Sc., Laval (Que.) U., 1951; M.Sc. in Meteorology, U. Toronto, 1953; Ph.D., McGill U., Montreal, 1965. Weather forecaster, then research scientist Govt. Can., Montreal, 1953-74; dir. Can. Meteorol. Centre, Montreal, 1974-81, sr. scientist, 1981—. Fellow Can. Meteorol. and Oceanographic Soc., Am. Meteorol. Soc. (2d Half Century award 1980); mem. Meteorol. Soc. Japan. Roman Catholic. Subspecialty: Meteorology. Current work: Development of efficient time integration schemes applicable to numerical models of the atmosphere. Home: 207 Desmarteau St Laval PQ Canada H7N 3N8 Office: 2121 Trans-Canada St Dorval PQ Canada H9P 1J3

ROBERTS, ALBERT SIDNEY, JR., mechanical engineer, educator; b. Washington, N.C., Sept. 16, 1935; s. Albert Sidney and Lorena Ruth (Jefferson) R.; m. Llewellyn Bowers, Sept. 21, 1957; children: Leigh, Sidney, Ellyn, Amanda. B.S. in Nuclear Engring, N.C. State U., 1957, Ph.D., 1965; M.S. in Mech. Engring, U. Pitts., 1959. Registered profl. engr., Va. Assoc. engr. Bettis Atomic Power Lab., Westinghouse Electric Corp., 1957-60; mem. faculty dept. mech. engring. Old Dominion U., Norfolk, Va., 1965-68, 70—, prof., 1973—; research engr. Swedish Atomic Energy Co., Studsvik, Sweden, 1968-69; mem. adv. com. Va. Ctr. Coal and Energy Research. Contbr. articles to profl. jours. NASA, Ford Found. grantee. Mem. ASME, AAAS, Am. Soc. Engring. Edn., Sigma Xi, Phi Kappa Phi, Tau Beta Pi. Episcopalian. Club: Torch (Norfolk). Subspecialties: Mechanical engineering; Nuclear engineering. Current work: Teaching and research in thermodynamics and heat transfer, active passive solar heating and nuclear reactors for power. Home: 5437 Glenhaven Crescent Norfolk VA 23508 Office: Dept Mech Engring Old Dominion U Norfolk VA 23508

ROBERTS, BRUCE ROGER, plant physiologist; b. Leonia, N.J., May 19, 1933; s. Clarence Roger and Florence (Kingsbury) R.; m. (div.); children: Amy V., Mark K. A.B., Gettysburg (Pa.) Coll., 1956; M.F., Duke U., 1960, Ph.D., 1963. Reseach plant physiologist Agrl. Research Service, U.S. Dept. Agr., Delaware, Ohio, 1963—. Contbr. chpts. to books, articles to profl. jours. Mem. Delaware City Park and Recreation Adv. Bd., 1982; mem. Delaware City Shade Tree Commn., 1978—. Mem. Am. Soc. Plant Physiologists, Internat. Soc. Arboriculture (research award 1975, author's citation award 1981, hon. life mem.), Am. Soc. Hort. Sci., Sigma Xi. Subspecialty: Plant physiology (agriculture). Current work: Physiological response of plants to environmental stress. Home: 72 W Winter St Delaware OH 43015 Office: 359 Main Rd Delaware OH 43015

ROBERTS, DANIEL ALTMAN, plant pathologist, educator; b. Micanopy, Fla., Jan. 8, 1922; s. Simon and Pearle (Altman) R.; m. Ruth Elizabeth (Remsen), Feb. 1, 1944; children: Peter R., Kathleen E. Roberts Veline, Stephen B. B.S., U. Fla., 1943, M.S., 1948; Ph.D, Cornell U., 1951. Asst. prof. Cornell U., Ithaca, 1951-55; assoc. prof. U. Fla., Gainesville, 1955-59, prof. plant pathology, 1959—. Author: (with C.W. Boothroyd) Fundamentals of Plant Pathology, 1972, 75, Fundamentals of Plant-Pest Control, 1978; contbr. articles to profl. jours.; cons. editor: McGraw-Hill Ency. Sci. and Tech., 1979—. Served to 1st lt. AUS, 1943-46. Recipient award of distinction Fla. Dept. Agr., 1976; Guggenheim fellow, 1958; NSF grantee, 1969. Mem. Am. Phytopath. Soc., Internat. Soc. Plant Pathologists, Soil and Crop Sci. Soc. Fla. Democrat. Presbyterian. Subspecialties: Plant pathology; Plant virology. Current work: Teaching plant pathology; research on physiology of plant viral infections and on diseases of forage legumes. Home: 204 SW 40th Terr Gainesville FL 32607 Office: Dept Plant Patholog U Fla Gainesville FL 32611

ROBERTS, DAVID CRAIG, chemistry educator; b. Madison, Wis., Feb. 21, 1948; s. Clifford Swain and Margaret) Lee (Haines) R.; m. Laura Beth Abbott, Mar. 17, 1979. B.A., U. Wis.-Madison, 1970; Ph.D., MIT, 1975. Postdoctoral assoc. dept. chemistry UCLA, 1976-77; asst. prof. dept. chemistry Rutgers U., New Brunswick, N.J., 1977—. NIH fellow, 1976-77, NIH research grantee, 1983—. Mem. Am. Chem. Soc. Subspecialties: Organic chemistry; Biophysical chemistry. Current work: Synthesis and reactions of peptides and related substances; peptide-protein interactions; biosynthesis and medicinal chemistry of peptides; chemistry/biology of human consciousness. Home: 129 Washington Rd Sayreville NJ 08872 Office: Dept Chemistry Rutgers U Busch Campus New Brunswick NJ 08903

ROBERTS, DONALD WILSON, biologist, insect pathologist; b. Phoenix, Jan. 20, 1933; s. Alpha Wilson and Rubye Clothilde (Finklea) R.; m. Mae Astrid Strand, June 17, 1959; children: Marc Donald, Sara Judith. B.S., Brigham Young U., 1957; M.S., Iowa State U., 1959; Ph.D., U. Calif.-Berkeley, 1964. NSF postdoctoral fellow Swiss Fed. Inst. Tech., Zurich, 1964-65; mem. staff Boyce Thompson Inst. for Plant Research, Ithaca, N.Y., 1965—, insect pathologist, 1974—; coordinator Insect Pathology Resource Center, 1979—. Editor: Pathogens of Medically Important Arthropods, 3d edit, 1982; mem. editorial bd.: Jour. Invertebrate Pathology, Intervirology; contbr. articles to profl. jours. USPHS fellow, 1959-64; NSF U.S./India exchange scientist awardee, 1978; grantee NIH, 1966-76, WHO, 1975—, Rockefeller Found., 1982, U.S. Dept. Agr., 1980—, AID, 1981—. Mem. Soc. for Invertebrate Pathology, AAAS, Entomol. Soc. Am., Entomol. Soc. Brazil, Am. Mosquito Control Assn., Internat. Orgn. for Biol. Control, Am. Soc. for Microbiology, Mycol. Soc. Am., Sigma Xi. Subspecialties: Insect pathology; Integrated pest management. Current work: Development of microorganisms (especially entomopathogenic fungi and viruses) for pest-insect control. Mode of action of insect pathogens (e.g., toxin production); utilization of microbial control in developing nations. Home: 9 Redwood Ln Ithaca NY 14850 Office: Boyce Thompson Inst Tower Rd Cornell U Ithaca NY 14853

ROBERTS, EDWARD BAER, technology management educator; b. Chelsea, Mass., Nov. 18, 1935; s. Nathan and Edna (Podradchik) R.; m. Nancy Helen Rosenthal, June 14, 1959; children: Valerie Jo, Mitchell Jonathan, Andrea Lynne. B.S., MIT, 1958, M.S., 1958, M.S. in Mgmt, 1960, Ph.D., 1962. Founding mem. systems dynamics program MIT, 1958—, instr., 1959-61, asst. prof., 1961-65, assoc. prof., 1965-70, prof., 1970—, David Sarnoff prof. mgmt. of tech., 1974—, assoc. dir. research program on mgmt. sci. and tech., 1963-73, chmn. tech. and health mgmt. group, 1973—; co-founder, pres. Pugh-Roberts Assocs., Inc., Cambridge, Mass., 1963—; also dir.; dir. MIT-Boston VA Joint Center on Health Care Mgmt., 1976-80; dir. MIT Joint Program on Mgmt. of Tech., 1980—, Med. Info. Tech. Inc., Cambridge, Bio Clin. Group, Inc., Zero Stage Capital Equity Fund; cons. Assn. Am. Med. Colls., also numerous corps.; mem. Task Force on Nuclear Medicine, ERDA, 1975-76, Nat. Research Council Com. on Ionizing Radiation Effects. Author: The Dynamics of Research and Development, 1964, (with others) Systems Simulation for Regional Analysis, 1969, The Persistent Poppy, 1975, The Dynamics of Human Service Delivery, 1976; prin. author, editor: Managerial Applications of System Dynamics, 1978; editor: Biomedical Innovation, 1981; editorial bd.: IEEE Trans. on Engring. Mgmt, 1968—, (with others) Indsl. Mktg. Mgmt, 1975—, Health Care Mgmt. Rev, 1976-78, Technol. Forecasting and Social Change, 1980—; contbr. articles to profl. jours. Mem. IEEE, Inst. Mgmt. Sci. (pres. Boston chpt. 1962-63), Sigma Xi, Tau Beta Pi, Eta Kappa Nu, Tau Kappa Alpha. Subspecialties: Management of technology; Systems engineering. Current work: Management of technical innovation; technical entrepreneurship. Home: 17 Fellsmere Rd Newton MA 02159 Office: 50 Memorial Dr Cambridge MA 02139

ROBERTS, EUGENE, neuroscientist, researcher, educator; b. Krasnodar, Russia, Jan. 19, 1920; s. Ruvim and Eva (Vassilevskaya) Rabinowitch; m. Ethel, June 22, 1941; children: Judith, Paul David, Miriam; m. Charlene Ruth, Dec. 23, 1977. B.S., Wayne State U., 1940; M.S., U. Mich., 1941, Ph.D, 1943. Asst. head inhalation sect. Manhattan Project, U. Rochester, N.Y., 1943-46; research assoc. Barnard Free Skin and Cancer Hosp. and Washington U. Sch. Medicine, St. Louis, 1946-54; chmn. dept. biochemistry City of Hope Research Inst., Duarte, Calif., 1954-68, chmn. div. neuroscis., 1968—; adj. prof. biochemistry U. So. Calif. Sch. Medicine, Los Angeles, 1970—; fellow Center Advanced Study in Behavioral Scis., Stanford, Calif., 1978; Mem. psychopharmacology study sect. NIH, 1961-66; mem. scientist exchange group to USSR NIMH, 1962; mem. profl. adv. sect. Scottish Rite Schizophrenia Research Program, La Jolla, Calif., 1966—; mem. com. brain scis., div. med. scis. NRC/Nat. Acad. Scis., 1966-70; mem. sci. adv. com. lab. behavior genetics U. Colo., Boulder, 1966-71; mem. research adv. com. Calif. Dept. Mental Hygiene, 1968-69; mem. bd. sci. counselors Nat. Inst. Neurol. Diseases and Stroke, 1969-73; mem. sci. adv. com. Found. Research in Hereditary Disease, Los Angeles, 1970—; mem. internat. adv. bd. Israeli Ctr. Psychobiology, Tel Aviv, 1973—; mem. med. adv. bd. A.L.S. Found., Inc., Los Angeles, 1973-74; mem. research group Huntington's Chorea, World Fedn. Neurology, 1974—; mem. epilepsy adv. com. Nat. Inst. Neurol. Disease and Stroke, 1974-78; mem. neuroscis. interdisciplinary cluster Pres.'s Biomed. Research Panel, 1975; mem. U.S. Nat. Com. for IBRO, 1976-79; mem. bd. sci. adv. La Jolla Cancer Research Found., 1979—. Contbr. numerous articles to sci. jours.; editorial bd.: Molecular Pharmacology. Chmn. bd. trustees Calif. Found. Biochem. Research, La Jolla, 1981-82; trustee, 1966—. Wayne State U. scholar, 1936-40; McGregor Fund scholar, 1940-41; U. Mich. fellow, 1941-43; recipient Disting. Service award Wayne State U., 1966, Alumni award, 1969. Fellow N.Y. Acad. Scis.; mem. Am. Chem. Soc., Am. Soc. Biol. Chemists, Soc Exptl. Biology and Medicine, Am. Assn. Cancer Research, World Fedn. Neurology, Council Internat. Orgn. Med. Scis., Internat. Soc. Neurochemistry, Am. Soc. Pharmacology and Exptl. Therapeutics, Am. Soc. Neurochemistry, Soc. Neurosci., Sigma Xi, Phi Lambda Upsilon, Phi Kappa Phi. Subspecialties: Neurochemistry; Neuropharmacology. Current work: Role of GABA in nervous system function. Office: 1450 E Duarte Rd Duarte CA 91010

ROBERTS, GEORGE ADAM, metallurgist; b. Uniontown, Pa., Feb. 18, 1919; s. Jacob Earle and Mary M. (Bower) R.; m. Betty E. Matthewson, May 31, 1941; children—George Thomas, William John, Mary Ellen; m. Jeanne Marie Polk. Student, U.S. Naval Acad., 1935-37; B.Sc., Carnegie Tech., 1939, M.Sc., 1941, D.Sc., 1942. Technician Bell Telephone Labs., N.Y.C., 1938; research dir. Vasco Metals Corp. (formerly Vanadium Alloys Steel Co.), Latrobe, Pa., 1940-45, chief metallurgist, 1945-53, v.p., 1953-61, pres., 1961-66; pres., dir. Teledyne, Inc. (merger with Vasco Metals Corp.), Los Angeles, 1966—; hon. lectr. Societe Francaise de Metallurgie, 1960. Author: Tool Steels, 1944, 62; Contbr. articles trade jours. Recipient silver medal from, Paris, 1955. Fellow Metall. Soc. Am. Inst. Mining, Metall. and Petroleum Engrs., Am. Soc. for Metals (chmn. Pitts. chpt. 1949-50, internat. pres. 1954-55, trustee Found. Edn. and Research 1954-59, 63-64, pres. Found. 1955-56, Gold medal 1977); mem. Nat. Acad. Engring., Metal Powder Industries Fedn. (dir. 1952-55, pres. 1957-61), Am. Soc. Metals, Am. Iron and Steel Inst., Soc. Mfg. Engrs., Tau Beta Pi; hon. life mem. several fgn. socs. Methodist. Subspecialty: Metallurgy. Office: 1901 Ave of the Stars Suite 1800 Los Angeles CA 90067

ROBERTS, GLENN DALE, microbiologist; b. Gilmer, Tex., Apr. 9, 1943; s. B. C. and Mary Fern (Baker) R.; m. Kathleen Louise Brackin, Oct. 13, 1973; children: Michael Glenn, Heather Michelle, Megan Louise. B.S., N. Tex. State U., 1967; M.S., U. Okla., 1969, Ph.D., 1972. Diplomate: Am. Bd. Med. Microbiology. Postdoctoral fellow dept. community medicine U. Ky., Lexington, 1971-72; cons. clin. microbiology Mayo Clinic, Rochester, Minn., 1972—; assoc. prof. lab. medicine and microbiology Mayo Med. Sch., Rochester, 1979—. Author: Practical Laboratory Mycology, 1978. Fellow Am. Acad. Microbiology; mem. Internat. Soc. Human and Animal Mycology, Am. Thoracic Soc., Am. Soc. Clin. Pathologists, Am. Soc. Microbiology. Lutheran. Subspecialties: Microbiology (medicine); Infectious diseases. Current work: Microbiology and immunology of fungal and mycobacterial diseases. Home: 1751 Walden Ln SW Rochester MN 55901 Office: Mayo Clin and Mayo Found 200 1st St SW Rochester MN 55901

ROBERTS, HYMAN JACOB, physician, researcher; b. Boston, May 29, 1924; s. Benjamin and Eva (Sherman) R.; m. Carol Antonia Klein, Aug. 9, 1953; children: David, Jonathan, Mark, Stephen, Scott, Pamela. M.D. cum laude, Tufts U., 1947. Diplomate: Am. Bd. Internal Medicine. Intern, resident Boston City Hosp., 1947-49; resident Mcpl. Hosp., Washington, 1949-50; research fellow, instr. med. Tufts Med. Sch., Boston, 1948-49, Georgetown Med. Sch., Washington, 1949-50; fellow in medicine Lahey Clinic, Boston, 1950-51; mem. sr. active staff Good Samaritan and St. Mary's Hosps., West Palm Beach, Fla., 1955—; dir. Palm Beach Inst. Med. Research, West Palm Beach, 1964—; U.S. rep. Council of Europe for Driving Standards, 1972. Author: Difficult Diagnosis, Spanish and Italian edits, 1958, The Causes, Ecology and Prevention of Traffic Accidents, 1971, Is Vasectomy Safe?, 1979; assoc. editor: Tufts Med. Alumni Bull, Boston, 1978—; contbr. sci. and med. articles to profl. jours. Pres. Jewish Community Day Sch., West Palm Beach, Fla., 1975-76; mem. pres. council U. Fla., Gainesville, 1974—; founder, dir. Jewish Fedn. Palm Beach County, West Palm Beach, 1960-72. Served to lt. USNR, 1951-54. Named Fla. Outstanding Young Man Jr. C. of C. Fla., 1958; recipient Gold Share cert. and silver certs. Inst. Agr. and Food Scis., U. Fla., 1974-78; Paul Harris fellow Rotary Found., 1980. Fellow Am. Coll. Chest Physicians, Am. Coll. Nutrition, Stroke Council; mem. AMA, ACP, Am. Soc. Internal Medicine, Endocrine Soc., Am. Diabetes Assn., Am. Heart Assn., Am. Fedn. Clin. Research, Am. Coll. Angiology (gov. 1981—), Pan Am. Med. Assn. (chmn. endocrinology 1982—), So. Med. Assn., N.Y. Acad. Scis., Am. Physicians Fellowship of Israel Med. Assn., Alpha Omega Alpha. Lodges: Rotary; B'nai B'rith. Subspecialties: Internal medicine; Endocrinology. Current work: Original and ongoing researches in medical diagnosis, diabetes, hypoglycemia, postvasectomy state, Vitamin E metabolism, pentacholophenol and heavy metal toxicity, narcolepsy, traffic accidents, thrombophlebitis, nutrition and bioethics.

Home: 6708 Pamela Ln West Palm Beach FL 33405 Office: Palm Beach Inst Med Research 300 27th St West Palm Beach FL 33407

ROBERTS, JOHN D., chemist, educator; b. Los Angeles, June 8, 1918; s. Allen Andrew and Flora (Dombrowski) R.; m. Edith Mary Johnson, July 11, 1942; children: Anne Christine, Donald William, John Paul, Allen Walter. A.B., UCLA, 1941, Ph.D., 1944; Dr. rer. nat. h.c., U. Munich, 1962; D.Sc., Temple U., 1964. Instr. chemistry U. Calif. at Los Angeles, 1944-45; NRC fellow chemistry Harvard, 1945-46, instr. chemistry, 1946, Mass. Inst. Tech., 1946, asst. prof., 1947-50, asso. prof., 1950-52; vis. prof. Ohio State U., 1952, Stanford U., 1973-74; prof. organic chemistry Calif. Inst. Tech., 1953-72, Inst. prof. chemistry, 1972—, dean of faculty, v.p., provost, 1980-83, chmn. div. chemistry and chem. engring., 1963-68, acting chmn., 1972-73; Foster lectr. U. Buffalo, 1956; Mack Meml. lectr. Ohio State U., 1957; Falk-Plaut lectr. Columbia U., 1957; Reynaud Found. lectr. Mich. State U., 1958; Bachmann Meml. lectr. U. Mich., 1958; vis. prof. Harvard, 1958-59, M. Tishler lectr., 1965; Reilly lectr. Notre Dame U., 1960; Am.-Swiss Found. lectr., 1960; O.M. Smith lectr. Okla. State U., 1962; M.S. Kharasch Meml. lectr. U. Chgo., 1962; K. Folkers lectr. U. Ill., 1962; Phillips lectr. Haverford Coll., 1963; vis. prof. U. Munich, 1962; Sloan lectr. U. Alaska, 1967; Disting. vis. prof. U. Iowa, 1967; Sprague lectr. U. Wis., 1967; Kilpatrick lectr. Ill. Inst. Tech., 1969; Pacific Northwest lectr., 1969; E.F. Smith lectr. U. Pa., 1970; vis. prof. chemistry Stanford U., 1973-74; S.C. Lind lectr. U. Tenn.; Arapahoe lectr. U. Colo., 1976; Mary E. Kapp lectr. Va. Commonwealth U., 1976; R.T. Major lectr. U. Conn., 1977; Nebr. lectr. Am. Chem. Soc., 1977; Leermakers lectr. Wesleyan U., 1980; Iddles Meml. lectr. U. N.H., 1981; Arapahoe lectr. Colo. State U., 1981; Winstein lectr. UCLA, 1981; Gilmar lectr. Iowa State U., 1982; Marvel lectr. U. Ill., 1982; dir., cons. editor W.A. Benjamin, Inc., 1961-67; cons. E.I. du Pont Co., 1950—; mem. adv. panel chemistry NSF, 1958-60, chmn., 1959-60, chmn. divisional com. math., phys. engring. scis., 1962-64, mem. math. and phys. sci. div. com., 1964-66; chemistry adv. panel Air Force Office Sci. Research, 1959-61; chmn. chemistry sect. Nat. Acad. Scis., 1968-71, chmn., 1976-78, councillor, 1980-83; dir. Organic Syntheses, Inc. Author: Basic Organic Chemistry, Part I, 1955, Nuclear Magnetic Resonance, 1958, Spin-Spin Splitting in High-Resolution Nuclear Magnetic Resonance Spectra, 1961, Molecular Orbital Calculations, 1961, (with M.C. Caserio) Basic Principles of Organic Chemistry, 1964, 2d edit., 1977, Modern Organic Chemistry, 1967, (with R. Stewart and M.C. Caserio) Organic Chemistry-Methane To Macromolecules, 1971; cons. editor: McGraw-Hill Series in Advanced Chemistry, 1957-60; editor-in-chief: Organic Syntheses, vol. 41; editorial bd.: Tetrahedron, Nouveau Chimie. Bd. dirs. L.S.B. Leakey Found. Recipient Alumni Profl. Achievement award UCLA, 1967; Guggenheim fellow, 1952-53, 55-56; recipient Am. Chem. Soc. award pure chemistry, 1954; Harrison Howe award, 1957; Roger Adams award in organic chemistry, 1967; Alumni Achievement award UCLA, 1967; Nichols medal, 1972; Tolman medal, 1975; Michelson-Morley award, 1976; Norris award, 1978; Pauling award, 1980; Theodore Wm. Richards medal, 1982; Willard Gibbs Gold medal, 1983. Mem. Am. Chem. Soc. (chmn. organic chemistry div. 1956-57, exec. com. organic div. 1953-57), Nat. Acad. Scis., Am. Philos. Soc., Am. Acad. Arts and Scis., Sigma Xi, Phi Lambda Upsilon, Alpha Chi Sigma. Subspecialties: Organic chemistry; Biochemistry (biology). Current work: Applications of nuclear magnetic resonance spectroscopy to organic chemistry; biology and medicine; computers and chemistry; theoretical organic chemistry. Office: Calif Inst Tech Pasadena CA 91125

ROBERTS, JOHN LEWIS, zoology educator, researcher; b. Waukesha, Wis., May 23, 1922; s. Frank Lewis and Leola (Bostwick) R.; m. Margaret Louise Ross, Apr. 23, 1950; children: Judith, David. B.S., U. Wis.-Madison, 1947, M.S., 1948; Ph.D., UCLA, 1953. Instr. physiology U. Mass., Amherst, 1952-54, asst. prof., 1954-61, assoc. prof. zoology, 1961-70, prof., 1970—, acting chmn., 1980-81; panelist regulatory biology program NSF, Washington, 1977-80. Assoc. editor: Physiol. Zoology, 1975—. Mem. Town Meeting, Amherst, 1956-59. Served to 1st lt. U. S. Army, 1943-46; ETO, PTO. NIH grantee, 1959-67; NSF grantee, 1968-72; NIH spl. fellow, 1966-67; NRC-NOAA sr. research assoc., 1973-74. Fellow AAAS; mem. Am. Soc. Zoologists (div. chmn. 1982-84), Am. Physiol. Soc., Am. Fisheries Soc., AAUP (pres. U. Mass. chpt. 1969-70). Subspecialties: Comparative physiology; Comparative neurobiology. Current work: Biomechanics and neural control of respiration and locomotion in fish. Office: Dept Zoology U Mass Amherst MA 01003

ROBERTS, JOSEPH, biochemist, researcher, educator; b. Bardejov, Czechoslovakia, May 24, 1936; came to U.S., 1963; s. Alexander and Anna (Stern) R.; m. Julie Roberts, Nov. 8, 1964; children: Lori Jennifer, Jeffrey Andrew. B.Sc., U. Toronto, 1959; M.S., U. Wis., 1961; Ph.D., McGill U., 1963. Research assoc. Johns Hopkins U., 1963-64; asst. prof. Baylor U., 1964-69, U. Wash., Seattle, 1969-73; assoc. prof. Cornell U., 1973—; assoc. mem. lab. head Sloan-Kettering inst. Cancer Research, Rye, N.Y., 1973—. Editor: Enzymes As Drugs, 1981. Leukemia Soc. Am. scholar, 1965-70; recipient Research Career Devel. award USPHS, 1971-73. Mem. Am. Assn. Cancer Research, Am. Chem. Soc. Subspecialties: Chemotherapy; Enzyme technology. Current work: Devel. of novel enzymes for use as therapeutic agents; enzyme engring.; devel. of new methods for targeting drugs to tumors. Home: 14 Homer Ave Larchmont NY 10538 Office: 145 Boston Post Rd Rye NY 10580

ROBERTS, MORTON SPITZ, astronomer; b. N.Y.C., Nov. 5, 1926; m. Josephine Taylor, Aug. 2, 1951; 1 dau., Elizabeth Mason. B.A., Pomona Coll., 1948; Sc.D. (hon.), 1979; M.Sc., Calif. Inst. Tech., 1950; Ph.D. (Lick Obs. fellow), U. Calif., Berkeley, 1958. Asst. prof. physics Occidental Coll., 1949-52; lectr. astronomy dept. U. Calif., Berkeley, 1959-60; lectr., research asso. Harvard Coll. Obs., Harvard U., 1960-64; scientist Nat. Radio Astronomy Obs., Charlottesville, Va., 1964-68, sr. scientist, dir., 1978—; Sigma Xi nat. lectr., 1970-71; vis. educator SUNY, Stony Brook, 1968, Cambridge U., 1972, U. Groningen, 1972. Bd. editors: Astronomy and Astrophysics, 1971-80; asso. editor: Astron. Jour, 1977-79. NSF postdoctoral fellow, 1958-59. Mem. Am. Astron. Soc. (vis. prof. program 1965-73, v.p. 1971-72, mem. council 1983—, publs. bd. 1979-80), AAAS (council mem. 1973-79), Internat. Astron. Union, Internat. Sci. Radio Union., Nat. Acad. Sci. Subspecialties: Radio and microwave astronomy. Current work: Spectroscopy of extra galactic systems; scientific administration. Home: 1826 Wayside Pl Charlottesville VA 22903 Office: Nat Radio Astronomy Obs Charlottesville VA 22901

ROBERTS, ROBERT MICHAEL, biochemistry and molecular biology educator; b. Ilkley, Yorkshire, Eng., Oct. 23, 1940; came to U.S., 1965; s. Charles Herbert and Muriel (Wilson) R.; m. Susan E. Wilson, Jan. 26, 1961; children: Samantha Clare, Mark Glyn. B.A., Oxford U., 1962, Ph.D., 1965. Postdoctoral assoc. SUNY-Buffalo, 1965-67, asst. prof., 1967-68; sr. research fellow U.S. AEC, Amersham, Eng., 1968-70; asst. prof. U. Fla., Gainesville, 1970-72, assoc. prof., 1972-76, prof. biochemistry and molecular biology, 1976—; pres. Coll Medicine Faculty Council, U. Fla., Gainesville, 1983-84. Co-author: Plant Structure and Metabolism, 1974, 2d edit., 1982. Recipient Research Career Devel. award USPHS U. Fla., 1972-77; 1st Prize Electrofocus Competition LKB Instruments, Inc., 1978; WHO fellow, 1977; NATO sr. fellow, 1977. Mem. Am. Soc. Biol. Chemists, Am. Soc. Cell Biology, Soc. Study Reprodn. Subspecialties: Membrane biology; Reproductive biology. Current work: Biochemical interactions between the conceptus and mother during pregnancy, control of secretory activity, transmembrane transport of proteins, membrane dynamics. Home: 600 NE 9th Ave Gainesville FL 32601 Office: Dept Biochemistry and Molecular Biology J Hillis Miller Health Ctr J-245 U Fla Gainesville FL 32610

ROBERTS, STEVEN KURT, writer, research cons.; b. Erie, Pa., Sept. 25, 1952; s. Edward H. and Phyllis M. Pres. Cybertronics, Inc., 1973-79; pres., research dir. The Info. Inst., Inc., Dublin, Ohio, 1979-83; writer; public speaker. Contbr. to mags.; contbr.: books include Micromatics, 1980; author: books including Creative Design with Microcomputers, 1983, Complete Guide to Microsystem Management, 1984. Mem. Am. Assn. Artificial Intelligence, Authors Guild, Mensa. Subspecialties: Artificial intelligence; Computer architecture. Current work: Artificial intelligence, industrial system design, robotics, natural language understanding, database research, human factors, software design. Home and Office: 116 Sheffield Rd Columbus OH 43214

ROBERTS, WILBUR EUGENE, orthodontics educator, bone researcher; b. Lubbock, Tex., Nov. 16, 1942; s. Wilbur Eugene and Elva Etna (Chance) Turnwall; m. Cheryl Ann Jones, June 6, 1967; children: Jeffery Alan, Carrie Jean. Student, U. Denver, 1961-63; D.D.S., Creighton U., 1967; Ph.D. in Anatomy, U. Utah, 1969. Cert. in orthodontics U. Conn.-Farmington, 1974. Journeyman roofer Arrow Roofing Co., Denver, 1960-65; postdoctoral fellow dept. anatomy U. Utah, 1967-69; resident in orthodontics U. Conn.-Farmington, 1971-74; asst. prof. orthodontics U. of Pacific, 1974-77, assoc. prof., 1977-82, prof., 1982—; pres., bd. dirs. Pacific Dental Research Found., San Francisco, 1977—. Research numerous publs. in field, 1974—. Precinct worker Republican Party Calif., Danville, 1978-82; supt. ch. sch. San Ramon Valley Methodist Ch., Alamo, Calif., 1980-82; asst. Boy Scout Troop 815 Diablo council Boy Scouts Am., 1980-82; head umpire Danville Little League Baseball, 1982. Served to lt. comdr. USN, 1969-71; Vietnam. Decorated Commendation medal U.S.; Tech. Service medal Republic of Vietnam, 1971; named Man of Yr. Lambda Chi Alpha, Denver, 1963. Fellow Internat. Coll. Dentists; mem. Medico-Dental Study Guild Calif. (pres. 1981-82), Internat. Assn. Dental Research, Am. Assn. Orthodontists, Am. Assn. Anatomists, Delta Sigma Delta (pres. San Francisco grad. chpt. 1982-83). Republican. Methodist. Subspecialty: Orthodontics. Home: 75 Saint Timothy Ct Danville CA 94526 Office: Sch Dentistry U of Pacific 2155 Webster St San Francisco CA 94115

ROBERTSON, DONALD SAGE, geneticist; b. Oakland, Calif., June 27, 1921; s. Milton Sage and Helen (Playter) R.; m. Roxana Ruth Robertson, Sept. 13, 1942; children: MarkThomas, Martha Leanne Alexander, William Paul. Student, San Jose State Coll., 1941-43; B.S., Stanford U., 1947; Ph.D., Calif. Inst. Tech., 1951. Chmn. sci. dept. Biola U., La Mirada, Calif., 1951-57; asst. prof. Iowa State U., Ames, 1957-60, assoc. prof., 1960-63, prof. genetics, 1963—, chmn. dept., 1975-80. Contbr. articles to profl. jours. Served with AUS, 1943-46. NSF grantee, 1961-68, 77—. Mem. AAAS, Genetics Soc. Am., Am. Genetics Assn., Am. Sci. Affiliation, Sigma Xi. Republican. Mem. Evangelical Free Ch. Subspecialties: Genetics and genetic engineering (biology); Genome organization. Current work: Research on maize genetics, especially genetic factors affecting mutation rates. Home: 1232 20th St Ames IA 50010 Office: Dept Genetics Iowa State U Ames IA 50011

ROBERTSON, HOWARD THOMAS, II, physician, educator; b. Abilene, Tex., Mar. 6, 1942; s. Howard Thomas and Mary Ellen (DeMotte) R.; m. Christine M. Johnston, June 13, 1970; children: Dana Marie, Marla Christine. B.A., Colgate U., 1964; M.D., Harvard U., 1968. Diplomate: Am. Bd. Internal Medicine. Instr. medicine U. Wash., Seattle, 1973-78, asst. prof., 1978-82, assoc. prof., 1982—. Author numerous sci. papers. Served to maj. AUS, 1970-72. NIH Young Investigator awardee, 1976-78; Research Career Devel. awardee, 1982-87. Fellow ACP; mem. Am. Physiol. Soc., Am. Thoracic Soc. Subspecialties: Pulmonary medicine; Physiology (medicine). Current work: research in pulmonary physiology; pulmonary gas exchange; lung injury responses and exercise physiology. Office: University Hospital Div Respiratory Diseases Room 12 Seattle WA 98195 Home: 4508 W Mercer Wy Mercer Island WA 98040

ROBERTSON, JAMES BRAGG, educator; b. Plainfield, N.J., Mar. 7, 1943; s. James Bragg and Lee (Pace) R.; m.; children: James, Andrea, George Denise. B.Sc., Miami U., Oxford, Ohio, 1969; M.S., Harvard U., 1974, Sc.D., 1976. Diplomate: Am. Bd. Toxicology. Research scientist Los Alamos (N.Mex.) Sci. Lab., 1976-77; research assoc. Harvard U., Sch. Pub. Health, Cambridge, Mass., 1977-80; asst. prof. Clarkson Coll. Tech., Potsdam, N.Y., 1980—; cons. Harvard U., 1980—, Canadian U. Ins. Co., Boston, 1983. Contbr. articles to profl. jours. Served to 1st lt. USMC, 1961-72. Recipient Culler prize Miami U., Ohio, 1968; AEC spl. fellow, 1972-74. Mem. Radiation Research Soc., Am. Forestry Assn., Sigma Xi. Subspecialties: Toxicology (medicine); Cancer research (medicine). Current work: Med. application of laser activated flow microfluorometry; detection of environmental mutagens/carcinogens; late health effects of nuclear power. Home: 104 Lawrence Ave Potsdam NY 13676 Office: Clarkson Coll Tech Dept Biology Potsdam NY 13676

ROBERTSON, JAMES SYDNOR, nuclear medicine physician, educator, researcher; b. Richmond, Va., Nov. 27, 1920; s. Paul Augustus and Beth O'Ferrall (Whitacre) R.; m. Ruth Elizabeth Henrici, Jan. 15, 1944; children: Kathleen, John, Marion. B.S., U. Minn., 1943, M.B., 1944, M.D., 1945; Ph.D. in Physiology, U. Calif.-Berkeley, 1949. Intern U.S. Naval Hosp., Annapolis, Md., 1944-45; physiologist U. Calif.-Berkeley, 1947-50; head med. physics Brookhaven Nat. Lab., Upton, N.Y., 1950-75; prof. lab. medicine Mayo Med. Sch., Rochester, Minn., 1975—; cons. nuclear medicine Mayo Clinic, Rochester, 1975—; adj. prof. biophysics SUNY-Stony Brook., 1968-75; prof. physics CCNY, N.Y.C., 1970-75; clin. assoc. nuclear medicine U. Wis.-La Crosse, 1976; cons. nuclear medicine M.D. Anderson Hosp., Houston, 1960-62. Author/editor: Compartmental Distribution of Radiotracers, 1983. Served to lt. comdr. USN, 1953-55. Fellow AAAS; mem. Am. Physiol. Soc., Soc. Nuclear Medicine, Health Physics Soc., Radiation Research Soc., Math. Assn. Am., Sigma Xi. Democrat. Methodist. Lodge: Masons. Subspecialties: Nuclear medicine; Physiology (medicine). Current work: Tracer theory, radiation disometry, imaging. Home: 1119 7th Ave SW Rochester MN 55901 Office: Mayo Clinic 200 1st St SW Rochester MN 55905

ROBERTSON, JAMES THOMAS, neurol. surgeon; b. McComb, Miss., Apr. 5, 1931; s. Clyde Aubrey and Roberta Darville R.; m., Nov. 26, 1952; children—James T., Elizabeth, Catherine, Clay, Roberta, Daniel. Student, Southwestern at Memphis, 1949-51; M.D., U. Tenn., 1951. Diplomate: Am. Be. Neurol. Surgery. Intern, resident in surgery Bapt. Meml. Hosp., City of Memphis Hosps., 1955-59; resident Peter Bent Bingham Hosp., Children's Hosp., Boston, 1959-60; mem. faculty dept. neurosurgery U. Tenn., Memphis. Editor: (with R.R. Smith) Subarachnoid Hemorrhage and Cerebrovascular Spasm, 1975. Fellow ACS; mem. AMA, Congress of Neurol. Surgeons, Am. Assn. Neurol. Surgeons, Am. Acad. Neurol. Surgery, Soc. Univ Neurosurgeons, Soc. Neurol. Surgeons, Tenn. Med. Assn., Memphis Med. Soc., Shelby County Med. Soc. Republican. Presbyterian. Subspecialty: Neurosurgery. Home: 628 N Trelevant Memphis TN 38112 Office: 956 Court Memphis TN 38163

ROBERTSON, LESLIE EARL, structural engineer; b. Los Angeles, Feb. 12, 1928; s. Garnet Roy and Tina (Grantham) R.; m. Saw-Teen See, Aug. 11, 1982; children: Jeanne, Christopher Alan, Sharon Miyuki. B.S., U. Calif., Berkeley, 1952. Structural engr. Kaiser Engrs., Oakland, Calif., 1952-54, John A. Blume, San Francisco, 1954-57, Raymond Internat. Co., N.Y.C., 1957-58; mng. partner Skilling, Helle, Christiansen, Robertson, N.Y.C., Seattle and Anchorage, 1958-82; chmn. Robertson, Fowler & Assocs., P.C., N.Y.C., 1982—; vice chmn. Council on Tall Bldgs. and Urban Habitat; dir. Wind Engring. Research Council.; mem. com. on natural disasters NRC. Author papers in field. Served with USNR, 1944-46. Fellow ASCE (Raymond C. Reese Research prize 1974); mem. Nat. Acad. Engring. Subspecialty: Structural engineering. Home: PO Box 8284 New Fairfield CT 06810 Office: 211 E 46th St New York NY 10017

ROBERTSON, ROBERT LAFON, entomology educator; b. Blountsville, Ala., July 20, 1925; s. Herbert Leon and Estelle (Lafon) R; m. Ruth Farmer, Mar. 13, 1965; 1 dau., Karen. B.S. in Agr, Auburn U., 1950, M.S. in Entomology, 1954. County agrl. agt. Ala. Extension Service, Moulton, 1950-52; research asst. Auburn U., 1952-54, asst. research entomologist, 1954-56; staff entomologist Am. Cyanamid Co., N.Y.C., 1956-58; extension entomologist U. Ga., 1958-60; prof. extension entomology N.C. State U., 1961—; Pesticide coordinator Office of Sec., Dept. Agr., Washington, 1979; survey coordinator Animal Plant Health Inspection Service, Dept. Agr., Hyattsville, Md., 1981-82. Served with USNR, 1943-46; PTO. Recipient Outstanding Extension award N.C. State U., 1976-77. Mem. Entomol. Soc. Am. (governing bd. 1979-80), N.C. Entomol. Soc. (pres. 1976, service award 1977), Turfgrass Council N.C. (service award 1981), Pesticide Assn. N.C. (adv. bd.). Democrat. Methodist. Subspecialty: Integrated pest management. Current work: Develop and implement integrated pest management programs in peanuts, turf, small grains, alfalfa and corn. Home: 409 Holly Circle Cary NC 27511 Office: North Carolina State U PO Box 5215 Raleigh NC 27650

ROBIN, EUGENE DEBS, biologist, physician; b. Detroit, Aug. 23, 1919; s. Benedict and Anna (Cooper) R.; m. Evelyn Cowen, Aug. 23, 1942; children—Anna Rosa, Donald Allan. A.A., U. Chgo., 1939; S.B., George Washington U., 1946, S.M., 1947, M.D., 1951. Intern Peter Bent Brigham Hosp., Boston, 1951-52, chief med. resident, 1954-55, asso. dir. cardiovascular tng., 1958-59; instr. medicine Harvard, 1955-58; asso. prof. medicine U. Pitts., 1959-63, prof. medicine, 1963-70; prof. medicine and physiology Stanford U. Sch. Medicine, 1970—; cons. to Gov. Pa. Tb care, 1968-70, Pneumoconiosis, 1964-66; vis. prof. medicine Washington U., St. Louis, 1964; vis. research prof. Cardiovascular Research Inst., San Francisco, 1964; vis. prof. Tufts U., 1968, Stanford, 1969; Towsley prof. medicine U. Mich., 1974; Farquarson prof. medicine U. Toronto, 1976; Eppinger prof. medicine Harvard U., 1977; vis. prof. medicine French Govt., 1978; Chmn., pulmonary adv. com. Nat. Heart and Lung Inst., NIH, 1971—, mem. cardiovascular and pulmonary study sect., 1976—; Amberson lectr., Boston, Washington, 1980, Rodbord lectr., 1980; McArthur lectr. U. Edinborough, 1980. Contbr. chpts. to books, articles to med. jours. Bd. dirs. Nat. Tb and Respiratory Disease Assn.; trustee Mt. Desert Island Biol. Lab. Mem. Am. Assn. Physicians, Am. Soc. Clin. Investigation, Am. Physiol. Soc., Am. Thoracic Soc. (pres. 1970—), Assn. Am. Physicians, Sigma Xi, Alpha Omega Alpha. Subspecialties: Solid state chemistry; Molecular biology. Current work: Genetic expression of slycolysis; condinate regulation of collage metabolism; mitochandrial biogenesis, heart-lung transplantaion. Office: Stanford U Med Center Stanford CA 94305

ROBIN, MITCHELL WOLFE, psychology educator; b. Bklyn., Apr. 30, 1944; s. Ben and Lee (White) R.; m. Regina Catherine Spires, Mar. 26, 1972; children: Elaine Dara, Abigail Alice. B.B.A., Baruch Sch. CCNY, 1965; M.A., New Sch. Social Research, 1969; Ph.D., NYU, 1983. Teaching fellow Baruch Sch. CCNY, 1965-67; lectr. N.Y.C. T.C., CUNY, 1968-70, asst. prof., 1970—; mem. faculty New Sch. for Social Research, 1970—; cons. in field. Contbr. articles to profl. jours.; co-author/contbr.: Cross Cultural Psychology at Issue, 1982; media cons.: jour. Psychology, 1971-72; contbr. articles to profl. jours. Mem. Am. Psychol. Assn., Internat. Orgn. for Study of Group Tensions (exec. sec.), Internat. Council Psychologists, Psi Chi. Democrat. Jewish. Subspecialty: Developmental psychology. Current work: Cross cultural approaches to child rearing and handling of deviance in childhood; psychological impact of video games. Office: Social Sci Dept CUNY 300 Jay St Brooklyn NY 11201

ROBINS, LEE NELKEN, sociology educator; b. New Orleans, Aug. 29, 1922; d. Abe and Leona (Reiman) Nelken; m. Eli Robins, Feb. 22, 1946; children: Paul, James, Thomas, Nicholas. Student, Newcomb Coll., 1938-40; B.A., Radcliffe Coll., 1942, M.A., 1943, Ph.D., 1951. Mem. faculty Washington U., St. Louis, 1954—, prof. sociology in psychiatry, 1968—, prof. sociology, 1969—; former mem. Nat. Adv. Council on Drug Abuse, Pres.'s Commn. on Mental Health task panels; expert adv. panel mental health WHO.; salmon lectr. N.Y. Acad. Medicine, 1983. Author 3 monographs; editor 5 books; contbr. articles to profl. jours. Recipient Research Scientist award USPHS, 1970—; Pacesetter Research award Nat. Inst. Drug Abuse, 1978; Radcliffe Coll. Grad. Soc. medal, 1979; Research grantee NIMH, Nat. Inst. on Drug Abuse, Nat. Inst. on Alcohol Abuse and Alcoholism. Fellow Am. Coll. Epidemiology; Mem. Am. Sociol. Assn., Internat. Sociol. Assn., Inst. of Medicine, Internat. Epidemiological Assn., Soc. Epidemiol. Research, Am. Psychopathol. Assn. (Paul Hoch award 1978), Am. Public Health Assn. (Rema Lapouse award 1979), Soc. Life History Research in Psychopathology, Am. Coll. Neuropsychopharmacology. Subspecialties: Epidemiology; Psychiatry. Current work: Risk and protein factors in major mental disorders. Office: Dept Psychiatry Med Sch Washington U Saint Louis MO 63110

ROBINS, MORRIS JOSEPH, organic chemist, educator; b. Nephi, Utah, Sept. 28, 1939; s. Waldo George and Mary Erda (Anderson) R.; m. Jerri Johnson, June 11, 1960 (div. 1972); children—Dayne M., Diane, Douglas W., Debra, Dale C.; m. Jackie Alene Robinson, Aug. 24, 1973; children: Mark K., Janetta A. B.A., U. Utah, 1961; Ph.D., Ariz. State U., 1965. Cancer research scientist Roswell Park Meml. Inst., Buffalo, 1965-66; research asso. U. Utah, Salt Lake City, 1966-69; asst. prof. dept. chemistry U. Alta. (Can.), Edmonton, 1969-71, asso. prof., 1971-78, prof., 1978—; cons. in organic biochemistry. Mem. editorial bd.: Nucleic Acids Research, 1980-83; contbr. articles to profl. jours. NSF fellow, 1963-64. Mem. Am. Chem. Soc., Am. Assn. for Cancer Research, Chem. Inst. Can., Phi Beta Kappa, Sigma Xi, Phi Eta Sigma, Phi Kappa Phi. Mormon. Subspecialties: Organic chemistry; Medicinal chemistry. Current work: Organic chemistry and biol. effects of nucleic acid related compounds, nucleoside and nucleotide chemistry; biol. effects of nucleosides and analogues; organic and bioorganic chemistry edn. Patentee in field. Office: Dept Chemistry U Alberta Edmonton AB Canada T6G 2G2

ROBINS, NORMAN ALAN, steel company executive; b. Chgo., Nov. 19, 1934; s. Irving and Sylvia (Robbin) R.; m. Sandra Ross, June 10, 1956; children: Lawrence Richard, Sherry Lynn. S.B. in Chem. Engring, MIT, 1955, S.M., 1956; Ph.D. in Math, Ill. Inst. Tech., 1972.

Metallurgist Inland Steel Co., Chgo., 1957-60, research metallurgist, 1960-62, asst. mgr., 1962-67, assoc. mgr., 1967-72, dir. process research, 1972-77, v.p. research, 1977—. Mem. Am. Iron and Steel Inst. (Regional Tech. Meeting award 1967, 72), AIME (Nat. Open Hearth Conf. award 1972), Am. Inst. Chem. Engrs., Indsl. Research Inst., Mathematics Assn. Am. Subspecialties: Chemical engineering; Metallurgical engineering. Current work: Process modelling and computer control. Patentee: (with F.H. Bugajski) control of coating thickness of hot-dip metal coating. Office: 3001 E Columbus Dr East Chicago IL 66312

ROBINSON, CASEY PERRY, pharmacologist, educator; b. Idabel, Okla., Oct. 10, 1932; s. Clarence Elbert and Susie Josephine R.; m. Eunice Vanda Bettes, Aug. 24, 1955; children: Cheryl, Kent, Brian. B.S., U. Okla, 1954, M.S., 1967; Ph.D., Vanderbilt U., 1970. Registered pharmacist. Pharmacist, then pharmacy mgr., then supr. chain of pharmacies, Calif., 1956-66; sr. sci. investigator Los Angeles County Heart Assn., UCLA, Med. Sch., 1970-71; faculty U. Okla., Oklahoma City, 1971—; prof. pharmacodynamics and toxicology Health Scis. Center, 1978—. Contbr. articles to sci. publs. Served with U.S. Army, 1954-56. NSF fellow, 1967-69; NIH fellow, 1969-70; Recipient teaching awards U. Okla., 1975, 76. Mem. Am. Soc. Pharmacology and Exptl. Therapeutics, Soc. Toxicology, AAAS, Soc. Exptl. Biology and Medicine, Rho Chi. Democrat. Mem. Ch. of Nazarene. Subspecialty: Pharmacology. Current work: Neuro-effector mechanisms in vascular tissue. Vascular tissue, norepinephrine metabolism, histamine receptors, insecticide effects. Home: 3204 Windsor Terr Oklahoma City OK 73122 Office: Coll Pharmacy U Okla Health Scis Center Oklahoma City OK 73190

ROBINSON, DANIEL NICHOLAS, psychology educator; b. Monticello, N.Y., Mar. 9, 1937; s. Henry Stoddard and Margaret Anne (Kimbiz) R.; m. Carol Wood, June, 1958 (div. Sept. 1961); children: Tracy Margaret, Kimberly; m. Francine S. Malasko, Sept. 18, 1967. B.A., Colgate U., 1958; M.A., Hofstra U., 1960; Ph.D., CUNY, 1965. Research psychologist elec. research labs. Columbia U., N.Y.C., 1960-64, sr. research psychologist, 1965-68; vis. lectr. Princeton (N.J.) U., 1965-68; asst. prof. to assoc. prof. Amherst (Mass.) Coll., 1968-71; assoc. prof. psychology Georgetown U., Washington, 1971-73, prof., 1973—; cons. NSF, 1970-74, NIH, 1971-73. Author: 10 books, including The Enlightened Machine, 1973, An Intellectual History of Psychology, 1976, Psychology and Law, 1981, Foundations of Psychobiology, 1983; editor numerous books; mem. editorial bd.: Jour. History of Behavioral Scis, 1979—. Research grantee NIH, 1970-73, NEH, 1982 (declined). Fellow Am. Psychol. Assn., Brit. Psychol. Soc. Roman Catholic. Subspecialties: Neuropsychology; Psychophysics. Current work: History and philosophy of science; neural sciences; human information processing. Office: Dept Psychology Georgetown U Washington DC 20057 Home: 6630 Carpenter Rd Frederick MD 21701

ROBINSON, DAVID LEE, research physiologist, educator; b. St. Louis, May 2, 1943; s. Harry and Minerva Ruth (James) R.; m. Lynne Scott; children: Kevin Andrew, Megan Lynn. B.S., Springfield (Mass.) Coll., 1965; M.S., Wake Forest U., 1968; Ph.D., U. Rochester, 1971. Research physiologist, postdoctoral fellow NIMH, 1971-74; research physiologist Armed Forces Radiobiology Research Inst., Bethesda, Md., 1974-78; Lab. Sensorimotor Research Nat. Eye Inst., Bethesda, Md., 1978—; assoc. prof. physiology and anatomy Uniformed Services U. of Health Scis., 1977—. Mem. Soc. Neurosci., Assn. Research in Vision and Ophthalmology. Episcopalian. Subspecialties: Neurophysiology; Neuropsychology. Current work: Visual attention; eye movement; visual processing in brain. Home: 9814 Fosbak Dr Vienna VA 22180 Offic: Nat Eye Inst 10/6C420 NI Bethesda MD 20205

ROBINSON, DEAN WENTWORTH, chemist, educator; b. Boston, July 22, 1929; s. Lawrence Dean and Doris Elizabeth (Prowse) R.; m.; children: Dean W., Amy E., Jonas W. B.S., U. N.H., 1951, M.S., 1952; Ph.D., M.I.T., 1955. Mem. faculty Johns Hopkins, 1955—, prof. chemistry, chmn. dept., 1976—. Author research papers. Guggenheim fellow, 1966; Fulbright fellow, 1966. Mem. Am. Chem. Soc., AAAS. Subspecialties: Laser photochemistry; Laser-induced chemistry. Current work: Polyatomic chemical lasers; gas-phase kinetics and energy transfer; laser induced physical chemistry. Home: 10 Sonachan Ct Baltimore MD 21204 Office: Chemistry Dept Johns Hopkins U Baltimore MD 21218

ROBINSON, EARL JAMES, management science educator; b. Wilmington, Del., Apr. 15, 1949; emigrated to Can., 1978; s. Harry and Minerva Ruth (James) R., m. Karen Frances Smith, July 5, 1980, 1 dau., Ruth Frances. A.B., Davidson Coll., 1971; M.S., Bucknell U., 1973; Ph.D., U. Ga., 1977. Registered psychologist, N.S. Testing services asst. dir. Central Susquehanna Intermediate Unit, Lewisburg, Pa., 1972-73; asst. dir. Instr. Evaluation Services, Athens, Ga., 1973-77; asst prof. U. Ga., Athens, 1977-78, St. Mary's U., Halifax, N.S., 1978-81, assc. prof. mgmt. sci., 1981—, chmn. dept., 1981—; cons. in field; bd. dirs. Internat. Assn. Students of Econs. and Commerce, Halifax, 1982—. Contbr. articles to profl. jours. Recipient Golden "M" award St. Mary's U., 1981; NSF grantee, 1978; FAA grantee, 1978; Ashland Oil Corp. grantee, 1978. Mem. Am. Psychol. Assn., Can. Psychol. Assn., Am. Statis. Assn., Ops. Research Soc. Am., Inst. Mgmt. Sci., Psychometric Soc., Sigma Xi, Sigma Phi Epsilon. Subspecialties: Software engineering; Statistics. Current work: Active in in-roads into applied computer tech. and edn., specifically in social scis., tech. in higher levels of statis. analysis. Home: 6183 Duncan St Halifax NS Canada B3L 1K1 Office: Dept Fin and Mgmt Sci Saint Mary's U Halifax NS Canada B3H 3C3

ROBINSON, EDWARD J., physicist, educator; b. N.Y.C., June 16, 1936; s. Irving L. and Fannie (Freeman); m. Toni Sandler, Dec. 26, 1959; children: Ian S., Gregory J. B.S., Queens Coll., 1957; Ph.D., NYU, 1964. Research assoc. Joint Inst. Lab. Astrophysics, Boulder, Colo., 1964-65; mem. faculty NYU, N.Y.C., 1965—, prof. physics, 1982—. Contbr. articles to profl. jours. Mem. Am. Phys. Soc. Subspecialties: Atomic and molecular physics; Theoretical physics. Current work: Theoretical laser-related atomic and molecular physics, especially the dynamics of atoms in strong fields and photon-aided collisions. Office: 4 Washington Pl New York NY 10003

ROBINSON, FARREL RICHARD, pathologist, toxicologist; b. Wellington, Kans., Mar. 23, 1927; s. Farrel Otis and Norine (Sloan) R.; m. Mimi Agatha Hathaway, June 5, 1949; children—Farrel Richard, Kelly S., E. Scott, Brian A. B.S., Kans. State U., 1950, D.V.M., M.S., 1958; Ph.D., Tex. A. and M. U., 1965. Diplomate: Am. Coll. Vet. Pathologists, Am. Bd. Vet. Toxicology (v.p. 1971-74, pres. 1976-79). Served with USN, 1945-46; commd. 2d lt. USAF, 1951, advanced through grades to lt. col., 1971; vet. pathologist Aerospace Med. Research Labs., Wright-Patterson AFB, Ohio, 1958-68; chief Vet. Pathology div. Armed Forces Inst. Pathology, Washington, 1968-74; ret., 1974; scientist asso. Univs. Assoc. for Research and Edn. in Pathology, Inc., 1972-74; asst. clin. prof. pathology George Washington U. Sch. Medicine, 1972-74; instr. NIH Grad. Program, 1973-74; prof. toxicology-pathology Sch. Vet. Medicine, Purdue U., 1974—; dir. Animal Disease Diagnostic Lab., 1978—, head dept. vet. sci., 1978—; cons. vet. pathology USAF surg. gen. and asst. surg. gen. for vet. services, 1970-74. Mem. editorial bd.: Human and Vet. Toxicology, 1976—; Contbr. sci. articles to profl. jours. Decorated USAF Commendation medal, Meritorious Service medal; recipient Aerospace Med. Research Labs. Scientist of Year award, 1967. Mem. AVMA, Am. Coll. Vet. Toxicologists, Am. Assn. Vet. Lab. Diagnosticians (gov.), Wildlife Disease Assn., Conf. Research Workers in Animal Disease, Soc. Toxicology, Am. Animal Health Assn., Sigma Xi, Phi Kappa Phi, Alpha Zeta, Phi Zeta. Democrat. Methodist. Subspecialties: Pathology (veterinary medicine); Toxicology (medicine). Current work: Pathologic changes caused by chemicals. Home: 201 W 600 N West Lafayette IN 47906 Office: Animal Disease Diagnostic Lab Purdue U West Lafayette IN 47907

ROBINSON, GEORGE HORINE, JR., psychology educator; b. Louisville, Feb. 19, 1939; s. George Horine and Frances (Ives) R.; m. Nancy Buford, Mar. 15, 1969. B.A., Millsaps Coll., Jackson, Miss., 1962; M.S., U. Miss., 1965, Ph.D., 1967. Research assoc. U. Ga., Athens, 1967-69, asst. prof., 1969-70, U. North Ala., Florence, 1970-73, assoc. prof., 1973-79, prof., 1979—. Mem. AAAS, Am. Psychol. Assn., Soc. Psychophysiol. Research, Optical Soc. Am. Subspecialties: Psychophysics; Sensory processes. Current work: Psychophysical theory and sensory functioning; psychophysiological relationships; neuropsychology; time perception. Home: 738 Dixie Ave Florence AL 35630 Office: Dept Psychology U North Ala Florence AL 35632

ROBINSON, HARRY JOHN, med. adminstr.; b. Elizabeth, N.J., July 30, 1913; s. Harry and Ann (McCourt) R.; m. Marion Neunert, Sept. 9, 1939; children: Linda Miele, Harry John, Raymond P. B.A. in Biology, N.Y. U., 1940; Ph.D. in Microbiology, Rutgers U., 1943; M.D., Columbia U., 1948; D.Sc. (hon.), Bucknell U., 1974. Asst. dir. research Merck Inst., Rahway, N.J., 1943-48, assoc. dir., 1948-56, dir., 1956-65; pres. Quinton Co. div. Merck & Co., Inc., Rahway, 1965-68; sr. v.p. Merck Sharp & Dohme Research Labs., 1968-70, v.p. sci. affairs, 1971-76; v.p. med. affairs Allied Corp., Morristown, N.J., 1976—; acting chmn. dept. bacteriology Coll. Medicine and Dentistry N.J., 1960-62, now clin. prof. medicine.; Mem. N.J. State Sci. Adv. Com., 1981—, N.J. Pub. Health Council, 1953-77. Contbr. numerous articles to sci. jours. Office Sci. Research and Devel. grantee, 1945. Mem. Am. Soc. Clin. Investigation, Am. Soc. Clin. Pharmacology and Therapeutics, Am. Soc. Microbiology, Am. Soc. Pharmacology and Exptl. Therapeutics, Infectious Diseases Soc., Genetic Toxicology Assn., Royal Soc. Medicine, Soc. Exptl. Biology and Medicine. Congregationalist. Clubs: Princeton (N.Y.C.); Baltusrol Country (Springfield, N.J.). Subspecialties: Microbiology (medicine); Molecular pharmacology. Current work: Diagnostic medicine, preventive medicine, therapeutics, toxicology, clinical chemistry, biotechnology, immunology, immunoassays applied to diagnosis and treatment of disease.

ROBINSON, JAMES ROBERT, engineering administrator, educator; b. Pitts., Apr. 18, 1927; s. Roger Robert and Marie (McDermott) R. B.S.M.E., U. Pitts., 1952; postgrad., San Diego State U., 1979. Supr. Devel. Lab., Reaction Motors Inc., Denville, N.J., 1952-54; flight test engr., asst. site mgr. Gen. Dynamics, Convair, San Diego, 1954-60, sr. project engr., 1960-69, site mgr., Quincy, Mass., 1969-75; lectr. in elec./computer engring. San Diego State U., 1976—; engring. mgr. Sci. Applications, Ft. Irwin Facility, La Jolla, Calif., 1982—; lectr. energy conversion systems; researcher, report co-author Satellite Receiver Sta., 1981. Editor-in-chief: Skyscraper Engr, 1950-51. Mem. Republican Presdl. Task Force, Washington. Mem. IEEE (exec. com. at large), IEEE Computer Soc. (chmn. San Diego chpt. 1982—), Soc. Photooptical Instrumentation Engrs., Air Force Assn., Amateur Radio Relay League, Pi Tau Sigma, Sigma Beta Sigma. Republican. Presbyterian. Subspecialties: Distributed systems and networks; Robotics. Current work: Distributed systems and networks: local area networks, assisting in establishing 1800 Port Packet switching network for USAF; fiber optics: development of fibre optic local area network in conjunction with distributed systems and networks: satellite studies. Office: Science Applications Inc 1200 Prospect La Jolla CA 92038 Office: San Diego State U San Diego CA 92182

ROBINSON, JOSEPH DOUGLASS, biochemist, educator; b. Asheville, N.C., Nov. 28, 1934; s. Joseph Douglass and Catherine Elva (Carrier) R.; m. Carol Jeannine Smith, Aug. 3, 1958; children: Karin Aileen, Lisa Jeannine. Student, Duke U., 1955; student, U. Wis.; M.D., Yale U., 1959. Diplomate: Nat. Bd. Med. Examiners. Intern Stanford U. Hosp., 1959-60; research asso. NIH, 1960-62; postdoctoral fellow Yale Sch. Medicine, 1962-64; NSF sr. fellow Cambridge (Eng.) U., 1971-72; from asst. prof. to prof. pharmacology SUNY, Syracuse, 1964—; mem. pharmacology study sect. NIH, Bethesda, Md., 1980-84. Mem. editorial bd.: Jour. Bioenergetics and Biomembranes, 1979—; Contbr. numerous articles and book chpts. to profl. lit. Bd. dirs. Syracuse Friends of Chamber Music, 1982—. Served as sr. asst. surgeon USPHS, 1960-62. NIH research grantee, 1964—. Mem. Biophys. Soc., Am. Soc. Nuerochemistry, Philosophy of Sci. Assn., History of Sci. Soc., Am. Soc. Pharmacology, Phi Beta Kappa, Sigma Xi. Subspecialties: Biochemistry (biology); Biophysics (biology). Current work: Biochemistry of membrane transport: reaction mechanism of transport ATPases. Modes of conflict resolution in science. Home: 109 Buffington Rd Syracuse NY 13224 Office: Pharmacology Dept Upstate Med Center Syracuse NY 13210

ROBINSON, JOSEPH EDWARD, geology educator; b. Regina, Sask, Can., June 25, 1925; came to U.S., 1976; s. Webb Gabriel Wilton and Blanche Marion (Schiefner) R.; m. Mary Corrine Maclaughlin, Nov. 1, 1952 (div.); children: Joseph Christopher, John Edward, Timothy Webb. B.Eng., McGill U., 1950, M.Sc., 1951; Ph.D., U. Alberta, 1968. Geophysicist Imperial Oil, Ltd., Can., 1951-68; sr. geologist Union Oil Co. of Can. Ltd., 1968-76; Cons. geologist J.E. Robinson Assn., Syracuse, N.Y., 1976—; prof. geology Syracuse U., 1976—; dir. Penerex Corp., Rochester, N.Y., Demco Corp., Columbus, Ohio. Author: Computer Applications in Petroleum Geology, 1982. Served with Can. Navy, 1943-46. Recipient Best Dissertation award Can. Assn. Petroleum Geologists, 1968. Mem. Am. Assn. Petroleum Geologists, Can. Assn. Petroleum Geologists, Soc. Exploration Geophysicists, Internatl. Assn. for Math. Geology (assoc. editor 1976-78). Subspecialties: Geology. Current work: Advanced techniques, including computer application, in petroleum exploration and production. Home: 837 Ackerman Ave Syracuse NY 13210 Office: Syracuse U Dept Geology Syracuse NY 13210

ROBINSON, RICHARD BRUCE, pharmacology educator; b. Chgo., Feb. 28, 1950; s. Norton and Charlotte (Patt) Sherey; m. Barbar Ann Dye, Aug. 12, 1973; children: Laura, Sandra. B.S., U. Ill.-Urbana, 1971, Ph.D., 1975. Research asst. dept. physiology and biophysics U. Ill.-Urbana, 1970-71, teaching asst., 1972-73; fellow Cardiovascular Ctr. U. Iowa, Iowa City, 1975-77; research assoc. dept pharmacology Columbia U., 1977-78, asst. prof., 1978—. Recipient Sr Investigatorship N.Y. Heart Assn., 1980; Young Investigator award NIH, 1981. Mem. Biophys. Soc., Am. Physiol. Soc., AAAS, Tissue Culture Assn. Subspecialties: Cellular pharmacology; Membrane biology. Current work: Cardiac cell culture; electrophysioloyg; developmental autonomic pharmacology; adrenergic receptor regulation. Office: Dept Pharmacology Columbia U 630 W 168th St New York NY 10032

ROBINSON, STEPHEN HOWARD, medical investigator, physician, educator; b. N.Y.C., Apr. 3, 1933; s. Nathan and Beatrice Leonia (Koen) R.; m. Carole Latter, June 25, 1956; children: Lisa, Susan, Michael. A.B., Harvard U., 1954; M.D., 1958. Diplomate: Am. Bd. Internal Medicine. Med. intern Harvard Med. Service, Boston, 1958-59, resident, 1959-61; asst. dir. hematology NIH, Bethesda, Md., 1961-63; asst. prof. medicine U. Chgo., 1963-65; from asst. prof. to assoc. prof. medicine Harvard Med. Sch., Boston, 1965-77, prof., 1977—; chief hematology Beth Israel Hosp., Boston, 1971—, clin. dir. dept. medicine, 1980—. Contbr. articles to prof. jours. Served to surgeon USPHS, 1963-65. Fellow ACP; mem. Assn. Am. Physicians, Am. Soc. Clin., Investigation, Am. Soc. Hematology (exec. sec. 1971-74), Am. Fedn. for Clin. Research. Democrat. Subspecialties: Hematology; Cell biology (medicine). Current work: Elucidating factors that regulate normal vs. leukemic cell growth in tissue culture, interactions between normal and leukemic cells, and the induction of differentiation in human leukemic cell lines by chemical and physiologic stimuli. Home: 35 Chatham Rd Newton Highlands MA 02161 Office: Dept Medicine Harvard Med Sch Beth Israel Hosp 330 Brookline Ave Boston MA 02215

ROBINSON, SUSAN ESTES, pharmacology educator; b. Radford, Va., Apr. 26, 1950; d. Cecil Bennett and Helen Elizabeth (Buckner) Estes; m. Robert McMurdo Gillespie, Nov. 7, 1981. B.A., Vanderbilt U., 1972, Ph.D. in Pharmacology, 1976. Staff fellow Lab. Preclin. Pharmacology, NIMH, St. Elizabeth's Hosp., Washington, 1976-79; asst. prof. dept. med. pharmacology Tex. A&M U., College Station, 1979-81; asst. prof. dept. pharmacology and toxicology Med. Coll. Va., Richmond, 1981—; researcher. Contbr. numerous articles to profl. jours. Recipient Lyndon Baines Johnson Research award Tex. affiliate Am. Heart Assn., 1980; NIMH grantee, 1980-83; Am. Parkinson Disease Assn. grantee, 1980-82; Am. Heart Assn. grantee, 1980-83. Mem. Am. Soc. Pharmacology and Exptl. Therapeutics, Soc. Neurosci., Am. Soc. Neurochemistry, Basic Scis. Council of Am. Heart Assn. Subspecialties: Neuropharmacology; Psychopharmacology. Current work: Interactions between neurotransmitters in brain areas relevant to blood pressure, schizophrenia, and Parkinsin's Disease.

ROBINSON, WALKER LEE, neurol. surgeon, cons.; b. Balt., Oct. 13, 1941; s. Edward and Wilma (Walker) R.; m. Mae Elizabeth Meads, Arp. 9, 1966; children: Kimberly, Walker. B.S., Morgan State Coll., 1962; M.D., U. Md., 1970. Diplomate: Am. Bd. Neurol. Surgery. Mgmt. trainee C & P Telephone Co. Md., 1964, office mgr. comml. div., 1965, growth forecaster, 1966; instr. U. Md. Med. Sch., Balt., 1976-77, asst. prof., 1977—; head pediatric neurosurg. sect. Univ. Hosp., 1978—; clin. asso. prof. Meharry Med. Sch., 1980; cons. Nat. Inst. Neurol. Disease and Stroke. Bd. dirs. East Balt. Community Corp.; mem. East Side Democratic Orgn. Served to capt. Spl. Forces USAR, 1965-70; Served to capt. Spl. Forces U.S. Army, 1962-64. Fellow ACS; mem. Am. Assn. Neurol. Sugeons (fellow stroke council), Congress Neurol. Surgeons. Subspecialties: Neurosurgery; Laser medicine. Current work: Use of laser in neurological surgery; hydrocephalus in children - research and evaluation. Home: 3701 Cedar Dr Baltimore MD 21207 Office: 1205 York Rd #18 Lutherville MD 21093

ROBINSON, WALTER LLOYD, radiation physicist; b. Lancaster, Pa., Jan. 13, 1946; s. Walter C. and Kathryn M. (Lloyd) R.; m. Donna M. Fetterhoff, July 4, 1968; children: Shane M., Todd A. B.A. in Biology, Millersville State Coll., 1968; M.S. in Radiol. Health, U. Mich., 1971. Cert. in nuclear instrumentation Am. Bd. Sci. in Nuclear Medicine. Biochem. analyst St. Joseph Hosp., Lancaster, 1968-69; radiation safety officer, radiation physicist St. John's Hosp., Detroit, 1971-72; v.p., dir. Pa. Services, Bionucleonics, Inc., Kenilworth, N.J., 1972-80; owner Walter L. Robinson & Assocs., cons. radiation and diagnostic imaging physicists, Lancaster, 1980—; program dir. Meml. Osteo. Hosp. Sch. Nuclear Medicine Tech., York, Pa.; part-time lectr. Harrisburg Area Community Coll. Author research papers. Active Cub Scouts and Boy Scouts Am., 1981-82. AEC grantee, 1970-71. Mem. Soc. Nuclear Medicine, Am. Assn. Physicists in Medicine (adj. mem. nuclear medicine com.). Republican. Lodge: East Hempfield (Pa.) Lions. Subspecialties: Nuclear medicine; Diagnostic radiology. Current work: Diagnostic medical imaging; nuclear medical physics; diagnostic radiology physics; seminars in nuclear medicine technology. Inventor in field. Home and Office: 2624 Spring Valley Rd Lancaster PA 17601

ROBINSON, WILLIAM ROBERT, chemist, educator; b. Longview, Tex., May 30, 1939; m. Phyllis J. Swart, July 11, 1962; children: Margaret, Kevin, Brian. B.S., Tex. Tech U., 1961, M.S., 1962; Ph.D., MIT, 1966. Asst. prof. Purdue U., West Lafayette, Ind., 1967-73, assoc. prof., 1973-79, prof. chemistry, 1979—. Author: General Chemistry, 1980; assoc. editor: Jour. Solid State Chemistry, 1981—. NSF postdoctoral fellow, 1966-67. Mem. Am. Chem. Soc., Am. Crystallographic Assn., AAAS. Subspecialties: Solid state chemistry; X-ray crystallography. Current work: Solid state synthetic chemistry, crystal structures, oxides, sulfides, phosphates. Home: 137 E Knox Dr West Lafayette IN 47906 Office: Dept Chemistry Purdue U West Lafayette IN 47907

ROBISON, G(EORGE) ALAN, pharmacology educator; b. Lethbridge, Alta., Can., Nov. 4, 1934; s. Douglas Charles and Margaret Elizabeth (Barr) R.; m. Jill Jeanine Seaman, Mar. 12, 1956; children: James Darcy, Amelia N'Orlean. B.Sc., U. Alta., 1957; M.S., Tulane U., 1960, Ph.D., 1962. Postdoctoral research fellow Western Res. U., Cleve., 1962-63; research assoc. Vanderbilt U., Nashville, 1963-64, instr. dept. pharmacology, 1964-66, asst. prof., 1966-70, assoc. prof. physiology and pharmacology, 1970-72; prof., chmn. dept. pharmacology U. Tex. Med. Sch.-Houston, 1972—; mem. research council Nelson Research and Devel. Co., Irvine, Calif., 1971—. Author: (with R.W. Butcher and E.W. Sutherland) Cyclic AMP, 1971; editor: (with P. Greengard) Advances in Cyclic Nucleotide Research, 1972—; contbr. articles to profl. jours. Recipient J. Murray Luck award Nat. Acad. Scis., 1979. Mem. Am. Chem. Soc., Am. Soc. Pharmacology and Exptl. Therapeutics, Endocrine Soc., Soc. Neurosci., Am. Humanist Assn. Subspecialties: Molecular pharmacology; Cellular pharmacology. Current work: Mechanism of drug action; cyclic nucleotide research. Home: 250 Stoney Creek Houston TX 77024 Office: PO Box 20708 Houston TX 77025

ROBSON, RICHARD MORRIS, biochemist, educator; b. Atlantic, Iowa, Dec. 7, 1941; s. Morris Henry and Eleanor Gwen (Livingston) R.; m. Susan Elaine Pelzer, June 4, 1964 (div.); children: Kristi Jo, Jeff Alan. B.S. in Animal Sci. with distinction, Iowa State U., Ames, 1964, M. Biochemistry, 1966, Ph.D. in Biochemistry (NIH predoctoral fellow), 1969. Asst. prof. biochemistry in animal sci. U. Ill., Urbana, 1969-72; assoc. prof. biochemistry and animal sci. Iowa State U., 1972-77, prof. biochemistry, animal sci. and food tech., 1977—. Contbr. articles to profl. jours. Served with USAR, 1960-64. Named to Outstanding Tchr. List U. Ill., 1972. Mem. Am. Soc. Biol. Chemists, Am. Soc. Animal Sci., Inst. Food Technologists, Biophys. Soc., Am. Soc. Cell Biology, Soc. Sigma Xi, Phi Kappa Phi, Gamma Sigma Delta. Club: Ames Town and Coll. Toastmasters (pres. 1982-83). Subspecialties: Biochemistry (biology); Cell and tissue culture. Current work: Biochemistry and structure of contractile and cytoskeletal proteins. Office: Muscle Biology Group Iowa State U Ames IA 50011

ROBYT, JOHN FRANCIS, biochemist, educator; b. Moline, Ill., Feb. 17, 1935; s. Frank August and Mary Margret (McFadyen) R.; m. Lois Tefft Kennedy, Apr. 12, 1958; children: Jan Clare, William John. B.S., St. Louis U., 1958; Ph.D., Iowa State U., 1962. Asst. prof. biochemistry La. State U., Baton Rouge, 1962-63; research assoc. Lister Inst. Medicine, London, 1963-64, Iowa State U., Ames, 1964-67, asst. prof. biochemistry, 1967-73, assoc. prof. biochemistry, 1973—; cons. E.I. DuPont Co., Wilmington, Del., 1972-83. Contbr. research writings to publs. in field. Pres. Ames Minor Hockey Assn., 1977-82; coach Ames High Sch. Hockey Club, 1978-79. Grantee U.S. Dept. Agr., 1965-70, Nat. Inst. Dental Research, NIH, 1971—. Mem. Am. Chem. Soc., Am. Soc. Biol. Chemists. Subspecialty: Biochemistry (biology). Current work: Enzyme mechanisms involved in polysaccharide hydrolysis and biosynthesis. Polysaccharide and eligosaccharide structure; enzyme purification and structure. Patentee inhibition of dental plaque. Home: RR 4 Ames IA 50010 Office: Dept Biochemistry and Biophysics Iowa State U Ames IA 50011

ROCCI, MARIO LOUIS, JR., pharmacokinetic and medical researcher, consultant; b. Utica, N.Y., Nov. 6, 1952; s. Mario Louis and Marie Joyce (Cuccaro) R.; m. Claudia Fiutak, Oct. 16, 1976. B.S., SUNY-Buffalo, 1976, Ph.D., 1981. Intern pharmacy Millard Fillmore Hosp., Buffalo, 1975-76, pharmacist, 1978-79; adj. instr. Thomas Jefferson U. Hosp., Phila., 1980-82, adj. asst. prof., 1982—; asst. prof. Phila. Coll. Pharmacy and Sci., 1980—; dir. Pharmacokinetic Research Lab., 1980—, chmn. research, pub. and staff devel. com., 1981—. Children's Hosp. Phila. Research Found. research grantee, 1981—; Sterling-Winthrop Labs. research grantee, 1982, 83; Glaxo Labs. Inc. research grantee, 1983. Mem. Am. Fedn. Clin. Research, Am. Soc. Clin. Pharmacology and Therapeutics, Am. Assn. Colls. Pharmacy, Tau Phi Zeta. Subspecialties: Pharmacokinetics; Pharmacology. Current work: Examination of the pharmacokinetics of drugs and their metabolism in health and disease and in animal models. Home: 7111 Brentwood Rd Philadelphia PA 19151 Office: Thomas Jefferson Univ Hosp M502 11th and Walnut Sts Philadelphia PA 19107

ROCKHOLD, ROBIN WILLIAM, pharmacologist; b. Dayton, Ohio, Sept. 29, 1951; s. William Thomas and Mildred (Davis) R.; m. Linda Raper, Dec. 23, 1978. A.B., Kenyon Coll., 1973; Ph.D., U. Tenn., 1978. Postdoctoral fellow U. Heidelberg (W. Ger.) Pharmacology Inst., 1978-79; U. Tenn. Ctr. Health Scis., Memphis, 1979-82, asst. prof., 1982-83; asst. prof. dept. pharmacology and toxicology U. Miss. Med. Ctr., Jackson, 1983—. Contbr. articles to profl. jours. Tenn. Heart Assn. fellow, 1975-76; NIH fellow, 1979-81, 81-82. Mem. Am. Heart Assn., Soc. Neurosci., Southeastern Regional Pharmacology Soc., Sigma Xi (assoc.). Subspecialties: Pharmacology; Neuroendocrinology. Current work: Cardiovascular regulation by peptide and monamine-containing neurons; hypertension, blood pressure, opioid peptides, vasopressin, monoamine neurons, cardiovascular reflexes. Office: Dept Pharmacology and Toxicology U Miss Med Center Jackson MS 39216

ROCKLAND, KATHLEEN SKIBA, neuroanatomist; b. Stamford, Conn., Oct. 10, 1947; d. Charles and Sophie (Markisz) Skiba; m. Charles Rockland, Aug. 2, 1970. B.A., Wellesley Coll., 1969; M.A., Princeton U., 1972; Ph.D., Boston U., 1979. Research fellow div. biology Calif. Inst. Tech., Pasadena, Calif., 1978-79; research fellow in neurology Children's Hosp. Med. Center, Boston, 1979-80; research assoc. dept. ophthalmology Med. U. S.C., Charleston, 1980-82; sr. research fellow E. K. Shriver Ctr., Waltham, Mass., 1982—; adj. asst. prof. dept. anatomy Boston U., 1983—. Contbr. articles to profl. jours. Woodrow Wilson fellow, 1969; Princeton Nat. fellow, 1969-72; NDEA fellow, 1969-72; NIH trainee, 1974-78; recipient Nat. Research Service award NIH, 1978-80. Mem. Assn. Research in Vision and Ophthalmology, Soc. Neurosci. Subspecialties: Neurobiology; Neurophysiology. Current work: Neuroanatomy; cerebral cortex, visual system primates.

ROCKWELL, SARA CAMPBELL, research scientist; b. Somerset, Pa., Sept. 8, 1943; d. W. Paul and Rebecca (Mostoller) Campbell; m. Charles Rockwell; children: Rebecca Jane, Karen Renee. B.S., Pa. State U., 1965; Ph.D., Stanford U., 1971. Engring., tech. asst. HRB-Singer, State Coll., Pa., 1963-64; postdoctoral fellow Stanford U., 1971-72, 74; attache de recherche Institut de Recherche de Radiobiologie Clinique, Institut Gustave Roussy, Villejuif, France, 1973; asst. prof. therapeutic radiology Yale U., 1974-78, asso. prof., 1978—. Editorial bd.: Cell and Tissue Kinetics; contbr. articles to jours. USPHS fellow, 1966-71; Damon Dunyon Meml Fund grantee, 1971-72, 74; Am. Cancer Soc. grantee; Nat. Cancer Inst. grantee. Mem. Cell Kinetics Soc. (pres. 1982-83), Radiation Research Soc., Am. Assn. Cancer Research, Bioelectromagnetics Soc., Tissue Culture Assn., AAAS, Phi Kappa Phi, Alpha Lambda Delta. Subspecialties: Cancer research (medicine); Radiology. Current work: Biology and treatment of cancer. Office: 333 Cedar St Room 284 New Haven CT 06510

ROCKWOOD, DONALD LEE, forestry educator, consultant; b. Rockford, Ill., Oct. 28, 1944; s. Claude Leslie and Agnes Anna (Krueger) R.; m. Virginia Joanne, Aug. 18, 1968; children: Kimberly, Brian. B.S., U. Ill., 1966, M.S., 1968; Ph.D., N.C. State U., 1972. Research assoc. U. Fla., Gainesville, 1972-73, asst. prof. forestry, 1973-80, assoc. prof., 1980—. Contbr. articles to profl. jours. Mem. AAAS, Soc. Am. Foresters, Sigma Xi, Gamma Sigma Delta. Subspecialties: Plant genetics; Biomass (energy science and technology). Current work: Research in tree improvement, woody biomass production and forest measurement. Office: University of Florida School of Forest Resources and Conservation Gainesville FL 32611

RODDIS, LOUIS HARRY, JR., retired naval officer, consulting engineer; b. Charleston, S.C., Sept. 9, 1918; s. Louis Harry and Winifred Emily (Stiles) R. B.S., U.S. Naval Acad., 1939; M.S., MIT, 1944. Registered profl. engr., Pa., D.C., N.Y., N.J., S.C.; chartered engr., U.K. Commd. ensign U.S. Navy, 1939, advanced through grades to capt., 1957; various assignments, sea duty, 1939-41, Pearl Harbor, 1941, assigned Phila. Naval Shipyard, 1944, staff of comdr. Joint Task Force I, atomic weapons tests, Bikini, 1946, assigned Clinton Labs., Manhattan Engring. Dist. (now Oak Ridge Nat. Lab.), 1946, staff bur. ships Dept. Navy, Washington; staff bur. ships, nuclear ship propulsion program AEC, assisted nuclear reactor design U.S.S. Nautilus Dept. Navy, 1947-55; dep. dir. div. reactor devel. AEC, 1955-58; pres., dir. Pa. Electric Co., Johnstown, 1958-67, chmn., 1967-69; dir. nuclear power activities Gen. Pub. Utilities Corp., N.Y.C., 1967-69; vice chmn. Consol. Edison Co. N.Y., 1969, 73-74, pres., 1969-73, also trustee; pres., chief exec. officer John J. McMullen Assos., Inc., N.Y.C., 1975-76; asso. co. Panero-Tizian Assos., Inc., 1975-76; cons. engr., 1976—; dir. Hammermill Paper Co., Detroit Edison Co., Inc., Gould Inc., Research-Cottrell, Inc.; mem. Pres.'s Adv. Council on Energy Research and Devel., 1973-75; cons. U.S. Dept. State, Disarmament Commn., 1960-61, U.S. Maritime Adminstrn., 1959-62; chmn. maritime research adv. com. Nat. Acad. Scis.-NRC, 1958-60; mem. Gov.'s Com. of 100 for Better Edn., 1962-64, Gov.'s Council Sci. and Tech., 1963-65, Pa. Indsl. Devel. Authority, 1961-68; pres. Atomic Indsl. Forum, 1962-64; mem. adv. com. Rockefeller U., 1972—; mem. energy research adv. bd. Dept. Energy, 1978—, chmn., 1981—. Author tech. articles on nuclear power and energy subjects. Bd. dirs. Mercy Hosp., Johnstown, 1959-67, Metal Properties Council, 1970-74. Recipient Outstanding Service award AEC, 1957; Arthur S. Flemming career award Washington C. of C., 1958; Outstanding Citizen award Johnstown Inter-Service Club Council, 1963. Fellow Royal Instn. Naval Architects; mem. ASME (Fellowship award), IEEE, Am. Nuclear Soc. (pres. 1969-70, Fellowship award 1970), Soc. Naval Engrs., Am. Soc. Naval Architects and Marine Engrs., Nat. Acad. Engring., Nat. Soc. Profl. Engrs., ASHRAE, N.Y. C. of C., Newcomen Soc. N.Am., Edison Electric Assn. (dir. 1969-73), Human Factor Soc., Am. Gas Assn. (dir. 1973-74), Commerce and Industry Assn. N.Y. (dir. 1969-74), Sigma Xi, Tau Beta Pi. Clubs: Army Navy (Washington); Army-Navy Country (Arlington, Va.); Rotary (Charleston); Chemists (N.Y.); Mantoloking (N.J.); Yacht. Subspecialty: Energy research and development. Current work: Energy policy and especially research and development of all energy forms, chairman research advisory board to U. S. Department of Energy. Home and Office: 46 State St Charleston SC 29401

RODENSKY, ROBERT L., research consultant; b. New London, Conn., Mar. 13, 1945; s. Abraham and Evelyn (Rich) R.; m. Catherine Anne Colby, Mar. 16, 1979. B.A., U. Pa., 1967; M.A., U. R.I., 1970; Ph.D., U. Western Ont., Can., 1977. Lectr. U. Western Ont., London, 1971-77; researcher Children's Aid Soc. Met. Toronto, 1977-80; research cons. program evaluation-child abuse, Kitchener, Ont., 1980—. NDEA grantee, 1963. Mem. Am. Psychol. Assn., Can. Psychol. Assn., Ont. Psychol. Assn. (communications com.), Can. Evaluation Soc. Club: Waterloo-Wellington Flyint (Breslau, Ont.). Subspecialties: Program Evaluation; Child Abuse. Current work: Implementation of methodologically sound research techniques to provide useful information regarding treatment and identification of child abuse. Home and office: 689 Doon Village Rd Kitchener ON Canada N2P 1A1

RODGERS, JOHN, educator, geologist; b. Albany, N.Y., July 11 1914; s. Henry D. and Louise W. (Allen) R. B.A., Cornell U., 1936, M.S., 1937; Ph.D., Yale U., 1944. Geologist, U.S. Geol. Survey, 1938-46, intermittently, 1946—; sci. cons. U.S. Army Engrs., 1944-46; instr. geology Yale U., 1946-47, asst. prof., 1947-52, asso. prof., 1952-59, prof., 1959-62, Silliman prof., 1962—; vis. lectr. Coll. de France, Paris, 1960; sec.-gen. commn. on stratigraphy Internat. Geol. Congress, 1952-60; commr. Conn. Geol. and Natural History Survey, 1961-71. Author: (with C.O. Dunbar) Principles of Stratigraphy, 1957, The Tectonics of the Appalachians, 1970; also articles in field.; editor: Symposium on the Cambrian System, 3 vols., 1956, 61; asst. editor: Am. Jour. Sci., 1948-54, editor, 1954—. NSF Sr. postdoctoral fellow, France, 1959-60; exchange visitor Geol. Inst., Acad. Scis. USSR, 1967; Guggenheim fellow, Australia, 1973-74; recipient medal of freedom U.S. Army, 1947. Fellow Geol. Soc. Am. (councillor 1962-65, pres. 1970, Penrose medal 1981), AAAS; mem. Am. Acad. Arts and Scis., Am. Assn. Petroleum Geologists, Conn. Acad. Sci. and Engring. (charter), Conn. Acad. Arts and Scis. (pres. 1969), Am. Geophys. Union, Nat. Acad. Scis., Geol. Soc. London (hon.), Société géologique de France (asso. mem., v.p. 1960), Acad. Scis. USSR (hon. fgn. mem.), Academia Real de Ciencias y Artes Barcelona (fgn. corr. mem.), Sigma Xi. Club: Elizabethan (New Haven). Subspecialty: Geology. Address: Dept Geology Yale U PO Box 6666 New Haven CT 06511

RODGERS, JOSEPH LEE, III, psychologist; b. Norman, Okla., Feb. 9, 1953; s. Joseph Lee and Mary Joyce (Norwood) R. B.S., U. Okla., 1975, B.A. in Math, 1975; M.A., U. N.C., 1979, Ph.D., 1981. Instr. U. N.C., Chapel Hill, 1980; asst. prof. U. Okla., Norman, 1981—. Tournament dir Norman Tennis Assn., 1982-83. Mem. Psychometric Soc., Am. Psychol. Assn., Population Assn. Am. Democrat. Methodist. Subspecialties: Psychometrics; Statistics. Current work: Scaling and measurement, evaluation methods, multivariate analysis, population psychology. Office: Dept Psycholog U Okla Norman OK 73014

RODKEY, LEO SCOTT, immunologist; b. Topeka, Jan. 18, 1941; s. Everett Marvin and Nola Ruth (Mitchell) R.; m. Dixie Dee Croft, June 2, 1963; 1 son, Travis Lincoln. A.B., U. Kans., 1964, Ph.D., 1968. Postdoctoral fellow U. Ill., Chgo., 1966-70; mem. faculty Kans. State U., 1970-82, prof., 1982; mem. faculty U. Tex. Med. Sch., Houston, 1982—; mem. Basel (Switzerland) Inst. Immunology, 1977-78, NIH Clin. Sci. I study sect., 1980—; mem., cons. Abbott Labs, 1983—. Contbr. articles to profl. jours. NIH research career devel. awardee, 1977-82. Mem. Am. Assn. Immunologists, Am. Soc. Microbiology. Subspecialty: Immunobiology and immunology. Current work: Regulation of immune response, idiotypic network. Office: Dept Pathology U Tex Med Sch PO Box 20708 Houston TX 77025

RODRIGUEZ, ELOY, biologist, phytochemist, educator, consultant; b. Edinburg, Tex., Jan. 7, 1947; s. Everardo and Hilaria (Calvillo) R. B.S. in Zoology, U. Tex., Austin, 1969, Ph.D., 1975. Postdoctoral scientist U. B.C. (Can.), Vancouver, 1976; asst. prof. U. Calif., Irvine, 1976-78, assoc. prof./research scientist, 1978-83, prof., research scientist, 1983—. Contbr. articles to sci. jours. Ford Found. Mexican-Am. fellow, 1972-74; Nat. Chicano Council fellow, 1978; Fulbright sr. scholar, 1978; NIH Research Career Devel. award, 1982—. Mem. Am. Chem. Soc., Phytochem. Soc., Am. Bot. Soc., Mexican Bot. Soc., Soc. Advancement of Chicanos and Native Ams. Subspecialties: Pharmacognosy; Medicinal chemistry. Current work: Phytochemistry, plant allergens, medicinal plant chemistry, chemical ecology. Office: Phytochem Lab Dept Ecology and Biology U Calif Irvine CA 92717

RODRIGUEZ, LUIS FELIPE, radio astronomer; b. Merida, Yucatan, Mexico, May 29, 1948; s. Vicente and Edith (Jorge) R.; m. Rosa Maria Ezquerro, Dec. 17, 1973. M.A., Harvard U., 1975, Ph.D. in Astronomy, 1978, diploma, 1979. Researcher Nat. U. Mexico, Mexico City, 1979—, dir., 1980—, also mem. tech. council. Contbr. articles to profl. jours. Recipient Robert J. Trumpler award Astron. Soc. Pacific, 1980. Mem. Internat. Astron. Union, Am. Astron. Soc., Mex. Soc. Physics. Subspecialty: Radio and microwave astronomy. Current work: Radio astronomical studies of interstellar medium. Microwave and milimiter spectroscopy, radio interferometry, very long baseline interferometry. Contributions to understanding of nature of nucleus of Milky Way and of phenomena related to star formation. Home: Villa Olimpica 11-102 Mexico DFMexico 14020 Office: Apartado Postal 70-264 Mexico DFMexico 04510

RODRIQUEZ-RIGAU, LUIS JOSE, physician, medical educator; b. Barcelona, Spain, June 16, 1951; came to U.S., 1975, naturalized, 1977; s. Jose A. and Dolores (Rigau) Rodriquez-R.; m. Carol M. Maycock, June 26, 1976; children: Robyn, Elena, Karen. B.Sc., U. Barcelona, Spain, 1967; M.D., Autonomous U. Med. Sch., Barcelona, 1974. Intern Univ. Hosp., Barcelona, 1973-74, resident in reproductive endocrinology, 1974-75; fellow U. Tex. Med. Sch., Houston, 1975-76, instr., 1976-77, asst. prof. reproductive medicine, 1977-81, assoc. prof., 1981—; sci. sec. 1st Internat. Symposium Reproductive Medicine, Houston, 1983-84. Contbr. numerous articles to profl. jours., chpts. to books; mem. editorial bd.: Jour. Andrology, 1981—. Bd. dirs. Houston chpt. March of Dimes, 1978—; mem. med. adv. bd. Infertility Network Inc., Houston, 1982—. March of Dimes grantee, 1978—; recipient AMA Physician Recognition award, 1977, 80, Sci. Exhibit award, 1981. Mem. Endocrine Soc., Am. Fertility Soc., Am. Soc. Andrology, Am. Fedn. Clin. Research, AMA. Subspecialties: Endocrinology; Reproductive endocrinology. Current work: Research, education and clinical practice in reproductive medicine in males and females. Home: 2002 Deer Springs Kingwood TX 77339 Office: Dept Reproductive Medicine U Tex Med Sch PO Box 20708 Houston TX 77025

ROE, BYRON PAUL, physicist, educator; b. St. Louis, Apr. 4, 1934; s. Sam S. and Gertrude H. (Claris) R.; m. Alice Susan Krauss, Sept. 27, 1961; children: Kenneth David, Diana Clare. B.A., Washington U., St. Louis, 1954; Ph.D., Cornell U., 1959. Instr. physics U. Mich., Ann Arbor., 1959-61, assoc. prof., 1961-64, 1964-69, prof. physics, 1969—; Brit. govt. Sci. Research Council fellow, Oxford, 1979-80; vis. scientist CERN, Geneva, Switzerland, 1967-68. Fellow Am. Phys. Soc. Subspecialty: Particle physics. Current work: Charmed particle production. Inventor Textcode translator, 1982. Home: 3610 Charter Pl Ann Arbor MI 48105 Office: Dept Physics U Mich Ann Arbor MI 48109

ROE, KENNETH ANDREW, engring. co. exec.; b. Perry, N.Y., Jan. 31, 1916; s. Ralph Coats and Esther Mae (Bishop) R.; m. Hazel Winifred Thropp, Feb. 22, 1942; children—Ralph Coats, Randall Brewster, Kenneth Keith, Hollace Lindsey, Willis Barton. B.A., Columbia U., 1938; B.S. in Chem. Engring, Mass. Inst. Tech., 1941; M.S. in Mech. Engring, U. Pa., 1946; D.Eng. (hon.), Stevens Inst. Tech., 1978. Registered profl. engr., 27 states; cert. Nat. Council Engring. Examiners. With Burns and Roe, Inc., N.Y.C., 1938—, exec. v.p., 1953-64, pres., 1963—, chmn. bd., 1971—; trustee, past chmn. exec. com. council engring. affairs, mem. com. acad. affairs of bd. trustees Manhattan Coll.; mem. bd. overseers U. Pa. Coll. Engring. and Applied Scis.; trustee, mem. exec. com. Stevens Inst. Tech.; mem. engring. council Columbia U.; founding mem. Columbia Engring. Affiliate; bd. dirs., investment adv. com. U.S. nat. com. World Energy Conf. Author numerous papers, articles in field. Trustee Round Hill Community Ch., Greenwich, Conn., 1972—; adv. council Bergen council Boy Scouts Am., 1974; past chmn. N.J. Conf. Promotion Better Govt.; past trustee Brit.-Am. Found. Served to lt. comdr. USNR, 1941-45. Recipient Carl Kayan award Columbia U., 1977; Named Engr. of Yr. N.J. Soc. Profl. Engrs. and Land Surveyors, 1980, Bergen County Soc. Profl. Engrs. Fellow AAAS, Inst. Mech. Engrs. (Eng.), Inst. Engrs. (Australia); hon. mem. ASME (past v.p., pres., Engring. award 1965, Edwin F. Church medal 1980); mem. Nat. Acad. Engring., Engrs. Joint Council (1st v.p. 1976-77, pres. 1978—), Am. Nuclear Soc., AIAA, IEEE, Nat. Soc. Profl. Engrs. (award 1981), N.J. Soc. Profl. Engrs., TAPPI (asso.), Am. Inst. Indsl. Engrs. (sr.), Soc. Petroleum Engrs., Electro-Chem. Soc., Am. Chem. Soc. (sr.), Am. Mgmt. Assn. (pres. assn.), Am. Soc. Quality Control, Am. Inst. Chem. Engrs., Pub. Utilities Assn. Virginias, Am. Assn. Engring. Socs. (chmn., acting pres. 1980), ASCE, Am. Soc. Engring. Edn. (hon.), Columbia U. Engring. Alumni Assn. (life; chmn. fund 1978-79), Brit. Nuclear Energy Soc. Subspecialties: Nuclear engineering; Electrical engineering. Current work: Advanced energy programs such as conversion, synthetic fuels,magnetohydrodynamics; fuel cells, solar, breeder reactors. Home: Baldwin Farms North Greenwich CT 06830 Office: 550 Kinderkamack Rd Oradell NJ 07649

ROE, KENNETH KEITH, engineering and construction company executive; b. Phila., oct. 17, 1945; s. Kenneth Andrew and Hazel (Thropp) R.; m. Elizabeth Eaton, June 28, 1975; children: Kenneth Andrew Roe II, Whitney Elizabeth, Edward Scott. B.S.N.E., Princeton U., 1968; M.E. in Nuclear Engring, MIT, 1974, M.S., 1974; postgrad., Harvard U., 1980. Lic. profl. engr., Calif., N.J., N.Y., Wash., P.R. Engr. Burns & Roe Inc., Hempstead, N.Y., 1971-75, asst. project mgr., 1975-77, project engr., resident project engr., Woodbury, N.Y., 1977-78, project engring. mgr., Oradell, 1978-79, asst. v.p., 1979-80, v.p., 1980-82, exec. v.p., 1982—, dir., 1971—, dir. indsl. services, Paramus, N.J., 1974—; dir. Burns and Roe Synthetic Fuels, 1980—, Fegles Power Service Corp., Mpls., Gen. Physics Corp., Columbia, Md. Mem. adv. council dept. mech. engring. Columbia U., N.Y.C., 1982—; chmn. pastor-parish relations com. First Ch. Round Hill, Greenwich, Conn., 1981—. Served to lt. USN, 1969-71. Sloan Found. fellow MIT, 1971. Mem. Am. Nuclear Soc., ASME (exec. com. net. sect. 1981—), Am. Inst. Chem. Engrs., AAAS, Water Pollution Control Assn., Colegio de Ingenieros. Republican. Clubs: Princeton of N.Y., Stanwich (Greenwich); Coral Beach (Paget, Bermuda); Sankaty Head (Siasconset, Mass). Subspecialties: Mechanical engineering; Nuclear engineering. Current work: Development of improved plant designs for both conventional fossil and nuclear fueled power plants and introduction of advanced power technologies, such as breeder reactors, fusion, synthetic fuels, and others. Office: Burns and Roe Inc 550 Kinderkamack Rd Oradell NJ 07649

ROEDER, ROBERT CHARLES, astrophysicist, educator; b. Stratford, Ont., Can., Oct. 7, 1937; s. Albert H.G. and Catharine A. R.; m. Dagmar F. Katterfeld, Dec. 30, 1961; children: Robert Charles, Thomas. B.Sc. in Physics, McMaster U., 1959, M.Sc., 1960; Ph.D. in Astronomy, U. Ill., 1963. Asst. prof. astronomy U. Ill., Champaign, 1962-63; asst. prof. physics Queen's U., 1963-64; asst. prof. astronomy U. Toronto, 1964-68, prof. astronomy, 1975—. Contbr. articles to profl. jours. Served to capt. Royal Can. Signals, 1955-63. Mem. Am. Astron. Soc., Can. Astron. Soc., Royal Astron. Soc. Can., Internat. Astron. Union, Gen. Relativity and Gravity Soc. Subspecialties: Cosmology; General relativity. Current work: Cosmology, quasars, general relativistic optics. Office: Phys Scis Scarborough Coll West Hill ON Canada M1C 1A4

ROEDER, STEPHEN BERNHARD WALTER, physical chemist, educator; s. Walter Martin and Katherine Elizabeth Ruth (Holtz) R.; m. Phoebe E. Barber, June 28, 1969; children: Adrienne H. K., Roland K. W. A.B. in Chemistry, Dartmouth Coll., 1961; Ph.D. in Phys. Chemistry, U. Wis., 1965. Postdoctoral fellow Bell Telephone Labs., 1965-66; lectr. in physics U. Oreg., 1966-68, postdoctoral researcher in chemistry, 1966-68; asst. prof. chemistry and physics San Diego State U., 1968-71, assoc. prof., 1972-75, prof., 1975—, chmn. dept. physics, 1975-78, chmn. dept. chemistry, 1979—; vis. assoc. prof. U. B.C. (Can.), Vancouver, 1974-75; mem. vis. staff Los Alamos Nat. Lab. Author: (with Eiichi Fukushima) Experimental Pulse NMR: Nuts and Bolts Approach, 1981; contbr. numerous articles on nuclear magnetic resonance to profl. jours. Recipient Outstanding Tchr. award San Diego State U., 1971; Research Corp. grantee, 1968-71; NSF grantee, 1978-79. Mem. AAAS, Am. Chem. Soc., Phi Kappa Phi. Subspecialties: Condensed matter physics; Physical chemistry. Current work: Nuclear magnetic resonance instrumentation; molecular motions; research in nuclear magnetic resonance instrumentation teaching. Home: 6789 Alamo Way La Mesa CA 92041 Office: Dept Chemistr San Diego State U San Diego CA 92182

ROEDERER, JUAN GUALTERIO, geophysics educator; b. Trieste, Italy, Sept. 2, 1929; came to U.S., 1967, naturalized, 1972; s. Ludwig Alexander and Anna Rafaela (Lohr) R.; m. Beatriz S. Cougnet, Dec. 20, 1952; children: Ernesto, Irene, Silvia, Mario. Ph.D., U. Buenos Aires, 1952. Research scientist Max Planck Inst., Gottingen, Germany, 1952-55; group leader Argentine AEC, 1953-59; prof. physics U. Buenos Aires, 1959-66, U. Denver, 1966-77; dir. Geophys. Inst., U. Alaska, Fairbanks, 1977—, dean coll. environ. scis., 1978-82; vis. staff Los Alamos Nat. Lab., 1969-81. Author: Dynamics of Geomagnetically Trapped Radiation, 1970, Physics and Psychophysics of Music, 1973, 3rd edit. 1979; contbr. articles to profl. jours. NAS/NASA Sr. research fellow, 1964-66. Fellow AAAS, Am. Geophys. Union; mem. Asociación Argentina de Geodestas y Geofísicos, Polar Research Bd., Internat. Council Sci. Unions (sci. com. on solar-terrestrial physics). Subspecialties: Space physics; Psychophysics. Current work: Study of plasma and energetic particles in the earth's magnetosphere; sci. policy issues for the Arctic; study of perception of music. Home: 105 Concordia Dr Fairbanks AK 99701 Office: Geophys Inst Univ Alaska Fairbanks AK 99701

ROEHRS, ROBERT JESSE, nondestructive testing specialist; b. Ellis Grove, Ill., Dec. 14, 1930; s. Sidney France and Eugenia Alma (Shea) R.; ; s. Elaine Renee and Stephen Todd.; m. Betty Evelyn Roehrs, Sept. 15, 1950. B.S. in Prodn. Mgmt, Washington U., St. Louis, 1964. Registered profl. engr., Calif. Equipment designer research and devel. uranium div. Mallinckrodt Chem. Co., Weldon Spring, Mo., 1956-61; mgr. quality assurance Nooter Corp., St. Louis, 1961-68; cons. nuclear surveys ASME, N.Y.C., 1968; chief mech. engr., mgr. quality control Killebrew Engring. Corp., St. Louis, 1968-69; sr. tech. specialist nondestructive testing material and process devel. McDonnell Aircraft Co., St. Louis, 1969—. Coordinator, contbg. author: Nondestructive Testing Handbook, 1982; contbr. numerous chpts. to books and articles to profl. jours. Deacon Tower Grove Baptist Ch., St. Louis, 1956-58, tchr. Sunday sch., 1954-71, chmn. recreation com., 1956-58. Served with U.S. Army, 1952-54. Fellow Am. Soc. Nondestructive Testing (dir. 1973-76, chmn. leak testing 1966-78, vice chmn. 1978—, Gold medal award, chmn. St. Louis sect. 1962-64, dir. 1964-68, chmn. 1964-65), ASTM (Charles W. Briggs award); mem. Am. Soc. Metals, ASME, Am. Vacuum Soc., Am. Welding Soc., Brit. Inst. Nondestructive Testing, Soc. Automotive Engrs. Subspecialties: Nondestructive testing; Materials (engineering). Current work: Computerized ultrasonics and radiography; nondestructive testing, inspection, quality assurance, materials evaluation. Home: 4762 Barroyal Dr Saint Louis MO 63128 Office: Department 357 Bldg 32/ PO Box 516 Saint Louis MO 6316

ROEL, LAWRENCE EDMUND, neurochemist; b. Bklyn., Aug. 19, 1949; s. Edmund Lawrence and Leslie Adele (Gonzales) R. A.B., Princeton U., 1971; Ph.D., M.I.T., 1976; Muscular Dystrophy Assn. postdoctoral fellow, U. Calif., San Diego, 1976-78; postgrad, U. Pa., 1981—. Asst. prof. anatomy Northwestern U. Med. Sch., Chgo., 1978-81; research assoc. U. Pa., Phila., 1981—; cons. Sunmark Research Co., Arlington, Va., Newsource Publs., Huntington, N.Y. Contbr. articles to profl. jours. Am. Cancer Soc. grantee, 1980. Mem. Soc. Neurosci., N.Y. Acad. Scis., AAAS, Am. Chem. Soc., Sigma Xi. Club: Princeton of Phila. Subspecialties: Neurochemistry; Neurology. Current work: Neurochemistry of monoamine, protein synthesis interacions. Home: 4310 Spruce St Philadelphia PA 19104

ROELFS, ALAN PAUL, plant pathologist; b. Stockton, Kans., Nov. 18, 1936; s. Paul Martin and Florence Lorene (Lewin) R.; m. Anita Faye Clark, Sept. 3, 1956 (div. 1981); children: David Alan, Judith Lynne, Lorene Mary.; m. LuAnne Beatrice, June 11, 1983. B.S., Kans. State U., 1959, M.S., 1964; Ph.D., U. Minn., 1970. Technician Agrl. Research Service, U.S. Dept. Agr., Manhattan, Kans., 1959-64, plant pathologist, St. Paul, 1965-70, Animal and Plant Health Inspection Service, 1971-75, Agrl. Research Service, 1975—; cons. Interam. Inst. for Agrl. Research, Interam. Inst. Coop. on Agr., 1980, 82. Contbr. articles to profl. jours. Served with U.S. Army, 1959-61. Dept. Agr. grantee, 1980, 82—. Mem. Am. Phytopathol. Soc., AAAS, Internat. Assn. for Aerobiology, U.S. Fedn. for Culture Collections, Internat. Soc. Plant Pathology, Sigma Xi, Gamma Sigma Delta. Subspecialty: Plant pathology. Current work: Epidemiology of the rusts of cereals; effects of the environment, pathogen genotype and host genotype on disease devel. and losses in the U.S. and Mexico; rust epidemiology worldwide. Home: 132 DeMont St Apt 238 Saint Paul MN 55117 Office: Cereal Rust Lab U Minn Saint Paul MN 55108

ROELOFS, WENDELL LEE, biochemistry educator, consultant; b. Orange City, Iowa, July 26, 1938; s. Edward and Edith (Beyers) R.; m. Marilyn Joyce Kuiken, Sept. 3, 1960; children: Brenda Jo, Caryn Jean, Jeffrey Lee, Kevin Jon. B.A., Central Coll., Pella, Iowa, 1960; Ph.D., Ind. U., 1964; postdoctoral fellow, MIT, 1965. Asst. prof. N.Y. State Agrl. Expt. Sta., Geneva, 1965-69, assoc. prof., 1969-76, prof., 1976—, Liberty Hyde Bailey prof. insect biochemistry, 1976—; cons. Albany Internat. Co. Contbr. over 200 articles to sci. jours. Recipient Alexander von Humboldt award in Agr., 1977; Outstanding Alumni award Central Coll., 1978; Wolf prize for Agr., 1982. Mem. Entomol. Soc. Am. (J. Everett Bussart Meml. award 1973, Founder's Meml. award 1980), Am. Chem. Soc., AAAS, Sigma Xi. Republican. Presbyterian. Subspecialties: Physiology (biology); Biochemistry (biology). Current work: Insect sex pheromones research: biosynthesis, behavior, genetics, identification of active components. Patentee in field (10). Home: 4 Crescence Dr Geneva NY 14456 Office: NY State Agrl Expt Sta Geneva NY 14456

ROENIGK, HENRY HERMAN, dermatologist, educator; b. Tuscalusa, Ala., July 21, 1934; s. Henry Herman and Irene L. (Rini) R.; m. Kathy Franck, July 27, 1973; children—Randall K., Ronald C., Scott S., Larry M. M.D., Northwestern U., 1960. Head dept. dermatology Cleve. Clin, 1967-77; prof., chmn. dept. dermatology Northwestern U. Med. Sch., Chgo., 1977—; dir. North Community Bank. Served with U.S. Army, 1960-63. Med. Am. Soc. Dermatologic Surgery (pres. elect), Am. Acad. Dermatology (dir.). Subspecialty: Dermatology. Office: 303 E Chicago Ave Chicago IL 60611

ROESING, TIMOTHY GEORGE, virologist; b. Abington, Pa., May 14, 1947; s. George A. and Dorothy E. (MacClaskey) R.; m. Charline Ann Korn, Aug. 23, 1969; children: Kristy Anne, Timothy George. B.S., U. Md., 1969; M.S., Hahnemann Med. Coll., 1973, Ph.D., 1977. Grad. teaching asst. dept. microbiology and immunology Hahnemann Med. Coll., 1971, research asst., 1971-76; postgrad. fellow Smith Kline Diagnostics, Phila., 1976-77, sr. scientist, 1977; sr. project devel. biologist, dept. biol. quality control tech. services Merck Sharp & Dohme, West Point, Pa., 1977-82, sr. project microbiologist, dept. biol. prodn. tech. services, 1982—. Contbr. articles to profl. jours. Mem. Edge Hill Fire Co., North Hills, Pa., 1969—; mem. Weldon Fire Co., Glenside, Pa., 1977—, treas., 1982. Mem. Am. Soc. Microbiology, Tissue Culture Soc., Soc. Gen. Microbiology. Subspecialties: Virology (medicine); Microbiology (medicine). Current work: Research and tech. support to biol. (vaccine) prodn. area where vaccines for measles, mumps, rubella, hepatitis B viruses are produced. Home: 508 Briarwood Rd Glenside PA 19038 Office: Merck Sharp & Dohme W28-231 Westpoint PA 19486

ROESSLER, DAVID MARTYN, physicist; b. London, Apr. 29, 1940; U.S., 1966; s. Alfred Ernest and Elizabeth Minnie (Cornish) R.; m. Linda Jean Ciupak, May 19, 1983. B.Sc. in Math. and Physics, King's Coll., U. London, 1961, Ph.D. in Ultraviolet Spectroscopy-Physics, 1966. Asso. King's Coll., 1961; postdoctoral research fellow physics dept. U. Calif., Santa Barbara, 1966-68; temporary mem. tech. staff Bell Labs., Murray Hill, N.J., 1968-70; staff research scientist physics dept. Gen. Motors Research Labs., Warren, Mich., 1970—. Contbr. articles on spectroscopy to profl. jours. Mem. Optical Soc. Am., Internat. Solar Energy Soc., Inst. Physics (U.K.), Sigma Xi. Subspecialties: Condensed matter physics; Spectroscopy. Current work: Interaction of light and matter; spectroscopy, including optical properties of solids and aerosols, luminescence, photoacoustic spectroscopy. Patentee photoacoustic spectroscopy field. Home: 22610 Oak Ct Hazel Park MI 48030 Office: Physics Dept Gen Motors Research Labs Warren MI 48090

ROFFMAN, MARK, clinical pharmacologist, educator; b. Boston, July 8, 1945; s. William and Florence A. Tobias (Ableman) R.; m. Ina Ellen Marritt, July 13, 1968; children: Gary William, Jeremy Michael. A.B., Boston U., 1967; M.S., U. R.I., 1971, Ph.D. in Behavioral Pharmacology, 1972. NIMH postdoctoral trainee dept. psychiatry N.Y. U., 1972-73; assoc. in psychiatry Harvard Med. Sch., Boston, 1973-75, asst. prof. dept. psychiatry, 1975-76; lectr. in psychology Emmanuel Coll., Boston, 1975-76; sr. scientist II dept. pharmacology, pharms. div. CIBA-GEIGY Corp., Summit, N.J., 1976-78, sr. assoc. clin. pharmacology, 1978-80, asst. dir. clin. pharmacology, 1980-81, assoc. dir. clin. pharmacology, 1981—, dir. clin. pharmacology, 1982, exec. dir. clin. research, 1983—; instr. Bloomfield (N.J.) Coll., 1977—; adj. asso. prof. psychiatry Coll. Medicine and Dentistry N.J.-Rutgers Med. Sch., Piscataway, 1977—; adj. asso. prof. grad. faculty Rutgers U., 1979—. Contbr. articles to profl. jours. Fellow Am. Coll. Clin. Pharmacology; mem. AAAS, Am. Soc. Pharmacology and Exptl. Therapeutics, Soc. Neuroscience, Soc. Stimulus Properties of Drugs, Am. Soc. Clin. Pharmacology and Therapeutics, N.Y. Acad. Sci., Soc. Clin. Trials, Sigma Xi, Rho Chi, Phi Sigma. Subspecialties: Neuropharmacology; Psychopharmacology. Current work: Preclinical and clinical pharmacology. Office: CIBA-GEIGY Corp 556 Morris Ave Summit NJ 07901

ROGAN, ELEANOR GROENIGER, cancer researcher; b. Cin., Nov. 25, 1942; d. Louis M. and Esther (Levinson) Groeniger; m. William John Rogan, June 12, 1965 (div.); 1 dau., Elizabeth Rebecca. A.B. cum laude, Mt. Holyoke Coll., 1963; Ph.D., Johns Hopkins U., 1968. Lectr. dept. biology Goucher Coll., Towson, Md., 1968-69; postdoctoral fellow U. Tenn., Knoxville, 1969-71, research assoc. dept. microbiology, 1971-73; research assoc. Eppley Inst., U. Nebr. Med. Center, Omaha, 1973-76, asst. prof., 1976-80, assoc. prof., 1980—. Contbr. articles to profl. jours. USPHS fellow, 1965-68; Nat. Cancer Inst. research grantee, 1979-83. Mem. Fedn. Am. Scientists, Common Cause, Am. Assn. Cancer Research, AAAS, Sigma Xi. Democrat. Roman Catholic. Subspecialties: Biochemistry (biology); Cancer research (medicine). Current work: Chemical carcinogenesis, enzymology. Home: 8210 Bowie Dr Omaha NE 68114 Office: Eppley Inst U Nebr Med Center Omaha NE 68105

ROGAWSKI, MICHAEL ANDREW, neuroscientist, physician; b. Los Angeles, Apr. 8, 1952; s. Alexander Simon and Elise (Berwin) R. B.A. magna cum laude, Amherst Coll., 1974; M.D., Yale U., 1980, Ph.D. in Pharmacology, 1980. Med. staff fellow Nat. Inst. Neurol and Communicative Disorders and Stroke, NIH, Bethesda, Md., 1981; resident and fellow in neurology Johns Hopkins Hosp., Balt., 1982—. Contbr. articles to profl. jours. John Woodruff Simpson fellow, 1974; USPHS fellow, 1977-80; Nat. Inst. Gen. Med. Scis. fellow, 1981-84. Mem. Am. Soc. Pharmacology and Exptl. Therapeutics, Soc. Neurosci., N.Y. Acad. Scis., Sigma Xi, Phi Beta Kappa. Subspecialties: Neuropharmacology; Neurology. Current work: Laboratory studies on mechanism of action of neuroactive drugs with particular emphasis on those used in the treatment of neurological or psychiatric disorders. Office: Lab Neurophysiology NINCDS NIH Bldg 36 Room 2C-02 Bethesda MD 20205

ROGER, DAVID FREEMAN, aerospace engineering, educator; b. Theresa, N.Y., Sept. 3, 1937; s. Lewis Freeman and Gladys Marian (Zoller) R.; m. Nancy Ann Nuttall, Sept. 5, 1959; children: Stephen D., Karen N., Ransom R. B.A.E., Rensselaer Poly. Inst., 1959, M.S.A.E., 1960, Ph.D., 1967. Jr. engr. Boeing Airplane Co., summer 1958; engr. Douglas Aircraft Co., summer 1959; research asst. Rensselaer Poly. Inst., 1962-64; asst. prof. U.S. Naval Acad., Annapolis, Md., 1964-67, assoc. prof., 1967-74, prof. aerospace engring., dir. computer aided design/interactive graphics, 1974—; hon. research scholar Univ. Coll., London, 1977-78; vis. prof. U. New South Wales, Sydney, Australia, 1982; ONR research prof., 1972. Editor: computers and edn. Peragamon Press, Oxford, Eng., 1977—; series editor: Butterworths Pub, 1983; Author: Procedural Elements for Computer Graphics, 1983, Mathematical Elements for Computer Graphics, 1976, Computer Aided Heat Transfer Analysis, 1972. Rensselaer Poly. Inst. fellow, 1960-61, 55-59. Fellow AIAA; mem. Assn. Computing Machinery, Soc. Naval Architects and Marine Engrs., Tau Beta Pi, Sigma Gamma Tau. Subspecialties: Graphics, image processing, and pattern recognition; Aeronautical engineering. Current work: Computer graphics, computer aided design, computer aided mfg., fluid dynamics, vehicle performance/stability and control, boundary layers and computational fluid dynamics. Home: 817 Holly Dr E Route 10 Annapolis MD 21401 Office: Aerospace Engring Dept US Naval Acad Annapolis MD 21402

ROGERS, CLAUDE MARVIN, biology educator; b. Bloomington, Ind., Sept. 22, 1919; s. Carl Homer and Olive Sprague (Blair) R.; m. Nancy Lee Duff, Aug. 26, 1944; children: Mary, James, Kenneth, Carolyn. A.B., Ind. U., 1940; M.A., U. Mich., 1947, Ph.D., 1950. Stockroom custodian U. Okla., Norman, 1943-44; instr. biology U. Tex.-Austin, 1948-49; asst. prof. dept. biol. scis. Wayne State U., Detroit, 1949-58, assoc. prof., 1958-70, prof., 1970—. Contbr. articles to profl. jours. Mem. Am. Soc. Plant Taxonomists, Internat. Assn. for Plant Taxonomy, Mich. Acad. Arts, Scis. and Letters, Sigma Xi. Subspecialties: Evolutionary biology; Taxonomy. Current work: Plant taxonomy, especially of the flax family (Linaceae). Office: Dept Biol Sci Wayne State U Detroit MI 48202

ROGERS, DAVID ELLIOTT, foundation administrator, physician, educator, author; b. N.Y.C., Mar. 17, 1926; s. Carl Ransom and Helen Martha (Elliott) R.; m. Cora Jane Baxter, Aug. 13, 1946 (dec. Feb. 1971); children—Anne Baxter, Gregory Baxter, Julia Cushing; m. Barbara Louise Lehan, Aug. 26, 1972. Student, Ohio State U., 1942-44, Miami U., Oxford, Ohio, 1944; M.D., Cornell U. Med. Coll., N.Y.C., 1948; Sc.D. (hon.), Thomas Jefferson U., 1973, Tufts U., 1982. Intern Johns Hopkins Hosp., 1948-49, asst. resident, 1949-50; USPHS postdoctoral fellow div. infectious disease N.Y. Hosp., N.Y.C., 1950-51, chief resident in medicine, 1951-52, attending physician, 1974—; pres. The Robert Wood Johnson Found., Princeton, N.J., 1972—; vis. investigator Rockefeller Inst. Med. Research, N.Y.C., 1954-55; Lowell M. Palmer Sr. fellow in medicine Rockefeller Inst. Med. Research and Cornell U. Med. Coll., 1955-57, asst. prof., 1954-56, chief div. infectious diseases, 1955-59; assoc. prof. Cornell U. Med. Coll., 1956-59; prof., chmn. dept. medicine Vanderbilt U., 1959-68; physician-in-chief Vanderbilt U. Hosp., 1959-68; prof., dean med. faculty Sch. Medicine, Johns Hopkins; v.p. univ., med. dir. Johns Hopkins Hosp., 1968-71; Cons. Surgeon Gen., 1958-68, HEW, 1969-72; chmn. sect. Streptococcal-Staphylococcal Disease Commn. Armed Forces Epidemiol. Bd., 1958-69; mem. sci. adv. bd. Mead Johnson Research Center, Evansville, Ind., 1961-68; mem. med. adv. bd. Nat. Bd. Med. Examiners, 1961-64; mem. Tenn. adv. com. U.S. Commn. Civil Rights, 1962-64; mem. adv. com. survey of research and edn. VA-NRC, 1966-68; chmn. nat. sci. adv. council Nat. Jewish Hosp. and Research Center, Denver, 1971-73; mem. sci. adv. bd. Scripps Clinic and Research Found., La Jolla, Cal., 1972-74; vis. prof. numerous univs. Author: American Medicine: Challenge for the 1980s; Editorial bd.: Clin. Research, 1958-61, Medicine, 1962—; editor: Year Book of Medicine, 1966—. Bd. visitors Charles R. Drew Postgrad. Med. Sch., 1971—. Served to lt. (s.g.) M.C. USNR, 1952-54. Recipient John Metcalf Polk prize, Alfred Moritz Michaelis prize Cornell U. Med. Sch., 1948; Distinguished Service award Nashville Jr. C. of C., 1960; One of Ten Outstanding Young Men of Year award U.S. Jr. C. of C., 1961; Centennial Achievement award Ohio State U., 1970; award of distinction Cornell Alumni Assn., 1976; decorated Royal Order of Cedar Govt. Lebanon, 1972. Fellow Johns Hopkins Soc. Scholars, AAAS; mem. Am. Fedn. Clin. Research, Am. Soc. Clin. Investigation, ACP (master), Assn. Am. Physicians (pres. 1974-75), Assn. Profs. Medicine (sec.-treas. 1965-67), AAUP, So. Soc. Clin. Research, Infectious Disease Soc. Am. (charter), Am. Clin. and Climatological Assn., Interurban Clin. Club, Council Fgn. Relations, Assn. Am. Med. Colls. (Distinguished Service award), Inst. Medicine, Alpha Omega Alpha. Club: Cosmos (Century Assn.). Subspecialty: Health services research. Home: RFD 1 Coppermine Rd Princeton NJ 08540 Office: The Robert Wood Johnson Found PO Box 2316 Princeton NJ 08540

ROGERS, DONALD PHILIP, botanist, educator; b. Toledo, Ohio, Feb. 5, 1908; s. Philip John and Ella Leona (Johnston) R.; m. Alpha Mae Looney, Dec. 25, 1934; 1 dau., Helen Patricia Keyt. Student, Toledo U., 1925-26; B.A., Oberlin Coll., 1929; postgrad., U. Nebr., 1929-30; Ph.D., U. Iowa, 1935. Instr. botany Oreg. State Coll., 1936-40, Brown U., 1941-42; assoc. prof. biology Am. Internat. Coll., Springfield, Mass., 1942-45; asst. prof. botany U. Hawaii, Honolulu, 1945-47; curator N.Y. Bot. Garden, N.Y.C., 1947-57; prof. botany U. Ill., Urbana, 1957-76, prof. emeritus, 1976—. Mem. Mycological Soc. Am. Democrat. Episcopalian. Subspecialties: Morphology; Taxonomy. Current work: Basidial Morphology and fungal taxonomy; interpretation of basidial septation, history of mycology. Home: 1809 20th St NE Auburn WA 98002

ROGERS, EDGAR STANFIELD, biochemist; b. Dyersburg, Tenn., Nov. 14, 1919; s. Charles Clifton and Floreta Duane (Stanfield) R.; m. Beatrice June Herzberg, May 8, 1946; children: Jane, Floreta, Clifton. B.S., Duke U., 1942, M.D., 1944; postgrad., Rockefeller Inst. Med. Research, 1947. Diplomate: Nat. Bd. Med. Examiners. Asst. Rockefeller Inst., 1947-52; assoc. prof. Duke U., Durham, N.C., 1952-58; dir. U. Tenn. Research Ctr., Knoxville, 1958-64; scientist Oak Ridge Nat. Lab., 1964-72; prof. biochemistry/microbiology U. Tenn. Ctr. Health Scis., Memphis, 1972—. Contbr. articles to profl. jours. Served with M.C. U.S. Army, 1945-47. Recipient Parke-Davis award, 1959. Mem. Pathologists and Cell Biologists in Cancer Research. Subspecialties: Pathology (medicine); Genetics and genetic engineering (medicine). Current work: Cancer and recombinant DNA, genetic engring/cancer research.

ROGERS, HUGO HOMER, JR., plant physiologist, educator, researcher; b. Atmore, Ala., Aug. 2, 1947; s. Hugo Homer and Katherine Monette (Brantley) R.; m. Mary Crystal Harmon, Mar. 21, 1970; 1 dau., Hannah Star. B.S., Auburn U., 1969; M.S., 1971; Ph.D., U.N.C., 1975. Environ. engr. Research Triangle Inst., N.C., 1975-67; research assoc. N.C. State U., Raleigh, 1975-76; plant physiologist Agrl. Research Service, U.S. Dept. Agr., 1976—; assoc. prof. N.C. State U., Raleigh, 1976—; cons. in field. Mem. Am. Soc. Plant Physiologists, Air Pollution Control Assn., Am. Soc. Agrl. Engring., Crop Sci. Soc. Am., Am. Soc. Agronomy, AAAS, Sigma Xi, Pi Kappa Phi. Lodge: Mason. Subspecialties: Integrated systems modelling and engineering; Ecosystems analysis. Current work: atmospheric/vegetation effects; elevated carbon dioxide effects on plants in the field. Office: Botany Dept North Carolina State U Raleigh NC 27650 Home: Route 6 Box 46 Apex NC 27502

ROGERS, JACK DAVID, plant pathologist, educator; b. Point Pleasant, W.Va., Sept. 3, 1937; s. Jack and Thelma Grace (Coon) R.; m. Belle Clay Spencer, June 7, 1958; children: Rebecca Ann, Barbara Lee. B.S., Davis & Elkins Coll., 1960; M.F., Duke U., 1960; Ph.D., U. Wis., 1963. Mem. faculty Wash. State U.-Pullman, 1963—. Mem. Mycol. Soc. Am., Bot. Soc. Am., Am. Phytopath. Soc. Subspecialties: Plant pathology; Evolutionary biology. Current work: Diseases of forest trees, genetics of fungi, cytology of fungi. Home: NW 1435 Kenny Dr Pullman WA 99163 Office: Dept Plant Pathology Wash State U Pullman WA 99164

ROGERS, MARLIN NORBERT, horticulture educator; b. Mexico, Mo., Dec. 18, 1923; s. Harold Eugene and Edna Helena (Shire) R.; m. Elizabeth Lucinda BonDurant, Dec. 29, 1956; children: Michael J., Mary E. B.S. in Agr, U. Mo., 1948, M.S., 1951; Ph.D., Cornell U., 1956. Asst. prof. horticulture dept. U. Mo., Columbia, 1956-62, assoc. prof., 1962-67, prof., 1967—; vis. prof. U. Fla., Gainesville, 1981-82. Assoc. editor: Horticultural Revs., Vol. 2, 1980; contbr. chpts. to books, articles to profl. jours. Served in USAAF, 1943-45; Okinawa. Mem. Soc. Am. Florists (Outstanding Research award 1964), Am. Soc. Hort. Sci. (L.M. Ware Disting. Teaching award 1979, assoc. editor 1976-80). Roman Catholic. Subspecialties: Plant physiology (agriculture); Plant pathology. Current work: Effects of mineral nutrition, chemical growth regulators and environmental physiology on greehouse flower crops; postharvest physiology of greenhouse flower crops. Home: 1011 Hickory Hill Dr Columbia MO 65201 Office: Dept Horticulture U MO 1-40 Agriculture Bldg Columbia MO 65211

ROGERS, OWEN MAURICE, plant scientist, educator, cons., researcher; b. Worcester, Mass., July 4, 1930; s. F(rancis) Wyman and Lona Harriet (Foss) R.; m., Apr. 6, 1956; children: Lucy Ann, Mary Frances. B.V.A., U. Mass., 1952; M.S., Cornell U., 1954; Ph.D., Pa. State U., 1959. Asst. prof. U. N.H., Durham, 1959-65, asso. prof., 1965-72, prof., 1972—, chmn. genetics program, 1975-80, chmn. dept. plant sci., 1978—. Contbr. editor: Syringa, Hortus III, 1976; contbr. articles to hort. jours. Served to maj. USAF, 1954-56. Recipient State Gold Seal N.H. Fedn. Garden Clubs, 1969, Lilac Festival award Parks Dept., Rochester, N.Y., 1980; Mercer fellow, 1967-68. Mem. Genetics Soc. Am., Am. Soc. Hort. Sci., Internat. Lilac Soc. (pres. 1978—, Dir.'s award 1979), AAAS, Royal Hort. Soc., Mass. Hort. Soc. (trustee 1979—). Subspecialties: Plant genetics; Plant cell and tissue culture. Current work: Genetics and breeding of ornamental plants, especially lilacs and geraniums. Office: Plant Sci Dept U NH Durham NH 03824

ROGERS, QUINTON RAY, nutritional biochemist; b. Palco, Kans., Nov. 24, 1936; s. Irwin Tollie and Margret Helen (Baldwin) R.; m. Deana Joyce Dykstra, Dec. 28, 1956; children: Katrina Lynn, Terrill Dean, Tamara Gayle, Kevin Andrew. B.S., U. Idaho, 1958; M.S., U. Wis., 1960, Ph.D., 1963. Asst. prof. dept. nutrition and food sci. M.I.T., 1964-66; from asst. prof. to assoc. prof. physiol. chemistry dept. phys., sci. Sch. Vet. Medicine, U. Calif.-Davis, 1966-69, prof., 1976—. Contbr. articles to profl. jours. Mem. Am. Inst. Nutrition, Am. Physiol. Soc., AAAS. Subspecialties: Nutrition (biology); Nutrition (medicine). Current work: Amino acid nutrition and metabolism, control of food intake, feline and canine nutrition. Home: 1918 Alpine Pl Davis CA 95616 Office: Dept Physiol Sci Veterinary Medicine U Calif Davis CA 95616

ROGERS, THOMAS FRANCIS, scientist, foundation executive; b. Providence, Aug. 11, 1923; s. Thomas Francis and H. Ann R.; m. Estelle E. Hunt, July 6, 1946; children: Clare, Judith Reynolds, Hope Grove. B.Sc., Providence Coll., 1945; A.M., Boston U., 1949. Research assoc. Harvard U., Cambridge, Mass., 1944-45; TV engr. Bell & Howell Co., Chgo., 1945-46; scientist A.F. Cambridge Research Ctr., 1946-54; scientist, head communications div., mem. steering com. MIT Lincoln Lab., 1951-64; dep. dir. research and engring. Office Sec. Def., Washington, 1964-67; dir. research and tech. Office of Sec. HUD, Washington, 1967-69; v.p. Mitre Corp., McLean, Va., 1969-71; cons., pvt. investor, 1971—; pres. Sophron Found., 1980—; dir. study of civil space stas. U.S. Congress, 1982—. Contbr. sci. and tech. articles to profl. jours. Recipient Outstanding Performance award U.S. Civil Service, 1957; Cert. of Commendation Sec. Navy, 1961; Meritorious Civilian Service award and medal Sec. Def., 1967; Constrn. Man of Yr. award Engring. News Record, 1969. Fellow IEEE; mem. Nat Acad. Scis. (mem. space applications bd.). Club: Cosmos (Washington). Subspecialties: Aerospace engineering and technology; Space communications. Current work: The considered, long-term development of space and its human occupancy; improving international communications; improving international generation and distribution of renewable energy; improving health service delivery. Office: Office of Tech Assessment Congress of United States Washington DC 20510

ROGERS, VERN CHILD, engineering consulting firm executive, consultant; b. Salt Lake City, Aug. 28, 1941; s. Vern S. and Ruth (Child) R.; m. Patricia Powell, Dec. 14, 1962; 6 children. B.S., U. Utah, 1965, M.S., 1965; Ph.D., M.I.T., 1969. Registered profl. engr., Calif. Assoc. prof. Brigham Young U., 1969-73, Lowell Tech. Inst., 1970-71; mgr. nuclear dept. IRT Corp, SAn Diego., 1973-76; v.p. Ford, Bacon & Davis Utah, Salt Lake City, 1976-80; pres. Rogers & Assocs. Engring. Corp., Salt Lake City, 1980—; cons. in field. Co-author: books, including Mechanisms of Radiation Effects in Electronic Materials, 1980; contbr. numerous articles to profl. publs. Mem. Am. Nuclear Soc., Am. Phys. Soc., Health Physics Soc., Am. Chem. Soc., Am. Soc. Profl. Engrs. Mormon. Subspecialties: Nuclear fission; Environmental engineering. Current work: Nuclear fuel cycle and decontamination of nuclear facilities; waste classification, disposal site selection and qualification; developed methodology for calculating LLW facility performance and cost benefit analysis, for cover effectivenss and cost-benefit analysis for uranium mill tailings impoundments; radioactive effluent transport and risk analysis. Office: Rogers & Assocs Engring Corp PO Box 330 Salt Lake City UT 84110

ROGERS, WILBUR FRANK, civil engineering educator, consulting engineer, researcher; b. Willow Lake, Nebr., July 23, 1916; s. Hubert Merrell and Bertha (Dumke) R.; m. Priscilla Frances Chain, June 12, 1941; children: John Chain, William Frank, Martha Davis. B.Sc., U. Nebr., 1939, M.Sc., 1941; Ph.D., Pa. State U., 1970. Registered profl. engr., Nebr., Wyo. Cons. engr. and geologist, contractor, Scottsbluff, Nebr., 1945-69; research asst. Pa. State U., University Park, 1969-70; prof. civil engring. U. Nebr., Omaha, 1970—. Contbr. articles in field to profl. jours. Served to lt. comdr. USN, 1941-45. Recipient Sorkin award U. Nebr., 1975. Mem. ASCE, Internat. Water Resources Assn., Am. Assn. Petroleum Geologists, Am. Geophys. Union, Sigma Xi. Republican. Lodges: Rotary; Elks. Subspecialties: Civil engineering; Ground water hydrology. Current work: Relation between peak discharge and runoff volume, drainage basin linearity and nonlinearity and its relation to the drainage basin geological and physical environment. Inventor portable infiltrometer, 1966. Home: 2415 S 100th St Omaha NE 68124 Office: U Nebr 60th and Dodge Sts Omaha NE 68182

ROGGLI, VICTOR LOUIS, pathologist, educator, pulmonary pathology consultant; b. Winchester, Tenn., Apr. 23, 1951; s. Albert Louis and Anna Brown (Patton) R.; m. Miriam Dianne Weddington, Apr. 11, 1971; 1 dau., Heather Dianne. B.A., Rice U., 1973; M.D., Baylor Coll. Medicine, 1976. Diplomate: Am. Bd. Pathology. Resident in tng. Baylor Affiliated Hosps., Houston, 1976-80; assoc. in pathology Duke U. Med. Ctr., Durham, N.C., 1980-81, asst. prof. pathology, 1981—, Durham VA Med. Ctr., 1981—. Mem. Internat. Acad. Pathology, Assn. Am. Pathologists, Am. Soc. Clin. Pathology. Subspecialties: Pathology (medicine); Environmental toxicology. Current work: Asbestos related diseases, scanning electron microscopy and energy dispersive X-ray analysis of lung dusts recovered from human subjects and experimental animals. Office: Dept Pathology Duke U Med Ctr PO Box 3712 Durham NC 27710

ROGOLSKY, MARVIN, microbiology educator; b. Passaic, N.J., Apr. 17, 1939; s. Reuben and Ruth R. B.A., Rutgers U.-Newark, 1960; M.S., Northwestern U., 1962; Ph.D., Syracuse U., 1965. Postdoctoral fellow Scripps Clinic and Research Found., 1965-67; asst. prof. microbiology U. Utah Coll. Medicine, Salt Lake City, 1967-76; assoc. prof. biology and medicine U. Mo.-Kansas City, 1976-81, prof. biology and medicine, 1981—; researcher. Contbr. articles to sci. jours. NIH grantee, 1968-71, 72-75, 76-79. Mem. Am. Soc. Microbiology. Subspecialties: Microbiology; Gene actions. Current work: Research on mechanism of action of genetic transfer and genetic regulation of toxin synthesis in Staphylococcus aureus. Home: 10109 W 93rd Overland Park KS 66212 Office: Biol Scis Bldg Univ Mo Kansas City MO 64110

ROHLES, FREDERICK HENRY, JR., univ. research adminstr., psychologist, researcher, educator; b. Chgo., Dec. 23, 1920; s. Frederick Henry and Anna (Kiefer) R.; m. Mertyce Bliss, Nov. 9, 1943; children: Nancy Rohles Denning, Frederick Henry, Susan Rohles Grapengater. B.S., Roosevelt U., 1942; M.S., U. Tex., Austin, 1950, Ph.D., 1956. Enlisted in U.S. Air Force, 1942, advanced through grades to lt. col., 1961; research psychologist (USAF Aviation Psychology Program), 1942-49, chief psychol. br., Fairbanks, Alaska, 1950-53, research psychologist, Randolph AFB, Tex., 1954-56, chief sect. unusual environments, Wright-Patterson AFB, Ohio, 1956-58, dir. research, chief comparative psychology, Holloman AFB, N.Mex., 1958-63, ret., 1963; dir. Inst. Environ. Research, Kans. State U., Manhattan, 1963—; cons. in field. Contbr. numerous articles to profl. jours. Chmn. aviation com. C. of C. Manhattan, 1975-80; mem. Manhattan Aviation Com., 1978-80. Fellow Am. Psychol. Assn., Aerospace Medicine Assn. (assoc.); mem. Human Factors Soc., ASHRAE, Psychonomic Soc. Presbyterian. Subspecialties: Human factors engineering; Physiological psychology. Current work: Human thermal comfort; environmental ergonomics. Home: 700 Harris Ave Manhattan KS 66502 Office: Inst Environ Research Kans State U Manhattan KS 66506

ROHLICH, GERARD ADDISON, civil and environmental engineer, educator; b. Bklyn., July 8, 1910; s. Henry Otto and Margaret Loretta (Burns) R.; m. Mary Elizabeth Murphy, Sept. 8, 1941; children: Mary Ellen, Gerard A., Thomas Henry, Karl Otto, Catherine Ann, Henry James, Virginia Jean, John Harold, Richard Joseph, James William. B.C.E., Cooper Union, 1934, U. Wis., 1936, M.C.E., 1937, Ph.D. in Civil Engring, 1940. Diplomate: Am. Acad. Environ Engrs. Instr. civil engring. Carnegie Inst. Tech., Pitts., 1937-39; asst. prof. civil engring. Pa. State Coll., State College, 1941-43, asso prof., 1945-46; sr. san. engr., office chief of engrs. Dept. War, Washington, 1943-44; chief project engr. Esna Corp., Union, N.J., 1944-45; prof. civil engring. U. Wis., Madison, 1946-72, asso. dean grad. sch., 1963-65, dir., 1963-71, 1967-70; C.W. Cook prof. environ. engring., prof. pub. affairs U. Tex., Austin, 1972—; vis. prof. U. Calif., Berkeley, 1963, U. Helsinki, Finland, 1970; Walker-Ames Prof. U. Wash., Seattle, 1972; mem. adv. bd. on hazardous wastes Office of Tech. Assessment, U.S. Congress; cons. WHO; commr. Madison Water Utility, 1961-70; mem. bd. Wis. Dept. Natural Resources, 1966-72. Contbr. articles in field to profl. jours. and books. Recipient Benjamin Smith Reynolds award Excellence Teaching Engring. U. Wis., 1962; named Wis. Water Man of Year Am. Water Works Assn., 1969; recipient Gordon Maskrow fair medal Water Pollution Control Fedn., 1980. Fellow ASCE (Karl Emil Hilgard Hydraulics award 1972); mem. Nat. Acad. Engring. Democrat. Roman Catholic. Subspecialties: Water supply and wastewater treatment; Resource management. Current work: Basis for establishing water quality standards; financing water resource development. Home: 2101 Pecos St Austin TX 78703 Office: 86 E Cockrell Hall U Tex Austin TX 78712

ROHSENOW, WARREN MAX, mech. engr., educator; b. Chgo., Feb. 12, 1921; s. Fred and Selma (Gorss) R.; m. Katharine Towneley Smith, Sept. 20, 1946; children—John, Brian, Damaris, Sandra, Anne. B.S., Northwestern U., 1941; M.Eng., Yale, 1943, D.Eng., 1944. Teaching asst., instr. mech. engring. Yale, 1941-44; mem. faculty Mass. Inst. Tech., 1946—, prof. mech. engring., 1955—, dir. heat transfer lab., 1954—; Chmn. bd. dirs. Dynatech Corp. Author: (with Choi) Heat Mass and Momentum Transfer, 1961; Editor: Developments in Heat Transfer, 1964, (with Hartnett) Handbook of Heat Transfer, 1973. Served as lt. (j.g.) USNR, 1944-46; mech. engr. gas turbine div. Engring. Expt. Sta.; Annapolis, Md. Recipient Pi Tau Sigma gold medal Am. Soc. M.E., 1951; award for advancement sci. Yale Engring. Assn., 1952; merit award Northwestern Alumni, 1955. Fellow Am. Acad. Arts and Scis., Nat. Acad. Engring., Am. Soc. M.E. (Heat Transfer Meml. award 1967, Max Jakob Meml. award 1970); mem. Sigma Xi, Tau Beta Pi, Pi Tau Sigma. Subspecialties: Mechanical engineering; High-temperature materials. Current work: Application of heat transfer; boiling condensation and heat exchanges. Home: 47 Windsor Rd Waban MA 02168 Office: Massachusetts Institute of Technology Cambridge MA 02138

ROISTACHER, CHESTER N., plant pathologist; b. Bklyn., June 17, 1924; s. Samuel and Minnie (Ainspan) R.; m. Roberta D. Lanning, July 4, 1950; children: Robin, Sandy, Mark, Leslee, Dawn; m. Jean M. Fairfield, Feb. 27, 1965. B.S., Cornell U., 1949. Plant pathologist U. Calif., Riverside, 1949—. Mem. Am. Phytopath. Soc., Internat. Orgn. Citrus Virologists. Subspecialties: Plant pathology; Plant virology. Current work: Specialist in virus diseases of citrus and their control, consultant to foreign countries in citrus virology. Home: 880 Navajo Dr Riverside CA 92507 Office: Dept Plant Pathology U Calif Riverside CA 92521

ROIZMAN, BERNARD, educator, microbiologist; b. Chisinau, Rumania, Apr. 17, 1929; came to U.S., 1947, naturalized, 1954; s. Abram and Liudmilla (Seinberg) R.; m. Betty Cohen, Aug. 26, 1950; children: Arthur, Niels. B.A., Temple U., 1952, M.S., 1954; Sc.D. in Microbiology, Johns Hopkins, 1956. From instr. microbiology to asst. prof. Johns Hopkins Med. Sch., 1956-65; mem. faculty div. biol. scis. U. Chgo., 1965—, prof. microbiology, 1969—, prof. biophysics, 1970—, chmn. com. virology, 1969—, Joseph Regenstein prof., 1981—; convener herpesvirus workshop, Cold Spring Harbor, N.Y., 1972—; lectr. Am. Found. for Microbiology, 1974-75; Mem. spl. virus cancer program, devel. research working group Nat. Cancer Inst., 1967-71, cons. inst., 1967-73; mem. steering com. human cell biology program NSF, 1971-74, cons. found., 1972-74; mem. adv. com. cell biology and virology Am. Cancer Soc., 1970-74; chmn. herpesvirus study group for Internat. Commn. Taxonomy of Viruses, 1971—; mem. Internat. Microbiol. Genetics Commn., Internat. Assn. Microbiol. Scis., 1974—; sci. adv. council N.Y. Cancer Inst., 1971—; med. adv. bd. Leukemia Research Found., 1972-77; mem. herpesvirus working team WHO/FOA, 1972-81; mem. bd. sci. consultants Sloan Kettering Inst., N.Y.C., 1975-81; mem. study sect. on exptl. virology NIH, 1976-80; mem. task force on virology Nat. Inst. Allergy and Infectious Disease, 1976-77; mem. external adv. com. Emory U. Cancer Center, 1973—, Northwestern U. Cancer Center, 1979—; cons. Institut Merieux, Lyon, France.; mem. sci. adv. com. Internat. Assn. for Study and Prevention Virus Assoc. Cancers, 1983—; mem. com. to establish vaccine priorities Nat. Inst. Medicine, 1983—; chmn. sci. adv. bd. Showa U. Inst. Biol. Scis., 1983—. Author sci. papers, chpts. in books.; Mem. editorial bd.: Jour. Hygiene, 1958-61; editor: Herpes viruses, Vol. 1, 1982, Vol. 2, 1983; Mem. editorial bd.: Infectious Diseases, 1965-69, Jour. Virology, 1970—, Jour. Intervirology, 1972—, Archives of Virology, 1975—, Virology, 1976-78, 83—, Microbiologica, 1978—, Cell, 1979-80; adv. editor: Progress in Surface Membrance Sci, 1972—. Trustee Goodwin Inst. for Cancer Research, 1977—. Recipient Lederle Med. Faculty award, 1960-61, Career Devel. award USPHS, 1963-65, Pasteur award Ill. Soc. Microbiology, 1972, Esther Langer award for achievement in cancer research, 1974; Am. Cancer Soc. scholar cancer research at Pasteur Inst., Paris, 1961-62; faculty research asso., 1966-71; traveling fellow Internat. Agy. Research Against Cancer, Karolinska Inst., Stockholm, Sweden, 1970; grantee USPHS/NIH, 1958—, Am. Cancer Soc., 1962—, NSF, 1962-79, Whitehall Found., 1966-74. Hon. fellow Pan Am. Cancer Soc.; mem. Nat. Acad. Scis., Am. Assn. Immunologists, Soc. Exptl. Biology and Medicine, Am. Soc. Microbiology, A.A.A.S., Am. Soc. Biol. Chemists, Brit. Soc. Gen. Microbiology. Club: Quadrangle (Chgo.). Subspecialty: Virology (medicine). Home: 5555 S Everett Ave Chicago IL 60637

ROLFE, STANLEY THEODORE, civil engineer, educator; b. Chgo., July 7, 1934; s. Stanley T. and Eunice (Fike) R.; m. Phyllis Williams, Aug. 11, 1956; children: David Stanley, Pamela Kay, Kathleen Ann. B.S., U. Ill., 1956, M.S., 1958, Ph.D., 1962. Registered profl. engr., Pa., Kans. Supr. structural-evaluation sect. ordnance products div. U.S. Steel Corp., 1962-69, div. chief mech. behavior of metals div., 1969; Ross H. Forney prof. civil engring. U. Kans., 1969—, chmn. civil engring. dept., 1975—; Chmn. metall. studies panel ship research com. Nat. Acad. Scis., 1967-70. Author: Fracture and Fatigue Control in Structures—Applications of Fracture Mechanics; co-author: textbook Strength of Materials; Contbr.: numerous articles to profl. jours. T.R. Higgins lectr., 1980; Recipient Sam Tour award Am. Soc. Testing Materials, 1971, H.E. Gould Distinguished Teaching award U. Kans., 1972, 75, AWS Adams Meml. Educator award, 1974. Mem. ASCE (chmn. task force on fracture), Am. Soc. Testing Materials, ASME, Soc. Exptl. Stress Analysis, Am. Soc. Engring. Edn., Nat. Acad. Engring., Chi Psi. Conglist. Club: Elk. Subspecialties: Fracture mechanics; Civil engineering. Current work: Application of fracture mechanics to fracture and fatigue control in structures. Home: 2001 Camelback Dr Lawrence KS 66044

ROLLINS, REED CLARK, emeritus botany educator; b. Lyman, Wyo., Dec. 7, 1911; s. William (Clarence) and Clara Rachel (Slade) R.; m. Alberta Fitz-Gerald, Sept. 23, 1939 (div. 1976); children: Linda Lee White, Richard Clark; m. Kathryn W. Roby, Apr. 2, 1978. A.B., U. Wyo., 1933; S.M., Wash. State Coll., 1936; Ph.D. (fellow Soc. of Fellows), Harvard U., 1941. Teaching fellow Wash. State Coll., 1934-36; teaching asst. summer sch. U. Wyo., 1935; teaching asst. biology Harvard U., 1936-37; instr. biology, asst. curator Dudley Herbarium, Stanford U., 1940-41, asst. prof., curator, 1941-47, assoc. prof., curator, 1947-48; asso. geneticist Guayule Research Project, Dept. Agr., 1943-45; prin. geneticist Stanford Research Inst., 1946-47; geneticist div. rubber plant investigations Dept. Agr., 1947-48; assoc. prof. botany Harvard U., 1948-54, Asa Gray prof. systematic botany, 1954-82, emeritus, 1982—, dir., 1948-78; chmn. Inst. Research Gen. Plant Morphology, 1933-63, Inst. of Plant Scis., 1963-69, supr. Bussey Instn., Harvard U., 1968-78; chmn. adminstrv. com. Farlow Library and Herbarium, 1974-78; v.p. nat. com. XI Internat. Bot. Congress, Seattle, 1969; v.p. XII, Leningrad, USSR, 1975; pres. sect. nomenclature XIII, Sydney, Australia, 1981. Author: (with E. Shaw) The Genus Lesquerella (Cruciferae) in North America; Past editor-in-chief: Rhodora; editor: Contributions from the Gray Herbarium, 1948-78; occasional papers Farlow Herbarium, 1974-78; Contbr. articles and tech. papers to profl. jours. Recipient Centenary medal French Bot. Soc., 1954; cert. of merit Bot. Soc. Am., 1960; Congress medal XI Internat. Bot. Congress, 1969, XII, 1975. Fellow AAAS, Linnean Soc. (London); mem. Orgn. Tropical Studies (pres., chmn. bd. 1964-65), Am. Soc. Naturalists (pres. 1966), Am. Acad. Arts and Scis., Am. Inst. Biol. Scis. (governing bd. 1961-63), Am. Soc. Plant Taxonomists (pres. 1951), Bot. Soc. Am. (v.p. 1961), Calif. Bot. Soc. (treas. 1945-48), Genetics Soc. Am., Internat. Assn. Plant Taxonomy (past pres., 25th Anniversary medal 1975), Nat. Acad. Scis., N.E. Bot. Club (past pres.), Soc. Study Evolution., Phi Beta Kappa, Sigma Xi, Sigma Chi, Phi Kappa Phi. Subspecialties: Systematics; Evolutionary biology. Current work: The role of interspecific hybridization in plant evolution; the evolution of seed-dispensing mechanisms. Home: 19 Chauncy St Cambridge MA 02138 Office: Gray Herbarium 22 Divinity Ave Cambridge MA 02138

ROLLO, F. DAVID, med. affairs exec., co. exec., educator; b. Endicott, N.Y., Apr. 15, 1939; s. Frank C. and Augustine R.; m. Deane M. Rollo, June 8, 1967; 1 dau., Mindanao. A. in Chem. Tech, Broome Tech. Community Coll., Binghamton, N.Y., 1957; diploma indsl. engring, Internat. Corr. Schs., Scranton, Pa., 1958; B.A., Harpur Coll., Binghamton, 1959; M.S. in Radiol. Physics, U. Miami, 1965; Ph.D. in Physics, Johns Hopkins U., 1968; M.D., SUNY-Upstate Med. Center, Syracuse, N.Y., 1972. Diplomate: Am. Bd. Nuclear Medicine; med. lic. Calif., Tenn., Ky. Devel. research physicist IBM Glendale Lab., Endicott, 1959-60; asso. prof. math., physics Broome Tech. Community Coll., 1960-64; cons. physicist Sinai Hosp., Balt., 1965-68, Md. Gen. Hosp., 1965-68, Greater Balt. Med. Center, 1965-68; research asst. Johns Hopkins Med. Instn., 1968, research cons., 1969—; research cons. dept. nuclear medicine Duke U., 1969-72; Intern U. Calif.-San Francisco, 1972-73, resident in radiology, 1972-76, resident nuclear medicine sect., 1973-74, hosp. radiation physicist, 1973-74, asst. prof. medicine, 1974-77, asst. prof. radiology, 1974-77; asst. chief nuclear medicine service VA Hosp., San Francisco, 1974-77; dir. radiol. services, 1977-81, acting dir. radiol. sci. div., 1977-78, asso. prof. radiology and radiol. scis., 1977-79, prof., 1979—, dir. med. services, 1979-81, asst. to dean hosp. affairs, 1979-81; chief nuclear medicine VA Hosp., Nashville, 1977-79, asst. chief, 1979-81; v.p. advanced med. tech. and med. affairs Humana, Inc., Louisville, 1980—; cons. Capintec, Hewlett Packard, New England Nuclear, Mediphysics/Hoffman La Roche. Editorial bd.: Computerized Tomography Jour, 1980—; editorial rev. bd.: Picker Jour. Nuclear Med. Instrumentation, 1980—; manuscript reviewer: Jour. Physics in Medicine and Biology, 1975—, Jour. Nuclear Medicine, 1975—, Am. Jour. Radiology, 1976—; contbr. numerous articles in field to profl. jours. Grantee in field. Fellow Am. Coll. Nuclear Physicians; mem. Soc. Nuclear Medicine (Calif. chpt. chmn. Ednl. com. 1975-77, chmn. quality assurance com. 1976-77, mem. exec. com. 1975-77, chmn. membership com. 1976-77, chmn. quality control com. instrumentation 1977—, pres. 1978-79, trustee 1979-82, chmn. info. subcom. 1980—), Am. Math. Soc., Assn. Physicists in Medicine and Biology, Health Physics Soc., Nat. Assn. Residents and Interns, Assn. Univ. Radiologists, Am. Coll. Radiologists, Radiol. Soc. N. Am., AMA, Jefferson County Med. Assn., Ky. Med. Assn., Louisville Radiol. Soc., Phi Theta Kappa. Subspecialties: Imaging technology; Radiology. Current work: To assess and implement appropriate technologies and methodologie. Home: 3717 Hillsdale Rd Louisville KY 40222 Office: PO Box 1438 1800 First National Tower Louisville KY 40201

ROMAN, ANN, virologist, educator; b. Tampa, Fla., Sept. 8, 1945; d. Herschel Lewis and Caryl (Kahn) R.; m. Richard Stephen Weiner, Jan. 24, 1974; 1 son, Aaron Roman. B.S. in Biology, Reed Coll., 1967, Ph.D., U. Calif.-San Diego, 1973. Postdoctoral fellow dept. microbiology and immunology Oreg. Health Scis. Ctr., Portland, 1973-75; asst. prof. microbiology and immunology Ind. U. Sch. Medicine, Indpls., 1975-82, assoc. prof., 1982—; vis. scientist dept. virology Weizmann Inst. Sci., Rehovot, Israel, 1982-83. NIH grantee, 1976-79, 80-82. Mem. AAAS, Am. Soc. Microbiology, Am. Soc. Virology. Subspecialties: Virology (biology); Molecular biology. Current work: Regulation of papovavirus replication; behavior of viruses in different cells; factors determining the intracellular fate of viral DNA. Office: Dept Microbiology and Immunology Ind U Med Sch Indianapolis IN 46220

ROMAN, HERSCHEL LEWIS, geneticist; b. Szumsk, Poland, Sept. 29, 1914; came to U.S., 1921, naturalized, 1927; s. Isadore and Anna R.; m. Caryl Kahn, Aug. 11, 1938; children—Linda, Ann. A.B., U. Mo., 1936, Ph.D., 1942. Instr. U. Wash., 1942-46, asst. prof., 1946-47, asso. prof., 1947-52, prof., 1952—, chmn. dept. genetics, 1959-80; vis. investigator Carlsberg Lab., Copenhagen, 1960; vis. prof. Australian Nat. U., Canberra, 1966; cons. in field. Editor: Ann. Rev. Genetics, 1965—. Served with AC U.S. Army, 1943-46. Guggenheim fellow, 1952; Fulbright fellow, 1956; recipient Emil Christian Hansen Found., Copenhagen, 1980. Mem. Am. Acad. Arts and Scis., Nat. Acad. Scis., Genetics Soc. Am. (pres. 1968). Subspecialty: Genetics and genetic engineering (biology). Current work: Studies of recombination in the yeast Saccharomyces cerevisiae. Research, publs. on maize genetics, yeast genetics. Home: 5619 NE 77th St Seattle WA 98115 Office: Dept Genetics U Wash Seattle WA 98195

ROMANO, JAMES ANTHONY, JR., psychologist, army officer; b. Jersey City, Nov. 12, 1944; s. James Anthony and Carmen (Beauchamp) R.; m. Candy Grimes, Apr. 3, 1971; children: Candace, Alicia. A.B. in Psychology, Coll. Holly Cross, 1966, M.A., Fordham U., 1968, Ph.D, 1975. Asst. prof. psychology Manhattan Coll., Bronx, N.Y., 1970-78; adj. asst. prof. psychology Coll. Mt. St. Vincent, Riverdale, N.Y., 1971-77, Hudson County Community Coll., Jersey City, 1976-78, Harford Community Coll., Bel Air, Md., 1980-83; commd. capt. U.S. Army, 1978; research psychologist Health Services U.S. Army Med. Research Inst. Chem. Def., Aberdeen Proving Ground, Md., 1978—. Contbr. numerous articles to profl. publs. Mem. mayor's adv. council Study Juvenile Criminal System, Jersey City, 1973-74. Mem. Soc. Neurosci., Del. Soc. Neurosci., Joppatowne (Md.) Jaycees (pres. 1980-82), Sigma Xi. Subspecialties:

Psychopharmacology; Learning. Current work: Opiate-cholinergic interactions in stressed animals; behavioral toxicology of anticholinesterase and anticholinergic compounds. Home: 25 Old Sound Dr Joppatowne MD 21085 Office: Comdr US Army Med Research Inst Chem Def Attn SGRD UV RB Capt Romano Aberdeen Proving Ground MD 21010

ROMANOWSKI, THOMAS ANDREW, educator; b. Warsaw, Poland, Apr. 17, 1925; came to U.S., 1946, naturalized, 1949; s. Bohdan and Alina (Sumowski) R.; m. Carmen des Rochers, Nov. 15, 1952; children—Alina, Dominique. B.S., Mass. Inst. Tech., 1952; M.S., Case Inst. Tech., 1956, Ph.D., 1957. Research asso. physics Carnegie Inst. Tech., 1956-60; asst. physicist high energy physics Argonne Nat. Lab., Ill., 1960-63, asso. physicist, 1963-72, physicist, 1972-78; prof. physics Ohio State U., Columbus, 1964—. Contbr. articles to profl. jours. and, papers to sci. meetings, seminars and workshops. Served with C.E. AUS, 1946-47. Fellow Am. Phys. Soc., AAAS; mem. Lambda Chi Alpha. Subspecialty: High Energy Physics. Current work: Experiments in high energy physics on neutrino properties; experimental instrumentation for subatomic particle detection. Research in nuclear and high energy physics. Patentee in field of electronics, cons. in physics. Home: 80 W Cooke Rd Columbus OH 43214 Office: Dept Physics Ohio State U 174 W 18th St Columbus OH 43210

ROMBOSKI, LAWRENCE DAVID, mathematical and computer science educator; b. Washington, Pa., Feb. 19, 1938; s. Lawrence Walter and Pauline (Popeck) R.; m. Joanne Lee Kapsi, Apr. 18, 1959; children: Joyce Ann, Lynn D'Anne, Lawrence David. B.A. in Math, Washington and Jefferson Coll., 1959, M.A., Rutgers U., 1965, M.S. in Stats, 1967, Ph.D., 1969. Instr. Washington and Jefferson Coll., Washington, Pa., 1962-64; teaching asst. Rutgers U., New Brunswick, N.J., 1964-67, instr., 1967-68, lectr., 1968-69; prof. math. California (Pa.) State Coll., 1969—; cons. McElrath & Assocs., Mpls., 1981—. Editor: Jour. Quality Tech, 1974-76. Vice pres. California Area Boys Baseball, 1972-76; mem. council St. Thomas Cath. Ch., California, 1976—. Served to 1st lt. U.S. Army, 1959-62. Recipient Exceptional Acad. Service award Pa. Dept. Edn., 1975, 76. Mem. Am. Soc. for Quality Control (Testimonial award 1977), Am. Statis. Assn., Assn. for Computing Machinery, Sigma Xi. Democrat. Subspecialties: Statistics; Mathematical software. Current work: Applications of statistical and computer methods to industrial quality control situations. Home: 175 W Ellsworth St California PA 15419 Office: Dept Math California State Coll California PA 15419

ROMERO, ALEJANDRO FRANCISCO, mechanical engineering educator, researcher; b. Mascota, Jalisco, Mexico, Nov. 16, 1937; s. Rafael and Andrea (López) R.; m. Marisa Llanes, Oct. 20, 1967; children: Berenice Ileana, Katia Lorena. B.S. in Engring, U. Mex., 1963, M. Engring., 1973, Dr. Engring., 1977; postgrad., Tech. U. Munich, W. Ger., 1978-80. Engr. trainee Combustion Engring. Inc., Windsor, Conn., 1964-65; propositions engring. chief Ce-rrey (S.A.), Monterrey, Mex., 1965-67; chief dept. boiler engring. Babcock & Wilcox de Mex., Cerro Gordo, 1967-71; head sect. thermoscis. Facultad de Ingeniería, Universidad Nacional Autónoma de México, Mexico City, 1971-78, head dept. mech. engring., 1980—. Author: Thermodynamic Properties of Water, 1975. Nat. Council Sci. and Tech. grantee, Munich, 1978-79; Deutscher Akademischer Austauschdienst grantee, 1978-82. Mem. Internat. Solar Energy Soc. Roman Catholic. Club: Terranova (Mexico City). Subspecialties: Mechanical engineering; Theoretical physics. Current work: Applied solar energy and thermodynamic properties of matter. Inventor domestic solar-powered air conditioner, 1977. Office: Facultad de Ingeniería UNAM Mexico DF Mexico 04510 Home: Ave Copilo 162-30-102 Coyoacán México DF México 04360

ROMESBERG, LAVERNE EUGENE, mech. engr.; b. York County, Pa., Dec. 12, 1941; s. Raymond Ralph and Velma Marie R.; m. Gaye Lee Romesberg, Jan 24, 1964; children: Arden V., John D. A. Engring., Pa. State U., York, 1959-61; B.M.E., U. N.Mex., Albuquerque, 1968, M.M.E., 1969. Sr. engr. Mechanics Research, Inc., Albuquerque, 1969-73; prin. investigator Civil Nuclear Systems, Inc., Albuquerque, 1973-77; staff mem., task leader Sandia Nat. Labs., Albuquerque, 1977—. Contbr. articles to profl. jours. Mem. ASME. Republican. Methodist. Subspecialties: Mechanical engineering; Solid mechanics. Current work: Devel. of transp. systems for def. radioactive waste. Home: 1070 Velvet Dr Bosque Farms NM 87068 Office: PO Box 5800 Div 9782 Albuquerque NM 87185

ROMNEY, CARL F., seismologist; b. Salt Lake City, June 5, 1924; m. Barbara Doughty; children: Carolyn Ann, Kim. B.S. in Meteorology, Calif. Inst. Tech.; Ph.D., U. Calif. Seismologist U.S. Dept. Air Force, 1955-58; asst. tech. dir. Air Force Tech. Applications Center, 1958-73; dep. dir. Nuclear Monitoring Research Office, Def. Advanced Research Projects Agy., 1973-75, dir., 1975-79; dep. dir. Def. Advanced Research Projects Agy., 1979—; tech. adviser U.S. reps. in negotiations Test Ban Treaty; mem. U.S. del. Geneva Conf. Experts, 1958, Conf. on Discontinuance Nuclear Weapons Tests, 1959, 60; negotiations on threshold Test Ban Treaty, Moscow, 1974; mem. U.S. del. Peaceful Nuclear Explosions Treaty, Moscow, 1974-75. Contbr. articles to tech. jours. Recipient Exceptional Civilian Service awards Air Force, 1959, Dept. Def., 1964, 79; Pres.'s award for Distinguished Fed. Civilian Service, for outstanding contbns. to devel. of control system for underground nuclear tests, 1967; Presdl. Rank of Meritorious Exec., 1980. Subspecialty: Geophysics. Current work: Generation, propagation; detection of Seismic Waves from underground nuclear explosives; discrimination between explosions and earthquakes; yield estimation. Research on earthquake mechanism, seismic noise; generation, propagation, detection seismic waves from underground explosions. Home: 4105 Sulgrave Dr Alexandria VA 22309 Office: 1400 Wilson Blvd Arlington VA 22209

ROONEY, LLOYD WILLIAMS, food science and technology educator; b. Atwood, Kans., July 17, 1939; s. Lloyd and Tamzan (Clement) R.; m. Maxine T. Barenberg, Aug. 31, 1963; children: William, Tammy, Marcille. B.S., Kans. State U.-Manhattan, 1961; postgrad., U. Minn., 1961-63; Ph.D., Kans. State U.-Manhattan, 1966. NIH trainee U. Minn., St. Paul, 1961-63; grad. asst. Kans. State U., Manhattan, 1963-65; asst. prof. Tex. A&M U., College Station, 1965-70, assoc. prof., 1970-75, prof. food sci. and tech., 1975—. Editor: Sorghum Food Quality, 1982. Mem. Am. Assn. Cereal Chemists, Internat. Assn. Cereal Chemistry, Inst. Food Technologists, Am. Soc. Agronomy, Council Agrl. Sci. and Tech. Roman Catholic. Subspecialty: Food science and technology. Current work: Cereal quality, technology of cereals, cereal quality improvement. Home: RFD 3 Box 336 Bryan TX 77801 Office: Cereal Lab Tex A&M U College Station TX 77843

ROOT, THOMAS MICHAEL, biologist, educator, researcher; b. Detroit, Mar. 6, 1952; s. Robert Leonard and Margaret Ludwina (Reese) R.; m. Lorraine Zack, May 22, 1976. B.S. in Biology, U. Detroit, 1974; M.S. in Zoology and Physiology, U. Wyo., 1976, Ph.D., 1979. Vis. lectr. U. Wyo., Laramie, 1979; asst. prof. biology Middlebury (Vt.) Coll., 1979—. Contbr. articles to sci. jours., chpt. to book. Mem. Soc. for Neurosci., AAAS, Am. Soc. Zoologists. Subspecialties: Comparative neurobiology; Ethology. Current work: Neural control of behavior; behavior of scorpions. Office: Biology Dept Middlebury College Middlebury VT 05753

ROPER, STEPHEN DAVID, medical researcher and educator; b. Rock Island, Ill., May 30, 1945; s. Wesley S. and Ruth M. (Tilley) R.; m. Kathleen Ann Herburger, June 23, 1945; children: Anna Lisbeth, Peter Wesley. B.A., Harvard Coll., 1967; Ph.D., U. London, 1970. Postdoctoral fellow Harvard Med. Sch., Boston, 1970-73, teaching fellow, 1971-73; asst. prof. U. Colo. Sch. Medicine, Denver, 1973-79, assoc. prof. dept. anatomy and dept. physiology, 1979—; cons. mem. neurobiology study sect. NIH, 1981—. Fulbright fellow, 1967-69; NSF predoctoral fellow, London, 1969-70; NIH postdoctoral fellow, Boston, 1972-73; research career devel. awardee NIH, 1978-82. Mem. Neurosci. Soc. (chpts. com. 1981—), Rocky Mountain Region Neurosci. Group (founder, chmn. 1979-82), Colo. Heart Assn. (research bd. 1979-82, dir. 1982—), Am. Heart Assn. (research bd. Dallas 1982—), Am. Assn. Anatomists, Am. Physiol. Soc., AAAS. Subspecialties: Neurobiology; Regeneration. Current work: Development and maintenance of neuronal connections, repair of injured nerves; biology of sensory receptor cells, particularly taste cells. Office: Univ Colo Health Scis Center 4200 E 9th Ave Denver CO 80262

ROSA, DAVID, plastics research engr.; b. N.Y.C., May 10, 1952; s. Luis and Guillermina (Cornier) R.; m. Susan Elizabeth Morrissey, July 24, 1976; children: David II, Amanda Veronica. B.E. in Mech. Engring, CCNY, 1975, M.S., Rutgers U., 1981. Tech. support engr. Monsanto Comml. Products, South Windsor, Conn., 1975-77; devel. engr. Celanese Plastics & Specialties, Summit, N.J., 1977-83; v.p. research and devel. Software Projections, Inc., Flourtown, Pa., 1983—. Mem. Soc. Plastics Engrs., ASME, Soc. Automotive Engrs. Subspecialties: Mechanical engineering; Plastics engineering. Current work: Plastics design and computer modeling of plastics processing; plastics research and devel.; new applications for computer tech. in plastics industry. Home: 123 Meadowbrook Rd Livingston NJ 07039

ROSAN, BURTON, microbiology educator; b. N.Y.C., Aug. 18, 1928; s. Harry A. and Rachel (Halpern) R.; m. Helen G. Mescon, Jan. 14, 1951; children: Rhea S., Felice B., Jonathan S. B.S., CCNY, 1950; D.D.S., U. Pa., 1957, M.Sc., 1962. Research assoc. U. Pa. Sch. Dental Medicine, Phila., 1959-63, asst. prof., 1963-67, assoc. prof., 1967-77, prof. microbiology, 1977—; vis. prof. Inst. Dental Research, Sydney, Australia, 1980; vis. assoc. prof. SUNY-Downstate Med. Ctr., Bklyn., 1971-72; vis. scientist Inst. Dental Research, Sydney, 1975. Mem. editorial bd.: Infection and Immunity, 1978—, Jour. Bacteriology, 1979-82. NIH postdoctoral fellow, 1957-59; NIH grantee, 1959—. Fellow Am. Acad. Microbiology; mem. AAAS, Am. Soc. Microbiology, Internat. Assn. Dental Research (pres. micro-immunology 1979-80), Am. Assn. Dental Research (chmn. fellowship com. 1980), Sigma Xi. Jewish. Subspecialties: Microbiology; Oral biology. Current work: Molecular basis of microbial adherence, dental plaque formation and maturation. Home: 113 Cambridge Rd Brodmall PA 19008 Office: U Pa Sch Dental Medicine 4001 Spruce St Philadelphia PA 19008

ROSE, ALBERT, retired physicist; b. N.Y.C., Mar. 30, 1910; s. Simon and Sarah (Cohen) R.; m. Lillian Loebel, Aug. 25, 1940; children: Mark Loebel, Jane Susan. A.B., Cornell U., 1931, Ph.D., 1935. Research RCA Labs., Princeton, N.J., 1935-75; dir. research Labs. RCA Ltd., Zurich, Switzerland, 1955-58; mem. planning com. Internat. Confs. on Semiconductors, Internat. Conf. on Photoconductivity, Internat. Conf. on Electrophotography; vis. prof. Stanford U., 1976, Hebrew U., Jerusalem, 1976-77, Boston U., 1977, 78, Poly. Inst., Mexico City, 1978, U. Del., 1979. Author: Concepts in Photoconductivity, 1963, Vision: Human and Electronic, 1974; editorial bd.: Phys. Rev, 1956-58, Advances in Electronics, 1948-75, Jour. Physics and Chemistry of Solids, 1958-75; contbr. articles to profl. jours. Recipient certificate of merit USN, 1946, TV Broadcasters award, 1945, David Sarnoff Gold medal Soc. Motion Picture and TV Engrs., 1958; Mary Shephard Upson distinguished prof. Cornell U., 1967; Fairchild distinguished scholar Calif. Inst. Tech., 1975; recipient Leo Friend award in chem. tech., 1982. Fellow Am. Phys. Soc., IEEE (Morris Liebman award 1945, Edison medal 1979); mem. Nat. Acad. Engring., Société Suisse de Physique., Soc. Photog. Scientists and Engrs. (hon.). Subspecialty: Electronics. Patentee in field. Home: 292 Stockton Rd Princeton NJ 08540

ROSE, ANN MARIE, molecular geneticist; b. Midale, Sask., Can., Mar. 22, 1946; d. Barnard Frank and Mary Ethel (Sinclair) Kuchinka; m. Leslie Alan Rose, Aug. 13, 1969. B.A. with honors, U. Sask., Saskatoon, 1970; Ph.D., Simon Fraser U., B.C., 1980. Research asst. U. Toronto, 1969-73; research technician Simon Fraser U., 1975; fellow dept. biochemistry U. B.S., Vancouver, 1980-82, asst. prof. dept. med. genetics, 1982—. Muscular Dystrophy Assn. fellow, 1980-82. Mem. Genetics Soc. Am., Can. Genetics Soc. Subspecialties: Genome organization; Molecular biology. Current work: Understanding genetic regulation during devel.; cielegans genetics; gene regulation; muscle genes; recombination frequency. Office: Dept Medical Genetics U BC Vancouver BC Canada V6T IW5

ROSE, FRANK EDWARD, physics educator; b. Junction city, Kans., Mar. 11, 1927; s. Julian Donovan and Bernice (Deichman) R.; m. Florence C. Reid, June 18, 1948; children: Val, Mark, Tom, Ralph. B.S., Greenville (Ill.) Coll., 1949; A.M., U. Mich., 1957; Ph.D., Cornell U., 1965. Tchr. Beecher High Sch., Flint, Mich., 1949-54; mem. faculty Gen. Motors Inst., Flint, 1954-58; research asst. Cornell U., Ithaca, N.Y., 1959-63; assoc. prof. physics U. Mich.-Flint, 1963—, chmn. dept. physics and astronomy, 1978-80. Contbr. articles to profl. jours.; editor's referee: Am. Jour. Physics. Chmn. elect Cable TV Com., Flint, 1978—. Served with USNR, 1945-46. NSF grantee. Mem. Am. Assn. Physics Tchrs., Am. Phys. Soc., Mich. Assn. Physics Tchrs. Methodist. Subspecialties: Condensed matter physics; Low temperature physics. Current work: Physics and astronomy education; electron properties of solids; helicon waves, cryogenis, computer graphics instruction, computer typography and editing. Home: 1014 Beard St Flint MI 48503 Office: Dept Physics U Mich Flint MI 48503

ROSE, GEORGE DAVID, biophysical chemistry research and educator; b. Chgo., Aug. 28, 1939. B.S., Bard Coll., 1963; M.S., Oreg. State U., 1972, Ph.D., 1976. Asst. prof., sr. research assoc. U. Del., Newark, 1975-80; assoc. prof. dept. biol. chemistry, dir. research computing Hershey Med. Ctr., Pa. State U., 1980—; mem. study sect. NIH, 1981. Research career devel. awardee NIH, 1980—. Mem. Am. Soc. Biol. Chemists, Am. Chem. Soc., Biophys. Soc., AAAS, N.Y. Acad. Scis. Subspecialties: Biophysical chemistry; Graphics, image processing, and pattern recognition. Current work: Structure, self-assembly, dynamics of macromolecules of biological interest. Home: 170 Woodbine Dr Hershey PA 17033 Office: PA State Univ Dept Biol Chemistry Hershey Med Center Hershey PA 17033

ROSE, GEORGE GIBSON, medical-dental researcher, educator; b. Liberty, Ind., Aug. 7, 1922; s. Joseph Sims and Dorothy (Gray) R.; m. Jeanne Whitney Roco, Dec. 24, 1945; children: Mary Dianne, George Glenn. B.A., U. Tex-Austin, 1947; M.D., U. Tex.-Galveston, 1951. Intern Hermann Hosp., Houston, 1951-52; research assoc. U. Tex. Med. Br., Galveston, 1954-60, Pasadena Found. Med. Research, Calif., 1960-65; asst. and assoc. biologist U. Tex., M.D. Anderson Hosp. and Tumor Inst., Houston, 1955-67, asst. prof. dental br., 1960-66, assoc. prof., 1966-72, prof., 1972—. Author, editor: Atlas of Vertebrate Cells in Tissue Culture, 1970; editor: Cineminography in Cell Biology, 1965. Served to lt. (j.g.) USNR, 1942-46; PTO. Mem. Tissue Culture Assn., Internat. Assn. Dental Research. Democrat. Methodist. Subspecialties: Tissue culture; Oral biology. Current work: Tissue culture, oral biology, periodontics, cancer research. Inventor multipurpose culture chamber, 1959, circumfusion system, 1965. Home: 5306 Rutherglenn St Houston TX 77096 Office: U Tex Health Sci Center Dental Br 6516 John Freeman Ave Houston TX 77025

ROSE, GLENN ROBERT, computer system consultant; b. Phila., Oct. 20, 1946; s. Victor Lamar and Edith Marion (Deakin) Rose T.; m. Barbara Siegel, June 1, 1969; children: Peter Aarron, Laura Yvonne. B.S.E.E., Drexel U., 1969; M.S.E.E., Stanford U., 1970. Mem. tech. staff Bell Telephone Labs., Holmdel, N.J., 1969-79, GTE Auto. Elec. Labs., Phoenix, 1979-80; pres. Computereze, Inc., Neshanic Station, N.J., 1980-82, Polymorphic Systems, Holmdel, 1982—. Mem. Assn. Computing Machinery, Tau Beta Pi, Phi Kappa Phi, Eta Kappa Nu. Republican. Subspecialty: Database systems. Current work: Database systems and distributed systems. Home: 501 Sheppard Ct Neshanic Station NJ 08853 Office: Polymorphic Systems Inc PO Box 237 Holmdel NJ 07733

ROSE, KATHLEEN MARY, biochemistry researcher, educator; b. St. Paul, Sept. 5, 1945; d. Merwin Anthony and Eileen (Brockman) Campbell; m. Richard C. Rose; 1 son, Clayton John. B.S., Mich. State U., 1966, M.S., 1969; Ph.D., Pa. State U., 1977. Research asst. Pa. State U., Hershey, 1973-78, asst. prof., 1978-82, assoc. prof. pharmacology, 1982—; cons. U.S. Dept. Edn., 1981-83, NSF, 1980-83. Research grantee Nat. Inst. Gen. Med. Sci., 1979—; Research Career Devel. awardee Nat. Cancer Inst., 1980—; Alcoa Found. awardee, 1980. Mem. Am. Soc. Biol. Chemists, Am. Soc. Cell Biology, Sigma Xi. Subspecialties: Biochemistry (medicine); Genetics and genetic engineering (medicine). Current work: Regulation of gene transcription/expression. Co-inventor diagnostic test for systemic lupus erythematosus. Office: Dept Pharmacolog Hershey Med Center Hershey PA 17033

ROSE, KENNETH DAVID, science educator; b. Newark, N.J., June 21, 1949; s. Victor William and Odette Adele (Messler) R.; m. Jennie Jerome Neumann, Apr. 25, 1981. B.S., Yale U., 1972; M.A., Harvard U., 1974; Ph.D., U. Mich-Ann Arbor, 1979. Post-doctoral fellow dept. paleobiology Smithsonian Instn., 1979-80; asst. prof. cell biology and anatomy Johns Hopkins U., Balt., 1980—; research collaborator dept. paleobiology Nat. Mus. Natural History Smithsonian Instn. Contbr. articles to profl. jours. Nat. Geographic Soc. grantee, 1981-82; Am. Philos. Soc. grantee, 1982-83; NSF grantee, 1983—. Mem. Soc. Vertebrate Paleontology, Paleontol. Soc., Am. Soc. Mammalogists (Shadle fellow 1978), Soc. Systematic Zoology, Sigma Xi. Subspecialties: Paleobiology; Anatomy and embryology. Current work: Systematics, evolution and comparative and functional anatomy of early Cenozoic mammals, chiefly North American. Home: 1101 Argonne Dr Baltimore MD 21218 Office: Johns Hopkins U Med Sch 107 Hunterian Baltimore MD 21218

ROSE, MICHAEL ROBERTSON, evolutionary biologist; b. Iserlohn, W.Ger., July 25, 1955; s. Barry S. and Julia (Horsey) R.; m. Frances Reay Wilson, Aug. 28, 1976. B.Sc. with 1st class honors, Queen's U., Kingston, Ont., Can., 1975, M.Sc. (Ont. grad. scholar), 1976; Ph.D. (Commonwealth scholar), U. Sussex, Eng., 1979. NATO postdoctoral sci. fellow U. Wis., 1979-81; asst.prof. biology, research scholar Dalhousie U., Halifax, N.S., 1981—. Contbr. articles to profl. jours. Natural Scis. and Engring. Research Council Can. grantee, 1981-84; Dalhousie Research Devel. Fund grantee, 1981-82. Mem. Am. Soc. Zoologists, Animal Behavior Soc., Genetics Soc. Am., Soc. Study Evolution. Club: Dalhousie Faculty. Subspecialties: Evolutionary biology; Genome organization. Current work: Evolution of life history, fitness and senescence in drosophila; role of selfish DNA in evolution; theoretical population biology. Office: Biology Dept Dalhousie U Halifax NS Canada B3H 4J1

ROSE, RONALD PALMER, nuclear reactor mfg. company executive; b. Boston, Sept. 20, 1930; s. Arthur Burnham Hatch and Ann Margaret (Palmer) R.; m. Marjorie Joann Staples, Aug. 6, 1955; children: Linda, Randolph, Pamela, Bryan. A.B., Dartmouth Coll., 1952, M.S., 1953; Ph.D., U. Pitts., 1960. Fellow engr. Bettis Atomic Power Lab., Pitts., 1954-64; sect. chief analysis Phillips Petroleum Co., Idaho Falls, 1964-67; mgr. systems analysis Westinghouse Astronuclear Lab., Pitts., 1967-71, mgr. systems analyst, 1971-73; mgr. fusion projects Nuclear Energy Systems, 1975—; vis. fellow Princeton Plasma Physics Lab., 1973-75, mem. lithium blanket module adv. group, 1980-81; mem. joint US/USSR group on fusion-fission Electric Power Research Inst./Kurchatov Inst., 1976-79. Mem. Peters Twp. Sch. Dist. Long-Range Planning Group, McMurray, Pa., 1977; mem. Peters Twp. Planning Commn., 1979-81. AEC fellow, 1955. Mem. Am. Nuclear Soc., ASME, Sigma Xi. Club: Sports Car Club Am. Lodge: Elks. Subspecialties: Nuclear fission; Nuclear fusion. Current work: Fusion energy, fusion-fission hybrids, aerospace nuclear power, breeder reactors, nuclear reactor safety, energy systems analysis and forecasts. Patentee inertial confinement fusion concept to produce a line source of neutrons. Home: 200 Grouse Dr Apt 1 Elizabeth PA 15037 Office: Westinghouse Advanced Reactors Div Madison PA 15663

ROSE, WILLIAM CARL, pharmaceutical research scientist; b. Bklyn., Nov. 15, 1946; s. Edward Alexander and Gladys (Korkin) R.; m. Cherie Lynn O'Neil, Apr. 7, 1969; children: Joshua, Rachael, Shayna. A.A.S., SUNY, Farmingdale; B.S., Cornell U., 1968; M.S., Med. Coll. Va., 1970, Ph.D., 1972. NIH postdoctoral research assoc. Wistar Inst., Phila., 1972-73; research assoc. Mt. Sinai Med. Ctr., Milw., 1973-75; sr. research scientist So. Research Inst., Birmingham, Ala., 1975-78; sr. research scientist, project leader Bristol Pharm. Research and Devel. Div., Syracuse, N.Y., 1978—. Author numerous papers, reports in field. Recipient Halco Chem. Co. award, 1964-65; N.Y. State Regents scholar, 1963-67; A.D. Williams scholar, 1968-72. Mem. Am. Soc. Microbiology, Am. Assn. Cancer Research, Sigma Xi. Jewish. Club: E. Syracuse Coin (sec. 1981-82). Subspecialties: Chemotherapy; Cancer research (medicine). Current work: Supr. in vivo evaluation of exptl. antitumor drugs. exptl. chemotherapy, in vivo screening, tumor model devel., drug evaluation. Office: Bristol Pharm Research and Devel Div PO Box 4755 Syracuse NY 13221

ROSE, WILLIAM KENNETH, astronomy educator; b. Ossining, N.Y., Aug. 10, 1935; s. Kenneth and Shirley Hazel (Near) R.; m. Sheila L. Tuchman, Apr. 3, 1961; children: Kenneth, Edward, Cindy. A.B., Columbia U., 1957, Ph.D., 1963. Research Staff Princeton (N.J.) U., 1963-67; asst. prof. M.I.T., Cambridge, 1967-71, assoc. prof., 1971; assoc. prof. astronomy U. Md., College Park, 1971-76, prof., 1976—. Contbr. articles to profl. jours. Recipient Washington Acad. Sci. ann. award for achievement in phys. sci., 1975. Mem. AAUP, AAAS, Washington Acad. Sci. (fellow), Internat. Astron. Union, Am. Astron. Soc. Club: Cosmos. Subspecialties: Theoretical astrophysics; High energy astrophysics. Current work: Theoretical astrophysics, high energy astrophysics, stellar evolution. Home: 10916 Picasso Ln Potomac MD 20854 Office: Astronomy Program U Md College Park MD 20742

ROSEKE, WILLIAM ROBERT, physician scientist; b. Evergreen Park, Ill., Aug. 7, 1941; s. William Frank and Louise Marie (Wachholz) R.; m. Eileen Mary, Dec. 8, 1965; children: Lisa, Laura. A.B., U.Calif.-Berkeley, 1963; M.D., Stanford U., 1970. Diplomate: Am. Bd. Internal Medicine. Intern, resident Case Western Res. U., Cleve.; fellow in cardiology U. Calif.-San Diego; asst. prof. internal medicine U. Ariz., Tucson, 1976-80, assoc. prof., 1980—, assoc. prof. pharmacology, 1982—; cons. NIH, Am. Heart Assn. Contbr. articles to profl. jours. NIH grantee. Mem. Am. Soc. Clin. Investigation., Am. Heart Assn., Am. Soc. Pharmacology and Exptl. Therapeutics. Subspecialties: Cardiology; Molecular pharmacology. Current work: Receptor pharmacology: cardiovascular pharmacology. Office: 1501 N Campbell Ave Tucson AZ 85724

ROSELLA, JOHN DANIEL, psychologist, educator; b. Phila., Sept. 12, 1938; s. Orazio and Angela (Cardone) R.; m. Rosemary T. Malloy, Nov. 14, 1964; children: Anne-Marie, John Daniel. B.S., Villanova U., 1961; cert., St. Joseph U., 1962-63; M. Ed., Temple U., 1966; Ph.D., Walden U., 1981. Lic. Psychologist, Pa. Tchr./counselor Fr. Judgh High Sch., Phila., 1962-67; counselor Bristol (Pa.) Twp. schs., 1967-69; clin. psychologist Abraham Perlman (M.D.), Phila., 1969-72, Newtown Psychol. Centre, Pa., 1973-79; prof. Bucks County Community Coll., Newtown, 1969—; dir. psychol. services Fairless Hills Med. Ctr., Pa., 1979—; cons. psychologist Bur. Vocat. Rehab., Rosemont, Pa., 1976—, Disability Determination, Wilkes-Barre, Pa., 1981—; project dir. Fairless Hills Hosp., 1982—. Contbr. articles to profl. jours. Bd. dirs. Bucks County Community Ctrs., 1980—, Valley Day Sch., Yardley, Pa., 1977-81; 1st vice chmn. Newtown Twp. Democratic party, 1978-80; advisor health Congressman James K. Coyne, 1981-82. Recipient Nat. Assn. to Advance Ethical Hypnosis Man of Yr. award, 1976; commendation Pa. Dept. Edn., 1967. Mem. Assn. to Advance Ethical Hypnosis (pres. chpt. 1976-77, dir. 1970-80), Am. Psychol. Assn., Pa. Psychol. Assn., Am. Assn. Marriage and Family Therapy, Pa. Assn. Marriage and Family Therapy, Am. Fedn. Tchrs. Lodge: KC. Subspecialties: Behavioral psychology. Current work: Research linking family communications and psychosomatic disorders; developing a free standing mental health hospital utilizing family therapy model; treatment of depression utilizing multi-modal approach. Home: 582 Sterling St Newtown PA 18940 Office: 515 S Olds Blvd Fairless Hills PA 19030

ROSEMAN, SAUL, biochemist, educator; b. Bklyn., Mar. 9, 1921; s. Emil and Rose (Markowitz) R.; m. Martha Ozrowitz, Sept. 9, 1941; children: Mark Alan, Dorinda Ann, Cynthia Bernice. B.S., City Coll. N.Y., 1941; M.S., U. Wis., 1944, Ph.D., 1947. From instr. to asst. prof. U. Chgo., 1948-53; from asst. prof. to prof. biol. chemistry, also Rackham Arthritis Research Unit, U. Mich., 1953-65; Ralph S. O'Connor prof. biology Johns Hopkins, 1965—, chmn. dept., 1969-73; dir. McCollum-Pratt Inst., 1969-73; cons. NIH, NSF, Am. Cancer Soc., Hosp. for Sick Children, Toronto; sci. counselor Nat. Cancer Inst. Author articles on metabolism of complex molecules containing carbohydrates and on solute transport.; former mem. editorial bd.: Biochemistry; mem. editorial bd.: Jour. Biol. Chemistry. Served with AUS, 1944-46. Recipient Sesquicentennial award U. Mich., 1967; T. Duckett Jones Meml. award Helen Hay Whitney Found., 1973; Rosenstiehl award Brandeis U., 1974; Internat. award Gairdner Found., 1981. Mem. Am. Soc. Biol. Chemists, Am. Soc. Cell Biology, Am. Acad. Arts and Scis., Nat. Acad. Scis., Am. Chem. Soc., Am. Soc. Microbiologists. Subspecialties: Membrane biology; Cell biology. Current work: Membrane transport; cell-cell recognition, adhesion. Home: 8206 Cranwood Ct Baltimore MD 21208

ROSEMARK, PETER JAY, radiation physicist, researcher; b. Los Angeles, July 28, 1955; s. Edward M. and Erika Annette (Happel) R. S.B. in Math, M.I.T., 1977; M.S., UCLA, 1979, Ph.D. in Med. Physics, 1982. Radiation physicist Cedars-Sinai Med. Ctr., Los Angeles, 1977—; researcher, educator. Contbr. articles to profl. publs. in field. UCLA travel grantee, 1980, 81. Mem. Am. Assn. Physicists in Medicine (grantee 1982), Hosp. Physicists Assn., Am. Endocurietherapy Soc. Democrat. Subspecialties: Radiology; Medical physics. Current work: Research in dose calculations involved in implantation of radioactive sources into diseased tissue for treatment of cancer. Office: Radiation Therapy Dept 8700 Beverly Blvd Los Angeles CA 90048

ROSEN, ARTHUR D., neurologist, neurophysiologist; b. Bklyn., Sept. 19, 1935; s. Elihu and Gertrude (Simonson) R.; m. Deborah Mandelberg, Sept. 29, 1958; children: Jody Lynn, Matthew Scot. B.A., Columbia U., 1956; M.D., SUNY, Bklyn., 1960. Diplomate: Am. Bd. Psychiatry and Neurology. Intern Bklyn. Jewish Hosp., 1960-61; instr. physiology SUNY Downstate Med. Center, Bklyn., 1961-64, asst. prof. neurology, 1966-73, dir. Sci. Computing Center, 1970-73; USPHS fellow in neurophysiology, 1961-62; resident Kings County Hosp., Bklyn., 1962-64; asso. prof. neurology SUNY, Stony Brook, 1973-80, prof., 1980—, dir. div., 1973-80; mem. adv. bd. Nat. Amyotrophic Lateral Sclerosis Found. Contbr. numerous articles on neurology and neurophysiology to sci. jours. Served to lt. comdr., M.C. USNR, 1964-66. Fellow Am. Acad. Neurology; mem. Soc. for Neurosci., Am. Epilepsy Soc. Subspecialties: Neurology; Neurophysiology. Current work: Neurophysiology of epilepsy, epilepsy, neurophysiology of multiple sclerosis. Office: Dept Neurology SUNY Stony Brook NY 11794

ROSEN, ARTHUR L(EONARD), physiologist, concultant; b. Chgo., Apr. 30, 1934; s. Victor and Edith (Gold) R.; m. Arlene Silver, Aug. 26, 1956; children: David, Laura. B.S., Roosevelt U., Chgo., 1957; M.S., U. Chgo., 1964, Ph.D., 1971. Research assoc. Michael Reese Hosp., Chgo., 1955-58, 60-64, physiologist, 1977—; asst. prof. biophysics U. Ill., Chgo., 1967-77; physiologist Hektoen Inst., Chgo., 1964-77; research assoc., asst. prof. U. Chgo., 1981—. Contbr. sci. articles to profl. publs. Served with U.S. Army, 1958-60. Mem. Am. Physiol. Soc., Biophysics Soc., IEEE, Am. Assn. Physicians in Medicine, Am. Heart Assn. Subspecialties: Artificial organs; Physiology (medicine). Current work: Development of temporary red cell substitutes, stroma free hemoglobin solution and fluorochemical emulsions. Office: Michael Reese Hosp 2900 S Ellis Chicago IL 60616

ROSEN, B. WALTER, engineering company executive; b. N.Y.C., Aug. 2, 1931; s. David B. and Alice B. (Blum) B.; m. Marcia Fay Melnick, Apr. 11, 1959; children: Lynn Diane, Adam Todd. B.C.E., Cooper Union, 1952; M.S., Va. Poly. Inst. and State U., 1955; C.E., Columbia U., 1959; Ph.D., U. Pa., 1968. Aero. research scientist NACA, Hampton, Va., 1952-56; dep. dir. reentry vehicles AVCO Corp., Wilmington, Mass., 1956-62; cons. engr. Gen. Electric Co., King of Prussia, Pa., 1962-70; pres. Materials Scis. Corp., Spring House, Pa., 1970—; cons. Nat. Acad. Scis., 1963-65, NATO, 1972, UN, 1977; vis. prof. Technion, Israel, 1968-69; adj. prof. U. Pa., 1983—; structures design lectr AIAA, 1971; Scala lectr. 2d Internat. Conf. on Composite Materials, 1978. Contbr. articles to profl. jours. Mem. Upper Dublin Sch. Bd., 1973-79; bd. dirs. Temple Sinai, 1970-72. Guggenheim fellow, 1955-56. Mem. ASME, Am. Soc. Metals, Soc. Automotive Engrs., Am. Mgmt. Assn., AIAA, Soc. Aerospace Materials and Process Engrs. Subspecialties: Theoretical and applied mechanics; Composite materials. Current work: Analysis, design and research on composite materials and structures. Office: Gwynedd Plaza II Bethlehem Pike Spring House PA 19477

ROSEN, GERALD A., psychologist; b. Phila., June 25, 1946; s. Irving and Mary (Lipshutz) R. A.B., Temple U., 1967; M.A., Queens Coll., 1969; postgrad., Temple U., 1968-70; Ed.D., U. Pa., 1981. Lic. psychologist, Pa. Dir. research Vocat. Research Inst., Phila., 1971-79; dir. acad. testing ctr. and prof. psychology Mercer Coll., Trenton, N.J., 1981; cons. psychologist in pvt. practice, Phila., 1981—; cons. Bd. Examiners-Bd. Edn., Bklyn., 1982—, Queens Coll., 1979—, Trenton State Prison, 1981—, Pa. Coll. Podiatric Medicine, Phila., 1978-79. Author: Views, 1977; contbr. articles to profl. jours. Mem. subcom. on evaluation of human potential Nat. Task Force on Welfare Reform, 1972. Mem. Am. Psychol. Assn., Nat. Council Measurement in Edn., Am. Ednl. Research Assn., Classification Soc., ACLU. Democrat. Jewish. Current work: Vocational evaluation of mentally retarded; the relationship of personality to academic and vocational choices. Office: 2375 Woodward St Suite 112N Philadelphia PA 19115

ROSEN, JOSEPH DAVID, food chemistry educator, consultant; b. N.Y.C., Feb. 26, 1935; s. Rubin and Tobie (Greenspan) R.; m. Doris Stieber, Sept. 8, 1962; children: Todd, Amy, Mark, Dayan. B.S., CCNY, 1956; Ph.D., Rutgers U., 1963. Research chemist E. I. duPont de Nemours & Co., Parlin, N.J., 1963-65; asst. prof. dept. food sci. Rutgers U., New Brunswick, N.J., 1965-68, assoc. prof., 1968-74, prof., 1974—; sci. adv. FDA, N.Y.C., 1974—. Editor: Sulfur in Pesticide Action and Metabolism, 1981; co-editor: Jour. Food Safety, 1977—. Mem. Am. Chem. Soc. (exec. com. pesticide div. 1983—), AAAS, Inst. Food Technologists, Soc. Toxicology. Subspecialties: Toxicology (medicine); Analytical chemistry. Current work: Mutagens and tumor promoters in food, analytical methodology development for mycotoxins, mass spectrometry. Office: Rutgers U Cook Coll Food Sci Dept New Brunswick NJ 08903

ROSEN, JUDAH BEN, computer scientist; b. Phila., May 5, 1922; s. Benjamin and Susan (Hurwich) R.; m.; children—Susan Beth, Lynn Ruth. B.S. in Elec. Engring, Johns Hopkins U., 1943; Ph.D. in Applied Math, Columbia U., 1952. Research asso. Princeton (N.J.) U., 1952-54; head applied math. dept. Shell Devel. Co., 1954-62; vis. prof. computer sci. dept. Stanford (Calif.) U., 1962-64; prof. computer sci. dept. and math. research center U. Wis., Madison, 1964-71; prof., head computer sci. dept. Inst. Tech., U. Minn., Mpls., 1971—; Fulbright prof. Technion, Israel, 1968-69; Lady Davis vis. prof., 1980; invited lectr. Chinese Acad. Sci., Peking, 1980; lectr.; cons. Argonne (Ill.) Nat. Lab.; mem. Nat. Computer Sci. Bd. Editor: Nonlinear Programming, 1970; asso. editor: Soc. Indsl. and Applied Math. Jour. on Control and Orgn, 1965-77, Jour. Computer System Scis, 1966—; contbr. articles to profl. jours. NSF grantee, 1969—. Mem. Assn. Computing Machinery, Soc. Indsl. and Applied Math., Math. Programming Soc. Subspecialties: Numerical analysis; Algorithms. Current work: Research on algorithms and computational methods for large-scale optimization problems. Home: 1904 W 49th St Minneapolis MN 55409 Office: Lind Hall U Minn Minneapolis MN 55455

ROSEN, LOUIS, physicist; b. N.Y.C., June 10, 1918; s. Jacob and Rose (Lipionski) R.; m. Mary Terry, Sept. 4, 1941; 1 son, Terry Leon. B.A., U. Ala., 1939, M.S., 1941; Ph.D., Pa. State U., 1944; D.Sc. (hon.), U. N.Mex., 1980. Instr. physics U. Ala., 1940-41, Pa. State U., 1943-44; mem. staff Los Alamos Sci. Lab., 1944—, group leader nuclear plate lab., 1949-65, alt. div. leader exptl. physics div., 1962-65, dir. meson physics facility, also div. leader medium energy physics div., 1965—; Sesquicentennial hon. prof. U. Ala., 1981. Author papers in nuclear sci. and applications of particle accelerators.; bd. editors: Applications of Nuclear Physics. Mem. Los Alamos Town Planning Bd., 1962-64; mem. Gov.'s Com. on Tech. Excellence in N.Mex., Nat. Acad. Panel on Nuclear Sci.; chmn. sub-panel on accelerators Nat. Acad. Panel on Nuclear Sci.; mem. N.Mex. Cancer Control Bd., 1976—, v.p., 1979—; mem. panel on future of nuclear sci. Nat. Research Council of Nat. Acad. Scis., 1976; mem. panel on instl. arrangements for orbiting space telescope NRC-Nat. Acad. Scis., 1976; mem. U.S.A.-USSR Joint Coordinating Com. on Fundamental Properties of Matter, 1976—; Co-chmn. Los Alamos Vols. for Stevenson, 1956; Democratic candidate for county commr., 1962; bd. dirs. Los Alamos Med. Center, 1977-83. Recipient E.O. Lawrence award AEC, 1963; Golden Plate award Am. Acad. Achievement, 1964; N.Mex. Disting. Public Service award, 1978; named Citizen of Year, N.Mex. Realtors Assn., 1973; Guggenheim fellow, 1959-60; alumni fellow Pa. State U., 1978. Fellow Am. Phys. Soc. (mem. council 1975-78, chmn. panel on public affairs 1980), AAAS; mem. Meson Facility Users (dir.; Louis Rosen Prize established). Subspecialties: Nuclear physics; Particle physics. Current work: Energy options; national security. Home: 1170 41st St Los Alamos NM 87544 Office: PO Box 1663 Los Alamos NM 87545 I have come to believe that only after one has learned to manage and set worthy goals for himself should he attempt to do so for others.

ROSEN, MARTIN HOWARD, biochemistry educator; b. Bklyn., July 29, 1942; s. Edward Morris and Gussie (Klar) R.; m. Beth Dee Werfel, June 9, 1967. B.S., Bklyn. Coll., 1964; M.A., CUNY, 1967; Ph.D., N.Y.U., 1974. Teaching asst. Bklyn. Coll., 1964-67, teaching fellow, 1966-67; lectr. N.Y.C. Community Coll., Bklyn., 1966-67, Queensborough Community Coll., Queens, N.Y., 1968, Kingsborough Community Coll., Bklyn., 1968; instr. Manhattan Community Coll., N.Y.C., 1969, 70; assoc. prof. Coll. Staten Island, CUNY, 1968—, dep. chmn., evening session supr. biology dept., 1978—; research asst. Beth Israel Hosp., N.Y.C., 1965-67; research assoc. N.Y.U., Washington Square, 1974-78, N.Y.U. Med. Center, dept. dermatology, N.Y.C., 1978-82. Contbr. articles in field to profl. jours. Bd. dirs. Met. Assn. Coll. and Univ. Biologists, 1977-79, Found. Research Against Disease, 1979—; bd. trustee Staten Island Zoo, 1981. Recipient numerous awards and grants in field. Mem. AAAS, Am. Inst. Biol. Scis., N.Y. Acad. Sci., Found. Research Against Disease, Sigma Xi, others. Subspecialties: Biochemistry (medicine); Cell biology (medicine). Current work: Biochemical research in enzymatic analysis of cancer cells. Office: Dept Biology Coll Staten Island CUNY 715 Ocean Terrace Staten Island NY 07748

ROSEN, MICHAEL ROBERT, electrophysiologist, pharmacology educator; b. N.Y.C., Oct. 8, 1938; s. Jacob Selig and Gertrude Harriet (Laibson) R.; m. Tove Smulovitz, June 14, 1964; children: Nadine Miriam, Jennifer Naomi, Rachel Susannah. B.A., Wesleyan U., Conn., 1960; M.D., SUNY, Bklyn., 1964. Diplomate: Am. Bd. Internal Medicine. Intern Montefiore Hosp., N.Y.C., 1964-65, resident, 1965-66, 68-70; fellow Columbia U., N.Y.C., 1970-72, asst. prof. pharmacology, 1973-75, asst. prof. pharmacology and pediatrics, 1975-76, assoc. prof., 1976-81, prof., 1981—; sr. investigator N.Y. Heart Assn., 1972-75; mem. cardiovascular-pulmonary study sect. NIH, 1977-81. Asso. editor: Circulation Research, 1975-81; mem. editorial bd.: Circulation, 1983—; contbr. over 150 articles, revs. and abstracts to profl. jours. Served to capt., M.C. USAF, 1966-68. Irma T. Hirschl Trust fellow, 1975-79; Nat. Heart, Lung and Blood Inst. grantee, 1975—. Mem. Am. Coll. Clin. Pharmacology (pres. 1982-84), Cardiac Electrophysiologic Soc. (pres. 1981-82), Am. Heart Assn., ACP, Am. Coll. Cardiology, Am. Soc. Pharmacology and Exptl. Therapeutics, Alpha Omega Alpha, others. Subspecialties: Pharmacology; Physiology (medicine). Current work: Research on mechanisms responsible for cardiac arrhythmias, on interactions of heart with autonomic nervous system, developmental changes in cardiac function. Office: Dept Pharmacology Columbia U 630 W 168th St New York NY 10032

ROSEN, WILLIAM G., mathematician; b. Portsmouth, N.H., May 13, 1921; s. Harry and Lena (Winebaum) R.; m. Meriam L., Aug. 17, 1947; children: Abigail Mindy, Rachel Sylvia; m. Lynn Brice Rooney, Nov. 29, 1979. B.S. in Physics, U. Ill., 1943, M.S., 1947, Ph.D. in Math, 1954. Instr. to asst. prof. math. U. Md., 1954-61; asst. program dir. undergrad. sci. edn. NSF, Washington, 1961-62, assoc. program dir. sci. devel. program, 1962-64, staff assoc. sci. devel. program, 1964-66; spl. asst. to dir., exec. sec. com. on acad. sci. and engring. Fed. Council for Sci. and Tech., 1966-70, program dir. modern analysis and probability math. scis. sect., 1970-79, head math. scis. sect., 1979—. Served to 1st lt. USAF, 1943-46. Mem. Am. Math. Soc., Math. Assn. Am., Assn. for Women in Math., Soc. for Indsl. and Applied Math. Current work: Administration of mathematics program. Office: 1800 G St Room 304 Washington DC 20550

ROSEN, WILLIAM MICHAEL, chemistry educator, researcher; b. Lynn, Mass., Dec. 27, 1941; s. Morris and Matilda (Chernin) R.; m. Sandra H. Bergman, Aug. 23, 1964; children: Maxine, Rachel. B.S., UCLA, 1963; Ph.D., U. Calif.-Riverside, 1967. Postdoctoral fellow Ohio State U., Columbus, 1967-69; asst. prof. chemistry Purdue U., West Lafayette, Ind., 1969-70; prof. chemistry U. R.I., Kingston, 1970—, dir. honors program, 1982-83, coordinator honors coll., 1980-81; research chemist Scripps Inst. Oceanography, La Jolla, Calif., 1976-77. Fellow Am. Inst. Chemists; mem. Am. Chem. Soc., Royal Soc. Britain. Subspecialty: Synthetic chemistry. Current work: Organic and inorganic synthesis-theoretically and naturally interesting molecules. Home: 7 Mark Glen Ct Kingston RI 02881 Office: Dept Chemistry U RI Kingston RI 02881

ROSENBERG, ERIC RONALD, physician, radiology educator; b. Morristown, N.J., Feb. 18, 1949; s. Alvin Abe and Evelyn Claire (Thaler) R.; m. Jean Lynn Defanti, May 21, 1977; 1 dau., Caroline. A.B., Rutgers U., 1971; M.D., N.Y. Med. Coll., 1975. Diplomate: Am. Bd. Radiology. Resident in radiology Duke U. Med. Ctr., 1975-79, fellow in radiology, 1979-80, asst. prof. radiology, 1980—. Mem. Am. Inst. Ultrasound in Medicine, Am. Coll. Radiology, Radiol. Soc. N.Am. Subspecialties: Diagnostic radiology; Diagnostic ultrasound. Current work: Diagnostic ultrasound. Home: 2833 Split Rail Pl Durham NC 27712 Office: Dept of Radiology Duke U Med Ctr Durham NC 27710

ROSENBERG, GARY DAVID, geology educator, researcher; b. Milw., Aug. 2, 1944; s. Leo F. and Phyllis (Weitzman) R. B.S., U. Wis.-Madison, 1966; Ph.D., UCLA, 1972. Sr. research assoc. Sch. Physics, U. Newcastle/Tyne, Eng., 1972-76; research assoc. Washington U., St. Louis, 1976-78; vis. asst. prof. Mich. State U., East Lansing, 1978-79; assoc. prof. dept. geology Ind./Purdue U., Indpls., 1979—. Editor: Growth Rhythms and the History of the Earth's Rotation, 1975. Bd. dirs., v.p. Garden Walk Condominiums, Indpls. NATO sr. fellow, 1973-74; NSF fellow, 1967-68; NDEA IV fellow, 1968-71; NSF grantee, 1982-84; Dept. Interior grantee, 1983-84. Mem. Paleontol. Soc., Paleontol. Assn. (U.K.), AAAS, Ind. Geologists Soc., Sigma Xi, Phi Kappa Phi. Subspecialties: Paleoecology; Chronobiology. Current work: Skeletal growth rhythms - geophysical, paleocological and medical applications. Office: 425 Agnes St Indianapolis IN 46202

ROSENBERG, GERSON, mechanical engineer, biomedical engineer; b. Phila., Aug. 20, 1944; s. Meyer and Grace (Falcone) R. A.A., Pa. State U., 1966, B.S., 1970, M.S., 1972, Ph.D. in Mech. Engring, 1975. Research asst. Pa. State U., University Park, 1975, instr. mech. engring., 1972-73, asst. prof. bioengring., 1981—, sr. project assoc., Hershey, 1976-78, research assoc., 1978—, asst. chief div. artificial organs, 1983—; cons. Vitamek, Inc., Houston, 1981, St. Louis U. Sch. Medicine, 1980, others. Contbr. chpts. to books, articles to profl. jours. Served to 1st lt. U.S. Army, 1970-78. Pa. Research Corp. grantee, 1978-80; R.J. and H.C. Kleberg Found. grantee, 1979-82; Whitaker Found. grantee, 1982—. Mem. ASME, Assn. for Advancement of Med. Instrumentation, AAAS, Am. Soc. for Artificial Internal Organs (membership com. 1981—), Internat. Soc. for Artificial Internal Organs, Soc. for Biomaterials. Club: Porsche of Am. (pres. Central Pa. 1983—). Subspecialties: Mechanical engineering; Biomedical engineering. Current work: Artificial organs. Home: 213 Lopax Rd Harrisburg PA 17112 Office: Pa State U MS Hershey Med Ctr Hershey PA 17033

ROSENBERG, HOWARD CHARLES, pharmacology educator; b. Atlantic City, Apr. 17, 1947; s. Leroy A. and Henrietta (Tenenbaum) R.; m. Ann R., June 22, 1969; children: Martin J., Lewis A. B.A., Ithaca Coll., 1969; Ph.D., Cornell U., 1975, M.D., 1976. Fellow in pharmacology Cornell U.-Ithaca, N.Y.C., 1976-77; asst. prof. Med. Coll. of Ohio, Toledo, 1977-82, assoc. prof., 1982—; cons. Chem. Dependency Ctr. Flower Hosp. Contbr. articles to profl. jours. Nat. Inst. on Drug Abuse grantee, 1979—. Mem. Am. Soc. Pharmacology and Exptl. Therapeutics, AAAS, Soc. for Neurosci., Sigma Xi. Subspecialties: Pharmacology; Neuropharmacology. Current work: Research on tolerance to and dependence on sedative and tranquilizer drugs. Office: Medical College of Ohio CS #10008 Toledo OH 43699

ROSENBERG, IRWIN HAROLD, physician; b. Madison, Wis., Jan. 6, 1935; s. Abraham Joseph and Celia (Mazursky) R.; m. Civia Muffs, May 24, 1964; children—Daniel, Ilana. B.S., U. Wis., 1956; M.D. cum laude, Harvard U., 1959. Intern Mass. Gen. Hosp., Boston, 1959-60, resident, 1960-61; research asso. NIH, 1961-64; fellow Thorndike Meml. Lab., Boston, 1964-65; instr. Harvard U. Med. Sch., 1965-67, asst. prof. medicine, 1967-70; asso. prof. U. Chgo., 1970-74, prof., 1974—, head sect. gastroenterology, 1971—, dir. Clin. Nutrition Research Ctr., 1979—; mem. gen. med. study sect. NIH; mem. food and nutrition bd. NRC/Nat. Acad. Scis., chmn. food and nutrition bd., 1981-83. Served with USPHS, 1961-64. Mem. Am. Gastroenterol. Assn., Am. Soc. Clin. Nutrition (Pres. 1983-84), Am. Soc. Clin. Investigation, AAAS. Subspecialty: Gastroenterology. Current work: Intenstinal absorption and malabsorption of vitamins including vitamin D; B vitamins; nutritional complication of gastrointestinal disease; intensive nutritional support. Office: 950 E 59th St Chicago IL 60637

ROSENBERG, JERRY C., thoracic surgeon; b. N.Y.C., Apr. 23, 1929; s. Abraham and Harriet Francis (Wendroff) R.; m. Corliss E. Bestland, July 3, 1955; children: David Rolf, Andrew Leon. B.S., Wagner Coll., Staten Island, N.Y., 1950; M.D., Chgo. Med. Sch., 1954; Ph.D., U. Minn., 1963; M.B.A., Mich. State U., 1983. Intern U. Minn., Mpls., 1954-55, resident in thoracic and gen. surgery, 1955-61; mem. staff Hutzel Hosp., Detroit, 1968—, chief surg. unit, 1978—; mem. faculty Wayne State U., Detroit, 1968—, prof. surgery, 1972—. Contbr. over 100 sci. articles to profl. publs. Fellow ACS; mem. Soc. Univ. Surgeons, Am. Assn. Immunologists, Am. Assn. Transplant Surgery, Soc. Surg. Oncology, Transplantation Soc. Mich. (pres. 1976-78). Jewish. Club: Detroit Athletic. Subspecialty: Surgery. Home: 8544 Huntington Rd Huntington Woods MI 48070 Office: Wayne State Univ 4707 Saint Antoine St Detroit MI 48021

ROSENBERG, MICHAEL J., cardiologist, educator; b. Chgo., Jan. 21, 1952; s. Sam and Yette (Schussler) R.; m. Sharon L. Rutenberg, Feb. 3, 1980. B.A., Northwestern U., 1973, M.D., 1976. Diplomate: Am. Bd. Internal Medicine, Nat. Bd. Med. Examiners. Med. intern Evanston Hosp./Northwestern U., Evanston, Ill., 1977, resident in medicine, 1978-79; fellow in cardiology Rush-Presbyn.-St. Luke's

Hosp., Chgo., 1980-82; attending cardiologist, dir. CCU, Luth. Gen. Hosp., Park Ridge, Ill., 1982—; instr. Rush. Med. Coll., Chgo., 1980-82, U. Ill-Chgo., 1982—; assoc.prof./lectr. U. Ill.-Chgo., 1982—. Author abstracts, articles, and chpts. in field of cardiology. Marcy scholar Northwestern U., 1973. Mem. Am. Coll. Cardiology, Am. Fedn. Clin. Research, ACP, Am. Heart Assn. Subspecialties: Cardiology; Internal medicine. Current work: Clinical research and teaching of medical students and house staff physicians. Office: Lutheran Gen Hosp 1775 Dempster Park Ridge IL 60068

ROSENBERG, NORMAN JACK, agricultural meteorologist; b. Bklyn., Feb. 22, 1930; s. Jacob and Rae (Dombrowitz) R.; m. Sarah Zacher, Dec. 30, 1950; children: Daniel Jonathon, Alyssa Yael. B.S., Mich. State U., 1951; M.S., Okla. State U., 1958; Ph.D., Rutgers U., 1961. Soil scientist Israel Soil Conservation Service, Haifa, 1953-55, Israel Water Authority, 1955-57; research asst. Okla. State U., 1957-58; research fellow Rutgers U., 1958-61; asst. prof. agrl. meteorology U. Nebr., Lincoln, 1961-64, asso. prof., 1964-67, prof., 1968—, prof. agrl. engring., 1975—, prof. agronomy, 1976—, George Holmes prof. agrl. meteorology, 1981—, leader sect. agrl. meteorology, 1974; dir. Center for Agrl. Meteorology and Climatology, 1979; cons. Dept. State AID, NOAA, Am. Public Health Assn.; mem. numerous ad hoc coms., and mem. standing com. on atmospheric sci. Nat. Acad. Scis./NRC, 1975-78, now mem. bd. on atmospheric sci. and climate; vis. prof. agrl. meteorology Israel Inst. Tech., Haifa, 1968; Lady Davis fellow Hebrew U., Jerusalem, 1977. Author: Microclimate: The Biological Environment, 1974, 2d edit., 1983, Chinese transl., 1983; also numerous articles in profl. jours.; editor: North American Droughts, 1978, Drought in the Great Plains: Research on Impacts and Strategies, 1980; tech. editor: Agronomy Jour, 1974—; cons. editor: Irrigation Sci., Agrl. Meteorology, Climatic Change, Jour. Climate and Applied Meteorology. NATO sr. fellow in sci., 1968; recipient Centennial medal Nat. Weather Service, 1970. Fellow AAAS; mem. Am. Soc. Agronomy, Am. Meteorol. Soc. (outstanding achievement in bioclimatology award 1978, councillor 1981—), Arid Zone Soc. India, Explorers Club, Sigma Xi, Alpha Zeta, Gamma Sigma Rho. Jewish. Clubs: Malib Poker Soc. of Lincoln.; Cosmos (Washington). Subspecialties: Micrometeorology; Climatology. Current work: Effects of increasing global carbon dioxide on climate and agriculture; drought management strategy. Home: 3145 S 31st St Lincoln NE 68502 Office: 243 LW Chase Hall U of Nebr Lincoln NE 68583

ROSENBERG, PAUL, physicist, consultant; b. N.Y.C., Mar. 31, 1910; s. Samuel and Evelyn (Abbey) R.; m. Marjorie S. Hillson, June 12, 1943; 1 dau., Gale B.E. A.B., Columbia U., 1930, M.A., 1933, postgrad., 1933-40. Chemist Hawthorne Paint & Varnish Corp., N.J., 1930-33; grad. asst. physics Columbia, 1934-39, lectr., 1939-41; instr. Hunter Coll., N.Y.C., 1939-41; research asso. elec. engring. Mass. Inst. Tech., Cambridge, 1941; staff mem. Radiation Lab., Nat. Def. Research Com., 1941-45; pres. Paul Rosenberg Assocs. (cons. physicists), Pelham, N.Y., 1945—, Inst. Nav., 1950-51; Mem. war com. radio Am. Standards Assn., 1942-44; gen. chmn. joint meeting Radio Tech. Commn. for Aeros., Radio Tech. Commn. for Marine Services and Inst. of Nav., 1950; co-chmn. Nat. Tech. Devel. Com. for upper atmostphere and interplanetary nav., 1947-50; Mem. maritime research Adv. com. Nat. Acad. Scis.-NRC, 1959-60; chmn. cartography panel space programs Earth resources survey NRC, 1973-76; chmn. panel on nav. and traffic control space applications study Nat. Acad. Scis., 1968; Bd. dirs. Center for Environment and Man, 1976-83. Contbr. sci., tech. articles to prof. jours.; Editorial com.: Jour. Aerospace Scis, 1952-60. Fellow AAAS (council 1961-73, v.p. 1966-69), IEEE, Am. Inst. Chemists, Explorers Club, AIAA (asso.); mem. Am. Phys. Soc., Am. Chem. Soc., Nat. Acad. Engring., Acoustical Soc. Am., Armed Forces Communication Assn., Optical Soc. Am., N.Y. Acad. Scis., Am. Soc. Photogrammetry (Talbert Abrams award 1955), Am. Assn. Physics Tchrs., Sigma Xi, Zeta Beta Tau. Clubs: Beach Point Yacht (Mamaroneck, N.Y.); Columbia U. Subspecialties: Remote sensing (geoscience); Solar energy. Current work: Photogrammetry, Mapping, Surveying; Solar heat collectors; robotics. Patentee in field. Home: 53 Fernwood Rd Larchmont NY 10538 Office: 330 5th Ave Pelham NY 10803

ROSENBERG, PHILIP, pharmacologist, educator, editor, researcher; b. Phila., July 28, 1931; s. Morris and Rose (Schwartz) R.; m. Sybil Edith Stepman, Oct. 21, 1956; children: Stuart Owen, Rachelle, Gail Linda. B.S., Temple U., 1953; M.S., U. Kans., 1955; Ph.D., Thomas Jefferson U., 1957. Registered pharmacist, Pa. Asst. prof. Jefferson U., Phila., 1957-58; research fellow, asst. prof. Columbia U. Coll. Physicians and Surgeons, N.Y.C., 1958-68; prof. pharmacology Pharmacy Sch., U. Conn., Storrs, 1968—, chmn. sect. pharmacology and toxicology, 1968—; cons. in field; Meyerhoff sr. fellow Weizmann Inst. Sci., Rehovoth, Israel, 1982-83. Editor: Toxins: Animal, Plant and Microbial, 1978; contbr. chpts. to books, articles to profl. jours. USPHS spl. postdoctoral fellow, 1960-62; USPHS career devel. award, 1964-68; NIH grantee, 1978—; U.S. Army Med. Research and Devel. Command grantee, 1982—. Fellow Acad. Pharm. Scis. (Redi award 1982); mem. AAAS, Am. Assn. Colls. Pharmacy, Am. Chem. Soc., Am. Pharm. Assn., Am. Profs. for Peace in Middle East (co-chmn. U. Conn. chpt.), Am. Soc. Neurochemistry, Am. Soc. Pharmacology and Exptl. Therapeutics, Corp. Marine Biol. Lab. (Woods Hole, Mass.), Council Biology Editors, Harvey Soc., Internat. Soc. on Toxinology, N.Y. Acad. Scis., Sigma Xi, Phi Kappa Phi, Rho Chi, Rho Pi Phi. Jewish. Subspecialties: Neurochemistry; Neuropharmacology. Current work: Cholinergic system and action of anticholinesterases; phospholipid function and action of phospholipases on nerve, muscle and synapse; use of toxins as tools in studying bioelectrical phenomenon. Home: 40 Middle Rd Ellington CT 06029 Office: Sch Pharmacy U Conn Storrs CT 06268

ROSENBERG, RICHARD ALLAN, research chemist; b. St. Louis, Sept. 13, 1948; s. William and Ruth (Mayer) R.; m. Mary Josephine Rosenberg, June 6, 1982. B.S. in Chem. Engring, Northwestern U., 1971; M.S. in Chemistry, U. Ill., 1974, Ph.D., U. Calif., Berkeley, 1979. Research and teaching asst. U. Calif. Lawrence Berkeley Lab., 1974-79; research chemist physics div. Michelson Labs., Naval Weapons Ctr., China Lake, Calif., 1979—. Contbr. articles to profl. jours. Mem. Am. Phys. Soc., Am. Chem. Soc., Sigma Xi. Democrat. Subspecialties: Condensed matter physics; Surface chemistry. Current work: Electronic structure of materials studied using synchrotron radiation; studies of the surface electronic structure of solids using synchrotron radiaiton, photon stimulated ion desorption and photoemission. Office: SSRL, SLAC Bin 69 PO Box 4349 Stanford CA 94305

ROSENBERG, ROBERT BRINKMANN, research organization executive; b. Chgo., Mar. 19, 1937; s. Sidney and Gertrude (Brinkmann) R.; m. Patricia Margaret Kane, Aug. 1, 1959; children: John Richard, Debra Ann. B.S. in Chem. Engring. with distinction, Ill. Inst. Tech., 1958, M.S. in Gas Tech, 1961, Ph.D., 1964. Registered profl. engr., Ill. Adj. asst. prof. Ill. Inst. Tech., 1965-69; mem. staff Inst. Gas Tech., Chgo., 1962-77, v.p. engring. research, 1973-77; v.p. research and devel. Gas Research Inst., Chgo., 1977-78, exec. v.p., 1978—. Author. Mem. Hinsdale (Ill.) Home Rule Ad Hoc Com., 1975-77; bd. dirs. Hinsdale Cultural Arts Soc., 1977—; pres. Triangle Frat. Edn. Found., 1974—; mem. vis. com. dept. chemistry U. Tex. Mem. Am. Inst. Chem. Engrs., Am. Gas Assn. (ex officio mem. mng. com. of operating sect.), Inst. Gas. Engrs., Combustion Inst. (past treas. bd. central states sect.), Atlantic Gas Research Exchange (chmn. mng. bd. 1980—), Internat. Gas Union (U.S. dir. subcom. F-2 1974-83), Gas Appliance Engrs. Soc. (past trustee), Air Pollution Control Assn. (past sect. com. residential pollution sources). Subspecialties: Fuels; Combustion processes. Current work: Research on production; distribution and utilization of natural gas and substitute gaseous fuels. Patentee in field. Home: 414 W 8th St Hinsdale IL 60521 Office: 8600 W Bryn Mawr Ave Chicago IL 60631

ROSENBERGER, DAVID A., plant pathologist; b. Quakertown, Pa., Sept. 14, 1947; s. William H. and Ada (Geissinger) R.; m. Carol Joyce Freeman, July 28, 1973; children: Sara Joy, Matthew David. B.A., Goshen Coll., 1969; Ph.D., Mich. State U., 1977. Assoc. prof. plant pathology Cornell U., N.Y. State Agrl. Expt. Sta., Geneva, 1977—. Contbr. articles to profl. jours. Mem. Am. Phytopath. Soc., AAAS. Democrat. Nazarene. Subspecialties: Plant pathology; Integrated pest management. Current work: Diseases of deciduous tree fruits; research on epidemiology and control of fungal and viral diseases of tree fruits; control of postharvest pathogens of apples. Office: Hudson Valley Lab Box 727 Highland NY 12528

ROSENBERGER, FRANZ ERNST, physics educator, consultant; b. Salzburg, Austria, May 31, 1933; s.; m. Renate H. Suessenbach, 1959; children: Uta, Bernd, Till. Diploma in physics, U. Stuttgart, 1964; Ph.D. in Physics, U. Utah, 1970. Asst. prof. physics U. Utah, 1973-77, assoc. prof., 1978-81, prof. physics, also adj. prof. materials sci. and engring., 1981—; dir. Crystal Growth Lab., 1967—. Author: Fundamentals of Crystal Growth, 1979; contbr. articles profl. jours. Named Outstanding Physics Tchr. U. Utah, 1978. Mem. Am. Phys. Soc., Am. Assn. Crystal Growth. Subspecialties: Condensed matter physics; Materials processing. Current work: Physico-chemical and fluid dynamics aspects of single crystal preparation. Office: Dept of Physics University of Utah Salt Lake City UT 84112

ROSENBERGER, JOHN KNOX, virologist, educator, cons.; b. Wilmington, Del., Dec. 8, 1942; s. George G. and Caroline F. (Russell) R.; m. Anne E. Ratledge, Aug. 8, 1964; children: Wendy, Jill. B.S., U. Del., 1964, M.S., 1966; Ph.D., U. Wis., 1972. Asst. prof. animal sci. U. Del., 1972-76, assoc. prof., 1976-81, prof., chmn. dept., 1976—; cons. to industry and govt. Contbr. articles to profl. jours. Served to capt. AUS, 1966-69. Recipient DPI medal of achievement, 1975, Outstanding Research award Poultry Sci., 1982; grantee Animal Health Competitive Grants, 1981-82. Mem. U.S. Animal Health Assn., Am. Assn. Avian Pathology, Am. Soc. Miocrobiology. Subspecialties: Animal virology; Immunopharmacology. Current work: Pathogenicity and immunology of selected animal reoviruses. Home: 29 Skycrest Landenberg PA 19350 Office: Dept Animal Science and Agriculture Biochemistry U Del Newark DE 19711

ROSENBLATT, MURRAY, mathematics educator; b. N.Y.C., Sept. 7, 1926; m. Adylin, July 1949; 2 children. B.S., CCNY, 1946; M.S., Cornell U., 1947, Ph.D. in Math., 1949. Asst. prof. U. Chgo., 1950-55; assoc. prof. math. Ind. U., 1956-59; prof. probability and stats. Brown U., 1959-64; prof. U. Calif.-San Diego, 1964—; vis. asst. prof. Columbia U., 1955; guest scientist Brookhaven Nat. Lab., 1959; vis. fellow U. Stockholm, 1953, Univ. Coll. London, 1965-66, Imperial Coll. and Univ. Coll. London, 1972-73, Australian Nat. U., 1976, 79; vis. scholar Stanford U., 1982. Author: (with U. Grenander) Statistical Analysis of Stationary Time Series, 1957, Random Processes, 1962, Edited Time Series Analysis, 1963, Markov Processes, Structure and Asympototic Behavior, 1971; editor: The North Holland Series in Probability and Statistics, 1981—, The Birkhauser Boston Inc. Progress in Probability and Stats. Series, 1982; mem. editorial bd.: Jour. Multivariate Analysis, 1970—, Ind. Jour. Math., 1957—, Jour. Time Series Analysis, 1981—; contbr. articles to profl. jours. Office Naval Research grantee, 1949-50; Guggenheim fellow, 1965-66, 71-72; overseas fellow Churchill Coll., Cambridge U., 1979; recipient Bronze medal U. Helsinki, 1978. Fellow Inst. Math. Stats., AAAS, Internat. Statis. Inst.; mem. Am. Math. Soc., Sigma Xi, Phi Beta Kappa. Subspecialty: Statistics. Current work: Probability theory; stochastic processes; time series analysis; turbulence. Office: Dept Math U Calif-San Diego La Jolla CA 92037

ROSENBLITH, WALTER ALTER, scientist, educator; b. Vienna, Austria, Sept. 21, 1913; came to U.S., 1939, naturalized, 1946; s. David A. and Gabriele (Roth) R.; m. Judy Olcott Francis, Sept. 27, 1941; children—Sandra Yvonne, Ronald Francis. Ingenieur Radiotelegraphiste, U. Bordeaux, 1936; Ing. Radioelectricien, Ecole Supérieure d'Electricité, Paris, 1937. Research engr., France, 1937-39; research asst. N.Y. U., 1939-40; grad. fellow, teaching fellow physics U. Calif. at Los Angeles, 1940-43; asst. prof., asso. prof., acting head dept. physics S.D. Sch. Mines and Tech., 1943-47; research fellow Psycho-Acoustic Lab., Harvard U., 1947-51; lectr. otolaryngology Harvard Med. Sch., also Mass. Eye and Ear Infirmary, 1957—; asso. prof. communications biophysics Mass. Inst. Tech., 1951-57, prof., 1957—, Inst. prof., 1975—; staff Research Lab. Electronics, 1951-69, chmn. faculty, 1967-69, asso. provost, 1969-71, provost, 1971-80; dir. Kaiser Industries, 1968-76; chmn. com. electronic computers in life scis. Nat. Acad. Scis.-NRC, 1960-64, mem. brain scis. com., 1965-68, chmn., 1966-67; mem. central council Internat. Brain Research Orgn., 1960-68, mem. exec. com., 1960-68, hon. treas., 1962-67; cons. life scis. panel Pres.'s Sci. Adv. Com., 1961-66; mem. council Internat. Union Pure and Applied Biophysics, 1961-69; inaugural lectr. Tata Inst. Fundamental Research, Bombay, 1962; Weizmann lectr. Weizmann Inst. Sci., Rehovoth, Israel, 1962; pres. Commn. on Biophysics of Communication and Control Processes, 1964-69; mem. Pres.'s Com. Urban Housing, 1967-68; cons. communications scis. WHO, 1964-65; mem. bd. medicine Nat. Acad. Sci., 1967-70; charter mem. Inst. Medicine, 1970—, mem. council, 1970-76; mem. adv. com. to dir. NIH, 1970-74; mem. governing bd. NRC, 1974-76; mem. adv. com. med. sci. AMA, 1972-74; chmn. sci. adv. council Callier Center for Communication Disorders, 1968—; mem. Com. on Scholarly Communication with People's Republic of China, 1977—, Bd. Fgn. Scholarships, 1978-81, chmn., 1980-81; co-chmn. NRC-IOM com. for study of saccharin and food safety policy, 1978-79. Contbr. articles and chpts. to profl. publs. Bd. govs. Weizmann Inst. Sci., 1973—; chmn. nat. com. on rehab. of physically handicapped NRC, 1975-77; trustee Brandeis U., 1979—. Fellow Acoustical Soc. Am., World Acad. Art and Sci., Am. Acad. Arts and Scis. (exec. bd. 1970-77), AAAS, IEEE; mem. Biophys. Soc. (council 1969-72, exec. bd. 1957-61), Nat. Acad. Engring., Nat. Acad. Scis., Soc. Exptl. Psychologists. Subspecialties: Neurophysiology; Biomedical engineering. Office: Mass Inst Tech Cambridge MA 02139

ROSENBLUETH, EMILIO, earthquake engineer; b. Mexico City, Mex., Apr. 8, 1926; s. Emilio and Charlotte (Deutsch) R.; m. Alicia Laguette, Feb. 20, 1954; children: David Arturo, Javier Fernando, Pablo Esteban, Monica Teresa. C.E., Nat. Autonomous U. Mex., 1947; M.S. in Civil Engring, U. Ill., 1949, Ph.D., 1951, U. Waterloo, Ont., Can. Surveyor and structural engr., 1945-47; soil mechanics asst. Ministry Hydraulic Resources, also U. Ill., 1947-50; structural engr. Fed. Electricity Commn., also Ministry Navy, 1951-55; prof. earthquake engring. Nat. Autonomous U. Mex., 1956—, regent, 1972-81; pres. DIRAC Group Cons., 1970-77; vice-minister Ministry Edn., 1977-82; pres. Reúnion Internationale des Laboratoires d'Essais des Materiaux (RILEM), 1965-66. Co-author: Fundamentals of Earthquake Engineering, 1971; Co-editor: Seismic Risk and Engineering Decisions, 1976; Contbr. to profl. publs. Trustee Autonomous Metropolitana U., 1974-77; mem. working group engring. seismology UNESCO, 1965. Mem. Mexican Acad. Sci. Research (pres. 1964-65), Mexican Soc. Earthquake Engring., Mex. Soc. Soil Mechanics (trustee), Internat. Assn. Earthquake Engring. (pres. 1973-77), Mexican Assn. Civil Engrs., Mexican Math. Soc., Mexican Geophys. Union, Coll. Engring. Iberoamerican U., Seismol. Soc. Am., Soc. Exptl. Stress Analysis, Latin Am. Assn. Seismology and Earthquake Engring., Assn. Computing Machinery, N.Z. Soc. Earthquake Engring., Sigma Xi; hon. mem. Am. Concrete Inst., ASCE, Internat. Assn. Earthquake Engring. (pres. 1972-76); fgn. assoc. Nat. Acad. Arts and Scis., Nat. Acad. Scis., Nat. Acad. Engring. Subspecialties: Civil engineering; Theoretical and applied mechanics. Current work: Structural reliability; decisions making; applied numerical methods; earthquake engineering. Office: Argentina 28 20 Piso of 318 Mexico 1 DF Mexico

ROSENBLUM, MARK LESTER, neurosurgeon, researcher; b. Bklyn., Feb. 6, 1944; s. Harry and Martha (Milstein) R.; m. Pamela Joyce Rosenblum, Sept. 20, 1970; children: Amy Dana, Scott David. B.S., Rensselaer Poly. Inst., Troy, N.Y., 1965; M.D., N.Y. Med. Coll., 1969. Diplomate: Am. Bd. Neurol. Surgery. Brain tumor researcher U. Calif.-San Francisco, 1970-72, resident in neurosurgery, 1973-79, mem. faculty, 1979—, asso. prof. neurosurgery, 1982—; researcher human tumor stem cells, cancer therapy. Editor: Progress in Experimental Tumor Research, 2 vols., in preparation; contbr. articles sci. jours. Served as surgeon USPHS, 1970-72. Recipient Tchr. Investigator Devel. award NIH, 1981—; grantee Nat. Cancer Inst., 1981—. Mem. Congress Neurol. Surgery, Am. Assn. Cancer Research, Am. Soc. Clin. Oncology, Alpha Omega Alpha. Subspecialties: Neurosurgery; Cancer research (medicine). Current work: Neurooncology, cancer biology, neurosurgery, brain tumors, brain infections.

ROSENBLUTH, MARSHALL NICHOLAS, physicist, educator; b. Albany, N.Y., Feb. 5, 1927; s. Robert and Margaret (Sondheim) R.; m. Sara Unger, Feb. 6, 1979; children by previous marriage—Alan Edward, Robin Ann, Mary Louise, Jean Pamela. B.A., Harvard, 1945; M.S., U. Chgo., 1947, Ph.D., 1949. Inst. Stanford, 1949-50; staff mem. Los Alamos Sci. Lab., 1950-56; sr. research adviser Gen. Atomic Corp., San Diego, 1956-67; prof. U. Calif., San Diego, 1960-67, Inst. for Advanced Study, Princeton, N.J., 1967-80; lectr. with rank prof. in astrophys. scis. Princeton U., also vis. sr. research physicist, 1967-80; dir. Inst. for Fusion Studies, U. Tex., 1980—; Andrew D. White vis. prof. Cornell U., 1976; cons. AEC, NASA, Inst. Def. Analysis. Served with USNR, 1944-46. Recipient E.O. Lawrence award, 1964, Albert Einstein award, 1967, Maxwell prize, 1976. Mem. Am. Phys. Soc., Nat. Acad. Sci., Am. Acad. Arts and Scis. Subspecialty: Plasma physics. Home: 4602 Balcones Austin TX 78731 Office: Dept Physics U Tex Austin TX 78712

ROSENBUSCH, RICARDO FRANCISCO, veterinary microbiologist, educator; b. Buenos Aires, Argentina, Oct. 15, 1942; came to U.S., 1977, naturalized, 1982; s. Carlos and Jeanette (Richardson) R.; m. Marcia K. Harmon, Dec. 21, 1967; children: Karina, Adrian. Sc.D. in Vet. Medicine, U. Buenos Aires, 1964; M.S., Iowa State U., 1966; Ph.D., 1969. Diplomate: Am. Coll. Vet. Microbiologist. Tech. dir. Instituto Rosenbusch, Inc., Buenos Aires, 1974-77; vis. asst. prof. Vet. Med. Research Inst., Iowa State U., 1977-78, asst. prof., 1978-80, assoc. prof. vet. microbiology, 1980—. Mem. Am. Coll. Vet. Microbiologists, Am. Soc. Microbiology, Internat. Orgn. Mycoplasmology. Subspecialties: Microbiology (veterinary medicine); Cell and tissue culture. Current work: Infectious bovine keratoconjunctivities; pathogenesis and immunity; bovine viral diarrhea, pathogenesis; mucosal and epidermal epithelial cells; interaction with pathogens. Office: Veterinary Med Research Inst Iowa State U Ames IA 50011

ROSENFELD, ISADORE, medical educator, cardiologist, lecturer; b. Montreal, Que., Can., Sept. 7, 1926; came to U.S., 1958; s. Morris and Vera (Friedman) R.; m. Camilla Master, Aug. 19, 1956; children: Arthur, Stephen, Hildi, Herbert. B.S., McGill U., 1947, M.D.C.M., 1951, diploma internal medicine, 1956. Intern Royal Victoria Hosp., Montreal; also resident Balt. City Hosp.; clin. asst. prof. medicine Cornell Med. Coll., N.Y.C., 1964-71, clin. assoc. prof., 1971-79, clin. prof., 1979—; attending N.Y. Hosp., N.Y.C., 1964—; pres. Rosenfeld Heart Found., N.Y.C., 1974—; juror Lasker Sci. Awards, 1972; pres. Found. Bio-Med. Research, N.Y.C., 1982; lectr., TV commentator.; vis. prof. Baylor U. Coll. Medicine, 1982. Author: ECG and X-Ray in Diseases of the Heart, 1963, The Complete Medical Exam, 1978, Second Opinion, 1981. Bd. dirs. N.Y. Heart Assn., 1979-82; mem. NTL adv. com. Harriman Inst. Advanced Study of Soviet Union, 1982; bd. overseers Cornell U. Med. Coll., 1980—. Recipient Vera award The Voice Found., 1981. Fellow ACP, Am. Coll. Chest Physicians, Am. Coll. Cardiology, Royal Coll. Physicians Can., N.Y. County Med. Soc. (bd. censors 1979-83, v.p. 1983-84). Jewish. Subspecialty: Cardiology. Current work: Hypertension, angina pectoris, sudden cardiac death, arteriosclerosis. Office: 125 E 72d St New York NY 10021

ROSENFELD, RON GERSHON, pediatric endocrinology educator; b. N.Y.C., June 22, 1946; s. Stanley Irving and Deborah (Levin) R.; m. Valerie Rae Spitz, June 16, 1968; children: Amy Lauren, Jeffrey Michael. B.A., Columbia U., 1968; M.D., Stanford U., 1973. Diplomate: Am. Bd. Pediatrics. Intern Stanford U., 1973-74, resident in pediatrics, 1974-75, chief resident in pediatrics, 1975-76; practice medicine specializing in pediatrics, Santa Barbara, Calif., 1976-77; postdoctoral fellow in pediatric endocrinology Stanford U., 1977-80, asst. prof. pediatrics, 1980—, assoc. dir. div. pediatric endocrinology, 1980—. Contbr. articles to profl. jours. Recipient New Investigator Research award NIH, 1981-84; Basil O'Connor Starter grant March of Dimes, 1982-84; Mellon Found. fellow, 1980-83. Mem. Am. Fedn. Clin. Research, Soc. Pediatric Research, Endocrine Soc., Lawson-Wilkins Pediatric Endocrine Soc. Subspecialties: Pediatrics; Endocrinology. Current work: Somatemedins and other growth factors, regulation of cell growth. Home: 582 Patricia Ln Palo Alto CA 94303 Office: Dept Pediatrics Stanford U Stanford CA 94303

ROSENKILDE, CARL EDWARD, physicist; b. Yakima, Wash., Mar. 16, 1937; s. Elmer Edward and Doris Edith (Fitzgerald) R.; m. Bernadine Doris Blumenstine, June 22, 1963; children: Karen Louise, Paul Eric. B.S. in Physics, Wash. State Coll., 1959, M.S., U. Chgo., 1960, Ph.D., 1966. Postdoctoral fellow Argonne (Ill.) Nat. Lab., 1966-68; asst. prof. math. NYU, 1968-70; asst. prof. physics Kans. State U., Manhattan, 1970-76, assoc. prof., 1976-79; physicist Lawrence Livermore (Calif.) Nat. Lab., 1979—, cons., 1974-79. Contbr. articles on physics to profl. jours. Woodrow Wilson fellow, 1959, 60. Mem. Am. Phys. Soc., Am. Astron. Soc., Soc. for Indsl. and Applied Math., Am. Geophys. Union, Phi Beta Kappa, Phi Kappa Phi, Phi Eta Sigma, Sigma Xi. Republican. Presbyterian. Club: Tubists Universal Brotherhood Assn. (TUBA). Subspecialties: Theoretical physics; Fluid dynamics. Current work: Nonlinear wave propagation, atmospheric scavenging. Office: Lawrence Livermore Nat Lab FO Box 808 Livermore CA 94550

ROSENKRANTZ, HARRIS, biochem. pharmacologist, educator, cons.; b. Bklyn., Mar. 23, 1922; s. Abraham and Miriam (Heller) E.; m. Natalee Faye Rosenkrantz, May 19, 1951; children: Elliot Dale,

Mark Steven. A.B., Bklyn. Coll., 1943; M.S., N.Y. U., 1946, Cornell U. Med. Sch., 1948; Ph.D., Tufts U. Med. Sch., 1952. Research fellow Worcester Found. Exptl. Biology, Shrewsbury, Mass., 1951-52, staff scientist, 1952-59; spl. lectr. in biochemistry Clark U., Worcester, Mass., 1955-58, asso. prof. biology, Worcester, 1959-62, prof. biochemistry, 1963—; adj. prof. toxicology dept. comparative medicine Tufts U. Sch. Vet. Medicine, 1981—; dir. biochemistry, dir. biochem. pharmacology EG&G Mason Research Inst., Worcester, 1959—, v.p., 1961—. Contbr. articles to sci. publs., chpts. to books. Recipient Admiral Earle award Worcester Engring. Soc., 1956. Mem. Am. Soc. Biol. Chemists, AAAS, Am. Chem. Soc., Am. Inst. Chemists, Am. Soc. Pharmacology and Exptl. Therapeutics, Coblentz Soc., Endocrine Soc., N.Y. Acad. Scis., Soc. Toxicology, Am. Physiol. Soc., Sigma Xi. Democrat. Jewish. Subspecialties: Biochemistry (medicine); Toxicology (medicine). Current work: Drugs of abuse; drugs to treat iron overload; contraceptive and fertility drugs. Home: 136 S Flagg St Worcester MA 01602

ROSENKRANZ, HERBERT S., microbiologist; b. Vienna, Austria, Sept. 27, 1933; came to U.S., 1948, naturalized, 1954; s. Samuel and Lea Rose (Marilles) R.; m. Deanna Eloise Green, Jan. 27, 1959; children: Pnina Gail, Eli Joshua, Marguerite E., Dara V., Jeremy Emil, Sara C., Naomi, Cynthia. B.S., CCNY., 1954; Ph.D., Cornell U., 1959. Predoctoral fellow Sloan-Kettering Inst. Cancer Research, N.Y.C., 1954-59, postdoctoral research fellow, 1959-60, research assoc., 1960; research assoc. dept. biochemistry Sch. Medicine, U. Pa., Phila., 1960-61; asst. prof. dept. microbiology Columbia U., N.Y.C., 1961-65, assoc. prof., 1965-69, prof., 1969-76; prof., chmn. dept. microbiology, prof. pediatrics N.Y. Med. Coll., Valhalla, 1976-81; cons. pathology Westchester County Med. Ctr., Valhalla, 1976-81; prof. epidemiology and community health, biochemistry and radiology, dir. Ctr. Environ. Health Scis., Case Western Res. U. Sch. Medicine, Cleve., 1981—; vis. prof. microbiology Hebrew U.-Hadassah Med. Sch., 1971-72; mem. panel on carcinogenicity and mutagenicity Nat. Cancer Inst., 1976—; cons. U.S. EPA, 1977—; mem. Internat. Commn. Protection Against Environ. Mutagens and Carcinogens, 1978—; mem. sci. rev. panel health research U.S. EPA, 1980—; cons. Internat. Agy. Research on Cancer WHO, 1980—. Mem. editorial bd.: Mutation Research, 1976—, Mutation Research Letters, 1980—, Environmental Mutagenesis, 1981—. Recipient Faculty Summer Research award Lalor Found., 1963, Career Devel. award USPHS, 1965-75, Spl. Corp. Recognition award Xerox Corp., 1980; named Aaron Bendich Meml. lectr. Cornell U. Med. Coll., 1980; Ochs-Adler scholar, 1954-56; Alfred P. Sloan Found. predoctoral fellow, 1956-59; Nat. Cancer Inst. postdoctoral fellow, 1959-60. Mem. AAAS, Am. Assn. Cancer Research, Am. Chem. Soc., Am. Soc. Biol. Chemists, Am. Soc. Microbiology, Corp. Marine Biol. Lab., Enzyme Club, Harvey Soc., Soc. Gen. Physiologists, Environ. Mutagen Soc., Am. Soc. Photobiology, Infectious Diseases Soc. Am., Genetic Soc. Am., Sigma Xi. Subspecialties: Cancer research (medicine); Molecular biology. Current work: Causes and prevention of cancer. Office: Case Western Res U Sch Medicine 2119 Abington Rd Cleveland OH 44106

ROSENKRANZ, ROBERTO PEDRO, pharmacologist; b. Mexico City, Mar. 30, 1950; s. George and Edith. A.B. in Psychology, Stanford U., 1971; Ph.D. in Comparative Pharmacology/Toxicology, U. Calif.-Davis, 1980. Neurobiologic researcher Instituto Nacional de Neruologia, Mex., 1971-72; Mexican del. Internat. Group on Drug Legis. and Programs, Geneva, 1971-73; dir. research Centro Mexicano de Estudios en Farmacodependencia, 1972-73; research fellow dept. medicine Stanford (Calif.) U., 1980-82; staff research Syntex Labs., Palo Alto, Calif., 1982—; cons.; pres. Lic. Luis Echeverria Alvarez. Contbr. articles on pharmacology to profl. jours. Mex. del. Joint U.S.-Mex. Exec. Conf. on Drug Abuse Planning, 1972; Med. del. UN Social Def. Research Inst., Rome, 1971-72. Mem. AAAS, N.Y. Acad. Scis., Soc. Neurosci., Am. Soc. Pharmacology and Exptl. Therapeutics, Western Pharmacology Soc., Internat. Soc. Study of Xenobiotics. Subspecialties: Pharmacology; Neuropharmacology. Current work: Pharmacology of conjugated vasoactive amines, calcium antagonists. Office: Syntex Research 3401 Hillview Palo Alto CA 94305

ROSENTHAL, HAROLD LESLIE, physiological chemist, educator; b. Elizabeth, N.J., Mar. 26, 1922; s. Isadore and Sophia (Shapiro) R.; m. Rose Schwartz, June 6, 1948; children: Jenifer Ann, Pamela Susan. B.S. in Biology and Chemistry, U. N.Mex., 1944; Ph.D., Rutgers U., 1951. Instr. Tulane U., New Orleans, 1951-53; chief biochemist Rochester (N.Y.) Gen. Hosp., 1953-58; prof. Washington U., St. Louis, 1958—; vis. prof. Minerva Found., Helsinki, Finland, 1966-67, Inst. Nutrition, Budapest, Hungary, 1973-74. Author: Protein Metabolism, 1960, The Vitamins, 1968, Radiation Biology, 1969, Stable Strontium, 1981. Served with USN, 1943-46. Minerva Found. fellow, Finland, 1966; Nat. Acad. Scis. vis. scientist, Hungary, 1973. Fellow Am. Inst. Chemists, AAAS; mem. Am. Soc. Biol. Chemists, Am. Chem. Soc., Soc. Exptl. Biol. Medicine, Am. Inst. Nutrition. Subspecialties: Biochemistry (biology); Biochemistry (medicine). Current work: Protein chemistry; insolation of proteins specifically for neural tissue, organ of corti and other tissues. Home: 7541 Teasdale St Louis MO 63130 Office: Washington U 4559 Scott Ave St Louis MO 63110

ROSENTHAL, KENNETH STEVEN, microbiology/immunology educator; b. Bklyn., July 23, 1951; s. Joseph and Muriel (Tirschler) R.; m. Judith E. Lindner, Dec. 27, 1975. B.S., U. Del., 1973; M.A., U. Ill., 1976, Ph.D. in Biochemistry, 1977. Research asst. U. Ill.-Urbana, 1974-77; postdoctoral fellow Sidney Farber Cancer Inst., Boston, 1977-79; asst. prof. microbiology/immunology N.E. Ohio U. Coll. Medicine, Rootstown, 1979—. Mem. Profl. edn. com., bd. dirs. dist. IV Am. Cancer Soc., Ohio, 1982. Grantee Am. Cancer Soc., 1981—, Nat. Cancer Inst., 1980-83; Am. Cancer Soc. fellow, 1977-79. Mem. N.Y. Acad. Sci., Am. Soc. Microbiology, Sigma Xi (nominating com. 1982—), Alpha Chi Sigma, Phi Lambda Upsilon (chpt. pres. 1976-77). Subspecialties: Virology (biology); Membrane biology. Current work: Research on interaction of Herpes Simplex Virus with its host: infection, immunity and cancer. Office: NE Ohio U Coll Medicine State Route 44 Rootstown OH 44272

ROSENZWEIG, MARK RICHARD, psychology educator; b. Rochester, N.Y., Sept. 12, 1922; s. Jacob Z. and Pearl (Grossman) R.; m. Janine S.A. Chappat, Aug. 1, 1947; children: Anne Janine, Suzanne Jacqueline, Philip Mark. B.A., U. Rochester, 1943, M.A., 1944; Ph.D., Harvard U., 1949; hon. doctorate, U. René Descartes, Sorbonne, 1980. Postdoctoral research fellow Harvard U., 1949-51; asst. prof. U. Calif., Berkeley, 1951-56, asso. prof., 1956-60, prof. psychology, 1960—, asso. research prof., 1958-59, research prof., 1965-66; vis. prof. biology Sorbonne, Paris, 1973-74; v.p. Internat. Union Psychol. Sci. Author: Biologie de la Mémoire, 1976, (with A.L. Leiman) Physiological Psychology, 1982; Editor: (with P. Mussen) Psychology: An Introduction, 1973, 2d edit., 1977, (with E.L. Bennett) Neural Mechanisms of Learning and Memory, 1976, (with L. Porter) Ann. Rev. of Psychology, 1968—; Contbr. articles to profl. jours. Served with USN, 1944-46. Fulbright research fellow; faculty research fellow Social Sci. Research Council, 1960-61; research grantee NSF, USPHS, Easter Seal Found. Fellow Am. Psychol. Assn. (Disting. Sci. Contbn. award 1982), AAAS; mem. Nat. Acad. Sci., Am. Physiol. Soc., Internat. Brain Research Orgn., Soc. Exptl. Psychologists, Soc. for Neuroscience, Société Française de Psychologie, NAACP (life), Common Cause, Phi Beta Kappa, Sigma Xi. Subspecialties: Learning; Neuropsychology. Office: Dept Psychology U Calif Berkeley CA 94720

ROSHKO, ANATOL, educator, aeronautical scientist; b. Bellevue, Alta., Can., July 15, 1923; came to U.S., 1950, naturalized, 1957; s. Peter and Helen (Macen) R.; m. Aydeth de Santa Ritta Seitz, Dec. 8, 1956; children: Peter, Tamara. B.Sc., U. Alta., Can., 1945; M.S., Calif. Inst. Tech., 1947, Ph.D., 1952. Instr. math. U. Alta., 1945-46, lectr. engring., 1949-50; research fellow Calif. Inst. Tech., Pasadena, 1952-55, asst. prof. aeronautics, 1955-58, assoc. prof., 1958-62, prof., 1962—; sci. liaison officer Office Naval Research, London, Eng., 1961-62, U.S.-India exchange scientist, Bangalore, 1969, cons. McDonnell-Douglas Corp., 1956-79, Rockwell Internat., 1981—; mem. exec. bd. Wind Engring. Research Council; mem. U.S. Nat. Commn. on Theoretical and Applied Mechanics. Author: (with H.W. Liepmann) Elements of Gasdynamics, 1957. Served with Canadian Army, 1945. Fellow AIAA, Canadian Aeros. and Space Inst., Am. Acad. Arts and Scis., Am. Phys. Soc.; mem. Arctic Inst. N.Am., Nat. Acad. Engring. Subspecialties: Aeronautical engineering; Fluid mechanics. Current work: Experimental research on turbulent mixing and physical structure of turbulent flows. Research in gasdynamics, separated flow and turbulence. Home: 3130 Maiden Ln Altadena CA 91001 Office: 1201 E California St Pasadena CA 91125

ROSI, FRED DAVID, materials scientist, educator; b. Meriden, Conn., Jan. 13, 1921; s. Umberto and Elvira (Tiezzi) R.; m. Frances DeBogory, Dec. 15, 1970; children: Catherine Bree Merrick, Wynne DeBogory Jillson, Deborah Horton Ekholm, Margaret Weaver Jillson. B.E., Yale U., 1942, M.E., 1947, Ph.D., 1949. Engkr. Sylvania Electric Products, Bayside, N.Y., 1949-54; staff v.p., dir. materials research lab. RCA Labs., Princeton, N.J., 1954-72; dir. research and devel. Am. Can Corp., Barrington, Ill., 1973-75; with Reynolds Metals Co., Richmond, Va., 1975-78; exec. dir. Energy Policy Studies Ctr., U. Va., Charlottesville, 1978—; prof. Sch. Engring. and Applied Sci., U. Va., 1978—; mem. research adv. com. NASA, 1960-65; mem. adv. com. dept. physics U. Ky., 1968-71; mem. adv. bd. St. Croix Exptl. Sta., Fairleigh Dickinson U., 1969-71; mem. Nat. Materials Adv. Bd., 1971-75; mem. com. electronic materials Nat. Acad. Sci., 1970-71; cons. NASA, CIA, ITT, 1979—; mem. sci. adv. com. Gould Inc., 1978—. Editorial bd.: IRI, 1968-71; mem. hon. adv. bd.: Jour. Materials Sci, 1965-72; contbr. articles to profl. jours. Chmn. bd. trustees Trenton State Coll., 1967-69; vice-chmn. N.J. Council State Colls., 1967-69; mem. N.J. Bd. Higher Edn., 1967-69, Va. Coal and Energy Commn., 1979—. Served to lt. USNR, 1946-49. Fellow Am. Phys. Soc., Am. Inst. Metall. Engrs. (David Sarnoff Gold medal 1963); mem. AAAS, Sigma Xi. Roman Catholic. Club: Westwood Racquet (Richmond). Subspecialties: Electronic materials; 3emiconductors. Current work: Materials characterization, management of industrial research and development, energy policy. Patentee electronic materials and devices (12). Home: 8903 Tolman Rd Richmond VA 23229 Office: U Va Charlottesville VA 22901

ROSKOSKI, ROBERT, JR., biochemistry educator; b. Elyria, Ohio, Dec. 10, 1939; d. Robert and Mary (Rudnicki) R.; m. Laura Martinsek, Aug. 27, 1974. B.S., Bowling Green State U., 1961; M.D., U. Chgo., 1964, Ph.D., 1968. Postdoctoral fellow U. Chgo., 1964-66, Rockefeller U., N.Y.C., 1969-72; asst. prof. biochemistry U. Iowa, Iowa City, 1972-75, assoc. prof., 1975-79; prof., head dept. biochemistry La. State U. Med. Ctr., New Orleans, 1979—; cons. test. com. Nat. Bd. Med. Examiners. Contbr. articles to profl. jours. Served to capt. USAF, 1966-69. Mem. Am. Chem. Soc., Am. Soc. Biol. Chemists, Am. Soc. Pharmacology and Exptl. Therapeutics, Am. Soc. Neurochemistry, Internat. Soc. Neurochemistry, N.Y. Acad. Sci. Democrat. Subspecialties: Neurochemistry; Biochemistry (medicine). Current work: neurochemistry and cardiovascular biochemistry. Home: 1206 Aline St New Orleans LA 70115 Office: 1901 Perdido St New Orleans LA 70112

ROSNER, ANTHONY LEOPOLD, research chemist, lab. director; b. Greensboro, N.C., Nov. 13, 1943; s. Albert Aaron and Elsie (Lincoln) R.; m. Ruth Frances Rosner, June 19, 1966; 1 dau., Rachel Elizabeth. B.S., Haverford Coll., 1966; Ph.D., Harvard U., 1972. NINCDS postdoctoral fellow NIH, Bethesda, Md., 1972-73, staff fellow, 1973-74; vis. investigator molecular biology CNRS, Gif sur Yvette, France, 1973; research fellow Cancer Research Center, Tufts U. Med. Sch., Boston, 1974-75; research fellow pathology Beth Israel Hosp., Boston, 1975-76; research assoc. in pathology, 1976—, asst. in pathology, 1981—, tech. dir. chemistry lab., 1981—, dir. estrogen receptor lab., 1976—. Andelot fellow, 1966-67; Haverford Coll. scholar, 1964-66. Mem. Am. Chem. Soc., AAAS, Am. Soc. Microbiology, Am. Soc. Cell Biology, Am. Assn. Clin. Chemistry, N.Y. Acad. Scis. Subspecialties: Clinical chemistry; Receptors. Current work: Biochemical control mechanisms in estrogen target tissue and in tumor cells arising in estrogen target tissue. Office: 330 Brookline Ave Boston MA 02215

ROSNER, JUDAH LEON, biologist; b. Bklyn., Oct. 18, 1939; s. Max Harry and Rose Ruth (Langbaum) R.; m. Grace Lockett, June 20, 1965; 1 son; Ari S. A.B., Columbia U., 1960; M.S., Yale U., 1964, Ph.D., 1967. Guest worker Bookhaven Nat. Lab., Upton, N.Y., 1961; teaching asst. dept. biology Yale U., 1960-62; lab. asst. Marine Biol. Lab., Woods Hole, Mass., summer 1964; guest lectr. biochem. genetics NIH, 1967; scientist USPHS Lab. Molecular Biology, NIH, Bethesda, Md., 1967-70, staff fellow, 1970-74, research biologist, 1974—. Mem. Am. Soc. Microbiology. Jewish. Subspecialties: Genome organization; Molecular biology. Current work: Genetics of bacteria and bacteriophage; transposons; bacteriophage Pl; bacterial mutants. Home: 4400 Illinois Ave NW Washington DC 20011 Office: NIH Bldg 2 Room 210 Bethesda MD 20205

ROSNOW, RALPH L(EON), psychology educator, researcher; b. Balt., Jan. 10, 1936; s. Irvin and Rebecca (Faber) R.; m. Marion Audrey Quin, Aug. 12, 1963. B.S., U. Md., 1957; M.A., George Washington U., 1958; Ph.D., Am. U., 1962. Asst. prof. Boston U., 1963-67; assoc. prof. Temple U., Phila., 1967-70, prof., 1970-82, Thaddeus L. Bolton prof., 1982—; Gen. series editor Oxford Univ. Press, N.Y.C., 1975-82; vis. prof. London Sch. Econs., 1973, Harvard U., 1974. Author: (with K. Craik et al) New Directions in Psychology IV, 1970; author: (with Robert E. Lana) Introduction to Contemporary Psychology, 1972, (with Robert Rosenthal) The Volunteer Subject, 1975, Primer of Methods for the Behavioral Sciences, 1975, Japanese edit., 1976, (with Gary Alan Fine) Rumor and Gossip: The Social Psychology of Hearsay, 1976, Japanese edit., 1982, Paradigms in Transition: The Methodology of Social Inquiry, 1981, (with Robert Rosenthal) Essentials of Behavioral Research: Methods and Data Analysis, 1984; editor: (with Edward J. Robinson) Experiments in Persuasion, 1967, (with Robert Rosenthal) Artifact in Behavioral Research, 1969, (with Robert E. Lana) Readings in Contemporary Psychology, 1972, Reconstruction of Society Series, 4 vols, 1975-77. Grantee NSF, 1966-73, NIH, 1964-66. Fellow AAAS, Am. Psychol. Assn.; mem. Soc. Exptl. Social Psychology, AAUP, Eastern Psychol. Assn. Subspecialty: Social psychology. Current work: Conceptual and empirical work on the social psychology of research and the scientific method. Other recent work has been on (a) the social psychology of rumor and (b) on the nature of behavior that is not what it purports to be, e.g., synthetic benevolence and malevolence. Home: 177 Biddulph Rd Radnor PA 19087 Office: Temple U Philadelphia PA 19122

ROSS, DENNIS KENT, physics educator; b. Hebron, Nebr., May 4, 1942; s. Gerald Murray and Ida Esther (Menke) R.; m. Roberta Graber, Dec. 26, 1965; 1 son: David. B.S., Calif. Inst. Tech., 1964; Ph.D., Stanford U., 1968. Instr. physics Iowa State U., Ames, 1968-69, asst. prof., 1969-72, assoc. prof., 1972-79, prof., 1979—. Contbr. articles to profl. jours. Woodrow Wilson fellow, 1964; Danforth Found. fellow, 1964-68. Mem. Am. Phys. Soc., Internat. Astron. Union. Subspecialties: Relativity and gravitation; Theoretical physics. Current work: Unified renomalizable gauge theories which include the gravitational field. Home: 1411 Curtiss St Ames IA 50010 Office: Dept Physics Iowa State Ame IA 50011

ROSS, DOUGLAS TAYLOR, software company executive, educator; b. Canton, China, Dec. 21, 1929; s. Robert Malcolm and Margaret (Taylor) R.; m. Patricia Mott, Jan. 24, 1951; children: Jane Louise, Kathryn Ross Chow, Margaret Ross Thrasher. A.B., Oberlin Coll., 1951; S.M., MIT, 1954; postgrad., 1954-58. Head computer applications group Electronic Systems Lab., MIT, Cambridge, 1952-69, lectr. elec. engring., 1960-69, lectr. elec. engring. and computer sci., 1983—; pres. SofTech, Inc., Waltham, Mass., 1969-75, chmn. bd., 1975—; chmn. evaluation panel for Inst. Computer Scis. and Tech. of Nat. Bur. Standards, 1978-83, chmn., 1981-83. Mem. editorial bd.: Software: Practice and Experience, 1971—, IEEE Transactions on Software Engring, 1975-79; mem. editorial adv. bd.: Computers in Industry, 1978—; contbr. articles to profl. jours. Mem. Lexington (Mass.) Town Meeting, 1960-70. Recipient Joseph Marie Jacquard award Numerical Control Soc., 1975; Disting. Contbn. award Soc. Mfg. Engrs., 1980; named Outstanding Young Man of Yr. Greater Boston Jaycees, 1959; Hon. Engr. of Yr. San Fernando Valley Engrs. Council, 1981. Mem. Assn. Computing Machinery, AAAS, Internat. Fedn. Info. Processing, Sigma Xi. Mem. United Ch. of Christ. Subspecialties: Systems analysis; Software engineering. Current work: PLEX philosophy of structure, structured analysis, user-oriented systems design and engineering. Home: 33 Dawes Rd Lexington MA 02173 Office: 460 Totten Pond Rd Waltham MA 02154

ROSS, ELLIOTT M(ORTON), biochemist, pharmacology educator; b. Stockton, Calif., Jan. 16, 1949; s. William D. and Hannah L. R.; m. Phyllis B., Aug. 11, 1973; 1 dau., Sharon M. B.S., U. Calif.-Davis, 1970; Ph.D., Cornell U., 1975. Postdoctoral fellow U. Va., Charlottesville, 1975-77, vis. asst. prof., 1977-78, asst. prof., 1978-81; assoc. prof. U. Tex. Health Sci. Ctr., Dallas, 1981—; Established investigator Am. Heart Assn., 1979—. Mem. Am. Soc. Biol. Chemists, Am. Soc. Pharmacology and Exptl. Therapeutics. Democrat. Jewish. Subspecialties: Biochemistry (biology); Molecular pharmacology. Current work: Research related to adenylate cyclase, cell surface receptors, function and interaction of membrane-bound proteins. Office: Univ Tex Health Sci Center Dept Pharmacology 5323 Harry Hines Blvd Dallas TX 75235

ROSS, HENRY A., horticulturist, park director; b. Cleve., Dec. 7, 1926; s. John and Frances (Dobin) R. B.Sc. in Horticulture, Ohio State U., 1949. Founder, dir. Gardenview Hort. Park, Strongsville, Ohio, 1949—. Served in U.S. Army, World War II. Recipient Conservation Achievement award Ohio Dept. Natural Resources, 1975; Pres.'s award Internat. Lilac Soc., 1980; profl. citation Am. Hort. Soc., 1981. Subspecialty: Resource conservation. Current work: Collection, introduction and display of rare and uncommon plants; plant breeding. Patentee flower varieties. Address: 16711 Pearl Rd Strongsville OH 44136

ROSS, HILDY S(HARON), psychology educator; b. Toronto, Ont., Can., Mar. 31, 1944; d. Jack and Kate (Kates) Strashin; m. Michael Arthur Ross, June 12, 1967; children: Corey, Jordan, Ilana. B.A., U. Toronto, 1967; Ph.D., U. N.C.-Chapel Hill, 1972. Asst. prof. psychology U. Waterloo, Ont., 1972-77, assoc. prof., 1977-83, prof., 1983—; lectr. features and functions of infant games Can. Psychol. Assn., 1976. Contbr. articles to profl. jours., 1972—; editor: Peer Relationships and Social Skills in Childhood, 1982. Can. Council grantee, 1972, 73, 76; Social Scis. and Humanities Research Council Can. grantee, 1979,83. Mem. Am. Psychol. Assn., Can. Psychol. Assn., Soc. Research in Child Devel. Jewish. Subspecialty: Developmental psychology. Current work: Peer social relations and social skills in infants and toddlers. Office: Dept Psychology U Waterloo Waterloo ON Canada N2L 3Gl

ROSS, HUGH COURTNEY, consulting electrical engineer; b. Turlock, Calif., Dec. 31, 1923; s. Clare W. and Jeanne (Pierson) R.; m. Sarah A., Dec. 16, 1950; children: John, James, Robert. Student, Calif. Inst. Tech., 1942; student, San Jose State Coll., 1946-47; B.S.E.E., Stanford U., 1950, postgrad. in high voltage, 1951. Registered profl. engr., Calif. Instr. San Benito High Sch. and Jr. Coll., 1950-51; chief engr. Jennings Radio Mfg. Corp., San Jose, Calif., 1951-62, ITT Jennings, San Jose, 1962-64; chief engr., owner Ross Engring. HV (high voltage) Cons., Saratoga, Calif., 1964—; pres., chief engr. Ross Engring. Corp., Campbell, Calif., 1964—. Served with USAAF, 1943-46. Fellow IEEE; mem. Am. Soc. Metals, Am. Vacuum Soc. Subspecialties: Electrical engineering; Electronics. Current work: Energy sources and generation; consulting, design, development and production of high voltage devices and energy sources. Patentee in field. Home: 11915 Shadybrook Saratoga CA 95070 Office: 559 Westchester Dr Campbell CA 95008

ROSS, IAN MUNRO, electrical engineer; b. Southport, Eng., Aug. 15, 1927; came to U.S., 1952, naturalized, 1960; m. Christina Leinberg Ross, Aug. 24, 1955; children: Timothy Ian, Nancy Lynn, Stina Marguerite. B.A., Gonville and Caius Coll., Cambridge (Eng.) U., 1948; M.A. in Elec. Engring, Cambridge U., 1952, Ph.D., 1952; D.Sc. (hon.), N.J. Inst. Tech., 1983, D.Engring., Stevens Inst. Tech., 1983. With Bell Telephone Labs., Inc. (and affiliates), 1952—, exec. dir. network planning div., 1971-73, v.p. network planning and customer services, 1973-76, exec. v.p. systems engring. and devel., Murray Hill, N.J., 1976-79, pres., 1979—; dir. Gulton Industries, Inc., Thomas & Betts Corp., B.F. Goodrich Co. Author. Recipient Liebmann Meml. prize IEEE, 1963; Pub. Service award NASA, 1969. Fellow IEEE, Am. Acad. Arts and Scis.; mem. Nat. Acad. Engring. Subspecialties: Systems engineering; Electrical engineering. Patentee in field. Home: 5 Blackpoint Horseshoe Rumson NJ 07760 Office: Crawfords Corner Rd Holmdel NJ 07733

ROSS, J(OHN) B(RANDON) ALEXANDER, biochemistry educator; b. Suffern, N.Y., Feb. 18, 1947; s. John and Alice Elizabeth (Hall) R.; m. Laurie Malia Franklin, Nov. 17, 1973. B.A., Antioch Coll., 1970; Ph.D., U. Wash., 1976. Research assoc. dept. chemistry U. Wash., 1976-78, sr. fellow dept. lab. medicine, 1980-82; sr. research scientist dept. biology Johns Hopkins U., 1978-80; asst. prof. biochemistry Mt. Sinai Sch. Medicine, N.Y.C., 1982—. Mem. Am. Chem. Soc., Am. Soc. Photobiology, Biophys. Soc., Am. Soc. Biol. Chemists, Sigma Xi. Subspecialties: Biophysical chemistry; Spectroscopy. Current work: Time-resolved, excited-state spectroscopy of biological molecules and membranes; molecular interactions and conformational dynamics. Office: Dept Biochemistry Mt Sinai Sch Medicine One Gustave L Levy Pl New York NY 10029

ROSS, JOHN MUNDER, psychologist; b. N.Y.C., June 20, 1945; s. Nathaniel and Barbara (Munder) R.; m. Katherine Wren Ball, Aug. 17, 1974; 1 son, Matthew Munder Ball. B.A. magna cum laude, Harvard U., 1967; M.A., NYU, 1973, Ph.D., 1974; postgrad., London Sch. Econs., 1969-70. Asst. prof. Ferkauf Grad. Sch., N.Y.C., 1976-78; vis. asst. prof. Albert Einstein Coll. Medicine, Bronx, 1977-78; clin. asst. prof. Downstate Med. Ctr., Bklyn., 1978-80; clin. instr. Cornell Med. Coll., N.Y.C., 1978-80, adj. asst. prof., 1980—; clin. assoc. prof. psychiatry Downstate Med. Ctr., 1980—; cons. Child Devel. Research, Port Washington, N.Y., 1977-79; mem. adv. bd. Bank St. Fatherhood Project, N.Y.C., 1981—; bd. dirs. Primary Prevention, Detroit, 1981—; pvt. practice psychotherapy, N.Y.C., 1974—. Contbr. articles to profl. jours.; co-editor: Father and Child Developmental and Clinical Perspectives, 1982. Recipient Detur prize, John Harvard and Harvard Coll. scholarships, 1963-67; NIMH fellow, 1967-68, 69-71; Leverhulme fellow Lond Sch. Econs., 1968-69; Disting. Tchr. award Downstate Med. Sch., 1979, 80; award for outstanding book in behavioral scis. Am. Assn. Pubs., 1982. Mem. Am. Psychol. Assn., Am. Psychoanalytic Assn., N.Y. Psychol. Assn., World Assn. for Infant Psychiatry. Democrat. Clubs: Harvard (N.Y.C.); Signet Soc. (Cambridge, Mass.). Current work: Research in parenthood and father's role in child development; sexuality in Eastern and Western literature. Home: 277 West End Ave New York NY 10023 Office: 243 West End Ave New York NY 10023

ROSS, MARY HARVEY, entomology educator; b. Albany, N.Y., Apr. 1, 1925; d. Roy Newman and Myrtle Adele (King) Harvey; m. Robert Donald Ross, Dec. 20, 1947; children: Mary Jane, Robert Douglas, Nancy Lee. B.A. in Geology with distinction, Cornell U., 1946, M.A., 1947, Ph.D., 1950. Biologist Oak Ridge Nat. Lab., 1950-51; instr. Va. Poly. Inst. and State U., 1959-70, asst. prof. entomology, 1970-74, assoc. prof., 1974-80, prof., 1980—; adj. prof. N.C. State U., 1978-80. Contbg.: author Handbook of Genetics, vol. 3, 1975; contbr. numerous articles to profl. publs. NSF grantee, 1971-74, 80-83; Office Naval Research grantee, 1975-83. Mem. Am. Genetics Assn. (council 1977, sec. 1978-82), Entomol. Soc. Am., Genetics Soc. Am., Washington Entomol. Soc., Phi Beta Kappa, Sigma Xi, Phi Kappa Phi, Phi Sigma. Republican. Presbyterian. Subspecialties: Animal genetics; Entomology. Current work: Research on cockroaches: genetics, cytogenetics, biology, behavior. Home: 614 Airport Rd Blacksburg VA 24060 Office: Dept Entomology Va Poly Inst and State U Blacksburg VA 24061

ROSS, MORRIS H., nutritionist, consultant; b. Phila., Jan. 3, 1917; s. Abraham and Jennie (Freedman) R.; m. Anita Luterman, Oct. 30, 1949; children: Steven, Kevin, Eric. V.M.D., U. Pa., 1943. Jr. biologist U.S. Dept. Agr., Angleton, Tex., 1943-44; chief biologist Biochem. Research Found., Newark, Del., 1944-67; sr. mem. Inst. Cancer Research, Phila., 1967—; lit. reviewer Franklin Inst., Phila., 1979-83. Contbr. articles to publs. in field. Recipient J.B. Allison award Rutgers U., 1970, research award Am. Aging Assn., 1975, Presdl. citation Manhattan Project, 1946. Fellow Gerontol. Soc.; Mem. Am. Inst. Nutrition. Subspecialties: Nutrition (biology); Cancer research (medicine). Current work: Relationship of nutrition to cancer susceptibility and to the aging process. Home: 4285 Thistlewood Rd Hatboro PA 19040 Office: Inst Cancer Research Fox Chase Philadelphia PA 19111

ROSS, NORTON MORRIS, research executive, dental pharmacology educator; b. Rockville, Conn., Sept. 29, 1925; m. Rosalyn Zamkow, June 19, 1949; children: Elaine, Carol, Kenneth, Neil. B.S., U. Conn., 1949; D.D.S., U. Md., 1954; M.A., Loyola Coll., Balt., 1967. Assoc. prof. pharmacology U. Md. Sch. Dentistry, Balt., 1954-67; asst. dir. med. research Squibb Inst. Med. Research, New Brunswick, N.J., 1957-69; assoc. prof. pharmacology Med. Coll. Ga., Augusta, 1970-73; prof., chmn. dept. pharmacology Fairleigh Dickinson U., Hackensack, N.J., 1973-78; dir. oral clin. research Warner-Lambert Co., Morris Plains, N.J., 1978—. Author: Lab Manual of Pharmacology, 1966; Contbr. chpts. to textbooks. Served with USN, 1943-46. Fellow Am. Acad. Oral Medicine (pres. 1971-72, Diamond Pin award 1975), Am. Coll. Dentists; mem. Internat. Assn. Dental Research (various offices 1973—), N.J. Health Sci. Group (chmn. 1980-82), AAAS. Subspecialties: Pharmacology; Preventive dentistry. Current work: Director of dental/oral clinical research to support safety and efficiacy of consumer oral health products. Office: Warner-Lambert Co 170 Tabor Rd Morris Plains NJ 07950

ROSS, RICHARD STARR, medical school dean, physician, cardiologist; b. Richmond, Ind., Jan. 18, 1924; s. Louis Francisco and Margaret (Starr) R.; m. Elizabeth McCracken, July 1, 1950; children: Deborah Starr, Margaret Casad, Richard McCracken. Student, Harvard U., 1942-44, M.D. cum laude, 1947; Sc.D. (hon.), Ind. U., 1981. Diplomate: Nat. Bd. Med. Examiners, Am. Bd. Internal Medicine. (subsplty. bd. cardiovascular disease). Successively intern, asst. resident, chief resident Osler Med. Service, Johns Hopkins Hosp., 1947-54; research fellow physiology Harvard Med. Sch., 1952-53; instr. medicine Johns Hopkins Med. Sch., 1954-56, asst. prof. medicine, 1956-59, asso. prof., 1959-65, asso. prof. radiology, 1960-71, prof. medicine, 1965—, Clayton prof. cardiovascular disease, 1969-75; dir. Wellcome Research Lab., Johns Hopkins; physician Johns Hopkins Hosp.; dir. cardiovascular div. dept. medicine, adult cardiac clinic Johns Hopkins Sch. Medicine and Hosp., dir. myocardial infarction research unit, 1967-75; dean med. faculty, v.p. medicine Johns Hopkins U., 1975—; Sir Thomas Lewis lectr. Brit. Cardiac Soc., 1969; John Kent Lewis lectr. Stanford U., 1972; Connor lectr. Am. Heart Assn., 1979; dir. Balt. Life Ins. Co., Waverly Press; mem. cardiovascular study sect. Nat. Heart and Lung Inst., 1965-69, chmn., 1966-69, mem. tng. grant com., 1971-73, chmn. heart panel, 1972-73, adv. council, 1974-78; bd. dirs. Union Meml. Hosp., 1972-76; mem. Inst. of Medicine, 1976—; chmn. vis. com. Harvard Med. and Dental Sch., 1979—; bd. overseers Harvard U., 1980—. Contbr. 120 articles on cardiovascular disease and physiology to med. jours.; editor: Modern Concepts Cardiovascular Disease, 1961-65, The Principles and Practice of Medicine, 17th-21th edits, 1968-83; editorial bd.: Circulation, 1968-74; editorial com.: Jour. Clin. Investigation, 1969-73. Served as capt. M.C. AUS, 1949-51. Master A.C.P.; fellow Am. Coll. Cardiology; mem. Boylston Med. Soc., Am. Fedn. Clin. Research, Am. Physiol. Soc., Assn. Am. Physicians (master 1979), Am. Soc. Clin. Investigation (councillor 1967-69), Am. Clin. and Climatol. Assn. (pres. 1978-79, councillor 1979-83), Assn. Univ. Cardiologists (councillor 1972-75), Am. Heart Assn. (chmn. sci. sessions program com. 1965-67, chmn. publs. com. 1970-73, pres. 1973-74, dir. 1974-77, Gold Heart award 1976, James B. Herrick award 1982), Heart Assn. Md. (pres. 1967-68), Sigma Xi, Alpha Omega Alpha; corr. mem. Brit. Cardiac Soc., Sociedad Peruana de Cardiologie, Cardiac Soc. Australia and New Zealand. Clubs: Peripatetic, Interurban (pres. 1978), Elkridge, 14 West Hamilton, Johns Hopkins. Subspecialty: Cardiology. Home: 214 Wendover Rd Baltimore MD 21218 Office: 720 Rutland Ave Baltimore MD 21205

ROSS, RUSSELL, pathologist, educator; b. St. Augustine, Fla., May 25, 1929; s. Samuel and Minnie (DuBoff) R.; m. Jean Long Teller, Feb. 22, 1956; children: Valerie Regina, Douglas Teller. A.B., Cornell U., 1951; D.D.S., Columbia U., 1955; Ph.D., U. Wash., 1962. Intern Columbia-Presbyn. Med. Center, 1955-56, USPHS Hosp., Seattle, 1956-58; spl. research fellow pathology U. Wash. Sch. Medicine, 1958-62; asst. prof. pathology and oral biology U. Wash. Sch. Medicine and Dentistry, 1962-65, asso. prof. pathology, 1965-69, prof., 1969—, adj. prof. biochemistry, 1978—; asso. dean for sci. affairs Sch. Medicine, 1971-78, chmn. dept. pathology, 1982—; vis. scientist Strangeways Research Lab., Cambridge, Eng.; mem. research com. Am. Heart Assn.; mem. adv. bd. Found. Cardiologique Princess Liliane, Brussels, Belgium; vis. fellow Clare Hall, Cambridge U.; Guggenheim fellow, 1966-67; mem. adv. council Nat. Heart, Lung and Blood Inst., NIH, 1978-81. Mem. editorial bd.: Proceedings Exptl. Biology and Medicine, 1971—; Mem. editorial bds.: Jour. Cell Biology, 1972-74; assoc. editor: Arteriosclerosis; contbr. articles in arteriosclerosis research and wound healing to profl. jours. Recipient Birnberg Research award Columbia U., 1975, Gordon Wilson medal Am. Clin. and Climatol. Assn., 1981. Mem. Am. Soc. Cell Biology, Tissue Culture Assn., Gerontol. Soc., Am. Assn. Pathologists, Internat. Soc. Cell Biology, Electron Microscope Soc. Am., Am. Heart Assn. (fellow Council on Arteriosclerosis), Royal Micros. Soc., AAAS, Am. Soc. Biol. Chemists, Belgian Acad. Medicine (fgn. corr. mem.), Sigma Xi. Subspecialties: Cell biology (medicine); Pathology (medicine). Current work: Cell and molecular biology, experimental pathology, atherscierosis, growth control of cells, connective tissue metabolism, inflammation, and cell proliferation. Home: 4811 NE 42d St Seattle WA 98105 Office: U of Wash Sch Medicine SM-30 Seattle WA 98195 I have always believed in open, direct communication in society as well as in science, softened with as much tact and sensitivity as possible, but uncompromising in so far as honesty is concerned. A modicum of intelligence combined with a great deal of hard work will go as far as a great deal of brilliance.

ROSS, WILLIAM NOEL, physiology educator, medical researcher; b. Bklyn., Mar. 18, 1942; s. Norman and Martha (Cohen) R.; m. Nechama Lasser, Oct. 24, 1978; children: Jonathan, Karen. B.A., Columbia U., 1962, M.A., 1964, Ph.D., 1968. Research assoc. dept. physics U. Calif., Berkeley, 1968-71; postdoctoral fellow dept. physiology Yale U. Med. Sch., New Haven, 1972-75; dept. neurobiology Harvard Med. Sch., Boston, 1975-78; asst. prof. physiology N.Y. Med. Coll., Valhalla, 1979—. Irma T. Hirschl career scientist awardee, 1980—. Subspecialties: Neurobiology; Biophysics (biology). Current work: Optical studies of neurons; properties of dendrites and presynaptic terminals. Office: Dept Physiology NY Med Coll Valhalla NY 10595

ROSSA, ROBERT FRANK, mathematics educator, researcher; b. Kankakee, Ill., Aug. 17, 1942; s. Frank Louis and Launa Elizabeth (Lovejoy) R.; m. Dean Shelton, May 31, 1969; 1 dau., Jennifer. B.A., U. Okla., 1963, M.A., 1966, Ph.D., 1971. Asst prof. Ark. State U., State University, 1969-71, assoc. prof., 1971—. Mem. Am. Math. Soc., Math. Assn. of Am., Assn. for Computing Machinery, IEEE, Phi Beta Kappa, Sigma Xi. Mem. Christian Ch. (Disciples of Christ). Current work: Radical theory in nonassociative rings. Home: 1901 Starling Dr Jonesboro AR 72401 Office: PO Box 151 State University AR 72467

ROSSEN, ROGER DOWNEY, immunologist, educator; b. Cleve., June 4, 1935; s. Joseph McKinley and Maguerite Harriet (Downey) R.; m. Katherine Mary Bosanquet, Apr. 15, 1961 (div.); children: Adam Charles, Roger Christopher, Justin Frederick; m. Holly Hyde Birdsall, Apr. 17, 1980. B.A., Yale U., 1957; M.D., Western Reserve U., 1961. Diplomate: Am. Bd. Allergy and Immunology. Intern in Internal Medicine Columbia U. Coll. Physicians and Surgeons, N.Y.C., 1961-63, resident in Internal Medicine, 1966-67; assoc. to clin. investigator lab. clin. investigations Nat. Inst. AID, NIH, Bethesda, Md., 1963-66; instr. Baylor Coll. Medicine, Houston, 1967-68, asst. prof., 1968-71, assoc. prof., 1971-73, prof., 1973—; mem. faculty Grad. Sch. Biomed. Scis., U. Tex., 1975-82; faculty cons. M.D. Anderson Hosp. and Tumor Inst., U. Tex.; chief clin. immunology VA Med. Ctr., Houston; cons. NIH. Grantee NIH, VA. Mem. Am. Assn. Immunologists, So., Central socs. clin. investigation, Am. Soc. Microbiology, Soc. Exptl. Biology and Medicine, Transplantation Soc., Infectious Diseases Soc. Am., Am. Soc. Pathologists. Subspecialties: Immunobiology and immunology; Allergy. Current work: immune compex diseases; factors mediating host resistance to infectious agents; tumor immunology especially the effect of antibodies and immune complexes on host defenses against tumors, transplantation immunology and allergic diseases. Home: 2624 Glen Haven Houston TX 77025 Office: Baylor College of Medicine 1200 Moursund Ave Houston TX 77030

ROSSI, BRUNO, physicist; b. Venice, Italy, Apr. 13, 1905; s. Rino and Lina (Minerbi) R.; m. Nora Lombroso, Apr. 10, 1938; children— Florence S., Frank R., Linda L. Student, U. Padua, 1923-25, U. Bologna, 1925-27; hon. doctorate, U. Palermo, 1964, U. Durham, Eng., 1974, U. Chgo., 1977. Asst. physics dept. U. Florence, 1928-32; prof. physics U. Padua, 1932-38; research asso. U. Manchester, Eng., 1939; research asso. in cosmic rays U. Chgo., 1939-40; asso. prof. physics Cornell U., 1940-43; prof. physics Mass. Inst. Tech., 1946—, Inst. prof., 1966-70, Inst. prof. emeritus, 1970—; Mem. staff Los Alamos Lab.; 1943-46, hon. fellow Tata Inst. Fundamental Research, Bombay, India, 1971; mem. physics com. NASA; hon. prof. U. Mayor, San Andres, La Paz, Bolivia. Author: Rayons Cosmiques, 1935, (with L. Pincherle) Lezioni di Fisica Sperimentale Elettrologia, 1936, Lezioni di Fisica Sperimentale Ottica, 1937, (with Staub) Ionization Chambers and Counters, 1949, High Energy Particles, 1952, Optics, 1957, Cosmic Rays, 1964, (with S. Olbert) Introduction to the Physics of Space, 1970. Recipient Cresson medal Franklin Inst., 1974; decorated Order of Merit, Republic of Italy). Mem. Am. Acad. Arts and Scis. (Rumford prize 1976), Nat. Acad. Sci. (space sci. bd., astronomy survey com.), Deutsche Akademieder Naturforscher Leopoldina, Am. Phys. Soc., Am. Inst. Physics, Accademia dei Lincel (Internat. Feltrinelli award 1971), Internat. Astron. Union, Am., Royal astron. socs., Accademia Patavina di Scienze, Letteree Arti, Accademia Ligure di Scienze e Lettere, Bolivian Acad. Scis. (corr.), A.A.A.S., Am. Philos. Soc., Italian Phys. Soc. (Gold medal 1970), Sigma Xi. Subspecialty: Cosmic ray high energy astrophysics. Address: 221 Mt Auburn St Cambridge MA 02138

ROSSI, JOHN JOSEPH, molecular geneticist; b. Washington, July 8, 1946; s. Oscar L. and Mina F. (Cappelletti) R.; m. Mary Jane, Sept. 30, 1969; children: Christina M., John M., Kathleen M. B.A. in Biology, U. N.H., 1969; M.S. in Genetics, U. Conn., Storrs, 1971; Ph.D. in Microbial Genetics, U. Conn., Storrs, 1976. NIH postdoctoral fellow div. biology and medicine Brown U., Providence, 1976-78, univ. fellow, 1978-80; asst. research scientist dept. molecular genetics City of Hope Research Inst., Duarte, Calif., 1981—. Contbr. chpts. to books, articles to sci. jours. Chmn. Powell Sch. Site Council, Azusa, Calif., 1982—. Served to 1st.lt. USAF, 1971-73. Mem. AAAS, Am. Soc. Microbiology, Phi Beta Kappa, Phi Kappa Phi, Sigma Xi. Democrat. Roman Catholic. Subspecialties: Gene actions; Genome organization. Current work: Mechanisms of gene regulation, prokaryotic promoters, eukaryotic promoters, splicing, transfer RNA gene orgn. and expression. Office: 1450 E Duarte Rd Kaplan Black Bldg Duarte CA 91010

ROSSI, RAYMOND PAUL, diagnostic medical physicist, educator; b. Chgo., Aug. 30, 1945; s. Frank and Nellie (Cecchi) R.; m. Harriet F., Aug. 7, 1967; children: Keith, Stefani. B.S., Loyola U., Chgo., 1967; M.S., DePaul U., 1969. Diplomate: Am. Bd. Radiology. Teaching asst. dept. physics DePaul U., Chgo., 1967-69; health physicist, dept. radiation therapy Michael Reese Hosp. & Med. Center, Chgo., 1969-72, instr. med. physics, 1971-75; instr. radiology med. physics U. Colo. Health Scis. Center, Denver, 1975-78, asst. prof., 1978—. Contbr. articles in field to profl. jours. Mem. Am. Assn. Physicists in Medicine, Am. Inst. Physics, Health Physics Soc., Radiol. Soc. N. Am., Am. Coll. Radiology, Soc. Radiol. Engrng. (dir. 1981—). Subspecialties: Imaging technology; Diagnostic radiology. Current work: Teaching image receptor development and evaluation. Office: University of Colorado 4200 E Ninth Ave C-278 Denver CO 80262

ROSSING, THOMAS HARRY, physician; b. Detroit, Feb. 18, 1950; s. Robert Grangaard and Dolores Evelyn (Christenson) R.; m. Jolyn Gayle Giese, Dec. 30, 1971. B.A., U. Tex., 1971; M.D., Harvard U., 1976. Diplomate: Am. Bd. Internal Medicine (Pulmonary Medicine). Intern Peter Bent Brigham Hosp., Boston, 1976-77, resident, 1977-79; dir. med. intesive care Brigham and Women's Hosp., Boston, 1981—. Mem. Am. Thoracic Soc., Am. Fedn. Clin. Research, ACP, Am. Coll. Chest Physicians, Soc. Critical Care Medicine. Subspecialties: Internal medicine; Pulmonary medicine. Current work: High frequency ventilation, asthma. Home: 96 Fair Oaks Ave Newton MA 02160 Office: Brigham and Women's Hosp 75 Francis St Boston MA 02115

ROSSINGTON, DAVID RALPH, engring. educator; b. London, July 13, 1932; s. George Leonard and Clara Fanny (Simmons) R.; m. Angela Mae Reynolds, Sept. 3, 1955; children: Andrew, Carolyn, Nicholas, Philip. B.Sc. with honors in Chemistry, U. Bristol, Eng., 1953, Ph.D. in Phys. Chemistry, 1956. Research fellow N.Y. State Coll. Ceramics, 1956-58; tech. officer Imperial Chem. Industries Ltd., Eng., 1958-60; asst. prof. N.Y. State Coll. Ceramics, 1960-63, assoc. prof., 1963-69, prof., 1969—, head div. engring. sci., 1982—. Contbr. articles to profl. jours. Justice Town of Alfred, N.Y., 1976—. Recipient SUNY Chancellors award for Excellence in Teaching, 1976. Mem. Am. Ceramic Soc., Nat. Inst. Ceramic Engrs., Am. Chem. Soc. Subspecialties: Ceramic engineering; Physical chemistry. Current work: Surface reactions; adsorption of gases on solids; catalysis; nuclear waste disposal. Home: 14 High St Alfred NY 14802 Office: NY State College of Ceramics Alfred U Alfred NY 14802

ROSSINI, FREDERICK DOMINIC, scientist, educator, consultant; b. Monongahela, Pa., July 18, 1899; s. Martino and Costanza Carra R.; m. Anne K. Landgraff, June 29, 1932 (dec. Dec. 1981); 1 son, Frederick Anthony. B.S., Carnegie Inst. Tech., (now Carnegie Mellon U.), 1925, M.S., 1926, D.Sc. (hon.), 1948; Ph.D., U. Calif.-Berkeley, 1928, ; D.Engr. Sci., Duquesne U., 1955; D.Sc., U. Notre Dame, 1959, Loyola U., Chgo., 1960, U. Portland, 1965; Litt.D., St. Francis Coll., Loretto, Pa., 1962; Ph.D., U. Lund, Sweden, 1974. Lab. asst. in physics Carnegie Inst. Tech., 1923-23, teaching asst. math., 1924-26, silliman prof., 1950-60, head dept. chemistry, 1950-60, dir., 1950-60; teaching fellow chemistry U. Calif.-Berkeley, 1926-28; phys. chemist Nat. Bur. Standards, 1928-36, chief sect. on thermochemistry and hydrocarbons, 1936-50; prof. chemistry U. Notre Dame, 1960-71, dean, 1960-67, assoc. dean grad. sch., 1960-67, v.p. research, 1967-71, prof. emeritus, 1971—; prof. chemistry Rice U., 1971-75, prof. emeritus in residence, 1975-78, prof. emeritus, 1978—; mem. panel on research materials and info. U.S.-Japan Coop. Sci. program, 1965; mem. adv. panel for chemistry NSF, 1951-54; mem. policy adv. bd. Argonne Nat. Lab., 1958-66; trustee State Ind. Ednl. Service Found., 1968-71; mem. tech. adv. com. State Ill. Bd. Higher Edn., 1970-71; chmn. environ. measurements adv. com. EPA, 1967-78, mem. sci. adv. bd. exec. com., 1976-78, cons., 1978-79; lectr. and cons. in field. Author: 11 books Including Thermochemistry of Chemical Substances, 1936, Hydrocarbons from Petroleum, 1953, Thermodynamics and Physics of Matter, 1955, Experimental Thermochemistry, 1956, Fundamental Measures and Constants for Science and Tehnology, 1974; contbr. numerous articles to profl. jours. Recipient Gold Medal exceptional Service award U.S. Dept. Commerce, 1950; award in chemistry Pitts. Jr. C. of C., 1957; Laetare medal U. Notre Dame, 1965; John Price Wetherill medal Franklin Inst., 1965; Carl Engler medal Deutsche Gesellschaft fur mineralolwissenschaft und Kohlechemie, Germany, 1976; Nat. Medal Sci. U.S.A., 1977. Fellow AAAS, Am. Inst. Chemists (life), Am. Phys. Soc., Washington Acad. Sci. (editor jour. 1937-40, sec. 1940-43, bd. mgrs. 1943-46, pres. 1948); mem. Nat. Acad. Sci., Am. Acad. Arts and Scis., Am. Chem. Soc. (mem. editorial bd. jour. 1946-56, chmn. com. constitution and bylaws 1949-50, mem. adv. bd. petroleum research fund 1954, 57-59), Am. Inst. Chem. Engrs., Am. Petroleum Inst., Am. Soc. Engring. Edn., Chem. Soc. London, Geochem. Soc., Philos. Soc. Washington, Albertus Magnus Guild, Cath. Assn. Internat Peace, Sigma Xi, Phi Kappa Theta (found. bd. trustees 1963-66, 73-76), Phi Lambda Upsilon, Tau Beta Pi. Republican. Roman Catholic. Club: Cosmos (Washington). Subspecialties: Physical chemistry; Fuels. Current work: Thermochemistry and thermodynamics; data for science and technology; physical chemistry of hydrocarbons and petroleum.

ROSSMAN, GEORGE ROBERT, mineralogist; b. LaCrosse, Wis., Aug. 3 1944; m. (m). Ph.D., Calif. Inst. Tech., 1971. Assoc. prof. mineralogy Calif. Inst. Tech., Pasadena, 1977—. Contbr. articles to profl. jours. Fellow Mineral. Soc. Am. Subspecialties: Mineralogy; Inorganic chemistry. Current work: Properties of minerals, optical, color; spectroscopy of minerals and related synthetic materials. Office: Dept Geol and Planetary Scis Calif Inst Tech Pasadena CA 91125

ROSSMAN, ISADORE, gerontologist; , 1943; 1 child. B.A., U. Wis., 1933; Ph.D. in Anatomy, U. Chgo., 1937, M.D., 1942. Med. dir. dept. home care and extended service Montefiore Hosp., N.Y.C., 1954—; assoc. prof. Albert Einstein Coll. Medicine, Bronx, N.Y., 1970-80, prof. community health, 1980—. Mem. Am. Geriatrics Soc. (pres.), Gerontol. Soc. Am. (Joseph Freeman award 1981). Subspecialty: Gerontology. Office: Montefiore Hosp 111 E 210th St New York NY 10467

ROSSMAN, TOBY G., genetic toxicologist, educator; b. Weekhawken, N.J., June 3, 1942; d. Norman N. and Sylvia B. (May) Natowitz; m. Neil I. Rossman, Sept. 16, 1962 (div. 1983). A.B. in Biology (univ. scholar), NYU, 1964, Ph.D. in Microbiology, 1968; postgrad. in Biochemistry, Brandeis U., 1964-65. Postdoctoral dept. pathology NYU Med. Center, N.Y.C., 1969-71, assoc. research scientist, dept. environ. medicine, 1971-73, asst. prof., 1974-78, assoc. prof., 1978—; cons. in field; reviewer tng. grants Nat. Inst. Environ. Health Scis. Contbr. articles to profl. publs. NIH predoctoral and postdoctoral trainee, 1964-71; research grantee EPA, 1981—, NIH, 1977—. Mem. Environ. Mutagen Soc., AAAS, Am. Soc. Microbiology, Am. Soc. Cell Biology, N.Y. Acad. Sci., Assn. Women in Sci., Beta Lambda Sigma. Subspecialties: Genetics and genetic engineering (medicine); Toxicology (medicine). Current work: Geneticseffects of chems. and radiation; genetic toxicology of metals, bisulfite, ascorbate; comutageness, systems and mechanisms; mutagenic specificity. Home: 102-19 66th Rd 16K Forest Hills NY 11375 Office: 550 First Ave New York NY 10016

ROST, THOMAS LOWELL, botany educator; b. St. Paul, Dec. 28, 1941; s. Lowell Henry Rost and Agnes Marie (Wojtowicz) Rost J.; m. Ann Marie Ruhland, Aug. 31, 1963; children: Christopher, Timothy, Jacquelyn. B.S., St. John's U., Collegeville, Minn., 1963; M.A., Mankato State U., 1965; Ph.D., Iowa State U., 1970. Postdoctoral fellow Brookhaven Nat. Lab., Upton, N.Y., 1970-72; asst. prof. botany U. Calif., Davis, 1979-80, faculty asst. to chancellor, 1982—, assoc. prof. botany, 1977—; vis. fellow Research Sch. Biol. Scis., Australian Nat. U., Canberra, 1979-80; FAO cons. Faculty Agronomy, U.

Uruguay, Montevideo, summer 1979. Co-author: Botany, A Brief Introduction to Plant Biology, 1979, Botany, An Introduction to Plant Biology, 1981, Laboratory Studies in Botany, 1982; contbr. articles to profl. jours.; co-editor: Mechanisms and Control of Cell Division, 1977. Served to capt. USAR, 1965-67. NSF grantee, 1978. Fellow Royal Micros. Soc.; mem. Bot. Soc. Am., Am. Inst. Biol. Sci., Sigma Xi, Gamma Sigma Delta. Subspecialties: Cell and tissue culture; Plant growth. Current work: Effects of stress on cell division in plant root meristems, aspects of morphogenesis in plants especially root growth and development, seed anatomy and histochemistry. Office: Dept Botany U Calif Davis CA 95616

ROTA, GIAN CARLO, educator, mathematician; b. Vigevano, Italy, Apr. 27, 1932; came to U.S., 1950, naturalized, 1961; s. Giovanni and Gina (Facoetti) R.; m. Teresa Rondón-Tarchetti, June 23, 1956 (div. 1979). B.A. summa cum laude, Princeton U., 1953; M.A., Yale U., 1954, Ph.D., 1956. Vis. fellow Courant Inst. Math. Scis., N.Y.U., 1956-57; B. Pierce instr. Harvard, 1957-59; asst. prof., then asso. prof. math. M.I.T., 1959-65, prof. math., 1967-74, prof. applied math. and philosophy, 1975—; prof. math. Rockefeller U., 1965-67; vis. prof. U. Calif. at Berkeley, 1961, Courant Inst. Math. Scis., 1964, U. Ill., 1965, Ind. U., 1964, U. Paris, 1972, U. Mex., 1973, Scuola Normale Superiore, 1975, U. Buenos Aires, 1975, U. Strasbourg, 1976, 78; Taft lectr. U. Cin., 1971; spl. vis. prof. U. Colo., 1969-82; Andre' Aisenstadt vis. prof. U. Montreal, 1971; Hardy lectr. London Math. Soc., 1973; professore linceo, Rome, 1979; cons. Rand Corp., 1965-71; Sigma Xi nat. lectr., 1980-81; fellow Los Alamos Sci. Lab., 1966—; cons. Brookhaven Nat. Lab., 1969-71. Author: (with G. Birkhoff) Ordinary Differential Equation, 1962, (with H. Crapo) Combinational Geometries, 1970, Finite Operator Calculus, 1975, MAA Survey in Combinatorics, 1978; also articles combinatorial theory, differential equations, probability, philosophy.; Editor: Jour. Combinatorial Theory, 1966—, Jour. Math. and Mechanics, 1965-71, Jour. Math. Analysis, 1966—, Utilitas Mathematica, 1973—; asso. editor: Am. Math. Monthly, 1966-73; (with H. Crapo) Procs. Royal Soc. Edinburgh, 1976—, Advances in Mathematics, 1967—; editor in chief: Advances in Applied Math, 1980—; editor: Bull. Am. Math. Soc, 1967-73, 79-84, (with H. Crapo) Ency. of Math, 1974—. Sloan fellow, 1962-64. Fellow Am. Acad. Arts and Scis., Academia Argentina de Ciencias, Inst. Math. Statistics, Nat. Acad. Scis., AAAS; mem. Am. Math. Assn. (Hedrick lectr. 1967), Am. Math Soc. (council 1967-72), London Math. Soc., Heidegger Circle, Soc. Indsl. and Applied Math. (v.p. 1975), Am. Philos. Assn., Soc. Phenomenology and Existential Philosophy, AAUP, Unione Mathematica Italiana, Phi Beta Kappa (pres. chpt. 1978-80). Roman Catholic. Subspecialties: Parasitology; Cognition. Current work: Combinatiorics, probability, logic. phenomenology, especially Husserl. Address: 2-351 Mass Inst Tech Cambridge MA 02139

ROTENBERG, A. DANIEL, cons. scientist; b. Toronto, Ont., Can., July 21, 1934; s. Meyer and Mattie (Levi) R.; m. Gita Leah Segal, Apr. 1, 1938; children: Miriam, Meir, Noam, Dahna. Ph.D., U. Toronto, 1962. Biophysicist Montreal (Que.) Gen. Hosp., 1963-69; physicist Jewish Gen. Hosp., Montreal, 1969-77; dir. research and devel. Coinamatic Inc., Montreal, 1977—; dir. BRRCM Inc. (Cons.), 1966. Fellow Can. Coll. Physicists in Medicine; mem. Can. Assn. Physicists, Am. Assn. Physicist in Medicine. Club: Town of Mt. Royal Curling. Subspecialties: Biophysics (physics); Physics consulting. Current work: Microprocessor devel., ionizing radiation safety. Home: 6616 Fleet Rd Montreal Canada PQ H4V 1A9

ROTH, DANIEL, pathologist; b. N.Y.C., Oct. 27, 1920; s. Samuel Joseph and Anna (James) R.; m. Roslyn Weintraub, Oct. 6, 1950; children: David Mark, Sally Michele. A.B., Columbia U., 1940; M.D., N.Y. Med. Coll., 1943. Diplomate: Am. Bd. Anat. Pathology, Am. Bd. Clin. Pathology. Intern Met. Hosp., N.Y.C., 1944; resident in pathology Mary Immaculate and Univ. hosps., N.Y.C. and; Grasslands Hosp., Valhalla, N.Y., 1947-50; asst. pathologist United Hosp., Port Chester, N.Y., 1951-56; dir. labs. Bergen Pines Hosp., Paramus, N.J., 1959-61, St. Barnabas Hosp., Bronx, N.Y., 1962-67; asst. prof. pathology NYU Med. Ctr., N.Y.C., 1956-59, assoc. prof., 1967-73; pathologist Goodwin Inst., Plantation, Fla., 1974—. Contbr. numerous sci. articles to profl. publs. Served to capt. AUS, 1944-46. Grantee Nat. Inst. Dental Health, 1967, N.Y.C. Health Dept., 1970. Mem. N.Y. Acad. Scis., Am. Assn. Pathologists, Am. Soc. Photobiology, Environ. Mutagen Soc., AAAS. Jewish. Subspecialties: Cancer research (medicine); Pathology (medicine). Current work: DNA denaturation due to free-radicals, its role in carcinogenesis, its reversal by antioxidants. Patentee cancer screening, 1970. Home: 14 Cornwall Ln Port Washington NY 11050 Office: Goodwin Inst 1850 NW 69th Ave Plantation FL 33313

ROTH, JACK ALAN, thoracic surgeon, oncologist; b. LaPorte, Ind., Jan. 29, 1945; s. Richard and Bernice (Saperstein) R.; m. Elizabeth Ann Grimm, Nov. 25, 1978. B.A. magna cum laude, Cornell U., 1967; M.D., Johns Hopkins U., 1971. Diplomate: Am. Bd. Surgery, Am. Bd. Thoracic Surgery. Intern, then resident in surgery Johns Hopkins Hosp., 1971-73; postgrad. research fellow UCLA Med. Sch., 1973-75; resident divs. gen. surgery and thoracic surgery, 1975-80; lectr. surgery Johns Hopkin Sch. Medicine, 1981—; sr. investigator surgery br. Nat. Cancer Inst., 1980—, head thoracic oncology sect., surgery bd., 1982—. Author papers, reports in field. Served to capt. M.C. USAR, 1972-78. Recipient Mead Johnson award, 1980; also various best paper awards. Mem. Am. Assn. Immunologists; mem. Am. Assn. Cancer Research; Mem. Assn. Acad. Surgeons; mem. Am. Soc. Clin. Oncology, Soc. Surg. Oncology; Mem. N.Y. Acad. Scis. Subspecialties: Cancer research (medicine); Cellular engineering. Current work: Devel. of hybridoma monoclonal antibodies to solid tumor-associated antigens, purification of tumor produced substances and growth factors that regulate cell division. Office: NIH Bldg 10 Room 10N116 Bethesda MD 20205

ROTH, JAMES FRANK, manufacturing company executive, chemist; b. Rahway, N.J., Dec. 7, 1925; s. Louis and Eleanor R.; m. Sharon E. Mattes, June 20, 1969; children by previous marriage: Lawrence, Edward, Sandra. B.A. in Chemistry, U. W.Va., 1947; Ph.D. in Phys. Chemistry, U. Md., 1951. Research chemist Franklin Inst., Phila., 1951-53, mgr. chemistry lab., 1958-60; chief chemist Lehigh Paints & Chems. Co., Allentown, Pa., 1953-55; research chemist GAF Corp., Easton, Pa., 1955-58; with Monsanto Co., St. Louis, 1960-80, dir. catalysis research, 1973-77, dir. process sci. research, 1977-80; corp. chief scientist Air Products and Chems., Inc., Allentown, 1980—. Contbr. articles to profl. jours.; editorial bd.: Jour. Catalysis, 1976—, Catalysis Revs, 1973—, Applied Catalysis, 1981—. Served with USN, 1943-46. Recipient Richard J. Kokes award Johns Hopkins U., 1977. Mem. Am. Chem. Soc. (St. Louis sect. St. Louis award 1975, E.V. Murphree nat. award 1976), Catalysis Soc. N. Am., Nat. Acad. Engring., Catalysis Club of Phila. (award 1981). Subspecialties: Catalysis chemistry; Synthetic chemistry. Inventor process biodegradable detergents, for acetic acid; U.S., fgn. patents in field. Office: PO Box 538 Allentown PA 18105

ROTH, JAMES LUTHER AUMONT, physician, educator; b. Milw., Mar. 8, 1917; s. Paul Wagner and Rose Marie (Schulzke) R.; m. Marion S. Main, June 7, 1938; children—Stephen Andrew, Kristina Marie, Lisa Kathryn. B.A., Carthage Coll., 1938, D.Sc. (hon.), 1957; M.A., U. Ill., 1939; M.D., Northwestern U., 1944, Ph.D., 1945. Intern Mass. Gen. Hosp., Boston, 1944-45; resident (Grad. Hosp.), 1945-46, 49-50, Hosp. U. Pa., Phila., 1948-49; practice medicine specializing in gastroenterology, Phila., 1950—; instr. physiology Northwestern U., 1942-44; instr. div. gastroenterology Grad. Sch. Medicine, U. Pa., 1950-52, asso., 1952-54, asst. prof., 1954-56, asso. prof., 1956-59, clin. prof. gastroenterology, 1959-68, dir. div., 1961-69, prof. clin. medicine, 1968—, chief gastroenterology clinic, 1953-67, chief gastroenterology service, 1961-69; dir. Inst. Gastroenterology, Presbyn. U. Pa. Med. Center, 1965—; cons. USN, Bethesda, Md., 1967—, USAF, 1946-48. Editorial bd.: Gastroenterology, 1960-67, Am. Jour. Gastroenterology, 1976-79, Current Therapy, 1972—, Current Concepts in Gastroenterology, 1976-81; editor: Bockus' 4th edit. Gastroenterology; Contbr. articles to med. jours. Recipient Bronze medal AMA, 1944; certificates of merit Am. Roentgen Ray Soc., AMA, 1958; Disting. Alumni award Carthage Coll., 1979. Fellow A.C.P.; mem. AMA, Pan Am. Med. Assn., N.Y. Acad. Sci., Am. Physiol. Soc., Am. Fedn. Clin. Research, Am. Gastroent. Assn. (chmn. admissions com.), Am. Coll. Gastroenterology (trustee 1977—, chmn. grad. edn. com. 1978-81, v.p. 1981), Bockus Internat. Soc. Gastroenterology (pres. 1973-75), Digestive Disease Fedn. (dir.), Union League Phila., Sigma Xi, Alpha Omega Alpha; hon. mem. Fla., Colombian, Venezuelian, Dominican Republic socs. gastroenterology. Republican. Lutheran. Subspecialty: Gastroenterology. Office: 51 N 39th St Philadelphia PA 19104

ROTH, JEROME ALLAN, research pharmacologist; b. N.Y.C., Aug. 20, 1943; s. Fred and Lillian (Klapwald) R.; m. Jane Reiss, June 22, 1969; children: Rachel Eva, Evan Mathew. M.Nutritional Sci., Cornell U., 1967, Ph.D. in Biochemistry, 1971. Asst. prof. anesthesiology Yale U., New Haven, 1972-76; asst. prof. pharmacology SUNY, Buffalo, 1976-79, assoc. prof., 1979—. Mem. Am. Soc. Pharmacology and Exptl. Therapeutics, Am. Soc. Neurochemistry, Internat. Soc. Neurochemistry. Subspecialties: Neurochemistry; Neuropharmacology. Current work: Disposition of biogenic amine in human brain; monoamine oxidase, catechol-o-methyltransferase, phenol sulfotrasnferase, properties and function of brain enzymes. Office: SUNY 127 Farber Hall Buffalo NY 14214

ROTH, JESSE, endocrinologist; b. N.Y.C., Aug. 5, 1934. B.A., Columbia U., 1955; M.D., Albert Einstein Coll. Medicine, Yeshiva U., 1959; hon. degree, U. Uppsala, Sweden, 1980. Intern Barnes Hosp., Washington U., St. Louis, 1959-60, resident, 1960-61; Am. Diabetes Assn. research fellow radioisotope service Bronx (N.Y.) VA Hosp., 1961-63, clin. asso., 1963-65, sr. investigator, 1965-66, chief diabetes sect., clin. endocrinology br., 1966-74; chief diabetes br. Nat. Inst. Arthritis, Diabetes, Digestive and Kidney Diseases, Bethesda, Md., 1974—, dir. div. research, 1983—; regents' lectr. U. Calif., 1977; G. Burroughs Mider lectr. NIH, 1978. Recipient Eli Lilly award Am. Diabetes Assn., 1974; Ernst Oppenheimer Meml. award Endocrine Soc., 1974; Spl. Achievement award HEW, 1974; David Rumbough Meml. award Juvenile Diabetes Found., 1977; Lita Annenberg Hazen award Mt. Sinai Sch. Medicine; Diaz Cristobal Found. prize Internat. Diabetes Found., 1979; Annual award Gairdner Found., 1980; Disting. Service medal Public Health Service, 1980. Subspecialties: Endocrinology; Evolutionary biology. Current work: Receptors; hormones, neurotransmitters, insulin, human disease, evolution. Office: Diabetes Br Nat Inst Arthritis, Diabetes, Digestive and Kidney Diseases 9000 Rockville Pike Bethesda MD 20205

ROTH, J(OHN) REECE, physics educator; b. Washington, Pa., Sept. 19, 1937; s. John Meyer and Ruth E. (Iams) R.; m. Helen Marie De Crane, Jan. 14, 1972; children: Nancy Ann, John Alexander. S.B. in Physics, Mass. Inst. Tech., 1959; Ph.D. in Engring. Physics, Cornell U., 1963. Aerospace research scientist NASA Lewis Research Center, Cleve., 1963-78, prin. investigator, 1967-78; vis. prof. elec. engring. U. Tenn., Knoxville, 1978-82, prof. physics, 1982—; cons. Westinghouse Corp., Pitts., 1980, TVA, Chattanooga, 1982—. Assoc. editor: IEEE Transactions on Plasma Sci, 1973—; contbr. numerous sci. articles to profl. publs. Recipient Awareness award NASA, 1976; Sloan scholar MIT, 1955-59; Ford fellow Cornell U., 1961-62. Fellow IEEE (mem. exec. com. 1974-77, 80-82, bd. dirs., mem. exec. com. East Tenn. chpt. 1982-85); mem. IEEE Nuclear and Plasma Scis. Soc. (sec. 1975), Am. Nuclear Soc., AIAA, Am. Phys. Soc., AAAS (life), Archaeol. Inst. Am., Am. Soc. Engring. Edn., Sigma Xi (life). Subspecialties: Plasma physics; Nuclear fusion. Current work: Electric field dominated plasmas; alternate magnetic confinement concepts; advanced fusion fuel cycles; public policy issues of fusion energy; plasma-wall interactions; fusion materials testing. Home: 4301 Hiawatha Dr Knoxville TN 37919 Office: U Tenn 409 Ferris Hall Knoxville TN 37996-2100

ROTH, LAWRENCE MAX, pathologist; b. McAlester, Okla., June 25, 1936; s. Herman Moe and Blanch (Brown) R.; m. Anna Berit Katarina Sundstrom, Apr. 3, 1965; children: Karen Esther, David Josef. B.A., Vanderbilt U., 1957; M.D., Harvard Med. Sch., 1960. Diplomate: Am. Bd. Pathology. Rotating intern U. Ill.-Chgo., 1960-61; resident anat. pathology Washington U., St. Louis, 1961-64; resident clin. pathology Calif. Med. Ctr., San Francisco, 1967-68; asst. prof. pathology Tulane U. Sch. Medicine, 1968-71, assoc. prof. pathology, 1971, Ind. U. Sch. Medicine, 1971-75, prof. pathology and dir. surg. pathology, 1975—. Series editor: Contemporary Issues in Surgical Pathology, 1981—; editorial bd.: Am. Jour. Surg. Pathology, 1977, Am. Jour. Clin. Pathology, 1981—. Mem. Am. Assn. Pathologists, Internat. Acad. Pathologists, Internat. Soc. Gynecol. Pathologists, So. Med. Assn. (chmn. pathology sec. 1980-81), Ind. Assn. Pathologists (chmn. seminar com. 1976—). Subspecialty: Pathology (medicine). Current work: Pathology and ultrastructure of ovarian neoplasia. Home: 7898 Ridge Rd Indianapolis IN 46240 Office: Ind U Med Ctr 926 W Michigan St Indianapolis IN 46223

ROTH, MITCHELL GODFREY, computer scientist; b. Bangor, Maine, Oct. 19, 1951; s. Fred Warner and Dora Maxine (Hart) R.; m. Diane Lee Suskind, Aug. 5, 1979. B.S., Mich. State U., 1973, M.S., 1973; Ph.D., U. Ill., 1981. Teaching asst. dept. computer sci. U. Ill., Urbana, 1975-76, research asst., 1976-79; sr. engr. Calif. Inst. Tech., Pasadena, 1979-82; dir. graphical information systems Setpoint/Alaska Inc., Anchorage, 1982-83; assoc. prof. math. scis. dept. U. Alaska, Fairbanks, 1983—. Mem. Nat. Computer Graphics Assn., Soc. Indsl. and Applied Math., Am. Geophys. Union. Subspecialties: Graphics, image processing, and pattern recognition; Numerical analysis. Current work: Computer graphics applications to geographical information systems, computer aided design and computer mapping; numerical methods for the solution of differential equations. Office: Math Scis Dept U Alaska Fairbanks AK 99701

ROTH, ROBERT ANDREW, JR., toxicologist, educator, cons.; b. McKeesport, Pa., Aug. 15, 1946; s. Robert and Jane (Cox) R.; m. Kathleen Johnson, June 12, 1970; children: Evan, Kelly. B.A., Duke U., 1968; Ph.D., Johns Hopkins U., 1975. Diplomate: Am. Bd. Toxicology. Research fellow Yale U. Sch. Medicine, New Haven, 1975-77; asso. prof. dept. pharmacology and toxicology Mich. State U., East Lansing, 1972-76, prof., 1976—; cons. Contbr. articles to profl. jours. Served in U.S. Army, 1969-71. NIH grantee, 1977—. Mem. Am. Soc. Pharmacology and Exptl. Therapeutics, Soc. Toxicology. Subspecialties: Toxicology (medicine); Environmental toxicology. Current work: Pulmonary and hepatic toxicology; mechanism of action of toxicants; monocrotaline induced cardiopulmonary injury; lung's role in metabolism of chemicals; carbon monoxide effects on pharmacokinetics. Home: 1465 Birchwood Dr Okemos MI 48864 Office: Mich State U East Lansing MI 48864

ROTHBERG, RICHARD MARTIN, medical educator, researcher; b. N.Y.C., July 15, 1933; s. Morris and Beatrice (Jacobs) R.; m. Laura Gail Clayman, June 26, 1955; children: Benjamin, Miriam, Jonathan. B.A., U. Rochester, 1955; M.D., U. Chgo., 1958. Diplomate: Am. Bd. Pediatrics, 1966. Resident in pediatrics U. Pitts., 1958-62; research fellow in immunology Scripps Clinic and Research Found., La Jolla, Calif., 1964-66; mem. faculty U. Chgo. Med. Sch., 1966—, prof. pediatrics and pathology, 1974—. Contbr. articles to sci. jours. Served to lt. comdr., M.C. USN, 1962-64. USPHS grantee, 1967-82. Mem. Ill. Soc. Allergy and Clin. Immunology (pres. 1981-82), Am. Assn. Immunologists, Am. Pediatric Soc., Soc. Pediatric Research. Democrat. Jewish. Subspecialties: Immunology (medicine); Immunobiology and immunology. Current work: Immune response of newborn (effects of maternal environmental antigens on infant responses); immunologic reconstitution of patients with immunologic deficiency diseases. Office: Box 133 950 E 59th St Chicago IL 60637

ROTHMAN, JOHN M., clinical pharmacologist, researcher; b. N.Y.C., June 11, 1948; s. Harold and Lenora (Reiss) R.; m. Patricia Murphy, Oct. 12, 1974; 1 son, Noah Christopher. B.A., Windham Coll., 1973; M.S. in Pharmacology, Tulane U. Sch. Medicine, 1979; Ph.D., City U. Los Angeles, 1981. Clin. research assoc. dept. clin. pharmacology Revlon Health Care Group, Tuckahoe, N.Y., 1980-82; med. research assoc. interferon clin. research Schering Corp., Bloomfield, N.J., 1982-83; clin. research scientist dept. med. oncology and immunology Hoffmann-LaRoche, Inc., Nutley, N.J., 1983—. Contbr. articles to profl. jours. NIH fellow, 1976-79; Sigma Xi grantee, 1978; Am. Fedn. Clin. Research grantee, 1979. Mem. Am. Fedn. Clin. Research, AAAS, N.Y. Acad. Scis., Neuroscience Soc., Sigma Xi, Phi Chi. Subspecialties: Cancer research (medicine); Immunopharmacology. Current work: Peptides, neuroendocrine and immune pharmacology. Office: 340 Kingsland St Nutley NJ 07110

ROTHROCK, RAY ALAN, software engineer; b. Ft. Worth, Dec. 1, 1954; s. Nate Paul and Wanda Sue (Brewer) R. B.S., Tex. A&M U., 1977; S.M., MIT, 1978. Assoc. engr. Yankee Atomic Elec. Co., Westboro, Mass., 1978-80; staff engr. Exxon Minerals, Houston, 1980-82; product mgr. Sagus Corp., Campbell, Calif., 1982—; mem. subcom. Atomic Indsl. Forum, Washington, 1978-80; cons. Sagus Corp., 1982—. Vol. Am. Cancer Soc., Westboro, Mass., 1980; mem. Westboro Community Chorus, 1980; asst. scoutmaster Boy Scouts Am. Mem. Am. Nuclear Soc., Nat. Soc. Profl. Engrs. Club: Am. Radio Relay League. Subspecialties: Software engineering; Nuclear engineering. Current work: Utilization of microprocessor based systems for large scale, integrated engineering applications. Home: 14973 Natalye Rd Monte Sereno CA 95030

ROTHSCHILD, BRIAN JAMES, biology educator; b. Newark, Aug. 14, 1934; s. Daniel Leslie and Dorothy (Goodman) R.; m.; children: Elise, Jessica. B.S., Rutgers U., 1957; M.S., U. Maine, 1959; Ph.D., Cornell U., 1962, postdoctoral vis. fellow in biometrics and stats, 1965-66. Chief Skipjack-Yellowfin Tuna Ecology program Bur. Comml. Fisheries, Biol. Lab., Honolulu, 1962-67, acting dep. dir., 1967-68, sr. sci., 1968; assoc. prof. U. Wash., Seattle, 1968-71, prof., 1971; dep. ctr. dir. Nothwest Fisheries Ctr., Nat. Marine Fisheries Service, Seattle, 1971-72; dir. Southwest Fisheries Ctr., La Jolla, Calif., 1972-76, acting dir. extended juris. planning staff div., Washington, 1976; dir. Office of Policy Devel. and Long Range Planning, 1976-78; sr. policy adv. and sr. sci., Office of Adminstr. NOAA, Washington, 1978-80; prof. U. Md. Ctr. Environ. and Estuarine Studies, Chesapeake Biol. Lab., Solomons, 1980—. Contbr. articles to profl. jours. Recipient first ann. Nautilus award Marine Tech. Soc., 1980; Spl. Achievement award U.S. Dept. Commerce, 1976; Achievement award cert. NOAA, and Local 2703. Fellow AAAS; mem. Am. Inst. Fishery Research Biologists, Am. Fisheries Soc. Subspecialties: Population biology; Fishery management. Current work: Population dynamics of fishes. Office: Ctr Environ and Estuarine Studies Chesapeake Biol Lab U Md Box 38 Solomons MD 20688

ROTHSCHILD, HENRY, acad. physician, molecular biologist; b. Horstein, Germany, June 5, 1932; came to U.S., 1939, naturalized, 1945; s. William Wolf and Fanny (Hahn) R.; m.; children: Shoshana Tamar, Jamin Kahlil. B.A., Cornell U., 1954; M.D., U. Chgo., 1958; Ph.D., Johns Hopkins U., 1968. Intern U. Chgo. Clinics, 1958-59; resident in internal medicine Univ. Hosp., Balt., 1959-62; instr. medicine Mass. Gen. Hosp., Boston, 1970-71; assoc. prof. medicine La. State U. Med. Center, New Orleans, 1971-75, assoc. prof. anatomy and research, 1972-75, assoc., 1973—, prof. medicine and anatomy, 1975—; med. dir. New Orleans Home and Rehab. Center, St. Margaret's Daus. Home; vis. prof. Universidad Autonoma de Nuevo Leon, Mex., 1981. Fellow ACP; mem. Am. Soc. Exptl. Pathology, Am. Fedn. Clin. Research, So. Soc. Clin. Investigation, Soc. Exptl. Biology and Medicine. Subspecialties: Genetics and genetic engineering (medicine); Gerontology. Current work: Molecular biology, genetics of lung cancer, gerontology. Home: 705 Pine St New Orleans LA 70118 Office: 1542 Tulane Ave New Orleans LA 70112

ROTHSTEIN, JEROME, computer science educator; b. N.Y.C., Dec. 14, 1918; s. Harry and Annetta (Simenhoff) R.; m. Charlotte Ella Weinrebe, Dec. 25, 1941; children: Louise, Judith, Deborah. B.S. cum laude, CCNY, 1938; M.A., Columbia U., 1940. Registered profl. engr., Mass. With U.S. Army Electronic Research and Devel. Labs., Red Bank, N.J., 1942-57; sr. sci. exec. EG&G Inc., Boston, 1957-62; v.p., chief scientist Maser Optics, Inc., Boston, 1962-63; sr. staff scientist LFE Inc., Boston, 1963-67; assoc. prof. Ohio State U., Columbus, 1967-69, prof. computer/info. sci., 1969—; cons. Contbr. articles to profl. jours. Recipient Best Paper award Internat. Conf. Parallel Processing, 1976, Most Original Paper award, 1977. Mem. Am. Phys. Soc., IEEE, AAAS, Biophys. Soc. Jewish. Subspecialties: Biophysics (physics); Statistical physics. Current work: Theoretical biophysics, distributed computation, development of theoretical computer models for biological systems, including cognitive functions. Patentee in field. Home: 2912 Zollinger Rd Upper Arlington OH 43221 Office: Ohio State 2035 Neil Ave Mall Columbus OH 43220

ROTTENBERG, DAVID ALLAN, neurologist, researcher; b. Detroit, Jan. 8, 1942; s. Leon and Adeline (Sax) R.; m. Rochelle E. Rottenberg, June 16, 1963; children: Elizabeth Grace, Catherine Anne. B.A., U. Mich., 1963; M.Sc., Cambridge (Eng.) U., 1967; M.D., Harvard Med. Sch., 1969. Diplomate: Am. Bd. Psychiatry and Neurology, (examiner, 1980). Intern in surgery Mass. Gen. Hosp., 1969-70; research asso. NIH, 1970-72; resident in neurology N.Y. Hosp., 1972-74, chief resident, 1974-75, asst. attending neurologist, 1975—; asst. prof. neurology Cornell U. Med. Coll., 1975-79, assoc. prof., 1979—; asst. attending neurologist Meml. Hosp., 1975-79, assoc. attending neurologist, 1979—; cons. NIH, also site reviewer. Reviewer med. and sci. jours.; Contbr. articles to profl. jours. Served to lt. comdr. USPHS, 1970-72. Grantee NINCDS. Mem. Am. Acad. Neurology, AAAS, Am. Neurol. Assn., Harvey Soc., Soc. Computerized Tomography and Neuro-Imaging, Soc. Neurosci., Phi Beta Kappa. Subspecialties: Physiology (medicine); Imaging technology. Current work: Positron emission tomography of the central nervous system, quantitative aspects of CT. Office: 1275 York Ave New York NY 10021

ROTTER, JEROME ISRAEL, geneticist, internist, educator; b. Los Angeles, Feb. 24, 1949; s. Leonard L. and Jeanette (Kronenfeld) R.; m. Deborah Tofield, July 14, 1970; children: Jonathan Moshe, Amy Esther. B.S., UCLA, 1970, M.D., 1973. Intern Harbor Gen. Hosp., Torrance, Calif., 1973-74; resident in internal medicine Wadsworth VA Hosp., Los Angeles, 1974-75; fellow in med. genetics Harbor-UCLA Med. Ctr., Torrance, 1975-78, asst. research pediatrician, 1978-79, asst. prof. medicine and pediatrics med. genetics div., 1979—; key investigator Ctr. Ulcer Research and Found., Los Angeles, 1980—. Editor: Genetics and Heterogeneity of Common Gastrointestinal Disorders, 1980; editorial bd.: Metabolism Jour, 1980—. NIH fellow, 1977-78; grantee, 1979-82; Basil O'Connor Starter Research award March of Dimes, 1979-82. Fellow ACP; mem. Am. Soc. Human Genetics, Am. Fedn. Clin. Research, Am. Gastroent. Assn., Am. Diabetes Assn., Soc. Pediatric Research. Jewish. Subspecialty: Genetics and genetic engineering (medicine). Current work: Genetics of common diseases, especially the genetics of peptic ulcer disease, gastrointestinal disorders, and diabetes mellitus. Office: Harbor-UCLA Med Center Med Genetics Div 1000 W Carson St Torrance CA 90509

ROTTMAN, FRITZ, M., molecular biology-microbiology educator; b. Muskegon, Mich., Mar. 29, 1937; s. John D. and Hattie R.; m. Carol Jean VandenBosch, June 9, 1959; children: Barbara, Douglas, Susan. B.A. in Chemistry, Calvin Coll., 1959; Ph.D. in Biochemistry, U. Mich., 1963. Postdoctoral fellow USPHS and Am. Cancer Soc. NIH, Bethesda, Md., 1963-66; asst. prof. biochemistry Mich. State U., East Lansing, 1966-70, assoc. prof., 1970-74, prof., 1974-81; prof. and chmn. molecular biology and microbiology Case Western Reserve U., Cleve., 1981—; vis. prof. biochemistry U. B.C., 1974-75; mem. biochemistry study sect. NIH, 1978-81. Contbr. articles to profl. jours. Am. Cancer Soc. scholar, 1974-75. Mem. AAAS, Am. Soc. Biol. Chemists, Am. Soc. for Microbiology, Am. Chem. Soc. Subspecialties: Molecular biology; Genetics and genetic engineering (medicine). Current work: Molecular biology-regulation of gene expression at the RNA level-recombinant DNA technology and the analysis of pituitary hormone genes. Home: 28500 Gates Mills Blvd Pepper Pike OH 44124 Office: Dept Molecular Biology and Microbiology Case Western Reserve U Cleveland OH 44106

ROTZ, C. ALAN, agricultural engineer, researcher; b. Chambersburg, Pa., Sept. 7, 1951; s. Robert Lester and Margaret Ann (Horst) R.; m. Robin McCartney, Nov. 12, 1976. B.A., Elizabethtown Coll., 1974; B.S. in Mech. Engring, Pa. State U., 1974; M.S. in Agrl. Engring, Pa. State U., 1975, Ph.D., 1977. Test technician Pratt & Whitney Aircraft, East Hartford, Conn., summer 1973; grad. research asst. argl. engring. dept. Pa. State U., University Park, 1974-77; asst. prof. agrl. engring. Mich. State U., East Lansing, 1977-80; agrl. engr. U.S. Dept. Agr., East Lansing, 1981—; cons. Mich. State U., 1981. Contbr. numerous articles to profl. jours. Mem. Am. Soc. Agrl. Engrs., Alpha Epsilon, Gamma Sigma Delta. Subspecialties: Agricultural engineering; Systems engineering. Current work: Research toward the improvement of dairy forage production through the development of better harvesting systems. Office: Mich State U Agrl Engring Bldg East Lansing MI 48824

ROUFA, DONALD JAY, molecular biologist, educator; b. St. Louis, Apr. 8, 1943; s. Jack H. and Edith Merriam (Feldman) R.; m. Eileen Joyce Weiner, Feb. 8, 1967; children: Karen Shani, Andrew Harris. A.B., Amherst Coll., 1965; Ph.D., Johns Hopkins U., 1970. Research scientist NIH, Bethesda, Md., 1967-70; research scientist N.Y. State Dept. Health, Albany, 1970-71; asst. prof. biochemistry and medicine Baylor Coll. Medicine, Houston, 1971-75; assoc. prof. biology Kans. State U., Manhattan, 1975-81, prof. biology, 1981—. Contbr. numerous articles to profl. jours. Recipient USPHS Research Career Devel. award, 1976-81; Am. Cancer Soc. research grantee, 1973—; NIH grantee, 1974—; NSF grantee, 1974. Mem. Am. Soc. Biol. Chemists, AAAS, N.Y. Acad. Sci., Am. Soc. Microbiology. Jewish. Club: Manhattan Country. Subspecialties: Molecular biology; Biochemistry (biology). Current work: Somatic cell genetics; ribosomes; DNA replication; recombinant DNA technology, tissue culture; genetics. Home: 1701 Denholm Dr Manhattan KS 66502 Office: Ackert Hall Kans State U Manhattan KS 66506

ROUHANI, SAYD ZIA, thermal-hydraulics researcher; b. Nahavand, Iran, Dec. 23, 1930; came to U.S., 1979; s. Sayd-Abdel-Hossein and Maassoomeh (Zarrab) R.; m. Hosnieh Khoskebarchi, July 12, 1961; children: Anita-Helena, Viola-Goli. Electro-mech. engring. diploma, U. Tehran, 1955; cert., Internat. Sch. Nuclear Sci. and Engring., Pa. State U. and Argonne Nat. Lab., 1959; M.Sc., U. Calif.-Berkeley, 1959; Dr. Tech., Royal Inst. Tech., Stockholm, 1979. Research engr. AB Atomenergi, Studsvik, Sweden, 1960-67, group leader, Stockholm, 1967-70; project mgr. Joint Scandinavian Research, Riso, Denmark, 1970-73; cons. engr. AB Atomenergi, Studsvik, Sweden, 1973-76; project engr. Studsvik Energiteknik AB, 1976-80; sr. engring. specialist E G & G Idaho, Inc., Idaho Falls, 1980—. Co-author: Two-Phase Flow and its Application to Nuclear Reactors, 1978. Mem. Am. Nuclear Soc., ASME. Subspecialties: Fluid mechanics; Nuclear engineering. Current work: Theoretical and experimental aspects of two-phase, cyrogenics, magneto hydrodynamics and fusion power systems. Patentee in field. Home: 3028 Hartert Dr Idaho Falls ID 83401 Office: EG&G Idaho Inc PO Box 1625 Idaho Falls ID 83415

ROUSE, ROY DENNIS, ednl. adminstr.; b. Andersonville, Ga., Sept. 20, 1920; s. Joseph B. and Janie (Wicker) R.; m. Madge Mathis, Mar. 6, 1946; children—David Benjamin, Sharon. Student, Ga. Southwestern Coll., 1937-39; B.S. in Agr, U. Ga., 1942, M.S., 1947; Ph.D., Purdue U., 1949. Asst. prof. agronomy and soils Auburn (Ala.) U., 1949-50, asso. prof., 1950-56, prof., 1956-66, asso. dir., asst. dean, 1966-72, dean, dir., 1972-81, emeritus, 1981—; mem. Com. of Nine, Dept. Agr., 1970-74. Contbr. articles to profl. jours. Served to capt. USNR, 1942-67; PTO. Recipient Leadership award Farm-City Com. Ala., 1975; Disting. Service award Catfish Farmers Am., 1976, Ala. Vocat.-Agrl. Tchrs. Assn., 1976; Man of Yr. in Agr. award Progressive Farmer, 1977; named Hon. State Farmer Future Farmers Am., 1976, Man of Yr. Crop Improvement Assn., 1981, Hon. County Agt., 1981. Fellow Am. Soc. Agronomy, Soil Sci. Soc. Am.; mem. So. Assn. Agrl. Scientists (pres. 1976), Assn. So. Agrl. Expt. Sta. Dirs. (chmn. 1974), Assn. Univs. and Land-Grant Colls. (chmn. expt. sta. com. on orgn. and policy 1977), Sigma Xi, Alpha Zeta, Phi Kappa Phi, Xi Phi Xi, Gamma Sigma Delta. Presbyterian. Clubs: Lions, Men's Camellia, Outing (Auburn). Subspecialties: Soil chemistry; Plant physiology (agriculture). Current work: Calibrating chemical analyses with crop response to addition of plant nutrients and to liming of soil. Home: 827 Salmon Dr Auburn AL 36830

ROUSEK, EDWIN JOSEPH, animal science educator; b. Burwell, Nebr., Sept. 8, 1917; s. Joseph and Anne (Chadek) R.; m. Anita Underwood, May 27, 1945; children: Kathy, Sally. B.Sc., U. Nebr., 1941; M.Sc., Cornell U., 1942. Instr. Cornell U., 1940-42; Pacific Stockman, 1946-48; faculty Calif. State U.-Fresno, 1948—, now chmn. animal sci. dept.; livestock judge. Contbr. articles to profl. jours.; author monographs. Mem. Fresno County Planning Commn., 1978—. Served with AUS, 1942-46. Recipient Salgo Outstanding Tchr. award, 1972; grantee in field. Mem. Nutrition Today Soc., Meat Sci. Assn., Phi Kappa Phi, Alpha Zeta, Alpha Phi Omega. Presbyterian. Club: San Joaquin. Lodge: Rotary. Subspecialty: Animal nutrition. Current work: Animal nutrition, beef cattle. Address: Calif State U Fresno CA 93740

ROUSH, WILLIAM RICHARD, chemistry educator, consultant; b. National City, Calif., Feb. 20, 1952; s. James Chester and Julia Mae (Martin) R.; m. Rosalie Broder, Mar. 21, 1982. B.S., U.C.L.A., 1974; Ph.D., Harvard U., 1977. Asst. prof. chemistry MIT, 1978-81, Firmenich asst. prof. natural products chemistry, 1981—, assoc. prof., 1983—; cons., prin. BioInformation Assocs., Boston, 1981—. Author: Reagents for Organic Synthesis, Vol. 9, 1981; Contbr. articles to profl. jours. Sloan Found. fellow, 1982; Eli Lilly Co. grantee, 1981. Mem. Am. Chem. Soc., Phi Beta Kappa. Subspecialty: Organic chemistry. Current work: Organic chemistry; organic synthesis; natural products chemistry. Office: MIT 77 Massachusetts Ave Cambridge MA 02139

ROUTTENBERG, ARYEH, neuroscience researcher and educator; b. Reading, Pa., Dec. 1, 1939. B.A., McGill U., 1961; M.Sc., Northwestern U., 1963; Ph.D., U. Mich., 1965. Asst. prof. behavioral neurosci. Northwestern U., Evanston, Ill., 1965-68, assoc. prof., 1968-71, prof., 1971—; Vis. scholar Hoffman-Laroche, Zurich, 1976; vis. prof. U. Tex., Austin, 1980. Editor: Facets of Reinforcement, 1980, Brain Phosphoproteins, 1982. Fellow AAAS, Am. Psychol. Assn.; mem. Soc. for Neurosci. Subspecialties: Neurochemistry; Neurobiology. Current work: Brain chemistry in relation to experience; reward, pleasure, memory and learning.

ROUX, KENNETH HENRY, immunologist; b. Phila., May 12, 1948; s. Henry J. and Ruth I. (Sieder) R.; m. Shirley E. Songer, June 13, 1970; 1 son, Kyle J. B.S., Delaware Valley Coll., Doylestown, Pa., 1970; M.S., Tulane U., 1972, Ph.D., 1974. Postdoctoral fellow U. Ill. Med. Ctr., Chgo., 1975-78; asst. prof., dept. biol. sci. Fla. State U., Tallahassee, 1978—. Contbr. articles to profl. jours. NIH grantee, 1979-82; Am. Cancer Soc. grantee, 1981-82. Mem. Am. Assn. Immunologists, Fedn. Am. Socs. Exptl. Biology, AAAS. Subspecialties: Immunobiology and immunology; Immunogenetics. Current work: Immunoglobulin genetics, regulation and biochemistry. Home: 232 Timberlane Rd Tallahassee FL 32312 Office: Dept Biol Sci Fla State U Tallahassee FL 32306

ROVETTO, MICHAEL JULIEN, physiologist; b. Challis, Idaho, Mar. 20, 1943; s. John and Hannah C. (Hill) R. B.S., Utah State U., 1965; M.S., U. Idaho, 1968; Ph.D., U. Va., 1970. Research assoc. Hershey Med. Coll., 1970-73, asst. prof., 1973-73, Jefferson Med. Coll., Phila., 1974-78, assoc. prof., 1978-80, U. Mo., Columbia, 1980—; ad hoc reviewer NIH, 1978-83. Mem. editorial bd.: Am. Jour. Physiology, 1981—, Circulation Research, 1983; co-editor: Patho Physiology and Therapeutics of Myocardial Ischemia, 1977; contbr. articles to profl. jours. NIH grantee, 1975; recipient Career Devel. award, 1977-83; Am. Heart Assn. grantee, 1978-82. Mem. Am. Physiol. Soc., Biophys. Soc., Cardiac Muscle Soc., Am. Heart Assn. (circulation council, basic sci. council). Subspecialties: Physiology (medicine); Biochemistry (medicine). Current work: Myocardial energy metabolism, control of cardiac muscle function, myocardial adenine nucleotide metabolism, effects of myocardial ischemia.

ROVICK, ALLAN ASHER, physiologist, medical educator; b. Chgo., Feb. 11, 1928; s. Max Israel and Tillie (Clausman) R.; m. Renah Adar Reinstein, Jan. 30, 1949; children: Sharon, Lynn, Joshua, Jonathan. B.S., Roosevelt U., 1951; M.S., U. Ill.-Chgo., 1954, Ph.D., 1958. Instr. Loyola U. Med. Sch., Chgo., 1957-60, asst. prof., 1960-66, assoc. prof., 1966-67; vis. prof. U. Ill. Project, Chiangmai, Thailand, 1967-69; assoc. prof. U. Ill. Med. Ctr., Chgo., 1970-71; exec. sec. Div. Research Grants, NIH, Bethesda, Md., 1971-72; chief cardiac diseases br. Nat. Heart, Lung Inst., 1972-73; prof. dept. physiology Rush Med. Sch., Chgo., 1978—. Contbr. articles in field to profl. jours. Grantee NIH, 1960-67, Chgo. Heart Assn., 1973-77, Am. Heart Assn., 1977-80. Mem. Am. Physiol. Soc., AAAS. Subspecialties: Physiology (medicine); Fluid mechanics. Current work: Computer-based education in physiology, blood flow characteristics, rheology. Home: 65 W Schiller St Chicago IL 60610 Office: Dept Physiology Rush Med Coll 1753 W Congress Pkwy Chicago IL 60612

ROWE, JOSEPH EVERETT, electrical engineer; b. Highland Park, Mich., June 4, 1927; s. Joseph and Lillian May (Osbourne) R.; m. Margaret Anne Prine, Sept. 1, 1950; children: Jonathan Dale, Carol Kay. B.S. in Engring, U. Mich., 1951, B.S. Engring. in Math, 1951, M.S. in Engring, 1952, Ph.D., 1955. Mem. faculty U. Mich., Ann Arbor, 1953-74, prof. elec. engring., 1960-74, dir. electron physics lab., 1958-68, chmn. dept. elec. and computer engring., 1968-74; vice provost, dean engring. Case Western Res. U., Cleve., 1974-76; provost Case Inst. Tech., 1976-78; v.p. tech. Harris Corp., Melbourne, Fla., 1978-81, v.p., gen. mgr., 1981-82; exec. v.p. research and def. Gould Inc., 1982, vice chmn., chief tech. officer, 1983—; cons. to industry. Mem. adv. group electron devices Dept. Def., 1966-78. Author: Nonlinear Electron-Wave Interaction Phenomena, 1965, also articles. Recipient Distinguished Faculty Achievement award U. Mich., 1970. Fellow IEEE (chmn. adminstrv. com. group electron devices 1968-69, editor proc. 1971-73), AAAS; mem. Am. Phys. Soc., Am. Soc. Engring. Edn. (Curtis McGraw research award 1964), Nat. Acad. Engring., Sigma Xi, Phi Kappa Phi, Tau Beta Pi, Eta Kappa Nu. Subspecialty: Electrical engineering. Address: Gould Inc 10 Gould Center Rolling Meadows IL 60521

ROWE, RANDALL CHARLES, plant pathologist; b. Balt., Sept. 26, 1945; s. Leonard Charles and Joy (Randall) R.; m. Sandra Margaret Campbell, June 24, 1967; children: Steven Randall, Mark David. B.S., Mich. State U., 1967; Ph.D., Oreg. State U., 1972. Research assoc. N.C. State U., Raleigh, 1972-74; asst. prof. Ohio State U. and Ohio Agrl. Research Devel. Ctr., Wooster, 1974-79, assoc. prof., 1979—. Contbr. articles in field to profl. jours. Recipient undergrad. scholastic award botany dept. Mich. State U., 1967; NDEA fellow Oreg. State U., 1967-70; U.S. Dept. Agr. research grantee, 1980-83. Mem. Am. Phytopath. Soc., Potato Assn. Am. Subspecialty: Plant pathology. Current work: Biology and control of soil-borne plant pathogens of vegetable crops. Office: Dept Plant Pathology Ohio Agr Research Devel Ctr Wooster OH 44691

ROWELL, PETER PUTNAM, research scientist, educator; b. St. Petersburg, Fla., July 24, 1946; s. John Putnam and Jeanne (Fontaine) R.; m. Kathryn Johnson, June 10, 1972; 1 dau., Julie Fontaine. B.A., Stetson U., 1964; Ph.D., U. Fla., 1975. Research asso. Vanderbilt U., 1975-77; asso. prof. dept. pharmacology and toxicology U. Louisville, 1977—. Served to 1st lt. U.S. Army, 1969-71. Decorated Bronze Star. Mem. Soc. for Neurosci., AAAS, Am. Assn. Pharmacology and Exptl. Therapists, Sigma Xi. Subspecialties: Neurochemistry; Neuropharmacology. Current work: Cholinergic, presynaptic regulation. Office: Dept Pharmacology U Louisville Sch Medicine Louisville KY 40292

ROWLAND, FRANK SHERWOOD, educator, chemist; b. Delaware, Ohio, June 28, 1927; s. Sidney and Margaret Lois (Drake) R.; m. Joan Evelyn Lundberg, June 7, 1952; children—Ingrid Drake, Jeffrey Sherwood. A.B., Ohio Wesleyan U., Delaware, 1948; M.S., U. Chgo., 1951, Ph.D. in Chemistry, 1952. Instr. Princeton U., 1952-56; mem. faculty U. Kans., Lawrence, 1956-64, prof. chemistry, 1963-64, U. Calif.-Irvine, 1964—, chmn. dept., 1964-70; Philips lectr. Haverford Coll., 1975; J.T. Donald lectr. McGill U., 1976; Voenable lectr. U. N.C., 1979; Snider vis. prof. U. Toronto, 1980; vis. sr. scientist Japan Soc. Promotion of Sci., 1980; Humboldt sr. scientist, W. Ger., 1980; cons. radiation chemistry IAEA, 1969, 74; Mem. vis. com. in chemistry Brookhaven Nat. Lab., 1972-75; mem. vis. com. in chemistry Argonne Nat. Lab., 1974-79, chmn., 1975; mem. Internat. Commn. on Atmos. Chemistry and Global Pollution, Internat. Assn. Meteorology and Atmos. Physics, 1979—, Ozone Commn., 1980—; mem. vis. com. Max Planck Inst. (W.Ger.), 1982—; chmn. sci. adv. com. High Altitude Pollution Program, FAA, 1982; Mem. acid rain peer rev. panel U.S. Office Sci. and Tech., 1982-83. Contbr. articles to profl. jours.; editorial bd.: Jour. Geophys. Research, 1963-65, Jour. Phys. Chemistry, 1968-77. Served with USNR, 1945-46. Recipient John Wiley Jones award Rochester Inst. Tech., 1975; Distinguished Faculty Research award U. Calif., Irvine, 1976; Profl. Achievement award U. Chgo., 1977; Gordon Billard award N.Y. Acad. Scis., 1977; Guggenheim fellow, 1962, 74; Erskine fellow U. Canterbury, N.Z., 1978. Fellow Am. Acad. Arts and Scis., Am. Geophys. Union, AAAS; mem. Nat. Acad. Scis. (com. on atmos. scis. 1979—, com. on solar-terrestrial research 1979—, CODATA com. 1977-82), Am. Chem. Soc. (chmn. div. nuclear chemistry and tech. 1973, chmn. phys. div. 1974, Orange County award 1975, Tolman medal 1976, E.F. Smith Meml. lectr. 1980, Zimmermann award 1980, Environ. Sci. and Tech. award 1983), Am. Phys. Soc. (Leo Szilard award 1979), Phi Beta Kappa, Sigma Xi (nat. lectr.). Club: Explorers. Subspecialty: Atmospheric chemistry. Co-discoverer theory that fluorocarbons deplete ozone layer of stratosphere. Home: 4807 Dorchester Rd Corona del Mar CA 92625 Office: Dept Chemistry U Calif Irvine CA 92717

ROWLAND, LEWIS PHILLIP, neurologist, educator; b. Bklyn., Aug. 3, 1925; s. Henry Alexander and Cecile (Coles) R.; m. Esther Edelman, Aug. 31, 1952; children: Andrew Simon, Steven Samuel, Judith Mora. B.S., Yale U., 1945, M.D., 1948. Diplomate: Am. Bd. Psychiatry and Neurology. Intern New Haven Hosp., 1949-50; asst. resident N.Y. Neurol. Inst., 1950-52, fellow, 1953; clin. asso. NIH, Bethesda, Md., 1953-54; practice research medicine, specializing in neurology, N.Y.C., 1954-67, Phila., 1967-73, N.Y.C., 1973—; asst. neurologist Montefiore Hosp., N.Y.C., 1954-57; vis. fellow Nat. Inst. Med. Research, London, Eng., 1956; from asst. prof. to prof. neurology Columbia Coll. Phys. and Surg., 1957-67; prof., chmn. dept. neurology U. Pa., Med. Sch., 1967-73, Columbia Coll. Phys. and Surg., 1973—; from asst. neurologist to attending neurologist Presbyn. Hosp., 1957-67; co-dir. Neurol. Clin. Research Center, 1961-67, dir. neurology service, 1973—; Mem. med. adv. bd. Myasthenia Gravis Found., pres., 1971-73; med. adv. bd. Muscular Dystrophy Assos., Nat. Multiple Sclerosis Soc., Com. to Combat Huntington's Disease; pres. Parkinson's Disease Found., 1979—; adv. Josiah Macy Jr. Found.; mem. tng. grants com. Nat. Inst. Neurol. Communicable Diseases and Stroke, NIH, 1971-73, bd. sci. counselors, 1978—, chmn., 1980—. Editorial bd.: Archives of Neurology, 1968-76, Advances in Neurology, 1969—, Italian Jour. Neurol. Sci., 1979—, Handbook of Clin. Neurology, 1982—; editor-in-chief: Neurology, 1977—. Served with USNR, 1942-44; with USPHS, 1953-54. Mem. Am. Neurol. Assn. (pres. 1980), Am. Acad. Neurology, Phila. Neurol. Soc. (pres. 1972), Assn. Research Nervous Mental Disease (pres. 1969, trustee 1976—, v.p. 1980), Assn. U. Profs. Neurology (sec. 1971-74, pres. 1978), Eastern Pa. Multiple Sclerosis Soc. (chmn. med. adv. bd. 1969-73), N.Y.C. Multiple Sclerosis Soc. (chmn. med. adv. bd. 1977—). Subspecialty: Neurology. Home: 404 Riverside Dr New York NY 10025 Office: Neurological Inst 710 W 168th St New York NY 10032

ROWLAND, WILLIAM JOSEPH, biology educator; b. Bklyn., Dec. 15, 1943; s. Ralph George and Ethel Rose (Baker) R.; m. Ineke Jacoba Charlotte Kamann, July 24, 1971; children: Marijke, Sylvia. B.A., Adelphi U., 1965; Ph.D., SUNY-Stony Brook, 1970. Lectr. biology SUNY-Stony Brook, 1969-70; sci. co-worker Zool. Lab., Groningen, Netherlands, 1970-71; asst. prof. biology Ind. U. Bloomington, 1971-77, assoc. prof., 1977—. Contbg. author: Behavioral Significance of Color, 1979. USPHS grantee, 1973-75, 80-82; Ind. U. grantee, 1973-76, 78-80. Mem. Ind. Acad. Sci., Animal Behavior Soc., AAAS, Internat. Assn. Fish Ethnologists, Sigma Xi. Subspecialties: Ethology; Sociobiology. Current work: Ethology, behavioral ecology and sociobiology of lower vertebrates, especially fishes. Home: 510 E First St Bloomington IN 47401 Office: Dept Biology Ind Univ Bloomington IN 47405

ROWLANDS, DAVID THOMAS, pathology educator; b. Wilkes-Barre, Pa., Mar. 22, 1930; s. David Thomas and Anna Jule (Morgan) R.; m. Gwendolyn Marie York, Mar. 1, 1958; children—Julie Marie, Carolyn Jane. M.D., U. Pa., 1955. Diplomate: Am. Bd. Pathology, Am. Bd. Allergy and Immunology. Intern Pa. Hosp., Phila., 1955-56; resident Cin. Gen. Hosp., 1956-60; asst. prof. U. Colo., 1962-64, Rockefeller U., 1964-66; asso. prof. Duke U., Durham, N.C., 1966-70; prof. pathology U. Pa., Phila., 1970-82, chmn. dept. pathology, 1973-78, prof. medicine, 1979-82; prof., chmn. dept. pathology U. So. Fla., Tampa, 1982—; assoc. dean U. Pa., 1983—. Mem. editorial bd.: Am. Jour. Pathology, 1971—, Developmental and Comparative Immunology, 1977-79. Served with USNR, 1960-62. Recipient Lederle Med. Faculty award U. Colo., 1964, Jacob Ehrenzeller award Pa. Hosp., 1976. Mem. Am. Assn. Pathologists, Internat. Acad. Pathology, Am. Soc. Clin. Pathology, Am. Assn. Immunologists. Presbyterian. Subspecialty: Pathology (medicine). Home: 13804 Cypress Village Circle Tampa FL 33624 Office: Dept Pathology Coll Medicine U South Fla Tampa FL 33612

ROWLANDS, JOHN ALAN, medical physicist; b. Altrincham, Cheshire, Eng., May 4, 1945; s. Alan and Esther Mary (Moore) R.; m. Cheryl Ann Goodfellow; children: Brock Jeffrey, Allison Jane. B.Sc. with first class honors in Physics, U. Leeds (Eng.), 1967, Ph.D., 1971. Postdoctoral fellow physics dept. U. Alta., 1971-73, research assoc., 1973-77, vis. asst. prof., 1977-78; vis. asst. prof. physics dept. Mich. State U., 1978-79; asst. prof. radiol. research labs., dept. radiology U. Toronto, 1979—; cons. physicist St. Joseph's Health Ctr. Contbr. articles to sci. jours. Recipient Izaak Walton Killam Fellowship, 1971-73. Mem. Can. Assn. Physicists, Am. Assn. Physicists in Medicine, Soc. Optical Engrs. Subspecialties: Diagnostic radiology; Imaging technology. Current work: Physics in Radiology; intravenous angiography, imaging. Home: 47 Hanley St Toronto ON Canada M6S 2H3 Office: Radiol Research Labs Med Sci Bld U Toronto Toronto ON Canada M5S 1A8

ROWLANDS, ROBERT EDWARD, engring. educator, cons.; b. Trail, C., Can., July 7, 1936; came to U.S., 1960; s. Edward Howell and Eda May (Randell) R.; m. Mary Roma Ranaghan, Nov. 14, 1959; children: Robert Philip, Edward Hugh. B.A.Sc., U. B.C., Vancouver, 1959; M.S., U. Ill., Urbana, 1964, Ph.D., 1967. Registered profl. engr., Wis. Mech. engr. MacMillan & Bloedel, Powell River, B.C., 1959-60; research engr. Ill. Inst. Tech. Research Inst., Chgo., 1967-71, sr. research engr., 1971-74; asst. prof. engring. U. Wis., 1974-76, assoc. prof., 1976-79, prof., 1979—; lectr. and cons. in field; Am. rep. to USSR-U.S.A. Advanced Composite Materials meeting, Riga, Latvia, 1978; mem. U.S.A. organizational com. U.S.A.-USSR Composite Materials Meeting, 1980. Contbr. articles profl. jours., chpts. in books. Active boys program YMCA, Park Forest, Ill., 1970-74; youth racing program Madison Ski Club, 1978—; mem. com. Boy Scouts Am., Madison, 1975—, Guardian mem. Mohawk dist. council, 1976—. Fellow Soc. Exptl. Stress Analysis (Hetenyi award 1971, 77); mem. ASME, N.Am. Photonics Assn., Am. Acad. Mechanics. Subspecialties:

Mechanical engineering; Solid mechanics. Current work: Stress analysis and materials, exptl. mechanics. Home: 5401 Russett Rd Madison WI 53711 Office: Mechanics Dept U Wis 1415 Johnson Dr Madison WI 53706

ROWLEY, DONALD A., pathologist-immunologist, educator; b. Owatonna, Minn., Feb. 4, 1923; m., 1948; 4 children. B.S., U. Chgo., 1945, M.S., 1950, M.D., 1950. Sr. asst. surgeon Nat. Inst. Allergy and Infectious Diseases, 1951-54; instr. pathology U. Chgo., 1954-57; asst. prof., 1957-61, asso. prof., 1961-68, prof. dept. pathology, 1969—, dept. pediatrics, 1973—; dir. research La Rabida-U. Chgo. Inst., 1973—, dir. inst., 1978—; dir. La Rabida Children's Hosp. and Research Center, 1978-82; sr. research fellow USPHS, 1959-69; vis. scientist Sir William Dunn Sch. Pathology, Oxford, Eng., 1961-62, 70-71. Mem. Soc. for Exptl. Pathology, Assn. Pathology and Bacteriology, Assn. Immunology. Subspecialty: Immunology (medicine). Office: LaRabida U of Chgo Inst 65th and Lake Michigan Chicago IL 60649

ROWLEY, JANET DAVISON, physician; b. N.Y.C., Apr. 5, 1925; d. Hurford Henry and Ethel Mary (Ballantyne) Davison; m. Donald A. Rowley, Dec. 18, 1948; children—Donald, David, Robert, Roger. B.S., U. Chgo., 1946, M.D., 1948. Research fellow Levinson Found., Cook County Hosp., 1955-61; intern Marine Hosp., USPHS, Chgo., 1950-51; research fellow Levinson Found., Cook County Hosp., Chgo., 1955-61; clin. instr. neurology U. Ill., 1957-61; research asso. dept medicine U. Chgo., 1962-71, asso. prof., 1971-78, prof., 1978—; current splty.: cytogenetic analysis of human hematologic malignant diseases; mem. Nat. Cancer Adv. Bd., 1979—. Contbr. chpts. to books, articles to profl. jours. Trustee Adler Planetarium, 1978—. Served with USPHS, 1950-51. Mem. Am. Soc. Human Genetics, Am. Soc. Hematology, Am. Assn. Cancer Research. Episcopalian. Subspecialties: Cancer research (medicine); Hematology. Current work: Identification of chromosome abnormalities in human leukemia and lymphoma. Home: 5310 University Ave Chicago IL 60615 Office: Box 420 950 E 59th St Chicago IL 60637

ROWLEY, PETER TEMPLETON, physician; b. Greenville, Pa., Apr. 29, 1929; s. George Hardy and Susan Mossman (Templeton) R.; m. Carol Stone, Mar. 19, 1967; children: Derek Stone, Jason Templeton. A.B. magna cum laude, Harvard U., 1951; M.D., Columbia U., 1955. Diplomate: Am. Bd. Internal Medicine. Intern med. service N.Y. Hosp.-Cornell Med. Center, 1955-56; clin. asso. Nat. Inst. Neurol. Disease and Blindness, NIH, 1956-58; asst. resident, then resident Harvard Med. Service, Boston City Hosp.; asst. in medicine Harvard U. Med. Sch. and researcher Thorndike Meml. Lab., 1958-60; hon. research asst. dept. eugenics, biometry and genetics Univ. Coll., U. London, 1960-61; postdoctoral fellow dept. microbiology N.Y. U. Sch. Medicine, 1961-63; asst. prof. medicine Stanford U., 1963-70; asso. prof. medicine pediatrics and genetics U. Rochester, 1970-75, prof. medicine, pediatrics, genetics and microbiology, 1975—, acting chmn. div. genetics, 1975—, dir. genetics and regulation postdoctoral tng. program, 1977—; mem. staff Highland Hosp.; physician, pediatrician Strong Meml. Hosp.; cons. NIH study, 1977-79; med. adv. bd. Ahepa Cooley's Anemia Found., 1978—; mem. N.Y. State Exec. and Adv. Coms. on Genetics Diseases, 1979—; WHO vis. scholar Inst. Biol. Chemistry, U. Ferrara, Italy, 1970. Editor: (with M. Lipkin Jr.) Genetic Responsibility: On Choosing Our Children's Genes, 1974. Served with USPHS, 1956-58. Recipient Excellence in Teaching award U. Rochester Class of 1976, 1973; NRC fellow, 1960-63; Buswell research fellow, 1970-71, 71-72. Fellow A.C.P.; mem. Am. Fedn. Clin. Research, Am. Soc. Hematology, Am. Soc. Human Genetics (social issues com. 1980—). Subspecialty: Genetics and genetic engineering (medicine). Office: Div Genetics PO Box 641 U Rochester Med Sch 601 Elmwood Ave Rochester NY 14642

ROWND, ROBERT HARVEY, biochemistry educator; b. Chgo., July 4, 1937; s. Walter Lemuel and Marie Francis (Joyce) R.; m. Rosalie Anne Lowery, June 13, 1959; children: Jennifer Rose, Robert Harvey, David Matthew. B.S. in Chemistry, St. Louis U., 1959; M.A. in Med. Scis, Harvard U., 1961; Ph.D. in Biophysics, Harvard U., 1963. Postdoctoral fellow Med. Research Council, NIH, Cambridge, Eng., 1963-65; postdoctoral fellow Nat. Acad. Scis.-NRC, Institut Pasteur, Paris, 1965-66; prof., chmn. molecular biology and biochemistry U. Wis., Madison, 1966-81; John G. Searle prof., chmn. molecular biology and biochemistry Med. and Dental Schs., Northwestern U., Chgo., 1981—; cons. NIH, NSF, Nat. Acad. Scis.-NRC. Contbr. numerous articles to sci. jours., books.; Mem. editorial bd.: Jour. of Bacteriology, 1975-81; editor, 1981—; assoc. editor plasmid, 1977—. Mem. troop com., treas. Four Lakes council Boy Scouts Am., Madison, 1973-77; mem. People to People Program del. of microbiologists to China, 1983. Fellow NSF, NIH, 1959-66; research grantee, 1966—; tng. grantee, 1970-79; USPHS Research Career Devel. awardee, 1968-73. Mem. Am. Soc. Microbiology, Assn. Harvard Chemists, Am. Soc. Biol. Chemists, Am. Acad. Microbiology, N.Y. Acad. Scis. Subspecialty: Molecular biology. Home: 506 Lake Ave Wilmette IL 60091 Office: Northwestern U Med and Dental Schs 303 E Chicago Ave Chicago IL 60611

ROWSE, ROBERT ALFRED, ceramic co. exec.; b. Worcester, Mass., Aug. 8, 1925; s. Alfred and Louise (Kelly) R.; m. Shirley Erikson; children: Paula Rowse Buonomo, JoAnn Rowse DiPilato. B.S. in Chem. Engring, Worcester Poly. Inst., 1949; grad. advanced mgmt. program, Harvard U., 1969. With Norton Co., Worcester, 1950—, v.p. div. ops., materials div., 1975-78, v.p. div. research and devel., materials div., 1978-80, div. v.p. research and new bus. devel., div. v.p. and gen. mgr. proppants, 1981—. Clk. and warden St. Matthew's Episcopal Ch., Worcester. Served to 2d lt. AC U.S. Army, 1943-46. Recipient Man of Yr. award Abrasive Engring. Soc., 1976. Mem. Am. Chem. Soc., Abrasive Grain Assn. Lodge: Masons. Subspecialties: Ceramics; Materials. Current work: Ceramic materials. Patentee abrasives. Home: 33 North St Shrewsbury MA 01545 Office: 1 New Bond St Worcester MA 01606

ROY, DENIS L., research physicist; b. Cap Chat, Que., Can., Nov. 21, 1946; s. Charles and Marguerite R.; m. Anne Marie Masson, July 6, 1968. B.Sc., Laval. U., 1970, M.Sc., 1971, D.Sc., 1974. Assoc. prof. Laval U., 1974-80, Nat. Sci. Engring. Research Council Can. research attache, 1980—. Contbr. articles to profl. jours. Mem. Assn. Canadienne-Francaise pour L'Avancement des Sciences, Can. Assn. Physicists, Am. Phys. Soc., Royal Astron. Soc. Can., Can. Nature Fedn. Subspecialty: Atomic and molecular physics. Current work: Research atomic and molecular physics, electron scattering spectroscopy. Office: Dept Physics Pav Vachon Laval U Quebec PQ Canada G1K 7P4

ROY, DIPAK, environmental engineer, consultant; b. Calcutta, West Bengal, India, Aug. 4, 1946; came to U.S., 1973; s. Jitendra Nath and Smriti Kana (Basu) R.; m. Bulbul Dutta, Nov. 30, 1975; 1 son, Neil. B.C.E., Jadaupur U., Calcutta, 1963-68; M.Tech., Indian Inst. Tech., Kanpur, 1969-71; Ph.D., U. Ill.-Urbana, 1979. Scientist Nat. Environ. Research Inst., Nagpur, India, 1971-73; engr. Catalytic, Inc., Phila., 1973-74, Johnson & March, Phila., 1974-75; asst. prof. dept. civil engring. La. State U., Baton Rouge, 1979—; cons. N.Y. Assocs., New Orleans, 1979. EPA research grantee, 1981-82; Dept. of Energy research grantee, 1981-82. Mem. Water Pollution Control Fedn., ASCE, Am. Water Works Assn., Assn. Environ. Engr. Profs., Internat. Assn. Water Pollution Research. Subspecialties: Water supply and wastewater treatment; Biomass (energy science and technology). Current work: Energy production from biomass, hazardous waste treatment, water supply and wastewater treatment. Home: 2023 General Beauregard Baton Rouge LA 70810 Office: La State U Dept Civil Engring Baton Rouge LA 70803

ROY, PETER ALAN, indsl. hygiene educator; b. New Bedford, Mass., July 9, 1947; s. Pierre Arthur and Helen Ann (Jata) R.; m. Mary Elizabeth Griffen, Sept. 16, 1972. B.S. in Biology, S.E. Mass. U., 1967-70; M.P.H., U. Minn., 1982. Cert. comprehensive practice of indsl. hygiene. Indsl. hygienist IHE, Hopkins, Minn., 1977; sr. corp. indsl. hygienist Medtronic, Inc., Mpls., 1978-80; asst. prof. indsl. hygiene U. Minn., 1980—; cons. in field. Served with USNR, 1970-73. NSF grantee, 1970; Nat. Inst. Occupational Health and Safety fellow, 1975-76. Mem. Am. Indsl. Hygiene Assn., Am. Conf. Govt. Indsl. Hygienists, Am. Acad. Indsl. Hygienists. Subspecialty: Contaminant exhaust systems. Current work: Analysis and control of toxic gases and vapors, control of gas and vapor hazards associated with health care industry and products. Home: 10 Village Dr #104 Proctor MN 55810 Office: U Minn 104 Voss-Kovach Hall Duluth MN 55812

ROY, RAJARSHI, laser physicist, educator; b. Calcutta, India, June 4, 1954; s. Tapash and Phullora (Ray) R. B.Sc. with honors, Delhi (India) U., 1973, M.Sc., 1975; M.A., U. Rochester, 1977, Ph.D., 1981. Postdoctoral research assoc. Joint Inst. for Lab. Astrophysics, U. Colo., Boulder; asst. prof. Sch. Physics, Ga. Inst. Tech., Atlanta. Contbr. articles to profl. jours. Mem. Optical Soc. Am., Am. Phys. Soc. Subspecialty: Spectroscopy. Office: Sch Physics Ga Inst Tech Atlanta GA 30332

ROY, RUSTUM, geochemistry educator; b. Ranchi, India, July 3. 1924; came to U.S., 1945, naturalized, 1961; s. Narendra Kumar and Rajkumari (Mukherjee) R.; m. Della M. Martin, June 8, 1948; children—Neill, Ronnen, Jeremy. B.Sc., Patna (India) U., 1942; M.Sc., Pa. State U., 1944, Ph.D., 1948. Research asst. Pa. State U., 1948-49, mem. faculty, 1950—, prof. geochemistry, 1957—, prof. solid state, 1968—, chmn. solid state tech. program, 1960-67, chmn. sci. tech. and soc. program, 1977—, dir. materials research lab., 1962—, Evan Pugh prof., 1981—; sr. sci. officer Nat. Ceramic Lab., India, 1950; mem. com. mineral sci. tech. Nat. Acad. Scis., 1967-69, com. survey materials sci. tech., 1970-74; exec. com. chem. div. NRC, 1967-70, nat. materials adv. bd., 1970-77, mem. com. radioactive waste mgmt., 1974-80, chmn. panel waste solidification, 1976—, chmn. com., USSR and Eastern Europe, 1976—; sci. policy fellow Brookings Instn., 1982-83; mem. Pa. Gov.'s Sci. Adv. Com.; chmn. materials adv. panel, 1965—; mem. adv. com. engring. NSF, 1968-72, adv. com. to ethical and human value inplications sci. and tech., 1974-76, adv. com. div. materials research, 1974-77; Hibbert lectr. U. London, 1979; dir. Kirkridge, Inc., Bangor, Pa.; cons. to industry. Author: Honest Sex, 1968, Crystal Chemistry of Non-metallic Materials, 1974, Experimenting with Truth, 1981, Radioactive Waste Disposal, Vol. 1, the Waste Package; also articles.; Editor-in-chief: Materials Research Bull, 1966—, Bull. Sci. Tech. and Soc. Chmn. bd. Dag Hammarskjold Coll., 1973-75; mem. ad hoc com. sci., tech. and ch. Nat. Council Chs., 1966-68. Mem. Nat. Acad. Engring., Mineral. Soc. Am. (award 1957), Am. Chem. Soc. (Petroleum Research Fund award 1960), Royal Swedish Acad. Engring. Scis. (fgn. mem.). Subspecialties: Ceramics; High-temperature materials. Current work: Novel materials preparation and synthesis; especially those involing solution-made ceramics via DMS and sol-gel techniques for zero expansion, high energy shortage, and radioactive waste solidification. Home: 528 S Pugh St State College PA 16801 Office: 202 Materials Research Lab University Park PA 16802 A continuing search for an accurate perception of the totality of "Reality," and a commitment both to sharing those (slightly changing) perceptions and to living by the highest values embedded in them, is central to my being. Integration of the insights of science and the vehicle of art into the highest religious value systems is the urgent need of our time. As a Christian I seek to increase society's commitment to proclaiming the centrality of other-centered love, and educating and training all citizens in it, even as a partial substitute for its commitment to science.

ROY-BURMAN, PRADIP, molecular biologist, biochemist, educator; b. Comilla, India, Nov. 12, 1938; came to U.S., 1963, naturalized, 1976; s. Prafulla Nath and Mrinalini (Barman) Roy-B.; m. Sumitra Roy-Burman, Nov. 26, 1963; children: Arup, Paula. B.Sc. in Chemistry with honors, Calcutta U., 1956, M.Sc., 1958, Ph.D., 1963. Asst. prof. biochemistry U. So. Calif., 1970, asst. prof. biochemistry and pathology, 1970-71, assoc. prof., 1972-78, prof., 1978—, chmn. grad. com. exptl. pathology, 1974—; prin. investigator viral oncology research Nat. Cancer Inst. and Am. Cancer Soc. Contbr. articles to profl. jours. Dernham Sr. Research fellow in oncology, 1966-71; Am. Cancer Soc. grantee, 1968-70; Nat. Cancer Inst. grantee, 1970—; So. Calif. Edison Co. grantee, 1979-82. Mem. Am. Soc. Biol. Chemists, Am. Soc. Microbiology, AAUP, Internat. Assn. Comparative Research on Leukemia and Related Diseases. Democrat. Subspecialties: Cancer research (medicine); Virology (medicine). Current work: Viral and cellular oncogenes, molecular oncology, oncodevelopmental genes feline leukemia virus, leukemogenesis, oncogenes and regulation of their expression in normal and leukemic hematopoietic cells, recombinant DNA technology. Office: U So Calif Sch Medicine Dept Pathology Los Angeles CA 90033

ROYDS, ROBERT B., clinical pharmacologist, physician; b. Harrogate, Yorks., Eng., Oct. 3, 1944; came to U.S., 1974; s. John Edmund and Ailsa Dorothea (Williams) R.; m. Marilyn Maria Valerio, Apr. 23, 1977; children: Elizabeth Caroline, Leslie Alexandria. M.B.B.S., St. Bartholomew's Hosp., London, 1967; M.R.C.S., L.R.C.P., 1967. Research fellow St. Bartholomew's Hosp., London, Eng., 1970-72, chief asst., 1972-74; assoc. dir. Merck, Sharp & Dohme Labs., Rahway, N.J., 1975-76; sr. research physician Hoffmann-La Roche Inc., Nutley, N.J., 1976-79; v.p. Besselaar Assocs., Princeton, N.J., 1979-82; chmn. Theradex Systems, Inc., Princeton, N.J., 1982—; adj. asst. prof. pharmacology U. Pa.-Phila., 1981—. Fellow Royal Soc. Medicine; mem. Royal Coll. Physicians of London, Am. Soc. Clin. Pharmacology and Therapeutics, Am. Fedn. Clin. Research. Subspecialties: Clinical Pharmacology; Bioinstrumentation. Current work: Computer assisted clinical study monitoring. Office: Theradex Systems Inc CN 5257 Princeton NJ 08540

ROYER, GARFIELD PAUL, biochemist; b. Waynesboro, Pa., Dec. 2, 1942; s. Paul Franklin and Dolores (Schnurr) R.; m. Alvilda Ann Hopcraft, Aug. 13, 1966; children: Thaddeus, Corynn, Paul. B.S., Juniata Coll., 1964; Ph.D., W.Va., U., 1968; postgrad., Northwestern U., 1968-70. Prof. biochemistry Ohio State U., Columbus, 1970-82; dir. biotech. div. Standard Oil (Ind.), Naperville, Ill., 1983—; cons. to various corps. Author: Fundamentals of Enzymology, 1982; mem. editorial bd.: Jour. Molecular Catalysis, 1977—; contbr. articles to profl. jours. NIH Research Career Devel. awardee, 1975. Mem. Am. Chem. Soc., Am. Soc. Biol. Chemists, AAAS. Republican. Lutheran. Subspecialties: Enzyme technology; Genetics and genetic engineering (agriculture). Current work: Research in areas of enzyme engineering, synthetic enzyme models, and genetic engineering. Patentee. Office: Amoco Research Center PO Box 400 Naperville IL 60566

ROYLANCE, DAVID KAYE, engineering educator; b. Salt Lake City, Sept. 22, 1940; s. Kaye Fautin and Mary Louise (Kimball) R.; m. Margaret Eileen Allen, Nov. 22, 1968; children: Stephen, Patricia, Michael. B.S. in Mech. Engring, U. Utah, 1964, Ph.D., 1968. Research assoc. U. Utah, Salt Lake City, 1964-68; group leader Army Materials Lab., Watertown, Mass., 1970-75; prof engring. MIT, Cambridge, Mass., 1975—. Contbr. articles to profl. jours. Served to capt. U.S. Army, 1968-70; Viet Nam. Mem. Am. Chem. Soc., Soc. Plastics Engrs., Soc. of Rheology. Subspecialties: Polymers; Solid mechanics. Current work: Mechanical properties of polymers and composites: deformation, fracture, degradation, processing. Home: 80 Morrill St Newton MA 02165 Office: MIT Room 6-202 Cambridge MA 02139

ROZEN, JEROME GEORGE, JR., research entomologist, mus. curator and adminstr.; b. Evanston, Ill., Mar. 19, 1928; s. Jerome George and Della (Kretchmar) R.; m. Barbara L. Lindner, Dec. 18, 1948; children—Steven George, Kenneth Charles, James Robert. Student, U. Pa., 1946-48; B.A., U. Kans., 1950; Ph.D., U. Calif. at Berkeley, 1955. Entomologist (taxonomy) U.S. Dept. Agr., 1956-58; asst. prof. entomology Ohio State U., 1958-60; asso. curator Hymenoptera, dept. entomology Am. Mus. Natural History, N.Y.C., 1960-65, curator of Hymenoptera, 1965—, chmn. dept., 1960-71, dep. dir. for research, 1972—; field expdns., U.S., Europe, Trinidad, Chile, Brazil, Morocco, So. Africa. Fellow AAAS; mem. Entomol. Soc. Am. (editor misc. publs. 1959-60), Soc. Study Evolution, Soc. Systematic Zoology, N.Y. Entomol. Soc. (pres. 1964-65), Washington Entomol. Soc. Club: Cosmos. Subspecialties: Evolutionary biology; Systematics. Current work: Evolutionary and systematic relationships of insects; especially bees. Research in evolutionary biology, especially systematics of bees and beetles. Home: 55 Haring St Closter NJ 07624 Office: Am Museum Natural History Central Park W at 79th St New York NY 10024

ROZEN, SIMON, internist, hematologist; b. Havana, Cuba, Oct. 26, 1928; came to U.S., 1960; s. Naun and Berta (Silberfarb) R.; m. Hilda Zaidman, Oct. 4, 1953; children: Griselle, Henry. M.D., U. Havana, 1953. Intern Michael Reese Hosp., Chgo., 1953-54, resident in hematology, 1955-56; resident in internal medicine VA Research Hosp., Chgo., 1961-62, Passavant Meml. Hosp., 1962-63; assoc. attending Mt. Sinai Hosp., Miami Beach, Fla., 1980—; attending Miami Heart Inst., U. Havana Club; instr. medicine Northwestern U., Chgo., 1964-65, clin. instr., 1965-68; assoc. prof. medicine, asst. prof. oncology U. Miami, 1982—. Recipient Ruth Reader award Hematology Research Found., 1955. Fellow ACP; mem. AMA, Am. Soc. Internal Medicine, Am. Soc. Hematology, Am. Soc. Clin. Oncology. Subspecialties: Hematology; Internal medicine. Current work: Clinical bleeding coagulation. Office: 605 Lincoln Rd Miami Beach FL 33139

ROZENBERGS, JANIS, physicist; b. Leeds, Eng., May 6, 1948; came to Can., naturalized, 1959; came to U.S., 1978; s. Peteris Arvids and Austra (Alers) R. B.Sc., Lakehead U., Thunder Bay, Can., 1970, 1971, M.Sc., 1974; DiplomIngenieur, Johannes Kepler U., Linz, Austria, 1975, Ph.D., 1978. Prin. engr. Electro-Optical Products Div. ITT, Roanoke, Va., 1978—. Contbr. articles to profl. jours. Mem. Am. Phys. Soc. Subspecialties: Semiconductors; Fiber optics. Current work: Semiconductor device research and development specializing in detectors for fiber optics and infrared sensing. Office: ITT Electro-Optical Products Division 7635 Plantation Rd Roanoke VA 24019

RUBAL, BERNARD J., physiologist; b. Johnstown, Pa., Sept. 21, 1949; s. Bernard R. and Helen (Korch) R.; m. Diane Rothacker, Sept. 14, 1971; children: Matthew, Michael, Melissa, Mark. B.A., Kent State U., 1971; M.S., 1973; Ph.D., 1976; Ph.D., Baylor U. Med. Coll., 1977. Temporary inst. Kent (Ohio) State U., 1973-74; grad. asst. Baylor Med. Sch., Houston, 1974-75; instr., asst. prof. Tex. Women's U., Denton, 1975-76, 79, assoc. prof., 1979-80; research physiologist; Brooke Army Med. Ctr., Fort Sam Houston, Tex., 1980—; adj. assoc. prof. Tex. Coll. Osteo. Medicine, Forth Worth, 1980-81; clin. assoc. prof. U. Tex. Health Sci. Center, San Antonio, 1982. Contbr. articles to profl. jours. Am. Heart Assn. grantee, 1979; Life Mark, Inc. grantee, 1979. Mem. Am. Physiol. Soc., Am. Coll. Sports Medicine, Soc. Exptl. Biology and Medicine, Sigma Xi. Subspecialties: Physiology (medicine); Biomedical engineering. Current work: Cardiovascular adaptations to stress, applied thermographic research. Office: Brooke Army Med Center Fort Sam Houston TX 78234

RUBENSTEIN, ARTHUR HAROLD, physician, educator; b. Johannesburg, South Africa, Dec. 28, 1937; came to U.S., 1967; s. Montague and Isabel (Nathanson) R.; m. Denise Hack, Aug. 19, 1962; children: Jeffrey Lawrence, Errol Charles. M.B., B.Ch., U. Witwatersrand, 1960. Fellow in endocrinology Postgrad. Med. Sch., London, 1965-66; fellow in medicine U. Chgo., 1967-68; asst. prof., 1968-70, asso. prof., 1970-74, prof., 1974—, Lowell T. Coggeshall prof. med. sci., 1981—, asso. chmn. dept. medicine, 1975-81, chmn., 1981—; attending physician Billings Hosp., U. Chgo., 1968—; mem. study sect. NIH, 1973-77; mem. adv. council Nat. Inst. Arthritis, Metabolism and Digestive Diseases, 1978-80; chmn. Nat. Diabetes Adv. Bd., 1982, mem., 1983. Editorial bd.: Diabetes, 1973-77, Endocrinology, 1973-77, Jour. Clin. Investigation, 1976-81, Am. Jour. Medicine, 1978—; contbr. articles to profl. jours. Recipient David Rumbough Meml. award Juvenile Diabetes Found., 1978. Fellow ACP, Coll. Physicians (S. Africa), Royal Coll. Physicians (London); mem. Am. Soc. for Clin. Investigation, Am. Diabetes Assn. (Eli Lilly award 1973, Banting medal award 1983), Endocrine Soc., Am. Fedn. for Clin. Research, Central Soc. for Clin. Research, Assn. Am. Physicians. Subspecialties: Endocrinology; Internal medicine. Current work: Etiology and pathogenesis of diabetes; mutant insulin formation improved therapy of diabetes with insulin. Home: 5517 S Kimbark Ave Chicago IL 60637 Office: 950 E 59th St Chicago IL 60637

RUBENSTEIN, EDWARD, physician, educator; b. Cin., Dec. 5, 1924; s. Louis and Nettie R.; m. Nancy Ellen Millman, June 20, 1954; children: John, William, James. M.D., U. Cin., 1947. Intern, jr. asst. resident, sr. asst. resident internal medicine Cin. Gen. Hosp., 1947-50; fellow May Inst., Cin., 1950; sr. asst. resident Ward Med. Service, Barnes Hosp., St. Louis, 1953-54; chief of medicine San Mateo County Hosp., Calif., 1960-70; asso. dean postgrad. med. edn., prof. clin. medicine Stanford U. Sch. Medicine, 1971—. Author: textbook Intensive Medical Care; editor-in-chief: Sci. Am. Medicine, 1978—. Served with USAF, 1950-52. Fellow ACP; mem. Inst. Medicine, Nat. Acad. Scis., Calif. Acad. Medicine, Western Assn. Physicians, Soc. Photo-Optical Engrs., Alpha Omega Alpha. Subspecialties: Imaging technology; Internal medicine. Current work: Non-invasive coronary angiography. Origin of chiral molecules. Research on synchrotron radiation. Office: TC 129 Stanford Med Center Stanford CA 94305

RUBERG, ROBERT LIONEL, physician, educator; b. Phila., July 22, 1941; s. Norman and Yetta (Wolfman) R.; m. Cynthia Lief, June 26, 1966; children: Frederick, Mark, Joshua. B.A., Haverford Coll., 1963; M.D., Harvard Med. Sch., 1967. Diplomate: Am. Bd. Surgery, Am. Bd. Plastic Surgery. Intern Hosp. U. Pa., 1967-68, resident, 1968-75; asst. instr. surgery U. Pa., 1967-72, instr. surgery, 1972-75; asst. prof. surgery Ohio State U., Columbus, 1975-81, assoc. prof. surgery, 1981—; dir. Nutrition Support Service, Ohio State U. Hosps., 1976—; co-dir. Burn Unit, 1975; bd. dirs. Am. Soc. for Parenteral and Enteral Nutrition, 1983—. Trustee Columbus Hebrew Sch., 1982. Research grantee Plastic Surgery Ednl. Found., 1976, 78; basic sci. prize, 1977. Fellow ACS; mem. Am. Assn. Plastic Surgeons, Central Surg. Assn., Plastic Surgery Research Council, Am. Soc. Plastic and Reconstructive

Surgeons. Club: Aesculapian (Boston). Subspecialties: Surgery; Nutrition (biology). Current work: Nutrition and drug effects on skin grafts and skin flaps, solutions and techniques for parenteral feeding. Home: 6243 Peach Tree Rd Columbus OH 43213 Office: Ohio State Univ Hosp Room N-809 410 W Tenth Ave Columbus OH 43213

RUBERT, MARY LOU, psychologist; b. San Juan, Oct. 5, 1951; d. Guillermo and Francisca (del Valle) R.; m.; 1 son, Guillermo Morales Rubert. B.A., U. P.R., 1972; M.S., Caribbean Center for Advanced Studies, 1976; Ph.D., Calif. Sch. Profl Psychology, 1980. Prof. psychology and clin. supr. Caribbean Center for Advanced Studies, Santurce, P.R., 1980—; dir. continuing edn. program, same, 1981—. Ford Found. fellow, 1978-80. Mem. Am. Psychol. Assn., AAAS. Subspecialties: Clinical psychology; Developmental psychology. Current work: Laboratory design for research on info. processing and developmental difference in individuals; behavioral dimension and decision-making models related to biomed. ethics. Home: Cond Pauque San Patricio 2 Apt 306 Caparra Heights PR 00922 Office: Caribbean Center for Advanced Studies 1409 Ponce de Leon Ave Santurce PR 00940

RUBIN, BERNARD, pharmacologist, researcher; b. N.Y.C., Feb. 15, 1919; s. Charles and Anna (Slutskin) R.; m. Betty Rose Schindler, June 17, 1945; children: Stefi Gail, Robert Henry. Ph.D. in Pharmacology, Yale U., 1951. Bacteriologist Bur. Labs., N.Y.C. Dept Health, 1940-43; med. lab. technician U.S. Marine Hosp., S.I., N.Y., 1944; health insp. Bur. Foods and Drugs, N.Y.C. Dept. Health, 1945; research biologist Nepera Chem. Co., Yonkers, N.Y., 1945-48; pharmacologist, sr. research group leader Squibb Inst. Med. Research, Princeton, N.J., 1950—. Contbr. articles to profl. jours. Served with AUS, 1942-43. AEC fellow, 1949-50; CIBA fellow, 1948. Mem. Am. Soc. Pharmacology and Exptl. Therapeutics, Internat. Soc. Hypertension, Soc. Exptl. Biology and Medicine, Am. Pharm. Assn., Am. Heart Assn. (council for high blood pressure research). Subspecialty: Pharmacology. Current work: Hypertension, novel types of antihypertensive agents (specific enzyme inhibitors), captopril. Office: PO Box 4000 Princeton NJ 08540

RUBIN, EDWARD STEPHEN, engineering educator; b. N.Y.C., Sept. 19, 1941; s. Hyman and Esther R.; m. Maria Carmen Seligra, Dec. 20, 1966; children: Denise, Lisa. B.E. cum laude, CCNY, 1964; M.S. in M.E, Stanford U., 1965, Ph.D., 1969. Asst. prof. mech. engrng. Carnegie-Mellon U., 1969-72, asst. prof. mech. engrng. and pub. affairs, 1972-74, assoc. prof. mech. engrng., engrng. and pub. policy, 1974-79, prof., 1979—, dir., 1978—; vis. mech. engr. Brookhaven Nat. Lab., Upton, N.Y., 1975, 77; vis. prof. dept. physics, U. Cambridge, Eng., 1979-80, vis. fellow, Churchill Coll., 197980, cons. in field. Contbr. articles to profl. jours. Mem. Allegheny County (Pa.) Air Pollution Control Adv. Bd., 1973-80; mem. Pa. Air and Water Quality Tech. Adv. com., 1977-79. NSF trainee, 1964-68; recipient Ralph R. Teetor Ednl. award Soc. Automotive Engrs., 1972. Mem. ASME (mem. exec. com. Pitts. sect. 1970-73, chmn. div. air pollution control 1983—), Air Pollution Control Assn., Sigma Xi, Tau Beta Pi, Pi Tua Sigma. Subspecialties: Environmental engineering; Mechanical engineering. Office: Carnegie Mellon U Pittsburgh PA 15213

RUBIN, GERALD M., molecular biologist, educator; b. Boston, Mar. 31, 1950; s. Benjamin H. and Edith R.; m. Lynn Suzanne Mastalir, May 7, 1978; 1 son, Alan F. B.S., M.I.T., 1971; Ph.D., Cambridge (Eng.) U., 1974. Helen Hay Whitney Found fellow dept. biochemistry Stanford U. Sch. Medicine, 1974-76; asst. prof. biol. chemistry Sidney Farber Cancer Inst./Harvard Med. Sch., Boston, 1977-80; staff mem. dept. embryology Carnegie Instn. of Washington, Balt., 1980-83; faculty dept. biochemistry U. Calif.-Berkeley, 1983—. Contbr. articles to profl. publs. NSF fellow, 1971-73; U.S. Churchill Found. fellow, 1974-76. Mem. Phi Beta Kappa, Phi Lambda Epsilon. Subspecialties: Genome organization; Molecular biology. Current work: Transposable elements in drosophila. Office: 115 W University Pkwy Baltimore MD 21210

RUBIN, LEWIS J., physician; b. N.Y.C., Aug. 5, 1950; s. Theodore and Erna (Kaufman) R.; m. Deborah F. Levine, Dec. 24, 1972. B.A., Yeshiva U., 1972, diploma Hebraic Studies, 1972; M.D., Albert Einstein Coll. Medicine, 1975. Diplomate: Am. Bd. Internal Medicine. Assoc. dept. medicine Duke U. Med. Ctr., Durham, N.C., 1979-80; asst. prof. medicine U. Tex. Health Sci. Ctr., Dallas, 1981—; chief pulmonary sect. VA Med. Ctr., Dallas, 1981—; cons. NHLBI, NIH, Bethesda, Md., 1982—. Editor: Pulmonary Heart Disease, 1983; contbr. chpt. to book, articles to profl. jours. E.L. Trudeau scholar Am. Lung Assn., 1980; NIH grantee, 1982; recipient A. Soyer Meml. award Yeshiva U., 1972; H. Schiff award Duke U., 1978. Fellow ACP; mem. Am. Thoracic Soc., Am. Heart Assn., Am. Fedn. Clin. Research. Jewish. Club: Brookhaven (Dallas). Subspecialties: Pulmonary medicine; Physiology (medicine). Current work: Pulmonary vascular physiology and pharmacology; cardiovascular pulmonary interrelationships and diseases of the pulmonary circulation. Home: 4341 Northview Ln Dallas TX 75229 Office: Dallas VA Med Ctr 111F 4500 S Lancaster Rd Dallas TX 75216

RUBIN, ROBERT HOWARD, astrophysicist, educator; b. Phila., Mar. 26, 1941; s. Abraham D. and Betty B. (Farber) R. B.S., Case Inst. Tech., 1963, Ph.D. in Astrophysics, 1967. Research assoc. Nat. Radio Astronomy Obs., Charlottesville, Va., 1967-69; research assoc. asst. prof. U. Ill., Urbana, 1969-72; assoc. prof. Calif. State U., Fullerton, 1972-81; sr. nat. research council assoc. NASA Ames Research Ctr., Moffett Field, Calif., 1981—. Contbr. writings to profl. publs. Stanford/NASA Ames faculty fellow, summers 1980, 81; Santa Clara U./NASA faculty fellow, summer 1979; NSF grantee, 1970-72. Mem. Am. Astron. Soc., Internat. Astron. Union, Union Radio Sci. Internat. Club: Pacific Wing and Rotor Flying. Subspecialties: Theoretical astrophysics; Radio and microwave astronomy. Current work: Studies of interstellar medium, theoretical modeling of gaseous nebulae with emphasis on predicting infrared line intensities. Home: 436 Sierra Vista Mountain View CA 94043 Office: NASA Ames Research Center Mail Stop 245-6 Moffett Field CA 94035

RUBIN, RONALD PHILIP, pharmacologist, educator; b. Newark, N.J., Jan. 4, 1933; s. Moe and Pearl R.; m. Lois Speyer, Aug. 21, 1955; children: Judith, Ellen, Lawrence. A.B., Harvard U., 1953, A.M.T., 1958; Ph.D., Albert Einstein Coll. Medicine, 1963. Instr. dept. pharmacology SUNY Downstate Med. Ctr., Bklyn., 1964-66, asst. prof., 1966-70, assoc. prof., 1970-74; prof., chief autonomic-cardiovascular div. Med. Coll. Va., Richmond, 1974—. Author: Calcium and the Secretory Process, 1974, Calcium and Cellular Secretion, 1982. Mem. Harvey Soc., Endocrine Soc., Am. Soc. Pharmacology and Exptl. Therapeutics. Subspecialties: Cellular pharmacology; Cell biology (medicine). Current work: Study secretory mechanisms. Home: 1607 Helmsdale Dr Richmond VA 23233 Office: Med Coll VA PO Box 613 MCV Station Richmond VA 23298

RUBIN, SHELDON, dynamics engineer, consultant; b. Chgo., July 19, 1932; s. George and Elsie (Braid) R.; m. Ann Renee Lustgarten, July 3, 1955; children: Geoffrey, Kenneth, Beth. B.S., Calif. Inst. Tech., 1953, M.S., 1954, Ph.D., 1956. Registered profl. engr., Calif. Research engr. sound and vibration Lockheed Aricraft Co., Burbank, Calif., 1956-58; sect. head electronic Packaging Hughes Aircraft Co., Culver City, Calif., 1958-62; sr. project engr. Aerospace Corp., El Segundo, Calif., 1962—; cons. NASA, U.S. Air Force. Contbr. articles on dynamics engrng. to profl. jours.; contbg. author: Shock and Vibration Handbook. Recipient Trustee's Disting. Achievement award Aerospace Corp., 1981; Shuttle Flight Cert. of Achievement NASA, 1981; Outstanding Accomplishment award Aerospace Corp., 1979. Assoc. fellow AIAA; mem. Soc. Automotive Engrs. (cert. of appreciation 1965), Am. Nat. Standards Inst. Subspecialties: Theoretical and applied mechanics; Shock and vibration. Current work: Dynamic system identification, emphasis on damage detection on fixed offshore oil platforms by vibration monitoring prevention of vibratory instability in liquid rockets, shock and vibration survivability. Patentee in field. Office: Po Box 92957 M4/899 Los Angeles CA 90009

RUBIN, ZICK, social psychologist, writer; b. N.Y.C., Apr. 29, 1944; s. Eli Hyman and Adena (Lipschitz) R.; m. Carol Lynn Moses, June 21, 1969; children; Elihu James, Noam Moses. B.A., Yale U., 1965; Ph.D., U. Mich., 1969. Asst. prof., assoc. prof. psychology Harvard U., Cambridge, Mass., 1969-76; Louis and Frances Salvage prof. social psychology Brandeis U., Waltham, Mass., 1976—; chmn. Yale U. Council Com. on Social Scis.-Behavioral, New Haven, 1981—. Author: Liking and Loving: An Invitation to Social Psychology, 1973, Children's Friendships, 1981 (Am. Psychol. Found. Nat. Media Award 3The Psychology of Being Human, 3d edit); contbg. editor: Psychology Today, 1980—. Recipient socio-psychol. prize AAAS, 1969; grantee Found. Child Devel., 1977-80; Ford Found. grantee, 1981-83. Fellow Am. Psychol. Assn., Soc. Psychol. Study of Social Issues (Council 1975-77); mem. Soc. Exptl. Social Psychology. Jewish. Club: Elihu (New Haven). Subspecialty: Social psychology. Current work: Research on father-son relationships, children's friendships, romantic love. Office: Dept Psychology Brandeis Univ Waltham MA 02159

RUBINOFF, IRA, biologist, conservationist; b. N.Y.C., Dec. 21, 1938; s. Jacob and Bessie (Rose) R.; m. Roberta Wolff, Mar. 19, 1961; children: Jason, Ana; m. Anabella Guardia, Feb. 10, 1978; 1 son, Andres. B.S., Queens Coll., 1959; A.M., Harvard U., 1960, Ph.D., 1963. Biologist, asst. dir. marine biology Smithsonian Tropical Research Inst., Balboa, Republic of Panama, 1964-70, asst. dir. sci., 1970-73, dir., 1973—; asso. in ichthyology Harvard U., 1965—; courtesy prof. Fla. State U., Tallahassee, 1976—; mem. sci. adv. bd. Gorgas Meml. Inst., 1964—; trustee Rare Animal Relief Effort, 1976—; bd. dirs. Charles Darwin Found. for Galapagos Islands, 1977—; chmn. bd. fellowships and grants Smithsonian Instn., 1978-79; vis. fellow Wolfson Coll., Oxford (Eng.) U., 1980-81. Author Strategy for Preservation of Moist Tropical Forests; Contbr. articles to profl. jours. Bd. dirs. Internat. Sch., Panama, 1983—. Fellow Linnean Soc. (London); mem. Am. Soc. Naturalists, Soc. Study of Evolution, N.Y. Acad. Scis., Orgn. Tropical Studies. Club: Cosmos (Washington). Subspecialties: Evolutionary biology; Behavioral ecology. Current work: Analysis of diving behavior and physiology of sea snakes using radiotelemtry; conservation strategy for tropical forests. Home: Box 2281 Balboa Republic of Panama Office: Smithsonian Tropical Research Inst APO Miami FL 34002

RUBY, LAWRENCE, nuclear engineering educator, consultant; b. Detroit, July 25, 1925; m. Judith, Apr. 8, 1951; children: Jill, Peter, Frederick. A.B., UCLA, 1945, M.A., 1947, Ph.D., 1950. Registered profl. engr., Calif. Physicist Lawrence Berkeley Lab., Calif., 1950—; assoc. prof. U. Calif.-Berkeley, 1962-67, prof. nuclear engrng., 1967—; cons. Mem. Am. Phys. Soc., Am. Assn. Physics Tchrs., Am. Nuclear Soc., U.S. Metric Assn. Subspecialties: Nuclear fission; Nuclear fusion. Current work: Fission reactor kinetics, radiation insrumentation, fusion reactor engineering. Home: 54 Cowper Ave Berkeley CA 94707 Office: U Calif Berkeley CA 94720

RUCH, RICHARD JULIUS, chemistry educator; b. Perryville, Mo., Jun 9, 1932; s. Julius Maurus and Zita Elizabeth (Boxdorfer) R.; m. Leola Sander, June 20, 1954; children: Stephen, David, Susan, Daniel. B.S., S.E. Mo. State U., 1954; M.S., Iowa State U., 1956, Ph.D., 1959. Asst. prof. chemistry State U. S.D., 1959-62; So. Ill. U., Carbondale, 1962-66; assoc. prof. chemistry Kent (Ohio) State U., 1966—. Author: (with T. Sato) Stabilization of Colloidal Dispersions by Polymer Adsorption, 1980; contbr. articles to profl. jours. Mem. Am. Chem Soc., Sigma Xi. Republican. Lutheran. Subspecialties: Physical chemistry; Surface chemistry. Current work: Properties of colloidal dispersions and film coatings, using dielectric, wetting, and rheological techniques. Home: 1955 Pineview Dr Kent OH 44240 Office: Dept Chemistry Kent State U 214B WMH Kent OH 44242

RUCKLE, WILLIAM HENRY, mathematics educator, researcher; b. Neptune, N.J., Oct. 29, 1936; s. Ernest George and Margret Elizabeth (Fallen) R.; m. Cynthia Ann Grill, Aug. 30, 1960; 1 dau., Marjorie Ann. A.B., Lincoln U., Lincoln University, Pa., 1960; Ph.D., Fla. State U., 1963. Assoc. prof. Lehigh U., 1963-69; prof. math. scis. Clemson (S.C.) U., 1969—; vis. prof. Goethe U., Frankfurt/Main, W.Ger., 1975-76, Western Wash. U., 1978-79. Author: Sequence Spaces, 1981, Geometric Games and Their Applications, 1982. Mem. Am. Math. Soc., Soc. Indsl. and Applied Math., Math Assn. Am., Soc. Risk Analysis, Sigma Xi. Subspecialties: Operations research (mathematics); Operations research (engineering). Current work: Two-person games with geometric content; the foundations or risk analysis; value theory; utility theory. Home: 106 Whippowill Dr Seneca SC 29678

RUCKMICK, JOHN CHRISTIAN, geologist; b. Iowa City, Iowa, Nov. 26, 1926; s. Christian Alban and Katherine T. R.; m. Jane E. Douglas, Sept. 10, 1955; children: Stephen C., Melissa K. B.A., Amherst Coll., 1952; M.S., Calif. Inst. Tech., 1954, Ph.D., 1957. Chief geologist Orinoco Mining Co., Ciudad Piar, Venezuela, 1957-67; exploration mgr. Homestake Mining Co., Kalgoorlie, Western Australia, 1967-70, Tucson, 1970-74, Texasgulf, Inc., Golden, Colo., 1974-82; exec. v.p. Exploration Ventures Co., Golden, 1982. Contbr. articles to profl. jours. Served with USNR, 1944-46. Fellow Geol. Soc. Am.; mem. Soc. Econ. Geologists, AIME, Am. Assn. Petroleum Geologists, Northwest Mining Assn. Republican. Subspecialties: Geology; Geochemistry. Current work: Research on soil gases as geochemical pathfinders to discovery of petroleum. Home: 2266 Pebble Beach Ct Evergreen CO 80439 Office: Exploration Ventures Co Inc 404 Violet St Golden CO 80401

RUDAVSKY, ALEXANDER BOHDAN, civil engr., educator; b. Poland, Jan. 17, 1925; s. Leo and Zenovia (Orlov) R.; m. Juanita Jean Enga, Nov. 5, 1955; 1 dau., Natica. Sc.D. (Dr.Ing.), Franzius Inst., Ger., 1966; B.S. and M.S., U. Minn., 1956. Civil engr. Justin & Courtney, Phila., 1956-57, Iran, 1957-58; mem. faculty San Jose (Calif.) State U., 1960—, prof., 1975—; dir., owner, pres. Hydro Research Sci., Santa Clara, Calif., 1964—; cons. Contbr. numerous articles to profl. jours. ASCE Freeman scholar, Europe, 1958. Mem. U.S. Com. Large Dams, Internat. Assn. Hydraulic Research, ASCE. Subspecialties: Fluid mechanics; Hydrology. Current work: Hydraulic research through model studies of engineering problems related to hydraulic structures, ports and harbors, coastal protection and thermal pollution. Office: 3334 Victor Ct Santa Clara CA 95050

RUDD, DALE FREDERICK, chem. engr.; b. Mpls., Mar. 2, 1935; s. Henry G. and Emmy E. (Grip) R.; m. Sandra Coryell, Aug. 1, 1964; children—Karen, David. Ph.D., U. Minn., 1959. Prof. chem. engring. U. Wis., Madison, 1961—; pres. Shanahan Valley Assos., 1979—; cons. Exxon Chem. Co. Author: (with Watson) Strategy of Process Engineering, 1968, (with Powers and Siirola) Process Synthesis, 1973, (with Berthouex) Strategy of Pollution Control, 1977, (with Fathi-Afshar, Treviño and Stadtherr) Petrochemical Technology Assessment, 1981. Am. Soc. Engring. Edn. lectr.; Guggenheim fellow, 1971-72. Mem. Am. Inst. Chem. Engring. (Colburn award 1972), Nat. Acad. Engring. Subspecialty: Chemical engineering. Office: Dept Chem Engring U Wis Madison WI 53706

RUDDAT, MANFRED, biology educator; b. Insterburg, Ger., Aug. 21, 1932; came to U.S., 1961; s. Otto and Helene (Naujoks) R.; m. Helga Kuntzel, Nov. 3, 1962; children: Michael, Monica. Ph.D., U. Tübingen, Ger., 1960. Sci. asst. botany U. Tübingen, 1960-61; NSF fellow Calif. Inst. Tech., Pasadena, 1961-64; asst. prof. botany U. Chgo., 1964-68, asst. prof., 1968-70, assoc. prof. biology, 1970—. Editor: Bot. Gazette, 1974. Recipient Quantrell award, 1969. Mem. Am. Soc. Plant Physiologists, Bot. Soc. Am., Japanese Soc. Plant Physiologists, AAAS, Internat. Plant Growth Substances Assn. Subspecialties: Plant physiology (biology); Plant growth. Current work: Developmental biology and biochemistry of plants and fungi, physiology and biochemistry of plant growth regulators. Office: Dept Biology U Chgo Barnes Lab 5630 S Ingleside Ave Chicago IL 60637

RUDDLE, FRANCIS HUGH, geneticist; b. West New York, N.Y., Aug. 19, 1929; s. Thomas Hugh and Mary Henley (Rodda) R.; m. Nancy Marion Hartman, Aug. 1, 1964; children—Kathlyn Gabrielle, Amy Elizabeth. B.A., Wayne State U., Detroit, 1953, M.S., 1956; Ph.D., U. Calif., Berkeley, 1960; hon. degree, Lawrence U., 1982, Weizmann Inst., Israel, 1983. Asst. prof. Yale U., New Haven, 1961-67, asso. prof., 1967-72, prof. biology and human genetics, 1972—, chmn. dept. biology, 1977—; adv. com. Am. Type Culture Collection, 1963-71; planning com. 1st Internat. Congress on Cell Biology, 1975; mem. cell study sect. NIH, 1965-70, chmn. mutant cell lines com., 1972-77; R.E. Dyer lectr., 1978; Merck lectr. Montreal Cancer Inst., 1979; Condon lectr. U. Oreg., 1981. Editorial bd.: Exptl. Cell Research, 1975—, Genetics, 1973—, In Vitro, 1970—, Somatic Cell Genetics, 1975—, Am. Jour. Cell Biology, Biochemistry. Served with USAAF, 1946-49. NIH postdoctoral fellow U. Glasgow, Scotland, 1960-61. Fellow N.Y. Acad. Scis.; Mem. Nat. Acad. Scis., Am. Acad. Arts and Scis., Conn. Acad. Sci. and Engring., Harvey Soc., Soc. Devel. Biology (pres. 1971-72, trustee 1970-73), AAAS, Am. Soc. Cell Biology, Am. Soc. Human Genetics (dir. 1972-75), Am. Soc. Naturalists, Am. Soc. Zoologists, Genetics Soc. Am., Am. Soc. Biochemistry, Pattern Recognition Soc., Tissue Culture Assn. (dir.), Am. Cancer Soc., Sigma Xi. Subspecialty: Cell biology. Office: Kline Biology Tower Yale U New Haven CT 06520

RUDIGER, CARL ERNEST, JR., electrical engineer; b. Mt. Kisco, N.Y., Oct. 24, 1939; s. Carl E. and Edna (Tanzer) R.; m. Catherine Childre, May 4, 1961 (div.); children: Carl, Catherine; m. Jeanne Kurtzon, Oct. 12, 1975; stepchildren: Lisa Cosmas, Jennifer Cosmas. B.S.E.E., Duke U., 1961; M.B.A., U. N.C., 1962. With Lockheed Missiles and Space Co., Palo Alto, Calif., 1962—, program mgr. ocean systems div., Sunnyvale, Calif., 1969-78, program mgr. biotech. div., Palo Alto, 1978—; dir. Alkco Mfg. Co., Franklin Park, Ill. Fellow Marine Tech. Soc. (chmn. San Francisco Bay region 1977-79, mem. First Group Fellows 1975, Cert. Appreciation 1980); mem. AIAA. Subspecialties: Aerospace engineering and technology; Space colonization. Current work: Program manager to develop life sciences research facilities for use on space stations and space shuttles. Home: 13060 La Vista Dr Saratoga CA 95070 Office: Lockheed Missiles and Space Co 3251 Hanover St Palo Alto CA 94304

RUDOFSKY, ULRICH HUBERT WALDEMAR, research scientist, microbiology and immunology educator; b. Prague, Czechoslovakia, Apr. 17, 1935; came to U.S., 1951, naturalized, 1965; s. Waldemar Ulrich and Edith Victoria (Reiss) R.; m. Barbara K. Kiernat, June 6, 1965; children: Frederick W.H., Christina M.S. Student, U. Minn., Mpls., 1953-56, U. Chgo., 1964; B.S., Roosevelt U., Chgo., 1970. Research asst. U. Chgo., 1957-58, 60-70; sr. research scientist N.Y. State Kidney Disease Inst.-N.Y. State Dept. Health, Albany, 1970—; research assoc. dermatology dept. medicine Albany Med. Coll., 1973—, asst. prof. dept. microbiology and immunology, 1982—. Contbr. articles to profl. jours. Pres. Tri-City Swim League, Albany, 1982-83. Served with U.S. Army, 1958-60. NIH grantee, 1972-83; Health Research Inc. grantee, 1982. Mem. Am. Assn. Immunologists, N.Y. Acad. Sci. Club: Delmar Dolfins Swim (Delmar, N.Y.). Subspecialties: Immunology (medicine); Pathology (medicine). Current work: Research on experimental and spontaneous kidney diseases, hypertension, autoimmune diseases, immunogenetics, tissue antigens, monoclonal auto-antibodies, dermatology. Home: 2 Brookside Dr Delmar NY 12054 Office: Kidney Disease Institute New York State Dept Health Empire State Plaza Lab Albany NY 12201

RUDOLPH, LUTHER DAY, computer and information science educator; b. Cleve., Aug. 10, 1930; s. George H. and Katherine (Day) R.; m. Janel S. Cole, May 1, 1954 (div.); children: Jennifer, Katherine, Elizabeth. B.S.E.E., Ohio State U., 1958; M.E.E., U. Okla., 1964; Ph.D., Syracuse U., 1968. Engr. Gen. Electric Co., 1958-64; research engr. Syracuse Research Corp., 1964-68; mem. faculty sch. computer and info. sci. Syracuse U., 1968—, prof., 1974—, dir. communications studies lab., 1979—. Contbr. articles to profl. jours. Served with USAF, 1951-54. Mem. IEEE (sr. mem., assoc. editor algebraic coding 1974-76), Parapsychol. Assn. (assoc.), Am. soc. Psychical Research (trustee 1981—), Eta Kappa Nu, Pi Mu Epsilon, Tau Beta Pi. Subspecialties: Electrical engineering; Applied mathematics. Current work: Theoretical and applied research in algebraic coding theory; application of communication theory to apparent anomalies in human communication. Patentee in field. Home: 105 Beach St Fayetteville NY 13066 Office: Sch Computer Info Sci Syracuse U 313 Link Hall Syracuse NY 13210

RUDY, YORAM, biomedical engineering educator; b. Tel-Aviv, Israel, Feb. 12, 1946; came to U.S., 1973, naturalized, 1981; s. Nahum and Yafa (Krinkin) R. B.Sc. in Physics, Technion, Israel Inst. Tech., 1971, M.Sc., 1973; Ph.D. in Biomed. Engring, Case Western Res. U., 1978. Research assoc. in biomed. engring. Case Western Res. U., Cleve., 1978-79, vis. asst. prof. biomed. engring., 1979-81, asst. prof. biomed. engring., 1981—; vis. prof. biomed. engring. Technion, Haifa, Israel, 1982-83. Guest editor: Annals of Biomed. Engring, 1983; contbr. sci. articles to profl. jours. NIH research grantee, 1978. Mem. Biomed. Engring. Soc. (sr.), Am. Physiol. Soc., IEEE Group on Engring. in Medicine and Biology, Am. Heart Assn. (basic sci. council), AAAS, Sigma Xi. Subspecialties: Biomedical engineering; Biophysics (physics). Current work: Bioelectric phenomena, cardiac electrophysiology, biophysical basis of electrocardiography, electrocardiographic body surface potential mapping. Office: Case Western Reserve University University Circle Cleveland Ohio 44106

RUE, ROLLAND RAY, chemistry educator; b. Marshfield, Wis., Apr. 25, 1935; s. Ray W. and Mrytle D. (Hemquist) R.; m. Donna Mae Erickson, Aug. 23, 1958; children: Shari, Julie. B.A., Macalester Coll., 1957; Ph.D., Iowa State U., 1962. Grad. asst. Iowa State U., Ames, 1957-62; faculty S.D. State U., Brookings, 1962—, assoc. prof. chemistry, 1968—; researcher in field. Treas. Lutheran Campus

Council, S.D. State U.; mem. state bd. S.D. State U. Mem. Am. Chem. Soc., S.D. Acad. Sci. Club: Rifle and Pistol. Lodge: Lions. Subspecialties: Physical chemistry; Soil chemistry. Current work: Thermodynamics, solutions, soil chemistry, modeling of salt and water movement through soils. Home: 2043 Elmwood Dr Brookings SD 57006 Office: Chemistry Dept SD State Univ Brookings SD 57007

RUECKERT, ROLAND RUDYARD, virology and biochemistry educator; b. Rhinelander, Wis., Nov. 24, 1931; s. George L. and Monica A. (Seiberlich) R.; m. Ruth Helen Ullrich, Sept. 5, 1959; 1 dau., Wanda Lynn. B.S. in Chemistry, U. Wis.-Madison, 1953, Ph.D. in Oncology, 1960. Postdoctoral fellow Max Planck Inst. for Biochemistry, Munich, W.Ger., 1960-61; postdoctoral researcher Max Planck Inst. Virology, Tubingen, W.Ger., 1961-62; asst. virologist Virus Lab., U. Calif., Berkeley, 1962-65; asst. prof. biophysics and biochemistry U. Wis.-Madison, 1965-69, assoc. prof., 1969-72, prof., 1972—. Served with U.S. Army, 1953-55. NIH grantee, 1965—; Am. Cancer Soc. grantee, 1969—. Mem. AAAS, Am. Assn. Biol. Chemistry, Am. Soc. Microbiology, Am. Soc. Virology, Soc. Gen. Microbiology (Eng.). Subspecialties: Virology (biology); Molecular biology. Current work: Picorna viruses, polio, common cold, nodaviruses, monoclonal antibodies, neutralization, immunochemistry, protein chemistry.

RUEGSEGGER, DONALD RAY, JR., radiol. physicist; b. Detroit, May 29, 1942; s. Donald Ray and Margaret Arlene (Elliot) R.; m. Judith Ann, Aug. 20, 1965; children: Steven, Susan, Mark, Ann. B.S., Wheaton Coll., 1964; M.S., Ariz. State U., Tempe, 1966, Ph.D., 1969. Diplomate: Am. Bd. Radiology. Radiol. physicist cons. VA Hosp., Dayton, Ohio, 1970-77; adj. asst. prof. physics Wright State U., Fairborn, Ohio, 1973—, clin. asst. prof. radiology, 1976-81, clin. assoc. prof. radiology, 1981—; radiol. physicist, chief med. physics sect. Miami Valley Hosp., Dayton, Ohio, 1969—; civilian cons. Wright Patterson AFB Med. Center, Fairborn, Ohio, 1982—. NDEA fellow, 1966-69. Mem. Am. Assn. Physicists in Medicine (chpt. pres. 1982-83), Am. Coll. Radiology, Am. Phys. Soc., Health Physics Soc., AAAS. Republican. Baptist. Subspecialties: Radiation therapy; Nuclear medicine. Current work: Applying physical principals and techniques in radiation therapy and nuclear medicine to develop better ways to treat cancer patients with radiation. Home: 2018 Washington Creek Ln Centerville OH 45459 Office: 1 Wyoming St Dayton OH 45409

RUEGSEGGER, PAUL MELCHIOR, physician, researcher; b. Berne, Switzerland, June 27, 1921; came to U.S., 1948, naturalized, 1954; s. Paul and Frieda Beatrice (Schmocker) R.; m. Freya Bundi Wipf, Sept. 6, 1948; children: Theodore Bernard, Christine Monica, Carole Suzanne. M.D., U. Zurich, Switzerland, 1946. Diplomate: Am. Bd. Internal Medicine. Intern Bellevue Hosp., N.Y.C., 1948-51, resident, 1951-52; resident in cardiology Meml. Sloan Kettering Cancer Center, N.Y.C., 1952-53, 55-56; asst. prof. clin. medicine Cornell U. Med. Sch., N.Y.C., 1959; research assoc. Sloan Kettering Inst., N.Y.C., 1959-67; attending physician Meml. Center for Cancer, N.Y.C., 1959-69, N.Y. Hosp., 1956-69; research dir. Med. Imaging Lab. (name now Biotronics Inst.), N.Y.C., 1970—; aero-med. cons. Swissair Lines, N.Y.C., 1956—; thermography cons. Trial Lawyers Assn., N.Y.C., 1982—; cons. med. imaging Hoffmann-LaRoche Corp., Nutley, N.J., 1969—. Author: Transaminase Tests, 1956, Coronary Thrombolysis, 1959, Walking EKG Stress Test, 1963, Thermography of Pain, 1969, 81. Served to capt. USAF, 1953-55; Japan. Mem. N.Y. Acad. Scis., Am. Fedn. Clin. Research, Harvey Soc. N.Y., N.Y. County Med. Soc., Am. Acad. Thermology, European Thermology Soc., Am. Soc. Internat. Medicine, Zool. Soc. N.Y., N.Y. Bot. Garden, Swiss Soc. N.Y. Subspecialties: Thermography of diseases; Biofeedback. Current work: Cybernetics of chronicity, recurrence in chronic pain, other disorders; infrared television imaging of altered physiological states. Office: Biotronics Inst 115 E 61st St New York NY 10021

RUESINK, ALBERT WILLIAM, biologist, educator; b. Adrian, Mich., Apr. 16, 1940; s. Lloyd William and Alberta May (Foltz) R.; m. Kathleen Joy Cramer, June 8, 1963; children: Jennifer Li, Adriana Eleanor. B.A., U. Mich., 1962; M.A., Harvard U., 1965, Ph.D., 1966. Postdoctoral Swiss Fed. Inst. Tech., Zurich, 1966-67; mem. faculty Ind. U., Bloomington, 1967—, prof. plant scis., 1980—. Contbr. articles to sci. jours. Recipient Disting. Teaching award Amoco, 1980. Mem. Am. Soc. Plant Physiology, Bot. Soc. Am., Am. Inst. Biol. Scis., Ind. Acad. Sci., Sigma Xi. Democrat. Mem. United Ch. Christ. Subspecialties: Plant physiology (biology); Plant cell and tissue culture. Current work: Plant physiology, especially cell wall-plasma membrane interactions; education. Home: 2605 E 5th St Bloomington IN 47401 Office: Ind U Dept Biology Bloomington IN 47405

RUFF, ROBERT LOUIS, physiologist, neurologist, educator; b. N.Y.C., Dec. 16, 1950; s. John Joseph and Rhoda (Alpert) R.; m. Louise Seymour Acheson, Apr. 26, 1980. B.S., Cooper Union U., 1971; M.D., U. Wash.-Seattle, 1976, Ph.D., 1976. Diplomate: Am. Bd. Neurology and Psychiatry. Intern U. Wash., Seattle, 1976-77; neurology resident Cornell Med. Coll., N.Y.C., 1977-79, neurology chief resident, 1979-80; asst. prof. dept. physiology, biophysics, neurology U. Wash., Seattle, 1980—. Recipient Tchr.-Investigator award NIH, 1980. Mem. Biophys. Soc., Am. Acad. Neurology, Soc. Neurosci., Stroke Counsel, AAAS, N.Y. Acad. Scis., Alpha Omega Alpha, Sigma Pi Sigma. Subspecialties: Physiology (medicine); Neurology. Current work: Endocrine myopathy, electrical studies of human muscle, animal models of stroke. Office: U Wash Dept Physiology and Biophysics SJ-40 Seattle WA 98195

RUFFOLO, ROBERT RICHARD, JR., pharmacologist; b. Yonkers, N.Y., Apr. 14, 1950; s. Robert Richard and Lorraine Regina (Varipapa) R.; m. Christine Bernice Nettleship, July 1, 1972. Ph.D., Ohio State U., 1976. Postdoctoral fellow Ohio State U., Columbus, 1976-77; staff fellow NIH, Bethesda, Md., 1977-78; sr. pharmacologist Lilly Research Labs., Indpls., 1978-82, research scientist, 1982—. Contbr. numerous articles to sci. publs. Mem. Am. Soc. Pharmacology and Exptl. Therapeutics. Subspecialty: Pharmacology. Current work: Adrenergic pharmacology and adrenergic receptors; hypertension, cardiovascular research. Home: 9903 Carefree Dr Indianapolis IN 46256 Office: Lilly Research Labs (MC-304) Indianapolis IN 46285

RUGGE, HENRY FERDINAND, high technology company executive; b. South San Francisco, Calif., Oct. 28, 1936; s. Hugo Heinrich and Marie Mathilde (Breiholz) R.; m. Sue, Dec. 29, 1967. A.B. in Physics, U. Calif.-Berkeley, 1958, Ph.D., 1963. Bus. mgr. Andros, Inc., Berkeley, 1969-72; v.p. Norse Systems, Inc., Hayward, Calif., 1972-75; v.p., gen. mgr. Rosor Assocs., Inc., Sunnyvale, Calif., 1975-80; now dir.; pres. Ultra Med Inc., Sunnyvale, 1980-81, Berliscan, Inc., 1981—; cons. in field. Contbr. articles to profl. jours. Mem. Am. Heart Assn., Phi Beta Kappa. Subspecialties: Biomedical engineering; CAT scan. Current work: Medical utlrasonic diagnostic agents and instrumentation. Home: 1626 Chestnut Berkeley CA 94702 Office: 253 Humboldt Ct Sunnyvale CA 94036

RUGH, JOHN DOUGLAS, dental psychology educator; b. Corvallis, Oreg., Feb. 20, 1940; s. Harold Kelly and Vida Esther (Toney) R.; m. Annie Louise Taylor, Dec. 30, 1960. A.A., Coll. San Mateo, 1965; B.A., U. Calif.-Santa Barbara, 1968, Ph.D., 1975. Dir. Psycho-tech. Lab., Claremont Grad. Sch., 1974-77; asst. prof. dept. psychiatry and dept. oral and maxillofacial surgery U. Tex. Health Sci. Ctr., San Antonio, 1977-80, assoc. prof., 1980—, dir., 1977-81; mem. Oral Biology and Medicine Study Sect., Nat. Inst. Dental Research, NIH, Bethesda, Md., 1982-86. Author: books, including Biofeedback in Dentistry, 1977, Oral Motor Behavior, 1979; contbr. numerous chpts., articles, abstracts to profl. publs. Served with USN, 1959-63. Mem. Biofeedback Soc. Am. (dir. 1980-82, pres. 1983-84), Am. assn. Dental Research (pres. San Antonio chpt. 1982-83), Soc. Behavioral Medicine, Am. Assn. Advancement Med. Instrumentation, Omicron Kappa Upsilon (hon.). Subspecialties: Physiological psychology; Bioinstrumentation. Current work: Investigations into etiology, diagnosis and treatment of stres-related oral disorders (bruxism, temporo mandibular joint disorders and soft-tissue lesions); also biomedical instrumentation development and application. Inventor, patentee bio-alarm security system, 1978. Office: U Tex Health Sci Ctr Dept Oral and Maxillofacial Surgery 7703 Floyd Curl Dr San Antonio TX 78284

RUIBAL, RODOLFO, biologist; b. Cuba, Oct. 27, 1927; s. Rodolfo and Antonia R.; m. Irene Shamu, Oct. 25, 1948; 1 son, Claude. B.A., Harvard U., 1950; Ph.D., Columbia U., 1955. Prof. biology U. Calif., Riverside, 1967—, chmn. dept. biology, 1979—. Served with U.S. Army, 1946-48. Guggenheim fellow, 1967-68. Subspecialties: Evolutionary biology; Ecology. Current work: Ecology of desert anurans; structure of amphibian skin. Office: Dept Biology U Calif Riverside CA 92521

RUMPEL, MAX LEONARD, chemist, educator; b. WaKeeney, Kans., Mar. 17, 1936; s. Philip A. and Rosa C. (Deines) R.; m. Joan C. Hubbell, Aug. 26, 1961; children: Craig, Karen. A.B., Fort Hays State U., 1957; Ph.D., U. Kans., 1962. Successively instr., asst. prof., assoc. prof. chemistry Fort Hays State U., 1961-67, prof. chemistry, 1968—, chmn. dept., 1972—. Contbr. articles to profl. jours. NSF grantee, 1965-67, 68-70, 1971. Mem. Am. Chem. Soc., Kans. Acad. Sci., Sigma Xi (treas. Fort Hays 1978, sec. 1979, pres.-elect 1980), Phi Kappa Phi, Phi Lambda Upsilon, Sigma Pi Sigma. Subspecialties: Analytical chemistry; Inorganic chemistry. Current work: Electrochemistry, unusual valences of metals, computer simulations. Home: 511 W 31st St Hays KS 67601 Office: Fort Hays State U 600 Park St Hays KS 67601

RUMSEY, VICTOR HENRY, electrical engineering educator; b. Devizes, Eng., Nov. 22, 1919; s. Albert Victor and Susan Mary (Norman) R.; m. Doris Herring, Apr. 2, 1942; children: John David, Peter Alan, Catherine Anne. B.A., Cambridge U., 1941, D.Sc. in Physics, 1972; D.Eng., Tohoku U., Japan, 1962. With U.K. Sci. Civil Service, 1941-48; asst. to asso. prof. Ohio State U., 1948-54; prof. U. Ill., 1954-57, U. Calif., Berkeley, 1957-66, prof. elec. engring. and computer scis., San Diego, 1966—, dept. chmn., 1977-81. Author 1 book in field; contbr. articles to profl. jours. Guggenheim fellow.; recipient George Sinclair award Ohio State U., 1982. Fellow IEEE (Morris Liebman prize), Union Radio Scientifique Internationale, Internat. Astron. Union; mem. Nat. Acad. Engring. Subspecialty: Computer engineering. Patentee in field. Home: 465 Hidden Pines Ln Del Mar CA 92014 Office: U Calif San Diego CA 92093

RUNKLE, JAMES READE, biology educator; b. Grove City, Pa., July 3, 1951; s. Irvin Lester and Beverly Aliene (Bell) R.; m. Janet Lynn Kreps, June 9, 1973; children: Benjamin, Matthew, William. B.A., Ohio Wesleyan U., 1973; Ph.D., Cornell U., 1979. Vis. asst. prof. biology U. Ill., Chgo., 1978-79; asst. prof. biol. scis. Wright State U., Dayton, Ohio, 1979—. Contbr. articles to profl. jours. Deacon First United Presbyn. Ch., Fairborn, Ohio, 1980—. NSF grantee, 1982—. Mem. AAAS, Am. Inst. Biol. Scis., Am. Soc. Naturalists, Brit. Ecol. Soc., Ecol. Soc. Am., Phi Beta Kappa, Pi Mu Epsilon. Presbyterian. Subspecialties: Ecology; Population biology. Current work: Effect of disturbance, especially small scale disturbance on forest vegetation, impact on species and communities; pattern analysis. Home: 240 Holmes Dr Fairborn OH 45324 Office: Dept Biol Scis Wright State Univ Colonel Glenn Hwy Dayton OH 45435

RUNYON, CAROLINE LOUISE, veterinary educator; b. Red Bank, N.J., May 16, 1948; d. James Edward and Louise Florence (Leckie) R.; m. Brian Lee Hill, Apr. 15, 1978. B.S., Marymount Coll., 1970; D.V.M., Colo. State U., 1977; M.S., Iowa State U., 1982. Gen. practice vet. medicine, Princeton, N.J., 1977-78; clin. intern Iowa State U., Ames, 1978-79, surg. resident, 1979-82, asst. prof. vet. orthopedic surgery, 1982—. Mem. Am. Vet. Med. Assn., Am. Animal Hosp. Assn., Am. Assn. Feline Practitioners, Am. Assn. Vet. Clinicians, Vet. Orthopedic Soc. Subspecialty: Surgery (veterinary medicine). Current work: Orthopedic surgery, oncology. Home: RR 1 Stagecoach Rd Ames IA 50010 Office: Iowa State U Ames IA 50011

RUPPEL, EARL GEORGE, research plant pathologist; b. Milw., Nov. 10, 1932; s. George Albert and Ida Elizabeth (Ptaschinski) R.; m. Joyce Ruth Port, Sept. 6, 1958; children: Susan T., Julia R., Michael R. B.S., U. Wis.-Milw., 1958; Ph.D., U. Wis.-Madison, 1962. Plant pathologist Agrl. Research Service, U.S. Dept. Agr., Mayagues, P.R., 1963-65, Mesa, Ariz., 1965-69, Ft. Collins, Colo., 1969—; faculty affiliate botany and plant pathology dept. Colo. State U., 1969—, mem. grad. faculty, 1971—. Contbr. numerous articles to sci. jours. Served with AUS, 1953-55. Am. Cancer Found. grantee, 1960, 62. Mem. Am. Phytopath. Soc., Internat. Soc. Plant Pathology, Mycol. Soc. Am., Am. Soc. Sugar Beet Technologists, Rocky Mountain Plant Protection Group, Sigma Xi, Gamma Sigma Delta. Democrat. Roman Catholic. Club: Fort Collins Camera. Subspecialties: Plant pathology; Plant virology. Current work: Epidemiology of sugarbeet diseases; nature of resistance to plant pathogens; breeding for resistance to plant pathogens. Office: USDA Crops Research Lab Colo State U Fort Collins CO 80523

RUSH, JAMES EDWARD, information scientist, educator; b. Warrensburg, Mo., July 18, 1935; m. Delores A. Lee, June 7, 1958; children: Susan, Pamela, Adam, Jennifer. B.S. in Chemistry and Math, Central Mo. State Coll., 1956, Ph.D., U. Mo., 1962. Asst. editor organic index editing dept. Chem. Abstracts Service, Columbus, Ohio, 1962-65, asst. head chem. info. procedures dept., 1965-67, tech. liason officer, 1967-68; asst. prof. computer and info. sci. Ohio State U., Columbus, 1968-69, assoc. prof., 1969-73, adj. assoc. prof., 1973-79, adj. prof., 1980—; pres. James E. Rush Assocs., Powell, Ohio, 1969—; dir. R&D, OCLC Inc., Dublin, Ohio, 1973-80, dir. research, 1980—; adj. prof. library sci. U. Ill., Champaign, 1980—; cons. in field. Contbr. articles to profl. jours. Served with USNG, 1952-60. Mem. Am. Soc. Info. Sci. (best paper award 1971), Am. Chem. Soc. (editor newsletter 1970-73), Sigma Zeta, Kappa Mu Epsilon, Phi Beta Pi, Sigma Xi. Home and Office: 2223 Carriage Rd Powell OH 43065

RUSH, JOHN EDWIN, JR., physicist, educator; b. Birmingham, Ala., Aug. 11, 1937; s. John Edwin and Sarah Martha (Hargrove) R.; m. Sharon Mayhall, June 8, 1963 (div.); children: Michael, Ray, Joanna, John; m. Donna Lee, Jan. 9, 1982; 1 dau.: Victoria. Student, Snead Jr. Coll., 1954-56; B.S., Birmingham-So. Coll., 1959; Ph.D. (NDEA fellow), Vanderbilt U., 1965. Asst. prof. physics U. of South, 1964-67; asst. prof. physics U. Ala., Huntsville, 1967-68, assoc. prof., 1968—, dean grad. studies and research, 1972-76; cons. NASA, Applied Research Corp. Contbr. articles on elem. particle theory to profl. jours. Bd. dirs., pres. The Key, Center for Creative Living, 1975-80; bd. dirs. Mental Health Assn., Huntsville, 1977—. NASA/Am. Soc. for Engring. Edn. summer faculty fellow, 1978, 79, 82. Mem. Am. Phys. Soc., Sigma Xi, Alpha Tau Omega. Democrat. Episcopalian. Subspecialties: Atomic and molecular physics; Materials processing. Current work: Atomic and molecular collision theory; containerless processing of materials; fluid mechanics; heat transfer; acoustics; nuclear weapons effects. Home: 11021 Louis Dr Huntsville AL 35803 Office: Kaman Scis Corp Huntsville AL 35899

RUSH, RICHARD WILLIAM, consulting geologist; b. Austin, Minn., July 14, 1921; s. James Francis and Irene (Peterson) R.; m. Florence Allison Rayman, Sept. 1945 (div. 1972); children: Richard William, Lucy E., Frederick J., Cynthia I. B.A., U. Iowa, 1945; M.A., Columbia U., 1948, Ph.D., 1954; diploma, Nat. Tech. Schs., Los Angeles, 1976. Registered profl. geologist, Calif. Instr. Colby Coll., Waterville, Maine, 1949-51; asst. prof. U. Tex.-Austin, 1952-57; assoc. prof. No. Ariz. U., Flagstaff, 1963-69; cons. Plateau Corp. River Products, Colo., Iowa, 1961-64; pvt. cons. geologist, Phoenix, 1969—; pres. ULC Opr., Los Angeles, 1974-75, WESGOE, Denver, 1982; cons. FMC Corp., White & Co., Austin, Tex., 1954-57, U.S. Steel Co., Que., 1960, and others. Mem. adv. bd. Republican Nat. Com., 1980—; supt. Episcopal Ch. Schs., Flagstaff, Ariz., 1965-67. Fellow Geol. Soc. Am., AAAS; mem. Am. Inst. Profl. Geologists (charter), N.Y. Acad. Sci., Am. Assn. Petroleum Geologists, Am. Geophys. Union, Sigma Gamma Epsilon. Subspecialties: Tectonics; Information systems, storage, and retrieval (computer science). Current work: Application of computer electronics to regional geologic tectonics.

RUSHMER, ROBERT FRAZER, bioengineer; b. Ogden, Utah, Nov. 30, 1914; s. John Todd and Emma (Osborn) R.; m. Estella Virginia Dix, Apr. 5, 1942; children—Donald Scott, Anne, Elizabeth. Student, Reed Coll., 1931-33; B.S., U. Chgo., 1939; M.D., Rush Med. Coll., 1939; Ph.D. (hon.), U. Linkoping, Sweden, 1977. Intern St. Luke's Hosp., San Francisco; fellow in pediatrics and aviation medicine Mayo Found., 1940-42; asst. prof. physiology U. So. Calif., 1946, U. Wash., Seattle, 1947-50, asso. prof., 1950-56, prof., 1956—, dir. Center Bioengring., 1967-75, asso. dean engring., 1980—, dir., 1978—; mem. heart council NIH, HEW, Nat. Inventors Council; sci. council Pacific Sci. Center. Author: Cardiovascular Dynamics, 4th edit, 1976, Medical Engineering, 1972, Humanizing Health Care, 1975, National Priorities for Health, 1980; editor: Methods of Medical Research, 1966; contbr. articles to profl. jours. Served with USAF, 1942-46. Recipient Disting. Service award U. Chgo., 1963; Outstanding Achievement award Mayo Found., 1964; Modern Medicine Disting. Achievement award, 1966; Alza lectr., 1978; Am. Assn. Med. Instrumentation Found. award, 1976. Fellow Am. Coll. Cardiology (hon.); Mem. IEEE; mem. Inst. Medicine of Nat. Acad. Sci.; Mem. ASTM (disting. lectr.), Alpha Omega Alpha (Ida B. Gould award 1958). Subspecialties: Biomedical engineering; Health services research. Current work: Health care systems analysis, particularly appropriate applications of health technologies. Home: 7050 56th Ave NE Seattle WA 98115 Office: Center Bioengring U Wash Seattle WA 98105

RUSSELL, ALLEN STEVENSON, retired aluminum company executive; b. Bedford, Pa., May 27, 1915; s. Arthur Stainton and Ruth (Stevenson) R.; m. Judith Pauline Sexauer, Apr. 5, 1941. B.S., Pa. State U., 1936, M.S., 1937, Ph.D., 1941. With Aluminum Co. Am., 1940—, asso. dir. research, 1973-74; v.p. Alcoa Labs., Alcoa Center, Pa., 1974-78, v.p. sci. and tech., Pitts., 1978-81; v.p., chief scientist, 1981-83; adj. prof. U. Pitts., 1981. Contbr. articles to profl. jours. Named IR-100 Scientist of Yr., 1979; Pa. State U. alumni fellow, 1980; K.J. Bayer medalist, 1981; recipient chem. Pioneer award Am. Inst. Chemists, 1983. Fellow Am. Soc. Metals (Gold medal 1982), AIME, Am. Inst. Chemists; mem. Am. Chem. Soc., Dirs. Indsl. Research, Nat. Materials Adv. Bd., Nat. Acad. Engring. (council), Sigma Xi. Republican. Presbyterian. Subspecialty: Physical chemistry. Patentee in field. Home: 929 Field Club Rd Pittsburgh PA 15238

RUSSELL, B. DON, engr., educator; b. Denison, Tex., May 25, 1948; s. Bill D. and Mickye R.; m. Rebecca Joan Crawford, Jan. 6, 1973; children: Christyn Joan, Jennifer Rebecca, John Paul. B.S. in Elec. Engring, Tex. A&M U., 1970, M.E., 1971; Ph.D., U. Okla., 1975. Registered profl. engr., Tex. Engr. asst. Tex. Instruments, Dallas, 1967-70; mem. faculty Abilene Christian Coll., 1971-73, U. Okla., 1973-74; design engr. Okla. Gas & Electric Co., Oklahoma City, 1974; cons. Electric Power Research Inst., Palo Alto, Calif., 1975; mem. faculty Tex. A&M U., College Station, 1976—; assoc. prof., research prin. investigator Tex. Engring. Experiment Sta., 1976—; pres. MICON Engring., Inc., 1978—. Editor: Power System Control and Protection, 1978; assoc. editor: Electric Power System Research, 1975—; contbr. articles to profl. jours. Recipient Outstanding patent submission award Electric Power Research Inst., 1975; Outstanding Young Engr. Braxos chpt. Tex. Soc. Profl. Engrs., 1978; Outstanding Engring. Achievement award Nat. Soc. Profl. Engrs. Mem. IEEE, Nat. Soc. Profl. Engrs., Tex. Soc. Profl. Engrs., Tex. Soc. Energy Auditors, Instrument Soc. Am., Power Engring. Soc. Republican. Mem. Ch. of Christ. Subspecialties: Electrical engineering; Power and energy transmission. Current work: Microcomputer applications to the automation, control and protection of process industry and power systems, electromagnetic interference in power systems. Home: Route 5 Box 1620 College Station TX 77840 Office: Dept Elec Engring Tex A&M U College Station TX 77843

RUSSELL, CHRISTOPHER THOMAS, geophysicist, educator; b. b, St. Albans, Eng., May 9, 1943; s. Thomas Daniel and Mary Teresa Ada Susan R.; m. Arlene Ann Thompson, June 25, 1966; children: Jennifer Ann, Danielle Suzanne. B.Sc., U. Toronto, 1964; Ph.D., UCLA, 1968. Research geophysicist Inst. Geophysics and Planetary Physics, UCLA, 1969-71, prof., 1982—. Author: The IMS Source Book, 1982, Solar Wind Three, 1974, Auroral Processes, 1978, Active Experiments in Space Plasmas, 1981; assoc. editor: Eos, 1978-82, Geophys. Research Letters, 1978-82; contbr. articles to profl. jours. Recipient Macelwane award Am. Geophys. Union, 1977. Fellow Am. Geophys. Union, AAAS; mem. European Geophys. Soc., Am. Astron. Soc. Subspecialties: Planetary science; Geophysics. Current work: Space plasma physics and planetary magnetism. Office: Inst Geophysics and Planetary Physics UCLA Los Angeles CA 90024

RUSSELL, DAVID ALLISON, aeronautical engineering educator; b. Saint John, N.B., Can., Apr. 25, 1935; came to U.S., 1954, naturalized, 1967; s. James Vener and Helen Ringen (Allison) R.; m. Hazel Anne Garnett, Mar. 22, 1957; children: Karen, Kristen, Kathryn. Student, U. N.B., Fredericton, 1950-54; B.Mech.Engring., U. So. Calif., 1956; M.Sc. in Aeros, Calif. Inst. Tech., 1957; Ph.D. in Aeros. and Physics, Calif. Inst. Tech., 1961. Sr. scientist Jet Propulsion Lab., Pasadena,

Calif., 1961-67; research asso. prof. aeros. and astronautics U. Wash., 1967-70, asso. prof., 1970-74, prof., 1974—, chmn. dept. aeros. and astronautics, 1977—; cons. in field. Contbr. articles on fluid mechanics and gas physics to profl. jours., especially on shock processes and laser fluid dynamics. Mem. Kirkland (Wash.) Planning Commn., 1971-79, chmn., 1977-79; chmn. Kirkland Land Use Policy Plan, 1977-79. NASA grantee, 1969-78; Dept. Def. grantee, 1969-71, 73-81. Fellow AIAA (assoc., Pacific N.W. Council 1974-76, Pacific N.W. sect. award for contbns. to aerospace tech. 1972), Am. Phys. Soc. (exec. council div. fluid dynamics 1977-78); mem. Sierra Club, Sigma Xi, Tau Beta Pi, Pi Tau Sigma. Subspecialty: Aeronautical engineering. Home: 4507 105th St NE Kirkland WA 98033 Office: 206 Guggenheim Hall U Wash FS-10 Seattle WA 98195

RUSSELL, DIANE HADDOCK, medical educator, researcher; b. Boise, Idaho, Sept. 9, 1935; d. Grove M. and Eileen F. (Gridley) Haddock; m. Kenneth Russell, May 27, 1953; children: Shauna, Keri. A.A., Boise Jr. Coll., 1961; B.S. in Zoology summa cum laude, Coll. Idaho, 1963; Ph.D. in Zoophysiology, Wash. State U., 1967; postdoctoral fellow Johns Hopkins U. Sch. Medicine, 1967-69. Research chemist Balt. Cancer Research Center, Nat. Cancer Inst., 1969-73; asso. prof. dept. pharmacology and research asso. dept. internal medicine U. Ariz. Health Scis. Center, Tucson, 1973-76, prof., 1976—; chmn. various symposia; NSF fellow NATO Advanced Study Inst. on Regulation of Function and Growth of Eukaryotic Cells, Belgium, 1974; Burroughs Wellcome fellow Internat. Symposium Clin. Pharmacology, W.Ger., 1974; resident scholar Rockefeller Found., Lake Como, Italy, 1977; Fogarty sr. internat. fellow Med. Research Council, Cambridge, Eng., 1980. Author: (with B.G. Durie) Polyamines as Markers of Normal and Malignant Growth, 1978; editor: Polyamines in Normal and Neoplastic Growth, 1973; contbr. chpts. to books, articles to profl. jours. Mem. Am. Soc. Pharmacology and Exptl. Therapeutics, Am. Soc. Biol. Chemists, Am. Physiol. Soc., Am. Assn. Cancer Research (dir. 1979-82), Am. Soc. Cell Biology, Soc. Developmental Biology, Am. Soc. Preventive Oncology, Sigma Xi; fellow AAAS. Subspecialties: Cellular pharmacology; Cancer research (medicine). Current work: Regulation of normal and neoplastic growth; polyamines as biochemical markets of cell kinetics; enzymology of ornithine decarboxylase and transglutaminase. Office: U Ariz Coll Medicine Tucson AZ 85724

RUSSELL, DONALD GLENN, oil company executive; b. Kansas City, Mo., Nov. 24, 1931; s. Virgil G. and Mae (Agey) R.; m. Norma J., Mar. 31, 1953; children: Karen L. Russell Gemmill, Steven G. B.S. in Math, Sam Houston State U., 1953; M.S. in Math. and Physics, U. Okla., 1955. With Shell Oil Co., 1974—, now v.p. prodn., Houston. Co-author: monograph Pressure Buildup and Flow Tests in Wells, 1967. Mem. exec. bd. Sam Houston Area council Boy Scouts Am. Mem. Soc. Petroleum Engrs. (Cedric K. Ferguson medal 1962, John Franklin Carll award 1980), Am. Petroleum Inst., Nat. Acad. Engring. Clubs: Coronado, Champions Golf (Houston). Subspecialty: Petroleum engineering. Office: Shell Oil Co Box 2463 Houston TX 77001

RUSSELL, ELBERT WINSLOW, neuropsychologist; b. Las Vegas, NM, June 4, 1929; s. Josiah Cox and Ruth (Annus) R.; m.; children: Gwendolyn Marie, Franklin Winslow, Kirsten Nash, Jonathan Nash. B.A., Earlham Coll., 1951; M.A., U. Ill., 3; M.S., Pa. State U., 1958; Ph.D., U. Kans., 1968. Lic. psychologist, Fla. Clin. psychologist Warnersville (Pa.) State Hosp., 1959-61; neuropsychologist Cin. VA Med. Ctr., 1968-71; dir. neuropsychology lab. Miami (Fla.) VA Med. Ctr., 1971—; adj. assoc. prof. U. Miami, 1979—, U. Miami Med Sch., 1980—; adj. prof. Nova U. Sch. Profl. Psychology, 1980—. Author: (with Charles Neuringer and Gerald Goldstein) Assessment of Brain Damage, 1970. Mem. Am. Psychol. Assn., Internat. Neuropsychol. Soc., Sigma Xi. Subspecialties: Neuropsychology; Cognition. Current work: Application of psychometric methods to the assessment of brain damage effects. Office: VA Med Center 1201 NW 16 St Miami FL 33125

RUSSELL, EUGENE A., mechanical engineer, corp. energy cons.; b. Centerville, Tenn., July 13, 1914; s. P.R. and Bessie I. (Mallory) R.; m. Jean C. Russell, July 13, 1952; children: David, Mark, Eric, Betsy, Cynthia. M.E., Columbia U., 1949; M.B.A., U. Calif.-Berkeley, 1950. Dir. research Electronic Devel. of Fla., 1954-58; dir research Automated Equipment Corp., Nashville; now dir. research Corp. Cons. (name changed to Corp. Energy Cons. 1980), Hermitage, Tenn. Served with USMC, 1945. Mem. Internat. Solar Energy Soc., ASHRAE. Lodges: Masons; Shriners. Subspecialties: Solar energy; Geothermal power. Current work: Research in field of energy alternates geothermal and solar. Patentee in field. Home: 6000 Panama Dr Hermitage TN 37076 Office: PO Box 332 Hermitage TN 37076

RUSSELL, GLEN ALLAN, chemistry educator, researcher; b. Johnsonville, N.Y., Aug. 23, 1925; s. John A. and Marion A. (Cottrell) R.; m. Martha Ellen Havill, June 6, 1953; children: Susan A., June E. B.Chem. Engring., Rensselaer Poly. Inst., 1947, M.S., 1948; Ph.D., Purdue U., 1951. Research assoc. Gen. Electric Research Lab., Schenectady, 1951-58; assoc. prof. chemistry Iowa State U., Ames, 1958-60, prof. chemistry, 1960—, Disting. prof., 1972—. Contbr. articles to profl. jours. Alfred P. Sloan fellow, 1959-63; John S. Guggenheim fellow, 1972; recipient James Flack Norvis award Am. Chem. Soc., 1983, Midwest award, 1975, Petroleum Chemistry award, 1965, Iowa medal Iowa City sect., 1971. Mem. Am. Chem. Soc., Royal Chem. Soc. (London), AAUP. Subspecialty: Organic chemistry. Current work: Physical organic chemistry of free radicals, electron spin resonance spectroscopy, electron transfer processes. Patentee in field. Office: Dept Chemistry Iowa State Univ Ames IA 50011

RUSSELL, RAY WILLIAM, astronomer, astrophysicist; b. Burlington, Vt., May 17, 1950; s. Edwin and Gloria Isabel (MaGill) R.; m. Andrea Marie, Jan. 8, 1972; 1 son, Brian James. B.S., SUNY, Stony Brook, 1972; M.S., U. Calif., La Jolla, 1974, Ph.D., 1978. Research physicist Cornell U., Ithaca, N.Y., 1978-81; mem. tech. staff Aerospace Corp., Los Angeles, 1981—. Contbr. articles to profl. jours. Mem. Am. Astron. Soc. Republican. Quaker. Subspecialty: Infrared optical astronomy. Current work: Astrophysical researcher in airborne infrared spectroscopy; development of far infrared photoconductive detectors, space-based detection systems. Office: Aerospace Corp M2-266 PO Box 92957 Los Angeles CA 90009

RUSSELL, ROBERT LEE, pharmacologist, educator; b. Independence, Mo., June 27, 1927; s. James Elijah and Kathryn Dorothea (Hiller) R.; m. Mary Frances Stewart, Nov. 25, 1950; children: Brett Vernon, Mark Stewart. A.B., U. Mo., 1950, M.A., 1952, Ph.D., 1954. Asst. in physiology and pharmacology U. Mo., Columbia, 1950-54, mem. faculty, 1954—, assoc. prof. pharmacology, 1957-65, prof., 1966—. Served with U.S. Army, 1945-46. Fellow AAAS; mem. Soc. for Exptl. Biology and Medicine, Am. Soc. for Pharmacology and Exptl. Therapeutics. Subspecialties: Pharmacology; Cellular pharmacology. Current work: Drugs affecting lipid metabolism. Home: Rural Route 4 Box 127 Columbia MO 65201 Office: U Mo M520 Med Center Columbia MO 65201

RUSSELL, ROBERT MITCHELL, medical educator; b. Boston, Apr. 9, 941; s. Stanley G. and Martha L. (Johnson) R.; m. Sharon E. Stanton, Aug. 28, 1965; children: Kimberly, Brooke. B.A., Harvard U., 1963; M.D., Columbia U., 1967. Diplomate: Am. Bd. Internal Medicine, Am. Bd. Gastroenterology. Intern U. Chgo., 1967-68, jr. asst. resident, 1968-69; resident epidemiology Water Reed Army Inst., 1969-71; sr. asst. resident Walter Reed Army Inst., 1971-72; vis. asst. prof. medicine Shiraz U., Iran, 1974-75; asst. prof. medicine U. Md., Balt., 1977-81, assoc. prof., 1981; assoc. prof. medicine Tufts U. Med. Sch., Boston, 1981—; assoc. dir. human studies Human Nutrition Research Ctr. on Aging, U.S. Dept. Agr., Boston, 1981—; clin. investigator VA Med. Ctr., Balt., 1979-81; cons. FDA, 1974-79; mem. Nat. Digestive Disease Adv. Bd., 1983—. Editorial bd.: Jour. Am. Coll. Nutrition, 1983—. Recipient Global Medicine award U.S. Army, 1971; VA grantee, 1975; NIH grantee, 1981; U.S. Dept. Agr. grantee, 1981. Mem. Am. Gastroenterology Assn. (council on aging), Am. Assn. Study of Liver Disease, Am. Soc. Clin. Nutrition, ACP, Am. Fedn. Clin. Research. Subspecialties: Gastroenterology; Internal medicine. Current work: Nutritional status in aging, gastrointestinal function in aging, vitamin A metabolism. Office: 711 Washington St Boston MA 02111

RUSSELL-HUNTER, W(ILLIAM) D(EVIGNE), zoology educator, research biologist, writer; b. Rutherglen, Scotland, May 3, 1926; came to U.S., 1963; s. Robert R. and Gwladys (Dew) R.-H.; m. Myra Porter Chapman, Mar. 22, 1951; 1 son, Peregrine D. B.Sc. with honors, U. Glasgow, 1946, Ph.D., 1953, D.Sc., 1961. Sci. officer Bisra/Brit. Admiralty, Millport, Scotland, 1946-48; asst. lectr. U. Glasgow, 1948-51, univ. lectr. zoology, 1951-63; examiner in biology Pharm. Soc. Gt. Britain, Edinburgh, 1957-63; chmn. dept. invertebrate zoology Marine Biol. Lab., Woods Hole, Mass., 1964-68; trustee, 1967-75, 77—; prof. zoology Syracuse (N.Y.) U., 1963—; cons. editor McGraw-Hill Encys., 1977—; dir. Upstate Freshwater Inst., Syracuse, 1981—. Author: Biology of Lower Invertebrates, 1967, Biology of Higher Invertebrates, 1968, Aquatic Productivity, 1970, A Life of Invertebrates, 1979; mng. editor: Biol. Bull, Woods Hole, Mass., 1968-70. Carnegie and Browne fellow, 1954; research grantee NIH, 1964-70, NSF, 1971-81. Fellow Linnean Soc. London, Royal Soc. Edinburgh, Inst. Biology U.K., AAAS; mem. Ecol. Soc. Am. Subspecialties: Behaviorism; Physiology (biology). Current work: Writing and editing of books and articles on biology; research on ecology and physiology of marine and freshwater invertebrates, actuarial bioenergetics, prevention of fouling and boring organisms. Home: 23 Hurd St Cazenovia NY 13035 Office: Syracuse University 027 Lyman Hall Syracuse NY 13210

RUSSO, IRMA HAYDEE ALVAREZ, physician, pathologist; b. San Rafael, Mendoza, Argentina, Feb. 28, 1942; d. Jose Maria and Maria Carmen (Martinez) Alvarez; m. Jose Russo, Feb. 8, 1969; 1 dau., Patricia Alexandra. B.A. in Edn, Escuela Normal M.T.S.M. de Balcarce, 1959; M.D., U. Nat. of Cuyo, Mendoza, Argentina, 1970. Diplomate: Am. Bd. Pathology, 1980. Intern Sch. of Medicine Hosps., Mendoza, 1969-70; resident in pathology Wayne State U. Sch. Medicine, Detroit, 1976-80; guest lectr. Sch. Medicine, U. Nat. of Cuyo, Mendoza, 1965-71, research asst., instr., 1963-71, assoc. prof. histology, 1970-72; research assoc. Inst. for Molecular and Cellular Evolution, U. Miami, Fla., 1972-73, Exptl. Pathology Lab., Mich. Cancer Found., Detroit, 1973-75, research scientist, 1975-76, vis. research scientist, 1976-80; pathologist Lab. Pathology, 1982—; resident physician Wayne State U. Sch. Medicine, Detroit, 1976-78, chief resident physician, 1978-80, asst. prof. pathology, 1980—; mem. med. staff in pathology Harper-Grace Hosps., Detroit, 1980-82. Contbr. numerous sci. articles and abstracts to profl. publs. Rockefeller Found. grantee, 1972-73; Nat. Cancer Inst. grantee, 1978-81. Mem. Am. Soc. Clin. Pathologists, Am. Assn. Cancer Research, Mich. Soc. Pathologists, AMA, Electron Microscopy Soc. Am., Mich. Electron Microscopy Forum, Sigma Xi. Roman Catholic. Subspecialties: Cancer research (medicine); Pathology (medicine). Current work: Studies on the pathogenesis and prevention of chemically induced breast carcinoma in experimental animals and pathogenesis and prevention of human breast cancer. Office: 110 E Warren Ave Detroit MI 48201

RUST, JAMES HAROLD, publisher, energy consultant; b. Peoria, Ill., Sept. 19, 1936; s. Harold Jacob and Gladys May (Laird) R. B.S.Ch.E., Purdue U., Lafayette, Ind., 1958; M.Sc., M.I.T., 1960; Ph.D., Purdue U., 1965. Asst. prof. U. Va., Charlottesville, 1964-67; asst. prof. Ga. Inst. Tech., Atlanta, 1967-69, assoc. prof., 1969-77, prof., 1977-81; pub. Haralson Pub. Co., Atlanta, 1979—; cons. Ga. Power Co., Atlanta, 1977. Author: (with others) Elements of Nuclear Reactor Design, 1977, Nuclear Power Plant Engineering, 1979; editor: Nuclear Power Safety, 1976. Named Engr. of Year Met. Atlanta E-Week, 1982. Mem. Nat. Soc. Profl. Engrs., Am. Nuclear Soc., ASME, Ga. Soc. Profl. Engrs. (pres. Atlanta chpt. 1983, Engr. of Year 1982). Methodist. Club: Bicker's (Atlanta). Subspecialty: Nuclear engineering. Current work: Energy conversion with nuclear power plants, heat transfer and fluid flow in nuclear reactor cores. Patentee in field. Home: 340 Garden Ln Atlanta GA 30309 Office: Haralson Pub Co PO Box 20366 Atlanta GA 30325

RUSTIGIAN, ROBERT CARNING, research virologist, educator; b. Boston, July 26, 1915; s. Nicholas and Rose (Avadanian) R.; m. Gloria Louise Florentino, May 27, 1956; children: Robert, Gabrielle, Christina. B.S., U. Mass., 1938; M.S., Brown U., 1940, Ph.D., 1943. NRC fellow Harvard Med. Sch., 1946-47, instr., 1947-49; asst. prof. U. Chgo., 1949-56, Tufts Med. Sch., Boston, 1956-62, assoc. prof., 1962-68; chief virology research lab. VA Med. Ctr., Brockton, Mass., 1968—; asst. prof. virology Harvard Med. Sch., 1968-72, assoc. prof., 1972—. Contbr. articles to profl. jours. Served to capt. AUS, 1943-46. Decorated Bronze Star. Fellow Am. Soc. Microbiologists; mem. Am. Assn. Immunologists. Subspecialties: Microbiology (medicine); Virology (medicine). Current work: Chronic human viral infections. Office: VA Med Ctr Brockton MA 02401

RUTAN, BURT, aircraft designer; b. 1943. Student, Calif. Poly. Inst. Civilian stability and control specialist Edwards AFB, 1972-75; aircraft designer J. Bede, Kans., 1975; founder Rutan Aircraft Factory, Mojave, Calif., 1975—. Designed Variviggen aircraft. Subspecialty: Aeronautical engineering. Designer of more than 90 aircraft, including VariEze, Quickie, and Solitaire sailplane. Office: Rutan Aircraft Factory Mojave CA 93501

RUTFORD, ROBERT HOXIE, university administrator; b. Duluth, Minn., Jan. 26, 1933; s. Skuli and Ruth (Hoxie) R.; m. Marjorie Ann, June 19, 1954; children: Gregory, Kristian, Barbara. B.A., U. Minn., 1954, M.A., 1963, Ph.D., 1969. Football and track coach Hamline U., 1958-62; research fellow U. Minn., 1963-66; asst. prof. geology U. S.D., 1967-70, asso. prof., 1970-72, chmn. dept. geology, 1968-72, chmn. dept. physics, 1971-72; dir. Ross Ice Shelf Project U. Nebr., Lincoln, 1972-75, vice chancellor for research and grad. studies, prof. geology, 1977-82, interim chancellor, 1980-81; pres. U. Tex., Dallas, 1982; dir. div. Polar Programs, NSF, Washington, 1975-77. Served to 1st lt. U.S. Army, 1954-56. Recipient Antarctic Service medal, 1964, Distinguished Service award NSF, 1977, Ernie Gunderson award for service to amateur athletics S.D. AAU, 1972. Fellow Geol. Soc. Am.; mem. Antarctican Soc., Arctic Inst. N.Am., Explorers Club, Am. Polar Soc., Nat. Council Univ. Research Adminstrs., Sigma Xi. Lutheran. Subspecialties: Geomorphology; Geology. Current work: Antarctic mineral resources and glacial history of antarctica. Home: 6809 Briar Cove Dr. Dallas TX 75240 Office: Univ. of Texas at Dallas PO Box 688 Richardson TX 75080

RUTLEDGE, CHARLES OZWIN, pharmacologist, pharm. cons., educator; b. Topeka, Oct. 1, 1937; s. Charles Ozwin and Alta (Seaman) R.; m. Jane Ellen Crow, Aug. 13, 1961; children: David Ozwin, Susan Harriett, Elizabeth Jane, Karen Ann. B.S. in Pharmacy, U. Kans., 1959, M.S. in Pharmacology, 1961, Ph.D., Harvard U., 1966. NATO postdoctoral fellow, Gothenburg, Sweden, 1966-67; asst. prof. pharmacology U. Colo. Med. Center, Denver, 1967-74, asso. prof., 1974-75; prof. pharmacology U. Kans., Lawrence, 1975—, chmn. dept., 1975—; cons. PharMat, Am. Heart Assn., NIH, NIMH, Am. Dysautomia Found., Kans. Pharmacy Found. Contbr. numerous articles to research publs. Kans. Heart Assn. grantee, 1977; NIH grantee, 1970—. Mem. Am. Soc. Pharmacology and Exptl. Therapeutics (specific field editor Jour. 1973—), Soc. Neurosci., Am. Soc. Neurochemistry, AAAS, Am. Assn. Colls. Pharmacy, Am. Pharm. Assn., Kans. Pharmacists Assn., Health Care Invesitgation Assn. Club: Cosmopolitan Internat. Subspecialties: Pharmacology; Neuropharmacology. Current work: Research involves the mechanisms by which drugs alter neurotransmitter function in both peripheral and central nervous systems. Home: 2620 Stratford Lawrence KS 66044 Office: Dept Pharmacology and Toxicology Sch Pharmacy U Kans Lawrence KS 66045

RUTMAN, ROBERT J., biochemist, educator; b. N.Y.C.; s. Leon and Anne (Porringer) R.; m.; children: Rose, Randy Rutman Allen, Stephen Johnson, Ellen Johnson, David Johnson. B.S., Pa. State U., 1940, M.S., 1975; Ph.D., U. Calif., Berkeley, 1950. Asst. prof. Jefferson Med. Coll., Phila., 1950-53; research assoc. in bioilogy U. Pa., 1954-56, research assoc. in chemistry, 1956-61, assoc. prof. chemistry, 1961-68; prof. biochemistry U. Pa. Sch. Vet. Medicine, 1968—, chair biochemistry, 1975-79; vis. prof. U. Ibadan, Nigeria, 1973-74, external examiner, 1977—, coordinator exchangeprogram, 1980—. Contbr. articles to profl. jours. Pres. Phila. Peace Council, 1968-70. Served to capt. AUS, 1946-50. Mem. Am. Assn. Sci. Workers (nat. sec.), Am. Soc. Biol. Chemistry, Am. Assn. Cancer Research, AAAS, AAUP, Am. Soc. Vet. Oncology (chmn. sci. com. chem.-biol. warfare). Subspecialties: Biochemistry (biology); Chemotherapy. Current work: Mechanism of anti-cancer DNA reactive drugs, monoclonal antibodies to modified DNA. Patentee in field. Home: Park Plaza Apt PH-P Philadelphia PA 19131 Office: Sch Vet Medicine U Pa Philadelphia PA 19104

RUWE, WILLIAM DAVID, medical neuroscientist; b. Lafayette, Ind., Feb. 18, 1953; s. Alfred Carl and Marcelline Emma (Warbelton) R. B.A., Wabash Coll., 1975; M.S., Purdue U., 1977; postgrad., U. N.C.-Chapel Hill, 1979-80; Ph.D., Purdue U., 1980. NIMH fellow Purdue U., West Lafayette, Ind., 1975-76, research asst., 1976-78, teaching asst., 1978-79; research fellow U. N.C., Chapel Hill, 1979-80; Alta. Heritage Found. Med. Research fellow U. Calgary, Alta., Can., 1981—. Lay tchr. Christian and Missionary Alliance Ch., Calgary, Alta., 1981—; lay leader Presbyn. Ch., West Lafayette, Ind., 1977-80; youth coordinator Babe Ruth Baseball Assn., 1971-75. NIMH predoctoral fellow Purdue U., 1975; Alta. Heritage Found. Med. Research research fellow U. Calgary, 1981—. Mem. Am. Physiol. Soc., Soc. Neurosci., Phi Beta Kappa, Sigma Xi, Psi Chi. Republican. Mem. Christian and Missionary Alliance Ch. Subspecialties: Physiology (medicine); Neuropharmacology. Current work: Neuropharmacological basis of basic physiological functions including thermoregulation, alcohol addiction, febrile convulsions. Home: Apt H206 1919 University Dr NW Calgary AB Canada T2N 4L1 Office: U Calgary 3330 Hospital Dr NW Calgary AB Canada T2N 1N4

RYAN, CHARLES LUCE, JR., polymer scientist; b. Cheverly, Md., May 29, 1953; s. Charles Luce and Lillian (Gaydos) R.; m. Nancy Dittig, June 28, 1975. B.S., Va. Poly. Inst. and State U., 1975; M.S., U. Mass.-Amherst, 1977, Ph.D., 1979. Research chemist Phillips Petroleum Co., Bartlesville, Okla., 1980—; research asst. U. Mass.-Amherst, 1975-79, postdoctoral fellow, 1979-80. Sci. fair judge, Bartlesville, Okla., 1981—; coach Pony League Baseball, Bartlesville, 1981—, Boys Club Basketball, Bartlesville, 1982—. Mem. Soc. Plastics Engrs (dir. 1982—), Am. Chem. Soc., Am. Phys. Soc., AAAS. Subspecialty: Polymer chemistry. Current work: Physical chemistry of macromolecular systems with particular emphasis on property/structure relationships. Office: Phillips Petroleum Co 320 PL PRC Bartlesville OK 74004 Home: 1121 S Osage St Bartlesville OK 74003

RYAN, JAMES BERNARD, agricultural chemical researcher; b. Greenville, S.C., Aug. 4, 1945; s. Edward Henry and Marjorie (Gowen) R.; m. Evelyn T. Nemeth, Nov. 22, 1969; children: Michael James, Macqueline E., Daniel E. B.S. in Plant Sci, Rutgers U., 1967, M.S. in Soils and Crops, 1969, Ph.D. in Weed Sci, 1975. Grad. research asst. aquatic weed control Rutgers U., 1967-69; field research rep. agrl. chems. Rohm & Haas Co., Phila., 1969-71, field research rep. Western US, Fresno, Calif., 1971-72, field research mgr. agrl. chems., Spring House, Pa., 1973-75, N.Am. regional mgr. research and product devel. agrl. chems., Phila., 1975—. Republican committeeman, 1982—. Mem. Weed Sci. Soc. Am., Council Agrl. Sci. and Tech., Am. Phytopath. Soc., So. Weed Sci. Soc., Plant Growth Regulators Soc. Am., Alpha Zeta. Roman Catholic. Club: Central Bucks Gymnastics. Subspecialties: Weed science; Agricultural chemicals development. Current work: Research, development and registration of products for improvement of yield and quality of agronomic and horticultural crops. Home: 439 Pine Run Rd Doylestown PA 19801 Office: Rohm and Haas Co Independence Mall W Philadelphia PA 19105

RYAN, KENNETH JOHN, physician, educator; b. N.Y.C., Aug. 26, 1926; s. Joseph M. R.; m. Marion Elizabeth Kinney, June 8, 1948; children: Alison Leigh, Kenneth John, Christopher Elliot. Student, Northwestern U., 1946-48; M.D., Harvard, 1952. Diplomate: Am. Bd. Obstetrics and Gynecology. Intern, then resident internal medicine Mass. Gen. Hosp., Boston, also Columbia-Presbyn. Med. Center, N.Y.C., 1952-54, 56-57; resident obstetrics and gynecology Boston Lying-in Hosp., also Free Hosp. for Women, Brookline, Mass., 1957-60; instr. obstetrics and gynecology Harvard U.; also dir. Harvard (Fearing Research Lab.), 1960-61; prof. obstetrics and gynecology, dir. dept. Western Res. U. Med. Sch., 1961- 70; prof. reproductive biology, dept. obstetrics and gynecology U. Calif. at San Diego, La Jolla, 1970-73; Kate Macy Ladd prof., chmn. dept. obstetrics and gynecology Harvard U. Med. Sch., 1973—; dir. Lab. Human Reprodn. and Reproductive Biology Harvard Med. Sch., 1974—; chief staff Boston Hosp. for Women, 1973—. Recipient Schering award Harvard Med. Sch., 1951, Soma Weis award, 1952, Bordon award, 1952; Ernst Oppenheimer award, 1964; Max Weinstein award, 1970; fellow Am. Cancer Soc., Mass. Gen. Hosp., 1954-56. Mem. Am. Coll. Obstetricians and Gynecologists, Am. Soc. Biol. Chemists, Endocrine Soc., Soc. Gynecol. Investigation, Mass. Med. Soc., Am. Soc. Clin. Investigation, Am. Gynecol. Soc., Alpha Omega Alpha. Subspecialties: Neuroendocrinology; Reproductive endocrinology. Current work: Reproductive endocrinology; steroid biochemistry. Home: 75 Francis St Boston MA 02115

RYAN, ROGER BAKER, research entomologist; b. Port Chester, N.Y., May 5, 1932; s. John Wynne and Adele (Baker) R.; m. Joan Marie Bennett, Nov. 26, 1955; children: James, Ann, Joseph, Patrick. B.S., N.Y. State Coll. Forestry, Syracuse, 1953; M.S., Oreg. State U., 1959, Ph.D., 1961. Research entomologist Dept. Agr. Forest Service, Corvallis, Oreg., 1961—. Served to lt. (j.g.) USNR, 1953-56. Mem. Entomol. Soc. Am., Entomol. Soc. Can., Internat. Orgn. Biol. Control. Roman Catholic. Subspecialties: Ecology; Integrated pest management. Current work: Biological control of forest insects with introduced parasites. Office: USDA Forest Service 3200 Jefferson Way Corvallis OR 97331 Home: 5427 NW Highland Dr Corvallis OR 97330

RYAN, STEWART RICHARD, physicist, educator; b. Schenectady, Jan. 26, 1942; s. August R. and Frances A. (Ruth) R.; m. Rita M. Sandman, July 9, 1966; children: Kathleen, Colleen, Ellen Mary. B.S. in Physics, U. Notre Dame, 1964; M.S., U. Mich., 1965, Ph.D., 1971. Research staff physicist, then instr. Yale U., New Haven, 1971-74; staff physicist U. Ariz., Tucson, 1974-77; asst. prof. physics U. Okla., Norman, 1977-82, asso. prof., 1982—, mem. engring. physics com. Contbr. numerous articles to profl. publs. Mem. Am. Phys. Soc., Am. Assn. Physics Tchrs., Am. Soc. Engring. Edn., Okla. Acad. Sci., Sigma Xi. Subspecialties: Atomic and molecular physics; Applied physics. Current work: Experimental atomic and molecular physics, applied physics and instrumentation, energy conservation, experimental low temperature physics. Home: 2711 Willow Creek Dr Norman OK 73071 Office: Dept Physics and Astronomy U Okla Norman OK 73019

RYAN, UNA SCULLY, med. researcher, cons.; b. Kuala Lumpur, Malaysia, Dec. 18, 1941; came to U.S., 1964; d. Henry and Amy (Yee) Scully; m. David Spencer Smith, July 18, 1964; 1 dau., Tamsin Spencer; m. James Walter Ryan, June 17, 1973; 1 dau., Amy Jean Susan. B.S., Bristol U., Eng., 1963; Ph.D., U. Cambridge, Eng., 1966. Fellow dept. biology U. Va., Charlottesville, 1964-66; fellow dept. medicine U. Miami, 1966-67, instr. medicine, 1967-72, adj. asst. prof. biology, 1968-71, asst. prof. medicine, 1972-77, assoc. prof., 1977-80, research prof., 1980—; vis. investigator Labs. Cardiovascular Research, Howard Hughes Med. Inst., Miami, 1967-71, dir., 1970-71; sr. scientist Papanicolaou Cancer Research Inst., Miami, 1972-77; chairperson heart lung and blood research rev. com. NIH, 1980-81. Editor: Tissue & Cell, 1981—; contbr. numerous articles to profl. jours. Mem. vestry St. Stephen's Ch., Miami; trustee Sch. Dept. Sci. and Indsl. Research; fellow, 1963. Ethel Sargant Research fellow, 1964, 65; Sci. Research Council fellow, 1966. Mem. Am. Soc. Cell Biology, Soc. Neurosci., Tissue Culture Assn., Am. Heart Assn., European Soc. Microcirculation, Am. Microcirculatory Soc., Am. Thoracic Soc., N.Y. Acad. Scis. Episcopalian. Subspecialties: Cell biology (medicine); Pulmonary medicine. Current work: Cell biology of pulmonary endothelium. Home: 3420 Poinciana Ave Miami FL 33133 Office: 1399 NW 17th Ave Miami FL 33125

RYDER, BENJAMIN MILLS, information sciences executive; b. Boston, March 2, 1945; s. Clayton Philip and Sarah Ann (Tewkes) R.; m. Deborah K. Ozaki, July 1, 1966; children: Philip, Tarah, Dana. B.A., Wake Forest U., 1966; M.S. in Computer Scis., U. Ill., 1968. Computer programmer Weston Lab., Chgo., 1970-74, dept. chmn., 1974-76; asst. mgr. info. scis. Time, Inc., Chgo., 1976-80, mgr. info. scis., 1981—; cons. to numerous pvt. cos., govt. agys. Contbg. editor: Computer World; contbr. articles to profl. jours. Trustee Oak Park Pub. Library. Mem. Assn. for Computing Machinery, Soc. for Indsl. and Applied Math. Club: Cliffdwellers (Chgo.). Lodge: Kiwanis. Subspecialties: Information systems, storage, and retrieval (computer science); Database systems. Office: Werik Bldg 24 N Wabash Ave Suite 823 Chicago IL 60602

RYDER, EDWARD JONAS, geneticist; b. N.Y.C., Oct. 6, 1929; s. Wilfred and Tillie (Brown) R.; m. Elouise Jones Viales, Mar. 10, 1962; children: Robert G. Viales, Lawrence D. Viales, Deborah L. B.A., Cornell U., 1951; Ph.D., U. Calif., Davis, 1954. Geneticist Agrl. Research Service, Dept. Agr., 1957—, research leader, 1972—, location leader, 1982—; cons. in field. Author: Leafy Salad Vegetables, 1979; contbr. articles to profl. jours. Pres. Y Men's Club, 1965. Served with AUS, 1954-56. Iceberg Lettuce Research Bd. grantee, 1974—; recipient The Grower award, 1981. Mem. Am. Soc. Hort. Sci., Genetics Soc. Am., Internat. Soc. Hort. Sci., Crops Sci. Soc. Subspecialties: Genetics and genetic engineering (agriculture); Plant breeding. Current work: Genetics and breeding of lettuce and other leafy vegetables. Developer 7 lettuce varieties. Home: 77 Paseo Hermoso Salinas CA 93908 Office: Box 5098 Salinas CA 93915

RYDER, STEVEN WILLIAM, clinical research laboratory administrator; b. Rockville Center, N.Y., Sept. 13, 1950; s. William Benjamin and Mary (Jeray) R.; m. Rita Claire Saturno, June 6, 1976; children: Patricia, William. M.D., Mt. Sinai Sch. Medicine, 1974. Med. intern, resident, fellow in endocrinology SUNY-Stony Brook, 1974-79; research assoc. Sol. A Berson Lab., Bronx, N.Y., 1979-81; asst. dir. clin. research Ayerst Labs., N.Y.C., 1982—. Fellow ACP; mem. Endocrine Soc., Am. Diabetes Assn. Subspecialties: Neuroendocrinology; Neuropharmacology. Current work: Aldose reductase, LHRH (neuropeptides), excitatory amino acid neurotransmitters. Home: PO Box 122 Saint James NY 11780 Office: Ayerst Labs 685 3d Ave New York NY 10021

RYKER, NORMAN JENKINS, JR., aerospace engineering and technology company executive; b. Seattle, Dec. 25, 1927; s. Norman Jenkins and Adelia Gustine (Macombee) R.; m. Kathleen Marie Crawford (div.); children: Jeanne Ryker Flores, Christina, Vickie Ryker Risley, Norman Jenkins, III, Cathy. B.S., U. Calif., 1949, M.S., 1951; M.S., Harvard U., 1973. Registered profl. engr., Calif. With North Am. Aviation (now Rockwell Internat.), El Segundo, Calif., 1951—, Apollo asst. chief engr., 1961-64, 67-68, v.p. research and test engring., 1968-70; v.p. research engring. and test newspaper press group, v.p. ops. newspaper press, 1973, v.p., gen. mgr., 1974, 1974-76, pres., 1976-83, sr. v.p. aerospace ops., Boston, 1983—; mem. NASA Manned Spacecraft Ctr. team which evaluated and recommended the lunar orbit rendezvous concept for landing men on the moon; dir. Warner Ctr. Bank. Recipient award of merit NASA, 1969, Disting. Pub. Service medal, 1981; Silver Knight award Nat. Mgmt. Assn., 1979; Indsl. Tech. Mgmt. award Calif. Soc. Profl. Engrs., 1979; Companion Instn. Prodn. Engrs., London, 1981; Engr. of Yr. award Inst. Advancement Engring. and Orange County Engring. Council, 1982. Fellow AIAA, Inst. Advancement Engring.; mem. ASCE, Am. Astron. Soc., Nat. Mgmt. Assn., Sigma Xi, Phi Epsilon. Subspecialty: Aerospace engineering and technology. Current work: Responsible for companies of Cleveland Pneumatic, National Water Lift. Office: 6633 Canoga Ave Canoga Park CA 91304

SAAD, SAMI MICHEL, botany and microbiology educator, plant pathologist, researcher; b. Beirut, June 10, 1939; U. S., 1965, naturalized, 1977; s. Michel K. and Mary G. (Hassoun) S. B.S. with distinction in Agr. and Ingenier Agricole diploma, Am. U. Beirut, 1961, M.S. in Agronomy-Seed Tech, 1965; Ph.D. In Plant Pathology, U. Wis., 1969. Postdoctoral fellow in plant pathology and bacteriology U Wis., Madison., 1969-71; vis. prof. botany U. Wis. Center, Washington County, West Bend, 1971-72, asso. prof. biol. scis., 1972—; plant industry specialist WI Dept. Agr., Trade and Consumer Protection, 1979-81; research specialist Libby, McNeil and Libby, Inc., Janesville, Wis., 1975-77; extension plant pathologist specialist, dept. plant pathology U. Wis., Madison, 1969, 70, 71, 73, 74; researcher and cons. plant pathology and entomology. Contbr. articles on plant pathology, botany and microbiology to profl. jours. Mem. Wis. Acad. Scis., Arts and Letters, Bot. Club Wis., Assn. Midwestern Coll. Biology Tchrs., Mycological Soc. Am., Washington County Agr. Assn., Horticulture Inspection Soc. (Central chpt.), Wis. Agrl. and Life Scis. Alumna Assn. Subspecialties: Plant pathology; Microbiology. Current work: Bacterial and viral diseases of vegetable crops, identification of wild mushrooms. Home: 734 A S 6th Ave West Bend WI 53095 Office: U Wis 100 University Dr West Bend WI 53095

SAADA, ADEL SELIM, civil engineer, educator; b. Heliopolis, Egypt, Oct. 24, 1934; came to U.S., 1959, naturalized, 1965; s. Selim N. and Marie (Chahyne) S.; m. Nancy Helen Hernan, June 5, 1960; children: Christiane Mona, Richard Adel. Ingénieur des Arts et Manufactures, École Centrale, Paris, France, 1958; M.S., U. Grenoble, France, 1959; Ph.D. in Civil Engring, Princeton U., 1961. Registered profl. engr., Ohio. Engr. Société Dumez, Paris, 1959; research asso. dept. civil engring. Princeton (N.J.) U., 1961-62; asst. prof. civil engring. Case Western Reserve U., Cleve., 1962-67, asso. prof., 1967-72, prof., 1973—, chmn. dept. civil engring., 1978—; cons., lectr. soil testing and properties Waterways Expt. Sta. (C.E.), Vicksburg, Miss., 1974-79; cons. to various firms, 1962—; pres. Adel Saada Corp., since 1979—. Author: Elasticity Theory and Applications, 1974; contbr. numerous articles on soil mechanics and foundation engring. to profl. jours. Fellow ASCE; Mem. Internat. Soc. Soil Mechanics, ASTM. Club: Executive. Subspecialties: Civil engineering; Fracture mechanics. Current work: Soil mechanics, foundations, mechanical behavior of particulate media; mechanics of solids and fracture. Inventor pneumatic analog computer and loading frame. Home: 3342 Braemar Rd Shaker Heights OH 44120 Office: Dept Civil Engring Case Inst Tech Case Western Reserve Univ Bingham Bldg Cleveland OH 44106

SAALFELD, FRED ERIC, chemist, research adminstr.; b. Joplin, Mo., Apr. 9, 1935; s. Eric A. and Milla E. (Kessler) S.; m. Elizabeth Renner, Nov. 22, 1958; 1 son, Fred E. B.S., S.E. Mo. State U., 1957; M.S., Iowa State U., 1959, Ph.D., 1961. Chemist Naval Research Lab., Washington, 1962, head mass spectrometry sect., 1963-73, head phys. chemistry br., 1974-75, supt. chemistry div., 1976-79, 1981-82; chief scientist and sci. dir. Officer Naval Research br. office, London, 1980; dir. research program Office Naval Research, Arlington, Va., 1982—. Contbr. numerous articles on chemistry to profl. jours. Sec. Lake Braddock High Sch. PTA, 1976. Farmers and Mchts. Bank scholar, 1956; recipient Navy Meritorious Civilian Service award, 1981. Mem. Am. Chem. Soc., Am. Soc. Mass Spectrometry, Soc. for Applied Spectroscopy, AAAS, Sigma Xi. Subspecialties: Physical chemistry; Analytical chemistry. Current work: Life support, contaminant ident, combustion, solid state chemistry. Office: Office of Naval Research 800 N Quincy St Arlington VA 22217

SABBAGHA, RUDY E., obstetrician, gynecologist, educator; b. Tel Aviv, Israel, Oct. 29, 1931; s. Elias C. and Sonia B. S.; m. Asma E. Sahyouny, Oct. 5, 1957; children: Elias, Randa. B.A., M.D., Am. U., Beirut, Lebanon. Sr. physician Tapline, Saudi Arabia, 1958-64, ob-gyn specialist, 1969-70; teaching fellow U. Pitts., 1965-68; asst. prof. ob-gyn, 1970-75; prof. Northwestern U., 1975—; obstetrician, gynecologist Prentice Women's Hosp., Chgo. Author: Ultrasound-High Risk Obstetrics, 1979; contbr. articles to profl. jours. Fellow Am. Coll. Obstetricians and Gynecologists; mem. Soc. Gynecologic Investigation, Central Assn. Obstetricians and Gynecologists, Assn. Profs. Ob-Gyn, Am. Inst. Ultrasound in Medicine. Subspecialty: Diagnostic ultrasound. Current work: Research in diagnostic ultrasound, obstetrics and gynecology. Home: 2415 Meadow Dr Wilmette IL 60091 Office: Prentice Women's Hosp Chicago IL 60611

SABBAHI, MOHAMED AHMED, neurophysiologist, consultant; b. Cairo, May 10, 1946; s. Ahmed Sabbahi and Hamida Abdel-Hamid (Salem) Awadalla; m. Nadra R. Metawey, July 28, 1977; children: Wesam, Rabab. B.Sc. in Phys. Therapy, Cairo U., 1966; Ph.D. in Physiology, U. Southampton, Eng., 1976. Lectr. Cairo U., 1977-78; cons. Nat. Magnet Lab., MIT, Cambridge, 1979-80; adj. asst. prof. Boston U., 1979-80; cons. Liberty Mut. Research Center, Hopkinton, Mass., 1979—; cons. in clin. neurophysiology, rehab. medicine. Contbr. articles to profl. publs. Served with Egyptian Air Force, 1969-71. Project HOPE fellow to U.S., 1978-79. Mem. Internat. Soc. Electrophysiol. Kinesiology, Soc. Neurosci., Brain Research Assn. U.K., European Brain and Behavior Soc., Egyptian Phys. Therapy Assn. (past pres.). Moslem. Subspecialties: Neurophysiology; Physical medicine and rehabilitation. Current work: Neural and muscle control by peripheral stimuli; testing and modification of physical rehabilitation methods; functional integrity of spinal and supraspinal reflexes; measuring local muscle fatigue. Home: 644 W 10th St Westwood MA 02090 Office: Neuromuscular Research Lab Children's Hosp Med Center 300 Longwood Ave Boston MA 02115

SABIN, JOHN ROGERS, physics educator; b. Springfield, Mass., Apr. 29, 1940; s. Henry Bowman and Elizabeth Rogers S.; m.; Children: Peter B., Amanda B. A.B., Williams Coll., 1962; Ph.D., U.N.H., 1966. NIH fellow Uppsala U., Sweden, 1966-67; research fellow Northwestern U., Evanston, Ill., 1967-68; asst. prof. U. Mo., Columbia, 1968-70; assoc. prof. U. Fla., Gainesville, 1970-77, prof. physics and chemistry, 1977—; NORDITA prof., Denmark, 1982-83. Editor: procs. of Sanibel Symposia, 1976-82; assoc. editor: Internat. Jour. Quantum Chemistry, 1973—. Fellow Am. Phys. Soc.; fellow Am. Chem. Soc.; Mem. Am. Inst. Chemists. Subspecialties: Theoretical physics; Atomic and molecular physics. Current work: Computational quantum mechanics of atoms, molecules, solids, stopping powers quantum physics. Home: 528 NW 34th Dr Gainesville FL 32607 Office: Quantum Theory Project Physics Dept U Fla Gainesville FL 32611

SABRI, MOHAMMAD IBRAHIM, neurochemist, neurotoxicologist, educator; b. India, Mar. 28, 1940; came to U.S., 1968; s. Abdul Karim and Bandi S.; m. Hashmi Begam, May 29, 1958; children: Mohammad Yunus, Mohd Yusuf, Armana. B.S., Bombay (India) U., 1959, Ph.D., 1967; M.S., Allahad (India) U., 1961. Jr. research fellow Central Drug Research Inst., Lucknow, India, 1961-63, sr. research fellow, 1964-67; research assoc. Ind. U. Med. Ctr., 1968-71; lectr. in neurochemistry U. London, 1971-77; asst. prof. neurology, asst. prof. neurosci. Albert Einstein Coll. Medicine, 1978—; cons. Enviroprobe Inc., Bronx, N.Y. Contbr. chpts., numerous articles and abstracts to profl. publs.; editorial bd.: Neurotoxicology; cons. editor, 1978—; editorial bd.: Neurochemistry Internat.; cons. editor, 1980—; reviewer various jours. Council Sci. and Indsl. Research sr. research fellow, 1964-67; named hon. citizen City of Indpls., 1969; recipient research travel award Royal Soc. Eng., 1976. Mem. Internat. Brain Research Orgn., Biochem. Soc. (U.K.), Am. Soc. Neurochemistry, Internat. Soc. Neurochemistry, Soc. Neurosci. Muslim. Subspecialties: Environmental toxicology; Neurochemistry. Current work: Neurochemistry with specific interest in determining biochemical and cellular mechanisms of action of toxic substances of environmental significance on the nervous system of experimental animals and axonal transport; axonal transport; myelination; brain development. Office: 1300 Morris Park Ave Bronx NY 10461

SACCO, WILLIAM P(ATRICK), psychology educator, consultant; b. Elizabeth, N.J., Aug. 8, 1951; s. Patrick James and Dorothy Evelyn (Brown) S. B.A. cum laude, U. Fla., 1973; M.S., Fla. State U., 1976, Ph.D., 1979. Lic. psychologist, Fla. Psychologist Community Mental Health Ctr., West Palm Beach, Fla., 1973-74; research asst. dept. psychology U. Fla., 1973; research asst. Fla. State U., 1974-78; clin. trainee VA Hosp., Biloxi-Gulfport, Miss., 1975; clin. intern Ctr. Cognitive Therapy dept. psychiatry U. Pa., 1978-79; asst. prof. U. South Fla., 1979—; pvt. practice psychology Associated Psychol. Services, Tampa, Fla., 1979—; clin. supr. U. South Fla. Psychol. Services Ctr., 1979—. Contbr. articles to profl. publs.; ad-hoc editor: Jour. Abnormal Psychology, 1980-82, Cognitive Therapy and Research, 1980-82, Jour. Personality and Social Psychology, 1981. USPHS fellow, 1976-78; U. South Fla. grantee, 1980, 81. Mem. Am. Psychol. Assn., Assn. Advancement Psychology, Southeastern Psychol. Assn., Phi Beta Kappa, Sigma Xi (Fla. State chpt. second Pl. Ann. Student Paper award 1977), Phi Kappa Phi, Psi Chi (faculty adv. chpt. 1979-81). Democrat. Subspecialty: Behavioral psychology. Current work: Etiological theories of depressive disorders; research methodology; program evaluation. Home: 3502 River Grove Dr Tampa FL 33610

SACHS, BENJAMIN DAVID, biopsychologist; b. Madrid, Spain, Mar. 4, 1936; s. Georg Eduard and Leonie Bernardine (Feiler) S.; m. Jacqueline Sachs, June 12, 1965; 1 dau.; Naomi. B.A., CCNY, 1957; M.S. Ed., 1961; Ph.D., U. Calif., Berkeley, 1966. Nat. Inst. Child Health and Human Devel. postdoctoral fellow Inst. Animal Behavior, Rutgers U., Newark, 1966-68; asst. prof. U. Conn., Storrs, 1968-72; assoc. prof., 1972-76, prof., 1976—; Vis. scholar Stanford Med. Sch., 1975-76. Editor: (with McGill and Dewsbury) Sex and Behavior, 1978; editorial adv. bd.: Current Contents/Life Scis, 1968—; contbr. articles to profl. jours. Served with AUS, 1958-60. USPHS grantee, 1969—. Fellow Am. Psychol. Assn.; mem. Eastern Psychol. Assn., Animal Behavior Soc., Internat. Soc. Psychoneuroendocrinology, Internat. Soc. Devel. Psychobiology, Soc. Neurosci., Sigma Xi, Psi Chi. Subspecialties: Neuropsychology; Psychobiology. Current work: Hormone-brain-behavior relations, especially regarding reproductive behavior. Office: Dept Psychology U Conn Storrs CT 06268

SACHS, HOWARD, neurologist; b. N.Y.C., July 12, 1926; m.; children: Susan, Linda, Nancy. Ph.D., Columbia U., 1953; M.D., Case Western Res. U., 1976. Intern Los Angeles County-U. So. Calif. Med. Ctr., 1976; resident U. Calif.-San Diego, 1977-80, UCLA, 1980-81; prof. physiology Case Western Res. U. Sch. Medicine, Cleve., 1969; sect. chief neurochemistry Roche Inst. Molecular Biology, Nutel, N.J., 1969-73; neurologist New Hope Pain Ctr., Los Angeles, 1981—. Contbr. articles to profl. jours. Served with inf. U.S. Army, 1944-46. Recipient Career Devel. award NIH, 1960-69; NSF grantee, 1958-69. Mem. Endocrine Soc., Fed. Soc. Biol. Chemists, Neurochem. Soc., Neurol Soc. Jewish. Subspecialties: Neurology; Neurochemistry. Current work: Neuropeptides, pain. Address: 100 S Raymond St Alhambra CA 91802

SACHS, ROBERT GREEN, physicist, educator, lab. adminstr.; b. Hagerstown, Md., May 4, 1916; s. Harry Maurice and Anna (Green) S.; m. Selma Solomon, Aug. 28, 1941; m. Jean K. Woolf, Dec. 17, 1950; children: Rebecca, Jennifer, Jeffrey, Judith, Joel; m. Carolyn L. Wolf, Aug. 21, 1968; stepchildren—Thomas Wolf, Jacqueline Wolf, Katherine Wolf. Ph.D., Johns Hopkins, 1939; D.Sc. (hon.), Purdue U., 1967, U. Ill., 1977. Research fellow George Washington U., 1939-41; instr. physics Purdue U., 1941-43; on leave as lectr., research fellow U. Calif. at Berkeley, 1941; 2sect. chief Ballistic Research Lab., Aberdeen (Md.) Proving Ground, 1943-45; dir. theoretical physics div. Argonne Nat. Lab., Ill., 81945-47; asso. prof. physics U. Wis., 1947-48, prof., 1948-64; asso. dir. Argonne Nat. Lab., 1964-68; prof. physics U. Chgo., 1964—; dir. Enrico Fermi Inst., 1968-73, Argonne Nat. Lab., 1973-79; Higgins vis. prof. Princeton U., 1955-56; vis. prof. U. Paris, 1959-60, Tohoku U., Japan, 1974; cons. Ballistic Research Labs., 1945-59, Argonne Nat. Lab., 1947-50, 60-64; cons. radiation lab. U. Calif. at Berkeley, 1955-59; adv. panel physics NSF, 1958-61; mem. physics survey com., chmn. elem. particle physics panel Nat. Acad. Scis., 1969-72; high energy physics adv. panel div. research AEC, 1966-69; mem. steering com. (Sci. and Tech., A Five Year Outlook), 1979. Author: Nuclear Theory, 1953; Chief editor: High Energy Nuclear Physics, 1957; editor: National Energy Issues: How Do We Decide?, 1979. Guggenheim fellow, 1959-60. Fellow Am. Acad. Arts and Scis. (v.p. Midwest Center); mem. Nat. Acad. Scis. (chmn. physics sect. 1977-80, chmn. Class I, Math. and Phys. Scis. 1980-83), Am. Phys. Soc. (council 1968-71, regional sec. Central States 1964-69), AAAS (v.p., chmn. physics sect. 1970-71), Am. Inst. Physics (mem. governing bd. 1969-71), Phi Beta Kappa, Sigma Xi. Subspecialties: Particle physics; Theoretical physics. Current work: Theory of weak and electro magnetic interactions of fundamental particles, violation of time-reversal and CP invariances. Research in theoretical nuclear and atomic physics, terminal ballistics, nuclear power reactors, theoretical particle physics. Home: 5490 South Shore Dr Chicago IL 60615 Office: Enrico Fermi Inst 5630 Ellis Ave U Chgo Chicago IL 60637

SACKETT, WILLIAM MALCOLM, chemist, marine science educator; b. St. Louis, Nov. 14, 1930; s. William E. and Hedvig M. (Sanuskar) S.; m. Ann Hughey, June 16, 1956; children: Susan Ann, Karen Lynn. B.A. in Chemistry, Wash. U.-St. Louis, 1953, Ph.D., 1958. Tech. group supr. Pan Am. Petroleum Corp., Tulsa, 1959-62; asst. prof. geology Columbia U., N.Y.C., 1962-64; research scientist Isotope Geochemistry Program, Jersey Production Research Co., 1964; vis. scientist Woods Hole Oceanographic Instn., summers 1965-66; assoc. prof. chemistry U. Tulsa, 1965-68; assoc. prof. oceanography Tex. A&M U., 1968-70, prof. oceanography, 1970-79, asst. dean, 1973-74; von Humboldt Found. sr. scientist, Hannover, Germany, 1976-77; prof. marine sci. U. So. Fla., St. Petersburg, 1979—, chmn. marine sci. dept., 1979-82; vis. scientist, Pau, France, summer 1967; Mem. adv. panel oceanography project support NSF, 1980-82; mem. U.S. Nat. Com. Geochemistry, Nat. Acad. Sci.-NRC, 1979-82; mem. com. ocean drilling, 1981-82, chmn. steering com. report on petrolium in marine environment, 1980-82, mem. ocean scis. bd., 1981-83; mem. adv. bd. Univ. Nat. Oceanographic Lab. Systems, 1980-83. Contbr. sci. articles to profl. publs. Grantee in field. Mem. AAAS, Am. Geophys. Union, Geochem. Soc. (chmn. organic geochem. div. 1976), Am. Quarternary Assn., Am. Geol. Inst., Internat. Assn. Geochemistry and Cosmochemistry, European Assn. Organic Geochemists. Democrat. Unitarian. Subspecialty: Marine Chemistry. Current work: Marine geochemistry; stable and unstable isotope variations in natural materials; radiochemistry; marine pollution; geochemical exploration; organic geochemistry. Home: 2500 Driftwood Rd SE Saint Petersburg FL 33705 Office: Univ So Fla Marine Sci Dept 140 Seventh Ave S Saint Petersburg FL 33701

SACKMANN, I-JULIANA, astrophysicist; b. Schoenau, Prussia, Feb. 8, 1942; d. Emil and Lilly (Stelter) S.; m. Robert Frederick Christy, Aug. 4, 1973; children: Ilia Christy, Alexa Christy. B.A. (Alumni scholar), U. Toronto, 1963, M.A. (NRC Can. scholar), 1965, Ph.D. (Univ. fellow), 1968. NRC Can. postdoctoral fellow U. Goettingen, Germany, 1968-69, Max Planck Inst. fuer Physik and Astrophysik, Munich, 1969-71; postdoctoral fellow U. Hamburg (W.Ger.) Obs., 1971; research fellow Calif. Inst. Tech., Pasadena, 1971-74; NRC research assoc. Jet Propulsion Lab., Pasadena, 1974-76; NRC sr. research fellow, 1976-81, faculty assoc. in physics, 1981—. Contbr.

articles to profl. jours. Recipient Alexander von Humboldt award, 1970-71. Mem. Am. Astron. Soc., Can. Astron. Soc. Subspecialty: Theoretical astrophysics. Current work: Thermonuclear run-away reactions in stars, observable consequences, late evolution of stars, creation of elements in stars. Office: Kellogg Lab #106-38 1201 E California Blvd Pasadena CA 91125

SACKS, WILLIAM, neurochemist; b. Phila, Feb. 17, 1924; s. Harry F. and Bessie (Schechter) S.; m. Shirley Sacks, Apr. 4, 1954; children: Harriet Sacks Cook, Roberta Sacks Malone, Stuart Barry. B.S., Pa. State U., 1947, M.S., 1948, Ph.D., 1951. Dir. chemistry lab. Einstein Med. Ct. So. Div., Phila., 1951-58; prin. research scientist Rockland Research Inst., Orangeburg, N.Y., 1958—, also dir.; research assoc. prof. psychiatry N.Y. U. Sch. Medicine, N.Y.C., 1979—. Contrb. articles to profl. jours. Served with U.S. Army, 1944-46; PTO. Mem. Internat. Soc. Neurochemistry, Am. Soc. Neurochemistry, Am. Chem. Soc., AAAS, Soc. Biol. Chemists, Soc. Neurosci., Soc. Biol. Psychiatry, Sigma Xi, Phi Eta Sigma, Phi Lambda Upsilon, Gamma Sigma Delta. Subspecialties: Neurochemistry; Biochemistry (medicine). Current work: Cerebral metabolism in humans in vivo using an original arteriovenous technique. Have demonstrated a difference in brain metabolism of chronic mental patients.

SADLER, PETER MICHAEL, research geologist, educator; b. London, Dec. 20, 1948; U.S., 1976; s. Victor Noel and Hilda (Ling) S.; m. Marilyn Ann Kooser, Nov. 20. 1978; children: Heidi Dawn Sadler, Kelly Lane Kooser. B.Sc., U. Bristol, Eng., 1970, Ph.D., 1973; postgrad., U. Goettingen, W.Ger., 1973-75. Research scientist U. Goettingen, 1973-75; asst. prof. geology U. Calif.-Riverside, 1976-83, assoc. prof., 1983—. Mem. Soc. Econ. Paleontologists and Mineralogists, Nat. Assn. Geology Tchrs. Subspecialties: Sedimentology; Tectonics. Current work: Development of methods to assess quantitatively completeness of stratigraphic record; documentation of influence of tectonics upon sedimentation during uplift of mountains. Office: Dept Earth Sci U Calif Riverside CA 92521

SADUN, ALFREDO ARRIGO UMBERTO, neuro-ophthalmologist; b. New Orleans, Oct. 23, 1950; s. Elvio H. and Lina (Ottolenghi) S.; m. Debra Rice, Mar. 18, 1978; 1 dau., Rebecca Eli. S.B., M.I.T., 1972; Ph.D., Albert Einstein Med. Sch., 1976, M.D., 1978. Intern Huntington Meml. Hosp., Pasadena, Calif., 1978-79; resident in ophthalmology Harvard U. Med. Sch., Cambridge, Mass., 1979-82, Heed Found. fellow in neuro-ophthalmology, 1982-83. Contbr. articles to profl. jours. Mem. Am. Assn. Anatomists, Soc. for Neurosci., Assn. for Research in Vision and Ophthalmology, Mass. Soc. Eye Physicians and Surgeons. Subspecialties: Ophthalmology; Neuroanatomy. Current work: Researcher in tracing fiber pathways in the human visual system. Home: 23 E Riverside Dr Dedham MA 02026 Office: Massachusetts Eye & Ear Infirmary 243 Charles St Boston MA 02114

SAETTLER, ALFRED WILLIAM, plant pathologist, educator; b. Peoria, Ill., Mar. 8, 1940; s. Erhard George and Opal Flossie (Coats) S.; m. Stephanie K., Apr. 24, 1976; children: Robert W., Elizabeth C., Melani. B.A., Beloit Coll., 1962; Ph.D., U. Wis., 1966. Research plant pathologist U.S Dept. Agr., East Lansing, Mich., 1966—; adj. assoc. prof. Mich. State U., East Lansing, 1967—. Contbr. articles to profl. jours. Mem. Am. Phytopath. Soc., Sigma Xi. Subspecialties: Plant pathology; Integrated pest management. Current work: Researcher in diagnosis and control of bean diseases and breeding disease resistance. Office: Dept Plant Pathlog Mich State U East Lansing MI 48824

SAFARJAN, WILLIAM ROBERT, psychologist; b. Visalia, Calif., Feb. 17, 1943; s. Robert and Alice Joy (Sharp) S.; m. Paula Ann Tinder, May 26, 1978. B.A. in Internat. Relations, U. Calif-Berkeley, 1966; A.B. in Psychology, San Diego State U., 1971, M.A., 1976; Ph.D. in Psychology, Rutgers U., 1980. Lic. psychologist, Calif. Research asst. San Diego State U., 1971-72, Naval Personnel Research and Devel. Ctr., San Diego, 1972-73, Bell Telephone Lab., Holmdel, N.J., 1977; teaching asst., instr. Rutgers U., New Brunswick, N.J., 1974-78, research intern, 1978-80; staff psychologist Porterville (Calif.) State Hosp., 1980—. Contbr. articles to sci. publs., also chpts. to books. Served to lt. USN, 1966-69. Mem. Assn. for Advancement of Psychology, Am. Psychol. Assn., AAAS, Western Psychol. Assn., San Joaquin Psychol. Assn., Delta Sigma Phi. Subspecialties: Developmental psychology; Learning. Office: Porterville State Hosp PO Box 2000 Porterville CA 93258

SAFER, MARTIN ALLEN, psychology educator; b. Milw., Oct. 26, 1946; s. Joseph and Rita (Friedman) S. B.S., U. Wis.-Madison, 1968, M.S., 1974, Ph.D., 1978. Program evaluator Milw. Health Dept., 1978-79; asst. prof. psychology Cath. U. Am., Washington, 1979—. Contbr. articles to profl. jours. Mem. Am. Psychol. Assn., Eastern Psychol. Assn. Subspecialties: Social psychology; Preventive medicine. Current work: Research on hemisphere specialization for emotion, health promotion, health care utilization. Home: 3149 Queens Chapel Rd Apt 302 Mount Rainier MD 20712 Office: Dept Psychology Cath Univ Am Washington DC 20064

SAFFRAN, JUDITH, clinical chemist, biochemistry researcher; b. Montreal, Que., Can., Nov. 5, 1923; d. Philip and Pauline (Wigdor) Cohen; m. Murray Saffran, June 8, 1947; children: David, Wilma, Arthur, Richard. B.S., McGill U., 1944, Ph.D., 1948. Cert. Nat. Registry Clin. Chemistry. Clin. chemist endocrinology Jewish Gen. Hosp., Montreal, Que., Can., 1955-69; research fellow Inst. Med. Research, Toledo Hosp., 1969-74; asst., then assoc. prof. Med. Coll. Ohio, Toledo, 1974-79, clin. chemist, 1980—. Contbr. sci. articles to profl. publs. Research grantee Med. Research Council Can., 1964-68; NIH, 1974-78. Mem. Am. Soc. Biol. Chemists, Endocrine Soc., Can. Biochem. Soc., Am. Assn. Clin. Chemistry. Subspecialties: Clinical chemistry; Receptors. Current work: General clinical chemistry. Home: 2331 Hempstead Toledo OH 43606 Office: Med Coll Ohio CS 10008 Toledo OH 43699

SAGALYN, RITA C., physicist, government research administrator; b. Lowell, Mass., Nov. 24, 1924; m., 1952; 2 children. B.S., U. Mich., 1948; M.S., Radcliffe Coll., 1950. Research physicist Air Force Cambridge Research Labs., 1948-58; research physicist Aeronomy and Ionospheric Physics Labs., 1958-69; br. chief space physics Ionospheric Physics Lab., 1969-75; research physicist U.S. Air Force Geophys. Lab., Hanscom AFB, Bedford, Mass., 1949—, chief elec. processes br., 1975—. Recipient Guenther Loeser award Air Force Cambridge Research Labs., 1958, Patricia Kayes Glass award USAF, 1966. Mem. Am. Geophys. Union, Sigma Xi. Subspecialty: Ionospheric physics. Office: Air Force Geophys Lab Hanscom AFB Bedford MA 01731

SAGAN, CARL EDWARD, astronomer, educator, author; b. N.Y.C., Nov. 9, 1934; s. Samuel and Rachel (Gruber) S.; m. Ann Druyan; children by previous marriages: Dorion Solomon, Jeremy Ethan, Nicholas; 1 dau., Alexandra. A.B. with gen. and spl. honors, U. Chgo., 1954, B.S., 1955, M.S., 1956, Ph.D., 1960; Sc.D. (hon.), Rensselaer Poly. Inst., 1975, Denison U., 1976, Clarkson Coll. Tech., 1977, Whittier Coll., 1978, Clark U., 1978, Am. U., 1980; D.H.L. (hon.), Skidmore Coll., 1976, Lewis and Clark Coll., 1980, Bklyn. Coll., CUNY, 1982; LL.D. (hon.), U. Wyo., 1978. Miller research fellow U. Calif.-Berkeley, 1960-62; vis. asst. prof. genetics Stanford Med. Sch., 1962-63; astrophysicist Smithsonian Astrophys. Obs., Cambridge, Mass., 1962-68; asst. prof. Harvard U., 1962-67; mem. faculty Cornell U., 1968—, prof. astronomy and space scis., 1970—, David Duncan prof., 1976—, dir. Lab. Planetary Studies, 1968—, asso. dir. Center for Radiophysics and Space Research, 1972-81; pres. Carl Sagan Prodns. (Cosmos TV series), 1977—; nonresident fellow Robotics Inst., Carnegie-Mellon U., 1982—; NSF-Am. Astron. Soc. vis. prof. various colls., 1963-67, Condon lectr., Oreg., 1967-68; Holiday lectr. AAAS, 1970; Vanuxem lectr. Princeton U., 1973; Smith lectr. Dartmouth Coll., 1974, 77; Wagner lectr. U. Pa., 1975; Bronowski lectr. U. Toronto, 1975; Philips lectr. Haverford Coll., 1975; Disting. scholar Am. U., 1976; Danz lectr. U. Wash., 1976; Clark Meml. lectr. U. Tex., 1976; Stahl lectr. Bowdoin Coll., 1977; Christmas lectr. Royal Instn., London, 1977; Menninger Meml. lectr. Am. Psychiat. Assn., 1978; Carver Meml. lectr. Tuskegee Inst., 1981; Feinstone lectr. U.S. Mil. Acad., 1981; Pal lectr. Motion Picture Acad. Arts and Scis., 1982; Dodge lectr. U. Ariz., 1982; other hon. lectureships. Author: Atmospheres of Mars and Venus, 1961, Planets, 1966, Intelligent Life in the Universe, 1966, Planetary Exploration, 1970, Mars and the Mind of Man, 1973, The Cosmic Connection, 1973, Other Worlds, 1975, The Dragons of Eden, 1977, Murmurs of Earth: The Voyager Interstellar Record, 1978, Broca's Brain, 1979, Cosmos, 1980; also numerous articles; editor: Icarus: Internat. Jour. Solar System Studies, 1968-79, Planetary Atmospheres, 1971, Space Research, 1971, UFO's: A Scientific Debate, 1972, Communication with Extraterrestrial Intelligence, 1973; editorial bd.: Origins of Life, 1974—, Icarus, 1962—, Climatic Change, 1976—, Science 80, 1979-82. Mem. various adv. groups NASA and Nat. Acad. Scis., 1959—; mem. council Smithsonian Instn., 1975—; vice chmn. working group moon and planets, space orgn. Internat. Council Sci. Unions, 1968-74; lectr. Apollo flight crews NASA, 1969-72; chmn. U.S. del. joint conf. U.S. Nat. and Soviet Acads. Sci. on Communication with Extraterrestrial Intelligence, 1971; responsible for Pioneer 10 and 11 and Voyager 1 and 2 interstellar messages; judge Nat. Book Awards, 1975; mem. fellowship panel Guggenheim Found., 1976—. Recipient Smith prize Harvard U., 1964; NASA medal for exceptional sci. achievement, 1972; Prix Galabert, 1973; John W. Campbell Meml. award, 1974; Klumpke-Roberts prize, 1974; Priestley award, 1975; NASA medal for disting. public service, 1977, 81; Pulitzer prize for lit., 1978; Washburn medal, 1978; Rittenhouse medal, 1980; Peabody award, 1981; Hugo award, 1981; Seaborg prize, 1981; Roe medal, 1981; NSF fellow, 1955-60; Sloan research fellow, 1963-67. Fellow AAAS (chmn. astronomy sect. 1975), Am. Acad. Arts and Scis., AIAA, Am. Geophys. Union (pres. planetology sect. 1980-82), Am. Astronautical Soc. (council 1976-81), Brit. Interplanetary Soc., Explorers Club (75th Anniversary award 1980); mem. Am. Phys. Soc., Am. Astron. Soc. (councillor, chmn. div. for planetary scis. 1975-76), Fedn. Am. Scientists (council 1977-81), Am. Com. on East-West Accord, Soc. Study of Evolution, Genetics Soc. Am., Internat. Astron. Union, Internat. Acad. Astronautics, Internat. Soc. Study Origin of Life (council 1980—), Planetary Soc. (pres. 1979—), Authors Guild, Phi Beta Kappa, Sigma Xi. Subspecialties: Cosmology; Planetary science. Current work: Research on physics and chemistry of planetary atmospheres and surfaces, origin of life, exobiology, spacecraft observations of planets. Research on physics and chemistry of planetary atmospheres and surfaces, origin of life, exobiology, Mariner, Viking and Voyager spacecraft observations of planets. Address: Space Sci Bldg Cornell Univ Ithaca NY 14853

SAGE, ANDREW PATRICK, JR., engineering educator; b. Charleston, S.C., Aug. 27, 1933; s. Andrew Patrick and Pearl Louise (Britt) S.; m. LaVerne Galhouse, Mar. 3, 1962; children: Theresa Annette, Karen Margaret, Philip Andrew. B.S. in Elec. Engring, The Citadel, 1955; S.M., MIT, 1956; Ph.D., Purdue U., 1960. Registered profl. engr., Tex. Instr. elec. engring. Purdue U., 1956-60; asso. prof. U. Ariz., 1960-63; mem. tech. staff Aerospace Corp., Los Angeles, 1963-64; prof. elec. engring. and nuclear engring. scis. U. Fla., 1964-67; prof., dir. Info. and Control Scis. Center, So. Methodist U., Dallas, 1967-74, head elec. engring. dept., 1973-74; Quarles prof. engring. sci. and systems U. Va., Charlottesville, 1974—, chmn. dept. chem. engring., 1974-75, chmn. dept. engring. sci. and systems, 1977—, asso. dean, 1974-80; cons. Martin Marietta, Collins Radio, Atlantic Richfield, Tex. Instruments, LTV Aerospace, Battelle Meml. Inst., TRW Systems, NSF; gen. chmn. Internat. Conf. on Systems, Man and Cybernetics, 1974; mem. spl. program panel on system sci. NATO, 1981-82. Author: Optimum Systems Control, 1968, 2d edit., 1977, Estimation Theory with Applications to Communications and Control, 1971, System Identification, 1971, An Introduction to Probability and Stochastic Processes, 1973, Methodology for Large Scale Systems, 1977, Systems Engineering: Methodology and Applications, 1977, Linear Systems Control, 1978, Economic Systems Analysis, 1983; asso. editor: IEEE Transactions on Systems Sci. and Cybernetics, 1968-72; editor: IEEE Transactions on Systems, Man and Cybernetics, 1972—; asso. editor: Automatica, 1968-81; editor, 1981—; editorial bd.: Systems Engring, 1968-72, IEEE Spectrum, 1972-73, Computers and Electrical Engineering, 1972—, Jour. Interdisciplinary Modeling and Simulation, 1976—; editor Elsevier North Holland textbook series in system sci. and engring., Matrix Press textbook series on circuits and systems, 1976—; co-editor-in-chief: Jour. Large Scale Systems: Theory and Applications, 1978—; contbr. articles on mgmt. sci. and systems engring. to profl. jours. Recipient Frederick Emmonds Terman award Am. Soc. for Engring. Edn., 1970; M. Barry Carlton award IEEE, 1970; also Norbert Wiener award; Case Centennial scholar, 1980. Fellow IEEE, AAAS; mem. Inst. Mgmt. Scis., Am. Soc. for Engring. Edn., Am. Inst. Decision Sci., Internat. Inst. Forecasting, Ops. Research Soc. Am., Sigma Xi, Eta Kappa Nu, Tau Beta Pi. Subspecialties: Systems engineering; Artificial intelligence. Current work: Design of decision support systems; expert system and intelligent data base research; management science. Home: 303 Ednam Dr Charlottesville VA 22901

SAGER, NAOMI, linguistic and information research scientist, educator; b. Chgo., Mar. 21, 1927; d. Leon B. and Hannah (Shulman) S. B.S.E.E., Columbia U., 1953; A.M., U. Pa., 1954, Ph.D., 1968. Elec. engr. biophysics dept. Sloan-Kettering Inst., N.Y.C., 1953-57; sons. in electronics dept. physiology, 1958-59; research assoc. U. Pa., Phila., 1958-66; sr. research scientist, dir. Linguistic String Project, NYU, N.Y.C., 1966—; adj. prof. Faculty of Arts and Scis. Author: Natural Language Information Processing: A Computer Grammar of English and Its Applications, 1981; contbr. articles to profl. jours.; editor, co-author String Program reports, 1966—; mem. editorial bd.: Am. Jour. Computational Linguistics, 1974-77; adv. bd.: Info. Systems, 1974—. NSF grantee, 1966—; NIH grantee, 1969—. Mem. Biophysical Soc., Am. Soc. Info. Sci., Assn. Computational Linguistics, Assn. for Computing Machinery, Assn. for Health Quality. Subspecialties: Automated language processing; Information systems, storage, and retrieval (computer science). Current work: Research in computer applications in medicine; patient records; finding computable info. structures in sci.; mapping language-borne info. into database. Office: NYU Warren Weaver Hall New York NY 10012

SAGER, RAY STUART, chemistry educator; b. Cuero, Tex., Feb. 24, 1942; s. Oscar H. and Edrie E. (Hartman) S.; m. Linda Gail, Dec. 22, 1962; children: Jason, Joel. B.S., Tex. Lutheran Coll., 1964; Ph.D., Tex. Christian U., Ft. Worth, 1968. Prof. chemistry Concordia Coll., Moorhead, Minn., 1968-69, Capital U., Columbus, Ohio, 1969-75, Pan Am. U., Edinburg, Tex., 1975—. Subspecialty: Inorganic chemistry. Current work: Synthesis and properties of copper (II) complexes. Home: Route 2 Box 351G Edinburg TX 78539 Office: Chemistry Dept Pan American U Edinburg TX 78539

SAGER, RUTH, geneticist; b. Chgo., Feb. 7, 1918; d. Leon S.; m. Arthur B. Pardee. B.S., U. Chgo., 1938; Ph.D., Columbia U., 1948. Merck fellow NRC, Rockefeller Inst., N.Y.C., 1949-51, mem. staff, 1951-55; research scientist Columbia U., 1955-65; prof. molecular biology Hunter Coll., N.Y.C., 1966-75; prof. cellular genetics, chief div. cancer genetics Sidney Farber Cancer Inst., Harvard Med. Sch., 1975—. Author: Cytoplasmic Genes and Organelles, 1972; co-author: Cell Heredity, 1961; Contbr. articles to profl. jours. Guggenheim fellow, 1972-73. Mem. Nat. Acad. Scis., Am. Acad. Arts and Scis., Genetics Soc. Am., Am. Soc. Naturalists, Am. Soc. Biol. Chemists, Am. Soc. Cell Biology. Subspecialties: Genetics and genetic engineering (biology); Cancer research (medicine). Current work: Cell and molecular genetic analysis of neoplasia; role of genome rearrangements; DNA methylation. Office: Dana-Farber Cancer Inst 44 Binney St Boston MA 02115

SAHA, SUBRATA, bioengineering educator; b. Kushtia, India, Nov. 2, 1942; came to U.S., 1968, naturalized, 1976; s. Jaladhar K. and Sushama S.; m. Pamela Sunday, Oct. 30, 1972; children: Sunil, Supriya. B.S., Calcutta U., 1963; M.S., Tenn. Tech. U., 1969; Ph.D., Stanford U., 1973. Engr. cons. firms, 1963-67; research and teaching asst. Tenn. Tech. U., Cookeville, 1968-69, Stanford U., 1969-74; asst. prof. Yale U., 1974-79; assoc. prof., coordinator bioengring. La. State U. Med. Center, Shreveport, 1979—. Contbr. numerous articles to nat. and internat. jours.; also abstracts. Recipient Research Career Devel. award NIH, 1978-83, U.S.-India Exchange Scientist award, 1978; Fulbright award, 1982. Mem. ASME, Soc. Exptl. Stress Analysis, Soc. Biomaterials, Orthopaedic Research Soc., Biomed. Engring. Soc., Am. Soc. Biologists, ASCE, Am. Acad. Mechanics, Sigma Xi. Subspecialties: Biomedical engineering; Biomedical engineering. Current work: Biomedical engineering, ultrasound in medicine, bioelectrical stimulation. Home: 7601 Old Spanish Trail Shreveport LA 71105 Office: Dept Orthopaedic Surgery La State U Med Ctr PO Box 33932 Shreveport LA 71130

SAHNI, OMESH, research scientist; b. Lahore, Jan. 15, 1940; U.S., 1969; s. Chaman and Vimla (Rathore) S.; m. Tiia Taks, June 17, 1970; 1 dau., Sarita Lea. M.S., Banaras Hindu U., Varanasi, India, 1960; M.E.E., Indian Inst. Tech., Bombay, 1962; Ph.D., Rennselaer Poly. Inst., 1972. Lectr. Indian Inst. Tech., Bombay, 1963-67; research assoc. Ctr. Nuclear Studies, AEC, Saclay, France, 1967-68; Lab. Plasma Physics, U. Paris, Orsay, 1968-69; research and teaching asst. electrophysics div. Rensnselaer Poly. Inst., Troy, N.Y., 1969-72; mem. research staff IBM, Watson Research Ctr., Yorktown Heights, N.Y., 1972—. Contbr. articles to sci. jours. Mem. IEEE, Am. Phys. Soc., Soc. Info. Disply, Sigma Xi (treas. 1978-79, v.p. 1979-80, pres. chpt. 1980-81). Current work: Physics and technology of information display devices and development of an ultra-thin film capacitor compatible with silicon technology. Patentee blue color gas discharge display panel; electro-optically matrix-addressed electroluminescence display. Home: Box 47 RFD Hanover St Yorktown Heights NY 10598 Office: IBM Watson Research Center Yorktown Heights NY 10598

SAHU, SAURA CHANDRA, biochemist, researcher, cons., educator; b. Cuttack, India, June 29, 1944; came to U.S., 1966, naturalized, 1978; s. Gopinath and Ichhamoni S.; m. Jharana Sahu, May 29, 1966; children: Meghamala, Sudhir, Subir. M.S. (faculty fellow), Columbia U., 1967; Ph.D., U. Pitts., 1971. Research asso. Mich. State U., East Lansing, 1971-72, C.F. Kettering Research Lab., Yellow Springs, Ohio, 1972-74; asst. research prof. Duke U. Med. Center, Durham, N.C., 1974-79; research biochemist Health Scis. Labs., Consumer Product Safety Commn., Washington, 1979—; cons., in field. Contbr. articles on lung biochemistry to sci. jours. Govt of India Atomic Energy fellow, 1964-66; India Found. Scholar, 1966. Mem. Am. Soc. Biol. Chemists, Am. Soc. Pharmacology and Exptl. Therapeutics, Soc. Complex Carbohydrates, Soc. Toxicology, N.Y. Acad. Scis. Subspecialty: Biochemistry (medicine). Current work: Lung biochemistry; inhalation toxicology; structure, function and metabolic activity of the lung. Home: 13321 Kurtz Rd Woodbridge VA 22193 Office: Health Scis Labs Consumer Product Safety Commn 200 C St SW Washington DC 20204

SAIER, MILTON HERMAN, JR., educator, scientist; b. Palo Alto, Calif., July 30, 1941; s. Milton Herman and Lucilia (Bates) S.; m. Jeanne Karen Woodhams, Apr. 13, 1963; children: Hans Herman, Anila Johanna, Amanda Lucelia. B.S., U. Calif., Berkeley, 1963, Ph.D., 1968. Prof. dept. biology U. Calif., San Diego, La Jolla, 1972—; cons. Warner-Lambert Corp. and Jet Propulsion Lab. Author: (with C. D. Stiles) Molecular Dynamics in Biological Membranes, 1975, (with G. R. Jacobson) The Molecular Basis of Sex and Differentiation, 1983; contbr. articles to sci. publs. Mem. Am. Soc. Microbiology, Am. Soc. Cell Biology, Am. Soc. Biol. Chemists. Subspecialties: Biochemistry (biology); Microbiology (medicine). Current work: Mechanisms and regulation of transmembrane solute transport; sugar and salt transport; bacteria; kidney cells; regulation; biol. membrane functions, tumorigenicity. Home: 666 Quail Gardens Dr Encinitas CA 92024 Office: Dept Biology Univ Calif San Diego La Jolla CA 92093

SAIGAL, ROMESH, mathematician, engineering educator; b. Rawalpindi, Pakistan, Mar. 6, 1940; U.S., 1964; s. David Dayal and Laxmi (Dhawan) S.; m. Veena Khullar, Dec. 11, 1966; children: Ashima, Shailesh. B.Tech., Indian Inst. Tech., Kharagpur, 1961, M.Tech., 1963; Ph.D., U. Calif.-Berkeley, 1968. Asst. prof. U. Calif.-Berkeley, 1967-73; mem. tech. staff Bell Telephone Labs., Holmdel, N.J., 1973-76; mem. faculty Northwestern U., Evanston, Ill., 1976—, assoc. prof. indsl. engring., 1976-79, prof., 1979—. Grantee NSF, 1978-82; U. Wis. fellow, 1979; CORE fellow, Belgium, 1972-73. Mem. Ops. Research Soc. Am., Math. Prog. Soc., AAAS. Subspecialties: Operations research (engineering); Applied mathematics. Current work: Forumlating and solving large, sparse and structured systems. Office: Dept Indsl Engring Northwestern Univ Evanston IL 60201

SAINI, RAVINDER KUMAR, pharmacologist; b. Hoshiarpur, Punjab, India, Jan. 28, 1946; s. Bishen Dass and Ram (Pyari) S.; m. Urmil Malhotra, Oct. 22, 1971; children: Sandro, Alvin. D.V.M., Coll. Vet. Medicine, Hissar, India., 1968; M.S.; postgrad., Med. Research Inst., Chandigarh, India, 1971; Ph.D., U. Naples, Italy, 1973. Research asso. U. Pa., 1973-74; research fellow U. Wis., Madison, 1974-76; postdoctoral fellow U. Miami, Fla., 1976-78; research pharmacologist Merrell Research Center, Cin., 1978-79; sr. research investigator Squibb Inst. Medicine, Princeton, N.J., 1979—. Contbr. articles to profl. jours., chpt. to book. Mem. Am. Soc. Pharmacology and Exptl. Therapeutics, Am. Heart Assn., Am. Coll. Angiology, Internat. Soc. Heart Research, Italian Pharmacol. Soc. Subspecialties: Pharmacology; Veterinary cardiology. Current work: Design and devel. antianginal, antiarrhythmic and antithrombotic drugs in animal models of cardiovascular and occlusive diseases. Home: 11 Lawnside Dr Lawrenceville NJ 08648 Office: PO Box 4000 Princeton NJ 08540

ST.CLAIR, LARRY LEE, botanist, educator, researcher, cons.; b. Roanoke, Va., June 8, 1950; s. Jack Eugene and Bonnie Lee (Ratcliffe) St. C.; m. Rieta Joan Cheney, Dec. 18, 1971; children: Samuel, Laura, Audra, Virginia, Katherine. B.S., Brigham Young U., 1974, M.S.,

1975; Ph.D. (NSF fellow), U. Colo., 1983. Mem. faculty dept. botany and range sci. Brigham Young U., Provo, Utah, 1976—; researcher. Contbr. articles in botany to profl. jours. Mem. Am. Bryological and Lichenological Soc., Brit. Lichenological Soc., Internat. Lichenological Soc., Sigma Xi. Republican. Mormon. Subspecialties: Taxonomy; Ecology. Current work: Lichenology, taxonomy, ecology, and pollution physiology. Home: 1076 W 550 S Orem UT 84057 Office: Brigham Young U 427 WIDB Provo UT 84602

ST. JOHN, ROBERT MAHARD, physicist, educator; b. Westmoreland, Kans., Mar. 20, 1927; s. Fred H. and Penelope S. (Mahard) St. J.; m. Phyllis M. Frank, June 25, 1951; children: Glenn, Carol. Student, Washburn U., 1946-47; B.S., Kans. State U., 1950, M.S., 1951; Ph.D., U. Wis., 1954. Asst., then assoc. prof. U. Okla., Norman, 1954-68, prof. physics, 1968—, dir. engring. physics, 1963-77. Contbr. articles to profl. jours. Served with USN, 1945-46. Fellow Am. Phys. Soc., Okla. Acad. Sci.; mem. Am. Assn. Physics Tchrs., Sigma Xi, Phi Kappa Phi. Methodist. Subspecialty: Atomic and molecular physics. Current work: Precise measurements of atomic oscillator strengths. Home: 1523 Rosemont St Norman OK 73069 Office: 440 W Brooks St Norman OK 73019

ST. OMER, VINCENT EDMUND VICTOR, vet. pharmacologist, educator; b. Castries, St. Lucia, W.I., Nov. 16, 1934; s. Victor and Josephine M. (Laurent) St. O.; m. Margaret Moran Muir, May 5, 1962; children: Ingrid, Denise, Jeffrey, Raymond. D.V.M., U. Toronto, Ont., Can., 1962; M.Sc., U. Man., Winnipeg, Can., 1965; Ph.D., U. Guelph, Ont., 1969. Pvt. vet. practice, 1962-63; research assoc. Bur. Child Research. U. Kans., 1968-71; asst. prof. vet. pharmacology Kans. State U., 1972-74; assoc. prof. Coll. Vet. Medicine, U. Mo., Columbia, 1974—; asst. prof. Sch. Medicine, U. Mo., Columbia, 1974—; adj. prof. physiology and cell biology U. Kans., 1970-83; cons., researcher neurotoxicology, devel. toxicology, behavioral teratology. Contbr. articles to profl. jours., chpts. in books. Recipient numerous research grants, 1970—. Fellow Am. Acad. Vet. Pharmacology and Therapeutics; mem. Soc. Neurosci., Research Workers in Animal Disease, Am. Animal Hosp. Assn. Roman Catholic. Club: Optimist. Subspecialties: Neurochemistry. Current work: Devel., neurobehavioral and neurochem. toxicology, devel. toxicology, behavioral toxicology, vet. pharmacology, neurotoxicology, neurochemistry, environ. toxicology, neuropharmacology. Home: 2504 Mallard Ct Columbia MO 65201 Office: Dept Vetinary Anatomy and Physiology U Mo Columbia MO 65211

SAKAGUCHI, RONALD LOUIS, dental researcher; b. Los Angeles, Aug. 3, 1955; s. Louis and Hatsy S. B.S., UCLA, 1976; D.D.S., Northwestern U. Chgo., 1980; grad. in prosthodontics, U. Minn., 1982—. Research asst. biol. chemistry UCLA, 1973-76; researcher Am. Dental Assn., Chgo., 1977—; staff dentist Chgo. Osteo. Med. Ctr., 1980-82; instr. prosthodontics U. Minn., 1982—; cons. Research Triangle Inst., 1981, Pediatric and Adolescent Comprehensive Care Program, Chgo., 1981-82; dental examiner, coordinator Little Village Health Fair, Chgo., 1980-82, Kosminski Sch., 1982, Chgo. Dental Soc., 1981. Research grantee Northwestern U., 1979. Mem. Am. Dental Assn., Am. Assn. Dental Research, Minn. Dental Assn., Internat. Assn. Dental Research, Am. Coll. Prosthodontists, Minn. Assn. Prosthodontists, Xi Psi Phi. Subspecialties: Prosthodontics; Biomedical engineering. Current work: Relationship of practice characteristics to urinary mercury levels in dentists, acoustic transmission and systems analysis of joint function. Office: School of Dentistry University of Minnesota 515 Delaware St SE Minneapolis MN 55455

SAKOL, MARVIN J(AY), physician, consultant; b. Pitts., Feb. 4, 1923; s. David and Bessie (Steele) S.; m. Ruth Eva Levy, Feb. 1, 1953; children: Deborah Kay, Jeffrey Neil. M.D., Ohio State U., 1950. Diplomate: Am. Bd. Internal Medicine. Intern U. Colo., Denver, 1950-51; resident U. Pitts., 1951-52, U. Louisville, 1952-53, Ohio State U., 1953-54; jr. staff Akron (Ohio) City Hosp., 1955-59, sr. staff, 1959—, chmn. dept. oncology, 1972—; assoc. prof. Northeastern Ohio Univs. Coll. Medicine, Rootstown, 1975—. Served to 1st lt. USAAF, 1942-46. Fellow Internat. Soc. Hematology; mem. Am. Soc. Clin. Oncology. Jewish. Subspecialties: Cancer research (medicine); Chemotherapy. Current work: Treatment malignant disease. Home: 1037 Bunker Dr Akron OH 44313 Office: Akron City Hosp 628 W Market St Akron OH 44303

SAKOVER, RAYMOND PAUL, radiologist; b. Chgo., Oct. 8, 1944; s. Max and Maria Adele (Berardi) S.; m. Patricia Ellyn Taylor, June 7, 1969; children: Shelley, Michael, David, Raymond. M.D., U. Ill.-Chgo., 1969. Diplomate: Am. Bd. Radiology, Nat. Bd. Med Examiners. Intern St. Francis Hosp., Evanston, Ill., 1969-70, resident in radiology, 1970-73; radiologist Riverside Radiology Med. Group, Calif., 1975—; pres. Computerized Diagnostic Med. Group, Riverside, 1979-82. Bd. dirs. Riverside Humane Soc., 1981-83. Served to lt. comdr. USNR, 1973-75. Mem. Am. Coll. Radiology, Soc. Nuclear Medicine, AMA, Calif. Med. Assn., Riverside County Med. Assn. (mediation com. 1982—), Am. Lung Assn. (dir. Riverside County chpt. 1976-83). Republican. Roman Catholic. Club: Rotary (Magnolia Center, Calif.). Subspecialties: Radiology; Nuclear medicine. Office: 6941 Brockton Ave Riverside CA 92506

SALADIN, KENNETH STANLEY, biology educator; b. Kalamazoo, Mich., May 6, 1949; s. Albert Roman Joseph and Jennie Louise (Wing) S.; m. C. Diane Campbell, Sept. 8, 1979; children: Emory Michael, Lisa Nicole. B.S., Mich. State U., 1971; Ph.D., Fla. State U., 1979. Asst. prof. biology Ga. Coll., Milledgeville, 1977-83, assoc. prof., 1983—. Bd. dirs. Big Bros./Big Sisters Am., Milledgeville, 1977—; bd. dirs. Nat. Ctr. Sci. Edn., Del., 1982—. Mem. Southeastern Soc. Parasitologists (Elon Byrd award 1978), Am. Soc. Parasitologists, Animal Behavior Soc., AAAS, Ga. Acad. Sci. Democrat. Subspecialties: Ethology; Parasitology. Current work: Ethology of host-finding by parasitic invertebrates; sensory physiology and behavioral ecology of parasites. Home: 112 Camellia Circle SW Milledgeville GA 31061 Office: Georgia College Milledgeville GA 31061

SALADINI, JOHN LOUIS, research biologist; b. Detroit, Feb. 13, 1946; s. John and Elaine (Bogardus) S. B.A. with high honors, W.Va. U., Morgantown, 1968; M.S., U. Fla., Gainesville, 1971; Ph.D., Ohio State U., Columbus, 1976. Research biologist E. I. duPont de Nemours & Co. (Expt. Sta.), Wilmington, Del., 1976-78, research and devel. rep., Denver, 1979—. Served to capt. Chem. Corps. USAR, 1971. Mem. Am. Phytopath. Soc., Entomol. Soc. Am., Weed Sci. Soc. Am., Phi Beta Kappa, Phi Kappa Phi. Republican. Presbyterian. Subspecialty: Agrichemicals. Current work: Research and development of commercial and experimental compounds for use by agricultural personnel. Home: 276 S Monaco Pkwy Apt B Denver CO 80224 Office: 7401 W Mansfield Ave Suite 300 Lakewood CO 80235

SALAMA, MAMDOUH M., mechanical engineer, researcher; b. Cairo, Feb. 11, 1945; s. Mohamed I. and Soria B. (Samour) S.; m. Nadia El-Khosht; children: Yasmine, Kareem, Mohammed. B.Sc., Ain shams U., Cairo, 1966, M.Sc., 1969; S.M., MIT, 1971, Mech.E., 1972, Sc.D., 1976. Research asst. MIT, 1970; with Stone & Webster Engring., Boston, 1972-77; research group leader materials group, prodn. research div. Conoco, Inc., Ponca City, Okla., 1977—. Contbr. articles to profl. jours. Mem. ASME, Am. Welding Soc., ASTM, Soc. Petroleum Engrs. Republican. Moslem. Subspecialties: Fracture mechanics; Metallurgy. Current work: Research activities related to oil production and offshore developments. Home: 400 Wren Dr Ponca City OK 74601 Office: Conoco Inc Ponca City OK 74603

SALAMONE, JOSEPH CHARLES, polymer chemist, cons. to plymer industries, univ. dean, educator; b. Bklyn., Dec. 27, 1939; s. Joseph John and Angela S. B.S. in Chemistry, Hofstra U., 1961, Ph.D., Poly. Inst. Bklyn., 1966. NIH postdoctoral fellow U. Liverpool, Eng., 1966-67; research assoc. U. Mich., 1967-70; asst. prof. dept. chemistry Lowell (Mass.) Technol. Inst., 1970-73, assoc. prof., 1973-76; assoc. prof. dept. chemistry U. Lowell, 1973-76, prof., 1976—, chmn. dept., 1975-78, dean, 1978—. Mem. editorial bd.: Polymer; mem. adv. bd.: Jour. Polymer Sci; contbr. numerous articles on polymer chemistry to sci. jours. Mem. Am. Chem. Soc. (divs. polymer chemistry, organic coatings and plastics chemistry, chmn. div. polymer chemistry 1982—). Subspecialty: Polymer chemistry. Current work: Synthesis of new monomers and polymers, particularly ionic systems; properties of polyelectrolytes. Home: Coll Pure and Applied Sci U Lowell North Campus Lowell MA 01854

SALAMUN, PETER J(OSEPH), botanist, educator, consultant; b. La Crosse, Wis., June 12, 1919; s. Peter and Melana (Hardi) S.; m. Lorraine Anne Saurman, June 6, 1946; children: Mary Janet Salamun Conrad, Elizabeth Alice Salamun Drake, Charles Peter, William Mark, Edward Joseph, David Robert, Lawrence George, Katherine Anne. B.S., Wis. State Tchrs. Coll., 1941; M.S., U. Wis., Madison, 1947, Ph.D., 1950. Instr. biology Wis. State Coll. (now U. Wis.), Milw., 1948-51, assoc. prof., 1951-56, prof. botany, 1957—, chmn dept., 1957-61, dir. herbarium, 1966—; cons. Contbr. articles to sci. jours. Served in USAF, 1941-45; CBI. Recipient AMOCO Teaching award, 1976. Mem. AAAS, Am. Inst. Biol. Scis., Am. Meteorol. Soc., Am. Soc. Plant Taxonomists, Bot. Soc. Am., Ecol. Soc. Am., Internat. Assn. Plant Taxonomy, Nature Conservancy, Soc. Study of Evolution, Wis. Acad. Sci., Sigma Xi. Roman Catholic. Subspecialties: Taxonomy; Ecology. Current work: Floristics of vascular plants of Wisconsin; vegetational analysis of the Lake Michigan shoreline. Home: 5013 N Elkhart Ave Whitefish Bay WI 53217 Office: U Wis LAP 450 Milwaukee WI 53201

SALCMAN, MICHAEL, Neurol. surgeon, educator, researcher; b. Pilsen, Czechoslovakia, Nov. 4, 1946; s. Arthur and Edith (Atlas) S.; m. Ilene, July 27, 1969; children: Joshua, Dara. B.A., M.D., Boston U., 1969. Diplomate: Am. Bd. Neurol. Surgeons, 1978, Nat. Bd. Med. Examiners, 1970. Intern in surgery Boston U. Med. Center, 1969-70; research assoc. Nat. Inst. Neurol. Diseases, NIH, Bethesda, Md., 1970-72; resident in neurosurgery Neurol. Inst. N.Y., Columbia U., N.Y.C., 1972-76; asst. prof. neurol. surgery U. Md. Sch. Medicine, Balt., 1976-79, assoc. prof., 1979—, chief neurooncology, 1978—; cons. Md. Inst. Emergency Med. Services; mem. exec. com. Congress Neurol. Surgeons. Author: Not Quite a Miracle, 1983; assoc. editor: Neurosurgery, 1982; Contbr. articles to sci. jours. Served with USPHS, 1970-72. Recipient award Lange Med. Pub., 1966; Nat. Eye Inst. grantee, 1972-78; Am. Cancer Soc. grantee, 1978-80; Nat. Inst. Neurol. and Communicative Disorders and Stroke grantee, 1980-82, 82-84. Mem. Washington Print Club, Friends of Modern Art, Balt. Mus. Art, Print and Drawing Soc. Balt. Mus. Art.; Fellow ACS; mem. Am. Assn. Neurol. Surgeons, AMA, Assn. Acad. Surgeons, Research Soc. Neurol. Surgery, Soc. Univ. Neurosurgeons, AAAS, Am. Physiol. Soc., ASME, Assn. Advancement Med. Instrumentation, IEEE, N.Y. Acad. Sci., Soc. Neurosci., Begg Soc., Phi Beta Kappa, Alpha Omega Alpha. Jewish. Subspecialties: Oncology; Neurobiology. Current work: Multimodality attack on malignant brain tumors using computer-controlled microwave hyperthermia, computer-guided radiation sources, polychemotherapy facilitated by drug opening of the blood-brain barrier, development of model brain tumors in animals and monoclonal antibodies to same; interaction of biomedical engineering and clinical neurosurgery (laser, etc). Patentee chronic microelectrode; co-developer microwave hyperthermia for brain; DMSO. Home: 5401 Springlake Way Baltimore MD 21212 Office: 22 S Greene St Baltimore MD 21201

SALEH, SHOUKRY DAWOOD, management sciences educator, consultant; b. Minia, Egypt, Jan. 17, 1930; emigrated to Can., 1962, naturalized, 1967; s. Dawood and Wahibah (Boulos) S.; m. Feb. 18, 1967; children: Krista, Marcie. B.A. in Sociology and Psychology, Cairo U., 1951; M.S. in Psychology, Case Western Res. U., 1960; Ph.D. in Indsl. Psychology, Case Western Res. U., 1963. Research asst. Psychol. Research Services, Case Western Res. U., 1961, research assoc., to 1963; exams. officer Ont. Dept. Civil Service, Toronto, 1963, testing supr., 1963-64, dir. personnel research, 1964-67; assoc. prof. mgmt. scis. U. Waterloot, Ont., 1967-69, prof., 1969—; chmn. dept. mgmt. scis., 1974-78; vis. prof. Faculty Commerce U. Nairobi, Kenya, 1979—; tng. lectr. various cos.; founding mem. Sanford Flemming Found.; reviewer Can. Council Research Grants Applications and various profl. jours.; cons. in field. Author: Administrative Behavriour, 1969; contbr. numerous articles to profl. jours. Recipient recognition for substantial contbn. to Dept. Civil Service and to Govt. Can. Hon. Charles MacNaughton, past. treas. Ont., 1967; Internat. Labour Orgn. fellow, U.S., 1959. Mem. Personnel Research Assn. (pres. 1964-65), Can. Assn. Organizational Behavior. Subspecialties: Social psychology; Organizational behavior. Home: 107 Longwood Dr Waterloo ON Canada N2L 4B6 Office: U Waterloo Dept Mgmt Scis Waterloo ON Canada N2L 3G1

SALEM, HARRY, toxicologist/pharmacologist; b. Ontario, Can., Mar. 21, 1929; s. S. Oscar and Bessie (Pierce) S.; m. Florence, June 30, 1957; 1 son, Jerome Sheldon. B.A., U. Western Ont., 1950; B.Sc., U. Mich., 1953; M.A., U. Toronto, 1955, Ph.D., 1958, postdoctoral, 1958-59; postgrad. in medicine, U. of East, Manila, 1977—. Pharmacologist Air Shields, Inc., 1959-62; sr. pharmacologist Smith, Kline & French Labs., 1962-65; dir. respiratory researchlabs. Nat. Drug Co., 1965-70; dir. pharmacology and toxicology Smith, Miller & Patch, Inc., New Brunswick, N.J., 1970-72, Cooper Labs., Inc., Cedar Knolls, N.J., 1972-76; pres. chief toxicologist Cannon Labs., Inc., Reading, Pa., 1977-79; pres. Cosmopolitan Safety Evaluations, Inc., Somerville, N.J., 1979-80, ToxiGenics, Inc., Decatur, Ill., 1980—; mem. faculty U. Pa., Phila., 1960—, assoc. prof. pharmacology, 1975—; adj. prof. environ. health Coll. Pharmacy, Temple U., Phila., 1965—; cons. pharmacologist/toxicologist Med. Documentation Service Coll. Physicians, Phila., 1968—, Franklin Research Ctr., 1978—. Mem. editorial adv. bd.: Drug Info. Jour, 1974—, Jour. Applied Toxicology, 1980, Internat. Ency. Pharmacology and Therapeutics, 1970—, Jour. Environ. Pathology and Toxicology, 1979—; contbr. articles to profl. jours. Fellow Am. Coll. Clin. Pharmacology, N.Y. Acad. Scis.; mem. Acad. Scis. Phila., AAAS, Am. Chem. Soc., Am. Coll. Toxicology, Am. Pharm. Assn., Am. Soc. Clin. Pharmacology and Therapeutics, Am. Soc. Pharmacology and Exptl. Therapeutics, Drug Info. Assn., Internat. Union Pharmacology, Pharmacology Soc. Can., Reticuloendothelial Soc., Soc. Comparative Ophthalmology (v.p.), Soc. Toxicology. Subspecialties: Toxicology (agriculture); Pharmacology. Patentee method of lowering intraocular pressure with antazoline. Office: Whittaker Toxigenics Inc 1800 Pershing Rd Decatur IL 62526

SALERNI, JOHN VINCENT, design engineer; b. Los Angeles, Jan. 30, 1940; s. John and Rose Clema (Longuevan) S.; m. Joy Annine, Jan. 20, 1962; children: Sheryl Annine, Nina Marie. B.S.M.E., Calif. State U.-Long Beach, 1965. Registered profl. engr., Calif. Prodn. scheduler Baker Oil Tools, Inc., Los Angeles, 1957-65, devel. engr., 1965-70; mfg. engring. supr. Baker Packers, Los Angeles, 1970-74, project engr., 1974-75; Mfg. engring. mgr., 1975-76, prodn. mgr., 1976-78; mgr. new product devel. Baker Sand Control, Houston, 1978—; instr. customer and employee tng. Served with U.S. Army, 1962-68. Mem. ASME, Soc. Mfg. Engrs., Soc. Petroleum Engrs. of AIME. Republican. Subspecialties: Mechanical engineering; Materials (engineering). Current work: Design mechanical and hydraulic devices for sub-surface use in oil and gas wells. Devices to function in a highly contaminated and corrosive environment with temperatures and pressures. Patentee in field. Office: PO Box 61486 Houston TX 77208

SALGANICOFF, LEON, pharmacologist, educator, cons.; b. Buenos Aires, Argentina, Sept. 11, 1924; came to U.S., 1964, naturalized, 1977; s. Marcos and Ana (Zelicson) S.; m. Matilde Saffier, Apr. 12, 1957; children: Alina, Marcos. D.Sc., U. Buenos Aires, 1955. Cert. Argentine Bd. Clin. Biochemistry. Vis. prof. Johnson Found., U. Pa., Phila., 1964-68; assoc. prof. dept. pharmacology Temple U. Sch. Medicine, Phila., 1968-76, pharmacology sect. leader, 1972-82, prof. pharmacology, 1976—; vis. prof. dept. gen. pathology U. Rome, 1977—, NATO vis. prof., 1982. Contbr. articles to profl. jours. Mem. Am. Soc. Pharmacology and Exptl. Therapeutics, AAAS. Subspecialties: Cellular pharmacology; Molecular pharmacology. Current work: Physiology and pharmacology of activated platelets; cyclic adenosin monophosphate and calcium control of contractile function. Home: 6409 N 11th St Philadelphia PA 19126 Office: 3F Pharmacology Temple U 3400 N Broad St Philadelphia PA 19140

SALINAS, FERNANDO A., oncologist, educator, researcher; b. Santiago, Chile, Oct. 30, 1939; emigrated to Can., 1976; s. Ramón Fernando and Mariá Teresa (Beà) S.; m. Alejandrina A. Marfful, Sept. 3, 1964; children: Claudia, Alejandro. B.Sc., U. Chile, 1959, D.V.M., 1963. Diplomate: in vet. medicine U. Chile, 1964. Instr. dept. biology U. Chile Sch. Vet. Medicine, 1959-63; asst. prof. dept. embryology Sch. Medicine, 1963-67, asst. prof. dept. exptl. medicine, 1967-70; Vis. investigator/prof. biology div. Oak Ridge Nat. Lab., 1971-73; asst. prof. dept. medicine U. So. Calif. Sch. Medicine, Los Angeles, 1974-76; assoc. prof. dept. pathology U. B.C., Can., Vancouver, 1976—; head lab. animal oncology Cancer Control Agy. B.C., Vancouver, 1976—. Contbr. articles to profl. jours. Mem. Internat. Soc. Exptl. Hematology, Am. Assn. Cancer Research, Am. Assn. Immunologists, Am. Soc. Clin. Oncology. Subspecialties: Oncology; Cancer research (medicine). Current work: Pathogenic and prognostic role of immune complexes in human cancer. Immunoregulatory significance of oncofetal antigens; biological response modifiers. Office: 2656 Heather St Vancouver BC Canada V5Z 3J3

SALK, JONAS EDWARD, physician, scientist; b. N.Y.C., Oct. 28, 1914; s. Daniel B. and Dora (Press) S.; m. Donna Lindsay, June 8, 1939; children: Peter Lindsay, Darrell John, Jonathan Daniel; m. Francoise Gilot, June 29, 1970. B.S., CCNY, 1934, LL.D. (hon.), 1955; M.D., NYU, 1939, Sc.D. (hon.), 1955, LL.D., U. Pitts., 1955, Ph.D., Hebrew U., 1959, LL.D., Roosevelt U., 1955, Sc.D., Turin U., 1957, U. Leeds, 1959, Hahnemann Med. Coll., 1959, Franklin and Marshall U., 1960, D.H.L., Yeshiva U., 1959, LL.D., Tuskegee Inst., 1964. Fellow in chemistry NYU, 1935-37, fellow in exptl. surgery, 1937-38, fellow in bacteriology, 1939-40; Intern Mt. Sinai Hosp., N.Y.C., 1940-42; NRC fellow Sch. Pub. Health, U. Mich., 1942-43, research fellow epidemiology, 1943-44, research asso., 1944-46, asst. prof. epidemiology, 1946-47; asso. research prof. bacteriology Sch. Medicine, U. Pitts., 1947-49, dir. virus research lab., 1947-63, research prof. bacteriology, 1949-55, Commonwealth prof. preventive medicine, 1955-57, Commonwealth prof. exptl. medicine, 1957-63; dir. Salk Inst. Biol. Studies, 1963-75, resident fellow, 1963—, founding dir., 1975—; developed vaccine, preventive of poliomyelitis, 1955, cons. epidemic diseases sec. war, 1944-47, sec. army, 1947-54; mem. commn. on influenza Army Epidemiol. Bd., 1944-54, acting dir. commn. on influenza, 1944; mem. expert adv. panel on virus diseases WHO; adj. prof. health scis., depts. psychiatry, community medicine and medicine U. Calif., San Diego, 1970—. Author: Man Unfolding, 1972, The Survival of the Wisest, 1973, (with Jonathan Salk) World Population and Human Values: A New Reality, 1981, Anatomy of Reality, 1983; Contbr. sci. articles to jours. Decorated chevalier Legion of Honor, France, 1955, officer, 1976; recipient Criss award, 1955, Lasker award, 1956, Gold medal of Congress and presdl. citation, 1955, Howard Ricketts award, 1957, Robert Koch medal, 1963, Mellon Inst. award, 1969; Presdl. medal of Freedom, 1977; Jawaharlal Nehru award for internat. understanding, 1976. Fellow A.A.A.S., Am. Pub. Health Assn.; asso. fellow Am. Acad. Pediatrics (hon.); mem. Am. Coll. Preventive Medicine, Am. Acad. Neurology, Assn. Am. Physicians., Soc. Exptl. Biology and Medicine, Am. Soc. Clin. Investigation, Am. Assn. Immunologists, Am. Epidemiol. Soc., Phi Beta Kappa, Alpha Omega Alpha, Delta Omega. Office: Salk Inst Biol Studies PO Box 85800 San Diego CA 92138

(SALK), SUNG HO, physicist, educator; b. Seoul, Korea, Apr. 14, 1939; came to U.S., 1964, naturalized, 1982; s. Chin S. and Kuk J. (Shin) Suck; m. Jung J. Yeon, Apr. 3, 1942; children: Tom T., Bob S. M.S., U. Houston, 1968; Ph.D., U. Tex.-Austin, 1972. Research assoc. U. Tex.-Austin, 1972-76; research asst. prof. U. Mo.-Rolla, 1977-82, research assoc. prof. physics and cloud physics, 1982—. Contbr. numerous articles to profl. jours. NSF fellow, 1980-82. Mem. Am. Phys. Soc., Sigma Xi, Phi Kappa Phi. Subspecialties: Atomic and molecular physics; Polymer physics. Current work: Molecular reaction dynamics (rearrangement collision theory); conductivity of organic crystals; atmospheric nucleation and radiation; air pollution. Home: 1114 Sycamore St Rolla MO 65401 Office: U Mo Rolla MO 65401

SALKOFF, LAWRENCE BENJAMIN, geneticist, neurobiologist; b. Bklyn., Mar. 3, 1944; s. Goodwin and Gladys (Bass) S.; m. Miranda Carey, Sept. 7, 1968; 1 dau., Leah Naomi. B.A., UCLA, 1967; Ph.D. in Genetics, U. Calif.-Berkeley, 1979. Vol. U.S. Peace Corps, Colombia, 1967-70; teaching assoc. U. Calif.-Berkeley, 1976-79; postdoctoral researcher Yale U., New Haven, 1979-82; asst. research physiologist U. Calif. Sch. Medicine, San Francisco, 1982—. Contbr. articles to profl. jours. Recipient John Belling prize in genetics U. Calif.-Berkeley, 1980. Mem. Soc. for Neurosci., Genetics Soc. Am., AAAS, Sierra Club. Subspecialties: Genetics and genetic engineering (biology); Neurobiology. Current work: Genetic determinants of neural function, especially regarding membrane electrical properties. Office: Dept Physiology U Calif Sch Medicine San Francisco CA 94143

SALOM, IRA LOUIS, physician, researcher; b. Bklyn., Sept. 13, 1952; s. Benjamin and Francine (Gratz) S. B.A., NYU, 1973; M.D., SUNY-Buffalo, 1977; M.Sc. (Med.), U. Minn., 1982. Intern Met. Hosp., N.Y.C., 1977-78; resident in internal medicine U. Minn., Mpls., 1978-80; in gastroenterology SUNY, 1980-82; asst. dir. clin. research Ayerst Labs., N.Y.C., 1982—. Hattie Strong scholar, 1976; recipient Florence Lee prize DAR, 1976. Fellow N.Y. Acad. Medicine; Mem. N.Y. Acad. Sci., Am. Coll. Nutrition, ACP, Am. Gastroent. Assn., Am. Fedn. Clin. Research, Am. Soc. Clin. Nutrition. Subspecialties: Gastroenterology; Internal medicine. Current work: Portal hypertension and its complications, gastric irritation, nutrition and host-disease relationships, general gastroenterology and internal medicine. Office: Ayerst Labs 685 3d Ave New York NY 10017

SALOMON, ROBERT EPHRAIM, chemistry educator, consultant; b. N.Y.C., June 8, 1933; s. Alexander R. and Claire (Meirowitz) Caine) S.; m. Ronaly Felice Greenstein, Mar. 25, 1961; children: Alexander R., Lawrence Paul. B.A., Bklyn. Coll., 1954; Ph.D., U. Oreg., 1960. Prof. chemistry Temple U., 1960—; cons. U.S. Nuclear Regulatory Commn. Mem. Am. Chem. Soc., AAUP, Electrochemical Soc., Sigma Xi. Subspecialties: Wave power; Solar energy. Current work: Developing system for the conversion of solar heat into electricity using metallic hydrides and hydrogen concentration cells. Patentee Conversion of ocean wave energy to electricity, solar thermal hydride conventer. Home: 1015 Lombard St Philadelphia PA 19147

SALPETER, EDWIN ERNEST, educator; b. Vienna, Austria, Dec. 3, 1924; came to U.S., 1949, naturalized, 1953; s. Jakob L. and Frieder (Horn) S.; m. Miriam Mark, June 11, 1950; children—Judy Gail, Shelley Ruth. M.S., Sydney U., 1946; Ph.D., Birmingham (Eng.) U., 1948; D.Sc., U. Chgo., 1969, Case-Western Reserve U., 1970. Research fellow Birmingham U., 1948-49; faculty Cornell U., Ithaca, N.Y., 1949—; now J.G. White prof. phys. scis.; mem. U.S. Nat. Sci. Bd., 1979—. Author: Quantum Mechanics, 1957, 77; Editorial bd.: Astrophys. Jour, 1966-69; asso. editor: Rev. Modern Physics, 1971—; Contbr. articles to profl. jours. Mem. AURA bd., 1970-72. Recipient gold medal Royal Astron. Soc., 1973; J.R. Oppenheimer Meml. prize, 1974. Mem. Am. Astron. Soc. (v.p. 1971-73), Am. Philos. Soc., Nat. Acad. Scis., Am. Acad. Arts and Scis., Deutsche Akademie Leopoldina. Subspecialties: Theoretical astrophysics; Radio and microwave astronomy. Home: 116 Westbourne Ln Ithaca NY 14850

SALSBURG, KEVYN ANNE, applied mathematician, engineer; b. Ilion, N.Y., July 7, 1951; d. Glenn William and Nyna Jeanne (Warburton) S. B.S. cum laude, St. Lawrence U., Canton, N.Y., 1972; M.S., U. R.I., Kingston, 1973; M.A., U. Md., College Park, 1978. Math. statistician U.S. Bur. Census, Suitland, Md., 1974-78; mathematician 2Nat. Bur. Standards, Gaithersburg, Md., 1978; programmer Gen. Tech. div. IBM, Essex Junction, Vt., 1978-82, staff engr., Manassas, Va., 1982—. Author tech. books; contbr. articles to profl. jours. Mem. Soc. for Indls. and Applied Math., Assn. Computing Machinery. Democrat. Roman Catholic. Subspecialties: Computer-aided design; Microchip technology (materials science). Current work: Development of design automation systems for very large scale integration; finite element and finite difference mathematical modeling in the development of CAD tools for process and device simulation. Home: 14016 Junter Hill Ln Nokesville VA 22123 Office: Fed Systems Div IBM 9500 Godwin Dr 120-024 Manassas VA 22110

SALSBURY, ROBERT LAWRENCE, animal nutrition educator, researcher; b. Vancouver, C., Can., July 4, 1916; came to U.S., 1945, naturalized, 1955; s. Herbert Edward and Jane Stenhouse (Watson) S.; m. Elizabeth M. Brown, July 29, 1944. B.A., U. B.C., 1942; Ph.D., Mich. State U., 1955. Jr. chemist Can. Govt., Ottawa, Ont., 1943-45; control chemist E.R. Squibb & Sons, New Brunswick, N.J., 1945-46; research assoc. N.J. Agr. Exptl. Sta., Sussex, N.J., 1946-47; bacteriologist Mich. State Dept. Health, Lansing, 1947-49, biochemist 11, 1949-50; asst. prof. Mich. State U., 1950-61; asso. prof. animal nutrition U. Del., Newark, 1961-73, prof., 1973—. Contbr. to profl. jours. Mem. Am. Chem. Soc., Am. Soc. Microbiology, Am. Soc. Animal Sci., Am. Dairy Sci. Assn., Poultry Sci. Assn., Del. Acad. Sci., N.Y. Acad. Sci. Subspecialties: Biochemistry (biology); Microbiology. Current work: Interaction of dietary ingredients with therapeutic and prophylactic agents animal nutrition. Office: Dept Animal Sci and Agr Biochemistry U Del Newark DE 19711

SALTERS, GRACE HEYWARD, botany educator; b. Florence, S.C., May 25, 1933; d. John Wayne and Wilhelmena (Wright) Heyward; m. Walter Leon Salters, Dec. 21, 1963; 1 dau., Damita Renee. B.S., Bennett Coll., Greensboro, N.C., 1955; M.S., Atlanta U., 1962, Ph.D., 1977. Tchr. sci. Alexander County (N.C.) Sch., 1956-59, Bonds-Wilson High Sch., North Charleston, S.C., 1960-68; instr. biology S.C. State Coll., 1968-72, asst. prof., 1972-81, assoc. prof., 1981—; assoc. dir. Minority Access to Research Careers/Honors Undergrad.; Research Tng. program, 1980—. Vice pres. bd. dirs. Boylan-Haven-Mather Acad., Camden, S.C., 1979—. NSF fellow, summers 1964-66; NSF COSIP fellow, 1973-74; NIH MARC fellow, 1974-77; NSF grantee, summer 1958, 59-60. Mem. S.C. State Employees Assn. (bd. dirs. 1980—), Assn. Southeastern Biologists, NAACP, Bennett Coll. Alumnae (pres. chpt. 1970—), Phi Delta Kappa, Delta Sigma Theta (treas. chpt. 1983), Beta Kappa Chi. Democrat. Methodist. Clubs: Shamrock Social (Orangeburg, S.C.) (pres. 1979-81); Alpha Wives (v.p. 1979-81, pres. 1983—. Subspecialties: Plant growth; Plant cell and tissue culture. Home: 211 Chestnut St NE Orangeburg SC 29115 Office: SC State Coll PO Box 2013 Orangeburg SC 29117

SALTZER, JEROME H(OWARD), computer science educator, researcher; b. Nampa, Idaho, Oct. 9, 1939; m., 1961; 3 children. S.B., MIT, 1961, S.M., 1963, Sc.D., 1966. From instr. to assoc. prof. elec. engring. MIT, Cambridge, 1963-76, prof. computer sci., 1976—; cons. in field. Mem. AAAS, Assn. Computing Machinery; Fellow IEEE (sr.). Office: MIT 545 Technology Sq Cambridge MA 02139

SALUJA, JAGDISH KUMAR, high technology energy company executive; b. Jhelum, West Punjab, Pakistan, Jan. 14, 1934; came to U.S., 1956, naturalized, 1967; s. Kirpa Ram and Raksha Devi (Ajmani) S.; m. Subhashini Guddie Bhalla, June 9, 1967; children: Sunil, Samir. B.S.E. in Elec. Engring, U. Mich., 1957, 1958, M.S.E., 1959; Ph.D., U. Fla., 1966. Nuclear engr. Argonne (Ill.) Nat. Lab., 1959-62; sr. nuclear engr. Westinghouse Co., Pitts., 1967-77; pres. Viking Energy Corp., Pitts., 1978—. Editor: Instrumentation and Controls Analysis Status Report for the 1137400 E Nerva Engine, 1972. Mem. Republican Presdl. Task Force, 1982. Mem. Am. Nuclear Soc. Democrat. Subspecialties: Nuclear fission; Solar energy. Current work: Nuclear power plant safety and development of alternate energy technologies including conservation. Office: Viking Energy Corp 121 N Highland Ave Suite 203 Pittsburgh PA 15215

SALUTSKY, MURRELL LEON, chemical company executive; b. Goodman, Miss., July 16, 1923; m. Mary Ellen Butler, Mar. 20, 1948 (dec. 1959); 1 dau., Laura; m. Phyllis Perlman, June 26, 1966; children: Susan, Ilene. Research chemist Mound Lab. div. Montsanto, Miamisburg, Ohio, 1950-55; sr. research chemist Monsanto Chem. Co., Dayton, Ohio, 1955-57; research supr. W.R. Grace & Co., Clarksville, Md., 1957-65; dir. research Dearborn Chem. Co., Lake Zurich, Ill., 1965-70, v.p. research, 1970-82, v.p. tech. ops., 1982—. Author: (with others) Precipitation from Homogeneous Solution, 1959, Comprehensive Analytical Chemistry, 1962, Treaties in Analytical Chemistry, 1959; Contbr.: articles to profl. jours. Treaties in Analytical Chemistry. Served with AUS, 1944-46. Mem. Am. Chem. Soc., Marine Tech. Soc., AAAS, Internat. Oceanographic Found., Research Dirs. Assn. Chgo. (pres. 1980-81). Current work: Water treatment chemicals, scale inhibitors, corrosion inhibitors, desalination, by-products from the sea, fertilizers, radiochemistry, rare earths. Home: 1950 Berkeley Rd Highland Park IL 60035 Office: Dearborn Chem Co 300 Genesee St Lake Zurich IL 60047

SALVADOR, RICHARD A., pharmaceutical company executive; b. Albany, N.Y., May 19, 1927; s. Domenico and Irma Ida (Vanetti) S.; m. Carole Ann Salvador, Sept. 17, 1966; children: Barbara A., Diana S. B.S., Siena Coll., 1951; A.M., Boston U., 1953; Ph.D., George Washington U., 1956. Pub. health research fellow NIH, Bethesda, Md., 1957-58; research instr. Washington U., St. Louis, 1958-60; sr. pharmacologist Burroughs Wellcome & Co., N.Y., 1960-69; with Hoffmann-LaRoche, Inc., Nutley, N.J., 1973—, asst. dir. dept., 1975-79, dir. exptl. therapeutics, 1979—, asst. v.p., 1982—. Served with U.S. Army, 1945-47. Mem. Am. Soc. Pharmacology and Exptl. Therapeutics, Am. Chem. Soc., AAAS, N.Y. Acad. Scis., Sigma Xi. Subspecialty: Pharmacology. Current work: Direct the departments of toxicology and pathology, pharmacy research and development, pharmacokinetics and biopharmaceutics, biochemistry and drug metabolism, clinical pharmacology. Home: 1 Birdseye Glen Verona NJ 07044 Office: Hoffmann-LaRoche Inc Kingsland St Nutley NJ 07110

SALVATORI, VINCENT LOUIS, research and consulting company executive; b. Phila., Apr. 22, 1932; s. Louis and Lydia (Tofani) S.; m. Enid Joan Dodd, Oct. 4, 1952; children: Leslie Ann, Robert Louis, Sandra Ann. B.S.E.E., Pa. State U., 1958. Electronic insp. automation systems Automation Timing & Control Corp., King of Prussia, Pa., 1952-54; head spl. detection group HRB-Singer, Inc., State College, Pa., 1959-60, head microwave techniques sect. and antennae lab., 1960-63, mgr. passive ECCM systems and DF technique programs, 1963-67; mgr. reconnaissance systems dept. Radiation Systems, Inc., McLean, Va., 1967-69, v.p. engring., 1969-73; v.p. tech. Quest Research Corp., McLean, 1973-81; now dir.; exec. v.p. tech. and planning, dir. QuesTech, Inc., McLean, 1981—; dir. DHR, Inc., Dynamic Engring., Inc., Engring. Resources, Inc. Contbr. articles to profl. jours. Served to sgt. USAF, 1948-52. Recipient 115-pound Championship Golden Gloves award, 1945. Mem. IEEE, Am. Optical Assn., Assn. Old Crows. Club: Annapolis (Md.) Yacht. Subspecialties: Systems engineering; Electronics. Current work: Application of corporate resources to technology exploitation. Patentee cubic function generation. Office: 6858 Old Dominion Dr McLean VA 22101

SALZBERG, BRIAN MATTHEW, neurobiologist, biophysicist, educator; b. N.Y.C., Sept. 4, 1942; s. Saul and Betty Bernice (Jacobs) S. B.S., Yale U., 1963; A.M. (Woodrow Wilson fellow), Harvard U., 1965, Ph.D., 1971; A.M. (hon.), U. Pa., 1982. Research asst. physics dept. Harvard U., 1964-71; research assoc. dept. physiology Yale U. Med. Sch., New Haven, 1971-75; asst. prof. physiology U. Pa., Phila., 1975-80, assoc. prof., 1980-82, prof., 1982—; Steps fellow Marine Biol. Lab., 1977, 78, trustee, 1980—. Contbr. numerous articles on biophysics and neurobiology to sci. publs. Recipient MBL prize Marine Biol. Lab., 1981. Mem. Biophys. Soc., Soc. Gen. Physiologists, Am. Phys. Soc., AAAS, Phi Beta Kappa, Sigma Xi. Subspecialties: Neurobiology; Biophysics (biology). Current work: Optical measurement of membrane potential; neurophysiology optical probes, membrane biophysics, potential dependent absorption and fluorescence changes; tissue culture of identified neurons. Co-discoverer merocyanine probes of membrane potential. Office: 4010 Locust St Philadelphia PA 19104

SALZENSTEIN, MARVIN A., consulting mechanical engineer; b. Chgo., May 12, 1929; m. Molla Zackler, Oct. 30, 1982; children: Lauren J., Alan N. B.S., Ill. Inst. Tech., 1951. Registered profl. engr., Ill., 1957, N.J., 1967. Asst. dir., assoc. engr. W.C. McCrone Assocs., Chgo., 1957-61; pres. Polytechnic, Inc., Chgo., 1961—. Served with AUS, 1951-53. Mem. ASME, Am. Soc. Safety Engrs., Air Pollution Control Assn., Systems Safety Soc., Nat. Fire Protection Assn. Subspecialties: Mechanical engineering; Human factors engineering. Current work: Machine design, safety, combustion. Office: 3740 Morse Ave Chicago IL 60645

SALZMAN, NORMAN POST, cell biologist, scientist; b. N.Y.C., Aug. 14, 1926; s. David K. and Pauline (Post) S.; m. Lenore R. Flaum, June 6, 1954; children: Ann Helen, David L., Nancy E. B.S., CCNY, 1948; M.S., U. Mich., 1949; Ph.D., U. Ill., 1953. With Nat. Heart Inst., 1953-55; with Nat. Inst. Allergy and Infectious Diseases, Bethesda, Md., 1955—, head sect. cell biology, 1961-67, chief lab. biology of viruses, 1967—; vis. prof. U. Turin, Italy, 1964-65; dept. molecular biology U. Geneva, Switzerland, 1973-74; professorial lectr. Georgetown U., 1967—; exchange scientist U.S.-French Cancer Program, 1977. Served with USNR, 1945-46. Recipient Superior Service award U.S. Govt., 1973. Mem. Am. Soc. Biol. Chemists, Am. Acad. Microbiology, Am. Soc. Microbiology, Am. Soc. Virology. Club: Old Georgetown (v.p. 1977-79). Subspecialty: Cell biology. Current work: Biomedical research virology; science administrator. Home: 10508 Gainsborough Rd Potomac MD 20854 Office: Nat Inst Health Bethesda MD 20205

SALZMAN, STEVEN KERRY, physiologist, pharmacologist, cons., researcher; b. N.Y.C., Feb. 19, 1952; s. Martin and Rose (Lakner) S.; m. Barbara Elaine Boutwell, June 24, 1974; children: Katherine Milann, Elana Rose. B.S. (scholar), U. Fla., Gainesville, 1974; Ph.D. in Pharmacology (NIMH fellow), U. Conn., 1979. Technician U. Fla., 1975, technologist, 1976; research asst. dept. biobehavioral sci. U. Conn., 1976-79; postdoctoral fellow Alfred I. DuPont Inst., Wilmington, Del., 1979-82; research scientist A.I. DuPont Inst. of Nemours Found., Wilmington, 1982—; guest lectr. U. Del.; cons. drug therapy and usage. Contbr. articles and abstracts to profl. jours. Recipient Outstanding Achievement award Miami Optimist Club, 1970. Mem. Soc. Neurosci. (treas. Del. chpt.). Subspecialties: Neuropharmacology; Neurophysiology. Current work: CNS control of thermoregulation, arousal, clin. neuropharmacology, neurophysiology, neuropharmacology, neurochemistry. Home: 105 Wentworth Dr Claymont DE 19703 Office: PO Box 269 Wilmington DE 19899

SALZMAN, WILLIAM RONALD, chemistry educator; b. Cutbank, Mont., Feb. 27, 1936; s. Ralph Irwin and Oleta Fern (Owens) S.; m. Virginia Ann Harbin; children: Suzanne, Sandra. B.S. in Chemistry, UCLA, 1959, M.S. in Physics, 1964, Ph.D. in Chemistry, 1967. Asst. prof. chemistry U. Ariz., Tucson, 1967-72, assoc. prof., 1972-79, prof., 1979—, head dept., 1977-83. Served with U.S. Army, 1959-61. Mem. AAUP, Am. Chem. Soc., Am. Phys. Soc., AAAS, Sigma Xi. Subspecialties: Theoretical chemistry; Physical chemistry. Current work: Theory of optical activity and interaction of light with matter. Office: Dept Chemistry U Ariz Tucson AZ 85721

SAMAAN, NAGUIB ABDELMALIK, endocrinologist; b. Girga, Egypt, Apr. 2, 1925; s. Abdelmalik and Amasil Hanna S.; m. Jean Moffatt, Nov. 18, 1961; children—Sarah Ann, Mary Elizabeth, Jane Susan, Catherine Thia, Michael James. M.B., Ch.B., Alexandria (Egypt) U., 1951; D.M. in Internal Medicine, 1953; Ph.D. in Medicine, U. London, 1964. Rotating intern Alexandria U. Hosp., 1951-52, resident, 1952-54, sr. med. resident, instr., 1954-55; sr. research fellow Chest Inst., Brompton Hosp., London, 1955-56, Neurology Inst., Queen Sq., 1956; clin. fellow Postgrad. Med. Sch., London, 1957; clin. asst. prof. dept. endocrinology and therapeutics, asst. physician Royal Infirmary, Edinburgh, Scotland, 1957; sr. med. resident North Cambridge (Eng.) Hosps., 1958-60; staff physician, asst. prof., sr. research fellow Royal Postgrad. Med. Sch., London, 1960-64; research asso., asst. physician and endocrinologist Case Western Res. U., Cleve., 1964-66; staff physician, asst. prof. dept. internal medicine U. Iowa Hosps., Iowa City, 1966-69; med. staff physician and endocrinologist VA Hosp., Iowa City, 1966-69; chief sect. endocrinology U. Tex. M.D. Anderson Hosp. and Tumor Inst., Houston, 1969—, asso. internist, asso. prof. medicine, 1969-72, internist, prof. medicine, 1972—; prof. medicine and physiology U. Tex. Grad. Sch. Biomed. Scis., Houston, 1969-72, prof., 1972—; prof. internal medicine U. Tex. M.D. Anderson Hosp. and Tumor Inst. and U. Tex. Med. Sch., Houston, 1973—; cons., attending physician dept. internal medicine Hermann Hosp., Houston, 1970—. Contbr. numerous articles to med. jours. Brit. Med. Research Council grantee, 1962-64; NIH grantee, 1969—; Am. Cancer Soc. grantee, 1971—. Fellow Royal Coll. Physicians (Scotland) Royal Coll. Physicians (Eng.), A.C.P.; mem. Brit. Med. Assn., AMA, Am. Endocrine Soc., Am. Fedn. Clin. Research, Am. Physiol. Soc., Fedn. Am. Socs. Exptl. Biology, Central Soc. Clin. Research, Soc. Gynecologic Investigation, N.Y. Acad. Sci., Harris County Med. Soc., Houston Soc. Internal Medicine, Am. Thyroid Assn., Am. Diabetes Assn. Club: Nottingham Forest (Houston). Subspecialties: Neuroendocrinology; Cancer research (medicine). Current work: Investigations of the mechanism of production of pituitary tumors, diagnosis and management; early diagnosis of thyroid and parathyroid tumors and managnment; investigations of possible mechanism of production of diabetes mellitus. Home: 14315 Heatherfield St Houston TX 77024 Office: MD Anderson Hosp and Tumor Inst 6723 Bertner Ave Houston TX 77030

SAMARAS, GEORGE MICHAEL, bioengr., physiologist, educator, cons.; b. Ottawa, Ont., Can., Jan. 6, 1948; came to U.S., 1949, naturalized, 1954; s. Demetrios George and Margaret (O'Connor) S.; m. Harrie Renee Stein, May 30, 1977. B.S.E.E., U. Md., 1972, M.S. in Physiology, 1974, Ph.D. in Neurophysiology/Neuropharmacology, 1976; postgrad. in engring. adminstrn, George Washington U., 1981—. Registered profl. engr., Md. Biochemistry lab. technician Bur. Radiol. Health, HEW, 1968-70, computer technician, 1970-71; biomed. engr. EPA, 1971-72; grad. teaching asst. dept. zoology U. Md., College Park, 1974-76, asst. prof. dept. radiation therapy, 1976-80, research asso. prof., 1980-82, assoc. prof., 1982—; dir. Neuro Oncology Research Labs., 1978-82; external reviewer NSF, 1978—, VA, 1981—; bioengring. cons. G.M.S. Engring. Corp. Contbr. articles to sci. jours., chpt. to book. Huntington's Chorea Found. grantee, 1973-77; Am. Cancer Soc. grantee, 1978-80; Whitaker Found. grantee, 1978-81; Nat. Inst. Neurol. Communicative Disease and Stroke grantee, 1979-83; U. Md. faculty research award, 1976-78. Mem. AAAS, Assn. for Advancement Med. Instrumentation (high frequency therapeutic device nat. standards com. 1977—), IEEE (sr.), Soc. for Neurosci., Nat. Soc. Profl. Engrs., N.Y. Acad. Scis. Subspecialties: Biomedical engineering; Physiology (biology). Current work: Microwave hyperthermia for brain tumor treatments; automated prostheses for the autonomic nervous system. Address: PO Box 2277 Columbia MD 21045 also: 10 S Pine St MSTF 6-34 Baltimore MD 21201

SAMEROFF, ARNOLD JOSHUA, developmental psychologist; b. N.Y.C., Apr. 20, 1937; s. Stanley and Zeena (Shapiro) S. B.S., U. Mich., 1961; Ph.D., Yale U., 1965. Postdoctoral clin. psychology trainee VA Hosp., West Haven, Conn., 1964-65; NIMH postdoctoral research fellow Inst. for Care of Mother and Child, Prague, Czechoslovakia, 1965-67; asst. prof. psychology, pediatrics and psychiatry U. Rochester, N.Y., 1967-70, assoc. prof., 1970-73, prof., 1973-78, dir. devel. psychology tng. program, 1975-78; vis. prof. U. London, 1974-75; vis. scientist Ctr. for Interdisciplinary Research U. Bielefeld, W. Ger., 1977-78; prof. psychology U. Ill.-Chgo., 1978—; assoc. dir. Inst. for Study of Devel. Disabilities, 1978—; dir. research Ill. Inst.for Devel. Disabilities, Ill. Dept. Mental Health Devel. Disabilities, 1978—. Contbr. numerous articles on devel. psychology to profl. jours.; mem. editorial bd.: Am. Jour. Mental Deficiency; author: Monographs of the Society for Research in Child Development, 1978, (with R. Seifer, M. Zax) Monographs of the Society for Research in Child Development, 1982. Mem. small grants adv. com. NIMH, 1977-81; mem. social and behavioral scis. research adv. com. March of Dimes Birth Defects Found., 1977—. Gen. Electric fellow, 1961; NIMH fellow, 1962. 1965-67; NIMH grantee, 1979—. Fellow Am. Psychol. Assn. (council rep. 1980-83); mem. AAAS, Soc. Research Child Devel., AAUP. Subspecialty: Developmental psychology. Current work: Theories of development, developmental disabilities, children-at-risk. Home: 1660 N LaSalle St Chicago IL 60614 Office: Inst for Study of Devel Disabilities U Ill at Chgo 1640 W Roosevelt Rd Chicago IL 60608

SAMIOS, NICHOLAS PETER, physicist; b. N.Y.C., Mar. 15, 1932; s. Peter and Niki (Vatick) S.; m. Mary Linakis, Jan. 12, 1958; children: Peter, Gregory, Alexandra. A.B., Columbia U., 1953, Ph.D., 1957. Instr. physics Columbia U., N.Y.C., 1956-59; asst. physicist Brookhaven Nat. Lab., Upton, N.Y., 1959-62, asso. physicist, 1962-64, physicist, 1964-68, sr. physicist, 1968—, group leader, 1965-75, chmn. dept. physics, 1975-81, dep. dir. for high energy and nuclear physics, 1981, dir., 1982—, adj. prof. Stevens Inst. Tech., 1969-73, Columbia U., 1970—. Contbr. articles in field to profl. jours. Recipient E.O. Lawrence Meml. award, 1980; award in phys. and math. scis. N.Y. Acad. Scis., 1980. Fellow Am. Phys. Soc., Am. Acad. Arts and Scis.; mem. Nat. Acad. Scis., Sigma Xi. Subspecialty: Particle physics. Expert field of particle spectroscopy and weak interactions. Office: Office of Director Brookhaven Nat Lab Upton NY 11973

SAMMET, JEAN E., computer scientist; b. N.Y.C.; d. Harry and Ruth S. B.A., Mount Holyoke Coll., Sc.D. (hon.), 1978; M.A., U. Ill. Group leader programming Sperry Gryoscope, Great Neck, N.Y., 1955-58; sect. head, staff cons. programming Sylvania Electric Products, Needham, Mass., 1958-61; with IBM, 1961—, Boston adv. program mgr., 1961-65, program lang. tech. mgr., 1965-68; programming tech. planning mgr. Fed. Systems div., 1968-74, programming lang. tech. mgr., 1974-79, software tech. mgr., 1979-81, div. software tech. mgr., 1981—; chmn. history of computing com. Am. Fedn. Info. Processing Socs., 1977-79. Author: Programming Languages: History and Fundamentals, 1969; editor-in-chief: Assn. Computing Machinery Computing Revs, 1979—; contbr. articles to profl. jours. Mem. Assn. Computing Machinery (pres. 1974-76), Math. Assn. Am., Nat. Acad. Engring., Upsilon Pi Epsilon. Subspecialty: Software engineering. Office: IBM Fed Systems Div 6600 Rockledge Dr Bethesda MD 20817

SAMMON, PETER, reservoir simulation scientist; b. Penticton, B.C., Can., Aug. 26, 1951; s. Andrew and Kathleen D. (Auton) S.; m. Christine E. Hellwig, Dec. 23, 1978. Sc.B., U. B.C., 1973, M.Sc., 1975; Ph.D., Cornell U., 1978. Asst. scientist Math Research Ctr., Madison, Wis., 1978-79; Dickson instr. U. Chgo., 1979-81; simulation scientist Computer Modelling Group, Calgary, Alta., Can., 1981-83, sr. simulation scientist, 1983—. Contbr. articles to profl. jours. NRC Can. postdoctoral scholar, 1976-79. Mem. Soc. Indsl. and Applied Math.; assoc. mem. Soc. Petroleum Engrs. Subspecialties: Applied mathematics; Petroleum engineering. Current work: Efficient numerical techniques for petroleum reservoir simulation. Office: Computer Modelling Group 3512 33d St NW Calgary AB Canada T2L 2A6

SAMPSON, HERSCHEL WAYNE, medical anatomy educator; b. Greenville, Tex., June 28, 1944; s. Clyde Edwin and Wanda Ruth (Brandon) S.; m. Patricia Jenell Hudson, Nov. 27, 1965; children: Nathan Paul, Susan Diane. Ph.D., Baylor U., 1970. Asst. prof. anatomy Creighton U. Med. Sch., Omaha, 1970-72; asst. prof. to assoc. prof. anatomy Baylor U. Dental Sch., Dallas, 1972-77; assoc. prof. anatomy Oral Roberts U. Med. Sch., Tulsa, 1977-78, Tex. A&M Med. Sch., College Station, 1979—. Contbr. articles to sci. publs. Bd.

dirs. Brazos Valley Rehab. Ctr., Bryan, Tex., Dallas Bible Coll., A&M Bapt. Student Ctr. NIH research grantee, 1981-84. Mem. Fedn. Am. Socs. Exptl. Biology, Am. Inst. Nutrition, Am. Soc. Cell Biology, Am. Assn. Anatomy, Electron Microscope Soc. Am. Subspecialties: Anatomy and embryology; Cell biology (medicine). Current work: Calcium localization, metabolism and homeostasis; presently working with the involvement of calcium in exocrine secretion and also gastric acid secretion. Home: 2901 Adrienne Dr College Station TX 77840 Office: Tex A&M U Med Sch College Station TX 77843

SAMUEL, ARYEH HERMANN ALBERT, research scientist; b. Hildesheim, West Germany, Feb. 19, 1924; came to U.S., 1941, naturalized, 1957; s. Rudolf and Erna (Ballheimer) S.; m. Betty Roth, Mar. 28, 1954; 1 son, Joshua Reuven. B.S., U. Ill.-Urbana, 1943; M.S., Northwestern U., 1946; Ph.D., U. Notre Dame, 1953. Research scientist Broadview Research, Burlingame, Calif., 1956-60, Stanford Research Inst., Menlo Park, Calif., 1960-65, 67-72; research mgr. Gen Precision Research Lab., Little Falls, N.J., 1965-67; ops. analyst Vector Research, Inc., Ann Arbor, Mich., 1974-77; prin. research scientist Battelle Meml. Inst., Washington, 1977—. Served with Israel Army, 1948-49. Mem. Ops. Research Soc. Am. (Lanchester prize 1962). Subspecialties: Operations research (engineering); Physical chemistry. Current work: Analysis of public sector systems (especially military, postal); radiation chemistry, mass spectrometry. Home: 10861 Bucknell Dr Wheaton MD 20902 Office: Battelle Memorial Institute 2030 M St NW Washington DC 20036

SAMUEL, CHARLES E., biochemistry and molecular biology educator; b. Portland, Oreg., Nov. 28, 1945; m., 1968; 2 children. B.S., Mont. State U., 1968; Ph.D., U. Calif.-Berkeley, 1972. Damon Runyon scholar Duke U. Med. Ctr., 1972-74; asst. prof. U. Calif.-Santa Barbara, 1974-79, assoc. prof. biochemistry and molecular biology, 1979-83, prof., 1983—. Mem. editorial bd. Virology, 1980—, Jour. Interferon Research, 1980—; contbr. articles to numerous publs. in field. Recipient Research Career Devel. award NIH, 1979-84; Am. Cancer Soc. grantee, 1975-84; NIH grantee, 1975-84. Mem. Am. Soc. Biol. Chemists, Am. Soc. Virology, Am. Soc. Microbiology. Subspecialties: Virology (biology); Molecular biology. Current work: Molecular virology, biochemistry of animal virus-cell interactions, mechanism of interferon action. Office: Dept Biol Scis U Calif Santa Barbara CA 93106

SAMUEL, MARK AARON, physicist, educator; b. Montreal, Que., Can., Jan. 26, 1944; s. Michael and Molly (Ofter) S.; m. Carol Anne, Dec. 23, 1965; children: Kenneth Brian, Tamara Sue. B.S., McGill U., 1964, M.S., 1966; Ph.D., U. Rochester, 1969. Asst. prof. physics Okla. State U., Stillwater, 1969-75, assoc. prof., 1975-81, prof. physics, 1981—; vis. scientist Aspen Ctr. for Physics, 1981, Niels Bohr Inst., 1977, Stanford Linear Accelerator Ctr., 1973, 75. Author: Group Theory Made Easy for Scientists and Engineers, 1979; contbr. articles to publs. Dept. Energy/ERDA research grantee, 1976—. Mem. Am. Phys. Soc., Am. Assn. Physics Tchrs., Can. Assn. Physicists. Subspecialties: Theoretical physics; Particle physics. Current work: Perturbative QED and QCD, properties of weak bosons, large order perturbation theory. Office: Physics Dept Okla State U Stillwater OK 74078

SAMUEL, PAUL, physician; b. Janoshaza, Hungary, Feb. 17, 1927; came to U.S., 1954, naturalized, 1960; s. Adolf and Magda (Zollner) S.; m. Gabriella R. Zeichner, Mar. 27, 1954; children: Robert Mark, Adrianne Jill. Baccalaureat, Kemeny Zsigmond Gymnasium, Budapest, 1945; M.D., U. Paris, 1953. Intern Queens Hosp. Ctr., N.Y.C., 1954-55; resident L.I. Jewish Med. Ctr., New Hyde Park, N.Y., 1959-61; adj. prof. chemistry Queens Coll., N.Y.C., 1969—; adj. prof. Rockefeller U., N.Y.C., 1971-81; adj. prof. medicine Cornell U., N.Y.C., 1979—; clin. prof. medicine Albert Einstein Coll. Medicine, Bronx, 1981—; dir. Arteriosclerosis Research Lab., L.I. Jewish-Hillside Med. Ctr., New Hyde Park, 1962—; chmn. N.Y. Lipid Research Club, Rockefeller U., 1977-78. Contbr. articles to profl. jours. Pres. Am. Heart Assn., Nassau County, 1980. Fellow Am. Coll. Cardiology; mem. Am. Heart Assn. (fellow council on arteriosclerosis, Disting. Achievement award 1975), Am. Fedn. Clin. Research, Harvey Soc., ACP. Subspecialties: Cardiology; Biochemistry (medicine). Current work: Cholesterol metabolism and kinetics; bile acid metabolism; arteriosclerosis and coronary heart disease. Home: 25 Nassau Dr Great Neck NY 11021 Office: Albert Einstein Coll Medicine 1300 Morris Park Ave Bronx NY 10461

SAMUELS, ROBERT, biologist, educator, protozoology, researcher; b. Phila., June 12, 1918; s. Irwin Louis and Anna (Weisberg) S.; m. Gloria Siegel, June 11, 1948; children: Deborah Samuels Freeman, Joel, Leslie Samuels Bodell. A.B., U. Pa., 1938, M.A., 1940; Ph.D., U. Calif., Berkeley, 1952. Med. entomologist USPHS, Atlanta, 1941-46; lectr. U. Calif., Berkeley, 1949-52; instr. Calif. Poly. State U., San Luis Obispo, 1953; instr., asst. prof. U. Colo. Med. Sch., Denver, 1953-63; prof. Meharry Med. Coll., Nashville, 1963-67, Ind.U.-Purdue U., Indpls., 1967-79; prof., chmn. biol. sci. E. Tenn. State U., Johnson City, 1979-82; cons. parasitologist Wind River Indian Reservation Health Survey, 1957. Contbr. articles to profl. jours. Exec. bd. Colo. Civil Liberties Union, 1960-63, Middle Tenn. Civil Liberties Union, 1966-67. Served with USPHS, 1943-46. NIH grantee, 1958-66; NSF grantee, 1958-62. Mem. Soc. Protozoologists, Am. Soc. Microbiology, Soc. Exptl. Biology and Medicine, Am. Micro. Soc., Am. Soc. Parasitologists, Am. Inst. Biol. Sci., AAAS, AAUP, Sigma Xi. Subspecialties: Cell biology; Biochemistry (biology). Current work: Protozoan karyotyping; regulation of carbohydrate metabolism. Office: E Tenn State Univ PO Box 23590 Johnson City TN 37614

SAMUELSON, DON ARTHUR, morphologist, mycologist, educator; b. Boston, Aug. 30, 1948; s. John Arthur and Laura Kotrina (Ornstedt) S.; m. Leslie Joyce Gilbert, Feb. 14, 1977; 1 son, Peter Andrew. B.A., Boston U., 1971; M.S., U. Fla., 1975, Ph.D., 1977, M.S., 1982. EPA researcher U. Fla., Gainesville, 1978-79, NIH research div. comparative ophthalmology, 1980-82, asst. prof. dept. comparative ophthalmology, 1982—. Contbr. articles to profl. jours. Mem. Assn. Southeastern Biologists, Mycological Soc. Am., Assn. Research in Vision and Ophthalmology. Congregationalist. Lodge: Lions. Subspecialties: Ophthalmology; Microbiology. Current work: Biochemical, physiologic and morphologic development of inherited cataract in a colony of dogs; biochemical, physiologic and morphologic progression of primary open angle glaucoma in a colony of dogs. Office: Coll Vet Medicine U Fla Box J-115 JHMHC Gainesville FL 32610

SAMUELSON, DOUGLAS ALAN, operations research analyst, consultant; b. Reno, July 27, 1948; s. Norman Harold and Shirley (Leder) S.; m. Francine Ruth Kimel, Jan. 7, 1979; 1 son, Andrew Neil. B.A., U. Calif.-Berkeley, 1969; M.S., George Washington U., 1981, postgrad., 1981—. Computer systems analyst Bank of Am., San Francisco, 1972-73; cons. computer systems, San Rafael, Calif., 1973-75; econ. statistician Dept. Interior, Washington, 1975-77; mathematician Fed. Preparedness Agy., 1977-78; ops. research analyst Dept. Energy, Washington, 1978-80, FAA, 1980-82, Evaluation Research Corp., Vienna, Va., 1982-83; cons. ops. research analyst, Falls Church, Va., 1983—; staff analyst White House Task Force on Inland Energy Impacts, Washington, 1977. Presdl. scholar, 1965. Mem. Washington Ops. Research Mgmt. Sci. Council (treas. 1982-83), Washington Statis. Soc. (membership chmn. 1981-83, natural resources program chmn. 1980-81), Am. Statis. Assn. (com. on sci. freedom 1980—), Ops. Research Soc. Am., Inst. Mgmt. Scis. Subspecialties: Operations research (mathematics); Probability. Current work: Probability modeling, reliability and risk analysis, statistical and operations research software for microcomputers, statistical pattern recognition, optimization modeling, forecasting. Home: 3443 Skyview Terr Falls Church VA 22042

SAMY, ANANTHA T. S., biochemist, researcher, educator; b. Hungenahally, Karnataka, India, Sept. 8, 1936; s. Suryanarayana T. and Rukminiamma Gundappa Iyer; m. Mangala Nuggehally, Nov. 3, 1965; children: Sanjay, Sharad. B.Sc., U. Mysore, India, 1958; M.S. in Agr. Chemistry, U. Poona, India, 1960; Ph.D. in Biochemistry, Indian Inst. Sci., 1966. Postdoctoral research assoc. hormone research lab. U. Calif. Med. Ctr., San Francisco, 1966-69; Ford Found. fellow Indian Inst. Sci., Bangalore, 1969-72; research assoc. Sidney Farber Cancer Inst., Boston, 1972-82; assoc. prof. dept. oncology U. Miami, Fla., 1982—. NIH grantee, 1974-82; Am. Cancer Soc. grantee, 1982-83. Mem. Biochem. Soc. (London), Am. Assn. Cancer Research, Am. Soc. Biol. Chemists, N.Y. Acad. Scis. Subspecialties: Biochemistry (biology); Chemotherapy. Current work: Structure, synthesis of proteins, antitumor proteins, mechanism of action of antitumor antibiotics on DNA; pharmacology and pharmacokinetic of antitumor drugs in experimental animals; understanding mechanism of drug resistance in cancer cells; targeting of drugs and efficient utilization of drugs in cancer chemotherapy. Office: Dept Oncology PO Box 016960 (R 71) Miami FL 33101

SANADI, D. RAO, research institute administrator, biochemistry educator, researcher; b. India, July 8, 1920; m., 1950; 2 children. Ph.D. in Biochemistry, U. Calif., 1949. Fellow Nat. Cancer Inst., 1949-52, research assoc., 1952-53; asst. prof. biochemistry U. Wis., 1953-55; asst. prof., U. Calif., 1955-58; chief sect. comparative biochemistry NIH, 1958-66; dir. dept. cell physiology Boston Biomed. Research Inst., 1966—, exec. dir., 1969-71, 75-77; assoc. prof. dep. biol. chemistry Harvard U. Med. Sch., Boston, 1975—; chmn. Gordon Research Conf. on Energy Coupling Mechanisms, 1969, 74, Gordon Research Conf. on Biology of Aging, 1974; mem. adult devel. and aging research and tng. com. Nat. Inst. Child Health and Human Devel., 1970-73; mem. adv. panel metabolic biology NSF, 1971-74. Editor: Jour. Bioenergetics Biomembrane, 1975—. Established investigator Am. Heart Assn., 1954-58. Fellow Gerontol. Soc.; mem. AAAS, Am. Chem. Soc., Am. Soc. Biol. Chemistry. Subspecialty: Cell biology. Office: Dept Cell Physiology Boston Biomed Research Inst 20 Staniford St Boston MA 02114

SANBERG, PAUL RONALD, neuroscientist; b. Coral Gables, Fla., Jan. 4, 1955; s. Bernard and Molly (Spector) S. B.Sc. with honors, York U., Can., 1976; M.Sc., U. B.C. (Can.), 1978; Ph.D., Australian Nat. U., 1981. Research asst. York U., Toronto, Ont., 1974-75; grad. research asst. U. B.C., Vancouver, 1976-78; postdoctoral research asst. Australian Nat. U., Canberra, 1981; postdoctoral fellow depts. neurosci. and psychiatry Johns Hopkins U., Balt., 1981-83-; asst. prof. psychology and biomed. sics. Ohio U., Athens, 1983—. Contbr. articles to profl. jours.; assoc. editor: Bird Behaviour. Recipient Maurice Klugman Meml. award Tourette Syndrome Assn., 1982, Sir. J.G. Crawford prize Australian Nat. U., 1981; NIH grantee, 1981-82. Mem. Soc. Neurosci., Internat. Brain Research Orgn., Psychonomic Soc., Am. Psychol. Assn. Subspecialties: Psychopharmacology; Psychobiology. Current work: Research into elucidating the role of the basal ganglia and specific neurotransmitter systems in motor and complex behavior. In addition, animal models of human movement disorders and diseases of dementia are being developed. Office: Dept Psychology Ohio U Athens OH 45701

SANBORN, TIMOTHY ALLEN, cardiologist, medical educator; b. N.Y.C., Aug. 23, 1947; s. Earl Boyce and Edna (Turner) S.; m. Ann Hill, Aug. 8, 1969; children: Meredith, Marnie. B.A., U. Wis.-Madison, 1969, M.S., 1976; M.D., Northwestern U., 1977. Diplomate: Am. Bd. Internal Medicine. Intern Boston City Hosp., 1977-78, resident in medicine, 1978-80, chief resident in medicine, 1979-80; clin. fellow in cardiology Univ. Hosp., Boston, 1980-82, asst. vis. prof. medicine, 1982—; asst. prof. medicine Boston U. Med. Sch., 1982—. Served with USAR, 1970-76. Biomed. Gen. Research Support grantee, 1982-83; NIH grantee, 1982-83. Mem. Am. Fedn. Clin. Research, Am. Heart Assn., ACP, Am. Coll. Cardiology, Mass. Med. Soc. Republican. Subspecialties: Cardiology; Laser medicine. Current work: Research on laser and balloon angioplasty in models of experimental atherosclerosis; lipoprotein cell interaction in atherosclerosis. Home: 15 Sylvan Rd Wellesley MA 02181 Office: University Hosp 75 E Newton St Boston MA 02118

SANCAKTAR, EROL, mech. engr., educator; b. Ankara, Turkey, July 13, 1952; came to U.S., 1974; s. Mehmet Ali and Ulker Mualla (Elveren) S.; m. Teresa Sue Davis, Feb. 16, 1979; 1 son, Orhan Ali. B.S. in Mech. Engring. with honors (Coll. scholar), Robert Coll., (now Bosphorus U.), Istanbul, Turkey, 1974, M.S., Va. Poly. Inst. and State U., 1975, Ph.D. in Engring. Mechanics, 1979. Teaching asst. dept. physics Robert Coll., 1971-74, dept. mech. engring., 1973-74; trainee engr. Chrysler Corp., Istanbul, summers 1972-73; research asst. dept. mech. engring. Va. Poly. Inst. and State U., 1974-75, dept. engring. sci. and mechanics, 1975-76, 77, research assoc., 1976-77, instr., 1977-78; research assoc. NASA Langley Research Ctr., Hampton, Va., 1976; instr. Clarkson Coll. Tech., 1978-79, asst. prof. mech. engring., 1979—; invited lectr./cons. NASA Langley Research Ctr. Contbr. articles to profl. jours., confs. Olin Corp. Trust grantee, 1979-80; Clarkson Coll. Tech. Research Award grantee, 1980, 81; NSF grantee, 1980-82; Gen. Electric Co. grantee, 1982; NASA Langley Research Ctr. grantee, 1982-83. Subspecialties: Solid mechanics; Polymers. Current work: viscoelasticity, adhesives, material characterization, rate and time dependence, stress analysis, experimental solid mechanics, plastics, creep, cure, fracture. Home: 65 Lawrence Ave Potsdam NY 13676 Office: Dept Mech and Indsl Engring Clarkson Coll Potsdam NY 13676

SANCHEZ, PEDRO ANTONIO, soil scientist; b. Habana, Cuba, Oct. 7, 1940; came to U.S., 1958, naturalized, 1968; s. Pedro A. and Georgina (San Martin) Sanchez-D.; m. Wendy Rise Levin, June 20, 1965; children: Jennifer, Evan, Juliana. B.S., Cornell U., 1962, M.S., 1964, Ph.D., 1968. Research asst. U. Philippines-Cornell Project, Los Baños, 1965-68; co-leader Nat. Rice Program, N.C. State U. Mission to Peru, Lambayeque, 1968-71; coordinator tropical soils program N.C. State U., Raleigh, 1971-76, 80—, asst., assoc. prof. and prof. soil sci., 1968—; coordinator tropical pastures program Centro Internacional de Agricultura Tropical, Cali, Colombia, 1976-79; chief N.C. Mission to Peru, 1982—; chmn. Univ. Consortium on Soils of Tropics, to 1974-75; cons., advisor AID, World Bank, FAO, Rockefeller Found., Nat. Acad. Scis. Author: Properties and Management of Soils in the Tropics, 1976, (with Pappendick and Tripplett) Multiple Cropping, 1976, (with L.E. Tergas) Pasture Production in Acid Soils of the Tropics, 1979; contbr. articles to profl. jours. Mem. Am. Soc. Agronomy (chmn. internat. agronomy div., dir.), Soil Sci. Soc. Am., Colombian Soil Sci. Soc. (hon.), Brazilian Soil Sci. Soc., Peruvian Soil Sci. Soc., N.C. Soil Sci. Soc. Subspecialties: Soil chemistry; Ecology. Current work: Management of tropical soils on an agronomically, economically and ecologically sound basis. Office: Soil Sci Dept NC State U Raleigh NC 27650

SANDAGE, ALLAN REX, astronomer; b. Iowa City, June 18, 1926; s. Charles Harold and Dorothy (Briggs) S.; m. Mary Lois Connelley, June 8, 1959; children—David Allan, John Howard. A.B., U. Ill., 1948, D.Sc., 1967; Ph.D., Calif. Inst. Tech., 1953; D.Sc., Yale, 1966, U. Chgo., 1967, Miami U., Oxford, Ohio, 1974; LL.D., U. So. Calif., 1971. Astronomer Mt. Wilson Obs., Palomar Obs., Carnegie Instn., Washington, 1952—; Peyton postdoctoral fellow Princeton U., 1952; asst. astronomer Hale Obs., Santa Barbara, Calif., 1952-56, astronomer, 1956—; vis. lectr. Harvard, 1957; cons. NSF, 1961-64; Sigma Xi nat. lectr., 1966; vis. prof. Mt. Stromlo Obs., Australian Nat. U., 1968-69. Mem. astron. expdn. to South Africa, 1958; Mem. permanent organizing com. Solvay Conf. in Physics. Served with USNR, 1944-45. Recipient gold medal Royal Astron. Soc., 1967, Pope Pius XI gold medal Pontifical Acad. Sci., 1966, Rittenhouse medal, 1968, Nat. Medal Sci., 1971; Fulbright-Hayes scholar, 1972. Mem. Am. Astron. Soc. (Helen Warner prize 1960, Russell prize 1973), Royal Astron. Soc. (Eddington medal 1963), Astron. Soc. Pacific (Gold medal 1975), Nat. Acad. Scis., Franklin Inst. (Elliott Cresson medal 1973), Phi Beta Kappa, Sigma Xi. Subspecialties: Cosmology; Optical astronomy. Current work: Research in observational cosmology and stellar evolution. Age dating of stars and star clusters. The size and age of the universe. Home: 8319 Josard Rd San Gabriel CA 91775 Office: 813 Santa Barbara St Pasadena CA 91101

SANDBERG, AVERY ABA, internist, educator, researcher; b. Poland, Jan. 29, 1921; m., 1943; 4 children. B.S., Wayne U., 1944, M.D., 1946. Diplomate: Am. Bd. Internal Medicine. Intern Receiving Hosp., Detroit, 1940-47; resident in cardiology Mt. Sinai Hosp., N.Y.C., 1949-50; resident in medicine VA Hosp., Salt Lake City, 1950-51; univ. fellow in medicine, NIH research fellow Coll. Medicine, U. Utah, 1951-52, resident instr. medicine, 1952-53, instr. in medicine, 1953-54; assoc. chief medicine Roswell Park Meml. Inst., 1954-57, chief medicine, 1957—; now also chief dept. genetics and endocrinology; asst. prof. Med. Sch. U. Buffalo, 1954—, asst. research prof. Grad. Sch., 1956-57, research prof., 1957—, Canisius Coll., 1969—, Niagara U., 1969—; cons. Med. Found. Buffalo. Dazian Found. Med. Research fellow, 1950, 52-53; Am. Cancer Soc. scholar, 1953-56. Mem. Am. Soc. Clin. Investigation, Endocrine Soc., Am. Soc. Hematology, Am. Assn. Cancer Research, Am. Fedn. Clin. Research. Subspecialties: Endocrinology; Cancer research (medicine). Office: Roswell park Meml Inst 666 Elm St Buffalo NY 14203

SANDBERG, CHARLES ALBERT, conodont biostratigrapher, petroleum geologist; b. Boston, June 12, 1929; s. Allan A. and Frances (Beres) S.; m. Dorothy Ann Taylor, Sept. 22, 1956; children: Susan Ann, Janet Lynn, William Allan. Geologist U.S. Geol. Survey, Denver, 1950—. Contbr. over 110 geol. articles to profl. publs. Served with U.S. Army, 1952-54; Korea. Fellow Geol. Soc. Am.; mem. Paleontol. Soc., Pander Soc., Am. Assn. Petroleum Geologists. Subspecialty: Paleoecology. Current work: Devonian and Mississippian conodont biostratigraphy, paleoecology, zonation, and taxonomy of North America and Europe. Home: 395 S Lee St Lakewood CO 80226 Office: US Geol Survey Box 25046 Mail Stop 940 Fed Center Denver CO 80225

SANDBERG, HERSHEL, physician; b. Sarny, Poland, Jan. 16, 1929; came to U.S., 1938; s. Jacob and Rebecca (Shapiro) S.; m. Lois Brown, June 12, 1951; children: Jacob, David, Sharon. B.S., Wayne State U., 1949, M.D., 1953. Intern Receiving Hosp. Detroit, 1953-54, resident in internal medicine, 1953-59; chief dept. endocrinology Sinai Hosp., Detroit, 1980—. Bd. dirs. Fedn. Arts; trustee Congregation Shaarey Zedek; pres. Wayne State U. Sch. Medicine Alumni Assn. Served as lt. USN, 1955-57. Subspecialties: Internal medicine; Endocrinology. Office: Sinai Hosp 17550 W 12 Mile Rd Southfield MI 48076

SANDBERG, IRWIN WALTER, research mathematician; b. N.Y.C., Jan. 23, 1934; s. Ben and Estelle (Hornick) S.; m. Barbara A. Zimmerman, June 15, 1958; 1 dau., Heidi L. Student, CCNY, 1951-53; B.E.E., Poly. Inst. Bklyn., 1955, M.E.E. (Westinghouse fellow), 1956, D.E.E. (Bell Telephone Labs. fellow), 1958. Tech. aid Bell Telephone Labs. Inc., Murray Hill, N.J., 1954, mem. tech. staff, 1958-67, head systems theory research dept., 1967-72, mem. math. and statis. research ctr., 1972—; engr. Wheeler Labs., Great Neck, N.Y., 1955; vis. prof. U. Calif.-Berkeley, 1965; lectr. study insts. NATO, Knokke, Belgium, 1966, Copenhagen, Denmark, 1970; disting. lectr. Asilomar Conf., 1973, 74. Del. Union Radio Scientifique Internationale, Munich, 1966; U.S. rep. NATO conf., 1972. Fellow IEEE (vice chmn. group circuit theory 1971-72); mem. Nat. Acad. Engring., AAAS, Eta Kappa Nu, Sigma Xi, Tau Beta Pi. Subspecialties: Electrical engineering; Applied mathematics. Current work: Research mathematician. Patentee in elec. engring. field. Home: 100 Lenape Ln Berkeley Heights NJ 07922 Office: Mountain Ave Murray Hill NJ 07974

SANDEL, BILL ROY, physicist; b. Brady, Tex., Nov. 19, 1945; s. Roy and Mary Lucretia (Collins) S.; m. Karen Kay DeLay, July 26, 1980; children: Aaron Francis Archer, Brody Steven Sandel. B.A. in Physics, Rice U., 1968, M.S. in Space Sci., 1971, Ph.D., 1972. Sr. research assoc. Kitt Peak Nat. Obs., Tucson, 1973-78; research assoc. Lunar and Planetary Lab., U. Ariz., 1978-79; research scientist Ctr. for Space Scis., U. So. Calif., 1979—; assoc, research scientist Lunar and Planetary Lab., U. Ariz., 1983—; co-investigator Voyager Ultraviolet Spectrometer Expt., 1978-. Contbr. articles on planetary sci., ultraviolet spectroscopy to profl. jours. Served to capt. USAR. Recipient Exceptional Sci. Achievement award NASA, 1981, Group Achievement award, 1981, 81. Mem. Am. Geophys. Union, Am. Astron. Soc., AAAS. Subspecialties: Planetary science; Planetary atmospheres. Current work: Investigation of bound and extended atmospheres of giant planets and their satellites; development of imaging detectors for ultraviolet rays. Home: 4442 E 6th St Tucson AZ 85711 Office: 3625 E Ajo Way Tucson AZ 85713

SANDERS, ARTHUR, project engineer, nuclear engineer; b. N.Y.C., Nov. 16, 1931; s. David and Rebecca (Manowitz) S.; m. Elaine Selman, July 5, 1955 (div. 1972); children: David, Richard. B.S. in Chem. Engring, CCNY, 1953, M.S., Columbia U., 1955; J.D., U. Conn., 1959. Profl. nuclear engr., Calif. Group leader Pratt & Whitney Aircraft, East Hartford, Conn., 1954-59; nuclear specialist Martin-Marietta Corp., Balt., 1962-63; nuclear engring. br. chief Douglas Aircraft Corp., Santa Monica, Calif., 1959-61, 63-66; pres. A.R.T. Research Corp., Los Angeles, 1966-72; nuclear group supr. Bechtel Power Corp., Norwalk, Calif., 1972-78, project engr., 1978—. Mem. Am. Nuclear Soc. Subspecialties: Nuclear fission; Nuclear engineering. Current work: Design of nuclear power plants in role of project engineer (responsible for engineering). Office: Bechtel Power Corp 12400 E Imperial Hwy Norwalk CA 90650

SANDERS, CHARLES ADDISON, pharmaceutical company executive, physician; b. Dallas, Feb. 10, 1932; s. Harold Barefoot and May Elizabeth (Forrester) S.; m. Elizabeth Ann Chipman, Mar. 6, 1956; children—Elizabeth, Charles Addison, Carlyn, Christopher. M.D., U. Tex., 1955. Intern, asst. resident Boston City Hosp., 1955-57, chief resident, 1957-58; clin. and research fellow in medicine Mass. Gen. Hosp., Boston, 1958-60, chief cardiac catheterization lab., 1962-

72, gen. dir., 1972-81, physician, 1973-81, program dir. myocardial infarction research unit, 1967-72, program dir., 1969-72; exec. v.p. Squibb Corp., 1981—; assoc. prof. medicine Harvard U. Med. Sch., 1969-80, prof., 1980-83; lectr. MIT, 1973—; dir. Bank of Boston Corp., New Eng. Life Ins. Co., Mohawk Rubber Co.; mem. Inst. Medicine, Nat. Acad. Scis.; chmn. Nat. Council Health Care Tech., 1980-81. Mem. editorial bd.: New Eng. Jour. Medicine, 1969-72. Past trustee Mass. Hosp. Assn.; dir. Charles Stark Draper Lab., Cambridge, Mass., Avco-Everett Research Lab. Served to capt. M.C. USAF, 1960-62. Mem. Am. Fedn. for Clin. Research, Am., Mass. heart assns., Mass. Med. Soc., A.C.P., Am. Physiol. Soc., Am. Clin. and Climatol. Soc., Am. Coll. Cardiology, Am. Soc. for Clin. Investigation, Soc. Hosp. Adminstrs., Greater Boston C. of C. (dir. 1977—). Unitarian. Club: Harvard. Subspecialty: Cardiology. Home: 70 Independence Dr Princeton NJ 08540 Office: ER Squibb's & Sons Inc Princeton NJ 08540

SANDERS, DONALD BENJAMIN, neurologist, neurophysiologist, educator; b. Sumter, S.C., Aug. 3, 1938; s. Colclough Eugene and Frances Ann (Humphries) S.; m. Polly Sandridge, Nov. 28, 1965; 1 dau., Colclough Allison; m. Lynda Frank, July 17, 1975; 1 dau., Kathleen Chatterton. B.S., U. of South, 1959; M.D., Harvard U., 1964. Diplomate: Am. Bd. Psychiatry and Neurology, 1972. From asst. prof. to asso. prof. neurology U. Va. Sch. Medicine, Charlottesville, 1972-80; prof. medicine Duke U. Med. Center, Durham, N.C., 1980—, also dir. electromyography lab. Contbr. numerous articles to sci. jours. Served to maj. USAF, 1969-72. USPHS fellow, 1961-62; Nat. Inst. Neurol. and Communicative Disorders and Stroke grantee, 1976-79; Muscular Dystrophy Assn. grantee, 1977-80, 82. Mem. Am. Acad. Neurology, Am. Neurol. Assn., Am. Assn. Electromyography and Electrodiagnosis, Soc. Neurosci., Myasthenia Gravis Found. Subspecialties: Neurology; Neurophysiology. Current work: Diagnosis and treatment of nerve, muscle and neuromuscular diseases. Office: Duke U Med Center Box 3403 Durham NC 27710

SANDERS, GILBERT OTIS, research psychologist, consultant, educator; b. Oklahoma City, Aug. 7, 1945; s. Richard Allen and Evelyn Wilmoth (Barker) S.; m. Marline Marie Lairmore, Nov. 1, 1969 (div. Oct. 1982); 1 dau., Lisa Dawn. A.S., Murray State Coll, 1965; B.A., Okla. State U., 1967; M.S., Troy State U., 1970; Ed.D., U. Tulsa, 1974. Research psychologist U.S. Army Research Inst., Ft. Hood, Tex., 1978-79; engring. psychologist U.S. Army Tng. and Doctrine Command Systems Analysis Activity, White Sands Missile Range, N. Mex., 1979-80; project dir./research psychologist Applied Sci. Assocs., Ft. Sill, Okla., 1980-81; research psychologist Res. Components Personnel and Adminstrn. Ctr., St. Louis, 1981—; cons. behavioral sci., St. Louis, 1981—; adj. prof. bus. and psychology Columbia Coll.-Buder Campus, St. Louis, 1982—; chmn. dept. computer sci., dir. computer services Calumet Coll., Whiting, Ind., 1975-78; dir. edn. Am. Humane Edn. Soc., Boston, 1975. Prin. editor: TRADOC Training Effectiveness Analysis Handbook, 1980; author research reports. Hon. col. Okla. Gov.'s Staff, Oklahoma City, 1972; hon. ambassador Gov. Okla., 1974. Recipient Kavanough Found. Community Builder award, 1967. Mem. Am. Psychol. Assn., Human Factors Soc., Am. Personnel and Guidance Assn., Assn. Ednl. Communications Tech., Tex. Psychol. Assn., Okla. Psychol. Assn., Mo. Psychol. Assn., Okla. Hist. Soc., Res. Officers Assn. Lodge: Masons. Subspecialties: Learning. Current work: Research in area of human-machine interface of developing military weapon systems and the development of cognitive skill test to predict training success; weapon systems include new battlefield computers (TACFIRE) and the Pershing II. Home: 184 Rue Grand Lake Saint Louis MO 63367 Office: Res Components Personnel and Adminstrn Ctr 9700 Page Blvd Saint Louis MO 63132

SANDERS, GLORIA TOLSON, physical therapy educator, researcher; b. Jacksonville, N.C., Nov. 17, 1944; d. William Mattocks and Zeta (Tolson) S. Student, N.C. Wesleyan Coll., 1962-65; B.S., Med. Coll. Va., 1967; M.S., East Carolina U., 1976. Cert. phys. therapist. Staff therapist Grady Meml. Hosp., Atlanta, 1967-69, Kennestone Hosp., Marietta, Ga., 1969-70; instr. East Carolina U., Greenville, N.C., 1970-71, assoc. prof. dept. phys. therapy, 1974—; dir. phys. therapy Univ. Hosp. of Jacksonville, Fla., 1971-74; mem. N.C. Bd. Phys. Therapy Examiners, 1982-83; mem., cons. State Arthritis Adv. Com., 1979-82; reviewer Profl. Exam. Service, N.J., 1980-83. Author Lower Limb Amputations: A Guide to Rehabilitation; manual, articles in field. 1983. Grantee N.C. Dept. Human Resources, 1980, 81, 82, East Carolina U., 1977, 78. Mem. Am. Phys. Therapy Assn., N.C. Phys. Therapy Assn. (peer reviewer 1975-83), Delta Kappa Gamma, Alpha Omicron Pi (chpt. advisor 1976-79). Democrat. Methodist. Subspecialties: Prosthetics; Physical medicine and rehabilitation. Current work: Study of barriers of handicapped information; research: nausea control during chemotherapy, pelvic tilt measurement, sacro-iliac joint mobility. Home: 1205 E Fifth St Greenville NC 27858 Office: East Carolina Univ Greenville NC 27834

SANDERS, ROBERT BURNETT, biochemist, educator; b. Augusta, Ga., Dec. 9, 1938; s. Robert and Lois Mabel (Jones) S.; m. Gladys Nealous, Dec. 23, 1961; children: Sylvia, William. B.S., Paine Coll., Augusta, 1959; M.S., U. Mich., 1961, Ph.D., 1964. Teaching fellow, trainee U. Mich., 1959-64; postdoctoral fellow U. Wis., Madison, 1964-66; asst., then asso. prof. biochemistry U. Kans., Lawrence, 1966—; vis. scientist Battelle Northwest, Richland, Wash., 1970-71; vis. asso. prof. U. Tex. Med. Sch., Houston, 1974-75; program dir. NSF, 1978-79; now cons.; cons. NIH. Contbr. articles to profl. jours. Bd. dirs. United Child Devel. Center, 1968—. Served with USAR, 1955-62. Paine Coll. Alumni Scholar, 1958-59; USPHS trainee, 1959-64; Am. Cancer Soc. postdoctoral fellow, 1964-66; Battelle Meml. Inst. fellow, 1970-71; NIH fellow, 1974-75; also research grantee. Mem. Am. Soc. Biol. Chemists, Am. Soc. Pharmacology and Exptl. Therapeutics, Sigma Xi, Alpha Kappa Mu. Methodist. Subspecialties: Biochemistry (biology); Biochemistry (medicine). Current work: Biochemistry of reproduction; biochemistry of the rat uterus. Office: Dept Biochemistry U Kans Lawrence KS 66045

SANDERS, WALTER L., astronomer; b. Evansville, Ind., Aug. 21, 1937; s. Walter I. and Dorothy L. (Ruckett) S.; m. Zinta Brunavs, June 17, 1960; 1 dau., Lia Paula. Dr.Sc., Gottingen (W.Ger.) U., 1970. Astron. asst. Lick Obs., U. Calif., 1963-66; research assoc. Yerkes Obs., U. Chgo., 1970-71, Kitt Peak Nat. Obs., Tucson, Ariz., 1971-72; asst. prof. N.Mex. State U., Las Cruces, 1972-77, assoc. prof. astronomy, 1978—. Served with USAF, 1954-58. Mem. Am. Astron. Soc., Internat. Astron. Union, Royal Astron. Soc., Astron. Soc. Pacific, Astronomische Gesellschaft. Subspecialties: Optical astronomy; Radio and microwave astronomy. Current work: Structure of the galaxy, distance scale galactic and extragalactic distance scale.

SANDERS, WILTON TURNER, III, physicist; b. Greenwood, Miss., Sept. 4, 1947; s. Wilton Turner Jr. and Virginia (Jones) S. B.A., Johns Hopkins U., 1969; M.S., U. Wis., 1972, Ph.D., 1976. Lectr., research assoc. dept. physics U., Wis.-Madison, 1976-77, asst. scientist, 1977-81, assoc. scientist, 1981—. Contbr. articles to profl. jours. Marie Christine Kohler fellow, 1972-74. Mem. AAAS, Am. Phys. Soc., Am. Astron. Soc., Internat. Astron. Union, Phi Beta Kappa, Sigma Xi. Subspecialty: 1-ray high energy astrophysics. Current work: Study of the low energy x-ray (0.1-2 kev) diffuse background, its spatial and spectral characteristics and the local interstellar medium which is thought to be origin of these x-rays; also stellar x-ray emission. Home: 2433 Fox Ave Madison WI 53711 Office: Dept Physics U Wis 1150 University Ave Madison WI 53706

SANDERSON, H. REED, wildlife biologist, range scientist; b. Houston, Idaho, July 8, 1932; s. Rol and Hazel Leanore (Stinson) Frost; m. Georgiana Weisgerber, Feb. 11, 1961; children: Christine Marie, Allen Reed. B.S., Humboldt State U., 1957; M.S., Colo. State U., 1959. Cert. wildlife biologist. With Forest Service Research, U.S. Dept. Agr., 1960—, involved in browse propagation, 1960-63, mountain meadow ecology and annual grass mgmt., Calif., 1963-66, eastern gray squirrel habitat mgmt., W.Va., 1966-76, integrated resource mgmt. research, La Grande, Oreg., 1976—. Contbr. articles to profl. jours. Served with U.S. Navy, 1950-54. Memm. Soc. Range Mgmt., Wildlife Soc. Subspecialties: Ecology; Range and wildlife habitat. Current work: Evaluation of the effects of range management activities on the other natural resources. Home: Route 4 2861 D St La Grande OR 97850 Office: Route 2 Box 2315 La Grande OR 97850

SANDERSON, IAN SCOTT, mechanical engineer; b. Luton, Eng., Mar. 28, 1939; came to U.S., 1964; s. Eric George and Marjorie Alice (White) S.; m.; children: Peter Brian, Mark Ian. B. Tech., U. Loughborough, Eng., 1961; M.S.M.E., U. Wash., 1967. Design engr. Brit. Aircraft Corp., 1955-64; product design engr. Ryan Aero. Co., San Diego, 1967-69; sr. devel. engr. Maremont Corp., Saco, Maine, 1973-78; sr. research engr. Grinnell Fire Protection Systems, Cranston, R.I., 1979—. Mem. ASME. Unitarian. Subspecialties: Mechanical engineering; Materials (engineering). Current work: New product design and development in fire protection system products. Patentee variable reluctance elec. generator, proportioning apparatus.

SANDHU, SHAHBEG SINGH, genetic toxicologist; b. Panjab, India; came to U.S., 1965, naturalized, 1974; s. Tara Singh and Ballwant (Kaur) S.; m. Surinder P., May 14, 1964; chidren: Amandeep K., Navdeep S., Ripple S. B.S., Punjab U., 1960, M.S., 1963; Ph.D., Purdue U., 1968. Research asst. Purdue U., West Lafayette Ind., 1965-68; assoc. prof. biology N.C. Central U., Durham, 1968-70, prof., 1970-76; research biologist div. genetic toxicology EPA, Research Triangle Park, N.C., 1977—. Mem. Environ. Mutagen Soc., Genetic Soc. Am., Soc. for Risk Analysis. Democrat. Lodge: Rotary. Subspecialties: Genetics and genetic engineering (agriculture); Plant genetics. Current work: Research in mutagenesis and carcinogenesis and genetic engineering. Office: EPA MD-68 Research Triangle Park NC 27711

SANDHU, TALJIT SINGH, physicist; b. Kot, India, June 25, 1949; came to U.S., 1970, naturalized, 1977; s. Jogindar Singh and Pritam Kaur (Jhaz) S.; m. Laurie A. Sandhu, Apr. 20, 1982. B.Sc. in Physics with honors, U. Delhi, India, 1968, M.Sc., 1970; Ph.D., SUNY-Buffalo, 1975. Cert. therapeutic radiol. physics Am. Bd. Radiology. Cancer research scientist Roswell Park Meml. Inst., Buffalo, 1975-77, med. physicist, 1977-78; sr. med. physicist Henry Ford Hosp., Detroit, 1978-81; adj. asst. prof. Wayne State U., 1979-81; asst. prof. radiology U. Utah, Salt Lake City, 1981—. Contbr. articles to profl. jours. Mem. Am. Assn. Physicists in Medicine, Radiation Research Soc., Bioelectromagnetics Soc., IEEE. Subspecialties: Biophysics (physics); Biomedical engineering. Current work: Interaction of ionizing and non-ionizing radiation with tissues; radiation, x-rays, gamma-rays, charged particles, microwaves, rf fields, hyperthermia, effects of ionizing and non-ionizing radiation on tumors and normal tissues, electromagnetic properties of tissues. Office: Dept Radiology U Utah Medical Center Salt Lake City UT 84132

SANDLER, SUMNER GERALD, medical administrator, laboratory director; b. Lawrence, Mass., Jan. 13, 1935; s. Maurice Lewis and Dorothy (Alman) S.; m. Katherine Rosenberg, May 6, 1969; children: Lisah, David, Jonathan, Joel. A.B., Princeton U., 1957; postgrad., Harvard Coll., 1957, Tufts U., 1957-58; M.D., NYU, 1962. Diplomate: Am. Bd. Internal Medicine; cert. blood banking, Am. Bd. Pathology. Intern Bellevue Hosp., N.Y.C., 1962-63; resident in internal medicine NYU-Bellevue Med. Ctr., N.Y.C., 1963-64, 66-68; asst. prof. medicine Georgetown U., Washington, 1968-72, clin. prof., 1978—; sr. lectr. medicine Hebrew U., Jerusalem, 1972-74, assoc. prof. hematology, 1974-78; assoc. dir. Blood Services Nat. Hdqrs. ARC, Washington, 1978—. Editor: Immunobiology of the Erythrocyte, 1980. Coordinator ARC. Registry Therapeutic Pheresis Registry, 1980—. Served with USPHS, 1964-66. Fellow ACP; mem. Internat. Soc. Blood Transfusion, Am. Soc. Hematology, Am. Fedn. Clin. Research, Am. Assn. Blood Banks (chmn. com. autologous tranfusion, com. standards 1980—). Jewish. Subspecialties: Hematology; Internal medicine. Current work: Direct American Red Cross Blood Services, National Headquarters r. Home: 5808 Ogden Ct Bethesda MD 20816 Office: Blood Services Nat Hdqrs ARC 1730 E St NW Washington DC 20006

SANDS, DONALD EDGAR, univ. adminstr.; b. Leominster, Mass., Feb. 25, 1929; s. George W. and Emily R. (Parker) S.; m. Elizabeth Stoll, July 28, 1956; children: Carolyn Sands Looff, Stephen Robert. B.S. in Chemistry, Worcester Poly. Inst., 1951; Ph.D. in Phys. Chemistry, Cornell U., 1955. Sr. chemist Lawrence Livermore Lab., 1956-62, asst., then assoc. prof., 1962-68; prof. chemistry U. Ky., Lexington, 1968—, dir. gen. chemistry, 1974-75, assoc. dean, 1975-81, acting dean arts and scis., 1978-81, assoc. v.p. acad. affairs, 1981-82, assoc. vice chancellor, 1982—. Author: Introduction to Crystallography, 1968, Vectors and Tensors in Crystallography, 1982; also articles, revs. Mem. Am. Chem. Soc., Am. Crystallographic Assn., N.Y. Acad. Scis., AAAS, Ky. Acad. Scis., AAUP, Sigma Xi. Democrat. Subspecialties: Physical chemistry; Crystallography. Current work: Mathematical methods in crystallography; foundations of thermodynamics; applications of tensor analysis to crystals. Home: 335 Cassidy Ave Lexington KY 40502 Office: 7 Adminstrn Bldg U Ky Lexington KY 40506

SANDSTEAD, HAROLD HILTON, physician, laboratory administrator; b. Omaha, May 25, 1932; s. Harold Russel and Florence (Hilton) S.; m. Kathryn Brownlee, June 6, 1959; children: Eleanor, James, William. B.A., Ohio Wesleyan U., 1954; M.D., Vanderbilt U., 1958. Diplomate: Am. Bd. Internal Medicine, Am. Bd. Clin. Nutrition, 1967. Intern Barnes Hosp., Washington U., St. Louis, 1958-59; asst. resident in medicine 1959-60; asst. resident in pathology Vanderbilt U. Hosp., Nashville, 1960-61, chief resident medicine, 1964-65, instr. medicine, 1965-68, asst. prof. medicine, 1968-71, asst. prof. biochemistry, 1965-70, assoc. prof. nutrition, 1970-71; dir. U.S. Dept. Agr.-Agrl. Research Service, Human Nutrition Research Ctr., Grand Forks, N.D., 1971—; adj. prof. biochemistry U. N.D. Sch. Medicine, 1971—, clin. prof. internal medicine, 1976—; cons. WHO. Contbr. numerous articles to sci. publs. Served as asst. surgeon USPHS, 1961-63; Egypt. Named Nutrition Found. Future Leader, 1968-70; recipient Am. Inst. Nutrition Mead Johnson award, 1971. Fellow ACP; mem. Am. Soc. Clin. Nutrition (pres. 1982-83), Am. Inst. Nutrition, Central Soc. Clin. Research, So. Soc. Clin. Investion. Subspecialties: Nutrition (medicine); Internal medicine. Current work: Nutritional biochemistry; essential trace elements and toxic elements in nutrition and physiologic function; nutrition and fetal development; nutrition and brain function and development; clinical nutrition, human requirements. Office: USDA-ARS Human Nutrition Research Ctr 2420 2d Ave N Grand Forks ND 58202

SANDWEISS, JACK, physicist, educator; b. Chgo., Aug. 19, 1930; s. Charles Ray and Florence (Hymovitz) S.; m. Letha Ann Boeck, Jan. 16, 1956; children: Daniel Howard, Anne Florence, Benjamin Lewis. Student, UCLA, 1948-50; B.S., U. Calif., Berkeley, 1952, Ph.D., 1957. Research assoc. Radiation Lab., U. Calif., Berkeley, 1957; instr. Yale U., New Haven, 1957-59, asst. prof., 1959-62, assoc. prof., 1962-64, prof. physics, 1964—, Donner prof. physics, 1979, also chmn. dept. physics; cons. Brookhaven Nat. Lab., Fermi Nat. Accelerator Lab.; chmn. high energy physics adv. panel Dept. Energy-NSF, 1982—. Contbr. articles to profl. jours. Fellow Am. Phys. Soc. (chmn. div. particles and fields 1980); mem. AAAS. Subspecialty: Particle physics. Home: 248 Ogden St New Haven CT 06511 Office: Sloane Physics Lab Yale U New Haven CT 06520

SANFORD, ALLAN ROBERT, geophysicist, educator; b. Pasadena, Calif., Apr. 25, 1927; s. Roscoe Frank and Mabel Aline (Dyer) S.; m. Alice Elaine Carlson, Aug. 31, 1956; children: Robert, Colleen. B.A., Pomona Coll., 1949; M.S., Calif. Inst. Tech., 1954, Ph.D., 1958. Geophysicist Western Geophys. Co., various locations, 1949-52; ordnance engr. Naval Ordnance Test Sta., Pasadena, Calif., 1952-53; asst. prof. geophysics N.Mex. Inst. Mining Tech., Socorro, 1957-64, assoc. prof., 1964-68, prof., 1968—. Served with USN, 1945-46. Grantee NSF Am. Petroleum Inst., Air Force Office Sci. Research, Water Resources Research Inst. Mem. Am. Geophys. Union, Seismological Soc. Am., AAAS, Soc. Exploration Geophysicists, Sigma Xi. Subspecialties: Geophysics; Tectonics. Current work: Detection of magma in the crust using microearthquakes sources, crustal exploration using microearthquake sources, seismicity of New Mexico. Home: 1302 North Dr Socorro NM 87801 Office: Dept Geosci NMex Inst Mining Tech Campus Station Socorro NM 87801

SANFORD, BARBARA HENDRICK, geneticist; b. Brockton, Mass., Oct. 17, 1927; d. Arthur A. and Grace E. (Brennan) Hendrick; m. (div.); children: Arthur, Jane, Brian, Paul. Ph.D., Brown U., 1963. Dir. Jackson Lab., Bar Harbor, Maine, 1981—. Mem. Genetics Soc. Am., Am. Assn. Immunologists, Am. Assn. Cancer Research. Subspecialties: Animal genetics; Immunology (agriculture). Current work: Mammalian genetics; administration. Office: Jackson Lab Bar Harbor ME 04609

SANFORD, JAY PHILIP, physician, govt. ofcl.; b. Madison, Wis., May 27, 1928; s. Joseph Arthur and Arlyn (Carlson) S.; m. Lorraine Burklund, Apr. 7, 1950; children—Jeb, Nancy, Sarah, Philip, Catherine. M.D., U. Mich., 1952. Intern Peter Bent Brigham Hosp., Boston, 1952-53; research fellow Harvard Med. Sch., Boston, 1953-54; resident Duke U. Hosp., Durham, N.C., 1956-57; practice medicine specializing in internal medicine, Dallas, 1957-75; mem. faculty U. Tex. Southwestern Med. Sch. at Dallas, 1957-75; prof. internal medicine, 1965-75; dean Sch. Medicine, Uniformed Services U. Health Scis., Bethesda, Md., 1975-81, pres., 1981—; chief microbiology lab. Parkland Meml. Hosp., Dallas, 1957-75, pres. med. staff, 1968-69; form mem. staff St. Paul Hosp., Dallas, Presbyn. Hosp., John Peter Smith Hosp., Ft. Worth; cons. Dallas VA Hosp., Wilford Hall USAF Hosp., Brooke Gen. Hosp., Ft. Sam Houston; mem. adv. council Dallas Health and Sci. Mus., 1968-75; chmn. Am. Bd. Internal Medicine, 1978-79; mem. Gov.'s Commn. Phys. Fitness, 1971-75. Contbr. articles to profl. jours. Served with M.C. U.S. Army, 1954-56. Recipient Cert. of award Div. Health Moblzn. USPHS, 1963, 64; Pfizer award for CD, 1965; Presdl. citation for Health Moblzn. Planning, 1970. Fellow Am. Acad. Microbiology, A.C.P.; mem. Assn. Am. Physicians, Nat. Inst. Allergy and Infectious Diseases (chmn. tng. grant com. 1971), Am. Fedn. Clin. Research (pres. 1968-69), Am. Soc. Microbiology, Central Soc. Clin. Research, Soc. Exptl. Biology and Medicine, Am. Soc. Clin. Investigation, Soc. Med. Consultants to Armed Forces (pres. 1976-77), Am. Thoracic Soc., Infectious Disease Soc. Am. (pres. 1978-79), Inst. of Medicine of Nat. Acad. Scis., Sigma Xi. Subspecialties: Internal medicine; Infectious diseases. Current work: Pharmacology of antimcrobial agents. Home: 10409 Windsor View Dr Potomac MD 20854 Office: Uniformed Services Univ of the Health Scis 4301 Jones Bridge Rd Bethesda MD 20814

SANGER, GREGORY M., optical engr.; b. Spokane, Feb. 2, 1946; s. Marvin M. and Maud F. (Taschereau) S.; m. Betsy Layne, Aug. 19, 1972; 1 dau., Elizabeth Ann. B.S. in Physics, Calif. State U., Northridge, 1968; M.S. in Optics, U. Ariz., 1971, Ph.D., 1976. Project engr., mgr. optics and optical support systems for multiple mirror telescope Lawrence Livermore Lab., 1971-76, group leader optics labs., 1976—. Contbr. articles to profl. jours. Subspecialties: Advanced optical manufacturing and testing; Laser fusion. Current work: Advanced high technology optical manufacturing and testing. Home: 5181 Diane Ct Livermore CA 94550 Office: PO Box 808L 432 Livermore CA 94550

SANGREY, DWIGHT A., civil engineer, educator; b. Lancaster, Pa., May 24, 1940; s. Abram W. and Dorothy L. (Herr) S.; m. Karla, June 24, 1978; children: William, Karla, Aline, Emily, Harlan. B.S., M.S.; Ph.D., Cornell U. Registered profl. engr. With H.L. Griswold Cons. Engrs., 1961-63; Shell Oil Co., 1964-65; faculty Queen's U., Kingston, 1967-70; Cornell U., Ithaca, N.Y., 1970-74; staff U.S. Geol. Survey, 1977-78; mem. faculty Carnegie-Mellon U., Pitts., 1980—, now prof. civil engring., head civil engring. dept. Contbr. over 70 sci. articles to profl. publs. Recipient Excellence in Teaching award Tau Beta Pi, Chi Epsilon. Mem. ASCE, ASTM, Internat. Soc. Soil Mechanics, and Found. Engring., Can. Geotech. Soc., Internat. Soc. Rock Mechanics, Am. Soc. Engring. Edn. Subspecialties: Semiconductors; Robotics. Current work: Geotechnical engineering; robotics in civil engineering and construction. Home: 7409 Richland Pl Pittsburgh PA 15208 Office: Carnegie Mellon Univ Civil Engring Dept Pittsburgh PA 15213

SANI, BRAHMA P., biochemist; b. Trichur, India, Sept. 13, 1937; s. Brahma L. and T. V. (Annam) Porinchu; m. Alice C. Sani, Nov. 27, 1967; children: Anita, Renju. B.S., St. Thomas Coll., Kerala, India, 1960; M.S., Holkar Coll., Indore, India, 1962; Ph.D., Indian Inst. Sci., 1967. Sr. research fellow Indian Inst. Sci., Bangalore, 1962-68; staff fellow Boston Biomed. Research Inst., 1968-71; research cons. Retina Found., Boston, 1970-71; research assoc. Inst. for Cancer Research, Fox Chase, Phila., 1971-74; sr. biochemist So. Research Inst., Birmingham, Ala., 1974-78, head protein biochem. sect., 1979—. NIH grantee, 1977—. Mem. Am. Assn. Cancer Research, Am. Soc. Biol. Chemists. Subspecialty: Biochemistry (biology). Current work: Molecular basis of carcinogenesis and anti carcinogensis; mechanism of action of retinoids and trace elements in growth and in chemoprevention of cancer. Office: 2000 9th Ave S Birmingham AL 35255

SAN JUAN, EDUARDO CARRION, aeronautical engineer, researcher, consultant; b. Manila, Mar. 18, 1924; U.S., 1955, naturalized, 1958; s. Fransico Luis San Juan and Elvira (Carrion) De San Juan; m. Patricia Parrott, Aug. 2, 1952; children: Elizabeth Anne San Juan Crawford, Elaine Jane, Edward Patrick, Elvyra Frances. B.S. in Mech. engring, Mapua Inst. Tech., Manila, 1948, D.Sc. in Engring. (hon.), 1976. Registered profl. mech. engr., Wash. Foreman machine shop San Juan Engring. Works, Manila, 1936-40; mech. draftsman Nat. Coconut Corp., Commonwealth Govt., Manila, 1940-44; plant supr. Philippine Indsl. Equipment Co., Manila, 1947-48; mech./erection engr. Ysmael Steel Co./Barredo Constrn. Co., Manila, 1949-50; base mgr./tech. adv. Adola Goldfields, Imperial Ethiopian Govt.

Dept. Mines, Ministry Fin., Addis Ababa, 1950-54; structural engr. Issacson Iron Works/Builder's Metals, Seattle, 1954-55; sr. design engr. Pacific Car and Foundry Co., Seattle, 1955-61; sr. project engr. Sverdrup and Parcel Cons. engrs., St. Louis, 1961-62; lead engr./designer A Boeing Aerospace Co., Huntsville, Ala., 1962-63; project engr./asst. mgr. Hayes Internat. Corp., Huntsville, 1963-64; sr. prin. research engr./br. mgr. Brown Engring. Co., Huntsville, 1964-66; design specialist/staff engr.-scientist Advanced System Div. advanced programs Lockheed Missile and Space Co., Inc., Sunnyvale, Calif., 1966—; cons. engring. Author various tech, publs, and reports. Speaker-mem. Industry Edn. Council Santa Clara County-Community Relations, San Jose, Calif., 1982; mem. People-To-People Citizen Ambassador Program, Spokane, Wash., 1982, U.S. Congressional Adv. Bd., Am. Security Council, Washington, 1982. Served with ROTC Philippine Army, 1949; with U.S. Army, 1945-47. Recipient Apollo 11 medallion NASA, 1971, Cert. award-Disting. Service award Filipino-Am. Jaycees, San Francisco, 1972, Most Outstanding Alumnus award Nat. Assn. Mapuai Alumni, Manila, 1972. Fellow AIAA (assoc., chmn. public relations S.E. Sect. 1965), Brit. Interplanetary Soc.; mem. Fluid Power Soc. (pres. Seattle chpt. 1958, organizer Denver chpt. 1959), Washington Profl. Engrs., Am. Rocket Soc., Am. Astronautical Soc. (sr., dir-at-large S.E. Sect. 1965), Air Force Assn., Lockheed Mgmt. Assn. (speaker's bur. 1966, sci. adv. group for environ. services 1968), Am. Def. Preparedness Assn., Nat. Space Club, Remotely Piloted Vehicle Soc., Nat. Space Inst. (charter), Philippine Soc. Mech. Engrs., Soc. Naval Architects and Marine Engrs. Democrat. Roman Catholic. Subspecialties: Aerospace engineering and technology; Mechanical engineering. Current work: Advanced science and engineering technology applications. Office: Lockheed Missile & Space Co PO Box 504 Sunnyvale CA 94086

SANNELLA, JOSEPH LEE, industrial company executive; b. Boston, July 27, 1933; s. Theodore and Anna (Barone) S.; m. Nancy Marshall, June 6, 1959; children: Joseph A., Sueanne E., Stephen J. A.B., Harvard U., 1955; M.S., U. Mass., 1957; Ph.D., Purdue U., 1963; M.B.A., U. Del., 1969. Product evaluation and tech. service supr. FMC Corp., Marcus Hook, Pa., 1962-67; with Ball Corp., Muncie, Ind., 1967—, supr. chem. research, 1971-74, dir. research, 1974—. Contbr. articles to profl. publs. Bd. dirs. Big Bros./Big Sisters, Delaware County, Ind., 1969-78, Isanogel Center, Delaware County, 1979, Ball Corp. Employees Credit Union, 1981—, Ball Corp. Polit. Action Com., 1978—. Mem. Am. Chem. Soc., Soc. Plastic Engrs., Nat. Metal Decorating Assn. Lodge: Kiwanis. Subspecialties: Polymers; Materials. Current work: Evaluation of new technologies related to plastic, metals and glass. Patentee in field. Home: 2803 W Woodbridge St Muncie IN 47304 Office: 1509 S Macedonia St Muncie IN 47302

SANNER, JOHN HARPER, pharmacologist; b. Anamosa, Iowa, Apr. 29, 1931; s. Lee Michael and Helen Grace (Smyth) S.; m. Marilyn Joan Eichorst, Dec. 28, 1958; children: Linda, Steven. B.S., U. Iowa, 1954, M.S., 1961, Ph.D., 1964. Lic. pharmacist, Iowa. Research investigator G. D. Searle & Co., Chgo., 1963-69, sr. research investigator, 1969-75, research fellow, 1975—. Contbr. articles to profl. jours. and books. Served to 1st lt. USAF, 1955-57. Mem. Am. Soc. Pharmacology and Exptl. Therapeutics, AAAS. Republican. Subspecialty: Pharmacology. Current work: Pharmacological anatgonists, pharmacology of prostaglandins and vasoactive peptides, smooth muscle pharmacology. Home: 959 Apple Tree Ln Deerfield IL 60015 Office: 4901 Searle Pkwy Skokie IL 60077

SANTELLA, REGINA MARIA, biochemist; b. Bklyn., Oct. 30, 1948; d. Joseph Anthony and Vinnie (Noto) Padronaggio; m. Dennis Santella, June 10, 1972; children: Nicholas, Anthony, Dennis. M.S., U. Mass., 1971; Ph.D., CUNY, 1976. Research assoc. pediatrics dept. Columbia U., N.Y.C., 1976-77; research assoc. Inst. Cancer Research, 1977-82, asst. prof. div. environ. sci., 1982—. Contbr. sci. articles to profl. pubs. Mem. Am. Assn. Cancer Research, Harvey Soc., Biophys. Soc., Am. Chem. Soc., Sigma Xi. Subspecialties: Cancer research (medicine); Biochemistry (biology). Home: 764 Queen Anne Rd Teaneck NJ 07666 Office: 701 W 168th St New York NY 10032

SANTI, GINO P., aerospace engineer; b. Bklyn., Feb. 5, 1916; s. Joseph and Emma (Grandi) S.; m. Dorothy Edna Hardy, May 23, 1948; children: Janice, Martha, Victor, Linda, Laura. B.C.E., CCNY, 1936, M.C.E., 1937; postgrad., Ohio State U., 1948-52, U. Dayton, 1956-60. Registered profl. engr., Ohio. Civil engr. Nat. Excavation Corp., N.Y.C., 1937-38, Johnson, Drake & Piper, Inc., Freeport, N.Y., 1938-39; naval architect Dept. Navy, Mare Island, Calif., 1939-40, NASA, Langley Field, Va., 1940-44; aerospace engr. Dept. Air Force, Wright-Patterson AFB, Ohio, 1947-80. Served to capt. USAAF, 1944-47; Served to capt. USAF, 1951-53. Recipient Meritorious Civilian Service award Dept. Air Force, 1977. Mem. AIAA, Survival and Flight Equipment Assn., Sigma Xi. Republican. Roman Catholic. Subspecialty: Aerospace engineering and technology. Current work: Conception, analysis, research, development and test of U.S. Air Force aircraft escape systems and the determination of research and development programs in all technical elements necessary to achieve performance requirements for advanced aircraft. Inventor automatic escape system release; designer 1st aircraft high-speed crew escape ejection seat. Home: 201 Enfield Rd Dayton OH 45459

SANTILLI, JOHN, JR., pediatrician, allergist, immunologist; b. Waterbury, Conn., Mar. 29, 1942; s. John and Caroline (Pranaiss) S.; m. Beverly M. McKee, July 2, 1966; children: Susan, Michael, Sandra. B.S., Villanova U., 1964; M.D., Georgetown U., 1968. Diplomate: Am. Bd. Pediatrics, Am. Bd. Allergy and Clin. Immunology. Intern in pediatrics Georgetown U. Hosp., Washington, 1968-69, resident in pediatrics, 1969-71; individual practice medicine specializing in clin. immunology and allergy, Bridgeport, Conn., 1975—; chief Allergy Clinic, Bridgeport Hosp., 19—. Contbr. numerous sci. articles to profl. publs. Served to maj. USAF, 1973-75. Mem. AMA, Am. Acad. Allergy. Roman Catholic. Subspecialty: Allergy. Current work: Molds as allergens and immunogenics in allergy; clinical allergist and researcher. Office: 4675 Main St Bridgeport CT 06606

SAPATINO, BRUNO VASILE, microbiologist, researcher; b. Bucharest, Romania, June 9, 1930; s. Stelian George and Paulina Chira (Negulescu) S. B.S. in Zoology, U. Bucharest, 1958, M.Sc., 1959; postgrad., Inst. Virology, Bucharest, 1970. With Inst. Oncology, Bucharest, 1960-63, Inst. Virology, 1963-70, Karolinska Inst., Stockholm, 1970, Inst. Anatomy, Zurich, Switzerland, 1970-72; faculty dept. biology Calif. Inst. Tech., 1972-73; with Organon, El Monte, Calif., 1973-75, Becton-Dickinson Labware, Oxnard, Calif., 1975—, now mgr. advanced biotech. research. Contbr. articles to profl. publs. Mem. Am. Soc. Microbiology, Tissue Culture Assn., N.Y. Acad. Scis. Republican. Eastern Orthodox. Subspecialties: Virology (biology); Tissue culture. Romanian, U.S. patentee in field. Home: 1401 S Port Dr Oxnard CA 93033 Office: 1950 Williams Dr Oxnard CA 93030

SAPERSTEIN, ALVIN MARTIN, physics, educator; b. Bronx, N.Y., June 3, 1930; s. Morris and Eva (Finkelstein) S.; m. Harriet Eve Brown, June 10, 1956; children: Shira, Rina. B.A., N.Y.U., 1951; M.S., Yale U., 1952, Ph.D., 1956. Research assoc. Woods Hole Oceanographic Instn., U. Mich., Brown U., 1957-59; asst. prof. U. Buffalo, 1959-62; assoc. prof. Wayne State U., Detroit, 1963-67, prof., 1967—, dir. program in environ. studies, 1977-80; mem. exec. bd. Ctr. Peace and Conflict Studies Wayne State U., 1979—. Author: Energy: Physics in the Environment, 1975; contbr. articles to profl. jours. Mem. exec. bd. Met. Detroit chpt. ACLU. NSF fellow, 1976; NSF grantee. Fellow Am. Phys. Soc., AAAS; mem. Am. Assn. Physics Tchrs., Fedn. Am. Scientists, Union Concerned Scientists, United Campuses to Prevent Nuclear War, Phi Beta Kappa, Sigma Xi. Jewish. Subspecialties: Theoretical physics; Nuclear physics. Current work: Theoretical physics, relations between science and society, impact of science and technology on environment, war and peace questions. Home: 1500 Chateaufort Pl Detroit MI 48207 Office: Dept Physics Wayne State U Detroit MI 48202

SAPERSTEIN, LEE WALDO, mining engineering educator; b. N.Y.C., July 14, 1943; s. Charles Levy and Freda Phyllis (Dornbush) S.; m. Priscilla Frances Hickson, Sept. 16, 1967; children: Adam Geoffrey, Clare Freda. B.S. in Mining Engring, Mont. Sch. Mines, 1964; D.Phil. in Engring. Sci. (Rhodes scholar), Oxford U., 1967. Registered profl. engr., Pa. Laborer, miner, engr. The Anaconda Co., Butte, Mont., and; N.Y.C., 1963-64; asst. prof. mining engring. Pa. State U., University Park, 1967-71, asso. prof., 1971-78, prof., 1978—, sect. chmn., 1974—; mem. engring. accreditation commn. Accreditation Bd. for Engring. and Tech.; cons. Visitor Accreditation Bd. Engring. and Tech. Contbr.: articles refereed jours. Mem. Soc. Mining Engrs., AIME, Am. Assn. Rhodes Scholars. Subspecialty: Mining engineering. Current work: Mining engineering education; surface mine design; reclamation and land use planning; training developments for miners. Home: 337 Ridge Ave State College PA 16801 Office: 118 Mineral Sciences Bldg University Park PA 16802

SAPICO, FRANCISCO LEJANO, physician, medical educator; b. Manila, Philippines, July 18, 1940; came to U.S., 1967; s. Urbano Loyola and Asuncion Limon (Lejano) S.; m. Margaret Mary Armstrong, Nov. 7, 1969; children: Erica Anne, Derek Armstrong. A.A., U. Philippines, 1957-60, M.D., 1965. Diplomate: Am. Bd. Internal Medicine (subcert. in infectious diseases). Rotating intern, resident in internal medicine Philippine Gen. Hosp., Manila, 1964-67; resident in internal medicine SUNY Med. Ctr.-Syracuse, 1967-69; fellow in infectious diseases UCLA Ctr. for Health Scis., 1969-71, adj. asst. prof. medicine, 1972-77, asst. prof., 1977-82, assoc. prof., 1982—; fellow in infectious diseases Wadsworth VA Hosp., Los Angeles, 1971-72, staff physician in internal medicine, 1972-77; physician specialist infectious diseases Rancho Los Amigos Hosp., Downey, Calif., 1977—. Contbr. articles to med. jours. Asst. coach Fullerton Rangers (youth soccer), 1981-83. Fellow ACP; mem. Am. Soc. Microbiology, Infectious Diseases Soc. Am., Am. Fedn. Clin. Research. Democrat. Subspecialties: Internal medicine; Infectious diseases. Current work: Clinical anaerobic microbiology and disease states; new antibiotic studies; staphylococcal infections; osteomyelitis; bacterial peritonitis. Office: Rancho los Amigos Hosp Infectious Diseases Div 7601 E Imperial Hwy Downey CA 90242

SAPORTA, SAMUEL, anatomist, researcher; b. Athens, Greece, Mar. 30, 1946; s. Daniel and Allegra (Ackrish) S.; m. Lucinda Marie Jensen, Oct. 17, 1970; 1 dau., Sara Rachelle. B.A., U. Calif., Davis, 1967; Ph.D., U. So. Calif., 1973; postdoctoral fellow, UCLA, 1973-76. Instr. UCLA, 1976, research asst., 1976-77; asst. prof. U. South Fla., 1977-82, asso. prof., 1982—. NIMH fellow, 1970-73; NIH fellow, 1973-76. Mem. Am. Assn. Anatomists, Soc. Neurosci., AAAS, Cajal Club, Sigma Xi. Subspecialties: Neurobiology; Anatomy and embryology. Current work: Neuranatomical and physiol. orgn. of somatosensory system. Office: 12901 N 39th St Tampa FL 33612

SARACHEK, ALVIN, geneticist; b. Pitts., July 29, 1927; s. Harry E. and Bertha (Balter) S.; m. Rosa Lee Ireland, Apr. 22, 1977. B.A., U. Mo., Kansas City, 1948, M.A., 1950; Ph.D., Kans. State U., 1957; postdoctorate, Inst. Microbiology, Rutgers U., 1957-58. Research microbiologist Am. Research Kitchens, Kansas City, Mo., 1948-50; research assoc., biol. research lab. So. Ill. U., 1951-54; asst. prof. biol. scis. Wichita State U., Kans., 1958-59, assoc. prof., 1959-61, prof., 1961—; program mgr. molecular genetics AEC, 1965-66; sr. profl. assoc. sci. edn. directorate NSF, Washington, 1977-78; Endowment Assn. disting. prof. natural scis. Wichita State U., 1974—; cons., mem. adv. com., panelist AEC, Dept. Energy, NSF, Am. Cancer Soc. Contbr. articles to profl. jours. Served with AUS, 1954-55. Office Naval Research grantee, 1958-65; Am. Cancer Soc. grantee, 1958-74; AEC grantee, 1968-73; recipient Teaching Excellence award Wichita State Regents, 1969. Fellow Am. Acad. Microbiology; mem. Am. Soc. Microbiology, Genetics Soc. Am., Environ. Mutagen Soc., Internat. Soc. Animal and Human Mycology, AAAS, Sigma Xi, Phi Kappa Phi. Subspecialties: Genetics and genetic engineering (biology); Microbiology. Current work: Researcher in devel. of artificial parasexual systems for genetic analysis of asexual pathogenic or industrially significant yeasts. Office: Dept Biol Scis Wichita State U Wichita KS 67208

SARACHMAN, THEODORE NICHOLAS, physics educator; b. W. Warwick, R.I., Feb. 26, 1932; s. Theodore and Tekla (Krawchuk) S.; m. Joann Marilyn McCulloch, July 24, 1965; children: Elisabeth Ann, Suzanne Alix. A.B., B.S., M.S., U. Chgo., 1954; Ph.D., Harvard U., 1961. NBS/NRC research assoc., 1961-63; asst. to prof. physics SUNY, Buffalo, 1963-70; assoc. prof. physics, chmn. dept. Whittier (Calif.) Coll., 1970—; chmn. bd. dirs. Western Obs., 1979-82. Contbr. articles to profl. jours. Mem. AAAS, Western Observatorium, Am. Phys. Soc., Am. Assn. Physics Tchrs. Subspecialty: Atomic and molecular physics. Current work: Laser spectroscopy, biophysics. Office: Dept Physics Whittier Coll Whittier CA 90608

SARAVIS, CALVIN ALBERT, immunochemist; b. Englewood, N.J., Feb. 27, 1930; s. Max and Eve Rachele (Adler) S.; m. Judith Alice Bloch, Sept. 12, 1954; children: Susan, Peter, Ellen, Joanne. A.B., Syracuse U., 1951; M.S., W.Va. U., 1955; Ph.D., Rutgers U., 1958. Head antiserum product and devel. Blood Grouping Lab., Boston, 1958-59; dir. immunochem. lab. Blood Research Inst., Boston, 1959-72; chief immunology div. Harvard Surg. Unit, Boston City Hosp., 1966-72; sr. research assoc. G.I. Research Lab., Mallory Inst. Pathology, Boston, 1971—; prin. assoc. in surgery Harvard Med. Sch., 1971—; sr. research assoc., dept. pathology Boston U. Med. Sch., 1974—; sr. research assoc. Cancer Research Inst., New Eng. Deaconess Hosp., 1979—, mem., 1982—. Contbr. articles to profl. jours. Mem. AAAS, Transplantation Soc., Am. Assn. Immunologists, Electrophoresis Soc. Subspecialties: Cancer research (medicine); Immunology (agriculture). Current work: Development of immunoassays systems to detect and characterize clinically important molecules; tumor markers; biomedical instrument development. Patentee in field. Home: 110 Evelyn Rd Waban MA 02168 Office: 784 Massachusetts Ave Boston MA 02118

SARAZIN, CRAIG LEIGH, astronomy educator; b. Milw., Aug. 11, 1950; s. Valley V. and Martha V. S.; m. Jane C. Curry, June 12, 1971; children: Stephen N., Andrew T. B.S. in Physics, Calif. Inst. Tech., 1972, M.A., Princeton U., 1973, Ph.D., 1975. Mem. Inst. for Advanced Study, Princeton, N.J., 1975-77; vis. prof. physics 1981-82; asst. prof. astronomy U. Va., Charlottesville, 1977-79, assoc. prof., 1979—; vis. asst. prof. astronomy U. Calif., Berkeley, 1979; vis. assoc. scientist Nat. Radio Astronomy Obs., Charlottesville, 1977-81. Contbr. articles to profl. jours. Nat. Merit scholar, 1968-72; NSF fellow, 1972-75; NSF grantee, 1981—. Mem. Am. Astron. Soc., Internat. Astron. Union. Subspecialties: Theoretical astrophysics; High energy astrophysics. Current work: Theoretical astrophysics; high energy astrophysics, interstellar medium, extragalactic astronomy. Office: Astronomy Dept PO Box 3818 Charlottesville VA 22903

SARETT, LEWIS HASTINGS, chemist; b. Champaign, Ill., Dec. 22, 1917; s. Lew and Margaret (Husted) S.; m. Mary Adams, Mar. 1, 1944 (div.); children: Mary Nicole, Katharine Wendy; m. Pamela Thorp, June 28, 1969; children: Will Hastings, Renée MacLeod. B.S., Northwestern U., 1939, D.Sc., 1972; Ph.D., Princeton U., 1942. With Merck & Co., 1942-82, asst. dir. organic and biol. research, 1948-52, dir. dept. med. chemistry, 1952-56, dir. dept. synthetic organic chemistry, 1956-62, exec. dir. fundamental research div., 1962-66, v.p. for basic research, 1966-69; pres. Merck Sharp & Dohme Research Labs., 1969-76, corporate v.p. sci. and tech., 1976-82; dir. Advanced Cardiovascular Systems Inc., Hybritech Inc., Queue Systems Inc., Vestar Inc., Zymark Inc.; Cons. chemotherapy of malaria and schistosomiasis Dept. Def., 1967-68; chmn. basic sci. adv. com. Nat. Cystic Fibrosis Research Found., 1966-69; rep. Indsl. Research Inst., 1968—, dir., 1974-77; chmn. vis. com. div. biology Calif. Inst. Tech., 1969-76; mem. indsl. adv. com. U. Calif., San Diego, 1971—; mem. overseers com. Sch. Public Health, Harvard U., 1979—; adv. com. U. Alexandria (Egypt) Research Center, 1979—; mem. sci. and tech. panel Reagan Transition Team, 1980-81. Editorial bd.: Chem. and Engring. News, 1969-71; Contbr. articles to profl. pubs. Trustee Cold Spring Harbor Lab., of Quantitative Biology, 1968-70, Med. Center at Princeton. Recipient Leo Hendrick Baekeland award, 1951, Award of Merit Northwestern Alumni Assn., 1951, East Union County (N.J.) C. of C. award, 1952, Merck Directors award, 1951, Julius W. Sturmer Meml. Lectr. award, 1959, award Synthetic Organic Chem. Mfrs. Assn., 1964; William Scheele Lectr. award Royal Pharm. Inst., Stockholm, Sweden, 1964; N.J. Patent award N.J. Council Research and Devel., 1966; Nat. Cystic Fibrosis Research Found. award, 1969; award for research mgmt. Indsl. Research Inst., 1980; James Madison medal Princeton U., 1982. Fellow Am. Inst. Chemists (chem. Pioneer award 1972, Inst. award 1981), Soc. Chem. Industry (Perkin medal 1976), Am. Chem. Soc. (award for creative work in synthetic organic chemistry 1964, Nat. medal Sci. 1975, mem. exec. com. organic chemistry div. 1967-68); mem. Nat. Acad. Scis. (Inst. Medicine, mem. sci. fin. com. 1977), Dirs. Indsl. Research, Pharm. Mfrs. Assn., Commn. on Drugs for Rare Diseases, Phi Beta Kappa, Sigma Xi, Phi Eta Sigma, Phi Kappa Psi. Clubs: Princeton, Bedens Brook Country. Subspecialties: Life sciences research and administration; Medicinal chemistry. Current work: Director technical companies. Home: Rolling Hill Rd Skillman NJ 08558 Office: Merck & Co Rahway NJ 07065

SARGENT, ANNEILA ISABEL, astronomer; b. Kirkcaldy, Fife, Scotland, June 30; d. Richard Anthony and Annie (Blaney) Cassells; m. Wallace Leslie William Sargent, Aug. 5, 1964; children: Lindsay Eleanor, Alison Clare. B.Sc. with honors in Physics, U. Edinburgh, 1963; M.S. in Astronomy, Calif. Inst. Tech., 1967, Ph.D., 1977. Research asst. Royal Greenwich Obs., 1963-64; sr. research asst. Calif. Inst. Tech., 1967-71, 72-74, postdoctoral research fellow, 1977-80; scientist, mem. profl. staff Downs Lab. Physics, 1980—. Mem. Royal Astron. Soc., Am. Astron. Soc. Subspecialties: Infrared optical astronomy; Millimeter/submillimeter wave astrophysics. Current work: Millimeter/submillimeter wave astrophysics infrared astronomy. Home: 400 S Berkeley Ave Pasadena CA 91107 Office: Downs Lab Physics Calif Inst Tech Pasadena CA 91125

SARGENT, ERNEST DOUGLAS, aerospace company executive; b. New Berlin, N.Y., Dec. 17, 1931; s. Jess Howard and Edna Ester (Strain) S.; m. Genevieve June Pettee, Feb. 13, 1982; children: Theresa, Mark, Doug. B.S.E.E., Heald Engring. Sch., 1957; M.S., U. So. Calif., 1975. Registered profl. engr., Calif.; cert. community coll. instr., Calif. Research engr. Lockheed Corp., Sunnyvale, Calif., 1964-66, supr. engring., 1966-68, mgr. systems engring., 1968-72, program mgr., 1972-76, dir. product assurnce, 1976-80, dir. engring. and tech., 1980-83, v.p. engring., 1983—; instr. Foothill Coll., 1976—; dir. Reliability and Maintanability Symposium, Orlando, Fla., 1979—. Chmn. div. fund raiser Jr. Achievement, Sunnyvale, Calif., 1979. Served with USN, 1950-55. Mem. Am. Soc. Quality Control (recipient Ben Lubelsky award 1979), AIAA, ASME, Am. Def. Preparedness Assn., Am. Secrit Indsl. Assn. Subspecialties: Systems engineering; Aerospace engineering and technology. Current work: General systems approach to implementing innovative concepts. Home: 2291 Via Madores Rd Los Altos CA 94022 Office: Lockheed Corp 1111 Lockheed Way Sunnyvale CA 94086

SARGENT, ROBERT GEORGE, engineering educator; b. Port Huron, Mich., June 14, 1937; s. George O. and Marie L. (Roome) S.; m. Dorothy M. Baum; 1 dau., Tiffany. A.S., Port Huron Jr. Coll., 1956; B.S.E., U. Mich., 1959, M.S., 1963, Ph.D., 1966. Electronic engr. Hughes Aircraft Co., Culver City, Calif., 1959-61; project engr. Gen. Foods, Battle Creek, Mich., 1963; mem. faculty dept. indsl. engring. Syracuse (N.Y.) U., 1966—, prof., 1982—, chmn. dept., 1982—. Co-editor: Winter Simulation Conf, 1976, 77. Bd. dirs. Winter Simulation Conf., 1975—; chmn., 1977, chmn. bd. dirs., 1980-82. Mem. Inst. Indsl. Engrs. (sr.), Ops. Research Soc. Am., Inst. Mgmt. Scis., Soc. Computer Simulation, Assn. Computing Machinery (dept. editor 1980—). Subspecialties: Operations research (engineering); Industrial engineering. Current work: Modelling and simulation, model validation, computer performance evaluation; applied operations research. Office: Dept Indsl Engring and Ops Research Syracuse U Syracuse NY 13210

SARGENT, WALLACE LESLIE WILLIAM, astronomer, educator; b. Elsham, Eng., Feb. 15, 1935; s. Leslie William and Eleanor (Dennis) S.; m. Anneila Isabel Cassells, Aug. 5, 1964; children: Lindsay Eleanor, Alison Clare. B.Sc., Manchester U., 1956, M.Sc., 1957, Ph.D., 1959. Research fellow Calif. Inst. Tech., 1959-62; sr. research fellow Royal Greenwich Obs., 1962-64; asst. prof. physics U. Calif., San Diego, 1964-66; mem. faculty dept. astronomy Calif. Inst. Tech., Pasadena, 1966—, prof., 1971-81, Ira S. Bowen prof. astronomy, 1981—. Contbr. articles to profl. jours. Alfred P. Sloan Found. fellow, 1968-70; recipient Helen B. Warner prize Am. Astron. Soc., 1969. Fellow Am. Acad. Arts and Scis., Royal Soc. (London); mem. Am. Astron. Soc., Royal Astron. Soc., Internat. Astron. Union. Club: Athenaeum (Pasadena). Subspecialties: Cosmology; Optical astronomy. Current work: Quasars, cosmology, evolution of galaxies. Home: 400 S Berkeley Ave Pasadena CA 91107 Office: Astronomy Dept 105-24 Calif Inst Tech Pasadena CA 91125

SARICH, VINCENT M., anthropologist, educator; b. Chgo., Dec. 13, 1934; s. Matt and Manda Saric; m. Jorjan Snyder; children: Kevin, Tamsin. B.S., Ill. Inst. Tech., 1955; Ph.D., U. Calif.-Berkeley, 1967. Instr. anthropology Stanford (Calif.) U., 1965; asst. prof. anthropology U. Calif.-Berkeley, 1967-80, assoc. prof., 1970-81, prof., 1981—. Mem. AAAS, Am. Assn. Phys. Anthropology, Am. Soc. Mammals. Subspecialties: Biochemistry (biology); Evolutionary biology. Current work: Research on application of biochemistry to problems in evolution. Office: Dept Anthropology U Calif Berkeley CA 94720

SARKAR, NURUL HAQUE, molecular virologist educator; b. Noroshinga Pur, West Bengal, India, Aug. 5, 1937; came to U.S., 1967, naturalized, 1982; s. Patana Uddin and Aleya (Mondal) S.; m. Rabeya Choudbury, Mar. 9, 1965; children: Atom, Tina. B.Sc. with honors (merit scholar), U. Calcutta, 1957, M.Sc., 1960, Ph.D., 1966. Lectr. Hooghly Moshin Coll., West Bengal, 1961-67; hon. research scholar biophysics div. Saha Inst. Nuclear Physics, Calcutta, India, 1961-67; research fellow Inst. for Med. Research, Camden, N.J., 1967-68, research assoc., 1968-69, assoc., 1969-71, assoc. mem., 1971-73, head div. electron microscopy, 1972-73; asst. prof. research pediatrics U. Pa. Sch. Medicine, Phila., 1972-75; assoc. prof. biology Sloan-Kettering div. Cornell U., N.Y.C., 1975-80, assoc. prof. genetics and molecular biology, 1981—; assoc. mem., head Lab. Molecular Virology, Sloan-Kettering Inst. for Cancer Research, N.Y.C., 1973—; mem. breast cancer task force exptl. biology com. NIH, 1976-79. Contbr. numerous articles on modification of low energy electron scattering theory for carbon and its application to electron microscopy, detection of virus-like particles assoc. with human breast cancer to sci. jours. Mem. Electron Microscopy Soc. Am. (bd. cert.), Am. Soc. for Microbiology, Am. Soc. for Virology, N.Y. Acad. Sics. Subspecialties: Molecular biology; Cancer research (medicine). Current work: To understand the mechanism by which virus causes breast cancer in mice and to extend such studies in discovering the etiology of human breast cancer. Home: 1161 York Ave Apt 5-L New York NY 10021 Office: 1275 York Ave Room 915-K New York NY 10021

SARMA, ABUL CHANDRA, research and development and production executive, educator; b. Mangaldoi, Assam, India, Feb. 1, 1939; came to U.S., 1965, naturalized, 1977; m. Delora J. Polk, Aug. 22, 1970 (dec. Aug. 1976); 1 son, Aryan; m. Eva Gohaun, Feb. 10, 1979; 1 son, Gaurab. B.Sc. with honors, Gauhati U., India, 1961; M.Sc., 1963; M.S., U. Minn., 1968; Ph.D., U. Louisville, 1971. Lectr. chemistry Cotton Coll., Gauhati, 1963-65; research assoc. U. Louisville, 1971-73, adj. assoc. prof., 1976—; exec. asst. to v.p. research and devel. Whip-Mix Corp., Louisville, 1973—. Contbr. articles to profl. jours. Pres. Assam Found. N. Am., 1981. Govt. India scholar, 1960-63; U. Minn. scholar, 1966; U. Louisville fellow, 1968-70. Mem. Am. Chem. Soc., Internat. Assn. Dental Research, Am. Nat. Standard Inst./ADA, Sigma Xi. Democrat. Hindu. Subspecialties: Solid state chemistry; High-temperature materials. Current work: In mechanism of using investments in foundry, dentistry, precision casting, high technology. Duplicating materials. Industrial waces, colloidal sclica, silica, and phosphate chemistry. Use of precious, semi-precious and non-precious metals and alloys. Studies in special cements with phosphate and silicate binding systems; Pyrolitic reactions. Patentee in field. Home: 2105 Merriwood Ct Louisville KY 40299 Office: Whip-Mix Corp 361 Farmington Ave Louisville KY 40217

SARMA, P.S. BALA, medical educator, psychiatrist; b. India, Aug. 8, 1941; came to U.S., 1964, naturalized, 1976. M.B., B.S., Stanley Med. Coll., 1963. Asst. prof. Univ. of Health Sci., Chgo. Med. Sch., North Chicago, Ill., 1972-80, assoc. prof., dir. child psychiatry, 1980—. Fellow Am. Acad. Child Psychiatry; mem. AMA, Am. Psychiat. Assn. Subspecialty: Psychiatry. Current work: Minor neurological signs in children; psychiatric evaluation of the elementary school child. Office: Department of Psychiatry University of Health Sciences Chicago Medical School 3333 Greenbay Rd North Chicago IL 60612

SARRAM, MEHDI, engineering company manager, nuclear engineer; b. Kerman, Iran, June 28, 1942; came to U.S., 1961, naturalized, 1982; s. Gholamreza and Legha (Akbarzadeh) S.; m.; children: Shiva, Bahman. B.S., U. Mich., 1965, M.S., 1966; Ph.D. equivalent, U. Tehran, 1969. Lic. sr. operator AEC. Reactor shift supr. U. Mich., Ann Arbor, 1963-67; asst. prof. Tehran (Iran) U., 1967-73; dir. safeguards Atomic Energy Orgn. of Iran, Tehran, 1973-81; safeguards cons. IAEA, Vienna, Austria, 1981-82; cons. United Engrs. Co., Phila., 1982—. Translator: (by John R. Lamarsh) Reactor Theory, 1973; contbr. sci. papers in nuclear field to profl. lit. Best Fgn. Student scholar, 1961-67; recipient Leadership award Iran Prime Minister's Office, 1977. Mem. Am. Nuclear Soc. (chmn. program com. Delaware Valley sect. 1982-83, mem. internat. com. 1979, 83—). Subspecialties: Nuclear engineering; Nuclear fission. Current work: Nuclear power generation, international safeguards, nuclear safety and radiation analysis. Home: 1225 Green Tree Ln Pennvalley PA 19072 Office: United Engrs Co 30 S 17th St Philadelphia PA 19101

SARTIANO, GEORGE P., oncologist, hematologist, educator; b. Bklyn., Nov. 16, 1934; s. George A. and Josephine (Tramantano) S.; m. Dianne M. E. Connolly, May 17, 1970; 3 children. B.A. magna cum laude, Blkyn. Coll., 1956; M.D., N.Y.U., 1960. Intern Bklyn. Jewish Hosp., 1960-61; resident in medicine N.Y. VA Hosp., 1961-63; clin. research trainee in hematology Meml. Hosp., 1963-64; fellow N.Y. Hosp. Cornell Med. Center, 1968-69, NIH fellow in hematology, 1969-70, asst. physician, 1968-70; chief hematology Meth. Hosp., Bklyn., 1964-66; asst. to dir. oncology Western Pa. Regional Med. Program, 1970-73; asst. prof. medicine U. Pitts., 1970-76; assoc. prof., dir. div. hematology and oncology U. S.C., 1976—, prof., 1979—; mem. staff VA Hosp., Columbia, S.C., 1976—, Richland Meml. Hosp., 1976—. Contbr. numerous articles in field to profl. jours. Served with MC U.S. Army, 1966-68. Recipient E. Lipson Siegel award, 1956; Sloan-Kettering Inst. fellow, 1963-64. Mem. Am. Assn. Blood Banks, AAAS, Am. Cancer Soc., Am. Fedn. Clin. Research, Am. Soc. Hematology, N.Y. Acad. Scis., Piedmont Oncology Assn., S.C. Oncology Assn., So. Soc. Clin. Investigation. Roman Catholic. Subspecialties: Chemotherapy; Cancer research (medicine). Current work: Mechanisms of action of chemotherapeutic agents; clin. cancer chemotherapy. Office: U SC Sch Medicine VA Enclave Columbia SC 29201

SARTORELLI, ALAN CLAYTON, pharmacology educator; b. Chelsea, Mass., Dec. 18, 1931; m. Alice C. Anderson, July 7, 1969. B.S., New Eng. Coll. Pharmacy Northeastern U., 1953; M.S., Middlebury (Vt.) Coll., 1955; Ph.D., U. Wis., 1958; M.A. (hon.), Yale U., 1967. Research chemist Samuel Roberts Noble Found., Ardmore, Okla., 1958-60, sr. research chemist, 1960-61; mem. faculty dept. pharmacology Yale Sch. Medicine, 1961—, prof., 1967—, head devel. therapeutics program Comprehensive Cancer Center, 1974-82, chmn. dept. pharmacology Comprehensive Cancer Center, 1977—, dep. dir. Comprehensive Cancer Ctr., 1982-84, dir. Comprehensive Cancer Ctr., 1984—, chmn. div. oncology, 1984—; Charles B. Smith vis. research prof. Meml. Sloan-Kettering Cancer Center, 1979; William N. Creasy vis. prof. clin. pharmacology Wayne State U., 1983; sci. adv. bd. ImmunoGen, Inc., 1981—; Mem. cancer clin. investigation rev. com. Nat. Cancer Inst., 1968-72, mem. com. to establish nat. coop. drug discovery groups, 1982-83; mem. instl. research grants com. Am. Cancer Soc., 1971-76; cons. in biochemistry U. Tex. M.D. Anderson Hosp. and Tumor Clinic, Houston, 1970-76; mem. exptl. therapeutics study sect. NIH, 1973-77, mem. working cadre nat. large bowel cancer project, 1973-76; mgmt. cons. to dir. div. cancer treatment Nat. Cancer Inst., 1975-77; cons. Sandoz Forschungs-Institut, Vienna, Austria, 1977-80; mem. council analysis and projection Am. Cancer Soc., 1978-79; mem. bd. sci. counselors, div. cancer treatment Nat. Cancer Inst., 1978-81; mem. adv. com. Cancer Research Center, Washington U. Sch. Medicine, 1971-75; mem. external adv. com. Wis. Clin. Cancer Center, 1978-79, Duke Comprehensive Cancer Ctr., 1983—; mem. sci. adv. com. U. Iowa Cancer Center, 1979-83; mem. external adv. bd. U. Ariz. Cancer Ctr., 1982—, U. So. Calif. Cancer Ctr., 1983—; mem. nat. program com. 13th Internat. Cancer Congress, 1979-81; mem. selection com. Bristol-Myers prize in cancer research, 1977—, chmn., 1979-81; mem. adv. bd. Drug and Vaccine Devel. Corp. (Center for Public Resources), 1980-81; mem. external adv. bd. Clin. Cancer Research Center, Brown U., 1980—; adv. bd. Specialized Cancer Center, Mt. Sinai Med. Center, 1981—; cons. Bristol-Myers Co., 1982—. Regional editor: Am. Continent Biochem. Pharmacology, 1968—; mem. editorial adv. bd.: Cancer Research, 1970-71; asso. editor, 1971-78; editorial bd.: Internat. Ency. Pharmacology and Therapeutics, 1972—; editor: Handbuch der experimentellen Pharmakologie vols. on antineoplastic and immunosuppressive agts; exec. editor: Pharmacology and Therapeutics, 1975—; editorial bd.: Seminars in Oncology, 1973—, Chemico-Biol. Interactions, 1975-78, Jour. Medicinal Chemistry, 1977-82, Cancer Drug Delivery, 1982—; adv. bd.: Advances in Chemistry Series, ACS Symposium Series, 1977-80; editor: series on cancer chemotherapy Am. Chem. Soc. Symposium, 1976; editorial cons.: Biol. Abstracts, 1984—; contbr. articles to profl. jours. Fellow N.Y. Acad. Scis.; mem. AAAS, Am. Assn. Cancer Research (dir. 1975-78, chmn. publs. com. 1981-82), Am. Chem. Soc., Am. Soc. Microbiology, Am. Soc. Biol. Chemists, Am. Soc. Cell Biology, Am. Soc. Pharmacology and Exptl. Therapeutics, Biochem. Soc. (London, Eng.). Subspecialties: Cancer research (medicine); Pharmacology. Current work: Mechanism of chemotherapeutic drug action; development of new chemotherapeutic agents; study of differentiation; study of metastic process; resistance mechanisms. Home: 4 Perkins Rd Woodbridge CT 06525 Office 333 Cedar St New Haven CT 06510

SARVA, RAJENDRA P., physician, researcher; b. Machilipatnam, India, Apr. 29, 1951; came to U.S., 1975; s. Madhusudana P. and Vedaniam S.; m. Sarada Sarva; 1 child, Suma. M.B.B.S., Guntur Med. Coll., India, 1973. Diplomate: Am. Bd. Internal Medicine (subcert. in gastroenterology). Asst. prof. medicine U. Pitts., 1981—; staff physician Oakland VA Med. Ctr., Pitts., 1981—. Contbr. chpt. to book. Recipient VA award, 1983; United Way award, 1983. Mem. ACP, Am. Fedn. Clin. Research, Am. Gastroenterological Assn. Hindu. Subspecialties: Internal medicine; Gastroenterology. Current work: Cholesterol and bile acid metabolism, gall bladder motility. Office: Oakland VA Hosp Univ Dr-C Pittsburgh PA 15240

SASLAW, LEONARD DAVID, chemist, toxicologist; b. Bklyn., Aug. 27, 1927; s. Issay and Sara S. B.S., CCNY, 1949; M.S. in Biochemistry, George Washington U., 1954; Ph.D. in Chemistry, Georgetown U., 1963. Lic. clin. lab. dir. N.Y.C. Chemist Nat. Cancer Inst., NIH, 1951-57, div. biophysics Sloan-Kettering Inst., 1957-58, biochem. br. Armed Forces Inst. Pathology, 1958-65; dir. div. biochem. pharmacology of cancer chemotherapy dept. Microbiol Assos., Inc., 1965-68; sr. biochemist Nat. Drug Co., 1968-69; chief Lab. Cellular Biochemistry, Albert Einstein Med. Center, 1969-70; clin. lab. dir. Med. Diagnostic Centers, Inc., 1970-71; lab. dir., research asso. Renal Lab., N.Y. Med. Coll., 1971-73; mgr. biochem. investigations Bio/dynamics Inc., N.J., 1973-74; profl. assoc. Smithsonian Sci. Info. Exchange, 1975-77; cons. Burton Parsons Co. Inc., Seat Pleasant, Md., 1977-78; physiologist div. toxicology Bur. Vet. Medicine, FDA, Washington, 1978—. Contbr. articles to sci. jours. Pres. Balt.-Washington Area chpt. B'nai B'rith Young Men, 1954. Served with USN, 1945-46. Recipient Meritorious Achievement award Armed Forces Inst. Pathology, 1964. Mem. AAAS, Am. Chem. Soc., Am. Soc. Pharmacology and Exptl. Therapeutics, Clin. Ligand Assay Soc., Am. Assn. Cancer Research, Am. Inst. Biol. Scis., Am. Coll. Toxicology, Soc. Toxicology, Internat. Platform Assn., Sigma Xi. Democrat. Jewish. Subspecialties: Toxicology (medicine); Cancer research (medicine). Current work: Metabolism and mode of action of drugs, analytical biochemistry and clinical chemistry. I am directly involved in regulatory area of the scientific assessment of safety of additives and drugs in animal feeds. Home: 425 G St SW Washington DC 20024 Office: 200 C St SW HFV-330 Washington DC 20204

SASSAMAN, CLAY A., biology educator; b. Washington, Dec. 27, 1948; s. Grant M. and Jemma A. (Stella) S. B.S., Coll. of William and Mary, 1970; Ph.D., Stanford U., 1976. Postdoctoral fellow Woods Hole (Mass.) Oceanographic Instn., 1975-76; asst. prof. biology U. Calif.-Riverside, 1976-82, assoc. prof., 1982—. Mem. Genetics Soc. Am., Am. Genetics Assn., AAAS, Crustacea Soc. Subspecialty: Population biology. Current work: Population structural analysis in invertebrates and fishes. Office: Dept Biology U Calif Riverside CA 92521

SASSEN, KENNETH, meterologist, educator, researcher, cons.; b. N.Y.C., Nov. 28, 1948; s. George and Olga (Horinovitch) S. B.S. in Meteorology and Oceanography, N.Y.U., 1970, M.S., 1973; Ph.D. in Atmospheric Sci, U. Wyo., 1976. Research instr. meteorology U. Utah, Salt Lake City, 1976-77, research asst. prof., 1977-81, research asso. prof., 1981—; cons. in field. Contbr. articles to profl. jours. Numerous research grants, including NSF, NASA, NOAA, Air Force Office Sci. Research. Mem. Am. Meteorol. Soc. (com. laser atmospheric studies 1982—), Optical Soc. Am., Meteorol. Soc. Japan, Weather Modification Assn., Fedn. Am. Scientists, Sigma Xi. Liberal Democrat. Subspecialty: Remote sensing (atmospheric science). Current work: Laser remote sensing of the atmosphere. Office: 819 W Browning Bldg U Utah Salt Lake City UT 84112

SATO, HIROSHI, materials engineering educator, researcher; b. Matsuzaka, Mie Prefecture, Japan, Aug. 31, 1918; came to U.S., 1954, naturalized, 1977; s. Masayoshi and Fusae (Ohara) S.; m. Kyoko A., Sept. 8, 1922; children: Norie M. Sato Berry, Nobuyuki Albert, Erika Michiko. B.S., Hokkaido (Japan) U., 1938, M.S., 1941; D.Sc., Tokyo U., 1951. Research asso. Hokkaido U., 1941-42, asst. prof., 1942-43; research physicist Inst. Phys. Chem. Research, Tokyo, 1943-47; prof. Tohoku U., Sendai, Japan, 1945-58; research physicist Westinghouse Research Labs., Pitts., 1954-56; prin. research scientist sci. lab. Ford Motor Co., Dearborn, Mich., 1956-74; prof. materials engring. Purdue U., 1974—; cons. in field. Contbr. numerous articles, chpts. to profl. publs.; editor: Single Crystal Films, 1964. Recipient prize of Merit in metal physics Japan Inst. Metals, 1951; Humboldt U.S. Sr. Scientist award, 1980; Guggenheim fellow, 1966; Japan Soc. for Promotion of Sci. Disting. fellow for sci. exchange with Japan, 1979. Fellow Am. Phys. Soc.; mem. Japan Phys. Soc., Am. Ceramic Soc., AIME, N.Y. Acad. Sci., Sigma Xi. Congregationalist. Subspecialties: Condensed matter physics; Solid state chemistry. Current work: Theoretical and exptl. study of solid state reactions and phase transitions; specific interest in irreversible stat. mechanics of transport phenomena, diffusion, superionic condrs., magnetism, transmission electron microscopy. Patentee in field. Home: 1601 Woodland Ave Lafayette IN 47906 Office: Sch Materials Engring Purdue U West Lafayette IN 47907

SATO, MAKIKO, planetary scientist; b. Nishinomiya, Japan, May 29, 1947; came to U.S., 1970; d. Masakazu and Yone (Takeichi) Hayashi; m. Makoto Sato, Dec. 27, 1969. B.S. in Physics, Osaka (Japan) U., 1970, M.A., Yeshiva U., N.Y.C., 1972, Ph.D., 1978. Research scientist Columbia U., 1978; research assoc. SUNY, Stony Brook, 1978-79; sci. analyst M/A-Com Sigma Data Services Corp., N.Y.C., 1980—; co-investigator Voyager photopolarimeter expt. Contbr. articles to profl. jours. Mem. Am. Astron. Soc. Subspecialty: Planetary atmospheres. Current work: Analyses of atmospheric spectra outer planets in visible and infrared ranges to deduce gas compositions, cloud haze structures, pressure-temperature profiles. Home: 240 Anderson Ave Closter NJ 07624 Office: 2880 Broadway New York NY 10025

SATTIN, ALBERT, psychiatrist; b. Cleve., Oct. 5, 1931; s. Sam and Edith (Stolarsky) S.; m. Renee Schnider, Dec. 16, 1962; children: Rebecca Lee, Michael Moshe. B.S., Western Res. U., 1953, M.D., 1957. Diplomate: Am. Bd. Psychiatry and Neurology. Intern Barnes Hosp., St. Louis, 1958-59; resident, teaching fellow Univ. Hosp., Cleve., 1959-62; fellow dept. biochemistry Inst. Psychiatry, London, 1965; sr. instr. psychiatry and pharmacology Case Western Res. U. Sch. Medicine, Cleve., 1965-70, asst. prof., 1970-77; assoc. prof. psychiatry Ind. U. Sch. Medicine, Indpls., 1977—; staff physician Indpls. VA Hosp., 1977—; cons. Hooverwood Home for Aged. Contbr. articles, abstracts and book chpts. to profl. lit. Fellow Am. Psychiat. Assn.; mem. Soc. Biol. Psychiatry, AAAS, Assn. Research in Nervous and Mental Disease, Soc. Neurosci., European Neurosci. Assn., Internat. Soc. Neurochemistry, Phi Beta Kappa. Subspecialties: Psychiatry; Neuropharmacology. Current work: Research on brain mechanisms of mental and nervous disease, including depression and Alzheimer's Disease. Office: Inst Psychiat Research Ind U Med Center Indianapolis IN 46223

SATYANARAYANA, T., research biochemist, educator; b. Gooty, Andhra Pradesh, India, Oct. 7, 1937; s. Tanjore and Suseela Bai Gopala Rao; m. Saraswathi T., Aug. 28, 1965. B.S. in Chemistry, Andhra U., Waltair, India, 1956; M.S. in Biochemistry, Indian Inst. Sci., Bangalore, 1960, Ph.D., 1965. Research assoc. Purdue U., West Lafayette, Ind., 1965-70; biochemist Wash. State U., Pullman, 1970-71; research scientist NASA-Ames Research Ctr., Moffett Field, Calif., 1973—; also prof. biochemistry U. San Francisco. Contbr. numerous research articles to profl. publs. Nat. Acad. Scis.-NRC fellow, 1971-73; NASA grantee, 1973—. Mem. Am. Soc. Microbiology, AAAS. Democrat. Hindu. Subspecialties: Biochemistry (biology); Evolutionary biology. Current work: Origin and evolution of life.

SAUDER, WILLIAM CONRAD, physicist, educator; b. Wheeling, W.Va., Jan. 3, 1934; s. Howard R. and Lodema I. (Burkholder) S.; m. Nanalou West, June 18, 1955; children: Anne E., William H.L. B.S., Va. Mil. Inst., 1955; Ph.D., Johns Hopkins U., 1963. Instr., asst. prof., assoc. prof., prof. physics Va. Mil. Inst., Lexington, 1955—, chmn. dept., 1979—; cons. Nat. Bur. Standards, 1965-82. Contbr. numerous articles to profl. jours. Vice chmn. Rockbridge County (Va.) Democratic Com., 1979—. Served to capt. USAF, 1955-57. Research Corp. grantee, 1967; Nat. Bur. Standards precision measurement grantee, 1979-82. Mem. Am. Phys. Soc., Am. Assn. Physics Tchrs., Va. Acad. Sci. Episcopalian. Subspecialties: Atomic and molecular physics; Acoustics. Current work: Research in fundamental constants, x-ray spectroscopy and positronium annihilation. Precision acoustic interferometry. Home: Route 1 Box 364 Lexington VA 24450 Office: Dept Physics Va Mil Inst Lexington VA 24450

SAUERBIER, WALTER, molecular geneticist, cons.; b. Butzbach, Germany, Aug. 8, 1932; s. Heinrich and Martha-Elisabeth (Lamp) S.; m. Sieglinde Herta Heil, 1958; 1 son, Arnim Walter. Ph.D. in Biophysics, J.W. Goethe U., Frankfurt, Ger., 1957. Asst. U. Cologne, Ger., 1958-61, 63-65; vis. prof. U. Oreg., 1961-62; asst.prof., assoc. prof., prof. U. Colo. Med. Sch., Denver, 1966-77, acting chmn. dept. biophysics and genetics, 1976-77; prof. U Minn., Mpls./St. Paul, 1978—, head dept. genetics and cell biology, 1978-80; industry cons.; tng. grant dir.; dir. NSF sponsored program; lectr. European Molecular Biology Orgn., 1975. Contbr. articles to profl. publs. Research grantee NIH and NSF. Mem. Am. Soc. Microbiology. Subspecialties: Genetics and genetic engineering (agriculture); Genome organization. Current work: Structural and functional orgn. of chromatin. Office: 1145 Gortner Ave Saint Paul MN 55108

SAUL, WILLIAM EDWARD, civil engineer, educator, cons; b. N.Y.C., May 15, 1934; s. George James and Fanny Ruth (Murokh) S.; m. J. Muriel Held-Eagleburger, June 11, 1976. B.S. with honors in Civil Engring, Mich. Tech. U., 1955, M.S. in Civil Engring, 1961; Ph.D. in Civil Engr. (Walter P. Murphy fellow, Royal E. Cabell fellow), Northwestern U., 1964. Registered profl. engr., Mich., Wis. Mech. engr. Shell Oil Co., New Orleans, 1955-59; teaching asst. Mich. Tech. U., Houghton, 1959-60, instr. engring. mechanics, 1960-62; research asst. Snow, Ice, and Permafrost Research Establishment, U.S. Army C.E., Houghton County Field Sta., 1960-62; asst. prof. civil engring. U. Wis. - Madison, 1964-67, assoc. prof., 1967-72, prof. civil and environ. engring., 1972—, chmn. dept., 1976-80; dean engring. U. Idaho-Moscow, 1984—; vis. prof. Institut fur Statik und Dynamik der Luft-und Raumfahrtkonstruktionen, U. Stuttgart, W.Ger., 1970-71; cons. in field. Contbr. articles to profl. jours. Fulbright scholar, 1970-71; Alexander von Humboldt fellow, 1970-71. Mem. ASCE (pres. Madison br. 1978-80, v.p. Wis. sect. 1982-83), Internat. Assn. for Bridge and Structural Engrs., Am. Concrete Inst., Am. Soc. for Engring. Edn., AAUP, Mensa, Sigma Xi, Phi Kappa Phi, Tau Beta Pi, Chi Epsilon. Subspecialties: Civil engineering; Theoretical and applied mechanics. Current work: Structural dynamics, earthquake and wind engineering; foundation engineering; computer applications. Home: 2817 Regent St Madison WI 53705 Office: Civil Engring Dept U Wis 1415 Johnson Dr Madison WI 53706

SAUNDERS, JAMES CHARLES, neurol. scientist; b. Elizabeth, N.J., May 8, 1941; s. Charles O. and Elizabeth (Drake) S.; m. Elaine P. Saunders, Oct. 1967; children: Breton M., Drew C. B.A., Ohio Wesleyan U., 1963; M.A., Conn. Coll., 1965, Princeton U., 1967, Ph.D., 1968; MA. (hon.), U. Pa., 1979. NIH postdoctoral fellow Princeton U., 1968; asst. prof. Monash U., Australia, 1969-72; research assoc. Central Inst. for Deaf, St. Louis, 1972-73; asst. prof. dept. otorhinolaryngology U. Pa., Phila., 1973-79, assoc. prof., 1979—, acting dir., 1981—; cons. NIH. Contbr. articles to profl. jours. Australian Research Grants Commn. fellow, 1970; recipient prize Am. Acad. Ophthalmology and Otolaryngology, 1978, Pa. Acad. Otolaryngoloy, 1982. Mem. Acoustical Soc. Am., Neuroscis. Soc., AAAS, N.Y. Acad. Sci., Assn. Research Otolaryngology. Subspecialties: Neurobiology. Current work: Auditory Neurobiology and physiol. acoustics, comparative bioacoustics, cochlear and middle ear mechanics, auditory devel., effects of noise. Home: 417 Bryn Mawr Ave Bala Cynwyd PA 19004 Office: 5-Silverstein ORL 3400 Spruce St Philadelphia PA 19104

SAVAGE, CARL RICHARD, JR., biochemist; b. Bloomsburg, Pa., Dec. 1, 1942; s. C. Richard and Martha J. (White) S. B.A., Gettysburg Coll., 1964; Ph.D., SUNY-Buffalo, 1971; postgrad., Vanderbilt U., 1971-73. Dir. research lab. VA Hosp., Albany, N.Y., 1973-75; asst. prof. biochemistry Temple U. Sch. Medicine, Phila., 1975-81, assoc. prof. biochemistry, 1981—. Mem. Fedn. Am. Soc. Biol. Chemists, Sigma Xi. Subspecialties: Biochemistry (biology); Receptors. Current work: Polypeptide hormones, receptors, tissue culture, hepatocytes. Home: 2035 Harts Ln Conshohocken PA 19428 Office: Biochemistry Temple U Sch Medicine Philadelphia PA 19140

SAVAGE, JOHN EDMUND, computer science educator, researcher; b. Lynn, Mass., Sept. 19, 1939; s. Edmund John and Eldora A. (Guay) S.; m. Patricia Landers, Sept. 13, 1945; children: Elizabeth, Kevin. Sc. B., MIT, Sc.M., 1962, Ph.D., 1965. Mem. tech. staff Bell Labs., Holmdel, N.J., 1965-67; mem. faculty Brown U., Providence, 1967—, prof. computer sci., 1974—; vis. prof. Tech. U. Eindhoven, Netherlands, 1973-74, U. Paris, 1980-81. Author: The Complexity of Computing, 1976. Mem. tech. resource network Wheaton Coll., 1982-83. Guggenheim fellow, 1973; recipient Fulbright-Hays award, 1973.

Mem. IEEE, Assn. Computing Machinery. Subspecialties: Theoretical computer science; Algorithms. Current work: Algorithms, analysis and design methodology for VLSI, space-time tradeoffs, applied theory of computation, coding and information theory. Patentee in field. Office: Dept Computer Sci Box 1910 Brown U Providence RI 02912

SAVAGE, WARREN FAIRBANK, educator; b. Harvard, Mass., Mar. 10, 1922; s. L. Kingston and Mildred (Fairbank) S.; m. Helen Agnes Lambert, Nov. 6, 1943; children—Sharon Anne (Mrs. David P. Mardon), Donald Fairbank. Grad., Lawrence Acad., 1939; B.Chem. Engring., Rensselaer Poly. Inst., 1942, M.Metall. Engring., 1949, Ph.D., 1954. Metall. engr. Adirondack Foundries & Steel Co., Watervliet, N.Y., 1942-44; research fellow Rensselaer Poly. Inst., 1945-48, mem. faculty, 1948—, prof. metall. engring., 1960—, dir. welding research, 1961—; Clarence Jackson honor lectr., 1976, cons. to govt. and industry, 1945—; mem. U.S./USSR Joint Commn. on Welding and Spl. Electrometallurgy, 1975—. Active local Boy Scouts Am. Fellow Am. Soc. Metals (pres. Eastern N.Y. chpt. 1952-53, Tchr. Recognition award Phila. chpt. 1965); hon. life mem. Am. Welding Soc. (Adams Meml. lectr. 1967, Houdremont lectr. 1980); mem. Am. Arbitration Bd., Nat. Acad. Engring., Rensselaer Soc. Engrs. (trustee, faculty adviser 1954-64), Sigma Xi, Phi Lambda Upsilon, Tau Beta Pi, Alpha Sigma Mu. Subspecialty: Metallurgy. Home: RD 2 Box 474 Averill Park NY 12018 Office: Materials Research Center Rensselaer Poly Inst Troy NY 12181

SAVAGE, WILLIAM FREDERICK, aero. engr.; b. Anchorage, May 23, 1923; s. Gordon Prescott and Josephine Isabelle S.; m. Mary Helen Carter, June 25, 1949; children: Kathleen C., William S. B.S. in Aero. Engring, Rensselaer Poly. Inst., 1943, M.S., Purdue U., 1949; student, Oak Ridge Sch. Reactor Tech., 1957-58. Registered profl. engr., Ohio. Aerodynamicist Convair, Ft. Worth, 1946-48; asst. prof. U. Ky., Lexington, 1948-52; chief engr. Kett Corp., Cin., 1952-55; mgr. tech. analysis Gen. Electric Co., Cin., 1956-60; dir. nuclear products Martin Co., Balt., 1960-67; asst. dir. Office of Saline Water, Dept of Interior, Washington, 1967-74; chief advanced systems Dept. Energy, Washington, 1974-82, mgr. instnl. and regulatory systems, 1982—. Contbr. articles to tech. jours. Served to capt. USAR and USAFR, 1948-58. Recipient Outstanding Performance award Dept. Interior, 1968, spl. achievement award ERDA, 1976. Mem. Am. Nuclear Soc., ASME. Methodist. Subspecialties: Nuclear engineering; Aeronautical engineering. Current work: Analysis of instnl. and regulatory issues in nuclear power systems applications. Home: 8025 Garlot Dr Annandale VA 22003

SAVAGEAU, MICHAEL ANTONIO, microbiology educator, researcher, cons.; b. Fargo, N.D., Dec. 3, 1940; s. Antonio D. and Jennie E. (Kaushagen) S.; m. Ann E., June 22, 1967; children: Mark E., Patrick D. B.S., U. Minn., 1962; M.S., U. Iowa, 1963; Ph.D. (Santa Clara Med. Soc. fellow, NIH fellow), Stanford U., 1967. NIH fellow UCLA, 1967-68, Stanford U., 1968-70, lectr., 1968; asst. prof. microbiology U. Mich., Ann Arbor, 1970-74, assoc. prof., 1974-78, prof., 1978—, acting chmn. dept. microbiology and immunology, 1979-80, 82-83; Fulbright sr. research fellow Max Planck Inst. Biophys. Chemistry, Gottingen, W.Ger., 1976-77; mem. spl. study sect. NIH; ad hoc reviewer NSF, Research Corp. and numerous sci. jours. Author: Biochemical Systems Analysis: A Study of Function and Design in Molecular Biology, 1976; contbr. numerous articles, chpts. to profl. publs.; mem. editorial bd.: Math. Bioscis, 1976—. Guggenheim fellow, 1976-77; NSF grantee, 1970-82; NIH grantee, 1982—. Mem. AAAS, Am. Chem. Soc., Am. Soc. Microbiology, Biophys. Soc., IEEE, N.Y. Acad. Sci., Soc. Gen. Physiologists, Soc. Indsl. and Applied Math., Soc. Math. Biology, Sigma Xi. Subspecialties: Microbiology; Applied mathematics. Current work: Function, design and evolution of biochemical and genetic regulatory systems; development and application of mathematical methods in biology; mathematical modeling in biochemical engineering. Home: 813 Berkshire Ann Arbor MI 48104 Office: Dept Microbiology U Mich Ann Arbor MI 48109

SAVILLE, THORNDIKE, JR., coastal engineer, consultant; b. Balt., Aug. 1, 1925; s. Thorndike and Edith Stedman (Wilson) S.; m. Janet Foster, Aug. 28, 1950; children: Sarah, Jennifer, Gordon. A.B., Harvard U., 1947; M.S., U. Calif.-Berkeley, 1949. Research asst. U. Calif., Berkeley, 1947-49; hydraulic engr. Beach Erosion Bd. and Coastal Engring. Research Center, Ft. Belvoir, Va., 1949-81, chief research div., 1964-71, tech. dir., 1971-81. Author numerous papers in engring. and sci. lit. Served with USAAF, 1943-46. Recipient Meritorious Civilian Service award Dept. Army, 1981. Fellow AAAS, Wash. Acad. Scis., ASCE (Moffatt-Nichol award 1979); mem. Am. Geophys. Union, Internat. Assn. for Hydraulic Research, Nat. Acad. Engring., Permanent Internat. Assn. Navigation Congresses. Subspecialties: Coastal, port, harbor engineering; Nearshore oceanography. Current work: Sediment transport on coasts. Home and Office: 5601 Albia Rd Bethesda MD 20816

SAVIN, SAMUEL MARVIN, geologist; b. Boston, Aug. 31, 1940; s. George and Sarah (Lewiton) S.; m. Norma Goulder, Nov. 4, 1978; children: Robert Goulder, Lisa Rebecca. B.A., Colgate U., 1961; Ph.D., Calif. Inst. Tech., 1967. Asst. prof. geol. scis. Case Western Res. U., Cleve., 1967-73, assoc. prof., 1973-76, prof., 1976—, dept. chmn., 1977-82; industry fellow Marathon Oil Co., Denver, 1976; mem. adv. panel earth scis. NSF, 1978-81. Asso. editor: Geochimica et Cosmochimica Acta, 1976-79, Marine Micropaleontology, 1979—; contbr articles to sci. jours. Fellow AAAS; mem. Geol. Soc. Am., Am. Assn. Petroleum Geology, Am. Geophys. Union, Geochem. Soc., Clay Minerals Soc. (councillor 1978-81), Nat. Assn. Geology Tchrs., No. Ohio Geol. Soc. Subspecialties: Geochemistry; Oceanography. Home: 2236 Demington Dr Cleveland Heights OH 44106 Office: Dept Geol Scis Case Western Res U Cleveland OH 44106

SAWYER, CHARLES HENRY, anatomist, educator; b. Ludlow, Vt., Jan. 24, 1915; s. John Guy and Edith Mabel (Morgan) S.; m. Ruth Eleanor Schaeffer, Aug. 23, 1941; 1 dau., Joan Eleanor. B.A., Middlebury Coll., 1937, D.Sc. (h.c.), 1975; student, Cambridge U., Eng., 1937-38; Ph.D., Yale, 1941. Instr. anatomy Stanford, 1941-44; asso., asst. prof., asso. prof., prof. anatomy Duke U., 1944-51; prof. anatomy U. Calif., Los Angeles, 1951—, chmn. dept., 1955-63, acting chmn., 1968-69, faculty research lectr., 1966-67. Editorial bd.: Endocrinology, 1955-59, Proc. Soc. Exptl. Biology and Medicine, 1959-63, Am. Jour. Physiology, 1972-75; Author papers on neuroendocrinology. Mem. Internat. Brain Research Orgn. (council 1964-68), AAAS, Am. Assn. Anatomists (v.p. 1969-70), Am. Physiol. Soc., Am. Zool. Soc., Neurosci. Soc., Endocrine Soc. (council 1968-70, Koch award 1973), Am. Acad. Arts and Scis., Nat. Acad. Scis., Soc. Exptl. Biology and Medicine, Soc. Study Reprodn. (dir. 1969-71, Hartman award 1977), Internat. Neuroendocrine Soc. (council 1972-76), Hungarian Soc. Endocrinology and Metabolism (hon.), Nat. Acad. Scis., Japan Endocrin Soc. (hon.), Phi Beta Kappa, Sigma Xi. Subspecialties: Neuroendocrinology; Anatomy and embryology. Current work: Mechanisms by which the brain controls endocrine secretions, especially the pituitary gonadotropins and ovarian steroids. Technics include stereotaxic surgery electrophysiology, neuropharmacology and radioimunoassay. Home: 466 Tuallitan Rd Los Angeles CA 90049 Office: U Calif Sch Medicine Los Angeles CA 90024

SAWYER, CONSTANCE BRAGDON, solar physicist, oceanographer; b. Lewiston, Maine, June 3, 1926; d. William Hayes, Jr. and Beatrice Goulding (Burr) S.; m. James Walter Warwick, Sept. 6, 1947; children: Sarah, David, Rachel Warwick, Joel McCulloch. A.B., Smith Coll., 1947; M.A., Radcliffe Coll., 1948, Ph.D., 1952. With Sacramento Peak Obs., Sunspot, N.Mex., 1952-55; with High Altitude Obs., Nat. Ctr. for Atmospheric Research, Boulder, Colo., 1955-58, staff scientist, 1979—; with Space Environ. Lab., NOAA, Boulder, 1958-74, Atlantic Oceanic and Meteorol. Lab., Miami, Fla., 1975-76, Pacific Marine Environ. Lab., Seattle, 1976-79. Contbr. research papers to profl. jours. Mem. Am. Astron. Soc., Am. Geophys. Union, AAAS, Internat. Astron. Union, Internat. Union Geology and Geophysics, Sigma Xi. Subspecialties: Solar physics; Remote sensing (atmospheric science). Current work: Solar activity, solar corona, spacecraft observations of sun and corona. Home: 850 20th St Apt 705 Boulder CO 80302 Office: Radiophysics 5475 Western Ave Boulder CO 80301

SAX, DANIEL SAUL, neurologist; b. Balt., Jan. 27, 1935; s. Benjamin Jacob and Miriam (Helfgott) S.; m. Joan Atherton Bond, Mar. 1962; children: Karen Bond, John Derek, Diana Liba-Atherton. B.A., Johns Hopkins U., 1955; M.D., U. Md., 1959. Diplomate: Am. Bd. Psychiatry and Neurology. Intern Boston City Hosp., 1959-60; resident in neurology N.E. Med. Ctr., Boston, 1960-61; resident in neuropathology Boston City Hosp., 1961-62, resident in neurology, 1962-64; teaching fellow in neurology Tufts U. Sch. Medicine, Boston, 1960-61; teaching and research fellow in neurology Harvard U. Sch. Medicine, Boston, 1961-64; asst. prof. neurology Northwestern U., Chgo., 1966-67, Albert Einstein Coll. Medicine, N.Y.C, 1967-69, vis. asst. prof. neuroanatomy, 1967-69; asso. prof. neurology Boston U. Sch. Medicine, 1969-77, prof. neurology, 1977—; chief neurology service VA Ct. St. Outpatienmt Clinic, Boston, 1974—; chief clin. neurophysiology lab. Boston VA Hosp., 1973—; sci. adv. Mass. chpt. Com to Combat Huntington's Disease; vis. neurologist Quincy (Mass.) Hosp., 1969—, Univ. Hosp., Boston, 1969—; cons. in neurology Cape Cod Hosp., Hyannis, Mass., 1969—, Brookline (Mass.) Hosp., 1969—; spl. cons. staff neurologist Brockton (Mass.) Hosp., 1969—. Contbr. chpts. to books, articles to profl. jours. Served to lt. comdr. M.C. USNR, 1964-66. VA Med. research grantee. Fellow Am. Acad. Neurology, N.Y. Acad. Medicine; mem. Boston Soc. Psychiatry and Neurology (sec.-treas. 1979-82), Am. Med. Electroencephalographic Assn., AMA, N.Y. Acad. Scis., Mass. Med. Soc., Am. Epilepsy Soc., Am. Electroencephalographic Soc., Stroke Council Am. Heart Assn., Am. Tree Farmers Assn., Am. Scotch Highland Breeders Assn. Subspecialties: Neurology; Neurophysiology. Current work: Research in neurology; movement disorders and the application of clinical neurophysiology to the study of such problems with special interest in Huntington's Disease, Multiple Sclerosis and neuro-urologic disorders and Polyneuropathies. Office: 720 Harrison Ave Boston MA 02118

SAX, GILBERT, educational psychologist, researcher, author, consultant; b. N.Y.C., July 13, 1930; s. and Belle (Wolfe) Carlin; m. Judith Iris Baker, Jan. 27, 1950 (div. Feb. 1967); children: Laurie, Kathy, Karen; m. Lois Mae Johnson, Sept. 2, 1968; stepchildren: Debra, Jennifer. B.A., UCLA, 1953, M.A., 1956; Ph.D., U. So. Calif., 1958. Asst. prof. ednl. psychology U. Hawaii, 1958-62, assoc. prof., 1962-66, chmn. dept. ednl. psychology, 1962-66; prof. edn. and psychology U. Wash., 1966—, chmn. dept. ednl. psychology, 1981—, chmn. behavioral scis. rev. com., 1971-73, 80—; mem. Bd. Psychology Examiners, Olympia, Wash., 1972-73. Author: Foundations of Educational Research, 1979, Principles of Educational and Psychological Measurement, 1981; contbr. numerous tech. articles, chpts. to profl. publs., encys. Served with USNR, 1947-52. Calif. Scholar Soc. scholar, 1947-49. Fellow Am. Psychol. Assn. (pres. Hawaii 1963-64); mem. Washington Ednl. Research Assn. (pres. 1969-70), Psychometric Soc., Am. Ednl. Research Assn., Nat. Council on Measurement in Edn., Psychometric Soc., AAUP (pres. Hawaii chpt. 1963-64), Sigma Xi (life), Phi Delta Kappa (pres. chpt. 1962-63, life mem). Current work: Test development and theory; statistics; experimental design; research methods. Home: 3733 47th Pl NE Seattle WA 98105 Office: U Wash 312 Miller Hall DQ-12 Seattle WA 98195

SAXE, GEOFFREY B., developmental psychology educator, researcher; b. Los Angeles, June 12, 1948; s. Carl H. and May L. S.; m. Maryl Gearhart, May 8, 1981; 1 son, Joshua Daniel Gearhart. B.A. in Psychology, U. Calif.-Berkeley, 1970, Ph.D., 1976. Research assoc. Edn. Research unit U. Papua, New Guinea, 1978-80; asst. prof. to assoc. prof. CUNY, 1977-83; assoc. prof. edn. UCLA, 1983—. Contbr. chpts. to books and articles to profl. jours. Nat. Inst. Edn. grantee, 1978—. Mem. Soc. Research in Child Devel., Am. Psychol. Assn., Sigma Xi. Subspecialties: Cognition; Developmental psychology. Current work: Research on cognitive development. Office: Dept Edn UCLA Los Angeles CA 90024

SAXENA, VINOD KUMAR, meteorology educator, consultant atmospheric sciences; b. Agra, India, May 23, 1944; came to U.S., 1968, naturalized, 1977; s. Kishori L. and Uma S.; m. Indra Nigam, Aug. 19, 1973; 1 dau.: Rita. Lectr. physics Agra Coll., 1963-64; asst. prof. U. Saugar, Sagar, India, 1967-68; fellow U. Mo.-Rolla, 1968-71; cloud physicist U. Denver, 1971-77; research assoc. prof. U. Utah, 1977-79; assoc. prof. meteorology N.C. State U., Raleigh, 1979—; invitational cons. State of N.C., Raleigh, 1980—, WRAL-TV, 1982—. Contbr. articles in field to profl. jours. Research grantee NSF, 1971—, U.S. Navy, 1972-75, Dept. Energy, 1977—; recipient Meteorol. award U. Utah, 1979. Mem. Am. Meteorol. Soc., Am. Geophys. Union, Am. Water Research Assn., Himalayan Internat. Inst., Brit. Royal Meteorol. Soc., Am. Platform Assn., Sigma Xi. Hindu. Subspecialties: Meteorologic instrumentation; Atmospheric chemistry. Current work: Cloud, aerosol and precipitation physics; particulate pollution; mechanisms involving the generation and scavenging of natural aerosols, development of instrumentation for cloud-active aerosols. Developer cloud condensation nucleus spectrometer. Home: 3616 Greywood Dr Raleigh NC 27604 Office: Dept Marine Earth and Atmospheric Scis N C State U Box 5068 Raleigh NC 27650

SAXON, GEORGE EDWARD, mfg. co. exec.; b. Pitts., July 17, 1932; s. George and Julia (Kavchak) S.; m. Frances J. Bartosiewicz, June 30, 1956; children: Regina, Edward, George, Gregory. B.S.M.E., U. Pitts., 1955. Foreman, mfg. engr. Gen. Electric Co., 1955-63; purchasing engr. atomic equipment-atomic fuel divs. Westinghouse Electric Corp., Cheswick, Pa., 1963-66; purchasing mgr. tubular products div. Babcock & Wilcox Co., Beaver Falls, Pa., 1967-70; gen. mgr. Universal Lubricating Systems, Inc., Oakmont, Pa., 1970-71; pres., chmn. bd. Conco Systems, Inc., Verona, Pa., 1971—; cons. engr. and researcher tube cleaning. Contbr. articles to prof. jours. Mem. ASME, Small Mfrs. Council Pitts. (past pres.), Small Bus. United (pres. elect 1982-83), Soc. Am. Valve Engrs., Am. Iron and Steel Engrs. Republican. Roman Catholic. Clubs: U. Pitts. Century, Golden Panthers. Lodges: Elks; Eagles. Subspecialty: Energy conservation. Current work: Conserving energy in power plants by making condenser tube cleaning tools. Patentee in field. Home: 665 9th St Oakmont PA 15139 Office: 135 Sylvan St Verona PA 15147

SAYRE, DAVID, crystallographer, microscopist; b. N.Y.C., Mar. 2, 1924; s. Ralph and Sylvia (Rosenbaum) S.; m. Anne Claire Bowns, Dec. 26, 1947. B.S. in Physics, Yale U., 1943; Ph.D. in Crystallography, Oxford (Eng.) U., 1951. Mem. staff radiation lab. MIT, 1943-46, Johnson Found., U. Pa., 1951-55; mem. research staff IBM, Yorktown Heights, N.Y., 1955—; mem. U.S. nat. com. on crystallography NRC. Contbr. articles to profl. jours. Bd. trustee Village of Head-of-the-Harbor, N.Y. Vis. fellow All Souls Coll., Oxford U., 1973-74. Mem. Am. Crystallographic Assn. (pres. 1983). Subspecialties: X-ray crystallography; X-ray microscopy. Current work: Imaging of matter at the atomic and molecular level; especially development of techniques for the use of low-energy X-ray photons for this purpose. Office: IBM Research Center Yorktown Heights NY 10598

SAYRE, RICHARD MARTIN, plant pathologist; b. Hillsboro, Oreg., Mar. 25, 1928; s. William Ontarius and Mary (Drasha) S.; m. Diane Pringle, June 30, 1962; 1 dau., Janet. Student, Pacific U., Forest Grove, Oreg., 1947-48; B.S., Oreg. State U., 1951, M.S., 1954; Ph.D., U. Nebr., Lincoln, 1958. Research officer plant pathology Canada Agr., Harrow, Ont., 1958-65; research plant pathologist U.S. Dept. Agr., Beltsville, Md., 1965—. Mem. Soc. Nematologists, Am. Phytopathol. Soc., Am. Inst. Biol. Scis., AAAS, Sigma Xi. Club: Toastmaster Internat. Subspecialties: Plant pathology; Integrated pest management. Current work: Diseases of vegetable crops; nematology; nematode diseases; biocontrol of plant nematodes using their parasites and predators. Home: 3111 Calverton Blvd Beltsville MD 20705 Office: Nematology Lab Bldg 011A BARC (W) Beltsville MD 20705

SAYRE, WILLIAM OLAF, geology educator; b. Fairfield, Iowa, Apr. 7, 1954; s. Ralph Mills and Lois Jean (Gregg) S.; m. Shirley Ann Hoelzen, May 29, 1982; 1 dau., Andria Jo Green. B.S., Western Wash. U., 1976; Ph.D., Southampton U., 1981. Research asst. Southampton (Eng.) U., 1976-79; asst. prof. geology Iowa State U., 1979—. Mem. Ames (Iowa) Water and Energy Adv. Bd., 1982—. Mem. Am. Geophys. Union, Geol. Soc. Am., Soc. Exploration Geophysicists. Democrat. Presbyterian. Lodge: Kiwanis. Subspecialty: Geophysics. Current work: Study of anisotropy of magnetic susceptibility of sedimentary rocks; use of magnetic polarity stratigraphy to establish temporal framework of evolution of man. Office: Dept Earth Scis Iowa State U Ames IA 50011

SBORDONE, ROBERT JOSEPH, neuropsychologist, educator; b. Boston, May 6, 1940; s. Saverio and Phylliss (Dellaria) Vella; m. Melinda Welles, June 30, 1972 (div. 1977). A.B., U. So. Calif., 1967; M.A., Calif. State U.-Los Angeles, 1969; Ph.D., UCLA, 1976, postdoctoral, 1977. Cert. psychologist Calif. Mem. staff psychology UCLA, 1977-78; sole practice psychology, Los Angeles, 1978-80; asst. prof. psychology U. Calif.-Irvine, 1980-82, asst. clin. prof., 1983—; pres. Robert Sbordone Inc., Garden Grove, Calif., 1982—. Contbr. chpt. to book and articles to profl. jours.; editor: Clinical Neoropsychology jour, 1979—. Mem. bd. dirs. So. Calif. Head Injury Found., 1982—; mem. adv. bd. Mardan Sch., Costa Mesa, Calif., 1982—. Served with USAF, 1962-66. NIMH grantee, 1973-77. Mem. Am. Psychol. Assn., Internat. Neuropsychol. Soc., Nat. Head Injury Found., Internat. Soc. Research in Aggression, N.Y. Acad. Scis. Subspecialties: Neuropsychology; Cognition. Current work: Neuropsychological assessment of brain injured; development of computer software for assessment and rehabilitation of brain injured patients. Developer of computer software, 1982. Home: 13412 Donegal Dr Garden Grove CA 92644 Office: Robert J Sbordone Inc 13412 Donegal Dr Garden Grove CA 92644

SCALIA, FRANK RICHARD, neurobiologist, educator; b. N.Y.C., Mar. 18, 1939; s. Frank F. and Isabelle (Plazza) S.; m. Patricia T. Raffaele, June 10, 1960; m. Julia R. Currie, Dec. 18, 1971; children: Jason, Russell, Katherine. B.A., NYU, Washington Sq. Coll., 1959; Ph.D., SUNY-Downstate Med. Ctr., 1964. Asst. prof. anatomy and cell biology SUNY-Downstate Med. Ctr., 1967-72, assoc. prof., 1972-77, prof., 1977—. Subspecialties: Neurobiology; Morphology. Current work: Studies on structure, function, development and regeneration of visual system in vertebrate species. Office: Dept Anatomy and Cell Biology SUNY-Downstate Med Ctr 450 Clarkson Ave Brooklyn NY 11203

SCANDALIOS, JOHN GEORGE, geneticist; b. Nisyros Isle, Greece, Nov. 1, 1934; s. George John and Calliope (Broujos) S.; m. Penelope Anne Lawrence, Jan. 18, 1961; children: Artemis Christina, Melissa Joan, Nikki Eleni. B.A., U. Va., 1957; M.S., Adelphi U., 1962; Ph.D., U. Hawaii, 1965. Asso. in bacterial genetics Cold Spring Harbor Labs., 1960-62; NIH postdoctoral fellow U. Hawaii Med. Sch., 1965; asst. prof. Mich. State U., East Lansing, 1965-70, asso. prof., 1970-72; prof., head dept. biology U. S.C., Columbia, 1973-75; prof., head dept. genetics N.C. State U., Raleigh, 1975—; vis. prof. genetics U. Calif., Davis, 1969; vis. prof. OAS, Argentina, Chile and Brazil, 1972. Author: Physiological Genetics, 1979; Editor: Developmental Genetics; co-editor: Isozymes, 4 vols., 1975, Monographs in Developmental Biology, 1968—. Served with USAF, 1957. Alexander von Humboldt travel fellow, 1976; mem. exchange program NAS, US/USSR. Mem. Genetics Soc. Am., Am. Soc. Human Genetics, Am. Genetic Assn. (pres.), AAAS, Soc. Devel. Biology (dir.), Am. Inst. Biol. Scis., Am. Soc. Plant Physiologists, Sigma Xi. Subspecialties: Genetics and genetic engineering (biology); Gene actions. Current work: Studing regulation of gene expression in eukaryotes (maize) during development at the molecular level; developmental-molecular genetics, use of recombinant-DNA to study gene expression in plants. Office: Dept Genetics NC State U PO Box 5487 Raleigh NC 27650

SCARL, DONALD, physicist, educator; b. Easton, Pa., Sept. 17, 1935; m. Barbara S. Cohen, Nov. 24, 1979; 1 dau., Judith Cohen. B.A., Lehigh U., 1957; Ph.D., Princeton U., 1962. Assoc. prof. physics Poly. Inst. N.Y., Farmingdale, 1966—. Contbr. articles to profl. jours. Mem. Am. Phys. Soc., Optical Soc. Am., AAAS, Phi Beta Kappa. Subspecialties: Optical signal processing; Atomic and molecular physics. Current work: Quantum optics, atomic state dynamics. Patentee in field. Home: 8 Woodland Rd Glen Cove NY 11542 Office: Polytechnic Institute of NY Farmingdale NY 11735

SCATTERGOOD, THOMAS W., planetologist, analytical chemist; b. Mt. Holly, N.J., Oct. 3, 1946; s. William E. and Grace (Paulin) S. B.S. in Chemistry, U. Del., Newark, 1968; M.S., SUNY-Stony Brook, 1972, Ph.D. in Chemistry, 1975. Teaching asst. SUNY-Stony Brook, 1970-72, research asst., 1972-75, research assoc., 1975-76, sr. research assoc., 1979—; research assoc. NRC, Nat. Acad. Scis., NASA Ames Research Ctr., 1977-78, Titan Probe Sci. Study Group, NASA Ames Research Ctr., Moffett Field, Calif., 1979-81. Contbr. articles to profl. publs. Pres. Cypress Point Lakes Homeowners Assn., Central Coast Counties Camera Club Council. Mem. AAAS, Am. Geophys. Union, Am. Astron. Soc. (div. planetary scis.), Planetary Soc., Photog. Soc. Am. Quaker. Club: Camaraderie Camera (Mountain View, Calif.). Subspecialties: Planetary atmospheres; Space chemistry. Current work: Study of photochemistry and energetic particle initiated chemistry in atmospheres of outer planets; determination of identigy (composition) and optical properties of clouds of Titan. Office: MX 239-12 NASA Ames Research Center Moffett Field CA 94035

SCHACHMAN, HOWARD KAPNEK, molecular biologist, educator; b. Phila., Dec. 5, 1918; s. Morris H. and Rose (Kapnek) S.; m. Ethel H. Lazarus, Oct. 20, 1945; children—Marc, David. B.S. in Chem. Engring, Mass. Inst. Tech., 1939; Ph.D. in Phys. Chemistry, Princeton, 1948; D.Sc. (hon.), Northwestern U., 1974. Fellow NIH, 1946-48; instr., asst. prof. U. Calif. at Berkeley, 1948-54, asso. prof.

biochemistry, 1954-59, prof. biochemistry and molecular biology, 1959—, chmn. dept. molecular biology, dir. virus lab., 1969-76. Author: Ultracentrifugation in Biochemistry, 1959. Served from ensign to lt. USNR, 1945-47. Guggenheim Meml. fellow, 1956; Recipient John Scott award, 1964, Warren Triennial prize Mass. Gen. Hosp., 1965. Mem. Nat. Acad. Sci., Am. Chem. Soc. (recipient award in Chem. Instrumentation 1962, Calif. sec. award 1958), AAAS, Am. Soc. Biol. Chemists, Am. Acad. Arts and Scis., Sigma Xi. Subspecialties: Biochemistry (medicine); Biophysical chemistry. Current work: Proteins, enzymes, nucleic acids. Macromolecular interactions. Office: Molecular Biology and Virus Lab U Calif Berkeley CA 94720

SCHACH VON WITTENAU, MANFRED EBERHARD, pharm. research exec.; b. Pennekow, Ger., June 19, 1930; s. Hans and Sibylle (von Below) S. von W.; m. Patricia Anita Mackey, Aug. 11, 1955; children: Hans, Alexis, Sibylle. B.S., U. Heidelberg, Ger., 1952, M.S., 1955, Ph.D., 1957. Postdoctoral fellow M.I.T., 1957-58; research chemist Pfizer, Inc., 1958-64, group supr., 1964-67, mgr. drug metabolism, 1967-71, asst. dir. drug metabolism, 1971-72, dir. drug metabolism, 1972-74, exec. dir. safety eval. and drug metabolism, 1974-81, v.p. safety eval., 1981—; vice chmn. drug safety steering com. Pharm. Mfrs. Assn., OSHA Task Force; mem. ad hoc com. health effects Conn. Acad. Sci. and Engring. Contbr. articles to profl. jours. Mem. Am. Soc. Pharmacology and Exptl. Therapeutics, Am. Chem. Soc., Gesellschaft Deutscher Chem., AAAS. Subspecialties: Medicinal chemistry; Toxicology (medicine). Current work: Mgmt. safety eval. and drug metabolism depts. Patentee in field. Office: Eastern Point Rd Groton CT 06340

SCHADE, ROBERT RICHARD, medical educator, researcher; b. Rockville Centre, N.Y., Jan. 5, 1948; s. Robert Richard and Loretta K. (McGovern) S.; m. Rosann Foster, Oct. 14, 1972; children: Danielle Nicole, Kimberly Anne, Allison Janine. A.B., Colgate U., 1969; M.D., George Washington U., 1973. Diplomate: Am. Bd. Internal Medicine, Nat. Bd. Med. Examiners. Intern Rush-Presbyn. Med. Ctr., Chgo., 1973-74, resident in internal medicine, 1974-76; fellow liver disease unit Yale U., New Haven, 1976-78, fellow in gastroenterology, 1978-80; asst. prof. medicine U. Pitts. Med. Sch., 1980—; dir. Clin. Gastrointestinal lab. Presbyn. U. Hosp., Pitts., 1982—; attending VA Hosp. of Oakland, Pitts., Children's Hosp. Pitts. Contbr. articles to profl. jours. Mem. sci. adv. bd. Pitts. chpt. Nat. Found. Ileitis and Colitis, 1981—. Mem. Am. Gastroent. Assn., N.Y. Acad. Sci., ACP, Am. Soc. Internal Medicine, Am. Coll. Gastroenterology, Am. Soc. Gastrointestinal Endoscopy, AAAS. Republican. Roman Catholic. Subspecialties: Gastroenterology; Internal medicine. Current work: Liver disease and liver transplantation; effects of alcohol on pituitary hormone secretion; studies in gastrointestinal motility. Office: U Pitts Pittsburgh PA 15261

SCHADLER, DANIEL LEO, biologist, educator; b. Dayton, Ky., Apr. 5, 1948; s. Alvin Peter and Irma Catherinne (Schack) S. A.B., Thomas More Coll., 1970; M.S., Cornell U., 1972, Ph.D., 1974. NSF grad. trainee Cornell U., Ithaca, N.Y., 1970-74; postdoctoral research assoc. U. Wis., Madison, 1974-75; asst. prof. biology Oglethorpe U., Atlanta, 1975-79, assoc. prof., 1979—. Mem. Am. Chem. Soc., Am. Phytopath. Soc., Internat. Soc. Plant Pathology, Ga. Chrysanthemum Soc. Roman Catholic. Subspecialties: Plant pathology; Microbiology. Current work: Phytotoxins; teaching introductory microbiology courses. Home: 4218 Admiral Dr Chamblee GA 30341 Office: 4484 Peachtree Rd NE Atlanta GA 30319

SCHAECHTER, MOSELIO, microbiologist, academic administrator; b. Milan, Italy, Apr. 26, 1928; s. Abraham Isaac and Victoria C. (Waksmann) S.; m. Barbara Ruth Thompson, Dec. 13, 1953; children: Judith A., John N. M.A., U. Kans., 1951; Ph.D., U. Pa., 1954. Postdoctoral State Serum Inst., Copenhagen, Denmark, 1956-58; instr. to assoc. prof. U. Fla., 1958-62; assoc. prof., prof. Tufts U., Boston, 1962—, chmn. dept. molecular biology, 1970—; chmn. NIH study sect. on bacteriology and mycology, 1978-79, mem., 1975-79; grant reviewer NSF. Contbr. books, articles to profl. publs. Served with U.S. Army, 1954-56. NIH research career devel. awardee, 1959-68. Mem. Am. Soc. Microbiology (chmn. com. on genetics, molecular and systematic microbiology 1979—), Soc. Genetic Microbiology, Boston Mycol. Club, Sigma Xi. Subspecialties: Microbiology; Molecular biology. Current work: DNA replication in bacteria. Home: 855 Commonwealth Newton MA 02159 Office: 136 Harrison Ave Boston MA 02111

SCHAEFER, ALBERT RUSSELL, physicist, educator; b. Oklahoma City, Oct. 13, 1944; s. Albert R. and Marcella (Russell) S.; m. Judith Ann Bracewell, Jan. 19, 1968; children: Amy, Brandon. B.S. in Physics with spl. distinction, U. Okla., 1966, Ph.D. in Atomic Physics, 1970. Physicist Nat. Bur. Standards, Washington, 1970—; adj. prof. physics Montgomery Coll., 1974—. Contbr. articles to profl. jours. Co-organizer, past bd. dirs. Greater Laytonville Area Citizens Assn. Recipient Outstanding Performance awards Nat. Bur. Standards, 1978, 80; Bronze medal Dept. Commerce, 1981; NSF fellow, 1966. Mem. Am. Phys. Soc., Optical Soc. Am., Soc. Photo-optical Instrumentation Engrs., Sigma Xi, Phi Beta Kappa. Subspecialties: Photo-optical instrumentation; Atomic and molecular physics. Current work: Radiometric physics, optical detectors, spectroscopy, silicon photodiodes, electro-optics, photometry, atomic lifetimes, astronomy. Office: Nat Bur Standards B-306-MET Washington DC 20234

SCHAEFER, CARL WALTER, II, biology educator; b. New Haven, Sept. 6, 1934; s. Jack Warner and Eugenia (Ives) S.; m.; children: Ann Jack Haislop, Madelyn Ives. B.A., Oberlin Coll., 1956; Ph.D., U. Conn., 1964. Instr. Bklyn. Coll., 1963-64, asst. prof., 1964-66, U. Conn., Storrs, 1966-68, assoc. prof., 1968-75, prof. biology, 1975—. Editor: Annals of Entomol. Soc. Am, 1973—, Heteropterists Newsletter, 1973—; editorial bd.: Entomologia Generalis, 1979-82, Uttar Pradesh Jour. Zoology, 1981—. Mem. Entomol. Soc. Am., Kans. Entomol. Soc., Conn. Entomol. Soc., Am. Soc. Naturalists, Soc. Study of Evolution, Soc. Systematic Zoology, N.Mex. Hist. Soc., Conn. Hist. Soc. Subspecialties: Systematics; Morphology. Current work: Morphology and phylogeny of Hemiptera (Insecta). Office: Bid Sci Group U Conn Storrs CT 06268

SCHAEFER, HENRY FREDERICK, III, chemist; b. Grand Rapids, Mich., June 8, 1944; s. Henry Frederick and Janice Christine (Trost) S.; m. Karen Regine Rasmussen, Sept. 2, 1966; children—Charlotte Ann, Pierre Edward, Theodore Christian. S.B., M.I.T., 1966; Ph.D., Stanford U., 1969. Faculty chemistry U. Calif., Berkeley, 1969—; faculty sr. scientist Lawrence Berkeley Lab., 1972—; Wilfred T. Doherty prof. U. Tex., Austin; also dir. Inst. for Theoretical Chemistry. Alfred P. Sloan fellow, 1972-74. Author: The Electronic Structure of Atoms and Molecules, 1972, Modern Theoretical Chemistry, 1977. John S. Guggenheim fellow, 1976-77. Mem. Am. Chem. Soc. (Pure Chemistry award 1979), Am. Phys. Soc., Sigma Xi, Phi Lambda Upsilon. Evangelical Christian. Subspecialties: Theoretical chemistry; Polymer chemistry. Current work: Quatum mechanics of molecular electronic structure. Office: Univ of Calif 714 Univ Hall Berkeley CA 94720

SCHAEFER, JACOB WERNLI, research laboratory executive; b. Paullina, Iowa, June 27, 1919; s. Louis B. and Minnie (Wernli) S.; m. Mary Snow Carter, July 26, 1941; children: Joanna, James, Scott. B.M.E., Ohio State U., 1941, D.Sc. (hon.), 1976. Mem. staff Bell Labs., 1941—; dir. Kwajalein (Marshall Islands) Field Sta., 1963-65, exec. dir., Holmdel, N.J., 1968-80, Murray Hill, N.J., 1980-81, Mil. Systems Div., 1981—. Author articles in field. Pres. Watchung (N.J.) Sch. Bd., 1954-63, Bancroft Sch., Haddonfield, N.J., 1978—; chmn. Watchung Planning Bd., 1967—; mem. Watchung Area council Boy Scouts Am., 1966-69. Served to capt. Ordnance Corps AUS, 1942-46. Decorated Commendation medal; recipient Disting. Alumnus award Ohio State U., 1966, Outstanding Civilian Service medal. Mem. Nat. Acad. Engring., ASME, IEEE (sr.), Army Ordnance Assn. Republican. Subspecialties: Systems engineering. Current work: Military systems: guided missile defense systems, underwater surveillance systems, and computer applications. Patentee command guidance for anti-aircraft missiles, optical tracking systems for anti-aircraft fire control. Home: 115 Century Ln Watchung NJ 07060 Office: 600 Mountain Ave Murray Hill NJ 07974

SCHAEFER, STEVEN DAVID, otorhinolaryngology educator; b. Los Angeles, Mar. 25, 1945; s. Glen Arthur and Alice Elizabeth (Malerstein) S.; m. Phyllis Lois Clark, July 1, 1967; 1 dau., Jessica Leigh. B.A., U. Calif.-Berkeley, 1967; M.D., U. Calif-Irvine, 1972. Diplomate: Am. Bd. Otolaryngology. Intern UCLA Med. Ctr., 1972-73, resident in gen. surgery, 1973-74; resident in otolaryngology Stanford U., 1974-77; asst. prof. surgery U. Tex., Dallas, 1977-81, assoc. prof. otolaryngology, 1982—. Contbr. chpts. to books and articles to profl. jours. Chmn. pub. edn. Am. Cancer Soc., Tex., 1980—. Investigator award NIH, 1978, 82. Fellow ACS, Am. Soc. Head and Neck Surgery, Am. Soc. Clin. Oncologists, Am. Laryngologic, Rhinologic and Otologic Soc. Subspecialties: Otorhinolaryngology; Oncology. Current work: Head and neck oncology and trauma. Home: 514 Durango Circle S Irving TX 75062 Office: U Tex Health Sci Ctr 5323 Harry Hines Blvd Dallas TX 75235

SCHAEFFER, BENSON, clinical psychologist; b. N.Y.C., Aug. 30, 1942; s. Charles and Gilda (Cwas) S.; m. Danna Wilner, Aug. 11, 1963; 1 dau., Rebecca. B.A., UCLA, 1962, M.A., 1964, Ph.D., 1967. Asst. prof. dept. psychology U. Oreg., Eugene, 1966-72, assoc. prof., 1972-82; pvt. practice clin. psychology, Portland, Oreg., 1982—. Author: Total Communication: A Signed Speech Program for Nonverbal Children, 1980. NIH grantee U. Oreg., 1975-76. Mem. Am. Psychol. Assn., Assn. Severely Handicapped. Subspecialties: Developmental psychology; Neuropsychology. Current work: Cognitive and linguistic development in and training for mentally handicapped individuals. Home: 3422 NE Pacific St Portland OR 97232 Office: 2330 NW Flanders St Suite 201 Portland OR 97210

SCHAEFFER, CHARLES DAVID, chemistry educator; b. Allentown, Pa., June 14, 1948. B.A., Franklin and Marshall Coll., 1970; Ph.D., SUNY-Albany, 1974. Postdoctoral fellow Yale U., New Haven, 1974-76; mem. faculty dept. chemistry Elizabethtown (Pa.) Coll., 1976—, assoc. prof., 1981—. Contbr. articles to profl. jours. Mem. Am. Chem. Soc., Royal Soc. Chemistry (London), AAAS, Sigma Xi. Subspecialties: Inorganic chemistry; Nuclear magnetic resonance (chemistry). Current work: Organometallic chemistry of the main group elements; NMR spectroscopy. Office: Dept Chemistry Elizabethtown Coll Elizabethtown PA 17022

SCHAEFFER, GIDEON W., plant physiologist, researcher; b. Menno, S.D., Nov. 5, 1929; m., 1953. B.A., Yankton Coll., 1952; M.Sc., U. Nebr., 1957, Ph.D. in Physiol. Genetics, 1960. AEC fellow, USPHS fellow, research assoc. in cellular biology Brookhaven Nat. Lab., 1960-62; fellow Phytotron Lab., Gif-sur-Yvette, France, 1962-63; research scientist Beltsville (Md.) Agrl. Research Ctr., Sci. and Edn. Adminstrn.-Agrl. Research, Dept. Agr., 1963—, chief lab., 1977. Mem. Am. Soc. Plant Physiology, AAAS, Tissue Culture Assn., Scandinavian Soc. Plant Physiology, Am. Soc. Agronomy. Subspecialty: Plant cell and tissue culture. Current work: Research in rice varieties, pollen culture. Office: Cell Culture and Nitrogen Fixation Lab Beltsville Agrl Research Ctr Beltsville MD 20705

SCHAEFFER, MORRIS, USPHS medical officer, educator; b. Berdichev, Ukraine, Russia, Dec. 31, 1907; s. Samuel Sim and Celia Zapora (Sellman) S.; m. Josephine Marie Wintzer, Nov. 5, 1943; children: Debora, Colin Sim, David Emil, Jessica. A.B., M.A., U. Ala., 1930; Ph.D., N.Y. U., 1935, M.D., 1944. Diplomate: Nat. Bd. Med. Examiners, Am. Bd. Preventive Medicine. Intern Boston City Hosp., 1944-45; resident in pediatrics Cleve. City Hosp., 1945-48; physician-in-chief Contagious Pavillion, 1945-49; asst. prof. pediatrics and microbiology Western Res. U., 1945-49; med. dir. virology br. Ctr. Disease Control, 1949-59, dir. pub. health labs., N.Y.C., 1959-71; asst. to dir. Bur. Biologics, FDA, Rockville, Md., 1972-82, Nat. Ctr. Drugs and Biologics, 1982—; adj. prof. medicine NYU, 1960—. Author: Experimental Poliomyelitis, 1940; editor: Viruses Affecting Man and Animals, 1971, Immunofluorescence and Related Staining Techniques, 1975, Federal Legislation and the Clinical Laboratory, 1981; contbr. articles profl. jours., chpts. to books. Recipient Disting. Service award N.Y. Health Dept., 1967; award of merit, N.Y.C., 1974; Commendable Service award FDA, 1978; Service citation NYU, 1981. Mem. Am. Pub. Health Assn., Soc. Exptl. Medicine, Am. Coll. Preventive Medicine, Am. Assn. Immunology, Infectious Disease Soc. Am., Washington Acad. Medicine, Am. Soc. Microbiology, Am. Acad. Microbiology, Sigma Xi, Alpha Omega Alpha, Delta Omega. Subspecialties: Infectious diseases; Preventive medicine. Current work: Safety and effectiveness of biological products, vaccine development, standardization of allergenic extracts, drug adv. com. coordination. Home: 8930 Bradmoor Dr Bethesda MD 20817 Office: HFN-6 Room 14B-26 5600 Fishers Ln Rockville Md 20857

SCHAEFFER, NORMAN M., physicist, cons.; b. Camden, Ark., Nov. 1, 1927; s. Sam and Lena (Sabludowski) S.; m. Cecille Marion Levinson, Aug. 14, 1949; children: Marc A., Jeanette A., Susan R. B.S., La. State U., 1947, M.S., 1949; Ph.D., U. Tex., 1953. With Gen. Dynamics, Ft. Worth, 1953-62, mgr. nuclear research, 1956-62; pres. Radiation Research Assocs., Ft. Worth, 1962—. Author: Reactor Shielding for Nuclear Engineers, 1973. Fellow Am. Nuclear Soc.; mem. Am. Phys. Soc. Subspecialties: Nuclear fission; Nuclear engineering. Current work: Reactor shielding, radiation transport; manage research co. specializing in radiation shielding and atomsopheric optics. Office: 3550 Hulen Fort Worth TX 76107

SCHAEFFER, ROBERT L., JR., biology educator; b. Allentown, Pa., Oct. 31, 1917; s. Robert L. and Millie Louisa (Ochs) S. B.S., Haverford Coll., 1940; Ph.D., U. Pa., 1948. Instr. botany U. Pa., Phila., 1946-47; asst. prof. biology Upsala Coll., East Orange, N.J., 1948-54, Muhlenberg Coll., Allentown, Pa., 1954-59, prof., 1960—. Author: Vascular Flora of Northampton County, Pa, 1948. Elder St. John's United Ch. of Christ, Allentown; active Lehigh County Hist. Soc. Served with USN, 1944-46. Recipient Lindbach Teaching award, 1963. Mem. Am. Fern Soc., Phila. Bot. Club, Torrey Bot., Am. Soc. Plant Taxonomists, Sigma Xi., Phi Beta Kappa. Subspecialties: Taxonomy; Systematics. Current work: Vascular flora of the Lehigh Valley. Home: 32 N 8th St Allentown PA 18101 Office: Muhlenberg Coll Allentown PA 18104

SCHAFER, ANDREW IMRE, medical educator, biomedical researcher; b. Budapest, Hungary, Mar. 20, 1948; came to U.S., 1961; s. Stephen and Lili (Reisner) S.; m. Pauline Santagate, June 3, 1973; children: Eric, Pamela. B.A., Northeastern U., 1969; M.D., U. Pa., 1973. Diplomate: Am. Bd. Internal Medicine (Hematology). Intern U. Chgo., 1973-74, resident in medicine, 1974-76; fellow in hematology Peter Bent Brigham Hosp., Boston, 1976-79; instr. medicine Harvard Med. Sch., Boston, 1979-81, asst. prof., 1981—; assoc. in medicine Brigham and Women's Hosp., Boston, 1982—; dir. hematology West Roxbury (Mass.) VA Hosp., 1979—. Contbr. articles to sci. jours. Med. Found. fellow, 1979; Am. Heart Assn. grantee, 1983. Fellow ACP, Boston Blood Club (pres. 1979-80); mem. Am. Soc. Hematology, Am. Fedn. Clin. Research. Subspecialties: Hematology; Biochemistry (medicine). Current work: Basic research in the biochemical control of blood platelet function and platelet-vessel wall interactions. Home: 153 Loker St Wayland MA 01778 Office: 75 Francis St Boston MA 02115

SCHAFER, JOHN FRANCIS, plant pathologist; b. Pullman, Wash., Feb. 17, 1921; s. Edwin George and Ella Frances (Miles) S.; m. Joyce A. Marcks, Aug. 16, 1947; children: Patricia, Janice, James. B.S., Wash. State U., 1942; Ph.D., U. Wis.-Madison, 1950. Mem. faculty dept. plant pathology Purdue U., 1949-68, prof., 1958-68; prof., head dept. plant pathology Kans. State U., 1968-72; prof., chmn. dept. plant pathology Wash. State U., 1972-80; integrated pest mgmt. coordinator sci. and edn. U.S. Dept. Agr., Beltsville, Md., 1980-81; acting nat. research program leader plant pathology and nematology Agrl. Research Service, 1981-82, dir. Cereal Rust Lab., St. Paul, 1982—; vis. research prof. Duquesne U., 1965-66; adj. prof. plant pathology U. Minn. Served with AUS, 1942-46. Fellow AAAS, Ind. Acad. Sci.; mem. Am. Phytopath. Soc. (past pres.), Am. Soc. Agronomy, Crop Sci. Soc. Am. Subspecialty: Plant pathology. Current work: Durability of plant disease resistance. Co-breeder of 30 cultivars of cereal crops including Arthur wheat. Home: 1753 Lindig St St Paul MN 55113 Office: 1551 Lindig St St Paul MN 55108

SCHAIRER, GEORGE SWIFT, aerodynamic engineer; b. Pitts., May 19, 1913; s. Otto Sorg and Elizabeth Blanche (Swift) S.; m. Mary Pauline Tarbox, June 20, 1935; children: Mary Elizabeth, George Edward, Sally Helen, John Otto. B.S., Swarthmore (Pa.) Coll., 1934, D.Eng. (hon.), 1958; M.S., MIT, 1935. With Bendix Aviation Corp., South Bend, Ind., 1935-37; cons. Vultee Aircraft Corp., San Diego, 1937-39; with Boeing Airplane Co., Seattle, 1939—, mem. aerodynamic staff, 1948-51, chief tech. staff, 1951-56, asst. chief engr., 1956-57, dir. research, 1957-59, v.p. research and devel., 1959-73, v.p. research, 1973-78; cons. in field; mem. aero. and space engring. bd. NRC, 1977-80. Contbr. articles to profl. jours. Trustee, contemporary Theatre Cornish Sch. Recipient Spirit of St. Louis medal ASME, 1959; Guggenheim medal, 1967. Fellow AIAA (Sylvanus Albert Reed award 1950, Wright Brothers lectr. 1964); mem. Nat. Acad. Engring., Nat. Acad. Scis., Internat. Acad. Aeronautics, Sigma Xi, Sigma Tau. Subspecialties: Aerospace engineering and technology; Astronautics. Current work: Consultant aerospace industry. Home: 4242 Hunts Point Rd Bellevue WA 98004 Office: Boeing Co Box 3707 Seattle WA 98124

SCHALLY, ANDREW VICTOR, medical research scientist; b. Poland, Nov. 30, 1926; came to U.S., 1957; s. Casimir Peter and Maria (Lacka) S.; m.; children: Karen, Gordon; m. Ana Maria Comaru, Aug. 1976. B.Sc., McGill U., Can., 1955, Ph.D. in Biochemistry, 1957, . Research asst. biochemistry Nat. Inst. Med. Research, London, 1949-52; dept. psychiatry McGill U., Montreal, Que., 1952-57; research asso., asst. prof. physiology and biochemistry Coll. Medicine, Baylor U., Houston, 1957-62; chief endocrine and polypeptide labs. VA Hosp., New Orleans, 1962—; asso. prof. Sch. Medicine, Tulane U., New Orleans, 1962-67, prof., 1967—; sr. med. investigator VA, 1973—. Author several books; contbr. articles to profl. jours. Recipient Dir.'s award for outstanding med. research VA Hosp., New Orleans, 1968; Van Meter prize Am. Thyroid Assn., 1969; Ayerst-Squibb award Endocrine Soc., 1970; William S. Middletown award VA, 1970; Ch. Mickle award U. Toronto, 1974; Gairdner Internat. award, 1974; Borden award Assn. Am. Med. Colls. and Borden Co. Found., 1975; Lasker Basic Research award, 1975; co-recipient Nobel prize for medicine, 1977; USPHS sr. research fellow, 1961-62. Mem. Endocrine Soc., Am. Physiol. Soc., Soc. Biol. Chemists, AAAS, Soc. Exptl. Biol. Medicine, Internat. Soc. Research Biology Reprodn., Soc. Study Reprodn., Soc. Internat. Brain Research Orgn., Mexican Acad. Medicine, Am. Soc. Animal Sci., Endocrine Soc. Madrid, Sigma Xi, others. Subspecialties: Neuroendocrinology; Oncology. Current work: Hypothalamic hormones, endocrine dependent cancer. Home: 5025 Kawanee Ave Metairie LA 70002 Office: 1601 Perdido St New Orleans LA 70146

SCHANFIELD, MOSES SAMUEL, geneticist, research director; b. Mpls., Sept. 7, 1944; s. Abraham and Fannie (Schwartz) S.; m. Nancy Bergren; 2 daus., Sara Abigail, Amanda Phylisa. B.A., U. Minn., 1966; M.A., Harvard U., 1969; Ph.D., U. Mich., 1971. Postdoctoral fellow dept. medicine U. Calif., San Francisco, 1971-74; asst. research geneticist dept. medicine, 1974-75; dir. reference lab. and transfusion service Milw. Blood Ctr., 1975-78; asst. dir. blood services, head immunohematology lab. ARC Blood Services, Bethesda, Md., 1978-83; dir. immunogenetics Serologicals, Inc., Atlanta, 1983—. Contbr. articles to profl. jours. NIH fellow, 1972-74; recipient gold medal 1st Latin Am. Congress Hemotherapy and Immunohematology, 1979. Mem. Am. Assn. Blood Banks, Am. Soc. Human Genetics, Am. Assn. Phys. Anthropologists, Soc. Study Human Biology, Human Biology Council, Am. Assn. Immunology, AAAS, Sigma Xi. Subspecialties: Immunogenetics; Evolutionary biology. Current work: Biological properties of genetic markers of antibodies. Office: Serologicals of Atlanta 401 W Peactree St NW Suite 1660 Atlanta GA 30308

SCHANK, ROGER CARL, computer science and psychology educator, business executive; b. N.Y.C., Mar. 12, 1946; s. Maxwell and Margaret (Rosenberg) S.; m. Diane Levine, Mar. 22, 1970; children: Hana, Joshua. B.S. in Math, Carnegie Inst. Tech., 1966; M.A. in Linguistics, U. Tex., 1967, Ph.D., 1969. Engr./scientist Semiotics group Tracor, Austin, Tex., 1966-68; research assoc. dept. computer sci. Stanford (Calif.) U., 1968-69, asst. prof. linguistics and computer sci., 1969-74; assoc. prof. computer sci. and psychology Yale U., New Haven, 1974-76, prof., 1976—; dir. cognitive sci. program, 1978—; dir. Artificial Intelligence Project, since 1974—, chmn. deptl computer sci., 1980—; pres. Cognitive Systems, Inc., New Haven, 1979—; research fellow Inst. Semantics and Cognition, Castagnola, Switzerland, 1973-74; cons. Bolt, Beranek, and Newman, 1974. Author: Conceptual Information Processing, 1975, (with R. Abelson) Scripts, Plans, Goals and Understanding: An Inquiry into Human Knowledge Structures, 1977, Reading and Understanding: Teaching from an Artificial Intelligence Perspective, 1981, Dynamic Memory: A theory of Learning in Computers and People, 1983; editor: (with K. Colby) Computer Models of Thought and Language, 1973, (with C. Riesbeck) Inside Computer Understanding: Five Programs Plus Miniatures, 1981; editorial bd.: Human Learning and Memory; contbr. numerous articles to profl. jours. Mem. Cognitive Sci. Soc. (bd. govs.). Subspecialties: Artificial intelligence; Cognition. Current work: Natural language processing; artificial intelligence. Office: 10 Hillhouse Ave New Haven CT 06520

SCHANKER, LEWIS STANLEY, pharmacologist, educator; b. Kansas City, Mo., Sept. 23, 1930; s. Herman H. and Florence (Fishman) S.; m. Joyce Ann Minkin, May 27, 1953; children: Neil B.,

Lynn S. Student, Kansas City Jr. Coll., 1947-48; B.S., U. Mo.-Kansas City, 1951, M.S., 1953; Ph.D., U. Wis., Madison, 1955. Registered pharmacist, Mo. Asst. instr. pharmacology U. Mo., Kansas City, 1951-53; research fellow in pharmacology U. Wis., Madison, 1953-55; asst. scientist, sr. asst. scientist USPHS, Bethesda, Md., 1955-57; pharmacologist, head sect. cellular pharmacology, head sect. biochemistry of drug action Lab. Chem. Pharmacology, Nat. Heart Inst., NIH, Bethesda, Md., 1957-66; trustee prof. pharmacology, prof. dentistry and medicine Schs. Pharmacy, Dentistry and Medicine, U. Mo., Kansas City, 1966—; mem. pharmacology panel U.S. Bd. Civil Service Examiners, 1959-66. Mem. editorial bd.: Jour. Pharmacology and Exptl Therapeutics, 1962-65, Proc. Soc. Exptl. Biology and Medicine, 1965-68, Internat. Jour. Clin. Pharmacology and Biopharmacy, 1967—, Jour. Toxicology and Environ. Health, 1980—; contbr. articles to profl. jours., chpts to books. Recipient John J. Abel prize Am. Soc. Pharmacology and Exptl. Therapeutics, 1966; Outstanding Alumni Achievement award U. Mo., Kansas City, 1967; Alumnus of Yr. award, 1969; N.T. Veatch award for disting. research, 1972; Research Achievement award in Pharmacology Am. Pharm. Assn., 1972. Fellow AAAS; mem. Am. Soc. Pharmacology and Exptl. Therapeutics, Am. Physiol. Soc., Soc. Exptl. Biology and Medicine, Sigma Xi. Subspecialties: Pharmacology; Physiology (medicine). Current work: Permeability of body membranes to drugs and other substances: gastrointestinal tract, lungs, liver, central nervous system. Office: Pharmacy Bldg City Kansas City MO 64110

SCHARFF, MATTHEW DANIEL, immunologist, cell biologist, educator; b. N.Y.C., Aug. 28, 1932; s. Harry and Constance S.; m. Carol Held, Dec. 19, 1954; children—Karen, Thomas, David. A.B., Brown U., 1954; M.D., N.Y. U., 1959. House officer II and IV med. service Boston City Hosp., 1959-61; research asso. NIH, 1961-63; asst. prof. Albert Einstein Coll. Medicine, Yeshiva U., Bronx, N.Y., 1963-67, asso. prof., 1967-71, prof. dept. cell biology, 1971—, chmn. dept., 1972-83, dir. div. biol. scis., 1975-81; asso. dir. Cancer Center, 1975—. Served with USPHS, 1961-63. Recipient Alumni Achievement award N.Y. U. Sch. Medicine, 1980. Mem. Am. Assn. Immunologists, Am. Soc. Clin. Investigation, Nat. Acad. Scis., Phi Beta Kappa, Sigma Xi, Alpha Omega Alpha. Subspecialties: Immunobiology and immunology; Cell biology. Current work: Studies on the genetic control of antibody production and diversity in cultured cells including hybridomas. Office: Dept Cell Biology Albert Einstein Coll Medicine 1300 Morris Park Ave Bronx NY 10461

SCHARRER, BERTA VOGEL, emeritus anatomy educator; b. Munich, Germany, Dec. 1, 1906; came to U.S., 1937, naturalized, 1944; d. Karl Phillip and Johanna (Greis) Vogel; m. Ernst Albert Scharrer, Mar 1, 1934 (dec. 1965). Ph.D., U. Munich, 1930; M.D. honoris causa, U. Giessen, W.Ger., 1976; Sc.D. (hon.), Northwestern U., 1977, U. N.C., 1978, Smith Coll., 1980, Harvard U., 1982, Yeshiva U., 1983. Research assoc. Research Inst. Psychiatry, Munich, 1931-34, Neurol. Inst., Frankfurt-am-Main, Germany, 1934-37; dept. anatomy U. Chgo., 1937-38, Rockefeller Inst. Med. Research, N.Y.C., 1938-40; sr. instr., fellow dept. anatomy Western Res U., Cleve., 1940-46; Guggenheim fellow dept. anatomy U. Colo. Sch. Medicine, Denver, 1947-48, USPHS spl. research fellow, 1948-50, asst. prof. research, 1950-55; prof. anatomy Albert Einstein Coll. Medicine, N.Y.C., from 1955, now prof. anatomy and neurosci. emeritus, acting chmn., 1965-66, 74-77. Co-editor: Cell and Tissue Research, 1957; editorial bd.: Jour. Gen. Comparative Endocrinology, 1968—, Biol. Bull. Recipient Kraepelin Gold medal, 1978, Koch medal, 1980, Henry Gray medal, 1982. Fellow AAAS; mem. Nat. Acad. Scis., Am. Acad. Arts and Scis., Deutsche Akademie der Naturforscher Leopoldina, Royal Netherlands Acad. Arts and Scis. (fgn.), Internat. Soc. Neuroendocrinology, Internat. Brain Research Orgn., NRC (com. brain scis.), Endocrine Soc. Subspecialties: Neuroendocrinology; Comparative neurobiology. Current work: Neuropeptide and neuropetide receptors in invertebrates. Researcher comparative neuroendocrinology, neurosecretion, fine structure neuroglandular elements, neurotrophic factors in tumor growth. Home: 1240 Neill Ave Bronx NY 10461 Office: 1300 Morris Park Ave Bronx NY 10461

SCHAUB, JOHN ROBERT, utilities company executive; b. Madison, Wis., Jan. 29, 1947; s. Donald Robert and Frances (Sullivan) S.; m. Jean Marie Stransky, June 7, 1969; children: Jennifer Ann, Karen Elizabeth. B.S.N.E., U. Wis.-Madison, 1969; M.B.A., Wayne State U., 1973. Registered profl. engr., Mich., Calif.; cert. health physicist. Project engr. Palisades, Consumers Power Co., Jackson, Mich., 1972-74, sr. engr. modifications, 1974-77, test engr., 1977-79, staff engr., 1979-80, project mgr. nuclear modifications, 1980-82, asst. project mgr., 1982—; mem. steering com. EPRI BWR Piping, 1980—. Chmn. Historic Dist. Commn., Jackson, Mich.; gen. chmn. Bicentennial Festivals Com., Jackson, 1973-76; active Big Bros., Jackson, 1975-79; com. chmn. Cursillo, Jackson. Recipient Outstanding Service award Jackson County Bicentennial, 1976; named Big Brother of Yr. Jackson County Big Bros., 1978. Mem. Am. Nuclear Soc. (chmn. Mich. 1970-83, governance award 1979), Health Physics Soc., Mich. Soc. Profl. Engrs. (v.p. 1973-83, com. chmn. 1973-83, Young Engr. of Yr. 1978). Roman Catholic. Subspecialties: Nuclear engineering; Nuclear fission. Current work: Project management of the planning, design and construction of nuclear power plants and modifications to nuclear power plants. Office: Consumers Power Co 1945 Parnall Rd Jackson MI 49201

SCHAWLOW, ARTHUR LEONARD, educator, physicist; b. Mt. Vernon, N.Y., May 5, 1921; s. Arthur and Helen (Mason) S.; m. Aurelia Keith Townes, May 19, 1951; children—Arthur Keith, Helen Aurelia, Edith Ellen. B.A., U. Toronto, 1941, M.A., 1942, Ph.D., 1949, LL.D. (hon.), 1970, D.Sc., U. Ghent, Belgium, 1968, U. Bradford, Eng., 1970. Postdoctoral fellow, research asso. Columbia, 1949-51, vis. asso. prof., 1960; research physicist Bell Telephone Labs., 1951-61, cons., 1961-62; prof. physics Stanford, 1961—, now J.G. Jackson-C.J. Wood prof. physics, exec. head dept., 1966-70, acting chmn. dept., 1973-74. Author: (with C.H. Townes) Microwave Spectroscopy, 1955; Co-inventor, optical laser on laser, 1958. Recipient Ballantine medal Franklin Inst., 1962, Thomas Young medal and prize Inst. Physics and Phys. Soc., London, 1963; Nobel prize in physics, 1981; named Calif. Scientist of Year, 1973, Marconi Internat. fellow, 1977. Fellow Am. Acad. Arts and Scis., Am. Phys. Soc. (council 1966-70, chmn. div. electron and atomic physics 1974, pres. 1981), Optical Soc. Am. (dir.-at-large 1966-68, pres. 1975, Frederick Ives medal 1976); mem. Nat. Acad. Scis., IEEE (Liebmann prize 1964), AAAS (chmn. physics sect. 1979). Subspecialties: Atomic and molecular physics; Spectroscopy. Current work: Radio frequency, opitcal and microwave spectroscopy, lasers and quantaum electronics. Home: 849 Esplanada Way Stanford CA 94305

SCHECHTER, ALAN NEIL, medical researcher and educator; b. N.Y.C., June 28, 1939; s. Sidney S. and Mildred L. S.; m. Geraldine Poppa, Feb. 6, 1965; children: Daniele, Andrew. A.B., Cornell U., 1959; M.D., Columbia U., 1963. Med. lic. N.Y., Calif. Intern Bronx-Mcpl. Hosp. Ctr., 1963-64, resident, 1964-65; research assoc. Nat. Inst. Arthritis, Diabetes, Digestive and Kidney Diseases NIH, Bethesda, 1965-67, research physician, 1967-72, sect. chief, 1972-82, chief, 1982—; mem. grad. faculty George Washington U., 1968—, Johns Hopkins U., 1976—; dir. Found. Advanced Edn. in Scis., Inc., 1972—. Editor: Current Topics in Biochemistry, 1972, 73, Molecular and Cellular Aspects of Sickle Cell Disease, 1976, Chemical Synthesis and Sequencing of Peptides and Proteins, 1981; contbr. articles to profl. jours. Bd. dirs. Green Acres Sch., Rockville, Md., 1981-83. Served with USPHS, 1965-67,83—. Recipient USPHS Superior Service award, 1982. Mem. Am. Soc. Biol. Chemists, Biophys. Soc., Am. Soc. Clin. Investigation, Am. Soc. Hematology, Am. Soc. Human Genetics. Subspecialties: Biochemistry (medicine); Genetics and genetic engineering (medicine). Current work: Genetic diseases; biochemistry and genetics, hemoglobin and erythroid cells. Office: National Institute Health Bldg 10 Room 9N-307 Bethesda MD 20205

SCHECHTER, MARTIN DAVID, pharmacologist; b. Bklyn., Feb. 28, 1946; s. Harry F. and Frances L. (Maybloom) S.; m. Audrey Ellen Freeman, June 27, 1968; 1 son, Jason Ben. B.S., Bklyn. Coll., 1965; Ph.D., SUNY Sch. Medicine, Buffalo, 1970. Research asso. Med. Coll. Va., Richmond, 1970-72; sr. research fellow U. Melbourne, Australia, 1972-74; asst. prof. Eastern Va. Med. Sch., 1974-76, assoc. prof., 1976-78, Northeastern Ohio U. Coll. Medicine, 1978-82, prof., 1982—; coordinator clin. pshychopharmacology courses Fallsview Psychiat. Clinic, Akron (Ohio) Gen. Hosp., 1979—. Served with USAF, 1965-67. NIMH grantee, 1980—; Nat. Inst. Environ. Health Inst. grantee, 1982—. Mem. Am. Soc. Pharmacology and Exptl. Therapeutics, Soc. Exptl. Biol. Medicine, Soc. Stimulus Prop. Drugs, AAAS, Sigma Xi. Subspecialties: Psychopharmacology; Toxicology (medicine). Current work: Psychopharmacology, behavioral toxicology, drug discrimination, hyperkinesis. Office: 4209 SR 44 Rootstown OH 44272

SCHECHTER, ROBERT SAMUEL, chem. engr., educator; b. Houston, Feb. 26, 1929; s. Morris Samuel and Helen Ruth S.; m. Mary Ethel Rosenberg, Feb. 15, 1953; children—Richard Martin, Alan Lawrence, Geoffrey Louis. B.S. in Chem. Engring, Tex. A. and M. U., 1950, Ph.D., U. Minn., 1956. Registered profl. engr., Tex. Asst. prof. chem. engring. U. Tex. at Austin, 1956-60, asso. prof., 1960-63, prof., 1963—; adminstrv. dir. Center Statis. Mechs. and Thermodynamics, 1968-72, chmn. dept. chem. engring., 1970-73, chmn. petroleum engring., 1975-78, E. J. Cockrell, Jr. prof. chem. and petroleum engring., 1975-81, Dula and Ernie Cockrell chair prof. engring., 1981—; vis. prof. U. Edinburgh, Scotland, 1965-66; Disting. vis. prof. U. Kans., spring 1968; vis. prof. U. Brussels, 1969; cons. Author: Variational Method in Engineering, 1967, (with G.S.G. Beveridge) Optimization—Theory and Practice, 1970, Adventures in Fortran Programming, 1975, (with B.B. Williams and J.L. Gidley) Acidizing Monograph, 1979, (with D.D. Shah) Enhanced Oil Recovery by Surfactants and Polymers, 1979; Contbr. numerous articles to profl. jours. Served to 1st lt., Chem. Corps AUS, 1951-53. Decorated chevalier Order Palmes Academique; recipient Outstanding Teaching award U. Tex., 1969, Outstanding Paper award, 1973, Katz lectr., 1979. Mem. Am. Inst. Chem. Engrs., Am. Chem. Soc., Soc. Petroleum Engrs., Nat. Acad. Engrs., Sigma Xi, Tau Beta Pi. Subspecialties: Chemical engineering; Surface chemistry. Current work: Application of the principles of surface and colloidal chemistry to the solution of resource recovery problems, in particular energy and metals from the earth. Developer methods of measuring surface viscosity and ultra low inter-facial tensions; discoverer instability of thermal diffusion. Home: 4700 Ridge Oak Dr Austin TX 78731

SCHECKLER, STEPHEN EDWARD, botany educator, researcher, consultant; b. Irvington, N.J., Mar. 17, 1944; s. Milton Ellisworth and Alice Jenevieve (McGoldrick) S.; m. Rebecca Klein, June 2, 1968; children: Charles W., Leah A., Nathaniel D. B.S., Cornell U., 1968, M.S., 1970, Ph.D., 1973. NRC Can. postdoctoral fellow dept. botany U. Alta., Can., Edmonton, 1973-75, 1976-77, asst. prof. (vis.) botany, 1975-76, 77; asst. prof. botany Va. Poly. Inst. and State U., Blacksburg, 1977-82, assoc. prof., 1982—; geol. cons. NRC Can. Contbr. articles on botany to profl. jours. Recipient Teaching Excellence award Va. Poly. Inst. and State U., 1981, 82; research honorarium N.Y. State Mus., 1967; Nat. Geog. Soc. grantee, 1981-82; Va. Ctr. Coal and Energy Research, 1983-84; NSF grantee, 1984-86. Mem. Bot. Soc. Am. (Dimond Fund award 1981), Can. Bot. Assn., Internat. Assn. Plant Taxonomy, Internat. Orgn. Paleobotany, Internat. Assn. Pteridologists, Internat. Assn. Angiosperm Paleobotany, Soc. Evolutionary Botanists, N.Y. Acad. Scis., Sigma Xi. Democrat. Jewish. Subspecialties: Evolutionary biology; Paleobiology. Current work: Adaptive radiation of early land plants, evolution of parts and tissues, reproductive biology of progymnosperm/gymnosperm transition, ecological interactions of early plant communities. Discovered new types of extinct plants. Office: Dept Biolog Va Poly Inst and State Uni Blacksburg VA 24061 Home: 1905 Gardenspring Dr Blacksburg VA 24060

SCHECTMAN, RICHARD MILTON, physics educator; b. Wilkes-Barre, Pa., Apr. 9, 1932; s. Isadore H. and Sadie (Coplan) S.; m. Devera H. Hillman, June 26, 1960; children: David, Jay. B.S. in Engring. Physics, Lehigh U., 1954; M.S., Pa. State U., 1956; Ph.D., Cornell U., 1961. Research asst. Brookhaven Nat. Lab., 1955, Los Alamos Sci. Lab., 1957; vis. scientist Lawrence Radiation Lab., summer 1963, cons., 1964; resident assoc. Argonne (Ill.) Nat. Lab., summer 1973, vis. scientist, 1975-76; teaching asst. Pa. State U., 1954-56, Cornell U., 1956-58, research asst., 1958-61; faculty U. Toledo, Ohio, 1961—, now prof. physics and astronomy; vis. assoc. prof. U. Ariz., fall 1968, Hebrew U., Jerusalem, 1969; vis. scientist Weizmann Inst. Sci., 1982-83. Contbr. articles to profl. jours. Mem. Am. Phys. Soc., Am. Assn. Physics Tchrs., Sigma Xi, Phi Beta Kappa, Tau Beta Pi. Subspecialty: Atomic and molecular physics. Current work: Properties of ions and atoms studied by atomic spectroscopy of fast ion beams and related techniques. Office: 2801 W Bancroft St Toledo OH 43606

SCHEELINE, ALEXANDER, chemistry educator; b. Hollidaysburg, Pa., June 6, 1952; s. Isaiah and Alice (Rauh) S. B.S., Mich. State U., 1974; Ph.D., U. Wis., 1978. NRC fellow Nat. Bur. Standards, Gaithersburg, Md., 1978-79; asst. prof. dept. chemistry U. Iowa, Iowa City, 1979-81; asst. prof. sch. chem. scis. U. Ill., Urbana, 1981—; cons. Nat. Bur. Standards, 1982—, Spectral Scis., Ind., 1983—. Contbr. articles to profl. jour. Petroleum Research Fund grantee, 1980; Research Corp. grantee, 1980; NSF grantee, 1980, 82. Mem. Am. Chem. Soc., Soc. Applied Spectroscopy, Optical Soc. Am., ASTM, Phi Kappa Phi. Republican. Jewish. Subspecialties: Analytical chemistry; Spectroscopy. Current work: Emission spectrochemical analysis of solids using sparks and theta pinches; laser spectroscopy; plasma diagnostics; dynamics of non-linear systems; computerization, electronics and instrumentation. Office: U Ill Sch Chem Scis 1209 W California Ave Urbana IL 61801

SCHEFF, STEPHEN WILLIAM, neurobiologist; b. St. Louis, June 1, 1948; s. Francis Benard and Olive Merle (Newell) S.; m. Mary Suzanne Lutz, Dec. 27, 1978; children: Michelle, Melissa. B.A., Washington U., St. Louis, 1970; M.A., U. Mo., 1972, Ph.D., 1974. Research fellow Dalton Research Ctr., Columbia, Mo., 1972-74; NIH postdoctoral research fellow U. Calif.-Irvine, 1974-77, asst. research psychobiologist, lectr. dept. psychobiology, 1977-80; asst. prof. dept. anatomy/neurology U. Ky., Lexington, 1980—. Mem. Am. Assn. Anatomists, Soc. for Neuroscis. Subspecialties: Neurobiology; Regeneration. Current work: Neuroanatomical plasticity in the CNS; aging and development in the CNS; recovery of function after damage to the CNS. Office: Dept Anatomy U Ky Lexington KY 40536

SCHEFFER, ROBERT PAUL, plant sci. researcher; b. Newton, N.C., Jan. 26, 1920; s. Paul and Mary Alice (Shuford) S.; m. Beulah Jennie Spoolman, June 12, 1951; children: Thomas Jay, Mary Karen. B.S., N.C. State U., 1947, M.S., 1949; Ph.D., U. Wis., 1952. Research asst. N.C. State U., Raleigh, 1947-49; research asst. U. Wis., 1949-52; research assoc., 1952-53; asst. prof. Mich. State U., 1953-58, prof. botany and plant pathology, 1963—; vis. researcher Rockefeller U., N.Y.C., 1960-61; cons. panel regulatory biology NSF, 1965-68. Contbr. numerous articles to profl. jours., chpts. to books. Served with USAAF, 1941-45. NSF grantee, 1958—. Fellow Am. Phytopath. Soc., Am. Soc. Plant Physiologists, AAUP. Subspecialties: Plant pathology; Plant physiology (biology). Current work: Research on the physiology and biochemistry of disease development in plants and in disease resistance. Research on host-selective determinants of disease. Home: 912 Gainsborough Dr East Lansing MI 48823 Office: Dept Botany and Plant Pathology Mich State U East Lansing MI 48824

SCHEIBEL, LEONARD WILLIAM, pharmacologist, educator; b. Hays, Kans., Jan. 18, 1938; s. Raymond P. and Thelma (Bane) S.; m. Melania Parada Valdes, May 1, 1976. B.S., Creighton U., 1960, M.S. in Chemistry, 1962; D.Sc., Johns Hopkins U., 1967; M.D., U. Fla., 1973. Diplomate: Am. Bd. Preventive Medicine, Nat. Bd. Med. Examiners. Rotating intern Gorgas Hosp., Balboa Heights, C.Z., 1973-74, resident in internal medicine, 1974-77; asst. prof. Lab. Parasitology, staff physician Rockefeller U. and Hosp., N.Y.C., 1977-81; assoc. prof. dept. preventive medicine and biometrics Sch. Medicine, Uniformed Services U. Health Scis., Bethesda, Md., 1981-82, assoc. prof., 1982—; assoc. dept. pathobiology Sch. Hygiene and Pub. Health, Johns Hopkins U., Balt., 1982—; adj. assoc. prof. U. Md. Sch. Medicine, Balt., 1982; diving medicine cons. Panama Canal Co., 1973-77; cons. Med. Service Cons., Inc., Arlington, Va., 1978—; guest prof. Gorgas Meml. Lab., Panama City, Panama, 1973-77; vis. asst. prof. tropical medicine Cornell U. Med. Coll., 1977-81; mem. U.S. Army Med. Research and Devel. Adv. Com., 1982—, U.S. Army Source Selection Bd. for Testing Antileishmanial Compounds, 1982—. Mem. editorial bd.: Jour. Parasitology, 1981—; ad hoc reviewer: Molecular and Biochem. Parasitology; contbr. articles to sci. jours. Served to capt. U.S Army, 1967-70. Recipient Igor I. Sikorsky Helicopter Rescue award, 1973; Physician's Recognition award AMA, 1979—; NIH fellow, 1962-67. Fellow Am. Coll. Preventive Medicine; mem. Fedn. Am. Sos. Exptl. Biology, Am. Soc. Pharmacology and Exptl. Therapeutics (div. clin. pharmacology and drug metabolism), Am. Soc. Tropical Medicine, N.Y. Soc. Tropical Medicine, Am. Soc. Parasitologists, Isthmian Med. Assn., Microbiol. Soc. Panama, Tropical Medicine Assn. Washington, Undersea Med. Soc., Inter-Am. Assn. San. Engring. Subspecialties: Molecular pharmacology; Preventive medicine. Current work: Research and teaching in tropical medicine with emphasis on metabolism and chemotherapy of protozoal and helminthic disease; rational design of new pharmacologic agents, their testing and employment in the field, especially in areas where drug resistant parasites have emerged. Office: 4301 Jones Bridge Rd Bethesda MD 20814

SCHEID, VERNON EDWARD, mineral resources educator; b. Balt., Sept. 5, 1906; s. Charles Christian and Blanche McLenny (Donaldson) S.; m. Martha Frances Helm, Aug. 17, 1934; children: Donald Edward, Margaret Kathryn. A.B. in Geology, Johns Hopkins U., 1928, Ph.D. in Econ. Geology, 1946; M.S. in Geol. Engring, U. Idaho, 1940. Registered profl. engr., Nev.; registered geologist, Calif. Instr. dept. geology Johns Hopkins U., Balt., 1931-34; prof. dept. geology U. Idaho, Moscow, 1934-42, 47-51; geologist U.S. Geol. Survey, Idaho, Mont., Wash., 1942-47; dean Mackay Sch. Mines, Reno, Nev., 1951-72, prof., 1972—; dir. Nev. Bur. Mines & Geology, Reno, 1951-72; cons. UN, 1973—. Author reports in field. Chmn., dir. Nev. Adv. Councils Mining, Maps, Water and Energy, 1953-72; chmn., dir. Nev. Oil & Gas Conservation Commn., 1953-72; mem. Nat. Adv. Com. Oceans & Atmosphere. Mem. Soc. Mining Engrs., AIME (McConnell award 1980), Soc. Econ. Geologists (leader del. to China 1982), Geol. Soc. Am., Assn. Am. State Geologists, Assn., Geoscientists Internat. Devel. Lodge: Rotary. Subspecialty: Economic geology. Current work: Mineral resources and economics; economic and engineering geology; national mineral policy. Office: Mackay Sch Mines U Nev Reno NV 89557 Home: 33 Rancho Manor Dr Reno NV 89509

SCHEIDT, W. ROBERT, chemistry educator. B.S., U. Mo.-Columbia, 1964; Ph.D., U. Mich., 1968. Asst. prof. chemistry U. Notre Dame, Ind., 1970-76, assoc. prof., 1976-80, prof., 1980—. Mem. Am. Chem. Soc., AAAS, AAUP, Sigma Xi. Subspecialties: X-ray crystallography; Inorganic chemistry. Current work: Structure and chemical characterization of metalloporphyrin compounds and relationship to the biological chemistry of hemas. Office: Dept Chemistry U Notre Dame Notre Dame IN 46556

SCHEIN, STANLEY JAY, neurobiologist; b. Balt., Jan. 27, 1950; s. Bernard and Sylvia (Selvin) S. S.B., MIT, 1970; Ph.D., Albert Einstein Coll. Medicine, 1976, M.D., 1977. Postdoctoral fellow MIT, Cambridge, 1977-78; postdoctoral fellow Nat. Eye Inst., Bethesda, Md., 1978-80, scientist, 1980—. Sloan Found. fellow, 1977-81; recipient NIH Nat. Research Service award, 1977-80. Mem. Soc. Neurosci., Assn. Research in Vision and Ophthalmology, AAAS, Am. Soc. Neurochemistry. Subspecialties: Neurobiology; Neurochemistry. Current work: Anatomy, physiology and functional architecture of primate vision at the level of the visual cortex and the retina. Office: Nat Eye Inst Bldg 9 Bethesda MD 20205

SCHELBERT, HEINRICH RUEDIGER, radiological sciences educator; b. Wuerzburg, Germany, Nov. 5, 1939; came to U.S., 1966; s. Heinrich Johannes and Hedwig (Fahnemann) S.; m. Barbara Wilde, Nov. 28, 1969; children: Kristina, Mark. M.D., U. Wuerzburg, 1964, Ph.D., 1965. Lic. physician, W.Ger., Calif. Intern Mercy Med. Ctr., Darby, Pa., 1966-67, resident in medicine, 1967-68, 70-71; resident in cardiology U. Duesseldorf, W. Ger., 1971-72; resident in physiology U. Calif., San Diego, 1968-69; assoc. prof. radiol. sci. UCLA Sch. Medicine, 1977-80, prof., 1980—; mem. VA merit rev. bd., Washington, 1982—, NIH cardiovascular study sect., Bethesda, Md., 1983—. Recipient Georg Von Hevesy Found. prize, 1978, 82. Fellow Am. Coll. Cardiology, Am. Heart Assn. (council circulation.); mem. Soc. Nuclear Medicine (silver medal 1977), Los Angeles Heart Assn. (vice-chmn. research com.). Subspecialties: Nuclear medicine; Cardiology. Current work: Noninvasive study of myocardial metabolism in health and disease with positron emission computed tomgraphy and metabolic tracers. Home: 521 Via De La Paz Pacific Palisades CA 90272 Office: Div Nuclear Medicine UCLA Sch Medicine Los Angeles CA 90024

SCHELLMAN, JOHN ANTHONY, chemistry educator; b. Phila., Oct. 24, 1924; s. John A. and Margaret Mary (Mason) S.; m. F. Charlotte, Mar. 16, 1922; children: Heidi, Lise. Ph.D., Princeton U., 1951. NIH fellow U. Utah, 1951-53, Garlsbergslab, Denmark, 1953-55; DuPont fellow U. Minn., Mpls., 1955, asst. prof. chemistry, 1956-58; prof. chemistry U. Oreg., Eugene, 1958—. Served with U.S. Army. Mem. Am. Chem. Soc., Am. Phys. Soc., Biophys. Soc., Am. Soc. Biol. Chemistry, Nat. Acad. Scis., Sigma Xi. Subspecialty: Physical chemistry. Current work: Biophysical chemistry. Office: Chemistry Dept U Oreg Eugene OR 97403

SCHENCK, HANS UWE, Chemical company executive; b. Munich, Germany, Aug. 24, 1942; s. Gerhard and Anna A. (Joeckel) S.; m. Gabriele Damm, May 5, 1967; children: Sandra B., Corinna E. M.S., Tech. U. Stuttgart, W. Ger., 1967; B.S., Free U. Berlin, 1964, Ph.D., 1969. Research asst. Free U. Berlin, 1969-70; mem. research staff BASF AG, Ludwigshafen, W. Ger., 1970-75, group leader polymeric auxiliaries research and devel., 1975-77, group leader research and devel., 1977-81; v.p. research and devel., BASF Wyandotte Corp., Parsippanny, N.J., 1981—. Contbr. articles to sci. publs. Mem. Am. Chem. Soc., Indsl. Research Inst., Soc. Chem. Industry, Comml. Devel. Assn., Sigma Xi. Subspecialties: Polymer chemistry; Organic chemistry. Current work: Urethane chemistry, functional fluids, nutritional products, nonionic surfactants, oil production chemicals. Holder 18 U.S. patents. Home: 21339 Masi Ct Grosse Ile MI 48138 Office: 1609 Biddle Ave Wyandotte MI 48192 Office: BASF Wyandotte Corp 100 Cherry Hill Rd Parsippanny NJ 07054

SCHENCK, ROBERT ROY, surgeon; b. Brimfield, Ill., Sept. 19, 1931; s. Isaac Barrett and Pearl Irene (Murnan) S.; m. Ruth Mary Helm, June 18, 1955 (div. Jan. 1977); children: Claudia, Lynn, Karen, Heidi, Robert Paul; m. Nanci Whitney, June 13, 1982. B.A., Taylor U., 1951; M.D., U. Ill.-Chgo., 1955. Intern Akron (Ohio) Gen. Hosp., 1955-57; med. missionary Ethiopia Bapt. Gen. Conf., Ambo, 1959-61; staff physician Centerville Clinic, Fredericktown, Pa., 1962-67; resident gen. surgery Western Pa. Hosp., Pitts., 1967-69; resident plastic surgery Columbia-Presbyn. Med. Center, N.Y.C., 1969-71; fellow hand surgery Roosevelt Hosp., N.Y.C., 1971-72; assoc. prof. depts. plastic and orthopedic surgery, dir. sect. hand surgery Rush-Presbyn.-St. Luke's Med. Center, Chgo., 1972—. Guest editor: Replantation and Reconstruction Microsurgery, 1977. Served with USPHS, 1957-59. Mem. Am. Assn. Plastic Surgeons, Am. Soc. Surgery for the Hand, Am. Soc. Plastic and Reconstructive Surgeons. Republican. Club: Union League (Chgo.). Subspecialties: Surgery; Microsurgery. Current work: Microsurgery, basic microvascular amastomotic techniques, replantation, clinical results. Office: Robert A Schenck & Assocs 1725 W Harrison St Suite 398 Chicago IL 60612

SCHENKMAN, JOHN BORIS, pharmacology educator; b. N.Y.C., Feb. 10, 1936; s. Abraham and Theresa Moses S.; m. Deanna Owen, June 5, 1960; children: Jeffrey Alan, Laura Ruth. B.A., Bklyn. Coll., 1960; Ph.D., SUNY-Upstate Med. Ctr.-Syracuse, 1964. Postdoctoral fellow Johnson Found., U. Pa., Phila., 1964-67; NSF vis. scientist Osaka U., Japan, 1967-68; research assoc. U. Tubingen, Germany, 1968; asst. prof. pharmacology Yale U. Sch. Medicine, 1968-71, assoc. prof., 1971-78; prof., head dept. pharmacology U. Conn. Health Ctr., Farmington, 1978—. Contbr. numerous articles to profl. jours., chpts. to books. Served with AUS, 1953-55. NSF grantee, 1967-68; recipient NIH Research Career Devel. award, 1971-76, grantee, 1968—. Mem. Am. Soc. Biol. Chemists, Am. Soc. Pharmacology and Exptl. Therapeutics, Brit. Biochem. Soc., German Pharmacol. Soc. Subspecialties: Biochemistry (medicine); Molecular pharmacology. Current work: Research on mechanism of action of hepatic cytochrome p-450 monooxygenase; research on active oxygen in lipid peroxidation. Office: Dept Pharmacology U Conn Health Center Farmington CT 06032

SCHENTAG, JEROME J., pharmacologist, educator; b. St. Clair, Mich., Jan. 25, 1950; s. John and Rose S.; m. Rita Sloan, June 26, 1976. B.S. in Pharmacy, U. Nebr., 1973; Pharm.D., Phila. Coll. Pharmacy and Scis., 1975. Postdoctoral fellow SUNY, Buffalo, 1975-76, asst. prof., 1976-79, assoc. prof. pharmaceutics, 1980—; dir. clin. pharmacokinetics lab. Millard Fillmore Hosp., Buffalo, 1981—. Mem. AAAS, N.Y. Acad. Scis., Am. Soc. Microbiology, Am. Soc. Clin. Pharmacology and Therapeutics, Am. Soc. Pharmacology and Exptl. Therapeutics, Am. Coll. Clin. Pharmacy, Am. Pharm. Assn., Am. Soc. Hosp. Pharmacists. Subspecialties: Pharmacokinetics; Pharmacology. Current work: Clinical pharmacokinetics; acute care medications. Office: Clin Pharmacokinetics Lab Millard Fillmore Hosp Buffalo NY 14209

SCHEPARTZ, SAUL ALEXANDER, biochemist; b. Nutley, N.J., May 18, 1929; s. Harry Hugo and Sadye (Rosenberg) S.; m. Marlyn Joan Rubenstein, June 17, 1956; children: Helen Leslie, Frederick Marc, Esther Diane (dec.), Ruth Adrianne. A.B. in Chemistry, Ind. U., 1951; M.S. in Biochemistry, U. Wis., 1953, Ph.D., 1955. Research asst. U. Wis., Madison, 1951-52, 54-55, USPHS fellow, 1952-54; research assoc. Wistar Inst., U. Pa., Phila., 1955-57; head dept. biochemistry Microbiol. Assocs. Inc., Bethesda, Md., 1957-58; with Nat. Cancer Inst., NIH, Bethesda, 1958—, assoc. dir. for drug research and devel. program, div. cancer treatment, 1972 74, acting dep. dir. div. cancer treatment, 1976-78, acting dir. div. cancer treatment, 1980—. Recipient Superior Service award HEW, 1972. Mem. Am. Chem. Soc., Am. Soc. for Microbiology, Am. Assn. for Cancer Research, AAAS, N.Y. Acad. Scis., Alpha Chi Sigma. Subspecialty: Cancer research (medicine). Patentee cell culture process. Home: 8603 Grimsby Ct Potomac MD 20854 Office: Nat Cancer Inst Bethesda MD 20205

SCHER, CHARLES DAVID, physician, researcher; b. Newark, July 25, 1939; s. Sidney Milton and Anne T. (Levin) S.; m. Reda Ellen Walsh, Dec. 27, 1964; children: Deborah Anne, Matthew Isaac. B.A., Brandeis U., 1961; M.D., U. Pa., 1965. Diplomate: Am. Bd. Pediatrics. Intern Bronx Mcpl. Hosp., 1965-66, resident, 1966-67, Children's Hosp. Med. Ctr., Boston, 1971-72; asst. prof. pediatrics Harvard U., 1974-78, assoc. prof., 1978-82; prof. U. Pa., 1982—. Contbr. articles in field to profl. jours. Served with USPHS, 1967-71. Life Ins. Med. Research Fund fellow, 1963-65; NIH fellow, 1972-74. Mem. Am. Soc. Microbiology, Am. Assn. Cancer Research, Am. Soc. Cell Biology, Am. Soc. Clin. Investigation. Jewish. Subspecialties: Cell biology (medicine); Cell study oncology. Current work: Cellular growth control. Office: Childrens Hosp Philadelphia 34th St and Civic Center Blvd Philadelphia PA 19104

SCHER, ROBERT SANDER, mech. engr.; b. Cin., May 24, 1934; s. Stanford Samuel and Eva (Ordan) S.; m. Audrey Erna Gordon, Oct. 21, 1961; children: Sarah E., Alexander B., Aaron Z. S.B., M.I.T., 1956, S.M., 1958, M.E., 1960, Sc.D., 1963. Research and teaching asst. M.I.T., 1957-62; control systems engr. RCA Astro Electronics Div., Hightstown, N.J., 1963-65; engring. mgr. Sequential Info. Systems, Elmsford, N.Y., 1965-71; tech. dir. Teledyne Gurley, Troy, N.Y., 1971—77, v.p. engring., 1977—. Bd. dirs. Lake George Opera Co., 1976-78. Mem. Optical Soc. Am. (pres. Hudson Mohawk chpt. 1977-78, sec. 1982—), ASME, Sigma Xi, Tau Beta Pi, Pi Tau Sigma. Jewish. Subspecialty: Mechanical engineering. Current work: Design of high accuracy optical encoding systems for industrial and aerospace application. Patentee in field. Home: 2 Laurel Oak Ln Clifton Park NY 12065 Office: 514 Fulton St Troy NY 12181

SCHER, WILLIAM, biochemist; b. Cleve., Feb. 23, 1933; s. William and Janet (Lewis) S.; m. Barbara Ann Messina, June 13, 1958; children: Gina Suzanne, William Arthur. B.S. in Microbiology, Yale U., 1955, M.S., 1957; M.D., U. Va., 1961. Bacteriology lab. asst. Yale U., 1956-57; intern in medecine, then resident Kings County Hosp., N.Y.C., 1961-63; guest investigator Rockefeller U., N.Y.C., 1963-65; research assoc. Population Council, N.Y.C., 1965-66; lt. comdr. USPHS, 1966-68; asst. prof. cell biology CUNY, 1968—; asst. prof. Ctr. Exptl. Cell Biology, Mt. Sinai Sch. Medecine, N.Y.C., 1968-78, asst. prof. medecine, div. med. oncology, 1978—. Author numerous papers, reports in field. Fellow NIH, 1963-65; grantee NIH, 1975-77, 78-81, Am. Cancer Soc., 1978; recipient Irving Albert Cancer Research award, 1981. Mem. AAAS., Am. Assn. Cancer Research, Am. Soc. Cell Biology, Am. Soc. Microbiology, Cell Kinetics Soc., Harvey Soc., N.Y. Acad. Sci., Soc. Exptl. Biology and Medicine, Soc. Study Blood, Tissue Culture Assn., Sigma Xi. Subspecialties: Biochemistry (biology); Cell biology. Current work: Control of gene expression, molecular control of cellular, erthroid differentiation, hemoglobin expression, relationship of differentiation to neoplasia.

SCHERAGA, HAROLD ABRAHAM, physical chemistry educator; b. Bklyn., Oct. 18, 1921; s. Samuel and Etta (Goldberg) S.; m. Miriam Kurnow, June 20, 1943; children: Judith Anne, Deborah Ruth, Daniel Michael. B.S., CCNY, 1941; A.M., Duke U., 1942, Ph.D., 1946, Sc.D. (hon.), 1961. Teaching, research asst. Duke U., 1941-46; fellow Harvard Med. Sch., 1946-47; instr. chemistry Cornell U., 1947-50, asst. prof., 1950-53, assoc. prof., 1953-58, prof., 1958—, Todd prof. chemistry, 1965—, chmn. dept., 1960-67; vis. assoc. biochemist Brookhaven Nat. Lab., summers 1950, 51, cons. biology dept., 1950-56; vis. lectr. div. protein chemistry Wool Research Labs., Melbourne, Australia, 1959; vis. prof. Soc. for Promotion Sci., Japan, Aug. 1977; mem. tech. adv. panel Xerox Corp., 1969-71, 74-79; Mem. biochemistry tng. com. NIH, 1963-65; mem. research career award com. NIGMS, 1967-71; commn. molecular biophysics Internat. Union for Pure and Applied Biophysics, 1965-69, mem. commn. macromolecular biophysics, 1969-75, pres., 1972-75, mem. commn. subcellular and macromolecular biophysics, 1975-81; adv. panel molecular biology NSF, 1960-62; Welch Found. lectr., 1962, Harvey lectr., 1968, Gallagher lectr., 1968, Lemieux lectr., 1973, Hill lectr., 1976, Venable lectr., 1981; co-chmn. Gordon Conf. on Proteins, 1963; mem. council Gordon Research Confs., 1969-71. Author: Protein Structure; Theory of Helix-Coil Transitions in Biopolymers; Co-editor: Molecular Biology, 1961—; Mem. editorial bd.: Physiol. Chemistry and Physics, 1969-75, Mechanochemistry and Motility, 1970-71, Thrombosis Research, 1972-76, Biophys. Jour, 1973-75, Macromolecules, 1973—, Computers and Chemistry, 1974—, Internat. Jour. Peptide and Protein Research, 1978—, Jour. Computational Chemistry, 1980—, Jour. Protein Chemistry, 1982—; corr.: PAABS Revista, 1971-73; editorial adv. bd.: Biopolymers, 1963—, Biochemistry, 1969-74. Mem. Ithaca Bd. Edn., 1958-59; Bd. govs. Weizmann Inst., Israel, 1970—; Mem. staff Naval Research Lab. Project, Air Force OSRD Project, World War II. Fulbright, Guggenheim fellow Carlsberg Lab., Denmark, 1956-57, Weizmann Inst., Israel, 1963; NIH Spl. fellow Weizmann Inst., 1970; Fogarty fellow NIH, 1984; recipient Townsend Harris medal CCNY, 1970, Chemistry Alumni Sci. Achievement award, 1977. Fellow AAAS; mem. Nat. Acad. Scis., Am. Chem. Soc. (chmn. Cornell sect. 1955-56, mem. exec. com. div. biol. chemistry 1966-69, vice chmn. div. biol. chemistry 1970, chmn. div. biol. chemistry 1971, Eli Lilly award 1957, Nichols medal 1974, Kendall award 1978, Linderstrøm-Lang medal 1983), Am. Soc. Biol. Chemists, Biophys. Soc. (council 1967-70), Am. Acad. Arts and Scis., Phi Beta Kappa, Sigma Xi, Phi Lambda Upsilon. Subspecialties: Biophysical chemistry; Physical chemistry. Current work: Physical chemistry of proteins and other macromolecules. Chemistry of blood clotting. Structure of water and dilute aqueous solutions. Home: 212 Homestead Terr Ithaca NY 14850

SCHERBERG, NEAL HARVEY, biochemist; b. Mpls., Nov. 10, 1939; s. Max and Goldie (Steinberg) S.; m. Sara Ann Greenwald, Sept. 28, 1975; children: Jacob, Hannah. A.B., Oberlin Coll., 1961; Ph.D., Tufts U., 1966. Postdoctoral fellow U. Chgo., 1966-71, research assoc., assoc. prof., 1971-83, dir. thyroid Lab., 1980—. Recipient Research Career Devel. award NIH, 1975. Mem. Am. Soc. Biol. Chemists. Jewish. Subspecialties: Biochemistry (biology); Endocrinology. Current work: Description of new hormones. Patentee in field. Office: U Chgo 950 E 59th St Chicago IL 60637

SCHEUERMANN, PETER I., computer scientist; b. Bucharest, Romania, Aug. 28, 1945; s. Salo and Gerda (Metsch) S.; m. Mona M. Scheuermann, Dec. 28, 1973. B.S. in Applied Math, Tel-Aviv U., 1969; Ph.D. in Computer Sci, SUNY-Stony Brook, 1976. Asst. prof. Coll. William and Mary, Williamsburg, Va., 1975-76; asst. prof. Northwestern U., Evanston, Ill., 1976-81, assoc. prof., 1981—; vis. prof. Free U., Amsterdam, 1983. Editor: (with M. Ouksel) Information Systems, Vol. 7, 1982, Improving Database Usability and Responsiveness, 1983, (with C.R. Carlson) Self-Assessment Procedure: Data Base Systems; assoc. editor: Internat. Jour. Policy Analysis and Info. Systems. Mem. Assn. Computing Machinery, IEEE, Simulation Council. Subspecialties: Database systems; Distributed systems and networks. Current work: Dynamic structures; distributed database systems, in particular, database machines and query processing. Office: Dept Elec Engring and Computer Sci Northwestern U Evanston IL 60201

SCHEY, JOHN ANTHONY, mechanical engineering educator, researcher, consultant; b. Sopron, Hungary, Dec. 19, 1922; came to U.S., 1962; s. Mihaly and Terez Hedvig (Topfel) S.; m. Margit Maria Sule, Sept. 25, 1948; 1 son, John Francis. Dip. Metall. Ing., Jozsef Nador Tech. U., Sopron, Hungary, 1946; Cand. Tech. Sci. in Metallurgy, Acad. Scis., Budapest, Hungary, 1953. Supt. metal works Steel and Metal Works, Csepel, Budapest, 1947-51; reader metals tech. Tech. U. Miskolc, Hungary, 1951-56; head dept. fabrication Research Labs., Brit. Aluminum Co., Chalfont Park, Eng., 1957-62; sr. metall. adv. Ill. Inst. Tech. Research Inst., Chgo., 1962-68; prof. metall. engring. U. Ill.-Chgo., 1968-74; prof. mech. engring. U. Waterloo, Ont., Can., 1974—; cons. in field; course dir. Forging Industry Assn. Author: Introduction to Manufacturing Processes, 1977, Tribology in Metalworking: Friction, Lubrication, and Wear, 1983; contbr. numerous articles to profl. publs. Recipient W.H.A. Robertson award Inst. Metals, London, 1966. Mem. Nat. Acad. Engring., Am. Soc. Metals, Soc. Mfg. Engrs. (Gold medal 1974), Sigma Xi. Subspecialties: Materials processing; Solid mechanics. Current work: Development of manufacturing processes; tribology of metalworking (friction, lubrication, and wear); interactions between material properties and process conditions in metalworking; social impact of technological advance. Patentee in field. Office: U Waterloo Waterloo On Canada N2L 3G1

SCHIEFER, HANS BRUNO, veterinarian, educator; b. Cologne, Germany, Aug. 25, 1929; s. Wilhelm and Therese (Meyer) S.; m. Elisabeth Hesse, Sept. 5, 1957; children: Bernhard, Barbara, Monica, Michael. D.V.M., U. Munich, 1956, M.S., 1958, Ph.D., 1965. Lic. veterinarian, Saskatchewan. Pvt. practice vet. medicine, Munich, Germany, 1956-58; instr., asst. prof., assoc. prof. dept. vet. medicine Pathology U. Munich, 1958-59; assoc. prof. to prof. dept. vet. pathology U. Saskatchewan, Saskatoon, 1969—, head dept. vet. pathology, 1974-77; vis. prof. U. Conn., 1966-67. Author numerous articles for profl. publs. Fulbright scholar, 1966-67. Mem. Can. Vet. Med. Assn., Can. Assn. Vet. Pathologists, Soc. Toxicology Can. Club: Rotary (Saskatoon-Nutana). Subspecialties: Toxicology (agriculture); Pathology (veterinary medicine). Current work: Mycotoxins, particularly trichothecene mycotoxins; toxicology problems related to agricultural chemicals. Office: Toxicology Research Centre University of Saskatchewan Saskatoon SK Canada S7N 0W0

SCHIFF, JEROME ARNOLD, biologist, educator; b. Bklyn., Feb. 20, 1931; s. Charles K. and Molly (Weinberg) S. B.A. in Biology and Chemistry, Bklyn. Coll., (summer scholar invertebrate zoology Woods Hole Marine Biol. Lab.), 1952; Ph.D. in Botany and Biochemistry, U. Pa., 1956. Predoctoral fellow USPHS, 1954-56; fellow Brookhaven Nat. Labs., summer 1956; research asso. biology Brandeis U., Waltham, Mass., 1956-57, mem. faculty, 1957—, prof. biology, 1966—, chmn. dept., 1972-75, Abraham and Etta Goodman chair biology, 1974—; dir. Inst. Photobiology of Cells and Organelles, 1975—; summer instr. Exptl. Marine Botany, Woods Hole, Mass., 1971, sr. investigator, 1972-74, dir. programs, 1974-79; Cons. developmental biology NSF, 1965-68; vis. prof. Tel Aviv U., 1972, Hebrew U., 1972, Weizmann Inst., Israel, 1977; Mem. biology grant rev. program U.S.-Israel Binat. Sci. Found., 1974—. Editorial bd.: Developmental Biology, 1971-74; editorial bd.: Plant Sci. Letters, 1972-78; asso. editor, 1978-81; chief co-editor, 1981—; asst. editor: Plant Physiology, 1964-69; adv. editor, 1969-79; mem. editorial com.: Ann. Rev. Plant Physiology, 1974-80. Carnegie Instn. fellow in plant biology, 1962-63; Recipient Disting. Alumni award Bklyn. Coll., 1972. Fellow Am. Acad. Arts and Scis., AAAS; mem. Soc. Developmental Biology (sec. 1964-66, exec. com.), Am. Soc. Biol. Chemists, Am. Soc. Plant Physiologists (exec. com. 1972—), Biophys. Soc., Can. Soc. Plant Physiologists, Internat. Phycol. Soc., Phycol. Soc. Am., Soc. Cell Biology, Soc. Gen. Microbiology, Soc. Protozoologists, Am. Soc. Microbiology, Internat. Soc. Developmental Biologists, Brit. Phycol. Soc., Am. Soc. Photobiology, Bot. Soc. Am., Sigma Xi (mem. corp. Marine Biol. Lab. 1972—). Subspecialties: Plant physiology (biology); Cell biology. Current work: Research in chloroplast development, replication and function, chiefly in Euglena; sulfate metabolism; metabolism of algae and higher plants. Research in metabolism of protista (algae), sulfate reduction by plants and microorganisms, pathway of biosynthesis of chlorophylls, carotenoids and anthocyanins, chloroplast devel., replication and function particularly in Euglena. Home: 37 Harland Rd Waltham MA 02154 Office: Inst for Photobiology of Cells and Organelles Brandeis Univ Waltham MA 02254

SCHIFFER, LEWIS MARTIN, physician, reseacher; b. N.Y.C., July 10, 1930; s. Milton and Minna S.; m. Lenore B. Herman, Dec. 11, 1955; children: Lawrence Carl, Carol Ann. B.A., N.Y.U., 1951; M.D., SUNY, Syracuse, 1955. Intern Grace New Haven Hosp., 1955-56; resident SUNY, 1958-61; scientist Brookhaven Nat. Lab., 1961-68; dir. cancer research labs. Allegheny Gen. Hosp., Pitts., 1968-82; chmn. dept. exptl. therapeutics AMC Cancer Research Center and Hosp., Lakewood, Colo., 1982—. Contbr. numerous articles to profl. jours. Served to capt., M.C. USAF, 1956-58. Mem. AAAS, Am. Fedn. Clin. Research, Am. Assn. Cancer Research, Am. Soc. Clin. Oncology, Cell Kinetics Soc., Am. Soc. Hematology. Subspecialties: Cancer research (medicine); Cell study oncology. Current work: Cell kinetics of tumors in animals and man. Office: 6401 W Colfax Ave Lakewood CO 80214

SCHIFFMAN, SUSAN STOLTE, psychologist, educator, cons., researcher; b. Chgo., Aug. 24, 1940; d. Paul Richard and Mildred Elizabeth (Glicksman) Stolte; m.; 1 dau., Amy. B.A. cum laude, Syracuse U., 1965; Ph.D., Duke U., 1970. Lic. practicing psychologist, N.C. Postdoctoral fellow Duke U. Med. Center, Durham, N.C., 1970-72, asst. prof. dept. psychiatry, 1972-78, assoc. prof., 1978-83, prof., 1983—; cons. in field. Contbr. numerous articles to sci. jours. Mem. Soc. for Neurosci., Am. Chem. Soc. Subspecialties: Neuropsychology; Psychobiology. Current work: Psysicochemical properties responsible for taste and smell; obesity aging. Office: Duke U Dept Psychology Durham NC 27706

SCHIFFMANN, ROBERT FRANZ, food processing consultant, researcher, food processing company executive; b. N.Y.C., Feb. 11, 1935; s. Franz and Sophie (Bohling) S.; m.; children: Carla Helen, Erica Margaret, Robert Franz. B.S., Columbia U., 1955; M.S., Purdue U., 1959. Research scientist and project mgr. DCA Food Industries, N.Y.C., 1959-62, 64-71; v.p. research Nucleonics Corp. Am., N.Y.C., 1963-64; ptnr. Bedrosian and Assocs., Alpine, N.J., 1971-78; v.p. tech. Natural Pak Systems, Alpine, 1976—; pres. R. F. Schiffmann Assocs., N.Y.C., 1979—; v.p. tech. Ranum Food Systems, N.Y.C., 1982; mem. faculty Center for Profl. Advancement. Contbr. numerous articles to profl. jours., chpt. to textbook. Recipient Putman award Putman Pub. Co., 1972. Mem. Internat. Microwave Power Inst. (pres. 1973-81, chmn. 1981-83), Soc. Plastics Industries, Inst. Food Technologists, Sigma Xi, Phi Lambda Upsilon, Rho Chi. Subspecialty: Food science and technology. Current work: Microwave heating applications; new product and process development; research and development in microwave heating applications for domestic, institutional and industrial use; development and research in new products and processes, primarily for food industry. Patentee, primarily in area of microwave heating, preservation of fruits and vegetables, and food processing. Home and Office: 149 W 88th St New York NY 10024

SCHIFTER, CATHERINE CRUTCHFIELD, dental hygiene educator, researcher; b. Abilene, Tex., Sept. 24, 1950; d. James Willard and Josephine (Palmer) Crutchfield; m. Stephan Clay Schifter, Aug. 17, 1974. B.S., Baylor Coll. Dentistry, 1972; M.Ed., U. Houston, 1973; post-grad., U. Pa., 1980—. Cert. dental hygienist. Research asst. Baylor Coll. Medicine, Houston, 1972-73; asso. in dental hygiene U. Pa., Phila., 1973-74; dental hygienist, West Chester, Pa., 1974-75, dental hygienist in periodontal practice, Phila., 1974-78; asst. prof. U. Pa., Phila., 1978—; speaker before profl. groups, 1981—. Mem. editorial bd.: R.D.H, 1981—; contbr. articles to dental jours. Recipient Acad. Effort award Pa. Dental Hygiene Assn., 1981, Earl Banks Hoyt award U. Pa., 1982. Mem. Am. Dental Hygiene Assn., Pa. Dental Hygiene Assn., Am. Assn. Dental Schs., Am. Assn. Dental Research, Internat. Assn. Dental Research, Phila. Orch. Assn., Alpha Phi Internat., Sigma Phi Alpha. Methodist. Current work: Use of dark field microscopy in monitoring patient disease/health. Use of dark field microscopy in evaluating effectiveness of antimicrobials. Inventor (with others) method of treating chronic inflammatory periodontal disease. Home: 1420 Locust St #32K Philadelphia PA 19102 Office: U Pa Sch Dental Medicine 4001 Spruce St Philadelphia PA 19104

SCHILD, RUDOLPH ERNEST, astronomer; b. Chgo., Jan. 10, 1940; s. Kasimir and Anneliese (Schuricht) S.; m. Jane Struss, July 31, 1982. B.S., U. Chgo., 1962, M.S., 1963, Ph.D., 1966. Postdoctoral fellow Calif. Inst. Tech., Pasadena, 1967-69; astronomer Harvard-Smithsonian Ctr. for Astrophysics, Cambridge, Mass., 1969—; Harlow Shapley vis. lectr. Am. Astron. Soc. Contbr. articles to profl. joursl. Smithsonian Research Found. grantee, 1976-77; NASA grantee, 1982—; Smithsonian scholar, 1982—. Mem. Am. Astron. Soc., Internat. Astron. Union. Democrat. Lutheran. Club: Antique Auto. Subspecialties: Optical astronomy; Cosmology. Current work: Application of CCD Camera electronic and computer technology to optical and near-infrared imaging of highly redshifted galaxies and QSO's. Patentee in field. Office: 60 Garden St Cambridge MA 02138

SCHILE, RICHARD DOUGLAS, mech. engr., cons.; b. New Haven, Apr. 3, 1931; s. George Emerson and Bernadette (Laurie) S.; m. Christine Ann Karajanis, Aug. 30, 1953; children: Theresa, Carol, Linda. B.Aero. Engring., Rensselaer Poly. Inst., 1953, M.S. in Mechanics, 1957, Ph.D., 1967. Sr. reserch scientist United Technologies Research Lab., East Hartford, Conn., 1957-69; assoc. prof. engring. Dartmouth Coll., Hanover, N.H., 1969-76; assoc. dir. research Ciba Geigy Corp., Ardsley, N.Y., 1976-82; ind. cons., 1982—. Contbr. articles to profl. jours. Served to lt. (j.g.) USNR, 1953-55.

Mem. ASME, Am. Acad. Mechanics. Subspecialties: Composite materials; Solid mechanics. Current work: Development of fiber-reinforced composite materials and structures including design, materials development and process development; stress analysis, and design of mechanical structures and systems; development of high performance materials for mechanical systems. Patentee in field (5). Home: 22 Bloomer Rd Ridgefield CT 06877

SCHILLING, EDWARD E., botany educator; b. Los Angeles, Sept. 23, 1953; s. Edward E. and Anne E. (Novicky) S. B.S., Mich. State U., 1974; Ph.D., Ind. U., 1978. Vis. instr. U. Tex., Austin, 1978-79; asst. prof. botany U. Tenn., Knoxville, 1979—. Contbr. sci. articles to profl. publs. NSF fellow, 1975; Am. Philos. Soc. grantee, 1980. Mem. Am. Soc. Plant Taxonomists, Torrey Bot. Club, Internat. Assn. Plant Taxonomy, AAAS. Subspecialties: Taxonomy; Evolutionary biology. Current work: Biology of sunflowers and black nightshades; taxonomy, systematics and phytochemistry. Office: Univ Tenn Botany Dept Knoxville TN 37996-1100

SCHILLING, JOHN ALBERT, surgeon; b. Kansas City, Mo., Nov. 5, 1917; s. Carl Fielding and Lottie Lee (Henderson) S.; m. Lucy West, June 8, 1957 (dec.); children: Christine Henderson, Katharine Ann, Jolyon David, John Jay; m. Helen R. Spelbrink, May 28, 1979. A.B. with honors, Dartmouth Coll., 1937; M.D., Harvard U., 1941. Diplomate: Am. Bd. Surgery (chmn. 1969). Intern, then resident in surgery Roosevelt Hosp., N.Y.C., 1941-44; mem. faculty U. Rochester (N.Y.) Med. Sch., 1945-53, asst. prof. surgery, 1955-56; prof. surgery, head dept. U. Okla. Med. Sch., 1956-74; prof. surgery U. Wash. Med. Sch., Seattle, 1974—, chmn. dept., 1975-83; mem. bd. sci. counselors Nat. Cancer Inst.; also mem. diagnosis subcom. breast cancer task force; chmn. adv. com. to surgeon gen. on metabolism of trauma Army Med. Research and Devel. Command; mem. surgery study sect., div. research grants NIH; chief surgery USAF Sch. Aviation Medicine; cons. Surgeon Gen. USAF, 1959-75. Author articles, chpts. in books, abstracts, reports.; Editorial bd.: Am. Jour. Surgery. Served to maj. M.C. USAF, 1953-55. Grantee Army Office Surgeon Gen., 1956—. Mem. A.C.S. (bd. govs., chmn. com. surg. edn. in med. schs.), Am., So., Western, Pan-Pacific, N. Pacific, Pacific Coast surg. assns., Soc. Univ. Surgeons, Am. Assn. Surgery Trauma, Surg. Biology Club, Am. Physiol. Soc., Soc. Surg. Chmn., Am. Trauma Soc., Seattle Surg. Soc., Soc. Exptl. Pathology, Soc. Surgery Alimentary Tract, Explorers Club, Alpha Omega Alpha. Clubs: Yacht, University (Seattle). Subspecialties: Surgery; Cell biology (medicine). Current work: Medical education; surgical research (wounds); clinical surgery (general, gastrointestinal). Home: 9807 Lake Washington Blvd NE Bellevue WA 98004 Office: Dept Surgery (RF-25 Univ Wash Medical Sch). Seattle WA 98195

SCHILLING, ROBERT FREDERICK, medical educator; b. Adell, Wis., Jan. 19; s. Edgar Frederick and Lillian Marie (Bollard) S.; m. Mariam Alan Hansen, Feb. 2, 1946; m. Marilyn Jean Carbon, July 14, 1973; children: Carla, Robert, Fredricka, Richard, & Anne. B.S., U. Wis., 1940, M.D., 1943. Diplomate: Am. Bd. Internal Medicine. Resident in medicine U. Wis.-Madison, 1946-48; research fellow in medicine Harvard U. Med. Sch., Boston, 1948-50; from asst. prof. to prof. medicine U. Wis.-Madison Med. Sch., 1951-83, Washburn prof., 1983—, also chmn. dept. Served with M.C. USMC, 1944-46. Mem. Assn. Am. Physicians, Am. Soc. Clin. Investigation, Central Soc. Clin. Research, Am. Soc. Hematology. Unitarian. Subspecialty: Hematology. Current work: Studies of the pathophysiology of hematologic disorders. Office: 600 Highland Ave H4-536 Madison WI 53792

SCHILMOELLER, NEIL HERMAN, engineering company executive; b. Granville, Iowa, July 9, 1934; s. Albin H.B. and Matilda (Ludwig) S.; m. Janice R. Gates, Aug. 30, 1956; children: Jean, James, Joellen, Jacqueline, Jennifer, Jane. B.S., Iowa State U., Ames, 1959, M.S., 1962, Ph.D., 1965. Asst. prof. U. Ill., Urbana, 1965-70; assoc. prof. U. Notre Dame, 1970-74; supr. licensing Stone & Webster Engrng. Corp., Cherry Hill, N.J., 1975-77, mktg. engr., 1977-79, mgr. mktg., Houston, 1980—. Served with U.S. Army, 1954-56. Mem. ASME, Am. Nuclear Soc., Soc. for Values in Higher Edn., Sigma Xi, Phi Theta Epsilon. Subspecialties: Nuclear engineering; Mechanical engineering. Current work: Managing a business development department; engineering ethics and decision-making in the technical fields and government. Home: 402 Commodore Way Houston TX 77079 Office: Stone & Webster Engrng Corp 16430 Park Ten Pl Houston TX 77084

SCHILSKY, RICHARD LEWIS, physician, researcher; b. N.Y.C., June 6, 1950; s. Murray and Shirley (Cohen) S.; m. Cynthis Schum, Sept. 24, 1977; children: Allison Michelle, Meredith Ann. B.A. cum laude, U. Pa., 1971; M.D. with honors, U. Chgo., 1975. Diplomate: Am. Bd. Internal Medicine. Intern Parkland Meml. Hosp., Dallas, 1975-76, resident in medicine, 1976-77; clin. assoc. medicine br. Nat. Cancer Inst., Bethesda, 1977-78, clin. assoc. clin. pharmacology br., 1978-80, cancer expert, 1980-81; asst. prof. med. oncology U. Mo.-Columbia, 1981—. Contbr. articles to profl. jours. Served with USPHS, 1977-80. VA grantee, 1982. Fellow ACP; mem. Am. Soc. Clin. Oncology, Am. Assn. Cancer Research, Am. Fedn. Clin. Research, AAAS. Democrat. Jewish. Subspecialties: Internal medicine; Chemotherapy. Current work: Biochemical pharmacology of antifolates; hormonal modulation of chemotherapy sensitivity; clinical toxicities of chemotherapy. Office: N408 Health Scis Ctr 1 Hospital Dr Columbia MO 65212

SCHIMA, FRANCIS JOSEPH, physicist; b. Chgo., Apr. 15, 1935; s. Frank J. and Marie (Jedlicka) S.; m. JoAnn Marie Haberman, Jan. 21, 1967; children: Francis Joseph III, Susan. B.S., Ill. Benedictine Coll., Lisle, Ill., 1957; Ph.D., U. Notre Dame, 1964. Research assoc. Ind. U., Bloomington, 1964-66; physicist Nat. Bur. Standards, U.S. Dept. Commerce, Washington, 1966—. Contbr. numerous articles to sci. jours. Mem. Am. Phys. Soc., Sigma Xi. Roman Catholic. Subspecialty: Nuclear physics. Current work: Nuclear structure studies, radioactivity metrology, radioisotope production. Home: 820 Jonker Ct Gaithersburg MD 20878 Office: Nat Bur Standards US Dept Commerce Radioactivity Group 532 Washington DC 20234

SCHINGOETHE, DAVID JOHN, dairy nutrition educator; b. Aurora, Ill., Feb. 15, 1942; s. John E. and Helen M. (Tesch) S.; m. Darlene Kay Wennlund, June 6, 1964; children: Darcy, Deanna. B.S., U. Ill., 1964, M.S., 1965; Ph.D., Mich. State U., 1968. Asst. prof. dairy sci. S.D. State U., Brookings, 1969-73, assoc. prof., 1973-80, prof., 1980—. Contbr. chpt. to book in field. Mem. Am. Dairy Sci. Assn. (editorial bd. 1973-78), Am. Soc. Animal Sci., Am. Inst. Nutrition. Republican. Lutheran. Lodge: Lions. Subspecialties: Animal nutrition; Nutrition (medicine). Current work: Dairy cattle nutrition, protein nutrition, whey utilization, sunflower products, vitamin E, protected proteins. Office: SD State U Box 2104 Brookings SD 57007

SCHIPPER, ARTHUR LOUIS, JR., plant pathologist; b. Bryan, Tex., Apr. 8, 1940; s. Arthur Louis and Mildred Helen (Johnson) S.; m. Sandra Lee Murphy, July 25, 1964; children: Aaron Michael, Amy Rebecca. B.S., U. of South, 1962; M.S., U. Minn., 1965, Ph.D., 1968. Research plant pathologist, research plant physiologist Forest Service, U.S. Dept. Agr., St. Paul, 1968-78, research adminstr., Washington, 1979—. Contbr. articles to profl. jours. Mem. AAAS, Am. Phytopath. Assn. Subspecialties: Plant pathology; Plant physiology (agriculture). Current work: Administration of forest pathology research. Office: US Dept Agriculture PO Box 2417 Washington DC 20013

SCHLACHTER, ALFRED SIMON, physicist; b. Cedar City, Utah, Feb. 18, 1942; s. Max and Rose (Rosenfeld) S. A.B., U. Calif. - Berkeley, 1963; M.A., U. Wis., 1965, Ph.D., 1969. Aerospace engr. Ames Research Ctr., NASA, summer 1963; research asst. U. Wis. - Madison, 1963-68; prin. research scientist Honeywell Corp. Research Ctr., Mpls., 1968-70; scientist U. Paris, Orsay, 1971-73, Saclay Nuclear Research Ctr., France, 1971-75; staff scientist Magentic Fusion Energy Group, Lawrence Berkeley Lab., U. Calif. - Berkeley, 1975—; vis. scientist Ctr. for Nuclear Studies, Fontenay-aux-Roses, France, 1977, Justus-Liebig U., Giessen, W.Ger., 1980-81. Contbr. articles to profl. jours. NSF fellow, 1964; Centre National de la Recherche Scientifique fellow U. Paris, 1971-72; Joliot&Curie fellow, Saclay, 1972-73; NATO travel grantee, 1980-81; Alexander von Humboldt Stiftung traveling fellow, 1980-81. Mem. Am. Phys. Soc. Subspecialties: Atomic and molecular physics; Fusion. Current work: Basic research in atomic physics, experimental atomic and molecular physics with applications to fusion energy and fusion plasma diagnostics. Patentee in field. Office: Lawrence Berkeley Lab Univ Calif Berkeley CA 94720

SCHLAGER, SEYMOUR IRVING, immunologist, cons.; b. Hannover, Germany, Apr. 20, 1949; came to U.S., 1956, naturalized, 1961; s. Conrad and Helen (Topol) S.; m. Diane R. Schlager, Dec. 17, 1971; children: Carin Stephanie, Jason Lee. B.S., U. Ill., Chgo., 1969, M.S., 1973, Ph.D., 1975. Research chemist DeSoto, Inc., Des Plains, Ill., 1969-71; successively staff fellow, sr. staff fellow, cancer expert Lab. of Immunobiology, Nat. Cancer Inst., NIH, Bethesda, Md., 1975-80; assoc. prof. microbiology U. Notre Dame, 1980—; cons. Contbr. articles to profl. jours. Recipient Milan V. Novak award in microbiology U. Ill. Med. Center, 1975, various research grants, 1980-82, 1st prize Sigma Xi Research Forums, 1974, 75. Mem. Am. Assn. Immunologists, Am. Assn. Cancer Research, N.Y. Acad. Scis., AAAS, NIH Alumni Assn. Republican. Jewish. Lodge: B'nai B'rith. Subspecialties: Immunology (medicine); Immunopharmacology. Current work: Cellular immunology, molecular interactions of tumor cells with immune system, tumor cell resistance to immune attack and pharmacol. intervention thereof. Office: Dept of Microbiology University of Notre Dame Notre Dame IN 46556

SCHLEGEL, DAVID EDWARD, political scientist, educator; b. Fresno, Calif., Sept. 3, 1927; s. Edward and Lydia (Denny) S.; m. Betty Carolyn Adkins, Sept. 12, 1948; children: Marsha Elaine, Linda Elise, Mark David, Susan Eileen. B.S., Oreg. State U., 1950; Ph.D., U. Calif. at Berkeley, 1954. Faculty dept. plant pathology U. Calif. at Berkeley, 1953—, prof. plant pathology, plant pathothologist, chmn. dept., 1970-76; assoc. dean research Coll. Natural Resources, 1976-79, dean, 1979—, Miller prof., 1966-67. Served with USNR, 1945-46. NIH grantee, 1960-73; NSF grantee, 1974-76. Mem. Am. Phytopath. Soc., Am. Inst. Biol. Sci., Soc. Gen. Microbiology (U.K.), Sigma Xi. Subspecialties: Plant virology; Plant pathology. Research on plant virus replication, structure function relationships, comparative virology. Home: 3995 Paseo Grande Moraga CA 94556 Office: Coll Natural Resources U Calif at Berkeley Berkeley CA 94720

SCHLEGEL, ROBERT ALLEN, molecular and cell biology educator, researcher; b. Chgo., Feb. 17, 1945; s. Conard and Miriam Geraldine (Calkins) S.; m. Peggy Lee Anfinson, Aug. 24, 1968. B.S. in Chemistry, U. Iowa, 1967; A.M., Harvard U., 1968, Ph.D. in Biochemistry and Molecular Biology, 1971. Postdoctoral fellow Walter and Eliza Hall Inst. Med. Research, Melbourne, Australia, 1971-74; research asst. prof. U. Utah, Salt Lake City, 1974-76; asst. prof. Pa. State U., University Park, 1976-82, assoc. prof., 1982—. Am. Heart Assn. investigator, 1983—; Jane Coffin Childs Meml. Fund fellow, 1971-74; NSF fellow, 1968-71; grantee NIH, Nat. Cancer Inst., Am. Cancer Soc. Mem. Am. Soc. Cell Biology, Internat. Cell Cycle Soc. Subspecialties: Cell biology; Membrane biology. Current work: Hematopoietic surfaces; cell fusion; membrane fusion; cell cycle control; chromatin and chromosome structure; sperm cell surfaces. Office: 101 S Frear Lab University Park PA 16802

SCHLEGEL, RONALD GENE, acoustical and mechanical engineer, aerospace company executive; b. Derby, Conn., Sept. 3, 1936; s. Walter Joseph and Dorothy Louise (Lyon) S.; m. Sandra Webster, Sept. 14, 1957; children: Kenneth Alan, Peter Ronald. B.S. with honors, U. Conn., 1957; M.E., Yale U., 1959. Registered profl. engr., Conn. With Sikorsky Aircraft div. United Technologies Corp., Stratford, Conn., 1957—, program mgr. civil helicopter program, 1974-76, chief acoustics, 1976—; Chmn. noise control com. Aerospace Industries, Inc., Washington, 1982-84; mem. ad hoc com. helicopter acoustics Nat. Acad. Engrng., Washington, 1968. Editor: Helicopter Manufacturers' Economic Impact Assessment, 1979; contbr. numerous articles to profl. publs. Mem. planning commn. City of Shelton, Conn., 1960-62, mem. sewer commn., 1964-66, mem. charter revision commn., 1973; lector, spl. minister St. Joseph's Ch., Shelton, 1981—. Mem. Am. Helicopter Soc. (mem. acoustics com. 1970-83), AIAA (mem. aeroacoustics com. 1977-78), Helicopter Assn. Internat. (mem. acoustics com. 1975-83), Soc. Aeromotive Engrng. Republican. Roman Catholic. Club: Seymour Fish and Game (jr. rifle instr. 1970-82). Subspecialties: Acoustical engineering; Mechanical engineering. Current work: Development of technology for the design of helicopters with low internal and external noise levels; work with U.S. and international civil authorities to develop acoustic certification standards for helicopters. Home: 43 Lady Slipper Dr Shelton CT 06484 Office: United Technologies Corp Sikorsky Aircraft Div North Main St Stratford CT 06601

SCHLENDER, KEITH KENDALL, pharmacology educator; b. Newton, Kans., Oct. 3, 1939; s. Milton and Edna (Nanninga) S.; m. Shirley L. Bonser, Sept. 28, 1963; children: Krisleah, Todd. B.A., Westmar Coll., 1961; M.S., Mich. State U., 1963, Ph.D., 1966. NIH research fellow U. Minn., Mpls., 1966-69; asst. prof. Med. Coll. Ohio, Toledo, 1969-74, assoc. prof., 1974-80, prof., 1980—. Recipient USPHS Career Devel. award, 1977-83. Mem. Am. Soc. Exptl. Pharmacology and Therapeutics, Am. Soc. Biol. Chemists, Am. Chem Soc., Sigma Xi (pres. 1982-83), Phi Kappa Phi. Subspecialties: Biochemistry (medicine); Molecular pharmacology. Current work: Enzymology - regulation of glycogen metabolism by protein phosphorylation. Home: 4156 Indian Rd Toledo OH 43606 Office: Med Coll Ohio CS 10008 Toledo OH 43699

SCHLENKER, BARRY RICHARD, psychologist, educator; b. Passaic, N.J., Feb. 21, 1947; s. Henry Walter and Ruth Stephanie (Gammelin) S.; m. Patricia Ann O'Rorke, July 22, 1972; 1 son, David Richard. A.B., U. Miami, Fla., 1969; M.A., U.S. Internat. U., 1970; Ph.D., SUNY-Albany, 1972. Asst. prof. U. Fla., Gainesville, 1972-76, assoc. prof., 1976-80, prof. psychology, 1980—; prof. mktg., 1977—, prof. clin. psychology, 1976—. Cons. editor: Jour. Personality and Social Psychology, 1977—; cons. editor: Social Psychology Quar, 1980—, Jour. Social and Clin. Psychology, 1982—; Author: Conflict, Power and Games, 1973, A Contemporary Introduction to Social Psychology, 1976, Impression Management, 1980. Recipient Research Scientist Devel. award NIMH, 1979; NSF predoctoral fellow, 1970. Fellow Am. Psychol. Assn., Soc. for Psychol. Study of Social Issues; mem. Southeastern Psychol. Assn. Subspecialties: Social psychology; Personality. Current work: Self Concept; interpersonal relations; self-presentation; social identity. Home: 3218 NW 46 Pl Gainesville FL 32605 Office: U Fla Dept Psychology Gainesville FL 32611

SCHLENKER, ROBERT ALISON, biophysicist; b. Rochester, N.Y., Oct. 25, 1940; s. Martin and Ruth K. (Isler) S.; m. Sara E. Law, June 8, 1968; children: Martin, Laura. S.B., M.I.T., 1962, Ph.D., 1968. Postdoctoral fellow M.I.T., 1968-69; asst. physicist Argonne (Ill.) Nat. Lab., 1970-75; biophysicist, 1975—; instr. Coll. DuPage, 1975-78; mem. Com. 57 Nat. Council Radiation Protection and Measurements, 1978—. Assoc. editor: Radiation Research, 1983—; Contbr. articles to profl. jours. Mem. AAAS, Health Physics Soc. Radiation Research Soc., Am. Assn. Physicists in Medicine, Soc. Risk Analysis. Subspecialties: Biophysics (physics); Cell and tissue culture. Current work: Radiation biophysics. Office: Argonne Nat Lab Bldg 203 Argonne IL 60439

SCHLESINGER, DAVID HARVEY, biochemist, consultant; b. N.Y.C., Apr. 28, 1939; s. Philip Theodore and Fay (Margolis) S.; m.; children: Sarah Jane, Karen Louise. B.A., Columbia U., 1962; M.S., Albany Med. Coll., 1965; Ph.D., CUNY, 1972. Med. research technician Brookhaven Nat. Labs., Upton, L.I., N.Y., 1965-68; research fellow Harvard Med. Sch. and Mass. Gen. Hosp., Boston, 1972-75, asst. in biochemistry, 1975-77; assoc. prof. U. Ill. Med. Ctr., Chgo., 1977-81; research prof. exptl. medicine and cell biology NYU Med. Ctr., N.Y.C., 1981—; cons. Ortho Pharm., Raritan, N.J., 1978—. Author: Neurohypophyseal Peptide Hormones and Other Biologically Active Reptides, 1981. Mem. Am. Physiol. Soc., Am. Soc. Biol. Chemistry, Am. Chem. Soc., N.Y. Acad. Scis. Subspecialties: Biochemistry (medicine); Neuroendocrinology. Current work: Protein sequencing, peptide synthesis, structure of calcium binding proteins, ubiquitin, thymopoetin, neurophysin, precursor of CRF. Patentee in field. Office: Dept Medicine NYU Med Ctr 550 1st Ave New York NY 10016

SCHLESINGER, EDWARD BRUCE, neurol. surgeon; b. Pitts., Sept. 6, 1913; s. Samuel B. and Sara Marie (Schlesinger) S.; m. Mary Eddy, Nov. 1941; children—Jane, Mary, Ralph, Prudence. B.A., U. Pa., 1934, M.D., 1938. Diplomate: Am. Bd. Neurosurgey. Mem. faculty Columbia Coll. Phys. and Surg., N.Y.C., 1946—, prof. clin. neurol. surgery, 1964—, Byron Stookey prof., chmn. dept. neurol. surgery, 1973-80, Byron Stookey prof. emeritus, 1980—; dir. neurol. surgery Columbia Presbyn. Hosp., 1973-80, pres. med. bd., 1976-79; cons. in neurosurgery Presbyn. Hosp., 1980—. Trustee Matheson Found. Fellow N.Y. Acad. Scis. (chmn. Elsberg fellowship com.); mem. Harvey Cushing Soc., Harvey Soc., AAAS, Assn. Research in Nervous and Mental Disease, Neurosurg. Soc. Am. (pres. 1970-71), Soc. Neurol. Surgeons, Am. Assn. Surgery of Trauma, Am. Rheumatism Soc., Am. Coll. Clin. Pharmacology and Chemotherapy, AMA, Eastern Assn. Electroencephalographers, Sigma Xi. Subspecialties: Neuropharmacology; Neurophysiology. Research, publs. on uses, effects of curare in neuromsucualr disease, lesions of central nervous system, localization of brain tumors using radioactive tagged isotopes. Home: Closter Dock Rd Alpine NJ 07620 Office: 710 W 168th St New York NY 10032

SCHLESINGER, MILTON JOSEPH, virologist, molecular biologist, biochemist, researcher, educator; b. Wheeling, W.Va., Nov. 26, 1927; m. Sondra, July 10, 1934. B.S., Yale U., 1951; M.S., U. Rochester, 1953; Ph.D., U. Mich., 1959. Research assoc. M.I.T., Cambridge, 1961-64; asst. prof. Washington U. Sch. Medicine, St. Louis, 1964-67, assoc. prof., 1967-72, prof. microbiology and immunology, 1972—, assoc. dir., 1977-83. Assoc. editor: Virology, 1975—; editor: Jour. Biol. Chemistry, 1982—; Contbr. numerous articles to profl. publs. Served with U.S. Army, 1946-47. AEC fellow, 1951-53; NIH fellow, 1959-60; NSF fellow, 1960-61. Mem Am. Chem. Soc., Am. Soc. Biol. Chemists, AAAS, AM. Soc. Microbiology, Am. Soc. Virologists, Phi Beta Kappa. Subspecialties: Molecular biology; Virology (medicine). Current work: Membrane proteins, gene expression, virus replication.

SCHLESSINGER, DAVID, microbiologist; b. Toronto, Ont., Can., Sspt. 20, 1936; s. Morris and Eleanor (Ciechanower) S.; m. Alice Ruth, Mar. 18, 1960; children: Lillian Rachel, Esther Susan. BA., U. Chgo., 1957; Ph.D., Harvard U., 1960. Postdoctoral fellow Pasteur Inst., Paris, 1960-62; mem. faculty Washington U. Med. Sch., St. Louis, 1962—, prof. microbiology in medicine and immunology, 1980—. Editor: ann. Microbiology, 1974—; contbr. articles to profl. jours. Recipient Eli Lilly award, 1969; Macy fellow, 1981. Mem. Am. Soc. Microbiology, Am. Chem. Soc., Am. Soc. Cell Biology, Am. Fedn. Biochemists, Biophys. Soc., Sigma Xi. Subspecialties: Microbiology (medicine); Genetics and genetic engineering (medicine). Current work: Study of protein synthesis, RNA and ribosome formation and metabolism.

SCHLOSSBERG, HARVEY, psychologist; b. N.Y.C., Jan. 27, 1936; s. Harry and Sally (Frankel) S.; m. Cynthia Marks, Sept. 5, 1964 (div. 1978); children: Mark, Alexander, James, Steven; m. Antoinette M. Collarini, Oct. 25, 1982. B.S., Bklyn. Coll., 1958; M.S., L.I. U., 1960; Ph.D., Yeshiva U., 1971. Lic. psychologist, N.J., N.Y. Patrol officer N.Y. City Police Dept., 1958-71, dir. psychol. services, 1971-78; psychoanalyst N.Y. Ctr. for Psychoanalytic Tng., 1971-76; pvt. practice clin. psychology, N.Y.C., 1972—; asst. prof. John Jay Coll., CUNY, 1974-81, L.I. U., 1977-79; cons. Internat. Law Enforcement Stress Assn. Author: novel Psychologist with Gun; Contbr. articles to prfl. jours. Recipient Sci. Contbns. to Police Work award Internat. Assn. Chiefs. Police, 1975; Alumni award L.I. U., 1979. Mem. Am. Psychol. Assn., N.Y. State Psychol. Assn., Soc. for Personality Assessment, N.Y. Acad. Scis., Internat. Council Psychologists, Internat. Acad. Forensic Psychology. Subspecialties: Clinical psychology; Forensic Psychology. Current work: Application of psychological principles to police work. Development, creation, implementation and training for hostage negotiations programs and techniques used internationally. Home: 67-39 108th St Forest Hills NY 11375

SCHLOSSER, ELIZABETH, museum director; b. Defiance, Ohio, Nov. 1, 1949; s. Charles William and Nancy (Bryson) S.; m. Charles G. Jordan, Apr. 1, 1980; children: Savannah Cass, Wallis Lanier. A.B., Stanford U., 1971; M.A., U. Colo., 1973. Community planner HUD, Denver, 1974-76; preservation planner Gage Davis & Assocs., Boulder, Colo., 1976-78; exec. dir. Historic, Inc., 1978—; cons. in field. Author: Chautauqua - A Future for the Past, 1977, Northwest Denver - A Case Study, 1977, The Tabor Grand - Leadville, Colorado, 1977. Mem. Curtis Park Neighborhood Assn. Home: 2662 Curtis St Denver CO 80205 Office: 770 Pennsylvania Denver CO 80203

SCHMANDT-BESSERAT, DENISE, educator; b. Ay, France, Aug. 10, 1933; came to U.S., 1965, naturalized, 1970; d. Victor and Jeanne (Crabit) Besserat; m. Jurgen Schmandt, Dec. 27, 1956; children—Alexander, Christopher, Phillip. Ancienne Elève, Ecole du Louvre, 1965. Research fellow in Near Eastern archaeology Peabody Mus., Harvard U., Cambridge, Mass., 1969-71; fellow Radcliffe Inst., Cambridge, 1969-71; asst. prof. Middle Eastern studies U. Tex., Austin, 1972-81, asso. prof., 1981—; acting chief curator U. Tex. Art Mus., 1978-79. Assn. Adv. editor: Tech. and Culture, 1978—; contbr. articles to profl. jours. Wenner-Gren Found. grantee, 1970-71; Nat. Endowment for Arts grantee, 1974-75, 77-78; Nat. Endowment for

Humanities fellow, 1979-80. Mem. Am. Oriental Soc., Archaeol. Inst. Am., Am. Anthropol. Subspecialties: Cognition; Information systems (information science). Current work: Reckoning device based on takens of many shapes used in the middle east 8000-3000 B.C., the origin of phonetic writing and arithmetic. Office: Univ of Tex Austin TX 78712

SCHMERLING, ERWIN ROBERT, government official; b. Vienna, Austria, July 28, 1929; came to U.S., 1955, naturalized, 1962; m., 1957. B.A., Cambridge U., 1950, M.A., 1954, Ph.D., 1958; grad., Advanced Mgmt. Program, Harvard, 1969, Fed. Exec. Inst., 1975. Vis. asst. prof. elec. engring. Pa. State U., 1955-57, asst. prof., 1957-60, assoc. prof., 1960-64; program chief ionospheric physics Office Space Sci., NASA Hdqrs., Washington, 1964-70, program chief magnetospheric physics, 1970-76, program chief space plasma physics, 1976—; mem. coms. III and IV Internat. Sci. Radio Union, 1958; sec. U.S. com. III, 1966-69, chmn., 1969-72; mem. Adv. Group Aerospace Research and Devel. Fellow IEEE (wave propagation standards com.); mem. Am. Geophys. Union, AAAS, Sigma Xi. Current work: Space plasma physics; use of computers for information processing, retrieval and communications. Office: NASA Hdqrs 400 Maryland Ave SW Washington DC 20546

SCHMID, GERHARD MARTIN, chemist, educator; b. Ravensburg, Ger., Oct. 26, 1929; s. Josef and Anna (Schenzle) S.; m. Waltraud Menges, May 26, 1958; children: K. Peter, Michael R. Ph.D., U. Innsbruck, Austria, 1958. Postdoctoral fellow, research assoc. U. Tex., Austin, 1958-62; sr. scientist Tracor, Inc., Austin, 1960-62; temporary asst. prof. U. Alta., Edmonton, 1962-64; asst. prof. U. Fla, Gainesville, 1964-74, assoc. prof. chemistry, 1974—, asst. chmn. dept., 1974—. Author research publs. Mem. Electrochem. Soc., Am. Chem. Soc., Deutsche Bunsengesellschaft, N.Y. Acad. Scis., Sigma Xi. Subspecialties: Analytical chemistry; Surface chemistry. Current work: Electrochemical reaction kinetics, adsorption; teaching and research in analytical chemistry. Office: Dept Chemistry U Fla Gainesville FL 32611

SCHMID, RUDI RUDOLF, physician, educator, researcher; b. Switzerland, May 2, 1922; came to U.S., 1948, naturalized, 1954; s. Rudolf and Bertha (Schiesser) S.; m. Sonja D. Wild, Sept. 17, 1949; children: Isabelle S., Peter R. B.S., U. Zurich, 1941, M.D., 1947; Ph.D., U. Minn., 1954. Intern U. Calif. Med. Center, San Francisco, 1948-49; resident medicine U. Minn., 1949-52, instr., 1952-54; research fellow biochemistry Columbia U., 1954-55; investigator NIH, Bethesda, Md., 1955-57; assoc. medicine Harvard U., 1957-59, asst. prof., 1959-62; prof. medicine U. Chgo., 1962-66, U. Calif., San Francisco, 1966—, dean Sch. Medicine, 1983—; Cons. U.S. Army Surgeon Gen., USPHS, VA. Mem. editorial bd.: Jour. Clin. Investigation, 1965-70, Jour. Lab. and Clin. Medicine, 1964-70, Blood, 1962-75, Gastroenterology, 1965-70, Jour. Investigative Dermatology, 1968-72, Annals Internal Medicine, 1975-79, Proc. Soc. Exptl. Biology and Medicine, 1976—. Served with Swiss Army, 1943-45. Fellow AAAS, N.Y. Acad. Scis.; mem. Nat. Acad. Scis., Am. Acad. Arts and Scis., Assn. Am. Physicians, Am. Soc. Clin. Investigation, A.C.P., Am. Soc. Biol. Chemists, Am. Soc. Exptl. Pathology, Am. Soc. Hematology, Am. Gastroenterol. Assn., Am. Assn. Study Liver Disease (pres. 1965), Internat. Assn. Study Liver (pres. 1980), Leopoldina. Subspecialty: Internal medicine. Research in metabolism of hemoglobin, heme, prophyrins, bile pigments and liver. Home: 211 Woodland Rd Kentfield CA 94904 Office: University of California Med Center San Francisco CA 94143

SCHMIDT, BERLIE LOUIS, agronomy educator; b. Treynor, Iowa, Oct. 2, 1932; s. Hans Frederick and Louisa Amalie (Guttau) S.; m. Joanne Doris Bruning, Sept. 4, 1954 (dec.); children: Brian, Luanne Schmidt Code, Kevin, Kimberly, Christy. B.S., Iowa State U., 1954, M.S., 1959, Ph.D., 1962. Soil scientist U.S. Soil Conservation Service, Council Bluffs, Iowa, 1954-57; grad. research assoc. agronomy dept. Iowa State U., Ames, 1957-62; mem. faculty Ohio Agrl. Research and Devel. Center, Ohio State U., Columbus, 1962—, prof. agronomy, 1969—, assoc. chmn. agronomy dept., 1969-75, chmn. agronomy dept., 1975—. Contbr. numerous sci. articles to profl. publs. Elder Worthington (Ohio) United Presbyn. Ch., 1983—. Served with U.S. Army, 1954-56. Fellow Ohio Acad. Sci.; mem. Am. Soc. Agronomy, Soil Sci. Soc. Am. (co-editor spl. publ. 1982), Soil Conservation Soc. Am. (Outstanding Chpt. Mem. award 1977), Council Agrl. Sci. and Tech., Worthington Civic Assn. (trustee 1982—). Republican. Subspecialties: Soil science; Agronomy. Current work: Soil conservation and soil erosion control; department administration of agronomy research; teaching. Office: 2021 Coffey Rd Columbus OH 43210

SCHMIDT, EDWARD GEORGE, astronomer, educator; b. Cut Bank, Mont., Dec. 13, 1942; s. Donald J. and Kathryn F. (Frevert) S.; m. Karen R., Sept. 21, 1963; children: Hope, Gwen, Sarah. B.S. in Physics, U. Chgo., 1965; Ph.D. in Astronomy, Australian Nat. U., Canberra, 1970. Research assoc. U. Ariz., 1970-72; sr. research fellow Royal Greenwich Obs., Herstmonceux, Sussex, Eng., 1972-74; asst. prof. physics U. Nebr., Lincoln, 1974-77, assoc. prof., 1977-82, prof. physics, 1982—; cons. NASA. Contbr. sci. papers to profl. publs. NSF grantee, 1978, 80, 83; NASA grantee, 1978, 79, 80, 81, 82; Research Corp. grantee, 1978. Mem. Internat. Astron. Union, Am. Astron. Soc., Royal Astron. Soc. Subspecialty: Optical astronomy. Current work: Research variable stars, stellar chromospheres, star clusters, space astronomy, astronomical instrumentation and techniques. Office: Dept Physics U Nebr Lincoln NE 68588

SCHMIDT, MAARTEN, educator, astronomer; b. Groningen, The Netherlands, Dec. 28, 1929; came to U.S., 1959; s. Wilhelm and Antje (Haringhuizen) S.; m. Cornelia Johanna Tom, Sept. 16, 1955; children—Elizabeth Tjimkje, Maryke Antje, Anne Wilhelmina. B.Sc., U. Groningen, 1949; Ph.D., Leiden (The Netherlands) U., 1956; Sc.D., Yale, 1966. Sci. officer Leiden (The Netherlands) Obs., 1953-59; postdoctoral Carnegie fellow Mt. Wilson Obs., Pasadena, Calif., 1956-58; mem. faculty Calif. Inst. Tech., 1959—, prof. astronomy, 1964—, exec. officer for astronomy, 1972-75, chmn. div. physics, math. and astronomy, 1975-78; Mem. staff Hale Obs., 1959-80, dir., 1978-80. Co-winner Calif. Scientist of Year award, 1964. Fellow Am. Acad. Arts and Scis. (Rumford award 1968); mem. Am. Astron. Soc. (Helen B. Warner prize 1964, Russell lecture award 1978), Nat. Acad. Scis. (fgn. asso.), Internat. Astron. Union, Royal Astron. Soc. (asso., Gold medal 1980), Spl. research dynamics of our galaxy, star formation, radio galaxies, quasi-stellar radio sources. Subspecialty: Radio and microwave astronomy. Office: Calif Inst Technology Pasadena CA 91125

SCHMIDT, MOSHE, physician; b. Breslau, Germany, July 17, 1929; came to U.S., 1972, naturalized, 1979; s. Ewald and Ruth (Tockuss) S.; m. Rina Jacoby, Jan. 14, 1951; children: Ehud, Joal. Student, U. Fribourg, U. Geneva, 1950-54; M.D., Hebrew U., Jerusalem, 1958. Diplomate: Am. Bd. Internal Medicine with subsplty. in gastroenterology. Staff physician Meir Hosp., 1966-70; cons. in medicine and gastroenterology Kupat-Holim Sick Fund, Petah-Tiqva and Tel-Aviv, Israel, 1969-72; research fellow Meml. Sloan-Kettering Cancer Ctr., N.Y.C., 1972-76; attending in gastroenterology Syosset (N.Y.) Hosp. and Nassau County (N.Y.) Med. Ctr., 1976-77; staff physician Bklyn. VA Med. Ctr., 1977—; asst. prof. medicine SUNY, Bklyn., 1978—. Contbr. articles to profl. jours. Served to lt. M.C. Israel Def. Force, 1948-49. Damon Runyon Research Fund fellow, 1972-73; Nat. Cancer Inst. grantee, 1974-76. Fellow Am. Coll. Gastroenterology; mem. Am. Assn. for Cancer Research, N.Y. Acad. Gastroenterology, Nat. Assn. VA Physicians, Bklyn. VA Med. Soc. (v.p. 1982—). Jewish. Subspecialties: Gastroenterology; Cancer research (medicine). Current work: Cancer epidemiology. Immunological markers in gastrointestinal neoplasia. Office: 9800 Poly Pl Brooklyn NY 11209

SCHMIDT, NATHALIE JOAN, virologist; b. Flagstaff, Ariz., Sept. 24, 1928; d. Joseph Francis and Gertrude Nathalie (Hill) S. B.A., U. Ariz., 1950; M.S, Northwestern U., 1952, Ph.D., 1953. Diplomate: Am. Bd. Med. Microbiology, 1965. Asst. instr. Northwestern U., 1950-53; microbiologist Evanston (Ill.) Hosp., 1953-54; research specialist Virus Lab., Calif. Dept. Health, Berkeley, 1954-81, research scientist, 1981—; lectr. Sch. Pub. Health, U. Calif. - Berkeley, 1971—; cons. NIH. Editor: Diagnostic Procedures for Viral, Rickettsial and Chlamydial Infections, 5th edit, 1979, Jour. Clin. Microbiology, 1975—; mem. editorial bd.: Intervirology, 1972—, Jour. Immunology, 1973-75, Proc. Soc. Exptl. Biology and Medicine, 1975—; contbr. articles to profl. jours. Recipient Kimble award in Lab. Methodology Kimble, Inc., 1977. Mem. Am. Acad. Microbiology, Am. Assn. Immunologists, Am. Public Health Assn., Am. Soc. Microbiology, N.Y. Acad. Scis., Soc. Exptl. Biology and Medicine. Subspecialty: Virology (medicine). Current work: Developmental viral diagnosis; immunology of viral diseases; immunology of coxsackievirus, herpes virus and rubella virus infections; immunoassays for viral antigens and antibodies. Office: 2151 Berkeley Way Berkeley CA 94704

SCHMIDT, RUTH A.M., geology educator; b. Bklyn., Apr. 22, 1916; d. Edward and Anna M. (Range) S. A.B., NYU, 1936; A.M., Columbia U., 1939, Ph.D., 1948. X-ray technician L.I. Coll. Hosp., Bklyn., 1936-38; researcher Am. Mus. Natural History, N.Y.C., 1941; geologist U.S. Geol. Survey, Washington, 1943-56, dist. geologist, Anchorage, 1956-63; prof., head dept. geology U. Alaska-Anchorage, Anchorage Community Coll., 1969—; cons. geologist, Anchorage, 1964—. Author: Geology Color Slide Sets, 1965-68; contbr. articles in field to profl. jours. Trustee Brooks Range Trust, 1973—; exec. bd. Alaska Ctr. Environment, 1981-83; bd. govs. Arctic Inst. N.Am. Fellow AAAS (past pres. local chpt.), Geol. Soc. Am.; mem. Fedn. Am. Scientists, Soc. Econ. Mineralogists and Paleontologists, Am. Assn. Petroleum Geologists, Alaska Geol. Soc. (past pres.), Am. Inst. Profl. Geologists (sect. v.p. 1968-69, cert. profl. geologist), Audubon Soc., Sierra Club, Nat. Parks and Conservation Assn., Sigma Xi (hon.)..). Subspecialties: Paleoecology; Geology. Current work: Quaternary micropaleontology Alaska; instructional television delivery of earth science to rural Alaska. Office: 1040-C-St Anchorage AK 99501

SCHMIDT-NIELSEN, KNUT, physiologist, educator; b. Norway, Sept. 24, 1915; U.S., 1946, naturalized, 1952; s. Sigval and Signe Torborg (Sturzen-Becker) Schmidt-N. Mag. Scient., U. Copenhagen, 1941, Dr. Phil., 1946. Research fellow Carlsberg Labs., Copenhagen, 1941-44, 1944-46; research assoc. zoology Swarthmore (Pa.) Coll., 1946-48; docent U. Oslo, Norway, 1947-49; research assoc. physiology Stanford, 1948-49; asst. prof. Coll. Medicine, U. Cin., 1949-52; prof. physiology Duke, Durham, N.C., 1952—, James B. Duke prof. physiology, 1963—; Harvey Soc. lectr., 1962; Regents' lectr. U. Calif. at Davis, 1963; Brody Meml. lectr. U. Mo., 1962; Hans Gadow lectr. Cambridge (Eng.) U., 1971; vis. Agassiz prof. Harvard, 1972; Mem. panel environmental biology NSF, 1957-61; mem. sci. adv. com. New Eng. Regional Primate Center, 1962-66; mem. nat. adv. bd. physiol. research lab. Scripps Instn. Oceanography, U. Calif. at San Diego, 1963-69, chmn., 1968-69; organizing com. 1st Internat. Conf. on Comparative Physiology, 1972-80; mem. U.S. nat. com. Internat. Union Physiol. Scis., 1966-78, vice chmn. U.S. nat. com., 1969-78, pres., 1980—; mem. subcom. on environmental physiology U.S. nat. com. Internat. Biol. Programme, 1965-67; mem. com. on research utilization uncommon animals, div. biology and agr. Nat. Acad. Scis., 1966-68; mem. animal resources adv. com. NIH, 1968; mem. adv. bd. Bio-Med. Scis., Inc., 1973-74; Chief scientist Scripps Instn. Amazon expdn., 1967. Author: Animal Physiology, 3d. edit, 1970, The Physiology of Desert Animals; Physiological Problems of Heat and Water, 1964, How Animals Work, 1972, Animal Physiology; Adaptation and Environment, 1975, 2d edit., 1979. Sect; editor: Am. Jour. Physiology, 1961-64, 70-76, Jour. Applied Physiology, 1961-64, 70-76; editorial bd.: Jour. Cellular and Comparative Physiology, 1961-66, Physiol. Zoology, 1959-70, Am. Jour. Physiology, 1971-76, Jour. Applied Physiology, 1971-76, Jour. Exptl. Biology, 1975-79; cons. editor: Annals of Arid Zone, 1962—; hon. editorial adv. bd.: Comparative Biochemistry and Physiology, 1962-63; Contbr. articles to sci. publs. Guggenheim fellow, 1953-54; grantee Office Naval Research, 1952-54, 58-61, UNESCO, 1953-54, Office Q.M. Gen., 1953-54, Office Surgeon Gen., 1953-54, NIH, 1955—, NSF, 1957-61, 59-60, 60-61, 61-63; recipient Research Career award USPHS, 1964. Fellow N.Y. Acad. Sci., AAAS, Am. Acad. Arts and Scis.; mem. Nat. Acad. Scis., N.C. Acad. Sci. (Poteat award 1957), Am. Physiol. Soc., Am. Soc. Zoologists (chmn. div. comparative physiology 1964), Soc. Exptl. Biology, Harvey Soc. (hon.), Royal Danish Acad., Académie des Sciences (France) (fgn. assoc.), Royal Norwegian Soc. Arts and Sci., Norwegian Acad. Scis., Physiol. Soc. London (assoc.). Subspecialty: Comparative physiology. Office: Dept Zoology Duke Univ Durham NC 27706

SCHMIEDER, ROBERT WILLIAM, physicist, marine scientist; b. Phoenix, July 10, 1941; s. Otto and Ruby Maybel (Harkey) S.; m. (div.); children: Robyn, Russell Otto, Robert Randall. A.B., Occidental Coll., 1963; B.S., Calif. Inst. Tech., 1963; M.A., Columbia U., 1965, Ph.D., 1968. Staff mem. Lawrence Berkeley Lab., 1969-71; instr. physics dept. U. Calif. - Berkeley, 1971-72; mem. tech. staff Sandia Nat. Lab., Livermore, Calif., 1972—. Contbr. articles to sci. jours. Nat. Geog. Soc. grantee, 1979; Explorers Club grantee, 1980; San Francisco Found. grantee, 1981; NOAA grantee, 1981-83. Mem. Am. Phys. Soc., Optical Soc. Am. Subspecialties: Combustion processes; Atomic and molecular physics. Current work: Research in combustion physics and chemistry; oceanic research expeditions. Patentee in field of nuclear instrumentation. Home: 4295 Walnut Blvd Walnut Creek CA 94596 Office: Div 8513 Sandia Nat Labs Livermore CA 94550

SCHMIT, LUCIEN ANDRÉ, JR., structural engineer; b. N.Y.C., May 5, 1928; s. Lucien Alexander and Eleanor Jessie (Donley) S.; m. Eleanor Constance Trabish, June 24, 1951; 1 son, Lucien Alexander, III. B.S., MIT, 1949, M.S., 1950. Structures engr. Grumman Aircraft Co., Bethpage, N.Y., 1951-53; research engr., aeroelastic and structures lab. MIT, 1954-58; asst. prof. engring. (Case Inst. Tech.), 1958-60, assoc. prof., 1961-63, prof., 1964-70; prof. engring. and applied sci. UCLA, 1970—; mem. sci. adv. bd. USAF. Contbr. numerous articles on analysis and synthesis of structural systems, finite element methods, design of fiber composite components to profl. jours. Fellow ASCE (Walter L. Huber Civil Engring. Research prize 1970), AIAA (assoc., Design Lecture award 1977, Structures, Structural Dynamics and Materials award 1979); mem. ASME, Sigma Xi. Subspecialties: Aeronautical engineering; Civil engineering. Current work: Methods for optimum and/or balanced design of structural systems subject to static and dynamic load condition. Home: 712 El Medio Ave Pacific Palisades CA 90272 Office: 6731K Boelter Hall UCLA Los Angeles CA 90024

SCHMITT, DONALD PETER, nematologist; b. New Hampton, Iowa, Oct. 29, 1941; s. Louis Stanley and Anna (Formanek) S.; m. MaryAnn Pualeialii, June 17, 1967; children: Julia, Peter, Anna, Cecilia. B.S., Iowa State U., 1967, M.S., 1969, Ph.D., 1971. Plant pathologist Tenn. Dept. Agr., Nashville, 1971-75; nematologist, plant pathologist N.C. State U., Raleigh, 1975—. Served with USAF, 1960-64. Mem. Soc. Nematologists, Am. Phytopathol. Soc., So. Soybean Disease Workers, Sigma Xi, Gamma Sigma Delta. Subspecialty: Plant pathology. Current work: Pesticide interaction effects on nematodes; research on effects of nematicides and herbicides on nematode behavior; damage threshholds. Office: 3127 Ligon St Raleigh NC 27607

SCHMITT, ROLAND WALTER, manufacturing company research executive; b. Seguin, Tex., July 24, 1923; s. Walter L. and Myrtle F. (Caldwell) S.; m. Claire Freeman Kunz, Sept. 19, 1957; children: Lorenz Allen, Brian Walter, Alice Elizabeth, Henry Caldwell. B.A. in Math, U. Tex., 1947, B.S. in Physics, 1947, M.A., 1948; Ph.D., Rice U., 1951. With Gen. Electric Co., 1951—, research and devel. mgr. phys. sci. and engring. Gen. Electric Corp. Research and Devel., Schenectady, 1967-74, research and devel. mgr. energy sci. and engring. Gen. Electric Corp. Research and Devel., 1974-78, v.p. corp. research and devel. Gen. Electric Corp. Research and Devel., 1978-82, sr. v.p. corp. research and devel. Gen. Electric Corp. Research and Devel., 1982—; mem. energy research adv. bd. Dept. Energy, 1977-83. Trustee Northeast Savs. Bank.; vice chmn. Bd. dirs., chmn. investment rev. com. N.Y. State Sci. and Tech. Found.; bd. govs. Albany Med. Center Hosp., 1979-82; trustee Union Coll., Schenectady, Argonne Univs. Assn., 1979-82, RPI; bd. dirs. Sunnyview Hosp. and Rehab. Center. Served with USAAF, 1943-46. Fellow Am. Phys. Soc., IEEE, AAAS; mem. Am. Inst. Physics (chmn. com. on corp. assocs., mem. governing bd.), Nat. Acad. Engring. (council), Nat. Sci. Bd., Dirs. Indsl. Research, Am. Nuclear Soc. Club: Cosmos. Subspecialty: Research management. Office: PO Box 8 Schenectady NY 12301

SCHMITZ, ROGER ANTHONY, chem. engr., univ. dean; b. Carlyle, Ill., Oct. 22, 1934; s. Alfred Bernard and Wilma Afra (Aarns) S.; m. Ruth M., Aug. 31, 1957; children: Jan, Joy, Joni. B.S., U. Ill., 1959; Ph.D., U. Minn., 1962. Prof. chem. engring. U. Ill.-Urbana, 1962-79; Keatin-Crawford prof. chem. engring. U. Notre Dame, 1979—, chmn. dept., 1979-81, dean Coll. Engring., 1981—. Contbr. articles to profl. jours. Served with U.S. Army, 1953-55. Recipient Allan P. Colburn award Am. Inst. Chem. Engrs., 1970, R.H. Wilhelm award, 1981; George Westinghouse award Am. Soc. Engring. Edn., 1977. Mem. Combustion Inst., Am. Inst. Chem. Engrs., Am. Chem. Soc., Am. Soc. Engring. Edn. Subspecialties: Chemical engineering; Catalysis chemistry. Current work: Dynamics, control and mathematical modeling of chemically reacting systems. Office: U Notre Dame 257 Fitzpatrick Hall Notre Dame IN 46556

SCHNAAR, RONALD LEE, research biochemist; b. Detroit, Nov. 1, 1950; s. Herbert N. and Faye T. (London) S.; m. Cynthia B. Roseman, June 24, 1972; children: Melissa, Stephen, Gregory. B.S., U. Mich., 1972; Ph.D., Johns Hopkins U., 1976. Postdoctoral fellow in biology Johns Hopkins U., 1977, Lab. Biochem. Genetics, NIH, 1978-79; asst. prof. dept. pharmacology and exptl. therapeutics Johns Hopkins U. Sch. Medicine, Balt., 1979—. Contbr. articles to profl. publs. Recipient Jr. Faculty Research award Am. Cancer Soc., 1981—; Muscular Dystrophy Soc. fellow. Mem. Soc. Neurosci. Subspecialties: Biochemistry (biology); Neurochemistry. Current work: The metabolism of cell surface complex carbohydrates and their role in intercellular interactions. Home: 9094 Goldamber Garth Columbia MD 21045 Office: 725 N Wolfe St Baltimore MD 21205

SCHNEBERGER, GERALD LEO, materials scientist, educational administrator; b. Flint, Mich., Sept. 2, 1936; s. Edward Leo and Alice Gertrude (Pickett) S.; m. Kathleen Ann, Aug. 24, 1939; children: David, Louise, Gregory, Steven, Janet, Keli. A.S., Flint Jr. Coll., 1956; A.B., U. Mich., 1959; Ph.D., W.Va. U., 1963. Registered profl. engr., Wis., 1982. Assoc. prof. chemistry West Liberty Coll., 1963-66; asst. prof. GM Engring. and Mgmt. Inst., 1966-68, assoc. prof., 1968-71, prof., 1971—, chmn. dept. matrials sci., 1974-75, chmn. dept. materials sci. and processes, 1975-77, chmn. dept. mech engring., 1977-79, head sci. and math., 1979-82, dir. research and profl. services, 1982—; vis. scientist/lectr. MIT, 1978-79, research affiliate, 1979-81; cons. paint and adhesives; speaker, lectr. in field. Contbr. articles to profl. publs; author: Understanding Adhesives, 1971, 2d edit., 1974, Understanding Paint and Painting Processes, 1975, 2d edit., 1980, Spray Paint Defects, 1983; Painting Forum columnist: Indsl. Finishing, 1977-80; editor, contbg. author: Adhesives in Manufacturing, 1983. Mem. Am. Soc. Engring. Edn., Am. Soc. Metals, Soc. Automotive Engrs. (Mid-Mich. Governing Bd. 1978), Soc. Mfg. Engrs. (cert mfg. engr.), ASME, Am. Chem. Soc. (chmn. Flint sub-sect., trustee Detroit sect. 1979—), Sigma Xi. Subspecialties: Materials (engineering); Polymers. Current work: Adhesives; paint consulting and training. Home: 2418 Nolen Dr Flint MI 48504

SCHNEER, CECIL J., geology educator, history of science; b. N.Y.C., Jan. 7, 1923; s. Jacob Bernard and Sid Leah (Glass) S.; m. Mary Barsam Temple, Jan. 25, 1943; children: Jean Schneer Silverman, David. A.B., Harvard Coll., 1943; A.M., Harvard U., 1949; Ph.D., Cornell U., 1954. Geologist Cerro de Pasco Copper Co., Morococha, Peru, 1943-44, U.S. Geol. Survey, 1948, 49; instr. Hamilton Coll., Clinton, N.Y., 1950-52; asst. prof. U. N.H., Durham, 1954-59, assoc. prof., 1959-64, prof. geology and history of sci., 1964—; acad. guest Swiss Fed. Inst. Tech., Zurich, 1960-61; guest prof. U. Milan, Italy, 1969-70; Sigma Xi nat. lectr., 1981-83; convenor N.H. Interdisciplinary Conf. on History of Geology, 1967, N.H. Bicentennial Conf. History of Geology in Am., 1976. Assoc. editor: Isis, 1977-80, Dictionary of Sci. Biography, 1969—; author: The Search for Order, 1960, Mind and Matter, 1970; editor: Toward a History of Geology, 1969, Crystal Form and Crystal Structure, 1977, Two Hundred Years of Geology in America, 1979. Served with USNR, 1944-46. Fellow Geol. Soc. Am. (chmn. history of geology div. 1978), Mineral. Soc. Am., Geol. Soc. London; mem. History of Sci. Soc. (council 1981—), Internat. Com. for History of Geol. Scis. (v.p. 1976—, corr. mem.). Subspecialties: Mineralogy; Crystallography. Current work: Phase thermodynamics, crystal morphology, history of geology and crystallography. Office: Dept Earth Scis U NH James Hall Durham NH 03824

SCHNEIDER, BARRY IRWIN, theoretical chemist, researcher; b. Bklyn., Nov. 16, 1940; s. Sidney and Beatrice (Berman) S.; m. Roberta Lain, Aug. 26, 1962; children: Cris Edward, Scot Mathew. B.S., Bklyn. Coll., 1962; M.S., Yale U., 1964; Ph.D. in Theoretical Chemistry, U. Chgo., 1968. Staff scientist GTE Labs., 1969-72; staff mem. T-Div. Los Alamos Nat. Lab., 1972—. Contbr. articles profl. jours. NATO grantee, 1979. Mem. Am. Phys. Soc., Sigma Xi. Subspecialties: Theoretical chemistry; Atomic and molecular physics. Current work: Scattering theory, photoionization, many-body theory, electron scattering from molecules. Office: Los Alamos Sci Lab MS J 569 Los Alamos NM 87645

SCHNEIDER, DONALD LEONARD, biochemistry educator; b. Muskegon, Mich., Jan. 15, 1941; s. Leonard and Hilma (Nichols) S.; m. Jean Chin, Dec. 29, 1979; 1 dau., Amy. B.A., Kalamazoo Coll., 1963; Ph.D., Mich. State U., 1969. Postdoctoral fellow Cornell U., Ithaca, N.Y., 1969-71; research assoc. Rockefeller U., N.Y.C., 1971-72, asst. prof., 1972-73, U. Mass., Amherst, 1973-76, Dartmouth Med. Sch., Hanover, N.H., 1977-82, assoc. prof. biochemistry, 1982—. Mem. Am. Soc. Cell Biology, Am. Soc. Biol. Chemists, AAAS, Am. Chem. Soc. Subspecialties: Biochemistry (medicine); Cell biology (medicine). Current work: Function of proton pump ATPase in lysosomes; microbicidal events in leukocytes. Office: Dept Biochemistry Dartmouth Med Sch Hanover NH 03756

SCHNEIDER, EDWARD LEE, botanist, researcher, educator; b. Portland, Oreg., Sept. 14, 1947; s. Edward John and Elizabeth (Matthews) S.; m. Sandra Lee, 2Aug. 2, 1968; children: Kenneth Lee, Cassandra Lee. B.S., Central Wash. U., 1969, M.S., 1971; Ph.D., U. Calif., Santa Barbara, 1974. Asst. prof. botany S.W. Tex. State U., 1974-79, assoc. prof., 1979—. Contbr. articles to profl. jours. NSF grantee, 1980-83. Mem. Bot. Soc. Am., Internat. Assn. Aquatic Vascular Plant Biology, Tex. Acad. Sci., Sigma Xi, Phi Sigma. Roman Catholic. Subspecialty: Evolutionary biology. Current work: Insect-flowering relationships; coevolution; aquatic plant mgmt.; ecological plant anatomy, morphology. Home: Route 1 Box SE 28 San Marcos TX 78666 Office: Dept Biology SW Tex State U San Marcos TX 78666

SCHNEIDER, FREDERICK H., banker. Grad., U. Pa., 1939; LL.B., St. John's U., 1947. Trustee Roosevelt Savs. Bank, Bklyn., 1951—, v.p., 1963-67, pres., 1967-83; dir. Instl. Securities Corp., 1968-74, 78-81, INSTLCORP, 1968-74, Savs. Banks Retirement System, 1973-79, Savs. Banks Trust Co., 1973-74, 81-82, Drayton Co. Ltd., 1972-75; mem. Nassau adv. bd. Mfrs. Hanover Trust, 1971-81. Mayor Village of Garden City, 1969-71, trustee, 1963-69; bd. dirs. Ottilie Home for Children, 1952-77, pres., 1961-63; bd. dirs. Swedenborg Found., Inc., 1981—. Served with USNR, World War II. Mem. Savs. Bank Assn. N.Y. State (pres. 1973-74, dir.), Nat. Assn. Mut. Savs. Banks (dir. 1973-79), L.I. Assn. (dir. 1975-80), Garden City C. of C., Nassau Bar Assn. Clubs: Rotary (Mineola-Garden City); Masons. Subspecialties: Cell biology (medicine); Pharmacology. Current work: In vitro assay development, medical uses of naturally occurring proteins, immunodiagnostics. Address: 1122 Franklin Ave Garden City NY 11530

SCHNEIDER, GERALD ELMORE, mechanical engineering educator, consultant; b. Waterloo, Ont., Can., Aug. 21, 1949; s. Almond Louis and Nelda (Jacobi) S.; m. Joan Diane Illig, June 26, 1968; children: Jeremy Glenn, Joshua James. B.A., U. Waterloo, Ont., Can., 1973, M.A., 1974, Ph.D., 1977. Research asst. U. Waterloo, Ont., Can., 1971-73, project engr., 1973-77, adj. lectr., 1976, asst. prof. mech. engring., 1977-81, assoc. prof., 1981—; cons. Cluff & Cluff Architects, Toronto, Ont., Can., 1980; on-site cons. Aerojet Electro Systems Inc., Azusa, Calif., 1980; cons. Valley Blades Ltd., Waterloo, Ont., 1980-81, Thermo Fluids Research and Devel. Inc., Waterloo, 1980—. Mem. AIAA, ASME. Subspecialties: Fluid mechanics; Numerical analysis. Current work: Heat transfer, numerical methods, tribology, solution procedures, incompressible fluid flow. Office: Dept Mech Engring U Waterloo University Ave W Waterloo ON Canada N2L 3G1 Home: 81 Culpepper Dr Waterloo ON Canada N2L 5K8

SCHNEIDER, MARILYN BETH, physicist; b. Bklyn., Mar. 11, 1952; d. Hyman Isadore and Clara (Feinstein) S.; m. Murdock G.D. Gilchriese, June 27, 1982. A.B., Columbia U., 1974; M.S. in Physics, Cornell U., 1978; Ph.D., 1983. Research asst. Columbia Astrophysics Lab., Columbia U., N.Y.C., 1972, teaching asst. dept. physics, 1974; teaching asst. dept. applied and engring. physics Cornell U., Ithaca, N.Y., 1974-77, research asst., 1977-83, postdoctoral assoc., 1983—. Mem. Am. Phys. Soc. Subspecialties: Condensed matter physics; Biophysics (physics). Current work: Liquid crystals, membrane mechanics, critical phenomena, biophysics, liquid vapor interface at the critical point with light scattering. Home: 107 Maplewood Dr Ithaca NY 14850 Office: Dept Applied & Engring Physics Clark Hall Cornell U Ithaca NY 14853

SCHNEIDER, NORMAN RICHARD, veterinarian, educator; b. Ellsworth, Kans., Mar. 28, 1943; s. Henry C. and Irene C. (Ney) S.; m. Karen Marjorie Nelson, July 1, 1968; 1 son, Nelson R. B.S., Kans. State U., 1967, D.V.M., 1968; M.S., Ohio State U., 1972. Diplomate: Am. Bd. Vet. Toxicology. Commd. capt. U.S. Air Force, 1968, advanced through grades to maj., 1976; base veterinarian Goose AB, Labrador, Can., 1968-70; veterinary scientist, toxicologist Armed Forces Radiobiology Research Inst., Bethesda, Md., 1972-76; veterinary toxicologist Aerospace Med. Research Lab., Wright-Patterson AFB, Dayton, 1976-79; assoc. prof., veterinary toxicologist dept. veterinary sci. U. Nebr., Lincoln, 1979—; adj. prof., dept. pharmacodynamics and toxicology U. Nebr. Med. Center, 1982—; chief environ. services Nebr. Air Nat. Guard, Lincoln, 1979—. Mem. Am. Bd. Vet. Toxicologists, Am. Coll. Vet. Toxicologists, Am. Vet. Med. Assn., Nebr. Vet. Med. Assn., Kans. Vet. Med. Assn., Am. Assn. Vet. Lab. Diagnosticians, Assn. Official Analytical Chemists, Nat. Guard Assn. U.S., Nat. Guard Assn. Nebr., Alliance Air Nat. Guard Flight Surgeons, Alliance Air Nat. Guard Veterinarians, Assn. Mil. Surgeons U.S., N.Y. Acad. Scis., Council Agrl. Sci. and Tech., Am. Legion, NRA, Alpha Zeta, Phi Zeta. Roman Catholic. Club: Farm House. Subspecialties: Toxicology (agriculture); Diagnostic veterinary toxicology. Current work: Mycotoxins and mycotoxicoses, nitrite/nitrate, pesticides, military munitions, natural toxicants. Home: Rt 1 Box 70 Ceresco NE 68017 Office: Vet Diagnostic Center Lincoln NE 68583

SCHNEIDER, RICHARD JOEL, neurobiologist, translator; b. Bklyn., July 25, 1944; s. Albert and Edith E. (Heltermann) S.; m. Diane J. Shannon, July 31, 1982; children: Kelley, Connie, Keith. Student, Colby Coll., Waterville, Maine, 1961-62; A.B., U. Chgo., 1966; Ph.D., U. Pitts., 1972. Instr. U. Md., Balt., 1971-72, asst. prof., 1972-78; head Lab. Neurosci., Md. Inst. Emergency Medicine, Balt., 1978-81; guest worker NIMH, Bethesda, 1981—; cons. Neurosci. Lab., Curtis Hand Ctr., Union Meml. Hosp., Balt. Contbr. articles to profl. jours. Nat. Multiple Sclerosis Soc. grantee, 1978-81; Bressler Research Found. grantee, 1976-80; Curtis Hand Found. grantee, 1980-81; Surgery Research award Plastic Surgery Ednl. Found., 1982; Nat. Media Library Teaching Tape award, 1974. Mem. AAAS, Am. Pain Soc., Balt. Neuroscience Soc., Md. Neurol. Soc., Fedn. Am. Scientists, Internat. Assn. Study Pain, Soc. Neurosci. Subspecialties: Neurobiology; Comparative physiology. Current work: Neurophysiology and neuroanatomy of somatic sensation, pain, demyelinating diseases (Mutiple Sclerosis), neurotransmitters in the somatosensory system; neural prostheses. Home: 5483 Endicott Ln Columbia MD 21044 Office: Bldg 89 Room INI07 Bethesda MD 20205

SCHNEIDER, ROBERT JAY, medical oncologist; b. Miami, Fla., May 31, 1949; s. Irving and Ethel (Pack) S.; m. Barbara Cunningham, June 1, 1974; children: Matthew Stephen, Kirsten Leigh. A.B., Boston U., 1972; M.D., Albert Einstein Coll. Medicine, 1975. Diplomate: Am. Bd. Internal Medicine. Intern Bronx Mcpl. Hosp. Ctr., 1975-76, resident, 1976-78; fellow in med. oncology Sloan-Kettering Meml. Ctr., N.Y.C., 1978-80; asst. prof. medicine N.Y. Med. Coll., Valhalla, 1980-81; adj. attending physician devel. chemotherapy service Meml. Sloan-Kettering Cancer Ctr., N.Y.C., 1981—; cons. cancer program No. Westchester Hosp. Ctr., Mt. Kisco, N.Y., 1981—; cons. TV documentary The Cancer Confrontation, 1982. Organizer and moderator Reader's Digest/No. Westchester Hosp. Ctr. Community Series on Cancer, Chappaqua, N.Y., 1981—. Am. Cancer Soc. fellow, 1978-79, 1980-81. Mem. Am. Soc. Clin. Oncology, Westchester County Med. Soc. Presbyterian. Club: Woodway Country (Darien, Conn.). Subspecialties: Cancer research (medicine); Oncology. Current work: Developmental chemotherapy, development of community cancer program and community education program in cancer-related issues. Office: No Westchester Hosp Ct 439 E Main St Mt Kisco NY 10549

SCHNEIDER, ROBERT WILLIAM, mechanical engineering consultant; b. S.I., N.Y., Dec. 30, 1925; s. Otto William and Anna Viola (Androvette) S.; m. Phyllis Mae Rantz, Dec. 24, 1946; children: Craig Robert, Dean Alan, David William. B.S. in Engring, Lehigh U., 1948, M.S., 1949. Registered profl. engr., Pa., Ont., Can. Asst. design and metall. engr. Linde Air Products Co., Tonawanda, N.Y., 1949-54; supr. ASME code shop inspection dept. Travelers Indemnity Co., Hartford, Conn., 1954-60; asst. supt. engring. and inspection dept. Oak Ridge Nat. Lab., 1960-68; mgr. engring. Bonney Forge, Allentown, Pa., 1968-81; pres. R.W. Schneider Assocs. (pressure vessels and piping cons.), Allentown, 1981—. Author: An Overview of the Structural Design of Piping Systems, 1978; contbr. articles on design and analysis of pressure vessels and piping to profl. jours. Served to ensign USN, 1943-46; PTO. Merck fellow, 1948-49. Fellow ASME; Mem. Welding Research Council, Nat. Soc. Profl. Engrs., Springhouse Farms Homeowners Assn. Presbyterian. Subspecialty: Mechanical engineering. Current work: Research and development to investigate stresses in nozzles of pressure vessels and piping due to external loads and internal pressure. Patentee in field. Home and Office: 3918 Lincoln Pkwy W Allentown PA 18104

SCHNEIDER, STEPHEN HENRY, climatologist; b. N.Y.C., Feb. 11, 1945; s. Samuel and Doris C. (Swarte) S.; m. Cheryl Kay Hatter, Aug. 19, 1978. B.S., Columbia U., 1966, M.S., 1967, Ph.D., 1971. Nat. Acad. Scis., NRC postdoctoral research asso. Goddard Inst. Space Studies NASA, N.Y.C., 1971-72; fellow advanced study program Nat. Center Atmospheric Research, Boulder, Colo., 1972-73, scientist, dep. head climate project, 1973-78, acting leader climate sensitivity group, 1978-80, head visitors program and dep. dir. advanced study program, 1980—; affiliate prof. U. Corp. Atmospheric Research Lamont-Doherty Geol. Obs., Columbia U.; mem. Carter-Mondale Sci. Policy Task Force, 1976—; mem. internat. sci. coms. climatic change, energy, food and pub. policy; expert witness congl. coms. Author: (with Lynne E. Mesirow) The Genesis Strategy: Climate and Global Survival, 1976, (with Lynne Morton) The Primordial Bond: Exploring Connections between Man and Nature through Humanities and Science, 1981, (with Randi S. Londer) The Coevolution of Climate and Life, 1984; sci. and popular articles on theory of climate, influence of climate on soc., relation of climatic change to world food, population, energy and environ. policy issues.; Editor: Climatic Change, 1976—, (with W. Bach) Interactions of Food and Climate, 1981, (with R.S. Chen and E. Boulding) Social Science Research and Climate Change: An Interdisciplinary Appraisal, 1983. Subspecialties: Climatology; Environmental Policy. Current work: Climatic modeling of paleoclimates and of human impacts on climate; e.g., carbon dioxide increases or aerosols generated by nuclear war. Also popular science writing. Office: Nat Center Atmospheric Research Box 3000 Boulder CO 80307

SCHNEIDERMAN, MARVIN ARTHUR, environmental toxicologist, statistician, educator; b. Bklyn., Dec. 25, 1918; s. Alexander and Mollie (Simpkins) S.; m. Irene Wolfson, Dec. 20, 1941; children: Jo Harte, Sarah, Susan. B.S., CCNY, 1939; postgrad., Ohio State U., 1944-45, Harvard U., 1944, London Sch. Hygiene and Tropical Medicine, 1959-60; Ph.D., Am. U., 1961. Clk. Nat. Container Corp., N.Y.C., 1939-40; jr. statistician Census Bur., Washington, 1940-41; statistician Office of Quartermaster Gen., Washington, 1941-42, Air Force, Dayton, Ohio, also Washington, 1946-48; with Nat. Cancer Inst., Bethesda, Md., 1948-80, assoc. dir., 1968-80; prof. biostats. in medicine Georgetown U., 1960—; prof. biostats. U. Pitts. Grad. Sch. Pub. Health, 1968-78; environ. cons. Clement Assocs., Arlington, Va., 1980—; prof. preventive medicine-biometrics Uniformed Services U. Health Scis., Bethesda, 1980—; cons. Nat. Acad. Scis. Contbr. articles to profl. jours. Served to 1st lt. USAAF, 1943-45. Recipient Disting. Service medal HEW, 1974; Outstanding Accomplishment in Govt. for Protection of Environ. award Environ. Def. Fund., 1975; Rockefeller Pub. Service fellow, 1959. Fellow Am. Statis. Assn., Royal Status. Soc., AAAS; mem. Biometrics Soc., Am. Assn. for Cancer Research, Internat. Epidemiol. Assn., Am. Soc. for Preventive Oncology, Soc. for Occupational and Environ. Health, Risk Assessment Soc., Phi Beta Kappa. Subspecialties: Environmental toxicology; Preventive medicine. Current work: Environmental protection-cancer prevention. Controlled clinical trials; mathematical modelling of biological systems, and risk analysis. Home: 6503 E Halbert Rd Bethesda MD 20817 Office: Clement Assocs 1515 Wilson Blvd Suite 700 Arlington VA 22209

SCHNEIDERWENT, MYRON OTTO, physicist, educator; b. Milw., Jan. 8, 1935; s. Richard R. and Anna C. S.; m. Marian L. Kolbeck, Dec. 13, 1954; children: Jean Erchul, Craig Schneiderwent. B.S., U. Wis. Stevens Point, 1960; M.A., Western Mich. U., 1963; M.S., U. Miss., 1964; Ed.D., U. No. Colo., 1970. Tchr. sci. and math. Muskegon (Mich.) Pub. Schs., 1960-63; instr. sci. and math. Interlochen (Mich.) Arts Acad., 1964-67; assoc. prof. physics U. Wis. Superior, 1967—; cons. Harvard Project Physics, Center Lake Superior Environ. Studies, U. Wis. Extension, Tech. Systems Inc., Combase. Author: Physical Science, 1979, Exploring Alternative Energy Systems, 1979, others. Served with U.S. Army, 1954-56. Recipient award Milw. Jour., 1976; Bendix award Soc. Physics Students, 1975. Mem. Nat. Sci. Tchrs. Assn., Am. Educators Tchrs. Sci., Am. Assn. Physics Tchrs., Lake Superior Sci. and Math Tchrs. Club, Assn. U. Wis. Faculties, Sigma Pi Sigma. Subspecialties: Theoretical physics; Solar energy. Current work: Solar energy and theoretical physics. Home: 1407 N 21st St Superior WI 54880 Office: U Wis Barstow 208 Superior WI 54880

SCHNIEDERJANS, MARC JAMES, management science educator; b. Pocahontas, Ark., Oct. 8, 1950; s. Oliver H. and Florence (Schutte) S.; m. Jill Marlene Schniederjans, Aug. 13, 1971; 1 son, Alexander J. B.S., U. Mo., 1972; M.B.A., St. Louis U., 1974, Ph.D., 1978. Program dir. St. Louis U., 1975-78; lectr. U. Mo.-St. Louis, 1976-78, asst. prof., 1979-80, U. Nebr.-Omaha, 1978-79, asst. prof. mgmt., Lincoln, 1981—; asst. prof. U. Hawaii, Honolulu, 1980-81; cons. Blue Hills Homes Corp., St. Louis, 1979—, Ralston Purina Corp., 1980—, Telex Corp., Lincoln, 1983—. Author: (with N.K. Kwak) Managerial Applications of Operations Research, 1982; contbr. articles to profl. jours. Mem. Am. Inst. Decision Scis., Inst. Mgmt. Scis., Ops. Research Soc. Am., Am. Prodn. and Inventory Soc. Subspecialties: Operations research (mathematics); Information systems, storage, and retrieval (computer science). Current work: Developing and applying operations research techniques in micro and macro computer information systems to aid in medical legal and business decision making. Home: 5220 S 66th Circle Lincoln NE 68516 Office: Dept of Management University of Nebraska Lincoln NE 68588

SCHNITZER, HOWARD JOEL, educator, physicist; b. Newark, Nov. 12, 1934; s. Albert and Helen (Ehrlich) S.; m. Phoebe Kazdin, May 22, 1966; children: Mark Jacob, Elizabeth Karen. B.S. in Mech. Engring., Newark Coll. Engring., 1955; Ph.D. in Physics, U. Rochester, 1960. Postdoctoral research asso. U. Rochester, 1960-61; mem. faculty Brandeis U., Waltham, Mass., 1961—, prof. physics, 1968—, dept. chmn., 1981-83; vis. prof. Rockefeller U., 1969-70; hon. research asso. Harvard U., 1974, 76-83. Asso. editor: Phys. Rev. Letters, 1978-80. Sloan fellow, 1965-67; Guggenheim fellow, 1983-84. Fellow Am. Phys. Soc.; mem. Sigma Xi. Subspecialty: Theoretical physics. Home: 397 Highland St Newtonville MA 02160 Office: Dept Physics Brandeis Univ Waltham MA 02254

SCHNURRENBERGER, PAUL ROBERT, veterinary educator, cons.; b. Youngstown, Ohio, Aug. 19, 1929; s. Gilbert Miles and Bernice Wannamaker (Parshall) S.; m. Marsha Blatt, Jan. 8, 1936; children: Jody Lynn, Gregory Paul. D.V.M., Ohio State U., 1953; M.P.H., U. Pitts., 1958; cert., Am. Coll. Vet. Preventive Medicine. Pvt. vet. practice, Otterbein, Ind., 1953-55; chief pub. health veterinarian Ohio Dept. Health, 1956-63, Ill. Dept. Pub. Health, 1963-72; prof. pub. health Auburn U., 1972-74, 76—; prof., head dept. vet. pub. health and preventive medicine Ahmadu Bello U., Zaria, Nigeria, 1974-76; cons. Ala. Bd. Health. Author: Diseases Transmitted from Animals to Man, 1975, Autotutorials on Preventive Medicine, 1976, Outline of the Zoonoses, 1981, Principles of Health Maintenance, 1982; contbr. articles to sci. jours. Served to 1st lt. USAF, 1953-55. Fulbright sr. scholar, 1982. Mem. AVMA, Wildlife Disease Assn., U.S. Animal Health Assn., Ala. Vet. Med. Assn. Subspecialties: Preventive medicine (veterinary medicine); Epidemiology. Current work: Health maintenance, epidemiology of the zoonoses. Home: 862 Cary Dr Auburn AL 36830 Office: Dept Microbiology Sch Vet Medicine Auburn U Auburn AL 36849

SCHOBER, ROBERT CHARLES, electrical engineer; b. Phila., Sept. 20, 1940; s. Rudolph Ernst and Kathryn Elizabeth (Ehrisman) S.; m. Mary Eve Kanuika, Jan. 14, 1961; children: Robert Charles, Stephen Scott, Susan Marya. B.S. in Engring. (Scott Award scholar), Widner U., 1965; postgrad., Bklyn. Poly. Extension at Gen. Electric Co., Valley Forge, Pa., 1965-67, U. Colo., 1968-69, Calif. State U.-Long Beach, 1969-75, U. So. Calif., 1983—. Engr. Gen. Electric Co., Valley Forge, 1965-68, Martin Marietta Corp., Denver, 1968-69; sr. engr. Jet Propulsion Lab., Pasadena, Calif., 1969-73; mem. tech. staff Hughes Semiconductor Co., Newport Beach, Calif., 1973-75; prin. engr. Am. Hosp. Supply Corp., Irvine, Calif., 1975-83; sr. staff engr. TRW Systems, Redondo Beach, Calif., 1983—; cons. Biomed. LSI, Huntington Beach, Calif. Mem. IEEE (student br. pres. 1963-65), Assn. for computing Machinery, Tau Bea Pi. Republican. Subspecialties: Integrated circuits; Cardiology. Current work: Develop large scale integrated circuits which employ and advance the leading edge of technology for biomedical applications including cardiac pacemakers and monitors, implantable and external hearing devices and electronic visual devices; also develop very high speed analog to digital converters and logic in the Gigaherz region. Patentee cardiac pacemakers. Home: 9411 Tiki Circle Huntington Beach CA 92646 Office: TRW Defense Systems Group Mail MSTS-1843 One Space Park Redondo Beach CA 90278

SCHOECH, WILLIAM JOSEPH, mechanical engineering educator; b. Clearwater, Fla., Mar. 14, 1944; s. Walter Frederick and Hilda Louise (Uhlig) S.; m. Susan Mary Sales, July 12, 1969; 1 son, Eric William. B.S., Valparaiso U., 1966; M.S., Pa. State U., 1968; Ph.D., Purdue U., 1971. Registered profl. engr., Ind. Asst. prof. mech. engring. Valparaiso U., 1971-78, assoc. prof., 1978—. Mem. ASME, Am. Soc. Metals, Soc. Mfs. Engrs., Tau Beta Pi. Lutheran. Subspecialties: Manufacturing engineering; Graphics, image processing, and pattern recognition. Current work: Computer graphics relating to manufacturing. Office: Valparaiso U Gellersen Engring Ctr Valparaiso IN 46383

SCHOEN, GEORGE JANSSEN, mechanical engineer; b. Danbury, Conn., May 7, 1938; s. Ernest George and Alma Lydia (Janssen) S.; m. Barbara Jean Smalley, Sept. 9, 1962; children: Susan E., Cindy-Anne. B.S.M.E., Worcester Poly. Inst., 1960. Application engr. Barden Corp., Danbury, Conn., 1961-74, supr. product engring., 1974-76, supr. product design, 1976-82, adminstr., 1982—. Served as 1st lt. Signal Corps U.S. Army, 1960-68. Mem. ASME, Sigma Xi. Lutheran. Subspecialty: Mechanical engineering. Current work: Design, use and testing of high precision ball bearings, principally for aerospace and machine tool use.

SCHOEN, MAX HOWARD, dentist, educator; b. N.Y.C., Feb. 4, 1922; s. Adolph and Ella (Grossman) S.; m. Beatrice Mildred Hoch, Feb. 5, 1950; children: Steven Charles, Karen Ruth. B.S., U. So. Calif., 1943, D.D.S., 1943; M.P.H., UCLA, 1962; D.P.H., U. Calif., Los Angeles, 1969. Diplomate: Am. Bd. Dental Pub. Health. Practice dentistry, Los Angeles, 1947-54, founding partner group dental practice, So. Los Angeles, 1954—; vis. prof. Sch. Dental Medicine, U. Conn., 1972; prof., dean pro-tem, asso. dean Sch. Dental Medicine, State U. N.Y., Stony Brook, 1973-76; prof. preventive dentistry and pub. health U. Calif. Sch. Dentistry, Los Angeles, 1976—, sect. chmn., 1976-82; mem. Com. for Nat. Health Ins.; chmn. dental adv. bd. Headstart, Los Angeles, 1966; cons. in field. Author papers in field. Served to capt. Dental Corps U.S. Army, 1943-46. Fellow Am. Pub. Health Assn.; mem. Inst. Medicine, Am. Dental Assn., Am. Assn. Pub. Health Dentists, Fedn. Dentaire Internat., Am. Assn. Dental Schs., Group Health Assn. Am. Subspecialty: Public health. Home: 5818 Sherbourne Dr Los Angeles CA 90056 Office: Sch Dentistry Univ Calif Los Angeles CA 90024

SCHOENFELD, MYRON PAUL, cardiologist; b. N.Y.C., Nov. 10, 1928; s. George and Rhoda (Kahn) S.; m. Gloria Toby Edis, June 14, 1959; children: Bradley Jon, Glenn Murray, Dawn Rhoda, Melody Lynn. B.A., NYU, 1948; M.A., Columbia U., 1949; M.D., Chgo. Med. Sch., 1953. Diplomate: Am. Bd. Internal Medicine. Intern Kings County Hosp., Bklyn., 1953-54; resident Kingsbridge VA Hosp., Bronx, N.Y., 1954-55, 58-59; research fellow Mt. Sinai Hosp., N.Y.C., 1959-60; practice medicine specializing in cardiology, sr. ptnr. Schoenfeld-Edis Med. Assoc., P.C., Scarsdale, N.Y., 1960—; founder, pres. Life-Line Spl. Med. Services, Scarsdale, 1978—, Royal Artcrafts, Inc., 1970—; cons. and attending numerous hosps. Editor-in-chief: Cardiovascular Ultrasonography, 1982-83; contbr. numerous articles in med. jours.; mem. editorial bd.: Investing Profl. Mag, 1975-76. Pres. Deer Hill Civic Assn., Scarsdale, 1965-83. Served to capt. USAF, 1955-57. Recipient Disting. Alumnus award Chgo. Med. Sch., 1974. Fellow ACP, N.Y. Cardiol. Soc., Am. Coll. Cardiology; mem. Soc. for Exptl. Biology and Medicine, Am. Fedn. For Clin. Research, Phi Beta Kappa. Democrat. Jewish. Subspecialties: Cardiology; Internal medicine. Current work: Echocardiography, lipid metabolism, clinical biochemistry, bio-medical instrumentation. Inventor infusion monitor, 1963. Office: Schoenfeld-Edis Med Assoc 2 Overhill Rd Scarsdale NY 10583

SCHOENIKE, ROLAND ERNEST, forestry educator; b. Watertown, Wis., May 9, 1925; s. Ernest Gustav and Erna (Hoyer) S. B.S., U. Minn., 1951, M.S., 1953, Ph.D., 1961. Forester U.S. Forest Service, Crossete, Ark., 1953-57; research asst. U. Minn., Mpls., 1957-62, asst. prof. dept. forestry, 1963; assoc. prof. Clemson(S.C.) U., 1963-77, prof., 1977—. Editor: Bibliography of Yellow Poplar, 1981. Served to sgt. U.S. Army, 1945-47. Mem. Soc. Am. Foresters, Am. Forestry Assn., Sigma Xi. Lutheran. Subspecialties: Plant genetics; Resource management. Current work: Genetics and breeding of forest trees, arboretum management, Christmas tree production and research. Office: Dept Forestry Clemson U Clemson SC 29630

SCHOENING, WILLIAM EDWARD, photographic astronomy researcher; b. St. Louis, Oct. 29, 1941; s. William F. and Catharine S. B.S., Central Meth. Coll., 1964; teaching cert., Central Mo. State Coll., 1965. Tchr. Maplewood (Mo.) Sch. System, 1966-67; photographic researcher Kitt Peak Nat. Obs., 1967—; head Photographic Research Lab., 1967—, sr. research assoc., 1970—. Contbr. articles to sci. jours. Mem. Am. Astron. Soc., Ariz. Paper and Photographic Conservation Group. Republican. Methodist. Subspecialty: Optical astronomy. Current work: Photographic Astronomical photography; emulsion hypersensitizing to reduce exposure times at the telescopes. Home: 542 N Country Club Dr Tucson AZ 85716 Office: 950 N Cherry St Tucson AZ 85726

SCHOENSTADT, ARTHUR LORING, mathematician, educator; b. N.Y.C., Nov. 8, 1942; s. Arthur Edmund and Shirely (Sinclair) S.; m. Elizabeth Ann Beardsley, Aug. 29, 1964; children: Michele Suzan, Arthur Lloyd. B.S. in Math, Rensselaer Poly. Inst., Troy, N.Y., 1964, M.S., 1965, Ph.D., 1968. Assoc. prof. math. Naval Postgrad. Sch., Monterey, Calif., 1970—; cons. INSIGHT, Inc., Washington, 1982—. Editor: Information Linkage Between Applied Math and Industry, vol. 1, 1979, vol. 2, 1980; Contbr. articles to profl. jours. Served to capt. U.S. Army, 1968-70. Decorated Army Commendatation medal with oak leaf cluster; recipient Excellence in Teaching award Naval Postgrad. Sch., 1981. Mem. Soc. for Indsl. and Applied Math., Armed Forces Communications-Electronics Assn., Sigma Xi (treas. local chpt. 1982-83). Subspecialties: Applied mathematics; Numerical analysis. Current work: Numerical methods in weather predictions; numerical solution of system of linear equations and of differential equations; simulation and modeling. Office: Dept Math Naval Postgrad Sch Code 53Zh Monterey CA 93940

SCHOENY, RITA SUE, environmental health educator, researcher; b. Cin., Sept. 16, 1949; d. Richard Thomas and Rita Catherine (Berg) S. B.S. in Biology magna cum laude (Pres.'s scholar), U. Dayton, 1970; Ph.D. in Microbiology, U. Cin., 1977. Med. lab. technician Our Lady of Mercy Hosp., Cin., 1971, Public Health Dept., 1971-72; teaching asst. U. Cin., 1973-76, postdoctoral fellow, 1977-80, asst. prof. environ. health, 1980—, mem. biohazards com., 1974-78, carcinogen safety com., 1978—, substantial risk com., 1979—, discussion meeting coordinator, 1981—; cons., lectr. in field. Contbr. articles to profl. pubIs. Active NOW; coordinator Pick Up Team Dance Co., 1980-82. Recipient Nat. Research Service award, 1979. Mem. Am. Soc., Microbiology, Environ. Mutagen Soc., Fedn. Am. Scientists, AAAS, Iota Sigma Pi. Subspecialties: Environmental toxicology; Genetics and genetic engineering (biology). Current work: Metabolism and mutagenicity of chemical carcinogens; induction by environmental agents of enzymes involved in xenobiotic metabolism of mutagenic agts.; biological activities of complex environmental mixtures; mutagenicity and modification of mutagenesis by chlorinated hydrocarbons; relationship of DNA damage and repair to mechanisms of teratogenesis; co-mutagenesis and promotion of mutagenesis by chemical carcinogens. Office: Kettering Lab 3223 Eden Ave Cincinnati OH 45213

SCHOERER, FRANK, mech. engr.; b. N.Y.C., Sept. 5, 1922; s. Louis Francis and Emily Louise (Roemer) S.; m. Lois Green, June 25, 1949; 1 son, John Arnold. B.S., Webb Inst. Naval Architecture, Glen Cove, N.Y., 1944; M.S., M.I.T., 1947. From engr. to mgr. advanced devel. Gas Turbine div. Westinghouse Aircraft, Phila. and Kansas City, Mo., 1947-57; from adv. engr. to mgr. reactor div. Westinghouse Bettis Atomic Power Lab., Pitts., 1957-64; from mgr. engrng. to tech. dir. NUS Corp., Washington, 1964-67; sr. assoc. Richard Lowe & Assoc., Washington, 1967-75; tech. dir. Nuclear Projects, Inc., Rockville, Md., 1975—; mem. nuclear safety rev. bd. Ginna Nuclear Power Plant, Callaway Nuclear Power Plant. Served to lt. U.S. Navy, 1943-46. Mem. Am. Nuclear Soc., ASME. Episcopalian. Clubs: Severn River Yacht (Annapolis, Md.); Chevy Chase (Md.) Recreation Assn. Subspecialties: Nuclear engineering; Mechanical engineering. Current work: Design, licensing and operation of nuclear power plants for electric power generation. Home: 7213 Rollingwood Dr Chevy Chase MD 20815 Office: 5 Choke Cherry Rd Rockville MD 20850

SCHOKNECHT, JEAN DONZE, mycologist, electron microscopist, educator, researcher; b. Urbana, Ill., Oct. 31, 1943; d. Joseph M. and Genevieve S. (Stanis) Donze. B.S. (Mathew Arnold scholar), U. Ill., 1965, M.S., 1967, Ph.D., 1972. Teaching and research asst. U. Ill., Champaign, Urbana, 1965-71, research assoc., 1972-74; asst. prof. microbiology Ind. State U., Terre Haute, 1974-78, assoc. prof., 1978—; mem. adj. faculty Ind. U. Med. Sch., 1974-78; assoc. mycologist sect. botany and plant pathology Ill. Natural History Survey, Champaign, 1982—; cons. to Western Paper & Mfg. Co.; cons. on fungi in foods Dept. Agr. Contbr. articles to sci. jours. Bd. dirs. Wabash Valley Audubon Soc., 1982-83. NSF grantee, 1980. Mem. Mycol. Soc. Am., Med. Mycol. Soc. Ams., Electroc Microscope Soc. Am., Am. Micros. Soc., Bot. Soc. Am., Brit. Mycol. Soc., Brit. Lichen Soc., Sigma Xi. Club: University (Terre Haute). Subspecialties: Microbiology; Microscopy. Current work: Ultrastructure, mineral accumulation, utilization, etc. by cells; teaching and research in cell biology, microbiology. Home: 119 Jackson Blvd Terre Haute IN 47803 Office: Life Scis Dept Ind State U Terre Haute IN 47809

SCHOLES, NORMAN WALLACE, pharmacology educator; b. Ogden, Utah, June 9, 1930; s. Wallace Burnham and Rozanna Lee S.; m. Adamina Badillo, June 30, 1950; 1 son, Norman Wallace; m. Marianne Schenkel, Mar. 27, 1974; children: Karl A. Armbrust, Walter T. Armbrust. B.S., U. Utah, 1953; M.S., UCLA, 1956, Ph.D., 1960; M.S. in Edn, U. So. Calif., 1973; B.S. in Pharmacy, Creighton U., 1982. Registered pharmacist, Nebr. With City of Hope Nat. Med. Center, Duarte, Calif., 1960-64; neuropharmacologist U. Calif.-Davis, 1964-68; mem. faculty Creighton U., Omaha, Nebr., 1968—, assoc. prof. pharmacology. Contbr. numerous neuropharm. articles to profl. publs. Active CAP, 1976—. Served with USAF, 1947-50. NIH fellow, 1955-58; Giannini Found. fellow, 1958-60. Mem. Am. Soc. Pharmacology and Exptl. Therapeutics, Soc. Exptl. Biology and Medicine, Nebr. Acad. Scis., Fedn. Am. Socs. Exptl. Biology, Sigma Xi. Subspecialties: Neuropharmacology; Psychopharmacology. Current work: Mechanism of action of psychtropic drugs. Home: 11615 Calhoun Rd Omaha NE 68112 Office: 2500 California St Omaha NE 68178

SCHONBECK, MARK WALTER, research crop scientist; b. Goshen, N.Y., Oct. 30, 1950; s. Rudolph George and Mary Elisabeth (Parsons) S. A.B., Oberlin Coll., 1972; postgrad., U. Calif.-Santa Barbara, 1973; Ph.D., U. Glasgow, Scotland, 1977. Research assoc. So. Weed Sci. Lab., Agrl. Research Service, Dept. Agr., Stoneville, Miss., 1978-79; postdoctoral research assoc. U. Calgary, Alta., Can., 1980-81; research plant physiologist Shell Devel. Co., Modesto, Calif., 1981—. Contbr. articles on intertidal ecology, weed seed germination to profl. jours. Mem. Am. Soc. Agronomy, Am. Soc. Plant Physiologists, Am. Inst. Biol. Scis., Sigma Xi. Scientologist. Subspecialties: Plant physiology (agriculture); Plant growth. Current work: Plant growth regulation, field physiology of crops, agronomy, plant growth regulator, field evaluation on major crops, yield enhancement. Home: 301 Standiford Ave Apt 134 Modesto CA 95350 Office: Shell Devel Co PO Box 4248 Modesto CA 95352

SCHONFELD, GUSTAV, medical educator, head research center; b. Mukacevo, Czechoslovakia, May 8, 1934; came to U.S., 1946; s. Alexander and Helena (Gottesmann) S.; m. Miriam Steinberg, May 28, 1961; children: Joshua, Julia, Jeremy. B.A., Washington U., St. Louis, 1956, M.D., 1960. Diplomate: Am. Bd. Internal Medicine. Intern. NYU-Bellevue Hosp., N.Y.C., 1960-61, resident, 1961-63; asst. prof. Med. Sch. Washington U., St. Louis, 1968-70, assoc. prof., 1972-77, prof. medicine, 1977—; head Lipid Research Ctr., St. Louis, 1972—; assoc. prof. MIT, Cambridge, Mass., 1970-72; mem. nutrition com. Am. Heart Assn., 1981—, grants rev. com., 1980-83; Mem. grants rev. com. Am. Egg Bd., 1980-83, Metabolism Drugs Evaluation Com. FDA, 1982—. Assoc. editor: Circulation, Jour. Am. Heart Assn, 1983. Served to capt. USAF, 1966-68. Grantee Nat. Heart Lung Blood Inst., 1972—. Fellow ACP; mem. Assn. Am. Physicians, Am. Soc. Clin. Investigation, Am. Soc. Biol. Chemists, Am. Soc. Clin. Nutrition. Democrat. Jewish. Subspecialties: Internal medicine; Preventive medicine. Current work: Hyperlipidemia; atherosclerosis; diabetes, lipoprotein structure, synthesis and metabolism; coronary heart disease prevention. Office: Sch Medicine Washington U 4566 Scott Ave Saint Louis MO 63110

SCHOOLAR, JOSEPH CLAYTON, pharmacologist, psychiatrist; b. Marks, Miss., Feb. 28, 1928; s. Adrian Taylor and Leah (Covington) S.; m. Betty Peck, Nov. 2, 1960; children—Johnathan Covington, Cynthia Jane, Geoffrey Michael, Catherine Elizabeth, Adrian Carson. A.B. (Univ. Faculty scholar), U. Tenn., 1950, M.S., 1952; Ph.D. (Merck scholar), U. Chgo., 1957, M.D., 1960. Diplomate: Am. Bd. Psychiatry and Neurology. Chief sect. drug abuse research Tex. Research Inst. Mental Scis., Houston, 1966-72, chief div. clin. research, 1967-69, chief clin. and sociol. research div., 1969-72, asst. dir. inst., 1968-72, dir., 1972—; prof. pharmacology Baylor Coll. Medicine, 1974—, prof. psychiatry, 1975—; chmn. central office research rev. com. Tex. Dept. Mental Health and Mental Retardation, 1975—; mem. drug abuse tng. rev. com. Nat. Inst. Drug Abuse, HEW, 1974-79. Editor: (with C.M. Gaitz) Research and the Psychiatric Patient, 1975, (with J.L. Claghorn) The Kinetics of Psychiatric Drugs, 1979. Recipient Eugen Kahn award Baylor Coll. Medicine, 1963. Fellow Am. Psychiat. Assn., Am. Coll. Psychiatrists; mem. Am. Soc. Clin. Pharmacology and Therapeutics, Am. Coll. Neuropsychopharmacology, Soc. Neurosci., Am. Acad. Clin. Toxicology, Internat. Council on Alcohol and Addictions, AAAS, Tex. Med. Assn. (chmn. spl. com. on alcoholism and drug abuse 1976—), Sigma Xi. Episcopalian. Subspecialty: Psychopharmacology. Home: 2222 Sunset Blvd Houston TX 77005 Office: 1300 Moursund Ave Houston TX 77030

SCHOONMAKER, RICHARD CLINTON, chemistry educator; b. Schenectady, N.Y., Dec. 21, 1930; s. James C. and Edna (Neuville) S.; m. Dina Bikerman, Feb. 13, 1956; children: Karen, Dirk, Timothy, Jonathan. B. Chem. Eng., Yale U., 1952; Ph.D., Cornell U., 1960. Research physicist Columbia U., 1959-60; mem. faculty Oberlin Coll., 1960—; prof. chemistry, 1967—. Author: Composition, Reaction, Equilibrium, 1970; contbr. articles to profl. jours. Served to lt. (j.g.) USNR, 1953-56. Gen. Electric predoctoral fellow, 1959; NSF faculty fellow, 1966-67; Brit. Sci. Research Coucil vis. sr. research fellow, 1973-74; NSF profl. devel. fellow U. Calif.-Berkeley, 1980-81. Mem. Am. Chem. Soc., Am. Phys. Soc., Sigma Xi. Subspecialties: Physical chemistry; Surface chemistry. Current work: Molecular beam scattering from surfaces; mechanism of condensation. Home: 270 E College St Oberlin OH 44074 Office: Dept of Chemistry Oberlin College Oberlin OH 44074

SCHOPF, THOMAS JOSEPH MORTON, geophysicist, researcher, educator; b. Urbana, Ill., Aug. 26, 1939. A.B., Oberlin Coll., 1960; Ph.D., Ohio State U., 1964. Teaching asst. Ohio State U., 1960-62; asst. Orton Geol. Mus., 1962-64; with U.S. Geol. Survey, 1961-76; postdoctoral fellow Marine Biol. Lab., Woods Hole, Mass., 1964-67, lectr., 1970-71, instr., 1972-76, cons., 1977; asst. prof. dept. geol. sci. Lehigh U., 1967-69; asst. prof. dept. geophys. sci. U. Chgo., 1969-72, assoc. prof., 1972-78, prof., 1978—, also mem. com. on evolutionary biology; faculty assoc. Center Advanced Studies Field Mus. Natural History, Chgo., 1969—, research assoc. dept. geology, 1980—; investigator Friday Harbor Labs. U. Wash., 1977, 78, 79; vis. prof. Inst. Geology and Paleontology, U. Hamburg, Germany, 1978-79; vis. research assoc. div. biology Calif. Inst. Tech., 1981-82, vis. research assoc. div. geol. and planetary sci., 1981-82. Mem. editorial bd.: Evolutionary Theory, 1973—, Paleobiology, 1975—, Biochem. Systematics and Ecology, 1976-77; contbr. chpts. to books and articles to profl. jours. Guggenheim fellow, 1981-82; NSF grantee; Office Naval Research grantee. Fellow Geol. Soc. Am.; mem. Am. Assn. Petroleum Geologists, AAAS, Am. Geol. Inst., Am. Soc. Zoologists, Atlantic Estuarine Research Soc., Bermuda Biol. Sta., Internat. Bryozoology Assn., Marine Biol. Lab., Pander Soc., Paleontol. Soc. (Schuchert award 1976), Palaeontol. Assn. Gt. Brit., Paleontol. Research Instn., Soc. Econ. Geologists, Soc. Systematic Zoology, Soc. Study Evolution, Systematics Assn., Internat. Paleontol. Union, Sigma Xi. Subspecialties: Evolutionary biology; Paleobiology. Office: 5734 S Ellis Ave Chicago IL 60637

SCHOPP, ROBERT THOMAS, physiologist, educator, researcher; b. Pontiac, Ill., Nov. 5, 1923; s. Thomas Henry and Rose Catherine (Tyler) S.; m. Kathleen Alice Ivens, June 15, 1950; children: Susan L. Schopp Wittenborn, Daniel Thomas, Jamie C. Schopp Lane, Lori Ann. Student, Lewis Coll., Lockport, Ill., 1946-47; B.S. in Biology, U. Ill., 1950, M.S., 1951; Ph.D. in Physiology, U. Wis., 1956. Instr. dept. biology St. Norbert Coll., DePere, Wis., 1951-52; teaching asst. dept. physiology U. Wis. Sch. Medicine, Madison, 1952-55; instr., 1955-56; instr. dept. physiology U. Colo. Sch. Medicine, Denver, 1956-59, asst. prof., 1959-67; asso. prof. dept. physiology Kirksville (Mo.) Coll. Osteopathy and Surgery, 1967-69; prof. physiology sect. So. Ill. U. Sch. Dental Medicine, Edwardsville, 1969—, head sect., 1969—. Contbr. articles to sci. jours. Served with USMC, 1943-46. NIH Heart Inst. fellow, 1966. Mem. Am. Physiol Soc., Am. Soc. Pharmacology and Exptl. Therpeutics, Western Pharmacology Soc., Nat. Fencing Coaches Assn. Am. Conservative Republican. Roman Catholic. Clubs: So. Ill. U. Fencing; North County Fencing (Florissant, Mo.). Lodge: KC. Subspecialty: Physiology (medicine). Current work: Exercise physiology. Office: Sch Dental Medicine So Ill U Edwardsville IL 62026

SCHOR, JOSEPH MARTIN, biochemist; b. Bklyn., Jan. 10, 1929; s. Aaron Jacob and Rhea Iress (Kay) S.; m. Sandra Anne Moshman, Sept. 6, 1953; children: Esther Helen, Joshua David, Gideon Alexander. B.S. magna cum laude, CCNY, 1951; Ph.D., Fla. State U., 1957. Cert. profl. chemist Am. Inst. Chemists. Sr. research chemist Armour Pharm. Co., Kankakee, Ill., 1957-59; Lederle Labs., Pearl River, N.Y., 1959-64; dir. biochemistry Endo Labs., Inc., E. I. DuPont de Nemours & Co., Inc., Garden City, N.Y., 1964-76; v.p. sci. affairs Forest Labs., N.Y.C., 1977—. Editor, contbg. author: Chemical Control of Fibrinolysis-Thrombolysis, 1970; contbr. articles to profl. jours. USPHS fellow, 1955-57. Fellow Am. Inst. Chemists, Internat. Soc. Hematology; mem. Internat. Soc. Thrombosis and Hemostasis, Am. Chem. Soc. (subsect. chmn. 1971-72, alt. councillor 1973-75), N.Y. Acad. Sci., AAAS, Phi Beta Kappa, Sigma Xi. Subspecialties: Biochemistry (medicine); Pharmacokinetics. Current work: Drug metabolism, analgetics, cardiovascular drugs and controlled-release technology. Patentee fibrinolytic agts. and sustained-release tech. Home: 28 Meleny Rd Locust Valley NY 11560 Office: 919 3d Ave New York NY 10022

SCHORRY, ROBERT ELMER, mech. engr.; b. Cin., Dec. 30, 1954; s. Elmer Daniel and Kathryn May (Neack) S.; m. Elizabeth Kolks, July 8, 1978. B.S. in Mech. Engring, U. Cin., 1977, M.S., Wayne State U., 1982. Automotive body engr. Chrysler Corp., Detroit, 1977-81; sr. project engr. Structural Mechanics Cons. Corp., Southfield, Mich., 1981-82; project engr. Cin. Milacron (applied sci. research and devel. dept.), 1982—. Recipient Student award ASTM, 1976. Mem. Soc. Automotive Engrs., ASME. Republican. Subspecialties: Theoretical and applied mechanics; Mechanical engineering. Current work: Stress analysis and design engring., computer aided engring. via finite element analysis. Office: 4701 Marburg Ave Cincinnati OH 45209

SCHOWALTER, WILLIAM RAYMOND, chemical engineering educator; b. Milw., Dec. 15, 1929; s. Raymond Philip and Martha (Kowalke) S.; m. Jane Ruth Gregg, Aug. 22, 1953; children: Katherine Ruth, Mary Patricia, David Gregg. B.S., U. Wis., 1951; postgrad., Inst. Paper Chemistry, 1951-52; M.S., U. Ill., 1953, Ph.D., 1957. Asst. prof. dept. chem. engring. Princeton U., 1957-63, asso. prof., 1963-66, prof., 1966—, acting chmn. dept. chem. engring., 1971, chmn. dept. chem. engring., 1978—, asso. dean Sch. Engring. and Applied Sci., 1972-77; Sherman Fairchild Disting. scholar Calif. Inst. Tech., 1977-78; vis. fellow U. Salford, Eng., 1974; vis. sr. fellow Sci. Research Council, U. Cambridge, Eng., 1970; cons. to chem. and petroleum cos.; mem. editorial adv. bd. McGraw-Hill Pub. Co., 1964—; co-chmn. Internat. Seminar for Heat and Mass Transfer, 1970; mem. vis. com. for chem. engring. MIT, 1979—, Lehigh U., 1980—; mem. evaluation panel Ctr. Chem. Engring., Nat. Bur. Standards, 1982—; mem. commn. engring. and tech. systems NRC, 1983—. Author: Mechanics of Non-Newtonian Fluids, 1978; mem. editorial com.: Ann. Rev. Fluid Mechanics, 1974-80; editorial bd.: Internat. Jour. Chem. Engring., 1974—, Indsl. and Engring. Chemistry Fundamentals, 1975-78, Jour. Non-Newtonian Fluid Mechanics, 1976—, Am. Inst. Chem. Engrs. Jour, 1979—; Contbr. articles to profl. jours. Served with AUS, 1953-55. Recipient Lectureship award Chem. Engring. div. Am. Soc. Engring. Edn., 1971, Disting. Service citation Coll. Engring., U. Wis.-Madison, 1983. Mem. Am. Inst. Chem. Engrs. (William H. Walker award 1982), Nat. Acad. Engring., Am. Chem. Soc., Soc. Rheology (exec. com. 1977-79, v.p. 1981-83, pres. 1983—), Sigma Xi, Tau Beta Pi, Phi Lambda Upsilon, Phi Eta Sigma. Subspecialties: Chemical engineering; Industrial engineering. Home: 106 Crestview Dr Princeton NJ 08540

SCHRADER, EDWARD LEON, geologist; b. Vicksburg, Miss., June 19, 1951; s. Edward and Dorothy (Chaney) S.; m. Myra Lee Ladner, June 2, 1973; children: Melanie Denise, Edward Austin. B.S., Millsaps Coll., 1973; M.S., U. Tenn., 1975; Ph.D., Duke U., 1977. Exploration geologist Chevron Minerals, Denver, 1975-76; geochemist Deep Sea Drilling Project, LaJolla, Calif., 1977, 79; project geologist Chevron Resources, Inc., Denver, 1977-78; asst. prof. econ. geology U. Ala., Tuscaloosa, 1978-80; adj. prof., 1980-82; chief geologist J.M. HuberCorp., Macon, Ga., 1981—; vis. lectr. Duke U., spring 1983; Co-chmn. UN sponsored conf., Athens, Greece, 1980. Editor, author: Geology of the Southeastern Clay Belt, 1983. Pres. Baptist Men's Orgn., Ingleside Baptist Ch., Macon, 1982-83; bd. visitors Duke U., 1982-83; mem. program adv. com. U. Ala., Tuscaloosa, 1979-80. HEW grantee, 1975; U.S. Dept. Interior grantee, 1978-80. Mem. Geol. Soc. Am., AIME (co-chmn. Ga. subsect. 1982), Clay Minerals Soc., Soc. Econ. Geologists, Internat. Assn. on the Genesis of Ore Deposits, Phi Beta Kappa, Sigma Xi. Democrat. Subspecialties: Geochemistry; Ore deposit origins, distributions. Current work: Geochemical and mineralogical analyses of modern and ancient ore forming environments; applications for exploration, also seafloor hotspring chemistry. Home: 390 Lokchapee Dr Macon GA 31210 Office: J M Huber Corp Route 4 Huber Macon GA 31298 Home: 390 Lokchapee Dr Macon GA 31210

SCHRAMM, DAVID NORMAN, astrophysicist, educator; b. St. Louis, Oct. 25, 1945; s. Marvin and Betty (Math) S.; m. Melinda Holzhauer, 1963 (div. 1979); children: Cary, Brett.; m. Colleen Rae, 1980. S.B. in Physics, M.I.T., 1967, Ph.D., Calif. Inst. Tech., 1971. Research fellow in physics Calif. Inst. Tech., Pasadena, 1971-72; asst. prof. astronomy and physics U. Tex., Austin, 1972-74; assoc. prof. astronomy, astrophysics and physics Enrico Fermi Inst. and the Coll., U. Chgo., 1974-77, prof., 1977—, Louis Block prof. phys. scis., 1982—, acting chmn. dept. astronomy and astrophysics, 1977, chmn., 1978—; resident cosmotologist Fermilab, 1982—; cons., lectr. Adler Planetarium; organizer numerous sci. confs.; frequent lectr. in field. Contbr. numerous articles to profl. jours.; co-editor: Explosive Nucleosynthesis, 1973; editor: Supernovae, 1977; assoc. editor: Am. Jour. Physics, 1978—; co-editor: Phys. Cosmology, 1980, Fundamental Problems in Stellar Evolution, 1980, Essays in Nucleosynthesis, 1981; editor: U. Chgo. series Theoretical Astrophysics; Physics Reports, 1981—; editorial bd.: Ann. Revs. Nuclear and Particle Sci, 1976-80; columnist: Outside mag; co-author: Advanced States of Stellar Evolution, 1977. Nat. Graeco-Roman wrestling champion, 1971; recipient Robert J. Trumpler award Astron. Soc. Pacific, 1974, Gravity Research Found. awards, 1974, 75, 76, 80. Fellow Am. Phys. Soc.; mem. Am. Astron. Soc. (Helen B. Warner prize 1978, exec. com. planetary sci. div. 1977-79, sec.-treas. high energy astrophysics div. 1979-81), Astron. Soc. Pacific, Meteoritical Soc., Internat. Astron. Union, Am. Alpine Club, Austrian Alpine Club, Sigma Xi. Club: Quadrangle. Subspecialties: Cosmology; Theoretical astrophysics. Current work: The interface of high energy particle physics, nuclear physics and astrophysics. Home: 19 Sauk Circle Fermilab Batavia IL 60510 Office: Astronomy and Astrophysics Center Univ Chicago 5640 S Ellis Ave Chicago IL 60637 One can learn new information about the universe if one does not allow oneself to be bound by the traditional boundaries of various disciplines but instead synthesizes.

SCHRECKENBERG, GERVASIA MARY, biology educator; b. Paderborn, Ger., Jan. 4, 1916; came to U.S., 1938, naturalized, 1962; d. Stephan and Maria (Dohmann) S. B.S., Cath. U. Am., 1952, M.S., 1954, Ph.D., 1956; postdoctoral, Columbia U., 1970, U. Tex., 1971. Asst. prof. biology Tombrock Coll., West Paterson, N.J., 1956-62, dean, 1956-60, pres., 1960-62; prof. neurobiology Fairleigh Dickinson U., Rutherford, N.J., 1972—; vis. prof. Nat. Cheng Kung U., Taiwan. Contbr. articles to profl. jours. Pres. bd. trustees Holy Family Residence. Mem. AAAS, N.Y. Acad. Sci., N.J. Acad. Sci., Soc. Neurosci., Developmental Biology Soc., Internat. Kirlian Research Assn., Sigma Xi. Democrat. Roman Catholic. Subspecialties: Neurobiology; Developmental biology. Current work: Neuropeptides (B-Endorphins); Sexual dimorphism in brain tissue of Ambystoma mexicanum; effects of disharmonious music on the mouse Hippocampus: a Golgi analysis; and IRA analysis in terms of B-endorphin levels. Office: Fairleigh Dickinson Univ Montross Ave Rutherford NJ 07070

SCHREIBER, HANS, pathologist, educator; b. Quedlinburg, Germany, Feb. 5, 1944; s. Wolfgang Gotthold and Dorothee Margarete (Zimmermann) S.; m. Karin Kugler, Oct. 28, 1969; children: Dorothee, Ute, Maya. M.D., U. Freiburg, W.Ger., 1969, D.M.Sc., 1969; Ph.D., U. Chgo., 1977. Research staff mem. biology div. Oak Ridge Nat. Lab., 1970-73; med. intern Moabit U. Hosp., Berlin, 1973-74; fellow immunology dept. pathology U. Chgo., 1974-77, asst. prof. pathology, 1977-81, assoc. prof., 1982—, mem. com. immunology, 1977—. Contbr. articles to profl. jours. Recipient Goedecke research prize Med. Faculty U. Freiburg, 1969; German Nat. Found. fellow, 1973; Nat. Cancer Inst. fellow, 1974-77; Nat. Cancer Inst. awardee, 1978—. Mem. Am. Assn. Cancer Research, Am. Assn. Pathologists, Am. Soc. Cytology, Am. Assn. Immunologists. Lutheran. Subspecialties: Cellular engineering; Oncology. Current work: Cancer immunology, specific regulation of immune responses. Home: 5467 S Dorchester Ave Chicago IL 60615 Office: La Rabida U Chgo Inst E 65th St at Lake Michigan Chicago IL 60649

SCHRIEBER, ROBERT ALAN, researcher, educator; b. Bklyn., Feb. 11, 1940; s. Benjamin and Syd (Hymowicz) S.; m. Patricia Lynn Schrieber, Dec. 26, 1966; children: Kay, Sam. B.A., U. N.C., 1965; M.A., U. Colo., 1969, Ph.D., 1970. Postdoctoral fellow Med. U. S.C., Charleston, 1970-74; asst. prof. U. Tenn. Ctr. Health Scis., 1974-80, assoc. prof. biochemistry, 1980—. Contbr. articles and abstracts to profl. lit. Served with U.S. Army, 1961-64. Fogarty Interat. fellow, 1982-83. Mem. Soc. Neurosci., Behavioral Genetics Assn., Internat. Soc. Developmental Psychobiology, Research Soc. Alcoholism, Sigma Xi, Alpha Chi Sigma. Democrat. Jewish. Subspecialties: Neurochemistry; Psychobiology. Current work: CNS reactivity; epilepsy; physical dependence. Animal model of epilepsy and physical dependence. Brain energy reserves. Home: 1444 Vance Ave Memphis TN 38104 Office: Dept Biochemistry 216 Nath Bldg 800 Madison Ave Memphis TN 38163

SCHRIEFFER, JOHN ROBERT, physicist; b. Oak Park, Ill., May 31, 1931; s. John Henry and Louise (Anderson) S.; m. Anne Grete Thomsen, Dec. 30, 1960; children—Anne Bolette, Paul Karsten, Anne Regina. B.S., Mass. Inst. Tech., 1953; M.S., U. Ill., 1954, Ph.D., 1957, Sc.D., 1974; Sc.D. (hon.), Tech. U., Munich, Germany, 1968, U. Geneva, 1968, U. Pa., 1973, U. Cin., 1977. NSF postdoctoral fellow U. Birmingham, Eng.), also; Niels Bohr Inst., Copenhagen, 1957-58; asst. prof. U. Chgo., 1957-59, U. Ill., 1959-60, assoc. prof., 1960-62; mem. faculty U. Pa., Phila., 1962—, Mary Amanda Wood prof. physics, 1964—; Andrew D. White prof. at large Cornell U., 1967-75. Author: Theory of Superconductivity, 1964. Guggenheim fellow, Copenhagen, 1967; Recipient Comstock prize Nat. Acad. Sci.; Nobel Prize for Physics, 1972; John Ericsson medal Am. Soc. Swedish Engrs., 1976; Alumni Achievement award U. Ill., 1979. Mem. Am. Phys. Soc. (Oliver E. Buckley solid state physics prize 1968), Nat. Acad. Sci., Am. Philos. Soc. Subspecialty: Theoretical physics. Office: Dept Physics U Calif Santa Barbara CA 93106

SCHRIER, BRUCE KENNETH, research neurobiologist; b. Kalamazoo, Mar. 26, 1938; s. Neil Mulder and Maurine (Niessink) S.; m. Shirley Fisher, Mar. 7, 1964; children: Jennifer, Peter, Katrin, David, Andrew, Timothy, Abigail. B.A., Coll. Wooster, 1960; M.D., Western Res. U., 1964; Ph.D., Tufts U. Med. Sch., 1967. Researcher in pharmacology Upjohn Co., Kalamazoo, summers 1960-63; NIH postdoctoral fellow Tufts U., Boston, 1964-67; commd. sr. surgeon USPHS, 1967; research assoc. Lab. Biochem. Genetics, Nat. Heart Inst., NIH, Bethesda, Md., 1967-70; research scientist, med. officer Lab. Devel. Neurobiology, Nat. Inst. Child Health and Human Devel., 1970—; fellow Pharmacology Research Assoc. Tng. Program, 1967-69; reviewer in molecular neurobiology NSF; sheep breeder; mem. bd. references Balt. Creation Fellowship. Contbr. chpts., numerous articles to profl. publs. Nat. Inst. Gen. Med. Scis. spl. fellow, 1969-70. Mem. AAAS, Am. Soc. Biol. Chemists, Am. Soc. Pharmacology and Exptl. Therapeutics, Md. Sheep Breeders Assn., Md. Hampshire Sheep Assn. Republican. Subspecialties: Genetics and genetic engineering (biology); Biomass (agriculture). Current work: Molecular biology with nervous system expressed genes, fuel and sludge production from poultry manure. Home: 1100 Whites Church Rd New Windsor MD 21776 Office: NIH Bldg 36 Room 2A21 Bethesda MD 20205

SCHRIER, ROBERT WILLIAM, physician; b. Indpls., Feb. 19, 1936; s. Arthur E. and Helen M. S.; m. Barbara Lindley, June 14, 1959; children: David, Debbie, Douglas, Derek, Denise. B.A., DePauw U., 1957; M.D., Ind. U., 1962. Intern Marion County (Ind.) Hosp., 1962-63; resident U. Wash., Seattle, 1963-65; practice medicine, specializing in nephrology, San Francisco, 1969-72, Denver, 1972—; asst. prof. U. Calif. Med. Center, San Francisco, 1969-72, asso. mem., 1970-72, asso. dir. renal div., 1971-72, asso. prof., 1972; prof., head renal disease U. Colo. Med. Center, 1972-76, prof., chmn. dept. medicine, 1976—. Served with U.S. Army, 1966-69. Mem. Am. Soc. Nephrology (sec.-treas. 1978-82, pres.-elect 1982, pres. 1983), Internat. Soc. Nephrology (former mem. council, treas. 1983), Western Soc. Clin. Research (pres. 1981), A.C.P., Am. Fedn. Clin. Research, Internat. Soc. Physiology, Am. Physiol. Soc., Am. Soc. Clin. Investigation (v.p. 1982), Am. Clin. and Climatol. Assn., Assn. Am. Physicians (rector 1982), Western Assn. Physicians (v.p. 1981, pres. 1982), Alpha Omega Alpha. Subspecialties: Nephrology; Internal medicine. Office: 4200 E 9th Ave Denver CO 80262

SCHROEDER, ALICE LOUISE, genetics educator; b. Knoxville, Tenn., June 22, 1941; m. Paul C. Schroeder, June 25, 1966. B.A. in Zoology, U. Colo., Boulder, 1963; Ph.D. in Biology, Stanford U., 1970. NIH fellow Wash. State U., Pullman, 1969-70, lectr. molecular genetics, 1971-74, asst. prof., 1975-80, assoc. prof., 1980—. Contbr. in field. Active Inland Empire Girl Scouts U.S.A., 1948—; active YWCA, 1970—. NIH grantee, 1971-74, 78-82; German Acad. Exchange Service grantee, 1978. Mem. Genetics Soc. Am., AAAS. Subspecialty: Genetics and genetic engineering (medicine). Current work: DNA repair, neurospora genetics. Office: Molecular Genetics Heald Hall Washington State U Pullman WA 99164

SCHROEDER, FRIEDHELM, pharmacologist, membrane researcher, educator; b. Kastorf, E. Ger., July 16, 1947; came to U.S., 1957, naturalized, 1965; s. Helmut R. and Irma Eva (Kaiser) S.; m. Ann Kier, Dec. 9, 1978. B.S. in Chemistry (Univ. scholar), U. Pitts., 1970; Ph.D. in Biochemistry (NSF grad. fellow), Mich. State U., 1974. Am. Cancer Soc. postdoctoral fellow Washington U. Sch. Medicine, St. Louis, 1974-76; asst. prof. pharmacology Sch. Medicine, U. Mo., Columbia, 1976-81, assoc. prof., 1982—; cons. hemotropic disease group (vet. microbiology), 1980—; cons. Miles Research Labs., 1976—; mem. adv. council on Huntington's disease U. Minn., 1980—; abstract presenter at nat. and internat. meetings. Contbr. numerous articles to profl. jours. Mem. Am. Soc. for Pharmacology and Exptl. Therapeutics, Am. Soc. Biol. Chemists, Soc. for Neurosci., Am. Oil Chemists Soc., Am. Chem. Soc., Am. Heart Assn., Soc. for Aquatic Vet. Medicine, Mus. Assocs., Friends of Music, Nat. Audubon Soc. Clubs: Spaulding Raquetball, Comic Art. Subspecialties: Membrane biology; Biophysical chemistry. Current work: Cancer, metastasis, membranes, lipids, fluorescence probes, lipoproteins, atherosclerosis, cholesterol asymmetry, toxicology. Home: Route 2 Carter School Rd Columbia MO 65201 Office: Dept Pharmacology U Mo Sch Medicine Columbia MO 65212

SCHROEDER, JAMES ERNEST, research psychologist; b. Independence, Iowa, Apr. 2, 1947; s. Earl Ernest and Lillian Gertrude (Pahl) S.; m. Elaine Kay Thompson, Dec. 16, 1967; children: Andrew Christopher, James William. B.S., U. Iowa-Iowa City, 1969; M.A., U. N.Mex.-Albuquerque, 1971, Ph.D., 1973. Cert. psychologist, Tex. Asst. prof. exptl. psychology Lamar U., Beaumont, Tex., 1973-78, assoc. prof., 1978-80; research psychologist Litton, Ft. Benning, Ga., 1980-81, U.S. Army Research Inst., Ft. Benning, 1980—, project mgr., 1980—, mgr., 1982—. Bd. dir. Fairway House, Inc., Beaumont, 1979-80; asst. leader Chattahoochee council Boy Scouts Am., 1982—. Recipient Regents Merit award for Excellence in Teaching Lamar U., 1975; U.S. Army Research Inst. award for Sustained Superior Performance, 1982. Mem. Am. Psychol. Assn., Southeastern Psychol. Assn., Southwestern Psychol. Assn., Soc. Applied Learning Technology, Sigma Xi, Beta Theata Pi. Subspecialties: Learning; Computer-assisted tranining and simulation. Current work: My interests are in mathematical models for learning and motivation, applying high technology products to training, education and simulation. Home: 1331 Peacock Ave Columbus GA 31906 Office: US Army Research Institute PO Box 2086 Fort Benning GA 21905

SCHROEDER, LELAND ROY, chemist, educator; b. Caledonia, Minn., June 26, 1938; m. J. Kathleen Ewers, Sept. 10, 1960; children: Todd, Michael. A.B. cum laude in Chemistry, Ripon Coll., 1960; M.S., Lawrence U., 1962; Ph.D. in Organic Chemistry, 1965. Faculty George Washington U., Washington, 1965-66; staff U.S. Army Fgn. Sci. and Tech. Ctr., Washington, 1965-67; faculty Inst. Paper Chemistry, Appleton, Wis., 1967—, prof. chemistry, 1977—, assoc. dean, 1980—. Co-editor: Jour. Wood Chemistry and Tech. Served to capt. U.S. Army, 1965-67. Mem. Am. Chem. Soc., Royal Chemistry Soc., TAPPI. Subspecialties: Organic chemistry; Carbohydrate chemistry. Office: 1042 E South River St Appleton WI 54911

SCHROER, BERNARD JON, university environmental and energy center director; b. Seymour, Ind., Oct. 11, 1941; s. Alvin J. and Selma A. (Mellencamp) S.; m. Kathleen Dittman, July 5, 1963; children: Shannon, Bradley. B.S. in Engring, Western Mich. U., 1964, M.S., U. Ala., 1967; Ph.D., Okla. State U., 1971. Registered profl. engr. Engr. Brown Engring., Huntsville, Ala., 1963-67, Boeing Corp., Huntsville, 1967-70, Computer Sci. Corp., 1970-72; dir. Kenneth E. Johnson Environ. and Energy Ctr., U. Ala., Huntsville, 1972—. Contbr. over 40 sci. articles to profl. publs. NSF grantee, 1971. Mem. Am. Inst. Indsl. Engring. (Outstanding Engr. award 1973, 77), Ala. Soc. Profl. Engrs., Nat. Soc. Profl. Engrs., Soc. Am. Mil. Engrs., Soc. Mfg. Engrs., Ala. Solar Energy Assn., Sigma Xi. Democrat. Lutheran. Subspecialties: Industrial engineering; Information systems, storage, and retrieval (computer science). Current work: Advanced manufacturing; robotics; simulations; information storage and retrieval. Home: 1710 Montdale St Huntsville AL 35801 Office: Kenneth E Johnson Environ and Energy Center Univ Ala Huntsville AL 35899

SCHROHENLOHER, RALPH EDWARD, biochemistry, educator; b. Cin., Aug. 6, 1933; s. Ralph Jacob and Delma Louise (Hagen) S.; m. Sandra Jean Welch, Oct. 30, 1960; children: Robin, John. B.S. in Pharmacy, U. Cin., 1955, Ph.D. in Biochemistry, 1959. Research chemist Nat. Cancer Inst., Hagerstown, Md., 1958-61; asst. prof. medicine U. Ala. Sch. Medicine, Birmingham, 1961-69, assoc. prof., 1969-73, assoc. prof. medicine and pathology, 1973—; guest investigator Rockefeller U., N.Y.C., 1963-64. Contbr. chpts. to books, articles to profl. jours. Nat. Inst. Arthritis and Metabolic Diseases/ Nat. Inst. Allergy and Infectious Diseaseds grantee, 1965-78; Nat. Cancer Inst. grantee, 1975-78. Mem. Am. Assn. Immunologists, Am. Rheumatism Assn., Am. Chem. Soc., AAAS, Sigma Xi. Presbyterian. Subspecialties: Biochemistry (medicine); Immunology (medicine). Current work: Immunoglobulin structure and function. Research on autoantibodies and immune complexes in connective tissue diseases. Home: 1125 Empire Ln Birmingham AL 35226 Office: U Ala in Birmingham Univ Station Birmingham AL 35294

SCHROTT, HELMUT GUNTHER, medical educator, endocrinologist; b. Vienna, Austria, Jan. 23, 1937; came to U.S., 1938, naturalized, 1955; s. Walter Quido Schrott and Anna (Klein) Pershing; m. Mara Llee Jones, Dec. 23, 1961; children: Heidi Lin, Dagon Gunther. B.A., Western Res. U., 1962; M.D., SUNY-Buffalo, 1966. Intern Buffalo Gen. Hosp., 1966-68; U. Utah Med. Ctr., Salt Lake City, 1968-70; fellow U. Wash., Seattle, 1970-72; assoc. cons. Mayo Clinic, Rochester, Minn., 1972-73; asst. prof. internal medicine U. Iowa, 1973-79, assoc. prof. internal medicine, 1979—, assoc. prof. preventive medicine, 1979—; co-dir. Lipid Research Clinic, Iowa City, 1976—; dir. Diabetes Control and Complications Trial, Iowa City, 1982—. Author: Nutrition Counseling, 1979. Pres. Johnson County (Iowa) Heart Assn., 1978. Nat. Heart, Lung and Blood Inst. grantee, 1972; Nat. Inst. Arthritis, Diabetes, Digestive and Kidney Disease grantee, 1982. Fellow Am. Heart Assn.; mem. Am. Fedn. Clin. Research, Soc. Clin. Trials, Am. Diabetes Assn. Democrat. Unitarian. Subspecialties: Endocrinology; Epidemiology. Current work: Genetics of coronary risk factors; nutrition counseling; preventive cardiology; clinical trials; diabetes control. Home: 3117 Alpine Ct Iowa City IA 52240 Office: Dept. Presentive Medicine U Iowa Westlawn S-212 Iowa City IA 52242

SCHUBEL, JERRY ROBERT, oceanographer, research ctr. adminstr.; b. Bad Axe, Mich., Jan. 16, 1936; s. Theodore Howard and Laura Alberta (Gobel) S.; m. Margaret Ann Hostetler, June 14, 1958; children: Susan, Elizabeth, Kathryn Ann. B.S., Alma Coll., 1957; M.A., Harvard U., 1959; Ph.D., Johns Hopkins U., 1968; post grad., U. Mich., 1958, Scripps Inst. Oceanography, 1964, Woods Holle Oceanographic Instn., 1962. Tchr. math. and physics high sch., Winchester, Mass., 1959-60; research asst. Chesapeake Bay Inst., Johns Hopkins U., Balt., 1962-65, head data processing, 1965-67, assoc. research scientist, 1968-69, research scientist, 1970-74, assoc. dir., 1973-74; now prof. oceanography, dir. Marine Scis. Research Ctr., SUNY, Stony Brook. Author: The Living Chesapeake, 1981; editor 2 books; contbr.: articles to profl. jours; editor: Coastal Ocean Pollution Assessment News. Mem. L.I. Environ. Council, 1975—, L.I. Marine Resources Council, 1974—. Alfred P. Sloan Fellow, 1958-59; Grantee NSF, NOAA, NASA. Mem. AAAS, Am. Shore and Beach Preservation Assn., Am. Soc. Limnology and Oceanography. Republican. Presbyterian. Subspecialties: Petroleum engineering; Resource management. Current work: Coastal oceanography; coastal sedimentation, coastal zone management, dredging and dredged material disposal; effects of power plants on aquatic systems. Office: Marine Scis Research Ctr SUNY Stony Brook NY 11794

SCHUBERT, WILLIAM KUENNETH, pediatric gastroenterologist; b. Cin., July 12, 1926; s. Wilfred S.; m. Mary Jane; children—Carol, Joanne, Barbara, Nancy. B.S., U. Cin., 1949, M.D., 1952. Diplomate: Am. Bd. Pediatrics. Intern Ind. U. Med. Center, 1952-53; resident in pediatrics Children's Hosp. Med. Center, Cin., 1953-55; research fellow pediatrics USPHS fellow, 1955-56; physician exec. dir., chief of staff; prof. pediatrics, chmn. dept. U. Cin. Med. Sch.; dir. Children's Hosp. Research Found. Author numerous articles in field. Served with USNR, 1944-46. Fellow Am. Acad. Pediatrics. Subspecialty: Pediatrics. Office: Children's Hosp Med Center Elland and Bethesda Aves Cincinnati OH 45220

SCHUDER, DONALD LLOYD, entomologist; b. Bartholomew County, Ind., Mar. 4, 1922; s. Henry Wilson and Pearl Estelle (Mounts) S.; m. Mary Ethel Ricketts, Mar. 31, 1945; children: Phillips R., David Lloyd. B.S., Purdue U., 1948, M.S., 1949, Ph.D., 1956. Prof. entomology Purdue U., West Lafayette, Ind., 1949—; exec. sec. Ind. Assn. Nurserymen, West Lafayette, 1955—. Author: Insects of Ornamental Plants, 1979. Served with USAF, 1942-45. Recipient Awards of merit Ind. Assn. Nurserymen, 1975, Photographic Soc. Am., 1972; named Spokesman of Yr. Farm Chems. Mag., 1972. Fellow Ind. Acad. Sci (chmn. Entomol. div.); mem. Entomol. Soc. Am. (chmn. insect photo salon 1973—, service award N. Central br. 1983), Internat. Soc. Arboriculture (bd. dirs. 1957-62), Ind. Jr. Hort. Soc. (dir.), Sigma Xi, Gamma Sigma Delta. Republican. Clubs: Wabash Valley Gem and Mineral Soc. (pres. 1955), Thomas Say Entomol. Soc. (pres. 1947-48), Greater Lafayette Aquarium Soc. (pres. 1949). Subspecialties: Population biology; Taxonomy. Current work: Biology of Zimmerman pine moth, history pouch gall, two-lined Japanese weevil, eastern pine shoot borer, Chionaspis heterophyllae and C. pinifoliae and major pine pests in midwest. Home: 2319 Sycamore Ln West Lafayette IN 47906 Office: Dept Entomology Purdue U West Lafayette IN 47907

SCHUEL, HERBERT, anatomist, educator, researcher; b. N.Y.C., Apr. 8, 1935; s. Irving and Frances (Pomerantz) S.; m. Regina Hirsch, July 8, 1962; children: Victor Marvin, Barbara Ellen. Research assoc. Fla. State U., 1960-61, biol. div; Oak Ridge Nat. Lab., 1961-63, dept. chemistry, Northwestern U., 1963-65; asst. prof. biology Oakland U., Rochester, Mich., 1965-68; asst. prof. anatomy Mt. Sinai Sch. Medicine, 1968-72, assoc. research prof., 1972-73; assoc. prof. biochemistry SUNY Downstate Med. Ctr., Bkln., 1973-76; assoc. prof. anatomy SUNY-Buffalo, 1977—; reviewer grant applications NSF, NIH, AEC; lectr. in field. Contbr. articles to profl. jours. NIH fellow, 1961-63; NSF grantee.; NIH grantee; Am. Cancer Soc. grantee, 1970-75; Population Council grantee, 1974-75; SUNY Research Found. grantee, 1975-76. Mem. N.Y. Acad. Sci., Am. Soc. Zoologists, Am. Inst. Biol. Scis., AAAS, Am. Soc. Cell Biology, Soc. Gen. Physiology, Biophysical Soc., Soc. Study Reproduction, Am. Physiol. Soc., Am. Assn. Anatomists, Sigma Xi. Subspecialty: Cell biology (medicine). Current work: Mechanisms responsible for the prevention of polyspermy during fertilization and the releases of stored secretary products by cells. Office: Dept Anatomical Scis SUNY Buffalo NY 14214

SCHUELEIN, MARIANNE, pediatric neurologist, educator; b. Stuttgart, Germany, Apr. 16, 1934; d. Curt Charles and Gertrude (Weil) S.; m. Ralph Mack Krause, June 26, 1960; children: Peter Carl, Steven Charles. B.A., Wellesley Coll., 1955; M.D., N.Y.U., 1959. Intern Yale U. Med. Ctr., New Haven, 1959-60; resident in pediatrics Michael Reese Hosp., Chgo., 1960-62; fellow Children's Hosp., Washington, 1962-63; resident in neurology Georgetown U., 1963-66; asst. prof. pediatrics and neurology, 1967—. Contbr. articles to profl. jours. Mem. D.C. Med. Soc., Women's Med. Assn., Child Neurology Soc., Am. Acad. Neurology, Am. Acad. Pediatrics, Internat. Child Neurology Soc., Soc. Developmental Medicine, Assn. for Research in Nervous and Mental Disease. Subspecialties: Neurology; Pediatrics. Current work: Diseases of the brain and nervous system. Home: 3208 44th St NW Washington DC 20016 Office: Georgetown U Med Ctr 3800 Reservoir Rd NW Washington DC 20007

SCHUETZ, CARY EDWARD, thermophysics engineer; b. San Diego, Dec. 6, 1953; s. Celestine Edward and Doris Marjorie (Berquist) S.; m. June Wong, Mar. 11, 1983. B.S. in Mech. Engring, U. Calif.-Santa Barbara, 1978. Engr. U. Calif.-Santa Barbara, 1976-78; engr., scientist McDonnell Douglas Corp., Long Beach, Calif., 1978-81; engr. Northrop Corp., Los Angeles, 1981—. Mem. AIAA, Nat. Soc. Profl. Engrs., Calif. Soc. Profl. Engrs. Subspecialties: Systems engineering; Thermophysics engineering. Current work: Product achievements in new function technologies through analytical system modeling with interactive hardware developments. Inventor thermoelectric generator to provide automobile electricity from wasted engine exhaust heat, 1978. Home: 10961 Roebling Ave Los Angeles CA 90024 Office: 1 Northrop Ave 3818/85 Hawthorne CA 90250

SCHUETZE, STEPHEN MARK, neurobiologist, educator; b. New Ulm, Minn., Apr. 26, 1951; s. Gerhard and Frieda (Polchow) S.; m. Roberta Rollene Pollock, July 15, 1978. B.S.E.E. with honors, Washington U., St. Louis, 1973, A.B. summa cum laude, 1973; M.A., Harvard U., 1975, Ph.D. in Cell and Developmental Biology, 1978. Postdoctoral fellow dept. anatomy Harvard U. Med. Sch., Boston, 1978-81; asst. prof. dept. biol. scis. Columbia U., N.Y.C., 1981—. Contbr. articles to sci. jours. NIH fellow, 1978; Sloan research fellow, 1982; grantee NIH, Muscular Dystrophy Assn., Sloan Found., March of Dimes Found.; recipient Joseph B. Butler prize Washington U., 1972, Century Electric award, 1972; Mo. Soc. Profl. Engrs. fellow, 1972-73. Mem. Soc. Neurosci., Phi Beta Kappa, Tua Beta Pi, Eta Kappa Nu. Subspecialties: Neurobiology; Developmental biology. Current work: Synapse formation; development of neuromuscular junction; synapse elimination. Home: 601 W 115th St Apt 26 New York NY 10025 Office: Columbia U Dept Biol Scis New York NY 10027

SCHUFLE, JOSEPH ALBERT, chemistry educator, historian of science; b. Akron, Ohio, Dec. 21, 1917; s. Albert Bernard and Daisy (Frick) S.; m. Lois Mytholar, May 31, 1942; children: Joseph Albert, Jean Ann (twins). B.S., U. Akron, 1938, M.S., 1942; Ph.D., Western Res. U., 1948. Accredited profl. chemist. Chemist, City of Akron, 1938-42; instr. Western Res. U., Cleve., 1946-48; asst. prof., assoc. prof., prof. chemistry N.Mex. Inst. Tech., Socorro, 1948—64; prof. chemistry N.Mex. Highlands U., Las Vegas, 1964—; vis. prof. Univ. Coll., Dublin, 1961-62; research prof. U. Uppsala, Sweden, 1977. Author: Bergman's Dissertation on Elective Attractions, 1968, Vacquero to Dominie, 1978, Juan Jose D'Elhuyar, 1981. Served to capt. U.S. Army, 1942-46. Named Outstanding Scientist N.Mex. Acad. Sci., 1972; recipient honor scroll N. Mex. Inst. Chemists, 1980. Mem. Am. Chem. Soc. (Clark medal 1982), AAAS. Democrat. Presbyterian. Lodge: Rotary. Subspecialties: Physical chemistry; Geochemistry. Current work: Structure of water; ion exchange dating of sediments; history of science. Patentee electrodialysis cell. Home: 1301 8th St Las Vegas NM 87701 Office: Highlands Univ Las Vegas NM 87701

SCHUG, KENNETH ROBERT, chemistry educator; b. Easton, Pa., Aug. 27, 1924; s. Howard Lester and Marion Henry (Hubert) S.; m. Miyoko Ishiyama, June 13, 1948; children: Carey Tyler, C(arson) Blake, Reed Porter. B.A. in Chemistry, Stanford U., 1945, Ph.D., U. So. Calif., 1955. Research assoc. U. Wis.-Madison, 1954-56; mem. faculty Ill. Inst. Tech., Chgo., 1956—, now prof. chemistry, chmn chemistry dept., 1976-82, dir., ; cons. Argonne (Ill.) Nat. Lab., 1960-62. Co-author book contbr. numerous sci. articles to profl. publs. Bd. dirs. Michael Reese Hosp. Health Plan, Chgo., Hyde Park Consumer Cooperative Soc., Chgo., COMDRAND, Inc. Fulbright Research fellow Kyushu (Japan) U., 1964-65. Mem. Am. Chem. Soc. (dir. Chgo. chpt.), AAAS. Subspecialties: Inorganic chemistry; Synthetic chemistry. Current work: Research on chemical reactions of coordination compounds of ruthenium and other transition metals. Home: 1466 E Park Pl Chicago IL 60637 Office: Ill Inst Tech Chicago IL 60616

SCHUH, G(EORGE) EDWARD, agricultural economics educator; b. Indpls., Sept. 13, 1930; s. George Edward and Viola (Lentz) S.; m. Maria Ignez, May 23, 1965; children: Audrey, Susan, Tanya. B.S. in Agrl. Edn, Purdue U., 1952; M.S. in Agrl. Econs., Mich. State U., 1954; M.A. in Econs, U. Chgo., 1958, Ph.D., 1961. From instr. to prof. agrl. econs. Purdue U., 1959-79; dir. Center for Public Policy and Public Affairs, 1977-78; dep. undersec. for internat. affairs and commodity programs Dept. Agr., Washington, 1978-79; prof. agrl. and applied econs., head dept. U. Minn., 1979—; program adv. Ford Found., 1966-72; sr. staff economist Pres.'s Council Econ. Adv's., 1974-75; dir. Nat. Bur. Econ. Research, 1977—; hon. prof. Fed. U. Vicosa, Brazil, 1965; bd. dirs. Econs. Inst., Boulder, Colo., 1979—, Mpls. Grain Exchange, 1980—. Author, editor profl. books; contbr. numerous articles to profl. publs. Served with U.S. Army, 1954-56. Fellow Am. Acad. Arts and Scis.; mem. Am. Agrl. Econs. Assn. (Thesis award 1962, Published Research award 1971, Article award 1975, Policy award 1979, dir. 1977-80, pres.-elect 1980-81, pres. 1981-82), Internat. Assn. Agrl. Econs., Am. Econ. Assn., Brazilian Soc. Agrl. Economists. Subspecialties: Agricultural economics; Resource management. Current work: Economics of technical change and rate of technical change. Office: 231 Classroom Office Bldg 1994 Buford Ave U Minn Saint Paul MN 55108

SCHUH, MERLYN DUANE, chemistry educator; b. Avon, S.D., Feb. 21, 1945; s. Edward Arthur and Amelia (Rueb) S.; m. Judy Anne Swigart, June 1, 1969. B.A. in Chemistry, U. S.D., 1967; Ph.D. in Phys. Chemistry, Ind. U., 1971. Asst. prof. chemistry Middlebury (Vt.) Coll., 1971-75; asst. prof. Davidson (N.C.) Coll., 1975-80, assoc. prof., 1980—. Contbr. articles on chemistry to profl. jours. Research Corp. grantee, 1972-83; Petroleum Research Fund grantee, 1976-83; NSF grantee, 1979-81; N.C. Bd. Sci. Tech. grantee, 1977-78. Mem. Am. Chem. Soc., Inter-Am. Photochem. Soc., Metrolina Assn. for the Blind (bd. dirs. 1977-82, pres. 1979-80), Sigma Xi. Democrat. Presbyterian. Club: Davidson Lions (pres. 1981-82). Subspecialties: Biophysical chemistry; Laser photochemistry. Current work: Gas-phase photochemistry and photophysics of aldehydes and their role in smog, photophysical energy transfer processes. Home: PO Box 704 Davidson NC 28036 Office: Davidson Coll G9 Martin Chemistry Davidson NC 28036

SCHULER, ROBERT HUGO, educator, chemist; b. Buffalo, Jan. 4, 1926; s. Robert H. and Mary J. (Mayer) S.; m. Florence J. Forrest, June 18, 1952; children: Mary A., Margaret A., Carol A., Robert E., Thomas C. B.S., Canisius Coll., Buffalo, 1946; Ph.D., U. Notre Dame, 1949. Asst. prof. chemistry Canisius Coll., 1949-53; asso. chemist, then chemist Brookhaven Nat. Lab., 1953-56; staff fellow, dir. radiation research lab. Mellon Inst., 1956-76, mem. adv. bd., 1962-76; prof. chemistry, dir. radiation research lab. Carnegie-Mellon U., 1967-76; prof. chemistry, dir. radiation lab. U. Notre Dame, Ind., 1976—. Author articles in field. Fellow AAAS; mem. Am. Chem. Soc., Am. Phys. Soc., Chem. Soc., Radiation Research Soc. (pres. 1975-76), Sigma Xi. Club: Cosmos. Subspecialties: Physical chemistry; Radiation chemistry. Current work: Radiation chemistry and its application to kinetic and structural problems. Office: Radiation Lab U Notre Dame Notre Dame IN 46556

SCHULTE, DANIEL HERMAN, optical designer; b. Osceola, Iowa, Aug. 3, 1929; s. Henry and Margaret Leona (Sherwood) S.; m. Sue Ellen Donnenwirth, June 29, 1955; children: Robert, Aden, Marta. A.B., Phillips U., 1951; Ph.D., U. Chgo., 1958. Optical designer, project engr. Perkin Elmer Corp., Norwalk, Conn., 1956-59; asst. astronomer, computer dept. mgr. Kitt Peak Nat. Obs., Tucson, 1959-65; staff physicist optical design dept. Itek Corp., Lexington, Mass., 1965-81; research scientist Lockheed Missile & Space Co., Palo Alto Research Lab., Calif., 1981—; cons. Haneman Corp., Tropel Inc. Contbr. articles to sci. jours. Mem. Am. Astron. Soc., Internat. Astron. Union, Optical Soc. Am. Current work: Optical systems design, astronomical and space optics. Home: 118 Mercy St Mountain View CA 94041 Office: 3251 Hanover St Palo Alto CA 94304

SCHULTES, RICHARD EVANS, botanist, museum executive, educator; b. Boston, Jan. 12, 1915; s. Otto Richard and Maude Beatrice (Bagley) S.; m. Dorothy Crawford McNeil, Mar. 26, 1959; children: Richard Evans II, Neil Parker and Alexandra Ames (twins). A.B., Harvard U., 1937, A.M., 1938, Ph.D., 1941; M.H. (hon.), Universidad Nacional de Colombia, Bogotá, 1953. Plant explorer, NRC fellow Harvard Bot. Mus., Cambridge, Mass., 1941-42, research asso., 1942-53; curator Orchid Herbarium of Oakes Ames, 1953-58, curator econ. botany, 1958—, exec. dir., 1967-70, dir., 1970—; Guggenheim Found. fellow, collaborator U.S. Dept. Agr., Amazon of Colombia, 1942-43; plant explorer in South Am. Bur. Plant Industry, 1944-54; prof. biology Harvard, 1970-72, Paul C. Mangelsdorf prof. natural scis., 1973-81, Edward C. Jeffrey prof. biology, 1981—; adj. prof. pharmacognosy U. Ill., Chgo., 1975—; Hubert Humphrey vis. prof. Macalaster Coll., 1979; field agt. Rubber Devel. Corp. of U.S. Govt., in S.Am., 1943-44; collaborator Instituto Agronômico Norte, Belem, Brazil, 1948-50; hon. prof. Universidad Nacional de Colombia, 1953—, prof. econ. botany, 1963; bot. cons. Smith, Kline & French Co., Phila., 1957-67; mem. NIH Adv. Panel, 1964; mem. selection com. for Latin Am. Guggenheim Found., 1964—; chmn. on-site visit U. Hawaii Natural Products Grant NIH, 1966, 67; Laura L. Barnes Annual lectr. Morris Arboretum, Phila., 1969; Koch lectr. Rho Chi Soc., Pitts., 1971, Chgo., 1974; vis. prof. econ. botany, plants in relation to man's progress Jardín Botánico, Medellín, Colombia, 1973; Cecil and Ida H. Green Vis. Lectr. U. B.C., Vancouver, Can., 1974; mem. adv. bd. Fitz Hugh Ludlow Library, San Francisco, 1974—. Author: (with P. A. Vestal) Economic Botany of the Kiowa Indians, 1941, Native Orchids of Trinidad and Tobago, 1960, (with A. F. Hill) Plants and Human Affairs, 1960, rev. edit., 1968, (with A. S. Pease) Generic Names of Orchids—their Origin and Meaning, 1963, (with A. Hofmann) The Botany and Chemistry of Hallucinogens, 1973, rev. edit., 1980, Plants of the Gods, 1979, Plant Hallucinogens, 1976; contbg. author: Ency. Biol. Scis, 1961, Ency. Brit, 1966, Ency. Biochemistry, 1967, McGraw-Hill Yearbook Sci., Tech, 1971; author numerous Harvard Bot. Mus. leaflets.; Asst. editor: Chronica Botanica, 1947-52; editor: Bot. Mus. Leaflets, 1957—, Econ. Botany, 1962-79; mem. editorial bd.: Lloydia, 1965-76, Altered States of Consciousness, 1973—, Jour. Psychedelic Drugs, 1974—; mem. adv. bd.: Cannabis Rx, Jour. Cannabis Research, 1973—, Horticulture, 1976-78, Jour. Ethnopharmacology, 1978—; Contbr. numerous articles to profl. jours. Mem. governing bd. Amazonas 2000, Bogotá; asso. in ethnobotany Museo del Oro, Bogota, 1974—; chmn. NRC panels, 1974, 75; mem. NRC Workshop on Natural Products, Sri Lanka, 1975; participant numerous sems., congresses, meetings. Decorated Orden de la Victoria Regia in recognition of work in Amazon Colombian Govt., 1969. Fellow Am. Acad. Arts, Scis., Am. Coll. Neuropsychopharmacology; mem. Nat. Acad. Sci., Linnean Soc., Academia Colombiana de Ciencias Exactas, Fisico-Quimicas y Naturales, Instituto Ecuatoriano de Ciencias Naturales, Sociedad Cientifica Antonio Alzate (Mexico), Argentine Acad. Scis., Am. Orchid Soc. (life hon.), Pan Am. Soc. New Eng. (gov.), Asociación de Amigos de Jardines Botánicos (life), Soc. Econ. Botany (organizer annual meeting 1961, Disting. Botanist of Yr. 1979), New Eng. Bot. Club (pres. 1954-60), Internat. Assn. Plant Taxonomy, Am. Soc. Pharmacognosy, Phytochem. Soc. N.Am., Socieded Colombiana de Orquideologia, Assn. Tropical Biology, Asociación Colombiana de Ingenieros Agrónomos, Sociedad Cubana de Botánica, Explorer's Club, Sigma Xi (pres. Harvard chpt. 1971-72), Beta Nu chpt. Phi Sigma (first hon.). Unitarian (vestryman Kings Chapel 1974-76, 82-85). Subspecialties: Taxonomy; Systematics. Current work: Plants employed as medicines in Amazon; rubber-yielding plants; flora of the northwest Amazon. Home: 78 Larchmont Rd Melrose MA 02176 Office: Bot Museum Harvard U Cambridge MA 02138

SCHULTZ, ALBERT BARRY, biomechanical engineer, educator, researcher; b. Phila., Oct. 10, 1933; s. George D. and Belle (Seidman) S.; m. Susan Resnikov, Aug. 25, 1955; children: Carl, Adam, Robin. B.S. (Navy ROTC scholar), U. Rochester, 1955; M. Engring, Yale U., 1959, Ph.D., 1962. Asst. prof. U. Del., 1962-65; asst. prof. mech. engring. U. Ill., Chgo., 1965, assoc. prof., 1966-71, prof., 1971-83, dir. biomechs.Research Lab., 1980-83; Vennema prof. engring. U. Mich., 1983—; chmn. U.S. Nat. Com. on Biomechs., 1982-85. Served with USN, 1955-58. Recipient Research Career award NIH, 1975-80; NIH spl. research fellow, 1971-72. Mem. ASME (chmn. div. bioengring. 1981-82), Am. Soc. Biomech. (pres. 1982-83), Internat. Soc. for Study Lumbar Spine (pres. 1981-82), Phi Beta Kappa, Phi Kappa Phi, Tau Beta Pi. Subspecialties: Biomedical engineering; Mechanical engineering. Current work: Orthopaedic biomechanics; biomechanics of human spine. Office: Mech Engring Dept U Mich Ann Arbor MI 48109

SCHULTZ, ALVIN LEROY, internist, med. educator; b. Mpls., July 27, 1921; s. Maurice Arthur and Elizabeth Leah (Gershin) S.; m. Martha Jean Graham, Aug. 14, 1947; children: Susan Kristine, David Matthew, Peter Jonathan, Michael Graham. B.A., U. Minn.-Mpls., 1943, M.B., 1946, M.D., 1947, M.S., 1952. Diplomate: Am. Bd. Internal Medicine. Instr. medicine U. Minn.-Mpls., 1952-54, asst. prof., 1954-59, assoc. prof., 1959-65, prof., 1965—; asst. chief medicine Mpls. VA Hosp., 1952-54; dir. endocrine clinic U. Minn., 1954-59; dir. medicine and research Mt. Sinai Hosp., Mpls., 1959-65; chief of medicine Hennepin County Med. Ctr., Mpls., 1965. Editor: Jour. Lab. and Clin. Medicine, 1966-69, Modern Medicine, 1960—; editorial bd.: Minn. Medicine, 1965—; contbr. articles profl. jours. Bd. dirs. Planned Parenthood of Minn., 1970-75, Hennepin County Med. Philanthropic Found., 1976—. Served to capt. AUS, 1947-49. Fellow ACP (Minn. gov. 1983-86); mem. Central Soc. Clin. Research, Am. Fedn. Clin. Research, Endocrine Soc., Am. Thyroid Assn., Minn. Med. Assn. (ho. of dels. 1980-85), Hennepin County Med. Soc. (dir. 1977-81). Republican. Jewish. Subspecialties: Internal medicine; Endocrinology. Current work: Thyroid disease, hormonal control lipid metabolism, training of medical students, general internists. Home: 5127 Irving Ave S Minneapolis MN 55419 Office: Hennepin County Med Center 701 Park Ave S Minneapolis MN 55415

SCHULTZ, FREDERICK H. C., physics educator, energy consultant; b. Hanks, N.D., June 11, 1921; s. Herman A. and Helvene G. (Ausl) S.; m. Lila Fay Gregory, Aug. 27, 1949; children: Michael F., Jane F., John F. Ph.B., U. N.D.-Grand Forks, 1942; M.S., U. Idaho-Moscow, 1950; Ph.D., Wash. State U.-Pullman, 1967. Instr. physics Mont. Sch. Mining, Butte, 1950-55; asst. prof. Mont. State U.-Bozeman, 1955-61; physicist U.S. Naval Ordnance Lab., Corona, Calif., summers 1957, 59, 61, 63; asst. prof. Minot (N.D.) State U., 1961-63, Wash. State U., 1963-68; prof. U. Wis.-Eau Claire, 1968—; dir. seismograph sta. U.S. Coast & Geol. Survery, Bozeman, 1955-61; dir. Summer Inst. NSF, Pullman, 1967-68; energy cons. W.C. Wis. Regional Planning Commn., Eau Claire, 1981—. Served with U.S. Navy, 1944-46. Mem. Wis. Assn. Physics Tchrs., Am. Assn. Physics Tchrs., N.Y. Acad. Sci., Optical Soc. Am., Sigma Xi. (local pres. 1979-81). Republican. Lutheran. Subspecialties: polarized infrared absorption and reflection; energy sources and uses. Current work: Profession-undergraduate physics teaching; consulting work in energy sources and uses with environmental effects such as acid rain. Office: Department Physics University Wisconsin Eau Claire WI 54701 Home: 3834 Nimitz St Eau Claire WI 54701

SCHULTZ, JANE SCHWARTZ, geneticist, educator; b. N.Y.C., July 28, 1932; d. Jacob and Helen (Rosenthal) Schwartz; m. Jerome S. Schultz, Sept. 1, 1955; children: Daniel S., Judith S., Kathryn A. B.A., Hunter Coll., 1953; M.S., Columbia U., 1955; Ph.D., U. Mich., 1970. Chemist Gen. Food Corp., Hoboken, N.J., 1954-55, USDA Forest Products Lab., Madison, Wis., 1955-58; tchr. sci. Pearl River (N.Y.) High Sch., 1958-64; immunogeneticist Univ. Hosp., Leiden, Holland, 1971-72; research assoc. dept. human genetics U. Mich., Ann Arbor, 1970-75, asst. prof., 1975-83; assodc. prof. CU. Mich., 1983; research adminstr. VA Central Office, Washington, 1977-80; geneticist VA Med. Center, Ann Arbor, Mich., 1972-83; asst. dean U. Mich. Med. Sch., 1979-83; research adminstr. NIH, 1983—. Contbr. articles to profl. jours. N.Y. State Regents Scholar, 1949. Mem. Genetics Soc. Am., Am. Assn. Immunologists, Am. Soc. Human Genetics, Am. Assn. Clin. Histocompatibility, Sigma Xi, Phi Beta Kappa. Subspecialties: Immunogenetics; Transplantation. Current work: Researcher in immunogenetics of transplantation. Home: 2743 Antietam Dr Ann Arbor MI 48105 Office: Med Sch Bldg 1301 E Catherine Rd Ann Arbor MI 48109

SCHULTZ, JEROME SAMSON, biochem. engr.; b. Bklyn., June 25, 1933; s. Henry Herman and Sally S.; m. Jane Paula Schwartz, Sept. 1, 1955; children—Daniel Stuart, Judith Susan, Kathryn Ann. B.S. in Chem. Engring, Columbia U., 1954, M.S., 1956; Ph.D. in Biochemistry, U. Wis., 1958. Group leader biochem. research Lederle Labs., N.Y.C., 1958-64; asst. prof. dept. chem. engring. U. Mich., Ann Arbor, 1964-67, asso. prof., 1967-70, prof., 1970—, chmn. dept., 1977—. Contbr. articles to profl. jours. NIH research career devel. awardee, 1970-75. Mem. Am. Chem. Soc. (past chmn. microbial chemistry div.), Am. Inst. Chem. Engrs. (past chmn. food and bioengring. div.), Am. Soc. Artifical Internal Organs, AAAS, Sigma Xi, Phi Lambda Upsilon, Tau Beta Pi. Subspecialties: Bioinstrumentation; Enzyme technology. Current work: Artifical hybrid organs; implantable sensors for metabolites; industrial use of genetic engineering; membranes for separations. Patentee in field. Home: 2743 Antietam Dr Ann Arbor MI 48105 Office: Dept Chem Engring Univ of Mich Ann Arbor MI 48109

SCHULTZ, K. DAVID, psychologist; b. Mt. Pleasant, Iowa, June 19, 1949; s. Kenneth Darrell and Virginia Lee (Rosa) S.; m. Deborah Boettger, Jan. 6, 1980. B.A. magna cum laude, U. Mich, 1971; M.S., Yale U., 1973, Ph.D., 1976. Diplomate: in clin. psychology, Am. Bd. Profl. Psychology.; Lic. clin. psychologist, Conn. Psychotherapist Highland Hts., New Haven, 1971-73; psychologist assoc. VA Med. Ctr., West Haven, Conn., 1973-76; teaching fellow Yale U., New Haven, 1974-76; instr. Quinnipiac Coll., Mt. Carmal, Conn., 1974-76; psychotherapist Psychotherapy Assocs., New Haven, 1976-79; psychologist Waterbury (Conn.) Hosp., 1976—; pvt. practice psychology, Woodbury, Conn., 1979—; bid. dirs. Human Resources Ctr., New Haven, 1976; Vice pres. Assn. Children Social and Learning Difficulties, Ann Arbor, Mich., 1970-71. Book reviewer: Div. 32 Am. Psychol. Assn, 1976—; reviewer for profl. jours.; Contbr. articles, chpts. to jours., books. Deacon First Congl. Ch., Woodbury, 1980-81; v.p. Woodlake Condominium II, Woodbury, 1980-82. Am. Psychol. Assn. travel scholar, 1976. Mem. Am. Psychol. Assn., Eastern Psychol. Assn., Conn. Psychol. Assn., Am. Orthopsychiatric Assn., Physicians Assn. Anthrop. Medicine, Phi Beta Kappa, Sigma Xi. Republican. Clubs: Yale Yacht, Yale Aviation. Current work: Research on schizophrenia, depression, cognitive dysfunction; int. Office: 64 Robbins St Waterbury CT 06720

SCHULTZ, RICHARD MICHAEL, tumor immunologist; b. Hillsboro, Wis., Mar. 11, 1948; s. Paul Raymond and Thelma Violet (Johns) S.; m. Erika Elisabeth Zander, June 18, 1976. B.S., U. Ill., 1971; postgrad., George Washington U., 1976-77. Research asst. Sch. Vet. Medicine U. Ill., 1971-74; research assoc. Frederick (Md.) Cancer Research Ctr., 1974; microbiologist Nat. Cancer Inst., NIH, Bethesda, Md., 1974-79; research scientist Eli Lilly and Co., Indpls., 1979—. Contbr. articles to profl. jours. Mem. Am. Assn. Immunologists., Am. Assn. Cancer Research, Reticuloendothelial Soc. Lutheran. Subspecialties: Immunopharmacology; Cancer research (medicine). Current work: Regulation of cytotoxic activity by interferons and prostaglandins. Discoverer autoregulation of macrophage functional activity by interferons and prostaglandin E. Office: 307 E McCarty St Dept M931 Indianapolis IN 46285 Home: 8700 Shelby St Indianapolis IN 46227

SCHULTZ, RICHARD MICHAEL, biochemist, educator; b. Phila., Oct. 28, 1942; s. William and Beatrice (Levine) S.; m. Rima Michele Lunin, Mar. 7, 1965; children: Carl Moses, Eli Joseph. B.A., Harpur Coll., SUNY-Binghamton, 1964; Ph.D., Brandeis U., 1969. Research fellow Harvard Med. Sch., Boston, 1969-71; asst. prof. Loyola U. Stritch Sch. Medicine, Maywood, Ill., 1971-78, assoc. prof. dept. biochemistry, 1978—. Contbr.: chapts. to books, sci. articles to jours. including Proc. Nat. Acad. Science. NIH grantee, 1973—; predoctoral fellow, 1967-69; postdoctoral fellow, 1969-71. Mem. Am. Chem. Soc. (Petroleum Research Found. fellow 1962-64), Am. Soc. Biol. Chemists, AAAS. Subspecialties: Biochemistry (medicine); Biophysical chemistry. Current work: Mechanism of enzyme catalysis, particularly serine proteases; inhibition of protease enzymes by peptidyl aldehyde transition-state analogy inhibitors; role of proteases in cancer cell metastases, coagulation and fibrinolysis. Office: Dept Biochemistry Loyole U Stritch Sch Medicine Maywood IL 60153

SCHULTZ, RONALD DAVID, pathobiological sciences educator, immunologist; b. Freeland, Pa., Apr. 21, 1944; s. Fred C. and Hilda A. (Biasi) S.; m. Carolyn D. Schultz, July 16, 1966; children: Stacey, Kim, Scott. B.S., M.S., Ph.D., Pa. State U. From research assoc. to asst. prof. N.Y. State Coll. Vet. Medicine, Cornell, 1971-73; from asst. to assoc. prof. J.A. Baker Inst., Cornell U., Ithaca, N.Y., 1973-78, assoc. dir. clin. labs., 1972-78; prof. Sch. Vet. Medicine, Auburn (Ala.) U., 1978-82, Sch. Vet. Medicine, U. Wis., Madison, 1982—, also chmn. dept., dir. Immunodiagnostic Lab.; cons. in field. Contbr. numerous sci. articles to profl. publs. Served to capt. U.S. Army, 1966-76. Mem. Am. Soc. Microbiology (mem. editorial bd.: Infection and Immunity), Am. Assn. Vet. Immunologists (pres. 1980), Am. Assn. Immunologists, U.S. Animal Health Assn., Conf. on Research Workers in Animal Diseases, Internat. Soc. Devel. and Comparative Immunology, Sigma Xi, Gamma Sigma Delta, Phi Sigma. Clubs: Antibody club, Immunology club. Subspecialty: Immunology. Current work: Developmental aspects of immune response, immunosupression, immunopathology, viral immunity, clinical immunology. Patentee in field. Home: 3619 Vickiann St Verona WI 53593 Office: Univ Wis Sch Veterinary Medicin Pathobiol Scis Dept Madison WI 53715

SCHULTZ, STANLEY GEORGE, physiologist, educator; b. Bayonne, N.J., Oct. 26, 1931; s. Aaron and Sylvia (Kaplan) S.; m. Harriet Taran, Dec. 25, 1960; children: Jeffery, Kenneth. A.B summa cum laude, Columbia U., 1952; M.D., N.Y. U., 1956. Intern Bellevue Hosp., N.Y.C., 1956-57, resident, 1957-59; research assoc. in biophysics Harvard U., 1959-62, instr. biophysics, 1964-67; assoc. prof. physiology U. Pitts., 1967-70, prof. physiology, 1970-79; prof., chmn. dept. physiology U. Tex. Med. Sch., Houston, 1979—; cons. USPHS, NIH, 1970—; mem. physiology test com. Nat. Bd. Med. Examiners, 1974-79, chmn., 1976-79. Editor: Jour. Applied Physiology, 1971-75, Physiol. Revs, 1979—; mem. editorial bd.: Jour. Gen. Physiology, 1969—; editorial bd.: Ann. Revs. Physiology, 1974-81; assoc. editor Ann. Revs. Physiology, 1977-81; editorial bd.: Current Topics in Membranes and Transport, 1975—, Jour. Membrane Biology, 1977—; Contbr. articles to profl. jours. Served to capt. M.C. USAF, 1962-64. Recipient Research Career award NIH, 1969-74; overseas fellow Churchill Coll., Cambridge U., 1975-76. Mem. Am. Heart Assn. (established investigator 1964-69), Am. Physiol. Soc., AAAS, Biophys. Soc., Soc. for Gen. Physiologists, Internat. Cell. Research Orgn., Internat. Union Physiol. Scis. (chmn. internat. com. gastrointestinal physiology 1977-80), Phi Beta Kappa, Alpha Omega Alpha. Subspecialties: Physiology (medicine); Cell biology (medicine). Current work: Solute transport by epithelial tissues. Membrane transport. Electrophysiology. Home: 4955 Heatherglen Dr Houston TX 77096

SCHUMACHER, HARRY, RALPH, rheumatologist, educator, researcher; b. Montreal, Que., Can., Feb. 14, 1933; came to U.S., 1934; s. H. Ralph and Dorothy (Shreiner) S.; m. Elizabeth Swisher, July 13, 1963; children: Heidi Ruth, Kaethe Beth. B.S., Ursinus Coll., 1955; M.D., U. Pa., 1959. Diplomate: Am. Bd. Internal Medicine, also Sub-bd. Rheumatology. Intern Denver Gen. Hosp., 1959-60; resident Wadsworth VA Hosp., Los Angeles, 1960-62, fellow in rheumatology, 1962-63, Robert B. Brigham Hosp., Boston, 1965-67; research assoc. Med. Sch., U. Calif.-San Francisco, 1964-65; asst. in medicine Peter B. Brigham Hosp., Boston, 1965-67; assoc. in medicine U. Pa., Phila., 1967-69, asst. prof. medicine, 1969-72, assoc. prof., 1972-79, prof., 1979—; dir. Rheumatology-Immunology Center, VA Med. Ctr., Phila., 1977—; acting chief arthritis sect. Hosp. of U. Pa., Phila., 1978-80. Author: Gout and Pseudogout, 1978, Essentials of A Differential Diagnosis of Rheumatoid Arthritis, 1981; editor: Primer on Rheumatic Diseases, 1983. Pres. Eastern Pa. Arthritis Found., Phila., 1980-81; bd. dirs. Am. Bd. Med. Advancement China, N.Y.C., 1982—. Fellow ACP; mem. Am. Rheumatism Assn. (nat. chmn. edn. com. 1978-81, pres. Southeastern region 1980-81), Electron Microscopy Soc. Am. (pres. Phila. chpt. 1975-76), Phila. Rheumatism Soc. (pres. 1980); hon. mem. Mexican Rheumatism Soc., Colombian Soc. Internal Medicine, Colombian Rheumatism Soc., Rheumatism Soc. Chile. Subspecialties: Internal medicine; Cell biology (medicine). Current work: Electron microscopic studies on pathogenesis of gout, rheumatoid arthritis and other rheumatic diseases; immuno electron microscopy, electron probe elemental analysis on synovial membrane and fluid. Home: 947 Longview Rd Gulph Mills PA 19406 Office: Hosp U Pa 24th and Spruce St Philadelphia PA 19104

SCHUMAN, LEONARD MICHAEL, physician, educator; b. Cleve., Mar. 4, 1913; s. Morris and Rebecca (Lazarowitz) S.; m. Marie Romich, Oct. 30, 1954; children by previous marriage—Lowell, Judith. A.B., Oberlin Coll., 1934; M.Sc. (fellow hygiene and bacteriology), Western Res. U., 1939, M.D., 1940; fellow nutrition, Hillman Hosp., Birmingham, Ala., Vanderbilt U., 1946. Diplomate: Am. Bd. Preventive Medicine. Intern U.S. Marine Hosp., Chgo., 1940-41; asst. epidemiologist Ill. Dept. Pub. Health, 1941-42, dist. health supt., 1943, asst. chief div. local health adminstrn., 1943-45, chief div. venereal disease control, 1947-50, dep. dir. div. preventive medicine, 1950-54; dir. S.E. Nutrition Research unit USPHS, 1945-47; epidemiologist Korean cold injury research team U.S. Dept. Def., 1951-53; asso. prof. U. Minn. Med. Sch., 1954-58, prof. epidemiology, 1958—; Cons. air pollution med. program Communicable Disease Center USPHS, Minn. Dept. Health; cons. pediatrics Hennepin County Gen. Hosp.;

mem. nat. adv. coms. on gamma globulin prophylaxis in poliomyelitis, polio vaccine field trials, also tech. adv. com. polio vaccine evaluation U. Mich.; sec. Nat. Subcom. Epidemic Intelligence; mem. nat. cancer control com. Nat. Cancer Inst., chmn. panel epidemiology and biometry, 1961-62, cons., 1958—; cons. div. radiol. health USPHS, 1961-69, cons. to chronic disease div., 1964-72, adv. com. on bio-effects radiation, 1966-69, task force on smoking and health, 1967-68, nat. adv. urban and indsl. health council, 1968-69; accident prevention research study sect. NIH, 1963-66; cons. tng. program in epidemiology U.S. Dept. Agr., 1961—; mem. Surgeon Gen.'s Adv. Com. Smoking and Health, 1962-64; field studies bd. Nat. Adv. Cancer Council, 1961-62; chmn. Conf. Chronic Disease Tng. Program Dirs., 1960-72; mem. Nat. Leukemia Study Com., 1959-63, mem. adv. com. on health protection and disease prevention to sec. health, edn. and welfare, 1969-71; mem. Nat. Adv. Environmental Control Council, 1969-71, Cancer Research Center rev. com. NIH, 1971-75, Task Group on Smoking, Task Group on Interstitial Lung Disease, Task Force on Epidemiology of Respiratory Diseases, Nat. Heart, Lung and Blood Disease Inst., 1978-79, Epidemiology and Disease Control Study Sect., NIH, 1981—; collaborative researcher Norwegian Cancer Registry, Oslo; cons. to WHO, Peru, 1967, India, 1971, Montserrat, 1973; vis. lectr. U. Valle Med. Sch., Cali, Colombia, 1975; AID cons., Ban-dung, Indonesia, 1977, Thailand and Sri Lanka, 1979; mem. steering com. AID World Research Needs in Nutrition. Contbr. sci. papers to profl. lit. Served as sr. surgeon USPHS, 1941-47; med. dir. USPHS Res., 1951-77. Mem. AMA, Royal Soc. Health, Public Health Cancer Assn., Middle States, Am. Public Health Assn. (governing council 1964-70, chmn. infectious disease monograph subcom. 1961-68, chmn. subcom. drugs 1964-65, chmn. epidemiology sect. 1967, chmn. program area com. communicable diseases 1966-70, mem. tech. devel. bd. 1966-70), Ill. Public Health Assn., Minn. Public Health Assn., Internat. Epidemiol. Soc., Internat. Soc. Cardiology, Internat. Soc. Thrombosis and Hemostasis, Am. Heart Assn. (fellow council epidemiology 1965), Am. Thoracic Soc., Am. Coll. Preventive Med. (chmn. council research), Am. Epidemiology Soc. (v.p. 1978), N.Y. Acad. Scis., Assn. Tchrs. Preventive Medicine, Am. Venereal Disease Assn., Am. Assn. Public Health Physicians, Assn. Mil. Surgeons U.S., AAAS, Soc. for Epidemiologic Research (exec. bd. 1979-80), Phi Beta Kappa, Sigma Xi, Alpha Omega Alpha, Phi Zeta Soc., Phi Kappa Phi. Subspecialty: Epidemiology. Discovered (with Dr. Thomas Lowry) silo-fillers' disease. Home: 17 Merilane S Edina MN 55436 Office: Health Scis-Unit A Bldg U Minn Minneapolis MN 55455

SCHUMAN, ROBERT PAUL, radiochemistry scientist; b. Milw., May 17, 1919; s. Ernest Scott and Ethel (Kanable) S.; m. Ellen Jean Bruner, June 28, 1949; children: Barbara, Nancy, Elizabeth. B.S., U. Denver, 1941; M.S., Ohio State U., 1944, Ph.D., 1946. Chemist U.Chgo, Metall. Lab., 1944-45, Knolls Atomic Power Lab., Gen. Electric Co., Schenectady, 1947-57; chemistry sect. head Nat. Reactor Testing Sta., Phillips Petroleum Corp and Idaho Nuclear Corp., Idaho Falls, Idaho, 1957-69; assoc. prof. Robert Coll., Bogazici U., Istanbul, Turkey, 1969-77; vis. prof. Iowa State U., Ames, 1977-78; chemist Idaho Nat. Engring. Lab., Exxon Nuclear Idaho Co. and EG & G Idaho, Inc., Idaho Falls, 1978—; mem. subcom. on radiochemistry Nat. Acad. Scis.-NRC, Washington, 1962-69. Fulbright lecture grantee radiochemistry, Tirupati, India, 1965-66. Mem. Am. Chem. Soc. (local sect. chmn. 1968-69), Am. Phys. Soc., Am. Nuclear Soc., Sigma Xi. Subspecialties: Physical chemistry; Nuclear engineering. Current work: Nuclear chemistry, nuclear waste management, actinide chemistry, cross section measurements, chemical hot cell separations. Home: 1766 Rainier St Idaho Falls ID 83402 Office: EG & G Idaho Inc Idaho Nat Engring Lab PO Box 1625 Idaho Falls ID 83402

SCHUMER, DOUGLAS BRIAN, applied physicist; b. Passaic, N.J., Mar. 22, 1951; s. William and Janet Ellen (Levine) S.; m. Barbara Lee Witte, May 23, 1976. B.S., Carnegie-Mellon U., 1973; M.S. in Measurement and Control, 1974; Ph.D., Rensselaer Poly Inst., 1977. Mem. tech. staff Bell Lab., Holmdel, N.J., 1977-80; mgr. research Chaus Scale Corp., Florham Park, N.J., 1980—. Contbr. articles to profl. jours. Mem. Am. Phys. Soc., Optical Soc. Am., Nat. Computer Graphics Assn. (charter mem.), IEEE, Soc. Photo-optical Instrumentation Engrs., Soc. Photographic Scientists and Engrs. Subspecialties: Optical signal processing; Ultrasonics. Current work: Application of coherent optics and physical acoustics to communications systems and to sensor technology. Home: 27 Iler Dr Middletown NJ 07748 Office: 28 Hanover Rd Florham Park NJ 07932

SCHUNK, DALE HANSEN, psychology educator, researcher; b. Chgo., Aug. 14, 1946; s. Elmer Charles and Milred Augusta (Hansen) S. B.S., U. Ill.-Urbana, 1968; M.Ed., Boston U., 1974; Ph.D., Stanford U., 1979. With NATO, Naples, Italy, 1970-74; teaching asst. Stanford U., 1975, research asst., 1975-59; prof. psychology U. Houston, 1979—; ednl. cons. Spring Branch Sch. Dist., Houston, 1981—. Contbr. articles to profl. jours. Served to capt. USAF, 1968-74. USPHS research grantee, 1980; NSF research grantee, 1980. Mem. Am. Psychol. Assn., Am. Ednl. Research Assn., Soc. Research in Child Devel., Southwestern Psychol. Assn., S.W. Ednl. Research Assn. Subspecialties: Cognition; Learning. Current work: Social learning theory; children's cognitive processes, learning and achievement; motivational processes. Home: 9001 S Braeswood St Houston TX 77074 Office: Coll Ed U Houston 4800 Calhoun St Houston TX 77004

SCHUNK, ROBERT WALTER, physics educator; b. N.Y.C., July 4, 1943; s. Walter and Anna (Weinkauff) S.; m. Ellen Veronica Morris, Dec. 11, 1965; children: Michael, Allison. B.S., NYU, 1965; Ph.D., Yale U., 1970. Postdoctoral fellow U. Mich., Ann Arbor, 1970-71; research assoc. Yale U., New Haven, 1971-73; assoc. research physicist U. Calif., San Diego, 1973-76; prof. physics Utah State U., Logan, 1976—. Contbr. over 100 sci. articles to profl. publs. Recipient William R. Bryans award NYU, 1965, Founders Day award, 1965; Gibbs Prize fellow Yale U., 1965-68. Mem. Am. Geophys. Union, AAAS, Sigma Pi Sigma. Subspecialties: Aeronomy; Planetary atmospheres. Current work: Develop computer models to describe the dynamic behavior of planetary ionospheres and atmospheres. Office: Utah State Univ Physics Dept Logan UT 84322

SCHUPLER, BRUCE RALPH, geophysicist; b. West Palm Beach, Fla., July 12, 1953; s. Solomon Simon and Gail (Horn) S. B.S., Calif. Inst. Tech., 1975; M.A., U. Tex., Austin, 1977. Research asst. U. Tex., Austin, 1975-77; mem. tech. staff Computer Scis. Corp., Goddard Space Flight Ctr., Greenbelt, Md., 1977—. Contbr. articles in field to profl. jours. Mem. Am. Astron. Soc., Am. Geophys. Union. Subspecialties: Radio and microwave astronomy; Geodesy. Current work: The rotation of the earth and the positions of radio sources, geodesy, radio astronomy, earth rotation, minicomputer techniques for scientific data analysis. Office: Code 974 Goddard Space Flight Center Greenbelt MD 20771

SCHURR, AVITAL, neuropharmacologist, educator; b. Jerusalem, Aug. 23, 1941; s. Areyeh and Esther (Glazner) S.; m. Dafna Safran, Aug. 9, 1965; children: Barak, Hila, Ori. B.Sc.Agr., Hebrew U., Jerusalem, 1967; M.Sc., Tel Aviv U., 1970; Ph.D., Ben Gurion U. of the Negev, Beer Sheva, Israel, 1977. Research asso. Inst. Arid Zone Research, Beer Sheva, 1970-72; research asso. Research and Devel. Authority, Beer Sheva, 1972-77; postdoctoral fellow Baylor Coll. Medicine, Houston, 1977-79; research asso. in neuroscience U. Tx. Med. Sch., Houston, 1979-81; asst. prof. anesthesiology U. Louisville Sch. Medicine, 1981—. Contbr. articles to profl. jours. Served with Israeli Army, 1959-62. Israeli Ministry of Health grantee, 1973-74; Israeli Ministry of Police grantee, 1974-75; Israeli Soc. Psychobiology grantee, 1975-76. Mem. Soc. Neurosci., N.Y. Acad. Sci., AAAS. Jewish. Current work: Mechanism of action of monoamine oxidase. Brain slice, its electrophysiology and the effects of drugs and neurotransmitters on it; the therapeutic potential of cannabis. Office: Dept Anesthesiology Louisville School Medicine Louisville KY 40292

SCHUSTER, MARVIN MEIER, physician, researcher, educator; b. Danville, Va., Aug. 30, 1929; s. Isaac and Rosel (Katzenstein) S.; m. Lois R. Bernstein, Feb. 19, 1961; children: Roberta, Nancy, Cathy. B.A., B.S., U. Chgo., 1951; M.D., 1955. Diplomate: Am. Bd. Internal Medicine. Intern Kings County Hosp., Bklyn., 1955-56; resident Balt. City Hosp., 1956-58, chief digestive disease div., resident, 1961—; Johns Hopkins Hosp., Balt., 1958-61; prof. medicine and psychiatry Johns Hopkins U. Sch. Medicine, Balt., 1976—. Editor: Gastrointestinal Motility Disorders, 1981; contbr. chpts. to textbooks, articles in field to profl. jours.; mem. editorial bd.: Gastroenterology, 1978-81, Gastrointestinal Endoscopy, 1979-81, Psychosomatics, 1979—. Bd. dirs. Am. Cancer Soc., 1975—, v.p., pres.-elect, 1982; chmn. med. adv. bd. Balt. Ostomy Assn., 1966—. Recipient St. George Disting. Service award Am. Cancer Soc., 1979. Fellow ACP, Am. Psychiat. Assn.; mem. Am. Gastroent. Assn. (chmn. audiovisual com. 1975-78), Am. Soc. Gastrointestinal Endoscopy (governing bd. 1975-78), Am. Physiol. Soc., AAUP. Democrat. Jewish. Subspecialties: Gastroenterology; Psychiatry. Current work: Teaching and research in gastroenterology and application of biofeedback to gastrointestinal control. Home: 3101 Northbrook Rd Baltimore MD 21208 Office: Baltimore City Hosp 4940 Eastern Ave Baltimore MD 21224

SCHUSTER, TODD MERVYN, biotechnology company executive; b. Mpls., June 27, 1933; s. David Theodore and Ann (Kaluser) S.; m. Nancy Joanne Mottashed, June 28, 1935; 1 dau., Lela Alexa. Ph.D., Washington U., 1963. Asst. prof. dept. biology SUNY, Buffalo, 1966-70; Assoc. prof. dept. biology U. Conn., Storrs, 1970-75, prof., 1975-82, head dept., 1977-81; pres., chief exec. officer Xenogen, Inc., Storrs, 1981—; vis. scientist Max Planck Inst., Goettingen, Germany, 1963-66, 67, 68; vis. prof. chemistry dept. Ind. U., 1976; McCollum-Pratt prof. biology dept. Johns Hopkins U., Balt., 1979-80; mem. NIH grant rev. panels, biophysics and biophys. chemistry panel, 1971-75, Biomed. Scis. Postdoctoral Fellowship Panel, 1976, Sickle Cell Disease Adv. Panel, 1977. USPHS fellow, 1959-63 63-66. Mem. Am. Chem. Soc., Biophys. Soc., AAAS, Am. Soc. Biol. Chemists. Subspecialties: Biophysical chemistry; Molecular biology. Current work: Biophysical chemistry, structure and functions of proteins, muscle protein structure, hemoglobin function, mechanisms of virus assembly, rapid reaction kinetics, spectroscopy of biopolymers, research management. Home: 557 Wormwood Hill Rd Mansfield CT 06250 Office: Xenogen Inc 1768 Storrs Rd Storrs CT 06268

SCHVARTZMAN, JORGE BERNARDO, cell biologist, researcher; b. Asunción, Paraguay, Dec. 14, 1948; s. Isaac and Olga (Blinder) S.; m. Dora B. Krimer, Jan. 7, 1971; children: Juan Manuel, Daniel Ignacio. Agrl. Engr., U. Nacional de Asunción, 1972; Ph.D., Universidad Politécnica de Madrid, 1978. Research asoc. Instituto de Biología Celular, Madrid, 1976-79, 82—, Brookhaven Nat. Lab., N.Y.C., 1980-82. Subspecialties: Cell biology; Genome organization. Current work: Plant genetics and cytogenetics, DNA replication, damage and repair in higher embryology.

SCHWAB, JEFFREY RICHARD, computer systems programmer; b. Cleve., Aug. 11, 1955; s. Richard Frank and Carol Joyce S. B.S. in Computer Sci, Purdue U., 1976, M.S., 1978. Assoc. programmer IBM Corp., Cape Canaveral, Fla., 1978-79; systems programmer Purdue U., West Lafayette, Ind., 1979—. Mem. ACM, IEEE Computer Soc. Subspecialties: Distributed systems and networks; Operating systems. Current work: Large scale time sharing systems on networks of distributed medium scale computers. Office: Purdue U Math Sci G169 West Lafayette IN 47907

SCHWAB, JOHN HARRIS, microbiology and immunology educator; b. St. Cloud, Minn., Nov. 20, 1927; s. John David and Katherine Palmer (Harris) S.; m. Ruth Graves, Sept. 1, 1951; children: Stewart, Thomas, Anna, Kellogg. B.S., U. Minn., 1949, M.S., 1950, Ph.D., 1953. Asst. prof. to prof. microbiology and immunology U. N.C.-Chapel Hill, 1953, Cary C. Boshamer prof. microbiology, 1982—; vis. scientist Lister Inst. Preventive Medicine, London, 1960-61, MRC Rheumatism Research Unit, Taplow, Eng., 1968-69, Radiobiol. Inst., Rijswick, Netherlands, 1975-76. Editor: Infection and Immunity, 1980; contbr. articles profl.jours., chpts. in books. NIH special fellow, 60-61, 68-69. Mem. Am. Assn. Immunologists, Am. Soc. Microbiology, AAAS. Subspecialties: Infectious diseases; Microbiology (medicine). Current work: Immunomodulation by bacteria, chronic inflammatory diseases induced by bacterial cell walls.

SCHWABEL, MARY JANE, virologist, nurse; b. Buffalo, Oct. 9, 1946; d. Albert Thomas and Doris Katherine (Schottin) S. B.S., Damen Coll., 1968; M.S., Canisius Coll., 1975; A.A.S. in Nursing, Trocaire Coll., 1983. Cert. in epidemiology, registered nurse, N.Y. Research asst. Erie County Virology Lab., Erie County Med. center, Buffalo, 1968-74, sr. serology technician/supr., 1974-79, chief virologist, 1980—. Author profl. papers. Mem. Western N.Y. Infection Control Soc., Am. Soc. Microbiology, Am. Public Health Assn., N.Y. State Assn. Public Health Labs., N.Y. State Public Health Assn., Erie County Soc. Prevention Cruelty to Animals, North Shroe Animal League, Nat. Antivivisection Soc., Internat. Wildlife Fedn., Nat. Wildlife Fedn., Am. Forestry Assn., Beta Beta Beta. Subspecialties: Virology (biology); Preventive medicine. Current work: Diagnostic virology and infection control. Office: 462 Grider St Buffalo NY 14215

SCHWABER, JERROLD, research immunologist, educator; b. Evanston, Ill., May 24, 1947. B.A., U. Chgo., 1969, Ph.D., 1974. Research asst. LaRabida-U. Chgo. Inst., 1971-74; research fellow Children's Hosp. and Harvard U. Med. Sch., Boston, 1974-78, instr. pediatrics, 1978-81; asst. prof. pathology Harvard U. Med. Sch., 1981—. Leukemia Soc. Am. scholar, 1980—; recipient Basil O'Connor award March of Dimes-Birth Defects Found., 1979-81. Mem. Am. Assn. Immunologists, Am. Soc. Cell Biology. Subspecialties: Immunobiology and immunology; Genetics and genetic engineering (medicine). Current work: Immune deficiency; B lymphocyte ontogeny and regulation; regulation of antibody production. Office: Children's Hosp 300 Longwood Ave Boston MA 02115

SCHWAN, HERMAN PAUL, educator, research scientist; b. Aachen, Germany, Aug. 7, 1915; came to U.S., 1947, naturalized, 1952; s. Wilhelm and Meta (Pattberg) S.; m. Anne Marie DelBorello, June 15, 1949; children—Barbara, Margaret, Steven, Carol, Cathryn. Student, U. Goettingen, 1934-37; Ph.D., U. Frankfurt, 1940; Doctor habil. in physics and biophysics, 1946. Research scientist, prof. Kaiser Wilhelm Inst. Biophysics, 1937-47; asst. dir. Kaiser Wilheim Inst. Biophysics, 1945-47; research sci. USN, 1947-50; prof. elec. engring., prof. elec. engring. in phys. medicine, assoc. prof. phys. medicine U. Pa., Phila., 1950—, Alfed F. Moore prof., 1983—, dir. electromed. div., 1952—, chmn. biomed. engring., 1961-73, program dir. biomed. engr. tng. program, 1960-77; vis. prof. U. Calif. at Berkeley, 1956, U. of Frankfurt, Germany, 1962; lectr. Johns Hopkins U., 1962-67; W.W. Clyde vis. prof. U. Utah, Salt Lake City, 1980; Fgn. sci. mem. Max Planck Soc. Adv. Research, Germany, 1962—; cons. NIH, 1962—; chmn. nat. and internat. meetings biomed. engring. and biophysics, 1959, 61, 65; mem. nat. adv. council environ. health HEW, 1969-71; mem. Nat. Acad. Scis.-NRC coms., 1968-77, Nat. Acad. Engring., 1975—. Co-author: Advances in Medical and Biological Physics, 1957, Therapeutic Heat, 1958, Physical Techniques in Medicine and Biology, 1963; Editor: Biol. Engring, 1969; Mem. editorial bd.: Environ. Biophysics; Contbr. articles to profl. jours. Recipient Citizenship award, Phila., 1952, 1st prize AIEE, 1953, Achievement award Phila. Inst. E.E. and Electronics, 1963, Rajewsky prize for biophysics, 1974; U.S. sr. scientist award Alexander von Humboldt Found., 1980-81; Biomed. Engring. Edn. award Am. Soc. Engring. Edn., 1983. Fellow IEE, IEEE (Morlock award 1967, Edison medal 1983, chmn. and vice chmn. nat. profl. group biomed. engring. 1955, 62-68), AAAS; mem. Am. Standards Assn. (chmn. com. 1961-64), Am. Phys. Soc., Biophys. Soc. (publicity com., council, constrn. com.), Soc. for Cryobiology, Internat. Fedn. Med., and Biol. Engring., Biomed. Engring. Soc. (dir.), Sigma Xi, Eta Kappa Nu. Subspecialties: Biomedical engineering; Biophysics (physics). Current work: Electrial and acoustic properties of biomatter; biological effects of nonionizing radiation; bioelectrodes, clinical applications of microwaves and ultrasound. Home: 99 Kynlyn Rd Radnor PA 19087 also 162 59th St Avalon NJ 08202 Office: Dept Bioengring D2 U Pa Philadelphia PA 19104

SCHWANHAUSSER, ROBERT ROWLAND, aerospace engineering company executive; b. Buffalo, N.Y., Sept. 15, 1930; s. Edwin Julius and Helen Putnam (Rowl) S.; m. Mary Lea Hunter, Oct. 17, 1953 (div. 1978); children: Robert Hunter, Mark Putnam; m. Beverly Bohn Allemann, Dec. 31, 1979. S.B. in Aero. Engring, MIT, 1952. Project engr. Continental Aviation & Engring. Co., Detroit, 1954; field service rep. Ryan Aero. Co., San Diego, 1954-56, project engr., 1956-59, program mgr., 1959-62, chief engr., 1962-64, dir. drone programs, 1964-66, v.p. aerospace, 1966-72, exec. v.p. programs, 1972-73, exec. v.p. internat., 1973-75; pres. Condur Aerospace/Condur Engring., San Diego/El Paso, Calif., 1975-77; v.p. bus. devel. All Am. Engring., Wilmington, Del., 1977-79; v.p. internat. and RPV programs Teledyne Ryan Aero., San Diego, 1979-82; v.p. advanced program Teledyne Brown Engring., Huntsville, Ala., 1982—. Bd. dirs. Cornerstone Found. Served to 1st lt. USAF, 1952-54. Fellow AIAA; mem. Am. Def. Preparedness Assn., Navy League, Nat. Rifle Assn., Nat. Assn. Remotely Piloted Vehicles (trustee 1972—), Theta Delta Chi. Club: Greenhead Hunting (founder, pres. San Diego 1937-77). Subspecialties: Aeronautical engineering; Aerospace engineering and technology. Current work: Unmanned and manned aircraft and missile systems engineering. Office: Teledyne Brown Engineering Cummings Research Park Huntsville AL 35807

SCHWARK, WAYNE STANLEY, veterinary educator, researcher; b. Vita, Man., Can, May 19, 1942; s. William and Bertha (Walters) S.; m. Donna Mae Laufersweiler, Dec. 27, 1963; children: Dwight Wayne, Lisa Lynn. D.V.M., U. Guelph, Ont., Can., 1965, M.Sc. in Pharmacology, 1967, Ph.D., U. Ottawa, 1970. Lectr. U. Guelph, 1965-67; biologist Food and Drug Directorate, Ottawa, Ont., 1967-70; vet. cons. U. Ottawa, 1969-70; research scientist Food and Drug Directorate, 1970-71; asst. prof. Cornell U., Ithaca, N.Y., 1972-77, assoc. prof. pharmacology, 1977—; cons. FDA, Poison Info. Center, Syracuse, N.Y. Contbr. articles to sci. jours. Grantee NIH, Epilepsy Found. Am., FDA. Mem. Soc. Neurosci., Can. Vet. Med. Assn., So. Tier Vet Assn. Lutheran. Subspecialties: Veterinary pharmacology; Neuropharmacology. Current work: Neurochemical and neuropharmacological basis of epilepsy; pharmacokinetic studies of drugs used in veterinary therapeutics. Home: 313 Winthrop Dr Ithaca NY 14850 Office: NY State Coll Vet Medicine Cornell U Ithaca NY 14853

SCHWARTZ, A(LBERT) TRUMAN, chemist, educator; b. Freeman, S.D., May 8, 1934; s. Albert and Edna (Kaufman) s.; m. Beverly Joan Beatty, Aug. 12, 1958; children: Ronald Eric, Katherine Marie. A.B., U. S.D., 1956; B.A., Oxford U., 1958, M.A., 1960, Ph.D., M.I.T., 1963. Teaching and research asst. M.I.T., Cambridge, 1958-63; research chemist Procter & Gamble Co., Cin., 1963-66; asst. prof. Macalester Coll., St. Paul., 1966-72, assoc. prof., 1972-78, dean faculty, 1974-76, prof., 1978-83, De Witt Wallace prof., 1983—, chmn. dept., 1980—; Arthur Lee Haines lectr., vis. scientist U. S.D., 1965; vis. prof. U. Wis., Madison, 1979-80; mem. state and regional coms. for selection of Rhodes Scholars, 1963—. Author: Chemistry: Imagination and Implication, 1973; contbr. articles on chemistry to profl. jours. Recipient Catalyst award for excellence in coll. chemistry teaching Chem. Mfrs. Assn., 1982; Rhodes scholar, 1956; NSF summer fellow, 1959; Macalester fgn. fellow, 1968; NSF-COSIP fellow, 1967, 72, 73; NSF grantee, 1979. Mem. Am. Chem. Soc. (mem. various coms. div. chem. edn.), AAAS, Assn. Am. Rhodes Scholars, Phi Beta Kappa, Sigma Xi. Subspecialties: Physical chemistry; Biophysical chemistry. Home: 68 Otis Ave St Paul MN 55104 Office: Dept Chemistry Macalester Coll St Paul MN 55105

SCHWARTZ, ALLEN DAVID, pediatrics educator; b. Balt., Feb. 26, 1938; s. Paul and Rose (Goldfinger) S.; m. Maria Gumbiwas, May 19, 1968; 1 son, Michael. B.A., Johns Hopkins U., 1960; M.D., U. Md., 1964. Intern Yale-New Haven Hosp., 1964-65, resident, 1966-68, Johns Hopkins Hosp., Balt., 1965-66; asst. prof. pediatrics Yale U., 1970-72; assoc. prof. Northwestern U., 1972-75, U. Md., Balt., 1975-78, prof., 1978—. Author: Malignant Diseases of Infancy, Childhood and Adolescence, 1978, 2d edit., 1982. Bd. Dirs. Md. Chpt. Leukemia Soc. Am., 1977—; bd. dirs. Hematology-Oncology Support Services Md., Balt., 1979-82, Md. chpt. Am. Cancer Soc., 1980—. Fellow Am. Acad. Pediatrics; mem. Am. Soc. Clin. Oncology, Am. Soc. Hematology, Am. Soc. Pediatric Hematology/Oncology, Soc. Pediatric Research. Subspecialties: Hematology; Oncology. Current work: Splenic function, platelet function disorders, malignant diseases of children. Office: Dept Pediatrics U Md Hosp Baltimore MD 21201

SCHWARTZ, ANTHONY, veterinary surgeon, educator, immunologist; b. Bklyn., July 30, 1940; s. Murray and Miriam Sarah (Wittes) S.; m. Claudia Rosenberg, July 21, 1963; children: Thomas Frederick, Eric Leigh. Student, Mich. State U., 1957-58; D.V.M., Cornell U., 1963; Ph.D., Ohio State U., 1972. Diplomate: Am. Coll. Vet. Surgeons, 1971. Pvt. practice vet. medicine, Huntington, N.Y., 1963-66; resident in surgery Animal Med. Ctr., N.Y.C., 1968-69, Ohio State U., 1969-70, asst. prof., head small animal surgery, Columbus, 1973; from asst. prof. to assoc. prof. Yale U. Sch. Medicine, 1973-79; from assoc. prof. to prof., chmn. dept. surgery Tufts U. Sch. Vet. Medicine, Boston, 1979—; cons. U.S. Surg. Corp., Norwalk, Conn. Assoc. editor: Yale Jour. Biology and Medicine, 1977-79; sect. editor: Textbook of Small Animal Surgery, 1983; contbr. chpts. to books, articles to profl. jours. Mem. adv. bd. Morris Animal Found., 1981—. Served to capt., vet. corps. U.S. Army, 1966-68. Recipient 1st prize N.Y. State Vet. Med. Soc., 1963; NIH grantee, 1975—. Mem. AVMA, AAAS, Am. Assn. Immunologists, Mass. Vet. Med. Assn., N.Y. Acad. Sci., Sigma Xi, Phi Kappa Phi, Phi Zeta. Democrat. Jewish. Subspecialties: Immunology (medicine); Immunopharmacology. Current work: T cell interactions in the immune response.

SCHWARTZ, ARTHUR GERALD, cancer researcher, gerontologist, educator; b. Balt., Mar. 13, 1941; s. Paul and Rose (Goldfinger) S. B.A., Johns Hopkins U., 1961; Ph.D., Harvard U., 1968. Postdoctoral fellow Albert Einstein Coll. Medicine, 1971-72; Asst. prof. microbiology, mem. Fels Research Inst., Temple U., 1972-77, assoc. prof., mem., 1977—; dir. gerontol. research Inst. on Aging, Temple U., 1982—. Contbr. articles to cancer research and gerontology jours. Jane Coffin Childs Meml. Fund fellow U. Oxford, 1968-71. Mem. Am. Assn. Cancer Research, Gerontol. Soc., Phi Beta Kappa. Subspecialties: Cancer research (medicine); Gerontology. Current work: Cancer prevention; gerontology; working with naturally occurring adrenal steroid, DHEA, which prevents many types of cancers in laboratory animals and appears to delay aging. Home: 220 Locust St Philadelphia PA 19106 Office: Fels Research Institute Temple Med Sch Philadelphia PA 19140

SCHWARTZ, BERNARD, ophthalmologist, educator; b. Toronto, Ont., Can., Nov. 12, 1927; s. Samuel and Gertrude S. (Levinsky) S.; m. Marcia Struhl, Nov. 30, 1980; children: Lawrence Frederick, Karen Lynne, Jennifer Carla, Ariane Samara. M.D., U. Toronto, 1951; M.S., State U. Iowa, 1953, Ph.D., 1959. Diplomate: Am. Bd. Ophthalmology. Intern Univ. Hosp., Iowa City, 1951-52, resident in ophthalmology, 1953-55; asst. and assoc. prof. ophthalmology SUNY Downstate Med. Ctr., Bklyn., 1958-68; assoc. vis. surgeon Kings County Hosp., Bklyn., 1958-68; cons. in ophthalmology Coney Island Hosp., 1954-68; attending ophthalmologist in charge Maimonides Hosp., Bklyn., 1963-68; attending ophthalmologist State U. Hosp. N.Y., 1966-68; ophthalmologist-in-chief Tufts-New Eng. Med. Ctr., Boston, 1968—; prof., chmn. dept. ophthalmology Tufts U., 19—; cons. ophthalmologist VA Hosp., 1970—. Editor: Corticosteroids and the Eye, 1966, Decision-Making in the Diagnosis and Therapy of the Glaucomas, 1969, Syphilis and the Eye, 1970; editor-in-chief: Survey of Ophthalmology; contbr. articles to profl. jours. Fellow ACS; mem. Am. Acad. Ophthalmology, Am. Soc. Cell Biology, Am. Soc. Clin. Pharmacology and Therapeutics, Assn. for Research in Vision in Ophthalmology, Council Biology Editors. Subspecialty: Ophthalmology. Current work: Application of computers to image analysis of photographs of the eye for diagnosis and therapy; development of techniques for measuring the circulation of blood in the eye; role of hormones in the development and treatment of ocular disease especially glaucoma. Office: 171 Harrison Ave Boston MA 02111

SCHWARTZ, DAVID MICHAEL, energy co. exec., architect, archtl. engr.; b. Pitts., Sept. 10, 1948; s. Meyer and Anne Lisa (Bernheim) S.; m. Shelby Hanson, May 30, 1976 (div.); m. Helen Glass, Dec. 8, 1979. B. Arch, Carnegie-Mellon U., 1972. Registered architect, D.C. Research assoc. ARK: Environ, Research, Inc., Pitts., 1971-73; pres. G.N.S. Co., Inc., Boston, 1974-77; dir. Energy Design & Analysis Co., Inc., Washington, 1977-78; dir. engring. Enercorp, Inc., Denver, 1978—; mem. faculty dept. mech. engring. U. Colo., Boulder, 1980; project leader Foam Plastic Housing Devel.; mem. energy com. Met. Washington Bd. Trade, 1979-80. HUD grantee, 1969. Mem. AIA (chmn. Met. Washington chpt. energy com. 1980), ASHRAE, Internat. Solar Energy Soc. Democrat. Jewish. Club: Effendi Billiard (Washington). Subspecialties: Solar energy; Environmental engineering. Current work: Energy research and devel; conservation in energy systems; solar energy applications; emerging energy tech. mgmt. Designer ABT Assoc. Inc. solar office bldg., Cambridge, Mass., 1976; patentee bldg. system, solar collector. Office: 666 Sherman St Denver CO 80203

SCHWARTZ, DREW, biology educator; b. Phila., Nov. 15, 1919; s. Isaac and Miriam (Bonn) S.; m. Pearl (Freeman), July 26, 1942; 1 dau., Rena Ann. B.S., Pa. State Coll., 1942; B.A., Columbia U., 1948, Ph.D., 1950. Research assoc. U. Ill., Urbana, 1950-51; sr. biologist Oak Ridge Nat. Lab., 1951-62; prof. Western Res. U., Cleve., 1962-64; prof. biology Ind. U., Bloomington, 1964—. Served with U.S. Army, 1942-46. Recipient Waterman research award Ind. U., 1979; Guggenheim fellow, 1970. Fellow AAAS, Genetics Soc. Am. (treas. 1960-62); mem. Sigma Xi. Subspecialties: Gene actions; Genetics and genetic engineering (biology). Current work: Regulation of gene action, analysis of transposable controlling elements in maize. Office: Dept Biology Ind U Bloomington IN 47401 Home: 4001 Saratoga Dr Bloomington IN 47401

SCHWARTZ, EDWARD, toxicologist; b. Phila., Dec. 25, 1932; s. Harry and Sylvia S.; m. Harriet Cohen, June 15, 1958; children: Lisa, Elaine, Daniel. B.Sc., Phila. Coll. Pharma. Sci., 1955; V.M.D., U. Pa., 1959; Ph.D., Jefferson Med. Coll., 1963; diploma in computer programming, LaSalle U., 1969. Sr. toxicologist Hoffman LaRoche Inc., 1962-65; sr. research assoc. Warner Lambert, Morris Plains, N.J., 1965-69, dir. toxicology, 1970-77; assoc. dir. toxicology Schering Corp., Lafayette, N.J., 1977, dir. pathology and toxicology, 1977-81, sr. dir. pathology and toxicology, 1981—. Contbr. articles to profl. publs. Mem. AVMA, Soc. Toxicology, Am. Soc. Vet. Clin. Pathologists, Am. Soc. Pharm. and Explt. Therapeutics, Can. Assn. Research in Toxicology, European Teratology Soc., Am. Coll. Toxicology, European Soc. Study drug Toxicology. Subspecialty: Toxicology (medicine). Home: 15 Walsh Way Morris Plains NJ 07950 Office: PO Box 32 Lafayette NJ 07848

SCHWARTZ, GEORGE JOHN, pediatric nephrologist, biomedical researcher; b. Jersey City, Feb. 18, 1947; s. B. Robert and Marilyn Thelma (Wolf) S. A.B., Colgate U., 1968; M.S., Case Western Res. U., 1972. Diplomate: Am. Bd. Pediatrics Am. Bd. Nephrology. Fellow pediatric nephrology Albert Einstein Coll. Medicine, Bronx, N.Y., 1974-76, asst. prof. pediatrics, 1979—; guest worker NIH, Bethesda, Md., 1976-78, sr. staff fellow, 1978-79; grants reviewer N.Y. Kidney Found., 1980-83. Contbr. articles to med. jours. Mem. Am. Soc. Nephrology, Am. Fedn. Clin. Research, Am. Physiol. Soc., N.Y. Acad. Scis., Soc. for Pediatric Research, Physicians for Social Responsibility. Jewish. Subspecialties: Pediatrics; Nephrology. Current work: Developmental renal physiology, maturation of transport in the isolated perfused proximal renal tubule, acidification by proximal tubule, salt handling by the immature cortical collecting duct. Office: Dept Pediatrics Albert Einstein Coll Medicine 1410 Pelham Pkwy S Bronx NY 10461

SCHWARTZ, HENRY GERARD, educator, surgeon; b. N.Y.C., Mar. 11, 1909; s. Nathan Theodore and Marie (Zagat) S.; m. Edith Courtenay Robinson, Sept. 13, 1934; children: Henry G., Michael R., Richard H. A.B., Princeton, 1928; M.D., Johns Hopkins, 1932. Diplomate: Am. Bd. Neurol. Surgery (chmn. 1968-70). Denison fellow with Prof. O. Forester, Breslau, Germany, 1931; surg. house officer Johns Hopkins Hosp., 1932-33; NRC fellow Harvard Med. Sch., 1933-35, instr. anatomy, 1935-36; fellow neurol. surgery Washington U. Med. Sch., St. Louis, 1936-37, instr., asst. prof., asso. prof. neurol. surgery, 1937-46, prof., 1946—; acting surgeon-in-chief Barnes and Allied hosps., 1965-67; chief neurosurgeon Barnes, St. Louis Children's hosps.; cons. neurosurgeon St. Louis City, Jewish, Los Alamos (N.M.) hosps.; cons. to surgeon gen. USPHS. to surgeon gen. U.S. Army.; Mem. subcom. neurosurgery NRC; del. World Fedn. Neurosurgery. Mem. editorial bd.: Jour. Neurosurgery (chmn, 1967-69; editor, 1975—. Served with AUS, 1942-45. Recipient ofcl. citation and commendation Brit. Army; decorated Legion of Merit; recipient Harvey Cushing medal, 1979. Fellow ACS (adv. council on neurosurgery 1950, 60, v.p. 1972-73); mem. Soc. Neurol. Surgeons (pres. 1968-69), Am. Acad. Neurol. Surgery (pres. 1951-52), Harvey Cushing Soc. (pres. 1967-68), Am. Neurol. Assn. (hon.), Assn. Research Nervous and Mental Disease, Central Neuropsychiat. Assn. Am., Assn. Anatomists, So. Neurosurg. Soc. (pres. 1953-54), Soc. Med. Cons. to Armed Forces, Am. Surg. Assn. (v.p. 1975-76), Soc. de Neuro-Chirurgie de Langue Francaise, Excelsior Surg. Soc., Johns Hopkins Soc. Scholars, Soc. Internat. de Chirugie. Subspecialty: Neurosurgery. Home: 2 Briar Oak St Ladue MO 63132 Office: Barnes Hospital Plaza St Louis MO 63110

SCHWARTZ, HERBERT, chemist; b. Limerick, Pa., Mar. 8, 1925; s. Edward and Nettie (Mamet) S.; m. Martha M. Scheepers, Oct. 16, 1958; children: Simone L., David. Student, Temple U., 1947-50; Dipl.-Chem., U. Freiburg, Germany, 1955; Ph.D., U. Utrecht, Netherlands, 1965. Research chemist Vineland Chem. Co., N.J., 1955-57; chemist FDA, 1957-58, Grad. Hosp., U. Pa., Phila., 1958; research coordinator Biovivan Research Inst., Vineland, 1958—; adj. prof. chemistry Camden County Coll., Blackwood, N.J., 1957; prof. organic chemistry Cumberland County Coll., Vineland, 1969-75. Vice chmn. Cumberland County Mosquito Control Study Commn., 1969. Served with U.S. Army, 1943-46; ETO. Mem. Am. Chem. Soc., AAAS, N.J. Acad. Sci., Am. Aging Assn., Del. Valley Translators Assn. Subspecialties: Organic chemistry; Cancer research (medicine). Current work: Removal of malodors and toxic pollutants from the atmospheres; development of theory to unify basic causes of hypertension, cardiovascular disease, and cancer; sociochronobiology. Patentee herbicides and fungicides. Office: Biovivan Research Institute PO Box 266 Vineland NJ 08360

SCHWARTZ, IRA S., biochemistry educator, researcher; b. N.Y.C., May 16, 1947; s. Max and Gussie (Gartner) S.; m. Arlene R. Schwartz Ebner, June 9, 1968; children: Kenneth, Aliza, Rina. B.S., CCNY, 1968; Ph.D., CUNY, 1973. Lectr. biochemistry CCNY, 1968-73; fellow Roche Inst. Molecular Biology, Nutley, N.J., 1973-75; asst. prof. U. Mass., 1975-80, N.Y. Med. Coll., Valhalla, 1980-82, assoc. prof., 1982—. Regents scholar, 1964-68; Sinsheimer scholar, 1981—. Mem. Am. Soc. Biol. Chemists, N.Y. Acad. Sci., AAAS, Am. Chem. Soc., Sigma Xi. Democrat. Jewish. Subspecialties: Biochemistry (biology); Molecular biology. Current work: Regulation of gene expression; molecular mechanism of protein biosynthesis. Home: 10 Dr Frank Rd Spring Valley NY 10977 Office: Dept Biochemistr New York Med Coll Valhalla NY 10595

SCHWARTZ, IRVING LEON, scientist, educator, physician; b. Cedarhurst, N.Y., Dec. 25, 1918; s. Abraham and Rose (Doniger) S.; m. Felice T. Nierenberg, Jan. 12, 1946; children: Cornelia Ann, Albert Anthony, James Oliver. A.B., Columbia U., 1939; M.D., N.Y.U., 1943. Diplomate: Am. Bd. Internal Medicine. Intern, asst. resident Bellevue Hosp., N.Y.C., 1943-44, 46-47; NIH fellow physiology N.Y.U. Coll. Medicine, 1947-50; Am. Physiol. Soc. Porter fellow, also Gibbs meml. fellow in clin. sci. Rockefeller Inst., N.Y.C., 1950-51, Am. Heart Assn. fellow, 1951-52, asst., then assoc., 1952-58; asst. physician, assoc. physician Rockefeller Inst. Hosp., 1950-58; sr. scientist Brookhaven Nat. Lab., Upton, L.I., N.Y., 1958-61; attending physician Brookhaven Nat. Lab. Hosp., 1958—; research collaborator Brookhaven Nat. Lab., 1961—; Joseph Eichberg prof. physiology, dir. dept. U. Cincinnati Coll. Medicine, 1961-65; dean grad. faculties Mt. Sinai Med. and Grad. Schs., 1965-81; prof. physiology and biophysics, chmn. dept. Mt. Sinai Med. and Grad. Schs. City U. N.Y., 1968-79; exec. officer biomed. scis. doctoral program City U. N.Y., 1969-72, Dr. Harold and Golden Lamport disting. prof., 1979—; dir. Ctr. Peptide and Membrane Research Mt. Sinai Med. Ctr., 1979—; chmn. emeritus dept. physiology and biophysics Mt. Sinai Sch. Medicine, 1979—; dean emeritus Mt. Sinai Grad. Sch. Biol. Scis., 1980—. Contbr. articles to sci. publs. Pres. Life Scis. Found., 1962—. Served from 1st lt. to capt., M.C. AUS, 1944-46. Fellow A.C.P.; mem. Am. Physiol. Soc., Soc. Exptl. Biology and Medicine, Am. Soc. Clin. Investigation, Am. Fedn. Clin. Research, Biophys. Soc., Endocrine Soc., Harvey Soc., Soc. for Neurosci., Am. Heart Assn., John Jay Assos. Columbia Coll., AAAS, N.Y. Acad. Sci., Sigma Xi, Alpha Omega Alpha. Subspecialties: Psychophysiology; Biophysics (physics). Current work: Conformation and mechanism of action of biologically-active peptides, particularly neurohypophyseal and other peptide hormones, membrane permeability and transport phenomena. Home: 1120 Fifth Ave New York NY 10028 also 9 Thorn Hedge Rd Bellport NY 11713 Office: Mount Sinai Med and Grad Schs City U NY 100th St and Fifth Ave New York NY 10029 also Med Research Center Brookhaven Nat Lab Upton NY 11973

SCHWARTZ, JACOB THEODORE, computer scientist; b. N.Y.C., Jan. 9, 1930; s. Harry and Hedwig (Kurzbartz) S.; m. Frances E. Allen, 1972; children: Rachel, Abby. B.S., CCNY, 1948; Ph.D., Yale U., 1951. Instr. Yale U., 1953-56, asst. prof., 1956-58; assoc. prof. computer sci. NYU, 1957-58; prof. computer sci. dept. Courant Inst. Math. Scis., 1958-59, dir. computer sci. div., 1982-83, chmn. dept., 1969-77; disting. lectr. math. and computer sci. U. Calif.-Santa Barbara, 1978—, MIT, 1980; chmn. computer sci. bd. NRC. Author: 11 books including Lectures on NonLinear Functional Analysis, 1968, Lectures on Differential Geometry and Topology, 1969; Editorial bd.: Jour. Computer and Systems Scis, 1980-83, Communications on Pure and Applied Math, 1975-83, Advances in Applied Math, 1980-83; assoc. editor: Jour. Programming Langs, 1979-83; Contbr. articles profl. jours. Sloane fellow, 1961-62; recipient Wilbur Cross medal Yale U., 1976, Townsend Harris medal CUNY, 1975, Steele prize Am. Math. Soc., 1981. Mem. Nat. Acad. Scis. Subspecialties: Algorithms; Computer architecture. Current work: Algorithms, computer architecture, robotics, programming languages, software engineering, theoretical computer science. Home: Finney Farm Rd Croton on Hudson NY 10520 Office: 251 Mercer St New York NY 10012

SCHWARTZ, JANICE BLUMENTHAL, medical educator; b. N.Y.C., Apr. 27, 1949; d. Martin and Frances Irene (Multach) Blumenthal; m. Stephen Andrew Schwartz, Dec. 23, 1971 (div.); m. Jerry C. Griffin, Apr. 25, 1981. B.S., Newcomb Coll., 1970; M.D., Tulane U., 1974. Diplomate: Am. Bd. Internal Medicine. Intern U. So. Calif. Hosp., Los angeles, 1974-75; resident in internal medicine Cedar-Sinai Med Ctr., Los Angeles, 1975-77, clin. fellow cardiology, 1977-78; fellow cardiology Stanford (Calif.) U., 1978-81; instr. medicine Baylor Coll., Houston, 1981-82, asst. prof. medicine, 1982—. Am. Heart Assn. grantee, 1982-84; recipient Pharm. Mfrs. Found. award, 1982; NIH awardee, 1982—. Fellow Council clin. Cardiology Am. Heart Assn., Am. Coll. Cardiology; mem. Am. Heart Assn., Am. Fedn. Clin. Research. Subspecialty: Cardiology. Current work: Investigating pharmacology of cardiovascular drugs. Home: 1736 Milford St Houston TX 77098 Office: Sect Cardiology Baylor Coll Medicine 6565 Fannin St MS-F905 Houston TX 77030

SCHWARTZ, KARLENE V., biologist; m. L. M. Schwartz. B.S., U. Wis.-Madison, M.S., 1963. Cert. U. Oslo, 1956. Biologist Boston U., 1968-72, U. Mass., Boston, 1972—; lectr. Newton (Mass.) Community Sch., 1981—. Author: (with L. Margulis) Five Kingdoms: A Guide to the Phyla of Life on Earth, 1981, Five Kingdoms Slides, 1982. Recipient Oslo Pubs. award U. Oslo, 1956; M. L. S. Clapp award Radcliffe Coll., Schlesinger Library, 1982-83; U. Mass. grantee, 1979. Clubs: Sierra; Squam Lake Sci. Ctr. (Holderness, N.H.); Appalachian Mountain (Boston). Subspecialties: Physiology (biology); Evolutionary biology. Current work: Edible wild plants of North America, comparative animal physiology. Office: Biology Dept U Mass Harbor Campus Boston MA 02125

SCHWARTZ, LESTER WILLIAM, veterinary pathologist, inhalation toxicologist-pathologist; b. Baraboo, Wis., Aug. 28, 1943; s. Lester Lee and Esther Lillian (Paepke) S.; m. Bonnie Ellen Wheeler, Sept. 8, 1962; children: Shelly Ann, Andrew John. D.V.M., Purdue U., 1967; Ph.D., U. Calif.-Davis, 1974. Diplomate: Am. Coll. Vet Pathologists. Asst. prof. U. Calif.-Davis, 1973-76, assoc. prof., 1977-81; assoc. dir. pathology Smith Kline & French Labs., Phila., 1982—; cons. Nat. Inst. Environ. Health Scis., Research Triangle Park, 1982, Health Effects Inst., Cambridge, Mass., 1982. Served to capt. U.S. Army, 1966-70. NIH trainee, 1970-73. Mem. Am. Coll. Vet. Pathologists (chmn. com. alternatives to animal testing 1981—), Am. Assn. Pathologists, Am. Thoracic Soc. Republican. Methodist. Subspecialties: Pathology (veterinary medicine); Immunotoxicology. Current work: Investigation of pathological and immunological effects of potential therapeutic agents using morphological and in vitro procedures. Home: PO Box 347 Chester Springs PA 19425 Office: Smith Kline and French Labs 1500 Spring Garden St L-60 Philadelphia PA 19101

SCHWARTZ, LYLE HOWARD, materials science educator; b. Chgo., Aug. 2, 1936; s. Joseph Kibbe and Helen (Shefsky) S.; m. Celsta Sue Jurkovich, Sept. 1, 1973; children by previous marriage: Ara, Justin. B.Sc. in Engring, Northwestern U., 1959, Ph.D., 1964. Postdoctoral fellow U. Paris, 1963-64; asst. prof. to prof. materials sci. and engring. Northwestern U., Evanston, Ill., 1964—; tech. staff Bell Labs., Murray Hill, N.J., 1972-73; dir. Materials Research Ctr., Northwestern U., 1979—. Contbr. articles to profl. jours.; author: (with J.B. Cohen) Diffraction from Materials, 1977. Predoctoral hon. fellow, 1959-60; Woodrow Wilson fellow, 1959-60; Allstate fellow, 1960-61; NSF fellow, 1963-64. Mem. Am. Inst. Physics, Am. Crystallography Soc., Am. Soc. Metals, AIME, AAAS, AAUP, Sigma Xi. Subspecialties: Alloys; X-ray crystallography. Current work: Study of structure and strength of spinodal alloys, characterization of heterogeneous catalysts, Mossbauer spectroscopy, x-ray and neutron diffraction. Office: 2145 N Sheridan Rd Evanston IL 60201

SCHWARTZ, MARSHALL ZANE, pediatric surgeon, educator; b. Mpls., Sept. 1, 1945; s. Sidney Shay and Peggy Belle (Lieberman) S.; m. Michele Carroll Walker, Oct. 16, 1971; children: Lisa, Jeffrey. B.S., U. Minn., 1968, M.D., 1970. Diplomate: Am. Bd. Surgery (cert. pediatric surgery). Instr., Harvard U., Boston, 1978-79; asst. in surgery Children's Hosp. Med. Ctr., Boston, 1978-79; asst. prof. surgery and pediatrics U. Tex. Med. Br., Galveston, 1979-81, assoc. prof., 1981-83, chief pediatric surgery, 1980-83, surgeon-in-chief Child Health Center, 1980-83; assoc. prof. surgery and pediatrics U. Calif.-Davis, 1983—. Author: (with others) Can We Influence Intestine Adaptation, 1983; contbr. articles to profl. jours. Recipient Basil O'Conner Research award March of Dimes Found., 1981; Young Investigator award NIH, 1982; Research award Found. for Children, 1982; James W. McLaughlin Fund award U. Tex., 1983. Fellow ACS, Am. Acad. Pediatrics; mem. AMA, Am. Pediatric Surgery Assn., Assn. Acad. Surgery, AAAS, N.Y. Acad. Sci., Soc. Surgery Alimentary Tract, Sigma Xi. Jewish. Subspecialties: Surgery; Physiology (medicine). Current work: Functional and immunologic development of the small intestine. The role of gastrointestinal hormones on the function of the small intestine. Office: U Calif Med Ctr 4301 X St Sacramento CA 95817

SCHWARTZ, MAURICE LEO, geology educator, consultant, researcher; b. Ft. Worth., Sept. 27, 1925; s. Simon and Bertha S.; m. Norma Sternberg, Jan. 1, 1950; children: Stephanie, Phebe, Philip, Howard, Ivan. B.S., Columbia U., 1963, M.A., 1964, Ph.D., 1966. Teaching asst. Columbia U., N.Y.C., 1963-66; instr. Bklyn. Coll., 1964-68; prof. dept. geology Western Wash. U., Bellingham, 1968—; pres. Coastal Cons., Inc., Bellingham, 1980—; adj. prof. Coastal Studies Inst., Ft. Lauderdale, Fla., 1980—. Editor: Spits and Bars, 1972, Barrier Islands, 1973, Encyclopedia Beaches and Coastal Enviroments, 1982. Fulbright-Hays scholar, Greece, 1973-74. Fellow Geol. Soc. Am.; mem. Coastal Soc., N.Y. Acad. Scis., Am. Shore and Beach Preservation Assn., Sigma Xi. Subspecialties: Geology; Oceanography. Current work: Coastal geology and processes, net shore-drift, coastal archaeology, beach erosion and restoration. Home: 423 N Garden Bellingham WA 98225 Office: Dept Geology Western Wash Univ High St Bellingham WA 98225

SCHWARTZ, MELVIN, educator, physicist; b. N.Y.C., Nov. 2, 1932; s. Harry and Hannah (Shulman) S.; m. Marilyn Fenster, Nov. 25, 1953; children—David N., Diane R., Betty Lynn. A.B., Columbia U., 1953, Ph.D., 1958. Asso. physicist Brookhaven Nat. Lab., 1956-58; mem. faculty Columbia U., 1958-66, prof. physics, 1963-66, Stanford U., 1966—; Pres. Digital Pathways, Inc. Bd. govs. Weizmann Inst. Sci. Guggenheim fellow, 1968. Fellow Am. Phys. Soc. (Hughes award 1964); mem. Nat. Acad. Scis. Subspecialty: Particle physics. Discoverer muon neutrino, 1962. Home: 770 Funston Ave San Francisco CA 94118 Office: Physics Dept Stanford U Stanford CA 94305

SCHWARTZ, MORTIMER LEONARD, medical educator, researcher; b. Newark, Jan. 12, 1915; s. Herman and Rose (Nusbaum) S.; m. Rene Kanengiser, Mar. 25, 1941; children: Gary, Jessica Schwartz Auerbach, Alison. M.D., Eclectic Med. Coll., Cin., 1938. Diplomate: Am. Bd. Internal Medicine, 1954, Cardiovascular Disease, 1962. Intern Alexian Bros. Hosp., Elizabeth, N.J., 1938-39; resident Jersey City Hosp., 1947-48; practice medicine specializing in internal medicine, and cardiovascular disease, N.J., 1940-42, 46-47, 47—; mem. faculty N.J. Med. Sch., Newark, 1958-72; prof. medicine Albert Einstein Coll. Medicine, Bronx, N.Y., 1973-77; chief cardiovascular sect. Bronx Lebanon Hosp., 1972-77; dir. medicine Mountainside Hosp., Montclair, N.J., 1977-80; prof. medicine U. Medicine and Dentistry/N.J. Med. Sch., Newark, 1966-72, 79—; dir. dept. medicine Bergen Pines County Hosp., Paramus, N.J., 1981—. Served to maj. U.S. Army, 1942-46. Recipient Harry Gold award, 1974. Fellow ACP Am. Coll. Cardiology, Am. Coll. Chest Physicians, Council on Clin. Cardiology of Am. Heart Assn., Am. Coll. Clin. Pharmacology. Subspecialties: Cardiology; Internal medicine. Current work: Clinical pharmacology. Home: 49 Sommer Ave Maplewood NJ 07040 Office: Bergen Pines County Hosp E Ridgewood Ave Paramus NJ 07652

SCHWARTZ, PAULINE MARY, pharmacologist; b. Phila., Aug. 8, 1947; d. Jack Paul and Marguerita (Slane) S.; m. Brad T. Garber, June 22 1975. B.S., Drexel U., 1970; M.S., U. Mich., Ph.D., 1975. Postdoctoral research scholar, assoc. U. So. Calif., Los Angeles, 1975-77; NIH postdoctoral fellow Yale U., 1977-80, research assoc. pharmacology, 1980-83; research pharmacologist DuPont Co., pharms. div., 1983—. Contbr. articles to profl. jours. A.T. Drexel scholar, 1967-70; Pharm. Mfrs. Assn. fellow, 1971-75; NIH young investigator's grantee, 1979-81; recipient Wilton E. Earle award Am. Tissue Culture Assn. Mem. Am. Assn. Cancer Research, Am. Chem. Soc., Am. Soc. Microbiology. Subspecialties: Molecular pharmacology; Chemotherapy. Current work: Design and development of antiviral and antineoplastic agent; metabolism and mechanics of action of antimetabolites; combination chemotherapy; purine and pyrimidine metabolism and regulation. Home: 101 Hammonasset Meadows Rd Madison CT 06443 Office: DuPont Co Pharm Div 500 S Ridgeway Ave Glenolden EPA 19036

SCHWARTZ, PHILIP RAYMOND, astrophysicist; b. Phila., Nov. 19, 1944; s. Raymond and Helen (Grabowski) S.; m. Nancy Lee

Schlosberg, Mar. 25, 1967; children: Eric, Michael. B.S. in Physics, M.I.T., 1966, Ph.D., 1971. Radio astronomer Naval Research Lab., Washington, 1971—. Contbr. numerous articles to sci. jours. Mem. Am. Astron. Soc., Internat. Astron. Union, Optical Soc. Am. Subspecialty: Radio and microwave astronomy. Current work: Galactic astrophysics, star formation and the interstellar medium, molecular spectroscopy of the earth's upper atmosphere. Office: Naval Research Lab Code 4138 Washington DC 20375

SCHWARTZ, RONALD HARRIS, research scientist; b. Buffalo, Apr. 19, 1943; s. Sidney and Roslyn (Liberman) S.; m. Joan P. Poyner, Jan. 19, 1967. A.B., Cornell U., 1966; M.D., Harvard U., 1970; Ph.D., Rutgers U., 1973. Staff assoc. immunology br. Nat. Cancer Inst., Bethesda, Md., 1972-74; research assoc. lab. immunology Nat. Inst. Allergy and Infectious Diseases, 1974-76, sr. investigator, 1976—. Contbr. numerous articles to profl. jours. Served with USPHS, 1972—. Recipient Soma Weiss Award Harvard U., 1970. Mem. Am. Assn. Immunologists, AAAS. Subspecialty: Immunobiology and immunology. Current work: Complementing immune response genes; T lymphocyte cloning; T cell activation. Office: Lab Immunology NIH Bldg 10 Rm 11N311 Bethesda MD 20205

SCHWARTZ, STANLEY ALLEN, immunologist, pediatrician, educator; b. Newark, N.J., July 20, 1941; s. Jack and Betty (Katz) S.; m. Diane I. Gottlieb, June 20, 1965. A.B., Rutgers U., 1963, M.S., 1965; Ph.D., U. Calif., San Diego, 1968; M.D., Albert Einstein Coll. Medicine, 1972. Diplomate: Am. Bd. Pediatrics, 1981. Asst. prof. pediatrics Cornell U. Med. Coll., 1977-78, asst. prof. biology, 1978; assoc. Sloan-Kettering Inst. Cancer Research, N.Y.C., 1977-78; assoc. prof. pediatrics U. Mich., Ann Arbor, 1978-83, assoc. prof. microbiology/immunology, 1978—, assoc. prof. epidemiology, 1981—, prof. pediatrics, 1983—. Contbr. articles to profl. jours. Recipient Meller award Meml. Sloan-Kettering Cancer Ctgr., 1977; Am. Cancer Soc. scholar, 1968-72; NIH Research Career Devel. award, 1983. Mem. Am. Assn. Immunologists, Am. Fedn. Clin. Research, Soc. Pediatric Research, Am. Acad. Allergy and Immunology, Midwest Soc. Pediatric Research. Subspecialties: Immunology (medicine); Pediatrics. Current work: Mechanisms of immunoregulation in man, immunnodeficiency diseases, tumor immunology. Office: Dept Epidemiology U Mich Ann Arbor MI 48109

SCHWARTZ, STEVEN, psychology educator, researcher, writer; b. N.Y.C., Nov. 5, 1946; emigrated to Australia, 1978, naturalized, 1982; s. Robert and Frances (Raiten) S.; m. Carolyn Susan Greenberg, June 23, 1968; children: Seth, Trica, Gregory. B.A., Bklyn. Coll., 1967; M.S., Syracuse U., 1970, Ph.D., 1971. Asst. prof. No. Ill. U., 1971-75; research psychologist U. Tex.-Galveston, 1975-78; sr. lectr. U. Western Australia, Perth, 1978-80; reader dept. psychology U. Queensland (Australia), St. Lucia, 1980—; cons. Commonwealth Sci. and Indsl. Research Orgn., Perth. Author: Complete Book of Gold, 1980, Psychopathology of Childhood, 1981, Measuring Reading Competence, 1982, Medical Decision Making, 1982. Bd. dirs. profl. bd. Fedn. Autistic Children's Assns., Autralia, Sydney, 1981—. NIMH grantee, 1976-78; Ednl. Research and Devel. Com. grantee, 1978-82; Australian Research Grants Com. grantee, 1981—. Mem. Am. Psychol. Assn., Australian Psychol. Assn., Sigma Xi. Anglican. Subspecialties: Cognition; Psychobiology. Current work: Brain behavior relationships, particularly those underlying language. Home: 71 Gem Rd Kenmore (Brisbane) Queensland Australia 4069 Office: Dept Psychology U Queensland Saint Lucia (Brisbane) Queensland Australia 4067

SCHWARTZ, WILLIAM BENJAMIN, educator, physician; b. Montgomery, Ala., May 16, 1922; s. William Benjamin and Molly (Vendruff) S.; m.; children: Eric A., Kenneth B., Laurie A. M.D., Duke U., 1945. Diplomate: Am. Bd. Internal Medicine (mem. test com. nephrology). Intern, then asst. resident medicine U. Chgo. Clinics, 1945-46; asst. medicine Peter Bent Brigham Hosp., Boston; also research fellow medicine Harvard Med. Sch., 1948-50; fellow medicine Children's Hosp., Boston, 1949-50; mem. faculty Tufts U. Sch. Medicine, 1950—, prof. medicine, 1958—, Endicott prof., 1975-76, Vannevar Bush Univ. prof., 1976—, chmn. dept. medicine, 1971-76; mem. staff New Eng. Center Hosps., 1950—, sr. physician, chief renal service, 1959-71, physician-in-chief, 1971-76; Established investigator Am. Heart Assn., 1956-61; chmn. gen. medicine study sect. NIH, 1965-69; mem. sci. adv. bd. USAF, 1965-68, Nat. Kidney Found., 1968—, chmn., 1970—; mem. tng. com. Nat. Heart Inst., 1969-70; chmn. health policy adv. bd. Rand Corp., 1970—, prin. adviser health scis. program, 1977—. Author numerous articles in field. Markle scholar med. scis., 1950-55. Mem. Inst. Medicine, Nat. Acad. Scis., Am. Soc. Nephrology (pres. 1974-75), Acad. Arts and Scis., A.C.P., Am. Fedn. Clin. Research, Am. Physiol. Soc., Am. Soc. Clin. Investigation, Assn. Am. Physicians, Phi Beta Kappa, Sigma Xi, Alpha Omega Alpha. Subspecialties: Laser medicine; Artificial intelligence. Current work: Development of computer programs which can simulate the performance of expert clinicians. Office: 65 East India Row Boston MA 02110

SCHWARTZ, WILLIAM JOSEPH, neurologist, neuroscientist; b. Phila., Mar. 28, 1950; s. Leon and Helene (Siris) S.; m. Randi Joy Eisner, July 1, 1979. Student, U. So. Calif., Los Angeles, 1967-68; B.S. summa cum laude, U. Calif., Irvine, 1971, M.D., 1974. Diplomate: Am. Bd. Psychiatry and Neurology, 1982. Intern U. Calif.-H.C. Moffitt Hosp., San Francisco, 1974-75, resident in neurology, 1978-81; research assoc. Lab. Neurophysiology NIMH, Bethesda, 1975-77, Lab. Cerebral Metabolism, 1977-78; instr. neurology Harvard U., 1981-82, asst. prof. neurology, 1982—; asst.in neurology Mass. Gen. Hosp., 1981-83, asst. neurologist, 1983—. Contbr. articles in field to profl. jours. Served with USPHS, 11975-78. Recipient Merck Manual award, 1974; Med. Found. Charles A. King Trust fellow, 1981-82; William F. Milton Fund of Harvard U. fellow, 1982; Charles H. Hood Found. fellow, 1982. Mem. Soc. Neuroscience, AAAS, Am. Acad. Neurology, Alpha Omega Alpha. Subspecialties: Neurology; Neuroendocrinology. Current work: Clinical neurology; anatomy physiology endocrinology, chemistry of circadian rhythms. Office: Neurology Research 4 Mass Gen Hosp Boston MA 02114

SCHWARTZKOPF, STEVEN HENRY, research scientist, consultant; b. Lincoln, Nebr., May 4, 1951; s. Leo Robert and Lydia Katherine (Filbert) S.; m. Johanna Witherspoon, Apr. 2, 1982. B.S., U. Nebr., 1973; M.A., U. Calif.-Davis, 1976, Ph.D., 1978. Teaching asst. U. Calif.-Davis, 1973-76, lectr., 1976-79; research scientist U. N.H., Durham, 1979—; cons. in field. Author: (with P.E. Stofan) A Chamber Design for Closed Ecological Systems Research, 1981. NASA grantee, 1982; U. Calif.-Davis fellow, 1977; Jessie Smith Noyes and Robert Sterling Clark scholar, 1976. Mem. AAAS, Ecol. Soc. Am. Democrat. Lutheran. Subspecialties: Space agriculture; Plant growth. Current work: Closed ecological life support systems, controlled environments, develop closed ecological life support systems (CELSS) for outer space life support applications and application to terrestrial agriculture. Inventor plant growth chamber for life support system experimentation-NASA. Office: NASA Ames Research Ctr MS 239-10 Moffett Field CA 94035

SCHWARTZMAN, BORIS, prosthodontist; b. Mexico City, Mex., June 27, 1954; s. Moses and Mira (Yasinovsky) S. B.S. Coll. Israelista de Mex, 1972; D.D.S. magna cum laude, Tech. U. Mex., 1976; postdoctoral cert. biomaterial scis., UCLA, 1976-77; splty. cert. in prosthetic dentistry, UCLA, 1979, UCLA, 1980. Practice dentistry specializing in prosthodontics, Mexico City, 1981—; faculty mem. Tech. U. Mex., 1981—; lectr. UCLA, 1981—; expert cons. Calif. Dental Assn., Los Angeles, 1981—. Contbr. chpt. to book, articles and abstracts to profl. jours. Bd. dirs. Jewish Sport Ctr., Mexico City, 1982—. Mem. Am. Acad. Maxillo Facial Prosthetics, Am. Coll. Prosthodontists (hon. mention 1979), Internat. Assn. Dental Research, Am. Acad. Crown and Bridge (Stanley D. Tylman award '80), Mexican Dental Assn., Atenco Nacional de Ciencias (hon.), Mex. Inst. Culture (hon.). Subspecialty: Prosthodontics. Current work: Clinical application of prosthetic dentistry, dental biomaterial sciences research in field. Office: Campos Eliseos 385 B7 piso Mexico City Mexico 11560

SCHWARTZMAN, ROBERT JAY, neurology educator; b. Washington, Nov. 28, 1939. B.A., Harvard U., 1961; M.D., U. Pa., 1965. Diplomate: Nat. Bd. Med. Examiners.; Am. Bd. Internal Medicine, 1972, Am. Bd. Neurology, 1974; Lic. physician, Fla., Calif., Pa., Tex. Intern in medicine Duke Hosp., Durham, N.C., 1965-66, jr. asst. resident, 1966-67; resident in neurology Hosp. U. Pa., Phila., 1967-69; clin. assoc. med. neurology br. NIH, Bethesda, Md., 1969-71; instr. Howard U., Washington, 1970-71; asst. prof. neurology U. Miami (Fla.) Sch. Medicine, 1971-74, assoc. prof., 1974-78; prof. neurology, chief div. U. Tex. Health Sci. Ctr., San Antonio, 1978—; staff Bexar County and Santa Rosa hosps.; cons. Audie Murphy VA Hosp., San Antonio. Editor: Continuing Med. Edn, 1981. Mem. sci. rev. com. NSF, 1975-77; mem. sci. council VA, 1975-77. Served to lt. comdr. USPHS, to 1971. Harvard Coll. scholar, 1957-61; recipient numerous Best Tchr. awards. Mem. Am. Neurol. Assn., Am. Acad. Neurology, Stroke Council, So. Neurol. Assn., AAAS, Soc. Neurosci., N.Y. Acad. Scis., Alpha Omega Alpha. Home: 820 Pennstone Rd Bryn Mawr PA 19010

SCHWARZ, ANTON J., research exec.; b. Munich, Ger., May 26, 1927; s. Joseph and Therese (Bauer) S.; m. Josephine F. Morrissey, Nov. 15, 1952; children: Theresa, Anton. M.D., Ludwigs Maximilian U., Munich, 1951; grad., Advanced Mgmt. Program, Harvard Bus. Sch., 1972. Intern, resident City Hosp., Munich, St. John's Hosp., L.I., N.Y., Dobbs Ferry Hosp., N.Y., 1951-54; sr. research assoc. U. Cin. Coll. Medicine, 1954-56; dir. virus research Pitman Medrelo div. Dow Chem. Co., 1965; dir. human health research and devel. Dow Chem. Co., 1965-71, dir. biol. research and devel., dir. biol. labs., 1971-75; med. dir. Dow Europe, 1975-77; dir. corp. med. research Schering-Plough Corp., Bloomfield, N.J., 1977-81, dir. med. sci., 1981—. Contbr. articles to profl. jours. Recipient Order of Merit of Medicine, Brazil; Wolverine Frontiersman award State of Mich., 1968. Mem. Soc. Exptl. Biology and Medicine, Am. Assn. Immunologists, N.Y. Acad. Scis., AAAS, AMA. Subspecialties: Internal medicine; Virology (medicine). Current work: Direction of research in new drugs and biologicals; evaluation of new products. Patentee in field; inventor, developer Schwarz strain further attenuated measles vaccine. Home: 1 Euclid Ave Apt 4A Summitt NJ 07901 Office: 60 Orange St Bloomfield NJ 07003

SCHWARZ, EMIL ARTHUR, retired manufacturing executive, mechanical engineer; b. St. Louis, Dec. 5, 1911; s. Michael and Maria Elizabeth (Nothum) S.; m. Stephanie Ostie, Aug. 3, 1940; children: Joanne Schwarz Becker, Richard, Stephen. S.B. in Mech. Engring. and Bus., Harvard U., 1934. Registered profl. engr., Mo. Successively mech. engr., supt., plant mgr., v.p. mfg. Crunden Martin Mfg. Co., St. Louis, 1934-83, also dir. Mem. ASME. Republican. Roman Catholic. Clubs: Harvard, KC, Engineers (St. Louis). Subspecialties: Mechanical engineering; Industrial engineering. Designer equipment to make kites, toy gliders and grills. Home: 6443 Devonshire Ave Saint Louis MO 63109

SCHWARZBECK, CHARLES, clinical psychologist, psychoanalyst, clinical infant researcher; b. N.Y.C., Dec. 21, 1945; s. Charles and June (West) S. A.B., Kenyon Coll., 1967; Ed.M., Boston U., 1970; Ph.D. summa cum laude, U. Tex., Austin, 1976. Intern adult psychiatry Menninger Found., Topeka, Kans., 1974-75; fellow dept. psychiatry Children's Hosp., Washington, 1975-76; chief clin. psychologist Psychiat. Inst., Washington, 1977-82; clin. instr. psychiatry George Washington Sch. Medicine, Washington, 1977-78, clin. prof., 1978-82; clin. investigator NIMH, Bethesda, Md., 1980—; cons. PRETERM Centre for Reproductive Health, Washington, 1980-82; chief cons. Chelsea Sch., Silver Spring, Md., 1977-82. Contbr. chpts., articles in adult, child psychology to research publs. NIMH Infant Study Sect. fellow, 1981-83; NIMH fellow, 1972-74; vis. fellow Tavistock Centre, London, 1976. Mem. Psychologists Interested in Psychoanalysis. Current work: Clinical infant research in neonates; stress reactions child and adult psychoanalysis; applied research. Office: 135 Wisconsin Circle Chevy Chase MD 20815

SCHWARZSCHILD, MARTIN, educator, astronomer; b. Potsdam, Germany, May 31, 1912; came to U.S., 1937, naturalized, 1942; s. Karl and Else (Rosenbach) S.; m. Barbara Cherry, Aug. 24, 1945. Ph.D., U. Goettingen, 1935; D.Sc. (hon.), Swarthmore Coll., 1960, Columbia U., 1973. Research fellow Inst. Astrophysics, Oslo (Norway) U., 1936-37, Harvard U. Obs., 1937-40; lectr., later asst. prof. Rutherfurd Obs., Columbia U., 1940-47; prof. Princeton U., 1947-50, Higgins prof. astronomy, 1950-79. Author: Structure and Evolution of the Stars. Served to 1st lt. AUS, 1942-45. Recipient Dannie Heineman prize Akademie der Wissenschaften zu Goettingen, Germany, 1967; Albert A. Michelson award Case Western Res. U., 1967; Newcomb Cleveland prize AAAS, 1957; Rittenhouse Silver medal, 1966; Prix Janssen Société astronomique de France, 1970. Fellow Am. Acad. Arts and Scis.; mem. Internat. Astron. Union (v.p. 1964-70), Akademie der Naturforscher Leopoldina, Royal Astron. Soc. (asso., Gold medal 1969, Eddington medal 1963), Royal Astron. Soc. Can. (hon.), Am. Astron. Soc. (pres. 1970-72), Nat. Acad. Scis. (Henry Draper medal 1961), Soc. Royale des Sciences de Liege (corr.), Royal Netherlands Acad. Sci. and Letters (fgn.), Royal Danish Acad. Sci. and Letters (fgn.), Norwegian Acad. Sci. and Letters, Astron. Soc. Pacific (Bruce medal 1965), Am. Philos. Soc., Sigma Xi. Subspecialty: Theoretical astrophysics.

SCHWEBEL, MILTON, psychologist, educator; b. Troy, N.Y., May 11, 1914; s. Frank and Sarah (Oxenhandler) S.; m. Bernice Lois Davison, Sept. 3, 1939; children—Andrew I., Robert S. A.B., Union Coll., 1934; M.A., SUNY at Albany, 1936; Ph.D., Columbia U., 1949; certificate in psychotherapy, Postgrad. Center for Mental Health, 1958. Diplomate: Am. Bd. Examiners Profl. Psychology; licensed psychologist, N.Y., N.J. Tchr. high schs., upstate N.Y., 1936-39; counselor Nat. Youth Adminstrn., White Plains, N.Y., 1939-41; employment counselor N.Y. State Employment Service, 1941-43; labor market analyst War Manpower Commn., 1943; psychometrist Coll. City N.Y., 1946; dir. guidance, asst. prof. psychology, chmn. philosophy-psychology dept. Mohawk Coll., 1946-48; asst. prof. psychology Champlain Coll., 1948-49; asst. to prof. edn. NYU, 1949-67, chmn. dept. guidance and personnel adminstrn., 1964-66, assoc. dean, 1965-67; dean, prof. edn. Grad. Sch. Edn., Rutgers U., New Brunswick, N.J., 1967-77; prof. Grad. Sch. Applied and Profl. Psychology, 1977—; vis. prof. U. So. Calif., summer 1953, U. Hawaii, summer 1965; lectr. psychologist Postgrad. Center for Mental Health, 1958-60; cons. NIMH, 1963-66, U.S. Office Edn., 1968-72, Hadassah-Wizo Research Inst., Israel, 1968-69, India, 1975, Rumania and Cypress, 1978, VA, 1978—, UNESCO, Paris, 1982—; pvt. cons. psychologist and psychotherapist, 1953—. Author: The Interests of Pharmacists, 1951; author: (with Ella Harris) Health Counseling, 1951, Resistance to Learning, 1963, Who Can Be Educated, 1968, (with Andrew, Bernice and Carol Schwebel) Student Teacher's Manual, 1979; (with Bernice Schwebel) film Why Some Children Don't Learn, 1962; co-author: State of the Art Report: Research on Cognitive Development and Its Facilitation, 1983; editor: Behavioral Science and Human Survival, 1965, (with Jane Raph) Piaget in the Classroom, 1973; rev. editor: Am. Jour. Orthopsychiatry, 1963-71; mem. editorial bd.: Jour. Contemporary Psychotherapy, 1965-73, Jour. Counseling Psychology, 1966-75; mem. editorial com.: Jour. Social Issues, 1965-70; editorial bd.: NYU Edn. Quar, 1969—, Rev. of Edn, 1974—, Rutgers Profl. Psychology Rev, 1981—; editorial advisory bd.: Change in Higher Edn, 1969-74. Trustee Edn. Law Center, 1973-81, Nat. Com. Employment Youth, Nat. Child Labor Com., 1967-75, Union Exptl. Colls. and Univs., 1976-78, Opera Theatre N.J., 1979—; pres. Opera Theatre N.J., 1981—; chmn. adv. com. Nat. Com. Edn. Migrant Children, 1970-75; exec. bd. Inst. and Mus. Fantasy and Play, 1977—; bd. dirs. Breast Disease Assn. Am., 1976-80; v.p. Nat. Orgn. for Migrant Children, 1976-81, pres., 1981—. Served with AUS, 1943-46; ETO. Postdoctoral fellow Postgrad. Center for Mental Health, 1954-56; Met. Applied Research Council fellow, 1970. Fellow Am. Psychol. Assn., Am. Orthopsychiatry Assn., Soc. Psychol. Study Social Issures; mem. Am. Personnel and Guidance Assn. (past pres. N.Y.), AAAS, AAUP, Jean Piaget Soc. (trustee), Am. Ednl. Research Assn., N.Y. Acad. Scis. Subspecialties: Cognition; Developmental psychology. Current work: The facilitation of cogintive development and cognitive functioning; utilization of cognitive psychology in education; psychological effects of nuclear dangers. Home: 1050 George St New Brunswick NJ 08901 Office: Grad Sch Applied and Prof Psychology Rutgers U PO Box 819 Piscataway NJ 08854

SCHWEIKER, GEORGE CHRISTIAN, chemical company executive; b. Phila., Feb. 17, 1924; s. William Frederick and Isabelle Lenox S.; m. Joyce E. Gilman, Feb. 14, 1950; children: G. Russell, Marguerite I., Susan J., Robert C., Laura K. A.B., Temple U., 1949, M.A., 1952, Ph.D., 1953. Research and devel. chemist, supr. Hooker Chem. Corp., Niagara Falls, N.Y., 1953-57; mgr. research and devel. Velsicol Chem. Corp., Chgo., 1957-60; mgr. plastics research and devel. Celanese Corp., Summit, N.J., 1960-65; dir. chems. and polymers Borg Warner Corp., Parkersburg, W.Va., 1965-71; dir. research and devel. PQ Corp. Research and Devel. Ctr., Lafayette Hill, Pa., 1971—; adj. prof. chemistry Drexel U., Phila., 1950-53. Chmn. bd. editors: Research Mgtm, 1979-81; contbr. articles to profl. jours. Dir. East Barrington Countryside Assn., 1958-60. Served with U.S. Maritime Service, 1942-46. Fellow AAAS, Chem. Soc. (London); mem. Am. Chem. Soc., Indsl. Research Inst. (chmn. pub. communications com. 1982-83), Soc. Chem. Industry, Assn. Research Dirs., N.Y. Acad. Scis., Phila. Research Mgmt. Group (pres. 1977-78), Am. Oil Chem. Soc., Soc. Plastics Engrs., Soap and Detergents Assn., Sigma Xi. Subspecialties: Chemical research administration; Materials science research administration. Current work: Direction of research and development activities concerned with industrial and specialty chemical products, processes and applications. Patentee in field. Home: 233 Shawnee Rd Ardmore PA 19003 Office: PO Box 258 Lafayette Hill PA 19444 Home: 233 Shawnee Rd Ardmore PA 19003

SCHWEIZER, FRANCOIS, astronomer; b. Switzerland, Aug. 16, 1942; s. Hans and Madeleine (Tobler) S.; m. Linda Y., May 25, 1975; children: Briana C., Maia K. Lizentiat, U. Bern, Switzerland, 1968; M.A. in Astronomy, U. Calif.-Berkeley, 1970, Ph.D., 1974. Carnegie postdoctoral fellow Hale Obs., Pasadena, Calif., 1975-75; staff astronomer Cerro Tololo Inter-Am. Obs., La Serena, Chile, 1976-81; staff astronomer, dept. terrestrial magnetism Carnegie Instn. of Washington, 1981—. Mem. Schweizerische Astronomische Gesellschaft, Am. Astron. Soc., Astron. Soc. of the Pacific, Internat. Astron. Union. Subspecialties: Optical astronomy; Graphics, image processing, and pattern recognition. Current work: Optical studies of colliding and merging galaxies; structure and formation of ellipticals; surface photometry and structure of spiral galaxies. Office: 5241 Broad Branch Rd NW Washington DC 20015

SCHWENSFEIR, ROBERT JAMES, JR., electronics manufacturing corporation executive; b. Hartford, Conn., June 27, 1934; s. Robert James and Elizabeth (Condron) S.; m. Margaret Gagosz, June 17, 1967; children: Thomas, Michael, Mary Elizabeth. B.A., Wesleyan U., Middletown, Conn., 1956; M.S., Trinity Coll., 1960; Ph.D., Brown U., 1966. Research assoc. Pratt & Whitney Aircraft Corp., Middletown, 1966-68; asst. prof. physics Bucknell U., Lewisburg, PA, 1968-74; nuclear criticality safety specialist UNC Naval Products, Uncasville, Conn., 1974-79, cons., 1979—; mgr. nuclear safety Tex. Instruments, Attleboro, Mass., 1979-82, mem. tech. staff, 1981—. Bucknell U. grantee, 1969, 71, 73; NSF grantee, 1970. Mem. Am. Nuclear Soc., N.Y. Acad. Scis., Am. Phys. Soc., Am. Assn. Phys. Tchrs., AAAS, Phi Beta Kappa, Sigma Xi. Roman Catholic. Lodge: KC. Subspecialties: Condensed matter physics; Nuclear physics. Current work: Research and development in structural materials. Home: 54 Marlise Dr Attleboro MA 02703 Office: Tex Instruments Inc. M/S 10-15 34 Forest St Attleboro MA 02703

SCHWER, ROGER EDWIN, metallurgist; b. Cin., June 14, 1932; s. M. Clayton and Olga Elizabeth (Christenson) S.; m. Margaret Dicken, Jan. 29, 1956; 1 dau., Susan Elizabeth. B.S., Mich. State U., 1954; M.B.A., Seidman Coll. Bus., 1980. Registered profl. engr., Calif. Tech. advisor Revere Copper & Brass, Los Angeles, 1955-56; application engr. Inco Inc., N.Y.C., 1966-72; dir. mktg. Cannon Muskegon Corp., Mich., 1972-80, pres., gen. mgr., 1980—; adj. prof. Seidman Grad. Coll. Bus., 1981—. Contbr. articles to profl. jours. Served to capt. USAF, 1954-57. Mem. Am. Soc. Metals, AIME, Metall. Soc., ASME, Nat. Assn. Corrosion Engrs. Club: Century. Subspecialties: Metallurgical engineering; High-temperature materials. Current work: Development and production of advanced high temperature superalloys; nickel base alloys; for critical gas turbine aerospace applications; single crystal alloys.

SCHWERZEL, ROBERT EDWARD, research scientist, photochemist, educator; b. Rockville Center, N.Y., Dec. 14, 1943; s. Robert August and Mary Louise (Burtis) S.; m. Sharon Edith Whidden, Apr. 24, 1967. B.S. in Chemistry, Va. Poly. Inst. and State U., 1965; Ph.D. in Phys. Organic Chemistry, Fla. State U., 1970. Univ. research fellow in photochemistry Cornell U., Ithaca, N.Y., 1970-71; Univ. research fellow in magnetic resonance Brown U., Providence, 1971-72; research chemist Syva Research Inst., Palo Alto, Calif., 1972-73; research scientist Columbus (Ohio) Labs., Battelle Meml. Inst., 1973-79, sr. research scientist, 1979—. Mem. editorial rev. bd.: Solar Energy Jour, 1977—; editor: Inter-Am. Photochem. Soc. Newsletter, 1980—; contbr. numerous sci. articles to profl. publs. Recipient I.R.-100 award Indsl. Research and Devel. mag., 1980; numerous research grants. Mem. Am. Solar Energy Soc. (dir. biotech. and chem. scis. div. 1978—, vice chmn. 1982, chmn. 1983), Electrochem. Soc. (chmn. Columbus sect. 1980), Am. Chem. Soc. (chmn. Columbus sect. 1981), Audubon Soc., Nat. Geog. Soc., Nature Conservancy, AAAS, European Photochemistry Assn., N.Y. Acad. Scis., Ohio Acad. Sci., Union Concerned Scientists, Sigma Xi. Clubs: Sierra, Columbus Ski. Subspecialties: Photochemistry; Laser photochemistry. Current work:

Research in solar energy conservation; spectroscopy, photochemistry, laser-induced chemistry, EXAFS spectroscopy with laser-produced x-rays. Holder four U.S. patents; co-inventor laser-EXAFS spectroscopy, 1980. Home: 1260 Olde Henderson Sq Columbus OH 43220 Office: Battelle Meml Inst Columbus Labs 505 King Ave Columbus OH 43201

SCHWETMAN, HERBERT DEWITT, physicist, former educator; b. Waco, Tex., Aug. 1, 1911; s. Henry William and Camilla Alice (Henderson) S.; m. Mary Jean Knight, July 29, 1939; children—Herbert Dewitt, John William, Rosemary. B.A., Baylor U., 1932; M.A., U. Tex., 1937, Ph.D., 1952; M.S., Harvard, 1934. Tchr. pub. schs., Brucevill-Eddy, Tex., also, Waco, 1933-41; instr. electrons, research asso. Harvard, 1941-47; asso. prof. physics Baylor U., 1947-52, prof., chmn. dept., 1952-81; cons. in field; sr. nuclear engr., nuclear analysis group Convair, Ft. Worth. Contbr. articles on analogue computers to jours. Fellow AAAS, Tex. Acad. Sci.; mem. AAUP, Am. Assn. Physics Tchrs. (pres. Tex. br.), Am. Phys. Soc., Am. Inst. Physics, Am. Inst. Radio Engrs., Am. Assn. Coll. Profs., Sigma Xi, Sigma Pi Sigma. Baptist. Subspecialty: Operating systems. Current work: Measurement, modeling and performance evaluation of computer systems. Home: 519 Edgewood Waco TX 76708

SCHWINGER, JULIAN, educator, physicist; b. N.Y.C., Feb. 12, 1918; s. Benjamin and Belle (Rosenfeld) S.; m. Clarice Carrol, 1947. A.B., Columbia U., 1936, Ph.D., 1939, D.Sc., 1966; D.Sc. (hon.), Purdue U., 1961, Harvard, 1962, Brandeis U., 1973, Gustavus Adolphus Coll., 1975; LL.D., CCNY, 1972. NRC fellow, 1939-40; research asso. U. Calif. at Berkeley, 1940-41; instr., then asst. prof. Purdue U., 1941-43; staff mem. Radiation Lab., Mass. Inst. Tech., 1943-46; staff Metall. Lab., U. Chgo., 1943; asso. prof. Harvard, 1945-47, prof., 1947-72, Higgins prof. physics, 1966-72; prof. physics UCLA, 1972-80, Univ. prof., 1980—; Mem. bd. sponsors Bull. Atomic Sci.; sponsor Fedn. Am. Scientists; J.W. Gibbs hon. lectr. Am. Math. Soc., 1960. Author: Particles and Sources, 1969, (with D. Saxon) Discontinuities in Wave Guides, 1968, Particles, Sources and Fields, 1970, Vol. II, 1973, Quantum Kinematics and Dynamics, 1970; Editor: Quantum Electrodynamics, 1958. Recipient C. L. Mayer nature of light award, 1949, univ. medal Columbia U., 1951, 1st Einstein prize award, 1951; Nat. Medal of Sci. award for physics, 1964; co-recipient Nobel prize in Physics, 1965; recipient Humboldt award, 1981; Guggenheim fellow, 1970. Mem. Nat. Acad. Scis., Am. Acad. Arts and Scis., Am. Phys. Soc., Royal Instn. Gt. Britain, ACLU, AAAS, N.Y. Acad. Scis. Subspecialty: Theoretical physics. Office: Dept Physics U Calif Los Angeles CA 90024

SCHWITTERS, ROY FREDERICK, physicist; b. Seattle, June 20, 1944; s. Walter Frederick and Margaret Lois (Boyer) S.; m. Karen Elizabeth Chrystal, June 18, 1965; children—Marc Frederick, Anne Elizabeth, Adam Thomas. S.B., M.I.T., 1966, Ph.D., 1971. Research asso. Stanford U. Linear Accelerator Center, 1971-74, asst. prof., then asso. prof., 1974-79; prof. physics Harvard U., 1979—. Author papers on high energy physics; asso. editor: Ann. Rev. Nuclear and Particle Sci; div. asso. editor: Phys. Rev. Letters. Recipient Alan T. Waterman award NSF, 1979. Mem. Am. Phys. Soc. Subspecialty: Particle physics. Current work: Experimental high energy physics involving colliding beams of protons and antiprotons. Construction of CDF detector for the fermilab tevatron collider. Home: 25 Central St Winchester MA 01890 Office: 235 Lyman Lab Harvard U Cambridge MA 02138

SCIAMMARELLA, CESAR AUGUSTO, mechanical engineer, educator, researcher; b. Buenos Aires, Argentina, Aug. 22, 1926; s. Emilio Silvio and Maria Belen (Mansilla) S.; m. Esther Elba Norbis; children: Alejandro, Eduardo, Federico. Diploma in Civil Engring, U. Buenos Aires, 1950; Ph.D., Ill. Inst. Tech., 1960. Prof. mech. engring. U. Buenos Aires, 1955-57, U. Fla., 1961-65, Poly. Inst. Bklyn., 1967-72, Ill. Inst. Tech., 1972—; vis. prof. Poly. Inst., Milan, Italy, 1979, U. Cagliari, Italy, 1979, Poly. Inst. Lausanne, Switzerland, 1979, U. Poitiers, France, 1980; cons. to govt., pvt. industry. Contbr. articles to profl. jours. Recipient Faculty Research award Sigma Xi, 1966; Outstanding Paper award Acad. Mechanics, 1970; Disting. Services award ASME, 1972; Frocht award Soc. for Exptl. Stress Analysis, 1980; Hetemy award, 1983. Fellow Soc. for Exptl. Stress Analysis, ASME; mem. ASTM, Gesellschaft für Angewandte Mathematik and Mechanik, Optical Soc. Am. Roman Catholic. Subspecialties: Theoretical and applied mechanics; Fracture mechanics. Current work: Optical techniques applied to stress analysis, mechanics of materials and fracture mechanics, fracture mechanics and fatigue. Patentee in field. Home: 175 E Delaware Pl Apt 5215 Chicago IL 60611 Office: Dept Mech Engring Ill Inst Tech Chicago IL 60616

SCISSON, SIDNEY E., engineering company executive, engineer; b. Danville, Ark., Feb. 4, 1917; s. Eugene and Arvie (Keathley) S.; m. Betti Shumaker, Sept. 8, 1942; children: Jane Scisson Grimshaw, Judith Scisson Fererri. Student, Ark. Tech. U., 1934-36; B.S. in Gen. Engring, Okla. State U., 1939. Registered profl. engr., Ill., Ky., Ohio, Okla., R.I. Civil engr. U.S. C.E., Tulsa, 1939-42; civil engr. Pate Engring. Co., Tulsa, 1945-48; a founder and with Fenix & Scisson, Inc. (engring. and constrn. services for govt. and industry), Tulsa, 1948—, formerly pres., now chmn. bd.; dir. Bank of Okla., Atlas Life Ins. Co. Treas.-pres. Tulsa Civic Ballet, 1965. Served with USNR, 1942-45. Named Disting. Alumnus Okla. State U., Stillwater, Ark. Tech. U., Russellville. Mem. Nat. Acad. Engring., AIME, ASCE, Okla. Soc. Civil Engrs., Nat. Soc. Profl. Engrs., Okla. Soc. Profl. Engrs., Am. Gas Assn., Natural Gas Processors Assn., Associated Gen. Contractors Am. Clubs: So. Hills Country; Tulsa (Tulsa). Subspecialties: Civil engineering; Mechanical engineering. Current work: Mined underground storage. Home: 4814 S Zunis St Tulsa OK 74105 Office: 1401 S Boulder Tulsa OK 74119

SCLAIR, MORTON H., biology educator, researcher, academic administrator; b. Bangor, Maine, Nov. 18, 1942; s. Samuel and Dora I. (Siegel) S.; m. Penny W. Weisberger, Aug. 8, 1966; children: Sara, Naomi. B.S. in Engring. Physics, U. Maine, 1964; M.S. in Biophysics, Pa. State U., 1965, Ph.D., 1969. Postdoctoral trainee in genetics U. N.C. Sch. Medicine-Chapel Hill, 1969-72; asst. prof. math./sci. Erie Community Coll.-City Campus, 1972-75; assoc. prof. biology, 1975-81, prof., 1981—, asst. curriculum coordinator, 1975-81, curriculum coordinator, 1981—; cancer research scientist Roswell Park Meml. Cancer Inst., Buffalo, 1973—. Contbr. articles to profl. jours. Treas. Young Israel of Greater Buffalo; trustee Kadimah Sch., Buffalo. NDEA fellow, 1964-69. Mem. Am. Soc. Microbiology, AAAS. Subspecialties: Molecular biology; Genetics and genetic engineering (biology). Current work: Study of viral genome and recombination in SV40 using DNA sequencing and gene splicing techniques. Office: Erie Community Coll-City Campus 121 Ellicott St Buffalo NY 14203

SCORA, RAINER WALTER, botanist, educator; b. Mokre, Silesia, Poland, Dec. 5, 1928; came to U.S., 1951, naturalized, 1956; s. Paul Wendelin and Helene Maria (Nester) S.; m. Christa Maria Fiala, Sept. 2, 1942; children: Georg-Alexander, Katharina-Monarda, Peter-Evans. B.S., DePaul U., 1955; M.S., U. Mich., 1958, Ph.D., 1964. Asst. prof. botany U. Calif., Riverside, 1964-70, assoc. prof., 1971-75, prof., 1975—. Author: Interspecific Relationships in the Genus Monarda, 1967; over 70 articles. Served with Signal Corps U.S. Army, 1955-57. Alfred P. Sloan fellow, 1959; recipient Cooley award Am. Inst. Biol. Scis., 1968; NSF fellow, 1963-64. Mem. Phytochem. Soc. N.Am., Internat. Soc. Plant Taxonomists, Internat. Orgn. Plant Biosystematists, Sigma Xi, Phi Sigma. Roman Catholic. Subspecialties: Evolutionary biology; Biochemistry (biology). Current work: Evolution of plants and their chemical components. Office: Botany and Plant Sci U Calif Riverside CA 92521

SCORDELIS, ALEXANDER COSTICAS, civil engineering educator; b. San Francisco, Sept. 27, 1923; s. Philip Kostas and Vasilica (Zois) S.; m. Georgia Gumas, May 9, 1948; children: Byron, Karen. B.S., U. Calif., Berkeley, 1948; M.S., M.I.T., 1949. Registered profl. engr., Calif. Structural designer Pacific Gas & Electric Co., San Francisco, 1948; engr. Bechtel Corp., San Francisco, summer 1951, 52, 53, 54; instr. civil engring. U. Calif., Berkeley, 1949-50, asst. prof., 1951-56, asso. prof., 1957-61, prof., 1962—, asst. dean, 1962-65, vice chmn. div. structural engring, structural mechanics, 1970-73; cons. engring. firms, govt. agys. Contbr. articles on analysis and design of complex structural systems, reinforced and prestressed concrete shell and bridge structures to profl. jours. Served to capt., C.E. U.S. Army, 1943-46; ETO. Decorated Bronze Star, Purple Heart; recipient Western Electric award Am. Soc. Engring. Edn., 1978; Axion award Hellenic Am. Profl. Soc., 1979; Best Paper award Canadian Soc. Civil Engring., 1982; K.B. Woods award Nat. Acad. Scis. Transp. Research Bd., 1983. Fellow ASCE (Moissieff award 1976, 81), Am. Concrete Inst.; mem. Internat. Assn. Shell and Spatial Structures, Structural Engrs. Assn. Calif., Nat. Acad. Engring. Subspecialty: Civil engineering. Home: 724 Gelston Pl El Cerrito CA 94530 Office: 729 Davis Hall U Calif Berkeley CA 94720

SCOTT, DAVID BYTOVETZSKI, dental research and forensic odontology consultant; b. Providence, May 8, 1919; m. (married); 3 children. A.B., Brown U., 1939; D.D.S., U. Md., 1943; M.S., U. Rochester, 1944; Sc.D. (hon.), Med. and Dental Coll. N.J., 1979, U. Louis Pasteur, Strasbourg, France, 1981. Staff Nat. Inst. Dental Research, NIH, Bethesda, Md., 1944-56, chief lab. histology and pathology, 1956-65, dir. inst., 1976-82; now pvt. cons. dental research and forensic odontology; faculty Case Western Res. U., Cleve., 1965-76, Thomas J. Hill Distinguished prof. phys. biology Sch. Dentistry, prof. anatomy Sch. Medicine, 1965-76, dean Sch. Dentistry, 1969-76. Recipient Arthur S. Flemming award, 1955; award for Research in Mineralization Internat. Assn. Dental Research, 1968; Research Achievement award Mass. Dental Soc., 1978; Fred Birnberg Dental Research medal Columbia U., 1978; decorated Order Rising Sun Japanese Govt., 1983. Mem. ADA, Am. Acad. Forensic Sci. (forensic odontology award 1981), Electron Micros. Soc. Am., Internat., Am. colls. dentists, Internat. Assn. Dental Research, Royal Soc. Medicine (hon.). Subspecialty: Oral biology. Current work: Consultant dental research and forensic odontology. Office: 10448 Wheatridge Dr Sun City AZ 85373

SCOTT, ELIZABETH LEONARD, statistics educator; b. Ft. Sill, Okla., Nov. 23, 1917; d. Richard C. and Elizabeth (Waterman) S. B.A., U. Calif., Berkeley, 1939, Ph.D., 1949. Research fellow U. Calif., Berkeley, 1939-49, mem. faculty, 1949—, assoc. prof., 1957-62, prof. stats., 1962—, chmn. dept. stats., 1968-73, asst. dean, 1965-67, co-chmn. group in biostats., 1972—; mem. Commn. on Nat. Stats. Nat. Acad. Scis., 1971-77, Commn. on Women in Sci., 1977-82, Commn. on Applied and Theoretical Stats., 1981—. Contbr. articles to sci. publs. Fellow Royal Statis. Soc. (hon.), Inst. Math. Stats. (pres. 1977-78, mem. council 1971-74, 76-79); mem. Biometric Soc. (council 1978-81), Am. Astron. Soc., Internat. Astron. Union, Internat. Stats. Inst. (v.p. 1981-83), Internat. Assn. Stats. in Phys. Sci. (sci. sec. 1960-72), Bernoulli Soc. (mem. council 1978-81, pres.-elect 1981-83, pres. 1983—), Astron. Soc. Pacific, AAAS (chmn. sect. U 1970-71, mem. council 1971-76). Subspecialty: Statistics. Research in math. stats. and applications. Home: 34 Tunnel Rd Berkeley CA 94705 Office: Dept Stats U Calif Berkeley CA 94720

SCOTT, FRANKLIN ROBERT, power research institute executive; b. Portland, Oreg., Aug. 23, 1922; s. Linden Douglas and Mabel Gay (Smith) S.; m. Christine Louise Golter, Aug. 10, 1950; children: Barbara Louise, Deborah Joanne, Kenneth Robert. B.A., Reed Coll., 1947; M.S., Ind. U.-Bloomington, 1949, Ph.D., 1952. Research staff mem. Los Alamos (N.Mex.) Sci. Lab., 1951-57; asst. dir. fusion div. Gen. Atomic, San Diego, 1957-67; cons. Oak Ridge Nat. Lab., 1967-73; prof. physics U. Tenn., Knoxville, 1967-73; chief open system br. Energy Research and Devel., Germantown, Md., 1973-75; program mgr. Electric Power Research Inst., Palo Alto, Calif., 1975—; cons. Dept. Energy, Washington, 1977—. Served with U.S. Army, 1943-46. Fellow Am. Phys. Soc.; mem. AAAS, Am. Vacuum Soc., IEEE, Sigma Xi. Club: Toastmasters (San Mateo, Calif.). Subspecialties: Nuclear fusion; Laser fusion. Current work: Advanced energy research mgmt. including research devel. planning. Patentee in field. Home: 2708 Wakefield Dr Belmont CA 94002 Office: Electric Power Research Inst 3412 Hillview Ave Palo Alto CA 94002

SCOTT, GERALD WILLIAM, surgeon, researcher; b. London, Jan. 12, 1931; s. Frederick William and Constance Ella (Burgess) S.; m. Beryl Elizabeth, May 14, 1955; children: Martin, Nigel, Elizabeth, Celia, Ian. M.B., B.S., U. London, 1955. Diplomate: Am. Bd. Surgery. Resident, fellow surgery Mayo Clinic and Grad. Sch. Medicine, Rochester, Minn., 1960-64; practice medicine specializing in surgery, Calgary, Alta., Can., after 1965—; assoc. prof. surgery U. Calgary, after 1968; prof. surgery U. Alta., Edmonton, after 1973; dir. Surg.-Med. Research Inst., 1978—. Served to lt. Royal Navy, 1957-59. Fulbright scholar, 1960; recipient E. Starr Judd award Mayo Found., 1964; Sir Peter Freyer Meml. lectr. U. Coll. Galway, 1983. Fellow Royal Coll. Surgeons Can. (exec. sec. surgery test com.), ACS; mem. Internat. Soc. Gastrointestinal Motility, British Pharmacol. Soc., Western Surg. Assn. Subspecialties: Artificial organs; Gastroenterology. Current work: Function of gallbladder gastrointestinal surgery, gastrointestinal surgery and research. Home: 5511 175th St Edmonton AB Canada T6M 1C3 Office: Surg Med Research Inst 1074B Dentistry/Pharmacy Bld U Alta Edmonto AB Canada T6G 2N8

SCOTT, IRENA MCCAMMON, physiologist, researcher technical writer; b. Delaware, Ohio, July, 1942; James Robert and Gay McCammon; m. John Watson Scott, Dec. 6, 1969. B.S., Ohio State U., 1965; M.S., U. Nev., 1972; Ph.D., U. Mo., 1976. Research assoc. Cornell U., Ithaca, N.Y., 1977-78; asst. prof. St. Bonaventure U., N.Y., 1978-79; research assoc. Ohio State U. Med. Sch., Columbus, 1980—; researcher Battelle Meml. Inst., Columbus, 1980—. Contbr. articles to sci. jours., poetry to mags. Vol. Ohio State U Radio Telescope, 1981—. Recipient Bausch & Lomb award, 1965. Mem. Mensa Writers Group, Ohio Archeol. Soc., Verse Writers Guild, Am. Physiol. Soc., AAAS, Am. Dairy Sci. Assn., Sigma Xi, Gamma Sigma Delta. Methodist. Club: Olentangy Poets (Delaware). Subspecialties: Neurochemistry; Neurophysiology. Current work: Electrophysiological study of the effects of hormones, neurotransmitters and fever producing substances on brain cells. Office: Ohio State U 310 Hamilton Hall Columbus OH 43210

SCOTT, JOHN W(ATTS), anatomy educator, neurobiologist; b. Birmingham, Ala., Oct. 5, 1938; s. John W. and Martha Frances (Herndon) S.; m. June Lynn Rothman, Aug. 22, 1966. A.B., Ala. Coll., 1961; Ph.D., U. Mich., 1965. Postdoctoral fellow Rockefeller U., 1965-67, asst. prof., 1967-69; asst. prof. anatomy Emory U., 1969-76, assoc. prof., 1976—. Nutrition Found. grantee, 1970; NIMH grantee, 1971; NSF grantee, 1972, 77-84; NIH grantee, 1977-85. Mem. Am. Assn. Anatomists, Soc. Neurosci. (pres. local chpt. 1980), Am. Chemoreception Soc. Subspecialty: Neurobiology. Current work: Neuroanatomy and neurophysiology of olfactory bulb and its connections to forebrain. Office: Dept Anatomy Emory U Atlanta GA 30322

SCOTT, JUNE ROTHMAN, microbiologist, educator, researcher; b. N.Y.C., Nov. 28, 1940; d. Samuel and Jennie (Pritz) Rothman; m. John W. Scott, Aug. 22, 1966. B.A., Swarthmore Coll., 1961; Ph.D., M.I.T., 1964. Faculty Rockefeller U., N.Y.C., 1964-69; mem. faculty Emory U., Atlanta, 1969—, assoc. prof.,then prof. microbiology. Mem. editorial bd.: Jour. Virology, Jour. Bacteriology; contbr. articles to sci. jours. Mem. DeKalb County Area Planning Council. Mem. Am. Soc. for Microbiology, Genetics Soc. Am., Am. Soc. for Virology. Subspecialties: Microbiology (medicine); Molecular biology. Current work: Molecular biology and genetics of bacteria, plasmids and bacteriophage.

SCOTT, LARRY DONALD, applied physics researcher; b. Tucson, Mar. 25, 1935; s. Melvin Rufus and Bonnie Irene (Hawes) S.; m. Enid Rollins Wylie, June 14, 1957; children: Yvette Patricia, Jean Diane, Brian William. B.S.E.E. with high distinction, U. Ariz., 1965; S.M., Harvard U., 1965, Ph.D., 1970. Field engr. Bendix Field Engring. Corp., Keflavik, Iceland, 1960-62; mem. tech. staff Bell Telephone Labs., North Andover, Mass., 1965-66; research fellow, lectr. Harvard U., 1970-72; mem. tech. staff Mission Research Corp., Albuquerque, 1972-77, div. mgr., 1977-82, mgr. and corp. tech. advisor for antenna devel., 1982—. Contbr. articles to profl. jours. Bd. dirs. Mission Research Corp. and Holiday Park Ch. of the Nazarene, 1978—; pres. Albuquerque chpt. Nazarene World Missionary Soc., 1982—. Served with USAF, 1956-59. Mem. Soc. Harvard Engrs. and Scientists, IEEE, Sigma Xi, Tau Beta Pi. Republican. Subspecialties: Electrical engineering; Remote sensing (geoscience). Current work: Antenna design and analysis, their use for communications and remote sensing in air, earth and water; buried antennas, subsurface communications, geophysical probing. Home: 4605 Oahu Dr NE Albuquerque NM 87111 Office: Mission Research Corp 1720 Randolph Rd SE Albuquerque NM 87106

SCOTT, NORMAN ROY, agricultural engineering educator; b. Spokane, Wash., Sept. 6, 1936; s. Roy Samuel and Agnes Sarafia (Lilljegren) S.; m. Sharon R. Cogley, June 17, 1961; children: Robin, Nanette, Shirlene. B.S. in Agrl. Engring, Wash. State U., 1958, Ph.D., Cornell U., 1962. Mem. faculty Cornell U., Ithaca, N.Y., 1962—, prof. agrl. engring., 1978—, chmn. agrl. engring. dept., 1978—; cons. in field. Contbr. numerous tech. articles to profl. publs. NIH grantee, 1974-81; NSF grantee, 1973-75. Mem. Am. Soc. Agrl. Engrs., Am. Soc. Engring. Edn., ASHRAE, Instrumentation Soc. Am., N.Y. Acad. Scis. Democrat. Methodist. Subspecialties: Agricultural engineering; Bioinstrumentation. Current work: Biomathematical modeling of animal systems; animal caliorimetry; environmental physiology; electronic instrumentation applied to physical and biological measurements. Patentee estrus detection probe, 1981. Home: 144 Burleigh Dr Ithaca NY 14850 Office: Cornell Univ Riley-Robb Hall Agrl Engring Dept Ithaca NY 14853

SCOTT, ROBERT BRADLEY, SR., medical educator; b. Petersburg, Va, Nov. 7, 1933; s. Charles Franklin and Hellen (Muse) S.; m. Harriet Wyche, June 21, 1958; children: Claiborne, Robert Bradley, Ann. B.S., U. Richmond, 1954; M.D., Va. Commonwealth U., 1958. Diplomate: Am. Bd. Internal Medicine. Intern, resident Bellevue and Meml. hosps., N.Y.C., 1958-61; research fellow dept. biology M.I.T., 1963-65; asst. prof. Med. Coll. Va., Va. Commonwealth U., Richmond, 1965-68, assoc. prof., 1968-75, prof. medicine, 1975—, assoc. dean clin. activities, 1979-82. Contbr. articles to profl. jours. Served with USPHS, 1961-63. Recipient Research Career Devel. award NIH, 1972-76. Fellow ACP; mem. Am. Soc. Hematology, Am. Fedn. Clin. Research, Am. Assn. Cancer Research, Am. Soc. Clin. Oncology, So. Soc. Clin. Investigation, Alpha Omega Alpha. Subspecialties: Hematology; Cell study oncology. Current work: Cell differentiation, leukemia, hemoglobin disorders, hematological research, hematology and oncology. Office: Med Coll Va Box 214 Richmond VA 23298

SCOTT, ROLAND BOYD, pediatrician; b. Houston, Apr. 18, 1909; s. Ernest John and Cordie (Clark) S.; m. Sarah Rosetta Weaver, June 24, 1935 (dec.); children—Roland Boyd, Venice Rosetta, Estelle Irene. B.S., Howard U., 1931, M.D., 1934; Gen. Edn. Bd. fellow, U. Chgo., 1936-39. Diplomate: Nat. Bd. Med. Examiners, Am. Bd. Pediatrics. Faculty Howard U., Washington, 1937—, prof. pediatrics, 1952-77, distinguished prof. pediatrics and child health, 1977—, chmn. dept. pediatrics, 1945-73; dir. Center for Sickle Cell Anemia, 1973—; chief pediatrician Freedmen's Hosp., 1947-73; professorial lectr. in child health and devel. George Washington U. Sch. Medicine, 1971—; staff Children's Hosp., Providence Hosp., Columbia Hosp., D.C. Gen. Hosp., Washington Hosp. Center; cons. in pediatrics to NIH, hosps.; Mem. com. Pub. Health Adv. Council, 1964—, U.S. Children's Bur., 1964—; mem. Nat. Com. for Children and Youth.; mem. sickle cell adv. com. NIH, 1983—. Author: (with Althea D. Kessler) Sickle Cell Anemia and Your Child, (with C.G. Uy) Guidelines For Care of Patients With Sickle Cell Disease; Editor: Procs. 1st Internat. Conf. on Sickle Cell Disease: A World Health Problem, 1979; Mem. editorial bd.: Advances in the Pathophysiology, Diagnosis and Treatment of Sickle Cell Disease, 1982, Clin. Pediatrics, 1962-80, Jour. Nat. Med. Assn, 1978; cons. editor: Medical Aspects of Human Sexuality; editorial bd.: Annals of Allergy, 1977. Recipient Sci. and Community award Medico-Chirurgical Med. Soc. D.C., 1971, Community Service award Med. Soc. D.C., 1972, award for contbns. to sickle cell research Delta Sigma Theta, 1973, 34 years Dedicated and Distinguished Service award in pediatrics dept. pediatrics Howard U., 1973, Faculty award for excellence in research, 1974, certificate of appreciation Sickle Cell Anemia Research and Edn., 1977; Mead Johnson award D.C. chpt. Am. Acad. Pediatrics, 1978; Percy L. Julian award We Do Care, Chgo., 1979; also plaques for work in sickle cell disease various orgns. including Elks, Nat. Assn. Med. Minority Educators, NIH, 1980-82. Mem. AMA, Am. Hematology Soc., Am. Pediatric Soc., Soc. Pediatric Research, Am. Acad. Allergy (v.p. 1966-67), Am. Acad. Pediatrics (mem. com. on children with handicaps, cons. head start program), Am. Fedn. Clin. Research, Nat. Med. Assn. (Distinguished Service medal 1966), AAAS, Internat. Corr. Soc. Allergists, Assn. Ambulatory Pediatric Services, AAUP, Internat. Congress Pediatrics, Am. Coll. Allergists (Distinguished Service award 1977), Can. Sickle Cell Soc. (hon. life), Phi Beta Kappa, Sigma Xi (Percy L. Julian award Howard U. chpt. 1977), Kappa Pi, Beta Kappa Chi, Alpha Omega Alpha, others. Subspecialties: Allergy; Hematology. Current work: Growth and development of children; sickler cell disease and allergic disorders in children. Research, publs. on sickle cell anemia, growth and devel. of infants and children, allergy in children. Home: 1723 Shepherd St NW Washington DC 20011 Office: 1114 Girard St NW Washington DC 20009

SCOTT, THOMAS RUSSELL, JR., neuroscientist; b. Ridley Park, Pa., Oct. 1,1944; s. Thomas Russell and Cathryn (Steciw) S.; m. Bonnie Kime, June 17, 1967; children: Heather Sheila, Ethan Kime, Heidi Cathryn Molly. B.A. cum laude, Princeton U., 1966; Ph.D., Duke U., 1970. Asst. prof. psychology U. Del., Newark, 1970-75, assoc. prof., 1975-83, prof., 1983—, chairperson dept., 1983—; adj. assoc. prof. physiol. psychology Rockefeller U., N.Y.C., 1980-82. Subspecialty: Neurophysiology. Current work: X. Office: Dept Psycholog U Del Newark DE 19711

SCOTT, WALTER NEIL, physiologist, educator; b. Evansville, Ind., Mar. 2, 1935; s. Paul Kruger and Pauline Virginia (Kimbley) S.; m. Margaret Ann Simon, Nov. 21, 1959; 1 son, Walter David Kimbley. B.S., Western Ky. State Coll., 1956; M.D., U. Louisville, 1960. Intern New Eng. Center Hosp., Boston, 1960-61, resident, 1961-62; NIH fellow medicine Mass. Meml. Hosps., Boston, 1962-63; USPHS fellow biophys. lab. Harvard Med. Sch., Brookline, Mass., 1963-65; spl. NIH fellow biochemistry Mass. Inst. Tech., Cambridge, 1965-66; biochemist Sch. Aerospace Medicine, San Antonio, 1966-68, acting chief biochem. pharmacology div., 1967-68; asst. prof. Mt. Sinai Grad. Sch., N.Y.C., 1968-71; mem. grad. faculty City U. N.Y., N.Y.C., 1968—; asst. prof. ophthalmology Mt. Sinai Med. Sch., N.Y.C., 1971-74, asso. prof. ophthalmology, 1974-79, research prof. ophthalmology, 1979—, asso. prof. physiology, 1974-79, prof., 1979—, asst. dean research, 1976-81, asso. dean research, 1981—; Lancaster vis. prof. Western Ky. U., 1980; mem. Nat. Eye Inst. Cornea Task Force, 1972, Vision Research Program Com., 1975-79; cons. metabolic biology program NSF, 1976—; established investigator Am. Heart Assn., 1971-76; Molly Berns sr. investigator N.Y. Heart Assn., 1976-80; chmn. Gordon Conf. Biology and Chemistry of Peptides, 1978; mem. organizing com. 3d Gordon Conf. on Peptides (6th Am. Peptide Symposium.). Mem. editorial bd.: Mt. Sinai Jour. Medicine; contbr. articles to sci. publs. Served to capt. USAF, 1966-68. Fellow N.Y. Acad. Scis. (gov. 1978—, v.p. 1981-82, pres.-elect 1982, vice-chmn. conf. organizing com. 1979, chmn. 1980); mem. Am. Physiol. Soc., Am. Soc. Biol. Chemists, Biophys. Soc., Soc. Exptl. Biology and Medicine (editorial bd. proceedings), Am. Heart Assn., AAAS, Am. Chem. Soc., Am. Soc. Nephrology, Endocrine Soc., Soc. Cell Biology, Sigma Xi, Alpha Omega Alpha. Subspecialty: Biophysics (physics). Home: 1095 Park Ave New York NY 10028 Calhoun KY 42327 Office: Mt Sinai Sch Medicine 100th St and Fifth Ave New York NY 10029

SCOTTI, VINCENT GUY, cons. nuclear engr.; b. Villamina, Avelino, Italy, Aug. 1, 1921; came to U.S., 1923; s. Nicholas and Giovanina (Santoro) S.; m. June Kupferschmid, Apr. 4, 1982; m. Hattie E. Tomlin, June 2, 1946 (dec. 1980); children: Vincent Guy, Nicholas, Michael, Richard, Kevin, Barbara. B.A., Boston U., 1950; postgrad., U. Calif., Berkeley, 1956, Ga. Inst. Tech., 1972, Kent State U., 1974. Sr. chemist Tracer Lab., Inc., Boston, 1950-55; asst. project engr. Pratt & Whitney Aircraft Co., East Hartford, Conn., 1955-65; assoc. chemist Argonne Nat. Lab., Ill., 1965; sect. mgr. Battelle Columbus, Ohio, 1969-74; biophysicist Swedish Hosp., Seattle, 1965-66; sr. cons. engr. Combustion Engring. Inc., Windsor, Conn., 1974—. Served with U.S. Navy, 1942-46; PTO. Mem. Am. Nuclear Soc. (exec. com. remote systems tech. div. 1980-82), Fusion Power Assn. Roman Catholic. Subspecialties: Nuclear engineering; Radiochemistry. Current work: Provide consulting services reactor design (fission) hotcell and fuel fab facilities; fusion reactor components. Office: Combustion Engring Inc 1000 Prospect Hill Rd Windsor CT 06095

SCOVELL, WILLIAM MARTIN, chemist, educator, researcher, cons.; b. Wilkes Barre, Pa., Jan. 16, 1944; s. Glenn W. and Anna M. (Gober) S.; m. Eleanor A. Krehely, July 10, 1965; children: Sherry Diane, William Martin, Jeffrey John. B.S. in Chemistry, Lebanon Valley Coll., 1961-65, Ph.D., U. Minn., Mpls., 1969. Research chemist E. I. duPont, 1965; postdoctoral fellow Princeton U., 1969-70, instr., 1970-72; asst. prof. chemistry SUNY - Buffalo, 1972-74; assoc. prof. Bowling Green State U., 1974-79, prof., 1979—; adj. prof. Med. Coll. Ohio at Toledo, 1980—. Contbr. articles to profl. jours. Active Pee Wee Baseball, Bowling Green, Ohio; chmn. fund dr. for Wood County (Ohio) Arthritis Found., N.W. Ohio chpt. NIH grantee, 1973-76; Am. Cancer Soc. grantee, 1981-83. Mem. Am. Chem. Soc., Sigma Xi. Methodist. Subspecialties: Biophysical chemistry; Inorganic chemistry. Current work: Drug and carcinogen interactions with DNA and chromatin; chromatin structure, effect of structural modification on biol functions; laser raman spectroscopy of DNA and DNA-protein interactions. Home: 1206 Bourgogne Ave Bowling Green OH 43402 Office: Dept Chemistr Bowling Green State U Hayes Hall Bowling Green OH 43403

SCRIBNER, JOHN DAVID, chemist, researcher; b. Portage, Wis., June 28, 1941; s. Charles Woodrow and Mary Louise (Tenley) S.; m. Nancy Carol Schmit, Dec. 22, 1941 (div.); m. Norma Katherine Naimy, May 14, 1942; 1 dau., Tara Lynn. B.S., U. Wis., Madison, 1963, Ph.D. (NSF predoctoral fellow, NIH predoctoral fellow), 1967, B.A., 1970. Postdoctoral fellow German Cancer Research Center, 1967-68, 1968-71; ind. investigator, program head chem. carcinogenesis Pacific N.W. Research Found., Seattle, 1971—; affiliate assoc. prof. pathology U. Wash.; assoc. mem. Fred Hutchinson Cancer Research Center; chem. pathology study sect. div. research grants NIH. Contbr. articles and revs. to profl. jours., chpts. in books. NSF postdoctoral fellow, 1967-69; NIH postdoctoral trainee, 1969-71; grantee Nat. Cancer Inst., 1972, 75, 78. Mem. Am. Chem. Soc., AAAS, Am. Assn. Cancer Research, N.Y. Acad. Sci. Methodist. Subspecialties: Cancer research (medicine); Toxicology (medicine). Current work: Chem. carcinogenesis, mechanism of action of tumor initiators and promoters, quantitative structure-activity relationships for chem. carcinogens. Office: 1102 Columbia St Seattle WA 98104

SCRIMA, LAWRENCE, sleep clinician, neuroscientist, physiological psychologist, educator; b. Tribes Hill, N.Y., May 6, 1950; s. Frank and Mary (Gitto) S.; m. Sonia Skakich, Sept. 27, 1979; 1 son, Xander Nicholas. B.A. (Wagner scholar), St. John Fisher Coll., 1972; M.A., Calif. State U.-Los Angeles, 1974; Ph.D., York U., 1979. Cert. clin. polysomnographer, Assn. Sleep Disorders Ctrs., 1982. Clin. coordinator Sleep Disorders Ctr., Mt. Sinai Med. Ctr., Miami Beach, Fla., 1979-82, dir., 1982—; asst. prof. psychology U. Miami, Coral Gables, Fla., 1980—; dir. Sleep Evaluation and Treatment Ctr., Sch. Medicine, Miami, 1982—, asst. prof. neurology, 1982—; assoc. Behavioral Medicine Inst., Inc., Miami, 1982—; cons. NASA. Contbr. articles to profl. jours. Founder and bd. dirs. Can. Assn. Narcolepsy, 1978—; founder and sci. cons. Disorders Excessive Sleepiness Club, Miami, 1979—. Mem. Soc. Neurosci., Soc. Psychophysiol. Research, Am. Psychol. Assn., Sleep Research Soc., European Brain and Behavior Soc (hon.), Brit. Brain Research Assn. (hon.). Subspecialties: Neurophysiology; Physiological psychology. Current work: Sleep disorders, functions; neuromechanisms of behavior, learning and memory; etiology/treatments of sleep disorders (esp, narcolepsy, cataplexy, nocturnal myoclonus/restless legs, sleep apnea, insomina); neuromechanisms of motivation, emotion, information processing, learning, memory and behavior, and neuromolecular correlates. Office: Sleep Evaluation Ctr U Miami Sch Med 1501 NW 9th Ave Miami FL 33136

SCRIMSHAW, NEVIN STEWART, food science educator, consultant; b. Milw., Jan. 20, 1918; s. Stewart and Harriet Fernwood (Smith) S.; m. Mary Ware Goodrich, Aug. 26, 1941; children: Susan March Norman Stewart, Nevin Baker, Steven Ware, Nathaniel Lewis. B.A., Ohio Wesleyan U., 1938, D.Pub. Service (hon.), 1961; M.A., Harvard U., 1939, Ph.D., 1941, M.P.H., 1959; M.D., U. Rochester, 1945; Ph.D. (hon.), U. Tokushima, 1979, Mahidol U., Bangkok, 1982. Diplomate: Nat. Bd. Med. Examiners, Am. Bd. Nutrition (sec.-treas. 1965-69). Teaching fellow comparative anatomy Harvard U., 1939-41; instr. embryology and comparative anatomy Ohio Wesleyan U., 1941-42; postdoctoral fellow nutrition and endocrinology U. Rochester, 1942-43, Rockefeller Found. postdoctoral fellow, 1946-47, Merck-NRC fellow, 1947- 49; rotating intern Gorgas Hosp., C.Z., 1945-46; asst. resident physician obstetrics and gynecology Strong Meml., Genesee hosps., Rochester, N.Y., 1948-49; cons. nutrition Pan Am. San. Bur., Office for Ams. WHO, 1948-49, chief nutrition sect., 1949-53, regional nutrition adviser, 1953-58; dir. Inst. Nutrition C.Am., Panama, Guatemala, 1949-61, cons. dir., 1961-65; cons. Inst. Nutrition C. Am., 1965—; vis. lectr. pub. tropical health Harvard U., 1968—; adj. prof. pub. health nutrition Columbia, 1959-61, vis. lectr., 1961-66; head dept. nutrition and food sci. Mass. Inst. Tech., 1961-79, Inst. prof., 1979—; Mem. coms. NIH, NRC, govt. depts.; sci. adv. com. Nutrition Found., 1960-68, Williams-Waterman Fund and Williams-Waterman Program Com. Research Corp., 1962-75; bd. govs. Pan. Am. Soc. New Eng., 1962-70; chmn. malnutrition panel U.S.-Japan Coop. Med. Sci. Program, 1964-74, mem. U.S. delegation of joint com., 1974—; chmn. internat. centers com. NIH, 1965-69; cons. med. and natural scis. program Rockefeller Found., 1966-71, trustee, 1971-83; mem. adv. com. med. research WHO, 1971-78, chmn., 1973-78, mem. expert adv. council on nutrition, 1975—; sr. adviser World Hunger Programme UN U., 1975-81; program dir. food, nutrition and poverty UN U., 1982—; mem. coms., del. confs. WHO, FAO, UNICEF, others. Author: (with Carl E. Taylor, John E. Gordon) Interactions of Nutrition and Infection, 1968; Editor: (with John E. Gordon) Malnutrition, Learning and Behavior, 1968, (with A.M. Altschul) Amino Acid Fortification of Protein Foods, 1971, (with A. Berg, D.L. Call) Nutrition, National Development and Planning, 1973, (with Steven R. Tamenbaum and Bruce R. Stillings) The Economics, Marketing, and Technology of Fish Protein Concentrate, 1974, (with M. Behar) Nutrition and Agricultural Development: Significance and Potential for the Tropics, 1976, (with others) Single-Cell Protein: Safety for Animal and Human Feeding, 1979, (with M.B. Wallerstein) Nutrition Policy Implementation: Issues and Experience, 1982; editorial bd.: (with L. Chen) Diarrhea and Malnutrition, 1983, (with others) Procs. Nat. Acad. Scis; contbr. numerous articles to profl. jours. Served with AUS, 1942-45. Recipient Mead-Johnson prize Rochester Acad. Medicine, 1947; Joseph Goldberger award AMA, 1969; Univ. alumni citation U. Rochester, 1969; 1st James R. Killian, Jr. Faculty Achievement award Mass. Inst. Tech., 1972; Bolton L. Corson medal Franklin Inst., 1976; medal of Honor Fundacion F. Cvenca Villoro, Spain, 1978; decorated Order Rodolfo Robles, Guatemala, 1961. Fellow Am. Pub. Health Assn. (chmn. food and nutrition sect. 1962-63, mem. council 1966—, award for excellence 1974), Royal Soc. Health, Am. Acad. Arts and Scis. (chmn. com. nutrition 1975—), Am. Coll. Preventive Medicine, Council on Fgn. Relations, Internat. Epidemiol. Assn.; mem. Am. Inst. Nutrition (Osborne-Mendel award 1960, Elvehjem award 1976), Nat. Acad. Scis. (mem. inst. medicine), Am. Soc. Clin. Nutrition (McCollum award 1975), Am. Physiol. Soc., Am. Chem. Soc., Inst. Food Technologists (Internat. award 1969), Am. Soc. Tropical Medicine and Hygiene, Genetics Soc. Am., Mass. Pub. Health Assn., Mass. Med. Soc., Asociación Pediátrica de Guatemala (hon.), Asociación Médica Nacional de Panamá (hon.), Royal Soc. Scis. Uppsala (Sweden), Czechoslovak Soc. for Gastroenterology and Nutrition (corr.), Asociación Argentina de Nutrición y Dietologiá (hon.), AAAS, Am. Epidemiological Soc., Asociación de Nutricionistas y Dietistas de América Central y Panamá (hon.), Sociedad Latino-americana de Nutrición, Internat. Union Nutritional Scis. (pres. 1978-81), Group European Nutritionists (corr.), Phi Beta Kappa, Sigma Xi (nat. lectr. 1971-72), Omicron Delta Kappa, Gamma Alpha, Phi Tau Sigma., Delta Omega, numerous others. Unitarian. Club: Cosmos (Washington). Subspecialties: Nutrition (medicine); Epidemiology. Current work: Clinical and public health nutrition; clinical evaluation of new protein sources; nutrient requirements. Home: Waterville Valley NH 03223 Office: Room 20A-201 MIT 18 Vassar St Cambridge MA 02139

SCRIVER, CHARLES ROBERT, physician; b. Montreal, Que., Can., Nov. 7, 1930; s. Walter deM. and Jessie (Boyd) S.; m. E.K. Peirce, Sept. 8, 1956; children: Dorothy, Peter, Julie, Paul. B.A. cum laude, McGill U., Montreal, 1951, M.D.C.M., 1955. Intern Royal Victoria Hosp., Montreal, 1955-56; resident Royal Victoria and Montreal Children's hosps., 1956-57, Children's Med. Center, Boston, 1957-58; McLaughlin travelling fellow Univ. Coll., London, 1958-60; chief resident pediatrics Montreal Children's Hosp., 1960-61; asst. prof. pediatrics McGill U., 1961; now prof. genetics (Human Genetics Center), prof. biology, prof. pediatrics. Co-author: Amino Acid Metabolism and Its Disorders, 1973; research publs. in field. Recipient Wood Gold medal McGill U., 1955, Borden award Nutrition Soc. Can., 1969, Gairdner Internat. award Gairdner Found., 1979; Markle scholar, 1962-67; Med. Research Council Can. asso., 1968—; Can. Rutherford lectr. Royal Soc. (Eng.), 1983. Fellow Royal Soc. Can. (McLaughlin medal 1981); mem. Can. Soc. Clin. Investigation (pres. 1974-75, G. Malcolm Brown award 1979), Soc. Pediatric Research (pres. 1975-76), Am. Soc. Human Genetics (dir. 1971-74, William Allan award 1978), Am. Soc. Clin. Investigation, Assn. Am. Physicians, Brit. Pediatric Assn. (hon.; 50th Anniversary lectr. 1978), Soc. Francaise de Pediat (hon.), Am. Acad. Pediatrics (Mead Johnson award 1968, Borden award 1973). Subspecialties: Genome organization; Genetics and genetic engineering (medicine). Current work: Mendelian and biochemical genetics of membrane transport functions. Prediction and prevention of genetic disease. Office: Montreal Children's Hosp Research Inst 2300 Tupper St Montreal PQ Canada H3H 1P3

SCUDDER, HARVEY ISRAEL, environmental and public health biologist, educator, consultant; b. Wellsburg, N.Y., Jan. 2, 1919; s. Henry Spaulding and Charlotte Evelyn (Draper) S.; m. Florence Viola Graff, June 16, 1945; children: Paul Harvey, Barbara Carol. B.S., Cornell U., 1939; postgrad., N.Y.U., 1939-42; Ph.D., Cornell U., 1953. Commd. scientist, officer USPHS, 1943-66; chief viruses and cancer program Nat. Cancer Inst., NIH, Bethesda, Md., 1959-62, chief research tng. grants br., 1962-65; chief health manpower program USPHS, Washington, 1965-66; head div. biol. and health sci. Calif. State-Hayward, 1967-70, prof. microbiology, 1967-80; research assoc. Calif. Acad. Scis., San Francisco, 1981—; trustee Marine Ecol. Inst., Redwood City, Calif., 1971—; chmn., 1974-78, 79-80, 82—; trustee Moss Landing Marine Labs., Moss Landings, Calif., 1967-70, chmn., 1969-70; chmn. health manpower Alameda County Comprehensive Health Planning Council, Oakland, Calif., 1973-76. Author: Malaria Control Manual, 1982; contbr. articles to profl. jours. Trustee St. Rose Hosp.; adv. com. Fairmont Hosp.; air conservation com. Lung Assn. Alameda County; trustee Alameda County Mosquito Abatement Dist., USPHS. Dupont Corp. fellow, 1939-41. Mem. Am. Pub. Health Assn., Am. Soc. Tropical Medicine and Hygiene, Am. Soc. Microbiology, Entomol. Soc. Am., Am. Mosquito Control Assn., Sigma Xi, Phi Kappa Phi. Republican. Mem. Christian Ch. Clubs: Pub. Health Service (pres. 1962, treas. 1961-63), Pub. Health Service (dir. 1960-64). Subspecialties: Paleoecology; Integrated systems modelling and engineering. Current work: Further studies on decision support systems and training for management of malaria control programs in underdeveloped countries for U.S. Agency for International Development; cooperative paleoecological studies of Middle Miocene in southwest Nevada. Inventor Scudder Fly Grill, 1947. Home: 7409 Hansen Dr Dublin CA 94568 Office: Dept Biol Scis Calif State U Hayward CA 94542

SCURA, LAWRENCE THOMAS, JR., agrl. engr.; b. Gowanda, N.Y., June 2, 1943; s. Lawrence Thomas and Irma Grace S. A.S., Jamestown Community Coll., 1967; B.S. in M.E, SUNY-Buffalo, 1971; M.S. in Agrl. Engring, Cornell U., 1977. Summer intern U.S. Naval Surface Weapons Ctr., Silver Spring, Md., 1970, engr., Dahlgren, Va., 1971-73; grad. research asst. Cornell U., 1973-76; engr. John Deere Product Engring. Ctr., Waterloo, Iowa, 1977—. Mem. Am. Soc. Automotive Engrs., ASME, Am. Soc. Metals. Methodist. Clubs: Blackhawk Creek Saddle, Pony Express Riders Iowa. Subspecialties: Agricultural engineering; Mechanical engineering. Current work: Scheduling and programming field test computers; assisting in field test planning; operation of field computers; interpretation of field test results. Home: 1130 Amherst Ave Waterloo IA 50702 Office: John Deere Product Engineering Center PO Box 8000 Waterloo IA 50704

SEABORG, GLENN THEODORE, chemistry educator; b. Ishpeming, Mich., Apr. 19, 1912; s. H. Theodore and Selma (Erickson) S.; m. Helen Griggs, June 6, 1942; children: Peter, Lynne Seaborg Cobb, David, Stephen, John Eric, Dianne. A.B., UCLA, 1934; Ph.D., U. Calif.-Berkeley, 1937; numerous hon. degrees; LL.D., U. Mich., 1958, Rutgers U., 1970; D.Sc., Northwestern U., 1954, U. Notre Dame, 1961, John Carroll U., Duquesne U., 1968, Ind. State U., 1969, U. Utah, 1970, Rockford Coll., 1975, Kent State U., 1975; L.H.D., No. Mich. Coll., 1962; D.P.S., George Washington U., 1962; D.P.A., U. Puget Sound, 1963; Litt.D., Lafayette Coll., 1966; D.Eng., Mich. Technol. U., 1970; Sc.D., U. Bucharest, 1971, Manhattan Coll., 1976. Research chemist U. Calif.-Berkeley, 1937-39, instr. dept. chemistry, 1939-41, asst. prof., 1941-45, prof., 1945-71, univ. prof., 1971, leave of absence, 1942-46, 61-71, dir. nuclear chem. research, 1946-58, 72-75, asso. dir. Lawrence Berkeley Lab., 1954-61, 71—; chancellor Univ. (U. Calif.-Berkeley), 1958-61, dir. Lawrence Hall of Sci., 1982—; sect. chief metall. lab. U. Chgo., 1942-46; chmn. AEC, 1961-71, gen. adv. com., 1946-50; research nuclear chemistry and physics, transuranium elements.; Chmn. bd. Kevex Corp., Burlingame, Calif., 1972—; Mem. Pres.'s Sci. Adv. Com., 1959-61; mem. nat. sci. bd. NSF, 1960-61; mem. Pres.'s Com. on Equal Employment Opportunity, 1961-65, Fed. Radiation Council, 1961-69, Nat. Aeros. and Space Council, 1961-71, Fed. Council Sci. and Tech., 1961-71, Nat. Com. Am.'s Goals and Resources, 1962-64, Pres.'s Com. Manpower, 1964-69, Nat. Council Marine Resources and Engring. Devel., 1966-71; chmn. Chem. Edn. Material Study, 1959-74, Nat. Programming Council for Pub. TV, 1970-72; dir. Ednl. TV and Radio Center, Ann Arbor, Mich., 1958-64, 67-70; pres. 4th UN Internat. Conf. Peaceful Uses Atomic Energy, Geneva, 1971, also chmn. U.S. del., 1964, 71; U.S. rep. 5th-15th gen. confs. IAEA, chmn., 1961-71, U.S. del. to USSR for signing Memorandum Cooperation Field Utilization Atomic Energy Peaceful Purposes, 1963; mem. U.S. del. for signing Limited Test Ban Treaty, 1963; mem. commn. on humanities Am. Council Learned Socs., 1962-65; mem. sci. adv. bd. Robert A. Welch Found., 1957—; mem. Internat. Orgn. for Chem. Scis. in Devel., UNESCO, 1980—, chmn., 1981; mem. Nat. Commn. on Excellence in Edn., Dept. Edn., 1981—. Author: (with Joseph J. Katz) The Actinide Elements, 1954, The Chemistry of the Actinide Elements, 1957, The Transuranium Elements, 1958, (with E.G. Valens) Elements of the Universe, 1958 (winner Thomas Alva Edison Found. award), Man-Made Transuranium Elements, 1963, (with D.M. Wilkes) Education and the Atom, 1964, (with E.K. Hyde, I. Perlman) Nuclear Properties of the Heavy Elements, 1964, (with others) Oppenheimer, 1969, (with W.R. Corliss) Man and Atom, 1971, Nuclear Milestones, 1972; editor: Transuranium Elements: Products of Modern Alchemy, 1978; asso. editor: Jour. Chem. Physics, 1948-50; editorial adv. bd.: Jour. Inorganic and Nuclear Chemistry, 1954-82, Indsl. Research, Inc, 1967-75; adv. bd.: Chem. and Engring. News, 1957-59; editorial bd.: Jour. Am. Chem. Soc, 1950-59, Ency. Chem. Tech., 1975—, Revs. in Inorganic Chemistry, 1977—; mem. hon. editorial adv. bd.: Internat. Ency. Phys. Chemistry and Chem. Physics, 1957—; mem. panel: Golden Picture Ency. for Children, 1957-61; mem. cons. and adv. bd.: Funk and Wagnells Universal Standard Ency, 1957-61; mem.: Am. Heritage Dictionary Panel Usage Cons, 1964—; contbr. articles to profl. jours. Trustee Pacific Sci. Center Found., 1962-77; trustee Sci. Service, 1965—, pres., 1966—; trustee Am.-Scandinavian Found., 1968—, Ednl. Broadcasting Corp., 1970-72, Swedish Council Am., 1976—; chmn. bd. dirs. Swedish Council Am., 1978-82; bd. dirs. World Future Soc., 1969—, Calif. Council for Environ. and Econ. Balance, 1974—; bd. govs. Am. Swedish Hist. Found., 1972—. Recipient John Ericsson Gold medal Am. Soc. Swedish Engrs., 1948; Nobel prize for Chemistry (with E.M. McMillan), 1951; John Scott award and medal City of Phila., 1953; Perkin medal Am. sect. Soc. Chem. Industry, 1957; U.S. AEC Enrico Fermi award, 1959; Joseph Priestley Meml. award Dickinson Coll., 1960; Sci. and Engring. award Fedn. Engring. Socs., Drexel Inst. Tech., Phila., 1962; named Swedish Am. of Year, Vasa Order of Am., 1962; Franklin medal Franklin Inst., 1963; 1st Spirit of St. Louis award, 1964; Leif Erikson Found. award, 1964; Washington award Western Soc. Engrs., 1965; Arches of Sci. award Pacific Sci. Center, 1968; Internat. Platform Assn. award, 1969; Prometheus award Nat. Elec. Mfrs. Assn., 1969; Nuclear Pioneer award Soc. Nuclear Medicine, 1971; Oliver Townsend award Atomic Indsl. Forum, 1971; Disting. Honor award U.S. Dept. State, 1971; Golden Plate award Am. Acad. Achievement, 1972; John R. Kuebler award Alpha Chi Sigma, 1978; Founders medal Hebrew U. Jerusalem, 1981; Henry DeWolf-Smyth award Am. Nuclear Soc., 1982; decorated officier Legion of Honor, France; Daniel Webster medal, 1976. Fellow Am. Phys. Soc., Am. Inst. Chemists (Pioneer award 1968, Gold medal award 1973), Chem. Soc. London (hon.), Royal Soc. Edinburgh (hon.), Am. Nuclear Soc., Calif., N.Y., Washington acads. scis., AAAS (pres. 1972, chmn. bd. 1973), Royal Soc. Arts (Eng.); mem. Am. Chem. Soc. (award in pure chemistry 1947, William H. Nichols medal N.Y. sect. 1948, Charles L. Parsons award 1964, Gibbs medal chgo. sect. 1966, Madison Marshall award No. Ala. sect. 1972, Priestley medal 1979, pres. 1976), Am. Philos. Soc., Royal Swedish Acad. Engring. Scis. (adv. council 1980), Am. Nat., Argentine Nat., Bavarian, Polish, Royal Swedish, USSR acads. scis., Royal Acad. Exact, Phys. and Natural Scis. Spain (acad. fgn. corr.), Soc. Nuclear Medicine (hon.), World Assn. World Federalists (v.p. 1980), Fedn. Am. Scientists (bd. sponsors 1980), Deutsche Akademie der Naturforscher Leopoldina (East Germany), Nat. Acad. Pub. Adminstrn., Internat. Platform Assn. (pres. 1981—), Am. Hiking Soc. (dir. 1979—, v.p. 1980), Phi Beta Kappa, Sigma Xi, Pi Mu Epsilon, Alpha Chi Sigma (John R. Kuebler award 1978), Phi Lambda Upsilon (hon.). Clubs: Bohemian (San Francisco); Chemists (N.Y.C.); Cosmos, University (Washington); Faculty (Berkeley). Subspecialty: Nuclear Chemistry. Co-discoverer elements 94-102, and 106: plutonium, 1940, americium, 1944-45, curium, 1944, berkelium, 1949, californium, 1950, einsteinium, 1952, fermium, 1953, mendelevium, 1955, nobelium, 1958, element 106, 1974; co-discoverer nuclear energy isotopes Pu-239, U-233, Np-237, other isotopes including I-131, Fe-59, Te-99m, Co-60; originator actinide concept for placing heaviest elements in periodic system. Office: Lawrence Berkeley Lab U Calif Berkeley CA 94720

SEAGO, JAMES LYNN, JR., biology educator; b. Alton, Ill., June 2, 1941; s. James L. and Dorothy F. (Watkins) S.; m. Katherine A. Brown Fanning, June 18, 1966; m. Jill P. Dabbs Arnold, Dec. 24, 1969; m. Marilyn Ann Meiss, Nov. 25, 1982; children: Kirstjan Erika, Robert Maclean. B.A., Knox Coll., 1963; M.A., Miami U., Oxford,

Ohio, 1966; Ph.D. in Botany, U. Ill., Urbana, 1969. Asst. prof. biology SUNY Coll.-Oswego, 1968-74, assoc. prof., 1974—, chmn. dept. biology, 1979—, also cross country coach; cons. biology. Contbr.: articles to Am. Jour. Botany; reviewer for jour. articles, books. N.Y. State grantee-in-aid, 1971; Nat. Soy Bean Crop Improvement Council grantee, 1976. Mem. Bot. Soc. Am., Torrey Bot. Club, Am. Soc. Agronomy, Sigma Xi. Subspecialty: Plant growth. Current work: Plant root development; cattail growth. Home: Box 316 Dumas Rd Minetto NY 13115 Office: SUNY Dept Biology Piez Hall Oswego NY 13126

SEALE, ROBERT LEWIS, nuclear and energy engineering educator, consultant; b. Rosenberg, Tex., Mar. 18, 1928; s. Benjamin Lewis and Olga Pauline (Zemanek) S.; m. Lina Glenn Jahn, Sept. 31, 1947; children: Katherine, Linda, James, Anne. B.S., U. Houston, 1947; M.A., U. Tex.-Austin, 1951, Ph.D., 1953. Registered profl. engr., Ariz. Geophysicist Gulf Research and Devel. Co., Houston, 1947-48; sr. engr. to chief nuclear ops. Gen. Dynamics, Ft. Worth, Tex., 1953-61; now prof., head dept. nuclear and energy engring. U. Ariz., Tucson; cons. in field. Fellow Am. Nuclear Soc.; mem. AAAS, Nat. Soc. Profl. Engrs., Sigma Xi. Methodist. Subspecialties: Nuclear engineering; Nuclear fission. Current work: Nuclear reactor systems design and operation safety. Office: U Ariz Tucson AZ 85721

SEAMAN, RONALD LEON, research engineer; b. Seaman, Ohio, Feb. 10, 1947; s. Thomas A. and Pauline A. (Black) S.; m. Patricia Ann Seaman, July 8, 1977; children: Lisa, Mike, Matthew, Emma. B.S.E.E. (RCA student scientist), U. Cin., 1970; Ph.D. in Biomed. Engring, Duke U., Durham, N.C., 1975. Research assoc. Duke U., 1974; research fellow U. Tex. Health Sci. Center, Dallas, 1975-76, instr., 1976-79; research engr. Ga. Tech. Engring. Expt. Sta., Atlanta, 1979-82, sr. research engr., 1982—; conf. presenter. Contbr. abstracts and articles to profl. jours. Recipient Student Br. award IEEE, 1970, Grad. Sch. Research award Duke U., 1972; grantee NIH. Mem. IEEE, AAAS, Soc. Neurosci., Internat. Microwave Power Inst., Bioelectromagnetics Soc., Tau Beta Pi, Eta Kappa Nu. Subspecialties: Biophysics (biology); Biomedical engineering. Current work: Biol. effects of electromagnetic waves at cellular and molecular levels, membrane biophysics, computer applications. Co-inventor open-ended coaxial exposure device. Office: Georgia Tech EES CTAB-ERB Atlanta GA 30332

SEAMAN, WILLIAM BERNARD, physician; b. Chgo., Jan. 5, 1917; s. Benjamin and Dorothy E. S.; m. Veryl Swick, February 26, 1944; children—Cheryl Dorothy, William David. Student, U. Mich., 1934-37; M.D., Harvard U., 1941. Diplomate: Am. Bd. Radiology. Intern Billings Hosp., U. Chgo., 1941-42; asst. radiology Yale U. Sch. Medicine, 1947-48, instr., 1948-49; instr. radiology Washington U. Sch. Medicine, St. Louis, 1949-51, assoc. prof., 1951-55, prof., 1955-56; prof. radiology, chmn. dept. Coll. Phys. and Surg., Columbia U., 1956—; James Picks prof. emeritus Columbia U., 1982—; dir. radiology service, trustee Presbyn. Hosp., N.Y.C. Served as maj. USAAF, 1942-46; flight surgeon. Recipient W.R. Cannon medal Soc. Gastro-intestinal Radiologists, 1979, Gold medal Am. Coll. Radiology, 1983. Mem. Radiol. Soc. N.A., Am. Roentgen Ray Soc. (pres. 1973-74), Am. Coll. Radiology (pres. 1980-81), Assn. U. Radiologists (pres. 1955-56, Gold medal 1979), N.Y. Roentgen Soc. (pres. 1961-62), N.Y. Gastroent. Soc. (pres. 1965-66), Soc. Chmn. Academic Radiology Depts. (pres. 1967-68). Presbyn. Subspecialties: Radiology; Diagnostic radiology. Current work: Radiology. Home: 261 Hickory St Tenafly NJ 07670 Office: 622 W 168th St New York City NY 10032

SEARLES, ARTHUR LANGLEY, chemistry educator; b. Nashua, N.H., Aug. 8, 1920; s. Arthur Langley and Georgine Johanna (Stemmerman) S. B.A., NYU, 1942, Ph.D., 1946. Chemist Squibb Inst. Med. Research, New Brunswick, N.J., 1944-45; instr. chemistry NYU, 1946-48, asst. prof., 1948-56; assoc. prof. Coll. Mt. St. Vincent, Bronx, N.Y., 1956-73, prof. chemistry, 1974-; chmn. dept. chemistry, 1970-72; cons. Roel-Cryston Corp., N.Y.C., 1950-52, FMC Corp., Princeton, N.J., 1955-57, U.S. Govt., 1956. Author: Laboratory Manual of Elementary Organic Chemistry, 1948, revised edit., 1950, 52; editor: Fantasy Commentator, 1943-53, 78—. Mem Bronx County Grand Jury, 1968—. Recipient Bene Merenti medal Coll. Mt. St. Vencent, 1981. Mem. Am. Chem. Soc., Chem. Soc. Gt. Britain, Sigma Xi, Phi Lambda Upsilon. Republican. Club: NYU Chemistry Alumni (pres. 1962-63). Subspecialties: Organic chemistry; Synthetic chemistry. Current work: Chemistry of beta-ketoanilides, synthesis of nitrogen heterocycles. Home: 7 E 235th St Bronx NY 10470 Office: Coll Mount St Vincent Riverdale Ave and W 263d St Bronx NY 10471

SEARLS, ROBERT LOUARN, biology educator; b. Madison, Wis., Oct. 26, 1931; s. Edward Marlborough and Annie Mary (Houghey) Searles; m. Ellen Donovan, June 10, 1961; children: Timothy, David, Paul Anne. B.S., U. Wis.-Madison, 1953; Ph.D., U. Calif.-Berkeley, 1960. Postdoctoral fellow Brandeis U., Waltham, Mass., 1960-63; asst. prof. U. Va., Charlottesville, 1963-68; assoc. prof. biology Temple U., Phila., 1968-74, prof., 1974—. Contbr. articles to profl. jours. Served to 1st lt. U.S. Army, 1954-56. NSF, NIH grantee. Mem. Am. Soc. Zoologist, Am. Soc. Cell Biologists, Am. Soc. Anatomists, Soc. Devel. Biology, Internat. Soc. Devel. Biologists. Subspecialties: Developmental biology; Biochemistry (biology). Current work: Limb development in the chick embryo, including morphogeneses of the limb and the differentiation of cartilage and muscle in the limb. Home: 1306 Jericho Rd Abington PA 19001 Office: Dept Biology Temple U Philadelphia PA 19122

SEARS, DUANE WILLIAM, immunologist, biochemist; b. Denver, Mar. 23, 1946; s. William A. and Florence E. (Harder) S.; m. Sheryn Elaine Rogers, Dec. 20, 1969; children—Rebecca Anne, David Allen. B.S., Colo. Coll., 1968; M.S., Columbia U., 1974, Ph.D., 1974. Postdoctoral fellow N.Y. Heart Assn.; NIH postdoctoral trainee Albert Einstein Coll. Medicine, 1974-77; asst. prof. U. Calif.-Santa Barbara, 1977—. Contbr. articles to profl. jours. Served as capt. U.S. Army, 1968-76. Nat. Cancer Inst. grantee, 1978—. Mem. Am. Assn. Immunologists, Am. Chem. Soc., AAAS, N.Y. Acad. Scis. Subspecialties: Immunogenetics; Transplantation. Current work: Immunogenetic and biochem. analysis of cytotoxic T lymphocytes.

SEARS, ERNEST ROBERT, geneticist, emeritus educator; b. Bethel, Oreg., Oct. 15, 1910; s. Jacob Perlonzo and Ada Estella (McKee) S.; m. Caroline F. Eichorn, July 5, 1936; 1 son, Michael Allan; m. Lotti Maria Steinitz, June 16, 1950; children: John, Barbara, Kathleen. B.S., Oreg. State Agrl. Coll., 1932; A.M., Harvard U., 1934, Ph.D., 1936; D.Sc., Goettingen U. Agt. Dept. Agr., 1936-41, geneticist, 1941-80; research assoc. U. Mo., Columbia, 1937-63, prof., 1963-80, prof. emeritus, 1980—. Contbr. articles to tech. jours. Fulbright research fellow, Germany, 1958; Recipient Stevenson award for research in agronomy, 1951, Hoblitzelle award for research in agrl. sci., 1958, Distinguished Service award Oreg. State Agrl. Coll., 1973. Fellow Agronomy Soc. Am., AAAS, Indian Soc. Genetics and Plant Breeding (hon.), Japanese Genetics Soc. (hon.); mem. Am. Acad. Arts and Scis., Nat. Acad. Scis., Bot. Soc. Am., Genetics Soc. Am. (pres. 1978-79), Am. Soc. Naturalists, Genetics Soc. Can. (Excellence award 1977), Am. Inst. Biol. Scis., Am. Assn. Cereal Chemists (hon.), Sigma Xi (chpt. research award 1970), Phi Kappa Phi, Alpha Zeta, Gamma Sigma Delta (Distinguished Service award 1958), Alpha Gamma Rho. Subspecialties: Plant genetics; Evolutionary biology. Current work: Cytogenetics and evolution of wheat. Transfer to wheat of useful characters from its relatives. Home: 2009 Mob Hill Columbia MO 65201

SEARS, HENRY FRANCIS, II, oncologic surgeon; b. N.Y.C., July 7, 1940; s. Henry and Mary (Pouch) S.; m. Sarah Day Storm, June 12, 1962; children: David, Nathaniel, H. Christopher. B.A., U. Pa., 1962; M.D., Columbia U., 1966. Diplomate: Am. Bd. Surgery, 1974. Asst. resident in surgery U. Va. Hosp., Charlottesville, 1968-71, chief resident, 1971-72; staff surgeon Bethesda (Md.) Naval Hosp., 1972-74; sr. investigator surgery br. Nat. Cancer Inst., 1974-77; dir. surg. research and edn. Fox Chase Cancer Ctr., Phila., 1977—; assoc. scientist Wistar Inst., Phila., 1979—; asst. clin. prof. surgery U. Pa., 1977—. Contbr. articles to profl. jours. Trustee Kent Sch., Chestertown, Md., 1982—. Served to lt. comdr. USN, 1972-74. Fellow ACS, Phila. Acad. Surgery; mem. Muller Surg. Soc., Assn. Acad. Surgeons, Clinico-Pathologic Soc. Washington, Soc. Surg. Oncology, Am. Assn. Cancer Research. Subspecialties: Cancer research (medicine); Surgery. Current work: Investigation of tumor immunology using monoclonal antibodies to define membrane antigens, circulating antigens and immunodiagnostic and therapeutic potential. Office: Am Oncologic Hosp Central and Shelmire Aves Philadelphia PA 19111 Home: 215 Sunrise Ln Philadelphia PA 19118

SEARS, WILLIAM REES, engr., educator; b. Mpls., Mar. 1, 1913; s. William Everett and Gertrude (Rees) S.; m. Mabel Jeannette Rhodes, Mar. 20, 1936; children—David William, Susan Carol. B.S. in Aero. Engring, U. Minn., 1934; Ph.D., Calif. Inst. Tech., 1938. Asst. prof. Calif. Inst. Tech., 1939-41; chief aerodynamics Northrop Aircraft, Inc., 1941-46; dir. Grad. Sch. Aero. Engring., Cornell U., Ithaca, N.Y., 1946-63, Center Applied Math., 1963-67, J.L. Given prof. engring., 1962-74; prof. aerospace and mech. engring. U. Ariz., Tucson, 1974—; Cons. aerodynamics. Author: The Airplane and its Components, 1941; Editor: Jet Propulsion and High-Speed Aerodynamics, vol. VI, 1954, Jour. Aerospace Scis, 1956-63, Ann. Revs. of Fluid Mechanics, Vol. I. Recipient Distinguished Alumnus award U. Minn., 1950; Vincent Bendix award Am. Soc. Engring. Edn., 1965; Prandtl Ring Deutsche Gesellschaft für Luft-und Raumfahrt, 1974; Von Karman lectr. Am. Inst. Aeros. and Astronautics, 1968; F.W. Lanchester lectr. Royal Aero. Soc., 1973, Von Karman medal AGARD, NATO, 1977. Fellow Internat. Acad. Astronautics, Am. Acad. Arts and Scis., Am. Inst. Aeros. and Astronautics (hon., G. Edward Pendray award 1975, S.A. Reed aeros. award 1981); mem. Nat. Acad. Engring., Nat. Acad. Engring. Mexico (fgn.), Nat. Acad. Scis., Sigma Xi. Subspecialties: Aeronautical engineering; Fluid mechanics. Current work: Wind tunnels. Home: 6560 Skyway Rd Tucson AZ 85718 Office: Aerospace and Mech Engring Dept U Ariz Tucson AZ 85721

SEATON, CRAIG EDWARD, psychology educator; b. Davenport, Iowa, Jan. 31, 1939; emigrated to Can., 1977; s. Leo Russell and Letha Delores (Kerr) S.; m. Marsha Joan Pellissier, Aug. 20, 1965; children: Jeffrey, Kathryne. B.A., Calif. State U., 1961, M.S., 1965; M.S., U. So. Claif., 1970, Ph.D., 1972. Lic. marriage/family/child counselor, Calif. Social worker Santa Clara County Walfare Dept., 1963-65; asst. prof. Biola Coll., 1968-69; research assoc. Calif. Coordinating Council for Higher Edn., 1972-73, Calif. Postsecondary Edn. Commn., 1974-77; chief tng. Calif. Dept. Social Welfare, 1973-74; faculty Rio Hondo Jr. Coll., 1968-69, Chapman Coll., 1973-74, Golden Gate U., San Francisco, 1974-76; pvt. practice marriage/family counseling, 1970-77; mem. faculty Trinity Western Coll., Langley, B.C., Can., 1977—, v.p., 1979-81, dean acad. affairs, 1977-80, assoc. prof. psychology and sociology, asst. to pres. for govt. relations, dir. Human Services Program, 1982—; dir. Fraser Valley Aging Resource Ctr., 1982—; cons. in field. Bd. Dirs. Langley Sr. Citizens Resource Ctr. Soc., 1980-82; Coach Little League Baseball, 1979-80. Mem. Delta Soc., Can. Guidance and Counseling Assn., Western Gerontal. Soc., Gerontology Assn. B.C. (bd. dirs. 1980-83), Religous Research Assn., Can. Assn. Gerontology, Can. Inst. Religion and Gerontology, Am. Sociol. Assn., Am. Psychol. Assn., Assn. for Advancement Psychology, Nat. Interfaith Coalition on Aging. Lodge: Rotary. Subspecialty: Developmental psychology. Current work: Teaching research supervison human services program for those desiring field placement experience; developmental issues especially mid-life problems, religious development across the life cycle. Office: 7600 Glover Rd Langley BC Canada V2S 1V5 Home: 33234 Farrant Crescent Abbortsford BC Canada V2S 1V5

SEAVER, PHILIP HENRY, chemical engineering company executive; b. Ridgewood, N.J., July 2, 1920; s. John Eliot and Helen Elizabeth (Benson) S.; m. Anne Lillian Laskowy, Apr. 12, 1947; children: John Benson, Scott Hall. B.Chem.E., Cornell U., 1943; grad. advanced mgmt. program, Harvard U., 1964. Registered profl. engr., Mass. Process engr. E.B. Badger & Sons Co., Boston, 1946-51; with Badger Co., Inc., Cambridge, Mass., 1951—, now pres., chief operating officer, dir. Bd. dirs. N.E. region NCCJ, 1980—; mem. corp. Boston Mus. Sci., Goodwill Industries. Served to lt. (j.g.) USNR, 1944-46. Mem. Nat. Soc. Profl. Engrs., Am. Inst. Chem. Engrs., Soc. Chem. Industry, Phi Kappa Sigma. Clubs: Chemists (N.Y.C.); Harvard, Algonquin (Boston); Eastern Yacht (Marblehead, Mass.). Subspecialty: Chemical engineering. Office: One Broadway Cambridge MA 02142

SEAVEY, STEVEN R., biology educator; b. Pueblo, Colo., Jan. 2, 1945; s. Boyd King and Dorothy Ruth (Downey) S.; m. Ann, Aug. 9, 1964; children: Justine Marie, Rebecca Pilar. B.S., San Diego State U., 1968; Ph.D., Stanford U., 1972. Mem. faculty Calif. State Coll., Bakersfield, 1972-76; assoc. prof. biology Lewis and Clark Coll., Portland, Oreg., 1976—; researcher. Contbr. articles to profl. jours. Mem. AAAS, Am. Inst. Biol. Scis., Bot. Soc. Am., Am. Soc. Plant Taxonomists, Soc. Study of Evolution. Subspecialties: Systematics; Evolutionary biology. Current work: Cytogenetics of hybrid plants, reproductive biology.

SEAY, THOMAS AUSTIN, psychologist, educator, consultant; b. Gunersville, Ala., Oct. 5, 1942; s. Barnard Austin and Mary (Croxton) S.; m. Mary Burt, Mar. 4, 1978; 1 son, Michael Alexander. M.A., Austin Peay State U., 1968; postgrad., U. Del., 1969-70; Ph.D., So. Ill. U., 1973, Lehigh U., 1974-75. Tchr. Christian County (Ky.) Sch. System, 1966-67; counselor Cecil County (Md.) Sch. System, 1968-69; asst. prof.psychology Cecil Community Coll., 1969-71; instr. So. Ill. U., Carbondale, 1971-73; researcher Kutztown State Coll., 1973—, asst. prof. counseling psychology, 1973-76, assoc. prof., 1976-80, prof., 1980—; psychol. cons. Community Psychol. Service Cons., Allentown, PA., 1980—; trainer Pa. Dept. Social Welfare, Harrisburg, 1978. Author: Systematic Eclectic Therapy, 1978, (with M. Braswell) Approaches to Counseling and Psychotherapy, 1980; author: test Counselor R Shale, 1976; editor: Pa. Jour. Counseling, 1978—, Jour. Counseling and Psychotherapy, 1982. Vice chmn. Allentown State Hosp. Patient Rights; bd. dirs. Rape Crisis Council Lehigh Valley, Allentown, Haven House, Allentown, Council on Drug and Alcohol Abuse, Allentown. Recipient Service award Rape Crisis Council Lehigh Valley, 1980; Md. Dept. Health grantee, 1970; Pa. Dept. Social Welfare Title XX grantee, 1978. Mem. Am. Assn. Marital and Family Therapists (clin.), Am. Psychol. Assn., Am. Ednl. Research Assn., Am. Personnel and Guidance Assn., Am. Acad. Psychologists in Marital, Family and Sex Therapy, Elkton Jaycees (v.p. 1968-71). Lodge: Elks. Current work: Verbal and nonverbal behavior; prototypes in marital perceptions; therapy styles; personality characteristics of adoptive parents; personality characteristics of alcohol abusers. Home: 325 S Fulton St Allentown PA 18102 Office: Kutztown State Coll Grad Ctr Kutztown PA 19530

SEBO, STEPHEN ANDREW, elec. engr., educator, researcher; b. Budapest, Hungary, June 10, 1934; came to U.S., 1967, naturalized, 1976; s. Emery and Elizabeth (Thieben) S.; m. Eva Agnes Vambery, May 25, 1968. M.S.E.E., Budapest Poly. U., 1957; Ph.D. in Elec. Engring, Hungarian Acad. Scis., 1966. Engr. Budapest Electric Co., 1957-61; asst. prof. Budapest Poly. U., 1961-66, assoc. prof., 1966-67; Ford Found. fellow Columbia U., N.Y.C., 1967-68; assoc. prof. elec. engring. Ohio State U., Columbus, 1968-74, prof., 1974—, Am. Electric Power prof. in power systems, 1982—; cons. engr. for electric utilities. Contbr. articles to profl. jours. Recipient MacQuigg award Ohio State U., 1978; Power Educator award Edison Electric Inst., 1981. Mem. IEEE (sr. mem., best paper award 1982), Internat. Conf. on Large High Voltage Electric Systems, Am. Power Conf. Subspecialties: Electrical engineering; High voltage. Current work: Performance of high voltage systems and electric power systems. Developer/organizer internat. short courses on electric power systems engring. Office: 2015 Neil Ave Columbus OH 43210

SEBRANEK, JOSEPH GEORGE, food technologists, researcher, educator, cons.; b. Richland Center, Wis., Feb. 22, 1948; s. Marvin Joseph and Elsie Margaret (Machovec) S.; m. Annette Kay, Jan. 17, 1970; children: Amy Christina, Abby Jo Anna. B.S., U. Wis., Platteville, 1970, M.S., 1971, Ph.D., 1974. NIH postdoctoral fellow U. Wis., 1974-75; asst. prof. animal sci. Iowa State U., 1975-76, asst. prof. animal sci. and food tech., 1976-79, assoc. prof. animal sci. and food tech., 1979—; cons. Dept. Agr. Author: Meat Science and Processing, 1978; contbr. articles to profl. publs. Mem. Gilbert (Iowa) Planning and Zoning Commn., 1978—. Recipient Livestock Service award Walnut Grove Co., 1981. Mem. Inst. Food Technologist (chmn.-elect Muscle Foods Div.), Am. Meat Sci. Assn., Am. Soc. Animal Sci. Roman Catholic. Subspecialty: Food science and technology. Current work: Toxicology and safety judgements concerning food additives, particulary those used in meat processing. Office: Iowa State U 215 Meat Lab Ames IA 50011

SECHZER, JERI ALTNEU, psychologist; b. N.Y.C., Nov. 1, 1930; d. Max M. and Sarah (Lefkowitz) Altneu; m. Philip H. Sechzer, Aug. 20, 1948. B.S., N.Y.U., 1956; M.A., U. Pa., 1961, Ph.D., 1962. Research fellow physiology U. Pa., 1962-63, mem. mission to Algeria, 1962, USPHS fellow, 1963-64; asst. prof. anatomy Baylor U. Med. Coll., 1964-66; research scientist in rehab. medicine and anatomy N.Y. U. Med. Sch., 1966-70; asst. prof. psychiatry Med. Coll. Cornell U., White Plains, N.Y., 1970-71, assoc. prof., 1971—; asst. attending psychologist N.Y. Hosp. Cornell Med. Ctr., 1970-71, assoc. attending psychologist, 1971-77. Contbr. articles to profl. jours. U. Pa. fellow, 1958-61; NIH grantee, 1964—; NSF grantee, 1973-75; Exxon Found. grantee, 1979. Fellow N.Y. Acad. Scis. (chmn. com. animal research 1977—, chmn. psychology sect. 1982-84, women in sci. com.), Am. Psychol. Assn. (chmn. com. animal research and experimentation 1982-83); mem. UN Assn., AAAS, Am. Assn. Anatomists, AAUP, Am. Physiol. Soc., Am. Psychol. Assn., Women in Sci., Psychonomics Soc., Soc. Neurosci., Sigma Xi. Subspecialties: Physiological psychology; Developmental psychology. Current work: Early devel. and behavior; neurobehavioral toxicology; ethical issues regarding the use of animals for experimentation. Office: Bourne Lab Cornell Med Center White Plains NY 10605

SEDLACEK, BLAHOSLAV JAN, polymer scientist; b. Drahotuse, Czechoslovakia, Oct. 10, 1925; s. Frantisek and Marie (Mahlova) S.; m. Jana Nejdlova, Dec. 22, 1951; children: Pavel, Milos. B.S. in Engring, Tech. U., 1948; Ph.D., Czechoslovak Acad. Scis., 1954, D.Sc., 1964. Lab. head Inst. Organic Chem. Biochemistry, Czechoslovak Acad. Sci., Prague, 1954-59; physics div. head Inst. Macromolecular Chemistry, 1960-72, lab. head, 1973—, dep. dir., 1960-72, organizer symposia and courses, 1957—. Editor: Coll. Czechoslovak Chemical Communs; contbr. articles to profl. jours. Recipient nat. price Czechoslovak, 1967. Mem. Czechoslovak Chem. Soc., Am. Phys. Soc., Optical Soc. Am. Subspecialties: Polymer physics; Physical optics. Current work: Physical optics and physical chemistry; polymers, colloids, biopolymers; solutions, dispersions, gels; structural order, phase transitions, molecular properties and interactions; dynamic phenomena; light scattering. Home: Na Pernikarce 37 Prague 6-Dejvice Czechoslovakia 16000 Office: Inst Macromolecular Chemistry Czechoslovak Acad Scis Heyrovsky sq 2 Prague 616-Petriny Czechoslovakia 16206

SEEBASS, ALFRED RICHARD, III, aerospace engr., educator, university dean; b. Denver, Mar. 27, 1936; s. Alfred Richard and Marie Estelle (Wright) S.; m. Nancy Jane Palm, June 20, 1958; children: Erik Peter, Scott Gregory. B.S.E. magna cum laude, Princeton, 1958, M.S.E. (Guggenheim fellow), 1961; Ph.D. (Woodrow Wilson fellow), Cornell U., 1962. Asst. prof. Cornell U., 1962-64, asso. prof., 1964-72, prof., 1972-75, asso. dean, 1972-75; prof. aerospace and mech. engring., prof. math. U. Ariz., Tucson, 1975-81; dean Coll. Engring. and Applied Sci., U. Colo., Boulder, 1981—; cons. in field. Mem. coms. Nat. Acad. Sci., NRC, NASA, Dept. Transp., 1970—; chmn. Aeros. and Space Engring. Bd., Air Force Studies Bd.; grant investigator NASA, Office Naval Research, Air Force Office Sci. Research, 1966—. Editor: Sonic Boom Research, 1967, Nonlinear Waves, 1974; editorial bd.: Ann. Rev. Fluid Mechanics, Phys. Fluids, AIAA Jour.; Contbg. author: Handbook of Applied Mathematics, 1974; Contbr. articles to profl. jours. Fellow AIAA, AAAS, mem., AAUP, Soc. Indsl. and Applied Math., Sigma Xi, Tau Beta Pi. Subspecialty: Applied mathematics. Office: Coll Engring and Applied Sci U Colo Boulder CO 80309

SEEBOLD, OTTO PAUL, JR., nuclear engr.; b. Granite City, Ill., Sept. 13, 1956; s. Otto Paul and Katheryn Paul S.; m. Debra Lee Forte, June 7, 1957; children: William, Genette. B.S., Northwestern U., 1978. Assoc. engr. Westinghouse Electric Co., West Mifflin, Pa., 1978-79; nuclear safety specialist Ill. Dept. Nuclear Safety, Springfield, 1982—. Mem. Am. Nuclear Soc., Tau Beta Phi. Subspecialties: Nuclear engineering; Fluid mechanics. Current work: Manager of computer system. Office: Ill Dept Nuclear Safety 1035 Outer Park Springfield IL 62704

SEED, HARRY BOLTON, civil engineering educator; b. Bolton, Eng., Aug. 19, 1922; came to U.S., 1947, naturalized, 1966; s. Arthur Bolton and Annie (Wood) S.; m. Muriel Johnson Evans, Dec. 29, 1953; children: Raymond Bolton, Jacqueline Carol. B.Sc. in Engring, Kings Coll., London U., 1944, Ph.D., 1947; S.M., Harvard, 1948. Asst. lectr. King's Coll., 1945-47; instr. Harvard, 1948-49; found. engr. Thomas Worcester Inc., Boston, 1949-50; mem. faculty U. Calif. at Berkeley, 1950—, prof. civil engring., 1960—; cons. U.S. Bur. Reclamation, 1965—, sec. interior, 1966-69, NASA, 1966-68, AEC, 1973—, C.E., U.S. Army, 1967—, Calif. Dept. Water Resources, 1963—, U.S. Navy, 1972, also engring. cos.; Tersaghi lectr., 1967, Rankins lectr., 1979; mem. Calif. Seismic Safety Commn., Presdl. Panel on Safety Fed. Dams. Recipient Croes medal ASCE, 1960, 62, 72, Middlebrooks award, 1958, 64, 66, 71, Wellington prize, 1968, Rowland prize, 1961, Research prize, 1962, Terzaghi lectr., 1967, Norman medal, 1968, 77, Karl Terzaghi award, 1973, Vincent Bendix award, 1976, T.K. Hsieh award, 1980, Lamme award, 1983, Disting. Engring. Achievement award, 1983. Mem. ASCE, Earthquake Engring.

Research Inst., Hwy. Research Bd., Nat. Acad. Engring., Seismol. Soc. Am., U.S. Com. on Large Dams, Structural Engrs. Assn. Calif. Subspecialty: Civil engineering. Home: 623 Crossridge Terr Orinda CA 94563

SEELEY, WILLIAM GLOVER, physicist, cons.; b. Hartford, Conn., Sept. 29, 1932; s. Walter E. and Helen C. (Glover) S.; m. Alice Jean Costa, Jan. 30, 1958; children: Wesley, David, Jeffrey. S.B.E.E., MIT, 1958; M.S. in Physics, Williams Coll., 1962, Ph.D., SUNY-Albany, 1971. Research and devel. engr. Sprague Electric Co., 1958-65; prof. physics, chmn. dept. North Adams (Mass.) State Coll., 1965—; cons. computers, math. physics, applied physics; dir. Theta Electronics Corp. Contbr. articles profl. jours. Chmn. North Adams Energy Adv. Commn. Served with USN, 1950-53. Mem. Internat. Solar Energy Soc. Subspecialty: Theoretical physics. Current work: Applied theoretical physics. Patentee in field. Home: 91 Summer St Williamstown MA 01267 Office: North Adams State College North Adams MA 01247

SEELY, JOHN HENRY, mech. engr., cons.; b. Pensacola, Fla., Sept. 24, 1921; s. John Henry and Eleanor Catherine (Gill) S.; m. Marcella Dennigan, July 4, 1945. M.E., Stevens Inst. Tech., 1949; M.M.E., Syracuse U., 1960; postgrad. (IBM fellow), Stanford U., 1964-65. Registered profl. engr., N.Y. State. Engr. IBM, 1949, engring. mgr., to, 1976; prof. mech. engring. Calif. State Poly. U., 1976-82; cons. mech. engring., Tallahassee, 1982—; lectr., condr. seminars at univs., profl. devel. schs., indsl. cos. Author: (with R. C. Chu) Heat Transfer in Microelectronic Equipment, 1972, Elements of Thermal Technology, 1981; contbr.: articles to profl. jours., mags., meetings. Elements of Thermal Technology. Served to lt. USN, 1942-46, 51-52. Recipient cert. for service to mech. design advs. com. Dutchess Community Coll., 1974, plaque for outstanding service Fla. A&M U. Coll. Sci. and Tech., 1975. Fellow ASME (7 certs. for various activities and com. chairmanships 1967-80, Meritorious Service Citation award Region II 1972, Centennial medallion and cert. 1980). Subspecialties: Mechanical engineering; Materials. Current work: Heat Transfer in microelectronic equipment. Home: 3045 Shamrock N Tallahassee FL 32308

SEETHARAM, BELLUR, biomedical researcher, educator; b. Channarayapatna, Mysore, India, June 3, 1943; came to U.S., 1972; s. Krishna and Gundamma Rao; m. Shakuntla, Mar. 3, 1971; children: Antariksh, Anil. B.S., Mysore U., 1961; M.S., Bangalore U., 1965, Ph.D., 1972. Postdoctoral fellow U. N.Mex., Albuquerque, 1972-74; instr. Washington U., St. Louis, 1974-77, asst. prof., 1977-83, assoc. prof., 1983—. NIH research grantee, 1979. Mem. Am. Gastroent. Assn., Am. Soc. Biochemists, Am. Fedn. Clin. Research. Hindu. Subspecialties: Biochemistry (medicine); Gastroenterology. Current work: Study of molecular organization of brush border membrane proteins. Vitamin B_{12} in health and disease. Home: 9034 Watsonia Olivette MO 63132 Office: Washington U Sch Medicine 660 S Euclid Saint Louis MO 63110

SEGAL, ALVIN, research organic chemist; b. N.Y.C., Mar. 21, 1929; s. Murray and Betty (Gordon) S.; m. Nanette R. Greenberg, June 1, 1958 (div. dec); children—Irene Jane, Richard. B.S. in Pharmacy, Bklyn. Coll. Pharmacy, 1958; Ph.D. in Organic Chemistry, N.Y. U., 1965. Registered pharmacist, N.J., N.Y. Pharmacist N.Y. U. Med. Sch. Hosp., 1958; Research organic chemist Ortho Pharm. Co., Raritan, N.J., 1959-61; asst. prof. pharmacognosy U. Tenn. Coll. Pharmacy, 1966-68; mem. faculty dept. environ. medicine N.Y. U. Med. Ctr., 1968—, research prof. environ. medicine, 1980—. Author articles in field. Recipient Founders Day award N.Y. U., 1966. Mem. Am. Assn. Cancer Research, Am. Chem. Soc., AAAS. Subspecialties: Inorganic chemistry; Cancer research (medicine). Current work: Mechanisms of chem. carcinogenesis, chem.-biol. interacios. Home: 6 Brevoort Ave Apt 1B Pomona NY 10970 Office: 550 First Ave New York NY 10016

SEGAL, BARRY M., internist, endocrinologist, educator; b. N.Y.C., Sept. 26, 1934; s. O. Saul and Doris (Eisenberg) S.; m. Susan Betty Grossman, June 17, 1961; children: Jeffrey Howard, Helaine Beth. B.A., Syracuse U., 1956; M.D., SUNY-Syracuse, 1959. Diplomate: Am. Bd. Internal Medicine, Am. Bd. Endocrinology, Am. Bd. Nuclear Medicine. Intern Bellevue Hosp., and Meml. Hosp. Center, N.Y.C., 1959-60, asst. resident in medicine, 1960-61; resident in medicine N.Y. Med. Coll., N.Y.C., 1962-63; vis. fellow in endocrinology Columbia U. Presbyn. Med. Center, N.Y.C., 1963-65; asst. physician Vanderbilt Clinic, Coll. of Physicians and Surgeons, 1965-71; asst. chief endocrinology sect. Westchester County Med. Center, Valhalla, N.Y., 1965; staff mem., cons. Phelps Meml. Hosp., Tarrytown, N.Y., 1965—; program chmn. medicine dept. 1981, chmn. continuing med. edn. com., 1982—; attending staff No. Westchester Hosp., Mt. Kisco, N.Y.; clin. asst. prof. medicine N.Y. Med. Coll., 1971-74, clin. assoc. prof., 1974—; cons. in endocrinology Peekskill (N.Y.) Community Hosp., 1965—. Contbr. sci. articles to profl. publs. Fellow ACP, Am. Coll. Nuclear Physicians; mem. Westchester Diabetes Assn. (mem. bd. dirs. 1965-70), Am. Diabetes Assn., AMA, Westchester County Med. Soc. (chmn. medicine sect. 1972-77, chief internal medicine sect. 1972-75), Endocrine Soc., Am. Thyroid Assn., Soc. Nuclear Medicine. Subspecialties: Endocrinology; Nuclear medicine. Current work: Endocrinology; nuclear medicine. Home: 520 Sleepy Hollow Rd Briarcliff NY 10510 Office: 316 Chappaqua Rd Briarcliff NY 10510

SEGAL, BERNARD LOUIS, physician, educator; b. Montreal, Que., Can., Feb. 13, 1929; came to U.S., 1961, naturalized, 1966; s. Irving and Fay (Schecter) S.; m. Idajane Fischman, Feb. 17, 1963; 1 dau., Jody Segal. B.Sc. cum laude, McGill U., 1950, postgrad., 1950-51, M.D., C.M. high standing, 1955. Diplomate: Am. Bd. Internal Medicine. Intern Jewish Gen Hosp., Montreal, 1955-56; resident Balt. City Hosp., 1956-57, Beth Israel Hosp., Boston, 1957-58, Georgetown Med. Center, Washington, 1958-59, St. George's Hosp., London, Eng., 1959-61; practice medicine specializing in internal medicine and cardiology, Phila., 1961—; prof., sr. attending physician medicine Hahnemann Med. Coll., Phila., 1964—, dir. William Likoff Cardiovascular Inst. Author: Auscultation of the Heart, 1965; Editor: Theory and Practice of Auscultation, 1964, Engineering in the Practice of Medicine, 1966, Your Heart, 1972, Arteriosclerosis and Coronary Heart Disease, 1972; Editorial bd.: Am. Jour. Cardiology, 1970—, Clin. Echocardiography, 1978; Contbr. numerous articles on cardiology to med. jours. Fellow A.C.P., Am. Coll. Cardiology (chmn. scholar-trainee com., trustee 1969-71), Am. Coll. Chest Physicians; mem. N.Y. Acad. Scis., Alpha Omega Alpha. Subspecialties: Cardiology; Internal medicine. Home: 1156 Red Rose Ln Villanova PA 19085 Office: 1320 Race St Philadelphia PA 19102

SEGAL, IRVING EZRA, mathematics educator; b. N.Y.C., Sept. 13, 1918; s. Aaron and Fannie (Weinstein) S.; m. Osa Skotting, Feb. 15, 1955; children: William, Andrew, Karen B. A.B., Princeton U., 1937; Ph.D., Yale U., 1940. Instr. Harvard U., 1941; research asst. Princeton U., 1941-42, assoc., 1942-43; asst. to O. Veblen, Inst. for Advanced Study, 1945-46; asst. prof. U. Chgo., 1948-53, assoc. prof., 1953-57, prof., 1957-60, Mass. Inst. Tech., Cambridge, 1960—; vis. assoc. prof. Columbia, 1953-54; vis. fellow Insts. Math. and Theoretical Physics, Copenhagen, 1958-59; vis. prof. Sorbonne, Paris, France, 1965, U. Lund, Sweden, 1971, Coll. de France, 1977. Author: Mathematical Problems of Relativistic Physics, 1963, (with R.A. Kunze) Integrals and Operators, 1968, Mathematical Cosmology and Extragalactic Astronomy, 1976; Editor: (with W.T. Martin) Analysis in Function Space, 1964, (with Roe Goodman) Mathematical Theory of Elementary Particles, 1966; Contbr. articles to profl. jours. Served with AUS, 1943-45. Guggenheim fellow, 1947, 51-52, 67-68. Mem. Am. Math. Soc., Am. Phys. Soc., Am. Acad. Arts and Sci., Royal Danish Acad. Scis., Am. Astron. Soc., Soc. Indsl. and Applied Math., Nat. Acad. Sci. Subspecialties: Cosmology; Theoretical physics. Current work: Quantum field and particle theory, especially in alternative space-times. Cosmological theory and statistics of galaxies, qasars, etc. Functional analysis and nonlinear partial differential equations. Home: 25 Moon Hill Rd Lexington MA 02173

SEGAL, JOSEPH, biochemist, educator; b. Shaar Ha-Amakim, Israel, Sept. 14, 1946, came to U.S., 1976; s. Mishael and Haya (Lerman) S.; m. Harriet Gildner, Sept. 3, 1975; 1 dau., Erin Michelle. B.Sch., Hebrew U., Israel, 1970, M.Sc., 1972, Ph.D., 1976. Research fellow in medicine Beth Israel Hosp., Boston, 1976-78; assoc. in medicine Harvard Med. Sch., Boston, 1976-78, instr., 1978-80, asst. prof., 1980—. Leah and Arthur Felix fellow, 1978; Nat. Inst. on Aging grantee, 1979. Mem. Am. Gerontol. Soc., Am. Fedn. for Clin. Research. Subspecialties: Biochemistry (biology); Receptors. Current work: Thyroid hormone: The mechanism of action of thyroid hormone at the plasma membrane level. Aging: The role of endocrine and immune systems in the process of aging. Office: U/Beth Israel Hosp 330 Brookline Ave Boston MA 02215

SEGAL, SHOSHANA, molecular biologist; b. Plovdive, Bulgaria, Feb. 27, 1941; came to U.S., 1969, naturalized, 1977; d. Manoach and Sara (Cohen) Ninio; m. David M. Segal, July 18, 1967; children: Ron, Ethan. B.Sc., Hebrew U., Jerusalem, 1965; Ph.D., Georgetown U., 1973. Lab. asst. Weizmann Inst. Sci., Israel, 1961-62, 65-68; fellow Nat. Cancer Inst., NIH, Bethesda, Md., 1973-76, sr. investigator, 1976-80, expert cons., 1980—; supr. students dept. biochemistry Georgetown U., 1981-82. Contbr. in field. Served with Israeli Army, 1959-61. Mem. Am. Soc. Microbiology, Sigma Xi. Subspecialties: Gene actions; Genetics and genetic engineering (biology). Current work: The use of genetic engineering for studying cellular differentiation and regulation of mammalian genes. Home: 5 Eton Overlook Rockville MD 20850 Office: National Cancer Institute National Institutes Health Bldg 37 Bethesda MD 20205

SEGALOFF, ALBERT, physician, researcher, educator; b. West Haven, Conn., July 25, 1916; s. Samuel and Lena Lilly (Merman) S.; m. Ann Esther Zaem, May 9, 1940; children: Ruth Tina, David Charles, Joel Paul. B.S. with distinction, Yale U., 1937; M.S. in Anatomy, Wayne State U., 1939, M.D., 1943. Diplomate: Am. Bd. Internal Medicine. Teaching fellow anatomy Wayne State U., Detroit, 1940-43; intern William J. Seymour Hosp., Eloise, Mich., 1943-44; dir. endocrine research Alton Ochsner Med. Found., New Orleans, 1945-67, head endocrine research sect. 1, 1967, dir. research, 1961-62, head oncology program, 1974—; asst. in medicine Vanderbilt U., 1944-45; cons. endocrinology USPHS Hosp., New Orleans, 1964-74; clin. prof. medicine, adj. prof. pharmacology Tulane U., 1963—; clin. prof. dept. pathology La. State U., 1977; cons. internal medicine Charity Hosp. La., New Orleans, 1965—. Author: Gouty Arthritis and Gout; editor: Breast Cancer, 1958, Current Concepts in Breast Cancer, 1967, Internat. Jour. Steroids, 1966; contbr. articles in field to profl. jours. Ciba fellow, 1940-43; recipient Ciba award Endocrine Soc., 1951; Disting. Service award Wayne State U. Coll. Medicine, 1966. Jewish. Subspecialties: Endocrinology; Cancer research (medicine). Current work: Oncology, endocrinology from organic chemical, biological, pharmacologic and clinical view points. Office: 1520 Jefferson Hwy New Orleans LA 70121

SEGAR, DOUGLAS ALLAN, tech. cons. co. exec.; b. Liverpool, Eng., Apr. 29, 1944; s. Thomas Harold and Mary (McEvoy) S.; m. Rene Jill, Sept. 10, 1965 (div.); children: Jennifer Kate, Anthony Allan. B.Sc., U. Liverpool, 1965, Ph.D. in Oceanography, 1969. Postdoctoral research fellow Marine Biol. Assn. U.K., Plymouth; asst. prof. Sch. Marine and Atmospheric Scis., U. Miami, Fla., 1970-74; oceanographer NOAA, Rockville, Md., 1974-77, Miami, 1974-77; profl. staff mem., cons. U.S. Congress Mcht. Marine and Fisheries Com., Washington, 1977-79; pres. SEAMOcean, Inc., Wheaton, Md., 1979—; mem. U.S. Adv. Com. to Internat. Conv. on Prevention of Marine Pollution by Dumping of Wastes and Other Materials, 1981—; congressional rep. U.S. del. Spl. Consultative Meeting Antarctic Treaty, Canberra, Australia, 1978, Buenos Aires, Argentina, 1978, Washington, 1978; cons. EPA, Govt. Poland, 1977. Contbr. articles to profl. jours. Mem. Am. Soc. Limnology and Oceanography (chmn. com. spl. symposia 1977-79), Am. Oceanic Orgn., Am. Geophys. Union, AAAS, Am. Chem. Soc., Optical Soc. Am., N.Y. Acad. Scis., Water Pollution Control Fedn. Current work: Pollution chemistry, optical analysis methods, environmental management.

SEGINER, ARNAN, aeronautical engineer; b. Poprad, Czechoslovakia, Apr. 13, 1936; s. David Bernard and Boriska (Korach) S.; m. Ora Targan, Oct. 15, 1959; children: Osnat, Vered. B.Sc., Israel Inst. Tech., Haifa, 1960, M.Sc., 1963, D.Sc., 1968. Lectr. Israel Inst. Tech., Haifa, 1965-68; research assoc. NASA, Ames Research Ctr., Moffett Field, Calif., 1968-70, 1975-77, 82—; sr. lectr. Israel Inst. Tech., 1970-75, assoc. prof., 1975-82; cons. Ministry Def., Tel Aviv, Israel, 1974-82; head Aerodynamics Lab. Technion, Haifa, 1977-82; mem. steering com. Ministry Energy, Jerusalem, 1980-82. Chmn. Israeli Student Orgn. U.S., N.Y.C., 1975-77. Served to lt. col. Israeli Army, 1954-56. Recipient Klein award Klein Fund, 1982; NRC Associateship award, 1968-75, 80. Mem. AIAA, Israel Soc. Aeros. and Astronautics. Subspecialties: Aeronautical engineering; Fluid mechanics. Current work: Development of computer codes for nonlinear aerodynamics of advanced flight vehicles and development of advanced wind tunnel techniques. Home: 127 Greenmeadow Way Palo Alto CA 94306 Office: NASA Ames Research Ctr MS 227-8 Moffett Field CA 94035

SEGRE, DIEGO, immunologist, educator; b. Milan, Italy, Feb. 3, 1922; s. Ulderico and Corinna (Corinaldi) S.; m. Mariangela Bertani, July 22, 1952; children: Carlo Segre, Alberto Segre. D.V.M., U. Milan, 1947; M.S., U. Nebr., 1954; Ph.D., U. Wis., 1957. Asst. prof. U. Milan, 1947-51; Fulbright scholar Purdue U., 1951-52; asst. U. Nebr., 1952-55; research asst. U. Wis., 1955-57, asst. prof., 1957-60; prof. immunology U. Ill., Urbana, 1960—; cons. NIH, NSF. Contbr. articles to profl. jours. Mem. Am. Assn. Immunologists, AAAS, Soc. Exptl. Biol. Medicine, Gerontol. Soc., AVMA, Conf. Research Workers Animal Diseases, Am. Coll. Vet. Microbiology. Subspecialty: Immunobiology and immunology. Current work: Regulation of immune response, immunogerontology. Home: 2010 Boudreau Dr Urbana IL 61801 Office: 2001 S Lincoln Ave Urbana IL 61801

SEGRÈ, EMILIO, physicist, educator; b. Tivoli, Rome, Italy, Feb. 1, 1905; came to U.S., 1938, naturalized, 1944; s. Giuseppe and Amelia (Treves) S.; m. Elfriede Spiro, Feb. 2, 1936 (dec. Oct. 1970); children: Claudio, Amelia, Fausta; m. Rosa Mines, Feb. 12, 1972. Ph.D., U. Rome, 1928; Dr. honoris causa, U. Palermo, Italy, Gustavus Adolphus Coll., St. Peter, Minn., Tel Aviv U., Israel, Hebrew Union Coll., Los Angeles. Asst. prof. U. Rome, 1932-36; dir. physics lab. U. Palermo, Italy, 1936-38; research asst. U. Calif. at Berkeley, 1938-43, prof. physics, 1945-72, emeritus, 1972—, group leader Los Alamos Sci. Lab., 1943-46; hon. prof. San Marcos U., Lima; vis. prof. U. Ill., Purdue U.; prof. physics U. Rome, 1974-75. Recipient Hofmann medal German Chem. Soc., Cannizzaro medal Accad. Lincei; Nobel prize in physics, 1959; decorated great cross merit Republic of Italy; Rockefeller Found. fellow, 1930-31; Guggenheim fellow, 1959; Fulbright fellow. Fellow Am. Phys. Soc.; mem. Nat. Acad. Scis., Am. Philos. Soc., Am. Acad. Arts and Scis., Heidelberg Akademie Wissenschaften, European Phys. Soc., Acad. Scis. Peru, Soc. Progress of Sci. (Uruguay), Società Italiana di fisica, Accad. Naz. Lincei (Italy), Accad. Naz. XL (Italy), Indian Acad. Scis. Bangalore, others. Subspecialty: Nuclear physics. Co-discoverer slow neutrons, elements technetium, astatine, plutonium and the antiproton. Home: 3802 Quail Ridge Rd Lafayette CA 94549 Office: Dept Physics U of California Berkeley CA 94720

SEGRE, MARIANGELA, immunology educator; b. Milan, Italy, Oct. 4, 1927; d. Carlo and Armida (Seveso) Bertani; m. Diego R. Segre, July 22, 1952; children: Carlo, Alberto. D.Sc., U. Milan, 1949. Research assoc. U. Milan, 1949-51; vis. investigator Animal Disease Research Lab., Weibridge, Eng., 1951; bacteriologist Montecatini Corp., Milan, 1951-52, Nebr. Dept. Health, 1953-54; research assoc. U. Ill., Urbana, 1963-73, asst. prof., 1973—. Contbr. articles to profl. jours. Mem. Am. Assn. Immunologists. Subspecialty: Immunobiology and immunology. Current work: Immunology of aging, cellular immunology. Home: 2010 Boudreau Dr Urbana IL 61801 Office: 1101 Peabody Dr Urbana IL 61801

SEGREST, JERE PALMER, pathology and biochemistry educator; b. Dothan, Ala., Aug. 16, 1940; s. Jere Palmer and Grace (Hudgins) Blevins) S.; m. Susan Freeman, Sept. 3, 1966; children: Stuart, Chamberlain, Austin. B.A., Vanderbilt U., 1962, M.D., 1967, Ph.D., 1969. Diplomate: Nat. Bd. Med. Examiners. Intern Vanderbilt U. Hosp., Nashville, 1969, resident, 1970; staff assoc. NIH, Bethesda, Md., 1970-74; resident George Washington U. Hosp., Washington, 1974; assoc. prof. pathology U Ala Med Ctr., Birmingham, 1974-80, prof. pathology, 1980—, prof. biochemistry, 1982—; mem. molecular cytology study sect. NIH, 1978-82. Served with USPHS, 1970-74. Am. Heart Assn. grantee, 1976-78; NIH grantee, 1976—; NSF grantee, 1976-78; Am. Egg Bd. grantee, 1981-84. Fellow Council on Atherosclerosis, Am. Heart Assn.; mem. Am. Chem. Soc. (com. on profl. tng. 1978-84), Am. Soc. Exptl. Pathology, Am. Soc. Biol. Chemists, Biophys. Soc., Am. Soc. Cell Biology. Democrat. Subspecialties: Biochemistry (medicine); Membrane biology. Current work: Studies of protein-lipid interactions in plasma lipoproteins and biological membranes, roles of plasma lipoprotein in pathophysiology of atherosclerosis. Inventor single vertical spin analysis of plasma lipoproteins. Home: 3709 Forest Run Rd Birmingham AL 35223 Office: U Ala Med Ctr Volker Hall Room G018 Birmingham AL 35294

SEHGAL, PRAVINKUMAR BHAGATRAM, physician, researcher, educator; b. Bombay, India, Sept. 11, 1949; came to U.S., 1973; m. Shashibala P., Dec. 2, 1972; 1 dau.: Neelima. M.B.B.S. (Nat. Merit Scholar India, Dorabji Tata spl. scholar, J. N. Tata scholar), Seth G. S. Med. Coll., Bombay, 1973; Ph.D., Rockefeller U., 1977. Intern King Edward Meml. Hosp., Bombay, 1971-72; grad. fellow Rockefeller U., 1973-77, postdoctoral fellow, 1977-79; asst. prof. virology, 1979—. Contbr. numerous articles to profl. jours.; editorial bd.: Jour. Interferon Research, 1980—, Virology, 1983—. Recipient Jr. Faculty Research award Am. Cancer Soc., 1980-82, Career Scientist award Irma T. Hirschl, 1982; Nat. Inst. Arthritis and Infectious Diseases grantee, 1979—; Am. Heart Assn. established investigator, 1983—. Mem. AAAS, Am. Soc. Microbiology, Am. Soc. Virology, N.Y. Acad. Sci., Internat. Soc. Interferon Research, Sigma Xi. Subspecialties: Animal virology; Molecular biology. Current work: Human interferons; halobenzimidazole ribosides; regulation gene expression. Office: Rockefeller U 1230 York New York NY 10021

SEIBERT, WARREN FREDERICK, applied research psychologist, research administrator; b. Carbondale, Ill., Mar. 19, 1927; s. Harry Harrison and Henrietta Caroline (Struckmeyer) S.; m. Nedra Alice Midjaas, Mar. 13, 1949; children: Diane Janice, Eric Larsen, Robin Suzanne. B.S. in Edn, So. Ill. U., 1950, M.S., 1951; Ph.D., Purdue U., 1956. Asst. prof. psychology Purdue U.-Calumet Campus, Hammond, Ind., 1955-56; assoc. prof. library sci. Purdue U., West Lafayette, Ind., 1956-63, prof., 1963-69, prof. edn. and engring. edn., 1972-80; chief ednl. research and evaluation br. Lister Hill Ctr., Nat. Library Medicine, Bethesda, Md., 1980—; dist. vis. prof. ednl. psychology Pa. State U., 1968; prof. psychology and dentistry U. Mich., 1969-72; observer European Broadcasting Union, Rome, 1961, animateur, Paris, 1968; chmn. research council Nat. Assn. Ednl. Broadcasters, Washington, 1965-68, 74-77; mem. adv. com. on tech. Nat. Acad. Engring., Washington, 1970-72. Joint author: monograph series Past and Likely Future of 58 Research Libraries, 1965-68; co-editor: Procs. Internat. Seminar on Instructional TV, 1962. Chmn. Wabash Democrats, West Lafayette, 1978-80. Served with USN, 1945-46; PTO. U.S. Office Edn. grantee, 1959-68, 63, 65; USPHS grantee, 1972; Lilly Endowment grantee, 1973. Mem. Am. Psychol. Assn., AAAS, Am. Ednl. Research Assn., Midwestern Psychol. Assn. Current work: Growth and development of research libraries; impact of information technology on libraries; design of instructional systems. Home: 5 Grason Ct Rockville MD 20850 Office: NMAC/LHNCBC Nat library of Medicine 8600 Rockville Pike Bethesda MD

SEIDELMANN, P. KENNETH, scientist; b. Cin., June 15, 1937; s. P. Emil and Esther Margaret (Momberg) S.; m. Roberta Jane Buck, Aug. 27, 1960; children: Holly, Alan. B.S. in E.E, U. Cin., 1960, M.S., 1962, Ph.D. in Dynamical Astronomy, 1968. Nautical almanac office astronomer U.S. Naval Obs., Washington, 1965-73, asst. dir., 1973-76, dir., 1976—; vis. assoc. prof. astronomy U. Md., 1973, 75, 77, 80. Contbr. numerous articles to sci. jours. Served with AUS, 1963-65. Recipient Disting. Alumni award Coll. Engring., U. Cin., 1975. Mem. Am. Astron. Soc., Internat. Astron. Union, Royal Astron. Soc., Inst. Nav., AIAA, AAAS, Sigma Xi. Lutheran. Subspecialties: Planetary science; Optical astronomy. Current work: Planetary and satellite motions, observations with charge coupled device for astrometry and solar system objects, very large array observations, space telescope. Co-discoverer satellite of Saturn, 1980; prepared star chart for Einstein Monument, Nat. Acad. Scis. grounds. Home: 6539 Windermere Circle Rockville MD 20852 Office: US Naval Observatory Washington DC 20390

SEIDL, MILOS, physicist, educator; b. Budapest, Hungary, May 24, 1923; came to U.S., 1968, naturalized, 1977; s. Ferdin and Therese (Studinka) S.; m. Vera Hnevsova, Sept. 14, 1960; 1 dau., Kathy. B.Sc., Tech. U. Prague, 1947, Ph.D., 1949, D.Sc., 1963; M.Eng. (hon.), Stevens Inst. Tech., 1979. Research staff Inst. Vacuum Electronics, Prague, Czechoslovakia, 1949-58; sr. scientist Czechoslovakian Acad. Sci., Prague, 1958-68; lectr. Tech. U. Prague, 1960-68; vis. scientist Stanford U., 1968-69; prof. physics Stevens Inst. Tech., 1969—. Recipient Stevens Inst. Tech. J. Davis Meml. Research award, 1978. Fellow Am. Phys. Soc., Am. Vacuum Soc., IEEE, N.Y. Acad. Sci. Presbyterian. Subspecialties: Plasma physics; Plasma engineering. Current work: Interaction of electron beams with plasma; plasma production; ion sources; sputtering of atomic particles from solid surfaces by ion bombardment. Home: 27 Clinton Ln Wayne NJ 07470 Office: Dept Physics Stevens Inst Tech Hoboken NJ 07030

SEIDLER, FREDERIC JOHN, pharmacologist; b. Newark, Mar. 24, 1951; s. Frederick B. and M. Elizabeth (Hill) S. B.S., Duke U., 1974.

Research technician dept. physiology and pharmacology Duke U. Med. Center, Durham, N.C., 1974-75, 76-78, sr. research technician dept. pharmacology, 1978-80, lab. research analyst, 1980—. Contbr. articles to profl. jours. Mem. Am. Soc. Pharmacology and Exptl. Therapeutics. Subspecialties: Neuropharmacology; Biochemistry (medicine). Current work: Developmental pharmacology, neuropharmacology, and polyamines. Office: Dept Pharmacology Duke U Med Center Box 3813 Durham NC 27710

SEIFER, ARNOLD DAVID, engineer; b. Newark, Apr. 22, 1940; s. Abe W. and Bessie R. (Coopersmith) S. B.S. in Math, Rensselaer Poly. Inst., 1962, M.S., 1964, Ph.D., 1968. Research specialist Gen. Dynamics Corp., Groton, Conn., 1967-73; sr. staff mathematician Applied Physics Lab., Laurel, MD., 1973-76; sr. staff engr. Emerson Electric Co., St. Louis, 1976-80; prin. engr. Raytheon Co., Wayland, Mass., 1980—. Contbr. sci. papers to profl. jours. Mem. IEEE (sr.), Soc. Indsl. and Applied Math., Sigma Xi. Subspecialties: Systems engineering; Applied mathematics. Current work: Radar systems analysis and design. Home: 66 Dinsmore Ave Apt 606 Framingham MA 01701 Office: Raytheon Co Equipment Devel Labs 430 Boston Post Rd Wayland MA 01778

SEIGER, MARVIN BARR, geneticist, educator; b. N.Y.C., Nov. 18, 1926; s. Isadore and Rose (Nicholas) S.; m. Haydee Corona; children: Leslie Anne, Andrea Michelle, Karen Elizabeth. B.S., Duquesne U., 1950; M.A., U. Tex., 1953; Ph.D., U. Toronto, Ont., Can., 1962. Vis. prof. Purdue U., 1962-63; research assoc. U. Notre Dame, 1963-64; Nat. Inst. Public Health fellow Behavior Genetics Inst., U. Calif., Berkeley, summer 1964; NIH postdoctoral trainee U. Rochester, 1964-65; assoc. prof. Wright State U., 1965—; NRC fellow, vis. prof. Fed. U. Rio de Janeiro, 1976-79. Pres., Miami Valley Orchid Soc.; chmn. Moon Soc., United Remnant Band, Shawnee Nation, 1970—; mem. Dayton Opera Chorus, 1979—; trustee Dayton Opera Co. Served with AUS, 1944-46; Served with USN, 1946. Mem. Am. Soc. Naturalists, Brazilian Genetics Assn., Ohio Acad. Sci., Sigma Xi. Subspecialties: Gene actions; Behaviorism. Current work: Behavior as a population phenomenon; behavior as evolutionary mechanism. Home: 6253 Leawood Dr Dayton OH 45424 Office: Dept Biol Sci Wright State U Dayton OH 45435

SEIGLER, DAVID STANLEY, botanist, educator, researcher; b. Wichita Falls, Tex., Sept. 11, 1940; s. Kenneth Ray and Floy Meleese (Wilkinson) S.; m. Janice Kay Cline, Jan. 20, 1961; children: Dava Kay Seigler Carson, Rebecca Joy. B.S., Southwestern State Coll., Weatherford, Okla., 1961; Ph.D., U. Okla., Norman, 1967. Postdoctoral assoc. Dept. Agr. No. Regional Lab., Peoria, Ill., 1967-68; postdoctoral fellow dept. botany U. Tex., Austin, 1968-70; asst. prof. botany U. Ill., Urbana, 1970-76, assoc. prof., 1976-79, prof., 1979—, assoc. head dept. botany, 1980-82; Fulbright Hays lectr., Argentina, 1976. Contbr. articles to profl. jours. Mem. Am. Chem. Soc., Bot. Soc. Am., Royal Chem. Soc., Phytochem. Soc., Soc. Econ. Botany, Southwestern Assn. Naturalists. Subspecialties: Systematics; Organic chemistry. Current work: Biochem. systematics; phytochemistry. Home: 510 W Vermont St Urbana IL 61801 Office: Dept Botany U Ill Urbana IL 61801

SEILER, DAVID GEORGE, physics educator, researcher; b. Green Bay, Wis., Dec. 17, 1940; s. George A. and Esther V. (Gustafson) S.; m. Nancy Sarah Cowdrick, June 15, 1963; children: Laurel E., Rebecca J. B.S. in Physics, Case Western Res. U., 1963; M.S., Purdue U., Lafayette, Ind., 1965, Ph.D. in Physics, 1969. Teaching and research asst. Purdue U., 1963-69; asst. prof. physics North Tex. State U., 1969-72, assoc. prof., 1973-79, prof., 1979—; physicist Nat. Bur. Standards, Boulder, Colo., 1972. Contbr. numerous articles to profl. jours. Case Western Res. scholar. Mem. Am. Phys. Soc., Optical Soc. Am. Subspecialty: Condensed matter physics. Current work: Semiconductor properties; interaction of laser radiation with semi-conductors. Office: Dept Physics North Tex State U Denton TX 76203

SEINFELD, JOHN HERSH, chemical engineering educator; b. Elmira, N.Y., Aug. 3, 1942; s. Ben B. and Minna (Johnson) S. B.S., U. Rochester (N.Y.), 1964; Ph.D., Princeton U., 1967. Asst. prof. chem. engring. Calif. Inst. Tech., Pasadena, 1967-70, assoc. prof., 1970-74, exec. officer for chem. engring., 1973—, prof., 1974—, Louis E. Nohl prof., 1980—; Allan P. Colburn meml. lectr. U. Del., 1976; Camille and Henry Dreyfus Found. lectr. MIT, 1979; Donald L. Katz lectr. U. Mich., 1981; Reilly lectr. U. Notre Dame, 1983; cons. Standard Oil of Calif.; mem. sci. adv. bd. EPA; Inst. lectr. Am. Inst. Chem. Engrs., 1980. Author: (with Leon Lapidus) Numerical Solution of Ordinary Differential Equations, 1971, Mathematical Methods in Chemical Engineering, Vol. III, Process Modeling, Estimation and Identification, 1974, Air Pollution: Physical and Chemical Fundamentals, 1975, Lectures in Atmospheric Chemistry, 1980; Assoc. editor: Environ. Sci., Tech, 1981—; mem. editorial bd.: Computers, Chem. Engring, 1974-76, (with Leon Lapidus) Jour. Colloid and Interface Sci, 1978—, Advances in Chem. Engring, 1980—, Revs. in Chem. Engring, 1980—; assoc. editor: Atmospheric Environment, 1976—. Recipient Donald P. Eckman award Am. Automatic Control Council, 1970; Curtis W. McGraw Research award Am. Soc. Engring. Edn., 1976; Allan P. Colburn award Am. Inst. Chem. Engrs., 1976; Pub. Service medal NASA, 1980; Camille and Henry Dreyfus Found. Tchr. Scholar grantee, 1972. Mem. Am. Inst. Chem. Engrs., Am. Soc. Engring. Edn., Nat. Acad. Engring., Air Pollution Control Assn., Am. Chem. Soc., Soc. Petroleum Engrs., Sigma Xi, Tau Beta Pi. Subspecialties: Chemical engineering; Atmospheric chemistry. Home: 121 S Wilson St Unit 101 Pasadena CA 91106 Office: Dept of Chem Engring Calif Inst of Tech Pasadena CA 91125

SEIREG, ALI A(BDEL HAY), mechanical engineer; b. Egypt, Oct. 26, 1927; came to U.S., 1951, naturalized, 1960; s. Abdel Hay and Aisha S.; m. Shirley Marachowsky, Dec. 24, 1954; children—Mirette Elizabeth, Pamela Aisha. B.Sc. M.E., U. Cairo, 1948; Ph.D., U. Wis., 1954. Lectr. Cairo U., 1954-56; staff adv. engr. Falk Corp., Milw., 1956-59; assoc. prof. theoretical and applied mechanics Marquette U., 1959-64, prof., 1964-65; prof. mech. engring. U. Wis., Madison, 1965—; cons. industry, ednl. and govt. agys.; chmn. U.S. council Internat. Fedn. Theory of Machines, 1974—; co-chmn. 5th World Congress of Theory of Machines, 1979. Author: Mechanical Systems Analysis, 1969; contbr. numerous articles to profl. jours. Fellow ASME (Richards Meml. award 1973, Machine Design award 1978, chmn. div. design engring. 1977-78, chmn. computer tech. 1978-81, mem. policy bd. communications 1978-80, mem. policy bd. gen. engring. 1979-80, chmn. Century II Internat. Computer Tech. Conf. 1980, founding chmn. computer engring. div. 1980-81, v.p. systems and design 1981—), Am. Soc. Engring. Edn. (George Westinghouse award 1970), Soc. Exptl. Stress Analysis, Am. Gear Mfg. Assn. (E. P. Connell award 1974), Automation Research Council. Subspecialties: Mechanical engineering; Biomedical engineering. Current work: Computers in mechanical engineering; biomechanics, tribology; ocean technology. Home: 219 DuRose Terr Madison WI 53705 Office: 1513 University Ave Madison WI 53706 I have always tried my best to look beyond what I hear, to think beyond what I see, to give more than I receive, and to do good as its own reward.

SEITZ, FREDERICK, university president emeritus; b. San Francisco, July 4, 1911; s. Frederick and Emily Charlotte (Hofman) S.; m. Elizabeth K. Marshall, May 18, 1935. A.B., Leland Stanford Jr. U., 1932; Ph.D., Princeton U., 1934; Doctorate Hon. Causa, U. Ghent, 1957; D.Sc., U. Reading, 1960, Rensselaer Poly. Inst., 1961, Marquette U., 1963, Carnegie Inst. Tech., 1963, Case Inst. Tech., 1964, Princeton U., 1964, Northwestern U., 1965, U. Del., 1966, Poly. Inst. Bklyn., 1967, U. Mich., 1967, U. Utah, 1968, Brown U., 1968, Duquesne U., 1968, St. Louis U., 1969, Nebr. Wesleyan U., 1970, U. Ill., 1972, Rockefeller U., 1981; LL.D., Lehigh U., 1966, U. Notre Dame, 1962, Mich. State U., 1965, Ill. Inst. Tech., 1968, N.Y. U., 1969; L.H.D., Davis and Elkins Coll., 1970, Rockefeller U., 1981. Instr. physics U. Rochester, 1935-36, asst. prof., 1936-37; physicist research labs. Gen. Electric Co., 1937-39; asst. prof. Randal Morgan Lab. Physics, U. Pa., 1939-41, asso. prof., 1941-42; prof. physics, head dept. Carnegie Inst. Tech., Pitts., 1942-49; prof. physics U. Ill., 1949-57, head dept., 1957-64, dir. control systems lab., 1951-52, dean Grad. Coll., v.p. research, 1964-65; exec. pres. Nat. Acad. Scis., 1962-69; pres. Rockefeller U., N.Y.C., 1968-78; trustee Ogden Corp., 1977—; dir. tng. program Clinton Labs., Oak Ridge, 1946-47; Chmn. Naval Research Adv. Com., 1960-62; vice chmn. Def. Sci. Bd., 1961-62, chmn., 1964-68; sci. adviser NATO, 1959-60; mem. nat. advisory com. Marine Biomed. Inst. U. Tex., Galveston, 1975-77; mem. adv. group White House Conf. Anticipated Advances in Sci. and Tech., 1975-76; mem. advisory bd. Desert Research Inst., 1975-79, Center Strategic and Internat. Studies, 1975—; mem. Nat. Cancer Advisory Bd., 1976-82; Dir. Akzona Inc., Tex. Instruments Inc. Author: Modern Theory of Solids, 1940; author: The Physics of Metals, 1943, Solid State Physics, 1955. Trustee Rockefeller Found., 1964-77, Princeton U., 1968-72, Lehigh U., 1970-81, Research Corp., 1966-82, Inst. Internat. Edn., 1971-78, Woodrow Wilson Nat. Fellowship Found., 1972—, Univ. Corp. Atmospheric Research, Am. Museum Natural History, 1975—; trustee John Simon Guggenheim Meml. Found., 1973—, chmn. bd., 1976—; mem. Belgian Am. Edn. Found.; bd. dirs. Richard Lounsbery Found., 1980—. Recipient Franklin medal Franklin Inst. Phila., 1965; Hoover medal Stanford U., 1968; Nat. Medal of Sci., 1973; James Madison award Princeton U., 1978; Edward R. Loveland Meml. award ACP, 1983; Vannevar Bush award Nat. Sci. Bd., 1983. Fellow Am. Phys. Soc. (pres. 1961); mem. Nat. Acad. Scis., Am. Acad. Arts and Scis., Am. Inst. Mining, Metall. and Petroleum Engrs., Am. Philos. Soc., Am. Inst. Physics (chmn. governing bd. 1954-59), Inst. for Def. Analysis, Finnish Acad. Sci. and Letters (fgn. mem.), Phi Beta Kappa Assos. Subspecialties: Theoretical physics; Materials. Address: Rockefeller U 1230 York Ave New York NY 10021

SEKANINA, ZDENEK, research scientist; b. M. Boleslav, Czechoslovakia, June 12, 1936; came to U.S., 1969, naturalized, 1974; s. Frantisek and Hedvika (Kolarikova) S.; m. Jana Soukupova, Apr. 1, 1966; 1 son, Sidney Jason. RNDr., Charles U., Prague, 1963. Fellow Institut d'Astrophysique, Liege, Belgium, 1968-69; physicist Smithsonian Astrophys. Obs., Cambridge, Mass., 1969-80; scientist Jet Propulsion Lab., Calif. Inst. Tech., 1980—. Contbr. numerous articles to profl. jours. Asteroid 1913 named Sekanina in his honor, 1976; Recipient Outstanding Sci. Performance cert. Smithsonian Astrophys. Obs., 1980. Mem. Am. Astron. Soc., Planetary Soc., Internat. Astron. Union (assoc. dir. Central Bur. for Astron. Telegrams 1970-80), Com. Space Research. Subspecialty: Planetary science. Current work: Physics of comets, especially of cometary nuclei; cometary dust and its relationship to interplanetary dust; dynamics of meteors and solid grains; physics of fireballs. Office: 4800 Oak Grove Dr MS 183-401 Pasadena CA 91109

SEKSARIA, DINESH CHAND, mechanical engineer, consultant; b. Agra, India, Apr. 20, 1944; came to U.S., 1963, naturalized, 1976; s. Gopal Das and Krishna Devi (Kejriwal) S.; m. Madhuri Raniwala, June 2, 1971; children: Shikha, Priti. B.Sc., Agra U., 1962; B.S.M.E., Chgo. Tech. Coll., 1964; M.S.E., U. Mich., 1966. Registered profl. engr., Ohio. Jr. designer M.P.L. Inc., Chgo., 1965; analysis engr. Nat. Waterlift Co. div. Pneumodynamics, Kalamazoo, 1967-71; chief engr. structures and materials Crane div. Clark Equipment Co., Lima, Ohio, 1971-78, tech. specialist II, 1971-78. Mem. ASME (vice chmn. Central Mich. sect. 1967), Soc. Automotive Engrs., India Assn. Kalamazoo. Hindu. Club: Lima Internat. (pres. 1973). Subspecialties: Theoretical and applied mechanics; Solid mechanics. Current work: Structural and material science. Structural and material evaluation of industrial products. Office: 525 N 24th St Battle Creek MI 49016

SELANDER, ROBERT KEITH, biologist; b. Garfield, Utah, July 21, 1927; s. Clyde Service and Genevieve S.; m. Bonnie Jean Slater, Sept. 21, 1952; children—David Martin, Jennifer Jean. B.S., U. Utah, 1950, M.S., 1951; Ph.D. in Zoology (Thompson Meml. fellow 1952, NSF fellow 1953, Univ. fellow 1955), U. Calif., Berkeley, 1956. Instr. U. Tex., Austin, 1956-58, asst. prof. zoology, 1959-63, asso. prof., 1964-67, prof., 1968-74; prof. biology U. Rochester, 1974—, chmn. dept. biology, 1976-78; research fellow Am. Mus. Natural History, 1961; Guggenheim fellow U. Calif., Berkeley, 1965; Herbert W. Rand fellow Marine Biol. Lab., Woods Hole, Mass., summer 1971. Contbr. numerous articles to profl. publs.; asso. editor: Evolution, 1972-74; editorial bd.: Evolutionary Theory, 1973—. Served with U.S. Army, 1946. Recipient Walker prize Mus. Sci., Boston, 1969, Painton award Cooper Ornithol. Soc., 1970; co-recipient Elliott Coues award Am. Ornithol. Soc., 1975; NSF grantee, 1956—; NIH grantee, 1968—; NATO grantee, 1977-78. Fellow Am. Ornithologists' Union, AAAS; mem. Am. Inst. Biol. Scis., Am. Soc. Human Genetics, Am. Soc. Mammalogists, Am. Soc. Naturalists, Animal Behavior Soc., Genetics Soc. Am., Soc. Study Evolution (pres. 1976), Soc. Systematic Zoology (editorial bd. 1973-76), Phi Beta Kappa. Subspecialty: Evolutionary biology. Home: 47 S Ridge Trail Fairport NY 14450 Office: U Rochester Dept Biology Rochester NY 14627

SELDIN, DONALD WAYNE, physician, educator; b. N.Y.C., Oct. 24, 1920; s. Abraham L. and Laura (Ueberal) S.; m. Muriel Goldberg, Apr. 1, 1943; children: Leslie Lynn, Donald Craig, Donna Leigh. B.A., NYU, 1940; M.D., Yale U., 1943; D.H.L., So. Meth. U., 1977; D.Sci., Med. Coll. Wis., 1980. Diplomate: Am. Bd. Internal Medicine (test com. on nephrology 1970-73). Intern New Haven Hosp., Yale U., 1943-44, resident, 1944-46, instr. medicine, 1948-50, asst. prof. internal medicine, 1950-51; mem. faculty U. Tex. Southwestern Med. Sch., Dallas, 1951—, William Buchanan prof. internal medicine, 1969—, chmn. dept. internal medicine, 1952—; chief med. service Parkland Meml. Hosp., Dallas, 1952—; chmn. dept. medicine Lisbon VA Hosp., Dallas; cons. Baylor Hosp., St. Paul's Hosp., Presbyn. Hosp., Dallas; Brooke Army Hosp., Fort Sam Houston, Walter Reed Army Hosp., Washington, also to Surgeon Gen. U.S., Surgeon Gen. USAF, and Eli Lilly Co., 1972—; mem. Bur. Budget, Exec. Office of Pres., 1966-67; chmn. dialysis and transplantation com. of sci. adv. bd. Nat. Kidney Found.; mem. bd. sci. councillors Nat. Inst. Arthritis and Metabolic Diseases, NIH, 1968-71; trustee Rand Corp., 1975—. Editorial bd.: Jour. Lab. and Clin. Medicine, 1958-60, Nephron, The Clinician, Medicine, Mineral and Electrolyte Metabolism, 1977-79; cons. editor: Am. Jour. Medicine; asso. editor: Kidney Internat., 1973-79; contbr. articles to profl. jours. Served as capt. U.S. Army, 1946-48. Recipient Disting. Achievement award Modern Medicine, 1977; Friedrich Von Muller hon. lectr. U. Munich, 1968. Fellow Royal Soc. Medicine, Am. Acad. Arts and Sci.; mem. Dallas County Med. Soc., Tex. Med. Assn., Dallas Diabetes Assn., So. Soc. Clin. Investigation (pres. 1964, Founders medal 1975), Central Soc. Clin. Research (pres. 1963), AMA, Am. Fedn. Clin. Research, Am. Soc. Clin. Investigation (pres. 1966), Assn. Profs. Medicine (pres. 1971, Robert H. Williams Disting. Chmn. Medicine award 1977), Assn. Am. Physicians (pres. 1980), ACP (master, disting. teaching award 1980), Am. Physiol. Soc., Am. Soc. Nephrology (pres. 1968), Nat. Kidney Found. (David Hume award 1981), Am. Heart Assn., Am. Clin. and Climatol. Assn., Soc. Med. Cons. to Armed Forces, Internat. Soc. Nephrology (councillor 1973-78, pres.-elect 1981), Australian Soc. Nephrology (hon.), Alpha Omega Alpha. Subspecialties: Internal medicine; Nephrology. Office: 5323 Harry Hines Blvd Dallas TX 75235

SELDNER, MICHAEL, systems engr.; b. Los Angeles, July 24, 1948; s. Abraham and Esther (Wachtel) S.; m. Elaine Carlson, Sept. 24, 1967; children: Eric, Jenny. A.B., Rutgers U., 1972; Ph.D., Princeton U., 1977. Research assoc. dept. physics Princeton (N.J.) U., 1977-81; mem. tech. staff Bell Telephone Labs., Holmdel, N.J., 1981—. Contbr. articles to profl. jours. Mem. Am. Astron. Soc., Phi Beta Kappa. Subspecialties: Software engineering; Cosmology. Current work: Remote equipment monitoring, distributed processing, data analysis, system definition. Office: Bell Labs Holmdel NJ 07733

SELF, HAZZLE LAFAYETTE, animal sci. educator; b. Erath County, Tex., Aug. 1, 1920; s. Hazzle K. and Ethel F. (Dowdy) S.; m. Martha Louise Smith, Oct. 16, 1943; children: Linda Lindell, Debra Parker, Ann Paterson, Michael Self. Cert. animal scientist. Asst. prof. animal husbandry Tarleton State U., 1948-52; asst. prof. U. Wis., 1954-59; prof. Iowa State U., 1961—, state extension swine specialist, 1959-61; dir. Outlying Research Ctr., 1961—. Contbr. to profl. jours. Served with U.S. Army, 1944-46. Recipient Iowa Cattlemen's Assn. Appreciation award, 1971. Mem. Am. Soc. Animal Sci. (Livestock Mgmt. award 1979), Am. Registry Cert. Animal Scientists AAAS, Sigma Xi, Alpha Gamma Rho, Sigma Phi. Methodist. Lodge: Masons. Subspecialties: Animal physiology; Integrated systems modelling and engineering. Current work: Reproductive management in beef cows; feedlot management of cattle; administrative research centers in Iowa; management beef cattle. Home: 2221 Clark Ave Ames IA 50010 Office: 20 Curtis Hall Iowa State U Ames IA 50011

SELIGER, HOWARD H., educator; b. Bklyn., Dec. 4, 1924; s. Morton and Essie (Datz) S.; m. Beatrice Semel, Jan. 15, 1944; children—Carol Ann, Susan Diane. B.S., Coll. City N.Y., 1943; M.S., Purdue U., 1948; Ph.D., U. Md., 1954. Supervisory physicist Nat. Bur. Standards, Washington, 1948-60; asso. prof. biology McCollum Pratt Inst., Johns Hopkins, Balt., 1962-68, prof. biology, 1968—, radiation safety officer, 1962—; Pres. Radiation Instrument Co., 1952-62; Mem. environmental research guidance com. Md. Acad. Sci. Author: (with W.D. McElroy) Light, Physical and Biological Action, 1965, Ecology of Bioluminescent Bays, Phytoplankton Ecology. Served to 1st lt. USAAF, 1943-46. Guggenheim fellow, 1958-59; recipient Meritorious Service award Dept. Commerce, 1958. Fellow Am. Phys. Soc.; mem. Am. Soc. Photobiology (pres. 1980-81). Subspecialty: Ecology. Spl. research devel. bonsai in artificial environment. Home: 1805 W Rogers Ave Baltimore MD 21209

SELIGSON, M. ROSS, psychologist; b. Balt., May 18, 1949; s. Joseph Jerome and Dorothy G. (Greenfeld) S. B.A., U. Md., 1971; Ed.M., Loyola Coll., Balt., 1975; Ph.D., Calif. Sch. Profl. Psychology, 1979. Lic. psychologist, Fla. Clin. psychology intern Long Beach (Calif.) Neuropsychiat. Inst. and Hosp., 1975-76, Orange Coast Coll. Student Health Ctr., Costa Mesa, Calif., 1976-77; psychology field trainee Sect. on Legal Psychiatry UCLA, 1977-78; clin. psychologist Logansport (Ind.) State Hosp. (Isaac Ray Unit), 1978-80, South Fla. State Hosp., Hollywood, 1980—, Counseling Affiliates, Inc., Ft. Lauderdale, 1981—; chmn. exec. com. Am. Med. Health Plan, Miami, 1982—; clin. dir. North Miami Community Mental Health Ctr.; staff writer Women's Issues, Ft. Lauderdale, 1982—; adj. faculty Barry U., 1982—, Nova U., 1982—. Contbr. articles to profl. jours. Mem. North Area Adv. Bd. of Regional Mental Health Team for Newport Beach, Costa Mesa and Irvine, Calif., 1976. Calif. Sch. Profl. Psychology scholar, 1976; recipient Certificate of Appreciation Am. Bus. Women's Assn., 1982. Mem. Wash. Area Counsel on Alcohol and Drug Abuse, Calif. Psychol. Assn., Am. Psychol. Assn., Fla. Assn. Practicing Psychologists. Democrat. Jewish. Subspecialties: Behavioral psychology. Current work: Psychological adjustment, social history factors and training performance measures as predictors of suicide prevention. Office: Counseling Affiliates Inc 3891 Stirling Rd #5&6 Fort Lauderdale FL 33312

SELIKOFF, IRVING JOHN, physician; b. Bklyn., Jan. 15, 1915; s. Abraham and Tilli (Katz) S.; m. Celia Shiffrin, Feb. 4, 1946. B.S., Columbia, 1935; M.D., Anderson Coll. Medicine, U. Melbourne, Australia, 1941; Sc.D. (hon.), Tufts U., 1976, Bucknell U., 1979, N.J. State Coll., 1980. Diplomate: Am. Bd. Preventive Medicine. Intern Newark Beth Israel Hosp., 1943-44; asst. morbid anatomy Mt. Sinai Hosp., N.Y.C., 1941, now attending physician community medicine; prof. medicine and community medicine Mt. Sinai Sch. Medicine, City U. N.Y.; prof. and chmn. div. environmental medicine, dir. environmental scis. lab.; resident Sea View Hosp., N.Y.C., 1944-46; cons. medicine Barnert Meml. Hosp., Paterson, N.J.; mem. Nat. Cancer Adv. Bd., 1979—; pres. Collegium Ramazzini, 1982—. Author: The Management of Tuberculosis, 1956, Biological Effects of Asbestos, 1965, Toxicity of Vinyl Chloride and Polyvinyl Chloride, (with D.H.K. Lee) Asbestos and Disease, 1978, Health Hazards of Asbestos Exposure, 1979; also articles profl. jours.; Sr. asso. editor: Mt. Sinai Jour. Medicine, N.Y.C.; editor-in-chief: Am. Jour. Indsl. Medicine, Environ. Research, 1970—. Recipient Lasker award in Medicine, 1955; Ann. Research award Am. Cancer Soc., 1976; Edwards medal Welsh Nat. Sch. Medicine, 1977. Fellow Am. Pub. Health Assn., Am. Coll. Chest Physicians (pres. N.J. chpt. 1954-55); mem. N.Y. Pub. Health Assn. (Haven Emerson award 1975), N.Y. Acad. Scis. (pres. 1969-70, bd. govs. 1970—, Poiley award and medal 1974), AMA, N.J., Passaic County med. socs., Inst. Medicine Nat. Acad. Scis., Soc. Occupational and Environ. Health (pres. 1973-74), Am. Coll. Toxicology (pres. 1980). Subspecialty: Environmental medicine. Research in treatment Tb, 1945; with Dr. Edward H. Robitzek introduced isoniazid and iproniazid, chemotherapy of Tb, 1951-52. Office: Mount Sinai Med Center 10 E 102d St New York NY 10029

SELING, THEODORE VICTOR, engineer; b. Lansing, Mich., Mar. 27, 1928; s. Ernest and Alice Helen (Venzke) S.; m. Gwendolyn Cleone Ludwig, Dec. 6, 1952; children: Stanley Ernest, Stuart Alan. B.S. in E.E, Mich. State U.; M.S.E. in Instrumentation, U. Mich., 1960; Ph.D. in E.E, U. Mich., 1969. With AC Spark Plug div. Gen. Motors Corp., Flint, Mich. and Milw., 1952-61; with Def. Systems Lab., Santa Barbara, Calif., 1961-62; research scientist radio astronomy U. Mich., 1962-82; chief engr. radio astronomy Calif. Inst. Tech., 1982—. Contbr. articles to profl. jours. Served with Signal Corps U.S. Army, 1950-52. Mem. IEEE, Am. Astron. Soc. Subspecialties: Radio and microwave astronomy; Electronics. Current work: Research in radio astronomy instrumentation; low noise receiving systems, interferometers very long baseline interferometry. Patentee in field. Home: 165 S Lima St Sierra Madre CA 91024 Office: Calif Inst Tech 105-24 Pasadena CA 91125

SELINGER, PATRICIA GRIFFITHS, computer scientist; b. Cleve., Oct. 15, 1949; d. Fred Robert and Olive Mae (Brewster) Priest; m. Robert David Selinger, July 22, 1978; 1 son, David Robert. A.B., Radcliffe Coll., 1971; S.M., Harvard U., 1972, Ph.D., 1975. Mem. research staff IBM, San Jose, Calif., 1975—, mgr. R* distributed data base, 1978-82; mgr. Office Systems Lab., 1983—; vice chmn. Spl.

Interest Group on Mgmt. of Data, Assn. Computing Machinery. Contbr. articles to profl. jours. Recipient Outstanding Contbn. award IBM, 1979. Mem. Assn. Computing Machinery, SIGMOD, Phi Beta Kappa. Subspecialties: Database systems; Office systems. Current work: Relational database management systems, research and development in office systems technology. Office: IBM Research Lab 5600 Cottle Rd San Jose CA 95193

SELKURT, EWALD ERDMAN, physiologist, emeritus educator; b. Edmonton, Alta., Can., Mar. 13, 1914; s. Ephraim and Amanda Olga (Stirle) S.; m. Ruth Marion Gesley, June 21, 1941; children: Claire Elaine, Sylvia Anne. B.A., U. Wis., 1937, M.A., 1939, Ph.D., 1941. Instr. physiology N.Y. U. Sch. Medicine, 1941-44; sr. instr. Western Res. U. Sch. Medicine, Cleve., 1944-46, asst. prof., 1946-49, asso. prof., 1949-58; prof., chmn. dept. physiology Ind. U. Sch. Medicine, Indpls., 1958-80, disting. prof., 1976-81, disting. prof. physiology emeritus, 1981—; Centennial vis. prof. physiology Ohio State U. Sch. Medicine, 1970; mem. Josiah Macey Conf. on Kidney, 1949-53, Conf. Shock and Circulatory Homeostasis, 1949-53; mem. subcom. on shock Com. Med. Scis., NRC, 1953-57; mem. cardiovascular study sect. Nat. Heart Inst., NIH, 1963-68; panel mem. evaluation of sci. faculty fellowships NSF, 1963-64; mem. kidney council Unitarian Service Com., Germany, 1954. Author: Textbook of Physiology, 1963, Basic Physiology for Health Sciences, 1975; cons. editor: Am. Jour. Physiology and Jour. Applied Physiology, 1963-67, 70-73, Am. Heart Jour, 1966-77; procs.: Soc. Exptl. Biology and Medicine, 1967—, Circulatory Shock, 1974—; Editorial com.: Ann. Rev. Physiology, 1965-69. Active Indpls. Symphonic Chorus. NSF fellow, Göttingen, W.Ger., 1964-65. Mem. Am. Soc. Nephrology, AAUP, Am. Heart Assn., Am. Physiol. Soc. (council 1971-74, pres. elect 1975), Harvey Soc., Assn. Chmn. Depts. Physiology (councilor 1967-70, pres. 1971-72, Service award 1979), Soc. Exptl. Biology and Medicine (council 1978—), Phi Beta Kappa, Sigma Xi, Phi Sigma, Gamma Alpha, Pi Kappa Delta. Lutheran. Club: Singer's (Cleve.). Subspecialties: Nephrology; Physiology (medicine). Current work: Renal blood flow and electrolyte handling; kidney function in hemorrhagic shock; renin production: role of histamine; prostaglandins and renal function. Home: 3269 W 42d St Indianapolis IN 46208

SELL, KENNETH WALTER, govt. adminstr., pediatrician; b. Valley City, N.D., Apr. 29, 1931; s. Walter Robert and Patricia Haldora (Gottskalkson) S.; m., Dec. 20, 1950; children—Gregory, Thomas, Barbette, Susan. B.A., U. N.D., 1953, B.S., 1954; M.D., Harvard U., 1956; Ph.D., Cambridge (Eng.) U., 1968. Diplomate: Am. Bd. Pediatrics. Commd. capt. M.C. U.S. Navy, 1956; intern (Nat. Naval Med. Center), Bethesda, Md., 1956-57, resident, pediatrician, 1956-65, dir., 1965-70, chmn. dept. clin. and exptl. immunology, 1970-74, inst. comdg. officer, 1974-77, ret., 1977; sci. dir. Nat. Inst. Allergy and Infectious Diseases, Bethesda, 1977—; clin. prof. pediatrics Georgetown U. Editor: Tissue Banking for Transplantation, 1976. Trustee Christ Lutheran Ch., Bethesda. Decorated Legion of Merit. Fellow Am. Acad. Pediatrics, Am. Acad. Pathology; mem. AMA, Brit. Soc. Immunology, Tissue Culture Assn., Soc. Cryobiology, Transplantation Soc., Am. Assn. Immunologists, Soc. Exptl. Hematology, Am. Assn. Tissue Banks (pres.), Am. Coll. Pathology, Am. Soc. Microbiology, Phi Beta Kappa, Sigma Xi, Phi Beta Pi. Subspecialty: Immunology (medicine). Home: 12311 Old Canal Rd Rockville MD 20854 Office: NIH Nat Inst Allergy and Infectious Diseases Bldg 5/137 Bethesda MD 20205

SELLERS, GREGORY JUDE, physicist; b. Far Rockaway, N.Y., June 20, 1947; s. Douglas L. and Rita R. (Dieringer) S. A.B. in Physics, Cornell U., 1968, M.S., U. Ill., 1970, Ph.D., 1975. Sr. scientist B-K Dynamics, Inc., Rockville, Md., 1974-76; with Allied Corp., Morristown, N.J., 1976—, applications physicist, 1977—; dir. Thermo-Tek, Inc., Madison, N.J. Mem. AAAS, Am. Phys. Soc., IEEE, Soc. Plastics Engrs. Subspecialties: Polymers; Materials. Current work: Development of applications for polymeric materials and glassy metals in the electrical and electronics area. Co-inventor adhesive bonding metallic glass, electromagnetic shielding, testing of thermal insulation, amorphous antipilferage marker, amorphous spring-shield. Home: PO Box 296 C Convent Station NJ 07961 Office: Allied Corp PO Box 2332 R Morristown NJ 07960

SELLIN, IVAN ARMAND, physicist, educator, researcher, cons.; b. Everett, Wash., Aug. 16, 1939; s. Petrus and Amelia Fanny (Josephson) S.; m. Helen K. Gill, June 16, 1962; children: Peter B., Frank E. Student, Harvard U., 1956-59; S.M., U. Chgo., 1960, Ph.D., 1964. Lectr., physicist U. Chgo., 1961-64, instr., research assoc., 1964-65; asst. prof. physics N.Y.U., 1965-67; research physicist Oak Ridge Nat. Lab., 1967-70, project dir., 1970—; assoc. prof. physics U. Tenn., Knoxville, 1970-74, prof., 1974—; cons., mem. coms. NRC. Editor 2 books; Contbr. numerous articles to physics jours., also encys., chpts. in books. Mem. adv. com. various pvt. and govt. agys.; Mgr. chamber music and concert series Oak Ridge Civil Music Assn., 1972-75. Recipient Chancellor's Research Scholar award U. Tenn., 1976; Alexander von Humboldt Found. Sr. U.S. Scientist award, W.Ger., 1977; sr. Fulbright-Hays grantee, 1977-78; 13 grants NSF, Dept. Energy, Office Naval Research, NASA, Oak Ridge Nat. Lab., 1972—. Fellow Am. Phys. Soc. (council 1979-83). Subspecialty: Atomic and molecular physics. Current work: Research in physics of highly ionized matter, high-velocity ion-atom and ion-solid collisions, and stored ion spectroscopy. Home: 1008 W Outer Dr Oak Ridge TN 37830 Office: Bldg 5500 PO Box X Oak Ridge TN 37830

SELLMYER, DAVID JULIAN, physics educator; b. Joliet, Ill., Sept. 28, 1938; s. Marcus Leo and Della Louise (Plumhoff) S.; m. Catherine Joyce Zakas, June 16, 1962; children: Rebecca Ann, Julia Maryn, Mark Anthony. B.S., U. Ill., 1960; Ph.D., Mich. State U., 1965. Asst. prof. MIT, Cambridge, Mass., 1965-72, assoc. prof., 1972; assoc. prof. physics U. Nebr., Lincoln, 1972-75, prof., 1975—, chmn. dept. physics and astronomy, 1978—; cons. Dale Electronics, Norfolk, Nebr., 1979—. Author: Solid State Physics, 1978, Landolt Bornstein, 1980. Recipient Tech. award NASA, 1972. Fellow Am. Phys. Soc.; mem. AAAS, Sigma Xi, Sigma Tau. Subspecialties: Condensed matter physics; Amorphous metals. Current work: Electronic structure and properties of metallic glasses, intermetallic compounds, low dimensional conductors, amorphous magnetism. Home: 2700 Rathbone Rd Lincoln NE 68502 Office: Dept Physics and Astronomy U Nebr Lincoln NE 68588

SELMANOWITZ, VICTOR JOEL, dermatologist; b. Bklyn., Mar. 30, 1938; s. Louis and Ethel (Simon) S. B.H.L., Yeshiva U., 1959; M.D., SUNY-Downstate Med. Ctr., 1962. Diplomate: Am. Bd. Dermatology. Intern Brookdale Hosp., 1962-63; resident NYU Med. Ctr., 1963-66; prof. and vice chmn. dept. dermatology N.Y. Med. Coll., Valhalla and Met. Hosp. Center, N.Y.C., 1978—; cons. Animal Med. Ctr., N.Y.C., 1968—; chmn. dept. dermatology Flower Hosp., N.Y.C., 1982—. Contbr. numerous articles on dermatology to med. jours.; co-author med. exam. rev. book in dermatology. Served to capt. USAF, 1966-68. Fellow ACP, AM.Acad. Dermatology, Am. Coll. Angiology; mem. Gerontol. Soc., Internat. Soc. Tropical Dermatology. Subspecialties: Dermatology; Physical chemistry. Current work: Genetics (genodermatology), gerontology, general dermatology. Office: NY Med Coll Met Hosp Center 1901 First Ave New York NY 10029

SELTZER, BENJAMIN, physician; b. Phila., Aug. 5, 1945; s. Albert and Sylvia S.; m. Natalie Ross, Oct. 13, 1924; children: Daniel, Jennifer. A.B., U. Pa., 1965; M.D., Jefferson Med. Coll., 1969. Diplomate: Am. Bd. Psychiatry and Neurology. Intern Boston City Hosp., 1969-70, resident in neurology, 1979-73; fellow, aphasia unit Boston VA Hosp., 1973; asst. attending neurologist Boston City Hosp., 1973-75; assoc. neurologist Beth Israel Hosp., Boston, 1975—; clin. fellow Harvard Med. Sch., 1970-73, lectr., 1978—; asst. prof. Boston U. Med. Sch., 1979—; neurologist geriatric research unit VA Hosp., Bedford, Mass., 1975—. Contbr. articles to profl. jours. Fellow Am. Acad. Neurology, Royal Soc. Medicine; mem. Soc. Neurosci., Internat. Neuropsychol. Soc., Phi Beta Kappa. Subspecialties: Neurology; Neuroanatomy. Current work: Connections of the cerebral cortex; human organic mental disorders. Home: 31 Page Rd Bedford MA 01730 Office: VA Hospital 200 Springs Rd Bedford MA 01730

SELTZER, RAYMOND EUGENE, agricultural economist, consultant; b. Elmwood, Ill., Apr. 29, 1918; s. Lester E. and Blanche (Laws) S.; m. Holly Jean Seltzer, July 5, 1943; children: Elizabeth, Stephen, Richard. B.S., U. Ill., Urbana, 1940; M.S. in Agrl. Econs., Kans. State U., Manhattan, 1942, Ph.D., U. Calif., Berkeley, 1954. Head dept. agrl. econs. U. Ariz., Tucson, 1947-61; chmn. bd., chief exec. officer Devel. Planning & Research Assocs., Inc., Manhattan, 1961-83; cons., 1983—; dir. Kans. State U. Research Found. Contbr. numerous articles to profl. jours. Served to maj. AUS, 1942-47. Decorated Purple Heart, Bronze Star. Mem. Am. Agrl. Econs. Assn., Internat. Assn. Agrl. Economists. Subspecialty: Agricultural economics. Current work: International economic development marketing, environmental economics. Home: 815 Wildcat Ridge Manhattan KS 66502

SELTZER, STEPHEN MICHAEL, physicist; b. N.Y.C., May 7, 1940; s. Morris and Henrietta (Seigel) S.; m. Francine Lynne Arthur, June 4, 1961; children: Lisa Norma, Holly Victoria. B.S., Va. Poly Inst., Blacksburg, 1962; M.S., U. Md.-College Park, 1973. Physicist Nat. Bur. Standards, Washington, 1962—. Contbr. numerous articles to profl. jours. Mem. Am. Phys. Soc., Am. Nuclear Soc., Sigma Pi Sigma, Phi Kappa Phi. Subspecialty: Theoretical physics. Current work: Transport of radiation (electrons, photons, protons) through matter; work includes development of cross section data and of analytical and monte carlo methods for transport calculations. Home: 1709 Glastonberry Rd Potomac MD 20854 Office: Nat Bur Standards Washington DC 20234

SEMBA, KAZUE, neuroscientist, educator; b. Tsuchiura, Japan, Jan. 13, 1949; d. Tadao and Tadako (Hirata) Numakura; m. Masayoshi Semba, Mar. 20, 1972. B.Ed., Tokyo U. Edn., 1971, M.A., 1973; Ph.D., Rutgers U., 1979. Postdoctoral fellow Iowa State U.-Ames, 1979-80; research specialist III Coll. Medicine and Dentistry of N.J.-Rutgers Med. Sch., Piscataway, N.J., 1980-81, instr., 1981—. Contbr. articles to profl. jours., conf. procs. Recipient Young Psychologist award Internat. Congress of Psychology, 1976; Japan Soc. of N.Y. grantee, 1978. Mem. Soc. Neurosci., Am. Psychol. Assn., AAUP, Sigma Xi. Subspecialties: Anatomy and embryology; Neurophysiology. Current work: Sensory mechanisms in spinal cord; sensorimotor control of the vibrissae. Office: Dept Anatomy UMDNJ-Rutgers Med Sch Piscataway NJ 08854

SEMKEN, HOLMES ALFORD, JR., geology educator, researcher; b. Knoxville, Tenn., Jan. 28s, 1935; s. Holmes Alford and Edith (Klinck) S.; m. Alma Elaine Friedrichs, Aug. 31, 1957; children: Steven Holmes, David Andrew. B.S. in Geology, U. Tex.-Austin, 1958, M.A., 1960, Ph.D., U. Mich., 1965. Mus. technician Tex. Meml. Mus., Austin, 1953-60; technician Smithsonian Instn., Washington, 1960-61; teaching asst. U. Mich., 1961-65; asst. prof. geology U. Iowa, 1965-69, assoc. prof., 1969-73, prof., 1973—; cons. in zooarchaeology, 1976—; mem. exec. com. Iowa Hall Project, Mus. Natural History exhibit hall, 1980—; co-chmn. Plains Conf., 1980. Contbr. articles to profl. jours.; editor, contbg. author: The Cherokee Excavations, 1980; assoc. dir.: film Iowa's Ancient Hunters, 1979. Served to 1st lt. C.E. U.S. Army, 1960. Recipient E. C. Case award U. Mich., 1974; prin. investigator NSF, 1969, 74, 76. Fellow Geol. Soc. Am.; mem. Internat. Union Quaternary Research, Am. Soc. Mammalogists, Soc.,Vertebrate Paleontology, Am. Quaternary Assn., Paleontol. Soc. Presbyterian. Clubs: Investment (Iowa City) (sec. 1976-81); Co. Mil Historians.). Subspecialties: Paleobiology; Paleoecology. Current work: Pleistocene and Holocene paleoecology and paleobiology of fossil vertebrates; zooarchaeology. Home: 1725 Winston Dr Iowa City IA 52240 Office: Dept Geology U Iowa Iowa City IA 52242

SEN, GANES C(HANDRA), molecular biologist, researcher, educator; b. Varanasi, India, Jan. 17, 1945; came to U.S., 1974; s. Tarapada and Kanak (Gupta) S.; m. Indira, Nov. 18, 1973; children: Srijan, Ritu. Ph.D., McMaster U., Hamilton, Ont., Can., 1974. Research assoc. Yale U., 1976-78; asst. mem. Sloan-Kettering Cancer Center, N.Y.C., 1978—; asst. prof. molecular biology Cornell U. Grad. Sch. Med. Schs., N.Y.C., 1979—. Contbr. numerous articles to profl. publs. Mem. Am. Soc. Microbiology, Am. Soc. Biol. Chemists, Am. Soc. Virology, N.Y. Acad. Scis. Subspecialties: Molecular biology; Virology (biology). Current work: Mechanism of action of interferons. Home: 504 E 81st St New York NY 10028 Office: 1275 York Ave New York NY 10021

SENDAX, VICTOR IRVEN, dentist, researcher, educator; b. N.Y.C., Sept. 14, 1930; s. Maurice and Molly R. S.; m. Deborah DeLand Cobb, Dec. 17, 1969 (div. June 1976); 1 dau., Jennifer Reiland. B.A., Washington Square Coll., NYU, 1951; D.D.S., Coll. Dentistry, 1955; postgrad., Harvard U. Sch. Dental Medicine, Boston, 1969-72. Commr. N.Y. State Dental Service Corp., N.Y.C., 1969-73; pres. Victor I. Sendax, D.D.S. (P.C.), N.Y.C., 1972—, Biodental Research Found., Inc., 1975—; adj. asst. prof. and dir. implant prosthodontics resident program Columbia U. Sch. Dental and Oral Surgery, 1974—; cons. in field. Author: Dental Implants and You, 1983; mem. editorial bd.: Oral Implantology, 1979—. Mem. adv. bd. Hist. Assn., L.I.; bd. dirs. Schola Cantorum, N.Y.C. Served to capt. Dental Corps USAF, 1955-57; Japan.. Recipient spl. recognition for significant contbrs. to dental health Am. Fund Dental Health, Chgo., 1982. Fellow Am. Coll. Dentists Internat. Coll. Dentists, Am. Acad. Implant Dentistry (pres. 1980-81, organizing chmn. 1st workshop-conf. on status of human oral implants 1976); mem. Am. Dental Assn. (ho. of dels. 1969), Am. Assn. Dental Schs. (chmn. spl. interest group 1982-83), Internat. Assn. Dental Research, Am. Assn. Dental Research, Sigma Epsilon Delta. Clubs: The Players (N.Y.C.); United Brothers. Subspecialties: Implantology; Prosthodontics. Current work: Research, teaching, and clinical practice related to implant prosthodontics. Co-inventor anti-calculus agt. Home: 70 E 77th St Apt 6A New York NY 10021 Office: Victor I Sendax DDS 30 Central Park S Suite 14B New York NY 10019

SEN GUPTA, BARUN LUMAR, educator, geologist; b. Jamshedpur, India, July 31, 1931; came to U.S., 1968; s. Tarapada and Sulata (Das Gupta) Sen. G.; m. Mandira Gupta, May 12, 1956; children: Sagaree, Upal. B.Sc. with honors, Calcutta (India) U., 1951, M.Sc., 1954; M.S., Cornell U., 1961; Ph.D., Indian Inst. Tech., Kharagpur, 1963. Registered prof. geologist, Ga. Asst. lectr. to lectr. Indian Inst. Tech., 1955-66; postdoctoral fellow Bedford Inst. Oceanography, Dartmouth, N.S., Can., 1966-68; asst. prof. to prof. U. Ga., Athens, 1969-79; prof. geology La. State U., Baton Rouge, 1979—. Assoc. editor: Jour. Foraminiferal Research, 1975—; contbr. articles to profl. jours. NSF grantee, 1969-72, 75-79, 82—. Fellow Geol. Soc. Am., Cushman Found. for Foraminiferal Research; mem. Paleontol. Soc. Soc. Econ. Paleontologists and Mineralosists, AAAS. Subspecialty: Paleoecology. Current work: Ecology and paleoecology of modern and ancient benthic foraminifera; paleoceanography. Office: Dept Geolog La State U Baton Rouge LA 7080 Home: 292 Baird Dr Baton Rouge LA 70808

SENICH, GEORGE A., chemist; b. Pitts., Jan. 5, 1953; s. George Albert and Lillian Catherine (Hoellein) S. B.S. Engring., Case Western Res. U., 1974; M.S., U. Mass., Amherst, 1976, Ph.D., 1979. Chemist Nat. Bur. Standards, Washington, 1978-80, research chemist, 1980—. Mem. rev. bd.: Polymer jour, 1979—, Polymer Engring. and Sci. jour, 1980—; contbr. articles to profl jours. Mem. Am. Chem. Soc., Am. Phys. Soc., Chem. Soc. Washington, Younger Chemists Com. D.C., Sigma Xi, Tau Beta Pi. Club: Tip Toppers. Subspecialties: Polymer chemistry; Polymers. Current work: Diffusion by chromatography with polymers; dynamic mechanical relaxation processes in linear and network polymers; radiation curing kinetics of resins; thermodynamics of oligomer/polymer systems. Office: Nat Bur Standards Bldg 224 Room B320 Washington DC 20234

SENIOR, DAVID FRANK, veterinarian; b. Melbourne, Australia, Aug. 16, 1946; came to U.S., 1975; s. Frank and Jean Helen (Houston) S.; m. Paulette Wilson, May 13, 1968; children: Darren, Justin. B.V.Sc., U. Melbourne, 1969. Diplomate: Am. Coll. Vet. Internal Medicine. Intern U. Melbourne, 1969-70; co-dir. Wetaskiwin (Alta., Can.) Vet. Hosp. Ltd., 1970-75; resident Sch. Vet. Medicine, U. Pa., Phila., 1975-77, instr., 1977-79; asst. prof. dept. med. scis. Coll. Vet. Medicine, U. Fla., Gainesville, 1979—. Mem. Am. Coll. Vet. Internal Medicine, AVMA. Subspecialties: Internal medicine (veterinary medicine); Urology. Current work: Veterinary urology. Office: U Fla J-126 JHMHC Gainesville FL 32610

SENSENY, PAUL EDWARD, cons. mech. engr.; b. Chambersburg, Pa., June 14, 1950; s. Paul Edward and Wilda Lucille (Kump) S.; m. Theresa Marie Hughes, Aug. 19, 1972; 1 dau., Jennifer Elaine. B.S.M.E., U. Pa., 1972; Ph.D., Brown U., 1976, Sc. M., 1974. Research engr. SRI Internat., Menlo Park, Calif., 1976-79; resident research engr. RE/SPEC Inc., Rapid City, S.D., 1979-80, mgr. materials lab., 1980—. Author numerous tech. reports; contbr. articles to profl. jours., symposia procs. Mem. Soc. Engring. Sci., Soc. Exptl. Stress Analysis, Am. Geophys. Union, ASME, ASCE. Republican. Lodge: Elks. Subspecialty: Solid mechanics. Current work: Constitutive modeling of rate-dependent deformation. Home: 3718 Brookside Dr Rapid City SD 57701 Office: PO Box 725 Rapid City SD 57709

SENTERFIT, LAURENCE BENFRED, microbiologist; b. Sarasota, Fla., July 30, 1920; s. Wallace L. and Eulalie V. (Beyer) S.; m. Catherine Anne Reed, Apr. 13, 1957; children: James, David, Reed. B.S., U. Fla., 1949, M.S., 1950; D.Sc., Johns Hopkins U., 1955. Diplomate: Am. Bd. Microbiology. Instr. Johns Hopkins U., Balt., 1955-58; chief microbiologist Charlotte (N.C.) Meml. Hosp., 1958-64; research scientist Pfizer Co., Terre Haute, Ind., 1964-65; assoc. prof. pathology St. Louis U., 1965-67; chief microbiologist St. Johns Mercy Hosp., St. Louis, 1967-80; dir. lab. microbiology, assoc. prof. Cornell U., N.Y. Hosp., N.Y.C., 1970—. Mem. Am. Assn. Immunologists, Am. Soc. Microbiology, Assn. Clin. Scientists, Venereal Disease Soc. Am., Soc. Explt. Biology and Medicine, Internat. Organ. Mycoplasmology. Democrat. Roman Catholic. Subspecialties: Microbiology (medicine); Immunology (medicine). Current work: Study of the antigens of mycoplasma pneumoniae, diagnostic uses of immunofluorescent technology in microbiology and virology. Office: Cornell U Med Coll 1300 York Ave New York NY 10021 Home: 174 Norman Dr Ramsey NJ 07446

SENTMAN, LEE HANLEY, aeronautical and astronautical engineering educator; b. Chgo., Jan. 27, 1937; s. Lee Hanley and Esther (Dore) S.; m. Mary Alice Lerch, June 6, 1959; children: Jeanne, Charles. B.S., U. Ill., 1958; Ph.D., Stanford U., 1965. Dynamics engr. Lockheed Missiles and Space Co., Sunnyvale, Calif., 1959-65; asst. prof. U. Ill., Urbana, 1965-69, assoc. prof., 1969-79; prof. aero. and astronautical engring., 1979—; vis. prof. aerospace engring. U. Ariz., Tucson, 1970-71; cons. in field. Contbr. articles to profl. jours. NRC assoc., 1972. Fellow AIAA (assoc.); mem. Optical Soc. Am., Am. Phys. Soc. Subspecialties: High energy lasers; Aerospace engineering and technology. Current work: Rotational nonequilibrium effects in chemical lasers; nonlinear interactions between chemical kinetics, fluid dynamics and optical resonator of fluid flow lasers. Home: Rural Route 1 Box 68 Dewey IL 61840 Office: Dept Aero and Astronautical Engring U Ill 113 Transp Bldg 104 S Mathews St Urbana IL 61801

SEPKOSKI, JOSEPH JOHN, JR., paleontology educator; b. Presque, Isle, Maine, July 26, 1948; s. Joseph John and Sally (Feuchtwanger) S.; m. Maureen Meter; 1 son, David Christopher. B.S., U. Notre Dame, 1970; Ph.D., Harvard U., 1977. Instr. U. Rochester, N.Y., 1974-77, asst. prof., 1977-78, U. Chgo., 1978-82, assoc. prof., 1982—. Editor Paleobiology, 1983—; Contbr. articles to profl. jours. Mem. AAAS, Paleontol. Soc., Sigma Xi. Subspecialty: Paleobiology. Current work: Study of large scale patterns in the history of life forms includes quantifying and modeling evolution of diversity and complexity in global ecological systems, identifying mass extinctions, documenting major changes in faunas. Office: Dept Geophys Scis U Chgo Chicago IL 60637

SEQUEIRA, LUIS, plant pathologist; b. San Jose, Costa Rica, Sept. 1, 1927; came to U.S., 1960, naturalized, 1967; s. Raul and Dora (Jenkins) S.; m. Elisabeth Steinvorth, May 27, 1954; children: Anabel, Marta, Roberto, Patricia. A.B., Harvard U., 1949, M.A., 1950, Ph.D.; Parker fellow, 1952. Plant pathologist, dir. Coto Research Sta., United Fruit Co., 1953-60; research asso. N.C. State U., Raleigh, 1960-61; prof. plant pathology and bacteriology U. Wis., Madison, 1961—, J.C. Walker prof. plant pathology and bacteriology, 1982—; cons. AID. Editor in chief: Phytopathology, 1979—. NSF sr. postdoctoral fellow, 1970-71. Fellow Am. Phytopath. Soc. (v.p. 1983); mem. Bot. Soc. Am., Am. Soc. Plant Physiologists, Council Biology Editors, Mycol. Soc. Am., Nat. Acad. Scis. Subspecialties: Plant pathology; Plant physiology (agriculture). Current work: Cell-cell interactions in host-parasite systems. Recognition phenomena and pathogenesis. Home: 10 Appomattox Ct Madison WI 53705 Office: 1630 Linden Dr Madison WI 53706

SERAVALLI, EGILDE PAOLA, immunologist, researcher; b. Florence, Italy, Sept. 26, 1937; d. Guido and Maria Pia (Nieri) S. D.Biol. Scis., Inst. Microbiology, U. Florence, 1963. Postdoctoral trainee Rockefeller U., 1965-68; research asst. N.Y. U., 1970-71; visitor investigator Sloan Kettering Meml. Cancer Ctr., N.Y.C., 1976-78; asst. prof. SUNY-Downstate Med. Sch., N.Y.C., 1978-82; research assoc., dir. research labs. anesthesia dept. Beth Israel Med. Ctr., N.Y.C., 1982—. Contbr. articles profl. jours., chpts. in books. NIH grantee; recipient Italian Lit. prize for articles on death and dying, 1983. Mem. N.Y. State Anesthesia Soc., Am. Soc. Anesthesiologists, Assn. Immunologists, AAAS. Subspecialties: Plant pathology; Cell and tissue culture. Current work: Effect of anesthetics on cel membranes and on cell Genomes; historical perspective on death and dying.

SERDULA, KENNETH JAMES, energy company executive, consultant; b. Esterhazy, Sask., Can., Apr. 29, 1936; s. John and Barbara (Tocher) S.; m. Florence Ann McLellan, Apr. 27, 1968; children: Claire, Jay. B.S. in Engring. Physics, U. Sask., 1959; M.S., U. Birmingham (Eng.), 1960, Ph.D., 1963. Registered profl. engr., Can. Research officer Chalk River (Ont.), Nuclear Labs., 1963-78; pres. Serdula Systems Ltd., Deep River, Ont., Can., 1978—, also chmn. Engring. Inst. Can. fellow, 1980; Atholone fellow U.K. Bd. Trade, 1959-61. Mem. 2Engring. Inst. Can. (regional exec. 1974-79, local exec. positions 1966-74), Can. Nuclear Soc., Am. Nuclear Soc. Clubs: Deep River Yacht and Tennis (officer 1966-69), Mt. Martin Ski (officer Deep River 1974-79). Subspecialties: Nuclear fission; Nuclear engineering. Current work: Research, design, development, commissioning and operation related to dynamics of nuclear systems with emphasis on control and safety, nuclear power feasibility and evaluation studies. Office: Serdula Systems Ltd PO Box 1808 Glendale Plaza Deep River ON Canada KOJ 1PO Home: 236 Thomas St Deep River ON Canada KOJ 1PO

SERENY, ARON, engring. scientist; b. Budapest, Hungary, Apr. 23, 1947; came to U.S., 1971, naturalized, 1978; s. Andrew and Elisabeth (Erdos) S.; m. Nelly Kwartler, Apr. 4, 1974; children: Carla, Andrea. B.Sc., Columbia U., 1973, M.Sc., 1974, D.Engring. Sci., 1977. Engring. cons., 1974—; mem. research staff N.Am. Philips Corp., Briarcliff Manor, N.Y., 1977-81; mem. research staff mech. engring. scis Palo Alto Research Center, Xerox Corp., North Tarrytown, N.Y., since 1981—; adj. prof. mech. engring. Columbia U., N.Y.C., 1980-81. Contbr. numerous articles on lubrication tech. to profl. jours. Mem. ASME, Sigma Xi, Tau Beta Pi. Subspecialties: Fluid mechanics; Numerical analysis. Current work: Research and consulting in field of mechanical engineering sciences. Office: 141 Webber Ave North Tarrytown NY 10591

SERLEMITSOS, PETER J., x-ray astronomer; b. Samos, Greece, Apr. 27 1933; came to U.S., 1951, naturalized, 1958; s. Jannetis P. and Theodora (Kavageorgiu) S.; m. Mary Xakellis, May 1, 1954 (div. 1977); children: John, Louann; m. Marigail Barcome, Mar. 27, 1978. B.S., Franklin and Marshall Coll., 1955; M.S., U. Md., 1964, Ph.D., 1966. Physicist Hamilton Watch Co., Lancaster, Pa., 1957-60; physicist Goddard Space Flight Ctr., Greenbelt, Md., 1961—. Author articles. Recipient John C. Lindsay award Goddard Space Flight Ctr., 1977. Mem. Am. Astron. Soc. Democrat. Greek Orthodox. Subspecialty: 1-ray high energy astrophysics. Current work: X-ray mirrors; high resolution spectrophotometry; Spacelab. Home: 212 Valleybrook Dr Silver Spring MD 20904 Office: Goddard Space Flight Ctr Code 661 Greenbelt MD 20771

SERRIN, JAMES BURTON, educator; b. Chgo., Nov. 1, 1926; s. James B. and Helen Elizabeth (Wingate) S.; m. Barbara West, Sept. 6, 1952; children—Martha Helen, Elizabeth Ruth, Janet Louise. Student, Northwestern U., 1944-46; B.A., Western Mich. U., 1947; M.A., Ind. U., Ph.D., 1951; D.Sc., U. Sussex, 1972. With Mass. Inst. Tech., 1952-54; faculty U. Minn., Mpls., 1955—, prof. math., 1959—, head, 1964-65, Regents prof., 1968—; Vis. prof. U. Chgo., 1964, 75, Johns Hopkins, 1966, Sussex U., 1967-68, 72, 76, U. Naples, 1979. Author: Mathematical Principles of Classical Fluid Mechanics, 1957; Co-editor: Archive for Rational Mechanics and Analysis. Mem. Met. Airport Sound Abatement Council, Mpls., 1969—. Recipient Disting. Alumni award Ind. U., 1979. Fellow AAAS; mem. Nat. Acad. Scis., Am. Math. Soc. (G.D. Birkhoff prize 1973), Math. Assn. Am., Soc. for Natural Philosophy (pres. 1969-70). Subspecialties: Applied mathematics; Thermodynamics. Current work: Foundations of continuum mechanics and of thermodynamics, with emphasis on the careful delineation of concepts and the theoretical roles of energy and entropy, continuum theory of phase transitions in fluids, both in equilibrium and in dynamic situations. Home: 4422 Dupont Ave S Minneapolis MN 55409

SERY, THEODORE WILSON, microbiologist; b. N.Y.C., Feb. 5, 1924; s. John and Rhoda (Saddler) S.; m. Doris Anne Dragner, Nov. 2, 1957; children: Paul, M. James, Elizabeth, Katherine. B.S., Columbia U., 1949; Ph.D., Cornell U. Med. Coll., 1954. Instr. Cornell U. Med. Coll., 1954-55; with research div. Wills Eye Hosp., Phila., 1955—, chief microbiology, sr. research scientist, 1965—, dir. research, 1971-80, dir. basic research, 1980—; research prof. ophthalmology Thomas Jefferson U., 1974—; Mem. study sect. Nat. Eye Inst., 1965-69, behavioral and neuroscis. study sect., 1981-85. Contbr. articles to profl. jours. Mem. N.J. Com. for Children with Spl. Learning Needs, 1981—. Served with AUS, 1943-46; ETO. Recipient Zentmeyer award Coll. Physicians, 1958; NIH grantee, 1958-63. Fellow Coll. Physicians Phila., Assn. Research in Vision and Ophthalmology; Mem. Am. Assn. Immunologists, Am. Soc. Microbiologists, N.Y. Acad. Sci., Am. Uveitis Soc., Tissue Culture Assn., Phila. Physiol. Soc., Am. Brittle Bone Soc. Subspecialties: Immunology (medicine); Cancer research (medicine). Current work: Immune mechanisms in the cornea, uvea and retina during virus and bacterial infections; photoradiation therapy of eye cancer. Home: 309 Warwick Rd Haddonfield NJ 08033 Office: Wills Eye Hosp 9th and Walnut Sts Philadelphia PA 19107

SESHADRI, SENGADU RANGASWAMY, electrical and computer engineering educator, researcher, writer; b. Madras, India, Oct. 25, 1928; came to U.S., 1955, naturalized, 1967; m. Susheela R. Chari, June 6, 1933. B.Sc. in Physics and Math with honors, Madras U., 1950, M.A., 1951; D.I.I.Sc. in Elec. Communication Engring, Indian Inst. Sci., Bangalore, India, 1952; Ph.D. in Engring. and Applied Physics, Harvard U., 1959. Lectr. Madras Inst. Tech., 1953-55; research asst. Harvard U., 1955-58, research fellow, 1959-60, 61-63; prin. sci. officer Electronics Research and Devel. Establishment, Bangalore, India, 1960-61; sr. engring. specialist Sylvania Applied Research Lab., Waltham, Mass., 1963-67; vis. prof. Toronto U., 1965-66; hon. research fellow Harvard U., 1963-67; prof. dept. elec. and computer engring. U. Wis., Madison, 1967—; vis. scientist GA Technologies, Inc., San Diego, 1982-83. Author: Fundamentals of Transmission Lines and Electromagnetic Fields, 1971, Fundamentals of Plasma Physics, 1973; contbr. articles to profl. jours. NSF Sr. postdoctoral fellow, 1970-71. Mem. Am. Phys. Soc., Optical Soc. Am., Electromagnetics Soc., IEEE. Subspecialties: Electrical engineering; Plasma. Current work: Research: surface acoustic waves, magnetic and integrated optical devices, electromagnetic theory and applied plasma physics; teaching: electromagnetic theory, plasma physics, solid state physics, linear and nonlinear systems. Home: 109 N Whitney Way Madison WI 53705 Office: Department Electrical and Computer Engineering The University of Wisconsin Madison WI 53706

SETHI, VIDYA SAGAR, cancer pharmacologist; b. Panjab, India, Dec. 1, 1937; came to U.S., 1971, naturalized, 1977; s. Dewan Ch and Ram Rakhi (Wadhwa) S.; m. Hema Lata Amarsingh, Jan. 19, 1969; children: Preeti Seema, Deepu Shikha. M.Pharmacy, Banaras Hindu U., India, 1960; Ph.D., U. Munich, W.Ger., 1964. Vis. scientist Max Planck Inst. Biochemistry, Munich, 1968-71; head sect. molecular pharmacology Litton Bionetics, Bethesda, Md., 1973-76; research assoc. prof. medicine Bowman Gray Sch. Medicine, Winston-Salem, N.C., 1977—; also dir. Clin. Pharmacology Lab., Oncology Research Center.; Pres. Indo-U.S. Cultural Assn., Winston-Salem, 1981; v.p. Unitarian Universalist Fellowship, Winston-Salem, 1982—. Contbr. articles to profl. jours. Mem. Am. Assn. Biol. Chemists, Am. Assn. Cancer Research, Am. Assn. Pharmacology and Exptl. Therapeutics, Am. Microbiol. Assn., Am. Chem. Soc., AAAS. Democrat. Hindu-Unitarian-Universalist. Subspecialties: Cancer research (medicine); Pharmacology. Current work: Biochemical, preclinical and clinical pharmacology and metabolism of anticancer drugs. Home: 636 Friar Tuck Rd Winston-Salem NC 27104 Office: Bowman Gray Sch Medicine Winston-Salem NC 27103

SETHNA, PATARASP RUSTOMJI, mechanics educator; b. Bombay, India, May 26, 1923; came to U.S., 1947, naturalized, 1965; s. R.E. and Naja (Engineer) S.; m. Shirley S. Smith, Sept. 4, 1954; children: James, Michael, Susan, John. B. Engring., U. Bombay, 1946; Ph.D., U. Mich., 1953. Sr. engr. Bendix Corp., 1953-56; faculty U. Minn., Mpls., 1956—, prof. mechanics, 1963—, head dept., 1966—; Vis. prof. applied math. Brown Univ., 1965-66; sr. vis. scholar U. Calif. at Berkeley, 1970; vis. prof. U. Warwick, Eng., 1972, Stanford U., 1977, Cornell U., 1983. Recipient Gold medal in applied mechanics Bombay U., 1946. Fellow ASME, Am. Acad. Mechanics, AIAA (asso.); mem. Am. Soc. Engring. Edn., Soc. Natural Philosophy, Soc. Indsl. and Applied Math. Subspecialty: Theoretical and applied mechanics. Home: 2147 W Hoyt St St Paul MN 55108 Office: Dept Aero and Engring Mechanics U Minn Minneapolis MN 55414

SETLER, PAULETTE ELIZABETH, biologist; b. Pitts., Jan. 1, 1938; d. Carl George and Mary Marguerite (Hynes) S. B.A. in Chemistry, Seton Hill Coll., Greensburg, Pa., 1959; Ph.D. in Physiology, U. Pa., 1970. Research asst. McNeil Labs., 1959-62; research asst. Smith Kline & French Labs., Phila., 1962-66, research scientist, 1972-81; postdoctoral fellow Cambridge (Eng.) U., 1970-72; research scientist on central nervous system McNeil Pharm. Co., Spring House, Pa., 1981-82, dir. biol. research, 1982—. Contbr. numerous articles to sci. jours. Mem. Am. Physiol. Soc., Soc. Neurosci., Am. Soc. Pharmacology and Exptl. Therapeutics, Alpha Lambda Delta (life). Subspecialties: Pharmacology; Neurobiology. Current work: Effects of drugs on the nervous system, mechanism of drug action, discovery of novel therapeutic agents. Patentee in field. Office: McNeil Pharm Co Spring House PA 19477

SETLIFF, EDSON CARMACK, mycologist; b. Indianola, Miss., Nov. 3, 1941; s. Abram Carmack and Martha (Baumann) S.; m. Dorene Dee Lyon, Aug. 7, 1939; children: Eric John, Alissa Ellen. B.S., N.C. State Coll., 1963; M.F., Yale U., 1964; Ph.D., SUNY-Syracuse, 1970. Vol. U.S. Peace Corps, Arusha, Tanzania, 1964-66; postdoctoral fellow U. Wis., Madison, 1970-73; forest pathologist Cary Arboretum, N.Y. Bot. Garden, Millbrook, N.Y., 1973-77; Coll. Environ. Sci. and Forestry, SUNY-Syracuse, 1977-80; research scientist Forintek Can. Corp., Vancouver, B.C., Can., 1980—; adj. assoc. prof. Vassar Coll., SUNY-Syracuse, U. B.C., Vancouver. Contbr. articles to profl. jours. Mem. Am. Phytopathol. Soc., Mycol. Soc. Am., Central N.Y. Mycol. Soc., Vancouver Mycol. Soc. Subspecialties: Microbiology; Biomass (agriculture). Current work: Fungal taxonomy, forest pathology, wood products pathology, bioconversion of wood wastes into food, alcohol, chemicals, identification of fungi important in forest and biotechnology. Patentee in field. Office: 6620 NW Marine Dr Vancouver BC Canada V6T 1X2

SETLOW, RICHARD BURTON, biophysicist; b. N.Y.C., Jan. 19, 1921; s. Charles Meyer and Elsie (Hurwitz) S.; m. Jane Kellock, June 6, 1942; children—Peter, Michael, Katherine, Charles. A.B., Swarthmore Coll., 1941; Ph.D., Yale U., 1947. Asso. prof. Yale U., 1956-61; biophysicist Oak Ridge Nat. Lab., 1961-74, sci. dir. biophysics and cell physiology, 1969-74; dir. U. Tenn.-Oak Ridge Grad. Sch. Biomed. Scis., 1972-74; sr. biophysicist Brookhaven Nat. Lab., Upton, N.Y., 1974—, chmn. biology dept., 1979—; prof. biomed. scis. U. Tenn., 1967-74; adj. prof. biochemistry State U. N.Y., Stony Brook, 1975—. Author: (with E.C. Pollard) Molecular Biophysics, 1962; editor: (with P.C. Hanawalt) Molecular Mechanisms for Repair of DNA, 1975. Mem. Nat. Acad. Scis., Am. Acad. Arts and Scis., Biophys. Soc. (pres. 1969-70), Comité Internat. de Photobiologie (pres. 1972-76), Radiation Research Soc., Am. Soc. Photobiology, Phi Beta Kappa. Subspecialties: Biophysics (biology); Cancer research (medicine). Current work: Damage to DNA and its repair. Repair deficient diseases. Ultraviolet photobiology. Home: 57 Valentine Rd Shoreham NY 11786 Office: Biology Dept Brookhaven Nat Lab Upton NY 11973

SETO, JOSEPH TOBEY, microbiologist, educator; b. Tacoma, Aug. 3, 1924; s. T. and K. (Morita) S.; m. Grace K., Aug. 9, 1959; children: Susan Lynn, Steven Fred. B.S., U. Minn., 1949; M.S., U. Wis.-Madison, 1955, Ph.D., 1957. Postdoctoral researcher UCLA Med. Center, 1958-59; asst. prof. San Francisco State U., 1959-60; prof. microbiology Calif. State U.-Los Angeles, 1960—. Served to sgt. U.S. Army, 1945-46. Recipient United Health Found. award, 1965; WHO award, 1972; Alexander von Humboldt award Humboldt Found., Germany, 1972; Sr. Scientist award NATO, 1979; Deutsche Forschungsgemeinschaft award, 1979. Fellow Am. Soc. Microbiology; mem. N.Y. Acad. Scis., Am. Soc. Microbiology, Soc. Gen. Microbiology, Electron Microscope Soc., AAAS, Japanese Am. Citizens League, Sigma Xi. Democrat. Subspecialties: Virology (medicine). Current work: Molecular biology of the pathogenesis of paramyxoviruses and influenza viruses; persistent infections of paramyxoviruses. Office: Dept Microbiology Calif State U Los Angeles CA 90032

SEVALL, JACK SANDERS, research biochemist, biologist; b. Atchison, Kans., Jan. 12, 1946; s. H. T. and M. T. S.; m. Carmen Elizabeth Castro, July 10, 1976. B.A., Willamette U., 1967; Ph.D., Purdue U., 1971. Damon Runyon research fellow Calif. Inst. Tech., Pasadena, 1971-74; asst. prof. Tex. Tech. U., Lubbock, 1974-80; sr. scientist II Wadley Insts., Dallas, 1980-83; assoc. scientist S.W. Found., San Antonio, 1983—; cons. Litton Industries, Fredrick, Md., 1975; minority lectr. Tougaloo (Miss.) Coll., 1982. Block capt. Dallas Police Dept., 1981—. NSF grantee, 1975; NIH grantee, 1976, 81. Mem. Biophys. Soc., AAAS, Soc. Cell Biology, Am. Fedn. Biol. Chemists. Subspecialties: Genetics and genetic engineering (agriculture); Biochemistry (biology). Current work: Protein-DNA interaction with respect to chromosome structure. Office: SW Found Research and Edn PO Box 28147 W Loop 410 at Military Dr San Antonio TX 78284 Home: 112 Wagon Trail Shavano Park TX 78284

SEVERN, DAVID JONES, biologist; b. Chgo., Feb. 11, 1921; s. Michael and Adele (Balzerit) S.; m. Theresa Mary McKee, Feb. 13, 1954. S.B., U. Chgo., 1940, S.M., 1942; M.L.S., Pratt Inst., 1968. Tech. writer Lederle Labs., N.Y.C., 1950-52; med. editor Physicians Publs., Inc., N.Y.C., 1952-70; asst. to pres. Sugar Assn., Inc., N.Y.C., 1970-75; tech. info. mgr. Am. Sugar div. Amstar Corp., N.Y.C., 1975-79; ind. researcher in population genetics, 1979—. Served with AUS, 1942-46; Served with USAF, 1950-51. Mem. Am. Inst. Biol. Scis., Bot. Soc. Am. Republican. Roman Catholic. Subspecialties: Genetics and genetic engineering (biology); Information systems (information science). Current work: Population genetics of partially lethal traits; population genetics, gene frequency, lethal traits. Home: 608 W Scott St Monett MO 65708

SEVERS, WALTER BRUCE, pharmacologist, researcher, educator; b. Pitts., June 10, 1938; s. Walter Bruce and Pauline M. (Sever) S.; m. Anne Elizabeth Daniels, Apr. 25, 1970; children—Mary, Jane, Steven, William, Katherine. B.S., U. Pitts., 1960, M.S., 1963, Ph.D., 1965. Cert. R.Ph., Pa. State Bd. Pharmacy, 1960. Asst. prof. pharmacology Ohio No. U., Ada, 1965-66; NIH fellow Nat. Heart Inst., 1966-68; asst. prof. pharmacology Pa. State U., 1968-71, asso. prof., 1971-77, prof., 1977—; cons. NASA, Life Systems, Inc. Editorial bd.: Pharmacology, 1978—, Am. Jour. Physiology, 1978—; Contbr. articles to profl. publs. Served as sgt. USAR, 1955-63. Named Disting. Alumnus U. Pitts., 1978; NIH and NASA research grantee; NSF summer faculty fellow, 1975. Mem. Am. Physiol. Soc. (cons.), Am. Soc. Pharmacology and Exptl. Therapeutics, Sigma Xi (pres. Pa. State U. chapt. 1981-82). Republican. Roman Catholic. Lodge: Kiwanis. Subspecialties: Neuropharmacology; Neuroendocrinology. Current work: Study of the properties of the brain renin-angiotensin system: relation to the pathophysiology of hypertension and diseases of salt/water balance. Study of the neuroendocrinology of vasopressin pathways in brain and interaction with other neurotransmitters. Home: 1011 Grubb Rd Palmyra PA 17078 Office: Dept Pharmacology Hershey Med Center Hershey PA 17033

SEWELL, KENNETH GLENN, physicist; b. Sherman, Tex., July 26, 1933; s. Arthur J. and Mary A. (Simmons) s.; m. Eudena Barfield, Nov. 20, 1952; children: Kadena, Kanette. M.S. in Aero. Engring, So. Meth. U., 1960; Ph.D. in Physics, Tex. Christian U., 1964. Research scientist LTV Research Center, 1957-60, sr. scientist, 1960-69; assoc. prof. physics Abilene Christian Coll., 1967-68; tech. dir. Isoray Internat. Corp., 1969-70; dir. research and devel. Integrated Systems div. Varo, Inc., Dallas, 1970-74, engring. mgr., 1974-78, gen. mgr., 1978—. Contbr. articles to tech. jours. Mem. IEEE, Optical Soc. Am., Am. Phys. Soc., Am. Mgmt. Assn. Subspecialty: Aerospace engineering and technology. Home: 7661 La Bolsa Dr Dallas TX 75240 Office: Varo Inc Integrated Systems Div 2201 Walnut Garlance TX 75080

SEYDEL, HORST GUNTER, physician, educator; b. Berlin, Germany, Aug. 5, 1929; came to U.S., 1955, naturalized, 1961; m. Emily Meginnity, 1958; children: Edgar, Charlotte, Carolyn. M.D., Goethe U., Frankfurt, Germany, 1955; M.S., Wayne State U., 1961. Intern Md. Gen. Hosp., Balt., 1955-56; resident in oncology Am. Oncologic Hosp., Phila., 1956-57, chief dept. radiation therapy, 1968-75; resident in radiology Wayne State U., Detroit, 1958-61, radiotherapist, 1961-64, U. Md. Sch. Medicine, Balt., 1964-68; cons. radiotherapy Perry Point VA Hosp., Perryville, Md., 1966-68, Prince George Gen. Hosp., Hyattsville, Md., 1966-68, Peninsular Gen. Hosp., Salisbury, Md., 1966-68, Cambridge (Md.) Hosp., 1966-68, St. Joseph Hosp., Balt., 1967-68; radiotherapist No. div. Albert Einstein Med. Ctr., Phila., 1975-77, Thomas Jefferson U. Med. Coll., 1977-79; cons. radiotherapy Nazareth Hosp., Phila., 1978-79, Chestnut Hill Hosp., 1978-79; dir. clin. radiation therapy service Harper-Grace Hosps., Detroit, 1979-82, vice chief radiation therapy, 1979-82; clin. prof. radiation oncology Wayne State Sch. Medicine, 1982—; chmn. therapeutic radiology Henry Ford Hosp., Detroit, 1982—. Contbr. articles to profl. jours. Fulbright scholar, 1955-56. Mem. Am. Assn. for Cancer Research, Am. Cancer Soc., Am. Coll. Radiology (accreditation com.), AMA, Am. Radium Soc., Am. Soc. Clin. Oncology, Am. Soc. Therapeutic Radiologists, Mich. State Med. Soc., Mich. State Radiol. Soc., Radiation Therapy Oncology Group (chmn. lung com.), Radiol. Soc. N.Am., Southwest Oncology Group (chmn. radiation therapy quality control. com.), Wayne County Med. Soc. Subspecialty: Radiology. Current work: Clinical cancer research.

SEYFRIED, THOMAS NEIL, neurogenetics educator; b. Flushing, N.Y., July 25, 1946; s. William Edward and Anne Marie (McGuire) S.; m. Karen Elizabeth Seyfried, Apr. 13, 1973; children: Melissa, Nicholas, Edward. B.A., St. Francis Coll., 1968; M.S., Ill. State U., 1973; Ph.D., U. Ill., 1976. Postdoctoral fellow dept. neurology Yale U., New Haven, 1976-79, asst. prof., 1979—. Contbr. articles to profl. jours. Served with U.S. Army, 1969-71; with Res., 1980—. Decorated Bronze Star medal with 2 oak leaf clusters, Air medal, Army Commendation medal; recipient Honored Student award Am. Oil Chemists Soc., 1975; Nat. Research Service fellow NIH, 1976-79; NIH grantee, 1981—; NIH research career devel. award, 1979—. Mem. Genetics Soc. Am., Soc. for Neurosci., Internat. Soc. Neurochemistry, AAAS. Subspecialties: Gene actions; Neurochemistry. Current work: Cellular localization and function of brain gangliosides; developmental genetics of inherited epilepsy in mice. Office: Dept Neurology Yale Univ Sch Medicine New Haven CT 06510 Home: 27 Laconia St Hamden CT 06514

SHA, GEORGE TZENG-TSUN, mech. engr., metallurgist; b. Kiangsu Province, China, Jan. 6, 1936; came to U.S., 1960, naturalized, 1972; s. Chin-Fung and Yu-Nei (Ge) S.; m. Annie L., Mar. 30, 1968; children: Michael C., Edward C. M.S., U. Idaho, 1963, Carnegie Inst. Tech., 1965; Ph.D., Lehigh U., 1971. Research metallurgist AMF Thermatool Corp., New Rochelle, N.Y., 1966-68; research scientist Bell Aerospace Co., Buffalo, 1971-73; sr. engr. Aluminum Co. Am., Pitts., 1973-75; engring. specialist Pratt & Whitney Aircraft, West Palm Beach, Fla., 1976-81; staff devel. engr. Detroit Diesel Allison div. Gen. Motors Corp., Indpls., 1981—. Contbr. tech. articles to profl. jours. Mem. ASME, Am. Soc. Metals, ASTM, Sigma Xi. Subspecialties: Fracture mechanics; Materials. Current work: Engineering fracture mechanics; damage tolerant design concept; component life prediction technologies; finite element stress analyses. Home: 11623 Valleybrook Pl N Carmel IN 46032 Office: PO Box 894 Indianapolis IN 46206

SHACK, WILLIAM JOHN, mechanical engineer; b. Pitts., Jan. 12, 1943; s. William and Anna (Mitriski) S.; m. Nancy Treeger, May 31, 1975; 1 dau., Emily. B.S.C.E., M.I.T., 1964; M.S. in Applied Mechanics, U. Calif.-Berkeley, Ph.D., 1968. Asst. prof. mech. engring. M.I.T., 1968-74, assoc. prof., 1974-75; research scientist Argonne Nat. Lab., Ill., 1975-79, group leader, metal properties group, materials sci. div., 1979—. Contbr. numerous articles to profl. jours. NSF fellow, 1964-67. Mem. ASME. Democrat. Subspecialties: Solid mechanics; Materials (engineering). Current work: Mechanics and materials engring. problems in energy systems. Office: Argonne Nat Lab Bldg. 212 Argonne IL 60437

SHADDUCK, JOHN ALLEN, vet. pathologist; b. Toledo, Ohio, Apr. 22, 1939; s. Hugh Allen and Martha Juliet (Niles) S.; m. Mary Lou Lambdin, Apr. 9, 1960; children—James Allen, Margaret Ann. D.V.M., Ohio State U., 1963, M.Sc., 1965, Ph.D., 1967. NIH postdoctoral fellow Ohio State U., Columbus, 1964-68, asst. prof., 1967-72, assoc. prof., 1972-73, U. Tex. Health Scis. Center, Dallas, 1973-80, prof., 1980-81; prof., head dept. vet. pathobiology U. Ill., Urbana-Champaign, 1981—. Editor: Vet. Pathology, 1979; contbr. articles in field to profl. jours. Grantee NIH, 1970; WHO, 1974; FDA, 1977-79. Mem. AAAS, Am. Coll. Vet. Pathologists, AVMA, Am. Assn. Patologists, Tissue Culture Assn. Subspecialties: Pathology (veterinary medicine); Microbiology (veterinary medicine). Current work: Comparative biology of ocular melanomas; inflammation; endothelial cells; immunology of photozoal diseases; in vitro alternatives to animals. Office: Dept Vet Pathology 1101 W Peabody St Urbana IL 61801

SHADE, BARBARA JEAN, psychology educator; b. Armstrong, Mo., Oct. 30, 1933; d. Murray Kenneth and Edna Rose (Bowman) Robinson; m. Oscar DePriest Shade, Mar. 22, 1954; children:

Christina Marie Shade Jones, Kenneth Eugene, Patricia Louise. B.S., Pittsburg (Kans.) State U., 1955; M.S., U. Wis.-Milw., 1967; Ph.D., U. Wis.-Madison, 1973. Lic. sch. psychologist, Wis.; lic. secondary tchr., Wis. Tchr. Milw. Pub. Schs., 1960-68; exec. dir. Dane County Head Start, Madison, Wis., 1969-71; postdoctoral fellow Nat. Endowment Humanities, 1973-74; urban edn. specialist Dept. Public Instrn., Madison, 1974-75; asst. prof. U. Wis., Madison, 1975-81; assoc. prof. ednl. psychology U. Wis.-Parkside, Kenosha, 1981—. Mem. Wis. Gov.'s Policy and Planning Task Force, 1971-72; bd. dirs. St. Mary's Hosp. Med. Center, Madison, 1973-81, chmn. bd., 1979-81; v.p. United Way of Dane County, Madison, 1979; mem. Wis. Humanities Com., 1980—. Recipient Chancellor's Service award U. Wis., Madison, 1980; Gov.'s Employment Tng. Office grantee, 1982; Wis. Center for Endl. Research grantee, 1980-82. Mem. Am. Psychol. Assn., Am. Ednl. Research Assn., AAAS, Delta Sigma Theta. Methodist. Subspecialty: Cognition. Current work: Study of cultural influences on Afro-American patterns of cognition. Home: 3110 Pelham Rd Madison WI 53141 Office: University of Wisconsin-Parkside PO Box 2000 Kenosha WI 53141

SHAFER, JULES ALAN, biological chemistry educator, consultant; b. N.Y.C., Nov. 21, 1937; s. Samuel Z. and Ada (Gams) S.; m. Marcia Anolik, June 21, 1959; children: Toby Beth, Howard Keith, Neil Evan. B. Chem. E., CUNY, 1959; Ph.D. in Polymer Chemistry, Poly. Inst. Bklyn., 1963. Postdoctoral fellow and tutor Harvard U., Boston, 1963-64; asst. prof. biochemistry U. Mich., Ann Arbor, 1964-69, assoc. prof. biochemistry, 1969-77, prof. biol. chemistry, 1977—. NIH fellow, 1964—. Mem. Am. Chem. Soc., N.Y. Acad. Sci., Am. Soc. Biol. Chemists, AAAS. Subspecialties: Enzyme technology; Biophysical chemistry. Current work: Mechanisms for the action of proteolytic enzymes and enzymes requiring vitamin B_6; interaction of polypeptide hormones with their receptors. Office: Biol Chemistry Dept Box 034 U Mich Med Sci Bldg I Ann Arbor MI 48109

SHAFFER, CLINTON JOHN, nuclear engineer, consultant; b. Preston, Idaho, Oct. 4, 1947; s. Dean and Delma (Swainston) S.; m. Tiffany Owens, Nov. 6, 1980; 1 stepdau., Melanie Lynn Robinson. B.S.M.E., U. Idaho, 1965-70; M.S.N.E., U. N.Mex., 1971, postgrad., 1977-78. Registered profl. engr., Idaho, N.Mex. Sr. engr. Aerojet Nuclear Co., Idaho Falls, Idaho, 1972-76; sr. engr. EG&G Idaho, Inc., Idaho Falls, 1976-77; grad. asst. U. N.Mex., Albuquerque, 1977-78; project mgr. Energy, Inc., Albuquerque, 1979-83; self-employed cons., Albuquerque, 1983—. Contbr. articles in field to profl. jours. Mem. Am. Nuclear Soc., ASME. Republican. Subspecialties: Nuclear fission; Nuclear fusion. Current work: Nuclear fusion/fission power systems, space propulsion and power systems, electromagnetic field theory and applications, alternate energy systems. Home: 7321 Gill St NE Albuquerque NM 87109 Office: 5345 Wyoming St NE Apt 101 Albuquerque NM 87109

SHAFFER, DAVID REED, psychology educator; b. Watsonville, Calif., Feb. 4, 1946; s. Duboise Herbert and Gerrie Doris (Cadlieux) S.; m. Garnett Sue Stokes, June 6, 1981. B.A., Humboldt State U., 1967, M.S., 1968; Ph.D., Kent State U., 1972. Asst. prof. Kent (Ohio) State U., 1972-73; asst. prof. U. Ga., Athens, 1973-76, assoc. prof., 1978-82, prof. psychology, 1982—, dir. undergrad. program in psychology, 1978-82, dir. devel. psychology program, 1982—. Author: Social and Personality Development, 1979; editor: Jour. Personality, 1981-82, Jour. Personality and Social Psychology, 1979-80; contbr. articles to profl. jours. Named Disting. Alumnus Humboldt State U., 1982. Mem. Am. Psychol. Assn., Soc. for Research in Child Devel., Soc. Exptl. Social Psychology, Southwestern Psychol. Assn., Sigma Xi. Democrat. Subspecialties: Social psychology; Developmental psychology. Current work: Attitudes and attitude changes, sex typing, sex role development, self-monitoring, self-focused attention. Home: 249 Ansley Dr Athens GA 30605 Office: Dept Psychology U Ga Athens GA 30602

SHAFFER, HOWARD JEFFREY, clinical psychologist, consultant, researcher, educator; b. Boston, Sept. 1, 1948; s. Milton and Ruth Ann (Weiner) S.; m. Linda Marie Andrews; 1 son, David Andres. B.A., U. N.H., 1970; M.S., U. Miami, Coral Gables, Fla., 1972, Ph.D., 1974. Lic. psychologist, Mass., N.H.; registered psychologist Nat. Register Health Service Providers in Psychology, 1982. Research dir. Psycho-Social Rehab. Ctr. Dade County (Fla.), Inc., 1974-75; clin. dir. Project Turnabout, Inc., Hingham, Mass., 1975-78, East Boston (Mass.) Drug Rehab. Clinic, 1976-77; dir. spl. consultation and treatment program for women Judge Gould Inst. Human Resources, Inc., Worcester, Mass., 1977-78; dir. narcotics treatment program Drug Problems Resource Center, dept. psychiatry Harvard Med. Sch. at Cambridge (Mass.) Hosp. and North Charles Found. for Tng. and Research, 1978-79, dir. tr., 1979-82; asst. dir. tng. dept. psychiatry Harvard Med. Sch. at Cambridge Hosp., 1982—; mem. adj. faculty U. Miami, 1974; mem. clin. faculty Barry Coll., 1974-75; mem. field faculty Lone Mountain Coll., 1974-75; teaching cons. U. Lowell, 1975-76; clin. assoc. prof. counseling psychology Boston U., 1976-78; instr. psychology dept. psychiatry Harvard Med. Sch. at Cambridge Hosp., 1978-81, asst. prof. psychology, 1982—; mem. faculty Mass. Psychol. Center, Boston, 1980—; mem. council on marijuana and health Nat. Orgn. for Reform Marijuana Laws, 1981—. Contbr. articles, revs., abstracts to profl. publs., presentations in field; editor: a book about drug issues Myths and Realities, 1977, (with M. E. Burglass) Classic Contributions in the Addictions, 1981 (alt. main selection Behavioral Sci. Book Club 1981); assoc. editor: Bull. of Psychologists in Addictive Behavior, 1982—; guest editor: spl. issue Advances in Alcohol and Substance Abuse, 1983; mem. editorial rev. bd.: Jour. Psychoactive Drugs, 1981—, Advances in Alcohol and Substance Abuse, 1982—. Mem. Andover (Mass.) Substance Abuse Com., 1982. Recipient 1st place award U. N.H. Undergrad. Conf. for Psychol. Research, 1969. Mem. Am. Psychol. Assn., AAAS, Soc. Psychologists in Addictive Behavior, Psi Chi, Phi Kappa Phi. Democrat. Jewish. Clubs: Harvard (Boston); Far Corner. Subspecialties: Substance use and abuse; Social psychology. Current work: Cognitive, behavioral, and psychodynamic factors associated with substance use and abuse; the social psychology of psychotherapy; philosophy of science. Home: 171 Summer St Andover MA 01810 Office: Dept Psychiatry Harvard Med Sch 1493 Cambridge St Cambridge MA 02139

SHAFI, MUHAMMAD IQBAL, biology educator; b. Kanpur, India, Aug. 23, 1943; came to U.S., 1980; s. Mohammad and Ashrafunnisa (Begum) S.; m. Fahmida Iqbal, Dec. 15, 1973; 1 child, Nadeem. B.Sc. with honors, U. Karachi, Pakistan, 1964, M.Sc., 1965; Ph.D., U. Toronto, Ont., Can., 1972. Research asst. Cereal Diseases Research, Murree, Pakistan, 1966-67; biologist Ontario Ministry of Natural Resources, Toronto, 1973-74; ecologist Andre Marsan and Assocs., Montreal, Que., Can., 1975-76; postdoctoral research asst. U. N.B. (Can.), Fredericton, 1977-79; asst. prof. dept. biology Rust Coll., Holly Springs, Miss., 1980—; councillor Karachi U. Student Council Soc., 1962-63. Contbr. articles and guide to profl. jours.; editor: The Botanist jour, 1964-65. Postdoctoral assistantship NRC Can., 1978; U.S. Dept. Edn. grantee, 1981—. Mem. Can. Botanical Soc., Am. Ecol. Soc., Brit. Ecol. Soc., Quebec Assn. Biologists, Internat. Assn. Ecologists. Muslim. Subspecialties: Plant cell and tissue culture; Cell and tissue culture. Current work: Fire ecology, fires and weeds, right of way and multiple uses. Home: 541 N Randolph St Holly Springs MS 38635 Office: Rust Coll 1 Rust Ave Holly Springs MS 38635

SHAFIZADEH, FRED, chemist, educator; b. Tehran, Iran, Jan. 26, 1924; s. Ali and Akhtar (Afshar) S.; m. Doreen Wigley, Mar. 2, 1954; 1 dau., Alexandera K. Shafizadeh Startin. Ph.D. in Organic Chemistry, U. Birmingham, Eng., 1950, D.Sc., 1972. Univ. research fellow U. Birmingham, 1946-50; research assoc. physics dept. Pa. State U., 1952-53; research assoc., assoc. supr. Ohio State U. Research Found., 1953-58; profl. specialist Weyerhaeuser Co., Longview, Wash., 1958-60, mgr. pioneering research, Seattle, 1960-66; prof., dir. wood chemistry lab. U. Mont., Missoula, 1966—. Contbr. articles to profl. jours. Recipient 1st Disting. Researcher award U. Mont., 1980. Mem. Am. Chem. Soc., Brit. Chem. Soc. Subspecialties: High temperature chemistry; Organic chemistry. Current work: Chemistry of carbohydrates and other natural products; thermanal conversion of biomass. Patentee in field. Office: Wood Chemistry Lab U Mont Missoula MT 59812

SHAFRITZ, DAVID ANDREW, internist, molecular biologist, researcher, educator; b. Phila., Oct. 5, 1940; s. Saul and Ethel (Kohn) S.; m. Sharon Klemow, Aug. 16, 1964; children: Gregory S., Adam B., Keith M. A.B. in Chemistry, U. Pa., 1962, M.D., 1966. Diplomate: Am. Bd. Internal Medicine. Research asst. U. Pa., 1962-66; intern U. Med. Hosp., 1966-67, resident, 1967-68; USPHS fellow, research assoc. NIH, Bethesda, Md., 1968-71; spl. research fellow Mass. Gen. Hosp., Boston, 1971-73; asst. prof. medicine, Harvard U., 1973; prof. medicine and cell biology Albert Einstein Coll. Medicine, 1973—, chief molecular hepatology unit, 1982—; temporary adv. WHO, Geneva, 1983. Contbr. numerous articles, abstracts to profl. publs.; editor: (with others) The Liver: Biology and Pathobiology, 1982; assoc. editor: Hepatology, 1981—; editorial bd.: Jour. Med. Virology, 1982—. Trustee Westchester Jewish Ctr., Mamaroneck, N.Y., 1980—. Served to lt. comdr. USPHS, 1968-71. Recipient Irma T. Herschl Career Scientist award, 1974-78, Research Career Devel. award NIH, 1975-80. Mem. Am. Soc. Biol. Chemists, Am. Soc. Clin. Investigation, Harvey Soc., N.Y. Acad. Scis., Am. Assn. Study Liver Disease (Merril lectr. 1978). Club: U. Pa. Alumni of Westchester County. Subspecialties: Gastroenterology; Genetics and genetic engineering (medicine). Current work: Molecular pathogenesis of liver disease; role of hepatitis B virus infection in primary liver cancer; regulation of gene expression in eukaryotic cells. Home: 4 Pheasant Run Larchmont NY 10538 Office: Albert Einstein Coll Medicine 1300 Morris Park Ave Bronx NY 10461

SHAH, HARSHAD, electrial engineer; b. Bombay, India, Nov. 28, 1942; s. Balubhai and Kanchenben S. B.S.E.E., Baroda U., 1964; M.S.E.E., Utah State U., 1966. Project engr. Honeywell, Inc., Mass., 1966-68; sr. engr. Raytheon Co., Mass., 1968-69; chief engr. Sy Sterling Assocs., Mich., 1969-72, Tenor Co., New Berlin, 1972-76; pres. MicroControl Systems, Milw., 1976—. Subspecialty: Electronics. Current work: Automated bldg. controls for energy mgmt. Office: 6579 N Sidney Pl Milwaukee WI 53209

SHAH, RAJIV RAJARAM, applied physicist; b. Maharashtra, India, July 11, 1953; came to U.S., 1972, naturalized, 1976; s. Rajaram Damodar and Kamala (Rajaram) S.; m. Shrida Rajiv, Dec. 18, 1980. B.Sc. in Physics, U. Poona, India, 1972; M.S. in E.E, Rice U., 1974, Ph.D., 1976. Research asst. Rice U., Houston, 1972-76; Dr. Chaim Weizmann postdoctoral research fellow Calif. Inst. Tech., Pasadena, 1976-79; mem. tech. staff Tex. Instruments, Dallas, 1979—. Contbr. articles to profl. jours. Govt. India Nat. Sci. Talent Search scholar, 1969-72. Mem. Am. Phys. Soc., IEEE, Optical Soc. Am., Electrochem. Soc., Am. Vaccum Soc., Materials Research Soc. Current work: Laser applications, device physics, device processing; laser processing of semiconductor devices, 3-D integration of semiconductor devices, device processing, device physics and device modeling. Developed laser assisted removal of saw damage in silicon slices; apparatus and method for photolithography with phase conjugate optics; reduction in temperature dependence and standard deviation of polysilicon resistors by laser annealing; large crystal growth in polisilicon on oxide or nitride by dual wavelength laser annealing for absorption optimized heating of oxide or nitride; laser processing of PSG, oxide and nitride via absorption optimized selective laser annealing. Home: 2116 Newcombe Dr Plano TX 75075 Office: Texas Instruments MS-82 PO Box 225012 Dallas TX 75265

SHAH, SAIYID MASROOR, med. physicist; b. Rampur, India, Jan. 8, 1938; came to U.S., 1974; s. Syed Maqsood and Noor Jehan (Ali) S.; m. Janice Moore, Aug. 8, 1970. Ph.D., Tex. Tech. U., 1970. Lectr. in physics SRA/DJ Colls., Karachi, 1958-64; asst. prof. physics U. Karachi, 1970-72; assoc. prof. physics Baluchistan U. Quetta, 1972-73; fgn. research fellow Sophia U., Tokyo, 1973-74; postdoctoral research fellow physics M.D. Anderson Hosp., Houston, 1976; med. physicist Rosewood Hosp., Houston, 1976—, cons., instr. continuing edn. program, 1976—. Contbr. articles to profl. jours. Japan Soc. for Promotion of Sci. Fgn. Research fellow, 1973; Fulbright Travel scholar, 1964; recipient Am. Friends of Middle East Hon. Mention award, 1969. Mem. Assn. Physicists in Medicine. Moslem. Subspecialties: Medical physics; Atomic and molecular physics. Current work: Radiation physics; cancer treatment planning; radiation safety; cancer treatment plan optimization for high energy radiation beams, quality control, dosimetry, radiation safety. Home: 15819 Ridgebar Circle Houston TX 77053 Office: 9200 Westheimer St Houston TX 77063

SHAH, SUDHIR AMRATLAL, biologist, cancer research scientist; b. Kampala, Uganda, Nov. 23, 1944; came to U.S., 1980, naturalized, 1981; s. Amratlal Tribhovandas and Chandramani (Amratlal) S.; m. Hema Sudhir, Mar. 7, 1972; children: Neha, Rina. B.Sc., U. Bombay, India, 1967; M.Phil., U. London, 1972; Ph.D., U. Newcastle upon Tyne, Eng., 1977; licentiate, Royal Inst. Chemistry, Slough Coll. Tech., Eng., 1969. Tech. officer Chester Beatty Research Inst., London, 1968-72; research assoc. U. Newcastle upon Tyne, 1972-76, research fellow, cancer research unit, 1976-80; guest worker-sr. research assoc. Cancer Research Lab., Carnegie Mellon U., Pitts., 1980-82; tumor biologist-immunopharmacologist Med. Diagnostic div. New Eng. Nuclear, North Billerica, Mass., 1983—. Contbr. numerous articles and papers to profl. jours. Brit. Empire Cancer Research Campaign postdoctoral research fellow, 1976-80. Mem. Am. Assn. for Cancer Research, AAAS, Sigma Xi. Subspecialties: Cancer research (medicine); Physiology (medicine). Current work: Tumor biology/physiology/immunology, hyperthermia, immunotherapy, chemotherapy and radiotherapy. Animal tumor models, human cancer cells in culture.

SHAHADY, EDWARD JOHN, physician; b. Fairmont, W.Va., June 2, 1938; s. Joseph E. and Mary V. (Mallamo) S.; m. Sandra Jean Kovach, Sept. 13, 1958; children—Mary, John, Thomas, Elizabeth, Edward, William. B.S. in Biology, Wheeling Coll., 1960; M.D., W.Va. U., 1964. Diplomate: Am. Bd. Family Practice. Rotating intern Akron (Ohio) City Hosp., 1964-65, resident, 1967-68; practice medicine specializing in family practice, Akron, 1968-70; dir. family practice residency Akron City Hosp., 1970-76; chmn. dept. family practice Northeastern Ohio Univs. Coll. Medicine, 1974-76; prof., chmn. dept. family medicine U. N.C., Chapel Hill, 1976—; football team physician Chapel Hill High Sch., 1979—. Contbr. articles on family medicine and med. edn. to profl. jours. Bd. dirs. Women's Health Counseling Service, Chapel Hill, 1979-80; trustee Child Guidance Center, Akron, 1971-76, Denton House, Akron, 1973-76. Served with USN, 1965-67. Recipient Eben J. Carey Phi Chi Meml. award, 1964; Mead Johnson scholar, 1968. Fellow Am. Acad. Family Physicians; mem. N.C. Acad. Family Physicians, Soc. of Tchrs. Family Medicine (chmn. edn. com. 1974-79, pres. 1980-81). Subspecialty: Family practice. Current work: Sports medicine. Home: 2515 Buxton Ct Chapel Hill NC 27514 Office: Dept of Family Medicine U NC Trailer 15 269H Chapel Hill NC 27514

SHAHANI, BHAGWAN TOPANDAS, physician, educator, researcher; b. Larkana, Pakistan, Dec. 16, 1936; came to U.S., 1963, naturalized, 1977; s. Topandas N. and Devi T. (Gidwani) S.; m. Maya B. Advani, Apr. 2 1937; 1 son, Robin. M.B.B.S., Bombay (India) U., 1962; D.Phil., Oxford (Eng.) U., 1970. Diplomate: Am. Bd. Phys. Medicine and Rehab. Intern B.Y.L. Nair Hosp., Bombay, 1962-63; resident NYU Med. Ctr., N.Y.C., 1963-66; med. officer dept. neurology Oxford U., 1966-70; dir. electromyography and motor control unit Clin. Neurophysiology Lab., Mass. Gen. Hosp., Boston, 1971—, assoc. neurologist, 1980—; assoc. prof. neurology Harvard U., 1980—; sci. dir. restorative medicine and rehab. Mass. Rehab. Hosp., Boston. Contbr. articles on electromyography and motor contol to sci. jours., books. Recipient Spl. award Am. Assn. Crippled Children, 1966-70; World Rehab. Fund fellow, 1963-66. Mem. Soc. Neurosci., Am. Acad. Neurology, Am. Assn. Electromyography and Electrodiagnosis (pres. 1981-82), Am. Acad. Phys. Medicine and Rehab. Subspecialties: Neurology; Physical medicine and rehabilitation. Current work: Clinical neurophysiology; electromyography; nerve conductor studies; reflex studies; studies of human motor control. Home: 10 Olde Village Dr Winchester MA 01890 Office: Mass Gen Hosp Boston MA 02114

SHAHIN, MICHAEL M., scientist, mgr.; b. Iran, Sept. 7, 1932; m. Andree Shahin; children: Francoise, Jean-Paul. B.Sc. with honors in Chemistry, U. Birmingham, Eng., 1955, Ph.D. in Phys. Chemistry, 1958. Research fellow NRC Can., Ottawa, 1958-60; research chemist E.I. duPont de Nemours, Wilmington, Del., 1960-61; with Yale U. Sch. Medicine, 1961-63, Xerox Corp., Webster, N.Y., 1967—, v.p., 1973-78, v.p. tech. planning, 1978—. Contbr. numerous articles tech. jours., chpt. in book. Mem. Am. Phys. Soc. (chmn. N.Y. chpt.), Am. Chem. Soc., Chem. Soc., Chem. Soc. London, Sigma Xi. Subspecialties: Atomic and molecular physics; Plasma physics. Current work: Physical chemistry, ionization phenomena, mass spectrometry. Home: 12 Widewaters Ln Pittsford NY 14534 Office: J C Wilson Center for Technology Webster NY 14580

SHAHINPOOR, MOHSEN, engineering educator; b. Tehran, Iran, Sept. 14, 1943; came to U.S., 1966, naturalized, 1981; s. Ali Asghar and Zarrin (Hajirajabali) S.; m. Jamileh Farahmand, Sept. 9, 1976; children: David, Parsa, Sheerin. B.Sc., Abadan Inst. Tech., 1966; M.Sc., U. Del., 1968, Ph.D., 1970. Assoc. prof. Pahlavi U., Shiraz, Iran, 1972-76, prof., 1976-78, Northwestern U., Evanston, Ill., 1978-79; assoc. prof. Clarkson Coll. Tech., Potsdam, N.Y., 1979-82, prof., chmn. dept. mech. and indsl. engring, 1982—, chmn. solid mechanics, 1982—. Editor-in-chief: Iran Jour. Sci. & Tech, 1975-78; book series: Adv. Mech. Flow Granular Mats, 1982—; contbr. over 150 articles to profl. jours. Named Researcher of Yr. Alborz Found., 1976; Inventor of Yr. Ministry of High Edn., Iran, 1977. Mem. ASME, Am. Acad. Mechanics, Soc. Engring. Sci., AAAS, U.S. Colloid and Surface Sci. Subspecialties: Mechanical engineering; Materials (engineering). Current work: Mechanics and flow granular materials, liquid crystals, materials with microstructure, systems analysis and machine design, design of industrial robots. Inventor in field. Office: Clarkson Coll Tech Potsdam NY 13676

SHAHRYAR, ISHAQ M., physical chemist; b. Kabul, Afghanistan, Jan. 10, 1936; s. Ahmad Ali and Zahra S. B.S. in Phys. Chemistry, U. Calif., 1961, M.A., 1968. Founder, pres. Solec Internat., Inc. (became subs. Pilkington Bros., P.L.C. of Eng.), Hawthorne, Calif., 1976—; Counsel to chancellor U. Calif., 1981—. Republican. Moslem. Subspecialty: Solar energy. Current work: Administration of technology, research and development, manufacturing future solar cell technolgy. Inventor low cost solar cell, 1972; patentee photovoltaics tech., 1977. Office: 12533 Chadron Ave Hawthorne CA 90250

SHAIKH, ZAHIR AHMAD, toxicologist, educator; b. Jullundur, India, Mar. 31, 1945; s. Zafer Ahmad and Mehmooda Beoum (Chohan) S.; m. Mary Butterfield, Aug. 23, 1975; children: Faraz, Kashan, Summur. B.Sc., U. Karachi, Pakistan, 1965, M.Sc., 1967; Ph.D., Dalhousie U., 1972. Research assoc. environ. health U. Okla., Oklahoma City, 1972-73; sr. postdoctoral fellow in toxicology U. Rochester, 1973-75, asst. prof., 1975-81, assoc. prof., 1981-82; assoc. prof. pharmacology and toxicology U. R.I., Kingston, 1982—; spl. reviewer toxicology study sect. NIH, 1983—. Contbr. articles to profl. jours. NIH fellow, 1973-75; NIH grantee, 1975—. Mem. Soc. Toxicology (pres. metals sect.), Soc. Exptl. Biology and Medicine, Am. Soc. Pharmacology and Exptl. Therapeutics, N.Y. Acad. Scis., AAAS. Subspecialties: Toxicology (medicine); Environmental toxicology. Current work: Heavy metal metabolism and toxicology; role of metallothionein in metal toxicology; detoxification processes; biol. indicators of metal exposure. Home: 75 Greenwood Dr Peace Dale RI 02879 Office: Dept Pharmacology and Toxicology U RI Kingston RI 02881

SHALLECK, ALAN BENNETT, energy control equipment co. exec.; b. N.Y.C., Nov. 14, 1938; s. Milton and Rosalyn (Baron) S. B.S., M.I.T., 1960; M.B.A., Harvard U., 1963. Bus. mgr. ASW programs Grumman Corp., N.Y.C., 1963-66; sr. cons. planning and fin. Diebold Group, N.Y.C., 1966-68; asst. v.p. corp. planning Reliance Group, Inc., N.Y.C., 1968-70; pres. Princeton (N.J.) Labs., Inc., 1970-75; v.p. corp. devel. Carter Wallace, Inc., N.Y.C., 1975-77; pres. Aegis Group, Inc., Pennington, N.J., 1977—, Aegis Energy Systems, Inc., Doylestown, Pa., 1977—; dir. Bioenergy Systems, Inc., Ellenville, N.Y. Served with Signal Corp. U.S. Army, 1960. Mem. IEEE, Solar Energy Industry, N.Y. Acad. Scis. Democrat. Clubs: Harvard, M.I.T. Subspecialties: Solar energy; Systems engineering. Home: 5 Park Ave Pennington NJ 08534 Office: 607 Airport Blvd Doylestown PA 18901

SHALVOY, RICHARD BARRY, research chemist; b. Norwalk, Conn., Apr. 26, 1949; s. Hugh Augustus and Eunice Eudora (Shufelt) S.; m. Karol Ann Mihailoff, July 15, 1972; 1 dau., Anastasia Estelle. B.S. in Physics, Rensselaer Poly. Inst., 1971, Ph.D., Brown U., 1976. Postdoctoral fellow Inst. Mining and Minerals Research, U. Ky., Lexington, 1976-78, sr. physicist, 1978-80; research chemist Stauffer Chem. Co., Dobbs Ferry, N.Y., 1980—. Contbr. articles to profl. jours. Mem. Am. Phys. Soc., Am. Vacuum Soc. Episcopalian. Club: Taconic Road Runners (Baldwin Place, N.Y.). Subspecialties: Analytical chemistry; Surface physics. Current work: Use of surface analytical instrumentation to solve problems with industrial chemicals and materials and in exploratory research. Office: Stauffer Chem Co Eastern Research Ctr Dobbs Ferry NY 10522

SHAMBAUGH, GEORGE ELMER, III, physician, educator, researcher; b. Boston, Dec. 21, 1931; s. George E. and Marietta Moss (Long) S.; m. Katharine Margaret Matthews, Dec. 29, 1956; children: George Elmer IV, Benjamin Albert, Daniel Frederick, James Bradley, Elizabeth Matthews. B.A., Oberlin Coll., 1954; M.D., Cornell U., 1958. Diplomate: Am. Bd. Internal Medicine. Intern Denver Gen. Hosp., 1958-59; resident in internal medicine Walter Reed Gen. Hosp., Washington, 1961-64; research internist Walter Reed Army Med. Ctr., Frederick, Md., 1964-67; fellow in physiologic chemistry U. Wis.,

Madison, 1967-69; mem. staff endocrinology Northwestern U. Med. Sch., Chgo., 1969—, asst. prof. medicine, 1969-74, assoc. prof. medicine, 1974-81, prof. medicine, 1981—; chief endocrinology and metabolism VA Lakeside Med. Ctr., Chgo., 1974—; attending physician Northwestern Meml. Hosp., Chgo., 1977—. Contbr. chpts. to textbooks, articles to profl. jours. Served to maj. U.S. Army, 1959-67. Schweppe Found. fellow, 1972-74; VA research grantee, Chgo., 1975—. Fellow ACP; mem. Am. Fedn. Clin. Research, Sci. Research Soc. Am., Endocrine Soc., Central Soc. for Clin. Research, Am. Thyroid Assn., Am. Inst. Nutrition, Am. Soc. Clin. Nutrition (vice chmn. com. on nutrition issues 1981—), Sigma Xi. Subspecialties: Internal medicine; Neuroendocrinology. Current work: Effects of altered nutrition on fetal and neonatal development, including brain, liver, placenta, cell replication, mechanism altered by changing nutrient composition. Home: 530 S Stone Ave La Grange IL 60525 Office: Va Lakeside Med Center 333 E Huron St Chicago IL 60611

SHAMBERGER, RAYMOND JOSEPH, JR., biochemist, clin. chemist, researcher; b. Munising, Mich., Aug. 23, 1934; s. Raymond Joseph and Kathryn Anna S.; m. Barbara Ann Walsh, Jan. 31, 1959; children: Chrissa, Erik, Monica, Kara, Michael, Shannon. B.S., Alma (Mich.) Coll., 1956; M.S., Oreg. State U., 1959; Ph.D., Miami (Fla.) U., 1963. Cert. Am. Bd. Clin. Chemistry, Nat. Registry Clin. Chemistry. Dir. research Sutton Research Corp., Santa Monica, Calif., 1963-64; with Roswell Park Meml. Inst., Buffalo, 1964-69; sect. head enzyme lab. Cleve. Clinic Found., 1969—; prof. clin. chemistry Cleve. State U. Contbr. articles profl. jours. Recipient award Cancer Research, 1981. Mem. Am. Inst. Biol. Sci., N.Y. Acad. Scis., European Assn. Cancer Research, Soc. Geochemistry and Health, Am. Chem. Soc., Am. Assn. Clin. Chemistry, Am. Assn. Cancer Research, Am. Soc. Clin. Pathology, Fedn. Am. Socs. Exptl. Biology, Sigma Xi. Roman Catholic. Subspecialties: Clinical chemistry; Nutrition (biology). Current work: Mechanisms of cancer formation, enzyme and trace metal chemistry. Patentee in field.

SHAMOIAN, CHARLES A., geriatric psychiatrist, educator, hospt. adminstr.; b. Worcester, Mass., Oct. 5, 1931; s. Garabed and Anna (Varjabedian) S.; m. Paula, Oct. 8, 1961; children: Paula Ann, Charles R. A.B., Clark U., 1950, M.A., 1954; Ph.D., Tufts U., 1961, M.D., 1966. Diplomate: Am. Bd. Psychiatry and Neurology, 1981. Asso. prof. clin. psychiatry and pharmacology Cornell U. Coll. Medicine, N.Y.C., 1979—; dir. geriatric services N.Y. Hosp.—Cornell Med. Center, White Plains, 1979—. Contbr. articles to sci. jours. Mem. Larchmont Little League. USPHS fellow, 1960-61. Fellow Am. Psychiat. Assn.; mem. Am. Psychopath. Assn., Am. Geriatric Soc., Am. Assn. Geriatric Psychiatry (dir.). Subspecialties: Psychiatry; Psychopharmacology. Current work: Geriatric psychiatry; geriatric clinical psychopharmacology, dementia, depression. Office: 21 Bloomingdale Rd White Plains NY 10605

SHAMOS, MORRIS HERBERT, educator; b. Cleve., Sept. 1, 1917; s. Max and Lillian (Wasser) S.; m. Marion Jean Cahn, Nov. 26, 1942; 1 son, Michael Ian. A.B., NYU, 1941, M.S., 1943, Ph.D., 1948; postgrad., MIT, 1941-42. Faculty NYU, 1942—, prof. physics, 1959—; chmn. dept. Washington Sq. Coll., 1957-70; sr. v.p. research and devel. Technicon Corp., 1970-75, chief sci. officer, 1975-83, also dir., prin. sci. cons., 1983—; pres. M.H. Shamos & Assocs., 1983—; Cons. pvt. industry; cons. Armament Center, USAF, 1955-57, Tung-Sol Electric, Inc., 1949-65, Office Pub. Information, UN, 1958, NBC, 1957-67, AEC, 1957-70, N.Y. Eye and Ear Infirmary, 1961-64, 79—, L.I. Jewish Hosp., 1962—, N.Y.C. Health Dept., 1961-70, Technicon Instruments Corp., 1964-70, U.S. Office Edn., 1964-72. Author: Great Experiments in Physics, 1959; Co-editor: Recent Advances in Science, 1956, Industrial and Safety Problems of Nuclear Technology, 1950; cons. editor, Addison-Wesley Pub. Co., 1965-69; Adv. bd.: Jour. Coll. Sci. Teaching, 1971-75, 76-80, Clin. Lab. Guide, Am. Chem. Soc, 1972-76. Dir. tng., N.Y.C. Office Civil Def., 1950-54; subscribing mem. N.Y. Philharmonic Soc.; mem. adv. council Pace U., 1971—, N.Y. Poly. Inst., 1980—; Trustee Hackley Sch., 1971-80, Westchester Arts Council. Fellow N.Y. Acad. Scis. (past chmn. phys. scis., bd. govs. 1977—, rec. sec. 1978-80, v.p. 1980-81, pres. 1982), AAAS; mem. Nat. Assn. Ednl. Broadcasters, Am. Phys. Soc., IEEE, AAUP, AFTRA, Am. Chem. Soc., Nat. Sci. Tchrs. Assn. Am. (pres. 1967), Assn. Physics Tchrs. Britain, Chemist's Club, Phi Beta Kappa, Sigma Xi, Pi Mu Epsilon, Sigma Pi Sigma. Club: Cosmos. Subspecialties: Biophysics (physics); Bioinstrumentation. Current work: Consultant in high technology to the investment community, particulary in health care field. Spl. research atomic and nuclear physics, biophysics. Home: 3515 Henry Hudson Pkwy New York NY 10463

SHAN, HSIN-TSAN GRACE, microbiologist; b. Taiwan, June 29, 1950; d. Tse-Wen and Chiung-Hwa (Lin) Lo; m. Ming-chien Shan, Aug. 3, 1974; children: Vivian Yu-Wen, Eric Yu-Sen. M.S., Rutgers U., 1974. Microbiologist SRI Internat., Menlo Park, Calif., 1975-80; mem. sci. staff BNR Inc., Mountain View, Calif., 1980—. Mem. Am. Microbiology Soc., Assn. Computing Machinery. Roman Catholic. Subspecialties: Microbiology. Current work: Computerized analysis methodology for microbiological studies; data base management system. Office: 685A E Middlefield Rd Mountain View CA 94043

SHANDS, HENRY LEE, plant geneticist, researcher; b. Madison, Wis., Aug. 30, 1935; s. Ruebush George and Elizabeth (Henry) S.; m. Catherine Miller, Nov. 20, 1962; children: Deborah A., Jeanne A., James L. B.S., U. Wis., 1957; M.S., Purdue U., 1961, Ph.D., 1963. NSF fellow Swedish Seed Assn., Svalov, Sweden, 1962-63; asst. prof. Purdue U., 1963-66; with Purdue U.-AID Project, Minas Gerais, Brazil, 1963-65; asst. prof. dept. botany and plant pathology Purdue U., Lafayette, Ind., 1965-66; research agronomist and eastern wheat project leader DEKALB Hybrid Wheat, Inc., Lafayette, 1966-79; research agronomist, dir. sunflower research DeKalb-Pfizer Genetics (formerly DeKalb AgResearch, Inc.), Glyndon, Minn., 1979—. Served as 1st lt. Chem. Corps U.S. Army, 1957-59. Mem. AAAS, Am. Soc. Agronomy, Crop Sci. Soc. Am., Am. Genetic Assn., Genetics Soc. Can., European Assn. for Research on Plant Breeding, Am. Phytopathol. Soc. Subspecialties: Plant genetics; Plant pathology. Current work: Plant improvement through genetic and cytogenetic modification and manipulation. Home: 2309 Fairway Dr Moorhead MN 56560 Office: DeKalb-Pfizer Genetics Sunflower Research Ctr Route 2 Box 8AA Glyndon MN 56547

SHANER, GREGORY ELLIS, botanist; b. Portland, Oreg., Dec. 19, 1942. B.S., Oreg. State U., 1964, Ph.D., 1968. Mem. faculty Purdue U., West Lafayette, Ind., 1968—, now prof. botany and plant pathology, head dept. Mem. Am. Phytopath. Soc., Am. Soc. Agronomy, Soc. Econ. Botany. Subspecialties: Plant pathology; Plant genetics. Current work: Disease resistance in wheat and oats. Office: Dept Botany and Plant Pathology Purdue U West Lafayette IN 47907

SHANI, URI, computer scientist; b. Hadera, Israel, Sept. 16, 1950; came to U.S., 1976; s. Arie and Fruma Ivri Scheininger; m. Zipi Zuker, Feb. 28, 1973; children: Guy, Erez. B.A., Technion of Israel, Haifa, 1975; M.Sc., U. Rochester, 1977, Ph.D., 1981. Teaching asst. Technion, Haifa, 1975-76; research asst. U. Rochester, N.Y., 1976-80; researcher Gen. Electric Corp., research and devel., Schenectady, N.Y., 1980—. Served with Israeli Army, 1968-71. Recipient A. Gutvirt scholar Technion, 1975. Mem. Assn. Computing Machinery. Jewish. Subspecialties: Graphics, image processing, and pattern recognition; Software engineering. Current work: Computer vision, computer graphics, geometric modeling, programming environment, artificial intelligence, expert systems, languages. Office: Gen Electric Corp Research and Devel 1 River Rd Schenectady NY 12345 Home: 1800 Hillside Ave Schenectady NY 12309

SHANK, CHARLES VERNON, electrical engineer; b. Mt. Holly, N.J., July 12, 1943. B.S., U. Calif.-Berkeley, 1965, M.S., 1966, Ph.D. in Elec. Engring, 1969. Mem. tech. staff Bell Labs., Holmdel, N.J., 1969—, head dept. quantum physics and electronics, 1976—. Recipient Edward Longstreth medal Franklin Inst., 1983; Morris E. Leeds award IEEE, 1983. Mem. Nat. Acad. Engring. Subspecialty: Electrical engineering. Office: Bell Labs Crawford Corners Rd Holmdel NJ 07733

SHANK, MAURICE EDWIN, aero. engr.; b. N.Y.C., Apr. 22, 1921; s. Edwin A. and Viola (Lewis) S.; m. Virginia Lee King, Sept. 25, 1948; children: Christopher King, Hilary Lee, Diana Lewis. B.S., Carnegie Mellon U., Pitts., 1942; Sc.D., M.I.T., 1944. Registered profl. engr., Mass. Assoc. prof. mech. engring. M.I.T., 1949-60; dir. Advanced Materials Research and Devel. Lab., Pratt & Whitney Aircraft, 1960-70, mgr. materials engring. and research, 1971-72, dir. engring. tech., 1972-80, dir. engine design and structures engring., 1980-81; dir. engring-tech., comml. engring. Pratt & Whitney Aircraft Group, United Technologies Corp., East Hartford, Conn., 1981—; mem. aeros. adv. com., mem. subcom. aero. propulsion NASA, 1978—, chmn., 1978-81; mem. vis. com. aeros. and astronautics M.I.T., 1979—; mem. vis. com. metallurgy and materials sci. U. Pa., 1970—. Cons. editor, McGraw-Hill Book Co., 1960—; Contbr. articles to profl. jours. Bd. dirs., mem. exec. com. Coordinating Research Council, 1976—. Served to maj. U.S. Army, 1942-46. Fellow ASME, Metall. Soc. of AIME; mem. AIAA, Conn. Acad. Sci. and Engring. Subspecialties: Mechanical engineering; Aeronautical engineering. Current work: Aeronautical propulsion, analytical and component activities to support present and future engine programs, research and devel. to assure a superior level future technology. Office: 400 Main St Mail Stop 162-31 East Hartford CT 06108

SHANK, PETER R., microbiologist, educator; b. Ithaca, N.Y., Feb. 17, 1946; s. Donald Jay and Ruth Ann (Rabe) S.; m. Kathleen Ryan, Aug. 15, 1970; 1 son, Jonathan Ryan. B.S. in Microbiology, Cornell U., 1968; Ph.D. in Microbiology and Biochemistry, U. N.C., Chapel Hill, 1973. Postdoctoral fellow U. Calif.-San Francisco, 1974-76; vis. asst. prof., 1977; asst. prof. med. sci. Brown U., Providence, 1977-83, assoc. prof., 1983—. Anna Fuller postdoctoral fellow, 1974; NIH trainee, 1975-76; NIH grantee, 1978—. Mem. Soc. Gen. Microbiology, Am. Soc. Microbiology, AAAS. Subspecialties: Virology (medicine); Gene actions. Current work: Research on the molecular biology of RNA tumor viruses and mechanism of cell transformation. Office: Div Biology and Medicine Brown Univ Providence RI 02912

SHANK, ROBERT ELY, physician, emeritus preventive medicine educator; b. Louisville, Sept. 2, 1914; s. Oliver Orlando and Isabel Thompson (Ely) S.; m. Eleanor Caswell, July 29, 1942; children: Jane, Robert Oliver, Bruce. A.B., Westminster Coll., 1935; M.D., Washington U., 1939. Diplomate: Am. Bd. Nutrition. Intern, house physician Barnes Hosp., 1939-41; asst. resident physician, asst. in research Hosp. Rockefeller Inst. Med. Research, 1941-46; research asso. div. nutrition and physiology Pub. Health Research Inst. City N.Y., 1946-48; prof. preventive medicine Washington U. Sch. Medicine, 1948-55, Danforth prof. preventive medicine, 1955-83, prof. emeritus preventive medicine, 1983—; Cutter lectr. preventive medicine Harvard, 1964; Mem. food and nutrition bd. NRC, 1949-69; spl. cons. nutrition USPHS, 1949-53; chmn. adv. bd. health and hosps., St. Louis County, 1949-54; med. adv. bd. St. Louis Vis. Nurses Assn., 1950—; mem. com. food and nutrition nat. adv. bd. health services A.R.C., 1950-53; mem. adv. com. metabolism Office Surgeon Gen., 1956-60, mem. adv. com. on nutrition, 1964-72; mem. Am. Bd Nutrition, 1955-64, sec., treas., 1958-64; mem. Nat. Bd. Med. Examiners, 1957-58; co-dir. nutrition survey NIH, Peru, 1959, N.E. Brazil, 1963; mem. sci. adv. bd. Nat. Vitamin Found., 1958-61; mem. nutrition study sect. NIH, 1964-68, chmn. sect., 1966-68; mem. gastroenterology and nutrition tng. com., 1968-69, mem. nat. adv. child health and human devel. council, 1969-73; mem. clin. application and prevention adv. com. Nat. Heart, Lung and Blood Inst., 1976-80. Author sects. in med. textbooks, sci. paper relating to nutritional, metabolic disorders.; Asso. editor: Nutrition Revs, 1948-58; editorial adv. bd.: Nutrition Today, 1966-76, Hepatology, 1980—. Served as lt. comdr. M.C. USNR, 1942-46. Recipient Alumni Achievement award Westminster Coll., 1970. Fellow Am. Pub. Health Assn. (governing council 1955-56); mem. N.Y. Acad. Scis., Harvey Soc., Am. Soc. Biol. Chemists, Soc. Exptl. Biology and Medicine (council 1952-54), Central Soc. Clin. Research, Am. Soc. Clin. Investigation, Assn. Tchrs. Preventive Medicine (v.p. 1955-57, pres. 1957-58), A.M.A. (council on foods and nutrition 1960-69, chmn. 1963-66), Gerontological Soc., Assn. Am. Physicians, Am. Soc. Clin. Nutrition (council 1963-65, pres. 1967-68), Am. Dietetic Assn. (hon.), Am. Soc. for Study Liver Diseases (council 1963-66, pres. 1966), Am. Heart Assn. (chmn. nutrition com. 1973-76, award of merit 1981), Am. Inst. Nutrition, Sigma Xi, Alpha Omega Alpha. Subspecialties: Nutrition (medicine); Preventive medicine. Current work: Nutrients and aging, cardiovascular disorders and liver disease. Home: 1325 Wilton Ln Kirkwood MO 63122 Office: 4566 Scott Ave St Louis MO 63110

SHANKER, ROY JAMES, natural resources consultant; b. Cleve., July 15, 1948; s. David and Jean (Swartz) S.; m. Rosemary P. Jackson, Mar. 17, 1976 (div. June 1980). S.B., Swarthmore (Pa.) Coll., 1970; M.S., Carnegie-Mellon U., 1972, Ph.D., 1975. Research staff Inst. Def. Analyses, Arlington, Va., 1973-76; prin., ptnr. Resource Planning Assocs., Washington, 1976-80, Hagler Bailly & Co., 1980-82; pvt. practice natural resources consulting, Washington, 1982—; advisor, cons., reviewer Dept. Energy, NSF, Nat. Bur. Standards. Contbr. articles to profl. jours. Mellon fellow, 1970-73. Mem. Ops. Research Soc. Am. (chmn. energy group 1979), Inst. Mgmt. Sci. Republican. Jewish. Subspecialties: Regulated utilities; Operations research (mathematics). Current work: Electric utility system planning (optimal system design and operating). Home: 1657 Park Rd NW Washington DC 20010

SHANKLIN, DOUGLAS RADFORD, medical educator; b. Camden, N.J., Nov. 25, 1930; s. John F. and Muriel K. (Morgan) S.; m. Virginia A. McClure, Apr. 7, 1956; children: Elizabeth, Leigh, Lois Virginia, John Carter, Eleanor. A.B. in Chemistry, Syracuse U., 1952; M.D., SUNY-Syracuse, 1955. Lic. physician N.Y., Fla., Ill., Tenn. Intern Duke Hosp., Durham, 1955-56, resident in pathology, 1958, SUNY, Syracuse, 1959; instr. to assoc. prof. U. Fla., Gainesville, 1960-67; prof. ob/gyn U. Chgo., 1967-78; exec. dir. Santa Fe Found., Gainesville, 1978-83; prof. pathology U. Tenn., Memphis, 1983—; sr. mem. Marine Biol. Lab., Woods Hole, Mass., 1970—. Author: Maternal Nutrition and Child Health, 1979, Preventive Obstetrics: The Motherwell Experience, 1983, Tumors of Placenta, 1983; chief editor: Jour. Reproductive Medicine, 1970-75. Trustee Coll. Light Opera Co., Falmouth, Mass., 1969—; trustee Hippodrome Theater, Gainesville, 1978—; chmn. Survey Adv. Com., USDA, 1979—. Served to lt. USNR, 1956-58. Named Freeman citizen City Council Glasgow, Scotland, 1981; Best Sci. Tchr. U. Fla., 1967. Mem. Am. Assn. Pathologists, Soc. Pediatric Research, N.Y. Acad. Scis., Sigma Xi. Democrat. Episcopalian. Clubs: Quadrangle (Chgo.); The Barclay. Subspecialties: Pathology (medicine); Nutrition (medicine). Current work: Cellular and molecular bases for various neonatal diseases. Home: 66 Monroe Ave #1004 Memphis TN 38103 Home: PO Box 3086 Memphis TN 38173 Office: Dept Pathology U Tenn 858 Madison Ave Memphis TN 38163

SHANKS, ROGER D., geneticist, educator; b. Libertyville, Ill., May 30, 1951; s. Douglas and Esther (Sage) S.; m. Wendy D. Shanks, Aug. 28, 1971. B.S., U. Ill., 1974; M.S., Iowa State U., 1977, Ph.D., 1979. Asst. prof. genetics dept. dairy sci U. Ill.-Urbana, 1979—. Contbr. articles to profl. and popular jours. Mem. Am. Agrl. Econs. Assn., AAAS, Am. Dairy Sci. Assn., Am. Genetic Assn., Genetics Soc. Am., Biometric Soc., Alpha Zeta, Gamma Sigma Delta, Phi Kappa Phi. Methodist. Subspecialties: Animal breeding and embryo transplants; Animal genetics. Current work: Quantitative genetics and applied animal breeding; genetic and economic aspects of dairy cattle improvement programs; sire and cow evaluations for multiple traits.

SHANMUGAM, KEELNATHAM THIRUNAVUKKARASU, microbiologist, educator; b. Keelnatham, India, Oct. 15, 1941; came to U.S., 1965; s. Kilnatham and Mangayarkarasu T.; m. Valli Narayanasamy, Aug. 27, 1972; 1 son, Nataraj. B.Sc., Annamalai U., 1963; M.Sc., U.P. Agr. U., 1965; Ph.D., U. Hawaii, 1969. Asst. research chemist U. Calif.-San Diego, 1973-75; asst. research agronomist U. Calif.-Davis, 1976-80; asst. research scientist U. Fla., Gainesville, 1980-81, asso. prof. microbiology, 1981—. Mem. Am. Soc. Microbiology. Subspecialties: Microbiology; Nitrogen fixation. Current work: Researcher in molecular biology of nitrogen fixation, energetics of nitrogen fixation, molecular biology of hydrogen metabolism and anaerobic metabolism and solar energy conversion. Office: U Fla 1059 McCarty Hall Gainesville FL 32611

SHANNON, JACK, engineer, aerodynamics consultant; b. Brenham, Tex., July 18, 1938; s. Aubrey Willys and Josephine Paul (Griffin) S.; m.; children: Mark, John. B.S., U. Tex. Registered profl. engr., Wash. Mem. aerodynamics staff Boeing Co., Seattle, 1961-74; owner, pres. Shannon Engring., Seattle, 1975—. Author test flight documents, and design studies. Mem. Soc. Automotive Engrs. Subspecialty: Aeronautical engineering. Current work: Low drag high performance airfoils. Aerodynamics and acoustic technology and small technical business growth and management. Office: Boeing Field 213 Terminal Bldg Seattle WA 98108

SHAO, CHENG-YUAN, astronomer; b. Luyi, Henan, China, Nov. 3, 1927; came to U.S., 1959, naturalized, 1973; s. Z.-C. and Chi (Jen) S.; m. Fen Fang, May 1, 1955. B.A., Nat. Taiwan U., 1952; M.A., Harvard U., 1962. Sr. research asst. Harvard Obs., Cambridge, Mass., 1962-66, astronomer, 1982—, Smithsonian Astrophys. Obs., Cambridge, 1966-82. Contbr. articles to profl. jours. Vice-pres. Boston chpt. Nat. Assn. Chinese Ams., 1980-82, pres., 1982—. Mem. Am. Astron. Soc., Internat. Astron. Union, Astron. Soc. Pacific, Am. Assn. Variable Star Observers. Subspecialties: Planetary science; Optical astronomy. Current work: Comet and asteroid astronomy; variable stars; optical study of X-ray sources. Home: 20 Bryant Rd Lexington MA 02173 Office: 60 Garden St Cambridge MA 02138

SHAPERO, DONALD CAMPBELL, physicist, government agency executive; b. Detroit, Apr. 17, 1942; s. Donald Mayer and Lillian Emily (Campbell) S.; m. Diana B. Berner, Dec. 17, 1969 (div.); 1 son, Stephen B. B.S., MIT, 1964, Ph.D., 1970. Thomas J. Watson fellow IBM Corp., Yorktown Heights, N.Y., 1970-72; asst. prof. physics Am. U., Washington, 1972-73, Catholic U., 1973-75; exec. dir. Energy Research Adv. Bd. U.S. Dept. Energy, Washington, 1978-79; sr. staff officer Nat. Acad. Scis., Washington, 1975-78, spl. asst. for program coordination, 1979-82, exec. dir., 1982—; Exec. sec. com. effects on multiple nuclear weapon detonations Nat. Acad. Scis.-NRC, Washington, 1975-76, exec. sec. geophys. data panel, 1976, exec. sec. panel to assess nat. need for facilities dedicated to prodn. synchrotron radiation, 1976, sr. staff officer geophys. study com., 1976-78; dir, com. sci. and pub. policy nuclear risk survey Nat. Acad. Sci., 1976-78, sr. staff officer for five yr. outlook sci. and tech., 1979-82, sr. staff officer workshop sci. instrumentation, 1982. Contbr. sci. articles to profl. publs. Vice pres. Bethesda Jewish Congregation, 1983-84. NSF fellow, 1964-68; Cottrell Research grantee, 1975. Mem. Am. Phys. Soc. Subspecialties: Theoretical physics; Particle physics. Current work: Broad interest in science policy, particularly physical sciences and computer science and technology. Home: 5821 Edson Ln Rockville MD 20852 Office: Nat Acad Scis 2101 Constitution Ave Washington DC 20418

SHAPIRA, RAYMOND, biochemist, educator; b. New Bedford, Mass., June 29, 1928; s. Maurice and Anna (Levine) S.; m. Doris B. Kohler, Jan. 12, 1956; children: Stuart K., Gary L., Nathan A. B.S., U. N. Mex., 1950; Ph.D., Fla. State U., 1954. Research biochemist Biology div. Oak Ridge Nat. Lab., 1954-57; asst., then assoc. prof. dept. biochemistry Emory U., Atlanta, 1958-71, prof., 1972—. Author. Served with F.A. U.S. Army, 1945-47. Fulbright grantee Weizmann Inst. Sch., 1967; recipient Sinsheimer award Emory U., 1963. Mem. AAAS, Am. Chem. Soc., Fedn. Am. Socs. Exptl. Biology, N.Y. Acad. Sci., Am. Soc. Biol. Chemists, Am. Soc. Neurochemistry, AAUP. Jewish. Subspecialties: Biochemistry (medicine); Neurochemistry. Current work: Structure, metabolism and function of myelin proteins; effects of ionizing radiation including chemical protection; neutrophil enzymes, structure and function. Patentee in field. Home: 954 Liawen Ct Atlanta GA 30329 Office: 159 Woodruff Meml Bldg Atlanta GA 30322

SHAPIRO, ASCHER HERMAN, mechanical engineer, educator; b. Bklyn., May 20, 1916; s. Bernard and Jennie (Kaplan) S.; m. Sylvia Charm, Dec. 24, 1939 (div. 1959); children: Peter Mark, Martha Ann, Bernett Mary; m. Regina Julia Lee, June 4, 1961 (div. 1972). Student, CCNY, 1932-35; S.B., MIT, 1938, Sc.D., 1946; D.Sc. (honoris causa), Salford U., Eng., 1978. Asst. mech. engring. MIT, 1938-40, faculty, 1940—, prof. mech. engring., 1952—, prof. charge fluid mechanics div., mech. engring. dept., 1954-65, Ford prof. engring., 1962-75, chmn. faculty, 1964-65, head dept. mech. engring., 1965-74, inst. prof., 1975—; vis. prof. applied thermodynamics U. Cambridge, Eng., 1955-56; Akroyd Stuart Meml. lectr. Nottingham (Eng.) U., 1956; editor Acad. Press, Inc., 1962-65; cons. United Aircraft Corp., M.W. Kellogg Co., Arthur D. Little, Inc., Hardie-Tynes Mfg. Co., Carbon & Carbide Chems. Corp., Oak Ridge, Rohm & Haas Co., Ultrasonic Corp., Jackson & Moreland (Engrs.), Stone & Webster, Bendix Aviation, Oak Ridge Nat. Lab., Acushnet Processing Co., Kennecott Copper Co., Welch Sci., Sargent-Welch, others; served on sub-coms. on turbines, internal flow, compressors and turbines NACA; mem. Lexington Project to study and report on nuclear powered flight to AEC, summer 1948; dir. Project Dynamo to study and report to AEC on technol. and econs. nuclear power for civilian use, 1953, Lamp Wick study Office Naval Research, 1955; mem. tech. adv. panel aeronautics Dept. Def.; cons. ops. evaluation group Navy Dept.; sci. adv. bd. USAF, 1964-66; mem. Nat. Com. for Fluid Mechanics Films, 1962—, chmn., 1962-65, 71—; chmn. com. on ednl. films Commn. on Engring. Edn., 1962-65. Author: The Dynamics and Thermodynamics of Compressible Fluid Flow, vol. 1, 1953, vol. 2, 1954, Shape and Flow, 1961; also ednl. films and articles, tech. jours.; Editorial board: Jour. Applied Mechanics, 1955-56; editorial com.: Ann. Rev. Fluid Mechanics, 1967-71; mem.

editorial bd.: M.I.T. Press, 1977—; chmn., 1982—. Mem. Town Meeting Arlington, Mass.; chmn. 1st Mass. chpt. Atlantic Union Com., 1951-52, mem. council, 1954—; dir. lab. for devel. power plants for use in torpedoes Navy Dept., 1943-45; mem. ad hoc med. devices com. FDA, HEW, 1970-72; mem. com. Nat. Council for Research and Devel., Israel, 1971—; mem. com. sci. and pub. policy Nat. Acad. Scis., 1970-74; bd. govs. Technion, Israel Inst. Tech., 1968—. Recipient Naval Ordnance Devel. award, 1945; joint certificate outstanding contbn. War and Navy depts., 1947; Richards Meml. award ASME, 1960; Worcester Reed Warner medal, 1965; Fluids Engring. award, 1981; Townsend Harris medal Coll. City N.Y., 1978. Fellow Am. Inst. Aeros. and Astronautics, Am. Acad. Arts and Scis. (councilor 1967-71), ASME; mem. Am. Sci. Films Assn., Nat. Acad. Scis. (com. on sci. and pub. policy 1973-77), Nat. Acad. Engring., Biomed. Engring. Soc. (charter mem. 1968), Am. Soc. Engring. Edn. (Lamme medal 1977, Lamme award o61977), AAAS, Sigma Xi, Tau Beta Pi, Pi Tau Sigma. Clubs: Mass. Inst. Tech. Faculty (Cambridge); Cavendish (Brookline, Mass.). Subspecialties: Biomedical engineering; Fluid mechanics. Current work: Diagnostic and therapeutic devices and systems for the cardiovascular and pulmonary systems. Patentee fluid metering equipment, combustion chambers, propulsion apparatus and gas turbine aux., magnetic disc, magnetic disc storage device, vacuum pump, low-density wind tunnel, recipe calculator. Office: Mass Inst Tech 77 Massachusetts Ave Cambridge MA 02139

SHAPIRO, BURTON LEONARD, educator; b. N.Y.C., Mar. 29, 1934; s. Nat Lazarus and Fay Rebecca (Gartenhouse) S.; m. Eileen Roman, Aug. 11, 1958; children—Norah Leah, Anne Rachael, Carla Faye. Student, Tufts U., 1951-54; D.D.S., N.Y. U., 1958; M.S., U. Minn., 1962, Ph.D., 1966. Faculty U. Minn. Sch. Dentistry, Mpls., 1962—, asso. prof. div. oral pathology, 1966-70, prof., chmn. div. oral biology, 1970—, prof., chmn. dept. oral biology, 1979, prof. dept. oral pathology and genetics, 1979, dir. grad. studies, mem. grad. faculty genetics, 1966—, mem. grad. faculty pathobiology, 1979, univ. senator, 1968—, also mem. med. staff, exec. com., chmn.; hon. research fellow Galton Lab. dept. human genetics Univ. Coll., London, 1974. Mem. adv. editorial bd.: Jour. Dental Research, 1971—; Contbr. articles to profl. jours. Served to lt. USNR, 1958-60. Am. Cancer Soc. postdoctoral fellow, 1960-62; advanced fellow, 1965-68. Fellow Am. Acad. Oral Pathology, AAAS, mem. Internat. Assn. Dental Research (councilor 1969—), Am. Soc. Human Genetics, Craniofacial Biology Soc. (pres. 1972), Sigma Xi, Omicron Kappa Upsilon. Subspecialties: Genetics and genetic engineering (medicine); Cell and tissue culture. Current work: Search for basic defect in cystic fibrosis; explanation of findings in Down syndrome; cell calcium and aging; genetics and human disease. Home: 76 Exeter Pl St Paul MN 55104 Office: Dept Oral Biology Sch Dentistry U Minn Minneapolis MN 55455

SHAPIRO, DONALD MICHAEL, physician; b. Chgo., Jan. 29, 1941; s. Jack Abraham and Helen (Grossman) S.; m. Martha Lee Nazor, Mar. 2, 1979; 1 son, Ryan; children from previous marriage: Steven, Michael, Daniel, Susan. B.S., Roosevelt U., 1963; D.O., Chgo. Coll. Osteopathy, 1968. Diplomate: Am.Bd. Internal Medicine. Intern Detroit Osteo. Hosp., 1968-69, resident in medicine, 1969-70, Wilford Hall Med. Ctr., San Antonio, 1970-73; fellow in hematology-oncology Baylor Coll. Medicine, Houston, 1973-75; practice medicine specializing in hematology, med. oncology, Dayton, Ohio, 1977—; assoc. clin. prof. medicine Wright State U. Medicine, Dayton, 1976—; mil. cons. U.S. Air Force, Dayton, 1975-78. Contbg. editor: Osteo. Physician, 1975-77; contbr. articles to profl. jours. Trustee Miami Valley unit Am. Cancer Soc., Dayton, 1978-79, Greene County unit, Xenia, Ohio, 1982-84; med. advisor Greene County unit, 1982-84. Served to maj. USAF, 1970-77. Fellow ACP; mem. Am. Soc. Hematology, Am. Soc. Clin. Oncology, Undersea Med. Soc., Am. Osteo. Assn., Ohio Osteo. Assn. (physicians effectiveness com. 1982-83). Subspecialties: Hematology; Chemotherapy. Current work: Clinical chemotherapy, undersea medicine, medical education. Home: 9031 Normandy Ln Dayton OH 45459 Office: 1141 N Monroe Dr Xenia OH 45385

SHAPIRO, HOWARD MAURICE, physician, researcher; b. Bklyn., Nov. 8, 1941; s. Alfred Lester and Jennie Geraldine (Epstein) S.; m. Leslie Leona Hochberg, June 21, 1964; children: Jill Elise, Peter Jay. A.B., Harvard U., 1961; M.D., N.Y.U., 1965. Asst. research scientist engring. math. N.Y.U., 1962-65; mem. surg. house staff, 1965-67; research assoc., sr. staff fellow Nat. Cancer Inst., NIH, Bethesda, Md., 1967-71; surg. house staff U. Ariz., 1971-72; asst. dir. to dir. clin. research diagnostic products G.D. Searle & Co., 1972-76; cons., inventor, pres. Howard M. Shapiro (M.D., P.C.), West Newton, Mass., 1976—; research assoc. Beth Israel Hosp., 1982—, Sidney Farber Cancer Inst., 1977—; lectr. pathology Harvard Med. Sch., 1976—. Served with USPHS, 1967-70. Mem. AAAS, IEEE, Assn. Advancement Med. Instrumentation, Am. Assn. Cancer Research, Am. Soc. Clin. Oncology, Am. Soc. Hematology, Cell Kinetics Soc., Histochem. Soc., Soc. Analytical Cytology. Subspecialties: Bioinstrumentation; Cancer research (medicine). Current work: Development of biomedical instrumentation for clinical and research laboratories; analytical and quantitative cell biology relating to cell growth control.

SHAPIRO, IRVING MEYER, biochemist; b. London, Oct. 28, 1937; U.S., 1969, naturalized, 1977; s. Syd and Betty (Silver) S.; m. Joan Poliner, July 4, 1965; 1 dau., Susan. B.Dental Surgery, U. London, 1961, Ph.D., 1968; M.Sc., U. Liverpool, Eng., 1964. Research fellow Royal Dental Hosp., London, 1966-69; mem. faculty U. Pa. Sch. Dental Medicine, Phila., 1969—, prof., chmn. dept. biochemistry, 1976—. Contbr. articles to profl. jours. Nuffield fellow, 1964-65. Mem. Internat. Assn. Dental Research (Basic Sci. award 1974), AAAS, Am. Soc. Biol. Chemists, Am. Chem. Soc. Subspecialties: Biochemistry (biology); Environmental toxicology. Current work: Studies of the mechanisms of biological mineralization. Effects of toxic elements on human populations. Home: 555 Haverford Rd Wynnewood PA 19096 Office: Dept Biochemistry Dental Med Sch U Pa Philadelphia PA 19104

SHAPIRO, IRWIN IRA, physicist, educator; b. N.Y.C., Oct. 10, 1929; s. Samuel and Esther (Feinberg) S.; m. Marian Helen Kaplun, Dec. 20, 1959; children: Steven, Nancy. A.B., Cornell U., 1950; A.M., Harvard U., 1951, Ph.D., 1955. Mem. staff Lincoln Lab. MIT, Lexington, 1954-70; Sherman Fairchild Distinguished scholar Calif. Inst. Tech., 1974; Morris Loeb lectr. physics Harvard, 1975; prof. geophysics and physics MIT, 1967-80, Schlumberger prof., 1980—; Paine prof. practical astronomy, prof. physics Harvard U., 1982—; sr. scientist Smithsonian Astrophys. Obs., 1982—; dir. Harvard-Smithsonian Ctr. for Astrophysics, 1983—; cons. NSF. Contbr. articles to profl. jours. Recipient Albert A. Michelson medal Franklin Inst., 1975, award in phys. and math. scis. N.Y. Acad. Scis., 1982; Guggenheim fellow, 1982. Fellow AAAS, Am. Geophys. Union, Am. Phys. Soc.; mem. Am. Acad. Arts and Scis., Nat. Acad. Scis. (Benjamin Apthorp Gould prize 1979), Am. Astron. Soc. (Dannie Heineman award 1983), Internat. Astron. Union, Phi Beta Kappa, Sigma Xi, Phi Kappa Phi. Subspecialty: Radio and microwave astronomy. Current work: Radio and radar techniques: applications to astrometry, astrophysics, geophysics, planetary physics, and tests of theories of gravitation. Home: 17 Lantern Ln Lexington MA 02173 Office: 60 Garden St Cambridge MA 02138

SHAPIRO, JAMES ALAN, microbial geneticist, educator; b. Chgo., May 18, 1943; s. Henry I. and Soretta (Baim) S.; m. Joan E. Shapiro, June 14, 1964; children: Jacob N., Danielle E. B.A. magna cum laude, in English Lit, Harvard U., 1964; Ph.D. in Genetics, U. Cambridge, Eng., 1968. Fellow Service de Genetique Cellulaire, Institut Pasteur, Paris, 1967-68; research fellow dept. bacteriology and immunology Harvard U., 1968-70; prof. genetics U. Havana, Cuba, 1970-72; fellow Rosenstiel Basic Med. Scis. Research Center, Waltham, Mass., 1972-73; asst. prof. microbiology U. Chgo., 1973-78, assoc. prof., 1978-82, prof., 1982—; vis. prof. dept. microbiology Tel Aviv (Israel) U., 1980; mem. NSF genetic biology panel, 1981—. Mem. editorial bd.: Jour. Bacteriology, 1976—, Enzyme and Microbial Tech, 1981—, Biotech. Series, 1981—; contbr. articles to profl. jours. Subspecialty: Genetics and genetic engineering (biology). Current work: Bacterial genetics, transposable elements, hydrocarbon oxidation. Office: 920 E 58th St Chicago IL 60637

SHAPIRO, LINDA GAIL, computer science educator, consultant; b. Chgo., June 10, 1949; d. Aaron W. and Sylvia (Goldstein) Rosenberg; m. Robert M. Haralick, Feb. 12, 1978. B.S., U. Ill.-Urbana, 1970; M.S., U. Iowa-Iowa City, 1972, Ph.D., 1974. Asst. prof. Kans. State U., Manhattan, 1974-78, Va. Poly. Inst., Blacksburg, 1978-81, assoc. prof. computer sci., 1981—. Author: (with R.J. Baron) Data Structures and Their Implementation, 1980; editor: Computer Vision, Graphics, and Image Processing. Mem. IEEE (tech. com. 1981—, newsletter co-editor), Assn. Computing Machinery, Pattern Recognition Soc., Am. Assn. Artificial Intelligence. Subspecialties: Graphics, image processing, and pattern recognition; Artificial intelligence. Current work: Recognition of images of three-dimensional objects. Home: Route 3 Box 231-D Blacksburg VA 24060 Office: Va Poly Inst Dept Computer Sci Blacksburg VA 24061

SHAPIRO, LUCILLE, molecular biology educator; b. N.Y.C., July 16, 1940; d. Philip and Yetta (Stein) Cohen; m. Roy Shapiro, Jan. 23, 1960; 1 son, Peter; m. Harley H. McAdams, July 28, 1978; stepchildren: Heather, Paul. A.B. cum laude, Bklyn. Coll., 1962; Ph.D. in Molecular Biology, Albert Einstein Coll. Medicine, 1966. Asst. prof. dept. molecular biology Albert Einstein Coll. Medicine, Bronx, N.Y., 1967-72, assoc. prof., 1972-77, Siegfried Ullmann Prof. molecular biology, 1977—, dept. chmn., 1977—, dir. div. biol. scis., 1981—; cons. Searle Adv. Bd., 1982—; NIH bd. sci. counselors Nat. Inst. Arthritis, Metabolism and Digestive Diseases, 1980—. Contbr. chpts. to books, sci. papers to profl. jours. Jane Coffin Childs fellow, 1966-67; recipient Am. Cancer Soc. Faculty Research award, 1968-76; Hirschl Career Scientist award, 1976; Spirit of Achievement award, 1978. Mem. Am. Soc. Biol. Chemists. Office: Dept Molecular Biology Albert Einstein Coll Medicine Bronx NY 10461

SHAPIRO, MAURICE MANDEL, astrophysicist; b. Jerusalem, Nov. 13, 1915; U.S., 1921; s. Asher and Miriam R. (Grunbaum) S.; m. Inez Weinfield, Feb. 8, 1942 (dec. Oct. 1964); children: Joel Nevin, Elana Shapiro Ashley, Raquel Tamar Shapiro Kislinger. B.S., U. Chgo., 1936, M.S., 1940, Ph.D., 1942. Instr. physics and math. Chgo. City Colls., 1937-41; chmn. dept. phys. and biol. scis. Austin Coll., 1938-41; instr. math. Gary Coll., 1942; physicist Dept. Navy, 1942-44; lectr. physics and math. George Washington U., 1943-44; group leader, mem. coordinating council of lab. Los Alamos Sci. Lab., U. Calif., 1944-46; sr. physicist, lectr. Oak Ridge Nat. Lab., Union Carbon and Carbide Corp., 1946-49; cons. div. nuclear energy for propulsion aircraft Fairchild Engine & Aircraft Corp., 1948-49; head cosmic ray br. nucleonics div. U.S. Naval Research Lab., Washington, 1949-65, supt. nucleonics div., 1953-65, chief scientist Lab. for Cosmic Ray Physics, 1965-82, apptd. to chair of cosmic ray physics, 1966-82, chief scientist emeritus, 1982—; lectr. U. Md., 1949-50, 52—, assoc. prof., 1950-51; vis. prof. physics and astronomy U. Iowa, 1981—; vis. prof. astrophysics U. Bonn, 1982-84; cons. Argonne Nat. Lab., 1949; cons. panel on cosmic rays U.S. nat. com. IGY; lectr. physics and engring. Nuclear Products-Erco div. ACF Industries, Inc., 1956-58; lectr. E. Fermi Internat. Sch. Physics, Varenna, Italy, 1962; vis. prof. Weizmann Inst. Sci., Rehovoth, Israel, 1962-63, Inst. Math. Scis., Madras, India, 1971; Inst. Astronomy and Geophysics Nat. U. Mex., 1976; vis. prof. physics and astronomy Northwestern U., Evanston, Ill., 1978; cons. space research in astronomy Space Sci. Bd., Nat. Acad. Scis., 1965; cons. Office Space Scis., NASA, 1965-66; prin. investigator Gemini S-9 Cosmic Ray Expts., NASA, 1964-69, Skylab, 1967-76, Long Duration Exposure Facility, 1977—; mem. Groupe de Travail de Biologie Spatiale, Council of Europe, 1970—; mem. steering com. DUMAND Consortium, 1976—, mem. exec. com., 1979—; lectr. Summer Space Inst., Deutsche Physikalische Gesellschaft, 1972; dir. Internat. Sch. Cosmic-Ray Astrophysics, Ettore Majorana Centre Sci. Culture, 1977—, also sr. corr., 1977—; chmn. U.S. IGY com. on interdisciplinary research, mem. nuclear emulsion panel space sci. bd.; Nat. Acad. Scis., 1959—; chief U.S. rep., steering com. Internat. Coop. Emulsion Flights for Cosmic Ray Research; cons. CREI Atomics, 1959—; vis. com. Bartol Research Found., Franklin Inst., 1967-74; mem. U.S. organizing com. 13th Internat. Conf. on Cosmic Rays; mem. sci. adv. com. Internat. Confs. on Nuclear Photography and Solid State Detectors, 1966—; mem. Com. of Honor for Einstein Centennial, Acad. Naz. Lincei, 1977—. Editorial bd.: Astrophysics and Space Sci, 1968-75; asso. editor: Phys. Rev. Letters, 1977—; contbr. to: Am. Inst. Physics Handbook. Recipient Distinguished Civilian Service award Dept. Navy, 1967; Guggenheim fellow, 1962-63; medal of Honor Société d'Encouragement au Progrès, 1978; Publs. award Naval Research Lab., 1970, 74, 76; Dir.'s Spl. award, 1974. Fellow Am. Phys. Soc. (chmn. organizing com. div. cosmic physics, chmn. 1971-72, com. on publs. 1977—), AAAS, Washington Acad. Scis. (past com. chmn.); mem. Am. Inst. Physics, Research Soc. Am., Am. Astron. Soc. (exec. com. div. high-energy astrophysics 1978—, chmn.-elect 1981), Philos. Soc. Washington (past pres.), Am. Technion Soc. (Washington bd.), Assn. Los Alamos Scientists (past chmn.), Assn. Oak Ridge Engrs. and Scientists (past chmn.), Fedn. Am. Scientists (past mem. exec. com., nat. council), Internat. Astron. Union (exec. com. commn. on high-energy astrophysics), Phi Beta Kappa, Sigma Xi. Subspecialties: High energy astrophysics; Cosmic ray high energy astrophysics. Current work: Composition, origin; propagation and nuclear transformations of cosmic rays; neutrino astrophysics. Research in cosmic radiation: composition, origin, propagation, and nuclear transformations; high-energy astrophysics; particles and fields; nuclear physics; neutron physics and fission reactors, neutrino astronomy. Office: Van Allen Hall U Iowa Iowa City IA 52240

SHAPIRO, RALPH, meteorologist; b. Malden, Mass., Nov. 9, 1922; s. Samuel and Sadie (Aberbalt) S.; m. Sylvia Olfson, Aug. 22, 1945; children: Susan, Paul, Nancy. B.S., Bridgewater (Mass.) State U., 1943; M.S., MIT, 1948, Sc.D., 1950. Mem. staff planetary atmospheres research Lowell Obs., Flagstaff, Ariz., 1950-52; chief atmospheric dynamics br. Air Force Cambridge Research Lab., Bedford, Mass., 1952-70; chief climatology and dynamics br. Air Force Geophysics Lab., Bedford, 1970-80; sr. scientist Systems and Applied Scis. Corp., Lexington, Mass., 1980—. Contbr. numerous sci. articles to profl. publs. Served with USNR, 1943-47. Recipient Gunter Loesser Sci. award Air Force Geophys. Lab., 1979. Fellow Am. Meteorol. Soc.; mem. Am. Geophys. Union, Royal Meteorol. Soc., Sigma Xi, AAAS. Subspecialties: Meteorology; Applied mathematics. Current work: Applied mathematical research on problems of atmospheric dynamics, statistics and climatology. Home: 30 Wayne Rd Needham MA 02194 Office: 109 Massachusetts Ave Lexington MA 02173

SHAPLEY, JOHN ROGER, chemist, educator; b. Manhattan, Kans., Apr. 15, 1946; s. Carl and Ruth Margaret (Wolf) S.; m. Laurie Beck, June 6, 1970 (div.); 1 dau., Rebecca Lynn. B.S. in Chemistry, U. Kans., Lawrence, 1967, Ph.D., Harvard U., 1972. Postdoctoral fellow Stanford U., 1971-72; asst. prof. chemistry U. Ill. Urbana, 1972-78, assoc. prof., 1978-79, prof., 1979—; cons. in field. Research numerous publs. in field. Recipient Dreyfus Tchr. Scholan award Camille and Henry Dreyfus Found., 1978, Fresenius award Phi Lambda Upsilon, 1980; Sloan research fellow, 1978. Mem. Am. Chem. Soc., Royal Soc. Chemistry. Subspecialties: Inorganic chemistry; Catalysis chemistry. Current work: Chemistry and catalysis with organotransition metal compounds; organometallic chemistry; homogeneous catalysis. Office: 505 S Mathews Box 20NL Urbana IL 61801

SHAPSHAK, PAUL, molecular virologist, educator; b. Johannesburg, South Africa, Oct. 3, 1942; s. Rene and Eugenie (Palca) S.; m. Solveig Anne-Marie Aberg, Apr. 10, 1972; children: Mans Karl-Johann, Helga, Dag. B.A., Harvard U., 1693; Ph.D. in Biochemistry, Princeton U., 1969. Research assoc. Cornell U., 1969-71, postdoctoral fellow applied physics, 1971-72; postdoctoral fellow biophysics U. Wis.-Madison, 1972-73; sr. scientist advanced systems lab. Frederick Cancer Research Ctr., Litton Bionetics, Inc., Ft. Detrick, Md., 1973; staff fellow Nat. Inst. Child Health and Human Devel., 1973-75; postdoctoral fellow microbiology, immunology and pathology UCLA, 1975-77; mgr. research and devel. biochemistry lab. Redken Labs., Inc., Van Nuys, Calif., 1977-78; adj. asst. prof. pediatrics Harbor-UCLA Med. Ctr., Torrance, Calif., 1979—; sr. research assoc. neurology VA Wadsworth Med. Ctr., Los Angeles, 1981—, co-dir. dept. neurology, 1981—; asst. research neurologist UCLA Sch. Medicine, Westwood, 1982—. Contbr. articles and abstracts to profl. jours. Counselor Boy Scouts Am., 1978—; trustee Temple Beth Sholom, 1982—; active polit. campaigns. Harvard Coll. scholar, 1959-63; N.Y. U. Med. Ctr. Honors Program scholar, 1963-64; NIH predoctoral fellow, 1964-69; NRC travel fellow, 1969; NIH grantee, 1979-81, 81-84; Kroc Found. grantee, 1982-84; NIH pediatrics grantee, 1982-87, VA, 1983-84. Mem. Am. Soc. Virology (founding), AAAS, Am. Soc. Microbiology, Tissue Culture Assn. So. Calif., N.Y. Acad. Scis., Am. Chem. Soc., Soc. Cosmetic Chemists. Democrat. Jewish. Subspecialties: Neuro-viral-immunology; Virology (biology). Current work: Molecular virology, cloning and sequencing viral CDNAs, multiple sclerosis, neurovirology, neuroimmunology, persistent viral infections of the central nervous system, virsu evolution. Home: 1137 23d St Apt 3 Santa Monica CA 90403 Office: VA Medical Center Neurology 691 127 Wadsworth Los Angeles CA 90073

SHARGEL, LEON DAVID, pharmacologist, educator; b. Balt., Nov. 18, 1941; s. Earl and Irene (Singer) S.; m. Sharon Lee Fine, Mar. 18, 1944; children: Deborah, Jeffrey. B.S. in Pharmacy cum laude, U. Md., 1963; Ph.D. in Pharmacology, George Washington U., 1969. Registered pharmacist. Assoc. research biologist, group leader dept. drug metabolism and disposition Sterling-Winthrop Research Inst., Rensselaer, N.Y., 1969-75; asst. prof. pharmacy and pharmacology Northeastern U. Coll. Pharmacy, Boston, 1975-78, assoc. prof. pharmacy and pharmacology, 1978-82, sect. leader pharmaceutics, 1980-82; assoc. prof. pharmacy and pharmacology, mgr. biopharm. analysis lab. Mass. Coll. Pharmacy and Allied Health Scis., Boston, 1982—; cons. Food Research, Boston, 1975-77, Appleton-Century Crofts, N.Y.C., 1977—, Dooner Labs., Inc., Haverhill, Mass., 1979—, Breon Labs., Inc., N.Y.C., 1981. Contbr. articles, research papers to sci. lit. Mem. AAAS, Am. Assn. Colls. Pharmacy, Am. Pharm. Assn., Acad. Pharm. Sci., Am. Soc. Pharmacology and Exptl. Therapeutics (div. drug metabolism), Internat. Soc. Study Xenobiotics, Sigma Xi, Rho Chi. Subspecialties: Pharmacology; Toxicology (medicine). Current work: Drug metabolism and disposition, pharmacokinetics, biopharmaceutics. Office: 179 Longwood Ave Boston MA 02115

SHARMA, JAGDEV MITTRA, veterinary microbiologist, researcher; b. India, June 28, 1941; came to U.S., 1962, naturalized, 1974; s. Hari Ram and Devki S.; m. Sylvia Ann Lemus, June 21, 1969; children: Dave, Susan. B.V.Sc., Punjab U., Hissar, India, 1961; M.S., U. Calif.-Davis, 1964, Ph.D., 1967. Instr. in vet. surgery Punjab U., 1961; jr. specialist U. Calif., Davis, 1962-66, Regents fellow in comparative pathology, 1966-67; poultry pathologist Wash. State U., 1967-71; vet. med. officer Regional Poultry Research Lab., U.S. Dept. Agr., East Lansing, Mich., 1971—. Contbr. numerous articles to sci. jours. Recipient Cert. of Merit Punjab U., 1961; Outstanding Performance award U.S. Dept. Agr., 1976-77; Achievement award Upjohn Co., 1982. Mem. AVMA, World Vet. Poultry Assn., Am. Soc. Microbiology, Am. Assn. Avian Pathologists, Sigma Xi. Subspecialty: Microbiology (veterinary medicine). Current work: Pathology and immunology of neoplastic and non-neoplastic diseases of chickens. Home: 4611 Sequoia Trail Okemos MI 48864 Office: 3606 E Mount Hope Rd East Lansing MI 48823

SHARMA, RAGHUBIR PRASAD, toxicologist, educator; b. India, Sept. 7, 1940; came to U.S., 1964, naturalized, 1979; s. Rammurti and Ramdulari S.; m. Lalita Shukla, Dec. 13, 1958; children: Rajesh, Sanjeev. D.V.M., U. Rajasthan, India, 1959; Ph.D., U. Minn., 1968. Diplomate: Am. Bd. Vet. Toxicology, Am. Bd. Toxicology; cert. indsl. hygienist. Clin. practice vet. medicine, 1959-60; instr. pharmacology; U.P. Agrl. U., India, 1960-61, asst. prof., 1961-64; research fellow U. Minn., 1964-69; asst. prof. toxicology Utah State U., Logan, 1969-75, asso. prof., 1975-79, prof., 1979—; cons. in field. Editor: Immunologic Considerations in Toxicology, Vol. I, 1981, Vol. II, 1981; contbr. over 150 articles and abstracts to sci. jours. Recipient various research awards USPHS, State of Utah, pvt. industry. Mem. Soc. Toxicology, Am. Soc. Pharmacology and Exptl. Therapeutics, AVMA, Am. Indsl. Hygiene Assn. Subspecialty: Toxicology (medicine). Current work: Immunotoxicology; biochemical mechanisms of toxic action. Home: 1375 North 1600 East Logan UT 84321 Office: Utah State U UMC 56 Logan UT 84322

SHARMA, SANSAR C., devel. neurobiologist, educator, researcher; b. Pirthipur, India, Mar. 10, 1938; came to U.S., 1968, naturalized, 1974; s. Ram Rattan and Hukmi Devi S.; m. Janet Phillips, Dec. 4, 1970; children: David, Nina. B.Sc., M.Sc. with honors, Panjab U., 1962; Ph.D. in Physiology, U. Edinburgh Med. Sch., 1967. Teaching asst. Panjab U., 1961-62; research scholar Council for Sci. and Indsl. Research, 1963-64; lectr. zoology, Ambala, India, 1962-63; research scholar U. Edinburgh Med. Sch., 1966-67, postdoctoral fellow, 1967-68; research assoc. Washington U., St. Louis, 1968-72; asst. prof. ophthalmology N.Y. Med. Coll., Valhalla, 1972-77, assoc. prof., 1977-81, prof., 1981—; adj. prof. anatomy, 1981—; summer fellow Marine Biol. Labs., Woods Hole, Mass., 1965. NSF research grantee, 1974-76; Nat. Eye Inst./NIH research grantee, 1975-83; NIH research career devel. grantee, 1978-83. Fellow Am. Soc. Exptl. Biology; mem. Soc. for Neurosci., Assn. for Research in Vision and Ophthalmology, Internat. Union Physiologists, Brit. Physiol. Soc., Physiol. Soc., Indian Physiol. Soc. Subspecialties: Comparative neurobiology; Neurophysiology. Current work: Genesis of visual pathways; retinotectal connections; study of development and regeneration of visual pathways between the retina and the brain in lower vertebrates. Office: Dept Opthalmology NY Med Coll Valhalla NY 10595

SHARMA, SUBHASH CHANDER, med. physicist; b. Bhiwani, India, Apr. 25, 1946; came to U.S., 1968, naturalized, 1981; s. Dhanpat

Rai and Asharfi (Devi) S.; m. Meena Tiwari, Nov. 13, 1971; children: Yashi, Akash. B.Sc., Panjab U., India, 1964; M.Sc., Birla Inst. Tech. and Sci., India, 1966, U. Nebr., Lincoln, 1970, Ph.D., 1971. Diplomate: Am. Bd. Radiology. Lectr. in physics Jat Coll., Rohtak, India, 1966-68; teaching asst. U. Nebr., Lincoln, 1968-70, research asst., 1970-71, instr., 1972, 75, postdoctoral research assoc. physics dept., 1971-75; tng. fellow med. radiation physics M.D. Anderson Hosp., Tex. Med. Cancer Ctr., Houston, 1975-76; instr. radiation physics div. radiation oncology Washington U. Sch. Medicine, St. Louis, 1976-78; asst. prof. dept. therapeutic radiology U. Minn. Hosp., Mpls., 1978-81; assoc. prof. dept. therapeutic radiology U. Louisville, James Graham Brown Cancer Ctr., Louisville, 1981—. Contbr. articles to profl. jours. Mem. Am. Assn. Physicists in Medicine, Am. Soc. Therapeutic Radiologist, Radiation Research Soc. Subspecialty: Radiology. Current work: Radiation physics related to clinical dosimetry of photons and electrons. Home: 1216 Westlynne Way Louisville KY 40222

SHARP, AARON JOHN, botanist, educator; b. Plain City, Ohio, July 29, 1904; s. Prentice Daniel and Maude Katharine (Herriott) S.; m. Cora Evelyn Bunch, July 25, 1929; children: Rosa Elizabeth, Maude Katharine, Mary Martha, Fred Prentice, Jennie Lou. A.B., Ohio Wesleyan U., 1923-27, D.Sc., 1952; M.S., U. Okla., 1927-29; Ph.D., Ohio State U., 1936-38. Botanist technician Nat. Park Service, Great Smoky Mountains Nat. Park, summer, 1934, ranger-naturalist, summers, 1939-41; acting editor The Bryologist, 1943- 44, asso. editor, 1938-42, 1945-53, Castanea, 1947-66; instr. summer sch. W.Va. U., 1939-41; instr. U. Tenn., 1929-37, asst. prof., 1937-40, asso. prof., 1940-46, prof., 1946-65, Distinguished Service prof., 1965-74, prof. emeritus, 1974—, curator herbarium, 1949-68, asso. curator, 1968—, head botany dept., 1951-61; mem. staff Hattori Bot. Lab., 1956—; vis. prof. Stanford U., 1951, U. Mich. Biol. Sta., 1954-57, 59-64, U. Minn. Biol. Sta., 1971, U. Mont. Biol. Sta., 1972, Nat. U. Taiwan, 1965, Instituto Universitario Pedagógico Experimental, Maracay, Venezuela, 1976, U. Va. Biol. Sta., 1980; Cecil Billington lectr. Cranbrook Inst. Sci., 1947; vis. lectr. Am. Inst. Biol. Scis., 1967-70; trustee Highlands (N.C.) Biol. Labs., 1934-38, 1948-64, bd. mgrs., 1946-52; sec. of sect. Inter-Am. Conf. on Conservation of Renewable Natural Resources, Denver, 1948; mem. nat. adv. bd. Ministry of Ecology, 1975-81. Contbr.: articles to sci. jours. Ency. Brit; asso. editor: Hattori Bot. Jour., Nichinan, Japan, 1961—. Bd. dirs. Gt. Smoky Mountains Conservation Assn., 1960—, Gt. Smoky Mountains Nat. Hist. Assn., 1979-81, U. Tenn. Arboretum Soc., 1979—. Guggenheim Found. fellow, 1944-46. Fellow AAAS (v.p. 1963); mem. New Eng. Bot. Club, Internat. Soc. Phytomorphologists, Internat. Assn. Plant Taxonomy, So. Appalachian Bot. Club (pres. 1946-47), Sullivant Moss Soc. (pres. 1935), Am. Bryol. and Lichen. Soc., Am. Fern Soc., Bot. Soc. Am. (editorial com. 1948-53, treas. 1957-62, v.p. 1963, pres. 1965, Merit award 1972), Soc. for Study Evolution, Soc. Botánica de México (hon.), Soc. Mexicana de Historia Natural, Tenn. Acad. Sci. (exec. com. 1943-44, v.p. 1952, pres. 1953), Am. Soc. Plant Taxonomists (pres. 1961), Assn. Southeastern Biologists (v.p. 1956, Meritorious Tchr. award 1972), Ecol. Soc. Am. (v.p. 1958-59), Torrey Bot. Club, Nature Conservancy (gov. 1955-61), Am. Soc. Naturalists, AAUP, Bot. Brit. Bryol. Soc., Internat. Soc. Tropical Ecology, Internat. Phycolog. Soc., Nat. Assn. Biology Tchrs., Palynolog. Soc. India, Phycolog. Soc. Am., Systematics Assn., Am. Assn. Stratigraphic Palynol., Soc. Latino-americand de Briologia (hon.), Phi Beta Kappa, Phi Kappa Phi, Sigma Xi, Phi Sigma, Phi Epsilon Phi, Sigma Delta Pi. Club: Explorers. Subspecialties: Taxonomy; Ecology. Current work: Plant geography in relation to geological and environmental factors. Home: 1201 Tobler Rd Knoxville TN 37919 Office: U Tenn Knoxville TN 37916 The Universe is so constructed that for every error committed someone must pay a penalty now or in the future.

SHARP, CHARLES WILLIAM, biochemist, government scientist; b. Winchester, Ind., June 27, 1933; s. John Edward and Gladys Pauline (Petty) S.; m. Betty Gertrude Gehlbac, Aug. 28, 1954; children: Jack Kevin, Steven Kary. B.S. in Pharmacy, Butler U., Indpls., 1957; Ph.D. in Biochemistry, Ind. U., 1966. Postdoctoral fellow dept. pediatrics U. Wash., 1966-69; staff fellow Nat. Inst. Environ. Health Scis., 1969-74; biochemist, project officer div. research Nat. Inst. Drug Abuse, Rockville, Md., 1974—. Author: Volatile Substances, 1981, also monographs; contbr. articles to profl. jours. Mem. Am. Soc. Neurochemistry, Soc. of Neurosci., AAAS. Subspecialties: Neurochemistry; Neuropharmacology. Current work: Monitor, develop and initiate research on drug abuse, primarily dealing with biochemical mechanisms of drug actions; with primary interests in volatile substances, opiates, and other psychoactive substances. Home: 14009 Arctic Ave Rockville MD 20853 Office: Research Div Nat Inst Drug Abuse 5600 Fishers Ln Rockville MD 20857

SHARP, ROBERT PHILLIP, geology educator, adminstr.; b. Oxnard, Calif., June 24, 1911; s. Julian Hebner and Alice (Darling) S.; m. Jean Todd, Sept. 7, 1938; children: Kristin Sharp Lytle, Bruce T. B.S., Calif. Inst. Tech., 1934; Ph.D., Harvard U., 1938. Inst., asst. prof. geology U. Ill., Urbana, 1938-43; assoc. prof., then prof. geology U. Minn., Mpls., 1945-47; prof. div. geol. scis. Calif. Inst. Tech., Pasadena, 1947—, chmn. div., 1952-67, prof., 1967—, prof. emeritus, 1979—. Contbr. numerous articles on geology to profl. jours. Served to capt. USAAF, 1943-45. Named One of Ten Outstanding U.S. Coll. Tchrs. Life Mag., 1950; recipient NASA Exceptional Sci. Achievement medal, 1971; Nat. Acad. Scis. GSA Penrose award, 1977. Fellow Geol. Soc. Am. (Kirk Bryan award 1964), AAAA; mem. Glaciological Soc., Am. Geophys. Union. Republican. Subspecialties: Geology; Geomorphology. Current work: Forms and processes on planetary surfaces, particularly earth and Mars. Office: 1201 California Blvd Pasadena CA 91125

SHARP, ROBERT RICHARD, biophysicist, educator; b. Newport News, Va., May 15, 1941; s. Edward Raymond and Elvira (Bona) S.; m. Maria Besan, July 25, 1966; 1 son, David. A.B., Case Western Reserve U., 1965, M.S., 1965, Ph.D., 1969; postdoctoral fellow, Oxford U., Eng., 1968-70. Prof. chemistry U. Mich., Ann Arbor, 1970—. Contbr. articles on biophysics to profl. jours. Mem. Am. Chem. Soc., AAAS, Am. Physics Assn., Sigma Xi. Subspecialties: Photosynthesis; Nuclear magnetic resonance (chemistry). Current work: Nuclear magnetic resonance studies of photosynthesis and neurochemistry. Home: 3047 Overridge Dr Ann Arbor MI 48104 Office: Chemistry Dep U Mich Ann Arbo MI 48109

SHARPE, LAWRENCE GRADY, drug addiction researcher; b. Macon, Ga., Feb. 2, 1935; s. Grady Benjamin and Kathleen (Postell) S.; m. Caryl Ann Bodine, Sept. 2, 1961; 1 dau., Lauren. Ph.D., Purdue U., 1968. Asst. prof. neuropsychology dept. psychiatry Washington U. Sch. Medicine, St. Louis, 1968-74; research psychologist Addiction Research Center, Nat. Inst. Drug Abuse, Lexington, Ky., 1974—. Contbr. articles to sci. jours. Served with U.S. Army, 1955-57. Mem. AAAS, Soc. Neurosci., Fedn. Am. Scientists. Subspecialties: Neuropharmacology; Neuropsychology. Current work: The role of brain biochemistry in the drug addiction process; chemical/electrical stimulation and recording of brain sites in animals to determine how drugs of addiction act on specific brain areas to cause addiction. Office: Addiction Research Center Nat Inst Drug Abuse PO Box 12390 Lexington KY 40583

SHARPLESS, NANSIE SUE, neurochemist, educator; b. West Chester, Pa., Oct. 11, 1932; d. George Roberts and Lorraine Eleanor (Way) S. B.A., Oberlin Coll., 1954; M.S., Wayne State U., 1956, Ph.D., 1970. Med. technologist Henry Ford Hosp., Detroit, 1956-67; research asst., research assoc., sr. research fellow Mayo Clinic/Mayo Found., Rochester, Minn., 1970-74; asst. prof. psychiatry Albert Einstein Coll. Medicine, Bronx, N.Y., 1975-81, assoc. prof. psychiatry and neurology, 1981—, lab. chief clin. neuropsycho pharmacology, 1983—. Bd. dirs. Alexander Graham Bell Assn. for Deaf, 1980-83; Pres. Found. Sci. and Handicapped, Inc., 1982-84. Recipient Disting. Alumni award Wayne State U., 1980. Mem. Soc. Neurosci., Internat. Soc. Neurochemistry, Am. Chem. Soc., N.Y. Acad. Scis., AAAS (com. office of opportunities in sci. 1977—), Am. Assn. Clin. Chemistry, Soc. Exptl. Biology and Medicine. Quaker. Subspecialties: Neurochemistry; Biochemistry (medicine). Current work: Basic and clinical research in neurotransmitter chemistry, biogenic amines and neurological and psychiatric disorders. Office: 1300 Morris Park Ave Bronx NY 10461

SHARPTON, FRANCIS ARTHUR, physics educator, researcher; b. New Vienna, Ohio, Aug. 2, 1936; m. Wanda Ruth Penland, Dec. 22, 1957; 1 son, Jeffrey. B.S., Coll. Ozarks, 1958; M.S., U. Ark., 1960; Ph.D., U. Okla., 1968. Asst. prof. physics Hendrix Coll., 1960-64; assoc. prof. Olivet Nazarene Coll., Kankakee, Ill., 1968-70; prof. physics Northwest Nazarene Coll., Nampa, Idaho, 1970—; researcher U. Ark., 1977; vis. prof. U. Wis., Madison, 1979, 80, 81, 82—. Contbr. articles to profl. jours. Research Corp. grantee, 1970. Mem. Am. Phys. Soc., Sigma Pi Sigma, Sigma Xi. Mem. Ch. of the Nazarene. Subspecialty: Atomic and molecular physics. Current work: Physics educator at undergrad. level and carrying on an active research program in atomic physics. Office: Dept Physics Northwest Nazarene College Nampa ID 83651

SHATKIN, AARON JEFFREY, biochemist; b. Providence, July 18, 1934; s. Morris and Doris S.; m. Joan A. Lynch, Nov. 30, 1957; 1 son, Gregory Martin. A.B., Bowdoin Coll., 1956, D.Sc. (hon.), 1979; Ph.D., Rockefeller Inst., 1961. Sr. asst. scientist NIH, Bethesda, Md., 1961-63, research chemist, 1963-68; vis. scientist Salk Inst., La Jolla, Callif., 1968-69; asso. mem. dept. cell biology Roche Inst. Molecular Biology, Nutley, N.J., 1968-73, full mem., 1973-77, head molecular virology lab., 1977—, head dept. cell biology, 1983—; adj. prof. cell biology Rockefeller U. Mem. editorial bd.: Jour. Viology, 1969-82, Archives of Biochemistry & Biophysics, 1972-82, Virology, 1973-76, Comprehensive Virology, 1974-82, Jour. Biol. Chemistry, 1977—; editor-in-chief: Molecular and Cellular Biology, 1980—. Served with USPHS, 1961-63. Recipient U.S. Steel Found. prize in molecular biology, 1977; Rockefeller fellow, 1956-61. Fellow N.Y. Acad. Scis.; Mem. Nat. Acad. Scis., Am. Soc. Microbiology, Am. Soc. Biol. Chemists, AAAS, N.Y. Acad. Scis., Harvey Soc. Subspecialties: Molecular biology; Biochemistry (biology). Current work: Mechanisms of Eukaryotic gene expression structure and function of animal viruses. Home: 4 Tanglewood Rd North Caldwell NJ 07006 Office: Roche Inst Molecular Biology Nutley NJ 07110

SHAW, CHORNG-GANG, radiological physicist, educator, cons.; b. Taichung, Taiwan, July 18, 1949; s. Ker-Chin and Jye-Shin (Jong) S.; m. Yue-Hua, Aug. 15, 1976; 1. Ph.D., U. Wis., Madison, 1981. Research asst. U. Wis., Madison, 1976-81; asst. prof. radiology, radiation physica SUNY Upstate Med. Center, Syracuse, 1981—; cons. med. imaging. Contbr. articles to profl. publs. Recipient gold medal for sci. exhibits Am. Roentgen Ray Soc., 1981. Mem. Am. Assn. Physicists in Medicine, Soc. Photo-Optical Instrumentation. Subspecialties: Imaging technology; Optical image processing. Current work: Physics and instrumentation of digital radiology, computerized tomography, nuclear magnetic resonance imaging. Office: SUNY Upstate Med Center 750 E Adams St Syracuse NY 13219

SHAW, DAVID ELLIOT, computer science educator; b. Chgo., Mar. 29, 1951; s. Charles Bergman and Marilyn (Baron) S. B.A., U. Calif.-San Diego, 1972; M.S., Stanford U., 1974, Ph.D., 1980. Pres. Stanford Systems Corp., Palo Alto, Calif., 1976-79; assoc. prof. dept. computer sci. Columbia U., 1980—, dir. NON-VON supercomputer project, 1980—; cons. IBM, Bell Labs., various other firms in computer, semi-condr. and info. systems industries; dir. Crofton Group, Inc., Computer Music Corp. Reviewer computer textbooks Prentice-Hall, various publs.; mem. editorial adv. bd.: Sci. Digest; contbr. articles to profl. jours. Advanced Research Projects Agy. grantee. Mem. Assn. for Computing Machinery, IEEE. Democrat. Jewish. Subspecialty: Parallel Supercomputers and VLSI Systems.. Current work: Principal architect of the NON-VON supercomputer, a highly parallel machine containing over a quarter million processing elements. Office: Dept Computer Sci Columbia U New York NY 10027

SHAW, EUGENE DOUGLAS, microbiologist, virologist; b. Burlington, Iowa, Aug. 3, 1925; s. William Douglas and Eunice Caroline (Sourwine) S.; m. Elizabeth Ann O'Brien, June 24, 1956; children: Douglas M., Gregory F., Kevin J., Daniel R. B.S., Northwestern U., 1949; M.S. in Zoology, Iowa State U., 1955; Ph.D. in Microbiology, Iowa State U., 1963. Commd. 2d lt. U.S. Army, 1951, advanced through grades to lt. col.; served in, Korea, 1952-53, 1953-55, Ft. McPherson, Ga., 1956-61, Washington, 1963-70, ret., 1970; prin. scientist Ortho Research Inst. Med. Sci., Raritan, N.J., 1970-78; dir. infectious diseases research and devel. Wampole Labs., Cranbury, N.J., 1978-82; pres. Shaw Systems, Inc., Princeton Junction, N.J., 1982—; engaged in indsl. research and devel. mgmt. Contbr. sci. articles to profl. jours. Sec. Sherbrooke Estates Homeowners Assn., 1980-82. Served with U.S. Army, 1944-46. Decorated Combat Infantryman badge; recipient Philip B. Hofmann award Johnson & Johnson, 1976. Mem. Am. Soc. Microbiology, Pan Am. Group for Rapid Viral Diagnosis, SAR, Assn. Vets. of Battle of Bulge, Sigma Xi. Republican. Methodist. Lodge: Lions. Subspecialties: Microbiology (medicine); Virology (medicine). Current work: Rapid diagnosis of infectious diseases. Patentee in field. Home: 35 Berkshire Dr Princeton Junction NJ 08550 Office: Shaw Systems Inc PO Box 140 Princeton Junction NJ 08550

SHAW, JAMES ELWOOD, microbiologist, researcher; b. Tulsa, Okla., Feb. 26, 1941; s. James France and Helen Rose (Smith) S.; m. Phyllis Ann Throckmorton, Aug. 8, 1964; 1 dau., Melanie Ann. B.A., N. Tex. State U., 1964, M.A., 1966; Ph.D., U. Okla., 1970. Postdoctoral fellow dept. biochemistry McGill U., 1970-73; research assoc. dept. bacteriology and immunology U. N.C., 1973-77, research asst. prof., 1977-82; asst. prof. dept. med. microbiology and immunology Ohio State U., 1982—. Contbr. articles to profl. jours. NIH trainee, 1967-70; Nat. Cancer Inst. Program Project awardee, 1975—. Mem. Fedn. Am. Scientists, Am. Soc. Microbiology, AAAS, Sigma Xi. Methodist. Subspecialties: Microbiology; Virology (biology). Current work: Teachning and basic research in molecular virology (persistent and latent viral infections). Office: Ohio State Dept Medical Microbiology and Immunolog 5072 Graves Hall 333 W 10th Ave Columbus OH 43210

SHAW, MARGERY WAYNE SCHLAMP, physician, lawyer; b. Evansville, Ind., Feb. 15, 1923; d. Arthur George and Louise (Meyer) Schlamp; m. Charles Raymond Shaw, May 31, 1942 (div. Nov. 1972); 1 dau., Barbara Rae. Student, Hanover Coll., 1940-41; A.B. magna cum laude, U. Ala., 1945; M.A., Columbia U., 1946; postgrad., Cornell U., 1947-48; M.D. cum laude, U. Mich., 1957; J.D., U. Houston, 1973; D.Sc. (hon.), U. Evansville, 1977. Intern St. Joseph Mercy Hosp., Ann Arbor, Mich., 1957-58; practice medicine specializing in human genetics, Ann Arbor, 1958-67, Houston, 1967—; instr. dept. human genetics Med. Sch. U. Mich., 1958-61, asst. prof., 1961-66, asso. prof., 1966-67; asso. prof. dept. biology Grad. Sch. Biomed. Scis., U. Tex., Houston, 1967-69, prof., 1969—, dir., 1971—, acting dean, 1976-78; mem. genetics study sect. NIH, Bethesda, Md., 1966-70, mem. genetics tng. com., 1970-74, adv. com. to dir., 1979-82; chromosome studies astronauts NASA, 1970-71; mem. med. adv. bd. Nat. Genetics Found., 1972—; research adv. bd. Planned Parenthood, Houston, 1972-79; vis. scholar Yale Law Sch., 1974; vis. prof. Cornell U., 1982—. Asso. editor: Am. Jour. Human Genetics, 1962-68; editorial bd.: Am. Jour. Med. Genetics, 1977—, Am. Jour. Law and Medicine, 1977—; contbr. articles to profl. jours. First aid instr. ARC, 1962-67; unit chmn. United Fund, 1966. Recipient Billings Silver medal AMA, 1966; Achievement award AAUW, 1970-71; Am. Jurisprudence award, 1973. Mem. Am. Soc. Human Genetics (past sec., dir., pres. 1982), Genetics Soc. Am. (sec. 1971-73, pres. 1977-78, Wilhelmene Key award 1977), Tissue Culture Assn. (trustee 1970-72), Environ. Mutagen Soc. (council), Am. Soc. Cell Biology, Phi Beta Kappa, Alpha Omega Alpha. Subspecialties: Genetics and genetic engineering (medicine); Health law. Home: 8100 Cambridge Dr 88 Houston TX 77054 Office: Health Law Inst Interfirst Bank Bldg Suite 600 Houston TX 77030

SHAW, MICHAEL TREVOR, physician, educator; b. Newcastle upon Tyne, Eng., Mar. 12, 1933; came to U.S., 1971, naturalized, 1978; s. Maurice and Sybil (Orlans) S.; m. Linda Black, June 25, 1957 (div. 1974); children: Gabrielle, Victoria, Kathryn; m. Melinda Kay Perry, Apr. 27, 1975. M.B., B.S., Durham (Eng.) U., 1956; M.D., U. Newcastle upon Tyne, Eng., 1971. Intern Royal Victoria Infirmary, Newcastle upon Tyne, resident, 1966-68; gen. practice medicine, Newcastle upon Tyne, 1957-65; assoc. prof. medicine U. Okla., 1972-75, Yale U., 1975-76; prof. medicine U. N.Mex., Albuquerque, 1976-79, 80—; prof. clin. oncology U. Newcastle upon Tyne, 1979-80. Editor: Chronic Granulocytic Leukemia, 1982. Chmn. pub. edn. com. Am. Cancer Soc., N.Mex. div., 1982—. Fellow Royal Coll. Physicians London, ACP. Democrat. Jewish. Subspecialties: Hematology; Chemotherapy. Current work: Research in cancer chemotherapy. Office: 2427 Telshor Blvd Las Cruces NM 88001

SHAW, MILTON CLAYTON, mechanical engineering educator; b. Phila., May 27, 1915; s. Milton Fredic and Nellie Edith (Clayton) S.; m. Mary Jane Greeninger, Sept. 6, 1939; children—Barbara Jane, Milton Stanley. B.S. in Mech. Engring, Drexel Inst. Tech., 1938; M.Eng. Sci., U. Cin., 1940, Sc.D., 1942; Dr. h.c., U. Louvain, Belgium, 1970. Research engr. Cin. Milling Machine Co., 1938-42; chief materials br. NACA, 1942-46; with Mass. Inst. Tech., 1946-61, prof. mech. engring., 1953-61, head materials processing div., 1952-61; prof., head dept. mech. engring. Carnegie Inst. Tech., Pitts., 1961-75, univ. prof., 1974-77; prof. engring. Ariz. State U., Tempe, 1978—; Cons. indsl. cos.; lectr. in, Europe, 1952; pres. Shaw Smith & Assos., Inc., Mass., 1951-61; Lucas prof. Birmingham (Eng.) U., 1961; Springer prof. U. Calif. at Berkeley, 1972; Distinguished guest prof. Ariz. State U., 1977; mem. Nat. Materials Adv. Bd., 1971-74; bd. dirs. Engring. Found., 1976, v.p. conf. com., 1976-78. Recipient gold medal Am. Soc. Tool Engrs., 1958; Outstanding Research award Ariz. State U., 1981; Guggenheim fellow, 1956; Fulbright lectr. Aachen T.H., Germany, 1957; OECD fellow to Europe, 1964—. Fellow Am. Acad. Arts and Scis., ASME (Hersey award 1967, Thurston lectr. 1971, meeting theme organizer 1977, hon. 1980); fellow Am. Soc. Lubrication Engrs., mem. (Nat. award 1964, hon. life); titular mem. Internat. Soc. Prodn. Engring. Research (pres. 1960-61, hon. mem. 1975), Am. Soc. Metals (Wilson award 1971); mem. Am. Soc. for Engring. Edn. (G. Westinghouse award 1956), Soc. Mfg. Engrs. (hon., P. McKenna award 1975, Gold medal 1958, Am. Machinist award 1973, internat. edn. award 1980), Nat. Acad. Engring., Polish Acad. Sci. Subspecialties: Mechanical engineering; Solid mechanics. Current work: Material processing, brittle fracture tribology; engineering design, materials behavior. Home: 3203 S College St Tempe AZ 85282 also 1540 Spanish Moss Way Flagstaff AZ 86001

SHAW, WILLIAM WEI-LIEN, microsurgeon, plastic surgeon; b. Kwei-Yang, Kwei-Chow, China, Mar. 12, 1942; came to U.S., 1957, naturalized, 1963; s. Emil and Rosemarie (Lam) S.; m. Margaret Pao, Oct. 19, 1975; children: Emily, Victor. B.A., UCLA, 1964, M.D., 1968. Diplomate: Am. Bd. Surgery, Am. Bd. Plastic Surgery. Surg. resident UCLA Med. Ctr., 1969-70, 73-74, Johns Hopkins Med. Ctr., 1972; plastic surgery resident Inst. Reconstructive Plastic Surgery, NYU Med. Ctr., N.Y.C., 1975-77, instr. surgery, 1977-79, asst. prof. surgery, 1979-83, assoc. prof., 1983—; dir. clin. microsurgery Bellevue (N.Y.) Hosp., 1977—, chief plastic surgery service, 1977—; cons. plastic surgeon VA Hosp., Castle Point, N.Y., 1977—; attending plastic surgeon Manhattan Eye, Ear and Throat Hosp., Univ. Hosp., N.Y. VA Hosp., N.Y. Infirmary/Beekman Downtown Hosp. Served to maj. U.S. Army, 1970-72. Regents scholar UCLA Sch. Medicine, 1963-68; recipient Golden Plate award Am. Acad. Achievement, 1980; Regents Disting. Service award Fairleigh Dickinson U., 1980; U.S. Army Meritorious Service award, 1972. Fellow ACS; mem. Am. Soc. Plastic and Reconstructive Surgeons, Internat. Soc. Reconstructive Microsurgery, Am. Assn. Surgery of Trauma, N.Y. County Med. Soc. Republican. Subspecialties: Microsurgery; Transplant surgery. Current work: Microsurgery in trauma, limb amputation and reconstructive plastic surgery, tissue preservation and transplantation, skin flap circulation. Office: 302 E 30th St New York NY 10016

SHAWHAN, STANLEY DEAN, physicist; b. Mpls., Feb. 7, 1941; s. Elbert Neil and Ena Maxine (Burdine) S.; m. Susan Jenkins, June 20, 1964; children—Peter Sven, Daniel Lloyd. B.A., Ohio Wesleyan U., 1963; M.S. (NASA trainee 1963-66), U. Iowa, 1965, Ph.D., 1966. Research asso. U. Iowa, 1966-68, mem. faculty, 1969—, prof. physics, 1978—; vis. scientist Royal Inst. Tech., Stockholm, 1968-69, 76-77, Danish Space Research Inst., Lyngby, summer 1969, 76; vis. asst. scientist Nat. Radio Astronomy Obs., Green Bank, W.Va., summer 1970; mem. working group ray tracing Internat. Sci. Radio Union-Internat. Assn. Geomagnetism and Aeronomy, 1976—; mem. joint bd. radio sci. Am. Geophys. Union-Internat. Sci. Radio Union, 1976-78; mem. space sci. bd., panel plasma processes Nat. Acad. Scis., 1976-77; mem. subcom. space sci., com. on radio frequencies NRC, 1977—; mem. subsatellite working group atmospheric, magnetospheric and plasmas in space program Goddard Space Flight Center, NASA, 1978-79; steering com. for dynamics Explorer Mission, 1976—; payload specialist selection com. Spacelab 2 Mission, 1978; del. internat. meetings, cons. in field. Contbr. articles to profl. jours. Grantee NSF, 1972—, NASA, 1975—. Mem. Am. Geophys. Union, Internat. Astron. Union, Am. Astron. Soc., Internat. Sci. Radio Union, Phi Beta Kappa, Sigma Xi, Pi Mu Epsilon, Chi Gamma Nu. Subspecialty: Space colonization. Home: 1147 E Court St Iowa City IA 52240 Office: Univ Iowa Dept Physics and Astronomy Iowa City IA 52242

SHAY, WILLIAM MCBRIDE, JR., mech. engr.; b. Kansas City, Mo., Jan. 13, 1936; s. William McBride and Grace Mae (McLemore) S.; m. Loretta Jean, July 3, 1955; children: Steven William, Kathy Jean. B.S.M.E., Wichita State U., 1965; postgrad., U. Ala., Huntsville, 1966-68. Registered profl. engr., Calif. Tech. mech. engr. Boeing Co., Wichita, Kans., 1958-65, Huntsville, 1965-68; mech. engr. Douglas Aircraft Co., Long Beach, Calif., 1965-66, Lawrence Livermore (Calif.) Nat. Lab., 1968—. Contbr. numerous articles to sci. jours. Served to sgt. USAF, 1954-58. Mem. ASME, Soc. Exptl. Stress Analysis. Democrat. Lodge: Masons. Current work: Experimental data on instrumentation. Office: PO Box 808 1-145 Livermore CA 94550

SHEA, DANIEL JOSEPH, engineering company executive, operations engineering consultant; b. Providence, Dec. 21, 1943; s. Daniel J. and Ann L. (Remillard) S. B.A., U. Louvain, Belgium, 1971, M.A., 1973. Lic. reactor operator, U.S. NRC. Dir. personnel analysis NUS/Halliburton, Rockville, Md., 1975-82; pres. Human Engring. Assocs., Inc., Houston, 1982—; cons. Houston Lighting & Power, La. Power & Ligh, Central Power & Light. Served in USN, 1960-65. Mem. Am. Nuclear Soc. Republican. Roman Catholic. Subspecialties: Human factors engineering; Operations research (engineering). Current work: Operations safety research relative to light water nuclear power plant operations; integration of hardware design, operating procedure design, job design and training design. Home: 1203 W Bell Houston TX 77019

SHEA, JOHN R., III, engineer, researcher; b. Balt., June 11, 1949; s. John R. and Patricia (Lindsay) S.; m. Virginia Fitz, June 13, 1981; 1 dau., Laura Ann. B.S.E., Princeton U., 1971; Ph.D. in Aeros, Calif. Inst. Tech., Pasadena, 1976. Asst. prof. aero. and mech. engring. Air Force Inst. Tech., Dayton, Ohio, 1976-79; mem. research staff Inst. for Def. Analyses, Alexandria, Va., 1979—. Contbr. articles to profl. jours. Served to capt. USAF, 1975-79. Fannie and John Hertz Found fellow, 1971-75. Mem. AIAA, Am. Phys. Soc., Mil. Ops. Research Soc., Am. Irish Hist. Soc. Roman Catholic. Subspecialties: Operations research (engineering); Aeronautical engineering. Current work: Devising and preparing analyses to reflect potential operational utility of tactical missile and aircraft systems being considered by the U.S. Department of Defense. Office: Inst for Def Analyses 1801 N Beauregard St Alexandria VA 22311

SHEA, JOSEPH FRANCIS, mech. engr.; b. N.Y.C., Sept. 5, 1926; s. Joseph Anthony and Mary Veronica (Tully) S.; m. Carol Dowd Manion; children from earlier marriage—Mary Linda Helt, Nancy Catherine, Patricia Ann Kelly, Elizabeth Carol Williams, Amy Virginia. B.S. in Math, U. Mich., 1949, M.S. in Engring. Mechanics, 1950, Ph.D., 1955. Instr. engring. mechanics U. Mich., 1948-50, 53-55; research mathematician Bell Telephone Labs., 1950-53, mil. devel. engr., 1955-59; dir. advanced system research and devel., also mgr. Titan inertial guidance program AC Spark Plug div. Gen. Motors Corp., 1959-61; space program dir. Space Tech. Labs., 1961-62; dep. dir. systems, manned space flight NASA, 1962-63; mgr. Apollo spacecraft program office Manned Spacecraft Center, 1963-67; v.p., dep. dir. engring. Polaroid Corp., Cambridge, Mass., 1967-68; v.p., gen. mgr. equipment div. Raytheon Co., Waltham, Mass., 1968—, sr. v.p., group exec., Lexington, Mass., 1975-81, sr. v.p. engring., 1981—. Served to ensign USNR, 1944-47. Recipient Arthur S. Flemming award U.S. Jr. C. of C., 1965. Fellow Am. Astronautical Soc., Am. Inst. Aeros. and Astronautics; mem. ASME, IEEE, Nat. Acad. Engring., Nat. Space Club. Subspecialties: Systems engineering; Aerospace engineering and technology. Home: 56 W Cedar St Boston MA 02114 Office: 141 Spring St Lexington MA 02173

SHEA, KEITH RAYMOND, forester; b. Greenfield, Iowa, Dec. 24, 1924; s. Raymond Edward and Helen Amy (Richards) S.; m. Dolores Jean Kinder, July 4, 1968; children by previous marriage: David E., Robin M., Julie A., J. Daniel, Paula J., C. Matthew. B.S., U. Minn., 1950; Ph.D., U. Wis., 1954. Asst. prof. U. Wis., 1954-56; research specialist Weyerhaeuser Co., Centralia, Wash., 1956-66; project leader U.S. Dept. Agr., Forest Service, Washington, 1966-71, chief forest disease research, 1971-73, dir. forest insect and disease research, 1973-74; staff officer Office Sec., U.S. Dept. Agr., 1974-76; dir. forest insect and disease research U.S. Dept. Agr., Forest Service, Washington, 1976-79; asst. dir. program mgmt. Sci. and Edn. Administrn., 1979-80, assoc. dep. chief research, 1980—. Served with USNR, 1943-46. Recipient Cert. of Merit U.S. Dept. Agr., 1974, 79, Superior Service award, 1977. Mem. Soc. Am. Foresters. Subspecialties: Plant pathology. Current work: Administration forest research programs—insects and diseases, timber management and forest environment, coordination and planning of research among federal agencies and state forestry schools. Office: US Dept Agr Forest Service PO Box 2417 Washington DC 20013

SHEAR, NATHANIEL, physicist; b. Bklyn., Dec. 20, 1908; s. Victor Jacob and Henrietta Leah (Robinson) S. A.B. with honors, Columbia U., 1930, M.A., 1932. Physicist U.S. Navy, Civil Service, Washington, Phila., Lakehurst, N.J., 1937-44; research physicist div. war research Columbia U., N.Y.C., 1944-46; ops. research analyst MIT Ops. Evaluation Group, U.S. Navy, Washington, 1946-48; physicist Bur. Ships, U.S. Navy, Washington, 1948-51; cons. physics, Alexandria, Va., 1951-60; physicist Emerson Research Lab., Silver Spring, Md., 1960-62; sr. physicist Johns Hopkins U. Applied Physics Lab., Laurel, Md., 1962-66; cons physics, Silver Spring, Md., 1966-69; ops. research analyst Def. Spl. Projects Group, Washington, 1969-72; retired, 1972—. Author sci. reports for, U.S. Navy. Mem. Va. Acad. Sci., Ops. Research Soc. Am., Am. Phys. Soc., U.S. Naval Inst. (assoc.), Phi Beta Kappa. Republican. Jewish. Subspecialties: Operations research (mathematics); Oceanography. Current work: Integrating and coordinating scientific papers prepared for the Navy. Home: 1401 Blair Mill Rd Apt 612 Silver Spring MD 20910

SHEARER, DAVID ROSS, behavioral medicine specialist, consultant in health psychology; b. Houston, Mar. 2, 1950; s. Hutton A. and Francile (Thompson) S.; m. Penelope Lynn Potter, May 18, 1972 (div. June 1977). B.A., Tex. Christian U., 1972, M.A., 1977; postgrad., Tex. Woman's U., 1975-76, Hawaii Sch. Profl. Psychology, 1983—. Lic. family therapist, Calif. Adolescent team coordinator Psychiat. Inst., Ft. Worth, 1974-75; psychol. asso. Ft. Worth Psychol. Center, 1974-77; psychologist Tex. Dept. Mental Health & Mental Retardation, San Angelo, 1977-78; flight service dir. Am. Airlines, Los Angeles, 1979-81; behavioral medicine specialist Straub Clinic, Honolulu, 1982—; adj. clin. prof. Antioch U. Hawaii, Honolulu, 1982—. Author: 12 weeks to a Healthier Lifestyle, 1982, Handling Stress at Home and on the Job, 1982, others. Pres. adv. bd. Diamond Head Mental Health Ctr., 1982; Clin research fellow Inst. Child Health, U. Benin, Nigeria, 1978. Mem. Am. Psychol. Assn., Am. Acad. Behavioral Medicine, Acad. Psychologists in Marital, Sex, and Family Therapy, Am. Orthopsychiat. Assn., Am. Soc. Tng. and Devel., Psi Chi. Republican. Episcopalian. Lodge: Masons. Subspecialties: Preventive medicine; Neuroimmunology. Current work: Behavioral medicine and health psychology, psychology of lifestyle change - psychoneuroimmunology, stress management, instructional systems design. Home: 2452 Tusitala #710 Honolulu HI 96815 Office: Straub Clinic & Hosp Inc 888 S King St Honolulu HI 96813

SHEARIN, NANCY LOUISE, physiology research educator; b. Meridian, Miss., May 17, 1938; d. Robert F. and Ossie (Wall) S. B.S., Millsaps Coll., 1960; M.S., Tenn. Tech. U., 1971; Ph.D., U. Wyo., 1974. Fellow Meml. U. Nfld., St. John's, Can., 1974-76; research assoc. U. Alta., Edmonton, Can., 1976-79; research assoc. dept. physiology U. Utah, Salt Lake City, 1979-82, research instr. gastrointestinal physiology, 1982—. Mem. Am. Physiol. Soc. Subspecialties: Physiology (medicine); Gastroenterology. Current work: Gastrointestinal motility, ionic flux in smooth muscle, role of prostaglandins in gastrointestinal physiology, direct intracellular measurement (ionselective microelectrode) of ionic activities in colonic smooth muscle. Office: Dept Surgery U Utah Sch Medicine 50 N Medical Dr Salt Lake City UT 84132

SHEEHY, RONALD J., biology educator; b. Jacksonville, Fla., May 29, 1945. B.Sc., Morehouse Coll., 1965; M.Sc., Atlanta U., 1967; postgrad., U. South Fla., 1969; Ph.D., U. Tenn., 1972. Postdoctoral fellow Pub. Health Research Inst., N.Y.C., 1973; Carnegie fellow Oak Ridge Nat. Lab., 1973; asst. prof. dept. biology Morehouse Coll., Atlanta, 1973-75, assoc. prof., 1975-76; interim dir. Office of Health Professions, 1978-79, prof., chmn. dept. biology, 1978-82, David E. Packard prof., chmn. dept. biology, 1982—. Contbr. articles to profl. jours. NIH fellow, 1969-72; Charles Merrill fgn. travel awardee, 1976; Max Planck Inst. grantee, 1978. Mem. Am. Soc. Microbiology, Nat. Assn. Black Newspaper Pubs., Mid-Atlantic Extrachromosomal Elements Group, Ga. Acad. Sci., Orgn. Black Scientists (sec.), Nat. Minority Health Affairs Assn., AAAS, N.Y. Acad. Scis., Beta Kappa Chi, Alpha Phi Alpha. Subspecialties: Behavioral ecology; Genome organization. Home: 125 Grande Ct College Park GA 20249 Office: Dept Biology Morehouse College Atlanta GA 30314

SHEEN, SHUH JI, biochemical geneticist; b. Woojiang, Jiangsu, China, Mar. 21, 1931; s. Chung Yi and Zhing Hwa (Chiang) S.; m. Rosetina You Hsu, Dec. 19, 1959; children: Vida Dee, Vernon Lyn, Vera Lo, Volney Leo, Vidal Ty. B.S., Taiwan Provincial Chung Hsing U., 1953; M.S., N.D. State U., 1958; Ph.D., U. Minn., 1962. Asst. prof. biology Hanover (Ind.) Coll., 1962-66; asst. prof. agronomy U. Ky., Lexington, 1966-72, assoc. prof., 1972-74, assoc. prof. plant pathology, 1974-79, prof. plant pathology, 1979—. Contbr. articles to profl. jours. U.S. Dept. Agr. grantee, 1966—. Mem. Am. Chem. Soc., AAAS, Am. Phytopathological Soc., Genetic Soc. Am., Am. Soc. Plant Physiologists, Phytochemical Soc. N.Am., Sigma Xi. Subspecialties: Plant pathology; Biomass (agriculture). Current work: Mechanism of plant resistance to virus and the utilization of tobacco biomass for protein production and safer smoking material. Home: 404 Lakeshore Dr Lexington KY 40502 Office: Dept Plant Pathology U Ky Lexington KY 40546

SHEFF, JAMES ROBERT, nuclear and energy engineering educator; b. Colorado Springs, Colo., Nov. 5, 1936; s. Robert Lee and Jessie Marie (Leathers) S.; m. Linda Anne Smith, June 7, 1959 (div. June 1973); children: Robert Benjamin, Nancy Elizabeth (dec.), Natalie Joy, Matthew Garner. B.S.Ch.E., U. Colo., Boulder, 1959; M.S., U. Wash.-Seattle, 1962, Ph.D., 1965; postgrad., Columbia Basin Coll., 1976. Registered profl. engr., Wash. Chem. engr. Gen. Electric Co., Richland, Wash., 1959-60; sr. research engr. Battelle N.W., Richland, 1965-70, WADCO (Westinghouse subs.), 1970-71; sr. research scientist Battelle-N.W., Richland, 1971-77; prof. nuclear and energy engring. U. Lowell, Mass., 1977—; pres., cons. Desert Ventures, Richland, Wash., 1973-77, Ventures in Energy, Lowell, Mass., 1977-81; cons. Gen. Pub. Utilities-Three Mile Island, Harrisburg, Pa., 1981—; pres., cons. Lowell (Mass.) Tech. Assocs., 1982—. Contbr. articles in field to profl. jours. Mem. Mass. Voice of Energy, Boston, 1977-80. U.S. Dept. Energy grantee, 1977. Mem. Am. Nuclear Soc., Soc. Profl. Well Log Analysts, N.Y. Acad. Scientists, Sigma Xi, Tau Beta Pi, Phi Lambda Upsilon. Club: Toastmasters (Richland, Wash.) (pres. 1965-70). Subspecialties: Nuclear engineering; Wind power. Current work: Nuclear, wind and geothermal engineering topics, rad waste disposal, reactor physics, windmill design theory, geothermal theory. Home: PO Box 811 Lowell MA 01853 Office: U Lowell 226 Engring Bldg Lowell MA 01854

SHEFFER, RICHARD DOUGLAS, cytogeneticist, educator; b. Portland, Ind., Apr. 19, 1942; s. Elwood Willard and Helen Irene (Hartzell) S.; m. Janet Elston, June 8, 1969; 1 son, Donald Douglas. Student, Ind. U., 1960-63; B.S. with distinction, Purdue U., 1970; Ph.D., U. Hawaii, 1974. Postdoctoral research asso. U. Hawaii, Honolulu, 1974-75; asst. prof. biology U. N.B. (Can.), Fredericton, 1975-76; insr. biology Montclair State Coll., Upper Montclair, N.J., 1976-77; asst. prof. biology Ind. U. N.W., Gary, 1977-81, assoc. prof., 1981—; research assoc. Mo. Bot. Garden, St. Louis, 1979—. Contbr. articles to sci. jours. Served with M.C. U.S. Army, 1964-66. NDEA fellow, 1970-72; Stanley Smith Hort. Trust asst., 1972-74. Mem. AAAS, Genetics Soc. Can., Genetics Soc. Am., Am. Genetics Assn., Am. Soc. Plant Taxonomists, Internat. Assn. Plant Taxonomists, Soc. Study of Evolution, Am. Inst. Biol. Scis., Am. Orchid Soc., Am. Soc. Hort. Sci., Internat. Aroid Soc., N.Y. Acad. Sci., Sigma Xi, Phi Kappa Phi, Gamma Sigma Delta. Subspecialties: Genome organization; Systematics. Current work: ITAnthurium cytogenetic and biosystematic studies including chromosome banding, karyotype analysis, chromosomal evolution and meiotic studies of Fone interspecific hybrides to determine interspecific relationships. Home: 208 Shorewood Dr Valparaiso IN 46383 Office: 3400 Broadway Gary IN 46408

SHEFFIELD, JOEL B., biology educator, consultant; b. Bklyn., Dec. 30, 1942; s. Reuben and Gertrude S.; m. Lucy Paige, June 20, 1965; 1 dau., Jennifer. A.B., Brandeis U., 1963; Ph.D., U. Chgo., 1969. Research assoc. Rockefeller U., 1963-64; postdoctoral fellow Netherlands Cancer Inst., 1969-71; asst. mem. Inst. Med. Research, Camden, N.J., 1971-77; asst. prof. biology Temple U., 1977-82, assoc. prof., 1982—. Subspecialties: Developmental biology; Neurobiology. Current work: Developmental cell biology of eye. Office: Dept Biology Temple U Philadelphia PA 19122

SHEFFIELD, JOHN, national laboratory administrator, researcher; b. Purley, Surrey, Eng., Dec. 15, 1936; came to U.S., 1966; s. Norman Everett and Ella Nancy (Dyson) S.; m. Dace Kancbergs, Aug. 29, 1966; children: Jason Everett, Nicholas John. B.Sc. with honors in Physics, Imperial Coll., London U., 1958; M.Sc., No. Poly. U., London, 1962; Ph.D., London U., 1966. Exptl. officer Harwell Lab., U.K. Atomic Energy Authority, 1958-61; expt. officer Culham Lab., 1961-66, prin. sci. officer, 1971-77; asst. prof. U. Tex., 1966-71; confinement sect. head fusion energy div. Oak Ridge Nat. Lab., 1977-80, assoc. dir., 1980—. Author: Plasma Scattering of Electro Magnetic Radiation, 1975. Fellow Am. Phys. Soc.; mem. Am. Nuclear Soc. Subspecialties: Fusion; Plasma physics. Current work: Fusion program management, assessment and planning of fusion. Office: Fusion Energy Division Oak Ridge National Laboratory PO Box Y Oak Ridge TN 37830

SHEINBEIN, MARC LESLIE, psychologist; b. Oklahoma City, Dec. 11, 1945; s. Isadore and Gloria (Davis) S.; m. Andrea M. Riff, Nov. 30, 1980; children by previous marriage: Amy Michelle, David Benjamin. B.A., Vanderbilt U., 1967; M.A., U. Tenn., 1969, Ph.D., 1972. Lic. psychologist, Tex. Doctoral intern U. Okla. Med. Sch., 1971-72; asst. prof. U. Tex. Health Scis. Ctr., Dallas, 1972-75; pvt. practice psychology, Dallas, 1973—; cons. Ctr. for Marriage and Family Therapy, Inc., Plano, Tex., 1981—. Author: Psychol. Test, Children's Sentence Completion Test, 1978; contbr. articles to profl. jours. Mem. Am. Psychol. Assn., Dallas Psychol. Assn., Am. Assn. Marriage and Family Therapists, Dallas Assn. Marriage and Family Therapists. Jewish. Club: AEPI Alumni. Current work: Test and case research. Home: 8734 Clover Meadow Dallas TX 75243 Office: 4255 LBJ Freeway Suite 276 Dallas TX 75234

SHELDON, JOSEPH KENNETH, biology educator; b. Ogden, Utah, Nov. 11, 1943; s. Donald David and Ina May (Kramp) S.; m. Donna Sherrie Zielaskowski, Aug. 21, 1965; children: Jodi Gwyn, Bret Eugene. B.S., Coll. Idaho, 1966; Ph.D., U. Ill., 1972. Prof. biology Eastern Coll., St. Davids, Pa., 1971—; research scientist Battelle NW, Richland, Wash., summer 1975; research cons., summer 1976. Contbr. articles to profl. jours. Sci. studies com. Valley Forge (Pa.) Audubon Soc., 1982-83. NDEA fellow, 1966-69; Am. Philos. Soc. grantee, 1973; recipient Bronze Medal Carnegie Hero Fund Com., 1973. Mem. Am. Entomol. Soc. (nat. v.p. 1981-83), AAAS, Ecol. Soc. Am., Cambridge Entomol. Soc., Sigma Zeta (nat. pres. 1979-80). Democrat. Subspecialty: Ecology. Current work: Insect ecology. Home: 172 Hillside Circle Villanova PA 19087 Office: Biology Dept Eastern Coll Saint Davids PA 19087

SHELLENBERGER, MELVIN KENT, pharmacologist, neurotoxicologist, educator; b. Pittsburg, Kans., Oct. 29, 1936; s. Chester Melvin and Gertrude Wilene (Thompson) S.; m. Carol L. Carpenter, July 4, 1959; children: William C., Stephen L., Daniel A., Karin D. B.S., Kans. State Tchrs. Coll., 1958; M.S., U. Wash., 1962, Ph.D., 1965. Postdoctoral fellow depts. pharmacology and physiology U. Mich., Ann Arbor, 1965-67; asst. prof. dept. pharmacology U. Kans., Kansas City, 1968-74, assoc. prof., 1974—; vis. asst. prof. dept. psychiatry UCLA, 1973-74; research assoc. Kans. Ctr. Mental Retardation, 1971—. Contbr. articles on pharmacology and neurotoxicology to profl. jours.; assoc. editor: Neurotoxicology; field editor: Neurochemistry Internat. NIMH grantee, 1967-68, 1972, 76; NIMH Research Career Devel. award, 1972-77. Mem. Soc. Neurosci., Am. Soc. Pharmacology and Exptl. Therapeutics, Internat. Soc. Neurochemistry. Subspecialties: Neuropharmacology; Neurochemistry. Current work: Correlation of chemical synaptic events with function, neurotransmitters, ontogeny, neurotoxicants, functional deficits, behavior. Office: Ralph L Smith Center for Mental Retardation Researc Kans U Med Center Kansas City KS 66103

SHELLEY, EDWIN FREEMAN, educational administrator, engineering and business consultant; b. N.Y.C., Feb. 19, 1921; s. Robert and Jessie Selma (Sinick) S.; m. Florence D. Dubroff, Aug. 31, 1941; children: Carolyn Shelley LeBel, William Edson. A.B., Columbia U., 1940, B.S.E.E., 1941; cert. advanced mgmt. program, Harvard U., 1957. Flight test engr. Curtiss Wright Corp., Caldwell, N.J., 1941-47; pres., chief engr. Am. Chronoscope Corp., Mt. Vernon, N.Y., 1948-50; v.p., chief engr. Bulova Research and Devel. Labs., Inc., Jackson Heights, N.Y., 1950-57; v.p., dir. advanced programs U.S. Industries, Inc., N.Y.C., 1957-64; pres., chmn. EF Shelley & Co., Inc., N.Y.C. and Washington, 1965-75; dir. center for energy policy and research N.Y. Inst. Tech., 1975—; dir. Solar-Micro, Inc., AquaSol, Inc.; trustee N.Y. Inst. Tech., Nova U. Mem. adv. bd. N.Y. State Legisl. Commn. on Sci. and Tech., 1977—; pres. Nat. Council on Aging, 1968-72, bd. dirs., 1962—. Mem. IEEE (sr.), AIAA, AAAS, Newcomen Soc. N.Am. Club: Harvard (N.Y.C.). Subspecialties: Systems engineering; Information systems (information science). Current work: Information systems, educational technology, flight systems, energy education, consulting for energy-related businesses, educational technology development, flight system development. Patentee in field. Home: 339 Oxford Rd New Rochelle NY 10804 Office: Center Energy Policy NY Inst Tech Old Westbury NY 11568

SHELLEY, WALTER BROWN, physician, educator; b. St. Paul, Feb. 6, 1917; s. Patrick K. and Alfaretta (Brown) S.; m. Marguerite H. Weber, 1942 (dec.); children: Peter B., Anne E. Kiselewich, Barbara A. (dec.); m. E. Dorinda Loeffel, 1980; children: Thomas R., Katharine D. B.S., U. Minn., 1940, Ph.D., 1941, M.D., 1943; M.A. honoris causa, U. Pa., 1971, M.D., U. Uppsala, Sweden, 1977. Diplomate: Am. Bd. Dermatology (pres. 1968-69, dir. 1960-69). Instr. physiology U. Pa., Phila., 1946-47, asst. instr. dermatology and syphilology, 1947-49, asst. prof. dermatology, 1950-53, asso. prof., 1953-57, prof., 1957-80, chmn. dept., 1965-80; prof. dermatology U. Ill. Peoria Sch. Medicine, 1980-83; prof. medicine (dermatology) Med. Coll. Ohio, 1983—; instr. dermatology Dartmouth Coll., 1949-50; Regional cons. dermatology VA, 1955-59; mem. com. on cutaneous system NRC, 1955-59, Commn. Cutaneous Diseases, Armed Forces Epidemiological Bd., 1958-61, dep. dir., 1959-61; cons. dermatology Surgeon Gen. USAF, 1958-61, U.S. Army, 1958-61; mem. NRC, 1961-64. Author: (with Crissey) Classics in Clinical Dermatology, 1953, (with Pillsbury, Kligman) Dermatology, 1956, Cutaneous Medicine, 1961, (with Hurley) The Human Apocrine Sweat Gland in Health and Disease, 1960, (with Botelho and Brooks) The Endocrine Glands, 1969, Consultations in Dermatology with Walter B. Shelley, 1972, Consultations II, 1974; Mem. editorial bd.: Jour. Investigative Dermatology, 1961-64, Archives of Dermatology, 1961-62, Skin and Allergy News, 1970—, Excerpta Medica Dermatologica, 1960—, Cutis, 1972—; asso. editor: Jour. Cutaneous Pathology, 1972-81; editorial cons.: Medcom, 1972—. Served as capt. M.C. AUS, 1944-46. Recipient Spl. award Soc. Cosmetic Chemists, 1955; Hellerstrom medal, 1971; Am. Med. Writers Assn. Best Med. Book award, 1973; Dohi medal, 1981. Master A.C.P.; fellow Assn. Am. Physicians, St. John's Dermatol. Soc. London (hon.); mem. AMA (chmn. residency rev. com. for dermatology 1963-67, chmn. sect. dermatology 1969-71), Assn. Profs. Dermatology (pres. 1972-73), Pacific Dermatol. Assn. (hon.), Am. Dermatol. Assn. (dir., pres. 1975-76), Soc. Investigative Dermatology (pres. 1961-62), Am., Phila. physiol. socs., Brit. Dermatol. Soc. (hon.), Phila. Dermatol. Soc. (pres. 1960-61), Iowa Dermatol. Soc., Ill. Dermatol. Soc., Chgo. Dermatol. Soc., Am. Acad. Dermatology (pres. 1971-72), Pa. Acad. Dermatology (pres. 1972-73), Am. Soc. for Dermatologic Surgery, Royal Soc. Medicine; corr. mem. Nederlandse Vereniging Van Dermatologen, Israeli Dermatol. Assn., Finnish Soc. Dermatology, Swedish Dermatol. Soc., French Dermatologic Soc.; fgn. hon. mem. Danish Dermatol. Assn., Japanese Dermatol. Assn., Dermatol. Soc. S.Africa. Subspecialty: Dermatology. Home: 24869 W River Rd Perrysburg OH 43551 Office: Med Coll Ohio CS 10008 Toledo OH 43699

SHELOKOV, ALEXIS, physician, epidemiology educator; b. Harbin, China, Oct. 18, 1919; came to U.S., 1938, naturalized, 1944; s. Ivan T. and Maria M. (Zolotov) S.; m. Paula Helbig, June 10, 1947; 1 son, Alexis Paul. A.B., Stanford U., 1943, M.D., 1948. Diplomate: Am. Bd. Microbiology, Am. Bd. Preventive Medicine. Physiologist climatic research lab. War Dept., 1943-44; research asst. Stanford, 1946-47; house officer Mass. Meml. Hosps., Boston, 1947-50; instr. medicine Boston U. Sch. Medicine, 1948-50; asst. in pediatrics Harvard Med. Sch., 1949-50; clin. instr. medicine Georgetown U., 1953-57; cons. D.C. Gen. Hosp., 1955-57, Gorgas, Coco Solo hosps., C.Z., 1958-61; dir. Middle Am. Research Unit, C.Z., 1957-61; mem. sci. adv. bd. Gorgas Meml. Inst., Panama, 1959-72, 76—; chief lab. tropical virology Nat. Inst. Allergy and Infectious Diseases, NIH, 1959-63; chief lab. virology and rickettsiology Div. Biologics Standards, NIH, 1963-68; officer (med. dir.) USPHS, 1950-68; prof., chmn. dept. microbiology U. Tex. Health Sci. Center, San Antonio, 1968-81; prof. dept. epidemiology Johns Hopkins U. Sch. Hygiene and Public Health, 1981—; dir. vaccine research Salk Inst., 1981—; mem. U.S. del. on virus diseases to USSR, 1961, chmn. U.S. del. on hemorrhagic fevers, USSR, 1965, 69; exec. council Am. Com. for Arthropod-borne Viruses, 1960-67; panel for arthropod-borne virus reagents. Nat. Inst. Allergy and Infectious Diseases, 1962-66, cons. geog. medicine br., 1972-76; mem. sci. adv. bd. WHO Serum Bank, Yale, 1964-68; cons. Pan Am. Health Orgn., 1958-63, 71, WHO, 1966; mem. commn. on viral infections Armed Forces Epidemiological Bd., 1967-68; mem. virology study sect. div. research grants NIH, 1968-70; mem. viral disease panel U.S.-Japan Coop. Med. Sci. Program, 1971-76; mem. Am. Tropical Medicine del., China; adv. bd. virology CRC Press, 1975-80. Mem. editorial bd.: Am. Jour. Tropical Medicine and Hygiene, 1973—, Jour. Clin. Microbiology, 1976—, Archives of Virology, 1979-81. Trustee Am. Type Culture Collection, 1969-72; bd. sci. counselors Nat. Inst. Dental Research, NIH, 1971-75. Decorated Order Rodolfo Robles (Guatemala). Mem. Am. Assn. Immunologists, Soc. Exptl. Biology and Medicine, Am. Soc. Tropical Medicine and Hygiene, Am. Epidemiologic Soc., A.M.A., Infectious Disease Soc. Am., Soc. Epidemiological Research, Am. Soc. Microbiology, Council Biology Editors. Subspecialties: Virology (medicine); Epidemiology. Current work: Research and development of viral vaccines. Home: 7233 Dockside Ln Columbia MD 21045 Office: Johns Hopkins U Sch Hygiene and Public Health Baltimore MD 21205

SHELUS, PETER J., astronomer; b. Phila., Aug. 2, 1942; s. Peter John and Helen Anna (Paulauskas) S.; m. Margaretann Stuhl, June 15, 1968; children: Peter John, Jonathan Andrew. A.B., U. Pa., 1964; M.A., U. Va., 1968, Ph.D., 1971. Postdoctoral fellow NASA Manned Spacecraft Center, Houston, 1969-71; research scientist McDonald Obs., U. Tex.-Austin, 1971—; lectr. Austin Community Coll., 1976—; cons. Ednl. Devel. Co., Austin. Mem. Am. Geophys. Union, Internat. Assn. Geodesy, Internat. Union Geodesy and Geophysics, Am. Astronom. Soc., Internat. Astronom. Union. Current work: Lunar laser ranging, lageos laser ranging, earth rotation, polar motion, plate tectonics, space telescope astrometry. Office: Astronomy Dept U Tex Austin TX 78712

SHEMIN, DAVID, biochemist, educator; b. N.Y.C., Mar. 18, 1911; s. Louis and Mary (Bush) S.; m. Mildred B. Sumpter (dec. 1962); children—Louise P., Elizabeth; m. Charlotte Norton, Mar. 1963. B.S., Coll. City N.Y., 1932; A.M., Columbia, 1933, Ph.D., 1938. Asst. prof. biochemistry Columbia, 1945-49, asso. prof., 1949-53, prof., 1953-68; chmn. dept. biochemistry and molecular biology Northwestern U., Evanston, Ill., 1968—, prof. biochemistry, 1968—. Contbr. articles to sci. publs. Recipient Pasteur medal, Stevens award.; Guggenheim fellow; Commonwealth fellow; Fogarty Internat. scholar, 1979—. Mem. Nat. Acad. Sci., Am. Acad. Arts and Scis. Subspecialties: Biochemistry (biology); Enzyme technology. Current work: Deputy director of Cancer Center at Northwestern University. Home: 902 Lincoln St Evanston IL 60201 Office: Dept Biochemistry and Molecular Biology Northwestern U Evanston IL 60201

SHEN, TSUNG YING, research pharmacologist; b. Peking, China, Sept. 28, 1924; s. Tsu-Wei and Sien-Wha (Nieu) S.; m. Amy Lin, 1953; children. B.Sc., Nat. Central U., China, 1946; D.I.C., Imperial Coll., London, 1948; Ph.D., U. Manchester, Eng., 1950, D.Sc. (hon.), 1978. Postdoctoral fellow Ohio State U., 1950; research assoc. MIT, 1952-56; sr. chemist Merck Sharp & Dohme Research Labs., Rahway, N.J., 1956-65, asst. dir. labs., 1965-69, dir., 1969-76, v.p. membrane chem. research, 1976-77, v.p. membrane and arthritis research, 1977—. Editorial bd.: Jour. Medicinal Chemistry; contbr. 120 articles to sci. jours. Recipient medal of Merit Giornate Mediche Internazionali del Collegium Biologicum Europa, 1977; Rene Descartes Silver medal U. Paris, 1977; Galileo medal U. Pisa, Italy, 1976; Dirs. Sci. award Merck & Co., Inc., 1976; Outstanding Patentee award N.J. R&D Council, 1975. Mem. Am. Chem. Soc. (Alfred Burger award in Medicinal Chemistry 1980), N.Y. Acad. Scis., AAAS, Internat. Soc. Immunopharmacology. Subspecialties: Synthetic chemistry; Medicinal chemistry. Current work: Medicinal chemistry and biomembrane research related to anti-inflammatory and immunopharmacological agents. Patentee in field (200). Home: 935 Minisink Way Westfield NJ 07090 Office: PO Box 2000 Rahway NJ 07065

SHEN, WEI-CHIANG, biochemist, educator; b. China, May 3, 1942; came to U.S., 1967, naturalized, 1976; s. Tze-Ping and Yi-Ching (Wen) S.; m. Daisy Teh-Chang, Jan. 13, 1968; children: Howard, Jerry. B.S., Tunghai U., Taichung, Taiwan, 1965; Ph.D., Boston U., 1972. Mass. Heart Assn. postdoctoral fellow dept. biol. chemistry Harvard U. Med. Sch., 1972-73; sr. research asso. dept. biochemistry Brandeis U., Waltham, Mass., 1973-76; asst. research prof. dept. pathology and pharmacology Boston U. Sch. Medicine, 1976-80, assoc. research prof., 1980-83, assoc. prof., 1983—. Contbr. articles to sci. jours. Am. Cancer Soc. cancer research scholar award, 1982—. Mem. Am. Soc. Cell Biology, Am. Soc. Pharmacology and Exptl. Therapeutics, N.Y. Acad. Scis. Subspecialties: Cell biology; Cancer research (medicine). Current work: Poly(basic amino acids) as drug carriers in cancer chemotherapy; conjugates of drugs and monoclonal antibodies as potential tumor-targeting agents; endocytosis and lysosomal functions. Home: 11 Dartmouth Ave Needham MA 02192 Office: Dept Pathology Boston U Sch Medicine Boston MA 02118

SHEN, YI-SHANG, computer systems/network designer, analyst, cons.; b. Taiwan, Nov. 23, 1947; came to U.S., 1970, naturalized, 1983; s. Ju-chang and Feng-chen (Chang) S.; m. Po-ching Yang, Nov. 27, 1976; children: Andrew Chun-yang, Jessica Chun-han. B.S. in Electronic Engring, Nat. Chiao-Tung U., Hsin-Chu, Taiwan, 1969; M.S. in Computer Sci, SUNY-Stony Brook, 1972; Ph.D. in Computer Scis, SUNY-Stony Brook, 1980. Quality control officer First Chinese Navy Shipyard, Kaohsiung, Taiwan, 1969-70; teaching, research asst. SUNY-Stony Brook, 1971-75; asst. prof. dept. computer scis. SUNY-Oswego, 1975; cons. specialist Gen. Electric Info. Services Co., Rockville, 1975-81; project dir. Gen. Systems Groups, Inc., Salem, N.H., 1981—; cons. Contbr. numerous articles to profl. jours. Mem. Assn. Computing Machinery, IEEE. Subspecialties: Distributed systems and networks; Software engineering. Current work: Telecommunications, computer system/network architecture, system analysis, capacity planning and implementation, software engring., project planning and management and solid-state electronic circuits. Office: 6849 Old Dominion Dr Suite 224 McLean VA 22101

SHEPARD, MARION LAVERNE, materials scientist, educator; b. Owosso, Mich., Dec. 20, 1937; s. Marion Elwin and Lola Margaret (Koivuniemi) S.; m. Cynthia Anne Swift, Jan. 6, 1962; 1 dau., Lori Anne. B.S. in Metall. Engring, Mich. Tech. U., 1959; M.S., Iowa State U., 1960, Ph.D., 1965. Registered profl. engr., N.C. Research asst. Ames Lab., Iowa State U., 1960; project metallurgist Pratt & Whitney Aircraft, East Hartford, Conn., 1961-62, 65-67; asst. prof. to prof. materials sci. Sch. Engring., Duke U., Durham, N.C., 1967-82, prof., assoc. dean, 1977—. Author: Introduction to Energy Technology, 1976; contbr. articles to profl. jours. Mem. Am. Soc. Engring. Edn., ASME. Republican. Subspecialties: Metallurgy; Materials. Current work: Thermodynamics of materials; failure analysis teaching, academic administration, research. Home: 3421 Pinafore Dr Durham NC 27705 Office: Sch Engring Duke Univ Durham NC 27706

SHEPARD, ROGER NEWLAND, psychologist, educator; b. Palo Alto, Calif., Jan. 30, 1929; s. Orson Cutler and Grace (Newland) S.; m. Barbaranne Bradley, Aug. 18, 1952; children—Newland Chenoweth, Todd David, Shenna Esther. B.A., Stanford U., 1951; Ph.D., Yale U., 1955; A.M. (hon.), Harvard U., 1966. Research asso. Naval Research Lab., 1955-56; research fellow Harvard, 1956-58; mem. tech. staff Bell Telephone Labs., 1958-66, dept. head, 1963-66; prof. psychology Harvard U., 1966-68, dir. psychol. labs., 1967-68; prof. psychology Stanford U., 1968—. Guggenheim fellow Center for Advanced Study in Behavioral Scis., 1971-72; recipient Howard Crosby Warren medal, 1981. Fellow AAAS, Am. Psychol. Assn. (pres. exptl. div. 1980-81, Disting. Sci. Contbn. award 1976); mem. Nat. Acad. Scis., Psychometric Soc. (pres. 1973-74), Psychonomic Soc., Soc. Exptl. Psychologists. Subspecialty: Experimental psychology. Office: Dept Psychology Stanford U Stanford CA 94305

SHEPHERD, ALEXANDER M.M., pharmacologist, physician, educator; b. Perth, Scotland, Apr. 26, 1945; came to U.S., 1978; m. Frances Mary Airtie, Apr. 6, 1971; children: Rebecca, Claire, Rachel. B.Sc., St. Andrews U., Fife, Scotland, 1965, M.B., Ch.B., 1969; Ph.D. in Clin. Pharmacology, U. Dundee, Scotland, 1978. House physician Dundee Royal Infirmary, 1969-70, sr. house physician, 1970-72, registrar in medicine, 1972-73; Med. Research Council fellow in clin. pharmacology U. Dundee, 1973-75, lectr., 1973-75; registrar in internal medicine Ninewells Hosp., Dundee, 1973-75, sr. registrar, 1975-77; asst. prof. pharmacology and medicine U. Tex. Health Sci. Ctr., San Antonio, 1978—, dir. hypertension service, 1981-, dir. clin. pharmacokinetics, 1981—, acting chief div. clin. pharmacology, 1982—. Contbr. articles to profl. jours. Lawrence Bequest fellow, 1976-77; grantee Am. Heart Assn., U. Tex. Health Sci. Ctr., NIH, Lilly Research Labs., Bristol Labs., Marion Labs., others. Mem. Royal Coll. Physicians (U.K.), Am. Fedn. Clin. Research, So. Soc. for Clin. Investigation, Am. Soc. for Pharmacology and Exptl. Therapeutics. Subspecialty: Pharmacology. Current work: Clinical pharmacology of Cardiovascular agents; effects of aging on response to cardio-vascular drugs. Office: Dept Pharmacology and Medicine U Tex Health Scis Center Room 212B 7703 Floyd Curl Dr San Antonio TX 78284

SHEPHERD, A.P., JR., physiology educator, consultant; b. Lexington, Miss., Dec. 29, 1943; s. A. P. and Virginia (Carter) S.; m. Melissa Ann Darnell, Dec. 18, 1965; 1 dau., Genevieve. B.S., Millsaps Coll., 1966; Ph.D., U. Miss., 1971. Asst. prof. U. Calif.-Irvine, 1973-74; asst. prof. physiology U. Tex.-San Antonio, 1974-76, assoc. prof., 1976—. Mem. Am. Physiol. Soc., Microcirculatory Soc., IEEE. Subspecialties: Physiology (biology); Biomedical engineering. Current work: Microcirculation, tissue oxygenation, instrumentation, laser Doppler velocimetry. Office: Dept Physiology U Tex Health Sci Ctr San Antonio TX 78284

SHEPHERD, HURLEY SIDNEY, botanist; b. Oxford, N.C., Dec. 17, 1950; s. Hurley Sidney and Anna Elizabeth (Keever) S. B.S., U. N.C., Chapel Hill, 1973; Ph.D., Duke U., 1978. Research assoc. U. Calif.-San Diego, 1978-81; asst. prof. botany and biochemistry U. Kans., Lawrence, 1981—. Duke U. fellow, 1973-76; NIH postdoctoral tng. grantee, 1979-81. Mem. Genetics Soc. Am., Plant Molecular Biology Assn., Am. Soc. Plant Physiologists. Subspecialties: Genome organization; Molecular biology. Current work: Researcher in control of gene expression in plants; developmentally regulated genes; nuclear-organelle genome interactions. Home: 1430 Louisiana St #5 Lawrence KS 66044 Office: Dept Botany U Kans Lawrence KS 66045

SHEPHERD, MARK, JR., electronics company executive; b. Dallas, Jan. 18, 1923; s. Mark and Louisa Florence (Daniell) S.; m. Mary Alice Murchland, Dec. 21, 1945; children: Debra Aline (Mrs. Rowland K. Robinson), MaryKay Theresa, Marc Blaine. B.S. in Elec. Engring, So. Meth. U., 1942, M.S., U. Ill., at Urbana, 1947. Registered profl. engr., Tex. With Gen. Elec. Co., 1942-43, Farnsworth Television and Radio Corp., 1947-48; with Tex. Instruments Inc., Dallas, 1948—, v.p., gen. mgr. semicondr.-components div., 1955-61, exec. v.p., chief operating officer, 1961-66, dir., 1963—, pres., chief operating officer, 1967-69, pres., chief exec. officer, 1969-76, chmn. bd., chief exec. officer, 1976—; dir. Republic Bank Dallas, Republic Bank Corp., U.S. Steel; mem. internat. council Morgan Guaranty Trust Co.; Mem. Adv. Council on Japan-U.S. Econ. Relations; bd. dirs. Found. for Sci. and Engring. So. Meth. U.; bd. govs., trustee So. Meth. U.; trustee Com. for Econ. Devel., Am. Enterprise Inst. Pub. Policy Research; trustee, councillor Conf. Bd.; mem. adv. council Am. Ditchley Found.; mem. Bus. Council, Dallas Citizens Council, Trilateral Commn.; mem. nat. bd. Com. on the Present Danger. Served to lt. (j.g.) USNR, 1943-46. Fellow IEEE; mem. Soc. Exploration Geophysicists, Newcomen Soc., Council on Fgn. Relations, U.S. Council Internat. Bus. (trustee), Nat. Acad. Engring., Sigma Xi, Eta Kappa Nu. Subspecialty: Electrical engineering. Office: Texas Instruments Inc PO Box 225474 MS 236 Dallas TX 75265

SHEPHERD, ROBERT JAMES, microbiologist, research plant virologist; b. Clinton, Okla., June 5, 1930; s. Lee Finis and Ruby (Gilleland) S.; m. Mary Ann Sall, Mar. 18, 1978; children by previous marriage: Steven, Eudora Deidre, David. B.S., Okla. State U., 1954, M.S., 1955; Ph.D., U. Wis., 1959. Asst. prof. plant pathology U. Wis., 1959-61; asst. prof. U. Calif., Davis, 1961-66, assoc. prof., 1966-72, prof., 1972—; cons. in field. Served with U.S. Army, 1950-52. Fulbright scholar, 1955-56. Fellow Am. Phytopath. Soc. (Ruth Allen award 1981); mem. Plant Molecular Biology Assn., Soc. Gen. Microbiology. Democrat. Subspecialties: Genetics and genetic engineering (biology); Plant virology. Current work: Biology of DNA plant viruses; genetic organization of DNA plant viruses, use of DNA plant viruses as recombinant DNA vector for plants. Patentee caulimoviruses as recombinant DNA vectors for plants. Home: 138 A St Davis CA 95616 Office: U Calif 560 Hutchison St Davis CA 95616

SHEPPARD, LOUIS CLARKE, biomed. engring. educator; b. Pine Bluff, Ark., May 28, 1933; s. Ellis Allen and Louise (Clarke) S.; m. Nancy Louise Mayer, Feb. 8, 1958; children: David, Susan, Lisa. B.S. in Chem. Engring, U. Ark., 1957; diploma, Imperial Coll. Sci. and Tech., London, 1976; Ph.D. in Elec. Engring, U. London, 1976. Devel. staff supr. Diamond Alkali Co., Deer Park, Tex., 1957-63; systems engr. IBM, Houston, 1963-64, staff engr., Rochester, Minn., 1964-66; assoc. prof. surgery U. Ala., Birmingham, 1966—, prof. chmn. dept. biomed. engring., 1979—; dir. acad. computing U. Ala. Med. Ctr., 1981—; mem. adv. bd. Pharmacontrol, Englewood Cliffs, N.J., 1982—; mem. med. adv. bd. Hewlett Packard, Waltham, Mass., 1980—; cons. IMED Devel. Corp., San Diego, 1982—, NIH, 1972—; dir. FBK Internat., Birmingham. Contbr. articles to profl. jours. Served with U.S. Army, 1958-66. Recipient various grants. Mem. Am. Inst. Chem. Engrs., IEEE (sr.), Nat. Soc. Profl. Engrs., Cardiovascular Dynamics Soc. (charter), Brit. Computer Soc. Subspecialties: Biomedical engineering; Health services research. Current work: Critical care systems, biomedical instrumentation, computer controlled therapy, infusion devices, drug delivery systems, pharmacodynamics, biological signal analysis, modeling and simulation. Home: 3644 Shamley Dr Birmingham AL 35223 Office: U Ala in Birmingham Univ Sta Birmingham AL 35294

SHEPS, CECIL GEORGE, physician; b. Winnipeg, Man., Can., July 24, 1913; came to U.S., 1946, naturalized, 1956; s. George and Polly (Lirenman) S.; m. Mindel Cherniack, May 29, 1937 (dec. Jan. 1973); 1 son, Samuel B.; m. Ann Shepherd. M.D., U. Man., 1936; M.P.H., Yale, 1947; D.Sc. (hon.), Chgo. Med. Sch., 1970; Ph.D. (hon.), Ben Gurion U. of the Negeu, 1983. Intern, St. Joseph's Hosp. Gen. Hosp., Winnipeg; resident Corbett Gen. Hosp., Stourbridge, Eng., Queen Mary's Hosp. for East End, London, Camp Shilo Mil. Hosp., Can., Ft. Osborne Mil. Hosp.; gen. dir. Beth Israel hosp., Boston; also clin. prof. preventive medicine Harvard Med. Sch., 1953-60; prof. med. and hosp. adminstrn. Grad. Sch. Pub. Health, U. Pitts., 1960-65; gen. dir. Beth Israel Med. Center, N.Y.C.; also prof. community medicine Mt. Sinai Sch. Medicine, 1965-68; prof. social medicine U. N.C., 1969-79, Taylor Grandy Disting. prof. social medicine, 1980—, dir. Health Services Research Center, 1969-71, vice chancellor health scis., 1971-76; cons. in field, 1947—. Author: Needed Research in Health and Medical Care-A Biosocial Approach, 1954, Community Organization-Action and Inaction, 1956, Medical Schools and Hospitals, 1965, The Sick Citadel: The American Academic Medical Center and the Public Interest, 1983. Chmn. health services research study sect. NIH, 1958-62; cons. med. affairs Welfare Adminstrn., HEW, 1964-67; mem. spl. commn. social scis. NSF, 1968-69; chmn. Milbank Meml. Fund. Commn. on Higher Edn. for Pub. Health, 1972-75. Served as capt. M.C., Royal Can. Army, 1943-46. Mem. Nat. Council Aging (v.p., dir.), Am. Nurses Found. (bd. mem. sec.-treas.), Inst. Medicine, Nat. Acad. Scis. Subspecialties: Preventive medicine; Epidemiology. Current work: Primary care; evoluation of health programs. Home: 1304 Arboretum Dr Chapel Hill NC 27514

SHER, FRANKLIN ALAN, immunologist; b. Nutley, N.J., Apr. 12, 1945; s. Seymour Cyrus and Roslyn (Lepow) S. A.B., Oberlin Coll., 1966; Ph.D., U. Calif., San Diego, 1972. Postdoctoral fellow Nat. Inst. Med. Research, London, 1972-74; prin. research assoc. dept. pathology Harvard U., 1974-77, asst. prof., 1978-81; head immunology and cell biology sect. Lab. Parasitic Disease, Nat. Inst. Allergy and Infectious Disease, NIH, Bethesda, MD, 1980—. Contbr. articles to profl. jours. Mem. Am. Assn. Immunologists, Am. Soc. Tropical Medicine and Hygiene, Brit. Soc. Parasitology. Subspecialties: Immunobiology and immunology; Parasitology. Current work: Immunology of parasitic infections. Office: NAID NIH Bldg 5 Room 114 Bethesda MD 20205

SHERALD, ALLEN FRANKLIN, III, geneticist, educator; b. Frederick, Md., Nov. 15, 1942; s. Allen Franklin Jr. and Betty Eileen (Harrop) S. B.S. cum laude, Frostburg (Md.) State Coll., 1964; Ph.D., U. Va., 1972. Sci. tchr., Montgomery County, Md., 1964-66; postdoctoral trainee Cornell U., Ithaca, N.Y., 1974-76; asst. prof. George Mason U., Fairfax, Va., 1976-80; asso. prof. biology, 1980—. Contbr. articles to profl. publs. NIH grantee, 1968, 69, 74; George Mason U. research grantee, 1978-81. Mem. AAAS, Genetics Soc. Am., Va. Acad. Scis., ACLU, Sigma Xi, Phi Sigma. Subspecialties: Gene actions; Developmental biology. Current work: Control of gene activity in development: sclerotization and coloration of cuticle of Drosophila. Home: 9451 Lee Hwy Apt 1209 Fairfax VA 22031 Office: Dept Biology George Mason U Fairfax VA 22030

SHERBON, JOHN WALTER, food science educator; b. Lewiston, Idaho, Oct. 31, 1933; s. Ollis W. and Agnes G. S.; m. Ruth Fern Poppens, Aug. 10, 1957; children—Barbara, William. B.S., Wash. State U., 1955; M.S., U Minn., 1958, Ph.D., 1963. Prof. food sci. Cornell U., Ithaca, N.Y., 1963—. Subspecialty: Food science and technology. Current work: Investigate rapid methods of food analysis, studies of physical state of fats in foods. Office: Cornell U 207 Stocking Hall Ithaca NY 14853

SHERIDAN, JOHN ROGER, physicist; b. Helena, Mont., Sept. 24, 1933; s. Maxwell Clark and Blanche Alta (Roberts) S.; m. Carol Marguerite Buckner, June 27, 1959; children: Lenita Lynn, Timothy Patrick. B.A., Reed Coll., 1955; Ph.D., U. Wash., 1964. Research engr. Boeing Co., 1955-60; research asst. U. Wash., Seattle, 1960-63, predoctoral assoc., 1963-64, research instr., 1964; asst. prof. physics U. Alaska Coll., Fairbanks, 1964-67, head dept., 1966-76, 78-81, assoc. prof., 1967-71, prof., 1971—; vis. physicist Stanford Research Inst., 1968-69; hon. prof. physics Queen's U., Belfast, 1975-76; mem. grad. and minority grad. and minority grad. panels NRC, 1980, 81, chmn. minority grad. fellowship panel, 1981. Contbr. articles to profl. jours. Precinct chmn. Democratic party, 1972-81; mem. Central Dist. Alaska Dem. Com., 1972-81, State Central Com. Alaska for Dems., 1981—. George F. Baker scholar George F. Baker Found., 1951-55. Fellow Am. Phys. Soc.; mem. Am. Assn. Physics Tchrs., AAAS, Sigma Xi. Democrat. Presbyterian. Club: Farthest North Bridge (Fairbanks, Alaska). Subspecialties: Atomic and molecular physics; Aeronomy. Current work: Use of photon-photon coincidence in measuring atomic reaction parameters and in calibrating optical systems; use of beam growth equations for measuring metastable reaction parameters and excitation cross sects. Home: Star Route Box 20260 Fairbanks AK 99701 Office: U Alaska Fairbanks AK 99701

SHERIDAN, ROBERT EDMUND, geology educator, consultant geophysicist; b. Hoboken, N.J., Oct. 11, 1940; s. Philip Michael and Mae Ann Miller S.; m. Maria Karen McCauley, May 28, 1966; children: Mary Jennifer, David Christopher, Steven Gregory. B.A., Rutgers U., 1962; M.A., Columbia U., 1965, Ph.D., 1968. Research scientist Lamont-Doherty Geol. Obs., Palisades, N.Y., 1968; instr. Columbia U., N.Y.C., 1968; asst. prof. U. Del., Newark, 1968-73, assoc. prof., 1973-81, prof. geology and geophysics, 1981—; vis. prof. Rutgers U., New Brunswick, N.J., 1974-79; chmn. Passive Margin Panel Joint Ocean, Deep Earth Sampling, 1979-81; pres. Monitor Research and Recovery Found., 1975-80; co-chief scientist Leg 44, Deep Sea Drilling Project, 1975, Leg 76, 1980. Author articles in field to profl. jours. Webelos leader Boy Scouts Am., 1971-73, com. chmn., 1977-79, treas., 1981-82. Recipient Antarctica medal Dept. Def., 1968. Fellow Geol. Soc. Am. (assoc. editor 1982—); mem. Am. Assn. Petroleum Geologists, Am. Geophys. Union. Republican. Roman Catholic. Lodge: KC. Subspecialties: Geophysics; Sea floor spreading. Current work: Atlantic continental margin, Bahama platform, shelf sedimentation, U.S.S. Monitor wreck. Office: Dept Geology U Del Newark DE 19711

SHERIDAN, ROBERT EMMETT, JR., computer scientist, consultant; b. Wilson, Pa., Jan. 18, 1943; s. Robert Emmett and Helen (Somogyi) S. B.S. in Aerospace Engring, Pa. State U., 1965, M.S., 1968, M.S. in Computer Sci, 1973. Assoc. research engr. Boeing Co., Seattle, 1968-70; sr. exptl. engr. Pratt & Whitney Fla. Research and Devel., West Palm Beach, 1970-71; grad. asst. Applied Research Lab., Pa. State U., 1971-73; program mgr. BDM Corp., Albuquerque, 1974-81, 83—; prin. scientist BDM Mgmt. Services Co., Killeen, Tex., 1981-83; instr. physics El Paso Community Coll., 1975-76; cons. Trainee NSF, 1965; proposal reviewer research grants, 1979. Mem. AIAA, Assn. Computing Machinery, IEEE Computer Soc., Digital Equipment Computer Users Soc. Democrat. Roman Catholic. Subspecialties: Mathematical software; Fluid mechanics. Current work: Modeling and simulation of mathematical and scientific systems, computational fluid mechanics. Home: 5528 Amistad NE Albuquerque NM 87111 Office: BDM Management Services Co Albuquerque NM 87106

SHERIFF, ROBERT EDWARD, geophysics educator; b. Mansfield, Ohio, Apr. 19, 1922; s. Charles Frederick and Marjorie (Norton) S.; m. Margaret Sites, Oct. 13, 1945; children: Anne Makowski, Richard K., Jeanne K., Susan Hunter, Barbara Barnes, Linda Barasch. A.B., Wittenberg U., 1943; M.Sc., Ohio State U., 1947, Ph.D., 1950. Chief geophysicist Standard Oil Co. (Calif.), various U.S., fgn. locations, 1950-75; sr. v.p. Seiscom Delta, Houston, 1975-80; adj. prof. dept. geoscis. U. Houston, 1972-79, prof., 1980—. Author: Dictionary Exploration Geophysics, 1973, First Course Geophysical Intrepretation, 1978, Seismic Stratigraphy, 1980, Exploration Seismology, 1982; co-author: Applied Geophysics, 1976. Mem. Am.

Assn. Petroleum Geologists, European Assn. Exploration Geophysicists, AAAS, Soc. Exploration Geophysicists (v.p., Kaufman gold medal 1969, hon.), Geophys. Soc. Houston (hon.). Democrat. Presbyterian. Subspecialties: Geophysics; Fuels. Current work: Seismic interpretaion techniques; seismic stratigraphy. Home: 12911 Butterfly Ln Houston TX 77024 Office: Geoscis Dept U Houston Houston TX 77004

SHERMAN, FRED, molecular biologist, educator; b. Mpls., May 21, 1932; s. Harry and Ann (Kaufman) S.; m. Revina Freeman, July 25, 1958; children: Aaron, Mark, Rhea. B.A., U. Minn., 1953; Ph.D., U. Calif., 1958. Postdoctoral fellow U. Wash., 1959-60, Lab. Genetique Physiol., France, 1960-61; sr. instr. U. Rochester, N.Y., 1961-62, asst. prof., 1962-66, assoc. prof., 1966-71, prof. radiation biology and biophysics, 1971—, prof. and chmn. dept. biochemistry, 1982—, Wilson prof., 1982-; A. Wander Meml. lectr., 1975. NIH fellow, 1959-61; NIH grantee, 1963—. Mem. Genetic Soc. Am., Environ. Mutagen Soc., AAAS, Am. Soc. Microbiology, Biophys. Soc. Subspecialties: Molecular biology; Gene actions. Current work: Yeast genetics, gene regulation, gene [protein relationship, mutagensis, recombinant DNA. Home: 340 Beresford Rd Rochester NY 14610 Office: Dept RBB U Rochester Med Sch Rochester NY 14642

SHERMAN, GORDON RAE, university computer center ofcl., educator; b. Menomonie, Mich., Feb. 24, 1928; s. Gordon Everett and Myrtle Harriet (Evensen) S.; children: Karen Rae, Gordon Thorstein. B.S., Iowa State U., 1953; Ph.D., Purdue U., 1960. Mem. faculty U. Tenn.-Knoxville, 1960—, prof. computer scis., 1969—, dir. Computing Center, 1960—; cons. U. Santiago (Chile), 1982; mem. People to People Computer Sci. del., USSR, Europe, 1982. Contbr. numerous sci. articles to profl. publs. Del. Knox County (Tenn.) Republican Com., 1968-69. Served with USAF, 1946-49, 50-51. Fellow Brit. Computer Soc.; mem. Data Processing Mgmt. Assn. (Profl. of Yr. award 1979), Assn. for Computing Machinery, Am. Statis. Assn., Operations Research Soc. Am. Republican. Lutheran. Subspecialties: Operating systems; Theoretical computer science. Current work: Discrete optimization; administration of University of Tennessee statewide computer system and network. Home: 301 Cheshire Dr Apt 105 Knoxville TN 37919 Office: Univ Tenn Computing Center 200 SMC Knoxville TN 37916

SHERMAN, JAMES HOWE, physiology educator; b. Detroit, Mar. 14, 1936; s. Willard H. and Esther M. (Redding) S.; m. Margrit E. Faulk, July 31, 1965; children: Diane, Douglas. B.S., U. Mich., 1957; Ph.D., Cornell U., 1963. Research asst. U. Mich., Ann Arbor, 1963-64, instr., 1964-66, asst. prof., 1966-70, assoc. prof., 1971—; asst. dir. Office Allied Health, 1972-77. Author: Human Physiology, 1970, Human Function and Structure, 1978. Mem. Am. Physiol. Soc., Biophys. Soc., N.Y. Acad. Sci. Subspecialties: Physiology (medicine); Cell biology (medicine). Current work: Membrane and muscle physiology. Address: Dept Physiology Univ Mich Med Sch Ann Arbor MI 48109

SHERMAN, JOHN FOORD, association official; b. Oneonta, N.Y., Sept. 4, 1919; s. Henry C. and Ruth (Foord) S.; m. Betsy Deane Murray, Feb. 8, 1944; children: Betsy Deane, Mary Ann. B.S., Albany Coll. Pharmacy of Union U., 1949, D.Sc., 1970; Ph.D., Yale U., 1953. With NIH, 1953-74; asso. dir. extramural programs Nat. Inst. Neurol. Diseases and Blindness, 1961-62, Nat. Inst. Arthritis and Metabolic Disease, 1962-63; asso. dir. for extramural programs Office Dir. NIH, 1964-68, dep. dir., 1968-74; v.p. Assn. Am. Med. Colls., Washington, 1974—. Asst. surgeon gen. USPHS, 1964-68; spl. research chemotheraphy and neuropharmacology; mem. Nat. Multiple Sclerosis Med. Adv. Bd.; mem. panel on data and studies NRC, 1976—; mem. biomed. library rev. com. NIH. Served with U.S. Army, 1941-46. Decorated Bronze Star; recipient Meritorious Service award USPHS, 1965, Disting. Service award HEW, 1971, Sec.'s Spl. Citation award, 1973, award Nat. Civil Service League, 1973; Disting. Alumnus award Union U.-Pharmacy Coll. Council, 1974. Mem. Nat. Soc. for Med. Research (pres.), AAAS, Inst. Medicine, Nat. Acad. Scis., Sigma Xi. Congregationalist. Club: Cosmos.

SHERMAN, LOUIS ALLEN, biophysicist; b. Chgo., Dec. 16, 1943; s. Stanley E. and Sarah S.; m. Debra Meddoff, June 15, 1969; children—Daniel, Jeffrey. B.S., U. Chgo., 1965, Ph.D., 1970. Postdoctoral fellow Cornell U., 1970-72; asst. prof. U. Mo., 1972-77, assoc. prof. div. biol. sci., 1977—. Contbr. articles to sci. jours. NIH postdoctoral fellow, 1970-72; NIH research grantee, 1975—; grantee U.S. Dept. Agr., Dept. Energy, 1980—. Mem. AAAS, Biophys. Soc., Am. Soc. Microbiology, Am. Soc. Plant Physiologists, Photobiology Soc. Democrat. Jewish. Subspecialties: Photosynthesis; Molecular biology. Current work: Study of the relationship of structure to function in photosynthetic membranes; cloning of photosynthesis genes in cyanobacteria. Office: U Mo 312 Tucker Hall Columbia MO 65211

SHERMAN, MERRY RUBIN, biochemist, educator; b. Bronx, N.Y., May 14, 1940; d. Geron T. and Edith (Williams) Margolish. B.A., Wellesley Coll., 1961; M.A., U. Calif., Berkeley, 1963, Ph.D. in Biophysics, 1966. Postdoctoral fellow Weizmann Inst., Rehovot, Israel, 1966-67; postdoctoral fellow Nat. Inst. Dental Research, NIH, 1967-70; successively research assoc., assoc., assoc. mem. Sloan-Kettering Inst., N.Y.C., 1970—, head endocrine biochemistry lab., 1975—; assoc. prof. biochemistry Cornell U. Grad. Sch. Med. Scis., N.Y.C., 1977—. Contbr. articles to profl. jours. Am. Cancer Soc. grantee, 1970; NIH grantee, 1974, 78, 80, 82. Mem. Am. Assn. Cancer Research, Endocrine Soc., Am. Assn. Biol. Chemists. Subspecialties: Cancer research (medicine); Cell study oncology. Current work: Steroid hormone receptors and mechanism of action. Office: 1275 York Ave New York NY 10021

SHERMAN, ROBERT GEORGE, zoologist, physiologist; b. Charlevoix, Mich., Mar. 22, 1942; s. George Robert and Alice (Bedell) S.; m. Barbara W. Sherman, June 12, 1965; children: Lisa, Amy. B.S., Alma (Mich.) Coll., 1964; M.S., Mich. State U., 1967, Ph.D., 1969. Postdoctoral fellow U. Toronto, Can., 1969-71; assoc. prof. physiology Clark U., Worcester, Mass., 1971-78; prof., chmn. dept. zoology Miami U., Oxford, Ohio, 1978—; corp. mem. Bermuda Biol. Sta. for Research. Contbr. articles to profl. jours., chpts. in books. Mem. Am. Soc. Zoology, N.Y. Acad. Scis., Electron Microscope Soc. Am., Ohio Acad. Sci., Soc. Gen. Physiology, Soc. Neurosci. Club: Oxford Country. Lodge: Oxford Rotary. Subspecialties: Comparative neurobiology; Morphology. Current work: Research, teaching and ednl. adminstrn. Office: Miami Univ 244 Upham Oxford OH 45056

SHERMAN, S. MURRAY, neurosciences educator; b. Pitts., Jan. 4, 1944; s. Julius Louis and Ida (Schooler) S.; m. Marjorie Jean Ebken, Feb. 1, 1969; children: Erika Kirsten, Benjamin William. B.S. in Biology, Calif. Inst. Tech., 1965; Ph.D. in Anatomy, U. Pa., 1969. Postdoctoral fellow Australian Nat. U., Canberra, 1970-72; asst. prof. physiology U. Va., Charlottesville, 1972-75, assoc. prof., 1975-78, prof., 1978-79; prof. neurobiology and behavior SUNY-Stony Brook, 1979—. Mem. editorial bd.: Physiol. Revs; contbr. chapts. to books, articles to profl. jours. Recipient numerous grants and fellowships Sloan Found., NIH, NSF. Mem. Soc. for Neurosci., Am. Assn. Anatomists, Assn. for Research in Vision and Ophthalmology, Cajal Club. Subspecialties: Neurobiology. Current work: Postnatal development of mammalian central visual pathways. Office: Dept Neurobiology SUNY Stony Brook NY 11794

SHERREN, ANNE TERRY, chemist; b. Atlanta, July 1, 1936; d. Edward Allison and Annie Ayres (Lewis) Terry; m. William Samuel Sherren, Aug. 13, 1966. B.A., Agnes Scott Coll., Decatur, Ga., 1957; Ph.D., U. Fla., Gainesville, 1961. Instr. to asst. prof. chemistry Tex. Woman's U., Denton, 1961-66; assoc. prof. chemistry North Central Coll., Naperville, Ill., 1966-76, prof., 1976—, chmn. dept., 1975-78, 81—; research assoc. Oak Ridge Nat. Lab., summers 1962, 63, Argonne Nat. Lab., summers 1968, 73-80. Mem. Am. Chem. Soc., Am. Inst. Chemists, Nat. Assn. Sci. Tchrs., AAUP, AAAS, Ill. Acad. Sci., Sigma Xi, Iota Sigma Pi (nat. pres. 1978-81), Delta Kappa Gamma. Presbyterian. Subspecialties: Analytical chemistry; Inorganic chemistry. Current work: Synthesis of inorganic complex ions and study of related spectra. Office: North Central Coll Naperville IL 60566

SHERRILL, THOMAS JOSEPH, space scientist; b. Cleve., Sept. 26, 1940; s. Otha Lee and Marie Catherine (Kontur) S. B.S. in Physics, Case Inst. Tech., 1962; M.A. in Astronomy, U. Calif., Berkeley, 1965, Ph.D., 1966. Research specialist Lockheed Missiles & Space Co., Sunnyvale, Calif., 1966—. Mem. AIAA, Am. Astron. Soc. Democrat. Subspecialties: Satellite studies; Aerospace engineering and technology. Current work: Space Telescope mission ops. planning; systems engrng. requirements analysis; satellite orbit perturbation studies; orbital tracking error analysis. Home: 226 Solana Dr Los Altos CA 94022 Office: Orgn 62-22 Bldg 579 Lockheed Missiles & Space Co 1111 Lockheed Way Sunnyvale CA 94088

SHERRIS, JOHN CHARLES, microbiology educator; b. Colchester, Eng., Mar. 8, 1921; came to U.S., 1959, naturalized, 1969; s. Cyril and Dorothy (Sherris); m. Elizabeth, July, 1944; children: Peter, Jacqueline. MRCS, LRCP, 1944; M.B., B.S., U. London, 1948, M.D., 1950; M.D. (hon.), Karolinska Inst., Stockholm, Sweden, 1975. Diplomate: Am. Bd. Pathology, Am. Bd. Med. Microbiology. House surgeon and physician King Edward VII Hosp., Windsor, Eng., 1944-45; trainee, asst. pathologist central lab. Sector V. Ministry Health, 1945-48; sr. registrar in pathology Aylesbury and Dist. Lab., Stoke, Mandeville, Eng., 1948-50, Radcliffe Infirmary, Oxford, Eng., 1950-52; lectr. in bacteriology U. Manchester (Eng.), 1953-56, sr. lectr., 1956-59; dir. clin. microbiology labs. U. Wash. Hosp., 1959-70; assoc. prof. microbiology U. Wash., 1959-63, prof., 1963—, chmn. dept. microbiology and immunology, 1970-80; bd. dirs. Nat. Found. Infectious Diseases. Editor: Cumitech Series, Am. Soc. Microbiology, 1974-80. Fellow Royal Coll. Pathologists, Am. Acad. Microbiology (gov. 1974-78, v.p. 1976-78); mem. Am. Soc. Microbiology (Becton-Dickinson award 1978, pres. 1982-83), Soc. Gen. Microbiology, Can. Soc. Microbiologists, Assn. Clin. Pathologists, Acad. Clin. Lab. Physicians, Scientists, Am. Bd. Med. Microbiology (chmn. 1971-73). Subspecialty: Microbiology. Research in antibiotic susceptibility and resistance of bacteria, automation in microbiology. Home: 10021 Lake Shore Blvd NE Seattle WA 98125 Office: Dept Microbiology and Immunology SC-42 U Wash Seattle WA 98195

SHERRY, SOL, physician, educator; b. N.Y.C., Dec. 8, 1916; s. Hyman and Ada (Greenman) S.; m. Dorothy Sitzman, Aug. 7, 1946; children—Judith Anne, Richard Leslie. A.B., NYU, 1935, M.D., 1939; D.Sc. (hon.), Temple U., 1980. Successively fellow, intern, resident 3d med. div. Bellevue Hosp., N.Y.C., 1939-42, 46; asst. prof. medicine N.Y.U. Coll. Medicine, 1946-51; dir. May Inst. Med. Research, Cin., 1951-54; dir. medicine Jewish Hosp., St. Louis, 1954-58; prof. medicine Washington U. Sch. Medicine, St. Louis, 1958-68, co-chmn. dept., 1964-68; chmn. dept. Temple U. Sch. Medicine, 1968—; dir. Thrombosis Research Center, 1970-79, dir. emeritus, 1979—; Cons. emeritus surgeon gen. army; past chmn. com. thrombolytic agts. USPHS, past chmn. com. on thrombosis; council on thrombosis Am. Heart Assn.; past mem. com. blood, past chmn. task force on thrombosis NRC; mem. Internat. Commn. on Hemostasis and Thrombosis; past chmn. and pres. Internat. Socs. on Thrombosis and Hemostasis; past mem. sci. adv. com. St. Louis Heart Assn., also St. Louis Multiple Sclerosis Soc. Contbr. articles to profl. jours. Mem. bd. Hillel, 1961-67; pres. S.E. Pa. chpt. Am. Heart Assn., 1978-79. Served as flight surgeon USAAF, 1942-46; ETO. Recipient medal for typhus control Lower Bavaria U.S. Army Typhus Commn., 1946; Research Career award USPHS, 1962; Distinghished Achievement award Modern Medicine, 1963, A.C.P., 1968, Am. Coll. Cardiology, 1971, Internat. Soc. Thrombosis and Haemostasis, 1978. Fellow Royal Coll. Physicians; mem. AMA (past mem. and chmn. council drugs, chmn. sect. exptl. medicine and therapeutics), Assn. Am. Physicians, Assn. Profs. Medicine (council 1973-75, pres. 1976), Am. Soc. Clin. Investigation, Am. Heart Assn., Central Soc. Clin. Research (council 1962-64), Am. Physiol. Soc., Soc. Exptl. Biology and Medicine, A.C.P. (master), Phi Beta Kappa, Sigma Xi (pres. Washington U. chpt. 1962-63), Alpha Omega Alpha. Subspecialties: Hematology; Cardiology. Current work: Primary developer of clot dissolving therapy and its newer applications. Active in organization and conduct of clinical trials evaluating agentsfor prevention or treatment of blood clots. Home: 408 Sprague Rd Narberth PA 19072

SHERTZER, HOWARD GRANT, toxicologist, educator, cons.; b. N.Y.C., Oct. 9, 1945; s. Sidney Maurice and Terry June (Rosenbaum) S.; m. Ellen Lea Asch, June 23, 1968; children—Kyle W., Kevin M. B.S. in Zoology, U. Mich., 1967; Ph.D. in Cell Biology, UCLA, 1973; postdoctoral student in biochemistry, Cornell U., 1975. Asst. prof. cell biology Tex. A&M U., College Station, 1975-79; assoc. prof. toxicology U. Cin. Med. Center, 1979—; also cons. Contbr. numerous articles to sci. jours. Grantee NIH, Nat. Cancer Inst., 1973—. Mem. AAUP, Am. Soc. Pharmacology and Exptl. Therapeutics, Soc. Toxicology, Internat. Soc. Study Xenobiotics. Subspecialties: Cancer research (medicine); Toxicology (medicine). Current work: Formation metabolism, toxicity and mechanism of action of chemical carcinogens. Metabolism of carcinogens, especially N-Nitrosamines.

SHETH, CHANDRAKANT HAKMICHAND, mechanical engineer; b. Gondal, Saurastra, India, Dec. 23, 1936; s. Hakmichand Anderji and Shantaben (Hakmich) S.; m. Manjulaben J. Shah, Jan. 31, 1940; 1 son, Sanjay. B.S.M.E., Gujarat U., India, 1960; M.S.I.E., NYU, 1962. Registered profl. engr., N.J., N.Y. Tech. staff engr. FEL Corp., Farmingdale, N.J., 1968-81; electronics engr. CECOM div. U.S. Army, Red Bank, N.J., 1981—. Mem. ASME. Subspecialties: Mechanical engineering; Electronics. Current work: Reliability and maintainability engineering in Dept of Defense communication and weapon systems, project management and design engineering activities in electronics systems. Home: 6 Old Bridge Dr Howell NJ 07731

SHETLER, STANWYN GERALD, botanist, museum curator, educator; b. Johnstown, Pa., Oct. 11, 1933; s. Sanford Grant and Florence Hazel (Young) S.; m. Elaine Marie Retburg, Feb. 2, 1963; children—Stephen Garth, Lara Suzanne. Student, Eastern Mennonite Coll., 1951-53; B.S. with distinction, Cornell U., 1955, M.S., 1958; Ph.D., U. Mich., 1979. Asst. curator phanerogam dept. botany Smithsonian Instn., Washington, 1962-63, assoc. curator phanerogams, 1963-81, curator phanerogams, 1981—; instr. Dept. Agr. Grad. Sch., 1962-64, 82—, 1966—; tour leader. Author: The Komarov Botanical Institute: 250 Years of Russian Research, 1967, Variation and Evolution of the Nearctic Harebells, 1982; contbr. numerous tech. and popular sci. articles, revs. to various publs.; editorial bd.: Systematic Botany Monographs, 1981—. Mem. land use citizens adv. com. Washington Met. Council Govts., 1977-79; mem. open space adv. com. Loudoun County, Va., 1980—. Recipient individual award Piedmont Environ. Council, 1981. Mem. AAAS, Am. Soc. Plant Taxonomists, Am. Inst. Biol. Scis., Bot. Soc. Am., Bot. Soc. Washington (v.p. 1982), Internat. Assn. Plant Taxonomy, Nat. Audubon Soc., Audubon Naturalist Soc. Central Atlantic States (dir. 1971-74, 80—, pres. 1974-77), Washington Biologists Field Club (v.p. 1981—), Sigma Xi. Mennonite. Subspecialties: Evolutionary biology; Systematics. Current work: Systematics and evolution of plant family Campanulaceae; floriistics of temperate N. Am., especially Alaska and central Atlantic region; history of Russian botany; use of info. retrieval in biol. data banking. Home: 142 Meadowland Ln E Sterling VA 22170 Office: Dept Botany Smithsonian Instn NHB 166 Washington DC 20560

SHEVACH, ETHAN MENAHEM, immunologist, physician; b. Brookline, Mass., Oct. 16, 1943; s. Benjamin Jacques and Anne (Pollack) S.; m. Ruth Schneider, May 30, 1967; children: Matthew, Seth. A.B., Boston U., 1967, M.D., 1967. Diplomate: Am. Bd. Internal Medicine. Intern and resident in medicine Bronx Municipal Hosp. Center, N.Y., 1967-69; clin. assoc. Lab. Clin. Investigation, Nat. Inst. Allergy and Infectious Diseases, NIH, Bethesda, Md., 1969-72, sr. staff fellow, 1972-73, sr. investigator, 1973—. Served to capt. USPHS, 1973—. Mem. Am. Assn. Immunologists, Am. Soc. Clin. Investigation, Am. Fedn. Clin. Research. Subspecialties: Immunogenetics; Cellular engineering. Current work: Immune response gene function; control of immunocompetent cell interactions.

SHEW, HARRY WAYNE, biologist, educator; b. Wilmington, N.C., Sept. 3, 1950; s. James Robert and Mary Viola (Mintz) S.; m. Rebecca Johnson, Dec. 16, 1972. B.A., U. N.C., 1971, M.A., 1974, Ph.D., 1977. Vis. lectr. biology U. N.C., Chapel Hill, 1978; asst. prof. biology Birmingham (Ala.) So. Coll., 1978—. Contbr. articles to profl. jours. Mellon Devel. grantee, 1980-81. Mem. Am Forestry Assn., Ala. Acad. Sci., Sigma Xi. Club: AED. Subspecialties: Genetics and genetic engineering (biology); Microbiology. Current work: Med. genetics - botany. Office: Birmingham So Coll 800 8th Ave W Birmingham AL 35254 Home: 1407 Sutherland Pl Birmingham AL 35209

SHEWMON, PAUL GRIFFITH, metall. engr., educator; b. Rochelle, Ill., Apr. 18, 1930; s. Joe Allen and Mildred Elizabeth (Griffith) S.; m. Dorothy E. Bond, Aug. 30, 1952; children—David A., Joan E. (dec.), Andrew B. B.S. in Metall. Engring, U. Ill., 1952; M.S., Carnegie Mellon U., 1954, Ph.D., 1955. Staff Westinghouse Research Lab., Pitts., 1955-58; prof. metall. engring. Carnegie Mellon U., Pitts., 1958-67; asso. dir. metal div., 1967-69, Exptl. Breeder Reactor II Project, Argonne Nat. Lab., 1969-70, dir. material sci. div., 1971-73; dir. div. materials research NSF, 1973-75; prof., head dept. Ohio State U., 1976-83; mem. adv. com. on reactor safeguards U.S. Nuclear Regulatory Commn., 1977—. Author: Diffusion in Metals, 1963-67, Phase Transformations, 1969. Dow fellow, 1952-55; NSF sci. faculty fellow, 1963-64. Fellow Am. Soc. Metals (Howe medal 1979); mem. Nat. Acad. Engring., Am. Nuclear Soc., Am. Inst. Metall. Engring. (asso. editor trans., R. Raymond award 1959, Noble award 1959). Subspecialty: Metallurgical engineering. Home: 2477 Lytham Rd Columbus OH 43220 Office: Dept of Metall Engring Ohio State Univ Columbus OH 43210

SHIBIB, MUHAMMED AYMAN, electrical engineer; b. Damascus, Syria, Feb. 14, 1953; came to U.S., 1975; s. Soubhi Taufik and Loutfieh (Shorkatli) S.; m. Reem Estwani, Mar. 30, 1982; 1 dau., Dena Reem. B.S., Am. U., Beirut, 1975; M.S., U. Fla.-Gainesville, 1976, Ph.D., 1979. Grad. research asst. U. Fla., Gainesville, 1976-79, research assoc., 1979, vis. asst. prof. elec. engring. dept., 1979-80; mem. tech. staff Bell Labs., Reading, Pa., 1980—. Contbr. articles to profl. jours. Mem. Am. Phys. Soc., Electrochem. Soc., IEEE, AAAS, Electron Device Soc., N.Y. Acad. Scis. Moslem. Subspecialties: 3emiconductors; Microchip technology (engineering). Current work: Research and development in semiconductor device physics and technology of integrated circuits. Home: 7 Tewkesbury Dr Wyomissing Hills PA 19610 Office: Bell Labs 2525 N 12th St Reading PA 19604

SHICHI, HITOSHI, biology educator and administrator; b. Nagoya, Japan, Dec. 20, 1932; came to U.S., 1967; s. Mitsuaki Takeichi and Tsuyako S.; m. Asae Elizabeth Nakagawa, Feb. 9, 1962; children: Yukari Grace, Mikaru Lucy. B.S., Nagoya U., 1955, M.S., 1957; Ph.D., U. Calif.-Berkeley, 1962. Asst. prof. Nagoya (Japan) U., 1962-63, Tokyo (Japan) U., 1963-67; sr. staff NIH, Bethesda, Md., 1967-81; prof. Inst. Biol. Scis., Oakland U., Rochester, Mich., 1981—. Author: Biochemistry of Vision, 1983; assoc. editor: jour. Photochemistry and Photobiology, 1977-81. Japan Scholarship Assn. scholar, 1951-57; Fulbright scholar, 1957-61. Mem. Am. Soc. Biol. Chemists, AAAS, Am. Soc. Photobiology, Biophys. Soc., Am. Chem. Soc., Assn. Research for Vision and Ophthalmology. Subspecialties: Biochemistry (biology); Biochemistry (medicine). Current work: Biochemistry of the visual process, ocular drug metabolism. Home: 4455 Pine Tree Trail Bloomfield Hills MI 48013 Office: Inst Biol Scis Oakland U Rochester MI 48063

SHIEH, PAULINUS SHEE-SHAN, nuclear engineer; b. Nanchang, Kiangsi, China, Apr. 5, 1931; came to U.S., 1956, naturalized, 1968; s. D.C. and C.H. (Chu) S.; m. Grace Y. Tung, Sept. 3, 1960; children: Jessie T., Patricia T., Victor T., Christopher T. B.S. in Physics, Taiwan U., Taipei, 1954, M.S., U. S.C., 1958; Ph.D. in Nuclear Engring, N.C. State U., 1969. Chmn. dept. physics King's Coll., Wilkes-Barre, Pa., 1960-63; instr. physics N.C. State U., 1963-68; assoc. prof. Miss. State U., 1969-79; sr. engr. Carolina Power & Light Co., Raleigh, N.C., 1979-81; supervising engr. Houston Lighting & Power Co., 1981—; cons. in field. Author: (with Inam-U-Rahman) Introduction to Nuclear Engineering, 1981. Mem. Am. Nuclear Soc., Am. Soc. Engring. Edn. Republican. Subspecialties: Nuclear engineering; Nuclear fusion. Current work: Nuclear analysis for LWR; fusion reactor design and neutronics. Home: 771 Seacliff Dr Houston TX 77062 Office: Houston Lighting and Power Company PO Box 1700 Houston TX 77001

SHIEH, YUCH-NING, geochemistry educator; b. Taiwan, China, Feb. 15, 1940; came to U.S., 1963; s. Min-chu and Hwei-tsau (Yang) S.; m. Tiee-Leou Ni, Sept. 10, 1966; children: Yi-shine, Yi-fong. B.S., Nat. Taiwan U., Taipei, 1962; Ph.D., Calif. Inst. Tech., 1969. Postdoctoral fellow McMaster U., Hamilton, Ont., Can., 1968-72; asst. prof. geoscis. Purdue U., 1972-79, assoc. prof., 1979—; vis. scientist Academia Sinica, Taipei, 1981. Mem. Am. Geophys. Union, Geochem. Soc. Subspecialties: Geochemistry; Petrology. Current work: Stable isotope geochemistry; oxygen, hydrogen and carbon isotopes in igneous and metamorphic petrology; sulfur isotopes in coal and ore deposits. Home: 4912 Hawthorne Ridge West Lafayette IN 47906 Office: Dept Geoscis Purdue U West Lafayette IN 47907

SHIELDS, GREGORY ALAN, astronomer, educator; b. Los Angeles, Oct. 22, 1946; s. Norman J. and Barbara (McVeigh) S.; m. Katharine M. Bowden, May 25, 1948; 1 son: Michael Gordon. B.S., Stanford U., 1968; M.S. in Astronomy, Calif. Inst. Tech., 1969, Ph.D., 1973. Research fellow Harvard Coll. Obs., 1972-74; asst. prof. astronomy U. Tex., Austin, 1974-79, assoc. prof., 1979—. Contrb. articles to profl. jours. Alfred P. Sloan research fellow, 1979. Mem. Astron. Soc. Pacific (editorial bd. publs. 1982—), Am. Astron. Soc., Royal Astron. Soc. Subspecialties: Theoretical astrophysics; Optical astronomy. Current work: Ionized nebulae, quasars, galaxies. Also emission lines, physical conditions, chemical abundances. Office: Dept Astronomy U Tex Austin TX 78712

SHIFFMAN, MAX, mathematician, consultant, researcher; b. N.Y.C., Oct. 30, 1914; s. Nathan and Eva (Krasilchick) S.; m. Bella Manel, June 1938 (div. 1957); children: Bernard, David. B.S., CCNY, 1935; M.S., NYU, 1936, Ph.D., 1938. Instr. in math. St. John's U., N.Y.C., 1938-39, CCNY, 1938-42; mathematician, assoc. prof. NYU, 1941-49; prof. math. Stanford U., 1949-66, Calif. State U.-Hayward, 1967-79; mathematician, owner Mathematico, San Francisco and Hayward, 1970—; cons. RAND Corp., Santa Monica, Calif., 1951; U.S. Govt., 1962-64; cons. math. logistics George Washington U., 1958-61. Contbr. articles to profl. jours. Recipient award of Merit U.S. Navy, 1945; Blumenthal fellow, 1935-38. Mem. Am. Math. Soc., Soc. Indsl. and Applied Math., Math. Assn. Am. Subspecialties: Applied mathematics. Current work: Mathematical aerodynamics for airplane wings; stability and instability in variational analysis; measure theory and non-measurable sets; generalized median; topics in algebraic systems. Home and Office: 16913 Meekland Ave Apt 7 Hayward CA 94541

SHIH, CHARLES CHIEN, nuclear generation engineer; b. Taipei, Taiwan, June 3, 1950; came to U.S., 1974, naturalized, 1983; s. Tsu-En and Chung-Chueh (Hsu) S.; m. Grace H. Cheng, June 3, 1978. B.S., Nat. Tsing Hua U., 1972; M.S., U. R.I.-Kingston, 1977. Registered profl. engr., Calif. Sr. nuclear engr. Kaiser Engrs., Inc., Oakland, Calif., 1977-80; nuclear generation engr. Pacific Gas & Electric Co., San Francisco, 1980—. Mem. Am. Nuclear Soc., Health Physics Soc. Subspecialties: Nuclear engineering; Software engineering. Current work: Radiological assessment, computer software engineering, nuclear criticality analysis, nuclear radiation analysis. Office: Pacific Gas & Electric Co 77 Beale St San Francisco CA 94106

SHIH, JASON CHIA-HSING, biotechnology educator, researcher; b. Chien-Cheng, Hunan, China, Oct. 8, 1939; came to U.S., 1969; s. Pang-Fang and Shue-Yin (Shen) S.; m. Jane Chu-Huei Chien, Aug. 31, 1966; children: Giles, Tim. B.S., Nat. Taiwan U., 1963, M.S., 1966; Ph.D., Cornell U., 1973. Lectr. Tunghai U., Taichung, Taiwan, 1966-69; research asst. Cornell U., Ithaca, N.Y., 1969-73, sr. research assoc., 1975-76; research assoc. U. Ill., Champaign-Urbana, 1973-75; asst. prof. N.C. State U., Raleigh, 1976-80, assoc. prof. biotechnology, 1980—; chmn., speaker Poultry Nutrition Conf., 1982; speaker Internat. Seminar Bio-Engergy, U.K., 1982; lectr. Ministry Agr., China, 1982. Pres. Triangle Area Chinese Am. Soc., 1982-83, exec. com., 1979-83; adv. bd. N.C. China Council, 1979—. Research grantee NIH, U.S. Dept. Energy, N.C. Energy Inst., 1977—. Mem. Am. Inst. Nutrition, Am. Chem. Soc., Am. Soc. Microbiology, Poultry Sci. Assn., Phi Kappa Phi. Subspecialties: Biochemistry (biology); Microbiology. Current work: Poultry biotechnology: anaerobic digestion of poultry waste for methane production, experimental atherosclerosis: quail model study for atherosclerosis disease. Office: Biotechnology Lab Dept Poultry Sci NC State Univ Raleigh NC 27650

SHIH, TSUNG-MING ANTHONY, Pharmacologist; b. Taipei, Taiwan, Oct. 8, 1944; came to U.S., 1968, naturalized, 1977; s. Chukun and Jean S.; m. Ming Lin, Sept. 5, 1970; children: Liane, Jason. B.S. in Pharmacy, Kaohsiung (Taiwan) Med. Coll., 1967; Ph.D. in Pharmacology, U. Pitts., 1974. Teaching asst. Columbia U., 1968-69; teaching asst. U. Pitts., 1969-71, research asst. 111, 1975-76, research assoc., 1976-78; pharmacologist U.S. Army Med. Research Inst. Chem. Def., Aberdeen Proving Ground, Md., 1978-80; Dean Chinese Lang. Sch. Balt., 1982—. Contbr. articles to profl. jours. NIH trainee, 1972-74, Western Psychiat. Inst. and Clinic fellow, 1974-76. Mem. N.Y. Acad. Sci., Soc. Neuroscio. Am. Chem. Soc., Am. Soc. Neurochemistry, Am. Soc. Mass Spectrometry, AAAS, Am. Mgmt. Assn., Sigma Xi. Democrat. Subspecialties: Neuropharmacology; Neurochemistry. Current work: Central neuropharmacological mechanisms of action of organophosphorus anticholinesterases and their treatment compound; central neurotransmitter system dynamics and interactions. Office: Bldg E-3100 Aberdeen Proving Ground MD 21010

SHIMAOKA, KATSUTARO, physician, researcher; b. Nara, Japan, Sept. 4, 1931; s. Shigeyuki and Shizue (Kishimoto) S.; m. Tomoko Suzuki, May 14, 1956; children: Julia E., Eva E. M.D., Keio U., Tokyo, 1955. Intern St. Luke's Hosp., Denver, 1956-57; resident in internal medicine Louisville Gen. Hosp., 1957-58; resident Roswell Park Meml. Inst., Buffalo, 1958-58, fellow in internal medicine, 1959-61; research asst. UCH Med. Sch., London, 1961-63; registrar UCH, 1961-63; sr. research assoc. Roswell Park Meml. Inst., 1963-65, sr. cancer research scientist, 1965-67, cancer research clinician, 1967-79, assoc. chief cancer research clinician, 1979—; research prof. SUNY-Buffalo, 1981—; cons. staff Erie County Med. Ctr., 1975—; research prof. Niagara U., 1977—; attending physician VA Hosp., 1979—. Contbr. numerous articles to profl. jours. Mem. Endocrine Soc., Am. Thyroid Assn., Japanese Soc. Internal Medicine, Japan Endocrine Soc., N.Y. Acad. Scis., Am. Assn. Cancer Research, Am. Soc. Clin. Oncology, Eastern Coop. Oncology Group, Am. Soc. Bone and Mineral Research, Am. Coll. Nuclear Medicine, others. Subspecialties: Oncology; Endocrinology. Current work: Endocrine tumors, consequences of cancer treatment.

SHIMOSATO, SHIRO, anesthesiologist; b. Matsumoto, Japan, June 14, 1929; s. Magojuro and Shizue (Sengoku) S.; m. Minako Tamamoto, Sept. 12, 1959; children: Keiko, Elliko, Yukiko. M.D., Shinshu U., 1955; Ph.D., Sapporo Med. Coll., 1977. Diplomate: Am. Bd. Anesthesiology. Practice medicine specializing in anesthesiology, prof. anesthesiology Tufts U. Med. Sch., Boston, 1972-74, U. Iowa Med. Coll., Iowa City, 1975-78, prof., 1978—; now dir. cardiovascular anesthesia research U. Iowa Hosps. and Clinics; prof., vice chmn. dept. U. Calif.-Davis, 1978-79. Contbr. articles to profl. jours. USPHS Research career devel. awardee; Charlton Fund fellow, 1959; Mass. Heart Assn. research fellow, 1960; Fulbright scholar, 1956. Fellow Am. Coll. Cardiology; mem. Assn. Univ. Anesthetists, Am. Soc. Anesthesiologists. Buddhist. Subspecialties: Anesthesiology; Health services research. Current work: Cardiocirculatory physiology and pharmacology as related to anesthesia. Home: 225 Gould St Iowa City IA 52242 Office: Dept Anesthesia U Iowa Hosps and Clinics Iowa City IA 52242

SHINEFIELD, HENRY ROBERT, pediatrician; b. Paterson, N.J., Oct. 11, 1925; s. Louis and Sarah (Kaplan) S.; m.; children—Jill, Michael. B.A., Columbia U., 1945, M.D., 1948. Diplomate: Am. Bd. Pediatrics (examiner, 1975—, bd. dirs., 1979—, v.p., 1981—). Rotating intern Mt. Sinai Hosp., N.Y.C., 1948-49; pediatric intern Duke Hosp., Durham, N.C., 1949-50; asst. resident pediatrician N.Y. Hosp. (Cornell), 1950-51, pediatrician to outpatients, 1953-59, instr. in pediatrics, 1959-60, asst. prof., 1960-64, asso. prof., 1964-65, asst. attending pediatrician, 1959-63, asso. attending pediatrician, 1963-65; pediatrician to outpatients Children's Hosp., Oakland, Calif., 1951-53; chief of pediatrics Kaiser-Permanente Med. Center, San Francisco, 1965—; asso. clin. prof. pediatrics Sch. Medicine U. Calif., 1966-68, clin. prof. pediatrics, 1968—, clin. prof. dermatology, 1970—; asso. attending pediatrician Paterson (N.J.) Gen. Hosp., 1955-59; chief of pediatrics Kaiser Found. Hosp., San Francisco, 1965—; attending Moffitt Hosp., San Francisco, 1967—; practice medicine specializing in pediatrics, Paterson, 1953-59; cons. San Francisco Gen. Hosp., 1967—, Childrens Hosp., San Francisco, 1970—, Mt. Zion Hosp., 1970—; mem. research grants rev. br. NIH, HEW, 1970-74; med. dir. USPHSR, 1969—; bd. dirs. San Francisco Peer Rev. Orgn., 1975—, sec., exec. com., 1976—; chmn. Calif. State Child Health Disability Bd., 1973—; sr. mem. Inst. of Medicine, Nat. Acad. Scis., 1980—; cons. Bur. Drugs FDA, 1970—, Nat. Acad. Scis., 1972—, NIH, HEW, 1974—. Editorial bds.: Western Jour. of Medicine, 1968—, American Jour. of Diseases of Children, 1970—; contbr. writings to profl. publs. Chmn. San Francisco Med. Adv. Com. Nat. Found. March of Dimes, 1969—. Served with USPHS, 1951-53. Fellow Am. Acad. Pediatrics (com. of fetus and newborn 1969-76, mem. com. on drugs 1978—); mem. AMA, Soc. Pediatric Research, Infectious Diseases Soc. Am., Western Pediatric Soc., Western Soc. Clin. Research, Am. Pediatric Soc., Phi Beta Kappa. Subspecialty: Pediatrics. Home: 2705 Larkin St San Francisco CA 94109 Office: 2200 O'Farrell St San Francisco CA 94115

SHING, YUH-HAN, material scientist, physicist; b. Tungtu, China, June 18, 1941; came to Can., 1965; s. Chih-An and Shin-I (Wong) S.; m. Jean Jen Chou, June 24, 1967; children: Mona, Sophia, Lara. B.Sc., Taiwan Normal U., 1963; M.Sc., U. Calgary, Alta., Can., 1969, Ph.D., 1972. Postdoctoral fellow McGill U., 1972-74; research assoc., reader, 1974-76, reader, 1976-79; mem. research staff Xerox Research Ctr., Mississauga, Ont., 1979-80; sr. research scientist ARCO Solar Industries, Woodland Hills, Calif., 1980-82; project leader Atlantic Richfield Co., Los Angeles, 1982—. Contbr. articles to sci. jours. Pres. Conejo Valley Chinese Assn., 1982. NRC of Can. grad. scholar, 1969-72; postdoctoral fellow NRC of Can., 1972-74; recipient grants, 1976-79. Mem. Am. Phys. Soc., IEEE, Electrochem. Soc., Am. Vacuum Soc. Subspecialties: Electronic materials; Solar energy. Current work: Photovoltaic energy conversion, thin film semiconductors, compound semiconductors, sputtering processes, semiconductor characterizations, vacuum evaporation, glow discharge, amorphous semiconductors. Home: 2134 Calle Riscoso Thousand Oaks CA 91362 Office: Materials Development Lab Atlantic Richfield Co 20717 Prairie St Chatsworth CA 91311

SHINNAR, REUEL, chemical engineer, educator; b. Vienna, Austria, 1923. B.S., M.S., Technion Haifa, Israel; Sc.D., Columbia U., 1957. Former mem. faculty Technion Haifa, Princeton U.; mem. faculty CCNY, 1964—, now prof. chem. engring., chmn. dept.; cons. chem. and petroleum industry. Contbr. articles to profl. jours. Mem. Am. Inst. Chem. Engrs. (Alpha Chi Sigma award 1979), Am. Chem. Soc., AIAA, N.Y. Acad. Scis., Ops. Research Soc. Am. Subspecialties: Chemical engineering; Fuels. Patentee in field. Office: Dept Chem Engring CCNY Convent Ave at 140th St New York NY 10031

SHINNICK, THOMAS MICHAEL, microbiologist, researcher; b. Madison, Wis., Apr. 1, 1953; s. Michael Grant and Cora Adele (Wilson) S.; m. Kathleen Marie McDowell, May 25, 1974. B.S., U. Wis., 1974; Ph.D. (Johnson & Johnson Industries predoctoral fellow, 1977-78), M.I.T., 1978. Research asst. U. Wis.-Madison, 1973-74; teaching asst. M.I.T., Cambridge, 1975-78; postdoctoral fellow Research Inst. of Scripps Clinic, La Jolla, Calif., 1978-80, asst. mem., 1980—. Contbr. sci. publs. to profl. jours. Helen Hay Whitney Found. fellow, 1978-81. Mem. Am. Soc. Microbiology. Subspecialties: Molecular biology; Developmental biology. Current work: Control of gene expression; biology of oncogenic viruses. Office: 10666 N Torrey Pines Rd La Jolla CA 92037

SHINOWARA, NANCY LEE, neurobiologist, cell biologist; b. Waynesboro, Va., Aug. 5, 1944; d. George Yukio and Alice Marie (Waterhouse) S. A.B., Mt. Holyoke Coll., 1966; M.A.T., Northwestern U., 1974, Ph.D., 1975. Postdoctoral research fellow div. biology Calif. Inst. Tech., Pasadena, 1974-78; staff fellow Lab. Neuroscis., Gerontology Research Center, Nat. Inst. on Aging, Balt., 1978-81, sr. staff fellow sect. exptl. morphology, 1981—. Contbr. articles on peripheral nerve and synapse, blood-nerve barrier, intercellular junctions in nerve and myelin, freeze-fracture methods, retinal morphology and glucose utilization during aging to profl. jours. Mem. Soc. for Neurosci., Am. Soc. Cell Biology, AAAS, Chesapeake Soc. Electron Microscopy (pres.). Subspecialties: Neurobiology; Cell biology (medicine). Current work: Functional morphology of the nervous system (peripheral nerve, blood -nerve and blood-ocular barriers, retina) during development and aging with emphasis on cell-cell interactions, cell surfaces, permeability and secretion. Office: Nat Inst Aging Gerontology Research Center Balt City Hosps Baltimore MD 21224

SHIPMAN, HARRY LONGFELLOW, astrophysicist, educator, researcher; b. Hartford, Conn., Feb. 20, 1948; s. Arthur Leffingwell and Mary Pepperell (Dana) S.; m. Editha Davidson, Apr. 10, 1970; children—Alice Elizabeth, Thomas Nathaniel. B.A. in Astronomy, Harvard U., 1969, M.S., Calif. Inst. Tech., 1970, Ph.D., 1971. Programmer Travelers Ins. Co., 1966; research asst. Smithsonian Astrophys. Obs., summers 1968, 69; teaching asst. Calif. Inst. Tech., 1969-71; J. W. Gibbs instr. in astronomy Yale U., 1971-73; asst. prof. physics U. Mo., St. Louis, 1973-74; astronomer McDonnel Planetarium, St. Louis, 1973-74; asst. prof. physics U. Del., 1974-77, assoc. prof., 1977-81, prof., 1981—; Harlow Shapley vis. lectr. Am. Astron. Soc., 1976—; trustee Mt. Cuba Astron. Obs., 1977—; cons. Author: The Restless Universe: An Introduction to Astronomy, 1978, Black Holes, Quasars, and the Universe, 2d edit, 1980; contbr. numerous articles to profl. jours. Guggenheim fellow, 1980-81; NSF grantee, 1974—; NASA grantee, 1976-79, 81. Mem. Am. Astron. Soc. (edn. officer 1979—), Am. Assn. Physics Tchrs., AAUP, AAAS, Fedn. Am. Scientists, Internat. Astron. Union, Astron. Soc. Pacific, Phi Beta Kappa, Sigma Xi (sec. U. Del. chpt. 1976-77, v.p. chpt. 1977-79, pres. chpt. 1979-80, chpt. Disting. Scientist award 1981). Subspecialties: Optical astronomy; Ultraviolet high energy astrophysics. Current work: Development numerical simulations of radiative transfer in outer layers of white-dwarf stars; application of these to observations of x-rays and ultraviolet radiation from these stars. Home: 346 Old Paper Mill Rd Newark DE 19711 Office: Dept Physics U Del Newark DE 19711

SHIPMAN, ROSS LOVELACE, university research executive, petroleum consultant; b. Jackson, Miss., Nov. 20, 1926; s. William Smylie and Jeanette Scott (Lovelace) S.; m. Lois Pegrim, June 6, 1948; 1 dau., Smylie Shipman Anderson. B.A., U. Miss., 1950. Geologist Humble Oil & Refining Co., West Tex., 1950-55; petroleum cons., Midland, Tex., 1955-60, Corpus Christi, Tex., 1960-67; asst. exec. dir. Am. Geol. Inst., Washington, 1967-71; assoc. dir. U. Tex. Marine Sci. Inst., Austin, 1971-79; assoc. v.p. for research U. Tex.-Austin, 1979—; prin. Petroleum Investments/Worldwide, 1975—; mem. Tex. Coastal and Marine Council, 1979—, U.S.-Mexico Boundary Water Study Program, 1978—; de. Argonne Univs. Assn., Chgo., 1982. Author numerous geol. reports and studies, 1955-; editor, pub.: The AGI Report newsletter, 1968-70; editor: Profl. Geologist, 1975-76. Served to cpl. U.S. Army, 1944-46; PTO. Fellow Geol. Soc. Am., AAAS, Geol. Soc. (London); mem. Am. Assn. Petroleum Geologists (cert. geologist), Am. Inst. Profl. Geologists (cert. profl. geologist, Tex. State pres. 1974, nat. editor 1975-76); Mem. Soc. Ind. Profl. Earth Scientists; mem. Nat. Council Univ. Research Adminstrs.; Mem. Soc. Research Adminstrs. Anglican. Club: Corpus Christi Yacht (rear commodore 1965-67). Subspecialties: Geology; Hydrogeology. Current work: Development, funding, administration of university research programs; petroleum and mineral exploration and enhanced oil recovery development; international water development. Home: Ambiente 1803 Great Oaks Dr Round Rock TX 78664 Office: U Tex Austin TX 78712

SHIPPEE, PAUL WESLEY, solar designer, consultant, builder; b. Providence, Nov. 15, 1937; s. Paul W. and Katherine (Carlson) S.; m. Mary Susan Newton, 1973 (div.); children: Jennifer, Elliott. B.S.C.E. with honors, U. Conn., 1959; postgrad., San Francisco State Coll., 1962-63. Ptnr. Colo. Mountain Builders, Livermore, 1972-74; civil engr. State of Calif., 1960-63; owner, operator Colo. Sunworks, Boulder, 1975—; cons. passive and active solar heating; conf. speaker; instr. Community Coll., Denver, 1978, U. Colo., 1976-77, Colo. State U., 1974-75. Contbr. articles to periodicals. Served with USAR, 1959-60. Recipient Solar award HUD, 1977, 79; Dept. Energy grantee, 1978-80. Mem. Internat. Solar Energy Soc., Colo. Solar Energy Assn. (founding pres. 1977, bd. dirs. 1978-79), N.Mex. Solar Energy Assn. Patentee in field. Office: PO Box 455 Boulder CO 80306

SHIRES, GEORGE THOMAS, physician, educator; b. Waco, Tex., Nov. 22, 1925; s. George Thomas and Donna Mae (Smith) S.; m. Robbie Jo Martin, Nov. 27, 1948; children—Donna (Mrs. James G. Blain), George Thomas III, Jo Ellen. M.D. (Life Ins. Med. Research fellow), U. Tex. at Dallas, 1947. Diplomate: Am. Bd. Surgery (dir. 1968-74, chmn. 1972-74). Intern Mass. Meml. Hosp., Boston, 1948-49; resident Parkland Meml. Hosp., Dallas, 1950-52; mem. faculty U. Tex. Southwestern Med. Sch. at Dallas, 1952-74, asso. prof. surgery, acting chmn. dept., 1960-61, prof., chmn. dept., 1961-74; surgeon in chief surg. services Parkland Meml. Hosp., 1960-74; prof., chmn. dept surgery U. Wash. Sch. Medicine, Seattle, 1974-75; chief of service Harborview Med. Center, Seattle, Univ. Hosp., 1974-75; chmn. dept. surgery N.Y. Hosp.-Cornell Med. Center, 1975—; attending surgeon Sloan-Kettering Cancer Center, 1979—; cons. Nat. Inst. Gen. Med. Sci., 1965—, to Surgeon Gen. Army, 1965-75; cons. to Surgeon Gen. Jamaica Hosp., 1978—; mem. com. metabolism and trauma Nat. Acad. Sci./NRC, 1964-71, com. trauma, 1964-71; mem. research program evaluation com., reviewer clin. investigation applications career devel. program VA, 1972-76; mem. gen. med. research program progress com. NIH, 1965-69; mem. Surgery A study sect., 1970-74, chmn., 1976-78; mem. Nat. Adv. Gen. Med. Scis. Council, 1980-84; cons. editorial bd. Jour. Trauma, 1968—, Tex. Medicine. Editorial bd.: Year Book Med. Publs, 1970—, Annals of Surgery, 1972—, Surgical Techniques Illustrated: An International Comparative Text, 1974-75, Am. Jour. Surgery, 1968—, Contemporary Surgery, 1973—; assoc. editor in chief: Infections in Surgery, 1981; mem. editorial council: Jour. Clin. Surgery, 1980-82; editor: Surgery, Gynecology and Obstetrics, 1982—. Served to lt. M.C. USNR, 1949-50, 53-55. Mem. Dallas Soc. Gen. Surgeons (pres.-elect, pres. 1972-74), Am. Assn. Surgery Trauma, A.C.S. (bd. regents 1971—, chmn. bd. regents 1978-80, pres. 1981-82), A.M.A., Am. Surg. Assn. (sec. 1969-74, pres. 1980), Digestive Disease Found (founding mem.), Halsted Soc., Internat. Soc. Burn Injuries, Internat. Surg. Soc. (sec. 1978-81), Pan-Am. Med. Assn. (surgery council 1971—), Pan Pacific Surg. Assn., Soc. Clin. Surgery, Soc. Surgery Alimentary Tract, Soc. Surg. Chairmen (pres. 1972-74), Soc. Univ. Surgeons (chmn. publs. com. 1969-71), So. Surg. Assn., Surg. Biology Club (sec. 1968-70), Western Surg. Assn., Allen O. Whipple Surg. Soc., James IV Assn. Surgeons, Alpha Omega Alpha, Alpha Pi Alpha, Phi Beta Pi. Subspecialty: Surgery. Current work: Trauma, shock and burns. Home: 450 E 63d St New York NY 10021 Office: 525 E 68th St New York NY 10021

SHIRES, THOMAS KAY, pharmacologist, educator; b. Buffalo, July 12, 1935; s. Frank Alexander and Bessie Cummings (Harder) S.; m. Ann Kirkpatrick, Aug. 27, 1961; children—Cynthia Louise, Heather Ann, Elliot Thomas. A.B., Colgate U., 1957; M.S., U. Okla., 1962, Ph.D., 1965. Instr. depts. anatomy and urology U. Okla. Sch. Medicine, 1965-67, asst. prof., 1967-68; asst. prof. dept. pharmacology U. Iowa Coll. Medicine, Iowa City, 1972-75, assoc. prof., 1975-81, prof., 1981—, assoc. prof. dept. pathology, 1975-77. Author 1 book in field; contbr. articles to profl. jours. Served with U.S. Army, 1958-60. Mem. Am. Soc. Pharmacology and Exptl. Therapeutics, Am. Soc. Cell Biology, Am. Assn. Pathologists, Am. Assn. Cancer Research, N.Y. Acad. Scis., Am. Ornithologists Union, Sigma Xi. Methodist. Subspecialties: Cell and tissue culture; Molecular pharmacology. Current work: Research into synthesis, degradation and function of the internal cellular membrane system with special concern for neoplastic and toxicologic alterations of the system. Office: Dept Pharmacology U Iowa Coll Medicine Iowa City IA 52242

SHIRKEY, HARRY CAMERON, pediatrician; b. Cin., July 2, 1911; s. Lewis Cameron and Pearl (Knight) S.; m. Mary Alice Brill; 3 children. B.S., U. Cin., 1939, M.D., 1945; D.Sc., Mass. Coll. Pharmacy, 1976. Diplomate: Am. Bd. Pediatrics, Am. Bd. Med. Toxicology. Pharmacist Children's Hosp., Cin., 1940-45; dir. out-patient dept., 1952-53; dir. pediatrics U. Cin., 1953-58, Children's Hosp., Birmingham, Ala., 1960-68, dir., 1968-77. Editor: Pediatric Therapy, 6th edit, 1980. Recipient Hugo Schaeffer medal Am. Pharmacy Assn., 1971. Mem. Am. Pharmacy Assn., Am. Acad. Pediatrics, AMA. Episcopalian. Subspecialty: Toxicology (medicine). Current work: Toxicology. Home: 1229 N Fort Thomas Ave Fort Thomas KY 41075

SHIRKEY, WILLIAM DAN, physicist, engineer, researcher; b. Roswell, N.Mex., Nov. 6, 1951; s. Robert Johnson and Joan (Savage) S. B.S. in Physics, SUNY-Brockport, 1973, M.S., Clarkson Coll. Tech.,

1980; postgrad., Cornell U. Grad. Sch. Mgmt., 1982-84. Quality assurance engr. Corning Glass Works, Canton, N.Y., 1974-76, sr. process engr., 1979-81, devel. engr., 1981, sr. process and product devel. engr., 1981-82; staff Cornell U. Grad. Sch., 1982—; research asst. Clarkson Coll. Tech., Potsdam, N.Y., 1977-79. Contbr. articles to profl. publs. Mem. Optical Soc. Am., Am. Inst. Physics, Nat. Peace Acad., Amnesty Internat., Sigma Pi Sigma. Subspecialties: Materials processing; Materials (engineering). Current work: Use of fused silica and low expansion glasses in the fabrication and frit sealing of lightweight mirrors, fusion sealing of lightweight mirrors, astronomical and space use mirror fabrication. Home: 15 Dart Dr Ithaca NY 14850 Office: Corning Glass Works PO Box 28 Canton NY 13617

SHIRLEY, BARBARA ANNE, physiology educator, researcher; b. Muskogee, Okla., Oct. 15, 1936; s. Harvey Burton and Nona Leone (Smith) S. B.A., Okla. Bapt. U., 1956; M.S., U. Okla.-Norman, 1961, Ph.D., 1964. Asst. prof. zoology U. Tulsa, 1964-70, assoc. prof., 1970-79, prof., 1979—. Author: Laboratory Manual of Mammalian Physiology, 1975, 2d edit., 1982; contbr. articles in field to profl. jours. Fellow Okla. Acad. Sci. (pres. 1975); mem. Am. Soc. Zoologists, Am. Physiol. Soc., AAAS, Southwestern Assn. Naturalists, Sigma Xi. Republican. Baptist. Subspecialties: Physiology (biology); Reproductive biology. Current work: Basic research relevant to human in vitro fertilization and embryo culture. Home: 4217 E 26th St Tulsa OK 75114 Office: Natural Scis Dept U Tulsa 600 S College St Tulsa OK 74104

SHIRLEY, DAVID ARTHUR, research laboratory administrator, chemist, educator; b. North Conway, N.H., Mar. 30, 1934; m. Virginia Schultz, June 23, 1956; children: David N., Diane, Michael, Eric, Gail. B.S. in Chemistry, U. Maine, 1955, Sc.D. (hon.), 1978; Ph.D. in Chemistry (NSF fellow), U. Calif., Berkeley, 1959. From lectr. to prof. chemistry U. Calif., Berkeley, 1960—, vice chmn. dept. chemistry, 1968-71, chmn. dept., 1971-75; assoc. dir. and head materials and molecular research div. Lawrence Berkeley Lab., 1975-80, dir. lab., 1980—. Recipient Ernest O. Lawrence award AEC, 1972; NSF sr. postdoctoral fellow, 1966-67, 70. Fellow Am. Phys. Soc.; mem. Am. Chem. Soc. (Calif. sect. award 1970), AAAS, Nat. Acad. Sci., Am. Acad. Arts and Sci. Subspecialties: Physical chemistry; Surface chemistry. Office: Lawrence Berkeley Lab One Cyclotron Rd Berkeley CA 94720

SHIVANANDAN, KANDIAH, research physicist; b. Parit Buntar, Malaysia, Aug. 22, 1929; s. Sangrapillai and Vallianaki K.; m. Mary Sheehy, Sept. 21, 1961; children: John, Marianne. B.Sc., U. Melbourne, Australia, 1957; M.S., U. Toronto, 1958; Ph.D., Cath. U. Am., 1966. Physicist M.I.T., Cambridge, Mass., 1957-61; physicist (Goddard Space Flight Ctr.), Greenbelt, Md., 1961-63, research physicist infrared sci. and tech., astronomy/asstrophysics, Washington, 1963—. Contbr. articles to profl. jours. Optical Soc. Am. fellow, 1965. Mem. Am. Inst. Physics, Indian Inst. Physics, Am. Astron. Soc., Royal Astron. Soc., Internat. Astron. Union. Subspecialties: Infrared optical astronomy; Acoustics. Current work: Infrared science and technology, astronomy and astrophysics. Home: 4711 Overbrook Rd Bethesda MD 20816

SHIVELY, JAMES NELSON, veterinary pathologist; b. Moran, Kans., Feb. 9, 1925; s. Carl Nelson and Clara Wanda (Keith) S.; m. Ann Webster, June 6, 1953; children: Elizabeth, Susan, Robert. D.V.M., Kans. State U., 1946; M.P.H., Johns Hopkins U., 1953; M.S., U. Rochester, 1956; Ph.D., Colo. State U., 1970. Prof. ultrastructural pathology dept. pathology N.Y. State Vet. Coll., Cornell U., Ithaca, N.Y., 1971-75; prof. vet. sci. dept. vet. sci. U. Ariz., Tucson, 1975—. Contbr. articles to profl. jours. Served with U.S. Army Vet. Corps, 1946-60; Served with USPHS, 1961-70. Mem. AVMA, Internat. Acad. Pathology, Electron Microscopy Soc. Am., Am. Coll. Vet. Toxicology, Sigma Xi. Subspecialties: Pathology (veterinary medicine); Morphology. Current work: Ultrastructural pathology, research and diagnostic. Office: Dept Vet Sci U Ariz Tucson AZ 85721

SHKAROFSKY, ISSIE PETER, physicist; b. Montreal, July 4, 1931; s. Frank and Sylvia (Alpert) S.; m. Agnes Spira, Mar. 10, 1957; children—Marvin David, Sema, Lou Aaron, Aviva Brocha. B.Sc., McGill U., 1952, M.Sc., 1953, Ph.D., 1957. Sr. mem. sci. staff RCA Ltd., Montreal, 1957-73, research/devel. fellow, 1973-76, MPB Technologies, Inc., Ste. Anne de Bellevue, Que., 1977—. Contbr. articles to profl. jours.; author: (with T.W. Johnston and M.P. Bachnyski) The Particle Kinetics of Plasmas, 1966. Fellow Am. Phys. Soc.; mem. Can. Assn. Physicists, Am. Geophys. Union, Assn. Orthodox Jewish Scientists. Subspecialties: Plasma physics; Fusion. Current work: Plasma transport, magnetic fusion, laser fusion, lasers, space physics, ionosphere, troposphere, others.

SHKEDI, ZVI, physicist; b. Munich, W. Ger., May 24, 1947; came to U.S., 1977, naturalized, 1982; s. Shaul and Fruma (Gerstner) S.; m. Chana S. Nafcha, Aug. 20, 1975; children—Ariel, Michal, Lior, Yael. B.Sc. cum laude, Hebrew U. Jerusalem, 1970; M.Sc., Weizmann Inst. Sci., Israel, 1972, Ph.D., 1977. Nuclear research asst. Weizmann Inst., 1970-77; research assoc. Brown U., Providence, 1977-79; mem. tech. staff Monogram Industries, Santa Monica, Calif., 1979-80; project engr. Garrett-AiResearch, Torrance, Calif., 1980—. Contbr. articles to profl. jours. Chaim Weizmann fellow, 1977. Mem. IEEE, Am. Phys. Soc. Subspecialties: Aerospace engineering and technology; Solar energy. Current work: Pressure and temperature transducers, thin films, micro-computer systems, numerical analysis, solar energy conversion, improving humane abilities, education. Office: AiResearch 2525 W 190th St Torrance CA 90509

SHNIDER, BRUCE I., physician, educator, researcher; b. Ludzk, Poland, Jan. 20, 1920; came to U.S., 1920, naturalized, 1926; s. Benjamin and Rose (Bornstein) S.; m. Doris Benjamin, June 22, 1942; children—Steven, Marc, Reed. B.S. in Edn, Wilson Tchrs. Coll., 1941; M.D. cum laude, Georgetown U., 1948. Cert. Am. Bd. Internal Medicine. Tchr. D.C. Public Schs., 1941-44; intern D.C. Gen. Hosp., 1948-49, resident in medicine, 1950-51, attending physician, 1958—, dir. tumor service, 1958—, chmn. subcom. ambulatory care, 1970-72, active staff, 1972—; mem. Tumor Bd., 1974-80; asst. resident Georgetown U. Hosp., 1949-50, chief resident, 1951-52; instr. Sch. Medicine, 1952-56, lectr. phys. diagnosis, 1953-61, asst. prof., 1956-61, asst. prof. pharmacology, 1961-63, prof. medicine and pharmacology, 1960-61, prof. medicine, 1966—, prof. pharmacology, 1966-79; asst. dean (Sch. Medicine), 1960-63, assoc. dean, 1963-73, assoc. dean acad. affairs, 1973-79; dir. Cancer Detection Ctr. Hosp., 1952-53; mem. Tumor Bd., 1962-65, cancer coordinator, 1960-65; chmn. cancer edn. com. Vincent T. Lombardi Cancer Ctr., 1971-75, coordinator teaching activities, 1961-79, dir. div. oncology, 1965-79, attending physician, 1958—; dir. Georgetown U. Chemotherapy Research program; VA Hosp., 1962-70; head dept. oncology Beilinson Hosp., Petach Tikva, Israel, 1972-74; vis. prof. Tel Aviv U., 1972-73, Hebrew U., 1979-80; cons. Contbr. articles to profl. jours. Served with USPHS, 1955-57. Fellow ACP, Am. Soc. Clin. Pharmacology and Therapeutics, Am. Coll. Clin. Pharmacology; mem. Med. Soc. D.C., N.Y. Acad. Scis., AAAS, Am. Fedn. Clin. Research, Assn. Am. Med. Colls., AAUP, Am. Assn. Cancer Edn., Georgetown U. Alumni Assn., Royal Soc. Medicine, Southeastern Assn. Cancer Research, Found. Thanatology, AMA, Am. Soc. Preventive Oncology. Subspecialties: Oncology; Chemotherapy. Office: DC General Hospital 19th St and Massachusetts Ave Suite 1505 Washington DC 20003

SHOCH, DAVID EUGENE, physician, educator; b. Warsaw, Poland, June 10, 1918; s. Henry and Hannah (Dembina) S.; m. Gertrude Amelia Weinstock, June 10, 1945; children—James, John. B.S., Coll. City N.Y., 1938; M.S., Northwestern U., 1939, Ph.D., 1943, M.D., 1946. Diplomate: Am. Bd. Ophthalmology (dir., vice chmn. 1978, chmn. 1979). Intern Cook County Hosp., Chgo., 1945-46, resident ophthalmology, 1948-52; practice medicine, specializing in ophthalmology, Chgo., 1952—; asst. prof. ophthalmology dept. Northwestern U., Chgo., 1952-66, prof. head ophthalmology dept., 1966—; head ophthalmology dept. Northwestern Meml. Hosp., Childrens Meml. Hosp., VA Lakeside Hosp. Editorial cons. in ophthalmology: Postgrad. Med; abstract editor: Am. Jour. Ophthalmology. Trustee Assn. U. Profs. (pres. 1973). Opthalmology; bd. dirs., sec. Heed Ophthalmic Found. Served to capt., M.C. AUS, 1946-48. Fellow A.C.S.; mem. A.M.A., AAAS, Am. Acad. Ophthalmology (sec. for instrn. 1972-78, pres. 1981), Assn. Research in Vision and Ophthalmology, Inst. Medicine Chgo., Chgo. Ophthalmol. Soc. (past pres.), Am. Ophthalmol. Soc., Pan-Am. Ophthalmol. Soc., French Ophthalmol. Soc., Sigma Xi. Subspecialty: Ophthalmology. Current work: Cataracts, ophthalmic optics; lens implants. Home: 1070 Hohlfelder Rd Glencoe IL 60022 Office: 303 E Chicago Ave Chicago IL 60611

SHOCHET, MELVYN JAY, physicist; b. Phila., Oct. 31, 1944; s. Abraham and Dorothy (Kaminsky) S.; m. Sheila Eileen Mazer, July 4, 1967; children: Stephen, Tara. B.A., U. Pa., 1966; M.A., Princeton U., 1972, Ph.D., 1972. Research assoc. U. Chgo., 1972-73, instr. physics, 1973-75, asst. prof., 1975-78, assoc., prof., 1978—. Alfred P. Sloan fellow, 1978-82. Mem. Am. Phys. Soc., AAAS. Subspecialty: Particle physics. Current work: Experimental studies of massive muon pair production, studies of very high energy interactions using proton anti-proton colliding beams. Home: 5749 S Dorchester Ave Chicago IL 60637 Office: U Chgo Enrico Fermi Inst 5640 S Ellis Ave Chicago IL 60637

SHOCK, D'ARCY ADRIANCE, engineering consultant; b. Fowler, Colo., June 13, 1911; s. Earl I. and Margaret (Adriance) S.; m. Barbara Beth Lounsbury, Feb. 11, 1954; children: Kathy Beth, David Christopher. B.S. in Chem. Engring, Colo. Coll., 1933; M.A. in Phys. Chemistry, U. Tex., 1946. Registered profl. engr., Okla. Chemist Dow Chem. Co., Midland, Mich., 1933-36; analytical lab. supr. McGean Chem. Co., Cleve., 1936-41; works chemist Internat. Mineras Magnesium Plant, Austin, Tex., 1941-44; Nat. Gasoline Assn. Am. research fellow U. Tex.-Austin, 1945-49; with Continental Oil Co., Ponca City, Okla., 1949-76; dir.-mgr. Corp. Research div. Conoco, Ponca City, 1956-73; mgr. Mining Research div., ret., 1976; cons. on slurry transport and solution mining, Ponca City, 1976—; mem. waste implacement pilot plant panel radioactive waste disposal com. Nat. Acad. Scis., 1976—. Contbr. sci. articles to profl. pubis. Mem. Okla. radiation adv. com. Okla. Health Dept., 1956-70. Mem. Nat. Assn. Corrosion Engring., Soc. Petroleum Engrs., AIME, Soc. Mining Engrs. of AIME. Subspecialties: Solution Mining; Radioactive Waste Disposal. Current work: Actively consulting in areas of slurry transport solution mining. Holder 27 U.S. patents. Home and Office: 233 Virginia St Ponca City OK 74601

SHOCKLEY, WILLIAM BRADFORD, physicist; b. London, Feb. 13, 1910; s. William Hillman and May (Bradford) S.; m. Jean A. Bailey, 1933 (div. 1955); children—Alison, William Alden, Richard Condit; m. Emmy Lanning, 1955. B.S., Calif. Inst. Tech., 1932; Ph.D., M.I.T., 1936; Sc.D. (hon.), Rutgers U., 1956, U. Pa., 1955, Gustavus Adolphus Coll., Minn., 1963. Teaching fellow M.I.T., 1932-36; mem. tech. staff Bell Telephone Labs., 1936-42, 45, became dir. transistor physics research, 1954; dir. Shockley Semicondr. Lab.; pres. Shockley Transistor Corp., 1958-60; cons. Shockley Transistor unit Clevite Transistor, 1960-65; lectr. Sanford U., 1958-63, Alexander M. Poniatoff prof. engring. sci. and applied sci., 1963-75, prof. emeritus, 1975—; exec. cons. Bell Telephone Labs., 1965-75; dep. dir. research, weapons systems evaluation group Dept. Def., 1954-55; expert cons. Office Sec. War, 1944-45; vis. lectr. Princeton U., 1946; vis. prof. Calif. Inst. Tech., 1954-55; sci. adv., policy council Joint Research and Devel. Bd., 1947-49; sr. cons. Army Sci. Adv. Panel.; Dir. research Anti-submarine Welfare Ops. Research Group USN, 1942-44. Author: Electrons and Holes in Semiconductors, 1950, (with W. A. Gong) Mechanics, 1966; editor: Imperfections of Nearly Perfect Crystals, 1952. Recipient medal for Merit; Air Force Assn. citation of honor, 1951; U.S. Army cert. of appreciation, 1953; co-winner (with John Bardeen and Walter H. Brattain) Nobel Prize in Physics, 1956; Wilhelm Exner medal Oesterreichischer Gewerbeverein, Austria, 1963; Holley medal ASME, 1963; Calif. Inst. Tech. Alumni Disting. Service award, 1966; NASA cert. of appreciation Apollo 8, 1969; Public Service Group Achievement award NASA, 1969; named to Inventor's Hall of Fame, 1974. Fellow AAAS; mem. Am. Phys. Soc. (O.E. Buckley prize 1953), Nat. Acad. Sci. (Comstock prize 1954), IEEE (Morris Liebmann prize 1952, Gold medal, 25th anniversary of transistor 1972, Medal of Honor 1980), Sigma Xi, Tau Beta Pi. Subspecialties: Condensed matter physics; 3emiconductors. Current work: Statistics of human quality, including I.Q. Holder over 90 patents. Inventor of junction transistor; research on energy bands of solids, ferromagnetic domains, plastic properties of metals; semicondr. theory applied to devices and device defects such as dislocations; fundamentals of electromagnetic energy and momentum; mental tools for sci. thinking; ops. research on human quality problems. Home: 797 Esplanada Way Stanford CA 94305 Office: Stanford Electronics Labs McCullough 202 Stanford U Stanford CA 94305

SHOEMAKER, DAVID POWELL, chemist, educator; b. Kooskia, Idaho, May 12, 1920; s. Roy Hopkins and Sarah (Anderson) S.; m. Clara Brink, Aug. 5, 1955; 1 son, Robert Brink. B.A., Reed Coll., 1942; Ph.D., Calif. Inst. Tech., 1947. Research asst. Calif. Inst. Tech., 1943-45, NRC fellow, 1945-47, sr. research fellow, 1948-51; fellow John Simon Guggenheim Meml. Found. Inst. Theoretical Physics, Copenhagen, Denmark, 1947-48; asst. prof. chemistry Mass. Inst. Tech., Cambridge, 1951-56, asso. prof., 1956-60, prof., 1960-70; prof. chemistry Oreg. State U., Corvallis, 1970—, chmn. dept., 1970-81; vis. scientist Lab. Cristallographie, CNRS, Grenoble, France, 1967, 78-79; vis. lectr. Kemisk Institut, Aarhus, Denmark, June 1979; cons. Exxon Co. U.S.A., Baton Rouge, 1957—; sec.-treas. U.S.A. Nat. Com. for Crystallography, 1962-64, chmn., 1967-69; mem. vis. com. chemistry dept. Brookhaven Nat. Lab., 1974-79; mem. evaluation panel material scis. div. Nat. Bur. Standards, 1977-79. Author: (with Carl W. Garland, Jeffrey I. Steinfeld, Joseph W. Nibler) Experiments in Physical Chemistry, 1962, 67, 74, 81; Am. co-editor: Acta Crystallographica, 1964-69. Mem. Am. Chem. Soc., Am. Phys. Soc., Am. Crystallographic Assn. (pres. 1970), Am. Acad. Arts and Scis., AAAS, Internat. Union Chrystallography (mem. exec. com. 1972-78), Phi Beta Kappa, Sigma Xi, Phi Lambda Upsilon, Phi Kappa Phi. Subspecialties: X-ray crystallography; Solid state chemistry. Current work: Transition metal alloys, alloy hydrides, zeolites, x-ray diffraction, neutron diffraction. Home: 3453 NW Hayes Ave Corvallis OR 97330 Office: Dept Chemistry Oreg State U Corvallis OR 97331

SHOEMAKER, EUGENE MERLE, geologist, educator; b. Los Angeles, Apr. 28, 1928; s. George Estel and Muriel May (Scott) S.; m. Carolyn Jean Spellmann, Aug. 18, 1951; children: Christine Carol, Patrick Gene, Linda Susan. B.S., Calif. Inst. Tech., 1947, M.S., 1948; M.A., Princeton U., 1954, Ph.D., 1960; Sc.D., Ariz. State Coll., 1965, Temple U., 1967. Exploration uranium deposits and investigation salt structures Colo. and Utah U.S. Geol. Survey, 1948-50, regional investigations geochemistry, vulcanology and structure Colorado Plateau, 1951-56, research structure and mechanics of meteorite impact and nuclear explosion craters, 1957-60, with E.C.T. Chao, discovered coesite, Meteor Crater, Ariz., 1960, investigation structure and history of moon, 1960-73, established lunar geol. time scale, methods of geol. mapping of moon, 1960, application TV systems to investigation extra-terrestrial geology, 1961—, geology and paleomagnetism; Colo. Plateau, 1969—, systematic search for planet-crossing asteroids, 1973—, geology of satellites of Jupiter and Saturn, 1978, organized br. of astrogeology U.S. Geol. Survey, 1961; co-investigator TV expt. Project Ranger, 1961-65; chief scientist, center of astrogeology U.S. Geol. Survey, 1966-68, research geologist, 1976—; prin. investigator geol. field investigations in Apollo lunar landing, 1965-70, also television expt. Project Surveyor, 1963-68, prof. geology, 1969—, Calif. Inst. Tech., chmn. div. geol. and planetary scis., 1969-72. Recipient (with E.C.T. Chao) Wetherill medal Franklin Inst., 1965; Arthur S. Flemming award, 1966; NASA medal for exceptional sci. achievement, 1967; honor award for meritorious service U.S. Dept. Interior, 1973; Disting. Service award, 1980. Mem. Nat. Acad. Sci., Geol. Soc. Am. (Day medal 1982), Mineral Soc. Am., Soc. Econ. Geologists, Geochem. Soc., Am. Assn. Petroleum Geologists, Am. Geophys. Union, Seismol. Soc. Am., Am. Astron. Soc., Internat. Astron. Union. Subspecialties: Geology; Planetary science. Home: PO Box 984 Flagstaff AZ 86002 Office: US Geol Survey 2255 N Gemini Dr Flagstaff AZ 86001

SHOEMAKER, PAUL BECK, III, plant pathologist; b. Bridgeton, N.J., Feb. 26, 1941; s. Paul Beck and Dorothy Matie (Ward) S.; m. Simone Marice Stad, Aug. 15, 1964; children: Lisa Anne, Laura Marie. B.S., Rutgers U., 1963, M.S., 1965; Ph.D., Cornell U., 1971. Asst. prof. plant pathology N.C. State U., 1970-73, assoc. prof., 1973-81, prof., 1981—, ext. plant pathology specialist, 1970—. Contbr. articles to profl. jours. Mem. Am. Phytopath. Soc., Sigma Xi, Phi Kappa Phi. Unitarian-Universalist. Subspecialty: Plant pathology. Current work: Epidemiology and control of diseases of vegetables and tobacco. Home: Box 31 Route 7 Hendersonville NC 28739 Office: Box 249 Route 2 Fletcher NC 28732

SHOFFNER, ROBERT NURMAN, geneticist, educator; b. Junction City, Kans., Mar. 3, 1916; s. Nurman A. and Hazel D. (Cunningham) S.; m. Gladys Marie Shoffner, Sept. 3, 1938; children: R. Kirk, Jane M., Patti I. B.S., Kans. State U., 1940; M.S., U. Minn., 1942, Ph.D., 1946. Instr. animal sci. U. Minn., 1942-46, asst. prof., 1946-49; assoc. prof., 1949-55, pres., 1955—; vis. prof. Iowa State U., 1957, U. Tex., 1969; FAO cons., 1976. Assoc. dir. St. Anthony Park, St. Paul. Fulbright scholar, 1962; recipient Merck Research award, 1982. Fellow Poultry Sci. Assn. (pres. 1966-67), AAAS; mem. Genetic Soc., Genetic Soc. Am., Sigma Xi, Gamma Sigma Delta. Subspecialties: Genetics and genetic engineering (agriculture); Animal breeding and embryo transplants. Current work: Animal cytogenetics, genetic modification. Home: 2066 Knapp St Saint Paul MN 55108 Office: Dept Animal Science U Minn 1404 Gortner Ave Saint Paul MN 55108

SHOFNER, FREDERICK MICHAEL, engineering company executive; b. Shelbyville, Tenn., Apr. 9, 1940; s. Joseph Cecil and Elva Hill (Perry) S.; m. Betty Jo Neese, Jan. 26, 1962; children: Michelle, Michael, Kyle. B.S., U. Tenn., 1962, M.S., 1964, Ph.D., 1966. Registered profl. engr., Tenn. Prof. elec. engring. U. Tenn. Space Inst., Tullahoma, 1966-71; v.p., dir. Environ. Systems, Knoxville, Tenn., 1971-76; pres., chmn. bd. ppm, Inc., Knoxville, 1976—; prof. elec. engring. U. Tenn., Knoxville, part-time 1971-81; cons. Contbr. articles to sci. jours.; patentee in field (6). Mem. Optical Soc. Am., Nat. Soc. Profl. Engrs., ASTM. Methodist. Subspecialties: Electrical engineering. Current work: Applications of lasers and electro-optics to aerosol measurements and control; invention, development and commercializing new instrumental methods.

SHOLTIS, JOSEPH ARNOLD, JR., physicist, air force officer; b. Monongahela, Pa., Nov. 28, 1948; s. Joseph and Gladys Virginia (Frye) S.; m. Cheryl Anita Senchur, Dec. 19, 1970; children: Christian Joseph, Carole Lynne. B.S. in Nuclear Engring, Pa. State U., 1970; M.S., U. N.Mex., 1977, postgrad., 1977-80. Lic. sr. reactor operator U.S. Nuclear Regulatory Commn. Mathematician, statistician U.S. Bur. Mines, Pitts., 1968-70; commd. officer U.S. Air Force, 1970; advanced through grades to maj., nuclear research officer Wright-Patterson AFB, Dayton, Ohio, 1970-74; space nuclear systems analyst Kirtland AFB, N.Mex., 1974-78; advanced nuclear systems safety analyst Sandia Nat. Lab., N.Mex., 1978-80; chief radiation sources, reactor physicist in charge Armed Forces Radiobiology Research Inst., Bethesda, Md., 1980—, instr., lectr., 1981—, mem. radiation safety com., 1981—. Author: handbook LMFBR Accident Delineation, 1980. Mem. N.Mex. gov.'s panel on energy resources, 1976-78; com. chmn. Ft. Detrick Catholic Community. Mem. Am. Nuclear Soc., AAAS, N.Y. Acad. Scis., Planetary Soc., Scientists and Engrs. for Source Energy. Republican. Roman Catholic. Clubs: Fort Detrick (Md.) Rod & Gun, Jefferson's Island. Subspecialties: Nuclear engineering; Nuclear physics. Current work: Nuclear reactor research, reactor operations, maintenance and upgrade, effects of ionizing radiation on matter, radiation sources development, design and use, reactor safety, risk assessments, reactor regulation. Office: Armed Forces Radiobiology Research Inst Bldg 42 Nat Naval Med Ctr Bethesda MD 20814

SHONS, ALAN RANCE, plastic surgeon; b. Freeport, Ill., Jan. 10, 1938; s. Ferral Caldwell and Margaret Zimmerman (Ziegler) S.; m. Mary Ella Misamore, Aug. 5, 1961; children: Lesley Margaret, Susan Campbell. A.B., Dartmouth Coll., 1960; M.D., Case Western Res. U., 1965; Ph.D., U. Minn., 1976. Diplomate: Am. Bd. Surgery, 1975, Am. Bd. Plastic Surgery, 1977. Intern Univ. Hosp., Cleve., 1965-66; surg. resident, 1966-67; research fellow U. Minn., Mpls., 1969-72, surg. resident, 1972-74, dir., 1976—; plastic surgery resident NYU, N.Y.C., 1974-76; chmn. Minn. Com. on Trauma. Author numerous sic. papers and book chpts. Served to capt. USAF, 1967-69. Dartmouth Coll. Wheelcock scholar, 1960. Fellow ACS; mem. Minn. Acad. Plastic Surgeons (pres. 1981-82), Am. Assn. Plastic Surgeons, Am. Assn. Surgery Trauma, Central Surg. Assn., Soc. Head and Neck Surgeons, Plastic Surgery Research Council. Republican. Presbyterian. Subspecialties: Microsurgery; Transplant surgery. Current work: Research on limb transplantation. Office: U Minn Hospital Minneapolis MN 55455

SHOORE, JOSEPH DAVID, physicist; b. Logan, Utah, Mar. 9, 1938; s. Isadore and Constance (Petersen) S.; m. Rita Fern Finn, June 30, 1960; children: Ginger Fern, David Joseph. B.S., Ariz. State U., 1960, M.S., 1963. Registered engr., Calif. Aeronutronic div. Philco-Ford Corp., Newport Beach, Calif., 1963-68; scientist Systems Div., Interstate Electronics Corp., Anaheim, Calif., 1968-69; prin. scientist, prin. test dir. aset facility. McDonnell Douglas Astronautics Corp., Huntington Beach, Calif., 1969—. Contbr. articles in field to profl. jours. Active Boy Scouts Am., 1970—; mem. Newport Beach Sch. Bd.

ednl. rev. com., 1981-82. Mem. IEEE, Optical Soc. Am., AAAS, Soc. Information Display. Mormon (high council 1973-82). Subspecialties: Infrared optical astronomy; Infrared spectroscopy. Current work: Infrared sensor test and evaluation for space applications; space science, astronomy, remote sensing, cyrogenics/IR detectors.

SHOPE, THOMAS CHARLES, pediatrician, medical virologist, educator, consultant; b. Princeton, N.J., Feb. 26, 1938; s. Richard Edwin and Helen (Ellis) S.; m. Jean Thatcher, Aug. 7, 1962; children: Timothy Russell, Jonathan Richard, Laura Jean. B.A., State U. Iowa, 1960; M.D., Cornell U., 1964. Diplomate: Am. Bd. Pediatrics. Intern Univ. Hosps., Mpls., 1964-65, resident in pediatrics, 1965-67; with Epidemic Intelligence Service, Centers for Disease Control, Atlanta, 1967-69; fellow in pediatrics (infectious diseases) Yale U., 1969-72, inst. pediatrics, 1972-73; asst. prof. pediatrics Wayne State U., Detroit, 1973-76, asoc. prof., 1976-82; assoc. prof. pediatrics and pathology U. Mich., Ann Arbor, 1982—; assoc. dir. pediatric infectious diseases Children's Hosp. Mich., Detroit, 1973-82; dir. pediatric infectious diseases C.S. Mott Children's Hosp., Ann Arbor, 1982—; cons. infectious diseases. Contbr. articles, chpts., numerous abstracts to profl. publs. Served to lt. comdr. USPHS, 1967-69. Fellow Am. Acad. Pediatrics, Infectious Diseases Soc. Am., Soc. Pediatric Research, Am. Soc. Microbiology. Republican. Unitarian. Subspecialties: Infectious diseases; Virology (medicine). Current work: Interactions between infected host and immune system, particularly virus infections and viral vaccines; evaluation of antiviral drugs. Office: U Mich Med Center F2815 Mott Hosp Box 066 Ann Arbor MI 48109

SHORE, KAREN FAY, quality assurance administrator; b. St. Paul, Aug. 17, 1948; d. Hobart Paul and Irene Susan (Baierl) Kern; m.; 1 dau. Kristen. B.S. in Chemistry, U. San Francisco, 1970; M.B.A., U. Santa Clara, 1983. Chemist Sondell Sci., Palo Alto, Calif., 1970-71; chemist Smith Corona, Palo Alto, 1972-75, Diamond Shamrock, Redwood City, Calif., 1975-76; quality assurance mgr. Beckman Instruments, Palo Alto, 1977—. Mem. Am. Soc. Quality Control, Regulatory Affairs Profls. Soc., Am. Mgmt. Assn. Democrat. Subspecialty: Analytical chemistry. Office: Beckman Instruments 1050 Page Mill Rd Palo Alto CA 94304

SHORTLE, WALTER CHARLES, research plant pathologist; b. Laconia, N.H., Apr. 26, 1945; s. Charles Walter and Merna (Reed) S.; m. Elizabeth Stickney, Apr. 30, 1966; children: Jennifer, Amy, Abigail, Emily. B.S., U. N.H., 1968, M.S., 1970; Ph.D., N.C. State U., 1974. Research plant pathologist U.S. Forest Service, Durham, N.H., 1974—; adj. prof. U. N.H., also; U. Maine. Contbr. articles to profl. jours. NSF Research fellow, 1966; recipient Alpha Zeta Academic Achievement award, 1964. Mem. Am. Phytopath. Soc., Sigma Xi. Republican. Mem. Christian Ch. Subspecialties: Plant physiology (agriculture); Biochemistry (biology). Current work: Biochemistry of plant responses to wounding and infection; electronic measurement of tree growth and wood decay. Office: US Forest Service Box 640 Durham NH 03824

SHORTLIFFE, EDWARD HANCE, medical computer educator, physician; b. Edmonton, Alta., Can., Aug. 28, 1947; came to U.S., 1954, naturalized, 1962; s. Ernest Carl and Elizabeth Joan (Rankin) S.; m. Linda Marie Dairiki, June 21, 1970; 1 dau., Lindsay Ann. A.B., Harvard U., 1970; Ph.D., Stanford U., 1975, M.D., 1976. Diplomate: Am. Bd. Internal Medicine, 1979. Intern Mass. Gen. Hosp., Boston, 1976-77; resident Stanford (Calif.) Hosp., 1977-79, asst. prof. medicine, 1979—, asst. prof. computer sci., 1979—; cons. and co-founder Teknowledge, Inc. Contbr. articles on med. computer sci. to profl. jours. Recipient Research Career Devel. award Nat. Library Medicine, 1979-84; Henry J. Kaiser Family Found. Faculty scholar, 1983—; NIH trainee, 1971-76. Fellow ACP; mem. Am. Fedn. Clin. Research, Assn. Computing Machinery (Grace Murray Hopper award 1976), Soc. Med. Decision Making, Am. Assn. Artificial Intelligence. Subspecialties: Internal medicine; Artificial intelligence. Current work: Medical computer science, applications of computers to clinical practice, research into development of computer-based clinical decision aids using artificial intelligence techniques. Developer Mycin and Oncocin Med. consultation computer programs. Office: TC-117 Stanford Med Sch Stanford CA 94305

SHOUP, WILLIAM DAVID, agricultural engineering educator, consultant; b. Lafayette, Ind., Feb. 7, 1951; s. William E. Shoup and Maxine Ann (Shelton) Brown; m. Roberta Eileen Sikora, July 26, 1981. B.S., Purdue U., 1973, M.S., 1974, Ph.D., 1980; postgrad., Mich. State U., 1974. Sales engr. Internat. Harvester, Chgo., 1974-75; grad. instr. Purdue U., 1975-76; dist. mgr. J. I. Case Co., Indpls., 1977-79, product planning mgr., Racine, Wis., 1979-80, v.p. D2M Corp., Gainesville, Fla., 1980—; asst. prof. agrl. engring. U. Fla., 1980—, cordinator for mechanized agr., 1980—; sr. cons. Harbor Assocs., Boston, 1981—; cons. CMF&Z Corp., Des Moines, 1982-83. Author: books, the most recent being Advanced Farm Machinery Systems, 1982, Technical Machine Sales Management, 1982. Leader 4-H Club, Ind., Fla., 1965—. Named to Outstanding Young Men Am. U.S. Jaycees, 1982. Mem. Am. Soc. Agrl. Engrs. (com. chmn., com. chmn. sect.), Am. Farm Bur. Fedn., Sigma Xi, Alpha Zeta (adv.), Gamma Sigma Delta, Alpha Mu. Club: U. Fla. Faculty. Subspecialties: Agricultural engineering; Systems engineering. Current work: Agricultural systems research, technology impact, robotic applications. Home: 4106 NW 19th Dr Gainesville FL 32605 Office: U Fla 3 Fazier Rogers Hall Gainesville FL 32611

SHREFFLER, DONALD CECIL, geneticist; b. Kankakee County, Ill., Apr. 29, 1933; s. Cecil LeRay and Laura Belle (Pearman) S.; m. Dorothy Ferne Kramer, Aug. 18, 1957; children—Douglas LeRay, David Kenneth. B.S., U. Ill., 1954, M.S., 1958; Ph.D. (NSF predoctoral fellow), Calif. Inst. Tech., 1962. Research asso. human genetics U. Mich., 1961-64, asst. prof., 1964-68, asso. prof., 1968-71, prof., 1971-75; prof. genetics Washington U., St. Louis, 1975—, James S. McDonnell prof., chmn. dept., 1977—. Editorial bd. 8 sci. jours.; contbr. numerous articles to sci. jours. Served with U.S. Army, 1954-56. Recipient Research Career Devel. award USPHS, 1966-75. Mem. Genetics Soc. Am., Am. Soc. Human Genetics, Am. Assn. Immunologists, AAAS, Inst. Medicine of Nat. Acad. Sci. Methodist. Subspecialties: Gene actions; Immunobiology and immunology. Current work: Immunogenetics and biochemical genetics of major histocompatibility complexes in mammals. Research in immunogenetics and biochem. genetics. Home: 4 Ricardo Ln Saint Louis MO 63124 Office: 660 S Euclid St Saint Louis MO 63110

SHRESTHA, BUDDHI MAN, scientist, cariologist, pathologist and pedodontist; b. Chainpur Bazar, Sankhuwa Sabha, Nepal, Sept. 28, 1936; came to U.S., 1973; s. Lok Man and Subhadra Devi Shresthea. B.D.S., Panjab (India) U., 1963; M.S. in Dental Sci, U. Rochester, 1970; Ph.D. in Pathology, U. Rochester, 1980. Cert. pedodontics, clin. intern. Dental surgeon Dept. Health Services, Govt. of Nepal, 1965-73; research assoc. Eastman Dental Ctr. Rochester, N.Y., 1973-74; postdoctoral fellow U. Rochester, 1975-80; scientist, cariologist Oral Health Research Ctr. Sch. Dentistry, Fairleigh Dickinson U., Hackensack, N.J., 1981-82, adj. assoc. prof. pathology, 1982—, dir. div. nutrition and cariology, 1982—; part-time clin. assoc. prof. oral medicine Coll. Dentistry NYU, N.Y.C., 1982—; mem. adv. com. on nutrition edn. in profl. schs. N.Y. Acad. Medicine, 1982—. Dental scholar Colombo Plan, 1959-63; research fellow Eastman Dental Ctr., 1968-70; postdoctoral fellow NIH, U. Rochester, 1975-78. Mem. Nepal Med. Assn. Biratnagar Br. (hon. joint sec. 1965-67), Nepal Dental Soc. (founding, hon. exec. sec. 1971-73), Internat. Assn. Dental Research, Am. Assn. Dental Research. Club: Genesee Golf (Rochester). Subspecialties: Cariology; Pathology (medicine). Current work: Effects of diet and nutrition on dental caries; fluoride and trace elements in prevention of dental caries; cariostatic effects of titanium tetrafluoride; development of prototype rate caries model with improved computerized UV caries scoring system; fluoride and trace elements in prevention and treatment of osteoporosis. Patentee method of coating teeth with a durable glaze; inventor UV method for detection and scoring of rat caries. Office: Sch Dentistry Fairleigh Dickinson U 110 Fuller Pl Hackensack NJ 07601

SHREVE, GREGORY MONROE, anthropology and computer educator; b. Munich, Bavaria, W.Ger., Aug. 3, 1950; came to U.S., 1964; s. J.L. and Rosa (Zerweiss) S.; m. Joan Marie Nelson, Dec. 29, 1971. B.A., Ohio State U., 1971, M.A., 1974, Ph.D., 1975; cert. advanced study in computer and info. sci, U. Pitts., 1980. Vis. asst. prof. Ohio State U., Columbus, 1975; asst. prof. Kent (Ohio) State U., 1975-80, assoc. prof., 1980—, acad. dean, 1981—, chmn. acad. computing adv. com., 1983; dir. computer ctr. Kent State U.-Liverpool, 1978-81; cons. BFG Inc., Burton, Ohio, 1982-83. Author: Genesis of Structures in Narrative (2 vols.), 1975, 2d edit., 1983. Pres. Geauga County Arts Council, Chardon, Ohio, 1981-83; mem. steering com. United Way Info. Line, 1981-83. Fellow Am. Anthrop. Assn., Am. Folklore Soc.; mem. Assn. Computing Machinery; installation del. Digital Equipment Users Soc. Subspecialties: Automated language processing; Distributed systems and networks. Current work: Relationship of software structures to natural language structures; general semiotic structures present in artificial languages and how they relate to natural language systems. Home: 14649 Evergreen Dr Burton OH 44021 Office: Kent State Univ 14111 Claridon-Troy Rd Burton OH 44021

SHRIVASTAV, BRIJ BHUSHAN, physiologist, biophysicist, pharmacologist; b. Guna, India, July 13, 1936; s. Gajanand Prashad and Kaushilya (Devi) S.; m. Shashi Beohar, Aug. 31, 1940; children—Pragya, Harsha. Ph.D., Western Ont., Can., 1968. Postdoctoral fellow Yale U. Med. Center, 1969-71; research asso. Harvard U. Sch. Medicine and Mass. Gen. Hosp., 1971-73; asst. prof. Duke U. Med. Center, Durham, N.C., 1973—. Contbr. articles to profl. jours. Mem. Am. Assn. Pharmacology and Exptl. Therapeutics, Am. Physiol. Assn., Biophys. Soc., Soc. Gen. Physiologists. Hindu. Subspecialties: Neuropharmacology; Neurophysiology. Current work: Electrophysiology, nueropharmacology, hyperbarienuerophysiology. Home: 5215 Russell Rd Durham NC 27712 Office: Duke U Med Center Box 3813 Durham NC 27710

SHRIVASTAVA, PRAKASH NARAYAN, med. physicist; b. Narsingpur, India, Sept. 5, 1940; s. Sunder Lal and Vidyavati S.; m. Uma Ravipati, Mar. 7, 1968; children: Anil, Rashmi, Anupam. B.S., Nagpur U., 1958, M.S., 1961; Ph.D., U. Tex., Austin, 1966. Cert. Am. Bd. Radiology, 1975, Am. Bd. Health Physics, 1975. Internat. research scholar Italian Nuclear Energy Com., Rome, 1961-62; research fellow U. Tex., Austin, 1966-68; research asso. Princess Margaret Hosp., Toronto, Ont., Can., 1968-69; asst., then assoc. attending radiol. physicist Allegheny Gen. Hosp., Pitts., 1969-74, dir. med. physics and engring., sr. attending radiol. physicist, 1974—; clin. prof. Community Coll. Allegheny County, 1980—; sr. lectr. Carnegie-Mellon U., 1982—; dir. Mideast Ctr., Radiol. Physics, 1974—. Contbr. articles to profl. jours. Pres. Hindu Temple, Pitts., 1979—. Recipient Gupta Gold medal, 1958, Paranjpe Gold meda, 1961. Mem. Am. Assn. Physicists in Medicine, Am. Coll. Radiology, Soc. Nuclear Medicine, Am. Pub. Health Assn., Health Physics Soc., Soc. Therapeutic Radiologists, N.T. Acad. Scis. Subspecialties: Cancer research (medicine); Medical physics. Office: 320 E North Ave Pittsburgh PA 15212

SHRIVER, BRUCE DOUGLAS, computer science educator, researcher; b. Buffalo, Oct. 18, 1940; s. Millard Douglas and Arlene J. (Schalk) S.; m. Beverly Connell, Aug. 17, 1963; children: Bruce D., Mark, Elizabeth, Matthew. B.S., Calif. State Poly. U., 1963; M.S., West Coast U., Los Angeles, 1968; Ph.D., SUNY-Buffalo, 1971. Vis. sr. staff mem. Aarhus, Denmark, 1971-73; prof. computer sci. U. Southwestern La., Lafayette, 1973—. Contbr. articles to profl. jours. NSF fellow, 1969-70. Mem. IEEE (editor-in-chief Sofeware mag 1983—, chmn. computational medicine 1981—, chmn, microprogramming 1981-83), Assn. Computing Machinery (vice chmn. SIGMICRO 1977-79), Soc. Indsl. and Applied Math., Am. Math. Soc., AAUP. Subspecialties: Computer architecture; Distributed systems and networks. Current work: Virtual computer systems, multi-level interpretive systems, firmware engineering, computer systems organization for distributed systems, graphics. Home: 402 Harwell Dr Lafayette LA 70503 Office: U Southwestern La PO Box 44330 Lafayette LA 70504

SHRIVER, DAVID ALLEN, gastronintestinal pharmacologist; b. Syracuse, N.Y., May 29, 1942; s. Harry J. and Elixabeth Jane (Allen) S.; m. Sharon L. Steinberg, Aug. 29, 1964; children—Amy E., Carrie J. Student, Union Coll., Cranford, N.J., 1960-62; B.S., Purdue U., 1966; M.S., U. Iowa, 1968, Ph.D., 1970. Research scientist Wyeth Labs., Inc., 1970-77; research mgr. GI/CNS/Autonomics Pharmacology Ortho Pharm. Corp., 1977—. Mem. Bd. Edn., Bridgewater-Rariton (N.J.) Sch. Dist. Mem. N.Y. Acad. Sci., Am. Soc. Pharmacology and Exptl. Therapeutics, Phila. Physiol. Soc. Subspecialties: Pharmacology; Gastroenterology. Current work: Gastrointestinal drug discovery. Patentee in field. Home: 2051 Lynne Way PO Box 8 Martinsville NJ 08836 Office: Ortho Pharm Corp Raritan NJ 08869

SHUB, DAVID A., biologist, educator, researcher; b. Bklyn., Sept. 25, 1939; s. Abraham and Bella (Bierstein) S.; m. Jeanne, June 10, 1962; 1 dau., Ellen. A.B., Columbia Coll., 1960; Ph.D., MIT, 1966. Research assoc. Inst. Molecular Biology, U. Geneva, 1968-70; asst. prof. dept. biology SUNY-Albany, 1970-76, assoc. prof. biol. scis., 1976—. Co-author articles to publs. in field. NIH grantee, 1973-80; NSF grantee, 1982—; N.Y. State Health Research Council grantee, 1981-82; N.Y. State Sci. and Tech. Found. grantee, 1983-84. Mem. Am. Soc. Microbiology. Subspecialties: Genetics and genetic engineering (biology); Gene actions. Current work: Use of bacteriophage T4 as a DNA cloning vector; viral regulation of host cell genetic activity; effect of altered DNA structure on protein-nucleic acid interaction. Home: 3 Parkwood St Albany NY 12203 Office: Biology Dept SUNY-Albany 1400 Washington Ave Albany NY 12222

SHUG, AUSTIN LEO, biochemist, researcher, educator, consultant; b. Paterson, N.J., July 23, 1925; s. Leo Austin and Alice (Fiederlein) S.; m. Kathryn Jean Snyder, Sept. 24, 1955; children: Barbara, Mary, Leo. B.S., U. Tenn., 1951, M.S., 1952; Ph.D., U. Wis., 1958. Postdoctoral fellow U. Ind., Bloomington, 1957-59; research assoc. Enzyme Inst. U. Wis., Madison, 1959-60; chemist, prof. neurology dept. U. Wis. and VA Hosp., Madison, 1961—; chemist NIH, Bethesda, Md., 1960-61; part time lab. dir. Metabolic Analysis Labs., Inc., Madison, 1981—; cons. Sigma-Tau Chem. Co., Rome, Italy, 1977—. Contbr. chpts. to books, articles to profl. jours. Served with USN, 1943-46. Grantee NIH, 1975-81, 81—, VA, 1961—, Muscular Dystrophy Research Assn., 1982-83. Fellow Am. Inst. Chemists; mem. Am. Soc. Biol. Chemists, Fedn. Am. Socs. for Exptl. Biology, Sigma Xi, Phi Lambda Upsilon. Democrat. Roman Catholic. Subspecialties: Cardiology; Cancer research (medicine). Current work: Metabolic control mechanisms, carnitine-linked metabolism, in various diseases such as Ischemia, cardiomyopathy, muscle myopathy, cancer. Office: VA Hosp 2500 Overlook Terr Madison WI 53705 Home: 1201 Shorewood Blvd Madison WI 53705

SHULER, MICHAEL LOUIS, chemical engineering educator, cons.; b. Joliet, Ill., Jan. 2, 1947; s. Louis D. and Mary C. (Boylan) S.; m. Karen J. Beck, June 24, 1972; children: Andrew, Kristin, Eric. B.S. in Chem. Engring, U. Notre Dame, 1969, Ph.D., U. Minn., 1973. Asst. prof. chem. engring. Cornell U., Ithaca, N.Y., 1974-79, assoc. prof., 1979—; vis. scholar U. Wash., Seattle, 1980-81; cons. in field. Contbr. articles on chem. engring. to profl. jours.; editor: Utilization and Recycle of Agricultural Wastesand Residues. Active Advs. for the Handicapped, Ithaca, 1978—, v.p., 1979-80. Served to capt. USAR, 1969-77. Named Colburn Lectr. U. Del., 1982; recipient Excellence in Teaching award Tau Beta Pi and Coll. Engring. Cornell U., 1977-78. Mem. Am. Inst. Chem. Engrs., Am. Soc. Pharmacognosy. Roman Catholic. Subspecialties: Chemical engineering; Plant cell and tissue culture. Current work: Biochemical engineering, plant cell tissue culture, immobilized cell reactors, mathematical models of cell growth, biological waste treatment. Office: Cornell U Sch Chem Engring Ithaca NY 14853

SHULL, CLIFFORD G., physicist, educator; b. Pitts., Sept. 23, 1915; s.David H and Daisy I. (Bistline) H.; m. Martha-Nuel Summer, June 19, 1941; children—John C., Robert D., William F. B.S., Carnegie Inst. Tech., 1937; Ph.D., N.Y.U., 1941. Research physicist Texas Co., 1941-46; chief physicist Oak Ridge Nat. Lab., 1946-55; prof. physics Mass. Inst. Tech., 1955—; Chmn. vis. com. Brookhaven Nat. Lab., 1961-62; chmn. vis. com. Nat. Bur. Standard reactor, 1972-73; chmn. vis. com. solid state div. Oak Ridge Nat. Lab., 1974-75; chmn. policy com. Nat. Small-Angle-Scattering Center, 1978-81. Contbr. articles to sci. jours. Recipient award of merit Alumni Assn. of Carnegie-Mellon U., 1968; Humboldt Sr. U.S. Scientist award, 1979. Fellow Am. Phys. Soc. (Buckley prize 1956, chmn. solid state physics div. 1962-63), A.A.A.S., Am. Acad. Arts and Scis., N.Y. Acad. Scis., Nat. Acad. Scis. (vice chmn. panel on neutron sci. 1977); mem. Am. Crystallographic Assn., Research Soc. Am., Sigma Xi, Tau Beta Pi, Phi Kappa Phi. Subspecialties: Condensed matter physics; Magnetic physics. Current work: Crystal physics; neutron interferometry. Home: 4 Wingate Rd Lexington MA 02173 Office: Dept Physics Mass Inst Tech Cambridge MA 02139

SHULMAN, ROBERT GERSON, biophysics educator; b. N.Y.C., Mar. 3, 1924; s. Joshua S. and Freda (Lipshay) S.; m. Saralee Deutsch, Aug., 1952; children: Joel, Mark, James. A.B., Columbia U., 1943, M.A., 1947, Ph.D., 1949. Research assoc. Columbia U. Radiation Lab., N.Y.C., 1949; AEC fellow in chemistry Calif. Inst. Tech., Pasadena, 1949-50; head semicondr. research sect. Hughes Aircraft Co., Culver City, Calif., 1950-53; mem. tech. staff Bell Labs., Murray Hill, N.J., 1953-66, head biophysics research dept., 1966-79; prof. molecular biophysics and biochemistry Yale U., 1979—; Rask Oersted lectr. U. Copenhagen, 1959; vis. prof. Ecole Normale Superieur, Paris, 1962; Appleton lectr. Brown U., 1965; vis. prof. physics U. Tokyo, 1965; Reilly lectr. U. Notre Dame, Ind., 1969; vis. prof. biophysics Princeton U., 1971-72; Regents lectr. UCLA, 1978. Guggenheim fellow in lab. molecular biology MRC Cambridge (Eng.) U., 1961-62; recipient Havinga medal Leiden U., 1983. Mem. Nat. Acad. Scis. Subspecialty: Biophysics (physics). Researcher spectroscopic techniques applied to physics, chemistry, biology. Home: 123 York St New Haven CT 06511 Office: JW Gibbs Research Lab Yale U New Haven CT

SHULTZ, LEONARD DONALD, immunologist; b. Boston, Apr. 16, 1945; s. Samson and Jean (Korim) S.; m. Kathryn L., Aug. 31, 1969; children: David, Sarah. B.A., Northeastern U., 1967; Ph.D., U. Mass., 1972. Research asst. Tufts U. Sch. Medicine, Cancer Research Lab., 1967-68; grad. teaching asst. U. Mass., 1968-70, lectr., 1970-71; predoctoral trainee U. Mass./USPHS, 1971-72; postdoctoral trainee Jackson Lab., Bar Harbor, Maine, 1972-74, research assoc. II, 1974-76, assoc. staff scientist, 1976-79, staff scientist, 1979—; cooperating prof. zoology U. Maine, Orono, 1981. Contbr. articles to profl. jours. Mem. Am. Soc. Microbiology, Am. Assn. Immunologists. Subspecialties: Immunology (medicine); Immunogenetics. Current work: Immunodeficiency and autoimmunity. Home: RFD 1 Bar Harbor ME 04609 Office: Jackson Lab Bar Harbor ME 04609

SHUM, RAYMOND HING-YAN, nuclear scientist, consultant; b. Un Long, Hong Kong, Oct. 24, 1936; U.S., 1960, naturalized, 1972; s. Ming Wing and King Fong (Lo) S.; m. Julia Hwa-Yueh Miao, Mar. 4, 1967; children: Wesley, Charlotte, Irene, Margaret. B.Sc. Nat. Taiwan U, Taipei, 1959; M.Sc., U. N.B., 1960; Ph.D., N.C. State U., 1966; M.B.A., U. Chgo., 1973. Registered profl. engr., Pa., N.J., Ill., Idaho. Power system engr. N.B. (Can.) Elec. Power Commn., 1960; teaching asst. N.C. State U., Raleigh, 1960-61, 62-65; asst. prof. Miss. State U., State College, 1960; lectr. Lynchburg (Va.) Coll., 1966-67; prin. engr. B & W Co., Lynchburg, 1966-67; sr. engr. Westinghouse Electric Co., Pitts., 1967-69; nuclear engr. Argonne Nat. Lab. (Ill.), 1969-73; prin. staff engr. P. S. E. & G. Co., Newark, 1973-79; prin. engr. Ebasco Services, Inc., N.Y.C., 1979—; Mem. exec. com. Chinese Am. Acad. and Profl. Assn., N.Y.C., 1979. Mem. Am. Nuclear Soc., N.Y. Acad. Scis., Nat. Taiwan U. Alumni Assn. (exec. adv. 1980). Club: Chgo. Bus. Sch. (N.Y.C.) (communication com. 1976). Subspecialties: Nuclear fission; Electrical engineering. Current work: Nuclear power reactor safety analyses, nuclear wate management and electrical power engineering. Home: 339 Walnut St Livingston NJ 07039 Office: Ebasco Services Inc 2 World Trade Ctr New York NY 10048

SHUMAN, FREDERICK GALE, meteorologist; b. South Bend, Ind., July 13, 1919; s. Fred William and Catherine (Grimm) S.; m. Elena Fragomeni, June 15, 1946; children: Frederick Gale, Marianne, Deborah Joan. B.S., Ball State U., 1941; M.S., Mass. Inst. Tech., 1948, Sc.D., 1951. Chief computation br. Nat. Meteorol. Center, U.S. Weather Bur., Washington, 1955-58, chief devel. br., 1958-64, dir. center, 1964-81, sr. research meteorologist, 1981—; mem. meteorology group Inst. Advanced Study, Princeton, N.J., 1952-54; vis. prof. N.Y.U., 1961. Served to maj. USAAF, 1941-45. Recipient Meritorious Service award U.S. Dept. Commerce, 1957, gold medal award, 1967; Alumni Distinguished Service award Ball State U., 1965. Mem. Am. Meteorol. Soc. (2d Half Century award 1980), Am. Geophys. Union, AAAS. Subspecialty: Meteorology. Home: 212 Inverness Ln Fort Washington MD 20744 Office: Nat Meteorol Center Nat Weather Service Washington DC 20233

SHUTER, WILLIAM LESLIE HAZLEWOOD, physicist, educator; b. Rangoon, Burma, Jan. 17, 1936; s. William F.S. and Marjory D. (Hazlewood) S.; m. Beverley Robin Smith, Dec. 14, 1963; children: William J., Edward M. B.Sc., Rhodes U., S. Africa, 1957, M.Sc., 1959; Ph.D., U. Manchester, Eng., 1963. Jr. Leverhulme fellow Jodrell Bank, U. Manchester, 1959-63; lectr. in physics Rhodes U., 1963-65; faculty physics dept. U. B.C., Vancouver, 1965—, prof., 1979; satellite communications studies for B.C. Govt. Contbr. articles to profl. jours. Mem. Internat. Astron. Union, Am. Astron. Soc., Can. Astron. Soc., Royal Astron. Soc. Subspecialty: Radio and microwave astronomy. Current work: Studies of interstellar medium and structure of galaxy using centimeter and millimeter wave radio astronomy techniques. Office: Dept Physics U BC Vancouver BC Canada V6T 1W5

SHVARTZ, ESAR, aerospace scientist, environmental physiologist; b. Tel-Aviv, Israel, Jan. 8, 1935; came to U.S., 1959, naturalized, 1982; s. Itzhak and Hinda (Levin) S.; m. Carmela Bojarsky, Jan. 24, 1971; children: Illan, Guy. B.S., UCLA, 1960, M.S., 1962; Ph.D., U. So. Calif., 1965. Asst. prof. Ind. State U., Terre Haute, 1966-67; dir. human physiol. lab. Negev Inst. Arid Zone Research, Beer-Sheva, Israel, 1967-70; vis. scientist Chamber of Mines S. Africa, Johannesburg, 1974; sr. scientist Heller Inst. Med. Research, Tel-Hashomer, Israel, 1970-76; Nat. Acad. Scis. sr. research assoc. NASA, Ames Research Ctr., Moffett Field, Calif., 1976-77; sr. engr., scientist Douglas Aircraft Co., Long Beach, Calif., 1978—; dir. phys. activity Israel Def. Forces, 1970-76; sr. lectr. Tel Aviv (Israel) U., 1971-73, Ben Gurion U., Beer-Sheva, Israel, 1971-73. Contbr. articles to profl. jours. Served to maj. M.C. Israel Def. Forces, 1970-76. Named Hon. Prof. U. Witwatersrand, 1974. Fellow Am. Coll. Sports Medicine, Aerospace Med. Assn. (assoc.; environ. sci. award 1982); mem. Human Factors Soc., Am. Physiol. Soc., Internat. Com. Phys. fitness Research. Subspecialties: Physiology (medicine); Space medicine. Current work: Effect of heat, cold, altitude, zero gravity and exercise on human cardiovascular and thermoregulatory responses, decompression problems in aviation. Home: 3580 Marna Ave Long Beach CA 90808

SIAKOTOS, ARISTOTLE NICHOLAS, biochemist, pathologist, educator; b. Dedham, Mass., July 19, 1928; s. Nicholas and Demetra (Brouma) S.; m. Janet Marie Sanders, June 5, 1972; children: Ellen, Kathleen, Christine, Michael, Catherine. B.S., U. Mass., 1952, M.S., 1954; Ph.D., Cornell U., 1958. Entomologist Med. Research Labs., U.S. Army Chem. Ctr., 1958-62, biochemist, 1962-68; asst. prof., then assoc. prof. pathology Ind. U.-Indpls., 1968-78, prof., 1978—. Served with AUS, 1946-48. Mem. Am. Chem. Soc., Am. Oil Chemists Soc., Am. Soc. Neurochemistry, Am. Soc. Neurosci., Am. Soc. Exptl. Pathology, Sigma Xi. Greek Orthodox. Lodge: Lions. Subspecialties: Biochemistry (medicine); Pathology (medicine). Current work: Biochemical pathology, inherited retinal degenerations, slow viruses, biochemistry of eye and brain, lipid biochemistry.

SIBLEY, CAROL HOPKINS, geneticist, educator; b. Freeport, N.Y.C, Oct. 9, 1943; d. Harlow Bliss and Helen Davidson (Hymers) Hopkins; m. Thomas Howard Sibley, July 2, 1966; children: David Hopkins, Sarah Hopkins. B.A., U. Rochester, 1965, M.A., 1966, M.S., 1969; Ph.D., U. Calif., San Francisco, 1974. Research fellow Calif. Inst. Tech., Pasadena, 1974-76; vis. faculty, 1981; asst. prof. U. Wash., 1976-83, assoc. prof., 1983—. Contbr. numerous articles to profl. jours. NIH grantee, 1976—; recipient faculty research award Am. Cancer Soc., 1978-83, grantee, 1981—. Mem. Am. Assn. Immunologists, AAAS, Assn. Women in Sci., NOW. Democrat. Presbyterian. Subspecialties: Genetics and genetic engineering (agriculture); Immunocytochemistry. Current work: Control of synthesis and processing of heavy chains of immunoglobulins, immunogenetics, protein chemistry, molecular biology, RNA processing. Office: Dept Genetics SK-50 U Wash Seattle WA 98195

SIBLEY, CHARLES GALD, biologist, educator; b. Fresno, Calif., Aug. 7, 1917; s. Charles Corydon and Ida (Gald) S.; m. Frances Louise Kelly, Feb. 7, 1942; children—Barbara Susanne, Dorothy Ellen, Carol Nadine. A.B., U. Calif. at Berkeley, 1940, Ph.D., 1948; M.A. (hon.), Yale U., 1965. Biologist USPHS, 1941-42; instr. zoology U. Kans., 1948-49; asst. prof. San Jose (Calif.) State Coll., 1949-53; asso. prof. ornithology Cornell U., 1953-59, prof. zoology, 1959-65; prof. biology Yale, 1965—, William Robertson Coe prof. ornithology, 1967—; dir. div. vertebrate zoology, curator birds Peabody Mus., 1965—, dir. mus., 1970-76; cons. systematic biology, 1963-65; Mem. adv. com. biol. medicine NSF, 1968—; exec. com. biol. agr. NRC, 1966-70. Contbr. profl. jours. Served to lt. USNR, 1943-45. Guggenheim fellow, 1959-60. Fellow AAAS; mem. Am. Inst. Biol. Scis., Soc. Study Evolution, Am. Soc. Zoology, Am. Soc. Naturalists, Soc. Systematic Biology, Am. Ornithol. Union, Royal Australian Ornithol. Union, Deutsche Ornithol. Gesellschaft, Internat. Ornithol. Congress (sec.-gen. 1962). Subspecialty: Taxonomy. Home: Old Quarry Rd Guilford CT 06437 Office: Peabody Museum Yale U New Haven CT 06520

SICILIANO, MICHAEL J., research biologist, educator, academic administrator; b. Bklyn., May 12, 1937; s. Michael J. and Marion S.; m. Jeanette Siciliano, June 9, 1961; children: Jeanne, Lorraine, Peter. B.S., St. Peter's Coll., 1959; M.S., L.I. U., 1962; Ph.D., NYU, 1970. Assoc. prof. biology L.I. U., 1970-72; asst. prof., then assoc. prof. U. Tex. M.D. Anderson Hosp. and Tumor Inst., Houston, 1972-82, prof., head dept. genetics, 1982—, also dir. genetics program; cons. in field. Contbr. numerous articles on genetics to profl. jours. Soccer coach Spring Br. Meml. Sch. Dist., Houston. Mem. Genetics Soc. Am., Environ. Mutagen Soc., AAAS, Am. Soc. Cell Biology. Subspecialties: Gene actions; Genome organization. Current work: Mutagenesis and gene control, mutagenesis, isozymes, mapping, regulation. Office: Dept Genetics MD Anderson Hosp and Tumor Inst Houston TX 77030

SIDBURY, JAMES BUREN, physician; b. Wilmington, N.C., Jan. 13, 1922; s. James Buren and Willie Wellington (Daniel) S.; m. Alice Lucas Rayle, Aug. 29, 1953; children—Anne, Mary, Patricia, James, Robert. B.S., Yale U., 1944; M.D., Columbia U., 1947. Intern, then asst. resident in internal medicine Roosevelt Hosp., N.Y.C., 1947-49; intern Johns Hopkins Hosp., 1949-50; asst. resident in pediatrics Univ. Hosp., Cleve., 1950-51; with Communicable Disease Center, USPHS, Atlanta, 1951-53; practice medicine specializing in pediatrics, Wilmington, 1953-54; research fellow Johns Hopkins Hosp., 1955-57; asst. prof. Johns Hopkins U. Med. Sch., 1957-61; asso. prof. pediatrics Duke U. Med. Center, 1961-65, prof., 1965—, dir. clin. research unit, 1961-75; sci. dir. intramural research program Nat. Inst. Child Health and Human Devel., Bethesda, Md., 1975—; cons. Lennox Baker Hosp., Durham; lectr. Johns Hopkins Hosp.; clin. prof. pediatrics Uniformed Services U. Health Scis., 1979—; prof. child health and devel. George Washington U. Med Sch., 1977—. Contbr. profl. jours. Served with USNR, 1943-45. Recipient Bicentennial medallion in pediatrics Columbia U. Coll. Physicians and Surg., 1967. Mem. Am. Acad. Pediatrics, Am., So. socs. pediatric research, AAAS, Am. Soc. Human Genetics, Am. Pediatric Soc. Subspecialty: Pediatrics. Home: 5200 W Cedar Ln Bethesda MD 20014 Office: NIH NICHD 31/2A-50 Bethesda MD 20014

SIDDIQI, SHAUKAT MAHMOOD, biology educator, taxonomy researcher; b. Gujrat, Pakistan, Mar. 10, 1936; s. Ghulam Haider and Khurshid (Khan) S.; m. Florence Alice Patterson, Sept. 24, 1969; children: John Haider, Javaid E., Jamil J., Jafary A. B.Sc., Peshawar U., Pakistan, 1959, M.Sc., 1961; M.S., Va. State U., 1970; Ph.D., U. S.C., 1980. Acct., Pakistan, 1955-56, lang. interpreter, 1958-59; lectr. Peshawar U., 1961-68; asst. prof. Va. State U., Ettrick, 1969-75; instr. U. S.C., Columbia, 1975-77; assoc. prof. biology Va. State U., 1977—; curator herbarium, 1980—. Sponsor Muslim Fellowship, 1977—. Recipient Belser award U. S.C., 1977; Merit award Peshawar U., 1959; HEW fellow, 1975-77. Mem. Bot. Soc. Am., Am. Soc. Plant Taxonomists, Internat. Assn. Plant Taxonomy, Sigma Xi. Democrat. Moslem. Subspecialties: Taxonomy; Species interaction. Home: 21215 Chesterfield Ave Ettrick VA 23803 Office: Va State U Petersburg VA 23803

SIDERIS, ANTONIOS GEORGE, nuclear engineer; b. Athens, Greece, Dec. 8, 1945; came to U.S., 1964; s. George Antonios and Anastasia (Moraki) S.; m. Sandra Vasa, July 4, 1973. B.S. Mech, Piraeus (Greece) Tech. Inst., 1964. Assoc. engr. TVA, Knoxville, 1967-70; engr. Parsons-Jurden Corp., N.Y.C., 1970-71; nuclear startup engr. Fluor-Pioneer, Kewaunee, Wis., 1971-74; cons. engr. Quadrex Corp., Campbell, Calif., 1974-76; mem. nuclear engring. staff Am. Nuclear Insurers, Farmington, Conn., 1976-79; engring. mgr. Engring. Cons. Inc., Atlanta, 1979-82; sr. program mgr. Inst. Nuclear Power Ops., Atlanta, 1982—; guest lectr., acad. adv. Rensselaer Polytech. Inst., Troy, N.Y., 1978; guest lectr. Ga. State U., Atlanta, 1982. Nuclear Regulatory Commn. grantee, 1977. Mem. Am. Nuclear Soc., ASME (chmn. systems). Subspecialties: Nuclear engineering; Nuclear fission. Current work: Nuclear risk assessment, nuclear safety, events analysis of nuclear power plant operations. Home: 2103 Mitchell Ct Marietta GA 30062 Office: Inst Nuclear Power Ops 1100 Circle 75 Parkway Atlanta GA 30339

SIDKY, YOUNAN ABDEL-MALIK, scientist; b. Khartoum, Sudan, Feb. 9, 1928; s. Abdel Malik and Iskandara (Abdullab) S.; m. Jan Fahmy, July 30, 1964; children: Emil, Sonya. B.Sc., Cairo U., 1950, M.Sc., 1955; Ph.D., Phillips U., 1956. Demonstrator Cairo U., 1950-60, lectr., 1960-65; research assoc. U. Wis.-Madison, 1965-69, project assoc., 1972-76, assoc. scientist, 1976—; research assoc. U. Alta., Can., 1969-72. Contbr. articles to profl. jours. Mem. Assn. Egyptian-Am. Scholars, Am. Assn. Immunologists. Coptic Christian. Subspecialties: Cancer research (medicine); Immunobiology and immunology. Current work: In vivo and in vitro study of lymphocyte and tumor-induced angiogenesis. Home: 10 Cheyenne Circle Madison WI 53705 Office: Clinical Science Ctr 600 Highland Ave #K4/414 Madison WI 53792

SIDLINGER, BRUCE DOUGLAS, software engineer; b. Dallas, Mar. 19, 1958; s. Bruce Chester and Joanne (Leonard) S.; m. Sarah Lynne Jennings. B.S., Trinity U., San Antonio, 1982, diploma cum laude, 1982. Sci. programmer Recognition Equipment, Irving, Tex., 1975-76; software engr. Tex. Instruments, Dallas, 1976-77; dir. new products Alcor Inc., San Antonio, 1977-80, research scientist, 1980-82, dir. research, 1982-83, v.p. research and engring., 1983—, also dir. Res. dep. Bexar County (Tex.) Sheriff's Dept., 1982. Trinity U. Pres.'s scholar, 1980. Mem. Assn. Computing Machinery (v.p. Trinity U. chpt. 1981-82), IEEE Computer Soc., Soc. Info. Display, Inst. Navigation, Upsilon Pi Epsilon (local pres. 1981-82), Aircraft Owners and Pilots Assn., Soaring Soc. Am., Seaplane Pilots assn., Porsche Club Am., Alpha Chi. Republican. Subspecialties: Artificial intelligence; Software engineering. Current work: Real-time expert systems for on-line control and fault diagnosis; software engineering for low-cost, human life-rated airborne computer systems. Patentee exercise trampoline; inventor stepper motor indicator, flight data recorder. Home: 7626 Callaghan Rd Suite 3006 San Antonio TX 78229 Office: Alcor Inc 10130 Jones Maltsberger San Antonio TX 78216

SIDOROWICZ, KENNETH JOSEPH, mech. engr., cons.; b. Topeka, Sept. 28, 1951; s. Norbert Joseph and Florence Matilda (Malnar) S.; m. Deborah Sue Hill, Nov. 24, 1979; 1 son, Andrew Joseph. B.Arch., Kans. State U., 1969, M.S. in Applied Mechanics/Mech. Engring, 1976; postgrad. in bus. adminstrn, U. Kans. Regents Ctr., Overland Park, 1980—. Registered profl. engr., Kans., Mo. Engr. Kansas City div. Bendix Corp., Mo., 1977-79, 80-82; mech. engr. Black & Veatch Cons. Engrs., Overland Park, 1979; engr. King Radio Corp., Olathe, Kans., 1979-80; assoc. cons. Wagner-Hohns-Inglis-Inc., Kansas City, Mo., 1982—. Mem. Westwood Hills Homeowners Assn. Mem. ASME (assoc.). Democrat. Roman Catholic. Subspecialties: Theoretical and applied mechanics; Laser welding stress analysis. Current work: Structural engineering, finite element method, thermal shock phenomenon, structural engineering consulting, investigative research for claims and litigation marketing and business administration. Home: 1908 W 49th St Westwood Hills KS 66205 Office: 4420 Madison Suite 215 Kansas City MO 64111

SIDWELL, ROBERT WILLIAM, scientist, educator; b. Huntington Park, Calif., Mar. 17, 1937; s. Robert Glen and Eva Amalie (Gordy) S.; m. Rhea Julander, May 31, 1957; children: Richard Dale, Jeanette Kathleen, David Eugene, Cynthia Diane, Michael Jason, Robert Odell. B.S., Brigham Young U., Provo, Utah, 1958; M.S., U. Utah, 1961, Ph.D., 1963. Head serology, ricketts and virus research Epizoology Lab., U. Utah, 1958-63; head virus div. So. Research Inst., Birmingham, Ala., 1963-69; head dept. virology ICN Nucleic Acid Research Inst., Irvine, Calif., 1969-72, head div. chemotherapy, 1972-75, dir. inst., 1975-77; prof. animal, dairy and vet. scis. Utah State U., Logan, 1977—; mem. faculty U. Ala. Med. Sch., 1968-69; speaker in field. Editorial bd.: Antimicrobial Agts. and Chemotherapy, 1972—, Chemotherapy, 1974, Jour. Antiviral research, 1980—; Contbr. articles to profl. jours. Mem. Nibley (Utah) City Planning and Zoning Commn., 1978—; mem. steering com. Irvine Sch. Bd., 1972, chmn. health edn. awareness forum, 1975. Scholar Order Eagles, 1954, Dept. Interior, 1954. Mem. AAAS, Am. Assn. Immunologists, Soc. Exptl. Biology and Medicine, Pan Am. Med. Assn., Internat. Soc. Chemotherapy, Inter-Am. Soc. Chemotherapy, Am. Soc. Microbiology, Am. Soc. Virology, Sigma Xi. Mormon. Subspecialties: Virology (biology); Virology (veterinary medicine). Current work: Chemotherapy and immunotherapy of viral diseases, effects of nutrition and specific nutritional factors on sensitivity to viral infection, major interest in the antiviral drug, ribavirin. Home: 162 Quarter Circle Dr Nibley UT 84321 Office: Dept Biology Utah State U Logan UT 84322

SIEDLER, ARTHUR JAMES, educator; b. Milw., Mar. 17, 1927; s. Arthur William and Margaret (Stadler) S.; m. Doris Jean Northrop, Feb. 23, 1976; children—William, Nancy Siedler Sowden, Sandra Siedler Lowman, Roxanne Rose Butler, Randy Rose, Rick Rose. B.S., U. Wis., 1951; M.S., U. Chgo., 1956, Ph.D., 1959. Chief div. biochemistry and nutrition Am. Meat Inst. Found., Chgo., 1959-64; group leader Norwich (N.Y.) Pharmacal Co., 1964-65, chief physiology sect., 1965-69, chief biochemistry sect., 1969-72; acting dir. div. nutritional scis. U. Ill., Urbana, 1978-81, prof., head dept. food sci., prof., 1972—. Served with USCG, 1945-46; PTO. NIH research grantee, 1960-63; Nat. Livestock and Meat Bd. grantee, 1959-64. Mem. Inst. Food Technologists, Am. Chem. Soc., Am. Inst. Nutrition, Am. Heart Assn., Assn. Vitamin Chemists (past pres. 1962), Am. Dairy Sci. Assn., Am. Assn. Cereal Chemists, Council for Agrl. Sci. and Tech., N.Y. Acad. Sci., Sigma Xi. Clubs: Rotary, Isaak Walton, Moose. Subspecialties: Food science and technology; Biochemistry (biology). Current work: Mode of anti bacterial activity of nitrate; food additives. Patentee in field. Home: 8 Stanford Pl Champaign IL 61820 Office: 567 Bevier Hall 905 S Goodwin Urbana IL 61801

SIEGAL, GENE PHILIP, surgical pathologist, educator; b. N.Y.C., Nov. 16, 1948; s. Murray Herman and Evelyne (Philips) S.; m. Sandra Helene Meyerowitz, Aug. 3, 1972; children: Gail Deborah, Rebecca Stacey. B.A., Adelphi U., 1970; M.D., U. Louisville, 1974; Ph.D., U. Minn., 1979. Diplomate: Nat. Bd. Med. Examiners, 1975, anat. pathology diplomate Am. Bd. Pathology, 1978. Intern to chief resident in pathology Mayo Grad. Sch. Medicine, Rochester, Minn., 1974-79, research fellow in pathology, 1976-77; research assoc. Nat. Cancer Inst., NIH, Bethesda, Md., 1979-81; med. specialist fellow U. Minn., Mpls., 1981-82; asst. prof. pathology U. N.C., 1982—; attending surg. pathologist N.C. Meml. Hosp., Chapel Hill, 1982—; Bd. editors Minn. Medicine, 1979-82. Contbr. numerous articles to sci. publs. Tutor Upward Bound Adelphi U., Garden City, N.Y., 1964-66. U. Minn. Am. Cancer Soc. clin. fellow, 1981-82. Fellow Am. Soc. Clin. Pathologists; mem. Am. Assn. Pathologists, Internat. Acad. Pathology, AAAS, N.Y. Acad. Scis., Sigma Xi, Omicron Delta Kappa, Beta Beta Beta, Phi Delta Epsilon. Democrat. Jewish. Club: Workmen's Circle (East Meadow, N.Y.). Subspecialties: Pathology (medicine); Cancer research (medicine). Current work: tumor invasion and metastasis; immunohistochemistry. Office: Dept Pathology U N C Bldg 228H Chapel Hill NC 27514

SIEGEL, ABRAHAM LAZARUS, biochemistry educator; b. N.Y.C., May 19, 1916; s. Philip and Rebecca (Shugerman) S.; m. Florence Rader., Jan. 7, 1942; children: Dorothy, David, Richard. B.Sc., CCNY, 1939; M.Sc., N.Y. U., 1949; Ph.D., U. Ala., 1958. Dir. chem. labs. Univ. Hosp., Birmingham, Ala., 1958-64; assoc. prof. clin. pathology U. Ala.-Birmingham Sch. Medicine, 1965-69, prof. dept. pathology, 1972—, dir. endocrinology lab., 1969—. Ala Heart Assn. fellow, 1956-58; NIH fellow, 1967-68. Mem. Am. Assn. Clin. Chemistry, AAAS, AAUP (past chpt. pres.), Physicians for Social Responsibility, Sigma Xi (past chpt. pres.). Subspecialties: Biochemistry (medicine); Endocrinology. Current work: Hormone markers for tumors of the endocrine system; steroids in amniotic fluid as markers for fetal viability. Home: 1640 Lakewood Dr Birmingham AL 35216 Office: U Ala-Birmingham Sch Medicine University Station Birmingham AL 35294

SIEGEL, ALBERT, molecular biologist, educator; b. N.Y.C., Aug. 20, 1924; s. Isaac and Jeanette (Kostinsky) S.; m. Betty Lois Knack, Aug. 21, 1947; children: Laurel Siegel Gord, Jacqueline Siegel Bartelt, Robin, Timothy. B.A., Cornell U., 1947; Ph.D., Calif. Inst. Tech., 1951. Asst. research botanist UCLA, 1951-59; prof. agrl. biochemistry U. Ariz., Tucson, 1959-72; program dir. biochemistry NSF, Washington, 1967-68; sr. postdoctoral fellow Virus Research Lab., Cambridge, Eng., 1965-66; prof. biol. scis. Wayne State U., Detroit, 1972—; vis. prof. dept. microbiology U. Mich., 1978-79. Contbr. articles to profl. jours. Served with U.S. Army, 1943-46; ETO. Recipient Disting. Research award Sigma Xi, 1982. Fellow Am. Phytopath. Soc.; mem. AAAS, Genetics Soc. Am., Am. Soc. Cell Biology, Am. Soc. Microbiology, Am. Virology Soc., Sigma Xi. Subspecialties: Plant virology; Genetics and genetic engineering (agriculture). Current work: Development of cloning vehicle for plants.

SIEGEL, BARRY ALAN, nuclear physician, diagnostic radiologist, educator; b. Nashville, Dec. 30, 1944; s. Walter Gross and Lillian (Tumbarello) S.; m. Pamela M. Mandel, Aug. 18, 1968 (div. Mar. 2, 1981); m. Marilyn J. Siegel, Jan. 29, 1983. A.B., Washington U., 1966, M.D., 1969. Diplomate: Am. Bd. Nuclear Medicine, Am. Bd. Radiology (Diagnostic Radiology and Nuclear Radiology). Intern in medicine Barnes Hosp., St. Louis, 1969-70; resident in radiology and nuclear medicine Washington U., St. Louis, 1970-73, asst. prof., 1973-76, assoc. prof., 1976-79, prof. radiology, 1979—, assoc. prof. medicine, 1980—, dir. div. nuclear medicine, 1973—; chief radiol. scis. div. Armed Forces Radiobiology Research Inst., Bethesda, 1974-76; radiologist Barnes Hosp./St. Louis Children's Hosp., St. Louis, 1973—; cons. FDA, Rockville, MD., 1972—, REMS Corp., Albuquerque, 1981—; mem. FDA radiopharm. drugs adv. com., 1974-77, 81—, chmn., 1982—. Served to maj. USAF, 1974-76. Fellow ACP, Am. Coll. Radiology, Am. Coll. Nuclear Physicians; mem. Soc. Nuclear Medicine (trustee 1981—), Radiol. Soc. N.Am. Jewish. Subspecialty: Nuclear medicine. Current work: Detection of thrombosis; positron emission tomography; cardiovascular nuclear medicine. Office: Mallinckrodt Inst Radiology Washington U Sch Medicine 510 S Kingshighway Blvd Saint Louis MO 63110

SIEGEL, GILBERT BYRON, public adminstration educator, consultant; b. Los Angeles, Apr. 19, 1930; s. Morris DeSagar and Rose (Vancott) S.; m. Darby Day Smith, Oct. 16, 1954; children: Clark Byron, Holly May. B.S., U. So. Calif., 1952, M.S., 1957; Ph.D., U. Pitts., 1964. Adminstrv. analyst County Los Angeles, 1954-57; vis. asst. prof. U. So. Calif. Faculty in Iran, Tehran, 1957-59; instr. U. Pitts., 1959-61; asst. prof. U. So. Calif., Los Angeles and Rio de Janeiro, Brazil, 1961-64, assoc. prof., 1964-75, prof. pub. adminstrn., 1975—, sector head air pollution control inst., 1964-67, 75-76, assoc. dean, 1976-79; cons. earthquake rehab. study U. Wis., Green Bay, 1982—; cons. instn. bldg. Served with U.S. Army, 1952-54. Medal of Merit Getulio Vargas Found., Brazil, 1975. Mem. Am. Soc. Pub. Adminstrn. (nat. conf. planning com. 1972), Internat. Personnel Mgmt. Assn. (nat. conf. planning com. 1982), Pi Sigma Alpha, Pi Alpha Alpha. Democrat. Mem. United Ch. Christ. Subspecialty: Resource management. Current work: Seismic hazard rehabilitation and prevention. Office: Sch Pub Adminstrn U So Calif University Park MC 0041 Los Angeles CA 90089-0041 Home: 208 N Poinsettia Ave Manhattan Beach CA 90266

SIEGEL, IVENS AARON, pharmacologist; b. Bay Shore, N.Y., Jan. 28, 1932; s. Nathan and Rose S.; m. Naomi Mary; children: Elizabeth, Sondra, Maria. B.S., Columbia U., 1953; M.S., U. Kans., 1958; Ph.D., U. Cin., 1962. Asst. prof. SUNY at Buffalo, 1962-68; asso. prof. U. Wash., 1968-72, prof., 1972-79, U. Ill., 1979—. Served with U.S. Army, 1954-56. Mem. Am. Soc. Pharmacology and Exptl. Therapeutics, Internat. Assn. Dental Research. Jewish. Subspecialties: Pharmacology; Oral biology. Current work: Transport in salivary glands and oral mucosa. Home: 910 W Park Champaign IL 61820 Office: 506 S Matthew Ave Champaign IL 61801

SIEGEL, JEROME, neuroscientist, educator; b. Bklyn., Sept. 1, 1929; s. David and Dora S.; m. Priscilla Wishnick, July 23, 1929; children: Peter, Bennett, Rebecca. A.B., U. Mich., 1955; Ph.D., Ohio state U., 1960. Postdoctoral fellow dept. physiology UCLA, 1960-62; mem. faculty Sch. Life and Health Sci., U. Del., Newark, 1962—, assoc. prof., 1968-73, prof., 1973—; NIH sr. postdoctoral fellow Inst. Animal Physiology, Babraham, Cambridge, Eng., 1968-69; vis. prof. dept. anatomy U. Rotterdam, Netherlands, 1975-76. Contbr. articles to profl. jours. Mem. Soc. for Neurosci., AAAS, Sigma Xi. Jewish. Subspecialties: Neurophysiology; Physiological psychology. Current work: The neural basis of arousal and inhibition of behavior. Home: 18 N Wynwyd Rd Newark DE 19711 Office: Inst for Neuroscience U Del Newark DE 19711

SIEGEL, JOEL MARVIN, nuclear engr.; b. Norwalk, Conn., Nov. 4, 1947; s. Robert L. and Ann (Rubin) S.; m. Robin Heather Alpert, June 4, 1972; children: Nathan Gabriell, Jonathan Adrian. B.A., U. Bridgeport, 1968; M.S., U. Miami, 1969, Ph.D., 1973; M.E., U. Fla., 1979. Sr. research scientist BDM Corp., McLean, Va., 1979-81; systems engr. Project Mgmt. Corp., Oak Ridge, 1981-82, Technology for Energy Corp., Knoxville, Tenn., 1982—. Contbr. articles to profl. jours. Violinist Oak Ridge Symphony Orch., 1981—. Served to lt. U.S. Navy, 1975-79. Recipient acad. awards and fellowships. Mem. Am. Nuclear Soc., Instrument Soc. Am., Health Physics Soc. Subspecialties: Nuclear engineering; Human factors engineering. Current work: Performance of PRAs for the electric utilities. Have developed innovative techniques. Human factors engineering applied to control rooms. Office: Technology for Energy Corp 1 Energy Center Pellissippi Pkwy Knoxville TN 37922

SIEGEL, MELVIN WALTER, physicist; b. N.Y.C., May 26, 1941; s. Moses and Mae (Sager) S.; m. Jane A.L. Licht, Aug. 18, 1968. B.A., Cornell U., 1962; M.S., U. Colo., 1967, Ph.D., 1970. Postdoctoral fellow U. Va., Charlottesville, 1970-72; asst. prof. SUNY-Buffalo, 1972-74; staff scientist Extranuclear Labs., Inc., Pitts., 1974-78, research devel. mgr., 1978-82; sr. research scientist The Robotics Inst., Carnegie-Mellon U., Pitts., 1982—; cons., patentee in field. Contbr. articles to profl. jours. Vol. Peace Corps, Ghana, 1962-64. Recipient I-R 100 awards, 1978, 79, Indsl. Research/Devel. Mag. Mem. AAAS, Am. Phys. Soc., ASTM, N.Y. Acad. Sci., Am. Soc. Mass Spectrometry, Soc. Analytical Chemists Pitts., Am. Assn. Artificial Intelligence, Spectroscopy Soc. Pitts. Subspecialties: Robotics; Artificial intelligence. Current work: Application of advanced sensor technology, microprocessor control, artificial intelligence to robotics and systems. Patentee in field. Home: 5232 Westminster Pl Pittsburgh PA 15232 Office: Robotics Inst Carnegie Mellon Univ Pittsburgh PA 15213

SIEGEL, MICHAEL ELLIOT, physician, educator, researcher; b. N.Y.C., May 13, 1942; s. Benjamin and Rose (Gilbert) S.; m. Marsha Rose Snower, Mar. 20, 1966; children: Herrick Jove, Meridith Ann. A.B., Cornell U., 1964; M.D., Chgo. Med. Sch., 1968. Diplomate: Nat. Bd. Med. Examiners. Intern Cedars-Sinai Med. Ctr., Los Angeles, 1968-69; resident in radiology, 1969-70; NIH fellow in radiology Temple U. Med. Ctr., Phila., 1970-71; NIH fellow in nuclear medicine Johns Hopkins Sch. Medicine, Balt., 1971-73, asst. prof. radiology, 1972-76; assoc. prof. radiology, medicine U. So. Calif., Los Angeles, 1976—, co-dir. nuclear cardiology, 1976—; dir. div. nuclear medicine, 1982—, vice-chmn. dept. radiology, 1983—; dir. div. nuclear medicine Kenneth Norris Cancer Hosp. and Research Ctr., Los Angeles, 1983—; dir. nuclear medicine, cons. Orthopaedic Hosp., Los Angeles, 1981—; cons. Intercommunity Hosp., Covina, Calif., 1981—, Rancho Los Amigos Hosp., Downey, Calif., 1976—. Author: Textbook of Nuclear Medicine, 1978, Vascular Surgery, 1983; editor: Nuclear Cardiology, 1981, Vascular Disease: Nuclear Medicine, 1983. Coach Little League, Beverly Hills; mem. Maple Ctr., Beverly Hills. Served as maj. USAF, 1974-76. Mem. Soc. Nuclear Medicine (sci. exhbn. com. 1978-79, program com. 1979-80, Silver medal 1975), Am. Coll. Nuclear Medicine (sci. investigator 1974, 76, nominations com. 1980, program com. 1983), Radiol. Soc. N.Am., Soc. Nuclear Magnetic Resonance Imaging. Lodge: Friars So. Calif. Subspecialties: Nuclear medicine; Radiology. Current work: Development of nuclear medicine techniques to: evaluate cardiovascular disease and diagnose and treat cancer, development of nuclear magnetic resonance (NMR) as a clinical imaging. Inventor pneumatic radiologic pressure system. Office: U So Calif Med Ctr PO Box 693 1200 N State St Los Angeles CA 90033

SIEGEL, MICHAEL JASON, astronomer, computer programmer; b. Boston, Dec. 16, 1951; s. Benjamin and Lena (Levovsky) S. A.B. in Astronomy, Boston U., 1973, M.S., U. Md., College Park, 1980. Obs. instr. dept. astronomy Boston U., 1974-75; teaching asst. astronomy program U. Md., College Park, 1976-80; tech. asst. Steward Obs., U. Ariz., Tucson, 1981-82; research asst. Kitt Peak Nat. Obs., Tucson, 1981—; researcher in field. Contbr. articles to profl. jours. Recipient Harvard Price Book, 1969. Mem. Astron. Soc. Pacific, Am. Astron. Soc. (assoc.). Jewish. Subspecialties: Optical astronomy; Data analysis programming. Current work: Photoelectric photometry, photometric determination of properties of stellar atmospheres; computer support of photometry. Office: Kitt Peak Nat Obs PO Box 26732 Tucson AZ 85726

SIEGEL, RONALD KEITH, psychopharmacologist, consultant; b. Little Falls, N.Y., Jan. 2, 1943; s. Harry and Freida (Rosen) S. B.A., Brandeis U., 1965; M.A., Dalhousie U., 1966, Ph.D., 1970. Postdoctoral fellow Albert Einstein Coll. Medicine, N.Y.C., 1970-73; adj. assoc. prof. UCLA, Los Angeles, 1974-76, assoc. research psychologist, 1975—; cons. LeDain Commn., Ottawa, Ont., Can., 1970-73, Shaeffer Commn., Washington, 1972-73; pres. Moksha Labs., Inc., Los Angeles, 1977—. Editor: Hallucinations, 1975; artist: comic book Mother's Oats No. 1, 1976; contbr. articles in field to profl. jours. Fellow Am. Psychol. Assn.; mem. Psychonomic Soc. Subspecialties: Behavioral psychology; Psychopharmacology. Current work: Self-administration of psychoactive plants by animals in natural habitats; effects of drugs on perception; cocaine and other drugs of abuse. Office: VA PO Box 84358 Los Angeles CA 90073

SIEGEL, STUART ELLIOTT, physician; b. Plainfield, N.J., July 16, 1943; s. Hyman and Charlotte (Freidberg) S.; m. Linda Wertkin, Jan. 20, 1968; 1 son, Joshua. B.A., Boston U., 1967, M.D., 1967. Diplomate: Am. Bd. Pediatrics, Am. Bd. Pediatric Hematology/Onbcology. Intern U. Minn. Hosp., 1967-68, resident, 1968-69; clin. assoc. NIH, Bethesda, Md., 1969-72; asst. prof. U. So. Calif., Los Angeles, 1972-76, assoc. prof., 1976-81, prof. pediatrics, 1981—; head div. hematology/oncology Children's Hosp., Los Angeles, 1976—; cons. in field. Editorial bd.: Current Concepts in Pediatric series, 1981—; reviewer: Jour. Pediatrics, 1980—, Med. and Pediatric Oncology, 1978—, Am. Jour. Pediatric Hematology-Oncology, 1981—, Jour. Nat. Cancer Inst, 1981—; contbr. numerous articles to profl. jours. Pres. So. Calif., Children's Cancer Services, Inc., 1978—; participant Nat. Leukemia Broadcast Council Radiothon, 1978-80, mem. adv. bd., 1978—; adv. bd. Leukemia Soc. Am., 1978—. Served with USPHS, 1969-72. NIH grantee, 1975-77; USPHS grantee, 1974—; Am. Cancer Soc. grantee, 1977-80; Nat. Cancer Inst. grantee, 1979-82; others. Mem. Los Angeles County Med. Soc., Los Angeles Pediatric Soc., Western Soc. Pediatric Research, AMA, AAAS, Am. Assn. Cancer Research, Am. Soc. Hematology, Am. Soc. Microbiology, Am. Soc. Clin. Oncology, Am. Soc. Pediatric Hematology-Oncology, Am. Assn Cancer Edn., Soc. Pediatric Research. Subspecialties: Pediatrics; Oncology. Current work: Clinical pediatric cancer research. Office: 4650 Sunset Blvd Los Angeles CA 90027

SIEGMAN, ANTHONY EDWARD, elec. engr.; b. Detroit, Nov. 23, 1931; s. Orra Leslie and Helen Salome (Winnie) S.; m. (married). A.B. summa cum laude, Harvard U., 1952; M.S., UCLA, 1954; Ph.D., Stanford U., 1957. Mem. faculty Stanford (Calif.) U., 1957—, asso. prof. elec. engring., 1960-65, prof., 1965—; dir. Edward L. Ginzton Lab., 1978—; cons. Lawrence Livermore Labs., United Technologies Corp., GTE; mem. Air Force Sci. Adv. Bd.; vis. prof. Harvard U., 1965. Author: Microwave Solid State Masers, 1964, An Introduction to Lasers and Masers, 1970; contbr. over 100 articles to profl. jours. Fellow IEEE (W.R.G. Baker award 1971, J.J. Ebers award 1977), Optical Soc. Am. (R.W. Wood prize 1980); mem. Nat. Acad. Engring., AAAS, AAUP, Phi Beta Kappa, Sigma Xi. Subspecialties: Spectroscopy; Laser-induced chemistry. Patentee microwave and optical devices and lasers, including the unstable optical resonator. Office: Ginzton Lab 35 Stanford U Stanford CA 94305

SIEGMAN, MARION JOYCE, physiologist, researcher; b. Bklyn., Sept. 7, 1933; s. C. Joseph and Helen Rachel (Wasserman) S. B.A., Tulane U., 1954; Ph.D., SUNY-Bklyn., 1966. Instr. Jefferson Med. Coll., Phila., 1967-68, asst. prof., 1968-72, assoc. prof., 1972-77, prof. physiology, 1977—; mem. physiology study sect. NIH, Bethesda, Md., 1979—83; cons. regulatory biology NSF, Washington, 1978-79. NIH research grantee, 1967-84. Mem. Am. Physiology Soc., Biophys. Soc., Soc. Gen. Physiologists, Physiology Soc. Phila. (pres. 1975-76), Sigma Xi (pres. Jefferson chpt. 1982-83). Subspecialties: Physiology (medicine); Biophysics (physics). Current work: Muscle mechanics and energetics, ultrastructure. Office: Thomas Jefferson U Jefferson Med Coll 1020 Locust St Philadelphia PA 19107

SIEKEVITZ, PHILIP, educator; b. Phila., Feb. 25, 1918; s. Joseph and Tillie (Kaplan) S.; m. Rebecca Burstein, Aug. 7, 1949; children—Ruth, Miriam. B.S. in Biology, Phila. Coll. Pharmacy and Scis., 1942, Ph.D. (hon.), 1972, U. Calif. at Berkeley, 1949, U. Stockholm, Sweden, 1974. USPHS fellow Mass. Gen. Hosp., 1949-51; fellow oncology McArdle Lab., U. Wis., 1951-54; mem. faculty Rockefeller U., 1954—, now prof. cell biology.; Mem. molecular biology panel NSF, 1964-67; panel Internat. Cell Research Orgn., 1963—; Bd. dirs., treas., chmn. Scientists Com. Pub. Information N.Y., 1962—; bd. dirs. N.Y. Univs. Com., 1965—; council Am. Fedn. Scientists, 1967—. Author: (with A. Loewy) Cell Structure and Function, 2d edit, 1969; Editor: Jour. Cell Biology, 1962-65, Jour. Cellular Physiology, 1970—, Biosci., Jour. Exptl. Zoology, 1969-73, Biochim. Biophysica Acta; mem. editorial bd.: Jour. Cell Sci. Served with USAAF, 1942-45. Mem. Am. Soc. Biol. Chemists, Am. Soc. Cell Biology (pres. 1966-67), A.A.A.S., Am. Inst. Biol. Scientists, N.Y. Acad. Scis. (governing bd. 1973—, pres. 1976), Nat. Acad. Scis., Sigma Xi. Subspecialty: Cell biology. Office: Dept Cell Biology Rockefeller U 1230 York Ave New York NY 10021

SIEM, LAURIE HELEN, energy/environmental engineer; b. N.Y.C., Apr. 17, 1954; d. John and Marianne (Boye) Kriss; m. John Edwin Siem, Aug. 29, 1981. B.S. cum laude in Earth Sci, Adelphia U., 1976; M.S. in Environ. Engring, SUNY-Stony Brook, 1980. Environ. engr. Pall Corp., Glen Cove, N.Y., 1977-81; energy engr. Pan Am, Subase Bangor, Bremerton, Wash., 1982—. Mass. Audubon Soc. grantee, 1975; recipient Dept. Earth Sci. award, 1976. Mem. Inst. Environ. Scis., Am. Soc. Quality Control, Marine Environ. Council. Subspecialties: Environmental engineering; Energy conservation. Current work: Hazardous waste management energy conservation plans, alternate energy sources, resource planning.

SIEMENS, ALBERT JOHN, biomedical research administrator, lecturer, consultant; b. Winnipeg, Man., Can., Dec. 7, 1943; m. Judith H. Horch, May 8, 1965; children James, Robert. B.Sc. in Pharmacy, U. Man., 1966; M.Sc., 1969; Ph.D., U. Toronto, 1973. Research sci. V N.Y. State Research Inst. Alcoholism, Buffalo, 1973-80; adj. asst. prof. pharmacology SUNY-Buffalo, 1974-80; asst. dir. Pfizer Pharms., Inc., N.Y.C., 1980-82, assoc. dir., 1982; assoc. dir. clin. research Pfizer Internat., Inc., 1982—. Contbr. articles to profl. jours. Mem. Am. Soc. Pharmacology and Exptl. Therapeutics, Am. Soc. Clin. Pharmacology and Therapeutics, Am. Pharmacology Assn., Research Soc. Alcoholism, Can. Pharm. Assn., Man. Pharm. Assn., AAAS. Subspecialties: Pharmacology; Pharmacokinetics. Current work: Management of clinical research; development of new drugs. Office: Pfizer Internat Inc 235 E 42d St New York NY 10017

SIENICKI, JAMES JOSEPH, nuclear engineer; b. Chgo., Aug. 27, 1948; s. Harry and Angeline S. B.S. in Physics, U. Ill., Chgo., 1969, M.S., 1970, Ph.D., 1975. Asst. nuclear engr. Argonne Nat. Lab., Ill., 1976-81, nuclear engr., 1981—. Contbr. articles to profl. jours. Mem. Am. Nuclear Soc. Subspecialties: Nuclear engineering; Mechanical engineering. Current work: Nuclear reactor safety analysis and experiments; developer of computer models for analysis of hypothetical accidents in nuclear reactors and for analysis of nuclear reactor safety experiments. Home: 5650 S Kilbourn Ave Chicago IL 60629 Office: Argonne Nat Lab 9700 S Cass Ave Argonne IL 60439

SIENKO, MICHELL J., chemist, educator; b. Bloomfield, N.J., May 15, 1923; s. Felix and Teofila (Kislova) S.; m. Carol Tanghe, Aug. 25, 1946; 1 dau., Tanya. A.B., Cornell U., 1943; Ph.D., U. Calif.-Berkeley, 1946. Research asso. Stanford U., 1946-47; instr. chemistry Cornell U., Ithaca, N.Y., 1947-50, asst. prof., 1950-53, asso. prof., 1953-58, prof., 1958—; Fulbright lectr. U. Toulouse, France, 1956-57; vis. prof. Am. Coll. in Paris, 1963-64; guest prof. U. Vienna, Austria, 1974-75, 82-83; vis. fellow Cambridge U., 1978-79. Author: (with Robert A. Plane) Chemistry, 5th edit., 1976, Experimental Chemistry, 6th edit., 1983, Physical Inorganic Chemistry, 1963, Stoichiometry and Structure, 1964, Equilibrium, 1964, Principles and Properties, 3d edit., 1979, Chemistry Problems, 2d edit., 1972; editor: (with Gerard Lepoutre) Solutions Metal-Ammoniac, 1964, (with Joseph Lagowski) Metal-Ammonia Colloque Weyl II, 1970, Jour. Solid State Chemistry. Recipient Sporn award Coll. Engring., Cornell U., 1963, Clark Disting. Teaching award Coll. Engring., Cornell U., 1982; Guggenheim fellow, Grenoble, France, 1970-71. Mem. Am. Phys. Soc., Am. Chem. Soc. (Chem. Edn. award 1983), Phi Beta Kappa, Sigma Xi, Phi Kappa Phi. Subspecialty: Solid state chemistry. Current work: Superconducting compounds; low temperature magnetic properties; metal-non-metal transition. Home: 493 Ellis Hollow Creek Rd Ithaca NY 14850 Office: Cornell U Dept Chemistry Ithaca NY 14853 I have always believed that man could do anything he set his mind to provided he was willing to work hard enough. However, no goal is worth pursuing if it means that human qualities are destroyed. Respect for one's fellowman comes first.

SIERACKI, LEONARD MARK, research/development company executive.; b. Hartford, Conn., Apr. 15, 1941; s. Joseph and Lillian (Golec) S.; m. Martha Elaine Shermeth, Sept. 12, 1964; children: Jeffrey Mark, Jennifer Elaine, Julie Elaine. B.M.E., Cath. U. Am., 1963, M.M.E., 1970; M. Adminstrv. Sci., Johns Hopkins U., 1977. Registered profl. engr., Md. Aerospace engr. Harry Diamond Labs., Washington, 1963-67; sr. project mgr. Hydrospace Research Corp., Rockville, Md., 1967-70; dir. applied mechanics div. Columbia Research Corp., Gaithersburg, Md., 1970-74; pres. Lorelei Corp., Gaithersburg, 1974-77, Tritec Inc., Columbia, Md., 1977—; cons. in field. Contbr. articles to profl. jours.; author: Handbook of Fluidic Sensors, 1977; prin. author: Submarine Cabled Systems Design and Planning Manual, 1976. Mem. Am. Assn. Small Research Cos., Am. Def. Preparedness Assn. Roman Catholic. Subspecialties: Fluid mechanics; Ocean engineering. Current work: Biomedical engineering as related to respiratory devices. Patentee in field. Office: Tritec Inc PO Box 56 Columbia MD 21045 Home: 9421 Kilimanjaro Rd Columbia MD 21045

SIESS, CHESTER PAUL, educator; b. Alexandria, La., July 28, 1916; s. Leo C. and Adele (Liebreich) S.; m. Helen Kranson, Oct. 5, 1941; 1 dau., Judith Ann. B.S., La. State U., 1936; M.S., U. Ill., 1939, Ph.D., 1948. Party chief La. Hwy. Commn., 1936-37; research asst. U. Ill., 1937-39; soil engr. Chgo. Subway Project, 1939-41; engr., draftsman N.Y.C. R.R. Co., 1941; mem. faculty U. Ill., 1941—, prof. civil engring., 1955-78, emeritus, head dept. civil engring., 1973-78; Mem. adv. com. on reactor safeguards Nuclear; Regulatory Commn., 1968—, chmn., 1972. Recipient award Concrete Reinforcing Steel Inst., 1956. Mem. ASCE (hon. mem., Research prize 1956, Howard medal 1968, Reese award 1970), Nat. Acad. Engring., Am. Concrete Inst. (pres. 1974—, Wason medal 1949, Turner medal 1964, hon. mem.), Reinforced Concrete Research Council (chmn. 1968-80, Boase award 1974), Internat. Assn. Bridge and Structural Engring., Sigma Xi, Tau Beta Pi, Phi Kappa Phi, Omicron Delta Kappa, Gamma Alpha, Chi Epsilon. Subspecialties: Civil engineering; Nuclear engineering. Current work: Pathology of structural concrete; safety of nuclear power plants. Research reinforced and prestressed concrete structures and hwy. bridges. Home: 805 Hamilton Dr Champaign IL 61820 Office: Newmark Lab 208 N Romine St Urbana IL 61801

SIEVER, RAYMOND, educator; b. Chgo., Sept. 14, 1923; s. Leo and Lillie (Katz) S.; m. Doris Fisher, Mar. 31, 1945; children—Larry Joseph, Michael David. B.S., U. Chgo., 1943, M.S., 1947, Ph.D., 1950; M.A. (hon.), Harvard U., 1960. With Ill. Geol. Survey, 1943-44, 47-56, geologist, 1953-56; research asso., NSF sr. postdoctoral fellow Harvard U., 1956-57, mem. faculty, 1957—, prof. geology, 1965—, chmn. dept. geol. scis., 1968-71, 76-81; assoc. geology Woods Hole (Mass.) Oceanographic Instn., 1957-65; cons. to industry and govt., 1957—. Author: (with others) Geology of Sandstones, 1965, Sand and Sandstone, 1972, Earth, 3d edit, 1982, Planet Earth, 1974, Energy and Environment, 1978; also numerous articles, papers. Served with USAAF, 1944-46. Recipient Best Paper award Soc. Econ. Paleontologists and Mineralogists, 1957; Pres.'s award Am. Assn. Petroleum Geologists, 1952. Fellow Am. Acad. Arts and Scis., AAAS, Geol. Soc. Am.; mem. Geochem. Soc. (pres. organic geochemistry group 1965), Am. Geophys. Union. Subspecialties: Sedimentology; Geochemistry. Current work: Diagenesis of sedimentary rock; geology of sandstones; geology and geochemistry of siliceous sediments and sedimentary geochemistry. Home: 38 Avon St Cambridge MA 02138 Office: Hoffman Lab Harvard Univ Cambridge MA 02138

SIEWIOREK, DANIEL PAUL, computer science researcher, consultant, educator; b. Cleve., June 2, 1946; s. Frank Stanley and Lena Mae (Pozum) S.; m. Karon Diane Walker, Aug. 5, 1972; children: Nora Ann, Gail Marie. B.S.E.E., U. Mich., 1968; M.S.E.E., Stanford U., 1969, Ph.D., 1972. Asst. prof. Carnegie-Mellon U., Pitts., 1972-76, assoc. prof., 1976-79, prof. computer sci. and elec. engring., 1979—; cons. engr. Digital Equipment Corp., Maynard, Mass., 1975—, Research Triangle Inst., Research Triangle Park, N.C., 1978—, Bendix Corp., Southfield, Mich., 1978—; mem. adv. com. United Technologies, Hartford, Conn., 1978—; reviewer Harper & Row Pubs., N.Y.C., 1982—. Author: (with H. Stone) Introduction to Computer Organization and Data Structures, 1975, (with G. Bell and A. Newell) Computer Structures: Principles and Examples, 1982, (with R. Swarz) The Theory and Practice of Reliable System Design, 1982, (with M. Barbacci) The Design and Analysis of Instruction Set Processors, 1982. Cons. Pitts. Pub. Schs. Computer Sci. Magnet Program, 1979; chmn. judging math. div. Pitts. Regional Sch. Sci. and Engring. Fair, Pitts., 1972—; judge Brashear High Sch. Super Bowl, Pitts., 1982—. John Huntington Found. scholar, 1964-65; Jonathan Logan scholar, 1964-68; Tektronix fellow, 1968; Scaife Found. faculty grantee, 1972. Fellow IEEE (chmn. tech. com. on fault tolerant computing 1981—, disting. visitor 1981, 82); mem. Assn. Computing Machinery (bd. dirs. spl. interest group on computer architecture 1974-78), Marine Tech. Soc., Assn. Computing Machinery Communications (editor computer systems dept. 1972-78), Sigma Xi, Tau Beta Pi, Eta Kappa Nu. Roman Catholic. Subspecialties: Computer architecture; Computer-aided design. Current work: Computer architecture, reliability modeling, fault-tolerant computing, modular design and design automation. Inventor iterative dell switch design for hybrid redundancy. Office: Carnegie-Mellon U Dept Computer Sci Pittsburgh PA 15213 Home: 1259 Bellerock St Pittsburgh PA 15217

SIGINER, AYDENIZ, engineering educator, researcher; b. Ankara, Turkey, July 10, 1943; came to U.S., 1976; s. Kazim Musa and Emine (Turkoz) S. Dr.Sc., Tech. U. Istanbul, Turkey, 1971; Ph.D., U. Minn., 1981. Asst. prof. civil engring. Tech. U. Istanbul, 1971-73; research and teaching assoc. U. Minn., Mpls., 1976-81; asst. prof. engring. mechanics U. Ala., Tuscaloosa, 1981—. Author: Handbook of Hydraulics, 1971, Problems in Hydraulics, 1978. Served to lt. C.E. Turkish Army, 1975. Sci. and Tech. Research Council Turkey fellow, 1969-71; recipient Henry Charles Ratcliff award U. Ala., 1983. Mem. ASME, NY Acad. Scis., Soc. Engring. Sci., Sigma Xi. Subspecialties: Fluid mechanics; Theoretical and applied mechanics. Current work: Research in mechanics of non-Newtonian fluids and fluid Mechanics in general. Home: 4317 Crabbe Rd Northport AL 35476 Office: Dept Engring Mechanics U Ala Box 2908 University AL 35486

SIKARSKIE, DAVID LAWRENCE, engineering educator; b. Marquette, Mich., Aug. 3, 1937; s. David Lawrence and Anna Dorothy (Price) S.; m. Marcia Gail Blum, June 22, 1957; children—David Lawrence, Anya Marie, Paul Benjamin. B.S. in Civil Engring, U. Pa., 1959; M.S. in Applied Mechanics, Columbia U., 1960, Sc.D., 1964. Registered profl. engr., Mich. Mem. tech. staff Ingersoll Rand Co., Princeton, N.J., 1963-66; vis. lectr. Princeton U., 1965-66; prof. aerospace engring. U. Mich., Ann Arbor, 1966-79; prof., chmn. dept. metallurgy, mechanics and materials sci. Mich. State U., East Lansing, 1979—; 1Vice pres., dir. William Ryder & Assocs., St. Clair Shores, Mich., 1966-73; cons. Tension Structures Inc., Plymouth, Mich., 1970-77; cons. various cos. Author, editor: Rock Mechanics Symposium, 1973; contbr. articles to profl. jours. Guggenheim Found. fellow, 1959; NDEA fellow, 1960. Mem. ASME, AAAS, Soc. Engring. Sci., Sigma Xi, Tau Beta Pi. Subspecialty: Theoretical and applied mechanics. Current work: Biomechanics, composite materials, rock mechanics. Office: Mich State U 330 Engring Bldg East Lansing MI 48824 Home: 4824 Buttercup Ln Okemos MI 48864

SIKOV, MELVIN RICHARD, developmental toxicologist; b. Detroit, July 8, 1928; s. Paul Merrill and Emma (Perlman) S.; m. Shirley Dressler, June 1, 1952; children: Peter H., Stacy J., Thomas R. B.S., Wayne State U., 1951; Ph.D., U. Rochester, 1955. Asst. prof. radiobiology Coll. Medicine, Wayne State U., Detroit, 1955-60, assoc. prof., 1961-65; sr. research scientist, biology dept. Battelle Pacific N.W. Lab., Richland, Wash., 1965-68, research assoc., 1968-78, mgr. devel. toxicology, 1978-81, sr. staff scientist, 1981—. Contbr. articles to profl. jours. Served with U.S. Army, 1946-47. Mem. Am. Inst. Ultrasound in Medicine, Am. Assn. for Cancer Research, Am. Assn. Pathologists, Health Physics Soc., Radiation Research Soc., Soc. for Exptl. Biology and Medicine, Soc. Toxicology, Teratology Soc. Subspecialties: Toxicology (medicine); Teratology. Current work: Age and environmental factors in radionuclide metabolism and effects; effects of ionizing and non-ionizing radiations and chemicals on embryonic and postnatal development and on tumorigenesis; biomedical imaging. Office: PO Box 999 Richland WA 99352

SILBERBERG, DONALD H., neurologist, educator; b. Washington, Mar. 2, 1934; s. William Aaron and Leslie Frances (Stone) S.; m. Marilyn Alice Damsky, June 7, 1959; children: Mark, Alan. M.D., U. Mich., 1958; M.A. hon.), U. Pa., 1971. Diplomate: Am. Bd. Neurology. Mem. faculty dept. neurology U. Pa. Sch. Medicine, Phila., 1963—, prof., 1971—, chmn. dept., 1982—; mem. numerous govtl. and pvt. non-profit health agy. coms. Contbr. articles to profl. jours.; mem. editorial bd.: Annals of Neurology, 1981—. Served with USPHS, 1951-61. Mem. Am. Acad. Neurology, Am. Neurol. Assn., Am. Assn. Neuropathologists, Assn. Research in Nervous and Mental Disease, Internat. Brain Research Orgn., Internat. Soc. Neurochemistry, Soc. Neurosci., Tissue Culture Assn. Subspecialties: Neurology; Neurobiology. Current work: Etiology and pathogenesis of demyelinating disease; normal development of and role of the oligodendrocyte in myelination, demyelination and remyelination; studies of myelination in vitro in organ cultures of mammalian CNS. Office: 3400 Spruce St Philadelphia PA 19104

SILBERBERG, REIN, research physicist, educational administrator; b. Tallinn, Estonia, Jan. 15, 1932; s. Juri and Elisabeth (Linkvest) S.; m. Ene-Liis, Aug. 28, 1965; children: Hugo, Ingrid. Ph.D. in Physics, U. Calif.-Berkeley, 1960. Grad. research asst. U. Calif.-Berkeley, 1957-60; NRC-NAS postdoctoral research assoc. Naval Research Lab., Washington, 1960-62, resarch physicist, 1962-81; acting chief scientist Lab. for Cosmic Ray Physics, 1981—; asst. dir. Internat. Sch. Cosmic Ray Astrophysics. Contbr. chpts to books, articles to sci. publs. Active Philos Soc., Washington. Recipient Naval Research Lab. publ. award, 1970, 75, 76; Meritorious Civil Service award U.S. Navy, 1980; Sigma Xi pure sci. award, 1981. Fellow Am. Phys. Soc.; mem. Am. Astron. Soc., Internat. Astron. Union, Am. Geophys. Union, Radiation Res. Soc., Sigma Xi. Subspecialties: Cosmic ray high energy astrophysics; Nuclear physics. Current work: Cosmic-ray transformations, nuclear astrophysics, neutrino astrophysics, nuclear spallation cross sections, active galactic nuclei. Home: 7507 Hamilton Spring Bethesda MD 20817 Office: Naval Research Lab Code 4154 Washington DC 20375

SILER, WILLIAM MACDOWELL, biomedical computer scientist, biomathematician; b. Houston, Aug. 5, 1920; s. Leon McDowell and Kathleen (Thompson) S.; m. Sylvia Parnell McDaniel, Nov. 9, 1974; children: James, Silliam, Sandra. M.S., Stevens Inst., Hoboken, N.J., 1949; Ph.D., Cuny, 1972. Asst. attending physicist Meml. Sloan Kettering Cancer Ctr., N.Y.C., 1959-65; Chmn. biomed. computer sci. dept. Downstate Med. Ctr., SUNY, Bklyn., 1965-72; prof. biomath. U. Ala., Birmingham, 1972-80, adj. prof., 1982—; dir. clin. computing Carraway Meth. Med. Ctr., Birmingham, 1980—; cons. NIH, mem. computer research study sect., 1969-74; mem. biostatics and epidemiology contract rev. com. Nat. Cancer Inst., 1982—. Editor: Computers in Life Science Research, 1975; contbr. articles in field to profl. jours. Served to 2d lt. AC U.S. Army, 1942-46. Mem. Biomed. Engrng. Soc., Soc. Math. Biology, IEEE Computer Soc., Am. Assn. Physicists in Medicine. Methodist. Subspecialties: Biomedical engineering; Graphics, image processing, and pattern recognition. Current work: Computers in biomedical research, biomathematics, computers, image processing, simulation, mathematical modelling. Home: 417 22d Ave S Birmingham AL 35205 Office: 1500 26th St N Birmingham AL 35234

SILL, WEBSTER HARRISON, JR., plant pathologist; b. Wheeling, W. Va., Dec. 4, 1916; s. W.H. and Grace (Clark); m. Charlyn A. Adams, July 25, 1943; children: Webster H. III, Susan C., Warren. B.S., W. Va. Wesleyan U., 1939; M.S., Pa. State U., 1942; M.A., Boston U., 1947; Ph.D., U. Wis.-Madison, 1951. Research assoc. U. Wis., 1951; faculty Kans. State U., 1952-60; prof., chmn. dept. biology U. S.D., 1969-81; sci. resident NSF, 1982—. Author: Plant Protection Discipline, 1978, Plant Protection, 1982; contbr. articles to profl. jours. Served with USMC, 1941-46. Mem. Am. Phytopathol. Soc., AAAS, Am. Assn. Biol. Scis., Am. Assn. Agrl. Cons., Sigma Xi, Phi Sigma, Alpha Delta Gamma, Phi Sigma. Republican. Methodist. Subspecialties: Plant virology; Integrated pest management. Current work: Virus diseases of small grains and grasses. Office: Biology Dep U SD Vermillion SD 57069 Home: 515 N Prentis Vermillion SD 57069

SILLMAN, ARNOLD JOEL, physiology educator, scientist; b. N.Y.C., Oct. 10, 1940; s. Philip and Anne L. (Pearlman) S.; m. Jean Fletcher Van Keuren, Sept. 26, 1969; children: Andrea Jose, Diana Van Keuren. B.A., UCLA, 1963, M.A., 1965, Ph.D., 1968. Asst. prof. ophthalmology UCLA, 1969-73; asst. prof. U. Calif., Davis, 1975-78, assoc. prof., 1978—; asst. prof. U. Pitts., 1973-75; cons. Bio-Med. Sci. Assocs., Davis. Contbr. articles to profl. jours. USPHS trainee, 1966-67; NSF fellow, 1967-68; Fight For Sight, Inc. fellow, 1968-69; NIH grantee, 1977—. Mem. Am. Physiol. Soc., Soc. Gen. Physiologists, Assn. Research in Vision and Ophthalmology, Am. Soc. Zoologists. Jewish. Subspecialties: Neurophysiology; Physiology (biology). Current work: Physiology, biochemistry and biology of the vertebrate visual system with special interest in the visual pigments, adaptation and transduction in photoreceptors; current research concerns the effects of heavy metals on photoreceptor function. Home: 1140 Los Robles St Davis CA 95616 Office: University of California Department of Animal Physiology Davis CA 95616

SILSBY, GRAHAM FORBES, mech. engr., govt. researcher; b. Richmond, Va., Apr. 7, 1943; s. Howard Wiswell II and Eleanor (Foltz) S.; m. Patricia Shields; m. Louise Rose, May 12, 1979; 1 son, Jeffrey Allen Shumate. B.S.M.E., Carnegie Inst. Tech., 1965; B.S.C.E., Johns Hopkins U., 1978. Jr. mech. engr. Farrington Electronics Inc., Springfield, Va., 1965-66; mech. engr. U.S. Army Nuclear Def. Lab., Edgewood Arsenal, Md., 1966-68, Dept. of Army—U.S. Army Nuclear Def. Lab., Edgewood Arsenal, 1968-72; research mech. engr. Dept. of Army—Ballistic Research Lab., Aberdeen Proving Ground, Md., 1972—. Mem. corp. bd. United Ch. of Christ Bd. for Homeland Ministries, N.Y.C., 1975-81. Served with U.S. Army, 1966-68. Recipient Am. Spirit Honor medal Citizen's Com. for Army Navy and Air Force, 1967. Mem. ASME. Subspecialties: Mechanical engineering. Current work: Mechanics of high speed impact; research into mechanics of interaction of kinetic energy penetrators and armors at increased velocities.

SILVA, J. ENRIQUE, physician, educator, researcher; b. Santiago, Chile, June 17, 1943; came to U.S., 1980; s. Enrique and Isabel (Solovera) S.; m. Eugenia M. Santelices, Mar. 5, 1966; children: Liliana, Enrique, Gonzalo. M.D., U. Chile, 1968. Diplomate: Am. Bd. Internal Medicine. Resident in internal medicine U. Chile Sch. Medicine, Santiago, 1968-71, instr. medicine, 1971-73, asst. prof. medicine, 1973-74; fellow in endocrinology Montefiore Hosp. Med. Ctr., Bronx, N.Y., 1974-76; research fellow in medicine Peter Bent Brigham Hosp./Harvard Med. Sch., Boston, 1976-77; asst. prof. medicine Harvard Med. Sch., Boston, 1980—; physician Brigham and Women's Hosp., Boston, 1980; assoc. investigator Howard Hughes Med. Inst., Boston, 1980—. Author research papers, chpts. and books. Mem. Endocrine Soc., N.Y. Acad. Sci., Am. Thyroid Assn., Latin Am. Thyroid Assn. Subspecialties: Receptors; Endocrinology. Current work: Regulation of thyroid function and mechanism of thyroid hormone action. Home: 30 Ferdinand St Melrose MA 02176 Office: Brigham and Women's Hosp Thyroid Diagnostic Ctr 75 Francis St Boston MA 02115

SILVER, DONALD, educator, surgeon; b. N.Y.C., Oct. 19, 1929; s. Herman and Cecilia (Meyer) S.; m. Helen Elizabeth Harnden, Aug. 9, 1959; children: Elizabeth Tyler, Donald Meyer, Stephanie Davies, William Paige. A.B., Duke U., 1950, B.S. in Medicine, M.D., 1955. Diplomate: Am. Bd. Surgery, Am. Bd. Thoracic Surgery. Intern Duke Med. Center, 1955-56, asst. resident, 1958-63, resident, 1963-64; mem. faculty Duke Med. Sch., 1964-75, prof. surgery, 1972-75; cons. Watts Hosp., Durham, 1965-75, VA Hosp., Durham, 1970-75, chief surgery, 1968-70; prof. surgery, chmn. dept. U. Mo. VA Med. Center, Columbia, 1975—; cons. Harry S. Truman Hosp., Columbia, 1975—; mem. bd. sci. advisers Cancer Research Center, Columbia, 1975—; mem. surg. study sect. A NIH. Contbr. articles to med. jours., chpts. to books. Served with USAF, 1956-58. James IV Surg. traveler, 1977. Fellow ACS, Deryl Hart Soc.; mem. Am., Mo. med. assns., Boone County Med. Soc., AAAS, Internat. Cardiovascular Soc., Soc. Univ. Surgeons, Am. Heart Assn. (Mo. affiliate research com.), Soc. Surgery Alimentary Tract, Assn. Acad. Surgery, So. Thoracic Surg. Assn., Internat. Soc. Surgery, Soc. Vascular Surgery, Am. Assn. Thoracic Surgery, Am., Central, Western surg. assns., Phi Beta Kappa, Alpha Omega Alpha. Subspecialties: Surgery; Physiology (medicine). Current work: Quantitation of peripheral blood flow; intravascular fibrinolysis; heparin-induced anti-platelet antibodies. Home: 1050 W Covered Bridge Rd RD 4 Columbia MO 65201 Office: Dept Surgery M580 Univ Mo Med Center Columbia MO 65212

SILVER, ELI ALFRED, geology educator; b. Worcester, Mass., June 3, 1942; s. Benjamin and Evelyn (Wellin) S.; m. Mary Wilcox, July 2, 1967; children: Monica Amy, Joel Benjamin. A.A., San Francisco City Coll., 1962; B.A., U. Calif.-Berkeley, 1964. Registered geophysicist, Calif. Editorial bd. Geology, 1979—. Research asst. Scripps Inst. Oceanography, LaJolla, Calif., 1964-69, postgrad researcher, 1969-70; geologist U.S. Geol. Survey, Menlo Park, Calif., 1970-73; asst. prof. U. Calif., Santa Cruz, 1973-75, assoc. prof., 1975-79, prof. geology, 1979—. Fellow Geol. Soc. Am.; mem. Am. Geophys. Union, Soc. Exploration Geophysicists, Seismol. Soc. Am., AAAS, Phi Beta Kappa. Subspecialties: Tectonics; Sea floor spreading. Current work: Marine geophysical studies of active subduction and collision zones; development of mountain belts. Home: 250 Sheldon Ave Santa Cruz CA 95060 Office: Earth Scis Bd Univ Calif Santa Cruz CA 95064

SILVER, HENRY K., pediatrician, educator; b. Phila., Apr. 22, 1918; s. Samuel and Dora (Kreitzer) S.; m. Harriet Ashkenas, June 15, 1941; children: Stephen, Andrew. B.A., U. Calif., Berkeley, 1938, M.D., 1942. Diplomate: Am. Bd. Pediatrics. Intern U. Calif. Hosp., San Francisco, 1941-42; resident Children's Hosp., Phila., 1942-43; instr., then asst. prof. pediatrics U. Calif. Med. Sch., San Francisco, 1946-52; asso. prof. Yale Med. Sch., 1952-57; prof. pediatrics U. Colo. Med. Sch., Denver, 1957—, asso. dean admissions, 1977—; clin. prof. nursing U. Colo. Sch. Nursing, 1976—. Co-author: Healthy Babies-Happy Parents, 1958, Current Pediatric Diagnosis and Treatment, 7th edit., 1982, Handbook of Pediatrics, 14th edit., 1982. Rosenberg Found. fellow, 1945-47; recipient George Armstrong award Ambulatory Pediatric Assn., 1972; Martha May Eliot award Am. Pub. Health Assn., 1974; Eleanor Roosevelt Humanitarian award Denver chpt. Hadassah, 1973. Mem. Inst. Medicine of Nat. Acad. Sci., Am. Acad. Pediatrics, Western Soc. Pediatric Research (Ross award in edn. 1962), Soc. Pediatric Research, Am. Pediatric Soc., Sigma Xi, Alpha Omega Alpha. Subspecialty: Pediatrics. Current work: Development of health manpower programs and health education. Developer of first nurse practitioner program and child health associate program in U.S.; first to describe Silver's Syndrome, 1953, Deprivation Dwarfism, 1967. Home: 135 S Ivy St Denver CO 80224 Office: 4200 E 9th Ave Denver CO 80262

SILVER, HULBERT KEYES BELFORD, physician; b. Montreal, Que., Can., July 14, 1941; s. Arthur Desraeli and Norah Joanna (Belford) S.; m. Susan Daphne Andrew, Oct. 8, 1967; children: Hulbert, Signe, William. B.Sc., Bishop's U., Lennoxville, Que., 1962; M.D., McGill U., Montreal, 1966, Ph.D., 1974. Asst. prof. UCLA Sch. Medicine, 1973-76; asst. prof. medicine U. B.C., Vancouver, 1976-79, assoc. prof., 1979—; med. oncologist Cancer Control Agy. B.C., 1976—. Contbr. articles to profl. jours. Served with Royal Can. Air Force, 1960-62. Recipient medal in medicine Royal Coll. Physicians and Surgeons Can., 1975. Fellow ACP, Royal Coll. Physicians and Surgeons Can.; mem. Am. Soc. Clin. Oncology, Am. Assn. Cancer Research, Can. Oncology Assn. Subspecialties: Cancer research (medicine); Oncology. Current work: Tumor markers, biologic response modifiers, immunology. Office: Cancer Control Agy British Columbia 2656 Heather St Vancouver BC Canada V5Z 3J3

SILVER, RICHARD TOBIAS, physician; b. N.Y.C., Jan. 18, 1929; s. Benjamin and Pauline (Nussbaum) S.; m. Barbara Polack, Jan. 18, 1963; 1 son, Adam Bennett. A.B., Cornell U., 1950, M.D., 1953. Diplomate: Am. Bd. Internal Medicine, Am. Bd. Clin. Oncology. Intern in medicine N.Y. Hosp.-Cornell U. Med. Ctr., 1953-54, asst. resident and resident in medicine and hematology, 1956-58; clin. assoc. medicine br. Nat. Cancer Inst., 1954-56; mem. faculty Cornell U. Med. Coll., 1956—, clin. prof. medicine, 1973—; vis. Fulbright prof. U. Bahia (Brazil) Sch. Medicine, 1958-59, chief oncology div., 1976—; attending physician N.Y. Hosp., 1973—; mem. rev. bds. NIH; cons., vis. lectr., prof. throughout world. Author numerous articles in field. Served with USPHS, 1954-56. Mem. ACP, Internat. Soc. Hematology, Am. Soc. Hematology, N.Y. Med. Soc. Study Blood, N.Y. Med.Soc., N.Y. County Med. Soc., Harvey Soc., Am. Fedn. Clin. Research, AMA, Am. Soc. Clin. Oncology, Am. Assn. Cancer Research, Explorers Club (dir. 1976-78, chmn. sci. adv. com. 1980), Sigma Xi. Subspecialties: Chemotherapy. Current work: Cancer research.

SILVER, ROBERT BENJAMIN, biochemistry educator, biomed. researcher; b. Chgo., Aug. 9, 1950; s. Seymour Martin and Mary (Brodsky) S. B.S. in Biology, Ill. Inst. Tech., 1973; Ph.D. in Zoology, U. Calif.-Berkeley, 1977. Am. Cancer Soc. postdoctoral research fellow dept. biochemistry U. Calif.-Berkeley, 1977-80, lectr., 1978-79; asst. prof. biochemistry dept. biochemistry Univ. Health Scis., Chgo. Med. Sch., 1981—, dir. electron microscope lab.; cons. in field. Contbr. articles on biochemistry to profl. jours. Am. Cancer Soc. grantee, 1981-83; Chgo. Heart Assn. grantee, 1981-83; STEPS fellow Marine Biol. Lab., Woods Hole, Mass., 1983. Mem. Am. Soc. Cell Biology, Soc. for Devel. Biology, Am. Soc. Zoologists, AAAS, Sigma Xi. Subspecialties: Cell and tissue culture; Cancer research (medicine). Current work: Role of membranes and calcium regulation in the assembly and functioning of the mitotic apparatus in cell division. Office: 3333 N Green Bay Rd North Chicago IL 60064

SILVER, SIMON DAVID, biochemist educator; b. Detroit, June 22, 1936; s. Paul and Molly (Chaenko) S.; m. Margaret Neave, Dec. 22, 1958; children: Andrew, Stephen. B.A., U. Mich., 1957; Ph.D., M.I.T., 1962; postgrad. (fellow), U. London, 1962-64. Asst. research biophysicist U. Calif.-Berkeley, 1964-66; asst. to full prof. dept. biology Washington U., St. Louis, 1966—, prof. biology and microbiology/immunology, 1976—; vis. fellow dept. biochemistry John Curtin Sch. Med. Research, Australian Nat. U., Canberra, 1974-75, 79-80. Editor, then editor-in-chief: Jour. Bacteriology, 1976—, NSF Metalbolic Biology Panel, 1980-83; contbr. numerous articles to profl. jours. Mem. Am. Soc. Microbiology, Am. Soc. Biol. Chemists, Genetics Soc. Am., Biophys. Soc., Soc. Gen. Microbiology, AAUP. Subspecialties: Biochemistry (biology); Microbiology. Current work: Mechanisms of transport systems in bacterial membranes, genetics and mechanisms of resistances to heavy metals in bacterial plasmids. Office: Biology Dept Washington U Saint Louis MO 63130

SILVERBERG, MICHAEL, research scientist, educator; b. London, Feb. 24, 1948; U.S., 1973; s. Morris and Rebecca (Weiner) S. B.A., U. Oxford, Eng., 1970, M.A., 1973, Ph.D., 1973. Research assoc. Yale U., 1973-76; research assoc. SUNY-Stony Brook, 1976-78, asst. prof. medicine, 1978—. Nat. Heart Lung and Blood Inst. grantee, 1983. Mem. Am. Heart Assn., Am. Assn. Pathologists, Am. Soc. Hematology, N.Y. Acad. Scis. Subspecialties: Biochemistry (medicine); Hematology. Current work: Molecular events associated with the initiation of blood clotting and inflammation. Office: SUNY Health Science Center Tl6-040 Stony Brook NY 11794

SILVERMAN, ALBERT JACK, psychiatrist, educator; b. Montreal, Que., Can., Jan. 27, 1925; came to U.S., 1950, naturalized, 1955; s. Norman and Molly (Cohen) S.; m. Halina Weinthal, June 22, 1947; children: Barry Evan, Marcy Lynn. B.Sc., McGill U., 1947, M.D., C.M., 1949; grad., Washington Psychoanalytic Inst., 1964. Diplomate: Am. Bd. Psychiatry and Neurology. Intern Jewish Gen. Hosp., Montreal, 1949-50; resident psychiatry Colo. U. Med. Center, 1950-53, instr., 1953; from assoc. to assoc. prof. psychiatry Duke Med. Center, 1953-63; prof. psychiatry, chmn. dept. Rutgers U. Med. Sch., 1964-70; prof. psychiatry U. Mich. Med. Sch., Ann Arbor, 1970—, chmn. dept., 1970-81; cons. to area hosps. Mem. biol. scis. tng. rev. com. NIMH, 1964-69, chmn., 1968-69, mem. research scientist devel. award com., 1970-75, chmn., 1973-75; mem. merit rev. bd. in behavioral scis. VA, 1975-78, chmn., 1976-78; Bd. mgrs. N.J. Neuropsychiat. Inst., 1965-69; trustee N.J. Fund Research and Devel. Nervous and Mental Diseases, 1965-67; bd. dirs. N.J. Mental Health Assn., 1964-69; mem. behavioral sci. com. Nat. Bd. Med. Examiners, 1978—. Cons. editor: Psychophysiology, 1970-74, Psychosomatic Medicine, 1972—; Contbr. articles in field. Served as capt. M.C. USAF, 1955-57. Fellow Am. Coll. Psychiatry (charter), Am. Psychiat. Assn. (chmn. council on med. edn. 1970-75), Am. Acad. Psychoanalysis, Am. Coll. Neuropsychopharmacology; mem. Am. Psychosomatic Soc. (council 1964-68, 70-74, pres. 1976-77), N.J. Psychoanalytic Soc. (trustee 1968-70), Assn. Research Nervous and Mental Diseases, N.J. Neuropsychiat. Assn. (council 1966-69), Group Advancement Psychiatry (chmn. com. psychopathology 1968-74), Soc. Psychophys. Research, Soc. Biol. Psychiatry, Mich. Psychiat. Soc. (council 1975-77), Mich. Psychoanalytic Soc. Subspecialties: Psychiatry; Psychophysiology. Current work: Physical differences in field dependency; brain lateralization; pre-clinical and clinical psychiatric studies. Home: 19 Regent Dr Ann Arbor MI 48104

SILVERMAN, BARRY GEORGE, engineering educator, researcher; b. Boston, July 5, 1952; s. Joseph M. Silverman and Matilda Falcon; m. Fern Linda Margolis, May 28, 1978; 1 dau., Rachel. B.S.E., U. Pa., 1975, M.S.E., 1977, Ph.D., 1977. Asst. prof. George Washington U., Washington, 1978-80, assoc. prof., 1980—; mem. adv. bd. Def. Systems Mgmt. Coll., Ft. Belvoir, Va., 1981—; tech. advisor Senator John Glenn, Washington, 1983—. Editor-in-chief: Triangle Mag, 1974-75; mem. editorial bd.: Jour. Tech. Mgmt, 1982—; assoc. editor procs.: Pitts. Conf. on Simulation, 1981; contbr. articles to profl. jours.; reviewer articles for profl. jours.; creator numerous software products. Bd. dirs. Holmes Run Civic Assn., Fairfax, Va., 1980-82; mem. Council of Leadership United Jewish Appeal, Washington, 1980—. U.S. Dept. Interior grantee, 1977; Office Sci. and Tech. Policy grantee, 1978; Office of Tech. Assessment grantee, 1980; NASA grantee, 1981—; Dept. Def. grantee, 1982—. Mem. Inst. Mgmt. Sci., IEEE, Assn. Computing Machinery, Inst. Indsl. Engrs., Sigma Xi. Subspecialties: Mathematical software; Systems engineering. Current work: Major focus on design, assessment and evaluation of computer models and software for decision makers; current focus on development of automated aids for analogical reasoning in systems engineering and technological innovation. Developed, tested, published theory on how experts use analogy in reasoning and problem solving. Office: George Washington U nlLibrary 636A Washington DC 20052

SILVERMAN, DAVID NORMAN, biochemist, educator; b. South Bend, Ind., July 22, 1942; s. Maurice and Edith (Coffman) S.; m. Carole Sue Katz, Aug. 21, 1966; children: Jennifer, Allison. B.S., Mich. State U., 1964; Ph.D., Columbia U., 1968. Research assoc., instr. Cornell U., Ithaca, N.Y., 1969-71; asst. prof. pharmacology U. Fla., Gainesville, 1971-76, assoc. prof., 1976-80, prof. pharmacology and biochemistry, 1980—. Mem. Am. Chem. Soc., Am. Soc. Biol. Chemists, Am. Soc. Pharmacology and Exptl. Therapeutics. Subspecialties: Biophysical chemistry; Molecular pharmacology. Current work: Catalytic mechanism of enzymes, nuclear magnetic resonance. Address: U Fla Box J267 Health Center Gainesville FL 32610

SILVERMAN, HIRSCH LAZAAR, clinical and forensic psychologist; b. N.Y.C., June 19, 1915; s. Herman Bear and Ida (Mackta) S.; m. Mildred Friedlander, Mar. 1, 1942; children: Hyla Susan. Morton Maier, Stuart Edward. B.Sc., CCNY, 1936, M.Sc., 1938; M.A., NYU, 1947; Ph.D., Yeshiva U., 1951; M.A., Seton Hall U., 1956; D.Sc. (hon.), Lane Coll., 1962; LL.D., Fla. Meml. Coll., 1965; L.H.D., Ohio Coll. Podiatric Medicine, 1972. Diplomate: Am. Bd. Profl. Psychology, Am. Bd. Forensic Psychology, Am. Acad. Behavioral Medicine., Am. Acad. Neuropsychology, Am. Assn. Clin. Counselors; cert. profl. psychotherapist; cert. family psychologist. Asst. prof. philosophy and psychology, chmn. dept. Philosophy Mohawk Coll., 1946-48; ednl. and vocational cons., psychol. lab. Stevens Inst. Tech., 1948-49; asst. prof. psychology Rutgers U., 1949-53; asst. to supt. schs., dir. psychol. services Nutley (N.J.) Bd. Edn., 1953-59; chmn. dept. ednl. and sch. psychology Yeshiva U. Grad. Sch., 1959-61, prof. psychology grad. div., 1959-65; prof. ednl. adminstrn., chmn. dept. Seton Hall U., 1965-80, prof. emeritus, 1980—; research clin. psychologist Columbus Hosp., Newark, 1963-72; psychol. cons. N.J. Rehab. Hosp., East Orange, N.J.; vis. prof. psychology Lane Coll., 1961—, Fla. Meml. Coll., 1961—; research clin. psychologist N.Y. Med. Coll., 1961-65; clin. psychologist, med. staff St. Vincent's Hosp., Montclair, N.J.; v.p. Am. Bd. Forensic Psychology, 1978-80. Author 15 books, numerous articles in field. Chmn. N.J. Bd. Marriage Counsel Examiners, 1968-74; mem. N.J. Mental Health Council, 1966-69, N.J. Bd. Psychol. Examiners, 1975—; chmn. N.J. Bd. Psychol. Examiners, 1979-82; mem. nat. adv. com. White House Conf. on Families.; Bd. dirs. Family Service Bur., Newark, 1961-82. Served with U.S. Army, 1942-46. Decorated Army Commendation medal; grand cross St. John of Jerusalem; Maltese cross Order St. John Denmark. Fellow Coll. Preceptors Eng., Philos. Soc. Eng. (hon. v.p.), Royal Soc. Health, Royal Soc. Arts, Gerontological Soc., AAAS, Assn. Advancement Psychotherapy, Am. Assn. Social Psychiatry, Am. Orthopsychiat. Assn., Assn. Advancement Psychotherapy, Am. Psychol. Assn., Am. Med. Writers Assn., Am. Pub. Health Assn., Am. Assn. Psychiat. Services for Children, Royal Soc. Medicine, World Acad. Art and Sci.; mem. N.J. Assn. Marriage and Family Counselors (pres. 1967-70), Acad. Psychology in Marital Therapy (pres. 1964-67), Essex County Psychol. Assn. (exec. com., sec., editor), Internat. Council Psychologists (treas. 1971-73), Am. Coll. Psychology (adv. bd.), Phi Beta Kappa, Sigma Xi, Psi Chi, Phi Sigma Tau, Kappa Delta Pi, Phi Delta Kappa. Subspecialties: Cognition; Neuropsychology. Current work: Psychotherapy;forensic psychology; humanities; psychopetry and psycho-diagnostic assessments: lecturing and teaching in psychological services. Home: 123 Gregory Ave West Orange NJ 07052 Office: McQuaid Hall Seton Hall Univ South Orange NJ 07079 One's standard ultimately is to be measured not in gold, but in growth; not in position, but in personal inner power; not in capital, but in character.

SILVERMAN, JOSEPH, educator, scientist; b. N.Y.C., Nov. 5, 1922; s. Jakob and Mary (Chechick) S.; m. Joan Aline Jacks, Jan. 14, 1951; children: Joshua Henry, David Avrom. B.A., Bklyn. Coll., 1944; A.M., Columbia U., 1948, Ph.D., 1951. Head research dept. Walter Kidde (nuclear labs.), Garden City, N.Y., 1952-54; v.p., tech. dir. RAI Research Corp., L.I. City, N.Y., 1954-59; asso. prof. chemistry State U. N.Y., Stony Brook, 1959-60; prof. dept. chem. engring. U. Md., College Park, 1960—; dir. Inst. Phys. Sci. and Tech., 1976—; cons. Danish AEC, Indsl. Research Inst., Japan, Boris Kidric Inst., Yugoslavia, Bechtel Co., IAEA, Vienna; disting. vis. prof. Tokyo U.,

1974; gen. chmn. 2d Internat. Meeting on Radiation Processing, Miami, Fla., 1978, 3d Tokyo, 1980. Editor: Internat. Jour. Applied Radiation and Isotopes, 1973-78, Internat. Jour. Radiation Engring, 1970-73, Trans. 1st Internat. Meetings on Radiation Processing, 1977, 2d, 1979, 3d, 1981; mem. editorial adv. bd.: Radiation Physics and Chemistry, 1978—. Served with AUS, 1944-46. Research fellow Brookhaven Nat. Lab., 1949-51; Guggenheim fellow, 1966-67. Fellow Nordic Soc. Radiation Chemistry and Tech., Am. Phys. Soc., Am. Nuclear Soc. (Radiation Industry award 1975); mem. Am. Inst. Chemists, Am. Chem. Soc., Sigma Xi. Club: Cosmos. Subspecialties: Polymer chemistry; Nuclear engineering. Current work: Applied radiation chemistry of vinyl monomers and polymers; radiation sources technology. Home: 320 Sisson St Silver Spring MD 20902 Office: Inst Phys Sci and Tech U Md College Park MD 20742

SILVERMAN, PAUL HYMAN, university president, zoologist; b. Mpls., Oct. 8, 1924; s. Adolph and Libbie (Idlekope) S.; m. Nancy Josephs, May 20, 1945; children: Daniel Joseph, Claire Storms. B.S., Roosevelt U., 1949; M.S. in Parasitology and Ecology, Northwestern U., 1951; D.Sc., U. Liverpool, 1968; Ph.D. in Parasitology and Epidemiology, U. Liverpool, 1955. Research fellow Malaria Research Sta., Hebrew U., Israel, 1951-53; sr. sci. officer dept. parasitology Moredun Inst., Edinburgh, Scotland, 1956-59; head dept. immunoparasitology Allen & Hanbury, Ltd., Ware, Eng., 1960-62; prof. zoology and vet. pathology and hygiene U. Ill., Urbana, 1963-72, chmn. dept. zoology, 1964-65, head dept. zoology, 1965-68; sr. staff mem. Ctr. for Zoonoses Research, 1964; prof. biology, head div. natural scis. Temple Buell Coll., Denver, 1970-71; prof. biology, chmn. dept. biology, acting v.p. research, v.p. research and grad. affairs, assoc. provost for acad. services and research U. N.Mex., 1972-77; provost for research and grad. studies SUNY, Central Adminstrn., Albany, 1977-79; pres. Research Found., SUNY, Albany, 1979-80, U. Maine, Orono, 1980—; adj. prof. U. Colo., Boulder, 1970-72; Fulbright prof. zoology Australian Nat. U., Canberra, 1969; faculty appointee Sandia Corp., Dept. Energy, Albuquerque, 1974-81; project dir. research in malaria immunology and vaccination AID, 1965-76; project dir. research in Helminth immunity USPHS, NIH, 1964-72; sr. cons. Ministry Edn. and Culture, Brazil, 1975—; bd. dirs. Inhalation Toxicology Research Inst., Lovelace Biomed. and Environ. Research Inst., Albuquerque, 1977—; mem. N.Y. State Gov.'s High Tech. Opportunities Task Force; chmn. research and rev. com. N.Y. State Sci. and Tech. Found.; bd. advs. Lovelace-Bataan Med. Ctr., Albuquerque, 1974-77; chmn. Commn. on Instns. Higher Edn. North Central Assn. Colls. and Secondary Schs., 1974-76, cons. examiner, 1964—. Contbr. chpts. to books, articles to profl. jours.; mem. editorial bd. profl. jours. Bd. dirs. Historic Albany Found. Served with AUS, 1943-46. Fellow Royal Soc. Tropical Medicine and Hygiene, N.Mex. Acad. Scis.; mem. Am. Soc. Parasitologists, Am. Soc. Immunologists, Brit. Soc. Parasitology (council), Brit. Soc. Immunologists, Soc. Gen. Microbiology, Soc. Parotozoologists, Am. Soc. Zoologists, Am. Inst. Biol. Scis., AAAS, N.Y. Acad. Scis., N.Y. Soc. Tropical Medicine, Sigma Xi, Phi Kappa Phi. Subspecialties: Immunology (agriculture); Parasitology. Home: U Maine President's House Orono ME 04469 Office: U Maine 209 Alumni Hall Orono ME 04469

SILVERMAN, RICHARD BRUCE, chemistry educator; b. Phila., May 12, 1946; s. Philip and S. Ruth (Simon) S.; m. Judith Tellerman, Nov. 27, 1971 (div. Dec. 1978); m. Barbara Kesner, Jan. 9, 1983; children: Matthew, Margaret. B.S., Pa. State U., 1968; A.M., Harvard U., 1972, Ph.D., 1974. NIH postdoctoral fellow Brandeis U., Waltham, Mass., 1974-76; asst. prof. Northwestern U., Evanston, Ill., 1976-82, assoc. prof. chemistry, 1982—. Contbr. articles to profl. jours. Served with U.S. Army, 1969-71. Recipient Research Career Devel. award NIH, 1982-87; Young Faculty award E. I. duPont de Nemours & Co., 1976; Alfred P. Sloan research fellow, 1981-83; NIH postdoctoral fellow, 1974-76. Mem. Am. Chem. Soc., AAAS. Subspecialties: Medicinal chemistry; Biochemistry (biology). Current work: Medicinal and bioorganic chemistry, mechanism of action and design of drugs, specific enzyme inhibition, enzyme mechanisms. Office: Dept Chemistry Northwestern U Evanston IL 60201

SILVERN, LEONARD CHARLES, Systems engineering company executive; b. N.Y.C., May 20, 1919; s. Ralph and Augusta (Thaler) S.; m. Gloria Marantz, June, 1948 (div. 1968); 1 son, Ronald; m. Elisabeth Beeny, Aug., 1969 (div. 1972). B.S., L.I.U., 1946; M.A., Columbia U., 1948, Ed.D., 1952. Registered profl. engr., Calif. Tech. tng. supr. U.S. Dept. Navy, N.Y.C., 1939-49; tech. tng. dir. State Div. Safety, Albany, N.Y., 1949-55; resident engring., psychologist Rand Corp., MIT Lincoln Lab., Lexington, Mass., 1955-56; research engr., dir. Hughes Aircraft Co., Culver City, Calif., 1956-62; dir. engring. lab. Northrop Corp., Hawthorne, Calif., 1962-64; prin. scientist, v.p., pres. Edn. and Tng. Consultants Co., Los Angeles, 1964-80; v.p. Systems Engrinmg. Labs. div. ETC Co., Sedona, Ariz., 1980—; adj. prof. U. So. Calif., 1957-65; vis. prof. UCLA, 1963-72; cons. Hdqrs. USAF Air Tng. Command, Randolph Air Force Base, 1964-68, U. Hawaii, 1970-74, Centro Nacional de Productividad, Mexico City, 1973-75. Author, editor: book series Systems Engineering of Education, 1964—; contbg. editor: Educational Technology mag, 1968-73, 81—; Contbr. articles to profl. jours., chpts. to books. Dist. ops. officer Los Angeles County Sheriff's Dept. Disaster Commn., 1973-75, dist. communications officer, 1975-76. Served with USNR, 1944-46. Mem. IEEE (sr.), Am. Psychol. Assn., Soc. Wireless Pioneers, Quarter Century Wireless Assn., Friendship Veterans Fir Engine Co., Am. Radio Relay League, Radio Amateur Satellite Corp., Ariz. Archeol. Soc., Verde Valley Archeol. Soc., Sierra Club. Club: Westerners. Subspecialties: Systems engineering; Computer engineering. Current work: Developing computer-controlled radio communications systems incorporation cartography, meteorological satellite data, and technical data bases including space science, geoscience, social and behavioral science, military and political science, energy science and electronics; communications worldwide. Office: Systems Engring Labs ETC Co Sedona AZ 86336 Home: Golden Eagle Dr Sedona AZ 86336

SILVERSTEIN, CHARLES I., psychologist; b. Chgo., July 27, 1951; s. Milton and Ruth (Pazley) S. B.A., Calif. State U., 1973; M.A., San Jose State U., 1974; Ph.D., U. Oreg., 1977. Lic. psychologist, Calif. Rearch asst. NASA, Moffett Field, Calif., 1973-74; psychologist Fairview Hosp., Salem, Oreg., 1974-76; research fellow CORBEH, Eugene, Oreg., 1976-77; coordinator N.W. Human Research Ctr., Williston, N.D., 1977-78; dir. psychology Grafton (N.D.) State Sch., 1978-79; psychologist Sonoma State Hosp., Eldridge, Calif., 1979—; pvt. practice psychology, 1981—. Contbr. articles to profl. jours. Mem. Am. Psychol. Assn., Am. Assn. Mental Deficiency, Western Psychol. Assn. Subspecialties: Behavioral psychology; Cognition. Current work: Time perception, vitamin and nutritional treatment of developmental disabilities. Address: Sonoma State Hosp PO Box 33 Eldridge CA 95431

SILVERSTEIN, ELLIOT MORTON, physicist; b. Chgo., Jan. 2, 1928; s. Joseph and Jeanette (Scudder) S.; m. Irma Arenz, Jan. 27, 1957; children: Marc, Julie, Frances. B.A., U. Chgo., 1950, M.S., 1953, Ph.D., 1958. Mem. tech. staff, ops. analyst Hughes Aircraft Co., 1958-61; sr. staff physicist Surveyor Spacecraft Lab., Hugh Aircraft Co., 1961-68; prin. engr./scientist Electro-Optics Lab., McDonnell Douglas Astronautics Co., Huntington Beach, Calif., 1968-81; sr. mem. tech. staff Aerospace Corp., El Segundo, Calif., 1981—. Contbr. articles on electro-optics and aerospace to profl. jours. Served with USN, 1946-48. Mem. Optical Soc. Am. (standards com., awards com.), Optical Soc. So. Calif. (past pres.), Am. Phys. Soc., IEEE, AAAS, Wilderness Soc., Phi Beta Kappa, Sigma Xi. Subspecialties: Aerospace engineering and technology; Optical engineering. Current work: Infrared and optical physics, optical and electro-optical imaging, image processing, detection and tracking; also, design and analysis of spacecraft instrumentation. Home: 8004 El Manor Ave Los Angeles CA 90045 Office: Aerospace Corp PO Box 92957 Los Angeles CA 90009

SILVESTRI, GEORGE JOSEPH, JR., mechanical engineer, systems engineer; b. Jessup, Pa., Aug. 3, 1927; s. George Joseph and Catherine (Garaventa) S.; m. Betty Ann Huber, Jan. 21, 1961; children: Mary Elizabeth, Janet Catherine. B.S. in Mech. Engring, Drexel U., 1953, M.S., 1956; cert. in bus. mgmt., U. Pa., 1965. Engr. Westinghouse, Phila., 1953-62, sr. engr., 1962-69, fellow engr., 1969-80, adv. engr., 1980—; project engr. EPRI, Palo Alto, Calif., 1973-74. Contbr. articles to profl. jours. Served with USN, 1945-46. Fellow ASME (author Steam Tables 1967); mem. Am. Nuclear Soc. Republican. Subspecialties: Mechanical engineering; Systems engineering. Patentee (16). Home: 1840 Cheryl Dr Winter Park FL 32792 Office: Westinghouse Electric Corp 6655 E Colonial Dr Orlando Fl 32807

SILVIDI, ANTHONY ALFRED, physics educator; b. Steubenville, Ohio, Jan. 17, 1920; s. Alfredo and Mary (DiMarzio) S.; m. Lillian Emily Ferrelli, June 7, 1947; children: Anita M., Gina E., Julius A., Anthony C. B.S., Ohio U., 1942, M.S., 1944; Ph.D., 1949. Research engr. Westinghouse Research Lab., East Pittsburgh, Pa., 1944-45; assoc. prof. physics Coll. Steubenville (Ohio), 1949-51, head dept. physics, 1949-51; research physicist Cornell Aero. Lab., Buffalo, 1951-52; asst. prof. physics Kent (Ohio) State U., 1952-56, assoc. prof., 1956-64, prof., 1964—, coordinator doctoral programs dept. physics, 1958-71, acting chmn. dept., 1969; research scientist Goodyear Aero. Corp. Akron, Ohio, 1952-58. Author: Physics in Speech Communiation, 1964, Biomechanics, 1977, (with G. Bush) The Atom, 1961. Precinct committeeman, Kent, 1975-79; mem., chmn. Planning Commn., Kent, 1976-79. Served to cpl. U.S. Army, 1942-44. NSF grantee, 1953-69; USAF grantee, 1957-60; Dept. Energy grantee, 1980, 82. Mem. Am. Phys. Soc. (Ohio chmn. 1975), Biophys. Soc., Sigma Xi (local chmn. 1961-62, 1971-72). Democrat. Roman Catholic. Club: Christopher Columbus. Subspecialties: Biophysics (biology); Nuclear magnetic resonance (biotechnology). Current work: Applying physical measurements to biological problems. Home: 311 Valleyview St Kent OH 44240 Office: Dept Physics 314 Smith Hall Kent State U Kent OH 44242

SIMHA, ROBERT, educator; b. Vienna, Aug. 4, 1912; s. Mercado and Mathilde S.; m. Genevieve Martha Cowling, June 7, 1941. Student, Poly. Inst., Vienna, 1930-31; Ph.D., U. Vienna, 1935. Research assoc. U. Vienna, 1935-38, Columbia U., 1939-41; mem. faculty Poly. Inst. Bklyn., 1941-42, Howard U., 1942-45; cons., coordinator Polymer Research, Nat. Bur. Standards, 1945-51; mem. faculty N.Y. U., 1951-58, U. So. Calif., 1958-67; prof. macromolecular sci. Case Western Res. U., 1968—; cons. in field. Contbr. articles in field to profl. jours. Lalor Found. fellow, 1940; J.F. Kennedy Meml. Found. fellow, 1966; British Sci. Research Council fellow, 1967; recipient Washington Acad. Sci. Disting. Service to Sci. award, 1946; Meritorious Service award U.S. Dept. Commerce, 1949. Fellow AAAS, Am. Inst. Chemistry, Am. Phys. Soc. (High Polymer Plasma prize 1981), N.Y. Acad. Sci., Washington Acad. Sci.; mem. AAUP, Am. Chem. Soc., Soc. Rheology (Bingham medal 1973), Sigma Xi. Subspecialties: Polymer chemistry; Polymer physics. Current work: Research on physics and physical chemistry of macromolecular systems. Office: Dept Macromolecular Science Case Western Res U Cleveland OH 44106

SIMITSES, GEORGE JOHN, mech. engr., aerospace engr., educator, researcher, cons.; b. Athens, Greece, July 31, 1932; s. John G. and Vasilike G. (Goutoufas) S.; m. Nena E., June 10, 1941; children: John G., William G., Alexandra. Student, U. Tampa, 1951-52; B.S. in Aero. Engring, Ga. Inst. Tech., 1955, M.S., 1956; Ph.D. in Aeros. and Astronautics, Stanford U., 1965. Research engr. Ga. Inst. Tech. Exptl. Sta., 1956; instr. Ga. Inst. Tech., 1957-59, asst. prof., 1959-66, assoc. prof., 1966-74, prof. engring. sci. and mechanics, 1974—; cons. to industry; summer faculty assoc. Lockheed-Ga. Co., 1958, 60, 61, 65. Author: Elastic Stability of Structures, 1976; contbr. numerous articles on structural analysis, stability and optimization. Fellow AIAA (assoc.); mem. ASME, Structural Stability Research Council, Hellenic Soc. Theoretical and Applied Mechanics (founding mem.), Am. Acad. Mechanics, Sigma Xi (Ga. Inst. Tech. chpt. Sustaining Research award 1980). Greek Orthodox. Subspecialties: Theoretical and applied mechanics; Aerospace engineering and technology. Current work: Structural stability under static and dynamic loads of arches, frames, plates and shells of various constrns. (metallic and fiber reinforced composites). Home: 1389 Spalding Dr NE Atlanta GA 30338 Office: care Ga Inst Tech ESM Atlanta GA 30332

SIMMON, VINCENT FOWLER, genetic engineering company executive; b. Los Angeles, Aug. 9, 1943; s. Vincent Joseph and Gertrude (Fowler) S.; m. Carol Ann Lamboy, Dec. 27, 1963 (div.); 1 dau., Stacy Anne; m. Berniece Irene Dower, Jan. 2, 1983. B.A., Amherst Coll., 1964; M.S., U. Toledo, 1966; Ph.D., Brown U., 1971. Chemist Gen. Tire and Rubber Co., Toledo, 1964-65; NIH fellow Stanford (Calif.) U., 1971-73; microbiologist toxicology dept. SRI Internat., Menlo Park, Calif., 1973-75, mgr. microbiology program, 1975-78, asst. dir. toxicology dept., 1978-79; with Genex Corp., Rockville, Md., 1979—, sr. v.p. research, 1982—. Contbr. articles to sci. publs. Mem. AAAS, Environ. Mutagen Soc., Am. Chem. Soc., Chem. Industry Assn. Subspecialties: Genetics and genetic engineering (biology); Environmental toxicology. Current work: Marketing research services; managing corporate research operations. Home: 10421 River Rd Potomac MD 20854 Office: 6110 Executive Blvd Suite 680 Rockville MD 20852

SIMMONS, GUSTAVUS JAMES, mathematician, researcher; b. Anstead, W.Va., Oct. 27, 1930; s. Earl Gustavus and Ida Mae (Skaggs) S.; m. Diana Alice Degenkolb, Mar. 12, 1950; 1 dau., Karen Simmons Hart. B.S. in Math. and Physics, U. N.Mex., 1955; M.S. in Physics and Math, U. Okla., 1958; Ph.D. in Math, U. N.Mex., 1969. Chmn. div. design and evaluation of command and control systems for U.S. nuclear weapons Sandia Nat. Labs., Albuquerque, 1962-69, mgr. dept. applied math., since 1972—; mem. Gov.'s Sci. Adv. Bd. Editor book on cryptography; contbr. articles to profl. jours. Bd. dirs. Southwest Coop. Edn. Lab., 1969-72. Served with USAF, 1948-53. Mem. IEEE, Am. Math. Soc., Math. Assn. Am. Subspecialties: Cryptography and data security; Discrete Mathematics.. Current work: Research primarily in the areas of combinatorics and graph theory and in the applied topics of information theory and cryptography especially as applied to message authentication and systems design to achieve this function. Patentee in teaching machines, computer devices, weapon systems. Office: Dept Math Sandia Nat Lab PO Box 5800 Albuquerque NM 87185

SIMMONS, HOWARD ENSIGN, JR., research adminstr., chemist; b. Norfolk, Va., June 17, 1929; s. Howard Ensign and Marie Magdalene (Weidenhammer) S.; m. Elizabeth Anne Warren, Sept. 1, 1951; children—Howard Ensign III, John W. B.S. in Chemistry, M.I.T., 1951, Ph.D. in Organic Chemistry, 1954. Mem. research staff E.I. duPont de Nemours & Co. (central research and devel. dept.), Wilmington, Del., 1954-59, research supr., 1959-70, asso. dir., 1970-74, dir. research, 1974-79, dir., 1979—; adj. prof. chemistry U. Del., 1974—; Sloan vis. prof. Harvard U., 1968; Kharasch vis. prof. U. Chgo., 1978. Author: (with A.G. Anastassiou) Theoretical Aspects of the Cyclobutadiene Problem in Cyclobutadiene, 1967; Editorial bd.: Jour. Organic Chemistry, 1969-74, Synthesis, 1969—, Chem. Revs, 1972-74. Trustee Gordon Research Confs., 1974—. Fellow N.Y. Acad. Scis.; mem. Am. Acad. Arts and Scis., Nat. Acad. Scis., Soc. Chem. Industry, Am. Chem. Soc., Indsl. Research Inst., AAAS, Delta Kappa Epsilon. Subspecialty: Organic chemistry. Home: PO Box 3874 Wilmington DE 19807 Office: Central Research and Development Dept EI DuPont deNemours & Co Wilmington DE 19898

SIMMONS, RALPH OLIVER, physicist, educator; b. Kensington, Kans., Feb. 19, 1928; s. Fred Charles and Nellie (Douglass) S.; m. Janet Lee Lull, Aug. 31, 1952; children—Katherine Ann, Bradley Alan, Jill Christine, Joy Diane. B.A., U. Kans., 1950, Oxford U., 1953; Ph.D., U. Ill., 1957. Research asso. U. Ill., Urbana, 1957-59, faculty physics, 1959—, asso. prof., 1961-65, prof. physics, 1965—, head physics dept., 1970—; mem. governing bd. Internat. Symposia on Thermal Expansion, 1970—; cons. Argonne Nat. Lab., 1978—; Chmn. Office of Phys. Scis., NRC, 1978—; mem. Assembly of Math. and Phys. Scis., 1978—, Geophysics Research Bd., 1978—; trustee Argonne Univs. Assn., 1979—. Mem. internat. adv. bd.: Jour. Physics C (Solid State Physics), 1971-76; mem. editorial bd.: Physical Review B, 1978-81. Sr. postdoctoral fellow NSF, 1965. Fellow Am. Phys. Soc. (vice chmn., chmn. div. solid state physics 1975-77); mem. AAAS, Am. Crystallographic Assn., Am. Assn. Physics Tchrs., Phi Beta Kappa, Sigma Xi, Pi Mu Epsilon. Subspecialties: Condensed matter physics; Low temperature physics. Current work: Condensed matter physics, low temperature physics. Research on atomic defects in solids, quantum solids, molecular crystals, crystal dynamics, radiation damage. Home: 1005 Foothill Dr Champaign IL 61820

SIMMONS, WILLIAM HOWARD, biochemistry and biophysics educator, researcher; b. Mansfield, Ohio, May 15, 1947; s. Harold Eugene and Mildred Eileen (Patterson) S. B.A., Wittenberg U., 1969; M.S., Bowling Green State U., 1973; postgrad., Ohio State U., 1975-76; Ph.D., U. Ill. Med. Ctr., 1979. Teaching asst. Bowling Green (Ohio) State U., 1971-75; research asst. Ohio State U., Columbus, 1975-76, U. Ill. Med. Ctr., Chgo., 1976-79, research assoc., 1979-81; asst. prof. biochemistry and biophysics Loyola U. Med. Ctr., Maywood, Ill., 1981—. Contbr. articles to profl. jours. NIH research grantee, 1981. Mem. Am. Assn. Biol. Chemists, Am. Chem. Soc., N.Y. Acad. Scis., AAAS, Sigma Xi. Lutheran. Subspecialties: Biochemistry (biology); Physiology (biology). Current work: Purification and characterization of enzymes which metabolize biologically active peptides. Purification of new brain peptides. Behavioral effects of peptides. Home: 1117 N Dearborn 806 Chicago IL 60610 Office: Dept Biochemistry and Biophysics Loyola Univ Med Center 2160 S First Ave Maywood IL 60153

SIMON, ALBERT, physicist, engineer, educator; b. N.Y.C., Dec. 27, 1924; s. Emanuel D. and Sarah (Leitner) S.; m. Harriet E. Rubinstein, Aug. 17, 1947 (dec. June 1970); children: Richard, Janet, David; m. Rita Shiffman, June 11, 1972. B.S., CCNY, 1947; Ph.D., U. Rochester, 1950. Registered profl. engr., N.Y. State. Physicist Oak Ridge Nat. Lab., 1950-54, asso. dir. neutron physics div., 1954-61; head plasma physics div. Gen. Atomic Co., San Diego, 1961-66; prof. dept. mech. engring. U. Rochester, N.Y., 1966—, prof. physics, 1968—, chmn. dept. mech. engring., 1977—; mem. Inst. for Advanced Study, Princeton, 1974-75; sr. vis. fellow U.K. Sci. Research Council, Oxford U., 1975. Author: An Introduction to Thermonuclear Research, 1959; contbr. to: Ency. Americana, 1964, 74; Editor: Advances in Plasma Physics, 1967—. John Simon Guggenheim fellow, 1964-65. Fellow Am. Phys. Soc. (chmn. plasma physics exec. com. 1963-64); mem. ASME, Am. Nuclear Soc. Subspecialties: Nuclear fusion; Nuclear engineering. Current work: Theoretical studies on plasma behavior in fusion devices; both magnet confinement and inerial confinement. Home: 263 Ashley Dr Rochester NY 14620

SIMON, BARRY, mathematician, physicist, educator; b. Bklyn., Apr. 16, 1946; s. Hy and Minnie (Landa) S.; m. Martha Katzin, Jan. 24, 1971; children: Rivka, Benjamin, Zvi. B.A., Harvard U., 1966; Ph.D., Princeton U., 1970. Instr. depts. math. and physics Princeton (N.J.) U., 1969-70, asst. prof., 1970-72, assoc. prof., 1972-76, prof., 1976-81; prof. math. and theoretical physics Calif. Inst. Tech., 1981—. Author 8 books in field; contbr. articles to profl. jours.; editor: Wadsworth Advanced Mathematics Book Series, 1980—, Mathematical Notes series, 1980 ; asso. editor: Jour. Operator Theory, 1979—, Communications in Math. Physics, 1979—, Duke Math. Jour, 1980—. Recipient medal Internat. Acad. Atomic and Molecular Sci., 1981, Stampacchia prize, 1982. Fellow Am. Phys. Soc.; mem. Am. Math. Soc. (council 1976-78, program com. for nat. meetings 1980-82). Subspecialties: Theoretical physics. Current work: Research in mathematical physics, especially the theory of Schrodinger operators and the statistical mechanics of lattice gases. Office: Calif Inst Tech 253-37 Pasadena CA 91125

SIMON, GARY LEONARD, physician, researcher, educator; b. Bklyn., Dec. 18, 1946; s. Bernard and Dorothy (Ligeti) S.; m. Vicki Ellen Thiessen, Aug. 29, 1970; children: Jason Richard, Jessica Ann. B.S., U. Md., 1968; M.S., U. Wis.-Madison, 1970, Ph.D., 1972; M.D. cum laude, U. Md., 1975. Diplomate: Am. Bd. Internal Medicine (Infectious Diseases). Intern U. Md. Hosp., 1975-76, resident, 1976-78; fellow Tufts-New Eng. Med. Ctr., Boston, 1978-80; asst. prof. medicine George Washington U., 1980—. Contbr. chpts., articles to profl. publs. Recipient Outstanding Faculty award George Washington U. Med. House Staff, 1980. Mem. ACP, AAAS, Am. Venereal Disease Assn., Am. Soc. Microbiology, Am. Fedn. Clin. Research, Alpha Omega Alpha. Democrat. Jewish. Subspecialties: Internal medicine; Infectious diseases. Current work: Host-defense mechanisms. Office: George Washington U Med Center 2150 Pennsylvania Ave NW Washington DC 20037

SIMON, MARCIA, clinical chemist, researcher; b. San Antonio, Apr. 12, 1946; d. Alton Carl and Lucy Jane (Young) Mielke; m. Robert Jeffrey Simon, July 1, 1968 (div.); 1 dau. B.A. in Chemistry, U. Colo., 1968; Ph.D. in Pharmacology, UCLA, 1974. Research asst. dept. biochemistry U. Colo., Boulder, 1966-68; mental health trainee dept. pharmacology UCLA, 1970-73, NIMH postdoctoral fellow dept. psychiatry, 1974-76; NIMH postdoctoral fellow Neurochemistry Lab., VA, Brentwood, Calif., 1974-76; research leader Bio. Sci. Labs., Van Nuys, Calif., 1976-83; asst. dir. Boston Med. Lab., Waltham, Mass., 1983—. Contbr. articles to sci. jours. Mem. Am. Assn. for Clin. Chemistry, Assn. for Women in Sci., Am. Soc. Pharmacology and Exptl. Therapeutics, Western Pharmacology Soc., NOW, ACLU. Subspecialty: Clinical chemistry. Current work: Parathyroid hormone, glycosylated hemoglobin, therapeutic drug monitoring; develop and validate new assays for diagnosis and monitoring human disease. Office: 15 Lunda St Waltham MA 02154

SIMON, RICHARD L., mechanical engineer, manufacturing company executive; b. Port Chester, N.Y., Feb. 28, 1948; s. Jack and Rebecca (Gross) S.; m. Anita G., Aug. 6, 1972; 1 dau., Laura Healey. B.S. in Mech. Engring, Rensselaer Poly. Inst., M.S. Mech. Engring, 1971;

M.B.A., Boston U., 1979. Analytical engr. Gen. Electric Knowles Atomic Power Lab., Schenectady, 1971-73; sr. engr. and mktg. specialist Digital Equipment Corp., Maynard, Mass., 1973-78; dir. cam product mgmt. and stratetic planning Computer-vision Corp., Bedford, Mass., 1978—. Contbr. articles on computer sci. to profl. jours. NASA scholar, 1970-71. Mem. Soc. Mfg. Engrs., Numerical Control Soc. Subspecialties: Mechanical engineering; Robotics. Current work: Cad-cam applied to manufacturing automation including numerical control, robotics, inspection, group technology, process planning and analysis. Office: 201 Burlington Rd Bedford MA 01730

SIMON, RICHARD MACEY, statistician; b. St. Louis, July 27, 1940; s. Leon and Rose (Lander) S.; m. Caroline Lebowitz, Aug. 9, 1970; 1 son: Jonathan Simon. B.S., Washington U., 1965, Ph.D., 1969. Statistician, clin. onocology program Nat. Cancer Inst., Bethesda, Md., 1971-78, chief biometric research br., 1978—. Assoc. editor: Jour. Nat. Cancer Inst, 1981—; contbr. articles to profl. jours. Pres. Springbrook Forest Citizens Assn., 1981-82. Served with USPHS, 1969-71. Mem. Am. Statis. Assn., Am. Assn. Cancer Research, Am. Soc. Clin. Oncology, Biometric Soc., Cell Kinetics Soc., Soc. for Clin. Trials. Subspecialties: Statistics; Cancer research (medicine). Current work: Biostatistics, clinical trials.

SIMONETTI, JOSEPH LAWRENCE, test engineer; b. Plainfield, N.J., Apr. 12, 1952; s. John Joseph and Philomena Rose (Mone) S. B.S.M.E., Rutgers Coll. Engring., 1976; B.A. in Psychology, Rutgers U., 1976. Test engr. Pratt and Whitney Aircraft, East Hartford, Conn., 1976, Sikorsky Aircraft, Stratford, Conn., 1976-77; test engr. level IV Avco Lycoming, Stratford, 1977—. NSF grantee, 1975. Mem. ASME, Soc. Automotive Engrs., Aircraft Owners and Pilots Assn., Pi Tau Sigma. Club: Flying Eagles (Stratford). Subspecialties: Mechanical engineering; Aerospace engineering and technology. Current work: Flight and ground testing of high by-pass ratio turbofan engines; aircraft powerplants. Office: 550 S Main St Stratford CT 06497

SIMONIAN, SIMON JOHN, surgeon, immunologist; b. Antioch, French Ter., Apr. 20, 1932; U.S., 1965, naturalized, 1976; s. John Simon and Marie Cecile (Tomboulian) S.; m. Arpi Ani Yeghiayan, July 11, 1965; children—Leonard Armen, Charles Haig, Andrew Hovig. M.D., U. London, 1957; B.A. with honors, U. Oxford, Eng., 1964, M.A., 1969; Sc.M., Harvard U., 1967, Sc.D., 1969. Diplomate: Am. Bd. Surgery, Brit. Bd. Surgery. Intern Edinburgh (Scotland) Royal Infirmary, 1957-58, resident, 1961-62, Edinburgh Western Gen. Hosp., 1958-59, Birmingham Accident and Burns Hosp., U. Birmingham, Eng., 1959-60; demonstrator dept. anatomy Edinburgh U., 1960-61; research fellow in pathology Harvard U., Cambridge, Mass., 1965-68; resident in surgery Boston City Hosp., 1970-74; dir. surg. immunology, asst. in surgery Peter Bent Brigham Hosp., Boston, 1968-70; attending surgeon in transplantation and gen. surgery services U. Chgo. Hosps. and Clinics, 1974-78; instr., asso. in surgery Harvard Med. Sch., 1968-70; asst. prof. surgery U. Chgo., 1974-78; head div. renal transplantation Hahnemann Med. Coll. and Hosp. of Phila., 1978—, prof. surgery, 1978—; surg. dir. End Stage Renal Disease Program, 1980—; lectr. in field. Contbr. articles in field to med. jours. Founder Armenian Youth Soc., Eng., 1953, pres., 1953-54; bd. govs. Friends Sch., London, 1963-65; Mass. del. Armenian Assembly, Washington, 1970-74. Nairn scholar, 1949-52; Middlesex scholar, 1952-57; recipient Suckling prize, 1957; Brit. Med. Research Council award, 1962-64; NIH award, 1965-70; Alt prize, 1973; Thompson award, 1974-77; Johnson award, 1975-77; named outstanding new citizen of 1976-77, Washington. Fellow Internat. Coll. Surgeons, Royal Coll. Surgeons Edinburgh, A.C.S.; mem. AAAS, Royal Coll. Surgeons of Eng., Nat. Assn. Armenian Studies and Research, Am. Armenian Med. Assn. (founder 1972, treas. 1972-74), Assn. Acad. Surgery, Transplantation Soc., Am. Fedn. Clin. Research, N.Y. Acad. Scis., Am. Soc. Transplant Surgeons (founding mem. 1974, chmn. immunosuppression study com. 1974-77), Phila. Acad. Scis. (co-chmn. membership com.), Assn. Immunologists Boston, Assn. Immunologists Chgo., Assn. Ill. Transplant Surgeons, Greater Delaware Valley Soc. Transplant Surgeons (pres. elect 1980-82), Assn. Cancer Research Boston, Phila. County Med. Soc., Pa. Med. Soc., AMA, Am. Technion Soc., Internat. Cardiovascular Spc., End Stage Renal Disease Network 24 (mem. med. rev. bd.), Sigma Xi. Clubs: Harvard, Med. (Phila.). Subspecialties: Transplant surgery; Transplantation. Current work: Organ transplantation surgery, biology and immunology. Office: 230 N Broad St Philadelphia PA 19102

SIMONIS, GEORGE JEROME, research physicist; b. Wisconsin Rapids, Wis., Nov. 9, 1946; s. Anselm William and Agnes (Olds) S.; m. Toni Simonis, Aug. 31, 1968. B.S. in Physics, Wis. State U., Plattville, 1968, Ph.D., Kans. State U., 1973. Research physicist Harry Diamond Labs., Adelphia, Md., 1974—. Served to 1st lt. U.S. Army, 1972-74. Mem. IEEE, Optical Soc. Am., Am. Phys. Soc. Subspecialties: Wave research; Condensed matter physics. Current work: Laser research, solid state spectroscopy, millimeter and far-infrared research. Home: 13609 Russett Terr Rockville MD 20853

SIMONS, ROY KENNETH, pomology educator, researcher; b. Kincheloe, W.Va., Dec. 26, 1920; s. Basil E. and Essie M. (Currey) S.; m. Frances Jensen, June 12, 1953; children: Janet Simons Orashan, Kenneth B. B.S., M.S., W.Va. U., 1947; Ph.D., Mich. State U., 1951. Instr. U. Del., Newark, 1947-48; grad. asst. Mich. State U., East Lansing, 1948-51; prof. horticulture U. Ill.-Urbana, 1951—. Contbr. numerous articles on pomology to profl. jours. Served with AUS, 1942-45. Rootstock Research Com. grantee, 1980-81; Hort. Research Inst. grantee, 1981. Fellow AAAS, Am. Soc. for Hort. Sci. (Stark award 1967, Hurov award 1977, 78); mem. Internat. Soc. Hort. Sci., Internat. Dwarf Fruit Tree Assn. (sec. rootstock research com.), Am. Pomological Soc., Ill. Hort. Soc., Mich. Hort. Soc. Subspecialties: Ecosystems analysis; Cell biology (medicine). Current work: Pomology, fruit growth as related to cultural practices, morphology anatomy of specific plant parts and dwarfing rootstock evaluations. Home: 1517 Alma Dr Champaign IL 61820 Office: 1707 S Orchard St Urbana IL 61801

SIMONS, SAMUEL STONEY, JR., research chemist; b. Phila., Sept. 13, 1945; s. S. Stoney and Virginia (Cooke) S.; m. Cyrena Scott Gouge, Apr. 25, 1970; children: C. Torrey, Caroline S. A.B., Princeton U., 1967; Ph.D., Harvard U., 1972. Postdoctoral fellow dept. biochemistry U. Calif.-San Francisco, 1972-75; staff fellow NIH, Bethesda, Md., 1975-80, research chemist, 1980—. Mem. Am. Soc. Biol. Chemists, Am. Chem. Soc., Phi Beta Kappa. Subspecialties: Receptors; Molecular biology. Current work: Steroid hormone action, glucocorticoid receptors, affinity labeling, anti-glucocorticoids. Patentee in field. Office: NIH Bldg 4 Room 132 NIADDK Bethesda MD 20205

SIMONSON, LLOYD GRANT, microbiology/biochemistry researcher; b. San Jose, Calif., Dec. 1, 1943; s. Rolfe Lau and Dorothy Fay (O'Scully) S.; m. Katherine Lenora Peck, Aug. 24, 1968. B.A., Western Ill. U., 1966; M.S., Ill. State U., Normal, 1968, Ph.D., 1974. Research cons. Chgo. Med. Sch., North Chicago, Ill., 1977-80; project mgr. Naval Dental Research Inst., Great Lakes, Ill., 1974—. Contbr. articles to profl. jours. Recipient Outstanding Performance award Naval Dental Research Inst. Mem. Am. Assn. Dental Research, Am. Soc. Microbiology, Ill. Soc. Microbiology, Internat. Assn. Dental Research, Phi Sigma, Beta Beta Beta. Subspecialties: Microbiology; Enzyme technology. Current work: Control of disease by enzymatic methods. Biochemistry of microbial proteins and carbohydrates. Hybridoma technology and immunoassays. Patentee in field. Office: US Naval Dental Research Inst Bldg 1-H Great Lakes IL 60088 Home: 1115 Knollwood Rd Deerfield IL 60015

SIMONSON, SIMON CHRISTIAN, III, nuclear engineering consultant; b. Fergus Falls, Minn., Dec. 20, 1938; s. Simon Christian and Flora Jane (Brown) S.; m. Jade Lin, Oct. 15, 1966; 1 son, Niko Christian. S.B., MIT, 1960; M.Sc., Ohio State U., 1965, Ph.D., 1967; M.Sc., U. Md., 1976. Registered profl. engr., Md., Calif. Sci. officer Leiden (Netherlands) Obs., 1967-69; asst. prof. astronomy U. Md., College Park, 1969-75; sr. project engr. Nuclear Assocs. Internat., Rockville, Md., 1976-78; sr. assoc., Sunnyvale, Calif., 1978—. Contbr. articles to profl. jours. Served to lt. USNR, 1960-63. Mem. Am. Nuclear Soc., ASME, Internat. Astron. Union, Internat. Union Radio Sci., Am. Astron. Soc., AAAS, Sigma Xi. Subspecialties: Nuclear engineering; Mechanical engineering. Current work: Analysis of nuclear reactor core physics and transient thermal hydraulics using computer models. Office: 215 Moffett Park Dr Sunnyvale CA 94086

SIMPSON, DENNIS DWAYNE, psychology educator; b. Lubbock, Tex., Nov. 9, 1943; s. Homer Arnold and Georgie Lee (Barrett) S.; m. Sherry Ann Johnson, Aug. 20, 1965; children: Jason Renn, Jeffrey Todd, Jennifer Lynn. B.A., U. Tex., Austin, 1966; Ph.D., Tex. Christian U., 1970. Asst. prof. Tex. Christian U., Ft. Worth, 1970-74, assoc. prof., 1974-79, prof., 1979-82; prof. psychology Tex. A&M U., College Station, 1982—, dir. behavioral research, 1982—; cons. in field. Editor: Effectiveness of Drug Abuse Treatment, 3 vols, 1976; contbr. articles to profl. jours. Mem. Am. Psychol. Assn., Evaluation Research Soc., Southwestern Psychol. Assn., Sigma Xi, Psi Chi. Ch. of Christ. Subspecialties: Behavioral psychology; Evaluation. Current work: Biofeedback and behavioral medicine; social service delivery systems; evaluation research; drug abuse, addiction, prevention. Home: 800 Hereford St College Station TX 77840 Office: Tex A and M Univ Dept Psychology College Station TX 77843

SIMPSON, EUGENE SIDNEY, hydrology educator, researcher; b. Schenectady, July 14, 1917; s. Irving and Dora (Mandel) S.; m. June Diana Angel, July 22, 1969. B.S. in Civil Engring, CCNY, 1944; M.A. in Geology, Columbia U., 1949, Ph.D., 1960. Geologist U.S. Geol. Survey, Washington, 1945-63; prof. hydrology and water resources U. Ariz., 1963—; cons. hydrogeology, Tucson, 1963—. Fellow Geol. Soc. Am.; mem. ASCE, Am. GEophys. Union. Subspecialties: Hydrogeology; Geology. Current work: Research in transport and dispersion of pollutants in aquifers; aquifer mechancis; teaching groundwater hydrology. Office: Dept Hydrology U Ariz Tucson AZ 85721

SIMPSON, FREDERICK JAMES, research administrator; b. Regina, Sask., Can., June 8, 1922; s. Ralph James and Lillian Mary (Anderson) S.; m. Margaret Christine Simpson, May 28, 1947; children: Christine Louise, Steven James, Leslie Coleen, Ralph Edwin, David Glen. B.Sc., U. Alta., Can., 1944, M.Sc. in Agr, 1946; Ph.D. in Bacteriology, U. Wis., 1952. With Nat. Research Council Can., 1946—; asst. dir. Atlantic Research Lab., Halifax, N.S., 1970-73, dir., 1973—; vis. scientist U. Ill., Urbana, 1955-56, vis. prof., 1964; mem. exec. council Atlantic Provinces Interuniv. Com. on Scis., 1976-79, chmn., 1981—; pres. Fed. Inst. Mgmt., Halifax, 1981-82. Contbr. numerous articles to profl. jours. Decorated Queen's Silver Anniversary medal. Mem. Can. Soc. Microbiologists (sec.-treas. 1969-70, v.p. 1971-72, pres. 1972-73), Am. Soc. Microbiology, Am. Soc. Biol. Chemists, Nova Scotian Inst. Sci. (v.p. 1975-76, pres. 1977-78), Atlantic Provinces Council on Scis. (chmn. 1981—), Internat. Phycological Soc., Sigma Xi. Mem. United Ch. of Canada. Subspecialties: Microbiology; Plant physiology (agriculture). Current work: Development of technology for growing marine plants under controlled conditions. Office: Atlantic Research Lab 1411 Oxford St Halifax NS Canada B3H 3Z1

SIMPSON, GEORGE GAYLORD, educator, vertebrate paleontologist; b. Chgo., June 16, 1902; s. Joseph Alexander and Helen Julia (Kinney) S.; m. Lydia Pedroja, Feb. 2, 1923 (div. Apr. 1938); children: Helen Frances, Patricia (dec.), Joan, Elizabeth; m. Anne Roe, May 27, 1938. Student, U. Colo., 1918-19, 20-22; Ph.B., Yale U., 1923, Ph.D., 1926, Sc.D., 1946; Sc.D., Princeton, 1947, U. Durham, 1951, Oxford U., 1951, U. N.Mex., 1954, U. Chgo., 1959, Cambridge, Eng., 1965, York U., 1966, Kenyon Coll., 1968, U. Colo., 1968, U. Ariz., 1982; LL.D., U. Glasgow, 1951; Dr. h.c., U. Paris, France, 1965; D. Honoris Causa, Universidad de La Plata, Argentina, 1977. Marsh fellow Peabody Mus. (Yale), research on Mesozoic mammals, 1924-26; field asst. Am. Mus. Natural History, N.Y.C., 1924, asst. curator of vertebrate paleontology, 1927, asso. curator, 1928-42, curator of fossil mammals, 1942-59, chmn. dept. geology and paleontology, 1944-58; prof. vertebrate paleontology Columbia, 1945-59; Agassiz prof. vertebrate paleontology Mus. Comparative Zoology, Harvard, 1959-70; prof. geoscis. U. Ariz, Tucson, 1967—; pres., trustee Simroe Found., 1968—; NRC and Internat. Edn. Bd. fellow in work on early fossil mammals, chiefly in Brit. Mus., London, also other instns., Eng., France and Germany, 1926-27; expdns. to collect fossil animals include-Northern, Tex., Mont., N.Mex., Fla. and S.E. states, Argentina, Venezuela, Brazil. Author books relating to field, 1928—; The Meaning of Evolution, rev. edit, 1967, Horses, 1951, Life of the Past, 1953, The Major Features of Evolution, 1953, Evolution and Geography, 1953, Life, 1957, Quantitative Zoology, 1960, Principles of Animal Taxonomy, 1961, This View of Life: The World of an Evolutionist, 1964, The Geography of Evolution, 1965, Biology and Man, 1969, Penguins, 1976, Concession to The Improbable, an Unconventional Autobiography, 1978, Splendid Isolation, The Curious History of South American Mammals, 1980, Why and How, Some Problems and Methods in Historical Biology, 1980, Fossils and the History of Life, 1983; Contbr. articles to profl. publs. Served from capt. to maj. U.S. Army, 1942-44. Recipient Cross medal Yale Grad. Assn., 1969, Lewis prize Am. Philos. Soc., 1942, Thompson medal Nat. Acad. Scis., 1943, Elliot medal, 1944, 65, Gaudry medal Geol. Soc. France, 1947; Hayden medal, 1951; Penrose medal Geol. Soc. Am., 1952; André H. Dumont medal Geol. Soc. Belgium, 1953; Darwin-Wallace medal Linnean Soc., 1958; Darwin Plakette Deutsche Akad. Naturforscher Leopoldina, 1959; Gold medal Linnean Soc., 1962; Darwin medal Royal Soc., 1962, Nat. Medal Sci., 1966; distinguished achievement medal Am. Mus., 1969; Paleontol. Soc. medal, 1973; Internat. award Nat. Mus. Natural History, 1976. Fellow Am. Acad. Arts and Scis., Nat. Acad. Scis., Am. Philos. Soc., Geol. Soc. Am., Paleontol. Soc., AAAS; mem. Soc. Vertebrate Paleontology (sec.-treas. 1940-41, pres. 1942), Soc. Study Evolution (pres. 1946), Am. Soc. Mammalogists, Soc. Systematic Zoology (pres. 1962), Am. Soc. Zoologists (pres. 1964), Academia Nazionale dei Lincei (Italy), Academia Nazionale dei XL (Italy), Sociedad Argentina de Estudios Geog. Gaea (hon. corr.), Zool. Soc. London (fgn. mem.), Royal Soc. (fgn. mem.), Linnean Soc. London (fgn. mem.), Academia de Ciencias (Venezuela, Brazil, Argentina), Asociacion Paleontologica Argentina (hon. mem.), Deutsche Gesellschaft für Saugetierkunde (hon.), Senkenbergische Naturforschende Gesellschaft, Phi Beta Kappa, Sigma Xi. Subspecialties: Evolutionary biology; Paleontology. Current work: Morphology; systematics; toxonomy; ecology; paleobiology; paleoecology. Home: 5151 E Holmes St Tucson AZ 85711

SIMPSON, LARRY DEAN, med. physicist; b. Satanta, Kans., Sept. 3, 1944; s. Melvin LeRoy and Dorothy Lorraine (Moody) S.; m. Sara Margaret Frandle, Aug. 14, 1965; 1 son: Rustin Dean. A.B., U. Kans., 1966, M.S. in Radiation Biophysics, 1969, Ph.D., 1971. Radiol. health traineeship div. radiol. health USPHS, Dept. Radiation Biophysics, U. Kans., Lawrence, 1967-71; postdoctoral fellow in med. physics Meml. Sloan-Kettering Cancer Ctr., N.Y.C., 1971, asst. physicist dept. med. physics, 1971-72, asst. attending physicist dept. med. physics, 1973-79, assoc. attending physicist, 1979-81; asst. prof. radiology Cornell U. Grad. Sch. Med. Scis., 1979-81; dir. med. physics sect. div. radiation oncology, assoc. prof. radiology, asst. prof. radiation, biology and biophysics U. Rochester Cancer Ctr., Rochester, N.Y., 1981—. Contbr. numerous articles to profl. jours. Mem. Health Physics Soc., Am. Assn. Physics Tchrs., AAAS, Am. Soc. Therapeutic Radiologists (mem. sci. program com., mem. com. on radiation therapy technologists' affairs), Am. Assn. Physicists in Medicine (bd. dirs., fin. com., publs. com., chmn. sci. program com., mem. sci. council, chmn. continuing edn. com.), Radiol. Soc. N.Am., Am. Coll. Radiology, Radiation Therapy Oncology Group (mem. med. physics com.), Sigma Xi. Subspecialties: Cancer research (medicine); Biophysics (physics). Current work: Optimization of radiation use in diagnosis, cure and research of cancer. Director radiation physics and radiation biophysics activities allied to cancer radiation treatments. Office: U Rochester Cancer Center 601 Elmwood Ave Box 647 Rochester NY 14642

SIMPSON, ORMAN ALLEN, JR., physicist, educator; b. Macon, Ga., June 15, 1952; s. Orman Allen and Nelle Jewelle (Wright) S. B.S. in Physics (Moody scholar, Boyd scholar, Physics scholar), Auburn U., 1974; M.S. in Physics (Pres.'s fellow), Ga. Inst. Tech., 1975, Ga. Inst. Tech., 1976; Ph.D. in Physics, Emory U., 1982. Research scientist I Ga. Inst. Tech., Atlanta, 1978-81, research scientist II, 1981-82; asst. prof. physics Emory U., Atlanta, 1982—; cons. PSI Inc. Contbr. articles to profl. jours. Mem. IEEE, Optical Soc. Am., Ga. Acad. Sci., Virginia-Highland Civic Assn., Sigma Xi, Sigma Pi Sigma. Current work: Molecular spectroscopy, IR-FIR laser/optical physics, MM-Wave systems and component development. Office: Dept Physics Emory U Atlanta GA 30322

SIMPSON, RICHARD ALLAN, planetary astronomer; b. Portsmouth, N.H., June 25, 1945; s. Allan Haines and Priscilla (Warren) S. S.B. in Elec. Sci. and Engring, M.I.T., 1967; M.S. in Elec. Engring, Stanford U., 1969; Ph.D., 1973. Engr. instrumentation lab. M.I.T., Cambridge, 1967; research asst. Ctr. for Radar Astronomy, Stanford (Calif.) U., 1967-73, research assoc., 1973-76, sr. research assoc., 1976—; mem. NASA Planetary Geology Rev. Panel, 1981-82. Contbr. articles to sci. publs. Active environ. and conservation groups including Calif. Wilderness Coalition, Sierra Club. Served to capt. USAR, 1967-75. Mem. IEEE (profl. groups Antennas and Propagation, Geosci. and Remote Sensing), Am. Geophys. Union, AAAS, Am. Astron. Soc. (div. planetary sci.), Internat. Sci. Radio Union (U.S. nat. com.), Eta Kappa Nu, Tau Beta Pi. Democrat. Subspecialties: Planetary science; Planetology. Current work: Radar astronomy; theoretical and experimental research on scattering of radio waves by planetary surfaces, atmospheres, and rings. Study electromagnetic wave propagation in inhomogeneous materials. Geophysical interpretation of experimental results. Home: 3326 Kipling St Palo Alto CA 94306 Office: Stanford Electronics Labs Durand 217 Stanford CA 94305

SIMPSON, RICHARD KENDALL, JR., physician, surgeon, researcher; b. Atlanta, Sept. 10, 1953; s. Richard Kendall and Juliet Hodges (Rowsey) S. B.A., Coker Coll., 1975; Ph.D., Med. U. S.C., 1980, M.D., 1982; postgrad., Warnborough Coll., Oxford, Eng., 1974. Diplomate: Nat. Bd. Med. Examiners. Teaching asst. dept. physiology Med. U. S.C., Charleston, 1976-80, research assoc., 1980—83, intern nuerology, 1982—83; resident neurosurgery dept. neurosurgery Baylor Coll. Medicine, Houston, 1983-89; cons. Spinal Cord Injury Research, Charleston, S.C., 1980-83; grant reviewer NSF, Charleston, 1981—. Author: Peripheral Nerve Fiber Group and Spinal Corp Pathway Contributions to the Somatosensory Evoked Potential, 1980. Mem. Am. Med. Polit. Action Com., 1983. Watson fellow Coker Coll., 1974. Mem. AMA, Am. Physiol. Soc., Soc. Neurosci., Digital Equipment Computer Users Soc., N.Y. Acad. Sci., Sigma Xi, Alpha Omega Alpha, S.C. Soc. Neurosci., S.C. Acad. Sci. Episcopalian. Subspecialties: Neurosurgery; Neurophysiology. Current work: Evaluation and management of patients with spinal cord and closed head injury utilizing somatosensory evoked potentials, the pathophysiology of central nervous system trauma, the neurophysiological mechanisms of evoked potential production. Home: 7209 Brompton Rd Apt 252-A Houston TX 77025 Office: Dept Neurosurgery Baylor Coll Medicine 1200 Moursund Ave Houston TX 77030

SIMPSON, ROBERT TODD, research scientist; b. Chgo., June 28, 1938; s. William Loyal and Evelyn T. (Berg) S.; m. Katherine Rupkey, Feb. 16, 1963; children: Todd Andrew, William Robert, Michael Scott, Brian David. B.A., Swarthmore Coll., 1959; M.D., Harvard U., 1963, Ph.D., 1969. Diplomate: Nat. Bd. Med. Examiners. Served as surgeon USPHS, 1969-72; sr. surgeon and chief sect. developmental biochemistry NIH, Bethesda, Md., 1972-79; chief lab. cellular and developmental biology, med. dir. USPHS, 1979—, also co-chmn. dept. biochemistry. Editorial bd.: Nucleic Acids Research; contbr. 80 articles to sci. jours. Recipient Commendation award USPHS, 1982; Soma Weiss award Harvard Med. Sch., 1963; Leon Resnick prize, 1963. Mem. Am. Soc. Biol. Chemists, Soc. Developmental Biology. Subspecialties: Biochemistry (biology); Developmental biology. Current work: Chromatin structure and gene regulation during development. Office: NIH 6/B1-38 Bethesda MD 20205

SIMRING, MARVIN, periodontist, educator; b. Bklyn., June 16, 1922; s. Julius Wolf and Fannie (Eisenberg) S.; m. Francine Robinson, Dec. 17, 1949; children: Elyse, James, June, Robert. B.A., Bklyn. Coll., 1942; D.D.S., N.Y. U., 1944. Diplomate: Am. Acad. Periodontology. Pvt. practice gen. dentistry, Bklyn., 1947-55, specializing in periodontics, 1955-79; clin. prof., dir. postdoctoral tng. dept. periodontics N.Y.U. Coll. Dentistry, N.Y.C., 1947-79; assoc. prof. periodontics U. Fla., Gainesville, 1979—; cons. in periodontics U.S. VA, Gainesville, 1979—; dir. dept. periodontics Jewish Hosp. and Med. Ctr., Bklyn., 1953-79; cons. in periodontics VA Hosp., Bklyn., 1970-79, U.S. Army 1st Army Hosp., 1962-69. Author: Manual of Clinical Periodontics, 1973, 2d edit, 1978; contbr. chpts. to books, articles to profl. jours. Served to capt. Dental Corps AUS, 1944-46; ETO. Recipient Isadore Hirschfeld award Northeastern Soc. Periodontists, 1975; Disting. Service award Am. Acad. Oral Medicine, 1982; Loos medal for excellence U. Frankfurt, Germany, 1980; Disting. Service award Jewish Hosp. and Med. Ctr., Bklyn., 1979. Fellow Am. Coll. Dentists, N.Y. Acad. Dentistry, Am. Acad. Oral Medicine, Internat. Coll. Dentists; mem. Am. Acad. Periodontology, Northeastern Soc. Periodontists (pres. 1968-70), Exotic Plant Soc., Omicron Kappa Upsilon. Subspecialties: Periodontics; Preventive dentistry. Current work: Occlusion relative to periodontics, temporomanibular joint. Tooth hypersensitivity. Endodontics. Oral Medicine. Implants. Tooth Movement. Home: 2501 NW 21st Ave Gainesville FL 32605 Office: 1817 NW 13th St Gainesville FL 32601

SINCLAIR, JOHN HENRY, zoology educator; b. Oakwood, Tex., Aug. 14, 1935; s. Leon R. and Leetie (Dorsett) S.; m. Sabina M. Wagner, Sept. 27, 1962; 1 son, Eric D. B.S., Tex. A&M U., 1958, M.S.,

1959; Ph.D., U. Chgo., 1966. Postdoctoral fellow Carnegie Inst. Washington, Balt., 1966-68; asst. prof. biology Ind. U., Bloomington, 1968-72, chmn. zoology, 1973-76, asso. prof., 1972—. NSF grantee; NIH grantee; U.S. Dept. Agr. grantee. Mem. Am. Soc. Cell Biology, Am. Genetics Soc., AAAS. Subspecialties: Genome organization; Plant genetics. Current work: Analysis of Genome organization of mitochondria of higher plants, especially zea mays. Home: 2511 N Dunn St Bloomington IN 47401 Office: Dept Biology Ind Univ Bloomington IN 47405

SINCOVEC, RICHARD F(RANK), academic adminsitrator, computer science educator; b. Pueblo, Colo., July 14, 1942; s. Frank and Kathryn (Zelnick) S.; m. Deanna Crossley; children: Mary, Jimmy. B.S., U. Colo.-Boulder, 1964; M.S., Iowa State U., 1967, Ph.D., 1968. Research mathematician Exxon Prodn. Research, Houston, 1968-70; asst. prof. Kans. State U., 1970-74, assoc. prof., 1974-77, prof., 1979-80; mgr. Boeing Computer Services, Seattle, 1977-79; prof. computer sci., chmn. dept. computer sci. U. Colo.-Colorado Springs, 1980; cons. Lawrence Livermore (Calif.) Lab., 1971-83. Author: Programming in ADA, 1983. Mem. Assn. Computing Machinery, Spl. Interest Group-Numerical Math (dir. 1980-83), Soc. Indsl. and Applied Math. Subspecialties: Mathematical software; Software engineering. Current work: Numerical analysis, mathematical modeling and model development, petroleum reservoir simulation, mathematical software. Office: U Colo Dept Computer Sci Austin Bluffs Pkwy Coloardo Springs CO 80933 Home: 655 Allegheny Dr Colorado Springs CO 80919

SINDELAR, WILLIAM FRANCIS, physician, researcher; b. Cleve., Mar. 13, 1945; s. William Frank and Josephine Ann (Storkan) S.; m. Aleta Beth Merkel, May 8, 1982. B.A., Western Res. U., 1967, M.A., 1968, Ph.D., 1970, M.D., 1971. Diplomate: Am. Bd. Surgery, 1979. Mem. faculty Western Res. U., Cleve., 1966-67, Marine Biol. Lab., Woods Hole, Mass., 1966-67; jr. house officer in surgery Johns Hopkins U., Balt., 1971-73; fellow in surgery NIH, Bethesda, Md., 1973-75; sr. house officer in surgery U. Md., Balt., 1975-77; sr. surgeon Nat. Cancer Inst., NIH, Bethesda, 1977—; cons. in surgery U. Md. Contbr. articles on cancer biology, immunobiology and clin. treatment to profl. jours. Commd. sr. surgeon USPHS, 1973—. Mem. ACS, Am. Assn. for Cancer Research, Am. Soc. Clin. Oncology, Am. Soc. for Cell Biology, Am. Radium Soc., Assn. for Acad. Surgery, Soc. of Surg. Oncology, Phi Beta Kappa, Alpha Omega Alpha. Subspecialties: Surgery; Immunobiology and immunology. Current work: Development of new clinical cancer treatment; basic research in cancer immunobiology, surgical oncology, cancer research, immunobiology. Office: Surgery Br Nat Cancer Inst NIH Bethesda MD 20205

SINFELT, JOHN HENRY, chemist; b. Munson, Pa., Feb. 18, 1931; s. Henry Gustave and June Lillian (McDonald) S.; m. Muriel Jean Vadersen, July 14, 1956; 1 son, Klaus Herbert. B.S., Pa. State U., 1951; Ph.D., U. Ill., 1954, D.Sc. (hon.), 1981. Research engr. Exxon Research Engrng. Co., Linden, N.J., 1954-57, sr. research engr., 1957-62, research asso., 1962-68, sr. research asso., 1968-72, sci. advisor, 1972-79, sr. sci. advisor, 1979—; vis. prof. chem. engrng. U. Minn., 1969; Lacey lectr. Calif. Inst. Tech., 1973; Reilly lectr. U. Notre Dame, 1974; Francois Gault lectr. catalysis Council Europe Research Group Catalysis, 1980; Mobay lectr. in chemistry U. Pitts., 1980; disting. vis. lectr. in chemistry U. Tex., 1981; Robert A. Welch Found. lectr. Confs. on Chem. Research, 1981; Camille and Henry Dreyfus lectr. UCLA, 1982. Contbr. articles to sci. jours. Recipient Dickson prize Carnegie-Mellon U., 1977; Internat. prize for new materials Am. Phys. Soc., 1978; Nat. Medal of Sci., 1979; Chem. Pioneer award Am. Inst. Chemists, 1981. Fellow Am. Acad. Arts and Scis.; mem. Am. Inst. Chem. Engrs. (Alpha Chi Sigma award 1971, Profl. Progress award 1975), Am. Chem. Soc. (Carothers lectr. Del. sect. 1982, Petroleum Chemistry award 1976), Catalysis Soc. (Emmett award 1973), Nat. Acad. Scis., Nat. Acad. Engrng. Methodist. Subspecialty: Catalysis chemistry. Current work: Heterogeneous catalysis; bimetallic cluster catalysts; catalysis of hydrocarbon reactions on metals; structure of metal catalysts. Inventor polymetallic cluster catalysts used commercially in petroleum reforming. Office: Exxon Research Engineering Co PO Box 45 Linden NJ 07036

SINGER, JACK WOLFE, physician, resaarcher; b. N.Y.C., Nov. 9, 1942; s. Leon Eugene and Sarah Betty (White) S.; m.; 1 son, Constantine Jeremiah. B.A., Columbia U., 1964; M.D., SUNY-Bklyn., 1968. Diplomate: Am. Bd. Internal Medicine. Guest investigator Rockefeller U., N.Y.C., 1967; intern and resident in medicine U. Chgo., 1968-70; chief hematology Center for Disease Control, Atlanta, 1970-72; fellow in hematology and oncology U. Wash., Seattle, 1972-73, assoc. prof. medicine, 1978—; chief med. oncology Seattle VA Med. Center, 1975—; asst. mem. Fred Hutchinson Cancer Center, Seattle, 1975—. Author: Cancer Care, 1980; contbr. sci. articles to profl. publs. Served to lt. comdr. USPHS, 1970-72. Mem. Am. Soc. Hematology, Internat. Soc. Exptl. Hematology, Am. Soc. Clin. Oncology, Western Soc. Clin. Research, Am. Fedn. Clin. Research. Subspecialties: Cancer research (medicine); Marrow transplant. Current work: Cell biology of hematopoletic neoplasic and aspects of hematopoletic reconstitution in marrow transplantation. Office: Seattle VA Med Center 4435 Beacon Ave S Seattle WA 98108

SINGER, MARCUS JOSEPH, educator, biologist; b. Pitts., Aug. 28, 1914; s. Benjamin and Rachel (Gershenson) S.; m. Leah Horelick, June 8, 1938; children—Robert H., Jon Fredric. B.S., U. Pitts., 1938; M.A., Harvard, 1940, Ph.D., 1942. Mem. faculty Harvard Med. Sch., 1942-51, asst. prof. anatomy, until 1951; vis. prof. anatomy L.I. Coll. Medicine, 1950; asso. prof., then prof. zoology and child devel. Cornell U., 1951-61; vis. fellow Dutch Brain Inst., Amsterdam; H.W. Payne prof. anatomy, dir. dept. Case Western Res. U. Sch. Medicine, Cleve., asso. dir., 1961-69, dir., 1969—; vis. prof. anatomy Hebrew U., Jerusalem, 1974; Zyskind hon. vis. prof. faculty health scis. Ben Gurion U. Negev, Beersheba, Israel, 1975-76; vis. prof. Gunma U., Japan, 1977; mem. cell biology study sect. NIH, 1971-74, neurology study sect., 1976—. Author: (with P. Yakovlev) Human Brain in Sagittal Section, 1954, Dog Brain in Section, 1962; Editor: Jour. of Morphology, 1965-70; asso. editor, 1970-72; Asso. editor: Jour. Exptl. Zoology, 1963-68, 70-71; editorial bd., 1970-74; Contbr. articles on nervous system, histochemistry regeneration, cytology to profl. publs. Guggenheim fellow, Rome, 1967. Fellow Am. Acad. Arts and Scis., Ohio Acad. Sci., AAAS; mem. Am. Neurol. Assn. (asso.), Am. Assn. Anatomists, Soc. Zoologists, Internat. Brain Research Orgn., Assn. Research Nervous and Mental Diseases, Soc. Devel. and Growth, Biol. Stain Commn., Sigma Xi. Subspecialty: Neurobiology. Office: Dept Anatomy Case Western Res U Cleveland OH 44106

SINGER, MAXINE FRANK, biochemist; b. N.Y.C., Feb. 15, 1931; d. Hyman S. and Henrietta (Perlowitz) Frank; m. Daniel Morris Singer, June 15, 1952; children: Amy Elizabeth, Ellen Ruth, David Byrd, Stephanie Frank. A.B., Swarthmore Coll., 1952, D.Sc. (hon.), 1978; Ph.D., Yale U., 1957; D.Sc., Wesleyan U., 1977. USPHS postdoctoral fellow NIH, Bethesda, Md., 1956-58, research chemist (biochemistry), 1958-74; head sect. on nucleic acid enzymology Nat. Cancer Inst., 1974-79; chief Lab. of Biochemistry, Nat. Cancer Inst., 1979—; Regents vis. lectr. U. Calif., Berkeley, 1981; Bd. dirs. Found. for Advanced Edn. in Scis., 1972-78; mem. sci. council Internat. Inst. Genetics and Biophysics, Naples, Italy, 1982—. Editor: Jour. Biol. Chemistry, 1968-74, Sci. mag, 1972-82; Contbr. articles to scholarly jours. Trustee Wesleyan U., Middletown, Conn., 1972-75, Yale Corp., New Haven, 1975—; bd. govs. Weizmann Inst. Sci., Rehovot, Israel, 1978—. Recipient award for achievement in biol. scis. Washington Acad. Scis., 1969, award for research in biol. scis. Yale Sci. and Engrng. Assn., 1974, Superior Service Honor award HEW, 1975, Dirs. award NIH, 1977, Disting. Service medal HHS, 1983. Fellow Am. Acad. Arts and Scis.; mem. AAAS, Am. Soc. Biol. Chemists, Am. Soc. Microbiologists, Am. Chem. Soc., Inst. Medicine (Nat. Acad. Scis.), Nat. Acad. Scis. Subspecialties: Genome organization; Molecular biology. Current work: Organization and function of highly-repeated DNA sequences; relation between host and viral genomes. Home: 5410 39th St NW Washington DC 20015 Office: Bldg 37 4E-28 Bethesda MD 20205

SINGER, ROBERT, biology educator; b. Chgo., July 13, 1949; s. Paul and Edith (Greenfield) S.; m. Lucinda Johnson; Dec. 30, 1974. B.S., U. Ill., 1971, M.S., 1973, Ph.D., 1977. Asst. prof. biology Colgate U., Hamilton, N.Y., 1977-83; adj. asst. prof. biology Rensselaer Poly. Inst., Troy, N.Y., 1983—; research assoc. dept. civil engrng. Syracuse U. (N.Y.), 1983—; cons. EPA, N.C. State U., U.S. Fish and Wildlife Service. Editor symposium procs., 1981. Subspecialty: Ecology. Current work: Lake ecology/limnology. Office: Rensselaer Poly Inst Troy NY 12181 Office: 150 Hinds Hall Syracuse U Syracuse NY 13210

SINGER, TIMOTHY JAMES, aviation psychologist, consultant; b. Champaign, Ill., June 3, 1947; s. Robert Roy and Earline Elizabeth (Harris) S.; m. Ann Kathleen Widmer, Jan. 14, 1978; children: Rachael Linn, Lindsay. B.A., Reed Coll., Portland, Oreg., 1973; M.S., Yale U., 1975, M. Phil., 1976, Ph.D., 1977. Commd. lt. comdr. USNR, 1980—; Clin. psychology resident Wilford Hall U.S. Air Force Med. Ctr., San Antonio, 1976-77; chief mental health service U.S. Air Force Hosp., Loring AFB, Jaine, 1977-80; flight surgeon trainee Naval Aerospace Med. Inst., Pensacola, Fla., 1980-81, spl. project officer operational psychology, 1981; aviation safety researcher COMNAVAIRPAC, San Diego, 1981—, aircraft accident investigator, 1981—; adj. asst. prof. George Washington U. Med. Sch., 1981—. Author U.S. Navy med. trg. manuals; contbr. articles in field to profl. jours. Decorated U.S. Air Force Commendation Medal, Armed Forces Expeditionary Medal. Mem. Aerospace Med. Assn., Am. Psychol. Assn., Assn. Aviation Psychologists, Am. Soc. Clin. Hypnosis, Human Factors Soc., Yale Sci. and Engrng. Assn., Phi Beta Kappa. Club: Yale (San Diego). Subspecialties: Aerospace engineering and technology; Human factors engineering. Current work: Research and development, testing and evaluation in the military aviation community; special interests in man-machine integration, human-computer interface, maximization of aircrew performance and safety, and the application of human factors research methods to advanced aviation technology projects. Home: 9939 Oviedo St San Diego CA 92129 Office: COMNAVAIRPAC(code 0143C) NAS North Island San Diego CA 92135

SINGH, ADYA PRASAD, botanist, researcher; b. Varanasi, India, Oct. 2, 1941; came to U.S., 1964; s. Ram Lakhan and Sone Deepa S.; m. Asha Singh, May 13, 1977; children: Monica, Neil. Ph.D., U. Ark., 1969. Postdoctoral fellow Simon Fraser U., Vancouver, B.C., Can., 1969-71; sci. pool officer G.B. Pant U. Agr. and Tech., India, 1972-73; postdoctoral fellow No. Ariz. U., Flagstaff, 1973-74; assoc. researcher U. Hawaii-Hilo, 1974-78; vis. asst. prof. Tex. A&M U., College Station, 1978-79; research scientist Forest Research Inst., Rotorua, N.Z., 1980—. Contbr. articles on botany to profl. jours. Subspecialty: Plant growth. Current work: Plant ultrastructure, cambial structure and activity, structure and development of xylem and phloem cells, microtubules, microfilaments.

SINGH, ARJUN, molecular biologist, geneticist; b. Gonda, India, Jan. 27, 1943; came to U.S., 1964, naturalized, 1978; s. Ram Awadh and Samkali S.; m. Maya, June 10, 1963; children: Rakesh, Dinesh, Deepika. B.S. with honors, Pantnagar U. Agr. and Tech., 1964; M.S., U. Ill.-Urbana-Champaign, 1969, Ph.D., 1969. Postdoctoral fellow Case Western Res. U., 1969-70; U. Calif.-Berkeley, 1970-71, U. Rochester, 1971-76, U. Mass., 1976-77; postdoctoral fellow U. Wis.-Madison, 1977-79, assoc., 1979-80; sr. scientist Nabisco Brands Research Ctr., Wilton, Conn., 1980-81; scientist Genentech, Inc., South San Francisco, Calif., 1981—. Contbr. sci. and tech. articles to profl. publs. Univ. merit scholar, 1963-64; U. Ill. fellow, 1964-66. Mem. Genetics Soc. Am., AAAS, Am. Soc. Microbiology. Subspecialties: Genetics and genetic engineering (agriculture); Molecular biology. Current work: Genetic engineering, heterologous gene expression in yeast, yeast genetics and molecular biology. Home: 735 Cape Breton Dr Pacifica CA 94044 Office: 460 Point San Bruno Blvd South San Francisco CA 94080

SINGH, BRAMAH NAND, cardiologist, researcher; b. Suva, Fiji, Mar. 3, 1938; U.S., 1975; s. Shri Ram and Janki S.; m. Roshni Ram, Dec. 11, 1964; children: Pramil, Nalini, Sanjiv. B.Med.Sc., U. Otago, N.Z., 1961, M.B.Ch.B., 1963; Ph.D., U. Oxford, Eng., 1971; M.D., U. Otago, 1975. Sr. lectr. medicine U. Auckland (N.Z.) Sch. Medicine, 1972-75; dir. inpatient cardiology Cedars-Sinai Med. Ctr., Los Angeles, 1976-78; assoc. prof. medicine UCLA, 1976-80, prof. medicine, 1980—; dir. cardiovascular research lab. VA Hosp., Los Angeles, 1978—; Mem. cardiology adv. com. Nat. Heart, Lung and Blood Inst., 1979-83. Contbr. numerous articles to profl. jours.; Editor: Calcium Antagonists, 1984; editorial bd. numerous med. jours. Am. Heart Assn. grantee, 1978—. Fellow Royal Coll. Physicians (London), Am. Coll. Cardiology; mem. Royal Australasian Coll. Physicians. Subspecialty: Cardiology. Current work: Cardiovascular research into the mechanisms and control of disturbances of cardiac rhythm and conduction and the biology of myocardial ischemia. Home: 16979 Encino Hills Dr Encino CA 91436 Office: UCLA Los Angeles CA 90073

SINGH, GURBAX, physics educator; b. Sargodha, W. Pakistan, June 5, 1933; came to U.S., 1963, naturalized, 1977; d. Jiwan and Kaushalia Dass; m. Balwant Singh; 1 dau., Kanwal. B.S. with honors in Physics, U. Delhi, India, 1957, M.Sc., 1959; Ph.D., U. Md., 1971. Tutorial fellow U. Delhi, 1959-61; sr. research asst. Inst. Nuclear Medicine and Allied Scis., Delhi, India, 1961-63; teaching asst. U. Md., 1963-64, research asst., 1964-70; asst. prof. physics U. Md. Eastern Shore, Princess Anne, Md., 1970-73, assoc. prof. physics, 1973—. Contbr. articles to profl. jours.; Mem. editorial bd.: Jour. Engrng. Prodn, India. Pres. India Assn. Eastern Shore, 1981—. Mem. Am. Phys. Soc., IEEE, Am. Solar Energy Soc., Phi Kappa Phi. Democrat. Sikh. Current work: Investigation of the possibility of drying grain with microwaves on a commercial scale. Collaborating with a nutritionist to improve the nutrition quality of soybean by deactivating "soybean trypsin inhibitor" with microwaves. Patentee multiple wire-cavity spark counter for alpha particle detection.

SINGH, HARPAL, biology educator; b. Pakistan, Aug. 16, 1941; came to U.S., 1966, naturalized, 1974; s. Prem and Harbans K. S.; m. Harbhajan Kaur, Dec. 29, 1969; children: Remma, Arvin, Amit. B.S., Punjab U., 1960, M.S., 1962; Ph.D., U. Tenn., 1970, M.P.H., 1974. Faculty Savannah (Ga.) State Coll., 1974—, prof. biology and coordinator, med. tech., 1980—. Contbr. articles to profl. jours. NIH grantee, 1975—. Mem. Environ. Health Soc. Am. Subspecialties: Environmental toxicology; Toxicology (agriculture). Current work: Chem. effects on germ cells. Office: Savannah State Coll Dept Biology Savannah GA 31404 Home: 602 Wild Turkey Rd Savannah GA 31406

SINGH, JARNALL, biologist, educator, environ. researcher; b. Amritsar, Punjab, India, Oct. 26, 1941; came to U.S., 1965, naturalized, 1978; s. Mihan and Basant (Kaur) S.; m. Brijindar Kaur, June 26, 1968; children: Harpreet, Vijaypreet. B.S. in Agr, Punjab U., Chandigarh, India, 1961; M.S., Punjab Agrl. U., Ludhiana, India, 1964; Ph.D., Kans. State U., 1968. Asso. prof. biology Stillman Coll., Tuscaloosa, Ala., 1969-81, prof., 1981—, chmn. div. math. and sci., 1970-73, prin. investigator minority biomed. research support program, 1974—. Author: Simple Laboratory Exercises for General Biology, 1982; contbr. articles to profl. jours. Mem. Tuscaloosa Clean Air Com., 1970-74. Recipient Dedication to Equal Opportunity in Edn. award United Negro Coll. Fund, 1973; Oak Ridge Assoc. Univs. fellow, 1973; United Negro Coll. Fund fellow, 1979; Nat. Health Research Service fellow, 1983-84. Mem. AAAS, AAUP, Inst. Environ. Scis. Democrat. Subspecialties: Environmental toxicology; Teratology. Current work: Research on the teratogenicity of protein deficient diets mycotoxins and air pollutant gases such as carbon monoxide, sulphur dioxide and nitrogen dioxide. Office: Stillman Coll 3700 15th St Tuscaloosa AL 35403

SINGH, RAJENDRA PRATAP, surgery educator; b. Allahbad, India; came to U.S., 1973, naturalized, 1981; s. Akbal Bahadur and Kamla S.; m. Sushma Singh, Feb. 16, 1971; children: Sonia, Jay Pal. B.S., Banaras (India) U., 1958; M.B.B.S., M.S., Agra (India) U., 1968; M.A.M.S., Indian Acad., 1971. Diplomate: Am. Bd. Surgery. Registrar in gen. and thoracic surgery, U.K., 1968-72, Beckley, N.Y., 1973-76; surg. resident Marshall U., W.Va., 1976-77, instr., 1977-78, asst. prof., 1978—; attending surgeon VA, Raleigh Gen., Beckley hosps. Author: Immediate Care of Sick and Injured Children, 1978; contbr. numerous articles to med. jours. Fellow ACS, Royal Coll. Surgeons Eng., Royal Coll. Surgeons Scotland; mem. Indian Acad. Med. Sci. Subspecialty: Surgery. Current work: Clinical surgical research; angioaccess for hemodialysis. Office: 201 Woodcrest Dr Beckley WV 25801

SINGH, SHIVA PUJAN, microbiologist, microbiology educator; b. Tulsipur Majha, Gonda, India, July 5, 1947; came to U.S., 1971, naturalized, 1977; s. Ram Bahadur and Dukhana (Devi) S.; m. Patricia Ann Gangloff, Sept. 12, 1973; children: Suman R., Raj K. B.S., Pant U., Pantnagar, India, 1969, M.S., 1971; Ph.D., Auburn U., 1976. Grad. asst. Pant U., 1969-71, Auburn U., 1971-75; research assoc. Tuskegee Inst., 1976; acting head gen. biology Ala. State U., 1978, asst. prof., 1976-82, assoc. prof. microbiology, 1982—. Rep. United Way Campaign, Montgomery, Ala., 1977. Argonne Nat. Lab. Faculty Research Fellowship awardee, 1979. Mem. AAAS, Am. Soc. Microbiology, Ala. Acad. Sci. Democrat. Hindu. Subspecialties: Microbiology; Virology (biology). Current work: Baculovirus serology and biochemistry; bacterial physiology and biochemistry. Home: 2186 W Aberdeen Dr Montgomery AL 36116 Office: Ala State U 915 S Jackson St Montgomery AL 36195

SINGHAKOWINTA, AMNUAY, physician, oncologist, educator; b. Thailand, Oct. 25, 1939; s. Prasong and Chu S.; m. Boonploog Tancharoenpol, June 20, 1965; children: Pearl, Ann. M.D., U. Med. Sci., Bangkok, Thailand, 1963. Diplomate: Am. Bd. Internal Medicine, Am. Bd. Med. Oncology. Instr. Wayne State U., Detroit, 1971; asst. prof. med. oncology, 1972-75, assoc. prof., 1976—; vice-chief oncology dept. Harper-Grace Hosp., Detroit, 1975—. Contbr. chpts. to books. Fellow ACP; mem. Am. Assn. Cancer Research, Am. Soc. Clin. Oncology. Buddhist. Subspecialties: Cancer research (medicine); Receptors. Current work: Breast cancer research and treatment; steroid receptors and cancer. Home: 2875 Homewood Troy MI 48098 Office: 785 N Lapeer Rd Lake Orion MI 48035

SINGHAL, ANIL KUMAR, mech. engr.; b. Gwalior, India, Oct. 17, 1947; came to U.S., 1969, naturalized, 1982; s. Uma Shanker and Shri Kumari (Goel) S.; m. Maria Da Graca Singhal, Dec. 1, 1980; 1 son: Neel Shanker. B.S.M.E. (Merit scholar), Birla Inst. Tech. and Sci., Pilani, India, 1969; M.S.I.E., U. Houston, 1971. Registered profl. engr., Tex., La. Project engr., supr. engrng. Graver Tank & Mfg. Co. div. Aerojet Gen. Corp., 1971-80; lead engr. Foster Wheeler Corp., 1980-81; sr. mech. engr. Bechtel Pet Corp., 1981-82; mgr. equipment design Sci. Design Corp., Houston, 1982-83. Mem. ASME, Cryogenic Soc. Am., Tex. Soc. Profl. Engrs., La. Soc. Profl. Engrs. Subspecialties: Mechanical engineering; Cryogenics. Current work: Process equipment design, study of various metals of construction, process equipment design and engineering management. Project dir. devel. of light weight aluminum floating roof for tanks storing volatile substances. Home: 11211 Wood Lodge Houston TX 77077

SINGHVI, SAMPAT MANAKCHAND, pharmacist, researcher; b. India, Oct. 14, 1947; came to U.S., 1968, naturalized, 1971; s. Manakchand R. and Jayanta (Balai) S.; m. Usha S. Singhvi, July 17, 1971; children: Nikhil, Nilima. B.Pharm., Birla Inst. Tech. and Sci., Pilani, India, 1967; M.S., Phila. Coll. Pharmacy and Sci., 1970; Ph.D., SUNY, Buffalo, 1974; M.B.A., Rider Coll., 1979. Research investigator Squibb Inst. Med. Research, New Brunswick, N.J., 1974-78, sr. research investigator, 1978-79, research group leader, 1979—. Contbr. articles on drug metabolism, biopharmaceutics, pharmacokinetics and pharm. analysis to profl. jours. Mem. Am. Pharm. Assn., Acad. Pharm. Scis., Am. Soc. Pharmacology and Exptl. Therapeutics, Drug Metabolism Discussion Group, Rho Chi. Subspecialties: Pharmacokinetics. Current work: Pharmacokinetics, pharmacodynamics, biopharmaceutics, and drug metabolism in animals and man. Office: PO Box 191 New Brunswick NJ 08903

SINGWI, KUNDAN SINGH, physics educator; b. Udaipur, India, Mar. 13, 1919; came to U.S., 1959, naturalized, 1973; s. Khubilal and Jatanbai (Khamesra) S.; m. Helga Clara Grêve, Apr. 5, 1953; children: Veena Astrid, Sonita Maya. B.Sc., Allahabad (India) U., 1938, D.Sc., 1949. Head theoretical physics Atomic Energy Establishment, Trombay, India, 1953-58; sr. scientist Argonne (Ill.) Nat. Lab., 1961-67; prof. physics Northwestern U., Evanston, Ill., 1968—, chmn. physics dept., 1980-82; cons. in field. Contbr. numerous sci. articles to profl. publs. Recipient Contbn. to Scis. award Govt. of India, 1979; Contbn. to Sci. award India League Am., 1980, Assn. Indians in Am., 1981; NSF grantee, 1968. Fellow Am. Phys. Soc.; mem. Indian Nat. Sci. Acad. Subspecialties: Condensed matter physics; Theoretical physics. Current work: Many-particle physics; electron correlations in metals; electron-hole liquid in semiconductors. Office: Northwestern Univ B750 Tech Bldg Evanston IL 60201

SINHA, NAVIN KUMAR, molecular biologist; b. Patna, India, Oct. 14, 1945; s. Ramananda and Shanti (Sahay) S.; m. Mary Shultz, Nov. 21, 1971; children: Tara, Neil. B.Sc., Patna U., 1962, M.Sc., 1964; Ph.D., U. Minn., 1972. Lectr. botany Patna U., 1964; research assoc. M.I.T., 1972; postdoctoral fellow Princeton U., 1973-76; asst. prof. Waksman Inst., Rutgers U., 1976-82, assoc. prof., 1982—. Recipient Sir Ali Imam Gold medal, Univ. Gold medal Patna U., 1965. Mem. Am. Soc. Microbiology, Genetics Soc. Am. Subspecialties: Genetics and genetic engineering (biology); Biochemistry (biology). Current work: Current research: Molecular mechanisms of mutation; DNA replication; mechanisms of action of carcinogens; DNA-protein interaction. Office: Waksman Inst Microbiology Rutgers U Piscataway NJ 08854 Home: 8 Evans Ct Somerset NJ 08873

SINIBALDI, RALPH MICHAEL, SR., research biologist; b. Elmhurst, Ill., Nov. 9, 1947; s. John and Cecile (Klima) S. B.S., U. Ill., Chgo., 1970, M.S., 1974, PH.D., 1978. Damon Runyon-Walter Winchell postdoctoral fellow in biochemistry U. Ill. Med. Center, Chgo., 1978-80, research assoc., 1980-81; NIH postdoctoral fellow in devel. biology U. Chgo., 1981-82; sr. research biologist Zoecon Corp., Palo Alto, Calif., 1982—. Contbr. writings to profl. publs. in field. Mem. Am. Soc. for Cell Biology, Genetics Soc. Am. Roman Catholic. Subspecialties: Genetics and genetic engineering (biology); Genetics and genetic engineering (agriculture). Current work: My goal is to understand the molecular basis of heat stress in plants and to perhaps genetically engineer a better stress response. Office: 975 California Ave Palo Alto CA 94304

SINKOVICS, JOSEPH G., oncologist, virologist, educator; b. Budapest, Hungary, June 17, 1924; came to U.S., 1956; s. Joseph and Maria (Rajnocha) S.; m. (div.); children: Geza, Eszter. M.D., Petrus Pazmany U. Hungary, 1948. Diplomate: Am. Bd. Internal Medicine. Prof. medicine U. Tex.-M.D. Anderson Hosp., Houston, 1972-79, chief sect. tumor virology and immunology, 1968-79, cons. oncology, 1979—; prof. virology Baylor Coll. Medicine, Houston, 1980—; pvt. practice med. oncology, hematology, infectious diseases and internal medicine. Author: Die Grundlagen der Virusforschung, 1956, Medical Oncology, 1979, (with J.E. Harris) The Immunology of Malignant Disease, 1976; contbr. articles to profl. jours. Fellow ACP; mem. Am. Soc. Clin. Oncology, Am. Assn. Cancer Research, Am. Soc. Microbiology, Infectious Diseases Soc. Am., others. Club: Doctors (Houston). Subspecialties: Cancer research (medicine); Infectious diseases. Current work: Cancer immunology and immunotherapy. Cancer cell culture. Chemoimmunotherapy of human cancer. Home: 2336 N Braeswood Blvd Houston TX 77030 Office: 909 Frostwood Suite 153 Houston TX 77024

SINKS, LUCIUS FREDERICK, pediatric oncologist; b. Newburyport, Mass., Mar. 14, 1931; s. Allen Thurman and Anna Chose (Batchelder) S.; m. Lillion Sharpless Taylor, Mar. 6, 1934; children: George W., Lillian F., Lucius Frederick. B.S., Yale U., 1953; M.Sc., Chgo. State U., 1963; M.D., Jefferson Med. Coll., Phila., 1957. Diplomate: Am. Bd. Pediatrics. Intern Grad. Hosp. of U. Pa., 1957-58; resident in pediatrics Children's Hosp., Ohio State U., Columbus, 1961-63, fellow in pediatric hematology/oncology, 1963-64; NIH fellow Cambridge (Eng.) U., 1964-66; chief dept. pediatrics Roswell Park Meml. Inst., Buffalo, 1966-76; prof. pediatrics Georgetown U. Med. Sch., 1976-81; prof. pediatrics, chief div. pediatrics and adolescent oncology Tufts-New Eng. Med. Ctr., Boston, 1981—. Served to capt. M.C. USAF, 1958-61. Mem. Am. Assn. Cancer Research, Am. Soc. Clin. Oncology. Subspecialties: Pediatrics; Chemotherapy. Current work: Pediatric oncology and cell biology. Address: 171 Harrison Ave Box 184 Boston MA 02111

SIOMOS, KONSTADINOS, physicist; b. Larissa, Greece, June 30, 1947; s. Grigorios and Maria (Skotiniotou) S.; m. Patricia Mary Macfie, Oct. 25, 1975; children: Maria Fiona, Christina Alexandra. Vordiplom, U. Heidelberg, W.Ger., 1969, diploma in physics, 1972; Ph.D. in Physics, U. Cologne, W.Ger., 1974. Asst. prof. physics U. Cologne, 1972-75; mem. staff, researcher Nuclear Research Ctr., Jülich, W.Ger., 1975-76; asst. prof. physics U. Tenn., Knoxville, 1980-81; staff mem., prin. investigator laser spectroscopy health and safety research div. Oak Ridge Nat. Lab., 1976—. Contbr. numerous articles to profl. jours. DAAD scholar, 1970-72. Mem. Optical Soc. Am., German Phys. Soc. Subspecialties: Spectroscopy; Atomic and molecular physics. Current work: Laser atomic and molecular spectroscopy; dye laser technology and development; laser induced ionization processes of molecules and negative ions in low and high pressure gases and liquids; non linear laser spectroscopy of molecules in gases and liquids. Office: Oak Ridge Nat Lab PO Box X Oak Ridge TN 37830

SIPE, HERBERT JAMES, JR., chemistry educator; b. Lewistown, Pa., Aug. 17, 1940; s. Herbert James and Esther Louise (Bossinger) S. B.S. (NSF grantee), Juniata Coll., 1962; Ph.D. in Phys. Chemistry, U. Wis.-Madison, 1969. Research Corp. grantee, 1969; asst. prof. chemistry Hampden-Sydney (Va.) Coll., 1968-74, assoc. prof., 1974-81, prof., 1981—, chmn. dept. chemistry, 1976-82; postdoctoral research assoc. EPR and ENDOR Spectroscopy, 1980-81. Contbr. articles on chemistry to profl. jours. NSF grantee, 1972, 78. Mem. Am. Chem. Soc., AAAS, AAUP, Alpha Chi Sigma, Sigma Xi. Subspecialties: Physical chemistry; Organic chemistry. Current work: Electron spin resonance and ENDOR spectroscopic studies of bonding and structure in organometallic analogs of pharmaceutically active compounds. Office: Dept Chemistry Hampden-Sydney Coll Hampden-Sydney VA 23943

SIRGY, MAGDY JOSEPH, psychology educator; b. Cairo, May 31, 1952; U.S., 1970, naturalized, 1972; s. Joseph Ibrahim and Odette Mikaheel (Hosni) S.; m. Karen Sue Cornett, Nov. 15, 1973; 1 dau., Melissa Jane. B.A. in Psychology, UCLA, 1974; M.A., Calif. State U., 1977; Ph.D., U. Mass., 1979. Interogator/interpreter U.S. Mil. Intelligence, Schofield, Hawaii, 1971-73; personnel analyst Calif. State Employment Dept., Los Angeles, 1975; teaching asst. Calif. State U.-Long Beach, 1975-77; teaching assoc. U. Mass., Amherst, 1977-79; asst. prof. Va. Poly. Inst. and State U., Blacksburg, 1979—; research cons. Mktg. & Mgmt. Research Assocs., Greensboro, 1980. Author: Social Cognition and Consumer Behavior, 1983; contbr. articles to profl. jours. Cons. Ctr. for Vol. Devel., Blacksburg, 1981-82, Montgomery County Recreation Dept., 1982, Va. Tech. Student Union, 1982. Served with U.S. Army, 1971-73. Va. Tech. Honors Program grantee, 1981; Va. Tech. Learning Resource Ctr. grantee, 1980-81; Calif. State U. scholar, 1979. Mem. Am. Psychol. Assn., Assn. Consumer Research, Acad. Mktg. Sci., Am. Mktg. Assn. Subspecialties: Social psychology; Consumer psychology. Current work: Examination of personality variables (e.g. self-concept, achievement motivation) and other social psychol. variables in relation to economic behavior. Home: 2102 Birch Leaf Ln Blacksburg VA 24060 Office: Va Poly Inst and State Univ Blacksburg VA 24061

SIRICA, ALPHONSE EUGENE, cancer researcher, educator; b. Waterbury, Conn., Jan. 16, 1944; s. Alphonse Eugene and Elena Virginia (Mascolo) S. B.A., St. Michaels Coll., Winooski Park, Vt., 1965; M.S., Fordham U., 1968; Ph.D. in Biomed. Scis, U. Conn., 1976. Research asso., supr. drug evaluation div. Microbiol. Assocs., Cancer Chemotherapy Research Lab., Bethesda, Md., 1969-71; postdoctoral trainee dept. oncology McArdle Lab. for Cancer Research, U. Wis., Madison, 1976-79; asst. prof. anatomy and hepatic pathology U. Wis. Med. Sch., 1979—. Contbr. articles to profl. jours. Nat. Cancer Inst. grantee, 1981-83; NIH grantee, 1981-84. Mem. AAAS, Am. Assn. Cancer Research, Am. Assn. Pathologists, N.Y. Acad. Scis., Tissue Culture Assn., Am. Soc. Cell Biology. Subspecialties: Cancer research (medicine); Pathology (medicine). Current work: Chemical hepatocarcinogenesis, pathobiology of liver neoplasms, regulatory mechanisms controlling hepatocyte differentiation and proliferation, liver cell culture. Office: 1300 University Ave Madison WI 53706

SIRIGNANO, WILLIAM ALFONSO, aerospace and mechanical scientist; b. Bronx, N.Y., Apr. 14, 1938; s. Anthony P and Lucy (Caruso) S.; m. Molly Van Leeuwen, Oct. 29, 1966 (div. 1975); 1 dau., Monica Ann; m. Lynn Haisfield, Nov. 26, 1977; 1 dau., Jacqueline Hope. B.Aero.Engring., Rensselaer Poly. Inst., 1959; Ph.D., Princeton U., 1964. Mem. research staff Guggenheim Labs., aerospace, mech. scis. dept. Princeton U., 1964-67, asst. prof. aerospace and mech. scis., 1967-69, asso. prof., 1969-73, prof., 1973-79, dept. dir. grad. studies, 1974-78; George Tallman Ladd prof., head dept. mech. engring. Carnegie-Mellon U., 1979—; cons. industry and govt., 1966—; mem. emissions control panel Nat. Acad. Scis., 1971-73; lectr. and cons. NATO adv. group on aero. research and devel., 1967, 75, 80; chmn. nat. and internat. tech. confs. Asso. editor: Combustion Sci. and Tech, 1969-70; contbr. articles to nat. and internat. profl. jours., also research monographs. United Aircraft research fellow, 1973-74. Mem. Combustion Inst. (treas. internat. orgn. and chmn. Eastern sect.), Soc. Indsl. Applied Math., AIAA, ASME. Subspecialties: Combustion processes; Fluid mechanics. Current work: Turbulent reacting flows; spray composition; fuel-droplet heatingand vaporization; ignition, combustion of pulverized coal; combustion instability, fire safety. Home: 5391 Northumberland St Pittsburgh PA 15217 Office: Carnegie-Mellon U Pittsburgh PA 15213

SISENWINE, SAMUEL FRED, pharmacologist, chemist; b. Phila., Dec. 30, 1940; s. Harry and Mary (Levin) S.; m. Phyllis Goldstein, June 10, 1962; children: Marnie, Joel, Jill. B.S., Phila. Coll. Pharmacology and Sci., 1962; Ph.D., U. Pa., 1966. Teaching asst. U. Pa., Phila., 1962-66; sr. scientist Wyeth Lab. Inc., Radnor, Pa., 1966-68, unit supr., 1969-78, mgr., 1978—, corp. radiation health safety officer, 1971—; editorial adv. bd. Drug Metabolism and Disposition, 1981-83. Contbr. sci. articles to publs. Mem. Am. Soc. Pharmacology and Exptl. Therapeutics, Delaware Valley Radiation Health Safety, Delaware Valley Drug Metabolism Discussion Group. Subspecialties: Pharmacokinetics; Drug metabolism. Current work: Studies in drug metabolism in various species. Home: 609 Galahad Rd Plymouth PA 19462 Office: Wyeth Lab Inc Radnor PA 19087

SISLER, HARRY HALL, chemist, educator; b. Ironton, Ohio, Mar. 13, 1917; s. Harry C. and Minta A. (Hall) S.; m. Helen E. Shaver, June 29, 1940; children: Elizabeth A., David F., Raymond K., Susan C.; m. Hannelore L. Wass, Apr. 13, 1978. B.Sc., Ohio State U., 1936; M.Sc., U. Ill., 1938, Ph.D., 1939; Doctorate honoris causa, U. Poznan, Poland, 1977. Vice chmn. Prudential-Bache Securities, Inc., 1939-41; from instr. to asso. prof. U. Kans., Lawrence, 1941-46; from asst. prof. to prof. chemistry Ohio State U., Columbus, 1946-56; Arthur and Ruth Sloan vis. prof. chemistry Harvard, fall, 1962-63; prof., chmn. dept. chemistry U. Fla., Gainesville, 1956-68, dean, 1968-70, exec. v.p., 1970-73, dean grad. sch., 1973-79, Disting. Service prof. chemistry, 1979—; indsl. cons. W.R. Grace & Co., Martin Marietta Aerospace, Naval Ordnance Lab., TVA; chemistry adv. panel, also vis. scientists panel NSF, 1959-62; cons. USAF Acad., Battelle Meml. Inst., chmn. interinstl. com. nuclear research, Fla., 1958-64; mem. Fla. Nuclear Devel. Commn. Teaching Sci. and Math., 1958; chemistry adv. panel Oak Ridge Nat. Lab., 1965-69. Author: Electronic Structure, Properties, and the Periodic Law, 2d edit, 1973, Starlight-A Book of Poems, 1976, Of Outer and Inner Space—A Book of Poems, 1981, (with others) Gen. Chemistry—A Systematic Approach, 2d edit, 1959, Coll. Chemistry—A Systematic Approach, 4th edit, 1980, Essentials of Chemistry, 2d edit, 1959, A Systematic Laboratory Course in Chemistry, 1950, Essentials of Experimental Chemistry, 2d edit, 1959, Semimicro Qualitative Analysis, 1958, rev. edit., 1965, Comprehensive Inorganic Chemistry, Vol. V, 1956, Chemistry in Non-Aqueous Solvents, 1961, The Chloramination Reaction, 1977; cons. editor: Dowden, Hutchinson & Ross, 1971-78; series editor: Phys. and Inorganic Textbook Series, Reinhold Pub. Corp, 1958-70; contbr. articles to profl. jours. Decorated Royal Order North Star, Sweden).; Named Outstanding Chemist in South, 1969, Outstanding Chemist in Southeast Am. Chem. Soc., 1960, James Flack Norris award, 1979; recipient Outstanding Centennial Achievement award Ohio State U., 1970. Mem. Am. Chem. Soc. (nat. chmn. div. chem. edn. 1957-58, exec. com. 1957-60, bd. publ. Jour. Chem. Edn. 1956-58), Phi Beta Kappa, Sigma Xi, Phi Delta Kappa, Phi Lambda Upsilon, Phi Kappa Phi, Alpha Chi Sigma. Methodist. Subspecialties: Inorganic chemistry; Synthetic chemistry. Current work: Reactions of chloramine with nitrogen and phosphorous bases; synthesis of hydrazine derivatives and related high energy fuels, molecular adducts of dinitrogen tetroxide. Patentee in field. Home: 6014 NW 54th Way Gainesville FL 32606

SISMANIS, ARISTIDES, otolaryngologist, educator; b. Athens, Greece, Nov. 6, 1949; came to U.S., 1973, naturalized, 1982; s. Dimitrios and Stella (Petropoulos) S.; m. Anna Rozaki, Mar. 10, 1974; children: Stamatina, Dimitrios. M.D., Med. Sch. Athens, 1973. Intern. in gen. surgery Maimonides Hosp., Bklyn., 1973-74; resident in surgery Bklyn. Jewish Hosp., 1973-75; resident in otolaryngology L.I. Coll. Hosp., Bklyn., 1975-78; also Univ. Hosp., Jamaica Plains (N.Y.) VA Hosp.; in head and neck surgery Boston U., 1978-79; fellow in otology and neurotology Otology Group, Nashville, 1979-80; asst. prof. otolaryngology Med. Coll. Va. Hosp., Va. Commonwealth U., Richmond, 1980—. Contbr. numerous med. articles to profl. publs. Fellow Am. Acad. Otolaryngology; mem. AMA (Physician's Recognition award 1980, 82). Subspecialty: Otorhinolaryngology. Current work: Otology, otorhinolaryngology. Home: 1917 Windingridge Dr Richmond VA 23233 Office: MCV Sta Box 146 Richmond VA 23298

SISSON, THOMAS RANDOLPH CLINTON, medical educator; b. Winnipeg, Man., Can., Jan. 22, 1920; s. Lorne Randolph Clinton and Edna (Wilson) S.; m. Shirley Anne Robson, May 5, 1945; children: Geoffrey R.L., Peter A.W., Paul C.R. A.B., Colgate U., 1941; M.D., Temple U., 1944. Diplomate: Am. Bd. Pediatrics. Intern St. John's Riverside Hosp., Yonkers, N.Y., 1944-45; resident U. Rochester Med. Center, 1946-48, sr. instr. pediatrics, obstetrics, 1953-59; asso. dir. clin. research Geigy Pharm., Ardsley, N.Y., 1960-64; asso. prof. pediatrics Albert Einstein Coll. Medicine, Bronx, 1959-65; assoc. prof. Loma Linda U., 1965-67; prof. pediatrics, dir. neonatal research Temple U., Phila., 1967-78; clin. prof. pediatrics Rutgers U., Piscataway, N.J., 1978—; chmn. dept. pediatrics Raritan Bay Health Service Corp., Perth Amboy, N.J., 1978—; mem. com. on photobiology NAS/NRC, 1972-76; cons. NIH, 1972—. Contbr. articles to profl. jours.; editorial bd.: Clin. Pediatrics, 1976—. Served to comdr. USNR, 1950-52, 45-46; PTO. Buswell Faculty fellow, 1956-59. Fellow Am. Acad. Pediatrics, Am. Coll. Nutrition; mem. Am. Soc. Photobiology (councillor 1972-76), Soc. Pediatric Research, Am. Inst. Nutrition, Internat. Soc. Chronobiology, Am. Fedn. Clin. Research, Alpha Omega Alpha. Anglican. Subspecialties: Neonatology; Nutrition (medicine). Current work: Photobiology, iron and bilirubin metabolism, neonatal nutrition chronobiology. Address: Dept Pediatric Raritan Bay Health Service Corp 530 New Brunswick Ave Perth Amboy NJ 08861

SISTERSON, JANET MARGOT, physicist; b. Edinburgh, Scotland, July 7, 1940; d. Thomas James and Lucy Margaret (Smith) Brownlee; m. Leonard Keith Sisterson, Oct. 23, 1965; children: James, Mark. B.S. in Physics, U. Durham, Eng., 1961; Ph.D. in High Energy Physics, Imperial Coll. U. London, Eng., 1965. Physicist London Hosp., 1964-66; sr. physicist Chelsea Hosp. for Women, London, 1966-68; research fellow Cambridge (Mass.) Electron Accelerator, 1968-73; research assoc. Cyclotron Lab., Harvard U., Cambridge, 1973—. Contbr. articles to profl. publs. Exec. bd. elem. sch. PTA, Lexington, Mass. Mem. Am. Phys. Soc., Am. Assn. Physicists in Medicine, Am. Women in Sci. Subspecialties: Nuclear physics; Medical physics. Current work: Proton activation analysis; application of proton beams in medicine. Developer sci. exhibits in field including Harvard Cyclotron Lab. Home: 36 Webb St Lexington MA 02173 Office: Harvard Cyclotron Lab 44 Oxford St Cambridge MA 02138

SITARZ, ANNELIESE LOTTE, pediatric hematologist/oncologist; b. Medellín, Colombia, Aug. 31, 1928; came to U.S., 1935, naturalized, 1946; d. Hans and Elisabeth (Noll) S. B.A. cum laude, Bryn Mawr Coll., 1950; M.D., Columbia U., 1954. Diplomate: Am. Bd. Pediatrics, Sub-bd. pediatric hematolegy/oncology. Intern Boston Children's Med. Ctr., 1954-55; resident Babies Hosp., Columbia-Presbyn. Med. Center, N.Y.C., 1955-57, vis. fellow, 1957-59, asst. pediatrician, 1959-64, asst. attending pediatrician, 1964-74, assoc. attending pediatrician, 1974-83, attending pediatrician, 1983—; assoc. prof. clin. pediatrics Columbia U., 1974-83, prof. clin. pediatrics, 1983—; cons. Overlook Hosp., Summit, N.J., 1975—. Contbr. articles to profl. jours. NIH fellow, 1957-61; Am. Cancer Soc. fellow, 1962-65. Fellow Am. Acad. Pediatrics; mem. Am. Soc. Hematology, Am. Soc. Clin. Oncology, Am. Assn. Cancer Research, Internat. Soc. Hematology, Harvey Soc., N.Y. Acad. Sci., N.Y. Soc. for Study of Blood. Subspecialties: Pediatrics; Oncology. Current work: Pediatric cancer treatment and pediatric blood disorders; co-principal investigator children's cancer study group. Office: Babies Hosp Broadway and 167th St New York NY 10032

SITTERLY, CHARLOTTE MOORE, physicist; b. Ercildoun, Pa., Sept. 24, 1898; m. (widowed). B.A., Swarthmore Coll., 1920, D.Sc. (hon.), 1962; Ph.D. in Astronomy, U. Calif., Berkeley, 1931; hon. doctorate, U. Kiel, W.Ger., 1968; D.Sc. (hon.), U. Mich., 1971. Computer Princeton U. Obs., 1920-25, research asst., 1931-36, research assoc., 1936-45; computer Mount Wilson Obs., 1925-28; physicist Atomic Physics div. Nat. Bur. Standards, Washington, 1945-68, Office of Standard Reference Data, 1968-70, ret., 1970, guest worker, 1971—; research physicist Naval Research Lab., Space Sci. Div., part-time, 1971-78, guest worker, 1978—; adv. bd. Office of Critical Tables, NRC, 1961-69, mem. com. on line spectra of elements, 1925-70. Author: (with others) Revision of Rowland's Preliminary Table of Solar Spectrum Wavelengths, 1928, Atomic Lines in the Sun-Spot Spectrum, 1933, (with Henry N. Russell) The Masses of the Stars, 1940, A Multiplet Table of Astrophysical Interest, 1945, (with Harold D. Babcock) The Solar Spectrum, 16600 to 113495, 1947, Atomic Energy Levels: Vol. I, 1949, Vol. II, 1952, Vol. III, 1958, An Ultraviolet Multiplet Table: Sect. I, 1950, Sect. 2, 1952, Sects. 3, 4, 5, 1962, (with Minnaert and Houtgast) The Solar Spectrum 2935Å to 8770Å, 1966, Selected Tables of Atomic Spectra, 10 Sects, 1965-83, Bibliography on the Analysis of Optical Atomic Spectra, 4 sects, 1968-69, Ionization Potentials and Ionization Limits, 1970; Contbr. numerous articles to profl. jours. Recipient Dept. Commerce Exceptional Service award, 1960; Fed. Woman's award, 1961; Annie Jump Cannon Centennial medal, 1963; Career Service award Nat. Civil Service League, 1966. Fellow Optical Soc. Am. (William F. Meggers award 1972), Am. Phys. Soc., Washington Acad. Scis.; mem. Internat. Astron. Union (pres. commn. on fundamental spectroscopic data 1961-67, also rep. to various joint commns. and coms.), Royal Astron. Soc. (fgn. assoc.), Philos. Soc. Washington, AAAS (v.p. sect. D on astronomy 1952), Am. Astron. Soc. (v.p. 1958-60, Annie J. Cannon prize 1937), Soc. Applied Spectroscopy (hon.), Société Royale des Sciences de Liege (Belgium) (corr.), Phi Beta Kappa, Sigma Xi. Subspecialties: Atomic and molecular physics; Solar physics. Current work: Work on identification of lines observed in solar and sun-spot spectra as to chemical origin. Studies of atoms and molecules present in solar atmosphere. Identifications of lines observed in far ultraviolet solar spectra from rockets and orbiting solar observatories. Critical evaluation of laboratory data on atomic spectra, of special interest to astrophysics for determination of abundances of chemical elements in sun and stars. Home: 3711 Brandywine St NW Washington DC 20016

SIURU, WILLIAM DENNIS, JR., air force officer, engineering administrator and educator; b. Detroit, Jan. 29, 1938; s. William D. and Bertha S. (Lindfors) S.; m. Nancy K. Watson, June 22, 1962; children: Brian, Andrea. B.S. in Mech. Engring, Wayne State U., 1960; M.S. in Aero. Engring, Air Force Inst. Tech., 1964; Ph.D., Ariz. State U., 1975. Registered profl. engr., Ohio, Colo. Commd. 2d lt. U.S. Air Force, 1960, advanced through grades to col., 1983; project engr. devel. planning (Air Force Space Div.), Los Angeles, 1964-68, chief space launch systems br. fgn. tech. div., Wright-Patterson AFB, Ohio, 1968-71; chief support tech br. Air Force Rocket Propulsion Lab., Edwards AFB, Calif., 1974-76; asst. prof. dept. engring. U.S. Mil. Acad., West Point, N.Y., 1976-79; comdr. Frank J Seiler Research Lab. (U.S. Air Force Acad.), Colo., 1979-83; also mem. faculty. (Frank J Seiler Research Lab., U.S. Air Force Acad.); div. flight systems engring. Wright-Patterson AFB, Ohio, 1983—. Contbr. numerous articles on engring. to profl. jours. Decorated Legion of Merit. Mem. AIAA. Subspecialties: Aeronautical engineering; Mechanical engineering. Current work: Management of miliary research and development. Office: ASD/ENF Wright-Patterson AFB OH 45433

SIUTA-MANGANO, PATRICIA, biochemist; b. Yonkers, N.Y., Aug. 18, 1953; d. Joseph and Beatrice (Stepanski) Siuta; m. Richard M. Mangano, Oct. 28, 1978. B.S., Fordham U., 1975; Ph.D., Johns Hopkins U., 1981. Assoc. research biochemist Stauffer Chem. Co., Dobbs Ferry, N.Y., 1981—. NIH Predoctoral Fellowship awardee, 1975-81. Mem. Am. Chem. Soc., N.Y. Acad. Scis., Phi Lambda Upsilon, Phi Beta Kappa. Subspecialty: Biochemistry (biology). Current work: Research and development in the area of biochemical technology. Office: Stauffer Chem Co Livingstone Ave Dobbs Ferry NY 10522

SIVAK, ANDREW, life scis. mgr., researcher; b. New Brunswick, N.J., May 31, 1931; s. Andrew and Isabelle (Paragh) S.; m. Mary Margaret Sivak, May 24, 1958; children: Paul, Thomas, Gustav. B.S., Rutgers U., 1952, M.S., 1957, Ph.D., 1960. Staff biochemist Arthur D. Little, Inc., Cambridge, Mass., 1961-63, sr. staff, 1975-77, sect. mgr. bio/med. research tech., 1977—, v.p., 1979—; research dir. Bio/Dynamics, Inc., East Millstone, N.J., 1963-64; research assoc. Inst. Environ. Medicine, N.Y. U. Med. Ctr., 1964-68, asst. prof., 1968-71, assoc. prof., 1971-74; vis. lectr. M.I.T., 1975—. Contbr. articles to profl. jours. Served with USNR, 1952-72. USPHS fellow, 1960-61. Mem. Am. Coll. Toxicology, Am. Assn. Cancer Research, AAAS, Am. Soc. Cell Biology, Am. Soc. Microbiology, Environ. Mutagen Soc., Soc. Toxicology, Tissue Culture Assn., Sigma Xi. Subspecialties: Cell biology; Toxicology (medicine). Current work: Carcinogenesis, tumor promotion, genetic toxicology, risk assessment. Office: Acorn Park Cambridge MA 02140

SIVERTSON, JOHN NEILOS, statistician; b. Everett, Mass., Dec. 28, 1915; s. John S. and Margaret S.; m. Annette Thurston, Apr. 20, 1939; children: Carol Sivertson Dehnert, John N. B.A., Northeastern U., 1938; M.S., Rutgers U., 1972. Food chemist First Nat. Stores, Inc., Somerville, Mass., 1938-39; chemist Lever Bros. Co., Cambridge, Mass. and Balt., 1939-40, Am. Smelting & Refining Co., Balt., 1940-42, So. Acid & Sulphur Co., Port Arthur, Tex., 1942-48, Memphis, 1942-48, Pasadena, Tex., 1942-48, Johnson & Johnson, New Brunswick, N.J., 1948-67, mgr. stats. and computer ops., 1967-79; cons., 1979—; assoc. J. A. Keane & Assocs., Princeton, N.J., 1979—; pres. Warren (N.J.) Statis. Services, 1982—. Pres. Edison Twp. Republican Club, 1960; v.p. Warren Twp. Bd. Health, 1963-67. Mem. Am. Chem. Soc., Am. Soc. Quality Control, Am. Statis. Assn., Biometric Soc.

Subspecialties: Statistics; Analytical chemistry. Current work: Experimental design, computers in quality control; biostatistics, non-parametric statistics. Address: 7 Katherine Dr RD 1 Warren NJ 07060

SIX, HOWARD RONALD, microbiology educator; b. Princeton, W.Va., Jan. 5, 1942; s. Howard Berkeley and June Joan (Pruitt) S.; m. Bobbye June Young, Feb. 14, 1964; children: Robert H., Kimberley G. B.A., David Lipscomb Coll., Nashville, 1963; Ph.D., Vanderbilt U., 1972. Postdoctoral fellow in pharmacology Washington U., St. Louis, 1972-74; research assoc. in molecular biology Vanderbilt U., Nashville, 1974-75; asst. prof. microbiology and immunology Baylor Coll. Medicine, Houston, 1975-82, assoc. prof., 1982—; adj. assoc. prof. immunology program U. Tex. Sch. Biomed. Scis., Houston, 1982—. Co-editor: Liposomes and Immunobiology, 1980; author articles. Mem. Am. Soc. for Microbiology, Am. Assn. Immunologists, Infectious Disease Soc. Am., Am. Assn. Virologists, Soc. for Exptl. Biology and Medicine. Subspecialties: Infectious diseases; Microbiology. Current work: Immunity to influenza virus; liposomes as adjuvants; antigenic variation, viral immunology, liposomes, immunobiology and vaccines.

SJOLUND, RICHARD DAVID, plant physiologist, educator; b. Iron River, Mich., Dec. 9, 1939; s. Runar Ray Sjölund and Helen Lucille (James) S. B.S., U. Wis., Milw., 1963; Ph.D., U. Calif.-Davis, 1968. Botanist NASA Ames Research Ctr., Moffett Field, Calif., 1963-64; research asst. U. Calif.-Davis, 1964-68; asst. prof. botany U. Iowa, Iowa City, 1968-73, assoc. prof., 1974—. Contbr. articles to profl. jours. Vol. Big Bros./Big Sisters Program, 1977—, bd. dirs., 1981—. Old Gold fellow, 1977; MUCIA grantee, 1982; Deutscher Akademischer Austauschdienst fellow, 1982. Mem. Am. Soc. Plant Physiologists, Bot. Soc. Am., Electron Microscopy Soc. Am., Iowa Microbeam Soc. (pres.). Subspecialties: Plant cell and tissue culture; Plant physiology (agriculture). Current work: Development of vascular tissue in cell cultures, function of phloem cells in tissue cultures, role of the sieve element membranes in transport of carbohydrates. Office: Dept Botany U Iowa Iowa City IA 52242

SKALAFURIS, ANGELO JAMES, applied mathematician, engr., researcher; b. Pitts., Dec. 9, 1931; s. James Thomas and Katherine (Marinos) S.; m. Marie Consuelo Guerrero, Aug. 12, 1967 (div.); 1 son: Christopher. B.S., Ill. Inst. Tech., 1954; M.S., U. Chgo., 1958; Ph.D., Brandeis U., 1963. Registered profl. engr., N.Y., Pa., D.C. Research physicist U. Chgo., 1954-55; nuclear engr. G.L. Martin, Balt., 1956-57; ops. research analyst Caywood Schiller, Chgo., 1958; cons. Allied Research Assocs., Boston, 1959-60; research physicist Harvard Obs., 1960-63; Fulbright prof. U. Athens, U. Paris, 1963-64; research assoc. Space Inst., NASA, N.Y.C., 1965-66; prof. physics CCNY, 1967-69; prof. math. Bartol Research Found., Swarthmore, Pa., 1970-72; sr. research assoc. SUNY, Albany, 1972-74; sr. staff researcher Naval Research Lab., Washington, 1974—; cons. Re-Energy Systems, Inc. Contbr. chpts. to books, articles to profl. jours. Nat. Acad. Scis. fellow, 1965-66. Mem. ASME, Am. Phys. Soc., Am. Astron. Soc. Subspecialties: Systems engineering; Applied mathematics. Current work: Infra-red image processing, computor simulations, artificial intelligence, cogeneration and solar energy systems. Home: 2401 Calvert St NW Washington DC 20008 Office: Naval Research Lab Code 7931 Washington DC 20375

SKARDA, ROMAN THOMAS, veterinary and anesthesiology educator; b. Bitterfeld, Germany, Mar. 9, 1944; s. Hubert Ludwig and Viktoria Catherina (Liska) S.; m. Nicole Coderey, Sept. 5, 1975; children: Michael, Amanda, Emmanuelle. Eidg Fachpruefung fuer Tieraerzte, U. Zurich, 1969, M., 1972. Asst. prof. vet. anesthesiology Ohio State U., Columbus, 1975—. Contbr. articles to profl. jours., chpt. in textbook. Served to capt., inf. Swiss Army, 1963-74. Recipient Rotary Found. award and postgrad. scholarship, 1974-75. Mem. Gruetli Soc., Internat. Orgn. Christian Businessmen, Phi Zeta., AVMA (autotutorial excellence award 1981), Am. Soc. Vet. Anesthesiology, Am. Assn. Equine Practitioners, Assn. Swiss Officers. Liberal. Roman Catholic. Club: Swiss Home Assn. Subspecialties: Vet. med. anesthesia; Anesthesiology. Current work: Local anesthesia in veterinary medicine, neuroanatomy, neurophysiology, pharmacology, spinal and epidural anesthesia in cows and horses, new techniques. Home: 16047 Hawn Rd Plain City OH 43064 Office: 1935 Coffey Rd Columbus OH 43210

SKEEN, LESLIE CARLISLE, neuroscientist, psychologist, educator; b. Dearborn, Mich., Feb. 28, 1942; s. Henry Alexander and Margaret Ann (Young) S.; m. Julie Taylor, Oct. 15, 1964; children: Christopher, Alicia, Caldwell. B.A., Fla. State U., 1966, M.A., 1968, M.S., 1969, Ph.D., 1973. Postdoctoral fellow Duke U. Med. Center, Durham, N.C., 1973-75, asst. prof. med. research, 1975-77; asst. prof. psychology Sch. Life and Health Scis. and Inst. Neurosci., U. Del., Newark, 1977—, dir., 1978—. Contbr. numerous articles on evolution of anthropoid intelligence and olfactory system to profl. jours. Served with USAF, 1958-62. NIMH fellow, 1975-76; NSF fellow, 1969-73; NIH grantee, 1978-81; U. Del. Research Found. grantee, 1978-79; NSF grantee, 1976-77. Mem. AAAS, Soc. Neurosci., Am. Anat. Assn., Del. Soc. Neurosci. (pres.), Am. Chem. Soc., Sigma Xi. Subspecialties: Comparative neurobiology; Neuropsychology. Current work: Apply neuroanatomical, neurochemical and behavioral methods to questions concerning the evolution of subdivisions in the frontal lobe, development and organization of olfactory system functions. Office: 220 Wolf Hall U Del Newark DE 19711

SKEIST, IRVING, polymer consultant; b. Worcester, Mass., Apr. 9, 1915; s. Samuel and Bertha (Glazer) S.; m. Dorothy Geist, June 18, 1939; children: Judith R. Skeist Goodman, Laurie Skeist Sewall, Robert J., Helen D. B.S., Worcester Poly. Inst., 1935; M.S., Bklyn. Poly. Inst., 1943, Ph.D., 1949. Research chemist Celanese Corp., 1941-51; tech. dir. Newark Parraffine Paper Co., 1951-53, Am. Molding Powder, 1953-54; mktg. specialist Gering Products, 1953-54; pres. Skeist Labs., Inc., Livingston, N.J., 1955—. Author and editor: also articles Handbook of Adhesives. Mem. Assn. Cons. Chemists and Chem. Engrs., Am. Chem. Soc., ASTM, Sigma Xi, Phi Lambda Upsilon. Subspecialties: Polymer chemistry; Coal. Current work: Techno-economic studies, laboratory research and consultant on polymers, plastics, adhesives, coatings and fibers. Patentee in field. Office: Skeist Labs Inc 112 Naylon Ave Livingston NJ 07039

SKELLEY, GEORGE CALVIN, JR., animal scientist, educator; b. Boise City, Okla., Jan. 28, 1937; s. George Calvin and Catherine Bell (May) S.; m. Aletha Clair Brown, June 22, 1958; children: Mary Laura, Martha Alice. B.S., Okla. Panhandle State U., 1958; M.A., U. Ky., 1960, Ph.D., 1963. Research assoc. U. Ky., 1958-62; asst. prof. animal sci. Clemson U., 1962-67, assoc. prof., 1967-72, prof., 1972—; cons. in field. Contbr. numerous articles, abstracts to profl. jours., 1962—; author numerous research series' publs., exptl. sta. circulars and bulls. Named to Honor Roll of Profs. Clemson U., 1981, 82. Mem. Am. Soc. Animal Sci. (cert. animal scientist), Am. Meat Sci. Assn., Inst. Food Technologists, Block and Bridle (hon.), Sigma Xi, Alpha Zeta (hon.), Gamma Sigma Delta. Republican. Methodist. Lodge: Kiwanis. Subspecialties: Animal nutrition; Food science and technology. Current work: Nutrition and treatment of beef and pork on meat. Home: 112 Knight Circle Clemson SC 29631 Office: Dept Animal Sci Clemson U Clemson SC 29631

SKELSKIE, STANLEY IRVIN, cons. co. exec.; b. Boston, May 10, 1922; s. Arthur Nathan and Ida Lee (Lipinski) S.; m. Marion Esther Thornton, June 5, 1948; children: Barbara, Judith, Cynthia, Arthur. B.S., M.I.T., 1946. Process engr. Welch Grape Jiuce Co., Westfield, N.Y., 1946-56; dir. quality control Loblaw's, Buffalo, 1963-69; dir. research and devel. Ocean Spray Co., Hanson, Mass., 1963-69; corp. tech. dir. Freezer Queen div. Nabisco, Buffalo, 1969-70; v.p. Herbert V. Shuster, Inc., Quincy, Mass., 1970—; cons., tchr. in field. Chmn. Westwood (Mass.) Bd. Health, 1972-76; mem. fin. com. Mass. Dept. Pub. Health, 1975-76. Served to 1st lt. USAAF, 1944-45. Decorated Air medal with seven oak leaf clusters. Mem. Am. Soc. for Qualtiy Control (cert., mem. editorial bd. food, drug and cosmetic div. 1976—), Internat. Assn. Sanitarians, Inst. Food Technologists, Am. Assn. Cereal Chemists. Jewish. Subspecialties: Food science and technology; Biomedical engineering. Current work: Consultant to food, drug, cosmetic, medical device, chemical speciality industries. Patentee food processing field. Home: 36 Woodridge Rd Westwood MA 02090 Office: 5 Hayward St Quincy MA 02171

SKELTON, DENNIS LEE, optical systems engr.; b. Pontiac, Mich., June 17, 1951; s. Walter Kenneth and Phyllis Elaine (Walter) S. B.S., U. Mich., 1973; M.S., U. Colo., 1978. Observer Radio Astronomy Obs., U. Mich., Ann Arbor, 1974; research asst. Lab. for Atmospheric and Space Physics, U. Colo., Boulder, 1975-78; systems engr., electro-optics design (Ball Aerospace Systems Div.), Boulder, 1978-82; staff optical engr. optical systems engring. Itek Optical Systems, Lexington, Mass., 1982—. Mem. Am. Astron. Soc., Soc. Photo-optical Instrumentation Engrs. Methodist. Subspecialties: Solar physics; Optical engineering. Current work: Solar physics, stellar atmospheres, radiative transfer, the design and operation of scientific instrumentation for aerospace applications.

SKILLING, DARROLL DEAN, research plant pathologist; b. Carson City, Mich., June 18, 1931; s. Gay and Thelma Estella (Howe) S.; m. Marie Lou Harper, June 10, 1951; children: Ann, James, Stephen. B.S., U. Mich., 1953, M.F., 1954; Ph.D., U. Minn., 1968. Research forester Lake States Forest Expt. Sta., St. Paul, 1954-61; research plant pathologist North Central Forest Expt. Sta., St. Paul, 1961-70, prin. plant pathologist, 1970—; adj. prof. plant pathology U. Minn., 1969—. Contbr. articles to profl. jours. Recipient Superior Service award Dept. Agr., 1976. Mem. Internat. Soc. Plant Pathologists, Am. Phytopath. Soc., Internat. Union Forest Research Orgns., Sigma Xi, Gamma Sigma Delta. Subspecialties: Plant pathology; Integrated pest management. Current work: Biological control of forest disease problems in nurseries and plantations. Office: 1992 Folwell Ave Saint Paul MN 55108

SKINNER, THOMAS PAUL, computer science educator, consultant; b. N.Y.C., May 17, 1944; s. Harold P. and Blanche A. (Schifferdecker) S.; m. Linda Joan Grillo, June 5, 1971; 1 dau., Kristin Lin. S.B., MIT, 1966; M.S., Worcester Poly. Inst., 1978; Ph.D., Boston U., 1982. Mem. research staff MIT Project MAC, Cambridge, Mass., 1966-71; sr. engr. Urgan Scis., Inc., Wellesley, Mass., 1971-72; treas., prin. engr. Systems of Security Inc., Boston, 1972-78; pres. Microcom Assocs., Framingham, Mass., 1978—; asst. prof. computer sci. Boston U., 1979—. Contbr. articles to profl. jours. Mem. IEEE, IEEE Computer Soc., Assn. Computing Machinery. Subspecialties: Operating systems; Distributed systems and networks. Current work: Microprocessor-based distributed systems; microprocessors; computing networks; distributed operating systems; personal computing; computer architecture systems. Home: 15 Rolling Dr Framingham MA 01701 Office: Boston U 755 Commonwealth Ave Room B4 Boston MA 02215

SKIPSKI, VLADIMIR P(AVLOVICH), research biochemist, educator; b. Ugrojedy, Ukraine, Russia, Oct. 15, 1913; s. Pavel G. and Anna I. (Novickova) S.; m. Irene A. Lysloff, Sept. 5, 1959 (div.); 1 son: Pavel V. M.S. equivalent, Kiev State U., 1938; Ph.D. equivalent, Inst. Exptl. Biology and Pathology, Kiev, 1941, U. So. Calif., 1956. Research assoc. Kavir Research Inst., Santa Barbara, Calif., 1950-51; research assoc. Sloan Kettering Inst. Cancer Research, N.Y.C., 1956-59, assoc., 1960-69, asoc. mem., 1970-82, head lab., 1979-82; asst. prof. biochemistry Cornell U. Grad. Sch. Med. Scis., 1960-70, assoc. prof., 1970-82. Contbr. to books, articles in field. Swift and Co. fellow, 1951-56; Damen Runyen Meml. Fund for Cancer Research grantee, 1971-73; NIH grantee, 1974-82. Mem. Am. Soc. Biol. Chemists, Am. Physiol. Soc., Biochem. Soc. Gt. Britain, Soc. Gen. Physiologists, Am. Assn. Cancer Research, Harvey Soc., N.Y. Lipid Research Club, Soc. Complex Carbohydrates, AAAS, Sigma Xi. Greek Orthodox. Subspecialties: Biochemistry (biology); Cancer research (medicine). Current work: Effect of malignant tumors upon glycosphingolipid protiles in host blood serum; possible role of glycosphingolipids and glycoproteins in formation of metastases. Home: 148 Wallace Ave Mount Verno NY 1055

SKOFRONICK, JAMES G., physicist; b. Merrill, Wis., Oct. 11, 1931; s. Gust Alfred and Esther Catherine (Franke) S.; m. Dorothy Nilles, Sept. 12, 1959; children: Gregory, Gail, Gary, Gretchen. B.S.E.E., U. Wis., 1959, M.S. in Physics, 1961, Ph.D., 1964. Research asst. U. Wis., Madison, 1958-64; vis. research asst. Fla. State U., Tallahassee, 1964-65, asst. prof. dept. physics, 1965-69, assoc. prof., 1969-74, prof., 1974—. Author: (with J.R. Cameron) Medical Physics, 1978. Served with USAF, 1951-55. Grantee, 1970-74; Max Planck grantee, summers 1980-82. Mem. Am. Phys. Soc., European Phys. Soc. Democrat. Roman Catholic. Subspecialties: Condensed matter physics; Medical physics. Current work: Dynamics of solid surfaces. Office: Dept Physic Fla State U Tallahassee FL 32306

SKOLNICK, PHIL, neuropharmacologist; b. N.Y.C., Feb. 26, 1947; s. David Louis and Gertrude (Gewirtzman) S. B.S. summa cum laude, L.I. U., 1968; Ph.D. in Pharmacology, George Washington U. Sch. Medicine, 1972. USPHS trainee in pharmacology/toxicology, 1968-71; teaching fellow George Washington U., 1971-72; staff fellow NIH, Bethesda, Md., 1972-75, sr. staff fellow, 1975-77, pharmacologist, sr. investigator, 1977—; nat. and internat. lectr. in field. Contbr. articles to profl. publs. Mem. Am. Soc. Pharmacology and Exptl. Therapeutics, Internat. Soc. Neurochemistry, Soc. Biol. Psychiatry (A.E. Bennett award 1980). Subspecialties: Neuropharmacology; Psychopharmacology. Current work: Investigation of neurochemical bases of anxiety and sleep disorders. Research on anxiety, sleep disorders, benzodiazepines, adenosine receptors, neurotransmitters. Patentee (with S. Paul) radioreceptor assay for benzodiazepines in plasma. Office: NIH Bldg 4 Room 212 Bethesda MD 20205

SKOLNIK, HERMAN, retired organic chemist; b. Harrisburg, Pa., Mar. 22, 1914; s. Morris and Edith (Locke) S.; m. Ida Kopp, Dec. 25, 1938 (dec.); children: Carol D. Skolnik Czitron, Susan J.; m. Emma-June H. Tillmanns, Apr. 20, 1981. B.S. in Chem. Engring, Pa. State U., 1937; M.S. in Organic Chemistry, U. Pa., 1941, Ph.D., 1943. With Hercules Research Ctr., Hercules, Inc., Wilmington, Del., 1942-79, research mgr., 1951-79, ret., 1979; cons. in field. Author: Multi-Sulfur and Oxygen Five and Six Numbered Heterocycles, 2 vols, 1966, A Century of Chemistry, 1976, The Literature Matrix of Chemistry, 1982; contbr. numerous sci. articles to profl. publs. Mem. bd. dirs. Del. Symphony, Wilmington, 1976-83. Recipient award Phila. Patent Assn., 1978. Mem. Am. Chem. Soc. (editor: Jour. Chem. Info. and Computer Scis. 1960-82, Crane-Patterson award Columbus/Dayton chpt. 1969, Div. Chem. Info. award for excellence 1976), AAAS, TAPPI. Subspecialties: Organic chemistry; Chemical information science. Current work: Chemical information science; history and philosophy of chemistry. Home: 239 Waverly Rd Wilmington DE 19803

SKORYNA, STANLEY CONSTANTINE, medical educator, researcher, consultant; b. Warsaw, Poland, Sept. 4, 1920; s. Constantine Gregory and Alexandra Lydia (Fabian) S.; m. Jane Marie Polud, Aug. 8, 1970 (dec.); children: Christopher, Richard, Elizabeth. M.D., U. Vienna, 1943; M.Sc., McGill U., 1950, Ph.D., 1963. With McGill U., 1947—; dir. Gastrointestinal Research Lab., 1959—, Rideau Inst. Advance Research Ctr. for Behavioral and Phys. Scis., Newboro, Ont., Can., 1969—; dir. med. expdn. to Easter Island, 1964-65. Author: Pathophysiology of Peptic Ulcer, 1964; author: Intestinal Absorption of Trace Elements, 1970, Handbook of Stable Strontium, 1981; creative works include Inside Mara, 1981, Rideau Lakes-A Ballet, 1982. Recipient Outstanding Citizenship award Montreal Citizenship Council, 1965, Outstanding Achievement award McGill Grad. Soc., 1965; medalist in surgery Royal Coll. Physicians and Surgeons Can., 1957. Fellow ACS; mem. Can. Physiol. Soc., Am. Physiol. Soc., Am. Assn. Cancer Research, Can. Med. Assn., Soc. Exptl. Biology, Am. Gastroent. Assn. Roman Catholic. Clubs: McGill Univ., Royal Montreal Curling. Subspecialties: Physiology (medicine); Surgery (veterinary medicine). Current work: Gastrointestinal pathophysiology; trace elements; alternate energy sources; cancer research. Office: 740 Penfield Ave Donner Bldg McGill U Montreal PQ Canada H3A 1A4

SKOTHEIM, TERJE ASBJORN, physicist; b. Molde, Norway, June 7, 1949; s. Ole and Liv Solveig (Rodal) S.; m. Ellen Dolsa, May 1, 1972; children: Jan Marcel, Suzanne. B.S., MIT, 1972; Ph.D., U. Calif.-Berkeley, 1979. Postdoctoral fellow CNRS-Meudon, France, 1979-80; vis. research scientist U. Linkoping, Sweden, 1980-81; physicist Brookhaven Nat. Lab., Upton, N.Y., 1981—. Contbr. articles to profl. jours.; editor: Handbook of Conducting Polymers, 1984. Mem. Am. Phys. Soc., Norwegian Phys. Soc., Am. Solar Energy Soc. Unitarian. Subspecialties: Polymers; Polymer physics. Current work: Synthesis and characterization of highly conducting and semiconducting polymers, applications of conducting polymers to electronic and energy conversion devices and catalysis. Patentee in field. Office: Brookhaven Nat Lab Bldg 815 Upton NY 11772

SKURLA, GEORGE MARTIN, aerospace company executive; b. Newark, July 2, 1921; m. Marie Brignoli; children: George Martin, Thomas Michael, Martin John, James Mathew. B.S. in Aero. Engring., U. Mich., 1944; postgrad., Harvard U. Bus. Sch., 1973. With Grumman Aerospace Corp., Bethpage, N.Y., 1944—, v.p., 1970-74, pres., chief operating officer, 1974—, chmn. bd., chief exec. officer, 1976—; dir. Grumman Corp.; chmn. bd. Grumman Houston Corp., Grumman St. Augustine Corp.; dir. LITCO Bancorp., L.I. Trust Bank. Mem. pres.'s adv. council for L.I., Poly. Inst. N.Y.; trustee Urban League of L.I., U.S.S. Intrepid Mus., Fla. Inst. Tech., Adelphi U., Nat. Aviation Mus. Found.; mem. nat. adv. council Ariz. Heart Inst.; bd. dirs. L.I. Assn., Air Force Acad. Found., Fla. Council of 100. Recipient Disting. Citizen award Suffolk County council Boy Scouts Am., 1977, Tree of Life award Jewish Nat. Fund, 1983; named Outstanding Man of Year in Aviation Mgmt. Roth Grad. Sch. Bus. Adminstrn., L.I. U., 1976; Disting. Citizen of Yr. Dowling Coll., 1980; recipient L.I. Tech. Leadership award Poly. Inst. N.Y., 1982. Fellow AIAA; mem. Soc. Logistics Engrs. (Founders medal 1977), Air Force Assn. (Ira Eakerfellow Iron Gate chpt. 1983, Man of Yr., H.H. Arnold chpt. 1975), Am. Def. Preparedness Assn. (exec. com.), Assn. U.S. Army, Nat. Space Club (bd. govs.), Assn. Nav. Aviation (dir.), Navy League U.S. (life). Subspecialty: Aerospace engineering and technology. Office: Grumman Aerospace Corp Bethpage NY 11714

SKUTNIK, BOLESH S. JOSEPH, phys. chemist; b. Passaic, N.J., Aug. 19, 1941; s. Boleslaus Stanley and Helen Marie (Dzierzynska) S.; m. Phyllis V. Wojciechowski, Sept. 2, 1967; children: Pamela, Janeen, Todd Jason. B.S., Seton Hall U., 1962; M.S., Yale U., 1964, Ph.D., 1967. Research assoc. phys. chemistry Brandeis U., 1967-69; sr. research scientist Firestone Radiation Research div. Firestone Tire & Rubber Co., Westbury, L.I., N.Y., 1969-73; asst. prof. chemistry Fairfield U., 1973-79; research scientist Ensign-Bickford Industries, Simsbury, Conn., 1979-81, supr. corp. research, 1981-82, supr. new product research and devel., 1982—; lectr. Brandeis U., 1968-69; abstractor Chem. Abstract Service, 1969—; cons. Acad. Press Inc., 1974-76, John Wiley & Sons, 1978-79, Ensign-Bickford Industries, 1979. Contbr. numerous articles to sci., math. jours. Mem. Commack (N.Y.) Community Council, 1972-73; co-chmn. Scholarship Com., Polish Heritage Soc. Bridgeport, 1974-76; pres. Fairfield U. Employees Fed. Credit Union, 1976-79; chmn. St. Mary's Sch. Endowment Fund, 1980-82. Yale U. fellow, 1965-66. Mem. Am. Chem. Soc., Am. Phys. Soc., Soc. Plastics Engrs., Assn. Finishing Processes of Soc. Mfg. Engrs., Conn. Rubber Group, AAUP (v.p., pres. Conn. State Conf. 1977-79). Republican. Roman Catholic. Club: Porsche Am. Subspecialties: Polymers; Fiber optics. Current work: Conductive polymers; coating for fiber optics; effects of radiation on matter; theoretical and applied; polymer structure-property correlations; inter and intramolecular energy transfer. Patentee irradiation of elastomers and coatings for fiber optics. Office: 660 Hopmeadow St Simsbury CT 06070

SKY-PECK, HOWARD H., biochemist, educator; b. London, July 24, 1923; s. Harry H. and Rebecca B. (Silverman) Sky-P.; m. Bernice Mogg, Dec. 20, 1952; children: Stephen, Kathryn, Constance. B.S., U. Calif., 1949; M.S., U. So. Calif., 1951, Ph.D., 1956. Asst. prof. U. Ill., 1958-62, assoc. prof., 1962-68, prof., 1968-71; dir. clin. chemistry Presbyn.-St. Luke's Hosp., Chgo., 1967-75, dir. cancer research lab., 1956-70; prof. biochemistry Rush Med. Coll., Chgo., 1971—, chmn. dept. biochemistry, 1971-80. Contbr. articles to profl. jours. Served to lt. col. U.S. Army, 1942-46. Am. Cancer Soc., Nat. Cancer Inst., NIH grantee. Mem. Nat. Acad. Clin. Biochemistry, Am. Assn. Cancer Research, AAAS, Internat. Union Against Cancer, Royal Soc. Medicine, Am. Assn. Clin. Chemistry, Sigma Xi. Democrat. Episcopalian. Subspecialties: Biochemistry (medicine); Cancer research (medicine). Current work: Distribution of trace elements in human cancer tissues and their effect on genetic control of nucleic acids; comparison between normal and tumor subcellular distribution of trace elements and kinetics of biological function. Home: 187 Olmsted Rd Riverside IL 60546 Office: Dept Biochemistry Rush Med Coll Chicago IL 60612

SLACK, DERALD ALLEN, plant pathologist, researcher, educator, academic administrator; b. Cedar City, Utah, Dec. 22, 1924; s. Fredrick Allen and Marcella (Perry) S.; m. Betty Lue Stevens, Dec. 5, 1944; children: Steven A., Bonnie Lue Slack Huens. B.S., Utah State U., 1949, M.S., 1950; Ph.D., U. Wis., 1952. Asst. prof. plant pathology U. Ark.-Fayetteville, 1952-54, assoc. prof., 1954-59, prof., 1959—, head dept. plant pathology, 1964—; mem. Ark. Plant Bd., 1964—. Author: (with others) Plant Pathology Laboratory Manual, 3d edit, 1968; contbr. numerous articles to profl. jours. Served to 2d lt. AC U.S. Army, 1943-45. Recipient Outstanding Instr. award Alpha Zeta, 1956, Disting. Faculty award U. Ark. Alumni Assn., 1961. Mem. Am. Phytopathol. Soc. (sec. 1977-80, councilor-at-large 1980-83), Soc. Nematologists, Ark. Hort. Soc., Sigma Xi, Alpha Zeta, Gamma Sigma

Delta. Lodge: Lions. Subspecialty: Plant pathology. Current work: Supervision teaching and research programs.

SLACK, NELSON HOSKING, biostatistician; b. Burlington, Vt., Feb. 7, 1935; s. Errol Carlton and Ivy (Hosking) S.; m. Patricia Jane Billow, Sept. 10, 1960; children: Gregory and Gordon (twins), Brian. B.S., U. Vt., 1957; M.S., Rutgers U., 1963, Ph.D., 1964. Cancer research scientist Roswell Park Meml. Inst., Buffalo, 1964-67, sr. cancer research scientist, 1967-70, asso., 1970—, dep. dir. clin. trials nat. prostatic cancer project, 1977—; asst. prof. SUNY, Buffalo, 1969—. Contbr. numerous articles to profl. jours. Served with U.S. Army, 1957-59. Recipient C. H. Walker prize U. Vt., 1957. Mem. Am. Statis. Assn., Am. Assn. Cancer Research, AAAS, Sigma Xi. Subspecialties: Statistics; Cancer research (medicine). Current work: Chemotherapy clin. trials; disease-related factors. Office: 666 Elm St Buffalo NY 14263

SLACK, STEPHEN THOMAS, med. physicist, radiation safety officer; b. Lawrence, Mass., Nov. 3, 1941; s. Kenneth Francis and Leona Gertrude (Hmurciak) S. B.A., Marist Coll., Poughkeepsie, N.Y., 1964; Ph.D., Pa. State U., 1974. Cert. therapeutic med. physics and diagnostic med. physics. Am. Bd. Radiology. Tchr. Central Catholic High Sch., Wheeling, W.Va., 1962-64; physics instr. Emmanuel Coll., Boston, 1972-74; postdoctoral fellow U. Calif., San Francisco, 1974-75; chief med. physics and radiation safety, assoc. prof. dept. radiology, radiation safety officer W.Va. U., Morgantown, 1977—. Active Common Cause. Mem. Am. Assn. Physicists in Medicine, Health Physics Soc., Am. Phys. Soc., Am. Assn. Physics Tchrs., AAAS. Democrat. Roman Catholic. Subspecialties: Medical physics; Radiation safety. Office: Dept Radiology WV Univ Med Center Morgantown WV 26506

SLADEK, CELIA DAVIS, neuroscientist, educator; b. Denver, Mar. 25, 1944; d. C. Willard Davis and Mildred Davis TeSelle; m. John R. Sladek, Jr., May 23, 1970; children: Jonathan, Stefan, Jessica. B.A., Hastings (Nebr.) Coll., 1966; M.S., Northwestern U., 1968, Ph.D., 1971. Asst. prof. physiology U. Ill., 1970-73; research assoc. U. Rochester, N.Y., 1974-76, asst. prof. neurology and anatomy, 1976-80, assoc. prof., 1980—. Editor: Brain Research Bull, 1980—; mem. editorial adv. bd.: Peptides, 1980—; contbr. articles in field to profl. jours. Nat. Inst. Arthritis Metabolism grantee, 1977—, Nat. Inst. Heart Lung, 1982—; recipient Research Career Devel. award Nat. Inst. Neurol. and Communicable Disorders and Stroke, 1977-82. Mem. N.Y. Acad. Sci., Neuroscience Soc., Am. Assn. Anatomists. Subspecialties: Neuroendocrinology; Neuropharmacology. Current work: Regulation of vasopressin and oxytocin release research using radioimmunoassay, organ and tissue culture and electrophysiological techniques. Office: 601 Elmwood Ave Rochester NY 14642

SLADEN, BERNARD JACOB, psychologist, educator; b. Chgo., Mar. 30, 1952; s. Mayer and Anne (Cohn) S. B.S., U. Ill., 1974; Ph.D., Washington U., 1979. Lic. psychologist, Ill. Intern U. Minn., Mpls., 1977; psychologist Mental Health Ctr., Ft. Wayne, Ind., 1978-80, Associated Cert. Psychologists, Chgo., 1982—; coordinator psychol. evaluation Ill. Inst. Profl. Psychology, Chgo., 1981—; faculty dept. psychiatry Loyola U. Med. Sch., Maywood, Ill., 1981—; asst. prof. clin. psychiatry Northwestern U. Med. Sch., Chgo., 1982—; psychologist Hines (Ill.) Hosp., 1980—; Speakers Bur. coordinator Ill. Psychol. Assn., 1982—. Contbr. articles to profl jours. VA trainee, 1974-76; NIMH fellow, 1977-78. Mem. Am. Psychol. Assn., Ill. Psychol. Assn. Subspecialty: Clinical psychology. Current work: Research in compliance to psychological, medical treatment; clinical activities include family therapy, psychological Rx of drug and alcohol patients. Home: 201 E Chestnut St Apt 20A Chicago IL 60611 Office: Hines Hosp 116B Hines IL 60141

SLAKEY, LINDA L., biochemistry educator; b. Oakland, Calif., Jan. 2, 1939. B.S., Siena Heights Coll., 1962; Ph.D., U. Mich., 1967. Instr. chemistry St. Dominic Coll., St. Charles, Ill., 1967-69; faculty research assoc. Argonne (Ill.) Nat. Lab., 1969-70; project assoc. U. Wis.-Madison, 1970-73; asst. prof. biochemistry U. Mass., Amherst, 1973-79, assoc. prof., 1979—. Recipient Glycerine Research award Am. Oil Chemists, 1968; NIH spl. fellow, 1970-73; NSF grad. fellow; 1963-67. Mem. Am. Heart Assn. (fellow council on arteriosclerosis), Am. Soc. Biol. Chemists, Sigma Xi, Sigma Delta Epsilon. Subspecialty: Biochemistry (medicine). Current work: Regulatory interactions between blood vessel walls and blood cells; turnover of cell surface proteins; lipid metabolism. Office: Dept Biochemistry U Mass Amherst MA 01003

SLAVKIN, HAROLD CHARLES, biochemistry and nutrition educator, laboratory chief; b. Chgo., Mar. 20, 1938; s. Samuel and Clara (Kahn) S.; m. Kay Miriam, June 9, 1957 (div.); children: Mark, Todd; m. Lois Eveloff, Dec. 8, 1982. B.A., U. So. Calif., 1963, D.D.S., 1965. Faculty cellular and molecular biology U. So. Calif., Los Angeles, 1968—, asst. prof., 1968-70, assoc. prof., 1971-73, prof., 1974—, chmn. grad. program craniofacial biology, 1977—, prin. investigator grants, chmn. grad. program, faculty thematic options, 1982—; chmn. oral biology and medicine study sect. NIH, Bethesda, Md., 1980-85; cons., bd. dirs. Nat. Inst. Dental Research, NIH, 1976-80; cons. U.S. News and World Report, 1980—. Author: Developmental Craniofacial Biology, 1979; editor: Current Advances in Skeletogenesis Development, 1982; co-editor: Jour. Craniofacial genetics and developmental biology, 1980—; editorial bd., Differentiation, 1981—. Served with Dental Corps U.S. Army, 1955-58. Hon. research scholar Univ. Coll., London, 1979; creative scholar; U. So. Calif., 1980; recipient Isaac Schour Meml. award Internat. Assn. Dental Research, 1976, USPHS research career devel. award, 1968-72. Fellow AAAS, Am. Assn. Anatomists, AAUP, Am. Insts. Biol. Scis., Am. Soc. Cell Biology. Democrat. Jewish. Subspecialties: Biochemistry (biology); Developmental biology. Current work: Normal and abnormal craniofacial morphogenesis, with special interests in epithelial-mesenchymal interactions, extracellular matrix influences on epithelial differentiation and genetic susceptibility to congenital anomalies. Office: Andrus Gerontology Center 314 USC University Park-MC 0191 Los Angeles CA 90089-0191

SLEESMAN, JOHN PAUL, chemical company executive, agricultural researcher; b. Wooster, Ohio, Sept. 14, 1946; s. Jay Petersen and Helen Elizabeth (Henshaw) S.; m. Jo Anne Bryant, Apr. 8, 1971; children: Jay Bryant, Melanie Anne, John Matthew. B.S., Coll. Wooster, 1968; Ph.D., Ohio State U., 1975. Field research rep. Agrl. Chems. div. Mobay Chem. Corp., Mich. and Ohio, 1975-78, research farm mgr., Howe, Ind., 1978-82, regional devel. mgr., Indpls., 1982—. Contbr. articles, chpt. to profl. publs. Served to 1st lt. U.S. Army, 1968-71. Mem. Am. Phytopathol. Soc., Weed Sci. Soc. Am., Sigma Xi, Gamma Sigma Delta. Subspecialty: Plant pathology. Current work: Agricultural pest control and high yield research. Home: Rural Route 2 Howe IN 96746 Office: PO Box 385 Howe IN 46746

SLEMON, GORDON RICHARD, electrical engineering educator; b. Bowmanville, Ont., Can., Aug. 15, 1924; s. Milton Everitt and Selena (Johns) S.; m. Margaret Jean Matheson, July 9, 1949; children: Sally, Stephen, Mark, Jane. B.A.Sc., U. Toronto, 1946, M.A.Sc., 1948; D.I.C., Imperial Coll. Sci., London (Eng.) U., 1951, Ph.D., 1952, D.Sc., 1968. Asst. prof. elec. engring. N.S. Tech. Coll., Can., 1953-55; assoc. prof. U. Toronto, Ont., Can., 1955-63, prof., 1964—, chmn. dept. elec. engring., 1966-76, dean, 1979—; Colombo plan adviser, India, 1963-64; pres. Elec. Engring. Consociates, 1976-79. Author: (with J.M. Ham) Scientific Basis of Electrical Engineering, Magnetoelectric Devices, (with A. Straughen) Electric Machinery; Contbr. articles to profl. jours. Chmn. bd. Innovations Found., 1980—. Recipient excellence in teaching award Western Electric, 1965. Fellow Engring. Inst. Can., Inst. Elec. Engrs., IEEE; mem. Am. Soc. for Engring. Edn., others. Subspecialties: Electrical engineering; Electrical Propulsion. Current work: Variable-speed drives for industry and transportation. Patentee in field. Home: 40 Chatfield Dr Don Mills ON Canada Office: Faculty Applied Sci and Engring Univ Toronto Toronto ON Canada

SLEPIAN, DAVID, mathematician, communications engr.; b. Pitts., June 30, 1923; s. Joseph and Rose Grace (Myerson) S.; m. Janice Dorothea Berek, Apr. 18, 1950; children—Steven Louis, Don Joseph, Anne Maria. Student, U. Mich., 1941-43; M.A., Harvard U., 1947, Ph.D., 1949. With Bell Labs., Murray Hill, N.J., 1950—, head math studies dept., 1970—; prof. elec. engring. U. Hawaii, Honolulu, 1970-81; McKay prof. elec. engring. U. Calif., Berkeley, 1957-58; regents lectr., 1977. Contbr. articles to profl. jours. Served with U.S. Army, 1943-46. Fellow IEEE (Alexander Graham Bell award 1981), Inst. Math. Statistics, AAAS; mem. Nat. Acad. Scis. Nat. Acad. Engring., Soc. Indsl. and Applied Math. Subspecialties: Probability; Electrical engineering. Patentee in field. Home: 212 Summit Ave Summit NJ 07901 Office: Bell Labs 600 Mountain Ave Murray Hill NJ 07974

SLETTEN, CARLYLE JOSEPH, electronics engineer, physicist, consultant; b. Chetek, Wis., Jan. 13, 1922; s. Oscar William and Adeline Myrtle (Watrud) S.; m. Ruth Barbara Lessard, Jan. 23, 1946; children: Helen, Joel, Joyce, Robert, Steven. B.S. in Physics, U. Wis.-Madison, 1947; M.A. in Applied Physics, Harvard U., 1949. Registered profl. engr., Mass. Dir. microwave physics lab. Air Force Cambridge Research Labs., Mass., 1958-76; pres. Solar Energy Tech., Inc., Bedford, Mass., 1976—; Fulbright prof. electronics Escuela TSE de Telecomunicacion, Madrid, 1963-64; UN cons. Cath. U., Rio de Janeiro, Brazil, 1975-78. Contbr. articles to profl. jours. Bd. dirs. Mental Health Assn. Concord, Mass. Served to 1st lt. USAF, 1943-46. Mem. URSI, Phi Beta Kappa, Sigma Xi. Democrat. Mem. United Ch. of Christ. Subspecialties: Electronics; Solar energy. Current work: Electromagnetics, numerical computer techniques; reflector antennas, EM scattering, array synthesis, non-Tracking solar collectors. Patentee in field. Home: 106 Nagog Hill Rd Acton MA 01720 Office: Civil Terminal Bldg Hanscom Field Bedford MA 01730

SLEVIN, JOHN THOMAS, neurologist, pharmacology educator; b. Parkersburg, W.Va., Dec. 15, 1948; s. John Marshall and Mary Belle (Kysor) S.; m. Barbara Nyere, June 26, 1971; children: John Robert, Amie Elizabeth. B.A., Johns Hopkins U., 1970; M.D., W.Va. U., 1975. Intern in medicine W.Va. U. Hosp., Morgantown, 1975-76; resident in neurology U. Va. Hosp., Charlottesville, 1976-79; fellow in neuropharmacology Johns Hopkins U., Balt., 1979-81; asst. prof. neurology U. Ky. Med. Ctr. Lexington, 1981, asst. prof. pharmacology, 1982—; staff neurologist VA Med. Ctr. Lexington, 1981; research assoc. Sanders Brown Multidisciplinary Ctr. Gerontology. NIH postdoctoral fellow, 1979-81; tchr.-investigator devel. awardee, 1982—; So. Med. Assn. med. student scholar, 1971-72. Mem. Am. Acad. Neurology, AAAS, Soc. Neurosci. Subspecialties: Neurology; Neuropharmacology. Current work: Aging and development of the nervous system, epilepsy, excitatory amino acid neurotransmitters, neuropharmacology of the basal ganglia. Home: 340 Angela Ct Lexington KY 40503 Office: Dept Neurology Univ Ky Med Center Lexington KY 40536

SLICHTER, CHARLES PENCE, physicist, educator; b. Ithaca, N.Y., Jan. 21, 1924; s. Sumner Huber and Ada (Pence) S.; m. Gertrude Thayer Almy, Aug. 23, 1952 (div. Sept. 1977); children: Sumner Pence, William Almy, Jacob Huber, Ann Thayer; m. Anne FitzGerald, June 7, 1980; 1 son, Daniel H. A.B., Harvard U., 1946, M.A., 1947, Ph.D., 1949. Research asst. Underwater Explosives Research Lab., Woods Hole, Mass., 1943-46; mem. faculty U. Ill. at Urbana, 1949—, prof. physics, 1955—, prof. Center for Advanced Study, 1968—; Morris Loeb lectr. Harvard U., 1961; dir. Polaroid Co.; mem. Pres.'s Sci. Adv. Com., 1964-69, Com. on Nat. Medal Sci., 1969-74, Nat. Sci. Bd., 1975—, Pres.'s Com. Sci. and Tech., 1976. Author: Principles of Magnetic Resonance, 1963, rev. edit., 1978; Contbr. articles to profl. jours. Mem. corp. Harvard U.; former trustee, mem. corp. Woods Hole Oceanographic Instn. Recipient Langmuir award Am. Phys. Soc., 1969; Alfred P. Sloan fellow, 1955-61. Mem. Nat. Acad. Scis., Am. Acad. Arts and Scis., Am. Philos. Soc. Subspecialties: Condensed matter physics; Magnetic physics. Current work: Nuclear magnetic resonance, electron spin resonance, experimental solid state physics. Home: 61 Chestnut Ct Champaign IL 61821

SLICHTER, WILLIAM PENCE, chemist; b. Ithaca, N.Y., Mar. 31, 1922; s. Sumner Huber and Ada (Pence) S.; m. Ruth Kaple, June 17, 1950; children—Nancy, Carol, Catherine, Margaret. B.A., Harvard U., 1944, M.A., 1949, Ph.D., 1950. Mem. tech. staff Bell Telephone Labs., Murray Hill, N.J., 1950—, head chem. physics research dept., 1958-67, chem. dir., 1967-73, exec. dir. research-materials sci. and engring. div., 1973—; mem. adv. coms. Nat. Bur. Standards, NSF, NASA, Harvard U., Northwestern U., Dartmouth Coll., Rutgers U., SUNY, Albany. Served to lt. U.S. Army, 1943-46. Fellow Am. Phys. Soc. (award in high polymer physics 1970); mem. Nat. Acad. Engring., Am. Chem. Soc., Electrochem. Soc., AAAS, ASTM (dir.), Sigma Xi. Subspecialties: Kinetics; Nuclear magnetic resonance (chemistry). Research on molecular motion and structure in high polymers, nuclear magnetic resonance spectrosocopy. Home: 55 Van Doren Ave Chatham NJ 07928 Office: Bell Telephone Labs Murray Hill NJ 07974

SLICK, GARY LEE, medical educator, clinician; b. Balt., Apr. 13, 1941; s. Roy M. and Mildred (Nelson) Krick; m. Marylyn Maxwell, July 8, 1968; children: Jacquelyn, Christopher. B.A., Eastern Nazarene Coll., 1963; M.A., U. Kans., 1969; D.O., Kansas City Coll. Osteo. Medicine, 1969. Diplomate: Am. Osteo. Bd. Internal Medicine, Am. Osteo. Bd. Nephrology. Assoc. clin. prof. medicine Mich. State U. Coll. Osteopathy, East Lansing, 1975-82; attending nephrologist Detroit Osteo. Hosp., 1974-82; prof. medicine Okla. Coll. Osteo. Medicine and Surgery, Tulsa, 1982—, chmn. dept. medicine, 1982—; vice chmn. Am. Osteo. Bd. Internal Medicine, 1982—. Contbr. articles on osteo. medicine and renal pathophysiology to profl. jours. Mem. sci. adv. bd. Mich. Kidney Found., Ann Arbor, 1979-82. NIH fellow, 1972-74. Mem. Am. Coll. Osteo. Internists, Am. Fedn. for Clin. Research, Am. Physiol. Soc., Am. Soc. Artificial Internal Organs, Am. Soc. Nephrology. Republican. Methodist. Subspecialties: Nephrology; Internal medicine. Current work: Role of renal nerves in sodium handling and congestive heart failure. Office: Okla Coll Osteo Medicine and Surgery PO Box 2280 Tulsa OK 74101

SLIKKER, WILLIAM, JR., pharmacologist, toxicologist; b. Bakersfield, Calif., Apr. 3, 1950; s. William and Hazel Marie (Marchant) S.; m. Cristine Blozan, July 4, 1975; 1 dau., Annamarie. B.A. in Biology, U. Calif., Santa Barbara, 1972, M.A. in Biol. Scis, 1974, Ph.D. in Pharmacology/Toxicology, 1978. Jr. staff fellow drug research and evaluation program and perinatal research program Nat. Center for Toxicol. Research, FDA, USPHS, Jefferson, Ark., 1978-79, supervisory pharmacologist pharmacodynamics br. teratogenesis research div., 1979—, acting br. chief pharmacodynamics br., 1979-80, chief, 1980—; adj. asst. prof. dept. pharmacology and interdisciplinary toxicology U. Ark. for Med. Sci., Little Rock, 1980—; adj. assoc. prof. dept. drug and material toxicology U. Tenn., Memphis, 1982—. Contbr. articles to profl. jours. Recipient award for outstanding achievement in study of estrogen metabolism FDA, 1979. Mem. AAAS, Am. Soc. for Study Xenobiotics (charter), Am. Soc. Pharmacology and Exptl. Therapeutics. Subspecialties: Pharmacology; Teratology. Current work: Understanding the influence of metabolism and disposition of chems. on their resultant teratogenicity; investigating estrogens, glucocorticoids and ethanol; surg. manipulation of appropriate animal models and analytical methods devel. are prerequisite to collection and analysis of transplacental pharmacokinetic data of interest. Office: Pharmacodynamics Br Div Teratogenesis Research Nat Center for Toxicological Research Jefferson AR 72079

SLINEY, DAVID HAMMOND, med. physicist; b. Washington, Feb. 21, 1941; s. David Xavier and Ida Lee (Echols) S.; m. Carol Ann Scott, Feb. 19, 1966; children: Sean Scott, David Scott, Stephen Paul. B.S. in Physics, Va. Poly. Inst. and State U., 1963; M.S. in Physics and Radiol. Health, Emory U., 1965. Chief laser br. laser microwave div. U.S. Army Environ. Hygiene Agy., Aberdeen Proving Ground, Md., 1965—; cons. to WHO, others. Author: (with M. Wolbarsht) Safety with Lasers and Other Optical Sources, 1980; editor: Health Physics Jour, 1976—; contbr. articles to profl. jours. Served to capt. Med. Service Corps U.S. Army, 1965-67. Fulbright fellow, Yugoslavia, 1976; recipient decoration for meritorious civilian service Dept. of Army, 1978. Mem. Optical Soc. Am., Health Physics Soc., Laser Inst. Am. (chmn. laser safety com., dir.), Soc. Photo-optical Instrumentation Engrs., Am. Conf. on Govtl. Hygienists, Assn. for Research in Vision and Ophthalmology, others. Unitarian. Subspecialties: Laser medicine; Biophysics (physics). Current work: Evaluation of health hazards from lasers; establishment of safety standards for lasers and other optical sources; vision research and ocular effects of lasers. Home: 406 Streamside Dr Fallston MD 21010 Office: Laser Microwave Div USAEHA Aberdeen Proving Ground MD 21010

SLOANE, CHRISTINE SCHEID, research scientist; b. Washington, May 1, 1945; d. Arthur M. and Polly (Notte) Scheid; m. Thompson M. Sloane, Aug. 30, 1945; children: Luke N., Derek W. B.S., Coll. William and Mary, 1967; Ph.D., MIT, 1971. Research assoc. dept. chemistry U. Calif.-Berkeley, 1972-73; asst. prof. chemistry Oakland U., Rochester, Mich., 1974-78, assoc. prof., 1978; sr. research scientist, staff research scientist environ. sci. dept. Gen. Motors Research Labs., Warren, Mich., 1978—. Contbr. articles to profl. jours. Research Corp. grantee, 1976; Oakland U. grantee, 1975, 77. Mem. Am. Chem. Soc., AAAS, Am. Meteorol. Soc., Sigma Xi, Phi Beta Kappa. Subspecialties: Theoretical chemistry; Environmental engineering. Current work: Modeling of atmospheric transport and chemistry; statistical analysis of environmental data. Office: Environmental Science Department General Motors Research Laboratory Warren MI 48090

SLOMIANY, BRONISLAW LESZEK, biochemist, researcher; b. Wlodzimierz Wolynski, Poland, Dec. 12, 1941; came to U.S., 1966; s. Feliks and Eugenia (Murdzia) S.; m. Amalia Niewieczerzal, May 27, 1967; children: Lee, Beatrix, Mark. M.S., U. Lodz, 1966; Ph.D., N.Y. Med Coll., 1971. Postdotoral fellow Yeshiva U., 1971-72; research asst. prof. biochemistry N.Y. Med. Coll., Valhalla, 1972-74, research assoc. prof., 1974-78, research prof., 1979—; head gastroenterology research lab. Met. Hosp. N.Y. Med. Coll., 1979—. Contbr. articles sci. jours., chpts. to books. NIH research grantee, 1978, 81. Mem. Am. Soc. Biol. Chemists, Soc. Complex Carbodydrates, N.Y. Acad. Scis., AAAS. Roman Catholic. Subspecialties: Gastroenterology; Oral biology. Current work: Lipids, glycolipds, glycoproteins, mucous secretions, gastrointestinal tract, salivary secretions in health and disease. Home: 130 Ridge Rd Rutherford NJ 07070 Office: New York Medical College Valhalla NY 10595

SLOTKIN, THEODORE ALAN, pharmacologist, educator; b. Bklyn., Feb. 17, 1947; s. Herman and Roselyn B. (Seplowitz) S.; m. Linda I. Yarowenko, July 16, 1967; children: Alexander P., Matthew C. B.S., Bklyn. Coll., 1967; Ph.D., U. Rochester, 1970. Postdoctoral fellow U. Rochester, 1970; research assoc. in biochemistry Duke U., 1970-71, asst. prof. physiology and pharmacology, 1971-75, assoc. prof. pharmacology, 1975-79, prof., 1979—; cons. Nat. Inst. Drug Abuse, Nat. Heart, Lung and Blood Inst., Nat. Inst. Neurol. and Communicative Diseases and Stroke, NSF. Contbr. articles to profl. lit. Named Outstanding Young Investigator N.C. Heart Assn., 1975; recipient faculty devel. award Pharm. Mfrs. Assn. Found., 1973-75, research scientist devel. award Nat. Inst. Drug Abuse, 1975—, John J. Abel award in pharmacology Am. Soc. Pharmacology and Exptl. Therapeutics, 1982. Mem. Am. Soc. Pharmacology and Exptl. Therapeutics, Soc. Neurosci., Am. Soc. Neurochemistry, Internat. Soc. Developmental Neurosci. Democrat. Jewish. Subspecialties: Pharmacology; Neuropharmacology. Current work: Developmental neuropharmacology, perinatal effects of drugs and toxins. Effects of drugs and toxic substances on development of nervous system, particularly drugs of abuse, environmental contaminants. Home: 604 Duluth St Durham NC 27705 Office: Duke U Med Center Box 3813 Durham NC 27710

SLUTSKY, ROBERT ALLEN, cardiology and radiology educator, researcher; b. Bklyn., Jan. 22, 1949; s. Abraham Morris and Evelyn (Morris) S.; m. Elizabeth Cass, June 9, 1974; children: Anna Elizabeth, Nicholas Robert. B.A., Tufts U., 1970; M.D., UCLA, 1974. Diplomate: Am. Bd. Internal Medicine. Inter, resident in medicine U. Calif.-San Diego, 1974-77, research fellow in medicine, 1977-79, fellow in cardiology, 1979-81, asst. prof. medicine and radiology, 1981—; cons. radiology and medicine San Diego VA Med. Ctr., 1981—. Contbr. numerous articles profl. jours. Mem. Uptown Democratic Club, San Diego, 1982—. Grantee Am. Heart Assn., Am. Lung Assn., Distilled Spirits Council U.S.A. Fellow ACP, Am. Coll. Cardiology; mem. Radiol. Soc. N.Am., Soc. Nuclear Medicine, Am. Fedn. Clin. Research. Jewish. Subspecialties: Cardiology; Imaging technology. Current work: Application of new imaging modalities to cardiopulmonary physiology. Home: 1913 Fort Stockton Dr San Diego CA 92103 Office: U Calif Med Sch 225 W Dickinson St San Diego CA 92103

SLYKHOUS, STEWART JAMES, physicist, consultant; b. Compton, Calif., Sept. 26, 1950; s. Merle Albert and Helen Louise (Lamb) S.; m. Janis Anne Milham, Aug. 3, 1974. B.A. in Physics, U. Calif.-Irvine, 1971; M.B.A. in Fin.-Mktg, UCLA, 1977. Electro-optical scientist KMS Tech., 1971-74; asst. dept. head Universal City Studios/Technicolor Inc., 1974-77; applications and mktg. specialist on acoustic emission Endevco, San Juan Capistrano, Calif., 1977-80; regional applications mgr. on acoustic emission Phys. Acoustics Corp., San Clemente, Calif., 1980—; tchr. acoustic emission for tng. certs. Mem. Am. Soc. for Nondestructive Testing, Soc. for Advancement Material and Process Engring., Soc. Plastic Engrs., Soc. Plastic Industries, Acoustic Emission Working Group, ASTM. Club: Irvine Athletic Assn. Subspecialty: Composite materials. Current work: Application of acoustic emission inspection techniques to structural integrity investigations of metal and composites structures. Home and office: 265 Ave La Cuesta San Clemente CA 92672

SMAGORINSKY, JOSEPH, meteorologist; b. N.Y.C., Jan. 29, 1924; s. Nathan and Dinah (Azaroff) S.; m. Margaret Knoepfel, May 29, 1948; children: Anne, Peter, Teresa, Julia, Frederick. B.S., N.Y. U., 1947, M.S., 1948, Ph.D., 1953; Sc.D. (hon.), U. Munich, Ger., 1972. Research asst., instr. meteorology N.Y. U., 1946-48; with U.S. Weather Bur., 1948-50, 53-65, chief gen. circulation research sect., 1955-63; meteorologist Inst. Advanced Study, Princeton, N.J., 1950-53; acting dir. Inst. Atmospheric Scis., Environ. Scis. Services Adminstrn., Washington, 1965-66; dir. Geophys. Fluid Dynamics Lab., Environ. Scis. Services Adminstrn.-NOAA, Washington and Princeton, 1964-83; Vice chmn. U.S. Com. Global Atmospheric Research Program, Nat. Acad. Sci., 1967-73, 80—, officer, 1974-77, mem. climate bd., 1978—, chmn. com. on internat. climate programs, 1979, bd. internat. orgns. and programs, 1979-83, chmn. climate research com., 1981—; chmn. joint organizing com. Global Atmospheric Research Program, Internat. Council Sci. Unions/World Meteorol. Orgn., 1976—, officer, 1967—; chmn. Joint Sci. Com. World Climate Research Program, 1980-81; chmn. climate coordinating forum Internat. Council Sci. Unions, 1980; vis. lectr. with rank of prof. Princeton U., 1968-83, vis. sr. fellow, 1983—. Contbr. to profl. publns. Served to 1st lt. USAAF, 1943-46. Decorated Air medal; recipient Gold medal Dept. Commerce, 1966; award for sci. research and achievement Environ. Sci. Services Adminstrn., 1970; Buys Ballot Gold medal Royal Netherlands Acad. Arts and Scis., 1973; IMO prize and Gold medal World Meteorol. Orgn., 1974. Fellow Am. Meteorol. Soc. (councilor 1974-77, asso. editor jour. 1965-74, Meisinger award 1967, Wexler Meml. lectr. 1969, Carl-Gustaf Rossby Research Gold medal 1972, Cleveland Abbe award for disting. service to atmospheric sci. 1980, presdl. award 1980, Symons meml. gold medal 1981); mem. Royal Meteorol. Soc. (Symons Meml. lectr. 1963). Subspecialties: Climatology; Meteorology. Current work: Geophysical fluid dynamics and thermodynamics; geophysical applications of high speed computers; atmospheric general circulation and theory of climate; atmospheric predictability. Home: 21 Duffield Pl Princeton NJ 08540 Office: Geol and Geophys Scis Guyot Hall Princeton Univ Princeton NJ 08540

SMALLEY, LARRY LEE, physics educator; b. Grand Island, Nebr., Aug. 7, 1937; s. Lionel Mullen and Grace B. (Reinecke) S.; m. Katherine F. Davenport, Nov. 16, 1957; children: Larry L, Daco S., Audrey M. B.S., U. Nebr., 1959, M.S., 1964, Ph.D., 1967. Grad. asst. U. Nebr., Lincoln, 1962-67; nuclear engr., Hallam Nuclear Power Facility, Nebr., 1964; faculty U. Ala., Huntsville, 1967—, prof. physics, 1980—, chmn. dept., 1972—; space scientist NASA-Marshall Space Flight Ctr., Huntsville, 1979—. Contbr. articles to profl. jours. Served to lt. USN, 1959-62. Nat. Acad. Sci./NRC sr. research assoc. NASA, 1974-75; Humboldt fellow, 1977-78. Mem. Am. Phys. Soc., Sigma Xi, Pi Mu Epsilon, Phi Sigma Iota. Subspecialty: Relativity and gravitation. Current work: Extensions of general relativity to microscopic spacetime; gauge theories of gravitation; discrete geometry; post-Newtonian approximations. Office: Dept Physics Univ Ala Huntsville AL 35807

SMALLEY, RICHARD ERRETT, chemistry educator, consultant; b. Akron, Ohio, June 6, 1943; s. Frank Dudley and Esther Virginia (Rhoads) S.; m. Mary Lynn Chapieski, July 10, 1980; m. Judith Grace Sampieri, May 4, 1968 (div. July 1978); 1 son, Chad Richard. B.S., U. Mich., 1965; M.A., Princeton U., 1971, Ph.D. (Harold W. Dodds fellow), 1973. Research chemist Shell Chem. Co., Woodbury, N.J., 1965-69; assoc. U. Chgo., 1973-76; asst. prof. chemistry Rice U., 1976-80, assoc. prof., 1980-81, prof., 1981-82, Hackerman prof. chemistry, 1982—; mem. steering com. Quantum Inst., 1979—. Mem. adv. bd.: Jour. Phys. Chemistry, 1980—, Chem. Phys. Letters, 1983—. Sloan fellow, 1978-80. Mem. Am. Chem. Soc., Am. Phys. Soc., Am. Inst. Physics, AAAS. Subspecialties: Physical chemistry; Laser-induced chemistry. Current work: Chemical physics, laser chemistry, catalysis, surface chemistry and physics. Home: 12306 Alston Dr Stafford TX 77477 Office: Dept Chemistry Rice U PO Box 1892 Houston TX 77251

SMALLRIDGE, ROBERT CHRISTIAN, endocrinologist, researcher; b. Charleston, W.Va., Dec. 28, 1944; s. Horace Hamilton and Isabel Maury (White) S.; m. Elizabeth Gray Cone, Aug. 20, 1966; children: Amy Brewster, Laura Fontaine. B.A., Yale U., 1966; M.D., Med. Coll. Va., 1970. Diplomate: Am. Bd. Internal Medicine. Intern Grady Meml. Hosp., Atlanta, 1970-71, resident, 1971-73; commd. 2d lt. U.S. Army, 1970, advanced through grades to lt. col., 1979; endocrinology fellow Walter Reed Army Med. Ctr, Washington, 1975-77, research internist, 1977—, chief dept. clin. physiology, 1980—; asst. prof. Uniformed Services Univ. of Health Scis., Bethesda, Md., 1978-82; assoc. prof., 1982—. Contbr. articles to profl. jours. Vice-pres. Greenwich Forest Citizens Assn., Bethesda, 1981-83; treas. Washington Antiquarian Book Fair, Rosslyn, Va., 1980-82. Fellow ACP; mem. Endocrine Soc., Am. Thyroid Assn., Am. Fedn. Clin. Research, Soc. Uniformed Endocrinologists, Alpha Omega Alpha, Sigma Zeta. Presbyterian. Subspecialties: Endocrinology; Physiology (biology). Current work: Thyroid hormone metabolism, hormone effects on the heart, pituitary tumors. Home: 7815 Overhill Rd Bethesda MD 20814 Office: Walter Reed Army Inst Research Washington DC 20012

SMEBY, ROBERT RUDOLPH, biochemist, educator; b. Chgo., Dec. 24, 1926; s. Rudolph P. and Nellie (Smith) S.; m. Patricia R. Dowling, June 10, 1950; children: Eric, Patrick, Alec, Darrick. B.S., U. Ill., Urbana, 1950; M.S., U. Wis.-Madison, 1952, Ph.D., 1954. Biochemist R.J. Reynolds Research, Winston-Salem, N.C., 1954-56, Miles-Ames Research, Elkhart, Ind., 1956-59; mem. research staff Cleve. Clinic, 1959—; adj. prof. biology Cleve. State U., 1970—, John Carroll U., University Heights, Ohio, 1966-70. Contbr. articles to sci. publs. Served with USN, 1944-46. Fellow AAAS, Council High Blood Pressure Research, Internat. Soc. Hypertension; mem. Am. Chem. Soc., Am. Physiol. Soc. Subspecialties: Biochemistry (medicine); Physiology (medicine). Current work: regulation of synthesis and release of renin by kidney; role of the renin-angiotensin system in brain. Inventor, holder 10 U.S. patents. Home: 36801 Riviera Rd Willoughby OH 44094 Office: Cleveland Clinic Foundation 9500 Euclid Ave Cleveland OH 44106

SMIDT, MARY LOUISE, plant pathologist. B.A. summa cum laude, Pacific Luth. U., 1974; Ph.D. in Plant Pathology, U. Calif.-Davis, 1978. Research assoc. dept. plant pathology U. Calif.-Davis, 1978-79; research assoc. dept. plant pathology U. Nebr., Lincoln, 1979—. Subspecialties: Plant pathology; Microbiology. Current work: Physiology of host-parasite interactions; biological control mechanisms of plant diseases. Office: Dept Plant Pathology U Nebr Lincoln NE 68583

SMILEY, CHARLES JACK, geologist, educator; b. Mt. Vernon, Wash., Dec. 2, 1924; s. Charley and Julia Carolina (Watson) S.; m. Marguerite Clayton, Aug. 24, 1954; children: John, Sharifah. B.A., Western Wash. Coll., 1951; Ph.D., U. Calif.-Berkeley, 1960, M.A., 1954. Asst. prof. geology Macalester Coll., St. Paul, 1956-61; fellow Harvard U., 1961-62; asst. prof. geology U. Idaho, Moscow, 1962-65, assoc. prof., 1965-67, prof., 1967—, assoc. dean, 1976-80, dir., 1980—; investigator in No. Alaska Arctic Inst. N.Am./U.S. Office Naval Research, 1956-67; lectr. Author: The Ellensburg Flora of Washington, 1963; Editor, contbr.: Later Cenozoic History of the Pacific North west, 1982; contbr. articles to sci. jours. Served in USN, 1943-47, 51-52. NSF fellow, 1954-55; Fulbright prof. U. Malaya, Kuala Lumpur, 1968-69; NSF vis. scientist, Japan, 1973. Mem. Am. Assn. Petroleum Geologists, Explorers Club, Bot. Soc. Am., Paleobot. Soc., Sigma Xi. Phi, Kappa Phi. Subspecialties: Paleontology; Evolutionary biology. Current work: Cretaceous and tertiary floras: taxonomy, systematics, biostratigraphy, paleoecology; research and consulting on Cretaceous floras of western North America; research on Miocene Clarkia compression flora of northern Idaho. Home: 2100 Robinson Lake Rd Moscow ID 83843 Office: U Idaho Moscow ID 83843

SMILEY, PARKER CLARK, mechanical engineer; b. Boone, Iowa, Oct. 17, 1931; s. Paul Parsons and Ruth Louise (Clark) S.; m. Dorothy Jean Strader, July 18, 1959; children: Scott Parker, Kevin David. B.S., U. Calif.-Berkeley, 1959, M. Engring., 1960. Mech. engr. Lawrence Livermore (Calif.) Nat. Lab., 1960-65; mech. engr. Physics Internat. Co., San Leandro, Calif., 1965-67, sr. mech. engr., 1967-81, sr. staff engr., 1981—. Contbr. articles to profl. jours., confs. Served with USN, 1951-55. Mem. ASME. Subspecialties: Biomedical engineering; Ceramic engineering. Current work: Piezoelectric actuators; ceramics; plastics; mechanical design and development of vacuum tubes for high voltage electron accelerators. Patentee temperature compensated hydraulic valve, lever motion multiplier driven by electroexpansive material, spring diaphragm, pump systems. Home: 6693 Saroni Dr Oakland CA 94611 Office: 2700 Merced St San Leandro CA 94577

SMILEY, TERAH LEROY, geosciences educator; b. Oak Hill, Kans., Aug. 21, 1914; s. Terah Edward and Frances Angelina (Huls) S.; m. Marie Lemley, July 1935; 1 dau., Terrie Lucille Scheele; m. Winifred Whiting Lindsay, June 10, 1947; children: John, Maureen, Kathlyn; 1 stepdau., Margaret Ann Taylor. Student, U. Kans., 1934-36; M.A., U. Ariz., 1949. With U.S. Nat. Park Service, 1939-41, U.S. Immigration Service, 1941-42; research Lab. of Tree-Ring, U. Ariz., Tucson, 1946-60, dir., 1958-60, dir. geochronology labs., 1957-67, head dept. geochronology, 1967-70, prof. geosciences, 1970—; Gen. chmn. Internat. Conf. on Forest Tree Growth, Tucson, 1960; vice chmn. U.S. Com. on Internat. Assn. for Quaternary Research, Nat. Acad. Sci., 1961-66; gen. chmn. First Internat. Conf. on Palynology, Tucson, 1962; mem. U.S. Com. on Internat. Hydrological Decade, Nat. Acad. Sci., 1964-66; gen. chmn. Internat. Conf. on Arid Lands, Tucson, 1969. Editor: (with James H. Zumberge) Polar Deserts and Modern Man, 1974, The Geological Story of the World's Deserts, 1982; Contbr. articles to profl. jours. Served with USNR, 1942-45. Research fellow Clare Coll., Cambridge, U., 1970; vis. prof. Kvartärgeologiska Institutionen, Uppsala (Sweden) U., 1970-71; hon. v.p. 2d Internat. Conf. on Palynology, Utrecht, 1966. Fellow Geol. Soc. Am., AAAS, Ariz. Acad. Sci.; mem. Am. Meteorol. Soc., Tree-Ring Soc., Ariz. Geol. Soc. (past pres.), Ariz. Archeol. Soc. (past pres.), Sigma Xi. Current work: Role of geology in past climates; geological story of world's deserts , formation of landscapes in Arizonas. Home: 2732 N Gill Ave Tucson AZ 85719

SMITH, ALBERT CHARLES, biologist; b. Springfield, Mass., Apr. 5, 1906; s. Henry Joseph and Jeanette Rose (Machol) S.; m. Nina Grönstrand, June 15, 1935; children—Katherine (Mrs. L.J. Campbell), Michael Alexis; m. Emma van Ginneken, Aug. 1, 1966. A.B., Columbia, 1926, Ph.D., 1933. Asst. curator N.Y. Bot. Garden, 1928-31, asso. curator, 1931-40; curator herbarium Arnold Arboretum of Harvard U., 1940-48; curator div. phanerogams U.S. Nat. Mus., Smithsonian Instn., 1948-56; program dir. systematic biology NSF, 1956-58; dir. Mus. of Natural History, Smithsonian Instn., 1958-62, asst. sec., 1962-63; prof. botany, dir. research U. Hawaii, Honolulu, 1963-65, Gerrit Parmile Wilder prof. botany, 1965-70, prof. emeritus, 1970—; Ray Ethan Torrey prof. botany U. Mass., Amherst, 1970-76, prof. emeritus, 1976—; editorial cons. Pacific Tropical Bot. Garden, Hawaii, 1977—; bot. expdns., Colombia, Peru, Brazil, Brit. Guiana, Fiji, West Indies, 1926-69; del. Internat. Bot. Congresses, Amsterdam, 1935, Stockholm, 1950; v.p. systematic sect., Montreal, 1959, Internat. Zool. Congress, London, 1958. Author: Flora Vitiensis Nova: a New Flora of Fiji, Vol. I, 1979, Vol. II, 1981; also tech. articles.; Editor: Brittonia, 1935-40, Jour. Arnold Arboretum, 1941-48, Sargentia, 1942-48, Allertonia, 1977—; editorial com.: International Code Botanical Nomenclature, 1954-64. Bishop Museum fellow Yale U., 1933-34; Guggenheim fellow, 1946-47; Robert Allerton award for excellence in tropical botany, 1979. Fellow Am. Acad. Arts and Scis., Linnean Soc. London; mem. Bot. Soc. Am., Assn. Tropical Biology (pres. 1967-68), Internat. Assn. Plant Taxonomy (v.p. 1959-64), Nat. Acad. Scis., Fiji Soc. (hon.). Club: Washington Biologists' Field (pres. 1962-64). Subspecialties: Evolutionary biology; Systematics. Current work: Angiosperm flora of southwestern pacific, evolution and phytogeography of angiosperms. Office: Dept of Botany Univ of Hawaii Honolulu HI 96822

SMITH, ALLAN LASLETT, chemistry educator; b. Newark, June 21, 1938; s. Arthur Laslett and Dorothy Opdyke (Priesler) SM.; m. Charity Fletcher Smith, Aug. 28, 1960; children: Fletcher, Nadja. B.A. in Chemistry and Physics magna cum laude, Harvard U., 1960; Ph.D. in Phys. Chemistry, M.I.T., 1965. Nat. Acad. Sci.-NRC fellow Nat. Bur. Standards, 1965-66; asst. prof. chemistry, Yale U., 1966-71, assoc. prof., 1971-75, Drexel U., 1975—. Alfred P. Sloan fellow, 1969-71; NATO fellow, 1971. Mem. Am. Chem. Soc., Am. Phys. Soc., IEEE, AAAS. Subspecialties: Spectroscopy; Photochemistry. Current work: Computer based methods in spectroscopy, kinetics, laser fluorescence detection of transients, spectral and kinetic analysis using computers. Office: Dept Chemistry Drexel U 32d St and Chestnut St Philadelphia PA 19104

SMITH, ANNA CLANTON, medical physicist; b. Thomasville, Ala., Nov. 27, 1957; s. Benjamin Lamar and Dovie Lee (Freeman) Clanton; m. Robert Lee Smith, Sept. 12, 1981. B.S., U. Ala.-Birmingham, 1980; M.S., Ga. Inst. Tech., 1981. Cert. nuclear medicine technologist. Med. physicist North Miss. Med. Ctr., Tupelo, Miss., 1981-83, Bapt. Med. Ctr.-Princeton, Birmingham, Ala., 1983—. Mem. Am. Assn. Physicists in Medicine, Soc. Nuclear Medicine, Health Physics Soc. Subspecialties: Nuclear physics; Nuclear medicine. Current work: Radiation therapy treatment planning; calibration of therapy equipment; nuclear medicine physics. Home: 25090 Mountain Lodge Circle Birmingham AL 35216 Office: Radiation Therapy Bapt Med Ctr-Princeton 801 Princeton Ave Birmingham AL 35211

SMITH, BARRY DECKER, psychology educator; b. Harford, Pa., June 12, 1940; s. Clinton T. and Gretchen (Decker) S.; m. Elizabeth W. Wormley, June 15, 1963; children: Douglas, Debra. B.S., Pa. State U., 1962; M.A., Bucknell U., 1964; Ph.D., U. Mass., 1967. Research assoc. U. Mass., Amherst, 1967; asst. prof. psychology U. Md., College Park, 1967-71, assoc. prof., 1971-80, prof., 1981—, asst. chmn., 1979-80, 81—; cons. GAO, 1977—, D. C. Pub. Schs., 1972—, Howard U. Coll. Medicine, 1977-79. Author: Motivation, 1967, Personality Theory, 1971, Theoretical Approaches to Personality, 1982; contbr. articles to profl. jours. Chmn. activity and funding com. Boy Scouts Am., Washington, 1978—; mem. exec. com. Eisenhower High Sch., Laurel, Md., 1978-80, Roosevelt High Sch., Greenbelt, Md., 1981—. Mem. Am. Psychol. Assn., Eastern Psychol. Assn., Soc. Psychophysiol. Research, AAAS. Subspecialties: Psychobiology; Neurophysiology. Current work: Biological foundations of human behavior. Psychophysiol and neurophysiological bases of personality functioning. Office: Dept Psychology Univ Md College Park MD 20742

SMITH, BARRY HAMILTON, neurosurgeon, neuro-oncologist, neuroscience researcher; b. Orange, N.J., Oct. 6, 1943; s. Kenneth Wright and Harriet Kathryn (Barr) S.; m. Carley Eldredge, Dec. 18, 1969; children: Christopher, Sara. B.A., Harvard U., 1965; Ph.D. (NIMH fellow), M.I.T., 1968; M.D. (Life Ins. Med. Research Fund med. scientist fellow), Cornell U., 1972. Intern N.Y. Hosp., N.Y.C., 1971-72, asst. resident and fellow in surgery, 1972-75, asst. resident and fellow in neurosurgery, 1973-75; staff scientist neuroscis. research program M.I.T., 1975, program dir., 1976-78; resident in neurosurgery Mass. Gen. Hosp., Boston, 1975-78; dep. chief surg. neurology br. Nat. Inst. Neurol. Communicative Disorders and Stroke, Bethesda, Md., 1976-77, dep. chief, 1978-83; neurosurgeon, mem. Sloan Kettering Cancer Ctr., 1983—; mem. med. adv. bd. Acoustic Neuroma Assn. Contbr. articles to sci. publs. Served with USPHS, 1978-82. Recipient EEO award Nat. Inst. Neurol. Communicative Disorders and Stroke, 1980; Commendation medal USPHS, 1982; ACS Schering scholar, 1975. Mem. AAAS, AMA, Soc. for Neurosci. (chmn. public info. com), Electron Microscopy Soc. Am., Internat. Brain Research Orgn., N.Y. Acad. Scis., Phi Beta Kappa, Sigma Xi, Alpha Omega Alpha. Club: Md. Capitol Yacht (Annapolis). Subspecialties: Neurosurgery; Cancer research (medicine). Current work: Neurooncology; cellular growth control and differentiation; image analysis; chemotherapy agent evaluation; neuronal regeneration. Home: 408 94th St New York NY 10128 Office: Sloan Kettering Cancer Ctr 1275 York Ave New York NY 10021

SMITH, BRUCE NEPHI, botanist, coll. dean; b. Logan, Utah, Apr. 3, 1934; s. Nephi Pratt and Laura (Peterson) S.; m. Ruth Olean Aamodt, Dec. 18, 1959; children: Rebecca, Trudy, Alan, Marilee, Edward, Samuel. B.S., U. Utah, 1959, M.S., 1962; Ph.D., U. Wash., 1964. Acting instr. U. Wash., 1962-63; postdoctoral fellow in plant biochemistry UCLA, 1964-65; research fellow in geochemistry Calif. Inst. Tech., 1965-68; asst. prof. botany U. Tex., Austin, 1968-74; prof. botany Brigham Young U., 1974—, chmn. dept. botany and range sci., 1976-79, dean, 1982—. Contbr. numerous articles, chpts. to profl. publs. Vice chmn. Orem (Utah) Beautification Com., 1980-82. Mem. AAAS, AAUP, Am. Soc. Plant Physiologists, Japanese Soc. Plant Physiologists, Geochem. Soc., Sixma Xi, Phi Kappa Phi. Democrat. Mormon. Club: SAR (pres. club). Subspecialties: Plant physiology (biology); Photosynthesis.

SMITH, C. WILLIAM, engineering educator; b. Christiansburg, Va., Jan. 1, 1926; s. Robert Floyd and Ollie May (Surface) S.; m. Doris Burton, Sept. 9, 1950; children: Terry J., David B. B.S. in Civil Engring, Va. Poly. Inst., 1946, M.S., 1949. Registered profl. engr., Va. Teaching fellow Va. Poly. Inst. and State U., Blacksburg, 1947, instr., 1948, asst. prof., 1949-53, assoc. prof., 1954-58, prof., 1958-81, alumni disting. prof., 1981—; cons. Kollmergen Corp., Redstone Arsenal, Western Electric Co., Brunswick Corp., Am.-Marietta Corp., Radford Arsenal, Litton Industries, Rubatex Inc., Masonite Corp., W.Va. Paper & Pulp Co. Contbr. numerous articles to tech. jours.; chpts. to books. NSF grantee, 1976—; NASA grantee, 1973-76; Dept. Def. grantee, 1968-78; Oak Ridge Nat. Lab. grantee, 1975-78; Delft U. Tech. grantee, 1976-78. Mem. ASME, Soc. Exptl. Stress Analysis, Soc. Engring. Sci., Am. Soc. Engring. Edn., ASTM, Internat. Assn. Structural Mechanics in Reactor Tech. Christian. Subspecialties: Fracture mechanics; Solid mechanics. Current work: Application of optical methods to three dimensional cracked body problems. Home: 107 College St Christiansburg VA 24073 Office: Va Poly Inst and State U Dept Engring Sci and Mechanics Blacksburg VA 24061

SMITH, CASSANDRA L., molecular biologist, researcher; b. N.Y.C., May 25, 1947; divorced. B.A. in Biology, W.Va. U., 1969, M.S. in Med. Microbiology, 1971; Ph.D. in Genetics, Tex. A&M U., 1974. Postdoctoral fellow Pub. Health Research Inst., NIH, 1974-78; research assoc. in chemistry and biology Columbia U., N.Y.C., 1978-82, research assoc. in human genetics and devel., 1982—. Mem. Am. Soc. Microbiology, Genetics Soc. Am. Subspecialties: Genetics and genetic engineering (biology); Molecular biology. Office: Dept Human Genetics and Devel Columbia U 701 W 168th St New York NY 10032

SMITH, CEDRIC MARTIN, physician, educator; b. Stillwater, Okla., Feb. 1, 1927; s. Otto Mitchel and Mary Catherine (Carr) S.; m. Mary E. Wylie, Dec. 21, 1948; children: Cristine Smith Riedel, Michael Cedric, Celia Louise. B.S., Okla. State U., 1950; M.D., M.S., U. Ill., 1953. Intern Phila. Gen. Hosp., 1953-54; successively instr., asst. prof., asso. prof. U. Ill., 1954-63, prof. pharmacology, 1963-66; staff scientist Inst. Def. Analyses, Arlington, Va., 1964-65; acting head dept. pharmacology U. Ill. Coll. Medicine, 1965-66; prof. dept. pharmacology and therapeutics Schs. Medicine and Dentistry, SUNY, Buffalo, 1966—, chmn. dept., 1966-73; founding dir. N.Y. State Research Inst. on Alcoholism, 1970-79; mem. med. staff Erie County Med. Center, 1976—; preceptor Family Practice Center, 1978—; mem. neuropharmacology adv. com. FDA, 1971-77, chmn., 1974-75; mem. drug abuse adv. com. N.Y. State Dept. Health, 1979—. Author 3 books in field; contbr. articles to profl. jours.; mem. editorial bd.: Jour. Pharmacology and Exptl. Therapeutics, 1959-65, Alcoholism: Clin. and Exptl. Research, 1977-81, Research Communications in Substance Abuse, 1979—. Bd. dirs. Narcotic and Drug Research, Inc., 1976—; mem. Erie County Profl. Adv. Com. on Alcoholism, 1974—; chmn. Lakes Area Emergency Med. Services Com. on Behavioral Emergencies, 1975-77; bd. dirs. Alcoholism Council Erie County, 1971-77, Research Found. Mental Hygiene, Inc., 1974-79. Served with U.S. Army, 1945-46. NIH, N.Y. State Health Research Council grantee. Mem. Am. Soc. Clin. Pharmacology and Exptl. Therapeutics (program com. 1970-76), Am. Soc. Clin. Pharmacology and Therapeutics, Am. Coll. Clin. Pharmacology, Am. Med. Soc. Alcoholism, AAAS, AMA, AAUP, Internat. Brain Research Orgn., Med. Soc. County of Erie (legis. com. 1980—), N.Y. State Med. Soc., Am. Coll. Neuropsychopharmacology, Collegium Internationale Neuropsychopharmacologicum, Research Soc. Alcoholism, Soc. Neuroscience, Nat. Council Alcoholism, Gerontol. Soc., Sigma Xi, Alpha Omega Alpha, Nu Sigma Nu. Subspecialties: Neuropharmacology; Pharmacology. Current work: Alcohol and drug consumption in man and animals; mechanism of drug action on brain and nervous system; neuropharmacology. Home: 220 Lakewood Pkwy Snyder NY 14226 Office: SUNY at Buffalo 127 Farber Hall Buffalo NY 14214

SMITH, CHARLES IRVEL, medicinal chemistry educator; b. Balt., Aug. 22, 1923; s. Louis Eldon and Lillian Marie (Akehurst) S.; m. Millicent Lois Yamin, Aug. 11, 1950; children: Carol Lois, Barbara Anne, Alan Craig. B.S., U. Md., 1944, Ph.D., 1950. Instr. physiol. chemistry dept. Sch. Medicine, Johns Hopkins U., Balt., 1950-52; sr. research scientist sect. drug metabolism pharmacology Squibb Inst. Med. Research, New Brunswick, N.J., 1952-60; assoc. prof. medicinal chemistry Coll. Pharmacy, U. R.I.-Kingston, 1960-74, prof., 1974—, chmn. dept., 1975-82. Contbr. articles to profl. jours. Served to lt. (j.g.) USNR, 1944-46. Recipient William Simon Gold medal Sch. Pharmacy, U. Md., 1944; U.S. AEC grantee, 1962-63. Fellow AAAS; mem. Am. Chem. Soc., Sigma Xi, Rho Chi, Kappa Psi. Subspecialties: Medicinal chemistry; Nuclear medicine. Current work: Synthesis, analysis, metabolism of drugs including radiopharms. Home: 6 Nichols Rd Kingston RI 02881 Office: Coll Pharmacy U RI Kingston RI 02881

SMITH, CHARLES ROGER, veterinary medicine educator; b. Hartville, Ohio, Mar. 31, 1918; s. Charles Roger and Ethel Olive (Seeman) S.; m. Genevieve Lorraine Taylor, Aug. 9, 1946; children—Ronald Roger, Debra Smith Beckstett, Eric William. Student, Ohio U., 1936-38, 40-41; D.V.M., Ohio State U., 1944, M.Sc., 1946, Ph.D., 1953. Diplomate: in cardiology Am. Coll. Veterinary Internal Medicine. Instr. veterinary medicine Ohio State U., 1944-53, asst. prof., 1953-55, asso. prof., chmn. dept. veterinary physiology and pharmacology, 1955-59, prof., 1957-69, research prof., chmn. dept. veterinary edn., 1969-71, acting dean, 1972-73, dean, 1973-80, asst., prof., dean emeritus, 1980—; research asso. dept. dairy sci. U. Minn., summer 1947; asst. research veterinarian Purdue U. Agrl. Expt. Sta., summers 1949, 50, 51; vis. scholar dept. physiology Coll. Medicine, U. Wash., summer 1960; cons. Morris Animal Found.; mem. nat. adv. com. Bur. Veterinary Medicine, FDA, 1968-74; mem. com. veterinary med. rev. NIH, 1970-73. Contbg. author: Duke's Physiology of Domestic Animals, 9th edit, 1977; reviewer: Am. Jour. Veterinary Research, 1959-73; contbr. numerous articles to profl. jours. Served with ASTP, 1943-44. Recipient Gamma award Omega Tau Sigma, 1964, Disting. Alumnus award Coll. Vet. Medicine, Ohio State U., 1981; named Ohio Veterinarian of Yr., 1980. Fellow Am. Coll. Vet. Pharmacology and Therapeutics (hon. mem. 1982); mem. Ohio Heart Assn., Am. Heart Assn. (chmn. affiliate com. research 1976-81, rev. and adv. com. 1975—, chmn. 1976—), AVMA (council on research 1959-69, chmn. 1967—), Am. Acad. Vet Physiologists and Pharmacologists (disting fellow 1982), Sigma Xi, Phi Zeta. Republican. Methodist. Club: Ohio State Faculty (Columbus Maennerchor). Lodge: Masons. Subspecialties: Comparative physiology; Cardiology. Current work: Veterinary physiology and pharmacology. Home: 4873 Chevy Chase Ave Columbus OH 43220 Office: Vet Medicine 1900 Coffey Rd Columbus OH 43210

SMITH, CLIFFORD WINSTON, botanist, educator; b. Hereford, Eng., Mar. 10, 1938; s. Clifford Richard Joseph and Elizabeth (Winstone (Smith)) S. B.Sc. with honors, Univ. Coll. North Wales, 1962; M.Sc., U. Manchester, Eng., 1963, Ph.D., 1965. Research assoc. Princeton U., 1966-67; from asst. prof. to assoc. prof. botany U. Hawaii-Manoa, 1967—; field assoc. Bishop Mus., Honolulu, 1978—; dir. coop. nat. parks resources studies unit U. Hawaii-Manoa, 1975—. Author sci. articles. Mem. Big Bros., Hawaii. Served as cpl. Royal Army Service Corps, 1956-58. Nat. Park Service grantee, 1975—; U.S. Fish and Wildlife grantee, 1979—; Nature Conservancy grantee, 1980-81. Mem. Bot. Soc. Am., Mycol. Soc. Am., Am. Bryological and Lichenological Soc., Audubon Soc., Am. Mus. Natural History. Roman Catholic. Subspecialties: Taxonomy; Ecology. Current work: Taxonomy and ecology of Hawaiian lichens; resource management of Hawaiian national parks. Home: 1205 Manulani St Kailua HI 96734 Office: 3190 Maile Way Honolulu HI 96822

SMITH, DALE METZ, botany educator, plant systematist; b. Portland, Ind., Dec. 23, 1928; s. Homer H. and Gertrude A. (Metz) S.; m. Ruth W., Aug. 12, 1950; children: Teresa, Gayle. B.S., Ind. U., 1950, Ph.D., 1957; M.S., Purdue U., 1952. Instr. U. Ariz., 1952; instr. to assoc. prof. U. Ky., 1955-60; assoc. prof. U. Ill., 1960-64; prof. botany U. Calif.-Santa Barbara, 1964—, chmn. dept. biology, 1979-81; also curator The Herbarium. Contbr. articles to profl. jours. Fellow AAAS; mem. Linnean Soc. London, Bot. Soc. Am., Calif. Bot. Soc., Am. Soc. Naturalists, Am. Soc. Plant Taxonomists (Cooley award 1963), Sigma Xi. Subspecialties: Systematics; Phytochemistry. Current work: Chemosystematics of vascular plants.

SMITH, DEAN F., astrophysicist; b. Los Angeles, July 25, 1942; s. Emmett F. and Mildred (Graveline) S.; m. Zdenka A. Kopal, Sept. 1, 1967; children: Helena D., Lara M. B.S., M.I.T., 1964; M.S. in Applied Physics, Stanford U., 1966; Ph.D. in Astrophysics, Stanford U., 1969. Postdoctoral fellow U.S.-USSR Cultural Exchange Program, 1969-70; mem. staff Nat. Ctr. for Atmospheric Research, Boulder, Colo., 1970-78; sr. research assoc., lectr. dept. astro-geophysics U. Colo., Boulder, 1978-82; staff scientist Berkeley Research Assoc., Berkeley, Calif., 1982—; cons. physics dept. U. Calif., Irvine.; Invited visitor Max-Planck Inst. fur Extraterrestrische Physik, Garching, W.Ger., 1975-76, Acad. Scis. USSR, 1980. Contbr. numerous articles to sci. jours. Mem. Internat. Astron. Union, Internat. Radio Sci. Union, Internat. Assn. Geomagnetism and Aeronomy, Am. Astron. Soc., Am. Phys. Soc. Club: American Alpine. Subspecialties: Theoretical astrophysics; Solar physics. Current work: Pulsar and planetary magnetospheres; solar flares and coronal loops; particle acceleration in astrophysics. Office: PO Box 241 Berkeley CA 94701

SMITH, DENNIS B., neurologist, hospital administrator, educator; b. Albany, N.Y., Sept. 6, 1939. B.A., Wesleyan U., 1961; M.D., Albany Med. Coll., 1965. Asst. prof. neurology U. Vt., Burlington, 1970-74; assoc. prof. neurology Med. Coll. Ga., Augusta, 1975-80, prof. neurology, 1980—; chief neurology services VA Med. Center, Augusta, 1975—. Contbr. numerous sci. and tech. articles to profl. publs. Bd. dirs. Summerville Assn., Augusta, 1982. Served with USAR, 1966-74. Westwood Pharms. Inc. grantee; VA Coop. Study grantee, 1978-82; Merrill-Dow Pharms. grantee; 1982. Mem. Am. Acad. Neurology, Montreal Neurol. Soc., AAAS, Am. EEG Soc., Eastern EEG Soc., Am. Epilepsy Soc., Soc. Neurosci., AAUP, Am. Med. EEG Soc., So. EEG Soc., Soc. Neurol. Edn., N.Y. Acad. Scis., Augusta Opera Assn. Subspecialties: Neurophysiology; Neuropharmacology. Current work: Field potentials and the EEG; new anticonvulsants. Office: VA Med Center Neurology Service 127 Augusta GA 30910

SMITH, ELVIN ESTUS, JR., physiology educator; b. Hattiesburg, Miss., Dec. 11, 1938; s. Elvin Estus and Ellen (Locke) S.; m. Irma Coward, Aug. 26, 1960; children: Elvin Estus, Forrest Chandler. B.S., William Carey Coll., 1960; Ph.D., U. Miss., 1964. Asst. prof. dept. physiology U. Miss. Sch. Medicine, Jackson, 1964-67, asso. prof., 1967-75, prof., 1975; prof. dept. med. physiology Tex. A&M U., College Station, 1975—, assoc. dean, 1975—. Editorial bd.: Jour. Applied Physiology, 1970-76, Am. Jour. Physiology, 1970-76. Bd. dirs. Am. Heart Assn., Austin, 1978—, v.p. Council VII, Longview, 1982—; adult leader 4H, College Station, 1979—. Recipient Disting. Service award Am. Heart Assn., 1978; Alumnus of Yr. William Carey Coll., 1965; Leadership award, 1960. Mem. Am. Physiol. Soc., Microcirculatory Soc., Sigma Xi, Phi Kappa Phi. Baptist. Subspecialty: Physiology (medicine). Current work: Cardiovascular control systems, hypoxia, cardiovascular response to exercise, cardiac hypertrophy, hemorragic shock. Office: Office of Dean College of Medicin Tex A and M Univ College Station TX 77843 Home: 1088 Rose Cir College Station TX 77840

SMITH, EMIL L., biochemist, consultant; b. N.Y.C., July 5, 1911; s. Abraham and Esther (Lubart) S.; m. Esther Press, Mar. 29, 1934. B.S., Columbia U., 1931, Ph.D., 1936. Guggenheim fellow, 1938-40; fellow Rockefeller Inst., N.Y.C., 1940-42; biochemist E. R. Squibb & Sons, New Brunswick, N.J., 1942-46; prof. biochemistry Sch. Medicine U. Utah, 1946-63; prof., chmn. dept. biochemistry UCLA Sch. Medicine, 1963-79, prof. emeritus, 1979—; cons. in field; cons. NIH, Nat. Acad. Scis. Co-author: Principles of Biochemistry, 7th edit, 1983; contbr. numerous articles on biochemistry to profl. jours. Mem. Nat. Acad. Scis. Subspecialties: Biochemistry (biology); Molecular biology. Current work: Protein chemistry and biology. Office: Dept Biochemistry UCLA Sch Medicine Los Angeles CA 90024

SMITH, FRANK ACKROYD, biochem. toxicologist, educator, cons.; b. Winnipeg, Man., Can., Feb. 14, 1919; s. Frank and Doris A. S.; m., Apr. 15, 1944; children: Susan J., Deborah A. B.A., Ohio State U., 1940, M.S., 1941, Ph.D., 1944. With Mellon Inst., Pitts., 1944; with Manhattan Project, U. Rochester, N.Y., 1944-46, with atomic energy project, 1946—, instr. toxicology, 1946-54, asst. prof., 1954-58, assoc. prof., 1958—; ad hoc cons. to govt. and industry. Author: (with Dr. Harold C. Hodge) Fluorine Chemistry, Vol. 4, 1965, (with Dr. Hodge and Dr. P.S. Chen) Fluorine Chemistry, Vol. 5, 1963; contbr. articles to profl. jours. Recipient (with Dr. Harold C. Hodge) Adolph Kammer award Am. Occupational Med. Assn., 1978. Mem. Am. Chem. Soc., Am. Indsl. Hygiene Assn., Am. Soc. Pharmacology and Exptl. Therapeutics, Soc. Toxicology (charter), AAAS, AAUP, Sigma Xi. Subspecialty: Toxicology (medicine). Current work: Biochemical toxicology of inorganic, organic fluorides, organic solvents especially as related to occupational exposure. Office: U Rochester Sch Medicine and Dentistry Dept Radiation Biology & Biophysics Rochester NY 14642

SMITH, GARY LEE, molecular biology educator; b. Rock Springs, Wyo., May 27, 1947; s. Richard and Blanche (Paulik) S.; m. Dianna Lynn Key, Dec. 16, 1967; 1 son, Benjamin. B.S., U. Wyo., 1969; Ph.D., Kan. State U., 1972. Postdoctoral fellow U. Wis., Madison, 1972-74; asst. prof. sch. life sci. U. Nebr., Lincoln, 1974-77, assoc. prof., 1977—, chmn. sect. genetics, cellular and molecular biology, 1982—. Contbr. articles to profl. jours. Nat. Cancer Inst. grantee, 1975—. Mem. Am. Soc. Cell Biology. Subspecialties: Cell biology; Cell and tissue culture. Current work: Hormonal regulation of animal cell growth - molecular and cellular endocrinology. Office: Sch Life Sci Univ Nebr Lincoln NE 68588

SMITH, GEORGE FRANKLIN, physician, researcher; b. N.Y.C., Sept. 5, 1924; s. George W. and May E. (Wolsiefer) S.; m. Patricia Norris, Apr. 30, 1944; children: G. Michael, Kathleen, Mark, Kevin, Mathew, Gregory. M.D., Georgetown U., 1953, U. Goteborg, Sweden, 1974. Diplomate: Am. Bd. Pediatrics. Asst. prof. U. Miami (Fla.) Sch. Medicine, 1960, assoc. prof., 1962-69; hon. research asst. Dept. Genetics, Biometry and Eugenics, Galton Lab., U. Coll., London, 1962-64; prof. pediatrics Loyola U. Stritch Sch. Medicine, Chgo., 1969-73; acting. chmn. Dept. Pediatrics, 1971-72; dir. genetics Rush Med. Sch., Chgo., 1973-75; chmn. Dept. Pediatrics Ill. Masonic Med. Center, Chgo., 1976—, chmn.; curator/med. dir. Internat. Coll. Surgeons Mus. and Hall of Fame, Chgo., 1979—. Contbr. articles to profl. jours. Served with USAF, 1942-46. Fellow Royal Soc. Medicine, Royal Microscopical Soc.; mem. Am. Pediatric Soc., Soc. Pediatric Research, Am. Med. Writers Assn., Am. Dermatoglyphics Assn., AAAS, others. Subspecialty: Gene actions. Current work: Work on human genetic problems and chromosomal and biochemical problems related to Downs Syndrome. Office: 836 W Wellington St Chicago IL 60657

SMITH, GERALD DUANE, chemist, college dean, educator; b. Cass City, Mich., Aug. 31, 1942; s. Clifford Ross and Mary Ellen (Smith) S.; m. Nancy Louise Myers, Aug. 19, 1967; 1 son, Michael L. B.S., Huntington Coll., 1964; Ph.D., Purdue U., 1972. AEC fellow U. Wash/Battelle N.W., Seattle, 1964-65; instr. chemistry Owosso Coll., 1965-67; asst. prof. chemistry Huntington Coll., 1967-73, assoc. prof., 1973-77, prof., 1977—, dean of coll., 1982—; cons. Ind. Emergency Radiation Response Team, 1973—. Mem. bd. adminstrn. College Park United Brethren Ch., Huntington, 1981-84, mem. exec. com., 1982; mem. Huntington Coll. Alumni Bd., 1981-84. Mem. Am. Chem. Soc. (N.E. Ind. Chemist of Yr. 1982, Petroleum Research Fund faculty research fellow 1982), AAUP, Am. Assn. Physics Tchrs., Ind. Acad. Sci. Democrat. Subspecialties: Nuclear medicine; Photochemistry. Current work: Radiation damage in crystals; photo-induced changes in electron energies in crystals. Home: 4190 N 580 W Huntington IN 46750 Office: 2303 College Ave Huntington IN 46750

SMITH, GERALD RALPH, microbiologist, researcher; b. Vandalia, Ill., Feb. 19, 1944. B.S., Cornell U., 1966; Ph.D., M.I.T., 1970. Postdoctoral fellow dept. biochemistry U. Calif.-Berkeley, 1970-72, U. Geneva, Switzerland, 1972-75; asst. prof. biology Inst. Molecular Biology, U. Oreg., Eugene, 1975-80, assoc. prof., 1980-82; assoc. mem. Fred Hutchinson Cancer Research Center, Seattle, 1982—. Contbr. articles to profl. jours. Helen Hay Whitney fellow, 1970-73; Swiss Nat. Sci. Found. fellow, 1973-75; recipient NIH Research Career Devel. award, 1980-84. Mem. AAAS, Genetics Soc. Am., Am. Soc. Microbiology. Subspecialties: Molecular biology; Gene actions. Current work: Molecular basis of genetic recombination; regulation of gene expression. Home: 7939 Montgomery Ave Elkins Park PA 19117 Address: 1124 Columbia St Seattle WA 98104 Office: 1015 Walnut St Philadelphia PA 19107

SMITH, HAMILTON OTHANEL, molecular biologist, educator; b. N.Y.C., Aug. 23, 1931; s. Bunnie Othanel and; ; s. Tommie Harkey and S.; m. Elizabeth Anne Bolton, May 25, 1957; children: Joel, Barry, Dirk, Bryan, Kirsten. Student, U. Ill., 1948-50; A.B. in Math, U. Calif., Berkeley, 1952; M.D., Johns Hopkins U., 1956. Intern Barnes Hosp., St. Louis, 1956-57; resident in medicine Henry Ford Hosp., Detroit, 1959-62; USPHS fellow dept. human genetics U. Mich., Ann Arbor, 1962-64, research asso., 1964-67; asst. prof. molecular biology and genetics Johns Hopkins U. Sch. Medicine, Balt., 1967-69, asso. prof., 1969-73, prof., 1973—; asso. Institut für Molekularbiologie der U. Zurich, Switzerland, 1975-76. Contbr. articles to profl. jours. Served to lt. M.C. USNR, 1957-59. Recipient Nobel Prize in medicine, 1978; Guggenheim fellow, 1975-76. Mem. Am. Soc. Microbiology, AAAS, Am. Soc. Biol. Chemists, Nat. Acad. Sci. Subspecialties: Immunobiology and immunology; Genetics and genetic engineering (biology). Current work: DNA transformation; restriction enzymes; DNA recombination. Office: 725 N Wolfe St Baltimore MD 21205

SMITH, HARDING EUGENE, astrophysicist; b. San Jose, Calif., May 10, 1947; s. Harding Eugene and Frances Bernice S.; m. Eileen O'Callahan, June 21, 1969. B.S., Calif. Inst. Tech., 1969; M.A., U. Calif., Berkeley, 1972, Ph.D., 1974. Research asst. Space Radiation Lab., Calif. Inst. Tech., Pasadena, 1967-69; research/teaching asst. U. Calif., Berkeley, 1969-74; asst. research physicist U. Calif. San Diego, La Jolla, 1974-77, asst. prof., 1977-80, assoc. prof. physics, 1980—. Mem. Internat. Astron. Union., Am. Astron. Soc., Astron. Soc. Pacific, AAAS. Subspecialties: Infrared optical astronomy; Cosmology. Current work: Observational extragalactic astronomy, active galaxies and quasi-stellar objects, evolution of galaxies observational cosmology. Office: Ctr Astrophysics and Space Scis U Calif San Diego C-011 La Jolla CA 92093

SMITH, HARLAN JAMES, educator, astronomer; b. Wheeling, W.Va., Aug. 25, 1924; s. Paul Elder and Anna Persis (McGregor) S.; m. Joan Greene, Dec. 21, 1950; children: Nathaniel, Sarah (dec.), Julia, Theodore, Hannah. A.B., Harvard, 1949, M.A., 1951, Ph.D., 1955; D.Phys. Sci. (hon.), Nicholas Copernicus U., Torun, Poland, 1973, Denison U., 1983. Research asst. astronomy, teaching fellow and research fellow Harvard, 1946-53; from instr. to asso. prof. astronomy Yale, 1953-63; prof. astronomy, chmn. dept., 1963-78; also dir. McDonald Obs., U. Tex., 1963—; mem. (Space Sci. Bd.), 1977-79. Co-editor: Astron. Jour, 1960-63. Served as weather observer USAAF, 1943-46. George R. Agassiz research fellow Harvard Obs., 1952-53. Fellow AAAS; Mem. Am. Astron. Soc. (acting sec. 1961-62, chmn. planetary div. 1974-75, council 1975-78, v.p. 1977-79), Royal Astron. Soc., Am. Geophys. Union, Asso. Univs. Research in Astronomy (chmn. bd. 1980—), AAAS, Internat. Astron. Union, Sigma Xi. Subspecialties: Optical astronomy; Planetary science. Current work: Planetary rotations; quasars; precise radial velocities, instrumentation; large telescopes. Home: 2705 Pecos St Austin TX 78703

SMITH, HARVEY ALVIN, mathematician, educator; b. Easton, Pa., Jan. 30, 1932; s. William Augustus and Ruth Carolyn (Krauth) S.; m. Ruth Kolb, Aug. 27, 1955; children: Deirdre Lynn, Kirsten Nadine, Brinton Averill. B.S., Lehigh U., 1952; M.S., U. Pa., 1955, A.M., 1958, Ph.D., 1964. Mem. tech. staff Inst. Def. Analyses, Arlington, Va., 1965-66; assoc. prof. math. Oakland U., Rochester, Mich., 1966-68; ops. research scientist Exec. Office of Pres., Washington, 1968-70; prof. math. Oakland U., 1970-77, Ariz. State U., Tempe, 1977—; cons. ACDA, Washington, 1973-79, Los Alamos Nat. Lab., N.Mex., 1980—. Author: Mathematical Foundations of Systems Analysis, 1969. Mem. Soc. Indsl. and Applied Math., Am. Math. Soc., AAAS, Sigma Xi. Subspecialties: Operations research (mathematics); Harmonic analysis. Current work: Harmonic analysis, operator algebras, integral operators, engineering analyses, mathematical physics, operations research, applications of mathematics to social sciences, strategic policy. Home: 18 E Concorda Dr Tempe AZ 85282 Office: Dept Mat Ariz State Univ Tempe AZ 85287

SMITH, HELENE SHEILA CARETTNAY, cancer cell biologist; b. Phila., Feb. 13, 1941; d. Joseph Herman and Jean (Abramson) Cohen; m. Allan-Lawrence Smith, June 10, 1962; 1 son, Joshua David. B.S. in Edn, U. Pa., 1962; Ph.D. in Microbiology, 1967. Postdoctoral fellow Princeton U., 1967-69, Nat. Cancer Inst., 1969-71; mem. staff Sch. Pub. Health, U. Calif., Berkeley, 1971-77, Lawrence Berkeley Lab., 1977-82; asst. dir. Peralta Cancer Research Inst., Oak, Calif., 1980—. Author numerous papers, reports in field. Mem. Am. Assn. Cancer Research, Am. Soc. Cell Biology, Tissue Culture Assn. Subspecialties: Cancer research (medicine); Cell study oncology. Current work: Biology of human mammary cells in vitro. Address: Peralta Cancer Research Inst 3023 Summit St Oak CA 94609

SMITH, HENRY JOHN PETER, physicist, software engr.; b. Bklyn., May 3, 1935; s. Henry J. and Catharine A. (Wheeler) S. B.S., Manhattan Coll., 1957; postgrad., U. Rochester, 1957-58; M.S., Tufts U., 1960; Ph.D., Northeatern U., Boston, 1966. Sr. scientist Am. Scientist and Engring., Cambridge, Mass., 1966-70, Visidyne Inc., Burlington, Mass., 1970—. Mem. Am. Phys. Soc., Am. Geophys. Union. Roman Catholic. Subspecialties: Aeronomy; Mathematical software. Current work: Simulation of upper atmospheric physics and chemistry. Use of micro- and mini- computers for scientific computation of large problems. Office: 5 Corporate P South Bedford St Burlington MA 01803

SMITH, HORACE ALDEN, astronomer; b. Willimantic, Conn., Apr. 5, 1952; s. Harlow and Sylvia (Adams) S. B.A., Wesleyan U., 1974; Ph.D., Yale U., 1980. Carnegie fellow Mt. Wilson and Las Campanas Obs., 1979-81; asst. prof. dept. physics and astronomy Mich. State U., East Lansing, 1981—. Contbr. articles to profl. jours. Mem. Am Astron. Soc. Subspecialty: Optical astronomy. Current work: Research on RR Lyrae stars, globular star clusters, chem. evolution of the galaxy and magellanic clouds.

SMITH, HORACE VERNON, JR., physicist; b. Rockford, Ill., July 23, 1942; s. Horace Vernon and Wanda Louise (Farley) S.; m. Susannah Eleanor Journeay, Aug. 14, 1964; 1 dau., Leah Catherine. B. Enging. Sci., U. Tex., Austin, 1964; M.S. in Physics, U. Ill., Urbana, 1965, Ph.D., U. Wis., Madison, 1971. Asst. prof. physics Prairie View A&M U., 1970-1971; research asst. Rice U., 1971; research assoc. U. Wis., Madison, 1971-74, asst. scientist, 1974-78; staff Los Alamos Nat. Lab., 1978—. Contbr. numerous articles to profl. jours. Mem. Am. Phys. Soc., IEEE, Tau Beta Pi, Sigma Pi Sigma. Subspecialty: Accelerator physics. Current work: Ion source physics for application in accelerator tech. Office: Mail Stop H818 PO Box 1663 Los Alamos NM 87545

SMITH, HOWARD ALAN, research astrophysicist; b. Boston, June 1, 1944; s. Edward and Sarah S. B.S. in Physics, M.I.T., 1966, 1966; Ph.D. in Physics, U. Calif., Berkeley, 1976. With U. Ariz., Tucson, 1977-81; astrophysicist Space Sci. div., Naval Research Lab., Washington, 1981—. Contbr. numerous articles to sci. jours. Mem. Am. Phys. Soc., Am. Astron. Soc., Internat. Astron. Union. Democrat. Jewish. Subspecialties: Infrared optical astronomy; Atomic and molecular physics. Current work: Infrared spectroscopy of the interstellar medium, particularly as reflecting star-formation processes. Home: 2700 Q St NW Apt 213 Washington DC 20007 Office: Space Sci Div Naval Research Lab Code 4138 Washington DC 20375

SMITH, IRA AUSTIN, clinical psychologist, psychological consultant; b. Wakefield, R.I., Jan. 28, 1945; s. Percy Austin and Ruth (Eastwood) S.; m. Kathryn Rigney Kelley, Aug. 13, 1977; children: Heather, Andrew, Rebecca. B.A., U. R.I., 1966, M.A., 1969. Lic. psychologist, R.I. Clin. psychologist Dr. Joseph H. Ladd Ctr., Exeter, R.I., 1968-72, chief clin. psychologist, 1972—; clin. psychologist Cranston Ctr. for Retarded Citizens, 1972—; psychologist Rocky Knoll Rest Home, Tiverton, R.I., 1978—; cons. clin. psychology; speaker profl. convs., lectr. in field on behavior treatment, techniques with retarded. Author papers in field. Vestryman Episcopal Ch., Wakefield, 1980—; coach Little League, Narragansett, 1980—, Narragansett Youth Basketball, 1980—. Mem. Am. Assn. Mental Deficiency (chmn. R.I. membership 1982-83), Council Exceptional Children, Am. Psychol. Assn. (assoc.). Subspecialties: Mental retardation/developmental disabilities; Behavioral psychology. Current work: Innovative behavioral/developmental therapeutic/programmatic techniques and clinical diagnosis with mentally retarded/developmentally disabled populations. Home: 4 Lawrence Dr Narragansett RI 02882 Office: Dr Joseph H Ladd Ctr Sch Land Rd Exeter RI 02852

SMITH, JACKSON BRUCE, physician, immunology researcher; b. Mt. Holly, N.J., Mar. 2, 1938; s. Jackson Burdette and Cynda Bruce (Hughes) S.; m. Penelope Lynne Prusa, June 7, 1963; children: Jackson Bruce, Joshua, Brian. B.S., Wake Forest U., 1960; M.D., Bowman Gray Sch. Medicine, 1965. Intern Pa. Hosp., Phila., 1965-66; resident, 1966-67, Hosp. of U. Pa., 1967-68; clin. research fellow Inst. Cancer Research, Phila., 1967-69; fellow Univ. Coll. London, 1972-74; research physician Inst. Cancer Research, Phila., 1974-81; assoc. prof. medicine Jefferson Med. Coll., Phila., 1981—. Contbr. articles to profl. jours. Served with U.S. Navy, 1969-72. NIH grantee, 1979—; Am. Cancer Soc. grantee, 1978-81. Mem. AAAS, Am. Assn. Immunologists, Am. Assn. Cancer Research, ACP. Democrat. Subspecialties: Internal medicine; Immunology (medicine). Current work: Lymphocyte interactions and immune system regulation; biological research; cell interactions immune regulation, autoimmunity, cancer. Home: 7939 Montgomery Ave Elkins Park PA 19117 Office: 1015 Walnut St Philadelphia PA 19107

SMITH, JAMES CECIL, JR., research lab. adminstr.; b. Little Orleans, Md., Jan. 17, 1934; s. James Cecil and A. Blanche (Brinkman) S.; m. Kay Lucille Plummer, Aug. 19, 1961; children: James Cecil, Michelle K., Deborah B. B.S., U. Md., 1956, M.S., 1959, Ph.D., 1964.

Predoctoral fellow U. Md., College Park, 1961-64; physiologist VA Hosp., Long Beach, Calif., 1964-66, lab. chief, Washington, 1971-77, Dept. Agr., Beltsville, Md., 1977—; professorial lectr. George Washington U. Sch. Medicine, 1975—. Contbr. articles on biochemistry and animal nutrition to profl. jours. Mem. com. Boy Scouts Am., Bowie, Md. Served to Lt. (j.g.) USPHS, 1959-61. Recipient Klaus Schwarz Commerative medal Internat. Assn. Bioinorganic Scientists, 1982. Mem. Am. Inst. Nutrition, Am. Soc. Bone and Mineral Research, Am. Assn. Gnotobiology, Soc. Environ. Geochemistry and Health. Methodist. Club: Methodist Men (Glenn Dale, Md.). Lodge: Masons. Subspecialties: Biochemistry (biology); Animal nutrition. Current work: Trace element (zinc, copper, selenium, chromium) metabolism in clinical nutrition, health and disease. Office: Human Nutrition Research Center Bldg 307 Baltsville MD 20769

SMITH, JAMES DOUGLAS, nephrology educator, researcher; b. Bardstown, Ky., Mar. 31, 1948; s. John Thomas and Gladys Lee (Hibbs) S. A.B., Harvard U., 1970; M.D., U. Ky., 1974. Diplomate: Am. Bd. Internal Medicine. Postgrad. in internal medicine Waterbury (Conn.) Hosp., 1974-77, chief resident in medicine, 1977-78; postdoctoral fellow in nephrology Yale U., New Haven, 1978-82, asst. prof. medicine, 1982—. Mem. Am. Soc. Nephrology, Internat. Soc. Nephrology, ACP, Council Kidney in Cardiovascular Disease (Am. Heart Assn.). Subspecialties: Internal medicine; Nephrology. Current work: Research interest is glucose and amino acid physiology, including inter-organ exchange and regulation by insulin, in normals and patients with uremia. Home: 223 Greene St New Haven CT 06511 Office: Yale U 2074 LMP 333 Cedar St New Haven CT 06510

SMITH, JESSE GRAHAM, JR., physician, educator; b. Winston-Salem, N.C., Nov. 22, 1928; s. Jesse Graham and Pauline Field (Griffith) S.; m. Dorothy Jean Butler, Dec. 28, 1950; children: Jesse Graham, Cynthia Lynn, Grant Butler. B.S., Duke U., 1962, M.D., 1951. Diplomate: Am. Bd. Dermatology (dir. 1974—, pres. 1980-81). Intern VA Hosp., Chamblee, Ga., 1951-52; resident in dermatology Duke U., 1954-56, asso. prof. dermatology, 1960-62, prof., 1962-67; resident U. Miami, 1956-57, asst. prof., 1957-60; prof. dermatology Med. Coll. Ga., 1967—, chmn. dept. dermatology, 1967—, acting chmn. dept. pathology, 1973-75; chief staff Talmadge Meml. Hosp., Augusta, Ga., 1970-72; mem. advisory council Nat. Inst. Arthritis, 1975-79. Contbr. chpts. in books.; Mem. editorial bd.: Archives of Dermatology, 1963-72, Jour. Investigative Dermatology, 1966-67, Jour. AMA, 1974-80, So. Med. Jour., 1976—; editor: Jour. Am. Acad. Dermatology, 1978—; Contbr. articles to profl. jours. Served with USPHS, 1952-54. Fellow A.C.P., Royal Soc. Medicine; mem. Am. Dermatol. Assn. (sec. 1976-81, pres. 1981-82), Soc. Investigative Dermatology (dir. 1964-69, pres. 1979-80), Am. Acad. Dermatology (dir. 1971-74, 78—), S.E. Dermatological Assn. (sec. 1970-71, pres. 1975-76), Ga. Soc. Dermatology (pres. 1979-80), So. Med. Assn. (chmn. sect. dermatology 1973-74), AMA (chmn. sect. dermatology 1981-84), Assn. Profs. Dermatology (dir. 1976-77, 80—, pres.-elect 1982-84), Med. Research Found. Ga. (dir. 1967—, pres. 1974-75), Alpha Omega Alpha. Subspecialty: Dermatology. Current work: Connective tissue and aging. Home: 606 Scotts Way Augusta GA 30909 Office: Medical College of Georgia Augusta GA 30912

SMITH, JOE MAUK, educator; b. Sterling, Colo., Feb. 14, 1916; s. Harold Rockwell and Mary Calista (Mauk) S.; m. Essie Johnstone McCutcheon, Dec. 23, 1943; children—Rebecca K., Marsha Mauk. B.S., Calif. Inst. Tech., 1937; Ph.D., Mass. Inst. Tech., 1943. Chem. engr. Texas Co., Standard Oil Co. of Calif., 1937-41; instr. chem. engring. Mass. Inst. Tech., 1943; asst. prof. chem. engring. U. Md., 1945; prof. chem. engring. Purdue U., 1945-56; dean Coll. Tech., U. N.H., 1956-57; prof. chem. engring. Northwestern U., 1957-59, Walter P. Murphy prof. chem. engring., 1959-61; prof. engring. U. Calif., 1961—, chmn. dept. chem. engring., 1964—; hon. prof. chem. engring. U. Buenos Aires, Argentina, 1964—; Fulbright lectr., Eng., Italy, Spain, 1965, and, Argentina, 1963, 65, Ecuador, 1970; Mudaliar Meml. lectr. U. Madras, India, 1967; UNESCO cons., Venezuela, 1972—. Author: Introduction to Chemical Engineering Thermodynamics, 3d edit, 1975, Chemical Engineering Kinetics, 1956, 3d edit., 1981. Guggenheim research award for study in, Holland; also Fulbright award, 1953-54. Mem. Am. Chem. Soc., Am. Inst. Chem. Engrs. 77 (Walker award 1960, Wilhelm award 1977), Nat. Acad. Engring., Sigma Xi, Tau Beta Pi. Subspecialties: Chemical engineering; Petroleum engineering. Home: 760 Elmwood Dr Davis CA 95616

SMITH, JOHN BARBER, utility company executive, mechanical electrical engineer; b. Florence, S.C., Feb. 11, 1930; s. John and Murry Kathern (Stephens) S.; m. Jewel Estell Dawson, Oct. 3, 1954; children: John Barber, David Douglas. B.S., Clemson U., 1952. Aircraft engr. Lockheed Aircraft Co., Marietta, Ga., 1954-57; project mgr. Ga. Power Co., Atlanta, 1957-71; gen. supt. Thomasville (Ga.) Utility Dept., 1971—. Mem. Republican nat. com., Washington, 1963, state del., Clarkston, Ga., 1964. Served to capt. USAF, 1952-54. Mem. Am. Pub. Gas Assn. (dir. 1974—), Am. Pub. Power Assn. (coms. 1971—), Am. Water Works Assn., Am. Nuclear Soc. Republican. Lodge: Rotary. Subspecialties: Electrical engineering; Mechanical engineering. Current work: Electric, gas, water, utilities, computerizing of utility activities. Home: 111 W Club Dr Thomasville GA 31792 Office: Thomasville Utilities Dept 411 W Jackson St Thomasville GA 31792

SMITH, JOHN BRYAN, pharmacologist, educator, researcher, cons.; b. Darlington, Eng., June 17, 1942; came to U.S., 1971; s. Robert Frederick and Phoebe (Mulhatton) S.; m. Angela Jane Fogg, Sept. 9, 1967; children: Timothy, Susanne. Ph.D. in Biochemistry, London U., 1971. Mem. staff Cardeza Found. for Hematologic Research, Thomas Jefferson U., Phila., 1971-82, mem. faculty dept. pharmacology, 1971-82, prof., 1981-82; prof. pharmacology and thrombosis Temple U., Phila., 1982—, asst. dir., 1982—. Contbr. over 100 articles to profl. jours. NIH grantee, 1982—. Mem. Am. Soc. Pharmacology and Exptl. Therapeutics, AAAS, Royal Inst. Chemistry, Chem. Soc., Sigma Xi. Subspecialties: Cellular pharmacology; Hematology. Current work: Inhibition of platelet involvement in thrombosis. Patentee in field. Office: SCOR Temple U Med Sch 3400 N Broad St Philadelphia PA 19115

SMITH, JOHN DOUGLAS, mechanical engineer; b. Detroit, Dec. 6, 1951; s. William and Pauline Agnes (Klein) S. B.S. M.E., Lawrence Inst. Tech. Design engr. Cameron Iron Works, Houston, 1976-78; sr. engr. N. L. Shaffer, Houston, 1979—. Mem. ASME, Soc. Automotive Engrs. Club: Gulf Coast. Subspecialties: Mechanical engineering; Offshore technology. Current work: Subsea electro-hydraulic control systems for offshore petroleum recovery. Patentee subsea electrohydraulic connector.

SMITH, JOHN ELVANS, chemistry educator, consultant; b. Washington, Sept. 6, 1929; s. Carl Bernard and Emily Thornton (Farr) S.; m. Eloise Emily Russell, Apr. 11, 1962; children: Katherine, Suzanne, Paul Stanley, Christopher John. Research chemist U.S. Naval Research Lab., Washington, 1960-62; head dept. chemistry U. So. Colo., Pueblo, 1964-82, prof. chemistry, 1982—. Author: Conserve Energy and Save Money, 1981 (Colo. Pacesetter's award 1982); contbr. articles to profl. jours. Mem. Pueblo County Task Force on Hazardous Waste Disposal, Pueblo, Colo., 1981-82; mem. Pueblo Energy Commn., 1981-82. Served to lt. USN, 1953-55. USPHS research fellow, 1956-60. Mem. Am. Chem. Soc., NEA, Am. Inst. Chemists, Am. Numismatic Assn., Nat. Rifle Assn., Colo. Edn. Assn. Episcopalian. Subspecialties: Energy conservation; Analytical chemistry. Current work: Dynamics and economics of energy conservation, chemistry's role in society, analytical chemistry of trace species. Home: 809 W Pitkin Ave Pueblo CO 81004 Office: U So Colo 2200 Bonforte Blvd Pueblo CO 81001

SMITH, JOHN HENRY, metallurgist; b. Rome, N.Y., July 3, 1937; s. Henry Behrens and Hester (Chapman) S. B.S., A.B., Lafayette Coll., Easton, Pa., 1958; M.S., U. Mo.-Rolla, 1959; Sc.D., MIT, 1964. Registered profl. engr., Pa. Metallurgist NASA Lewis Research Labs., Cleve., 1964-76, U.S. Steel Corp., Monroeville, Pa., 1966-75, Nat. Bur. Standards, 1975—. Contbr. articles to profl. jours. Served to capt. U.S. Army, 1964-66. Dept. Commerce fellow, 1980-81. Mem. Am. Soc. Metals, AIME, ASTM, ASME, Am. Welding Soc., AAAS, Sigma Xi. Subspecialties: Materials (engineering); Fracture mechanics. Current work: Fracture mechanics, mechanical properties of materials. Home: 8174 Inverness Ridge Rd Potomac MD 20854 Office: B 266 Materials Bld Nat Bur Standards Washington DC 20234

SMITH, JOHN PHILIP, psychologist; b. N.Y.C., Dec. 14, 1933; s. Philip and Anna Josephine (Burke) S.; m. Sylvia Ann Esposito, Dec. 18, 1965; children: Richard, Kevin, Amy. A.B., Fordham U., 1954, M.A., 1958, Ph.D., 964. Lic. psychologist, N.Y. State. Psychologist in pvt. practice, Eastchester, N.Y., 1971—. Served to 1st lt. U.S. Army, 1954-56. Mem. Am. Psychol. Assn., Sigma Xi. Subspecialties: Behavioral psychology; Social psychology. Current work: Elaboration of natural, absolute ethics; establishment of scientific basis for law. Office: 33 Deerfield Ave Eastchester NY 10707

SMITH, JOSEPH LORENZO, physician, educator, cons.; b. Green River, Wyo., Oct. 15, 1946; s. Joseph Franklin and Vera (Robinson) S.; m. Judy Peterson, Aug. 8, 1969; children: Joseph Lorenzo II, A. Theodore, Brett Lowell, Heather, Megan. B.S. magna cum laude, U. Utah, 1970; M.D., Cornell U., 1972. Diplomate: Am. Bd. Internal Medicine, also Sub-Bd. Infectious Diseases, Am. Bd. Med. Examiners. Internal medicine intern Pa. State U. Hershey Med. Ctr., 1972-73, resident in internal medicine, 1973-74; epidemiologist Phoenix labs. div. Bur. Epidemiology, Ctr. for Disease Control, USPHS, 1974-76; postdoctoral fellow div. infectious diseases Stanford (Calif.) U. Sch. Medicine, 1976-78; practice medicine specializing in infectious diseases, Ogden, Utah, 1979-81; clin. instr. medicine U. Utah, Salt Lake City, 1979-81, postdoctoral fellow div. respiratory, critical care and occupational medicine (respiratory), 1981—; Tb cons. Utah Dept. Health, 1979—; cons. in infection control McKay Dee Hosp., Ogden, 1981—. Contbr. articles and abstracts to sci. jours. Fellow ACP; mem. Am. Soc. Internal Medicine, Am. Soc. Microbiology, Infectious Disease Soc. Am., Am. Coll. Chest Physicians, Am. Thoracic Soc., Utah Med. Assn., Weber County Med. Soc., Alpha Epsilon Delta, Phi Kappa Phi. Subspecialties: Critical care; Pulmonary medicine. Current work: Sepsis, septic shock. Office: LDS Hosp 325 8th Ave Salt Lake City UT 84143

SMITH, KENDALL OWEN, microbiologist, educator; b. Wilson, N.C., Sept. 5, 1928; s. Walter R. and Ellen L. (Owens) S.; m. Lillian Irene Smith, Apr. 17, 1927; 1 son, David M. B.A., George Washington U., 1952; M.S., U. N.C., 1957, Ph.D., 1959. Assoc. prof. Baylor U. Coll. Medicine, Houston, 1960-65; research scientist NIH, Bethesda, Md., 1965-69; prof. microbiology U. Tex. Health Sci. Center, San Antonio, 1969—. Contbr. over 100 articles to sci. jours. Served in U.S. Army, 1947-54. Recipient Research Career Devel. award NIH, 1964. Mem. Am. Assn. Immunologists, Am. Soc. Microbiologists. Baptist. Subspecialties: Virology (medicine); Immunology (medicine). Current work: Viral latency, chemotherapy, herpes viruses, tumor immunology. Patentee in field (4). Home: 133 Trillium Ln San Antonio TX 78213 Office: U Tex Health Sci Ctr San Antonio TX 78284

SMITH, KENDRIC CHARLES, PhotoGiologist, radiation biology educator; b. Oakwood, Ill., Oct. 13, 1926; s. Russell Wilson and Virginia Frances (Mozley) S.; m. Marion Edmonds, Feb. 5, 1955; children: Nancy Carol, Martha Ellen. Student, Berea Coll., 1944-46; B.S., Stanford U., 1947; Ph.D., U. Calif.-Berkeley, 1952. Research asst. in radiology U. Calif. Med. Sch., San Francisco, 1954-56; research assoc. in radiology Stanford U. Sch. Medicine, 1956-62, asst. prof. radiology, 1962-65, assoc. prof., 1965-73, prof. radiology (radiation biology), 1973—; Mem. U.S. Nat. Com. for Photobiology, NAS/NRC, 1964-74, chmn., 1970-74; mem. U.S. Nat. Com. of Internat. Union Biol. Scis., 1967-73, sec., 1971-73; tech. adv. council GTE Labs., 1970-74. Author: (with P.C. Hanawalt) Molecular Photobiology, 1969; author, editor: Aging, Carcinogenesis and Radiation Biology, 1976, (with P.C. Hanawalt) The Science of Photobiology, 1977, Biological Impacts of Increased Intensities of Solar Untraviolet Radiation, 1973; editor: Photochem. and Photobiol. Revs. Vol. 1-7, 1976-83; exec. editor: Photochemistry and Photobiology, 1962-72; editorial bd.: Carcinogenesis, 1979—. USPHS fellow, 1952-54; recipient Research Career Devel. award, 1966-71. Mem. Am. Chem. Soc., Am. Soc. Microbiology, Brit. Photobiology Soc., Am. Soc. Photobiology (pres. 1972-74, disting. service award 1974), Am. Inst. Biol. Scis. (pres. 1982-83), Sigma Xi. Subspecialties: Genetics and genetic engineering (biology); Cancer research (medicine). Current work: Genetic control and biochem. basis of repair of damaged deoxyribonucleic acid (DNA), the role of DNA repair in mutagenesis (radiation and chemicals), and in spontaneous mutagenesis. Home: 927 Mears Ct Stanford CA 94305 Office: Dept Radiology Stanford Sch Medicine Stanford CA 94305

SMITH, KEVIN RICHARD, mech. engr.; b. Kansas City, Mo., Oct. 7, 1953; s. Richard Frazier and Betty Jean (Waters) S.; m. Cathy Lynne Nelson, Aug. 8, 1976. B.S., U. Kans., Lawrence, 1975. Designer Burnett Instruments Co., Lawrence, Kans., 1973-75; product design engr. Dazey Products Co., Kansas City, Mo., 1975-77; mech. project engr., engring. computer system mgr. King Radio Corp., Olathe, Kans., 1977—. Recipient Mech. Design award Kans. U. Endowment Assn., 1973. Mem. ASME. Presbyterian. Subspecialties: Mechanical engineering; Computer-aided design. Current work: Computer aided engineering design and management in electronic/electromechanical industry, computer graphics, engineering data base applications. Patentee in field. Office: 400 N Rogers Rd Olathe KS 66062

SMITH, LEONARD CHARLES, chemist, educator; b. Spokane, Wash., Jan. 31, 1921; s. Leonard Charles and Edith L. (McLellan) S.; m. Mary Elaine Rush, Oct. 1, 1945; children: David E., L. Frederick, Steven M., Peter D., Andrew I. A.B., U. Mont., Missoula, 1943; Ph.D., U. Ill., Urbana, 1949. Instr. in exptl. medicine Northwestern U. Med. Sch., 1950-51, instr. in biochemistry, 1952-56; research biochemist Hines (Ill.) VA Hosp., 1949-56; asst. prof. biochemistry U. S.D. Sch. Medicine, 1956-57, assoc. prof., 1957-62, prof., 1962-66; prof. chemistry Ind. State U., 1966; vis. lectr. Glasgow (Scotland) U., 1961-62; cons. to pub. cos. Contbr. articles to profl. publs. Mem. Chem. Soc. (U.K.), Soc. Exptl. Biology and Medicine, AMA, Am. Chem. Soc., Sigma Xi, Phi Sigma, Pi Mu Epsilon. Lodges: Masons; Shriners. Subspecialties: Biochemistry (biology); Nutrition (biology). Current work: Research in amino acid metabolism.

SMITH, LEWIS DENNIS, biological sciences educator; b. Muncie, Ind., Jan. 18, 1938; s. Thurman Lewis and Dorothy A. (Dennis) S.; m. Suzanne F. Metcalfe, June 10, 1961; children: Lauren Kaye, Raymond Bradley. A.B. in Zoology and Chemistry, Ind. U., 1959, Ph.D. in Exptl. Embryology, 1963. Research assoc. zoology Ind. U., Bloomington, 1963-64; asst. embryologist to assoc. embryologist Argonne Nat. Lab., 1964-69; assoc. prof. biology Purdue U., West Lafayette, Ind., 1969-73, prof., 1973—, assoc. head dept. biol. scis., 1978-80, head dept., 1980—; Mem. space biology panel NASA. Contbr. articles to profl. jours. Recipient NIH research career devel. award, 1969-74; NIH grantee, 1969—. Mem. Soc. Developmental Biology, Internat. Soc. Developmental Biology, AAAS, Am. Inst. Biol. Scis., Am. Soc. Biol. Chemists. Subspecialties: Developmental biology; Molecular biology. Current work: Molecular aspects of nucleocytoplasmic interactions during oogenesis and early development of amphibians. Office: Dept Biol Scis Purdue U West Lafayette IN 47907

SMITH, LINDA CHERYL, information science educator; b. Rochester, N.Y., Jan. 27, 1949. B.S. summa cum laude, Allegheny Coll., 1971; M.S. in Library Sci, U. Ill., 1972; cert. in systems and data processing, Washington U., St. Louis, 1974; M.S. in Info. and Computer Sci, Ga. Inst. Tech., 1975; Ph.D., Syracuse U., 1979. Trainee med. library Washington U., St. Louis, 1972-73; lit. searcher Info. Exchange Ctr., Ga. Inst. Tech., Atlanta, 1974-75; asst. prof. Grad. Sch. Library and Info. Sci., U. Ill., Urbana, 1977- 83, assoc. prof., 1983—; research asst. prof. Info. Retrieval Research Lab., Coordinated Sci. Lab., 1979—, mem. publs. com., 1979—, mem. grad. faculty, 1980—; lectr. in field. Contbr. chpts. to books, articles, reports and revs. to profl. jours. Mem. Am. Assn. Artificial Intelligence, Assn. Computing Machinery, Am. Soc. Info. Sci. (mem. numerous coms.), Assn. Am. Library Schs. (program com.), Med. Library Assn. (library com.), Spl. Libraries Assn. (faculty advisor), On-line Users Group (Urbana-Champaign), Phi Beta Kappa, Beta Phi Mu. Subspecialties: Information systems, storage, and retrieval (computer science); Information systems (information science). Current work: Representations in information retrieval system design; compatibility of information systems. Office: Grad Sch Library & Info Sci U Ill Urbana IL 61801

SMITH, LYNWOOD HERBERT, nephrologist, educator; b. Kansas City, Mo., Aug. 2, 1929; s. Lynwood Herbert and Arline Estel (Chandler) S.; m. Margery Davis Waddell, Dec. 15, 1951; children: Michael Chandler, Katherine Ann, Philip Waddell, Martha Lynn. B.S., U. Kans., 1951, M.D., 1960. Intern Met. Gen. Hosp., Cleve., 1961-62; resident internal medicine Mayo Grad. Sch., Rochester, Minn., 1962-64; research fellow Johns Hopkins U., Balt., 1964-65; cons. Mephrology/urology Mayo Clinic, Rochester, Minn., 1965—, dir. urulithiasis research, 1966—, prof. medicine, 1977—. Contbr. articles to profl. jours. Mem. vestry St. Luke's Episcopal Ch., Rochester, 1972-75. Served to lt. (j.g.) USN, 1952-56. Mayo Grad. Sch. travel grantee, 1965; named Tchr. of Yr., 1982. Mem. Am. Soc. Nephrology, Internat. Soc. Nephrology, ACP, Central Soc. Clin. Research, Am. Clin. and Climatol. Assn., Alpha Omega Alpha, Omicron Delta Kappa, Beta Theta Pi, Sigma Xi. Republican. Subspecialties: Nephrology; Internal medicine. Current work: Basic and clinical research relating to formation of urinary calculi, crystal formation and inhibition and calcium matabolism. Home: 4912 Weatherhills Dr SW Rochester MN 55901 Office: Mayo Clinic 200 First St SW Rochester MN 55905

SMITH, MYRON ARTHUR, astronomer; b. Kosciusko, Miss., Dec. 25, 1944; s. Myron Bausman and Eleanor (McMillan) S.; m.; children: Hilary Wills, Zachary Andrew. B.S. in Astronomy, Case Inst. Tech., 1966, Ph.D., U. Ariz., 1970. Postdoctoral fellow U. Calif.-Santa Cruz, 1971-72; asst. prof. astronomy U. Toledo, 1972-73, U. Tex.-Austin, 1973-80; assoc. astronomer Sacramento Peak Obs., Sunspot, N. Mex., 1980—. Grumman Aerospace scholar, 1962-66. Mem. Am. Inst. Physics, Am. Astron. Soc., Internat. Astron. Union. Republican. Presbyterian. Subspecialty: Optical astronomy. Current work: High resolution stellar spectroscopy; velocity fields as diagnostics of the atmospheric and internal structure of stars.

SMITH, OLIVER HUGH, biology educator; b. Rochester, N.Y., Sept. 13, 1929; s. George Oliver and Stephanie (Grocki) S.; m. Sarah Haynam, Sept. 16, 1968; children: Stephen, Jessica. B.S., St. Louis U., 1951; M.S., Syracuse U., 1953; postgrad., Case Western Res. U., 1955-57; Ph.D., Stanford U., 1961. Postdoctoral fellow Nat. Found., Oak Ridge Nat. Labs., 1961-63; asst. prof. dept. biology Marquette U., Milw., 1963-68, assoc. prof., 1968-76, prof., 1972—; vis. scientist Lab. Chem. Biology, NIH, Washington, 1972, Institut Pasteur, Paris, 1976; microbiologist Div. Biol. Energy Research, U.S. Dept. Energy, 1981–82. Contbr. articles to profl. jours. Served with U.S. Army, 1953-55. Nat. Found. fellow, 1961-63; NIH fellow, 1972. Mem. Am. Soc. Microbiology, Genetics Soc. Am., AAAS, Sigma Xi. Subspecialties: Microbiology; Genetics and genetic engineering (biology). Current work: Genetic control mechanisms in bacteria, genetic control of virulence, mutational alterations of protein structure. Office: Dept Biology Marquette U Milwaukee WI 53233

SMITH, ORVILLE LEE, physicist, educator; b. Pittsburg, Kans., Mar. 19, 1935; s. Orville Lesley and Ruth (Miller) S.; m. Nancy Rose, June 25, 1963; 1 dau., Christy. Ph.D., U. Mo., 1962. Group leader reactor div. Oak Ridge Nat. Lab., 1961-70, research staff mem. environ. scis. div., 1971-79, sr. staff mem. instrumentation and controls div., 1980-82, sr. staff mem. and group leader, 1982—; assoc. prof. nuclear engring. Univ. Tenn., Knoxville, 1966—. Author: Soil Microbiology: A Model of Decomposition and Nutrient Cycling, 1981; contbr. articles to profl. jours. Recipient Albert S. Eisenstein award U. Mo., 1961. Mem. Sigma Xi. Subspecialties: Nuclear physics; Biophysics (physics). Current work: Electric power plant design and operation, nuclear power safety, mathematical analysis of ecosystems, including plant and animal interactions, mathematical modeling of human behavior, particularly as applied to man-machine interactions. Patentee in field. Office: PO Box X Oak Ridge TN 37830

SMITH, OTTO J.M., electrical engineering educator; b. Urbana, Ill., Aug. 6, 1917; s. Otto Mitchell and Mary Catherine (Carr) S.; m. Phyllis Pearl Sterling, Sept. 3, 1941; children: Candace Smith Shock, Otto J.A., Sterling Barton, Stanford. B.S. in Chemistry, Okla. State U., 1938, U. Okla., 1938, Ph.D., Stanford U., 1941. Registered profl. engr., Calif. Research asst. Stanford U., 1938-41; instr. elec. engring. Tufts U., Medford, Mass., 1941-43; asst. prof. elec. engring. U. Denver, 1943-44; prof. electronics Instituto Tecnologico de Aeronautica, Sao Jose dos Campos, Sao Paulo, Brazil, 1954-56; sr. research fellow in econs. and engring. Monash U., Melbourne, Australia, 1966-67; prof. elec. engring. and computer scis. U. Calif.-Berkeley, 1947—; dir. Wind Electric Systems, Inc. Author: Feedback Control Systems, 1958; contbr. articles to profl. jours. Dist. commr. Boy Scouts Am., 1949-53. Guggenheim fellow Tech. Hoch. Darmstadt, W.Ger., 1960. Fellow AAAS, IEEE; mem. Am. Soc. for Engring. Edn., Soc. for Social Responsibility in Sci., Soc. for Social Responsibility in Engring., AAUP, Fedn. Am. Scientists, Union Concerned Scientists, Internat. Solar Energy Soc., No. Calif. Solar Energy Assn. Democrat. Methodist. Club: City Commons (Berkeley). Subspecialties: Solar energy; Wind power. Current work: Design of solar-electric plants and wind turbines. System optimization, economic planning, appropriate technology, power system stabilization, automatic control, measurements. Patentee solar power, wind power, variable speed generators, stepping motors, heliostats, others. Home: 612 Euclid Ave Berkeley CA 94708 Office: Dept EECS U Calif Berkeley CA 94720

SMITH, PETER LLOYD, physicist; b. Victoria, B.C., Can., Apr. 28, 1944; s. Lloyd Wood and Joan Mary (Mercer) S.; m. Lois Elaine Hodgson, July 20, 1968. B.Sc., U. B.C., 1965; Ph.D., Calif. Inst. Tech., 1972. Research fellow Calif. Inst. Tech., 1972; asst. prof. Harvey Mudd Coll., Claremont, Calif., 1972-73; research fellow Harvard Coll. Obs., Cambridge, Mass., 1973-75, research assoc., 1975—, lectr. astronomy, 1983—. Mem. Am. Phys. Soc., Optical Soc. Am., Internat. Astron. Union, Am. Astron. Soc. Subspecialties: Ultraviolet high energy astrophysics; Atomic and molecular physics. Current work: Measurement of atomic and molecular parameters needed for the diagnosis of astrophysical plasmas. Development of space-borne astronomical instrumentation. Office: Harvard Coll Obs P-243 60 Garden St Cambridge MA 02138

SMITH, PETER WILLIAM, quantum electronics scientist; b. London, Nov. 3, 1937; U.S., 1963, naturalized, 1964; m. Jacqueline Marie Smith, June 18, 1966; children: Christal, Dawn Noelle. B.Sc., McGill U., Montreal, Que., Can., 1958, M.Sc., 1961, Ph.D., 1963. Mem. tech. staff Bell Labs., Holmdel, N.J., 1963—; vis. MacKay prof. U. Calif., Berkeley, 1970; vis. research scientist Laboratoire D'Optique Quantique, Ecole Polytechnique, Palaiseau, France, 1978-79. Contbr. numerous articles on quantum electronics to profl. jours.; editorial bd.: Comtex Sci, 1982—. Bd. dirs. Monmouth Arts Found., Red Bank, N.J. Recipient Sr. Scientist award NATO, 1979. Fellow IEEE, Optical Soc. Am.; mem. Am. Phys. Soc. Subspecialties: Laser research; Optical signal processing. Current work: Quantum electronics research in optical switching and optical signal processing. Patentee in field.

SMITH, PHILIP LAWRENCE, physiology educator, researcher; b. Brunswick, Maine, June 19, 1949; s. Oscar Samuel and Mary Anna (Utecht) S.; m. Bonnie Lee McPhail, Aug. 21, 1972; children: Pamela Anne, Matthew Philip. B.A., U. Maine, 1972; M.S., Northeastern U., 1975, Ph.D., 1978. USPHS fellow U. Chgo., 1978-79, U. Tex., Houston, 1979-81; asst. prof. physiology U. Kans., Kansas City, 1981—. Contbr. articles to sci. jours. Served with U.S. Army, 1970-76. Grantee USPHS, 1982, Kans. Lung Assn., 1982, Kans. Heart Assn., 1982. Mem. Am. Physiol. Soc., Soc. Gen. Physiologists, N.Y. Acad. Scis., Mt. Desert Island Biol. Lab. Methodist. Subspecialties: Gastroenterology; Physiology (medicine). Current work: Mechanisms and regulation of intestinal ion transport. Home: 6921 W 81st St Overland Park KS 66204 Office: Dept Physiology Univ Kans 39th and Rainbow Blvd Kansas City KS 66103

SMITH, PHILLIP JOSEPH, atmospheric science educator; b. Muncie, Ind., Oct. 2, 1938; s. and Marie I. (Perry) S.; m. Linda A. Whitney, Aug. 19, 1960; children: Amy, Andrew, Allyson. B.S., Ball State U., 1960; M.S., U. Wis., 1964, Ph.D., 1967. Research & teaching asst. dept. meteorology U. Wis.-Madison, 1963-67; asst. prof. to prof. atmospheric science Purdue U., West Lafayette, Ind., 1967—, prof., 1981—; sr. postdoctoral fellow Nat. Ctr. Atmospheric Research, Boulder, Colo., 1972-73. Contbr. articles to profl. jours. Served to lt USAF, 1960-63. Mem. Am. Meteorol. Soc., Am. Geophs. Union, Ind. Acad. Sci., Sigma Xi. Subspecialties: Meteorology; Synoptic meteorology. Current work: Dynamic and synoptic meteorology, kinematics and energetics of extratropical cyclone systems; role of latent heat release in cyclone development. Home: 3000 Georgton Rd West Lafayette IN 47906 Office: Purdue U Dept Geoscis West Lafayette IN 47907

SMITH, PHYLLIS STERLING, writer, translator, wind energy research co. exec.; b. Berkeley, Calif., Aug. 27, 1921; d. Allen and Pearl (Sitzler) Sterling; m. Otto Joseph Mitchell Smith, Sept. 3, 1941; children: Candace Barbara, Otto Joseph Allen, Sterling Mitchell, Stanford Douglas. Student in art and psychology, Stanford U., 1938-41, in psychology, Tufts U., 1941-43. Survey adminstr. Am. Psychology Corp., Boston, 1941-43; instr. in oil painting and ceramics Art Studio Scranton, Pa., 1945-47; pres. bd., 1972, free-lance Portuguese translator, Berkeley, 1975—; research asst., writer Wind Systems, Inc., Berkeley, 1978—, pres., 1980—. Free-lance author, Berkeley, 1968-71; author sci. fiction, travel articles, poetry. Bd. dirs. Berkeley Food Project; mem. social action com. Ch. Women United, Berkeley. Recipient Grand prize Poets Dinner, Berkeley, 1964, 71. Mem. Calif. Writers Club, Internat. Solar Energy Soc., No. Calif. Solar Energy Assn. Democrat. Methodist. Subspecialties: Wind power; Solar energy. Current work: Appropriate tech. for developing countries' energy, wind energy, solar power, solar-electricity, solar-heating, solar pasteurization, biomass conversion. Patentee apparatus for providing radiative heat rejection from a working fluid used in a rankine cycle tape system.

SMITH, RALPH EARL, microbiologist; b. Yuma, Colo., May 10, 1940; s. Robert Conrad and Esther Clara (Schwarz) S.; m. Sheila Lee Kondy, Aug. 26, 1961; 1 dau., Andrea Denise. B.S., Colo. State U., 1961; Ph.D., U. Colo. Med. Sch., 1968. USPHS postdoctoral fellow Duke U. Med. Ctr., Durham, N.C., 1968-70, asst. prof. dept. microbiology and immunology, 1970-74, assoc. prof., 1974-80, prof., 1980-82; prof., chmn. dept. microbiology and environ. health Colo. State U., Ft. Collins, 1983—; cons. Bellco Glass, Inc., Vineland, N.J., 1977-80. Contbr. articles to profl. jours. Asst. scoutmaster Boy Scouts Am., 1972—. Eleanor Roosevelt fellow Internat. Union Against Cancer, 1978-79. Mem. Am. Soc. Microbiology, N.Y. Acad. Scis., Am. Soc. Virology. Democrat. Methodist. Subspecialties: Virology (medicine); Microbiology (medicine). Current work: Isolated avian leukosis viruses which had limited oncogenic spectra; characterized virus causing bone tumors in chickens; isolated viruses causing obesity and lung tumors in chickens. Patentee in field. Office: Dept Microbiology and Environ Health Colo State U Fort Collins CO 80523

SMITH, RENE JOSE, med. physicist; b. Havana, Cuba, Mar. 19, 1938; s. Rene Eustasio and Dolores Eladia (Cuervo) S.; m. Maria Elena Sanchez, Apr. 1, 1967; children: Ana Lucia, James David. B.S. in Physics, Loyola U., New Orleans, 1962; M.S., Fordham U., Bronx, N.Y., 1964; Ph.D. in Biophysics, Va. Commonwealth U., Richmond, 1973. Instr. physics Richmond (Va.) Profl. Inst., 1964-68, acting head physics dept., 1966-68; sr. physicist Meml. Sloan-Kettering Cancer Ctr., N.Y.C., 1973-75; med. physicist, radiation safety officer VA Med. Ctr., East Orange, N.J., 1975—; cons. therapeutic radiol. physics. Contbr articles to profl. jours. Recipient Health Scis. Div. Va. Commonwealth U. grants, 1971-72, 72-73. Mem. Am. Assn. Physicists in Medicine, N.J. Med Physics Soc. (pres. 1982), Health Physics Soc., Am. Physics Soc., Biophysics Soc. Roman Catholic. Club: Century Club Loyola U. Subspecialties: Medical physics; Biophysics (physics). Current work: Optimization of radiation therapy treatment planning by means of CT scanning; determination of dose to different organs from radiation therapy treatments. Office: Radiation Therapy Service VA Med Ctr East Orange NJ 07019

SMITH, RICHARD JAMES, mfg. co. exec.; b. Oregon, Ill., Dec. 2, 1921; s. Henry Adelmon and Jessie Gertrude (Salzman) S.; m. Avalene Snodgrass, Apr. 2, 1945; children: George Victor, Jeffrey Halleck. B.S. in Mech. Engring, Purdue U., 1944. Design engr. Torrington Co., South Bend, Ind., 1945-48; sales engr. Jack & Heinz, Inc., Cleve., 1948-51; design engr., chief engr. McGill Mfg. Co., Valparaiso, Ind., 1951—, v.p. engring., 1962—. Mem. ASME, Soc. Automotive Engrs. Club: Union League. Lodge: Elks. Subspecialty: Mechanical engineering. Current work: Design, application and failure analysis of anti-friction ball and roller bearings. Patentee in field. Home: 106 Sheffield Dr Valparaiso IN 46383 Office: 909 Lafayette St Valparaiso IN 46383

SMITH, RICHARD PETRI, psychology educator; b. Middletown, N.Y., Aug. 22, 1926; s. Charles William and Lela (Percival) S.; m. Sara Ann Winges, June 6, 1964; 1 dau.: Eileen. A.B., Hofstra Coll., 1951; Ph.D., Emory U., 1961. Sr. ops. research analyst Lockheed Ga. Co., Marietta, 1958-61; asst. prof. psychology U. Louisville, 1961-64, assoc. prof., 1964-66, prof., 1970—. Contbr. articles to sci. jours. Served with USAAF, 1944-46. Mem. Psychonomic Soc., So. Soc. Philosophy and Psychology, AAAS, Human Factors Soc., Soc. Neurosci., Am. Orchid Soc., Ky. Orchid Soc. Republican. Episcopalian. Subspecialties: Psychobiology; Neuropharmacology. Current work: Behavioral effects of drugs; psychology of boredom; research on boredom and monotony. Home: 1318 Ridgeway Ave New Albany IN 47150 Office: 353 Life Scis Bldg U Louisville Belknap Campus Louisville KY 40292

SMITH, RICHARD STANLEY, JR., plant pathologist; b. Somerville, N.J., Mar. 2, 1930; s. Richard Stanley and Suzanna Evangeline (Drummond) S.; m. Virginia Lee Dahl, Sept. 15, 1956; children: Brian Marshall, Mark Drummond, Matthew Scott, Karyn Marie. B.S., Utah State U., 1958; Ph.D., U. Calif.-Berkeley, 1963. Research plant pathologist Forest Service, U.S. Dept. Agr., Berkeley, Calif., 1961-75, Honolulu, 1975-76, supervisory plant pathologist, San Francisco, 1976—. Contbr. numerous articles to sci. jours., chpts. to books. Served with USNR, 1950-54. Mem. Am. Phytopath. Soc. Subspecialties: Plant pathology; Integrated pest management. Current work: Consultant to forest managers on the control and prevention of forest diseases. Home: 152 Nova Dr Piedmont CA 94610 Office: 630 Sansome St Room 343 San Francisco CA 94111

SMITH, ROBERT CHARLES, biochemist, biochemistry educator; b. Chgo., Sept. 15, 1932; s. Theodore Earl and Mary Katherine (Scheidler) S.; m. Katherine Ann Kopacz, June 15, 1957; children: Robert Steven, Laura Mary, Jean Ann. B.S., Elmhurst (Ill) Coll., 1954; M.S., U. Ill., 1958, Ph.D., 1960. Postdoctoral fellow Univ. Inst. Microbiology, Copenhagen, 1959-61; asst. prof. dept. animal and dairy sci. Auburn U., 1916-63, assoc. prof., 1963-69, prof., 1969—. Contbr. to profl. jours. Mem. Am. Soc. Biol. Chemists, Am. Soc. Animal Sci., Soc. Exptl. Biology and Medicine., N.Y. Acad. Scis. Subspecialties: Biochemistry (biology); Molecular biology. Current work: Nucleoside and nucleotide composition of bovine erythrocytes; possible role of uric acid and ribosyluric acid as antioxidants. Home: 212 Dogwood Dr Auburn AL 36830 Office: Dept Animal Sci Auburn U Auburn AL 36849

SMITH, ROBERT DAVID, computer science educator, consultant, researcher; b. Erie, Pa., Dec. 7, 1937; s. Ira E. and Rozella (Zimmerman) S.; m. Vilma J. Monserrate, Aug. 28, 1966; children: Roger, Victor, Elena. B.S., Gannon Coll., 1959; M.S., Pa. State U., 1964, Ph.D., 1966. Instr. Pa. State U., University Park, 1962-66, asst. prof., 1966-68; assoc. prof. computer sci. Kent State U., Ohio, 1968-72, prof., 1972—; researcher, cons. McDermott/Babock, Alliance, Ohio, 1968—; research and devel. Goodyear Aerospace, Akron, Ohio, 1974—; mgmt. devel. exec. Akron City Hosp., 1980—; with Rehab. Services Adminstrn., 1979—; faculty Ohio Bankers Assn., Columbus, 1971—. Co-author: Personnel Management: A Computer Approach, 1970, Systems Analysis and Design, 1976, Personnel Management: A Human Resource System, 1982. Mem. Diocesan Bd. Edn., Youngstown, Ohio, 1982—. Served to lt. U.S. Army, 1960-62. Ford Found. grantee, 1974; Bur. Vocat. Rehab. Ohio grantee, 1981. Mem. Am. Soc. Personnel Adminstrn. (adv. 1977, Nat. Merit award 1980). Roman Catholic. Subspecialties: Information systems, storage, and retrieval (computer science); Human resource information systems. Current work: Computer systems in banking; research and development management; management of technological innovation. Home: 1587 Morris Rd Kent OH 44240 Office: Kent State U Kent OH 44242

SMITH, ROBERT L., auditory neurophysiologist, educator; b. N.Y.C., Mar. 29, 1941; s. Abe J. and Bertha (Rosenthal) S.; m. Carolee Smith, Nov. 16, 1968; children: Jana, Shayna, Marnina. B.E.E., CCNY, 1962; M.S.E.E., N.Y.U., 1966; Ph.D., Syracuse U., 1973. Devel. engr. Wheeler Labs., Great Neck, N.Y., 1962-64; lectr. elec. engring. CCNY, N.Y.C., 1964-66; instr. elec. and computer engring. Syracuse (N.Y.) U., 1970-74, asst. prof. sensory scis., 1974-80, assoc. prof., 1980—. Contbr. articles to profl. publs. Recipient Career devel. award NIH, 1979—; NSF grantee, 1981—. Mem. Accoustical Soc. Am., Soc. Neurosci., Assn. Research in Otolaryngology, N.Y. Acad. Scis., Sigma Xi, Eta Kappa Nu, Tau Beta Pi. Jewish. Subspecialties: Neurophysiology; Sensory processes. Current work: Investigation the encoding and transmission of information in auditory nervous system. Recording and analyzing responses to sound of single cells in cochlea, cochlear nerve and nucleus, and modelling of results. Home: 3 Haverhill Pl Dewitt NY 13214 Office: Inst Sensory Research Syracuse U Syracuse NY 13210

SMITH, ROBERTS ANGUS, biochemist, educator, pharm. co. exec.; b. Vancouver, C., Can., Dec. 22, 1928; came to U.S., 1953, naturalized, 1963; s. Alvin Roberts and Hazel (Mather) S.; m. Adela Marriott, Aug. 22, 1953; children: Roberts H.A., James A.D., Eric J.M., Richard I.F. B.S.A. with honors, U. B.C., 1952, M.Sc., 1953; Ph.D., U. Ill., 1957. Instr. U. Ill., Champaign-Urbana, 1957-58; asst. prof. UCLA, 1958-62, prof. biochemistry, 1962—; pres. Viratek, Inc., Covina, Calif., 1980—; also dir.; dir. ICN Pharms., Inc. Editor: Ribavirin: A Broad Spectrum Antiviral Agent, 1980; contbr. articles to profl. jours. Guggenheim fellow, 1963-64. Mem. Am. Chem. Soc., AAAS, Am. Soc. Biol. Chemists. Subspecialties: Biochemistry (medicine); Molecular pharmacology. Current work: Viral chemotherapy; protein phosphorylation. Home: 221 17th St Santa Monica CA 90402 Office: 222 N Vincent Ave Covina CA 91722

SMITH, ROGER POWELL, pharmacologist, toxicologist, educator, researcher, cons.; b. Hokuchin, Korea, July 16, 1932; s. Burton Powell and Mary Hannah (McMullen) S.; m. Rena Joan Pointer, Nov. 1, 1956; children: Sam F., Joan B., Ben H. B.S., Purdue U., 1953, M.S., 1955, Ph.D., 1957; M.A. (hon.), Dartmouth Coll., 1975. Lic. pharmacist, N.H. Instr. dept. pharmacology and toxicology Dartmouth Med. Sch., Hanover, N.H., 1960-63, asst. prof., 1963-68, assoc. prof., 1968-73, prof., 1973—, chmn. dept., 1976—; adj. prof. Vt. Law Sch., South Royalton, 1981—. Author: (with others) Clinical Toxicology of Commercial Products, 1983; contbr. articles to profl. jours. Committeeman Cub Scouts, Boy Scouts Am., Hanover Youth Hockey Assn.; deacon Ch. of Christ, Hanover. Served with Med. Service Corps. U.S. Army, 1957-60. USPHS career devel. award, 1966-71; Nat. Air Pollution Control Adminstrn. grantee, 1962-70; Nat. Heart, Lung and Blood Inst. grantee, 1970—. Mem. AAAS, Soc. Toxicology, Am. Soc. Pharmacology and Exptl. Therapeutics, Soc. Exptl. Biology and Medicine, Sigma Xi. Republican. Club: Hanover Country. Subspecialties: Toxicology (medicine); Pharmacology. Current work: Biochemical toxicology of cytochrome oxidase inhibitors, chemically induced damage to red blood cells, toxicology of vasodilator agents, pharmacology of vascular smooth muscle and platelet aggregation. Home: 12 Kingsford Rd Hanover NH 03755 Office: Dept Pharmacology and Toxicology Dartmouth Med Sch Hanover NH 03756

SMITH, RONALD GENE, research chemist; b. Woodland, Wash., Jan. 24, 1944; s. Claude Allen and Eileen Bogarta (Smearman) S.; m. Deloris Ann Caldwell, Aug. 22, 1970; children Alison Ann, Brian Jay. B.S., Whitworth Coll., 1966; Ph.D., Purdue U., 1972. Postdoctoral research assoc. Oreg. Grad. Ctr., Beavteron, 1972-74, instr., 1974-77; asst. prof. organic chemistry and asst. organic chemist U. Tex. M.D. Anderson Hosp. and Tumor Inst., Houston, 1977-80, assoc. prof. and assoc. organic chemist, 1980—; mem. faculty U. Tex. Grad. Sch. Biomed. Scis., 1978—. Contbr. articles to profl. jours. Mem. Am. Chem. Soc., Am. Soc. Mass Spectrometry, Am. Assn. Cancer Research, Phi Lambda Upsilon. Presbyterian. Subspecialties: Mass spectrometry; Pharmacokinetics. Current work: Applications of mass spectrometry to drug metabolism, pharmacokinetics and mechanism of drug action and toxicity in cancer research. Home: 6003 Arboles St Houston TX 77035 Office: 6723 Bertner Ave Houston TX 77030

SMITH, RONALD WILLIAM, materials engr.; b. Salem, Mass., Dec. 4, 1953; s. Harland Payson and Claire Arlene S.; m. Patricia Ann, Mar. 12, 1977; children: Erika Marie, Ryan. B.S.M.E., Northeastern U., 1976, M.S.M.E., 1976; postgrad., Drexel U., 1981—. Registered profl. engr., N.Y. State, 1981. Asst. engr. Ledgemont Lab., Kennecott Copper Corp., Lexington, Mass., 1975; research asst. Northeastern U., 1976; materials devel. engr. gas turbine div. Gen. Electric Co., Schenectady, 1976—. Mem. ASME, Am. Soc. Metals, Am. Powder Metallurgy Inst. Subspecialties: Materials processing; High-temperature materials. Current work: Plasma deposition of high temperature metals; powder meatls; melting and solidifcation processing. Home: 1416 Via Del Mar Schenectady NY 12309 Office: 1 River Rd 53-316 Schenectady NY 12345

SMITH, SAMUEL HOWARD, plant pathologist, university administrator; b. Salinas, Calif., Feb. 4, 1940; s. Adrian Reed and Elsa Rose (Jacop) S.; m. Patricia Ann Walter, July 8, 1960; children: Samuel Howard, Linda M. B.S., U. Calif.-Berkeley, 1961, Ph.D., 1964. NATO postdoctoral fellow Glasshouse Crops Research Inst., Sussex, Eng., 1964-65; asst. prof. plant pathology U. Calif.-Berkeley, 1965-69; assoc. prof. plant pathology Pa. State U.-Arendtsville, 1969-71, assoc. prof., University Park, 1971-74, prof., 1974—, head dept. plant pathology, 1976-81, dean Coll. Agr., 1981—, dir. Agrl. Expt. Sta., and dir. Coop. Extension Service, 1981—. Contbr. articles to profl. jours. Mem. Am. Phytopathol. Soc., AAAS, Am. Inst. Biol. Scis., Assn. Applied Biologists, State Hort. Assn. Pa., Pa. Flower Growers Assn., Latin Am. Phytopathol. Soc., Brit. Fedn. Plant Pathologists, Am. Mushroom Inst. Subspecialties: Plant pathology; Plant virology. Home: 1108 Kay St Boalsburg PA 16827 Office: Pa State U 201 Agr Adminstrn Bldg University Park PA 16802

SMITH, SHELDON MAGILL, physicist, astronomer; b. St. Paul, Apr. 19, 1931; s. Sheldon Holloway and Sarah Alice (Matteson) S.; m.; children: Sheldon, Jennifer, Christina; m. Carolyn Joan Rounds, July 14, 1978; stepchildren: Timothy, Coleen, Cathleen. B.S., U. Calif.-Berkeley, 1953; M.A. in Physics, U. Calif.-Davis, 1962; postgrad., Stanford U., 1965-66. Research scientist NASA Ames Research Center, Moffett Field, Calif., 1962—, sr. research scientist, 1979—. Contbr. articles to sci. jours. Vice pres. Stanford NACCP, 1972-73; asst. scoutmaster Boy Scouts Am., 1976-77. Served to lt. USN, 1954-64. Mem. Am. Astron. Soc., AAAS, Soc. Photo-Optical Instrumentation Engrs., Sigma Pi. Democrat. Episcopalian. Clubs: Toastmasters, Almaden Cycle Touring (v.p. 1981—). Lodge: Masons. Subspecialties: Satellite studies; Infrared spectroscopy. Current work: Space shuttle infrared telescope facility; study and design of far-infrared attenuating coatings to reduce stray light in the SIRTF telescope. Far-infrared reflectance; scattering; optical constants; photometry. Designer spl. instrumentation for several solar eclipse expdns.; developer (with Peter Sturrock) magnetic field configuration of solar streamers. Home: 10665 Martinwood Way Cupertino CA 95014 Office: NASA-Ames Research Center MS 245-6 Moffett Field CA 94036

SMITH, THOMAS GRAVES, JR., neurophysiol. researcher, cons.; b. Winnsboro, S.C., Mar. 22, 1931; s. Thomas Graves and Mary Eula (Mungo) S.; m. Jo Ann Horn, June 16, 1956. B.A., Emory U., 1953, Oxford (Eng.) U., 1956; M.D., Columbia U., 1960. Intern in medicine Bronx (N.Y.) Mcpl. Hosp. Ctr., 1960-61; research assoc. USPHS, NIH, Bethesda, Md., 1961-64; med. officer Lab. Neurophysiology, 1966-68; sect. chief Lab. Neurophysiology, 1968—; research assoc. M.I.T., 1964-66; cons. in field; grants reviewer. Contbr. articles to profl. jours.; editor: (with J. L. Barker) The Role of Peptides in Neuronal Function, 1980. Served with USPHS, 1961-64. Mem. Am. Physiol. Soc., Soc. for Neurosci. Subspecialties: Neurophysiology; Biophysics (physics). Current work: Interneuronal communication, neuronal electrical and chemical excitability, voltage clamping technology. Home: 5512 Oakmont Ave Bethesda MD 20817 Office: NIH Bldg 36 Room 2C02 Bethesda MD 20205

SMITH, THOMAS LLOYD, horticulturist, educator; b. Cin., July 29, 1946; s. Lloyd Damon and Aline Elizabeth (Rust) S.; m. Susan Sprow, Mar. 1, 1969; children: David Thomas Maguire, Lisa Suzanne Rust, Kathleen Elizabeth Rust. B.S., U. Cin., 1968; M. in Forest Sci, Yale U., 1970, postgrad., 1971. Grad. asst. Yale U., 1970-71; assoc. supt. horticulturist Spring Grove Cemetery, Cin., 1972—; lectr. horticulture U. Cin., 1973—; pres. Flora Therapy, Inc., Cin., 1979-81. Bd. dirs. Civic Garden Ctr. Greater Cin., 1980-83, Urban Forestry Dept. Parks, Cin., 1981—. Mem. Profl. Grounds Mgmt. Soc. (chmn. nat. conv. 1983), So. Ohio Profl. Grounds Mgmt. Soc., Ohio Forestry Assn., Ohio Nurseryman's Assn., Beta Theta Pi. Republican. Episcopalian. Clubs: Diogenes, Yale Alumni (Cin.). Subspecialties: Plant growth; Urban forest resource management. Current work: Involved with urban forest resource management of both developed and undeveloped grounds within the largest nonprofit cemetery in the United States; specific interests have included tree physiology and air pollution/photosynthetic tolerance of certain plant materials. Office: Spring Grove Cemetery 4521 Spring Grove Ave Cincinnati OH 45232

SMITH, THOMAS WOODWARD, cardiologist, educator; b. Akron, Ohio, Mar. 29, 1936; s. Luther David and Beatrice Pearl (Woodward) S.; m. Sherley Goodwin, Sept. 13, 1958; children: Julia Goodwin, Geoffrey Woodward, Allison Lloyd. A.B., Harvard U., 1958, M.D. magna cum laude, 1965. Diplomate: Am. Bd. Internal Medicine, 1971. Intern, then resident in medicine Mass. Gen. Hosp., Boston, 1965-67, clin. fellow in cardiology, 1967-69, Nat. Heart and Lung asso. program dir. myocardial infarction research unit, 1972-74; Inst. spl. fellow, 1969-71; instr. medicine Harvard Med. Sch., 1969-71, asst. prof. medicine, 1971-73, assoc. prof. medicine, 1973-79, prof. medicine, 1979—; chief of cardiology Peter Bent Brigham Hosp., 1974—; cons. Children's Hosp. Med. Center, 1975—, Mass. Gen. Hosp., 1977—, Sidney Farber Cancer Inst., 1978—. Served to lt. (j.g.) USN, 1958-61. Mem. Fellow Am. Coll. Cardiology, ACP; mem. Am. Heart Assn. (council clin. cardiology, council basic sci., council circulation), Am. Fedn. Clin. Research, Paul Dudley White Soc., AAAS, Am. Soc. Pharmacology and Exptl. Therapeutics, Am. Soc. Clin. Investigation, Assn. Univ. Cardiologists, Am. Physiol. Soc., Assn. Am. Physicians, Soc. Gen. Physiologists. Subspecialties: Cardiology; Internal medicine. Current work: Investigative cardiology and cardiovascular pharmacology. Cardiac glycoside mechanisms of action; antibody reversal of toxicity; receptor and inotropic state control in cardiac muscle. Office: Cardiovascular Div Brigham and Women's Hosp 75 Francis St Boston MA 02115

SMITH, WALTER LAWS, statistics educator, researcher; b. London, Nov. 12, 1926; U.S., 1954; s. Thomas and Dorothy Emily (Bush) S.; m. Mary Ramsden, Redman, Sept. 4, 1950; children: Caroline, Simon. B.A., Cambridge (Eng.) U., 1947, M.A., 1951, Ph.D., 1953. Statistician Med. Sch., Cambridge U., 1953-54, lectr. in math., 1956-58; asst. prof. stats. U. N.C.-Chapel Hill, 1954-56, assoc. prof., 1958-61, prof., 1961—, chmn. dept. stats., 1981—. Author: (with D.R. Cox) Queues, 1961; editor: (with W.E. Wilkinson) Procs. Symposium on Congestion theory, 1965; research numerous publs. in field. Served to lt. Royal Navy, 1947-50. Recipient Adams prize Cambridge U., 1961; Guggenheim fellow, 1974-75. Fellow Inst. Math. Stats., Am. Statis. Soc., Royal Statis. Soc., Cambridge Philos. Soc.; mem. Internat. Statis. Inst., Ops. Research Soc. Am. Subspecialties: Statistics; Probability. Current work: Random processes arising in various congestion systems.

SMITH, WAYNE D., electrical engineer; b. Ingram, Tex., Apr. 14, 1933; children: Michelle, Leisa, Mathew; m. Beverly Smith, July 3, 1975; children: Drew, D.R., Dana. B.A. in Psychology, Tex. Christian U., 1962. Photographer Swayze Studio, Kerrville, Tex., 1949-52; Humrro pilot adviser Dept. Def. Washington, 1957-61; engr. Ling Temco Vought Co., Dallas, 1961-65; engr. mgr. Boeing Co., Seattle, 1965—; mgr. crew systems research Boeing Comml. Airplane Co., Seattle, 1967—. Served with U.S. Army, 1953-57. Fellow AIAA. Democrat. Baptist. Lodge: Moose. Subspecialties: Electronics; Human factors engineering. Current work: Controls and displays systems development, crew interface performance test and evaluation, crew workload evaluation. Home: 2929 76th SE Apt 412 Mercer Island WA 98040 Office: Boeing Co PO Box 3707 Seattle WA

SMITH, WILBUR STEVENSON, consulting engineer; b. Columbia, S.C., Sept. 6, 1911; s. George W. and Margaret Rebecca (Stevenson) S.; m. Sarah E. Bolick, Dec. 22, 1934; children: Sarah, Margaret, Stephanie. B.S., U. S.C., 1932, M.S. magna cum laude, 1933, LL.D., 1963; postgrad., Harvard U., 1936-37; L.H.D., Lander Coll., 1975. Registered profl. engr., 50 states, D.C. Electric hwy. engr. S.C. Hwy. Dept., 1933-34, asst. traffic engr., 1935-37, traffic engr., 1937-42; traffic cons. FBI, 1942-63; chmn., dir. Freeman, Fox, Wilbur Smith & Assos. (cons. engrs.), London, 1964-71; transp. cons. U.S. Office CD, 1942-43; assoc. dir. Yale U. Bur. Hwy. Traffic, 1943-57, research assoc., 1957-68; chmn. bd. Eno Found. for Transp., Westport, Conn., Wilbur Smith & Assos., Columbia, S.C. and New Haven, 1946—, also Richmond and Falls Church, Va., N.Y.C., Washington, San Francisco, Los Angeles Houston, Pitts., Miami, Fla., Toronto, Melbourne; dir. Koger Properties, Inc., Am. Exec. Life Ins. Co., Columbia. Author: (with N. Hebden) State City Relationships in Highway Affairs, 1950, (with T. Matson, F. Hurd) Traffic Engineering, 1955; author tech. bulls., reports, traffic surveys. Former bd. dirs. Transp. Assn. Am.; bd. dirs. Hwy. Users Fedn. for Safety and Mobility; trustee U. S.C.-Bus. Partnership Found.; former trustee Presbyn. Coll.; chmn. U. S.C. Chair Endowment Club; former bd. visitors Sch. Engring., Duke U.; former mem. citizens adv. com. Comprehensive Cancer Duke U. Med. Center. Recipient Disting. Alumni award U. S.C., 1968. Mem. Internat. Road Fedn., Inst. Engrs. Australia, Inst. Civil Engrs. (U.K.), N.Z. Instn. Engrs., Internat. Bridge, Tunnel and Turnpike Assn., Am. Inst. Cons. Engrs., ASCE (hon., past chmn. hwy. div. exec. com., past chmn. transp. planning com., past chmn. nat. transp. policy com.), Transp. Research Bd. (Roy W. Crum award 1980), Inst. Transp. Engrs. (hon., past nat. pres.), Nat. Acad. Engring., Am. Rd. and Transp. Builders Assn. (past pres., dir.), Hong Kong Instn. Engrs., Am. Soc. Safety Engrs., IEEE, Nat. Soc. Profl. Engrs., S.C. Soc. Profl. Engrs. (Engr. of Yr., 1964), N.Y. Soc. Profl. Engrs., Am. Pub. Works Assn. (hon. mem. Inst. Transp.), Phi Beta Kappa, Tau Beta Pi, Phi Sigma Kappa, Chi Epsilon (hon.), Blue Key. Clubs: Forest Lake Country, Palmetto, Summit (Columbia, S.C.); Grads. (New Haven); Cosmos (Washington); Miami; Yale (N.Y.C.); St. Stephens (London); Wild Dunes (S.C.) Beach and Racquet. Subspecialty: Transportation Engineering. Home: 1630 Kathwood Dr Columbia SC 29206 Office: Bankers Trust Tower PO Box 92 Columbia SC 29202 Office: 135 College St New Haven CT 06507

SMITH, WILLIAM BOYCE, statistics educator, consultant; b. Port Arthur, Tex., Sept. 7, 1938; s. Benjamin Thomas and Eula (Wactor) S.; m. Patricia Rutherford, Apr. 12, 1963; children: Leah Ann, Scott Andrew, Angela Rae. B.S. in Math., Lamar U., 1959, M.S., Tex. A&M U., 1960, Ph.D. in Stats. (NDEA fellow), 1967; postgrad., Rice U., 1960-62. Asst. prof. math. Lamar U., 1962-64; mathematician Sandia Corp., summers 1961-62; asst. prof. to prof. stats. Tex. A&M U., 1967—, asst. deanColl. Sci., 1972-77; head dept. stats. (Tex. A&M U.), 1977—, legal cons., researcher. Contbr. articles profl. jours. Baseball coach, football ofcl.; mem. S.W. Football Ofcls. Assn. Ctr. Energy and Mineral Resources grantee, 1979-82; grantee NASA, 1969-71, 82—. Fellow Japanese Soc. Promotion of Sci.; mem. Am. Statis. Assn, Math. Assn. Am., Biometric Soc., Sigma Xi. Methodist. Subspecialty: Statistics. Current work: Multivariate statistical methods — development and application. Home: 1040 Rose Circle College Station TX 77840 Office: Dept Statistics Tex A&M U College Station TX 7784

SMITH, WILLIAM LEO, meteorologist, educator; b. Detroit, May 13, 1942; s. Rolland Francis and Ruth Mary (Gerhardstein) S.; m. Marcia Jean Simmerer, Aug. 20, 1962; children: William Leo, Jeanne, Steven, Julie, Joanna, Jonathon, Sarah, Kiara. B.S. in Meteorology, St. Louis U., 1963, M.S., U. Wis., 1964, Ph.D., 1966. Chief mesoscale applications br. NOAA, Nat. Earth Satellite Service, 1966-81; prof. meteorology U. Wis., Madison, 1978—, assoc. dir., 1977—; dir. devel. lab. NOAA/Nat. Earth Satellite and Data Info. Service, 1981—. Editor: Jour. Climate and Applied Meteorology, 1982-84; Contbr. articles to profl. jours. Recipient Spl. Achievement award Dept. of Commerce, 1970, Gold medal award, 1973, cert. of recognition, 1978, performance awards NOAA, 1967-83. Fellow Am. Meteorol. Soc. (Clarence Leroy Meisenger award 1970); mem. Sigma Xi. Subspecialties: Satellite meteorology; Remote sensing (atmospheric science). Current work: Development of remote sensing radiometers, algorithms of satellite data processing and techniques for data applications to weather forecasting. Office: 1225 W Dayton St Madison WI 53706

SMOLENSKY, MICHAEL HALE, physiology educator, chronobiology researcher; b. Chgo., May 10, 1942; s. Louis Rosenzweig and Lottie (Benditzon) Smolensky-R.; m. Michele Tauber, Aug. 17, 19637 (div. Apr. 1978); 1 dau., Susan; m. Nita Beth Gathing, Oct. 18, 1980; children: Brian Borroughs, Melissa. B.S. in Zoology, U. Ill., 1964, M.S., 1964, Ph.D., 1971. Asst. prof. U. Tex. Sch. Pub. Heatlh, Houston, 1970-74, assoc. prof. environ. physiology, 1974—; research assoc. McGovern Allergy Clinic, Houston, 1972—; research cons. Tex. Allergy Research Found., Houston, 1974—. Co-author: Biological Rhythms and Medicine, 1983; co-editor: Chronobiology in Allergy and Immunology, 1977, Recent Advances in the Chronobiology of Allergy and Immunology, 1980, Internat. Jour. Chronobiology, 1983—. Active Cypress Creek Emergency Med. Service, Houston, 1974-77. Grantee NASA, Nat. Inst. Occupational Safety and Health. Mem. Internat. Soc. Chronobiology (dir. 1982—), Soc. Menstrual Cycle Research (dir. 1976-77), AAAS, Sigma Xi. Subspecialties: Physiology (biology); Chronobiology. Current work: Environmental physiology, environmental and occupational heatlh, public health, pharmacology, toxicology. Office: U Tex Sch Pub Health Houston TX 77030

SMYTH, KERMIT CAMPBELL, chemist; b. Portland, Maine, Jan. 16, 1946; s. Duncan and Jean (Corthell) S.; m.; 1 dau., Kirsten. B.A., Amherst Coll., 1968; Ph.D., Stanford U., 1972. Mem. tech. staff Bell Labs., Murray Hill, N.J., 1972-74; research chemist, head exploratory fire research group Nat. Bur. Standards, Washington, 1974—. Contbr. articles to profl. jours. Recipient Bronze medal Dept. Commerce, 1981. Mem. Am. Phys. Soc., Sigma Xi, Phi Beta Kappa. Subspecialties: Spectroscopy; Atomic and molecular physics. Current work: Application of laser-based optical diagnostic methods to study the chemistry of soot formation in flames. Office: Nat Bur Standards Bldg 224 Rm B260 Washington DC 20234

SNAPE, WILLIAM JOHN, JR., medical educator, researcher; b. Camden, N.J., Aug. 24, 1943; s. William John and Barbara (Fleischman) S.; m. Margaret Frye, Mar. 7, 1982; children: William John, Rebecca J. B.A., Princeton U., 1965; M.D., Jefferson Med. Coll., 1969. Intern Bronx (N.Y.) Mcpl. Hosp., 1969-70, resident, 1970-71; asst. prof. medicine U. Pa., Phila., 1975-80, assoc. prof., 1980-81; prof. medicine UCLA, 1982—; chief gastroenterology Harbor Med. Ctr.-UCLA, Torrance, 1982—. Served to lt. comdr. USN, 1971-73. NIH grantee, 1979—. Fellow ACP; mem. Am. Gastroenterology Assn., Am. Physiology Soc., Western Soc. Clin. Investigation. Presbyterian. Subspecialties: Gastroenterology; Neurophysiology. Current work: Smooth Muscle physiology. Office: Harbor UCLA Med Center Gastrointestinal Sect 1000 W Carson St Torrance CA 90509

SNAVELY, BENJAMIN BRENEMAN, physicist; b. Lancaster, Pa., Jan. 6, 1936; s. Benjamin Lichty and Anne Espenshade (Breneman) S.; m. Sabine Maria, Aug. 19, 1961; children—Judith M., Eric B. B.S.E.E., Swarthmore Coll., 1957; M.S.E.E., Princeton U., 1959; Ph.D. in Engring. Physics (McMullen fellow, 1960, RCA fellow, 1951), Cornell U., 1962. Sr. research physicist, physics div. Kodak Research Labs., Rochester, N.Y., 1962-65, research asso., sr. staff physics div., 1965-68, lab. head, solid state physics lab., 1969-73, asst. dir. physics div., 1975-81, tech. asst. to dir., 1981—; vis. prof. U. Marburg, Ger., 1968-69; assoc. div. leader laser div. Lawrence Livermore (Calif.) Lab., 1973-75; assoc. prof. part time Inst. of Optics, U. Rochester, 1970-73. Fellow Optical Soc. Am.; mem. Am. Phys. Soc., AAAS, IEEE, Nat. Acad. Sci.-NRC Com. Atomic and Molecular Sci., Sigma Xi, Phi Kappa Phi, Sigma Tau. Club: Cosmos (Washington). Subspecialties: Condensed matter physics; Tunable lasers. Current work: Physics of tunable lasers. Office: Kodak Research Labs Bldg 83 Eastman Kodak Co Rochester NY 14650

SNELL, GEORGE DAVIS, geneticist; b. Bradford, Mass., Dec. 19, 1903; s. Cullen Bryant and Katharine (Davis) S.; m. Rhoda Carson, July 28, 1937; children: Thomas Carleton, Roy Carson, Peter Garland. B.S., Dartmouth Coll., 1926; M.S., Harvard U., 1928, Sc.D., 1930; M.D. (hon.), Charles U., Prague, 1967; LL.D. (hon.), Colby Coll., 1982; Sc.D., Dartmouth Coll., 1974, Gustavus Adolphus Coll., 1981, U. Maine, 1981, Bates Coll., 1982. Instr. zoology Dartmouth, 1929-30, Brown U., 1930-31; asst. prof. Washington U., St. Louis, 1933-34; research asso. Jackson Lab., 1935-56, sr. staff scientist, 1957—, emeritus, 1969—, sci. adminstr., 1949-50. Co-author: Histocompatibility, 1976; also sci. papers in field; editor: The Biology of the Laboratory Mouse, 1941. NRC fellow U. Tex., 1931-33; NIH health research grantee for study genetics and immunology of tissue transplantation, 1950-73 (allergy and immunology study sect. 1958-62); Guggenheim fellow, 1953-54; Recipient Bertner Found. award in field cancer research, 1962; Griffin award Animal Care Panel, 1962; career award Nat. Cancer Inst., 1964-68; Gregor Mendel medal Czechoslovak Acad. Scis., 1967; Internat. award Gairdner Found., 1976; Wolf Found. prize in medicine, 1978; award Nat. Inst. Arthritis and Infectious Disease-Nat. Cancer Inst., 1978; Nobel prize in physiology or medicine, 1980. Mem. Nat. Acad. Scis., Transplantation Soc., Am. Acad. Arts and Sci., French Acad. Scis. (fgn. asso.), Am. Philos. Soc., Brit. Transplantation Soc. (hon.), Phi Beta Kappa. Subspecialties: Immunogenetics; Transplantation. Current work: Study of histocompatibility genes, especially of the major histocompatibillity complex, in the mouse. Study of cell surface alloantigens of lymphocytes. Home: 21 Atlantic Ave Bar Harbor ME 04609

SNIDER, BARRY BERNARD, chemistry educator, consultant; b. Chgo., Jan. 13, 1950; s. Gordon L. and Ruth (Tobias) S.; m. Katalin Boros, July 12, 1975; 1 dau., Emily. B.S. in Chemistry, U. Mich., 1970; Ph.D., Harvard U., 1973. Research assoc. Columbia U., 1973-75; asst. prof. Princeton U., 1975-81; assoc. prof. chemistry Brandeis U., 1981—. Contbr. numerous articles to profl. jours. Sloan fellow, 1979-83; Camille and Henry Dreyfus Tchr.-Scholar grantee, 1982—. Mem. Am. Chem. Soc. Subspecialties: Organic chemistry; Synthetic chemistry. Current work: Development of new synthetic methods; total synthesis of biologically active natural products. Home: 45 Lafayette Rd Newton MA 02162 Office: Dept Chemistry Brandeis U Waltham MA 02254

SNIDER, JERRY ALLEN, biological sciences educator; b. Danville, Ill., Feb. 17, 1937; s. Norman and Thelma (Jones) S.; m. Linda Kathryn Upchurch, May 24, 1967; children: Steven Michael, Jeffrey Andrew. B.A. in Botany, So. Ill. U., 1967, M.A., U. N.C., 1969, Ph.D., Duke U., 1973. Asst. prof. biology Baylor U., 1973-74; asst. prof. biol. scis. U. Cin., 1974-79, assoc. prof., 1979—. Contbr. articles to profl. jours. Mem. Internat. Assn. Bryologists, Internat. Assn. Plant Taxonomists, Am. Bryological and Lichenological Soc., Brit. Bryological Soc., Norwegian Bryological Soc. Subspecialties: Morphology; Taxonomy. Current work: Cytology, morphology, devel. and taxonomy of bryophytes. Home: 5804 Cedaridge Dr Cincinnati OH 45247 Office: Dept Biol Sci U Cin Cincinnati OH 45221

SNIDER, JOSEPH LYONS, physicist; b. Boston, June 10, 1934; s. Joseph Lyons and Greta (Wood) S.; m. Ann Lackritz Fuller, June 27, 1981; children: Karen, Benjamin. B.A., Amherst Coll., 1956; Ph.D., Princeton U., 1961. Instr., research fellow in physics Harvard U., 1961-64, asst. prof. physics, 1964-69; assoc. prof. physics Oberlin (Ohio) Coll., 1969-75, prof., 1975—. Mem. AAAS, Am. Phys. Soc., Am. Assn. Physics Tchrs., AAUP. Subspecialties: Solar physics; Relativity and gravitation. Current work: Solar rotation and oscillation, teaching relativity and gen. physics. Office: Dept of Physics Oberlin College Oberlin OH 44074

SNOW, ANNE EVELYN, pharmacologist, researcher, consultant; b. Spokane, Wash., Apr. 9, 1943; d. Frank and Gertrude Alma (Krueger) S.; m. Ellis Branson Ridgway, June 7, 1980. B.A., U. Oreg., 1966; Ph.D., U. Wash., 1977. Undergrad. research asst. U. Oreg., 1963; research technician U. Wash., 1966-73, research asst., 1973-74, NIH predoctoral trainee, 1974-77; postdoctoral research assoc. Med. Coll. Va., 1977-80, research assoc., 1980—; cons. A.H. Robins. Contbr. articles to profl. publs. Recipient Biomed. Research Fund award U. Wash., 1977. Mem. Soc. Neurosci., Intersci. Research Found., AAAS. Clubs: Briarwood, Nautilus (Richmond, Va.). Subspecialties: Neuropharmacology; Neurochemistry. Current work: Stress and neuroendocine perturbations; antinociception-mechanisms of; antinociceptive responses to stress; neurochem. alterations following stress. Office: Med Coll Va Box 551 Richmond VA 23298

SNOW, JAMES BYRON, JR., physician, medical educator; b. Oklahoma City, Mar. 12, 1932; s. James B. and Charlotte Louise (Andersen) S.; m. Sallie Lee Ricker, July 16, 1954; children: James B., John Andrew, Sallie Lee Louise. B.S., U. Okla., 1953; M.D. cum laude, Harvard U., 1956; M.A. (hon.), U. Pa., 1973. Diplomate: Am. Bd. Otolaryngology (dir. 1972—). Intern Johns Hopkins Hosp., Balt., 1956-57; resident Mass. Eye and Ear Infirmary, Boston, 1957-60; prof., head dept. otorhinolaryngology Sch. Medicine U. Okla., Oklahoma City, 1962-72; prof., chmn. dept. otorhinolaryngology and human communication U. Pa. at Phila., 1972—; Mem. nat. adv. council neurol. and communicative disorders and stroke NIH, 1972-76, 82—; chmn. Nat. Com. Research Neurol. and Communicative Disorders, 1979-80. Editor: Am. Jour. Otolaryngology, 1979-83; Contbr. articles to sci. and profl. jours. Served with M.C. AUS, 1960-62. Recipient Regents award for superior teaching U. Okla., 1970. Mem. Soc. Univ. Otolaryngologists (pres. 1975), Am. Acad. Otolaryngology-Head and Neck Surgery, Assn. Acad. Depts. Otolaryngology (pres. 1981-82), A.C.S. (regent 1982—), Am. Laryngol. Rhinol., Otol. Soc., Am. Otol. Soc., AMA (council on sci. affairs 1975—), Am. Laryngol. Assn. (editor 1983—), Am. Broncho-esophagol. Assn. (editor trans. 1973-77, pres. 1979), Collegium Otorhinolaryngologicum, Phi Beta Kappa, Alpha Omega Alpha. Subspecialty: Otorhinolaryngology. Home: 708 Governors Circle Newtown Square PA 19073 Office: 3400 Spruce St Philadelphia PA 19104

SNOW, WILLIAM ROSEBROOK, physicist, technical adminstrator; b. N.Y.C., Jan. 6, 1930; s. William Boring and Genevieve (Rosebrook) S.; m. Phyllis Dodson, Aug. 25, 1951; children: William Dodson, David James. B.S. in Physics, Stanford U., 1952, M.S., U. Wash., 1964, Ph.D., 1966. Reactor physicist Gen. Electric Co., Hanford, Wash., 1952-54; physicist Precision Tech., Inc., Calif., 1954-58; staff assoc. Gen. Atomic, 1958-62; research staff Aerospace Corp., Calif., 1966-68; asst. prof. physics U. Mo., Rolla, 1968-73, assoc. prof., 1973-78; staff scientist Pacific Western Systems, Inc., Mountain View, Calif., 1978-81, tech. dir., 1981—. Contbr. articles in field to profl. jours. Mem. Am. Phys. Soc., Am. Vacuum Soc., Electrochemical Soc., Sigma Xi. Subspecialties: Atomic and molecular physics; Plasma physics. Current work: Negative ion charge transfer, molecular beams, plasma enhanced chem. vapor deposition. Office: 505 E Evelyn Ave Mountain View CA 94041

SNYDER, FRED LEONARD, biochemist; b. New Ulm, Minn., Nov. 22, 1931; s. George B. and Lillian R. (Meyer) S.; m. Joy D. Synder, Nov. 22, 1978; children: Vicki, David, Jon. Ph.D., U. N.D., 1958. Chief scientist med. and health sci. div. Oak Ridge Assos. Univs., Tenn., 1958-75; faculty U. Tenn.—Oak Ridge Grad. Sch. Biomed. Scis., 1972—, asst. chmn. med. and health sci. div., 1975-79, assoc. chmn., 1979—; prof. biochemistry U. Tenn. Center for Health Scis., Memphis, 1964—; prof. medicinal chemistry U. N.C., Chapel Hill, 1966—; chmn., 1981-82; mem. Am. Cancer Soc. adv. com. on biochemistry and chem. carcinogenesis, 1979-82. Editorial bd.: Reviews on Cancer, 1973-78, Biochimica Biophys. Acta, 1972-78, Archives of Biochemistry and Biophysics, 1972-78, Jour. Lipid Research, 1966—; asso. editor: Cancer Research, 1971-78; contbr. articles to profl. jours. & Lipid biochemistry, membranes, biol. chem. mediators. Nat. Heart Inst. fellow, 1955-58. Mem. Am. Soc. Biol. Chemists, Am. Assn. Cancer Research, Soc. Exptl. Biology and Medicine, Sigma Xi. Subspecialty: Biochemistry (biology). Office: ORAU PO Box 117 Oak Ridge TN 37830

SNYDER, GLENN JACOB, mech. engr.; b. Akron, Ohio, Aug. 1, 1923; s. George and Henrietta Louise (Ley) S.; m. Donna Lou Jones, Oct. 11, 1927; children: Gavin Jay, Leslie Marie, Adrianne Lee, Jeffrey David. B.S.M.E., Ohio U., 1949; postgrad. in thermodynamics, Akron U., 1950, in mech. engring, U. Va., 1961-64, in bus. adminstrn, Lynchburg Coll., 1965-69. Registered profl. engr., Ohio, Va. Plant engr. Ohio Boxboard Co., Rittman, 1950-52; sr. plant engr. Goodyear Aerospace Corp., Akron, 1952-55; group supr. remote handling and maintenance equipment design Nuclear Power Generation div. Babcock and Wilcox Co., Lynchburg, Va., 1955-61, unit mgr. reactor vessel, structural internals and rod drive mechanisms design, 1961-79, prin. engr. advanced reactor unit on liquid metal fast breeder reactor, large devel. plant, 1979—; pub. speaker, trainer on nuclear power generation and other subjects. Deacon, elder Rivermont Presbyterian Ch., 1959—; Meml. Bible Plan rep. Gideons Internat., Lynchburg North Camp, 1976—. Served as staff sgt. U.S. Army, 1943-46; PTO. Mem. ASME (chmn. Va. sect. 1979-80, 1982—), Nat. Soc. Profl. Engrs., Soc. Mfg. Engrs., Am. Nuclear Soc. Republican. Clubs: Lynchburg, Babcock & Wilcox, Toastmasters. (area gov. 1978-79). Subspecialties: Mechanical engineering; Systems engineering. Current work: Liquid metal fast breeder reactor commercial electric power development and conceptual design; large Development Plant nuclear island components. Home: 3300 Woodridge Pl Lynchburg VA 23503 Office: PO Box 1260 Lynchburg VA 24505

SNYDER, J. HERBERT, educator; b. McCloud, Calif., May 5, 1926; s. Clyde Herbert and Grace Altheda (Peaslee) S.; m. Ruth Margaret Forsyth, Feb. 14, 1948; children—Craig H., Neal R., Roy R. B.S., U. Calif., 1949, Ph.D., 1954. Asst. prof. to prof. agrl. econs. U. Calif. at Davis, 1953—, chmn. dept., 1966-70, chmn. div. resource scis., 1970-71, asst. dean environ. studies, 1971-72, prof., 1971—, dir., 1972—; Chmn. Gov.'s Adv. Com. Regional Land Use Info. Systems, 1964-67; com. cons. to Calif. Joint Legis. Com. on Open Space Land, 1967-70; mem. citizens adv. com. to Calif. Assembly Com. on Agrl. Land Use, 1964-67; citizens tech. adv. com. open space land to Calif. Legislature, 1967-71; mem. exec. bd. Univ. Council on Water Resources, 1976-82. Contbr. articles profl. jours. Mem. Davis City Planning Commn., 1968-70. Served with USNR, 1944-46. Fellow Soil Conservation Soc. Am.; mem. Am., Western agrl. econs. assns., AAAS, Calif. Soil Conservation Soc., Sigma Alpha Epsilon, Alpha Zeta. Club: Mason. Subspecialties: Agricultural economics; Resource management. Current work: Land and water resource management; economics of natural resources; natural resources policy and conservation; research administration. Home: 1106 Maple Ln Davis CA 95616

SNYDER, JAMES NEWTON, physicist, computer scientist, educator; b. Akron, Ohio, Feb. 17, 1923; s. Louis Emery and Mary (Sullivan) S.; m. Betty Jane Cooper, July 28, 1944; 1 son, James Newton. B.S., Harvard, 1945, M.A., 1947, Ph.D., 1949. Mem. faculty physics dept. U. Ill. at Urbana, 1949—; asso. prof. Digital Computer Lab., 1957-58, prof., 1958-70, prof. computer sci. and physics, head dept. computer sci., 1970—; head computer div. Midwestern U. Research Assn., Madison, Wis., 1956-57. Contbr. articles to profl. jours. Mem. Am. Phys. Soc., A.A.A.S., Assn. Computing Machinery, I.E.E.E., Phi Beta Kappa, Sigma Xi. Subspecialty: Administration. Home: 2304 S Vine St Urbana IL 61801

SNYDER, THOMA MEES VAN'T HOFF, consulting scientist/engineer; b. Balt., May 21, 1916; s. Charles David and Aleida Jenny (van't Hoff) S.; m. Charlotta Untiedt, Nov. 22, 1958 (dec. Dec. 1981); children: James A., Charles E. Ph.D., Johns Hopkins U., 1940. Registered profl. engr., Calif. Instr. physics Princeton (N.J.) U., 1940-43; research assoc. Los Alamos Sci. Lab., 1943-45; mgr. exptl. physics

Knolls Atomic Power Lab., Schenectady, 1946-51, mgr. advanced reactors, 1951-56, mgr. research, 1956-57; mgr. physics Vallecitors lab. Gen. Electric Co., Pleasanton, Calif., 1957-64; cons. scientist/engr. nuclear energy operation, San Jose, Calif., 1964—, tech. hazards council, 1957-67; tech. adv. AEC, Washington, 1950-72; gen. office rev. bd. GPU Nuclear Corp., Morristown, Pa., 1967—; ops. rev. com. Cin. Gas & Electric Co., 1980—. Pres., bd. dirs. El Camino Youth Symphony, 1964-67; bd. dirs. Crippled Children's Soc., Santa Clara County, 1969-71. Recipient citations Sec. War, Sec. Navy, 1945; Steinmetz award Gen. Electric Co., 1979. Fellow Am. Nuclear Soc. (chmn. univ./industry com. 1968-70, chmn. honors and awards com. 1970-73), Am. Phys. Soc.; mem. AIAA, Sigma Xi, Phi Beta Kappa. Republican. Subspecialties: Nuclear engineering; Nuclear fission. Current work: Nuclear science, technology and engineering; materials; nuclear safety. Home: 702 Meadow Ln Los Altos CA 94022 Office: Gen Electric Co 175 Curtner Ave M/C 120 San Jose CA 95125

SOBIN, LESLIE HOWARD, pathologist; b. N.Y.C., Feb. 10, 1934; s. Martin L. and Catherine N. (Soloway) S.; m. Margareta E.D. Ahlstrom, Dec. 21, 1962; 1 dau., Annika Dorothea. B.S., Union Coll., Schenectady, 1955; M.D., SUNY-N.Y.C., 1959. Cert. Nat. Bd. Med. Examiners, Am. Bd. Pathology. Intern N.Y. Hosp., N.Y.C., 1959-60; resident Cornell Med. Ctr., N.Y.C., 1960-65, asst. prof. pathology, 1965, assoc. prof., 1968-70, adj. prof., 1977—; WHO vis. prof., 1977—; U. Kabul, Afghanistan, 1965-68; pathologist WHO, Geneva, 1970-81; staff pathologist Armed Forces Inst. Pathology, Washington, 1981—; mem. expert adv. panel cancer WHO, Geneva, 1981—; chmn. TNM project Internat. Union Against Cancer, Geneva, 1982—; clin. prof. pathology Uniformed Services U. of Health Scis., 1981—. Editor: Internat. Histological Classification of Tumors, 1967-81; author: A Pathology Primer, 1978. Fellow Royal Coll. Pathologists; mem. Internat. Acad. Pathology (sec. 1982—), Am. Assn. Pathologists. Subspecialties: Pathology (medicine); Oncology. Current work: Standardization of tumor classification; research in gastrointestinal pathology. Office: Dept Gastrointestinal Pathology Armed Forces Inst Pathology Alaska St and 14th St Washington DC 20306

SOCIE, DARRELL FREDERICK, mechanical engineering educator; b. Toledo, Oct. 29, 1948; s. Frederick James and Emerence Marie (Lupinski) S.; m. Pamela Sue Doll, Apr. 15, 1951; children: Benjamin, Bethany. B.S., U. Cin., 1971, M.S. in Metall. Engring, 1973; Ph.D. in Theoretical and Applied Mechanics, U. Ill., 1977. Registered profl. engr., Ill., Ohio. Project engr. Structural Dynamics Research Corp., Cin., 1972-74; mem. faculty U. Ill.-Urbana, 1974—, assoc. prof. mech. engring., 1981—. Contbr. articles to profl. publs. Mem. Nat. Soc. Profl. Engrs., ASTM, Am. Soc. Metals. Subspecialties: Fracture mechanics; Metallurgy. Current work: Mechanical behavior of materials: fatigue, fracture, creep, cyclic deformation. Home: RR 1 Box 4 Saint Joseph IL 61873 Office: Univ Ill 1206 W Green St Urbana IL 61801

SOCOLOW, ROBERT HARRY, scientist, engineering educator; b. N.Y.C., Dec. 27, 1937; s. A. Walter and Edith (Gutman) S.; m. Elizabeth Anne Sussman, June 10, 1962 (div. Apr. 27, 1982); children: David, Seth. B.A., Harvard U., 1959, M.A., 1961, Ph.D., 1964. Asst. prof. physics Yale U., New Haven, 1966-71; assoc. prof. mech. and aerospace engring. Princeton U. (N.J.), 1971-77, prof. mech. and aerospace engring., 1977—; mem. Inst. Advanced Study, Princeton, 1971; dir. Center for Energy and Environmental Studies, Princeton, 1978—. Author: (with John Harte) Patient Earth, 1971, (with K. Ford, G. Rochlin, M. Ross) Efficient Use of Energy, 1975, (with H.A. Feiveson, F.W. Sinden) Boundaries of Analysis: An Inquiry into the Tocks Island Dam Controversy, 1976, Saving Energy in the Home: Princeton's Experiments at Twin Rivers, 1979. Mem. Council Fedn. Am. Scientists, 1981—. John Simon Guggenheim fellow, 1976-77; German Marshall Fund fellow, 1976-77; Yale Jr. Faculty fellow, 1970-71; NSF Postdoctoral fellow, 1964-66; NSF Predoctoral fellow, 1960-64. Mem. Am. Phys. Soc., Fedn. Am. Scientists, AAUP, AAAS. Jewish. Current work: Energy use in buildings, energy policy, global resources. Home: 37 Laurel Rd Princeton NJ 08540 Office: H102 Engineering Quad Princeton Univ Princeton NJ 08544

SODD, VINCENT JOSEPH, nuclear medicine administrator, radiology educator; b. Toledo, Nov. 20, 1934; s. Abraham and Sarah (Hamway) S.; m. Dorothy Langenderfer, Oct. 20, 1956; children: Vincent Joseph, Anthony Newman, Joseph William, Anne Marie. B.S., Xavier U., 1956, M.S., 1958; Ph.D., U. Pitts., 1964. Commd. health service officer USPHS, 1957; chemist Robert A. Taft San. Engring. Ctr., Cin., 1958-60, nuclear chemist, 1963-66, dep. chief nuclear medicine lab., 1966-71, acting chief nuclear and medicine lab. and chief radiopharm. devel. sect., 1971-72; prof. radiology U. Cin., 1977—; dir. nuclear medicine lab. FDA, Univ. Hosp., Cin., 1972—. Contbr. articles on nuclear medicine to profl. jours. Recipient Disting. Service award USPHS, 1972; Dorst Chemistry Key, 1959; Silver award ann. meeting Ohio State Med. Assn., 1968; NASA award, 1968. Mem. Cin. Radiation Soc., Soc. Nuclear Medicine (pres. southeastern chpt. 1979-80, trustee 1980-81, bd. dirs. radiopharm. sci. council 1978-80, bd. govs. instrumentation council 1982-84, mem. council southeastern chpt. 1981-85). Roman Catholic. Subspecialties: Inorganic chemistry; Nuclear medicine. Current work: Cyclotron chemistry and physics, charged-particle reactions, semi-conductor detector systems. Patentee in field. Office: FDA Nuclear Medicine Lab Univ Hosp Cincinnati OH 45267

SODERBLOM, DAVID ROBERT, research scientist; b. Oakland, Calif., July 8, 1948; s. Robert John and Helen Lorraine (Umbson) S.; m. Judith Anne Stoffer, July 22, 1978. A.B., U. Calif., Berkeley, 1969, Ph.D., 1980. Vis. scientist High Altitude Obs. Nat. Center for Atmos. Research, Boulder, Colo., 1980-81. Served with U.S. Army, 1970-72. Langley-Abbot fellow Harvard-Smithsonian Center for Astrophysics, Cambridge, Mass., 1981—. Mem. Am. Astron. Soc. Subspecialty: Optical astronomy. Current work: Solar-type stars; stellar rotation; high resolution spectroscopy; stellar chromospheres; evolution of low mass stars. Office: 60 Garden St Cambridge MA 02138

SOELDNER, JOHN STUART, diabetes researcher, medical educator; b. Boston, Sept. 22, 1932; s. Frank and Mary Amelia (Stuart) S.; m. Elsie Irene Harnish, Aug. 25, 1962; children: Judith M., Elizabeth A., Stephen J.D. B.S., Tufts U., 1954; M.D., Dalhousie U., Halifax, N.S., Can., 1959. Postgrad.in medicine Harvard Med. Sch., Boston, 1961-64, instr. medicine, 1964-67, assoc. in medicine, 1967-69, asst. prof. medicine, 1969-72; assoc. prof. medicine, sr. investigator Joslin Diabetes Ctr., 1972—. Author over 200 sci. publs. Recipient Sci. award Juvenile Diabetes Found., 1973; Sr. Scientist award Humboldt Found., W.Ger., 1975; Banting lecture awardee U. Toronto, Ont., Can., 1980. Mem. Am. Diabetes Assn. (citation 1981), Am. Soc. Clin. Investigation, Am. Physiol. Soc., Am. Soc. Artificial Organs, Internat. Soc. Artificial Organs. Democrat. Roman Catholic. Subspecialties: Endocrinology; Artificial organs. Current work: Diabetes-beta cell studies, genetics, metabolism, insulin, artificial beta cell, glucose sensor. Patentee glucose sensor. Home: 109 Oak Hill St Newton MA 02159 Office: Joslin Research Lab One Joslin Pl Boston MA 02215

SOFFER, BERNARD H., physicist; b. Bklyn., Mar. 2, 1931; s. Meyer and Sabina (Adlersheim) S.; m. Reba Nussbaum; 1 son: Roger. B.S., Bklyn. Coll., 1953; M.S., M.I.T., 1958. Staff mem. Lab. for Insulation Research, M.I.T., 1958-59; mem. tech. staff Hughes Research Lab., Malibu, Calif., 1959-61; sr. scientist Korad Corp., Santa Monica, Calif., 1961-69; sr. staff physicist, 1969—. Contbr. articles to profl. jours. Served with U.S. Army, 1953-55. Fellow Optical Soc. Am.; mem. IEEE (sr.), Am. Phys. Soc., Sigma Xi. Subspecialties: Optical signal processing. Current work: Research and development in optical signal and image processing and devices. Patentee in field. Home: 665 Bienveneda Ave Pacific Palisades CA 90272 Office: 3011 Malibu Canyon Rd Malibu CA 90265

SOFIA, ROBERT DUANE, pharm. co. exec., pharmacologist; b. Ellwood City, Pa., Oct. 8, 1942; s. Mario and Clara (Mancini) S.; m. Shirley Ann., Sept. 11, 1965; children: Robert, Maria, Tricia, Joella. B.S., Geneva Coll., 1964; M.S., Fairleigh-Dickinson U., 1969; Ph.D. (NIH predoctoral trainee), U. Pitts., 1971. Research biologist dept. pharmacology and exptl. therapeutics Lederle Labs., Inc., Pearl River, N.Y., 1964-67; research asso. dept. pharmacology Union Carbide Corp., Tuxedo, N.Y., 1967-69; investigator Pharmakon Labs., Scranton, Pa., 1969, supr., 1969-71, cons., 1971; sr. pharmacologist Wallace Labs. div. Carter-Wallace, Inc., Cranbury, N.J., 1971-73, dir. dept. pharmacology and toxicology, 1973-76, v.p. biol. research, 1976-80, v.p. research and devel., 1980-82; v.p. preclin. research, 1982—; co-adj. lectr. continuing edn. Rutgers U. Coll. Pharmacy, New Brunswick, N.J., 1973, 77, vis. asso. prof. pharmacology, 1979—; mem. tech., merit rev. panel Nat. Inst. Neurol. and Communicative Disorders and Stroke, 1979. Contbr. numerous abstracts, articles to profl. jours. Mem. AAAS, N.Y. Acad. Scis. (Pulmonary Research Group), Soc. Neurosci., Inflammation Research Assn. (pres. 1975-78), Internat. Assn. for Study Pain (founding), Am. Soc. Pharmacology and Exptl. Therapeutics, Soc. Toxicology, Eastern Pain Assn. (charter), Physiol. Soc. Phila., Am. Chem. Soc., Am. Rheumatism Assn., Am. Assn. Lab. Animal Sci. (Wallace Labs. rep. 1977—), Research and Devel. Council N.J. (Carter-Wallace rep. 1977—), European Biol. Research Assn. (assoc.), Am. Pain Soc. (charter), Soc. Research Administrs., Internat. Soc. for Study Xenobiotics (charter), Sigma Xi. Subspecialties: Pharmacology; Toxicology (medicine). Current work: Research adminstrn. Office: Wallace Labs Half Acre Rd Cranbury NJ 08512

SOGANDARES-BERNAL, FRANKLIN, biology educator, researcher, consultant; b. Ancon, C.A., Panama, May 12, 1931; came to U.S., 1951; s. Anastasio and Blanca Helena Bernal-Almillategui; m. Lucy Ann McAlister, 1960 (div. 1982); children: Franklin McAlister, Maria Helena, John Francis Marion. B.S., Tulane U., 1954; M.S., U. Nebr., Lincoln, 1955, Ph.D., 1958. Instr., asst. prof., prof. Tulane U., New Orleans, 1959-71, coordinator sci. planning, 1965-68; prof., chmn. zoology U. Mont., Missoula, 1971-72, prof. microbiology, 1972-74; prof. biology So. Meth. U., Dallas, 1974—, chmn. biology, 1974-76; cons. in pathology Baylor U. Med. Center, Dallas, 1974—, med. staff affiliate, 1978—; research affiliate Nebr. State Mus., Lincoln, 1972—; pvt. practice cons. immunologist, Dallas, 1978—; sci. adv. bd. EPA, Washington, 1980-82. Contbr. articles to profl. jours. Mem. Am. Soc. Parasitologists (council 1970-73, editorial bd 1964-67, H.B. Ward medal 1969), Am. Assn. Pathologists, Soc. Wildlife Diseases, Conf. Biol. Editors, N.Y. Acad. Scis., Am. Soc. Zoologists, Nat. Rifle Assn. (life). Democrat. Subspecialties: Parasitology; Pathology (medicine). Current work: Consultant in parasitology and immunopathology, pathology; research parasitic disease and tropical medicine, antigen fractionation, Chagas Disease, cerebral cysticercosis diagnosis. Home: 10622 Royal Chapel Dr Dallas TX 75229 Office: Southern Meth Univ Dallas TX 75275

SOHMER, MARCUS FRANK, JR., physician; b. Danville, Va., June 15, 1924; s. Marcus Frank and Ila (Bullington) S.; m. Betty C. Crawford, Dec. 16, 1950 (div. 1970); children: Robert Marcus, Frances Elizabeth, Susan Paige, Louisa Jeanette. M.D., Bowman Gray Sch. Medicine, 1952. Diplomate: Am. Bd. Internal Medicine. Asst. prof. medicine Bowman Gray Sch. Medicine, Winston-Salem, N.C., 1974—; attending physician Forsyth Meml. Hosp., Winston-Salem, 1964—, N.C. Bapt. Hosp., 1956—; pres. Forsyth Med. Specialists, Winston-Salem, 1982—. Contbr. articles to profl. jours. Served with U.S. Navy, 1943-46. Mem. Forsyth County Med. Soc. (pres. 1974-75), N.C. Med. Soc. (v.p. 1975-76, pres. 1979-80), Piedmont Med. Found. (pres. 1972-74), AMA; fellow ACP, So. Med. Assn. Republican. Methodist. Subspecialty: Internal medicine. Office: Forsyth Med Specialists PA 2240 Cloverdale Ave NC 27103 Home: 9808 Reynolda Rd Tobaccoville NC 27050

SOHMER, SEYMOUR HANS, botanist; b. Bronx, N.Y., Feb. 27, 1941; s. Ralph and Tanya S.; m. Sara Harrison, Aug. 5, 1967; children: Rebecca Rose, Rachel Adrienne. B.S., CCNY, 1963; M.S., U. Tenn., 1966; Ph.D., U. Hawaii, 1971. Instr. U. Wis.-LaCrosse, 1967, asst. prof., dir. herbarium, 1971-76, assoc. prof., 1976-80; chmn. dept. botany Bishop Mus., Honolulu, 1980—; staff assoc. NSF, Washington, 1977-78; research botanist Office of Forests, Lae, Papua, New Guinea, 1979. Contbr. articles to sci. jours. Smithsonian research fellow, 1975-76. Mem. Am. Soc. Plant Taxonomists, Internat. Assn. Plant Taxonomists, Am. Inst. Biol. Scis., Hawaiian Bot. Soc., Sigma Xi. Subspecialties: Systematics; Taxonomy. Current work: Relationships and identification of species in complex higher plant groups; administration of department of botany, with pursuit of research in systematic botany. Office: Bishop Museum PO Box 19000-A Honolulu HI 96822

SOHN, JOHN EDWIN, research chemist; b. Tokyo, Sept. 12, 1952; s. Edwin Carl and Janice (Baak) S.; m. Catherine Ann Angell, Aug. 10, 1974; 1 dau. Karen Elizabeth. B.S., U. Calif.-Davis, 1974; M.S., Berkeley, 1976, Ph.D., 1981. Research assoc. U. Pa., 1978-79; mem. research staff Western Electric Co., Princeton, N.J., 1979-82, sr. mem. research staff, 1982—. Contbr. articles to profl. jours. Mem. Am. Chem. Soc., Am. Phys. Soc., AAAS, sigma Xi. Subspecialties: Organic chemistry; Polymer chemistry. Current work: Organic chemistry, polymer chemistry, applied to communication systems, optical processes, interfacial properties and chemistry. Office: Western Electric Company PO Box 900 Princeton NJ 08540

SOIFER, DAVID, research scientist, cell biologist, educator; b. N.Y.C., Sept. 16, 1937; s. Israel and Margaret (Krenzler) S.; m. Lea Yitshaki, June 2, 1960; children Hillel, Boaz. Student, Swarthmore Coll., 1954-57; B.S., Columbia U., 1961; Ph.D., Cornell U., 1969. Research assoc. AMA Inst. Biomed. Research, Chgo., 1968-70; research scientist Inst. Basic Research on Devel. Disabilities, S.I., N.Y., 1970—; guest investigator Cambridge (Eng.) U., 1971, Max Planck Inst. Biophys. Chemistry, Gottingen, Germany, 1974; vis. prof. Hebrew U., Jerusalem, 1976; assoc. prof. anatomy and cell biology SUNY Downstate Med. Center, Bklyn., 1979—. Contr. articles to profl. jours. Mem. adv. com. U.S. Com. for Israel Environment; mem. sci. adv. council Am. Fedn. Aging Research; trustee S.I. Zool. Soc. Mem. Am. Soc. Cell Biology, Am. Soc. Neurochemistry, Internat. Soc. Neurochemistry, Soc. for Neurosci., AAAS, N.Y. Acad. Scis. Subspecialties: Cell biology (medicine); Neurochemistry. Current work: Neuronal cytoskeleton, especially microtubules and neurofilaments; application of molecular biological techniques to the study of the neuronal cytoskeleton and axoplasmic transport. Office: 1050 Forest Hill Rd Staten Island NY 10314

SOIKE, KENNETH FIEROE, JR., microbiologist, researcher, educator; b. Mpls., July 8, 1927; s. Kenneth Fieroe and Viola Mabel S.; m. Marian L. Laursen, June 11, 1950; children: Dianna, David, Kevin, Kathryn; m. Mary M. Morris, Feb. 14, 1978. B.A., U. Minn., 1949; Ph.D., Oreg. State U., 1955. Microbiologist Sterling Winthrop Research Inst., Rensselear, N.Y., 1955-61; assoc. prof. Albany (N.Y.) Med. Coll., 1961-75; with Delta Regional Primate Research Ctr., Tulane U., Covington, La., 1975—, head dept. microbiology, 1975—, adj. prof. dept. microbiology, 1975—. Served with USN, 1945-46. Mem. Am. Soc. Microbiology, Sigma Xi. Subspecialties: Virology (medicine); Virology (biology). Current work: Viral pathogenesis, antiviral drugs. Home: 17 Beth Dr Covington LA 70433 Office: Delta Regional Primate Research Ctr Covington LA 70433

SOKOL, ROBERT JAMES, researcher, medical educator, obstetrician-gynecologist; b. Rochester, N.Y., Nov. 18, 1941; s. Eli and Mildred (Levine) S.; m. Roberta Sue Kahn, July 26, 1964; children: Melissa Anne, Eric Russell, Andrew Ian. B.A. with highest distinction, U. Rochester, 1963, M.D. with honors, 1966. Diplomate: Am. Bd. Ob-Gyn, 1972, Maternal-Fetal Medicine, 1975. Resident in ob-gyn Barnes Hosp./Washington U., St. Louis, 1966-70; Buswell fellow in maternal-fetal medicine Strong Meml. Hosp./U. Rochester, 1972-73; asst. prof. Cleve. Met. Gen. Hosp./Case Western Res. U., 1973-77, co-program dir. perinatal clin. research ctr., 1973-81, assoc. prof., 1977-81, prof., 1981-83, assoc. dir. dept. ob-gyn, 1981-83, program dir. perinatal clin. research ctr., 1982-83; prof., chmn. dept. ob-gyn Wayne State U., Detroit, 1983—; chief ob-gyn Hutzel Hosp., 1983—; cons. Nat. Inst. Child Health and Human Devel., Nat. Inst. Alcohol Abuse and Alcoholism, Ctr. Disease Control, NIH; mem. alcohol psychosocial research rev. com. Nat. Inst. Alcohol Abuse and Alcoholism. Reviewer profl. jours.; contbr. articles to sci. jours., chpts. to books. Served from capt. to maj. USAF, 1970-72. Grantee NIH, Nat. Inst. Alcohol Abuse and Alcoholism, others. Mem. Soc. Perinatal Obstetricians, Perinatal Research Soc., Soc. Gynecologic Investigation, Research Soc. on Alcoholism, Assn. Profs. Gynecology and Obstetrics, Assn. Dirs. Gen. Clin. Research Centers, Royal Soc. Medicine, Central Assn. Obstetricians and Gynecologists, Am. Coll. Obstetricians and Gynecologists, Phi Beta Kappa, Alpha Omega Alpha. Subspecialties: Maternal and fetal medicine; Database systems. Current work: Perinatal risk assessment database management and statistical analysis; alcohol-related birth defects; low birth weight risks and outcomes; algorithmic diagnosis and management. Home: 5200 Rector Ct Bloomfield Hills MI 48013 Office: 4707 Saint Antoine Detroit MI 48201

SOKOLL, MARTIN DAVID, anesthesiologist, educator; b. Harrisville, Ohio, Oct. 14, 1932; s. Frank and Lottie (Pesta) S. B.S., Coll. of Steubenville, Ohio, 1954; M.D., U. Pitts., 1958. Bar: diplomate Am. Bd. Anesthesiology. Intern St. Margaret Hosp., 1958-59; resident Mercy Hosp., Pitts., 1959-61; faculty dept. anesthesiology U. Iowa Coll. Medicine, Iowa City, 1963-67, 68—, prof. anesthesiology, dir. neurosurg. anesthesia, 1974—; prof. pharmacology U. Lund, Sweden, 1967-68. Contbr. papers to profl. lit. Served with USAF, 1961-63. USPHS postdoctoral fellow, 1966. Mem. AMA, Am. Soc. Anesthesiologists, Am. Soc. Pharmacology and Exptl. Therapeutics, N.Y. Acad. Scis., AAAS. Republican. Roman Catholic. Subspecialty: Anesthesiology. Current work: Neuromuscular pharmacology and neurosurgical anesthesia. Home: 818 20th Ave Coralville IA 52241 Office: Dept Anesthesia U Iowa Hosp Iowa City IA 52242

SOKOLOFF, ALEXANDER, biology educator; b. Tokyo, May 16, 1920; s. Dimitri Fyodorovitch and Sofia (Alexandrovna (Solovieff)) S.; m. Barbara B. Bryant, June 24, 1956; children: Nadine A., Elaine A., Michael A. B.A., UCLA, 1948; Ph.D., U. Chgo., 1954. Research assoc. U. Chgo., 1954, instr., 1955; asst. prof. Hofstra U., 1955-58; geneticist W. H. Miner Agr. Research Inst., Chazy, N.Y., 1958-60; assoc. research botanist UCLA, 1960-61; assoc. research geneticist U. Calif.-Berkeley, 1961-66; assoc. prof. Calif. State Coll., San Bernardino, 1965-66, prof. zoology, 1966—. Author: The Genetics of Tribolium and Related Species, 1966, The Biology of Tribolium with Special Emphasis on Genetic Aspects, Vol. 1, 1972, Vol. 2, 1975, Vol. 3, 1977; editor: Tribolium Info. Bull, 1960—; mem. editorial bd.: Jour. Stored Products Research, 1965—; assoc. editor, mem. adv. bd.: Jour. Advanced Zoology, 1980—; assoc. editor: Evolution, 1972-74; contbr. articles to profl. jours. Served with USAAF, 1942-46. NSF, USPHS, U.S. Army Research Office grantee. Mem. Soc. for Study of Evolution, Genetics Soc. Am., Am. Genetic Assn., Am. Soc. Naturalists, Am. Soc. Zoologists, Genetics Soc. Can., Japanese Soc. Population Ecology, Entomol. Soc. Am., Ga. Entomol. Soc., Soc. Western Naturalists, Sigma Xi. Subspecialties: Animal genetics; Population biology. Current work: Ecological genetics of flour beetles, population biology of natural and experimental populations of tribolium. Home: 3324 Sepulveda San Bernardino CA 92404 Office: Dept Biology Calif State Coll San Bernardino 5500 State College Pkwy San Bernardino CA 92407

SOKOLOFF, LOUIS, physiologist, neurochemist; b. Phila., Oct. 14, 1921; m. (married); 2 children. B.A., U. Pa., 1943, M.D., 1946. Intern Phila. Gen. Hosp., 1946-47; research fellow in physiology U. Pa. Grad. Sch. Medicine, 1949-51, instr., then asso., 1951-56; asso. chief, then chief sect. cerebral metabolism NIMH, Bethesda, Md., 1953-68, chief lab. cerebral metabolism, 1968—. Chief editor: Jour. Neurochemistry, 1974-78. Served to capt. M.C. U.S. Army, 1947-49. Mem. Am. Physiol. Soc., Assn. Research Nervous and Mental Diseases, Am. Biophys. Soc., Am. Acad. Neurology, Am. Neurol. Assn., Am. Soc. Biol. Chemists, Am. Soc. Neurochemistry, U.S. Nat. Acad. Scis. Subspecialty: Physiology (biology). Address: NIMH Bethesda MD 20205

SOKOLOWSKI, MARLA BERGER, biologist, educator, researcher; b. Toronto, Ont, Can., July 20, 1955; d. Ernest and Ruth (Mendrofsky) Berger; m. Allen Bernard Sokolowski, July 1975. B.S., U. Toronto, 1977, Ph.D. in Zoology, 1981. Nat. Sci. and Engring. Research Council Can. postdoctoral fellow York U., Downsview, Ont., 1981-82, lectr. in animal behavior dept. biology, 1981-82, Univ. research fellow, 1982—. Recipient Theodosius Dobzhansky award Behavior Genetics Assn., 1979; Ramsey Wright award U. Toronto, 1980. Mem. Behavior Genetics Assn., Soc. for Study of Evolution, Am. Naturalists Soc., Genetics Soc. Am., Drosophila Info. Service. Subspecialties: Genome organization; Evolutionary biology. Current work: Behavioral genetics and evolutionary biology of the fruit fly Drosophila. Office: Dept Biology York Univ Downsview ON Canada M3J 1P3

SOLBERG, RUELL FLOYD, JR., mech. engr.; b. Norse, Tex., July 27, 1939; s. Ruel Floyd and Ruby Mae (Rogstad) S.; m. Laquetta Jane Massey, Oct. 3, 1959; children: Chandra Dawn, Marla Gaye. B.S., U. Tex., 1962, M.S., 1967; M.B.A., Trinity U., San Antonio, 1977. Registered profl. engr., Tex. Research engr. assoc. II, acoustics div. Def. Research Lab., Austin, Tex., 1962-65, research engr. assoc. III, 1965-67, asst. supr. mech. engring. sect., 1966-67; research engr., dept. applied electromagnetics S.W. Research Inst., San Antonio, 1967-70, sr. research engr., 1970-74, sr. research engr. electromagnetics div., 1974-75, sr. research engr. dept. electromagnetic engring., 1975—. Co-author: The Solbergs from Norway to Texas, 1979; contbr. articles to profl. jours. Charter mem. Norwegian Soc. Tex., Nordland Heritage Found., Bosque Meml. Mus. Served with U.S. Army. Howell Instruments scholar, 1961. Mem. ASME (Charles E. Balleisen award 1976, council cert. 1977, 79, 80, 81, Centennial medal 1980, Centennial award 1980), Nat. Soc. Profl. Engrs., Tex. Soc. Profl. Engrs., Consumer Products Tech. Group, Human Factors Soc., Sigma Xi, Tau Beta Pi, Pi Tau Sigma, Sigma Iota Epsilon. Subspecialties: Mechanical engineering; Theoretical and applied mechanics. Current work: Measurement of weight (mass) in space flight; zero gravity effects in space flight; electromechanical research and development; structural optimization; environmental effects; theoretical and experimental analysis; response of structures to periodic and impulsive loading. Patentee in field. Home: 5906 Forest Cove San Antonio TX 78240 Office: PO Drawer 28510 San Antonio TX 79284

SOLIMAN, KARAM FARAG ATTIA, pharmacology educator; b. Cairo, Oct. 15, 1944; U.S., 1968, naturalized, 1979; s. Farah Attia and Elaine (Kallini) S.; m. Samia Sidhom, Aug. 30, 1973; children: John K., Gina K., Mark K. B.S., Cairo U., 1964; M.S., U. Ga., 1971, Ph.D., 1972. Research asst. U. Ga., Athens, 1968-71, teaching asst., 1971-72; asst. prof. Tuskegee (Ala.) Inst. Sch. Vet. Medicine, 1972-75; assoc. prof. Fla. A&M U. Coll. Pharmacy, Tallahassee, 1975-79, dir. basic sci., 1982—, prof. pharmacology, 1979—. Editor: Practical Clinical Pharmacology, 1977, Chronopharmacology, 1981; contbr. articles to profl. jours. NASA grantee, 1977—; NIH grantee, 1976-78, 79-81, 82-84. Mem. Am. Physiol. Soc., Endocrine Soc., Neurosci. Soc., Sigma Xi. Democrat. Christian Orthodox. Subspecialties: Neuropharmacology; Neuroendocrinology. Current work: Establishment of new basic concept of drug therapy related to time of the day (chronotherapy) and helping to understand the neuroendocrinological basis of hormonal circadian rhythm. Home: 2414 Blarney Dr Tallahassee FL 32308 Office: Coll Pharmacy and Pharm Sci Fla A and M U Tallahassee FL 32308

SOLINGER, ALAN MICHAEL, physician, rheumatologist/ immunologist; b. Cin., Nov. 27, 1948; s. Frank and Ruth (Schuhmann) S. B.A. in Chem. Biology, Columbia U., 1970; M.D., U. Cin., 1974. Diplomate: Am. Bd. Internal Medicine. Intern in internal medicine U. Mo., 1974-75, resident in internal medicine, 1975-76; research fellow div. rheumatology-clin. immunology U. Calif.-San Francisco, 1978-81; asst. prof. medicine, div. immunology U. Cin., 1981—; travel scholar Internat. League Against Rheumatism, Paris, 1981. Contbr. articles, chpts. to profl. publs. Served with USPHS, 1976-78. Recipient Arthritis Investigator award Arthritis Found., 1982—; Nat. Arthritis Found. clin. fellowship grantee, 1978-81; No. Calif. Arthritis Found. project grantee, 1980-81. Fellow ACP; mem. AAAS, Am. Assn. Immunologists, Am. Fedn. Clin. Research, AAAS, AMA (Calif. housestaff 1978-81, Am. Rheumatism Assn., Sr. Rheumatology Scholar award 1981), Fedn. Advancement of Sci., N.Y. Acad. Scis., Ohio Rheumatism Soc. (exec. com.), Sigma Xi. Democrat. Jewish. Subspecialties: Immunogenetics; Rheumatology. Office: 7462 MSB U Cin Med Center 231 Bethesda Ave Cincinnati OH 45267

SOLLER, R. WILLIAM, pharmacologist, consultant; b. Bronxville, N.Y., Nov. 18, 1946; s. William Henry and Barbara (Bryde) S.; m. Phyllis Hess, Apr. 21, 1979; children: Adam K., Eric C. B.A., Colby Coll., 1968; Ph.D., Cornell U., 1973. Asst. in anatomy Cornell U. Grad. Sch. Medicine, N.Y.C., 1968-73; research assoc. U. Pa. Sch. Medicine, Phila., 1973-76, asst. prof. pharmacology, 1977-79; sci. assoc. Glenbrook Labs./Sterling Drug Inc., N.Y.C., 1979-80, dir. sci. affairs, 1980-81, v.p., dir. sci. affairs, 1981—. Contbr. articles to profl. pubs. Bd. dirs. Aspirin Found. NATO fellow; Pharm. Mfrs. Assn. fellow; Penna Plan scholar. Mem. Soc. for Neurosci., Assn. Clin. Pharmacology, N.Y. Acad. Scis., Sigma Xi. Republican. Subspecialties: Epidemiology; Neurobiology. Current work: Pharmaceutical research and development, clinical trial management, regulatory affairs.

SOLLID, JON ERIK, physicist; b. Denver, Oct. 1, 1939; s. Erik J. and Faye Sollid (Eising) S.; m. Margaret Louise Bashant Augustson, Dec. 23, 1940; m. Lynnette Madelon Mawson Sollid Little, Sept. 1, 1945; children—Sonje, Shije, Erika; m.; stepchildren—Erik, Scott, Leif, Dayn, Brin. B.S., U. Mich., Ann Arbor, 1961; M.S., N. Mex. State U., Las Cruces, 1965, Ph.D., 1967. Project leader Los Alamos (N. Mex.) Nat. Lab., 1974-82; leader optics group Ford Motor Co. Sci. Research Lab., Dearborn, Mich., 1972-74; sr. research scientist Gen. Dynamics, Fort Worth, 1967-72; adj. prof. physics Tex. Christian U., Fort Worth, 1968-72, No. N. Mex. Community Coll., 1978-79, U. N. Mex., Los Alamos, 1980—; diagnostic system mgr. Antares Laser Fusion Project, Los Alamos Sci. Lab., 1979—. Contbr. articles in field to profl. jours. Served with USMC, 1958-64. Mem. Am. Phys. Soc., Optical Soc. Am., Soc. Photo-optical Instrumentation Engrs., Sigma Xi, Sigma Pi Sigma. Subspecialties: Laser fusion; Optical engineering. Current work: Free electron laser, laser fusion, optical instrumentation, radiant energy measurement, holography, exptl. mechanics, plasma physics. Home: 89 Encino St Los Alamos NM 87544 Office: Los Alamos Nat Lab PO Box 1663 MS E533 Los Alamos NM 87545

SOLLINGER, HANS WERNER, physician, educator; b. Munich, Germany, Aug. 30, 1946; came to U.S., 1976, naturalized, 1979; s. Johann and Maria S.; m. Mary Lou Lang, Mar. 16, 1976; children: Nicola, Christina. M.D., U. Munich, 1973, Ph.D., 1975. Intern U. Munich Hosp., 1973-74; resident U. Wis. Hosp.-Madison, 1976-80; research assoc. U. Munich, 1974-75; postdoctoral fellow U. Wis.-Madison, 1975-77; asst. prof. U. Wis. Hosp., 1980—; dir. Tissue Typing Labs., 1981—. Recipient Max-Kade Found. research award, 1975; Deutsche Forschungsgemeinschaft research award, 1976. Mem. Internat. Transplantation Soc., Am. Soc. Transplant Surgeons, Madison Surg. Soc., Assn. Acad. Surgery, Soc. Univ. Surgeons. Subspecialties: Surgery; Transplant surgery. Current work: Transplantation immunology, pancreas transplantation, tissue typing. Office: Dept Surgery U Wis Hosp 600 Highland Ave Madison WI 53792

SOLLNER, TRAUGOTT CARL LUDWIG GERHARD, electrical engineering educator; b. Nashville, Oct. 14, 1943; s. Traugott Carl Ludwig Gerhard and Araminta Louise (Henrichsen) S.; m. Patricia Mae Roberson, Mar. 21, 1947. B.S., Ga. Inst. Tech., 1967; Ph.D., U. Colo., 1974. Postdoctoral fellow U. London, 1974-76; faculty research assoc. U. Mass., 1976-80, asst. prof., 1980—; cons. Lincoln Lab., M.I.T., 1977-82, staff, 1982—. Contbr. articles to prof. jours. NASA grantee, 1980-81; Raytheon grantee, 1980—. Mem. IEEE, Am. Inst. Physics, Internat. Astron. Union. Subspecialties: Condensed matter physics; Semiconductors. Current work: Quantum physics; birth of stars. Home: 68 Willard Grant Rd Sudbury MA 01776 Office: MIT Lincoln Lab Lexington MA 02173

SOLLNER-WEBB, BARBARA THEA, molecular biologist, educator; b. Washington, Dec. 21, 1948; d. Karl and Helen (Rosenberg) Sollner; m. Denis Conrad Webb, Oct. 21, 1973; 1 dau.: Lisa Ellen. B.S. in Physics, M.I.T., 1970; Ph.D. in Biology, Stanford U., 1976. NIH postodoctoral fellow, 1976-77; postdoctoral fellow Carnegie Inst., 1977-79; asst. prof. physiol. chemistry dept. Johns Hopkins U. Sch. Medicine, Balt., 1980—. Subspecialties: Molecular biology; Gene actions. Current work: Transcription of rRNA genes of mouse and frog. Home: 7508 Citadel Dr College Park MD 20740 Office: 725 N Wolfe St Baltimore MD 21205

SOLOMON, DAVID HARRIS, educator, physician; b. Cambridge, Mass., Mar. 7, 1923; s. Frank and Rose (Roud) S.; m. Ronda L. Markson, June 23, 1946; children—Patricia Jean (Mrs. Richard E. Sinaiko), Nancy Ellen. A.B., Brown U., 1944; M.D., Harvard U., 1946. Intern Peter Bent Brigham Hosp., Boston, 1946-47, resident, 1947-48, 50-51; fellow endocrinology New Eng. Center Hosp., Boston, 1951-52; faculty UCLA Sch. Medicine, 1952—, prof. medicine, 1966—, vice chmn. dept. medicine, 1968-71, chmn. dept., 1971-81; chief med. service Harbor Gen. Hosp., Torrance, Calif., 1966-71; cons. Wadsworth VA Hosp., Los Angeles, 1952—, Sepulveda VA Hosp., 1971—; cons. metabolism tng. com. USPHS, 1960-64, endocrinology study sect., 1970-73; mem. dean's com. Wadsworth, Sepulveda VA hosps. Contbr. numerous articles to profl. jours. Mem. Assn. Am. Physicians, Am. Soc. Clin. Investigation, Am. Fedn. Clin. Research, Am. Physiol. Soc., Western Soc. Clin. Research (councillor 1963-65, pres. 1983-84), Endocrine Soc., Am. Diabetes Assn., Am. Thyroid Assn. (pres. 1973-74), Soc. Exptl. Biology and Medicine, ACP, Los Angeles Soc. Internal Medicine (exec. council 1960-62), Inst. Medicine Nat. Acad. Scis., AAAS, Assn. Profs. Medicine (pres. 1980-81), Western Assn. Physicians (councillor 1972-75), Phi Beta Kappa, Sigma Xi, Alpha Omega Alpha. Subspecialties: Endocrinology; Gerontology. Current work: Treatment of hyperthyroidism; thyroid hormone methabolism in sickness, health and aging. Home: 863 Woodacres Rd Santa Monica CA 90402 Office: Dept Medicine UCLA Sch Medicine Los Angeles CA 90024

SOLOMON, NATHAN A., physician, educator, researcher; b. Bklyn., Oct. 31, 1922; s. Barnett W. and Gertrude (Rosen) S.; m. Florence Schneider, Mar. 15, 1944; children: William, Robert, Judith. B.S., CCNY, 1942; M.S., NYU, 1947, Ph.D., 1954; M.D., SUNY-Bklyn., 1959. Diplomate: Am. Bd. Nuclear Medicine, 1972. Chief chemistry Maimonides Med. Ctr., Bklyn., 1945-51, chief surg. research, 1951-55, dir. nuclear medicine, 1965-70; prof. radiology SUNY Downstate Med. Ctr., Bklyn., 1970—; cons. VA, Cath. hosps. Editor: Nuclear Med. Rev, 1975, 79, 83. Vice-pres. Flatbush chpt. Am. Jewish Congress; adult edn. East Midwood Jewish Ctr., Bklyn. Mem. N.Y. Acad. Sci. (A. Cressy Morrison award 1954), AMA, Soc. Nuclear Medicine, Am. Assn. Clin. Chemists, Sigma Xi, Alpha Omega Alpha, Phi Lambda Upsilon. Jewish. Subspecialties: Nuclear medicine; Pharmacology. Current work: Radiopharmacology and clinical applications. Patentee radiopharms. (2). Home: 1308 E 21st St Brooklyn NY 11210 Office: 450 Clarkson Ave Brooklyn NY 11210

SOLOMON, PAUL ROBERT, psychologist, educator; b. Bklyn., Aug. 27, 1948; s. Maynard and Norma H. (Rubin) S.; m. Suellen Zablow, Aug. 16, 1972; 1 son, Todd M. B.A., SUNY-New Paltz, 1970, M.A., 1972; Ph.D., U. Mass., 1975. Lic. psychologist, Mass. Teaching assoc. U. Mass., 1973-75; assoc. prof. psychology Williams Coll, Williamstown, Mass., 1976—; vis. research scientist U. Calif.-Irvine, 1979-80. Guest editor: Physiol. Psychology, June 1980, also articles. Recipient Disting. Teaching award. U. Mass., 1975; grantee NIMH, NSF. Mem. AAAS, Soc. Neurosci., Am. Psychol. Assn., Eastern Psychol. Assn. Subspecialties: Physiological psychology; Neurophysiology. Current work: Neural aspects of learning and memory; animal models of behavioral disorders; neuropharmacology. Home: 100 Hoxsey St Williamstown MA 01267 Office: Dept Psychology Bronfman Sci Ctr Williams Coll Williamstown MA 01267

SOLOMON, RICHARD LESTER, educator; b. Boston, Oct. 2, 1918; s. Frank and Rose (Roud) S.; m.; children by previous marriage—Janet Ellen, Elizabeth Grace. A.B., Brown U., 1940, M.Sc., 1942, Ph.D., 1947. Instr. psychology Brown U., 1946; asst. prof. Harvard U., 1947-50, asso. prof., 1950-57, prof. social psychology, 1957-60; prof. psychology U. Pa., Phila., 1960-74, James M. Skinner Univ. prof. sci., 1975—; Staff OSRD, 1942-45. Mem. Am. Psychol. Assn., Eastern Psychol. Assn. (pres.), Psychonomic Soc. (chmn. governing bd.), Am. Acad. Arts and Scis., Soc. Exptl. Psychologists, AAAS, Nat. Acad. Sci., Phi Beta Kappa, Sigma Xi. Subspecialty: Experimental psychology. Office: 3815 Walnut St Philadelphia PA 19104

SOLOMONS, HOPE COWEN, psychology educator; b. Providence, Sept. 20, 1932; d. Morris Louis and Anna Emma (Wunsch) Cowen; m. Gerald Solomons, June 12, 1955; children: Nan Martha, Mary Lorna. B.A., Clark U., 1952; student, Harvard U., summer 1951, Boston U., summer 1952; M.A., Wellesley Coll., 1954; Ed.D., Boston U., 1957. Lic. psychologist, Iowa. Pvt. practice psychology, Providence, 1952-62; cons. psychologist Community Workshops, Providence, 1959; instr. to asst. prof. R.I. Sch. Design, Providence, 1956-58, 58-60; asst. prof. psychology R.I. Coll., Providence, 1960-62; research assoc. neurology U. Iowa, Iowa City, 1962-63; sch. psychologist Iowa City Schs., 1963-67; assoc. prof. to prof. psychology Coll. Nursing, U. Iowa, Iowa City, 1967-82, 82—. Editorial bd.: Developmental Medicine & Child Neurology, 1978-83, Jour. Pediatric Psychology, 1980-82; editorial cons. various book pubs.; contbr. articles to profl. jours. Pres. exec. com. Iowa Wellesley Club, 1974-81, 81—. Fellow Am. Acad. Cerebral Palsy (exec. bd.); mem. Am. Psychol. Assn., Internat. Council-Psychologists (Iowa rep. 1978—), Soc. Pediatric Psychology, Iowa Psychol. Assn. (pres. 1975-76), Iowa-Yucatan Ptnrs. of the Ams. (chmn. rehab. com. 1974-76), Sigma Xi, Phi Beta Kappa, Pi Lambda Theta. Clubs: Pan Am., Spanish Conversation. Subspecialty: Developmental psychology. Current work: Cross cultural standarization of infant tests; infant motor development; health and accidents in day care centers; literature for parents of handicapped children. Home: 319 Mullin Ave Iowa City IA 52240 Office: Coll Nursing Univ of Iowa Iowa City IA 52242

SOLOMONS, NOEL W., nutrition educator, researcher, physician; b. Boston, Dec. 31, 1944; s. Gustave Martinez and Olivia Mae (Stead) S. Grad., Harvard U., 1966, M.D., 1970. Diplomate: Am. Bd. Internal Medicine. Fellow in gastroenterology and clin. nutrition U. Chgo., 1973, instr. in medicine, 1974, asst. prof. medicine, 1976; tutorial fellow Inst. Nutrition Central Am. and Panama, 1975, research assoc., 1976-77, affiliated investigator, Guatemala City, 1978—; asst. prof. clin. nutrition MIT, 1977-80, assoc. prof., 1980—; cons. in field. Recipient Faculty Fellowship award Josiah Macy Jr. Found., 1974-76, Future Leader award Nutrition Found., 1974-77, Clin. Investigator award NIH, 1979-82. Mem. Am. Gastroent. Assn., Am. Inst. Nutrition, Am. Soc. Clin. Nutrition, Latin Am. Nutrition Soc., Am. Coll. Nutrition, Com. Nutritional Anthropology, Am. Fedn. Clin. Research. Subspecialties: Gastroenterology; Nutrition (biology). Current work: Trace mineral nutrition; carbohydrate absorption; mineral bioavailability; gastrointestinal physiology. Office: 18 Vassar St Room 20B-213 Cambridge MA 02139

SOLON, LEONARD R(AYMOND), radiation physicist; b. White Plains, N.Y., Sept. 11, 1925; s. Morris and Rebecca (Bobrov) S.; m. Charlotte Rothman, June 30, 1946; children: Miriam, Matthew, Emily. B.A., Hamilton Coll., 1947; M.S., Rutgers U., 1949; Ph.D., N.Y.U., 1960. Cert. health physicist Am. Bd. Health Physics. Physicist Nuclear Devel. Assocs., 1950-52; asst. chief, then chief radiation br. AEC Health and Safety Lab., N.Y.C., 1952-60; dir. applied nuclear tech. Tech. Research Group, Inc., 1960-62; cons. Burns & Roe, Servo Corp. of Am., 1962-64; mgr. research and devel. Del Electronics Corp., Mount Vernon, N.Y., 1964-67; founder, exec. v.p., tech. dir. Hadron, Inc., 1967-75; dir. Bur. Radiation Control, N.Y.C. Dept. Health, 1975—; successively lectr., adj. asst. prof., adj. assoc. prof. dept. environ. medicine N.Y.U. Med. Ctr., 1955—; prof. health physics Mcht. Marine Acad., Kings Point, N.Y., 1963. Contbr. articles to profl. jours. Served with inf. AUS, 1944-46; ETO. Decorated Combat Infantryman's badge; recipient Southworth prize in physics Hamilton Coll., 1947; cert. of appreciation EPA, 1979. Mem. Conf. Radiation Control Program Dirs., AAAS, Am. Nuclear Soc., Health Physics Soc., Am. Phys. Soc., N.Y. Acad. Scis., Radiol. and Med. Physics Soc. N.Y., Phi Beta Kappa. Subspecialties: Health physics; Laser medicine. Current work: Environmental standards and public health control for ionizing and non-ionizing radiation, biomedical laser applications; radiation emergency management; nuclear disarmament and proliferation. Home: 28 Pilgrim Ave Yonkers NY 10710 Office: 65 Worth St New York NY 10013

SOLORZANO, ROBERT F., veterinary microbiologist, educator; b. N.Y.C., May 21, 1929; s. Frank and Jean Solorzano; m. (div.); children: Jean Louise, Carolyn Ann, Robert Stephen. B.S., Georgetown U., 1951; M.S., Pa. State U., 1956, Ph.D., 1962. Bacteriologist, Montefiore Hosp. for Chronic Diseases, N.Y.C., 1953-55; research asst. Children's Hosp., Phila., 1956-58; grad. asst. Pa. State U., University Park, 1958-62; asst. prof. U. Ga., Tifton, 1962-68; assoc. prof. vet. microbiology U. Mo., Columbia, 1968-78, prof., 1978—. Contbr. articles to profl. jours. Served to cpl. U.S. Army, 1951-53. Recipient Outstanding Research award Diamond Labs., 1973. Mem. Am. Soc. Microbiology, Am. Assn. Vet. Lab. Diagnosticians, AAAS, Am. Leptospirosis Research Conf., Phi Zeta. Roman Catholic. Subspecialties: Animal virology; Immunology (agriculture). Current work: Diagnostic virology and serology, enteric virus infections of animals, pseudorabies, parvoviruses, bluetongue virus. Home: 1021 Ashland Gravel Rd Apt 1006 Columbia MO 65201 Office: Vet Med Diagnostic La U Mo Columbia MO 65211

SOLOTOROVSKY, MORRIS, microbiology educator's consultant; b. N.Y.C., Oct. 10, 1913; s. Samuel and Ida (Feinstein) S.; m. Mary Louise Whitty, Aug. 27, 1945; children: Peter, Nina, Emilie, Julian. B.S., U. Va., 1934; Ph.D., Columbia U., 1946. Research asst., instr. Coll. Physicians and Surgeons, Columbia U., 1937-43; research assoc. Research Labs., Merck Sharp and Dohme, Rahway, N.J., 1946-58; prof. microbiology Rutgers U., 1958—, vis. scientist, 1965-66, 73-74; cons. Centre for Applied Microbiology and Research-Public Health Lab. Service Gt. Brit. Contbr. numerous articles to profl. jours. Served to maj., Med. Services Corps U.S. Army, 1945-46. Fellow Am. Acad. Microbiology, AAAS; mem. Am. Assn. Immunologists, Am. Assn. Microbiology, Am. Thoracic Soc. Subspecialties: Infectious diseases; Immunobiology and immunology. Current work: Subcellular vaccines for haemophilus and pasteurella infections; mechanisms of host-parasite interaction. Home: 165 Springhill Rd Skillman NJ 08558 Office: Rutgers U New Brunswick NJ 08903

SOLOWAY, ALBERT HERMAN, chemist, college dean; b. Worcester, Mass., May 29, 1925; s. Bernard and Mollie (Raphaelson) S.; m. Barbara Berkowicz, Nov. 29, 1953; children—Madeleine Rae, Paul Daniel, Renee Ellen. Student, U.S. Naval Acad., 1945-46; B.S., Worcester Poly. Inst., 1948; Ph.D., U. Rochester, 1951. Postdoctoral fellow Nat. Cancer Inst. at Sloan-Kettering Inst., N.Y.C., 1951-53; research chemist Eastman Kodak Co., Rochester, N.Y., 1953-56; asst. chemist Mass. Gen. Hosp., Boston, 1956-61, asso. chemist, 1961-73; asso. prof. med. chemistry Northeastern U., Boston, 1966-68, prof. medicinal chemistry, chmn. dept., 1968-71, prof. medicinal chemistry and chemistry, chmn. dept. medicinal chemistry and pharmacology, 1971-74; dean Coll. Pharmacy and Allied Health Professions, 1975-77; dean Coll. Pharmacy, prof. medicinal chemistry Ohio State U., Columbus, 1977—; pres. Cambridge (Mass.) Nuclear Corp., 1960-61, v.p., 1961-74, dir., 1960-74. Author research medicinal chemistry. Served with USNR, 1944-45. Fellow AAAS, Acad. Pharm. Soc.; Mem. Am. Chem. Soc., Am. Pharm. Assn., Am. Assn. Coll. Pharmacy, Ohio Pharm. Assn., Am. Assn. Cancer Research. Jewish (trustee temple 1969-77, v.p., 1970-73). Subspecialties: Medicinal chemistry; Cancer research (medicine). Current work: Design of new structures and new approaches to chemoradiotherapy and chemoimmunotherapy of cancer. Office: 500 W 12th Ave Columbus OH 43210

SOMANI, PITAMBAR, physician, educator; b. Chirawah, India, Oct. 31, 1937; s. Narendra K. and Sarla (Maheshwari) S.; m. Kamlesh Somani, May 5, 1960; children: Anita, Alok, Jyoti. M.D., G. R. Med. Coll., 1960; Ph.D., Marquette U., 1965. Instr. dept. pharmacology Marquette U., Milw., 1965-66, asst. prof., 1966-69; asso. prof. pharmacology Med. Coll. Wis., 1969-74; mgr. pharmacology Abbott Labs., North Chicago, Ill., 1971-74; prof. pharmacology and medicine U. Miami, 1974-80; prof. pharmacology and medicine, dir. clin. pharmacology Med. Coll. Ohio, Toledo, 1980—; also cons. Author articles, monographs and book chpts. in field. Pres. India Assn. of Greater Miami, 1977-78. Wis. Heart Assn. fellow, 1963-65; NIH grantee, 1967-78; Fla. Heart Assn. grantee, 1977-80. Mem. Am. Soc. Pharmacology and Exptl. Therapeutics, Am. Soc. Clin. Pharmacology, Am. Fedn. Clin. Research, Am. Coll. Clin. Pharmacology, Soc. Exptl. Biology and Medicine., AMA. Subspecialties: Pharmacology; Internal medicine. Current work: Clinical pharmacology, antiarrhythmic drugs; cardiovascular pharmacology; coronary circulation; peritoneal dialysis. Home: 2437 Knights Hill Ln Toledo OH 43614 Office: Med Coll Ohio CS 10008 Toledo OH 43699

SOMBERG, JOHN CHARIN, medical scientist, cardiologist, cardiovascular pharmacologist, educator; b. N.Y.C., Oct. 8, 1948. B.A., Univ. Hights-NYU, 1970; M.D., N.Y. Med. Sch., 1974. Intern N.Y. Med. Coll., 1974-75, resident, 1975-77; fellow in cardiology Peter Bent Brigham Hosp.-Harvard U. Med. Sch., Boston, 1977-79, instr. in medicine and cardiology, 1979-80; asst. prof. medicine and molecular pharmacology Albert Einstein Coll. Medicine, 1980—, dir. cardiac arrhythmia service, 1981—. Contbr. 50 articles to profl. jours. Established investigator Am. Heart Assn., 1980-85. Fellow Am. Coll. Clin. Pharmacology; mem. Fedn. Am. Socs. Exptl. Biology, Fedn. Clin. Research, Basic Sci. Council of Am. Heart Assn. Subspecialties: Cardiology; Molecular pharmacology. Current work: Anti-arrhythmic pharmacology; role of nervous system in facilitating cardiac arrhythmias; neural control of circulation; clinical anti-arrhythmic assessment of anti-arrhythmic agents in prevention sudden death. Office: Albert Einstein Coll Medicine Forchheimer Bldg Room 216 Eastchester Rd/Morris Park Ave Bronx NY 10461

SOMERVILLE, RONALD LAMONT, biochemist, educator; b. Vancouver, C., Can., Feb. 27, 1935; s. Thomas Lennox and Vivian May (Code) S.; m. Joyce Elizabeth Crowe, May 4, 1955; children: Gregory M., Kenneth G., Gordon P., Victoria L., Daniel E. B.A., U. B.C., 1956, M.S., 1957; Ph.D., U. Mich., 1961. Cert. Am. Soc. Biol. Chemists, 1972. Grad. research asst. U. Mich., 1957-61, asst. prof., 1964-67; vis. scholar Stanford U., 1961-64; assoc. prof., prof. biochemistry Purdue U., West Lafayette, Ind., 1967—; adj. prof. Ind. U. Sch. Medicine, 1972—; lectr. in field; cons. to biotechnology industry. Contbr. research articles to sci. publs. Jane Coffin Childs Meml. Fund for Med. Research grantee, 1961-63; NIH grantee, 1964—. Mem. Am. Soc. Biol. Chemists, Genetics Soc. Am. Subspecialties: Biochemistry (biology); Genetics and genetic engineering (biology). Current work: Molecular basis of DNA-protein interaction regulatory mechanisms; repressor proteins; microbial viral genetics; tryptophan biosynthesis. Office: Dept Biochemistry Purdue Univ West Lafayette IN 47907

SOMLYO, AVRIL VIRGINIA, physiologist, researcher; b. Sask., Can., Apr. 9, 1939; d. Frederick William and Miriam Virginia (MacDougal) Russell; m. Andrew Paul Somlyo, May 25, 1961; 1 son, Andrew Paul. B.A., U. Sask., 1959, M.Sc., 1961; Ph.D., U. Pa., 1976. Co-prin. investigator dept. pathology and Pa. Muscle Inst. Presbyn. U. Pa. Med. Center, Phila., 1968-79, asso. research prof., 1978-82, research prof. physiology, 1982—. Contbr. articles to profl. jours. Mem. Biophys. Soc., Soc. Gen. Physiologists, Am. Soc. Pharmacology and Exptl. Therapeutics. Anglican. Subspecialties: Cell biology (medicine); Biophysics (physics). Current work: Muscle physiology, blood vessels, excitation-contraction coupling, electron optical techniques, electron probe analysis, in situ elemental localization. Office: Pa Muscle Inst Univ Pa Med Sch Philadelphia PA 19104

SOMMER, NOEL FREDERICK, plant pathologist; b. Scio, Oreg., Jan. 21, 1920; s. John Frederick and Anna Effie (Holt) S.; m. Connie Inez Truxillo, May 1, 1946; 1 son, Gary Frederick. B.S., Oreg., State U., 1941; M.S., U., Calif.-Davis, 1952, Ph.D., 1955. Plant pathologist USDA, Raleigh, N.C., 1955-56; pomologist U. Calif.-Davis, 1956-80, chmn. dept. pomology, 1975-81, postharvest pathologist, 1981—; cons. Hawaii Dept. Agr., 1970, TRC Corp., 1973, Alexander and Baldwin Corp., 1975-76, Banco de Mexico, 1980, Bakki Steamship Co., 1981, UN Devel. Program-India, 1982; participant AID Devel. Project-Egypt, 1977-81. Contbr. articles to profl. jours. Served to maj. F.A. AUS, 1941-46. Fellow Institut Nationale de la Recherche Agronomique French Ministry of Agr., 1968-69; recipient Bronze medal Chambre d'Agriculture du Vaucluse, France, 1978. Mem. Mycol. Soc. Am., Am. Soc. Microbiology, Am. Phytopath. Soc., Am. Soc. Hort. Sci. Subspecialties: Plant pathology; Postharvest pathology. Current work: Prevention of losses in perishable raw food commodities through research in the teachnology of handling, storage and transportation and the suppression of microbial activity. Office: Dept Pomology Univ Calif Davis CA 95616

SOMMERS, SHELDON CHARLES, pathologist; b. Indpls., July 7, 1916; s. Charles Birk and Leonore (Dickey) S.; m. Edith, Nov. 9, 1943. S.B., Harvard U., 1937, M.D., 1941. Diplomate: Am. Bd. Pathology. Intern U. Chgo. Clinics, 1941-42; resident on pathology New Eng. Deaconess Hosp., Boston, 1946-48, Free Hosp., also Boston Lying-In Hosp., 1948-49, Henry Ford Hosp., Detroit, 1949-50; staff pathologist New Eng. Deaconess Hosp., Boston, 1950-53; lab. dir. Mass. Meml. Hosps., Boston, 1953-61, Scripps Meml. Hosp., LaJolla, Calif., 1961-63; asso. dir., lab. dir. Delafield Hosp., N.Y.C., 1963-68; lab. dir. Lenox Hill Hosp., N.Y.C., 1968-81; sci. dir. Council Tobacco Research U.S., 1981—; clin. prof. pathology Columbia U. Coll. Phys. and Surg., U. So. Calif. Med. Sch.; cons. in field. Author numerous papers in field.; Co-editor: Pathology Annual. Chmn. N.Y. State Mental Hygiene Med. Rev. Bd., 1976—. Served to capt. M.C. AUS, 194-46. Decorated Silver Star, Bronze Star; Croix de Guerre. Mem. Am. Assn. Pathologists, Coll. Am. Pathologists, Am. Soc. Clin. Pathologists, Internat. Acad. Pathology, Arthur Purdy Stout Soc. Surg. Pathology (pres. elect 1981-82), N.Y. County Med. Soc., N.Y. Pathol. Soc. (pres. 1976). Clubs: Harvard (N.Y.C.); Knickerbocker Country (Tenafly, N.J.). Subspecialties: Pathology (medicine); Cancer research (medicine). Current work: Exptl. carcinogenesis, endrotine, gynecology and G-1 pathology, biomed. research fuding.

SOMORJAI, GABOR A(RPAD), chemist, educator; b. Hungary, May 4, 1935; came to U.S., 1957, naturalized, 1962; s. Charles and Livia (Ormos) S.; m. Judith K., Sept. 2, 1957; children: Nicole, John. Ph.D. in Chemistry, U. Calif.-Berkeley, 1960. Mem. research staff IBM Research, 1960-64; prin. investigator materials research div. Lawrence Berkeley Lab., 1964—; asst. prof. chemistry U. Calif.-Berkeley, 1964-67, assoc. prof., 1967-72, prof., 1972—, Miller prof., 1971, 78; cons. in field. Author: Principles of Surface Chemistry, 1972, Chemistry in Two Dimensional Surface, 1981; contbr. numerous articles to profl. jours. Recipient Kokes award Johns Hopkins U., 1976, Emmett award Am. Catalysis Soc., 1977; Guggenheim fellow, 1969. Fellow Am. Phys. Soc., AAAS; mem. Nat. Acad. Scis., Am. Chem. Soc. (Colloid and Surface Chemistry award 1981). Subspecialties: Catalysis chemistry; Surface chemistry. Current work: Surface chemistry; the surface science of heterogeneous catalysis. Home: 665 San Luis Rd Berkeley CA 95707 Office: Dept Chemistry U Calif Berkeley CA 94720

SOMSEL, JOSEPH KENT, nuclear engineer; b. Kokomo, Ind., Dec. 29, 1950; s. Richard M. and Louise (Hughes) S.; m. Ellen Joan Sellinger, Nov. 18, 1977; children: Francis, Walter, Isaac. B.S. in Nuclear Engring, U. Fla., Gainesville, 1977. Registered nuclear engr., Calif. Engr. EDS Nuclear, San Francisco, 1977-78; nuclear group leader Bechtel Power Corp., San Francisco, 1978—. Mem. Am. Nuclear Soc., Atomic Indsl. Forum, Health Physics Soc., Engrs. Club San Francisco. Republican. Subspecialty: Nuclear engineering. Current work: Safety analysis of nuclear power plants; radiation protection; radioactive decontamination; public information about nuclear power. Office: Bechtel Power Corp PO Box 3965 San Francisco CA 94119

SONDEL, PAUL M., immunologist, pediatric hematologist-oncologist; b. Milw., Aug. 14, 1950; s. Robert F. and Audrey J. (Dworkus) S.; m. Sherie A. Katz, Jan. 1, 1973; children: Jesse Adam, Beth Leah. B.S. with honors, U. Wis.-Madison, 1971, Ph.D. in Genetics, 1975; M.D. magna cum laude, Harvard U., 1977. Diplomate: Am. Bd. Pediatrics. Postdoctoral research fellow Sidney Farber Cancer Ctr., Boston, 1975-77; intern U. Minn. Hosp., Mpls., 1977-78; resident in pediatrics U. Wis. Hosp., Madison, 1978-80; asst. prof. depts. pediatrics, human oncology and genetics U. Wis. Med. Sch., Madison, 1980—. Contbr. articles to profl. jours. Co-chmn. Madison Gan Ha Yeled Nursery Sch. Com. J.A. and G.L. Hartford Found. fellow, 1981-84; Leukemia Soc. Am. scholar, 1981-76; Nat. Cancer Inst. grantee, 1982-85; Am. Cancer Soc. grantee, 1982-84. Mem. Internat. Transplantation Soc., Am. Assn. Immunologists, Am. Fedn. Clin. Research, Soc. Pediatric Research. Jewish. Subspecialties: Immunogenetics; Oncology. Current work: Using in vitro techniques to study the immune responses of fresh and cloned human T lymphocytes to histocompatible tumor cells. Home: 4329 Bagley Pkwy Madison WI 53705 Office: 600 Highland Ave Madison WI 53792

SONEIRA, RAYMOND M., astrophysicist; b. N.Y.C., July 10, 1949; s. Ramon Mario and Amelia Jane (Rodriguez) S. A.B., Columbia U., 1972; Ph.D., Princeton U., 1978. Engring. asst., engring. and devel. dept. CBS-TV Network, N.Y.C., 1967-70; research asst. dept. physics Princeton U., 1972-78; mem. Inst. for Advanced Study, Princeton, 1978—. Contbr. articles to profl. jours. Mem. Am. Phys. Soc., Am. Astron. Soc., Sigma Xi. Subspecialties: Cosmology; Theoretical astrophysics. Current work: In cosmology, the clustering and superclustering of galaxies using correlation function techniques. In galactic structure, the star distribution in our galaxy using computer models. Home: 86 Einstein Dr Princeton NJ 08540 Office: Inst for Advanced Study Princeton NJ 08540

SONENSHEIN, ABRAHAM LINCOLN, microbiologist, molecular biologist, educator; b. Paterson, N.J., Jan. 13, 1944; s. Israel Louis and Celia (Rabinowitz) S.; m. Gail Entner, Jan. 29, 1967; children: Dina Miriam, Adam Israel. A.B., Princeton U., 1965; Ph.D., M.I.T., 1970. Am. Cancer Soc. postdoctoral fellow Institut de Microbiologie, Universite de Paris-Sud, Orsay, France, 1970-72; asst. prof. molecular biology and microbiology Sch. Medicine Tufts U., Boston, 1972-78, assoc. prof., 1978—; cons. in field. Contbr. writings to profl. publs. in field. Mem. Am. Soc. Microbiology, AAAS, Fedn. Am. Scientists, AAUP. Subspecialties: Molecular biology; Microbiology. Current work: Control of gene expression during bacterial differentiation. Office: 136 Harrison Ave Boston MA 02111

SONG, CHANG WON, radiobiology educator, consultant; b. Korea, Apr. 10, 1932; came to U.S., 1959, naturalized, 1978; s. Jong Joo and So Soon (Kim) S.; m. Jai Kang, May 12, 1934; 3 children. B.S. in Chemistry, Seoul Nat. U., 1957; M.S. in Biochemistry, Korea U., Seoul, 1959; Ph.D. in Radiobiology, U. Iowa, 1964. Radiobiologist Albert Einstein Med. Center, Phila., 1964-69; asst. prof. Med. Coll. Va., 1969-71; asst. prof. radiobiology U. Minn., Mpls., 1971-75, assoc. prof., 1975-79, prof., 1979—; dir. Radiobiology Lab., 1971—; vis. prof. grantee Korean Dept. Sci. and Tech., 1972. Contbr. numerous articles to profl. jours. Nat. Cancer Inst. grantee, 1972—; Am. Cancer Soc. grantee, 1974-76. Mem. Radiation Research Soc., Am. Assn. Cancer Research, AAAS, Cell Kinetic Soc. Subspecialties: Cancer research (medicine); Biophysics (biology). Current work: Role of blood flow, po2 and pH in treatment of cancer with radiation or hyperthermia; chemical sensitization of tumor cells to radiation and hyperthermia. Home: 9359 Hyland Creek Circle Bloomington MN 55437 Office: U Minn Box 494 Med Sch Minneapolis MN 55455

SONG, JOSEPH, pathologist, cancer researcher; b. Pyong Yang, Korea, May 11, 1927; s. Ha Chu and Who Som (Koh) S.; m. Kumsan Ryu, Mar, 2, 1958; children: Patrick, Michael, Jeff. M.D., Seoul Nat. U., 1950; M.S., U. Tenn., 1956; M.D., U. Ark., 1965. Diplomate: Am. Bd. Pathology. Pathologist-dir. R.I. State Cancer Detection Project, 1956-59; assoc. pathologist Providence Lying-in Hosp., 1959-61; assoc. prof. U. Ark., 1961-64; pathologist, dir. Mercy Hosp., Des Moines, 1965—; prof. Creighton U., Omaha, 1969—; cons. in field. Author: Human Uterus, 1964, Pathology of Sickle Cell Disease, 1971; contbr. articles to profl. jours. Served to maj. Korean Army, 1950-52. Recipient Martin Luther King award. Mem. AMA, Am. Assn. Pathologists, Internat. Acad. Pathology, Coll. Am. Pathologists, ACP. Presbyterian. Club: Wakonda (Des Moines). Subspecialties: Pathology (medicine); Cancer research (medicine). Current work: Cancer research. Home: 2345 Park Ave Des Moines IA 50321 Office: Mercy Hosp 6th and University Des Moines IA 50314

SONG, LEILA SHIA, chemistry educator; b. Chen-kiang, China, May 27, 1947; d. Chien-Kuang and Yu-Lien Tu Shia; m. Yih H-Song, Sept. 6, 1974; children: Priscilla, Berwin. B.S. in Chem. Engring, Nat. Taiwan U., Taipei, 1970; M.A. in Chemistry, SUNY-Plattsburgh, 1973. Teaching asst. Nat. Taiwan U., 1970-71; research asst., lab. instr. SUNY-Plattsburgh, 1971-73, Tufts U., Medford, Mass., 1973-75; research chemist Thermo Electron Co., Waltham, Mass., 1976-78; adj. prof. N.J. Inst. Tech., Newark, 1979—. Contbr. articles to profl. jours. Subspecialties: Environmental Chemistry; Chemical engineering. Current work: Environmental carcinogenic research. Office: Dept Chemistry NJ Inst Tech 323 High St Newark NJ 07102

SONNENFELD, GERALD, microbiologist, educator; b. N.Y.C., Oct. 14, 1949; s. Otto and Ann (Perelman) S.; m. Elaine Marie Budd, May 6, 1978. B.S., CCNY, 1970; Ph.D., U. Pitts. Sch. Medicine, 1975. Postdoctoral fellow Stanford (Calif.) U. Sch. Medicine, 1976-78; assoc. guest worker biomed. research div. NASA-Ames Research Ctr., Moffett Field, Calif., 1976-78; asst. prof. dept. microbiology and immunology U. Louisville Sch. Medicine, 1978—. Sect. editor immunology, mem. editorial bd.: Jour. Interferon Research; contbr. articles to sci. jours. Trustee, head profl. edn. com. Ky. chpt. Leukemia Soc. Am. Recipient Esther Teplitz award U. Pitts. Sch. Medicine, 1974; NIH trainee, 1970-75; grantee NASA, EPA, Ky. Tobacco and Health Research Inst. Mem. Am. Assn. Immunologists, Am. Soc. Microbiology, Sigma Xi. Subspecialties: Immunology (medicine); Immunobiology and immunology. Current work: Regulation of immune response by interferon; interactions between carcinogens and interferon. Office: U Louisville Sch Medicine Louisville KY 40292

SONTAG, EDUARDO DANIEL, mathematics educator, researcher, consultant; b. Buenos Aires, Argentina, Apr. 16, 1951; came to U.S., 1972, naturalized, 1981; s. Bruno and Sofia (Pietruszka) S.; m. Frances David, Mar. 9, 1980; 1 dau., Laura Beth. Licenciado, U. Buenos Aires, 1972; Ph.D. in Math, U. Fla., 1976. Systems analyst Nat. Geohelio Phys. Obs., San Miguel, Argentina, 1970; research asst. to Prof. R. E. Kalman, dept. elec. engring. U. Fla., 1972-76, postdoctoral fellow, 1976-77; asst. prof. math. Rutgers-The State U., New Brunswick, N.J., 1977-82, assoc. prof., 1982—; vis. mem. tech. staff Bell Telephone Labs., Piscataway, N.J., summer 1980; cons. in field; grant reviewer U.S. Army, USAF, NSF. Author: Temas de Inteligencia Artificial, 1972, Polynomial Response Maps, 1977; contbr. numerous articles to profl. jours. USAF grantee, 1979—; U.S. Army grantee, 1974-76; NSF grantee, 1974-76. Mem. Am. Math. Soc. (chmn. com. human rights of mathematicians 1982—), Soc. Indsl. and Applied Math. (v.p. N.J. chpt. 1982—), Phi Beta Kappa. Democrat. Jewish. Subspecialty: Applied mathematics. Current work: Control system theory and applications; relations with artificial intelligence and other areas of computer science and operations research. Home: 60 Seymour Terr Piscataway NJ 08854 Office: Dept Math Rutgers U New Brunswick NJ 08903

SONTAG, MARC ROBERT, med. physicist; b. Bklyn., June 1, 1950; s. Howard Monroe and Marjorie Louise (Seldowitz) S.; m. Robin Joy Sontag, July 3, 1977; 1 dau.: Lisa Michelle. B.S., Bucknell U., 1972, M.S., 1974; Ph.D., U. Toronto, 1979. Med. physicist, instr. radiation therapy Thomas Jefferson U., Phila., 1979; med. physicist, asst. prof. U. Pa. Med. Sch., Phila., 1980—; pres. Micro Med. Physics and Engring. Co., Cherry Hill, N.J., 1981—. Contbr. articles to profl. jours. Mem. Am. Assn. Physicists in Medicine, Am. Soc. Therapeutic Radiologists, AAAS, N.Y. Acad. Sci. Subspecialties: Radiology. Current work: Methods of radiation dose calculations. Office: Dept Radiation Therapy Hosp U Pa Box 522 34th and Spruce Sts Philadelphia PA 19104

SOO, SHAO LEE, mechanical engineer, educator; b. Peking, China, Mar. 1, 1922; came to U.S., 1947, naturalized, 1962; s. Hsi Yi and Yun Chuan (Chin) S.; m. Hermia G. Dan, June 7, 1952; children—Shirley A. Soo Benford, Lydia M., David D. B.S., Nat. Chiaotung U., 1945; M.S., Ga. Inst. Tech., 1948; Sc.D., Harvard U., 1951. Engr. China Nat. Aviation, Calcutta and Shanghai, 1945-47; lectr. Princeton (N.J.) U., 1951-54, asst. prof., 1954-57, assoc. prof. mech. engring., 1957-59; prof. mech. engring. U. Ill., Urbana, 1959—; dir. S. L. Soo Assocs., Ltd., Urbana, 1980—; cons. NASA, NIH, Dept. Energy, EPA, NATO; mem. sci. adv. bd. EPA, 1976-78; adv. energy transp. World Bank, 1979; dir. Internat. Powder Inst., 1976—; NATO AGARD lectr.; Fulbright-Hays disting. lectr., 1974—, lectr. Chinese Acad. Sci., 1980. Author 5 books on thermodynamics, energy conversion and multiphase flow; mem. editorial bd.: Internat. Jour. Multiphase Flow, 1972—, Jour. Pipelines, 1980—, Internat. Jour. Sci. and Engring., 1983; contbr. numerous articles to profl. jours. Recipient Applied Mechanics Rev. award, 1972, Disting. Lecture award Internat. Pipeline Assn., 1981. Fellow ASME; mem. ASEE, Combustion Inst., Fine Particle Soc. (chmn. fluidized beds com.), Sigma Xi, Pi Tau Sigma (hon.). Methodist. Subspecialties: Fluid mechanics. Current work: Basic differential-integral equations of multiphase flow; pneumatic transport systems; and slurry pipeline systems; computer modeling for nuclear reactor safety and energy policy. Patentee in field (5). Home: 2020 Cureton Dr Urbana IL 61801 Office: 1206 W Green St Urbana IL 61801

SOOST, ROBERT KENNETH, geneticist; b. Sacramento, Nov. 12, 1920; s. John W. and Edna M. (Olsen) S.; m. Jean C. Carter, Nov. 26, 1949; children: Anita, Janet, Elaine. A.B., U. Calif.-Berkeley, 1942; Ph.D., U. Calif.-Davis, 1949. With U. Calif.-Riverside, 1949—, geneticist, prof. Served with AUS, 1943-46. Fellow Am. Soc. Hort. Sci.; mem. AAAS, Genetics Soc. Am., Am. Inst. Biol. Sci., Am. Genetics Assn., Phi Beta Kappa, Sigma Xi. Subspecialty: Plant genetics. Current work: Current research: Genetics and breeding of citrus fruit. Patentee in field.

SOPER, HENRY VICTOR, scientist, educator; b. Glen Ridge, N.J., Mar. 10, 1945; s. Kenneth L. and Sylvia (Caldwell) S. B.A., Yale U., 1966; M.A., U. Conn., 1972, Ph.D., 1974. Neurophysiologist Brain Research Inst., UCLA, 1974-76, NIH fellow psychology dept., 1976-78; intern, fellow in neuropsychology Camarillo (Calif.) State Hosp. Neuropsychiat. UCLA, 1982—; research assoc., lectr. U. Ill.-Chgo., 1978-82, dir., 1978-81; dir. Exptl. Epilepsy Lab., UCLA, Brain Research Inst., Los Angeles, 1974-78; manuscript reviewer Am. Psychol. Assn., Washington, 1974-75. Contbr. articles to profl. jours.; author: Behind the Laws of Rugby, 1979. Served to 1st lt. U.S. Army, 1966-68; Vietnam. Alcohol, Drug Abuse and Mental Health Adminstrn. fellow, 1976. Mem. AAAS, Internat. Primatological Soc., Am. Assn. Primatologists, Am. Psychol. Assn., Soc. Neurosci., N.Y. Acad. Sci., Sigma Xi, U.S.A. Rugby Football Union (referee nat. com. 1979-82). Subspecialties: Neuropsychology; Physiological psychology. Current work: Functions of the cerebral association cortices in man and monkeys; assessment of higher cortical dysfunction in man and subsequent rehabilitation, neurological bases of autism, mental retardation, other disorders, human cerebral organization. Home: 625 Mariposa El Segundo CA 90245 Office: UCAL-NPI Research Program PO Box A Camarillo CA 93010

SORENSEN, DANNY CHRIS, numerical analyst, mathematician, computer scientist; b. Concord, Calif., Mar. 6, 1946; s. Golden E. and Genevieve P. (Stevens) S.; m. Susan F. Shapiro, Dec. 15, 1976; 1 dau., Lisa Marie. B.S., U. Calif.-Davis, 1972; M.A., U. Calif.-San Diego, La Jolla, 1975, Ph.D., 1977. Asst. prof. math. U. Ky., 1977-80; cons. Argonne (Ill.) Nat. Lab., 1977-80, asst. computer scientist, 1980—; vis. assoc. prof. ops. research Stanford U., summer 1980; vis. assoc. prof math. U. Calif.-San Diego, La Jolla, spring 1982. Editor: (with R. J. B. Wets) Nondifferential and Variational Techniques in Optimization, 1982, Algorithms and Theory in Filtering and Control, 1982; assoc. editor: Soc. Indsl. and Applied Math. Jour. Sci. and Statis. Computing, 1982—. Served to sgt. U.S. Army, 1965-67; Vietnam. NSF grantee, 1979-80. Mem. Soc. Indsl. and Applied Math. Subspecialties: Numerical analysis; Mathematical software. Current work: Primary research in numerical methods for optimization (Newton, quasi-Newton and Trust Region methods); also work in numerical linear algebra, mathematical software. Home: 7757 Deerfield Woodridge IL 60517 Office: MCSD Argonne Nat Lab 9700 S Cass Ave Argonne IL 60439

SORENSEN, RAYMOND ANDREW, physics educator; b. Pitts., Feb. 27, 1931; s. Andrew J. and Dora (Thuesen) S.; m. Audrey Nickols, Apr. 2, 1953; 1 dau., Lisa Kirsten. B.S., Carnegie Inst. Tech., 1953, M.S., 1955, Ph.D., 1958. Mem. faculty Columbia, 1959-61; asst. prof. Carnegie-Mellon U., Pitts., 1961-65, assoc. prof., 1965-68, prof. physics, 1968—, chmn. dept., 1980—. NSF sr. postdoctoral fellow, 1965-66. Mem. Am. Phys. Soc. Subspecialties: Nuclear physics; Theoretical physics. Current work: Theory of the structure of nuclei; collective and independent particle excitations. Home: 1235 Murdoch Rd Pittsburgh PA 15217

SORENSON, HAROLD WAYNE, engring. scientist, educator; b. Omaha, Aug. 28, 1936; s. Harold Henry and Dorothy Marion (Wilson) S.; m. Sara Glee Flagor, Sept. 27, 1958 (div. Dec. 1974); children: Shelley G., Sara G., Eric R; m. Ruth McFarland, June 26, 1976. B.S. in Aero. Engring, Iowa State U., 1957; M.S. in Engring. (Control Systems), UCLA, 1963, Ph.D., 1966. Sr. research engr. Gen. Dynamics/Astronautics, San Diego, 1957-62; head space systems group AC Electronics div. GM, El Segundo, Calif., 1963-66; guest scientist Institut fur Steuer-und Regeltechnik, Oberpfaffenhoffen, W.Ger., 1966-67; prof. engring. scis. U. Calif. San Diego, La Jolla, 1968—; founder, pres. Orincon Corp., La Jolla, 1973-81; mem. Air Force Sci. Adv. Bd., 1981—. Author: Parameter Estimation: Principles and Problems, 1980; contbr. chpts. to books, articles to profl. jours. Recipient numerous research grants. Fellow IEEE; mem. AIAA, Ops. Research Soc. Am. Subspecialties: Systems engineering; Control/Feedback Systems. Current work: Control and estimation/signal processing for stochastic dynamic systems. Office: Dept AMES Univ Calif San Diego La Jolla CA 92093

SOROF, SAM, cancer researcher, biochemist; b. N.Y.C., Jan. 24, 1922; s. Morris S.; m. Phyllis F. Sanders, Oct. 24, 1967; children: Jonathan Michael, Lauren Hope. B.S., CCNY, 1944; Ph.D., U. Wis., Madison, 1950. Sr. mem. staff Inst. for Cancer Research, Fox Chase Cancer Center, Phila., 1952—; adj. prof. dept. pathology and lab. medicine U. Pa., 1981—; editor Differentiation, Cancer Biochemistry and Biophysics. Subspecialty: Cancer research (medicine). Office: Inst for Cancer Research 7701 Burholme Ave Philadelphia PA 19111

SOROKIN, PETER PITIRIMOVICH, physicist; b. Boston, July 10, 1931; s. Pitirim Alexandrovich and Elena Petrovna (Baratynskaya) S.; m. Anita J. Schell, Oct. 1, 1977. A.B., Harvard U., 1952, M.S., 1953, Ph.D., 1958. Research physicist IBM Watson Research Center, Yorktown Heights, N.Y., 1957—. Contbr. articles in quantum electronics to profl. jours. Recipient Michelson medal Franklin Soc., 1974; R.W. Wood award Optical Soc. Am., 1978; IBM fellow, 1968—. Mem. Nat. Acad. Sci. (Comstock award 1983), Am. Acad. Sci., N.Y. Acad. Sci. Subspecialties: Atomic and molecular physics; Spectroscopy. Current work: Atomic and molecular physics; laser spectroscopy. Patentee laser devices. Home: PO Box 225 Millwood NY 10546 Office: IBM Watson Research Center PO Box 218 Yorktown Heights NY 10598

SOSKEL, NORMAN TERRY, physician; b. Norfolk, Va., Sept. 1, 1948; s. Fred and Ruth (Chapel) S.; m. Judith Anne Barrie, Apr. 9, 1980; 1 son, Daniel Aaron. B.A., U. Va., 1970, M.D., 1974. Med. intern Hosp. of St. Raphael, New Haven, 1974-75; resident in medicine Salem (Va.) Hosp., 1975-76, U. Va. Hosp., Charlottesville, 1976-77; pulmonary fellow U. Utah Sch. Medicine, Salt Lake City, 1977-80, asst. prof. medicine, 1982—. Recipient Paderewski medal Am. Fedn. Music Tchrs., 1967; NIH grantee, 1980—. Mem. Am. Thoracic Soc., Am. Fedn. Clin. Investigation, Nat. Speleological Soc., AAAS, N.Y. Acad. Sci., Sigma Xi. Subspecialties: Internal medicine; Pulmonary medicine. Current work: Mechanisms of producing lung injury with specific interest in connective tissue macromolecules. Mechanisms of modulating lung injury. Office: Div Respiratory Critical Care and Occupational Medicine Room 3E544 U Utah Med Center 50 N Medical Dr Salt Lake City UT 84132

SOTER, NICHOLAS ARTHUR, physician, educator; b. Niagara Falls, N.Y., Oct. 26, 1939; s. Nicholas Anthony and Mildred Elizabeth (Brady) S. B.A., Tex. Christian U., 1961; M.D., U. Tex.-Dallas, 1965. Diplomate: Nat. Bd. Med. Examiners, Am. Bd. Dermatology. Med. intern Boston City Hosp., 1965-66; resident in dermatology Baylor U. Affiliated Hosps., Houston, 1966-68, Mass. Gen. Hosp., Boston, 1968-69; research fellow in immunology Harvard U. Med. Sch., Boston, 1971-73, instr. dermatology, 1973-75, asst. prof., 1975-79, assoc. prof.,

1979—; asst. physician Robert B. Brigham Hosp., Boston, 1973-76; physician Robert B. Brigham div. Brigham and Women's Hosp., Boston, 1976—. Contbr. articles to profl. jours. Served to maj., M.C. U.S. Army, 1969-71. Mem. Soc. for Investigative Dermatology, Am. Acad. Dermatology, Am. Fedn. for Clin. Research, Am. Assn. Immunologists, Am. Acad. Allergy and Immunology. Club: Harvard (Boston). Subspecialties: Dermatology; Immunology (medicine). Current work: Immunodermatology and pathogenesis of inflammatory and immunologic skin diseases. Home: 10 Commercial Wharf Boston MA 02110 Office: Sealy Mudd Bldg Room 610 250 Longwood Ave Boston MA 02115

SOUKUP, RODNEY JOSEPH, electrical engineering educator; b. Faribault, Minn., Mar. 9, 1939; s. Joseph Edward and Lillian (Fierst) S.; m. Theresa Marie Rockers, Dec. 8, 1965; children: Richard Joseph, Michael Harold, Stephen Rodney. B.S. with high distinction, U. Minn., Mpls., 1961, M.S.E.E., 1964, Ph.D., 1969. Prin. devel. engr. Univac, St. Paul, 1969-71; instr. U. Minn.-Mpls., 1972; asst. prof. elec. engring. U. Iowa, Iowa City, 1972-76; mem. faculty U. Nebr.-Lincoln, 1976—, prof. elec. engring., 1979—, chmn. elec. engring. dept., 1978—. Contbr. numerous solid state device articles to profl. publs. Mem. IEEE (sr.), Am. Vacuum Soc., Am. Soc. Engring. Edn., Tau Beta Pi, Eta Kappa Nu. Subspecialties: Semiconductors; Microelectronics. Current work: Measurements of semiconductor materials and devices, solar cells, and scanning electron microscopy. Home: 2000 Pacific Dr Lincoln NE 68506 Office: Univ Nebr W194 NH Lincoln NE 68588-0511

SOULE, DOROTHY FISHER, marine scientist; b. Lakewood, Ohio, Oct. 8, 1923; d. Eugene N. and Hermion Fisher; m. John D. Soule, 1943; 2 daus. B.A., Miami U., Oxford, Ohio, 1945; M.A. Occidental Coll, Los Angeles, 1963; Ph.D., Claremont Grad. Sch., Calif., 1969. Curator of Bryozoa Allan Hancock Found., U. So. Calif., Los Angeles, 1964—, adj. prof. environ. engring., 1974—, sr. research scientist, 1976—, dir., 1970—, assoc. dir., 1974-76, dir., 1977—; dir. Las Cruces Marine Biol. Sta., Baja California, 1970; vis. lectr. Calif. Coll. Medicine, 1965-66; asst. prof. zoology Calif. State U.-Los Angeles, 1963-65; cons. marine biology, 1960—; mem. marine fisheries adv. council U.S. Dept. Commerce, 1977-80; mem. sci. adv. bd. EPA, 1978—, marine ecosystem monitoring task group, 1980—; mem. offshore marine mining environ. safety panel Nat. Acad. Engring., 1973-75; mem. adv. com. invertebrate systematics resources NSF, 1974—; mem. subcoms. Pacific Fishery Mgmt. Council; mem. harbor liaison group U.S. Army C.E., 1973; mem. comml. fisheries adv. group Port of Los Angeles, 1981, fish harbor planning subcom., 1981; field research throughout world. Sec.-treas. Irene McCulloch Found. Research and Edn. in Marine Biology, 1968—. Contbr. numerous articles to sci. jours. NSF grantee, 1966-68, 78-80; Sea Grant Program grantee, 1972-75, 76-78; others. Mem. So. Calif. Acad. Scis., Internat. Bryozoology Assn., Los Angeles Mus. Alliance, AAAS, Marine Tech. Soc., Western Soc. Naturalists, Am. Inst. Biol. Sci., Pacific Sci. Assn., Am. Soc. Zoologists, Assn. Environ. Profls., Calif. Water Pollution Control Assn., Phi Beta Kappa, Sigma Xi, Phi Sigma, Sigma Delta Epsilon, Phi Delta Gamma. Current work: Marine environmental management. Office: Inst Marine Studies Allan Hancock Found U So Calif Los Angeles CA

SOURKES, THEODORE LIONEL, biochemistry educator; b. Montreal, Que., Can., Feb. 21, 1919; s. Irving and Fannie (Golt) S.; m. Shena Rosenblatt, Jan. 17, 1943; children: Barbara, Myra. B.Sc., McGill U., 1939, M.Sc. magna cum laude, 1946; Ph.D., Cornell U., 1948. Asst. prof. pharmacology Georgetown U. Med. Sch., 1948-50; research asso. dept. enzyme chemistry Merck Inst. Therapeutic Research, Rahway, N.J., 1950-53; sr. research biochemist Allan Meml. Inst., Montreal, 1953-65, dir. lab. chem. neurobiology, 1965—; mem. faculty McGill U., Montreal, 1954—, prof. biochemistry, 1965—, asso. dean for research, 1972-75; Mem. Que. Med. Research Council, 1971-77; sr. fellow Parkinson's Disease Found., N.Y.C., 1963-66. Author: Biochemistry of Mental Disease, 1962, Nobel Prize Winners in Medicine and Physiology, 1901-1965, 1967; Mem. editorial bd.: Essays in Neurochemistry and Neuropharmacology. Fellow Royal Soc. Can.; mem. Canadian Biochem. Soc., Pharmacol. Soc. Can., Canadian Coll. Neuropsychopharmacology (Heinz Lehmann award), Am. Soc. Biol. Chemists, Am. Soc. Pharmacology and Exptl. Therapeutics, Am. Soc. Neurochemistry, Internat. Soc. Neurochemistry, Internat. Brain Research Orgn., Canadian Soc. Study History and Philosophy Sci., Sigma Xi. Subspecialties: Neurochemistry; Neuropharmacology. Research and publs. on drugs for treatment high blood pressure; 1st basic research on methyldopa; elucidation of role of dopamine and other monamines in nervous system; biochemical and histological bases of Parkinson's disease, biochemistry of mental depression, pathways of stress in the nervous system. Home: 4645 Montclair Ave Montreal PQ Canada H4B 2J8

SOUTAS-LITTLE, ROBERT WILLIAM, mechanical engineer, educator; b. Oklahoma City, Feb. 25, 1933; s. Harry Glenn and Mary Evelyn (Miller) Little; m. Patricia Soutas, Sept. 3, 1982; children: Deborah, Catherine, Colleen, Jennifer, Karen. B.S. in Mech. Engring, Duke U., 1955; M.S., U. Wis., 1959, Ph.D., 1962. Design engr. Allis Chalmers Mfg. Co., Milw., 1955-57; instr. mech. engring. Marquette U., 1957-59; instr. U. Wis., Madison, 1959-62, asst. prof., 1962-63, Okla. State U., 1963-65; prof. Mich. State U., 1965—, chmn. dept. mech. engring., 1972-77, chmn. dept. biomechanics, 1977—; Cons. A. C. Electronics Co., Ford Motor Co., CBS Research Lab., B. F. Goodrich Co.; lectr. AID, India, 1965. Author: Elasticity, 1973; Contbr. articles to profl. jours. Vice pres. Okomos (Mich.) Sch. Bd., 1967-72; mem. Meridian Twp. (Mich.) Charter Commn., 1969-70, Meridian Twp. Zoning Bd. Appeals, 1969-71. NSF grantee, 1964-69, 79; recipient award for excellence in instrn. engring. students Western Electric Co., 1970-71; NIH grantee, 1973-75, 79—. Mem. Soc. Engring. Sci., ASME, Am. Soc. Biomechanics, Sigma Xi, Pi Tau Sigma, Tau Beta Pi. Subspecialties: Biomedical engineering; Theoretical and applied mechanics. Current work: Biomechanics studies of the muscoloskeletal system and tissue biomechanics. Home: 2402 Hulett Rd Okemos MI 48864 Office: Dept of Biomechanics Mich State U East Lansing MI 48824

SOUTH, HUGH MILES, engineering administrator; b. Houston, Nov. 10, 1947; s. Hugh Wilson and Lala (Miles) S.; m. Marilyn Morrell, Oct. 1, 1976. B.A., Rice U., 1971; Ph.D., Johns Hopkins U., 1981. Instr. engring. Johns Hopkins Evening Coll., Balt., 1972-76; sr. engr. Johns Hopkins Applied Physics Lab., Laurel, Md., 1976-79, Span Lab. supr., 1979—. Contbr. articles to profl. jours. Mem. IEEE, European Soc. for Signal Processing, Acoustical Soc. Am. (assoc.), Phi Beta Kappa, Sigma Xi, Tau Beta Pi. Democrat. Subspecialties: Electrical engineering; Acoustics. Current work: Signal processing hardware/software systems, signal processing algorithms, ultrasonic waves. Home: 7242 Lasting Light Way Columbia MD 21045 Office: Johns Hopkins Applied Physics Lab Johns Hopkins Rd Laurel MD 20707

SOUTH, MARY ANN, pediatric immunologist, educator; b. Portales, N.Mex., May 23, 1933; s. John Anderson and Carrie (Schumpert) S. Student, Baylor U., 1951-53; B.A., Eastern N.Mex. U., 1955; M.D., Baylor Coll. Medicine, 1959. Diplomate: Am. Bd. Pediatrics. Instr. U. Minn., 1964-66; asst. prof. pediatrics Baylor Coll. Medicine, Houston, 1966-70, assoc. prof., 1970-73; assoc. prof. dept. pediatrics U. Pa., 1973-77; dir. pediatric immunology Children's Hosp., Phila., 1973-77; prof., chmn. dept. pediatrics Tex. Tech. U. Health Sci. Ctr., Lubbock, 1977-79, research prof., chief div. immunology dept. pediatrics, 1979—. Contbr. articles to profl. jours. USPHS awardee, 1968-73. Mem. Am. Assn. Immunologists, AAUP, Am. Pediatric Soc., Infectious Diseases Soc. Am., Internat. Soc. Exptl. Hematology, Soc. Pediatric Research, So. Soc. Pediatric Research, Western Soc. Pediatric Research, Reticuloendothelial Soc. Subspecialties: Immunology (medicine); Infectious diseases. Office: Dept Pediatrics Tex Tech U Health Sci Ct Sch Medicine Lubbock TX 79430

SOWERS, JOSEPH LOUIS, computer scientist, educator; b. Phila., Nov. 20, 1946; s. Joseph Louis and Edith Frances (Ober) S.; m. Louise Ann Schissler, Aug. 18, 1949; children: Matthew Sean, Kerri Leigh. A.B., Rutgers U., 1968; M.A., Temple U., 1971, Ph.D., 1978. Lab. instr. Temple U., Phila., 1968-69; instr. physics Rutgers U.-Camden, N.J., 1969-74, programmer analyst (CCIS), 1974-82, asst. prof. computer scis., 1982—. Contbr. articles to profl. jours. Mem. Am. Astron. Soc., Am. Phys. Soc., U.S. Coast Guard Aux., Am. Assn. Physics Tchrs., Soc. Computer Simulation. Republican. Roman Catholic. Subspecialties: Simulation; Programming languages. Current work: Researcher in computer simulation. Office: Dept Computer Sci Rutgers U Camden NJ 08102

SOYKA, LESTER FRANK, physician, pharmaceutical researcher; b. Chgo., Mar. 12, 1931; s. Anton Frank and Anna (Buchholz) S.; m.; children: Peter A., Gregory A., David L., Leslie A. B.S., U. Wis., 1952; M.S. in Pharmacology, U. Ill., 1960, M.D., 1961. House officer Children's service Mass. Gen. Hosp., Boston, 1961-62, asst. resident, 1962-63, clin. and research fellow, 1963-65, asst. in pediatrics, 1965-67; instr. pediatrics Harvard Med. Sch., Boston, 1965-67; asst. prof. pediatrics Stanford U. Sch. Medicine, 1967-70; assoc. prof. pharmacology and pediatrics U. Ill. Coll. Medicine, 1970-72, prof., 1973; prof. pharmacology and pediatrics U. Vt. Coll. Medicine, 1973-81, chmn. dept. pharmacology, 1975-81; with Devel. Pharmacology br. Nat. Inst. Child Health and Human Devel., NIH, Bethesda, Md., 1980-81; dir. clin. pharmacology Mead Johnson & Co., Evansville, Ind., 1981-82; dir. cardiovascular clin. research, pharm. research and devel. div. Bristol-Myers Co., Evansville, Ind., 1982—. Contbr. articles to profl. jours. Served to 1st lt. Med. Service Corps U.S. Army, 1952-54. Wyeth Pediatrics fellow, 1962-64; Pharm. Mfrs. Assn. Found. grantee. Mem. Am. Acad. Pediatrics, Am. Fedn. for Clin. Research, Am. Pediatric Soc., Am. Soc. for Clin. Pharmacology and Therapeutics, Am. Soc. for Pharmacology and Exptl. Therapeutics, Lawson Wilkins Pediatric Endocrine Soc., Mass. Med. Soc., Midwest Soc. for Pediatric Research, Soc. for Pediatric Research, Endocrine Soc., Vt. Diabetes Assn., Vt. State Med. Soc., Western Soc. for Pediatric Research, Drug Info. Assn. Subspecialties: Pediatrics; Pharmacology. Current work: Clinical pharmacology; treatment of hypertension; treatment of arrhythmias, clinical trials methodology.

SOZEN, METE AVNI, civil engineering educator; b. Istanbul, Turkey, May 22, 1930; came to U.S., 1951, naturalized, 1960; s. Huseyin Avni and Ayse Saliha S.; m. Joan E. Bates, Nov. 22, 1971; children: Tim, Adria, Ayse. B.S. in Civil Engring, Robert Coll., Istanbul, Turkey, 1951, Ph.D., U. Ill., 1956. Asst. prof. civil engring. U. Ill., 1957-59, assoc. prof., 1959-62, prof., 1962—; cons. in field. Mem. ASCE (Research award 1963, Reese award 1971, Moisseif award 1972), Am. Concrete Inst. (Kelly award 1976), Seismological Soc. Am., Nat. Acad. Engring. Subspecialty: Civil engineering. Current work: Earthquake response of structures. Home: 503 W Michigan Ave Urbana IL 61801 Office: 3112 Newmark CE Research Lab University of Illinois Urbana IL 61801

SPALDING, DONALD HOOD, plant pathologist, researcher; b. Pawtucket, R.I., Dec. 1, 1925; s. John Davis and Mildred Idella (Hood) S.; m. Marion Elizabeth Headley, Jan. 30, 1953; children: Darcy Elizabeth, Donna Jane. A.B. in Biology, Brown U., 1950; M.A. in Bacteriology, U. Kans., 1953; Ph.D. in Plant Pathology, Wash. State U., 1960. Microbiologist Dow Chem. Co., Midland, Mich., 1953-57; research asst. Wash. State U., Pullman, 1957-60; research plant pathologist Agr. Research Service, U.S. Dept. Agr., Wenatchee, Wash., 1960-61, Beltsville, Md., 1961-71, Subtropical Hort. Research Sta., Miami, Fla., 1971—. Served with USN, 1943-46; PTO. Mem. Am. Phytopathol. Soc., Am. Soc. for Hort. Sci., Fla. Hort. Soc. Presbyterian. Subspecialty: Plant pathology. Current work: Head agriculture research station for plant introduction and evaluation, research on biology and control of insects and market quality; research on postharvest technology for fruits and vegetables. Home: 17500 SW 89th Ct Miami FL 33157 Office: 13601 Old Cutler Rd Miami FL 33158

SPARACINO, JACK ROBERT, human resources researcher and consultant, psychologist; b. N.Y.C., Nov. 4, 1950; s. Robert R. and Marguerite (Riff) S. B.A., Syracuse U., 1972; Ph.D., U. Chgo., 1978. Lectr. U. Chgo., 1977-78; instr. Roosevelt U., Chgo., 1974-78; sr. research assoc. Ohio State U., 1980-81; assoc. project dir. Yankelovich, Skelly and White, Inc., Stamford, Conn., 1982—; cons. Columbus (Ohio) Quality of Working Life Program, 1979-81; adj. asst. prof. Ohio State U., 1980-81. Contbr. articles to profl. jours. Mem. Am. Psychol. Assn. Subspecialties: Social psychology. Current work: Currently working with Yankelovich, Skelly, and White's human resource research and consulting group; organizational personality psychology and the psychology of work behavior are my primary research areas; my firm now offers full service research and consulting capabilities and continues to conduct groundbreaking studies. Home: 73 Willuwbrook Ave Stamford CT 06902 Office: Yankelovich Skelly and White Inc 969 High Ridge Rd Stamford CT 06905

SPARKS, DAVID LEE, neuroscientist, educator; b. Guntersville, Ala., Dec. 22, 1937; s. Houston Lee and Ruth Bertha (Mooney) S.; m. Betty Ann Ellis, Aug. 31, 1963; children: Steven Edward, Robert Gregory, Michael Scott. B.A., U. Ala., 1959, M.A., 1962, Ph.D., 1963. USPHS postdoctoral research fellow U. Miss. Med. Sch., 1963-65; instr. psychiatry U. Ala., Birmingham, 1965-67, asst. prof., 1967-69, assoc. prof. psychology, 1969-72, prof., 1972-73, chmn. dept., 1969-72, prof. psychology and neurosci., 1973-81, prof. physiology and biophysics, 1981—, prof. neurosci., 1981—, Univ. scholar, 1974-81. Mem. editorial bd.: Revs. Oculomotor Research; contbr. articles to sci. jours. NIMH grantee, 1966-71; Nat. Eye Inst. grantee, 1973—. Mem. AAAS, Assn. Research in Vision and Ophthalmology, Soc. Neurosci. Subspecialties: Physiological psychology; Neurophysiology. Current work: Neural control of eye movements; motor physiology; sensory physiology; teaching medical and dental students; research; graduate education. Home: 3501 Old Leeds Ct Birmingham AL 35213 Office: Dept Physiology U Ala Birmingham AL 35294

SPARKS, HARVEY VISE, JR., physiologist; b. Flint, Mich., June 22, 1938; s. Harvey Vise and Ellen Louise (Paschall) S.; m. Barbara M. Taylor, Jan. 17, 1969; children—Matthew Taylor, Catherine Elliott, Wendy Sue, Harvey Vise. Student, U. Mich., 1956-59, M.D., 1963. Postdoctoral fellow dept. physiology Harvard Med. Sch., Boston and; U. Goteborg, Sweden; instr. U. Mich., 1966-67, asst. prof. physiology, 1967-70, asso. prof., 1970-74, prof., 1974-78, asst. to dean, 1970-71, asst. dean, 1971-72; prof., chmn. dept. physiology Mich. State U., 1979—; mem. survey team, liaison com. on med. edn. AMA and; Am. Assn. Med. Colls.; mem. rev. teams NIH. Author: Casebook of Physiology, 1973; contbr. numerous articles to profl. jours.; editor: (with others) Handbook of Physiology, 1979. Recipient Meritorious Service award Mich. Heart Assn., 1962; Borden award for med. student research, 1963; Mich. Heart Assn. student fellow, 1962-63; John and Mary Markle scholar, 1967-72; USPHS postdoctoral fellow, 1963-66; U. Mich. student research fellow, 1960-61; USPHS grantee, 1963—. Mem. Am. Physiol. Soc. (editorial bd. Am. Jour. Physiology 1974—), AAAS, Microcirculatory Soc., Soc. Exptl. Biology and Medicine, Am. Heart Assn. (council circulation), Mich. Heart Assn., Am. Coll. Sports Medicine, Victor Vaughn Soc., Alpha Omega Alpha, Phi Kappa Phi. Subspecialties: Physiology (medicine); Cardiology. Current work: Cononary blood flow, capillary transport, exercise myocardial ischemia. Home: 7996 Lovejoy Rd Perry MI 48872 Office: Dept Physiology Giltner Hall Mich State U East Lansing MI 48824

SPARKS, MORGAN, physicist; b. Pagosa Springs, Colo., July 6, 1916; s. Harry Lysinger and Pearl (Morgan) S.; m. Elizabeth MacEvoy, Apr. 30, 1949; children: Margaret Ellen, Patricia Rae, Morgan MacEvoy, Gordon K. B.A., Rice U., 1938, M.A., 1940; Ph.D., U. Ill., 1943. With Bell Telephone Labs., Murray Hill, N.J., 1943-72, exec. dir. semicondr. components, 1968-69, v.p. tech. info. and personnel, 1969-71, v.p. electronic tech., 1971-72; pres. Sandia Nat. Labs., Albuquerque, 1972-81; dean Anderson Sch. Mgmt., U. N.Mex., Albuquerque, 1981—. Contbr. articles on transistors, pn junctions and properties of semicondrs. to profl. jours. Rockefeller Found. fellow U. Ill., 1940-43. Fellow Am. Phys. Soc., IEEE (Jack A. Morton award 1977), Am. Inst. Chemists; mem. Am. Chem. Soc., Nat. Acad. Engring., Phi Beta Kappa, Sigma Xi, Phi Lambda Upsilon. Subspecialties: Semiconductors; Theoretical chemistry. Patentee semicondr. electronics. Home: 904 Lamp Post Circle SE Albuquerque NM 87123 Office: U NMex Albuquerque NM 87131

SPARKS, RONALD WAYNE, physicist; b. Wilmington, N.C., Apr. 14, 1955; s. William Marcellis and Sarah Lois S.; m. Cheryl Hawes, Nov. 24, 1979; children: Xylia Maxine, Adam Tobias. B.S. in Physics, U. N.C.-Wilmington, 1979. Salesman Sears Roebuck & Co., Wilmington, N.C., 1975-79; physicist Inst. for Marine Bio-Med. Research, Wrightsville Beach, N.C., 1979; tchr. chemistry, physics, math, electronics New Hanover County Bd. Edn., Wilmington, 1979-80; analyst Talbert, Cox & Assocs. Inc. (Cons. Engrs.), Wilmington, 1980-81; application engr. Climate Control, Wilmington, 1982—; energy cons., cons. physicist. Mem. Am. Phys. Soc., Forum of Physics and Soc. Seventh-day Adventist. Subspecialty: Enhanced heat transfer.

SPARLING, DONALD WESLEY, statistician; b. Chgo., Sept. 20, 1949; s. Donald W. and Lorraine H. (Ehlen) S.; m. Paulette Sittler, Mar. 20, 1971; children: Justin, Jessica. B.S., So. Ill. U., 1971, M.A., 1974; Ph.D., U. N.D., 1979. Instr. U. Minn. Tech. Coll., Crookston, 1974-78; asst. dir. Coop. Wildlife Research Lab., So. Ill. U., Carbondale, 1978-79; asst. prof. dept. biology Ball State U., Muncie, Ind., 1979-82, asst. prof. mgmt. sci. dept., 1982; statistician U.S. Fish and Wildlife Service, Jamestown, N.D., 1982—; cons. Middle River (Minn.) Sch., 1975-78. Mem. Wilson Ornithol. Soc., Am. Ornithologists Union, Animal Behavior Soc., N.D. Am. Statis. Soc., Sigma Xi. Roman Catholic. Subspecialties: Ecology; Biostatistics. Current work: Application of statistics and data processing to wildlife biology and areas of behavioral ecology. Home: 925 8th Ave NE Jamestown ND 58401 Office: US Fish and Wildlife Service Box 1747 Jamestown ND 58401

SPARROW, JANET RUTHE, anatomy educator, neurobiologist, researcher; b. Toronto, Ont., Can., Jan. 6, 1952; d. Robert John and Helen A. (Mathieson) S.; m. Alan D. Miller, July 22, 1978. B.Sc., U. Western Ont., 1974, Ph.D., 1980. Postdoctoral fellow dept. physiology Cornell U. Med. Coll., N.Y.C., 1980-82, research assoc., 1982—. Author sci. articles. Province of Ont. scholar; recipient Collip award U. Western Ont., 1980; Nat. Multiple Sclerosis Soc. fellow, 1980-82. Mem. Soc. Neurosci., N.Y. Acad. Sci. Subspecialties: Regeneration; Neurobiology. Current work: Research interests include axonal regeneration, neuronal response to injury, axonal transport, establishment of connections between neurons during development and regeneration. Office: 1300 York Ave New York NY 10021

SPAZIANI, EUGENE, biologist, educator; b. Detroit, July 22, 1930; s. Edward and Mary (Cataldi) S.; m. Carol Lynn, Nov. 12, 1953; children: Andrew, Daniel. Student, Chaffey Coll., 1948-50; B.A., UCLA, 1952, M.A. (Arthritis and Rheumatism Found. fellow), 1954, Ph.D., 1958; postgrad., Univ. Coll., U. London, 1958-59. Instr. U. Iowa, 1959-60, asst. prof. zoology, 1960-64, assoc. prof., 1964-68, prof., 1968—, chmn. dept. zoology, 1977-80; vis. prof. physiology U. Calif., San Francisco, 1981; vis. investigator AMA Inst. for Biomed. Research, Chgo., 1967. Research publs. in field. Endocrine Soc. Squibb fellow U. London, 1958; Lalor Found. fellow, 1960; Am. Cancer Soc. grantee, 1960-63; NIH grantee, 1960-61; NSF grantee, 1973-76. Mem. Am. Physiol. Soc., Endocrine Soc., Am. Soc. Zoologists, Soc. Study Reprodn., AAAS, AAUP. Subspecialties: Physiology (biology); Endocrinology. Current work: Endocrinology of reproduction; mechanism of steroid hormone action; hormone biosynthesis; comparative invertebrate endocrinology. Office: Dept Zoology U Iowa Iowa City IA 52242

SPEAR, NORMAN EBERMAN, psychology educator; b. Canton, Ohio, Oct. 15, 1937; s. Jesse B. and Florence (Eberman) S.; m. Linda Patia, June 14, 1977; children: Amanda Teri, Jennifer Lee. B.S., Bowling Green State U., 1959, M.S., 1961; Ph.D., Northwestern U., 1963. Asst. prof. psychology Rutgers U., 1963-65, assoc. prof., 1965-69, prof., 1969-74; disting. prof. psychology SUNY-Binghamton, 1974—, chmn. dept., 1976-78; vis. cons. U. So. Calif., 1971; vis. prof. Oxford U., 1973-74. Author: (with others) Learning and Memory, 1977, Processing Memories, 1978; editor: (with B.A. Campbell) Ontogeny of Learning and Memory, 1979, (with R.R. Miller) Information Processing in Animals, 1981, (with R. Isaacson) The Expression of Knowledge, 1982; contbr. articles to profl. jours. NIMH career devel. awardee, 1970-74; MINH, NSF grantee, 1963—. Mem. Am. Psychol. Assn., Soc. for Neurosci., Psychonomic Sci., Midwestern Psychol. Assn., Eastern Psychol. Assn., Internat. Soc. Devel. Psychology, Soc. Stimulus Properties of Drugs, Sigma Xi. Subspecialties: Learning; Developmental psychology. Home: 201 African Rd Vestal NY 13850 Office: Dept Psychology SUNY Binghamton NY 13901

SPEAS, ROBERT DIXON, aviation consultant; b. Davis County, N.C., Apr. 14, 1916; s. William Paul and Nora Estelle (Dixon) S.; m. Manette Lansing Hollingsworth, Mar. 4, 1944; children: Robert Dixon, Jay Hollingsworth. B.S., MIT, 1940; grad., Boeing Sch. Aero., 1938. Aviation reporter Winston Salem Jour., 1934; sales rep. Trans World Airlines, 1937-38; engr. Am. Airlines, 1940-44, asst. to v.p., 1944-47, dir. maintenance and engring., cargo div., 1947-48, spl. asst. to pres., 1948-50; U.S. rep. A.V. Roe Can., Ltd., 1950-51; pres., chmn. bd. R. Dixon Speas Assos., Inc. (aviation cons.), 1951-76; chmn., chief exec. officer Speas-Harris Airport Devel., Inc., 1974-76; chmn. bd., pres. Aviation Consulting, Inc., 1976—; Mem. aeros. and space engring. bd. Nat. Research Council, 1980—. Author: Airplane Performance and Operations, 1945, Pilots' Technical Manual, 1946, Airline Operation, 1949, Technical Aspects of Air Transport Management, 1955. Recipient 1st award Ann. Nat. Boeing Thesis Competition, 1937; research award; Am. Air Transport Assn., 1942. Fellow AIAA (treas. 1963-64, council 1963-64), Royal Aero. Soc.;

mem. Soc. Automotive Engrs. (v.p. 1955, mem. council 1964-66), Flight Safety Found. (bd. govs. 1958-66, 78—, exec. com. 1979—), Inst. Aero. Scis. (past treas.; council 1959-62, exec. com. 1962), L.I. Assn. Commerce and Industry (mem. bd.), Manhasset C. of C. (pres. 1962), ASME. Clubs: Wings (pres. 1968-69, council 1966-71, 73—), Port Washington Yacht). Subspecialty: Aeronautical engineering. Home: 591 Park Ave Manhasset NY 11030 Office: 58 Hillside Ave Manhasset NY 11030

SPECIALE, ROSS ALDO, scientist, researcher; b. Palermo, Sicily, Italy, July 24, 1927; came to U.S., 1970; s. Pietro Sebastiano and Rosalia (Trizzino) S.; m. Louisa Maria Robijns, Mar. 6, 1962; children: Paul, Claudia, Alexander. Ph.D., Politecnico di Milano, Italy, 1955. Microwave engr. Magneti Marelli, Milan, 1955-58; elec. engr. Laben, Milan, 1958-62; sr. scientist Philips N.V., Eindhoven-Holland, 1962-70, Tektronix, Inc., Beaverton, Oreg., 1970-77, TRW Elec. Systems Group, Redondo Beach, Calif., 1977—. Sr. mem. IEEE; mem. Am. Math. Soc., Soc. Indsl. and Applied Math., Assn. Computing Machinery. Subspecialties: Electronics; Computer-aided design. Home: 639 Camino de Encanto Redondo Beach CA 90277 Office: TRW Electronics and Defense Sector One Space Park Dr Redondo Beach CA 90278

SPECTER, STEVEN CARL, immunology educator; b. Phila., June 4, 1947; s. Herbert Louis and Miriam Selma (Rubin) S.; m. Randie Yeuson, June 1, 1969; children: Ross, Rachel. A.B., Temple U., 1969, Ph.D., 1975. Postdoctoral fellow Albert Einstein Med. Ctr., Phila., 1974-76; asst. dir. microbiology, 1976-79; asst. prof. immunology U. South Fla. Coll. Medicine, Tampa, 1979—, dir. virology lab., 1982—; cons. Tampa Gen. Hosp. Virus Lab., 1980—. Editorial bd.: Jour. Clin. Microbiology, 1981—; editor: Diagnosis of Viral Infections, 1979, Biological Response Modifiers in Human Oncology, 1983; contbr. articles to profl. jours. Mem. Am. Soc. Microbiology (br. newsletter editor 1977-79, 82—), AAAS, N.Y. Acad. Sci., Reticuloendothelial Soc. Republican. Jewish. Subspecialties: Immunobiology and immunology; Virology (biology). Current work: Alteration of immunity by tumors and viruses; immunopharmacology, development of rapid methods for viral diagnosis. Home: 2008 Chickwood Ct Tampa FL 33618 Office: Univ South Fla Coll Medicine 12901 N 30th St Tampa FL 33612

SPECTOR, REYNOLD, pharmacologist, physician, educator; b. Stoneham, Mass., Nov. 3, 1940; s. Asher and Esther (Karelitz) S.; m. Michiko, Mar. 1, 1973; children: Regine Amy, June Thalia. B.A., Harvard U., 1962; M.D., Yale U., 1966. Diplomate: Am. Bd. Internal Medicine, 1973. Intern and resident Peter Bent Brigham Hosp., Boston, 1966-68, 70-71; instr. Harvard U. Med. Sch., 1971-74, asst. prof., 1974-78, assoc. prof., 1978; prof. medicine and pharmacology U. Iowa, 1978—, dir. divs. clin. pharmacology and gen. medicine, 1978—. Contbr. articles to profl. jours. Served to maj. M.C. U.S. Army, 1968-70. NIH grantee, 1974—. Mem. Am. Soc. Clin. Investigation, Am. Soc. Pharmacology and Exptl. Therapeutics, Am. Coll. Clin. Pharmacology, Central Soc. Clin. Research, Am. Fedn. Clin. Research. Club: Hasty Pudding (Cambridge, Mass.). Subspecialties: Neurochemistry; Pharmacology. Current work: Blood-brain barrier; vitamin and drug transport in brain; DNA synthesis in brain; treatment of poisoned patients. Patentee in field. Office: Univ Hosps/clinics Univ Iowa E331 B4 GH Iowa City IA 52242

SPEHLMANN, RAINER, neurology educator; b. Mitau, Latvia, Apr. 18, 1931; came to U.S., 1962, naturalized, 1966; s. Felix and Elisabeth (Flaschel) S.; m. Phyllis Mattei, Sept. 22, 1934; children: Marc, John, Ben. M.D., U. Heidelberg, 1957. Diplomate: Am. Bd. Neurology. Intern Mercer Hosp., Trenton, N.J., 1957-58; fellow in neurophysiology Inst. Clin. Neurophysiology, Freiburg, Ger., 1960-62; resident in neurology Mayo Clinic, Rochester, Minn., 1962-65; asst. chief neurol. service VA Lakeside Med. Ctr., Chgo., 1965—; asst. prof. neurology Northwestern U. Med. Sch., Chgo., 1965-68, assoc. prof., 1968-73, prof., 1973—, prof. pharmacology, 1974—. Author: EEG Primer, 1982; contbr. articles to profl. jours. NIH grantee, 1967-80; VA grantee, 1965—. Fellow Am. Acad. Neurology; mem. Am. EEG Soc., Am. Neurol. Assn., Am. Physiol. Soc., Am. Epilepsy Soc. Subspecialties: Neurology; Neurophysiology. Current work: Physiology and pharmacology of single nerve cells in the brain of animals prepared as models of human diseases (epilepsy); evoked potentials in normal human and neurological disorders. Office: VA Lakeside Med Center 333 E Huron St Chicago IL 60611

SPEIS, THEMIS P., government official; b. Pa., Nov. 17, 1930; s. Philip T. and Athena (Katinos) S.; m. Mary Kanacopoulos, Nov. 17, 1971. B.S. Ch.E., U. Pitts., 1956, M.S.M.E., 1966; Ph.D. in Nuclear Engring, U. Md., 1975. Engr. Blaw-Knox Chem. Co., Pitts., 1956-58; sr. engr. Westinghouse Electric Co., Pitts., 1958-67, Atomic Energy Commn., Washington, 1967-74, branch chief, 1974-81; asst. dir. NRC, Washington, 1981-83, dir. div. safety tech., 1983—. Contbr. articles in field to profl. jours. Mem. Am. Nuclear Soc., AAAS. Greek Orthodox. Subspecialties: Nuclear engineering; Nuclear power plant risk assessment.. Current work: Manage and direct activities of the NRC division responsible for the identification and solution of technical issues relating to nuclear power plant safety and the assessment of nuclear power plant risks using probabalistic risk assessment techniques and methods. Home: 14 Watchwater Way Rockville MD 20850 Office: US NRC Washington DC 20555

SPELLMAN, DONALD JEROME, consulting engineer; b. Galesburg, Ill., Sept. 9, 1941; s. Walter Anthony and Verla Mae (Lorance) S.; m. Sharon A. McMahan, Mar. 3, 1962; children: Mark Raymond, Lisa Mae. B.S. in E.E, Purdue U., 1967; M.S. in Ocean Engring, U. R.I., 1974. Enlisted in U.S. Navy, advanced through grades to lt. comdr., 1980; ops. Officer Submarine Group 8, Naples, Italy, 1978-80; ret., 1980; project mgr. Gas Cooled Reactor Assoc., La Jolla, Calif., 1980-82; sr. cons. engr. Advanced Sci. and Tech. Assocs., Solana Beach, Calif., 1982—. Contbr. in field. Vol. Pete Wilson Senatorial Campaign, San Diego, 1982. Recipient Navy Achievement medal Sec. Navy, 1980. Mem. Am. Nuclear Soc. (exec. com. 1982), San Diego Voice Energy. Republican. Episcopalian. Club: San Diego Yacht. Subspecialties: Nuclear engineering; Electronics. Current work: Completed two years research engineering on hi temperature gas cooled reactors, currently with consulting firm at San Onofre nuclear generating station. Home: 3751 Jennings St San Diego CA 92106 Office: Advanced Science and Technology Associates 337 S Cedros Ave Suite J Solana Beach CA 92075

SPELLMAN, MITCHELL WRIGHT, physician, dean, educator; b. Alexandria, La., Dec. 1, 1919; s. Frank Jackson and Altonette Beulah (Mitchell) S.; m. Billie Rita Rhodes, June 27, 1947; children: Frank A., Michael A., Mitchell A., Maria A., Melva A., Mark A., Manly A., Rita A. A.B. magna cum laude, Dillard U., 1940, LL.D. (hon.), 1983; M.D., Howard U., 1944; Ph.D. in Surgery (Commonwealth Fund fellow), U. Minn., Mpls., 1955; D.Sc. (hon.), Georgetown U., 1974, U. Fla., 1977. Intern Cleve. Met. Gen. Hosp., 1944-45, asst. resident in surgery, 1945-46, Howard U. and Freedmen's Hosp., Washington, 1946-47, chief resident in thoracic surgery, 1947-48, teaching asst. in physiology, 1948-49, chief resident in surgery, 1949-50, teaching asst. in surgery, 1950-51; asst. prof. surgery Howard U., 1954-56, assoc. prof., 1956-60, prof., 1960-68; dir. Howard surgery service at D.C. Gen. Hosp., 1961-68; fellow in surgery U. Minn., 1951-54; sr. resident in surgery U. Minn. Med. Sch. and Hosp., 1953-54; dean Charles R. Drew Postgrad. Med. Sch., Los Angeles, 1969-77, prof. surgery, 1969-78; asst. dean, prof. surgery Sch. Medicine, U. Calif. at Los Angeles, 1969-78; clin. prof. surgery Sch. Med., U. So. Calif., 1969-78; dean for Med. Services, prof. surgery Harvard Med. Sch., Boston, 1978—; exec. v.p. Harvard Med. Center, 1978—; dir. State Mut. Life Ins. Co. Am.; fellow Center for Advanced Study in Behavioral Scis.; vis. prof. Stanford, 1975-76; Bd. dirs. Kaiser Found. Hosps., Kaiser Found. Health Plan; Mem. D.C. Bd. Examiners in Medicine and Osteopathy, 1955-68; mem. Nat. Rev. Com. for Regional Med. Programs, 1968-70; mem. spl. med. adv. group, nat. surg. cons. VA, 1969-73; mem. Commn. for Study Accreditation of Selected Health Edni. Programs, 1970-72; chmn. adv. com. br. med. devices Nat. Heart and Lung Inst., 1972; Am. health del. to visit, People's Republic of China, 1973. Mem. editorial bd.: Jour. Medicine and Philosophy, 1977—; Contbr. articles on cardiovascular physiology and surgery, measurement of blood volume, and radiation biology to profl. jours. Past bd. dirs. Sun Valley Forum on Nat. Health; former trustee Occidental Coll.; former bd. overseers com. to visit univ. health service Harvard; former regent Georgetown U.; former vis. com. U. Mass. Med. Center. Markle scholar in med. scis., 1954-59; recipient Distinguished Alumnus award Dillard U., 1963; Distinguished Postgrad. Achievement award Howard U., 1974; Outstanding Achievement award U. Minn., 1979. Mem. AMA, Nat. Med. Assn. (William A. Sinkler Surgery award 1968), AAAS, AAUP, A.C.S., Soc. Univ. Surgeons, Am. Coll. Cardiology, Am. Surg. Assn., Inst. of Medicine of Nat. Acad. Scis. (chmn. program com. 1976-78, governing council 1971-78), NRC, Nat. Acad. Scis. Roman Catholic. Subspecialty: Surgery. Current work: Dean medical school, educator. Office: Office of Dean Harvard U Med Sch 25 Shattuck St Boston MA 02115

SPENCE, MARY ANNE, geneticist; b. Tulsa, Sept. 8, 1944; d. John and Mary (Queen) Campbell; m. Arthur George Spence, Sept. 2, 1972. B.A., Grinnell Coll., 1966; Ph.D., U. Hawaii, 1969. NIH postdoctoral fellow U. N.C., Chapel Hill, 1969-70; asst. prof. UCLA, 1970-75, assoc. prof., 1975-80, prof. dept. psychiatry and biomath., 1980—; cons. NIH, NIMH. Contbr. articles to profl. jours. Mem. Genetics Soc. Am., Am. Soc. Human Genetics, Genetics Behavior Soc., AAAS, Sigma Xi. Subspecialties: Genetics and genetic engineering (medicine); Epidemiology. Current work: Human genetics, inherited diseases, family studies, gene linkage; developed methodology and analyzed data on inherited diseases. Office: MR Unit NPI 760 Westwood Plaza Los Angeles CA 90024

SPENCER, DWAIN FRANK, electric power research institute administrator, electrical engineer; b. South Bend, Ind., Oct. 6, 1934; s. Frank E. and Florence M. (VandeWalle) S.; m. Mary Jane, Feb. 5, 1955; children: Deborah Anne, Darrin Andrew, Daniel Arthur. B.S. in Chem. Engring, U. Notre Dame, 1956; M.S. in Engring, Purdue U., West Lafayette, Ind., 1958; postgrad., UCLA, 1960-65. Design engr. Babcock & Wilcox AED, Lynchburg, Va., 1956-57; research engr., project mgr. Caltech-Jet Propulsion Lab., Pasadena, Calif., 1958-72; program mgr. IGPA, NSF, Washington, 1972-74; dir. advanced power systems Electric Power Research Inst., Palo Alto, Calif., 1974—; cons. Am. Technol. U., 1975-76. Contbr. over 60 tech. articles to profl. pubs. Mem. fossils energy adv. com. U.S. Dept. Energy; mem. adv. bd. Solar Energy Research Inst.; Mem. Com. Pasadena Tournament of Roses, 1972—. Mem. AAAS, Am. Inst. Chem. Engrs., Palo Alto C. of C., Sigma Xi, Tau Beta Pi. Republican. Roman Catholic. Club: Notre Dame Alumni (Los Angeles). Subspecialties: Energy Sciences Technology; Chemical engineering. Current work: Demonstration of advanced coal conversion and renewable resource technologies; maintain key role in bringing advanced modular electric power generation systems to commercial development. Home: 1046 Covington Rd Los Altos CA 94022 Office: 3412 Hillview Ave Palo Alto CA 94303

SPENCER, JAMES ALPHUS, plant pathologist; b. Clayton, Okla., Nov. 5, 1930; s. James E. and Lola Lea (Booth) S.; m.; children from previous marriage: James Timothy, Jeannie Alexander, Jay Barton. M.S., U. Ark., 1962; Ph.D., N.C. State U., 1966. Agrl. research technician U. Ark., 1957-62; agrl. research technician USDA, N.C. State U., Raleigh, 1962-66; plant pathologist Miss. State U., State College, 1967—. Contbr. articles to profl. jours. Served with AUS, 1953-55. Mem. Am. Phytopath. Soc., Am. Rose Soc., Gamma Sigma Delta. Mem. Ch. of Christ. Club: Kiwanis. Subspecialty: Plant pathology. Current work: Woody ornamental diseases; especially rose blackspot. Office: Drawer PG Mississippi State MS 29762

SPENCER, JOHN BROCKETT, chemistry educator, educational administrator; b. Horton, Kans., Sept. 25, 1939; s. John Sheldon and Ellen Louise (Brockett) S.; m. Barbara Kathleen McBride, July 11, 1964; children: Scott, Brett. B.A. magna cum laude, Carleton Coll., 1961; Ph.D., U. Calif.-Berkeley, 1965. Instr. Beloit Coll., 1965-66, asst. prof. chemistry, 1966-71, assoc. prof., 1971-78, prof., 1978—, chmn. dept. chemistry, 1979—; vis. prof. Case Western Res. U., 1967-68, Uppsala (Sweden) U., 1971-72, U. Calif.-Berkeley, 1979. Recipient Tchr. of Yr. award Beloit Coll., 1976; Woodrow Wilson fellow, 1961-62; NSF grad. fellow, 1961-64. Mem. Am. Chem. Soc., AAAS, AAUP, ACLU, Phi Beta Kappa, Sigma Xi. Subspecialties: Physical chemistry; Inorganic chemistry. Current work: Crystallographic and spectroscopic studies of transition metal, rare earth, and hydrogen-bonded materials. Office: Beloit Coll Beloit WI 53511

SPENCER, LORRAINE BARNEY, biologist, educator; b. Ogden, N.Y., Jan. 26, 1924; d. Elmer Cecil and Edna Justine (Zinter) Barney; m. Richard Earl Spencer, Sept. 12, 1942 (div.); children: Linda, Susan, Deborah, Nancy. B.S. (Dana fellow), Guilford Coll., 1966; M.A. (scholar), Wake Forest U., 1970, Ph.D., 1973. Instr. in biology lab. Wake Forest U., 1968-72, research fellow, 1972-73; adj. prof. Guilford Coll., 1974; asst. prof. biology St. Augustine's Coll., Raleigh, N.C., 1974—. Contbr. articles to Jour. Phycology. Mem. So. Appalachian Bot. Club, Phycological Soc. Am., Bot. Soc. Am., Assn. S.E. Biologists, Am. Inst. Biol. Scis., Sigma Xi. Subspecialties: Systematics; Taxonomy. Current work: Preparation monograph on genus zephyranthes north of S.Am. (with others). Home: 315 White Oak Dr Cary NC 27511 Office: Saint Augustine's Coll Raleigh NC 27611

SPENCER, MARY JOSEPHINE MASON, physician, cons., research, educator; b. Joliet, Ill., Oct. 19; d. Ray Miller and Marjorie Elizabeth (Tedens) Mason; m. Donald J. Spencer, June 13, 1960; children: Ken, Marjorie, Katherine, Christine. B.A., U. Colo., 1958; M.D., UCLA, 1964. Diplomate: Am. Bd. Pediatrics. Intern Los Angeles County Gen. Hosp., 1964-65; resident in pediatrics Harbor Gen. Hosp.-UCLA, Torrance, 1969-70; practice medicine specializing in pediatrics, Los Angeles, 1971-73; fellow in pediatric infectious diseases UCLA Sch. Medicine, 1973-75; asst. prof. pediatrics, 1975-82; chief div. ambulatory pediatrics, dir. Marion Davies Children's Clinic, UCLA, 1975-82. Contbr. articles to med. jours. Named Santa Monica Woman of Yr. Santa Monica YWCA, 1982. Mem. Am. Soc. for Microbiology, Am. Fedn. for Clin. Research, Am. Acad. Pediatrics, Am. Pub. Health Assn., Am. Fedn. Clin. Research, Western Soc. for Pediatric Research, Infectious Disease Soc. Am. Republican. Episcopalian. Subspecialties: Pediatrics; Infectious diseases. Current work: Parasitology; epidemiologic research on parasitology (protozoan). Home: 18675 Avenidas Cordillera San Diego Ca 92128 Office: 910 E Ohio St Escondido CA San Diego Children's Hosp 7500 Frost St San Diego CA

SPENCER, RANDALL S(COTT), geologist, academic administrator, educator; b. Milw., Sept. 29, 1937; m. Linda Joyce Horton, Sept. 3, 1966; children: Scott Taylor, Anne Cameron. B.S., U. Wis.-Madison, 1960; M.S., U. Kans.- Lawrence, 1962, Ph.D., 1968. Asst. prof. geophys. scis. Old Dominion U., 1966-70, assoc. prof., 1970-78, prof., 1978—, asst. chmm. dept. geophys. scis., 1974-77, chmn. dept., 1981—, assoc. dean, 1977-81. Mem. exec., com., trustee Sci. Museum Assn., Eastern Va., Norfolk, 198;—. Mem. Va. Acad. Sci. (exec. com. 1980-82), Paleontol. Soc. (v.p. S.E. 1982-83, pres. S.E. 1983-84), Sigma Xi, Phi Kappa Phi. Subspecialties: Paleoecology; Paleontology. Current work: Micropaleontology, foraminifera, paleoecology, biostratigraphy, Cenozoic, Holocene. Office: Dept Geophys Scis Old Dominion U Norfolk VA 23508

SPENCER, ROBERT FREDERICK, neuroanatomist; b. Jamaica, N.Y., Oct. 20, 1949; s. Lincoln Ryder and Roberta Florence (Johnson) S.; m. Patricia Caldwell Ducker, July 9, 1977. B.Sc., Boston U., 1971; Ph.D., U. Rochester Med. Sch., 1974. USPHS postdoctoral research fellow Inst. Neurol. Scis., U. Pa. Med. Sch., Phila., 1974-77; asst. prof. anatomy Med. Coll. Va., Richmond, 1977-82, assoc. prof., 1982—; vis. assoc. prof. Uniformed Services U. Health Scis. Med. Sch., Bethesda, Md., 1982—. Contbr. articles to profl. jours. Nat. Eye Inst. grantee, 1978—. Mem. Am. Assn. Anatomists, Soc. Neurosci., Assn. Research in Vision and Ophthalmology, N.Y. Acad. Scis., Internat. Brain Research Orgn., Richmond Engrs., Cajal Club, Sigma Xi. Methodist. Subspecialties: Anatomy and embryology; Neurobiology. Current work: Anatomy and physiology of oculomotor system; synaptic transmission; axoplasmic transport; neurotransmitters, excitatory and inhibitory synaptic mechanisms, innervation and structure of extraocular muscle. Home: 5203 Devonshire Rd Richmond VA 23225 Office: Dept Anatomy Med Coll VA PO Box 709 Richmond VA 23298

SPENCER, WILLIAM ALBERT, physician; b. Oklahoma City, Feb. 16, 1922; s. Arthur M. and Lillian (Wehremeyer) S.; m. Helen Margaret Hart, Mar. 29, 1945; children—William Albert, Susan Hart. B.S. cum laude, Georgetown U., 1942; M.D., Johns Hopkins, 1946. Diplomate: Am. Bd. Pediatrics. Intern. asst. resident pediatrics Harriet Lane Home, Johns Hopkins Hosp., 1946-48; instr. pediatrics, asst. prof. physiology Baylor U. Coll., Medicine, 1950—; med. dir. Southwestern Poliomyelitis Respiratory Center, Jefferson Davis Hosp., Houston, 1950-58; dir. regional respirator center Nat. Found. Infantile Paralysis; chmn. medical research study section Vocational Rehabilitation Adminstrn.; asst. prof., depts. physiology and pediatrics Baylor U. Coll. Medicine, 1950-57, prof., chmn. dept. of rehabilitation, 1957—; pres. Inst. Rehab. and Research, Tex. Med. Center, 1958—; cons. spl. med. adv. group VA; cons. Nat. Inst. for Handicapped Research, 1979—, dep. dir., acting dir., 1979; pres. Assn. Research and Tng. Centers. Author and editor: Treatment of Acute Poliomyelitis, 1954. Served as capt. M.C. AUS, 1948-50. Named One of Ten Outstanding Young Men of Am. U.S. Jr. C. of C., 1955; Physician of Year Pres.'s Com. Employment Handicapped, 1964. Fellow Am. Acad. Pediatrics; mem. Am., Tex med. assns., Harris County Med. Soc., Houston Pediatric Soc., Am. Congress Rehab. Medicine (chmn. legis. com.), Phi Beta Kappa, Sigma Xi, also, Phi Beta Pi, also mem., Alpha Omega Alpha. Episcopalian. Subspecialty: Physical medicine and rehabilitation. Home: 2836 Bellefontaine Houston TX 77025 Office: 1333 Moursund Ave Houston TX 77025

SPERELAKIS, NICK, physiology educator, researcher, consultant, editor; b. Joliet, Ill., Mar. 3, 1930; s. James and Arestea (Kayaidakis) S.; m. Dolores Martinis, Jan. 28, 1960; children: Nicholas, Mark Demetri, Christine Marie, Sophia Ann, Thomas Andreas, Anthony James. B.S., U. Ill., 1954, M.S., 1955, Ph.D., 1957. Assoc. prof. physiology Case Western Res. U., 1957-66; prof. U. Va., 1966—; cons. in field. Contbr. numerous articles to profl. publs.; Assoc. editor: Circulation Research Jour, 1970-75; co-editor: Handbook of Physiology on the Heart, 1979. Served as sgt. USMC; Korea. NIH grantee, 1958-83; NSF grantee, 1978; Muscular Dystrophy Assn. grantee, 1976-82; Am. Heart Assn. grantee, 1970-75; established investigator Am. Heart Assn., 1961-66. Mem. Am. Physiol. Soc., Am. Soc. Pharmacology and Exptl. Therapeutics, Biophys. Soc., Soc. Gen. Physiologists, Sierra Club, Greenpace. Greek Orthodox. Club: Ahepa. Subspecialties: Physiology (medicine); Cellular pharmacology. Current work: Electrophysiology of muscle and excitation-contraction coupling. Home: 1615 Cedar Hill Rd Charlottesville VA 22901 Office: U Va Sch Medicine Dept Physiology Charlottesville VA 22908

SPERGEL, PHILIP, technical executive; b. N.Y.C., Mar. 5, 1926; s. Julius and Frieda (Mann) S.; m. Fay Ruth Spergel, Dec. 19, 1948; children: Gale A., Joan. B.E.E., CCNY, 1948; M.E.E., NYU, 1951. Engr. Sperry Gyroscope Co., Great Neck, N.Y., 1948-54; chief engr. Indsl. Nucleonics, Inc., Columbus, Ohio, 1954-58, Epsco., Inc., Cambridge, Mass., 1958-61; tech. staff Mitre Corp., Bedford, Mass., 1962; dir. engring. Baird Atomic, Inc., Cambridge, 1962-67; v.p. corporate quality assurance Instrumentation Lab., Inc., Lexington, Mass., 1967—. Served with USN, 1944-46. Mem. IEEE, Am. Assn. Quality Control. Subspecialties: Bioinstrumentation; Quality assurance. Current work: Bioinstrumentation, biomedical application. Patentee in field. Home: 19 Holmes Rd Lexington MA 02173 Office: 113 Hartwell Ave Lexington MA 02173 Home: 19 Holmes Rd Lexington MA 02173

SPERRY, ROGER WOLCOTT, neurobiologist, educator; b. Hartford, Conn., Aug. 20, 1913; s. Francis B. and Florence (Kraemer) S.; m. Norma G. Deupree, Dec. 28, 1949; children: Glenn Tad, Janeth Hope. B.A., Oberlin Coll., 1935, M.A., 1937, D.Sc. (hon.), 1982; Ph.D., U. Chgo., 1941, D.Sc. (hon.), 1977, Cambridge U., 1972, Kenyon Coll., 1979, Rockefeller U., 1980. Research fellow Harvard and Yerkes Labs., 1941-46; asst. prof. anatomy U. Chgo., 1946-52, sect. chief Nat. Inst. Neurol. Diseases of NIH, also asso. prof. psychology, 1952-53; Hixon prof. psychobiology Calif. Inst. Tech., 1954—; research brain orgn. and neural mechanism. Contbr. articles to profl. jours., chpts. to books.; Editorial bd.: Behavioral Biology. Recipient Oberlin Coll. Alumni citation, 1954; Howard Crosby Warren medal Soc. Exptl. Psychologists, 1969; Calif. Scientist of Year award Calif. Mus. Sci. and Industry, 1972; award Passano Found., 1973; Albert Lasker Basic Med. Research award, 1979; co-recipient William Thomas Wakeman Research award Nat. Paraplegia Found., 1972, Claude Bernard sci. journalism award, 1975, Disting. research award Internat. Visual Literacy Assn., 1979; Wolf Found. prize in medicine, 1979; Nobel prize in physiology or medicine, 1981. Fellow AAAS, Am. Acad. Arts and Scis., Am. Psychol. Assn. (recipient Distinguished Sci. Contbn. award 1971); mem. Royal Acad. (fgn. mem.), Nat. Acad. Scis., Am. Physiol. Soc., Am. Assn. Anatomists, Internat. Brain Research Orgn., Soc. for Study of Devel. and Growth, Psychonomic Soc., Am. Soc. Naturalists, Am. Zool. Soc., Soc. Developmental Biology, Am. Philos. Soc. (Lashley prize 1976), Am. Neurol. Assn. (hon.), Soc. for Neurosci., Internat. Soc. Devel. Biologists, AAUP, Pontifical Acad. Scis., Sigma Xi. Subspecialty: Psychobiology. Home: 3625 Lombardy Rd Pasadena CA 91107 Office: 1201 E California St Pasadena CA 91125

SPESSARD, GARY OLIVER, chemistry educator, researcher; b. Orange, Calif., Sept. 27, 1944; s. Everett L. and Elaine (Vincent) S.; m.

Carol A. Hagen, Apr. 13, 1968; 1 dau., Sarah. B.S., Harvey Mudd Coll., 1966; M.S., U. Wis.-Madison, 1968; Ph.D., Wesleyan U., 1971. Chemist Union Oil Co. Calif., Brea, 1966; postdoctoral fellow U. Alta., Edmonton, 1970-72; vis. research assoc. Ohio State U., Columbus, 1972-73; asst. prof. St. Olaf Coll., Northfield, Minn., 1973-79; vis. assoc. prof. U. Utah, Salt Lake City, 1979-80; assoc. prof. chemistry dept. St. Olaf Coll., Northfield, Minn., 1979—; cons. Circuit Services, Inc., Mpls., 1981—. Chmn. Environ. Quality Commn. Northfield, Minn., 1980—. Research Corp. grantee, 1977-79, 83-84; NSF trainee, 1966-68. Mem. Am. Chem. Soc., Midwest Assn. Chemistry Tchrs. Liberal Arts Colls., Phi Lambda Upsilon, Pi Mu Epsilon. Lutheran. Subspecialties: Organic chemistry; Synthetic chemistry. Current work: Synthesis of enzyme substrate analogs, ring opening of strained cyclic ethers by use of organosilicon compounds, ergot alkaloid biosynthesis. Home: 703 Greenvale Ave Northfield MN 55057 Office: Chemistry Dept St Olaf Coll Northfield MN 55057

SPICER, WILLIAM EDWARD, III, educator, physicist; b. Baton Rouge, Sept. 7, 1929; s. William Edward II and Kate Crystal (Watkins) S.; m. Cynthia Stanley, June 12, 1951 (div. 1969); children: William Edward IV, Sally Ann; m. Diane Lubarsky, Apr. 24, 1969; 1 dau., Jacqueline Kate. B.S., Coll. William and Mary, 1949, MIT, 1951; M.A., U. Mo., 1953, Ph.D., 1955; D.Tech. (hon.), U. Linköping, Sweden, 1975. Scientist, RCA Labs., Lawrence Radiation Lab., U. Calif.-Livermore, 1961-62; mem. faculty Stanford U., 1962—, prof. elec. engring. and materials sci. engring., 1965—, prof. by courtesy applied physics, 1976—, Stanford Ascherman prof. engring., 1978—; dir. Acad. Skills, Inc., Los Altos, Calif., 1971-73; dep. dir. Stanford Synchrotron Radiation Lab., 1973-75, cons. dir., 1975—; cons. to govt. and industry, 1962—; mem. solid state scis. panel Nat. Acad. Sci.-NRC, 1965-73; cons., lectr. Chinese univ. devel. project World Bank-Fudan U., 1983; mem. panel 222.00 atomic and molecular physics div. Nat. Bur. Standards, 1966-73, chmn., 1971-73; mem. adv. group election devices Dept. Def., 1975-82; fellow Churchill Coll., Cambridge U., Eng., 1979. Editorial bd.: Jour. Crystal Growth, 1981—; author publs. theory and experiment solid state and surface physics and chemistry, photoemission, optical properties solids, electronic structure metals, semiconductors, insulators. Bd. dirs. Princeton (N.J.) YMCA, 1960-62. Recipient Achievement award RCA, 1957, 60; named Scientist of Yr., Indsl. Research and Devel. mag., 1981; Guggenheim fellow, 1978-79. Fellow IEEE, Am. Phys. Soc. (Oliver Buckley Solid State Physics prize 1980); mem. Am. Vacuum Soc. (chmn. electronics material div. 1978-79, dir. 1978-80, trustee 1981—), Phi Beta Kappa. Subspecialties: Condensed matter physics; Microchip technology (engineering). Current work: Fundamentals of surface and interface applications to microchip technology; applications of synchrotron radiation. Home: 785 Mayfield Rd Stanford CA 94305 Office: McCullough Bldg Stanford Univ Stanford CA 94305

SPIEGEL, STANLEY LAWRENCE, mathematics educator, researcher; b. N.Y.C., Oct. 27, 1935; s. Sidney Daniel and Gertrude (Milsky) S.; m. Margery Joan Spiegel, July 1, 1962 (div. July 1979); 1 dau., Stephanie Berit; m. Diana Lees, Aug. 13, 1972; children: David Solomon, Sarah Caren. B.S., NYU, 1957; A.M., Harvard U., 1959, Ph.D., 1966. Postdoctoral fellow meteorology dept. MIT, Cambridge, 1966-68; research assoc. math. dept. Northeastern U., Boston, 1969-73; sr. scientist EG&G Environ. Cons., Waltham, Mass., 1978-79, cons., 1978—; prof. math. U. Lowell, Mass., 1973—; cons. Tri-Con Assocs., Cambridge, Mass., 1980—. Contbr. articles to profl. jours. Mem. Town Meeting, Brookline, Mass., 1981—. Summer faculty research fellow; Air Force Office Sci. Research, 1981, 82; research grantee, 1982. Mem. Am. Geophys. Union, Am. Meteorol. Soc., N.Y. Acad. Scis., Sigma Xi, Pi Mu Epsilon. Jewish. Subspecialties: Satellite studies. Current work: Development and implementation of a computer algorithm to detect charging of spacecraft at geosynchronous orbit in real time, error analysis of a multi layer, spectral global numerical weather prediction model. Home: 39 Stetson St Brookline MA 02146 Office: Math Dept U Lowell 1 University Ave Lowell MA 01854

SPIEGEL, ZANE, consulting ground-water hydrologist, hydrology educator; b. Middletown, N.Y., Nov. 6, 1926; s. Nathan and Anna Rebecca (Mayer) S.; m. Maryanne Geissler, Dec. 19, 1959; children: Austin Gregory, Evan Nathaniel. Student, Lafayette Coll., 1944-46; B.S. in Gen. Sci, U. Chgo., 1949; M.S. in Geology, U. Chgo., 1952; Ph.D. in Earth Sci, N.Mex. Inst. Mining and Tech., 1962. Registered profl. engr., N.Mex., N.Y., Mass.,Ohio. Geologist U.S. Geol. Survey, Albuquerque, 1949-53; water resource engr. N.Mex. State Engr. Office, Sante Fe, 1954-58, 66-71; Fulbright Commn. lectr., Arequipa, Peru, 1958-59; postdoctoral fellow Harvard U., 1962-63; vis. prof. Imperial Coll. Tech., London, 1963-64; project mgr. UN Spl. Fund, San Juan, Argentina, 1964-66; prin. Zane Spiegel, 1971—; cons. hydrologist, Santa Fe, 1971—; vis. lectr. U. Minn.-Mpls., 1967-68, N.Mex. Inst. Mining and Tech., 1971; course coordinator Coll. Sante Fe, 1974-77; assoc. prof. Ohio State U., 1980-82; hydrology advisor City/County Planning Commn., Sante Fe, 1970-71; extramural reviewer of research proposals EPA, 1973-77. Author: Geology and Groundwater of Northeast Socorro County, New Mexico, 1958, Hydraulics of Certain Stream-Connected Aquifer Systems, 1962, (with B. Baldwin) Geology and Water of Sante Fe Area, New Mexico, 1963; hydrology advisor to: film When the Rivers Run Dry, 1979. Served with AUS, 1945-46. Lafayette Coll. scholar, 1944-46; NSF coop. fellow, 1960-61. Fellow Geol. Soc. Am.; mem. Am. Geophys. Union, Nat. Water Well Assn., Assn. Geologists for Internat. Devel., ASCE. Republican. Unitarian. Subspecialties: Ground water hydrology; Civil engineering. Current work: Rehabilitation of brine-contaminated aquifers; diversion of stream flow by wells; aquifer-stream modelling and management; waste spray irrigation; equilibrium theory. Home and Office: Zane Spiegel PO Box 1541 Sante Fe NM 87501

SPIELVOGEL, LAWRENCE GEORGE, consulting engineer; b. Newark, N.J., June 2, 1938; s. Joseph and Fanny (Ravitz) S. B.S. in Mech. Engring, Drexel U., 1962. Registered profl. engr. in 49 states. Asst. post engr. Walter Reed Army Med. Ctr., 1963-65; engr. Utility Survey Corp., Chgo., 1965-66; assoc. Robert G. Werden & Assocs., Inc., Jenkintown, Pa., 1959-70; pres. Lawrence G. Spielvogel, Inc., Wyncote, Pa., 1970—; vis. lectr. Yale U., 1975-81. Author: Energy Management Handbook, 1982. Served to 1st lt. U.S. Army, 1963-65. Mem. ASHRAE (Disting. Service award 1975, award of merit 1976, best jour. award 1980, 81, Crosby Field award 1981), Illuminating Engring. Soc., Am. Cons. Engrs. Council, ASME, Internat. Solar Energy Soc., Chartered Instn. Bldg. Services (Eng.), Bldg. Services Research and Info. Assn. (Eng.). Subspecialties: Mechanical engineering. Current work: Energy use in buildings. Address: Wyncote House Wyncote PA 19095

SPIES, HAROLD GLEN, research institution administrator, biomedical researcher; b. Mountain View, Okla., Mar. 30, 1934; s. Chester Charles and Tessie Elisabeth (Hawkins) S.; m. Emma Jean Burkes, Dec. 22, 1955 (div. 1963); children: Russell Lee, Terry Wayne; m. Peggy Diane Wilp, Aug. 6, 1967 (div. 1980). B.S. in Animal Scis, Okla. State U., 1956; M.S. in Genetics, U. Wis.-Madison, 1957; Ph.D. in Physiology, U. Wis.-Madison, 1959. Research asst. U. Wis.-Madison, 1956-59; asst. prof. animal physiology Kans. State U., Manhattan, 1959-64, assoc. prof., 1964-66; NIH spl. fellow Brain Research Inst., UCLA, 1966-67, research assoc. anatomist, 1967-68; research scientist, assoc. prof. Delta Regional Primate Research Ctr. and Tulane U., New Orleans, 1968-72; prof. anatomy Oreg. Health Scis. U., Portland, 1973—; sr. scientist, chmn. reproductive physiology Oreg. Regional Primate Research Ctr., Beaverton, 1972—, interim dir., 1982, assoc. dir. research, 1983—, chmn., 1982—; invited lectr. various univs. and profl. confs.; cons. WHO, Geneva, 1977-79, G.D. Searle and Co., 1974; mem. various NIH study sects. Editorial bd.: Biology of Reproduction, 1974-78, Endocrinology, 1975-79; reviewer profl. jours.; contbr. chpts. to books, articles to publs. Mem. Am. Physiol. Soc., Neurosci. Soc., Internat. Neuroendocrine Soc., Endocrine Soc., Soc. Study of Reproduction (dir. 1978-79, pres. 1980-81), Internat. Soc. Biochem. Endocrinology (research council 1978—), Am. Assn. Anatomists, Soc. Exptl. Biology and Medicine, Am. Soc. Animal Sci., Sigma Xi, Phi Kappa Phi, Phi Eta Sigma, Alpha Zeta, Gamma Sigma Delta. Democrat. Subspecialties: Reproductive biology; Neuroendocrinology. Current work: Biomedical research in primates; central nervous system function and the endocrine system of primates. Office: Oreg Regional Primate Research Center 505 NW 185th Ave Beaverton OR 97006

SPIESS, ELIOT BRUCE, biological sciences educator, geneticist; b. Boston, Oct. 13, 1921; s. George N. and Rena (Bunce) S.; m. Luretta F. Davis, June 23, 1951; children: Arthur E., Bruce D. A.B., Harvard U., 1943, M.A., 1947, Ph.D., 1949. Asst. prof. U. Pitts., 1952-56, assoc. prof., 1956-65, prof., 1965-66; prof. biol. scis. U. Ill., Chgo., 1966—. Author: Genes in Populations, 1977. NSF grantee, 1972—. Fellow AAAS; mem. Am. Soc. Naturalists (pres. 1981), Genetics Soc. Am., Behavior Genetics Assn., Soc. Study Evolution (editor Evolution 1975-78). Subspecialty: Evolutionary biology. Current work: Genotype control over Darwinian fitness properties in populations, particularaly in Drosophila. Mating activity control by chromosomal and genic polymorphs with resulting sexual selection is of principal interest. Home: 1153 Asbury Ave Winnetka IL 60093 Office: Univ Ill Biol Scis Dept Chicago IL 60680

SPIESS, FRED NOEL, oceanographer, educator; b. Oakland, Calif., Dec. 25, 1919; s. Fred Henry and Elva Josephine (Monck) S.; m. Sarah Scott Whitton, July 25, 1942; children: Katherine Spiess Dallaire, Mary Elizabeth Spiess DeJong, John Frederick, Helen Spiess Shamble, Margaret Josephine. A.B., U. Calif., Berkeley, 1941, Ph.D., 1951; M.S., Harvard U., 1947. With Marine Phys. Lab., U. Calif., San Diego, 1952—, dir., 1958-80, U. Calif. Inst. Marine Resources, 1980—, Scripps Inst. Oceanography, La Jolla, 1964-65, prof. oceanography, 1961—; Mem. Naval Research Adv. Commn., 1978—; mem. com. on geodesy Nat. Acad. Scis., 1980—. Served with USNR, 1941-46, 53. Decorated Silver Star medal, Bronze Star medal; recipient John Prince Wetherill medal Franklin Inst., 1965; Compass Disting. Scientist award Marine Technol. Soc., 1971; Robert Dexter Conrad award U.S. Sec. of Navy, 1974; Newcomb Cleveland prize AAAS, 1981. Fellow Acoustical Soc. Am.; mem. Am. Geophys. Union (Maurice Ewing award 1983), Marine Tech. Soc., Phi Beta Kappa, Sigma Xi. Subspecialties: Oceanography; Ocean engineering. Current work: Studies of the sea floor; development of seafloor work technology. Home: 9450 La Jolla Shore Dr La Jolla CA 92037 Office: Scripps Inst Oceanography U Calif San Diego La Jolla CA 92093

SPIESS, LURETTA D(AVIS), biologist, educator; b. Chgo., Feb. 3, 1927; d. Arthur G. and Luretta E. (Lindefield) Davis; m. Eliot B., June 23, 1951; children: Arthur Eliot, Bruce Davis. A.B. cum laude, Radcliffe Coll., 1949; Ph.D., Harvard U., 1953. Instr. Northwestern U., 1968-70, sr. research asso. in biology, 1978—, lectr. in biology, 1978—; asst. prof. Lake Forest Coll., 1970-78. Contbr. numerous articles to profl. jours. NSF grantee, 1979, 82. Mem. Bot. Soc. Am., Am. Soc. Plant Physiologists, Bryological Soc. Am. Subspecialties: Developmental biology; Plant growth. Current work: Physiology and devel. of moss; role of bacterial assns. Home: 1153 Asbury Ave Winnetka IL 60093 Office: Northwestern U Hogan Hall Evanston IL 60201

SPILBERG, ISAIAS, rheumatologist, medical educator; b. Trujillo, Peru, July 18, 1936; s. Salik and Rosa (Kolker) S.; m. Fradi Goldstein, June 14, 1962; 1 son, Mark. M.D., U. San Marcos Sch. Medicine, Lima, Peru, 1963. Intern U. Louisville Med. Sch., 1963-64, resident in medicine, 1964-66; fellow in rheumatology N.Y.U. Sch. Medicine, 1966-68; instr. Washington U. Sch. Medicine, St. Louis, 1969-70, asst. prof., 1970-77, assoc. prof., 1978—; head rheumatology sect. St. Louis City Hosp., 1970-77; dir. rheumatology VA Med. Ctr., St. Louis, 1977—; mem. grants com. Arthritis Found. Contbr. articles to profl. jours. NIH grantee; Arthritis Found. grantee, 1975—. Mem. Am. Rheumatism Assn., Am. Fedn. Clin. Research, Central Soc. Clin. Research, N.Y. Acad. Scis., Am. Soc. Clin. Investigation. Subspecialties: Immunology (medicine); Cell biology (medicine). Current work: Research in arthritis, acute inflammation chemotaxis. Home: 8144 Pershing Clayton MO 63105 Office: 15 N Grand Ave 151Jc St Louis Mo 63125

SPILKER, BERT, pharmacologist, researcher, educator, physician; b. Washington, July 3, 1941; s. Victor and Sara (Robbins) S.; m. Arlene Titow, July 27, 1967; children: Adam, Karen. Ph.D. in Pharmacology, SUNY Downstate Med. Center, Bklyn., 1967; M.D., U. Miami, 1977. Resident in internal medicine Brown U. Med. Sch., Providence, 1977-78; with Pfizer Ltd., Kent, Eng., 1969-70, Phillips-Duphar B.V., Weesp, Netherlands, 1970-72, Sterling-Winthrop Research Inst., Rensselaer, N.Y., 1972-75, JRB Assos., Inc., McLean, Va., 1978-79; pvt. practice medicine specializing in internal medicine, Reston, Va., 1978-79; sr. clin. research scientist Burroughs-Wellcome Co., Research Triangle Park, N.C., 1979—; clin. asst. prof. medicine, adj. asso. prof. pharmacology, researcher U. N.C. Med. Sch., Chapel Hill, 1979—. Contbr. articles to profl. jours. Bd. dirs. Common Cause. Mem. Am. Soc. Pharmacology and Exptl. Therapeutics, Am. Epilepsy Soc., Am. Soc. Clin. Pharmacology and Therapeutics, Am. Pain Soc. Subspecialties: Internal medicine; Pharmacology. Current work: Clinical pharmacology and development of new drugs in various areas, especially neurology. Home: 2556 Booker Creek Rd Chapel Hill NC 27514 Office: 3030 Cornwallis Rd Research Triangle Park NC 27709

SPILLER, EBERHARD, physicist; b. Halbendorf, Germany, Apr. 16, 1933; s. Walter and Ruth S.; m. Marga Dietz, Dec. 18, 1964; children: Michael, Bettina. Diploma in Physics, U. Frankfurt, 1960, Ph.D., 1964. Faculty U. Frankfurt, Ger., 1959-68; research staff IBM T. J. Watson Research Center, Yorktown Heights, N.Y., 1968—. Contbr. chpts. to books, articles to sci. jours.; editor: High Resolution X-Ray Optic, 1981. Fellow Optical Soc. Am.; mem. German Phys. Soc., AAAS. Subspecialty: Optics of X-rays. Current work: Thin film coatings, X-ray microscopy, X-ray astronomy. Office: IBM Research Center PO Box 218 Yorktown Heights NY 10598

SPILLER, GENE ALAN, nutritionist, clinical human nutrition research consultant; b. Milan, Italy, Feb. 19, 1927; came to U.S., 1950, naturalized, 1962; s. Silvio and Beatrice (Galli) S. D.Chemistry, U. Milan, 1949; M.S., U. Calif.-Berkeley, 1968, Ph.D. in Nutrition, 1972. Cons. nutrition research and edn., Los Angeles, 1952-65; research chemist U. Calif., Berkeley, 1966-67, assoc. specialist physiology dept., 1968-72; prin. scientist, head nutritional physiology Syntex Research, Palo Alto, Calif., 1972-80; cons. clin. nutrition research, Los Altos, Calif., 1981—; lectr. Mills Coll., Oakland, Calif., 1971-81, Foothill Coll., Los Altos, 1974—. Editor: Fiber in Human Nutrition, 1976, Topics in Dietary Fiber, 1978, Medical Aspects of Dietary Fiber, 1980, Nutritional Pharmacology, 1981; reviewer papers: Am. Jour. Clin. Nutrition, 1976-83. Mem. Am. Inst. Nutrition, Am. Soc. Clin. Nutrition, Brit. Nutrition Soc., Am. Assn. Cereal Chemists. Club: Alpine Hills. Subspecialties: Nutrition (medicine); Physiology (medicine). Current work: Research in human nutrition; principal investigator in human nutrition studies; dietary fiber and carbohydrates effect on human health; role of lesser known food components in nutrition; non-human primates as models for human nutrition. Office: PO Box 123 Los Altos CA 94022

SPINDEL, WILLIAM, scientist, administrator; b. N.Y.C., Sept. 9, 1922; s. Joseph and Esther (Goldstein) S.; m. Louise Phyllis Hoodenpyl, July 30, 1967; children: Robert Andrew, Lawrence Marshall. B.A., Bklyn. Coll., 1944; M.A., Columbia U., 1947, Ph.D., 1950. Jr. scientist Los Alamos Lab, Manhattan Dist., 1944-45; instr. Poly. Inst., Bklyn., 1949-50; asso. prof. State U. N.Y., 1950-54; research asso., vis. prof. Columbia, 1954-57, vis. prof., sr. lectr., 1962-74; asso. prof. then prof. Rutgers U., 1957-64; prof., chmn. dept. chemistry Belfer Grad. Sch. Sci., Yeshiva U., 1964-74; exec. sec., office chemistry and chem. tech. Nat. Acad. Scis.-NRC, 1974—, also exec. sec. bd. on chem. scis. and tech., 1974—; vis. Am. scientist, Yugoslavia, 1971-72. Contbr. articles to profl. jours. Served with AUS, 1943-46. Guggenheim fellow, 1961-62; Fulbright Research scholar, 1961-62. Fellow AAAS; mem. Am. Chem. Soc., Am. Phys. Soc. Club: Cosmos. Subspecialties: Physical chemistry; Science policy, chemical science administration. Current work: Science policy studies in the chemical sciences; needs and scientific opportunities in chemical sciences, and potential of chemical sciences for solution of social needs. Patentee in field. Home: 6503 Dearborn Dr Falls Church VA 22044 Office: Nat Acad Scis 2101 Constitution Ave Washington DC 20418

SPIRO, THOMAS GEORGE, chemistry educator; b. Aruba, Netherlands Antilles, Nov. 7, 1935; s. Andor and Ilona S.; m. Helen Hendin, Aug. 21, 1959; children—Peter, Michael. B.S., UCLA, 1956; Ph.D., M.I.T., 1960. Fulbright researcher U. Copenhagen, Denmark, 1960-61; NIH fellow Royal Inst. Tech., Stockholm, 1962-63; research chemist Calif. Research Corp., LaHabra, 1961-62; mem. faculty Princeton U., 1963—, prof. chemistry, 1974—, head dept., 1979—, Eugene Higgins prof., 1981—; NATO sr. fellow, 1972. Author: (with William M. Stigliani) Environmental Issues in Chemical Perspective, 1980, Environmental Science in Perspective, 1980; contbr. articles to profl. jours. Mem. Am. Chem. Soc., AAAS, Phi Beta Kappa, Sigma Xi. Subspecialties: Biophysical chemistry; Inorganic chemistry. Current work: Biomolecular structure, metalloproteins studied by Raman and X-ray absorption spectroscopy; photoelectrochemistry. Office: Dept Chemistry Princeton Univ Princeton NJ 08544

SPISAK, JOHN FRANCIS, metallurgist, minerals company executive; b. Cleve., Mar. 27, 1950; s. Ernest L. and Adele M. (Chipko) S.; m. Barbara Ann Heisman, June 10, 1972; 1 son, John Stefan. B.S. in Chemistry and Biology with honors, Purdue U., 1972. Research engr. Anaconda Co., Tucson, 1972-79; chief metallurgist Fed. Am. Uranium, Riverton, Wyo., 1979-80, Anschutz Mining Corp., Denver, 1980—; lectr. Colo. Sch. Mines, Golden, 1982—. Author: Metallurgical Effluents-Growing Challenges for Second Generation Treatment, 1979, Solvent Extraction of Copper from Smelter Dust Treatment Liquirs with ACORGA P-5100, 1981, Recovery of Cobalt, Nickel and Copper from the Madison Mine, 1983. Recipient Presdl. Service award Purdue U., 1972. Mem. AIME, Am. Inst. Biol. Scis., Am. Chem. Soc., Purdue U. All-Am. Club. Republican. Roman Catholic. Lodge: Purdue Elks. Subspecialties: Metallurgical engineering; Integrated systems modelling and engineering. Current work: Metallurgical process design and development; bioprocessing of heavy metals; process design management; financial analyses; biometallurgical process development; biotechnology company formation. Home: 8224 Everett St Arvada CO 80005 Office 555 17th St Suite 2400 Denver CO 80202

SPITAL, ROBIN DAVID, physicist; b. N.Y.C., Oct. 29, 1948; s. Max and Elaine (Steinberg) S. A.B., Harvard U., 1969; M.S., Cornell U., 1972, Ph.D., 1974. Asst. prof. physics Ill. State U., Normal, 1974-75; cons. scientist Pfizer Med. Systems, Columbia, MD., 1976-81; prin. devel. engr. AAI Corp., Cockeysville, Md., 1981—. Contbr. articles to profl. jours. Mem. Am. Phys. Soc., Am. Assn. Physicists in Medicine, Phi Beta Kappa. Club: Marshall Chess (N.Y.C.). Subspecialties: Theoretical physics; Software engineering. Current work: Researcher in computer simulation of tactical environment for military training, medical and theoretical physics. Office: AAI Corp PO Box 6767 Baltimore MD 21204

SPITLER, LYNN E(LLEN), monoclonal antibody company executive, researcher, physician; b. Grand Rapids, Mich., sept. 28, 1938; d. Orrie Jay and Cornelia Ellen Dykman; m. Harley J. Spitler, Nov. 17, 1967; children: Diane, Paul. M.D., U. Mich., 1963. Diplomate: Am. Bd. Internal Medicine, Am. Bd. Allergy and Immunology. Intern Highland-Alameda County Hosp., Oakland, Calif., 1963-64; resident U. Calif. Med. Ctr., San Francisco, 1964-66; instr. in medicine, 1970-71, asst. prof. medicine, 1971—; research assoc. U. Calif.-San Francisco Cancer Research Inst., 1971-78; dir. research Children's Hosp., San Francisco, 1978-81; sr. v.p. XOMA Corp. at Children's Hosp., San Francisco, 1981—; cons. NIH, Bethesda, Md., 1970—; mem. alergy and immunology research com. Nat. Inst. Allergy and Infectious Diseases, NIH, USPHS, Bethesda, 1976-80; mem. merit rev. bd. VA, Washington, 1976-80. Contbr. numerous articles to profl. jours.; editorial bd.: Jour. Immunology, 1975-78, Internat. Jour. Immunopharmacology, 1979—; manuscript reviewer med. jours. Recipient Research CAreer Devel. award NIH, 1971-76; Am. Cancer Soc. Calif. Div. Dernham sr. fellow, 1969-71. Mem. Am. Assn. Immunologists, AAAS, Western Soc. Clin. Research, Am. Fedn. Clin. Research. Subspecialties: Internal medicine; Immunopharmacology. Current work: Development of monoclonal antibodies for therapeutic use in patients; development of improved means for diagnosis and therapy of melanoma. Office: XOMA Corp Children's Hosp 3700 California St San Francisco CA 94118

SPITSBERG, VITALY LEV, biochemistry educator; b. Kazatin, USSR, July 27, 1938; came to U.S., 1975, naturalized, 1981; s. Lev Israel and Anna Efim (Rubchinsky) S.; m. Margarita Bakaeva, Sept., 1960 (div. 1970); 1 son, Vladimir; m. Natalia Krasnova, Nov. 19, 1971; 1 son, Andrey. M.D., Moscow First Med. Inst., 1961; Ph.D., Inst. Biophysics, Moscow, 1966. Jr. scientist Inst. Med. Chemistry, Moscow, 1961-62; sr. scientist Inst. Molecular Biology, Moscow, 1966-71, Inst. Devel. Biology, 1971-74; asst. research prof. St. Louis U., 1981-82; vis. research prof. U. Nebr.-Lincoln, 1982—; project assoc. Northwestern U. Med. Sch., 1975, Inst. Enzyme Research, U. Wis.-Madison, 1976, research assoc. dept. pharmacology, 1977-78; research assoc. Syracuse U., 1978-79; sr. research assoc. St. Louis U., 1979-81. NSF grantee, 1982. Mem. Am. Soc. Biol. Chemists, N.Y. Acad. Scis. Subspecialties: Biochemistry (biology); Biochemical evolution. Current work: Oxidative phosphorylation, photosynthesis, biochemical evolution, cell biology. Home: 1205 S 20th St Lincoln NE 68502 Office: U Nebr-Lincoln 705 Hamilton Hall Lincoln NE 68588

SPITZER, ADRIAN, pediatrics educator, pediatrician; b. Bucharest, Rumania, Dec. 21, 1927; came to U.S., 1963, naturalized, 1968; s.

Osias and Sophia S.; m. Carole Zelter, Oct. 30, 1951; 1 son, Vlad. B.S., Matei Basarab Lyceum, Bucharest, 1946; M.D., Med. Sch. Bucharest, 1952. Intern White Plains (N.Y.) Hosp., 1964-65; resident in pediatrics Med. Coll. Pa., Phila., 1965-66; asst. prof. pediatrics Albert Einstein Coll. Medicine, 1968-72, assoc. prof., 1972-76, prof., dir. div. nephrology, 1976—; mem. gen. medicine B study sect. NIH, 1976-80; mem. sub-bd. in pediatric nephrology Am. Bd. Pediatrics, 1977-83, chmn., 1981-82; mem. com. health and sci. affairs Nat. Kidney Found., 1978-80, mem. sci. adv. bd., 1982. Assoc. editor: Pediatric Kidney Disease, 1980; editor: The Kidney During Development, 1982. NIH spl fellow, 1968; NIH Fogarty sr. internat. fellow, 1982; NIH grantee, 1968—; Health Research Council N.Y. grantee, 1978—; Nat. Kidney Found. grantee, 1970—. Fellow Am. Acad. Pediatrics; mem. Am. Physiol. Soc., Am. Soc. Nephrology, Am. Soc. Pediatric Nephrology (council 1976-83, pres. 1981-82), Soc. Pediatric Research, Am. Pediatric Soc. Subspecialties: Pediatrics; Pediatric nephrology. Current work: Developmental renal physiology; kidney disease in children. Home: 27 Sycamore Ln Irvington NY 10533 Office: Albert Einstein Coll Medicine 1410 Pelhm Pkwy S Bronx NY 10461

SPITZER, FRANK LUDWIG, mathematician, educator; b. Vienna, Austria, July 24, 1926; came to U.S., 1945, naturalized, 1951; s. Gustav and Margaret (Herzog) S.; m. Jean Wallach, Aug. 11, 1951; children—Karen, Timothy; m. Ingeborg Wald; 1975. Ph.D. in Math, U. Mich., 1952. Instr. Cal. Inst. Tech., 1953-58; asso. prof. U. Minn., 1959-60; NSF fellow Princeton, 1960-61; prof. Cornell U., Ithaca, N.Y., 1961—. Author: Principles of Random Walk, 1964. Guggenheim fellow, 1965-66. Mem. Nat. Acad. Scis. Subspecialty: Probability. Current work: Stochastic processes. Home: 114 Cascadilla Park Ithaca NY 14850

SPITZER, LYMAN, JR., astronomer; b. Toledo, June 26, 1914; s. Lyman and Blanche C. (Brumback) S.; m. Doreen D. Canaday, June 29, 1940; children: Nicholas, Dionis, Lutetia, Lydia. A.B., Yale U., 1935, D.Sc., 1958; postgrad., Cambridge (Eng.) U., 1935-36; Ph.D., Princeton U., 1938; D.Sc., Case Inst. Tech., 1961, Harvard U., 1975; LL.D., Toledo U., 1963. Nat. research fellow Harvard U., 1938-39; instr. physics and astronomy Yale U., 1939-42; scientist spl. studies group div. war research Columbia U., 1942-44, dir. sonar analysis group, 1944-46; assoc. prof. astrophysics Yale U., 1946-47; prof. astronomy, chmn. dept. and dir. obs. Princeton U., 1947-79, Charles A. Young prof. astronomy, 1952-82, chmn. research bd., 1967-72, dir., 1953-61; chmn. exec. com. (Plasma Physics Lab.), 1961-66; prin. investigator Princeton telescope on Copernicus satellite; trustee Woods Hole Oceanographic Inst., 1946-51; mem. com. on undersea warfare NRC, 1948-51; mem. Yale U. Council, 1948-51; chmn. Scientists Com. on Loyalty Problems, 1948-51. Author: Physics of Fully Ionized Gases, 1956, 62, Physical Processes in the Interstellar Medium, 1978, Searching between the Stars, 1982; editor: Physics of Sound in the Sea, 1946; contbr. articles to profl. jours. Recipient Rittenhouse medal, 1957; Exceptional Sci. Achievement medal NASA, 1972; Bruce Gold medal, 1973; Henry Draper gold medal, 1974; James C. Maxwell prize, 1975; Disting. Pub. Service medal NASA, 1976; Nat. Medal of Sci., 1980; Janssen medal, 1980; Franklin medal, 1980. Fellow Am. Phys. Soc.; mem. Nat. Acad. Sci., Am. Acad. Arts and Scis., Am. Philos. Soc., Am. Astron. Soc. (past pres.), Royal Astron. Soc. (assoc., gold medal 1978), Am. Geophys. Union, Astron. Soc. Pacific, Royal Soc. Scis. (Liege) (fgn. corr. mem.). Unitarian. Club: Am. Alpine. Subspecialties: Optical astronomy; Theoretical astrophysics. Current work: Chairman of AURA's Space Telescope Institute Council, with administrative oversight of Space Telescope Science Institute; member of various committees for Space Telescope. Home: 659 Lake Dr Princeton NJ 08540

SPITZER, NICHOLAS CANADAY, biology educator; b. N.Y.C., Nov. 8, 1942; s. Lyman, Jr. and Doreen D. (Canaday) S.; m. Janet Edith Lamborghini, Aug. 26, 1967; 1 son, Julian Elliott. B.A., Harvard U., 1964, Ph.D., 1969. Asst. prof. biology U. Calif.-San Diego, LaJolla, 1973-77, assoc. prof., 1977-82, prof., 1982—. Editor: Neuronal Development, 1982. Subspecialties: Neurobiology; Developmental biology. Current work: Developmental neurobiology. Home: 2526 Lozana Rd Del Mar CA 92014 Office: Dept Biology B 022 U Calif San Diego La Jolla CA 92093

SPITZER, ROGER EARL, physician, medical educator; b. Washington, June 20, 1935; s. Ronald Heller and Mildred Edith (Jaffee) S.; m. Rosalie J. Gutride, June 10, 1962; children: Scott, Neal, Amy. B.S., George Washington U., 1958; M.D., Howard U., 1962. Diplomate: Am. Bd. Pediatrics. Instr. chemistry George Washington U., Washington, 1956-58; asst. prof. pediatrics U. Cin., 1969-73; assoc. prof. pediatrics SUNY Upstate Med. Ctr., Syracuse, 1973-77, prof., 1977—, dir. pediatric nephrology, 1978—; adv. Central N.Y. Kidney Disease Soc. Contbr. articles to profl. jours. NIH fellow; USPHS grantee; Am. Cancer Soc. grantee. Mem. Am. Soc. Nephrology, Soc. Pediatric Research, Am. Pediatric Soc., Am. Assn. Immunologists. Democrat. Jewish. Subspecialties: Nephrology; Immunology (medicine). Current work: Glomerulonephritis and the complement system; pediatric kidney disease, mechanisms of complement activity and the role of complement in leukemia. Home: 104 E Genesee Pkwy Syracuse NY 13214 Office: 750 E Adams St Syracuse NY 13210

SPIVEY, HOWARD OLIN, biochemistry educator; b. Gainesville, Fla., Dec. 10, 1931; s. Herman Everette and Havens Edna (Taylor) S.; m. Dorothy Eleanor Luke, July 19, 1959; children: Bruce Allen, Curt Olin, Diane Elizabeth. B.S., U. Ky., 1954; Ph.D., Harvard U., 1962. Research assoc. Rockefeller U., N.Y.C., 1962-64; NIH fellow MIT, Cambridge, 1964-65; asst. prof. chemistry U. Md., College Park, 1965-67; asst. prof. Okla. State U., Stillwater, 1967-69, assoc. prof., 1969-75, prof. biochemistry, 1975—. NSF grantee, 1969-74; NIH grantee, 1969-72, 75-78, 79-83. Mem. Common Cause, Am. Chem. Soc., Am. Soc. Biol. Chemists, Sigma Xi, Phi Beta Kappa, Phi Kappa Phi, Phi Lambda Upsilon. Democrat. Presbyterian. Subspecialties: Biochemistry (biology); Biophysical chemistry. Current work: Evaluate extents and effects of enzyme associations in vivo. Home: 2222 W 11th St Stillwater OK 74078 Office: Biochemistry Dept Okla State U Stillwater OK 74078

SPLINTER, WILLIAM ELDON, agricultural engineer; b. North Platte, Nebr., Nov. 24, 1925; s. William John and Minnie (Calhoun) S.; m. Eleanor Love Peterson, Jan. 5, 1953; children: Kathryn Love, William John, Karen Ann, Robert Marvin. B.Sc., U. Nebr., 1950; M.Sc., Mich. State U., 1951, Ph.D., 1955. Instr. agrl. engring. Mich. State U., 1953-54; asso. prof. biology and agrl. engring. N.C. State U., 1954-61, prof., 1961-68; prof., chmn. dept. agrl. engring. U. Nebr., 1968—; cons. engr. Mem. exec. bd.; Am. Assn. Engring. Socs. Contbr. articles tech. jours. Served with USNR, 1946-51. Recipient Massey Ferguson Ednl. award; named to Nebr. Hall of Agrl. Achievement. Fellow Am. Soc. Agrl. Engrs. (pres., adminstrv. council), AAAS; mem. Soc. Automotive Engrs., Am. Soc. Engring. Edn., Nat. Soc. Profl. Engrs., Sigma Xi, Sigma Tau, Sigma Pi Sigma, Pi Mu Epsilon, Gamma Sigma Delta, Phi Kappa Phi, Beta Sigma Psi. Subspecialty: Agricultural engineering. Current work: Measurement and computer modeling of plant growth. Patentee in field. Home: 7105 N Hampton St Lincoln NE 68520

SPOELHOF, CHARLES P., photographic company executive; b. Hackensack, N.J., Aug. 6, 1930; s. Charles and Elizabeth (Keegstra) S.; m. Kay, June 11, 1953; children: Beth, Philip, Gordon, Ronald. Student in engring, Calvin Coll., 1948-51; B.S. in Engring. Physics, U. Mich., 1953, U. Mich., 1953; M.S. in Physics, U. Mich., 1954; postgrad. in optics, U. Rochester, 1954-55, in advanced engring. studies, MIT, 1965-66. With Apparatus Div., Eastman Kodak Co., Rochester, N.Y., 1954—, dir. research and engring., 1973-75, mgr. bus. and profl. products, 1975-82; v.p., asst. gen. mgr. Kodak Apparatus Div., 1982—. Bd. dirs. N.Y. State Epilepsy Assn. Recipient Apollo Achievement award NASA, 1970. Mem. Nat. Acad. Engring., Optical Soc. Am., Soc. Photog. Scientists and Engrs., Rochester C. of C., Rochester Acad. Sci., Phi Kappa Phi. Mem. Christian Reformed Ch. Subspecialties: Optical engineering; Systems engineering. Current work: Management of development and production of photographic apparatus. Patentee wide angle optical system. Office: 901 Elmgrove Rd Rochester NY 14650

SPOERI, RANDALL KEITH, statistics and operations research educator, consultant, administrator; b. Cleve., June 12, 1946; s. Theodore Warren and Marion (Barrick) S.; m. Kathleen Loma Bryden, Aug. 31, 1968 (div. 1981); 1 dau., Jennifer Anne; m. Deborah Jean Hammett, June 20, 1981. B.S. in Math, Calif. Poly. State U., 1968; M.S. in Stats, Tex. A&M U., 1970, Ph.D., 1976. Spl. asst., math. statistician Bur. of Census, Washington, 1976-80, quantitative methods research br. chief, 1980; assoc. prof. stats. and ops. research U.S. Naval Acad., 1980-83; statis. cons. Md. Energy Adminstrn.; adj. prof. U. Md., 1978—; statis. cons. Ctr. Mgmt. and Policy Research, Inc., Washington, 1980-83, Trident Engring., Inc., Annapolis, Md., 1982-83; assoc. exec. dir. Am. Statis. Assn., Washington, 1983—. Author, editor: Operations Research, 1972. Served to 1st lt. U.S. Army, 1970-72. Decorated Army Commendation Medal; recipient Outstanding Performance rating Bur. of Census, 1978. Mem. Am. Statis. Assn., Ops. Research Soc. Am., Inst. Mgmt. Scis., Mil. Ops. Research Soc., Phi Kappa Phi, Omega Rho, Mu Sigma Rho. Republican. Subspecialties: Statistics; Operations research (mathematics). Current work: Development and innovative application of statistical and operations research methodologies; computer technology to support diverse problem areas such as military tactics, educational and association administration, census data use, ecology and engineering. Home: 929 Burnett Ave Arnold MD 21012 Office: 806 15th St NW Washington DC 20005

SPOHN, HERBERT EMIL, psychologist; b. Berlin, Germany, June 10, 1923; s. Herbert F. and Bertha S.; m. Billie M. Powell, July 28, 1973; children—Jessica, Madeleine. B.S.S., CCNY, 1949; Ph.D., Columbia U., 1955. Research psychologist VA Hosp., Montrose, N.Y., 1955-60, chief research sect., 1960-64; sr. research psychologist Menninger Found., Topeka, 1965-80, dir. hosp. research, 1979—, dir. research dept., 1981—; mem. mental health small grant com. NIMH, 1972-76, mem. treatment assessment rev. com., 1983—. Author: (with Gardner Murphy) Encounter with Reality, 1968; contbr. articles to profl. jours; assoc. editor: Schizophrenia Bull, 1970—. Served with AUS, World War II. USPHS grantee, 1964—. Mem. Am. Psychol. Assn., AAAS, N.Y. Acad. Sci., Phi Beta Kappa, Sigma Xi. Subspecialties: Psychopharmacology; Experimental psychopathology. Current work: Experimental psychopathology-schisophrena; treatment evalation research. Office: Menninger Found Box 829 Topeka KS 66601

SPOLSKY, CHRISTINA MARIA, molecular biologist; b. Reutte, Austria, Mar. 3, 1945; came to U.S., 1967, naturalized, 1977; d. Yaroslaw Stephen and Maria Anna (Ivanytsky) S.; m. Thomas Uzzell, June 26, 1975; children: Stephan Thomas, Renata Christina. B.Sc., U. Toronto, Ont., Can., 1967; M. Phil., Yale U., 1971, Ph.D., 1973. Postdoctoral fellow U. Pa., 1973-75; research assoc. Wistar Inst. Anatomy and Biology, Phila., 1975-76, Acad. Natural Scis., 1978—. Contbr. articles to profl. jours. Province Ont. scholar, 1962; St. Michael's Coll. scholar, 1963; Am. Cancer Soc. fellow, 1974-75; Am. Philos. Soc. grantee, 1976; NSF grantee, 1979; Whitehall Found. grantee, 1981-84. Mem. Am. Soc. Microbiology, Am. Soc. Cell Biology, Tissue Culture Assn., Sigma Xi. Subspecialties: Molecular biology; Evolutionary biology. Current work: Genetics and evolution of mitochondrial DNA; rates of evolution of mitochondrial genome; use of mitochondrial DNA to determine phylogenetic relationships. Home: 2424 Golf Rd Philadelphia PA 19131 Office: Acad Natural Scis 19th and Pkwy Philadelphia PA 19103

SPRAGUE, CHARLES CAMERON, college president; b. Dallas, Nov. 14, 1916; s. George Able and Minna (Schwartz) S.; m. Margaret Frederica Dickson, Sept. 7, 1943; 1 dau., Cynthia Cameron. B.B.A., B.S., D.Sc., So. Meth. U., D.Sc. (hon.), 1966; M.D., U. Tex., 1943. Intern U.S. Naval Med. Center, Bethesda, Md., 1943-44; resident Charity Hosp., New Orleans, 1947-48, Tulane U. Med. Sch., 1948-50; Commonwealth research fellow in hematology Washington U. Sch. Medicine, St. Louis, also Oxford (Eng.) U., 1950-52; mem. faculty Med. Sch. Tulane U., 1952-67, prof. medicine, 1959-67, dean, 1963-67; prof., dean U. Tex. Southwestern Med. Sch., Dallas, 1967-72; pres. U. Tex. Health Sci. Center, Dallas, 1972—; Mem. Nat. Adv. Council, 1966-70; mem. adv. com. to dir. NIH, 1973—; chmn. Gov.'s Task Force Health Manpower, 1981, Gov.'s Med. Edn. Mgmt. Effectiveness Com.; chmn. allied health edn. adv. com., coordinating bd. Tex. Coll. and Univ. System. Served with USNR, 1943-47. Recipient Ashbel Smith Distinguished Alumnus award U. Tex. Med. Br., 1967; Distinguished Alumnus award So. Meth. U., 1965; recipient Sports Illustrated Silver Anniversary award, 1963. Mem. Assn. Am. Med. Colls. (chmn. council deans 1970, chmn. exec. council and assembly 1972-73), Am. Soc. Hematology (pres. 1968), Assn. Acad. Health Ctrs. (bd. dirs. 1982—). Subspecialties: Medical administration; Hematology. Office: U Tex Health Sci Center 5323 Harry Hines Blvd Dallas TX 75235

SPRAGUE, GEORGE FREDERICK, geneticist; b. Crete, Nebr., Sept. 3, 1902; s. Elmer Ellsworth and Emily Kent (Manville) S. B.S., U. Nebr., 1924, M.S., 1926, D.Sc., 1958; Ph.D., Cornell U., 1930. With Dept. Agr., 1924-72, leader corn and sorghum investigations, 1958-72; mem. faculty U. Ill., Urbana, 1973—, now prof. genetics and plant breeding. Editor: Corn and Corn Improvement, 2d edit, 1977; Contbr. articles to profl. jours. Recipient Superior Service award Dept. Agr., 1960, Distinguished Service award, 1970. Fellow AAAS, Washington Acad. Scis., Am. Soc. Agronomy (pres. 1960, Crops Research award 1957); mem. Nat. Acad. Scis., Crops Sci. Soc. (pres. 1951), Am. Genetics Assn., Genetics Soc. Am., Am. Soc. Plant Physiologists, Am. Naturalists, Biometrics Soc. Subspecialty: Plant genetics. Home: 2212 S Lynn St Urbana IL 61801 Office: Dept Agronomy Univ Ill Urbana IL 61801

SPRAGUE, JAMES MATHER, medical scientist, educator; b. Kansas City, Mo., Aug. 31, 1916; s. James P. and Lelia (Mather) S.; m. Dolores Marie Eberhart, Nov. 25, 1959; 1 son, James B. B.S., U. Kans., 1938, M.A., 1940; Ph.D., Harvard U., 1942; A.M. (hon.), U. Pa., 1971. From asst. to asst. prof. anatomy Hopkins Med. Sch., 1942-50; asst. prof. to prof. anatomy U. Pa. Med. Sch., Phila., 1950—, chmn. dept., 1967-76, Joseph Leidy prof. anatomy, 1973—, dir. Inst. Neurol. Sci., 1973-80, chmn. faculty senate, 1963; vis. prof. Northwestern U., 1948, Rockefeller U., 1955, Cambridge U., 1956, U. Pisa, 1966, 74-75; sci. cons. NIH, 1957-60. Co-editor: Progress in Psychobiology and Physiological Psychology, 1966—; asso. editor: Acta Neurobiol. Exper., 1976; contbr. articles to profl. jours. Recipient Macy faculty award, 1974-75; Guggenheim fellow, 1948-49. Mem. Am. Assn. Anatomists (v.p. 1976-78), Japanese Assn. Anatomists (hon.), Soc. Neurosci. Democrat. Subspecialties: Neurobiology; Anatomy and embryology. Current work: Neutral mechanisms of visually-guided behavior- attentions; pereption; discrimination activity. Home: 631 Moreno Rd Narberth PA 19072 Office: Sch Med U Pa Philadelphia PA 19104

SPRATT, JAMES LEO, pharmacologist; b. Chgo., Jan. 27, 1932; s. William and Margaret (Callahan) S.; m.; children: James, Sheila. A.B., U. Chgo., 1953, Ph.D., 1957, M.D., 1961. Asst. prof. pharmacology U. Iowa, Iowa City, 1961-65, asso. prof., 1965-71, prof., 1971—. Contbr. articles to sci. jours. Recipient Research Career Devel. award USPHS, 1963-68; Markle scholar, 1963-68. Mem. Am. Chem. Soc., AAAS, Am. Soc. Exptl. Pharmacology and Therapeutics. Subspecialty: Pharmacology. Current work: Metabolism and actions of cardiac and neuropharmacological agents.

SPRATT, JOHN STRICKLIN, surgeon; b. San Angelo, Tex., Jan. 3, 1929; s. John Stricklin and Nannie Lee (Morgan) S.; m. Beverly Jane Winfiele, Dec. 27, 1951; children: John Arthur, Shelley Winfiele, Robert Stricklin. B.S., So. Meth. U., 1948; M.S.P.H., U. Mo., Columbia, 1970; M.D., U. Tex. Southwestern Med. Sch., 1952. Diplomate: Am. Bd. Surgery. Intern, resident in surgery Barnes Hosp., Washington U., St. Louis, 1952-59; faculty U. Mo., 1961-76, U. Louisville, 1976—; prof. surgery, adj. prof. community heals, asso. Systems Sci. Inst., J. Graham Brown Cancer Center, 1979—; chief surgeon Ellis Fischel State Cancer Hosp., Columbia, 1961-76. Author numerous books and articles. Served to capt. USNR, 1952—. Nat. Cancer Inst. grantee, 1958-76. Mem. Am. Surg. Assn., A.C.S. (fellow), SAR, Naval Order. Baptist. Clubs: Rotary (Louisville); Cosmos (Washington). Subspecialties: Surgery; Cancer research (medicine). Current work: Cancer, surgery, epdiemiology, med. edn., med. econs. Office: J Graham Brown Cancer Center 529 S Jackson St Louisville KY 40202

SPRAWLS, PERRY, JR., physicist, educator; b. Williston, S.C., Mar. 2, 1934; s. Perry and Neva (Mathis) S.; m. Charlotte Williams, Dec. 16, 1961; 1 son, Charles. B.S., Clemson U., 1956, M.S., 1961, Ph.D., 1968. Cert. radiol. physicist Am. Bd. Radiology, 1974; cert. clin. engr. Bd. Examiners for Clin. Engring., 1974, Am. Bd. Clin. Engring., 1976; registered profl. engr., Ga. Devel. engr. Bell Labs., Winston-Salem, N.C., 1956-58; physicist Savannah River Plant, AEC, Aiken, S.C., 1959; instr. to prof. radiology Emory U., Atlanta, 1959—, dir. div. radiol. scis. and edn., 1981--; researcher. Author: The Physical Principles of Diagnostic Radiology, 1977, The Physics and Instrumentation of Nuclear Medicine, 1981. Served to 2d lt. Signal Corps AUS, 1957. Mem. Am. Assn. Physicists in Medicine, Am. Coll. Radiology. Subspecialties: Atomic and molecular physics; Biomedical engineering. Current work: Med. imaging, radiation effects, teaching methodology. Office: Emory University School of Medicine 402 Woodruff Memorial Bldg Atlanta GA 30322

SPREITZER, ROBERT JOSEPH, geneticist; b. Cleve., Apr. 12, 1952; s. Charles Joseph and Sophia Amelia (Zelasko) S.; m. Nancy Jean Pitts, June 3, 1978. B.S. cum laude, Cleve. State U., 1974; Ph.D. (NIH fellow), Case Western Res. U., 1979. Research assoc. in agronomy U. Ill., Urbana, 1979-80, vis. scholar, 1980-82; research assoc. molecular biology U. Geneva, 1982—. Rockefeller Found. postdoctoral fellow, 1980-82; EMBO fellow, 1982; Swiss NSF fellow, 1982—. Mem. AAAS, Am. Soc. Plant Physiologists, Bot. Soc. Am. Subspecialties: Genetics and genetic engineering (biology); Photosynthesis. Current work: Research in chloroplast genetics, genetics of photosynthesis, Chlamydomonas genetics, molecular biology RUBP carboxylase. Office: Dept Biologie Moléculaire Univ Genève 30 Quai Ernest-Ansermet CH-1211 Genève 4 Switzerland

SPREMULLI, LINDA LUCY, chemistry educator, consultant; b. Corning, N.Y., Sept. 6, 1947; d. Paul Francis and Gertrude Lollabella (Haspeslaph) S. B.A. in Chemistry, U. Rochester, 1969; Ph.D. in Biochemistry, M.I.T., 1973. Postdoctoral fellow Clayton Found. Biochem. Inst., U. Tex., Austin, 1973-76; asst. prof. dept. chemistry U. N.C., Chapel Hill, 1976-81, assoc. prof., 1981—; cons. Contbr. research papers to profl. publs. NIH. Research grantee Eli Lilly Corp., 1980-82. Mem. AAAS, Assn. Women in Sci., Am. Chem. Soc., Am. Soc. Microbiology, Am. Soc. Biol. Chemists. Subspecialties: Biochemistry (biology); Molecular biology. Current work: Mechanism of protein biosynthesis.

SPRINGER, ALAN DAVID, anatomist, educator; b. Linz, Austria, Jan. 6, 1948; s. Charles and Shiela (Lacher) S.; m. Linda Mona Springer, Mar. 22, 1969; children: Adam, Eric. B.S., Bklyn. Coll., 1969; Ph.D., CUNY, 1973; postdoctoral student, U. Mich., 1973-77. Asst. prof. phsyiology U. Ill. Med. Ctr., Chgo., 1977-79; assoc. prof. anatomy N.Y. Med. Coll., 1979—. Contbr. articles to profl. publs. NSF, Nat. Inst. Aging, Nat. Eye Inst. grantee, 1981—. Mem. Soc. Neurosci., Assn. Research in Vision and Ophthalmology, N.Y. Acad. Scis., AAAS, Am. Assn. Anatomists. Subspecialties: Regeneration; Developmental biology. Current work: Optic nerve regeneration, vision, neuroplasticity. Office: Dept Anatomy NY Med Coll Valhalla NY 10595

SPRINGER, ALLAN MATTHEW, paper science and engineering educator; b. Baraboo, Wis., Oct. 2, 1944; s. Lester Warren and Tressia Pauline (Dischler) S.; m. Sandra Jean Prothero, June 24, 1967. B.S. in Chem. Engring, U. Wis-Madison, 1966; M.S., Lawrence U., 1969, Ph.D., 1972. Process engr. Olin Mathieson Chem. Corp., Baraboo, Wis., 1966-67; research engr. Nat. Council Paper Industry for Air and Stream Improvement, Kalamazoo, 1972-76; asst. prof. Miami U., Oxford, Ohio, 1976-81, assoc. prof. paper sci. and engring., 1981—; cons. Distbrs. Processing Porterville, Calif., 1977-83, Black-Clawson, 1979, Tripoli Cons., Ltd., Kuala Lumpur, Malaysia, 1980, Eastman Kodak, 1981, Ethyl Corp., 1982. Contbr. articles to profl. jours. Fulbright-Hayes sr. lectr., Malaysia, 1979-80. Mem. TAPPI, Assn. Environ. Engring. Educators, Am. Inst. Chem. Engrs., Sigma Xi. Clubs: South Haven (Mich.) Yacht, Great Lakes (Ill.) Cruising. Subspecialties: Chemical engineering; Water supply and wastewater treatment. Current work: Air and water pollution abatement through process modification; wet-end chemistry of papermaking systems; wastewater treatment plant optimization. Home: 109 McKee Ave Oxford OH 45056 Office: Dept Paper Sci and Engring Miami Univ Oxford OH 45056

SPRINGER, CHARLES SINCLAIR, JR., chemist, educator; b. Houston, Nov. 2, 1940; s. Charles Sinclair and Claire Francis (Butler) S.; m. Karen Elizabeth Thayer, Dec. 28, 1963; children: Linda Kay, Amy Lee. B.Sc., St. Louis U., 1962; M.Sc., Ohio State U., 1964, Ph.D., 1967. Research chemist Aerospace Research Labs., Wright-Patterson AFB, Ohio, 1965-68; asst. prof. chemistry SUNY, Stony Brook, 1968-74, assoc. prof., 1974—; vis. assoc. Calif. Inst. Tech., 1976-77; vis. assoc. prof. Harvard Med. Sch., 1983-84. Contbr. numerous articles to profl. jours. Served to 1st lt. USAF, 1965-68. Recipient Research and Devel. award USAF, 1967. Mem. Am. Chem. Soc., AAAS, Internat. Soc. Magnetic Resonance, N.Y. Acad. Sci., Biophys. Soc., Sigma Xi. Subspecialties: Biophysical chemistry; Nuclear magnetic resonance (biotechnology). Current work: Magnetic resonance studies of biol. membranes, especially cation transport. Office: Dept Chemistry SUNY Stony Brook NY 11794

SPRINGER, DONALD HAROLD, software engr.; b. Rochester, Pa., Jan. 2, 1942; s. Harold L. and Anna (Fatula) S.; m. Judy Clarissa, Apr. 10, 1970; children: Clarissa, Jonathan, Joel, Stephen. B.A., San Diego State U., 1966; M.S., Calif. State U., Hayward, 1973; M.B.A., U. Santa Clara, 1976, postgrad., 1977. Assoc. engr. Advanced Memory Systems, Sunnyvale, Calif., 1971-75; systems programmer Diablo Systems, Hayward, Calif., 1975-76; software engr. Anderson-Jacobson, San Jose, Calif., 1976-78; advanced tech. and applications div. Boeing Computer Service, Seattle, 1978; mem. adj. faculty dept. software engring. Seattle U. Grad. Sch. Engring., 1981—. Served with USN, 1966-70. Mem. Assn. Computing Machinery (nat. recognition service), IEEE. Subspecialties: Software engineering; Distributed systems and networks. Current work: Local network design and protocol software, distributed microprocessor based systems, architecture and performance.

SPRINGER, GEORGE STEPHEN, mechanical engineering educator; b. Budapest, Hungary, Dec. 12, 1933; came to U.S., 1959, naturalized, 1966; s. Josef and Susan (Grausz) S.; m. Susan M. Flory, Sept. 15, 1963; children: Elizabeth, Mary. B.E., U. Sydney, 1959, M.Eng., 1960, M.S., 1961; Ph.D., Yale U., 1962. Registered profl. engr., Mass. Mem. faculty mech. engring. M.I.T., 1962-67; prof. mech. engring. U. Mich., Ann Arbor, 1967—; cons. Author: Environmental Effects on Composite Materials, 1981, (with Duderstadt and Knoll) Principles of Engineering, 1982; others.; Contbr. articles to profl. publs. Fellow ASME, AIAA (assoc.); mem. Soc. Automotive Engrs. (R. Teetor award 1978), Am. Phys. Soc., Sigma Xi, Tau Beta Pi. Subspecialty: Mechanical engineering. Current work: Composite materials, environmental effects. Office: Fluid Dynamics Lab U Mich 2282 GG Brown Lab Ann Arbor MI 48109

SPRINGER, JOSEPH TUCKER, biology educator; b. Lo-Ping, Kiang Si, China, Aug. 9, 1949; s. Charles Oliver and Marion E. (Tucker) S.; m. Elaine Castle, Oct. 15, 1978. B.A., Knox Coll., 1971; M.S., Wash. State U., 1976, Ph.D., 1977. Cert. wildlife biologist, Nebr. Wildlife ecologist Wyo. Game and Fish Dept., Laramie, 1977-79; asst. prof. biology Kearney (Nebr.) State Coll., 1979—; research ecologist/cons. Nature Conservancy, 1980; research program dir. NSF, Kearney, 1981; research cons. Nat. Audubon Soc., 1982. Author: Interactions Between and Some Ecological Aspects of Coyotes and Mule Deer, 1982; contbr. articles to profl. jours.; editor: Jour. Wildlife Mgmt, 1980—, Wildlife Soc. Bull, 1980—. Coordinator Recycle for Wildlife, Kearney, 1981, 82; instr. Elder Hostel, Minden, 1982—. NSF grantee, 1969; Sloan Found. grantee, 1970; others. Mem. Am. Soc. Mammalogists, Nat. Audubon Soc. (chpt. v.p., conservation chmn.), Nat. Wildlife Fedn., Wildlife Soc. (chpt. exec. bd.), Pacific N.W. Bird and Mammal Soc. Democrat. Subspecialties: Resource management; Species interaction. Current work: Behavioral ecology of coyotes; effects of prairie fires on small mammal populations. Office: Dept Biology Kearney State Coll Kearney NE 68849-0531

SPROULL, ROBERT FLETCHER, computer scientist; b. Ithaca, N.Y., June 6, 1947; s. Robert L. and Mary (Knickerbocker) S.; m. Lee S. Sonastine, June 26, 1971. A.B. in Physics, Harvard U., 1968; Ph.D., Stanford U., 1976. Computer specialist NIH, Bethesda, Md., 1970-72; mem. research staff Xerox Palo Alto Research Ctr., Calif., 1972-77; assoc. prof. computer sci. Carnegie-Mellon U., Pitts., 1977—; v.p. Sutherland, Sproull & Assocs., Inc., Pitts., 1979—; mem. tech. adv. council R.R. Donelley & Sons, Inc. Author: (with W.M. Newman) Principles of Interactive Computer Graphics, 2d edit. 1979; Contbr. articles to profl. jours. Served with USPHS, 1970-72. Mem. Assn. for Computing Machinery, IEEE. Subspecialties: Graphics, image processing, and pattern recognition; Computer engineering. Current work: Interactive computer graphics, graphics hardware design, integrated circuit design. Patentee in field. Home: 4419 Schenley Farms Ter Pittsburgh PA 15213

SPRUNG, CHARLES LEON, internist, medical center administrator; b. Bklyn., Sept. 4, 1949; s. Milton and Miriam (Rubin) S.; m. Rebecca Ann Levine, June 3, 1973; children: Elliot Jeffrey, Nina Brynn, Eric Benjamin. B.A., Yeshiva U., 1971; M.D., SUNY-Downstate Med Ctr., Bklyn., 1974. Intern Kings County Med. Ctr., Bklyn., 1974-75, resident in internal medicine, 1975-76, fellow in critical care medicine, 1976-77, chief resident in internal medicine, 1977-78; chief med. ICU VA Med. Ctr., Miami, Fla., 1978—, chief respiratory therapy, 1978—; asst. prof. medicine U. Miami (Fla.) Sch. Medicine, 1978-82, assoc. prof., 1982—, dir. critical care fellowships, 1981—, dir. sect. critical care medicine. Author and editor: The Pulmonary Artery Catheter, 1983; contbr. sci. articles to profl. publs. Recipient Merit Review award VA, Miami, 1981. Fellow ACP; mem. Am. Fedn. Clin. Research, Soc. Critical Care Medicine, Am. Heart Assn., Am. Thoracic Soc. Jewish. Subspecialties: Internal medicine. Current work: Pulmonary edema; shock; catheterization; heat stroke. Office: VA Med Center 1201 NW 16th St Miami FL 33125

SPRUNG, DONALD WHITFIELD LOYAL, physicist, university dean; b. Kitchener, Ont., Can., June 6, 1934; s. Lyall McCauley and Doreen Bishop (Price) S.; m. Hannah Sueko Nagai, Dec. 12, 1958; children—Anne Elizabeth, Carol Hanako. B.A., U. Toronto, 1957; Ph.D., U. Birmingham, Eng., 1961, D.Sc., 1977. Instr. Cornell U., 1961-62; mem. research staff Mass. Inst. Tech., 1964-65; asst. prof. physics McMaster U., Hamilton, Ont., 1962-66, asso. prof., 1966-71, prof., 1971—, dean Faculty of Sci., 1975—; vis. prof. U. Tuebingen, Ger., 1980-81. Contbr. numerous articles to physics jours.; mem. editorial bd.: Canadian Jour. Physics, 1975-80. C.D. Howe Meml. fellow Institut de Physique Nucleaire, Orsay, France, 1969-70. Fellow Royal Soc. Can.; mem. Canadian Assn. Physicists (Herzberg medal 1972), Am. Phys. Soc., Inst. Physics (London). Subspecialties: Nuclear physics; Theoretical physics. Current work: Models of the nucleon-nucleon interaction; Hartree-Fock calculations of nuclear structure, electron scattering from nuclei, structure of the deuteron. Home: 15 Little John Rd Dundas ON Canada L9H 4G5 Office: Faculty of Sci GSB-114 McMaster Univ Hamilton ON Canada L8S 4K1

SPURR, ARTHUR RICHARD, botany educator, researcher; b. Glendale, Calif., July 21, 1915; s. John Oliver and Frieda Helen (Biendara) S.; m. Winifred Mayo Blair, Aug. 31, 1942; children: Jeffrey, John, Douglas, Pamela Spurr Seager. B.S., UCLA, 1938, M.A., 1940; M.A., Harvard U., 1942, Ph.D., 1947. Instr. biology Harvard U., Cambridge, Mass., 1947-48; mem. faculty U. Calif., Davis, 1948—, prof. vegetable crops, 1973—. Contbr. articles on botany to profl. jours. Pres. Friends of Davis Arboretum, 1974-76. Served to lt. col. AUS, 1942-46. Recipient Battelle Meml. Inst. award, 1973; Australian Acad. Sci. award, 1974; NIH grantee, 1961; NSF grantee, 1971; Ministre des Affaires Etrangères (France) grantee, 1978. Mem. Bot. Soc. Am., Am. Soc. Hort. Sci., Microbeam Analysis Soc., Am. Soc. Plant Physiologists, Electron Microscopy Soc. Am., Internat. Soc. Plant Morphologists, Soc. Econ. Botany, AAAS, Am. Inst. Biol. Scis., No. Calif. Soc. Electron Microscopy (pres. 1974-75), Sigma Xi, Alpha Zeta. Republican. Unitarian. Subspecialties: Plant physiology (biology); Morphology. Current work: Microanalysis in relation to salinity, boron toxicity, ultrastructure, plant anatomy. Patentee in field. Office: Dept Vegetable Crops U Calif Davis CA 95616 Home: 617 Elmwood Dr Davis CA 95616

SQUIRE, LARRY RYAN, neuroscientist, psychologist, educator; b. Cherokee, Iowa, May 4, 1941; s. Harold Walter and Jean (Ryan) S. B.A., Oberlin Coll., 1963; postgrad., Stanford U., 1963-64; Ph.D. in Psychology, MIT, 1968. Asst. prof. dept. psychiatry U. Calif.-San Diego, 1973-76, assoc. prof., 1976-81, prof., 1981—; psychologist VA Med. Ctr., San Diego, 1976-78, neuropsychology cons., 1978-80, research career scientist, 1980—; lectr. Univ. fellow Stanford U., 1963-64; NIMH predoctoral fellow, 1964-68, NIMH Interdisciplinary fellow, 1968-70; research career scientist VA, 1980—. Mem. editorial adv. bd.: Jour. Clin. Neuropsychology; reviewer numerous profl. jours.; contbr. abstracts and articles to profl. jours., chpts. to books. Fellow Am. Psychol. Assn.; mem. Soc. Neurosci., Internat. Neuropsychol. Soc., Psychonomic Soc., AAAS. Subspecialties: Neuropsychology; Psychobiology. Current work: Neuroscience, neuropsychology, memory and the brain. Office: VA Medical Center 3350 LaJolla Village Dr San Diego CA 92161

SQUIRE, PHIL G., biochemistry educator; b. Ephraim, Utah, Sept. 11, 1922; s. George E. and Utah M. (Jensen) S.; m. Gwendolyn B., Sept. 5, 1946; children: Dale P., Karen, Kenneth B. B.S., Brigham Young U., 1948; M.S., U. Wis.-Madison, 1951; Ph.D., U. Calif.-Berkeley, 1957. Mem. faculty U. Calif.-Berkeley, 1957-67, asst. prof. exptl. endocrinology, 1961-64, assoc. prof. biochemistry, San Francisco, 1966-67, Colo. State U., Ft. Collins, 1967, now prof. Contbr. numerous articles to profl. publs. Served with U.S. Army, 1941-45. Am. Cancer Soc. scholar Inst. Biochemistry, U. Uppsala, Sweden, 1962-63; NSF fellow, 1964-76. Mem. Am. Chem. Soc., Am. Soc. Biol. Chemists, Am. Soc. Microbiology, AAAS, Sigma Xi. Subspecialties: Biochemistry (medicine); Pathology (veterinary medicine). Current work: Role of bacterial macromolecules in the infections disease process, specifically, the role of Pasteurella macromolecules in shipping fever pneumonia. Office: Colo State Univ Biochemistry Dept Fort Collins CO 80523

SQUIRES, RICHARD FELT, neuropharmacologist, neurochemist, biochemist; b. Sparta, Mich., Jan. 15, 1933; s. Monas Nathan and Dorothy Lois (Felt) S.; m. Else Saederup; 1 child: Iben Saederup. B.S. in Chemistry, Mich. State U., 1958. Dir. dept. biochemistry A/S Ferrosan, Soeborg, Denmark, 1963-78; group leader CNS Research Lederle Labs., Pearl River, N.Y., 1978-79; prin. investigator Rockland Research Inst., Orangeburg, N.Y., 1979—. Contbr. articles in field to profl. jours. Mem. Soc. Neurosci., Internat. Soc. Neurochemistry, Internat. Soc. Psychoneuroendocrinology, European Neurosci. Assn., Collegium Internationale Neuro-Psychopharmacologicum, Am. Soc. Neurochemistry. Subspecialties: Neuropharmacology; Neurochemistry. Current work: Benzodiazepine, GABA, Picrotoxin receptors in brain, characterization of receptors in CNS, devel. of novel psychotropic agts., neurol. and psychiat. disorders. Home: 10 Termakay Dr New City NY 10956 Office: Rockland Research Institute Orangeburg NY 10962

SREENIVASAN, S. RANGA, physicist; b. Mysore, India, Oct. 20, 1933; emigrated to Can., 1966, naturalized, 1974; s. H. Sreenivasachari and Alamelammal (Rangaswami) S.; m. Claire Selma Julie de Reineck, Oct. 16, 1963; children: Gopal, Govind, Gauri, Gayatri, Aravind. B.Sc. with honors in Physics, U. Mysore, 1950, 1952, Ph.D. in Theoret. Physics, 1958. Research fellow Harvard U., 1958-61; research asso. NASA, Goddard Inst. Space Studies, N.Y.C., 1961-64; vis. scientist Max Planck Inst, Munich, W.Ger., 1964-66; research fellow U. Calgary, Alta., Can., 1966-67, asst. prof. physics, 1967-68, asso. prof., 1968-74, prof., 1974—; vis. prof. Royal Inst. Tech., Stockholm, Sweden, 1974-75. Mem. Internat. Astronom. Union, Am. Geophys. Union, Am. Phys. Soc., Am. Astron. Soc., Am. Meteorol. Soc. Subspecialties: Theoretical astrophysics; Plasma physics. Current work: Stellar structure and evolution; solar physics; stellar atmospheres; interstellar medium; plasma astrophysics; controlled thermonuclear fusion. Home: 2110-30 Ave SW Calgary AB Canada T2T 1R4 Office: 2500 University Dr NW Calgary AB Canada T2N 1N4

SREEVALSAN, THAZEPADATH, microbiology educator, researcher; b. Kanjiramattom, India, Jan. 25, 1935; came to U.S., 1961, naturalized, 1982; s. Achuthan E. and Narayani (Narayni) Panickar. B.Sc., U. Travancore, India, 1953; M.Sc., 1956; Ph.D., U. Tex., 1964. Tchr. St. Ignatious High Sch., Kanjiramattom, 1953-54; research asst. Pasteur Inst., Coonoor, India, 1956-61; research assoc. U. Tex., 1964-66; chemist E.I. DuPont de Nemours, Wilmington, Del., 1966-69; asst. prof. microbiology Georgetown U., 1969-73, assoc. prof., 1973-80, prof., 1980—. Contbr. articles to profl. jours. Fogarty Internat. fellow NIH, 1978; recipient Margaret McKinley award U. Tex., 1963, O.B. Williams award Am. Soc. Microbiology, 1964, William Peck award Potgrad. Med. Assn., 1972. Mem. Am. Soc. Microbiology, AAAS, N.Y. Acad. Sci. Subspecialties: Microbiology, Cell biology. Current work: Growth control in animal cells as studied using interferon to inhibit proliferation of both normal and neoplastic cells. Home: 10710 Muirfield Dr Potomac MD 20854 Office: Department of Microbiology Schools of Medicine and Dentistry Washington DC 20007

SRIDARAN, RAJAGOPALA, reproductive endocrinologist, researcher, educator; b. Papireddipatti, India, Feb. 22, 1950; came to U.S., 1973; s. Rajagopal and Chellmmal (Viswanathan) R.; m. Geetha Kothandaram. B.S., U. Madras, 1970, M.S., 1972; Ph.D. in Physiology (fellow), U. Health Scis., North Chicago, Ill., 1977. Lectr. endocrine physiology Chgo. Med. Sch., 1976; postdoctoral research asso. in reproductive neuroendocrinology U. Nebr., Omaha, 1977-78; postdoctoral research asso. in endocrinology dept. physiology and biophysics U. Ill. at Med. Center, Chgo., 1978-81; asst. prof. dept. physiology Morehouse Sch. Medicine, Atlanta, 1981—. Reviewer: Neurosci. and Biobehavioral Revs; contbr. articles and abstracts to sci. jours. Mem. Am. Physiol. Soc., Endocrine Soc., Soc. for Neurosci., Soc. for Study Reprodn., AAAAS. Subspecialties: Neuroendocrinology; Receptors. Current work: Corpus luteum function and maintenance of pregnancy in the rat, abortifacient effects of dihydrotestosterone in the pregnant rat, neuroendocrine control of gondotropin secretion and ovulation, circadian rhythms in reproductive endocrinology. Home: 2069 Tidwell Trail Stone Mountain GA 30088 Office: Dept Physiology Morehouse Sch Medicine 720 Westview Dr SW Atlanta GA 30310

SRIVASTAVA, SATISH KUMAR, geologist; b. Sitapur, Uttar Pradesh, India, June 28, 1935; came to U.S., 1970, naturalized, 1979; s. Hazari Lal and Sheopiari S.; m. Rosalind Ann Catterall, Apr. 17, 1945. Ph.D. in Geology, U. Alta. (Can.), Edmonton, 1968. Research asst. Forest Research Inst. Dehradun, India, 1954-57; sr. tech. asst. Oil and Gas Commn., Dehradun, 1957-68; Killam postdoctoral fellow dept. botany U. B.C. (Can.), Vancouver, 1968-70; research geologist Chevron Oil Field Research Co., La Habra, Calif., 1970-80, sr. research geologist, 1980—. Contbr. numerous articles on palynology to nat., internat. profl. jours. Fellow Geol. Assn. Can., Linnean Soc. London; mem. Am. Assn. Stratigraphic Palynologists, Am. Assn. Petroleum Geologists, Soc. Econ. Paleontologists and Mineralogists, Am. Bot. Soc., Internat. Orgn. Palaeobotany (life), Palynological Soc. India, Indian Soc. Palynostratigraphers. Subspecialties: Paleontology; Paleoecology. Current work: Fossil spore-pollen taxonomy; biostratigraphy; paleoecology; evolution; study of palynology applied to oil exploration. Home: 3054 S Blandford Dr Rowland Heights CA 91748 Office: 3282 Beach Blvd La Habra CA 90631

STAATS, JOAN, medical librarian; b. Chgo., July 31, 1921; d. Walter J. and Betty (Pischel) S. B.S. in Psychology, U. Ill., 1943, M.S. in Zoology, 1947. Asst. in labs. Wilson & Co., 1943-44; exptl. pharmacologist Armour Pharm. Co., 1944-45; teaching asst. U. Ill.-Urbana, 1945-46, acting instr., 1946-47; med. librarian Jackson Lab., Bar Harbor, Maine, 1949—, staff scientist, 1955-76, sr. staff scientist, 1976—. Contbr. articles to sci. jours. Corporator, bd. dirs. Jesup Meml. Library, Bar Harbor, 1964—. Mem. AAAS, Genetics Soc. Am., Med. Library Assn., AAUW, Health Sci. Libraries and Info. Coop. (founding), North Atlantic Health Sci. Libraries (founding). Episcopalian. Subspecialty: Biofeedback. Current work: Mammalian genetics, bibliographic research. Home: Lower Main St Bar Harbor ME 04609 Office: Jackson Lab Bar Harbor ME 04609

STAATS, WILLIAM RICHARD, research adminstr., chem. engr.; b. Chgo., Sept. 10, 1935; s. William Jakob and Agnes Julia (Kasid) S.; m. Ann Cecille Staats, June 29, 1980. B.S. in Chem. Engring, Ill. Inst. Tech., Chgo., 1957, M.S. in Gas Engring, 1960, Ph.D. in Chem. Engring, 1970. Registered profl. engr., Ill. Mgr. applied combustion research Inst. Gas Tech., Chgo., 1957-58, 62-69; v.p. Poly., Inc., Lincolnwood, Ill., 1970-75; assoc. dir. process devel. Inst. Gas Tech., Chgo., 1975-79; dir. basic research Gas Research Inst., Chgo., 1979—. Served to 1st lt. USAF, 1959-62. Mem. Am. Inst. Chem. Engrs., Combusion Inst., Am. Chem. Soc., AAAS, Sigma Xi. Subspecialty: Chemical engineering. Patentee in field. Office: 8600 W Bryn Mawr St Chicago IL 60631

STABA, EMIL JOHN, pharmacognosy educator; b. N.Y.C., May 16, 1928; s. Frank and Marianna T. (Mack) P.; m. Joyce Elizabeth Ellert, June 19, 1954; children—Marianna, Joanna, Sarah Jane, John, Mark. B.S. cum laude, St. John's U., 1952; M.S., Duquesne U., 1954; Ph.D., U. Conn., 1957. Asst. prof. U. Nebr., 1957-60, prof., chmn. dept., 1968; prof., chmn. dept. pharmacognosy U. Minn., 1968—; Cons. econ. plants, plant tissue culture U.S. Army Q.M. Corps; Cons. drug plants, plant tissue culture NASA; cons. Korean govt., pharm. industry.; Con. internat. vis. prof. Dalhousie U., 1983; mem. natural products revision com. U.S. Pharmacopeia, 1980—. Mem. editorial bd.: Jour. Plant Cell, Tissue and Organ Culture, 1980—. Served with USNR, 1945-46; PTO. Sr. fgn. fellow NSF, Poland, 1969; Fulbright fellow, Germany, 1970; CSIR-NSF fellow, India, 1973; PCSIR-NSF fellow, Pakistan, 1978. Fellow AAAS; mem. Am. Soc. Pharmacognosy (pres. 1971-72), Am. Assn. Colls. Pharmacy (chmn. tchrs. sect. 1972-73, dir. 1976-77), Tissue Culture Assn. (pres. plant sect. 1972-74), Am. Pharm. Assn. and Acad. (chmn. pharmacognosy and nat. products 1977—), Soc. Econ. Botany. Subspecialties: Plant cell and tissue culture; Pharmacognosy. Current work: Medicenal and economic plant cell and organ culture; quatic plant chemistry and herbs. Home: 2840 Stinson Blvd NE Minneapolis MN 55418

STABLER, TIMOTHY ALLEN, biology educator; b. Port Jervis, N.Y., Sept. 27, 1940; s. Ralph Allen and Marjorie (Hoyt) S. B.A., Drew U., 1962; M.A., DePauw U., 1964; postgrad., Albany Med. Coll., 1964-65; Ph.D., U. Vt., 1969. Postdoctoral fellow U. Minn., Mpls., 1968-69; asst. prof. Hope Coll., Holland, Mich., 1969-71; postdoctoral fellow Boston U. Med. Sch., 1971-73; asst. prof. Ind. U. N.W., Gary, 1973-76, assoc. prof. biology, 1976—, chmn. dept., 1983—, health profl. adv., 1973—. Mem. Am. Soc. Zoologists, Nat. Assn. Advs. Health Profession (dir. 1980—), Central Assn. Advs. Health Profession (treas. 1978—), Ind. Acad. Sci., N.Y. Acad. Sci. Republican. Methodist. Subspecialties: Cell and tissue culture; Developmental biology. Current work: Cell culture of endocrine cells and histochemical cell indentification. Home: 405 E Institute Valparaiso IN 46383 Office: Ind U NW 3400 Broadway Gary IN 46408

STACH, ROBERT WILLIAM, biochemist; b. Chgo., Feb. 12, 1945; s. Edward Edwin and Chrystal Julia S. (Schwieger) S.; m. Bettie Marie Jones, Jan. 29, 1966; 1 dau., Jeannette Lynn. B.A., Ill. Wesleyan U., 1967; Ph.D. in Organic Chemistry, U. Wis.-Madison, 1972. Postdoctoral trainee Stanford U. Sch. Medicine, 1972-74; asst. prof. dept. biochemistry SUNY Upstate Med. Center, Syracuse, 1974-80; asso. mem. faculty Center for Neurobehavioral Scis., 1979—, adj. asso. prof. dept. anatomy, 1980—, asso. prof. dept. biochemistry, 1980—; adj. asso. prof. U. Tex. Med. Br., Galveston, 1982-83. Contbr. articles to profl. jours. NIH grantee, 1975—. Mem. Fedn. Am. Scientists, N.Y. Acad. Scis., AAAS, Am. Chem. Soc., Soc. Neurosci., Am. Soc. Biol. Chemists, Internat. Soc. Neurochemistry, Am. Soc. Neurochemistry, Sigma Xi. Republican. Methodist. Subspecialties: Neurobiology; Neurochemistry. Current work: Developmental neurobiochemistry growth and development, sensory and sympathetic nervous systems, receptors, peptide hormones. Home: 10 Juniper Liverpool NY 13088 Office: 766 Irving Ave Syracuse NY 13210

STACHNIK, ROBERT VICTOR, research scientist; b. Yonkers, N.Y., Mar. 30, 1947; s. Victor A. and Catherine V. S. B.S., Villanova U., 1969; M.S., SUNY at Stony Brook, 1971, Ph.D., 1975. Sr. electro-optical scientist in physics Itek Corp., Lexington, Mass., 1975-78; research fellow Ctr. for Earth and Planetary Physics, Harvard U., 1978; research assoc. Harvard-Smithsonian Ctr. for Astrophysics, 1979—. Author papers on astrophysics, optics and space science. Mem. Am. Astron. Soc. Democrat. Subspecialties: Optical astronomy; Optical image processing. Current work: Speckle imaging and interferometry; high optical resolution astronomy; space optical VLBI; digital image processing; astrophysics or stellar structure. Office: Ctr for Astrophysics 60 Garden St Cambridge MA 02138

STACPOOLE, PETER WALLACE, physician, medical educator and researcher; b. Torrance, Calif., Feb. 24, 1945; s. Albert Charles and Joyce Phyllis (Wallace) S.; m. Lee Ann Dodson, June 29, 1980. B.A. in Chemistry, U. of the South, 1967, M.S., U. San Francisco, 1968; Ph.D. in Pharmacology, U. Calif.-San Francisco, 1972; M.D., Vanderbilt U., 1976. Diplomate: Am. Bd. Internal Medicine. Postdoctoral fellow U.Calif.-San Francisco, 1971-72; intern in internal medicine Vanderbilt U., Nashville, 1976-77, resident in internal medicine, 1977-78, fellow in endocrinology, 1978-80; asst. prof. medicine U. Fla., Gainesville, 1980—; dir. Lipid Clinic Shands Hosp., Gainesville, 1981—, assoc. dir. Clin. Research Ctr., 1983—; dir. Lipid Clinic, Shands Hosp., Gainesville, 1981—. Author 55 sci. publs. Research grantee NIH, 1981-84, 82-85. Mem. Am. Assn. Clin. Research, Am. Diabetes Assn. (grantee 1982-84), Endocrine Soc., Am. Heart Assn. (grantee 1981-82), European Diabetes Assn., AAAS, Sigma Xi. Subspecialties: Endocrinology; Biochemistry (medicine). Current work: Pharmacological and nutritional control of cholesterol metabolism; treatment of lipid disorders; clinical pharmacology. Patent use of novel lipid-lowering drugs. Home: 6205 NW 143d St Gainesville FL 32606 Office: U Fla Coll Medicine Gainesville FL 32610

STADELMAN, WILLIAM JACOB, food science educator; b. Vancouver, Wash., Aug. 8, 1971; s. William Henry and Eda Wilhelmena (Huber) S.; m. Margaret Jane Lloyd, Apr. 3, 1942; children: Ralph Lindsay, Paula Gardner Jurgonski. B.S., Wash. State U., 1940; M.S., Pa. State U., 1942, Ph.D., 1948. Asst. prof. poultry sci. Wash. State U., Pullman, 1948-53, assoc. prof., 1953-55; assoc. prof. poultry sci. Purdue U., West Lafayette, Ind., 1955-58, prof. food sci., 1958—. Co-editor: Egg Science and Technology, 1977; contbr. articles to profl. jours. Served with USNR, 1942-67. Recipient Christie award

Poultry and Egg Nat. Bd., 1955. Fellow AAAS, Inst. Food Technologists (Scientist of Yr. Phila. sec. 1977), Poultry Sci. Assn.; mem. World Poultry Sci. Assn., Am. Meat Sci. Assn., Internat. Inst. Refrigeration, Research and Devel. Assocs. Republican. Subspecialties: Food science and technology; Animal nutrition. Current work: Preservation and quality evaluation of poultry meat and eggs; development of products from meat and eggs;factors influencing tenderness and flavor of poultry meat. Patentee in field. Home: 1429 N Salisbury St West Lafayette IN 47906 Office: Purdue U Smith Hall West Lafayette IN 47907

STADTMAN, EARL REECE, biochemist; b. Carrizozo, N.Mex., Nov. 15, 1919; s. Walter William and Minnie Ethyl (Reece) S.; m. Thressa Campbell, Oct. 19, 1943. B.S., U. Calif., Berkeley, 1942, Ph.D, 1949. With Alcan Hwy. survey Public Rds. Adminstrn., 1942-43; research asst. U. Calif., Berkeley, 1938-49, sr. lab. technican, 1949; AEC fellow Mass. Gen. Hosp., Boston, 1949-50; chemist lab. cellular physiology Nat. Heart Inst., 1950-58, chief enzyme sect., 1958-62, chief lab. biochemistry, 1962—; Biochemist Max Planck Inst., Munich, Germany, Pasteur Inst., Paris, 1959-60; faculty dept. microbiology U. Md.; prof. biochemistry grad. program dept. biology Johns Hopkins U.; adv. com. Life Scis. Research Office, Am. Fedn. Biol. Sci., 1974-77; Bd. dirs. Found. Advanced Edn. Scis., 1966-70, chmn. dept. biochemistry, 1966-68; biochem. study sect. research grants NIH, 1959-63. Editor: Jour. Biol. Chemistry, 1960-65, Current Topics in Cellular Regulation, 1968—, Circulation Research, 1968-70; exec. editor: Archives Biochemistry and Biophysics, 1960—, Life Scis, 1973-75, Procs. Nat. Acad. Sci, 1975-81, Trends in Biochem. Research, 1975-78; editorial adv. bd.: Biochemistry, 1969-76, 81—. Recipient medallion Soc. de Chemie Biologique, 1955, U. Pisa, 1966, Presdl. rank award as Disting. Sr. Exec., 1981. Mem. Am. Chem. Soc. (Paul Lewis Lab. award in enzyme chemistry 1952, exec. com. biol. div. 1959-64, chmn. div. 1963-64, Hillebrand award 1969), Am. Soc. Biol. Chemists (publs. com. 1966-70, council 1974—, Merck award 1983), Nat. Acad. Scis. (award in microbiology 1970), Am. Acad. Arts and Scis., Am. Soc. Microbiology, Washington Acad. Scis. (award biol. chemistry 1957, nat. medal sci. 1979, meritorious exec. award 1980). Subspecialties: Biochemistry (biology); Microbiology. Current work: Regulation of cellular metabolilsm; enzyme chemistry. Home: 16907 Redland Rd Derwood MD 20855 Office: Nat Heart and Lung Inst Bethesda MD 20014

STADTMAN, THRESSA CAMPBELL, biochemist, microbiologist, researcher; b. Sterling, N.Y., Feb. 12, 1920; d. Earl and Bessie (Waldron) Campbell; m. Earl Reece Stadtman, Oct. 19, 1943. B.S., Cornell U., 1940, M.S., 1942; Ph.D., U. Calif.-Berkeley, 1949. Research assoc. U. Calif.-Berkeley, 1943-46; research asst. Harvard U. Med. Sch., 1949-50; biochemist Nat. Heart Inst., NIH, Bethesda, Md., 1950—; sect. chief lab. biochemistry Nat. Heart, Lung, Blood Inst., 1974—; lectr. in field; mem. U.S. nat. com. Internat. Union Biochemistry, 1976-82; mem. U.S. nat. com. Internat. Union Microbiology, 1982—. Contbr. numerous articles, revs. on amino acid metabolism, vitamin B12-dependent enzymes, methane biosynthesis, selenium dependent enzymes and seleno tRNAs to profl. pubis. Recipient Hillebrand award Chem. Soc. Washington, 1979, Superior Service award USPHS, 1980; Helen Hay Whitney fellow Oxford (Eng.) U., 1954-55; Rockefeller Found. fellow U. Munich, W.Ger., 1959. Mem. Nat. Acad. Scis., Am. Soc. Biol. Chemists (sec. and program chmn. 1979-81), Am. Chem. Soc., Am. Soc. Microbiology, Am. Acad. Arts and Scis. Subspecialties: Biochemistry (biology); Microbiology. Current work: Selenium biochemistry. Office: NIH Bldg 3 Room 108 Bethesda MD 20205

STAEHLE, CHARLES MICHAEL, marine/ocean engineer, consultant; b. Lovell, Wyo., Oct. 4, 1938; s. Eddie Leroy and Frances Lenora (Glenn) S.; m. Margaret Elizabeth Allen, Jan. 21, 1963 (div. July 1982); children: Cynthia Marie, Mark Allen. B.S. in Physics, U. Okla., 1963. Project engr. oceanic div. Westinghouse Corp., Annapolis, Md., 1970-77; tech. mgr. NOAA, Washington, 1977-82; mgr. spl. programs Perry Offshore, Inc., Riviera Beach, Fla., 1982—. Pres. Chase Creek Civic Assn., Arnold, Md., 1975. Served to lt. comdr. USN, 1958-70; to capt. Res., 1970—. Mem. Soc. Naval Architects and Marine Engrs., Marine Tech. Soc. (underwater physics com. 1981—, marine mining and minerals com. 1983—), Deep Submersible Pilots Assn., Naval Res. Assn., Navy Submarine League. Republican. Episcopalian. Lodges: Masons; Shriners. Subspecialties: Ocean engineering; Systems engineering. Current work: Ocean engineering and marine technology, particularly as applied to the deep ocean and sea bed; research on marine minerals survey and recovery, nuclear waste disposal and related marine ecological technology. Home: 8716 Satalite Terr Lake Park FL 33403 Office: Perry Offshore Inc PO Box 10297 275 W 10th St Riviera Beach FL 33404

STAFFORD, FRED E., physicial science administrator; b. N.Y.C., Mar. 30, 1935; s. Frank H. and Eva S.; m. Barbara M. A.B., Cornell U., 1956; Ph.D. in Chemistry, U. Calif.-Berkeley, 1959. NSF postdoctoral fellow U. Brussels, Belgium, 1959-61; faculty chemistry Northwestern U., Evanston, Ill., 1961-73; spl. asst. NSF, Washington, 1973-75, now program dir. solid state chemistry. Contbr. sci. articles to profl. jours. Mem. Am. Chem. Soc., Am. Phys. Soc., AAAS, Fed. Exec. Inst. Alumni Assn. Subspecialties: Solid state chemistry; Surface chemistry. Office: Solid State Chemistry NSF 1800 G St NW Washington DC 20550

STAFFORD, HELEN ADELE, biology educator; b. Phila., Oct. 9, 1922; s. Morton Ogden and Ethel (Scherer) S. B.S., Wellesley Coll., 1944; M.A., Conn. Coll., 1948; Ph.D., U. Pa., 1951. Research assoc. instr. biochemistry and botany U. Chgo., 1951-54; asst. prof. biology Reed Coll., Portland, Oreg., 1954-58, assoc. prof., 1959-65, prof., 1965—. Contbr. articles to profl. jours. Guggenheim fellow, 1958-59; NSF postdoctoral fellow, 1963-64. Subspecialty: Plant physiology (biology). Current work: Regulation and metabolism of condensed tannins. Office: 3203 SE Woodstock Blvd 102 Portland OR 97202

STAHL, CHARLES DREW, engineering educator, consultant; b. Altoona, Pa., Aug. 28, 1923; s. C. Asher and Anna M. (Leinhoff) S.; m. Barbara M. Morrison, Aug. 14, 1948; 1 son, Kevin M. B.S., Pa. State U., 1947, M.S., 1950, Ph.D., 1953. Registered profl. engr., Pa. Research assoc. Pa. State U., 1947-49; reservoir engr. Mobil Internat., 1958-59; assoc. prof. Pa. State U., University Park, 1959-61, prof. dept. petroleum and natural gas engring., 1962-83, head dept., 1962—; cons. to major oil cos., also Commonwealth of Pa. Contbr. articles to profl. jours. Recipient Wilson award Pa. State U., 1981. Mem. Soc. Petroleum Engrs. of AIME, Am. Petroleum Inst. Republican. Subspecialties: Petroleum engineering; Fluid mechanics. Current work: Tertiary recovery of petroleum; flow and displacement of oil in porous media by use of micellar and other soluble fluids. Home: 209 Norle St State College PA 16801 Office: 101 Mineral Sci Bldg University Park PA 16802

STAHL, JOEL S., plastics company executive; b. Youngstown, Ohio, June 10, 1918; s. John Charles and Anna (Nadler) S.; m. Jane Elizabeth Anglin, June 23, 1950; 1 son, John A. B.Ch.E., Ohio State U., 1939. Mgr. spl. products, coordinator sales, mfg. and transp. Ashland Oil, Inc., Ky., 1939-50; pres., gen. mgr. Cool Ray Co., 1950-51; pres. Stahl Industries, Inc., Youngstown, Ohio, 1951—. Bd. dirs. Boardman Civic Assn. Mem. Soc. Plastics Engrs. (chmn. plastics in bldg. sect.), Soc. Plastics Industry, ASTM. Republican. Christian Scientist. Clubs: Circumnavigators (N.Y.C.); Berlin Yacht (North Benton, Ohio). Lodge: Shriners. Subspecialties: Polymer chemistry; Polymers. Current work: Plastics in building construction, plastic foam core panels and fire resistant plastic products for building construction. Patentee in field. Home: 746 Golfview Ave Youngstown OH 44512 Office: 9th Floor Dollar Bank Bldg Youngstown OH 44503

STAHL, RAYMOND EARL, chemist, researcher; b. Chgo., Feb. 21, 1936; s. Arthur Daniel and Gladys Hazel (Lockwood) S. Ph.B. in Chemistry and Math, Northwestern U., 1971. Group leader Morton Chem. Co., Elk Grove Village, Ill., 1962-65; tech. dir. Am. Indsl. Finishes, Chgo., 1965-67, formulator DeSoto, Inc., Chgo., 1960-62, research assoc., 1967-73, Midland div. Dexter Corp., Waukegan, Ill., 1973—; cons. superior system devel. Fellow Am. Inst. Chemists; mem. Am. Chem. Soc., Am. Inst. Physics, Am. Phys. Soc., AAAS, Fedn. Socs. for Coating Tech., Math. Assn. Am., Am. Statis Assn., Assn. Computer Users, Am. Mgmt. Assn., Ill. State Acad. Sci., Nat. Rifle Assn., Ill. State Rifle Assn. Republican. Subspecialties: Mathematical software; Statistics. Current work: Increasing productivity in the laboratory; maximization of computer utilization where ever applicable; statistical design of experiments and analysis of data. Home: 2207 Rolling Ridge Ln Lindenhurst IL 60046 Office: Dexter Corp Midland Div E Water St Waukegan IL 60085

STAHLY, TIM SCOTT, nutritionist, educator; b. Huron, S.D., May 10, 1949; s. Vernon A. and Francis V. (Mines) S.; m. Sharon K. Stahly; children: Lisa, Charles, Benjamin. B.S., S.D. State U., 1970, M.S., 1972; Ph.D., U. Nebr., 1975. Assoc. prof. dept. animal scis. U. Ky., Lexington, 1975—. Mem. Am. Soc. Animal Sci., Brit. Soc. Animal Prodn., Council Agrl. Sci. and Tech. Republican. Presbyterian. Subspecialty: Animal nutrition. Current work: Nutrition of the gravid and lactating dam and its effect on the subsequent development of the offspring and the reproductive capacity of the dam. Office: U Ky 610 Agr Sci S Lexington KY 40546

STAHR, HENRY MICHAEL, chemist, educator; b. White, S.D., Dec. 10, 1931; s. George Conrad and Kathryn Evelyn (Smith) S.; m. Irene Frances Sondey, July 27, 1952; children: Michael, John, Mary, Patrick, Matthew. B.S., S.D. State U., 1956; M.S., Union Coll., 1961; postgrad., U. S.C., 1961-63, U. Richmond, 1965-68; Ph.D., Iowa State U., 1976. Analyst, sr. devel. chemist Gen. Electric, 1956-65; research chemist, sr. scientist Philip Morris Research, Richmond, Va., 1965-69; asst. prof. chemistry Iowa State U., 1969-76, assoc. prof., 1976-82, prof., 1982—, chief chemist, 1969—. Contbr. to profl. jours. Mem. Ogden Sch. Bd., 1977—. Served with USMC, 1949-53. Mem. Am. Chem. Soc., Microchem. Soc., Electrochem. Soc., Am. Soc. Mass Spectroscopy, Am. Coll. Vet. Toxicology, Am. Assn. Vet. Lab. Diagnosticians. Democrat. Roman Catholic. Lodges: Lions; KC; Am. Legion. Subspecialties: Analytical chemistry; Toxicology (agriculture). Current work: Method development for toxic substances, utilization of modern instrumentation for objective biomedical information; research on toxic substances. Home: Route 1 Box 180 Ogden IA 50212 Office: Iowa State U 1636 Coll Vet Medicine Ames IA 50011

STAKGOLD, IVAR, educator; b. Oslo, Norway, Dec. 13, 1925; came to U.S., 1941, naturalized, 1947; s. Henri and Rose (Wishengrad) S.; m. Alice Calvert O'Keefe, Nov. 27, 1964; 1 dau., Alissa Dent. B.Mech. Engring., Cornell U., 1945, M.Mech. Engring., 1946; Ph.D., Harvard U., 1949. Instr., then asst. prof. Harvard, 1949-56; head math. and logistics brs. Office Naval Research, 1956-59; faculty Northwestern U., 1960-75, prof. math. and engring. scis., 1964-75, chmn. engring. scis., 1969-75; prof., chmn. math. U. Del., 1975—; mem. U.S. Army basic research com. NRC, 1977-80; mem. com. on applications of math. Nat. Acad. Scis.-NRC, 1982—; vis. faculty Math. Inst., Oxford (Eng.) U., 1973, Univ. Coll., London, 1978, Victoria U., Wellington, N.Z., 1981, Ecole Polytechnique Federale de Lausanne, Switzerland, 1981. Author: Boundary Value Problems of Mathematical Physics, vols. I and II, 1967, Green's Functions and Boundary Value Problems, 1978; asso. editor: Am. Math. Monthly, 1975-80, Jour, Applicable Analysis, 1977—, Internat. Jour. Engring. Sci, 1977—, Jour. Integral Equations, 1978—. Mem. Soc. Indsl. and Applied Math. (trustee 1976—, chmn. 1979—). Subspecialty: Applied mathematics. Current work: Nonlinear boundary valve problems. U.S. rep. World Bridge Championships, 1959, 60; holder 7 nat. bridge championships. Home: 13 Fairfield Dr Newark DE 19711

STALEY, FREDERICK JOSEPH, research mathematician; b. Terre Haute, Ind., June 28, 1944; s. Harold Joseph and Catherine (Roetker) S.; m. Pamela Kay Pierce, Nov. 17, 1966 (div. 1972); m. Carol Heineman, Jan. 2, 1980. B.S., SUNY, Albany, 1978; M.S. in Physics, Colo. State U., 1980. Self-employed as research mathematician, Ft. Collins, Colo., 1980—. Author monographs. Mem. Am. Phys. Soc. Subspecialties: Foundations of mathematics; Psychophysics. Current work: Research on the foundations of mathematics and on the applications of mathematics to new fields; catastrophe physics. Home and Office: 600 Blevins Ct Fort Collins CO 80521

STALEY, JOHN MERRILL, plant pathologist; b. Three Rivers, Mich., Sept. 12, 1929; s. Jesse H. and Mary M. (Muirhead) S.; m. Bonnie B. Barton, Feb. 6, 1953; children: Mary, Anna, Patricia, Jeffrey; m. Janice Cheryl Conner, Mar. 16, 1980. B.S. in Forestry, U. Mon., 1951; M.S., W.Va. U., 1953; Ph.D. in Plant Pathology, Cornell U., 1962. Plant pathologist N.E. Forest Expt. Sta., U.S. Forest Service, 1956-62; plant pathologist Rocky Mountain Forest and Range Expt. Sta., Ft. Collins, Colo., 1962-82; v.p., dir. Eastglade Corp., Ft. Collins, 1982—; affiliate prof. Colo. State U., 1967—. Contbr. articles to profl. jours. Served with U.S. Army, 1954-56. Mem. Am. Phytopathol. Soc., Mycol. Soc. Am. Congregationalist. Subspecialties: Plant pathology; Taxonomy. Current work: Diagnosis of plant disease and plant disease control. Home and Office: 1406 Peterson St Fort Collins CO 80524

STALEY, RALPH HORTON, chemist; b. Boston, Mar. 15, 1945; s. Carroll Hallowell and Patricia Lewis (Horton) S. A.B., Dartmouth Coll., 1967; Ph.D., Calif. Inst. Tech., 1976. Asst. prof. chemistry M.I.T., 1975-81; research leader, central research and devel. dept. E. I. DuPont de Nemours & Co., Wilmington, Del., 1981—. Contbr. revs., numerous articles to profl. jours. Mem. Am. Chem. Soc. Subspecialties: Physical chemistry; Catalysis chemistry. Home: 7 Stage Rd Newark DE 19711 Office: DuPont Co E356 Wilmington DE 19898

STALKER, KENNETH WALTER, mech. engr.; b. St. John, Kans., Oct. 3, 1918; s. Walter Richard and Bertha (Bisset) S.; m. Eva Leona Stalker, Feb. 7, 1947. B.A., U. Colo., 1941; LL.B., LaSalle U., 1955; postgrad., Coll. Engring. U. Ark., 1964-66. Prodn. mgr. Wright Mfg. Co., Erie, Pa., 1942-45; process engr. DeLaval Steam Turbine Co., Trenton, 1946-51; mfr. mfg. engring and process devel. Gen. Electric Co., Cin., 1950-59; mgr. engring. Goodman Mfg. Co., Chgo., 1959-64; cons. engr. Gen. Electric Co., Cin., 1966-81, Pratt & Whitney Aircraft Co., West Palm Beach, Fla., 1981—. Chmn. Republican Precinct Com., Chgo., 1963-64. Recipient William Badger award Gen. Electric Co., 1970. Mem. ASME. Presbyterian. Lodges: Masons; Shriners. Subspecialties: Mechanical engineering; Metallurgical engineering. Current work: Devel. of mfg. processes for future products to improve design and achieve lower cost production, including mining machinery, aircraft engines, missiles, space vehicles. Patentee hot isostatic pressing of castings, inertia welding, using plasma guns to fabricate components, coldformed components for aircraft engines.

STALL, ROBERT EUGENE, plant pathologist, educator; b. Leipsic, Ohio, Dec. 11, 1931; s. Charles Alvin and Blanche (Frayer) S.; m. Laura Elizabeth Burkholder, Aug. 9, 1931; children: Ronald Dean, David Allen. B.S., Ohio State U., 1953, Ph.D., 1957. Asst. prof. plant pathology U. Fla., 1957-64; assoc. prof., 1964-69, prof., 1969—. Fellow Am. Phytopath. Soc.; mem. Fla. State Hort. Soc. Methodist. Subspecialty: Plant pathology. Current work: Research with bacteria that cause plant disease. Home: 1507 NW 36th Way Gainesville FL 32605 Office: Plant Pathology Dept U Fla Gainesville FL 32611

STALLARD, RICHARD ELGIN, dentist, health administrator; b. Eau Claire, Wis., May 30, 1934; s. Elgin Gale and Caroline Francis (Betz) S.; m. Norma Ann Woock, Oct. 15, 1956 (dec. 1973); children: Rondi Lynn, Alison Judith; m. Jaxon Shirley Sandlin, May 2, 1974; 1 son, Elgin Sandlin. B.S., U. Minn., 1956, D.D.S., 1958, M.S., 1959, Ph.D., 1962. Co-dir. periodontal research Eastman Dental Center, Rochester, N.Y., 1962-65; prof., head dept. periodontology Sch. Dentistry, U. Minn., Mpls., 1965-68, adj. prof. public health, 1976—; asst. dir. (Eastman Dental Center), 1968-70; prof. anatomy Boston U. Sch. Medicine, 1970-74; asst. dean Sch. Grad. Dentistry, dir. clin. research center Boston U. Sch. Grad. Dentistry, 1970-74; dental dir., head dept. periodontology Group Health Plan, Inc., St. Paul, 1974-79; exec. v.p., dental dir. Minndent, Inc., Mpls., 1980-82; pres. R.E. Stallard & Assocs., Mpls., 1982—; cons. USAF, 1968—, U.S. Navy, 1971-75; mem. tng. grant com. NIH/Nat. Inst. Dental Research, 1969-72; edn. cons. Project Vietnam, AID, Saigon, 1969-74; mem. grants and allocations com. Am. Fund for Dental Health, 1976—. Author preventive dentistry textbook, articles in profl. jours. Recipient Meritorious Achievement citation for dental research and edn. Boston U., 1970. Fellow Am., Internat. colls. dentists, Acad. Gen. Dentistry, Internat. Coll. Oral Implantology (pres. 1980-81); mem. Am. Acad. Periodontology (pres. 1974), Am. Acad. Dental Group Practice, Fedn. Dentaire Internationale, Royal Soc. Medicine (Eng.), Am. Acad. Dental Radiology, Det Danske Akademi Oral Implantologi (Denmark), Am. Acad. Dental Spltys. (pres. 1971-75), Am. Pub. Health Assn., Omicron Kappa Upsilon, Sigma Xi. Club: Alumni (Mpls.). Subspecialties: Implantology; Preventive dentistry. Current work: Research and education into the cause and prevention of dental disease and the replacement of portions of the dental apparatus lost through disease or accident by means of transplants or implants. Home: 4200 W 44th St Edina MN 55424 Office: 7645 Metro Blvd Minneapolis MN 55435

STALLWORTH, CHARLES DOROTHEA, JR., psychologist; b. Riderwood, Ala., July 4, 1940; s. Charles D. and Annie (Horn) S. B.S., Tenn. State U., Nashville, 1963, M.S., 1966; postgrad, Calif. Sch. Profl. Psychology, 1977-79, U. Ky., 1980, Internat. Coll., 1981-82, U. South Ala., summer 1967, Tuskegee Inst., summer 1968, Auburn U., summer 1969, Harvard U., summer 1975. Psychiat. asst. Hubbard Hosp., Nashville, 1964-66; counselor, math tchr. North Central High Sch., Chatom, Ala., 1969-70; supr. adult edn. Washington County Bd. Edn., Chatom, 1968-70; dir. counseling ctr. Albany State Coll., Ga., 1970—; cons. Peace Corps, 1979-82. Bd. dirs. Dougherty County CODAC, Inc., Albany, 1973-77, Albany Area Council V.D. Control, 1975-77. Recipient grants HEW, 1970-77, U.S. Office Edn., 1972. Mem. Am. Psychol. Assn. (assoc.), Alpha Phi Alpha. Democrat. Baptist. Subspecialties: Behavioral psychology; Cognition. Current work: Impact of the affective domain on learning outcomes and on the application of cognative therapies as a means of controlling negative effects. Home: 805 E 4th Ave Albany GA 31705 Office: Dept Psychology Albany State Coll Albany GA 31705

STALNAKER, JOHN MARSHALL, emeritus educational corporation administrator; b. Duluth, Minn., Aug. 17, 1903; s. William Edward and Sara (Tatham) S.; m. Ruth Elizabeth Culp, July 29, 1933 (dec. Apr. 1968); children—John Culp, Robert Culp, Judith S. Weikel; m. Edna Remmers, Aug. 21, 1969. B.S. with honors, U. Chgo., 1925; A.M. in Psychology, 1928; LL.D., Purdue U., 1956, Centre Coll., 1960. Tchr. rural sch., Hardisty, Alta., Can., 1922; tchr. math., sci. Harvard Sch. for Boys, 1925-26; instr. psychology, spl. research asst. to pres., 1926-30; asst. prof. edn. and psychology (on leave) Purdue U., 1930-31; dir. attitude measurement, athletic survey U. Minn., 1930-31; examiner (instr.) bd. exams. U. Chgo., 1931-36, asst. prof., 1936-37, asso. prof., 1937-44; prof. Princeton, 1944-45; research asso. Coll. Entrance Exam Bd., 1936-37, cons. examiner, 1937-42; asso. sec., 1942-45, dir. Navy test research unit, 1942-45; contractor's tech. rep. for N.D.R.C. project N-106, 1942-45; dir. Army-Navy Coll. Qualifying Test, 1943-45; dean students, prof. psychology Stanford, 1945-49; prof. psychology, coordinator psychol. scis. and services Ill. Inst. Tech., 1949-51; cons. Fund for Advancement Edn., 1952-55, NSF, 1952-56; dir. studies Assn. Am. Med. Colls., 1949-55; pres. Nat. Merit Scholarship Corp., 1955-69, pres. emeritus, hon. dir., 1969—; mem. bd. North Shore Mental Health Assn., 1967-70; mem. Northfield Twp. Mental Health Adv. Bd., 1978-79, Ill. Bd. Higher Edn., 1969-75; trustee, dir. Pepsi-Cola Scholarship Bd., 1945-54; mem. adv. com. Fgn. Service Exam. Dept. State, 1941-51; sci. adv. bd. to Chief of Staff, USAF, 1950-53; mem. bd. fgn. scholarships Dept. State, 1962-67, chmn., 1962-65. Contbr. articles to edn., psychol. jours. Recipient Certificate Merit Pres. U.S., 1948, Distinguished Civilian Service award Sec. Navy, 1946; citation for outstanding contbn. to edn. Nat. Assn. Secondary Sch. Prins., 1970; Distinguished Service medal Coll. Entrance Exam. Bd., 1976. Fellow Am. Psychol. Assn.; mem. Psychometric Soc., Am. Edn. Research Assn., Phi Beta Kappa Assos., Sigma Xi, Tau Kappa Epsilon. Clubs: Univ. (Chgo.); Cosmos (Washington). Current work: Discovery and stimulation of gifted and talented. Home: 3839 Tangier Terr Sarasota FL 33579

STALTER, RICHARD, educator; b. Montvale, N.J., Jan. 16, 1942; s. Lester Clifford and Betty (Rivell) S.; m. Elaine Preston, Aug. 25, 1968; children: Laurie. B.S., Rutgers U., 1963; M.S., U. R.I., 1966; Ph.D., U. S.C., 1968; postgrad., High Point Coll., 1968-69, Pfeiffer Coll., 1969-70. Asst. prof. S.C. Pollution Control Authority, Columbia, 1970-71; with St. John's U., Jamaica, N.Y., 1971—, dir. environ. studies program, 1974—. NSF grantee, 1975—; Danforth Found. grantee, 1976; Eastern Parks and Monuments Assn. grantee, 1980. Mem. Assn. Souteastern Biologists, S.C. Acad. Sci., Ecol. Soc. Am., So. Appalachian Bot. Club, Northeastern Weed Sci. Soc., Assn. Souteastern Biolgosits, S.C. Acad. Sci. Skull and Circle Honor Soc., Sigmz Xi, Phi Sigma. Republican. Episcopalian. Club: Rugby. Subspecialty: Ecology. Current work: The ecology of salt marsh, sand dune, and maritime forests, floristic studies of barrier islands. Office: St Johns U Jamaica NY 11439

STAMBAUGH, JOHN EDGAR, JR., physician, researcher; b. Everrett, Pa., Apr. 30, 1940; s. John Edgar and Rhoda Irene (Becker) S.; m. Shirley Fultz, Sept. 15, 1940; children: Bambi, Michele, Michael, Heather. B.S., Dickinson Coll., 1962; M.D., Jefferson Med. Coll., 1966, Ph.D. in Pharmacology, 1968. Intern Thomas Jefferson Univ. Hosp., 1966-67, resident in medicine, 1968-70, fellow in oncology, 1970-72; asst. prof. pharmacology Jefferson Med. Coll., Phila., 1968-70, assoc. prof., 1970-82, prof., 1982—, asst. prof. medicine, 1972—; practice medicine specializing in oncology and hematology, Woodbury, N.J., 1972—; research clin. pharmacology Chronic Pain Ctr., Haddon Heights, N.J., 1982—. Contbr. articles on drug kinetics, analgesic trials, Phase I and II trials of antineoplastics, antibiotics, antiemetics and analgesics to profl. jours. Fellow Am. Soc. Exptl. Biology; mem. AMA, Am. Soc. Clin. Pharmacology, Am. Soc. Clin. Oncology, Am. Assn. Cancer Research, Am. Pain Soc., Internat. Pain Soc., Eastern U.S. Pain Soc., Am. Assn. Clin. Research, Sigma Xi. Subspecialties: Internal medicine; Pharmacokinetics. Current work: Cancer chemotherapy, chronic pain therapy, drug kinetics. Office: 119 White Horse Pike Haddon Heights NJ 08035

STAMPER, HUGH BLAIR, JR., health science adminstr., microbiologist, immunologist; b. Warren, Ohio, Dec. 13, 1943; s. Hugh Blair and Berniece Josephine (Polaski) S.; m. Gwen Lee Demshok, July 14, 1972; children: Lucy Demshok, Kevin Demshok. Student, St. Joseph's Coll., Rensselaer, Ind., 1961-62, Youngstown State U., 1963-65; B.Sc., Ohio State U., 1967, M.Sc., 1968, Ph.D., 1972. Tchr. Newark (Ohio) City Schs., 1968-69; bacteriologist Licking County Meml. Hosp., Newark, 1968-69; asst. prof. biology Old Dominion U., Norfolk, Va., 1972-75; research assoc. SUNY Downstate Med. Center, Bklyn., 1975-77; health scientist adminstr., div. lung diseases Nat. Heart, Lung and Blood Inst., NIH, Bethesda, Md., 1977—. Contbr. articles to profl. jours. Old Dominion U. Research Found. grantee, 1972, 73. Mem. Am. Soc. for Microbiology, AAAS, Am. Assn. Immunologists, Sigma Xi. Club: TSS Flying (Gaithersburg, Md.). Subspecialties: Pulmonary medicine; Immunobiology and immunology. Current work: The fibrotic process in the lungs. Pulmonary, fibrosis, immunology. Office: NHLBI/NIH Westwood Bldg Room 6A05 Bethesda MD 20205

STANBRO, WILLIAM DAVID, chemist; b. St. Louis, Nov. 29, 1946; s. William Woodrow and Rosemary Muriel (Conners) S.; m. Helen Frances deChabert, June 20, 1969; children: Jennifer Margaret, Elizabeth Marie, Patrick William. B.S., George Washington U., 1968, Ph.D., 1972. Research asst. George Washington U., Washington, 1967-68, teaching asst., 1971-72; research asst. Carnegie Inst. Washington, 1969-70; NSF intern applied physics lab. Johns Hopkins U., Laurel, Md., 1972-73, chemist, 1973—. Contbr. articles to profl. jours. Mem. Am. Chem. Soc., AAAS. Roman Catholic. Subspecialties: Kinetics; Laser-induced chemistry. Current work: Chemical kinetics in aqueous solution, photochemistry of singlet delta oxygen. Office: Applied Physics Lab Johns Hopkins U Johns Hopkins Rd Laurel MD 20707

STANCEL, GEORGE MICHAEL, pharmacology educator; b. Chgo., Dec. 29, 1944; s. George M. and Josephine (Zavislak) S.; m. Mary Lee, Aug. 19, 1972; children: George Michael, Gregory Alan, Emily Ann. B.S. in Chemistry, Coll. St. Thomas, 1966; Ph.D. in Biochemistry, Mich. State U., 1970. Postdoctoral fellow U. Ill.-Urbana, 1970-72; asst. prof. pharmacology U. Tex. Med. Sch., Houston, 1972-77, assoc. prof., 1977-82, prof., 1982—. Subspecialty: Pharmacology. Office: Dept Pharmacology U Tex Med Sch PO Box 20708 Houston TX 77025

STANDISH, ERLAND MYLES, JR., astronomer; b. Hartford, Conn., Mar. 5, 1939; s. Erland Myles and Hilda Chaffee (Crosby) S.; m. Jeannine Patricia Roy, July 20, 1968; children: Erland Myles III, Brian Roy, Kevin James. B.A., Wesleyan U., 1960, M.A., 1962; Ph.D., Yale U., 1967. Asst. prof. Yale U., New Haven, Conn., 1968-72; mem. tech. staff Jet Propulsion Lab., Pasadena, Calif., 1972—. Mem. editorial bd.: Celestial Mechanics Jour, 1977; contbr. articles in field to profl. jours. Recipient Group Achievement awards NASA, 1979, 81. Mem. Internat. Astron. Union, Am. Astron. Soc. Subspecialty: Planetary science. Current work: Devel. accurate planetary ephemerides spacecraft navigation. Home: 5180 Princess Anne Rd La Canada CA 91011 Office: Jet Propulsion Lab 264-664 CA 91109

STANEK, KAREN ANN, physiologist, researcher; b. Orofino, Idaho, Oct. 19, 1950; s. Francis James and Emma Elizabeth (Elmshaeuser) S. B.S., U. Idaho-Moscow, 1973; M.S., U. Wis.-Madison, 1975, Ph.D., 1978. Postdoctoral fellow U. Miss. Med. Ctr., Jackson, 1978-79, instr., 1979-80, asst. prof. physiology, 1980—, cons. microsphere expts., 1980—. Counselor Young Servants for Christ, Jackson, Miss., 1981. Recipient Nat. Research Service award NIH, 1979-80, grantee, 1980-83. Mem. Phi Beta Kappa, Sigma Delta Epsilon, Phi Sigma. Republican. Lutheran. Subspecialty: Physiology (medicine). Current work: Quantitating changes in flow and resistance patterns observed in the development and treatment of hypertension in conscious rats. This is accomplished by innovative radioactive microsphere techniques which I have developed and verified. Home: 1233 Melwood Pl Jackson MS 39206 Office: U Miss Med Center 2500 N State St Jackson MS 39216

STANFORD, LAURENCE RALPH, neuroscientist; b. New Haven, Nov. 20, 1949; s. Ralph G. and Anne A. (Moore) S. B.A., So. Conn. State Coll., 1971, M.S., 1973; Ph.D., U. Ill., 1979. Lectr. vet. bioscis. U. Ill.-Urbana, 1975-79; fellow in neurobiology and behavior SUNY-Stony Brook, 1979-81, research assoc., 1981, research asst. prof., 1981-82; asst. prof. dept. structural and functional scis. Coll. Vet. Medicine, U. Wis.-Madison, 1982—; prin. investigator, project dir. NIH Grant, 1982—. Contbr. articles to profl. jours. Mem. AAAS, Soc. for Neurosci., N.Y. Acad. Scis. Subspecialties: Neurophysiology; Neurobiology. Current work: Researcher in structure/function correlations and their development in the mammalian visual system using intracellular anatomical and physiological microelectrode recording techniques. Office: Waisman Ctr for Mental Retardation and Human Devel Biomed Unit U Wis Madison WI

STANFORD, ROBERT ERNEST, operations research analyst; b. Montgomery, Ala., May 26, 1944; s. Robert Ernest and Carolyn Ann (Wagner) Stanford Van R.; m. Janet Lee Zicarelli, Dec. 4, 1982. B.A., U. of South, Sewanee, Tenn., 1965; M.S., Ga. Inst. Tech., 1967; Ph.D., U. Calif.-Berkeley, 1971. Lectr. Calif. State U., Hayward, 1971-73; U. Calif.-Berkeley, 1973-74, U. Calif.-Riverside, 1974-76; asst. prof. Auburn U., Ala., 1976-79, assoc. prof., 1980-81; assoc. prof. ops. research U. Ala., Birmingham, 1982—; research assoc. U.S. Naval Postgrad. Sch., Monterey, Calif., 1979. Contbr. articles to profl. jours. NASA/Am. Inst. Indsl. Engring. fellow, 1977. Mem. Ops. Research Soc. Am., Inst. Mgmt. Sci., Sigma Pi Sigma. Subspecialties: Operations research (engineering); Operations research (mathematics). Current work: Manpower system models, applied stochastic processes, decision support systems, productivity of white collar labor, quality control analysis. Home: 2615 Montevallo Rd Birmingham AL 35223 Office: Dept Econs Univ Ala Birmingham AL 35294

STANG, PETER JOHN, chemistry educator; b. Nurnberg, Germany, Nov. 17, 1941; came to U.S., 1956, naturalized, 1963; s. John and Margaret (Pollman) S.; m. Christine S. Schirmer, June 12, 1969; children: Antonia, Alexandra. B.S., DePaul U., 1963; Ph.D., U. Calif.-Berkeley, 1966. Instr. Princeton U., N.J., 1968-69; asst. prof. U. Utah, Salt Lake City, 1969-75, assoc. prof., 1975-79, prof. chemistry, 1979—. Editorial adviser: Acad. Press; Author: Vinyl Cations, 1979; assoc. editor: Jour. Am. Chem. Soc, 1982—. Recipient Alexander von Humboldt Sr. U.S. Scientist award, 1977. Mem. Am. Chem. Soc., Chem. Soc. (London), AAAS. Subspecialty: Organic chemistry. Current work: Mechanistic organic chemistry: unsaturated reactive intermediates; vinyl cations, carbenes, dication ether salts; antitumor agents, organometallic chemistry. Office: Chemistry Department U. of Utah Salt Lake City UT 84112

STANGE, JAMES HENRY, architect; b. Davenport, Iowa, May 25, 1930; s. Henry Claus and Norma Strange; m. Mary Suanne Peterson, Dec. 12, 1954; children: Wade Weston, Drew Dayton, Grant Owen. B.Arch., Iowa State U., 1954. Lic. architect, Nebr., Iowa, Kans., Mo., Okla. Designer Davis & Wilson, Lincoln, Nebr., 1954-62, v.p., 1962-68; v.p., sec. Davis/Fenton/Stange/Darling, Lincoln, 1968-76, pres., 1976-78, pres., chmn. bd., 1978—. Archtl. works include Lincoln Hosp, 1974, Hastings YMCA, 1980, Nebr. Wesleyan U. Theatre, 1980, Bryan Hosp, 1982, others. Pres. Lincoln Ctr. Assn., 1979; pres. Capitol Assn. Retarded Citizens, 1972. Recipient numerous design awards AIA. Mem. AIA, Nebr. Soc. AIA (pres. 1968, dir. 1964-65), Am. Assn. Health Planners, Lincoln C. of C. (dir.), Interfaith Forum on Religion, Art, Architecture. Republican. Presbyterian. Clubs: Crucible, Hillcrest Country (past pres.), Executive (past pres.). Subspecialties: Solar energy; Environmental engineering. Current work: Passive solar system integration and application to existing or new structure; integration of structures into existing environmental conditions. Home: 3545 Calvert Lincoln NE 68506 Office: 226 Sutart Bldg Lincoln NE 68508

STANGE, MORTON DOUGLAS, geophysical sciences educator; b. Corpus Christi, Tex., July 5, 1945; s. Curtis Fenwick and Martha Evelyn (Bowker) S.; m. Sharon Irene Cunningham, Dec. 1, 1969; children: Valerie, Marc. B.S. in Chemistry, U. Chgo., 1966, M.S. in Geology, 1968; Ph.D., Yale U., 1972. Postdoctoral fellow dept. geology and geophysics Yale U., New Haven, 1972-74; asst. prof. geology U. Ariz., Tucson, 1974-78, assoc. prof., 1978-82, prof. geology, 1982—; vis. staff U.S. Geol. Survey, Denver, 1981. Sloan research fellow, 1978; grantee Dreyfuss Found., 1983. Mem. Am. Geophys. Union, AAAS, Geochem. Soc., Am. Soc. Limnology and Oceanography, Phi Beta Kappa, Sigma Xi. Subspecialties: Oceanography; Geochemistry. Office: Werik Bldg 5399 E 29th St Tucson AZ 85711

STANLEY, STEVEN MITCHELL, paleobiologist, educator; b. Detroit, Nov. 2, 1941; s. William Thomas and Mildred Elizabeth (Baker) S.; m. Nell Williams Gilmore, Oct. 11, 1969. A.B. with highest honors, Princeton U., 1963; Ph.D., Yale U., 1968. Asst. prof. U. Rochester, 1967-69; asst. prof. paleobiology Johns Hopkins U., 1969-71, assoc. prof., 1971-74, prof., 1974—; assoc. in research Smithsonian Instn., 1972—. Author: Relation of Shell Form to Life Habits in the Bivalvia, 1970, (with D.M. Raup) Principles of Paleontology, 1971, Macroevolution: Pattern and Process, 1979, The New Evolutionary Timetable: Fossils, Genes, and the Origin of Species, 1981; editorial bd.: Am. Jour. Sci, 1975—, Paleobiology, 1975—, Evolutionary Theory, 1973—. Recipient Outstanding Paper award Jour. Paleontology, 1968, Allan C. Davis medal Md. Acad. Scis., 1973; Guggenheim fellow, 1981. Fellow Geol. Soc. Am. (chmn. Penrose com. 1978); mem. Paleontol. Soc. (councilor 1976-77, Charles Schuchert award 1977), Soc. for Study Evolution. Subspecialties: Paleobiology; Evolutionary biology. Current work: Macroevolution as documented in the fossil record; extinction; functional morphology. Home: 1110 Bellemore Rd Baltimore MD 21210 Office: Dept Earth and Planetary Scis Johns Hopkins U Baltimore MD 21218

STANNARD, JAN GREGORY, dental materials educator, chemurgist; b. Detroit, Dec. 30, 1953; s. Frank Kempster and Edith (Olmsted) S.; m. Roberta Anne Salay, Sept. 5, 1981. B.S. in Med. Chemistry, U. Mich., 1976, M.S. in Polymer Chemistry, 1978, Ph.D. in Macromolecular Sci. and Engring. and Dental Materials, 1981. Chemist Dow Chem. Co. U.S.A., Midland, Mich., 1972-74; praktikant Hoffman-LaRoche, Basel, Switzerland, 1974-75; research assoc. U. Mich., 1974-78; asst. prof. dental materials U. Nebr.-Lincoln, 1981—; co-owner Design Interactions, Lincoln, 1981—; cons. in field. Contbr. articles to profl. jours.; author 2 patent disclosure documents. Recipient Dept. Commerce award, 1971, USN Sci. award, 1971, Internat. Sci. and Engring. award Sci. Service, 1971, Mich. Soc. Profl. Engrs. award, 1972, award of Excellence Gen. Motors, 1972, Key to City of Flint, Mich., 1972; Westinghouse Sci. Talent Search Scholar, 1972; Welch sci. scholar, 1972; Nat. Research Service award Nat. Inst. Dental Research, 1977-81. Mem. Internat. Assn. Dental Research, N.Y. Acad. Sci., Nebr. Acad. Sci., Mich. Sci. Research Club, Assn. Study of Higher Edn., Sigma Xi. Subspecialties: Biomaterials. Current work: Biomedical research; development and testing of new biocompatible materials and devices for internal and external interfacing. Home: 5523 S 31-12 Lincoln NE 68516 Office: Dental Materials U Nebr Lincoln NE 68583

STANSFIELD, WILLIAM DONALD, biology educator; b. Los Angeles, Feb. 7, 1930; s. William F. and Myrtle E. (Hall) S.; m. Janis R. Doty, July 23, 1933; children: Lorraine, Lynn Dee, William Donald. B.S. in Agr, Calif. Poly. State U., 1952; M.A. in Agrl. Edn, 1960; M.S. in Genetics, U. Calif.-Davis, 1962, Ph.D., 1963. Instr. vocat. agr. Fortuna (Calif.) Union High Sch., 1957-59; instr. biol. scis. Calif. Poly. State U., 1963—. Author: Genetics—Schaum's Outline Series, 1969-83, The Science of Evolution, 1977, Serology and Immunology—A Clinical Approach, 1981. Served as line officer USNR, 1953-57. Mem. Am. Genetic Assn., Genetics Soc. Am., Soc. Study of Evolution, Sigma Xi. Subspecialty: Immunogenetics. Current work: Genetics, biological sciences. Home: 653 Stanford Dr San Luis Obispo CA 93401 Office: Biological Sciences Dept California Polytechnic State University San Luis Obispo CA 93407

STANTON, GLENNON JOHN, microbiologist, educator, researcher; b. Rock Springs, Wyo., Nov. 18, 1931; s. Glennon George and Nan (Wortheim) S.; m. Beverly M. Doak; children: James, Lori, Barbara, Michael, Shauna. B.A., Stanford U., 1956; Ph.D., U. Utah, 1967. Asst. prof. microbiology U. Tex. Med. Br., Galveston, 1967-77, assoc. prof., 1977-81, prof., 1981—; cons. in field. Contbr. to profl. jours., books, 1964—. USPHS grantee, 1976—. Mem. Am. Soc. Microbiology, Ocular Microbiology and Immunology Group, Assn. for Research and Vision in Ophthalmology. Subspecialties: Immunobiology and immunology; Virology (biology). Current work: Host defenses and viral pathogenesis; interferons; lymphokines; lymphokine induced cell communication; leukocyte defenses against viral disease and cancer. Home: 9214 San Jose Texas City TX 77590 Office: Dept Microbiology U Tex Med Br Galveston TX 77550

STANTON, RALPH, mathematics and computer science educator, researcher; b. Lambeth, Ont., Can., Oct. 21, 1925; s. Gordon Wyman and Ida Maude (Robertson) S. B.A., U. Western Ont., 1944; M.A., U. Toronto, 1945, Ph.D., 1948; C.L.P., Rio de Janeiro, 1946; D.Sc., U. Newcastle, Australia, 1979. Instr. math. U. Mich., Ann Arbor, 1948-49; asst. prof. math. U. Toronto, 1949-57; dean grad. studies U. Waterloo, Ont., Can., 1957-66, head dept. math, 1957-66; prof. math. and computer sci. York U., Toronto, 1967-69; head dept. computer sci. U. Man., Winnipeg, 1970—; pres. Charles Babbage Research Centre, St. Pierre, Man., 1973—, Utilitas Mathematica Publ. Inc., Winnipeg, 1972—. Author: Numerical Analysis, 1961, Algebra and Vector Geometry, 1965; editor books. Recipient Founders medal U. Waterloo, 1982. Fellow Inst. Math and Its Applications, Royal Statis. Soc.; mem. Internat. Statis. Inst., Combinatorial Math. Soc. Australia. Subspecialties: Algorithms; Statistics. Current work: Algorithmic methods in combinatorics and graph theory. Office: Dept Computer Sci U Man Winnipeg MB Canada R3T 2N2 Home: 50 Lord Ave Winnipeg MB Canada R3V 1G5

STAPLES, GEORGE EMMETT, research veterinarian; b. Kanosh, Utah, Nov. 2, 1918; s. Grant DeVere and Grace E. (Whitaker) S.; m. Katherine Elna Stanford, Dec. 29, 1948; children: Fred S., James P., Gayla M., David G., Joseph G., Daniel R., Pter M. B.S., Utah State Agrl. Coll., 1947; M.S., S.D. State U., 1949; D.V.M., Colo. State U., 1954. Asst. prof. animal sci. S.D. State U., Brookings, 1948-50; pvt. practice vet. medicine, Afton, Wyo., 1954-57; area veterinarian U.S. Dept. Agr., Utah, 1957-60; research veterinarian Morris Research Labs., Inc., Topeka, Kans., 1960-64; assoc. prof. vet. sci. N.D. State U., Fargo, 1965—. Contbr. articles to profl. jours. Justice of peace, Meriden, Kans., 1963-64. Served to lt. USN, 1942-45. Mem. N.D. Vet. Med. Assn., Sigma Xi. Mormon. Subspecialties: Microbiology (veterinary medicine); Dietary therapeutics for animals. Current work: Research with infectious bovine keratoconjunctivitis and with neonatal diarrhea in farm animals; also surgical innovations. Patentee in field. Home: Route 3 Box 72 Fargo ND 58103 Office: ND State U Fargo ND 58105

STAPLES, RICHARD CROMWELL, plant biochemist; b. Hinsdale, Ill., Jan. 29, 1926; s. George Allen and Ruth (Larken) S.; m. Mildred J. Durdik, Aug. 7, 1954; children: Cynthia, Laura, Robert. B.S., Colo. State U., 1950; A.M., Columbia U., 1954, Ph.D., 1957. With Boyce Thompson Inst. Plant Research, Cornell U., Ithaca, N.Y., 1952—, program dir. plant stress, 1966—; vis. scientist dept. biochemistry U. Wis.-Madison, 1961-62; policy analyst Policy Research and Analysis Div., NSF, 1975-76. Editor: (with G.A. Toenniessen) 3 books in field including Plant Disease Control, 1981; contbr. articles to profl. jours. Served with USN, 1944-46, 50-52. Alexander von Humboldt Found. sr. U.S. scientist awardee, 1980; Humboldt awardee Institut für Biologie III, Aachen, W.Ger., 1981-82. Mem. Am. Soc. Plant Physiologists, Am. Phytopath. Soc. Subspecialties: Plant pathology; Plant physiology (biology). Current work: Basic research on sensing in fungi especially development of infection structures in the rust fungi. Office: Boyce Thompson Inst Tower Rd Ithaca NY 14850

STAPLETON, LEROY EARL, therapist, clinical psychologist, foundation executive; b. Harlan, Ky., Nov. 9, 1923; s. William Jesse and Joyce Mae (Belcher) S.; m. Mary Lillian Wynn, May 8, 1945; children: Teresa Dwen, Earl Patrick, Carrie Ellen. B.E., U. Nebr.-Omaha, 1957; M.A., Stanford U., 1959; Ph.D., Fla. State U., 1973. Lic. psychologist, Fla. Commd. 2d lt. U.S. Air Force, 1942, advanced through grades to lt. col., 1961, served as pilot, navigator, radar operator, bombardier, gunner, mechanic, performance engr., Eng., Africa, Pacific Islands, ret., 1966; prof. Troy State U., Ft. Walton Beach, Fla., 1973-75; dir. curriculum Air U., USAF, Montgomery, Ala., 1964-67; dir. psychol. services Okaloosa Bd. Edn., Ft. Walton Beach, 1968-77; psychologist, Ft. Walton Beach, 1979-81; pres. Child Family Research Found., Ft. Walton Beach, 1978—; pres., chmn. bd. Fla. Research Found., Ft. Walton Beach, 1978-83; clin. psychologist Fla. Regional Found., Ft. Walton Beach, 1975—; devel. and learning psychologist Okaloosa Bd. Edn., 1968-75; bio-astronautics psychologist USAF Proving Ground, Ft. Walton Beach, 1963-64. Author, creator gravitational bomb table, USAF, 1952; author, designer: Learning Disability Mental Analysis Profile for Children, 1975 (bd. edn. recognition 1975, spl. proclamation Bd. Edn. Okaloosa County 1975). Founder Fla. Regional Family Life Child Devel. Research Found., Ft. Walton Beach, 1978; bd. dirs. Mental Health Clinic, 1970-75, Fla. Crime Commn., Panhandle, 1973-75. Recipient Disting. Flying award USAF, 1952-53, Disting. Educator award Okaloosa County Bd. Edn., 1975. Mem. Fla. Assn. Practicing Psychologists (pres. 1979-81), Am. Psychol. Assn., Fla. Psychol. Assn., Fla. Assn. Sch. Psychologist Ethics (exec. bd. 1970-72), Am. Contract Bridge League (advanced sr. sr. 1963-83), Psi Chi. Methodist. Lodges: Masons; Shriners. Subspecialties: Behavioral psychology; Neuropsychology. Current work: Deep sleep hypnotherapy; hypnotherapy as a functional study of retardation, regression study in behavioral trauma. Featherweight boxing champion USAF, 1943. Home: 583 Pocahontas Dr Fort Walton Beach FL 32548 Office: Fla Regional Family-Life Child Devel Research Found 2000 Lewis Turner Blvd Fort Walton Beach FL 32548

STARFIELD, BARBARA HELEN, physician, educator; b. Bklyn., Dec. 18, 1932; d. Martin and Eva (Illions) S.; m. Neil A. Holtzman, June 12, 1955; children—Robert, Jon, Steven, Deborah. A.B., Swarthmore Coll., 1954; M.D., SUNY, 1959; M.P.H., Johns Hopkins U., 1963. Teaching asst. in anatomy Downstate Med. Center, N.Y.C., 1955-57; intern in pediatrics Johns Hopkins U., 1959-60, resident, 1960-62, dir. pediatric med. care clinic, 1963-66, dir. community staff comprehensive child care project, 1966-67, dir. pediatric clin. scholars program, 1971-76, asso. prof. pediatrics, 1973—, prof. health care orgn., 1975—; cons. HEW. Contbr. articles to profl. jours., Mem. editorial bd.: Med. Care, 1977-79, Pediatrics, 1977—, Internat. Jour. Health Services, 1978—, Med. Care Rev, 1980—. Recipient Dave Luckman Meml. award, 1958; HEW Career Devel. award, 1970-75. Mem. Nat. Acad. Sci. Inst. Medicine (governing council 1981—), Am. Pediatric Soc., Soc. Pediatric Research, Internat. Epidemiologic Assn., Ambulatory Pediatric Assn. (pres. 1980), Am. Public Health Assn., Sigma Xi, Alpha Omega Alpha., Nat. Acad. Sci. Inst. Medicine (governing bd. NRC 1983). Subspecialties: Health services research; Epidemiology. Current work: Cost effectiveness of care; health needs-care of population subgroups; primary care delivery; expidemiology of child health problems. Office: 615 N Wolfe St Baltimore MD 21205

STARK, JACK ALAN, clinical psychologist, educator; b. Hastings, Nebr., Sept. 20, 1946; s. Arlen Odale and Mary Virginia (Dryden) S.; m. Shirley A. Theis, Aug. 1, 1970; children: John, Nick, Suzanne. B.A., St. Francis Coll., Milw., 1968; M.A., U. Nebr., 1970, Ph.D., 1973. Lic. psychologist and rehab. counselor, Nebr.; vocat. expert Social Security Adminstrn. Secondary tchr. Lincoln Parochial Schs., 1968-70, sch. psychologist, 1970-71, counseling psychologist, 1971-73; assoc. prof. med. psychology, dir. family rehab. U. Nebr. Med. Ctr., Omaha, 1973—, dir., 1978-82, 1982—. Editor: Curative Aspects of Mental Retardation—Biomedical and Behavioral Aspects, 1983, Mental Illness in Mental Retardation, 1983; Contbr. articles to jours., presentations to confs., chpts. to books. Tng. grantee HEW, 1968-70; recipient Outstanding Profl. Employee award Greater Omaha Assn. for Retarded Citizens, 1979. Mem. Am. Assn. Mental Deficiency (v.p. vocat. rehab. 1982-84), Assn. for Advancement Community Services (v.p. 1979—). Democrat. Roman Catholic. Subspecialties: Behavioral psychology; Health services research. Current work: Psychology, mental retardation, behavioral medicine. Home: 306 Heavenly Dr Omaha NE 68154 Office: U Nebr Med Ctr 444 S 41st St Omaha NE 68131

STARKEY, WALTER LEROY, mechanical engineer, consultant; b. Mpls., Oct. 5, 1920; s. Harry N. and Rena B. (Towne) S.; m. Bonna B. Preston; children: David H., John M. B.M.E., U. Louisville, 1943; M.Sc., Ohio State U., 1947, Ph.D., 1950. Registered profl. engr., Ohio. Instr. U. Louisville, 1943-46; asst. prof. mech. engring. Ohio State U., 1950-54, asso. prof., 1954-58, prof., 1958-78, prof. emeritus, 1978—; research supr. Ohio State U. Research Found., 1950-78; cons. to industry, 1944—, cons. litigation, 1960—. Author: Motorhome Facts, 1971, 72, 73, 74; contbr. articles on theories of mech. failure, dynamics of machinery, and machine design to profl. jours. Fellow ASME (Machine Design medal 1971). Republican. Subspecialties: Mechanical engineering; Solid mechanics. Current work: Modes of mech. failure; mech. invention. Home and Office: 7000 Coffman Rd Dublin OH 43017

STARKSCHALL, GEORGE, medical physicist; b. Weiden, Germany, Sept. 11, 1946; came to U.S., 1947, naturalized, 1953; 1953; s. Nathan and Rosa (Ginsberg) S.; m. Carol Eisenberg, Apr. 20, 1974; 1 dau., Jessica. B.S., MIT, 1967; A.M., Harvard U., 1968, Ph.D., 1972. Cert. therapeutic radiol. physics Am. Bd. Radiology. Research asso. U. Chgo., 1972-74; physicist VA Hosp., Hines, Ill., 1974-75; radiol. physicist St. Francis Hosp., Evanston, Ill., 1975-77; asst. prof. Chgo. Med. Sch., North Chicago, Ill., 1977-80, U. Kans. Med. Center, Kansas City, 1980—. Mem. Am. Assn. Physicists in Medicine, Am. Coll. Radiology, [illegible], Am. Phys. Soc. Jewish. Subspecialty: Medical physics. Current work: Radiation therapy treatment planning, clin. radiation dosimetry. Home: 9809 Craig Dr Overland Park KS 66212 Office: 39th and Rainbow St Kansas City KS 66103

STARKWEATHER, HOWARD WARNER, JR., chemist; b. Cambridge, Mass., July 20, 1926; s. Howard Warner and Matilda (Golding) S.; m. Elizabeth Muir Focardi, June 5, 1948. A.B., Haverford Coll., 1948; A.M., Harvard U., 1950; Ph.D., Poly. Inst. Bklyn., 1952. Chemist central research and devel. depts. E.I. duPont de Nemours and Co., Wilmington, Del., 1952—. Subspecialties: Polymer chemistry; Physical chemistry. Home: 3931 Heather Dr Wilmington DE 19807 Office: Central Research and Devel Dept Exptl Station E I duPont de Nemours and Co Wilmington DE 19898

STARR, B(ARRY) JAMES, psychology educator; b. Phila., July 19, 1941; s. Oscar and Margarethe Ruth (Weinheimer) S.; m.; children: Andrew Cory, Jamie Lyn. B.S. with high distinction and deptl. honors, Pa. State U., 1965; student, Millersville State Coll., 1961-63; Ph.D., SUNY-Buffalo, 1970. Assoc. research scientist Ctr. for Social Orgn. of Schs., Johns Hopkins U., Balt., 1971-73; Fulbright grantee Lady Irwin Coll., New Delhi, 1974-75, U. Allahabad, India, 1974-75; assoc. prof. dept. psychology Howard U., Washington, 1973-76; profl. assoc. Culture Learning Inst., East-West Ctr., Honolulu, 1977; cons. Lawrence Johnson & Assocs., Washington, 1979—; grad. assoc. prof. dept. psychology Howard U., 1979—; cons. Geomet Inc., 1972, Koba Assocs., 1976-77, Litigation Support Services, Inc., 1980. Referee/reviewer various pub. houses, jours.; Contbr. articles to profl. jours.; editor: spl. edit. jour Topics In Culture Learning, 1977. Sec.-treas. New Democratic Coalition, Erie County, N.Y., 1969. U.S. Dept. Edn. grantee, 1982—; recipient Participant award NSF, 1972, 74, 76. Mem. Am. Psychol. Assn., Soc. Cross Cultural Research (sec.-treas.), Assn. Computing Machinery, N.Y. Acad. Scis., Internat. Assn. Cross Cultural Psychology. Democrat. Current work: Method and measurement in multicultural settings; application of computing technology in research and training settings. Home: 417 Browning Ct Takoma Park MD 20912 Office: Dept Psychology Howard Univ 2401 6th St NW Washington DC 20059

STARR, CHAUNCEY, research inst. exec.; b. Newark, Apr. 14, 1912; s. Rubin and Rose (Dropkin) S.; m. Doris Evelyn Debel, Mar. 20, 1938; children—Ross M., Ariel E. E.E., Rensselaer Poly. Inst., 1932, Ph.D., 1935, D.Engring. (hon.), 1964, Swiss ETH, 1980. Research fellow physics Harvard, 1935-37; research asso. Mass. Inst. Tech., 1938-41; research physicist D.W. Taylor Model Basin, Bur. Ships, 1941-42; staff radiation lab. U. Calif., 1942-43, Tenn. Eastman Corp., Oak Ridge, 1943-46, 1946; chief spl. research N. Am. Aviation, Inc., Downey, Calif., 1946-49, dir. atomic energy research dept., 1949-55, v.p., 1955-66, gen. mgr., 1955-60, pres. div., 1960-66; dean engring. U. Calif. at Los Angeles, 1966-73; cons. prof. Stanford, 1974—; pres. Electric Power Research Inst., 1973-78, vice chmn., 1978—; Dir. Atomic Indsl. Forum. Contbr. sci. articles to profl. jours. Decorated Legion of Honor, France). Fellow Am. Nuclear Soc. (past pres.), Am. Phys. Soc., AAAS (dir.); mem. Am. Inst. Aeros. and Astronautics (sr.), Am. Power Conf., Nat. Acad. Engring., Am. Soc. Engring. Edn., Royal Swedish Acad. for Engring. Scis., Eta Kappa Nu, Sigma Xi. Subspecialties: Fuels and sources; Nuclear fusion. Current work: Risk assessment; impact of energy supply on society. Home: 95 Stern Ln Atherton CA 94025

STARR, RICHARD CAWTHON, educator; b. Greensboro, Ga., Aug. 24, 1924; s. Richard Neal and Ida Wynn (Cawthon) S. B.S. in Secondary Edn, Ga. So. Coll., 1944; M.A., George Peabody Coll., 1947; postgrad. (Fulbright scholar), Cambridge (Eng.) U., 1950-51; Ph.D., Vanderbilt U., 1952. Faculty, Ind. U., 1952—, prof. botany, 1960-76, founder, head culture collection algae, 1953-76; prof. botany U. Tex. at Austin, 1976—; Head course marine botany Marine Biol. Lab., Woods Hole, Mass., 1959-63. Algae sect. editor: Biol. Abstracts, 1959—; editorial bd.: Jour. Phycology, 1965-68, 76—, Archiv für Protistenkunde; asso. editor: Phycologia, 1963-69; Contbr. articles to profl. jours. Trustee Am. Type Culture Collection, 1962-68, 80—. Guggenheim fellow, 1959; sr. fellow Alexander von Humboldt-Stiftung, 1972-73. Fellow AAAS, Ind. Acad. Sci.; mem. Nat. Acad. Scis., Am. Inst. Biol. Scis. (governing bd. 1976-77, exec. com. 1980—), Bot. Soc. Am. (sec. 1965-69, v.p. 1970, pres. 1971, Darbaker prize 1955), Phycological Soc. Am. (past pres., v.p., treas.), Soc. Protozoologists, Internat. Phycological Soc. (sec. 1964-68), Brit. Phycological Soc., Sigma Xi. Current work: Cultivation of algae; sexual reproduction and cell differentiation in colonial green algae; viral infections in algae. Office: Dept Botany Univ Texas Austin TX 78712

STARRFIELD, SUMNER GROSBY, astronomer; b. Los Angeles, Dec. 29, 1940; s. Harold Earnest and Eve (Grosby) S.; m. Susan Lee Hutt, Aug. 6, 1966; children: Barry, Brian, Sara. B.A., U. Calif., Berkeley, 1962; M.A., UCLA, 1966, Ph.D., 1969. Lectr. Yale U., 1967-69, asst. prof., 1969-71; fellow IBM Watson Research Center, 1971-72; asst. prof. physics and astronomy Ariz. State U. Tempe, 1972-75, assoc. prof., 1975-80, prof., 1981—; vis. staff mem. Los Alamos Nat. Lab., 1975—. Contbr. articles to profl. jours. NSF grantee, 1975—; NASA grantee, 1979—; U.S. Israel Binat. Found. grantee, 1981-84. Mem. Internat. Astron. Union, Am. Astron. Soc., Royal Astron. Soc., Astron. Soc. Pacific. Democrat. Club: Puli. Subspecialties: Theoretical astrophysics; High energy astrophysics. Current work: Hydrodynamic stellar evolution, evolution of nova outburst, nucleo synthesis in stars, white dwarf pulsations. Office: Dept Physics Ariz State U Tempe AZ 85281

STARY, FRANK EDWARD, chemistry educator; b. St. Paul, Minn., Jan. 3, 1941; s. Frank C. and Elaine E. (Anderson) S.; m. Sonja G. Dalsbo, Aug. 30, 1964. B.Chem., U. Minn., 1963; Ph.D., U. Cin., 1969. Fellow, instr. U. Calif., Irvine, 1969-72; research assoc. U. Mo., 1972-74; asst. prof. chemistry Maryville Coll., St. Louis, 1974-78, assoc. prof., 1978-81, prof., 1982—; cons in field. Contbr. articles to profl. jours. Recipient Disting. Teaching award Maryville Coll., 1981. Mem. Am. Chem. Soc., Sigma Xi, Phi Lambda Upsilon, St. Louis Audio Soc. Subspecialties: Nuclear magnetic resonance (chemistry); Polymer chemistry. Current work: Pulsed magnetic resonance on solids, chem. lasers, chemiluminescence of polymers. Office: 13550 Conway Rd Creve Coeur MO 63141

STARZAK, MICHAEL EDWARD, chemist; b. Woonsocket, R.I., Apr. 21, 1942; s. Michael and Ida Dolores (Bielagus) S.; m. Anndrea Lee Zahorak, July 22, 1967; children: Jocelyn Ann, Alissa Michelle. Sc.B., Brown U.; Ph.D., Northwestern U., 1968. Acting instr. U. Calif.-Santa Cruz, 1968-69, acting asst. prof., 1969-70; vis. lectr. U. Calif.-Berkeley, 1970, asst. prof., 1970-76; assoc. prof. chemistry SUNY-Binghamton, 1976—. Author: The Physical Chemistry of Membranes. Mem. Biophys. Soc., Am. Chem. Soc., Am. Phys. Soc., Sigma Xi. Roman Catholic. Subspecialties: Biophysics (physics); Biophysical chemistry. Current work: Kinetics of ionic channels in membranes, ion interactions with channels, photochem. studies of ionic channels, calorimetry and proteins, laser photochemistry. Office: Chemistry Dept SUNY Binghamton NY 13901

STASIAK, ROGER STANLEY, environmental health educator, software consultant; b. Oak Park, Ill., May 29, 1943; s. Henry and Irene S.; m. Ann Renee Goodbout, Aug. 6, 1966; children: Christopher Jon, Ann-Marie. B.A., St. Mary's Coll., Winona, Minn., 1965; M.S., U. Ariz., 1968; M.S.P.H., U. N.C., 1972, Ph.D., 1975. Registered sanitarian. Assoc. prof. environ. and occupational health Calif. State U.-Northridge, 1975-79; chmn. dept. environ. health sci. Eastern Ky. U., Richmond, 1979-81; prof. environ. health U. Ga., Athens, 1981—. Served to comdr. Med. Service Corps USN, 1969-71. Decorated Bronze Star. Mem. AAAS, Am. Pub. Health Assn. Subspecialties: Preventive medicine; Environmental toxicology. Current work: Teaching/research in environmental toxicology.

STASZESKY, FRANCIS MYRON, consultant; b. Wilmington, Del., Apr. 16, 1918; s. Frank J. and Ruth (Jones) S.; m. Barbara F. Kearney, May 30, 1943; children—Francis Myron, John B., Barbara J., Faith A., Paul D. B.S. in Mech. Engring, Mass. Inst. Tech., 1943, M.S., 1943. Mech. engr. Union Oil Co. Calif., Los Angeles, 1943-45; with E.I. duPont de Nemours Co., Wilmington, Del., 1946-48; joined Boston Edison Co., 1948, supervising engr. design and constrn., 1948-57, supt. engring. and constrn. dept., 1957-64, v.p., asst. to pres., 1964-67, exec. v.p., 1967-79, pres., chief operating officer, 1979-83; cons., 1983—; dir. Boston Edison Co., 1968-83; dir. Shawmut Corp., Shawmut Bank Boston N.A. Fellow ASME; mem. IEEE, Nat. Acad. Engring., Engring. Soc. New Eng. (pres. 1961-62). Clubs: Engineers, Algonquin (Boston); Brae Burn Country. Subspecialties: Electrical engineering; Energy consulting. Current work: Independent consultant to corporate management in capital intensive industry and on issues of future energy supply. Address: 144 Chestnut Circle Lincoln MA 01773

STAUBUS, JOHN REGINALD, dairy sci. educator; b. Cissna Park, Ill., Mar. 21, 1926; s. Raymond Reginald and Anna Evelyn (Newell) S.; m. Lorene Lawrence, Sept. 8, 1951; 1 dau., Anna Marie Staubus Jones. B.S., U. Ill., 1950, M.S., 1956, Ph.D., 1959. Mortgage rep. Phoenix Mut. Life Ins. Co., 1951-54; research assoc. dept. dairy sci. U. Ill., 1954-60; asst. prof. dairy sci. Ohio State U., 1960-64, assoc. prof., 1964-69, prof., 1969—. Served with C.E. U.S. Army, 1945-47. Mem. Am. Dairy Sci. Assn., Am. Soc. Animal Sci., AAAS. Methodist. Subspecialties: Animal nutrition; Microbiology. Current work: Ruminant nutrition and management teaching.

STAVELY, JOSEPH RENNIE, plant pathologist; b. Wilmington, Del., May 28, 1939; s. Joseph Glover and Susan Frances S.; m. Nancy Carol Gall, Aug. 15, 1965; 1 son, Joseph Carl. B.S., U. Del., 1961; M.S., U. Wis., 1963, Ph.D., 1965. Grad. research asst. U. Wis., Madison, 1961-65, postdoctoral NIH research fellow, 1965-66; research plant pathologist tobacco Agrl. Research Services, U.S. Dept. Agr., Beltsville, Md., 1966-80, research plant pathologist beans, 1980—. Contbr. articles to profl. jours. Vestryman St. Philip's Episcopal Ch., Laurel, Md., 1977-80; Webelos leader Cub Pack 1349, Silver Spring, Md., 1981-82. NIH postdoctoral research grantee, 1965-66; U.S. Dept. Agr. research grantee, 1980—. Mem. Am. Phytopath. Soc. (pres. Potomac div. 1980), Am. Genetic Assn., Am. Soc. Hort. Sci., Crop Sci. Soc. Am. Episcopalian. Subspecialties: Plant pathology; Plant genetics. Current work: Research on pathogen variability, host-pathogen relationships, and disease resistance in beans, including breeding for disease resistance. Office: Applied Plant Pathology Lab Bldg 004 BARC-W US Dept Agr Beltsville MD 20705

STAVINOHA, WILLIAM B., pharmacology and toxicology educator; b. Temple, Tex., June 11, 1928; s. Adolph Alfonso and Lillian Ruby (Schiller) S.; m. Karen L. Peterson, June 20, 1956; children: Anna Christa, Elizabeth Shannon, Elena Siobhan, Rose Catelin; m. Marilyn Rose Cole, Feb. 8, 1965. B.S., U. Tex.-Austin, 1950, M.S., 1953; Ph.D., U. Tex.-Galveston, 1960. Chief div. toxicology C.A.M.I., FAA, 1960-64, chief toxicology lab., 1964-68; assoc. prof. pharmacology med. br. U. Tex.-San Antonio, 1968-72, prof. pharmacology, 1973—, also chief toxicology div.; cons. Contbr. over 50 sci. articles to profl. pubis. Served with U.S. Army, 1950-52. Recipient Outstanding Alumnus award U. Tex.-Galveston, 1980; NIH; NIMH; Consumer Protection Agy.; Klebry Found.; Morrison Found. Mem. Soc. Toxicology, Soc. Pharmacology and Exptl. Therapeutics, Internat. Soc. Neurochemistry, Am. Soc. Neurochemistry, Soc. Exptl. Biology and Medicine, Neurosci. Soc. Subspecialties: Bioinstrumentation; Toxicology (medicine). Current work: Instrumentation for the study of rapid reactions in the central nervous system; toxicology; neurochemical identification of depression. Office: Univ Tex Health Sci Center 7703 Floyd Curl Dr San Antonio TX 78284

STEADMAN, JAMES ROBERT, plant pathologist, researcher, educator; b. Cleve., Feb. 7, 1942; s. Jerry E. and Theresa H. (Dunne) S.; m. Sharon A., Mar. 21, 1964; children: Cindy, Lesley, Tracey, Jason. B.A., Hiram Coll., 1964; M.S., U. Wis., Madison, 1968, Ph.D. in Plant Pathology, 1969. Wis. Alumni Research Found. fellow U. Wis., 1964-65; NIH predoctoral fellow, 1965-69; assoc. prof. plant pathology U. Nebr., Lincoln, 1969—; cons. on bean diseases, vegetable prodn. problems, Australia, U.S., Latin Am. Contbr. numerous articles to sci. jours. Chmn. bd. trustees Nebr. San. and Improvement Dist. 2. Dept. Interior Water Resources grantee, 1972-77; Dept. Agr. Regional Research grantee, 1977-83; Union Pacific Found. grantee, 1980-83; Dept. Agr.-AID Title VII Internat. Program grantee, 1981-83. Mem. Am. Phytopath. Soc., Internat. Soc. Plant Pathology, Bean Improvement Coop., Sigma Xi. Subspecialty: Plant pathology. Current work: Epidemiology and control of bean and other vegetable diseases; research, some teaching on epidemiology of plant diseases, primarily beans and cucurbits. Office: 406 H Plant Sci Hall U Nebr Lincoln NE 68583

STEARNS, CHARLES RICHARD, meteorology educator; b. McKeesport, Pa., May 21, 1925; s. Fenton and Lois Annette (Sellers) S.; m. Anita Bernice Dahlke, June 21, 1950; children: James William, Laura Ann; m. Nancy Ray Williams, Jan. 19, 1962. B.S., U. Wis., 1950, M.S. in Meteorology, 1952, Ph.D., 1965. Research asst. U. Wis., Madison, 1950-52; farmer, 1952-55; physicist Winzen Research, 1956-57; research assoc. U. Wis., Madison, 1957-60, asst. prof., 1965-70, assoc. prof., 1970-78, prof. meteorology, 1978—; cons. in field. Chmn. Town of Oregon (Wis.) Planning Commn. Served with U.S. Army, 1943-45. Decorated Bronze Star medal. Mem. AAAS, Am. Meteorology Soc. Subspecialties: Meteorology; Meteorologic instrumentation. Current work: Antarctic meteorology, automatic weather stations. Home: 4890 CTHA Oregon WI 53575 Office: 1225 W Dayton St Madison WI 53706

STEARNS, MARY BETH, physicist; b. Mpls.; m. Martin Stearns; children: Daniel, Richard. Ph.D., Cornell U., 1952. Research assoc. Carnegie Inst. Tech., 1952-57; scientist Gen. Atomic, 1957-60; staff scientist Ford Motor Co., Dearborn, Mich., 1960-81; prof. physics Ariz. State U., Tempe, 1981—; cons. in field. Contbr. articles to profl. jours. Fellow Am. Phys. Soc. Subspecialties: Condensed matter physics; Magnetic physics. Current work: Electronic structure of transition metals compositionally modulated thin metallic films exafs analysis structure, ferromagnetism. Office: Dept Physics Arizona State U Tempe AZ 85287

STEARNS, STEPHEN CURTIS, population biologist, educator; b. Kapaau, Hawaii, Dec. 12, 1946; s. Alvan Curtis and Ruth (Musgrove) S.; m. Beverly Peterson, Dec. 11, 1971; children: Justin K., Jason K. B.A., Yale U., 1967; M.S., U. Wis.-Madison, 1971; Ph.D., U. B.C., Vancouver, 1975. Asst. prof. biology Reed Coll., Portland, 1970—. Contbr. articles to profl. jours. Council mem. Malheur Field Sta., Burns, Oreg., 1978-83. Miller fellow, 1975-78. Mem. Ecol. Soc. Am., Am. Soc. Naturalists, Soc. Study Evolution, Am. Soc. Zoologists, AAAS. Democrat. Subspecialties: Evolutionary biology; Theoretical ecology. Current work: The evolution of life-history traits; the evolution of developmental plasticity; the analysis of the process of selection and the definition of fitness; the role of the whole organism in the evolutionary process. Home: 4155 SE Ogden St Portland OR 97202 Office: Reed College 3203 SE Woodstock Blvd Portland OR 97202

STEBBINS, GEORGE LEDYARD, emeritus genetics educator; b. Lawrence, N.Y., Jan. 6, 1906; s. George Ledyard and Edith Alden (Candler) S.; m. Margaret Chamberlaine, June 14, 1931 (div. 1958); children: Edith Candler, Robert Lloyd, George Ledyard (dec.); m. Barbara Jean Brumley, July 27, 1958. A.B., Harvard U., 1928, A.M., 1939, Ph.D., 1931. Instr. biology Colgate U., 1931-35; jr. geneticist U. Calif., Berkeley, 1935-39, asst. prof. genetics, 1939-40, asso. prof., 1940-47, prof., 1947-50, prof. genetics, Davis, 1950-73, prof. emeritus, 1973—. Author: Processes of Organic Evolution, 1966, The Basis of Progressive Evolution, 1969, Chromosomal Evolution in Higher Plants, 1971, Flowering Plants: Evolution above the Species Level, 1974, others. Guggenheim fellow, 19S4, 60-61; recipient Verrill medal Yale U., 1968; Gold medal Linnean Soc. London, 1973; Nat. Medal of Sci., 1980. Mem. Soc. Study of Evolution (pres. 1950), Bot. Soc. Am. (pres. 1962), Am. Soc. Naturalists (pres. 1969), Calif. Native Plant Soc. (pres. 1966-72), Western Soc. Naturalists (pres. 1976), Internat. Union Biol. Scis. (sec. gen. 19S9-64), Nat. Acad. Scis., Am. Philos. Soc., Am. Acad. Arts and Scis., Genetics Soc. Am., Royal Swedish Acad. Scis., German Leopoldina Acad., Phi Beta Kappa, Sigma Xi. Democrat. Unitarian. Subspecialties: Evolutionary biology; Plant physiology (agriculture). Current work: Evolution of higher plants, especially origin of species and of asexual reproduction. Home: 1435 Wake Forest Dr #1 Davis CA 95616 Office: Dept Genetics U Calif Davis CA 95616

STEBLAY, RAYMOND WILLIAM, research physician; b. Chgo., Mar. 28, 1922; s. Joseph Bernard and Margare (Kobadich) S. A.B. magna cum laude, Princeton U., 1947; M.D., U. Chgo., 1952. Asst. prof. U. Chgo. Med. Sch., 1959-69; research physician N.Y. State Kidney Disease Inst., Albany, 1969—; cons. in medicine Schering Corp., Bloomfield, N.J., 1969-72. Served with U.S. Army, 1942-43. Mem. Phi Beta Kappa. Subspecialties: Immunology (medicine); Nephrology. Current work: Immunologic basis of kidney diseases; discoverer two models that explain two different kidney diseases. Home: 107 Heritage Rd Apt 11 Guilderland NY 12084 Office: NY Kidney Disease Inst Empire State Plaza Labs Albany NY 12201

STECK, THEODORE LYLE, physician; b. Chgo., May 3, 1939; s. Irving E. and Mary L. S.; children: David B., Oliver M. B.S. in Chemistry, Lawrence Coll., 1960; M.D., Harvard U., 1964. Intern Beth Israel Hosp., Boston, 1964-65, fellow, 1965-66; research asso. Nat. Cancer Inst., NIH, Bethesda, Md., 1966-68, Harvard U. Med. Sch., Boston, 1968-70; asst. prof. medicine U. Chgo., 1970-74, asst. prof. biochemistry and medicine, 1973-74, asso. prof., 1974-77, prof., 1977—, chmn. dept. biochemistry, 1979—. Subspecialties: Biochemistry (biology); Membrane biology. Current work: Structure and function of biological mechanisms. Office: 920 E 58th St Chicago IL 60637

STECKEL, RICHARD J., physician, educator; b. Scranton, Pa., Apr. 17, 1936; s. Morris Leo and Lucille (Yellin) S.; m. Julie Raskin, June 16, 1960; children—Jan Marie, David Matthew. B.S. magna cum laude, Harvard U., 1957, M.D. cum laude, 1961. Diplomate: Am. Bd. Radiology. Intern UCLA Hosp., 1961-62; resident in radiology Mass. Gen. Hosp., Boston, 1962-65; clin./research asso. Nat. Cancer Inst., 1965-67; mem. faculty UCLA Med. Sch., 1967—, prof. radiol. scis. and radiation oncology, dir. Jonsson Comprehensive Cancer Center, 1974—; pres. Assn. Am. Cancer Insts., 1981. Author two books, over seventy articles in field of radiology and cancer diagnosis. Fellow Am. Coll. Radiology; mem. Radiol. Soc. N. Am., Am. Roentgen Ray Soc. Subspecialties: Diagnostic radiology; Cancer research (medicine). Current work: New diagnostic methods for cancer. Home: 248 24th St Santa Monica CA 90402 Office: Louis Factor Health Scis Bldg UCLA Center Health Scis Los Angeles CA 90024

STECKER, FLOYD WILLIAM, astrophysicist; b. N.Y.C., Aug. 12, 1942; s. Norman and Helen Lilian (Stern) S.; m. Dorothy Ruth Bick, July 4, 1965; children: Benjamin, Jonathan. B.S., M.I.T., 1963; M.A., Harvard U., 1965, Ph.D., 1968. Research asst. Lab. for Nuclear Sci., M.I.T., 1962-63, teaching asst., 1963; physicist Harvard-Smithsonian Center for Astrophysics, Cambridge, 1963-66, predoctoral intern, 1966-67; Nat. Acad. Sci. postdoctoral research asso. NASA Goddard Space Flight Center, 1967-68, astrophysicist, 1968-77; sr. astrophysicist Lab. High Energy Astrophysics, 1977—. Author: Cosmic Gamma Rays; contbr. articles to profl. jours. Recipient NASA Medal for exceptional sci. achievement, 1973. Fellow Am. Phys. Soc.; mem. Internat. Astron. Union (Commn. on Galactic Structure), Am. Astron. Soc., Sigma Xi. Subspecialties: Theoretical astrophysics; Cosmology. Current work: High energy astrophysics, cosmic ray astrophysics, gamma ray astrophysics, galactic structure, cosmology. Address: Code 660 NASA Goddard Space Flight Center Greenbelt MD 20771

STEELE, GLENN DANIEL, JR., surg. oncologist, immunologist; b. Balt., June 23, 1944; s. Glenn Daniel and Alice E. S.; m. Lisa LaBoissiere, Nov. 19, 1977; children: Joshua, Kirsten. B.A., Harvard U., 1966; M.D., N.Y.U., 1970; Ph.D., U. Lund, Sweden, 1975. Diplomate: Am. Bd. Surgery, 1977. Instr. surgery Harvard Med. Sch., 1976-78, asst. prof. surgery, 1978-81, assoc. prof. surgery, 1981—; clin. assoc. surg. oncology Sidney Farber Cancer Inst., 1978-79, asst. physician in surg.oncology, 1979—; jr. assoc. in surgery Brigham and Women's Hosp., Boston, 1976—. Contbr. articles to profl. jours. Nat. Cancer Inst. grantee. Fellow ACS; mem. Am. Assn. Immunologists, Boston Surg. Soc., Fedn. Am. Socs. Exptl. Biology, Soc. Surg. Oncology, Soc. Univ. Surgeons, Phi Beta Kappa, Alpha Omega Alpha. Clubs: Denver Athletic, Boston Tennis and Racquet. Subspecialties: Surgical oncology; Cellular engineering. Current work: Clinical oncology and research in tumor biology; clinical specialist in surgical oncology and research in carcinogenesis and immunologic manipulation in attempts to develop effective immunotherapy. Office: 75 Francis St Boston MA 02115

STEELE, JOHN HYSLOP, marine scientist, oceanographic institute administrator; b. Edinburgh, U.K., Nov. 15, 1926; s. Adam and Annie

H.; m. Margaret Evelyn Travis, Mar. 2, 1956; 1 son, Hugh. B.Sc., Univ. Coll., London U., 1946, D.Sc., 1964. Marine scientist Marine Lab., Aberdeen, Scotland, 1951-66, sr. prin. sci. officer, 1966-73, dep. dir., 1973-77; dir. Woods Hole Oceanographic Instn., Mass., 1977—, vis. research fellow, 1958, lectr. marine biol. lab., 1967; vis. prof. U. Miami, 1961; mem. NSF panel for internat. Decade of Ocean Exploration, 1972-73, Council Marine Biol. Assn. of U.K., 1974-76, Council Scottish Marine Biol. Assn., 1974-78; mem. Bd. Ocean Sci. and Policy Nat. Research Council; corp. mem. Marine Biol. Lab., Woods Hole, Mass. Author: The Structure of Marine Ecosystems, 1974; Contbr. articles to profl. jours. Served with Brit. Royal Air Force, 1947-49. Recipient Alexander Agassiz medal Nat. Acad. Sci., 1973. Fellow Royal Soc. Edinburgh, Royal Soc. London, Am. Acad. Arts and Scis. Subspecialty: Oceanography. Home: Meteor House Woods Hole MA 02543 Office: Woods Hole Oceanographic Instn Woods Hole MA 02543

STEELE, OLIVER LEON, plant geneticist, consultant; b. Bloomington, Ill., Apr. 8, 1915; s. Bhondee Wood and Mary Vance (Eagle) S.; m. Ruth Holbert, June 21, 1941; children: David, Dennis, Nancy Steele Brokaw. B.S., Ill. Wesleyan U., 1940, Sc.D. (hon.), 1967. Mgr. research dept. Funk Seeds Internat., Bloomington, Ill., 1940-52, assoc. research dir., 1952-57, research dir., 1957-78, v.p., 1963-78, research cons., 1978—; cons. in field. Elder 2d Presbyterian Ch, Bloomington. Mem. AAAS, Am. Soc. Agronomy, Genetics Soc., Am. Genetic Assn. Republican. Lodge: Rotary. Subspecialty: Plant genetics. Current work: Cytoplasm interactions; seed improvement in South Asia.

STEELMAN, ROBERT JOE, oral and maxillofacial surgeon; b. Richland, Wash., Apr. 11, 1949; s. Earl and Betty Catherine (Young) S.; m. Marie Carol Hobson, Dec. 20, 1980. A.A.S., Columbia Basin Coll., Pasco, Wash., 1969; B.A., U. Wash., 1972; A.M., Washington U., St. Louis, 1974, D.M.D., 1982. Ecology researcher Washington U., 1974-76, med. researcher, 1976-78; resident in oral and maxillofacial surgery Emory U., Atlanta, 1982—. Contbr. articles to profl. jours. Mem. ADA, Am. Assn. Dental Research, Sigma Xi, Omicron Kappa Upsilon. Subspecialty: Oral and maxillofacial surgery. Current work: Long term effects of therapeutic radiation to the temporomandibular joint and associated structures. Home: 2410 F Dunwoody Crossing Atlanta GA 30338 Office: Emory U 1462 Clifton Rd Atlanta GA 30322

STEEVES, RICHARD ALLISON, radiation oncologist; b. Fredericksburg, Va., Feb. 2, 1938; s. William Horace and Doris Calkin (Cole) S.; m. Eliane Monique Brunet, Sept. 29, 1965; children: Pascal, Colin, Rachel. M.D., U. Western Ont., 1961; Ph.D., U. Toronto, Ont., Can., 1966. Diplomate: Am. Bd. Radiology. Postdoctoral fellow McMaster U., Hamilton, Ont., 1966-67; assoc. cancer research scientist Roswell Park Meml. Inst., Buffalo, 1967-72; assoc. prof. Albert Einstein Coll. Medicine, Bronx, N.Y., 1972-80; asst. prof. U. Wis., Madison, 1980-82, dir. div. radiation oncology, 1982—; dep. dir. Wis. Clin. Cancer Center, 1982—. Contbr. numerous articles to profl. jours. Recipient Rowntree prize U. Western Ont., 1961; Nat. Cancer Inst. grantee, 1972-82. Mem. AAAS, Am. Assn. Cancer Research, Am. Soc. Microbiology, Am. Soc. Therapeutic Radiologists, Exptl. Aircraft Assn. Unitarian. Subspecialties: Oncology; Cancer research (medicine). Current work: Role of hyperthermia in treatment of cancer. Office: 600 Highland Ave K4/B100 Madison WI 53792

STEFANINI, MARIO, pathologist; b. Chieri, Piedmont, Italy, June 11, 1920; s. Eleuterio and Marie Therese (Trivereau) S.; m. Elizabeth Shields Just, Feb. 12, 1949; children: Marie Therese, Virginia Elizabeth. M.D., U. Rome, 1939; M.Sc., Marquette U., 1947. Diplomate: Am. Bd. Pathology. Asst. physician New Eng. Med. Ctr., Boston, 1949-55; dir. research St. Elizabeth Hosp., Boston, 1956-61; assoc. pathologist St. Joseph Hosp., Chgo., 1962-65; pathologist St. Elizabeth Hosp., Danville, Ill., 1966-79, Clinch Valley Hosp., Richlands, Va., 1979—; cons. VA Hosp., Boston, 1950-61; asst. prof. Tufts U. Sch. Medicine, 1950-52, prof. medicine, 1952-61. Author: The Hemorrhagic Disorders, 1955, 2nd edit. 1962; editor series: Progress in Clin. Pathology, 1966-81. Served to capt. USAF, 1941-45. Recipient Piccinini prize, 1946; Cert. of Merit Nat. Gastroent. Assn., 1948, AMA, 1952; 1st prize Am. Assn. Blood Banks, 1952, 56; Nuffield medal Royal Soc. London, 1979; USPHS sr. research fellow, 1947-49; Damon Runyon clin. Cancer fellow, 1950-52; established investigator Am. Heart Assn., 1952-58. Fellow Internat. Soc. Hematology, Coll. Am. Pathologists, Am. Coll. Nuclear Medicine, Am. Soc. Exptl. Pathology; mem. Coll. Nuclear Physicians, Am. Soc. Clin. Investigation. Roman Catholic. Subspecialties: Pathology (medicine); Chemotherapy. Current work: Techniques for preparation and transfusion of human platelets; introduction of proteolytic enzymes; discovery of the syndrome of DIC (disseminated intravascular coagulation); present interest: treatment of malignancies through immune technology. Home: 723 Cresswood Dr Richlands VA 24641 Office: Clinch Valley Community Hosp 2949 W Front St Richlands VA 24641

STEFANO, GEORGE B., neurobiologist, researcher; b. N.Y.C., Sept. 11, 1945; s. George and Agnes (Hendrickson) S.; m. Judith Mary Stefano, Aug. 24, 1968; 1 dau., Michelle Laura. Ph.D., Fordham U., 1973. Mem. faculty N.Y.C. Community Coll., 1972-79, Medgar Evers Coll., CUNY, 1979-82; assoc. prof. cell biology, chmn. dept. biol. sci. SUNY-Old Westbury, 1982—; pres., dir. East Coast Neurosci. Found., Dix Hills, N.Y., 1977—; research coordinator dept. anesthesiology St. Joseph Hosp. and Med. Ctr., Paterson, N.J., 1979—. Contbr. articles to sci. jours. Nat. Acad. Scis. grantee, 1978, 80; NIMH grantee, 1979—. Mem. Soc. Neurosci., N.Y. Acad. Sci., Gerontol. Soc. Am., AAAS. Subspecialties: Neurobiology; Molecular pharmacology. Current work: Opioid mechanisms in neural tissues; characterization and demonstration of opiate binding sites and pharmacological effects in neural tissues.

STEGER, RICHARD WARREN, endocrinologist, educator; b. Richmond, Calif., Aug. 4, 1948; s. Richard Warren and Lois (Burger) Davis; m. Marlene M. Vincent, Apr. 17, 1979; children: Kyle, Kimberly. B.A., U. Wyo., 1970, Ph.D., 1974. Postdoctoral fellow Wayne State U., Detroit, 1974-77, Mich. State U., East Lansing, 1977-79; asst. prof. U. Tex. Health Sci. Ctr., San Antonio, 1980—. Editor: Pharmacologic Methodology for the Study of the Neuroendocrine System, 1984; editorial bd.: Exptl. Aging Research. Recipient U. Wyo. Standard Oil grad. teaching award, 1974; Wayne State U. Ford Found. fellow, 1974-76; Mich. State U., NIH fellow, 1977-79. Mem. Endocrine Soc., Am. Physiol. Soc., Soc. Study of Reproduction, AAAS, Sigma Xi, Phi Kappa Phi. Subspecialties: Neuroendocrinology; Gerontology. Current work: aging of neuroendocrine system; reproductive neuroendocrinology. Office: Ob-Gyn Dept U Tex Health Science Center 7703 Floyd Curl San Antonio TX 78284

STEIGMAN, GARY, physicist; b. N.Y.C., Feb. 23, 1941; E2Charles and Pearl (Platzer) S. B.S., CCNY, 1961; postgrad., Cornell U., 1961-62; M.S., N.Y.U., 1963, Ph.D., 1968. Vis. fellow Inst. Theoretical Astronomy, Cambridge, Eng., 1968-70; research fellow Kellogg Lab., Calif. Inst. Tech., Pasadena, Calif., 1970-72; asst. prof. astronomy dept. Yale U., New Haven, 1972-78, asso. prof., 1978-80; prof. physics Bartol Research Found., U. Del., Newark, 1980—; vis. prof. Stanford U., 1978, U. Calif., Santa Barbara, 1981. Recipient Gravity Prize Gravity Research Found., 1980; named Del. Scientist of Year Sigma Xi, 1980. Mem. Am. Astron. Soc. Subspecialties: Cosmology; Theoretical astrophysics. Current work: Cosmology - The Physics of the Early Universe. Office: Bartol Research Found U Del Newark DE 19711

STEIHAUG, TROND, mathematical science educator; b. Lillehammer, Norway, Aug. 6, 1950; came to U.S., 1977, naturalized, 1977; s. Arne and Signe S.; m. Nina Naess, Aug. 19, 1972; children: Ole Martin, Espen. Candidatus Magisterii, U. Oslo, 1973, Candidatus Realium, 1975; M.A., Yale U., 1980, Ph.D., 1981. Asst. prof. math. sci. U. Oslo, 1976-77; research assoc. Chalmers U. Tech., Gothenburg, Sweden, 1977; asst. prof. math. sci. Rice U., 1981—; cons. Mobil Co., Dallas. Mem. Soc. Indsl. and Applied Math., ACM, Ops. Research Soc. Am., Math. Programming Soc. Subspecialties: Numerical analysis; Operations research (mathematics). Current work: Large scale nonlinear programming. Office: Rice U Math Sci Dept PO Box 1892 Houston TX 77251

STEIN, BARRY EDWARD, neurophysiologist, educator; b. N.Y.C., Dec. 3, 1944; s. Allen E. and Naomi (Schnittman) S.; m. Nancy London, Aug. 11, 1968. B.A., Queens Coll., 1966, M.A., 1969; Ph.D., CUNY, 1970. Postdoctoral fellow dept. anatomy and Brain Research Inst., UCLA, 1970-72, asst. research anatomist, 1972-75; asst. prof. physiology Med. Coll. Va., 1975-78, assoc. prof., 1978-82, prof., 1982—. Contbr. articles to profl. publs. Grantee A.D. Williams Found., 1975-76, NIMH, 1976-77, 76-79, NIH, 1979-82, 82-85, NSF, 1981—, Jeffess Found., 1982-85. Mem. Internat. Brain Research Orgn., Soc. Neurosci., AAAS, Am. Mus. Natural History, Smithsonian Instn., Sigma Xi. Jewish. Subspecialties: Neurophysiology; Neurobiology. Current work: Development and organization of sensory and motor representation in brain, sensorimotor transduction, developmental plasticity, multisensory integration, pain. Office: Dept Physiology Med Coll Va Richmond VA 23298

STEIN, DIANA B., molecular botanist, educator; b. N.Y.C., July 5, 1937; d. James J. and Genya P. (Becker) Borut; m. Otto L. Stein, June 15, 1958; children: Deborah Lee, Judith Ann and Suzanne Beth (twins), Jonathan Henri Richard. A.B., Barnard Coll., 1958; M.A., U. Mont., 1961; Ph.D., U. Mass., 1976. Research technician U. Mont., Missoula, 1958-59; instr. U. Mass., Amherst, 1964-69, postdoctoral assoc., 1976-80; asst. prof. molecular biology Mt. Holyoke Coll., South Hadley, Mass., 1980—. Contbr. articles to profl. jours. William and Flora Hewlett Found. grantee, 1980, 82. Mem. Bot. Soc. Am., Am. Fern Soc., Soc. for Devel. Biology, Sigma Xi. Subspecialties: Molecular biology; Genome organization. Current work: Use of DNA to study evolutionary relationships. Home: 140 Red Gate Ln Amherst MA 01002 Office: Dept Biol Scis South Hadley MA 01075

STEIN, DONALD GERALD, psychologist, educator, researcher; b. N.Y.C., Jan. 27, 1939; s. Frank and Elizabeth (Bernstein) S.; m. Darel Hammer, June 18, 1941; children: Jessica, Matthew. B.A., Mich. State U., 1960, M.A., 1962; Ph.D., U. Oreg., 1965. USPHS fellow M.I.T., Cambridge, 1965-66; prof. psychology Clark U., Worcester, Mass., 1966—; prof. neurology U. Mass. Med. Sch., 1978—; cons. psychologist sleep disorder ctr. U. Mass. Med. Sch.; chmn. Nat. Coalition for Sci. and Tech., 1981—. Contbr. numerous articles on psychology and neurology to profl. jours.; editor 3 vols. on recovery from brain damage and aging; co-author: Brain Damage and Recovery, 1982; editor, writer: Nat. Coalition for Sci. and Tech. Newsletter. USPHS fellow, 1963-65; Fulbright fellow, 1971-72; NIH fellow, 1973-78; Inserm (France) fellow, 1975-76; AAAS fellow, 1980-81. Mem. Am. Psychol. Assn., Soc. Neurosci., AAAS, European Brain and Behavioral Soc., Internat. Neuropsychology Soc., Am. Aging Assn. Democrat. Subspecialties: Regeneration; Physiological psychology. Current work: Recovery from brain damage, neural organization and development psychology of aging and neuroplasticity.

STEIN, GARY STEPHEN, cell biologist, educator; b. Bklyn., July 30, 1943; s. Abraham and Nesbit (Edkiss) S.; m. Janet Lee Swinehart, Feb. 2, 1974. B.A., Hofstra U., 1966; M.A., 1968; Ph.D., U. Vt., 1969. Research assoc. dept. pathology Temple U. Sch. Medicine, Phila., 1971-72; asst. prof. Coll. Medicine, U. Fla., Gainesville, 1972-75, assoc. prof., 1975-79, prof., 1979—, assoc. chmn. dept. biochemistry and molecular biology, 1981—. Editor: (with L.J. Kleinsmith) Chromosomal Proteins and Their Role in the Regulation of Gene Expression), 1975, (with J.L. Stein and L.J. Kleinsmith) Methods in Cell Biology: Chromatin and Chromosomal Protein Research, 4 vols, 1977, (with J.L. Stein) Recombinant DNA Approaches to Studying Control of Cell Proliferation, 1984, (with J.L. Stein and W. Marzluff) Histone Genes and Histone Gene Expression, 1984; mem. editorial adv. bd.: Growth and Cancer, Jour. Cell Physiology, Cell and Tissue Kinetics, Cell Biophysics, Anticancer Research, Jour. Nutrition. Grantee NSF, NIH, Am. Cancer Soc.; Damon Runyon-Walter Winchell grantee for cancer research. Mem. Am. Assn. for Cancer Research, Am. Chem. Soc., Am. Soc. for Cell Biology, AAAS, Southeastern Cancer Research Assn. (pres., chmn. bd. dirs.), Am. Soc. Biol. Chemists, N.Y. Acad. Scis. Subspecialties: Cell biology; Molecular biology. Current work: Regulation of gene expression and control of cell proliferation. Home: 5511 NW 55th Ln Gainesville FL 32606 Office: JHM Health Ctr U Fla Coll Medicine Box J-245 Gainesville FL 32610

STEIN, GEORGE NATHAN, radiologist, educator; b. Phila., Aug. 11, 1917; s. Samuel M. and Marie B. S.; m. Hazel Gloria Gould, Aug. 18, 1948; children—Stephen G., Eric J., James M. B.A., U. Pa., 1938; M.D., Jefferson Med. Coll., 1942; postgrad., Sch. Medicine, U. Pa., 1946. Asso. radiologist Grad. Hosp. U. Pa., 1948-60, asst. dir. dept. radiology, 1960-71; dir. dept. radiology Presbyn. U. Pa. Med. Center and prof. radiology U. Pa., 1971—; Prof. extraordinaire Javeriana U., Bogota, Colombia, 1960; cons. VA Hosp., Wilmington, Del., 1968-70. Author 3 books.; Contbr. numerous articles to profl. jours. Served with M.C. AUS, 1943-46. Fellow Am. Coll. Radiology; mem. Phila. Roentgen Ray Soc. (pres. 1968-69), Am. Roentgen Ray Soc., Tex. Radiol. Soc. (hon.), Philadelphia County, Pa. State med. socs., Radiol. Soc. N.Am., AMA. Subspecialties: Gastroenterology; Diagnostic radiology. Home: 544 Howe Rd Merion Station PA 19066 Office: 51 N 39th St Philadelphia PA 19104

STEIN, IRA DAVID, physician; b. N.Y.C., Mar. 13, 1955; s. Harry S. and Freida (Lagman) S.; m. Shirley B. Miller, Aug. 16, 1958; children: Joel, Tamara Judith, Shira Robin. B.S. in Biology, Rutgers U., 1956; M.D., George Washington U., 1960. Intern USPHS Hosp., San Francisco, 1960-61; resident in internal medicine VA Hosp., Coral Gables, Fla., 1963-66; USPHS fellow Mt. Sinai Hosp. Services, City Hosp. Ctr., Elmhurst, N.Y., 1966-68, clin. asst. hematology, 1968-69; staff physician VA Hosp., East Orange, N.J., 1969-81, chief of medicine, Erie, Pa., 1981—; asst. prof. medicine Coll. Medicine and Dentistry N.J., Newark, 1969-75, assoc. prof. medicine, 1975-81. Contbr. chpt. to book. Served with USPHS, 1961-63. Fellow ACP; mem. AAAS, Am. Soc. Clin. Research, Am. Soc. Hematology, Sigma Xi. Democrat. Jewish. Subspecialty: Hematology. Current work: Metabolic diseases of bones and their structural and mechanical consequences. Home: 135 E 38th St Blvd Erie PA 16501 Office: VA Med Ctr 135 E 38th St Blvd Erie PA 16501

STEIN, ROBERT FOSTER, astrophysicist; b. N.Y.C., Mar. 4, 1935; s. Arthur H. and Louise (Halpern) S.; m. Laura Cooper, Dec. 21, 1958; children: Karen, Tamara. B.S., U. Chgo., 1957; Ph.D., Columbia U., 1966. Cons. Smithsonian Astrophys. Obs., Cambridge, Mass., 1969-78; asst. prof. astrophysics Brandeis U., Waltham, Mass., 1969-76; assoc. prof. astronomy and astrophysics Mich. State U., East Lansing, 1956-91, prof., 1981-; vis. fellow Joint Inst. lab. Astrophysics, Boulder, Colo., 1973-74; vis. scientist Observatoire de Nice (France), 1981. Contbr. articles to profl. jours. Grantee NSF, 1974—, NASA, 1975—. Mem. Am. Astron. Soc., Internat. Astron. Union. Subspecialties: Theoretical astrophysics; Solar physics. Current work: Astrophysical fluid dynamics. Office: Dept Physics and Astronomy Michigan State U East Lansing MI 48824

STEIN, SCOTT ALLEN, systems programmer; b. Detroit, June 12, 1953; s. Wesley Herman and Rachel Lenore (Allen) S. B.S., U. Mich., 1977, M.S., 1982. Asso. systems programmer Burroughs Corp., Plymouth, Mich., 1978-80, systems programmer, Paoli, Pa., 1980-81, sr. systems programmer, 1981-82, project systems programmer, 1982—. Mem. Fedn. Am. Scientists, Assn. Computing Machinery, IEEE. Subspecialties: Distributed systems and networks; Operating systems. Current work: Networking architectures, distributed processing, communications protocols, protocol representation, distributed operating systems. Office: PO Box 235 Downingtown PA 19335

STEINACKER, ANTOINETTE, neurophysiologist; b. Balt., May 27, 1938; d. Clayton E. and Anna (Frederick) S.; m. (divorced). Ph.D., U. Pacific, 1972. Postdoctoral fellow U. Calif.-San Diego, 1973-75, Albert Einstein Coll. Medicine, Bronx, N.Y., 1975-78; research assoc. Rockefeller U., N.Y.C., 1979-82; asst. prof. Albert Einstein Coll. Medicine, 1982—. Contbr. articles to profl. jours. Mem. Soc. Neurosci., Biophys. Soc., N.Y. Acad. Scis. Subspecialty: Learning. Current work: Acetylcholine receptor, electrophysiology receptor structure function relationships. Office: 1410 Pelham Pkwy Bronx NY 10461

STEINBERG, ALFRED DAVID, immunologist; b. N.Y.C., Nov. 4, 1940; s. Philip M. and Sylvia S.; m. Susan Connor; children: Bonnie Jean, Robert T., Ellen Beth, Kathleen Triem. A.B., Princeton U., 1962; M.D. cum laude, Harvard U., 1966. Diplomate: Am. Bd. Internal Medicine. USPHS trainee Huntington Meml. Labs., Mass. Gen. Hosp., Boston, 1963-64; intern and resident in internal medicine Bronx (N.Y.) Mcpl. Hosp. Ctr., 1966-68; clin. assoc. arthritis and rheumatism br. Nat. Inst. Arthritis, Metabolism and Digestive Diseases, NIH, Bethesda, Md., 1968-70, sr. staff fellow, 1970-71, sr. investigator, 1971—, chief sect. cellular immunology, 1981—; capt. USPHS, 1968—; mem. clin. research com. NIH, 1971, 72, peer rev. com., 1972-74; USA-USSR Coop. in Rheumatology and Immunology, 1974—. Contbr. numerous articles on immunology and rheumatology to profl. jours.; assoc. editor: Jour. Immunology, 1976-78, 1978-80; editorial bd.: Jour. Immunopharmacology, 1977—, African Jour. Clin. Immunology, 1981—. Recipient Philip Hench award, 1974; Washington Acad. Scis. award, 1978; Commendation medal Dept. HHS, 1981. Fellow ACP; mem. Am. Fedn. Clin. Research, Am. Rheumatism Assn. (publs. com. 1979-83, fellowship com. 1981), Am. Assn. Immunologists (membership com. 1979-82), Am. Soc. Exptl. Pathology, Transplantation Soc., Soc. Exptl. Biology and Medicine, Am. Soc. Clin. Investigation, N.Y. Acad. Sci., Washington Acad. Scis., Phi Beta Kappa, Sigma Xi. Subspecialties: Immunology (medicine); Immunogenetics. Current work: Authority on autoimmune diseases, especially systemic lupus; studies of pathogenesis of autoimmune diseases as well as to basic immunology which are leading to improved therapy. Home: 8814 Bells Mill Rd Potomac MD 20854 Office: NIH Bldg 10 Room 9N-218 9000 Rockville Pike Bethesda MD 20205

STEINBERG, FREDERICK, mathematician, engineering research analyst; b. Bklyn., Feb. 20, 1924; s. Ephriam and Helen (Steinberg) S. B.S. in Sci, L.I. U., 1949; M.A. in Math, NYU, 1951; diploma in local govt., U. Calif.-Berkeley, 1971. Cert. secondary sch. teaching, N.Y., Calif. Mathematician Aerojet-Gen. Corp., Azusa and Sacramento, Calif., 1951-58; data systems analyst TRW Systems, Inc., Los Angeles, 1958-62; research analyst N.Am. Rockwell/Space, Downey, Calif., 1962-66; ops. researcher Lockheed Missiles & Space Co., Sunnyvale, Calif., 1966-70; data systems analyst Space Group Hughes Aircraft Co., Los Angeles, 1973-81, El Segundo, Calif., 1981—; tchr., cons. Santa Clara Unifed Sch. Dist., 1970-71; mgmt. systems analyst City Adminstrv. Office, City of Los Angeles, 1972. Served in USAAF, 1943-46. N.Y. State Bd. Edn. scholar, 1946. Mem. AIAA, IEEE (sr.). Republican. Jewish. Clubs: Mulholland (Los Angeles); NYU (N.Y.C.). Subspecialties: Operations research (engineering); Systems engineering. Current work: Electro-optical (laser) engineering applications to tactical weapon systems, computer reporting systems for management visibility and control. Home: 933 N Cedar St 13 Inglewood CA 90302 Office: Hughes Aircraft Co Bldg El 2000 El Segundo Blvd El Segundo CA 90245

STEINBERG, JACOB JONAH, pathologist, researcher; b. N.Y.C., June 1, 1951; s. Moses Aaron and Clara (Junowitz) S.; m. Sari Sue Miller, Aug. 7, 1977. B.S., CCNY, 1973; M.D., Loyola U., Chgo., 1976; postgrad., Columbia U., 1982—. Diplomate: Nat. Bd. Examiners. Intern Hosp. Univ. Pa., Phila., 1976-77, jr. resident, 1977-79, sr. resident, 1979-81; resident NYU Med. Ctr., 1981-83, fellow, 1983—; fellow Harrison Dept. Surg. Research, U. Pa., Phila., 1979-81. Contbr. articles to med. jours.; mem. editorial bd.: Jour. Surg. Pathology Quar. Index, 1982—. Runyon-Winchell Cancer Fund fellow and grantee, 1979-81; Am. Cancer Soc. fellow, 1979-80. Mem. Am. Fedn. Clin. Research, European Assn. Cancer Research, Soc. Controlled Clin. Trials, Harvard-MIT Metabolism Group. Subspecialties: Pathology (medicine); Cancer research (medicine). Current work: Basic research, mechanisms of DNA damage and repair, hybridoma technique for probe of DNA damage and quantity, clinical research, cancer cachexia. Office: NYU Med Center MSB#611 550 1st Ave New York NY 10016

STEINBERG, MARSHALL, toxicologist, educator; b. Pitts., Sept. 18, 1932; s. Harry L. and Eva (Goldstein) S.; m. Patricia Louise Zobac, Nov. 3, 1962; children: Leslie R., Michael A., Maureen S. B.S., Georgetown U., 1954; M.S., U. Pitts., 1956; Ph.D., U. Tex., Galveston, 1966. Cert. Acad. Toxicol. Scis., Nat. Registry in Clin. Chemistry. Sr. v.p., dir. life scis. Hazleton Lab. Am. Inc., Vienna, Va., 1978—; adj. prof. Am. U.; mem. threshold limit value com. Am. Conf. Govtl. Indsl. Hygienists, 1971—. Author chpts. in books, also articles. Bd. dirs. Toxicology Lab. Accreditation Bd. Served to col. U.S. Army, 1956-76. Decorated Legion of Merit. Mem. Acad. Toxicol. Scis., Am. Soc. Pharmacology and Exptl. Therapeutics, Am. Conf. Govt. Indsl. Hygienists, Am. Indsl. Hygiene Assn., Am. Coll. Toxicology, Soc. Toxicology. Jewish. Subspecialty: Toxicology (medicine). Current work: Dermal toxicology and laboratory management.

STEINBERG, MARTIN H., med. research adminstr., educator; b. N.Y.C., July 2, 1936; s. Meyer and Anne (Palatnik) S.; m. Susan Elizabeth McDaniel, Nov. 24, 1973; 1 dau., Elizabeth Anne. A.B., Cornell U., 1958; M.D., Tufts U., 1962. Intern Bellevue Hosp., N.Y.C., 1962-63; fellow in hematology New Eng. Med. Center, Boston, 1968-70; asst. prof., assoc. prof. U. Miss., Jackson, 1970-77, prof., 1977—; dir. research VA Med. Center, Jackson, 1973—; cons. Am. Heart

Assn., 1980—, VA Research Service, Washington, 1976—. Contbr. numerous articles to med. jours.; editorial bd.: Am. Jour. Hematology, 1978—. Served to capt. USAF, 1963-66. NIH grantee, 1978—; VA grantee, 1970—. Fellow ACP; mem. Am. Soc. Clin. Investigation, Central Soc. Clin. Research, So. Soc. Clin. Investigation, Am. Fedn. Clin. Research (councilor 1976-78), Alpha Omega Alpha. Subspecialties: Hematology; Molecular biology. Current work: Disorders of hemoglobin structure and synthesis especially related to sickle cell disease and thalassemia. Home: 4506 Meadowhill Rd Jackson MS 39206 Office: VA Med Center Jackson MS 39216

STEINBERG, MORRIS ALBERT, metallurgist; b. Hartford, Conn., Sept. 24, 1920; s. Abraham and Rose (Bellow) S.; m. Natlee Jean Haas, Apr. 29, 1951; children: Laurie Beth, Cathy, James J. B.S. (Competitive scholar), M.I.T., 1942, M.S. in Metallurgy (S.K. Wellman fellow), 1946, D.Sc., 1948; postgrad., Harvard Bus. Sch., 1970. Registered profl. engr., Calif. Instr. dept. metallurgy M.I.T., Cambridge, Mass., 1946-48; head and dir. Horizons Inc., Cleve., 1948-58; officer and dir. Micrometric Instrument Corp., Cleve., 1949-51; sec., dir. Diwolfram Corp., Cleve., 1950-53; chief metallurgist Horizons Titanium Corp., Cleve., 1951-54; dir. material sci. lab. Lockheed Missiles & Space Co., Palo Alto, Calif., 1958-65; dep. chief scientist Lockheed Corp., Burbank, Calif., 1965-74, dir. tech. applications, 1974-83, v.p. sci., 1983—; adj. prof. materials dept. Sch. Engring. and Applied Sci., UCLA, 1978—; mem. structures tech. panel for Chief of Staff, USAF, 1971-72; mem. Air Force studies bd. NRC, 1982—; mem. ad hoc com. of B-1 structures, sci. adv. bd., 1972—; mem. Nat. Materials Adv. Bd., NRC, Nat. Acad. Scis., 1976—; expert NASA Shuttle Flight Cert. Com., 1979-81. Contbr. numerous articles on metallurgy and aerospace materials to engring. jours. Served to capt. USAF, 1942-46; ETO. Fellow Am. Soc. Metals (mem. public affairs com. 1973-75), Inst. for Advancement of Engring., AAAS, Am. Inst. Chemists, AIAA; mem. Soc. Mfg. Engring. (chmn. engring. materials com. 1969-72), Soc. Automotive Engrs., N.Y. Acad. Scis., Aerospace Industries Assn. (chmn. materials and structures com. 1973-74), Nat. Acad. Engring. (panel laser mirror reliability 1982—), Soc. Mfg. Engrs. (mem. tech. council 1969-72), AIAA (chmn. materials com. 1962-64, chmn. honors and awards com. Los Angeles sect. 1977-79). Subspecialties: Materials (engineering); Metallurgy. Current work: Materials applications for aircraft, missiles and spacecraft, with interest in the relationship of material structure to properties. Home: 348 Homewood Rd Los Angeles CA 90049 Office: Lockheed Corp PO Box 551 Burbank CA 91520

STEINBRUEGGE, KENNETH BRIAN, electronics research company executive; b. St. Louis, Dec. 9, 1939; s. Theodore Watson and Ruth Maxine (Weiner) S.; m. Eileen Kelly, Aug. 27, 1966; children: Terri Lynn, Criag Allen, Steven Brian. B.S. in Engring. Physics, U. Mo.-Rolla, 1962, M.S., 1963. Assoc. engr., quantum electronics dept. Westinghouse Research Labs., Pitts., 1963-65, engr., 1965-69, sr. engr., 1969-82, fellow scientist, applied physics dept., 1982, mgr. sensor applications, electronic tech. div., 1982—. Contbr. articles to profl. jours. Mem. Republican Com., South Murrysville, Pa. Co-recipient IR-100 award, 1966; recipient cert. of recognition NASA, 1982. Mem. Optical Soc. Am. (pres. Pitts. sect. 1977-79), Spectroscopy Soc. Pitts. Methodist. Subspecialties: Fiber optics; Laser materials, systems, applications. Current work: Development of new optical and fiber optic sensor, communication and vision systems as well as application of acoustic and magnetic sensors for automated status monitoring. Patentee. Home: 3496 Ivy Ln Murrysville PA 15668 Office: Westinghouse Research Labs 1310 Beulah Rd Pittsburgh PA 15235

STEINER, BRUCE WATSON, materials scientist; b. Oberlin, Ohio, May 14, 1931; s. Luke Eby and Helen Annette (Watson) S.; m. Ruth Piette, Jan. 1, 1960; children: Jonathan, Miriam. A.B., Oberlin Coll., 1953; Ph.D., Princeton U., 1956. Research assoc. U. Chgo., 1958-61; research scientist Nat. Bur. Standards, 1961-69, chief Interdiv., 1969-71, research scientist, 1971-74, program analyst office of the dir., 1974-76; program officer atomic and material physics NSF, 1976-78; chmn. energy standards planning task force Nat. Bur. Standards, 1977; spl. asst. for long range planning Center for Material Sci., 1978—; mem. U.S. Nat. Com., Internat. Com. for Optics. Contbr. articles to profl. jours. Fellow Optical Soc. Am., Am. Inst. Chemists; mem. Am. Phys. Soc., Am. Chem. Soc., AAAS, N.Y. Acad. Scis. Current work: Development of programs in advanced materials science of innovative materials especially in advanced optics. Office: Nat Bur Standards Washington DC 20234 Home: 6624 Barnaby St NW Washington DC 20015

STEINER, ROBERT FRANK, physical biochemist; b. Manila, Philippines, Sept. 29, 1926; came to U.S., 1933; s. Frank and Clara Nell (Weems) S.; m. Ethel Mae Fisher, Nov. 3, 1956; children—Victoria, Laura. A.B., Princeton U., 1947; Ph.D., Harvard U., 1950. Chemist Naval Med. Research Inst., Bethesda, Md., 1950-70, chief lab. phys. biochemistry, 1965-70; prof. chemistry U. Md., Balt., 1970—, chmn. dept. chemistry, 1974—; mem. biophysics study sect. NIH, 1976. Author: Life Chemistry, 1968; author: Excited States of Proteins and Nucleic Acids, 1971, The Chemistry of Living Systems, 1981; editor: Jour. Biophys. Chemistry, 1972—. Served with AUS, 1945-47. Recipient Superior Civilian Achievement award Dept. Def., 1966; NSF research grantee, 1971; NIH research grantee, 1973. Fellow Washington Acad. Sci.; mem. Am. Soc. Biol. Chemists. Club: Princeton (Washington). Subspecialties: Biophysical chemistry; Biochemistry (biology). Current work: Physical properties of proteins and nucleic acids; fluorescence; allosteric enzymes. Home: 2609 Turf Valley Rd Ellicott City MD 21043 Office: 5401 Wilkens Ave Baltimore MD 21228

STEINER, WILLIAM WALLACE MOKAHI, population geneticist, consultant; b. Honolulu, Nov. 16, 1942; s. Charles William and Florence Wahine O'Mau Haleakala (Vera Cruz) S.; m. Judith Ann Matthews, June 26, 1964; children: Angela June, Shawna Lynne. A.S., Boise State Coll., 1964; B.A., U. Hawaii, 1970; Ph.D., 1974. Research asst. Hawaii Internat. Biol. project, Honolulu, 1972-74; asst. prof. dept. genetics U. Ill., Urbana, 1974-81; assoc. research scientist Ill. Natural History Survey, Champaign, 1981—; cons. S.E. Asian Ministers of Edn. Orgn., 1981—. Lead editor: Recent Developments in the Genetics of Insect Disease Vectors, 1982; contbr. articles to profl. jours. Grantee Hasselblad Found., Sweden, 1981-83, Dept. Energy and Natural Resources, 1982-84, Rockefeller Found., 1980, NIH, NSF, 1976-78. Mem. Genetics Soc. Am., Am. Genetics Assn., AAAS, Soc. Study Evolution, Am. Mosquito Control Assn. Roman Catholic. Subspecialties: Genetics and genetic engineering (biology); Systematics. Current work: Conducting research in the areas of genetics of insect disease vectors and conservation of genetic resources in natural populations; major interests ecological genetics, genetic biochemical polymorphisms, environment-genotype interaction. Office: Ill Natural History Survey 607 E Peabody St Champaign IL 61820

STEINFEIS, GEORGE FRANCIS, pharmacologist; b. Newark, June 22, 1954; s. George G. and Helen (Skutack) S.; m. Alison B. Barnat, Aug. 20, 1978. B.A. in Biology, Johns Hopkins U., 1976; Ph.D., U. Md., Balt., 1980. Instr. pharmacology, research asst. U. Md. Sch. Pharmacy, Balt., 1977-80; vis. research fellow dept. psychology Princeton U., 1980—. Contbr. articles to profl. jours. Issac E. Emerson fellow, 1979; NIH fellow, 1981; Am. Phils. Soc. grantee, 1981. Mem. Am. Soc. Pharmacology and Exptl. Therapeutics, Soc. Neurosci., AAAS, N.Y. Acad. Scis., Brit. Brain Research Assn., European Brain and Behavior Soc., Rho Chi. Club: River Rd. Rangers' Soccer (New Brunswick, N.J.). Subspecialties: Pharmacology; Neuropharmacology. Current work: Effects of psychoactive drugs on brain and behavior; neurophysiological recording of single cells in behaving animals; drug addiction in experimental animals/neurophysiology and pharmacology of sleep. Office: Princeton U Dept Psychology Green Hall Princeton NJ 08544 Home: Faculty Rd Hibben Apt 5F Princeton NJ 08540

STEINGRAEBER, DAVID ALLEN, botanist, researcher, educator; b. Waukesha, Wis., May 16, 1953; s. Joseph Anthony and Elaine Emily (Kalman) S.; m. Ita Marie Lindquist, Aug. 17, 1974. B.S., U. Wis.-Milw., 1974; Ph.D., U. Wis.-Madison, 1980. Research assoc. Fairchild Tropical Garden, Miami, Fla., 1980-81; asst. prof. botany Colo. State U., Ft. Collins, 1981—. Mem. Bot. Soc. Am.; mem. Ecol. Soc. Am., AAAS; Mem. Assn. Tropical Biology; mem. Phi Beta Kappa; Mem. Sigma Xi. Subspecialties: Plant morphology; Ecology. Current work: Ecol. plant morphology; plant form in relation to environment; patterns of shoot devel, branching and leaf placement in different environments; heterophylly; tropical botany; evolutionary ecology. Office: Dept Botany and Plant Pathology Colo State U Fort Collins CO 80523

STEINHERZ, PETER GUSTAV, pediatric hematologist oncologist; b. Budapest, Hungary, Apr. 5, 1942; came to U.S., 1957, naturalized, 1962; s. Simon and Lenke (Lowinger) S.; m. Laurel Weinberger, July 4, 1967; children: Jennifer Ann, Jonathan Aaron, Daniel Evan, David Ashely. B.S., Bklyn. Coll., 1964; M.D., Albert Einstein Coll. Medicine, 1968. Intern N.Y. Hosp.-Cornell Med. Ctr., N.Y.C., 1968-69, 1969-70; assoc. attending Meml. Sloan Kettering Cancer Ctr., N.Y.C., 1980—, research assoc., 1980—; assoc. attending N.Y. Hosp., 1980—; assoc. prof. Cornell U., 1980—; chmn. study com. Children's Cancer Study Group, 1980—. Contbr. articles to profl. jours. Served to maj. USAF, 1971-73. Fabian scholar Children's Blood Found., N.Y.C., 1973-75. Mem. Am. Assn. Cancer Research, Am. Soc. Clin. Oncology, Am. Soc. Hematology, Am. Soc. Pediatric Hematology-Oncology, Internat. Soc. Pediatric Oncology. Subspecialties: Pediatrics; Hematology. Current work: Treatment of pediatric molignance. Office: Memorial Sloan Kettering Cancer Center 1275 York Ave New York NY 10021

STEINITZ, LOUIS JOSEPH, mech. engr.; b. Balt., Sept. 17, 1920; s. Louis Adam and Elizabeth Mary (Benda) S.; m. Helen Mary, June 12, 1948; children: Mary Ellen, Kathleen, Paul, Louis, Anne. B.S.M.E., U. Md., 1950; B.S.E.E., Johns Hopkins U., 1960. Registered profl. engr., Md. Mech. engr. Indsl. Research Labs., Balt., 1950-55, Aeronca Mfg. Corp., 1955-65; sr. staff engr. Communications div. Bendix, Balt., 1965—. Adv. Balt. council Boy Scouts Am. Served with USN, 1943-46. Mem. ASME, Soc. Automotive Engrs. Club: KC. Subspecialty: Mechanical engineering. Home: 1229 Gleneagle Rd Baltimore MD 21239 Office: 1300 E Joppa Rd Baltimore MD 21204

STEINKER, DON COOPER, geology educator, researcher; b. Seymour, Ind., Oct. 6, 1936; s. Virgil and Velma (Cooper) S.; m. Paula Jean Dziak, Mar. 6, 1971. B.S., Ind. U., 1959; M.S., U. Kans., 1961; Ph.D., U. Calif.-Berkeley, 1969. Asst. prof. San Jose (Calif.) State Coll., 1965-66; instr. Bodega Marine Lab., Calif., 1966; prof. Bowling Green (Ohio) State U., 1967—. Editor: The Compass, 1975-80; contbr. articles to sci. jours. Mem. Paleontol. Soc., Fla. Acad. Scis., Nat. Assn. Geology Tchrs., Am. Assn. Earth Sci. Editors, Ohio Acad. Sci. Subspecialties: Paleobiology; Paleoecology. Current work: Foraminiferal ecology and paleoecology. Office: Dept Geology Bowling Green Mtate U Bowling Green OH 43403 Home: 311 Ordway Bowling Green OH 43402

STEINMETZ, WAYNE EDWARD, chemistry educator; b. Huron, Ohio, Feb. 16, 1945; s. Ralph Freeman and Helen Therese (Rossman) S. A.B., Oberlin Coll., 1967; A.M., Harvard U., 1968, Ph.D., 1973. Instr. St. Peter's Boys' High Sch., Gloucester, Mass., 1969-70; lab. instr. Oberlin (Ohio) Coll., 1970-71; asst. prof. chemistry Pomona Coll., Claremont, Calif., 1973-79, assoc. prof., 1979—. Scoutmaster Boy Scouts Am., Claremont, 1973—. Wig fellow, 1979; recipient Dist. award of merit Boy Scouts Am., 1978. Mem. Am. Chem. Soc., AAUP, Phi Beta Kappa. Democrat. Roman Catholic. Subspecialties: Physical chemistry; Nuclear magnetic resonance (chemistry). Current work: Applications of molecular spectroscopy to conformationalanalysis, two dimensional NMR spectroscopy. Home: 1081 Cascade Pl Claremont CA 91711 Office: Dept Chemistry Pomona Coll Claremont CA 91711

STEITZ, JOAN ARGETSINGER, molecular biophysics and biochemistry educator; b. Mpls., Jan. 26, 1941; d. Glenn D. and Elaine A. (Magnusson) Argetsinger; m. Thomas A. Steitz, Aug. 20, 1966; 1 son, Jonathan. B.S., Antioch Coll., Yellow Springs, Ohio, 1963; Ph.D., Harvard U., 1967. Asst. prof. molecular biophysics and biochemistry Yale U., New Haven, 1970-74, assoc. prof., 1974-78, prof., 1978—. Editorial bd.: Molecular/Cell Biology; contbr. numerous articles to profl. jours. NSF fellow, 1967-69; Jane Coffin Childs fellow, 1969-70; NIH grantee, 1979—; recipient Passano Found. Young Scientist award, 1975, Eli Lilly award, 1976, U.S. Steel Found. award in molecular biology, 1982. Fellow AAAS, Am. Acad. Arts and Scis.; mem. Nat. Acad. Scis., Am. Soc. Biol. Chemists. Subspecialties: Molecular biology; Gene actions. Current work: Structure and function of small ribonucleoprotein complexes from Eukaryotes, control of transcription and translation, RNA processing. Office: Molecular Biophysics and Biochemistry Yale U PO Box 3333 New Haven CT 06510

STEKOLL, MICHAEL STEVEN, biochemistry educator; b. Tulsa, May 7, 1947; s. Marion and Virginia (Bell) S.; m. Deborah Hansen, Apr. 25, 1976; children: Justin, Skye, Spencer. B.S., Stanford U., 1971; Ph.D., UCLA, 1976. Teaching asst. UCLA, 1971-76, research asst., 1971-76; postdoctoral fellow U. Alaska, Fairbanks, 1976-78; research biochemist Nat. Marine Fisheries Service, Juneau, Alaska, 1979; asst. prof. Sch. Fisheries and Sci., U. Alaska, Juneau, 1978—, acting dean, 1981-82. Contbr. articles to profl. jours. Pres. Auke Bay Pre-Sch., 1980-81. NSF scholar, 1971. Mem. Am. Chem. Soc. (treas. Alaska 1983), AAAS, Phycological Soc. Am. Subspecialties: Biochemistry (biology); Plant physiology (biology). Current work: Marine Algal physiology and ecology. Fisheries biochemistry. Pollution chemistry. Office: Sch Fisheries and Sci U Alaska 11120 Glacier Hwy Juneau AK 99801

STELLAR, ELIOT, psychologist; b. Boston, Nov. 1, 1919; s. Samuel B. and Bella S.; m. Betty E. Housel, Nov. 10, 1945; children: James R., Elizabeth Stellar Fallon. A.B., Harvard U., 1941; M.Sc., Brown U., 1942, Ph.D., 1947; M.A. (hon.), U. Pa., 1972, D.Sc., Ursinus Coll., 1978. Asst. prof. psychology Johns Hopkins U., 1947-54; assoc. prof. physiol. psychology in anatomy Inst. Neurol. Scis., U. Pa., Phila., 1954-60, prof., 1960—, provost, 1973-78; mem. adv. bd. Nat. Primate Ctr. U. Wash.; mem. Sloan Fellowship Com., Rockefeller Hosp. Com. on Research. Author: (with Morgan) Physiological Psychology, 1950, (with Dethier) Animal Behavior, 1961,64,70, (with Sprague) Progress in Physiological Psychology, 1966-73; contbr. numerous articles to profl. jours. Mem. vis. com. Emory U., Lehigh U., Yale U.; bd. dirs. Ursinus Coll. Served with USAF, 1942-44. NRC fellow, 1946-57; NIH grantee; recipient Warren medal Soc. Exptl. Psychology, 1967. Mem. Am. Psychol. Assn., Soc. Neurosci. Nat. Acad. Scis., Am. Philos. Soc. Democrat. Methodist. Subspecialties: Physiological psychology; Neuropsychology. Current work: Physiology of motivation; learning and memory. Home: 172 Cedarbrook Rd Ardmore PA 19003 Office: Dept Anatomy Univ Pa Philadelphia PA 19104

STELLINGWERF, ROBERT F., astrophysicist; b. N.J., Apr. 22, 1947; s. Jacob and Germaine (Bonte) S.; m. Marge Stellingwerf, May 17, 1981. B.A., Rice U., 1969; M.S., U. Colo., 1972, Ph.D., 1974. Postdoctoral trainee Columbia U., 1974-77; asst. prof. astronomy Rutgers U.-New Brunswick, N.J., 1977-80; research scientist Mission Research Corp., Albuquerque, 1980—. Mem. Am. Astron. Soc., Internat. Astron. Union, Am. Phys. Soc. Christian. Subspecialties: Theoretical astrophysics; Laser fusion. Current work: Astrophysical hydrodynamic stability, pulsating stars, winds and accretion flow. Leaser/plasma hydrodynamics, plasma stability. Office: 1720 Randolph Rd SE Albuquerque NM 87106

STEMMLER, EDWARD JOSEPH, physician, university dean; b. Phila., Feb. 15, 1929; s. Edward C. and Josephine (Heitzmann) S.; m. Joan C. Koster, Dec. 27, 1958; children: Elizabeth, Margaret, Edward C., Catherine, Joan. B.A., La Salle Coll., Phila., 1950; M.D., U. Pa., 1960; Sc.D. (hon.), Ursinus Coll., 1977. Diplomate: Am. Bd. Internal Medicine. Intern U. Pa. Hosp., 1960-61, med. resident, 1961-63, fellow in cardiology, 1963-64, chief med. resident, 1964-65, chief med. outpatient dept., 1966-67; chief of medicine U. Pa. Med. Service, VA Hosp., Phila., 1967-73; cons. pulmonary disease VA Hosp., Phila., 1973—, mem. deans com., 1974—; cons. pulmonary disease Phila. Gen. Hosp., 1973-76; NIH postdoctoral research trainee, dept. physiology Grad. Div. Medicine, U. Pa., 1965-67, instr. medicine, 1964-66, asso. in medicine, 1966-67; asso. in physiology Grad. Div. Medicine, 1967-72, asst. prof. medicine, 1967-70, asso. prof., 1970-74, prof., 1974—, Robert G. Dunlop prof., 1981—; asso. dean Univ. Hosp. (Sch. Medicine), 1973, asso. dean student affairs, 1973-75, acting dean, 1974-75, dean, 1975—; dir. Rorer Group, Inc.; Mem. Gov.'s Com. on Health Edn., 1974-76; mem. Nat. Bd. Med. Examiners, 1974-76; mem. ednl. policy com. Nat. Fund for Med. Edn., 1975-77; mem. policy governing bd. Advanced Tech. Ctr. Southeastern Pa., 1975-77. Contbr. articles to med. jours. Chmn. Pa. Dean's Com., 1976—; bd. govs. Mid-Eastern Regional Med. Library Services, 1977—, chmn., 1978-81; mem. bd. visitors U. Pitts. Sch. Medicine, 1980—. Served with Chem. Corps U.S. Army, 1951-53. Decorated Commendation medal; recipient Frederick A. Packard award, 1960, Albert Einstein Med. Center staff award, 1960, Roche award, 1960. Fellow ACP (treas., chmn. investment com. 1975-80); mem. AMA (health policy agenda), Phila. County Med. Soc., Am. Heart Assn.., Am., Pa. thoracic socs., Am. Fedn. for Clin. Research, Assn. Am. Med. Colls. (ad hoc external exam. rev. com. 1980—, exec. council 1980—, council of deans adminstrv. bd. 1980—, chmn. elect council of deans), Coll. of Physicians of Phila. (bd. censors 1979—, council 1979—), Laennec Soc., Am. Clin. and Climatological Soc., John Morgan Soc., Alpha Omega Alpha. Republican. Mem. Christian Ch. Subspecialties: Internal medicine; Pulmonary medicine. Home: 139 E Wynnewood Rd Merion PA 19066 Office: 36th and Hamilton Walk U Pa Philadelphia PA 19104

STENBACK, WAYNE ALBERT, electron microscoptist, clin. pathologist; b. Brush, Colo., June 12, 1929; s. Christian Rasmus and Mabel Eleanor (Christensen) S.; m. Lillian Marie Cacciatore, June 6, 1954; 1 son, Peter (dec.). B.S., U. Colo., 1955; M.S., U. Denver, 1957; Ph.D., U. Mo., 1962. Asst. instr. microbiology, then fellow microbiology U. Mo., Columbia, 1957-61; mem. faculty Baylor Coll. Medicine, Houston, 1962—, adj. asst. prof. pathology, 1976—; electron microscopist dept. pathology Tex. Children's Hosp., Houston, 1975—, med. staff affiliate, 1980—, dir. electron microscope lab., 1975—. Contbr. articles to profl. jours., chpts. to books. Mem. Houston-Harris County Civil Def., 1975—. Recipient Phi Sigma award, 1957; grantee USPHS. Mem. Electron Microscopy Soc., Am. Assn. Cancer Research, AAAS, N.Y. Acad. Scis., Houston Soc. Clin. Pathologists, Sigma Xi. Club: Quarter Century Wireless Assn. Subspecialties: Pathology (medicine); Microscopy. Current work: Ultrastructural pathology, especially pediatric, ultrastructure viruses. Home: 5518 Lymbar Dr Houston TX 77096 Office: Dept Pathology Tex Children's Hosp Houston TX 77030

STENCHEVER, MORTON ALBERT, physician, educator; b. Paterson, N.J., Jan. 25, 1931; s. Harold and Lena (Suresky) S.; m. Diane Bilsky, June 19, 1955; children: Michael A., Marc R., Douglas A. A.B., NYU, 1951; M.D., U. Buffalo, 1956. Intern Mt. Sinai Hosp., 1956-57; resident obstetrics and gynecology Columbia-Presbyn. Med. Center, N.Y.C., 1957-60; asst. prof., Oglebey Research fellow Case-Western Res. U., Cleve., 1962-66, asso. prof. dept. reproductive biology, 1967-70, dir., 1965-70, coordinator, 1969-70; prof., chmn. dept. obstetrics-gynecology U. Utah Med. Sch., Salt Lake City, 1970-77, U. Wash. Sch. Medicine, Seattle, 1977—; test com. chmn. for Ob-Gyn Nat. Bd. Med. Examiners, 1979-82. Author: Labor: Workbook in Obstetrics, 1968, Human Sexual Behavior: A Workbook in Reproductive Biology, 1970, Human Cytogenics: A Workbook in Reproductive Biology, 1973, Introductory Gynecology: A Workbook in Reproductive Biology, 1974; Contbr. articles to profl. jours. Served to capt. USAF, 1960-62. Fellow Am. Coll. Obstetricians and Gynecologists, Am. Assn. Obstetricians and Gynecologists, Am. Gynecol. Soc., Pacific Coast Ob-Gyn Soc.; mem. Assn. Profs. Gynecology and Obstetrics (chmn. steering com. teaching methods in obstetrics and gynecology 1970-79, v.p. 1975—, pres. 1983-84), AMA, AAAS, Pacific Northwest Obstetrics-Gynecology Soc., Wash. State Med. Assn., Seattle Gynec. Soc. (v.p. 1981, pres.-elect 1982, pres. 1982-83), Pacific Coast Obstet.-Gynecol. Soc., Am. Soc. Human Genetics, Central Assn. Obstetrics and Gynecology, Soc. Gynecologic Investigation, Wash. State Obstet. Soc., Tissue Culture Assn., N.Y. Acad. Sci., Utah Obstetrics-Gynecology Soc., Utah State Med. Assn., Teratology Soc. Subspecialties: Obstetrics and gynecology; Perinatal diagnosis and therapy. Current work: Reproductive genetics including male fertility problems; sperm penetration assay. Home: 8301 SE 83d St Mercer Island WA 98040 Office: Dept Obstetrics and Gynecology U Wash Sch Medicine Seattle WA 98105

STENESH, JOCHANAN, chemistry educator; b. Magdeburg, Germany, Dec. 19, 1927; came to U.S., 1950, naturalized, 1965; m. Alfred and Edith (Sturmthal) S.; m. Mabel Holstein, July 26, 1957; children: Ilan, Oron. B.S., U. Oreg.-Eugene, 1953; postgrad., Cornell U., 1953-54; Ph.D., U. Calif.-Berkeley, 1958. Research assoc. Weizmann Inst. Sci., Israel, 1958-60; sr. research assoc. Purdue U., West Lafayette, Ind., 1960-63; asst. prof. chemistry Western Mich. U.,

Kalamazoo, 1963-66, assoc. prof., 1966-71, prof., 1971—. NIH grantee, 1964-70; Am. Cancer Soc. grantee, 1964-66; Faculty Research Fund grantee Western Mich. U., 1968-71. Mem. AAUP, AAAS, Am. Soc. Microbiology, Am. Soc. Biol. Chemists, Am. Chem. Soc. Subspecialties: Biochemistry (biology); Molecular biology. Current work: Molecular biology of proteins and mucleic acids, DNA replication, protein synthesis. Office: Chemistry Dept Western Mich U Kalamazoo MI 49008

STENT, GUNTHER SIEGMUND, molecular biologist, educator; b. Berlin, Germany, Mar. 28, 1924; came to U.S., 1940, naturalized, 1945; s. George and Elizabeth (Karfunkelstein) S.; m. Inga Loftsdottir, Oct. 27, 1951; 1 son, Stefan Loftur. B.S., U. Ill., 1945, Ph.D., 1948. Research asst. U. Ill., 1945-48; research fellow Calif. Inst. Tech., 1948-50, U. Copenhagen, Denmark, 1950-51, Pasteur Inst., Paris, France, 1951-52; asst. research biochemist U. Calif., Berkeley, 1952-56, faculty, 1956—, prof. molecular biology, 1959—, prof. arts and scis., 1967-68, chmn. molecular biology, 1980—, dir. virus lab., 1980—; Document analyst U.S. Field Intelligence Agy. Tech., 1946-47; mem. genetics panel NIH, 1959-64, NSF, 1965-68. Author: Papers On Bacterial Viruses, 2d edit, 1966, Molecular Biology of Bacterial Viruses, 1963, Phage and the Origin of Molecular Biology, 1966, The Coming of the Golden Age, 1969, Function and Formation of Neural Systems, 1977, Morality as a Biological Phenomenon, 1978, Paradoxes of Progress, 1978, Molecular Genetics, 2d edit, 1978; Mem. editorial bd.: Jour. Molecular Biology, 1965-68, Genetics, 1963-68, Zeitschrift für Vererbungslehre, 1962-68, Ann. Revs. Genetics, 1965-69, Ann. Revs. Microbiology, 1966-70; Contbr. numerous sci. papers to profl. lit. Merck fellow NRC, 1948-54; sr. fellow NSF, 1960-61; Guggenheim fellow, 1969-70. Fellow Am. Acad. Arts and Scis.; mem. Soc. Neurosci., Nat. Acad. Scis. Subspecialties: Neurobiology; Molecular biology. Current work: Developmental neurobiology; history and philosophy of science. Home: 145 Purdue Ave Berkeley CA 94708

STEPH, NICK CHARLES, physics educator; b. Tulsa, Okla., Apr. 23, 1947; s. Claude M. and Alice (Gere) S.; m. Linda McDonald, Jan. 20, 1978; 1 son, Derek Michael. B.S. in Biology, U. So. Colo., 1969; Ph.D. in Physics, U. Okla., 1979. Research asst. U. Okla., 1976-79, research assoc., 1979-82, asst. prof. physics and astronomy, 1982—. Contbr. articles to profl. jours. Served with U.S. Army, 1969-72. Recipient J. Rud Nielson prize U. Okla., 1979. Mem. Am. Phys. Soc., AAAS, Sigma Xi. Subspecialties: Atomic and molecular physics; Developmental biology. Current work: Coincidence studies of collisional excitation of atoms and molecules by impact with electrons and ions. Office: 440 W Brooks St Normal OK 73019

STEPHENS, CATHY LAMAR, cancer research; b. Gadsden, Ala., Aug. 6, 1940; d. William and Lola Lamar (Smith) Blanton. B.A., U. N.Mex., 1963; Ph.D., UCLA, 1972. Postdoctoral researcher UCLA, 1972-76; research physiologist Nat. Cancer Inst. NIH, 1976-81; dir. research West Fla. Cancer Research Found., Inc., Pensacola, 1981—. Contbr. articles to profl. jours. Mem. Neuroscience Soc., N.Y. Acad. Sci., AAAS. Democrat. Episcopalian. Subspecialties: Cancer research (medicine); Membrane biology. Current work: Immunological targeting of antitumor drugs, membrane transport, mechanisms of selective cytotoxicity. Home: 1902 Copley Dr Pensacola FL 32503 Office: 2266 La Vista St Pensacola FL 32504

STEPHENS, JAMES FRANCIS, college dean, chemistry educator; b. Walker, W.VA., Aug. 2, 1939; s. Cecil Francis and Dora Lee (Jenkins) S.; m. Nancy Ruth Harman, July 6, 1967; 1 dau., Emily Ruth. B.S. in Chemistry, W.Va. U., 1963, M.S., 1965, Ph.D., U. Tenn.-Knoxville, 1969. Asst. prof. chemistry St Andrews Presbyterian Coll., 1969—, registrar, 1974—, assoc. acad. dean, 1976—. Contbr. articles to profl. jours. Mem. Am. Chem. Soc., Am. Assn. Registrars and Admissions Officers. Democrat. Methodist. Subspecialties: Inorganic chemistry; Nuclear magnetic resonance (chemistry). Office: Saint Andrews Presbyn Coll Laurinburg NC 28352

STEPHENS, PHILIP JOHN, biologist, educator, researcher; b. London, Apr. 24, 1951; s. Gerwyn Lougher and Elizabeth (Morgan) S.; m. Margaret Lesley Stephens, Sept. 26, 1975; children: Charles Andrew, David Philip. B.Sc., London U., 1972; Ph.D., Aberdeen U., Scotland, 1977. Alfred P. Sloan research fellow U. Va., Charlottesville, 1975-78, postdoctoral fellow, 1978-79; postdoctoral research fellow U. Toronto, Ont., Can., 1979-80; Grass Found. research fellow Woods Hole (Mass.) Biol. Lab., Mass., 1980; asst. prof. dept biology Villanova (Pa.) U., 1980—. Contbr. articles to sci. jours. Whitehall Found. grantee, 1981-84; NSF grantee, 1982-84. Mem. Soc. Neurosci. Subspecialty: Neurophysiology. Current work: Nerve and muscle physiology; temperature effects on neuromuscular physiology; nerve-muscle interactions during development; axon firing mechanisms. Office: Villanova U Villanova PA 19085

STEPHENS, TRENT DEE, anatomy educator; b. Wendell, Idaho, Aug. 14, 1948; s. Herbert Raymond and Phyllis (Behunin) S.; m. Kathleen Rae Brown, Sept. 4, 1971; children: Summer, Rhett Dee, Brittani, Derek Ray, Blake Christopher. B.S., Brigham Young U., 1973, M.S., 1974; Ph.D., U. Pa., 1977. Sr. fellow U. Wash., Seattle, 1977-79, research assoc., 1979-81; asst. prof. anatomy Idaho State U., Pocatello, 1981—. Author: Atlas of Human Embryology, 1980. Rotary Found. grantee, 1973-74; NIH fellow, 1974-77; Poncin scholar, 1977-79. Mem. Teratology Soc., Am. Assn. Anatomists, Idaho Acad. Sci., AAAS. Mormon. Subspecialties: Anatomy and embryology; Teratology. Current work: Morphogenesis, pattern formation, comparative embryology, limb field induction and placement, comparative and evolutionary morphogenesis. Home: Route 5 S Williamsburg Ln Pocatello ID 83204 Office: Dept Biology Idaho State U Pocatello ID 83209

STEPHENSON, EDWARD HAYES, veterinary microbiologist, army officer; b. Ada, Okla., Jan. 7, 1937; s. William Hayes and Beryl Rae (Moody) S.; m. Rita Mae Elms, Sept. 4, 1957; m. Barbara June Thompson, May 31, 1968; children: Scherri Lynn Stephenson Allen, Alisa Monnette, Edward Hayes; m.; stepchildren: Melody Kay Manley, Gary S. Manley. D.V.M., Tex. A&M U., 1961; M.S., U. Wis., 1966; Ph.D., Colo. State U., 1973. Clin. practice vet. medicine, 1961-62; commd. 1st lt. U.S. Army, 1962, advanced through grades to col., 1981; meat and dairy products inspection officer (Boston Army Base), Mass., 1962-64, asst. chief dept. vet. medicine, Vietnam, 1966-67, chief dept. vet. medicine, Ft. Sam Houston, Tex., 1967-70; vet. microbiologist Walter Reed Army Inst. Research, Washington, 1973-81; chief aerobiology div. U.S. Army Med. Research Inst. of Infectious Diseases, Frederick, Md., 1981—. Contbr. articles to profl. jours. Decorated Army Commendation medal. Mem. Am. Assn. Vet. Lab. Diagnosticians, Am. Coll. Vet. Microbiologists (diplomate 1975), Am. Soc. Microbiology, AVMA, Conf. Research Workers in Animal Disease, D.C. Vet. Med. Assn., Sigma Xi. Democrat. Mem. Christian Ch. (Disciples of Christ). Subspecialties: Microbiology (medicine); Microbiology (veterinary medicine). Current work: Rickettsiology-virology, aerobiology; respiratory infections—immunopathology, serodiagnosis, prophylaxis, and vaccine production. Home: 18820 Stoneyhurst St Olney MD 20832 Office: Aerobiology Div US Army Med Research Inst of Infectious Diseases Frederick MD 21701

STEPHENSON, MARY LOUISE, biochemist; b. Brookline, Mass., Feb. 23, 1921; d. George Eustis and Louise (Dixon) S. B.A., Conn. Coll., 1943; Ph.D., Radcliffe Coll., 1956. Assoc. biochemist Mass. Gen. Hosp., Boston, 1969—; prin. research assoc. Harvard U. Med. Sch., Boston, 1969—. Mem. Am. Soc. Biol. Chemists, Am. Assn. Cancer Research, Am. Soc. Cell Biology. Subspecialties: Biochemistry (biology); Molecular biology. Current work: Protein biosynthesis. Home: 308 Ocean Ave Marblehead MA 01945 Office: Arthritis Unit Mass Gen Hosp Fruit St Boston MA 02114

STERGAKOS, ELIAS P(ANAGIOTES), nuclear engineer, energy resources consultant; b. Potamia, Laconia, Greece, Apr. 7, 1942; came to U.S., 1956, naturalized, 1961; s. Panagiotes and Katina (Exarhakos) S.; m. Eleonor Ipliktsiades, Aug. 16, 1970; children: Katina, Miltiades, Panagiotes. B.A., Adelphia U., 1966; M.S., Va. Poly. Inst., 1967, Ph.D., 1970. Grad. asst. Va. Poly. Inst., 1966-70; lead nuclear engr. Duke Power Co., Charlotte, N.C., 1970-72; nuclear engring. specialist Burns & Roe, Inc., Oradell, N.J., 1972—. Pres. Assn. Potamiton, Bklyn., 1976—. Mem. Am. Nuclear Soc., Assn. Laconon. Greek Orthodox. Subspecialties: Nuclear engineering; Fuels and sources. Current work: Supervise and direct nuclear and nuclear instrumentation analyses and design performed in construction of nuclear power plants; evaluate availability of future energy resources. Office: Burns & Roe 185 Crossways Park Dr Woodbury NY 11797

STERLING, RAYMOND LESLIE, civil engr., educator, researcher, cons.; b. London, Apr. 19, 1949; U.S., 1966; s. Richard Howard and Joan Valeria (Skinner) S.; m. Linda Lee Lundquist, Aug. 7, 1970 (div. Sept. 1982); children: Paul Nathan, Juliet Paige, Erika Joy. B. Civil and Structural Engring., U. Sheffield, Eng., 1970; M.S. in Geol. Engring, U. Minn., 1975; Ph.D. in Civil Engring, U. Minn., 1977. Registered profl. engr., Minn. Civil engr. Egil Wefald & Assocs., Mpls., 1969-71; structural engr. Husband & Co., Sheffield, 1971-73, Setter, Leach & Lindstrom, Inc., Mpls., 1976-77; asst. prof. civil and mineral engring. dept. U. Minn., Mpls., 1977—; dir. Underground Space Ctr., 1977—; prin. cons. Itasca Cons. Group, Inc., Mpls., 1981—. Author: Earth Sheltered Housing Design, 1979, Earth Sheltered Housing: Code, Zoning and Financing Issues, 1980, Earth Sheltered Community Design, 1981 (Assn. Am. Pubs. architecture and urban planning book of yr. 1981), Earth Sheltered Residential Design Manual, 1982. Mem. Mpls. Energy Futures Com., 1981-82. Mem. ASCE (Young Engr. of Yr. award Minn. sect. 1982), Instn. Structural Engrs. (U.K.), Instn. Civil Engrs. (U.K.) (assoc. mem.), Nat. Soc. Profl. Engrs., Internat. Solar Energy Soc., Am. Underground Space Assn. Club: Mpls. Engrs. Subspecialties: Underground engineering; Earth-sheltered buildings. Current work: Use of underground space for land and energy conservation research, underground space use, earth sheltered buildings, passive energy conservation building techniques. Home: 4933 Dupont Ave S Minneapolis MN 55409 Office: 128 Pleasant St SE Minneapolis MN 55455

STERN, DANIEL HENRY, biology educator; b. Richmond, Va., June 18, 1934; s. Henry S. and Adele S. (Lewit) S.; m.; 1 son, Alexander. B.S. in Chemistry, Richmond Coll., 1955; M.S. in Biology, U. Richmond, 1959; Ph.D. in Zoology, U. Ill., 1964. Asst. prof. biology Tenn. Tech. U., Cookeville, 1964-66, La. State U., New Orleans, 1966-69; mem. faculty U. Mo.-Kansas City, 1969—, prof. biology, 1975—; cons. Midwest Research Inst., 1972—. Contbr. articles to profl. jours.; reviewer jours., books. Mem. polit. action com. NEA, Western Mo., 1982-83; bd. dirs. Maple Woods Bd., Gladstone, Mo., 1980—. Named Outstanding Tchr. La. State U., 1968. Mem. Am. Soc. Limnology and Oceanography, Ecol. Soc. Am., N.Am. Benthological Soc., Phycol. Soc. Am., Am. Micros. Soc., Am. Water Resources Assn., Am. Inst. Biol. Scis., Am. Soc. Zoologists, Sigma Xi, Beta Beta Beta, Phi Sigma. Democrat. Jewish. Subspecialties: Ecology; Environmental toxicology. Current work: Environmental impact analysis; toxicity of heavy metals in aquatic systems; human ecology; limnology, including water qulaity, plankton, macrobenthos, periphyton, land-water interactions. Home: 10114 Locust St Kansas City MO 64131 Office: Dept Biology Univ Mo Kansas City 109 Biol Scis Bldg Kansas City MO 64110

STERN, KINGSLEY ROWLAND, botany educator; b. Port Elizabeth, S. Africa, Oct. 30, 1927; s. Julius Charles and Vera Grace (Estment) S.; m. Janet Elaine McLeland, June 9, 1956; children: Kevin Douglas, Sharon Maureen. B.S., Wheaton (Ill.) Coll., 1949; M.A., U. Mich., 1950; Ph.D., U. Minn., 1959. Mem. faculty Calif. State U., Chico, 1959—, prof. botany, 1968—; cons. in field. Author: Introductory Plant Biology, 1979, 2d. ed., 1982; contbr. sci. articles to profl. publs. Recipient Profl. Achievement award Calif. State U., 1981; NSF grantee, 1963-70; Conway Macmillan Research fellow U. Minn., 1956-57. Mem. Am. Soc. Plant Taxonomists, Bot. Soc. Am., Calif. Bot. Soc., Calif. Native Plant Soc. Club: No. Calif. Center for Sports Medicine (Chico). Subspecialties: Evolutionary biology; Taxonomy. Current work: Biosystematics of vascular plants. Home: 1430 Arcadian Ave Chico CA 95926 Office: Calif. State Univ Biol Scis Dept Chico CA 95929

STERN, ROBERT MORRIS, educator; b. N.Y.C., June 18, 1937; s. Irving Dan and Nellie (Wachstetter) S.; m. Wilma Olch, June 19, 1960; children—Jessica Leigh, Alison Rachel. A.B., Franklin and Marshall Coll., 1958; M.S., Tufts U., 1960; Ph.D., Ind. U., 1963. Research assoc. dept. psychology Ind. U., 1963-65; asst. prof. psychology Pa. State U., 1965-68, asso. prof., 1968-73, prof., 1973—, head dept. psychology, 1978—. Author: (with W.J. Ray) Biofeedback, 1977, (with W.J. Ray and C.M. Davis) Psychophysiological Recording, 1980; Contbr. articles to profl. jours. Recipient Nat. Media award Am. Psychol. Found., 1978. Mem. Am. Psychol. Assn., Eastern Psychol. Assn., Soc. Psychophysiol. Research. Subspecialties: Physiological psychology; Gastroenterology. Current work: Gastrointestinal functioning in humans and health, disease and behavior. Home: 1360 Greenwood Circle State College PA 16801 Office: 417 Moore Bldg Pennsylvania State Univ University Park PA 16802

STERN, WARREN CHARLES, pharm. co. exec., psychopharmacologist, cons.; b. Bronx, N.Y., June 1, 1944; s. Julius and Eleanor (Fox) S.; m. Carol Joy, June 13, 1965; children: Andrew, Douglas, Gregory. B.S., Bklyn. Coll., 1965; Ph.D., Ind. U., Bloomington, 1969. Postdoctoral fellow Boston State Hosp., 1969; postdoctoral fellow Worcester (Mass.) Found. Exptl. Biology, 1970, staff scientist, 1971-75; neuropharmacologist Squibb Inst. Med. Research, Princeton, N.J., 1976; head sect. psychiatry dept. clin. research Burroughs Wellcome Co., Research Triangle Park, N.C., 1976—; research scientist D. Dix Hosp.; adj. asso. prof. Sch. Pharmacy, U. N.C. Contbr. numerous articles on neurosci., psychopharmacology, biol. psychiatry to profl. jours.; regional editor: Pharmacology, Biochemistry and Behavior, 1975—, Drug Devel. Research, 1981—. Recipient First prize Psychopharmacology div. Am. Psychol. Assn., 1971. Mem. Am. Soc. Pharm. and Exptl. Therapeutics, Biol. Psychiatry, AAAS, New Clin. Drug Eval. Unit. Subspecialties: Neuropharmacology; Psychopharmacology. Current work: Clin. devel. new psychiat. drugs; animal neuropharmacology; human and animal psychopharmacology. Home: 8904 Willow Wood Ct Raleigh NC 27612 Office: Burroughs Wellcome Co Research Triangle Park NC 27709

STERNBERG, DAVID EDWARD, psychiatrist, psychopharmacology researcher; b. Norfolk, Va., Jan. 18, 1946; s. Theodore and Bella (Rosenblatt) S.; m. Frances T. Glazer, Aug. 23, 1970; 2 sons, Jonathan Theodore, Daniel Alexander. B.A. in Biopsychology, U. Chgo., 1967; M.D., Tufts U., 1971. Diplomate: Am. Bd. Psychiatry and Neurology. Postdoctoral fellow in psychiatry Yale U., New Haven, 1972-75; staff psychiatrist and dir. alcohol rehab. unit Nat. Naval Med. Ctr., Bethesda, Md., 1975-77; research coordinator and staff psychiatrist sect. neuropsychopharmacology (schizophrenia studies), biol. psychiatry br. NIMH, Bethesda, 1977-79; chief clin. research unit and asst. prof. dept. psychiatry Yale U. Sch. Medicine, New Haven, 1979-83; lectr. dept. psychiatry Nat. Naval Med. Ctr., Bethesda, Md., 1983—; clin. dir. Falkirk Hosp., Central Valley, N.Y. Contbr. numerous articles to profl. jours. Served to lt. comdr. USN, 1975-77; to comdr. USPHS, 1977-79. Recipient Research Honors award Tufts U. Sch. Medicine, 1971; Seymour Lustman Research award Yale U. Dept. Psychiatry, 1975. Mem. Am. Psychiat. Assn., Soc. Biol. Psychiatry, Soc. Neurosci., AAAS. Subspecialties: Psychiatry; Psychopharmacology. Current work: Research into the etiology and pharmacotherapy of schizophrenia, mania and depression, focussing on neurotransmitter function by analysis of body fluids and pharmacologic challenges to brain receptor systems of specific neurotransmitters. Office: Falkirk Hosp Central Valley NY 10917

STERNHAGEN, CHARLES JAMES, therapeutic radiologist; b. Glasgow, Mont., Oct. 15, 1933; s. Joseph Peter and Mary Catherine (Carignan) S.; m. Marlene Linda Keubler, Dec. 19, 1952; children: Charlene, Charles, Linda, Joseph, William, Marc, Donald, Bernard, Mary, Scott, Catherine, Marleen. B.A., Carroll Coll., Helena, Mont., 1956; M.D., Loyola U., 1959. Intern St. Joseph Hosp., South Bend, Ind., 1959-60; resident U. Okla., Oklahoma City, 1969-72; instr. environ. health U. Okla. Health Scis. Ctr., Oklahoma City, 1968-72, instr. radiation oncology, 1972; asst. prof. radiology Cancer Research and Treatment Ctr., U. N.Mex.-Los Alamos Sci. Lab., 1971-75, assoc. prof., 1976-78; chief radiation oncology Lovelace Bataan Med. Center, Albuquerque, 1972-78; dir. Cancer Therapy Ctr., Providence Hosp., Anchorage, 1978—; Pres.-elect State of Alaska div. Am. Cancer Soc., 1982; state surgeon N.Mex. N.G., 1973-76. Contbr. articles to profl. jours. Served to col. USAR, 1964-82. Lederly research fellow, 1956; recipient various grants. Fellow Am. Pub. Health Assn., Royal Soc. Health (London); mem. Am. Endocurietherapy Soc., Am. Soc. Preventive Oncology, Okla. Acad. Environ. Health Scis., Alaska State Med. Assn. (councillor, conv. chmn.), Am. Soc. Therapeutic Radiologists, Am. Coll. Radiology, Am. Soc. Clin. Oncology, Am. Roentgen Ray Soc., Am. Assn. for Cancer Research, Health Physics Soc. (ethics com.), Internat. Radiation Protection Assn., Radiation Research Soc., Radiol. Soc. N.Am., Am. Radium Soc., N.Mex. Soc. Radiology, AMA, Anchorage Med. Soc., AAAS. Subspecialties: Oncology; Cancer research (medicine). Current work: Particles and radiation therapy; the negative Pi-meson research efforts in cancer; hyperthermia and cancer research; nasopharyngeal cancer research in Alaska natives. Home: 8421 Pioneer Dr Anchorage AK 99504 Office: 3200 Providence Blvd Anchorage AL 99504

STERNICK, EDWARD SELBY, med. physicist; b. Cambridge, Mass., Feb. 10, 1939; s. Charles and Adele N. (Stengel) S.; m.; children: Heidi, Jennifer, Peter. B.S., Tufts U., 1960; M.A., Boston U., 1963; Ph.D., UCLA, 1968. Diplomate: Am. Bd. Radiology. Research scientist NASA, Ames Research Center, Moffett Field, Calif., 1962-63; NIH fellow UCLA, Los Angeles, 1964-67; dir. med. physics Dartmouth-Hitchcock Med. Center, Hanover, N.H., 1968-78; dir. med. physics div. Tufts-New Eng. Med. Center, Boston, 1978—; assoc. prof. therapeutic radiology and diagnostic radiology Tufts U. Sch. Medicine, Boston, 1978—. Mem. Am. Assn. Physicists in Medicine, Health Physics Soc., Am. Coll. Radiology, Radiol. Soc. N.Am., Soc. Nuclear Medicine, Sigma Xi. Unitarian. Current work: Applications of computer tech. to medicine. Office: Med Phys Div 171 Harrison Ave Boston MA 02111

STERN-TOMLINSON, WENDY BARBARA, neurophysiologist; b. Bronx, N.Y., Jan. 12, 1943; d. Aaron and Hannah (Katz) Stern; m. Charles D. Yingling, July 3, 1964; 1 dau., Arden; m. Christopher J. Tomlinson, June 28, 1975; 1 dau., Stephanie. Student, Barnard Coll., 1960-61; B.A., Rice U., 1964, Ph.D., 1976. Research asst. N.Y. U. Med. Sch., N.Y.C., 1966-69; research assoc. dept. biology Tufts U., Medford, Mass., 1974-76; postdoctoral fellow dept. physiology U. Pa. Sch. Medicine, Phila., 1978-82; postgrad. research biologist, dept. biology U. Calif.-San Diego, La Jolla, 1982—. Contbr. articles to sci. jours. NIH fellow, 1978-81; NSF fellow, 1977; Whitehall Found. grantee, 1982—. Mem. AAAS, Soc. Neurosci., Assn. Women in Sci., Women in Neurosci. Subspecialties: Neurobiology; Neurophysiology. Current work: Neural circuitry underlying behavior in invertebrates. Office: U Calif La Jolla CA 92093

STETLER, DWIGHT L(AWRENCE) (BOB STETLER), mechanical engineer, consultant; b. Falls City, Nebr., Mar. 8, 1926; s. Dwight Irl and Verna Maude (Flowers) S.; m. Natalie Mary Gill, Apr. 29, 1950. Sr. draftsman Mobil Oil Co., Los Angeles, 1951-62; v.p. Charles W. Jones Engring., Los Angeles, 1962-71; project mgr. Ameron Corp., Monterey Park, Calif., 1971-77; chief engr. Wilbur Curtis Co., Los Angeles, 1977-83, Safe T Jack Corp., Santa Fe Springs, Calif., 1983—. Served with AC U.S. Army, 1944-45. Mem. ASME, Nat. Soc. Profl. Engrs., Aircraft Owners and Pilots Assn. Republican. Baptist. Club: Long Beach (Calif.) Flyers. Subspecialties: Solar energy; Fuels and sources. Current work: Conducting fundamental research in solar energy, fuels, fuel sources. Patentee in field. Office: Safe T Jack Corp 11823 E Slauson Ave Suite 34 Santa Fe Springs CA 90670

STETTEN, DEWITT, JR., biochemist; b. N.Y.C., May 31, 1909; s. DeWitt and Magdalen (Ernst) S.; m. Marjorie Roloff, Feb. 7, 1941; children—Gail (Mrs. Peter C. Maloney), Nancy (Mrs. Frank Einstein), Mary (Mrs. Michael Carson), George. B.A., Harvard, 1930; M.D., Columbia, 1934, Ph.D., 1940; D.Sc. (hon.), Washington U., St. Louis, 1974, Coll. Medicine and Dentistry N.J., 1976, Worcester Found. for Exptl. Biology, 1979. Intern, 3d med. div. Bellevue Hosp., N.Y.C., 1934-37; instr., asst. prof. biochemistry Coll. Phys. and Surgs., Columbia, 1940-47; asso. medicine Peter Bent Brigham Hosp., Boston, 1947-49; asst. prof. biol. chemistry Harvard Med. Sch., 1947-48; chief div. nutrition and physiology Pub. Health Research Inst. City N.Y., 1948-54; asso. dir. charge research Nat. Inst. Arthritis and Metabolic Diseases, NIH, 1954-62; dean Rutgers Med. Sch., New Brunswick, N.J., 1962-70; dir. Nat. Inst. Gen. Med. Scis., NIH, Bethesda, Md., 1970-74; dep. dir. for sci. NIH, 1974-79, sr. sci. adv. to dir., 1979—; mem. subcom. liver diseases NRC, 1948-49, panel intermediary metabolism, com. on growth, 1948-52, chmn. panel, 1950, mem. exec. bd. div. med. scis., 1965-69; cons. Walter Reed Army Med. Center, 1948-53; chmn. research com. N.Y. Diabetes Assn., 1951-52; cons., study sect. mem. NIH, 1952-53; mem. adv. com. health research facilities and resources; mem. adv. council dept. biology Princeton;

chmn. nat. adv. com. Okla. Found. Med. Research, 1966; chmn. nat. sci. adv. com. Roche Inst. Molecular Biology, 1966-70. Author: (with others) Principles of Biochemistry, 1954, 59; Mem. editorial bd.: Sci. and Perspectives in Biology and Medicine; Contbr. articles to profl. jours. Trustee N.J. Mental Health Research and Devel. Bd.; bd. dirs. Found. for Advanced Edn. in Scis., pres., 1973; mem. vis. com. Oklahoma City campus U. Okla.; bd. visitors Grad. Sch. Pub. Health, U. Pitts. Recipient Joseph Mather Smith prize Columbia, 1943; Alvarenga prize Phila. Coll. Physicians, 1954; Banting medal Am. Diabetes Assn., 1957; Superior Service award HEW, 1973; gold medal for distinguished achievement in medicine Columbia U. Coll. Phys. and Surg., 1974; Disting. Service award HEW, 1976; Woodrow Wilson fellow, 1979—. Mem. Nat. Acad. Scis., AAAS (chmn. sect. on medicine and v.p. 1962), Am. Acad. Arts and Scis., Nat. Acad. Scis. (council 1976-79), Am. Chem. Soc., Am. Soc. Biol. Chemists, N.Y. Acad. Scis., Harvey Soc., Soc. Exptl. Biology and Medicine (pres. 1977-79), Washington Acad. Medicine, Med. Soc. N.J., Phi Beta Kappa, Sigma Xi, Alpha Omega Alpha. Subspecialty: Medical research administration. Current work: Scientific and medical adminstration. Home: 2 West Dr Bethesda MD 20814 Office: NIH Bldg 16 Rm 118 Bethesda MD 20205

STEUER, ANTON FRANCIS, cell biologist; b. Reading, Pa., Oct. 23, 1947; s. Anton and Rose (Schroeder) S.; m. Mary Kathryn Steuer, Dec. 18, 1971; children: Andre, Shaun. B.S., Allentown Coll., Center Valley, Pa., 1969; M.S., Cath. U. Am., Washington, 1972, Ph.D., 1974. Teaching asst. Cath. U. Am., 1971-73; dir. dept. cell biology Biotech Research Labs., Inc., Rockville, Md., 1974—; lectr. gen. biology Dunbarton Coll., 1972. Contbr. articles and abstracts to profl. jours. Mem. Tissue Culture Assn., Am. Assn. Cancer Research, Am. Soc. Microbiology, Sigma Xi. Subspecialty: Cell and tissue culture. Current work: Hybridoma research, devel. of biol. detection systems. Office: Biotech Research Labs 1600 E Gude Dr Rockville MD 20850

STEVENS, CHARLES F., neurobiologist, educator; b. Chgo., Sept. 1, 1934; m. Jane Robinson; 3 children. B.A., Harvard U., 1956; M.D., Yale U., 1960; Ph.D., Rockefeller U., 1964. Asst. prof. dept. physiology and biophysics U. Wash. Sch. Medicine, 1963-68, asso. prof., 1968-72; guest investigator Lorentz Inst. Theoretical Physics, Leiden (Neitherlands) U., 1969-70; prof. dept. physiology Yale U. Sch. Medicine, New Haven, 1975—. Author: Neurophysiology: A Primer, 1966; contbr. articles to profl. jours. Subspecialties: Neurophysiology; Biophysics (biology). Office: Department Physiology Yale School Medicine 333 Cedar St PO Box 3333 New Haven CT 06510

STEVENS, GWENDOLYN RUTH, psychology educator; b. Los Angeles, Feb. 29, 1944; d. Oscar and Alice (Whalen) S.; m. David Nichols, Feb. 24, 1962 (div.); children: Lorin Ann, Stephen Forrest; m. Sheldon Gardner, Oct. 24, 1972. B.A., Calif. State U.-Los Angeles, 1973, M.A., 1974; Ph.D., U. Calif.-Riverside, 1978. Instr. Tri-Community Nursery, Coving, Calif., 1964-67; instr. psychology East Los Angeles Community Coll., 1975, Cypress Community Coll., Calif., 1975, Whittier Coll., 1976; asst. prof. psychology Southeast Mo. State U., Cape Girardeau, 1978-82; asst. prof. psychology dept. humanities U.S. Coast Guard Acad., New London, Conn., 1982—; cons. Calif. Grad. Inst., Orange, 1975-78; research dir. Client Assistance, Downey, Calif., 1975-77; research asst. Rancho Los Amigos Hosp., Downey, 1973-75; counselor Luth. Family Service, Cape Girardeau, 1978-81, 1977-80. Author: Women in Psychology, vol. I and II, 1981, Care and Cultivation of Parents, 1979; contbr. articles to profl. jours. Southeast Mo. State U. scholar, 1981. Mem. Am. Psychol. Assn., Internat. Council Psychology, Nat. Women's Studies Assn., Internat. Assn. for History of Social Scis. Libertarian. Subspecialties: Social psychology. Current work: Gender roles, attribution process, psychometric properties of attitude scales, history of women in psychology. Home: 75 Hilltop Rd Mystic CT 06355 Office: Dept Humanities US Coast Guard Acad New London CT 06320

STEVENS, JAMES THOMAS, toxicologist, pharmacologist; b. Wellsboro, Pa., June 23, 1946; s. William and Jeannette (Hatherill) S.; m. Brenda Joyce Davis, Sept. 23, 1967; children: Trent Douglas, Kelly Anne, Andrew James. B.A., Pa. State U., 1967; M.S. in Reproductive Physiology, W.Va. U., 1970; Ph.D. in Pharmacology, W.Va. U., 1972. Asst. prof. pharmacology Hershey (Pa.) Med. Sch., 1972-74; sect. chief inhalation toxicology EPA, Research Triangle Park, N.C., 1974-77; toxicologist, then sr. toxicologist Ciba-Geigy Corp., Greensboro, N.C., 1977-80, group leader inhalation toxicology and asst. to dir., Switzerland, 1980-81, mgr. toxicology, 1981—; vis. prof. Bowman-Grey Med. Sch., Winston-Salem, N.C., 1978; vis. lectr. Nat. Inst. Environ. Health Sci., 1981. Contbr. articles to profl. jours. Mem. Am. Soc. Pharmacology and Exptl. Therapeutics, Soc.Toxicology, Am. Indsl. Hygiene Assn., Jaycees, Sigma Xi. Republican. Methodist. Lodge: Elks. Subspecialties: Toxicology (medicine); Pharmacology. Current work: Inhalation toxicology and biochemical pharmacology. Home: 4107 Oak Hollow Dr High Point NC 27260 Office: PO Box 1830 Greensboro NC 27419

STEVENS, JOHN GEHRET, info. specialist, educator; b. Mount Holly, N.J., Dec. 16, 1941; s. Robert B. and Helen V. (Gehret) S.; m. Virginia C. Entwistle, June 14, 1963; children: Sybil M., Robert J., John G. B.S. in Chemistry, N.C. State U., 1964; Ph.D. in Phys. Chemistry, 1968. Asst., then assoc. prof. U. N.C., Asheville, 1968-79, prof. chemistry, 1979—; dir Mossbauer Effect Data Ctr., 1970—; dir. Computer Center, 1982—; research assoc. Argonne Nat. Lab., summers 1969, 70, 71, Max-Planck Inst. Solid State Physics, 1973; research prof. U. Nijmegen, Netherlands, 1976-77, summers 1978, 79, 80, 81. Contbr. articles to profl. lit., chpts. to books. Coordinator Advocates for Nuclear Arms Freeze; scoutmaster local troop Boy Scouts Am. Recipient Outstanding Young Am. award Sigma Pi Sigma; Outstanding Vol. award State of N.C. Mem. Am. Chem. Soc., Am. Phys. Soc., Fedn. Am. Scientists, AAAS, Sigma Xi. Presbyterian. Subspecialties: Physical chemistry; Solid state chemistry. Current work: Electronic structure of antimony molecules, the handling of scientific data and information. Antimony, Mossbauer, data evaluation, information handling, chemical bonding, powdered x-ray defraction. Home: 71 Woodbury Rd Asheville NC 28804 Office: Chemistry Dept U NC Asheville NC 28814

STEVENS, KARL KENT, engring. educator, cons.; b. Topeka, Jan. 24, 1939; s. Virgil Leo and Ida Janice S.; m. Diana Doherty, June 2, 1960; children: Robin, Scott, Michael; m. Josephine H. French, Sept. 21, 1973. B.S., Kans. State U., 1961; postgrad., U. N.Mex., 1961-62; M.S., U. Ill. Urbana, 1963, Ph.D., 1965. Registered profl. engr., Fla., Ohio. Mem. staff Sandia Corp., Albuquerque, 1960-62; prof. engring. mechanics Ohio State U., Columbus, 1965-78; prof. ocean engring. Fla. Atlantic U., Boca Raton, 1978-83, chmn. dept., 1981-83, prof. mech. engring., 1983—; cons. in field. Author: Statics and Strength of Materials, 1979; contbr. articles to profl. jours. Recipient Alumni award for Disting. Teaching Ohio State U., 1975; Outstanding Achievement award Ohio Soc. Profl. Engrs., 1972; NSF fellow, 1967. Mem. Am. Soc. Engring. Edn., ASME, Soc. Naval Architects and Marine Engrs., Am. Acad. Mechanics. Subspecialties: Theoretical and applied mechanics; Ocean engineering. Current work: Teaching and research in engineering design and vibrations. Home: 3533 NE 6th Dr Boca Raton FL 33431 Office: Fla Atlantic U Boca Raton FL 33431

STEVENS, REGGIE HARRISON, radiation biology educator, researcher; b. Iowa City, Aug. 7, 1941; s. Charles and Doris S.; m. Sylvia A.; 1 son, Reggie C. B.S., U. Iowa, 1968, M.S., 1970, Ph.D., 1972. Research biochemist VA, Iowa City, 1972-74; research investigator U. Iowa, Iowa City, 1974-75, instr., 1975-76, asst. prof., 1976-80, assoc. prof. radiation biology, 1980—. Served in USAF, 1959-63. Mem. Am. Coll. Toxicology, Radiation Research Soc., Am. Assn. Clin. Chemists, Am. Chem. Soc., Sigma Xi, Phi Lambda Upsilon. Subspecialties: Radiation biology; Environmental toxicology. Current work: Identification and quantification of radiation and chemicals' potential for inducing cancer. Home: 1850 Friendship Iowa City IA 52240 Office: 14 Medical Lab U Iowa Iowa City IA 52242

STEVENS, ROY HARRIS, microbiologist, dentist; b. N.Y.C., Jan. 8, 1948; s. Daniel and Gladys (Sporn) S.; m. Jeanne Marie Connors, Sept. 14, 1971; 1 dau., Jocelyn Natalie. B.A., Adelphi U., 1969; M.S., Rutgers U., 1972; D.D.S., Columbia U., 1976. Resident dental service Beth Israel Med. Ctr., N.Y.C., 1976-77; postdoctoral fellow dept. microbiology Dental Sch., U. Pa., Phila., 1977-79, research assoc., 1979-80, research asst. prof., 1980—. Author: Host-Parasite Interactions in Periodontal Disease, 1982; contbr. articles to profl. jours. Nat. Inst. Dental Research/NIH grantee, 1978—. Mem. Am. Soc. Microbiology, Internat. Assn. Dental Research. Jewish. Subspecialties: Oral biology; Microbiology. Current work: Oral microbiology, interaction of oral microorganisms on host tissues. Home: 4 Winfield Dr Berlin NJ 08009 Office: Dept Microbiology Sch Dental Medicine Univ Pa 4001 Spruce St Philadelphia PA 19104

STEVENSON, DAVID JOHN, planetary physicist; b. Wellington, New Zealand, Sept. 2, 1948; came to U.S., 1971; s. Ian McIvor and Gwenyth (Carroll) S. B. Sc., Victoria U., Wellington, 1971, M.Sc., 1972; Ph.D., Cornell U., 1976. Research fellow Australian Nat. U., Canberra, 1976-78; asst. prof. planetary physics UCLA, 1978-80; asso. prof. planetary sci. Calif. Inst. Tech., Pasadena, 1980—; mem. com. planetary and lunar exploration Nat. Acad. Scis., 1982—. Contbr. articles to profl. jours. Fulbright scholar, 1971-76; NASA grantee, 1979—; NSF grantee, 1982. Mem. Am. Astron. Soc., Am. Geophys. Union, AAAS. Club: Sierra. Subspecialties: Planetary science; Geophysics. Current work: Researcher working with structure and evolution of planets, including the earth with emphasis on behavior of materials at high pressure and temperature. Home: 2670 N Lake Ave Apt 3 Altadena CA 91001 Office: California Institute of Technology Div Geological and Planetary Science Pasadena CA 91125

STEVENSON, HARLAN QUINN, biology educator; b. Waynesboro, Pa., Apr. 1, 1927; s. Wilbur Harlan and E. Roberta (Quinn) S.; m. Katharine Gebhard, Aug. 27, 1960; children: Pamela Jean, Heather Ann. B.S., Pa. State U., 1950; postgrad., Cornell U., 1951-56; Ph.D., U. Fla., 1963. Research assoc. dept. biology Brookhaven Nat. Lab., Upton, N.Y., 1956-60; asst. prof. zoology U. Fla., Gainesville, 1963-64; asst. prof. dept biology So. Conn. State U., New Haven, 1964-68, assoc. prof., 1968-73, prof., 1973—. Contbr. articles to profl. jours. Trustee Bethany Conservation Trust, 1971-73; mem. Conservation commn. Town of Bethany, 1971-74. Served with USNR, 1945-46; ATO; Served with USNR; PTO. U. Fla. fellow, 1960-61; Nuclear Sci. fellow, 1961-63. Mem. AAAS, Am. Inst. Biol. Sci., Genetics Soc. Am., Am. Genetics Assn., Soc. for Study of Evolution, Am. Soc. Human Genetics, N.Y. Acad. Sci., Inst. Soc., Ethics and Life Scis., Conn. Acad. Arts & Scis., Phi Sigma, Gamma Alpha. Subspecialties: Genetics and genetic engineering (biology); Cell and tissue culture. Current work: Cytogenetics, bioethics. Home: 71 Doolittle Dr Bethany CT 06525 Office: So Conn State U New Haven CT 06515

STEWARD, OSWALD, research scientist, educator; b. Bermuda, Sept. 12, 1948; s. Oswald and Ann (Griffiths) S.; m. Kathy P. Steward; children: Jessica, Oswald IV. B.A. magna cum laude, U. Colo., 1970; Ph.D. in Psychobiology, U. Calif. - Irvine, 1974. Asst. prof. neurosurgery and physiology U. Va. Sch. Medicine, Charlottesville, 1974-79, assoc. prof., 1979—. Contbr. articles to profl. jours. Recipient Research Career Devel. award NIH, 1978—; NIH, NSF research grantee. Mem. Soc. Neurosci., Am. Assn. Anatomists, Am. Soc. Cell Biology, Sigma Xi. Subspecialties: Neurobiology; Neuropsychology. Current work: Molecular and cellular neurobiology, regeneration, neuronal plasticity, synapse formation, regulation of cell form, neurophysiology, cytoskeleton of neurons.

STEWART, DAVID HARRY, management consulting firm executive; b. Detroit, Oct. 16, 1939; s. Versile Harry and Alice Louise (Jackson) S.; m. Donna O.T. Lee, Jan. 5, 1980; 1 son, Eric Edward. B.A. in Philosophy, Calif. State U.-Long Beach, 1962. Computer programmer Aeronutronics, Newport Beach, Calif., 1959-63, Iowa State U., Ames, 1963-65; computer scientist Sch. Medicine U. So. Calif., Los Angeles, 1965-69; dept. head Rand Corp., Santa Monica, Calif., 1969-80; pvt. practice cons., Alexandria, Va., 1980-82; v.p. Viar and Co., Inc., Alexandria, 1982—; vis. fellow U. Copenhagen, Denmark, 1971. Bd. dirs. Los Angeles Regional Family Planning, 1976-77; mem. fiscal adv. com. Santa Monica Sch. Bd., 1980. Fellow Inst. for Advancement of Engring. (life), Phi Sigma Tau (life); mem. Assn. Computing Machinery, IEEE Computer Soc. (chpt. pres. 1978-80). Democrat. Club: Palos Verdes Yacht. Subspecialties: Information systems, storage, and retrieval (computer science); Database systems. Current work: Decision support systems, office automation, complex textual data bases, chemical data bases, structured office practices. Office: Viar and Co Inc 300 N Lee Alexandria VA 22314 Home: 324 S Pitt St Alexandria VA 22314

STEWART, DONALD MARTIN, plant pathologist; b. Rembrandt, Iowa, Jan. 20, 1908; s. Alexander Porter and Nellie Louise (Martin) S.; m. Marion Grace Christiansen, May 14, 1938; children: Margo Jeanne Gardner, Bonnie Ann. B.S., U. Minn., 1931, Ph.D., 1953. Jr. forester Bur Entomology and Plant Quarantine, U.S. Dept. Agr., 1933-35; dist. leader White Pine Blister Rust Control, Duluth, Minn., 1935-51; research plant pathologist U. Minn., St. Paul, 1951-70; project mgr. field crops productivity FAO/UN Devel. Project, Cairo, Egypt, 1970-74; profl. cons. plant scis. dept. U. Ariz., Tucson, 1977—; agronomist USAID, North Yemen, 1977-78. Contbr. articles in field to profl. jours. Recipient Cert. of Merit U.S. Dept Agr., 1958, cert. of tng., 1963, cert. of appreciation, 1974; cert. of appreciation Mpls. Pub. Schs., 1968-69; Fulbright-Hays research and lectr. grantee, Romania, 1965. Mem. Am. Phytopath. Soc., Minn. Archaeol. Soc. (pres. 1965), Sigma Xi. Presbyterian. Lodge: Masons. Subspecialties: Plant genetics; Plant cell and tissue culture. Current work: Development of new field crops. Home: 9476 E Shiloh St Tucson AZ 85710

STEWART, HORACE FLOYD, clinical psychologist, neuropsychologist; b. Daytona Beach, Fla., Apr. 20, 1928; s. Horace Floyd Stewart and Ruth Irma (Dawson) Cooper; m. Elizabeth Dunn, Sept., 1954 (div. Sept. 1977); children: Dona, Bonnie, Pamela, Terence; m. Billie Cundiff, Dec. 17, 1977. Ph.D., U. Fla., 1962. Lic. psychologist, Ga. Clin. psychologist state hosp., Milledgeville, Ga., 1958-65; assoc. prof. psychology West Ga. Coll., 1967-72, prof., 1972-80; dir. psychol. services Anneewakee Treatment Ctr., Douglasville, Ga., 1974—. Contbr. articles to profl. publs.; editor: (with J. Thomas) Introduction to Experiential Psychology, 1972. Served to cpl. U.S. Army, 1946-49; MTO. NIMH grantee, 1968. Mem. Am. Psychol. Assn., Southeastern Psychol. Assn., Ga. Psychol. Assn., Assn. Transpersonal Psychology, Nat. Acad. Neuropsychology. Episcopalian. Subspecialties: Transpersonal psychology. Current work: Psychotherapy, transpersonal orientation, neuropsychology, psychological assessment and treatment of children and youth. Home: 3090 E Hwy 166 Carrollton GA 30117 Office: Anneewakee Treatment Ctr 4771 Anneewakee Rd Douglasville GA 30135

STEWART, JAMES EDMUND, III, nuclear engineer, consultant; b. Bristol, Tenn., Aug. 17, 1943; s. James Edmund and Jane Reeve (Booher) S.; m. Peggy Elizabeth Adams, Dec. 30, 1966; children: Margaret Ann, James E. B.S., U. Tenn., 1966; M.E., U. Va., 1970, Ph.D., 1974. Assoc. and lead engr. Babcock & Wilcox Co., Lynchburg, Va., 1966-69; teaching asst. U. Va., Charlottesville, 1970-71; reactor engr. AEC, Washington, 1972-73; staff member and project leader Los Alamos Nat. Lab., 1974—; cons. Los Alamos Tech. Assocs., 1976—; Principal investigator neutron monitoring for centrifuge enrichment plant inspections, 1979-83. Author: (with T. D. Reilly, et al) Passive Nondestructive Assay of Fissionable Material, 1982-83. Mem. Am. Nuclear Soc., Inst. Nuclear Materials Mgmt., AAAS, Sigma Xi. Subspecialties: Nuclear fission; Nuclear engineering. Current work: Monte Carlo modeling of particle transport processes, non-destructive assay instrumentation development and training for international safeguards and nuclear materials accountability, radiation shielding. Office: Los Alamos Nat Lab PO Box 1663 MS #540 Los Alamos NM 87545

STEWART, JOHN ALVIN, public health physician, educator; b. Hamden, N.Y., July 25, 1934; s. Alvin W. and Pearl M. (McGague) S.; m. Madeline J. Stewart, Dec. 28, 1956; children: Bruce, Rebecca, Steven, Gary. A.B., Houghton Coll., 1956; M.D., U. Rochester, 1961, M.S., 1968. Diplomate: Am. Bd. Pediatrics. Intern Cleve. Met. Gen. Hosp., 1961-62, resident in pediatrics, 1962-64; instr. pediatrics Emory U., Atlanta, 1964—; acting chief, virus reference unit Center for Disease Control, Atlanta, 1966-67, asst. chief, virology sect., 1967-68, chief perinatal virology unit, 1968-73, asst. chief, perinatal virology br., 1973—. Served to lt. comdr. USPHS, 1964-67. Mem. Am. Soc. Microbiology, Ga. Acad. Pediatrics. Subspecialties: Microbiology (medicine); Preventive medicine. Current work: Researcher in rapid diagnosis of perinatal viral infection. Office: Center for Disease Control Bldg 7-240 Atlanta GA 30333

STEWART, LARRY GENE, psychologist, consultant; b. San Angelo, Tex., Oct. 9, 1937; s. John Summers and Bertha Irene (Barnes) S.; m. Shirley Hosephine Hanrahan, Dec. 30, 1958 (div. 1974); children: Lamar Gregory, Lee Garrett. B.S., Gallaudet Coll., 1957; M.Ed., U. Mo., 1963; Ed.D., U. Ariz., 1970. Lic. psychologist, Calif., Ariz., Tex., sch. psychologist, Ariz. Assoc. prof. U. Ariz., Tucson, 1972-75, research specialist, 1976-79; pvt. practice psychology, Tucson, 1973-79; exec. dir. Tex. Comm. for Deaf, Austin, 1979-80; pvt. practice psychology, Huntington Beach, Calif., 1980—; forensic psychologist Los Angeles, Orange County, Superior Cts. So. Calif., 1980—; cons. psychologist Dayle McIntosh Center, Garden Grove, Calif., 1980—, Calif. State U.-Northridge, 1981—, Drs. Hosp. Lakewood, Calif., 1983—; rehab. cons. various state and fed. agys., 1970—. Author: Hearing Impaired Developmentally Disabled Children and Adults, 1980, Guide on Hearing Impaired Developmentally Disabled for State Development Disability Councils, 1978; contbr. chpt. to book in field. Co-innovator Ariz. Council for the Deaf, Phoenix, 1974-75, Greater Kansas City Adv. Council for the Deaf, 1965, Greater Tucson Adv. Council for the Deaf, 1972-73. HEW grantee, 1970-71, 74, 74-76, 76-79. Mem. Profl. Rehab. Workers with Adult Deaf (pres. 1973-75), Am. Psychol. Assn., Calif. Psychol. Assn. Current work: Research: forensic psychological assessment methods, techniques, instruments with deaf individuals, neuropsychological assessment of deaf individuals having additional impairments affecting learning, communication, and adjustment. Office: 5200 Warner Ave Suite 109 Park Pl Huntington Beach CA 92649

STEWART, PATRICK BRIAN, physician, corporation executive; b. Champion Reef, India, Aug. 7, 1922; s. James and Lily (Wallace) S.; m. Eunice Sixsmith, Jan. 1, 1949; children: Jennifer, Duncan, Katherine. M.B.B.S., Middlesex Hosp. Med. Sch., U. London, 1950. Intern Middlesex Hosp., 1951; resident in pulmonary diseases Brompton Hosp. for Diseases of Chest, London, 1952; med. dir. Geigy Pharms., Can., 1957-62; dir. research Pharma-Research Can., Ltd., Pointe Claire, Montreal, 1962-77; sr. v.p. research and devel. Boehringer Ingelheim, Ltd., Ridgefield, Conn., 1977—; assoc. prof. medicine and clin. medicine McGill U., 1958-77; assoc. physician Royal Victoria Hosp., Montreal, 1969-77. Contbr. articles and abstracts to profl. jours. Co. rep. Am. Indsl. Health Council, Conn. Acad. Sci. and Engring., Indsl. Research Inst. Served with RAF, 1941-46. Decorated D.F.C.; research assoc. NRC, Can., 1956-58. Mem. Royal Coll. Physicians, AAAS, Am. Assn. Immunologists, Am. Physiol. Soc., Brit. Med. Assn., Can. Med. Assn., Conn. Med. Assn., Litchfield County Med. Assn., N.Y. Acad. Scis., Ont. Coll. Physicians and Surgeons, Profl. Corp. Physicians of Que., Transplantation Soc. Subspecialties: Immunopharmacology; Allergy. Current work: Clinical immunology, drug development. Home: Turner Rd Washington Depot CT 06794 Office: PO Box 368 90 E Ridge St Ridgefield CT 06877

STEWART, ROBERT MURRAY, JR., educator; b. Washington, May 6, 1924; s. Robert Murray and Emily (Smith) S.; m. Patricia Mary Alberding, June 27, 1945; children—Martha Murray, Scott Robert. Student, U. Utah, 1941-43; B.S. in Elec. Engring, Iowa State Coll., 1945; Ph.D. in Physics, Iowa State Coll., 1954. Faculty Iowa State U., 1946—, research asso., 1948-53, asst. prof. physics, 1953, asso. prof. physics, elec. engring., 1956-60, prof. physics, elec. engring., 1960—; research scientist AEC, Ames Lab., 1963-73, chief engr. cyclone digital computer, 1956-66; asso. dir. Iowa State U. Computation Center, 1963—, prof. and chmn. computer sci. dept., 1969—; pres. Computer Sci. Bd., 1976-77; cons. Midwest Research Inst., Kansas City, Mo., Collins Radio Corp., Cedar Rapids, Iowa, Dept. Meteorology and oceanography Tex. A. and M. Coll., Nat. Acad. Sci., Electronic Assos., Inc.; vis. scientist NSF, Assn. for Computing Machinery, Coll. Cons. Service. Mgmt. bd. YMCA, Snow Mountain Ranch, 1969-77; Bd. dirs. Ames Town and Gown Chamber Music Soc., 1973-77; pres. bd. dirs. Story County Youth and Shelter Service, 1975—. Served as ensign USNR, 1945-46. Mem. Assn. Computing Machinery, Am. Phys. Soc., IEEE (adminstrv. com. profl. group on electronic computers 1962-64), Iowa Acad. Sci., Osborn Research Club (chmn. 1971), Am. Fedn. Information Processing Socs. (mem. edn. com.), YMCA of Rockies (dir. 1977—), Ames Soc. for Arts (pres. 1970), UN Assn. (bd. dirs. Ames chpt.), Sigma Xi (pres. Iowa State U. chpt. 1974-75). Subspecialties: Computer architecture; Distributed systems and networks. Current work: Distributed microprocessor systems. Chief investigator NSF sponsored time sharing computer research, 1968-78; designer, implementer computer control system for expts. Ames Lab. Research Reactor AEC, 1963-73. Home: 3416 Oakland Ames IA 50010

STICHT, FRANK DAVIS, med. and dental educator, pharmacologist; b. Plattsburg, Miss., June 14, 1919; s. Frank and Nannie Pearl (Davis) S.; m. Sarah Catherine Tabor, May 6, 1941; children: Sarah Elizabeth, Rebekah Lois. B.S. in Pharmacy, U. Miss., 1948; D.D.S., Baylor U., 1956; M.S., U. Tenn., 1965. Lic. dentist and pharmacist, Miss. Pvt. practice dentistry, 1956-58; USPHS postdoctoral research fellow U. Tenn. Center for Health Scis., 1959-61, faculty, 1961—, prof.

pharmacology, 1979—, researcher in cardiovascular pharmacology, 1959—. Served with USN, 1942-45; to 1st lt. U.S. Army, 1950-52. Mem. Am. Soc. Pharmacology and Exptl. Therapeutics, Sigma Xi. Baptist. Subspecialty: Pharmacology. Current work: Research in pharmacology of antihypertensive drugs. Teacher of pharmacology to dental and medical students. Research in cardiovascular pharmacology. Home: 4802 Rocky Knob Dr Memphis TN 38116 Office: Dept Pharmacology Coll Medicine U Tenn Memphis TN 38163

STICKNEY, ROBERT ROY, educator; b. Mpls., July 2, 1941; s. Roy E. and Helen Doris (Nelson) S.; m. LuVerne C. Whiteley, Dec. 29, 1961; children: Robert Roy, Marolan Margaret. B.S., U. Nebr., 1967; M.A., U. Mo., 1968; Ph.D., Fla. State U., 1971. Cert. fisheries scientist. Research assoc. Skidaway Inst. Oceanography, Savannah, Ga., 1971-73, asst. prof., 1973-75, Tex. A&M U., College Station, 1975-78, assoc. prof., 1978-83, prof., 1983—; chmn. S-168 Com., So. Regional Coop. Research Project, 1981—. Author: Principles of Warmwater Aquaculture, 1979; contbr. articles to profl. jours. Served with USAF, 1959-63. Named Tex. Aquaculturist of Yr. Am. Fisheries Soc., 1979. Mem. Am. Fisheries Soc. (pres. fish culture sect. 1983-84), Am. Inst. Fish Research Biologists (Tex. div. dir.), Am. Inst. Nutrition, AAAS, Tex. Acad. Sci., Am. Soc. Limnology and Oceanography. Current work: Nutritional/environ. requirements of freshwater/marine fishes. Emphasis on lipid and fatty acid requirements of fish. Home: 1812 Hondo St College Station TX 77840 Office: Tex A&M U Dept Wildlife and Fisheries College Station TX 77843

STIDD, BENTON MAURICE, biology educator; b. Bloomington, Ill., June 30, 1936; s. Benjamin David and Alma Mae (Selzer) S. Ph.D., U. Ill.-Urbana, 1968. Asst. prof. botany U. Minn., St. Paul, 1968-70; asst. prof. biology Western Ill. U., Macomb, 1970-74, assoc. prof., 1974-79, prof., 1981—, chmn. biol. scis. dept., 1980—. Contbr. numerous articles on carbonferous fossil plants to profl. publs. Western Ill. U. Research Council grantee, 1971, 80; NSF grantee, 1974, 76; Nat. Endowment Humanities grantee, summer 1982. Mem. Bot. Soc. Am., Philosophy Sci. Assn., Internat. Orgn. Paleobotany, Soc. Systematic Zoology, Am. Fern Soc., Ill. Acad. Sci., Sigma Xi. Subspecialties: Evolutionary biology; Paleobiology. Current work: Carboniferous fossil plants-seed ferns; philosophy of biology. Home: 20 Grandview Dr Macomb IL 61455 Office: Western Ill Uni Biol Scis Dept Macomb IL 62455

STIEFEL, JOHN T., nuclear consultant; b. Pitts., Jan. 15, 1921; s. Ira Brokaw and Helen Mary (Jones) S.; 1 son, Reid; m. Joan L. Lico, Oct. 11, 1974. B.S. Ch.E., U. Ill.-Urbana, 1942. Lab. mgr. Gen. Electric Co., Bloomfield, N.J., 1945-49; v.p. Westinghouse Co., Pitts., 1949-72; sr. v.p. Fluor Co., Los Angeles, 1972-74; pres. Stiefel Assocs., Sarasota, Fla., 1974—, Bonacker Ltd., N.Y.C., 1981—; dir. Scottish Inns, Knoxville, Tenn., 1974-78, Vulcan Internat. & Assocs., Annapolis, Md., 1978—. Served to maj. Chem. Corps U.S. Army, 1942-45. Recipient Order of Merit Westinghouse Co., 1958. Mem. Am. Nuclear Soc. Republican. Presbyterian. Club: Westchester Country (golf com. 1978—). Subspecialties: Nuclear engineering; Nuclear fission. Current work: Management consultant and expert witness in the nuclear field to utilities and industry. Inventor S/G design, 1954. Home: 185 Country Ridge Rd Scarsdale NY 10583 also: 7001 Gulf of Mexico Dr # 14 Sarasota FL 33577 Office: Stiefel Assocs Inc Suite 810 1605 Main St Sarasota FL 33577

STILLE, JOHN KENNETH, chemistry educator; b. Tucson, May 8, 1930; s. John Rudolph and Margaret Victoria (Sakrison) S.; m. Dolores Jean Engelking, June 7, 1958; children: John Robert, James Kenneth. B.S., U. Ariz., 1952, M.S., 1953; Ph.D., U. Ill., 1957. Instr. organic chemistry U. Iowa, Iowa City, 1956-59, asst. prof., 1959-63, asso. prof., 1963-65, prof., 1965-77, Colo. State U., Ft. Collins, 1977—; guest prof. Royal Inst. Tech., Stockholm, 1969; cons. E.I. duPont de Nemours & Co., Inc. Author: Introduction to Polymer Chemistry, 1962, Industrial Organic Chemistry, 1968; Editor: Condensation Monomers (vol. 27 High Polymers), Jour. Am. Chem. Soc., 1982—; Asso. editor: Macromolecules, 1968-82; editorial bd.: Jour. Macromolecular Sci; adv. bd.: Jour. Polymer Sci. Served to lt. comdr. USNR, 1953-55. Fellow Chem. Soc. London; mem. Am. Chem. Soc. (chmn. polymer div. 1975, award in polymer chemistry 1982), Phi Lambda Upsilon, Alpha Chi Sigma, Soc. Sigma Xi. Subspecialties: Organic chemistry; Polymer chemistry. Current work: Organometallic chemistry; mechanisms and organic synthesis catalyzed by transition metals; polymer synthesis; structure property relationships. Home: 1523 Miramont St Fort Collins CO 80524 Office: Dept Chemistry Colo State U Fort Collins CO 80523

STILLER, CALVIN RALPH, medical scientist-clinician, educator, business executive; b. Naicam, Sask., Can., Feb. 12, 1941; s. Carl Hilmer and Mildred Ruth (Parson) S.; m. Marlene Catherine Onn, Sept. 1, 1962; children: Cynthia, Robert, Denise, Troy, Debra, Timothy. Student, U. Sask., 1958-60, M.D., 1965; postgrad, U. Western Ont., Can., 1965-71, U. Alta., Can., 1971-72. Chief Nephrology and transplantation Univ. Hosp., London, Ont., 1972—; dir. transplantation immunology, asst. prof. medicine U. Western Ontario, London, 1972-81, prof., 1982; chmn. Centre for Transplant Studies, London, 1981; chmn. bd. Diversicare Corp., London, 1982; vis. prof., Europe, Africa, N.Am. Mem. editorial bd.: Exec. Congress Immunology, 1983—; contbr. articles to profl. jours; editor and co-editor three books. Bd. dirs. Kidney Found. Can.; active United Way, Can. Diabetes Assn., Borrd-Mission Services of London, Child Care Internat. Recipient Med. Research Council of Can. award, 1973-83; Optimists Internat. Humanitarian award, 1983. Fellow Royal Coll. Physicians (Can.); mem. Royal Coll. Physicians and Surgeons (Can.), Can. Soc. Nephrology (bd. dirs.), Internat. Soc. Nephrology, Internat. Transplantation Soc. (chmn. com.), Can. Med. Assn., Can. Soc. Clin. Investigators. Club: London. Subspecialties: Transplantation; Nephrology. Current work: Immune monitoring, immunogenetics and immunosuppression; induction of tolerance and immunosuppression by biologic and pharmocologic (cyclospurine) means in transplantation (renal, cardiac, hepatic, pancreatic); immune probes in autoimmune disease; lymphokines and control of immune response. Office: University Hospital STE 2R34 London ON Canada N6G 2K3

STILLER, RICHARD LOUIS, clin. pharmacologist; b. N.Y.C., Feb. 15, 1933; s. Anthony J. and Marie M. (Urban) S.; m. Pamela R., Oct. 25, 1972; children: Adria M., Nicole E. A.B., Hunter Coll., 1959; A.B.T., Bklyn. Coll., 1962; M.S., St. John's U., 1968, Ph.D., 1972. Research chemist N.Y. State Psychol. Inst., N.Y.C., 1961-74, sr. research scientist in neurotoxicology, 1974-79; asst. prof. psychiatry U. Pitts., 1979—; cons. clin. pharmacology. Mem. Am. Chem. Soc., AAAS, N.Y. Acad. Scis., Am. Soc. Pharmacology and Exptl. Therapeutics. Subspecialties: Psychopharmacology; Toxicology (medicine). Current work: Pharmacokinetics and pharmodynamics of drugs especially psychotropic, anesthetic cardiovascular and other drugs that effect the central nervous system. Office: 3811 O'Hara St Room 1218 Pittsburgh PA 15213

STILLWAGON, GARY BOULDIN, radiation protection physicist; b. Memphis., Dec. 30, 1951; s. Jack Wright and Ida Jean (Bouldin) S.; m. Leta Fern Miller, Jan. 20, 1979. B.S. in Physics, Ga. Inst. Tech., 1974, M.S. in Nuclear Engring, 1975, Ph.D., 1978. Cert. Health Physics Bd., Part I, Nat. Bd. Med. Examiners, Parts I and II. Med. physicist Meth. Hosp., Memphis, 1974; research asst. Ga. Inst. Tech., Atlanta, 1975-78; radiation safety officer, and physicist VA Med. Center, Memphis, 1978-80, cons. radiation safety, 1980—; cons. in radiation safety to various area hosps. Contbr. articles to profl. jours. Active Boy Scouts Am., Bapt. Ch. Sunday Sch. Dept. Energy fellow, 1976-78. Mem. Health Physics Soc., Am. Assn. Physicists in Medicine, Am. Nuclear Soc., AAAS, AMA, Sigma Xi. Republican. Current work: In-situ dosimetry of internally deposited alpha emitters. Internal dosimetry. Radiation protection. Home: 772 Metcalf Pl Memphis TN 38104 Office: 1030 Jefferson Ave Memphis TN 38104

STIMPFLING, JACK HERMAN, immunogeneticist, research institution exec.; b. Denver, June 11, 1924; s. Herman Joseph and Clara Belle (Espy) S.; m. Bertha Helene Unruh, Apr. 20, 1950 (dec.); children: Lynn, Karen, Kurt, Lise. B.S., Denver U., 1949, M.S., 1950; Ph.D., U. Wis.-Madison, 1958. Assoc. staff scientist Jackson Lab., Bar Harbor, Maine, 1957-64; dir. McLaughlin Research Inst., Great Falls, Mont., 1964—; cons. in field. Contbr. over 50 articles to profl. publs. Pres. chpt. ACLU, 1981—; state v.p. Common Cause, 1982—. Served with U.S. Army, 1943-46. Mem. AAAS, Genetic Soc. Am., Am. Assn. Immunologists. Subspecialties: Animal genetics; Immunology (agriculture). Current work: The genetic and immunological properties of mammalian cell antigen systems. Office: 1625 3d Ave N Great Falls MT 59401

STINCHCOMB, THOMAS GLENN, physicist; b. Tiffin, Ohio, Sept. 12, 1922; s. George Alfred and Ruth Elise (Br) S.; m. Maxine Orr Kohler, Nov. 22, 1945; children: James A., William J., David G., Dan T. B.S., Heidelberg Coll., 1944; S.M. (univ. fellow, 1946-47), U. Chgo., 1948, Ph.D., 1951. Grad. teaching asst. U. Chgo., 1947-48; research asst. U. Chgo. High Altitude Cosmic-Ray Lab., Climax, Colo., 1948-51; full time vis. prof. radiology dept. U. Chgo., 1976-77, vis. research asso. part time radiology dept., 1977—; asst. prof. physics dept. State U. Wash., Pullman, 1951-54; prof. physics dept. Heidelberg Coll., Tiffin, Ohio, 1954-61; sr. physucist Ill. Inst. Tech. Research Inst., Chgo., 1961-68; prof. physics dept. DePaul U., Chgo., 1968—, chmn., 1968-76. Contbr.: articles to profl. publs. including Radiation Research. Pres. Lincoln Sch. PTA, Tiffin, 1959-60; lay adv. council Evanston Twp. High Sch., 1966-69; bd. deacons St. Paul's United Ch. of Christ, 1976-82; bd. dirs. Com. for Nuclear Overkill Moratorium, 1976—; treas. Chgo. Area Faculty for a Freeze on Nuclear Arms Race, 1981—. Served to lt. j.g. USNR, 1943-46. Recipient Admiral's award U.S. Naval Acad., 1944; mem. honor soc. Heidelberg Coll., 1943. Mem. Am. Assn. Physics Tchrs., Am. Assn. Physicists in Medicine, Am. Nuclear Soc., Physics Club Chgo. (dir. 1976—, pres. 1974-76), Chgo. Assn. Technol. Socs. (treas. 1981—), Sigma Xi, Sigma Pi Sigma. Democrat. Current work: Radiation physics (cosmic-rays, neutrons, gamma-rays, applications to medicine, nuclear-weapon effects). Neutron beam therapy for cancer, dosimetry, neutron and gamma-ray microdosimetry, exptl. and theoretical, radiation protection against neutrons. Home: 429 Grant Pl Apt D Chicago IL 60614 Office: 2219 N Kenmore St Chicago IL 60614

STINI, WILLIAM ARTHUR, anthropology educator; b. Oshkosh, Wis., Oct. 9, 1930; s. Louis Alois and Clara (Larsen) S.; m. Mary Ruth Kalous, Feb. 11, 1950; children: Patricia L., Paulette A., Suzanne K. B.B.A., U. Wis.-Madison, 1960; M.S., 1967; Ph.D., 1969. Asst. prof. Cornell U., Ithaca, N.Y., 1968-71, assoc. prof., 1971-73, U. Kans.-Lawrence, 1973-76; prof. U. Ariz., Tucson, 1976—; mem. anthropology rev. panel NSF, Washington, 1976-78; cons. NIH, Bethesda, Md., 1974-82. Author: Ecology and Human Adaptation, 1975, (with David Greenwood) Nature, Culture and Human History, 1977; editor: Physiological and Morphological Adaptation and Evolution, 1979. Mem. Gov's. Adv. Council on Aging, Phoenix, 1980-83. Recipient Cert. of Merit Office of Gov., 1983. Fellow AAAS, Am. Anthrop. Assn.; mem. Am. Assn. Phys. Anthropology (exec. com. 1978-81), Human Biology Council (exec. com. 1978-81), Am. Inst. Nutrition. Democrat. Subspecialties: Nutrition (biology); Evolutionary biology. Current work: Assessment of changes in bone mineral content with increasing age and under differing nutritional regimes, human adaptation to nutritional stress. Home: 6240 N Camino Miraval Tucson AZ 85718 Office: U Ariz Dept Anthropology Tucson AZ 85721

STINNETT, JIMMY DWIGHT, immunologist, educator, researcher; b. Athens, Ala., Dec. 11, 1949; s. Billy Joe and Hazel Willowdean (Lovell) S.; m. Mary Virginia Smith, May 30, 1970; 1 dau.: Suzanne Michelle. B.S. in Microbiology, U. Ala., 1971; Ph.D., U. Ga., 1974. Postdoctoral research fellow Med. Coll. Va., Richmond, 1974-75; mem. faculty U. Cin. Med. Ctr., 1975—, assoc. prof. research surgery and microbiology, 1979—, assoc. dir. surg. immunobiology lab., 1975—; dir. basic research Cin. unit Shriners Burns Inst., 1979—. Mem. editorial bd.: Jour. Bacteriology, 1977-79, Jour. Reticuloendothelial Soc, 1982—; contbr. chpts. to books, articles to profl. jours. Served to maj. USAF, 1973-82. USPHS grantee. Mem. Am. Soc. Microbiology, Reticuloendothelial Soc., Am. Assn. Immunologists, Internat. Soc. Immunopharmacology, N.Y. Acad. Scis., Sigma Xi. Baptist. Subspecialties: Immunopharmacology; Nutrition (medicine). Current work: Infections in immunocompromised patients, modification by nutrition and immunomodulators; infection, immunology, immunocompromised, immunomodulators, nutrition, burns, trauma. Office: 231 Bethesda Ave Cincinnati OH 45267

STINSKI, MARK FRANCIS, virologist, educator; b. Appleton, Wis., Jan. 6, 1941; s. Harold H. and Naomi M. (LaBerge) S.; m. Mary Ellen Lester, Nov. 30, 1946; children: Mark Kurtis, Brent Fritgerald. B.S., Mich. State U., 1964, M.S., 1966, Ph.D., 1969. Postdoctoral fellow U. Pa., 1971-73; asst. prof. microbiology U. Iowa, 1973-78, assoc. prof., 1978—. Editorial bd.: Jour. Virology, 1980—. Served with Chem. Corps U.S. Army, 1969-71. Recipient Research Career Devel. award NIH, 1980. Mem. Am. Soc. Microbiology, AAAS, Am. Soc. Virology. Subspecialties: Virology (medicine); Microbiology (medicine). Current work: Herpes viruses; Cytomegaloviruses. Office: Dept Microbiology U Iowa Iowa City IA 52240

STINSON, JOSEPH MCLESTER, medical educator, physician; b. Hartwell, Ga., July 27, 1939; s. James Isham and Julia Mae (Martin) S.; m. Elizabeth P. Lunceford, Dec. 26, 1964; children: Joseph, Jeffrey, Julie. B.S., Paine Coll., 1960; M.D., Meharry Med. Coll., 1964. Intern Hubbard Hosp., Nashville, 1964-65; instr. Meharry Med. Coll., Nashville, 1965-66, assoc. prof., 1972—, chmn. physiology, 1981—; research fellow Harvard Med. Sch., Boston, 1966-68; clin. fellow Vandervilt Med. Sch., Nashville, 1974-76; cons. VA Hosp., Murfreesboro, Tenn., 1976—. Contbr. to book in field, articles to profl. jours. Sec. to bd. St. Luke Geriatric Ctr., Nashville, 1977; advisor Community Health Services, Nashville, 1979; bd. dirs. Am. Lung Assn. Tenn., 1976—. Served to maj. M.C. USAF, 1968-72. Recipient Pulmonary Acad. award NIH, 1977. Mem. Am. Physiol. Soc., Am. Thoracic Soc. (reviewer 1978-79), Nat. Med. Assn., Am. Coll. Chest Physicians, Tenn. Thoracic Soc. (pres.-elect 1982—), Alpha Omega Alpha, Alpha Kappa Mu. Methodist. Club: Jack and Jill. Subspecialties: Physiology (medicine); Pulmonary medicine. Current work: Determination of lung function standards in selected populations. Home: 6616 Valley Dr Brentwood TN 37027 Office: Meharry Med Coll 1005 18th Ave N Nashville TN 37208

STITH, JAMES HERMAN, physicist, educator; b. Brunswick County, Va., July 17, 1941; s. Pierpoint and Ruth (Stith) Morgan; m. Alberta Juanita Hill, Oct. 2, 1965; children: Adrienne Yvette, Andrea Lynn, Alyssa Joy. B.S., Va. State U., 1963, M.S., 1964; D.Ed., Pa. State U., 1972. Asst. instr. Va. State U., Petersburg, 1964-65; assoc. engr. RCA, Lancaster, Pa., 1967-69; teaching asst. Pa. State U., 1969-72; assoc. prof. physics U.S. Mil. Acad., West Point, N.Y., 1972—; commd. capt. U.S. Army, 1967, advanced through grades to maj., 1979. Contbr. articles to profl. jours. NSF fellow. Mem. Am. Phys. Soc., Am. Assn. Physics Tchrs., Nat. Soc. Black Physicists, Sigma Pi Sigman, Phi Kappa Phi, Alpha Phi Alpha (Man of Yr. 1980). Baptist. Club: West Point Bowling. Subspecialties: Physics education; Theoretical physics. Current work: Computer assisted instruction. Home: 155D Gardinier Rd West Point NY 10996 Office: Dept Physics US Mil Acad West Point NY 10996

STITH, WILLIAM JOSEPH, biomedical engineer; b. Oklahoma City, Feb. 7, 1942; s. Joseph B. and Lera O. (Hall) S.; m. Linda Elizabeth Back, July 30, 1966; children: Debra E., William Joseph, David V. B.A. in Chemistry, Phillips U., 1964; Ph.D. in Biochemistry, 1972. Sr. cell physiologist Fenwal div. Baxter Travenol Labs., Round Lake, Ill., 1973-77, sect. mgr., 1976-77; with Med. Engring. Corp., Racine, Wis., 1977-81, v.p. sci. affairs, 1980-81; gen. mgr. bioengring. dept. Lord Corp., Erie, Pa., 1981—. Contbr. articles to profl. jours. Served to lt. comdr. Med. Service Corps U.S. Navy, 1965-69. Mem. Am. Soc. Quality Control, Sigma Xi. Baptist. Subspecialties: Biomaterials; Orthopedics. Current work: Development of new biomedical devices for orthopedic health care field. Home: 590 Hawthorne Trace Fairview PA 16415 Office: 2101 Peninsula Dr Erie PA 16514

STITZEL, ROBERT ELI, educator; b. N.Y.C., Feb. 22, 1937; s. Louis H. and Betty (Podell) S.; m. Judith Gold, June 4, 1961; 1 son, David. B.S., Columbia U., 1959, M.S., 1961; Ph.D., U. Minn., 1964. Asst. prof. dept. pharmacology and toxicology W.Va. U., 1965-69, asso. prof., 1969-73, prof., 1973—, asst. chmn. dept., 1977-79, asso. chmn., 1979—; cons., research in field. Author: Modern Pharmacology, 1982; contbr. articles to profl. jours. NIH career devel. grantee, 1970-75. Mem. Am. Soc. Pharmacology and Exptl. Therapeutics, AAAS. Subspecialties: Neuropharmacology; Cellular pharmacology. Current work: Hypertension, autonomic pharmacology, biochem. pharmacology. Home: 449 Devon Rd Morgantown WV 26505 Office: Dept Pharmacology WVa U Med Sch Morgantown WV 26506

STOCK, DAVID ALLEN, microbiologist, geneticist, educator; b. Elyria, Ohio, Feb. 8, 1941; s. Donald L. and Grace E. (Uthe) S.; m. Joyce A. Gardner, Sept. 5, 1964; children: Michelle, James. B.S. in Forestry, Mich. State U., 1963; M.S. in Genetics, N.C. State U., 1966; Ph.D. in Forestry and Genetics, N.C. State U., 1968. Instr. microbiology U. Miss. Sch. Medicine, Jackson, 1967-68; postdoctoral fellow in microbiology and dermatology Baylor Coll. Medicine, Houston, 1968-70; asst. prof. biology Stetson U., DeLand, Fla., 1970-77, assoc. prof., 1977—; also cons. Author publs. in field. Mem. Environ. Bd. of Volusia County, Fla., 1979—; bd. dirs. West Volusia br. Am. Cancer Soc., 1981—, West Volusia Christmas Bird Count, 1975—; chmn. Central Fla. Council for Clean Air, 1975-76. NASA and NIH predoctoral fellow, 1963-67; NIH and U.S. Dept. Agr. postdoctoral fellow, 1967-70; grantee Fla. div. Am. Cancer Soc., 1978. Mem. Am. Soc. Microbiology, Am. Phytopath. Soc., Fla. Ornithol. Soc., West Volusia Audubon Soc. (pres. 1978-81). Subspecialties: Gene actions; Microbiology. Current work: DNA metabolism of Eukaryotes (yeast); biology of limpkins (a bird); mechanisms of disease causation by Candida albicans (a yeast).

STOCK, RODNEY DENNIS, computer engineer; b. N.Y.C., July 21, 1947; s. Dennis and Norma Elizabeth (Waymouth) S. Student, Northeastern U., Boston, 1969. Jr. engr. NASA, Boston, 1967; technician Adage Corp., Boston, 1967-68; test engr. Electronic Image Systems, Boston, 1969; design engr. automated material handling Mobility Systems, Santa Clara, Calif., 1970-72; programmer Teleterminal Analysis Center, Memorex Corp., Santa Clara, 1972-73; project engr. flight simulators Evans and Sutherland Computer Corp., Salt Lake City, 1973-76; sr. engr. Ampex Video Art System, Ampex Corp., Redwood City, Calif., 1978-80; graphics engr., mgr. advanced image synthesis computer for movies Pixar) Lucasfilm, Ltd., San Rafael, Calif., 1980—. Mem. IEEE, Assn. Computing Machinery, Soc. Motion Picture and TV Engrs. Club: Sierra (San Francisco). Subspecialties: Graphics, image processing, and pattern recognition; Computer architecture. Current work: Computer graphics hardware, computer graphics algorithms, 3-D computer graphics, computer architecture, image processing, digital video, parallel processing, local area networks. Patentee in field. Office: PO Box 2009 San Rafael CA 94912

STOCKDALE, JOHN ALEXANDER DOUGLAS, sci. instruments mfr., painter, photographer, researcher, cons.; b. Ipswich, Queensland, Australia, Mar. 15, 1936; came to U.S., 1966, naturalized, 1971; s. Reginald Ian Henry and Catriona Mary Caulfield (Cameron) S.; m. Helen Margaret Sutton, July 31, 1957; children: Helen; Alexander; Shane. B.Sc. with honors, U. Sydney, Australia, 1958, M.Sc. in Physics, 1960, Ph.D., U. Tenn., 1969. Research scientist Australian Atomic Energy Comm., 1958-66; mem. research staff Oak Ridge Nat. Lab., 1966-82; pres. Comstock, Inc., Oak Ridge, 1980—; cons. Atomic Beams Lab. of N.Y. U.; chem. physics sect. Oak Ridge Nat. Lab.; painter, photographer. Represented in permanent collections, New South Wales Art Gallery; represented, Nat. Gallery of Victoria, Mus. Modern Art of San Francisco, Corcoran Gallery, Univ. Art Mus., Berkeley, Calif., Hunter Mus., Chattanooga, U. Melbourne.; Contbr. articles to profl. jours. Guggenheim fellow in painting, 1970-71. Mem. Am. Phys. Soc. Subspecialty: Atomic and molecular physics. Current work: Devel. of instrumentation for electrostatic energy analysis of changed particles, mass spectrometry with energy analysis, basic research in atomic and molecular physics. Home: 11323 Berry Hill Dr Knoxville TN 37921 Office: Comstock Inc PO Box 199 Oak Ridge TN 37830

STOCKINGER, SIEGFRIED LUDWIG, consultant; b. Yokohama, Japan, Feb. 3, 1943; came to U.S., 1952; s. Alois and Maria (Brokamp) S.; m. Bernadette Kulesz, Oct. 16, 1965; 1 son, Trevor. B.S. in Mech. Engring, Stevens Inst. Tech., 1965; M.S. in Environ. Engring, Loyola U., Los Angeles, 1973. Registered profl. engr., Calif., N.J., N.Y., N.C. Engr. electric boat div. Gen. Dynamics Corp., Groton, Conn., 1965-68; research and devel. labs. Mobil Chems., Edison, N.J., 1968-69; sect. mgr. Litton Ship Systems, Los Angeles, 1969-73; chief engr. Cosmodyne, Torrance, Calif., 1973-76; cons., Los Angeles, 1976-77; sr. supr. Ebasco Services, Inc., N.Y.C., 1977—. Contbr. writings in field to profl. publs. Active Republican Club, Short Hills, N.J., 1980—, USCG Aux., Redondo Beach, Calif., 1970-72, Eagle Scout. Mem. ASME, Am. Nuclear Soc., Alpha Sigma Phi (v.p. 1964). Roman Catholic. Club: Racquets (Short Hills, N.J.). Subspecialties: Environmental engineering; Chemical engineering. Current work: Chemical engineering and advanced technology applications in radioactive waste management and chemical process applications for treating effluents from nuclear operations.

STOCKLAND, WAYNE LUVERN, director animal nutrition research, researcher; b. Lake Lillian, Minn., May 4, 1942; s. Gaylord

Luvern and Betty (Springstein) S.; m. Jerrilyn Jean Herje, June 5, 1971; children: Lisa, Renee, Eric. B.S., U. Minn.-St. Paul, 1964, Ph.D., 1969. Cert. nutritionist. Research asst. U. Minn.-St. Paul, 1969, research fellow, 1969-70; research nutritionist Internat. Multifoods, Courtland, Minn., 1970-75, research nutritionist, statis. mgr., 1975-76, dir. animal nutrition research, 1976—. Contbr. articles to publs. in field. Pres. Courtland Vol. Fire Dept., 1980—, Businessmen's Assn., 1979—; mem. City Council, Courtland, 1978-82. Mem. Am. Soc. Animal Sci., Am. Dairy Sci. Assn., Poultry Sci. Assn., Am. Inst. Nutrition, N.Y. Acad. Scis., Assn. Ofcl. Analytical Chemists. Republican. Lutheran. Lodge: Lions. Subspecialties: Animal nutrition; Nutrition (biology). Current work: Animal nutrition research specializing in amino acid requirements and factors affecting amino acid requirements. Office: Internat Multifoods Supersweet Research Farm Courtland MN 56021 Home: 421 Riverview Dr Courtland MN 56021

STOCKMAYER, WALTER HUGO, educator; b. Rutherford, N.J., Apr. 7, 1914; s. Hugo Paul and Dagmar (Bostroem) S.; m. Sylvia Kleist Bergen, Aug. 12, 1938; children—Ralph, Hugh. S.B., Mass. Inst. Tech., 1935, Ph.D., 1940; B.Sc. (Rhodes scholar), Oxford U., 1937; hon. doctorate, U. Louis-Pasteur, Strasbourg, France, 1972. Instr. Mass. Inst. Tech., 1939-41, asst. prof., 1943-46, asso. prof., 1946-52, prof., 1952-61; prof. chemistry Dartmouth, 1961-79, prof. emeritus, 1979—; instr. Columbia, 1941-43; Cons. E.I. duPont de Nemours & Co., Inc., 1945—; vis. com. Nat. Bur. Standards, 1979-84. Contbr. articles on phys. and macromolecular chemistry to sci. jours. Guggenheim fellow, 1954-55; hon. fellow Jesus Coll., Oxford, Eng., 1976; Alexander von Humboldt fellow, W. Ger., 1978-79; Recipient MCA Coll. Chemistry Tchr. award, 1960. Fellow Am. Acad. Arts and Scis., Am. Phys. Soc. (Polymer Physics prize 1975); mem. Am. Chem. Soc. (asso. editor Macromolecules 1968-74, 76—, chmn. polymer chem. div. 1968, Polymer Chemistry award 1965, Peter Debye award 1974), Nat. Acad. Scis. (exec. com. assembly math. phys. sci. 1974-77), Sigma Xi. Club: Appalachian Mountain. Subspecialties: Physical chemistry; Polymer chemistry. Current work: Polymer chemistry, polymer physics. Home: Norwich VT 05055

STOCKWELL, CHARLES WARREN, neurophysiologist; b. Port Angeles, Wash., Dec. 31, 1940; s. Frank Edward and Esther Marie S.; m. Marsha Jo Stockwell, Aug. 23, 1966; children: Laura, Frank. B.A., Western Wash. U., 1964; M.A., U. Ill., 1966, Ph.D., 1968. Research psychologist Naval Aerospace Med. Inst., Pensacola, Fla., 1968-72; asst. prof. Ohio State U., Columbus, 1972-77, assoc. prof. otolaryngology, 1977—. Author: Manual of Electronystagmograph, 1976, 2d edit., 1980, ENG Workbook, 1983; Contbr. articles to profl. jours. Served with AUS, 1968-72. Mem. Assn. Research in Otolaryngology, Barany Soc., Soc. Neurosci. Subspecialties: Neurophysiology; Otorhinolaryngology. Current work: Vestibular physiology. Home: 1652 Cardiff Rd Columbus OH 43221 Office: 4024 University Hosps Clinic 456 Clinic Dr Columbus OH 43210

STODDARD, PATRICIA ANN, medical technologist, chemistry educator; b. Albert Lea, Minn., Apr. 5, 1930; d. Armond William and Lois Roberta (Remo) Olson Armstrong; m. Charles Gilbert Stoddard, Jan. 7, 1966; m. William Anton Hoogendijk, July 16, 1949 (div. Dec. 1964); 1 son, Christopher John. A.Sc., Clackamas Community Coll., 1973; B.A., Linfield Coll., 1976. Med. technologist King County Blood Bank, Seattle, 1962-63; surg. research technician Bishop Eye Research, Seattle, 1964-65; gen. supr. Willamette Falls Hosp., Oregon City, Oreg., 1965-68; adj. asst prof. U. Portland, Oreg., 1977—; dir. Willamette Animal Lab., Corbett, Oreg., 1969—. Author: Veterinarian Medicine for Small Animal Clinician, 1976, 77. Instr. ARC, Vancouver, Wash., 1979—; flotilla vice comdr. Coast Guard Aux., Vancouver, 1983. Mem. Am. Med. Technologists (Nat. Achievement award 1972, 74), Am. Soc. Microbiologists, Am. Inst. Biol. Sci., Am. Chem. Soc. (exec. bd. Portland sect. 1982—), Iota Sigma Phi. Republican. Episcopalian. Club: Dolphin Yacht (Camas, Wash.) (fleet surgeon 1981—). Subspecialties: Microbiology (veterinary medicine); Immunology (agriculture). Current work: Developed the first effect fungus vaccine to treat as well as prevent microsporim canis in cats and dogs and aspergillus fumigatus for birds, producer of many veterinarian vaccines for domestic animals. Home: 17317 SE Evergreen Hwy Camas WA 98607 Office: Willamette Animal Lab 36610 E Crown Point Hwy Corbett OR 97019

STOECKLEY, THOMAS ROBERT, astronomer; b. Ft. Wayne, Ind., Dec. 6, 1942; s. Robert George and Beatrice Mary (Kuckuck) S.; m. Kathleen Viola Stoeckley, Dec. 15, 1973; children: Andrew James, Heidi Nicole. B.S., Mich. State U., 1964; Ph.D. in Astronomy, U. Cambridge, Eng., 1967. Asst. prof. astronomy and astrophysics Mich. State U., East Lansing, 1967-73, assoc. prof., 1973-81, assoc. prof. dept. physics and astronomy, 1981—, chmn, dept. astronomy and astrophysics, 1976-81; researcher in field. Contbr. articles to astron. jours. Mem. Am. Astron. Soc., Royal Astron. Soc. Subspecialty: Optical astronomy. Current work: Steller spectroscopy, stellar rotation. Home: 1017 Whittier Dr East Lansing MI 48823 Office: Dept of Physics and Astronomy Michigan State University East Lansing MI 48824

STOFFER, DONALD CARL, mechanical engineering educator; b. Phila., May 15, 1938; s. Elmer Charles and Edna Ruth (McGraw) S.; m. Joan Clair Dreyer, June 30, 1962; children: Brett K., Scott C. B.S. in M.E, Drexel U., 1961, M.S., 1965; Ph.D., U. Mich., 1968. Registered Profl. Engr., Ohio. Engr. Philco-Ford Corp., Phila., 1961-63; engr. Westinghouse Electric Corp., Balt., 1963-65; lectr. U. Mich., Ann Arbor, 1965-69; mem. faculty U. Cin., 1969—, prof. aerospace engring. and applied mechanics, 1981—; vis. scientist Wright Patterson AFB, Dayton, summers 1975, 77, 78; U.S.-Australian Cooperative Sci. fellow Melbourne U., 1976. Contbr. articles to profl. jours. Tecumseh Products fellow, 1965-66; NSF grantee, 1974—; U.S. Air Force Systems Command grantee, 1977—. Mem. Soc. Rheology, ASME, Soc. Engring. Sco., Am. Acad. Mechanics, Internat. Assn. for Structural Mechanics in Reactor Tech. Subspecialties: Solid mechanics; Theoretical and applied mechanics. Current work: Development of mathematical models for the physical behavior of solid materials under various loading and environmental histories. Office: U Cin Mail Loc 70 Cinnati OH 45221

STOFFER, SHELDON SAUL, endocrinologist, educator; b. Detroit, May 14, 1940; s. Harry and Mary (Sherman) S.; m. Phyllis Elaine Raub, July 8, 1962; children: Amy, Lori. M.D., Wayne State U., 1965. Diplomate: Am. Bd. Internal Medicine. Intern Detroit Gen. Hosp., 1965-66; resident Wayne County Gen. Hosp., Eloise, Mich., 1968-69, Henry Ford Hosp., Detroit, 1969-70; endocrine fellow Mayo Clinic, Rochester, Minn., 1970-72; asst. prof. medicine Wayne State U., Detroit, 1972-74; cons. Associated Endocrinologists, Southfield, Mich., 1974—; mem. staff Detroit Gen. Hosp., 1972—, Harper Hosp. of Detroit, 1972—, Grace Hosp. of Detroit, 1974—, Sinai Hosp. of Detroit, 1980—. Contbr. numerous articles to med. jours. Served to lt. commdr. USPHS/NIMH, 1966-68. Fellow ACP; mem. Endocrine Soc., Am. Thyroid Assn., Am. Fedn. for Clin. Research, AMA, Am. Fertility Soc., Internat. Fedn. Fertility Socs., Am. Joggers Assn., Am. Coll. Sports Medicine. Subspecialties: Endocrinology; Reproductive endocrinology. Current work: Thyroid function tests, hyperparathyroidism, effects of thyroid dysfunction on mentruation. Home: 5375 Pocono West Bloomfield MI 48033 Office: 4400 Prudential #275 Southfield MI 48075

STOHS, SIDNEY JOHN, pharmacologist, educator; b. Ludell, Kans., May 24, 1939; s. John and Lydia (Holtz) S.; m. Susan Joan Stehl, Sept. 4, 1960; children: Sarah E., Timothy W. B.S. with distinction, U. Nebr., 1962, M.S., 1964; Ph.D., U. Wis., 1967; fellow, Karolinska Inst., Stockholm, 1975-76. Registered pharmacist, Nebr. Asst. prof. U. Nebr., 1967-71, asso. prof., chmn. dept. medicinal chemistry, 1971-74, prof., chmn. dept. medicinal chemistry, 1974-78, prof., chmn. dept. biomedical chemistry, 1978—. Contbr. over 130 articles to profl. jours. Pres. First Lutheran Ch., 1979-81. NIH fellow, 1962-64, 64-67; recipient Outstanding Tchr. award U. Nebr., 981. Mem. Am. Pharm. Assn., Acad. Pharm. Sci., Am. Chem. Soc., Am. Soc. Pharmacognosy, AAAS, Am. Assn. Colls. Pharmacy, Drug Metabolism Group, Am. Soc. Pharmacology and Exptl. Therapeutics, Internat. Union Pharmacology, Internat. Soc. Study of Xenobiotics, Sigma Xi, Rho Chi. Republican. Subspecialties: Medicinal chemistry; Molecular pharmacology. Current work: Drug metabolism; molecular mechanisms of aging; chemical carcinogenesis and xenobiotic toxicity. Office: U Nebr Med Center Omaha NE 68105

STOKES, JOHN BISPHAM, III, medical investigator, educator, physician; b. Temple, Tex., Mar. 26, 1944; s. John B. and Joy (Dumas) S.; m. Jackquelyn Joy Goodwin, June 22, 1968; children: Jennifer Elizabeth, Josephine Lynne. B.S., Gettysburg (Pa.) Coll., 1966; M.D., Temple U., 1971. Diplomate: Am. Bd. Internal Medicine. Chemist E.I. DuPont, Phila., 1966-67; intern Washington U., St. Louis, 1971-72; med. officer NIH, Bethesda, Md., 1972-74; resident in medicine U. Tex. Southwestern Med. Sch., Dallas, 1974-75, fellow in nephrology, 1975-78; asst. prof. medicine U. Iowa, Iowa City, 1978-82, assoc. prof., 1982—, dir. div. nephrology, 1982—. Served to lt. comdr. USPHS, 1972-73. NIH grantee, 1978—; Research Service awardee USPHS, 1976-78; established investigator Am. Heart Assn., 1983—. Fellow ACP; mem. Am. Heart Assn., Am. Fedn. Clin. Research (councillor Midwest sect. 1981—), Central Soc. Clin. Research, Am. Soc. Nephrology. Subspecialties: Nephrology; Membrane biology. Current work: Mechanisms of ion transport across epithelia, mechanisms of hormone regulation of transport, cellular regulation of renal countercurrent system. Home: 148 Penfro St Iowa City IA 52240 Office: Dept Internal Medicine U Iowa Hosps Iowa City IA 52242

STOLBERG, MARVIN A., chemist, high tech. co. exec.; b. N.Y.C., Oct. 29, 1925; s. William H. and Edna (Kostick) S.; m. Arlene Baumrind, Jan. 23, 1949; children—Susan Howe, Carol Stolberg Engellenner. B.S., Columbia U., 1950; M.S., U. Del., 1954, Ph.D., 1956. Asst. br. chief chemotherapy br. U.S. Army Chem. Center, Edgewood, Md., 1950-56; mgr. chemistry dept. Tracerlab, Inc., Waltham, Mass., 1956-60; with New Eng. Nuclear Corp., Boston, 1970—; now pres., chief exec. officer, also dir.; pres. South End Tech. Sq. Assos., Inc.; dir. Asso. Industries Mass., Mass. High Tech. Council. Trustee Univ. Hosp. Served with U.S. Army, 1944-46, 50-51. Decorated Purple Heart. Mem. Am. Chem. Soc., Soc. Nuclear Medicine, Sigma Xi. Subspecialties: Organic chemistry; Radiochemistry. Office: New England Nuclear Corp 549 Albany St Boston MA 02118

STOLINSKY, DAVID C., physician; b. Fargo, N.D., Nov. 19, 1934; s. Aaron and Rose Charlotte (Meblin) S.; m. Stefanie Auerbach, May 17, 1966. A.B. with honors, U. Calif., Berkeley, 1955, M.D., 1958. Intern U. Calif., San Francisco, 1958-59; resident Mt. Zion and San Francisco Gen. Hosps., 1959-61; fellow in hematology U. Calif., San Francisco, 1961-63; asst. research physician Cancer Research Inst., 1965-66; fellow in med. oncology U. So. Calif., Los Angeles, 1963-65, asst. prof. medicine, 1966—. Contbr. chpts. to books, articles to profl. jours. Served to lt. comdr. USPHS, 1963-65. Mem. Am. Assn. Cancer Research, Am. Soc. Clin. Oncology, Am. Fedn. Clin. Research, Western Cancer Study Group (sec. 1968-76), Phi Beta Kappa, Phi Delta Epsilon. Subspecialties: Cancer research (medicine); Chemotherapy. Current work: Clinical trials of chemo- and immunotherapy in patients with cancer. Office: 1200 N State St Los Angeles CA 90033

STOLLAR, BERNARD DAVID, biochemistry educator; b. Saskatoon, Sask., Can., Aug. 11, 1936; s. Percival and Rose (Direnfeld) S.; m. Carol Ann Singer, Oct. 7, 1956; children: Lawrence Benjamin, Michael Brent, Suzanne Naomi. B.A., U. Sask., 1958, M.D., 1959. Intern U. Sask. Hosp., Saskatoon, 1959-60; postdoctoral fellow Brandeis U., Waltham, Mass., 1960-62; dep. chief biol. sci. div. U.S. Air Force Office Sci. Research, Washington, 1962-64; asst. prof. Tufts U Sch. Medicine, Boston, 1964-68, assoc. prof., 1968-74, prof., 1974—; vis. prof. Internat. Course in Immunology and Immunochemistry, Mexico City, 1971; sr. fellow Weizmann Inst. Sci., Rehovot, Israel, 1971-72; vis. prof. Wellesley (Mass.) Coll., 1976, U. Tromsö (Norway) Sch. medicine, 1981; cons. Cetus, Inc., Palo Alto, Calif., 1982—. Mem. editorial bd.: Jour. Immunology, 1981—, Analytical Biochemistry, 1981—, Molecular Immunology, 1982—; contbr. over 130 articles to sci. jours., chpts. to books. Chmn. adult edn. Temple Reyim, Newton, Mass., 1974-77. Served to capt. USAF, 1962-64. Recipient Gold medal U. Sask. Coll. Medicine, 1959; grantee NSF, 1964—, NIH, 1976—. Mem. Am. Assn. Immunologists, Am. Soc. Biol. Chemists, AAAS. Jewish. Subspecialties: Immunology (medicine); Biochemistry (medicine). Current work: Autoimmune diseases and antibodies to nucleic acids, histones and nuclear enzymes; studies of chromatin organization. Home: 158 Clark St Newton MA 02159 Office: 136 Harrison Ave Boston MA 02111

STOLLAR, VICTOR, microbiologist, educator; b. Saskatoon, Sask., Can., Dec. 6, 1933; came to U.S., 1965, naturalized, 1980; s. Percy and Rose S.; m. Eva Claudine Svigelj, June 25, 1967; children: Lisa, Miriam, Anna. M.D., C.M., Queens U., Kingston, Ont., Can., 1956. Intern Montreal (Que., Can.) Gen. Hosp., 1956-57; postdoctoral fellow dept. biochemistry Brandeis U., Waltham, Mass., 1958-60, 61-62; research fellow Weizmann Inst. Sci., Rehovot, Israel, 1962-65; asst. prof. microbiology Rutgers Med. Sch. Univ. Medicine and Dentistry N.J., Piscataway, 1965-70, assoc. prof., 1970-75, prof., 1975—; mem. virology study sect. NIH, 1980-83. Mem. AAAS, Am. Soc. Microbiology, Am. Assn. Immunologists, Am. Soc. for Virology. Subspecialties: Virology (medicine); Cell biology (medicine). Current work: Arthropod-borne viruses; cell biology; somatic cell genetics of cultured mosquito cells. Home: 35 Brookside Dr Warren NJ 07060 Office: Dept Microbiology Rutgers Med Sch Univ Medicine and Dentistry NJ Piscataway NJ 08854

STOLLEMAN, GENE HOWARD, internal medicine educator, medical writer and editor; b. N.Y.C., Dec. 6, 1920; s. Maurice William and Sarah Dorothy (Mezz) S.; m. Corynne Miller, Jan. 21, 1945; children: Lee Denise Stollreman Meyburg, Anne Barbara, John Eliot. A.B. summa cum laude, Dartmouth Coll., 1941; M.D., Columbia U., 1944. Diplomate: Am. Bd. Internal Medicine (chmn. certifying exam. com. 1969-73) mem. exec. com. 1971-73). Intern Mt. Sinai Hosp., N.Y.C., 1944-45, asst. resident in medicine, 1945-46, chief resident in medicine, 1948-49; research fellow dept. microbiology N.Y.U. Coll. Medicine, 1949-50; med. dir. Irvington House for children with heart disease, Irvington-on-Hudson, N.Y., 1951-55; asst. prof. medicine Northwestern U., 1955-57, assoc. prof., 1957-62, prof., 1962-64; dir. Samuel J. Sackett Lab. for Research in Rheumatic and Infectious Diseases, 1955-64; prof., chmn. dept. medicine U. Tenn., 1965-81, Goodman prof., 1977-81; prof. Boston U. Sch. Medicine, 1981—; attending physician Univ. Hosp., Boston, 1981—; cons. in field; mem. Commn. on Streptococcal and Staphyloccal Diseases, Armed Forces Epidemiol. Bd., 1956-73; mem. arthritis delegation to Soviet Union, Cultural Exchange Program U.S.-USSR, State Dept., 1964; mem. expert ad. panel on cardiovascular diseases WHO, 1966-81; mem. various coms., panels USPHS; chmn. rev. panel of bacterial vaccines and toxoids Bur. Biologics, FDA, 1973—; mem. nat. adv. council Nat. Inst. Allergy and Infectious Diseases, 1979—. Author: Rheumatic Fever and Streptoccal Disease, 1975; contbr. numerous articles, editorials, chpts. to med. publs.; editorial bd.: Jour. Lab. and Clin. Medicine, 1959-65, Ann. Rev. Medicine, 1963-68, Antibiotic Agts. and Chemotherapy, 1965-78; editor: Advances in Internal Medicine, 1967—, Capsule and Comment, 1977—. Served to capt. U.S. Army, 1946-48. Recipient Bicentennial award Columbia U., 1967, Maimonides award State of Israel, 1978. Fellow A.C.P. (master region v.p. 1976—); mem. Assn. Am. Physicians (pres.), Am. Soc. Clin. Investigation, Profs. Medicine (assoc.), Phi Beta Kappa, Alpha Omega Alpha. Jewish. Subspecialties: Internal medicine; Microbiology (medicine). Current work: Rheumatic fever and streptococcal infections; teaching, research, and patient care. Home: 30 Rutgers Rd Wellesley MA 02181 Office: 720 Harrison Ave 1108 Boston MA 02118

STOLLNITZ, FRED, psychologist, science foundation official; b. N.Y.C., Apr. 13, 1939; s. Henry Sande and Helen Cecile (Bessemer) S.; m. Janet Louise Gabar, Aug. 6, 1961; children: Nancy Beth, Eric Joel. B.A. with high honors, Swarthmore Coll., 1959; Sc.M., Brown U., 1961, Ph.D., 1963. Research assoc. Brown U., Providence, 1963-64, asst. prof. psychology, 1964-66, Cornell U., Ithaca, N.Y., 1966-71; asst. program dir. for psychobiology NSF, Washington, 1971-72, assoc. program dir. for psychobiology, 1972-76, program dir. for psychobiology, 1976—. Co-editor: Behavior of Nonhuman Primates, 5 vols, 1965-74; contbr. articles to profl. jours., chpt. to book. USPHS fellow, 1960-62; NSF fellow, 1962-63. Mem. AAAS, Am. Psychol. Assn., Animal Behavior Soc., Internat. Primatological Soc., Psychonomic Soc., Phi Beta Kappa, Sigma Xi. Subspecialties: Psychobiology; Behaviorism. Current work: Administration of a program that awards grants in aid of research on animal behavior and its biological bases, including genetic, environmental, hormonal and neural factors in learning and memory, foraging and feeding, orientation and migration, reproductive and social behavior, and communication. Office: Psychobiology Program Nat Sci Found Washington DC 20550

STOLOFF, DAVID ROBERT, veterinary surgeon; b. Hartford, Conn., Aug. 25, 1943; s. William Benjamin and Kathryn S.; m. Lynne Abrams, Dec. 30, 1974; children: Allison, Rebecca, Matthew. B.S., U. Conn., 1966; D.V.M., Mich. State U., 1969; M.S., Kans. State U., 1977. Diplomate: Am. Coll. Vet. Surgeons, 1982. Intern Kans. State U., Manhattan, 1974-75, resident in surgery, 1975-77, asst. prof. surgery, 1977-78, Sch. Vet. Medicine, La. State U., Baton Rouge, 1978-81; staff surgeon Rowley Meml. Animal Hosp., Springfield, Mass., 1981—. Contbr. articles to profl. jours. Mem. Am. Animal Hosp. Assn., Am., Coll. Vet. Surgeons, AVMA, Conn. Vet. Med. Assn., Alpha Zeta, Gamma Sigma Delta, Phi Kappa Phi, Phi Zeta. Subspecialties: Surgery (veterinary medicine); Pathology (veterinary medicine). Current work: Comparison of healing following colonic anastomosis in the canine. Home: 212 Ridgewood Rd West Hartford CT 06107 Office: 53 Bliss St Springfield MA 01105

STOMMEL, HENRY MELSON, oceanographer; b. Wilmington, Del., Sept. 27, 1920. B.S., Yale U., 1942, D.Sc. (hon.), 1970, Gothenburg U., 1964, U. Chgo., 1970. Instr. math. and astronomy Yale U., 1942-44; research assoc. phys. oceanography Oceanographic Inst., Woods Hole, Mass., 1944-59, oceanographer, 1978—; prof. oceanography MIT, 1959-60, prof., 1963-78, Harvard U., 1960-63. Recipient Crafoord prize, 1983. Mem. Nat. Acad. Sci., Am. Astron. Soc., Am. Soc. Limnology and Oceanography, Am. Acad. Arts and Scis., Am. Geophys. Union. Subspecialty: Oceanography. Office: Woods Hole Oceanographic Inst Woods Hole MA 02543

STONE, EDWARD CARROLL, JR., physicist; b. Knoxville, Iowa, Jan. 23, 1936; s. Edward Carroll and Ferne Elizabeth (Baber) S.; m. Alice Trabue Wickliffe, Aug. 4, 1962; children—Susan, Janet. A.A., Burlington Community Coll., 1956; M.S. (NASA fellow), U. Chgo., 1959, Ph.D., 1964. Research fellow in physics Calif. Inst. Tech., Pasadena, 1964-66, sr. research fellow, 1967, mem. faculty, 1967—, prof. physics, 1976—, chmn. div. physics, math. and astronomy, 1983—, Voyager project scientist, 1972—; cons. Office of Space Scis., NASA, 1969—, mem. adv. com. outer planets, 1972-73, editorial bd., 1975-81, Space Sci. Revs., 1982—, Astrophysics and Space Sci., 1982—; mem. high energy astrophysics mgmt. operating working group Office of Space Scis., NASA (Space Sci. Instrumentation), 1976—; mem. com. on space astronomy and astrophysics Space Sci. Bd., 1979—. Recipient medal for exceptional sci. achievement NASA, 1980, also; Disting. Service medal, 1981; Am. Edn. award, 1981; Dryden award, 1983; Sloan Found. fellow, 1971-73. Fellow Am. Phys. Soc. (chmn. cosmic physics div. 1979-80, exec. com. 1974-76), Am. Geophys. Union; mem. Am. Astron. Soc. (com. mem. div. planetary scis. 1981-84), Am. Assn. Physics Tchrs., AAAS., AIAA; corres. mem. Internat. Acad. Astronautics. Subspecialties: Cosmic ray high energy astrophysics; Planetary science. Current work: Experimental studies of galactic and solar cosmic rays and planetary magnetospheres. Scientific coordination of outer planet exploration. Office: Dept Physics Calif Inst Tech 1201 E California Blvd Pasadena CA 91125

STONE, ERIC ANDREW, psychiatry educator; b. N.Y.C., Jan. 14, 1941; s. Philip and Ruth Steinbaum; m. Elaine Friedman, Aug. 1, 1970; 1 dau., Allison. B.A., NYU, 1962; Ph.D., Yale U., 1967. Asst. prof. Rockefeller U., N.Y.C., 1967-70; assoc. prof. dept. psychiatry NYU Sch. Medicine, N.Y.C., 1980—. USPHS, NIMH research grantee, 1970—. Mem. AAAS, Soc. Neurosci., Am. Physiol. Soc., Am. Psychol. Assn. Subspecialties: Neuropharmacology; Neurochemistry. Current work: Research. Office: NYU Sch Medicine 550 1st Ave New York NY 10016

STONE, GREGORY MICHAEL, research engineer, business executive; b. Hartford, Conn., July 31, 1959; s. George William and Patricia Gertrude (Fitton) S. A.B., Loyola U., Chgo., 1981; postgrad., Pacific Western U., 1982. Electronic system design engr. Barrett Electronics Corp., Northbrook, Ill., 1977-80; cons. engr. Tyndale-Dausen, Ltd., Mundelein, Ill., 1977-80, Sashs/Freeman Assocs., Inc., Northbrook, 1980—; mng. dir., chief scientist Stone Industries Inc., Mundelein, 1982—, also dir.; dir. Consol. News Service Ltd. Mem. IEEE, Soc. Am. Mil. Engrs., Soc. Cable TV Engrs., Amateur Radio Satellite Corp., Soc. Automotive Engrs., Radio Club Am., Midwest Intersystem Repeater Assn., John Birch Soc., SAR, Mayflower Soc. Republican. Subspecialties: Electronics; Robotics. Current work: Audio frequency bandwidth compression; magnetic anomolie detection communications systems design; radio frequency propagation empirical research. Home: 1112 Regency Ln Libertyville IL 60048 Office: PO Box 485 Mundelein IL 60060

STONE, JULIAN, physicist; b. N.Y.C., Apr. 12, 1929; s. Harry and Etta (Fishman) S.; m. Florence Pistreich, June 16, 1951; children: David, Hillary, Robert. B.S., CCNY, 1950, M.S., 1951; Ph.D., N.Y.U.,

1958. Material scientist Naval Material Lab., 1952; tutor CCNY, 1952-53, 1956-57; assoc. dir. Hudson Labs., Columbia U., 1953-69; mem. tech. staff Bell Labs., Holmdel, N.J., 1969—. Contbr. numerous articles on optics and acoustics to tech. jours. Mem. Optical Soc. Am. Subspecialties: Fiber optics; Spectroscopy. Current work: Fiber optics and lasers.

STONE, LAWRENCE ALLAN, physician, educator; b. Boise, Idaho, June 28, 1946; s. H. Joel and Sylvia (Sackt) S.; m. Brenda Fox, Jan. 30, 1949; 1 son, Jason Kenneth. A.B., Ind. U., Bloomington, 1968; M.D., Northwestern U., 1972. Diplomate: Am. Bd. Internal Medicine, Nat. Bd. Med. Examiners.; Lic. physician, Ill. Intern Northwestern U./McGaw Med Ctr., Chgo., 1972-73, resident, 1973-75; fellow in developmental therapeutics M.D. Anderson Hosp., Houston, 1975-76; fellow in med. oncology Rush Med. Coll., Chgo., 1976-77; attending physician Luth. Gen. Hosp., Park Ridge, Ill., 1977—; asst. prof. U. Ill. Col. Medicine, Chgo., 1978—; cons. Bristol Labs., Syracuse, N.Y., Ill. Cancer Council. Bd. dirs. Am. Cancer Soc., Palatine, Ill., 1977—; trustee Ill. Cancer Soc., 1982—. Mem. AMA, Ill. State Med. Soc., Chgo. Med. Soc., Am. Soc. Clin. Oncology, Eastern Coop. Oncology Group, Alpha Epsilon Pi, Phi Rho Sigma. Subspecialties: Internal medicine; Chemotherapy. Current work: Clinical trial development of new choretherapy and immunotherapy . Office: 9301 W Golf Rd #205 Des Plaines IL 60016

STONE, MARSHALL HARVEY, emeritus mathematics educator; b. N.Y.C., Apr. 8, 1903; s. Harlan Fiske and Agnes (Harvey) S.; m. Emmy Portman, June 15, 1927 (div. July 1962); children: Doris Portman, Cynthia Harvey, Phoebe G.; m. Raviojla Perendija Kostic, Aug. 8, 1962; 1 stepdau., Svetlana Kostic-Stone. A.B., Harvard U., 1922, A.M., 1924, Ph.D., 1926; postgrad., U. Paris, France, 1924-25; Sc.D., Kenyon Coll., 1939; hon. Dr., Universidad de San Marcos, Lima, Peru, 1943, Universidad de Buenos Aires, 1947, U. Athens, 1954; Sc.D., Amherst Coll., 1954, Colby Coll., 1959, U. Mass., 1966. Part-time instr. Harvard, 1922-23, instr., 1927-28, asst. prof. math., 1928-31, asso. prof., 1933-37; instr. Columbia, 1925-27; asso. prof. math. Yale, 1931-33; acting asso. prof. Stanford, summer 1933; prof. math. Harvard, 1937-46, chmn. dept., 1942; Andrew MacLeish Distinguished Service prof. math. U. Chgo., 1946-68, chmn. dept. math., 1946-52, mem. com. on social thought, 1962-68, prof. emeritus, 1968—; George David Birkhoff prof. math. U. Mass., 1968-73, prof., 1973-80, emeritus, 1980—; Walker Ames lectr. U. Wash., summer 1942; vis. lectr. Facultad de Ingenieria, U. Buenos Aires, 1943, U. do Brasil, Rio de Janeiro, 1947, Tata Inst. Fundamental Research, Bombay, 1949-50, Am. Math. Soc., 1951-52, College de France, 1953, Australian Math. Soc., 1959, Middle East Tech. U., Ankara, 1963, Pakistan Acad. Sci., 1964, U. Geneva, 1964; 1st Ramanujan vis. prof. Inst. Math Scis. Madras, 1963; vis. prof. Research Inst. Math. Scis., Kyoto, Japan, 1965; vis. scientist C.E.R.N., Geneva, 1966; Fulbright lectr. Australian Nat. U., 1967; hon. prof. Madurai Kamaraj U., South India, 1967—; hon. Disting. prof. Tchrs. Coll., Columbia U., 1978—; vis. lectr. U. Campinas (Brazil), 1979, U. Tócnica, Santiago, Chile, 1979, Academia Sinica, China, 1979; Fulbright travel grantee, vis. prof. Tata Inst., Bombay, 1980-81; cons. to USN Dept. Bur. Ordnance and Office Vice Chief Naval Operations, 1942-43; Civil Service employee U.S. War Dept. office chief of staff MIS, spl. br., 1944-45; with overseas service U.S. War Dept. office chief of staff, CBI, ETO; Cons. State Dept., India, 1961, NSF-AID, 1968-69, 73; vis. prof. U. Islamabad, Pakistan, 1968-69; pres. Inter-Am. Com. Math. Edn., 1961-72, hon. pres., 1972—; vice chmn. div. math. NRC, 1951-52. Author: Linear Transformations in Hilbert Space and Their Applications to Analysis, 1932; Contbr. articles on math. and math. edn. to profl. jours. Pres. Internat. Math. Union, 1952-54, Internat. Commn. on Math. Instrn., 1959-62, Inter Union Com. Teaching Sci., Internat. Council Sci. Unions, 1962-65. Recipient Nat. Medal for Sci., 1983. Mem. Nat. Acad. Scis., Am. Philos. Soc., Union Mathematica Argentine (hon.), Indian Math. Soc. (hon.), Centre International de Mathematiques Pures et Appliquees (adminstrv. and sci. councils), Acad. Brasileira de Ciencias, Lund (Sweden) Physiographical Soc., Bologna (Italy) Acad. Club: Explorers (N.Y.C.). Subspecialties: Linear transformations; Theoretical physics. Current work: Coordination and unification of symbolic (descriptive), logical (analytic), and conceptual (axiomatic) foundations of mathematics and physics. Home: 260 Lincoln Ave Amherst MA 01002

STONE, WILLIAM HAROLD, geneticist, educator; b. Boston, Dec. 15, 1924; s. Robert and Rita (Scheinberg) S.; m. Elaine Morein, Nov. 24, 1947; children: Susan Joy, Debra M.; m. Carmen Maqueda, Dec. 22, 1971; 1 son, Alexander R.M. A.B., Brown U., 1948; M.S., U. Maine, 1949; Ph.D., U. Wis., 1953. Research asst. Jackson Meml. Lab., Bar Harbor, Maine, 1947-48; faculty dept. genetics U. Wis., Madison, 1949-83, prof., 1961-83, prof. med. genetics, 1964-83; Cowles Disting. prof. dept. biology Trinity U., San Antonio, 1983—; staff scientist S.W. Found. for Research and Edn., San Antonio, 1983—; NIH fellow Calif. Inst. Tech., 1960-61; Mem. panel blood group experts FAO, 1962-67, program dir. immunogenetics research, Spain, 1971-74. Author: Immunogenetics, 1967; Contbr. articles to profl. jours. Recipient I.I. Ivanov medal USSR, 1974. Mem. Am. Inst. Biol. Scis., NRC, Assembly Life Scis., Nat. Acad. Scis., AAAS, Am. Soc. Immunologists, Am. Genetics Assn., Am. Aging Soc., Genetics Soc., Am. Soc. Human Genetics, Research Soc. Am., Internat. Soc. Transplant, Am. Soc. Animal Sci., Internat. Primatological Soc., Sigma Xi, Gamma Alpha. Subspecialties: Animal genetics; Immunology (agriculture). Current work: Immuno genetics in mammals. Home: 210 Pike Rd San Antonio TX 78209 Office: Dept Biology Trinity U San Antonio TX 78284 Some things are better never than late. We need more education and less legislation. I can't believe I get paid for something I have so much fun doing. Goodness with strength is a virtue, goodness with weakness is stupidity. Quite often, bad luck is really good luck gone unrecognized.

STONE, WILLIAM LAWRENCE, biomedical sciences educator; b. N.Y.C., Oct. 26, 1944; s. William and Dorothy (Healy) S.; m. Virginia C. Richardson, Dec. 16, 1967; children: Nora M., Isaac W. B.S., SUNY-Stony Brook, 1966, Ph.D., 1972; M.S., Marshall U., 1968. Phys. chemist Dow Chem. Co., Midland, Mich., 1966-67; research assoc. biochemistry dept. Duke U., Durham, N.C., 1973-75; asst. research chemist U. Calif.-Santa Cruz, 1975-79; asst. prof. div. biomed. scis. Meharry Med. Coll., Nashville, 1979—, grad. program dir. div. biomed. sci., 1982-83. Am. Heart Assn. grantee-in-aid, 1980; NIH research grantee, 1981-84; Office Naval Research grantee, 1981-82. Mem. Am. Inst. Nutrition, Biophys. Soc. Subspecialties: Animal nutrition; Toxicology (agriculture). Current work: Research is related to the pathology resulting from dietary deficiencies of vitamin E and selenium in animals. Co-inventor electronic analog-digital converter for microcomputer. Home: 2803 Meadow Rose Dr Nashville TN 37206 Office: Meharry Med Coll 18th Ave N Nashville TN 37208

STONE, WILLIAM ROSS, physicist, high technology research and development co. executive; b. San Diego, Aug. 26, 1947; s. William Jack and Winifred (Beckcom) S.; m. Susan L. Lane, Aug. 8, 1970; 1 dau., Ann Michele. B.A., U. Calif.-San Diego, La Jolla, 1967, M.S., 1973, Ph.C., 1978, Ph.D., 1978. Research asst. U. Calif.-San Diego, 1967-69; sr. physicist Gulf Gen. Atomic, La Jolla, 1969-73; sr. scientist Megatek Corp., San Diego, 1973-80; prin. physicist, leader inverse scattering group IRT Corp., San Diego, 1980—; pres. Stoneware Ltd., La Jolla, 1976—. Contbr. numerous articles to profl. publs.; editor: New Methods for Optical, Quasi-Optical, Acoustic, and Electromagnetic Synthesis, 1981; assoc. editor, columnist: IEEE Antennas and Propagation Soc, 1980—. Mem. adminstrv. bd. First United Methodist Ch., San Diego, 1964—, chmn. fin. com., 1975-80. Recipient medal San Diego Soc. Tech. Writers and Pubs., 1962. Mem. Internat. Radio Sci. Union (mem. U.S. nat. com. Commn. B. 1979—), U.S. Nat. Com. of Union Radio Scientifique Internationale (assoc. mem. Commn. G. 1979—), IEEE (profl. activities coordinator 1969—), Optical Soc. Am., Acoustical Soc. Am., Soc. Exploration Geophysicists, AAUP. Democrat. Subspecialties: Theoretical physics; Electronics. Current work: Fundamental research in inverse scattering theory, its application to seismic exploration, radar, sonar, and optical remote probing; basic and applied research in electromagnetic field theory. Home: 1446 Vista Claridad La Jolla CA 92037 Office: IRT Corp PO Box 80817 San Diego CA 92138

STONEHILL, ELLIOTT H., sci. adminstr., geneticist, cell biologist; b. Bklyn., Sept. 22, 1928; m. Harriett M., Jan. 28, 1951; children: Brian A., Eve Stonehill Kadar. B.S., CCNY, 1950; M.A., Bklyn. Coll., 1956; Ph.D., Grad. Sch. Med. Scis. Cornell U., 1965. Predoctoral researcher: antibiotic research S.B. Penick & Co., 1953-56, microbiology dept. Washington U., St. Louis, 1956-58, Sloan-Kettering Inst., 1959-62; Am. Cancer Soc.; postdoctoral fellow for fgn. studies, Villejuif, France, and; U. Sussex (Eng.), 1965-67; assoc. scientist Sloan-Kettering Inst.; asst. prof. microbiology Cornell U. Med. Coll., 1967-74; research planning officer, geneticist, study sect. exec. sec. NIH, Bethesda, Md., 1975-81; asst. dir. Nat. Cancer Inst., 1981—. Contbr. writings to publs. in field, papers to profl. confs. Served to 2d lt. USAF, 1951-53. Decorated Am. Spirit Honor medal. Mem. Am. Soc. Microbiology, Am. Soc. Cell Biology, N.Y. Acad. Scis., AAAS, Am. Inst. Biol. Scis., Am. Soc. Preventive Oncology, Am. Assn. Cancer Research, Sigma Xi. Subspecialties: Gene actions; Cancer research (medicine). Current work: Genetic expression and genetic control mechanisms. Office: Nat Cancer Inst NIH Bethesda MD 20205

STONER, CLINTON DALE, biochemistry educator; b. Mellette, S.D., Feb. 8, 1933; s. Homer H. and Della L. (Smith) S.; m. Elaine Blatt, Apr. 18, 1965; children: Robert Daniel, Michael David. B.S., S.D. State U., 1957, M.S., 1960; Ph.D., U. Ill., 1965. Research asst. U. Ill.-Urbana, 1961-64; postdoctoral trainee U. Wis.-Madison, 1964-65; asst. prof. Ohio State U., Columbus, 1965-69, assoc. prof. biochemistry, 1969—. Served in U.S. Army, 1953-55. NIH grantee, 1969-73, 75-79. Mem. AAAS. Unitarian. Subspecialties: Biochemistry (biology); Cell biology. Current work: Structure and function of mitochondria; steady-state kinetics of multienzyme reactions. Home: 1014 Kenway Ct Columbus OH 43220 Office: 1645 Neil Ave Columbus OH 43210

STONER, MARTIN FRANKLIN, plant pathology educator, cons.; b. Pasadena, Calif., Jan. 19, 1942; s. Robert Chester and Genevieve Virginia (Perrin) S.; m. Darleen Kay Roberts, June 13, 1963. B.S., Calif. State Poly. Coll., 1963; Ph.D., Washington State U., 1967. Research asst. Wash. State U., Pullman, 1963-67; asst. prof. Calif. State Poly. U., Pomona, 1967-70, assoc. prof., 1970-75, prof. plant pathology and mycology, 1975—, plant pathologist, 1967—; vis. prof., researcher plant pathology U. Hawaii, Honolulu, 1980-81. Named Disting. Alumnus in Sci. Calif. State Poly. U., 1982; Hawaii Ecosystems Analysis Project, Internat. Biol. Program, NSF grantee, 1972-75. Mem. Am. Phytopath. Soc. (pres. Pacific div. 1980-81), AAAS, Bot. Soc. Am., Mycological Soc. Am., Am. Theater Organ Soc. Subspecialties: Plant pathology; Integrated systems modelling and engineering. Current work: Applications of soil mycology and plant pathology in agrotech. transfer systems; innovative methods for assessment of fungi in soil ecosystems; integrated mgmt. plant diseases with spl. emphasis on cultural factors. Office: 3801 W Temple Ave Pomona CA 91768

STOOKEY, GEORGE K., university oral health research institute executive, dentistry educator; b. Waterloo, Ind., Nov. 6, 1935; s. Emra Gladison and Mary Catherine (Anglin) S.; m. Nola Jean Meek, Jan. 15, 1955; children: Lynda Marie, Lisa Ann, Laura Jean, Kenneth Ray. A.B., Ind. U.-Bloomington, 1957, M.S.D., 1961, Ph.D., 1972. Mem. faculty Ind. U., Indpls., 1953—, prof. preventive dentistry, 1978—, dir., 1981—. Contbr. over 120 sci. articles to profl. publs. Mem. Internat. Assn. Dental Research, ADA. Methodist. Subspecialties: Cariology; Analytical chemistry. Current work: Dental caries etiology and prevention; fluoride-cariostasis, metabolism; direct and administer research programs in various aspects of preventive dentistry. Patentee in field. Office: 415 Lansing St Indianapolis IN 46202

STORB, URSULA, molecular immunologist, educator; b. Stuttgart, Germany, July 6, 1936; d. Walter and Marianne (Kaemmerer) Stemmer. M.D., U. Tubingen, Ger., 1960. Lic. physician, Germany. Postdoctoral fellow Pasteur Inst., Paris, 1963-64; postdoctoral fellow dept. microbiology U. Wash., Seattle, 1966-69, mem. faculty, 1969—, prof. microbiology and immunology, 1981—; vis. prof. Deutsches Krebsforschungszentrum, Heidelberg, Germany, 1979-80. Contbr. articles to sci. jours. NIH grantee, 1972—; NSF grantee, 1978—. Mem. Am. Assn. Immunologists, Am. Soc. Cell Biology, Assn. Women in Sci., NOW. Subspecialties: Immunobiology and immunology; Molecular biology. Current work: Organization and expression of immunoglobulin genes; molecular basis of immunity. Office: U Wash Seattle WA 98195

STOREY, ARTHUR THOMAS, preventive dental science educator; b. Sarnia, Ont., Can., July 8, 1929; s. Edwin Arthur and Kathleen Harriet (Robinson) S.; m. Jaquelaine Muriel Holt, May 2, 1964; children: Eric Edwin, Patricia Ann, Lynne Alison. D.D.S., U. Toronto, 1953; M.S., U. Mich.-Ann Arbor, 1960, Ph.D., 1964. Instr. U. Mich.-Ann Arbor, 1962-65, asst. prof., 1965-66, U. Toronto, 1966-69, assoc. prof., 1966-77, prof., 1971-77; prof., head dept. preventive dental sci. U. Man., 1977—. Author: The Neural Basis of Oral and Facial Function, 1981. Chmn. children's dental health rev. com. Province of Man., 1979-81. Mem. Can. Dental Assn. (Grieve Meml. lectr. 1982), Can. Assn. Orthodontists, Am. Assn. Orthodontists, Internat. Assn. Dental Research, Soc. Neurosci., Sigma Xi, Phi Kappa Phi, Omicron Kappa Upsilon. Subspecialties: Orthodontics; Neurophysiology. Current work: Swallowing, jaw mechanics and reflexes, articular adaptation in temporomandibular joint and dentition. Home: 656 Oak St Winnipeg MB Canada R3M 3R7 Office: U Man Faculty of Dentistry 780 Bannatyne Ave Winnipeg MB Canada R3E OW3

STORK, GILBERT (JOSSE), chemistry educator; b. Brussels, Belgium, Dec. 31, 1921; s. Jacques and Simone (Weil) S.; m. Winifred Stewart, June 9, 1944; children: Diana, Linda, Janet, Philip. B.S., U. Fla., 1942; Ph.D., U. Wis., 1945; D.Sc. (hon.), Lawrence Coll., 1961, U. Paris, 1979, U. Rochester, 1982. Sr. research chemist Lakeside Labs., 1945-46; instr. chemistry Harvard, 1946-48, asst. prof., 1948-53; asso. prof. Columbia, 1953-55, prof., 1955-67, Eugene Higgins prof., 1967—, chmn. dept., 1973-76; plenary lectr. numerous internat. symposia. Vis. lectr. Swiss Am. Found. Sci. Exchange, 1959; Coover lectr., 1958, Folkers lectr.; Bachmann lectr., 1962, Treat B. Johnson lectr.; Frank Burnet Dains lectr., 1964; Kharasch vis. prof. U. Chgo.; Foster lectr. U. Buffalo, 1966; vis. scholar Fisk U., 1967; T. Dale Stewart lectr.; distinguished vis. prof. U. Iowa, 1968; Seydell Wooley lectr. Ga. Inst. Tech.; Karl Pfister lectr. Mass. Inst. Tech.; Robert Gnehm lectr. ETH, Zurich, 1969; Univ. lectr. U. Western Ont., 1974; Phi Lambda Upsilon lectr. Johns Hopkins, 1974; Centenary lectr. Brit. Chem. Soc., 1975; Tishler lectr. Harvard, 1975; John Howard Appleton lectr. Brown U.; Distinguished vis. lectr. U. Rochester; Benjamin Rush lectr. U. Pa., 1976; Royal Australian Chem. Inst. lectr., 1977; McElvain vis. scholar U. Wis., 1977; Debye lectr. Cornell U., 1977; Alexander Todd vis. prof. Cambridge (Eng.) U., 1979; Sandin lectr. U. Alta., 1980; Frank Mathers lectr. U. Ind., 1981; Greater Manchester lectr., 1982; Lemieux lectr. U. Ottawa, 1982; H. Martin Friedman lectr. Rutgers U., 1982; cons. Syntex, Internat. Flavors and Fragrances; chmn. Gordon Steroid Conf., 1958-59; bd. editors Jour. Organic Chemistry, 1955-61. Hon. adv. editor: Tetrahedron Letters, Nouveau Jour. de Chimie, Heterocycles; Editorial bd.: Accounts of Chem. Research, 1968-71. Recipient Baekeland medal, 1961; Harrison Howe award, 1962; Edward Curtis Franklin Meml. award Stanford, 1966; award for creative work in synthetic organic chemistry Am. Chem. Soc., 1967; Gold medal Synthetic Chems. Mfrs. Assn., 1971; Nebr. award, 1973; Roussel prize in steroid chemistry, 1978; Recipient Edgar Fahs Smith award, 1982, Willard Gilbs medal, 1982, Nat. medal of sci., 1982; Guggenheim fellow, 1959. Fellow Am. Acad. Arts and Scis., Nat. Acad. Scis. (award in chem scis. 1982); mem. Chemist Club (hon.), Am. Chem. Soc. (award in pure chemistry 1957, Nichols medal 1980, Arthur C. Cope award 1980, chmn. organic chemistry div. 1967), Chem. Soc. London. Subspecialty: Organic chemistry. Home: 459 Next Day Hill Englewood NJ 07631 Office: Columbia U New York NY 10027

STORM, DAVID ANTHONY, chemist; b. Bridgeton, N.J., Nov. 8, 1938; s. Francis Anthony and Gertrude (Robbinson) S.; m. Joanne Runk, Sept. 9, 1961; children: John, Elaine. B.S., Lehigh U., 1960; Ph.D. in Chemistry, Poly. Inst. Bklyn., 1970, Stanford U., 1978. Engr. Celanese Corp. Am., Houston, 1960-65; research assoc. U. Tex., Dallas, 1971-75; sr. research engr. Halcon Research and Devel., Montavle, N.J., 1978-81; group leader Texaco, Inc., Beacon, N.Y., 1981-83, research coordinator, 1983—. Contbr. articles to profl. jours. NASA grad. trainee Poly. Inst. Bklyn., 1967-69; NSF grad. trainee, 1969-70; Robert A. Welch Found. fellow U. Tex., 1971-75. Mem. Am. Chem. Soc., Am. Inst. Chem. Engrs., Am. Phys. Soc., ASTM, N.Y. Acad. Scis., Sigma Xi. Presbyterian. Subspecialties: Catalysis chemistry; Chemical engineering. Current work: Surface chemistry, catalysis chemistry, inorganic chemistry, chemical engineering, solid state chemistry. Patentee in field. Home: 35 Valley View Terrace Montvale NJ 07645 Office: PO Box 509 New Tech Research Beacon NY 12508

STORMONT, CLYDE JUNIOR, immunogeneticist, retired business executive; b. Viola, Wis., June 25, 1916; s. Clyde James and Lulu Elizabeth (Mathews) S.; m. Marguerite Butzen, Aug. 31, 1940; children: Bonnie Lu, Michael Clyde, Robert Thomas, Charles James, Janet Jean. B.A. in Zoology, U. Wis., 1938, Ph.D. in Genetics, 1947. Instr., lectr., then asst. prof. U. Wis., 1946-50; asst., then asso., then prof. dept. vet. microbiology U. Calif., Davis, 1950-60, prof. dept. reprodn., 1973-82, ret., 1982; dir. serology lab., 1952-81; chmn. Stormont Labs., Inc., Woodland, Calif., 1981—. Contbr. numerous articles to sci. jours. Served with USNR, 1944-46; PTO. Fulbright fellow, N.Z., 1949-50; Ellen B. Scripps fellow San Diego Zool. Soc., 1957-58, 64-65. Mem. AAAS, Am. Genetics Assn., Am. Assn. Immunologists, Am. Soc. Naturalists, Genetics Soc. Am., Am. Soc. Human Genetics, Soc. Exptl. Biology and Medicine, Internat. Soc. Animal Bloodgroup Research, Nat. Buffalo Assn.; mem. N.Y. Acad. Scis.; Mem. Sigma Xi. Subspecialties: Immunogenetics; Animal genetics. Current work: Immunogenetic analysis of antigenic determinants which characterize the membrane of blood cells (erythrocytes, lymphocytes and platelets). Home: Route 1 Box 264 Winters CA 95694 Office: 1237 E Beamer St Suite D Woodland CA 95695

STORRS, CHARLES L., physicist; b. Shaowu Fukien, China, Oct. 25, 1925; s. Charles L. and Mary Merrick (Goodwin) S.; m. Betty L. Wood, Nov. 23, 1957; children: Charles, Alexander, Richard. B.S., M.I.T., 1949, Ph.D., 1952. Various positions Gen. Elecrric Co., Oak Ridge, Tenn., 1952-55, aircraft nuclear propulsion dept., Evendale, Ohio, 1955, Idaho Falls, Idaho, 1955-65; dir. HWOCR program office Atomics Internat./Combustion Engring. Joint Venture, Canoga Park, Calif., 1965-67; dir. advanced devel. Combustion Engring., Windsor, Conn., 1967—. Mem. Am. Nuclear Soc. (bd. dirs. 1969-72), Am. Phys. Soc., Conn. Acad. Sci. and Engring. (council 1980—), Sigma Xi. Club: Civitan (Bloomfield, Conn.) (treas. 1982-84). Subspecialties: Nuclear fission; Nuclear engineering. Current work: Design and development of advanced energy sources, principally nuclear and fusion. Home: 76 Adams Rd Bloomfield CT 06002 Office: Combustion Engring Inc 1000 Prospect Hill Rd Windsor CT 06095

STOSKOPF, MICHAEL KERRY, aquatic veterinary medical educator, aquatic toxicologist; b. Garden City, Kans., Mar. 21, 1950; s. Cleve W. and Doris J. (Griffis) S.; m. Suzanne Kennedy, May 30, 1981. B.S., Colo. State U., 1973, D.V.M., 1975. Staff veterinarian Overton Park Zoo, Memphis, 1975-77; instr. Johns Hopkins U., Balt., 1977-79; asst. prof. pathology U. Md., Balt., 1981—; chief veterinarian Nat. Aquarium, Balt., 1981—; asst. prof. comparative medicine Johns Hopkins U., Balt., 1979—; staff veterinarian Balt. Zool. Soc., 1976-81; cons. veterinarian Md. Consortium Herpetol. Research, Balt., 1977—; cons. Instituto Roberto Franco, Meta, Colombia, 1976-78. Assoc. editor: Jour. Cardiovascular Ultrasonography, 1982, Jour. Zoo. Animal Medicine, 1976—. Mem. Am. Assn. Zoo Veterinarians (officer), Acad. Zoo Medicine (chmn. 1979—), Wildlife Disease Assn., Internat. Assn. Aquatic Animal Medicine, Phi Zeta, Phi Kappa Phi. Current work: Xenobiotic metabolism, comparative hematology, computer applications, diagnostic ultrasound, comparative anesthesiology, environmental kinetic modeling. Home: 2742 N Calvert St Baltimore MD 21218 Office: Johns Hopkins Sch Medicine 720 Rutland Baltimore MD 21205

STOTT, DONALD FRANKLIN, geologist; b. Reston, Man., Can., Apr. 30, 1928; s. Franklin Brisbin and Catherine Alice (Parker) S.; m. Margaret Elinor Hutton, Oct. 8, 1960; children: Glenn, David, Donald. B.Sc. with honors, U. Man., 1953, M.Sc., 1954; M.Sc., Princeton U., 1956, Ph.D., 1958. Jr. geologist Consol. Mining and Smelter Co., 1952, Calif. Standard Oil Co., summer 1953; research geologist Geol. Survey Can., 1957-66, head mesozoic stratigraphy sect., 1967-72, head regional geology subdiv., 1972-73; dir. Inst. Sedimentary and Petroleum Geology, Calgary, Alta., 1973-80, research scientist, 1980—. Contbr. articles to profl. jours. Recipient Willet G. Miller medal Royal Soc. Can., 1983. Fellow Geol. Assn. Can., Geol. Soc. Am.; mem. Can. Soc. Petroleum Geologists (v.p. 1977, pres. 1978), Soc. Econ. Paleontologists, Sigma Xi. Subspecialty: Geology. Office: 3303 33d St NW Calgary AB Canada T2L 2A7

STOTTS, LARRY BRUCE, physicist, electronics engr.; b. Pickstown, S.D., Sept. 21, 1949; s. Lawrence Zenith and Georgia Anne (Buresh) S. B.A., U. Calif., San Diego, 1972. Physicist Naval Electronics Lab., San Diego, 1971-77; staff scientist Naval Ocean System Center, San Diego, 1977—, chief scientist submarine laser communication program, 1979-81. Contbr. articles to profl. jours. Mem. Optical Soc. Am., Assn. Old Crows. Club: Coronado (Calif.) Yacht. Subspecialties: Laser communication; Optics research. Current work: Optics and signal processing. Patentee in field. Office: Code 7301 Naval Ocean System Center San Diego CA 92152

STOTZKY, GUENTHER, microbiologist, educator, cons., researcher; b. Leipzig, Germany, May 24, 1931; came to U.S., 1939, naturalized, 1944; s. Morris and Erna (Angres) S.; m. Kayla Baker, Mar. 23, 1958; children: Jay, Martha, Deborah. B.S., Calif. State Poly. U., 1952; M.S., Ohio State U., 1954, Ph.D., 1956. Research asst. agronomy dept. Ohio State U., 1953-56; research assoc. botany dept. U. Mich., 1956-58; head soil microbiology Central Research Labs., United Fruit Co., Norwood, Mass., 1958-63; chmn. Kitchawan Research Labs., Bklyn. Botanic Garden, 1963-66, assoc. prof. biology N.Y.U., N.Y.C., 1967-70, prof., 1970—, chmn. dept., 1970-77; cons. to numerous indsl. and govtl. orgns.; trustee N.Y. Ocean Sci. Lab., 1973—. Assoc. editor: Can. Jour. Microbiology, 1971-74, Applied and Environ. Microbiology, 1971-78; regional editor: Soil Biology and Biochemistry, 1969—; contbr. numerous articles to profl. jours., 35 chpts. to books; ad hoc reviewer for jours. and pubs. Served with USCG, 1957-58. Grantee USPHS, EPA, U.S. Dept. Agr., Am. Cancer Soc., NSF, NATO, N.Y. State Found. for Sci. and Tech., N.Y. State Health Research Council. Fellow AAAS, Am. Acad. Microbiology, Soil Sci. Soc. Am., Am. Soc. Agronomy; mem. Am. Soc. for Microbiology, Can. Soc. for Microbiology, Bot. Soc. Am., Am. Inst. Biol. Scis., Air Pollution Control Assn., Soc. for Environ. Geochemistry and Health, Internat. Assn. for Ecology, Internat. Soil Sci. Soc., Med. Mycol. Soc. Ams., Internat. Symposium on Environ. Biogeochemistry, Sigma Xi. Subspecialties: Microbiology; Environmental toxicology. Current work: Microbial ecology; surface interactions; soil, water and air pollution; environmental microbiology, virology and immunology; ecotoxicology; clinical microbiology. Home: 110 Bleecker St Apt 2D New York NY 10012 Office: Biology Dept NYU New York NY 10003

STOUT, ROBERT DANIEL, educator, univ. dean emeritus; b. Reading, Pa., Jan. 2, 1915; s. Harry Herbert and Anna (Guldin) S.; m. Elizabeth Allwein, Aug. 16, 1939; 1 dau., Elizabeth Ann. B.S., Pa. State U., 1935; M.S., Lehigh U., 1941, Ph.D., 1944; D.Sc., Albright Coll., 1967. Metall. engr. Carpenter Steel Co., 1935-39; faculty Lehigh U., Bethlehem, Pa., 1939—, prof. metallurgy, 1950—, dean, 1960-80, dean emeritus, 1980—; Metall. cons. Author: Weldability of Steels, 1953, 71; Contbr. articles to profl. pubis. Fellow Am. Soc. Metals (Stoughton Teaching award 1952, White Disting. Teaching award 1974); mem. Am. Welding Soc. (pres. 1972-73, Lincoln gold medal 1943, Adams lectr. 1960, Sparagen award 1964, Thomas award 1973, Jennings award 1974, Am. del. Internat. Inst. Welding 1955—, Houdremont lectr. 1970), Am. Soc. Engring. Edn., Sigma Xi. Subspecialties: Metallurgical engineering; Fracture mechanics. Current work: Metallurgical effects of welding; mechanical properties of welded alloy steels used for pressure vessels and cryogenic applications. Home: 135 E Market St Bethlehem PA 18018

STOVER, BETSY JONES, chemist, educator; b. Salt Lake City, May 13, 1926; d. Richard Hugh and Bessie Ina (Miers) Jones; m. (div.); children: Susan, Steven Nathan. B.A. with high honors, U. Utah, 1947; Ph.D. in Chemistry, U. Calif., Berkeley, 1950. Teaching asst., research assoc. U. Calif., Berkeley, 1947-50; asst. research prof. dept. chemistry U. Utah, Salt Lake City, 1950-58, assoc. research prof., 1958-70; chemistry group leader Radiobiology Lab., 1950-70, adj. assoc. research prof. dept. anatomy, 1970-75, adj. research prof., 1975-79, adj. prof. dept. pharmacology, 1979—, cons., 1970—; assoc. prof. dept. pharmacology U. N.C., Chapel Hill, 1970-76, prof., 1976—. Co-editor: Some Aspects of Internal Irradiation, 1962, Delayed Effects of Bone-Seeking Radionuclides, 1969; co-editor: Radiobiology of Plutonium, 1972; co-author: Statistical Mechanics and Dynamics, 1964, The Theory of Rate Processes in Biology and Medicine, 1974, Statistical Mechanics and Dynamics, 2d edit, 1982; also articles. Mem. AAAS, AAUP, Am. Chem. Soc., Am. Phys. Soc., Am. Soc. for Bone and Mineral research, Am. Soc. for Pharmacology and Exptl. Therapeutics, Assn. for Women in Sci., Internat. Soc. Quantum Biology, Radiation Research Soc., Soc. for Expt. Biology and Medicine, Soc. toxicology, ACLU, NOW, Friends of Kennedy Ctr., Phi Beta Kappa, Sigma Xi, Alpha Lambda Delta, Phi Kappa Phi. Subspecialties: Toxicology (medicine); Theoretical chemistry. Current work: Toxicology of radionuclides, theoretical chemistry of biology. Home: 103 Marion Way Robin's Wood Chapel Hill NC 27514 Office: Pharmacology Sch Medicine U NC 1128 FLOB 231H Chapel Hill NC 27514

STOVER, JANET, veterinarian, reproductive biology researcher; b. Pitts., Jan. 23, 1952; d. William Lloyd and Elizabeth (Gibson) S. Student, Colo. State U., 1969-72; V.M.D., U. Pa., 1976. Lic. vet., N.Y., Pa., Calif. Pathology intern Sch. of Vet. Medicine U. Pa., Phila., 1976-77; postgrad. fellow in reprodn. U. Pa. and Phila. Zool. Garden, 1977-78; clin. vet. Phila. Zool. Garden, 1978-79; resident in vet. medicine N.Y. Zool. Soc., 1982—, now assoc. vet.; vis. lectr. Sch. Vet. Medicine U. Pa. Muskiwinni Found. grantee, 1980-82. Mem. Am. Assn. Zoo Vets., Am. Assn. Zool. Parks and Aquariums, AVMA, Wildlife Disease Assn., Internat. Assn. Aquatic Animal Medicine, Internat. Embryo Transfer Soc., Am. Assn. Theriogenology, Am. Fertility Soc. Methodist. Subspecialties: Reproductive biology; Embryo transplants. Current work: Research in gene- banking for endangered species, working with frozen oocytes, embryo and sperm banks. Home and Office: N Y Zool Soc 185th and Southern Blvd Bronx NY 10460

STOWE, BRUCE BERNOT, biologist, educator, plant biochemist; b. Neuilly-sur-Seine, France, Dec. 9, 1927; s. Lel and Ruth Florida (Bernot) S.; m. Elizabeth Louise Kwasny, June 23, 1951; children: Mark Kwasny, Eric Bernot. B.Sc., Calif. Inst. Tech., 1950; A.M., Harvard U., 1951, Ph.D., 1954; M.A. (hon.), Yale U., 1971. NSF fellow Univ. Coll. North Wales, 1954-55; instr. biology Harvard U., Cambridge, Mass., 1955-58, tutor biochem. scis., 1956-58, lectr. botany, 1958-59; asst. prof. botany Yale U., New Haven, 1959-63, assoc. prof. biology, 1963-71, prof. biology, 1971—, prof. forestry, 1974—; dir. Marsh Bot. Garden, 1975-78, 82; vis. prof. U. Osaka, Japan, 1972, 73, Waite Agrl. Research Inst., U. Adelaide, South Australia, 1972-73; external examiner various Can. and Indian univs.; cons. plant biochemistry to industry. Contbr. articles to sci. jours.; editorial bd.: Plant Physiology, 1965—; editorial com.: Ann. Rev. Plant Physiology, 1968-73. Served with Signal Corps U.S. Army, 1946-47. AEC fellow, 1951-53; Lalor Found. fellow, 1953-54; Guggenheim fellow, 1965-66. Fellow AAAS; mem. Am. Soc. Plant Physiologists (exec. com. 1960-65, sec. 1963-65, trustee 1970-72, 75-77), AAUP (chpt. exec. com. 1963-75, chpt. sec.-treas. 1969-75, chpt. pres. 1966-68), Am. Soc. Biol. Chemists, Am. Soc. Plant Physiology, Bot. Soc. Am., Phytochem. Soc., Soc. Plant Growth Regulation, Soc. Developmental Biology, Soc. Gen. Physiology, Fedn. Am. Scientists, Am. Inst. Biol. Scis., Biochem. Soc. (U.K.), Soc. Exptl. Biology (U.K.), Soc. Physiol. Végétale (France), Soc. Botanique de France, Scandinavian Soc. Plant Physiology, Japanese Soc. Plant Physiology, Sigma Xi, Australian Soc. Plant Physiology, Sigma Xi. Democrat. Subspecialties: Plant physiology (biology); Biochemistry (biology). Current work: Biochemistry and physiology of plant hormones and membranes; physiologically active lipids; membrane modelling by bioassay; computer correlations of biological data and molecular models; computer aids to handicapped. Home: 161 Grand View Ave Hamden CT 06514 Office: 46A Kline Biology Tower Yale U Box 6666 New Haven CT 06511

STOWELL, ROBERT EUGENE, pathologist, med. educator, adminstr., researcher, cons.; b. Cashmere, Wash., Dec. 25, 1914; s. Eugene Frances and Mary (Wilson) S.; m. Eva Mae Chambers, Dec. 1, 1945; children: Susan Jane, Robert Eugene. Student, Whitman Coll., 1932-33; A.B. with honors, Stanford U., 1936, M.D., 1941; Ph.D. (Univ. fellow), Washington U., St. Louis, 1944. Diplomate: Am. Bd. Pathology. Asst. resident in pathology Barnes, McMillan, Children's hosps., St. Louis, 1942-43, resident, 1943-44; research assoc. Barnard Free Skin and Cancer Hosp., St. Louis, 1942-48; instr. pathology Washington U. Sch. Medicine, St. Louis, 1943-45, asst. prof., 1945-48, assoc. prof., 1948; chmn. dept. oncology U. Kans. Med. Center, 1948-51, prof. pathology and oncology, 1948-59, dir. cancer research, 1948-59, chmn. dept. pathology and oncology, 1951-59, pathologist-in-chief, 1951-59; sci. dir. Armed Forces Inst. Pathology, Washington, 1959-67; vis. prof. pathology U. Md., 1960-67; chmn. dept. pathology U. Calif. Sch. Medicine-Davis, 1967-69, asst. dean, 1967-72, prof. pathology, 1967-82, emeritus prof. pathology, 1982—; acting dir. Nat. Ctr. Primate Biology, 1968-69, dir., 1969-71; dir. div. pathology Sacramento Med. Center, 1967-69; mem. numerous adv. coms. and cons. to govt. Contbr. articles in pathology to med. jours. Recipient Meritorious Service award Dept. Army, 1963, Exceptional Civilian Service award, 1965. Mem. AAAS, Am. Assn. Cancer Research, Am. Assn. Pathologists and Bacteriologists (pres. 1970-71), AMA, Am. Soc. Clin. Pathologists, Am. Soc. Exptl. Pathology (pres. 1964-65), Assn. Am. Med. Colls., Calif. Med. Soc., Calif. Soc. Pathologists, Coll. Am. Pathologists, Histochem. Soc., Internat. Acad. Pathology (pres. 1959-60), Radiation Research Soc., Soc. Cryobiology (bd. govs. 1968-71), Soc. Exptl. Biology and Medicine, Volo County Med. Soc., Phi Beta Kappa, Sigma Xi, Alpha Omega Alpha. Subspecialties: Pathology (medicine); Cancer research (medicine). Current work: Human and experimental pathology; pathology education and administration; cancer research; consultant. Home: PO Box 3061 El Macero CA 95618 Office: Dept Pathology Sch Medicine U Calif Davis CA 95616

STRADER, HERMAN LEE, biology educator; b. Danville, Va., Aug. 9, 1920; s. Wesley Samuel and Margaret (Clark) S.; m. Edna Mae Mitchell, Nov. 27, 1948. B.S. in Biology, Va. Union U., 1946; M.S., Columbia U., 1952, Ed.D., 1965. Lab. asst. dept. biology Va. Union U., Richmond, 1945-46, instr., 1948-50, asst. prof., 1952-60, assoc. prof., 1960-72, prof., 1972—; sci. cons. Richmond Pub. Schs., Va., 1972-80. Author: A Program of Conservation Education, 1970. Mem. Falls of the James Environ. Com., Richmond, 1978-82. Danforth Found. fellow, 1958; United Negro Coll. Fund fellow, 1963; Nat. Wildlife Found. fellow, 1965; Macy Found. fellow, 1972. Mem. Am. Inst. Biol. Scis., Bot. Soc. Am., AAAS, Nat. Sci. Tchrs. Assn., Ecol. Soc. Am., Beta Kappa Chi. Baptist. Subspecialties: Ecology; Plant growth. Current work: General biology, ecology and comparative anatomy. Home: 2209 Northumberland Ave Richmond VA 23220 Office: Va Union U Dept Biology 1500 N Lombardy St Richmond VA 23220

STRAIN, JOHN WILLARD, aerospace company executive, consultant; b. Ottumwa, Iowa, Dec. 31, 1929; s. John Wells and Agnes Gertrude (Keaans) S.; m. Elizabeth LaVonne Moment, Dec. 27, 1952 (dec. 1969); children: James Anthony, Mary Therese, Michael Douglas, Meagan Kathaleen. B.A., U. No. Iowa, 1952. Rocket power plant engr. White Sands Proving Ground, Las Cruces, N.Mex., 1954-55; mgr. Lockheed Missile & Space Co., Santa Cruz, Calif., 1960-63, Sunnyvale, Calif., 1967-73, RPV Aquila chief test engr., 1975-79, factory test mgr., 1979-82, div. mgr., 1982—; owner Indsl. Systems Co., Sunnyvale, 1976—. Assoc. editor: Missile Away Mag, 1954-55. Bd. dirs. San Jose Civic Light Opera, 1972; leader Boy Scouts Am., San Jose, 1964. Served with U.S. Army, 1952-54. Recipient Alumni Service award U. No. Iowa, 1981. Fellow AIAA (assoc.); mem. Inst. Environ. Scis. (sr.), Assn. Unmanned Vehicle Systems (treas. 1982—), AAAS, Nat. Mgmt. Assn. Republican. Roman Catholic. Subspecialties: Aerospace engineering and technology; Combustion processes. Current work: Development and test of remotely piloted vehicles with integrated datalink, video sensor, laser capabilities. Home: 626 Oneida Dr Sunnyvale CA 94087 Office: Lockheed Missiles and Space Co 1111 Lockheed Way Sunnyvale CA 94086

STRAITON, ARCHIE WAUGH, electrical engineering educator; b. Arlington, Tex., Aug. 27, 1907; s. John and Jeannie (Waugh) S.; m. Esther McDonald, Dec. 28, 1932; children: Janelle (Mrs. Thomas Henry Holman), Carolyn (Mrs. John Erlinger). B.S. in Elec. Engring, U. Tex., 1929, M.A., 1931, Ph.D., 1939. Engr. Bell Telephone Labs., N.Y.C., 1929-30; from instr. to assoc. prof. Tex. Coll. Arts and Industries, 1931-41, prof., 1941-43, head dept. engring., 1941-43; faculty U. Tex., Austin, 1943—, prof., 1948—, dir. elec. engring. research lab., 1947-72, Ashbel Smith prof. elec. engring., 1963—, chmn. dept., 1966-71, acting v.p., grad. dean, 1972-73. Contbr. articles to profl. jours. Fellow IEEE; mem. Nat. Acad. Engring., Sigma Xi, Tau Beta Pi, Eta Kappa Nu. Subspecialties: Electrical engineering; Applied magnetics. Current work: Transmissions of millimeter radio wave through earth's atmosphere. Home: 4212 Far West Blvd Austin TX 78731

STRAKA, WILLIAM CHARLES, physicist, astronomer; b. Phoenix, Oct. 21, 1940; s. William Charles and Martha Nadene (Marshall) S.; m. Barbara Ellen Thayer, Jan. 29, 1966; 1 son, William Charles. B.S., Calif. Inst. Tech., 1962; M.S., UCLA, 1975, Ph.D., 1969. Tchr. Long Beach City Coll., 1966-70; asst. prof. Boston U., 1970-74; asst. prof. physics and astronomy Jackson State U., 1974-76, assoc. prof., 1976-82, prof. physics, 1982—; vis. staff mem. Los Alamos Nat. Lab., 1976—; exec. sec. Astron Adv. Com., NSF, 1978-79; summer fellow ASEE-NASA, Goddard Space Flight Ctr., 1982. NSF Research grantee, 1978—. Mem. Am. Astron Soc., Miss. Acad. Sci., Miss. Assn. Physicists (pres. 1981, v.p. 1982), Nat. Sci. Tchrs. Assn. Subspecialties: Theoretical astrophysics; High energy astrophysics. Current work: Supernova remnant nebulae; numerical models of hydrodynamics of supernova remnant nebulae.0. Office: Dept Physics Jackson State Jackson MS 39217

STRASEN, STEPHEN M., software engr.; b. Pitts., Feb. 17, 1941; s. Philip E. and Barbara E. (Griefen) Callanen; m. Barbara E. Ehrlich, June 12, 1963. B.S., Carnegie-Mellon U., 1962; Ph.D., U. Calif., Berkeley, 1968. Asst. prof. math. U. Santa Clara, Calif., 1968-71; programmer NCR Corp., Rancho Bernardo, Calif., 1972-79; staff programmer Gen. Automation, Anaheim, Calif., 1979-80, Tech. Mktg., Inc., Irvine, Calif., 1980—. NSF fellow, 1962-65. Mem. IEEE, Phi Kappa Phi. Subspecialties: Computer architecture; Software engineering. Office: 17862 Fitch St Irvine CA 92714 Home: 353 Ocean View Ave Encinitas CA 92024

STRASSER, ALFRED ANTHONY, consulting engineer; b. Budapest, Hungary, Jan. 21, 1927; came to U.S., 1939; s. Paul and and Margaret (Fuerth) S.; m. Barbara Schoenbrodt, Aug. 1, 1970; 1 son, Christopher. B.S. in Metall. Engring, Purdue U., 1948; M.S. in Metallurgy, Stevens Inst. Tech., 1952. Metallurgist M.W. Kellogg Co., Jersey City, 1948-51, U.S. Air Force, Dayton, Ohio, 1951-54; mgr. plutonium fuels dept. United Nuclear Corp., Elmsford, N.Y., 1954-72; v.p. reactor tech. S.M. Stoller Corp., N.Y.C., 1972—. Contbr. writings in field to profl. publs. Served with USN, 1945-46. Mem. Am. Soc. Metals, Am. Nuclear Soc. Club: Appalachian Mountain. Subspecialties: Nuclear engineering; Materials. Current work: Evaluation of materials, their performance and design on reactor component and system performance. Office: S M Stoller Corp 1250 Broadway New York NY 10001

STRATT, RICHARD MARK, chemistry educator; b. Phila., Feb. 21, 1954; s. Stanford Lloyd and Florence Claire (Sussman) S. S.B., MIT, 1975; Ph.D., U. Calif.-Berkeley, 1979. Postdoctoral research assoc. U. Ill.-Champaign/Urbana, 1979-80, NSF postdoctoral fellow, 1980; asst. prof. chemistry Brown U., Providence, 1981—. Contbr. articles in field to profl. jours. NSF postdoctoral fellow, 1980. Mem. Am. Phys. Soc., Sigma Xi, Phi Lambda Upsilon. Subspecialties: Theoretical chemistry; Statistical mechanics. Current work: The behavior of the internal degrees of freedom of molecules and the ordering of molecules, in condensed phases. Office: Dept Chemistry Brown U Providence RI 02912

STRATTON, JAMES FORREST, geology educator; b. Chicago Heights, Ill., Nov. 29, 1943; s. James Westbrook and Evelyn Juanita (Hasty) S.; m. Patrice Areta Fanuko, Feb. 23, 1980. B.S., Ind. State U., 1965; M.A.T., Ind. U., 1967, A.M. in Geology, 1972, Ph.D., 1975. Instr. geology Shippensburg (Pa.) State Coll., 1967-70; assoc. prof. geology Eastern Ill U., Charleston, 1975—, dir. summer field geology, 1978—. NSF fellow, 1971. Mem. Am. Assn. Petroleum Geologists, Soc. Econ. Paleontologists and Mineralogists, Paleontol. Soc., Ind. Acad. Sci., Internat. Bryozoology Assn. Methodist. Subspecialty: Paleobiology. Current work: Functional morphology and taxonomy of paleozoic fenestrate bryozoans. Office: Dept Geology Eastern Ill U Charleston IL 61920 Home: Rural Route 4 Box 65 Charleston IL 61920

STRAUB, RICHARD OTTO, psychology educator, experimental psychologist; b. Renton, Wash., May 10, 1954; s. Lee Fred and Phyllis Joan (Sullivan) S.; m. Pamela Lee Hanlon, July 31, 1977; 1 son, Jeremy Alan. B.S., Fla. So. U., 1975; M.A., Columbia U., 1976, M.Phil., 1978, Ph.D., 1979. Faculty fellow Columbia U., 1975-79; research asst. N.Y. State Psychiat. Inst., N.Y.C., 1976-77; asst. prof. psychology U. Mich., Dearborn, 1979—; cons., reviewer Worth Pubs., N.Y.C. Editor: Readings in the Psychobiology of Motivation, 1982; contbr. in field. NIMH fellow, 1977; recipient Disting. Tchr. award U. Mich., 1982. Mem. Am. Psychol. Assn., Mich. Acad. Sci., Arts and Letters, Psi Chi, Omicron Delta Kappa. Subspecialties: Psychobiology; Learning. Current work: The development of an animal model of coronary-prone behavior and selectively-bred substrains of coronary-prone and coronary-resistant animals as a vehicle for behavioral, genetic and pathophysiological research into the etiology of type A behavior and precursors to coronary heart disease. Home: 8740 Honeycomb Circle Bldg 5 Apr 159 Danton MI 48187 Office: U Mich 4901 Evergreen Rd Dearborn MI 48128

STRAUS, DAVID JEREMY, physician; b. Urbana, Ill., Apr. 22, 1944; s. Gerhard D. and Lois M. (Marin) S.; m. Karen B. Straus, Aug. 20, 1966; children: Jennifer, Emily. A.B., U. Chgo., 1965; M.D., Marquette U., 1969. Intern, then resident Montefiore Hosp., N.Y.C., 1969-70; resident Beth Israel Hosp., Boston, 1972-73, Meml. Hosp., 1975-77; practice medicine, specializing in hematology and oncology; assoc. attending physician Meml. Sloan Kettering Cancer Ctr., N.Y.C., 1982—; assoc. prof. dept. medicine Cornell U., Ithaca, N.Y., 1981—. Served to maj. M.C. U.S. Army, 1973-75. Fellow ACP; mem. Am. Soc. Hematology, Am. Soc. Clin. Oncology, Am. Assn. Cancer Research, Am. Fedn. Clin. Research. Subspecialties: Oncology; Hematology. Current work: Malignant lymphoma and other hemopoactic malignancies. Office: 1275 York Ave New York NY 10021

STRAUS, FRANCIS HOWE, II, surgical pathology educator; b. Chgo., Mar. 16, 1932; s. Francis Howe and Elizabeth (Kales) S.; m. Helen Lorna Puttkammer, June 11, 1955; children: Francis Howe III, Helen Elizabeth, Christopher Monroe, Michael Wilfred. A.B., Harvard Coll., 1953; M.D., U. Chgo., 1957, M.S., 1964. Diplomate: Am. Bd. Pathology. Instr. pathology U. Chgo., 1962-65, asst. prof., 1965-71, assoc. prof., 1971-78, prof. pathology, 1978—, assoc. dir., 1976—; pres. Chgo. unit Am. Cancer Soc., Chgo., 1979-83. Author: Hyperparathyroidism, 1973, Essentials of Surgical Pathology, 1974. Mem. Internat. Acad. Pathology, Am. Soc. Exptl. Pathologists, Am. Assn. Pathologists, Inst. Medicine Chgo. Clubs: Chgo. Literary, Cliff Dwellers (Chgo.). Subspecialty: Pathology (medicine). Current work: All aspects of endocrine and breast pathology, surface staining of urinary tract tumors, morphologic aspects of testicular infertility. Home: 5642 S Kimbark Ave Chicago IL 60637 Office: U Chgo 950 E 59th St Chicago IL 60637

STRAUSS, BERNARD S., educator, geneticist; b. N.Y.C., Apr. 18, 1927; s. Joseph and Kate (Silk) S.; m. Carol Maxine Dunham, Sept. 8, 1949; children: Leslie Joan Travis, David Wilson, Paul Leonard. B.S., CCNY, 1947; Ph.D., Calif. Inst. Tech., 1950; postdoctoral fellow, U. Tex., 1950-52. Teaching asst. Calif. Inst. Tech., 1947-48; asst. prof. Syracuse U., 1952-56, asso. prof., 1956-60; research asso. Brookhaven Nat. Lab., 1954-55; asso. prof. U. Chgo., 1960, prof., 1965—, chmn. com. genetics, 1962-76, chmn. dept. microbiology, 1969—; vis. prof. U. Sydney, 1967, Hadassah Med. Sch., Hebrew U., Jerusalem, 1975, 81; Mem. genetics tng. com. NIH, 1962-66, 70-73. Editorial bd.: Cancer Research, Mutation Research. Served with U.S. Mcht. Marine, 1945-47. Fulbright, Guggenheim fellow Osaka U., Japan, 1958-59. Mem. Am. Soc. Biol. Chemists, Am. Assn. Cancer Research, Genetics Soc. Am., Am. Soc. Microbiology, Environ. Mutagen Soc. (councilor 1975-78), Phi Beta Kappa, Sigma Xi. Subspecialties: Molecular biology; Cancer research (medicine). Current work: Molecular mode of action of mutagenic and carcinogenic agents. Home: 5431 S Ridgewood Ct Chicago IL 60615

STRAUSS, JOHN STEINERT, dermatologist, academic administrator, educator; b. New Haven, July 15, 1926; s. Maurice J. and Carolyn M. (Ullman) S.; m. Susan Thalheimer, Aug. 19, 1950; children: Joan Sue, Mary Lynn. B.S., Yale U., 1946, M.D., 1950. Diplomate: Am. Bd. Dermatology. Intern U. Chgo. Clinics, 1950-51; resident in dermatology Hosp. U. Pa., 1951-52, fellow in dermatology, 1954-56, inst. dermatology, 1954-57; asst. prof. dermatology Boston U., 1957-61, assoc. prof., 1961-66, prof., 1966-78; prof., head dept. dermatology U. Iowa, 1978—. Contbr. numerous articles to profl. jours. Recipient Heller medal Dept. Dermatology Karolinska Inst., Sweden, 1978. Fellow Am. Acad. Dermatology (dir. 1977-80, pres. 1982-83); mem. Soc. Investigative Dermatology (dir. 1963-68, pres. 1975-76), Dermatology Found. (trustee 1969-74, pres. 1972-74). Subspecialty: Dermatology. Current work: Pathogenesis of acne; physiology of sebaceous gland function; biochemistry of sebaceous and epidermal lipids. Office: Dept Dermatology Univ Hosps 2BT Iowa City IA 52242

STRAUSS, MONTY JOSEPH, mathematics educator, consultant; b. Tyler, Tex., Aug. 26, 1945; s. Milton and Ann (Schloss) Salfield; m. Linda Joyce Sank, June 16, 1968 (div. Aug. 1973); m. Jane Louise Winer, Nov. 4, 1978. B.A. magna cum laude, Rice U., 1967; Ph.D., Courant Inst., NYU, 1971. Asst. prof. math. Tex. Tech U., 1971-75, assoc. prof., 1975—, dept. dir. grad. studies, 1980—; cons. in field, 1975—. Contbr. numerous articles on research math., including oil exploration, partial differential equations, and computer literacy, to profl. jours. Faculty adv., founder Hillel Counselorship, Tex. Tech U., Lubbock. NSF grantee, 1975—. Mem. Am. Math. Soc., Math. Assn. Am., Soc. Indsl. and Applied Math., AAUP, Tex. Assn. Coll. Tchrs.

(state dir. 1975-76), Assn. Computers in Math. and Sci. Teaching, Phi Beta Kappa, Kappa Mu Epsilon (faculty sponsor 1979-80). Democrat. Jewish. Lodge: B'nai B'rith. Subspecialties: Applied mathematics. Current work: Partial differential equations; oil exploration; computer literacy. Home: 7010 Nashville Dr Lubbock TX 79413 Office: Dept Math Tex Tech Univ PO Box 4319 Lubbock TX 79409

STRAUSS, WALTER ALEXANDER, mathematician, educator; b. Aachen, Germany, Oct. 28, 1937; came to U.S., 1940; s. Charles and Johanna (Goldschmidt) S.; m. Phyllis Romanoff; children: Charles, Nathaniel. A.B., Columbia U., 1958; M.S., U. Chgo., 1960; Ph.D., MIT, 1962. NSF postdoctoral fellow U. Paris, 1962-63; vis. asst. prof. Stanford (Calif.) U., 1963-66; assoc. prof. Brown U., Providence, 1966-71, prof. math., 1971—. Mem. editorial bd.: Jour. Differential Equations, 1978—; contbr. articles to profl. jours. Fulbright lectr. U.S. State Dept., Rio de Janeiro, 1967; Guggenheim fellow, 1971; Japan Soc. Promotion Sci. fellow, 1972. Mem. Am. Math. Soc., Soc. Indsl. and Applied Math. Subspecialty: Applied mathematics. Current work: Nonlinear wave equations;scattering theory. Office: Brown U Providence RI 02912

STRAUSSNER, JOEL HARVEY, psychologist, researcher; b. Bklyn., Apr. 7, 1947; s. William Gabriel and Henrietta Beth (Edelson) S.; m. Shulamith Lala Ashenberg, Dec. 28, 1969; 1 son, Adam Ashenberg. M.S., CCNY, 1972, 75, Yeshiva U., 1977, cert., 1977, Ph.D., 1982. Cert. sch. psychologist, N.Y.; cert. tchr. K-6, N.Y. State. Tchr. N.Y.C. Bd. Edn., 1969-72, spl. edn., 1972-73; clin. psychology intern N.J. Dept. Human Services, Trenton, 1976-77; sch. psychologist Bur. Child Guidance N.Y.C., 1978-80; sch. psychologist, com. handicapped N.Y.C. Bd. Edn., 1980—; reading cons. N.Y. League for Hard of Hearing, N.Y.C., 1972-73; learning disabilities cons. W. H. Solan, Teaneck, N.J., 1973-75, St. Joseph's Home, Peekskill, N.Y., 1978; dir. child and adolescent services Whitestone (N.Y.) Counseling Ctr., 1981—. Contbr. articles to profl. jours. N.J. Dept. Human Services jr. fellow, 1977. Mem. Am. Psychol. Assn. (award of research competency div. 16 1982), N.Y. Soc. Clin. Psychologists, Internat. Reading Assn. Subspecialties: Cognition; Educational psychology. Current work: Investigation of differences in adult problem solving styles and subsequent results on adult's behavior, relationship of problem solving behavior and its effect susceptibility to the expectancy phenomenon; also investigating cross-cultural educational issues of newly-arrived immigrants. Home: 124 W 79th St New York NY 10024 Office: NYC Bd Edn 801 Park Pl Brooklyn NY 11216

STRAWDERMAN, WILLIAM EDWARD, statistics educator, researcher; b. Westerly, R.I., Apr. 25, 1941; s. Robert Lee and Alida Browning (Dow) S.; m. Eileen Carol Young, Sept. 5, 1964; children: Robert Lee, William Edward, Heather Lynne. B.S. in Engring, U. R.I., 1963; M.S. in Stats, Rutgers U., 1967, Ph.D., 1969; M.S. in Math, Cornell U., 1965. Mem. tech. staff Bell Labs., Holmdell, N.J., 1965-67; mem. faculty Rutgers U., New Brunswick, N.J., 1967—, prof. stats., 1976—, also chmn. stats. dept.; vis. asst. prof. stats. Stanford (Calif.) U., 1969-70; cons. in field. Contbr. articles to profl. publs. NSF grantee, 1971—. Fellow Inst. Math. Statis.; mem. Am. Statis. Assn. Subspecialty: Statistics. Current work: Statistical decision theory; mathematical statistics; applied statistics; regression theory. Office: Statistics Dept Rutgers U New Brunswick NJ 08903

STRAYER, DAVID SHELDON, pathologist, educator; b. N.Y.C., Apr. 25, 1949; s. Robert and Julia Blossom (Aboulafia) S.; m. Frances Anne Novarr, June 27, 1971; children: Reuben Jesse, Rebecca Miriam, Rachel Anna. A.B., Cornell U., 1970; Ph.D., U. Chgo., 1974, M.D., 1976. Diplomate: Am. Bd. Pathology. Intern Washington U., St. Louis, 1976-77, resident, 1977-79; instr., 1979-80; asst. prof. U. Calif., La Jolla., 1980—. Cancer Research Coordinating Com. grantee, 1981—; Am. Cancer Soc. fellow, 1981—. Mem. Am. Assn. Pathologists, Pediatric Oncology Group, Phi Beta Kappa. Subspecialty: Pathology (medicine). Current work: Research interests in the role of viruses, both oncogenic and nononcogenic, in the suppression of the immune response, secondary interest in the role of exogenously administered human surfactant in the therapy of neonatal lung disease. Office: Dept Pathology U Calif M-012 La Jolla CA 92093

STREET, ROBERT LYNNWOOD, civil and mechanical engineer; b. Honolulu, Dec. 18, 1934; s. Evelyn Mansel and Dorothy Heather (Brook) S.; m. Norma Jeanette Ensminger, Feb. 6, 1959; children: Brian Clarke (dec.), Deborah Lynne, Kimberley Anne. M.S., Stanford U., 1957, Ph.D. (NSF grad. fellow 1960-62), 1963. Mem. faculty Stanford U. Sch. Engring., 1962—, prof. civil engring., asso. chmn. dept., 1970-72, chmn. dept., 1972-80, prof. fluid mechanics and applied math., 1972—; asso. dean research Sch. Engring., 1971-83, acting vice provost, dean research, 1979, vice provost for acad. computing and info. systems, 1983; vis. prof. U. Liverpool, Eng., 1970-71; trustee Univ. Corp. Atmospheric Research, 1983, chmn. sci. programs evaluation com., 1981, chmn. audit com.; cons. in field. Author: The Analysis and Solution of Partial Differential Equations, 1973; co-author: Elementary Fluid Mechanics, 6th edit, 1982; asso. editor: Jour. Fluids Engring, 1978-81; author articles in field; mem. editorial bds. profl. jours. Served as officer C.E., USN, 1957-60. Sr. postdoctoral fellow Nat. Center Atmospheric Research, 1978-79; fellow N.E. Asia-U.S. Forum on Internat. Policy Stanford U. Mem. Am. Soc. Engring. Edn., ASCE (chmn. publs. com. hydraulics div. 1978-80, Walter Huber prize 1972), ASME, Am. Geophys. Union, Phi Beta Kappa, Sigma Xi, Tau Beta Pi. Subspecialties: Fluid mechanics; Ground water hydrology. Current work: Fluid mechanics; experimental studies and numerical simulation of turbulent stratified flows in nature, ground water hydrology; numerical simulation modeling of hazardous waste transport. Office: Dept Civil Engring Stanford Univ Stanford CA 94305

STREETEN, BARBARA ANNE WIARD, opthalmologist, educator, researcher; b. Candia, N.H., Mar. 3, 1925; d. Robert Campbell and Gertrude Sarah (Matheson) Wiard; m. David Henry Palmer Streeten, Aug. 2, 1952; children: Robert, Elizabeth, John. A.B., Tufts U., 1945, M.D., 1950. Diplomate: Am. Bd. Ophthalmology. Instr. U. Mich. Med. Sch., Ann Arbor, 1956-60; from asst. prof. to prof. ophthalmology and pathology Upstate Med. Ctr., Syracuse, N.Y., 1964—, dir. eye pathology lab., 1966—; mem. study sect. on vision Nat. Eye Inst., NIH, 1977-80, bd. sci. counselors, 1982-86. Editorial bd.: Investigative Ophthalmology and Visual Sci, 1979-82; editorial adv.: Opthalmology, 1982—; contbr. articles to sci. jours., chpts. to books. NIH grantee, 1975—. Fellow Am. Acad. Ophthalmology; mem. Am. Ophthal. Soc., Am. Assn. Ophthalmic Pathologists (charter), Verhoeff Soc. (pres.-elect), Eastern Ophthalmic Pathology Soc., Assn. Research in Vision and Ophthalmology (sect. chair 1976), Phi Beta Kappa, Alpha Omega Alpha. Episcopalian. Subspecialties: Ophthalmology; Pathology (medicine). Current work: Ultrastructure, immunology and biochemistry of extracellular matrix related to ophthalmic disease. Home: 334 Berkeley Dr Syracuse NY 13210 Office: 766 Irving Ave Syracuse NY 13210

STREGOWSKI, THOMAS JOHN, mech. engr., cons.; b. New Britain, Conn., Jan. 31, 1948; s. Stanley J. and Catherine (Holyst) S.; m. Elva G. Stregowski, Apr. 22, 1972; 1 son: Christopher. B.S.M.E., U. Conn., 1971, postgrad., 1972-77. Registered profl. engr., Conn. Advanced mfg. engr. Colt Firearms; mfg. engr. Arrow-Hart; engr. Supr. Plessey Ltd.; project engr. Union Carbide Corp., East Hartford, Conn., 1978—; cons. engring.; pres. Conn. Joint Fedn. Recipient Lyman Johnson Meml. award Am. Soc. Non-Destructive Testing, 1971. Mem. Nat. Soc. Profl. Engrs., Conn. Soc. Profl. Engrs., Soc. Am. Value Engrs. (pres. chpt. 1979-81), ASME. Democrat. Roman Catholic. Club: Timberlin Men's Golf. Subspecialties: Materials processing; Materials (engineering). Current work: Process capabilities of LLDPE versus LDPE resins and their relative strengths. Home: 189 Patterson Way Berlin CT 06037 Office: 88 Long Hill St East Hartford CT 06108

STREICHER, EUGENE, scientist, administrator; b. N.Y.C., Oct. 25, 1926; s. Samuel and Lucy (Carlin) S.; m. Janet Reid Hutcheson, Aug. 9, 1979; 1 son, William Leigh. B.A., Cornell U., 1947, M.A., 1948; Ph.D., U. Chgo., 1953. Physiologist med. div. Army Chem. Ctr., Edgewood, Md., 1948-50; physiologist, health scientist adminstrn. NIH, Bethesda, Md., 1954—, program dir. fundamental neuroscis., 1979—. Served with USN, 1945-46. Mem. Soc. for Neurosci., Am. Neuropath. Assn., AAAS, Soc. Exptl. Biology and Medicine. Subspecialties: Neurochemistry; Neurobiology. Office: NIH Federal Bldg Bethesda MD 20205

STRELKAUSKAS, ANTHONY JAMES, cellular immunologist; b. Newark, N.J., Apr. 25, 1944; s. Anthony and Stephanie (Nowselski) S.; m. Jennifer Dianne Snow, July 11, 1970; children: Jennifer E., Daniel A. B.A., U. Calif.-Riverside, 1971; M.S., U. Wyo., 1973; Ph.D., U. Ill., 1976. Postdoctoral fellow Harvard Med. Sch., 1976-78; research assoc. Farber Cancer Ctr., 1978-79; asst. prof. Med. U. S.C., Charleston, 1979-81, assoc. prof. immunology and microbiology, 1981—; assoc. prof. pediatrics Med. U.S.C., 1982—. Served with USMC, 1962-67. Mem. Am. Assn. Immunologists, Am. Soc. Microbiologists, Soc. Exptl. Biology and Medicine, AAAS. Subspecialties: Immunobiology and immunology; Genetics and genetic engineering (biology). Current work: Molecular mechanisms of cellular interactions and immunoregulation; construction of human hybridoma clones for dissection of human immune response thereby facilitating understanding of immunological diseases. Home: 26 27th Ave Isle of Palms SC 29451 Office: Med U SC 171 Ashley Ave Charleston SC 29425

STRICKLER, STEWART JEFFERY, chemist; b. Mussoorie, U.P., India, July 12, 1934; s. Herbert J. and Martha T. (Steward); m., June 27, 1959; children: Janet Carol, Peter Herbert. B.A., Coll. of Wooster, 1956; Ph.D., Fla. State U., 1961. Lectr. Rice U., Houston, 1962-63; asst. prof. U. Colo., Boulder, 1963-68, assoc. prof., 1968-73, prof., 1973, chmn. dept. chemistry, 1974-77; hon. fellow Australian Nat. U., Canberra, 1972-73. Contbr. articles to profl. jours. Mem. Am. Chem. Soc., Am. Phys. Soc., AAAS. Republican. Presbyterian. Subspecialties: Physical chemistry; Photochemistry. Current work: Research in molecular spectroscopy, quantum chemistry, photochemistry, solar energy conversion. Home: 1690 Wilson Ct Boulder CO 80302 Office: Chemistry Dept U Colo Boulder CO 80309

STRIEDER, WILLIAM CHRISTIAN, chemical engineering educator; b. Erie, Pa., Jan. 19, 1938; s. William A. and Virginia (Parmenter) S.; m. Joyce K. Panchot, Dec. 27, 1967; children: John, Katherine, Tracy, Joseph. B.S., Pa. State U., 1959; Ph.D., Case Inst. Tech., 1963. Postdoctoral fellow U. Brussels, Belgium, 1963-65, U. Minn.-Mpls., 1965-66; prof. chem. engring. U. Notre Dame, Ind., 1966—. Contbr. articles to profl. jours.; author: (with R. Aris) Variational Methods Applied to Problems of Diffusion and Reaction, 1973. Union Carbide fellow, 1962; Belgian Govt. fellow, 1964; NSF fellow, 1961; grantee, 1967-69, 1977-79, 80-83; Whirlpool grantee, 1977; AFOSR grantee, 1978-79; Dept. of Transp. grantee, 1980-83. Mem. Am. Phys. Soc., Am. Chem. Soc., Am. Inst. Chem. Engrs., Am. Soc. Engring. Edn, Sigma Xi. Roman Catholic. Subspecialties: Chemical engineering; Statistical mechanics. Current work: Thermodynamics porous media, heat and mass transport, statistical mechanics. Address: Dept Chem Engring U Notre Dame Notre Dame IN 46556

STRIKE, TERRY LEE, research scientist; b. Sacramento, July 11, 1947; s. Orlo Robert and Edith Emma (Smith) S.; m. Randon Lee Smith, Dec. 18, 1970; children: Charles Howard, Daniel Randon. B.A., Sacramento State Coll., 1970; M.S., Calif. State U., Sacramento, 1972; Ph.D., U. Calif.-Davis, 1978. Grad. asst. Calif. State U., Sacramento, 1970-72, lectr., 1972-73; research asst. U. Calif., Davis, 1973-78, postdoctoral scholar, 1978-79, staff research scientist, 1979—. Sr. warden Trinity Cathedral Ch., 1981-82. Mem. AAAS, Genetics Soc. Am., Sigma Xi. Democrat. Episcopalian. Subspecialties: Genetics and genetic engineering (biology); Cell biology. Current work: Genetic engineering—recombinant DNA. Home: 4914 Bowman Oaks Way Carmichael CA 95608 Office: Dept Genetics Univ Calif Davis CA 95616

STRINGFELLOW, DALE ALAN, microbiologist, virologist; b. Salt Lake City, Sept. 13, 1944; s. Paul Bennion and Jean (Barton) S.; m. Jean Racker, June 17, 1965; children: Jennifer, Wendy, Ashley. A.S., Weber State Coll., 1964; B.S. in Microbiology, U. Utah, 1967, M.S., 1970, Ph.D., 1972. Culture curator U. Utah, 1967-72, teaching asst., 1967-70, grad. research asst., 1968-72, instr. microbiology, 1972-73; Sr. research scientist Upjohn co., Kalamazoo, 1973-82, research head, cancer and virus research, 1979-82; dir. cancer research Bristol-Myers Corp., Syracuse, N.Y., 1982--; cons. in field. Contbr. articles to profl. publs.; editor: books, including Interferons and Interferon Inducers, 1983. NIH postdoctoral research fellow, 1972-73. Mem. Am. Assn. Cancer Research, AAAS, Am. Soc. Microbiology. Subspecialties: Virology (medicine); Cancer research (medicine). Current work: Antiviral and antireoplastic chemotherapy, interferon, regulation cell function, host def. mechanism, cellular communication, differentiation, transformation. Office: Cancer Research Bristol-Myers Corp Syracuse NY

STROBER, SAMUEL, immunologist, educator; b. N.Y.C., May 8, 1940; s. Julius and Lee (Lander) S.; m. Myra Hoffenberg, June 23, 1963; children: Jason, Elizabeth. A.B. in Liberal Arts, Columbia U., 1961; M.D. magna cum laude, Harvard U., 1966. Intern Mass. Gen. Hosp., Boston, 1966-67; resident in internal medicine Stanford (Calif.) U. Hosp., 1970-71; research fellow Peter Bent Brigham Hosp., Boston, 1962-63, 65-66, Oxford (Eng.) U., 1963-64; research assoc. Lab. Cell Biology, Nat. Cancer Inst., NIH, Bethesda, Md., 1967-70; instr. medicine Stanford U., 1971-72, asst. prof., 1972-78, assoc. prof. medicine, 1978-82, chief div. immunology, 1978—, prof. medicine, 1982—; investigator Howard Hughes Med. Inst., Miami, Fla., 1976-81. Assoc. editor: Jour. Immunology, 1981—, Transplantation, 1981—; contbr. articles to profl. jours. Served with USPHS, 1967-70. Recipient Leon Riznick Meml. Research prize Harvard U., 1966, Career Devel. award Nat. Insts. Allergy and Infectious Diseases, 1971-76. Mem. Am. Assn. Immunology, Am. Soc. Clin. Investigation, Am. Rheumatism Assn., Alpha Omega. Subspecialties: Immunobiology and immunology; Immunology (medicine). Current work: Immunobiology and immunology, transplantation, cellular engineering. Office: Stanford University School of Medicine Stanford CA 94305

STROM, JOEL ANDREW, cardiologist, educator; b. N.Y.C., Sept. 14, 1943; s. Irving and Sophie (Shreefter) S.; m. Jane Rosemarie Golin, June 18, 1967; children: Rebecca Nan, Jessica Marie. B.S. in Engring, Cornell U., 1965, M.S. in Engring. Physiology, 1966; M.D., SUNY, 1970. Diplomate: Am. Bd. Internal Medicine, 1973, Am. Bd. Cardiovascular Disease, 1977. Intern in medicine Upstate Med. Ctr., Syracuse, N.Y., 1970-71, resident, 1971-73; fellow in cardiology Cornell Med. Ctr., N.Y.C., 1973-74; research med. officer Brooks AFB, San Antonio, 1974-76; fellow in cardiology Albert Einstein Coll. Medicine, Bronx, N.Y., 1976-77, asst. prof. medicine and radiology, 1978—, dir. clin. cardiac labs., assoc. attending physician. Contbr. articles to sci. publs. Frieda and Herman Saporta cardiology fellow, 1980. Fellow N.Y. Cardiologic Soc.; mem. Am. Heart Assn., N.Y. Acad. Sci., AAAS, Am. Fedn. Clin. Research. Jewish. Subspecialties: Cardiology; Imaging technology. Current work: Cardiac ultrasound. Home: 24 Wildwood Dr Dix Hills NY 11746 Office: Hosp Albert Einstein College Medicine 1600 Tenbroeck Ave Bronx NY 10461

STROM, ROBERT GREGSON, planetary scientist; b. Long Beach, Calif., Oct. 1, 1933; s. John Irving and Ruth (Gregson) S.; m.; 1 son, Eric Gregson. B.S., U. Redlands, 1955; M.S., Stanford U., 1957. Geologist Standard-Vacuum Oil Co., White Plains, N.Y., 1957-60; asst. research geologist Space Scis. Lab., U. Calif., Berkeley, 1961-63; asst. prof. Lunar and Planetary Lab., U. Ariz., Tucson, 1963-72, assoc. prof. planetary scis. dept., 1972-81, prof., 1981—; cons. NASA, 1968—. Contbr. articles to profl. jours. NASA Research grantee, 1973—; recipient Group Achievement awards NASA, 1974, 81. Mem. Internat. Astron. Union, Am. Geophys. Union, Am. Astron. Soc. Democrat. Subspecialties: Planetary science; Planetology. Current work: Planetary geology, planetary geology research and planetary sci. edn. Home: 5331 Camino de la Cumbre Tucson AZ 85715 Office: Dept Planetary Sci U Ariz Tucson AZ 85721

STROM, TERRY BARTON, physician, educator; b. Chgo., Nov. 30, 1941; s. David and Sylvia D. (Abelson) S.; m. Margot Stern, Aug. 2, 1964; children: Adam Frederick, Rachel Fan. M.D., U. Ill., 1966. Diplomate: Am. Bd. Internal Medicine, 1970. Intern U. Ill. Hosp., Chgo., 1966-67; resident Beth Israel Hosp., Boston, 1970-71; instr. medicine Harvard U. Med. Sch., Boston, 1973-75, asst. prof., 1978, assoc. prof., 1978—; med. dir. renal transplant service, co-dir. lab. immunogenetics and transplantation Brigham and Women's Hosp., Boston, 1973—. Contbr. over 140 articles to sci. jours. Mem. Union Concerned Scientists; active local polit. campaigns. Served to capt. USAF, 1968-70. NIH awardee, 1976; also grantee. Mem. Am. Soc. Clin. Investigation, Transplantation Soc., Am. Assn. Immunologists, Am. Soc. Nephrology, Internat. Soc. Nephrology. Democrat. Jewish. Subspecialties: Cellular engineering; Transplantation. Current work: Immunology (cellular and transplantation). Home: 22 Kennard Rd Brookline MA 02146 Office: 75 Francis St Boston MA 02115

STROMBERG, BERT EDWIN, JR., parasitologist; b. Trenton, N.J., May 19, 1944; s. Bert Edwin and Ruth Evelyn (Eklund) S.; m. JoAnn Earling, Oct. 10, 1946; children: B. Erik, Kristin. B.A., Lafayette Coll., 1966; M.A., U. Mass., 1968; Ph.D., U. Pa., 1973. Instr. Trenton State Coll., 1968-70; NIH trainee U. Pa., 1971-73, asst. prof., 1973-79; assoc. prof. Coll. Vet. Medicine U. Minn., 1979—. Contbr. articles to profl. jours. Mem. Am. Soc. Parasitologists, Am. Assn. Vet. Parasitologists, Am. Assn. Immunologists. Subspecialties: Parasitology; Immunology (medicine). Current work: Evaluating the immune response of animals to the parasites which infect them. Home: 4219 Oakdale Ave S Edina MN 55416 Office: 1971 Commonwealth Ave Saint Paul MN 55108

STROMER, MARVIN HENRY, cell biologist, researcher, educator; b. Readlyn, Iowa, Sept. 1, 1936; s. Roy Henry and Luella (Klemp) S.; m. Shirley Louene Roepke, Apr. 3, 1960; 1 son, Craig. B.S. in Animal Sci. (Nat. Merit scholar, George Gund scholar), Iowa State U., Ames, 1959, 1959, Ph.D. in Cell Biology, 1966. With product devel. dept. George A. Hormel & Co., Austin, Minn., 1959-62; research in biochemistry Iowa State U., 1962-66, assoc. prof., 1968-76, prof., 1976—; postdoctoral fellow biochemistry Carnegie-Mellon U., 1966-68; vis. lectr. Tex. A&M U., 1971; vis. prof. U. Ariz., 1979-80; Humboldt fellow, vis. scientist Max Planck Inst. Med. Research, Heidelberg, W.Ger., 1974-75. Contbr. articles, revs. and abstracts to profl. jours., chpts. in books. Served with Army NG, 1959-65. Mem. Am. Soc. Cell Biology, Electron Microscopy Soc. Am., Biophys. Soc., Am. Heart Assn., Sigma Xi, Alpha Zeta. Republican. Lutheran. Subspecialties: Cell biology; Biochemistry (biology). Current work: Ultrastructure and biochemistry of striated and smooth muscle and other movement systems. Office: Muscle Biology Group Iowa State University Ames IA 50011

STRONG, DOUGLAS MICHAEL, researcher; b. Newport, Wash., Sept. 4, 1941; s. George and Dorothea (Rednour) S.; m. Geraldine O'Melveny, Jan. 30, 1965; children: Michael Phillip, David Richard, Patricia Anne. B.S., Gonzaga U., 1963; Ph.D., Med. Coll. Wis., 1973. Med. technologist Am. Soc. Clin. Pathology.; Supr. hematology Blood Bank U.S. Naval Hosp., Phila., 1965-67; clin. chemist Med. Tech. Sch., U.S. Naval Hosp., Gt. Lakes, Ill., 1967-69; chief clihn. immunology-histocompatability lab. Naval Med. Research Inst., Bethesda, Md., 1973-78, dep. chmn., 1978-79, head transplant research br., 1979-82; program mgr., fleet health care systems Naval Med. Research and Devel. Command, Bethesda, 1982—; asso. prof. Uniformed Services U., 1979—. Contbr. numerous articles to profl. jours. Soccer coach Wheaton Boys Club, 1973-78, Montgomery and Rockville United Soccer Clubs, 1978—. Served to comdr. USN, 1965—. Mem. Naval Med. Research Inst., Soc. Cryobiology (editorial bd. 1980—), Am. Soc. Clin. Histocompatability Testing, Assn. Immunology, Transplantation Soc., Am. Assn. Tissue Banks. Roman Catholic. Subspecialties: Transplantation; Marrow transplant. Current work: Cell and tissue antigens, hybridoma tech., genetic engring. Office: Code 45 Naval Med R & D Command Betheda MD 20814 Home: 8 Atwell Ct Potomac MD 20854

STRONG, KEITH TEMPLE, solar physicist; b. Torquay, Devonshire, Eng., Apr. 28, 1951; came to U.S., 1979; s. Clarence Hugh and Ivy and Rose (Foster) S.; m. Yvonne Margaret Parker, Sept. 22, 1979. B.Sc., Univ. Coll., London, 1973; Ph.D., Mullard Space Sci. Lab., Holmbury St. Mary, Eng., 1973-78. Sci. cons. Lockheed Palo Alto (Calif.) Research Lab., 1978; research asst. Mullard Space Sci. Lab., Holmburg St. Mary, Eng., 1978-79; research scientist Lockheed Palo Alto (Calif.) Research Lab., 1979—. Contbr. articles to profl. jours. Fellow Royal Astron. Soc.; mem. Am. Astron. Soc., Hampshire Astron. Group, Nat. Coalition to Ban Handguna. Subspecialty: Solar physics. Current work: Study of X-ray emission from solar corona, sun. corona, X-ray spectroscopy. Home: 682 16th Ave Menlo Park CA 94025 Office: 3251 Hanover St 52-12/255 Palo Alto CA 94304

STROSBERG, ARTHUR MARTIN, cardiovascular pharmacologist; b. Albany, N.Y., Sept. 16, 1940; s. Benjamin and Dorothy (Pearlman) S.; m. Sheila Landan, Aug. 26, 1973; children: Darryl, Mark. B.S. cum laude, Siena Coll., 1962; Ph.D., U. Calif., San Francisco, 1970. Pharmacologist Syntex Research, Palo Alto, Calif., 1970-71, sect. head cardiovascular pharmacology, 1972-75, prin. scientist, 1975—. Contbr. articles to profl. jours. NIH fellow, 1968-70. Mem. Am. Soc. Pharmacology and Exptl. Therapeutics, Western Pharmacology Soc., Am. Heart Assn., AAAS, Sigma Xi. Democrat. Jewish. Subspecialty: Pharmacology. Current work: Hypertension, heart failue, angina, arrhythmias; design and cardiovascular eval. antihypertensive drugs, cardiotonic agts. Patentee in field. Office: 3401 Hillview Ave Palo Alto CA 94304

STROUD, ROBERT MICHAEL, biochemistry educator, researcher. B.A., U. Cambridge, Eng., 1964, M.A., 1968; M.S., U. London, 1965, Ph.D., 1968. Postdoctoral fellOw Calif. Inst. Tech., Pasadena, 1968-71, asst. prof. chemistry, 1971-75, assoc. prof., 1975-77; assoc. prof. biochemisry U. Calif.-San Francisco, 1977-79, prof., 1979—; cons. NIH, 1977—, NSF, 1971—. Contbr. chpts. to books, articles to sci. jours. State scholar, 1961-64; Moulton scholar, 1965-68; Sloan fellow, 1973-75; recipient Career award NIH, 1972-77. Mem. N.Y. Acad. Sci., Biophys. Soc., Brit. Biophys. Soc. Subspecialties: Biophysical chemistry; Membrane biology. Current work: Membrane, structure and function; structural basis of biological function; protein folding; protein structure and function; x ray crystallography; genetic engineering to make new proteins. Office: S-960 Biochemistry Dept U Calif San Francisco CA 94143

STRUCK, ROBERT FREDERICK, medicinal chemist, pharmacologist, cons.; b. Pensacola, Fla., Jan. 9, 1932; s. Carl Herman and Hilda (Ropke) S.; m. Ruby Richardson, June 8, 1963; children: Lesley Dianne, Bert Richardson. B.S., Auburn U., 1953, M.S., 1957, Ph.D., 1961. Assoc. chemist So. Research Inst., Birmingham, Ala., 1957-58, research chemist, 1961-64, sr. chemist, 1964-80, head metabolism sect., 1980—; organic chemist U.S. Dept. Agr., Winter Haven, Fla., 1961; cons. Nat. Cancer Inst., 1976—. Contbr. numerous articles to sci. jours., also chpts. to books on metabolism of anticancer drugs and medicinal chemistry of anticancer drugs. Served as lt. USAF, 1954-56. Nat. Cancer Inst. grantee, 1979—. Mem. Am. Chem. Soc., Am. Assn. Cancer Research, Am. Soc. Pharmacology and Exptl. Therapeutics, Gideons Internat. Democrat. Presbyterian. Subspecialties: Cancer research (medicine); Medicinal chemistry. Current work: Medicinal chemistry, drug metabolism, pharmacology, and pharmacokinetics of anticancer drugs and carcinogens; mechanism of action of anticancer drugs and carcinogens. Home: 3533 Laurel View Ln Birmingham AL 35216 Office: 2000 9th Ave S Birmingham AL 35255

STRUNK, ROBERT CHARLES, pediatrician, educator; b. Evanston, Ill., May 29, 1942; s. Norman Wesley and Marion Mildred (Ree) S.; m. Alison Leigh Gans, Apr. 3, 1971; children: Christopher Robert, Alix Elizabeth. B.A., Northwestern U., 1964, M.D., 1968, M.S., 1968. Diplomate: Am. Bd. Pediatrics, Am. Bd. Allergy and Immunology. Intern Cin. Children's Hosp., 1968-69, resident, 1969-70; pediatrician Newport (R.I.) Naval Hosp., 1970-72; research fellow in pediatrics Harvard Med. Sch., Boston, 1972-74; fellow in medicine, immunology and allergy Children's Hosp. Med. Ctr., Boston, 1972-74; asst. prof. pediatrics Ariz. Health Scis. Ctr., 1974-78; assoc. prof. pediatrics Nat. Jewish Hosp. and Research Ctr. and U. Colo. Med. Ctr., 1978—, dir. clin. services, dept. pediatrics, 1979—, vice chmn. pediatrics, 1982—. Contbr. articles to profl. jours. Served to lt. USN, 1970-72. Recipient Borden Research award, 1968. Mem. Soc. for Pediatric Research, Am. Assn. Immunologists, Western Soc. for Pediatric Research. Democrat. Subspecialties: Pediatrics; Allergy. Current work: Role of complement in inflammation—regulation of complement synthesis by mononuclear phagocytes, clinical allergy, asthma. Office: 3800 E Colfax Ave Denver CO 80206

STUART, ANN ELIZABETH, neurophysiologist, educator; b. Harrisburg, Pa., Oct. 5, 1943; s. James Allan and Grace Sellers (Snyder) S.; m. John W. Moore, Apr. 2, 1978. B.A., Swarthmore Coll., 1965; Ph.D., Yale U., 1969. Research fellow in neurobiology Harvard Med. Sch., 1969-71; research fellow UCLA, 1971-73; asst. prof. neurobiology Harvard Med. Sch., Boston, 1973-78, assoc. prof., 1978; assoc. prof. physiology U. N.C., Chapel Hill, 1978—; mem. visual disorders Study sect. NIH, 1980—. Contbr. articles to profl. jours. Trustee Marine Biol. Labs., Woods Hole, Mass., 1979—. Recipient Nat. Eye Inst. research career devel. award, 1973-78; R.J. Reynolds jr. faculty devel. award U. N.C., 1980. Mem. Soc. Neurosci. (councillor N.C.), Assn. Research in Vision and Ophthalmology, Soc. Gen. Physiologists, Am. Women in Sci. Subspecialties: Neurophysiology; Neurobiology. Current work: Processing of sensory info. by single nerve cells in small nervous systems; vision in invertebrates; integration; info. processing by interneurons. Home: 1818 N Lakeshore Dr Chapel Hill NC 27514 Office: Dept Physiology U NC Chapel Hill NC 27514

STUART, ROBERT KENNETH, physician, scientist, educator; b. Baton Rouge, July 6, 1948; s. Walter Bynum, III and Rita Bess (Kleinpeter) S.; m. Gail Elaine Wiscarz, June 12, 1971; children: Robert Morgan, Elaine Catherine. B.S., Georgetown U., 1970; M.D., Johns Hopkins U., 1974. Diplomate: Am. Bd. Internal Medicine. Resident in medicine Johns Hopkins U. Sch. Medicine, Balt., 1974-76, fellow in medicine and oncology, 1976-79, asst. prof. medicine and oncology, 1979—; vis. research fellow Meml. Sloan Kettering Cancer Ctr., N.Y.C., 1978-79; jr. faculty clin. fellow Am. Cancer Soc., 1979; spl. fellow Leukemia Soc. Am., 1980-81. Mem. Internat. Soc. Exptl. Hematology, Am. Soc. Hematology, Am. Fedn. Clin. Research, Am. Soc. Clin. Oncology, Am. Assn. Cancer Research. Democrat. Roman Catholic. Subspecialties: Marrow transplant; Oncology. Current work: Experimental hematology, bone marrow transplantation, aplastic anemia research and treatment, clinical nutrition. Office: Johns Hopkins Oncology Ctr 600 N Wolfe St Baltimore MD 21205

STUART, WILLIAM DORSEY, JR., geneticist, educator; b. St. Louis, Mar. 28, 1939; s. William Dorsey and Alice Margueritte (Holleman) S. B.A. in Chemistry, Fla. State U., 1969, M.S. in Genetics; Univ. fellow, 1970; Ph.D. in Biochem. Genetics, 1973. Technologist Virology Lab., Mt. Sinai Hosp., N.Y.C., 1960-64; NIH predoctoral trainee Fla. State U., 1970-73; NIH postdoctoral fellow dept. biology Stanford (Calif.) U., 1973-76; research fellow in human genetics depts. pediatrics and genetics U. Melbourne, Australia, 1976-78; asst. prof. genetics Sch. Medicine, U. Hawaii, Honolulu, 1978—; cons. CETUS Corp., Berkeley, Calif., 1980—, Biotech. Group, Honolulu, 1982—. Contbr. articles to sci. jours. Mem. Genetics Soc. Am., Am. Soc. Microbiology, Genetics Soc. Australia, Sigma Xi, Phi Eta Sigma. Subspecialties: Genetics and genetic engineering (biology); Immunocytochemistry. Current work: Human biochemical and cytological genetics. Office: Dept Genetics U Hawaii at Manoa Honolulu HI 96822 Home: 1002-A Prospect Apt 29 Honolulu HI

STUCKY, GARY LEE, chemist, educator; b. Murdock, Kans., May 18, 1941; s. Martin and Ruby (Stucky) S. A.B., Bethel Coll., Kans., 1963; Ph.D., Kans. State U., 1967. With Miles Labs., Elkhart, Ind., 1967-70, Kivuvu Labs., Kimpere, Zaire, 1971; mem. faculty Eastern Mennonite Coll., Harrisonburg, Va., 1972—, prof. chemistry, 1979—; vis. prof. U. Rochester, N.Y., 1981-82. Mem. Am. Chem. Soc., Chem. Soc. London, AAAS. Mennonite. Subspecialties: Inorganic chemistry; Physical chemistry. Current work: Bioinorganic chemistry; metallo proteins; coordination chemistry. Home: PO Box 1053 Harrisonburg VA 22801 Office: Dept Chemistry Eastern Mennonite Coll Harrisonburg VA 22801

STUDY, ROBERT EDWARD, JR., research neurophysiologist; b. Tokyo, Japan, Aug. 9, 1949; s. Robert Edward and Barbara June (Smith) S. B.S., Union (N.Y.) Coll., 1971; Ph.D., Yale U., 1978. Research assoc. Yale U. Sch. Medicine, 1978; pharmacology research assoc. Nat. Inst. Gen. Med. Scis., NIH, Bethesda, Md., 1978-80; staff fellow Nat. Inst. Neurol. and Communicative Disorders and Stroke, 1980—. Contbr. articles to profl. jours. Mem. Soc. for Neurosci., Common Cause, ACLU. Subspecialties: Neuropharmacology; Neurophysiology. Current work: Mechanism of action of drugs on central nervous system. Cellular physiology/anticonvulsant drugs/neurotransmission/epilepsy sedative drugs/neurophysiology/biophysics.

STUESSY, TOD FALOR, plant systematist; b. Pitts., Nov. 18, 1943; s. Haydn and Mary Louise (Falor) S.; m. Carol Liebe, June 15, 1968 (div.); children: Mary Elizabeth, Alan Briscoe. B.A., DePauw U., 1965; Ph.D., U. Tex., Austin, 1968. Asst. prof. botany Ohio State U., 1968-71, 72-74, assoc. prof., 1974-78, prof., 1979—; Maria Moors Cabot postdoctoral fellow Harvard U., 1971-72; assoc. dir. systematic Biology Program, NSF, Washington, 1978-79; dir. Ohio State U. Herbarium, 1980—. Contbr. numerous articles on plant systematics to profl. jours. NSF grantee, 74-82; Sigma Xi grantee, 1968; Nat. Geog. Soc. grantee, 1975; Am. Philos. Soc. grantee, 1969. Mem. AAAS, Am. Soc. Plant Taxonomy, Assn. Tropical Biology, Bot. Soc. Am., Calif. Bot. Soc., Classification Soc., Internat. Assn. Plant Taxonomy, Soc. Study Evolution, Southwestern Assn. Naturalists, Soc. Systematic Zoology. Methodist. Subspecialty: Systematics. Current work: Systematic and evolutionary studies of flowering plants; compositae, herbarium, Latin Am., Juan Fernandez Islands, cladistics, Chile. Office: Dept Botany Ohio State U 1735 Neil Ave Columbus OH 43210

STUMPE, WARREN ROBERT, manufacturing company executive; b. Bronx, N.Y., July 15, 1925; s. William A. and Emma J. (Mann) S.; m. Jean Marie Mannion, June 5, 1952; children: Jeffrey, Kathy, William. B.S., U.S. Mil. Acad., 1945; M.S., Cornell U., 1949, N.Y. U., 1965. Registered profl. engr., N.Y., Fla., Wis. Commd. 2d lt., C.E. U.S. Army, 1945, advanced through grades to capt., 1954; with (65th Engr. Bn.), 1945-48; asst. prof. mechanics U.S. Mil. Acad., 1951-54; resigned, 1954; from capt. to col. Res., 1958-79; dep. gen. mgr., gen. engring. div. AMF, Stamford, Conn., 1954-63; exec. v.p. Dortech, Inc., Stamford, 1963-69; dir. systems mgmt. group Mathews Conveyor div. REX, Darien, Conn., 1969-71; dir. research and devel. Rexnord, Inc., Milw., 1971-73, v.p. corp. research and tech., 1973—, v.p. bus. devel. sector, 1981—; civilian aide to sec. army for, State of Wis. Contbr. articles to profl. jours. Founder, pres. No. Little League, Stamford, 1965-69; pres. Turn of River Jr. High Sh. PTA, Stamford, 1967-68; vice chmn. for Wis. Dept. Def., Nat. Com. Employer Support Guard and Res.; bd. regents Milw. Sch. Engring.; mem. liaison council Coll. Engring., U. Wis., also mem. indsl. adv. council; mem. Wis. Gov.'s Task Force on Energy. Mem. Am. Water Pollution Control Fedn., Process Equipment Mfrs. Assn., Indsl. Research Inst. (v.p., dir.), West Point Soc. N.Y. (career adv. bd.), West Point Soc. Wis., Tau Beta Pi, Phi Kappa Phi. Clubs: Wisconsin, Ozaukee Country. Subspecialties: Mechanical engineering; Systems engineering. Current work: Responsible for all company-wide technology activities which includes bringing to marketplace a product line of imaging recognition equipment. Office: PO Box 2022 Milwaukee WI 53201

STUMPF, WALTER ERICH, anatomist, pharmacologist, educator; b. Oelsnitz/Vogtland, Ger., Jan. 10, 1927; came to U.S., 1963; m. Ursula Emily Schwinge, May 20, 1961; children: Andrea, Martin, Carolin, Silva. Student, U. Leipzig, Germany, 1946-50; M.D. summa cum laude, Humboldt U., Berlin, 1952; Ph.D. in Pharmacology, U. Chgo., 1967. Intern Charite Hosp., Humboldt U., 1952-53, resident in neurology and psychiatry, 1954-57; trainee in psychotherapy and psychoanalysis Inst. Psychotherapy, Berlin, 1954-56; teaching asst. in clin. neurology Humboldt U., 1956-57; resident dept. neurology and psychiatry U. Marburg, W.Ger., 1958-60, Lab. Isotope Research and Radiobiology, U. Marburg, 1961-62; research assoc. dept. pharmacology U. Chgo., 1963-67, asst. prof. pharmacology, 1963-64, 67-70; assoc. prof. anatomy and pharmacology U. N.C., Chapel Hill, 1970-73, prof., 1973—; vis. psychiatrist Maudsley Hosp., London, 1959; vis. prof. Max-Planck Inst. Cell Biology, Wilhelmshaven, Germany, 1975; vis. prof. clin. morphology U. Ulm, Ger., 1981; research scientist, child devel. Biol. Research Center, N.C. Meml. Hosp.; assoc. mem. Carolina Population Center, Chapel Hill; cons. in field; mem. council, exec. com. Inst. Lab. Animal Resources, NRC, Nat. Acad. Scis., 1978-81. Editorial bd.: Jour. Histochemistry and Cytochemistry; coor. editor: Cell and Tissue Research; Contbr. numerous articles chpts. to profl. jours. Berlin Labs. fellow, 1968-69. Mem. AAAS, Am. Assn. Anatomists, N.Y. Acad. Scis., Soc. Exptl. Biology and Medicine, Internat. Brain Research Orgn., Am. Soc. Zoologists, Histochem. Soc. (council 1977-81), Internat. Soc. Psychoneuroendocrinology, Am. Chem. Soc., Soc. Neurosci., Internat. Soc. Xenobiotics (charter). Subspecialties: Neuroendocrinology; Cytology and histology. Current work: Steroid hormone receptors; brain; pituitary; autoradiorgraphy; immunohistochemistry; peptide hormones. Design cryosption pump, Harris wide range cryostat. Home: Route 5 Box 380 Chapel Hill NC 27514 Office: U NC Dept Anatomy 111 Swing Bldg Chapel Hill NC 27514

STURDEVANT, EUGENE J., elec. engr.; b. Newton, Kans., Dec. 27, 1930; s. Jesse Jackson and Lillian (David) S.; m. Ruth Jane Moore, Jan. 4, 1958 (dec.); children: Eugene J., Tiffany Mark. B.S.E.E., U. Calif., Berkeley, 1963. Registered profl. engr., Del. Sr. research technician Lawrence Berkeley (Calif.) Lab., 1961-63; research engr. Engring.-Physics Lab., E.I. DuPont de Nemours, Inc., Wilmington, Del., 1963-68; devel. engr. Holotron Corp., Newark, Del., 1968-71; cons. optics and electronics, Wilmington, Del., 1971-76; advance research engr. SCM Corp., Proctor-Silex Group, King of Prussia, Pa., 1976-80, sr. scientist, 1980—. Served to capt. USAF, 1951-58. Mem. Optical Soc. Am., Soc. Photo-Optical Instrumentation Engrs., Nat. Soc. Profl. Engrs. Subspecialties: Electrohydrodynamics in heat transfer; Electronics. Current work: Heat and mass transfer with corona discharges and electropydrodynamics, investigation in corona and EHD effects on heat and mass transfer; development and design of electronic and optical systems used in heat-mass transfer experiments. Patentee in field. Home: 2306 Centerville Rd Wilmington DE 19808 Office: 1016 W 9th Ave King of Prussia PA 19406

STURGES, FRANKLIN WRIGHT, biology educator; b. Santa Monica, Calif., June 19, 1930; s. Robert W. and Merl (McClees) S.; m. Patricia M. Patterson, June 15, 1952; children: Sheryl, Karen. B.A., San Jose State U., 1952; M.A., Oreg. State U., 1955, Ph.D., 1957. Instr. to assoc. prof. So. Oreg. State Coll., 1957-66; prof., chmn. dept. biology Beaver Coll., Glenside, Pa., 1966-72; prof., chmn. div. sci. and math. Shepherd Coll., Shepherdstown, W.Va., 1972—. Mem. Ecol. Soc. Am., Am. Ornithologists Union, Am. Soc. Mammalogists, Animal Behavior Soc. Subspecialties: Behavioral ecology; Population biology. Current work: Avian population and community ecology and behavior. Home: RD 1 Box 722 Shepherdstown WV 25443 Office: Shepherd Coll Shepherdstown WV 25443

STURM, JAMES EDWARD, chemistry educator, researcher; b. New Ulm, Minn., Mar. 28, 1930; s. Bernard Joseph and Magdalene Jeanette (Forerster) S.; m. Margaret Ruth Adams, Apr. 30, 1955; children: Johanna, Karl, Madeleine, Eric, Gretchen, Hans, Kristen. B.S., St. John's U., Collegeville, Minn., 1951; Ph.D., U. Notre Dame, 1957. Part-time instr. St. Mary's Coll., South Bend, Ind., 1954; research assoc. U. Wis., 1955-56; asst. prof. Lehigh U., Bethlehem, Pa., 1956-62, assoc. prof., 1962-72, prof. dept. chemistry, 1972—; vis. research assoc. Brookhaven Nat. Lab., summer 1957, Argonne Nat. Lab., summer 1958; cons. Edgewood Arsenal, summers 1968, 69. Bicentennial Com. Lower Saucon Twp., 1974-76. Mem. Am. Chem. Soc. (petroleum research fund 1958-60), Faraday Soc., Sigma Xi. Democrat. Roman Catholic. Subspecialties: Photochemistry; Kinetics. Current work: Photochemical kinetics; reactions of high-velocity atoms; vacuum ultraviolet photochemistry. Home: 1343 Wassergass Rd Hellertown PA 18055 Office: Lehigh Univ Dept Chemistry #6 Bethlehem PA 18015

STURTEVANT, FRANK MILTON, pharm. co. exec.; b. Evanston, Ill., Mar. 8, 1927; s. Frank Milton and Marguerite Marie (Walsh) S.; m. Ruthann Patterson, Mar. 18, 1950; children: Jill Diane Sturtevant Rovani, Jan Kimberly Sturtevant Cassidy. B.A. cum laude, Lake Forest Coll., 1948; M.S. (fellow), Northwestern U., 1950, Ph.D., 1951. Grad. asst. in biol. scis. Northwestern U., Evanston, 1949-51; sr. investigator div. biol. research G.D. Searle & Co., Chgo., 1951-58, asso. dir. research and devel., 1972-80; dir. Office Sci. Affairs, 1980—; sr. pharmacologist Research and Devel. div. Smith, Kline and French Labs., Phila., 1958-60; sr. research fellow in biochemorphology Mead Johnson Research Center, Evansville, Ind., 1960-64, dir. dept. neuro- and psychopharmacology, 1963-64, dir. sci. and regulatory affairs, 1963-72, with govtl. liaison dept, 1964-72; lectr. genetics U. Evansville, 1972. Cons. editor: Internat. Jour. Chronobiology; contbr. over 100 articles to profl. jours. Bd. govs. Lake Forest Coll. Served with U.S. Army, 1945-46. Fellow AAAS; mem. Am. Soc. Pharmacology and Exptl. Therapeutics, Soc. Exptl. Biology and Medicine, Am. Fertility Soc., Pacific Coast Fertility Soc., Internat. Soc. Reproductive Medicine, Drug Info. Assn. (charter), Internat. Soc. Chronobiology, Am. Coll. Toxicology (charter), Internat. Union Pharmacology (sect. toxicology), N.Y. Acad. Scis., Sigma Xi. Subspecialties: Reproductive biology (medicine); Chronobiology. Current work: Biometrics, chronopharmacokinetics. Home: 1868 Mission Hills Ln Northbrook IL 60062 Office: GD Searle & Co 4901 Searle Pkwy Skokie IL 60077

STUTEVILLE, DONALD LEE, plant pathologist, educator; b. Okeene, Okla., Sept. 7, 1930; s. Martin Harry and Mayme Lee (Herriman) S.; m. Lorene Julaine Dringenberg, Sept. 7, 1952; children: Susan Ann Stuteville Carlson, Donald Brian, Robert Vincent. B.S., Kans. State U., 1959, M.S., 1961; Ph.D., U Wis., 1964. Plant pathologist Kans. State U., Manhattan, 1964—, prof., 1979—. Served with U.S. Army, 1953-55. Mem. Am. Phytopath. Soc., Sigma Xi, Phi Kappa Phi, Alpha Zeta, Gamma Sigma Delta. Mem. Disciples of Christ Ch.. Subspecialty: Plant pathology. Current work: Forage crop pathology, particularly alfalfa. Home: 2006 Stillman Dr Manhattan KS 66502 Office: Dept Plant Pathology Kans State U Manhattan KS 66506

STWALLEY, WILLIAM CALVIN, chemistry and physics educator; b. Glendale, Calif., Oct. 7, 1942; s. Calvin Murdoch and Diette Clarice (Hanson) S.; m. Mauricette Frisius, June 14, 1963; children: Kenneth William, Steven Edward. B.S., Calif. Inst. Tech., 1964; Ph.D., Harvard U., 1969. Asst. prof. chemistry U. Iowa, Iowa City, 1968-72, assoc. prof., 1972-75, prof., 1975—, prof. physics, 1977—, dir. laser facility, 1979—. Editor: Metal Bonding and Interactions in High Temperature Systems, 1982. Leeds and Northrop Found. fellow, 1964-65; NSF fellow, 1965-68; Alfred P. Sloan Found. fellow, 1970-72; Japan Soc. Promotion Sci. fellow, 1982. Mem. Am. Chem. Soc., Am. Phys. Soc.; fellow AAAS. Democrat. Subspecialties: Spectroscopy; Atomic and molecular physics. Current work: Atomic interactions, optically pumped lasers, alkali metal vapors, spin-polarized atoms, laser-produced plasmas, long-range molecules, energy transfer and ionization processes. Home: 141 Green Mountain Dr Iowa City IA 52240 Office: Dept Chemistry U Iowa Iowa City IA 52242

SU, CHE, pharmacologist; b. Taipei, Taiwan, June 12, 1932; came to U.S., 1961, naturalized, 1971. Ph.D., UCLA, 1965. Asst. to assoc. prof. UCLA, 1967-78; prof. So. Ill. U. Sch. Medicine, Springfield, 1978—. Mem. Am. Pharmacol. Soc. Subspecialty: Pharmacology. Current work: Presynaptic receptors in hypertension; purinergic mechanisms in blood vessels. Office: So Ill U Sch Medicine 801 N Rutledge Springfield IL 62702 Home: 66 W Hazel Dell Ln Springfield IL 62707

SUBBIAH, RAVI MANDEPANDA THIMMIAH, biochemistry educator, researcher; b. Mercara, Mysore, India, June 30, 1942; came to U.S., 1970; s. Mandepanda C. and Bolamma (Kanda) Thimmiah; m. Deachu Muddayya, Jan. 15, 1972; children: Jeevan, Rekha. B.S., U. Mysore, Mercara, India, 1961; M.S., U. Baroda, India, 1964; Ph.D., U. Toronto, 1970. Research fellow Mayo Clinic, Rochester, Minn., 1970-72, research assoc., 1972-74, cons., 1974-78, asst. prof., 1975-78; assoc. prof. U. Cin. Med. tr., 1978-81, prof. exptl. medicine, 1981—, dir. lipid biochemistry, 1978—, dir. lipids-nutrition tng. program, 1980—. Editorial bd.: Proc. Soc. Exptl. Biology and Medicine, Atherogenese, 1983—; contbr. articles to profl. jours. Recipient First prize Govt. Coll., Mercara, India, 1959-61; NIH research grantee, 1974-83. Fellow Council Arteriosclerosis; mem. Am. Inst. Nutrition, Am. Chem. Soc., Am. Oil Chemists Soc., Soc. Exptl. Biology and Medicine. Subspecialty: Biochemistry (medicine). Current work: Atherosclerosis, cholesterol metabolism, bileacids, prostaglandins. Home: 4036 Ridgedale Dr Cincinnati OH 45247 Office: U Cin Med Center 234 Goodman St Cincinnati OH 45267

SUBUDHI, MANOMOHAN, mech. engr.; b. Daspalla, India, Sept. 27, 1946; s. Brindaban and Paluni (Sahoo) S.; m. Shantilata Subudhi, July 2, 1971; children: Sumit, Mili. B.SC., Banaras Hindu U., 1969; S.M., M.I.T., 1971; Ph.D., Ploy. Inst. N.Y., 1974. Sr. stress analyst Nuclear Power Services, N.Y.C., 1973-75; sr. engr. Bechtel Power Corp., Gaithersburg, Md., 1975-76; assoc. engr. Brookhaven Nat. Lab., Upton, N.Y., 1976-79, engr., 1979—; mem. adj. faculty various univs.; cons. Mem. ASME (Assoc.), Sigma Xi. Subspecialties: Solid mechanics; Theoretical and applied mechanics. Current work: Finite elements, piping stress, solid mechanics, equipment qualification, seismic design of structures. Home: 4 Somerset In East Setauket NY 11733 Office: Brookhaven Nat Lab Bldg 129 Upton NY 11973

SUCHORA, DANIEL HENRY, engineer, educator; b. Youngstown, Ohio, Dec. 2, 1945; s. Stanley and Stella (Tocicki) S.; m. Patricia Ann Sanders, Sept. 2, 1968; children: Kevin, Sherri, Matthew. B.E. in Mech. Engring, Youngstown State U., 1968, M.S., 1970, Ph.D., Case Western Res. U., Cleve., 1973. Registered profl. engr., Ohio. Project engr. Ajax Magnethermic Corp., Warren, Ohio, 1968-70; mem. faculty chmn. dept. engring. Central Ohio Tech. Coll., Newark, 1973-75; assoc. prof. mech. engring. Youngstown State U.; cons. in field. Contbr. articles and papers to profl. jours., 1975—. Bd. dirs., coach Boardman Little League. Recipient Outstanding Jr. in Mech. Engring. at Youngstown State U. award Dow Chem. Corp., 1967; Outstanding Mech. Engring. Grad. at Youngstown State U. award ASME, 1968; Henry Roemer prize in mech. engring., 1968; NSF fellow, 1970. Mem. ASME, Soc. for Exptl. Stress Analysis, Scientific Research Soc. N.Am., Eastern Ohio Conservation Club. Roman Catholic. Subspecialties: Mechanical engineering; Theoretical and applied mechanics. Current work: Kinematics, engineering education and mechanical engineering consulting in stress analysis, kinematics and computer aided design. Home: 617 Oakridge Dr Boardman Ohio 44512 Office: Dept Mechanical Engring Youngstown State Univ 410 Wick Ave Youngstown OH 44555

SUCIU, SPIRIDON N., electric co. ofcl.; b. Flint, Mich., Dec. 11, 1921; s. Nicholas and Mary (Moian) S.; m. Jean E. Suciu, Aug. 27, 1949; children: Nancy Susan, Barbara Jean, Richard Spiridon, James

Nicholas, Ronald Edward. B.S.M.E., Purdue U., 1944, M.S.M.E., 1949, Ph.D., 1951. Registered profl. engr., Ohio. Mgr. applied research and aerodynamic design Aircraft Engine Bus. div. Gen. Electric Co., Cin., 1958-67; mgr. design tech. operation Aircraft Engine div., Cin., 1967-71; gen. mgr. GTPD Engring. dept., Gas Turbine Products div., Schenectady, N.Y., 1971-76; mgr. energy tech. operation Energy Systems and Tech. div., Schenectady, 1976-78; gen. mgr. Neutron Devices dept. Aerospace Bus. Group, St. Petersburg, Fla., 1978—; chmn, air breathing propulsion com. NASA; mem. Air Force Sci. Adv. Bd. Mem. Suncoast Chamber Adv. Council, Com. of 100. Served to ensign USNR, 1944. Recipient Akroyd Stuart award Royal Aero. Sco., 1972. Mem. ASME, AIAA, Am. Mgmt. Assn. Club: Eastlake Woodlands Golf and Racquet. Subspecialties: Mechanical engineering; Electrical engineering. Current work: Management of high technology businesses, general manager research, development, manufacturing advanced electro-mechanical devices. Research on aircraft engine design, heavy duty gas turbine design, properties and chem. composition of cumbustion products, residual fuels, solubility of gases.

SUDAN, RAVINDRA NATH, physicist, educator; b. Chineni, Kashmir, India, June 8, 1931; came to U.S., 1958, naturalized, 1971; s. Brahm Nath and Shanti Devi (Mehta) S.; m. Dipali Ray, July 3, 1959; children: Rajani, Ranjeet. B.A. with first class honors, U. Punjab, 1948; diploma, Indian Inst. Sci., 1952, Imperial Coll., London, 1953; Ph.D., U. London, 1955. Engr., Brit. Thomson-Houston Co., Rugby, Eng, 1955-57; Engr. Imperial Chem. Industries, Calcutta, India, 1957-58; research asso. Cornell U., Ithaca, N.Y., 1958-59, asst. prof. elec. engring., 1959-63, asso. prof., 1963-68, prof., 1968-75, IBM prof. engring., 1975—, dir. Lab. Plasma Studies, 1975—; cons. Lawrence Livermore Lab., Los Alamos Sci. Lab., Sci. Applications Inc., Physics Internat. Co.; vis. research asso. Stanford U., summer 1963; cons. U.K. Atomic Energy Authority, Culham Lab., summer 1965; vis. scientist Internat. Center Theoretical Physics, Trieste, Italy, 1965-66, summers 1970, 73, Plasma Physics Lab. Princeton U., 1966-67, Inst. for Advanced Study, Princeton, N.J., spring 1975; head theoretical plasma physics group U.S. Naval Research Lab., 1970-71, sci. adviser to dir., 1974-75; chmn. Ann. Conf. on Theoretical Aspects of Controlled Fusion, 1975, 2d Internat. Conf. on High Power Electron and Ion Beam Research and Tech., 1977. Mem. editorial bd.: Nuclear Fusion, 1976—; mem. editorial bd.: Physics of Fluids, 1973-76, Comments on Plasma Physics, 1973—; co-editor: Handbook of Plasma Physics; contbr. over 100 articles to sci. jours. Fellow Am. Phys. Soc., IEEE; mem. Sigma Xi. Subspecialties: Plasma physics; Nuclear fusion. Current work: Physics and technology of high powered charged particle beams; high energy particle accelerators; plasma confinement for nuclear fusion and spaceplasma physics. Patentee (with S. Humphries, Jr) intense ion beam generator. Office: 290 Grumman Hall Cornell Univ Ithaca NY 14853

SUDDATH, FRED LEROY, JR., biochemist; b. Macon, Ga., May 6, 1942; s. Fred L. and Betty Rose (Parker) S.; m. Lee Gafford, Sept. 4, 1965; children: Brantley V., Suzanne V. B.S. in Chemistry, Ga. Inst. Tech., 1965, Ph.D., 1970. NIH fellow M.I.T., 1970-72, Am. Cancer Soc. fellow, 1972-75; asst. prof. biochemistry U. Ala., 1975-78, assoc. prof., 1978—. Mem. AAAS, Am. Crystallographic Soc., Biophys. Soc., Sigma Xi. Subspecialties: Biophysics (biology); Biochemistry (medicine). Current work: X-ray crystallography of proteins, nucleic acids rapid data collection methods, instrumentation, protein structure/function, computer graphics. Office: University Alabama Birmingham AL 35294

SUDIA, THEODORE WILLIAM, govt. sci. exec.; b. Ambridge, Pa., Oct. 10, 1925; s. Frank and Paraskevia (Storowska) S.; m. Cecelia Elson, Nov. 24, 1949; children: Frank, Rachael, Norah. B.S., Kent State U., 1950; M.S., Ohio State U., 1951, Ph.D., 1954. Asst. prof. biology Winona (Minn.) State Coll., 1955-58; research fellow dept. plant pathology U. Minn., St. Paul, 1958-59, research assoc., 1959-61, asst. prof., 1961-63, assoc. prof., 1963-66; assoc. dir. Am. Inst. Biol. Sci., Washington, 1967-69; chief ecol. services Nat. Park Service, Dept. Interior, Washington, 1969-74, chief scientist, 1974-78, assoc. dir. sci. and tech., 1978-80, sr. scientist, 1981-82; dept. dir. Office Trust Responsibilities, Bur. Indian Affairs, 1982—; dep. sci. adv. to Sec. Interior, Dept. Interior, Washington, 1980-81; mem. nat. com. UNESCO Man and the Biosphere Com. Contbr. numerous articles to profl. jours. Served with USNR, 1943-46; PTO. AEC grantee, 1958, 59-62, 66-68; NSF grantee, 1959-62; Coop. States Research Service-Dept. Interior grantee, 1963-66; Coop. States Research Service-Dept. Agr. grantee, 1964-66. Fellow AAAS; mem. Am. Soc. Plant Physiology, Ecol. Soc. Am., Bot. Soc. Am., N.Y. Acad. Sci., George Wright Soc. (dir.), Sigma xi, Phi Epsilon Phi, Phi Sigma Xi, Gamma Sigma Delta. Unitarian. Subspecialties: Ecology; Resource management. Current work: Site-specific resources management planning for land in public ownership and trust. Co-patentee catalytic coating to directly generate heat upon surface of heat dome. Home: 1117 E Capitol St SE Washington DC 20003 Office: Bur Indian Affairs Washington DC 20240

SUDILOVSKY, OSCAR, pathologist, educator; b. Rosario, Argentina, Nov. 8, 1933; s. Malquiel and Esther (Busel) S. M.D., U. Littoral, Rosario, Argentina, 1959; Ph.D., Case Western Res. U., Cleve., 1972. Chief tissue culture lab. U. Littoral, 1959-62; hon. fellow McArdle Lab. Cancer Research, U. Wis., Madison, 1969-71; asst. prof. pathology, also dir. Tissue Culture Lab. and Hybridoma Facility Case Western Res. U., Cleve., 1970-76, assoc. prof., div. Tissue Culture Lab. and Hybridoma Facility, 1976—; dir. autopsy service Univ. Hosps. of Cleve., 1970-76; mem. pathology B study sect. NIH. Contbr.; articles to profl. jours. Nat. Council Sci. and Tech. Research (Argentina) fellow, 1959; NIH spl. research fellow, 1967-71. Mem. Am. Assn. Cancer Research, Tissue Culture Assn., Am. Assn. Pathologists, AAAS. Subspecialties: Cancer research (medicine); Pathology (medicine). Current work: Hepatocarcinogenesis, hybridomas and monoclonal antibodies; somatic cell genetics. Office: 2085 Adelbert Rd Cleveland OH 44106

SUEDFELD, PETER, educator, psychologist; b. Budapest, Hungary, Aug. 30, 1935; emigrated to U.S., 1948, naturalized, 1952; s. Leslie John and Jolan (Eichenbaum) Field; m. Gabrielle Debra Guterman, June 11, 1961 (div. 1980); children: Michael Thomas, Joanne Ruth, David Lee. Student, U. Philippines, 1956-57; B.A., Queens Coll., 1960; M.A., Princeton U., 1962, Ph.D., 1963. Research assoc. Princeton; lectr. Trenton State Coll., 1963-64; vis. asst. prof. psychology U. Ill., 1964-65; asst. prof. psychology Univ. Coll. Rutgers U., 1965-67, assoc. prof., 1967-71, prof., 1971-72, chmn. dept., 1967-72; prof., head dept. psychology U. B.C., Vancouver, 1972—; cons. in field. Author: Restricted Environmental Stimulation: Research and Clinical Applications, 1980; editor: Attitude Change: The Competing Views, 1971, Personality Theory and Information Processing, 1971, The Behavioral Basis of Design, 1976, Jour. Applied Social Psychology, 1975-82; Contbr. articles to profl. jours. Served with U.S. Army, 1955-58. NIMH grantee, 1970-72; Can. Council grantee, 1973—; Nat. Research Council Can. grantee, 1973—; NIH grantee, 1980—. Fellow Can., Am. psychol. assns.; mem. N.Y. Acad. Scis., AAAS, Psychonomic Soc., Soc. Exptl. Social Psychology, Acad. Behavioral Medicine Research, Phi Beta Kappa, Sigma Xi. Subspecialties: Behavioral psychology; Cognition. Current work: Behavioral, cognitive, attitudinal and biological effects and uses of reduced stimulation (habit modification, isolated areas, solitary confinement); cognitive complexity of information processing and decision-making as a function of international crisis, governmental disruption, and personal life experiences. Office: Dept Psychology U BC Vancouver BC Canada

SUESS, HANS EDUARD, geochemistry educator; b. Vienna, Austria, Dec. 16, 1909; came to U.S., 1950, naturalized, 1955; s. Franz Eduard and Olga (Frenzl) S.; m. Ruth Viola Teutenberg, Dec. 30, 1940; children: Beate Maria Suess Lapayre, Stephen E. Ph.D., U. Vienna, 1935; Sc.D. (hon.), Queens U., Belfast, 1980. Prof. chemistry U. Hamburg, 1937-50; research assoc. U. Chgo., 1951; research chemist U.S. Geol. Survey, 1952-55; geochemist Scripps Instn. Oceanography, La Jolla, Calif., 1955-58; prof. geochemistry U. Calif.-San Diego, 1958—. Contbr. articles to profl. jours. Recipient Humboldt prize, 1977-78; Guggenheim fellow, 1966; NSF grantee; AEC grantee; NASA grantee. Mem. Nat. Acad. Sci., Am. Acad. Arts and Sci., Max Planck Soc., Heidelberg Acad. Sci., Austrian Acad. Sci. Subspecialties: Space chemistry; Climatology. Current work: The CO2 problem, geologic age determinations, solar activity, abundances of the elements. Office: Department of Chemistry B-017 University of California San Diego La Jolla CA 92093

SUGAI, IWAO, computer engineer; b. Tokyo, Oct. 19, 1928; U.S., 1952; s. Yonekichi and Tomeo (Nabatame) S.; m. Noriko Obata, Sept. 4, 1958; children: Edward K., Emily Y., Eileen T. Assoc. B.S., Tohoku U., Sendai, Japan, 1952; B.S., UCLA, 1955; M.S.E.E., Calif. Inst. Tech., 1956; D.Sc., George Washington U., 1971. Assoc. engr. IBM Research Ctr., Yorktown Heights, N.Y., 1958-59; tech. specialist ITT Labs., Nutley, N.J., 1959-63; computer scientist Computer Scis. Corp., Falls Church, Va., 1963-72; corp. dir. sci. application Old Dominion Systems, Inc., Gaithersburg, Md., 1972-73; sr. staff engr. Applied Physics Lab., Johns Hopkins U., Laurel, Md., 1973—; chmn. publicity com. Summer Computer Simulation Conf. Bd., 1977—. Contbr. articles in field to profl. jours. Calif. Intercollegiate Nisei Orgn. scholar, 1954; Microwave Research Inst., Poly. Inst. Bklyn. sr. research fellow, 1956-57. Mem. IEEE, Soc. Computer Simulation, Sigma Xi, Tau Beta Pi. Presbyterian. Subspecialties: Applied magnetics; Applied mathematics. Current work: Electromagnetic wave propagation, mathematical analysis of fluid dynamics and ballistic missile trajectory calculations. Home: 14637 Stonewall Dr Silver Spring MD 20904 Office: Applied Physics Lab Johns Hopkins U Johns Hopkins Rd Laurel MD 20707

SUGARBAKER, EVERETT VAN DYKE, surgeon/oncologist; b. N.Y.C., Aug. 6, 1940; s. Everett D. and Geneva I. (Van Dyke) S.; m. Catherine M. Mongiello, Sept. 16, 1968; children: Everett M., Kathryn A. B.S., Wheaton Coll., 1962; M.D., Cornell U., 1966. Resident in gen./thoracic surgery Mass. Gen. Hosp., Boston, 1966-68, 70-74; fellow surg. oncology M.D. Anderson Hosp., Houston, 1974-75; assoc. prof. surgery/oncology U. Miami (Fla.) Sch. Medicine, 1978-80; dir. surg. oncology Miami Cancer Inst., 1980—; pres. Surg. Oncology Assocs., Inc., Miami, 1980—; dir. dept. pathology Oncology Labs. div. Cedar Health Care Ctr., Miami; staff Mercy Hosp., Miami North Shore Hosp., Miami, Victoria Hosp., Miami. Contbr. articles to profl. jours. Served to lt. comdr. USPHS, 1968-70. Recipient Mo. Med. Soc. Sci. award, 1958; Wheaton Coll. scholar, 1960; Polk prize in surgery Cornell Med. Coll., 1966; ACS fellow, 1979; cert. of merit. Am. Cancer Soc., 1979-82; others. Fellow ACS; mem. Mass. Med. Soc., Assn. for Acad. Surgery, Dade County Med. Assn., Fla. Med. Assn., Am. Assn. Cancer Research, Am. Soc. Clin. Oncology, Am. Radium Soc., AMA, Soc. Surg. Oncology, Surg. Hist. Soc., Soc. Univ. Surgeons, Soc. Head and Neck Surgeons, AAAS, Miami Cancer Inst., Northwestern Med. Assn., Internat. Assn. Breast Cancer Research, others. Republican. Subspecialties: Surgery; Oncology. Current work: Cancer biology and metastasis research. Home: 6916 Sunrise Terr Coral Gables FL 33133 Office: 1399 SW 1st Ave Miami FL 33131

SUGG, LARRY CORBITT, automotive company executive, electrical engineer; b. Dyersburg, Tenn., May 5, 1942; s. Albert Glen and Delia Irene (Corbitt) S.; m. Martha Kay Wisner, Apr. 17, 1970. B.S. in Elec. Engring, Detroit Inst. Tech., 1965; M.A. in Engring, Chrysler Inst. Engring., 1967; M.B.A., Wayne State U., 1975. With Chrysler Corp., Highland Park, Mich., impact systems computer specialist, 1972-73, supr. product devel., 1973-75, supr. software systems devel., 1975-76, mgr. engring. systems devel., 1976—; tchr. microcomputer course Chrysler Inst. Counselor Jr. Achievement, 1970. Mem. alumni fund-raising com. Detroit Inst. Tech. Mem. Soc. Automotive Engrs., Enging. Soc. Detroit, Tau Phi. Baptist. Office: Chrysler Corporation 12800 Lynn Townsend Dr Highland Park MI 48288

SUGIURA, MASAHISA, geophysicist; b. Tokyo, Dec. 8, 1925; U.S., 1952, naturalized, 1961; s. Shotaro and Mitsuye (Matsumura) S.; m. Keiko Adachi, June 12, 1962; 1 child, Michi. M.S., U. Tokyo, 1949; Ph.D., U. Alaska, 1955. Asst. prof. geophys. research Geophys. Inst., U. Alaska, 1955-57; assoc. prof. geophysics, 1957-62, prof., 1962; Nat. Acad. Scis. sr. assoc. NASA, Greenbelt, Md., 1962-64; mem. staff NASA Goddard Space Flight Ctr., Greenbelt, 1964—, astrophysicist, 1964—; prof. atmospheric scis. U. Wash., Seattle, 1966-67. Contbr. articles to sci. jours. Recipient Exceptional Performance award NASA Goddard Space Flight Ctr., 1982; Guggenheim fellow, 1959. Mem. Am. Geophys. Union, Soc. Terrestrial Magnetism and Electricity Japan (Tanakadate prize 1950), Internat. Assn. Geomagnetism and Aeronomy. Subspecialties: Space plasma physics; Geophysics. Current work: Study of electrodynamics in the earth's plasma environment; magnetometer experiments on the Dynamics Explorer satellites. Home: 9408 Pin Oak Dr Silver Spring MD 20910 Office: Code 696 Goddard Space Flight Center Greenbelt MD 20771

SUH, NAM PYO, educator; b. Seoul, Korea, Apr. 22, 1936; came to U.S., 1954, naturalized, 1963; s. Doo Soo and Joon Joo (Lee) S.; m. Young Ja Surh, June 24, 1961; children—Mary M., Helen H., Grace J., Carolina Y. S.B., MIT, 1959, S.M., 1961; Ph.D., Carnegie-Mellon U., 1964. Devel. engr. Guild Plastics Inc., Cambridge, Mass., 1958-60; sr. research engr., project mgr. USM Corp., Beverly, Mass., 1961-65; asst. prof. U. S.C., Columbia, 1965-68, assoc. prof., 1968-69; assoc. prof. mech. engring. MIT, Cambridge, 1970-75, prof., 1975—, dir., 1977—, 1973—; chmn. bd. Axiomatics Corp., Sudbury, Mass.; dir. Surftech Corp., Hollis, N.H., Intelitec Corp., Billerica, Mass. Author: (with A.P.L. Turner) Mechanical Behavior of Solids, 1975; editor: (with N. Saka) Fundamentals of Tribology, 1980, (with N. Sung) Science and Technology of Polymer Procs, 1979, The Delamination Theory of Wear, 1977. Chmn. bd. Korean-Am. Soc. New Eng., 1979. Recipient Best Paper award Soc. Plastics Engrs., 1981; Citation Classic Inst. for Sci. Info., 1981; USM Corp. fellow, 1962-63. Mem. ASME (Gustus L. Larson Meml. award 1976, Blackall award 1982), Sigma Xi, Pi Tau Sigma, Phi Kappa Phi. Methodist. Subspecialty: Mechanical engineering. Patentee in field. Home: 34 Maynard Farm Rd Sudbury MA 01776 Office: Room 35-136 Mass Inst Tech Cambridge MA 02139

SUIT, HERMAN DAY, radiation therapist; b. Houston, Feb. 8, 1929; m. Joan Lucia Countryman, Nov. 11, 1960. B.A. in Biology, U. Houston, 1948; M.S. in Biochemistry, Baylor U., 1952, M.D., 1952; D.Phil. in Radiation Biology, Oxford U., Eng., 1956. Diplomate: Am. Bd. Radiology. Intern Jefferson Davis Hosp., Houston, 1952-53, resident in radiology, 1953; research fellow radiobiology lab. Churchill Hosp., Oxford, Eng., 1954-56, registrar in radiotherapy, 1956-57; practice medicine specializing in radiation therapy, Houston, 1959-70, Boston, 1970—; sr. asst. surgeon Nat. Cancer Inst., 1957-59; asst. radiotherapist M.D. Anderson Hosp. and Tumor Inst., Houston, 1959-63, asso. radiotherapist, 1963-68; asso. gen. faculty U. Tex. Grad. Sch. Biomed. Scis., Houston, 1965-70; prof. radiotherapy U. Tex., Houston, 1968-71; chief dept. radiation medicine Mass. Gen. Hosp., Boston, 1970—; prof. radiation therapy Harvard Med. Sch., Boston, 1971—; mem. cons. med. staff Mount Auburn Hosp., Cambridge, Mass., 1972—, Waltham (Mass.) Hosp., 1974—; Malden (Mass.) Hosp., 1975—, Roswell Park Meml. Inst., Buffalo, 1975—; mem. staff Cape Cod Hosp., Hyannis, Mass., 1980—; cons. AEC, 1969-71, Brookhaven Nat. Lab., 1974-77, Am. Cancer Soc., 1972—; med. adv. com. Fermi Nat. Accelerator Lab., Batavia, Ill., 1976—. Contbr. numerous articles on radiation research, cancer research and radiotherapy to profl. jours.; editorial bd.: Radiation Research, 1969-74, Cancer, 1974—, Cancer Research, 1969-74. Mem. Am. Coll. Radiology, Am. Assn. Cancer Research, Am. Soc. Therapeutic Radiologists (pres. 1981), Internat. Union Against Cancer, Radiation Research Soc. (mem. council 1979—), Mass. Radiol. Soc., Am. Radium Soc., New Eng. Soc. for Radiation Oncology, New Eng. Cancer Soc., Mass. Med. Soc., AMA, Radiol. Soc. N. Am., AAAS. Subspecialty: Cancer research (medicine). Current work: Sarcoma of soft tissue and bone; portion beam radiation therapy; metabolic determinants of radiation response. Home: 165 Merriam St Weston MA 02193 Office: Dept Radiation Medicine Mass General Hosp Boston MA 02114

SULLIVAN, ANN CLARE, biochemistry educator; b. Tillamook, Oreg., June 3, 1943; d. Edward A. and Esther K. (Makari) Supple. B.A. in Biology, Coll. Notre Dame, Balt., 1965; M.S. in Biol. Sci, Northwestern U., 1967; Ph.D. in Biochemistry, NYU, 1973. Research assoc., grad. teaching asst. Northwestern U., 1965-66; research assoc. Sci. and Engring., Inc., Waltham, Mass., 1966-68; asst. biochemist Hoffmann-LaRoche, Inc., Nutley, N.J., 1969-71, assoc. biochemist, 1971-72, biochemist, 1972-73, sr. biochemist, 1973-75, assoc. dir. biochem. nutrition, 1978-79, dir. pharmacology dept. II, 1979-81, dir. depts. pharmacology I and II, assoc. dir. exptl. biology, 1981—; adj. prof. Inst. Human Nutrition, Columbia U., 1976—; exec. com. Obesity Center, St. Luke's Hosp. Ctr., N.Y.C., 1978—. Contbr. articles profl. jours. Recipient Tribute to Women and Industry award YWCA, 1979. Mem. Am. Soc. Pharmacology and Exptl. Theapeutics, Am. Inst. Nutrition (Bio Serv award 1983), AAAS, Am. Chem. Soc., Am. Oil Chemists Soc. (Gold Medal Bond award 1976), N.Y. Acad. Scis., Assn. Women in Sci. (Women Scientist award 1977), Assn. Study of Obesity, Soc. Exptl. Biology and Medicine, Columbia U. Seminar on Appetitive Behavior, Sigma Xi, Delta Epsilon Sigma, Kappa Gamma Pi, Beta Beta Beta. Subspecialties: Biochemistry (biology); Pharmacology. Current work: Regulation of lipid and carbohydrate metabolism, control of appetite and energy balance, metabolic aspects of obesity and hyperlipidemia, development of pharmacological agents for treatment of obesity and hyperlipidemia. Patentee in field. Home: 48 Lakewood Ave Cedar Grove NJ 07009 Office: Hoffmann LaRoche Inc 340 Kingsland St Nutley NJ 07110

SULLIVAN, CAROLE A., radiation therapist, educator; b. Chgo., Nov. 16, 1941; d. John Norman and Arlene Mary (Narducy) S. M.Ed., U. Okla., 1982. Cert. in radiation therapy U. Chgo. Hosps. Chief technologist Radiation Therapy Ctr., U. Chgo. Hosps., 1963-67; dosimetrist, chief technologist Radiation Therapy Ctr. Northwestern Meml. Hosp., Chgo., 1968-73; tech. dir. Radiation Therapy Ctr., Univ. Hosp., Okla. U. Health Scis. Center, Oklahoma City, 1973—; assoc. prof., dir. radiation therapy program Coll. Allied Health, U. Okla., Oklahoma City, 1973-78, prof. chmn. dept. radiologic tech., 1980—; cons. ednl. accreditation, adminstrn. of health care facilities. Contbr. articles to profl. jours. Bd. dirs. Children's Tomorrow House Okla. Fellow Am. Soc. Radiologic Tech.; mem. Am. Assn. Physicists in Medicine, Okla. Acad. Scis., AAUP, Assn. Univ. Radiologic Technologists, Phi Delta Kappa. Roman Catholic. Subspecialties: Radiation oncology; Medical physics. Current work: Radiation biology-radiation exposure. Specific interest in the long term effects of chronic long term radiation exposure to radiation workers such as physicians and technologists. Home: 1118 Bedford Dr Oklahoma City OK 73116 Office: PO Box 26901 Dept Radiologic Tech Coll Allied Health Room 441 Oklahoma City OK 73190

SULLIVAN, HUGH RICHARD, JR., chemist, pharmacologist; b. Indpls., Apr. 8, 1926; s. Hugh Richard and Josephine Cecelia (Gill) S.; m. Betty Catherine Smith, Oct. 5, 1924; children: Hugh R., Kathleen, Mark K., Marianne, Kevin J. B.S. in Organic Chemistry, Notre Dame U., 1948; M.S., Temple U., 1954. Assoc. research chemist Mobil Oil Co., Paulsboro, N.J., 1948-51; assoc. research chemist Lilly Research Labs., Indpls., 1951-58, research chemist, 1958-67, sr. research chemist, 1967-69, research scientist, 1969-72, research assoc., group leader, 1972—. Contbr. articles to profl. jours. Served with USN, 1944-46. Mem. AAAS, Am. Soc. Mass Spectrometry, N.Y. Acad. Scis., Am. Soc. Pharmacology and Exptl. Therapeutics. Democrat. Roman Catholic. Club: KC. Subspecialties: Pharmacokinetics. Current work: Drug Metabolism and pharmacokinetic research. Patentee in field. Home: 7135 Kingswood Circle Indianapolis IN 46256 Office: Lilly Research Labs 307 E McCarty St Indianapolis IN 46285

SULLIVAN, JAY MICHAEL, medical educator; b. Brockton, Mass., Aug. 3, 1936; s. William Dennis and Wanda Nancy (Kelpsh) S.; m. Mary Suzanne Baxter, Dec. 30, 1964; children: Elizabeth, Suzanne, Christopher. B.S. cum laude, Georgetown U., 1958, M.D. magna cum laude, 1962. Diplomate: Nat. Bd. Med. Examiners, Am. Bd. Internal Medicine. Nat. Heart Inst. fellow, 1964; Med. Found. research fellow, 1967. Med. intern Peter Bent Brigham Hosp., Boston, 1962-63, resident, 1963-64, 66-67, chief resident, 1969-70, fellow in cardiology, 1964-66, dir. hypertnesion unit, 1970-74; preceptorship in biol. chemistry Harvard Med. Sch., Boston, 1967-69, asst. prof. medicine, 1970-74; dir. med. services Boston Hosp. for Women, 1973-74; prof. medicine, chief div. cardiovascular diseases U. Tenn. Coll. Medicine, Memphis, 1974—, vice-chmn. dept. medicine, 1982—; mem. staff City of Memphis, VA, Bapt. Meml. hosps., U. Tenn. Medical Center-Wm. F. Bowld Hosp.; mem. med. adv. bd. Council for High Blood Pressure Research; cons. Nat. Heart, Lung and Blood Inst., 1974—, U.S. VA, 1983—. Contbr. articles to sci. jours. Served with M.C., U.S. Army, 1963-70. Fellow ACP, Am. Coll. Cardiology, Council on Circulation of Am. Heart Assn.; mem. Assn. Univ. Cardiologists, Internat. Soc. Hypertension, Am. Fedn. Clin. Research, Am. Heart Assn. (pres. chpt. 1982-83), AAAS, Alpha Omega Alpha, Sigma Xi, Alpha Sigma Nu. Roman Catholic. Club: Racquet of Memphis. Subspecialty: Cardiology. Current work: Pathophysiology and hemodynamics of hypertension. Home: 6077 Maiden Ln Memphis TN 38117 Office: 951 Court Ave Room 353D Memphis TN 38163

SULLIVAN, LOUIS WADE, med. sch. dean, physician; b. Atlanta, Nov. 3, 1933; s. Walter Wade and Lubirda Elizabeth (Priester) S.; m. Eve Williamson, Sept. 30, 1955; children—Paul, Shanta, Halsted. B.S., Morehouse Coll., Atlanta, 1954; M.D. cum laude, Boston U., 1958. Diplomate: Am. Bd. Internal Medicine. Intern N.Y. Hosp.-Cornell Med. Center, N.Y.C., 1958-59, resident in internal medicine, 1959-60; fellow in pathology Mass. Gen. Hosp., Boston, 1960-61; research fellow Thorndike Meml. Lab. Harvard Med. Sch., Boston, 1961-63; instr. medicine Harvard Med. Sch., 1963-64; asst. prof. medicine N.J. Coll. Medicine, 1964-66; co-dir. hematology Boston U. Med. Center, 1966; asso. prof. medicine Boston U., 1968-74; dir. hematology Boston

City Hosp., 1973-75; also prof. medicine and physiology Boston U., 1974-75; dean Sch. Medicine, Morehouse Coll., 1975—; mem. sickle cell anemia adv. com. NIH, 1974-75; ad hoc panel on blood diseases Nat. Heart, Lung Blood Disease Bur., 1973, Nat. Adv. Research Council, 1977; mem. med. adv. bd. Nat. Leukemia Assn., 1968-70, chmn., 1970. John Hay Whitney Found. Opportunity fellow, 1960-61. Mem. Am. Soc. Hematology, Am. Soc. Clin. Investigation, Inst. Medicine, Phi Beta Kappa, Alpha Omega Alpha. Episcopalian. Subspecialty: Medical education administration.. Research on suppression of hematopoiesis by ethanol, pernicious anemia in childhood, folates in human nutrition. Home: 1525 New Hope Rd Atlanta GA 30331 Office: 223 Chestnut St Atlanta GA 30314

SULLIVAN, WOODRUFF TURNER, III, astronomy educator; b. Colorado Springs, Colo., June 17, 1944; s. Woodruff Turner and Virginia Lucille (Ward) S.; m. Barbara Phillips, June 8, 1968; 2 daus., Rachel, Sarah. B.S. in Physics, M.I.T., 1966; Ph.D. in Astronomy, U. Md., 1971. Astronomer Naval Research Lab., Washington, 1969-71; postdoctoral fellow U. Groningen, Holland, 1971-73; assoc. prof. astronomy U. Wash., Seattle, 1973—; NASA cons. Author: Classics in Radio Astronomy, 1982; editor: The Early Years of Radio Astronomy; Contbr. articles to profl. jours. Vis. fellow Cambridge U., Eng., 1980-81; NATO sr. fellow Groningen U., 1978; NSF grantee, 1976—; NASA grantee, 1980—. Mem. Internat. Astron. Union, Internat. Union of Radio Sci., History of Sci. Soc., Am. Astron. Soc., Astron. Unit. Subspecialty: Radio and microwave astronomy. Current work: Radio and optical properties of galaxies; history of astronomy; search for extraterrestrial intelligence. Home: 6532 Palatine Ave N Seattle WA 98103 Office: Dept Astronomy FM-20 Univ Wash Seattle WA 98195

SULZBACH, DANIEL SCOTT, computer consultant; b. Iowa City, Feb. 6, 1949; s. John F. and Elizabeth S. B.A., St. John's U., 1971; M.Sc., Queen's U., 1975; Ph.D., Ind. U., 1979. Research fellow in statis. genetics Wesleyan U., Middletown, Conn., 1979-80; research fellow in insect population biology U. Calif.-San Diego, 1980-82; instructional computing cons. Univ. Computer Ctr., 1982—. Contbr. articles to profl. jours. NIH fellow, 1980-82. Mem. Genetics Soc. Am., Soc. for Study of Evolution, Behavior Genetics Assn., Ecol. Soc. Am. Subspecialties: Evolutionary biology. Current work: Application of computer technology to research and instruction; quantitative genetics computers.

SUMMERFELT, ROBERT CLAR, aquatic ecologist, research adminstrator; b. Chgo., Aug. 2, 1935; s. Clarence Glen and Tina Clara (Henschel) S.; m. Deanne E. Walsh, Dec. 5, 1938; children: Scott R., Steven T., Sloan M. B.S., U. Wis.-Stevens Point, 1957; M.S., So. Ill. U., 1959, Ph.D., 1964; postgrad., Marine Lab. Duke U., 1962. Lectr. So. Ill. U., 1962-64; asst. prof. Kans. State U., 1964-66; assoc. prof. dept. zoology Okla. State U., 1966-71, prof., 1971-76; leader Okla. Coop. Fishery Research Unit U.S. Fish and Wildlife Service, 1966-76; prof. animal ecology Iowa State U., 1976—, chmn. dept., 1976—; vis. prof. So. Ill. U., summer 1963, Oreg. State U. Inst. Marine Biology, 1975, Utah State U., winter 1983; cons. in field. Contbr. articles to profl. jours. Bd. dirs. Ill. River Conservation Council. Recipient Bur. Sport Fisheries and Wildlife Spl. Achievement awards, 1969, 70, 78, U.S. Fish and Wildlife Service Citation for Outstanding Performance, 1976. Fellow Am. Inst. Fisheries Research Biologists; mem. Am. Fisheries Soc., Ecol. Soc. Am., Ill. Acad. Sci., Kans. Acad. Sci., Okla. Acad. Sci., Iowa Acad. Sci., Wildlife Disease Assn., Sigma Xi, Sigma Zeta, Phi Kappa Phi, Gamma Sigma Delta. Democrat. Lutheran. Clubs: Izaak Walton League Am., Wilderness Soc., Sierra. Lodge: Lions. Subspecialties: Ecology; Resource management. Current work: Biology of fishes, reservoir ecology, fish culture. Home: 2021 Greenbriar Circle Ames IA 50010 Office: Dept Animal Ecology Iowa State U Ames IA 50011

SUMMERS, DAVID ARCHIBOLD, mining educator, research center administrator; b. Newcastle on Tyne, Eng., Feb. 2, 1944; came to U.S., 1968; s. William Archibold and Margaret Kilpatrick (Little) S.; m. Barbara Lois Muchnick, July 30, 1972; children: Daniel Archibald, Joseph Andrew. B.Sc. in Mining, U. Leeds, U.K., 1965, Ph.D., 1968. Chartered engr., 1971. Asst. prof. mining U. Mo., Rolla, 1968-74, assoc. prof., 1974-77, prof., 1977-80, Curators prof., 1980—, dir. rock mechanics, 1976—; cons. in field. Recipient Brit. Ropes Mining prize, 1965, Alumni Merit award for research, 1974, G.C. Greenwell medal, 1978, NASA cert. of recognition, 1980. Fellow Instn. Mining Engrs.; mem. Instn. Mining and Metallurgy, AIME, ASME, Am. Soc. Engring. Edn., Rolla C. of C., Tau Beta Pi. Anglican. Subspecialty: Mining engineering. Current work: Mining, water jet applications, cavitation, jet cleaning, rock mechanics. Patentee longwall coal mining machine. Home: 808 Cypress Dr Rolla MO 65401 Office: Rock Mechanics Univ Missouri Rolla MO 65401

SUMMERS, GEORGE DONALD, technical company executive; b. Eldorado, Ill., Jan. 16, 1927; s. Arthur W. and Georgia Pearl (Horn) S.; m. Margot Gene Sturken, Dec. 28, 1949; children: Emmy L., Susan H.; m. Sachiko Orui, Aug. 1, 1977. B.S., U.S. Mil. Acad., 1949; postgrad., Army schs., 1949-50, 54-55. Registered profl. engr. Project engr. Am. Bosch Arma Corp., Garden City, N.Y., 1956-58; program mgr. Fairchild Industries (formerly Republic Aviation), Mineola and Farmingdale, N.Y., 1958-72; with Atlantic Research Corp., Alexandria, Va., 1972—, v.p., 1980—. Served to capt. F.A. U.S. Army, 1949-56. Recipient 100 award Indsl. Research publ., 1974. Mem. IEEE, AIAA, AAAS, Am. Soc. Artificial Internal Organs, Armed Forces Communications and Electronics Assn., Assn. U.S. Army, Air Force Assn., U.S. Naval Inst., Assn. Unmanned Vehicles, Am. Def. Preparedness Assn., Soc. Info. Display, Mensa Internat. Subspecialty: Systems engineering. Current work: Application of science and technology to solve problems such as systems engring. Patents, publs. in biomed., aerospace systems, and communications fields. Home: 10402 Hollyoak Pl Fairfax VA 22032 Office: Atlantic Research Corp 5390 Cherokee Ave Alexandria VA 22314

SUMMERS, ROBERT WENDELL, physician, educator; b. Lansing, Mich., July 28, 1938; s. Raymond Willard and Edna June (Cook) S.; m. Edith Marlene Brenneman, Aug. 26, 1961; children: Kristine Anne, Rebecca Marie, Rachel Karin. B.S., Mich. State U., 1961; M.D., U. Iowa, 1965. Diplomate: Am. Bd. Internal Medicine, Am. Bd. Gastroenterology. Intern, resident Cleve. Met. Hosp., 1965-67; intern, resident U. Iowa Hosp. Clinics, Iowa City, 1967-68, gastroenterology fellow, 1968-70, asst. prof., 1970-74, assoc. prof., 1974-83, prof., 1983—; cons. VA Hosp., Iowa City, 1970-76, staff physician, 1976—. Contbr. articles to sci. jours. NIH grantee, 1972-77; USPHS, NIH fellow, 1980; VA Adminstrn. merit rev. grantee, 1978-81, 81-84. Fellow ACP; mem. Am. Gastroent. Assn., Am. Physiol. Assn., Central Soc. Clin. Research, Am. Assn. Study Liver Disease. Mennonite. Subspecialties: Gastroenterology; Internal medicine. Current work: intestinal motility; inflamatory bowel diseases. Office: Div Gastroenterology-Hepatology U Hospitals and Clinics Iowa City IA 52242

SUMMERS, WILLIAM KOOPMANS, neuropsychiatrist, educator; b. Jefferson City, Mo., Apr. 14, 1944; s. Joseph Steward and Amy (Koopmans) S.; m. Angela Forbes Taveras, Oct. 24, 1972; children: Lawrence Pierce, Elisabeth Stuart, Wilhelmina Derek. B.A., U. Mo., 1966; M.D., Washington U.-St. Louis, 1971. Intern Barnes Hosp.-Washington U., St. Louis, 1971-72, resident in medicine and psychiatry, 1970-76; instr. dept. psychiatry and medicine Washington U., St. Louis, 1971-76; asst. prof. psychiatry, internal medicine U. Pitts., 1976-77, asst. dir. disorder clinic, 1976-77; asst. prof. psychiatry, internal medicine U. So. Calif., Los Angeles, 1977-82, ward chief med. center, 1977-81; asst. prof. research dept. psychiatry UCLA, 1982—. Contbr. articles to profl. jours. Markle Found. fellow Washington U., 1971. Mem. ACP, Am. Fedn. Clin. Research, Soc. Biol. Psychiatry. Republican. Episcopalian. Subspecialties: Neuropharmacology; Internal medicine. Current work: General studies of neuropsychiatry and specific study of clinical central nervous system acetylcholine mechanisms. Office: 624 W Duarte Rd Arcadia CA 91006

SUMMIT, ROGER KENT, information systems and services company executive; b. Detroit, Oct. 14, 1930; s. Paul Maurice and Mildred Suzanne S.; m. Virginia Buckhorn, Aug. 8, 1964; children: Jennifer Lee, Scott Wesley. A.B., Stanford U., 1952, M.B.A., 1957, Ph.D., 1965. Research scientist Lockheed Research Lab., Palo Alto, Calif., 1965-72; mgr. DIALOG info. retrieval service Lockheed Corp., Palo Alto, 1972-77, pres.; (subs.), 1981—; dir. Lockheed Info. Systems, Palo Alto, 1977—; mem. sci. and tech. info. working group Dept. Commerce and M.I.T.; mem. public-pvt. sector relations task force Nat. Commn. on Libraries and Info. Sci.; trustee Engring. Index, Inc. Contbr. to profl. publs. Served to lt. (j.g.) USN, 1952-55. Recipient Spl. Invention award Lockheed Corp. Mem. Info. Industry Assn. (dir., Info. Product of Year award 1975, Hall of Fame award 1982), Assn. Info. and Dissemination Centers (v.p. 1975-76), Am. Soc. Info. Sci., IEEE. Subspecialties: Information systems, storage, and retrieval (computer science); Information systems (information science). Inventor aerospace bus. environment simulator, DIALOG info. retrieval lang. Office: 3460 Hillview Ave Palo Alto CA 94304

SUMMITT, (WILLIAM) ROBERT, chemist, educator; b. Flint, Mich., Dec. 6, 1935; s. William Fletcher and Jessie Louise (Tilson) S.; m. Nancy Jo Holland, Apr. 2, 1956; children—Elizabeth Louise, David Stanley. A.S., Flint Jr. Coll., 1955; B.S. in Chemistry, U. Mich., 1957; Ph.D. (Union Carbide Chems. fellow), Purdue U., 1961. Research asso., instr. chemistry Mich. State U., 1961-62, asst. prof. metallurgy, mechs. and materials, 1965-68, asso. prof., 1968-73, chmn. dept. metallurgy mechs. and materials sci., 1972-78, prof., 1973—; research chemist Corning Glass Works, 1962-65; cons. in field. NRC Sr. Research asso. Air Force Materials Lab., Fairborn, Ohio, 1974-75. Leader Chief Okemos council Boy Scouts Am., 1973-76. Mem. Am. Chem. Soc., Am. Phys. Soc., Am. Soc. Metals, Nat. Assn. Corrosion Engrs., Inter-Soc. Color Council, Detroit Soc. Coatings Tech., Sigma Xi. Subspecialties: Physical chemistry; Corrosion. Current work: Corrosion control, prevention in aerospace systems; failure analysis consultant. Research, publs. in optical properties of materials, spectroscopy, corrosion and color sci. Home: 5424 Blue Haven Dr East Lansing MI 48823

SUMNERS, CAROLYN TAYLOR, museum administrator, investigator; b. Chattanooga, Mar. 7, 1948; d. Robert Armstrong and Eunice (Kenney) T.; m. Robert William Sumners, Dec. 27, 1970 (div.); children: Robert William, Jonathan Taylor. B.A. in Physics Astronomy magna cum laude, Vanderbilt U., 1969; Ed.D., U. Houston, 1979. Lectr. astronomy Burke Baker Planetarium, Houston, 1970-73, curator, 1973-78; curator astronomy Houston Mus. Natural Sci., 1978-81, dir. astronomy and physics, 1982—; co-prin. investigator Informal Sci. Study U. Houston, 1980—; tchr., cons. Kinkaid/Houston Sci. and Math. Inst. for Gifted Minority Students. Author articles, physics textbooks, elem. sci. textbooks. Woodrow Wilson fellow, 1978; recipient citation for excellence in energy edn. Tex. Edn. Agy., 1982. Mem. Met. Assn. Tchrs. Sci., Am. Assn. Physics Tchrs., Am. Astron. Soc., Nat. Sci. Tchrs. Assn., Internat. Solar Energy Soc., Planetary Soc., Houston Astron. Soc. Subspecialties: Optical astronomy; Solar energy. Current work: Development of astronomy data center mini-computer software package; implementation of an astronomy bulletin board system; development of a radio remote-computer controlled robot. Home: 9238 Westwood Village Dr Houston TX 77036 Office: 1 Hermann Circle Dr Houston TX 77030

SUMNEY, LARRY W., research administrator; b. Washington, Pa., Aug. 8, 1940; s. Raymond E. and Virginia E. (Hinerman) S.; m. Barbara A. Sumney, July 14, 1962; 1 son, Michael. B.A. with honors in Physics, Washington and Jefferson Coll., 1962; postgrad., U. Md., 1963-64; M.E.A., George Washington U., 1969, 1970-72. Research physicis Naval Research Lab., 1962-67, electronics engr., 1967-72; with Naval Electronics Systems Command, 1972-78, research dir., 1977-78; staff specialist for electron devices and integrated circuit tech. Office of Under Sec. of Def., 1978-80; dir. very high speed integrated circuit program office Office of Under Sec. Def. for Research and Engring., 1980-82; exec. dir. Semiconductor Research Corp. (subs. Semiconductor Industry Assn.), Durham, N.C., 1982—. Naval Research Lab. Edison fellow, 1970-72. Mem. IEEE. Methodist. Subspecialties: Microchip technology (engineering); Systems engineering. Office: 300 Park Research Triangle Durham NC 27709

SUN, CHANG-TSEN, engineering educator; b. Shenyang, China, Feb. 20, 1928; came to U.S., 1958, naturalized, 1969; s. Yao-Tsung and Chung-Shuan (Lee) S.; m. Jenny M. Lin, Aug. 17, 1963; children: Barry I-Lung, Karen I-Teh, Nancy I-Huei. B.S., Taiwan U., 1953; M.S., Stevens Inst. Tech., 1960; D. Engring., Yale U., 1964. With Iowa State U., Ames, 1965-77; with Gen. Motors Research Labs., Warren, Mich., 1977-79; prof. dept. engring. scis. U. Fla., Gainesville, 1979—. Contbr. articles to sci. jours. Mem. ASTM, ASME. Republican. Subspecialties: Solid mechanics; Fracture mechanics. Office: Dept Engring Scis U Fla Gainesville FL 32611

SUN, GRACE YAN CHI, educator, researcher; b. Hong Kong, Oct. 4, 1939; came to U.S., 1957; d. Paul and Mei-Sin (Lau) Cheung; m. Albert Y. Sun, May 9, 1964; 1 dau., Aggie. B.S., Seattle Pacific U., 1961; Ph.D., Oreg. State U., Corvallis, 1966. Research scientist Cleve. Psychiat. Inst., 1966-74; asst. prof. dept. chemistry U. Mo., Kansas City, 1974-75, assoc. prof. dept. biochemistry, Columbia, 1978—; research prof. Sinclair Research Farm, Columbia, 1975—. Mem. Am. Soc. Neurochemists, Internat. Soc. for Neurochemists, Neurosci. Soc., Fedn. Am. Socs. for Exptl. Biology, Internat. Soc. for Biomed. Research in Alcoholism. Subspecialties: Neurochemistry; Biochemistry (medicine). Current work: Plasmapheresis, Plasma exchange therapy, human autoimmunity, antibody production, immuno suppressive therapy. Home: Box 335 Route 12 Columbia MO 65201 Office: Sinclair Research Farm Univ Missouri Columbia MO 65201

SUN, JOHN, aerodynamics engineer, researcher, educator, consultant; b. Shanghai, China, Mar. 20, 1942; came to U.S., 1965; s. Cheung and Wan-Ching (Chang) S.; m. Michele T. Tollie, Sept. 1, 1973; children: Stephanie Francis, Valerie Michele, Gregory William. B.S., Provincial Taiwan Cheng Kung U., 1964; M.S. in Aerospace Engring, Va. Poly. Inst. and State U., 1969; Ph.D. in Engring. Mechanics, Va. Poly. Inst. and State U., 1976. Grad. teaching asst. Va. Poly. Inst. and State U., Blacksburg, 1965-66; aerospace engr. Naval Surface Weapons Ctr., Dahlgren, Va., 1967-80; sr. staff engr. Missile Systems Group, Hughes Aircraft Co., Canoga Park, Calif., 1980—. Contbr. articles to profl. publs. Recipient awards Naval Surface Weapons Ctr., 1976, 77, 78, 81, Hughes Aircraft Co., 1981, 82. Mem. AIAA, Nat. Honor Soc. Aerospace Engring., Naval Aeroballistic Com. Navy/Dept. Def. Subspecialty: Aeronautical engineering. Current work: researcher, developer state-of-the-art missile aerodynamics predicition computer code for generalized missiles configuration, radar cross-sections, designs for new missiles. Patentee controlled store separation system. Home: 664 San Telmo Circle Newbury Park CA 91320 Office: Hughes Aircraft Co Canoga Park CA 91304

SUN, MIKE, potato specialist, plant pathologist; b. Taiwan, Feb. 2, 1938; s. A. T. and Y. T. (Lieu) S.; m. Anna Wei, Oct. 7, 1967; children: Joannie, Theodore. Ph.D. in Plant Pathology, N.C. State U., 1971. Postdoctoral fellow N.C. State U., 1971-73; assoc. plant pathologist Asian Vegetable Research and Devel. Ctr., Taiwan, 1973-75; research assoc. Mich. State U., 1976-78; potato specialist, extension plant pathologist Mont. State U., 1978—. Subspecialties: Plant virology; Plant cell and tissue culture. Current work: Research on potato diseases and potato seed improvement; certify potato seed. Home: 1710 Park View Pl Bozeman MT 59715 Office: Potato Laborator Montana State University Bozeman MT 59717

SUND, ELDON H(AROLD), chemist, educator, cons.; b. Plentywood, Mont., June 6, 1930; s. Lawrence and Ethel May (Andersen) S.; m. Roberta Faulkner, July 13, 1933; children: Sharon Ellen, Elizabeth Dianne, Phillip Lawrence, Nancy Annemarie. B.S., U. Ill., Urbana, 1952; Ph.D., U. Tex., Austin, 1960. Research chemist E. I. duPont de Nemours & Co., Wilmington, Del., 1959-66; asst. prof. chemistry Ohio No. U., 1966-67; prof. Midwestern State U., Wichita Falls, Tex., 1967—, Hardin prof., 1975. Contbr. articles to profl. jours., 1970—. Served to 1st lt. USAF, 1952-56. NSF grantee, 1968-70; Robert A. Welch Found. grantee, 1970—. Mem. Am. Chem. Soc., Internat. Soc. Heterocyclic Chemistry, AAAS, Tex. Acad. Sci., Sigma Xi. Lutheran. Subspecialties: Organic chemistry; Synthetic chemistry. Current work: Synthesis of heterocyclic ketones capable of undergoing enol-keto tautomerism. Patentee in field. Office: 3400 Taft Blvd Wichita Falls TX 76308

SUNDSMO, JOHN SIEVERT, biomed. researcher; b. Adlington Hall, Cheshire, Eng., June 1, 1945; s. Oliver Frederick and Joan Mary (Holl) S.; m. Mary Patricia Wiley, Aug. 31, 1969; children: Amanda, Andrew. B.A., U. Calif.-Irvine, 1967, M.S., 1969; Ph.D., U. Wash., 1974. Postdoctoral fellow Fred Hutchinson Cancer Research Ctr., 1973-75; postdoctoral fellow Scripps Clinic and Research Found., La Jolla, Calif., 1975-77, research assoc., 1977-78, asst. mem. research inst., 1978-79, asst. mem. II, 1980—; asst. mem. Trudeau Inst., 1979-80; cons. Eli Lilly & Co. Contbr. articles to profl. jours. Nat. Cancer Inst. grantee, 1979—; grantee Nat. Inst. Allergy and Infectious Diseases, 1982—. Mem. Am. Assn. Immunologists, N.Y. Acad. Sci., Reticuloendothelial Soc., AAAS. Ind. Democrat. Episcopalian. Club: Mission Bay Yacht (San Diego). Subspecialties: Immunology (agriculture); Biochemistry (medicine). Current work: Complement, mechanisms of activation of lymphocytes and mononuclear phagocytes. Office: Dept Molecular Immunology 10666 N Torrey Pines Rd La Jolla CA 92037

SUNG, ZINMAY RENEE, plant cell genetics educator; b. Shanghai, China, Feb. 14, 1947; came to U.S., 1968; d. Feng-en and DiLee (Fan) S.; m. Nelson N.H. Teng, Mar. 12, 1974. B.S., Nat. Taiwan U., 1967; Ph.D., U. Calif.-Berkeley, 1973. Research asst. Max Planck Inst. Cell Physiology, Berlin, 1967-68; research asst. U. Calif.-Berkeley, 1968-73; research assoc. MIT, Cambridge, 1973-76; asst. prof. dept. genetics and dept. plant pathology U. Calif.-Berkeley, 1976-82, assoc. prof., 1982—; cons. Cetus Corp., Berkeley, 1982—. Mem. Am. Soc. Plant Pathologists, Am. Soc. Developmental Biologists, Am. Soc. Plant Physiologists. Subspecialty: Plant physiology (agriculture). Current work: Plant somatic genetics, developmental biology of plants, genetic engineering of plant via tissue culture manipulations. Office: Dept Genetics and Dept Plant Pathology U Calif Berkeley CA 994720

SUNKARA, SAI PRASAD, tumor biologist; b. Valivarthi Padu, India, June 18, 1948; came to U.S., 1975, naturalized, 1979; s. Sanyasi Rao and Nancheramma (Ambati) S.; m. Kusuma A. Sunkara, Aug. 8, 1974; children: Haritha, Srinivas. B.S., A.P. Agrl. U., Bapatla, India, 1969; M.S., U.P. Agrl. U., Pantnagar, India, 1971; Ph.D., Indian Inst. Sci., Bangalore, 1975. Fellow M.D. Anderson Hosp. and Tumor Inst., Houston, 1976-78, asst. prof., 1978-80; sr. research biochemist Merrell Dow Research Ctr., Cin., 1980-82, research leader, 1982—. Contbr. articles to profl. jours. Recipient Indian Nat. Sci. Acad. Young Scientist award, 1976; Indian Inst. Sci. Hanumantha Rao Meml. award, 1976. Mem. Am. Soc. Cell Biology, Am. Assn. Cancer Research. Subspecialties: Cancer research (medicine); Cell biology (medicine). Current work: To understand the biochemical differences between normal and tumor cells in order to design effective anti cancer drugs; tumor cell biology; biochemical regulation of DNA synthesis and mitosis; interferon; polyamine metabolism in normal and tumor cells. Office: 2110 E Galbraith Rd Cincinnati OH 45215

SUOMI, STEPHEN JOHN, psychology researcher, educator; b. Chgo., Dec. 16, 1945; s. Verner E. and Paula A. (Meyer) S.; m. Karen Francis Basele, Nov. 1, 1975. B.A., Stanford U., 1968; M.A., U. Wis.-Madison, 1969, Ph.D., 1971. Research assoc. U. Wis.-Madison, 1971-75, asst. prof. psychology, 1975-79, assoc. prof., 1979—; cons. BBC, Italian Nat. TV. Contbr. chpts. to books and articles to profl. jours. Mem. gov. selection com. Nat. Youth Sci. Camp, 1981. NIH fellow, 1970; Effron lectr. Am. Coll. Neuropsychopharmacology, 1981; recipient Excellence in Teaching award U. Wis., 1982. Mem. Internat. Primatol. Soc., Soc. Research in Child Devel., Am. Soc. Primatologists (editorial bd. jour. 1981—), Am. Psychol. Assn., Animal Behavior Soc. Lutheran. Subspecialties: Psychobiology; Developmental psychology. Current work: Study of primate psychobiological and social development, focusing on genetic environmental transactions, developmental continuities and individual differences, creation of primate models of human development. Office: Primate Lab 22 N Charter St Wis Madison WI 53706

SURKO, CLIFFORD MICHAEL, physicist; b. Sacramento, Oct. 11, 1941; s. Vlaho John and Alma Beatrice (Horan) S.; m. Pamela T. Hansen, Jan. 30, 1965; children: Michael, Leslie. A.B. in Physics and Math, U. Calif.-Berkeley, 1964, Ph.D., 1968. Research asst. U. Calif.-Berkeley, 1968-69; mem. tech. staff Bell Labs., Murray Hill, N.J., 1969-82, head semicondr. and chem. physics research dept., 1982—; vis. research scientist Ecole Polytechnique, Palaiseau, France, 1979, Plasma Fusion Center, M.I.T., Cambridge, 1978—. Contbr. articles to profl. jours. Mem. Am. Phys. Soc., AAAS, Phi Beta Kappa, Sigma Xi. Episcopalian. Subspecialties: Plasma physics; Condensed matter physics. Current work: Light scattering from waves and fluctuations in tokamak plasmas and fluid turbulence. Patentee in field. Office: Bell Labs Room 1D432 Murray Hill NJ 07974

SURMACZ, JOSEPH GEORGE, indsl. engr., researcher, cons.; b. Pitts.; m. Margaret Edmundson. B.S. in Elec. Engring, Carnegie-Mellon U., 1934, 1940. Registered profl. engr., Pa. Wis. Elec. engr. Duquesne Light Co., Pitts., 1934-40; indsl. engr. U.S. Steel (Am. Steel & Wire Div.), Pitts., 1940-48; v.p. mfg. P & H Harnishfeger Corp., Milw., 1948-67; exec. v.p., gen. mgr. Taylor Machines, Inc., Louisville, Miss., 1967-72; cons. and researcher in indsl. software, Diamond Bar, Calif., 1972—. Mem. ASME, Ops. Research Soc. Am., Am. Inst. Indsl. Engrs., Soc. Mfg. Engrs. Subspecialties: Graphics, image processing, and pattern recognition; Computer architecture. Current work: Computer-aided industrial design and manufacturing integration; computer-aided industrial productivity enhancement.

SUSSKIND, HERBERT, biomedical engineer, researcher; b. Ratibor, Germany, Mar. 23, 1929; came to U.S., 1938; s. Alex and Hertha (Loewy) S.; m. E. Suzanne Lieberman, June 18, 1961; children: Helen J., Alex M., David A. B.Ch.E. cum laude, CCNY, 1950; M.Ch.E., NYU, 1961. Registered profl. engr., N.Y. Engr. sect. supr. Brookhaven Nat. Lab., Upton, N.Y., 1950-77, biomed. engr., 1977—; assoc. prof. medicine dept. medicine SUNY-Stony Brook, 1979—. Co-founder, 1st pres. Huntington Twp. Jewish Forum, Huntington, N.Y., 1970-73; trustee Huntington Hebrew Congregation, 1970-78. Mem. Biomed. Engring. Soc., Soc. Nuclear Medicine, Am. Thoracic Soc., Am. Nuclear Soc. (treas., mem. exec. com. L.I. sect. 1978—), Am. Inst. Chem. Engrs., CCNY Alumni Assn. (pres. 1982-84), CCNY Engring. and Architecture Soc. (pres. 1963-65). Jewish. Subspecialties: Pulmonary medicine; Physiology (medicine). Current work: Application of chemical engineering to pulmonary medicine and physiology. Patentee in field. Home: 12 Hadland Dr Huntington NY 11743 Office: Brookhaven Nat Lab Med Dept Upton NY 11973

SUSSMAN, IRA ISRAEL, physician, researcher; b. Passaic, N.J., Oct. 25, 1943; s. William and Ceil (Grossman) S.; m. Nancy A. Antell, Apr. 7, 1968; children: Stephen, Carl, Rachel. B.A., Rutgers U., 1965; M.D., Albert Einstein Coll. Medicine, 1969. Diplomate: Am. Bd. Internal Medicine. Intern Lincoln Hosp., Bronx, N.Y., 1969-70, resident, 1970-71, N.Y. Hosp., N.Y.C., 1971-72; head coagulation lab. Queens (N.Y.) Hosp. Ctr., 1974-76; asst. head hematology Montefiore Med. Ctr., Bronx, N.Y., 1976—. Mem. Am. Physiol. Soc., Soc. Exptl. Medicine and Biology, Am. Soc. Hematology, Internat. Soc. Thrombosis and Hemostasis. Subspecialties: Hematology; Biochemistry (medicine). Current work: Perform research in area of hemostasis; primary work in hemophilia and von Willebrand's disease; also work in area of experimental arteriosclerosis and platelet vessel wall interactions. Office: Montefiore Hospital 111 E 210th St Bronx NY 10467

SUTTON, GEORGE WALTER, research laboratory executive; b. Bklyn., Aug. 3, 1927; s. Jack and Pauline (Aaron) S.; m. Evelyn D. Kunnes, Dec. 25, 1952; children—James E., Charles S., Richard E., Stewart A. B. Mech. Engring., Cornell U., 1952; M.S., Calif. Inst. Tech., 1953, Ph.D. magna cum laude, 1955. Research scientist Lockheed Missle Co., 1955; research engr. Space Sci. Lab., Gen. Elec. Co., 1955-61, mgr. magnetohydrodynamic power generation, 1962-63; vis. Ford prof. Mass. Inst. Tech., 1961-62; sci. adviser directorate devel. plans Hdqrs. USAF, 1963-65; with Avco Research Lab., 1965—, dir. laser devel., 1971—; v.p. Avco Everett Research Lab., Everett, Mass., 1972-80, mgr. govt. mktg. Helionetics Laser div., San Diego, 1983—; spl. cons. Energy Agy., 1977-79; lectr. magnetohydrodynamics U. Pa., 1960-63, Stanford, 1964. Author: Proceedings 4th Symposium Engineering Aspects of Magnetohydrodynamics, 1964, (with A. Sherman) Engineering Magnetohydrodynamics, 1965, Direct Energy Conversion, 1966; Editor-in-chief: Jour. Am. Inst. Aeros. and Astronautics, 1967—; Contbr. 70 articles to profl. jours. Served with USAAF, 1945-47. Recipient Arthur Flemming award for outstanding govt. service, 1965. Fellow AIAA (chmn. plasmadynamics tech. com., Thermophysics award 1980), ASME; mem. Symposium Engring. Aspects Magnetohydrodynamics (pres.), AAAS. Subspecialties: Solar energy; Laser-induced chemistry. Spl. research on ablation of heat protection for ICBM re-entry and high energy lasers. Home: 3268-37 Via Marin La Jolla CA 92037 Office: 3878 Ruffin Rd San Diego CA 92123

SUTTON, HARRY ELDON, geneticist, educator; b. Cameron, Tex., Mar. 5, 1927; s. Grant Edwin and Myrtle Dovie (Fowler) S.; m. Beverly Earlene Jewell, July 7, 1962; children: Susan Elaine, Caroline Virginia. B.S. in Chemistry, U. Tex., Austin, 1948, M.A., 1949, Ph.D. in Biochemistry, 1953. Biologist U. Mich., 1952-56, instr., 1956-57, asst. prof. human genetics, 1957-60; asso. prof. zoology U. Tex., Austin, 1960-64, prof., 1964—, chmn. dept. zoology, 1970-73, asso. dean Grad. Sch., 1967-70, 73-75, v.p. for research, 1975-79; mem. adv. council Nat. Inst. Environ. Health Scis., 1968-72, council sci. advs., 1972-76; mem. various coms. Nat. Acad. Scis.-NRC; cons. in field; bd. dirs. Associated Univs. for Research in Astronomy, 1975-79, Argonne Univs. Assn., 1975-79, Univ. Corp. for Atmospheric Research, 1975-79, Associated Western Univs., 1978-79. Author: Genes, Enzymes, and Inherited Disease, 1961, An Introduction to Human Genetics, 1965, Genetics: A Human Concern, 1984; editor: First Macy Conference on Genetics, 1960, Mutagenic Effects of Environmental Contaminants, 1972, Am. Jour. Human Genetics, 1964-69. Trustee S.W. Tex. Corp. Public Broadcasting, 1977-80, sec., 1979-80; bd. dirs. Austin Civic Ballet, 1978—. Served with U.S. Army, 1945-46. Mem. Am. Soc. Human Genetics (dir. 1961-69, pres. 1979), Genetics Soc. Am., Am. Soc. Biol. Chemists, Am. Chem. Soc., Tex. Genetics Soc. (pres. 1979), Environ. Mutagen Soc., AAAS, Am. Genetic Assn. Club: Headliners (Austin). Subspecialties: Gene actions; Environmental toxicology. Current work: Envirnmental mutagens and human somatic mutation; effects of mutation on enzyme function. Research, publs. in human genetics. Home: 1103 Gaston Ave Austin TX 78703 Office: Dept Zoology U Tex Austin TX 78712

SUZUKI, JON BYRON, periodontist, microbiologist, immunologist, educator; b. San Antonio, July 22, 1947; s. George K. and Ruby K. (Kanaya) S. B.A., Ill. Wesleyan U., 1968; Ph.D., Ill. Inst. Tech., 1972; S.M., Am. Acad. Microbiology, 1973; D.D.S., Loyola U.-Chgo., Maywood, Ill., 1978; cert., U. Md.-Balt., 1980. Diplomate: Am. Acd. Microbiology. Instr. Columbia Coll. Physicians and Surgeons, N.Y.C., 1971-73; dir. med. tech. U. Hawaii, Honolulu, 1973-74; instr. microbiology Loyola U.-Chgo., Maywood, 1974-78; NIH fellow periodontics U. Wash., Seattle, 1978-80; mem. adv. council NASA, Houston, 1976-82; assoc. prof. microbiology and periodontics U. Md., Balt., 1980—; lectr. periodontics and oral pathology, instr. pre-clin. endodontics, adv. grad. oral biology program Loyola U., Maywood, Ill., 1974—; pvt. practice dentistry, Ill., Wash., Md., 1978-82, pvt. practice periodontics, Towson, Md., 1982—. Author: Clinical Laboratory Methods, 1974, Soft Tissue Curettage, 1979; contbr. articles to profl. jours. Instr. cardiopulmonary resuscitation Am. Heart Assn.; water safety instr., sr. lifesaver ARC. Recipient Outstanding Alumnus of Year award Ill. Wesleyan U., 1977; Pres. medallion St. Apollonia Guild, Loyola U. Chgo., 1978; Gold Key medallion, 1978; Dean's Spl. Recognition award Loyola U.-Chgo., 1978; Disting. Service award St. Apollonia Jesuit Nat. Bd., 1978; Research award NASA, 1980; Excellence in Dentistry award Bd. Trustees, Loyola U.-Chgo., 1981; Achievement award Dentsply Internat. Clinicians, 1982; Omicron Kappa Upsilon-Am. Fund for Dental Health Charles Craig scholar, 1980-82. Mem. Internat. Assn. Dental Research, Internat. Soc. Biophysicists, Internat. Soc. Endocrinologists, Am. Acad. Microbiology, Am. Acad. Periodontology (First place Orban prize 1982), Am. Dental Assn. (Henry M. Thornton SCADA fellow 1978), Am. Acad. Oral Pathology (Oral Pathology award 1977), AAAS, AAUP, Am. Inst. Biol. Scis., Am. Soc. Microbiology, Soc. Indsl. Microbiology, Western Soc. Periodontology, N.Y. Acad. Scis., Sigma Xi, Blue Key, Omicron Kappa Upsilon. Subspecialties: Periodontics; Immunology (medicine). Current work: Biology of immunodeficient patients, periodontal disease-microbiology and immunology. Office: U Md 666 W Baltimore St Baltimore MD 21201

SUZUKI, TATEYUKI, engineering educator; b. Kawagoe, Saitama, Japan, Apr. 28, 1945; s. Sentaro and Michiyo (Akiyama) S. B.Engring., U. Tokyo, 1968, M.Engring., 1970, D.Engring., 1973. Research fellow dept. aeros. U. Tokyo, 1973-74, lectr. dept. mechanics, Kawagoe, Japan, 1974-76; lectr. U. Saitama Urawa, Japan, 1974—; research assoc. NASA-Ames Research Center, Moffett Field, Calif., 1976-77; assoc. prof. dept. mechanics Saitama Inst. Tech., Okabe, Japan, 1978—. Author: Characteristics of a Blast Wave over Dust Deposit, 1982; contbr. articles to profl. jours. Recipient research prizes Japan Pvt. Sch. Promotion Found., 1981, 82, Japan Securities Found., 1982. Mem. Kanto Inst. Engring. Edn. Buddhist. Subspecialties: Fluid mechanics; Combustion processes. Current work: Interaction of a shock wave with dust deposit; ignition of hydrogen injected into a stagnant region behind a reflected shock; ignition and combustion of coal particles. Home: 15-12 Minami-Torimachi Kawagoe Japan 350 Office: Saitama Inst Tech 1690 Fusaiji Okabe Japan 369-02

SUZUKI, TUSNEO, immunology educator; b. Nagoya, Japan, Nov. 23, 1931; s. Morichika and Toshiko (Kita) S.; m. Denise Judith, Jan. 20, 1970; children: Riichiro, Aijiro, Yozo. M.D., U. Tokyo, 1958; Ph.D., U. Hokkaido, 1969. Lic. physician, Japan. Research assoc. dept. physiol. chemistry U. Wis.-Madison, 1976-77; asst. prof. U. Kans. Med. Ctr., 1970-79, assoc. prof., 1979—; vis. asst. prof. U. Tex. Southwestern Med. Sch., 1976-77. Contbr. articles to profl. jours. Mem. Am. Assn. Immunologists. Subspecialties: Immunology (medicine); Biochemistry (biology). Current work: Mechanism of gamma receptor-mediated regulation of cell functions. Office: Ranbow Blvd at 39th Kansas City KS 66103 Home: 3620 W 73d St Prairie Village KS 66208

SVANUM, SOREN, psychologist, educator; b. San Rafael, Calif., July 22, 1945; s. Soren and Theresa (Estes) S.; m. Haya Ascher, Aug. 22, 1980; 1 son, Erez. A.B. in Psychology, San Francisco State U., 1971, M.A., U. Mont., 1973, Ph.D. in Clin. Psychology, 1976. Teaching asst. U. Mont., 1972-73, 73-74, 75-76; intern in health care psychology div. health care psychology U. Minn. Hosps., Mpls., 1974-75; asst. prof. psychology Purdue U. Sch. Sci., Indpls., 1976-82, assoc. prof. psychology, 1982—; cons. psychologist Fairbanks Hosp., Indpls., 1978—; mem. smoking task force com. Ind. chpt. Am. Cancer Soc., 1978-80; cons. in field. Contbr. articles to profl. publs.; author profl. papers; article reviewer profl. jours.; book rev. editor: Internat. Jour. Intercultural Relations, 1978-80. NIMH trainee, 1973-74; fellow, 1974-75. Mem. Am. Psychol. Assn., Midwestern Psychol. Assn., Soc. Personality Assessment, Central Ind. Psychol. Assn., Sigma Xi, Psi Chi. Current work: Clinical psychology, health care psychology, substance abuse. Office: Dept Psychology Purdue U Sch Sci at Indpls 1201 E 38th St PO Box 647 Indianapolis IN 46223 Home: 4360 Washington Blvd Indianapolis IN 46205

SVARRER, ROBERT W., research engineer, engineering educator; b. Bklyn., Mar. 3, 1932; s. Viggo A. and Emily E. (Marr) S.; m. Rita Rose Bruno, May 4, 1958; children: Donna Rose, Scott Christopher. A.A.S., N.Y. Inst. Tech., 1963, B.S., 1967. Product engr. Reeves Instrument Co., Garden City, N.Y., 1959-64; sr. engr. Bendix Corp., Teterboro, N.J., 1964-75; faculty mem. Tech. Career Inst., N.Y.C., 1975-81; asst. prof. County Coll. Morris, Randolph, N.J., 1981-82; chmn. sci. and tech. Hudson County Community Coll., Jersey City, N.J., 1982—; adj. faculty mem. Stevens Inst. Tech., Hoboken, N.J., 1979—; research electronics engr. Office Naval Research, Arlington, Va., 1980—; Aux. police lt. N.Y.C. Police Emergency Div., 1956-70. Served to sgt. USMC, 1949-55; to lt. col. USMCR, 1956—. Mem. AIAA, Am. Soc. Engring. Edn., IEEE, Assn. Computing Machinery, N.Y. Acad. Scis., Marine Corps Res. Officers Assn., Marine Corps Aviation Assn. Subspecialties: Computer engineering; Computer architecture. Current work: Digital avionic systems, computer systems, electronics, computer technology, and computer science education and research. Designer and co-designer of avionic systems for helicopters and fighter/attack aircraft, automatic flight systems for DC-10s. Home: 5 Martin Pl Fairfield NJ 07006 Office: Hudson County Community Coll Jersey City NJ 07306

SWAIMAN, KENNETH FRED, pediatric neurologist, educator; b. St. Paul, Nov. 19, 1931; s. Lester J. and Shirley (Ryan) S.; m.; children: Lisa, Jerrold, Barbara, Dana. B.A. magna cum laude, U. Minn., 1952, B.S., 1953, M.D., 1955. Diplomate: Am. Bd. Psychiatry and Neurology. Intern Mpls. Gen. Hosp., 1955-56; resident in pediatrics U. Minn., 1956-58, neurology, 1960-63; asst. prof. pediatrics, neurology U. Minn. Med. Sch., Mpls., 1963-66, assoc. prof., 1966-69, prof., dir. pediatric neurology, 1969—, exec. officer dept. neurology, 1977 , mem. internship adv. council exec. faculty, 1966-70; cons pediatric neurology Hennepin County Gen. Hosp., Mpls., St. Paul-Ramsey Hosp., St. Paul Children's Hosp., Mpls. Children's Hosp.; mem. human devel. study sect. NIH, 1976-79, guest worker, 1978-81. Author: (with Francis S. Wright) Neuromuscular Diseases in Infancy and Childhood, 1969, Pediatric Neuromuscular Diseases, 1979, (with Stephen Ashwal) Pediatric Neurology Case Studies, 1978; editor: (with John A. Anderson) Penylketonuria and Allied Metabolic Diseases, 1966, (with Francis S. Wright) Practice Pediatric Neurology, 1975, 2d edit., 1982; mem. editorial bd.: Annals of Neurology, 1977-82, Neurology Update, 1977-82, Pediatric Update, 1977—, Brain and Devel. (jour. Japanese Soc. Child Neurology), 1980—, Neuropediatrics, 1982—; contbr. articles to profl. jours. Nat. Inst. Neurologic Diseases and Blindness fellow, 1960-63. Fellow Am. Acad. Pediatrics, Am. Acad. Neurology (rep. to nat. council Nat. Soc. Med. Research); mem. Soc. Pediatric Research, Central Soc. Clin. Research, Central Soc. Neurol. Research, Internat. Soc. Neurochemistry, Am. Neurol. Assn., Minn. Neurol. Soc., AAAS, Midwest Pediatric Soc., Am. Soc. Neurochemistry, Child Neurology Soc. (1st pres. 1972-73, Hower award 1981), Internat. Assn. Child Neurologists (exec. com. 1975-79), Profs. Child Neurology (1st pres. 1978-80), Phi Beta Kappa, Sigma Xi. Subspecialties: Neurology; Developmental neuroscience. Current work: Metabolism of developing brain; cerebral iron metabolism; childhood neuromovement disorders. Office: U Minn Med Sch Dept Pediatric Neurology Minneapolis MN 55455

SWAKON, DOREEN H.D., animal scientist, educator; b. Berwyn, Ill., Oct. 9, 1953; d. Darrell L. and Elaine D. (Jensen) Downer; m. Lawrence W. Swakon, Nov. 30, 1974; 1 dau., Casey Laine. B.S. with high honors in Agr, U. Ill., 1975; M.S., U. Fla., 1977, Ph.D. in Animal Sci.-Nutrition, 1980. Grad. research asst. U. Fla., Gainesville, 1975-80; asst. prof. animal sci. Tex. A&I U., Kingsville, 1980—; dir. Forage Testing Lab., 1981—. Contbr. articles to profl. jours. Recipient Am. Forage and Grassland Council Young Scientist award, 1980. Mem. Am. Soc. Animal Sci., Range Mgmt. Soc., Am. Forage and Grassland Council, Tex. Forage and Grassland Council. Lutheran. Subspecialties: Animal nutrition; Agronomy-Forage Crops. Current work: Forage quality evaluation and utilization by ruminants and other herbivores. Home: 11121 Timbergrove Corpus Christi TX 78410 Office: Tex A&I U Coll Agr Box 156 Kingsville TX 78363

SWALIN, RICHARD ARTHUR, scientist, company executive; b. Mpls., Mar. 18, 1929; s. Arthur and Mae (Hurley) S.; m. Helen Marguerite Van Wagenen, June 28, 1952; children: Karen, Kent, Kristin. B.S. with distinction, U. Minn., 1951, Ph.D., 1954. Research assoc. Gen. Electric Co., 1954-56; mem. faculty U. Minn., Mpls., 1956-77, prof., head Sch. Mineral and Metall Engring., 1962-68, assoc. dean Inst. Tech., 1968-71, dean Inst. Tech., 1971-77; acting dir. Space Sci. Center, 1965; v.p. tech. Eltra Corp., N.Y.C., 1977-80; v.p. research and devel. Allied Corp., Morristown, N.J., 1980—; guest scientist Max Planck Inst. für Phys. Chemie, Göttingen, Germany, 1963, Lawrence Radiation Lab., Livermore, Cal., 1967; Cons. to govt. and industry. Dir. Medtronic Corp.; dir. BMC Industries. Author: Thermodynamics of Solids, 2d edit, 1972; Contbr. articles to profl. jours. Trustee Midwest Research Inst., 1975-78, Sci. Mus. Minn., 1973-77. Recipient Disting. Teaching award Inst. Tech., U. Minn., 1967; NATO sr. fellow in sci., 1971. Mem. AAAS, Am. Phys. Soc., Sigma Xi, Tau Beta Pi, Phi Delta Theta, Gamma Alpha. Subspecialties: Metallurgy; Amorphous metals. Current work: Management of diverse research activities in order to yield new products and develop new technologies which could form the basis of new business. Home: 264 Oak Ridge Ave Summit NJ 07901 Office: PO Box 3000R Allied Corp Morristown NJ 07960

SWALLOM, DANIEL WARREN, research laboratory power generation manager; b. Sioux Falls, S.D., Oct. 13, 1946; s. Maurice Leo and Mary Alice (Jacobs) S.; m. Elizabeth Ruth Pederson, June 7, 1969; children: Bradley Jay, Jeffrey Dean. B.S. in Mech. Engring, U. Iowa, 1969, M.S., 1970, Ph.D., 1972. Research asst. U. Iowa, Iowa City, 1971-72; mech. engr. Argonne (Ill.) Nat. Lab., 1975-76; mgr. ops. Maxwell Labs., Inc., Woburn, Mass., 1976-78; pres. Odin Internat. Corp., Woburn, 1978-79; mgr. mil. power programs Avco Everett Research Lab., Everett, Mass., 1979—; mem. indsl. adv. panel Lincoln Coll., Northeastern U., Boston, 1980-82. Served to capt. USAF, 1972-75. Merit scholar U. Iowa, 1964; recipient Mil. Engr. award USAF, 1968; NDEA Title IV grad. fellow, 1969-71. Assoc. fellow AIAA (plasma scis. tech. com. 1981—); mem. Am. Phys. Soc., Sigma Xi, Tau Beta Pi, Pi Tau Sigma. Congregationalist. Club: Meadow Brook Golf (Reading, Mass.). Subspecialties: Mechanical engineering; Magnetohydrodynamic energy conversion. Current work: Development of magnetohydrodynamic energy conversion systems for utility and military applications, devices which are compact, have high efficiency and high performance. Office: Avco Everett Research Lab 2385 Revere Beach Pkwy Everett MA 02149

SWANSON, DAVID HENRY, economist, educator; b. Anoka, Minn., Nov. 1, 1930; s. Henry Otto and Louise Isabell (Holiday) S.; m. Suzanne Nash, Jan. 19, 1952; children: Matthew David, Christopher James. B.A. in Econs, St. Cloud State Coll., 1953, M.A., U. Minn., 1955. Economist area devel. dept. No. States Power Co., Mpls., 1955-56, staff asst., v.p. sales, 1956-57, economist indsl. devel. dept., 1957-63; dir. area devel. dept. Iowa So. Utilities Co., Centerville, 1963-67, dir. econ. devel. and research, 1967-70; dir. New Orleans Econ. Devel. Council, 1970-72; div. mgr. Kaiser Aetna, New Orleans, 1972-73; dir. corp. research United Services Automobile Assn., San Antonio, 1973-76; pres. Lantern Corp., San Antonio, 1974-79; administr. bus. devel. State of Wis., Madison, 1976-78; dir. Center Indsl. Research and Service, Iowa State U., Ames, 1978—, mem. mktg. faculty, 1979—, mem. adv. council, 1978—; dir. Applied Strategies Internat. Ltd., 1983—; Mem. adv. council Center Indsl. Research and Service, 1967-70. Contbr. numerous tech. articles to profl. publs. Vice chmn. Planning Commn. Roseville, Minn., 1961; mem. Iowa Airport Planning Council, 1968-70; mem. adv. council Office Comprehensive Health Planning, 1967-70; mem. Dist. Export Council, 1978—; mem. region 7 adv. council SBA, 1978—; dir. Mid-Continent Research and Devel. Council, 1980—, Iowa Devel. Council, 1982—; chmn. Iowa del. White House Conf. on Small Bus., 1980, Gov.'s Task Force on High Tech., 1982, Gov.'s High Tech. Commn., 1983—; mem. adv. com. U. New Orleans, 1972-73; county fin. chmn. Republican Party, 1966-67; mem. bd. dirs. Greater New Orleans Urban League, 1970-73. Served with USAF, 1951-52. Mem. Small Bus. Inst. Dirs. Assn., Iowa Profl. Developers, Nat. Assn. Mgmt. Tng. Assistance Centers, Internat. Council Small Bus., Nat. Univ. Edn. Assn. Republican. Episcopalian. Lodges: Rotary; Toastmasters (past pres.). Subspecialties: Technology transfer administration. Home: 1007 Kennedy Dr Ames IA 50010 Office: Center Indsl Research and Service Iowa State Univ Ames IA 50011

SWANSON, DON RICHARD, univ. dean; b. Los Angeles, Oct. 10, 1924; s. Harry Windfield and Grace Clara (Sandstrom) S.; m. Patricia Elizabeth Klick, Aug. 22, 1976; children—Douglas Alan, Richard Brian, Judith Ann. B.S., Calif. Inst. Tech., 1945; M.A., Rice U., 1947; Ph.D., U. Calif., Berkeley, 1952. Physicist U. Calif. Radiation Lab., Berkeley, 1947-52, Hughes Research and devel. Labs., Culver City, Calif., 1952-55; research scientist TRW, Inc., Canoga Park, Calif., 1955-63; prof. Grad. Library Sch., U. Chgo., 1963—, dean, 1963-72, 77-79; mem. Sci. Info. Council, NSF, 1960-65; mem. toxicology info. panel Pres.'s Sci. Advisory Com., 1964-66; mem. library vis. com. Mass. Inst. Tech., 1966-71, mem. com. on sci. and tech. communication Nat. Acad. Scis., 1966-69. Editor: The Intellectual Founds. of Library Education, 1965, The Role of Libraries in the Growth of Knowledge, 1980; co-editor: Operations Research: Implications for Libraries, 1972, Management Education: Implications for Libraries and Library Schools, 1974; mem. editorial bd.: Library Quarterly, 1963—; Contbr.: chpt. to Ency. Brit, 1968—; sci. articles to profl. jours. Trustee Nat. Opinion Research Center, 1964-73; Research fellow Chgo. Inst. for Psychoanalysis, 1972-76. Served with USNR, 1943-46. Mem. Assn. Am. Library Schs., Am. Soc. for Info. Sci. Current work: Communication to public of science information; microcomputer applications to library problems. Home: 5550 Dorchester Ave Apt 809 Chicago IL 60637 Office: Grad Library Sch U Chgo 1100 E 57th St Chicago IL 60637 Problems I wish I could solve: how the universe got started, how volition arose in a heap of molecules, how to find information in a library, and how to sum up life in a sentence or two.

SWANSON, MARGARET MACMORRIS, curator genetics center, researcher; b. Ames, Iowa, Sept. 4, 1948; d. David and Jeanne Alice (Peruse) MacMorris; m. Richard Carl Swanson, Sept. 4, 1971; children: Abigail Beth, Julia Rees. B.A. in Math. with distinction, U. Colo., 1970; M.S. in biology, Calif. Inst. Tech., 1972; Ph.D., U. Calif.-Santa Cruz, 1977. Instr., researcher U. Mo., 1977-78; curator Caenorhabditis Genetics Center, Columbia, Mo., 1980—. Contbr. articles to profl. jours. Bd. dirs. Planned Parenthood Central Mo., 1978-82, pres., 1982. NIH fellow, 1970-71, 78-80. Mem. AAAS, Genetics Soc. Am., Phi Beta Kappa. Subspecialties: Genome organization; Developmental biology. Current work: Genetic regulation of early development; supervision and organization of genetic stock center for C. elegans and computerized information retrieval system for data bank genetic, strain. Office: University of Missouri 110 Tucker Hall Columbia MO 65211

SWANSON, PAUL NORMAN, physicist; b. San Mateo, Calif., June 29, 1936; s. Arnold C. and Velma G. (Dubois) S.; m. Sandra J. Berube, Dec. 28, 1960; children: Kyle, Brian. B.S., Calif. State Poly. U., 1962; Ph.D., Pa. State U., 1968. Faculty dept astronomy Pa. State U., State College, 1968-75; mem. tech. staff, group supr. Jet Propulsion Lab., Calif. Inst. Tech., Pasadena, 1975—. Contbr. articles to profl. publs. Served with USN, 1954-58. NASA grantee, 1965-68. Mem. AAAS, Am. Astron. Soc., IEEE, Sigma Xi, Aircraft Owners and Pilots Assn. Subspecialties: Aerospace engineering and technology; Radio and microwave astronomy. Current work: Submillimeter astronomy from space; space science and technology, astronomy, project management. Home: 3338 Los Olivos La Crescente CA 91214 Office: 4800 Oak Grove Dr Pasadena CA 91103

SWANSON, ROBERT A., genetic engineering company executive; b. N.Y.C., Nov. 29, 1947; s. Arthur John and Arline (Baker) S.; m. Judy Church, Sept. 2, 1980. S.B., M.I.T., 1970, S.M. in Mgmt, 1970. Asst. treas. Citicorp Venture Capital Ltd., N.Y.C., 1970-74; partner Kleiner & Perkins Venture Capital Partnership, San Francisco, 1975; pres. Genentech, Inc., South San Francisco, Calif., 1976—. Named Entrepreneur of Year Chgo. Research Dirs. Assn., 1981. Mem. Am. Chem. Soc., Am. Soc. Microbiology. Subspecialty: Genetics and genetic engineering (biology). Office: 460 Point San Bruno Blvd South San Francisco CA 94080

SWANSON, WILLIAM MASON, aluminum company executive; b. Chgo., Jan. 12, 1932; s. C. William and Lois A. (Rieff) S.; m. Joan Emily Krause, Aug. 2, 1952; children: Roger, Jill Champion, Daniel, Jeff, Paul. B.S. in Chem. Engring., Purdue U., West Lafayette, Ind., 1953; M.B.A., U. Chgo., 1969. Petrochem. process devel. ofcl. UOP, Inc., Des Plaines, Ill., 1961-66; mgr. mktg. Process div. Japan UOP, Inc., 1966-69; dir. mktg. Process div. Far East UOP, Inc., 1969-72; v.p. Mgmt. Services div. UOP, Inc., Des Plaines, 1972-74, v.p., gen. mgr. Minerals Scis. div., Tucson, 1974-82; pres. Toth Aluminum Corp., New Orleans, 1982—; seminar participant. Contbr. articles to profl. jours. Leader, Boy Scouts Am.; pres. bd. Adlai E. Stevenson High Sch., Prairie View, Ill.; active United Way, Tucson, 1980-82. Named Exec. of Week, Sta. WGSO News Radio, New Orleans, 1982. Mem. AIME, Am. Chem. Soc., Am. Petroleum Inst., Can. Inst. Mining, Beta Gamma Sigma. Subspecialties: Chemical engineering; Metallurgy. Current work: Production of aluminum trichloride and silicon tetrachloride by carbo-chlorination from clay, recovery of high purity nickel and cobalt through ammoniacal leaching and solvent extraction. Patentee in field. Home: 645 Carmenere S Kenner LA 70065 Office: 3101 W Napoleon Ave 200 Metairie LA 70001

SWARM, RICHARD LEE, physician; b. St. Louis, June 9, 1927; s. Clarence Lee and Elsie Viola (Parker) S.; m. Pauline Kirksey Alexander, Dec. 28, 1950; children: Lee Ann, Robert Alexander. B.A., Washington U., 1949, B.S., 1950, M.D., 1950. Diplomate: Am. Bd. Pathology. Intern, resident, chief resident and instr. pathology Washington U., St. Louis, 1950-54; assoc. prof. pathology U. Cin., 1965-68; pathologist Hoffmann LaRoche Inc. (Research Div.), Nutley, N.J., 1968-82, dir. exptl. pathology and toxicology, 1968-75, assoc. dir. exptl. therapeutics, 1975-82; sr. cons., dir Health Sci. Assocs., Ridgewood, N.J., 1982—; assoc. prof. pathology Coll. Physicians and Surgeons, Columbia U., N.Y.C., 1970—. Contbr. articles to profl. jours. Served with USPHS, 1954-65. Am. Cancer Soc. clin. fellow, 1953; grantee Am. Cancer Soc., 1965-68, USPHS, 1965-68. Mem. Am. Assn. Pathologists, AAAS, Am. Assn. for Cancer Research, AMA, Am. Soc. Clin. Pathologists, Coll. Am. Pathologists, Internat. Acad. Pathology, N.Y. Acad. Sci., Radiation Research Soc., Soc. Cryobiology, Soc. Toxicology, Soc. Toxicologic Pathologists, Sigma Xi, Phi Delta Theta. Subspecialties: Pathology (medicine); Toxicology (medicine). Current work: Study of neoplasms in man and animals, particularly bone tumors; human and experimental cancer pathology; chemotherapy and drug development; data management by computer. Office: PO Box 808 Ridgewood NJ 07451

SWARTZ, WILLIAM EDWARD, JR., chemistry educator, consultant; b. Braddock, Pa., Aug. 16, 1944; s. William Edward and Catherine (Brockschmidt) S.; m. Sandra Youngk, June 10, 1967; children: Jennifer E., Edward R. B.S., Juniata Coll., 1966; Ph.D., MIT, 1971. Research asst. U.S. Steel Co., Monroeville, Pa., 1965-67; postdoctoral research, assoc. U. Ga., Athens, 1971, U. Md., College Park, 1972; asst. prof. U. South Fla., Tampa, 1972-77, assoc. prof., 1977-82, prof. chemistry, chmn., 1982—. NIH predoctoral fellow, 1967-71. Mem. Am. Chem. Soc., Am. Vacuum Soc. (chmn. Fla. chpt. 1980), ASTM, Soc. Applied Spectroscopy (chmn. Fla. sect. 1980-81). Subspecialties: Analytical chemistry; Surface chemistry. Current work: Involved in the application of X-ray photoelectron spectroscopy and Auger electron spectroscopy to chemical problems, current interests centered on heterogeneous catalysis and microelectronic reliability. Office: U South Fla Dept Chemistry Tampa FL 33620

SWARZ, JEFFREY ROBERT, genetic engring. co. exec., neuroscientist; b. Newark, Nov. 9, 1949; s. Irvin Brad and Blanche S. (Marcus) S.; m. Kathy Helen Kafer, June 20, 1976. B.S. with honors, U. Calif.-Irvine, 1971; Ph.D. (NIMH trainee 1971-74, NIH fellow 1975-76), U. Rochester, 1976. Postdoctoral fellow in neurovirology Johns Hopkins U. Sch. Medicine, 1976-79; staff fellow Infectious Disease br. NIH, Bethesda, Md., 1979-80; dir. biotech. group Teknekron Research Inc., McLean, Va., 1980-81; pres. AgroBiotics, Inc., Balt., 1981-82, Urbana, Ill., 1981-82; sr. scientist Pall Corp., Glen Cove, N.Y., 1982—; cons. U.S. Senate Subcom. on Sci., Tech. and Space, 1979-80. Author: (with others) Genetic Engineering: Issues and Trends, 1982; contbr. numerous articles to profl. jours. Mem. Common Cause. Recipient Undergrad. Research award Bank of Am., 1970-71, Nat. Research Service award, 1976-79. Mem. Am. Assn. Neuropathologists, Neuroscis. Soc., Am. Chem. Soc., Soc. Indsl. Microbiology. Democrat. Jewish. Club: Bowley's Quarters Marina and Yacht (Balt.). Subspecialties: Genetics and genetic engineering (agriculture); Neurobiology. Current work: Development of new plant strains through biotechnological techniques, including genetic engineering, tissue culture, chemical mutagenesis, protoplast fusion, development of new analytical research techniques for genetic engineering.

SWAZEY, JUDITH POUND, college president; b. Bronxville, N.Y., Apr. 21, 1939; d. Robert Earl and Louise Titus (Hanson) Pound; m. Peter Woodman Swazey, Nov. 28, 1964; children: Elizabeth, Peter. A.B. (scholar), Wellesley Coll., 1961; Ph.D. (Wellesley Coll. Alumnae fellow, NIH predoctoral fellow, Radcliffe Coll. grad. fellow), Harvard U., 1966. Research asso. Harvard U., 1966-71, lectr., 1969-71, research fellow, 1971-72; cons. com. brain scis. Nat. Research Council, 1971-73; staff scientist M.I.T. Neuroscis. Research Program, 1973-74; asso. prof. history Boston U., 1974-80, asso. prof. dept. socio-med. scis. and community medicine, 1974-77, prof., 1977-80, adj. prof., 1980—; exec. dir. Medicine in the Public Interest, Inc., Boston and Washington, 1979-82; pres. Coll. of the Atlantic, Bar Harbor, Maine, 1982—; dir. Public Responsibility in Medicine and Research, Ctr. Drug Devel. Author: Reflexes and Motor Integration, the Development of Sherrington's Integrative Action Concept, 1969, (with others) Human Aspects of Biomedical Innovation, 1971, (with R.C. Fox) The Courage to Fail, a Social View of Organ Transplants and Hemodialysis, 1975, rev. edit., 1978 (hon. mention Am. Med. Writers Assn., C. Wright Mills award Am. Sociol. Assn.), Chlorpromazine in Psychiatry, a Study of Therapeutic Innovation, 1974, (with K. Reeds) Today's Medicine, Tomorrow's Science, Essays on Paths of Discovery in the Biomedical Sciences, 1978; editor: (with C. Wong) Dilemmas of Dying, Policies and Procedures for Decisions Not to Treat, 1981, (with F. Worden and G. Adelman) The Neurosciences: Paths of Discovery, 1975; contbr. articles to profl. jours.; asso. editor: IRB: A Jour. of Human Subjects Research, 1979—; editorial bd.: Bioethics Quar., 1981—, Bioethics Reporter, 1982—. Fellow Inst. Soc., Ethics and Life Scis. (v.p. 1980); mem. Inst. Medicine, Nat. Acad. Scis., AAAS, Sherrington Soc., Soc. Health and Human Values, Phi Beta Kappa, Sigma Xi. Subspecialty: Social and ethical aspects of biomedicine. Current work: Study of social, ethical, local and policy issues in biomedical research and medical care. Office: Coll of the Atlantic Bar Harbor ME 04609

SWEADNER, KATHLEEN J., neurobiochemist; b. Pitts., Oct. 17, 1949. B.A., U. Calif.-Santa Barbara, 1971; Ph.D., Harvard U., 1977. Instr. physiology Harvard Med. Sch., 1980-81, asst. prof., 1981—. Established investigator Am. Heart Assn. Mem. Soc. Neurosci. Subspecialties: Membrane biology; Neurobiology. Current work: Membrane proteins of the nervous system: development, cellular regulation, and ion transport. Office: Warren 4 Mass Gen Hosp Fruit St Boston MA 02114

SWEDLOW, JEROLD LINDSAY, mech. engr.; b. Denver, Aug. 31, 1935; s. Jack and Evelyn Lilian (Weinstein) S.; m. Patricia L. Lauer, Mar. 23, 1959; children: Jason, Pamela, Kathryn. B.S., Cal. Inst. Tech., 1957; M.S., Stanford U., 1960; Ph.D., Calif. Inst. Tech., 1965. Research fellow Calif. Inst. Tech., 1965-66; research scientist U.S. Steel Co., 1966; mem. faculty Carnegie Mellon U., Pitts., 1966—, prof. mech. engring., 1973—, assoc. dean engring., 1977-79; sr. vis. fellow Imperial Coll., London, 1973-74. Contbr. articles to profl. jours.; editor: reports of current research Internat. Jour. Fracture, 1969—. Trustee, 1st v.p. First Unitarian Ch. Pitts., 1979-82. Recipient Philip M. McKenna Meml. award Kennametal, Inc., 1978, Ralph Coats Roe award Am. Soc. Engring. Edn., 1981. Mem. Am. Acad. Mechanics, ASME, ASTM, Internat. Congress Fracture (founding); Fellow AAAS, AIAA (assoc.). Democrat. Subspecialties: Solid mechanics; Fracture mechanics. Current work: Elastic-plastic fracture of metals. Office: Dept Mech Engring Carnegie Mellon U Pittsburgh PA 15213

SWEENEY, MARY ANN, plasma physicist; b. Hagerstown, Md., Sept. 25, 1945; d. Daniel Joseph and Dorothy Virginia (Daub) S.; m. Edward Robert Ricco, Feb. 2, 1974; 1 dau.: Alanna Catherine Sweeney. B.A., Mt. Holyoke Coll., 1967; M.Ph., Columbia U., 1973, Ph.D. in Astronomy, 1974. Instr. astronomy Fairleigh Dickinson U., 1972-73; instr. astronomy William Paterson Coll. of N.J., Paterson, 1972-73; postdoctoral fellow Sandia Nat. Labs., Albuquerque, 1974-76, mem. tech. staff, 1976—, mem.; mem. exec. bd. Plasma Sci. and Applications, IEEE, 1982—. Editor: Surveying Your Future: Nontraditional Careers for Young Women, 1981. Bd. dirs. Downtown Neighborhoods Assn. Albuquerque, 1979-81, sec., 1980-81. Mem. Am. Phys. Soc., Am. Nuclear Soc., Am. Astron. Soc., Astron. Soc. Pacific, IEEE, Phi Beta Kappa, Sigma Xi. Subspecialties: Nuclear fusion; Plasma physics. Current work: Particle beam fusion research: Diode performance, modeling target experiments, reactor design. Weapons research: Storage and transport of nuclear weapons. Home: 312 Keleher NW Albuquerque NM 87102 Office: Sandia Nat Labs Div 1265 Albuquerque NM 87185

SWEENEY, PATRICK J., computer software consultant; b. N.Y.C., Feb. 8, 1954; s. Thomas John and Eileen (McCaffrey) S.; m. Karin Kathleen A'Hearn, Mar. 17, 1979; 1 son, Timothy. B.S. in Computer Sci, SUNY, Stony Brook, 1974, M.S., Columbia U., 1980; M.B.A., NYU, 1984. Computer software cons. Digital Equipment Corp., N.Y.C., 1975—. Vol. U.S. Peace Corps, Togo, West Africa, 1974-75. Mem. IEEE Computer Soc., Assn. Computing Machinery. Subspecialties: Distributed systems and networks; Software engineering. Home: 39-59 59th St Woodside NY 11377 Office: Digital Equipment Corp 1 Pennsylvania Plaza New York NY 10119

SWEENEY, URBAN JOSEPH, corporation librarian, information services consultant; b. v. St. John, N.B., Can., Jan. 18, 1922; came to U.S., 1927; s. Urban James and Dorothy Elizabeth (Murray) S.; m. Margaret Stretz, Jan. 12, 1952; children: Dennis, Steven, Edward, Mark, Barbara. B.S., NYU, 1955; M.L.S., Pratt Inst., 1956. Cert. profl. librarian, N.Y.; cert. coll. tchr., Calif. Asst. librarian Sperry Gyroscope Corp., Lake Success, N.Y., 1956-58; chief librarian Republic Aviation Corp., Farmingdale, N.Y., 1958-66, Gen. Electronics, Rochester, N.Y., 1966-71, Convair div. Gen. Dynamics, San Diego, 1971—; library cons., San Diego, 1976—. Author: Initialisms of Science and Technology Organizations, 1978. Served with USAAF, 1941-45. Mem. Spl. Libraries Assn. (pres. San Diego chpt. 1973-74, chmn. aerospace div. 1978-80, award 1980), AIAA, Assn. Computing Machinery. Republican. Roman Catholic. Subspecialties: Information systems, storage, and retrieval (computer science); Information systems (information science). Current work: On-line information, retrieval and microcomputer applications. Home: 7311 Borla Pl Carlsbad CA 92008 Office: Gen Dynamics Convair Div PO Box 85386 San Diego CA 92138

SWEET, HAVEN COLBY, biology educator, computer programmer; b. Boston, Mar. 1, 1942; s. Harman Royden and Ester Wilkins (Colby) S.; m. Gail Lever, Aug. 17, 1963; children: Kristen Lee, Jeffrey Stephen. B.S., Tufts U., 1963; Ph.D., Syracuse U., 1967. Postdoctoral fellow Brookhaven Nat. Lab., Upton, N.Y., 1967-68; research analyst Brown & Root-Northrop, Houston, 1968-71; asst. prof. dept. biol. sci. U. Central Fla., Orlando, 1971-77, assoc. prof., 1978—. Contbr. articles to sci. jours. Grantee NASA, 1971-74, 76-77, Fla. Dept. Agr., 1979-81, NSF, 1981-83; NDEA fellow, 1963-67; ASEE-NASA fellow, 1976-78. Mem. Bot. Soc. Am., Am. Assn. Plant Physiologists, AAAS, Linnean Soc. London, Fla. Acad. Sci., Sigma Xi, Omicron Delta Kappa, Fla. Defenders of Environ., Environ. Def. Fund, Sierra Club. Subspecialties: Plant physiology (biology); Remote sensing (atmospheric science). Current work: Developing computer techniques for data analysis; teaching, research, software development. Office: U Central Fla Orlando FL 32816

SWEET, RAY DOUGLAS, mech. engr.; b. Bellows Falls, Vt., Sept. 27, 1923; s. Preston Merritt and Florence (Ray) S.; m. Louise Marion, May 9, 1959. B.S. with distinction in Mech. Engring, U. Conn., 1945. With Singer Co., Bridgeport, Conn., 1947-60, Central Research Lab., Singer Co., Denville, N.J., 1960-67, sr. project engr., 1963-67; with Pratt & Whitney Aircraft Div., United Technologies Corp., East Hartford, Conn., 1967—, now sr. engr. instrumentation group. Served with USN, 1944-46. Mem. ASME, Tau Beta Pi. Republican. Mem. Ch. of Christ. Subspecialties: Electronics; Mechanical engineering. Current work: Dynamic measurements data reduction stress, vibration, acoustical noise, pressure. Patentee sewing machines.

SWEET, ROBERT MAHLON, biophys. chemist; b. Omaha, Sept. 21, 1943; s. Mahlon and Elizabeth K. S.; m. Diana D., Aug. 20, 1966; children: Anna, Charles, Joseph. B.S., Calif. Inst. Tech., 1965; Ph.D., U. Wis.-Madison, 1970. Research fellow Med. Research Council, Cambridge, Eng., 1970-73; asst. prof. chemistry UCLA, 1973-81, specialist in molecular biology, 1981—. Mem. Am. Crystallographic Assn., Sigma Xi. Subspecialties: Biophysical chemistry; X-ray crystallography. Current work: Study of protein structure and function; x-ray diffraction studies of light-harvesting systems from photosynthetic organisms. Office: Dept Chemistry UCLA 405 Hilgard Ave Los Angeles CA 90024

SWENSEN, CLIFFORD HENRIK, JR., psychology educator; b. Welch, W.Va., Nov. 25, 1926; s. Clifford Henrik and Cora Edith (Clovis) S.; m. Doris Ann Gaines, June 16, 1948; children: Betsy, Susan, Lisa, Timothy, Barbara. B.S., U. Pitts., 1949, M.S., 1950, Ph.D., 1952. Diplomate: in clin. psychology Am. Bd. Profl. Psychology. Clin. psychologist VA, Marion, Ind. and Knoxville, 1952-54; asst. to assoc. prof. U. Tenn., Knoxville, 1954-62; vis. prof. U. Fla., Gainesville, 1968-69, U. Bergen, Norway, 1976-77; prof. psychology Purdue U., West Lafayette, Ind., 1962—; cons. VA, Indpls., Danville, Ill., 1962—; allied profl. staff St. Elizabeth Hosp., Lafayette, 1982—. Author: An Approach to Case Conceptualization, 1968, Interpersonal Relations, 1973; contbr. articles to profl. jours. Served with USN, 1944-46. Am. Psychol. Assn./NSF disting. sci. lectr., 1969; Fulbright lectr./researcher, 1968-69. Fellow Am. Psychol. Assn. (pres. Div. 13 1976-77), Soc. for Personality Assessment; mem. Gerontol. Soc., Midwestern Psychol. Assn. Current work: Interpersonal relations; psychotherapy; gerontological psychology; ego development. Home: 611 Hillcrest Rd West Lafayette IN 47906 Office: Purdue U Dept Psychol Scis West Lafayette IN 47907

SWENSON, CLAYTON A., physicist, educator; b. Mpls., Nov. 11, 1923; s. Nels and Anna (Roth) S.; m. Heather M.F. Gell, Sept. 2, 1950 (dec. 1977); children—Anna, Paul, Wendy; m. Ruth B. Wildman, Jan. 1, 1980. B.S., Harvard, 1944; D.Phil., Oxford U., 1949. Mem. staff Los Alamos Sci. Lab., 1944-46; instr. Harvard, 1949-52; Div. Indsl. Cooperation staff mem. Mass. Inst. Tech., 1952-55; prof. physics, disting. prof. scis. and humanities, sr. physicist Ames Lab., U.S. Dept. Energy, Iowa State U., 1955—; chmn. dept. physics Iowa State U., 1975-82; mem. cons. com. on thermometry Internat. Com. on Weights and Measures. Fellow Am. Phys. Soc.; mem. Am. Assn. Physics Tchrs., AAAS, Phi Beta Kappa, Sigma Xi. Subspecialties: Condensed matter physics; Low temperature physics. Research, publs. in solid state physics with emphasis on low temperature and high pressure and combinations of these; publs. on understanding of elementary solids (inert gases and alkali metals) at low temperatures, devel. temperature scales below 30K. Home: 2102 Kildee St Ames IA 50010

SWENSON, DONALD OTIS, mech. engr.; b. Manhattan, Kans., Feb. 19, 1937; s. Donald D. and Florence (Knapp) S.; m. Harriett Swenson; 4 children. B.S. in Mech. Engring, U. Kans., 1963; M.S., 1965; Ph.D., 1967. Registered profl. engr., Calif., Fla., Kans., Minn., Mo., Nebr., Nev., N.D., Utah, Wis. Sr. research assoc. Pratt and Whitney div. United Aircraft Corp., East Hartford, Conn., 1967-71; cons. engr. Black & Veatch Cons. Engrs., Kansas City, Mo., 1971—. Contbr. articles to profl. jours. Served with USN, 1955-59. Summerfield scholar, 1961-63; U. Kans. fellow, 1966-67. Mem. ASME, ASTM, Air Pollution Control Assn., Sigma Xi, Tau Beta Pi, Pi Tau Sigma, Sigma Tau. Roman Catholic. Subspecialties: Mechanical engineering; Gas cleaning systems. Current work: Air pollution control systems, fabric filters and electrostatic precipitators.

SWENSON, GEORGE WARNER, JR., electronics engineer, radio astronomer, educator; b. Mpls., Sept. 22, 1922; s. George Warner and Vernie (Larson) S.; m. Virginia Laura Savard, June 26, 1943 (div. 1970); children: George Warner III, Vernie Laura, Julie Loretta, Donna Joan; m. Joy Janice Locke, July 2, 1971. B.S., Mich. Coll. Mining and Tech., 1944, E.E., 1950; M.S., Mass. Inst. Tech., 1948; Ph.D., U. Wis., 1951. Asso. prof. elec. engring. Washington U., St. Louis, 1952-53; prof. U. Alaska, 1953-54; asso. prof. Mich. State U., 1954-56; faculty U. Ill., Urbana, 1956—, prof. elec. engring. and astronomy, 1958—, acting head dept. astronomy, 1970-72, head dept. elec. engring., 1979—; dir. Vermilion River Obs., 1968-81; Vis. scientist Nat. Radio Astronomy Obs., 1964-68; cons. to govt. agys. and other sci. bodies. Author: Principles of Modern Acoustics, 1953, An Amateur Radio Telescope, 1980; Contbr. articles to profl. jours. Fellow IEEE, AAAS; mem. Nat. Acad. Engring., Internat. Sci. Radio Union (mem. U.S. nat. com. 1965-67, 80-82), Internat. Astron. Union, Sigma Xi, Eta Kappa Nu, Tau Beta Pi, Phi Kappa Phi. Subspecialties: Electronics; Planetary science. Current work: Radio astronomy instruments, antennas, optics administration of electrical engineering education and research. Home: 1107 Kenwood Rd Champaign IL 61821 Office: U Ill Urbana IL 61801

SWINEHART, JAMES STEPHEN, educator; b. Cleve., July 27, 1929; s. James Franklin and Bertha Bogniard (Brunk) S.; m. Ann Fundis, Sept. 12, 1963; 1 dau., Susan. B.S., Western Res. U., 1950; M.S., U. Cin., 1951; Ph.D., N.Y.U., 1959. Asst. prof. Wagner Coll., S.I., N.Y., 1957-61; assoc. prof. Am. U., 1961-65; sr. scientist Atlantic Research, Alexandria, Va., 1965-67; sr. spectroscopist Perkin Elmer Co., Norwalk, Conn., 1967-69, Digilab div. Block Engring., Silver Spring, Md., 1969-70; chmn., prof. chemistry SUNY, Cortland, 1970—; cons. CEIR, Inc., 1962-64, Atlantic Research, 1964-65, Dept. Energy, 1978-81. Author: Experiments in Techniques of Infrared Spectroscopy, 1967, Organic Chemistry and Experimental Approach, 1969, Interpretation of Spectra of Organic Compos, 1983. Coordinator Citizens Energy Workshops, 1978-81. NSF equipment grantee, 1974; Petroleum Research Found. grantee, 1964; Research Found. SUNY grantee, 1975, 77. Mem. Am. Chem. Soc., AAUP, Sigma Xi, Phi Beta Kappa. Subspecialties: Organic chemistry; Nuclear magnetic resonance (chemistry). Current work: Analytical organic chemistry, spectra structure correlation, structure-property correlations. Patentee in field. Home: 11 Gwen Ln RD 3 Cortland NY 13045 Office: Chemistry Dept SUNY Box 2000 Cortland NY 13045

SWINGLE, KARL FREDERICK, pharmacologist; b. Richland Center, Wis., Feb. 16, 1935; s. John Frederick Swingle and Lillian (Klein) Jacobson; m. Rosemary E. McSherry, July 12, 1958 (div. 1972); children: Kari Ann, Joseph F., Jane E., Katherine J.; m. Linda Rae Skog, Jan. 3, 1980. B.A., U. Wis.-Madison, 1958; Ph.D., U. Minn., 1968. Research specialist 3M Co., St. Paul, 1971-75, sr. pharmacologist, 1968-70, sr. research specialist, 1975. Contbg. author: Anti-Inflammatory Agents, 1974, Inflammation and Anti-Inflammatory Drugs, 1978, Modern Pharmacology, 1982; contbr. articles in pharmacology to profl. jours. Served with U.S. Army, 1959-61. NIH fellow, 1964. Mem. Am. Soc. for Pharmacology and Exptl. Therapeutics, Soc. for Exptl. Biology and Medicine. Democrat. Subspecialty: Pharmacology. Current work: Pulmonary pharmacology, anti-inflammatory drugs, dermatopharmacology. Home: 2244 E Minnehaha Ave St Paul MN 55119 Office: Riker Labs 3M Co 3M Center-Bldg 270-2S St Paul MN 55144

SWISHER, JOSEPH VINCENT, chemist; b. Kansas City, Mo., Jan. 12, 1932; s. Joe K. and Dorothy M. (De Ponds) S.; m. Mary M. Redfield, Aug. 29, 1960 (div.); children: Catherine, Margaret, William. A.B., Central Coll., Fayette, Mo., 1956; Ph.D., U. Mo., Columbia, 1960. Postdoctoral researcher Purdue U., Lafayette, Ind., 1960-61; mem. faculty U. Detroit, 1961—, assoc. prof. chemistry, 1969—. Vol. fireman, Huntington Woods, Mich., 1972-80. Mem. Am. Chem. Soc., Sigma Xi. Subspecialties: Organic chemistry. Current work: Organosilicon chemistry and stereochemistry. Home: 13328 Talbot St Huntington Woods MI 48070 Office: U Detroit 4001 W McNichols St Detroit MI 48221

SWISHER, SCOTT NEIL, physician, educator; b. LeCenter, Minn., July 30, 1918; s. Scott Neil and Edna (Stenlund) S.; m. Edna Lenora Briese, Sept. 5, 1944; children: Scott Neil III, Edward Allan. B.S., U. Minn., 1943, M.D., 1945, postgrad., 1948-49. Intern, resident U. Rochester Med. Center, 1945, 47-48; practice medicine, specializing in internal medicine, Rochester, N.Y., 1949-67, Lansing, Mich., 1967—; from instr. to prof. medicine, head hematology unit U. Rochester Sch. Medicine and Dentistry, 1950-67; prof., chmn. dept. medicine Coll. Human Medicine, Mich. State U., 1967-77, asso. dean research, 1978—; cons. NIH, Dept. Def. research and engring. div., NASA Manned Spacecraft Center and Hdqrs.; chmn. med. adv. com. ARC Nat. Blood Program, 1973-76, bd. govs., 1976-83; chmn. NRC Com. on Blood, 1958-63; chmn. panel 6 Bur. Biologics, FDA, 1975-79. Contbr. articles to profl. jours. Served from 1st lt. to capt. AUS, 1945-

47. Mem. Am. Soc. Clin Investigation, Assn. Am. Physicians, Am. Bd. Internal Medicine, Alpha Omega Alpha. Subspecialties: Hematology; Internal medicine. Current work: Autoimmunity, autoimmune, acquired hemolytic anernia, blood transfusion, immunology, comporative immunogenetics and immunohematology. Home: 1290 Whittier Dr East Lansing MI 48823

SWITZER, GEORGE LESTER, forestry educator, researcher; b. Chester, W.Va., Nov. 5, 1924; s. Francis John and Eleanora Stella (Seitz) S.; m. Meadie Exum Montgomery, Sept. 8, 1956; 1 son Francis M. B.S., W.Va. U., Morgantown, 1949; M.F., Yale U., 1950; Ph.D., SUNY-Syracuse, 1962. Asst. forester, forester Miss. Agrl. and Forestry Expt. Sta., Mississippi State, Miss., 1950—; prof. forestry Miss. State U. Contbr. articles to profl. jours. Bd. dirs. Crosby Arboretum Found., Hattiesburg, Miss., 1981—. Served with USN, 1942-46. Recipient award Miss. State U. Alumni Assn., 1975. Mem. Soil Sci. Soc. Am. (assoc. editor jour. 1979-82), AAAS, Am. Naturalist Assn. Republican. Episcopalian. Subspecialties: Ecosystems analysis; Plant growth. Current work: Nutrition and associated dynamics of forest ecosystems, biosystamatics and genecology of forest plants. Home: PO Box 385 Mississippi State MS 39762 Office: Dept Forestry Miss State U Drawer FR Mississippi State MS 39762

SYBERS, HARLEY DUANE, pathology educator, consultant; b. Tony, Wis., June 18, 1933; s. Henry and Dorothy S. B.S., U. Wis.-Madison, 1956, M.S., 1963, M.D., 1963, Ph.D., 1969. Med. intern USPHS Hosps., Balt., 1963-64; resident in pathology U. Wis.-Madison, 1964-68, postdoctoral fellow dept. physiology, 1968-69; asst. prof. pathology U. Calif.-San Diego, 1969-75; assoc. prof. pathology Baylor Coll. Medicine, Houston, 1975—; mem. cardiovascular renal study sect. NIH, Washington, 1980-84; cons. NIH Heart, Lund and Blood Inst., NIH, 1975—. Served to 1st lt. U.S. Army, 1956-58. NIH research grantee, 1961-75, 75-83. Mem. Internat. Soc. Heart Research (exec. com. Am. sect. 1975-80), Internat. Acad. Pathology, Am. Physiol. Soc., Am. Assn. Pathologists, Electron Microsconic Soc. Am. Subspecialties: Pathology (medicine); Morphology. Current work: Cardiovascular pathophysiology. Home: 5223 Valkeith St Houston TX 77096 Office: Dept Pathology Baylor Coll Medicine 1200 Moursund St Houston TX 77030

SYBERTZ, EDMUND J., JR., pharmacologist; b. Boston, Feb. 3, 1951. B.S., Fairfield U., 1972; Ph.D., U. Minn., 1977. Postdoctoral fellow U. Va., 1978; sr. scientist Schering Corp., Bloomfield, N.J., 1979-81; prin. scientist, 1982—; adj. assoc. prof. Fairleigh Dickinson U., 1981—. Contbr. articles to profl. jours. USPHS fellow, 1973-77, 78. Mem. Am. Soc. Pharmacology and Exptl. Therapeutics, AAAS. Subspecialties: Pharmacology; Physiology (medicine). Current work: Pathophysiology of hypertension; mechanisms of action of cardiovascular drugs.

SYDORAK, JAROSLAVA KUZMYCZ, quality control exec.; b. Mittenwald, Germany, Oct. 14, 1946; came to U.S., 1949, naturalized, 1959; d. George C. and Valentine P. (Dziwak) Kuzmycz; m. Mark Zenobius Sydorak, June 15, 1968; children: Larissa, Darya. B.S., Queens Coll., CUNY, 1967; M.A., CUNY, 1971. Import-export liaison dir. Podarogift, USSR, 1968-69; researcher in immunologic reactions Rockefeller U., N.Y.C., 1970-77; quality assurance mgr. Diagnostic Tech., Hauppauge, N.Y., 1977-81, product mgr., 1981—. Mem. Am. Soc. for Quality Control, N.Y. Acad. Scis., Ukrainian Inst. Am. Subspecialties: Hematology; Neuroimmunology. Current work: Introduction of monoclonal antibody techniques into clinical laboratory testing. Home: 66 Richards Rd Port Washington NY 11050 Office: 240 Vanderbilt Pkwy Hauppauge NY 11788

SYED, IBRAHIM BIJLI, medical physicist, educator; b. Bellary, India, Mar. 16, 1939; came to U.S., 1969, naturalized, 1975; s. Ahmed Bijli and Mumtaz Begum (Maniyar) S.; m. Sajida Shariff, Nov. 29, 1964; children: Mubin, Zafrin. B.S., Mysore U., 1960, M.S., 1962; D.Sc., Johns Hopkins U., 1972. Diplomate: Am. Bd. Radiology, Am. Bd. Health Physics. Med. physicist, radiation safety officer Victoria Hosp., India, 1964-67, Halifax (N.S., Can.) Infirmary, 1967-69; dir. med. physics, radiation safety officer Baystate Med. Center, Springfield, Mass., 1973-79; med. physicist, radiation safety officer VA Med. Center, Louisville, 1979—; prof. medicine and nuclear medicine tech. U. Louisville Sch. Medicine, 1979—; guest examiner Am. Bd. Radiology; mem. panel of examiners Am. Bd. Health Physics; Ph.D. thesis examiner U. Delhi; Pres. Springfield Islamic Center, 1973-79, India Assn., Louisville, 1980-81; v.p. Islamic Cultural Assn., Louisville, 1979-80, trustee, 1980—; vice chmn. bd., 1980—. Author: Radiation Safety for Allied Health Professionals; mem. editorial bd.: Jour. Islamic Med. Assn, 1981—; contbr. 80 articles to sci. jours. Trustee India Community Found. Louisville, 1980—; bd. dirs. Child Guidance Clinic, Springfield, 1973-79, Heritage Corp., Louisville, 1981—, others; active Am. Cancer Soc., Heart Fund. Recipient Disting. Community Service award India Community Found., 1982. Fellow Inst. Physics (U.K.), Am. Inst. Chemists, Royal Soc. Health, Am. Coll. Radiology; mem. Am. Coll. Nuclear Medicine, Health Physics Soc., Am. Assn. Physicists in Medicine, Soc. Nuclear Medicine, Nat. Assn. Ams. of Asian Indian Descent (chmn. state pub. relations com. 1982—). Islamic. Office: 800 Zorn Ave Louisville KY 40202 Home: 7102 Shefford Ln Louisville KY 40222

SYLVIA, DAVID MARTIN, plant pathologist; b. Ankara, Turkey, Nov. 12, 1951; s. Irving W. and Mary Anne (Hare) S.; m. Jeanne Gardiner, Mar. 25, 1972. B.S. summa cum laude, U. Mass., 1975; M.S. in Plant Pathology, 1977, Ph.D., Cornell U., 1981. Research asst. U. Mass., Amherst, 1975-77, Cornell U., Ithaca, N.Y., 1977-81; research assoc. dept. plant pathology U. Fla., Gainesville, 1981—. Contbr. articles to profl. jours. Mem. Am. Phytopath. Soc., Inst. Biol. Scis., Soc. Am. Foresters, Mycol. Soc. Am., Northeastern Forest Pest Council, Xi Sigma Pi. Subspecialties: Plant pathology. Current work: Ecology and physiology of ecto- and endomycorrhizae. Application of mycorrhizal technology to agriculture. Office: Dept Plant Pathology U Fla Gainesville FL 32611

SYMMES, DAVID, neurobiologist, researcher; b. N.Y.C., Sept. 4, 1929; s. William Rittle and Gladys Dwight (Jones) S.; m. Jean Sinclair Symmes, Apr. 20, 1957; children: Brian D., Deborah J., Patrick W. B.A., Harvard U., 1952; Ph.D., U. Chgo., 1956. Postdoctoral fellow Yale Med. Sch., 1956-60, asst. prof. physiology, 1960-67; chief sect. brain and behavior NIH, Bethesda, Md., 1967—. Contbr. numerous articles to profl. jours. Mem. Soc. Neurosci., N.Y. Acad. Sci., Am. Physiol. Soc. Democrat. Subspecialty: Neuropsychology. Office: NIH Bldg T-18 Bethesda MD 20205

SYMON, KEITH RANDOLPH, physics educator, researcher, computing center administrator; b. Ft. Wayne, Ind., Mar. 25, 1920; s. James Jefferson Keith and Claribel (Crego) S.; m. Mary Louise Reinhardt, Apr. 7, 1922; children: Judith Elizabeth, Keith Joseph, James Randolph, Rowena Louise. S.B., Harvard U., 1942, A.M., 1943, Ph.D., 1948. Instr. Wayne U., 1947-48, asst. prof. physics, 1948-53, assoc. prof., 1953-55; asst. prof. U. Wis.-Madison, 1955-56, assoc. prof., 1956-57, prof., 1957—; acting dir. Madison Acad. Computing Center, 1982—; cons. theoretical plasma physics; head advanced research group to tech. dir. Midwestern Univs. Research Assn., 1956-69. Contbr. numerous articles to profl. publs.; author: Mechanics, 1953, 3d edit., 1971. Subspecialties: Theoretical physics; Plasma physics. Current work: Theoretical plasma physics, particle orbit theory, research and teaching of physics. Patentee aircraft radio communications equipment, fixed-field alternating gradient particle accelerators. Home: 1816 Vilas Ave Madison WI 53711 Office: Dept Physics U Wis Madison WI 53706

SYNDER, FREEMAN WOODROW, plant physiologist; b. Phila., Dec. 6, 1917; s. Freeman and Mary Anna (Shaffer) S.; m. Elizabeth Fink, Aug. 6, 1938; children: Robert Gordon, Barbara Naneen Synder St. John. B.S., U. Idaho, 1938; Ph.D., Cornell U., 1950. Asst. agronomist U. Ark., Fayetteville, 1950-53; adj. asst. prof. adj. assoc. prof. crop and soil scis. Mich. State U., East Lansing, 1957-75; plant physiologist U.S. Dept. Agr., East Lansing, 1953-75, Beltsville, Md., 1975—. Contbr. article to profl. jours. Served with AUS, 1943-46. Mem. AAAS, Am. Inst. Biol. Scis., Bot. Soc. Am., Crop Sci. Soc. Am., Am. Soc. Plant Physiologists, Am. Soc. Sugarbeet Technologists. Subspecialties: Plant physiology (biology); Plant physiology (agriculture). Current work: Use of physiological and morphological characteristics to increase yield of alfalfa, sugarbeet and other crops; partitioning of the photosynthate. Office: US Dept Agr Bldg 046A BARC West Beltsville MD 20705

SYNDER, JAMES NEVIN, mechanical engineer, researcher; b. Phillipsburg, N.J., Nov. 3, 1923; s. George William and Estella (Hagerman) S.; m. Mary Elizabeth, Aug. 16, 1925. B.S.E., Princeton U., 1948, M.S.E., 1949. Lic. profl. engr., Pa. Trainee Bethlehem Steel Corp., Pa., 1949-51, foreman, 1951-61, engr., 1961-64, research engr., 1964-83; ret., 1983. Served with USMC, 1943-46. Mem. ASME, Am. Soc. Metals. Mem. United Ch. of Christ. Subspecialty: Mechanical engineering. Current work: Physical modeling of hot-rolling process. Patentee in field. Home: 1330 Stafore Dr S Bethlehem PA 18017

SYPERT, GEORGE WALTER, neurosurgeon; b. Marlin, Tex., Sept. 25, 1941; s. Claude Carl and Ruth Helen (Brown) S.; m. Nancy Susan Rojo, Dec. 10, 1971; children: Kirsten Diane, Shannon Ruth. B.A., U. Washington, 1963, M.D. with highest honors, 1967. Diplomate: Am. Bd. Neurol. Surgery. Intern Barnes Hosp., Washington U., St. Louis, 1967-68; asst. resident in neurol. surgery U. Wash. Sch. Medicine, Seattle, 1968, 70-72, chief resident in neurol. surgery, 1973-74, instr., 1973-74; asst. prof. neurosurgery and neuroscience Grad. Faculty U. Fla. Coll. Medicine, Gainesville, 1974-77, asso. prof., 1977-80, prof., 1980—; attending staff neuol. surgery Shands Teaching Hosp., U. Fla., 1974—; staff neurosurgeon Ft. Gordon Army Hosp., Augusta, Ga., 1968-69; asst. chief div. neurosurgery Fitzsimmons Gen. Hosp., Denver, 1969-70; chief neurosurgery sect. VA Hosp., Gainesville, 1974—; vis. prof. Rush-Presbyn.-St. Luke's Med. Center, Chgo., 1981; mem. merit rev. bd. neurobiology med. research service VA, 1979—; reviewer NIH, USPHS. Mem. editorial bd.: Neurosurgery, 1978—; contbr. articles to profl. jours. Served to capt. M.C. U.S. Army, 1968-70. Fellow ACS, Internat. Coll. Surgeons; mem. Research So. Neurol. Surgeons (pres. 1980), Congress Neurol. Surgeons (exec. com. 1980—), Am. Soc. Stereotactic and Functional Neurosurgery (v.p. 1981—), Soc. Neurosci., AAAS, Am. Assn. Neurol. Surgeons, Am. Physiol. Soc., So. Neurosurg. Soc., Fla. Neurosurg. Soc., Med. Research Soc., Fla. State Med. Assn., Alachua County Med. Soc., AMA, Am. Epilepsy Soc., Epilepsy Found. Am., Internat. Neurosurg. Forum, Brain Surgery Soc., Internat. Soc. Physiol. Soc., Internat. Congress Physiol. Scis., Internat. Soc. Stereotactic and Functional Surgery, Sigma Xi, Alpha Omega Alpha, Phi Sigma. Subspecialties: Neurosurgery; Otorhinolaryngology. Current work: Mammalian synaptic mechanisms; segmental motor control; spinal cord regeneration. Office: U Fla Health Center Neurosurgery Box J 265 Gainesville FL 32610

SYTSMA, LOUIS F., chemistry educator, consultant; b. Chgo., July 20, 1946; s. Frederick Louis and Bernice F. (Vander Ploeg) S.; m. Charlene Myroup, Aug. 19, 1967; children: Erick, Anne. Student, Trinity Christian Coll., 1963-65; B.A., Calvin Coll., 1967; Ph.D., Ohio U., 1971. Chemist Anderson Devel. Co., Adrian, Mich., 1971-77; asst. prof. Siena Heights Coll., Adrian, 1973-76; assoc. prof. Trinity Christian Coll., Palos Heights, Ill., 1977—; cons. Anderson Devel. Co., Gary, Ind., 1977—; summer instr. AuSable Trails Inst. Environ. Studies, Mancelona, Mich., 1982—. Contbr. articles in field to profl. jours. Mem. Am.Chem. Soc., Mid-Am. Chemistry Tchrs. at Liberal Arts Colls., Sigma Xi. Christian Reformed. Subspecialties: Organic chemistry. Current work: Teaching general, organic, analytical and environmental chemistry, consulting in organic synthesis, environmental monitoring, and analyses of industrial processes, research in environmental chemistry. Home: 4910 Forest Ct Oak Forest IL 60452 Office: Trinity Christian Coll 6601 W College Dr Palos Heights IL 60463

SZABO, SANDOR, research pathologist; b. Ada, Yougslavia, Feb. 9, 1944; s. Gyorgy and Ilona (Komlos) S.; m. Ildiko Mecs, Feb. 19, 1972; children: Peter, David. M.D., U. Belgrade, Yugoslavia, 1968; M.Sc., U. Montreal, Que., Can., 1971, Ph.D., 1973. Intern U. Belgrade Med. Sch. and Med. Center, Senta, Yugoslavia, 1968-69; vis. scientist Inst. Exptl. Medicine and Surgery, U. Montreal, 1969-70; resident in pathology Peter Bent Brigham Hosp.-Harvard U. Med. Sch., Boston, 1973-77; research fellow Harvard U. Med. Sch., 1975-77, asst. prof. pathology, 1977-81, assoc. prof., 1981—. Contbr. over 200 articles to sci. jours. Recipient Physician's Recognition award AMA, 1976; Milton Fund award Harvard U., 1978; NIH award, 1980. Mem. Am. Assn. Pathologists, Am. Soc. Pharmacology and Exptl. Therapeutics, Soc. Exptl. Biology and Medicine, Am. Gastroent. Assn., Endocrine Soc., N.Y. Acad. Scis., others. Roman Catholic. Subspecialties: Pathology (medicine); Gastroenterology. Current work: Studying pathogenesis of chemically-induced diseases. Home: 46 Clearwater Rd Brookline MA 02167 Office: 75 Francis St Boston MA 02115

SZAL, MARCEL MICHAEL, radiological physicist, biologist; b. McKees Rocks, Pa., Nov. 2l, 1954; s. Valerian F. and Florence (Drost) S.; m. Kathleen Ann Koczur, June 20, 1981. M.S., U. Pitts., 1981. Radiol. physicist Mid-East Center for Radiol. Physics, Allegheny Gen. Hosp., Pitts., 1980—. Mem. Nuclear Medicine Soc., Health Physics Soc., Am. Assn. Physicists in Medicine. Subspecialties: Cancer research (medicine); Radiology. Current work: Radiation oncology. Home: 238 Helen St McKees Rocks PA 15136 Office: 320 E North Ave Pittsburgh PA 15212

SZALAI, IMRE A., physicist, educator; b. Gyula, Hungary, Sept. 20, 1936; s. Istvan J. and Veron A. (Galbatz) S.; m. Beatrice Benner; children: Veronika, Brandon, Temi, Diana. B.A. in Physics; M.S. in Applied Physics. Instr. physics Bell and Howell Inst., 1962-63; mem. faculty, chmn. dept. physics Spring Garden Coll., Phila., from 1963; prof. Delaware County Community Coll., Media, Pa., 1968—. Vice-pres. Montgomery County Pub. Library Friends. Mem. Am. Assn. Physics Tchrs. Subspecialties: Solar energy; Nuclear physics. Current work: Solar energy; investigation of solar ponds using microprocessors. Office: Delaware County Community Coll Media Line Rd Media PA 19063

SZALDA, DAVID JOSEPH, chemistry educator, researcher; b. Buffalo, May 25, 1950; s. Aloysius and Florence (Lewandowski) S.; m. Victorai Anne Galgano, June 2, 1974; 1 dau., Dava Elizabeth. B.S., Manhattan Coll., 1972; M.A., Johns Hopkins U., 1974, Ph.D., 1976. Postdoctoral research assoc. Columbia U., N.Y.C., 1976-78; assoc. prof. chemistry Bernard M. Baruch Coll., N.Y.C., 1978—; research collaborator Brookhaven Nat. Lab., Upton, N.Y., 1981—. Contbr. articles to profl. jours. NIH fellow, 1976-78; Koscuiszo Found. fellow, 1972-73. Mem. Am. Chem. Soc., Sigma Xi, Phi Beta Kappa. Democrat. Roman Catholic. Subspecialties: Inorganic chemistry; X-ray crystallography. Current work: Study of the effects of electron transfer reactions on bond lengths in metal complexes. Study of metal binding to nucleic acid bases. Home: 17 Robin Ln Kings Park NY 11754 Office: Baruch Coll 17 Lexington Ave New York NY 10010

SZARA, STEPHEN, psychopharmacologist; b. Budapest, Hungary, Mar. 21, 1923; came to U.S., 1957, naturalized, 1963; s. Janos and Maria (Katona) S.; m. Madeline Gadanyi, Sept. 5, 1959 (div.); 1 son, Christopher. D.Sc. in Chemistry, Pazmany U., Budapest, 1950; M.D., Med. U. Budapest, 1951. Asst. prof. Biochem. Inst. Med. U., Budapest, 1950-53; chief biochem. lab. Central State Inst. Mental Disease, Budapest, 1953-56; chief sect. on pscyhopharmacology NIMH, Washington, 1957-71, chief clin. drug studies sect., Rockville, Md., 1971-74; chief biomed. br. div. research Nat. Inst. Drug Abuse, Rockville, 1974—. Author: (with M. Meszaros) Organic Chemistry, 1954, (with H. Weil, Malherbe) Biochemistry of Functional and Experimental Psychoses, 1971; editor: (with M.C. Braude) The Pharmacology of Marihuana, 1976; contbr. numerous articles on psychopharmacology to profl. jours. Fellow Am. Coll. Neuropsychopharmacology; mem. AAAS, Am. Soc. Pharmacology and Exptl. Therapeutics, Internat. Narcotic Research Conf., Collegium Internationale Neuropsychopharmacologicum, Computer Soc., IEEE. Subspecialties: Psychopharmacology; Neuropharmacology. Current work: Biomedical research on drugs of abuse (opiates, marijuana); program director for grant support in this area. Office: 5600 Fishers Ln 10A-31 Rockville Md 20857

SZE, PAUL YI LING, neurobiologist; b. China, June 7, 1938; s. Kung-Meng and Yuet-Fong (Wong) S. Ph.D., U. Chgo., 1969. Asst. prof. biobehavioral scis. dept. U. Conn., Storrs, 1969-72, assoc. prof., 1972-80, prof., 1980—; mem. neurology A study sect. NIH. Contbr. articles to profl. jours. Mem. Soc. Neurosci., Am. Soc. Neurochemistry, Internat. Soc. Neurochemistry. Subspecialties: Comparative neurobiology; Neurochemistry. Current work: biochemical regulatory functions of hormones in the mammalian central nervous system, particularly those involved in neural and behavioral activities. Office: Dept Biobehavioral Scis U Conn Storrs CT 06268

SZEBENYI, EMIL STEVEN, cancer researcher, corporation president; b. Budapest, Hungary, June 9, 1920; came to U.S., 1957, naturalized, 1962; s. Edmund and Franciska (Jalovits) S.; m. Clasa Goots de Jaszo, Apr. 21, 1944; children: Thomas, Andrew, Steve. Diploma agr., zootechnic, Poly. U., Budapest, 1942, doctorate in zoo-genetics, 1943. Asst. prof. animal genetics Agrl. U., Godollo, Hungary, 1950-51, assoc. prof., 1952-56; asst. prof. Fairleigh Dickinson U., Rutherford, N.J., 1962-69, assoc. prof., 1969-73, prof., 1973—, chmn. dept. biol. scis., 1971-80; pres. Alfacell Corp., Bloomfield, N.J., 1982—. Author: Atlas of Macaca Mulatta, 1969, Atlas of Developmental Embryology, 1979, Anatomy of Squalus Acautias, 1982, Anatomy of Felis domestica, 1982. Mem. presdl. task force Republican Party, 1980—. Mem. Soc. Developmental Biology, Am. Soc. Zoology, Am. Soc. Morphogenesis, Am. Assn. Lab. Animal Sci., AAAS, Sigma Xi. Subspecialties: Cancer research (veterinary medicine); Animal breeding and embryo transplants. Current work: Cancer; antitumor agents. Home: 5 Stephen Pl Little Falls NJ 07424 Office: Alfacell Corp 225 Belleville Ave Bloomfield NJ 07003

SZEGO, GEORGE CHARLES, engr.; b. Budapest, Hungary, Aug. 10, 1919; s. Paul S. and Helen E. (Elek) S.; m.; children: Deirdre Alexandra Wind, Luisa, Viviana Flor de Maria, Thea Thu, Vivian Rose, Isobel, Camilla Tuyet. B.S., U. Denver, 1947, M.S., 1950; Ph.D., U. Wash., 1956. Registered profl. engr. 19 states. Prof., head dept. chem. engring. Seattle U., 1948-56; pres. G.C. Szego & Assocs., Seattle, 1951-56; mgr. space power and space propulsion Gen. Electric Co., 1956-59; sr. staff mem. TRW Space Tech. Labs., Los Angeles, 1959-61, Inst. Def. Analyses, 1961-70; pres., chmn InterTech. -Solar Corp., Warrenton, Va., 1970-81; pres. Dr. George C. Szego, P.E. & Assocs., Inc., Arlington, Va., 1981—; dir. ITC/Solar Do Brasil Ltda., ITC/Solar, Italia, Korea; founding chmn. Intersoc. Energy Conversion Engring. Confs., Solar Energy Research Found. Contbr. chpts. to books. Served with AUS, 1943-46. Recipient 9th Ann. award for energy efficient design Owens-Corning, 1980. Fellow Am. Inst. Chem. Engrs. Subspecialties: Chemical engineering; Solar energy. Current work: Renewable, cost-competitive, CO2-free energy; synfuels from biomass. Patentee in field. Office: PO Box 4070 Annapolis MD 21403

SZEWCZYK, ALBIN ANTHONY, aerospace and mech. engr., educator, researcher, cons.; b. Chgo., Feb. 26, 1935; s. Andrew Aloysius and Jean Cecelia (Wojcik) S.; m. Barbara Valerie, June 16, 1956; children: Karen Marie Szewczyk Finkenbinder, Lisa Anne, Andrea Jean Szewczyk Harmon, Terese Helene. B.S.M.E., U. Notre Dame, 1956, M.S.M.E., 1958; Ph.D., U. Md., 1961. Mem. tech. staff Aerospace Corp., El Segundo, Calif., summer 1962; asst. prof. aerospace and mech. engring. U. Notre Dame, 1962-65, assoc. prof., 1965-67, prof., 1967—, chmn. dept. aerospace and mech. engring., 1978—; cons. Argonne Nat. Lab.; cons. as expert witness. Research numerous publs. in field. Office Naval Research grantee, 1967-72; NSF grantee, 1971-82. Mem. Am. Phys. Soc., ASME, AIAA. Roman Catholic. Clubs: Univ. (Notre Dame, Ind.); South Bend (Ind.) Country. Subspecialties: Mechanical engineering; Fluid mechanics. Current work: Experimental fluid mechanics; bluff body flows. Office: Dept Aerospace and Mech Engring Notre Dame Notre Dame IN 46556

SZILAGYI, MIKE NICHOLAS, engineer, physicist, educator; b. Budapest, Hungary, Feb. 4, 1936; came to U.S., 1981; s. Karoly and Ilona (Abraham) S.; m. Larissa Dorner, Feb. 23, 1957; m. Julia Levai, May 31, 1975; children: Gabor, Zoltan Charles. Diploma in Engring.-Physics, Poly. U. Leningrad, 1954-60; Ph.D., 1965; D.Tech., Tech. U., Budapest, 1965; D.Sc., Hungarian Acad. Sci., 1979. Research asst. dept. phys. electronics Poly U. Leningrad, 1958-60; research assoc. Research Inst. Tech. Physics, Hungarian Acad. Sci., Budapest, 1960-66; sci. adviser Nat. Research Inst. Neurosurgery, Budapest, 1966-70; chief research scientist, head Lab. Electron Optics, Tech. U., Budapest, 1966-71; prof. physics head dept. phys. scis. K. Kando Coll. Elec. Engring., Budapest, 1971-79, rector, 1971-74; guest prof. Aarhus U., Denmark, 1979-81; vis. sr. research assoc. engring. physics Cornell U., Ithaca, N.Y., 1981-82; prof. elec. engring. U. Ariz., Tucson, 1982—. Author: Introduction to the Theory of Space Charge Optics, 1974, Fachlexikon Physik, 1977, others; Contbr. articles to profl. jours. Mem. Am. Phys. Soc., IEEE (sr.), Danish Phys. Soc., Danish Engring. Soc., European Soc. Sterotactic and Functional Neurosurgery, J. Neumann Soc. Computer Sci., L. Eötvös Phys. Soc. (Brody prize 1964). Subspecialties: Computer-aided design. Current work: Teaching and research in computer-aided design of various engineering devices used in the field of microelectronics, ion-beam lithography and physical electronics. Office: University of Arizona Dept Electrical Engineering Tucson AZ 85721

TAAFFE, GORDON, psychologist; b. Atlanta, Aug. 22, 1916; s. Roderick Arthur and Susan Stewart (Fairbanks) T. B.S., Tulane, U., 1948; M.A., U. So. Calif., 1952; Ed.D., Wayne State U., 1968.

Research assoc. John Tracy Clinic, Los Angeles, 1956-61; project dir. Los Angeles State Coll., 1961-63; counselor of students U. Detroit, 1963-68, Health Care Research, Blue Cross & Blue Shield, Detroit, 1968—; bd. dirs. Health Service Adv. Com., Ann Arbor, Mich., 1978-79; exec. dir. Mich. Health Data Corp., Detroit, 1980-82; adj. prof. Oakland Community Coll., Farmington Hills, Mich., 1979—. Bd. dirs. Native Am. Vocat. Orgn., Pontiac, Mich., 1982. Served with U.S. Army, 1941-44; PTO. Mem. Am. Psychol. Assn., Mich. Psychol. Assn. Democrat. Club: Civil War (Dearborn, Mich.). Subspecialties: Information systems, storage, and retrieval (computer science); Statistics. Current work: Studies of patterns and profiles of health care utilization in all medical environments. Office: Blue Cross and Blue Shield Mich 600 Lafayette E Detroit MI 48226

TABACHNICK, WALTER JAY, geneticist; b. N.Y.C., June 14, 1947; s. Jack and Sylvia (Breidbart) T. B.S., CUNY, 1968; M.S., Rutgers U., 1971, Ph.D., 1974. Asst. prof. U. Wis.–Parkside, Kenosha, 1973-75; postdoctoral fellow Yale U., New Haven, 1975-78, research assoc., 1979-82; asst. prof. Loyola U., Chgo., 1982—; lectr. UCLA, 1978-79. Recipient Research Individual Service award NIH, 1975-78, Young Investigator award NIH, 1980-83; NDEA Title IV fellow, 1970-73. Mem. AAAS, Genetics Soc. Am., Am. Genetic Assn., Soc. Study Evolution, Am. Soc. Tropical Medicine and Hygiene, Entomol. Soc. Am., Sigma Xi. Subspecialty: Evolutionary biology. Current work: Genetics and evolution of arthropod vectors of human diseases, particularly the yellow fever mosquito Aedes Aegypti; genetic basis of the ability of mosquitoes to vector human disease. Office: Dept Biology Loyola U Chicago IL 60606

TABER, DOUGLASS FLEMING, chemistry educator; b. Berkeley, Calif., Nov. 11, 1948; s. Richard Douglas and Barbara (Fleming) T.; m. Susan Buhler, Dec. 30, 1969; children: John, Alan, Emma, Christina, Abigail. B.S., Stanford U., 1970; Ph.D., Columbia U., 1974. Research instr. dept. pharmacology Vanderbilt U., Nashville, 1975-77, asst. prof., 1977-82; asst. prof. dept. chemistry U. Del., Newark, 1982—; cons. dept. pharmacology Vanderbilt U., 1982—. Petroleum Research Fund grantee, 1976-83; NIH grantee, 1977—; Alfred P. Sloan Found. fellow, 1983—. Mem. Am. Chem. Soc. Mormon. Subspecialties: Organic chemistry; Synthetic chemistry. Current work: New methods in synthetic organic chemistry; natural product synthesis; synthesis with stable isotopes; drug metabolites; medicinal chemistry. Home: 717 Harvard Ln Newark DE 19711 Office: Dept Chemistry U Del Newark DE 19711

TABER, HARRY WARREN, molecular biologist, educator; b. Longview, Wash., Oct. 30, 1935; s. Russell James and Mildred Lucile (Schulstad) T.; m. Jeroo S. Kotval, Apr. 29, 1976; 1 dau., Shahnaz Kirsten. B.A. in Chemistry, Reed Coll., 1957; Ph.D. in Biochemistry, U. Rochester, 1963. Guest worker NIH, Bethesda, MD., 1964-66; research staff Centre National de la Recherche Scientifique, Gif-sur-Yvette, France, 1966-67; asst. prof. microbiology and radiation biology/biophysics U. Rochester, N.Y., 1968-73, assoc. prof., 1973-79; assoc. prof. microbiology and immunology Albany (N.Y.) Med. Coll., 1979-80, prof., 1980—; vis. prof. biochemistry U. Calif., Berkeley, 1975-76; mem. ad hoc rev. group NIH, 1979—. Contbr. articles to sci. jours. Active Planned Parenthood. Recipient NIH Career Devel. award, 1974-79. Mem. Am. Soc. Microbiology, Am. Soc. Photobiology, Soc. Gen. Microbiology, AAAS, Genetics Soc. Am., N.Y. Acad. Scis., Common Cause, Sierra Club, Nature Conservancy. Democrat. Unitarian. Subspecialties: Genetics and genetic engineering (biology); Molecular biology. Current work: Research in molecular biology of membrane function. Home: 135 Hackett Blvd Albany NY 12208 Office: Albany Medical College Albany NY 12208

TABER, ROBERT IRVING, pharmacologist; b. Perth Amboy, N.J., June 28, 1936; m. (married), 1960; children: Scott and Stacy (twins), Jennifer. B.S., Rutgers U., 1958; Ph.D., Med. Coll. Va., 1962. Pharmacologist research div. Schering Plough Pharm., Bloomfield, N.J., 1962-66, sr. pharmacologist, 1967-71, mgr. pharmacology, 1971-74, assoc. dir. biol. research, 1974-77, dir. biol. research, 1977-82; dir. pharm. research DuPont Pharms., Glenolden, Pa., 1982—. Mem. AAAS, Am. Coll. Neuropsychopharmacology, Am. Soc. Pharmacology and Exptl. Therapeutics, Acad. Pharm. Sci., Am. Pharm. Assn. Subspecialty: Pharmacology. Home: 120 Guernsey Rd Swarthmore PA 19081 Office: Du Pont Co Glenolden Lab Glenolden PA

TABER-PIERCE, ELIZABETH, neuroscientist; b. Paterson, N.J., Sept. 24, 1926; d. Leslie Ray and Orrell Margaret (Bond) Taber; m. Elleson C. Pierce, Aug. 16, 1957; children: Charles A., Cynthia Carol. B.A., Mt. Holyoke Coll., 1948; Ph.D., Columbia U., 1958. Postdoctoral fellow U. Oslo, 1959; with Harvard U. Med. Sch., Boston, 1960—, prin. assoc. dept. anatomy, 1976—. Author: (with Sidman and Angevine) Atlas of the Mouse Brain and Spinal Cord, 1971. Mem. AAAS, Am. Assn. Anatomists, Am. Soc. Neurologists, Cajal Club, World Affairs Council. Subspecialty: Neurophysiology. Current work: Study of development in the nervous system. Office: Dept Psychiatry Mass Gen Hosp Fruits Boston MA

TABOR, HERBERT, biochemist; b. N.Y.C., Nov. 28, 1918; s. Edward and Henrietta (Tally) T.; m. Celia White, Apr. 8, 1946; children: Edward, Marilyn, Richard, Stanley. A.B., Harvard U., 1937, M.D., 1941. Intern Yale U. and New Haven Hosp., 1942; chief Lab. Biochem. Pharmacology Nat. Inst. Arthritis, Metabolism and Digestive Diseases, Bethesda, Md.; med. dir. USPHS. Editor-in-chief: Jour. Biol. Chemistry; Contbr. articles to profl. jours. Mem. Am. Soc. Biol. Chemistry, Nat. Acad. Sci., Am. Acad. Arts and Scis., Am. Soc. Pharm. and Exptl. Therapeutics, Am. Chem. Soc. Subspecialty: Biochemistry (medicine). Office: Bldg 4 Room 110 NIH Bethesda MD 20205

TABOREK, JERRY, design engr.; b. Klaster, Czechoslovakia, Oct. 29, 1922; s. Jaroslav and Jana T.; m. (widowed); children: Peter, Eric. Dr.Eng., Techn. U., Dresden (W.Ger.), 1947. Design engr. Massey-Ferguson, Toronto, Ont., Can., 1951-55; hydraulic engr. Tow Motor Corp., Cleve., 1955-56; heat transfer/computer engr. Phillips Petroleum Co., Bartlesville, Okla., 1956-62; tech. dir. Heat Transfer Reserach Inc., Alhambra, Calif., 1962—. Mem. editorial bd.: Heat Transfer Engineering; author articles. Recipient Transenter medal U. Liege, Belgium, 1982. Mem. Am. Inst. Chem. Engrs. (D.Q. Kern award 1976, Tech. Achievement award 1974). Subspecialties: Chemical engineering; Database systems. Current work: Heat transfer research, heat exchanger technology and computer program for heat exchanger design. Home: 2165 Canyon Rd Arcadia CA 91006 Office: 1000 S Fremont Alhambra CA 91802

TACHE, YVETTE FRANCE, neuro-gastroenterologist; b. Lyon, Rhone, France, Feb. 1, 1945; came to U.S., 1982; d. Lucien Joseph and Jeanne Marthe (Fouillat) Laurent; m. Jean Arthur Tache, June 20, 1970 (dec. 1979); children: Stephanie, Veronique. Baccalaureat, Lycée Tarare, 1965; Maitrise, Faculty Scis. U. Claude Bernard, Lyon, France, 1968, D.E.A., 1969; Ph.D., Faculty Medicine U. Montreal, 1974. Asst. research prof. U. Montreal, Que., Can., 1977-78, asst. research prof., 1980-81, assoc. research prof., 1981-82; vis. scientist Salk Inst., La Jolla, Calif., 1978-80; assoc. prof. in residence UCLA, 1982—; external referee Med. Research Council Que., 1981—, mem. selection com., 1982. External referee specialized sci. jours., U.S., 1977—; contbr. writings to publs. Fellow Med. Research Council Que., 1974-78; fellow Med. Research Council, 1974-78; Centennial fellow, 1978-80; scholar, 1982; research grantee Med. Research Council Que., Med. Research Council Can., NIH, 1977. Mem. Internat. Soc. Psychoneuroendocrinology, Endocrine Soc., Soc. Neurosci., Am. Physiol. Soc., Am. Gastroent. Soc., Brain Research Inst. Subspecialties: Gastroenterology; Neurophysiology. Current work: Independent research on brain control of gastric function. Home: 1180 McClellan Dr Los Angeles CA 90049 Office: CURE VA Wadsworth Bldg 115 Room 217 Los Angeles CA 90073

TAFTLE, RAYMOND, physician; b. Washington, 1947; s. Herman and Shirley T.; m. Laurie Daniels, 1969; children: Julie, Sarah. B.S., U. Mich., 1969; M.D., Northwestern U., 1973. Intern U. Calif.-San Diego, 1973-74, resident in medicine, 1974-76, fellow, 1976—. Subspecialties: Hematology; Oncology. Office: Univ Hosp 225 W Dickinson St San Diego CA 92103

TAGGART, GEORGE BRUCE, physicist; b. Phila., Apr. 8, 1942; s. Robert Henry and Rachel Elizabeth (Burtt) T.; m. Grace Elizabeth Beamon, Feb. 2, 1979. B.S., Coll. William and Mary, 1964; postgrad., U. Pa., 1964-65; Ph.D., Temple U., 1971. Vis. asst. prof. dept. physics Temple U., Phila., 1970-71; prof. physics Va. Commonwealth U., Richmond, 1971-83; sr. staff mem. in advanced tech. BDM Corp., McLean, Va., 1983—; vis. prof. Oxford (Eng.) U., 1978; guest worker Nat. Bur. Standards, Gaithersburg, Md., 1978—; vis. assoc. prof. U. Ill.-Urbana, 1978-79; vis. prof. Fed. U. Pernambuco, Recife, Brazil, 1980; vis. scientist Naval Research Lab., Washington, 1982-83. Contbr. articles in field to profl. jours. Ford Found. fellow U. Pa., 1964; Temple U. fellow, 1968-70; Am. Soc. Engrng. Edn. grantee, 1982. Mem. Am. Ceramic Soc., Materials Research Soc., Fedn. Am. Scientists; mem. Am. Soc. Metals; Mem. AAAS.; mem. Am. Carbon Soc., Metall. Soc. of AIME. Subspecialties: Condensed matter physics; Alloys. Current work: Laser-Material interactions, composite materials, materials science. Home: 116622 Newbridge Ct Reston VA 22091 Office: BDM Corp 7915 Jones Branch Dr McLean VA 22102

TAGGART, ROBERT THOMAS, geneticist educator; b. Long Beach, Calif., July 4, 1951; s. Robert Thomas and Elizabeth Ann (DeKay) T. B.S., U. Denver, 1973; Ph.D., Ind. U., 1978. NIH fellow, dept. human genetics Yale U., 1978-80; asst. prof. medicine UCLA, 1980—; chief human genetics research VA Med. Center, Sepulveda, Calif., 1981—. Contbr. articles to profl. jours. Mem. AAAS, Am. Soc. Human Genetics, Genetics Soc. Am., Sierra Club, Phi Beta Kappa, Sigma Xi. Episcopalian. Subspecialties: Immunogenetics; Genetics and genetic engineering (medicine). Current work: Hybridoma research in cellular and biochemical immunology. Office: VA Medical Center Human Genetics Research 16111 Plummer St Sepulveda CA 91343

TAI, CHEN-YU, physicist; b. Hofei, An-Whi, China, Sept. 4, 1945; s. Tseng-Hsi and Wen-Min (Chang) T.; m. Ping-Kuang Ku, Dec. 29, 1971; children: Yunsian Tai, Jan-Sian Tai. B.S., Nat. Taiwan U., 1968; M.A., Columbia U., 1971, M.Phil., 1973, Ph.D., 1974. Postdoctoral fellow U. B.C., Vancouver, 1974-76, research assoc., 1976-79; asst. prof. physics U. Toledo, 1979-83, assoc. prof. physics, 1983—. Contbr. articles to profl. jours. NSF grantee, 1981—. Mem. Am. Phys. Soc., Sigma Xi. Subspecialty: Atomic and molecular physics. Current work: Coherent spectroscopy, nonlinear optics, atomic and molecular structure. Office: Dept Physics and Astronomy Univ Toldeo 2801 W Bancroft St Toledo OH 43606

TAI, DOUGLAS LEUNG-TAK, radiol. physicist; b. Hong Kong, Nov. 6, 1940; s. Kam Yue and Sau Lam (Ching) T.; m. Christine Yiu-Kam, June 2, 1943; children: Stephanie, Oliver. Ph.D., Cornell U., 1969. Postdoctoral research fellow Cornell U., 1969-71; faculty research fellow Ariz. State U., Tempe, 1971-72, vis. asst. prof., 1972-73; instr. U. Ky., Lexington, 1973-77, PHS fellow, 1975-77; instr. U. Miss. Med. Center, Jackson, 1977-79, asst. prof., 1979-81, U. Tenn. Center for Health Scis., Memphis, 1981—; radiol. physicist City of Memphis Hosp., 1981—. Contbr. articles to profl. jours. PHS Fellow, 1975-77. Mem. Am. Assn. Physicists in Medicine, Soc. Photo-optical Instrumentation Engrs., Health Physics Soc. Subspecialties: Radiological physics; Imaging technology. Current work: Computer treatment planning for radiotherapy; radiation dosimetry optimization technique for radiotherapy. Office: Dept Radiology U Tenn Center for the Health Scis 865 Jefferson Ave Memphis TN 38163

TAI, TSZE CHENG, aerodynamicist, research aerospace engineer; b. Shaoxing, Checkiag, China, Apr. 29, 1933; came to U.S., 1963, naturalized, 1972; m. Shih Lin Sun, Aug. 27, 1965; children: Kuangheng, Kuangkai, Kuangshin. M.S., Clemson U., 1965; Ph.D., Va. Poly. Inst., 1968. Aircraft insp. Taoyuan Airbase, Taoyuan, Taiwan, 1958-63; research asst. Clemson U., 1963-65; grad. asst. Va. Poly. Inst., 1965-67, instr., 1967-68; research engr. Naval Ship Research and Devel. Ctr., Bethesda, Md., 1968—; chmn. panel U.S. Navy Aeroballistics Com., Washington, 1978-81; lectr. von Karman Inst. Fluid Dynamics, Belgium, 1980; Prin. Potomac (Md.) Chinese Sch., 1981-82. Recipient Eugene Brooks award Naval Ship Research and Devel. Center, 1979. Assoc. fellow AIAA; mem. Sigma Xi (chmn. awards com. 1979-80). Subspecialties: Fluid mechanics; Aerospace engineering and technology. Current work: Research in transonic aerodynamics and three-dimensional flow separation. Home: 10705 Tara Rd Potomac MD 20854 Office: David Taylor Naval Ship Research and Devel Center Code 1606 Bethesda MD 20084

TAI, YUAN-HENG, research physiologist; b. Quinming, Yunnan, China, Aug. 14, 1941; came to U.S., 1965; s. Pei-Chih and I-Yun (Chou) T.; m. Chung-Yui Betty Li, Mar. 25, 1972; 1 dau., Katherine Chi. B.S., Nat. Taiwan U., 1963; M.Phil., Yale U., 1969, Ph.D., 1971. Postdoctoral assoc. Yale U., New Haven, 1971-74, research assoc., 1974-76; research physiologist Walter Reed Army Research Inst., Washington, 1976—. Contbr.: articles to profl. jours. including Gastroenterology. Mem. Am. Physiol. Soc., Biophys. Soc., N.Y. Acad. Scis. Subspecialties: Physiology (medicine); Biophysics (physics). Current work: Membrane physiology and biophysics, particularly the transport mechanisms of electrolytes and non-electrolytes across biological membranes. Home: 1 Shanandale Ct Silver Spring MD 20904 Office: Walter Reed Army Inst Research Washington DC 20307

TAINTOR, JERRY FRANK, dental educator, researcher; b. Trenton, Aug. 10, 1942; s. Aubra Ray and Frances Lilly (Marek) Nooncaster; m. Mary Jane McComas, Feb. 1, 1965. B.S., Okla. State U., 1965; D.D.S., U. Mo.-Kansas City, 1967; M.S., U. Iowa, 1975. Asst. prof. Med. Coll. Ga., 1975-76, U. Nebr.-Lincoln, 1976-79; assoc. prof. endodontics, chmn. dept. UCLA, 1979—; cons. Sepulveda VA Hosp., Los Angeles, 1982—. Author: (with J.I. Ingle) Endodontics, 1982; contbr. numerous articles to profl. jours. Bd. dirs. Am. Cancer Soc., Alaska, 1971. Med. Coll. Ga. grantee, 1975-76; U. Nebr. grantee, 1977-78. Fellow So. Calif. Acad. Endodontics; mem. AAUP (v.p. UCLA chpt. 1981-82, pres. 1982-83), Am. Assn. Endodontics (chmn. com. 1979—), ADA, Clyde Davis Endodontic Study Club (sec. 1977-80). Democrat. Presbyterian. Subspecialty: Endodontics. Current work: Research on pulpal biology. Home: 5021 Tilden Ave Sherman Oaks CA 91423 Office: UCLA Sch Dentistry Center Health Scis Los Angeles CA 90024

TAKAHASHI, ELLEN SHIZUKO, optometrist, physiol. opticist, educator, researcher; b. Berkeley, Calif., May 4, 1931; d. Henry and Barbara (Yammamoto) T.; m. Clyde William Oyster, May 27, 1967. M.Opt., U. Calif., Berkeley, 1953, Ph.D., 1968. Cert. Calif. State Bd. Examiners in Optometry. Pvt. practice optometry, Berkeley, Calif., 1953-56; optometrist Stanford Univ. Hosps., San Francisco, 1956-62; from clin. instr. to clin. prof. optometry U. Calif., Berkeley, 1961-67; research fellow Australian Nat. U., 1968-70; sr. research fellow Med. Faculty Rotterdam, Netherlands, 1970; from asst. prof. to prof. physiol. optics U. Ala., Birmingham, 1970—; mem. visual scis. study sect. NIH, 1978-82, mem. adv. council health professions edn., 1972-75. Contbr. articles to sci. jours. NIH grantee, 1968—. Fellow Am. Acad. Optometry; mem. Assn. Research in Vision and Ophthalmology, Soc. Neurosci. Subspecialties: Sensory processes; Optometry. Current work: Vision research: anatomy and physiology of vertebrate retina and visual pathways. Home: 3517 Belle Meade Way Birmingham AL 35223 Office: U Ala Birmingham AL 35294

TAKASHIMA, SHIRO, bioengineering educator; b. Tokyo, May 12, 1923; s. Tokuji and Yoshie (Miyoshi) T.; m. Yuki Morita, June 26, 1953; children: Nozomi L., Makoto D. B.S., U. Tokyo, 1947, Ph.D., 1953. Postdoctoral fellow in chemistry U. Minn., 1955-57; research assoc. U. Pa. Sch. Enring., 1957-59; prof. Protein Research Inst., Osaka (Japan) U., 1959-63; vis. scientist Walter Reed Army Med. Ctr., Washington, 1963-64; asst. prof. U. Pa. Sch. Engrng., Phila., 1964-70, assoc. prof., 1970-76, prof., 1976—. Contbr. chpts. to books, articles to sci. jours. Grantee NIH, NSF, Office Naval Research. Mem. Biophys. Soc., N.Y. Acad. Sci., IEEE, Bioelectric Magnetism Soc., Marine Biol. Lab. Democrat. Methodist. Subspecialties: Biophysics (biology); Biomedical engineering. Current work: Electrical properties of biological membranes and macromolecules; research on sickle cell anemia; interaction of biological systems with electromagnetic energy. Home: 263 Beechwood Ave Springfield PA 19064 Office: Sch Engring U Pa Philadelphia PA 19104

TAKEMORI, AKIRA EDDIE, pharmacologist, educator, researcher, cons.; b. Stockton, Calif., Dec. 9, 1929; s. Matsutaro and Haruko (Teshima) T.; m. Valerie Williams, June 22, 1958; children: Tensho, Rima. A.B. in Physiology, U. Calif., Berkeley, 1951, M.S. in Comparative Pharmacology and Toxicology, 1953; Ph.D. in Pharmacology, U. Wis., 1958. Instr., asst. prof. SUNY Upstate Med. Center, Syracuse, 1959-63; asst. prof. pharmacology U. Minn., Mpls., 1963-65, assoc. prof., 1965-69, prof., 1969—; researcher; cons. Contbr. numerous articles to profl. jours. Bd. dirs. Minnehon Arts Center, 1973-79, pres., 1975; mem. Edina (Minn.) Baseball Bd., 1978-81; precinct del. Democratic-Farm Labor Party, 1972. Served with U.S. Army, 1953-55. Am. Cancer Soc. postdoctoral fellow, 1958-59; Nat. Acad. Scis. internat. travel award, 1962, 65; China Med. Bd. N.Y. Alan Gregg fellow in med. edn., 1971; recipient Scientist award Japan Soc. for Promotion of Sci., 1971; Basic Sci. Tchr. award U. Minn., 1979. Mem. Am. Soc. for Pharmacology and Exptl. Therapeutics, AAAS, Am. Chem. Soc., Am. Coll. Neuropsychopharmacology, Soc. for Exptl. Biology and Medicine, AAUP, Minn. Assn. for Retarded Citizens, Sigma Xi. Buddhist. Subspecialties: Pharmacology; Neuropharmacology. Current work: Mechanism of action of narcotic analgesics; mechanism of narcotic tolerance and dependence. Home: 5237 Wooddale Ave Edina MN 55424 Office: 3-260 Millard Hall U Minn 435 Delaware St. SE Minneapolis MN 55455

TAKEMOTO, KENNETH K., virologist; b. Kauai, Hawaii, Sept. 26, 1920; s. Yutaro and Sumi (Hironaka) T.; m. Alice Imamoto, Jan. 28, 1951; children: Ruth E., Pual H. B.S., George Washington U., 1948, M.S., 1950, Ph.D., 1953. Postdoctoral fellow NIH, Bethesda, Md., 1953-54, virologist, 1958-68, 1968-78, chief sect. on viral biology, 1978—. Contbr. chpts. to books, articles to profl. jours. Served with U.S. Army, 1943-45. Decorated Purple Heart, Bronze Star.; Recipient Commendation medal USPHS, 1971, Meritorious Service medal, 1978. Mem. Am. Soc. Microbiologists, Am. Soc. Virologists. Democrat. Subspecialties: Virology (biology); Cell and tissue culture. Current work: DNA-containing tumor viruses; tumor cell biology. Home: 11308 Mitscher St Kensington MD 20895 Office: NIH Bethesda MD 20205

TAKESUE, EDWARD I., clin. researcher; b. Honolulu, Dec. 31, 1927; s. Buichi and Tazu (Harada) T.; m. Edna S. Takesue, Aug. 1, 1957; children: Blaine, Valerie, Evelyn, Burt. B.S. in Pharmacy, Phila. Coll. Pharmacy and Sci., 1952; M.S., Purdue U., 1954, Ph.D., 1955. Registered pharmacist, Pa., 1952. Pharmacologist Columbus-Pharmacal, 1955-58, Lederle Labs., Pearl River, N.Y., 1958-62; assoc. sect. head Sandoz Pharm. Co., Hanover, N.J., 1962-72, fellow clin. research., 1972-75; dir. biomed. pharmacology Purdue Frederick Co., Norwalk, Conn., 1975—. Contbr. numerous articles to profl. jours. Served with U.S. Army, 1946-47. Mem. Am. Soc. Pharmacology and Exptl. Therapeutics, N.Y. Acad. Sci., Am. Rheumatism Assn., AAAS. Subspecialties: Pharmacology; Clinical research, pharmacology. Current work: Conduct and design trials of pre-clinical and clinical nature to determine safety and efficiency of drugs. Patentee in field. Home: 109 Wilson Rd Easton CT 06612 Office: 50 Washington St Norwalk CT 06856

TALAMO, BARBARA RUTH, neurobiologist, educator; b. Washington, May 30, 1939; d. Isaac and Tessie (Silverman) Lisann; m. Richard Charles Talamo, June 22, 1958 (dec.); children: Jonathan H., David A., Anna B. A.B., Radcliffe Coll., 1960; Ph.D., Harvard U., 1972. Asst. prof. neurology, asst. prof. physiol. chemistry Johns Hopkins U. Sch. Medicine, Balt., 1974-80; assoc. prof. neurology, asst. prof physiology Tufts U. Sch. Medicine, Boston, 1980—, dir. neuroscis. program. NIH fellow, 1972-74; Grass fellow, 1973; NIH grantee, 1976—. Mem. Soc. for Neuroscis., Am. Soc. Neurochemistry, Internat. Soc. Neurochemistry. Subspecialties: Neurochemistry; Neurobiology. Current work: Neural regulation of secretory development, receptor mediated sensitivity. Office: 136 Harrison Ave S Cove 603 Boston MA 02111

TALAPATRA, DIPAK CHANDRA, mech. engr.; b. Bangladesh, Jan. 20, 1942; U.S., 1971, naturalized, 1977; s. Upendra Chandra and Jalada Sundari (Banik) T.; m. Brigitte Hildegard Talapatra, Dec. 14, 1973; children: Idrani, Anika. B.Tech. with honors, Indian Inst. Tech., Kharagpur, India, 1963; M.Eng.; NRC research asst., McGill U., 1968; Ph.D.M.E., U. B.C., 1972. Sr. scientist Ensco, Inc., Springfield, Va., 1972-77; sr. research engr. Gen. Tire & Rubber Co., Akron, Ohio, 1977-80; mech. engr. Dept. of Navy Naval Ordnance Sta., Indian Head, Md., 1980; aerospace engr. NASA Goddard Space Flight Center, Greenbelt, Md., 1981—. Contbr. articles to profl. jours. Mem. ASME, Soc. Exptl. Stress Analysis. Subspecialties: Theoretical and applied mechanics; Aerospace engineering and technology. Current work: Finite element methods, structural dynamics, composites, fracture mechanics, environmental testing of space shuttle payloads. Patentee in field. Office: NASA Goddard Space Flight Center Greenbelt MD 20771

TALMAGE, DAVID WILSON, physician; b. Kwangju, Korea, Sept. 15, 1919; s. John Van Neste and Eliza (Emerson) T.; m. LaVeryn Marie Hunicke, June 23, 1944; children: Janet, Marilyn, David, Mark, Carol. Student, Maryville (Tenn.) Coll., 1937-38; B.S., Davidson (N.C.) Coll., 1941; M.D., Washington U., St. Louis, 1944. Intern Ga. Baptist Hosp., 1944-45; resident medicine Barnes Hosp., St. Louis, 1948-50, fellow medicine, 1950-51; asst. prof. pathology U. Pitts., 1951-52; asst. prof., then assoc. prof. medicine U. Chgo., 1952-59;

prof. medicine U. Colo., 1959—, prof. microbiology, 1960—, chmn. dept., 1963-65, assoc. dean, 1966-68, dean, 1969-71; dir. Webb-Waring Lung Inst., 1973-83; mem. nat. council Nat. Inst. Allergy and Infectious Diseases, NIH, 1963-66, 73-77. Author: (with John Cann) Chemistry of Immunity in Health and Disease; editor: (with M. Samter) Jour. Allergy, 1963-67, Immunological Diseases. Served with M.C. AUS, 1945-48. Markle scholar, 1955-60. Mem. Am. Acad. Allergy (pres.), Am. Assn. Immunologists (pres.), Am. Acad. Arts and Scis., Nat. Acad. Scis., Inst. Medicine, Phi Beta Kappa. Subspecialties: Immunobiology and immunology; Allergy. Current work: Transplantation and tumor immunology. Office: U Colo Sch Medicine Denver CO 80262

TALMAN, JAMES DAVIS, applied mathematician, educator; b. Toronto, Ont., Can., July 24, 1931; s. James John and Ruth Helen (Davis) T.; m. Ragnhild Bruun Nilssen, Feb. 2, 1957; children: Liv Elizabeth, Stephen James, Marianne Ruth, Eric Andreas. B.A., U. Western Ont., 1953, M.Sc., 1954; Ph.D., Princeton U., 1959. Instr. Princeton U., 1957-59; asst. prof. Am. U. of Beirut, 1959-60; asst. prof. applied math. U. Western Ont., London, 1960-63, assoc. prof., 1963-65, prof., 1965—, dir., 1982—. Author: Special Functions: A Group Theoretic Approach, 1968. Mem. Canadian Assn. Physicists, Am. Phys. Soc., Soc. Indsl. and Applied Math., AAAS. Club: London Squash Racquets. Subspecialties: Atomic and molecular physics; Theoretical physics. Current work: Atomic and molecular structure theory, computational methods in physics and chemistry.

TAMARIN, ROBERT HARVEY, population biologist, educator; b. Bklyn., Dec. 14, 1942; s. Leon and Zelda R. (Hirsch) T.; m. Virginia M. Londy, May 31, 1968; children: David L., Bonnie. B.S., CUNY, 1963; Ph.D., Ind. U., 1968. Postdoctoral fellow U. Hawaii, 1968-70; research assoc. Princeton U., 1970-71; asst. prof. Boston U., 1971-77, assoc. prof. biology, 1977-83, prof. biology, 1983—. Author: Population Regulation, 1978, Principles of Genetics, 1982, also articles, abstracts, book revs. NIH grantee, 1972-76, 78-81; NSF grantee, 1981-84. Mem. AAAS, Am. Soc. Mammalogists, Am. Soc. Naturalists. Subspecialties: Population biology; Evolutionary biology. Current work: Population regulation in small animals; behavior, genetics, demography, reproductive physiology of island and mainland voles; sociobiology using radionuclides. Office: Dept Biolog Boston Boston MA 02215

TAMM, IGOR, biologist, educator; b. Tapa, Estonia, Apr. 27, 1922; came to U.S., 1945, naturalized, 1954; s. Alexander and Olga T.; m. Olive Emma Pitkin, May 9, 1953; children—Carol, Eric, Ellen. Student, Tartu (Estonia) U., 1942-43, Karolinska Inst., Stockholm, 1944-45; M.D. cum laude, Yale U., 1947. Intern, then asst. resident in medicine Grace-New Haven Community Hosp. Univ. Service, 1947-49; also asst. in medicine Yale U. Med. Sch.; mem staff and faculty Rockefeller Inst., N.Y.C., 1949—, prof., sr. physician, 1964—; chmn. bd. sci. cons. Sloan-Kettering Inst., 1972-73; chmn. task force virology Nat. Inst. Allergy and Infectious Diseases, 1976-78; Centennial lectr. U. Ill., 1968. Author, editor papers and jours. in field. Recipient Alfred Benzon Found. prize, 1967, Sarah L. Poiley Meml. award N.Y. Acad. Scis., 1977. Fellow N.Y. Acad. Scis.; mem. AAAS, Am. Soc. Microbiology, Am. Assn. Immunology, Soc. Exptl. Biology and Medicine, Am. Soc. Clin. Investigation, Am. Acad. Microbiology, Am. Soc. Cell Biology, Soc. Gen. Microbiology, Assn. Am. Physicians, Nat. Acad. Scis., Harvey Soc. (pres. 1974-75), Sigma Xi, Alpha Omega Alpha. Subspecialties: Virology (biology); Cell biology. Current work: Interferon action; cell cy and motility; cytoskeleton; effects of transforming viruses and growth factors on cells; RNA transcription; DNA replication. Office: 1230 York Ave New York NY 10021

TAN, ENG MENG, biomedical scientist; b. Seremban, Maalysia, Aug. 26, 1926; came to U.S., 1950; s. Ming Kee and Chooi Eng (Ang) T.; m. Liselotte Filippi, June 30, 1962; children: Philip, Peter. B.A., Johns Hopkins U., 1952, M.D., 1956. Asst. prof. Washington U. Sch. Medicine, St. Louis, 1965-67; assoc. mem. Scripps Clinic and Research Found., LaJolla, Calif., 1967-70, mem., 1970-77, dir. autoimmune disease ctr., 1982—; prof. U. Colo. Sch. Medicine, Denver, 1977-82; chmn. allergy and immunology research com. NIH, Bethesda, Md., 1980—; mem. nat. arthritis adv. bd. HHS, Washington, 1981—. Contbr. chpts. to books, articles to profl. jours. Mem. Arthritis Found. (pres. San Diego chpt. 1974-75), Am. Rheumatism Assn. (chmn. lupus com. 1980-82), United Scleroderma Found. (ann. award 1982), Assn. Am. Physicians, Am. Soc. Clin. Investigation, Western Assn. Physicians (v.p. 1980-81), Am. Assn. Immunologists (hon.), Argentina Rheumatism Assn. (hon.), Australian Rheumatism Assn. Subspecialties: Immunology (medicine); Internal medicine. Current work: Characterization of autoantibodies in autoimmune diseases; systemic lupus erythematosus, scleroderma, sjogren's syndrome, myositis and mixed connective tissue disease; relationship of autoantibodies to pathogenesis. Patentee in field. Home: 8303 Sugarman Dr La Jolla CA 92037 Office: Scripps Clinic and Research Found 10666 N Torrey Pines Rd La Jolla CA 92037

TANANBAUM, HARVEY DALE, astrophysicist, educator; b. Buffalo, July 17, 1942; m. Rona, 1964; children. B.A. summa cum laude, Yale U., 1964; Ph.D., MIT, 1968. Research asst. MIT, 1964-68; sr. staff scientist Am. Sci. and Engring., Inc., 1968-73; assoc. Harvard Coll. Obs., 1973—; also astrophysicist Smithsonian Astrophys. Obs., 1973—; assoc. dir. high energy astrophysics Harvard/Smithsonian Center Astrophysics, 1981—; project scientist UHURU Satellite, 1969-73; sci. program mgr. Einstein (HEAO-B) Obs., 1972—, prin. scientist, 1978—, prin. investigator, 1981—; mem. com. data mgmt. and computation Space Sci. Bd., 1978-82, com. space astronomy and astrophysics, 1981—; research briefing panel astronomy and astrophysics Nat. Acad. Scis., 1982. Contbr. articles profl. jours.. Nat. Merit scholar; recipient Chauncey Brewster Tinker award Yale U., Anthony Stanley award, Davenport Coll. award, 1980, Exceptional Sci. Achievement award NASA, 1980. Mem. Am. Astron. Soc., AAAS, Phi Beta Kappa. Subspecialty: 1-ray high energy astrophysics. Current work: X-ray astronomy, especially discrete cosmic X-ray sources with satellite payloads. Office: 60 Garden St Cambridge MA 02138

TANCRELL, ROGER HENRY, physicist, elec. engr.; b. Whitinsville, Mass., Feb. 17, 1935; s. Walter Henry and Eugenie Marie (Rioux) T. B.S. in Elec. Engring, Worcester Poly. Inst., 1956; M.S., M.I.T., 1958; Ph.D., Harvard U., 1968. Staff engr. M.I.T., Lexington, 1956-60; teaching fellow, researcher Harvard U., Cambridge, Mass., 1960-68; with Raytheon Research Div., Lexington, Mass., 1968—, prin. scientist, group leader, 1974—. Contbr. articles in field to profl. jours. Incorporator, bd. dirs. Adult Literacy Program, Inc., 1969—. Mem. Am. Inst. Ultrasound in Medicine, IEEE, Am. Assn. Physicists in Medicine, Acoustical Soc. Am., Sigma Xi, Tau Beta Pi, Eta Kappa Nu. Subspecialties: Acoustics; Biomedical engineering. Current work: Developer of ultrasonic instruments for medical imaging and underwater sonar. Patentee in field. Office: Raytheon Research Div 131 Spring St Lexington MA 02173

TANDANAND, SATHIT, mining engr.; b. Songkhla, Thailand, May 11, 1922; came to U.S., 1958, naturalized, 1971; m. (div.); children—Sathirut, Sornnudda, Lynda Tritip. B.S., Chulalongkorn U., Thailand, 1942; M.Sc., Colo. Sch. Mines, 1957; Ph.D., Pa. State U., 1962. Mining engr., field engr., project engr. in, Thailand, 1943-54, 56-58; research fellow Pa. State U., 1959-62; mining engr. U.S. Bur. Mines, 1962—, group supr., Mpls., 1962—. Author articles. Fulbright scholar, 1954-55; Nat. Acad. Sci. fellow, 1958-60. Mem. Soc. Mining Engrs., Am. Inst. Mining Engrs., ASME, Nat. Soc. Profl. Engrs., Minn. Soc. Profl. Engrs. Buddhist. Clubs: U.S. Tennis Assn., Eagandale Tennis (Eagan, Minn.). Subspecialty: Solid mechanics. Current work: Applied rock mechanics in mining engineering. Rock fragmentation; stability of rock mass; strata mechanics, and strata control engineering. Patentee in field. Home: 1481 Woodview E Eagan MN 55122 Office: 5629 Minnehaha Ave S Minneapolis MN 55417

TANDBERG-HANSSEN, EINAR ANDREAS, astronomer, physicist; b. Bergen, Norway, Aug. 6, 1921; s. Birger and Mona (Meier) T.-H.; m. Erna, June 22, 1951; children: Else Bartels, Karin Willoughby. Ph.D., U. Oslo, 1960. Research assoc. U. Oslo, 1954-57, lectr., 1959-61; research assoc. High Altitude Obs., Boulder, Colo., 1957-59, mem. sr. research staff, 1961-74; sr. research scientist NASA/Marshall Space Flight Center, Huntsville, Ala., 1974—; adj. prof. physics U. Ala., Huntsville, 1976—. Author: Radioastronomy, 1961, Solar Activity, 1967, Solar Prominences, 1974; contbr. numerous articles on astronomy, solar physics to sci. jours. Recipient Exceptional Service medal NASA, 1979. Mem. Norwegian Acad. Sci., Geophysics Research Bd., Internat. Astron. Union, Am. Astron. Soc. Lutheran. Subspecialties: Solar physics; Optical astronomy. Current work: Nature of solar activity, solar flares prominences, solar magnetic fields. Home: 4010 Granada Dr Huntsville AL 35802 Office: ES01 NASA Marshall Space Flight Center Huntsville AL 35812

TANENBAUM, MORRIS, telephone company executive; b. Huntington, W.Va., Nov. 10, 1928; s. Reuben S. and Mollie (Kadensky) T.; m. Charlotte Silver, June 4, 1950; children: Robin Sue, Michael Alan. B.A. in Chemistry, Johns Hopkins U., 1949, M.A., Princeton U., 1950, Ph.D. in Phys. Chemistry, 1952. Mem. tech. staff Bell Telephone Labs., Murray Hill, N.J., 1952-55, dept. head, 1955-60; asst. dir. Metall. Research Lab., 1960-62, dir. solid state device lab., 1962-64, exec. v.p., 1975-76; also dir.; with Western Electric Co., N.Y.C., 1964-74, v.p. engring., 1971-72, v.p. transmission equipment, 1972-75; v.p. engring. and network services AT&T, Basking Ridge, N.J., 1976-78; pres. N.J. Bell Telephone Co., Newark, 1978-80; exec. v.p. AT&T, N.Y.C., 1980—; dir. Am. Cyanamid Co., Cabot Corp., Southwestern Bell Telephone Co., AT&T Long Lines., Bell Telephone Labs. Contbr. numerous articles to profl. jours. Trustee Battelle Meml. Inst., Brookings Instn.; trustee Johns Hopkins U., Ednl. Broadcasting Corp. Fellow IEEE, Am. Phys. Soc.; mem. Nat. Acad. Engring., Am. Chem. Soc., Am. Inst. Metall., Mining and Petroleum Engrs. Patentee in field. Office: 295 N Maple Ave Basking Ridge NJ 07920

TANFORD, CHARLES, educator; b. Halle, Germany, Dec. 29, 1921; came to U.S., 1939, naturalized, 1947; s. Max and Charlotte (Eisenbruch) T.; m. Lucia Lander Brown, Apr. 3, 1948 (div. Feb. 1969); children—Victoria, James Alexander, Sarah Lander. B.A., N.Y. U., 1943; M.A., Princeton, 1944, Ph.D., 1947. Postdoctoral fellow Harvard, 1947-49, vis. prof. chemistry, 1966; asst. prof. chemistry U. Iowa, 1949-54, asso. prof., 1954-59, prof., 1956-60; prof. biochemistry Duke, 1960-70, James B. Duke prof. biochemistry, 1970-80, James B. Duke prof. physiology, 1980—; George Eastman vis. prof. Oxford U., 1977-78; Walker-Ames prof. U. Wash., 1979; Reilly lectr. U. Notre Dame, 1979. Author: Physical Chemistry of Macromolecules, 1961, The Hydrophobic Effect, 1973, 2d edit., 1980; Contbr. numerous articles to profl. jour. Guggenheim fellow, 1956. Mem. Am. Acad. Arts, Scis., Nat. Acad. Scis. Unitarian (pres. Unitarian-Universalist Fellowship 1968). Subspecialties: Membrane biology; Biophysics (biology). Current work: Active transport of ions. Home: 1430 N Mangum St Durham NC 27701 Office: Dept Physiology Box 3709 Duke Univ Med Center Durham NC 27710

TANG, RUEN C., educator; b. Chenkiang, China, Oct. 31, 1934; came to U.S., 1963; s. Ping H. and I-Chen (Shen) T.; m. Anna C.Y. Huang, Dec. 25, 1960; children: Gina, Sophia, Jayne. B.S., Nat. Chung-Hsing U., 1957; Ph.D., N.C. State U., 1968. Forester Taiwan Forest Bur., China, 1959-63; research asst. N.C. State U., Raleigh, 1963-67; research assoc. U. Ky., Lexington, 1968-69, asst. prof., 1970-74, assoc. prof., 1974-77; prof. Auburn U., Ala., 1978—; cons. Forest Products Industries, Ky., Ala., 1968—; adv. weed sci. Miss. State U., Starkville, 1979—. Contbr. numerous articles to profl. jours. Mem. Soc. Wood Sci. and Tech., Soc. Exptl. Stress Analysis, Soc. Am. Foresters, Forest Products Research Soc. (chmn. Physics sect. 1977-78), Am. Forest Assn., ASTM, AAAS, Internat. Assn. Math. Modeling, Soc. Computer Stimulation, Sigma Xi, Gamma Sigma Delta, Xi Sigma Pi. Confusianism. Subspecialties: Theoretical and applied mechanics; Composite materials. Current work: Anisotropic elasticity, composite materials, fiber mechanisms, wood engineering, timber physics, computer simulation and modeling. Home: 687 Longweed Rd Lexington KY 40503 Office: Dept Forestry Auburn U Auburn AL 36849

TANNOCK, IAN FREDERICK, physician, cancer research scientist; b. Hatfield, Hertshire, Eng., Nov. 22, 1943; emigrated to Can., 1974, naturalized, 1979; s. Archibald A. and Freda A.G. (Rickels) T.; m. Rosemary Tannock, May 20, 1967; children: Stuart, Lisa, Steven. B.A., Cambridge (Eng.) U., 1965; Ph.D., Inst. Cancer Research, London, 1968; M.D., U. Pa., 1974. Postdoctoral fellow M.D. Anderson Hosp., Houston, 1968-70; vis. scientist Radiobiol. Inst. T.N.O., Rijswijk, Netherlands, 1971; resident in internal medicine and med. oncology U. Toronto, 1974-78; staff physician, sr. scientist Ont. Cancer Inst., Princess Margaret Hosp., Toronto, 1978—; mem. Nat. Cancer Inst. Can. Grants panel, 1982-85; mem. exptl. therapeutics study sect. NIH, 1980-84. Contbr. articles to profl. jours. Mem. Am. Soc. Clin. Oncology, Am. Assn. Cancer Research, Radiation Research Soc., Cell Kinetics Soc. Subspecialties: Cancer research (medicine); Chemotherapy. Current work: Clinical and experimental studies of drug treatment of solid tumors, biology of human tumors. Office: Ontario Cancer Institute Department Medicine 500 Sherbourne St Toronto ON Canada M4X 1K9

TANSEY, MICHAEL ANSELME, clinical psychologist, biofeedback clinician; b. Frankfurt, Germany, Feb. 4, 1948; naturalized, 1948; s. Frank Vincent and Ginette (Tommassini) T.; m. Mary Loretta Glod, June 19, 1971; children: Jennifer, Michael, Matthew. B.A. in Psychology, Seton Hall U., South Orange, N.J., 1970, M.A., Fairleigh Dickinson U., Teaneck, N.J., 1972; Ph.D. in Clin. Psychology, Calif. Sch. Profl. Psychology, Fresno, Calif., 1976. Lic. clin. psychologist, N.J. Pres. Michael A. Tansey (P.A.), Union, N.J., 1978—; co-dir., owner T.H.E. Biofeedback and Counselling Assocs., Hightstown, N.J., 1981—; teaching fellow Fairleigh Dickinson U., 1971. Mem. Am. Psychol. Assn., N.J., Psychol. Assn., Biofeedback Soc. N.J. Lodge: K.C. Subspecialties: Neuropsychology; Biofeedback. Current work: The remediation of learning disabilities via the direct enhancement of discrete neural discharge over the Rolandic cortex of the brain. Home: 2297 Camplain Rd Somerville NJ 08876 Office: 2810 Morris Ave Union NJ 07083

TAO, LI-CHUNG, research scientist, mechanical engineer; b. WuChen, China, Dec. 20, 1932; came to U.S., 1957, naturalized, 1969; s. Yao-Kai and Kuang-Show (Hu) T.; m. Cecilia Chiang, June 17, 1961; children: Jeffrey, Angela. Ph.D., Yale U., 1968. Sr. research engr. Gen. Tire & Rubber Co., Akron, Ohio, 1968-71, research scientist, 1971-77; sr. engr. Bechtel Power Corp., Gaithersburg, Md., 1977—. Contbr. articles to profl. jours. Fellow Am. Orthopsychiat. Assn., Am. Coll. Psychology; Mem. ASME. Subspecialties: Solid mechanics; Composite materials. Current work: Research in polymer mechanics for the understanding of cord-rubber composites. Basic stress analysis of pressure vessels and piping. Home: 709 Hope Ln Gaithersburg MD 20878 Office: 15740 Shady Grove Rd Gaithersburg MD 20877

TAPLEY, BYRON DEAN, aerospace engineer; b. Charleston, Miss., Jan. 16, 1933; s. Ebbie Byron and Myrtle (Myers) T.; m. Sophia Philen, Aug. 28, 1959; children—Mark Byron, Craig Philen. B.S., U. Tex., 1956, M.S., 1958, Ph.D., 1960. Research engr. Structural Mechanics Research Lab., U. Tex., Austin, 1954-58, instr. dept. mech. engring., 1958, prof. dept. aerospace engring. and engring. mechanics, 1960—, chmn. dept., 1966-77, adv. scientist, Sunnyvale, Calif., 1961, W.R. Woolrich prof. engring., 1974—; cons. NASA Manned Spacecraft Center, 1965—, mem. adv. com. on guidance control and nav., 1966-67, com. on space research, panel I, 1974—, chmn. region IV, engring. council on profl. devel., 1974—. Editor: Celestial Mech. Jour, 1976-79; assoc. editor: Jour. Guidance and Control, 1978-79; asso. editor: Geophys. Revs, 1979—. Mem. ASME, Am. Acad. Mechanics, Am. Astronautics Soc., Soc. Engring. Sci., Am. Inst. Aeros. and Astronautics (chmn. com. on astrodynamics 1976-78), IEEE, Am. Geophys. Union, Am. Astron. Soc., AAAS, Internat. Astron. Union (chmn. 1981—), Sigma Xi, Pi Tau Sigma, Sigma Gamma Tau, Phi Kappa Phi, Tau Beta Pi. Subspecialties: Aerospace engineering and technology; Satellite studies. Current work: Satellite geodesy, satellite oceanography and remote sensing; application of satellite remote sensed data to study of earth. Home: 3100 Perry Ln Austin TX 78731

TARAMAN, KHALIL SHOWKY, manufacturing engineering educator, consultant; b. Cairo, Egypt, July 10, 1939; came to U.S., 1967, naturalized, 1975; s. Showky K. and Saadat M.A. (Ghany) T.; m. Sanaa R. Taraman, July 4, 1968; children: Shaoky, Sharief. B.S.M.E., Ain Shams U., Egypt, 1964, M.S.M.E., 1967; M.S. (WARF fellow), U. Wis., 1969; Ph.D. (fellow), Tex. Tech U., 1971. Registered profl. engr., Mich., Calif.; cert. mfg. engr. Instr., research fellow Tex. Tech U., 1969-70; asst. prof. U. Detroit, 1970-73, assoc. prof., 1973-77, dir. mfg. engring. isnt., 1975, prof., chmn. dept. mech. engring., 1977—; sr. tech. cons., research supr. corps. Author: Computer Aided Design/ Computer Aided Manufacturing, 1980; contbr. articles profl. jours. Named Eminent Engr. Tau Beta Pi, 1982. Mem. Soc. Mfg. Engrs. (internat. dir., chpt. chmn. 1975-76, resolution of appreciation 1977, 79, 80), ASME, Am. Soc. Engring. Edn., Am. Egyptian Scholars, Pi Tau Sigma, Alpha Pi Mu. Subspecialties: Mechanical engineering; Materials processing. Current work: Manufacturing engineering systems and its productivity, material removal. Office: 4001 W McNichols Rd Detroit MI 48221

TARBELL, DEAN STANLEY, chemistry educator; b. Hancock, N.H., Oct. 19, 1913; s. Sanford and Ethel (Millikan) T.; m. Ann Hoar Tracy, Aug. 15, 1942; children: William Sanford, Linda Tracy, Theodore Dean. A.B., Harvard U., 1934, M.A., 1935, Ph.D., 1937. Postdoctoral fellow U. Ill., 1937; mem. faculty U. Rochester, 1938—, successively instr., asst. prof., asso. prof., 1938-48, prof. chemistry, 1948-62, Charles Frederick Houghton prof. chemistry, 1960—, chmn. dept., 1964—; Disting. prof. chemistry Vanderbilt U., 1967—, Branscom disting. prof., 1975-76, disting. prof. emeritus, 1981—; Guggenheim fellow and vis. lectr. chemistry Stanford U., 1961-62; Fuson lectr., 1972; cons. USPHS, Army Q.M.C.; mem. various sci. adv. bds. to govt. agencies. Author research papers in field, history of chemistry; author: (with Ann T. Tarbell) Roger Adams, Scientist and Statesman, 1981. Recipient Herty medal, 1973; Guggenheim fellow, 1946-47. Mem. Nat. Acad. Sci., Am. Chem. Soc. (chmn. div. history of chemistry 1980-81), Chem. Soc. London, Am. Acad. Arts and Scis., History of Sci. Soc. Subspecialty: Organic chemistry. Current work: Structure of antibiotics; mechanism of organic reactions, history of chemistry. Home: 6033 Sherwood Dr Nashville TN 37215

TARDIFF, ROBERT GEORGE, toxicologist, adminstr.; b. Lowell, Mass., Feb. 1, 1942; s. George and Eugenie (St. Jean) T.; m. Kathleen Wuersig, Jan. 10, 1970; children: Rebecca, Brian, Rachel. B.A., Merrimack Coll., 1964; Ph.D., U. Chgo., 1968. Research toxicologist USPHS, Cin., 1968-70; chief toxicol. assessment br. EPA, Cin., 1970-77; exec. dir. bd. toxicology and environ. health hazards Nat. Acad. Sci., Washington, 1979—. Editor: Aquatic Pollutants and Biologic Effects with Emphasis on Neoplasia, 1977; sect. editor: Jour. Risk Analysis. USPHS fellow, 1964-68; recipient Sci. and Tech. Achievenemt award EPA, 1981. Subspecialty: Toxicology (medicine). Current work: Risk assessment of chemicals (extrapolation from high to low doses, extrapolation from laboratory animals to humans), toxicological evaluation of multiple substances. Office: Bd Toxicology and Environ Health Hazards Nat Acad Sci 2101 Constitution Ave NW Washington DC 20418

TARJAN, ROBERT ENDRE, computer scientist, educator; b. Pomona, Calif., Apr. 30, 1948. B.S. in Math, Calif. Inst. Tech., 1969; M.S. in Computer Sci, Stanford U., 1971, Ph.D., 1972. Asst. prof. computer sci. Cornell U., 1972-74; Miller research fellow U. Calif.-Berkeley, 1973-75; from asst. prof. to assoc. prof. computer sci. Stanford U., 1974-81; mem. tech. staff Bell Labs., Murray Hill, N.J., 1980—; adj. prof. NYU, 1981—. Recipient Nevanlinna prize Internat. Math. Union, 1983. Subspecialties: Algorithms; Group theory, Data structures. Office: Bell Labs Murray Hill NJ

TARR, DONALD ARTHUR, chemist, educator, researcher; b. Norfolk, Nebr., Aug. 1, 1932; s. Arthur A. and Bessie E. (Craven) T.; m. Marjorie Jean Hooper, Aug. 26, 1957; children: Stephen, Elizabeth. B.A., Doane Coll., 1954; M.S., Yale U., 1956, Ph.D., 1959. Instr. to asst. prof. Coll. Wooster, Ohio, 1958-65; asst. prof. St. Olaf Coll., Northfield, Minn., 1965-67, assoc. prof., 1967—. Contbr. articles to chem. jours. Danforth fellow, 1954-58. Mem. AAAS, AAUP, Am. Chem. Soc. Subspecialties: Inorganic chemistry; Kinetics. Current work: Kinetics and mechanisms of cobalt (III) substitution reactions. Home: 905 Ivanhoe Dr Northfield MN 55057 Office: St Olaf Coll Northfield MN 55057

TARR, MELINDA JEAN, immunotoxicologist, veterinary pathologist; b. San Francisco, Aug. 18, 1948; d. Cedric Winship and Janis Arlene (White) T. B.S., U. Calif., Davis, 1971, D.V.M., 1973; M.S., Ohio State U., 1976; Ph.D., 1979. Cert. Am. Coll. Vet. Pathology. Pvt. practice vet. medicine, Sebastopol, Calif., 1973-74; resident in equine medicine and surgery Ohio State U., Columbus, 1974-76, NIH postdoctoral trainee, fellow, 1976-79, research assoc., dept. vet. pathobiology, 1979-81, asst. prof., 1981—. Contbr. articles to profl. jours. Mem. Internat. Soc. for Immunopharmacology, Am. Assn. Vet. Immunologists, AVMA, Calif. Vet. Med. Assn. Subspecialties: Immunotoxicology; Immunopharmacology. Current work: Immunotoxicology, immunopharmacology, immunopathology, comparative pathobiology. Home: 1780 Bowtown Rd Delaware OH 43015 Office: Dept Veterinary Pathobiology Ohio State U 1925 Coffey Rd Columbus OH 43210

TARR, RICHARD ROBERT, orthopaedics researcher, cons.; b. Springfield, Ill., June 13, 1948; s. Robert William and Margaret Helen (Kobialka) T.; m. Joan Elaine Marcks, Aug. 15, 1970; 1 dau.: Julianne Janell. B.S., Northwestern U., Evanston, Ill., 1970; diploma, Northwestern Sch. Radiol. Tech., 1975; M.S., Tex. A&M U., 1977.

Cert. engr. in tng., 1970; registered radiol. technician, 1975. Design engr. rehab. engring. program Northwestern U., Chgo., 1973-75; research assoc. dept. orthpaedics and rehab. U. Miami, Fla., 1977-78; instr. dept. orthopaedics U. So. Calif., 1978—, chmn. animal utilization com., 1980-83; orthopaedic researcher. Contbr. articles to profl. jours., chpt. in book. Sec. bd. elders St. John's Lutheran Ch., Orange, Calif., 1981-82, chmn. bd. elders, Orange, Calif., 1983-84. Served with AUS, 1970-73. Recipient award Lowman Club, 1981, 82, 83. Mem. Nat. Assn. Bioengrs., ASME, Vet. Orthopedic Soc., Orthopaedic Research Soc., Am. Register Radiol. Tech., Soc. Biomaterials. Subspecialties: Biomedical engineering; Orthopedics. Current work: Artificial joint design and analysis, bone fracture management analysis, experimental mechanics, finite element analysis, animal modeling for orthopedic problems and orthopedic appliances. Home: 15022 Fleming St Westminster CA 92683 Office: Orthopaedic Hospital 2400 S Flower St Los Angeles CA 90007

TARTER, C. BRUCE, physicist; b. Louisville, Sept. 26, 1939; s. Curtis B. and Marian Turner (Cundiff) T.; m. Jill Cornell Tarter, June 6, 1964; 1 dau.: Shana. B.S., M.I.T., 1961; Ph.D., Cornell U., 1967. Sr. scientist aero. div. Philco Ford, Newport Beach, Calif., 1967; mem. staff Lawrence Livermore Nat. Lab., 1967-69, group leader, 1969-74, dep. head theoretical physics div., 1974-78, head theoretical physics div., 1978—; lectr. dept. applied sci. U. Calif., Davis, 1970—. Contbr. articles to profl. jours. Mem. Am. Phys. Soc., Am. Astron. Soc., Internat. Astron. Union, Sigma Xi. Subspecialties: Theoretical astrophysics; Statistical physics. Current work: Theoretical astrophysics, properties of matter at high temperatures and densities.

TASK, HARRY LEE, research physicist; b. Chgo., May 31, 1946; s. Harry John and Christine Virginia (Rozell) T.; m. Marjorie F. Howard, Apr. 4, 1970; 1 dau.: Christine Marie. B.S. in Physics, Ohio U., 1968, M.S., Purdue U., 1971, U. Ariz., 1978; Ph.D. in Optical Scis, U. Ariz., 1978. Optical scis. researcher human engring. div. Air Force Aerospace Med. Research Lab., Wright-Patterson AFB, Ohio, 1971—. Contbr. articles to profl. jours. Pres. Multi-Metrics Inc., Dayton, Ohio, 1982—. Served with USAF, 1969-70. Mem. Optical Soc. Am., Human Factors Soc. Current work: Display image quality, night vision and night lighting, night vision goggles, aircraft transparency evaluation. Patentee in optico-visual devices field. Office: AFAMRL/HEF Wright-Patterson AFB OH 45433

TASSOUL, JEAN-LOUIS, physicist, educator; b. Brussels, Nov. 1, 1938; s. Louis and Melanie (Brans) T.; m. Monique Cretin, May 7, 1966. B.S., U. Libre de Bruxelles, 1961, D.Sc., 1964. Research assoc. U. Libre de Bruxelles, 1965-66; research assoc. U. Chgo., 1966-67, Princeton U. Obs., 1967-68; asst. prof. physics U. Montreal, 1968-70, assoc. prof., 1970-75, prof., 1975—. Contbr. articles to sci. jours.; Author: Theory of Rotating Stars, 1978. Served with Belgian Army, 1964-65. NRC Can. grantee. Mem. Internat. Astron. Union, Am. Astron. Soc. Subspecialty: Theoretical astrophysics. Current work: Stellar structure; rotation, meridional circulation, magnetic fields, and instabilities in stars. Office: Dept Physics U Montreal Box 6128 Sta A Montreal PQ Canada H3C 3J7

TATINA, ROBERT EDWARD, biologist, educator; b. Chgo., May 18, 1942; s. Edward and Gertrude (Dase) T.; m. Geraldyne Sansone, May 26, 1978; children: Tomas Murphy, Heather Murphy. B.S.Ed., No. Ill. U., 1965; M.A. in Zoology, So. Ill. U., 1971; Ph.D. in Botany, So. Ill. U., 1982. Sci. and math. tchr. Trewyn Jr. High Sch., Peoria, Ill., 1965-66; biology tchr. Evergreen Park (Ill.) High Sch., 1966-69; assoc. prof. biology Dakota Wesleyan U., Mitchell, S.D., 1975—. Author: (with W. Houk) A Guide to the Natural History of the Quarry Bridge Outdoor Classroom, 1980. Mem. Am. Soc. Plant Physiologists, Bot. Soc. Am., Nat. Assn. Biology Tchrs., S.D. Acad. Sci., Phi Sigma, Kappa Delta Pi, Sigma Zeta, Phi Kappa Phi., Sigma Xi. Democrat. Lutheran. Subspecialty: Plant physiology (biology). Current work: Flora of South Dakota. Office: Dakota Wesleyan U Mitchell SD 57301

TATOM, FRANK BUCK, engineering company executive; b. Montgomery, Ala., Nov. 27, 1934; s. Thomas Athor and Elizabeth Hargrove (Buck) T.; m. Roberta Wood, Sept., 1959; children: Frank Thomas, John Wood, Briana Claire. B.S., U.S. Naval Acad., 1956; M.S.M.E., Auburn U., 1962; Ph.D. in M.E, Ga. Tech., 1971. Registered profl. engr. Ala., Tenn. Commd. ensign U.S. Navy, 1952, advanced through grades to lt. served to, 1960; now capt. USNR; process engr. DuPont, Camden, S.C., 1960-61; grad. assoc., asst. prof. dept. mech. engring. Auburn (Ala.) U., 1961-63; chief Aerothermodynamics br. Northrop Space Lab., Huntsville, Ala., 1963-68; instr., fellow Ga. Inst. Tech., Atlanta, 1968-71; mem. tech. staff Tex. Instruments Inc., Dallas, 1971-74; chief scientist Sci. Applications Inc., Huntsville, 1974-78; pres., chief engr. Engring. Analysis Inc., Huntsville, 1978—; guest lectr. U. Ala.-Huntsville, 1974-78; chmn. Engring. Socs. Secretarial Service, 1981—; cons. in field. Contbr. articles to profl. jours. Mem. Nat. Soc. Profl. Engrs., ASME, Soc. Am. Mil. Engrs., U.S. Naval Acad. Alumni Assn., Auburn U. Alumni Assn. Republican. So. Baptist. Subspecialties: Fluid mechanics; Environmental engineering. Current work: Jets, plumes, wakes, turbulence, immersed particle motion, cloud electrostatics, river hydraulics, curvilinear coordinates. Home: 3818 F Westwind Circle Huntsville AL 35805 Office: 2109 Clinton Ave Suite 432 Huntsville AL 35805

TATSCH, JAMES HENDERSON, research scientist, cons., lectr., writer; b. Eldorado, Tex., Dec. 7, 1916; s. Henry Charles and Ellen Mae (Specht) T.; m. Helen Gailis, July 23, 1946; children: James Alexis Wolfgang, Karyn Tatsch Brady. B.S.E.E., U.S. Naval Acad., 1940; postgrad., U. Ariz., Tucson, 1963. Commd. 2d lt. U.S. Marine Corps, 1940, advanced through grades to lt. col., 1960, ret., 1960; staff research scientist Collins Radio Co., Cedar Rapids, Iowa, 1963-64; dir. advance econ. planning Ingersoll Milling Machine Co., Rockford, Ill., 1965-67; prin. engr. Raytheon Co., Sudbury, Mass., 1967-70; chief scientist Tatsch Assocs., Sudbury, Mass., 1970—; cons. energy and minerals communities. Author: 11 books, including Coal Deposits: Origin, Evolution and Present Characteristics, 1980, The Earth's Tectonosphere: Past Development and Present Behavior, 1977; contbr. more than 200 articles on astronomy and geology to profl. jours. Decorated Bronze Star with combat disting. device. Mem. AAAS, Brit. Assn. Advancement Sci., Am. Geophys. Union, Geochem. Soc., Seismol. Soc. Am., Soc. Exploration Geophysicists, European Assn. Exploration Geophysicists. Current work: The origin, evolution and present characteristics of the Earth and its energy and mineral resources, the research results of the past 18 years in 11 hardcover books.

TATTAR, TERRY ALAN, plant pathology educator; b. Port Chester, N.Y., May 9, 1943; s. John and Emma (Hinlicky) T.; m. Donna San Juan, June 7, 1969. B.A., Northeastern Y., 1967; Ph.D., U. N.H., 1971. Plant pathologist U.S. Forest Service, Portsmouth, N.H., 1971-73; prof. plant pathology U. Mass., Amherst, 1973—. Author: Diseases of Shade Trees, 1978, Laboratory and Field Guide to Tree Pathology, 1981. Mem. Conservation Commn., Pelham, Mass., 1981—. Mem. Am. Phytopathol. Soc. Lutheran. Subspecialty: Plant pathology. Current work: Disease of shade trees. Home: 6 Pine Tree Circle Pelham MA 01002 Office: Shade Tree Lab U Mass Amherst MA 01003

TATTERSALL, IAN MICHAEL, physical anthropologist, museum curator; b. Paignton, Devon, Eng., May 10, 1945. B.A., Cambridge (Eng.) U., 1967, M.A., 1970; M.Phil., Yale U., 1970, Ph.D. in Geology, 1971. Asst. curator Am. Mus. Natural History, N.Y.C., 1971-76, assoc. curator, 1976-81, curator phys. anthropology, 1981—; vis. lectr. Grad. Faculty, New Sch. Social Research, 1971-72; adj. asst. prof. Lehman Coll., CUNY, 1971-74; adj. assoc. prof. Columbia U., 1978-79. Mem. Assn. Phys. Anthropologists, Soc. Vertebrate Paleontology, Am. Soc. Primatology, Internat. Primatology Soc., AAAS. Subspecialties: Evolutionary biology; Primatology. Office: Dept Anthropology Am Mus Natural History Central Park W at 79th St New York NY 10024

TAUB, EDWARD, research physiological psychologist, educator; b. Bklyn., Oct. 22, 1931; s. Samuel Hart and Ida Pearl (Kimmel) T.; m. Mildred Allen. B.A., Bklyn. Coll., 1953; M.A., Columbia U., 1959; Ph.D., NYU, 1970. Research asst. dept. psychology Columbia U., N.Y.C., 1956; research asst. dept. exptl. neurology Kingsbrook Jewish Med. Center, Bklyn., 1957-60, research assoc., 1960-69; asst. prof. Johns Hopkins U. Sch. Medicine, Balt., 1970—; dir. Behavioral Biology Ctr. Inst. Behavioral Research, Silver Spring, Md., 1969—; bd. dirs. Biofeedback Certification Inst. Am.; cons. in field. Contbr. articles to profl. jours. NIH grantee; Dept. Def. grantee. Fellow Am. Psychol. Assn.; mem. Biofeedback Soc. Am. (past pres.), Biofeedback Soc. Washington (past pres.), Fedn. Behavioral Medicine Socs. (founding mem.), Eastern Psychol. Assn., Psychonomic Soc., Soc. Neuroscience, Soc. Psychophysiological Research, Sigma Xi. Subspecialties: Physiological psychology; Neurophysiology. Current work: Somatosensory deafferentation, motor control, biofeedback, behavioral medicine. Co-invented technique of thermal biofeedback. Home: 1812 Metzerott Rd Adelphi MD 20783 Office: 1816 Metzeroff Rd Adelphi MD 20783

TAUBE, HENRY, chemist; b. Sask., Can., Nov. 30, 1915; came to U.S., 1937, naturalized, 1942; s. Samuel and Albertina (Tiledetski) T.; m. Mary Alice Wesche, Nov. 27, 1952; children—Linda, Marianna, Heinrich, Karl. B.S., U. Sask., 1935, M.S., 1937, LL.D., 1973; Ph.D., U. Calif., 1940, Hebrew U. of Jerusalem, 1979, D.Sc., U. Chgo., 1983. Instr. U. Calif., 1940-41; instr., asst. prof. Cornell U., 1941-46; faculty U. Chgo., 1946-62, prof., 1952-62, chmn. dept. chemistry, 1955-59; prof. chemistry Stanford U., 1962—, chmn. dept., 1971-74; Baker lectr. Cornell U., 1965. Recipient Harrison Howe award, 1961; Chandler medal Columbia U., 1964; F.P. Dwyer medal U. N.S.W., Australia, 1973; Nat. medal of Sci., 1976, 77; Allied Chem. award for Excellence in Grad. Teaching and Innovative Sci., 1979, Nobel prize, 1983; Guggenheim fellow, 1949, 55. Mem. Am. Acad. Arts and Scis., Nat. Acad. Scis. (award in chem. scis. 1983), Am. Chem. Soc. (Kirkwood award New Haven sect. 1965, award for nuclear application in chemistry 1955, Nichols medal N.Y. sect. 1971, Willard Gibbs medal Chgo. sect. 1971, Distinguished Service in Advancement Inorganic Chemistry award 1967, T.W. Richards medal NE sect. 1980, Monsanto Co. award in inorganic chemistry 1981, Linus Pauling award Puget Sound sect. 1981), Royal Physiographical Soc. of Lund, Phi Beta Kappa, Sigma Xi, Phi Lambda Upsilon (hon.). Subspecialties: Inorganic chemistry; Kinetics. Current work: Reactivity of inorganic substances; electron transfer reactions; mixed-valence molecules; systematic study of back-bonding.

TAUC, JAN, physics educator; b. Pardubice, Czechoslovakia, Apr. 15, 1922; came to U.S., 1969, naturalized, 1978; s. Jan and Josefa (Semonska) T.; m. Vera Koubelova, Oct. 18, 1947; children: Elena (Mrs. Milan Kokta), Jan. Ing.Dr. in Elec. Engring, Czech Inst. Tech., Prague, 1949; RNDr., Charles U., Prague, 1956; Dr.Sc. in Physics, Czechoslovak Acad. Scis., 1956. Jan. Scientist microwave research Sci. and Tech. Research Inst., Tanvald and Prague, 1949-52; head semiconductor dept. Inst. Solid State Physics, Czechoslovak Acad. Scis., 1953-69; prof. exptl. physics Inst. Physics, Charles U., 1964-69, dir., 1968-69; mem. tech. staff Bell Telephone Labs., Murray Hill, N.J., 1969-70; prof. engring. and physics Brown U., 1970—, L. Herbert Ballou prof. engring. and physics, 1983—; dir. E. Fermi Summer Sch., Varenna, Italy, 1965; vis. prof. U. Paris, 1969, Stanford U., 1977, Max Planck Inst. Solid State Research, Stuttgart, Germany, 1982—; UNESCO fellow, Harvard, 1961-62. Author: Photo and Thermoelectric Effects in Semiconductors, 1962, also numerous articles.; Editor: The Optical Properties of Solids, 1966, Amorphous and Liquid Semiconductors, 1974; co-editor: Solid State Communications, 1963—. Recipient Nat. prize Czechoslovak Govt., 1955, 69; Sr. U.S. Scientist award Humboldt Found., 1981. Fellow Am. Phys. Soc. (Frank Isakson prize 1982), AAAS; founding mem. European Phys. Soc.; corr. mem. Czechoslovak Acad. Scis., 1963-71. Subspecialties: Condensed matter physics; 3emiconductors. Current work: Electronic properties of amorphous semiconductors; picosecond relaxation processes. Office: Div Engring Brown Univ Providence RI 02912

TAULBEE, EARL SELDON, psychologist; b. Lee City, Ky., Mar. 22, 1923; s. Courtney F. and Ida (Wilson) T.; m. Lucille R. Resch, Oct. 21, 1961. B.A., Georgetown Coll., 1948; Ph.D., U. Nebr., 1953. Lic. psychologist, Ala. Chief psychology service Norfolk (Nebr.) State Hosp., 1958-59, VA Med. Ctr., Lincoln, Nebr., 1959-63, Tuscaloosa, Ala., 1963-68, Bay Pines, Fla., 1968-80, Asheville, N.C., 1980—; adj. prof. U. So. Fla., Tampa, 1974—, Fla. State U., Tallahassee, 1969-72, U. Miami, 1970-82; cons. psychologist Community Mental Health Ctr., Sylacauga, Ala., 1966-76. Author: MMPI Bibliography, 1977, Bibliography of Selected Psychology Tests, 1982; contbr. articles to profl. jours. Served with USAAF, 1942-45. Fellow Am. Psychol. Assn., Soc. for Personality Assessment (treas. 1965-71, staff editor jours. 1965—), Sigma Xi. Current work: Psychological assessment, psychotherapy, teaching, applied psychological research. Office: VA Med Ctr Asheville NC 28805

TAVASSOLI, MEHDI, physician, cancer researcher; b. Tehran, Iran, Mar. 30, 1933; came to U.S., 1961, naturalized, 1975; s. Hassan and Sakineh (Aghdasi) Farid; m. Marie Ellen DeLuca, Apr. 30, 1966; children: Ali, Javad. Cherine. M.D., Tehran U., 1961. Diplomate: Am. Bd. Internal Medicine. Intern Cambridge (Mass.) City Hosp., 1962-63; resident Cook County Hosp., Chgo., 1963-64, Carney Hosp., Boston, 1964-66; fellow in hematology, then asst. prof. medicine Tufts U. Med. Sch., 1966-72; staff hematologist Scripps Clinic and Research Found., LaJolla, Calif., 1972-81; prof. medicine U. Miss. Med. Center, Jackson, 1981—; chief hematology-oncology VA Hosp., Jackson, 1981—. Author numerous papers in field. Recipient John Larkin award Guild St. Luke, 1966, Career Devel. award NIH, 1974-79; Charleston fellow, 1968-69; research fellow Med. Found., Boston, 1970-71. Mem. Am. Soc. Hematology, ACP, Internat. Soc. Exptl. Hematology, Am. Assn. Cancer Research, Am. Soc. Clin. Oncology, Soc. Exptl. Biology and Medicine, Am. Soc. Clin. Pathology, Am. Assn. Pathologists. Subspecialties: Hematology; Cell biology (medicine). Current work: Hemopoiesis, bone marrow structure and function, membrane structure and function, cell surface recptors, post translational recognition.

TAVES, DONALD R., toxicology and pharmacology educator; b. American Falls, Idaho, July 22, 1926; s. John L. and Lois (Kendrick) Toevs; m. Ellen Myers, Mar. 30, 1951; children: Ann, Bennett C., Karen M., John B. B.S., U. Wash., 1949, M.D., 1953; M.P.H., U. Calif., 1957; Ph.D. in Radiation Biology, U. Rochester, 1963. Health officer Shasta County, Calif., 1957-60; asst. prof. radiation biology U. Rochester, 1963-67, assoc. prof. toxicology and pharmacology, 1967—. Served with USNR. NIH postdoctoral fellow, 1969-63. Mem. Soc. Pharmacology, Soc. Toxicology. Subspecialties: Toxicology (medicine); Pharmacology. Current work: Toxicology and pharmacology of fluoride and fluorocarbons. Office: Rochester Medical Center 4 7523 Rochester NY 14642

TAYLOR, BERNARD FRANKLIN, laboratory administrator, educator; b. Charlestown, W.Va., Mar. 21, 1930; s. Beverly Douglas and Harriet Elizabeth (Dotson) T.; m. Sylvia Adora Spriggs, Jan. 28, 1957; children: Bernard Franklin, Michael Lensen. B.S. in Biology cum laude, Storer Coll., 1952; postgrad., Bluefield State Coll., 1951; M.S. in Microbiology and pub. health, Mich. State U., 1959, Rider Coll., 1961, Trenton Jr. Coll., 1964; Ph.D. in Microbiology, Rutgers U., 1972; M.A. in Adminstrn, Rider Coll., 1980. Cert. Inst. Med. Research, Camden, N.J. 1974. Bacteriologist Bur. Virology, Dept. Health, State of Mich., Lansing, 1954-56, virologist, 1956-59; instr. sci., coach football Elizabeth City (N.C.) State Tchrs. Coll., 1959-60; virologist div. labs. Dept. Health State of N.J., 1960-61, sr. virologist, 1961-62, prin. virologist, 1962-67, chief virologist, 1967-79, dir. pub. health lab. service div. pub. health and environ. labs., 1979—; med. technologist Helene Fuld Hosp., Trenton, 1961-64; co-adj. dept. biology Trenton State Coll., 1972—; co-adj. Mercer County Community Coll., 1981—. Contbr. articles to profl. jours. Mem. juvenile cof. com. County of Mercer (N.J.); chmn. United Way campaign N.J. Dept. Health, 1973; asst. scoutmaster troop 31 Boy Scouts Am., Charlestown, 1949. Recipient Ella P. Stewart Biology award, 1952. Mem. Am. Soc. Microbiologists, Am. Acad. Microbiology, Am. Assn. for Lab. Animal Sci., Found. Infectious Disease, Assn. State and Territorial Pub. Health Lab. Dirs., Am. Soc. Pub. Adminstrs., Am. Pub. Health Assn., N.Y. Acad. Scis., Sigma Xi, Beta Kappa Chi, Alpha Phi Alpha. Democrat. Lodge: Masons. Subspecialties: Virology (biology); Microbiology. Current work: Immunology as a sero-diagnostic tool. Home: 438 Walnut Ave Trenton NJ 08609 Office: CN 360 New Jersey State Department Health Trenton NJ 08625

TAYLOR, CHARLES ELLETT, biologist; b. Chgo., Sept. 9, 1945; s. Stewart Ferguson and Barbara (Ellett) T.; m. Minna Taylor. A.B., U. Calif., Berkeley, 1968; Ph.D., SUNY, Stony Brook, 1973. Asst. prof. U. Calif., Riverside, 1974-80, assoc. prof. biology, Los Angeles, 1980—. Mem. Genetics Soc. Am., Soc. Study Evolution, Am. Soc. Naturalists. Subspecialties: Evolutionary biology; Population biology. Office: Department of Biology University of California Los Angeles CA 90024

TAYLOR, DAVID JOHN, physicist; b. Syracuse, N.Y., Dec. 29, 1944; s. John Edward and Marie Elizabeth (Uhrig) T.; m. Beverly Michele, May 27, 1967; children: Karen Michele, Jennifer Dawn. B.S. in Elec. Engring, MIT, 1966, M.S., 1967; Ph.D. in Laser Physics, Stanford U., 1972. MIT coop. program student-staff mem. IBM, East Fishkill, N.Y., 1964, 65, 66; integrated-circuit designer Hewlett-Packard, Palo Alto, Calif., 1967-69; mem. staff Gen. Electric Research and Devel. Ctr., Schenectady, N.Y., 1972-74, Los Alamos Nat. Lab., 1974—. Mem. Optical Soc. Am., IEEE. Democrat. Subspecialties: Laser-induced chemistry; Coal. Current work: Laser diagnostics in coal gasification. Home: 305 Kilby Los Alamos NM 87544 Office: Los Alamos Nat Lab MS J567 Los Alamos NM 87545

TAYLOR, GEORGE WILLIAM, electrical engineer, electronics company executive; b. Perth, Australia, June 16, 1934; came to U.S., 1962, naturalized, 1968; s. George William and Myrtle (Spigl) T.; m. Cynthia Hatch, Aug. 24, 1957; children: Susan, George William, Deborah, Felicity. B.E.E. with honors, U. Western Australia, 1957, D.Eng., 1981; Ph.D., U. London, 1961. Registered profl. engr., Australia, Eng. Lectr. U. Sydney, Australia, 1961-62; mem. tech. staff RCA Labs., Princeton, N.J., 1962-70; exec. v.p. research and engring. Princeton Materials Research, 1970-75; pres. Princeton Resources, Inc., 1975—, dir., 1975—; dir. Cottesloe Corp., Perth, Aberdare Co., Princeton, Abbotsford Co., Launceston, Tasmania, Australia. Author: Polar Dielectrics and Their Applications, 1979; editor: Ferroelectrics Jour, 1970—, Ferroelectrics Letters, 1981—, Display and Imaging Technology Jour, 1982—; contbr. over 50 sci. articles to profl. publs. Recipient RCA Labs. Achievement award, 1966; Hackett Overseas scholar, 1958-59; CSIRO Overseas scholar, 1960; NSF fellow, 1976. Fellow Instr. Elec. Engrs. (London), Inst. Engrs. (Australia); mem. IEEE (sr.), Am. Phys. Soc., Soc. for Info. Display, Sigma Xi. Episcopalian. Subspecialties: Electronics; Electronic materials. Current work: Piezoelectronics, ferroelectrics; electronic displays; liquid crystals. Holder 15 patents. Office: Box 211 Princeton NJ 08540

TAYLOR, JOHN EDWARD, researcher; b. Montpelier, Idaho, Jan. 30, 1947; s. John Woodrow and Enid Maureen (Peterson) T.; m. Barbara Strecper, June 6, 1969; children: Janna, Robyn, Leigh, Stephanie. B.S., Brigham Young U., 1971; M.S., U. Pacific, Stockton, Calif., 1974, Ph.D., 1977. Fellow Mayo Clinic, Rochester, Minn., 1978-81; asst. prof. pharmacology Ind. U., Evansville, 1981-82; group leader molecular-pharmacology Biomeasure Inc., Hopkinton, Mass., 1982—. Contbr. articles to profl. jours. Recipient Nat. Research Service award Alcohol Drug Abuse Mental Health Adminstrn., 1979-81. Mem. AAAS, N.Y. Acad. Sci., Am. Soc. Neurochemistry, Soc. Neuroscience. Mormon. Subspecialties: Molecular pharmacology; Psychopharmacology. Current work: Biochemical aspects of psychopharmacology, rouvonal tissue culture. Home: 74 Fisk Mill Rd Upton MA 01568 Office: 11-15 E Ave Hopkinton MA 01748

TAYLOR, JOHN JOSEPH, electric co. exec.; b. Hackensack, N.J., Feb. 27, 1922; s. John Daniel and Johanna Frances (Thomas) T.; m. Lorraine Antoinette Crowley, Feb. 5, 1943; children—John Brian, Nancy Mary, Susan Mary. A.B. (N.Y. Regents scholar, Univ. scholar), St. Johns U., Bklyn., 1942, D.Sc. (hon.), 1974; M.S. (tchg. fellow), Notre Dame U., 1946. Mathematician Bendix Aviation Corp., Teterboro, N.J., 1946-47; scientist Kellex Corp., N.Y.C., 1947-50; mgr. devel. Westinghouse Bettis Atomic Power Lab., Pitts., 1950-67; engring. mgr. Westinghouse Nuclear Energy Systems, Pitts., 1967-70; v.p., gen. mgr. Westinghouse Advanced Nuclear Divs., Pitts., 1970-76, Westinghouse Water Reactor Divs., 1976-81; dir. Nuclear Power Electric Power Research Inst., Palo Alto, Calif., 1981—. Co-author: (with T. Rockwell) Reactor Shielding Manual, 1956; Editor: (with A. Radkowsky) Naval Reactor Physics Manual, 1964; Contbr. articles to profl. jours. Served with USNR, 1942-45; PTO. Recipient Westinghouse Order of Merit, 1955. Fellow A.A.A.S. (mem. council), Am. Nuclear Soc. (dir.); mem. Am. Phys. Soc., Nat. Acad. Engring. (Internat. Platform Assn.). Subspecialty: Nuclear fusion. Office: PO Box 355 Pittsburgh PA 15230

TAYLOR, LARRY THOMAS, chemist, educator, researcher; b. Woodruff, S.C., Dec. 31, 1939; s. and Elberta (Boozer) T.; ; s. and Elberta (Boozer) Adams; m. Gail Moss, Oct. 27, 1939; children: Judith Michele, Marc Thomas. B.S., Clemson U., 1962, Ph.D., 1965. Research assoc. Ohio State U., 1965-67; asst. prof. to prof. chemistry Va. Poly. Inst., 1967—. Contbr. numerous articles to profl. publs.; editor: (with others) Bioinorganic Chemistry, 1971. Pres. F-M-R Shelter Home Inc. Recipient J. Sheldon Horsley award Va. Acad. Scis., 1973; Philip Sporn award, 1977; acad. Teaching Excellence award Va. Poly. Inst. and State U., 1977-80; grantee Va. Poly Inst. and State U.; Sigma Si; NASA; Research Corp.; Gen. Instrument Corp.; ERDA; Dept. Energy; Dept. Interior; Electric Power Research Inst. Mem. Am.

Chem. Soc., Sigma Xi, Phi Lambda Upsilon, Sigma Tau Epsilon, Phi Eta Sigma (hon.), Phi Kappa Phi. Baptist. Subspecialties: Inorganic chemistry; Analytical chemistry. Current work: Synthesis and characterization of metal containing compounds which reversibly absorb oxygen and other small molecules; modification of polymer properties by metal ion addition; development and application of analytical techniques to coal conversion products. Patentee in field. Office: Dept Chemistry Va Poly Inst Blacksburg VA 24061

TAYLOR, MALCOLM ERNEST, mechanical engineer, cons; b. Liverpool, Eng., May 3, 1934; came to U.S., 1967, naturalized, 1978; s. Ernest and Ida (Geddes) T.; m. Jean Margaret Taylor, June 3, 1961; children: Deborah, Michael, Dawn, Jonathan. B.S.M.E., Liverpool Coll. Tech., 1957. From apprentice to design engr. in product design, various cos., U.K., 1949-65; sr. machine engr. Gillette Industries, Reading, Berkshire, Eng., 1965-67; resident engr. N.J. Machine Corp., Lebanon, N.H., 1967-72; sr. project engr. Schick Safety Razor, Milford, Conn., 1972-75; equipment mgr. McNeil Labs., Inc., Ft. Washington, Pa., 1975-77; div. programs dir. Foster Miller Assos., Waltham, Mass., 1977—; cons. to various industries. Contbr. articles to profl. jours. Mem. Instn. Mech. Engrs. (London), ASME. Republican. Baptis. Subspecialties: Mechanical engineering; Systems engineering. Current work: Mechanization automation for food, pharm., electronics and med. device industries, integrating all fields of advanced engrng. tech. Patentee mech. assembly and packaging machinery. Home: 2 Vose Rd Westford MA 01886 Office: 350 Second Ave Waltham MA 02154

TAYLOR, MILTON WILLIAM, microbiologist educator; b. Glasgow, Scotland, Dec. 10, 1931; came to U.S., 1958; s. Hyman and Jessie (Mitchell) T.; m. Miriam Reifer, Feb. 24, 1958; children: Yuval, Jonathan. B.S., Cornell U., 1961; Ph.D., Stanford U., 1966. Asst. prof. biology Ind. U., 1967-70, assoc. prof., 1970-76, prof., 1976—. Contbr. numerous articles to sci. jours. USPHS grantee, 1970-72; Nat. March of Dimes grantee, 1977-79. Mem. Am. Soc. Microbiology, Fedn. Am. Soc. Exptl. Biology. Subspecialties: Gene actions; Virology (biology). Current work: Mechanism of mutation in mammalian cells; cloning of eukaryotic gene. Office: Jordan Hall 341 Bloomington IN 47402

TAYLOR, ROBERT LEE, statistics and mathematics educator; b. Anderson County, Tenn., July 23, 1943; s. Carl Steven and Cora Lee T.; m. Mary Ann Findley, July 3, 1968; children: Kevin Lee, Keri Ann, Amy Elizabeth. B.S., U. Tenn., 1966; M.S., Fla. State U., 1969, Ph.D. in Stats, 1971. Statistician Oak Ridge Nat. Lab., summer 1968; asst. prof. U.S.C., Columbia, 1971-74, assoc. prof., 1974-80, prof., 1980—; vis. assoc. prof. Fla. State U., Tallahassee, 1977. Contbr. articles to profl. jours. U.S. Bur. Mines grantee, 1973-75; HEW grantee, 1977-78; Air Force Office Sci. Research grantee, 1979, 81; other grants. Mem. Am. Statis. Assn. (council 1978-79, pres. S.C. chpt. 1978-79), Inst. Math. Stats., Math. Assn. Am. Episcopalian. Subspecialties: Statistics; Probability. Current work: Research in laws of large numbers, stochastic convergence, and estimation of probability density functions. Statistical consulting in industrial and physiological studies. Home: 847 Gardendale Dr Columbia SC 29210 Office: Dept Math and Stas U SC Columbia SC 29208

TAYLOR, ROGER NORRIS, immunologist, cons., researcher, writer, educator; b. Farmington, Utah, Oct. 23, 1941; s. Norris John and Josephine (Hardy) T.; m. Sydney Moulton, Apr. 1, 1965; children: Michael, Stephen, Reuben, Melissa, Marcus, Benjamin. B.S., U. Utah, 1969, M.S., 1971, Ph.D., 1974. Clin. and research microbiologist VA Hosp., Salt Lake City, 1969-72; teaching assoc. U. Utah Coll. Medicine, 1970-74; microbiologist, lab improvement Utah Div. Health, Salt Lake City, 1972-74; chief diagnostic immunology Ctrs. for Diseases Control, Atlanta, 1975—; cons., tchr. in field. Author monographs, also numerous articles. Scouting coordinator, mem. dist. scout com., mem. dist. scout tng. com. and tng. staff DeKalb dist. Atlanta Area Council Boy Scouts Am.; Tucker (Ga.) stake primary scouting dir. Mormon Ch. Served to cpl. USMC, 1960-64. Recipient Order of Arrow award Boy Scouts Am., 1981, Scouters' Tng. award, 1981, Woodbadge, 1981, On My Honor award, 1981, Dist. award of merit, 1982. Mem. Am. Acad. Microbiology, Am. Bd. Bioanalysis, Nat. Com. Clin. Lab. Standards, N.Y. Acad. Sci., Sigma Xi. Subspecialties: Immunology (medicine); Health services research. Current work: Measurement and improvement of laboratory proficiency with immunologic tests; proficiency testing; immunology; quality control; laboratory improvement; serology; biometrics. Home: 308 Westwind Dr Lilburn GA 30247 Office: 1600 Clifton Rd Bldg 6 Room 319 Atlanta GA 30333

TAYLOR, RONALD J., biology educator, environmental consultant; b. Victor, Idaho, Oct. 16, 1932; s. George G. and Elva A. (Drake) T.; m. Gloria M. Wood, Apr. 26, 1955; children: Ryan, Rhonda. B.S., Idaho State U., 1956; M.S., U. Wyo., 1960; Ph.D., Wash. State U., 1964. Mem. faculty Western Wash. U., Bellingham, 1964—, prof. biology, 1972—; environ. cons. Author: Sagebrush Country, 1974, Mountain Wildflower Pacific Northwest, 1975, Mosses of North America, 1980, Rocky Mountain Wildflowers-The Mountaineers, 1981; contbr. articles on biology to profl. jours. Served to 1st lt. USAF, 1954-59. Mem. AAAS, Bot. Soc. Am., Northwest Sci. Assn., Am. Soc. Plant Taxonomists, Sigma Xi. Subspecialties: Evolutionary biology; Systematics. Current work: Biosystematic of vascular plants, chemosystematics. Home: 4241 Northwest Rd Bellingham WA 98226 Office: Western Wash U H H 337 High St Bellingham WA 98225

TAYLOR, ROSS LAWTON, dentist, educator; b. Sydney, Australia, Feb. 14, 1923; s. Donald Alexander and Christina Considine (Lawton) T.; m. Margaret Jean McKinlay, Nov. 26, 1948; children: Jane Rosemary Taylor Griffiths, Ian Alexander. B.D.S., U. Sydney, 1943, M.D.S., 1946; M.D.Sc., U. Western Australia, 1955. Lectr. in prosthetic dentistry U. Sydney, 1943-47; assoc. prof. in prosthodontics U. Western Australia, 1947-71; Fulbright Research fellow U. Minn., 1959-60; vis. prof. posthetic denistry Northwestern U., Evanston, Ill., 1968-69, prof., chmn. dept. removable prosthodontics, 1971—; cons., lectr. in field. Author dental health program for Western Australia for Minister of Health, Govt. of Western Australia, 1966; Contbr. articles to profl. jours. Served with Australian Mil. Forces, 1940-41. Recipient Outstanding Instr. award Northwestern U., 1975. Fellow Internat. Coll. Dentists, Royal Australian Coll. Dental Surgeons, Am. Coll. Dentists; mem. Dental Study Group Western Australia (hon. life), Minn. Acad. Restorative Denistry (hon. life), Australian Dental Assn. (hon. life mem. Western Australia br.), Am. Prosthodontic Soc., Federation Dentaire Internationale, Fedn. Prosthodontic Orgns., Internat. Assn. Dental Research, Midwest Acad. Prosthodontics, Australian Soc. Prosthodontists, Odontographic Soc. Chgo., Chgo. Council Fgn. Relations, Omicron Kappa Upsilon. Club: Australian Am. Wine (Chgo.). Subspecialties: Prosthodontics; Implantology. Current work: Prosthodontics, materials, techniques, partial denture design, osseointegration. Office: Northwestern Univ Dental Sch 240 E Huron St Room 2466 Chicago IL 60611

TAYLOR, SAMUEL EDWIN, pharmacologist, educator, researcher; b. Tuskegee, Ala., Oct. 19, 1941; s. Grady Clifton and Edna (Crapps) T.; m. Ouida Faye Oswalt, Aug. 9, 1961; children: Samuel Edwin, Leslie Ann. B.S., U. Ala.-Tuscaloosa, 1963, Ph.D., 1971. Postdoctoral trainee U. Tenn., Memphis, 1971-72; asst. prof. pharmacology U. Oreg. Sch. Dentistry, Portland, 1972-77; asst. prof. Baylor Coll. Dentistry, 1977-78, assoc. prof., 1978—. Contbr. articles to profl. jours. Mem. Am. Assn. Dental Schs., Internat. Assn. Dental Research, Sigma Xi. Baptist. Subspecialty: Pharmacology. Current work: Autonomic/cardiovascular pharmacology, beta-adrenergic receptors smooth muscle and cardiac. Home: 1512 Stoneham Pl Richardson TX 75081 Office: Baylor Coll Dentistry 3302 Gaston Ave Dallas TX 75246

TAYLOR, STEPHEN KEITH, chemistry educator; b. Los Angeles, Mar. 28, 1944; s. Jolly Joe and Inez (Moore) T.; m. Nancy Carol Thomson, Aug. 22, 1969; children: Daniel, Melissa. B.A., Pasadena Coll., 1969; Ph.D., U. Nev., Reno, 1974. Research chemist DuPont Co., Wilmington, Del., 1973-78; assoc. prof. chemistry Olivet Nazarene Coll., Kankakee, Ill., 1978—; cons. Miles Biochems., Kankakee, 1981-82. Contbr. articles to profl. jours. Served with USAF, 1962-64. Recipient Alumni award Pasadena Coll., 1978. Mem. Am. Chem. Soc., Soc. Photog. Scientists and Engrs. (Outstanding Publ. award 1979), Sigma Xi, Pi Kappa Delta. Republican. Nazarene. Subspecialty: Organic chemistry. Current work: Reactions of epoxides, protein reactions, and organometallic chemistry. Home: 340 E Charles St Bourbonnais IL 60914 Office: Olivet Nazarene Coll Kankakee IL 60901

TAYLOR, STEVE LLOYD, food safety research scientist, educator; b. Portland, Oreg., July 19, 1946; s. Lloyd E. and Frances H. (Hanson) T.; m. Susan Kerns, June 23, 1973; children: Amanda, Andrew. B.S., Oreg. State U.-Corvallis, 1968, M.S., 1969; Ph.D., U. Calif.-Davis, 1973. Research assoc. U. Calif., Davis, 1973-74, Nat. Inst. Environ.-Health Scis. fellow, 1974-75; research chemist, chief Food Toxicology Lab, Letterman Army Inst. Research, San Francisco, 1976-78; asst. prof. food microbiology and toxicology Food Research Inst., U. Wis., Madison, 1978—; cons. WHO, Geneva, Switzerland, 1982—. Contbr. over 40 sci. articles to profl. publs. Assn. Official Analytical Chemists scholar, 1967-68; Stayton Canning Co. scholar, 1964-65. Mem. Inst. Food Technologists (chmn. toxicology and safety evaluation div. 1981-82, mem. expert panel on food safety and nutrition 1982-85, sci. letcr. 1983-86), Am. Acad. Allergy and Immunology (mem. adverse reactions to foods com. 1981—), Am. Chem. Co., Am. Peanut Research and Edn. Soc. Presbyterian. Subspecialties: Food science and technology; Toxicology (agriculture). Current work: Food allergies; scombroid fish poisoning or histamine poisoning; naturally-occurring toxicants in foods; effects of foodborne and environmental toxicants; general foodborne disease. Office: Univ Wis 202 Food Research Inst Madison WI 53706

TEAGUE, HOWARD STANLEY, research coordinator; b. Rockville, Ind., Jan. 16, 1922; s. Archie Grant and Versa Elizabeth (Witlock) T.; m. Marlyn Joy Spencer, Dec. 12, 1943; children: Douglas, Richard, Edward, Allen. B.S., U. Nebr., 1948, M.S., 1949; Ph.D., U. Minn., 1952. Asst. scientist Hormel Inst., Austin, Minn., 1950; prof. animal sci. Ohio State U. and Ohio Agrl. Exptl. Sta., Wooster, 1952-72; research leader U.S. Meat Animal Research Ctr., Clay Center, Nebr., 1972-76; nutritionist Sci. and Edn. Adminstrn., U.S. Dept Agr., Washington, 1976—. Served as capt. USAF, 1941-44. Mem. Am. Soc. Animal Sci. (pres. Midwestern sect. 1975-76), Soc. Study Reprodn., Am. Inst. Nutrition. Democrat. Presbyterian. Lodge: Kiwanis Internat. Subspecialties: Animal nutrition; Animal physiology. Current work: Coordination and review of agriculture-related research at Land Grant institutions and federal agencies and program management for specific areas of special grants programs. Office: Sci and Edn US Dept Agr Washington DC 20250 Home: 9207 Leamington Ct Fairfax VA 22031

TEASDALL, ROBERT DOUGLAS, physician, researcher; b. London, Ont., Can., Dec. 9, 1920; came to U.S., 1950; s. Douglas Rupert and Marjorie (Irvine) T.; m. Veronica Publow, July 17, 1948; children: Susan Teasdall Oldendorp, Robert D. M.D., U. Western Ont., 1946, Ph.D., 1950. Intern St. Michael's Hosp., Toronto, 1946-47; resident in neurology Balt. City Hosp., Johns Hopkins Hosp., 1950-54; instr. neurology Johns Hopkins Hosp., Balt., from 1954, assoc. prof. neurology, until 1974; neurologist Henry Ford Hosp., Detroit, 1974—. Contbr. numerous articles to profl. jours. Fellow Am. Acad. Neurology; mem. Am. Neurol. Assn., Am. Physiol. Soc. Subspecialties: Neurology; Comparative physiology. Current work: Clinical neurology. Home: 16240 N Park Dr Southfield MI 48075 Office: Henry Ford Hosp 2799 W Grand Blvd Detroit MI 48202

TEDDLIE, CHARLES BENTON, education research consultant, social psychologist; b. Winnfield, La., July 30, 1949; s. Charles Ray and Blanche (Wilson) T.; m. Susan Elizabeth Kochan, July 4, 1982; m. Karen Antoinette Lafontaine, Dec. 1972 (div. 1974). B.S. cum laude, La. State U., 1972; M.A. in Social Psychology, U.N.C.-Chapel Hill, 1977, Ph.D., 1979. Vis. asst. prof. La. State U., Baton Rouge, 1979, adj. asst. prof., 1980—; internal research cons. La. State Dept. Edn., Baton Rouge, 1980—; pres. La Data, Baton Rouge, 1980—. Contbr. articles to profl. jours.; editorial cons.: Personality and Social Psychology Bull, 1980, Population: Behavioral, Social and Environ. Issues, 1977-79, Representative Research in Social Psychology, 1973-79. Mem. Democratic Socialist Organizing Com., 1980—. Recipient Paul C. Young award La. State U., 1971-72. Mem. Am. Ednl. Research Assn., Am. Psychol. Assn., Assn. Instl. Research, SAS Users Group Internat., Phi Kappa Phi. Democrat. Subspecialties: Social psychology; Information systems (information science). Current work: Forecasting educational enrollment and policy trends, school efficiency and effectiveness studies, racial differences in social perception, management information systems, communications research. Home: 9065 Redbud Baton Rouge LA 70815 Office: PO Box 44064 Baton Rouge LA 70804

TEDESCHI, DAVID HENRY, biosciences research director.; b. Newark, N.J., Feb. 20, 1930; s. Edward D. and Iris C. (Beltrani) T.; children: Paul David, Douglas James, Linda Diane; m. Carol Mahoney. B.Sc. in Pharmacy, Rutgers U., 1952; Ph.D. in Pharmacology, U. Utah, 1955. Assoc. dir. pharmacology Smith Kline French Labs., 1955-68; dep. dir. biosci. CIBA-Geigy Co., 1968-72; dir. central nervous system/cardiovascular research Lederle Labs., 1972-78; dir. bioscis. 3M Co., St. Paul, 1978—; adj. prof. pharmacology Coll. Pharmacy, U. Minn., 1978—. Editor: Importance of Fundamental Principles in Drug Evaluation, 1969; contbr. over 80 articles on pharmacology, physiology, biochemistry, biotech. Recipient award in pharmacodynamics Am. Pharm. Assn. Found., 1968. Fellow Am. Coll. Neuropsychopharmacology; mem. Am. Soc. Pharmacology and Exptl. Therapeutics, Collegium International Neuropsychopharmacology. Subspecialties: PET scan; Enzyme technology. Current work: Bioinstrumentation; genetic engineering; biomaterials; enzymology; immunology;neurophysiology; pharmaceutics;pharmacology; wound management.

TEDESCHI, HENRY, biology educator; b. Novara, Italy, Feb. 3, 1930; s. Edoardo Vittorio and Berta (Minerbi) T.; m. Terry Kershner, Nov. 29, 1957; children: Alexander, Devorah, David. B.S., U. Pitts., 1950; Ph.D., U. Chgo., 1955. Research assoc. U. Chgo., 1955-57, asst. prof., 1957-60, U. Ill., 1960-65, assoc. prof., 1965; prof. SUNY-Albany, 1965—, chmn. dept. biol. sci., 1983—. Author: Cell Physiology: Molecular Dynamics, 1974, Mitochondria: Structure, Biogenesis and Transducing Function, 1976. Mem. Am. Physiol. Soc., Biophys. Soc., Am. Soc. Cell Biology, Soc. Neurosci, Am. Soc. Biol. Chemists, Soc. Gen. Physiologists. Subspecialties: Membrane biology; Cell and tissue culture. Current work: Structure and function of biological membranes. Home: 10 Westlyn Pl Albany NY 12203 Office: 1400 Washington Ave Albany NY 12222

TEDESCO, EDWARD FRANCIS, astronomer; b. Bklyn., Feb. 6, 1947; s. Anthony and Eileen Ann (Currier) T.; m. Gina Gale Quintero, May 24, 1980; 1 son: Preston. B.S., St. John's U., 1969; M.S., Fordham U., 1974, N. Mex. State U., 1978, Ph.D., 1979. NDEA fellow Fordham U., Bronx, N.Y., 1972-74; research asst. N.Mex. State U., Las Cruces, 1974-78; research asst. II Lunar and Planetary Lab., U. Ariz., Tucson, 1978-79, research assoc., 1979-81; NRC resident research assoc. Jet Propulsion Lab., Pasadena, Calif., 1981—. Contbr. articles to profl. jours. and encys. N.Y. state regents coll. scholar, 1965-69; NDEA fellow, 1972-74; NRC resident research assoc. fellow, 1981—. Mem. Internat. Astron. Union, Am. Astron. Soc., Astron. Soc. Pacific, Sigma Xi. Subspecialties: Optical astronomy; Planetary science. Current work: Solar system origin and evolution, visual and infrared ground based observations of asteroids, comets and planetary satellites. Home: 2601 E La Cienega Dr Tucson AZ 85716 Office: 4800 Oak Grove Dr MS 183-501 Pasadena CA 91109

TEED, DOUGLAS EARLE, electrical equipment manufacturing company executive; b. Truro, N.S., Can., Dec. 22, 1938; s. Willard Earle and Lucy (Fletcher) T.; m. Marie Maxine Taylor, Nov. 22, 1961; children: Michael, Donna, Kathryn. Engrng. diploma, Dalhousie U., 1959; B. Engrng. with honors, Tech. U.N.S., 1961; M.Sc., Northwestern U., 1963. With Can. Gen. Electric Co., Peterborough, Ont., Can., supr. fuel devel., 1975-80, mgr. engrng., 1981—. Mem. Am. Nuclear Soc., Can. Nuclear Soc. Subspecialty: Nuclear engineering. Current work: Fuel performance, fuel fabrication techniques. Home: 1609 Champlin Dr Peterborough ON Canada K9L 1N5 Office: Can Gen Electric Co 107 Park St N Peterborough ON Canada K9J 7B5

TEETER, JAMES WALLIS, geology educator; b. Hamilton, Ont., Can., Mar. 14, 1937; came to U.S., 1962; s. James C. and Dorothy Elizabeth (Wallis) T.; m. Gladys Wanda Wrobel, October 14, 1960; children: Laurie, Katherine, Judith. B. Sc., McMaster U., 1960, M.Sc., 1962; Ph. D., Rise U., 1966. Asst. geologist Geol. Survey of Can., Ottawa, Ont., 1958-60; field geologist Shell Can., Calgary and Edmonton, Alta., summers 1961-63; geologist Shell Oil, Houston Exploration Div., 1964; mem.faculty dept. geology U. Akron, Ohio, 1965—, prof., 1977—. Contbr. articles to profl. jours. Named Outstanding Grad. Student Dept. Geology, Rice U., 1965. Mem. No. Ohio Geol. Soc. (pres. 1972-73), Soc. of Econ. Paleontologists and Mineralogists (treas. 1976-77), Sigma Xi (pres. elect U. Akron chpt.). Subspecialties: Paleobiology; Paleoecology. Current work: Holocene sealevel fluctuation - Bahamas and climatic change. Home: 2437 Grapevine Circle Stow OH 44224 Office: Dept Geology U Akron Akron OH 44325

TEEVAN, RICHARD COLLIER, psychology educator; b. Shelton, Conn., June 12, 1919; s. Daniel Joseph and Elizabeth (Hallowell) T.; m. Virginia Agnes Stehle, July 28, 1945; children—Jan Elizabeth, Kim Ellen, Clay Collier, Allison Tracy. B.A., Wesleyan U., Middletown, Conn., 1951; M.A., U. Mich., 1952, Ph.D., 1955. Rubber buffer Sponge Rubber Product Co., Derby, Conn., 1939-41; with U. Mich., 1951-57, teaching fellow, 1951-53, instr., 1953-57; asst. prof. Smith Coll., 1957-60; assoc. prof. Bucknell U., 1960-64, prof., 1964-69; chmn. psychology, prof. SUNY-Albany, 1969—. Author: Reinforcement, 1961; author: Instinct, 1961, Color Vision, 1961, Measuring Human Motivation, 1962, Theories of Motivation in Learning, 1964, Theories of Motivation in Personality and Social Psychology, 1964, Motivation, 1967, Fear of Failure, 1969, Readings in Elementary Psychology, 1973; contbr. articles to sci. jours. Served to capt. AUS, 1941-47; prisoner of war, 1943-45; Ger. Office Naval Research grantee, 1958-72; recipient Lindbach award Bucknell U., 1966. Mem. Am. Psychol. Assn., Eastern Psychol. Assn., AAUP, AAAS, Phi Beta Kappa, Sigma Xi. Subspecialties: Social psychology; Cognition. Current work: Fear of failure in achievement situations, in women, need for achievement, projective measurement of motivation. Address: 45 Pine St Delmar NY 12054 Office: Dept Psychology SUNY 1400 Washington Ave Albany NY 12222

TEITELBAUM, PHILIP, psychologist; b. Bklyn., Oct. 9, 1928; s. Bernard and Betty (Schechter) T.; m. Evelyn Satinoff, Dec. 26, 1963; children: Benjamin, Daniel, David. B.S., Coll. City N.Y., 1950; M.A., Johns Hopkins U., 1952, Ph.D., 1954. Instr., asst. prof. physiol. psychology Harvard U., 1956-59; assoc. prof. psychology U. Pa., Phila., 1959-63, prof., 1963-73; prof. psychology U. Ill., Urbana-Champaign, 1973—, prof., 1979—, Disting. prof. Ctr. Advanced Studies, 1980—; fellow Center for Advanced Study in Behavioral Scis., Stanford U., Palo Alto, Calif., 1975-76. Author: Fundamental Principles of Physiological Psychology, 1967; Contbr. chpts. to books, articles to profl. jours. Fulbright fellow Tel Aviv U., 1978-79. Fellow Am. Psychol. Assn. (pres. div. physiol. psychology, disting. sci. contbn. award 1978); mem. Nat. Acad. Scis., AAAS, Am. Physiol. Soc., Soc. Neuroscis. Subspecialty: Physiological psychology. Current work: Animal models of Parkinsonism; recovery from hypothalmic damage; movement disorders produced by brain damage, use of movement-rotation. Office: Dept Psychology U Ill Urbana Champaign IL 61820

TEITELBAUM, RAY (TIM TEITELBAUM), computer scientist, educator, researcher; b. N.Y.C., Apr. 12, 1943; s. David and Sylvia (Lowenthal) T.; m. Ann Pitkin, Aug. 1964 (div.); children: Benjamin, Felix. S.B., MIT, 1964; Ph.D. in Computer Sci, Carnegie-Mellon U., Pitts., 1973. Programmer Western Electric, N.Y.C., 1964-65; sr. programmer dept. physics Columbia U., 1965-68; mem. faculty dept. computer sci. Cornell U., 1973—, now assoc. prof. Mem. Assn. Computing Machinery. Subspecialties: Programming languages; Software engineering. Current work: Design and implementation of syntax-directed systems and programming environments. Office: Upson Hall Cornell U Ithaca NY 14853

TEJADA, FRANCISCO, physician, med. researcher; b. Moyobamba, Peru, July 25, 1942; s. Francisco Tejada Rojas and Semiramis Reategui Tuesta; m. Barbara Ann Kotowski, Feb. 1, 1970; children: Ana Maria, Semiramis, Barbara Lee, Francisco, James. B.S., Universidad Nacional Mayor de San Marcos, Lima, 1961; M.D., Peruvian U. Cayetano Heredia, 1967. Diplomate: Am. Bd. Internal Medicine. Sr. cancer research internist Nat. Cancer Inst., NIH, 1973-75; asst. prof. George Washington U., 1974-75; asst. prof. medicine and oncology U. Miami, 1975-80, asst. to dir., 1975-80, assoc. prof. oncology and otolaryngology, 1980—; head med. oncology div. Miami Cancer Inst., 1980—; cons. in field. Contbr. articles to profl. jours. Mem. community health assn., 1980—. Recipient Hipolito Unanue Inst. award, 1969; grantee NIH, 1976-79, 78—, Lilly Research Lab., 1978-79. Mem. Peruvian Coll. Physicians, ACP (fellow), Am. Assn. Cancer Research, AM. Soc. Hematology, Am. Soc. Clin. Oncology, Cell Kinetic Soc., AAAS. Roman Catholic. Subspecialties: Cancer research (medicine); Cell study oncology. Current work: Cell kinetics of solid tumors and its pertubation by drugs and radiation. Office: 1339 SW 1st Ave Miami FL 33130 Home: 1550 SW 132d St Miami FL 33156

TELESCO, CHARLES MICHAEL, astronomer; b. Trenton, N.J., Oct. 18, 1946; s. Charles Edward and Aileen Mary (Harle) Watters; m. Patricia Gaynor Telesco, Mar. 6, 1982. A.S. in Physics, Palm Beach Jr. Coll., 1966, B.S., Case Western Res. U., 1969, M.S., Purdue U., 1971, U. Chgo., 1975, Ph.D., 1977. Postdoctoral research fellow Ctr. for Space Research, M.I.T., 1977-78; asst. astronomer U. Hawaii; also staff astronomer NASA Infrared Telescope Facility, Honolulu, 1979-82; NRC research assoc. Space Sci. div. Astrophys. Expts. br. NASA Ames Research Ctr., Moffett Field, Calif., 1982-83; space scientist Space Sci. Lab NASA Marshall Space Flight Ctr., Huntsville, Ala., 1983—. Contbr. articles to sci. jours. Mem. Am. Astron. Soc., Astron. Soc. of Pacific. Subspecialty: Infrared optical astronomy. Current work: Extragalactic infrared emission; infrared instrumentation; star formation. Home: 13027 Camelot Dr Huntsville AL 35803 Office: Space Sci Lab NASA Marshall Space Flight Mail Stop ES-63 Huntsville AL 35812

TELESHAK, STEPHEN, metall. engr., cons.; b. Monessen, Pa., May 10, 1922; s. Konstantin and Tekla (Kostura) Telischak; m. Alice Marie Feldman; 1 dau.: Tekla. B.S. in Metall. Engring, U. Pitts., 1949. Registered profl. engr., Pa., Tex., La. Mgr. metall. dept. Pitts. Testing Lab., 1952-74; owner, prin. Teleshak Metall. Lab., New Orleans, 1974—. Served with AUS, 1942-46. Mem. Am. Soc. Metals, ASME, Am. Welding Soc., ASTM, Am. Soc. Nondestructive Testing, Electron Microscope Soc. Am., Nat. Assn. Corrosion Engrs, VFW. Subspecialties: Metallurgy; Failure anaylsis. Home: 113 Eden Isles Dr Slidell LA 70458 Office: 4315 Royal St New Orleans LA 70117

TELKES, MARIA, scientist, engr., educator; b. Budapest, Hungary, Dec. 12, 1900; came to U.S., 1925, naturalized, 1937; d. Aladar and Maria (Laban) T. Ph.D., U. Budapest, 1924. Research asst. U. Budapest, 1923-24; biophysicist Cleve. Clinic, 1925-37; research engr. Westinghouse Research Labs., 1937-39; research asso. Mass. Inst. Tech., 1939-53; project dir. Coll. Engring., N.Y.U., 1953-58; research dir. solar energy lab. Princeton (div. Curtiss-Wright Co.), 1958-60; dir. research and devel. Cryo-Therm Co., 1961-64; head solar energy applications lab. MELPAR, Inc. (subsidiary Westinghouse Air-Brake Co.), 1965-69; chief scientist Inst. Direct Energy Conversion, U. Pa., Phila., 1970-72; adj. prof., cons. energy conversion Inst. of Energy Conversion, U. Del., 1972-77; dir. solarthermal storage devel. Am. Technol. U., Killeen, Tex., 1977-80; cons., 1980—. Contbr. numerous articles to profl. jours., chpts. in books. Mem. Solar Energy Soc. (dir., C.G. Abbott award), Am. Chem. Soc., Soc. Women Engrs. (hon. life, recipient 1st award 1952), Hellenic Soc. Solar Energy (hon., Quarter Century award), Nat. Acad. Sci./NRC. Subspecialties: Inorganic chemistry; Physical chemistry. Current work: Solar energy conversion, thermal energy storage. Specializes in solar energy research. Developed solar stills for life rafts, thermoelectric materials and solar thermoelectric generators, designed solar heating equipment for bldgs.; developed heat storage materials used in accessory equipment in spacecraft and solar heated buildings. Office: NAHB Research Found PO Box 1627 Rockville MD 20850

TELLER, CECIL MARTIN, II, research institute executive; b. Galveston, Tex., Oct. 25, 1939; s. Cecil Martin and Laura Mary (Adascheck) T.; m. Valerie Diana Klossner, June 18, 1966; children: Cecil Martin III, Diana Lynn. B.S.M.E., U. Tex.-Austin, 1964, M.S.M.E., 1966, Ph.D., 1971. Registered profl. engr., Tex. Engr., scientist Def. Research Lab., U. Tex., Austin, 1965-66, research engr. dept. mech. engring., 1970-71; mgr. materials tech. sect. dept. applied sci. Tracor, Inc., Austin, 1972-74; dep. br. chief U.S. Govt., Washington, 1974-77; mgr. nondestructive evaluation research instrumentation research div. Southwest Research Inst., San Antonio, Tex., 1977-83; tech. dir. Tex. Research Inst., Austin, 1983—. Contbr. articles to profl. jours. Served to lt. Civil Engr. Corps USNR, 1966-69. Cameron Iron Works fellow, 1969-71. Mem. ASME, Am. Soc. Metals, Am. Soc. Nondestructive Testing, Tau Beta Pi, Pi Tau Sigma, Phi Kappa Phi. Mem. Ch. of Christ. Subspecialties: Mechanical engineering; Materials (engineering). Current work: Materials testing and characterization, nondestructive evaluation; technical direction and management of a wide variety of research and development contracts in materials, chemistry and mechanical engineering. Office: 9063 Bee Caves RD Austin TX 78746

TELLER, EDWARD, physicist; b. Budapest, Hungary, Jan. 15, 1908; naturalized, 1941; s. Max and Ilona (Deutch) T.; m. Augusta Harkanyi, Feb. 26, 1934; children—Paul, Susan Wendy. Student, Inst. Tech., Karlsruhe, Germany, 1926-28, U. Munich, 1928-29; Ph.D., U. Leipzig, Germany, 1930; D.Sc. (hon.), Yale, 1954, U. Alaska, 1959, Fordham U., 1960, George Washington U., 1960, U. So. Calif., 1960, St. Louis U., 1960, Rochester Inst. Tech., 1962, PMC Colls., 1963, U. Detroit, 1964, Clemson U., 1966, Clarkson Coll., 1969, U. Md. at Heidelberg, 1977; LL.D., Boston Coll., 1961, Seattle U., U. Cin., 1962, U. Pitts., 1963, Pepperdine Coll., 1973; D.Sc., L.H.D., Mt. Mary Coll., 1964; Ph.D., Tel Aviv U., 1972; D.Natural Sci., DeLaSalle U., Manila, 1981. Research asso., Leipzig, 1929-31, Gottingen, Germany, 1931-33, Rockefeller fellow, Copenhagen, Denmark, 1934; lectr. U. London, 1934-35; prof. physics George Washington U., Washington, 1935-41, Columbia, 1941-42; physicist Manhattan Engr. Dist., U. Chgo., 1942-43, Los Alamos Sci. Lab., 1943-46; prof. physics U. Chgo., 1946-52, U. Calif., 1953-60, prof. physics-at-large, 1960-70, Univ. prof., 1970-75, prof. emeritus, chmn. dept. applied sci., Davis and Livermore, 1963-66; asst. dir. Los Alamos Sci. Lab., 1949-52; cons. Livermore br. U. Calif. Radiation Lab., 1952-53; asso. dir. Lawrence Livermore Radiation Lab., U. Calif., 1954-58, 60-75, dir., 1958-60, now dir. emeritus, cons.; concerned with planning and prediction function atomic bomb and hydrogen bomb, Manhattan Dist. of Columbia, 1942-46; also Metall. and Lab. of Argonne Nat. Lab., U. Chgo., 1942-43, 46-52, and Los Alamos, N.Mex., 1943-46, Radiation Lab., Livermore, Calif., 1952-75; sr. research fellow Hoover Instn. War, Revolution and Peace, Stanford, 1975—; Bd. mem. Thermo Electron Corp.; Mem. sci. adv. bd. USAF.; Mem. bd. Fed. Union; bd. govs. Am. Friends of Tel Aviv U.; sponsor Atlantic Union, Atlantic Council of U.S., Univ. Centers for Rational Alternatives; past mem. gen. adv. com. AEC; mem. Com. to Unite Am., Inc.; former mem. President's Fgn. Intelligence Adv. Bd. Author: (with Francis Owen Rice) The Structure of Matter, 1949, (with A. L. Latter) Our Nuclear Future, 1958, (with Allen Brown) The Legacy of Hiroshima, 1962, The Reluctant Revolutionary, 1964, (with G.W. Johnson, W.K. Talley, G.H. Higgins) The Constructive Uses of Nuclear Explosives, 1968, (with Segre, Kaplan and Schiff) Great Men of Physics, 1969, The Miracle of Freedom, 1972, Energy: A Plan for Action, 1975, Nuclear Energy in the Developing World, 1977, Energy from Heaven and The Earth, 1979, The Pursuit of Simplicity, 1980. Bd. visitors, past bd. dirs. Def. Intelligence Sch., Naval War Coll. Recipient Joseph Priestley Meml. award Dickinson Coll., 1957; Albert Einstein award, 1958; Gen. Donovan Meml. award, 1959; Midwest Research Inst. award, 1960; Research Inst. Am. Living History award, 1960; Golden Plate award, 1961; Thomas E. White and Enrico Fermi awards, 1962; Robins award of Am., 1963; Leslie R. Groves Gold medal, 1974; Harvey prize in sci. and tech. Technion-Israel Inst., 1975; Semmelweiss medal, 1977; Albert Einstein award Technion Inst., 1977; Henry T. Heald award Ill. Inst. Tech., 1978; Gold medal Am. Coll. Nuclear Medicine, 1980; A.C. Eringen award, 1980; named ARCS Man of Yr., 1980, Disting. Scientist Nat. Sci. Devel. Bd., 1981. Fellow Am. Nuclear Soc., Am. Phys. Soc.; mem. Am. Acad. Arts and Scis., Am. Ordnance Assn., Nat. Acad. Scis., Am. Geophys. Union, Soc. Engring. Scis. Subspecialties: Nuclear physics; Theoretical physics. Current work: Research on chemistry; molecular and nuclear physics, quantum mechanics, thermonuclear reactions, applications of nuclear energy, astrophysics, spectroscopy of polymatomic molecules, theory of atomic nuclei. Research on chem., molecular and nuclear physics, quantum mechanics, thermonuclear reactions, applications of nuclear energy, astrophysics, spectroscopy of polyatomic molecules, theory of atomic nuclei. Office: Hoover Instn Stanford CA 94305

TEMES, GABOR CHARLES, electrical engineering educator; b. Budapest, Hungary, Oct. 14, 1929; s. Erno and Rozsa (Angyal) Wohl-Temes; m. Ibi Kutasi-Temes, Feb. 6, 1954; children: Roy Thomas, Carla Andrea. Dipl.Ing., Tech. U. Budapest, 1952; Dipl.Phys., Eotvos U., Budapest, 1954; Ph.D., U. Ottawa, Ont., Can., 1961. Asst. prof. Tech. U. Budapest, 1952-56; project engr. Measurement Engring. Ltd., 1956-59; dept. head No. Electric Co. Ltd., 1959-64; group leader Stanford Linear Accelerator Center, 1964-66; corporate cons. Ampex Corp., 1966-69; prof. elec. engring. UCLA, 1969—, chmn. dept., 1975—; Cons. Ampex Corp., Collins Radio Co. Author: (with others) Introduction to Circuit Synthesis and Design, 1977; Asso. editor: Jour. Franklin Inst, 1971—; Co-editor, contbg. author: Modern Filter Theory and Design, 1973. Recipient Best Paper award IEEE, 1969, 81, Western Electric Fund award Am. Soc. Engring. Edn., 1982; NSF grantee, 1970—. Fellow IEEE (editor Transactions on Circuit Theory 1969-71). Subspecialties: Computer-aided design; Integrated circuits. Current work: Analog MOS integrated circuits; filters; signal theory and processing. Home: 2015 Stradella Rd Los Angeles CA 90024

TEMIN, HOWARD MARTIN, scientist, educator; b. Phila., Dec. 10, 1934; s. Henry and Annette (Lehman) T.; m. Rayla Greenberg, May 27, 1962; children: Sarah Beth, Miriam Judith. B.A., Swarthmore Coll., 1955, D.Sc. (hon.), 1972; Ph.D., Calif. Inst. Tech., 1959; D.Sc. (hon.), N.Y. Med. Coll., 1972, U. Pa., 1976, Hahnemann Med. Coll., 1976, Lawrence U., 1976, Temple U., 1979, Med. Coll. Wis., 1981. Postdoctoral fellow Calif. Inst. Tech., 1959-60; asst. prof. oncology U. Wis., 1960-64, asso. prof., 1964-69, prof., 1969—, Wis. Alumni Research Found. prof. cancer research, 1971-80, Am. Cancer Soc. prof. viral oncology and cell biology, 1974—, H.P. Rusch prof. cancer research, 1980—, Steenbock prof. biol. scis., 1982—; mem. NIH (virology study sect.), 1971-74, mem. dir.'s adv. com., 1979-83; mem. Nat. Cancer Inst. (spl. virus cancer program tumor virus detection segment working group), 1972-73; sponsor Fedn. Am. Scientists, 1976—; sci. adv. Stehlin Found., Houston, 1972—; mem. Waksman award com. Nat. Acad. Sci., 1976—; mem. U.S. Steel award Com., 1980—, chmn., 1982. Assoc. editor: Jour. Cellular physiology, 1966-77, Cancer Research, 1971-74; mem. editorial bd.: Jour. Virology, 1971—, Intervirology, 1972-75, Proc. Nat. Acad. Scis, 1975-80, Archives of Virology, 1975-77. Co-recipient Warren Triennial prize Mass. Gen. Hosp., 1971, Gairdner Found. Internat. award, 1974, Nobel Prize in medicine, 1975; recipient Med. Soc. Wis. Spl. commendation, 1971; Papanicolaou Inst. PAP award, 1972; M.D. Anderson Hosp. and Tumor Inst. Bertner award, 1972; U.S. Steel Found. award in Molecular Biology, 1972; Theobald Smith Soc. Waksman award, 1972; Am. Chem. Soc. award in Enzyme Chemistry, 1973; Modern Medicine award for Distinguished Achievement, 1973; Harry Shay Meml. lectr. Fels Research Inst., 1973; Griffuel prize Assn. Devel. Recherche Cancer, Villejuif, 1972; G.H.A. Clowes lectr. award Assn. Cancer Research, 1974; NIH Dyer lectr. award, 1974; Harvey lectr., 1974; Charlton lectr. Tufts U., 1976; Hoffman-LaRoche lectr. Rutgers U., 1979; Albert Lasker award in basic med. sci., 1974; Lucy Wortham James award Soc. Surg. Oncologists, 1976; Alumni Disting. Service award Calif. Inst. Tech., 1976; Gruber award Am. Acad. Dermatology, 1981; mem. Central High Sch. Hall of Fame, Phila., 1976; Pub. Health Service Research Career Devel. awardee Nat. Cancer Inst., 1964-74. Fellow Am. Acad. Arts and Scis.; fellow Wis. Acad. Sci., Arts and Letters; mem. Nat. Acad. Scis., Am. Philos. Soc. Subspecialties: Animal virology; Genetics and genetic engineering (biology). Current work: Molecular biology and genetics; virus evolution and variation. Office: McArdle Lab 450 N Randall St U Wis Madison WI 53706

TEMPLE, WALLEY JOHN, general and oncological surgeon, educator; b. Ann Arbor, Mich., May 8, 1946; s. Victor Clarence and Marna (Walley) T.; m. Doreen H. Farley, Sept. 3, 1966; children: Lara, Claire, Philip, Martha. M.D., Queen's U., Kingston, Ont., Can., 1970; postgrad., U. Man. (Can.), Winnipeg, 1976. Intern Health Scis. Center, 1972-76; fellow in surg. oncology U. Miami, Fla., 1976-78, asst. prof. surgery, 1979-82, assoc. prof., 1982—; mem. surg. staff VA Hosp., Miami, 1979—, Jackson Meml. Hosp., 1981—; mem. surg. com. Southeastern Cancer Study Group, 1981—. Am. Cancer Soc. fellow, 1978-79, 80-83. Fellow Royal Coll. Physicians and Surgeons Can., ACS; mem. Soc. Surg. Oncology, Soc. Head and Neck Surgeons. Methodist. Subspecialties: Surgery; Oncology. Current work: Tumor immunology cell cycle kinetics; clinical trials melanoma and colon cancer. Office: Univ Miami Sch Medicine 1600 NW 10th Ave Miami FL 33101 Home: 7301 SW 113th St Miami FL 33156

TENEICK, ROBERT EDWIN, cardiac electrophysiologist, consultant; b. Portchester, N.Y., Oct. 14, 1937; s. Arthur and Viola (Spence) TenE.; m. Mary Louise Costa, July 1, 1962; children: Matthew, Andrew. B.S., Columbia U., 1963, Ph.D., 1968. Guest investigator Rockefeller U., N.Y.C., 1968; vis. sr. scientist U. Saarlandes, W.Ger., 1974-75; asst. prof. pharmacology Northwestern U., Chgo., 1968-75, asso. prof., 1974-81, prof., 1981—; cons. Contbr. articles to profl. jours. Active Boy Scouts Am.; mem. exec. com. basic sci. council Am. Heart Assn., 1980-82. USPHS awardee, 1974-80; recpient Research Achievement award Chgo. Heart Assn., 1978. Mem. Am. Physiol. Soc., Am. Soc. Pharmacology and Exptl. Therapeutics, Cardiac Muscle Soc., Am. Heart Assn. Subspecialties: Cellular pharmacology; Membrane biology. Current work: Cellular electrical activity of the heart and cardiac excitation-contraction coupling; cardiac muscle research. Office: 320 E Superior St Chicago IL 60611

TENGERDY, CATHERINE ELIZABETH, biochemist, radiochemist; b. Budapest, Hungary, Nov. 14, 1931; d. Lajos and Sophia (Wartha) K.; m. Robert P. Tengerdy, Nov. 14, 1953; children: Thomas, Peter. B.S., Tech. U., Budapest, 1956; M.S., Colo. State U., 1966, Ph.D., 1973. Head technologist St. Luke's Hosp., N.Y.C., 1957-61; research assoc., radiol. biology Colo. State U., Ft. Collins, 1973—. Mem. Health Physics Soc., Phi Kappa Phi. Subspecialties: Neurochemistry; Radiology. Current work: Environmental Surveillance. Home: 1236 Country Club Rd Fort Collins CO 80524 Office: Colo State U Fort Collins CO 80523

TENGERDY, ROBERT PAUL, microbiologist; b. Budapest, Hungary, Dec. 17, 1930; s. Ferenc and Aranka (Garay) T.; m. Catherine E. Kökény, Nov. 14, 1953; children: Thomas, Peter. Dipl. Chem. Eng., Tech. U. Budapest, 1953; Ph.D. In Microbiology, St. John's U., N.Y., 1961. Asst. prof. bioengring. Tech. U. Budapest, 1953-56; research biochemist Chas. Pfizer & Co., Bklyn., 1957-61; prof. microbiology Colo. State U., Ft. Collins, 1961—. Contbr. articles to sci. jours. Alexander von Humboldt fellow, 1968. Mem. Am. Soc. Microbiology, Am. Assn. Immunologists, Soc. Indsl. Microbiology, AAAS. Subspecialties: Biomass (agriculture); Infectious diseases. Current work: Conversion of lignocellulose to single cell protein; immunoenhancement with Vitamin E. Office: Colo State U Fort Collins CO 80523

TENISON, ROBERT BLAKE, oil co. exec.; b. Houston, Jan. 26, 1924; s. Jack R. and Auban (Blake) Pope T.; m. (married); children: William B., Robert B., John Thomas, Susan. Student, U. Tex., 1942-45. With ind. oil and gas cos., 1946-61; v.p. Consol. Oil & Gas, Denver, 1961-67; pres. and chief exec. officer Worldwide Energy Corp., Denver, Worldwide One, Inc., Worldwide Energy Co., Ltd., Calgary, Alta., Can., Semco Gas, Inc., Cold Lake Transmission, Ltd., Bonnyville, Alta. Served to lt. (j.g.) AC USN, 1943-46. Mem. Ind. Petroleum Assn. Am., Am. Petroleum Inst. Republican. Episcopalian. Clubs: Elephant, Petroleum, Denver, Tennis World (Denver); Garden of Gods (Colorado Springs & Vail); Calgary Golf and Country. Subspecialties: Fuels; Coal. Office: Worldwide Energy Corporation 1700 Broadway Suite 1600 Denver CO 80290

TENNENBAUM, JAMES IRVING, physician; b. Cin., Aug. 21, 1932; s. Louis and Virginia (Klein) T.; m. Carole Elaine Pittler, Feb. 1, 1959; children: Charles, William, Craig, Ginny. B.S., U. Cin., 1954, M.D., 1958. Diplomate: Am. Bd. Internal Medicine. Intern, asst. med. resident Jewish Hosp., Cin., 1958-60; resident Bellevue Hosp., N.Y.C., 1960-61; sr. med. resident VA Research Hosp., Chgo., 1961-62; fellow in allergy Northwestern U. Med. Sch., Chgo., 1962-63; asst. chief allergy sect. Wilford Hall Hosp., Lackland AFB, Tex., 1963-65; mem. faculty dept. medicine Ohio State U. Coll. Medicine, Columbus, 1965—, assoc. prof. medicine, 1970-74, clin. assoc. prof., 1974-75, clin. prof., 1975—, acting dir. div. allergy dept. medicine, 1970-71, dir. div. allergy, 1971—; mem. staff Univ. Hosp., Columbus, 1965—; cons. VA Hosp., Dayton, 1965-74, Wright-Patterson AFD Hosp., 1965-74, Nat. Jewish Hosp., Denver, Allergy Rehab. Found. Mem. editorial bd.: Contemporary Therapy, 1974—; assoc. editor: Jour. Immunology and Allergy Practice, 1980—; contbr. articles to profl. jours. Served with USAF, 1963-65. Fellow ACP; mem. Ohio Soc. Allergy and Immunology (pres. 1973-75), Ohio State Med. Assn. (chmn. sect. allergy 1973—), Am. Acad. Allergy, Am. Fedn. Clin. Research, AAAS, Am. Assn. Immunologists, AMA, Columbus Acad. Medicine, Ohio State Univ. Hosps. Med. Soc., Phi Beta Kappa, Alpha Omega Alpha. Subspecialties: Allergy; Immunology (medicine). Current work: Pharmacodynamics of aminophyllines. Home: 45 S Merkle Rd Columbus OH 43209 Office: 3341 S Livingston Columbus OH 43227

TENNEY, STEPHEN MARSH, physiologist, educator; b. Bloomington, Ill., Oct. 22, 1922; s. Harry Houser and Caroline (Marsh) T.; m. Carolyn Cartwright, Oct. 18, 1947; children: Joyce B., Karen M., Stephen M. A.B., Dartmouth; M.D., Cornell U. Instr. medicine U. Rochester Sch. Medicine, 1951-54, instr. physiology, 1953-54, asst. prof. physiology and medicine, 1954-56, asso. prof., 1956; prof. physiology Dartmouth Med. Sch., Hanover, N.H., 1956—, dean, 1960-62, acting dean, 1966, 73—, dir. med. scis., 1957-59, chmn. dept. physiology, 1956-77, Nathan Smith prof. physiology, 1974—; med. dir. Parker B. Francis Found., 1975—; Chmn. physiology study sect. NIH, 1962-65; tng. com. Nat. Heart Inst., 1968-71; mem. exec. com. NRC; mem. physiology panel NIH study Office Sci. and Tech.; mem. regulatory biology panel NSF, 1971-75; chmn. bd. sci. counselors Nat. Heart and Lung Inst., 1974-78; chmn. Commn. Respiratory Physiology Internat. Union Physiol. Scis. Asso. editor: Jour. Applied Physiology, 1976—, Handbook of Physiology; editorial bd.: Physiol. Revs; Contbr. articles to sci. jours. Served with USNR, 1947-49; sr. med. officer; Shanghai. Markle scholar in med. sci., 1954-59. Fellow Am. Acad. Arts and Scis., AAAS; mem. Inst. Medicine of Nat. Acad. Scis., Am. Physiol. Soc., Am. Soc. Clin. Investigation, N.Y. Acad. Scis., Gerontol. Soc., Am. Heart Assn., Assn. Am. Med. Colls., Sigma Xi. Subspecialty: Physiology (biology). Current work: Control of breathing; comparative respiratory physiology; high altitude. Home: 18 Rope Ferry Rd Hanover NH 03755

TEPHLY, THOMAS ROBERT, pharmacologist, educator; b. Norwich, Conn., Feb. 1, 1936; s. Samuel and Anna (Pieniadz) T.; m. Joan Bernice Clifcorn, Dec. 17, 1960; children: Susan Lynn, Linda Ann, Annette Michele. B.S. in Pharmacy, U. Conn., 1957; Ph.D. in Pharmacology, U. Wis., 1962; M.D., U. Minn., 1965. Research asst. U. Wis., Madison, 1957-62, instr. dept. pharmacology and toxicology, 1962; asst. prof. dept. pharmacology U. Mich., 1965-69, assoc. prof., 1969-71; prof., dir. toxicology Center, dept. pharmacology U. Iowa, 1971—. Contbr. articles to profl. jours. NIH grantee; Am. Cancer Soc. scholar, 1962-65; Fogarty Sr. Internat. fellow, 1978. Mem. Am. Soc. Pharmacology and Exptl. Therapeutics, Soc. Toxicology, AAAS, Am. Soc. Biol. Chemists, Research Soc. on Alcoholism, Sigma Xi. Subspecialties: Molecular pharmacology; Toxicology (medicine). Current work: Biochemical pharmacology and toxicology. Home: 6 Lakeview Dr Iowa City IA 52240 Office: Toxicology Center Dept Pharmacology U Iowa Iowa City IA 52242

TEPLEY, NORMAN, physicist, educator; b. Denver, Dec. 14, 1935; s. David Jack and Ida Elizabeth (Cohan) T.; m. Elaine Tepley, Nov. 29, 1939; children: Jamina Esther, Philip Scot, Alan Joseph. B.S. in Physics, M.I.T., 1957; postgrad., Columbia U., 1957-59; Ph.D. in Physics, M.I.T., 1963. Asst. prof. physics Wayne State U., Detroit, 1963-69; assoc. prof. physics Oakland U.-Rochester, Mich., 1969-77, prof., 1977—, acting chmn. dept. physics, 1981—; vis. prof. U. Lancaster (U.K.), 1970; faculty research participant Argonne (Ill.) Nat. Lab., 1971. Contbr. articles to profl. jours. Faculty Research fellow Wayne State U., 1965; AFOSR Research grantee, 1964-69; Research Corp. grantee, 1971; Mich. Heart Assn. research grantee, 1975-79. Mem. Am. Phys. Soc., AAAS, Sigma Xi. Current work: Biomagnetism including magnetocardiography, magnetoplethysmography, cell magnetism, low temperature physics including superconductivity, Fermi surfaces. Office: Dept of Physics Oakland University Rochester MI 48063

TEPOORTEN, BERNARD A., osteo. medicine educator; b. Pontiac, Mich., Apr. 28, 1927; s. Bernard Angus and Ada Lucile (Thurman) TeP.; m. Nancy Louise Fish, Jan. 18, 1959 (div. Jan. 1969); m. Elizabeth Ann Klock, Nov. 21, 1969; children: Michael, Leslie. B.S., Mich. State U., 1952; D.O., Kirksville Coll. Osteo. Medicine, 1956. Pvt. practice osteo. medicine and surgery, Kezar Falls, Maine, 1957-59, Tucson, 1959-75; prof., div. chmn. U. Osteo. Medicine and Health Scis., Des Moines, 1975—. Contbr. articles to profl. jours. Served with U.S. Army, 1945-47; ETO. Fellow Am. Acad. Osteopathy; mem. Ariz.Acad. Applied Osteopathy (pres. 1960-62), Am. Osteo Assn., Polk County Osteo. Assn. Republican. Roman Catholic. Subspecialties: Family practice; Osteopathy. Home: 3200 John Lynde Rd Des Moines IA 50312 Office: U Osteo Medicine and Health Scis 3200 Grand Ave Des Moines IA 50312

TEPPER, LLOYD BARTON, physician, corporate executive, environmental researcher, educator; b. Los Angeles, Dec. 21, 1931; m., 1957; 2 children. A.B., Dartmouth Coll., 1954; M.D., Harvard U., 1957, M.I.H., 1960, Sc.D. in Occupational Medicine, 1962. Diplomate: Am. Bd. Preventive Medicine. Fellow Mass. Gen. Hosp., Boston, 1958-60, MIT, Cambridge, 1959-61; physician Eastman Kodak Co., 1961-62, AEC, 1962-65; assoc. dir. occupational medicine and inst. environ. health Kettering Lab., U. Cin., 1965-72; assoc. prof. environ. health U. Cin., 1965-71, prof., 1971-72, assoc. prof. medicine, 1969-72; assoc. commr. sci. FDA, 1972-76; corp. med. dir. Air Products & Chem., Inc., Allentown, Pa., 1976—; adj. prof. environ. medicine U. Pa., 1977—. Editor: Jour. Occupational Medicine, 1979—. Mem. Am. Acad. Occupational Medicine (pres. 1980-81), Am. Occupational Medicine Assn. Subspecialties: Toxicology (medicine); Environmental medicine. Office: Air Products & Chem Inc PO Box 538 Allentown PA 18105

TEREBA, ALLAN MICHAEL, cancer researcher, virology educator; b. Wichita, Kans., Feb. 6, 1947; s. Louis Carl and Annette Jane (Miller) T.; m. Barbara Mae Miles, Dec. 7, 1974; children: Christina, Daniel, Jonathan, Elizabeth. B.S. cum laude in Chemistry, Ind. U., 1969; Ph.D. in Biochemistry, U. Wash., 1973. Asst. mem. St. Jude Children's Research Hosp., Memphis, 1975-80, assoc. mem., 1980—; assoc. prof. virology U. Tenn., 1982—. Contbr. articles to profl. jours. NSF fellow, 1967-69; NIH trainee, 1969-73; Damon Runyon fellow, 1973-75; Am. Cancer Soc. grantee, 1975-77; NSF grantee, 1978-80; NIH grantee, 1980—. Mem. Am. Soc. Microbiology, Phi Lambda Upsilon, Alpha Chi Sigma. Subspecialties: Cancer research (medicine); Gene actions. Current work: Analysis of oncogene expression in childhood leukemia tissue and chromosomal localization and characterization of oncogenes by insituhybridization and DNA cloning techniques. Office: 332 N Lauderdale PO Box 318 Memphis TN 38101

TERMAN, DAVID STEPHEN, physician, researcher, educator; b. N.Y.C., Oct. 23, 1940; s. Joseph and Pearl (Scharfman) T.; m. Naomi Sue Auerbach, Sept. 17, 1967; children: Erica, Jennifer. B.A., Syracuse U., 1962; M.D., Georgetown U., 1966. Intern in internal medicine U. Ala. Med. Center, Birmingham, 1966-67, resident in internal medicine, 1967-68; fellow in nephrology U. Colo. Med. Center, Denver, 1968-69, fellow in immunology, 1969-71, asst. prof. medicine, 1973-76; assoc. prof. Baylor U. Coll. Medicine, Houston, 1978—, dir. cancer biology program, 1982—; cons. biologic response modifiers program NIH, Washington, 1982—; dir. cancer immunology program Methodist Hosp., Houston, 1980—. Assoc. editor: Internat. Jour. Artificial Organs, 1977; assoc. editor: Plasma Therapy, 1979—; contbr. articles to profl. jours. Served to maj. USAF, 1971-73. Recipient Research Career Devel. award NIH, 1978-83; clin. investigator VA, 1973-76. Mem. Am. Soc. Clin. Investigation, So. Soc. Clin. Investigation, AAAS, Am. Fedn. Clin. Research, AMA. Jewish. Subspecialties: Cancer research (medicine); Immunology (medicine). Office: Baylor Coll Medicine 1200 Moursund Ave Houston TX 77030

TERNER, CHARLES, biology educator; b. Lublin, Poland, Apr. 30, 1916; came to U.S., 1955, naturalized, 1966; s. Isidore and Fanny (Schachter) T.; m. Ruth Hilde Cohn, Aug. 26, 1945; children: James, Michael, Anne S. B.S., U. London, 1944; D.Sc., 1969; Ph.D., U. Sheffield, Eng., 1949. Mem. Med. Research Council Unit for Research in Cell Metabolism, Sheffield, 1947-51; sr. sci. officer Nat. Inst. for Research in Dairying, Reading, Eng., 1951-55; mem. staff Worcester Found. for Exptl. Biology, Shrewsburg, Mass., 1955-59; prof. Boston U., 1959—. Contbr. articles on biology to profl. jours. Mem. Biochem. Soc. (Gt. Brit.), Am. Soc. Biol. Chemists, Soc. for Study of Reproduction, Soc. for Exptl. Biology and Medicine, AAAS. Jewish. Subspecialties: Biochemistry (biology); Reproductive biology. Current work: Biochemistry of the male reproductive system, spermatogenesis and metabolism of spermatozoa, regulation of growth and function of prostate, male contraception. Home: 19 Avalon Rd Newton MA 02168 Office: Boston U 2 Cummington St Boston MA 02215

TERRELL, JAMES, (JR.) (NELSON TERRELL), physicist, astrophysicist; b. Houston, Aug. 15, 1923; s. Nelson James and Gladys Delphine (Stevens) T.; m. Elizabeth Anne Pearson, June 9, 1945; children: Anne, Barbara, Jean. B.A. (Graham Baker scholar), Rice U., 1944, M.A. (Rice U. fellow 1946-48), 1947, Ph.D. in Physics (AEC fellow), 1950. Research asst. in physics Rice U., Houston, 1950; asst. prof. physics Western Res. U., Cleve., 1950-51; staff mem. Los Alamos Nat. Lab., 1951—; vis. prof. N. Mex. Highlands U., Las Vegas, summer 1959; vis. staff mem. Lawrence Radiation Lab., U. Calif., Berkeley, summer 1963. Contbr. articles to sci. jours.; co-producer computer-generated movie on x-ray sky, 1982. Served from pvt. to 1st lt., Signal Corps AUS, 1944-46. Fellow AAAS, Am. Phys. Soc.; mem. Internat. Astron. Union, Am. Astron. Soc., Los Alamos Choral Soc., Phi Beta Kappa, Sigma Xi. Club: Los Alamos Ski. Subspecialties: High energy astrophysics; Theoretical astrophysics. Current work: X-ray astronomy, Gamma-burst astronomy, quasar theory, Fourier analysis, special and general relativity. Home: 85 Obsidian Loop Los Alamos NM 87544 Office: Los Alamos Nat Lab Mail Stop D436 Los Alamos NM 87545

TERRELL, ROSS CLARK, chemist; b. Oneonta, N.Y., Sept. 22, 1925; s. Ralph Leslie and Esther (Clark) T. B.S., Hartwick Coll., 1950; Ph.D., Columbia U., 1955; postgrad. (fellow), U. Birmingham, Eng., 1966-67. Research chemist Shulton Inc., 1955-59, Airco Inc., 1959-67; mgr. Ohio Med. Anesthetics, Murray Hill, N.J., 1967-82, dir. research, 1982. Contbr. articles to profl. jours. Mem. Gov.'s Sci. Adv. Com. of N.J., 1981—. Served with USMC, 1941-45. Subspecialties: Organic chemistry; Synthetic chemistry. Current work: Medicinal chemistry synthesis of pharmaceuticals. Patentee in field. Home: 615 Goodmans Crossing Clark NJ 07066 Office: 100 Mountain Ave Murray Hill NJ 07974

TERSHAK, DANIEL RICHARD, microbiology educator, cons., researcher; b. Wilkes-Barre, Pa., Nov. 19, 1936; s. Andrew and Catherine (Gay) T.; m. Mary Jane Prischak, Aug. 5, 1967; children: Daniel, Suzanne. B.S., Kings Coll., 1958; Ph.D., Yale U., 1962. Mem. faculty Pa. State U., University Park, 1964—, assoc. prof. molecular biology and microbiology, 1969, chmn. dept. microbiology and cell biology, 1978-79, program dir., 1979-80; vis. scientist in biochemistry Virus Research Inst., Pirbright, Eng., 1982. Contbr. articles on virology to profl. jours. NIH grantee, 1965—. Mem. N.Y. Acad. Scis., Am. Soc. Microbiology, Am. Soc. Virology. Republican. Subspecialties: Virology (medicine); Molecular biology. Current work: Poliovirus: therapy-chemo, synthesis viral macromolecules, vaccines.

TERZIAN, YERVANT, educator; b. Alexandria, Egypt, Feb. 9, 1939; came to U.S., 1960, naturalized, 1971; s. Bedros and Maria (Kiriakaki) T.; m. Araxy Hovsepian, Apr. 16, 1966; children: Sevan, Tamar. B.Sc., Am. U. Cairo, 1960; M.Sc., Ind. U., 1963; Ph.D., 1965. Staff Nat. Radio Astronomy Obs., 1963-65, Cornell U. Arecibo Obs., Ithaca, N.Y., 1965-67, prof. astronomy, chmn. dept., 1967—; vis. prof. astronomy U. Montreal, 1973, U. Salonica, 1974. Contbr. articles to profl. jours.; editor: Interstellar Ionized Hydrogen, 1968, Planetary Nebulae Observations and Theory, 1978, Cosmology and Astrophysics, 1982. Mem. Internat. Astron. Union, Internat. Union Radio Sci., Am. Astron. Soc. Subspecialties: Radio and microwave astronomy; High energy astrophysics. Current work: Radio astronomy, specialist in physics of interstellar medium. Office: Space Scis Bldg Cornell Univ Ithaca NY 14853 Home: 109 Brandywine Dr Ithaca NY 14850

TESKE, RICHARD GLENN, astronomy educator; b. Cleve., Aug. 16, 1930; s. William Frederick and Elsie Wilhelmina (Zornow) T.; m. Yvonne Lorene Russell, Oct. 25, 1975. B.S., Bowling Green U., 1952; M.S., Ohio State U., 1956; Ph.D., Harvard U., 1961. Instr. dept. astronomy U. Mich., Ann Arbor, 1960-62, asst. prof., 1962-67, assoc. prof., 1967-74, prof., 1974—. Served with U.S. Army, 1953-54. Mem. Am. Astron. Soc., Internat. Astron. Union. Subspecialties: Solar physics; Optical astronomy. Current work: Solar corona, solar photosphere, stellar photospheres, coronal emission lines.

TESSIER, BRUCE JOSEPH, plant pathologist; b. Holyoke, Mass., Mar. 20, 1954; s. Henry Raymond and Lillian (Plouffe) T.; m. Claire Jean Lancourt, June 10, 1978; children: William Joseph, Matthew Bruce. B.A., St. Anselm Coll., 1976; M.S., U. R.I., 1980, Ph.D. candidate, since 1980—. Mem. Am. Phytopathol. Soc., Pisum Genetic Assn. Subspecialties: Plant pathology; Plant physiology (agriculture). Current work: Ultrastructural characterization of disease resistance in plants. Home: Kingstown Rd Grad Village 514 Kingston RI 02881 Office: Dept Plant Pathology Univ RI Kingston RI 02881

TESTARDI, LOUIS RICHARD, metallurgist, research scientist; b. Phila., Sept. 23, 1930; s. Louis and Alice (Petracca) T.; m. Fulvia Pieraccini, Aug. 3, 1957; children: Stephen L., Mark R., David A., Richard P. A.B., U. Calif.-Berkeley, 1955; M.S., U. Pa., 1961, Ph.D., 1963. Mem. tech. staff Bell Labs., Murray Hill, N.J., 1963-80; dir. materials processing in space NASA, Washington, 1980-82; now chief metallurgy div. Nat. Bur. Standards, Gaithersburg, Md. Contbr. articles to profl. jours. Served with U.S. Army, 1948-50, 50-51. Fellow Am. Phys. Soc. Subspecialties: Condensed matter physics; Materials. Current work: Structural instabilisites; high temperature superconductivity; ultrasonics; elastic, electrical magnetic and optical properties of solids. Home: 20312 Aspenwood Ln Gaithersbug MD 20879 Office: Nat Bur Standards Washington DC 20234

TEUKOLSKY, SAUL ARNO, physicist, educator; b. Johannesburg, South Africa, Aug. 2, 1947; came to U.S., 1970, naturalized, 1981; s. Isaac and Ethel (Cramer) T.; m. Roselyn Siew, June 27, 1971; children: Rachel, Lauren. B.Sc. with honors, U. Witwatersrand, South Africa, 1970; Ph. D., Calif. Inst. Tech., 1973. Richard Chace Tolman research fellow Calif. Inst. Tech., Pasadena, 1973-74; asst. prof. physics Cornell U., Ithaca, N.Y., 1974-77, assoc. prof., 1977-83, prof., 1983—. Co-author 2 books; author articles. U.S. Steel Found. fellow, 1972-73; Alfred P. Sloan Found. fellow, 1975-77; John Simon Guggenheim Meml. Found. fellow, 1981-82. Mem. Am. Phys. Soc., Am. Astron. Soc. Subspecialties: General relativity; Theoretical astrophysics. Current work: Applications of general relativity to theoretical astrophysics. Office: Newman Lab Cornell U Ithaca NY 14853

TEUSCHER, MICHAEL COOK, mechanical engineer, consultant, designer; b. Montpelier, Idaho, Apr. 29, 1941; s. Sid J. and Marie (Cook) T.; m. Karen Hodges, July 6, 1962; children: Phillip, Rebecca, David, Stephen, Richard, Carrie. B.S.M.E., Utah State U., 1964, M.S.M.E., 1974. Assoc. engr. IBM, Vestal, N.Y., 1964-65, sr. assoc. engr., Essex Junction, Vt., 1966-71; research asst. Utah State U. Water Research Lab., 1973-74; environ. systems engr. TenEck, Louisville, 1975-76; lead engr. Thiokol, Brigham City, Utah, 1976—; pres. Rocky Mountain Design, Inc., Logan, Utah. Recipient Silver Snoopy award Space Shuttle NASA astronauts, 1979, cert. NASA, 1982. Mem. ASME. Mormon. Subspecialties: Mechanical engineering; Water supply and wastewater treatment. Current work: Lateral buckling of lifting beams and robotics, tools and equipment for manufacturing solid fuel rocket motor cases and nozzles. Home: 750 N 1st St W Logan UT 84321 Office: PO Box 524 MS 552D Brigham City UT 84302

TEW, KENNETH DAVID, molecular pharmacologist, researcher; b. Dumbarton, Scotland, Apr. 20, 1952; came to U.S., 1977; s. Kenneth William and Britalena (Jamieson) T. B.Sc., U. Wales, Swansea, 1973; Ph.D., U. London, 1977. Research assoc. depts. medicine and biochemistry Georgetown U., 1977-79; instr. medicine, 1979-80, asst. prof., medicine and biochemistry, 1980—. Contbr. to books, also articles to profl. jours. Active Am. Cancer Soc. Nat. Cancer Inst. grantee, 1980—; head basic pharmacology program Vincent T. Lombardi Cancer Research Ctr. Mem. Am. Assn. Cancer Research, Am. Soc. Cell Biology, AAAS., Am. Soc. Pharmacology and Exptl. Therapeutics. Club: Old Red Rugby Football (Washington). Subspecialties: Molecular pharmacology; Cancer research (medicine). Current work: Nuclear structure and drug mechanisms of action; application of molecular biology to pharmacology; anticancer drugs, drug resistance, carcinogenisis, European cancer research. Office: Lombardi Cancer Center Georgetown University Hospital 3800 Reservoir Rd NW Washington DC 20007

TEYLER, TIMOTHY JAMES, neurobiology educator, consultant; b. Portland, Oreg., Nov. 25, 1942; s. Otto F. and Ione M. (Erickson) T.; m. Lisbeth Nilsen, June 17, 1964; 1 son, Erik. M.S., U. Oreg. Med. Sch., 1967, Ph.D., 1968. Asst. prof. psychology U. So. Calif., 1968-69; assoc. research psychobiologist U. Calif., Irvine, 1969-72; NATO/Fulbright scholar U. Oslo Inst. Neurophysiology, 1972-73; assoc. prof. psychology and social relations Harvard U., 1973-77; prof. neurobiology N.E. Ohio U. Coll. Medicine, 1978—, chmn. neurobiology area com., 1978—. Contbr. articles to profl. publs. NIMH predoctoral fellow, 1965-69; postdoctoral fellow, 1970-72. Mem. Soc. for Neursci. Subspecialties: Neurobiology; Neurophysiology. Current work: Brain mechanisms of learning and memory. Office: NE Ohio Coll Medicine Rootstown OH 44272

THACORE, HARSHAD RAI, microbiology educator; b. India, Dec. 1, 1939; s. Cyril M. and Josephine M. T.; m. Premleela Thacore, Sept. 18, 1965; 1 child, Harshad R. B.Sc., Lucknow (India) Christian Coll., 1958; M.Sc., Lucknow U., 1960; Ph.D., Duke U., 1965. Postdoctoral fellow Ohio State U., 1965-67; research assoc. U. Pitts., 1967-69, instr., 1969-72, asst. research prof. dept. microbiology, 1971-74; asst. prof. SUNY-Buffalo, 1974-79, assoc. prof. microbiology, 1979—. Contbr. articles to profl. jours. Mem. Am. Soc. Microbiology, Am. Soc. Virology, Sigma Xi. Methodist. Subspecialties: Microbiology; Virology (biology). Current work: Virus-Cell interactions, viral interference mechanisms and the study of the interferon system in human cell cultures. Office: Dept Microbiology Sch Medicine SUNY-Buffalo Buffalo NY 14214

THAELER, CHARLES SCHROPP, JR., biology educator, researcher; b. Phila., Jan. 9, 1932; s. Charles Schropp and Luch Barnard (Kummel) T.; m. Eunice Adelle Gothard, Dec. 19, 1980; m. Marianne Merriam Hobbs, Nov. 30, 1957 (div. Dec. 1980); children: Charles Schropp, Kent H., Bret K. A.B., Earham Coll., 1954; M.A., U. Calif.-Berkeley, 1960, Ph.D., 1964. Acting asst. prof. and acting asst. curator mammals U. Calif.-Berkeley, 1963-64; asst. prof. Ind. U., South Bend, 1964-66; from asst. prof. to prof. biology N.Mex. State U., Las Cruces, 1966—; cons. Ariz. State U., 1981—. Served in U.S. Army, 1954-56. NSF grantee, 1968-70, 72-74. Fellow AAAS; mem. Am. Soc. Mammalogists (com. chmn. 1979—), Soc. Study of Evolution, Soc. Systematic Zoologists. Democrat. Subspecialties: Systematics; Evolutionary biology. Current work: Speciation, cytotaxonomy and systematics in the Geomyidae especially the genus Thomomys. Home: PO Box 4308 University Park Las Cruces NM 88003 Office: NMex State U Las Cruces NM 88003

THAKKAR, BHARATKUMAR S., mech. engr., educator; b. Magodi, Gujarat State, India, Oct. 26, 1941; came to U.S., 1964, naturalized, 1967; s. Shantilal S. and Lalita S. T.; m. Indira R. Thakkar, Jan. 22, 1968; children: Akhil, Virat. Ph.D. in Mech. Engring, Ill. Inst. Tech., 1976. Devel. engr. Foster Wheeler Corp., Livingston, N.J., 1965-69; sr. engr. Continental Can Co., Chgo., 1969-72; prin. engr. Nat. Can Co., Chgo., 1972-77; mem. tech. staff Bell Labs., Naperville, Ill., 1977—; chmn. mech. engring. dept Midwest Coll. Engring., Lombard, Ill.; instr. Ill. Inst. Tech., Chgo., 1979—. Mem. ASME, Am. Soc. for Metals, Soc. Plastics Engrs. Subspecialties: Materials processing; Solid mechanics. Current work: Residual stresses in polymers; processing of materials. Patentee on explosive forming. Home: 1418 Wesley Ct Westmont IL 60559 Office: 1100 E Warrenville Rd Naperville IL 60566

THAMMAVARAM, N. RAO, computer science and electrical engineering educator; b. Andhra Pradesh, India, June 5, 1933; s. Ramakrishnaish and Laxmamma (Maddipati) T.; m. R. Mala Thammavaram, Aug. 27, 1981; children from previous marriage: Krishna, Chandrika, Radhika. B. Sc., Andhra U., 1952; D.I.I.Sc., Indian Inst. Sci., Bangalore, 1955; M.S., U. Mich., 1961, Ph.D., 1964. Registered profl. engr., Tex. Assoc. prof. elec. engring. U. Md., College Park, 1967-75; prof. computer sci. and elec. engring. So. Meth. U., Dallas, 1975-80, U. Southwestern La., Lafayette, 1980—. Author: Error Coding for Arithmetic Processors, 1974; contbr. articles to profl. jours. Mem. IEEE, Assn. Computing Machinery. Subspecialties: Computer architecture; Computer engineering. Current work: Algebraic coding, fault-tolerant computing, computer arithmetic, cryptography and data security. Office: U Southwestern La Box 44330 Lafayette LA 70504

THAWLEY, DAVID GORDON, veterinary epidemiologist, educator; b. Hastings, N.Z., Oct. 4, 1946; s. Harold Gordon and Edna (Blanche) T.; m. Helen Margaret McGregor, Apr. 2, 1971; children: Shannon Margaret, David McGregor. B.V.Sc., Massey U., N.Z., 1969; Ph.D., U. Guelph, Ont., Can., 1974; Cert., Am. Coll. Vet. Preventive Medicine, 1978, Am. Bd. Vet. Pub. Health, 1978. Practice vet. medicine, N.Z., 1970-71; lectr. Sch. Vet. Medicine, Palmerston North, N.Z., 1975; asst. prof. dept. vet. microbiology Coll. Vet. Medicine, U. Mo., Columbia, 1976-79, assoc. prof., 1980—. Contbr. articles to profl. jours. Fed. govt. and producer grantee. Mem. AVMA, U.S. Animal Health Assn., Assn. Tchrs. Vet. Preventive Medicine. Presbyterian. Subspecialties: Preventive medicine (veterinary medicine); Virology (veterinary medicine). Current work: Research on the epidemiology and control of pseudorabies in swine; epidemiology of microbial agents. Home: RR 4 Box 76A Columbia MO 65211 Office: Coll Vet Medicine U Mo Columbia MO 65211

THEDFORD, THOMAS RAY, veterinarian, veterinary medicine and surgery educator; b. Tyler, Tex., Jan. 28, 1936; s. Ray and Sara Lee (Selman) T.; m. Nancy Jane Martin, June 26, 1958; children: Rebecca Ann, Miriam Clare. B.S., D.V.M., Tex. A&M U., 1959. Pvt. practice vet. medicine, 1959-60, Floydada, Tex., 1960-65; instr. vet. medicine and surgery Okla. State U., 1965-66, asst. prof., 1967-70, assoc. prof., 1970-77, prof., ext. veterinarian, 1981—; vis. assoc. prof. U. Nairobi, Kenya, 1974-76; cons. Winrock Internat., Morrilton, Ark., 1982—, Wildlife Health Cons., Stillwater, Okla., 1980—. Sarkey Found. grantee, 1982—. Mem. Okla. Vet. Med. Assn., AVMA, Am. Assn. Swine Practitioners, Am. Assn. Sheep and Goat Practitioners, Am. Assn. Ext. Veterinarians, Am. Assn. Vet. Clinicians. Presbyterian. Lodge: Masons. Subspecialties: Preventive medicine (veterinary medicine); wildlife diseases-foreign animal diseases. Current work: Educational programming/ multi media preventive medicine; research wildlife disease. Office: Teaching Hosp Okla State U Stillwater OK 74078 Home: Route 3 Tantara #8 Stillwater OK 74074

THEISEN, JEFFREY A., solar energy designer; b. Appleton, Wis., May 25, 1951; s. Donald W. and Lorrain (Wilhelm) W.; m. Marica Clair, Aug. 23, 1973; 1 dau., Roslyn. B.S. in Environ. Sci, U. Wis.-Green Bay. Analytical chemist Inst. Paper Chemistry, Appleton, Wis.; research scientist consumer products Kimberly Clark Corp., Neenah, Wis.; owner Solsorce Research, Appleton. Mem. Am. Solar Energy Soc., Solar Energy Resource Assn. Subspecialty: Solar energy. Current work: Effective passive product design for northern climates currently developing movable insulation, special glazing and solar storage systems. Office: PO Box 2732 Appleton WI 54913

THEOHARIDES, THEOHARIS CONSTANTIN, pharmacologist; b. Thessaloniki, Greece, Feb. 11, 1950; s. Constantin A. and Marika (Krava) T.; m. Efthalia I. Triarchou, July 10, 1981. Diploma with honors, Anatolia Coll., 1968, Bach., Yale U., 1972, M.S., 1975, M.Phil., 1975, Ph.D. in Pharmacology, 1978, M.D., 1983. Asst. in research biology Yale U., 1968-71, asst. in research pharmacology, 1973-78, research assoc. faculty clin. immunology, 1978-83; asst. prof. biochemistry and pharmacology Tufts U., 1983—; spl. instr. modern Greek Yale U., 1974, 77; vis. faculty Aristotelian U. Sch. Medicine, Thessaloniki, 1979. Author book on pharmacology; contbr. numerous articles to profl. jours. Bd. dirs., v.p. for relations with Greece, Krikos, 1978-79. Recipient Theodore Cuyler award Yale U., 1972; George Papanicoalou Grad. award, 1977; Med. award Hellenic Med. Soc. N.Y., 1979; M.C. Winternitz prize in pathology Yale U.; others. Mem. Hellenic Biochem. and Biophys. Soc., AMA, AAUP, N.Y. Acad. Scis., Am. Inst. History Pharmacy, AAAS, Soc. Health and Human Values, Am. Assn. History Medicine, Am. Soc. Cell Biology, Soc. Neurosci., Am. Fedn. Clin. Research, Conn. Acad. Arts and Scis., Am. Soc. Pharmacology and Exptl. Therapeutics, Hellenic Soc. Cancer Research, Hellenic Soc. Med. Chemistry, Internat. Soc. Immunopharmacology, Am. Soc. Microbiology, Am. Assn. Immunologists, Internat. Soc. History of Medicine, Sigma Xi. Subspecialties: Endocrinology; Cellular pharmacology. Current work: Mechanisms of release of secretory products; hormonal induction of ornithine decarboxylase and membrane functions of polyamines; pathophysiology of mast cells. Office: Dept Biochemistry and Pharmacology Tufts U 136 Harrison Ave Boston MA 02111 Home: 61 N Washington St Apt 5A Boston MA 02114

THEOLOGIDES, ATHANASIOS, physician; b. Ptolwmais, Greece, Feb. 5, 1931; came to U.S., 1956, naturalized, 1971; s. Stylianos and Stergia (Hatjinota) T.; m. Maria Mystakidou, June 19, 1965; children: Stergios, Evangelia. M.D., Aristotle U., 1955; Ph.D., U. Minn. Mem. faculty U. Minn. Med. Sch., 1967—, prof. medicine, 1974—. Author numerous papers in field. Served with Greek Army, 1963-65. Mem. Greek Orthodox Ch. Subspecialties: Internal medicine; Oncology. Current work: Biochemistry of cancer host. Home: 138 Windsor Ct New Beighton MN 55112 Office: U Minn Med Center Minneapolis MN 55455

THEON, JOHN SPERIDON, meteorologist; b. Washington, Dec. 12, 1934; s. Lewis and Merope (Xydias) T.; m. Joanne Edens, July 31, 1965; children—Christopher James, Catherine. B.S. in Aero. Engring, U. Md., 1957, Pa. State U., 1959, M.S., 1962. Aero. engr. Douglas Aircraft Co., Santa Monica, Calif., 1957-58; engr. U.S. Naval Ordnance Lab., White Oak, Md., 1962; research meteorologist NASA Goddard Space Flight Center, Greenbelt, Md., 1962-74, head meteorology br., 1974-77; asst. chief Lab. for Atmospheric Scis., 1977-78, Nimbus project scientist, 1972-78, Landsat discipline leader meteorol. investigations, 1974-78; mgr. global weather research program NASA Hdqrs., Washington, 1978-82, chief Atmospheric Dynamics and Radiation br., 1982—, Spacelab 3 program scientist, 1979—. Contbr. articles to profl. jours. Served with USAF, 1958-60. Recipient Nimbus F Instrument Team award NASA-Goddard, 1976, Exceptional Performance award, 1978. Mem. Am. Meteorol. Soc., Am.

Geophys. Union, AAAS. Presbyterian. Subspecialties: Remote sensing (atmospheric science); Meteorology. Current work: Responsible for synthesizing, planning, executing and evaluating a broad research and development program in remote sensing of the atmosphere and applying resulting satellite data to problems in meteorology and climatology. Home: 6801 Lupine Ln McLean VA 22101 Office: 600 Independence Ave Washington DC 20546

THEURER, J. CLAIR, research geneticist; b. Logan, Utah, Sept. 4, 1928; s. John Jessop and Rhoda-Durfey T.; m. Carma Lake, Apr. 24, 1953; children: Michael, Scott, Bruce, David. B.S., Utah State U., 1953; M.S., 1956; Ph.D., U. Minn., 1962. Research asst. dept. plant genetics U. Minn., 1957-61, postdoctoral fellow, 1961-62; research agronomist Agrl. Research Service Dept. Agr., Logan, Utah, 1962-63, research plant geneticist, 1963-83, West Range Plant Service Greenhouse, East Lansing, Mich., 1983—. Contbr. articles to profl. jours. Served to 1st lt. USAF, 1953-55. Recipient Superior Service award Dept. Agr., 1978. Mem. Am. Soc. Agronomy, Crop Sci. Soc., Am. Soc. Sugarbeet Technologists, Utah Farm Bu., Sigma Xi, Gamma Sigma Delta, Alpha Zeta. Mormon. Subspecialties: Plant genetics; Plant growth. Current work: Sugarbeet breeding and genetics, plant growth, photosynthate partitioning, biomass for ethanol production. Office: Utah State U UMC 63 Logan UT 84322

THIEL, LEO ALBERT, mech. engr., cons.; b. Covington, Ky., Sept. 30, 1952; s. Peter Edward and Hilda Bertha (Mueller) T.; m. Susan Leah Spaulding, Apr. 25, 1981; 1 dau., Carrie Sue. B.S.M.E., U. Cin., 1978. Project engr. Structural Dynamics Research Corp., Milford, Ohio, 1978—. Mem. ASME, Soc. Automotive Engrs. Republican. Roman Catholic. Subspecialties: Computer-aided design; Robotics. Current work: Applications of solids modeling, weight optimization and finite element modeling, artificial intelligence, software engineering, distribution systems; packaging; visibility; evaluation of software; application training; control design; computer programming; optimization. Home: 1287 Deblin Dr Milford OH 45150 Office: 2000 Eastman Dr Milford OH 45150

THIEMAN, JAMES RICHARD, astrophysicist; b. Dayton, Ohio, Aug. 8, 1947; s. Richard Frank and Agnes Marie (Heckman) T.; m. Barbara Eve Brandes, Sept. 29, 1973; children: Mark Edward, Cheryl Marie. B.S., U. Dayton, 1969; Ph.D., U. Fla., 1977. Teaching, research asst. U. Fla., 1969-77; research assoc. Nat. Acad. Sci./NRC, Goddard Space Flight Ctr., 1977-79; scientist, analyst ORI Inc., Silver Spring, Md., 1979—; tchr. George Mason U., 1980. Contbr. articles to profl. jours. Mem. Am. Geophys. Union, Am. Astron. Soc., Phi Beta Kappa, Sigma Pi Sigma. Subspecialties: Satellite studies; Radio and microwave astronomy. Current work: Plasma instrument data on Dynamics Explorer satellites. Home: 1203 Pennington Ln Bowie MD 20716 Office: 1400 Spring St Silver Spring MD 20910

THIEMENS, MARK H., chemist, educator; b. St. Louis, Jan. 6, 1950; s. Artur A. and Justine C. (Alt) T.; m. Candra C. Busse, Dec. 25, 1952. B.S. in Chemistry, U. Miami, Fla., 1972; M.S. in Chem. Oceanography, Old Dominion U., 1974; postgrad., Fla. Inst. Tech., 1974-75; Ph.D. in Geol. Oceanography, Fla. State U., 1977. Grad. research asst. Old Dominion U., Norfolk, Va., 1972-74, Fla. Inst. Tech., Melbourne, 1974-75; NIH fellow Fla. State U., Tallahassee, 1975-76; staff scientist Brookhaven Nat. Lab., Upton, N.Y., 1976-77; research asso. Enrico Fermi Inst., U. Chgo., 1977-80; vis. asst. prof., 1981—; researcher in field. Contbr. papers to profl. jours. Mem. AAAS, Am. Phys. Soc., Am. Geochem. Soc., Am. Geophys. Union, Am. Chem. Soc., Am. Astron. Soc., Am. Radio Relay League. Subspecialties: Cosmology; Space chemistry. Current work: Cosmochemistry research; evolution of sun and solar system.

THIER, SAMUEL OSIAH, physician, educator; b. Bklyn., June 23, 1937; s. Sidney and May Henrietta (Kanner) T.; m. Paula Dell Finkelstein, June 28, 1958; children: Audrey Lauren, Stephanie Ellen, Sara Leslie. Student, Cornell U., 1953-56; M.D., State U. N.Y., Syracuse, 1960. Diplomate: Am. Bd. Internal Medicine (dir. 1977—, exec. com. 1981—, chmn.-elect 1983-84). Intern Mass. Gen. Hosp., Boston, 1960-61, asst. resident, 1961-62, sr. resident, 1964-65, clin. and research fellow, 1965, chief resident, 1966; clin. asso. Nat. Inst. Arthritis and Metabolic Diseases, 1962-64; from instr. to asst. prof. medicine Harvard U. Med. Sch., 1967-69; asst. in medicine, chief renal unit Mass. Gen. Hosp., Boston, 1967-69; asso. prof., then prof. medicine U. Pa. Med. Sch., 1969-72, vice chmn. dept., 1971-74; asso. dir. med. services Hosp. U. Pa., 1969-74; David Paige Smith prof. internal medicine, 1978-81, Sterling prof. medicine, 1981—; chmn. dept. Yale U. Sch. Medicine, 1975—; chief medicine Yale-New Haven Hosp., 1975—, bd. dirs., 1978—, Hospice, Inc., 1976-82. Mem. editorial bd.: New Eng. Jour. Medicine, 1978-81; Contbr. articles to med. jours. Mem. adv. com. to the dir. NIH, 1980—. Served with USPHS, 1962-64. Recipient Christian R. and Mary F. Lindback Found. Distinguished Teaching award, 1971. Fellow ACP (bd. regents 1982—); mem. Assn. Am. Med. Colls. (adminstrv. bd. council acad. socs.), John Morgan Soc., Am. Fedn. Clin. Research (pres. 1976-77), Am. Soc. Nephrology, Am. Physiol. Soc., Inst. Medicine, Nat. Acad. Scis., Internat. Soc. Nephrology, Assn. Profs. Medicine, Assn. Am. Physicians, Interurban Clin. Club, Alpha Omega Alpha. Subspecialties: Internal medicine; Nephrology. Home: 8 Spector Rd Woodbridge CT 06525 Office: PO Box 3333 New Haven CT 06510

THIERSTEIN, HANS RUDOLF, geologist, educator; b. Zurich, Switzerland, May 27, 1944; came to U.S., 1973; s. Walter E.O.R. and Martha (Gege) T.; m. Verena M. Handschin, Sept. 26, 1969; children: Franziska Susanna, Stephanie Barbara. Ph.d., U. Zurich, 1972. Research asst. Swiss Inst. Tech., Zurich, 1972-73; postdoctoral fellow Swiss Nat. Sci. Found., 1973-76; asst. prof. geology Scripps Instn. Oceanography, U. Calif.-San Diego, 1976-80, assoc. prof., 1980—. NSF grantee, 1976—. Mem. Geol. Soc. Am., Am. Geophys. Union, Swiss Geol. Soc. Subspecialties: Geology; Oceanography. Current work: History of ocean circulation and climate; evolution of marine plankton. Office: Scripps Instn Oceanography U Calif-San Diego La Jolla CA 92093

THIGPEN, JAMES TATE, physician, oncology educator; b. Columbia, Miss., June 6, 1944; m. Louisa Berdie Kessler, June 14, 1969; children: Monroe Tate, James Howard, Samuel Calvin, Richard Allen. B.S., U. Miss., 1964, M.D., 1969. Intern Strong Meml. Hosp., U. Rochester, N.Y., 1969-70; resident U. Miss. Sch. Medicine, 1970-71, fellow dept. hematology, 1971-73, assoc. prof., dir. div. med. oncology dept. internal medicine, 1973—. Fellow ACP; mem. AMA, Miss. Med. Assn., Central Med. Soc., Jackson Acad. Medicine, Miss. Acad. Scis., S.W. Oncology Group, Gynecologic Oncology Group, Am. Fedn. Clin. Research, Am. Assn. Cancer Edn., Am. Soc. Clin. Oncology, Am. Assn. Cancer Research, Am. Soc. Hematology. Baptist (deacon 1978—, Sunday sch. tchr. 1979—). Subspecialty: Gynecological oncology. Home: 1135 Briarwood Dr Jackson MS 39211 Office: 2500 N State St Jackson MS 39216

THIMANN, KENNETH VIVIAN, educator, biologist; b. Ashford, Eng., Aug. 5, 1904; came to U.S., 1930; s. Israel Phoebus and Muriel Kate (Harding) T.; m. Ann Mary Bateman, Mar. 20, 1929; children—Vivianne (Mrs. J. Nachmias), Karen Thimann Romer, Linda Thimann Dewing. Student, Caterham Sch., Eng., 1915-21; B.Sc., Imperial Coll. Sci. and Tech., London, Royal Coll. Sci., 1924, A.R.C.S., 1924, Ph.D., 1928; A.M. (hon.), Harvard, 1938, Ph.D., U. Basel, Switzerland, 1960, U. Clermont-Ferrand, France, 1961. Demonstrator bacteriology King's Coll., London, 1927-29; instr. biochemistry and bacteriology Calif. Inst. Tech., Pasadena, 1930-35; lectr. botany Harvard, 1935-36, asst. prof. plant physiology, 1936-39, asso., 1939-46, prof., 1946-62, Higgins prof. biology, 1962-65, emeritus, 1965—; prof. biology U. Calif. at Santa Cruz, 1965—; provost Crown Coll., 1965-72; dir. Biol. Labs. Harvard, 1946-50; tutor in biology Eliot House, 1936-52, asso., 1952-65; master East House (Radcliffe Coll.), 1962-65; exchange prof. U. Paris, 1954-55; vis. prof. U. Mass., 1974, U. Tex., 1976; pres. XI Internat. Bot. Congress, 1969, 2d Nat. Biol. Congress, 1971, Internat. Plant Growth Substance Assn. Congress, Tokyo, 1973. Author: (with F. W. Went) Phytohormones, 1937, The Life of Bacteria, 2d edit, 1963, The Natural Plant Hormones, 1972, Hormones in the Whole Life of Plants, 1977; author (with others); editor: Senescence in Plants, 1981, (with R. S. Harris) Vitamins and Hormones (annual) Vol. 1, 1943, to Vol. 20, 1962, (with G. Pincus) The Hormones, 5 vols, 1948, 55, 63; Editorial bd.: Archives of Biochemistry and Biophysics, 1949-70, Canadian Jour. Botany, 1966-73, Plant Physiology, 1974—; Contbr. numerous articles to tech. jours. Bd. dirs. Found. Microbiology, Biol. Scis. Info. Services. Served as civilian scientist USN, 1942-45. Recipient Stephen Hales prize research Am. Soc. Plant Physiologists, 1936; Guggenheim fellow Eng., 1950-51, Italy, 1958; medallist Internat. Plant Growth Substance Assn., 1976. Fgn. mem. Royal Soc. (London), Soc. Nazionale dei Lincei (Rome), Akad. Leopoldina (Halle), Acad. Nat. de Roumanie (Bucharest), Acad. des Sci. (Paris), Bot. Soc. Netherlands, Bot. Soc. Japan; mem. Am. Soc. Biol. Chemists, Am. Philos. Soc. (council 1973-76), Am. Acad. Arts and Scis., Nat. Acad. Scis. (chmn. botany sect. 1962-65, mem. council 1967-71, exec. com. assembly life scis. 1972-76), Bot. Soc. Am. (pres. 1960), AAAS (dir. 1968-72), Am. Soc. Plant Physiologists (pres. 1950-51), Soc. Gen. Physiologists (pres. 1949-50), Biochem. Soc., Am. Soc. Naturalists (pres. 1954-55), Am. Inst. Biol. Scis. (pres. 1965), Soc. Study Devel. and Growth (pres. 1955-56). Subspecialties: Cell and tissue culture; Plant physiology (biology). Current work: Plant senescence; role of hormones in leaf physiology. Home: 36 Pasatiempo Dr Santa Cruz CA 95060 Office: Thimann Labs Univ Calif Santa Cruz CA 95064

THIRION, JEAN-PAUL JOSEPH, geneticist, educator; b. Metz, France, July 30, 1939; emigrated to U.S., 1963, naturalized, 1972; s. Paul Roger and Anne-Marie Josephine (Averlant) T.; m. Nancy Ouei, Oct. 12, 1967; children: Daniel, Philippe. Diplome d'Ingénieur, ENSIC, Nancy, France, 1963; Ph.D., U. Wis., 1966; Doctorat-ès-Sciences, U. Paris, 1969. Chargé de recherce CNRS, France, 1967-76; asst. prof. dept. microbiology U. Sherbrooke, Que., Can., 1972-76, assoc. prof., 1976-82, prof., 1982—. Fulbright fellow, 1962-66; MRC-Can. scholar, 1972-77; Chercheur scholar, 1977—. Subspecialties: Genetics and genetic engineering (agriculture); Genetics and genetic engineering (medicine). Current work: Somatic cell genetics, DNA recombinant.

THODE, HENRY GEORGE, scientist, educator; b. Dundurn, Sask., Can., Sept. 10, 1910; s. Charles Herman and Zelma Ann (Jacoby) T.; m. Sadie Alicia Patrick, Feb. 1, 1935; children: John Charles, Henry Patrick, Richard Lee. B.Sc., U. Sask., 1930, M.Sc., 1932, LL.D., 1958; Ph.D., U. Chgo., 1934; D.Sc. (hon.), U. Toronto, 1955, U. B.C., 1960, Acadia U., 1960, Laval U., 1963, Royal Mil. Coll. Can., 1964, McGill U., 1966, Queen's U., 1967, York U., 1972, McMaster U., 1973, LL.D., U. Regina, 1983. Research asst. chemistry Columbia, 1936-38; research chemist U.S. Rubber Co., Passaic, N.J., 1938-39; asst. prof. chemistry McMaster U., 1939-42, asso. prof. chemistry, 1942-44, prof. chemistry, 1944-79, prof. emeritus, 1979—; head dept. chemistry, 1948-52, dir. research, 1947-61; prin. Hamilton Coll., 1949-63, v.p., 1957-61, pres. and vice chancellor, 1961-72; Research asso. atomic energy project NRC, 1943-46; Sr. Fgn. Scientist fellow NSF, 1970; mem. commn. on atomic weights Internat. Union Pure and Applied Chemistry, 1963-79; mem. Can. nat. com. to internat. union, 1975-80; dir. Atomic Energy Can. Ltd., 1966-81, Stelco Inc. Mem. editorial adv. bd.: Jour. Inorganic and Nuclear Chemistry, 1954—, Earth and Planetary Sci. Letters, 1965—; Contbr. numerous articles in field. Bd. govs. Ont. Research Found., 1955-82; bd. dirs. Western N.Y. Nuclear Research Centre, 1965-73, Royal Bot. Gardens, 1961-72. Decorated mem. Order Brit. Empire, companion Order Can.; recipient medal Chem. Inst. Can., 1957; Shell Can. merit fellow, 1974. Fellow Royal Soc. Can. (pres. 1959-60, Tory medal 1959, Centenary medal 1982), Chem. Inst. Can. (hon., pres. 1951-52), Royal Soc. (London), Geol. Soc. Am. (Day medal 1980). Office: Nuclear Research Bldg McMaster Univ 1280 Main St W Hamilton ON Canada L8S 4K1

THOMAS, CARLTON EUGENE, elec. engr.; b. Cleve., Dec. 16, 1939; s. Clyde and Laura Bernice (Miles) T.; m. Anne Edna Todd, Jan. 28, 1961; children: Scott, Todd, Julie, Penny. B.S.E.E., U. Mich., 1961, M.S.E.E., 1963, Ph.D., 1971. Research scientist Conductron Corp., Ann Arbor, Mich., 1962-67; head advanced optics KMS Industries, Ann Arbor and Van Nuys, Calif., 1967-72; dir. laser and optics div. KMS Fusion, Ann Arbor, 1972-81; group leader photoengring. Standard Oil Co. (Ohio), Cleve., 1981—. Contbr. articles to profl. jours. Bd. dirs. Ann Arbor YMCA, 1979-80; mem. Interfaith Council for Peace, Ann Arbor, 1976-81. Mem. Internat. Solar Energy Soc., Fedn. Am. Scientists, AAAS, IEEE, Optical Soc. Am., Sigma Xi. Methodist. Subspecialty: Solar energy. Current work: Development of inexpensive photovoltaic solar cell systems based on amorphous silicon technology. Patentee in lasers and optics fields. Home: 2681 Rochester St Shaker Heights OH 44122 Office: 3092 Broadway Cleveland OH 44115

THOMAS, CHARLES SAMUEL, physical therapist, educator; b. Nazareth, South India, June 25, 1920; came to U.S., 1946, naturalized, 1954; s. Edward and Sellammal (Monicham) T.; m. Virginia Mae Learned, June 18, 1940; children: Dale, Carol. B.A., Pacific Union Coll., 1949; B.S., Loma Linda U., 1952; M.A., Stanford U., 1959; Ph.D., Clairemont Grad. Sch., 1966. Staff therapist White Meml. Hosp., 1952-54; instr. Sch. Phys. Therapy, Loma Linda U., Calif., 1954-58, asst. phys. medicine, 1958-63, developer home rehab. program, dept. phys. medicine and rehab., 1962-64, instr. div. pub. health, 1966, asst. prof. pub. health practice, 1967-69, asst. prof. health edn., 1970-71, asst. prof. preventive care, 1971-74, asst. prof. health sci., 1974-75, assoc. prof., 1975-82, assoc. prof. emeritus, 1982—; rehab. cons. Thailand Refugees UN, 1982. Seventh-day Adventist. Subspecialty: Health services research. Current work: Developing hydrotherapy for home and simple treatments for home. Office: Loma Linda U Sch Health Loma Linda CA 92220

THOMAS, CLAUDE EARLE, plant pathologist, researcher; b. Spartanburg, S.C., Dec. 4, 1940; s. John E. and Bertha Leola (Holder) T.; m. June Gilliam Oakman, Aug. 27, 1960; children: Christopher Lee, Matthew Earle, Andrew Beauregard. A.B., Wofford Coll., 1962; M.S., Clemson Coll., 1964; Ph.D., Clemson U., 1966. NDEA grad. fellow Clemson U., 1962-65, sr. grad. teaching asst., 1965-66; research plant pathologist Subtropical Research Lab., Agrl. Research Service, Dept. Agr., Weslaco, Tex., 1966-82, U.S. Vegetable Lab., Charleston, S.C., 1982—; vis. mem. grad. faculty Tex. A&M U., 1978—. Contbr. articles to profl. jours., 5 germ plasm releases, 4 vegetable cultivar releases. Mem. Am. Phytopath. Soc., Am. Soc. Hort. Sci., Cucurbit Genetics Coop., Tex. Vegetable Assn. Baptist. Lodge: Rotary. Subspecialties: Plant pathology; Integrated pest management. Current work: Diseases of vegetable crops (especially cucurbits); germ plasm enhancement for disease resistance; nature of genetic disease resistance; integrated management of foliar diseases-vegetables. Office: 2875 Savannah Hwy Charleston SC 29407

THOMAS, DONALD CHARLES, univ. dean, microbiologist, educator; b. Cin., Sept. 26, 1935; s. Howard G. and Elsie M. (Sack) T.; m. Barbara J., Sept. 2, 1957; children: Mark, Matthew, Michael. B.S., Xavier U., Cin., 1957; M.S., U. Cin., 1959; Ph.D., St. Louis U., 1968. Asst. dir. Surg. Bacteriology Labs. dept. surgery U. Cin., 1959-61; instr. in biology Villa Madonna Coll., 1961-63; instr. depts. med. microbiology and pediatrics Coll. Medicine, Ohio State U., 1968-69, asst. prof. med. microbiology and pediatrics, 1969-72, asst. prof. pediatrics, adj. asst. prof. med. microbiology, and asso. dir. div. program devel. assistance, 1972-77; dir. contracts and grants mgmt. Sch. Medicine, adj. assoc. prof. microbiology and immunology Sch. Medicine and Coll. Sci. and Engring., Wright State U., 1977-78, assoc. prof. pathology, 1978—, assoc. prof. microbiology and immunology, 1978—, dir. univ. research services, 1978—, asst. dean for research, 1979-80, assoc. dean for research, 1980—, acting dean, 1980—; mem. vis. faculty Princeton U., 1979; mem. bd. advisers St. Leonard Coll., 1980—; 2d v.p., mem. exec. com. Trustee Hospice Dayton (Ohio), Inc., 1981—; cons. Econ. Task Force for Social Justice, sponsored by Archdiocese Cin., Office Social Action and World Peace, 1981—. Author: (with others) Molecular Basis of Viral Carcinogenesis in Exploitable Molecular Mechanisms and Neoplasia, 1969; contbr. chpt., articles to profl. publs. Mem. steering com. Consortium for Cancer Control in Ohio, 1978—, mem. evaluation com., 1980—; mem. pilot research com. Ohio div. Am. Cancer Soc., 1979-82; mem. regents adv. com. for grad. studies Ohio Bd. Regents, 1980—; mem. Gov.'s Tech. Task Force, Ohio Dept. Econ. and Community Devel., 1982. Grantee Nat. Inst. Allergy and Infectious Disease, 1968-72, others. Mem. Am. Soc. Microbiology, AAAS, Fedn. Am. Scientists, Soc. Research Adminstrs., Nat. Council Univ. Research Adminstrs., Soc. Univ. Pat. Adminstrs., Nat. Soc. Med. Research, Licensing Execs. Soc. Club: Dayton Execs. Subspecialties: Virology (biology); Microbiology. Current work: diagnostic virology; molecular aspects of virus replication; pathogenesis of virus diseases; devels. in tumor viruses.

THOMAS, EDWARD DONNALL, physician, educator; b. Mart, Tex., Mar. 15, 1920; s. Edward E. and Angie (Hill) T.; m. Dorothy Martin, Dec. 20, 1942; children—Edward Donnall, Jeffery A., Elaine. B.A., U. Tex., 1941, M.A., 1943; M.D., Harvard, 1946. Diplomate: Am. Bd. Internal Medicine. NRC fellow medicine dept. biology Mass. Inst. Tech., 1950-51; instr. medicine Harvard Med. Sch., Boston; also hematologist Peter Bent Brigham Hosp., 1953-55; research asso. Cancer Research Found., Children Med. Center, Boston, 1953-55; physician in chief Mary Imogene Bassett Hosp., also asso. clin. prof. medicine Coll. Phys. and Surg., Columbia, 1955-63; prof. U. Wash. Sch. Medicine, Seattle, 1963—. Mem. Am. Soc. Clin. Investigation, Assn. Am. Physicians, Am. Soc. Hematology, Am. Fedn. Clin. Research, Internat. Soc. Hematology, Am. Assn. for Cancer Research, Western Assn. Physicians, Am. Soc. Clin. Oncology, Transplantation Soc. Subspecialties: Marrow transplant; Oncology. Research and numerous publs. on hematology, marrow transplantation, biochemistry and irradiation biology. Home: 1920 92d Ave NE Bellevue WA 98004 Office: Fred Hutchinson Cancer Research Center 1124 Columbia St Seattle WA 98104

THOMAS, FRANCIS T., transplant surgeon, cardiac surgeon, immunology researcher, cardiac consultant; b. Hibbing, Minn., June 24, 1939; s. Gerald M. and Patricia Ellen (Thornton) T.; m. Judith Marie, June 20, 1969; children: Francis, Scott, David Randolph, Jason Hunter. B.S., U. Minn., 1962, M.D., 1964. Diplomate: Am. Bd. Surgery, Am. Bd. Thoracic Surgery. Asst. prof. Med. Coll. Va., from 1971, assoc. prof., to 1979; prof. surgery, chief transplantation surgery East Carolina, U., 1979—; cons. health, immunology NIH; cardiac cons. to corps. Contbr. numerous articles on transplantation, cardiac surgery and immunobiology to profl. jours. Served with USAR, 1964-69. Am. Cancer Soc. fellow, 1966-69. Mem. Societe de Churgurie Internationale, Soc. Univ. Surgeons, Am. Assn. Immunologists, Royal Soc. Medicine. Democrat. Episcopalian. Subspecialties: Transplant surgery; Immunology (agriculture). Current work: Immunobiology of organ transplantation. Office: East Carolina U Sch Medicine Greenville NC 27834

THOMAS, FRANK JOSEPH, research exec., elec. engr., phys. scientist; b. Pocatello, Idaho, Apr. 15, 1930; s. Emil and Jennie Ruth (Jones) T.; m. Carol Jones, Feb. 4, 1949; children: Dale, Wayne, Keith, Ralph. B.S.E.E., U. Idaho, 1952; M.S., U. Calif., Berkeley, 1957. Registered profl. engr., Calif., 1959. Staff mem. Sandia Corp., 1952-56; teaching asst. U. Calif., 1956-57, lectr., 1959-60; div. mgr. Aerojet Gen. Nucleonics, San Ramon, Calif., 1957-64; asst. dir. def. research and engring. Office Sec. Def., Dept. Def., Washington, 1964-67; Staff Rand Corp., Santa Monica, Calif., 1967-71; pres., dir. research Pacific-Sierra Research, Los Angeles, 1971—; cons. to govt. agys. Author: Evasive Foreign Nuclear Testing, 1971; contbr. articles to profl. jours. Recipient Master Design award Product Engr. Mag., 1963, Meritorious Civilian Service medal Sec. Def., 1967. Mem. AAAS, AIAA. Club: So. Calif. Striders. Subspecialties: Nuclear fission; Electrical engineering. Current work: Tech. verification of arms control treaties. Designer ML-1 Nuclear Power Plant, Idaho, 1963; developed black jack strategy. Home: 21442 Paseo Portola Malibu CA 90265 Office: 12340 Santa Monica Blvd Los Angeles CA 90025

THOMAS, FRANKLIN AUGUSTINE, foundation executive; b. Bklyn., May 27, 1934; s. James and Viola (Atherley) T.; m. (div.); children: Keith, Hillary, Kerrie, Kyle. B.A., Columbia U., 1956, LL.B., 1963; LL.D. (hon.), Yale U., 1970, Fordham U., 1972, Pratt Inst., 1974, Pace U., 1977, Columbia U., 1979. Bar: N.Y. 1964. Atty. Fed. Housing and Home Finance Agy., N.Y.C., 1963-64; asst. U.S. atty. for So. Dist. N.Y., 1964-65; dep. police commr. charge legal matters, N.Y.C., 1965-67; pres., chief exec. officer Bedford Stuyvesant Restoration Corp., Bklyn., 1967-77; pres. The Ford Found., 1979—; dir. Citicorp./Citibank, CBS, Inc., Aluminum Co. Am., Allied Stores Corp., Cummins Engine Co. Trustee J.H. Whitney Found., Columbia U., 1969-75. Served with USAF, 1956-60. Recipient LBJ Found. award for contbn. to betterment of urban life, 1974, medal of excellence Columbia U., 1976.

THOMAS, GEORGE JOSEPH, JR., chemistry educator, research scientist; b. New Bedford, Mass., Dec. 24, 1941; m. Martha Ann Sheehan, July 2, 1966; children: Elizabeth Ann, George Joseph, Jeanine Marie. B.A., Boston Coll., 1963; Ph.D., MIT, 1967. Research fellow King's Coll., London, 1967-68; asst. prof. Southeastern Mass. U., North Dartmouth, 1968-71, assoc. prof., 1971-74, prof. chemistry, 1974—; vis. scientist dept. biology Osaka (Japan) U., 1975-76, MIT, Cambridge, 1982-83; chmn. biophys. chemistry study sect. NIH, Bethesda, Md., 1979-83. Editorial bd.: Biophys. Jour, 1982—; contbr. articles to sci. publs. Mem. Sch. Com. Westport, Mass., 1978—, chmn., 1982-83; trustee Westport Free Public Library, 1975-78. Recipient Coblentz award Coblentz Soc., 1976. Mem. AAAS, Biophys. Soc. Subspecialties: Biophysical chemistry; Spectroscopy. Current work: Structure and assembly of viruses and nucleoproteins; vibrational spectroscopy of nucleic acids and proteins. Office: Southeastern Mass U Dept Chemistry North Dartmouth MA 02747

THOMAS, H. RONALD, chemist; b. Auburn, Ind., June 9, 1942; s. Herbert Ronald and Margaret Lonise (Sheely) T.; m. Louise; children: Jason, Morganna. M.Sc., U. Durham, Eng., 1975, Ph.D., 1977. Mem. sci. staff Xerox Corp., 1967-82; sr. research scientist Pfizer, Inc., 1982—; adj. prof. chem. engring. U. Wash., 1982—; instr. Am. Chem. Soc. courses, 1979—; cons. Center Adhesion, Va. Inst. Tech., 1982—. Co-editor: Characterizations of Polymer Molecular Structure by Photon, Electron and Ion Probes, 1981; contbr. articles to profl. jours. Inst. Petroleum grantee, 1975-76; Xerox Corp. grantee, 1974-77. Mem. Am. Chem. Soc., Am. Phys. Soc. Subspecialties: Surface chemistry; Polymer physics. Current work: Surface chemistry and surface physics on polymer and inorganic surfaces. Home: 4105 N Delaware Dr Easton PA 18042 Office: 640 N 13th St Easton PA 18042

THOMAS, JOHN ALVA, biochemist, educator; b. Berwyn, Ill., May 9, 1940; s. Alva and Marion Louise (Jacobs) T.; m. Loretta Marie, June 19, 1965; children: Richard Allen, Wandy Allen. A.B., DePauw U., 1962; Ph.D. (NIH fellow), U. Ill., Urbana, 1968. NIH postdoctoral fellow U. Pa., 1968-70; asst. prof. biochemistry U. S.D., 1970-77, assoc. prof., 1979—; vis. prof. Cornell U., 1977-78. Mem. Am. Soc. Biol. Chemistry, Biophysics Soc., N.Y. Acad. Sci., S.D. Acad. Sci., AAAS, Sigma Xi. Subspecialties: Biochemistry (medicine); Biophysics (biology). Current work: Measurement and regulation of intracellular pH; bioenergetics; oxidative phosphorylation. Home: 1210 Valley View Dr Vermillion SD 57069 Office: Dept Biochemistry U SD Vermillion SD 57069

THOMAS, JOHN ARLEN, scientific corporation executive, pharmacologist, toxicologist; b. LaCrosse, Wis., Apr. 6, 1933; s. John M. and Eva Hazel (Nelson) T.; m. Barbara Ann Fisler, June 22, 1957; children: Michael, Jane. B.S., U. Wis.-LaCrosse, 1956; M.A., U. Iowa, 1958, Ph.D., 1961. Instr. physiology U. Iowa, Iowa City, 1961; asst. prof. pharmacology U. Va., Charlottesville, 1961-64; assoc. prof. pharmacology Creighton U., Omaha, 1964-67; prof. pharmacology and toxicology W Va. U., Morgantown, 1968-82; asst. dean, assoc. dean medicine W.Va. U., 1973-82; v.p. Travenol Labs., Inc., Morton Grove, Ill., 1982—; mem. study sect. Nat. Cancer Inst., Bethesda, Md., 1975-79, EPA, 1970-75; mem. rev. panels Nat. Toxicology Program, Research Triangle Park, N.C., 1978-79, Dept. Def., 1982-83. Author, co-author books in field of endocrine pharmacology, sci. articles; author films and audio-visual materials; editor: Advanced Sex Hormone Research; editorial bd.: Jour. Environ. Toxicology and Pathology, Jour. Toxicology and Applied Pharmacology, 1970-83. NIH grantee, 1968-80; recipient cert. sci. service EPA, 1977; Disting. Alumni award U. Wis.-LaCrosse, 1978; named outstanding tchr. W.Va. U. Sch. Medicine, 1971, 73, 77. Fellow Am. Sch. Health Assn.; mem. Endocrine Soc., Am. Soc. Pharmacology and Exptl. Therapeutics, Pharm. Mfg. Assn., Soc. Toxicology (com. chmn. 1978-83), Teratology Soc. Republican. Subspecialties: Pharmacology; Cytology and histology. Current work: Endocrine pharmacology/toxicology; prostate biochemistry and pathology; phthalate acid esters; genetic engineering. Home: 1304 Woodland Dr Deerfield IL 60015 Office: Travenol Labs Inc 6301 Lincoln Ave Morton Grove IL 60053

THOMAS, JOHN HOWARD, engring. educator, astrophysicist; b. Chgo., Apr. 9, 1941; s. William Whitney and Dorothy L. T.; m. Lois Moffit, Aug. 11, 1962; children: Jeffrey, Laura. B.S., Purdue U., 1962, M.S., 1964, Ph.D. in Engring. Scis, 1966. Registered profl. engr., N.Y. NATO postdoctoral fellow U. Cambridge, Eng., 1966-67; mem. faculty U. Rochester, N.Y., 1967—, prof. mech. and aerospace scis., 1981—, assoc. dean grad. studies, 1981—; vis. scientist Max-Planck Inst. for Physics and Astrophysics, Munich, W. Ger., 1973-74. Contbr. articles to profl. jours. NSF, NASA, U.S. Air Force grantee. Mem. Am. Astron. Soc., Internat. Astron. Union, Am. Phys. Soc., Am. Geophys. Union, ASME, Am. Acad. Mechanics, Am. Soc. Engring. Edn. Subspecialties: Solar physics; Fluid mechanics. Current work: Theoretical research in solar physics, especially sunspots; geophysical and astrophysical fluid dynamics; applied mathematics. Office: Coll Engring and Applied Sci U Rochester Rochester NY 14627

THOMAS, JOHN WILLIAM, animal agriculture educator, nutrition researcher; b. Spanish Fork, Utah, Mar. 25, 1918; s. John Banks and Kate (Tolhurst) T.; m. Carolyn M. Palmer, Jan. 6, 1945; children: Linda K., John P., Barbara E., Christopher. B.S., Utah State U., 1940; Ph.D., Cornell U., 1946. Grad. asst. U. Wis., Madison, 1940-41, Cornell U., Ithaca, N.Y., 1941-42, 45-46; research assoc. Nat. Def. Research Orgn., Northwestern U., Evanston, Ill., 1942-46; dairy husbandman and biochemist U.S. Dept. Agr., Beltsville, Md., 1946-60; prof. nutrition Mich. State U., East Lansing, 1960—; cons., Emprapa, Brazil, 1982. Contbr. numerous articles to profl. jours. Recipient Am. Feed. Mfr's. award Am. Dairy Sci. Assn., 1953; superior service award U.S. Dept. Agr., 1959; outstanding specialist award Mich. State U., 1982. Fellow AAAS; mem. Am. Inst. Nutrition, Am. Assn. Animal Sci., Am. Dairy Sci. Assn. (Borden award 1973), Soc. Exptl. Biology and Medicine, N.Y. Acad. Sci., Sigma Xi. Subspecialties: Nutrition (biology); Animal physiology. Current work: Research and write on forage harvesting, preservation, evaluation and feeding; supervise graduate students on nutrition; teach practical dairymen feeding, nutrition and management and diagnose problems and desirable practices. Home: 316 John R St East Lansing MI 48823

THOMAS, LEO JOHN, mfg. co. exec.; b. St. Paul, Oct. 30, 1936; s. Leo John and Christal (Dietrich) T.; m. Joanne Juliani, Dec. 27, 1958; children—Christopher, Gregory, Cynthia, Jeffrey. B.S., U. Minn., 1958; M.S., U. Ill., 1960, Ph.D., 1962. With Eastman Kodak Co., Rochester, N.Y., 1961—, asst. dir. research labs., 1975-77, dir., 1977—, v.p., 1977-78, sr. v.p., 1978—; bd. dirs. N.Y. State Sci. and Tech. Found., Indsl. Research, Indsl. Research Inst.; mem. founding com. Chem. Research Council; dir. Security Trust Co., Rochester. Mem. Am. Inst. Chem. Engrs., Soc. Photog. Scientists and Engrs., Soc. Motion Picture and TV Engrs., N.Y. Acad. Scis., Rochester C. of C. Subspecialty: Chemical engineering. Office: Eastman Kodak Co Research Labs Rochester NY 14560

THOMAS, LEWIS, physician, educator, med. adminstr.; b. Flushing, N.Y., Nov. 25, 1913; s. Joseph S. and Grace Emma (Peck) T.; m. Beryl Dawson, Jan. 1, 1941; children—Abigail Luttinger, Judith, Eliza. B.S., Princeton, 1933, Sc.D. (hon.), 1976; M.D., Harvard, 1937; M.A., Yale, 1969; Sc.D. (hon.), U. Rochester, 1974, U. Ohio at Toledo, 1976, Columbia U., 1978, Meml. U. Nfld., 1978, U. N.C., 1979, Worcester Found., 1979, LL.D., Johns Hopkins U., 1976, Trinity Coll., 1980, L.H.D., Duke U., 1976, Reed Coll., 1978, Litt.D., Dickinson Coll., 1980, Ursinus Coll., 1981. Intern Boston City Hosp., 1937-39, Neurol. Inst., N.Y.C., 1939-41; Tilney Meml. fellow Thorndike Lab., Boston City Hosp., 1941-42; vis. investigator Rockefeller Inst., 1942-46; asst. prof. pediatrics Med. Sch. Johns Hopkins, 1946-48; asso. prof. medicine Med. Sch. Tulane, 1948-50, prof. medicine, 1950; prof. pediatrics and medicine, dir. pediatric research labs. Heart Hosp., U. Minn., 1950-54; prof., chmn. dept. pathology N.Y. U. Sch. Medicine, 1954-58, prof., chmn. dept. medicine, 1958-66, dean, 1966-69; prof., chmn. dept. pathology Yale, 1969-72, dean, 1972-73; prof. medicine, pathology Med. Sch. Cornell U., N.Y.C., 1973—, prof. biology (SKI) div., 1973—, co-dir., 1974-80; adj. prof. Rockefeller U., 1975—; pres., chief exec. officer Meml. Sloan-Kettering Cancer Center, N.Y.C., 1973-80, chancellor, 1980—; dir. 3d and 4th med. divs. Bellevue Hosp., 1958-66, pres. med. bd., 1963-66; cons. Manhattan VA Hosp., 1954-69; cons. to surgeon gen. Dept. Army, surgeon gen. USPHS; mem. pathology study sect. NIH, 1954-58, nat. adv. health council, 1958-62, nat. adv. child health and human devel. council, 1963-67; mem. commn. on streptococcal disease Armed Forces Epidemiological Bd., 1950-62; mem. Pres.'s Sci. Adv. Com., 1967-70, Inst. Medicine, 1971, Nat. Acad. Scis., 1972—, mem. council and governing bd., 1979—; chmn. overview cluster subcom. Pres.'s Biomed. Research Panel, 1975-76; mem. Tech. Assessment Adv. Council, 1980—; Dir. Squibb Corp. Mem. Bd. Health, N.Y.C., 1956-69; mem. bd. sci. consultants Sloan-Kettering Inst. Cancer Research, 1966-72; mem. Sloan-Kettering Inst., 1973—; bd. dirs. Josiah Macy Jr. Found., 1975—; bd. sci. advisers Mass. Gen. Hosp., 1970-73, Scripps Clinic and Research Found., 1969—; bd. dirs., research council Public Health Research Inst. of City N.Y., 1964-69; bd. overseers Harvard Coll., 1976—; mem. sci. adv. com. Sidney Farber Cancer Inst., 1978—; mem. council Grad. Sch. Bus. and Public Adminstrn., Cornell U., 1978—; mem. awards assembly Gen. Motors Cancer Research Found., 1978—; Asso. fellow Ezra Stiles Coll. Yale U. Author: Lives of a Cell, 1974, Medusa and the Snail, 1979; Editorial bd.: Inflammation. Trustee N.Y.C.-Rand Inst., 1967-71, The Rockefeller U., 1975—, Draper Lab., 1975-81, John Simon Guggenheim Meml. Found., 1975—, Mt. Sinai Sch. Medicine, 1979—, Menninger Found., 1980—. Served as lt. comdr. M.C. USNR, 1941-46. Recipient Distinguished Achievement award Modern Medicine, 1975; Nat. Book award for arts and letters, 1975; Honor award Am. Med. Writers Assn., 1978; Med. Edn. award AMA, 1979; Bard award in medicine and sci. Bard Coll., 1979; Am. Book award, 1981. Fellow Am. Acad. Arts and Scis., Am. Rheumatism Assn.; mem. Nat. Acad. Scis., Am. Philos. Assn., Am. Soc. Exptl. Pathology, Practitioners Soc., Am. Acad. Microbiology, Peripatetic Clin. Soc., Am. Soc. Clin. Investigation, Am. Assn. Immunologists, Soc. Am. Bacteriologist, Assn. Am. Physicians, Am. Pediatric Soc., N.Y. Acad. Scis., Harvey Soc. (councillor), AAUP, Soc. Exptl. Biology and Medicine, Am. Soc. Clin. Oncology, Interurban Clin. Club, Phi Beta Kappa, Alpha Omega Alpha. Clubs: Harvard (N.Y.C.); Century Assn. Subspecialty: Internal medicine. Office: Memorial-Sloan Kettering Cancer Center 1275 York Ave New York NY 10021

THOMAS, NORMAN GENE, astronomer; b. Alamosa, Colo., May 1, 1930; s. Vivian Russell and Sonia Louise (Moeller) T.; m. Maryanna Ruth Brown, June 3, 1953; children: Kathryn Gail Hazelton, Carol Louise Baltutis, Brian Douglas, Bruce Gregory. Student, Adams State Coll., 1948-49; B.A., Wesleyan U., Middletown, Conn., 1952; postgrad., U. Calif., Berkeley, 1955-59; M.A., No. Ariz. U., Flagstaff, 1974. Astronomer U.S. Naval Obs., Washington, 1952-53; astronomer, party chief U.S. Army Map Service, 1953-55; research asst. various labs. U. Calif., Berkeley, 1955-59; research assoc. Lowell Obs., Flagstaff, Ariz., 1959—; astronomer-tchr. Community Coll.-Pub. Schs., Flagstaff, 1972—. Served with U.S. Army, 1953-55. Recipient Littell prize Wesleyan U., 1952. Mem. AAAS, Am. Astron. Soc., Astron. Soc. Pacific, Ariz. Acad. Sci., Sigma Xi. Lodge: Elks. Subspecialties: Optical astronomy; Planetary science. Current work: Astrometry: studies of star and QSO distribution in space, comet and asteroid statistics. Office: PO Box 1269 Mars Hill Rd Flagstaff AZ 86002

THOMAS, PAUL ELBERT, research biochemist, educator; b. Columbus, Ohio, Oct. 20, 1942; s. Ralph Elbert and Esther (Kaiser) T.; m. Marcia Ann Nunnikhoven, July 11, 1964; children: Mark, Ruth, Karen. B.S., Otterbein Coll., 1965; Ph.D., Ohio State U., 1970. Postdoctoral fellow Roche Inst. of Molecular Biology, Nutley, N.J., 1970-72; sr. scientist Hoffmann-LaRoche Inc., Nutley, 1972-79, research fellow, 1980—; adj. faculty King's Coll., Briarcliff Manor, N.Y., 1978-81. Contbr. numerous articles on biochemistry to profl. jours. Mem. Am. Soc. for Pharmacology and Exptl. Therapeutics, Am. Soc. Biol. Chemists, Am. Chem. Soc. Subspecialties: Molecular pharmacology; Biochemistry (biology). Current work: Drug metabolism and toxicity, cytochrome P450. Home: 28 Oakcrest Rd W Orange NJ 07052 Office: 340 Kingsland Rd Nutley NJ 07110

THOMAS, RICHARD EUGENE, university administrator, aerospace engineering educator; b. Logan, Ohio, Dec. 29, 1925; s. William Henry and Helen Marguerite (McGhee) T.; m. Jeannette Belle Que, July 2, 1950; children: Christine, Paula, Eric. B.Aero.E., Ohio State U., 1951, B.A. in Math, 1953, M.S., 1956, Ph.D., 1964. Registered profl. engr., Tex. Prof. aerospace engring. Tex. A&M U., 1966-69, 71-83, spl. asst. to dean engring., 1971, dir., 1972-77, assoc. dean engring., 1977-78, interim dean engring., 1978-79, dir., 1979—; mem. computer aided mfg. adv. group USAF Integrated Computer Aided Mfg. Program; mem. So. Assn. Univs. Accreditation Team. Author: (with Mark Gamache) reports Cobalt: Strategic Alternatives for Supply, 1981, (with Charles L. Smith) Natural Gas in Western Siberia, 1982. Mem. Hilliard (Ohio) City Council, 1958-62. Served with USAAF, 1944-45. Recipient award for disting. achievement in teaching Standard Oil Found., 1968. Fellow AIAA (assoc.); mem. Am. Soc. Engring. Edn., Aerospace Dept. Chairman's Assn., Tau Beta Pi, Phi Kappa Phi, Sigma Gamma Tau. Methodist. Club: Army and Navy (Washington). Subspecialties: Aerospace engineering and technology; Aircraft aerodynamics. Current work: U.S./foreign military technology; advanced manufacturing systems; space technology. Home: Route 1 Box 488 Bryan TX 77801 Office: Tex A&M U Box 83 FM College Station TX 77843

THOMAS, ROBERT JAMES, biologist, educator; b. Flint, Mich., July 5, 1949; s. Allan James and Ruth Pauline (Brandt) T. B.A., U. Mich., 1971; Ph.D., U. Calif., Santa Cruz, 1975. Asst. prof. biology Bates Coll., Lewiston, Maine, 1975-82, assoc. prof., since 1982—. Contbr. articles to sci. jours. NSF grantee, 1979-81; Cottrell Coll. Sci. grantee, 1980-82. Mem. Bot. Soc. Am., Am. Soc. Plant Physiologists, Am. Bryol. and Lichenol. Soc., Internat. Assn. Bryologists. Subspecialties: Plant physiology (biology); Plant growth. Current work: Physiology and development of bryophytes, including culture, hormone effects on growth, tropisms, protoplast isolation and transport studies.

THOMAS, WILLIAM ERIC, biochemist; b. Nashville, Aug. 2, 1951; s. Andrew Johnson and Alphonsa Lucille (Williams) T.; m. Linda Love, Aug. 25, 1973; 1 dau., Kimberly Monique. B.S., Tenn. State U., 1973, M.S., 1975; Ph.D., Meharry Med. Sch., 1980. Postdoctoral research fellow Harvard Med. Sch., Boston, 1980—. Contbr. articles to profl. jours. Mem. Soc. for Neurosci., AAAS. Subspecialties: Neurochemistry; Biochemistry (biology). Current work: Study of cerbral cortical neurotransmitters, cerebral cortical tissue culture, neurotransmitter metabolism. Home: 15 Pasadena Rd Dorchester MA 02121 Office: Dept Neurobiology Harvard Med Sch Boston MA 02115

THOMAS, WILLIAM WAYT, plant systematist; b. Durham, N.C., Apr. 11, 1951; s. William Wayt and Sara (Tillett) T.; m. Marie Antoinette Sick, June 29, 1974. B.A., U. N.C., 1973; M.S., U. Mich., 1976, Ph.D., 1982. Hon. research assoc. N.Y. Bot. Garden, N.Y.C., 1981; collection mgr. Herbarium Carnegie Mus. Natural History, Pitts., 1982—. Contbr. articles to profl. jours. NSF grantee, 1979-80. Mem. Am. Assn. Plant Taxonomists, Assn. Tropical Biology, Bot. Soc. Am., Internat. Assn. Plant Taxonomists, Mich. Bot. Club, So. Appalachian Bot. Club. Subspecialties: Systematics; Taxonomy. Current work: Systematics of the Cyperaceae; systematics of Rhynchospora for the southeastern U.S. and Central America. Home: 431 Shady Ave Pittsburgh PA 15206 Office: Sect Botany Carnegie Mus Natural History 4400 Forbes Ave Pittsburgh PA 15213

THOMPSON, ANTHONY RICHARD, elec. engr.; b. Hull, Yorkshire, Eng., Apr. 7, 1931; s. George and Ada Mary (Laybourn) T.; m. Sheila Margaret Press, Oct. 12, 1963; 1 dau., Sarah Louise. B.Sc. with honors, U. Manchester, Eng., 1952, Ph.D., 1956. Engr. E.M.I. Electronics Ltd., Feltham, Eng., 1956-57; research assoc., research fellow Harvard Coll. Obs., Cambridge, Mass., 1957-62; radio astronomer Stanford (Calif.) U., 1962-70; sr. research assoc., 1970-72; dep. project mgr. VLA project Nat. Radio Astronomy Obs., Socorro, N.Mex., 1973-78, systems engr., radio frequency coordinator, 1978—; chmn. radio astronomy subcom. of com. on radio frequencies Nat. Acad. Scis.; chmn. radio astronomy subgroup U.S. Study Group 2 Internat. Radio Consultative Com. Contbr. articles to profl. jours. Mem. IEEE, Am. Astron. Soc., Internat. Astron. Union (commn. 40), Internat. Union Radio Sci. Subspecialties: Radio and microwave astronomy. Current work: Design of radio astronomy instrumentation in particular interferometers and arrays for synthesis mapping; electromagnetic compatibility and frequency protection and radio astronomy. Office: PO Box O Socorro NM 87801

THOMPSON, BENNY LOUIS, energy engineering co. exec.; b. Seminole, Okla., July 1, 1928; s. Kermit Louis and Pauline (Forbes) T.; m. Joanne Crowell, Mar. 23, 1957 (div.); children— Pamela, Dianne, Janice, Steven. Student, San Jose State U., 1947-48, Pierce Coll., Woodland Hills, Calif., 1957-58. Pres. Thomco Equipment Engring. Co., San Fernando, Calif., 1952-65; owner B&K Auto Parts, Marysville, Calif., 1965-71; pres. Ben Thompson Design Corp., Yuba City, Calif., 1971—, Biomass Corp., Yuba City, 1978—; cons. for low BTU gas engine conversion systems. Vice chmn. Sutter County Bicentennial Commn., 1976; mem. Sutter County Republican Central Com. Served with USN, 1948-52. Recipient Sutter County Clean Air award, 1974. Lodges: Lions; Shriners; Masons. Subspecialties: Gasification; Gas cleaning systems. Current work: Broadest application of producer gas to power mechanical systems. Patentee automatic plowing systems, specialized boring systems for problem soils, compacting system for problem soils. Home and Office: 10921 Live Oak Hwy Live Oak CA 95953

THOMPSON, BRIAN JOHN, college dean, optics scientist; b. Glossop, Derbyshire, Eng., June 10, 1932; came to U.S., 1962; s. Alexander William and Edna May (Gould) T.; m. Joyce Emily Cheshire, Mar. 31, 1956; children: Karen Joyce, Andrew Derrick. B.Sc.Tech., U. Manchester, Eng., 1955, Ph.D., 1959. Demonstrator in physics Faculty of Tech., U. Manchester, 1955-56, asst. lectr., 1957-59; lectr. physics U. Leeds, Eng., 1959-62; sr. physicist Tech. Ops., Inc., Burlington, Mass., 1963-65, dir. optics dept., 1965-67, mgr. West, tech. dir., 1967-68; prof. Inst. Optics, U. Rochester, N.Y., 1968—, Wm. F. May prof. engring., 1982—, dir., 1968-75, dean, 1974—. Am. editor: Optica Acta, 1974—; Asso. editor: Optics Communications, 1974—, Optical Engring., 1973—; editorial adv. bd.: Laser Focus, 1967—; internat. editor: Optics and Laser Tech., 1970—; contbr. articles to profl. jours. Served with Brit. Army, 1950-52. Fellow Optical Soc. Am. (dir. 1969-72, exec. com. 1970-73, asso. editor Jour. 1966-77), Inst. Physics and Phys. Soc. (Gt. Britain), Soc. Photo-Optical Instrumentation Engrs. (pres. 1974, 75-76, Pres.'s award 1967, Pezzuto award 1978, Kingslake medal 1978); mem. Am. Phys. Soc., AAAS, N.Y. Acad. Scis. Mem. Christ Ch. Subspecialties: Optical engineering; Holography. Current work: Holographic particle size and velocity measurements; phase microscopy, image processing; design of coherent optical systems. Home: 9 Esternay Ln Pittsford NY 14534 Office: U Rochester Coll Engring and Applied Sci Rochester NY 14627

THOMPSON, DAVID JEROME, chemical company executive; b. Sand Creek, Wis., July 21, 1937; s. Marshall T. and Bernice (Severson) T.; m. Virginia Ruth Williams, Aug. 11, 1962; children: Keith D., Craig M. B.S., U. Wis.-Madison, 1960, M.S., 1961, Ph.D., 1963; M.B.A., U. Chgo., 1975. Research biochemist Internat. Minerals & Chem. Corp., Libertyville, Ill., 1964-68, supr. animal research, 1968-69, mgr. tech. service, Mundelein, Ill., 1969-78, dir. tech. service, 1978-79, sales mgr., 1979-81, v.p. sci. and tech., Northbook, 1981—; cons. Am. Assn. Feed Control Ofcls., 1974—, NRC, Nat. Acad. Sci., 1976-80; chmn. Nutrition Council of Am. Feed Mfrs. Assn. Co-author: Mineral Tolerance of Domestic Animals, 1980; contbr. articles to jours. and papers to internat. confs. Mem. AAAS, N.Y. Acad. Sci., Am. Inst. Nutrition, Poultry Sci. Assn., Am. Dairy Sci. Assn., Am. Soc. Animal Sci., Nutrition Today Soc., Am. Chem. Soc., Council for Agrl. Sci. and Tech., Sigma Xi, Gamma Alpha. Subspecialties: Animal nutrition; Biochemistry (biology). Current work: Business aspects of science activities. Home: 826 Fairway Libertyville IL 60048 Office: Internat Minerals & Chem Corp 2315 Sanders Rd Northbrook IL 60062

THOMPSON, DAVID JOHN, astrophysicist; b. Cin., Jan. 11, 1945; s. Martin Monroe and Jean Mildred (Beckert) T.; m. Carlynn Jean Grumbles, May 27, 1972; children: Lessa Anne, Kira Jo. B.A., Johns Hopkins U., 1967; Ph.D., U. Md., College Park, 1973. Research assoc. U. Md., College Park, 1973; astrophysicist NASA/Goddard Space Flight Center, Greenbelt, Md., 1973—. Contbr. articles to profl. jours. Recipient 3d prize Griffith Observer writing contest, 1977. Mem. Am. Phys. Soc., Am. Astron. Soc. Baptist. Subspecialty: Gamma ray high energy astrophysics. Current work: Research into high energy processes in universe, particularly as revealed by study gamma rays from cosmic sources, data analysis and interpretation, detector development gamma ray observatory. Home: 2901 Stoneybrook Dr Bowie MD 20715 Office: Code 662 NASA/Goddard Space Flight Center Greenbelt MD 20771

THOMPSON, DUDLEY, energy consultant; b. Kansas City, Mo, Aug. 25, 1929; s. John Randall and Eleanor (Griggs) Levy T.; m. Linda Giannone, June 6, 1951; children: John, Ann. B.S., U.S. Mil. Acad., 1951; M.S.E., Purdue U., 1956. Registered profl. engr., Calif. Group leader reactor ops. Brookhaven Nat. Lab., Upton, N.Y., 1960-67; br. chief, asst. dir., dir. AEC, NRC, Washington, 1967-82; pres. Thompson Assoc., Potomac, Md., 1982—. Served to capt. U.S. Army, 1951-60. Mem. Am. Nuclear Soc., Tau Beta Pi, Eta Kappa Nu. Republican. Episcopalian. Subspecialties: Nuclear engineering; Nuclear fission. Current work: Consultant on nuclear industry concerns, emphasis on quality assurance, probabilistic risk assessment, reactor operations and response to government enforcement actions. Home: 2197 Stratton Dr Potomac MD 20854 Office: Thompson Assocs 2197 Stratton Dr Potomac MD 20854

THOMPSON, EDWARD IVINS BRADBRIDGE, cell and molecular biologist; b. Burlington, Iowa, Dec. 20, 1933; s. Edward Bills and Lois Elizabeth (Bradbridge) T.; m. Lynn Taylor Parsons, June 27, 1957; children: Elizabeth Lynn, Edward Ernest Bradbridge. B.A. with distinction, Rice U., 1955; postgrad., Cambridge (Eng.) U., 1957-58; M.D., Harvard U., 1960. Diplomate: Nat. Bd. Med. Examiners. Intern then resident in internal medicine Presbyn. Hosp., Columbia U. Med. Ctr., N.Y.C.; commd. USPHS, 1962; research asst. NIMH, 1962-64; research scientist Nat. Inst. Arthritis and Metabolic Diseases, 1964-68, Nat. Cancer Inst., 1968—, chief sect. biochemistry gene expression, 1973—; attending physician endocrine service Nat. Naval Med. Ctr., 1978-80; chmn. Task Force Hormones and Cancer, 1978-80; mem. faculty Found. Advanced Edn. in Scis. Author articles in field; editor:

Regulation of Gene Expression, 1973, Gene Expression and Carcinogenesis in Cultured Liver, 1975, Steroid Receptors and the Management of Cancer, 1979; editorial bd.: Jour. Steroid Biochemistry; assoc. editor: Cancer Research. Active local Girl Scouts, Rice U. Alumni Fund. Hohenthal scholar, 1955; Harvard U. Med. Sch. scholar, 1955-57; Maco Stewart fellow, 1955-57; John Parker fellow, 1957-58. Mem. Am. Chem. Soc., Am. Soc. Cell Biology, Am. Assn. Cancer Research, Am. Soc. Biol. Chemists, Endorine Soc., Phi Beta Kappa, Alpha Omega Alpha. Club: Aspen Hill Racquet (Silver Spring, Md). Subspecialties: Cell and tissue culture; Genetics and genetic engineering (medicine). Current work: Steroid hormone effects on cells and gene expression. Office: NIH Bldg 37 Room 1C09 Bethesda MD 20205

THOMPSON, ELIZABETH BARNES, research biologist; b. San Antonio, June 15, 1942; d. Virgil Everett and Mildred Louise (Adlof) Barnes; m. Hugh Walter Thompson, June 18, 1964; 1 dau., Victoria Mireille. B.A., Radcliffe Coll., 1964; Ph.D., Cornell U. Grad. Sch. Med. Scis., 1971. Instr. dept. physiology NYU Med. Sch., 1971-74; asst. dept. neurobiology and behavior Pub. Health Research Inst. City N.Y., 1971-74; instr. physiology Coll. Physicians and Surgeons, Columbia U., N.Y.C., 1974-77, asst. prof., 1977-78, asst. prof. physiology and asst. prof. anatomy, 1978-80; vis. scientist Population Council Ctr. for Biomed. Research, Rockefeller U., N.Y.C., 1980—. Contbr. articles to profl. jours. NIH grantee, 1975-81; Career Devel. award, 1975-80; NSF grantee, 1979—. Mem. Am. Soc. Cell Biology, Soc. Neurosci., N.Y. Soc. Electron Microscopists. Subspecialties: Neurobiology; Cell biology. Current work: Ultrastructure of identified synapses in Aplysia which exhibit behaviorally relevant plasticity, development of techniques to mark uniquely identified synaptic contacts for study with light and electron microscopes. Office: 1230 York Ave New York NY 10021

THOMPSON, ERIC DOUGLAS, electrical engineering educator, physicist; b. Buffalo, Mar. 24, 1934; s. Milton Leslie and Nellie Evelyn (Throop) T.; m. Gail Michelle O'Brien, July 2, 1960; children: Aileen Jo, Eric Douglas, Ronald Brien. B.S.E.E., M.I.T., 1956, M.S.E.E., 1956, Ph.D. in Physics, 1960. Mem. tech. staff Bell Telephone Labs., Murray Hill, N.J., 1953-55; staff asso. Lincoln Labs., Lexington, Mass., 1957-60; asst. prof. elec. engring. and applied physics Case Western Res. U., 1963-66, asso. prof., 1966-70, prof., 1970-83, NSF program mgr., 1981-82; prof., chmn. elec. engring. and computer sci. Lehigh U., Bethlehem, Pa., 1983—. Served with USAF, 1960-62. NSF postdoctoral fellow, 1962-63; NRC sr. resident research asso., 1972-73. Fellow Am. Phys. Soc.; mem. IEEE (sr.), Sigma Xi, Eta Kappa Nu, Tau Beta Pi. Subspecialties: Microchip technology (engineering); Condensed matter physics. Current work: Josephson Junction integrated circuits; atomic scattering from solids; ultra-high speed electronic circuits. Office: Elec Engring and Computer Scis Dept Lehigh U Bethlehem PA 18015

THOMPSON, GEORGE ALBERT, geophysics educator; b. Swissvale, Pa., June 5, 1919; s. George A. and Maude A. (Harkness) T.; m. Anita Kimmell, July 20, 1944; children: Albert J., Dan A., David C. B.S., Pa. State U., 1941; M.S., MIT, 1942; Ph.D., Stanford U., 1949. Geologist, geophysicist U.S. Geol. Survey, 1942-76; asst. prof. Stanford, 1949-55, assoc. prof., 1955-60; prof. geophysics Stanford U., 1960—, Otto N. Miller prof. earth scis., 1980—, chmn. geophysics dept., 1967—, geology dept., 1979-82. Served from ensign to lt. (j.g.) USNR, 1944-46. NSF postdoctoral fellow, 1956-57; Guggenheim fellow, 1963-64; Recipient G.K. Gilbert award in seismic geology, 1964. Fellow AAAS, Geol. Soc. Am., Geophys. Union; mem. Seismol. Soc. Am., Soc. Exploration Geophysicists, Soc. Econ. Geologists. Subspecialties: Geophysics; Tectonics. Current work: Geophysics of the continental lithosphere(crust and upper mantle)based on reflection, seismic and gravity data. Home: 421 Adobe Pl Palo Alto CA 94306 Office: Geophysics Dept Stanford U Stanford CA 94305

THOMPSON, H. BRADFORD, chemist, educator; b. Detroit, Apr. 22, 1927; s. Herbert B. and Margaret Ann (Gilbert) T.; m. Jane Elizabeth Lang, June 13, 1949. B.S., Olivet (Mich.) Coll., 1948; A.M., Oberlin Coll., 1950; Ph.D., Mich. State U., 1953. Research assoc. Mich. State U., 1953-55; asst. prof. chemistry Gustavus Adolphus Coll., 1955-57, assoc. prof., 1957-63; sr. research assoc. Iowa State U. Ames Lab., 1963-65, U. Mich., 1965-67; prof. U. Toledo, 1967—. Mem. Common Cause, ACLU. Served with U.S. Army, 1945-46. Mem. Am. Chem. Soc., Am. Phys. Soc., AAAS, Sigma Xi. Presbyterian. Subspecialties: Physical chemistry; Theoretical chemistry. Current work: Molecular structure and geometry; concepts of chemical bond. Office: Dept Chemistry U Toledo Toledo OH 43606

THOMPSON, JACK MANSFIELD, JR., mechanical engineer; b. Hartford, Conn., Mar. 30, 1951; s. Jack Mansfield and Sarah Gertrude (Matthews) T. B.S. in Engring, Cornell U., 1973, M.M.E., 1974. Engr. Gleason Works, Rochester, N.Y., 1973; project mgr., sr. tech. cons. Structural Dynamics Research Corp., Milford, Ohio, 1974—. Contbr. articles to profl. jours. Recipient award Lincoln Arc Welding Found., 1973, 1978. Mem. ASME, Soc. Automotive Engring., Delta Tau Delta. Subspecialties: Mechanical engineering; Robotics. Current work: State of the art mechanical and structural design, specializing in composite materials and robotics.

THOMPSON, JAMES BURLEIGH, JR., geologist, educator; b. Calais, Maine, Nov. 20, 1921; s. James Burleigh and Edith (Peabody) T.; m. Eleanora Mairs, Aug. 3, 1957; 1 son, Michael A. A.B., Dartmouth, 1942, D.Sc. (hon.), 1975; Ph.D., Mass. Inst. Tech., 1950. Instr. geology Dartmouth, 1942; research asst. Mass. Inst. Tech., 1946-47, instr., 1947-49; instr. petrology Harvard, 1949-50, asst. prof., 1950-55, asso. prof. mineralogy, 1955-60, prof., 1960-77, Sturgis Hooper prof. geology, 1977—; guest prof. Swiss Fed. Inst. Tech., 1977-78. Served to 1st lt. USAAF, 1942-46. Guggenheim fellow, 1963; Sherman Fairchild distinguished scholar Calif. Inst. Tech., 1976. Fellow Mineral. Soc. Am. (pres. 1967-68, recipient Roebling medal 1978), Geol. Soc. Am. (A.L. Day medal 1964), Am. Acad. Arts and Scis.; mem. AAAS, Am. Geophys. Union, Nat. Acad. Scis., Geochem. Soc. (pres. 1968-69), Sigma Xi. Subspecialties: Petrology; Mineralogy. Current work: Metamorphic petrology; regional geology of appalachians. Office: Harvard U Cambridge MA 02138

THOMPSON, JAMES ELTON, electrical engineering educator; b. Lubbock, Tex., Jan. 24, 1946; s. James Elton and Marie Labelle (Wild) T.; m. Elizabeth Ann McCaleb, Sept. 8, 1967; children: Matthew, Julie. B.S.E.E., Tex. Tech. U., 1968, MSEE, 1970, Ph.D. in Elec. Engring, 1974. Scientist Laser Lab., Northrop Corp., Los Angeles, 1974-76; prof. engring. U. S.C., Columbia, 1976—, Tamper prof., 1980-83; prof., chmn. elec. engring. U. Tex., Arlington, 1983—. Contbr. articles to tech. jours. Mem. IEEE, Optical Soc. Am. Subspecialties: Electrical engineering; Optical engineering. Current work: Pulsed power, high voltage, fast electrical and optical diagnostics, high voltage and high current materials and devices. Office: Dept Elec Engring U Tex. Arlington TX 76019

THOMPSON, JAMES JARRARD, microbiologist, immunologist; b. Des Moines, May 21, 1944; s. Lloyd Howard and Elmina Ruth (Jarrard) T.; m. Harriet Crawford Willis, Nov. 19, 1969; children: Mark Lloyd, Neil Crawford, Julliette Willis. B.A., U. Iowa, 1965, M.S., 1968, Ph.D., 1970. Instr. U. Iowa, Iowa City, 1970-71; asst. prof. Temple U., Phila., 1972-74; mem. faculty La. State U. Med. Ctr., New Orleans, 1974—, assoc. prof., 1979—, acting head dept. microbiology and immunology, 1981—. Contbr. articles to profl. jours. NSF grantee, 1975-77; Am. Heart Assn. grantee, 1977-78; Arthritis Found. grantee, 1979; NIH grantee, 1980—. Mem. Am. Soc. Microbiology (sec.-treas. S. Central br. 1980—), AAAS, Am. Assn. Immunologists, Sigma Xi. Subspecialties: Microbiology (medicine); Immunology (medicine). Current work: Hybridoma technology; immunoassays; structure and function of apolipoproteins; immune responses in peridental diseases. Office: Dept Microbiology and Immunology La State U Med Ctr 1901 Perdido St New Orleans LA 70112

THOMPSON, LARRY CLARK, chemistry educator; b. Hoquiam, Wash., June 13, 1935; s. Lester Clark and Elizabeth (Young) T.; m. Frances Eula Dressel, Sept. 11, 1955; children: Martha Elaine, Whitney Kathryn. B.S., Willamette U., 1957; M.S., U. Ill., Ph.D., 1960. Asst. prof. chemistry U. Minn., Duluth, 1960-63, assoc. prof., 1963-68, prof., 1968—, head dept., 1972—; vis. prof. U. Sao Paulo, Brazil, 1969, Fed. U. Geara, Fortuleza, Brazil, 1973, 74, Fed. U. Pernambuco, Recife, Brazil, 1977. Contbr. articles to profl. jours. Grantee NSF, NIH, OAS. Mem. Am. Chem. Soc. (local chmn.), Sigma Xi. Subspecialty: Inorganic chemistry. Current work: Chemistry of rare earth elements. Home: 301 W Oxford St Duluth MN 55803 Office: Dept Chemistry U Minn Duluth MN 55812

THOMPSON, LOREN B., nuclear engineer, researcher; b. Portland, Mar. 9, 1941; s. L.B. and Dorothy (Ellison) T. B.S., Harvey Mudd Coll., 1964; Ph.D., U. Utah, 1974. Sr. engr. Idaho Nat. Engring. Lab., Idaho Falls, 1974-77; project mgr. U.S. Nuclear Regulatory Comm., Washington, 1977-80, Electric Power Research Inst., Palo Alto, 1980—. Contbr. articles to profl. jours. Mem. Am. Nuclear Soc. Subspecialty: Nuclear engineering. Office: Electric Power Research Inst 3412 Hillview Ave Palo Alto CA 94303

THOMPSON, MARY EILEEN, chemist, educator; b. Mpls., Dec. 21, 1928; d. Albert Charles and Mary Blanche (McAvoy) T. B.A., Coll. St. Catherine, 1953; M.S., U. Minn., Mpls., 1958; Ph.D., U. Calif., Berkeley, 1964. Joined Sisters of St. Joseph of Carondolet, Roman Catholic Ch., 1950; tchr. chemistry and math. Derham Hall High Sch., St. Paul, 1953-56, 58-9; instr. dept. chemistry lab. Coll. St. Catherine, 1953-56, instr., 1964-65, asst. prof. chemistry, 1965-69, assoc. prof., 1969-78, prof., 1978—, chairperson dept. chemistry, 1969—; cons. Center Ednl. Affairs, Argonne Nat. Lab., 1968-71. Research Corp. grantee, 1981. Mem. Am. Chem. Soc., Chem. Soc. London, AAAS, Midwestern Assn. Chemistry Teaching in Liberal Arts Colls., N.Y. Acad. Scis., Nat. Sci. Tchrs. Assn., Phi Beta Kappa, Sigma Xi. Subspecialties: Inorganic chemistry; Physical chemistry. Current work: Separation and characterizations of Cr(III) hydolytic polymers; oxidation-reduction kinetics of inorganic complexes; superoxo complexes of Co(III): synthesis, characterizations. Home: 1132 Grand Ave Saint Paul MN 55105 Office: 2004 Randolph Ave Saint Paul MN 55105

THOMPSON, RICHARD CLAUDE, chemistry educator; b. Kansas City, Mo., Mar. 12, 1939; s. Claude S. T.; m. Jeannette E. Thompson, Mar. 30, 1969. B.S., U. Chgo., 1961; Ph.D., U. Md., 1965. Resident research assoc. Argonne Nat. Lab. (Ill.), 1965-66; asst. prof. (Ill. Inst. Tech.), 1966-67; with U. Mo., 1967—, prof. chemistry, 1977—. Contbr. articles to profl. jours. Mem. Am. Chem. Soc., Sigma Xi, Alpha Chi Sigma. Subspecialties: Inorganic chemistry; Kinetics. Current work: Mechanisms of inorganic reactions.

THOMPSON, RICHARD FREDERICK, psychobiology educator; b. Portland, Oreg., Sept. 6, 1930; s. Frederick Albert and Margaret (St. Clair) T.; m. Judith K. Pedersen, June 22, 1960; children: Kathryn M., Elizabeth K., Virginia, St. Clair. B.A., Reed Coll., 1952; M.S., U. Wis.-Madison, 1953, Ph.D., 1956. Faculty mem. U. Oreg. Med. Sch., Portland, 1959-67, prof. med. psychology, 1965-67; prof. psychobiology U. Calif.-Irvine, 1967-73, 75-80; prof. psychology Harvard U., Cambridge, Mass., 1973-74; prof. psychology, human biology Stanford (Calif.) U., 1980—; mem. exec. com. Assembly of Behavioral and Social Sci., NRC, 1977—. Author: Foundations of Physiological Psychology, 1967; contbr. articles in field to profl. lit.; editor: Jour. Comparative and Physiological Psychology, 1981—; regional editor: Behavioral Brain Research, 1979—; assoc. editor: Ann. Revs. of Neurosci, 1980—. Recipient Research Scientist Career award NIMH, 1967-77; fellow Center for Advanced Studies in Behavioral Sci., Stanford U., 1978-79. Fellow Am. Psychol. Assn. (pres. Div. 6 1972, Disting. Sci. Contbn. award 1977); mem. Nat. Acad. Scis., Soc. Exptl. Psychologists, Psychonomic Soc. (gov. bd. 1972-77, chmn. 1976). Subspecialties: Psychobiology; Neurophysiology. Current work: Brain mechanisms of learning and memory. Office: Psychology Dept Stanford U Stanford CA 94305 Home: 1097 Cathcart Way Stanford CA 94305

THOMPSON, RODGER IRWIN, astrophysicist, educator, cons.; b. Texarkana, Tex., Aug. 9, 1944; s. William B. and Pearle J. (Goodman) R. S.B. in Physics, M.I.T., 1966, Ph.D., 1970. Asst. prof. optical sci. U. Ariz., Tucson, 1970-71, asst., then assoc. prof. astronomy, 1971-81, prof. astronomy, astronomer, 1981—. Contbr. articles and papers to sci. lit. Mem. curriculum com. Tucson Unified Sch. Dist., 1979—. NSF research grantee. Mem. Internat. Astron. Union, Am. Astron. Soc., Am. Phys. Soc. Subspecialties: Infrared optical astronomy; Theoretical astrophysics. Current work: Infrared astronomy applied to star formation; accretion disks in star formation. Office: Steward Obs U Ariz Tucson AZ 85721

THOMPSON, SAMUEL STANLEY, JR., plant pathologist; b. Monroe, La., Nov. 25, 1936; s. Samuel Stanley and Minnie Francis (Gossett) T.; m. Coleta Belinda Loper, Aug. 24, 1963; children: Belinda, Sharon, Theron. B.S. in Botany, La. Tech. U., 1959; M.S., Purdue U., 1961, Ph.D. in Plant Pathology, 1965. Extension plant pathology U. Ga., Tiffon, 1962—. Contbr. articles to profl. jours. Pres. Len Lastinger PTA, 1977-78. Mem. Am. Phytopath. Soc., Ga. Assn. Plant Pathologists, Nat. Assn. County Agts., Ga. Assn. County Agts. Club: Sertoma (Tifton) (pres. 1981-82). Subspecialty: Plant pathology. Current work: Extension education in diseases of peanuts and small grain. Office: PO Box 1209 Tifton GA 31794

THOMPSON, SYLVESTER, industrial mathematician; b. Osceola, Ark., Apr. 9, 1948; s. Raymond Alton and Ola Francis (Maynard) T.; m. Judy Faye Finch, Dec. 23, 1977; children: Michael Finch, David Eric. A.B. in Math, U. Mo.-Columbia, 1970, A.M., 1971, Ph.D., 1974. Asst. prof. U.S.C., 1974-76; prin. mathematician Babcock and Wilcox, Lynchburg, Va., 1976—. Contbr. articles to profl. publs. NSF fellow, 1970-74. Mem. Soc. Indsl. and Applied Math. (jour. referee 1978-79), Assn. Computing Machinery, Internat. Assn. Math. and Computers in Simulation, Internat. Assn. Math. Modelling. Unitarian. Subspecialties: Applied mathematics; Mathematical software. Current work: Numerical solution of ordinary differential equations; continuous simulation; mathematical modelling; numerical solution of hyperbolic partial differential equations. Home: Route 6 Box 132 Madison Heights VA 24572 Office: Babcock and Wilcox Co 3315 Old Forest Rd Lynchburg VA 24501

THOMPSON, THOMAS WILLIAM, astronomer; b. Canton, Ohio, May 25, 1936; s. Clifford Earl and Doris Marie (Flickinger) T.; m. Alicia Kathleen, July 16, 1966; children: Kimberly Robin, Marisa Lynn. B.S., Case Inst. Tech., 1958; M.S., Yale U., 1959; Ph.D., Cornell U., 1966. Research assoc. Arecibo Obs., Arecibo, P.R., 1966-69; mem. tech. staff Jet Propulsion Lab., Pasadena, Calif., 1969-76, scientist, 1982—; staff scientist Planetary Sci. Inst., Pasadena, 1977-81. Contbr. articles to profl. jours. Mem. Internat. Astron. Union, Am. Geophys. Union, IEEE, Sigma Xi. Subspecialties: Planetology; Remote sensing (geoscience). Current work: Radar astronomy; radar remote sensing. Home: 3043 Cloudcrest Rd La Crescanta CA 91214

THOMPSON, VINTON, biologist, educator, researcher; b. Mount Holly, N.J., July 24, 1947; s. Vinton and Marie (Coville) T.; m. Ruth Moscovitch, Oct. 5, 1975; 1 son, Isaiah. A.B., Harvard U., 1969; Ph.D., U. Chgo., 1974. Indsl. hygienist U.S. Dept. Labor, Chgo., 1975-77; instr. Loyola U., Chgo., 1979; asst. prof. biology Roosevelt U., Chgo., 1980—. Mem. Soc. Study Evolution, Genetics Soc. Am. Subspecialties: Evolutionary biology; Genome organization. Current work: Evolutionary genetics of drosophila, ecological genetics of spittlebugs. Home: 1747 W 21st St Chicago IL 60608 Office: Dept Biology Roosevelt U Chicago IL 60605

THOMPSON, WILLIAM JOSEPH, pharmacologist, educator; b. Indpls., Oct. 28, 1943; s. W. Ralph and Jane (Shelley) T.; m. Maryellen S. Thompson, Dec. 23, 1971; children: Kathryn A., Emily A. Ph.D., U. So. Calif., 1971. Postdoctoral fellow dept. medicine U. Wash., 1971-73; asst., then assoc. prof. dept. pharmacology U. Tex. Med. Sch., Houston, 1973-82, prof., 1982—. Contbr. numerous articles to profl. jours. Recipient award for paper in biochemistry, 1971. Subspecialties: Molecular pharmacology; Biochemistry (medicine). Current work: Regulation of cellular metabolism by Hommes insulin, cyclic AMP, catecholamines, phosphodiesterases, adenylyl cyclase, histamine, acid secretion. Home: 5506 Caversham St Houston TX 77096 Office: Dept Pharmacology U Tex Med Sch PO Box 20708 Houston TX 77030

THOMPSON, WILLIAM WARREN, educator, educationalpsychology consultant; b. Londonderry, No. Ireland, Nov. 8, 1944; emigrated to Can., 1977; s. John and Ivy (Fulton) T.; m. Rebecca Louise Chapman, June 29, 1974; 1 dau., Stephanie Sara. B.A., Trinity Coll., Dublin, 1966, M.A., 1969; M.Ed., Queens U., Belfast, No. Ireland, 1968, Ph.D., 1971. Registered psychologist, N.S., N.B. Sch. psychologist County Derry Schs., Londonderry, 1971-72, Argyllshire Schs., Dunoon, 1972-73, Isle of Wight (U.K.) Schs., 1973-74; sr. clin. psychologist Wessex Unit, Portsmouth, Eng., 1974-77; dir. pupil evaluation Sch. Dist. 20, Saint John, N.B., Can., 1977-80; assoc. prof. Mt. Saint Vincent U., Halifax, N.S., Can., 1980—, coordinator sch. psychology, 1982—. Author: Children and Their Problems, 1977; Contbr. chpts. to books, articles to profl. jours. Mem. com. Saint John Family Services, 1978. Mem. Brit. Psychol. Soc., Can. Psychol. Assn., Am. Psychol. Assn., Nat. Assn. Sch. Psychologists, Assn. Psychologists of N.S. (exec. sec. 1980—). Conservative. Presbyterian. Subspecialties: Behavioral psychology; School psychology. Current work: Environmental influences on educational performance; effective school psychology delivery systems; behavioral approaches to the treatment of teenage obesity. Home: Box 505 Waverley NS Canada BON 2SO Office: Mount Saint Vincent Univ Bedford Hwy Halifax NS Canada B3M 2S6

THOMPSON, WILMER LEIGH, pharmacologist, critical care physician; b. Shreveport, La., June 25, 1938; s. Wilmer Leigh and Mary Bissell (McIver) T.; m. Maurice Eugenie Horne, Mar 29, 1957; 1 dau., Mary Linton Bounetheau. B.S., Coll. Charleston, S.C., 1958; M.S., Med. U. S.C., 1960, Ph.D. in Pharmacology, 1963; M.D., Johns Hopkins U., 1965. Diplomate: Am. Bd. Internal Medicine. Osler med. resident Johns Hopkins U., Balt., 1965-70; staff assoc. NIH, Bethesda, Md., 1967-69; asst. prof. medicine and pharmacology Johns Hopkins U., 1970-74; also dir. med. critical care Johns Hopkins Hosp.; prof. medicine, assoc. prof. pharmacology, dir. clin. pharmacology, dir. critical care and drug info. Case Western Res. U and Univ. Hosps. Cleve., 1974-82; exec. dir. Lilly Research Labs., Eli Lilly & Co., Indpls., 1982—. Contbr. 165 articles to profl. jours. Served as sr. surgeon USPHS, 1967-69. Burroughs Wellcome scholar, 1975-80; Pharm. Mfrs. Assn. faculty awardee, 1971-74; Sir Henry Hallett Dale lectr. Johns Hopkins U., 1981; Frohlich vis. prof. Royal Soc. Medicine Found., London, 1981. Fellow ACP; mem. Soc. Critical Care Medicine (pres. 1982), Central Soc. Clin. Research Am. Soc. Pharmacology and Exptl. Therapeutics, Am. Soc. Clin. Pharmacology and Therapeutics, Council Clin. Pharmacology, Univ. Assn. Emergency Physicians, Phi Beta Kappa, Alpha Omega Alpha. Episcopalian. Subspecialties: Pharmacology; Internal medicine. Current work: Drug development. Developed hydroxyethyl starch, an artificial blood; tapmot, an emetic for poisoning. Home: 1044 Indian Pipe Circle Carmel IN 46032 Office: 307 E McCarty St Indianapolis IN 46285

THOMSEN, JOHN STEARNS, physicist; b. Balt., June 10, 1921; s. Herman Ivah and Alice Wellington (Sawyer) T.; m. Helen Calvert Steuart, Feb. 2, 1952; children: Mary H. Thomsen Davisson, Steuart H., Alice Thomsen Bockman, J. Marshall. B.E., Johns Hopkins U., 1943, Ph.D. in Physics, 1952. Asst. prof. Stevens Inst. Tech., 1953-54; research scientist Johns Hopkins U., 1954-55, asst. prof. mechanics, 1955-61, research scientist, 1961-69, fellow-by-courtesy, 1970—; com. fundamental constants Nat. Acad. Scis.-NRC, 1961-72, chmn., 1970. Contbr. articles profl. jours., chpt. in book. Alt. Republican Nat. Conv., 1976; treas. Balt. City Rep. Central Com., 1978-82; mem. Balt. City Community Relations Commn., 1965-75, chmn., 1973-75; pres. Balt. Streetcar Mus., 1966-70. Fellow Am. Phys. Soc.; mem. Am. Assn. Physics Tchrs. Episcopalian. Club: Johns Hopkins. Subspecialty: Atomic and molecular physics. Current work: Study of x-ray line widths and wavelengths. Home: 4710 Keswick Rd Baltimore MD 21210 Office: Dept Physic Johns Hopkins U Baltimore MD 21218

THOMSON, DAVID MARSHALL PARKS, physician, researcher, educator; b. Windsor, Ont., Can., Nov. 4, 1939; s. Marshall Clinton and Doris Catherine (Parks) T.; m. Janice Elaine, Dec. 2, 1938; children: Tracey Jane, Sally Doris. M.D., U. Western Ont., London, 1964; Ph.D., U. London, 1973. Intern Montreal (Que., Can.) Gen. Hosp., 1964-65; trainee in clin. immunology and medicine McGill U.-Montreal Gen. Hosp., 1965-69; Med. Research Council Can. fellow Chester Beatty Cancer Inst., Sutton, Eng., 1969-72; asst. prof. medicine McGill U., Montreal, 1972-77, assoc. prof, 1977-81, prof., 1981—; cons. physician, tchr., cancer researcher Montreal Gen. Hosp., 1972—. Med. Research Council Can. scholar, 1972-77. Fellow Royal Coll. Physicians and Surgeons Can. (medalist in medicine 1971); mem. Am. Assn. Clin. Immunology and Allergy, Am. Assn. Cancer Research, Am. Assn. Immunology, Can. Soc. Clin. Investigation. Subspecialties: Allergy; Cancer research (medicine). Current work: Cancer Immunology; patentee carcinoembryonic antigen. Patentee radioimmunoassay, tube leukocyte adherence inhibition assay for breast cancer. Office: 1650 Cedar Ave Room 7113 Montreal PQ Canada H3G 1A4

THOMSON, DONALD ARTHUR, ecology and evolutionary biology educator, curator; b. Detroit, Apr. 9, 1932; s. Arthur and Theresa Rita (Stasin) T.; m. Jenean Gruner, Apr. 6, 1957; children: Erin, Kurt, Lisa, Madelon. B.S., U. Mich., 1955, M.S., 1957; Ph.D., U. Hawaii, 1963. Fishery technician Inst. Fishery Research, Ann Arbor, 1954-55;

research asst. U. Mich., 1956-57; fishery research biologist U.S. Bur. Comml. Fisheries, Honolulu, 1957-61; research asst. U. Hawaii Marine Lab., Honolulu, 1962-63; asst. prof., then assoc. prof. ecology and evolutionary biology U. Ariz., Tucson, 1963-77, prof., 1978—; chief scientist R/V Te Vega, Stanford U., 1967; chmn. marine scis. program. U.-Ariz., 1973—. Author: Reef Fishes of the Sea of Cortez, 1979, Tide Calendar for the Northern Gulf of California, 1967—. Vice pres. Ariz. Consumers Council, 1968-70. Recipient research contracts Office Naval Research, 1965-69; research grantee NSF, 1968-70. Mem. Am. Soc. Icthyologists and Herpetologists, Assn. Systematics Collections (U. Ariz. rep.). Democrat. Subspecialties: Ecology; Behavioral ecology. Current work: Community ecology of Gulf of California fishes. Home: 3150 W Tucana St Tucson AZ 85745 Office: Dept of Ecology and Evolutionary Biology U Ariz Tucson AZ 85721

THOMSON, GEORGE WILLIS, forestry educator, researcher; b. Seward, Ill., July 10, 1921; m., 1945; 3 children. B.S., Iowa State U., 1943, M.S., 1947, Ph.D. in Silviculture and Soils, 1956. Instr. in gen forestry Iowa State U., 1947-52, asst. prof. forest mgmt., 1952-56; assoc. prof. 1956-60, prof. mensuration and photogrammetry, 1960—, acting head dept. forestry, 1967, 75, chmn. dept., 1975—. Fellow Soc. Am. Foresters; mem. Soc. Range Mgmt., Am. Soc. Photogrammetry. Subspecialties: Remote sensing (geoscience); Forestry. Office: Dept Forestry Iowa State U Ames IA 50011

THONNARD, NORBERT, physicist; b. Berlin, Jan. 22, 1943; U.S., 1958, naturalized, 1964; s. Ernst and Costanza (Marotti) T.; m. Roslyn Dee Oglesby, Apr. 19, 1964; children: Stefan, Paul, Janeen, Deanna. B.A., Fla. State U., 1964; M.S., U. Ky., 1969, Ph.D., 1971. Physicist U.S. Army Engr. Research/Devel. Labs., Ft. Belvoir, Va., 1964-66; teaching asst. dept. physics/astronomy U. Ky., Lexington, 1966-67, research assoc., 1967-70; postdoctoral fellow Carnegie Instn. Washington, 1970-72, staff mem. dept. terrestial magnetism, 1972—; cons. Battelle Pacific N.W. Labs., Richland, Wash., 1972-75; mem. users com. Nat. Radio Astronomy Obs., 1978-81. Contbr. articles to profl. jours. Mem. Am. Phys. Soc., Am. Astron. Soc., Internat. Astron. Union, AAAS. Democrat. Roman Catholic. Subspecialties: Radio and microwave astronomy; Optical astronomy. Current work: Study of systematic properties of spiral galaxies as a function of Hubble type and luminosity using both radio and optical astron. techniques. Office: 5241 Broad Branch Rd NW Washington DC 20015 Home: 1909 Gold Mine Rd Brookeville MD 20833

THONNEY, MICHAEL LARRY, animal science educator; b. Moscow, Idaho, June 2, 1949; s. Larry R. and Phyllis (Webster) T.; m. Patricia E. Foster, June 2, 1979. B.S., Wash. State U., 1971; M.S., U. Minn., 1973, Ph.D., 1975. Asst. prof. animal sci. Cornell U., 1975-81, assoc. prof., 1981—. Mem. Am. Soc. Animal Sci., Nutrition Soc., Brit. Soc. Animal Prodn., Am. Meat Sci. Assn., Phi Kappa Phi. Subspecialties: Animal nutrition; Developmental biology. Current work: Control and quantifying growth of domestic animals; growth, cattle, sheep, animal efficiency. Office: Cornell U 114 Morrison Hall Ithaca NY 14853

THORMAR, HALLDOR, virologist; b. Iceland, Mar. 9, 1929; came to U.S., 1967; s. Thorvardur and Olina Marta Jonsdottir; m. Lilja Asdis, Dec. 25, 1962; children: Sigridur, Asdis Birna, Olina Marta. Mag. Scient. in Cell Physiology, U. Copenhagen, 1956, Dr. phil. in Virology, 1966. Research scientist Inst. Exptl. Pathology, U. Iceland, Reykjavik, 1958-60, 62-65, 66-67, Statens Serum Inst., Copenhagen, 1960-62, Centro do Virologia, IVIC, Caracas, Venezuela, 1965-66; chief research scientist Inst. Basic Research in Mental Retardation, S.I., N.Y., 1967—. Contbr. numerous articles on cell biology and virology to profl. jours. Mem. Am. Soc. Microbiology, N.Y. Acad. Scis. Subspecialty: Virology (biology). Current work: Pathogenesis of viral infections of the central nervous system. Home: 8 Galewood Dr Holmdel NJ 07733 Office: 1050 Forest Hill Rd Staten Island NY 10314

THORNBER, JAMES PHILIP, biochemist; b. Yorkshire, Eng., Dec. 22, 1934; came to U.S., 1967; s. James Robert and Ethel (Greenwood) T.; m.; children: Karen, Emma. B.A., U. Cambridge, Eng., 1958, M.A., 1961, Ph.D., 1962. Sci. officer plant biochemistry sect. Twyford Labs. (subs. A. Guinness & Sons, Ltd.), London, 1961-67; from research assoc. to asst. scientist biology dept. Brookhaven Nat. Lab., Upton, N.Y., 1967-70; asst. prof. biology UCLA, 1970-72, assoc. prof., 1972-75, prof., 1975—, chmn. dept. biology. Contbr. articles on biochemistry to profl. jours. NSF grantee, 1971—; Guggenheim Found. fellow, 1976-77; Dept. Agr. grantee, 1978—; NSF panel mem., 1982—. Mem. Am. Soc. Biol. Chemists, Am. Soc. Photobiology, Am. Soc. Plant Physiologists. Subspecialties: Photosynthesis; Biochemistry (biology). Current work: The composition, structure and biosynthesis of the photosynthetic unit in higher plants, algae and bacteria, the molecular organization of chlorophyll in vivo, the mechanism of the primary photochemical event in photosynthesis. Home: 2337 Veteran Ave Los Angeles CA 90064 Office: Dept Biolog UCLA 2203 Life Scis Los Angeles CA 90024

THORNE, KIP STEPHEN, physicist, educator; b. Logan, Utah, June 1, 1940; s. David Wynne and Alison (Comish) T.; m. Linda Jeanne Peterson, Sept. 12, 1960 (div. 1977); children: Kares Anne, Bret Carter. B.S. in Physics, Calif. Inst. Tech., 1962; A.M. in Physics (Woodrow Wilson fellow, Danforth Found. fellow), Princeton U., 1963; Ph.D. in Physics (Danforth Found. fellow, NSF fellow), Princeton U., 1965; postgrad. (NSF postdoctoral fellow), Princeton U., 1965-66; D.Sc. (hon.), Ill. Coll., 1979; Dr.h.c., Moscow U., 1981. Research fellow Calif. Inst. Tech., 1966-67, asso. prof. theoretical physics, 1967-70, prof., 1970—, William R. Kenan, Jr. prof., 1981—; Fulbright lectr., France, 1966; vis. asso. prof. U. Chgo., 1968; vis. prof. Moscow U., 1969, 75, 78, 81, 83; vis. sr. research asso. Cornell U., 1977; adj. prof. U. Utah, 1971—; mem. Internat. Com. on Gen. Relativity and Gravitation, 1971-80, Com. on U.S.-USSR Coop. in Physics, 1978-79, Space Sci. Bd., NASA, 1980-83. Co-author: Gravitation Theory and Gravitational Collapse, 1965, Gravitation, 1973. Alfred P. Sloan Found. Research fellow, 1966-68; John Simon Guggenheim fellow, 1967; recipient Sci. Writing award in physics and astronomy Am. Inst. Physics-U.S. Steel Found., 1969. Mem. Nat. Acad. Scis., Am. Acad. Arts and Scis., Am. Astron. Soc., Am. Phys. Soc., Internat. Astron. Union, AAAS, Sigma Xi, Tau Beta Pi. Subspecialties: Theoretical physics; Theoretical astrophysics. Current work: Research in theoretical physics and astrophysics with emphasis on gravitation, black holes, gravitational waves, experimental tests of general relativity and quantum mechanical aspects of high-precision measurements. Research in theoretical physics and astrophysics. Office: 130-33 Calif Inst Tech Pasadena CA 91125

THORNTON, GUNNAR, engineering consultant, nuclear engineer; b. Oppegaard, Norway, Apr. 8, 1920; came to U.S., 1922, naturalized, 1937; s. Jens and Ingeborg (Herstad) T.; m. Nancy Alice Williamson, Nov. 22, 1948; children: Leslie Ann, Eleanor Jean, John Lauritz, Douglas Gunnar. B.S. in Engring, Tufts U., 1947; M.A. in Physics, Harvard U., 1948. Engr. U. Calif., Los Alamos, N.Mex., 1944-46; engr. physicist Fairchild Co., Oak Ridge, 1948-51; corp. engring. cons. Gen. Electric Co., Schenectady, N.Y., 1965—, engring. mgr., Evendale, Ohio, 1951-65. Author: (with D.C. Layman) Remote Handling of Mobile Nuclear Systems, 1965. Served with C.E. U.S. Army, 1942-46. Mem. Am. Nuclear Soc. Current work: Corporate staff review of major technical projects; aerospace nuclear propulsion. Office: 1 River Rd Bldg 36 517 Schenectady NY 12345 Home: 1017 Onondaga Rd Schenectady NY 12309

THORNTON, H. RICHARD, mech. engr., educator, cons., researcher; b. Van Etten, N.Y., Nov. 15, 1932; s. Hubert Fred and Gertrude Alvina T.; m. Shirley Ann, June 27, 1959; children: Cheryl Lynn, Scott Richard. B.S., Alfred U., 1954, M.S. in Glass Tech, 1957; Ph.D. in Ceramic Engring, U. Ill., 1963. Registered profl. engr., Tex., 1970. Research engr. Nat. Bur. Standards, 1956-59; research assoc. U. Ill., 1959-63; project structures engr. Gen. Dynamics, Ft. Worth, 1963-67; assoc. prof. mech. engring. Tex. A&M U., 1967-77, prof., 1977—; cons. Contbr. articles to profl. publs. Mgr., pres. Little League, Bryan, Tex.; pres. PTA, Bryan; chmn. Forum for Christian Studies, Bryan. Recipient Adams award Am. Welding Soc., 1974. Fellow Am. Inst. Chemists; mem. Nat. Soc. Profl. Engrs., Tex. Soc. Profl. Engrs., Am. Ceramic Soc., ASME (assoc. editor transactions 1974-78), Am. Soc. Metals, Soc. Materials and Process Engrs., Keramos, Sigma Xi, Phi Kappa Phi, Pi Tau Sigma. Mem. United Ch. of Christ. Club: Briarcrest Country (Bryan). Subspecialties: Ceramics; Composite materials. Current work: Research in friction and wear of threaded joints, acoustic emission within aluminum during deformation, hydrogen-metal systems. Home: 2505 Willow Bend Dr Bryan TX 77801 Office: Tex A&M U College Station TX 77843

THORNTON, J(OSEPH) SCOTT, JR., materials scientist, research inst. exec.; b. Sewickley, Pa., Feb. 6, 1936; s. Joseph Scott and Evelyn G. (Miller) T.; m. Terry Ray, Mar. 12, 1980; children: J. Scott III, Chris P. B.S.M.E., U. Tex.-Austin, 1957, Ph.D. (Alcoa fellow 1964, R.C. Baker Found. fellow 1967), 1969; M.S., Carnegie-Mellon U., 1962. Stress analyst Walworth Valve Co., Boston, 1958; research asst. Carnegie Inst. Tech., Pitts., 1958-61; materials engr. Westinghouse Astronuclear Lab., Pitts., 1961-64; instr. U. Tex., Austin, 1964-67; mgr. sect. advanced materials TRACOR Inc., Austin, 1967-69, dir. applied scis., 1973-75; mgr. dept. materials Horizons Research Inc., Cleve., 1963-73; pres., tech. dir. Tex. Research Inst. Inc., Austin, 1975—. Contbr. numerous articles to profl. jours. Mem. ASTM, ASME, Am. Soc. Metals. Subspecialties: Materials; Materials (engineering). Current work: Accelerated Testing and Life Prediction; relating fundamental degradation mechanisms in materials to reliability of components in service. Patentee in field. Office: 9063 Bee Cave Rd Austin TX 78746

THORSON, ROBERT MARK, geology educator, geologist; b. Edgerton, Wis., Oct. 6, 1951; s. Theodore W. and Margaret (Andersen) T.; m. Kristine Hoy, Aug. 21, 1977; children: Karsten, Adam. B.S. summa cum laude, Bemidji State U., 1973; M.S., U. Alaska-Fairbanks, 1975; Ph.D., U. Wash., Seattle, 1979. Geologist Arctic Environ. Project, U.S. Geol. Survey, Menlo Park, Calif., 1975-76, Puget Sound ESA Project, Seattle, 1976-81; asst. prof. geology U. Wis., Oshkosh, 1979-80; asst. prof. geology, geology/geophysics program U. Alaska, Fairbanks, 1980—, chmn. office quaternary studies, 1981—, dir., 1982—; head surficial geology U. Alaska Mus., Fairbanks, 1982—. Pres. Unitarian-Universalist Assn. Fairbanks, 1982. Alaska Council on Sci. and Tech. grantee, 1982. Mem. AAAS, Geol. Soc. Am. (Penrose Bequest grant 1974), Am. Quarternary Assn., Soc. Archeol. Scis., Alaska Geol. Soc., Alaska Quarternary Group. Unitarian/Universalist. Club: Scandinavian Fraternity Am. Subspecialties: Geology; Geological hazards. Current work: Arctic and subarctic geomorphology, quaternary paleogeography, paleoclimatology and paleoenvironments, volcanic hazards research, environmental problems associated with permafrost, early man chronology and stratigraphy, glacier processes and hazards. Home: Ithaca Rd University Heights Fairbanks AK 99701 Office: Geology/Geophysics Program U Alaska Fairbanks AK 99701

THOURET, WOLFGANG EMERY, physicist; b. Berlin, Germany, Aug. 27, 1914; s. Nicholas Leonard and Julie Emily (Schwab) T.; m. Marianne Margaret Conrad, Dec. 20, 1962. M.S. in Physics, Tech. U. Berlin, 1936; Ph.D., Tech. U. Karlsruhe, W.Ger., 1952. Physicist research dept. Osram Corp., Berlin, Germany, 1936-48; physicist Quartzlamp Co., Hanau, W.Ger., 1942-52; light source devel. engr. Westinghouse Electric Corp., Bloomfield, N.J., 1952-57; with Duro-Test Corp., North Bergen, N.J., 1957—, dir. engring. and research, 1962—; mem. com. C78 on electric lamps Nat. Standards Inst. Contbr. articles to profl. jours. Fellow Illuminating Engring. Soc. N.Am.; mem. Illuminating Engring. Soc. Gt. Britain, Illuminating Engring. Soc. Germany, Am. Phys. Soc., Optical Soc. Am., IEEE, Inst. Environ. Scis., Soc. Motion Picture and TV Engrs., AAAS, N.Y. Acad. Scis. Subspecialties: Electrical engineering; Plasma physics. Current work: Technology of light and radiation sources. Patentee in field of light and radiation sources. Home: Claridge House I Verona NJ 07044 Office: 2321 Kennedy Blvd North Bergen NJ 07047

THRELFALL, WALTER RONALD, veterinarian, theriogenologist; b. Marion, Ohio, Oct. 12, 1944; s. Walter and Ruth Alma (Fox) T.; m. Barbara Ann James, June 25, 1966; children: Ronda Ann, Deanna Dawn, Andrea Jo. D.V.M., Ohio State U., Columbus, 1968; M.S., U. Mo., Columbia, 1970, Ph.D., 1972. Cert. Am. Coll. Theriogenologists. Instr. U. Mo., 1968-70, resident, 1970-72; asst. prof. vet. medicine Ohio State U., 1972-77, assoc. prof., 1977—; cons., researcher, clinician. Subspecialties: Embryo transplants; Endocrinology. Current work: Embryo, hormone, endocrines, antibiotics, uterine infections, contraceptives, infertility, testicular biopsies, vasectomies. Home: 7012 Liberty Rd Powell OH 43065 Office: 1935 Coffey Rd Columbus OH 43210

THRODAHL, MONTE CORDEN, chemical company executive; b. Mpls., Mar. 25, 1919; s. Monte Conrad and Hilda (Larson) T.; m. Josephine Crandall, Nov. 6, 1948; children: Mark Crandall, Peter Douglas. B.S., Iowa State U., 1941. With Monsanto Co., St. Louis, 1941—, gen. mgr. internat. div., 1964-66, v.p., 1965—, dir., 1966—, group v.p. tech., 1974—, sr. v.p., 1979—; v.p., dir. Monsanto Research Corp.; dir. Boatmen's Nat. Bank. Bd. dirs., mem. exec. com. Webster Coll. Fellow Am. Inst. Chemists, Am. Inst. Chem. Engrs.; mem. Am. Chem. Soc., Am. Inst. Chem. Engrs., Nat. Acad. Engring., Comml. Devel. Assn., AAAS, Soc. Chem. Industry, Alpha Chi Sigma. Clubs: Old Warson Country, St. Louis (St. Louis). Subspecialty: Chemical research manangement. Home: No 36 Briarcliff Ladue MO 63124 Office: 800 N Lindbergh Blvd Saint Louis MO 63167

THRONER, GUY CHARLES, ordnance technology research administrator; b. Mpls., Sept. 14, 1919; m., 1943; 3 children. A.B., Oberlin Coll., 1943; postgrad. Grad. Sch. Mgmt., UCLA, 1962. Head explosive ORD, explosive ballistics U.S. Naval Weapon Ctr., 1937-53; Mgr. ordnance div. Aerojet Gen. Corp., 1953-63, mgr. research and devel. tactical weapon systems div., 1963-64; v.p., gen. mgr. Def. Tech. Labs. and Steel Products div. FMC Corp., 1964-74; dir. research and devel. Vacu-Blast & TronicCorp., 1976-78; v.p. engring. and mfg. Dahlman Inc., 1978-79; sect. mgr. research and devel. Columbus (Ohio) Labs, Battelle Meml. Inst., beginning 1979, now sect. mgr. ordnance systems and tech.; exec. v.p. research and devel. Am. Vidionetics Corp., 1976-78. Mem. Air Armament Bd., Bomb and Warhead Steering Com., Underwater Weapons Steering Com. Mem. Am. Def. Preparedness Assn., Sigma Xi. Subspecialty: Ordnance technology. Office: Columbus Labs Battelle Meml Inst 505 King Ave Columbus OH 43201

THUAN, TRINH XUAN, astrophysicist; b. Hanoi, Vietnam, Aug. 20, 1948; s. Trinh Xuan and Le Thi (Nghia) Ngan. B.S. in Physics, Calif. Inst. Teh., Pasadena, 1970; Ph.D. in Astrophysics, Princeton U., 1974. Research fellow in astronomy Calif. Inst. Tech. and Mt. Wilson and Las Campanas Obs., 1974-76; assoc. prof. astronomy U. Va., Charlottesville, 1976—; vis. lectr. Institut d'Astrophysique, Paris, 1978, Centre d'Etudes Nucleaires de Saclay, France, 1981-82. Contbr. articles profl. jours. Mem. Am. Astro. Soc., Internat. Astron. Union. Subspecialties: Cosmology; Infrared optical astronomy. Current work: formation and evolution of galaxies, observational cosmology. Office: PO Box 3818 University Station Charlottesville VA 22903

THUESON, DAVID OREL, med. researcher, educator; b. Twin Falls, Idaho, May 9, 1947; s. Orel Grover and Shirley Jean (Archer) T.; m. Sherrie Linn Lowe, June 14, 1969; children: Sean David, Kirsten Marie, Eric Michael, Ryan Paul, Todd Alan. B.S., Brigham Young U., 1971; Ph.D., U. Utah, 1976. Asst. prof. medicine and pharmacology U. Tex. Med. Br., Galveston, 1978—. Contbr. articles to profl. jours. NIH awardee, 1978-81. Mem. Am. Acad. Allergy, Am. Assn. Immunology. Subspecialties: Immunopharmacology; Allergy. Current work: Immunopharmacology, cellular immunity, delayed hypersensitivity. Home: 7306 Jones Dr Galveston TX 77551 Office: Dept Internal Medicin U Tex Med Br Galveston TX 77550

THURBER, ROBERT EUGENE, physiology educator, researcher, cons.; b. Bay Shore, N.Y., Oct. 11, 1932; s. Hallett Eliot and Mary Jean (Winkler) T.; m. Barbara Meyer, July 27, 1953 (div. July 1982); children: Robert, Joseph, Karl, Michael. B.S., Holy Cross Coll., 1954; M.S., Adelphi Coll, 1961; Ph.D., U. Kans., 1964. Research assoc. Brookhaven Nat. Lab., Upton, N.Y., 1956-61; asst. prof. Med. Coll. Va., Richmond, 1964-69; assoc. prof. Jefferson Med. Coll., Phila., 1969-70; prof. and chmn. dept. physiology East Carolina U., Greenville, 1970—; cons. NASA, Langley, Va., 1966-70, Va. Bd. Med. Examiners, Portsmouth, 1966-70, Psychol. Cons., Inc., Richmond, 1968-71; Cherry Hosp., Goldsboro, N.C., 1972-75. Pres. Am. Heart Assn., N.C. affiliate, Chapel Hill, 1978, chmn. Mid-Atlantic research rev. com., 1982-83, mem. regional nat. research com., Dallas, 1982-83. Served with U.S. Army, 1954-56; Korea. NIH fellow, 1962; NASA fellow, 1967; NEH fellow, 1978. Mem. Am. Physiol. Soc., Assn. Chmn. Depts. Physiology, Soc. Health and Human Values, N.Y. Acad. Scis., Sigma Xi. Democrat. Roman Catholic. Club: Greenville (N.C.) Country. Subspecialties: Physiology (medicine); Biophysics (physics). Current work: Non-equilibrium thermodynamics applied to material transfer across cell membranes. Office: Physiology Dept East Carolina U Sch Medicine Brody Bldg Greenville NC 27834

THURMAN, LLOY DUANE, biology educator; b. Oconto, Nebr., Sept 3, 1933; s. Jesse Lloy and Mable Ione (Parrish) T.; m. Barbara Joan Patterson, May 19, 1957; children: Sonya, Daniel, Linda, Bryan. B.S., U. Nebr., 1959, M.S., 1961; Ph.D., U. Calif., Berkeley, 1966. Asst. prof. biology So. Calif. Coll., Costa Mesa, 1965-67; mem. faculty Oral Roberts U., Tulsa, 1967—, prof. biology, 1973—, chmn. dept. natural scis., 1969-73, chief health professions advisor, 1969—; cons. in field. Author: How To Think About Evolution, 1978; contbr. articles to profl. publs. Served with USAF, 1953-57. Arthur W. Sampson fellow, 1959-61. Mem. Ecol. Soc. Am., Am. Sci. Affiliation, Assn. for Computing in Math and Sci., Soc. Coll. Sci. Tchrs. Republican. Methodist. Subspecialties: Nutrition (biology); Psychobiology. Current work: Human nutrition as affected by beliefs, personality, culture and amount of science training. Office: Oral Roberts Univ Natural Scis Dept Tulsa OK 74171

THURMAN, RONALD GLENN, pharmacologist, consultant, educator; b. Carbondale, Ill., Nov. 25, 1941; s. Glenn and Melba T. T. B.S. in Pharmacy, St. Louis Coll. Pharmacy, 1963; Ph.D. in Pharmacology, U. Ill., 1968. Postdoctoral fellow dept. biophysics and phys. biochemistry Johnson Research Found., U. Pa., 1968-69, asst. prof. biochemistry and biophysics, 1972-77; postdoctoral fellow Inst. Physiol. Chemistry and Phys. Biochemistry, U. Munich, W.Ger., 1969-72; assoc. prof. pharmacology U. N.C., Chapel Hill, 1977-82, prof., 1982—; reviewer, cons. Editor: (with Williamson, Drott and Chance) Alcohol and Aldehyde Metabolizing Systems, Vol. I, 1974, Vol. II, 1977, Vol. III, 1977, Vol. IV, 1980; contbr. numerous articles to sci. jours. Recipient Research Career Devel. award Nat. Inst. Alcohol Abuse and Alcoholism, 1973-83; Alexander von Humboldt postdoctoral fellow, 1970-71. Mem. Soc. Toxicology, Research Soc. on Alcoholism, Internat. Soc. Biomed. Research on Alcoholism, N.Y. Acad. Scis., Am. Soc. Biol. Chemists, Am. Soc. Pharmacology and Exptl. Therapeutics, Internat. Council Alcohol and Addictions, Argentine Soc. Pathology (life). Subspecialty: Pharmacology. Current work: Regulation of hepatic mixed-function oxidation, methods development for carcinogen metabolites, control of hepatic ethanol metabolism, drug and alcohol metabolism. Home: Route 2 Box 735A Chapel Hill NC 27514 Office: University of North Carolina 1124 Faculty Lab Office Bldg Chapel Hill NC 27514

THURSTON, WILLIAM P., mathematician, educator; b. Washington, Oct. 30, 1946; m. (married); 3 children. B.A., New Coll., Sarasota, Fla., 1967; Ph.D., U. Calif.-Berkeley. With Inst. Advanced Study, 1972-73; asst. prof. MIT, Cambridge, 1973-74; prof. math. Princeton (N.J.) U., 1974—. Recipient Veblen prize in geometry, 1976, Waterman award, 1979, Fields medal, 1982; Alfred P. Sloan Found. fellow, 1974-75. Office: Dept Math Princeton U Princeton NJ 08544

TIDWELL, TIMOTHY EUGENE, plant pathologist, researcher; b. Alameda, Calif., Dec. 30, 1949; s. Eugene Delbert and Golda May (Coldwell) T.; m. Lynette Claire Petersen, July 10, 1976; children: Andrea Lynn, Alissa Kay. A.B., Calif. State U.-Sacramento, 1973; M.S., U. Calif.-Davis, 1977. Postgrad. research plant pathology dept. plant pathology U. Calif.-Davis, 1977; clin. plant pathologist Calif. Dept. Food and Agr., Sacramento, 1977—. Contbr. articles to sci., popular publs.; plant pathology editor Calif. Plant Pest and Disease Report, 1980—. Recipient cert. outstanding achievement Calif. Dept. Food and Agr., 1980. Mem. Am. Phytopathol. Soc., Am. Hort. Soc. Democrat. Subspecialties: Plant pathology; Integrated pest management. Current work: Diseases of Cucurbits, shade trees and Conifers; plant disease diagnosis and research. Office: 1220 N St Room 340 Sacramento CA 95814

TIEMAN, SUZANNAH BLISS, neurobiologist; b. Washington, Oct. 10, 1943; d. John Alden and Winifred Texas (Bell) Bliss; m. David

George Tieman, Dec. 19, 1969. A.B., Cornell U., 1965; postgrad., MIT, 1966-67, Calif. Inst. Tech., 1971-72; Ph.D., Stanford U., 1974. Nat. Eye Inst. postdoctoral fellow U. Calif., San Francisco, 1974-77; research assoc., adj. asst. prof. Neurobiology Research Ctr., SUNY-Albany, 1977—. Contbr. articles to profl. jours. NSF fellow, 1970-73; NIH fellow, 1973-74; Nat. Eye Inst. grantee, 1979-82. Mem. AAAS, Am. Assn. Aatomists, Assn. Research in Vision and Ophthalmology, Assn. Women in Sci., Soc. Neurosci. Women in Eye Research, Women in Neurosci. Subspecialties: Neurobiology; Sensory processes. Current work: anatomical and behavioral studies of visual system development. Office: Neurobiology Research Center 1400 Washington Ave Albany NY 12222

TIEN, CHANG LIN, engineering educator; b. Wuhan, China, July 24, 1935; came to U.S., 1956, naturalized, 1969; s. Yun Chien and Yun Di (Lee) T.; m. Di Hwa Liu, July 25, 1959; children: Norman Chihnan, Phyllis Chihping, Christine Chihyih. B.S., Nat. Taiwan U., 1955; M.M.E., U. Louisville, 1957; M.A., Ph.D., Princeton U., 1959. Acting asst. prof. dept. mech. engring. U. Calif.—Berkeley, Berkeley, 1959-60, asst. prof., 1960-64, asso. prof., 1964-68, prof., 1968—, dept. chmn., 1974-81, also vice chancellor for research; tech. cons. Lockheed Missiles & Space Co., Gen. Electric Co. Contbr. articles to profl jours. Guggenheim fellow, 1965. Fellow ASME (Max Jakob Heml. award ASME—Am. Inst. Chem. Engrs. 1981, Heat Transfer Meml. award 1974, Larson Meml. award 1975), AIAA (assoc., Thermophysics award 1977); mem. Nat. Acad. Engring. Subspecialties: Heat transfer; Thermophysics. Current work: Thermal radiation transport; heat transport in porous media; heat transfer in multiphase systems. Home: 1451 Olympus Ave Berkeley CA 94708 Office: Dept Mech Engring U Calif Berkeley CA 94720

TIEN, HO TI, biophysicist, educator; b. Peking, China, Feb. 1, 1938; m. Joseleyne Slade, Jan. 31, 1953; children—Stephen, Robbins, Adrienne, Jennifer. B.S. in Chem. Engring, U. Nebr.; Ph.D. in Chemistry, Temple U., 1963. Chem. engr. Allied Chem. Corp., 1953-57; med. scientist Ea. Pa. Psychiat. Inst., Phila.; prof. biophysics, chmn. dept. Mich. State U., East Lansing, 1978—. Author: Bilayer Lipid Membranes: Theory and Practice, 1974; contbr. numerous articles internat. jours. Grantee NIH, 1964—, NSF, 1977-79, Dept. Energy, 1980-82. Mem. Am. Chem. Soc., AAAS, Biophys. Soc. Subspecialties: Solar energy; Biophysics (biology). Current work: Membrane biophysics, electrochem. solar cells, PGV cells, energy transduction including solar, primary events in photosynthesis and vision. Office: Biophysics Dept Michigan State U East Lansing MI 48824

TIEN, PING KING, electronic engineer; b. Chekieng, China, Aug. 2, 1919; came to U.S., 1947; s. N.S. and Y.S. (Chao) T.; m. Nancy N.Y. Chen, Apr. 19, 1952; children: Emily Ju-Psia, Julia Ju-Wen. M.S., Stanford U., 1948, Ph.D., 1951. Head dept. electron physics Bell Telephone Labs., Holmdel, N.J., 1952—. Contbr. sci. and tech. articles to profl. jours. Recipient Achievement award Chinese Inst. Engrs., 1966. Fellow IEEE (Morris N. Liebmann award 1979), Optical Soc. Am.; mem. Nat. Acad. Sci., Nat. Acad. Engring. Subspecialties: Fiber optics; Microelectronics. Current work: Advanced research in opto electronics for optical communication. Patentee in field. Home: Lisa Dr Apt 19 Chatham NJ 07928 Office: Bell Labs Holmdel NJ 07733

TIFFANY, LOIS HATTERY, botanist, educator; b. Story County, Iowa, Mar. 8, 1924; d. Charles R. and Emma Blanche (Brown) Hattery; m. F. Henry Tiffany, May 14, 1945; children: Charles Ray, Jean Marie, David Henry. B.S., Iowa State U., 1945, M.S., 1947, Ph.D., 1950. Mem. faculty dept. botany Iowa State U., 1950—, prof., 1965—. Recipient Weston Teaching excellence award Mycol. Soc. Am., 1980; Outstanding Teaching award Iowa State U., 1981; Disting. Iowa Scientist award Iowa Acad. Sci., 1981; Iowa Gov.'s Sci. medal, 1982. Mem. Am. Phytopath. Soc., Mycol. Soc. Am., Bot. Soc. Am., Iowa Acad. Sci. Subspecialty: Mycology. Current work: Ascomycete fungi. Office: Dept Botany Iowa State U Ames IA 50011

TIFFNEY, BRUCE HAYNES, research biologist, educator; b. Sharon, Mass., July 3, 1949; s. Wesley Newell and Sarah (Cousins) T. B.A. cum laude, Boston U., 1971; Ph.D., Harvard U., 1977. Asst. prof. dept. biology Yale U., New Haven, 1977-82, assoc. prof., 1982—; curator paleobot. Collections Peabody Mus. Natural History, 1977—. Contbr. articles to sci. jours. Mem. Bot. Soc. Am., Geol. Soc. Am., Internat. Assn. Plant Taxonomists, Assn. Tropical Biology, Internat. Orgn. Paleobotany, New Eng. Bot. Club, Tertiary Research Group. Subspecialties: Paleobiology; Evolutionary biology. Current work: Study of evolution of vascular land plants, especially angiosperms, with emphasis on contribution of paleobotany to general evolutionary theory. Office: Dept Biology Yale U Box 6666 New Haven CT 06511

TIFFNEY, WESLEY NEWELL, JR., environ. cons.; b. Springfield, Mass., June 10, 1940; s. Wesley Newell and Sarah Margaret (Cousins) T. B.S. Ed., Boston U., 1963; M.S. in Botany, U. N.H., 1965; Ph.D. in Bot. Ecology, U. N.H., 1972. Instr. biology U. Mass., Boston, 1967-72; developed teaching and research field sta., Nantucket Island, 1969-74, dir., 1974—, asst. prof. biology, 1972-74; chmn. Internat. Conf. Georges Bank Oil, 1982; environ. cons. Contbr. articles to sci. jours. Mem. Am. Soc. Environ. Edn. (trustee), Am. Bot. Soc., Am. Ecol. Soc., Brit. Ecol. Soc., Torrey Bot. Club, New Eng. Bot. Club, Nantucket Hist. Assn. (life), Nantucket Maria Mitchell Assn. (life, chmn. natural scis. com.), Sigma Xi. Subspecialties: Nitrogen fixation; Ecology. Current work: Nitrogen fixation, coastal ecology, health ecology. Office: U Mass Nantucket Field Sta Box 756 Nantucket MA 02554

TIFFT, WILLIAM GRANT, astronomer, educator; b. Derby, Conn., Apr. 5, 1932; s. William Charles and Marguerite Howe (Hubbell) T.; m. Carol Ruth Nordquist, June 1, 1957 (div.); children: Jennifer, William; m. Janet Ann (Lindner) Homewood, June 2, 1965; 1 dau., Amy; m.; stepchildren: -Patricia, Susan, Hollis. A.B., Harvard U., 1954; Ph.D., Calif. Inst. Tech., 1958. Hon. research fellow Mt. Stromolo Obs., Australian Nat. U., Canberra, 1958-60; research assoc. Vanderbilt U., Nashville, 1960-61; astronomer Lowell Obs., Flagstaff, Ariz., 1961-64; prof. astronomy U. Ariz., Tucson, 1964—. Author: Revised New General Catalog, 1973; contbr. articles to tech. jours. NSF predoctoral fellow, 1954-58; postdoctoral fellow, 1958-60; NASA grantee, 1964-72; NSF grantee, 1983—. Mem. Am. Astron. Soc., Internat. Astron. Union. Episcopalian. Subspecialties: Optical astronomy; Cosmology. Current work: Galaxies, the Redshift, large-scale structure and forces; galaxies, redshift, superclusters, gravitation. Office: U Ariz Steward Observatory Tucson AZ 85721

TIHON, CLAUDE, biochemist; b. Shanghai, China, June 12, 1944; came to U.S., 1961, naturalized, 1972; m. Berengere M. Baluda, July 27, 1969. B.A., U. Colo., 1965; Ph.D., Columbia U., 1971. Scientist Frederick Cancer Research Ctr., Md., 1974-76; sr. investigator Nat. Jewish Hosp., Denver, 1976-79; sr. scientist Bristol Labs., Syracuse, N.Y., 1979-80, asst. dir. clin. cancer research, 1980—. Mem. Am. Soc. Cell Biologists, Am. Assn. Cancer Research. Subspecialties: Cell and tissue culture; Cancer research (medicine). Current work: Anti-cancer drug development, cell biology, cancer research. Address: PO Box 657 Syracuse NY 13201

TILL, GERD OSKAR, pathologist; b. Neudek, Sudentenland, Germany, Dec. 12, 1939; came to U.S., 1980; s. Oskar and Maria (Rieger) T.; m. Anita Zimmer, Sept. 30, 1966; children: Joachim, Christopher. M.D., U. Freiberg, 1967; Dr. med. habil., U. Heidelberg, 1978. Postdoctoral fellow U. Conn., Farmington, 1972-73; wissenschaftlicher angestellter U. Heidelberg, W. Ger., 1973-78, privat-dozent, 1978-80; assoc. prof. pathology U. Mich., Ann Arbor, 1980—. Author: (with K. Rother) Komplement: Biochemie und Pathologie, 1974; editor: (with H.U. Keller) Leukocyte Locomotion and Chemotaxis, 1983. Mem. Am. Assn. Pathologists, Am. Assn. Immunologists, Am. Burn Assn., German Soc. Immunology, German Soc. Cell Biology. Subspecialties: Pathology (medicine); Immunobiology and immunology. Current work: Regulatory processes in inflamation; complement biology; leukocyte functions; thermal injury. Home: 2009 Shadford Rd Ann Arbor MI 48104 Office: Dept Pathology U Mich 1315 Catherine Rd Ann Arbor MI 48109

TIMBERLAKE, WILLIAM EDWARD, biologist, educator; b. Washington, May 2, 1948; s. George Taylor and Helen Doris (Nelson) T.; m. Sally Price, Aug. 23, 1969; children: Martha Anne, Nathan William. B.S., SUNY-Syracuse, 1970, Ph.D., 1973. Asst. prof. dept. biology Wayne State U., Detroit, 1974-79, assoc. prof., 1979-81; assoc. prof. dept. plant pathology U. Calif.-Davis, 1981—. Mem. Mycol. Soc. Am., Genetics Soc. Am., AAAS. Subspecialties: Developmental biology; Gene actions. Current work: Gene control during fungal and plant development. Office: Dept Plant Pathology U Calif Davis CA 95616

TIMIAN, ROLAND GUSTAV, plant pathologist; b. Langdon, N.D., Mar. 5, 1920; s. Gustav Albert and Adelena (Sueltz) T.; m. Frances Anna Newman, Mar. 17, 1949; children: James, Carol, Yvonne, Steven, Paulette. B.S., N.D. State U., 1949, M.S., 1950; Ph.D., Iowa State U., 1953. Research plant pathologist Agr. Research Service, USDA, Fargo, 1953—; adj. prof. N.D. State U., Fargo, 1960—. Contbr. articles to profl. jours.; editor: Barley Newsletter, 1977-80. Mem. Nat. Barley Improvement Com., 1979—. Served with USN, 1942-45. Decorated D.F.C., Air medal (5). Mem. Am. Phytopath. Soc., N.D. Acad. Sci., Sigma Xi, Gamma Sigma Delta. Lutheran. Subspecialties: Plant virology; Plant pathology. Current work: Host-virus interactions. Home: 2305 10th St Fargo ND 58102 Office: ND State U Fargo ND 58105

TIMIRAS, PAOLA SILVESTRI, physiologist; b. Rome, Italy, July 21, 1923; d. Mario and Maria (Varni) Silvestri; m. Nicholas Timaras; children: Mary-Letitia, Paul. M.D., U. Rome, 1947; Ph.D., U. Montreal, 1953. Prof. physiology, chmn. dept. physiology-anatomy U. Calif., Berkeley, 1978—; cons. space biology program NASA, 1976; panel sudden infant death syndrome Nat. Inst. Child Health and Human Devel., 1973, Biology and Medicine div. Med. Research br. AEC, 1967. Author: Stereotoxic Developing Rat Brain, 1970, Development and Physiology Aging, 1972, (with A. Vernadkais) Hormones Development and Aging, 1981; contbr. articles to profl. jours. USPHS grantee, 1978—; Am. Heart Assn. grantee, 1982—. Mem. Am. Physiol. Soc., Internat. Soc. Devel. Neurosci., Assn. Women Chemists, Internat. Soc. Psychoneuroendocrinology (v.p.), Endocrine Soc., Am. Soc. Pharmacology and Explt. Therapeutics, AAAS, Soc. Research in Child Devel., Gerontol. Soc., Internat. Soc. Neurochemistry, Internat. Soc. Biometerology, Sigma Xi, Iota Sigma Pi (pres. 1970). Subspecialties: Comparative physiology; Neuroendocrinology. Current work: Neuroendocrinology of development and aging. Office: Dept Physiology U Calif Berkeley CA 94720

TIMMERHAUS, KLAUS DIETER, coll. dean; b. Mpls., Sept. 10, 1924; s. Paul P. and Elsa L. (Bever) T.; m. Jean L. Mevis, Aug. 3, 1952; 1 dau., Carol Jane. B.S. in Chem. Engring, U. Ill., 1948, M.S., 1949, Ph.D., 1951. Diplomate: Registered profl. engr., Colo. Process design engr. Calif. Research Corp., Richmond, 1952-53; extension lectr. U. Calif. at Berkeley, 1952; mem. faculty U. Colo., 1953—, prof. chem. engring., 1961—, asso. dean engring., 1963—, dir. engring. research center, 1963—, chmn. aeorspace dept., 1979—; chem. engr. cryogenics lab. Nat. Bur. Standards, Boulder, summers 1955,57,59,61; lectr. U. Calif. at Los Angeles, 1961-62; sect. head Engring. div. NSF, 1972-73; cons. in field. Bd. dirs. Colo. Engring. Expt. Sta., Inc., Engring. Measurements Co, both Boulder. Editor: Advances in Cryogenic Engineering, vols. 1-25, 1954—; co-editor: Internat. Cryogenic Monograph Series, 1965—. Served with USNR, 1944-46. Recipient Distinguished Service award Dept. Commerce, 1957; Samuel C. Collins award outstanding contbns. to cryogenic tech., 1967; George Westinghouse award, 1968; Alpha Chi Sigma award chem. engring. research, 1968; Meritorious Service award Cryogenic Engring. Conf., 1967. Mem. Nat. Acad. Engring., AAAS, Am. Astronautical Soc., Am. Inst. Chem. Engrs. (pres. 1976, Founders award 1978), Am. Soc. for Engring. Edn. (3M Chem Engring. div. award 1980), Internat. Inst. Refrigeration (v.p. 1979—), Cryogenic Engring. Conf. (chmn. 1956-67), Sigma Xi (dir. 1981—), Sigma Tau, Tau Beta Pi, Phi Lambda Upsilon. Subspecialties: Chemical engineering; Cryogenics. Current work: Cryogenic refrigeration; heat transfer and thermodynamic properties. Home: 905 Brooklawn Dr Boulder CO 80303

TIMMERMAN, ROBERT WILSON, mech. engr. and researcher; b. Abington, Pa., July 27, 1944; s. Clarence Arthur and Mildred Wilson (Slack) T.; m. Nancy Jean Spinka, Sept. 28, 1974. Student, Wesleyan U., 1961-62; B.S., Cornell U., 1965, M.E., 1966; postgrad., Northwestern U., 1971-72, U. Pa., 1972-74. Registered profl. engr., Mass., Pa. Project engr. Monsanto Co., Springfield, Mass., 1966-68; engr. Stone & Webster Engring. Co., Boston, 1968-71, United Engrs. & Constructors, 1974-75; sr. engr. R.W. Beck & Assocs., Wellesley, Mass., 1975-77; founder, prin. R.W. Timmerman & Assocs., Boston, 1977—. Mem. ASME (chmn. Boston sect. 1979-80), ASHRAE (assoc.), Assn. Energy Engrs. (cert. energy mgr.), Internat. Dist. Heating Assn. Presbyterian. Subspecialties: Mechanical engineering. Current work: District heating, cogeneration, waste heat utilization. Patentee in heating field. Home and Office: 25 Upton St Boston MA 02118

TIMO, DOMINIC PETER, electrical engineer; b. Seattle, Aug. 27, 1923; s. Vincent and Fortunata (Baima) T.; m. Frances Ames Garvin, Mar. 31, 1951; children: Barbara, Michael, Daniel, Nancy, Christopher. B.S.E.E., U. Wash., 1947; student, Gen. Electric Advanced Engring. Program, 1949-51. Registered profl. engr., N.Y. Analytical engr. Knolls Atomic Power Lab., Schenectady, 1948-52; electromech. engr. M & P Lab., Gen. Electric Co., 1952-53, mgr. electromech. engring., 1954-60, mgr. structural engring. large steam turbine dept., 1961-67, mgr. structural devel. engring., 1967—; lectr. Contbr. articles to tech. jours. Bd. dirs. Niskayuna Babe Ruth Baseball League. Served to staff sgt. AUS, 1943-46. Fellow ASME; mem. Elfun Soc., Schenectady Gen. Electric Engrs. Assn. Roman Catholic. Subspecialties: Mechanical engineering; High-temperature materials. Current work: Analysis and evaluation of turbine components, defining inspection intervals for prevention of catastrophic failures, developing material, stress, thermal, fracture information. Patentee in field. Home: 872 Hereford Way Schenectady NY 12309 Office: General Electric Co 1 River Rd Bldg 40 Schenectady NY 12345

TING, CHOU-CHIK, physician; b. Eu-Yang, An-Hwei, China, July 2, 1937; came to U.S., 1963, naturalized, 1970; s. Sing-Wu and Shu-Ying Fu T.; m. Kai-Li Hsia, June 23, 1968; children—Ray Tung, Raynan. M.D., Nat. Taiwan U., 1962. Intern Nat. Taiwan Univ. Hosp., 1961-62; pathology resident N.Y. Med. Coll., 1963-64, Albert Einstein Coll. Medicine, 1964-68, instr. pathology, 1966-68; spl. fellow Nat. Cancer Inst., NIH, Bethesda, Md., 1968-70, staff fellow, 1970-72, sr. investigator, 1972—. Served with Chinese Air Force, 1962-63. Nat. Acad. Sci. research scholar, 1981. Mem. Am. Assn. Immunologists, Am. Assn. Cancer Research. Seventh-day Adventist. Subspecialty: Immunology (agriculture). Current work: To study the mechanisms for induction of specific cell-mediated tumor immunity. Home: 8221 Scotch Bend Way Potomac MD 20854 Office: Lab Cell Biology Nat Cancer Inst NIH Bethesda MD 20205

TING, FRANCIS TA-CHUAN, geologist, educator; b. Tsingtao, China, Apr. 26, 1934; came to U.S., 1959, naturalized, 1971; s. Teh Hsien and Hwei Po (Wu) T.; m. Susan Tong, July 22, 1966; children: Eric, Tracy. B.S., Nat. Taiwan U., 1957; M.S., U. Minn., 1962; Ph.D., Pa. State U., 1967. Postdoctoral fellow Pa. State U., University Park, 1967-68, research assoc., 1969-70; asst. prof. Macalester Coll., St. Paul, 1968-69, U. N.D., Grand Forks, 1970-73, assoc. prof., 1973-74, W.Va. U., Morgantown, 1974-79, prof., 1979—. Contbr. articles to profl. jours. NATO sr. fellow, summer 1973. Mem. Geol. Soc. Am., Am. Assn. Petroleum Geologists, Soc. Mining Engrs. of AIME, Bot. Soc. Am., Soc. Econ. Paleontologists and Mineralogists. Club: Kiwanis (Morgantown). Subspecialties: Geology; Organic geochemistry. Current work: Coal petrology, comparative coal petrology, petrology and geochemistry of organic materials in sediments. Office: Dept Geology and Geography W Va U Morgantown WV 26506

TING, YU-CHEN, geneticist; b. Henan, China, Oct. 3, 1920; s. Chin-Yung and Yi-Yung (Wang) T.; m. Jovina Y.H. Chen, June 25, 1960; children: Andrew, Claire. B.S., Henan U., China, 1944; M.S.A., Cornell U., 1951; Ph.D., La. State U., 1954. Postdoctoral fellow Harvard U., 1955-62; asst. prof. genetics Boston Coll., 1962-64, assoc. prof., 1964-67, prof., 1967—. Contbr. articles to profl. jours. Bd. dirs. Nat. Assn. Chinese Ams., Boston. Research collaborator Brookhaven Nat. Lab., 1963-65; Nat. Acad. Scientists sr. fellow, 1979. Mem. Am. Genetics Assn., Genetics Soc. Am., AAAS, Bot. Soc. Am., New Eng. Bot. Club. Subspecialties: Cell and tissue culture; Genetics and genetic engineering (agriculture). Current work: Researcher in cytogenesis of maize and its relatives, fine structure of meiotic chromosomes, plant cell and tissue culture. Home: 230 Bonad Rd Brookline MA 02167 Office: Dept Biology Boston Coll Chestnut Hill MA 02167

TINKHAM, MICHAEL, physicist, educator; b. Green Lake County, Wis., Feb. 23, 1928; s. Clayton Harold and LaVerna (Krause) T.; m. Mary Stephanie Merin, June 24, 1961; children: Jeffrey Michael, Christopher Gillespie. A.B., Ripon (Wis.) Coll., 1951, Sc.D. (hon.), 1976; M.S., MIT, 1951; Ph.D., 1954; M.A. (hon.), Harvard, 1966. NSF postdoctoral fellow at Clarendon Lab., Oxford (Eng.) U., 1954-55; successively research physicist, lectr., asst. prof., asso. prof., prof. physics U. Calif. at Berkeley, 1955-66; Gordon McKay prof. applied physics Harvard U., 1966—, prof. physics, 1966-80, Rumford prof. physics, 1980—, chmn. physics dept., 1975-78; cons. to industry, 1958—; participant internat. seminars and confs. Mem. commn. on very low temperatures Internat. Union Pure and Applied Physics, 1972-78. Author: Group Theory and Quantum Mechanics, 1964, Superconductivity, 1965, Introduction to Superconductivity, 1975; also numerous articles. Served USNR, 1945-46. Recipient award Alexander von Humboldt Found. U. Karlsruhe, W. Ger., 1978-79; NSF sr. postdoctoral fellow Cavendish lab.; vis. fellow Clare Hall Cambridge (Eng.) U., 1971-72; Guggenheim fellow, 1963-64. Fellow Am. Phys. Soc. (chmn. div. solid state physics 1966-67, Buckley prize 1974, Richtmyer lectr. 1977), AAAS; mem. Am. Acad. Arts and Scis., Nat. Acad. Scis. Subspecialties: Low temperature physics; Superconductors. Current work: Superconducting Josephson junctions and nonequilibrium superconductivity. Home: 98 Rutledge Rd Belmont MA 02178 Office: Physics Dept Harvard Univ Cambridge MA 02138

TIPLER, FRANK JENNINGS, III, mathematical physicist, educator; b. Andalusia, Ala., Feb. 1, 1947; s. Frank Jennings Jr. and Anne (Kearley) T. S.B., MIT, 1969; Ph.D., U. Md., 1976. NSF research mathematician U. Calif.-Berkeley, 1976-79; sr. research fellow Oxford (Eng.) U., 1979; research assoc. U. Tex.-Austin, 1979-81; assoc. prof. math. Tulane U.-New Orleans, 1981—. Editor: Essays in General Relativity: A Festschrift for Abraham H. Taub, 1980; contbr. numerous articles to sci. publs. Mem. Am. Phys. Soc., Royal Astron. Soc. Subspecialties: General relativity; Cosmology. Current work: Research on topics in global general relativity: quantum cosmology, physics of black holes, and long-time evolution of the Universe. Home: 3915 St Charles Ave Apt 313 New Orleans LA 70115 Office: Tulane Univ 312 Gibson Hall New Orleans LA 70118

TIRAS, HERBERT GERALD, engineering company executive; b. Houston, Aug. 11, 1924; s. Samuel Louis and Rose (Seibel) T.; m. Aileen Wilkenfeld, Dec. 14, 1955; children: Sheryle, Leslie. Student, Tex. A&M U., 1941-42; B.S., U. Houston, 1965. Registered prof. engr., Calif. Mfg. engr. Reed Roller Bit, Houston, 1942-60; pres. Tex-Truss, Houston, 1960-77; chief exec. officer OMNICO, Houston, 1977—; nat. dir. Coll. and Univ. Mfg. Ednl. Council, 1978—; resources dir., Region VI Fed. Emergency Mgmt. Agy., 1982—. Mem. Soc. Mfg. Engrs. (cert. of accomplishment, cert. in mfg. mgmt. and robotics), Robot Inst. Am., Robotics Internat., Marine Tech. Soc., Nat. Soc. Profl. Engrs. Lodge: Masons, Shriners. Subspecialties: Robotics; Remote sensing (geoscience). Current work: Application of robotics and remote sensing to sub-sea mining. Patentee in field. Home: 9703 Runnymeade Houston TX 77096

TISCHLER, ARTHUR S., exptl. pathologist; b. N.Y.C., July 10, 1946; s. Louis N. and Bertie (Brumberger) T.; m. Joanne L. Hager, July 13, 1980. B.S., Pa State U., 1966; M.D., Thomas Jefferson U., 1971. Diplomate: Am. Bd. Pathology. Intern, resident, chief resident, then research fellow in pathology Beth Israel Hosp.-Harvard Med. Sch., Boston, 1971-76; staff pathologist Walter Reed Army Med. Ctr., Washington, 1976-78; asst. prof. pathology Tufts U. Sch. Medicine, Boston, 1978—; asst pathologist New Eng. Med. Ctr., Boston, 1978—. Served to maj. M.C. U.S. Army, 1976-78. Grantee Nat. Cancer Inst., 1979, Am. Cancer Soc., 1980, Charles A. King Trust, 1980. Mem. Internat. Acad. Pathologists, New Eng. Soc. Pathologists, Soc. Neurosci., Tissue Culture Assn. Subspecialties: Pathology (medicine); Neuroendocrinology. Current work: Environmental influences on differentiation and function of neuroendocrine cells and neuroendocrine tumors. Office: Dept Pathology Tufts U Sch Medicine 136 Harrison Ave Boston MA 02111

TISCHLER, MARC ELIOT, biochemistry educator, researcher; b. N.Y.C., Nov. 10, 1949; s. Henry M. and Harriet (Green) T.; m. Meryl Green, Aug. 5, 1979; 1 dau., Rebecca. B.A., Boston U., 1971; M.S., U. S.C., 1973; Ph.D., U. Pa., 1977. Postdoctoral fellow Harvard Med. Sch., Boston, 1977-79; asst. prof. U. Ariz. Sch. Medicine, Tucson, 1979—. Mem. phys. edn. com. Jewish Community Ctr., Tucson. Am. Heart Assn. established investigator, 1982—; Muscular Dystrophy Assn. fellow, 1978-79; Am. Heart Assn. fellow, 1977-78. Mem. Biophys. Soc., Am. Physiol. Soc., Am. Heart Assn., Am. Soc. Biol. Chemists. Jewish. Subspecialties: Biochemistry (biology); Biochemistry (medicine). Current work: Regulation of protein degradation in muscle by hormones and the response of this process and other aspects of muscle metabolism to fasting, diabetes, trauma and muscle disuse. Office: U Ariz Sch Medicine Tucson AZ 85724

TISHLER, MAX, chemist; b. Boston, Oct. 30, 1906; s. Samuel and Anna (Gray) T.; m. Elizabeth M. Verveer, June 17, 1934; children—Peter Verveer, Carl Lewis. B.S., Tufts Coll., 1928; M.A., Harvard U., 1933, Ph.D., 1934; D.Sc., Tufts U., 1956, Bucknell U., 1962, Phila. Coll. Pharmacy, 1966, U. Strathlyde, Glasgow, 1969, Rider Coll., 1970, Upsala Coll., 1972, Fairfield U., 1972, Wesleyan U., 1981; D.Eng., Stevens Inst. Tech., 1966. Teaching asst. Tufts Coll., 1924; Austin teaching fellow Harvard, 1930-34, research asso., 1934-36, instr. chemistry, 1936-37; research chemist Merck & Co., Inc., 1937-41, head sect. process devel., 1941-44, dir. devel. research, 1944-51, asso. dir. research and devel., 1951-54, v.p. sci. activities, chem. div., 1954-56, pres., 1956-69, sr. v.p. research and devel., 1969-70; Rennebohm lectr. sch. pharmacy U. Wis., 1963; prof. chemistry Wesleyan U., Middletown, Conn., 1970-72, Univ. prof. scis., 1972-75, Univ. prof. scis. emeritus, 1975—, chmn. dept. chemistry, 1973-75; Mem. vis. com. dept. chemistry Harvard U., 1965-71, Sch. Pub. Health, 1963-69; mem. vis. com. dept. chem. engring. M.I.T.; trustee, chmn. vis. com. dept. chemistry Tufts U.; trustee Union (N.J.) Coll., 1965-70; asso. trustee sci. U. Pa., 1960-65, 70; adv. council Newark Coll. Engring., 1956-59; adminstrv. bd. Tufts New Eng. Med. Center, adv. bd. Coll., 1964-68; adv. com. biol. scis. Princeton U., 1972-75. Author: (with J.B. Conant) Chemistry of Organic Compounds, 1937, Streptomycin, (with S. A. Waksman), 1949; co-chmn. study group, author: Chemistry in the Economy, 1973; Editorial bd.: Organic Syntheses, 1953-61, Separation Science, Clin. Pharmacology and Therapeutics; editor: Organic Syntheses, 1960-61. Bd. dirs. Merck & Co. Found.; bd. govs. Weizmann Inst.; mem. nat. adv. council Hampshire Coll., 1964-68; bd. visitors Faculty Health Scis., SUNY, Buffalo, 1965-68; trustee Royal Soc. Medicine Found., 1965-69; bd. sci. advisers Sloan Kettering Inst., 1974—. Recipient Merck & Co., Inc. Bd. Dirs. Sci. award (resulting in establishment Max Tishler Vis. Lectureship at Harvard and Max Tishler Scholarship (ann.) at Tufts Coll., 1951; medalist Indsl. Research Inst., 1961, Soc. Chem. Industry, 1963; Julius W. Sturmer Meml. Lecture award Phila. Coll. Pharmacy, 1964; Chemistry lectr. Royal Swedish Acad. Engring. Scis., 1964; Kauffman Meml. lectr. Ohio State U., 1967; Chem. Pioneer award Am. Inst. Chemists, 1968; Gold medal, 1977; DuPont lectr. Dartmouth; Priestley medalist Am. Chem. Soc., 1970; Found. Patent award, 1970; Eli Whitney award Conn. Patent Law Assn., 1974; New Brunswick lectr. Am. Soc. Microbiology, 1974. Fellow N.Y. Acad. Sci., Royal Soc. Chemistry (hon.), Chem. Soc. London, Am. Inst. Chemists, Soc. Chem. Industry (chmn. Am. sect. 1966, hon. v.p. 1968), AAAS, Acad. Pharm. Scis. of Am. Pharm. Assn. (hon. mem.), Conn. Acad. Sci. and Engring.; mem. Am. Acad. Arts and Scis., Indsl. Research Inst., Nat. Acad. Scis., Am. Chem. Soc. (chmn. dir. organic chemistry 1951, pres. 1972, bd. dirs. 1973), Swiss Chem. Soc., Conn. Acad. Scis. and Engring., Harvard Assn. Chemists, Société Chimique de France (hon.), Phi Beta Kappa, Sigma Xi, Pi Lambda Phi (Big Pi award 1951). Club: Chemists (N.Y.C.) (hon.). Subspecialties: Organic chemistry; Medicinal chemistry. Current work: Synthesis of bioisoteres of l-amino acids and of small peptides containing these bioisoteres and to test them for possible therapeutic or enzyme inhibiting activities. Patentee in field. Home: 6 Red-Orange Rd Middletown CT 06457

TISSERAT, BRENT HOWARD, research geneticist; b. Berkeley, Calif., Oct. 2, 1951; s. William H. and Tina (Rose) T. A.A., San Bernardino Valley Coll., 1971; B.S., Calif. State U.-Pomona, 1973; Ph.D., U. Calif.-Riverside, 1976. Research assoc. U. Calif.-Riverside, 1973-76; Cabot fellow Harvard U., Cambridge, Mass., 1976-77; research geneticist U.S. Date and Citrus Sta., Dept. Agr. Research Service, Indio, Calif., 1977-80, 1980—; cons. on date palm tissue culture and microculture. Contbr. articles on genetics and plant cell and tissue culture to profl. jours. Mem. Palm Soc., Crop Sci. Soc. Am. Subspecialties: Plant genetics; Plant cell and tissue culture. Current work: Micro-breeding, cryogenetics, mass plant propagation, plant morphogenesis, especially in the palm family. Office: U S Dept Agr Agrl Research Service Fruit and Vegetable Chemistry Lab 263 S Chester Ave Pasadena CA 91106

TIVIN, FRED, analytical chemist, research executive; b. Chgo., Sept. 28, 1937; s. Edward and Min (Stone) T.; m. Sandra Sue Tivin, Dec. 12, 1941; children: David Scott, Mark Ingram, Brian Joseph. B.A., Kalamazoo Coll., 1958; M.S., U. Ill., 1960, Ph.D., 1963. Research chemist Exxon Co., 1963-66; with Procter & Gamble Co., Cin., 1966—, dir. analytical devel. and quality assurance, 1979—. Contbr. articles to profl. jours. Mem. Am. Chem. Soc., AAAS, Am. Soc. Quality Control, Sigma Xi, Phi Lambda Upsilon. Subspecialties: Analytical chemistry; Pharmaceutical quality assurance. Current work: Separations, pharmaceutical product development, director of analytical development and quality assurance for pharmaceutical product development. Office: PO Box 39175 Cincinnati OH 45247

TJIOE, SARAH ARCHAMBAULT, pharmacologist, educator; b. Phila., Oct. 12, 1944; d. Alfred and Joanna (Pomicter) Archambault; m. Gim Beng Tjioe, Aug. 19, 1967; children: Susan, James. B.A., U. Pa., 1966, Ph.D., 1971. Postdoctoral fellow Ohio State U., 1971-72, instr., 1972-75, asst. prof., 1975—. Pharm. Mfrs. Assn. Found. fellow, 1972-74. Mem. Neurosci. Soc., Am. Soc. Pharmacology and Exptl. Therapeutics. Subspecialties: Neurochemistry; Neuropharmacology. Current work: Developing brain and effect of drugs on developing nerve. Office: 333 W 10th Ave Columbus OH 43210

TOBE, STEPHEN SOLOMON, educator, physiologist; b. Niagara-on-the-Lake, Ont., Can., Oct. 11, 1944; s. John Harold and Rose (Bolter) T.; m. Martha, Oct. 19, 1969. B.Sc., Queen's U., Kingston, Ont., 1967; M.Sc., York U., Downsview, Ont., 1969; Ph.D., McGill U., Montreal, Que., Can., 1972. Research fellow Agrl. Research Council, U. Sussex, U.K., 1972-74; asst. prof. dept. zoology U. Toronto, 1974-78, assoc. prof., 1978-82, prof., 1982—. Contbr. articles to profl. jours. NRC of Can. bursary, 1968-70; scholar, 1970-72; postdoctoral fellow, 1972-74; E.W.R. Steacie Meml. fellow Natural Scis. and Engring. Research Council, 1982—; recipient C. Gordon Hewitt award Entomol. Soc. Can., 1982. Fellow Royal Entomol. Soc. London; mem. AAAS, Am. Soc. Zoologists, Biochem. Soc., Can. Soc. Zoologists, Entomol. Soc. Can., Entomol. Soc. Ont., Soc. for Endocrinology, Soc. for Exptl. Biology, Sigma Xi. Subspecialties: Physiology (biology); Reproductive biology. Current work: Invertebrate endocrinology. Regulation of hormone biosynthesis and control of hormone titre in insects. Role of hormones in reproduction of invertebrates. Hormonal regulation of metamorphosis in sects. Mode of action of hormone agonists/antagonists. Home: 55 Marmot St Toronto ON Canada M4S 2T4 Office: Dept Zoology U Toronto 25 Harbord St Room 537 Toronto ON Canada M5S 1A1

TOBEY, ROBERT A., microbiologist; b. Owosso, Mich., May 26, 1937; m. Eileen Massallek, July 9, 1960; children: Karen, Kevin. B.S., Mich. State U., 1959; Ph.D. in Microbiology, U. Ill., 1963. Staff mem. toxicology group Los Alamos Nat. Lab., 1964--, group leader toxicology group life sci. div., 1979-82. Contbr. articles to profl. jours. Mem. Am. Assn. Cancer Research, Am. Soc. Biol. Chemists, Am. Soc. Cell Biology. Subspecialties: Cell biology; Chemotherapy. Current work: Development of methodology of cell synchronization with emphasis on investigation of effects of physical, chemical and biological agents upon metabolic operations at specific stages of the cell cycle. Home: 102 Rover Blvd Los Alamos NM 87544 Office: Los Alamos Nat Lab PO Box 1663 Los Alamos NM 87545

TOBIN, GORDON ROSS, plastic surgeon; b. Twin Falls, Idaho, Jan. 6, 1943; s. Gordon Ross and Garnett Othalia (Peterson) T.; m. Elisabeth Ann Pelcher, Dec. 21, 1968; children: Christopher Ross, Anne Elise. A.B., Whitman Coll., 1965; M.D., U. Calif.-San Francisco, 1969. Diplomate: Am. Bd. Surgery, Am. Bd. Plastic Surgery. Intern San Francisco Gen. Hosp., 1969-70; resident U. Ariz. Affiliated Hosps., Tucson, 1970-76; practice medicine, specializing in plastic surgery, Louisville, 1977—; asst. prof. surgery U. Calif.-Irvine, 1976-77; assoc. prof. surgery U. Louisville, 1977—, dir., 1977—. Paralized Vets of Am. grantee, 1975; Am. Cancer Soc. grantee, 1980. Fellow ACS; mem. Plastic Surgery Research Council, Am. Soc. Plastic and Reconstructive Surgeons, Southeastern Soc. Plastic and Reconstructive Surgeons, Ky. Soc. Plastic and Reconstructive Surgery (pres.), Delta Tau Delta. Democrat. Presbyterian. Subspecialties: Surgery; Anatomy and embryology. Current work: Myology, microneurovascular anatomy, collagen biology, soft tissue microcirculatory physiology, central nervous system wound healing, tissue transfer, burn physiology, implantable biomaterials, medical education. Office: Dept Surgery Univ Louisville 530 W Jackson St Louisville KY 40292 Home: 2413 Tavener St Louisville KY 40222

TOBIN, THOMAS, veterinarian, educator; b. Dublin, Ireland, Aug. 7, 1941; came to U.S., 1970, naturalized, 1977; s. Nicholas and Mary Bridget (Ryan) T.; m. Maria Jolanta Pilacinska, Dec. 20, 1970. M.V.B., Univ. Coll., Dublin, 1964; M.Sc., U. Guelph, Ont., Can., 1966; Ph.D., U. Toronto, Ont., 1970. Cert. toxicologist. Research assoc. U. Toronto, 1966-70; asst. prof. Mich. State U., East Lansing, 1970-75; assoc. prof. vet. sci. and toxicology U. Ky., Lexington, 1975-78, prof., 1978—; cons. equine forensic toxicology. Author: Drugs and the Performance Horse, 1981, also articles. NIH grantee, 1972-75; NSF grantee, 1974-76; Ky. Equine Drug Research Program grantee, 1975-82. Mem. Royal Coll. Vet. Surgeons, Soc. Toxicology, Royal Dublin Soc. Club: Thoroughbred of Am. (Lexington). Subspecialties: Pharmacology; Toxicology (medicine). Current work: Forensic pharmacology and chemistry, especially as it relates to racehorses. Home: 423 Henry Clay Blvd Lexington KY 40502 Office: Dept Vet Sci U Ky Lexington KY 40506

TOCCO, DOMINICK JOSEPH, drug metabolism researcher; b. N.Y.C., Jan. 25, 1930; s. Frank Paul and Rose (D'Ambra) T.; m. Yvonne V. Cali, June 29, 1952; children: Donald, Gregory, Douglas, Brian. B.S., St. John's U., 1951, M.S., 1953; Ph.D., Georgetown U., 1960. Mem. staff NIH, Bethesda, Md., 1955-60; research fellow Merck Inst. for Therapeutic Research, Rahway, N.J., 1960-66, Shell Devel. Co., Modesto, Calif., 1966-70; sr. research fellow drug metabolism Merck Sharp & Dohme, West Point, Pa., 1970—. Served with U.S. Army, 1953-55. Mem. Fedn. Am. Socs. for Exptl. Biology. Subspecialty: Pharmacokinetics. Current work: Disposition and metabolism of drugs in animals and men. Office: Merck Sharp & Dohme West Point PA 19486

TODARO, GEORGE JOSEPH, pathologist; b. N.Y.C., July 1, 1937; s. George J. and Antoinette (Piccinni) T.; m. Jane Lehv, Aug. 12, 1962; children: Wendy C., Thomas M., Anthony A. B.S., Swarthmore Coll., 1958; M.D., N.Y. U., 1963. Intern N.Y. U. Sch. Medicine, N.Y.C., 1963-64, fellow in pathology, 1964-65, asst. prof. pathology, 1965-67; staff assoc. Viral Carcinogenesis br. Nat. Cancer Inst., Bethesda, Md., 1967-70, head molecular biology sect., 1969-70, chief, 1970-83; sci. dir. Oncogen, Seattle, 1983—; faculty mem. Genetics Program, George Washington U.; prof. pathology U. Wash., Seattle, 1983—. Editor: Cancer Research, 1973—, Archives of Virology, 1976—, Jour. Biol. Chemistry, 1979—; contbr. articles to profl. jours. Served as med. officer USPHS, 1967-69. Recipient Borden Undergrad. Research award, 1963, USPHS Career Devel. award, 1967, HEW Superior Service award, 1971, Gustav Stern award for virology, 1972, Parke-Davis award in exptl. pathology, 1975; Walter Hubert lectr. Brit. Cancer Soc., 1977. Mem. Am. Soc. Microbiology, Am. Assn. Cancer Research, Soc. Exptl. Biology and Medicine, Am. Soc. Biol. Chemists, Am. Soc. Clin. Investigation. Subspecialty: Cancer research (medicine). Home: 1940 15th Ave E Seattle WA 98112 Office: Oncogen 3005 1st Ave Seattle WA 98121

TODD, DEBORAH ANN, engineer, educator; b. Long Island City, N.Y., June 18, 1955; d. Ralph and Mae W. (Milligan) T. B.S. in Mech. Engring, Union Coll., 1976, M.S., Rensselaer Poly. Inst., 1978; postgrad., Harvard Bus. Sch. Registered profl. engr., N.Y. Mech. design engr. Large Steam Turbine div. Gen. Electric Co., Schenectady, N.Y., 1976-78, mech. engr. projects engring ops., 1978-79, project engr. dept. internat. projects, N.Y.C., 1980—; mech. engr. Gibbs & Hill, Inc., N.Y.C., 1979-80; lectr. in field. Recipient Gen. Electric Co. Top Young Engrs award, 1982. Mem Nat Soc Profl Engrs, ASME, Tau Beta Phi. Republican. Clubs: Appalachian Mountains, Appalachian Trail Conf. Subspecialties: Power Engineering; Systems engineering. Current work: Optimization of power plant cycles; waterhammer; water treatment control and supervision of architect-engr. firms performing proposal/contract engring. on internat. and domestic power plant projects. Home: 39 Netto Ln Plainview NY 11803 Office: Harvad Bus Sch Chase C3-6 HBS Boston MA 02163

TŐKÉS, ZOLTÁN ANDRÁS, biochemistry educator; b. Budapest, Hungary, May 14, 1940; s. Elemér and Marianne (Tilzer) T.; m. Dorcas-May Vanian; children: Krisztina, Géza. B.Sc., U. So. Calif., 1964; Ph.D. in Biochemistry, Calif. Inst. Tech., 1971. Lectr. U. Malaya Sch. Medicine, Kuala Lumpur, Malaysia, 1971-72; ind. investigator Basel (Switzerland) Sch. Immunology, 1972-74; asst. prof. biochemistry U. So. Calif. Sch. Medicine, Los Angeles, 1974-79, assoc. prof., 1980—, dir. cell culture and tumor marker labs., 1977—. Contbr. articles to profl. jours. Nat. Inst. Cancer grantee. Mem. Am. Soc. Biol. Chemists, Am. Assn. Cancer Research, Am. Assn. Pathologists, Am. Soc. Cell Biology, N.Y. Acad. Scis. Subspecialties: Biochemistry (biology); Cancer research (medicine). Current work: Detection of cell surface changes due to malignancy and other pathological conditions; molecular basis of cell-cell recognition. Patentee in field. Office: 1303 N Mission Rd Los Angeles CA 90033

TOKUNAGA, ALAN TAKASHI, astronomer; b. Puunene, Hawaii, Dec. 17, 1949; s. Hiroshi and Carol (Takemoto) T.; m. Cheryl Yip, Jan. 4, 1975. Student, Pacific U., 1967-69; B.A., Pomona Coll., 1971; Ph.D., SUNY, Stony Brook, 1976. Research assoc. NASA Ames Research Center, Moffet Field, Calif., 1976-77, Steward Observatory, Tucson, 1977-79; asst. astronomer U. Hawaii, Honolulu, 1979-82, assoc. astronomer, 1982—; mem. Kitt Peak Users Com., 1978-81, NASA Infrared Expts. Working Group, 1980, NASA Large Deployable Reflector Workshop, 1982. Contbr. articles to sci. jours. Mem. Am. Astron. Soc. Subspecialties: Infrared optical astronomy; Planetary science. Current work: Infrared spectroscopy of the outer solar system; infrared spectroscopy; infrared photometry; astron. instrumentation. Office: Inst Astronomy 2680 Woodlawn Dr Honolulu HI 96822

TOLBERT, DONALD DEAN, med. physicist; b. Beloit, Kans., Jan. 11, 1940; s. Ivan Floyd and Marjorie Ruth (Grecian) T.; m. Marie Julie Adams, Aug. 20, 1960 (div.); children: Jay, Jodi, Janese; m. Carol Margret Boff, Aug. 17, 1976; children: Cari, Michelle. B.A., Kans. Wesleyan U., 1962; Ph.D., U. Kans., 1968. Research assoc. Fla. State U., 1968-70; postdoctoral trainee radiol. physics U. Wis. Med. Sch., Madison, 1970-71; dir. Wis. Radiol. Physics Lab., 1971-77, dir. radiotherapy physics sect., 1971-77, asst. prof. human oncology, 1971-77; head radiol. physics div. Cancer Center Hawaii, Honolulu, 1977—; mng. partner Mid-Pacific Med. Physics, Honolulu, 1981—. Contbr. articles to sci. jours. Mem. Am. Soc. Therapeutic Radiology, Am. Assn. Physicists in Medicine, Health Physics Soc., Sigma Xi. Subspecialty: Medical physics. Current work: Application of physics and computers to radiation oncology treatment planning. Home: 47-010F Hui Iwa Pl Kaneohe HI 96744 Office: 1301 Punchbowl St Suite 307 Honolulu HI 96813

TOLBERT, LELLAND CLYDE, neurochemistry educator; b. Canton, Ill., July 14, 1943; s. Walter Clyde and Hazel (Stanger) T.; m. Rebecca Jane Bankhead, Sept. 16, 1967; 1 dau., Jennifer Rebecca. Lic. Practical Nurse, Anchorage Community Coll., 1964; B.A., Alaska Meth. U., 1968; Ph.D., Med. U. S.C., 1978. Psychiat. practical nurse Alaska Psychiat. Inst., Anchorage, 1964-68; project chemist TriBorough Air Resources Mgmt. Dist., Anchorage, 1968-70; postdoctoral fellow in neurosci. U. Ala., Birmingham, 1978-80, research assoc., 1980-81, asst. prof. depts. psychiatry and psychology, 1981—. Contbr. articles to profl. jours. Mem. Soc. Neurosci., Sigma Xi, Phi Alpha Theta. Subspecialties: Neurochemistry; Neuropharmacology. Current work: Biochemistry of mental illness; regulation of biogenic amines; one-carbon metabolism; schizophrenia; major affective disorders; psychopharmacology of biogenic amines. Home: 4177 Winston Way Birmingham AL 35213 Office: U Ala Birmingham AL 35294

TOLCHINSKY, PAUL D., psychologist; b. Cleve., Sept. 30, 1946; s. Sanford M. and Frances (Klein) T.; m. Laura S. Schermer, Nov. 2, 1968; children: Heidi E., Dana M. B.A., Bowling Green State U., 1971; Ph.D., Purdue U., 1978. Trainer Detroit Bank & Trust Co., 1971-73; cons. Babock & Wilcox, Barberton, Ohio, 1973-75, Gen. Foods Corp., West Lafayette, Ind., 1975-77; prof. psychology U. Akron, Ohio, 1978-81; pres. Creative Work-Life, Shaker Heights, Ohio, 1979—; cons. in field; prof. Bowling Green (Ohio) State U., 1980-82. Contbr. articles to profl. jours. Served with U.S. Army, 1966-69. Bowling Green State U. research grantee, 1971; David Ross grantee Purdue Found., 1977-78. Mem. Acad. Mgmt., Am. Psychol. Assn., Am. Mgmt. Assn. Democrat. Jewish. Subspecialty: Behavioral psychology. Current work: Improving orgnl. effectiveness and efficiency thru use of behavioral and socio-tech. sci. applications. Office: 3310 Warrensville Rd Shaker Heights OH 44122

TOLIN, SUE ANN, plant pathology educator; b. Montezuma, Ind., Nov. 29, 1938; d. Roy Willard and Ina Mae (Spaw) T. B.S., Purdue U., 1960; M.S., U. Nebr., 1962, Ph.D., 1965. Research asst. U. Nebr., 1960-65; research assoc. Purdue U., West Lafayette, Ind., 1965-66; asst. prof. plant pathology Va. Poly. Inst. and State U., Blacksburg, 1966-71, assoc. prof., 1971—; plant pathologist U.S. Dept. Agr., 1978-79; liaison mem. recombinant DNA adv. com. NIH, 1979—. Mem. Am. Phytopath. Soc., Am. Soc. Virology, Am. Soc. Microbiology, AAAS, Plant Molecular Biology Assn. Subspecialties: Plant virology; Genetics and genetic engineering (agriculture). Current work: Molecular biological basis for varibility of RNA viruses and relation to genetic basis of response of host plants, including corn, soybeans and other legumes. Office: Dept Plant Pathology and Physiology Va Poly Inst and State U Blacksburg VA 24061

TOLLER, GARY NEIL, astronomer; b. Phila., Dec. 13, 1950; s. Harry and Beatrice (Schaffer) T. B.S. in Physics and Math, Dickinson Coll., 1972; M.S. in Astronomy, SUNY-Albany, 1975, Ph.D., SUNY-Stony Brook, 1981. Astronomer Space Astronomy Lab., Albany, N.Y., 1973-80, Gainesville, Fla., 1981—; asst. research scientist U. Fla., Gainesville, 1981—. Mem. Am. Astron. Soc. Subspecialty: Optical astronomy. Current work: Galactic structure, space-based experimentation, integrated starlight, diffuse galactic light, cosmic light. Office: 1810 NW 6th St Gainesville FL 32601

TOLLES, WILLIAM MARSHALL, physical chemist, college dean; b. New Britain, Conn., June 30, 1937; s. Marshall H. and Lily V. (Calmback) T.; m. Elizabeth Blackman, July 31, 1937; children: Christopher, Laurie. B.A., U. Conn., 1958; Ph.D., U. Calif., 1962. Postdoctoral fellow Rice U., 1962; successively asst. prof., assoc. prof., prof. chemistry Naval Postgrad. Sch., Monterey, Calif., 1962-78, dean research, dean sci. and engring., 1978—. Author: Introduction to Non-linear Phenomena, 1981, also articles. Mem. Am. Phys. Soc., Am. Chem. Soc., Am. Optical Soc., Am. Soc. Engring. Edn. Subspecialties: Atomic and molecular physics; Spectroscopy. Current work: Non-linear spectroscopy, Coherent Anti-Stokes Raman Spectroscopy, Raman Induced Kerr Effect. Office: Naval Postgrad Sch Monterey CA 93940

TOLMAN, EDWARD LAURIA, biochem. pharmacologist; b. Chelsea, Mass., Oct. 9, 1942; s. Max and Frances (Baker) T.; m. Anita Young, June 25, 1967; 1 dau., Jennifer. B.A., U. Mass., 1964, M.A., 1965; Ph.D., SUNY, Syracuse, 1970. Postdoctoral fellow Hershey Med. Center, 1969-72; sr. scientist Lederle Labs., 1972-76, project leader, 1977-78, group leader, 1978-80; sect. head biochem. pharmacology Ortho Pharm. Corp., Raritan, N.J., 1980—; adj. asst. prof. human biochemistry Fairleigh Dickinson U. Sch. Dentistry, 1979—. Contbr. articles to profl. jours. Mem. Am. Soc. Pharmacology and Exptl. Therapeutics, Am. Diabetes Assn., Soc. Exptl. Biology and Medicine, Am. Chem. Soc., N.J. Acad. Sci., Sigma Xi. Subspecialties: Cellular pharmacology; Biochemistry (medicine). Current work: Prostaglandins, inflammation, diabetes drug discovery; research administration. Office: Biochem Research Ortho Pharm Corp Raritan NJ 08869

TOM, BALDWIN H., immunologist, conference developer and consultant; b. San Francisco, Sept. 19, 1940; s. Fred and Lily (Wong) T.; m. Madeline R. Nobori, June 13, 1964; children: Darren, Alyson. B.A. in Biochemistry, U. Calif.-Berkeley, 1963; M.S., U. Ariz., 1967, Ph.D. in Microbiology and Immunology, 1970. Research fellow in immunology Stanford U. Med. Sch., 1970-73; asst. prof. Northwestern U. Med. Sch., 1973-77; asst. prof., supr. Human Tumor Immunobiology Lab., U. Tex. Med. Sch.-Houston, 1977-83, assoc. prof. biochemistry, molecular biology, surgery, 1983—; chmn./organizer Nat. Symposium on Liposomes and Immunobiology, 1980, Nat. Symposium on Hybridomas and Cellular Immortality, 1981, Internat. Beijing Symposium on Interaction Traditional Chinese Medicine and Western Medicine: Impact on Immunology, 1983. Author: books, including Hybridomas and Cellular Immortality; contbr. numerous articles, chpts. to profl. publs. Recipient Research Career Devel. award Nat. Cancer Inst., 1979; Stanford U. Dean's fellow, 1970; Internat. Union Against Cancer exchange fellow, 1981. Mem. Am. Assn. Immunologists, Am. Assn. Cancer Research, Soc. Exptl. Biology and Medicine, N.Y. Acad. Scis., Fedn. Am. Scientists, Sigma Xi, Beta Beta Beta. Presbyterian. Subspecialties: Cancer research (medicine); Cellular engineering. Current work: Molecular immunology of human cancers. Patentee cellular prodn. of carcinoembryonic antigen. Office: U Tex Med Sch MSMB 6240 Houston TX 77030

TOMA, RAMSES BARSOUM, food science and nutrition educator; b. Heliopolis, Cairo, Egypt, Nov. 9, 1938; came to U.S., 1968, naturalized, 1973; s. Barsoum Toma and Fieka Ibrahim (Ghobriel) Khalil; m. Rosette R. Habib, Sept. 7, 1969; 1 son, Narmer. B.Sc., Ain Shams U., Cairo, 1959, M.Sc., 1965; Ph.D., La. State U., Baton Rouge,

1971; M.P.H., U. Minn., Mpls., 1980. Chemist, sr. chemist Ministry Food Supplies, Cairo, 1961-67; chemist quality assurance Crystal Foods, Inc., New Orleans, 1968; grad. research asst. La. State U., Baton Rouge, 1969-71; dir. tech. services Evangeline Foods, St. Martinsville, La., 1972; asst., assoc. prof., chmn. U. N.D., Grand Forks, 1972-80, prof. food sci./nutrition, 1980—; vis. assoc. prof. Mansoura U., Egypt, 1978-79. Contbr. articles in field to profl. jours. Vice-pres. Greek Orthodox Ch., N.D. chpt., 1978-80; adv./translator Fellowship of ISA, Bread for the World, Washington, 1982; mem. Trade Mission to Middle East countries State of N.D., Bismarck, 1976. Mem. Inst. Food Tech., Am. Dietetic Assn., Am. Chemist Soc., Am. Home Econs. Assn., Egyptian Am. Scholars Assn., Am. Pub. Health Assn., Am. Inst. Nutrition, Am. Assn. Cereal Chemists. Republican. Coptic Orthodox. Subspecialties: Food science and technology; Nutrition (medicine). Current work: Chemical analyses and nutrition evaluation of foods, food product development, basic and applied research in cereal chemistry. Home: 3640 9th Ave N Grand Forks ND 58201 Office: Dept Home Econs and Nutrition Box 8273 University Sta Grand Forks ND 58202

TOMASZ, ALEXANDER, microbiology educator, biochemistry and cell biology researcher; b. Budapest, Hungary, Dec. 23, 1930; m., 1956; 1 child. Diploma, Pazmany Peter U., Budapest, 1953; Ph.D. in Biochemistry, Columbia U., 1961. Research assoc. in cytochemistry Inst. Genetics, Hungarian Nat. Acad., 1953-56; Am. Cancer Soc. fellow, guest investigator genetics Rockefeller U., N.Y.C., 1961-63, from asst. prof. to assoc. prof. genetics and biochemistry, 1963-77, prof. microbiology, chmn. dept. microbiology, 1977—, now also dir. microbiology research lab. Mem. AAAS, Am. Soc. Microbiology, Am. Soc. Cell Biology, Harvey Soc. Subspecialty: Cell biology. Office: Dept Microbiology Rockefeller U New York NY 10021

TOME, RICHARD EARLE, mechanical engineer; b. Berea, Ohio, Mar. 1, 1936; s. Lloyd Edward and Ruth Helen (Blazek) T.; m. Eileen Costolo, July 1, 1961; children: Brian, Kevin, Kari. B.A., Baldwin Wallace Coll., 1959; B.S.M.E., Columbia U., 1959; M.S.M.E., Case Inst. Tech., 1961. Registered profl. engr., Pa. Grad. asst. dept. mech. engring. Case Inst. Tech., 1959-61; mech. engr. water reactor divs. Westinghouse Electric Corp., Pitts., 1961—, fellow engr. in primary components engring., nuclear tech. div., 1976—. Author numerous company publs. Ruling elder First Presbyterian Ch., Murrysville, Pa., 1978—; adult committeman Troop 205 Westmoreland-Fayette council Boy Scouts Am., 1981—. Mem. ASME (cert. for services in devel. standards and codes 1978, mem. Working Group on Vessel Design of Sect. III Nuclear Power Plant Components of Boiler and Pressure Vessel Code 1963—). Republican. Subspecialty: Mechanical engineering. Current work: Design, analysis and fabrication of pressurized water nuclear power plant reactor vessels and reactor coolant loop piping. Inventor self adjustable reactor vessel head insulation, improved reactor vessel head removable insulation. Home: 3710 Gleneagle Dr Murrysville PA 15668 Office: Westinghouse Electric Corp Nuclear Tech Div PO Box 355 Pittsburgh PA 15230

TOMEI, L. DAVID, cell physiologist; b. Williamsport, Pa., Apr. 27, 1945; s. Louis J. and Florence V. (Orbanac) T.; m. Angela M. Huber, Sept. 3, 1966; m. Louise M. Bartels, July 27, 1978; children: Annette, Monica, Maria, David, Victoria. B.S., Canisius Coll., 1968; M.S., SUNY, Roswell Park Meml. Inst., Buffalo, 1970, Ph.D., 1974. Research chemist Agrl. Research Service, U.S. Dept. Agr., Plum Island Animal Disease Ctr., 1974-75; assoc. dir. research Don Monti Leukemia Found.; career scientist dept. medicine North Shore Univ. Hosp., L.I., N.Y., 1975-76; sr. cancer research scientist Roswell Park Meml. Inst., Buffalo., 1976-80; Comprehensive research scientist Ohio State U. Cancer Ctr., Columbus, 1980—. Contbr. articles to profl. jours. Mem. Am. Assn. Cancer Research., Cell Kinetics Soc. Subspecialties: Cell study oncology; Optical image processing. Current work: Regulation of the phenotypic expression of malignancy cell cycle control; information transfer, cell culture, image analysis, laser optics, high speed scanning. Office: Ohio State U Comprehensive Cancer Center Suite 302 410 W 12th Ave Columbus OH 43210

TOMITA, NOBUYA, engineer; b. Japan, Feb. 25, 1941; s. Niuemon and Tsune T.; m. Taeko, Aug. 12, 1966; children: Aoy V., Ryu W. B.Engring. Sci., U. Tokyo; M.S.M.E., Ph.D., U. Ill. Registered profl. engr., Ill. Project mgr. U.S. Indsl. Chem. Co., Cin., 1981—. Contbr. articles to profl. jours. Mem. ASME, ASTM, Nat. Soc. Profl. Engrs. Baptist. Subspecialties: Theoretical and applied mechanics; Mechanical engineering. Current work: Crack initation, facture, gatigue.

TOMIZAWA, JUNICHI, molecular biologist; b. Tokyo, June 24, 1924; U.S., 1971; s. Zenjiro and Haruno T.; m. Keiko Ohomura, Apr. 1, 1954. B.S., Tokyo U., 1947, Ph.D., 1957. Mem. research staff Nat. Inst. Health of Japan, 1947-62, head dept. chemistry, 1962-69; prof. biology Osaka (Japan) U., 1969-71; chief sect. molecular genetics lab. molecular biology Nat. Inst. Arthritis, Diabetes and Digestive and Kidney Diseases, NIH, Bethesda, Md., 1971-. Contbr. articles to profl. jours. Recipient Japanese Genetics Soc. award, 1962; Matsuanaga award, 1967; Superior Service award USPHS, HEW, 1976. Mem. Am. Soc. Biol. Chemist (hon.), Am. Acad. Arts and Scis. (hon.). Subspecialty: Molecular biology. Current work: Molecular biology of nucleic acids, DNA synthesis, chromosome replication. Office: NIH 9000 Rockville Pike Bldg 2 Room 304 Bethesda MD 20205

TOMLINSON, RICHARD LEE, nuclear physicist; b. Davenport, Iowa, July 16, 1929; s. Thell Dodson and Ayliffe Hazel (Lewis) T.; m. Carol Jeanette Nyre, July 11, 1959; children: Kari Lee, Kyle Marie. B.S., Lake Forest Coll., 1951; postgrad, U. Calif., Berkeley, 1955-56; cert. bus., Diablo Valley Coll., Pleasant Hill, Calif., 1976. Physicist Gen. Electric Co., Richland, Wash., 1951-55, N. Am. Aviation Co., Canoga Park, Calif., 1955-62; physicist Aerojet-Gen. Corp., San Ramon, Calif., 1962-74; mgr. reactor ops. Aerotest Ops. Inc., San Ramon, 1974-78, Westinghouse Hanford Co., Richland, Wash., 1978—. Author: Nuclear Reactor Operator Training Manual, 1983. Mem. Am. Nuclear Soc., Am. Soc. Non-Destructive Testing, ASTM. Subspecialties: Nuclear engineering; Nuclear fission. Current work: Design, construction, operation of reactor facilities for commercial and government use for neutron radiography applications. Office: Westinghouse Hanford Co PO Box 1970 Richland WA 99352

TOMLINSON, W. JOHN, physicist; b. Phila., Apr. 3, 1938; s. W. John and Olive (Greatorex) T.; m. Barbara Kellog, June 10, 1963; 1 child, Robin B. S.B., M.I.T., 1960, Ph.D. in Physics, 1963. Mem. tech. research staff Bell Telephone Labs., Holmdel, N.J., 1965-81, supr. optical disk rec. group, 1981—. Served to capt. U.S. Army, 1963-65. Mem. Optical Soc. Am., Am. Phys. Soc. Subspecialties: Optics research; Laser data storage and reproduction. Current work: Optical disk recording technology and media; nonlinear optics. Home: 22 Indian Creek Rd Holmdel NJ 07733 Office: Bell Labs 555 Union Blvd Allentown PA 18103

TOMONTO, JAMES ROBERT, consultant; b. White Plains, N.Y., Apr. 14, 1932; s. James and Gladys (Hammond) T.; m. Irene Terenzio, Apr. 6, 1946; children: Robert J., Charles V., Patrice I., Kristin E., Melissa A. B.S. in Physics, Villanova U., 1954; M.S., Rensselaer Poly, Inst., 1959; postgrad., NYU, 1967-70. Elec. engr. Airborne Inst. Lab., Thornwood, N.Y., 1957-58; nuclear engr. Alco Products Inc., Schenectady, 1958-59; exptl. physicist Knolls Atomic Power Lab., Schenectady, 1959-64; mgr. nuclear engring. United Nuclear Corp., Elmsford, N.Y., 1964-74; sr. cons. Fla. Power & Light Co., Miami, 1974—; v.p. HTH Assocs., Miami, 1978—. Contbr. numerous articles, reports to profl. lit. Nat. pres. Christian Family Movement, Ames, Iowa, 1977-81; exec. couple Internat. Confedn. Christian Family Movements N.Am., 1980—. Served to lt. USNR, 1954-77. Mem. Am. Nuclear Soc.; mem. ASME, Sigma Xi. Subspecialties: Nuclear physics; Nuclear fission. Home: 14311 S W 74th Ct Miami FL 33158 Office: Fla Power & Light Co 9200 W Flagler St Miami Fl 33152

TOMPKINS, LAURIE, biology educator; b. N.Y.C., Mar. 29, 1950; d. Bruce and Jean (Murray) T. B.A., Swarthmore Coll., 1972; Ph.D., Princeton U., 1977. Asst. prof. biology Temple U., 1981—. Contbr. articles profl. jours. NIH grantee, 1981-84. Mem. AAAS, Genetics Soc. Am., Behavior Genetics Assn., Animal Behavior Soc., Sigma Xi. Subspecialties: Gene actions; Neurobiology. Current work: Analyzing genetic control of chemoreception and reproductive behavior in Drosophila melanogaster. Office: Dept of Biology Temple University Philadelphia PA 19122

TOMREN, DOUGLAS ROY, optical physicist; b. Buffalo, July 16, 1945; s. Raymond Henry and Dorothy Gertrude (Berg) T.; m. Linda Sue Kenne, Aug. 24, 1971; children: Holly Ann, Erik Roy, Michael Douglas. B.S., UCLA, 1967; postgrad., Optical Scis. Center, U. Ariz., 1971. Mem. tech. staff RCA, Burlington, Mass., 1967-69; laser engr. Martin Marietta Co., Orlando, Fla., 1969; holographer TRW, Redondo Beach, Calif., 1972-76, laser effects physicist, 1976—, project mgr., 1978—. Mem. 4th dist. adv. council City of Long Beach, Calif., 1979—. Mem. Optical Soc. Am., AIAA. Democrat. Methodist. Subspecialties: Laser research; Holography. Current work: Laser effects testing, laser research and development; applied research in nonlinear optical phenomena, development and testing of laser hardened materials. Office: TRW One Space Park R1/1112 Redondo Beach CA 90278

TONDREAU, SUK-PAN, (SUE), immunologist/microbiologist, cell biologist; b. Seremban, Malaysia, Sept. 21, 1935; b. May 24, 1974.; d. Ping-Kee and Ha (Sz-Tho) Chan; m. Cyril Vincent Tondreau. B.S. in Math. Sci, Cambridge (Eng.) U., 1955; M.Sc. in Biology and Chemistry, King George Instn., 1957. Research asst. U.S. Army Med. Unit, Kuala Lumpur, Malaysia, 1958-62, N.Y. U. Med. Ctr., N.Y.C., 1963-66; supr. immunology sect. Microbiol. Assocs., Bethesda, Md., 1966-69; immunologist/biologist biomed. research div. Litton Bionetics, Inc., Kensington, Md., 1969-75, prin. investigaotr, dept. cell biology, 1975—. Contbr. articles to med. jours. Mem. Am. Assn. Immunologists, Am. Chem. Soc., Am. Soc. Microbiology, N.Y. Acad. Scis. Roman Catholic. Subspecialties: Cancer research (medicine); Cellular engineering. Current work: Lymphokines: production, quantitation and separation. Home: 4716 Bel Pre Rd Rockville MD 20853 Office: 5516 Nicholson Ln C-200 Kensington MD 20895

TONNA, EDGAR ANTHONY, histology educator, university research administrator, cell physiology and chemistry researcher; b. Malta, May 10, 1928; m., 1951; 4 children. B.S., St. John's U., N.Y., 1951; M.S., N.Y. U., 1953, Ph.D. in Biology, 1956. Research collaborator div. exptl. pathology Med. Research Ctr., Brookhaven Nat. Lab., 1956-69, head histochem. and cytochem. research lab., 1959-67; prof. histology Grad. Sch. Basic Med. Sch., N.Y. U., 1967—; dir. Inst. Dental Research, Coll. Dentistry, 1971—, dir. lab. cellular research, 1967—; research biochemist Hosp. Spl. Surgery, N.Y.C., 1953-56, head histochem. and cytochem. research lab., 1956-59; adj. assoc. prof. Grad. Sch., L.I.U., 1956-62; cons. radiobiology Inst. Dental Research, N.Y. U., 1964-67. Editor in chief: Gerodontology, 1981—. Recipient R. Morton cert. and G. Mendel award St. John's U. Fellow Gerontology Soc., N.Y. Acad. Sci., Royal Microscopy Soc.; mem. Histochem. Soc., Sigma Xi. Subspecialties: Dental growth and development; Physiology (biology). Office: Inst Dental Research NY U Coll Dentistry 345 E 24th St New York NY 10010

TOOMRE, ALAR, mathematics educator, astronomer; b. Rakvere, Estonia, Feb. 5, 1937; came to U.S., 1949, naturalized, 1955; s. Elmar and Linda (Aghen) T.; m. Joyce Stetson, June 14, 1958; children: Lars, Erik, Anya. B.S. in Aero. Engring, MIT, 1957, 1957; Ph.D., Manchester (Eng.) U., 1960. Instr. math. MIT, Cambridge, 1960-62, asst. prof., 1963-65, assoc. prof., 1965-70, prof. applied math., 1970—; fellow Inst. Advanced Study, Princeton, N.J., 1962-63; vis. assoc. Calif. Inst. Tech., Pasadena, 1969-70. Guggenheim fellow, 1969-70; Fairchild scholar Calif. Inst. Tech., 1975. Mem. Am. Astron. Soc., Internat. Astron. Union, Am. Acad. Arts and Scis., Nat. Acad. Scis. Subspecialties: Theoretical astrophysics; Applied mathematics. Current work: Research in dynamics of galaxies. Office: Room 2-371 MIT Cambridge MA 02139

TOON, OWEN BRIAN, research geophysicist; b. Bethesda, Md., May 26, 1947; s. Owen Russel and Adrienne Joan (Van Burk) T.; m. Teresa Eileen Hand, Sept. 6, 1968. Student, UCLA, 1965-67, U. Edinburgh, Scotland, 1967-68; A.B., U. Calif.-Berkeley, 1969; Ph.D., Cornell U., 1975. Research asst. Cornell U., Ithaca, N.Y., 1973-75, research assoc., 1977-78; NRC resident assoc. Nat. Acad. Scis., 1975-79; research scientist NASA Ames Research Ctr., Moffett Field, Calif., 1978—. Contbr. articles to profl. jours. Recipient Exceptional Sci. Achievement medal NASA, 1983. Mem. Am. Geophys. Union, Am. Meteorol. Soc., Am. Astron. Soc. Subspecialties: Planetary atmospheres; Climatology. Office: M S 245-3 NASA Ames Moffett Field CA 94035

TOPPER, YALE J., biochemist; b. Chgo., Aug. 11, 1916; s. Ben and Aida T.; m. Hildegrad P. Pokorny, Oct. 12, 1956; children—David, Nina, James, Ethan. B.S., Northwestern U., 1942; M.A., Harvard U., 1943, Ph.D., 1947. Research fellow dept. biochemistry Harvard U., 1946-48; asso. Public Health Research Inst., City of N.Y., 1948-53; fellow Am. Heart Assn., Mass. Gen. Hosp., Boston, 1953-54; with Nat. Inst. Arthritis, Diabetes, Digestive and Kidney Diseases, NIH, Bethesda, Md., 1954—, chief sect. intermediary metabolism, 1963—. Contbr. articles to profl. jours. Mem. Am. Soc. Biol. Chemists, Endocrine Soc. Subspecialties: Developmental biology; Cell and tissue culture. Current work: Hormoedependent gene expression. Home: 11608 Danville Dr Rockville MD 20852 Office: NIH Bethesda MD 20205

TORAN-ALLERAND, C(LAUDE) DOMINIQUE, neurosciences educator; b. Paris, Oct. 19, 1933; U.S., 1942, naturalized, 1948; d. Jacques G. and Catherine G. (Elias) Aller; m. Edward A. Toran, Sept. 23, 1972. A.B. summa cum laude, Smith Coll., 1955; M.D., Albany Med. Coll., 1959. Diplomate: Am. Bd. Med. Examiners, Am. Bd. Psychiatry and Neurology. Intern in medicine Albany (N.Y.) Med. Ctr. Hosp., 1959-60, asst. resident, 1960-61; asst. resident dept. neurology Columbia U. Coll. Phys. and Surgs. and Columbia-Presbyn. Med. Ctr., 1961-63, fellow, 1963-66, asst., 1966-68; instr. dept. neurology N.Y.U. Sch. Medicine, N.Y.C., 1968-69, asst. prof., 1969-73; asst. vis. neurologist Bellevue Hosp., N.Y.C., 1968-73; staff neurologist VA Hosp., N.Y.C., 1968-73; asst. attending neurologist Columbia-Presbyn. Med. Center, N.Y.C., 1974-81, assoc. attending neurologist, 1981—; research assoc. Internat. Inst. for Study of Human Reprodn., Columbia U. Coll. Phys. and Surg., 1973-74, asst. prof., 1974-81, assoc. prof. clin. neurology, 1981—. Contbr. articles to profl. jours. Nat. Multiple Sclerosis Soc. fellow, 1963-65; 70-72; USPHS fellow, 1965-68; Rockefeller Found. grantee, 1974-77; NIH grantee, 1974—; NSF grantee, 1977—; Nat. Found./March of Dimes grantee, 1971-73, 77-81; W.T. Grant Found. grantee, 1979-80; Whitehall Found. grantee, 1982—; NIMH grantee, 1978—. Mem. AAAS, Tissue Culture Assn., Soc. for Neurosci., Am. Assn. Anatomists, Internat. Soc. for Devel. Neurosci., European Neurosci. Assn., N.Y. Acad. Scis., Phi Beta Kappa, Sigma XI. Subspecialties: Neurobiology; Cell and tissue culture. Current work: Developmental neurobiology; hormones and brain development; tissue culture of developing nervous tissue; ontogeny of sexual differentiation of the brain; steroid receptor ontogeny; neurobiology of alpha-fetoprotein. Office: Columbia U Coll Phys and Surg 630 W 168th St New York NY 10032

TORIGOE, RODNEY YOSHITO, clinical psychologist; b. Honolulu, Feb. 1, 1945; s. Samuel Yoshio and Sueko Thelma (Inakazu) T.; m. Bess Misao, Aug. 7, 1971; 1 dau., Tiffany Kikue Nakamura. B.A., U. Hawaii, 1968; M.A., U. Colo., 1973, Ph.D., 1976. Diplomate: Am. Acad. Behavioral Medicine. Affirmative action officer Western Interstate Commn. for Higher Edn., Boulder, Colo., 1973-74; psychology assoc. Ariz. State Hosp., Phoenix, 1975-76; clin. psychologist VA, Phoenix, 1976-78, chief, Honolulu, 1978—, chief psychology service, 1980—; asst. clin. prof. psychology U. Hawaii, 1980—, asst. clin. prof. psychiatry, 1981—. Pres. bd. dirs. House, Inc., 1981-82; mem. Neighborhood Bd. Hawaii, 1981-82. NIMH fellow, 1971-72; recipient recognition award Western Interstate Commn. for Higher Edn., 1974. Mem. Am. Psychol. Assn., Hawaii Psychol. Assn. (treas. Honolulu 1981-82), Am. Assn. Sex Educators, Counselors, and Therapists, Biofeedback Soc. Am., Am. Assn. Marriage and Family Therapy, Internat. Council Psychologists. Democrat. Buddhist. Current work: Sexual therapy, paradoxical maxims, sexuality for the aged, respiration and biofeedback, medical psychology. Office: VA OPC PO Box 50188 Honolulu HI 96850

TORMEY, DOUGLASS C., oncology researcher; b. Madison, Wis., Sept. 2, 1938; s. Weston C. and Marion D. (Douglass) T.; m. Patricia Bevington, Jan. 25, 1964; children: Bruce, Paula, Marc. B.S., U. Wis., 1960, M.D., 1964, M.S., 1964, Ph.D., 1969. Intern U. Calif. Med. Ctr., San Francisco, 1964-65, resident, 1965-66; postdoctoral fellow U. Wis. Med. Ctr., Madison, 1966-69; fellow in oncology Roswell Park Meml. Inst., Buffalo, 1969-70; staff oncologist Walter Reed Gen. Hosp., Washington, 1970-72; with NIH, Bethesda, Md., 1972-76; cons. in research, cell physiology, dept. biol. scis. George Washington U., Washington, 1974-76; assoc. prof. dept. human oncology U. Wis. Med. Sch., Madison, 1976-82, assoc. prof. dept. medicine, 1976-82, prof. depts. human oncology and medicine, 1982—; cons. William S. Middleton Meml. VA Hosp., Madison, 1979—. Contbr. articles to profl. jours. Served to maj. U.S. Army, 1970-72; to sr. surgeon USPHS, 1972-76. Recipient Borden award, 1964; Am. Cancer Soc. postdoctoral fellow, 1967-69; grantee NIH, Armour Pharm., ICI Americas, Melay Labs., Eli Lilly Co., others. Mem. AAAS, Am. Assn. for Cancer Research, Am. Soc. Hematology, Am. Soc. Clin. Oncology, Am. Fedn. Clin. Oncologic Socs., Cell Kinetic Soc., Am. Fedn. for Clin. Research, Internat. Assn. for Breast Cancer Research, Sigma Xi, Nu Sigma Nu. Subspecialties: Oncology; Cancer research (medicine). Current work: Therapeutic research in clinical oncology. Office: 600 Highland Ave K4/632 CSC Madison WI 53792

TORNECK, CALVIN DAVID, endodontist; b. Toronto, Ont., Can., June 9, 1935; s. Maxwell and Ruth (Kuchar) T.; m. Fay Rotman, June 1, 1958; children: Sandra Gail, Lynda Beth, Paula Teresa. D.D.S., U. Toronto, 1958; M.S., U. Mich., (1959.). Diplomate: Am. Bd. Endodontists. Clin. endodontist, Toronto, 1959—; clin. demonstrator U. Toronto Faculty Dentistry, 1959-61, assoc. dentistry, 1961-76, asst. prof. faculty dentistry, 1976—; cons. dentistry Mt. Sinai Hosp., Toronto, 1960—, Standards Council Can., 1976—, U.S. Naval Dental Ctr., Bethesda, Md., 1978-81. Fellow Internat. Coll. Dentists, Royal Coll. Dentistry; mem. Internat. Assn. Dental Research (pres. elect pulp biology group 1981-82), Can. Acad. Endodontics (pres. 1966—), Am. Assn. Endodontists (exec. council 1969-72), Fedn. Dentair Internationale (chmn. com. on biol. testing-endodontics sect. 1972-74), Omicron Kappa Upsilon. Jewish. Subspecialties: Endodontics; Oral biology. Office: U Toronto Faculty Dentistry 124 Edward St Toronto ON Canada M5G 1G6

TORR, MARSHA RUSSELL, space science physicist, educator; b. Pretoria, South Africa, Dec. 4, 1942; d. John Russell and Joan Ereina Marshall (Vercueil) Harding; m. Douglas G. Torr, Dec. 15, 1965. B.Sc. in Physics with distinction, Rhodes U., Grahamstown, S. Africa, 1963, 1964, M.Sc. in Physics with distinction, 1966, Ph.D., 1969. Jr. lectr. Rhodes U., 1965; lectr. physics U. Witwatersrand, S.Africa, 1966-67; sr. chief research officer South African Council Sci. and Indsl. Research, 1968-81; vis. fellow Yale U., 1973-74; assoc. research scientist U. Mich., 1974-80; prof. physics Utah State U., 1980—; mem. various coms. Spacelab 1. Assoc. editor: Jour. Geophys. Research-Space Physics, 1982—; contbr. numerous articles profl. jours. NASA grantee; NSF grantee. Mem. Am. Geophys. Union, Optical Soc. Am., South African Inst. Physics, Am. Astron. Union, AAAS. Subspecialties: Aeronomy; Satellite studies. Current work: Study of physics and chemistry of near-earth environment (atmosphere/ionosphere/magnetosphere) by theoretical and experimental techniques; development of remote sensing instrumentation with which to probe this region. Office: Dept of Physics UMC 41 Utah State U Logan UT 84322

TORRENCE, PAUL FREDERICK, biochemist, researcher; b. New Brighton, Pa., Apr. 22, 1943; s. Orville Theodore and Ada Ruth (Gordon) T.; m. Glenda Ann Margeson, June 3, 1967; children: Ian James, Jessica Renee. B.S. in Chemistry magna cum laude with honors, Geneva Coll., 1961-65; Ph.D. (NSF trainee), SUNY-Buffalo, 1969. Staff fellow Lab. Chemistry, Nat. Inst. Arthritis and Metabolic Diseases, NIH, Bethesda, 1969-71, sr. staff fellow, 1971-74; research chemist. Lab. Chemistry, Nat. Inst. Arthritis, Diabetes, Digestive and Kidney Diseases, NIH, Bethesda, 1974—; ad hoc cons. Nat. Inst. Allergy and Infectious Diseases. Contbr. numerous articles, chpts. to profl. jours.; bd. editors: Antiviral Research, 1981—. Recipient EEO award NIH, 1981. Mem. Sierra Club, Nat. Conservation and Parks Assn., Wilderness Soc., Nat. Audubon Soc., Am. Chem. Soc., Am. Assn. Biol. Chemists, Soc. Exptl. Biology and Medicine, Am. Soc. Microbiology, AAAS. Subspecialties: Biochemistry (medicine); Medicinal chemistry. Current work: Interferon induction and action, nucleic acids, nucleosides, antiviral and antitumor agents. Home: 106 E Deer Park Dr Gaithersburg MD 20877 Office: NIH Bldg 4 Room 126 Bethesda MD 20205

TORRES-MEDINA, ALFONSO, veterinary science educator; b. Colombia, Aug. 28, 1945; came to U.S., 1978; s. Alfonso Torres-Rudas and Maria Elena Medina de Torres; m. Maria Cecilia Cuervo, Aug. 26, 1972; children: Julian, Marcela. D.V.M., Nat. U. Colombia, 1968; M.S., U. Nebr., 1971, Ph.D., 1973. Asst. instr. Sch. Vet. Medicine, Nat. U. Colombia, Bogota, 1969; asst. prof. dept. vet. sci. U. Nebr.-Lincoln, 1973-75, 78-81, assoc. prof., 1981—; new products mgr. for Latin Am. Ames Co. div. Miles Labs. Inc., 1976-78. Contbr. articles to profl. jours. FAO-UN postgrad. fellow, 1971-73. Mem. Conf. Research Workers in Animal Diseases, Sigma Xi, Gamma Sigma Delta. Roman Catholic. Subspecialties: Virology (veterinary medicine); Pathology (veterinary medicine). Current work: Viral enteric infections of newborn animals (calves, piglets, and human). Office: U Nebr 126 VBS Dept Veterinary Sci Lincoln NE 68583

TORRIANI, ANNAMARIA GORINI, educator; b. Milan, Italy, Dec. 19, 1918; came to U.S., 1955, naturalized, 1959; d. Carlo and Ada (Forti) T.; m. Luigi Gorini, Dec. 6, 1959; 1 son: Daniel. Ph.D., U. Milan, 1942. Research assoc. Instituto Chimica e Biochimica, G. Ronzoni, Milan, Italy, 1942-48; charge de recherche dept. cellular physiology Institut Pasteur, Paris, 1948-56; Fulbright postdoctoral fellow dept. microbiology N.Y.U., 1956-58; research assoc. Harvard Med. Sch., Boston, 1958-59; research assoc. biology dept. Harvard U., 1959-60; research assoc. M.I.T., Cambridge, 1960-71, assoc. prof. microbiology, 1971-76, prof., 1976—. Contbr. articles to profl. jours. Mem. Am. Soc. Microbiology. Subspecialties: Microbiology; Molecular biology. Home: 115 Longwood Ave Brookline MA 02146 Office: Dept Biology Mass Inst Tec Cambridge MA 02114

TORTORELLA, MICHAEL J., communications engineer; b. N.Y.C., June 15, 1947; s. Salvatore L. and Carmela (Barile) T.; m. Elizabeth A. Garafola, Dec. 28, 1969; 1 son, Matthew. A.B. in Math, Fordham U., 1968, M.S., Purdue U., 1970, Ph.D., 1973. Lectr. in math. U. Wis.-Milw., 1973-75; mem. tech. staff Bell Telephone Labs., Holmdel, N.J., 1975—. Contbr. articles to profl. jours. Served to capt. USAR, 1973. Mem. Inst. Math. Statistics, Soc. for Indsl. and Applied Math. Subspecialties: Operations research (mathematics); Probability. Current work: Stochastic processes and applications to teletraffic theory and reliability of communications systems, including undersea fiber-optic systems. Office: Bell Telephone Labs Crawfords Corner Rd Holmdel NJ 07733

TORTORICI, MARIANNE RITA, radiological scientist, educator; b. Waterbury, Conn., May 22, 1947; d. Anthony and Carmela Emily (DiNapoli) T. B.S., Incarnate Word Coll., San Antonio, 1972; M.S., U. Nev., Las Vegas, 1975; Ed.D., U. Houston, 1979. Staff radiographer St. Raphael's Hosp., New Haven, Conn., 1968-72, Santa Rosa Med. Ctr., San Antonio, 1972; dir. Sch. Radiol. Tech., North Country Community Coll., Saranac Lake, N.Y., 1972-74; prof., chmn. dept. radiol. scis. U. Nev., Las Vegas, 1974—; lectr. U. Calif.-Santa Barbara Extension, Barstow, Calif., 1980, Western Intercollegiate Consortium on Edn. in Radiol. Tech., Reno, 1981, v.p., 1980-82; accrediting visitor AMA, Chgo., 1982—. Author: Fundamentals of Angiography, 1983, also lab. manuals.; Contbr. articles to profl. jours. Mem. Nev. Soc. Radiologic Tech. (pres. 1976-77, dir. 1977-78), Am. Soc. Radiologic Technologists (counselor 1974-76), Phi Kappa Phi, Alpha Beta Gamma. Subspecialty: Diagnostic radiology. Current work: Radiographic techniques and quality control, angiography, education. Home: 4139 Newcastle Las Vegas NV 89103 Office: Dept Radiologic Scis U Nevada 4505 Maryland Pkwy Las Vegas NV 89154

TOSTESON, DANIEL CHARLES, physiologist, dean medical school; b. Milw., Feb. 5, 1925; s. Alexis H. and Dilys (Bodycombe) T.; m. Penelope Kinsley, Dec. 17, 1949 (div. 1969); children: Carrie Marias, Heather Reich, Tor, Zoe Losada; m. Magdalena Tieffenberg, July 8, 1969; children: Joshua, Ingrid. Student, Harvard U., 1942-44, M.D., 1949; Dr. hon. causa, U. Liege, 1983. Fellow physiology Harvard Med. Sch., 1947-48; intern, then asst. resident medicine Presbyn. Hosp., N.Y.C., 1949-51; research fellow medicine Brookhaven Nat. Lab., 1951-53; lab. kidney and electrolyte metabolism Nat. Heart Inst., 1953-55, 57, research fellow biol. isotope research lab., Copenhagen, 1955-56; research fellow Physiol. Lab., Cambridge, Eng., 1956-57; assoc. prof. physiology Washington U. Sch. Medicine, St. Louis, 1958-61; prof., chmn. dept. physiology and pharmacology Duke U. Sch. Medicine, 1961-75, James B. Duke Distinguished prof., 1971-75; dean div. biol. scis., dean Pritzker Sch. Medicine U. Chgo., Lowell T. Coggleshall prof. med. scis., v.p. for Med. Center, 1975-77; dean and Caroline Shields Walker prof. physiology Harvard Med. Sch., Boston, 1977—, pres. Med. Center, 1977; mem. molecular biology panel NSF, 1959-62; cons. sci. rev. com. NIH, 1964-67, nat. adv. gen. med. scis. council, 1982—; mem. U.S. Office Tech. Assessment, 1976; ethics adv. bd. HEW, 1977—; nat. adv. gen. med. scis. council NIH, 1982—; mem. governing bd. NRC, 1977; founding mem. Nat. Found. for Depression, 1982—; mem. sci. com. Fondation pour l'Etude du Systeme Nerveux Central et Peripherique, 1982—. Mem. Am. Physiol. Soc. (council 1967-75, pres. 1973-74), Soc. Gen. Physiologists (pres. 1968-69), Nat. Found. for Depression (founding 1982), Biophys. Soc. (council 1970-73), Mem. Found. pour L'Etude du Systeme Nerveux Central et Peripherique (sci. com. 1982—), AAAS, Assn. Am. Med. Colls. (chmn. council acad. socs. 1969-70, chmn. assembly 1973-74), Assn. Am. Physicians, Inst. Medicine, Nat. Acad. Sci. (council 1975-78); mem. Red Cell Club, Soc. Health and Human Values; Mem. Danish Royal Soc. (fellow), Alpha Omega Alpha. Subspecialty: Membrane biology. Current work: Research directed toward understanding the cellular functions and molecular mechanisms of ion transport across membranes. Spl. research cellular transport processes, red cell membranes. Office: Office of Dean Harvard Medical School 25 Shattuck St Boston MA 02115

TOTH, ATTILA, obstetrician/gynecologist; b. Szekszard, Hungary, Mar. 19, 1940; came to U.S., 1968, naturalized, 1977; s. Sandor and Ibolya (Tarjani) T.; m. Constance Wesley Brooks, Mar. 20, 1980. M.D., U. Budapest, 1964. Diplomate: Am. Bd. Pathology, Am. Bd. Obstetrics and Gynecology. Research fellow Cleve. Clinic, 1968-69; intern Mt. Sinai Hosp., N.Y.C., 1969-70, resident in pathology, 1970-73; asst. prof. Mt. Sinai Med. Sch., N.Y.C., 1973-74; resident obstetrics/gynecology N.Y. Hosp./Cornell, N.Y.C., 1974-77, asst. prof., attending, 1977—; dir. MacLeod Lab. Infertility, N.Y.C., 1977—. Contbr. articles to profl. jours. Mem. Am. Fertility Soc., Am. Soc. Andrology. Republican. Mem. Dutch Reform Ch. Subspecialties: Obstetrics and gynecology; Pathology (medicine). Current work: Role of infection in male/female reproductive performance. Home: 436 E 69th St Apt 8C New York NY 10021 Office: New York Hosp Cornell Med Center 525 E 68th St New York NY 10021

TOTH, BELA, exptl. pathologist; b. Pecs, Hungary, Oct. 26, 1931; s. Bela and Maria (Csicsek) T.; m. Anna Rozsa, Sept. 2, 1963; children—Agnes, Reka, Attila, Bela. D.V.M., U. Vet. Sci., Budapest, 1956. Research asst. div. oncology Chgo. Med. Sch., 1959-61, research assoc., 1961-63, asst. prof., 1963-66; Eleanor Roosevelt Internat. Cancer Research fellow, Rehovoth, Israel, 1967; assoc. prof. Eppley Inst. Research in Cancer, U. Nebr.-Omaha, 1968-72, prof., 1972—. Editorial bd.: Anticancer Research; contbr. numerous articles to profl. jours. Research award UPHS, 1969-74; U.S. Nat. Com. Internat. Union against Cancer, 1972—. Mem. Am. Assn. Cancer Research, European Assn. Cancer Research, Am. Assn. Pathologists, Internat. Acad. Pathologists. Subspecialties: Cancer research (medicine); Pathology (medicine). Current work: Chem. carcinogenesis, exptl. pathology. Office: 24th & Dwey Ave Omaha NE 68105 Home: 11839 Cedar St Omaha NE 68144

TOU, JULIUS T., computer science educator; b. Shanghai, China, Aug. 15, 1925; m. Lisa Tou; children: Albert, Fred, Ivan, Sylvia. B.S., Nat. Chiao-Tung U., 1947; M.S., Harvard U., 1950; D.Eng., Yale U., 1952. Project engr. Philco Corp., 1952-54; asst. prof. U. Pa., 1954-57; assoc. prof. Purdue U., 1957-61; now cons.; prof. elec. engring., dir. Computer Sci. Lab. Northwestern U., 1961-64; dir. research Battelle Meml. Inst., Columbus, Ohio, 1964-67; grad. research prof., dir. Center for Info. Research, U. Fla., Gainesville, 1967—; cons. N.Am. Aviation, Dept. Army, IBM, Gen. Electric Co., Martin Marietta, McDonnell Douglas. Author 18 books on computer and info. sci.; editor-in-chief: Internat. Jour. Computer and Info. Sci., 1970—; editor: Internat. Series on Advances in Info. System Sci, 1968—. Fellow IEEE, Academia Sinica. Subspecialties: Graphics, image processing, and pattern recognition; Information systems (information science). Current work: Advanced automation; knowledge-based septems; image processing; CAD/CAM. Office: 339 Larsen Hall U of Fla Gainesville FL 32611

TOUGH, ALLEN MACNEILL, psychology educator; b. Montreal, Que., Can, Jan. 6, 1936; s. David Lloyd and Margaret Phyllis (Allen) T.; m. Elaine Posluns, June 10, 1981; children: (by previous marriage) Susan Anne, Paul Allen. B.A., U. Toronto, 1958, M.A., 1962; Ph.D., U. Chgo., 1965. Tchr. Scarborough (Ont.) Bd. Edn., 1959-61; asst. prof. U. Toronto, Ont., 1964-66; asst. to full prof. psychology and adult learning Ont. Inst. Studies in Edn. and U. Toronto, 1966—; cons. editor Adult Edn. Jour., 1967-73; conf. chmn. Nat. Seminar Adult Edn. Research, Toronto, 1969; v.p. UNESCO Meeting, Paris, 1979; cons. Nat. Inst. Edn., 1978-79. Author: Learning Without a Teacher, 1967, Adult's Learning Projects, 2d edit, 1979, Expand Your Life, 1980, Intentional Changes, 1982. Kellogg Found. fellow, 1963; Can. Council scholar, 1964; Ont. Inst. grantee, 1966-67. Mem. Am. Psychol. Assn., World Futures Studies Fedn., World Future Soc., Commn. Profs. of Adult Edn. (exec. com.). Subspecialties: Learning; Developmental psychology. Current work: Research on highly intentional change processes and major self-planned learning projects during adulthood. Office: Ont Inst Studies in Edn 252 Bloor St W Toronto ON Canada M5S 1V6

TOURTELLOTTE, WALLACE WILLIAM, neurologist, educator; b. Great Falls, Mont., Sept. 13, 1924; s. Nathaniel Mills and Frances Victoria (Charlton) T.; m. Jean Esther Toncray, Feb. 14, 1953; children: Wallace William, George Mills, James Millard, Warren Gerard. B.S. in Anatomy, U. Chgo., 1945, Ph.B., 1945, Ph.D. in Neurochem. Pharmacology, 1948, M.D., 1951. Diplomate: Am. Bd. Psychiatry and Neurology; lic. physician, N.Y., Mich., Calif.; cert. lab. technician Commn. Lab Accreditation. Intern Strong Meml. Hosp., U. Rochester (N.Y.) Sch. Medicine and Dentistry, 1951-52; resident in neurology U. Mich. Med. Ctr. Hosp., Ann Arbor, 1954-57; postdoctoral research dept. pharmacology U. Chgo. Toxicity Lab., 1948-51, instr. pharmacology, 1948-51; from asst. prof. to prof. neurology U. Mich., Ann Arbor, 1957-71; prof. in residence dept. neurology UCLA, 1971—, vice chmn. dept. neurology, 1971—; chief neurology service VA Wadsworth Med. Ctr., Los Angeles, 1971—; dir. neurology resident tng. program, dir. multiple sclerosis treatment ctr., co-dir. memory enhancement clinic, dir. Parkinson's disease ctr., dir. neurofunction lab., dir. Nat. Neurol. Research Bank, 1971—; dir. Human Speciman Bank of Am., 1976—; mem. adv. com. Life Inst., 1979—; cons. dept. consumer affairs Bd. Med. Quality Assurance, 1980—; cons. and lectr. in medicine; mem. numerous research coms. and orgns. Co-author: (with numerous co-authors) Post-Lumbar Puncture Headaches, 1964, Clinical Aspects of Cerebrospinal Fluid and Relevant Basic Science Information, 1973, A Selected Chronological, Annotated Bibliography of Measurements in Clinical Neurology, 1974, Measurements in Clinical Neurology, Vol. I & II, 1974, Antibodies IncorporatedElectro-Immuno Diffusion Kit, 1976, Dynamics of the Cerebrospinal Fluid System in Health and Disease, 1977, Cerebral, Brain Stem, and Spinal Cord Sensory Evoked Responses in Multiple Sclerosis and the Effects of Core Temperature, 1978, The VA Wadsworth Hospital Memory Enhancement Clinic Manual for Older Adults, 1980, others; co-editor: Multiple Sclerosis: Pathology, Diagnosis and Management, 1983; contbr. chpts. to books, articles to profl. jours. Mem. com. Internat. Maltese Mus. Fine Art. Served with USN, 1952-54. Recipient Disting. Alumni Service award U. Chgo., 1982; nominated Nobel Prize in physiology and medicine. Fellow Am. Acad. Neurology (sect. neuropharmacy 1981—, S. Weir Mitchell Neurology Research award 1959), ACP; mem. AAAS, World Fedn. Neurology (founding mem., multiple sclerosis research com. 1970—), Life Found. of Medicine and Law (adv. com., Life Bank subcom. 1981—), Family Survival Project (mem. sci. adv. council 1982—), Am. Neurol. Assn. (council 1982—), Am. Bd. Neurology and Psychiatry (asst. examiner 1965—), Internat. Soc. Neurochemistry (founding mem.), World Assn. Neurol. Commns. (founding mem.; neurochemistry commn. 1965—), Am. Soc. Pharmacology and Exptl. Therapeutics, Am. Therapeutic Soc., Am. Soc. Neurochemistry (founding mem.; exec. com. 1968), Fedn. Western Socs. of Neurol. Sci. (program chmn. 1973), Assn. Am. Med. Colls., AAUP (VAcom. 1980—), Nat. Multiple Sclerosis Soc. (nat. med. adv. bd. 1968—, So. Calif. chpt. 1976—, Hope Chest award 1982), Internat. Fedn. Multiple Sclerosis Socs. (adv. bd. 1972—), Soc. Neurosci., Am. Assn. Neuropathologists (assoc. mem.). Clubs: Pasadena Wine and Food Soc., Soc. Med. Friends of Wine, Beverly Hills Wine and Food Soc., Confrerie de la Chaine des Rotisseur (chevalier), Argentier du Baillage de Los Angeles (vice chancelier), La Chaine, Les Amis du Vin, Vigneron d'Honnear du Cru Morgan, Academie des Vins du Mts. Daphne. Lodge: Moose. Subspecialties: Neurology; Neurophysiology. Current work: Organic neurology; establishment of a neuro-function lab. to quan. Home: 1140 Tellem Rd Pacific Palisades CA 90272 Office: VA Med Ctr- West Los Angeles 691/W127 Wilshire and Sawtelle Blvds Los Angeles CA 90073

TOUSEY, RICHARD, physicist; b. Somerville, Mass., May 18, 1908; s. Coleman and Adella Richards (Hill) T.; m. Ruth Lowe, June 29, 1932; 1 dau., Joanna. A.B., Tufts U., 1928, Sc.D. (hon.), 1961; A.M., Harvard, 1929; Ph.D., 1933. Instr. physics Harvard, 1933-36, tutor div. phys. scis., 1934-36; research instr. Tufts U., 1936-41; physicist U.S. Naval Research Lab. optics div., 1941-58, head instrument sect., 1942-45, head micron waves br., 1945-58, head rocket spectroscopy br., atmosphere and astrophysics div., 1958-67, space sci. div., 1967-78, cons., 1978—; Mem. com. vision Armed Forces-NRC, 1944—; line spectra of elements com. NRC, 1960-72; mem. Rocket and Satellite Research Panel, 1958—; mem. astronomy subcom. space sci. steering com. NASA, 1960-62, mem. solar physics subcom., 1969-71; prin. investigator expts. including Skylab; mem. com. aeronomy Internat. Union Geodesy and Geophysics, 1958—; U.S. nat. com. Internat. Commn. Optics, 1960-66; mem. sci. steering com. Project Vanguard, 1956-58; mem. adv. com. to office sci. personnel Nat. Acad. Scis.-NRC, 1969-72. Contbr. articles to sci. jours. and books. Bayard Cutting fellow Harvard, 1931-33, 35-36; recipient Meritorious Civilian Service award U.S. Navy, 1945; E.O. Hulburt award Naval Research Labs., 1958; Progress medal photog. Soc. Am., 1959; Prix Ancel Soc. Francaise de Photographie, 1962; Henry Draper medal Nat. Acad. Scis., 1963; Navy award for distinguished achievement in sci, 1963; Eddington medal, 1964; NASA medal for exceptional sci. achievement, 1974; George Darwin lectr. Royal Astron. Soc., 1963. Fellow Am. Acad. Arts and Scis., Am. Phys. Soc., Optical Soc. Am. (dir. 1953-57, Frederic Ives medal 1960), Am. Geophys. Union; mem. Internat. Acad. Astronautics, Nat., Washington acads. scis., Am. Astron. Soc. (v.p. 1964-66, Henry Norris Russell lectr. 1966), Soc. Applied Spectroscopy, AAAS, Am. Geophys. Union, Philos. Soc. Washington, Internat. Astron. Union, Nuttall Ornithol. Club, Audubon Naturalists Soc., Phi Beta Kappa, Sigma Xi, Theta Delta Chi. Subspecialties: Satellite studies; Solar physics. Current work: Consultant in space research; solar physics, optics, spectroscopy. Home: 7725 Oxon Hill Rd Oxon Hill MD 20745 Office: US Naval Research Lab Washington DC 20375

TOWBIN, ABRAHAM, researcher, medi. legal cons.; b. Cripple Creek, Colo., Apr. 26, 1916; s. Mike and Esther (Jaffee) T.; m. Margret Mary Towbin, Sept. 17, 1942; 1 dau.: Mary Kathryn Towbin Potere. B.S., U. Denver, 1936; M.D., U. Colo., 1940. Cert. Am. Bd. Pathology, Am. Bd. Neuropathology. Intern N. Hudson Hosp., Weehawken, N.J., 1941-42; resident N.Y.C. Hosp., 1945-48; instr. pathology Ohio State U., 1948-51, asst. prof., 1951-54, assoc. prof., 1954-55; assoc. prof. pathology SUNY, Bklyn., 1955-57; Fulbright research scholar Max Planck Inst., Munich, W.Ger., 1957-58; prof. pathology Chgo. Med. Sch., 1959-62; pathologist NIH collaborative perinatal project, dept. neuropathology Harvard U., 1962-65; pathologist, clinician Mass. Dept. Mental Health, 1965-80; med. dir. Mental Retardation Research Inst., Waltham, Mass., 1980—. Contbr. numerous articles to profl. jours. Served to col., M.C. U.S. Army, 1942-65. Mem. Am. Assn. Pathology, Am. Assn. Neuropathologists. Jewish. Subspecialties: Pathology (medicine); Maternal and fetal medicine. Current work: Research in perinal brain damage and its sequels: Mental retardation, cerebral palsy, epilepsy, psychopathy; fetal-neonatal brain damage; psychopathy. Office: Mental Retardation Research Inst 775 Trapelo Rd Waltham MA 02154 Home: 18 Inis Circle West Newton MA 02165

TOWER, RONI BETH, psychologist; b. Akron, Ohio, Dec. 11, 1943; d. Arnold Edward and Elva Hermoine (Gross) Weinstein; m. Stephen Edward Tower, June 11, 1964 (div. 1980); children: Jennifer, Daniel. B.A., Barnard Coll., 1964; M.S., Yale U., 1977, Ph.D., 1980. Lic. psychologist, Conn. Spl. edn. tutor Town of Fairfield, Conn., 1969-74; research asst. Yale U., New Haven, 1976-77, teaching asst., 1976-79; psychologist Silver Hill Found., New Canaan, Conn., 1979-82; lectr. Yale U., 1981-82; pvt. practice clin. psychology, Westport, Conn., 1982—; cons. in field. Mem. editorial bd. Jour. Imagination, Cognition & Personality, 1982—; contbr. articles and chpts. to jours., books. USPHS trainee, 1978-79. Mem. Am. Psychol. Assn., Eastern Psychol. Assn., Conn. Psychol. Assn., Am. Assn. for Study of Mental Imagery (dir.), LWV. Republican. Jewish. Club: Jr. League Greater Bridgeport. Current work: Special interest in imagery, adult values and influences of parents on children. Home: 363 Mount Laurel Rd Fairfield CT 06430 Office: 10 Bay St Westport CT 06880

TOWN, DONALD EARL, systems analyst, consultant; b. Forestville, N.Y., Dec. 5, 1949; s. Earl Grover and Violet Grace (Kear) T. B.A., DePauw U., 1971; M.S., Ohio State U., 1973; Ph.D., Brown U., 1978. Teaching asst. Ohio State U., 1971-73; research asst. Brown U., 1975-78; asst. prof. Wellesley Coll., 1980; research mathematician Thomte & Co., Inc., Boston, 1978-80; staff systems analyst Dynamics Research Corp., Wilmington, Mass., 1980—; cons. Thomte & Co., Inc. Coach Wellesley Coll., 1979—. Mem. IEEE, Am. Math. Soc.,., Soc. Indsl. and Applied Math., Am. Statis. Assn., Math. Assn. Am., Phi Beta Kappa, Sigma Xi, Sigma Pi Sigma. Democrat. Methodist. Club: Providence Turners. Subspecialties: Statistics; Software engineering. Current work: Software quality metrics using applied mathematical techniques from statistics and pattern theory. Home: 15 Leamington Rd Brighton MA 02135 Office: Dynamics Research Corp 60 Concord St Wilmington MA 01887

TOWNES, CHARLES HARD, physics educator; b. Greenville, S.C., July 28, 1915; s. Henry Keith and Ellen Sumter (Hard) T.; m. Frances H. Brown, May 4, 1941; children: Linda Lewis, Ellen Screven, Carla Keith, Holly Robinson. B.A., B.S., Furman U., 1935; M.A., Duke U., 1937; Ph.D., Calif. Inst. Tech., 1939. Mem. tech. staff Bell Telephone Lab., 1939-47; asso. prof. physics Columbia U., 1948-50, prof. physics, 1950-61; exec. dir. Columbia Radiation Lab., 1950-52, chmn. physics dept., 1952-55; provost and prof. physics Mass. Inst. Tech., 1961-66, Inst. prof., 1966-67; v.p., dir. research Inst. Def. Analyses, Washington, 1959-61; Univ. prof. U. Calif. at Berkeley, 1967—; Guggenheim fellow, 1955-56; Fulbright lectr. U. Paris, 1955-56, U. Tokyo, 1956; lectr., 1955, 60,; dir. Enrico Fermi Internat. Sch. Physics, 1963; Scott lectr. U. Cambridge, 1963; Centennial lectr. U. Toronto, 1967; Lincoln lectr., 1972-73, Halley lectr., 1976; dir. Perkin-Elmer Corp., Gen. Motors Corp.; mem. Pres.'s Sci. Adv. Com., 1966-69, vice chmn., 1967-69; chmn. sci. and tech. adv. com. for manned space flight NASA, 1964-69; mem. Pres.'s Com. on Sci. and Tech., 1976. Author: (with A.L. Schawlow) Microwave Spectroscopy, 1955; author, co-editor: Quantum Electronics, 1960, Quantum Electronics and Coherent Light, 1964; editorial bd.: (with A.L. Schawlow) Rev. Sci. Instrument, 1950-52, Phys. Rev, 1951-53; bd.: Phys., Rev, 1951-53, Jour. Molecular Spectroscopy, 1957-60, Procs. Nat. Acad. Scis, 1978 ; contbr. articles to sci. publs. Trustee Calif. Inst. Tech., Carnegie Instn. of Washington; mem. corp. Woods Hole Oceanographic Instn. Recipient numerous hon. degrees and awards, including; Nobel prize for physics, 1964; Stuart Ballantine medal Franklin Inst., 1959, 62; Thomas Young medal and prize Inst. Physics and Phys. Soc., Eng., 1963; Disting. Public Service medal NASA, 1969; Wilhelm Exner award, Austria, 1970; Niels Bohr Internat. Gold medal, 1979; Nat. Sci. medal, 1983; named to Nat. Inventors Hall of Fame, 1976. Fellow Am. Phys. Soc. (council 1959-62, 65-71, pres. 1967, Plyler prize 1977), Optical Soc. Am. (hon., Mees medal 1968), IEEE (medal of honor 1967), Calif. Acad. Scis.; mem. Am. Philos. Soc., Am. Astron. Soc., Am. Acad. Arts and Scis., Nat. Acad. Scis. (council 1969-72, 78-81, chmn. space sci. bd. 1970-73, Comstock award 1959), Société Française de Physique (council 1956-58), Royal Soc. (fgn.), Pontifical Acad. Scis. Subspecialties: Infrared optical astronomy; Atomic and molecular physics. Current work: Microwave physics, laser physics and quantum electronics, nuclear and molecular structure, and astrophysics paricularly infrared and micowave astronomy. Patentee masers and lasers; research nuclear and molecular structure, quantum electronics, interstellar molecules, radio and infrared astrophysics. Office: Dept Physics U Calif at Berkeley Berkeley CA 94720

TOWNSEND, ALDEN MILLER, research geneticist; b. Tulsa, Mar. 4, 1942; s. Albert MacMillan and Mary Margaret (Miller) T.; m. Anne Bushfield, June 8, 1968; children—Jeffrey Bushfield, David Alden. B.S., Pa. State U., 1964; M.F., Yale U., 1966; Ph.D., Mich. State U., 1969. Supr. genetics project U.S. Forest Service, Cottage Grove, Oreg., summers, 1965,66; research geneticist Agrl. Research Service, Nursery Crops Research Lab., U.S. Dept. Agr., 1970—; mem. exec. council N. Central Tree Improvement Conf.; sec.-treas. Met. Tree Improvement Alliance, 1976-80, pres., 1982—. Contbr. articles to sci., tech. jours. Served with AUS, 1964-67. Mem. Am. Phytopath. Soc., Internat. Soc. Arboriculture (Research award 1982), Soc. Am. Foresters, Metria, Xi Sigma Pi, Gamma Sigma Delta. Presbyterian. Club: Optimist Club of Del. Subspecialty: Plant genetics. Current work: Development of elms resistant to Dutch elm disease; genetic variation and improvement of red maple, Colorado Blue Spruce, European alder. Selection of trees resistant to deicing salts. Home: 245 Grandview Ave Delaware OH 43015 Office: USDA ARS 359 Main Rd Delaware OH 43015

TOWNSEND, JOHN WILLIAM, JR., physicist, aerospace company executive; b. Washington, Mar. 19, 1924; s. John William and Elenore (Eby) T.; m. Mary Irene Lewis, Feb. 7, 1948; children: Bruce Alan, Nancy Dewitt, John William III, Megan Lewis. B.A., Williams Coll., 1947, M.A., 1949, Sc.D., 1961. With Naval Research Lab., 1949-55, br. head, 1955-58; with NASA, 1958-68, dep. dir. Goddard Space Flight

Ctr., 1965-68; dep. adminstrn. Environmental Scis. Services Adminstrn., 1968-70; asso. adminstr. Nat. Oceanic and Atmospheric Adminstrn., 1970-77; pres. Fairchild Space and Electronics Co., 1977-82; v.p. Fairchild Industries, 1979—; pres. Fairchild Space Co., 1983—; Mem. U.S. Rocket, Satellite Research Panel, 1950—. Author numerous papers, reports in field. Pres. town council, Forest Heights, Md., 1951-55. Served with USAAF, 1943-46. Recipient Profl. Achievement award Engrs. and Architects Day, 1957; Meritorious Civilian Service award Navy Dept., 1957; Outstanding Leadership medal NASA, 1962; Distinguished Service medal, 1971; recipient Arthur S. Fleming award Fed. Govt., 1963. Fellow Am. Meteorol. Soc., AIAA, AAAS; mem. Am. Phys. Soc., Nat. Acad. Engring., Am. Geophys. Union. Internat. Astronautical Fedn. (mem. internat., acad. astronautics), Sigma Xi. Subspecialties: Aerospace engineering and technology; Remote Sensing. Home: 15810 Comus Rd Clarksburg MD 20734

TOWNSEND, MARJORIE RHODES, aerospace engineer, business executive; b. Washington, Mar. 12, 1930; d. Lewis Boling and Marjorie Olive (Trees) Rhodes; m. Charles Eby Townsend, June 7, 1948; children: Charles Eby Jr., Lewis Rhodes, John Cunningham, Richard Leo. B.E.E., George Washington U., 1951. Registered profl. engr., D.C. Electronic scientist Naval Research Lab., Washington, 1951-59; research engr. to sect. head NASA-Goddard Space Flight Ctr., Greenbelt, Md., 1959-65, tech, asst. to chief systems div., 1965-66, project mgr. small astronomy satellites, 1966-75, project mgr. applications explorer missions, 1975-76, mgr. preliminary systems design group, 1976-80; aerospace and electronics cons., Washington, 1980-83; v.p. systems devel. Am. Sci. and Tech. Corp., 1983—. Decorated knight Order Italian Republic, 1972; recipient Fed. Woman's award, 1973; award Culture Assn. EUR, Rome, 1974; Engr. Alumni Achievement award George Washington U., 1975; Gen. Alumni Assn. Achievement award, 1976; Exceptional Service medal NASA, 1971; Outstanding Leadership medal, 1980. Fellow IEEE (chmn. Washington sect. 1974-75, program chmn. 1971), AIAA (nat. capitol sect. council 1973-75, 79-83), Washington Acad. Sci. (pres. 1980-81); mem. AAAS, N.Y. Acad. Scis., Am. Geophys. Union, Soc. Women Engrs., Sigma Kappa, DAR, Daus. Colonial Wars, Daughters 1812, Mensa. Republican. Episcopalian. Subspecialties: Aerospace engineering and technology; Electronics. Current work: In process of designing a spacecraft mission to be privately funded. Patentee digital telemetry system. Home: 3529 Tilden St NW Washington DC 20008

TRABER, DANIEL LEE, anesthesiologist, physiologist; b. Victoria, Tex., Apr. 28, 1938; s. Bruno James and Margaret Louise T.; m. Lillian Dee, Apr. 2, 1959; children: Daniel Stephen, Curt Anderson. B.A., St. Mary's U., San Antonio, 1959; M.A., U. Tex.-Galveston, 1962, Ph.D., 1965. Postdoctoral fellow Ohio State U., 1966; Mem. faculty U. Tex. Med. Br., Galveston, 1966—, assoc. prof. anesthesiology and physiology, now prof. Contbr. articles to profl. jours. Mem. choir, bd. religious edn., tchr. confraternity classes St. Peter the Apostle Ch., Galveston. NIH fellow, 1961-65; recipient Disting. Teaching award Grad. Sch. Biomed. Scis. U. Tex. Med. Br., 1976, Arabia Crown Jewels award Shriners N.A., 1976. Mem. Shock Soc. (charter), AAAS, AAUP, N.Y. Acad. Scis., Tex. Acad. Scis., Am. Soc. Anesthesiologists (assoc.), Tex. Soc. Anesthesiologists (assoc.), Internat. Anesthesia Research Soc., Am. Burn Assn., Am. Physiol. Soc., Soc. Exptl. Biology and Medicine (chmn. S. W. sect.), Am. Soc. Pharmacology and Exptl. Therapeutics, Am. Heart Assn., Bay Area Heart Assn., Internat. Soc. Burn Injury, Am. Assn. Lab. Animal Sci., Sigma Xi. Subspecialties: Anesthesiology; Cardiology. Current work: Cardiopulmonary responses to burn injury, smoke inhalation and sepsis using a chronic sheep model. Office: Anesthesia Research 610 Texas Ave Galveston TX 77550

TRACHTENBERG, EDWARD NORMAN, chemistry educator; b. N.Y.C., Dec. 8, 1927; s. Jacob M. and Eva (Adwokat) T.; m. Victoria A. Gotsky, Aug. 21, 1954; children: Ellen C. Gilbert, Judith Ann, Richard Bruce. A.B., N.Y.U., 1949; A.M., Harvard U., 1951, Ph.D., 1953. Postdoctoral fellow U. Colo., 1952-53; instr., then asst. prof. Columbia U., 1953-58; successively asst. prof., assoc. prof., prof. chemistry Clark U., Worcester, Mass., 1958—; cons. Radiation Applications, Inc., 1956-61, E.F. Drew and Co., 1958-62. Contbr. articles to profl. jours. Mem. Worcester County Democratic Com., 1972—, Mass. Dem. Platform Com., 1972-76; mem. exec. com. Citizens for Participation in Polit. Action, 1972—, treas., 1976—. Served with USAAF, 1946-47. NSF sci. faculty fellow, 1967-68. Mem. Am. Chem. Soc., Am. Inst. Chemists, Chem. Soc. London, Civil Liberties Union Mass. (exec. com. 1968-74, 81—). Subspecialty: Organic chemistry. Current work: Mechanisms of organic reactions. Home: 28 S Lenox St Worcester MA 01602 Office: Chemistry Dept Clark U Worcester MA 01610

TRAFTON, LAURENCE MUNRO, astronomer; b. Boston, July 31, 1938; s. Hurbert and Vesta Estelle (Trafton) Meara. B.S., Calif. Inst. Tech., 1960, M.S., 1961, Ph.D., 1965. Research scientist U. Tex., Austin, 1973—, spl. research assoc., 1969-73; physicist GS-13 Air Force Weapons Lab., Albuquerque, 1968-69. Served to 1st lt. USAF, 1965-68. Mem. Internat. Astron. Union, Am. Astron. Soc., AAAS. Congregationalist. Club: Northcross Figure Skating (Austin, Tex.). Subspecialty: Planetary atmospheres. Current work: Atmospheres of the major planets, observational astronomy, spectroscopy, planetary atmospheres, solar system studies. Home: 1208 A Elm St Austin TX 78703 Office: Astronomy Dept U Tex RLM 16 342 Austin TX 78712

TRAGER, WILLIAM, biologist, educator; b. Newark, Mar. 20, 1910; s. Leon and Anna (Emilfork) T.; m. Ida Sosnow, June 16, 1935; children—Leslie, Carolyn, Lillian. B.S., Rutgers U., 1930, Sc.D. (hon.), 1965; M.A., Harvard, 1931, Ph.D., 1933. Fellow Rockefeller U., N.Y.C., 1934-35, mem. faculty, 1935—, asso. prof., 1950-64, prof. biology, 1964—; guest investigator West African Inst. Trypanosomiasis Research, 1958-59; vis. prof. Fla. State U., 1962, U. P.R. Med. Sch., 1963, U. Mexico Med. Sch., 1965; guest investigator Nigerian Inst. Trypanosomiasis Research, 1973-74; Mem. study sect. parasitology and tropical medicine Nat. Inst. Allergy and Infectious Diseases, 1954-58, 67-70, tng. grant com., 1961-64, microbiology and infectious diseases adv. com., 1978-79; mem. malaria commn. Armed Forces Epidemiol. Bd., 1965-73; mem. study group parasitic diseases Walter Reed Army Inst. Research, 1977-79; chmn. sci. adv. council Liberian Inst. Tropical Medicine, 1965-66; rapporteur 6th, 7th Congresses Tropical Medicine; pres. Am. Found. for Tropical Medicine, 1966-69; mem. steering com. Malaria Immunology Group, WHO, 1977-80; cons. WHO, Bangkok, 1978, Panama, 1979, Shanghai, 1979. Author: Symbiosis, 1970; Editor: Jour. Protozoology, 1954-65; Contbr. articles on insect physiology and exptl. parasitology to profl. jours. Served to capt. AUS, 1943-45. NRC fellow, 1933-34; Guggenheim Found. fellow, 1973-74; recipient Darling medal WHO, 1980. Fellow AAAS, N.Y. Acad. Scis.; mem. Nat. Acad. Sci., Am. Soc. Parasitologists (council 1956-57, v.p. 1973, pres. 1974), Soc. Protozoologists (pres. 1960-61), Am. Soc. Tropical Medicine and Hygiene (pres. 1978-79). Subspecialty: Parasitology. Home: 89 Lee Rd Scarsdale NY 10583 Office: Rockefeller Univ York Ave at 66th St New York NY 10021

TRAIFOROS, SPYROS ANTHONY, engineering administrator; b. Piraeus, Greece, May 1, 1947; came to U.S., 1971; s. Anthony and Urania (Karnadakis) T.; m. Angela Katsipis, Aug. 29, 1971; 1 dau., Urania Anastasia. Diploma in mech. and elec. engring, Nat. Tech. U., Athens; M.S. in Nuclear Engring, U. Lowell, Ph.D., U. Md. Registered profl. engr., Ill., Greece. Nuclear Analyst Sargent and Lundy Engrs., Chgo., 1975-77; sr. engr. Bechtel Power Corp., Gaithersburg, Md., 1978-79, group leader, 1980—. Active Soc. Preservation Green Heritage, Athens Club. Mem. Am. Nuclear Soc., ASME, Am. Soc. Profl. Engrs. Greek Orthodox. Club: Toastmasters (pres. 11980). Subspecialties: Nuclear engineering; Mechanical engineering. Current work: Thermal Hydraulics of nuclear power plants with emphasis on consequences of postulated pipe breaks (subcompartment pressurization, water hammer analysis, jet impingement effects). Home: 7224 Selkirk Dr Bethesda MD 20817 Office: Bechtel Power Corp 15740 Shady Grove Rd Bethesda MD 20877

TRAJMAR, SANDOR, scientist; b. Bogacs, Hungary, Sept. 7, 1931; came to U.S., 1957, naturalized, 1962; s. Gyula and Maria (Balazs) T.; m. Magdolna Csanak, Feb. 1, 1930; 1 son, Peter. Dipl. Phys. Chemistry, L. Kossuth U., Hungary, 1955; Ph.D. in Phys. Chemistry, U. Calif., Berkeley, 1961. Research chemist N. Hungarian Chem. Works, Miskolc, 1955-57, Stauffer Chem. Co., Richmond, Calif., 1957-58; research asst. Lawrence Radiation Lab., U. Calif., Berkeley, 1959-61; sr. scientist Jet Propulsion Lab., Calif. Inst. Tech., Pasadena, 1961-80, research fellow div. chemistry, 1965-69, sr. research fellow, 1970-72, sr. research scientist, head Electron Collision group, 1970—. Contbr. articles to profl. jours. Recipient NASA medal for exceptional sci. achievement, 1973. Fellow Am. Phys. Soc.; mem. Am. Chem. Soc. Subspecialties: Atomic and molecular physics; Physical chemistry. Current work: Research and supervision of research in electron collision physics and spectroscopy. Home: 400 S Parkwood Ave Pasadena CA 91107 Office: 4800 Oak Grove Dr Pasadena CA 91109

TRAMILL, JAMES LOUIS, psychology educator; b. Clarksville, Tenn., July 25, 1945; s. Louis Howell and Mable (Clark) T.; m. P. Jeannie Kleinhammer, May 19, 1982. B.S., Austin Peay State U., 1967, M.A., 1977; Ph.D., U. So. Miss., 1981. Instr. psychology Austin Peay State U., Clarksville, 1977-78; asst. prof. ednl. psychology Wichita (Kans.) State U., 1980—. Contbr. articles to profl. jours. Mem. Am. Psychol. Assn., Soc. Research in Child Devel., Am. Ednl. Research Assn., Assn. Psychology and Edn. Research in Kans. (pres. 1981-83), Kans. Acad. Sci. (jour. editor 1980-83). Subspecialties: Developmental psychology; Physiological psychology. Current work: Behavioral and physiological effects of ethanol challenges, cognitive development, sex roles, ego development, life span. Office: Wichita State U Instructional Services 123 Wichita KS 67208

TRAPASSO, LOUIS MICHAEL, geography, geology educator; b. Niagara Falls, N.Y., Apr. 25, 1953; s. Domenick and Maria (Minervini) T. B.A., SUNY-Buffalo, 1975; M.A., Ind. StateU., 1977, Ph.D., 1980. Teaching asst. dept. geography Ind. U., Bloomington, 1975-77; research asst. geography/geology dept. Ind. State U., Terre Haute, 1977-80; asst. prof. dept. geography/geology Western Ky. U., Bowling Green, 1980—; advisor Holcomb Research Inst., Indpls., 1981. Contbr. Ency. of Climatology and lab. exercise manural, chpt. to book and articles in field to profl. jours. Dames and Moore project grantee, 1983. Mem. Assn. Am. Geographers, Am. Meteorol. Soc., AAAS, Ky. Acad. Sci., Internat. Graphoanalysis Soc., Sigma Xi, Gama Theta Upsilon (v.p. 1979-80). Democrat. Roman Catholic. Subspecialties: Climatology; Remote sensing (atmospheric science). Current work: Climate and the human body, climatic changes, climate and crime, environmental perception and environmental systems, thunderstorm trigger mechanisms. Office: Dept Geography/Geology Western Ky U Bowling Green KY 42101

TRAUB, ROGER DENNIS, neurologist; b. Washington, Feb. 26, 1946; s. Robert and Renee Charlotte (Gluck) T.; m. Stephanie Kinter, May 4, 1974; 1 son: Matthew. A.B., Princeton U., 1967; M.D., U. Pa., 1972. Research assoc. NIH, Bethesda, Md., 1973-75; research staff mem. IBM Watson Research Lab., Yorktown Heights, N.Y., 1975—; resident in neurology Neurol. Inst. N.Y., 1978-81; asst. prof. clin. neurology Columbia U., 1981—. Contbr. articles to profl. jours. Mem. Soc. Neurosci., Am. Acad. Neurology. Subspecialties: Neurophysiology; Neurology. Current work: Cellular mechanisms of epilepsy, neurophysiology of hippocampus, computer simulation, dementia. Office: IBM Watson Research Center Yorktown Heights NY 10598

TRAUB, WESLEY ARTHUR, physicist; b. Milw., Sept. 25, 1940; s. Carl A. and Garnet (Busch) T.; m. Esther G. Traub, June 22, 1963; 1 son: Jeremy C. B.S. in Applied Math., Engring. Physics, U. Wis., Milw., 1962, Ph.D. in Physics, 1968. Physicist Smithsonian Astrophys. Obs., Cambridge, Mass., 1968—; lectr. astronomy Harvard U., 1976—; cons. Contbr. articles to profl. jours. Mem. AAAS, Am. Astron. Soc., Am. Phys. Soc., Optical Soc. Am., Sigma Xi. Subspecialties: Infrared optical astronomy; Aeronomy. Current work: Infrared astronomy, stratospheric composition, astron. instrumentation. Office: 60 Garden St Cambridge MA 02138

TRAVIS, LARRY DEAN, space scientist; b. Burlington, Iowa, July 29, 1943; s. Dean Frank and Martha Virginia (Logsdon) T. B.A., U. Iowa, 1965, M.S., 1967; Ph.D., Pa. State U., 1971. Asst. prof. physics Pa. State U., University Park, 1971-73; sr. sci. analyst Computer Scis. Corp., N.Y.C., 1973-74, GTE Info. Systems, 1974-77, Sigma Data Services Corp., 1977-78; space scientist NASA Goddard Inst. for Space Studies, N.Y.C., 1978—. Contbr. articles to profl. jours. Recipient Exceptional Sci. Achievement medal NASA, 1980. Mem. Am. Astron. Soc., Am. Geophys. Union, AAAS, Sigma Xi. Subspecialties: Planetary atmospheres; Satellite studies. Current work: Remote sensing and analysis of planetary atmospheres. Office: 2880 Broadway New York NY 10025

TRAYNHAM, RICHARD NEVILLE, clin. psychologist; b. Seattle, Oct. 14, 1947; s. David James and Merceile Lois (Neville) T.; m. Billie J. Holmquist, June 19, 1967; children: William, Amy. B.A., Western Wash. U., 1971; M.A., U. Ark., 1974, Ph.D., 1977. Lic. clin. psychologist, Mont. Psychology technician VA Hosp., Hampton, Va., 1971-72; psychologist Warm Springs (Mont.) State Hosp., 1976-77, research dir., 1977-78; dir. clin. GTU, 1978-79; pvt. practice clin. psychology, Bozeman, Mont., 1978—; cons. Gov.'s Office, Mont. Mental Disabilities Bd. of Visitors, 1978—, Mont. State Prison, 1978-80; clin. psychologist Mont. State U., 1980-81. Contbr. articles to profl. jours. Team cons. Gallatin County Child Abuse, Bozeman, 1979—. Mem. Am. Psychol. Assn., Biofeedback Soc. Am., Assn. for Transpersonal Psychology, Assn. Rural Mental Health, Internat. Ctr. for Social Gerontology. Current work: Provide evaluation, treatment and consulting to nontraditional mental health delivery system in rural catchment area; research in the integration of transpersonal approaches with biofeedback and health psychology areas. Home: 504 Henderson St Bozeman MT 59715 Office: 111 S Tracy Ave Bozeman MT 59715

TREDER, JOHN DAVID, mech. engr.; b. Oakland, Calif., Jan. 19, 1940; s. David Louis and Edna Elizabeth (Finkle) T. B.S.M.E., Santa Clara U., 1962. Staff engr. Owens Corning Fiberglas, Santa Clara, Calif., 1962-63; with IBM, San Jose, Calif., 1965-81, staff engr., 1974-79, adv. engr., 1979-81, Evotek, Fremont, Calif, 1981—. Served with U.S. Army, 1963-65. Mem. ASME. Clubs: Racing Drivers (Fremont); Sports Car Am. Subspecialty: Mechanical engineering. Current work: Machine design specialist in computer peripherals, especially disk drives, head positioners, automatic teller machines. Patentee continuously movable info storage and retrieval systems, document cartridge and mounting apparatus. Office: 1220 Page Ave Fremont CA 94538

TREHUB, ARNOLD, psychologist; b. Malden, Mass., Oct. 19, 1923; s. Clarence and Rose (Issner) T.; m. Elaine Dorothy Epstein, Aug. 12, 1950; children: Craig, Aaron, Lorna. B.A., Northeastern U., 1949; M.A., Boston U., 1951, Ph.D., 1954. Research psychologist Mass. Gen. Hosp., Boston, 1953-54, VA Med. Center, Northampton, Mass., 1954—; adj. prof. U. Mass., Amherst. Served with AC U.S. Army, 1943-46. Mem. Soc. Neurosci., N.Y. Acad. Scis., AAAS. Subspecialties: Cognition; Neurophysiology. Current work: Neronal mechanisms for cognitive processes, brain, cognition, synapse, perception, learning, memory, vision, imagination. Home: 145 Farview Way Amherst MA 01002 Office: VA Med Center Northampton MA 01060

TREIMAN, SAM BARD, educator; b. Chgo., May 27, 1925; s. Abraham and Sarah (Bard) T.; m. Joan Little, Dec. 27, 1952; children—Rebecca, Katherine, Thomas. Student, Northwestern U., 1942-44; S.B., U. Chgo., 1948, S.M., 1949, Ph.D., 1952. Mem. faculty Princeton, 1952—, instr., 1952-54, asst. prof., 1954-58, asso. prof., 1958-63, prof. physics, 1963—. Author: (with M. Grossjean) Formal Scattering Theory, 1960, (with R. Jackiw and D.J. Gross) Current Algebra and Its Applications, 1972; Contbr. articles to profl. jours. Served with USNR, 1944-46. Mem. Am. Phys. Soc., Nat. Acad. Sci., Am. Acad. Arts and Scis. Subspecialties: Particle physics; Theoretical physics. Home: 60 McCosh Circle Princeton NJ 08540

TREMAINE, SCOTT DUNCAN, physics educator; b. Toronto, Can., May 25, 1950; s. Vincent Joseph and Beatrice Delphine (Sharp). B.Sc. with honors, McMaster U., 1971; M.A., Princeton U., 1973, Ph.D., 1975. Research fellow Calif Inst. Tech., Pasadena, 1975-77; research assoc. Inst. Astronomy, Cambridge, Eng., 1977-78; mem. Inst. Advanced Study, Princeton, N.J., 1978-81; assoc. prof. physics MIT, Cambridge, 1981—. Contbr. articles to profl. jours. Woodrow Wilson fellow, 1971-72; Josephine de Karman fellow, 1972-73; NRC Can. fellow, 1975-77; Alfred P. Sloan fellow, 1982—. Mem. Am. Astron. Soc. Subspecialty: Theoretical astrophysics. Office: MIT Room 6-211 77 Massachusetts Ave Cambridge MA 02139

TRENT, DENNIS WAYNE, virologist; b. Bend, Oreg., Oct. 17, 1935; s. Wayne Elbert and Ila Blanche (Stroke) T.; m. Joyce Taylor, Mar. 18, 1955; children: Jonathan, Leslie, Robert, Peter, Heidi, Patrick, James. B.S., Brigham Young U., 1959, M.S., 1961; Ph.D. (NIH fellow), U. Okla. Sch. Medicine, 1964. Asst. prof. microbiology Brigham Young U., 1964-66, assoc. prof., 1966-69; asst. prof. U. Tex. Med. Sch., San Antonio, 1969-70, assoc. prof., 1970-74; chief immunochemistry br., vector-borne diseases div. Ctr. for Disease Control, Ft. Collins, Colo., 1974—; mem. affiliate faculty depts. microbiology and biochemistry Colo. State U. Contbr. numerous articles to profl. jours. NIH grantee, 1964-74; Army Research and Devel. Command grantee, 1973-74. Mem. Am. Soc. Microbiology, Am. Soc. Virology, Am. Soc. Tropical Medicine and Hygiene, Sigma Xi. Democrat. Mormon. Subspecialties: Virology (medicine); Genetics and genetic engineering (medicine). Current work: Virology; viral biochemistry; molecular epidemiology; gene structure; genetic engineering. Office: PO Box 2087b Fort Collins CO 80522

TRETTER, MARIETTA JOAN, business statistics and computer science educator, researcher; b. Pueblo, Colo., Dec. 16, 1944; d. Vincent Joseph and Lena (Oberto) T. A.A., Pueblo Jr. Coll., 1964; B.S., U. So. Colo., 1965; M.S., U. Wis.-Madison, 1969, Ph.D., 1973. Programmer 1st Nat. Bank of Pueblo, 1965-66; instr. data processing Albuquerque Tech. Vocat. Inst., 1966-67; research asst. U. Wis.-Madison, 1969-75; asst. prof. mgmt. sci. Pa. State U., 1975-81; assoc. prof. bus. analysis Tex. A&M U., 1981—; cons. load research. Author: Software Interval Arithmetic, 1981. Mem. Soc. Indsl. and Applied Math. (vis. lectr. 1981), Inst. Math. Stats., Am. Statis. Assn. Subspecialties: Mathematical software; Applied mathematics. Current work: Distribution functions in applied statistics, interval arithmetic, symbolic manipulation, continued fractions, special functions, graphics. Office: Dept Business Analysis and Research Texas A&M U College Station TX 77843

TREVES, SAMUEL BLAIN, educator, geologist; b. Detroit, Sept. 11, 1925; s. Samuel and Stella (Stork) T.; m. Jane Patricia Mitoray, Nov. 24, 1960; children—John Samuel, David Samuel. B.S., Mich. Technol. U., 1951; postgrad. (Fulbright scholar), U. Otago, New Zealand, 1953-54; M.S., U. Idaho, 1953; Ph.D., Ohio State U., 1959. Geologist Ford Motor Co., 1951, Idaho Bur. Mines and Geology, 1952, Otago Catchment Bd., N.Z., 1953-54; mem. faculty U. Nebr., Lincoln, 1958—, prof. geology, 1966—, chmn. dept., 1964-70, 74—; curator geology Nebr. State Mus., 1964—; expdns. to, Antarctica and Greenland, 1960, 61, 63, 65, 70, 72, 73, 74, 75, 76. Fellow Geol. Soc. Am.; mem. Am. Mineral. Soc., Royal Soc. New Zealand, Nat. Assn. Geology Tchrs., Am. Polar Soc., AAAS, Sigma Xi, Tau Beta Pi, Sigma Gamma Epsilon. Club: Explorers. Subspecialties: Petrology; Tectonics. Current work: Investigation of mineral phases of alkaline basalts from Antarctica and precambrian of Nebraska. Research and publs. on geology of igneous and metamorphic rocks of Idaho, N.Z., Mich., Antarctica, Greenland, emphasis on origin of Precambrian granite complexes and volcanology. Home: 1710 B St Lincoln NE 68502

TRIBBLE, LELAND FLOYD, animal science educator; b. Oxnard, Calif., July 12, 1923; s. Floyd A. and Pearl I. (Adkins) T.; m. Betty L. Oberdiek., Aug. 16, 1952; children: Kent, Alan, Karen, Greg. B.S., U. Mo., Columbia, 1949, M.S., 1950, Ph.D., 1956. Prof. animal sci., swine nutrition prodn. and mgmt. U. Mo., Columbia, 1949-67; vis. prof. Kans. State U., 1965-66; prof. animal sci. Tex. Tech U., Lubbock, 1967—; vis. scientist non-ruminant nutritionist Coop. State Research Service, Dept. Agr., Washington. Contbr. articles to profl. jours. Served with USAAF, 1943-45. Mem. Am. Soc. Animal Sci., Sigma Xi, Gamma Sigma Delta. Subspecialty: Animal nutrition. Current work: Animal nutrition, swine nutrition production and management.

Home: 6613 Norfolk Lubbock TX 79413 Office: Tex Tech U Animal Sci Dept Lubbock TX 79409

TRIFFET, TERRY, research engineer, educator; b. Enid, Okla., June 10, 1922; m. Millicent McMaster, May 26, 1946; children: Patricia A., Terrence P., Melanie K. B.A., U. Okla., 1945; B.S., U. Colo., 1948, M.S., 1950; Ph.D. in Structural Mechanics, Stanford U., 1957. Instr. engring. U. Colo., Boulder, 1947-50; gen. engr. rocket and guided missile research U.S. Naval Ordnance Test Sta., 1950-55; gen. engr. radiol. research, head radiol. effects br. U.S. Naval Radiol. Def. Lab., 1955-59; assoc. prof. applied mechanics Mich. State U., East Lansing, 1959-63, prof. mechanics and material sci., 1963-76; assoc. dean research Coll. Engring., U. Ariz., Tucson, 1976—; cons. U.S. Dept. Def., 1959-65, Battelle Meml. Inst., 1965-68, Lear-Siegler, Inc., 1965—. Australian Research Com. grantee, 1966-67, 72-73. Mem. Am. Phys. Soc., Am. Math. Soc., Soc. Engring. Sci., Soc. Indsl. and Applied Math, IEEE. Subspecialties: Theoretical and applied mechanics; Artificial intelligence. Current work: Mathematical mechanics, software engineering, neural modeling. Office: Coll Engring U Ariz Dean's Office Tucson AZ 85721

TRIGG, MICHAEL EDWARD, pediatric oncologist, pediatric hematologist; b. Hartford, Conn., Aug. 18, 1949; s. Vincent Paul and Jean (Greenberg) T. B.A. with honors, Trinity Coll., 1971; M.D. with distinction, George Washington U., 1975. Diplomate: Nat. Bd. Med. Examiners. Student scientist lab. gen. and comparative biochemistry NIMH, Bethesda, Md., 1972-75; resident in pediatrics Northwestern U., Children's Mem. Hosp.-Prentice Maternity Ctr., Evanston, (Ill.) Hosp., 1975-78; cons. Prentice Maternity Ctr., Chgo., 1977-78; clin. assoc. pediatric oncology br. Nat. Cancer Inst., Bethesda, 1978-80; pediatrician Washington Adventist Hosp., Takoma Park, Md., 1978-80; guest worker Nat. Cancer Inst., 1980—; asst. prof. pediatrics U. Wis.-Madison, 1980—; assoc. mem. Wis. Clin. Cancer Ctr., Madison, 1982—; sci. dir. Molecular Medicine, Bethesda, 1980—, Inst. Immuno-oncology and Genetics, 1981—. Contbr. numerous articles to profl. jours. Served with USPHS, 1978-80. NIH tng. grantee, 1972; recipient William Beaumont Med. Research award, 1974; Ronald M. Ferguson prize, 1971; Am. Cancer Soc: Jr. Faculty fellow. Jr. fellow Am. Acad. Pediatrics; mem. William Beaumont Med. Soc., So. Med. Assn., Am. Soc. Clin. Oncology, Am. Soc. Pediatric Hematology/Oncology, Am. Assn. Cancer Research, Midwest Soc. Pediatric Research, Children's Cancer Study Group, Phi Beta Kappa, Alpha Omega Alpha. Subspecialties: Marrow transplant; Immunopharmacology. Current work: Serotherapy, malignant tumors, allogeneic and autologous bone marrow transplantation; allogeneic transplantation with T-lymphocyte depletion. Home: 7359 Tree Ln Madison WI 53717 Office: U Wis Dept Pediatrics 600 Highland Ave Madison WI 54792

TRIM, CYNTHIA MARY, vet. scientist, educator; b. Uxbridge, Eng., Apr. 12, 1947; d. Ettrick and Katherine Theodocia (Bradburne) T.; m. James Neil Moore, Sept. 2, 1978. B.V.Sc., U. Liverpool, Eng., 1970; diploma in vet. anesthesia, Royal Coll. Vet. Surgeons. Diplomate: Am. Coll. Vet. Anesthesiologists. Research asst. dept. clin. studies U. Cambridge, 1970-72; gen. practice vet. medicine Nixon & Partners, Bury St. Edmunds, Eng., 1972-74; asst. prof. anesthesiology, dept. clin. studies Ont. (Can.) Vet. Coll., Guelph, 1974-76; asst. prof. anesthesiology, dept. vet. clin. medicine Coll. Vet. Medicine, U. Ill., Urbana, 1976-77; assoc. prof. anesthesiology, dept. vet. medicine and surgery U. Mo., Columbia, 1977-80; assoc. prof. anesthesiology, dept. large animal medicine U. Ga., Athens, 1981—. Contbr. articles to profl. jours. Recipient prize in pathology U. Liverpool, 1969; Alumnae Anniversary award for teaching excellence U. Mo., 1980; also various grants. Mem. Royal Coll. Vet. Surgeons, Brit. Vet. Assn., Assn. Vet. Anesthetists Gt. Britain and Ireland, Am. Coll. Vet. Anesthesiologists, AVMA, Internat. Anesthesia Research Soc., Women's Vet. Med. Assn., Vet. Critical Care Soc. Episcopalian. Subspecialty: Veterinary Anesthesiology. Current work: Anesthesiology, cardiopulmonary physiology, pathophysiology of shock, pharmacology of anesthesia-related drugs. Office: Dept Large Animal Medicine Coll Vet Medicine Athens Ga 30602

TRIMBLE, ROBERT BOGUE, research biochemist; b. Balt., July 2, 1943; s. George Simpson and Janet Anna (Bogue) T.; m. Kathleen Marie Davis, May 17, 1969; 1 dau., Alison Bogue. B.Sc., Rensselaer Poly. Inst., 1965, M.Sc., 1967, Ph.D., 1969. Postdoctoral fellow Health Research Inc., Albany, N.Y., 1969-70; research scientist N.Y. State Dept. Health, Albany, 1971-76, sr. research scientist, 1977-80, assoc. research scientist, 1981—; cons. Contbr. articles to profl. jours. Active YMCA Youth Devel. Programs, 1979—. NIH grantee, 1976—. Mem. AAAS, Am. Soc. Microbiology, Am. Soc. Biol. Chemists. Subspecialties: Biochemistry (biology); Cell biology. Current work: Study of the biochemistry of cellular glycoproteins, including their synthesis, modification, transport and secretion. Office: NY State Dept Health Empire State Plaza Albany NY 12201

TRIMBLE, VIRGINIA LOUISE, physicist, astronomer, educator; b. Los Angeles, Nov. 15, 1943; d. Lyne Starling and Virginia Frances (Farmer) T.; m. Joseph Weber, May 17, 1919. B.A., UCLA, 1964; M.S., Calif. Inst. Tech., 1965, Ph.D., 1968; M.A., Cambridge (Eng.) U., 1969. Asst. prof. Smith Coll., Northampton, Mass., 1968-69; research fellow Inst. Theoretical Astronomy, Cambridge, Eng., 1969-71; asst. prof., then assoc. prof. U. Calif., Irvine, 1971-80, prof., 1980—; asst. prof., then assoc. prof. U. Md., College Park, 1973-80, prof., 1980—. Contbr. research papers and articles on astronomy and astrophysics to profl. lit. Named Outstanding Young Scientist Md. Acad. Scis., 1976; NSF predoctoral research fellow; NATO postdoctoral research fellow; Sloan Found. research fellow. Fellow AAAS; mem. Internat. Astron. Union, European Phys. Soc., Am. Astron. Soc., Royal Astron. Soc., Internat. Soc. Gen. Relativity and Gravitation, Astron. Soc. Pacific, Internat. Assn. math. Physics. Non-libertarian Socialist. Jewish. Subspecialties: Optical astronomy; Theoretical astrophysics. Current work: Structure and evolution of stars and galaxies; cosmology. Office: U Calif Dept Physics Irvine CA 92717

TRIPATHI, ANAND VARDHAN RAGHUNANDAN, computer scientist; b. Indore, India, June 17, 1949; came to U.S., 1976, naturalized, 1981; s. Raghunandan and Suman (Sule) T. B.Tech., Indian Inst. Tech., 1972; Ph.D., U. Tex., Austin, 1980. Prin. research scientist Honeywell, Inc., Bloomington, Minn., 1981—; sci. officer Bhabha Atomic Research Center, Bombay, India, 1972-76; research asst. dept. elec. engring. U. Tex., Austin, 1976-80. Contbr. articles to profl. jours. Mem. IEEE, Assn. Computing Machinery. Subspecialties: Distributed systems and networks; Operating systems. Current work: Development of concepts related to designing of highly reliable distributed systems; formal analysis of system properties such as fault-tolerance, functional correctness, performance and security. Office: 10701 Lyndale Ave S Bloomington MN 55420 Home: 200 W 97 St #116 Bloomington MN 55420

TRIPATHI, RAMESH CHANDRA, ophthalmology educator, ophthalmologist and pathologist; b. Jamira, India, July 1, 1936; came to U.S., 1977; s. Arjun and Gandhari T.; m. Brenda Jennifer Lane, May 20, 1969; children: Anita, Paul. I.Sc., Lucknow (India) Christian Coll., 1954; M.B., B.S., Argra Med. Coll., 1959; M.S. in Ophthalmology, Lucknow U., 1963; M.Surgery; Ph.D., U. London, 1970. Diplomate: ophthalmology Exam. Bd. Eng., Royal Coll. Pathologist Eng. Intern GSVM Med. Coll. Hosps., Kanpur, 1959-60, Lucknow U., 1959-60; resident, 1960-63; fellow Univ. Eye Clinic, Ghent, Belgium, 1964; registrar S.W. Middlesex Hosps., Isleworth, Middlesex, Eng., 1965-67; lectr. U. London, 1967-70, sr. lectr., 1970-77; cons. pathologist and ophthalmologist Moorfields Eye Hosp., London, 1972-77; prof. ophthalmology U. Chgo., 1977—, sec. dept., 1977—, mem. med. staff, 1977—, cons. pediatric tumor bd., 1978-80, prof., 1978—. Exec. editor: Exptl. Eye Research, 1973—; editorial bd.: Ophthalmic Research, 1974—, Lens Research, 1983—; sect. editor: Cornea, 1981—; assoc. editor: Afro-Asian Jour. Ophthalmology, 1981—; contbr. chpts. to books and articles to profl. jours. Council mem. Friends of India Soc. Internat., Chgo., 1982. Med. Research Council grantee, 1972-75; Nat. Eye Inst. grantee, 1977—; recipient Ophthalmology prize Royal Soc. Medicine, 1971, Royal Eye London prize Ophtalmol. Soc., 1976. Fellow Royal Soc. Medicine, Am. Acad. Opthalmology, Internat. Coll. Surgeons, Royal Coll. Physicians and Surgeons; mem. AMA, Physiol. Soc. London, Fedn. Am. Soc. Exptl. Biologists, Assn. Indians in Am. (co-chmn. med. council Chgo. 1983). Club: Quadrangle (Chgo.). Subspecialties: Ophthalmology; Pathology (medicine). Current work: Authored more than 170 scientific publications and 20 book chapters encompassing medical and biological fields of morphology, pathophysiology, clinical and experimental ophthalmology, genetics, cell biology, tissue culture, electron microscopy, immunopathology. Home: 5545 S Harper Ave Chicago IL 60637 Office: U Chicago 939 E 57th St Chicago IL 60637

TRIPATHY, SUKANT KISHORE, physicist; b. Chakradharpur, Bihar, India, Aug. 4, 1952; came to U.S., 1976, naturalized, 1981; s. Jyotish Chandra and Usha Rani (Pani) T.; m. Susan Jean Thomson, Sept. 5, 1981. B.S., Indian Inst. Tech., 1972, M.S., 1974; Ph.D., Case Western Res. U., 1980. Research fellow Indian Inst. Tech., Kharagpur, 1974-76; grad. fellow Case Western Res U., Cleve., 1976-80; mem. tech. staff GTE Labs., Inc., Waltham, Mass., 1981—. Contbr. articles to profl. jours. Recipient Nat. Sci. award Nat. Council India, 1969, Talent Search award, 1969. Mem. Am. Phys. Soc., Am. Chem. Soc., Sigma Xi. Subspecialties: Polymer physics. Current work: Structural studies in polymeric and composite systems having interesting electrical electronic and optical properties; theoretical modelling and statistical mechanics of chain molecules. Home: 34 Dartmouth St Arlington MA 02174 Office: 40 Sylvan Rd Waltham MA 02254

TRIPLETT, GLOVER BROWN, JR., agronomy educator, researcher; b. Miss., June 2, 1930; m., 1951; 1 child. B.S., Miss. State U., 1951, M.S., 1955; Ph.D. in Farm Crops, Mich. State U., 1959. From asst. prof. to assoc. prof. Ohio Agrl. Research and Devel. Ctr., Wooster, 1959-67, prof. agronomy, 1967—, agronomist, 1959—. Mem. Am. Soc. Agronomy, Weed Sci. Soc. Am. Subspecialties: Resource conservation; Agronomy. Office: Ohio Agrl Research and Devel Ctr Wooster OH 44691

TRIPODI, DANIEL, immuno-chemist; b. Cliffside Park, N.J., Mar. 13, 1939; s. Charles and Carrie (Pitetti) T.; m. Sharon, June 22, 1963; children: Daniel, Thomas. B.S., U. Del., 1961, M.S., 1963; Ph.D., Temple U., 1966. Dir. Immuno Diagnostics, Raritan, N.J., 1966-74; gen. mgr. biomed. div. New Eng. Nuclear, North Billerica, Mass., 1974-76; dir. tech. planning B-D Corp., Paramus, N.J., 1976-79; assoc. dir. Cambridge Research Labs., tech. devel. Ortho Diagnostics, Raritan, 1979-83; dir. biotech. Corp. Sci. and Tech., Johnson and Johnson, New Brunswick, N.J., 1983—; assoc. prof. microbiology U. Pa., Phila., 1979—. Contbr. articles on antibody affinity, immunochem. methodology, pharm. research and other topics to profl. jours. Fellow Am. Soc. Exptl. Biology and Medicine. Subspecialties: Immunology (medicine); Microbiology (medicine). Current work: New technical screening and development; introduce new technologies to corporations and guide the development into new business. Patentee in field. Home: RD4 Box 35A Lebanon NJ 08833 Office: COSAT Johnson and Johnson New Brunswick NJ 08933

TRIPPODO, NICK CHARLES, physiologist, researcher, consultant; b. Galveston, Tex., Sept. 27, 1945; s. Pete and Rena (Menotti) T.; m. Linda Sue Evers, Aug. 26, 1967; children: Joseph Brent, Julie Robin. B.S., Stephen F. Austin State U., 1968, M.S., 1969; postgrad., Col. State U.-Ft. Collins, 1969-70; Ph.D., U. Tex.-Med. Br., Galveston, 1974. Research assoc. U. Miss. Sch. Medicine, Jackson, 1974-75, instr., 1975-76; staff mem. Alton Ochsner Med. Found., New Orleans, 1976-79; asst. prof. La. State U. Med. Sch., New Orleans, 1976; research coordinator Alton Ochsner Med. Found., New Orleans, 1979—; cons. Travenol Labs., Chgo., 1978-79, Ciba-Geigy Corp., Summit, N.J., 1983. Contbr. articles in field to profl. jours. Active Assn. for Gifted and Talented Students, Metairie, La. Recipient Excellence of Research award Bay Area Heart Assn., Galveston, Tex., 1973; Am. Heart Assn. Fellow, 1975-76; grantee, 1977-78; USPHS research grantee, 1978-84. Mem. Am. Physiol. Soc., Am. Heart Assn, Internat. Soc. Hypertension, Soc. Exptl. Biology and Medicine. Subspecialty: Physiology (medicine). Current work: Cardiovascular phsiology, hypertension, salt water regulation, cardiovascular capacity and blood volume, chemical and physiological characterization of endogenous diuretic substance found in human and other mammalian heart tissue. Home: 5239 Trenton St Metairie LA 70002 Office: Alton Ochsner Med Found 1516 Jefferson Hwy New Orleans LA 70121

TRITSCH, GEORGE LEOPOLD, research biochemist; b. Vienna, Austria, Apr. 8, 1929; came to U.S., 1940; s. Robert James and Edith Mary (Halporn) T.; m. Norma Elsie, June 16, 1951; children: George Leopold, Margaret Ellen, Douglas Evan. B.A., N.Y.U., 1948; M.S., U. Md., 1951; Ph.D., Purdue U., 1954. Research assoc. Cornell U. Med. Coll., N.Y.C., 1954-56; research assoc. Rockefeller U., 1956-59; cancer research scientist Roswell Park Meml. Inst., Buffalo, 1959; assoc. grad. dept. biochemistry Roswell Park div. SUNY, Buffalo, 1960—, Niagara U., 1971—. Editor: Axenic Mammalian Cell Reaction, 1969; contbr. numerous articles to profl. jours. USPHS grantee, 1960—; Am. Cancer Soc. grantee, 1960—; Damon Runyan Cancer Fund grantee, 1962—. Mem. Am. Soc. Biol. Chemists, Am. Inst. Nutrition, Am. Soc. Pharmacology and Exptl. Therapeutics, Soc. Exptl. Biology and Medicine, Am. Assn. Cancer Research, Harvey Soc., N.Y. Acad. Sci., Sigma Xi, Phi Lambda Upsilon, Alpha Chi Sigma. Club: Buffalo Athletic. Subspecialties: Biochemistry (medicine); Cancer research (medicine). Current work: Metabolic control of purine salvage pathway enzymology by metabolites and antimetabolites in normal and tumor cells and in macrophages. Patentee in field. Office: Roswell Park Meml Inst Buffalo NY 14263

TRIVELPIECE, ALVIN WILLIAM, physicist, government official; b. Stockton, Calif., Mar. 15, 1931; s. Alvin Stevens and Mae (Hughes) T.; m. Shirley Ann Ross, Mar. 23, 1953; children: Craig Evan, Steve Edward, Keith Eric. B.S., Calif. Poly. Coll., San Luis Obispo, 1953; M.S., Calif. Inst. Tech., 1955, Ph.D., 1958. Fulbright scholar Delft (Netherlands) U., 1958-59; asst. prof., then asso. prof. U. Calif. at Berkeley, 1959-66; prof. physics U. Md., 1966-76; on leave as asst. dir. for research div. controlled thermonuclear research AEC, Washington, 1973-75; v.p. Maxwell Labs. Inc., San Diego, 1976-78; corp. v.p. Sci. Applications, Inc., La Jolla, Calif., 1978-81; dir. Office of Energy Research, U.S. Dept. Energy, Washington, 1981—, Fusion Power Assos.; cons. to govt. and industry. Author: Slow Wave Propagation in Plasma Wave Guides, 1966, (with N.A. Krall) Principles of Plasma Physics, 1973; also articles. Named Disting. Alumnus Calif. Poly. State U., 1978; Guggenheim fellow, 1966. Fellow Am. Phys. Soc., AAAS, IEEE; mem. Am. Nuclear Soc., Ams. for Energy Independence, AAUP, N.Y. Acad. Scis., Am. Assn. Physics Tchrs., Sigma Xi. Club: Cosmos. Subspecialties: Plasma physics; Plasma. Current work: Properties of plasmas, waves-experimental/theoretical, fusion-confinement, heating, stability, transport, nonneutral plasmas-experimental, accelerators: cyclotron wave. Patentee in field. Home: 3001 Veazey Terr NW 1410 Washington DC 20008 Office: US Dept Energy 1000 Independence Ave SW Washington DC 20585

TROJANOWSKI, JOHN QUINN, neuropathologist; b. Bridgeport, Conn., Dec. 17, 1946; s. Maurice John and Margaret (Quinn) T.; m. Virginia Trojanowski. B.A., King's Coll., 1970; postgrad, U. Vienna, 1967-69, Erasmus U., Rotterdam, 1973-74; M.D. (Charlton Fund fellow), Tufts U., 1976, Ph.D., 1976. Diplomate: Am. Bd. Pathology and Neuropathology. Clin. fellow in medicine Harvard Med. Sch., 1976-77; intern Mt. Auburn Hosp., Cambridge, 1976-77; clin. fellow in pathology Harvard Med. Sch., 1977-78, resident in neuropathology, Boston, 1978-79; resident in pathology and neuropathology Hosp. U. Pa., Phila., 1979-80; asst. prof. pathology U. Pa. Med. Sch., Phila., 1981—. Contbr. articles to profl. jours. Recipient Nat. Inst. Neurol. Communicative Disorders and Stroke award, 1978-79. Mem. Am. Assn. Neuropathologists, Am. Assn. Anatomists, Am. Assn. Pathologists, Soc. Neurosci. Subspecialties: Neurobiology; Pathology (medicine). Current work: Studies of axoplasmic transport and endocytosis in neurons of mammalian nervous system; studies of human neoplasms with monoclonal antibodies against intermediate filament proteins.

TROPF, WILLIAM JACOB, physicist; b. Chgo., Jan. 14, 1947; s. William Jacob and Ardith Shirley (Clausen) T.; m. Cheryl Lynn Griffiths, Aug. 31, 1968; 1 son, Andrew Zachary. B.S., Coll. William and Mary, 1968; Ph.D., U. Va., 1973. Project dir. B-K Dynamics, Inc., Rockville, Md., 1973-76; prin. staff physicist, asst. supr. dynamics analysis group Applied Physics Lab., Johns Hopkins U., Laurel, Md., 1977—. Served with USAR, 1968-78. Mem. Am. Phys. Soc., Optical Soc. Am., Sigma Xi, Sigma Pi Sigma. Home: 13060 Saint Patricks Ct Highland MD 20707 Office: Johns Hopkins Rd Laurel MD 20707

TROSKO, JAMES EDWARD, educator; b. Muskegon, Mich., Apr. 2, 1938; s. Andrew and Christina (Nemeth) T.; m. Beverly Kay Dowell, Sept. 3, 1960; 1 son, Philip Randal. B.A., Central Mich. U., 1970; M.S., Mich. State U., 1962, Ph.D., 1963. Postdoctoral fellow Oak Ridge Nat. Lab., 1963-66; asst. prof. Mich. State U., East Lansing, 1966-71, assoc. prof., 1971-75, prof., 1975—; vis. prof. McArdle Lab. Cancer Research U. Wis., 1972-73. Contbr. numerous articles to profl. jours. NDEA fellow, 1960-63; Nat. Cancer Inst. research career devel. award, 1972-77; Searle award U.K. Environ. Mutagen Soc., 1979. Mem. Am. Assn. Cancer Research, Genetic Soc. Am., AAAS, Toxicology Soc., Environ. Mutagen Soc., Tissue Culture Assn., Sigma Xi. Subspecialties: Cancer research (medicine); Genetics and genetic engineering (medicine). Current work: Genetic causes of chronic diseases.

TROST, BARRY MARTIN, chemist, educator; b. Phila., June 13, 1941; s. Joseph and Esther T.; m. Susan Paula Shapiro, Nov. 25, 1967; children: Aaron David, Carey Daniel. B.A., U. Pa., 1962; Ph.D., MIT, 1965. Mem. faculty U. Wis., Madison, 1965—, prof., chemistry, 1969—, Evan P. and Marion Helfaer prof. chemistry, from 1976, now Vilas research prof. chemistry; cons. Merck, Sharp & Dohme, E.I. duPont de Nemours.; Chem. Soc. centenary lectr., 1982. Author: Problems in Spectroscopy, 1967, Sulfur Ylides, 1975; editor: Structure and Reactivity Concepts in Organic Chemistry series, 1972—; asso. editor: Jour. Am. Chem. Soc, 1974-80; editorial bd.: Organic Reactions Series, 1971—; contbr. numerous articles to profl. jours. NSF fellow, 1963-65; Sloan Found. fellow, 1967-69; Am. Swiss Found. fellow, 1975—; recipient Dreyfus Found. tchr.-scholar, award, 1970; Am. Chem. Soc. pure chemistry award, 1977; Creative work in synthetic organic chemistry award, 1981; Baekeland medal, 1981; named Chem. Pioneer Am. Inst. Chemists, 1983. Mem. Am. Chem. Soc., Nat. Acad. Scis., Am. Acad. Arts and Scis., AAAS, Chem. soc. London. Subspecialties: Organic chemistry; Synthetic chemistry. Home: 209 N Whitney Way Madison WI 53705 Office: Dept Chemistry U Wis Madison WI 53706

TROWBRIDGE, RICHARD STUART, research scientist; b. Cambridge, Mass., Apr. 3, 1942; s. Walter Henry and Lamia Andree (Sanderson) T.; m. Sue Hitchcock Trowbridge, June 12, 1965; 1 son, John Richard. B.S., U. Mass., 1964, M.S., 1966, Ph.D., 1971. Research scientist dept. virology N.Y. State Inst. Basic Research in Mental Retardation, S.I., 1970-72, sr. research scientist, 1972-77, research scientist III, 1980-81, research scientist IV, 1981—; grants adminstr., dir. Research Found. Mental Hygiene, Inc., S.I., 1977-80. Contbr. articles to profl. jours. Mem. exec. bd. S.I. Council Boy Scouts Am., 1981—; v.p. Meals on Wheels S.I., Inc., 1981-82, pres., 1982—. NIH grantee, 1974-80. Mem. Am. Soc. Microbiology, Tissue Culture Assn., N.Y. Acad. Sci. Lodge: Rotary. Subspecialties: Cell and tissue culture; Virology (biology). Current work: Virus-host cell interactions which culminate in persistent infections of the central nervous system. Home: 180 Woodward Ave Staten Island NY 10314 Office: 1050 Forest Hill Rd Staten Island NY 10314

TROWER, WILLIAM PETER, physics educator; b. Rapid City, S.D., May 25, 1935; s. Wendell Phillips and Florence Evelyn (Hagen) T.; m. Katherine Ann Bache, Feb. 15, 1961 (div.); children: Alexandra Christine, Andrea Carroll. A.B., U. Calif.-Berkeley, 1957; M.S., U. Ill., 1964, Ph.D., 1966. Physicist Lawrence-Berkeley (Calif.) Nat. Lab., 1960-62; physicist U. Ill., Urbana, 1962-66; prof. physics Va. Poly. Inst., Blacksburg, 1966—; cons. Stanford (Calif.) U., Brookhaven Nat. Lab., Fermi Nat. Accelerator Lab., Va. Associated Research Ctr., Johnson & Humphrey, 1966—; dir. Council on Municipal Performance, N.Y.C., 1972-74. Author: Society of Technical Writers and Publishers, 1965; contbr. articles to profl. jours.; lectr. to profl. confs; founder: Gordon Conf. Series on High Energy Physics, 1973; co-founder: Physics in Collision Conf. Series, 1981—. Served with USMC, 1957-60. Fellow AAAS; mem. Am. Phys. Soc., Va. Acad. Sci., European Phys. Soc., Italian Phys. Soc. Democrat. Subspecialties: Particle physics; Graphics, image processing, and pattern recognition. Current work: Elementary particle physics experiments; magnetic monopole; high mass resources; computer communications. Home: 1105 Highland Circle SE Blacksburg VA 24060 Office: Dept Physics Va Tech U Blacksburg VA 24061

TROWN, PATRICK WILLOUGHBY, pharmaceutical company research executive, immunobiologist; b. Birmingham, Eng., Mar. 17, 1937; came to U.S., 1962; s. Ronald Hugh and Evelyn Mary (Willoughby) T.; m. Marie-Claire Allain Labbe, Aug. 18, 1962; children: Christopher, Nicolas. B.A. with honors, Oxford (Eng.) U. Oriel Coll., 1960, M.A., 1962, D.Phil., 1962. Research scientist Lederle Labs., Pearl River, N.Y., 1964-69; chief research group Hoffmann-La Roche, Inc., Nutley, N.J., 1969-76, asst. dir. dept. chemotherapy, 1976-78, asso. dir. dept. chemotherapy, 1978-80, dir. dept. immunotherapy, 1980—; vis. asst. prof. molecular biology Albert Einstein Coll. Medicine. Contbr. articles to profl. jours. Mem. Am. Assn. Cancer Research, Internat. Soc. Immunopharmacology. Subspecialties: Immunobiology and immunology; Virology (biology). Current work: Biology of interferons and other lymphokines; antitumor chemotherapy and immunotherapy. Office: Hoffmann-La Roche Inc Kingsland Ave Nutley NJ 07110

TROY, FREDERIC ARTHUR, II, medical biochemistry, educator, enologist; b. Evanston, Ill., Feb. 16, 1937; s. Charles McGregor and Virginia (Minto) T.; m. Linda Ann, Mar. 23, 1959; children: Karen M., Janet R. B.S., Washington U., St. Louis, 1961; Ph.D., Purdue U., 1966. Am. Cancer Soc. postdoctoral fellow dept. physiol. chemistry Johns Hopkins U. Sch. Medicine, 1966-68; asst. prof. biol. chemistry U. Calif. Sch. Medicine, Davis, 1968-74, assoc. prof., 1974-80, prof., 1980—, co-prin. investigator tumor biology tng. program, 1972; vis. research prof. dept. tumor biology Karolinska Inst. Med. Sch., Stockholm, 1976-77; cons. NIH, NSF, Damon Runyon-Walter Winchell Cancer Fund; cons. enology; tchr. wine chemistry. Contbr. rev. article, numerous research articles to profl. jours. Recipient Research Career Devel. award NIH, 1975-80; Am. Cancer Soc.-Eleanor Roosevelt-Internat. Cancer fellow awarded by Union International Contre le Cancer, Geneva, 1976-77; NIH grantee, 1968—. Mem. Am. Soc. Biol. Chemists, Am. Soc. Cancer Research, Biochem. Soc. (London), Am. Chem. Soc., Biophys. Soc., Am. Soc. Microbiology, Am. Soc. Enologists, Soc. Complex Carbohydrates, N.Y. Acad. Scis., AAAS, Sigma Xi. Subspecialties: Cancer research (medicine); Membrane biology. Current work: Research on molecular structure of cell membranes of human and mouse tumor cells for understanding relevance of cell surface changes to metastatic and invasive potential of tumor in vivo.

TROZZOLO, ANTHONY MARION, chemist; b. Chgo., Jan. 11, 1930; s. Pasquale and Francesca (Vercillo) T.; m. Doris C. Stoffregen, Oct. 8, 1955; children: Thomas, Susan (Mrs. Bruce Hecklinski), Patricia, Michael, Lisa, Laura. B.S., Ill. Inst. Tech., 1950; M.S., U. Chgo., 1957, Ph.D., 1960. Asst. chemist Chgo. Midway Labs., 1952-53; asso. chemist Armour Research Found., Chgo., 1953-56; mem. tech. staff Bell Labs., Murray Hill, N.J., 1959-75; Charles L. Huisking prof. chemistry U. Notre Dame, 1975—; vis. prof. Columbia U., N.Y.C., 1971, U. Colo., 1981; AEC fellow, 1951, NSF fellow, 1957-59; Phillips lectr. U. Okla., 1971; P.C. Reilly lectr. U. Notre Dame, 1972; C.L. Brown lectr. Rutgers U., 1975; Sigma Xi lectr. Bowling Green U., 1976, Abbott Labs., 1978; M. Faraday lectr. No. Ill. U., 1976; F.O. Butler lectr. S.D. State U., 1978. Asso. editor: Jour. Am. Chem. Soc, 1975-76; editor: Chem. Reviews, 1977—; editorial adv. bd.: Accounts of Chem. Research, 1977—; contbr. articles to profl. jours. Fellow N.Y. Acad. Scis. (Halpern award in Photochemistry 1980), AAAS, Am. Inst. Chemists; mem. Am. Chem. Soc. (Disting. Service award St. Joseph Valley sect. 1979, Coronado lectr. 1980), AAUP, Sigma Xi. Roman Catholic. Subspecialties: Photochemistry; Solid state chemistry. Current work: Physical organic chemistry: detection and characterization of reactive intermediaties: laser photochemistry: photodegradation of polymers. Patentee. Home: 1329 E Washington St South Bend IN 46617 Office: U Notre Dame Notre Dame IN 46556

TRUBATCH, JANETT, govt. ofcl., researcher, cons.; b. N.Y.C., Oct. 13, 1942; d. Louis L. and Lee J. (Jacobs) Rosenberg; m. Sheldon L. Trubatch, Aug. 26, 1962; children: A. David, Beth, Ruth Shoshana, Joel. B.Sc., Poly. Inst. N.Y., 1962; Ph.D., Brandeis U., 1968. Asst. prof. physics Calif. State U., Los Angeles, 1967-68; postdoctoral fellow Calif. Inst. Tech., 1968-73; asst. prof.-physiology N.Y. Med. Coll., 1973-77; program dir. neurobiology NSF, Washington, 1977-79; assoc. research prof. physiology George Washington U., 1979-80; health sci. adminstr. neurol. disorders program Nat. Inst. Neurol. and Communicative Disorders and Stroke, Bethesda, Md., 1980—, acting dep. dir. program, 1981—; guest speaker Nobel Symposium on Chem. Neurotransmission, Stockholm, 1980. Contbr. chpts. to books. Recipient Outstanding Performance award NSF, 1979; Whitehall Found. grantee, 1976-80; NSF grantee, 1976-81. Mem. Am. Phys. Soc., Soc. Neurosci., AAAS, Sigma Xi, Sigma Pi Sigma. Subspecialties: Neurobiology; Neurophysiology. Current work: Mechanisms of synaptic transmission; synaptic plasticity and correlates of learning; basic mechanisms underlying neurological disorders. Home: 3838 Garrison St NW Washington DC 20016 Office: NINCDS/NDP Fed Bldg Room 704 Bethesda MD 20550

TRUMP, DONALD LYNN, med. oncologist, clin. cancer researcher; b. Greencastle, Ind., July 31, 1945; s. Donald Liller and Virginia Elizabeth (Alexander) T.; m.; 1 son, Christopher Liller. B.A., Johns Hopkins U., 1967, M.D., 1970. Diplomate: Am. Bd. Internal Medicine. Intern, resident in internal medicine and med. oncology Johns Hopkins U., 1970-75; chief resident in medicine Johns Hopkins Hosp., 1975; oncologist U.S. Navy, Phila., 1975-77; asst. prof. Johns Hopkins U., 1977-81; asst. prof. human oncology and medicine U. Wis., Madison, 1981—; chief oncology Middleton VA Hosp., Madison, 1981—. Contbr. articles profl. jours. Served to lt. comdr. USNR, 1975-77. Mem. Am. Soc. Clin. Oncology, Am. Assn. Cancer Research, Am. Fed. Clin. Research. Subspecialties: Oncology; Cancer research (medicine). Current work: Clin. cancer research, primarily genitourinary cancer. Office: Wisconsin Clinical Cancer Center K 4 662 600 Highland Ave Madison WI 53792

TRUMPLER, WILLIAM E., mechanical engineer; b. East Orange, N.J., Aug. 14, 1916; s. William E. and Louise (Jelinek) T.; m. Irma Steinert, Nov. 13, 1940; children: Johanna, Albert, Maria. B.S.M.E., Lehigh U., 1937. Registered profl. engr., Pa. Design engr. steam turbines Westinghouse Corp., Phila., 1937-50, mgr. mech. design, large steam turbines, 1950-70, cons. engr., 1970-75; cons. engr., v.p. Trumpler Assocs., Inc., West Chester, Pa., 1975—. Contbr. articles to profl. jours. Fellow ASME (life, Council award 1971); mem. ASTM. Presbyterian. Clubs: Commodore Delaware Valley Sail, Swarthmore Tennis. Subspecialties: Mechanical engineering; Fracture mechanics. Current work: Failure analysis. Turbine failures both rotors and blading. Design of pressure vessels material evaluation. Patentee in field. Office: West Chester PA

TRUNE, DENNIS ROYAL, biomedical educator and researcher; b. Flint, Mich., Apr. 20, 1950; s. Walter Royal and Carol Mae (Wilcox) T. B.A. in Biology, U. Mich.-Flint, 1972, M.S., No. Ariz. U., 1974; Ph.D. in Anatomy, La. State U. Med. Ctr., 1979. NIH post doctoral research fellow dept. otolaryngology Ohio State U. Coll. Medicine, 1979-83; research assoc. dept. anatomy U. Kans. Med. Ctr., Kansas City, 1983—. Contbr. articles profl. jours. Recipient NIH Nat. Research Service award, 1979-82. Mem. Assn. Research in Otolaryngology, Centurions of Deafness Research Found., Soc. Neurosci., Am. Assn. Anatomists. Mem. Ch. of Christ. Subspecialties: Neurobiology; Anatomy and embryology. Current work: Inner ear and brain development. Office: U. Kans Med Ctr 39th and Rainbow Blvd Kansas City KS 66103

TRYBUS, RAYMOND J., research administrator, clinical psychologist; b. Chgo., Jan. 9, 1944; s. Fred and Cecilia (Liszka) T.; m. Sandra A. Noone, Aug. 19, 1967; children: David, Nicole. B.S., St. Louis U., 1965, M.S., 1970, Ph.D., 1971. Lic. psychologist, Md., D.C. Clin. psychologist Jewish Vocat. Service, St. Louis, 1968-71; clin. psychologist Gallaudet Coll., Washington, 1971-72, research psychologist, 1972-74, dir. demographic studies, 1974-78, dir., 1982—; dean Gallaudet Research Inst., 1978—; cons. Mental Health Ctr. for Deaf, Laurel, Md., 1982—, Congl. Research Service, 1982—. Editor, contbg. author: The Future of Mental Health Services for the Deaf, 1978. Grantee NIMH, Spencer Found., Tex. Edn. Agy., W.K. Kellogg Found. Mem. Internat. Assn. Study of Interdisciplinary Research, Am. Psychol. Assn., Soc. Research Adminstrs., AAAS, von Bekesy Soc. (sect. chair 1982—). Roman Catholic. Current work: Development and management of interdisciplinary research on deafness, prevention of deafness, restoration of hearing, accommodation to deafness. Home: 8806 Altimont Ln Chevy Chase MD 20815 Office: 800 Florida Ave NE Washington DC 20002

TRYON, EDWARD POLK, physics educator; b. Terre Haute, Ind., Sept. 4, 1940; s. Philip Freeland and Elizabeth Marsh (Banker) T. A.B., Cornell, 1962; Ph.D., U. Calif., Berkeley, 1967. Research asso. Columbia U., N.Y.C., 1967-68, asst. prof., 1968-71; asst. prof. physics Hunter Coll. City U., N.Y.C., 1971-73, asso. prof., 1974-79, prof., 1979— (on leave 1977-78); vis. mem. Inst. Advanced Study Princeton, 1977-78. Contbr. articles to profl. jours. Hon. Woodrow Wilson fellow, 1962-64; NSF fellow, 1962-64; CUNY scholar, 1982-83. Mem. Am. Phys. Soc., N.Y. Acad. Scis., Phi Beta Kappa, Sigma Xi (nat. lectr. 1982-84), Phi Kappa Phi. Subspecialties: Cosmology; Particle physics. Current work: Creation of universe from nothing as quantum fluctuation, particle theory. Originator quantum fluctuation theory for creation of universe from nothing, 1973. Office: Dept Physics Hunter Coll New York NY 10021

TRZASKO, JOSEPH ANTHONY, psychologist, educator; b. Jamaica, N.Y., June 4, 1946; s. Joseph Anthony and Lottie Marion (Nadraus) T.; m. Ann Elizabeth Kidd, June 26, 1971; 1 son, Joshua Damon. B.A. cum laude, U. N.H., 1967; M.A., U. Vt., 1969, Ph.D., 1972. Dir. instl. testing and research Mercy Coll., Dobbs Ferry, N.Y., 1969-80, prof. psychology, 1969—; staff psychologist St. Dominic's Intermediate Care Facility, Blauvelt, N.Y., 1980—; cons. staff psychologist J.G.B. Newman Intermediate Care Facility, Yonkers, N.Y., 1982; postdoctoral intern Colo. Dept. Instns., Wheat Ridge, 1980; adj. prof. L.I. U., Dobbs Ferry, N.Y., 1978—; project dir. U.S. Office Edn. Career Edn. Grant, 1976-77; cons. psychologist, Hartsdale, N.Y., 1982—. NDEA fellow, 1967-69; NSF grantee Edn. Commn. of States, 1976. Mem. Internat. Council Psychologists, Am. Psychol. Assn., Am. Ednl. Research Assn., AAUP, Westchester County Psychol. Assn., Psi Chi, Pi Gamma Mu, Kappa Delta Pi. Subspecialties: Behavioral psychology. Current work: Development of procedures for the establishment of intermediate care facilities for mentally retarded and developmentally disabled. Home: 18 Dunham Rd Hartsdale NY 10530 Office: Mercy Coll 555 Broadway Dobbs Ferry NY 10522

TRZYNA, THADDEUS CHARLES, political scientist, conservationist, research institute administrator; b. Chgo., Oct. 26, 1929; s. Thaddeus Stephen and Irene Mary (Giese) T.; m. 1 dau., Jennifer. B.A., U. So. Calif., 1961; Ph.D., Claremont (Calif.) Grad. Sch., 1975. U.S. Fgn. Service Officer U.S. Dept. State, Washington, 1962-69, Africa, 1962-69; pres. Calif. Inst. Pub. Affairs, Claremont, 1969—; cons. to fed. and Calif. state agys. Author: The California Environmental Quality Act, 1974, The California Handbook, 4th edit, 1981. Dir. Calif. Farmlands Project, 1981—; mem. commn. environ. planning Internat. Union for Conservation of Nature and Natural Resources, 1982—. Mem. Am. Fgn. Service Assn., Am. Planning Assn., Authors Guild, Royal Geog. Soc. (London), Delta Phi Epsilon, Sierra Club (v.p. 1975-77). Club: Univ. (Claremont). Subspecialty: Resource management. Current work: Land use, particularly in California; resource management in developing countries; organizations concerned with resource policy and management. Office: PO Box 10 Claremont CA 91711

TSAI, JAMES HSI-CHO, entomology educator; b. Fuzhou, Fujian, China, June 10, 1934; came to U.S., 1964, naturalized, 1973; s. Chuan Li and Chu Yin (Chen) T.; m. Sue Cheng, June 15, 1959; children: Cynthia H., Julie C. M.S., Mich. State U., 1967, Ph.D., 1969. Entomologist Internat. Inst. Tropical Agr., Ibadan, Nigeria, 1969-70; research assoc. dept. entomology Mich. State U., East Lansing, 1970-72; asst. prof. entomology U. Fla., Ft. Lauderdale, 1973-78, assoc. prof., 1978—. Contbr. articles to profl. jours. Hort. Research Inst. grantee, 1975, 76, 79; NSF grantee, 1976-80; Nat. Acad. Scis. grantee, 1980. Mem. Internat. Orgn. Citrus Virologists, Am. Phytopathol. Soc., Entomol. Soc. Am. Republican. Subspecialties: Plant virology; Plant pathology. Current work: Conducting research on plant mycoplasmal and virus diseases. Home: 6491 Plantation Rd Plantation FL 33317 Office: University of Florida 3205 SW College Ave Fort Lauderdale FL 33314

TSANG, ALFRED KWONG-Y, immunologist; b. Hong Kong, June 23, 1945; s. Chi Keung and Shui Hing (Lee) T.; m. Susan J. Tsang, May 13, 1972. B.S., Chinese U. Hong Kong, 1966; M.S., Bowling Green State U., 1969, Ph.D. (univ. fellow), 1974. Research asst. biol. sci. Bowling Green (Ohio) State U., 1973-74; postdoctoral research assoc. dept. surgery Med. Coll. Ohio, 1974-75, instr., 1975-77, asst. prof., 1977-79; instr. immunology Med. U. S.C., Charleston, 1979-81, asst. prof., 1981—. Recipient Merrill Chase prize in cellular immunology, 1982; Nat. Cancer Inst. grantee, 1981-84. Mem. Am. Acad. Microbiology, Am. Soc. Microbiologists, Electron Microscopy Soc. Northwestern Ohio, AAAS, Am. Assn. Immunologists, Sigma Xi. Subspecialties: Cancer research (medicine); Immunology (medicine). Current work: Tumor immunology, clinical immunology. Office: Dept Immunology Med U SC Charleston SC 29425

TSANG, REGINALD CHUN-NAU, ob-gyn and pediatrics educator, researcher; b. Hong Kong, Sept. 20, 1940; came to U.S., 1966; m. Esther Saewen, June 7, 1966; children: Trevor, Olivia. M.B.B.S., Hong Kong U. Med. Sch., 1964. Intern Queen Mary Hosp., Hong Kong U., 1964-65, resident in pediatrics, 1966; resident Hong Kong Psychiat. Hosp., 1965; resident in pediatrics Michael Reese Hosp., Chgo., 1967-68; instr. pediatircs U. Cin. Med. Coll., 1969-71, Fels asst. prof. pediatrics, 1971-75, dir., 1974-77, Fels assoc. prof. pediatrics and assoc. prof. ob-gyn, 1975-77, assoc. prof. pediatrics, ob-gyn, 1977-79, prof., 1979—; attending pediatrician U. Cin. Med. Ctr., Children's and Jewish Hosp., Cin., 1971—; Pediatric dir. lipid research clinic Children's Hosp., Cin., 1973; dir. neonatology tng. program U. Cin. Med. Coll, 1978, dir. NIH program project grant diabetes in pregnancy, 1978—, dir. newborn div., 1983. Contbr. numerous articles to med. jours. NIH grantee. Fellow Am. Coll. Nutrition (sec.-treas. 1979—); mem. M.W. Soc. for Pediatric Research (pres. 1978-79), Am. Soc. for Bone Mineral Research, Soc. for Pediatric Research, Perinatal Research So., Central Soc. Clin. Research. Subspecialties: Neonatology; Nutrition (medicine). Current work: Calcium, vitamin D, parathyroid hormone calcitonin metabolish, lipid metabolism, diabetes in pregnancy, neonatal nutrition. Office: Pediatris Dept U Cin Coll Medicine Cincinnati OH 45267

TSAO, GEORGE T., chemical engineer, microbiologist, educator, researcher; b. Nanking, China, Dec. 4, 1931; m., 1960: 3 children. B.Sc., Nat. Taiwan U., 1953; M.Sc., U. Fla., 1956; Ph.D. in Chem. Enging, U. Mich., 1960. Asst. prof. physics Olivet Coll., 1959-60; chem. engr. Merck & Co., Inc., 1960-61; research chemist TVA, 1961-62; sect. leader hydroloysis and fermentation, research dept. Union Starch & Refining Co., Inc. (div. Miles Labs., Inc.), 1962-65, asst. research dir., 1965-66; from assoc. prof. to prof. chem. enging. Iowa State U., 1966-77; prof. chem. enging. Purdue U., West Lafayette, Ind., 1977—, also now dir. Mem. Am. Chem. Soc., Am. Inst. Chem. Engrs., Am. Soc. Enging. Edn. Subspecialty: Chemical engineering. Office: Dept Chem Enging Purdue U West Lafayette IN 47907

TSAO, NAI-KUAN, computer science educator; b. Shanghai, China, June 25, 1939; s. Hung-Hsi and Yen (Wen) T.; m. Chung Long Wong, Mar. 25, 1965; 1 child, Eve Da-Yu. B.E.E., Nat. Taiwan U.-Taipei, 1961; M.E.E., Nat. Chiao Tung U., Hsinchu, Taiwan; M.S., U. Hawaii-Honolulu, 1966, Ph.D., 1970. Research assoc. Aerospace Research Labs., Dayton, Ohio, 1971-73; research analyst 1973-74; mem. faculty Wayne State U., Detroit, 1974—, assoc. prof. computer sci., 1980—. Co-author: Linear Circuits and Computations, 1973; contbr. articles to profl. jours. Mem. Math. Assn. Am., Assn. Computing Machinery, Soc. Indsl. and Applied Math., IEEE Computer Soc., U.S. Table Tennis Assn. Subspecialties: Numerical analysis; Mathematical software. Current work: Research in the implementation and analysis of numerical and non-numerical algorithms. Home: 20670 Secluded Ln Southfield MI 48075 Office: Wayne State Univ Computer Sci Dept Detroit MI 48202

TSE, HARLEY Y., immunologist, educator; b. China, July 17, 1947; s. Ton-cheuk and Hou-Ying (Choy) T.; m. Kwai-Fong Chui, Jan. 13, 1979; 1 son: Kevin Y. B.S. with honors, Calif. Inst. Tech., 1972; Ph.D., U. Calif.-San Diego, 1977. Fellow Arthritis Found., NIH, Bethesda, Md., 1977-80; sr. research immunologist Merck Sharp & Dohme Research Lab., Rahway, N.J., 1980—; adj. asst. prof. Columbia U., 1981—. Contbr. articles to profl. jours. Bd. dirs. Chinese Social Service Center, San Diego, 1975. Calif. Biochem. Research fellow, 1975; Arthritis Found. fellow, 1977-80. Mem. Am. Assn. Immunologists. Roman Catholic. Subspecialty: Immunobiology and immunology. Current work: Study of the nature and mechanisms of immune regulation. Office: MSDRL PO Box 2000 Rahway NJ 07065 Home: 11 Dana Ct Princeton NJ 08540

TSENG, AMPERE AN-PEI, mechanical engineer; b. Kiangsi, China, Jan. 21, 1946; came to U.S., 1971, naturalized, 1982; s. Chi-Kung and Ai-Chung; m. Maggie Shih-Ying Yang, Aug. 9, 1975; children: Claire, Karen, Miles. Coll. Diploma, Taipei (Taiwan) Inst. Tech., 1966; M.S., U. Ill., Urbana, 1974; Ph.D., Ga. Inst. Tech., 1978. Mech. engr. Taitan (Taiwan) Industries Pty. Ltd., 1967-71; structural engr. Westinghouse Electric Corp., Tampa, Fla., 1977-79; staff engr. Martin Marietta Labs., Balt., 1979—. Contbr. articles to profl. jours., confs.; reviewer: Nuclear Enging. and Design, ASME Transactions, Internat. Jour. Fracture. Mem. ASME, Soc. Enging. Sci. Subspecialties: Solid mechanics; Fracture mechanics. Current work: Numerical simulation of metal forming processes; computer-aided design and manufacturing; fracture mechanics; finite element method; nuclear pressure vessel analysis. Home: 5275 Five Fingers Way Columbis MD 21045 Office: 1450 S Rolling Rd Baltimore MD 21227

TSENG, CHARLES C(HIAO), biology educator, researcher; b. Fuchow, Fukien, China, Dec. 20, 1932; came to U.S., 1960, naturalized, 1977; s. Peter P. and Shu C. (Lin) T.; m. Ming Hwa Fu, June 26, 1965; children: Ernest, Anne. B.S., Taiwan Normal U., Taipei, 1955; M.S., Nat. Taiwan U., Taipei, 1957; Ph.D., UCLA, 1965. Asst. prof. biology Windham Coll., 1965-67, assoc. prof., 1967-75, Purdue U.-Calumet, Hammond, Ind., 1975-79, prof., 1979—. Contbr. articles to profl. jours. Named Outstanding Sci. Faculty Mem. Purdue U.-Calumet, 1980; NSF grantee, 1969-71, 71-74. Mem. Bot. Soc. Am., AAAS, Am. Inst. Biol. Sci., Sigma Xi. Subspecialties: Morphology; Evolutionary biology. Current work: Palynology, enzymology, systematics. Home: 3018 44th St Highland IN 46322 Office: 2233 171st St Hammond IN 46323

TSENG, JEENAN, immunologist, researcher; b. Taipei, Taiwan, China, Oct. 24, 1940; came to U.S., 1970, naturalized, 1978; s. Hualin and Wen Chang T.; m. Leeying Hsu, Dec. 24, 1974; children: Ann, Sharon. B.S., Kaohsiung Med. Coll., Taiwan, 1965; M.S., Nat. Taiwan U., 1967; Ph.D., U. Ill.-Chgo., 1974. Instr. bacteriology and immunology Nat. Taiwan U., 1966-67; asst. investigator Vets. Gen. Hosp., Taipei, 1968-70; instr. U. Ill., Chgo., 1972-74; fellow Johns Hopkins U., 1975-79; sect. head Walter Reed Army Inst. Research, Washington, 1979-83, acting chmn., 1981-83, dept. chmn., 1983—; dir. immunopathology NRC Associateship, 1983—. Ad hoc reviewer: Jour. Immunology, 1982; contbr. in field. NIAID fellow, 1978; recipient Mayor's award Taipei City, 1954. Mem. Chinese Assn. Microbiologists; mem. Am. Soc. Microbiology; Mem. Am. Assn. Immunologists. Subspecialties: Immunobiology and immunology; Microbiology (medicine). Current work: Migration and differentiation of lymphoid and nonlymphoid cells; development of mucosal immunity to microbial infections. Office: Dept Exptl Pathology Walter Reed Army Inst Research Dahlia St and 14th St Washington DC 20307

TSENG, LEON, pharmacologist, educator; b. Taiwan, Nov. 20, 1937; s. An-Chueng and Che T.; m. Grace Y. Tseng, Jan. 7, 1943; children: Joshua, Daniel, Joseph, Esther. B.S., Nat. Taiwan U., 1961; M.S., U. Kans., 1964, Ph.D., 1970. Asst. Research pharmacologist U. Calif., San Francisco, 1972-75, lectr. dept. psychiatry and pharmacology, 1973-75, adj. asst. prof. pharmacology, 1975-78; asst. prof. pharmacology Med. Coll. Wis., Milw., 1978-80, assoc. prof., 1980—. Contbr. articles to profl. lit. Mem. Am. Soc. Pharmacology and Exptl. Therapeutics, Soc. Neurosci. Subspecialties: Pharmacology; Neuropharmacology. Current work: Mechanism of actions of opioids and endorphins. Home: 12660 W Eden Trail New Berlin WI 53151 Office: Med Coll Wis Dept Pharmacology and Toxicology Milwaukee WI 53226

TSICHRITZIS, DENNIS, computer science educator; b. May 29, 1943. Diploma in Elec. Enging, Athens (Greece) Tech. U., 1965; M.A., Princeton U., 1967, Ph.D., 1968. Asst. prof. computer sci. U. Toronto, Ont., Can., 1968-71, assoc. prof., 1971-76, prof., 1976—; cons. in field. Mem. ACM, IEEE, Can. Info. Processing Soc., Can. Computer Sci. Assn. Current work: Office information systems; data base management systems, small business systems; computational complexity. Office: Dept Computer Sci U Toronto ON Canada M5S 1A1

TS'O, PAUL ON-PONG, biophysical chemist, educator; b. July 17, 1929. B.S., Lingnan U., 1949; M.S., Mich. State U., 1951; Ph.D., Calif. Inst. Tech., 1955. Teaching asst. Calif. Inst. Tech., 1952-55, research fellow biology div., 1955-61, sr. research fellow, 1961-62; asso. prof. biophys. chemistry dept. radiol. scis. Johns Hopkins U., Balt., 1962-67, prof., 1967-73, prof., dir. div. biophysics, 1973—, prof. dept. environ. health scis. div. environ. health biology, 1980—; cons. Nat. Cancer Inst., 1972-75; mem. study sect. A on biophysics and biophys. chemistry NIH, 1976-80; mem. Clearinghouse on Environ. Carcinogens, Nat. Cancer Inst., 1976-80; mem. European expert com. on biophysics UNESCO. Editor: Basic Principles in Nucleic Acid Chemistry, Vol. I and II, 1974, The Molecular Biology of the Mammalian Genetic Apparatus, Vol. I and II, 1977; co-editor: The Nucleohistones, 1964, Chemical Carcinogensis, Part A and Part B, 1974, Polycyclic Hydrocarbons and Cancer: Environment, Chemistry and Metabolism; and Molecular and Cell Biology, Vol. 1 and 2, 1978, Vol. 3, 1981, Carcinogensis: Fundamental Mechanisms and Environmental Effects, 1980; mem. editorial bd.: Molecular Pharmacology, 1964—, Biophys. Jour, 1969-72, Biochimica et Biophysica Acta, 1971-81, Cancer Rev, 1973—, Jour. Environ. Health Scis, 1976-81; asso. editor: Cancer Research, 1975—; mem. editorial adv. bd.: Biochemistry, 1966-74, Biopolymers, 1979—; contbr. over 240 articles and revs. to profl. jours. Named Md. Chemist of Yr., 1981. Fellow AAAS; mem. Biophys. Soc. (chmn. public sci. policy com.

1972-76, council mem. 1975-78, exec. bd. 1975), Am. Soc. Biol. Chemists, Am. Soc. Microbiology, Am. Soc. Cell Biology, Biology Alliance for Public Affairs (chmn. organizing com. 1973-76), Am. Assn. Cancer Research, Am. Chem. Soc., Academia Sinica, European Acad. Arts, Scis. and Humanities, Sigma Xi. Subspecialty: Biophysics (biology). Current work: Nucleic acid chemistry and biology; nuclear magnetic resonance in biochemical research; chemical and viral carcinogenesis, interferon research,cellular research on aging and differentiation; application of recombinant DNA techniques in cell biology. Office: Div Biophysics Johns Hopkins U Sch Hygiene and Public Health Baltimore MD 21205

TSOI, MANG-SO, immunologist; b. Hong Kong, Dec. 13, 1934; s. Sui-Po and King-Chong (Chan) Leung; m.; children— Douglas, Kenneth. B.S., Whitworth Coll., Spokane, 1961; M.S., U. Wash., 1963, Ph.D., 1966. Tchr. sub-insp. schs. Edn. Dept., Hong Kong, 1956-59; research assoc. dept. microbiology U. Wash., 1966-67; research assoc. Virginia Mason Research Ctr., Seattle, 1971-72; research assoc. dept. medicine U. Wash., 1972-75, research asst. prof., 1975-79, research assoc. prof. dept. medicine, 1979—; assoc. mem. Fred Hutchinson Cancer Research Ctr., 1979—. Contbr. articles to profl. jours. Am. Cancer Soc. grantee; Nat. Cancer Inst. grantee. Mem. Am. Assn. Immunologists, Reticuloendothelial Soc., N.Y. Acad. Scis., AAUP, Sigma Xi. Subspecialties: Transplantation; Cellular engineering. Current work: Study of immunological mechnaisms of graft-versus-host disease and graft host tolerance; immunology of cellular interactions in patients after bone marrow transplantation. Office: 1124 Columbia St Seattle WA 98104

TU, ANTHONY TSUCHIEN, biochemistry educator; b. Taipei, Taiwan, Aug. 12, 1930; came to U.S., 1954, naturalized, 1960; s. Tsungming and Songsui (Lin) T.; m. Kazuko Yamamoto, May 3, 1957; children: Marcia, Janice, Caroline, Kenneth, Alan. B.S., Nat. Taiwan U., 1953; M.S., U. Notre Dame, 1956; Ph.D., Stanford U., 1961. Research assoc. Stanford (Calif.) U., 1960-61, Yale U., New Haven, 1961-62; asst. prof. Utah State U., Logan, 1962-67; assoc. prof. Colo. State U., Ft. Collins, 1967-70, prof. biochemistry, 1970—. Author: Venoms: Chemistry and Molecules, 1977, Raman Spectroscopy in Biology, 1982; editor: Rattlesnake Venoms, 1982. Recipient Career Devel. award NIH, 1969. Mem. Am. Soc. Biol. Chemists, Am. Chem. Soc. Subspecialties: Toxicology (agriculture); Biophysical chemistry. Current work: Raman spectroscopy, venoms and toxins. Office: Dept Biochemistry Colo State U Fort Collins CO 80523

TU, YIH-O, mathematician; b. Jiangxi, China, Jan. 8, 1920; came to U.S., 1952, naturalized, 1968; s. Yin Su and Yih Chun (Tsou) T.; m. Frances Lucy Chang, July 3, 1960; 1 dau., Sharon Olivia. M.S., Carnegie Inst. Tech., 1955; Ph.D., Rensselaer Poly. Inst., 1959. Mem. research staff research div. IBM, San Jose, Calif., 1959—. Mem. Am. Math. Soc., Am. Phys. Soc., Soc. Indsl. and Applied Math., ASME. Subspecialties: Applied mathematics; Theoretical physics. Current work: Mathematical modelling of electro-chemico-mechanical devices. Home: 6716 Bret Harte Dr San Jose CA 95120 Office: 5600 Cottle Rd San Jose CA 95193

TUAN, DEBBIE FU-TAI, educator; b. Kiangsu, China, Feb. 2, 1930; d. Shian-gien and Chen Lee T. B.S., Nat. Taiwan U., 1954, M.S., 1958; M.S., Yale U., 1960, Ph.D., 1961. Postdoctoral research fellow dept. chemistry Yale U., 1961-64; postdoctoral research assoc. Theore Chem. Inst., U. Wis., 1964-65; asst. prof. chemistry dept. Kent (Ohio) State U., 1965-70, assoc. prof. chemistry, 1970-73, prof., 1973—; research fellow dept. chemistry Harvard U., 1969-70; vis. scientist SRI Internat., Menlo Park, Calif., 1981. Contbr. articles to profl. jours. Univ. faculty research fellow Kent State U., 1966, 68, 71. Mem. Am. Chem. Soc., Am. Phys. Soc., N.Y. Acad. Scis., Sigma Xi. Subspecialties: Theoretical chemistry; Physical chemistry. Current work: Research involves use of quantum mech. methods to study atomic and molecular problems such as electronic structures calculated by the self-consistent field X scattered wave method, calculation of phys. properties by perturbation methods, error bounds for theoretical predictions of phys. properties, exam. of electronic correlations by means of many-electron theory. Office: Chemistry Dept Kent State U Kent OH 44242

TUBBS, DAVID LEE, physicist, cons.; b. Clearfield, Pa., May 22, 1951; s. Dudley Earl and Donna Mae (Cruikshank) T. B.A., U. Wis., Madison, 1973, postgrad., U. Tex., Austin, 1973-74; Ph.D., U. Chgo., 1977. Research fellow Kellogg Radiation Lab., Calif. Inst. Tech., Pasadena, 1977-79; research assoc. U. Chgo., 1979-80; staff physicist Los Alamos Nat. Lab., 1980—; cons. Lawrence Livermore (Calif.) Nat. Lab., 1977—. Contbr. articles to profl. jours. Recipient Marc Perry Galler award U. Chgo., 1978. Mem. Am. Phys. Soc., Am. Astron. Soc., Phi Beta Kappa. Subspecialties: Theoretical astrophysics; Applied physics. Current work: Supernova physics; radiation and particle transport. Office: Los Alamos Nat Lab X-2 MSB220 Los Alamos NM 87545

TUCKER, ARTHUR OLIVER, botanist, researcher; b. Allentown, Pa., June 22, 1945; s. Arthur Oliver and Clara Althea (Funk) T.; m. Sharon Smith, June 12, 1971; children: Melissa, Angelica, Arthur Oliver IV. B.A., Kutztown State Coll., 1967; M.S., Rutgers U., 1970, Ph.D., 1975. Research assoc. Del. State Coll., 1976—. Mem. Bot. Soc. Am., Am. Soc. Plant Taxonomists, Internat. Assn. Plant Taxonomy, Torrey Bot. Club, Herb Soc. Am., Soc. Natural History Del., AAAS, Phila. Bot. Club, Washington Bot. Club, Am. soc. Hort. Soc., Pa. Hort. Soc., Sigma Xi, Kappa Delta Pi. Subspecialty: Taxonomy. Office: Dept Agrl and Natural Resources Del State Coll Dover DE 19901

TUCKER, MARC STEPHEN, computer research administrator, consultant; b. Boston, Nov. 15, 1939; s. David Jones and Natalie (Croman) T.; m. Linda Beth Hepler, Sept. 27, 1964 (div. 1973); children: Matthew, Joshua. A.B., Brown U., 1961; M.S., George Washington U., 1982. Lighting dir., camera, sta. WGBH-TV, Boston, 1962-64, asst. dir. edn. div., 1964-66; asst. to pres. Edn. Devel. Ctr., Newton, Mass., 1966-71; asst. dir. NWREL, Portland, Oreg., 1971-72; assoc. dir. Nat. Inst. Edn., Washington, 1972-81; dir. Project on Info. Tech. and Edn., Washington, 1981—; cons. Cresap, McCormick, Washington, 1982, Vanderbilt U., 1982, BRS, Albany, N.Y., 1982, U. Pa., 1982. Chmn., pres. Brass Chamber Music Soc. Annapolis, 1980-81; mem. sch. bd., Bedford, Mass., 1971. Mem. AAAS, Assn. Computing Machinery, Assn. for Pub. Policy Analysis and Mgmt. Democrat. Subspecialties: Information systems (information science); Information systems, storage, and retrieval (computer science). Current work: Public policy on use of computers and telecommunications technology in schools, colleges and universities. Development of improved hardware, software and management systems for education applications. Home: 510 2d St Annapolis MD 21403 Office: Project on Info Tech and Edn Suite 301 1001 Connecticut Ave NW Washington DC 20036

TUCKER, WALLACE HAMPTON, astrophysicist; b. McAlester, Okla., Nov. 4, 1939; s. Charles B. and Josephine E. T.; m. Karen A. Slagle; children: Kerry, Stuart. B.S. in Math, U. Okla., 1961, M.S., 1962; Ph.D. in Physics, U. Calif., San Diego, 1966. Instr., research assoc. Cornell U., 1966-67; asst. prof. Rice U., Houston, 1967-69; sr. staff scientist Am. Sci. and Engring., Inc., 1969-72, cons., 1972-76; astrophysicist Smithsonian Astrophys. Obs., Cambridge, Mass., 1976—; lectr. U.S. Internat. U., 1979-80, U. Calif., Irvine, 1980—. Mem. Am. Astron. Soc., Astron. Soc. Pacific, Internat. Astron. Union. Subspecialties: High energy astrophysics; 1-ray high energy astrophysics. Home: PO Box 266 Bonsall CA 92003

TULLOCH, MICHAEL VERN, computer technology research development consultant; b. Rolla, Mo., June 14, 1945; s. Stewart Bear and Ruth Ellen (Fisher) T.; m. Judith M. Brown, June 13, 1969; children: Michael Vern, Scott McIntosh. B.S. in Physics, Mo. Sch. Mines, 1968; postgrad., U. Ala., 1965-68; Ph.D., U. Tenn., 1973. Elec. engr. J.M. Cone Lab., U.S. Army, Redstone Arsenal, Ala., 1966-68; sr. human factors engr. Gen. Dynamics, Pomona, Calif., 1973-76; sr. scientist and project mgr. Systems Research Labs., Elgin AFB, Fla., 1976-80; project mgr. Inst. Nuclear Power Ops., Atlanta, 1980—; pres. Intelligent Homes Systems, Roswell, Ga., 1983—; cons. Sci. Assoc., 1978-80, The Computer Store, 1978-80. Contbr. articles to profl. jours; assoc. editor: Instant Software, 1978-80. Mem. nat. adv. bd. Am. Security Council, 1980—. NSF trainee, 1968-72. Mem. Am. Psychol. (mil. psychogists div.), Am. Def. Preparedness Assn., Soc. Engring. Psychologists, AAAS, Nat. Mgmt. Assn., Mensa, Phi Beta Iota, Phi Kappa Pi. Republican. Methodist. Current work: Application of computer technology to man-machine interface, intelligent controllers, displays; consultant engineering psychology to the aerospace and nuclear power industry. Home: 1500 Song Sparrow Ct Marietta GA 30062 Office: PO Box 0858 Roswell GA 30077

TULLY, RICHARD BRENT, astronomer; b. Toronto, Ont., Can., Mar. 9, 1943; came to U.S., 1975; s. William M. and Margaret J. (Eaton) T.; m. Janine Alicia Alvarez, June 30, 1972; children: Vincent Munro, Gina Amelia, Tanya Leilani. B.Sc., U. B.C., 1964; Ph.D., U. Md., 1972. Postdoctoral fellow David Dunlap Obs., Toronto, Ont., 1973, Observatoire de Marseille, France, 1973-75; asst. astronomer U. Hawaii, Honolulu, 1975-78, assoc. astronomer, 1978-83, astronomer, 1983—; vis. sr. scientist Cerro Tololo, Interam. Obs., La Serena, Chile, 1983; mem. sci. adv. com. Canada France Hawaii Telescope. Contbr. articles to profl. jours. NSF grantee, 1976—; NATO grantee, 1983. Mem. Am. Astron. Soc., Internat. Astron. Union. Subspecialties: Optical astronomy; Cosmology. Current work: Formation and dynamics of galaxies and clusters of galaxies; extragalactic distance scale; clustering properties of galaxies; formation of galaxies and large scale structure; density of the universe. Address: Inst for Astronomy 2680 Woodlawn Dr Honolulu HI 69822

TUMA, SAMIR NAIF, physician, researcher; b. Acco, Israel, Feb. 17, 1940; came to U.S., 1975; s. Naif J. and Miriam I. T.; m. Grace I. Anfous, May 9, 1970; children: Mona, Noha. M.D., Hebrew U., Jerusalem, 1968. Intern Rambam Univ. Hosp., Haifa, Israel, 1968-69, resident internal medicine, 1969-72, fellow in nephrology, 1972-74, sr. nephrologist, 1974-75; cons. physician Rotchild U. Hosp., Haifa, 1974-75; instr. medicine U So. Calif. Med. Ctr., Los Angeles, 1975-78; asst. prof. Medicine Baylor Coll. Medicine, Houston, 1978—; Co-establisher dialysis units Evangilican Mission, Nablis, West Bank, Israel, 1972-75, Nicosea Gen. Hosp., Nicosea, Cypress, 1973-74, Ramallah New Hosp., West Bank, Israel, 1973-74; cons. physician Nat. Israeli Radio, Jerusalem, 1972-75. Founder, 1st and 2d pres. Arab Am. Med. Assn., Houston, 1979; mem. Nat. Council High Edn., Tel Aviv, 1974. Recipient Tavori award Hebrew U., 1968. Mem. Internat. Soc. Nephrology, Am. Fedn. Clin. Research, Am. Soc. Nephrology, AAUP, N.Y. Acad. Sci. Club: Eliaho Hanovi (warden 1974-75). Subspecialties: Nephrology; Internal medicine. Current work: Clinical nephrology, parathyroid hormone and uremia toxicity. Home: 10915 Paulwood St Houston TX 77071 Office: Baylor Coll Medicine 1200 Moursund Ave Houston TX 77030

TUMBLESON, M.E., physiological chemist, educator; b. Mountain Lake, Minn., Mar. 13, 1937; s. Leonard Orville and Marie Kathryn (Meyer) T.; m.; children: Elise Marie, Eric Jon, Ellen Rae. B.S., U. Minn., 1958, M.S., 1961, Ph.D., 1964. Asst. prof. vet. physiology U. Mo., Columbia, 1966-69, assoc. prof., 1969-80; prof. vet. anatomy and physiology, research prof. Sinclair Research Farm, 1980—. Contbr. numerous articles and sci. abstracts to profl. jours. Nat. bd. dirs. Alpha Gamma Rho. Recipient Disting. award in research Gamma Sigma Delta, 1980. Fellow Gerontol. Soc.; mem. Am. Inst. Nutrition, Am. Soc. Animal Sci., Am. Soc. Biol. Chemists, Am. Soc. Clin. Nutrition, Am. Soc. Neurochemistry, Am. Soc. Vet. Physiologists and Pharmacologists, Nat. Council on Alcoholism, Soc. Exptl. Biology and Medicine, Sigma Xi. Subspecialties: Biochemistry (medicine); Nutrition (biology). Current work: Alcohol metabolism in mammals during devel. and aging. Home: Route 10 Box 41 Columbia MO 65202 Office: Coll Vet Medicine U MO Columbia MO 65211

TUMMALA, V. M. RAO, operations research and information systems educator; b. Pedalimgala, India, Sept. 10, 1937; came to U.S., 1961; s. Veeraiah and Subbamma T.; m. Parvati, May 12, 1955; children: Chandrasekhar, Prabhakhar, Sreenidhi. B.A., Andhra U., Waltair, India, 1957; M.A., Gujarat U., Ahmedabad, India, 1959; M.S., Mich. State U., 1962, Ph.D., 1968. Asst. prof. U. Detroit, 1967-70, assoc. prof., 1970-75, prof., area coordinator, 1977-81; prof. ops. research and info. systems Coll. Bus., Eastern Mich. U., 1976-77, 81—, head dept. ops. research and info. systems, 1981—. Author: Decision Analysis with Business Applications, 1974, Modern Decision Models, 1975; editor.: Procs. of 1981 Midwest conf, Am. Inst. Decision Scis. Pres. Telugu Assn. N. Am., Detroit, 1979-81; v.p. India League Am., Mich. chpt., Detroit, 1980—. Mem. Am. Stats. Assn., Inst. Mgmt. Scis., Am. Inst. Decision Scis., Soc. Risk Analysis. Subspecialties: Statistics; Information systems (information science). Current work: Multi-attribute utility; subjective probability; risk; uncertainy; hierarchical decision making; Bayesian inference. Home: 1825 Huntingwood Ln Bloomfield Hills MI 48013 Office: Dept Ops Research and Info Systems Coll Bus Eastern Mich U Ypsilanti MI 48197

TUNDERMANN, JOHN HAYES, metallurgist, alloy corporation executive; b. Newcastle-on-Tyne, U.K., July 17, 1940; s. Werner O. and Eve (Pickering) T.; m. Ann Pauline Trusiak, Jan. 10, 1940. B.M.E., Ga. Inst. Tech., 1963, M.M.E., 1964; Ph.D., U. Wales, Swansea, 1967. Research metallurgist Paul D. Merica Research Lab., Inc., Suffern, N.Y., 1969-72; sr. project mgr. devel. dept. Internat. Nickel Co., Inc., N.Y.C., 1972-77; mgr. engring. dept. Inco Research and Devel., Suffern, 1977-79, mgr. research, 1980-82; planning mgr. Huntington Alloys, Inc., W.Va., 1982, dir. tech., 1982-83, v.p. tech., 1983—. Mem. Am. Powder Metallurgy Inst., Nat. Mgmt. Assn., Nat. Assn. Corrosion Engrs., Am. Soc. Metals, AIME, ASTM, Nat. Acad. Sci. Episcopalian. Subspecialties: Metallurgy; Alloys. Current work: Hi-nickel alloys, powder metallurgy; managing and directing alloy corporation research and development, plant metallurgy and testing activities. Holder 7 U.S. patents. Home: 67 Copper Glen Dr Huntington WV 25701 Office: PO Box 1958 Riverside Dr Huntington WV 25720

TURCHAN, OTTO CHARLES, engineering physicist; b. Ostrava, Czechoslovakia, Dec. 30, 1925; s. Karl and Felicia (Szymanski) T.; m. Irene Nairi, June 7, 1952; children: Carl Michael, Gary Stephan. Diploma in Engring, German. Inst. Tech., 1945; D.Sc., U. Prague, 1947; B.S., U. Detroit, 1950, M.S., 1952. Cert. profl. engr., mech. engr., nuclear engr. Calif. Engr. Junkers Flugzeugwerke, Magdeburg, Germany, 1943; teaching fellow U. Detroit, 1950-53; mem. tech. staff Research and Devel. Labs., Hughes Aircraft Co., 1955-61; sr. tech. staff Space Tech. Labs. Inc. (Research Lab.), Redondo Beach, Calif., 1961-62; sr. staff Spl. Studies Directorate, Aerospace Corp., El Segundo, Calif., 1962-66; sr. tech. specialist Strategic Missiles Systems div. Rockwell Internat., Anaheim, Calif., 1966-70; prin. engr. Raytheon Co., 1971-72; engr. Bechtel Power Corp., 1972-77; prin. cons. Energy Systems Engring., Beverly Hills, Calif., 1978—; cons. in field. Mem. N.Y. Acad. Scis., Am. Phys. Soc., AIAA, Am. Geophys. Union, Am. Nuclear Soc. Subspecialties: Aerospace engineering and technology; Fusion. Developer electro-magnetic radiation induced fusion reactor; advanced inertial guidance, control and navigation system; electrohydraulic servo control system. Office: PO Box 3093 Beverly Hills CA 90212

TURCHIN, VALENTIN FEDOROVICH, computer science educator; b. Podolsk, USSR, Feb. 14, 1931; came to U.S., 1978; s. Fedor Vassilievich and Lubov Dmitrievna (Bagler) T.; m. Tatiana I. Novikova, Sept. 26, 1956; children: Peter, Dimitri. B.S., Moscow U., 1952; Ph.D., Inst. for Physics and Energy, Obninsk, USSR, 1958. Sr. scientist Inst. Physics and Energy, 1953-64, Inst. Applied Math., Moscow, 1964-73; assoc. research scientist Courant Inst. Math. Scis., N.Y.C., 1978-79; prof. computer sci. CCNY, CUNY, 1979—. Author: Slow Neutrons, 1965, The Phenomenon of Science, 1977, The Inertia of Fear and the Scientific Worldview, 1981. Chmn. Moscow chpt. Amnesty Internat., 1974-77. NSF grantee, 1980-84. Mem. Assn. Computing Machinery. Subspecialties: Artificial intelligence; Programming languages. Current work: Computers, artificial intelligence, philosophy of science, foundations of mathematics. Home: 75-34 113th St Forest Hills NY 11375 Office: Computer Sci Dept CCNY New York NY 10031

TURCO, SALVATORE JOSEPH, biochemistry educator; b. New Kensington, Pa., Sept. 16, 1950; s. Antonio Salvatore and Elvira (Pate) T.; m. Cathy Jane Hummel, Dec. 22, 1973; children: Jason Christopher, Joseph Matthew. B.S., Ind. U., 1972; Ph.D., U. Pitts., 1972-76. Postdoctoral fellow M.I.T., Cambridge, 1976-78; asst. prof. biochemistry U. Ky., Lexington, 1978—. NIH award, 1983-88; grantee, 1980-87; Am. Cancer Soc. grantee, 1980-81; Leukemia Soc. Am. fellow, 1977-78. Mem. N.Y. Acad. Scis., Soc. Complex Carbohydrates, AAAS, Am. Soc. Biol. Chemists. Republican. Methodist. Subspecialties: Biochemistry (medicine); Biochemistry (biology). Current work: Structure, function and biosynthesis of complex carbohydrate-containing macromolecules and the involvement of these substances in abnormal states. Home: 2908 E Hills Dr Lexington KY 40502 Office: U Ky Dept Biochemistry Lexington KY 40536

TURCOTTE, DONALD LAWSON, educator; b. Bellingham, Wash., Apr. 22, 1932; s. Lawson Phillip and Eva (Pearson) T.; m. Joan Meredith Luecke, May 17, 1957; children—Phillip Lawson, Stephen Bradford. B.S., Calif. Inst. Tech., 1954, Ph.D., 1958; M.Aero.Engring., Cornell U., 1955. Asst. prof. aero. engring. U.S. Naval Postgrad. Sch., Monterey, Calif., 1958-59; asst. prof. aero. engring. Cornell U., Ithaca, N.Y., 1959-63, asso. prof., 1963-67, prof., 1967-73, prof. geol. scis., 1973—, chmn., 1981—. Author: (with others) Statistical Thermodynamics, 1963, Space Propulsion, 1965, Geodynamics, 1982. Trustee U. Space Research Assn., 1975-79. NSF sr. postdoctoral research fellow, 1965-66; Guggenheim fellow, 1972-73. Mem. Am. Geophys. Union, Am. Phys. Soc., Geol. Soc. Am. (Day medal 1982), Seismol. Soc. Am. Club: Ithaca Country. Subspecialties: Geophysics; Tectonics. Current work: Mantle convection, sedimentary basins, stress in earth, faulting. Home: 703 Cayuga Heights Rd Ithaca NY 14850 Office: Kimball Hall Cornell U Ithaca NY 14853

TURCOTTE, JEREMIAH GEORGE, physician, educator; b. Detroit., Jan. 20, 1933; s. Vincent Joseph and Margaret Campau (Meldrum) T.; m. Claire Mary Lenz, July 5, 1958; children—Elizabeth Margaret, John Jeremiah, Sara Lenz, Claire Meldrum. B.S. with high distinction, U. Mich., 1955, M.D. cum laude, 1957. Diplomate: Am. Bd. Surgery. Intern U. Mich. Med. Center, 1957-58, resident in surgery, 1958-60, 61-63; research asst. USPHS grant U. Mich. surgery dept., 1960-61; mem. faculty U. Mich. Med. Sch., 1963—, prof. surgery, 1971—, chmn. dept., head sect. gen. surgery, 1974-81; sci. adv. bd. Mich. Kidney Found. Contbr. to med. jours. Recipient Henry Russell award U. Mich., 1970. Fellow A.C.S.; mem. Transplantation Soc. Mich. (pres. 1973-75), Assn. Academic Surgeons, Am. Surg. Assn., Soc. Univ. Surgeons. Internat. Transplantation Soc., Central Surg. Assn., Western Surg. Assn., Soc. Surgery Alimentary Tract, Am. Soc. Transplant Surgeons (pres. 1979-80), Frederick A. Collor Soc. (pres. 1982-83), Am. Trauma Soc. (a founder). Roman Catholic. Subspecialties: Surgery; Transplant surgery. Home: 769 Heatherway St Ann Arbor MI 48104 Office: Univ Hosp B3912 Box 17 1405 E Ann St Ann Arbor MI 48109

TUREK, JEFFERY LEE, sytems engineer; b. Toledo, Oct. 3, 1946; s. Robert Otis and Eve (Stribrny) T.; m. Patricia Jean Skilinchar, Feb. 8, 1969 (div. Oct. 1982). B.A., Washington and Jefferson Coll., 1968; M.S., Va. Poly. Inst. and State U., 1977, Ph.D., 1980. Research assoc. Va. Poly. Inst. and State U., Blacksburg, 1978-80, mgr. research and devel., 1981—; assoc. Systems Research & Applications, Arlington, Va., 1980-81. Served in USN, 1968-74. Assoc. mem. Am. Inst. Indsl. Engrs., Ops. Research Soc. Am. Subspecialty: Information systems (information science). Current work: Fuzzy sets, optimal stopping theory, risk analyses, human factors and their integration into a decisionsupport system aimed at practical, effective management of large research and development programs. Home: Rt 1 Box 57-F Newport VA 24128 Office: 106 Faculty St Blacksburg VA 24061

TUREKIAN, KARL KAREKIN, geology educator; b. N.Y.C., Oct. 25, 1927; s. Vaughan Thomas and Victoria (Guleserian) T.; m. Arax Roxanne Hagopian, Apr. 22, 1962; children: Karla Ann, Vaughan Charles. A.B., Wheaton (Ill.) Coll., 1949; M.A., Columbia U., 1951, Ph.D., 1955. Lectr. geology Columbia U., 1953-54, research asso., 1954-56; faculty Yale U., 1956—, prof. geology and geophysics, 1965-72, Henry Barnard Davis prof. geology and geophysics, 1972—, chmn. dept., 1982—; curator meteorites Peabody Mus., 1964—; cons. Pres.'s Commn. Marine Sci. Engring. and Resources, 1967-68; oceanography panel NSF, 1968-70; NASA exobiology panel NASA, Am. Inst. Biol. Scientists, 1966-69; U.S. nat. com. for geochemistry NAS-NRC, 1970-73, mem. climate research bd., 1977-80, ocean sci. bd., 1979-82; mem. group experts sci. aspects Marine Pollution UN, 1971-73. Author: Oceans, 1968, 2d edit., 1976, Chemistry of the Earth, 1972, (with B.J. Skinner) Man and the Ocean, 1973, (with C.L. Drake, J. Imbrie and J.A. Knauss) Oceanography, 1978; editor: Late Cenozoic Glacial Ages, 1971, Jour. Geophys. Research, 1969-75, Earth Planet Sci. Letters, 1975—; asso. editor: Geochim. et Cosmochim. Acta, 1967-70, Jour. Geophys. Research, 1967-69; editor: Discovery, Yale Peabody Mus., 1978-81; contbr. articles to profl. jours. Served with USNR, 1945-46. Guggenheim fellow, 1962-63. Fellow Geol. Soc. Am., Meteoritical Soc., Am. Geophys. Union, AAAS; mem. Am. Chem. Soc., Geochem. Soc. (pres. 1975-76), Sigma Xi. Subspecialty: Geochemistry. Current work: Marine and surficial geochemistry, geochemistry of planetary evolution. Home: 555 Skiff St North Haven CT 06473 Office: Dept Geology and Geophysics Yale U Box 6666 New Haven CT 06511

TURINSKY, PAUL JOSEF, nuclear engineering educator; b. Hoboken, N.J., Oct. 20, 1944; s. Paul Josef and Wilma Ann (Budig) T.; m. Karen Ann DeLuca, Aug. 20, 1966; children: Grant Dean, Beth Noelle. B.S. in Chem. Engring, U. R.I., 1966; M.S. in Nuclear Engring, U. Mich., 1967, Ph.D., 1970; M.B.A., U. Pitts., 1979. Asst. prof. nuclear engring. and sci. Rensselear Poly. Inst., Troy, N.Y., 1971-73; sr. nuclear design engr. Nuclear Fuel div. Westinghouse Electric Corp., Pitts., 1973-75, lead nuclear design engr., 1975-76, fellow nuclear design engr., 1976, mgr. nuclear design group, 1976-78, mgr. core devel., 1978-80; prof. nuclear engring. N.C. State U., Raleigh, 1980—, head nuclear engring. dept., 1980—; cons. in field. Contbr. sci. articles to profl. publs. Recipient Meritorious Service citation Westinghouse Electric Corp., 1974; AEC fellow, 1966-69. Mem. Am. Nuclear Soc., Am. Power Conf., Am. Soc. Engring. Edn., Sigma Xi., Tau Beta Pi, Phi Kappa Phi. Subspecialties: Nuclear fission; Fluid mechanics. Current work: Applications of mathematical optimization techniques to nuclear fuel cycle management; real time computer simulation of nuclear power plant's neutronic and thermal-hydraulic behaviors. Office: NC State Univ 1110 Burlington Engring Labs Raleigh NC 27650

TURK, DONALD EARLE, food science educator, researcher; b. Dryden, N.Y., Sept.4, 1931; s. Dewey Dolph and Laura (Carpenter) T. B.S., Cornell U., 1953, M.N.S., 1957; Ph.D., U. Wis.-Madison, 1960. From asst. prof. to assoc. prof. poultry sci. Clemson (S.C.) U., 1960-74, prof. food sci., 1974—. Reviewer: Poultry Sci, 1978—; contbr. articles and revs. to sci. jours. Mem. Am. Inst. Nutrition, Poultry Sci. Assn., Am. Chem. Soc., Am. Inst. Biol. Scis., AAAS. Lodges: Kiwanis; Lions. Subspecialties: Animal nutrition; Nutrition (medicine). Current work: Parasitism as related to nutrient absorption, gut physiology and nutrient utilization, trace element nutrition. Home: Route 1 Townville SC 29689 Office: Clemson U Clemson SC 29631

TURKANIS, STUART ALLEN, pharmacologist, neurophysiologist, researcher, educator; b. Everett, Mass., Dec. 15, 1936; s. Edward and Charlotte (Barr) T.; m.; children: Jonathan, Rebecca. B.S., Mass. Coll. Pharmacy, 1958, M.S., 1960; Ph.D., U. Utah, 1967. Registered pharmacist, Mass., Utah. Research assoc. dept. biophysics Univ. coll., London U., 1967-69; asst. prof. pharmacology U. Utah Sch. Medicine, 1969-76, assoc. prof., 1976—. Contbr. numerous chpts., articles to profl. publs. Served to capt. USAR, 1960-67. NIH grantee, 1969-75; Nat. Inst. Drug Abuse grantee, 1974-77, 77-80, 80-83, 83-87; Utah Heart Assn. grantee, 1974-76. Mem. Am. Soc. Pharmacology and Exptl. Therapeutics, Soc. Neurosci, AAUP. Subspecialties: Neuropharmacology; Neurophysiology. Current work: Effects and mechanisms of action of centrally acting drugs; neuropharmacology of cannabinoids. Office: U Utah Sch Medicine 2C202 Med Ctr Salt Lake City UT 84132

TURKAT, DAVID MARK, psychologist, consultant; b. N.Y.C., Apr. 7, 1952; s. Michael M. and Phyllis (Schiff) T. B.A., Brandeis U., 1973; M.A., La. State U., 1974, Ph.D., 1978. Counselor Hillside Hosp., N.Y.C., 1972; trainer So. U., Baton Rouge, 1973-74; psychologist La. State U., 1975-77, Ga. Mental Health Inst., Atlanta, 1977-78; tng. officer Div. Mental Health, Atlanta, 1978-80; pvt. practice psychology, prin. Atlanta Psychol. Assocs., 1979—; cons. in field. Contbr. articles to profl. jours, 1975—; author: radio reports Psychol. Issues program, 1979-83; producer: cable TV show Your Health News, 1981-83. Recipient Caber award Cable Atlanta, 1982. Fellow Menninger Found.; mem. Am. Psychol. Assn., Assn. Media Psychology (v.p. 1982-84), Mental Health Assn., Southeastern Psychol. Assn. Subspecialty: Media psychology. Current work: Media psychology, eating disorders, chronic schizophrenia. Home: 3221-I Post Woods Dr Atlanta GA 30339 Office: Atlanta Psychol Assocs 3390 Peachtree Rd Atlanta GA 30326

TURKEVICH, ANTHONY LEONID, chemist, educator; b. N.Y.C., July 23, 1916; s. Leonid Jerome and Anna (Chervinsky) T.; m. Ireene Podlesak, Sept. 20, 1948; children—Leonid, Darya. B.A., Dartmouth Coll., 1937, D.Sc., 1971; Ph.D., Princeton U., 1940. Research asso. spectroscopy physics dept. U. Chgo., 1940-41; asst. prof., research on nuclear transformations Enrico Fermi Inst. and chemistry dept., 1946-48, asso. prof., 1948-53, prof., 1953—, James Franck prof. chemistry, 1965-70, Distinguished Ser. prof., 1970—; war research Manhattan Project, Columbia U., 1942-43, U. Chgo., 1943-45, Los Alamos Sci. Lab., 1945-46; Participant test first nuclear bomb, Alamagordo, N.Mex., 1945, in theoretical work on and test of thermonuclear reactions, 1945—, chem. analysis of moon, 1967—; cons. to AEC Labs.; fellow Los Alamos Sci. Lab., 1972—. Del. Geneva Conf. on Nuclear Test Suspension, 1958, 59. Recipient E.O. Lawrence Meml. award AEC, 1962; Atoms for Peace award, 1969. Fellow Am. Phys. Soc.; mem. Am. Chem. Soc. (nuclear applications award 1972), AAAS, Nat. Acad. Scis., Am. Acad. Arts and Scis., Royal Soc. Arts (London), Phi Beta Kappa. Mem. Russian Orthodox Greek Cath. Ch. Clubs: Quadrangle, Cosmos. Subspecialties: Nuclear Chemistry; Space chemistry. Home: 175 Briarwood Loop Briarwood Lakes Oak Brook IL 60521 Office: 5640 S Ellis Ave Chicago IL 60637

TURLAPATY, PRASAD DURGA MALLIKHARJUNA VARA, pharmacologist, clin. researcher; b. Vijayawada, India, June 1, 1942; came to U.S., 1967; s. Sangameswara Rao and Visalakshmi (Nidumolu) T.; m. Snehalatha D. Kurapati, Sept. 3, 1972; 1 dau., Neelima. M. Pharmacy, Andhra (India) U., 1965; Ph.D. in Pharmacology, U. Hawaii, 1971. Postdoctoral fellow dept pharmacology U. Tex. Health Sci. Ctr., 1972-74; Sci. officer Jipmer, Pondicherry, India, 1975-77; instr. dept. physiology Downstate Med. Center N.Y., Bklyn., 1977-78, asst. prof., 1978-80; clin. investigative assoc. Ives Labs., N.Y.C., 1980-81, sr. clin. investigative assoc., 1982-83; asst. dir. clin. research Am. Critical Care, Chgo., 1983—. Contbr. articles to profl. publs. Recipient Gold medals Andhra U., 1964, 65; awards Dept. state, East-West Ctr., Honolulu, 1967-69. Mem. Am. Physiol. Soc., Am. Soc. Pharmacology and Exptl. Therapeutics, N.Y. Acad. Scis. Subspecialties: Pharmacology; Physiology (medicine). Current work: Physiology and pharmacology of vascular smooth muscle in hypertension and diabetes; role of magnesium deficiency; clinical research on new drugs in hypertension, angina, arrhythmia. Home: 1102 Whitman Rd Vernon Hills IL 60061 Office: 1600 Waukegan Rd McGraw Park IL 60085

TURNBULL, DAVID, educator, physical chemist; b. Elmira, Ill., Feb. 18, 1915; s. David and Luzetta Agnes (Murray) T.; m. Carol May Cornell, Aug. 3, 1946; children: Lowell D., Murray M., Joyce M. B.S., Monmouth (Ill.) Coll., 1936, Sc.D. (hon.), 1958; Ph.D., U. Ill., 1939; A.M. (hon.), Harvard U., 1962. Tchr. research Case Sch. Applied Sci., 1939-46; scientist Gen. Electric Co. Research Lab., 1946-62, mgr. chem. metallurgy sect., 1950-58; adj. prof. metallurgy Rennselaer Poly. Inst., 1954-62; Gordon McKay prof. applied physics Harvard U., 1962—; chmn. Gordon Conf. Physics and Chemistry Metals, 1952; Internat. Conf. Crystal Growth, 1958; Internat. Conf. Chem. Physics of Non-metallic Crystals, 1961; Office Naval Research panel study growth and morphology crystals, 1959-60. Editor: (with Seitz, Ehrenreich) Solid State Physics, 35 vols., 1955—, (with Doremus, Roberts) Growth and Perfection of Crystals, 1958; assoc. editor: Jour. Chem. Physics, 1961-63; editorial adv. bd.: Jour. Physics and Chemistry Solids, 1955—. Chmn. citizens curriculum com. Niskayuna (N.Y.) Sr. High Sch., 1954. Recipient von Hippel award, 1979. Fellow Am. Phys. Soc. (Internat. prize new materials 1983), N.Y. Acad. Scis., Am. Soc. Metals (chmn. seminar com. 1954), Am. Inst. Mining and Metall. Engrs. (lectr. Inst. Metals div. 1961); mem. Am. Chem. Soc., Nat. Acad. Scis., Am. Acad. Arts and Scis. Subspecialties: Materials; Condensed matter physics. Current work: Glass state; crystal growth; crystal defects. Home: 77 Summer St Weston MA 02193 Office: Applied Physics Dept Harvard Univ Cambridge MA 02138

TURNER, BRUCE JAY, biologist, educator; b. Bklyn., Sept. 19, 1945; s. Frederic Alexander and Lillian (Lindenberg) T.; m. Barbara Anne Bush, Dec. 18, 1972; children: Jonathan, Mikaila. B.S., Bklyn. Coll., 1966; M.A., UCLA, 1967, Ph.D., 1972. Postdoctoral fellow Neuropsychiatric Inst., UCLA, 1972-74; research assoc. Rockefeller U., N.Y.C., 1974-76; vis. research scientist Mus. Zoology, U. Mich., Ann Arbor, 1976-78; asst. prof. zoology dept. biology Va. Poly. Inst. and State U., Blacksburg, 1978—; cons. in field. Contbr. articles to profl. jours. N.Y. State Regents scholar, 1962-66; USPHS fellow, 1968-70; NSF grantee, 1976, 79; Nat. Geog. Soc. grantee, 1980. Mem. Am. Soc. Naturalists, Soc. for Study Evolution, Soc. Ichthyologists and Herpetologists, Soc. for Systematic Zoology, Genetics Soc. Am., Am. Fisheries Soc., AAUP. Subspecialties: Evolutionary biology; Systematics. Current work: Elucidation of factors determining genetic differentiation of fish populations. Office: Va Poly Inst and State Univ Dept Biology Blacksburg VA 24061

TURNER, DEREK TERENCE, materials science educator, researcher; b. London, Dec. 19, 1926; U.S., 1962, naturalized, 1973; s. Henry Godfrey and (Ellen Grant) T.; m. Anais Quevedo, Aug. 6, 1954; children; Luis, Christine. B.Sc., London U., 1951; A.I.R.I., Nat. Coll. Rubber Tech., London, 1952; Ph.D., London U., 1957. Research chemist Brit. Insulated Callenders Cables, London, 1952-55; sr. chemist British Rubber Producers Research Assn., Welwyn Garden City, 1955-62, Camille Dreyfus Lab., Research Triangle Inst., N.C., 1962-68; prof. metall. engring. Drexel U., Phila., 1968-71; prof. oral biology U. N.C., Chapel Hill, 1971-80, prof. operative dentistry, 1980—. Contbr. articles to profl. jours. Served to lt. U.S. Army, 1945-48. Mem. Am. Phys. Soc., Soc. Rheology, Am. Chem. Soc., Soc. Biomaterials, Internat. Assn. Dental Research. Subspecialties: Polymer chemistry; Materials. Current work: Design of composite materials to restore teeth, work on coupling agents and on satisfactory service performance in an aqueous environment, structure and properties of highly crosslinked glassy polymers. Home: 230 Wild Turkey Trail Chapel Hill NC 27514

TURNER, EARL JAMES, civil engineer; b. Marine, Ill., Feb. 11, 1918; s. James Robert and Emma (Suter) T.; m. Marjorie M. Hunziker, Feb. 22, 1944 (dec.); children: James Allen, John Earl, Gary Lee. B.S. in Civil Engring, Washington U., St. Louis, 1943. Registered profl. engr., Mo. Archtl. draftsman Mo. Pacific R.R., 1946-48; structural engr. sect. spl. structures Sverdrup & Parcel & Assocs., Inc., St. Louis, 1948—, project engr., 1955—. Served to 1st lt. inf. U.S. Army, 1944-46. Decorated Bronze Star. Mem. ASCE (chmn. com. aerodynamics Aerospace Div.), Nat. Soc. Profl. Engrs., Am. Concrete Inst., ASME, Res. Officers Assn. Mem. United Ch. of Christ. Subspecialties: Civil engineering; Aerospace engineering and technology. Current work: Project engineer. Office: 801 N 11th Blvd Saint Louis MO 63101

TURNER, EDWIN LEWIS, astronomer; b. Knoxville, Tenn., May 3, 1949; s. George Lewis and Gladys Love (Gregory) T.; m. Joyce Beldon, Aug. 15, 1971; 1 son, Alexander Ross. S.B., M.I.T., 1971; Ph.D., Calif. Inst. Tech., 1975. Postdoctoral fellow Inst. Advanced Study, Princeton, N.J., 1975-76; asst. prof. dept. astronomy Harvard U., Cambridge, Mass., 1977-78; asst. prof. dept. astrophs. scis. Princeton U., N.J., 1978-81, assoc. prof., 1981—; bd. dirs. Assn. Univs. Research in Astronomy, Inc., 1980—. Contbr. articles to astron. jours. Nat. merit scholar, 1967-71; NSF grad. fellow, 1971-74; Alfred P. Sloan basic research fellow, 1980—. Mem. Internat. Astron. Union, Am. Astron. Soc. Subspecialties: Optical astronomy; Cosmology. Current work: Extragalactic astronomy and cosmology research and teaching. Office: Princeton U Obs Peyton Hall Princeton NJ 08544

TURNER, HOWARD SINCLAIR, retired construction company executive; b. Jenkintown, Pa., Nov. 27, 1911; s. Joseph Archer and Helen (Carre) T.; m. Katharine Swett, 1936; children: Susan, Helen Christine, Barbara Jean. A.B., Swarthmore Coll., 1933, LL.D. (hon.), 1977; Ph.D., M.I.T., 1936. Research chemist, technologist, supr. E.I. duPont de Nemours & Co., Inc., 1936-47; dir. research and devel. Pitts. Consolidation Coal Co., 1947-54; v.p. research and devel. Jones & Laughlin Steel Corp., 1954-65, dir., 1965-71; pres. Turner Constrn. Co., N.Y.C., 1965-70, chief exec. officer, 1968-76, chmn. bd., 1971-78, chmn. exec. com., 1978-82, dir., 1952-82; dir. Ingersoll-Rand, 1969-83, Asarco, 1971-81; trustee Dime Savs. Bank of N.Y., 1973-83; Adviser to Q.M. Gen. on research, 1942-44, Office Indsl. Applications, NASA, 1962-65; mem. vis. com. dept. chem. engring. Carnegie Inst. Tech., 1962-65; mem. tech. adv. bd. Dept. Commerce, 1963-67; mem. Viet Nam task force HEW, 1966; mem. research and engring. adv. council P.O. Dept., 1967-68; exec. com. div.. engring. NRC, 1967-71; mem. Pres.'s Sci. Advisory Com., 1972; mem. adv. com. Center for Bldg. Tech., Nat. Bur. Standards, 1973-75; mem. vis. com. dept. civil engring. MIT, 1974; mem. adv. council Sch. Architecture and Urban Planning, Princeton U., 1974. Mem. Swarthmore Borough Council, 1944-47; Bd. mgrs. Swarthmore Coll., 1952-64, 68-72; trustee Am. Acad. Ednl. Devel., C.F. Kettering Found., 1979—; pres. Indsl. Research Inst., 1962; mem. council Rockefeller U.; bd. visitors, gov. Washington Coll., 1983—. Mem. Am. Chem. Soc., Dirs. Indsl. Research (emeritus), Com. for Econ. Devel. (hon. trustee), Nat. Acad. Engring., Econ. Club N.Y. Phi Beta Kappa, Sigma Xi, Phi Kappa Psi. Club: University. Subspecialty: Chemical engineering. Home: 870 UN Plaza New York NY 10017 Office: 633 3d Ave New York NY 10017

TURNER, JAMES ELDRIDGE, neuroscientist, educator, biomedical researcher; b. Richmond, Va., Oct. 1, 1942; s. Shelley Ivy and Aylease Frances (Moore) T.; m. Virginia Tucker Clark, May 27, 1967; children: Catherine Aylease, Annamarie Frances, James Eldridge. B.A., Va. Mil. Inst., 1965; M.S., U. Richmond, 1967; Ph.D., U. Tenn. Knoxville, 1970. Postdoctoral fellow Case Western Res. U. Sch. Medicine, 1970-71, NIH postdoctoral fellow, 1972-74; asst. prof. biology Va. Mil. Inst., 1971-72; asst. prof. anatomy Bowman Gray Sch. Medicine, Wake Forest U., 1974-78, assoc. prof., 1978-83, prof., 1983—; vis. prof. Max-Planck Inst. Psychiatry, Munich, W.Ger., 1980-81. Contbr. articles to profl. jours. Pres. ch. council Augsburg Lutheran Ch., Winston-Salem, N.C., 1983. Recipient Research Career Devel. award NIH, 1978-83; NIH fellow, 1967-69; March of Dime Basel O'Connor fellow, 1975-78. Mem. N.C. Soc Neurosci. (councillor 1982-85), Soc. Neurosci., Internat. Soc. Neurochemistry, Am. Assn. Anatomists, Am. Soc. Zoologists. Subspecialties: Regeneration; Neurobiology. Current work: Neurotrophic activity in regeneration and development of nervous system. Home: 450 Lynn Ave Winston-Salem NC 27104 Office: Dept Anatomy Bowman Gray Sch Medicine Wake Forest U 300 Hawthorne Rd Winston-Salem NC 27104

TURNER, MALCOLM ELIJAH, JR., educator, biomathematician; b. Atlanta, May 27, 1929; s. Malcolm Elijah and Margaret (Parker) T.; m. Ann Clay Bowers, Sept. 16, 1948; children: Malcolm Elijah IV, Allison Ann, Clay Shumate, Margaret Jean; m. Rachel Patricia Farmer, Feb. 1, 1968. Student, Emory U., 1947-48; B.A., Duke U., 1952; M.Exptl. Stats., N.C. State U., 1955, Ph.D., 1959. Analytical statistician Communicable Disease Center, USPHS, Atlanta, 1953; research asso. U. Cin., 1955, asst. prof., 1955-58; asst. statistician N.C. State U., Raleigh, 1957-58; asso. prof. Med. Coll. Va., Richmond, 1958-63, chmn. div. biometry, 1959-63; prof., chmn. dept. statistics and biometry Emory U., Atlanta, 1963-69; prof., chmn. dept. biomath., prof. biostats. and math. U. Ala., Birmingham, 1970—; instr. summers Yale U., 1966, U. Calif. at Berkeley, 1971, Vanderbilt U., 1975; prof. U. Kans., 1968-69; vis. prof. Atlanta U., 1969; cons. to industry. Contbr. articles to profl. jours. Fellow Ala. Acad. Sci.; hon. fellow Am. Statis. Assn., AAAS, mem. Biometric Soc. (mng. editor Biometrics 1962-69), N.Y. Acad. Sci., Soc. for Math. Biology, Soc. for Indsl. and Applied Math., AMA (affiliate), Mensa, Sigma Xi, Phi Kappa Phi, Phi Delta Theta, Phi Sigma. Subspecialties: Applied mathematics; Statistics. Current work: Math. biology, inductive inference. Home: 1734 Tecumseh Trail Pelham AL 35124 The logic of induction is the quest.

TURNER, ROBERT ALEXANDER, JR., rheumatologist, educator; b. Englewood, N.J., Oct. 12, 1937; s. Robert Alexander and Marie Antoinette (Fensterer) T.; m. Florence Elizabeth McGowan, June 25, 1960; children: John Alexander, Katheryn Elizabeth, Robert Andrew. A.B., U.N.C., Chapel Hill, 1959; M.D., U. Ala. in Birmingham, 1966. Diplomate: Nat. Bd. Med. Examiners, Am. Bd. Internal Medicine with subsplty. in rheumatology; lic., N.C. Resident in internal medicine N.C. Bapt. Hosp., 1966-69; fellow in rheumatology Hosp. U. Pa., 1969-71; asst. prof. Bowman Gray Sch. Medicine, Winston-Salem, N.C., 1971-75, assoc. prof., 1975-81, prof., 1981—, chief sect. rheumatology, 1979—; mem. N.C. Arthritis Program Com., 1979—. Contbr. articles to profl. jours. Bd. dirs. N.C. chpt. Arthritis Found., 1975—, pres., 1981-82. Served with USMC, 1959-62. Fellow ACP; mem. Am. Assn. Immunolgists, Am. Fedn. Clin. Research, Am. Rheumatism Assn., Am. Soc. Clin. Pharmacology and Therapeutics, Forsyth County Med. Soc., N.C. Med. Soc., Soc. Exptl. Biology and Medicine, So. Soc. Clin. Investigation, N.Y. Acad. Scis. Republican. Episcopalian. Current work: Etiopathogenesis, clin. pathology, treatment of rheumatoid arthristis. Home: 2801 Robinhood Rd Winston-Salem NC 27106 Office: Rheumatology Sect Dept Medicine Bowman Gray Sch Medicine 300 S Hawthorne Rd Winston-Salem NC 27103

TURNQUIST, PAUL KENNETH, agricultural engineering educator, academic administrator; b. Lindsborg, Kans., Jan. 3, 1935; s. Leonard O. and Myrtle E. (Ryding) T.; m. Peggy A. James, Dec. 22, 1962; children: Todd, Scott, Greg. B.S., Kans. State U., 1957; M.S., Okla. State U., 1961, Ph.D., 1965. Registered profl. engr., Okla. Research engr. Caterpillar Tractor Co., 1957; instr. Okla. State U., 1958-61, asst. prof., 1961-62; assoc. prof. S.D. State U., 1964-71, prof., 1971-76; prof. agrl. engring., head dept. Auburn U., 1977—. Author: (with others) Tractors and Their Power Units, 3d edit, 1979; contbr. numerous tech. articles to profl. jours. Mem. Am. Soc. Agrl. Engrs. (A. W. Farrell Young Educator award 1972), Am. Soc. Engring. Edn. (Outstanding Young Faculty award 1970), Nat. Soc. Profl. Engrs. Methodist. Lodge: Kiwanis. Subspecialty: Agricultural engineering. Current work: Administration; power and machinery. Office: Dept Agrl Engring Auburn U Auburn AL 36849

TURNQUIST, TRUMAN DALE, chemistry educator; b. Kipling, Sask., Can., Apr. 8, 1940; came to U.S., 1940; s. Leonard Ray and Muriel Amanda (Bjorklund) T.; m. Anna Linnea Linden, Aug. 22, 1964; children: David, Dale, Eric. B.A., Bethel Coll., St. Paul, 1961; Ph.D., U. Minn., 1965. Prof. chemistry Mt. Union Coll., Alliance, Ohio, 1965—; cons., researcher Babcock & Wilcox Co., Alliance, 1978—. NASA fellow, 1962-65. Mem. Am. Chem. Soc. Subspecialty: Analytical chemistry. Current work: Trace metal analysis. Office: Dept Chemistry Mt Union Coll Alliance OH 44601

TURRO, NICHOLAS JOHN, chemistry educator; b. Middletown, Conn., May 18, 1938; s. Nicholas John and Philomena (Russo) T.; m. Sandra Jean Misenti, Aug. 6, 1960; children—Cynthia Suzanne, Claire Melinda. B.A., Wesleyan U., 1960; Ph.D., Calif. Inst. Tech., 1963. Instr. Columbia U., N.Y.C., 1964-65, asst. prof., 1965-67, assoc. prof., 1967-69, prof. chemistry, 1969—, William P. Schweitzer prof. chemistry, 1982—, chmn. chemistry dept., 1981-84; Cons. E.I. duPont de Nemours and Co., Inc. Author: Molecular Photochemistry, 1965, (with G.S. Hammond, J.N. Pitts, D.H. Valentine) Survey of Photochemistry, vol. 1, 1968, vol. 2, 1970, vol. 3, 1971, (with A.A. Lamola) Energy Transfer and Organic Photochemistry, 1971, Modern Molecular Photochemistry, 1978; Editorial bd.: Jour. Organic Chemistry, 1974—. NSF fellow; Alfred P. Sloan Found. fellow.; Recipient Eastman Kodak award for excellence in grad. research; award for pure chemistry, 1973. Mem. Nat. Acad. Sci., Am. Chem. Soc., Am. Acad. Arts and Scis., Am. Chem. Soc. (Fresenius award 1973, award for pure chemistry 1974), Chem. Soc. (London), N.Y. Acad. Scis. (Freda and Gregory Halpern award in photochemistry 1977), Am. Soc. Photochemistry and Photobiology, Internat. Solar Energy Soc., Phi Beta Kappa, Sigma Xi. Subspecialty: Photochemistry. Current work: Molecular photochemistry; chemiluminis-cent organic reactions; small ring compounds; energy transfer mechanisms; laser application to photochemical problems; models for solar energy capture and storage; fluorescence probes of micelles and biological membranes; photochemistry in polymers; reactions of molecular oxygen. Home: 125 Downey Dr Tenafly NJ 07670

TUTTE, WILLIAM THOMAS, mathematics educator, researcher; b. Newmarket, Eng., May 14, 1917; m., 1949. Ph.D. in Math, Cambridge (Eng.) U., 1948. Lectr. U. Toronto, Ont., Can., 1948-52, from asst. prof. to assoc. prof., 1952-62; prof. math. U. Waterloo, Ont., 1962—. Fellow Royal Soc. Can. (Henry Marshall Tory medal 1975); mem. Am. Math. Soc., Math. Assn. Am., Can. Math. Soc., London Math. Soc. Office: Faculty of Math U Waterloo Waterloo ON Canada N2L 3G1

TUTTLE, JEREMY BALLOU, neuroscientist, researcher, educator; b. N.Y.C., Oct. 9, 1947; s. John Bauman and Charlotte Marion (Root) T.; m. Sara Jane Stasko, Mar. 26, 1971. A.B., U. Rochester, 1969; Ph.D. in Physiology (fellow), Johns Hopkins U., 1977. Research asst. Sloan-Kettering Inst., N.Y.C., summers 1966-68; postdoctoral fellow physiology sect. biol. sci. group U. Conn., Storrs, 1976-79, Spinal Cord Injury Found. fellow, 1980, asst. prof. in residence physiology sect. biol. sci. group, 1980—. Contbr. articles to sci. jours. Nat. Inst. Neurol. Communicative Disease and Stroke nat. research service awardee, 1976-79; research career devel. awardee, 1981—. Mem. Soc. for Neurosci., AAAS, N.Y. Acad. Scis., Johns Hopkins U. Med. and Surg. Assn., Sigma Xi. Subspecialties: Neurobiology; Neurophysiology. Current work: Cellular neurophysiology; trophic interactions of neural tissue in cell culture; neuronal cell culture; neural circuit formation; pharmacology of excitable cells. Home: 268 Warrenville Rd Mansfield Center CT 06250 Office: U Conn U-42 Physiology Dept Storrs CT 06268

TUTTLE, RONALD RALPH, pharmacologist; b. Colorado Springs, Colo., July 10, 1936; s. Ralph A. and Susan C. (McAlroy) T.; m. Arlene Elaine Casselman, Sept. 28, 1963; 1 son, Jeff Douglas. B.A. in Zoology, Colo. Coll., 1960; M.S. in Pharmacology, U. Man., Can., 1964, Ph.D., 1966; postdoctoral, Emory U., 1966-67. With Eli Lilly & Co., Indpls., 1967-80; dir. new drug devel. Key Pharms., Inc., Miami, Fla., 1980-81, v.p. and dir. new drug devel., 1981—. 437 Contbr. articles to profl. jours. Edward L. Drewry scholar, 1962. Fellow Am.

Heart Assn. Circulation Council; mem. Am. Soc. Pharmacology and Exptl. Therapeutics, Internat. Soc. Heart Research, N.Y. Acad. Scis., Sigma Xi. Subspecialty: Pharmacology. Office: 18425 NW 2d Ave Miami FL 33169

TUUL, JOHANNES, physics educator, researcher; b. Tarvastu, Viljandi, Estonia, May 23, 1922; came to U.S., 1956, naturalized, 1962; s. Johan and Emilie (Tulf) T.; m. Marjatta Murtoniemi, July 14, 1957 (div. Aug. 1971); children: Melinda, Melissa; m. Sonia Esmeralda Manosalva, Sept. 15, 1976; 1 son, Johannes. B.S., U. Stockholm, 1955, M.A., 1956; Sc.M., Brown U., 1957, Ph.D., 1960. Research physicist Am. Cyanamid Co., Stamford, Conn., 1960-62; sr. research physicist Bell & Howell Research Center, Pasadena, Calif., 1962-65; asst. prof., assoc. prof. Calif. State Poly. U., Pomona, 1965-68; vis. prof. Pahlavi U., Shiraz, Iran, 1968-70; chmn. phys. earth sci. Calif. State Poly. U., Pomona, 1971-75, prof. physics, 1975—; cons. Bell & Howell Research Center, Pasadena, Calif., 1965, Teledyne Co., Pasadena, 1968; guest researcher Naval Weapons Center, China Lake, Calif., 1967, 72. Author: Physics Made Easy, 1974; contbr. articles in field to profl. jours. Pres. Group Against Smoking Pollution, Pomona Valley, Calif., 1976; foster parent Foster Parents Plan, Inc., Warwick, R.I., 1964-83; block capt. Neighborhood Watch, West Covina, Calif., 1982. Brown U. fellow, 1957; U. Namur (Belgium) research grantee, 1978; Centre Nat. de la Recherche Scientifique research grantee, 1979; recipient Humanitarian Fellowship award Save the Children Fedn., 1968. Mem. Am. Phys. Soc., AAAS (life), Am. Assn. Physics Tchrs., N.Y. Acad Scis. Republican. Roman Catholic. Subspecialties: Condensed matter physics; Surface chemistry. Current work: Research in the area of energy conservation and new energy technologies. Office: Calif State Poly U 3801 W Temple Ave Pomona CA 91768

TWEEDY, BILLY G., biochemist, plant pathologist, chemical company executive; b. Cobden, Ill., Dec. 31, 1934; s. Amos Glenn and Lona Clara (Klughart) T.; m. Patsy Ann Glasco, Dec. 28, 1957; children: Patricia Lynn, Glenna Jean, Carol Jane. B.S., So. Ill. U., 1956; M.S., U. Ill., 1959, Ph.D., 1961. Asst. plant pathologist Boyce Thompson Inst., Yonkers, N.Y., 1961-65; mem. faculty dept. plant pathology U. Mo., Columbia, 1965-73, prof., 1973; prin. plant pathologist U.S. Dept. Agr., Washington, 1971-72; mgr. residue investigations CIBA-Geigy Corp., Greensboro, N.C., 1973-78, dir. biochemistry, 1978—; project leader U.S.-USSR Bilateral Agreement on Integrated Pest Mgmt. Contbr. chpts. to books, articles to profl. jours. Mem. Am. Phytopathol. Soc., Am. Chem. Soc., Weed Sci. Soc., AAAS, Internat. Congress Plant Protection (standing com.), Sigma Xi, Gamma Sigma Delta. Baptist. Subspecialties: Biochemistry (biology); Plant pathology. Current work: Administrator in area of biochemistry of pesticide in plants, soil and animals. Home: 111 Crest Hill Rd Jamestown NC 27282 Office: PO Box 18300 Greensboro NC 27419

TWIGGS, LEO BROOKHART, physician, educator; b. Benham, Ky., Dec. 31, 1946; m. Martha Esparza. B.S., U. Mich., 1968, M.D., 1972. Diplomate: Am. Bd. Ob-Gyn. (subcert. in gynecologic oncology). Intern Los Angeles County Hosp.- U. So. Calif. Med. Ctr., 1972-73, resident, 1973-76, fellow gynecologic oncology, 1976-78, clin. instr. ob-gyn, 1976-78; asst. prof. U. Minn., 1978-82, assoc. prof., 1982—; dir. Colposcopy Clinic Hosp., 1978—; gynecologic oncology cons. Hennepin County Med. Ctr., 1978-81; mem. tumor bd. Meth. Hosp., Mpls., 1978—; co-dir. Upper Midwest Trophoblastic Disease Ctr., 1981—. Contbr. chpts. to books and articles to profl. jours. Am. Cancer Soc. grantee. Fellow Am. Coll. Obstetricians and Gynecologists; mem. Soc. Ob-Gyn, Internat. Soc. Gynecological Pathologists, Am. Soc. Cervical Pathology and Colposcopy, Soc. Gynecologic Oncology, Am. Soc. Clin. Oncology. Subspecialties: Gynecological oncology; Oncology. Current work: Tumor markers in trophoblastic disease, hormone receptors in cancer. Home: 4200 Fremont Ave S Minneapolis MN 55409 Office: U Minn Hosps Box 395 Mayo Memorial Bldg 420 Delaware St SE Minneapolis MN 55455

TWISS, PAGE CHARLES, geology educator; b. Columbus, Ohio, Jan. 2, 1929; s. George Ransom and Blanche (Olin) T.; m. Nancy Homer Hubbard, Aug. 29, 1954; children—Stephen Ransom, Catherine Grace, Thomas Stuart. B.S. in Geology, Kans. State U., 1950, M.S., 1955; Ph.D., U. Tex. at Austin, 1959. Mem. faculty dept. geology Kans. State U., Manhattan, 1959—, assoc. prof., 1964-69, prof., 1969—, also head dept., 1968-77; geologist agrl. research service U.S. Dept. Agr., 1966-68; research scientist U. Tex., Austin, 1966-67. Contbr. articles to profl. jours. Chmn. Manhattan Council Human Relations, 1960-61; vice pres. Riley County Democratic Club, 1970-71; mem. Dem. Precinct Com., Manhattan, 1970-72, 74-80. Served with USAAF, 1951-53. Fellow Pan Am. Petroleum Found., 1957-58; Shell Found. fellow, 1958-59. Fellow Geol. Soc. Am. (chmn. South Central sect. 1972-73, sec.-treas. 1980—); mem. Am. Assn. Petroleum Geologists (geologic maps com. 1968-70), Soc. Econ. Paleontologists and Mineralogists, Kans. Acad. Sci. (mem. research awards com. 1966—, assoc. editor 1977—), AAAS, Clay Minerals Soc., Kans. Geol. Soc., W. Tex. Geol. Soc., Am. Soc. Agronomy, Soil Sci. Soc. Am., Internat. Soc. Soil Sci., Internat. Assn. Sedimentologists, Assn. Internationale pour l'Etude des Argiles, Am. Quaternary Assn., AAUP (chpt. v.p. 1971-72, chpt. pres. 1972-73), Nat. Assn. Geology Tchrs., Mineral. Soc. Am., Sigma Xi, Sigma Gamma Epsilon, Gamma Sigma Delta. Club: Mason. Subspecialties: Geology; Sedimentology. Current work: Origin, distribution, and classification of silica bodies (phytoliths)in plants, atmospheric dust deposition and composition, pertology and geochemistry of sandstone, carbonate rocks, and ignimbrites. Home: 2327 Bailey Dr Manhattan KS 66502

TWIST, E. MICHAEL, microbiologist, educator; b. Louisville, Apr. 19, 1944. B.S., Fla. Atlantic U., 1966, M.S., 1968; Ph.D., Nova U., 1976. Tchr. sci. Nova High Sch., Ft. Lauderdale, Fla., 1968-71; instr. chemistry Broward Community Coll., Ft. Lauderdale, 1971-76; postdoctoral student Inst. for Cancer Research, Phila., 1976-78; asst. prof. Wistar Inst., Phila., 1978—. Named Fla. High Sch. Tchr. of Yr. Am. Chem. Soc., 1971. Mem. Am. Soc. for Microbiology. Current work: Molecular biology of rotavirus pathogenesis; transcriptional mapping of hepatitis B virus. Home: 510 W Harvey St Philadelphia PA 19144 Office: Children's Hosp Philadelphia PA 19104

TYE, BIK-KWOON, biochemistry educator; b. Hong Kong, Jan. 7, 1947; d. James Ngo and Ching-Yue (Ma) Yeung; m. Sze-Hoi Henry Tye, June 25, 1971; 1 dau. Kay. B.A., Wellesley Coll., 1969; M.Sc., U. Calif.-San Francisco, 1971; Ph.D., MIT, 1974. Postdoctoral fellow Stanford U. Med. Sch., 1974-77; asst. prof. biochemistry Cornell U., Ithaca, N.Y., 1977—. Mem. Genetics Soc. Am. Subspecialties: Genome organization; Genetics and genetic engineering (biology). Current work: Structure and function of chromosomes, DNA replication; replication origins, centromeres, telomeres, chromosome construction, gene regulation. Office: Dept Biochemistry Cornell U W327 Wing Hall Ithaca NY 14853

TYLER, GEORGE LEONARD, electrical engineer, radar astronomer; b. Bartow, Fla., Oct. 18, 1940; m.; 2 children. B.S., Ga. Inst. Tech., 1963; M.S., Stanford U., 1964, Ph.D. in Elec. Engring, 1967. Research assoc. Stanford U., 1967-79, research engr., 1969-71, sr. research assoc., 1971-74, adj. prof. elec. engring., 1974-82, prof. elec. engring. (research), 1982—; team leader Voyager Radio Sci. Team, NASA, 1979-82. Recipient NASA Exceptional Sci. Achievement medal, 1977, 81. Mem. IEEE, Am. Astron. Soc., Am. Geophys. Union, Internat. Astron. Union, Internat. Union Radio Sci. Subspecialty: Radio and microwave astronomy. Office: Electronic Lab Ctr Radar Astronomy Stanford U Stanford CA 94305

TYLER, KENNETH LAURENCE, neurologist; b. Boston, May 6, 1953; s. H. Richard and Joyce (Colby) T.; m. Lisa Johnson, Oct. 27, 1979. A.B., Harvard U., 1974; M.D., Johns Hopkins U., 1978. Resident Peter Bent Brigham Hosp., Boston, 1978-80; resident in neurology Mass. Gen. Hosp., Boston, 1980-83, chief resident in neurology, 1983—; clin. fellow in medicine Harvard Med. Sch., Boston, 1978-80, clin. fellow in neurology, 1980—; cons. in neurology Mass. Rehab. Com. Contbr. articles on neurology to med. jours. Mem. ACP (assoc.), Am. Fedn. Clin. Research, Am. Acad. Neurology (jr.), Am. Soc. Neurol. Investigation, Countway Assocs. Harvard Med. Sch., Phi Beta Kappa, Alpha Omega Alpha. Clubs: Harvard, Spee, Pithotomy. Subspecialties: Neurology; Virology (biology). Current work: Neurovirology: study of the molecular biology of viral infection of the nervous system. Office: Dept Neurology Mass Gen Hosp/ Harvard Med Sch Fruit St Boston MA 02215 Home: 36 Russell St Brookline MA 02148

TYLER, LORAINE LYON, housing educator; b. Bath., N.Y., Feb. 13, 1947; d. William J. and Doris B. (Johnson) Lyon; m. Richard F. Tyler, Oct. 4, 1969; 1 dau., Jessica Lynne. B.S., State U. Coll., Oneonta, N.Y., 1969; M.S., Cornell U., 1972; Ph.D., Va. Poly. Inst., 1982. Tchr. home econs. secondary sch. Ellenville (N.Y.) Central Sch., 1969-70; teaching asst. Cornell U., 1970-72; instr. State U. Coll., Oneonta, 1972-79, asst. prof., 1981—; instr. Va. Poly. Inst., Blacksburg, 1979-81; cons., mem. pub. adv. com. Town of Blacksburg Planning Dept., 1980-81. Dept. Energy grantee, 1980; Am. Home Econs. Assn. grantee, 1981. Mem. Am. Assn. Housing Educators, Am. Home Econs. Assn., Omicron Nu, Phi Upsilon Omicron, Kappa Delta Pi, Phi Kappa Phi. Subspecialty: Energy Education. Current work: Energy education, housing alternatives. Home: RD 2 Box 478 Oneonta NY 13820 Office: State Univ Coll 215 Dept Home Econs Oneonta NY 13820

TYLER, TIPTON RANSOM, toxicologist; b. Milw., Jan. 3, 1941; s. Ransom and Margerite (Struble) T.; m. Barbara Ann Manville, Apr. 21, 1962; children: Timothy, Gregory, Jennifer. B.S., Colo. State U., 1963, Ph.D., 1968; M.S., N.C. State U., 1965. Diplomate: Am. Bd. Toxicology. Sr. research chemist Merck Sharp & Dohme Research Labs., Rahway, N.J., 1968-73; asst. prof. animal sci. U. Ill., Urbana, 1974-75; sr. scientist, chem, hygiene fellow Bushy Run Research Lab., Carnegie-Mellon U., Pitts., 1976-81; asst. corp. dir. applied toxicology Union Carbide Corp., South Charleston, W.Va., 1981—; adj. assoc. prof. toxicology dept. pharmacology Sch. Pharmacy, U. Pitts., 1979—. Contbr. articles to sci. jours. Mem. Soc. Toxicology, Am. Soc. Pharmacology and Exptl. Therapeutics, Am. Coll. Toxicology, Am. Chem. Soc. Subspecialties: Toxicology (medicine); Pharmacokinetics. Current work: Metabolism of chemicals and foreign compounds. Office: PO Box 8361 South Charleston WV 25303

TYNDALL, BRUCE MAPES, mathematics educator, automotive analysis consultant and researcher; b. Iowa City, June 16, 1930; s. Edward Philip Theodore and Irene Beatrice (Mapes) T.; m. Antoinette van der Berg, Aug. 22, 1957; children: Suzanne, Juliet. B.A., U. Iowa, 1955, M.S., 1956; student, Roosevelt U., 1950-51; Ph.D. equivalent Faculty Com. on Tenure and Promotion, U. New Haven, 1976. Cons. actuary Huggins & Co.; Phila., 1956-57; instr. Elizabethtown Coll., 1957-59, asst. prof. math., 1959-61, assoc. prof., 1961; instr. in mechanics Johns Hopkins U., 1961-63; cons. actuary Monumental Life Ins. Co. and O. Herschman Assocs., Balt., 1963-65; assoc. prof. math. U. New Haven, 1965-76, prof., 1976—; cons. automotive analysis Bruce Tyndall Assocs., Guilford, Conn., 1963—; pres. Tyndall Motor Car Co., Guilford, 1973—. Co-founder, co-editor: Essays in Arts and Scis, 1970—; inventor new forms of math. analysis, aluminum frame for impact absorption. Concert oboist, numerous concerts, East Coast, 1960-81; pianist, tchr. U. N.H. Served to cpl. U.S. Army, 1952-54. Recipient Author's award Automotive Industries jour., 1964; Outstanding Tchr. award U. New Haven, 1975. Mem. Soc. Indsl. and Applied Math., Soc. Automotive Engrs., Soc. Logistics Engrs. (chmn. chpt. 1982-84), Soc. Automotive Historians, Internat. Double Reed Soc. Subspecialties: Applied mathematics; Materials. Current work: Nonlinear differential equations with new forms of analytical solutions and approximation methods; numerous applications to impact, mortality, music, automotive and neuroscience. Office: Dept Math U New Haven West Haven CT 06516

TYRE, TIMOTHY EDWARD, psychologist, hospital program administrator; b. Oak Park, Ill., Mar. 12, 1947; s. Edward William and Shirley (Litton) T.; m. Marilyn Yanchar, Dec. 28, 1969; children: Emily, Lisa, Kateri, Sean. B.A., St. Mary's Coll., Winona, Minn., 1969; M.A., Xavier U., Cin., 1971; Ph.D., U. Wis.-Milw., 1973. Dir. Children's Clinic, Archdiocese Milw., 1973-75; asst. dir. dept. psychol. services Norris Health Center, U. Wis.-Milw., 1975-77; pvt. clin. practice psychology, Waukesha, Wis., 1977—; dir. Pain Clinic, Waukesha Meml. Hosp., 1979—; dir. cancer research program West Allis (Wis.) Hosp., 1981—; cons. psychology dept. phys. medicine St. Luke's Hosp., Milw., 1977—; chmn. Wis. Bd. Examiners in Psychology, Madison, 1980—. Research publs. in field. Mem. Soc. Clin./Cons. Psychology State Wis. (pres. 1978-79), Am. Psychol. Assn., Wis. Psychol. Assn. (pres. Div. I 1978-79), Soc. Behavioral Medicine, Soc. Clin. and Exptl. Hypnosis. Republican. Roman Catholic. Subspecialties: Cancer research (medicine); Behavioral psychology. Current work: Management of chronic pain-both benign and malignant; psychological variables active in neoplastic disease. Home and Office: 1519 Summit Ave Waukesha WI 53186

TYRRELL, JAMES, chemistry educator; b. Kilsyth, Scotland, Apr. 19, 1938; came to U.S., 1967, naturalized, 1977; s. Thomas and Jean (Bauld) T.; m. Karine Babrowski, Nov. 4, 1940; 1 son: Dalton; m. Neena Lynne Summers, Sept. 9, 1955. B.Sc. with honors, U. Glasgow, Scotland, 1960; Ph.D., 1963. Teaching postdoctoral fellow dept. chemistry McMaster U., Hamilton, Ont., Can., 1963-65; postdoctoral fellow div. pure physics NRC, Ottawa, Ont., Can., 1965-67; asst. prof. dept. chemistry So. Ill. U., Carbondale, 1967-72, assoc. prof., 1972-80, prof., 1980—, chmn. dept., 1982—. Mem. Am. Chem. Soc., Sigma Xi. Democrat. Subspecialties: Theoretical chemistry; Physical chemistry. Current work: AB initio molecular orbital calculations on complexes of transition metals and small gaseous molecules. Home: 1433 E Walnut St 7 C Carbondale IL 62901 Office: So Ill U Dept Chemistry Carbondale IL 62901

TYTELL, MICHAEL, anatomist, educator; b. N.Y.C., Jan. 3, 1948; s. Samuel and Ida Rachel (Smoller) T.; m. Frances Wilke, Mar. 3, 1973; children: Eric Daniel, Alison Rebecca. B.A., Queens Coll., 1969; M.S., Purdue U., 1973; Ph.D., Baylor Coll. Medicine, 1977. NIH postdoctoral fellow Case Western Res. U., Cleve., 1977-79, sr. research assoc., 1979-80; asst. prof. dept. anatomy Bowman Gray Sch. Medicine, Wake Forest U., Winston-Salem, N.C., 1980—. Contbr. articles to profl. jours. NSF grantee, 1982-84. Mem. AAAS, Soc. Neurosci., Am. Soc. Neurochemistry, Am. Soc. Cell Biology, N.Y. Acad. Scis., Planetary Soc., Sigma Xi. Jewish. Subspecialties: Cell biology; Neurobiology. Current work: Transport of proteins within axons of nerve cells; research on cell biology of neurons. Office: Dept Anatomy Bowman Gray Sch Medicine Winston Salem NC 27103

TYZNIK, WILLIAM JOHN, animal scientist, educator; b. Milw., Apr. 26, 1927; s. John and Anna (Pliska) T.; m. Elizabeth Ann Coughlin, June 21, 1950; children: Melissa, John, Lori, Patricia, James. B.S., U. Wis., 1948, M.S., 1949, Ph.D., 1951. Asst. prof. animal sci. and vet. preventive medicine Ohio State U., Columbus, 1951-54, assoc. prof., 1954-57, prof., 1957—; cons. animal nutrition; pres. Tizco Inc. Mem. Grandview Heights (Ohio) City Council, 1969-74. Recipient Alumni Teaching award Ohio State U., 1970. Mem. Am. Assn. Animal Sci. Roman Catholic. Subspecialty: Animal nutrition. Current work: Equine research. Patentee Fido Freeze treat for dogs. Office: 2121 Fyffe Rd Columbus OH 43212

TZENG, OLIVER CHUN SHUN, psychology educator; b. Taiwan, Dec. 1, 1939; s. Pen-Yeo and Jin-Mei T.; m. Diana Yu-Maan, Sept. 3, 1969; children: Bertrand JePing, Sophia SeaPing. B.Ed., Taiwan Normal U., 1966; M.S., U. Wis.-Menomonie, 1969; Ph.D., U. Ill., 1972. Asst. prof. U. Wis.-Menomonie, 1972; research asst. prof. U. Ill., 1972-76; asst. prof. Purdue U., Indpls., 1976-77, assoc. prof., 1978-81, prof., 1981—, dir. grad. studies, 1980—; advisor Taiwan Ministry of Edn., 1982-83; research advisor Taiwan Inst. Tchrs. Inservice, 1981—; dir. Osgood Inst. Cross-cultural Psychosemantics Research, 1982—. Contbr. chpts. to books, articles to profl. jours. U.S. Army grantee, 1977-79; Spencer Found. grantee, 1977-79; Conrail grantee, 1978-81. Mem. Am. Psychol. Assn., Soc. Intercultural Edn., Tng. and Research, Psychometric Soc., Soc. for Cross-cultural Research, Sigma Xi. Subspecialties: Cross-cultural Psychology; Psychosemantics.. Current work: Cross-cultural social and personality research, inter-group relations, inter-cultural training, psychosemantics, quantitative methodologies, and applied social psychology. Home: 3019 Lehigh Ct Indianapolis IN 46268 Office: Purdue Univ 1201 E 38th St Indianapolis IN 46223

TZIANABOS, THEODORE, microbiologist, researcher; b. Manchester, N.H., Feb. 12, 1933; s. Arthur and Marika (Tsouhidon) T.; m. Irene Davis, Aug. 11, 1962; children: Suzanne Tresselt, Peter T. B.A., U. N.H., 1955, M.S., 1959; Ph.D. (Lotta Crabtree fellow), U. Mass., 1965. Researcher U. N.H., 1955-57; microbiologist virus serology Mass. Dept. Pub. Health, Boston, 1958-59; research instr. U. Mass., 1961-65; postdoctoral resident Ctr. Disease Control, Atlanta, 1965-67, rickettsiologist, viral and rickettsial zoonoses br., viral disease div., 1972—; research microbiologist research and devel. U.S. Army Labs., Ft. Detrick, Md., 1967-70, Beckman Instruments, Atlanta, 1970-72. Contbr. numerous articles to profl. jours. Mem. Am. Soc. Microbiology, Am. Soc. Tropical Medicine and Hygiene (trustee Am. Type Culture Collection), Am. Acad. Microbiology, Atlanta Photog. Soc. (pres. 1981-82), Am. Soc. Rickettsiology, Sigma Xi. Greek Orthodox. Subspecialties: Microbiology; Virology (biology). Current work: Research in rickettsiology—reagents, structure and function. Office: 1600 Clifton Rd Atlanta GA 30333

UDENFRIEND, SIDNEY, biochemist; b. Bklyn., Apr. 5, 1918; s. Max and Esther (Tabak) U.; m. Shirley Reidel, June 20, 1943; children: Aliza, Elliot. Ph.D., N.Y.U., 1948. Chief lab. clin. biochemistry Nat. Heart Inst., NIH, Bethesda, Md., 1956-68; dir. Roche Inst. Molecular Biology, Nutley, N.J., 1968—; adj. prof. biochemistry Cornell Med. Ctr., N.Y.C., 1982—. Contbr. articles to profl. publs.; editorial bds. numerous jours. Mem. N.Y. Acad. Scis. (trustee), Am. Coll. Neuropsychopharmacology. Subspecialties: Biochemistry (biology); Molecular biology. Current work: Microprotein and peptide chemistry and their applications to neurobiology. Home: 22 Maple Dr North Caldwell NJ 07006 Office: Roche Inst Molecular Biology Nutley NJ 07110

UDIN, SUSAN BOYMEL, neurobiology researcher, educator; b. Phila., Aug. 11, 1947; d. Jules and Pauline (Friedman) Boymel; m. David Udin, June 3, 1967. B.S., M.I.T., 1969, Ph.D. (NSF grad. fellow), 1975. Sr. sci. staff mem. Nat. Inst. for Med. Research, Mill Hill, Eng., 1978-79; asst. prof. neurobiology SUNY, Buffalo, 1979—. Contbr. articles to sci. jours. NIH fellow, 1975-77; NIH grantee, 1977-78; Nat. Eye Inst. grantee, 1980—; Nat. Acad. Scis. travel grantee, 1980. Mem. Soc. for Neurosci., AAAS, Assn. for Women in Sci. (pres. Buffalo chpt. 1981-82), NOW, Nat. Abortion Rights Action League, Sigma Xi. Jewish. Subspecialties: Neurobiology; Developmental biology. Current work: Effects of sensory input on development of connections in the brain; developmental neurobiology; neuroanatomy of connections in developing brain. Office: SUNY 327 Cary Hall Buffalo NY 14214

UEDA, TETSUFUMI, research scientist, educator; b. Osaka, Japan, July 11, 1940; came to U.S., 1966, naturalized, 19; s. Ryuji and Takao (Hamaguchi) U.; m. Yasuko Amano, Jan. 11, 1970; children: Jane Kiyoko, Judy Yoshiko. B.S. in Chemistry, Kyoto (Japan) U., 1966; Ph.D. in Biol. Chemistry, U. Mich., 1971. Mem. faculty Yale U., 1971-78, NIH fellow, 1974-76, research assoc., 1976-78; asst. prof. U. Mich., Ann Arbor, 1978-81, assoc. prof., 1981—, asst. research scientist, 1978-81, assoc. research scientist, 1981—. Contbr. numerous articles to profl. publs. NIH grantee, 1979-82; NSF grantee, 1982-85; U. Mich. Rackham Grad. Sch. grantee, 1981. Mem. Am. Soc. Biol. Chemists, Am. Soc. Neurochemistry, Internat. Soc. Neurochemistry, Soc. Neurosci., N.Y. Acad. Sci. Subspecialties: Neurochemistry; Neurobiology. Current work: To understand the role of cyclic AMP and protein phosphorylation in the synapse, and to define the first messenger which lends to an increase in the phosphorylation of synaptic Protein I through initial activation of adenylate cyclase. Home: 3474 Richmond Ct Ann Arbor MI 48105 Office: Univ Mich Mental Health Research Inst 205 Washtenaw Pl Ann Arbor MI 48109

UETZ, GEORGE WILLIAM, zoologist, educator; b. Phila., Dec. 8, 1946; s. George Preston and Bertha (Swayne) U.; m. Mary Katherine Walters, Oct. 6, 1979. B.A., Albion (Mich.) Coll., 1968; M.S., U. Del., 1970; Ph.D., U. Ill.-Champaign, 1976. Research asst. U. Del., 1968-70; tchr. biology Sanford Prep. Sch., Hockessin, Del., 1970-72; research asst. U. Ill.-Champaign, 1972-75, teaching asst., 1975-76; asst. prof. U. Cin., 1976-81, assoc. prof., 1981—; assoc. curator Cin. Mus. Natural History, 1977—. Contbr. articles to profl. jours. Nat. Geog. Soc. grantee, 1978-80; Am. Philos. Soc. grantee, 1979, 82. Mem. Am. Arachnalogical Soc., Brit. Arachnalogical Soc., Animal Behavior Soc., Ecol. Soc., Entomol. Soc., AAAS. Subspecialties: Behavioral ecology; Sociobiology. Current work: Ecology and behavior of arthropods, especially spiders; social behavior courtship, communication, foraging. Office: U Cin Dept Biol Scis 006 Cincinnati OH 54221

UGELOW, ALBERT JAY, mechanical engineer; b. Bklyn., Mar. 16, 1950; s. Seymour Joseph and Lea (Payenson) U.; m. Paula Statsky, Apr. 29, 1973; children: Sharon Heather, Brianne Melissa. B.S., Columbia U., 1971; M.Engring., Stevens Inst. Tech., 1976. Registered profl. engr., N.Y., Fla. Asst. engr. Ebasco Services, N.Y.C., 1971-73, assoc. engr., 1973-74, engr., 1974-77, lead nuclear steam supply system engr., 1977-80, prin. engr., 1980-81, lead mech. discipline engr., Jensen Beach, Fla., 1981—. Named Young Engr. of Yr. N.Y. State Soc. Profl. Engrs., 1982. Mem. ASME, Nat. Soc. Profl. Engrs., Am. Nuclear Soc. Democrat. Lodge: KP. Subspecialties: Mechanical engineering; Nuclear fission. Current work: Mech. design of nuclear power plants, storage of spent nuclear fuel. Office: Ebasco Services PO Box 1117 Jensen Beach FL 33457

UHL, CHARLES H., botanical cytologist; b. Schenectady, May 28, 1918; s. Harry C. and Florence H. U.; m. Natalie Whitford, Aug. 15, 1945; children: Jean, Mary, Charles H., Elizabeth. B.A., Emory U., 1939, M.S., 1941; Ph.D., Cornell U., 1947. Assoc. prof. botany Cornell U., Ithaca, N.Y., 1952—. Served to lt. U.S. Navy, 1942-45. Mem. AAAS, Bot. Soc. Am., Genetics Soc., Soc. Study of Evolution, Internat. Assn. Plant Taxonomy. Subspecialty: Evolutionary biology. Current work: Hybrids and cytotaxonomy of American crassulaceae; chromosome pairing. Home: 1504 Hanshaw Rd Ithaca NY 14850 Office: Sect Plant Biology Cornell U Ithaca NY 14853

UHLENBECK, KAREN K., mathematician, educator; b. Cleve., Aug. 24, 1942; m., 1965. B.A., U. Mich., 1964; Ph.D. in Math, Brandeis U., 1968. Instr. math. MIT, 1968-69; lectr. U. Calif.-Berkeley, 1969-71; asst. prof. math. U. Ill., Urbana, 1971-76; assoc. prof. U. Ill.-Chgo., 1976-78, prof., 1978—; mem. faculty U. Chgo., 1983—. Sloan fellow, 1974-76; MacArthur fellow, 1983. Mem. Am. Math. Soc., Assn. Women in Math. Office: Dept Math U Ill PO Box 4348 Chicago IL 60680

UHRIG, ROBERT EUGENE, engineer, utility company executive; b. Raymond, Ill., Aug. 6, 1928; s. John Matthew and Anna LaDonna (Fireman) U.; m. Paula Margaret Schnepf, Nov. 27, 1954; children: Robert John, Joseph Charles, Mary Catherine, Charles William, Jean Marie, Thomas Paul, Frederick James. B.S. with honors, U. Ill., 1948; M.S. Iowa State U, 1950; Ph.D., 1954; grad. Advanced Mgmt. Program, Harvard U., 1976. Registered profl. engr., Iowa, Fla. Instr. engring. mechanics Iowa State U., 1948-51; asso. engr., research asst. Inst. Atomic Research (at univ.), 1951-54, asso. prof. engring. mechanics and nuclear engring., also group leader, 1956-60; prof. nuclear engring., chmn. dept. U. Fla., Gainesville, 1960-68, on leave, 1967-68; dean Coll. Engring., 1968-73; dep. asst. dir. research Dept. Def., Washington, 1967-68; dir. nuclear affairs Fla. Power & Light Co., Miami, 1973-74, v.p. for nuclear affairs, 1974-75, v.p. nuclear and gen. engring., 1976-78, v.p. advanced systems and tech., 1978—; Miami. Rep. Dept. Def. to com. on acad. sci. and engring. Fed. Council Sci. and Tech., 1967; chmn. engring. adv. com. NSF, 1972-73; bd. dirs. Engring. Council Profl. Devel., 1968-72; mem. commn. edn. for engring. profession Nat. Assn. State Univs. and Land Grant Colls., 1969-72. Author: Random Noise Techniques in Nuclear Reactor Systems, 1970, trans. into Russian, 1974. Served to 1st lt. USAF; instr. engring. mechanics U.S. Mil. Acad., 1954-56. Recipient Sec. of Def. Civilian Service award, 1968, Outstanding Alumni award U. Ill. Coll. Engring., 1970, Alumni Profl. Achievement award Iowa State U., 1972, President's medallion U. Fla., 1973; Disting. Achievement citation Iowa State U. Alumni Assn., 1980. Fellow Am. Nuclear Soc. (chmn. edn. com. 1962-64, chmn. tech. group for edn. 1964-66, dir. 1965-68, exec. com. of bd. 1966-68), ASME (Richards Meml. award 1969), AAAS; mem. Am. Soc. Engring. Edn. (pres. S.E. sect. 1972-73, chmn. nuclear engring. div. 1966-67, research award S.E. sect. 1962), John Henry Newman Honor Soc., Sigma Xi, Tau Beta Pi, Pi Mu Epsilon, Pi Tau Sigma, Phi Kappa Phi. Subspecialty: Nuclear engineering. Current work: Primarily concerned with utility research and development (coal-liquid mixtures, load management etc.), nuclear licensing of power plants, environmental affairs and quality assurance. Home: 164 SE Turtle Creek Dr Tequesta FL 33458 Office: Fla Power & Light Co PO Box 14000 Juno Beach FL 33408

ULABY, FAWWAZ TAYSSIR, electrical engineering educator; b. Damascus, Syria, Feb. 4, 1943; s. Tayssir and Moukarram U.; m. Mary Ann Hammond, Aug. 28, 1968; children: Neda Elizabeth, Aziza Marie, Laith Arthur. B.S. in Physics, Am. U. Beirut, 1964; M.S.E.E., U. Tex.-Austin, 1966, Ph.D., 1968. Assoc. research engr. U. Tex., 1966-68, coordinator millimeter wave sci. lab., 1968; asst. prof. elec. engring. U. Kans., Lawrence, 1968-69, assoc. prof., 1971-76, prof., 1976—, assoc. dir. remote sensing lab., 1969-71, dir. lab., 1971—, J.L. Constant Disting. prof., 1980—. Contbr. over 120 articles to sci./tech. jours.; co-author: Microwave Remote Sensing: Active and Passive, 3 vols; assoc. editor: Manual of Remote Sensing, 1983. Recipient Chancellor's award U. Kans., 1980. Fellow IEEE; mem. IEEE Geosci. and Remote Sensing Soc. (Outstanding Service award 1982, pres. 1979-81), URSI, Internat. Soc. Photogrammetry, Am. Soc. Photogrammetry, Am. Soc. Engring. Edn., Am. Geophys. Union, AIAA, Eta Kappa Nu (Holmes MacDonald award 1975), Sigma Xi, Tau Beta Pi. Subspecialties: Electrical engineering; Remote sensing (geoscience). Current work: Radar remote sensing; microwave radiometry; antennas; microwave communications systems; radar systems. Home: 2523 Arkansas Lawrence KS 66045 Office: 2291 Irving Hill Dr Lawrence KS 66045

ULLIMAN, JOSEPH JAMES, forestry educator; b. Springfield, Ohio, July 19, 1935; s. Joseph James and Iola Mae (Roth) U.; m. Barbara Gish, Apr. 29, 1961; children: Kathryn Michele, Barbara Anne, Mark Joseph. B.A. in English, U. Dayton, 1958; M.F. in Forest Mgmt, U. Minn., 1968, Ph.D., 1971. Research and teaching asst. U. Minn., St. Paul, 1966-68, instr., 1968-71, asst. prof., 1971-74; mem. land use planning team Willamette Nat. Forest, Eugene, Oreg., 1973; assoc. prof. U. Idaho, Moscow, 1974-79, prof., 1979—; instr. workshop on aerial photography and aerial photo interpretation; trainer vis. scientists in remote sensing Remote Sensing Inst., S.D. State U., Brookings; instr. workshop in remote sensing, Nairobi, Kenya. Contbr. articles to profl. jours. Mem. Environ. Commn., St. Paul, 1972-74, chmn., 1974. Served to capt. U.S. Army, 1959-65. Recipient Floyd Bartlett award Am. Soc. Photogrammetry, 1981. Mem. Soc. Am. Foresters, Am. Soc. Photogrammetry, Moscow C. of C. (chmn. natural resources com.), Sigma Xi, Xi Sigma Pi, Gamma Sigma Delta. Democrat. Roman Catholic. Subspecialties: Remote sensing (geoscience); Resource management. Current work: Aerial photo interpretation, remote sensing, mapping, teaching and research. Home: 2226 Weymouth St Moscow ID 83843 Office: Coll of Forestry U Idaha Moscow ID 83843

ULLMAN, JEFFREY DAVID, computer science educator; b. N.Y.C., Nov. 22, 1942; s. Seymour and Nedra Lucille (Hart) U.; m. Holly Elizabeth Scharf, Nov. 19, 1967; children: Peter, Scott. B.S., Columbia U., 1963; Ph.D., Princeton U., 1966, U. Brussels, Belgium, 1975. Mem. tech. staff Bell Labs., Murray Hill, N.J., 1966-69; prof. elec. engring. and computer sci. Princeton U., N.J., 1969-79; prof. computer sci. Stanford U., Calif., 1979—; cons. Bell Labs., Murray Hill, 1969—; mem. examination com. for computer sci. grad. record examination Ednl. Testing Service, Princeton, 1978—; cons. editor Computer Sci. Press, Rockville, Md., 1982—. Author: (with J.E. Hopcroft) Formal Languages and Their Relation to Automata, 1969, Introduction to Automata Theory, Languages, and Computation, 1979, (with A.V. Aho) The Theory of Parsing, Translation and Compiling, 1972, Principles of Compiler Design, 1977, Fundamental Concepts of Programming Systems, 1976, (with A.V. Aho and J.E. Hopcroft) The Design and Analysis of Computer Algorithms, 1974, Data Structures and Algorithms, 1982, Principles of Database Systems, 1980. Mem. Assn. for Computing Machinery (council 1978-80, sec.-treas. spl. interest group on automata and computability theory 1973-77). Jewish. Subspecialties: Theoretical computer science; Database systems. Current work: Design of algorithms; development of simple user interfaces for database systems; development of compilers for high-level languages into integrated circuit layout. Home: 1023 Cathcart Way Stanford CA 94305 Office: Computer Sci Dept Stanford U Stanford CA 94305

ULLREY, DUANE, animal nutrition educator; b. Niles, Mich., May 27, 1928; s. Ebon Earl and Jennie Verneda (Knott) U.; m. C. Lee Ingram, Dec. 16, 1961 (div. 1976); m. Susanne C. Traver, May 8, 1976; children: Henry J. Custer, Kay Lynn Custer, Kim Marie Custer. B.S., Mich. State U., 1950, M.S., 1951; Ph.D., U. Ill.-Urbana, 1954. Instr. Okla. State U., Stillwater, 1954-56; asst. prof. Mich. State U., East Lansing, 1956-61, assoc. prof., 1961-68, prof., 1968—; vis. prof. U. Calif.-San Diego, La Jolla, 1978; research assoc. San Diego Zoo, 1978, 81; chmn. nat. research council commn. on animal nutrition Nat. Acad. Sci., Washington, 1982—. Contbr. numerous articles on animal nutrition to profl. jours. Mem. Am. Inst. Nutrition, Am. Soc. Animal Sci. (Bohstedt award 1969), Equine Nutrition and Physiol. Soc. (bd. dirs. 1980—), Am. Assn. Zoo Veterinarians. Club: Meridian Optimist (founding mem. Okemos, Mich. 1982). Subspecialty: Animal nutrition. Current work: Comparative animal nutrition, mineral and vitamin metabolism. Home: 2090 Tamarack Okemos MI 48864 Office: Animal Sci Dept Mich State U East Lansing MI 48824

ULLRICH, ROBERT CARL, geneticist; b. Dumont, N.J., Aug. 4, 1940; s. Charles and Elsa (Dambach) U.; m. Sonja Ullrich, July 4, 1964; children: Jonathan Mark, Max Benjamin. B.S., U. Minn., 1968; M.A., Harvard U., 1969, Ph.D., 1973. Lectr. Harvard U., 1973-74; asst. prof. botany U. Vt., Burlington, 1974-79, assoc. prof., 1979—; founding mem. Genetics Resources, Inc., Burlington, 1981—. Contbr. articles to profl. jours. Cub Scout master; coach Little League. NSF grantee; Dept. Agr. grantee. Mem. AAAS, Genetics Soc. Am., Mycol. Soc. Am. Subspecialties: Genetics and genetic engineering (agriculture); Plant pathology. Current work: Researcher in mating type, incompatibility and molecular genetics of fungi. Office: U Vt Dept Botany Burlington VT 05405

ULMER, MELVILLE PAUL, physics and astronomy educator, researcher; b. Washington, Mar. 12, 1943; s. Melville Jack and Naiomi Louise (Zinken) U.; m. Patricia Elifson, Dec. 12, 1947; children: Andrew Todd, Jeremy John, Rachel Ann. B.A., Johns Hopkins U., 1965; Ph.D., U. Wis., Madison, 1970. Asst. research physicist U. Calif., San Diego, 1970-74; astrophysicist Harvard-Smithsonian Center for Astrophysics, 1974-76; asst. prof. dept. physics and astronomy Northwestern U., 1976-82, assoc. prof., 1982—. Contbr. articles to profl. jours. NASA grantee, 1976. Mem. Am. Astron. Soc., Roya Astron. Soc., Internat. Astron. Union, Am. Physical Soc. Subspecialties: 1-ray high energy astrophysics; Gamma ray high energy astrophysics. Current work: X-ray astronomy, clusters of galaxies, x-ray instrumentation, gamma ray astronomy. Office: Dept Physics and Astronomy Northwestern U Evanston IL 60201 Home: 2021 Noyes St Evanston IL 60201

UMEK, ANTHONY M., company administrator; b. New Kensington, Pa., Apr. 17, 1947; s. Anthony and Veronica (Yugovich) U.; m. Kristy S. Kocher, July 9, 1977. B.S. in Mech. Engring, Carnegie-Mellon Inst., 1969; M.B.A., U. Pitts., 1971. Engr. Westinghouse Electric Corp., Madison, Pa., 1970-74; mgr. line Westinghouse Hanford Co., Richland, Wash., 1974-80, dept. mgr., 1981—. Author: Non-Lethal Weapon, 1969; Contbr. articles to profl. jours. Karate instr. YMCA; pres. Tri-City Estates Water Dist. Mem. ASME (pub. affairs com.), Am. Nuclear Soc., Am. Mgmt. Assn. Roman Catholic. Subspecialties: Mechanical engineering; Nuclear fission. Current work: Manager of an engineering and startup testing department, in charge of nuclear reactor related facilities. Office: Westinghouse Hanford Co Box 1970 Richland WA 99352

UMMINGER, BRUCE LYNN, science administrator, biology educator; b. Dayton, Ohio, Apr. 10, 1941; s. Frederick William and Elnora Mae (Waltemathe) U.; m. Judith Lackey Bryant, Dec. 17, 1966; children: Alison Grace, April Lynn. B.S., Yale U., 1963, M.S., 1966, M.Phil., 1968, Ph.D., 1969. Asst. prof. biol. sci. U. Cin., 1969-73, assoc. prof. biol. sci., 1973-75, acting head biol. sci., 1973-75, prof. biol. sci., 1975-81, dir. grad. affairs biol. sci., 1978-79; program dir. NSF, Washington, 1979—; mem. Space Shuttle Proposal Rev. Panel, NASA, 1978; mem. adv. screening com. in life scis. Council Internat. Exchange of Scholars, 1978-81; liaison rep., adv. council Nat. Heart, Lung and Blood Inst., Bethesda, Md., 1979—; adv. bd. Campbell Comml. Coll., Cin., 1977-79. Assoc. editor: Jour. Exptl. Zoology, 1977-79; editorial adv. bd.: Gen. and Comparative Endocrinology, 1982. Mem. world mission com. Ch. of the Redeemer, New Haven, 1967-68; mem. Sunday sch. steering com. Calvary Episcopal Ch., Cin., 1972-73, sr. acolyte, 1972-77. Recipient George Rieveschl, Jr. univ. research award U. Cin., 1973, fellow grad. sch., 1977—; NSF grantee, 1971, 73, 76; fellow, 1964; Nat. Acad. Scis. travel grantee, 1974. Fellow AAAS (council 1980-83), N.Y. Acad. Scis; mem. Am. Physiol. Soc. (comparative physiology sect. program officer 1978-81, program exec. com. 1983—), Am. Soc. Zoologists (sec. 1979-81), Sigma Xi (pres. chpt. 1977-79, Disting. Research award 1973). Clubs: Yale of Washington, Mory's Assn. (New Haven). Lodge: Masons. Subspecialties: Comparative physiology; Endocrinology. Current work: Comparative physiology, endocrinology, biochemistry; research into fish systems, and administration of federal funding programs to assure health of science in these areas. Home: 4087-B S Four Mile Run Dr Arlington VA 22204 Office: Regulatory Biology Program NSF 1800 G St NW Washington DC 20550

UMSAWASDI, THEERA, physician, educator; b. Bangkok, Thailand, Oct. 13, 1942; s. Manit and Priub (Jaiprasart) U.; m. Chantana Wisoopakan, Jan. 20, 1967; children: Alyssa, Charlie, Marisa. M.D., Faculty of Medicine-Siriraj Hosp., Bangkok, Thailand, 1965. Diplomate: Medicine, Thailand, 1965. Intern Faculty of Medicine-Siriraj Hosp., 1965-66, instr., 1972-77; resident in medicine Bangkok Sanatorium Hosp., 1966-67; intern St. John Hosp., Detroit, 1967-68; resident in medicine Sinai Hosp., Detroit, 1968-69; fellow in med. oncology U. Miami (Fla.) Sch. Medicine, 1977-79; project investigator U. Tex.-M.D. Anderson Hosp. and Tumor Inst., Houston, 1969-72, asst. prof. internal medicine, 1979—. Contbr. numerous sci. articles and abstracts to profl. publs. Mem. AMA, Tex. Med. Assn., Harris County Med. Assn., Am. Assn. Cancer Research, Am. Soc. Clin. Oncology., N.Y. Acad. Scis. Subspecialty: Oncology. Current work: Chemotherapy and combined modality in treatment of lung cancer. Office: 6723 Bertner Ave Houston TX 77030

UNBEHAUN, LARAINE MARIE, educator; b. Kearney, Nebr., May 4, 1940; d. Vern Edgar and Olive Lucille (Trivelpiece) Clel; m. John Robert Unbehaun, Aug. 21, 1965. B.A., Kearney State Coll., 1961; M.A., U. No. Colo., 1964; Ph.D., Va. Poly. Inst. and State U., 1969. Tchr. biology, gen. sci. Ragan (Nebr.) Consol. Schs., 1961-63; teaching assoc. biology dept. U. Colo., Boulder, 1964-65; research asst. plant pathology and physiology Va. Poly. Inst. and State U., Blacksburg, 1965-66; prof. biology U. Wis., La Crosse, 1969—. Contbr. numerous articles to profl. jours. NSF fellow, 1966-69. Mem. Am. Phytopath. Soc., Sigma Xi. Subspecialties: Cell and tissue culture; Plant physiology (biology). Current work: Cell wall degrading enzymes produced by plant pathogens research; pectic enzymes. Home: Route 1 West Salem WI 54669 Office: U Wis 1707 Pine St La Crosse WI 54601

UNDERDAHL, NORMAN RUSSELL, microbiologist, educator; b. Freeborn County, Minn., June 5, 1918; s. Knut O. and Maria (Stoa) U.; m. Bernice Eleanor Nagle, Aug. 29, 1948; 1 dau., Kimbra. B.A., St. Olaf Coll., Northfield, Minn., 1941; M.S., U. Minn., Mpls., 1948. With Hormel Inst., U. Minn., Austin, 1946-55; asst. prof. vet. sci. U. Nebr., 1955-61, assoc. prof., 1961-68, prof., 1968—; cons. pharm. cos. Author: Specific Pathogen-Free Swine, 1973; contbr. numerous articles profl. jours. Served with USNR, 1942-46; PTO. Mem. Am. Legion., Nebr. Vet. Med. Assn. (hon.), Nebr. Specific—Pathogen-Free Accrediting Assn. (service award), Nat. Specific-Pathogen—Free Accrediting Assn. (service award), AVMA (assoc.), Am. Soc. Microbiology, Nat. Swine Repopulation Assn., Assn. Gnotobiology, Sigma Xi, Gamma Sigma Delta. Subspecialties: Microbiology; Preventive medicine (veterinary medicine). Current work: Use of probiotics as a preventive measure of enteric diseases, effect of combination of bacterial organisms on pneumonia in swine. Patentee breeder for raising surgically obtained pigs; co-developer specific-pathogen-free methods for obtaining and rearing surgically obtained pigs. Home: 935 N 67th St Lincoln NE 68505 Office: Dept Veterinary Science U Nebr Lincoln NE 68583

UNDERHILL, ANNE BARBARA, astronomer, astrophysicist; b. Vancouver, B.C., Can., June 12, 1920; d. Frederic Clare and Irene Anna (Creery) U. B.A., U. B.C., 1942, M.A., 1944; Ph.D., U. Chgo., 1948; D.Sc. (hon.), York U., 1969. Research scientist Dominion Astrophys. Obs., Victoria, B.C., 1949-62; prof. astrophysics State U. at Utrecht, Netherlands, 1962-70; chief Lab. for Optical Astronomy, NASA Goddard Space Flight Center, Greenbelt, Md., 1970-77, sr. scientist, 1977—. Author: The Early Type Stars, 1966, (with V. Doazan) B Stars with and without Emission Lines, 1982; also articles. Can. Fedn. Univ. Women sr. fellow, 1947; U.S. Nat. Research fellow, 1948. Mem. Am. Astron. Soc., Royal Astron. Soc., Can. Astron. Soc., Internat. Astron. Union, Royal Astron. Soc. Can., Astron. Soc. Pacific. Episcopalian. Subspecialties: Optical astronomy; Theoretical astrophysics. Current work: Interpretation of the spectra of hot stars; physical state of hot stars; interpretation of the ultraviolet spectra of early-type stars. Office: Code 680 Goddard Space Flight Center Greenbelt MD 20771

UNDERWOOD, ARTHUR LOUIS, JR., chemist, educator; b. Rochester, N.Y., May 18, 1924; s. Arthur Louis and Grace (Porter) U.; m. Elizabeth Emery, June 30, 1948; children: Paul W., Robert E., Susan E. B.S. in Chemistry, U. Rochester, 1944, Ph.D. in Biochemistry, 1951. Research assoc. U. Rochester Atomic Energy Project, 1946-51; research assoc. in chemistry M.I.T., 1951-52; asst. prof. chemistry Emory U., 1952-58, assoc. prof., 1958-62, prof., 1962—; research assoc. in chemistry Cornell U., 1959-60; vis. prof. Mont. State U., summers 1979-82. Author: (with R.A. Day, Jr.) Quantitative Analysis, 1980; research, numerous publs. in field. Served with USN, 1944-46. Mem. AAAS, Am. Chem. Soc., Phi Beta Kappa, Sigma Xi. Subspecialties: Biochemistry (biology); Analytical chemistry. Current work: Studies on Micelles, particularly the effects of organic counterions on micellar parameters. Home: 1354 Springdale Rd NE atlanta GA 30306 Office: Dept Chemistry Emory U Atlanta GA 30322

UNGAR, IRWIN A., botany educator; b. N.Y.C., Jan. 21, 1934; s. Isadore and Gertrude (Feigeles) U.; m. Ana del Cid, Aug. 10, 1959; children: Steven, Sandra, Sharon. B.S., CCNY, 1955; M.S., U. Kans., 1957, Ph.D., 1961. Instr. dept. botany U. R.I., Kingston, 1961; asst. prof. dept. biology Quincy (Ill.) Coll., 1962-66; prof. dept. botany Ohio U., Athens, 1975—; research assoc. Centre National de la Recherche Scientifique, 1972-73. Contbr. articles to profl. jours. NSF grantee, 1963-65, 67-69, 74-76, 76-78, 80—. Mem. Bot. Soc. Am., Ecol. Soc. Am., AAAS, Sigma Xi. Jewish. Subspecialty: Ecology. Current work: Ecology of halophytes. Office: Dept Botany Ohio U Athens OH 45701

UNO, HIDEO, pathologist, researcher; b. Tokyo, Nov. 28, 1929; U.S., 1970; s. Yoshinori and Hana U.; m. Shoko Ohashi, Apr. 1, 1956; children: Takeshi, Yayoi. M.D., Yokohama (Japan) Med. Coll., 1955, Ph.D., 1961. Asst. prof. Yokohama City U., 1961-64, assoc. prof., 1968-70; instr. Jefferson Med. Coll., Phila., 1964-66; vis. scientist Oreg. Primate Research Center, Beaverton, 1966-68, scientist, 1970-79; sr. scientist Wis. Primate Research Center, Madison, 1979—. Grantee Oreg. Med. Found., 1976; Upjohn Co., 1982. Mem. Am. Assn. Pathologists, Am. Assn. Anatomists, Soc. Investigative Dermatology, Internat. Acad. Pathology, Japanese Soc. Histochemistry and Cytochemistry (counselor). Buddhist. Club: Internat. House of Japan (Tokyo). Subspecialties: Neuroendocrinology; Gerontology. Current work: Comparative pathology of nonhuman primates, peptides in sensory autonomic nerve systems, aging changes in captive rhesus monkeys. Home: 3722 Ross St Madison WI 53705 Office: Wis Regional Primate Research Center U Wis 1223 Capitol Ct Madison WI 53715-1299

UNRUH, WILLIAM GEORGE, physics educator; b. Winnipeg, Man., Can., Aug. 28, 1945. B.Sc with honors, U. Man., 1967; M.A., Princeton U., 1969, Ph.D., 1971. NRC Can. fellow in physics U. London, 1971-72; Miller fellow in physics U. Calif.-Berkeley, 1973-74; asst. prof. applied math. McMaster U., Hamilton, Ont., Can., 1974-76; asst. prof. physics U. B.C., Vancouver, 1976-80, assoc. prof., 1980-82, prof., 1982—. Recipient Rutherford medal Royal Soc. Can., 1982, Herzberg medal Can. Assn. Physics, 1983; Sloan fellow, 1978. Subspecialties: Relativity and gravitation; Theoretical physics. Current work: The relation of quantum mechanics and gravity on each other. Office: Physics Dept Univ BC Vancouver BC Canada V6T 2A6

UNTERSTEINER, NORBERT, science adminstrator, educator; b. Meran, Italy, Feb. 24, 1926; s. Raimund and Anna (Sperk) U.; m. Krystyna Untersteiner, Aug. 11, 1980. Dr.phil., U. Innsbruck, Austria, 1950; Dozent, U. Vienna, 1960. Asst. prof. U. Vienna, 1951-56; research meteorologist Central Inst. Meteorology and Geodynamics, Vienna, 1957-62; research assoc. prof. U. Wash., 1962-67, prof. atmospheric scis. and geophysics, 1967—; dir. Polar Sci. Center. Contbr. articles to profl. jours. Decorated Hon. Cross for Arts and Scis., Austria, 1959; recipient Antarctic Service Medal U.S. Govt., 1967, numerous research grants. Mem. AAAS, Am. Geophys. Union, Internat. Glaciol. Soc., Comite Arctique International, German Polar Soc., Norsk Polar Club. Subspecialties: Oceanography; Climatology. Current work: Air-sea interaction at high latitudes, climatic change, research planning and coordination including field work and theoretical analysis. Home: 5536 E Greenlake Way N Seattle WA 98103 Office: U Wash Applied Physics Lab Seattle WA 98105

UNWIN, STEPHEN CHARLES, research scientist; b. Bromley, Kent, Eng., Sept. 8, 1953; s. Thomas Eric and Barbara Jean (Herrington) U. B.A. with honors in Physics and Theoretical Physics, Cambridge U., Eng., 1976, Ph.D. in Radio Astronomy, 1979, M.A., 1980. Staff scientist radio astronomy Owens Valley Radio Obs., Calif. Inst. Tech., 1979—. Contbr. articles to profl. jours. Mem. Am. Astron. Soc., Royal Astron. Soc. Subspecialty: Radio and microwave astronomy. Current work: Research in radio astronomy using interferometric methods. Mapping of compact radio sources; study of kinematics and spectra of variable sources. Office: Owens Valley Radio Calif Inst Tech Mail Code 105-24 Pasadena CA 91125

UPATNIEKS, JURIS, research engr.; b. Riga, Latvia, May 7, 1936; s. Karlis and Elenora (Jegers) U.; m. Ilze Induss, July 13, 1978; children: Ivars, Ansis. B.E.E., U. Akron, 1960; M.S.E.E., U. Mich., 1965. Jr. engr. Goodyear Aircraft Corp., Akron, 1957-59; research engr. Environ. Research Inst. Mich., Ann Arbor, 1960—; adj. assoc. prof. elec. and computing engring. dept. U. Mich., 1974—. Contbr. articles to profl. jours. Served to lt. U.S. Army, 1961-62. Recipient Holley medal ASME, 1976; R.W. Wood prize Optical Soc. Am., 1976. Fellow Optical Soc. Am., Soc. Photog. Instrument Engrs.; mem. IEEE, Am. Latvian Assn. Subspecialties: Holography; Optical signal processing. Current work: Coherent optics, holography, optical data processing research. Patentee in field. Office: PO Box 8618 Ann Arbor MI 48107

UPGREN, ARTHUR REINHOLD, JR., astronomer, educator; b. Mpls., Feb. 21, 1933; s. Arthur Reinhold and Marion Elizabeth (Andrews) U.; m. Joan Josephine Koswoski, Jan. 7, 1967; 1 dau., Amy Joan. B.A., U. Minn., 1955; M.S., U. Mich., 1958; Ph.D., Case Inst. Tech., 1961. Research assoc. Swarthmore Coll., 1961-63; astronomer U.S. Naval Obs., Washington, 1963-66; asst. prof. astronomy Conn. Wesleyan U., Middletown, 1966-73, assoc. prof., 1973-81, Van Vleck prof. astronomy, dir., 1981—; vis. prof. Yale U., 1979-80. Contbr. numerous articles to profl. jours. Mem. Internat. Astronom. Union, Am. Astronom. Soc., Royal Astronom. Soc., Sigma Xi. Democrat. Subspecialty: Optical astronomy. Current work: Galactic structure, astrometry. Office: Van Vleck Observatory Conn Wesleyan U Middletown CT 06457

UPTON, ARTHUR CANFIELD, pathologist; b. Ann Arbor, Mich., Feb. 27, 1923; s. Herbert Hawkes and Ellen (Canfield) U.; m. Elizabeth Bache Perry, Mar. 1, 1946; children—Rebecca A., Melissa P., Bradley C. Grad., Phillips Acad., Andover, Mass., 1941; B.A., U. Mich., 1944, M.D., 1946. Intern Univ. Hosp., Ann Arbor, 1947, resident, 1948-49; instr. pathology U. Mich. Med. Sch., 1950-51; pathologist Oak Ridge Nat. Lab., 1951-54, chief pathology-physiology sect., 1954-69; prof. pathology SUNY Med. Sch. at Stony Brook, 1969-77, chmn. dept. pathology, 1969-70; dean Sch. Basic Health Scis., 1970-75; dir. Nat. Cancer Inst., Bethesda, Md., 1977-79; prof., chmn. dept. environ. medicine N.Y. U. Med. Sch., N.Y.C., 1979—; Mem. various coms. nat. and internat. orgns. Asso. editor: Cancer Research; mem. editorial bd.: Internat. Union Against Cancer. Served with AUS, 1943-46. Recipient Ernest Orlando Lawrence award for atomic field, 1965; Comfort-Crookshank award for cancer research, 1979. Mem. Am. Assn. Pathologists and Bacteriologists, Internat. Acad. Pathology, Radiation Research Soc. (councilor 1963-64, pres. elect 1964-65, pres. 1965-66), Am. Assn. Cancer Research (pres. 1963-64), Am. Soc. Exptl. Pathology (pres 1967-68), AAAS, Gerontol. Soc., Sci. Research Soc. Am., Soc. Exptl. Biology and Medicine, Phi Beta Kappa, Phi Gamma Delta, Alpha Omega Alpha, Nu Sigma Nu. Subspecialty: Pathology (medicine). Home: 3 Washington Square Village New York NY 10012 Office: NY U Sch Medicine 550 1st Ave New York NY 10016

URALIL, FRANCIS STEPHEN, physicist; b. Monippally, India, June 3, 1950; came to U.S., 1970; s. Chacko Stephen and Chechamma (Joseph) U.; m. Annie Francis Kuruvilla, Dec. 15, 1980; 1 dau., Sherene Elizabeth. B.S., U. Kerala, India, 1969; M.S., Marquette U., Milw., 1972; Ph.D., U. Del., Newark, 1976. Research assoc. Case Inst. Tech., Cleve., 1976-78; mem. profl. staff Schlumberger, Richfield, Conn., 1978-79; research scientist Battelle Labs., Columbus, Ohio, 1979—. Mem. Am. Phys. Soc. Subspecialties: Polymers; Polymer physics. Current work: Mechanical properties of polymers, structure-property relationships, fracture mechanics. Home: 4503 Mobile Dr Columbus OH 43220 Office: Battelle Labs 505 King Ave Columbus OH 43220

URANO, MUNEYASU, radiation biologist, educator; b. Osaka, Japan, Apr. 21, 1936; came to U.S., 1977, naturalized, 1978; s. Iekazu and Shizuyo (Hirata) U.; m. Michiyo, Mar. 2, 1963; children: Shin-ichi, Jun. M.D., Kyoto (Japan) Prefectural U. Medicine, 1961, Ph.D., 1968. Intern Kyoto Prefectural U. Medicine, 1961-62, resident, 1962-66, asst. prof., 1968-71; sr. researcher and staff radiologist Nat. Inst. Radiol. Scis., Chiba, Japan, 1971-77; asst. radiation biologist Mass. Gen. Hosp., Boston, 1977-81, assoc. radiation biologist, 1982—; asst. prof. radiation biology Harvard U. Med. Sch., Boston, 1977-82, assoc. prof., 1983—. Contbr. articles to profl. jours. Nat. Cancer Soc. grantee, 1979—; Internat. Union against Cancer fellow, 1982. Mem. Radiation Research Soc., Am. Soc. Therapeutic Radiologists, N.Am. Hyperthermia Group, Cancer Research, AAAS. Subspecialties: Oncology; Cancer research (medicine). Current work: Hyperthermia; thermal effect on animal malignant and normal tissues; hyperthermia combined with radiation or chemotherapy; proton radiation biology. Home: 3 Laurel Hill Ln Winchester MA 01890 Office: Mass Gen Hosp Cox 7 Fruit St Boston MA 02114

URBACH, JOHN CHARLES, optical scientist; b. Vienna, Austria, Feb. 18, 1934; came to U.S., 1939, naturalized, 1947; s. Franz and Annie U.; m. Mary Trevor, July 7, 1956; children: Thomas Paul, Michael Karl, Katherine Joanne. B.S., U. Rochester, 1955, Ph.D. in Optics, 1962; M.S. in Physics, MIT, 1957. Assoc. physicist IBM, Ossining, N.Y., 1957-58; NATO postdoctoral fellow in sci. Royal Inst. Tech., Stockholm, 1961-62; scientist Xerox Research Ctr., Rochester, N.Y., 1963-67, area mgr., Rochester, Palo Alto, Calif., 1967-75; mgr. optical sci. lab. Xerox Palo Alto Research Ctrs., 1975—. Contbr. articles to profl. jours. Mem. conservation com. Town of Portola Valley, Calif., 1979-80. N.Y. State Regents scholar, 1951-55. Fellow Optical Soc. Am.; mem. Soc. Photo-optical Instrumentation Engrs., Soc. Photog. Scientists and Engrs., Phi Beta Kappa, Sigma Xi. Subspecialties: Laser data storage and reproduction; Holography. Current work: Management of research programs in several areas of optics and electro-optics, including high density optical data storage, laser scanning for high quality printing, and fiber optical communications for local area networks. Office: 3333 Coyote Hill Rd Palo Alto CA 94304

URBAN, JAMES EDWARD, microbiologist, researcher; b. Dime Box, Tex., Jan. 5, 1942; s. Ermo and Esther M. (Marburger) U.; m. Dianne Kieke, Aug. 31, 1963; children: Jill, Amy. B.A., U. Tex.-Austin, 1965, Ph.D., 1968. NSF post-doctoral fellow Kans. State U., 1968-70, asst. prof. microbiology, 1970-77, assoc. prof., 1977—. Mem. Am. Soc. Microbiology. Subspecialties: Microbiology (medicine); Nitrogen fixation. Current work: Analysis of induction of bacteroid formation in Rhizobium sp. Office: Kansas State University Division of Biology Ackert Hall Manhattan KS 66506

URQUILLA, PEDRO RAMON, clin. pharmacologist; b. San Miguel, El Salvador, July 28, 1939; came to U.S., 1966; s. Ruben Hector Cabezas and Maria Luisa U.; m. Maria A. Flores, Dec. 19, 1964; children: Marian, Marta, Pedro. M.D., U. El Salvador, 1965. Pan Am. Health Orgn. fellow, 1966-68; USPHS postdoctoral fellow in pharmacology U. W. Va., 1968-69; instr. pharmacology U. El Salvador, 1965-66, assoc. prof. dept. physiology and pharmacology, 1969-72, acting chmn. dept., 1969-71; assoc. prof. dept. physiol. scis. Faculty Medicine, U. Autonoma, Madrid, 1972-73; asst. prof. dept. pharmacology W.Va. U. Med. Center, Morgantown, 1973-76, assoc. prof., 1976-79; assoc. dir. med. research Miles Pharmaceuticals, West Haven, Conn., 1979-81; assoc. dir. clin. research Pfizer Inc., Groton, Conn., 1981—; lectr. clin. pharmacology VA Hosp., Clarksburg, W.Va., 1974; mem. research com. W.Va. Heart Assn., 1974-76. Contbr. articles to profl. jours. Co-recipient MacLachlan award for outstanding teaching in basic scis. W.Va. U. Med. Center, 1975; recipient same award, 1976. Mem. Am. Soc. Pharmacology and Exptl. Therapeutics, Am. Soc. for Clin. Pharmacology and Therapeutics, AMA. Roman Catholic. Subspecialty: Pharmacology. Current work: Clinical evaluation of thromboxane synthesis inhibitors. Home: 7 E Creek Circle Gilford CT 06437 Office: Pfizer Inc Groton CT

URRY, VERN WILLIAM, research psychologist; b. Salt Lake City, Sept. 20, 1931; s. Herbert William and Emma Irene (Swaner) U.; m. Billie Jeanne Nevius, Sept. 24, 1957; 1 dau., Gloria Jeanne. B.A., U. Utah, 1955, M.S., 1962; Ph.D., Purdue U., (1970.). Research psychologist U.S. Army Enlisted Evaluation Ctr., Ft. Benjamin Harrison, Ind., 1961-67; head systems and programming Measurement and Research Ctr., Purdue U., West Lafayette, 1967-70; asst. dir. Bur. Testing U. Wash., 1970-72; personnel research psychologist U.S. Office Personnel Mgmt., Washington, 1972—. Author: Tailored Testing: Its Theory and Practice, Part I, 1983. Served with U.S. Army, 1952-54. Recipient cert. achievement U.S. Army Enlisted Evaluation Ctr., 1967. Mem. AAAS, Am. Psychol. Assn., Md. Psychol. Assn., Psychometric Soc., Phi Kappa Phi. Subspecialties: Test theory; Algorithms. Current work: Research in tailored testing, a computerized mode of personnel or psychological testing, including the development of statistical models and the derivation of algorithms. Home: 3301 Accolade Dr Clinton MD 20735 Office: Office Personnel Research and Devel US Office Personnel Mgmt 1900 E St NW Washington DC 20415

URTHALER, FERDINAND, cardiologist, educator; b. Delemont, Jura, Switzerland, Jan. 31, 1938; came to U.S., 1970; s. Ferdin and Frieda Sophie (Kohler) U.; m. Josianne Yvette Fluckiger, June 11, 1964; children: Dominique Aude, Damien Jerome, Amanda Florence. B.S., Gymnase Cantonal, 1957; M.D., U. Zurich, 1963. Asst. pathology Pathology U., Zurich, Switzerland, 1964-65; intern in medicine Univ. Clinic, Zurich, 1965-67, resident in medicine, 1967-69, fellow cardiology, 1969-70; research fellow in cardiology U. Ala., Birmingham, 1970-73, asst. prof. medicine, 1973-75, assoc. prof. medicine, 1975-81, prof. medicine, 1981—. Contbr. numerous articles to profl. publs. Recipient Swiss Soc. Cardiologists award, 1976. Fellow Am. Health Assn. Council on Circulation; Mem. N.Y. Acad. Scis., So. Soc. Clin. Investigation, Am. Physiol. Soc., Am. Fedn. Classical Research. Subspecialties: Cardiology; Comparative physiology. Current work: Electrophysiology; disruption of electrical stability, sudden death, autonomic control of sinus node and AV junction. Home: 3609 River Ridge Rd Birmingham AL 35243 Office: U Ala Birmingham AL 35294

USSELMAN, THOMAS MICHAEL, geophysicist; b. Bismarck, N.D., Aug. 9, 1947; s. Bernard P. and Helen B. (Grembos) U.; m. Jonelle M. Gaus, Mar. 10, 1973. B.A., Franklin and Marshall Coll., Lancaster, Pa., 1969; M.S., Lehigh U., Bethlehem, Pa., 1971, Ph.D., 1973. NRC research assoc. NASA Johnson Space Center, Houston, 1973-75, Lunar Sci. Inst., 1975-76; asst. prof. geology SUNY-Buffalo, 1976-78; staff geophysicist com. on solar terrestrial research NRC, Washington, 1978—. Contbr. articles to profl. jours. Mem. Am. Geophys. Union, Geol. Soc. Am. Subspecialty: Geophysics. Office: NRC 2101 Constitution Ave Washington DC 20418

UTECH, FREDERICK HERBERT, curator, research botanist; b. Merrill, Wis., Apr. 19, 1943; s. Herbert Arthur and Priscilla Margaret (Pickrum) U.; m. Carla L. Shagass, July 20, 1974; m. Susan I. Haggerson, May 10, 1980; 1 dau., Rebecca Jane. B.S. with honors, U. Wis., 1966, M.S., 1968; Ph.D., Washington U., St. Louis, 1973. Vis. scientist U.S.-Japan Coop. Sci. Program, Toyama (Japan) U., 1974-75; fellow Japanese Soc. for Promotion Sci., 1975-76; curator sect. botany Carnegie Mus. Natural History, Pitts., 1976—; adj. scientist Hunt Inst. for Bot. Documentation, Carnegie-Mellon U., Pitts., 1977—. Mem. Bot. Soc. Am., Internat. Assn. Plant Taxonomy, So. Appalachian Bot. Club, Torrey Bot. Club. Subspecialties: Systematics; Evolutionary biology. Current work: Floral vascular anatomy of Liliaceae; cytotaxonomy of Liliaceae. Office: 4400 Forbes Ave Pittsburgh PA 15213

UTLAUT, WILLIAM FREDERICK, elec. engr.; b. Sterling, Colo., July 26, 1922; s. Frederick Ernst and Francis Ruth Hanna U.; m. Jeanne Elizabeth Pomeroy, Aug. 4, 1946; children—Mark William, Niles Frederick, Paige Elizabeth. Utlaut Hodges. B.S.E.E., U. Colo., 1944, M.S.E.E., 1950, Ph.D. in Elec. Engring, 1966; diploma, Naval Radar Sch., 1945. Engr. Gen. Electric Co., Schenectady, 1946-48, Nat. Bur. Standards, Boulder, Colo., 1952-53; instr. U. Colo., 1948-52, 53-54; dir. Nat. Telecom and Info. Adminstrn., Inst. for Telecommunication Scis., U.S. Dept. Commerce, Boulder, 1954—; chmn. U.S. study group 1, Internat. Radio Consultative Com., 1975-81, mem. U.S. nat. com., 1970-81; mem. electromagnetic wave propagation panel, adv. group aerospace research and devel. NATO, 1978-81. Guest co-editor spl. joint issue: IEEE Trans. on Spectrum Mgmt, 1981, IEEE Trans. on Communications, 1975; guest editor spl. issue: Radio Sci, 1974; contbr. numerous articles to profl. jours. Bd. dirs. YMCA, 1955—; mem. bd. mgmt. 1st Congl. Ch., 1960-66, 78—; mem. engring. devel. council U. Colo., 1969-81. Served in USN, 1943-46. Recipient Gold medal U.S. Dept. Commerce, 1971; Disting. Engring. Alumnus award U. Colo., 1973. Fellow IEEE (policy bd. Communications Soc.), Internat. Sci. Radio Union. Subspecialties: Radio spectrum; Telecommunication systems. Current work: Radio spectrum utilization and management improvement; radio propagation research; telecommunication system international standards development. Office: 325 Broadway Boulder CO 80303

UTLEY, PHILIP RAY, animal science educator, researcher; b. Albion, Ill., Dec. 18, 1941; s. Philip Everett and Mildred (Hocking) U.; m. Dorothy Lathrop, Dec. 22, 1962; children: Scott, Shawn, Sam. B.S., So. Ill. U., 1964; M.S., U. Mo.-Columbia, 1967; Ph.D., U. Ky., 1969. Asst. prof. Coastal Plain Sta., U. Ga., Tifton, 1970-74, assoc. prof., 1974-79, prof., 1979—, acting dept. head, 1982—. Contbr. articles to profl. jours. Mem. Am. Soc. Animal Sci., Assn. So. Agrl. Workers, Tift County Cattlemen's Assn., Ga. Cattlemen's Assn., Gamma Sigma Delta. Methodist. Subspecialty: Animal nutrition. Current work: Large animal nutrition; evaluating pasture management systems, growing-finishing systems and forage varieties for feeding beef cattle. Home: Route 3 Tifton GA 31794 Office: Coastal Plain Station U Ga PO Box 748 Tifton GA 31793

UTTING, KENNETH, software engineer; b. Riverhead, N.Y., Mar. 6, 1959; s. Walter Arthur and Nancy Rita (Ballirano) U. B.S.E., Princeton U., 1981. Software engr. Sanders Assoc., Nashua, N.H., 1981—. Mem. Assn. Computing Machinery. Subspecialties: Graphics, image processing, and pattern recognition; Artificial intelligence. Current work: Computer graphics, image processing, computer animation. Home: 160 Central St Hudson NH 03051 Office: Sanders Assoc 95 Canal St Nashua NH 03061

UTZ, WINFIELD ROY, mathematician, educator; b. Boonville, Mo., Nov. 17, 1919; s. W. Roy and Flora (Holloman) U.; m.; children: David, Charles, John. B.A., Central Meth. Coll., Fayette, Mo., 1941; M.A., U. Mo.-Columbia, 1942; Ph.D., U. Va., 1948. Instr. in math. U. Notre Dame, South Bend, Ind., 1943-44; prof. math. U. Mo.-Columbia, 1949—; vis. scholar U. Calif.-Berkeley, 1962-63; prof. math. U. Iowa, Iowa City, 1966; vis. prof. Brown U., Providence, 1969; mem. Inst. for Advanced Study, Princeton, N.J., 1955-56; vis. fellow U. Warwick, Eng., 1976; hon. fellow Math. Research Ctr., Madison, Wis., 1980. Author: Differential Equations, 1967; Contbr. articles to profl. jours. Recipient Faculty Alumni award U. Mo., 1969; Disting. Alumni award Central Meth. Coll., 1974. Mem. Am. Math. Soc., Math. Assn. Am., London Math. Soc., Soc. for Indsl. and Applied Math., Sigma Xi, Phi Kappa Phi, Pi Mu Epsilon. Subspecialty: Applied mathematics. Current work: Differential equations, topological dynamics. Home: 600 Manor Dr Columbia MO 65201 Office: U Missouri Columbia MO 65202

UYENO, EDWARD TEISO, psychopharmacologist; b. Vancouver, B.C., Can., Mar. 31, 1921; came to U.S., 1958; s. Ritsuichi and Kuye (Matsumiya) U.; m. Dorothy Hill, Apr. 27, 1969. B.A., U. Toronto, 1947, M.A., 1952, Ph.D., 1958. Research asst. U. Toronto, 1955-57; research assoc. Stanford (Calif.) U., 1958-61; psychopharmacologist Stanford Research Inst., 1961-76; S.R.I. Internat., 1977—. Author numerous sci. articles, also chpts. in books. NIMH grantee, 1963-66, 68-70, 71-73, 71-74. Mem. Am. Soc. Pharmacology and Exptl. Therapeutics, Western Pharmacology Soc., Am. Psychol. Assn., Can. Psychol. Assn., Psychonomic Soc., Internat. Primatol. Soc., Behavior Genetics Assn. Subspecialties: Psychopharmacology; Behavioral psychology. Current work: Evaluation of new analgesics, antidepressants and antianxiety agents. Behavioral toxicology, behavioral pharmacology. Bioassay of anesthetics, stimulants, sedatives, neurotoxicants and neuropeptides. Home: 535 Everett Palo Alto CA 94301 Office: SRI Internat Life Sci Div 333 Ravenswood Ave Menlo Park CA 94025

UZMAN, BETTY GEREN, pathologist; b. Fort Smith, Ark., Nov. 17, 1922; d. Benton Asbury and Myra Estelle (Petty) Geren; m. L. Lahut Uzman, Dec. 17, 1955 (dec.); 1 dau., Betty Tuba. Student, Fort Smith Jr. Coll., 1939-40; B.S., U. Ark., 1942; M.D., Washington U., 1945; postgrad., M.I.T., 1948-49; M.A. (hon.), Harvard U., 1967. Intern Childrens Hosp., Boston, 1945-46; resident in pathology Barnes Hosp., St. Louis, 1946-48; Am. Cancer Soc. research fellow M.I.T., Cambridge, Mass., 1948-50; chief biol. ultrastructure and exptl. pathology Children's Cancer Research Found., Boston, 1950-71; instr. Harvard Med. Sch., Boston, 1949-53, assoc., 1953-56, research assoc., 1956-67, assoc. prof., 1967-71, prof., 1971-72; head research dept. Sparks Regional Med. Center, Fort Smith, 1972-74; prof. pathology La. State U., Shreveport, 1972-74, U. Tenn., Memphis, 1977—; assoc. chief staff research VA, Shreveport, 1974-77, staff pathologist, Memphis, 1977—; chief field ops., spl. asst. to dir. VA Central Office, Washington, 1978-79, dir. med. research services, 1979-80; chmn. pathology A Study sect. NIH, 1973-76; cons. to sci. dir. Children's Cancer Research Found., Boston, 1971-73; mem. adv. com. on prevention, diagnosis and treatment Am. Cancer Soc., 1970-73, 77-80; disting. vis. investigator Inst. Venezolano Investigation Cientificas, Caracas, 1972-74. Decorated Order of Andres Bello 1st class, Venezuela; recipient Weinstein award United Cerebral Palsy, 1964; Am. Cancer Soc. research fellow, 1948-50. Mem. AAAS, Am. Soc. Cell Biology, Soc. Devel. Biology, Am. Acad. Neurology (asso.), Am. Soc. Neurochemistry (asso. mem.), Electron Microscope Soc., Am., Internat. Acad. Pathology, Am. Assn. Neuropathology (assoc.), Soc. Neuroscience, Am. Assn. Cancer Research. Subspecialties: Pathology (medicine); Regeneration. Current work: Nerve structure; peripheral nerve regeneration. Home: Apt 2102 99 N Main St Memphis TN 38103 Office: 1030 Jefferson Ave Memphis TN 38104 Constantly re-examine your assumptions; tell yourself the truth; do not be afraid of the innovative or untried path; most of all, persevere.

VACCA, LINDA L., research scientist, educator; b. Paterson, N.J., Mar. 10, 1947; d. Samuel J. and Rose M. (Felice) V. B.S., Coll. William and Mary, 1968; M.S., Tulane U., 1971, Ph.D., 1973. NIMH postdoctoral fellow Coll. Physicians and Surgeons, N.Y.C., 1973-74; instr. dept. physiology NYU Med. Ctr., N.Y.C., 1974-76; asst. prof. pathology, dir. histopath lab. Med. Coll. Ga., Augusta, 1976-82; asst. prof. anatomy U. Kans. Med. Ctr., Kansas City, Kans., 1982—. Author: Laboratory Manual of Histochemistry. Grantee Com. to Combat Huntington's Disease. Mem. AAAS, Soc. Neurosci., Am. Assn. Anatomists, Histochem. Soc., Sigma Xi. Democrat. Roman Catholic. Subspecialties: Neurobiology; Immunocytochemistry. Current work: Peptides in spinal pain processing and in Huntington's Disease. Home: 720 W 48th St Apt 405 Kansas City MO 64123 Office: U Kans Med Ctr Dept Anatomy Kansas City KS 66108

VADLAMUDI, SRI KRISHNA, pharmacologist, immunologist; b. Moparru, Tenali, Andhra, Pradesh, India, Aug. 15, 1927; s. Venkataratnam and Pitchamma (Yelavarti) V.; m. Jamuna Bai Narra, Oct. 14, 1955; children: Anula D. Nagarjuna, Venkata Ratnam, Gautam Kishore. B.V.Sc., Madras U., 1952, D.V.M., 1973; M.S., U. Wis., 1959, Ph.D., 1963. Sr. investigator Infectious Diseases div. Abbott Labs., North Chicago, Ill., 1963-65; chief Cancer Chemotherapy div. Microbiol. Assoc., Inc., Bethesda, Md., 1965-75; exec. sec. immunology panel div. clin. lab. devices Nat. Ctr. for Devices and Radiol. Health, FDA, Dept. Health and Human Services, Silver Spring, Md., 1975-79, br. chief immunology, 1979—. Contbr. articles to profl. jours. Bd. dirs. PTA, Telugu Assn. N. Am., Greater Washington Telugu Cultural Soc. Madras U. scholar, 1947-52. Mem. Am. Assn. Cancer Research, Internat. Union Against Cancer, AVMA, Am. Soc. Microbiology, Am. Assn. Pathology, Fedn. Am. Socs. Biology, Sigma Xi. Hindu. Subspecialties: Toxicology (medicine); Preventive medicine (veterinary medicine). Current work: Immunology, immunodiagnostics, tumormarkers, inborn metabolic errors, evaluation of commercial products before marketing in clinical chemistry, toxicology, and immunodiagnostics areas. Office: 8757 Georgia Ave Silver Spring MD 20910

VAGELOS, PINDAROS ROY, biomedical research executive; b. Westfield, N.J., Oct. 8, 1929; s. Roy John and Marianthi (Lambrinides) V.; m. Diana Touliatos, July 10, 1955; children: Randall, Cynthia, Andrew, Ellen. A.B., U. Pa., 1950; M.D., Columbia U., 1954; D.Sc. (hon.), Washington U., 1980, Brown U., 1982. Intern medicine Mass. Gen. Hosp., 1954-55, asst. resident medicine, 1955-56; surgeon Lab. Cellular Physiology, NIH, 1956-59, Lab. Biochemistry, 1959-64, head sect. comparative biochemistry, 1964-66; prof. biochemistry, chmn. dept. biol. chemistry Washington U. Sch. Medicine, St. Louis, 1966-75, dir. div. biology and biomed. scis., 1973-75; sr. v.p. research Merck, Sharp & Dohme Research Labs., 1975-76, pres., 1976—; corp. sr. v.p. Merck & Co., Inc., 1982—; mem. molecular biology study sect. NIH, 1967-71, mem. physiol. chemistry study sect., 1973-75; mem. commn. on human resources NRC, 1973-76, Inst. Medicine, Nat. Acad. Scis., 1974—. Mem. editorial bds. jours. in field. Mem. Am. Chem. Soc. (chmn. div. biol. chemistry 1973, award enzyme chemistry 1967), Am. Soc. Biol. Chemists, A.A.A.S., Nat. Acad. Scis., Am. Acad. Arts and Scis. Subspecialty: Biomedical research management. Discoverer of acyl-carrier protein. Home: 10 Canterbury Ln Watchung NJ 07060

VAIDYA, AKHIL BABUBHAI, molecular biologist; b. Gondal, India, Oct. 24, 1947; s. Babubhai Pranjivan and Dayaben Vajeshankar (Dave) V.; m. Sheila Rao Vaidya, May 6, 1973; 1 child, Ashish. B.Sc., Bhavan's Coll., U. Bombay, 1967, Ph.D., 1972. Research fellow Cancer Research inst., Bombay, 1967-70, sci. asst., 1970-72; assoc. Inst. Med. Research, Camden, N.J., 1972-77; asst. prof. dept. microbiology Hahnemann Med. Coll., 1977-82, assoc. prof., 1982—; mem. spl. study sect. NIH. Contbr. articles to profl. jours. Recipient Gold medal for acad. achievement Bhavan's Coll., 1967. Mem. Am. Soc. Microbiology. Subspecialties: Gene actions; Molecular biology. Current work: Research interests deal with studying mechanisms underlying eukaryotic gene regulation with mouse mammary tumor virus as a model; also conducting molecular biological studies on malarial parasites using recombinant DNA techniques. Office: Dept Microbiology Hahnemann Medical College Broad and Vine Philadelphia PA 19102

VAIL, PETER ROBBINS, geologist; b. N.Y.C., Jan. 13, 1930; s. Donald Bain and Eleanor (Robbins) V.; m. Carolyn Flesher, Sept. 15, 1956; children: Andrea, Susan, Timothy Edward. A.B., Dartmouth Coll., 1952; M.S., Northwestern U., 1955, Ph.D. (Shell fellow), 1959. Asst. geologist U.S. Geol. Survey, Spokane, Wash., 1952-56, Evanston, Ill.; research geologist Carter Oil Co., Tulsa, 1956-58, Jersey Prodn. Research Co., 1958-62, sr. research geologist, 1962-65; sr. research specialist Exxon Prodn. Research Co., Houston, 1965-66, research assoc., 1966, research supr., 1966-70, sr. research assoc., 1970-72, sr. research adv., 1972-75, research scientist, 1975-81, sr. research scientist, 1981—; mem. Consortium for Continental Reflection Profiling Site Selection Com., 1974-81; mem. internat. subcom. on stratigraphic classification Internat. Union Geol. Seismology com. on Stratigraphy, 1976—; mem. ocean sci. bd. U.S. Nat. Acad. Scis., 1979-82; mem. Joint Oceanography Insts. Deep Earth Sampling Passive Margin Panel, 1978—; William Smith lectr. Geol. Soc. London, 1978; Bullerwell lectr. U.K. Geophys. Soc., 1983; mem. Am. Petroleum Inst., 1978—. Contbr. to profl. jours. articles on seismic stratigraphy, global changes of sea level, tectonics. Recipient Offshore Tech. Conf. Individual Disting. Achievement Award, 1983. Fellow Geol. Soc. Am. (research grants com. 1977-79, chmn. 1979, councilor 1979-82), AAAS; mem. Am. Assn. Petroleum Geologists (disting. lectr. 1975-76, marine geology com. 1977, research com. 1978-84, co-recipient Pres.'s award for best pub. paper 1979, co-recipient Matson award for author of best paper 1980 ann. conv.), Soc. Exploration Geophysicists (Virgil Kaufmann Gold medal for advancement of science of geophys. exploration 1976), Soc. Econ. Paleontologists and Mineralogists, Houston Geophys. Soc. (hon. mem.), Houston Geol. Soc. (Best Paper award 1982-83), Mayflower Soc., Sigma Xi. Subspecialties: Geophysics; Sedimentology. Current work: Seismic interpretation of stratigraphy and structure. Home: 3745 Del Monte Dr Houston TX 77019 Office: Exxon Production Research Co Box 2189 Houston TX 77001

VALA, MARTIN THORVALD, JR., chemistry educator; b. Bklyn., Mar. 28, 1938; s. Martin Thorvald and Norun (Nilsen) V.; m. Vibeke Wilken-Jensen, June 23, 1966; children: Lars, Carsten, Steffen. B.A., St. Olaf Coll., 1960; M.S., U. Chgo., 1962, Ph.D., 1964. Postdoctoral fellow U. Copenhagen, 1965-66, U. Nagoya, Japan, 1966-67; asst. prof. U. Fla., Gainesville, 1968-72, assoc. prof., 1972-78, prof. chemistry, 1978—; vis. scientist U.S.-India Exchange, 1977. Contbr. articles on chemistry to profl. jours. Merck Found. fellow, 1970-71; George Marshall Found. fellow, 1971; NATO fellow, Paris, 1973; Fulbright sr. fellow, Paris, 1974. Mem. Am. Chem. Soc., Am. Phys. Soc., Inter-Am. Photochem. Soc. Democrat. Lutheran. Subspecialties: High temperature chemistry; Spectroscopy. Current work: Spectroscopy of matrix-isolated inorganic and organic systems. Home: 3432 NW 11th Ave Gainesville FL 32605 Office: U Fla Dept Chemistry Gainesville FL 32611

VALDES, JAMES JOHN, neurotoxicologist, cons.; b. San Antonio, Apr. 25, 1951; s. Fernando and Barbara Marie (Sachtleben) V.; m. Leslie Elizabeth Valdes, June 6, 1981. B.S., Loyola U., Chgo., 1973; M.S., Trinity U., 1976; Ph.D. (chemistry of Behavior fellow), Tex. Chirstian U., 1979; postdoctoral fellow, Johns Hopkins U., 1979-81. Instr. Tex. Christian U., Ft. Worth, 1978-79; phys. scientist toxicology br. Aberdeen Proving Ground, Md., 1982—; instr. environ. psychobiology, cons. neurochemistry and biostats. Johns Hopkins U.; instr. physiol. psychology and pharmacology Hood Coll., 1982-83. Assoc. editor: Neurobehavioral Toxicology and Teratology, 1982; contbr. articles profl. jours. Grantee U.S. Army Research Office, 1982—, Dept. Def. Chem. Systems Lab., 1982-83. Mem. Soc. Neurosci., AAAS, Brit. Brain Research Assn., European Brain and Behavior Soc. Subspecialties: Neuropharmacology; Toxicology (medicine). Current work: Neurotransmitter compensatory mechanisms and recovery of central nervous system function after toxic insult, neurochemical effects of mycotoxins. Office: Toxicology Branch CSL Aberdeen Proving Ground MD 21010

VALENSTEIN, ELLIOT SPIRO, psychology educator; b. N.Y.C., Dec. 9, 1923; s. Louis and Helen (Spiro) V.; m. Thelma Lewis, June 15, 1947; children: Paul, Carl. B.S., CCNY, 1949; M.A., U. Kans.-Lawrence, 1953, Ph.D., 1954. Chief neuropsychology sect. Walter Reed Inst. Research, Washington, 1957-61; sr. research assoc. Fels Research Inst., Yellow Springs, Ohio, 1961-71; prof. psychology Antioch Coll., 1963-71, U. Mich., 1970—, chmn, psychobiology area, 1979—; fellow Ctr. Advanced Studies Behavioral Sci., Stanford, Calif., 1976-77; vis. prof. Hebrew U., Jerusalem, 1980; mem. chmn. various sci. rev. panels NIH, NSF, 1964—; chmn, adv. bd. Wis. Regional Primate Ctr., 1972—. Author: Brain Stimulation and Motivation, 1973, Brain Control, 1973; editor, contbg. author: Psychosurgery Debate, 1980. Recipient Kenneth Craik Research award Cambridge (Eng.) U., 1980-81; Fulbright-Hays fellow, Israel, 1980. Fellow Am. Psychol. Assn. (pres. Physiol. and Comparative Div. 1976-77); mem. Internat. Brain Research Orgn., Internat. Union Physiol. Sci., Colegio de Sciencias de la Conducta Inst. Mexicano de Coltura (corr.). Subspecialties: Psychobiology; Neuropsychology. Current work: physiology of motivation. Home: 260 Indian River Pl Ann Arbor MI 48104 Office: U Mich Neurosci Lab Bldg 1103 E Huron Ann Arbor MI 48109

VALENTINE, MARTIN DOUGLAS, physician, researcher, educator; b. Greenwich, Conn., Apr. 13, 1935; s. Emanuel Henriques and Betty (Resnick) V.; m. Leah Helen David, June 16, 1957; children: Mark D., Daniel S., Rachel L., Joshua R. B.S., Union Coll. Schenectady, 1956; M.D., Tufts U., 1960. Diplomate: Am. Bd. Internal Medicine, Am. Bd. Allergy and Immunology. Research fellow Harvard U.-Mass. Gen. Hosp., 1965-66, Peter Bent Brigham Hosp., 1968-68, Robert Breck Brigham Hosp., 1966-68; jr. assoc. in medicine Peter Bent Brigham Hosp.-Harvard U., 1968-70; asst. prof. medicine Johns Hopkins U., 1970-77, assoc. prof., 1977—. Mem. editorial bd.: Jour. Allergy and Clin. Immunology, 1980—; contbr. articles to profl. jours. Served to capt. USAF, 1962-64. Nat. Inst. Allergy and Infectious Disease grantee, 1978—. Mem. Am. Acad. Allergy, Am. Assn. Immunologists, Am. Thoracic Soc. Subspecialties: Allergy; Immunology (medicine). Current work: New methods in treatment and prevention of insect allergy. Office: 5601 Loch Raven Blvd Baltimore MD 21239

VALFER, ERNST SIEGMAR, management scientist, psychologist, consultant; b. Frankfurt, Ger., July 4, 1925; came to U.S., 1941; s. Hermann Heinrich and Frieda (Kahn) V.; m. Lois Brandwynne, July 8, 1961; children: Rachel, Lilah. A.A., San Francisco City Coll., 1948; B.S., U. Calif.-Berkeley, 1950, M.S., 1952, Ph.D., 1965. Diplomate: Am. Bd. Profl. Psychology; Lic. psychologist, Calif.; registered profl. engr., Calif. Supt. indsl. planning U.S. Navy, Alameda, Calif., 1952-57; research scientist, dir. NRC-Nat. Acad. Scis., San Francisco and Washington, 1957-62; assoc. research engr. U. Calif.-Berkeley, 1961-64, sr. lectr., research engr., 1965-67; cons. in pvt. practice, Berkeley, 1981; adj. sr. fellow UCLA, 1973—; dir. mgmt. sci. staff. U.S. Dept. Agr.-Forest Service, Berkeley, 1962—; Western bd. mem. Am. Bd. Profl. Psychology, 1980—; cons. numerous pvt., govt. and acad. orgns.; examiner Calif. State Bd. Profl. Engrs., Sacramento, 1967-69. Contbr. articles to profl. jours., chpts. to books. Chmn. Agy. Jewish Edn., 1982—; bd. dirs. Jewish Fedn., Oakland, 1983—, Tehiyah Sch., Berkeley, 1979-83, various charitable orgns. in, San Francisco Bay area, 1970-80. Served to lt. AUS, 1944-46; ETO. Recipient citation Pres. of Nat. Acad. Scis.-NRC, 1961-62, recognition Am. Inst. Indsl. Engrs., 1958, various awards Dept Agr.-Forest Service, 1970s. Mem. Inst. Indsl. Engrs. (sr.), Am. Psychol. Assn., Inst. Mgmt. Scis., Sigma Xi. Jewish. Subspecialties: Social psychology; Human factors engineering. Current work: Multi-disciplinary approach to complex decision making in large o. Home: 2621 Rose St Berkeley CA 94708 Office: Dept Agriculture Forest Service PO Box 245 Berkeley CA 94701

VALLEE, BERT LESTER, biochemist, physician, educator; b. Hemer, Westphalia, Germany, June 1, 1919; came to U.S., 1938, naturalized, 1948; s. Joseph and Rosa (Kronenberger) V.; m. Natalie T. Kugris, May 29, 1947. Sc.B., U. Berne, Switzerland, 1938; M.D., N.Y. U. Coll. Medicine, 1943; A.M. (hon.), Harvard, 1960. Research fellow Harvard Med. Sch., Boston, 1946-49, research asso., 1949-51, asso., 1951-56, asst. prof. medicine, 1956-60, asso. prof., 1960-64, prof. biol. chemistry, 1964-65, Paul C. Cabot prof. biol. chemistry, 1965-80, Paul C. Cabot prof. biochem. scis., 1980—; research asso. dept. biology Mass. Inst. Tech., Cambridge, 1948—; physician Peter Bent Brigham Hosp., Boston, 1961-80; biochemist-in-chief Brigham & Women's Hosp., Boston, 1980—; sci. dir. Biophysics Research Lab., Harvard Med. Sch., Peter Bent Brigham Hosp., 1954-80; head Center for Biochem. and Biophys. Scis. and Medicine, Harvard Med. Sch. and Brigham & Women's Hosp., 1980—. Author book.; Contbr. articles and chpts. to sci. publs. Founder, trustee Boston Biophysics Research Found., 1957—; founder, pres. Endowment for Research in Human Biology, Inc., 1980—. Recipient Buchman Meml. award Calif. Inst. Tech., 1976; Linderstrøm-Lang award and gold medal, 1980; William C. Rose award in biochemistry, 1982. Fellow AAAS, Nat. Acad. Scis., Am. Acad. Arts and Scis., N.Y. Acad. Scis.; mem. Am. Soc. Biol. Chemists, Am. Chem. Soc. (Willard Gibbs gold medal 1981), Optical Soc. Am., Biophys. Soc., Swiss Biochem. Soc. (hon. fgn. mem.), Royal Danish Acad. Scis. and Letters, Alpha Omega Alpha. Subspecialty: Biophysical chemistry. Home: 56 Browne St Brookline MA 02146

VALLERY, STAFFORD JEAN, II, mech. engr.; b. Longview, Tex., Nov. 13, 1940; s. Stafford Jean and Cleo Warren V.; m. Janie Tanner, Sept. 1, 1962; 1 son, Gregory Lawrence. B.S., La. Poly. Inst., 1965. Field engr. atomic energy plant DuPont Co., Aiken, S.C., 1965-66, staff engr. central engring. offices, Wilmington, Del., 1966-68; plant engr., Victoria, Tex., 1968-82; corp. dir. tech. services Armstrong Internat., Three Rivers, Mich., 1982—. Mem. home selection chmn. Am. Field Services, Fgn. Student Exchange, 1980-82; mem. DeLeon dist. com. Boy Scouts Am., 1975-82; leader Cub Scouts, 1971-74. Served with USMC, 1960-65. Mem. ASME (chmn. 1981-82, region X del. to nat. agenda bd. 1982-83). Republican. Methodist. Clubs: Antique Automobile, Victoria Wings, CAP (Victoria). Subspecialties: Energy conservation; Mechanical engineering. Current work: Energy reductions in plant process systems, steam trap applications and development, condensate recovery, desuperheating, humidification, and steam systems training. Home: 110 Nottingham Dr Victoria TX 77904 Office: 900 Maple St Three Rivrs MI 49093

VALYI, EMERY I., machinery company executive; b. Murska Sobota, Yugoslavia, July 14, 1911; came to U.S., 1940, naturalized, 1946; s. Alexander and Elizabeth (Arvay) V.; m. Ilsabe von Behr; children: Katherine, Thomas. M.E., Swiss Fed. Inst. Tech., 1933, D.Sc., 1937. Research engr. Swiss Fed. Inst. Testing Materials, 1934-37; metall. engr. Injecta, Ltd., Switzerland, 1937-40; mgr. die casting machine div. Press Mfg. Co., Mt. Gilead, Ohio, 1940-42; v.p. Sam Tour Co., N.Y.C., 1942-45; pres. ARD Corp., Yonkers, N.Y., 1945-62; cons. maj. U.S., fgn. corps., 1972-80; pres. tpT Machinery Corp., Norwalk, Conn., 1974—. Contbr. numerous articles to profl. jours. Mem. Am. Inst. Mech. Engrs., Soc. Plastics Industry, Soc. Plastics Engrs., ASME, Am. Inst. Chemists. Subspecialties: Materials processing; Polymers. Current work: Conversion process for polymers. Patentee in field. Office: 3 Eversley Ave Norwalk CT 06851

VAN ALFEN, NEAL K., biologist, educator, cons.; b. Ogden, Utah, July 17, 1943; s. Gerrit Johan and Marguerite (Noorda) Van A.; m. Susan Duffin, Dec. 18, 1965; children: Peter, Anne, David, Christina. Ph.D., U. Calif.-Davis, 1972. Asst. plant pathologist Conn. Agrl. Expt. Sta., New Haven, 1972-75; asst. prof. biology Utah State U., 1975-78, assoc. prof., 1978-82, prof., 1982—; cons. Kennecott Corp., 1976—. Research, publs. in field. NSF grantee, 1976-78, 78-79, 81-84; Dept. Agr. Competitive Research Grants Office grantee, 1978-80, 80-82, 82-84; Dept. Agr. Forest Service grantee, 1979-82, 1981-83; Dept. Interior Park Service grantee, 1979-81. Mem. Am. Phytopath. Soc., Am. Soc. Plant Physiologists. Subspecialties: Plant pathology; Genetics and genetic engineering (biology). Current work: Mechanisms of pathogen virulence; biochemistry and genetics of virulence expression by plant pathogens. Office: Dept Biology UMC 45 Utah State U Logan UT 84322

VAN ALLEN, JAMES ALFRED, physicist, educator; b. Mt. Pleasant, Iowa, Sept. 7, 1914; s. Alfred Morris and Alma E. (Olney) Van A.; m. Abigail Fithian Halsey, Oct. 13, 1945; children: Cynthia Olney Van Allen (Schaffner), Margot Isham, Sarah Halsey, Thomas Halsey, Peter Cornelius. B.S., Iowa Wesleyan Coll., 1935, Sc.D., 1951; M.S., U. Iowa, 1936, Ph.D., 1939; Sc.D., Grinnell Coll., 1957, Coe Coll., 1958, Cornell Coll., 1959, U. Dubuque, 1960, U. Mich., 1961, Northwestern U., 1961, Ill. Coll., 1963, Butler U., 1966, Boston Coll., 1966, Southampton Coll., 1967, Augustana Coll., 1969, St. Ambrose Coll., 1981. Research fellow, physicist dept. terrestrial magnetism Carnegie Instn., Washington, 1939-42; physicist, group and unit supr., applied physics lab. Johns Hopkins, 1942, 1946-50; organizer, leader sci. expdns. study cosmic radiation, Peru, 1949, Gulf of Alaska, 1950, Greenland, 1952, 57, Antarctica, 1957; Carver prof. physics, head dept. U. Iowa, 1951—; Regents fellow Smithsonian Instn., 1981; research asso. Princeton, 1953-54; dir. Iowa Electric Light and Power Co., 1st Nat. Bank Iowa City; Devel. radio proximity fuze Nat. Def. Research Council, OSRD; pioneer high altitude research with rockets, satellites and space probes. Contbg. author: Physics and Med. of Upper Atmosphere, 1952, Rocket Exploration of the Upper Atmosphere; author: Origins of Magnetospheric Physics, 1983; Editor: Scientific Uses of Earth Satellites, 1956; asso. editor: Jour. Geophysical Research, 1959-64, Physics of Fluids, 1958-62; Contbr. numerous articles to sci. jours. Served as lt. comdr. U.S. Navy, 1942-46; ordnance and gunnery specialist, combat observer. Received C.N. Hickman medal for devel. Aerobee rocket Am. Rocket Soc., 1949; physics award Washington Acad. Sci., 1949; Guggenheim Meml. Found. research fellow, 1951; space flight award Am. Astronautical Soc., 1958; Louis W. Hill space transp. award Inst. Aero. Scis., 1959; Elliot Cresson medal Franklin Inst., 1961; John A. Fleming award Am. Geophys. Union, 1963, 64; Golden Omega award Elec. Insulation Conf., 1963; comdr. Order du Merit Pour la Recherche et L'Invention, 1964; Iowa Broadcasters Assn. award, 1964; Fellows award of merit Am. Cons. Engrs. Council, 1978. Fellow Am. Rocket Soc., IEEE, Am. Phys. Soc., Am. Geophys. Union (pres. 1982-84, William Bowie medal 1977); mem. Iowa Acad. Sci., Nat. Acad. Scis., Internat. Acad. Astronautics (founding mem.), Am. Philos. Soc., Am. Astron. Soc., Royal Astron. Soc. (U.K.) (gold medal 1978), Am. Acad. Arts and Scis., Sigma Xi, Gamma Alpha. Presbyn. Club: Cosmos (Washington). Subspecialties: Cosmic ray high energy astrophysics; Satellite studies. Discoverer radiation belts around earth. Office: 203 Physics Bldg U Iowa Iowa City IA 52242

VANAMAN, THOMAS C., biochemist, educator; b. Louisville, Ky., Aug. 12, 1941; s. Sherman Benton and Thelma Cecilia (Clark) V.; m. Linda Leigh Bell, Mar. 17, 1962; children: Thomas Randolph, John Tyler. B.S. in Chemistry, U. Ky., 1964; Ph.D., Duke U., 1968; postgrad., Stanford U., 1968-70. Asst. prof. microbiology and immunology Duke U., 1970-75, assoc. prof., 1975-79, prof., 1979-83, dir. Basic Research Comprehensive Cancer Center, 1981-83; prof., chmn. biochemistry U. Ky. Med. Ctr., Lexington, 1983—. Mem. editorial bd.: Jour. Biol. Chemistry; Contbr. in field. Served with USAR, 1961-67. Recipient Josiah Macy faculty scholar award, 1977-78; grantee in field. Mem. Am. Soc. Biol. Chemistry, Am. Soc. Microbiology, N.Y. Acad. Sci., AAAS. Subspecialties: Biochemistry (medicine); Biochemistry (biology). Current work: Protein chemistry, cell regulation, 2d messenger signalling systems, viral proteins. Office: Dept Biochemistry U Ky Med Ctr Lexington KY 40536

VAN ATTA, CHARLES WILLIAM, engineering sciences, educator; b. New London, Conn., Feb. 24, 1934; s. George William and Marguerite Ethyl (Myrmel) Van A.; m. Ann Louise Lyle, June 7, 1958; 1 dau., Pamela. B.S. in Aero. Engring, U. Mich., 1957, 1957, M.S. in Aero. Engring, 1958; Ph.D., Calif. Inst. Tech., 1964. Prof. engring. sci. and oceanography U. Calif., San Diego, La Jolla, 1965—. Contbr. articles to profl. jours. Guggenheim fellow, 1972-73; Nat. Acad. Scis. exchange scientist, USSR, 1973. Mem. Am. Phys. Soc. (exec. com. div. fluid dynamics 1982-83), Am. Geophys. Union, Am. Acad. Mechanics, Sierra Club. Subspecialties: Fluid mechanics; Oceanography. Current work: Experimental and theoretical research on the structure of turbulent flows in nature and the laboratory. Office: U Calif San Diego Ames B-010 La Jolla CA 92093

VAN CAMPEN, DARRELL R., nutritionist, laboratory administrator; b. Two Buttes, Colo., July 15, 1935; s. Robert L. and Emma Pauline (Comer) Van C.; m. Orlene Crone, Sept. 8, 1958; children: Anthony, Bryan; m. Judith Anne Gorsky, June 27, 1977; 1 son, John Tassone. B.S., Colo. State U., 1957; M.S., N.C. State U., 1960, Ph.D., 1962. USPHS fellow Cornell U., Ithaca, N.Y., 1962-63; research chemist U.S. Plant, Soil & Nutrition Lab., Ithaca, 1964-79, dir. lab., 1979—. Contbr. numerous sci. articles to profl. publs. Mem. Am. Inst. Nutrition, Soc. Exptl. Biol. Medicine, AAAS, N.Y. Acad. Sci., Fedn. Am. Soc. Exptl. Biology, Sigma Xi, Alpha Zeta, Phi Kappa Phi. Subspecialties: Animal nutrition; Nutrition (biology). Current work: Nutritional quality of plant foods and feeds; absorption and utilization of mineral elements in man and animals. Home: 117 Simsbury Dr Ithaca NY 14850 Office: US Plant Soil and Nutrition Lab Tower Rd Ithaca NY 14853

VANCE, B(ENJAMIN) DWAIN, botany educator; b. Cave City, Ark., May 7, 1932; s. William Huston and Eva Derenda (Bray) V.; m. Barbara Gloff, June 7, 1952; children: Valerie, Michael, Michelle, Melissa. B.S., Tex. Tech. U., 1958; A.M., U. Mo.-Columbia, 1959, Ph.D., 1962. Asst. prof. botany Tex. Tech. U., Lubbock, 1962-63; asst. prof. N. Tex. State U., Denton, 1963-69; assoc. prof. botany, 1969—; owner Greenway Lawn & Turf, Denton, 1973—; dir. Cyclic Energies, Inc., Denton, 1980—. Contbr. articles to profl. jours. Trustee, sec. Denton Ind. Sch. Dist., 1980—. Served with U.S. Army, 1952-54. NIH fellow, 1959-62; NIH grantee; Tex. Acad. Scis. fellow, 1969. Mem. Pyschological Soc. Am. Democrat. Lutheran. Subspecialties: Plant physiology (biology); Ecology. Current work: Carbon metabolism in algae in acid waters. Home: 2124 Glen Garden Dr Denton TX 76201 Office: Dept Biol Sci N Tex State U Denton TX 76203

VAN DE MARK, MICHAEL ROY, chemistry educator; b. Pigeon, Mich., May 21, 1950; s. Roy Donald and Alice Elizabeth (Gremel) Van De M.; m. Suzan Elaine Yelek, Nov. 26, 1976. B.S., Saginaw Valley State Coll., 1972; Ph.D., Texas A&M U., 1976. Assoc. prof., dir. grad. studies in chemistry U. Miami, Coral Gables, Fla., 1978—. Contbr. numerous articles to profl. jours. Mem. Am. Chem. Soc. (Miami subsect. sec./treas. 1979, chmn.-elect 1980, chmn. 1981), Electrochem. Soc., Phi Lambda Upsilon, Sigma Xi. Subspecialties: Polymer chemistry; Surface chemistry. Current work: Polymer adsorption on surfaces for electrode modification and corrosion inhibition through chelating adsorbed polymers. In addition am involved in synthetic organic electrochemistry and polymer synthesis. Home: 9800 SW 80 Dr Miami FL 33173 Office: U Miami Dept Chemistry Coral Gables FL 33134

VANDEMARK, NOLAND LEROY, physiologist, educator; b. Columbus Grove, Ohio, July 6, 1919; s. Daniel Leroy and Mary Frances (Bogart) V; m. Beda Alta Basinger, Aug. 18, 1940; children: Gary Lee, Judy Beth, Linda Kay. B.S., Ohio State U., 1941, M.S., 1942; Ph.D., Cornell U., 1948. Asst. in animal husbandry Ohio State U., 1941-42; vitamin chemist div. plant industry Ohio Dept. Agr., 1942; livestock specialist U.S. Allied Commn., Austria, 1946-47; asst. in animal husbandry Cornell U., 1942-44, 48; asst. prof. physiology dept. dairy sci. U. Ill., 1948-51, assoc. prof., 1951-55, prof., 19 -64; prof. dairy sci., chmn. dept. dairy sci. Ohio State U. and Ohio Agrl. Research and Devel. Ctr., 1964-73; dir. research N.Y. State Coll. Agr. and Life Scis., 1974-81; dir. Cornell U. Agrl. Expt. Sta., Ithaca, 1974-81, prof. animal sci., 1974—; cons. reproductive biology study sect. Nat. Inst. Child Health, 1966-70. Author: (with G. W. Salisbury) Physiology of Reproduction and Artifical Insemination of Cattle, 1961, 78; co-editor, contbr.: The Testis, 3 vols, 1970. Served to 2d lt. inf. and CIC AUS, 1944-46. Recipient award in dairy sci. Borden, 1959; Alpha Zeta Outstanding Agr. Tchr. award U. Ill., 1960; Italian Master Pioneer Gold Medal award 5th Internat. Congress Reprodn., 1964; Ivanov Centennial medal, Russia, 1975. Fellow AAAS; mem. Am. Physiol. Soc., Soc. Study Fertility, Am. Dairy Sci. Assn. (chmn. prodn. sect. 1969-70, dir. 1971-74), Soc. Study Reprodn. (v.p. 1968-69, pres. 1969-70, dir. 1970-71), Am. Soc. Animal Sci., Sigma Xi, Gamma Sigma Delta. Subspecialties: Animal physiology; Research administration. Office: Cornell U 320 Morrison Hall Ithaca NY 14853

VANDENBARK, ARTHUR ALLEN, immunology educator, researcher; b. Twin Falls, Idaho, Feb. 9, 1946; s. James Adams and Ruth Lillian (Dankenbring) V.; m. Halina Offner, Apr. 23, 1983. A.B., Stanford U., 1968; M.S., Wash. State U., 1971, Ph.D., 1973. Research immunologist VA Hosp., Portland, Oreg., 1973-82; asst. prof. Oreg. Health Scis. U., Portland, 1976-82, assoc. prof., 1982—. Recipient Fulbright-Hayes award Council Internat. Exchange for Scholars, 1979; Paul J. Cams Found. fellow Dr. L. Willems-Instituut, Diepenbeek, Belgium, 1981-83; VA Med. Center grantee, 1975-83; NIH grantee,

1979-83; Nat. Cancer Inst. grantee, 1978-80; Wetenschappelijk Onderzoek Multiple Sclerose grantee, 1981-83. Mem. Am. Assn. Immunologists, Soc. Exptl. Biology and Medicine, Reticuloendothelial Soc., N.Y. Acad. Scis., Sigma Xi. Democrat. Subspecialties: Immunology (medicine); Neuroimmunology. Current work: Selective stimulation or suppression of immune response to autoantigens; established and original assays of immune capability; tissue culture; T-cell-monocyte interaction; autoantigen identification. Office: VA Med Center Portland OR 97201

VANDENBERG, EDWIN JAMES, polymer chemist; b. Hawthorne, N.J., Sept. 13, 1918; m., 1950; 2 children. M.E., Stevens Inst. Tech., 1939, D.Eng. (hon.), 1965. Research chemist Sunflower Ordnance Works Hercules Inc., Wilmington, Kans., 1939-44, asst. shift supr. Sunflower Ordnance Works, Kans., 1944-45, from research chemist to sr. research chemist Research Ctr., Wilmington, 1945-65, research assoc., 1965-78, sr. research assoc., 1978—. Mem. adv. bd.: Jour. Polymer Sci, 1966—, Macromolecules, 1977-80. Recipient Indsl. Research 100 award, 1965. Mem. Am. Chem. Soc. Subspecialty: Polymer chemistry. Office: Hercules Inc Research Ctr Wilmington DE 19899

VANDENBERG, JOHN LEE, geneticist; b. Appleton, Wis., June 14, 1947; s. Gale LeRoy and Zona Idell (Raine) VandeB.; m. Jane Frances Barr, Mar. 29, 1975; children: Jason Cash, James Robert. B.S., U. Wis., 1969; B.Sc. with honors, La Trobe U., Australia, 1970; Ph.D., Macquarie U., (1975.), Australia. Research assoc., postdoctoral fellow lab. genetics U. Wis., Madison, 1975-79, asst. scientist lab. genetics and Wis. Regional Primate Research Ctr., 1979-80; dir., assoc. scientist dept. genetics S.W. Found. Research and Edn., San Antonio, 1980—; asst. prof. dept. anatomy and pathology U. Tex. Health Sci. Ctr., San Antonio, 1980—82, assoc. prof., 1982—; mem. internat. awards rev. group NIH, 1981—; mem. com. animal models and genetic stocks NRC, 1982—. Contbr. articles to profl. jours. Recipient award of Merit Gamma Sigma Delta, 1967; Fulbright fellow, 1969-70. Mem. AAAS, Tex. Genetics Soc., Gentics Soc. Am. Subspecialty: Genetics and genetic engineering (biology). Current work: Biochemical genetics of mammals; genetic control of heart disease. Office: PO Box 28147 San Antonio TX 78284

VAN DEN BERGH, SIDNEY, astronomer; b. Wassenaar, Netherlands, May 20, 1929; emigrated to U.S., 1956; s. Sidney J. and Mieke (van den Berg) vandenB.; m. Gretchen Krause; children by previous marriage: Peter, Mieke, Sabine. Student, Leiden (Netherlands) U., 1947-48; A.B., Princeton U., 1950; M.Sc., Ohio State U., 1952; Dr. rer. nat., Goettingen U., 1956. Asst. prof. Perkins Obs., Ohio State U., Columbus, 1956-58; research asso. Mt. Wilson Obs., Palomar Obs., Pasadena, Calif., 1968-69; prof. astronomy David Dunlap Obs., U. Toronto, Ont., Can., 1958-77; dir. Dominion Astrophys. Obs., Victoria, B.C., 1977—; dir. Can.-France-Hawaii Telescope Corp. Mem. Am., Royal astron. socs., Royal Soc. Can. Subspecialty: Optical astronomy. Current work: Galaxies, super noval; star clusters. Home: 418 Lands End Rd Sidney BC Canada Office: Dominion Astrophysic Observatory Victoria BC Canada

VANDENBOSCH, ROBERT, chemistry educator, researcher; b. Lexington, Ky., Dec. 12, 1932; m., 1956; 2 children. A.B., Calvin Coll., 1954; Ph.D. in Chemistry, U. Calif., 1957. From asst. chemist to assoc. chemist Argonne Nat. Lab., 1957-63; prof. chemistry U. Wash., Seattle, 1963—. Recipient award for nuclear chemistry Am. Chem. Soc., 1981. Fellow Am. Phys. Soc. Subspecialty: Nuclear chemistry. Office: Dept Chemistry U Wash Seattle WA 98195

VANDENDORPE, MARY MOORE, psychology educator; b. Chgo., June 2, 1947; d. Era William and Mary (Dobis) Moore; m. James Edward Vandendorpe, Aug. 16, 1969; 1 dau., Laura. A.B., St. Louis U., 1969; M.S., Ill. Inst. Tech., 1975, Ph.D., 1980. Adj. instr. Ill. Inst. Tech., Chgo., 1975; adj. instr. Lewis U., Romeoville, Ill., 1975-80, asst. prof., 1980—; adult officer, asst. historian Ill. Jr. Acad. Sci., Champaign, 1973-75. Author: History of the Illinois Junior Academy of Science, 1974, Faculty Student Interaction and Undergraduate Careers, 1975, Student Nurses and the Aged, 1980. Mem. Legis. Ednl. Network of DuPage, Naperville, Ill., 1982—; active Naperville Heritage Soc., 1982—, Naperville Humane Soc., 1980—; mem. Naper Carriage Homeowners; bd. dirs. Music Theatre Chgo., 1972-74. Mem. Am. Psychol. Assn., Gerontol. Soc., Chgo. Psychol. Assn. (mem. council 1979-82, pres. 1982-83), Psi Chi, Gamma Pi Epsilon. Subspecialties: Developmental psychology; Cognition. Current work: Development of work and family attitudes; human information processing; attitudes towards aging and death. Office: Lewis Univ Romeoville IL 60441

VANDERGRAFT, JAMES SAUL, computer scientist, consultant; b. Gooding, Idaho, Apr. 29, 1937; s. Fred and May Salome (Lewis) V. B.S., Stanford U., 1959, M.S, 1963; Ph.D., U. Md., 1966. Programmer Lawrence Radiation Lab., Livermore, Calif., 1959-61; staff scientist Bellcomm, Inc., Washington, 1963-64; NASA trainee U Md., 1964-66, assoc. prof. computer sci., 1966-79; asst. dir. Automated Sci. Group, Silver Spring, Md., 1979—; guest prof. Eidgenossische Technische Hochschule, Zurich, Switzerland, 1974. Author: Introduction to Numerical Computations, 1978. Mem. Assn. Computing Machinery, Soc. Indsl. and Applied Math. Subspecialties: Numerical analysis; Mathematical software. Current work: Numerical solution of linear and nonlinear system; development of mathematical software. Home: 772 11th St SE Washington DC 20003 Office: Automated Scis Group Inc 700 Roeder Rd Silver Spring MD 20910

VANDERHOEF, LARRY NEIL, biology educator, university administrator; b. Perham, Minn., Mar. 20, 1941; s. Wilmar James and Ida Lucille (Wothe) V.; m. Rosalie Suzanne Slifka, Aug. 31, 1963; children: Susan Marie, Jonathan Lee. B.S., U. Wis., Milw., 1964, M.S., 1965; Ph.D., Purdue U., 1969. Postdoctorate U. Wis., Madison, 1969-70, research assoc., summers 1970-72; asst. prof. biology U. Ill., Urbana, 1970-74, assoc. prof., 1974-77, prof., 1977—, head dept. botany, 1977-80; provost Agrl. and Life Scis., U. Md., College Park, 1980—; vis. investigator Carnegie Inst., 1976-77, Edinburgh (Scotland) U., 1978; cons. in field. NRC postdoctoral fellow, 1969-70; Dimond travel grantee, 1975; NSF grantee, 1972, 74, 76, 77, 78, 79; NATO grantee, 1980. Mem. AAAS, Am. Soc. Plant Physiology (bd. editors Plant Physiology 1977-82, trustee, mem. exec. com., treas. 1983—), Nat. Assn. State Univ. and Land Grant Colls. Subspecialties: Plant physiology (agriculture); Plant growth. Home: 7010 Eversfield Dr Hyattsville MD 20782 Office: 1104 Symons Hall U Md College Park MD 20742

VAN DER MEER, JOHN PETER, geneticist, researcher, educator; b. Netherlands, June 25, 1943; s. Edward and Sophie (Alkema) van der M.; m. Nellie Slingerland, July 2, 1966; children: Lawrence, Deborah. B.Sc. in Botany, U. Western Ont., 1966; Ph.D. in Genetics, Cornell U., 1971. Postdoctoral fellow Best Inst., U. Toronto (Ont., Can.), 1971-74; asst. research officer Atlantic Research Lab., NRC Can., Halifax, N.S., 1974-81, assoc. research officer, 1981—; mem. faculty Dalhousie U., 1979—. Contbr. articles to sci. jours. Recipient Bd. Govs. Gold medal for Botany U. Western Ont., 1966; research grants. Mem. Genetics Soc. Can., Genetics Soc. Am., Am. Genetics Assn., Internat. Physiol. Soc. Subspecialties: Genetics and genetic engineering (biology); Plant growth. Current work: Genetic modification of marine red algae; studies on life cycles and reproduction, mutation induction and characterization, breeding of novel strains, biochemistry. Home: 2730 Connaught Ave Halifax NS Canada B3L 2Z7 Office: 1411 Oxford St Halifax NS Canada B3H 3Z1

VANDERSLICE, THOMAS AQUINAS, electrical manufacturing company executive; b. Phila., Jan. 8, 1932; s. Joseph R. and Mae (Daly) V.; m. Margaret Hurley, June 9, 1956; children: Thomas Aquinas, Paul Thomas Aquinas, John Thomas Aquinas, Peter Thomas Aquinas. B.S. in Chemistry and Philosophy, Boston Coll., 1953; Ph.D. in Chemistry and Physics, Catholic U. Am., 1956. With Gen. Electric Co., 1956—, gen. mgr. electronic components bus. div., 1970-72, v.p., 1970—, group exec. spl. systems and products group, Fairfield, Conn., 1972-77, sr. v.p., sector exec., 1977-79, exec. v.p., sector exec., 1979; pres., chief operating officer, dir. Gen. Telephone & Electronics Corp., Stamford, Conn., 1979—; dir. Texaco Inc., Emery Worldwide; mem. com. on energy Aspen Inst. for Humanistic Studies; mem. Oxford Energy Policy Club, Oxford U. Co-author: Ultra High Vacuum and Its Applications, 1963; reviser: Scientific Foundations of Vacuum Technique, 1960; Contbr. to profl. jours. Trustee Boston Coll., Com. Econ. Devel.; bd. dirs. Southwestern Area Commerce and Industry Assn. of Conn., Inc.; mem. Conn. Crime Commn.; mem. Econ. Policy Council UNA; mem. Bus.-Higher Edn. Forum, Nat. Commn. on Indsl. Innovation. Fulbright scholar, 1953-56; recipient Golden Plate award Acad. Achievement, 1963; Bicentennial medallion Boston Coll., 1976. Mem. Nat. Acad. Engring., Conn. Acad. Sci. and Engring., Am. Vacuum Soc., ASTM, Am. Chem. Soc., Am. Inst. Physics, Sigma Xi, Tau Beta Pi, Alpha Sigma Nu, Sigma Pi Sigma. Clubs: Conn. Golf (Easton, Conn.); Patterson (Fairfield). Subspecialty: Electrical research; manufacturing management. Patentee low pressure gas measurements and analysis, gas surface interactions and elec. discharge. Office: Gen Telephone & Electronics Corp 1 Stamford Forum Stamford CT 06904

VANDER SLUIS, KENNETH L., physicist; b. Holland, Mich., Dec. 19, 1925; s. Leonard and Jennie (Vander Woude) Vander S.; m. Joan Harvie, June 14, 1962; children: Lisa Joan, Stephen Harvie, David Kenneth. Student, U. Notre Dame, 1945-46; B.S., Baldwin Wallace Coll., 1947; M.S., Pa. State U., 1950, Ph.D. in Physics, 1952. Physicist Oak Ridge Nat. Lab., 1952—. Subspecialties: Laser systems. Address: 954 West Outer Dr Oak Ridge TN 37830

VANDERVOORT, PETER OLIVER, astronomer, educator; b. Detroit, Apr. 25, 1935; s. William Bernard and Geraldine Catherine (Case) V.; m. Frances Sheridan, Sept. 23, 1956; children: William Frederick, Dirk Sheridan. A.B., U. Chgo., 1954, S.B. in Physics, 1955, M.S., 1956, Ph.D. in Physics, 1960. Research assoc. Nat. Radio Astronomy Obs., Green Bank, W.Va., 1960; postdoctoral fellow Princeton U. Obs., 1960-61; asst. prof. astronomy and astrophysics U. Chgo., 1961-65, assoc. prof., 1965-80, prof., 1980—; sr. postdoctoral fellow Leiden (Netherlands) Obs., 1967-68. Contbr. articles in field of interstellar gas-dynamics and on the structure and dynamics of galaxies to profl. jours. Trustee Adler Planetarium, Chgo., 1974—. NSF fellow, 1960-61, 67-68. Mem. Am. Astron. Soc., Royal Astron. Soc. (U.K.), Am. Phys. Soc., Internat. Astron. Union. Subspecialties: Theoretical astrophysics; Dynamical astronomy. Current work: Stellar orbits in galaxies, dynamics of galaxies; teaching astronomy and astrophysics. Office: 5640 Ellis Ave Chicago IL 60637

VANDERWEIL, RAIMUND GERHARD, JR., consulting engineering firm executive, mechanical engineer; b. Neptune, N.J., Nov. 21, 1940; s. Raimund Gerhard and Janet Stelle (Letson) V.; m. Anne Stuart Hinshaw, Oct. 10, 1970; children: Alexander Raimund, Shelley McMillan, Stefan Gerhard. A.B. in Engring. and Applied Physics, Harvard U., 1961; M.S. in Mech. Engring, MIT, 1963. Registered profl. engr., N.Y., Calif., Mass., and other states. Engr. Lockheed Missiles & Space Co., Sunnyvale, Calif., 1963-65; atomic power equipment dept. Gen. Electric, San Jose, Calif., 1965-67; engr. R. G. Vanderweil Engrs. Inc., Boston, 1967—, pres., 1973—; lectr. in archtl. tech., dept. architecture Harvard U. Grad. Sch. Design, 1974-82. Recipient Young Engr. of Yr. award Mass. Soc. Profl. Engrs., 1976. Mem. ASME, ASHRAE, Nat. Soc. Profl. Engrs. Subspecialties: Mechanical engineering; Electrical engineering. Current work: Mech. and elec. systems for bldgs.; indsl. scale power. Office: 38 Chauncey St Boston MA 02111

VAN DER ZIEL, ALDERT, elec. engr.; b. Zandeweer, Netherlands, Dec. 12, 1910; naturalized, 1955; s. Jan and Hendrika (Kiel) Van der Z.; m. Jantina J. deWit, Nov. 22, 1935; children—Jan P., Cornelia H., Joanna C. B.A., U. Groningen, The Netherlands, 1930, M.A., 1933, Ph.D., 1934. Research physicist Philips Labs., Eindhoven, Netherlands, 1934-47; asso. prof. physics U. B.C., Vancouver, Can., 1947-50; prof. dept. elec. engring. U. Minn., Mpls., 1950—; grad. research prof. U. Fla., Gainesville, 1968—. Author 10 books in field; contbr. numerous articles to profl. jours. Recipient Vincent Bendix award Am. Soc. Engring. Edn., 1975. Fellow IEEE (Edn. medal 1980); mem. Am. Phys. Soc., Sigma Xi. Lutheran. Subspecialties: Statistical physics; Semiconductors. Current work: Noise in solid state devices and circuits. Patentee in field. Office: Elec Engring Dept Univ of Minn Minneapolis MN 55455

VAN DEVENDER, ROBERT WAYNE, ecologist, educator, herpetologist; b. Roswell, N.Mex., July 16, 1947; s. Fred Roger and Winifred Allen (Drawhorn) Van D.; m. Amy Shrader, Dec. 29, 1973; children: Martha Noelle, Eve Amanda, Barnabas Bennett. B.S., Yale U., 1969; M.S., U. Mich., 1970, Ph.D., 1975. Curatorial asst. Yale U., New Haven, 1968-69; research assoc. Mus. Zoology, U. Mich., Ann Arbor, 1975-76; vis. asst. prof. Okla. State U., Stillwater, 1977; postdoctoral scholar U. Mich., 1977-78; asst. prof. Appalachian State U., Boone, N.C., 1978-82, assoc. prof., 1982—; adj. asst. prof. Western Carolina U., Cullohwhee, N.C., 1980. Mem. Herpetologists League, Am. Soc. Ichthyologists and Herpetologists, Ecol. Soc. Am., Soc. Study of Evolution, Soc. Study of Amphibians and Reptiles. Subspecialties: Evolutionary biology; Theoretical ecology. Current work: Ecology and evolution of local life history phenomena of amphibians and reptiles; applied photography; evolutionary theory; populations. Home: Route 4 Box 441 Boone NC 28607 Office: Appalachian State U Boone NC 28608

VAN DE WATER, THOMAS ROGER, scientist, laboratory administrator, medical educator; b. Oceanside, N.Y., Dec. 6, 1939; s. Lynn Patterson and Leonora (Winterson) Van De W.; m. Jeanette Adele Vilece, July 11, 1964; children: Ann Marie, Thomas Scott, Christopher Lynlee, Elizabeth Adele. A.A.S. in Forestry, Pauls Smiths Coll., 1959; B.S. in Biology, Western Carolina U., 1961; M.S., Hofstra U., 1965; Ph.D. in Biology, N.Y. U., 1976. Research assoc. Yale U. Sch. Medicine, 1964-65; research scientist N.Y. U. Med. Sch., 1966-68; research assoc., instr. Albert Einstein Coll. Medicine, 1968-75, asst. prof. otolaryngology and neurosci., 1976-81, assoc. prof., 1981—; vis. prof. Karolinska Inst., Stockholm, 1976-77; cons. communications disorders panel NIH, Washington; dir. Devel. Otobiology Lab., Bronx, N.Y., 1976—; spl. reviewer grant study sect. NIH, Bethesda, Md., 1982-83; invited lectr. Chaba sect. Nat. Acad. Sci., 1979. Regional rep. Catholic Charismatic Renewal, Nausau County, N.Y., 1980-82; field cons. Nassau County council Boy Scouts Am., 1981; bd. dirs. Cath. Fellowship, 1977-82. Served with U.S. Army, 1962. Swedish Med. Research Council grantee, 1976-77; NIH grantee, 1980-85; March of Dimes Birth Defects Found. grantee, 1983; Am. Otol. Found. grantee, 1983. Mem. N.Y. Acad. Sci., Am. Assn. Anatomists (pres. Morphogensis Club 1983-84), Assn. Research in Otolaryngology, Internat. Soc. Devel. Neurosci., Soc. Developmental Biology. Subspecialties: Cell and tissue culture; Developmental biology. Current work: Investigation of tissue interactions during development of inner ear that result in normal and abnormal morphogenesis employing techniques of organ culture, ultrastructure, biochemistry, teratology, genetic mutations and immunocyotchemistry. Home: 262 Pennsylvania Ave Freeport NY 11520 Office: Albert Einstein Coll Medicine 1300 Morris Park Ave Bronx NY 10461

VANDE WOUDE, GEORGE F., JR., molecular biologist; b. Bklyn., Dec. 25, 1935; s. George F. and Alice B. Vande W.; m. Dorothy H., Apr. 5, 1959; children: Susan, Gail, Cynthia, Alice. B.A., Hofstra U., 1959; M.S., Rutgers U., 1963, Ph.D., 1964. Postdoctoral research assoc. Nat. Acad. Sci., 1964-65; research chemist Dept. Agr., Greenport, N.Y., 1965-72; chief sect. tumor biochemistry Nat. Cancer Inst., NIH, Bethesda, Md., 1972-81, chief lab. molecular oncology, 1983—; dir. basic research program Frederick (Md.) Cancer Research Facility, 1983—. Contbr. articles to profl. jours. Served with U.S. Army, 1954-56. Mem. Am. Chem. Soc., Am. Soc. Microbiologists, Am. Soc. Virologists, AAAS, Sigma Xi. Subspecialties: Cancer research (medicine); Molecular biology. Current work: Molecular mechanisms of oncogenesis. Office: Nat Cancer Inst Bldg 4 Suite 100 Bethesda MD 20205

VAN DRIEL, HENRY MARTIN, physicist, educator; b. Breda, Netherlands, Dec. 27, 1946; s. John Cornelis and Anna Marie (Marijnissen) Van D.; m. Christine Marie Dent, June 27, 1970; children: Martin, Katherine, Peter. B.Sc., U. Toronto, 1970, M.Sc., 1971, Ph.D., 1975. Research assoc. U. Ariz., 1975-76; asst. prof. physics U. Toronto, 1976-81, assoc. prof. physics, 1982—; cons. optics cos. Contbr. articles to profl. jours. Mem. Can. Assn. Physics, Optical Soc. Am., Am. Phys. Soc. Roman Catholic. Subspecialty: Condensed matter physics. Current work: Laser interactions with semiconductors. Home: 386 Clarksville Ct Mississauga ON Canada L5A 1G8 Office: Dept Physics U Toronto Toronto ON Canada M5A 1A7

VAN DYKE, CECIL GERALD, botany educator, plant pathologist; b. Effingham, Ill., Feb. 4, 1941; s. Cecil Garrett and Vera Alice (Madden) Van D.; m. Susan Davis, Apr. 2, 1969; children: Anna Kate, Amanda Jane, Emily Lynn, Sallie Elaine. B.S. in Edn, Eastern Ill. U., 1963; M.S., U. Ill., 1966, Ph.D., 1968. Research asst. and assoc. U. Ill., 1963-68; research assoc., instr. N.C. State U., Raleigh, 1968-69, asst. prof., 1969-78, assoc. prof., 1978—; lectr. on creation. Contbr. articles on fungi to profl. jours. Active Pro-Family Leadership. Ciba-Geigy Corp. grantee, 1982. Mem. Mycological Soc. Am., Am. Phypath. Soc., Southeastern Biol. Control Working Group, Creation Research Soc. Republican. Mem. Assemblies of God Ch. Subspecialties: Plant pathology; Biological control. Current work: Biological control of weeds with fungal pathogens, ultrastructure of host-parasite fungus-plant interactions with electron microscopy. Home: 1612 Lorraine Rd Raleigh NC 27607 Office: Botany Dept NC State U 4205 Gardner Hall Raleigh NC 27650

VAN DYKE, KNOX, pharmacologist, toxicologist, educator, researcher, consultant; b. Chgo., June 23, 1939; s. Peter Alexander and Marjory Eleanor (Horter) Van D.; m. Anita Takahashi, July 15, 1961 (div. 1971); children: Cynthia Joyce, Christopher Jon, Richard Kenneth, Teri Lynn, Mark Alexander; m. Cynthia Ann Rogers, May 7, 1978; children: Teri Lynn, Mark Alexander. A.B. in Chemistry, Knox Coll., 1961; Ph.D. in Biochemistry (NIH fellow), St. Louis U., 1966. Postdoctoral fellow W.Va. U. Med. Sch., Morgantown, 1966-69, asst. prof. pharmacology, 1969-73, assoc. prof., 1973-77, prof. pharmacology-toxicology, 1977—; cons. in chemiluminescence tech.; W.Va. U. rep. Oak Ridge Asso. Univs., 1972-75. Contbr. over 150 articles to profl. publs., chpts to books. Mem. Am. Soc. Pharmacology and Exptl. Therapeutics, Am. Photobiology Soc., Sigma Xi. Methodist. Subspecialties: Cellular pharmacology; Toxicology (medicine). Current work: Inflammation, free radicals, chemiluminescence, cancer inflammation, malaria, glycoproteins. Home: 106 Morgan Dr Morgantown WV 26506 Office: W Va U Med Center Morgantown WV 26506

VAN DYKE, MILTON DENMAN, educator; b. Chgo., Aug. 1, 1922; s. James Richard and Ruth (Barr) Van D.; m. Sylvia Jean Agard Adams, June 16, 1962; children: Russell B., Eric J., Nina A., Brooke A. and Byron J. and Christopher M. (triplets). B.S., Harvard, 1943; M.S., Calif. Inst. Tech., 1947, Ph.D., 1949. Research engr. NACA, 1943-46, 50-54, 55-58; vis. prof. U. Paris, France, 1958- 59; prof. aero. Stanford, 1959—; cons. aerospace industry, 1949—; pres. Parabolic Press. Author: Perturbation Methods in Fluid Mechanics, 1964, An Album of Fluid Motion, 1982. Served with USNR, 1944-46. Guggenheim and Fulbright fellow, 1954-55. Mem. Am. Acad. Arts and Scis., Nat. Acad. Engring., Am. Phys. Soc., Phi Beta Kappa, Sigma Xi, Sierra Club. Subspecialties: Fluid mechanics; Aeronautical engineering. Current work: Perturbation methods in fluid mechanics. Home: 506 Campus Dr Stanford CA 94305 Office: Div Applied Mechanics Stanford Univ Stanford CA 94305

VAN DYKE, RUSSELL AUSTIN, pharmacologist, biochemist, consultant; b. Rochester, N.Y., Feb. 8, 1930; s. Russell H. and Geneva A. Van D.; m. Sally L. Olsen, July 28, 1956; children: Ceclia, Linda. B.S., Hope Coll., 1951; M.S., U. Mich., 1953; Ph.D., U. Ill., 1960. Research assoc. U. Colo., Denver, 1960-61; biochemist DOW Chem. Co., Midland, Mich., 1961-68; instr. biochemistry Saginaw Valley Coll., Saginaw, Mich., 1964-68; cons. Mayo Clinic, Rochester, Minn., 1968—; cons. in field. Author: Metabolism of Volatile Anesthetics: Implications for Toxicity, 1977. Served with U.S. Army, 1954-56. NIH grantee. Mem. Am. Soc. Pharmacology and Exptl. Therapeutics, Soc. Toxicology, Am. Soc. Anesthesiologists, Assn. Univ. Anaesthetists. Subspecialties: Biochemistry (medicine); Anesthesiology. Current work: Drug metabolism and microsomal enzyme systems. Home: 1735 Walden Ln SW Rochester MN 55901 Office: Mayo Clinic 200 1st St SW Rochester MN 55905

VAN ETTEN, HANS D., plant pathology educator, researcher; b. Peoria, Ill., Sept. 16, 1941; s. Cecil Herman and Freeda (Byl) Van E.; m. Janet Susan, Aug. 18, 1941; children: Erica Lynn, Laura Nadine. B.A., Wabash Coll., 1963; M.S., Cornell U., 1966, Ph.D., 1970. Tech.

assoc. dept. plant pathology Cornell U., Ithaca, N.Y., 1967-68, asst. prof. plant pathology, 1970-76, assoc. prof., 1976—. Alexander von Humboldt fellow, 1978-79; Fulbright-Hays fellow, 1978-79. Mem. Am. Phytopathology Soc., Phytochemistry Soc. N.Am., AAAS, Am. Soc. Microbiology. Subspecialty: Plant pathology. Current work: Disease physiology; molecular mechanism of pathogenesis. Home: 147 Starks Rd Newfield NY 14867 Office: Dept Plant Pathology Cornell U Ithaca NY 14853

VAN ETTEN, JAMES LEE, microbiologist; b. Cherrydale, Va., Jan. 7, 1938; s. Cecil H. and Frieda (Byl) Van E.; m. Arvalyn R. Van Etten, June 12, 1960 (div.); 1 son, Robert James. B.A., Carleton Coll., 1960; M.S., U. Ill., Urbana, 1963, Ph.D., 1965. NSF postdoctoral fellow dept. genetics U. Pavia, Italy, 1966; asst. prof. dept. plant pathology U. Nebr., Lincoln, 1966-69, assoc. prof., 1969-73, prof., 1973—. Contbr. articles to sci. jours. Mem. AAAS, Am. Phytopath. Soc., Am. Microbiol. Soc., Soc. Gen. Microbiology. Subspecialties: Developmental biology; Virology (biology). Current work: Microbiological biochemistry. Home: 5820 Pawnee St Lincoln NE 68506 Office: Dept Plant Pathology U Nebr Lincoln NE 68583

VAN HARTESVELDT, CAROL JEAN, psychology and neuroscience educator; b. Upper Darby, Pa., Oct. 1, 1940; d. Carroll Henry and Margaret (Str) Van H. B.A., Oberlin Coll., 1962; M.A., U. Mich., 1963; Ph.D., U. Rochester, 1968. Psychology research assoc. VA Hosp., Denver, 1968-70; asst. prof. psychology and neurosci. U. Fla., 1970-74, assoc. prof., 1974—, asst. dean, 1977-78, co-dir., 1984—. Author: Study Guide for Psychology: The Science of Behavior, 1971; contbr. articles, chpts. to profl. publs. Woodrow Wilson fellow, 1963. Mem. Soc. Neurosci., Internat. Soc. Devel. Psychobiology, Phi Beta Kappa, Sigma Xi. Subspecialties: Physiological psychology; Psychobiology. Current work: Devel. neurotransmitters and behavior; role of neurotransmitters and hormones in basal ganglia and behavior. Home: 60 NW 44th Terr Gainesville FL 32607 Office: Dept Psychology U Fla Gainesville FL 32601

VAN HECKE, GERALD RAYMOND, chemist; b. Evanston, Ill., Nov. 1, 1939; s. Joseph Michael and Sylvia Alina (Brygarri) Van H. B.S., Harvey Mudd Coll., 1961; A.M., Princeton U., 1963, Ph.D., 1966. Chemist Shell Devel. Co., Emeryville, Calif., 1966-70; mem. faculty dept. chemistry Harvey Mudd Coll., Claremont, Calif., 1970—, prof. chemistry, 1980—; Nat. Acad. Scis. exchange scientist, Poland, 1980. Contbr. articles to profl. jours. NASA summer faculty fellow, 1982. Fellow Am. Inst. Chemists; mem. Am. Chem. Soc., AAAS, Royal Inst. Chemistry (London), N.Y. Acad. Scis., Sigma Xi. Subspecialties: Thermodynamics; Condensed matter physics. Current work: Physical properties especially thermodynamics of liquid crystals. Home: 3667 N Walnut Grove Rosemead CA 91770 Office: 12th and Dartmouth Claremont CA 91711

VAN HORN, HUGH MOODY, theoretical astrophysicist; b. Williamsport, Pa., Mar. 5, 1938; s. Robert Dix and Virginia Elizabeth (Moody) Van H.; m. Mary Sue Boon, Sept. 17, 1960; children: Kathleen, Mary, Michael. B.Sc., Case Inst. Tech., 1960; Ph.D., Cornell U., 1965. Research assoc. U. Rochester, N.Y., 1965-67, asst. prof. physics and astronomy, 1967-72, assoc. prof., 1972-77, prof., 1977—, dept. chmn., 1980—. Contbr. articles in field to profl. jours. Mem. Am. Astron. Soc., Internat. Astron. Union, AAAS. Republican. Subspecialty: Theoretical astrophysics. Current work: Teaching and research in structure and properties of white dwarfs, neutron stars and accretion disks. Office: University of Rochester Dept Physics and Astronomy U Rochester Rochester NY 14627

VAN HOUTEN, JUDITH LEE, geneticist, educator; b. Paterson, N.J., Apr. 26, 1948; d. Charles D. and Jean T. Post; m. John C. Van Houten, Dec. 23, 1967; 1 dau., June H. Student, Elizabethtown Coll., 1965-68, U. Del., 1969-70; B.S., Pacific Luth. U., 1972; Ph.D., U. Calif., Santa Barbara, 1976. NIH postdoctoral researcher dept. pharmacology U. B.C., 1977-79; vis. asst. prof. U. Iowa, Iowa City, 1979-80; asst. prof. dept. zoology U. Vt., Burlington, 1980—. Contbr. articles to profl. publs. Grantee NIH, NSF. Mem. Soc. Neurosci., Am. Soc. Cell Biology, European Chemoreception Research Orgn., Assn. Chemoreception Scis., N.Y. Acad Scis., AAAS, Soc. Protozoologists, Sigma Xi. Subspecialties: Developmental biology; Membrane biology. Current work: Chemoreception and related membrane phenomena in normal and mutant paramecia. Research in genetic dissection, membrane biology, ciliary motility, sensory transduction im microorganisms; teaching of general, neurobehavioral, cell cycle and developmental genetics. Office: Dept Zoology U Vt Burlington VT 05405

VAN HOVEN, GERARD, physics educator; b. Los Angeles, Nov. 23, 1932; s. Aaron and Josephine (Hoppenjans) Van H; m. Barbara Hoxie, Dec. 29, 1956; children: Enid, Ian. B.S., Calif. Inst. Tech., 1954; Ph.D., Stanford U., 1963. Mem. tech. staff Bell Telephone Labs., 1954-56; electron physicist Gen. Electric Co., Palo Alto, Calif., 1956-63; research assoc. W.W. Hansen Labs. Physics, Stanford U., 1963-65; research physicist Inst. Plasma Research, 1965-68; asst. prof. to assoc. prof. U. Calif.-Irvine, 1968-79, prof. physics, 1979—; Fulbright fellow Vienna Tech. U., 1963-64; cons. Gen. Electric Co., 1963-65, Varian Assocs., 1965-68, Smithsonian Astrophys. Obs., 1975, Aerospace Corp., 1975-77; Langley-Abbot vis. scientist Ctr. Astrophysics, Harvard U., 1975; vis. astrophysicist Oss Astrofiscio Arectri, 1976, 82. Contbr. articles to profl. jours. Mem. Am. Phys. Soc., Am. Astron. Soc., Internat. Astron. Union. Subspecialties: Plasma physics; Theoretical astrophysics. Current work: Plasma instabilities in solar physics. Office: Dept Physics U Calif Irvine CA 92717

VAN KAMMEN, DANIEL PAUL, psychiatrist; b. Dordrecht, Netherlands, Aug. 26, 1943; s. Daniel Paul and Christina Petronella (Rinse) Van K.; m. Welmoet Bok, Apr. 4, 1970; 1 dau., Marleen. M.D., Utrecht (Netherlands) Med. Sch., 1966, Ph.D., 1978. Diplomate: in psychiatry Royal Dutch Specialty Bd. Unit chief sect. of neuropsychopharmacology Biol. Psychiatry br., NIMH, Bethesda, Md., 1975—. Contbr. articles to profl. jours. Mem. Am. Psychiat. Assn., Biol. Psychiatry, AAAS, Soc. Neuroscience, Am. Coll. Neuropsychopharmacology, Internat. Soc. Psycho Neuroendocrinology, Collegium Internationale Neuropsychopharmacologicum., Psychiat. Soc. (hon. mem.). Subspecialties: Psychiatry; Psychopharmacology. Current work: Schizophrenia, biochemistry, pharmacology, drug response. Office: 9000 Rockville Pike NIMH Bldg 10 Room 4N214 Bethesda MD 20205

VAN KLEY, HAROLD JAMES, biochemist; b. Chgo., Mar. 7, 1932; s. Jacob and Dora (Dekker) Van K.; m. Helen Priscilla Hawks, Sept. 12, 1959; children: Cynthia, Michael. Student, U. Ill., Chgo., 1949-50; A.B., Calvin Coll., 1953; M.S., U. Wis., Madison, 1955, Ph.D., 1958. Instr., Edward A. Doisy dept. biochemistry St. Louis U. Sch. Medicine, 1958-61, asst. prof., 1961-82; dir. biochem. research St. Mary's Health Center, St. Louis, 1967-82; prof. chemistry Trinity Christian Coll., Palos Heights, Ill., 1982—. Contbr. articles to profl. jours. Trustee Covenant Theol. Sem., 1962—, sec., 1963—. NIH predoctoral fellow, 1956-57. Mem. Am. Chem. Soc., Am. Soc. Biol. Chemists, AAAS, Am. Pancreatic Assn., N.Y. Acad. Sci., Sigma Xi. Presbyterian. Subspecialties: Biochemistry (biology); Clinical chemistry. Current work: Investigations of biochem. changes as markers for disease, especially altered structure and function of enzymes and proteins. Office: 6601 W College Dr Palos Heights IL 60463

VAN LOON, GLEN RICHARD, physician, scientist; b. Petrolia, Ont., Can., Mar. 1, 1940; came to U.S., 1983; s. John E. and Margaret B. (Thomson) Van L.; m. Joye L. Gerry, June 30, 1962. B.S., McMaster U., Hamilton, Ont., 1961; M.D., U. Toronto, 1965; Ph.D., U. Calif.-San Francisco, 1970. Intern Toronto Gen. Hosp., 1965-66, resident, 1971-72; jr. asst. resident Royal Victoria Hosp., Montreal, 1966-67; fellow in endocrinology Vanderbilt U., 1970-71; asst. prof. medicine and physiology U. Toronto, 1972-79, assoc prof., 1979-82; prof. medicine U. Ky., Lexington, 1982—; staff physician Toronto Gen. Hosp., 1972-80, U. Ky. Med. Ctr., Lexington, 1983—, VA Med. Ctr., 1983—. Contbr. articles to profl. jours. and books. Fellow Royal Coll. Physicians Can.; mem. Am. Fedn. Clin. Research, Am. Soc. Pharmacology and Exptl. Therapeutics, Am. Physiol. Soc., Endocrine Soc., Soc. Neuroscience, Can. Soc. Clin. Investigation, Can. Physiol. Soc., Can. Soc. Endocrinology Metabolism, Central Soc. Clin. Research, Pharmacol. Soc. Can., Internat. Soc. Neuroendocrinology. Subspecialties: Neuroendocrinology; Neuropharmacology. Current work: interest: Endogenous opioid peptides, catecholamines, Acth, regulation of responses to stress. Home: Route 4 Keene Pike Nicholasville KY 40356 Office: VA Medical Center (111) Lexington KY 40511

VAN NAGELL, JOHN RENSSELAER, JR., physician, educator; b. N.Y.C., Sept. 16, 1939; s. John Rensselaer and Rosamond Porter (Musgrave) van N.; m. Elizabeth Clemens Gay, June 10, 1965; children: John, Lucy, Elizabeth Knox. B.A., Harvard Coll., 1961; M.D., U. Pa., 1967. Diplomate: Am. Bd. Ob-Gyn. Am. Cancer Soc. fellow in gynecologic oncology, 1971-73; asst. prof. U. Ky. Med. Center, 1971-74, assoc. prof. gynecologic oncology, 1974-76, prof., dir. gynecologic oncology div., 1978—, Am. Cancer Soc. prof. clin. oncology, 1980—. Contbr. articles to profl. jours.; Author: Modern Concepts of Gynecologic Oncology, 1982. Bd. dirs. Ephraim McDowell Cancer Center, Lexington; trustee U. Ky. Hosp., 1978—. Served with USNR, 1968-75. Mem. Am. Assn. Cancer Research, Am. Assn. Clin. Oncology, Soc. Pelvic Surgeons, Soc. Gynecologic Oncology, Internat. Soc. Pathologists, Alpha Omega Alpha. Republican. Episcopalian. Clubs: Fly, Idle Hour Country, Iroquois Hunt (Lexington). Subspecialties: Gynecological oncology; Cancer research (medicine). Current work: Tumor markers in gynecologic malignancies; in vitro chemotherapy testing. Home: 226 Henry Clay Blvd Lexington KY 40502 Office: 800 Rose St Lexington KY 40536

VANONI, VITO AUGUST, hydraulic engr.; b. Calif., Aug. 30, 1904; s. Battitta and Mariana V.; m. Edith Maria Fulcinella, June 23, 1934. B.S. in Civil Engring, Calif. Inst. Tech., 1926, M.S., 1932, Ph.D., 1940. Supr. research lab. U.S. Soil Conservation Service, Pasadena, Calif., 1935-47; asst. prof. hydraulics Calif. Inst. Tech., 1942-49, asso. prof., 1949-55, prof., 1955-74, prof. emeritus, 1974—. Contbr. numerous articles on hydraulics and sedimentation to profl. jours.; editor: Sedimentation Engineering, 1975. Mem. ASCE (hon.), Internat. Assn. Hydraulic Research, Nat. Acad. Engring., Sigma Xi. Subspecialty: Civil engineering. Current work: River mechanics. Home: 3545 Lombardy Rd Pasadena CA 91107 Office: 1201 E California St Pasadena CA 91125

VAN PEENEN, PETER F(RANZ) D(IRK), physician, epidemiologist; b. Pensacola, Fla., Sept. 18, 1932; s. Hubert J. and Mavis M. (Warner) Van P.; m. Eleanor J., Apr. 26, 1958; children: Martha, Dirk, Peter, Charles. A.B. (Regional scholar), Princeton U., 1953; M.S. (Gov. Pardee scholar), U. Calif., San Francisco, 1956, M.D., 1957; M.P.H., Johns Hopkins U., 1959, Dr.P.H., 1960. Diplomate: Am. Bd. Preventive Medicine; Am. Bd. Occupational Medicine. Intern U.S. Naval Hosp., San Diego, 1957-58; commd. ensign U.S. Navy, 1956, advanced through grades to capt., 1972; served in, Vietnam, 1965-66, Indonesia, 1970-74, Taiwan, 1974-76, ret., 1978; prof. preventive medicine, chmn. dept. Uniformed Services U. Health Scis., 1976-79; epidemiologist Mich. div. Dow Chem. Co., Midland, 1979-82; dir. epidemiology Standard Oil Co. (Ind.), Chgo., 1982—. Author monographs; contbr. numerous articles to profl. jours. Decorated Legion of Merit, Bronze Star with combat V. Mem. Am. Pub. Health Assn., AMA, Am. Occupational Medicine Assn., Soc. Epidemiol. Research. Democrat. Congregationalist. Club: Dial. Subspecialties: Epidemiology; Preventive medicine. Current work: Epidemiology, tropical medicine, industrial epidemiology. Home: 233 E Wacker Dr Apt 1013 Chicago IL 60601 Office: 200 E Randolph Dr Box 5910-A Chicago IL 60601

VAN SCHMUS, WILLIAM RANDALL, geology educator; b. Aurora, Ill., Oct. 4, 1938; s. William George and Laura Jean (McKinstry) Schmus; m. Edna Jean Evison, June 30, 1961; children: Brian R., Derek R., Jennifer L. B.S., Calif. Inst. Tech., 1960; Ph.D., UCLA, 1964. Asst. prof. geology U. Kans., Lawrence, 1967-70, assoc. prof., 1970-75, prof., 1975—. Assoc. editor: Geol. Soc. Am. Bull, 1976—; contbr. articles to profl. jours. Active Boy Scouts Am., 1972-82. Served to 1st lt. USAF, 1964-67. Decorated Air Force Commendation medal; NSF grantee, 1967—. Mem. Geol. Soc. Am., Geochem. Soc., Am. Geophys. Union, Geol. Assn. Can. Subspecialties: Geochemistry; Geology. Current work: Application of geochronology and isotyope geochemistry to understanding continental evolution during Precambrian times. Office: Dept Geology U Kans Lawrence KS 66045 Home: 813 W 27th Terr Lawrence KS 66044

VANSPEYBROECK, LEON PAUL, astrophysicist; b. Wichita, Kans., Aug. 27, 1935; m., 1959; 3 children. B.S., MIT, 1957, Ph.D., 1965. Research assoc. in high energy physics MIT, 1965-67; staff scientist x-ray astronomy Am. Sci. & Engring., Inc., Mass., 1967-74; staff scientist Center Astrophysics, Cambridge, Mass., 1974—. Subspecialty: 1-ray high energy astrophysics. Office: Center for Astrophysics 60 Garden St Cambridge MA 02138

VAN TAMELEN, EUGENE EARLE, chemist, educator; b. Zeeland, Mich., July 20, 1925; s. Gerrit and Henrietta (Vanden Bosch) van T.; m. Mary Ruth Houtman, June 16, 1951; children—Jane Elizabeth, Carey Catherine, Peter Gerrit. A.B., Hope Coll., 1947, D.Sc., 1970; M.S., Harvard, 1949, Ph.D., 1950; D.Sc., Bucknell U., 1970. Instr. U. Wis., 1950-52, from asst. to asso. prof., 1952-59, prof., 1959-61, Homer Adkins prof. chemistry, 1961-62; prof. chemistry Stanford, 1962—, chmn. dept., 1974-78; Am.-Swiss Found. lectr., 1964. Mem. editorial adv. bd.: Chem. and Engring. News, 1968-70, Synthesis, 1969—, Accounts of Chem. Research, 1970-73; editor: Bioorganic Chemistry, 1971—. Recipient A.T. Godfrey award, 1947; G. Haight traveling fellow, 1957; Guggenheim fellow, 1965, 73; Leo Hendrik Baekeland award, 1965; Prof. Extraordinarius, Netherlands, 1967-73. Mem. Am. Chem. Soc. (Pure Chemistry award 1961, Creative Work in Synthetic Organic Chemistry award 1970), Am. Acad. Arts and Scis., Nat. Acad. Sci. Subspecialties: Organic chemistry; Synthetic chemistry. Current work: Total synthesis of natural products; new reactions; bioorganic chemistry. Home: 23570 Camino Hermoso Los Altos Hills CA 94022 also Office: 1 Smugglers Cove Cap Estate Castries St Lucia West Indies also Sealodge Princeville Hanalei Kauai HI 96714 Office: Dept Chemistry Stanford U Stanford CA 94305

VAN'T HOF, JACK, research cytologist; b. Grand Rapids, Mich., Apr. 11, 1932; s. Jacob and Martha (Folkersma) Van't H.; m. Nancy Elizabeth Hyma, Dec. 1, 1932; children: Thomas, Steven. B.A., Calvin Coll., 1957; Ph.D., Mich. State U., 1961. Biologist Hanford Labs., Richland, Wash., 1961-62; research assoc. Brookhaven Nat. Lab., Upton, N.Y., 1962-64, asst. cytologist, 1964-65, cytologist, 1966-80, sr. cytologist, 1980—; asst. prof. U. Minn., Mpls., 1965-66. Served with U.S. Army, 1953-55. Mem. Am. Soc. Cell Biology, Genetics Soc. Am., Radiation Research Soc., N.Y. Acad. Scis. Subspecialties: Cell and tissue culture; Plant cell and tissue culture. Current work: Molecular biology of plant cell division, chromosome structure and DNA replication. Home: 6 Trout Ponds Ct Brookhaven NY 11719 Office: Biology Dept Brookhaven Nat Lab Upton NY 11973

VANT-HULL, LORIN LEE, research physicist, educator, consultant; b. Matlock, Iowa, June 26, 1932; s. John A. and Bessie Anita (Visser) Vant-H.; m. Mary Evelyn Prunty, Feb. 4, 1955; children: Julia Maureen, Barry P. J., Brian L.C. B.S. in Physics, U. Minn., 1954, M.S., UCLA, 1955, Ph.D., Calif. Inst. Tech., 1966. Scientist Hughes Research Lab., Malibu, Calif., 1954-59; sr. research scientist Ford Sci. Lab., Newport Beach, Calif., 1966-69; assoc. prof. physics U. Houston, 1969-77, solar div. head energy lab., prof.physics, 1977—; program mgr. Dept. Energy sponsored Solar Thermal Advanced Research Ctr. Contbr. articles to profl. jours. Asst. scoutmaster Boy Scounts Am. NSF grantee, 1973-77. Mem. Am. Phys. Soc., Internat. Solar Energy Soc., Sigma Xi, Phi Kappa Phi, Phi Beta Kappa. Subspecialties: Solar energy; Low temperature physics. Current work: Co-originator of modern solar central receiver system for which extensive computer codes are developed, maintained and applied to the system design optimization, and performance analysis of heliostat fields and the heliostat-receiver interaction. Patentee in field. Office: 4800 Calhoun 113 SPA Houston TX 77004

VAN TIENHOVEN, ARI, animal physiology educator; b. The Hague, Netherlands, Apr. 22, 1922; s. Adrianus Baltus and Wilhelmina Hendrika (Mulder) van T.; m. Annie van Haselen, Jan. 26, 1923. Landbouwkundig Ingenieur, Agrl. U., Wageningen, The Netherlands, 1949; M.S., U. Ill., 1951, Ph.D., 1953. Asst. prof. animal physiology Miss. State Coll., 1953-55; asst. prof. Cornell U., 1955-61, assoc. prof., 1961-67, prof., 1967—. Author: Reproductive Physiology of Vertebrates, 1968, 82. Mem. Bd. Edn., Ithaca City Sch. Dist., 1969-72, pres., 1971-72; bd. drs. Econ. Opportunity Corp. of Tompkins County, 1957-59, 82—. NATO fellow, 1961-62. Fellow AAAS; mem. Soc. Study Reproduction, Am. Assn. Anatomy, Animal Soc. Zoology, Poultry Sci. Assn., Societe de Biologie Bordeaux. Subspecialties: Reproductive biology; Animal physiology. Current work: Endocrinology of ovulation in birds; thermoregulation in birds; biological rhythm of chickens. Home: 9 Hudson Pl Ithaca NY 14850 Office: Cornell U Rice Hall Ithaca NY 14853

VAN UITERT, LEGRAND GERARD, chemist; b. Salt Lake City, May 6, 1922; s. Antone and Lambertha Maria (Groeneveld) Van U.; m. Marion Emma Woolley, June 8, 1945; children: Robert, Bonnie, Craig. B.S., George Washington U., 1949; M.S., Pa. State U., 1951, Ph.D., 1952. Union Carbide fellow Pa. State U., 1951-52; materials scientist Bell Telephone Labs., Murray Hill, N.J., 1952—; cons. to materials adv. bd. Nat. Acad. Sci. Served with USN, 1940-46. Co-recipient W.R.G. Baker award IEEE, 1971; recipient H. N. Potts award Franklin Inst., 1975; award Indsl. Research Inst., 1976; internat. award for new materials Am. Phys. Soc., 1981. Mem. Nat. Acad. Engring., Am. Chem. Soc. (award for creative invention 1978), Sigma Xi. Subspecialties: Materials; Theoretical chemistry. Current work: Inovative semiconductor procesing, optical fibers; luminescence and the correlation of physical properties of crystals and glasses. Research on microwave ferrites, lasers, bubble domain memory, electro, non-linear and acusto-optic materials, luminescence, optical fibers, crystal growth, passive displays, semicondr. for processing and correlation of properties of matter. Patentee in field. Home: 2 Terry Dr Morristown NJ 07960 Office: Bell Labs Murray Hill NJ 07974

VAN VALKENBURG, MAC ELWYN, engineering educator; b. Union, Utah, Oct. 5, 1921; s. Charles Mac and Nora (Walker) Van V.; m. Evelyn J. Pate, Aug. 27, 1943; children: Charles Mac II, JoLynne, Kaye, David R., Nancy J., Susan L. B.S. in Elec. Engring, U. Utah, 1943; M.S., Mass. Inst. Tech., 1946; Ph.D., Stanford, 1952. With Radiation Lab., Mass. Inst. Tech., 1943-45, Research Lab. Electronics, 1945-46; mem. faculty U. Utah, 1946-55, U. Ill., 1955-66; prof. elec. engring. Princeton, 1966-74, chmn. dept., 1966-72; prof. elec. engring. U. Ill., 1974—; vis. prof. Stanford, U. Colo., U. Calif., Berkeley, U. Hawaii, Manoa, 1978-79, U. Ariz., 1982-83. Author: Network Analysis, 3d edit, 1974, Introduction to Modern Network Synthesis, 1960, Introductory Signals and Circuits, 1967, Signals in Linear Circuits, 1974, Circuit Theory: Foundations and Classical Contributions, 1974, Linear Circuits, 1982, Analog Filter Design, 1982; editor-in-chief: IEEE Press, 1983—. Mem. IEEE (v.p., dir. 1969-73, editor transactions 1960-63, proc. 1966-69, Edn. medal 1972), Am. Soc. Engring. Edn. (George Westinghouse award 1963, Benjamin Garver Lamme award 1978, Guillemin prize 1978), Nat. Acad. Engring., Sigma Xi, Tau Beta Pi, Phi Kappa Phi. Subspecialties: Computer-aided design; Computer engineering. Home: 902 S Lincoln Ave Urbana IL 61801 Office: U Ill Dept Elec Engring 1406 W Green St Urbana IL 61801

VAN VLIET, CAROLYN MARINA, educator, physicist; b. Dordrecht, Netherlands, Dec. 27, 1929; came to U.S., 1960, naturalized, 1967; d. Marinus and Jacoba (de Lange) Van V.; divorced; children: Elsa Marianne, Mark Edward, Cynthia Joyce, Renata Annette Carolina. B.S., Free U. Amsterdam, Netherlands, 1949, M.A., 1953, Ph.D. in Physics, 1956. Research fellow Free U. Amsterdam, 1950-54, research asso., 1954-56, asst. dir., 1958-60; postdoctoral fellow U. Minn., Mpls., 1956-57, mem. faculty, 1957-58, 60-70, prof. elec. engring. and physics, 1965-70; sr. research mem., prof., centre recherches math. U. Montreal, Que., 1969—; vis. prof. U. Fla., 1974, 4-6 months annually, 78—. Contbg. author: Fluctuation Phenomena in Solids, 1965; author numerous articles. Research grantee NSF, Air Force OSR, Nat. Sci. and Engring.; Research Council, Ottawa. Fellow Am. Sci. Affiliation; mem. Am., European, Dutch, Canadian phys. socs., IEEE (sr.). Subspecialties: 3emiconductors; Statistical physics. Current work: Nonequilibrium statistical mechanics; theoretical solid state physics; fluctuations and stochastic processes; solid state device theory. Home: 514-4 SW 34th St Gainesville FL 32607 The purpose of life is to honor God and to serve mankind.

VAN WART, HAROLD EDGAR, chemist; b. Bay Shore, N.Y., Oct. 29, 1947; s. Harold Edgar and Anna (Prygocki) Van W.; m. Jane Menges, Nov. 30, 1975; children: Sarah, Laura, Andrew. B.A., SUNY-Binghamton, 1969; M.S., Cornell U., 1971, Ph.D., 1974. Temporary asst. prof. Cornell U., Ithaca, N.Y., 1974-75; postdoctoral fellow Harvard Med. Sch., Boston, 1975-78; asst. prof. Fla. State U., Tallahassee, 1978-82, assoc. prof., biochemistry, 1982—, dir. molecular biophysics Ph.D. program, 1982—. NIH trainee Cornell U., 1971-74; NIH postdoctoral fellow Harvard U., 1975-78; recipient Career Devel. award NIH, 1982-87. Mem. Am. Chem. Soc., Am. Soc. Biol. Chemists. Subspecialties: Biophysical chemistry; Biochemistry (biology). Current work: Enzymology, protein chemistry, Raman spectroscopy. Office: Dept Chemistry Fla State U Tallahassee FL 32306

VAN WOERT, MELVIN HOLMES, neuropharmacologist; b. Bklyn., Nov. 3, 1929; s. Melvin and Lottie (Kramer) Van W.; m. Wanda Mae Stoker, Nov. 21, 1933; children: Peter, David. B.A., Columbia U., N.Y.C., 1951; M.D., N.Y. Med. Coll., 1956. Intern, then resident in internal medicine U. Chgo., 1956-60; research asst. in gastroenterology, 1962-63; asst. scientist Brookhaven Nat. Lab., Upton, N.Y., 1963-66, assoc. scientist, 1966-67; asst. prof. Yale U. Med. Sch., New Haven, 1967-71, assoc. prof., 1971-74; prof. medicine and pharmacology Mt. Sinai Med. Sch., N.Y.C., 1974-78, prof. neurology and pharmacology, 1978—; med. adviser Nat. Myoclonus Found.; adv. com. Tourette Syndrome Assn.; med. dir. Nat. Orgn. for Rare Disorders. Contbr. articles to profl. jours.; editorial bd.: Jour. Clin. Neuropharmacology, 1981—. Served with AUS, 1960-62. NIH grantee. Fellow ACP; mem. Am. Soc. Pharmacology and Exptl. Therapeutics, AAAS, Soc. Neurosci., Soc. Neuroschemistry, N.Y. Acad. Sci. Subspecialty: Neuropharmacology. Home: 252 Highbrook Ave Pelham NY 10803 Office: Dept Neurology Mt Sinai Med Sch 1 Gustave L Levy Pl New York NY 10029

VANZANT, KENT LEE, palynologist; b. Humboldt, Nebr., July 5, 1947; s. Evan S. and Ruth (Gibson) Van Z.; m. Nancy K. Patton, June 19, 1971. B.A., Earlham Coll., 1969; M.S., U. Iowa, 1973, Ph.D., 1976. Tchr. sci. Scattergood Sch., West Branch, Iowa, 1969-71; asst. prof. geology Beloit (Wis.) Coll., 1976-78, Earlham Coll., Richmond, Ind., 1978-81; petroleum paleontologist Amoco Prodn. Co., Denver, 1981—. Contbr. articles in field to profl. jours. Research Corp. grantee, 1977, 1979; Nat. Acad. Sci. grantee, 1976, 80. Mem. Geol. Soc. Am., Am. Assn. Stratigraphic Palynologists, Internat. Orgn. Paleonotanists, Iowa Acad. Sci. Quaker. Subspecialties: Paleontology; Paleoecology. Current work: Cenozoic palynology. Office: Amoco Prodn Co 1670 Broadway Denver CO 80202

VARGA, JANOS M., biochemist, researcher; b. Hungary, June 19, 1935; s. Janos and Maria (Toth) V.; m. Eva Pierrou, Oct. 15, 1969; children: Daniel, Paul, Elisabet. B.S., U. Tech., Budapest, 1959; Ph.D., U. Sci., Budapest, 1966. Research assoc. Inst. Pharmacology, Budapest, 1960-67; research asso. Royal Inst. Tech., Stockholm, 1967-69; sect. head dept. biochemistry, Uppsala, Sweden, 1970; postdoctoral asso. Yale U., 1971-74, asst. prof., 1974-76, asso. prof., 1976—; vis. scientist Inst. Microbiol. Chemistry, Rome, 1963-64. Contbr. numerous articles to profl. jours. Postdoctoral fellow USPHS, 1971-74; Research Career Devel. award, 1976-81. Mem. Am. Assn. Immunologists, Internat. Pigment Cell Soc. Subspecialties: Receptors; Immunology (agriculture). Current work: Chemistry, specificity and cell cycle dependence on cell surface receptors in eukaryotic cells. Patentee in field.

VARGAS, JOHN DAVID, mathematics educator, text reviewer and writer; b. Bronx, N.Y., Nov. 9, 1947; s. Paul and Elsie (Withenshaw) V.; m. Kathy Mary Capasso, June 8, 1974. B.S., Hunter Coll., 1969; M.S., N.Y. U., 1971, Adelphi U., 1982, Ph.D., 1983. Instr. Hostos Community Coll. CUNY, Bronx, N.Y., 1972-77, asst. prof., 1977-80; asst. prof. math. Mercy Coll., Dobbs Ferry, N.Y., 1980—; text reviewer Prentice Hall, Englewood Cliffs, N.J., 1982, Acad. Press, N.Y.C., 1980-82. Co-author: Solutions Manual - Calculus, 1984. Served with Army N.G., 1969-75. Teaching fellow N.Y. U., 1969-71; Sloan Found. fellow Adelphi U., 1975-77. Mem. Soc. Indsl. and Applied Math., Math. Assn. Am., Am. Math. Soc., N.Y. Acad. Scis., N.Y. State Math. Assn. Two-Yr. Colls. (campus rep.), Pi Mu Epsilon. Democrat. Roman Catholic. Subspecialties: Applied mathematics; Solid mechanics. Current work: Application of integral transform techniques to solve problems in wave propagation. Home: 36 Irving Ave Floral Park NY 11001 Office: Mercy Coll 555 Broadway Dobbs Ferry NY 10522

VARGHESE, SANKOORIKAL LONAPPAN, physicist, educator, cons.; b. Narakal, India, Mar. 13, 1943; came to U.S., 1966; s. Lonappan C. and Anna L. Kanjooparampil S.; m. Leela Varghese, Jan. 21, 1971; children: Teena-Ann, Emma-Betty, Geena-Mary, Binu-John. B.Sc. in Physics, Kerala U., 1963, M.Sc., 1965, M.S., U. Louisville, 1967, Ph.D., Yale U., 1974. Postdoctoral research staff physicist Yale U., New Haven, 1974; research assoc. Kans. State U., 1974-76; vis. asst. prof. U. Okla., 1976-77, East Carolina U., 1977-80; asst. prof. physics U. South Ala., Mobile, 1980-81, assoc. prof., 1981—; cons. Burroughs Wellcome Co., Greenville, N.C. Contbr. articles and abstracts on physics and automation to profl. jours. Oak Ridge Assoc. Univ. travel grantee, 1981—; also various univ. grants. Mem. Am. Phys. Soc., Sigma Xi, Sigma Pi Sigma. Subspecialties: Atomic and molecular physics; Integrated circuits. Current work: Experimental atomic physics; automation. Patentee in field. Home: 5722 Long Meadow Rd Mobile AL 36609 Office: Physics Dept U South Ala Mobile AL 36688

VARKEY, ALEXANDER, biology educator; b. Tiruvalla, Kerala, India, July 27, 1934; came to U.S., 1966, naturalized, 1982; s. Puthenparampil and Karukakkalathil (Aley) V.; m. Mercy Alexander Varkey, Nov. 26, 1962; 1 son, Daniel. B.Sc., St. Berchmans' Coll., India, 1954; M.Sc., Agra Coll., India, 1958; Ph.D., La. State U., 1973. Lectr. U. Kerala, 1958-59; head biology Cuttington Coll., Monrovia, Liberia, 1961-66; instr. St. Thomas Coll., St. Paul., 1967-68; teaching asst. La. State U., Baton Rouge, 1968-73; asst. prof. W.Va. Wesleyan Coll., Buckhannon, 1973-77; prof. Liberty Bapt. Coll., Lynchburg, Va., 1977—; research asst. Govt. India Dept. Fisheries, Mandapam, 1959-61. Author: A Laboratory Study of Vertebrate Zoology, 1982, Comparative Cranial Myology of North American Natricine Snakes, 1979. Recipient Merit cert. Cuttington Coll., 1966. Mem. Am. Soc. Zoologists, Herpetol. League, Soc. Study of Amphibians and Reptiles. Baptist. Subspecialties: Morphology; Systematics. Current work: Cranial myology of snake genus Helicops, Conophanes. Home: 103 Hillview Dr Lynchburg VA 24502 Office: PO Box 2000 C Lynchburg VA 24506

VARLASHKIN, PAUL GREGORY, physics educator; b. San Antonio, Aug. 28, 1931; s. Peter and Sara Lottie (Riggs) V.; m. Charlotte Duke, Aug. 8, 1953; children: Peter G., Paula A., Charlotte M., Carol S. B.S., U. Tex., 1952, M.A., 1954, Ph.D., 1963. Chief research and devel. Electro-Mechanics Co., Austin, Tex., 1963-64; research assoc. dept. physics U. N.C., Chapel Hill, 1964-66; prof. physics La. State U., Baton Rouge, 1966-72, East Carolina U., Greenville, N.C., 1972—. Contbr. articles to profl. jours. Tex. Instruments Found. fellow, 1963. Mem. N.C. Acad. Sci. (co-dir. students chpt.), Am. Phys. Soc., Sigma Xi. Subspecialties: Atomic and molecular physics; Condensed matter physics. Current work: Pedagogy, chemical physics. Home: 305 Prince Rd Greenville NC 27834 Office: Dept Physics East Carolina U Greenville NC 27834

VARLEY, RONALD ARTHUR, nuclear engineer; b. Buffalo, Aug. 26, 1949; s. Arthur and Jeanette Louise (Black) V.; m. Rejenia Ann McKinney, Sept. 28, 1969; children: Heather Lynn, Dawn Marie. Student, Naval Nuclear Power Sch., 1970-71. Tng. engr. Westinghouse Hanford Co., Richland, Wash., 1977-81; sr. emergency planning specialist Energy Cons., Inc., Harrisburg, Pa., 1981-83; sr. engr. EDS Nuclear, Melville, N.Y., 1983—; Adv., bd. dirs. Columbia Basin Coll., Pasco, Wash., 1977-81. Served with USN, 1969-77. Mem. Am. Nuclear Soc., Scientists and Engrs. for Secure Energy. Subspecialties: Nuclear engineering; Nuclear fission. Current work: Developing emergency planning concepts, programs, procedures and facilities for the nuclear energy. Home: 20 Wilson St Port Jefferson Station NY 11776 Office: EDS Nuclear 225 Broad Hollow Rd Melville NY 11747

VARMA, ARVIND, chemical engineering educator, researcher; b. Ferozabad, India, Oct. 13, 1947; came to U.S., 1968, naturalized, 1979; s. Hans Raj and Vijay L. (Jhanjhee) V.; m. Karen K. Guse, Aug. 7, 1971; children: Anita, Sophia. B.S. in Chem. Engring, Panjab U., Chandigarh, India, 1966, M.S., U. N.B. (Can.), Fredericton, 1968, Ph.D., U. Minn.-Mpls., 1972. Asst. prof. U. Minn.-Mpls., 1972-73; sr. research engr. Union Carbide Corp., Tarrytown, N.Y., 1973-75; asst. prof. chem. engring. U. Notre Dame, 1975-77, assoc. prof., 1977-80, prof., 1980—, chmn. dept. chem. engring., 1983—; vis. prof. U. Fla., 1980, U. Wis.-Madison, 1981; Chevron vis. prof. Calif. Inst. Tech., 1982. Research numerous publs. in field; editor: (with R. Aris) The Mathematical Understanding of Chemical Engineering Systems, 1980. Nalco Found. grantee, 1976-81; various indsl. orgns. grantee, 1977-82; NSF grantee, 1979-80; Dept. Energy grantee, 1978-82. Mem. Am. Inst. Chem. Engrs., Am. Chem. Soc., Soc. Indsl. and Applied Math., Sigma Xi. Subspecialty: Chemical engineering. Current work: Chemical and catalytic reaction engineering; kinetics and catalysis; mathematical modeling. Home: 421 Tonti South Bend IN 46617 Office: Dept Chem Engring U Notre Dame Notre Dame IN 46556

VARNER, REED WILLIAM, chemical company executive; b. Columbus, Ohio, Oct. 29, 1916; s. Charles Oscar and Lulu (Hildebr) V.; m. Helen Olga Prockiw, May 4, 1946; children: Sandra L., Elaine M., Nancy B. Student, Ohio State U., 1934-37; B.S.F., U. Mich., 1939, M.F., 1941, Ph.D., 1951. Biologist E. I. duPont de Nemours, Wilmington, Del., 1948-51, tech. field specialist, 1951-53, field research investigator, 1953-55, research supr., 1955-67, field research mgr., 1968—. Served to lt. USN, 1943-46. Mem. Am. Phytopathology Soc., Weed Soc. Am. Republican. Presbyterian. Club: Del. Camera. Lodge: Masons (Wilmington). Subspecialties: Integrated pest management; Plant pathology. Current work: Manager research and development in crop protection chems. Patentee selective tropical crop herbicides. Home: 109 Warwick Dr Wilmington DE 19803 Office: Barley Mill Plaza Wilmington DE 19898

VARNOS, HAROLD ELIOT, physician, educator; b. Oceanside, N.Y., Dec. 18, 1939; s. Frank and Beatrice (Barasch) V.; m. Constance Louise Casey, Oct. 25, 1969; children: Jacob, Christopher. B.A., Amherst Coll., 1961; M.A., Harvard U., 1962; M.D., Columbia U., 1966. Intern, resident dept. medicine Presbyn. Hosp., N.Y.C., 1966-68; clin. assoc. NIH, Bethesda, Md., 1968-70; lectr. dept. microbiology and immunology U. Calif.-San Francisco, 1970-72, asst. prof., 1972-74, assoc. prof., 1974-79, prof., 1979; cons. Chiron Corp., Emoryville, Calif. Editor/contbr.: Molecular Biology and Tumor Viruses, 1982, Selected Readings in Tumor Virology, 1983; assoc. editor: Cell. Jour, Virology Jour. Macy Found. Josiah Macy fellow, 1978-79; recipient Lasker Found. award, N.Y.C., 1982, Passano Found. award, Balt., 1983; named Mus. Sci. and Tech. Calif. Scientist of Yr. Los Angeles, 1982. Mem. AAAS, Am. Soc. Microbiology, Am. Soc. Virology. Subspecialties: Virology (biology); Cancer research (medicine). Current work: Replication of animal viruses; tumor induction by viruses and cellular oncogenes. Home: 956 Ashbury St San Franciso CA 94117 Office: Dept Microbiology and Immunology U Calif Med Sch San Francisco CA 94143

VARUGHESE, POTHEN, chemistry educator, researcher; b. Thiruvalla, Kerala, India, Aug. 4, 1936; s. Varughese and Mariamma (Pothen) Thomas; m. Keiko Agnes Matsumoto, Aug. 9, 1968; children: Lily Jean, Michelle Marie. B.Sc., U. Kerala, 1956; M.Sc., U. Saugar, Madhya Prades, India, 1959; Ph.D., Kent State U., 1974. Lectr. Union Christian Coll., Alwaye, Kerala, 1959-61; lectr. Cuttington Coll., Suacoco, Liberia, 1962-68; teaching fellow Kent State U., 1968-73, instr. chemistry, 1973-74; asst. prof. Indiana U. of Pa., 1974-80, assoc. prof., 1980—. Mem. Am. Chem. Soc., Pa. Acad. Sci., Sigma Xi. Presbyterian. Lodge: Kiwanis. Subspecialties: Organic chemistry; Medicinal chemistry. Current work: Design and synthesis of antimicrobial compounds; study of organic reactions using polymer-supported reagents, catalysts and under phase-transfer catalytic conditions. Home: 145 Concord St Indiana PA 15701 Office: Indiana U of Pennsylvania Indiana PA 15705

VASEEN, VESPER ALBERT, cons. san engr., inventor; b. Denver, Sept. 13, 1917; s. Albert and Ruby Cornelia (Weisz) V.; m. June Lily Novak, Feb. 2, 1941; children. P.E., Colo. Sch. Mines, 1939; tchrs. cert., Denver U., 1941; postgrad. in san engring. and public health, U. Mich., 1941; student, Denver U., 1941-43, 57-58; postgrad. in civil engring. U. Colo., 1946-48; student in Arctic engring. U. Alaska, 1977-78, also other spl. courses; D.Sc. (hon.), U. Del Norte, Chile, 1981. Registered profl. engr., Colo., Kans., Utah, Pa., Alaska. Asst. state san. engr. Colo. Bd. Health, 1941-43; partner Ripple and Howe, Inc. (Cons. Engrs. (Utilities)), 1946-55, pres., 1955-66, Ogallilla Inc., Colo., 1962-66; v.p. engring. Chart Devel. Corp., Jefferson County, Colo., 1963-66, Tyrol-Devco Corp., Jefferson County, 1963-66; with Indsl. Environ. Co., 1966-71; v.p. Weiss Bakery Inc., Denver, 1966-78; process/project engr. Steirns Roger Corp., Denver, 1966-80; pres. AVASCO, Wheat Ridge, Colo., 1978—, Technometrics Inc., Wheat Ridge, 1982—; spl. cons. Insitu Tech. Inc., 1980, 81; mem. Nat. Def. Exec. Res., Dept. Commerce, 1961-78. Contbr. articles to profl. jours. Served as capt. AUS, 1941-46. Recipient cert. accomplishment Denver C. of C., 1959, cert. merit, 1960, 61 (2), 62; cert. accomplishment Internat. Graphoanalysis Soc., Inc., 1962; master in graphoanalysis cert., 1964; cert. appreciation U. Colo. Water and Sewage Plant Operators Sch. Council, 1968; cert. award Evergreen (Colo.) High Sch., 1974. Mem. Am. Water Works Assn. (pres. Rocky Mountain sect. 1953, dir. sect. 1954-57, life mem.), Water Pollution Control Fedn. (trustee Rocky Mountain sect. 1956-58, life mem.), Inter Am. Assn. San. Engring. Republican. Lodges: Masons; Rosicrucians. Subspecialties: Civil engineering; Environmental engineering. Numerous patents in environ. improvement, solar energy, public health, plant tissue culture.

VASEK, FRANK C(HARLES), botany educator, researcher; b. Maple Heights, Ohio, May 9, 1927; s. Frank John and Florence Marie (Kraus) V.; m. G. Maxine McClelen, Aug. 28, 1954; children: Cheryl Denise, Cynthia Louise. B.S., Ohio U., 1950; Ph.D., UCLA, 1955. Instr. botany U. Calif.-Riverside, 1954-56, asst. prof. botany, 1956-63, assoc. prof., 1963-69, prof., 1969—; cons. in field. Contbr. tech. articles to nat., regional, fgn. profl. jours. Served to pfc. USMC, 1945-46. Mem. Bot. Soc. Am., Ecol. Soc. Am., Soc. Study Evolution, Calif. Bot. Soc., Am. Soc. Plant Taxonomists, Sierra Club, Nature Conservancy, Solar Lobby, Common Cause, Friends of Earth, Wilderness Soc., Environ. Def. Fund, Environ. Task Force, Nat. Parks Conservation Assn. Subspecialties: Systematics; Evolutionary biology. Current work: Evolution and adaptation in Claria; age and development of clones in Larrea; vegetation of California and Western U.S. Office: Dept Botany and Plant Sci U Calif Riverside CA 92521

VASIL, INDRA KUMAR, research biologist, educator, cons.; b. Basti, U.P., India, Aug. 31, 1932; came to U.S., 1962, naturalized, 1974; s. Lal Ch and Pushpalata (Abrol) V.; m. Vimla Negi, May 15, 1959; children: Kavita, Charu. B.S., Banaras Hindu U., 1952; M.S., U. Delhi, 1954, Ph.D., 1958. Assoc. prof., prof. botany U. Fla., Gainesville, 1967-79, grad. research prof. botany, 1979—; cons. Contbr. over 150 articles on biology to profl. jours. Recipient Sr. U.S. Scientist award (Humboldt award) Fed. Republic Germany, 1975-76. Mem. Bot. Soc. Am., Internat. Assn. Plant Tissue and Cell Culture, Internat. Assn. Plant Morphologists, Nat. Geog. Soc., AAAS. Subspecialties: Genetics and genetic engineering (agriculture); Plant cell and tissue culture. Current work: Plant cell and tissue culture for genetic improvement of crop plants. Home: 4901 NW 19th Pl Gainesville FL 32605 Office: Dept Botany U Fla Gainesville FL 32611

VASILAKIS, JOHN DIMITRI, mech. engr., researcher, educator; b. Somerville, Mass., Sept. 23, 1938; s. Dimitrios John and Penelope (Vasilakis); m. Anne Ruth, Aug. 25, 1963; children: Dimitri John, Damon William, Penne Chunae. B.S.M.E., Northeastern U., 1961; M.S. in Mechanics, Rensselaer Poly. Inst., 1963; Ph.D. in Theoretical Mechanics, Rensselaer Poly. Inst., 1967. Engr. Gen. Electric, Fitchburg, Mass., summer 1961, Raytheon Co., Bedford, Mass., summer 1962; postdoctoral researcher Rensselaer Poly. Inst., 1967-69, adj. prof. mech. engring., 1976—; engr. Watervliet Arsenal, N.Y., 1969—. Contbr. articles to profl. jours. Pres. council St. Basil Greek Orthodox Ch., Troy, 1982. Mem. ASME, Soc. Exptl. Stress Analysis, Sigma Xi, Tau Beta Pi, Pi Tau Sigma. Subspecialties: Theoretical and applied mechanics; Numerical analysis. Current work: Thermo-elastic-plastic stress analysis as applied to response of multilayered gun tubes and to determination of residual stress states after quencing. Office: Watervliet Arsenal Watervliet NY 12189

VASINGTON, PAUL JOHN, microbiologist, bus. exec.; b. Norwich, Conn., June 6, 1927; s. Berardino and Angela (Romano) V.; m. Shirley M. Vasington, Sept. 11, 1954; children: Ann Marie, Paul Bernard, Carol Lynn. B.S., U. Conn., 1956; M.S., U. Md., 1959, Ph.D., 1961. Asst. prof. microbiology St. John's U., 1960-61; dir. research and devel. Flow Labs., 1961-64; head dept. virus testing Lederle Labs., Pearl River, N.Y., 1964-65, mgr. biol. quality control, 1965-68, mgr. biol. mfg., 1968-70, gen. mgr., 1970-75; v.p., gen. mgr. Damon Biotech, Needham Heights, Mass., 1975—; pres. Karyon Tech. Inc., Norwood, Mass., 1983—. Contbr. articles in field to profl. jours. Mem. Sch. Bd., Pearl River, N.Y., 1973-75; bd. dirs. YMCA, Nyack, N.Y., 1970. Served with U.S. Army, 1945-47; Served with USAF, 1950-51. Mem. Am. Soc. Microbiology, Am. Assn. Clin. Chemists, N.Y. Acad. Sci. Republican. Roman Catholic. Club: Hatherly Country (Scituate, Mass.). Subspecialties: Artificial organs; Virology (biology). Current work: Biotechnology, virology and microencapsulation of living tissues for genetic engring., transplantation, fermentation. Office: 115 4th Ave Needham Heights MA 02194

VASSALLE, MARIO, physiology educator, physician; b. Viareggio, Lucca, Italy, May 26, 1928; came to U.S., 1958, naturalized, 1973; s. Giuseppe and Antonietta V.; m. Anna Maria Petrucci, Sept. 7, 1959; children: Andrew G., Alessandra A., Massimo B., Roberto M., Francesca A. M.D. cum laude, U. Pisa, 1953, diploma in cardiology cum laude, 1955. Diplomate: in medicine, Italy. Intern dept. medicine, cardiology U. Pisa Med. Sch., 1953-55, asst. dept. med. pathology, 1956-58; acting chief resident in medicine French Hosp., N.Y.C., 1958-59; trainee cardiovascular research and tng. program, dept. physiology Med. Coll. Ga., Augusta, 1959-60; fellow physiology SUNY-Downstate Med. Ctr., Bklyn., 1960-61, N.Y. Heart Assn. fellow, 1961-62, faculty mem., 1964—, prof. physiology, 1971—; NIH fellow U. Bern, Switzerland, 1962-64; vis. prof. U. Ferrara, Italy, 1971, U. Vt., 1978; bd. dirs. N.Y. Heart Assn., N.Y.C., 1980—; mem. study sect. NIH, Bethesda, Md., 1981—. Editor: Research in Physiology, 1971, Cardiac Physiology for the Clinician, 1976, Excitation and Neural Control of the Heart, 1982; assoc. editor: Am. Jour. Physiology - Heart and Circulatory Physiology, 1976-80; editorial bd.: Circulation Research, 1974-80; editorial cons.: European Jour. Pharmacology, 1979—; contbr. chpts. to books, numerous articles, abstracts, revs. in field to profl. jours. Fulbright travel grantee, 1958; recipient French Hosp. Alumni award, 1959, A. and A. Sinsheimer Fund award, 1966-71, N.Y. Health Research Council award, 1972-75. Mem. Am. Physiol. Soc., N.Y. Acad. Scis., Am. Heart Assn. (bd. dirs.), AAAS, Harvey Soc., Cardiac Muscle Soc., Cardiac Electrophysiol. Group (pres. 1972-73), Internat. Study Group for Research in Cardiac Metabolism, Sigma Xi. Roman Catholic. Subspecialty: Physiology (medicine). Current work: Study of normal and abnormal cardiac automaticity with various techniques. Home: 104 Huntington Rd Port Washington NY 11050 Office: Dept Physiology SUNY-Downstate Med Ctr 450 Clarkson Ave Brooklyn NY 11203

VAUGHAN, DAVID SHERWOOD, mechanical engineer; b. Attleboro, Mass., Jan. 2, 1923; s. John George and Golda Faye (Sherwood) V.; m. Carol Brining, Dec. 22, 1946; children: Barbara Vaughan Levy, William Stewart. B. Engring. M.E., Yale U., 1947; M.S., Case Inst. Tech., 1953. Registered profl. engr., Ohio, N.Y. Engring. draftsman Fairbanks Morse & Co., Three Rivers, Mich., 1947-48; engring. technician White Motor Co., Cleve., 1948-51; sr. engr. Goodyear Aircraft Corp., Akron, Ohio, 1951-54; project engr. Paragon Gear Works, Taunton, Mass., 1954-55, Facet Enterprises, Elmira, N.Y., 1955—. Chmn. Erin (N.Y.) Zoning Bd. Appeals, 1971-81. Served to tech. sgt. USAAF, 1943-45. Decorated D.F.C., Air Medal. Mem. Soc. Automotive Engrs., ASME. Republican. Methodist. Subspecialties: Mechanical engineering; Theoretical and applied mechanics. Current work: Design and development electromagnetic clutches and brakes. Patentee electromagnetic clutches. Home: Box 253 RD 1 Erin NY 14838 Office: Facet Enterprises Elmira NY 14903

VAUGHAN, JOHN HEATH, physician, medical researcher; b. Richmond, Va., Nov. 7, 1921; s. Warren Taylor and Emma (Heath) V.; m. Marjorie Seybold (div.); children: John, Nancy, David, Margaret. A.B. cum laude, Harvard U., 1942, M.D., 1945. Diplomate: Am. Bd. Internal Medicine. Intern Peter Bent Brigham Hosp., Harvard U. Med. Sch., Boston, 1945-46, research fellow, 1948-50, sr. asst. resident in medicine, 1950-51; asst. prof. medicine Med. Coll. Va., Richmond, 1953-58; assoc. prof. medicine, asst. prof. bacteriology and immunology U. Rochester (N.Y.) Med. Sch., 1958-63, prof. medicine, head immunological and infectious diseases, 1963-70; adj. prof. medicine U. Calif.-San Diego, La Jolla, 1970—; chmn. clin. divs. Scripps Clinic and Research Found., La Jolla, 1970-74, chmn. dept. clin. research, 1974-77, head div. clin. immunology, 1977—. Editor: Immunological Diseases, 3d edit, 1978, Dermatology in General Medicine, 1971; contbr. articles to profl. jours. Served with U.S. Army, 1946-48. NRC fellow, 1951-53. Mem. Am. Acad. Allergy (pres. 1966-67), Am. Assn. Immunologists, Am. Clin. and Climatological Assn., ACP, Am. Fedn. for Clin. Research, Am. Rheumatism Assn. (pres. 1970-71), Am. Soc. for Clin. Investigation, Assn. Am. Physicians, Infectious Diseases Soc., San Diego County Med. Soc., Western Assn. Physicians, Western Soc. for Clin. Research, Alpha Omega Alpha. Subspecialties: Allergy; Immunology (medicine). Current work: Emphasis on etiology and pathogenesis of rheumatoid arthritis, a disease apparently involving a partial deficiency in the immune system which controls infection with certain viruses, especially the Epstein-Barr virus. Office: 10666 N Torrey Pines Rd La Jolla CA 92037

VAUGHAN, WILLIAM WALTON, atmospheric scientist; b. Clearwater, Fla., Sept. 7, 1930; s. William Walton and Ella Vermelle (Warr) V.; m. Wilma Geraldine Stapleton, Dec. 23, 1951; children: Stephen W., David A., William D., Robert T. B.S., U. Fla., 1951; Ph.D., U. Tenn., 1976. Sci. asst. Air Force Armament Center, Eglin AFB, Fla., 1955-58; Army Ballistic Missile Agy., Huntsville, Ala., 1958-60; chief aerospace environ. div. Marshall Space Flight Center, NASA, Huntsville, 1960-76, chief atmospheric scis. div., 1976—.

Contbr. articles to profl. jours. Served with USAF, 1951-55. Recipient Exceptional Service medal NASA, 1971. Mem. Am. Meteorol. Soc., AIAA (Losey Atmospheric Scis. award 1980), Am. Geophys. Union, AAAS, Sigma Xi. Subspecialties: Meteorology; Aerospace engineering and technology. Current work: Atmospheric dynamics and space technology applications to atmospheric science problems; natural environment inputs for aerospace engineering applications. Office: Marshall Space Flight Center NASA Huntsville AL 35812

VAUGHN, CLARENCE BENJAMIN, physician, educator; b. Phila., Dec. 14, 1928; s. Albert and Aretha (Johnson) V.; m. Sarah Campbell, Sept. 18, 1953; children: Stephen, Annette, Carl, Ronald. B.S., Benedict Coll., 1951; M.S., Howard U., 1955, M.D., 1957; Ph.D., Wayne State U., 1965. Intern D.C. Gen. Hosp., Washington, 1957-58; resident Freedman's Hosp., Washington, 1958-59; lab. asst. in chemistry Benedict Coll., Columbia, S.C., 1945-51; grad. asst. chemistry dept. Howard U., Washington, 1951-53, research asst. biochemistry dept., 1954-56; chemist geology and petrology br. Dept. Interior, Washington, 1953-55; NIH spl. research fellow Wayne State U. Sch. Medicine, Detroit, 1962-64, lab. instr. dept. physiol. chemistry, 1963-67, asst. prof. oncology, 1967-78, clin. assoc. prof., 1967—, assoc. dept. biochemistry, 1967-81; research physician Milton A. Darling Meml. Center, Mich. Cancer Found., 1964-70, clin. dir., 1970-72; mem. HEW public adv. com., Washington, 1976—; med. dir. oncology service Providence Hosp., 1973—; cons. med. staff dept. medicine Kirwood Gen. Hosp., 1967—, Oakwood Hosp., 1968—, Detroit Meml. Hosp., 1970—, Crittenton Hosp., 1972—; quadrangle affiliate staff dept. medicine Sinai Hosp. Detroit, 1977—; courtesy med. staff dept. medicine Grace Hosp., 1973—, Hutzel Hosp., 1975—, jr. attending physician dept. medicine, 1975—, Southfield Rehab. Ctr., 1976—. Contbr. articles in field to profl. jours. Pres. Wayne County unit; bd. dirs. Am. Cancer Soc. Served with USAF, 1959-61. Fellow Am. Coll. Clin. Pharmacology; mem. AAAS, Am. Chem. Soc. ACP, AAUP, AMA, Am. Soc. Clin. Oncology, Am. Soc. Preventive Oncology, Nat. Med. Assn. (nat. chmn. aerospace and mil. medicine sect. 1980), S.W. Oncology Study (prin. investigator), Mich. Assn. Med. Edn., N.Y. Acad. Scis., Oakland County Med. Soc. (constitution and bylaws com.), Wayne County Med. Soc., Res. Officers Assn. (v.p.). Episcopalian. Subspecialty: Chemotherapy. Current work: Chemotherapy, hormonal therapy, organic acid metabolism, tissue ferritin.

VAUGHN, JAMES LLOYD, microbiologist; b. Marshfield, Wis., Mar. 2, 1934; s. Lloyd Robert and Irene Opal (Tague) V.; m. Jeanette Darlene Lustig, June 11, 1955; children: Susan, Katherine, Michael, David. B.S. in Bacteriology, U. Wis., Madison, 1957, M.S., 1959, Ph.D., 1962. Research asso. U. Wis., Madison, 1960-61; research officer Can. Dept. Forestry, Sault Ste. Marie, Ont., 1961-65; research microbiologist Dept. Agr., Beltsville, Md., 1965-74, supervisory microbiologist, 1974-79, lab. chief, 1979—; mem. adj. faculty W. Alton Jones Cell Sci. Center, Lake Placid, N.Y. Mem. Am. Soc. Microbiology, Soc. Invertebrate Pathology, Tissue Culture Assn. Subspecialties: Tissue culture; Integrated pest management. Current work: The study of replication of viruses pathogenic to insects and the devel. of insect cell culture tech. for use in basic research and as prototypes for comml. prodn. of insect viruses for use in integrated pest mgmt. Office: Dept Agr Room 214 Bldg 011A BARC-W Beltsville MD 20705

VAUPEL, DONALD BRUCE, pharmacologist; b. Hackensack, N.J., Aug. 30, 1942; s. Donald Fackert and Ruth Irma (Guenther) V.; m. Susan Skinner, Aug. 12, 1967; children: Lisa Marie, Jonathan Bruce. B.A., Wittenberg U., 1964; M.S., U. Ky., 1970, Ph.D., 1974. Chemist quality control sect. Am. Cyanamid Co., Lederle Labs., Pearl River, N.Y., 1964-66; pharmacologist clin. neuropharmacology and whole animal pharmacology sects. Nat. Inst. on Drug Abuse, Addiction Research Ctr., Lexington, Ky., 1972—; asst. adj. prof. dept. pharmacology U. Ky. Coll. Medicine. Contbr. articles on pharmacology to profl. jours. Mem. Am. Soc. Pharmacology and Exptl. Therapeutics, Sigma Xi. Episcopalian. Subspecialty: Neuropharmacology. Current work: Evaluating the pharmacology of drugs of abuse, especially hallucinogens, opiates, and amphetamines, and determining their abuse liability on the basis of pharmacologic equivalence. Home: 125 N Arcadia Park Lexington KY 40503 Office: PO Box 12390 Lexington KY 40583

VAURIO, JUSSI KALERVO, nuclear engineer, research administrator; b. Lapua, Finland, Mar. 4, 1940; came to U.S., 1975; s. Juho Kullervo and Rauha Anni (Hietala) V.; m. Eija K. Hyotylainen, June 27, 1963; children: Sari, Katja, Lena. Diploma engr., Tech. U., Helsinki, 1967, licentiate of tech., 1970, D.S. in Tech, 1971. Tchr., lectr. Tech. U., Espoo, Finland, 1965-72; operating engr. Reactor Lab., Espoo, 1966-68; researcher, engr. Tech. Research Ctr. of Finland, Espoo, 1968-72; nuclear engr. Imatra Power Co., Helsinki, 1972-75, Argonne (Ill.) Nat. Lab., 1975-80, program mgr., 1980—; alternate mem. Radioation Protection Commn., Helsinki, 1967-69. Editor: Dictionary Nuclear Engineering Terms, 1972; contbr. articles to profl. jours. Bd. dirs. Student Congregation, Espoo, 1965, League Finnish Am. Socs., Chgo., 1978-82; mem. engring. bd. City of Espoo, 1973-75. Served to lt. Finnish Army, 1959-60. Finnish Cultural Found. fellow, 1969; Emil Aaltonen Found. fellow, 1970; grantee, 1968; Jenny and Antti Wihuri Found. fellow, 1971. Mem. Finnish Nuclear Soc., Am. Nuclear Soc., Soc. Risk Analysis. Lutheran. Club: Coalition Party (Espoo). Subspecialties: Nuclear engineering; Probability. Current work: Reliability analysis models and methods, probabilistic risk assessment, man-machine interface technology methodology. Office: Argonne National Laboratory Bldg 207 9700 S Cass Ave Argonne IL 60439

VAVRA, TERRY GWYN, advertising agency executive; b. Los Angeles, Mar. 22, 1941; s. Marvin Joseph and Gwen Charlotte (Filipy) V.; m. Linda Faye Dallas, Dec. 19, 1970; children: Stacy Dallas, Kerry Lynn, Tammy Gwen. B.S., UCLA, 1964, M.S., 1967; Ph.D. in Mktg, U. Ill., Champaign, 1973. Dir. news audience research NBC, N.Y.C., 1972-77; group research dir. Kenyon & Eckhardt Advt., N.Y.C., 1977-79; dir. research Batten, Barton, Durstine & Osborn, Internat., N.Y.C., 1979-81; dir. mktg./research Levine, Huntley, Schmidt & Beaver, N.Y.C., 1981—; cons. Rolls-Royce Motors, Lyndhurst, N.J. Commr.-soccer Allendale (N.J.) Athletic Assn., 1982—; treas. Bergen Highland United Methodist Ch., Upper Saddle River, N.J., 1979-81. Recipient Pub. Service award Retinitis Pigmentosa Found., 1981. Mem. Am. Psychol. Assn., Am. Mktg. Assn., Assn. for Consumer Research, Psychometric Soc. Republican. Current work: Quantification of the impact of brand image on brand selection in consumer purchasing of package goods; micro computers for collecting survey data. Home: 4 Michele Ct Allendale NJ 07401

VAZIRI, NOSTRATOLA DABIR, internist, nephrologist, educator; b. Tehran, Iran, Oct. 13, 1939; came to U.S., 1969, naturalized, 1977; s. Abbas and Tahera V. M.D., Tehran U., 1966. Diplomate: Am. Bd. Internal Medicine, Am. Bd. Nephrology. Intern Cook County Hosp., Chgo., 1969-70; resident Berkshire Med. Ctr., Pittsfield, Mass., 1970-71, Wadsworth VA Med. Ctr., 1971-72, UCLA Med. Ctr., 1972-74; prof. medicine U. Calif.-Irvine, 1979—, chief nephrology div., 1977—, dir. hemodialysis unit, 1977—, dir., 1980, vice chmn. dept. medicine, 1982—; mem. sci. adv. council Nat. Kidney Found., 1977—. Contbr. numerous articles to med. jours. Recipient Golden Apple award, 1977; named outstanding tchr. U. Calif-Irvine, 1975, 78, 79, 80, 82. Fellow ACP; mem. Am. Soc. Nephrology, Am. Paraplegia Soc., Alpha Omega Alpha. Subspecialties: Internal medicine; Nephrology. Current work: Pathophysiology of end-stage renal disease particularly in spinal cord injured patients, active in field acid-base metabolism and development of new dialysis modalities. Home: 66 Balboa Coves Newport Beach CA 92663 Office: Div Nephrology Dept Medicine Room C351 Med Sci I U Calif Irvine CA 92717

VEBER, DANIEL FRANK, research organic chemist; b. New Brunswick, N.J., Sept. 9, 1939; s. Frank B. and Agnes (Olam) V.; m. Marilyn Franck, Sept. 12, 1959; children: Paul D., David F. B.A., Yale U., 1961, M.S., 1962, Ph.D., 1964. With Merck Sharp & Dohme Research Labs., Rahway, N.J., 1964—, research fellow, 1966-72, sr. research fellow, 1972-73, West Point, Pa., 1973-75, sr. dir., 1980—, head. medicinal chemistry dept., 1982—; Mem. pharmacological scis. study sect. NIH, Bethesda, Md., 1981—, cons. contraceptive devel. br., 1981-82, mem. ad hoc bioorganic study sect., 1981; mem. planning com. Am. Peptide Symposium. Contbr. over 100 sci. articles to profl. publs. NIH fellow, 1962-64. Fellow N.Y. Acad. Scis.; mem. Am. Chem. Soc., AAAS, Sigma Xi. Lutheran. Subspecialties: Medicinal chemistry; Organic chemistry. Current work: Medicinal chemistry: synthesis of bioactive peptides, peptide conformation, bioactive peptides, peptide synthesis, diabetes research. Patentee in field. Office: Merck Sharp & Dohme Research Labs West Point PA 19486

VEBLEN, DAVID RODLI, earth sciences educator; b. Mpls., Apr. 27, 1947; s. Paul Frederick and Alic Irma (Hankey) Benton) V.; m. Sarah Elizabeth Clebsch, Dec. 11, 1976; 1 dau., Annie Elizabeth. B.A., Harvard U., 1969, M.A., 1974, Ph.D., 1976. Faculty research assoc. Ariz. State U.-Tempe, 1976-79, asst. prof., 1979-81, Johns Hopkins U., Balt., 1981-82, assoc. prof. dept. earth and planetary sci., 1982—. Editor: Amphiboles, 1981-82; Contbr. articles to profl. jours. Recipient award Mineral. Soc. Am., 1983; grantee NSF, 1979—. Fellow Mineral. Soc. Am.; mem. Am. Geophys. Union, AAAS, Mineral. Soc. Can., Electron Microscopy Soc. Am. Club: Johns Hopkins (Balt.). Subspecialties: Mineralogy; Crystallography. Current work: Solid-state reactions and structural disorder in rockforming minerals, high-resolution transmission electron microscopy and analytical electron microscopy. Office: Dept Earth and Planetary Scis Johns Hopkins U Baltimore MD 21218

VEGORS, STANLEY HENRY, JR., physics educator; b. Detroit, Jan. 5, 1929; s. Stanley Henry and Esther Eloise (Sharpe) V.; m. Ann Hope, Aug. 11, 1951; children: Eric, Susan, Heidi. B.A., Middlebury Coll., 1951; B.S., MIT, 1951; M.S., U. Ill., 1952, Ph.D., 1955. Research assoc. U. Ill.-Champaign, 1955-56; physicist Phillips Petroleum Co., Idaho Falls, Idaho, 1956-58; assoc. prof. physics Idaho State U., 1958-61, prof., 1961—, head and chmn. dept. physics, 1958-65; prof. U. Petroleum and Minerals, Dhahran, Saudi Arabia, 1982—; cons. EG&G Idaho, Idaho Falls, 1976-81. Contbr. numerous articles to profl. jours.; author profl. papers. Mem. Am. Phys. Soc., Internat. Solar Energy Soc., Phi Beta Kappa. Republican. Methodist. Clubs: Seagull Bay Yacht, Univ. Racquet (Pocatello, Idaho). Subspecialties: Nuclear physics; Solar physics. Current work: Low energy nuclear physics, especially nuclear waste disposal; solar energy, especially as applied to households; household energy conservation. Home: 59 Drake St Pocatello ID 83201 Office: Dept Physics U Petroleum and Minerals PO Box 144 Box 656 Dhahran Internat Airport Dhahran Saudi Arabia

VEIS, ARTHUR, biochemistry educator; b. Pitts., Dec. 23, 1925; s. Fred M. and Sarah (Landis) V.; m. Eve Zenner, June 24, 1951; children: Judith, Sharon, Deborah. B.S., U. Okla., 1947; Ph.D., Northwestern U., 1951. Instr. phys. chemistry U. Okla., 1951-52; research chemist Armour & Co., Chgo., 1952-60, head phys. chemistry dept., 1959-60; spl. instr. Crane Jr. Coll., Chgo., 1955-56, Loyola U., 1957-58; mem. faculty Northwestern U. Med. Sch., Chgo., 1960—, prof. biochemistry, 1965—, asst. dean, 1968-70, asso. dean, 1970-76, chmn. dept. oral biology, 1977—; Disting. vis. prof. U. Adelaide, Australia, 1980; chmn. Gordon Conf. on Structural Macromolecules: Collagen, 1981; chmn. dental insts. and spl. projects adv. com. Nat. Inst. Dental Research, 1974-78. Author: Macromolecular Chemistry of Gelatin, 1964; Editor: Biological Polyelectrolytes, 1970, Chemistry and Biology of Mineralized Connective Tissues, 1981; also articles. Served with USNR, 1943-46. Recipient Fogarty Sr. Internat. Scholar award, 1977; Guggenheim fellow, 1967; Case Centennial scholar, 1980; award Internat. Assn. Dental Research, 1981. Mem. Am. Chem. Soc., Am. Soc. Biol. Chemists, Biophys. Soc., N.Y. Acad. Scis., Sigma Xi, Phi Lambda Upsilon. Subspecialties: Biochemistry (biology); Biophysical chemistry. Current work: Research in connective tissue biology, particulary collagen structure and biomineralization. Patentee in field. Home: 7633 Lowell St Skokie IL 60076 Office: 303 E Chicago Ave Chicago IL 60611

VEITH, FRANK JAMES, surgeon, educator; b. N.Y.C., Aug. 29, 1931; m., 1954; 4 children. A.B., Cornell U., 1952, M.D., 1955. NIH fellow Harvard U. Med. Sch., 1963-64; asst. prof. surgery Cornell U., 1964-67; assoc. prof. Albert Einstein Coll. Medicine, 1967-71, prof. surgery, 1971—; co-dir. kidney transplant unit Montefiore Hosp., N.Y.C., 1967—, assoc. attending surgeon, 1967-71, attending surgeon and chief vascular surgery, 1972—; cons. Heart-Lung Project Com., 1971—. Recipient Career Scientist award Health Research Council, City of N.Y. and Montefiore Hosp., 1965-72; Markle scholar in acad. medicine Cornell U., Albert Einstein Coll. Medicine and Montefiore Hosp., 1964-69. Mem. Soc. Univ. Surgeons, Soc. Vascular Surgery, Am. Assn. Thoracic Surgery, Am. Surg. Assn., Transplantation Soc. Subspecialty: Transplant surgery. Office: Dept Surgery Montefiore Hosp and Med Ctr New York NY 10467

VELAMAKANNI, KRISHNAMURTY SEETARAMA, pharmacologist; b. India, Aug. 25, 1939; s. Lakshminaryana and Bhaskaramma V.; m. Vijaya, Dec. 21, 1975; children: Shona, Priya. B.Sc., Andhra U., 1961; M.Sc., M.S., U. Baroda, 1968, Ph.D., 1974. Postdoctoral fellow in cardiology, 1974-76; instr. dept. medicine Southwestern Med. Sch., U. Tex. Health Scis. Center, Dallas, 1976-78; Am. Heart Assn. Advanced fellow UCLA, 1978-79; asst. prof. pharmacology Tulane U., New Orleans, 1979—. Contbr. articles to profl. jours. Grantee Am. Heart Assn., 1976-78, NIH, 1979-82; La. Heart Assn., 1982-83. Mem. Am. Heart Assn., Soc. Pharmacology and Exptl. Therapeutics, N.Y. Acad. Scis. Hindu. Subspecialties: Pharmacology; Comparative physiology. Current work: Vascular research on ionic regulation, mechanical function and hormonal interactions. Home: 676 Wall Blvd Gretna LA 70053 Office: 1430 Tulane Ave New Orleans LA 70112

VELARDO, JOSEPH THOMAS, molecular biologist, endocrinologist, curricula consultant; b. Newark, Jan. 27, 1923; s. Michael Arthur and Antoinette (Iacullo) V.; m. Forresta M. Monica Power, Aug. 12, 1948 (dec. July 1976). A.B., No. Colo. U., 1948; S.M., Miami U., 1949; Ph.D., Harvard U., 1952. Teaching research fellow Harvard, 1949-52, research fellow in biology and endocrinology, 1952-53; research asso. in pathology Sch. Medicine, 1953-54; research asso. in surgery, 1954-55; asst. in surgery Peter Bent Brigham Hosp., Boston, 1954-55; asst. prof. anatomy and endocrinology Yale Sch. Medicine, 1955-61; prof. anatomy, chmn. dept. N.Y. Med. Coll., N.Y.C., 1961-62; cons. N.Y. Fertility Inst., 1961-62; dir. Inst. for Study Human Reprodn., Cleve., 1962-67; prof. biology John Carroll U., Cleve., 1962-67; mem. research and edn. divs. St. Ann Obstetric and Gynecologic Hosp., Cleve., 1962-67, head dept. research, 1964-67; prof. anatomy dept. Stritch Sch. Medicine Loyola U., Chgo., 1967-79; chmn., 1967-73; mem. med. adv. bd. Barren Found., 1973—; cons. internat. basic and biomed. curricula, 1973—. Editor, contbr. to: Endocrinology of Reproduction, 1958; editor, contbr. to: Essentials of Human Reproduction, 1958; cons. editor, co-author: The Uterus, 1959; contbg. author: The Ovary, 1963, The Ureter, 1967, rev. edit., 1981; co-editor, contbr.: Biology of Reproduction, Basic and Clinical Studies, 1973; Co-author: Histochemistry of Enzymes in the Female Genital System, 1963. Served with USAAF, 1943-45. Recipient award Lederle Med. Fac. Awards Com., 1955-58; named hon. citizen of Sao Paulo, Brazil, 1972; U.S. del. Vatican 6th Internat. Congress on Animal Reprodn., 1964. Fellow AAAS, N.Y. Acad. Scis., Gerontol. Soc., Pacific Coast Fertility Soc. (hon.); mem. Am. Assn. Anatomists, Am. Soc. Zoologists, Am. Physiol. Soc., Endocrine Soc., Soc. Endocrinology (Gt. Britain), Soc. Exptl. Biology and Medicine, Am. Soc. Study Sterility (Rubin award 1954), Internat. Fertility Assn., Pan Am. Assn. Anatomy, Midwestern Soc. Anatomists (pres. 1973-74), Mexican Soc. Anatomy (hon.), Sigma Xi, Kappa Delta Pi, Phi Sigma, Gamma Alpha, Alpha Epsilon Delta. Club: Harvard (Chgo.). Subspecialties: Cell biology (medicine); Neuroendocrinology. Current work: Hormonal effects on cellular mechanisms, effects of hormones on reproductive mechanisms. Home: E Wilson Ave and Cherry Ln Old Grove East Lombard IL 60148 Office: 607 E Wilson Rd Lombard IL 60148 Success is best highlighted by the invincible instruments of truth, hard work, thinking, running the extra mile, leading or giving help where no other help seems forthcoming, recognizing the talents of our fellow man and lady, and above all, practicing of the Golden Rule.

VELASCO-SUAREZ, MANUAL M., physician, surgeon; b. San Cristobal las Casas, Chiapas, Mex., Dec. 28, 1915; s. Jose Manual Velasco Balboa and Maria (Suarez) Velasco-S.; m. Elvira Siles, Mar. 1, 1946; children: Jose Manual, Maria Cristina, Guadalupe, Jesus Agustin, Francisco Javier, Juan Antonio, Maria de Lourdes, Elvira, Lucia Angelica, Teresita, Agnet. M.C., Universidad Nacional Autonomous Mex., 1939. Resident U. Iowa, 1940; resident in neurology, neurol. surgery Harvard U., 1941-42; resident in neuropathology, neurol. surgery; Mass. Gen. Hosp.; neurosurgeon Hosp. Juarez, Mexico City, 1947-58; head neuropsychiat. dept. S.S.A., nat. health agy., Mexico City, 1953-59; head neurology and neurosurgery Hosp. Juarez, from 1958; dir. Inst. Nacional de Neurologia, 1959—; prof. Universidad Nacional Autonomous Mex., 1944—; chief prof. neurology and neurosurg. group; now pres. Academia Nacional de Ciencias Mex. Fellow ACS, Am. Psychiat. Assn.; mem. Liga Mex. contra la Epilepsia (founding pres.), Mex. Soc. Neurology and Psychiatry (pres.), World Group of Experts on Epilepsy (pres. 1968), other Mex., fgn. and internat. orgns. Subspecialty: Neurosurgery. Office: Academia Nacional de Ciencias Apdo M 77-98 Mexico 1 DF

VELTRI, ROBERT WILLIAM, microbiologist; b. McKeesport, Pa., Dec. 1, 1941; s. Anthony and Desdemona (D'Innocenzo) V.; m. Suzanne Jones, Apr. 10, 1961; children: Anthony J., Katherine M. (dec.). A.B., Youngstown State U., 1963; M.S. in Microbiology, W.Va., U., 1965, Ph.D., 1968. Asst. prof., dir. research depts. otolaryngology and microbiology W.Va. U. Med. Ctr., Morgantown, 1968-72, assoc. prof., dir. research, 1972-76, prof., dir. research, 1976-81; dir. research and devel. immunology-serology Diagnostics div. Cooper Biomed., Inc., Malvern, Pa., 1982—; regional lab. dir. Nat. Found. for Cancer Research, 1980—. Contbr. articles to profl. jours. Bd. dirs. Monongahela County (W.Va.) unit Am. Cancer Soc., 1979-81, bd. dirs. W.Va. state unit, 1980-81. Grantee Deafness Research Found., 1971-74, Nat. Cancer Inst., 1974-78, 80-83. Fellow Am. Acad. Microbiology; mem. AAAS, Am. Assn. Immunologists, Am. Assn. Cancer Research, Soc. Infectious Diseases, Am. Soc. Microbiology, Sigma Xi. Republican. Roman Catholic. Subspecialties: Immunology (medicine); Cancer research (medicine). Current work: Development of new in vitro diagnostic tests for cancer and infectious diseases; cancer research on human tumor markers; ivestigations on new immune modulatory and anti-cancer agents produced by participating scientists of the National Foundation for Cancer Research. Home: 1025 Goodwin Dr West Chester PA 19380 Office: One Technology Ct Malvern PA 19355

VENETTE, JAMES RAYMOND, plant pathologist, consultant, educator; b. Boulder, Colo., Sept. 24, 1945; s. Elmer Robert and Toots Elizabeth (Schreiter) V.; m. Patricia Ann Wiggett, June 11, 1966; children: Robert Charles, Steven James. B.S., Colo. State U., 1967, M.S., 1969; Ph.D., U. Minn., 1975. Plant pathologist N.D. State U., Fargo, 1976—, instr.; cons. in field. Contbr. articles on plant pathology to profl. jours. Judge sci. fairs; regional coordinator for U.S. Navy Sci. awards program. Served to lt. (jg) USN, 1969-71; to lt. comdr. USNR, 1979—. Mem. Am. Phytopath. Soc., Crop Sci. Soc., N.D. Acad. Sci., Bean Improvement Coop., Sigma Xi, Gamma Sigma Delta. Subspecialties: Plant pathology; Integrated systems modelling and engineering. Current work: Tissue culture, aerosol transport of bacteria, research on epidemiology of bacterial and fungal diseases of phaseolus beans. Home: 1422 W Gateway Circle Fargo ND 58103 Office: Dept Plant Pathology ND State U Fargo ND 58105

VENEZIA, WILLIAM ALBERT, scientist; b. Portchester, N.Y., Jan. 22, 1947; s. Caesar A. and Alice (Schrader) V.; m. Ingrid Marie Johnson, Mar. 21, 1970; children: Anthony, Kristine. A.S., Broward Community Coll., 1967; B.S., Fla. Atlantic U., 1970; M.S., Clemson U., 1971, Ph.D., 1975. Sr. staff engr. Johns Hopkins U., Laurel, Md., 1975-80; assoc. prof. ocean engring. Fla. Atlantic U., Boca Raton, 1980-81; prin. scientist Gen. Offshore Corp., Fort Lauderdale, Fla., 1981—; adj. prof. Nova U., Fort Lauderdale, 1981—; cons. Johns Hopkins U. Applied Physics Lab., Laurel, Md., 1980-81; mem. tech. com. Naval Sea Systems Command, Washington, 1981—; dir. master's research Fla. Atlantic U., Boca Raton, 1980-81; cons. Naval Surface Weapons Center, Fort Lauderdale, Fla., 1980—. Mem. Allview Arrowhead Civic Assn., Columbia, Md., 1975-80. NSF trainee, 1970-75; Fla. Atlantic U. research grantee, 1981; Fla. Inst. Oceanography research grantee, 1981; Johns Hopkins U. research grantee, 1981. Mem. Marine Tech. Soc. (chmn. student chpt. 1968-70). Republican. Roman Catholic. Subspecialties: Ocean engineering; Offshore technology. Current work: Applied ocean science, working in areas of fiber optics in ocean systems, ocean surveillance, and state of the art ocean system development. Home: 375 NW 35th Ln Boca Raton FL 33431 Office: Gen Offshore Corp 2605 Stirling Rd Fort Lauderdale FL 33312

VENKATARAMIAH, AMARANENI, environ. physiologist; b. Atmakur, India, Aug. 16, 1928; came to U.S., 1969; s. A. and A. (Lakshvamma) Rangaiah; m. Swarajyam Kurra, June 9, 1949; children: Sulochana, Rao, Bharadwaj, Kumar, Sujatha. B.S. in Biology and Chemistry, Andhra U., Waltair, India, 1955; M.A. in Zoology, Sri Venkateswara U., Tirupati, India, 1957; Ph.D. in Eco-Physiology, Sri Venkateswara U., Tirupati, India, 1965. Lectr. Andhra Loyola Coll., Vijayawada, India, 1957-61, Sri Venkateswara U., 1965-66; research fellow Council of Sci. and Indsl. Research, New Delhi, India, 1966-67; head physiology sect. Gulf Research Lab., Ocean Springs, 1969—; adj. prof. biology U. So. Miss., Hattiesburg, 1978—. Contbr. articles to profl. jours. Dept. Army grantee, 1970-77; Dept. Energy grantee, 1978-

82; other grants. Mem. AAAS, Am. Soc. Zoologists, Gulf Estuarine Research Soc., Miss. Acad. Scis., Inc., World Mariculture Soc. Subspecialties: Comparative physiology; Ocean thermal energy conversion. Current work: Physiological ecology of marine animals with emphasis on osmoregulation and metabolic problems, crustacean aquaculture, toxicological effects of discharges from the ocean thermal energy conversion plants on marine animals. Patentee in field. Home: 219-1/2 Halstead Rd PO Drawer AG Ocean Springs MS 39564 Office: Gulf Coast Research Lab East Beach Dr Ocean Springs MS 39564

VENKATESAN, T., physicist; b. Madras, India, June 19, 1949; came to U.S., 1971, naturalized, 1979; d. T. N. and Veda Ramanujam; m. Lakshmi V. Venkatesan, Jan. 17, 1977. B.S., Indian Inst. Tech., Kharagpur, India, 1969; M.S., 1971; Ph.D., CUNY, 1977. Mem. tech. staff Bell Labs., Crawford Hill, N.J., 1977-79, Murray Hill, N.J., 1979-83, Research mgr., 1983—. Contbr. articles to sci. jours. Recipient Govt. of India Nat. Sci. Talent award, 1966-71. Mem. Am. Phys. Soc.; mem. Opital Soc. Am.; Mem. Sigma Xi. Subspecialties: Atomic and molecular physics; Microchip technology (materials science). Current work: Beam solid interactions, ion beam lithography; material science and optical bistability. Patentee in field. Office: IE-347 Bell Labs 600 Mountain Ave Murray Hill NJ 07974

VENTER, J. CRAIG, research scientist, biochemist, educator; b. Salt Lake City, Oct. 14, 1946; s. John and Elizabeth (Wisdom) V.; m. Claire M. Fraser, Oct. 10, 1981; 1 son: Christopher. B.A., U. Calif., San Diego, 1972, Ph.D. in Physiology and Pharmacology, 1975. Postgrad. research pharmacologist and chemist U. Calif., San Diego, 1975-76; asst. prof. pharmacology and therpaeutics SUNY, Buffalo, 1976-81, assoc. prof. biochemistry, 1981-82; assoc. chief molecular immunology Roswell Park Meml. Inst., Buffalo, 1982—; adj. prof. biochem. pharmacology SUNY-Buffalo, 1983—; indsl. cons. Contbr. articles to sci. jours. Served with MC USN, 1965-68; Vietnam. NIH grantee, 1977—; Am. Heart Assn. grantee, 1977-82; Pharm. Mfrs. Assn. grantee, 1977. Mem. Am. Soc. Exptl. Biology. Subspecialties: Biochemistry (medicine); Receptors. Current work: Neurotransmitter receptor purification; monoclonal antibodies; autoimmune receptor diseases (asthma), radiation inactivation. Office: Roswell Park Meml Inst NY State Dept Health 666 Elm St Buffalo NY 14263

VERAY, FRANCISCO X., cardiology educator, history of medicine and philosophy researcher; b. Yauco, P.R., Mar. 9, 1933; s. Francisco Veray-Marin and Margarita Torregrosa; m. Maria del Pilar Mazo Franco, Sept. 13, 1959; children: Carmen, Francisco X III, Maria-Jose, Carlos-Jaime. B.S., U. P.R.-Rio Piedras, 1953, Instituto Balmes, Barcelona, Spain, 1956; M.D., U. Barcelona, 1959. Rotating intern Damas Hosp., Ponce, P.R., 1959-60; resident in internal medicine VA Hosp., Ft. Howard, Md., 1962-64; pres. elect., dir., exec. com. P.R. Heart Assn., 1983; fellow in cardiology dept. medicine Univ. Hosp., Rio Piedras, P.R., 1964-66; assoc. in medicine dept. medicine Recinto de Ciencias Medicas, Rio Piedras, 1966-68, asst. prof. medicine, 1968-70, assoc. prof., 1977; med. dir. coronary care unit Univ. Hosp., Recinto de Ciencias Medicas, Rio Piedras, 1972—. Author: Betances, El Medico, 1969. Sec. P.R. Heart Assn., 1969, exec. com., 1969-73, pres. elect, dir., exec. com., 1983; pres. com. history and cultural Asociacion Medica de Puerto Rico, 1974-80. Served to 1st lt. USAR, 1953-60. Fellow Am. Coll. Chest Physicians, Am. Coll. Angiology (assoc.); mem. N.Y. Acad. Scis., Sociedad Cardiologia P.R. (sec. 1969-70), Hastings Ctr., AAAS, Am. Soc. Internal Medicine, Am. Soc. Clin. Research, AAUP. Roman Catholic. Subspecialties: Cardiology; Internal medicine. Current work: History of medicine and philosophy, heart disease in pregnancy, beta blockers chronic use after acute myocardial infarction. Office: Dept of Medicine Recinto de Ciencias Medicas University of Puerto Rico Rio Piedras PR 00936

VERCELLOTTI, JOHN RAYMOND, laboratory research executive; b. Joliet, Ill., May 2, 1933; s. Joseph Francis and Mary Therese (Walowski) V.; m. Sharon Cecile Vergez, Sept. 3, 1966; children: Ellen, Paul. B.A. with distinction, St. Bonaventure U., Olean, N.Y., 1955; postgrad., Cath. U. Am., 1955-58, U. Pitts., 1958; M.Sc., Marquette U., 1960; Ph.D., Ohio State U., 1963. Resident research assoc. fellow Ohio State U., Columbus, 1963-64; asst. prof. chemistry Marquette U., Milw., 1965-67; assoc. prof. U. Tenn., Knoxville, 1969-70, Va. Poly. Inst. and State U., Blacksburg, 1970-73, prof., 1973-79; research dir. V-Labs., Inc., Covington, La., 1979—. Contbr. articles in chemistry to profl. jours. NSF grantee, 1971-74; U.S. Dept. Agr. grantee, 1973-76; NATO grantee, 1978-80; Crinos Pharm. grantee, 1980-83. Fellow Am. Inst. Chemists, Royal Soc. Chemistry London; mem. Am. Soc. Biol. Chemists, Am. Chem. Soc., Sigma Xi. Democrat. Roman Catholic. Subspecialties: Biochemistry (medicine); Organic chemistry. Current work: Research on chemistry and biochemistry of the carbohydrates, enzymology of glycosidic linkages, fermentations and organic synthesis. Home: 215 E 4th Ave Covington LA 70433 Office: V-Labs Inc 423 N Theard St Covington LA 70433

VEREBEY, KARL G., clin. pharmacologist, educator; b. Budapest, Hungary, Mar. 12, 1938; came to U.S., 1956, naturalized, 1963; s. Karoly and Etelka (Szabo) V.; m. Debra M. Adler, Feb. 22, 1962; children: Todd, Marc. A.A. in Humanities, Eotvos J. Gimnazium, 1956; B.A. in Physiology, Hunter Coll., 1965; M.A. in Molecular Biology, CUNY, 1968; Ph.D. in Pharmacology and Biochemistry, Cornell U., 1972. Cert. clin. lab. dir. N.Y.C. Dept. Health. Research assoc. in neurology and pharmacology Cornell U. Med. Coll., N.Y.C., 1972-73; dir. clin. pharmacology N.Y. State Div. Substance Abuse Services, Bklyn., 1973—; research prof. psychiatry N.Y. Med. Coll., 1977—; assoc. prof. psychiatry SUNY Downstate Med. Sch., Bklyn., 1979—; clin. lab. dir. Psychiat. Diagnostic Labs. Am., Summit, N.J., 1982—; mem. exec. com. Council Research Scientists, N.Y. State, 1980—. Contbr. over 50 articles to sci. publs. Served with U.S. Army, 1961-63. USPHS fellow, 1972-73; Nat. Inst. Drug Abuse grantee, 1974-77, 79-81. Mem. Am. Soc. Pharmacology and Exptl. Therapeutics, N.Y. Acad. Scis. Subspecialties: Pharmacology; Psychopharmacology. Current work: Development of sensitive analytic methods for drugs in body fluids, evaluate plasma levels vs. behavioral responses; experiments in pharmacodynamics correlated with psychopharmacology. Office: 80 Hanson Pl Brooklyn NY 11217

VERINK, ELLIS DANIEL, JR., metallurgical engineering educator; b. Peking, Feb. 9, 1920; s. Ellis Daniel and Phoebe Elizabeth (Smith) V.; m. Martha Eulala Owens, July 4, 1942; children: Barbara Ann, Wendy Susan. B.S. in Metall. Engring, Purdue U., 1941, M.S., Ohio State U., 1963, Ph.D., 1965. Registered profl. engr., Fla., Pa., Calif. Engr., mgr. chem. sect. Alcoa, New Kensington, Pa., 1946-59, mgr. chem. and petroleum industry sales, Pitts., 1959-62; mem. faculty U. Fla., Gainesville, 1965—, prof. metall. materials engring, 1973—, chmn. materials sci. and engring. dept., 1973—; pres. Materials Cons., Inc., Gainesville, 1970—; cons. in field; chmn. Gordon Research Conf. on Corrosion, New London, N.H., 1978. Contbr. over 70 articles to tech. publs. Pres. bd. dirs. YMCA, Gainesville. Served to comdr. USNR, 1941-46. Recipient Tchr.-Scholar of Yr. award U. Fla., 1979, Disting. Alumnus award Ohio State U., 1982. Mem. AIME, ASTM (Sam Tour award 1978), Am. Welding Soc., Electrochem. Soc., Am. Soc. for Metals, Nat. Assn. Corrosion Engrs. (Willis Rodney Whitney award 1982). Republican. Presbyterian. Lodges: Masons; Shiners; Kiwanis. Subspecialties: Corrosion; Materials (engineering). Current work: Improving the ability to predict corrosion behavior of alloys through electrochemical and complementary methods, thereby increasing cost effectiveness of research. Home: 4401 NW 18th Pl Gainesville FL 32605 Office: Materials Sci and Engring Dept Univ Fla Gainesville FL 32611

VERMA, OM PRAKASH, physiology and pharmacology educator, researcher; b. Mardan, Pakistan, Jan. 5, 1931; came to U.S., 1961, naturalized, 1974; s. Ram Rakha and Kaushaliya (Maniktals) V.; m. Sushma Luthra, Dec. 12, 1961; children: Randy, Aarti. B.V.Sc., Agrl U., India, 1956; M.V.Sc., IVRI, Izatnagar, India, 1960; Ph.D., U. Wis.-Madison, 1965. State veterinarian State Dept. Animal Husbandry, Lucknow, India, 1956-58; assoc. prof. Agrl. U., Ludhiana, India, 1965-68; research assoc. McDonald Coll., Ste. Ann DeBellevue, Can., 1968-69; asst. prof. Tuskegee (Ala.) Inst., 1969-76, assoc. prof., 1976-78, prof., head dept. physiology and pharmacology, 1978—; vet. med. officer U.S. Dept. Agr., 1977-78. Fellow Am. Coll. Vet. Pharmacology and Therapeutics; mem. Am. Physiology Soc., AVMA, Tuskegee Vet. Med. Assn. (V.P. 1981—). Subspecialties: Animal physiology; Reproductive biology. Current work: Control of reproduction in goats, teaching administration, research. Home: 109 Coosada Dr Montgomery AL 36117 Office: Sch Vet Medicine Tuskegee Inst Tuskegee Institute AL 36117

VERMA, RAM SAGAR, geneticist, educator; b. India, Mar. 3, 1946; s. Gaya Prasad and Moonga Devi V.; m. Shakuntala Devi, May 4, 1962; 1 child, Harendra Kumar. B.Sc. in Agr, Agra U., India, 1965, M.Sc., 1967; Ph.D. in Cytogenetics, U. Western Ont., Can., 1972. Plant breeder dept. genetics and plant breeding Govt. Agrl. Coll., Janpur, India, 1965-67; teaching asst. U. Western Ont., 1967-73, research asst. dept. plant scis., 1967-73; research assoc. dept. pediatrics U. Colo., Denver, 1973-74, fellow, 1974-76; instr. dept. medicine SUNY-Downstate Med. Ctr., Bklyn., 1976, asst. prof., 1976-79, assoc. prof., 1979—; assoc. dir. cytogenetics Jewish Hosp. and Med. Ctr., Bklyn., 1976-78, chief cytogenetics, 1978-79, chief div. cytogenetics, 1980—; mem. cytogenetic adv. com. for prenatal diagnosis lab. N.Y.C. Dept. Health, 1978—. Contbr. numerous articles to profl. jours. Fellow N.Y. Acad. Scis.; mem. Am. Soc. Cell Biology, AAAS, Am. Fedn. Clin. Research, Am. Genetic Assn., Am. Soc. Human Genetics, European Soc. Human Genetics, Genetic Soc. Am., Genetic Soc. Can., Internat. Assn. Human Biologists, Indian Soc. Human Genetics, Soc. Exptl. Biology and Medicine. Subspecialties: Animal genetics; Biofeedback. Home: 42-70 65th Pl Woodside NY 11377 Office: 555 Prospect Pl Brooklyn NY 11238

VERNADAKIS, ANTONIA (MRS. H. L. OCKERMAN), developmental neurobiologist, educator, researcher; b. Canea, Crete, Greece, May 11, 1930; m. H. L. Ockerman, 1961. B.A., U. Utah, 1955, M.S., 1957, Ph.D. in Anat. Pharmacology, 1961. Research assoc., research instr. in anatomy and pharmacology U. Utah Coll. Medicine, 1961-64; interdisciplinary tng. program fellow in pharmacology U. Calif.-San Francisco Med. Center, 1964-65; asst. research physiologist U. Calif.-Berkeley, 1965-67; asst. prof. U. Colo. Sch. Medicine-Denver, 1967-70, assoc. prof., 1970-78, prof. psychiatry and pharmacology, 1978—. Recipient Research Scientist Devel. award NIMH, 1969-79. Mem. Am. Soc. Pharmacology and Exptl. Therapeutics, Am. Physiol. Soc., Am. Neurochem. Soc., Internat. Soc. Neurochemistry, Internat. Soc. Psychoneuroendocrinology. Subspecialty: Neurobiology. Office: U Colo Sch Medicine 4200 E Ninth Ave Denver CO 80220

VERNAZZA, JORGE ENRIQUE, physicist; b. Buenos Aires, Jan. 16, 1943; U.S., 1967, naturalized, 1978; s. Enrique and Mercedez (Gómez) V.; m. Lucina K. Vernazza; children: Daniel R., Diana J. B.S., U. Buenos Aires, 1967; Ph.D., Harvard U., 1972. Research fellow, Harvard U., 1972-74, research assoc., 1974-78; sr. scientist Atmospheric 2nd Environ. Research Ctr., Cambridge, Mass., 1978-79; physicist Lawrence Livermore Lab., Livermore, Calif., 1979—. Contbr. articles in field to profl. jours. Mem. Am. Phys. Soc., Am. Astron. Soc. Subspecialties: Plasma physics; Database systems. Current work: Radiation transfer and non-LTE physics; development of computer codes for radiation transfer and non-LTE physics; also interested in database systems and their application to data analysis. Office: PO Box 808 Livermore CA 94550

VERRILLO, RONALD THOMAS, sensory scientist, researcher; b. Hartford, Conn., July 31, 1927; s. Francesco Paul and Angela (Forte) V.; m. Violet Verrillo, June 3, 1950; children: Erica, Dan, Thomas. B.A., Syracuse U., 1952; Ph.D., U. Rochester, 1958. Asst. prof. Syracuse U., 1957-62, research assoc., 1959-63, research fellow, 1963-67, assoc. prof., 1967-74, prof., 1977—, assoc. dir., 1980—; vis. prof. Karolinska Inst., Stockholm, 1977. Contbr. numerous articles in field to profl. jours. Served with USN, 1945-46. Am. Found. Blind fellow, 1956; NATO fellow Oxford U., 1970-71; NSF grantee, 1967-72; NIH grantee, 1972—; recipient research award Am. Personnel and Guidance Assn., 1961. Mem. Acoustical Soc. Am., Psychonomic Soc., Eastern Psychol. Assn., AAAS, Soc. Neurosci., N.Y. Acad. Sci., Internat. Assn. Study Pain, Am. Pain Soc., Sigma Xi (recipient Research award 1982). Subspecialties: Sensory processes; Psychophysics. Current work: Cutaneous sensory systems; sensory, cutaneous, mechanoreceptor, psychophysics. Home: 312 Berkeley Dr Syracuse NY 13210 Office: Inst Sensory Research Syracuse U Syracuse NY 13210

VERRY, WILLIAM ROBERT, systems engineer; b. Portland, Oreg., July 11, 1933; s. William Richard and Maurine Houser (Braden) V.; m. Bette Lee Ronspiess, Nov. 20, 1955 (div. 1981); children: William David, Sandra Kay Verry Londregan, Steven Bruce, Kenneth Scott; m. Jean Elizabeth Morrison, Oct. 16, 1982; step-children: Lucinda Jean Hale, Christine Carol Hale Fortner, Martha Jean Johnson, Brian Kenneth Lackey, Robert Morrison Lackey. B.A., Reed Coll., 1955; B.S., Portland State U., 1957; M.A., Fresno State U., 1960; Ph.D., Ohio State U.-Columbus, 1972. Instr. chemistry Reedley (Calif.) Coll., 1957-60; ops. research analyst Naval Weapons Center, China Lake, Calif., 1960-63; ordnance engr. Honeywell Ordnance, Hopkins, Minn., 1963-64; sr. scientist Litton Industries, St. Paul., 1964-67; project mgr. Tech. Ops., Inc., Alexandria, Va., 1967-70; research assoc. Ohio State U., Columbus, 1970-72; prin. engr. Computer Sci. Corp., Falls Church, Va., 1972-77; mem. tech. staff MITRE Corp., Albuquerque, 1977—. Mem. Ops. Research Soc. Am. Subspecialties: Operations research (engineering); Systems engineering. Current work: Systems engineering of simulations on a Joint Test Force designing and implementing a state-of-the-art simulation of command control and communications (C3) in the airland battle using current combat and C3 simulations as components. Home: 10224 Cielito Lindo NE Albuquerque NM 87111 Office: MITRE Corp PO Box 5520 Kirtland AFB NM 87185

VERSNYDER, FRANCIS LOUIS, research executive; b. Utica, N.Y., May 27, 1925; s. Frederick P. and Bessie (Becker) VerS.; m. Katherine Ann Kelly, May 30, 1948; children: Constance, Christine, Kelly. B.S., U. Notre Dame, 1950. Registered profl. engr., Conn. Analytical chemist metallurgist Sibley Machine & Foundry Co., 1948-50; tech. engr. Gen. Electric Co., Lynn, Mass., 1950-54, supr. metall. and high temperature testing, 1954-55, research assoc. metall. and ceramic research dept., 1955-61; asst. dir. alloy and materials devel. AMRDL-Pratt & Whitney Aircraft Group, 1961-64, assoc. dir. advanced materials, 1964-70, mgr. materials engring. and research lab., 1970-74, mgr. materials engring. and research dept., 1974-77; m; gr. materials tech. United Tech. Research Ctr., 1977-81; asst. dir. research for materials tech. United Techs. Research Ctr., East Hartford, Conn., 1977—; cons. in field. Contbr. articles to profl. jours. Bd. dirs. Acta Metallurgica. Served with U.S. Army, 1943-45. Decorated Purple Heart with oak leaf cluster.; Recipient George Mead Gold medal United Aircraft Corp., 1965; Arch T. Colwell award Soc. Automotive Engrs., 1971; Dickson Prize award and medal Carnegie Mellon U., 1972; Francis J. Clamer medal Franklin Inst., 1973; Engring. Honor award U. Notre Dame, 1975. Fellow Am. Soc. Metals (Henry Marion Howe medal 1954, engring. materials achievement 1975); mem. Nat. Acad. Engring., N.Y. Acad. Sci., Conn. Acad. Sci. and Engring., AIME, AAAS, Am. Foundrymen's Soc., Planetary Soc., Am. Vacuum Soc., Alpha Sigma Mu. Republican. Club: Farms Country. Subspecialties: Materials; High-temperature materials. Current work: Materials technology; research management. Patentee in field.

VESEL, RICHARD WARREN, engineering products company executive; b. Wickliffe, Ohio, Sept. 1, 1954; s. Clarence George and Wanda Jacqueline (Purpura) V.; m. Nancy Hallam, Aug. 13, 1982. B.S.E.E., Case Western Res. U., 1976, M.S.E.E., 1978. Registered profl. engr. Ohio. Pres. Method Systems Inc., Euclid, Ohio, 1976—; supr. engring. Bailey Controls Co., Wickliffe, 1978—. Mem. IEEE, Am. Nuclear Soc. Republican. Subspecialties: Graphics, image processing, and pattern recognition; Robotics. Current work: Empirically derived algorithms for vision based decisions. Developer vortex shedding Flowmeter with optical transducer, 1982, optical communications loop terminal module, 1982. Home: 38405 N Lane Apt E-207 Willoughby OH 44094 Office: Method Systems Inc 19751 S Lakeshore Dr Euclid OH 44119

VESELL, ELLIOT SAUL, pharmacologist, educator; b. N.Y.C., Dec. 24, 1933; s. Harry and Evelyn (Jaffe) V.; m. Kristen Peery, Mar. 24, 1968; children: Liane, Hilary. A.B. magna cum laude, Harvard Coll., 1955, M.D., Harvard U., 1959. Intern in pediatrics Mass. Gen. Hosp., 1959-60; research assoc. in human genetics, asst. physician Rockefeller Inst., 1960-62; asst. resident in medicine Peter Bent Brigham Hosp., 1962-63; clin. assoc. Nat. Inst. Arthritis and Metabolic Disease, 1963-65; head. sect. pharmacogenetics Lab. Chem. Pharmacology, Nat. Heart Inst., 1965-69; prof. pharmacology, genetics and medicine, also chmn. dept. pharmacology Coll. Medicine, Pa. State U., Hershey, 1969—, Evan Pugh prof., 1981—; Julius W. Sturmer Meml. lectr. Phila. Coll. Pharmacy and Sci., 1980. Contbr. articles to profl. publs. Mem. Am. Soc. Pharmacology and Exptl. Therapeutics (Therapeutics award 1971), Soc. Exptl. Biology and Medicine (Samuel J. Meltzer award 1967), Am. Soc. Clin. Investigation, Am. Coll. Clin. Pharmacology, Am. Soc. Clin. Pharmacology and Therapeutics. Jewish. Clubs: Cosmos (Washington); Century Assn. (N.Y.C.). Subspecialties: Pharmacology; Genetics and genetic engineering (medicine). Current work: Molecular and clinical pharmacology; effects of heredity and environmental factors on the disposition of drugs. Office: 500 University Dr Hershey PA 17033

VESSELLA, ROBERT LOUIS, JR., immunologist, educator; b. New Britain, Conn., Apr. 10, 1948; s. Robert Louis and Wyvonne Florence (Skibinski) V.; m. Ann Elizabeth Knouft, May 15, 1976; children: Thomas Robert, James Henry. B.A., U. Conn., 1970; Ph.D., U. Miss. Med. Ctr., 1974. Postdoctoral fellow dept. surgery U. Kans. Med. Center, Kansas City, 1974-76; asst. prof., research immunologist dept. urologic surgery U. Minn. and VA Med. Ctrs., Mpls., 1976—. Mem. Am. Assn. Immunologists, Am. Soc. Microbiology. Republican. Subspecialties: Cancer research (medicine); Immunology (medicine). Current work: Immunobiology of genitourinary tumors with an emphasis on development of monoclonal antibodies; immunodiagnostic assays and basic cellular immunology. Home: 13026 Euclid Ave Apple Valley MN 55124 Office: VA Med Center Research Bldg 31 Minneapolis MN 55417

VEST, ANTHONY LEON, nuclear engineering services executive; b. Kingsport, Tenn., Aug. 26, 1947; s. Landon Leon and Helen Maige (Payne) V.; m. Nancy Ann Morton, Dec. 3, 1970; 1 dau., Shannon Tamralyn. B.S. in Mech. Engring, U. Tenn., Knoxville, 1969; M.B.A., 1980. Field engr. Gen. Electric Co., Atlanta, 1969-70, requisition engr., Knoxville, 1970-71, nuclear test engr., Cordova, Ill., 1971-72, shift engr., Morris, Ill., 1972-73, outage coordinator, Waterford, Conn., 1972-73, pre-op test coordinator, Auburn, Nebr., 1973-74, project mgr., Vidalia, Ga., 1974-75, mgr. nuclear operating plant, Chattanooga, 1975-80, mgr. nuclear plant services, Atlanta, 1980—. Speaker Scientists and Engrs. for Secure Energy, N.Y.C., 1981; mem. Save the Ocoee River Com., Chattanooga, 1981, Citizens Tax Relief Com., Atlanta, 1981, Feed the Hungry, Atlanta, 1980. Recipient Managerial award Gen. Electric Co., 1974, Brass Ring award, 1973. Mem. Am. Nuclear Soc., Elfur Soc. Republican. Methodist. Clubs: Atlanta Track; Tennessee Valley Canoe (Chattanooga) (editor 1979-80). Subspecialty: Nuclear fission. Current work: Improvement in the construction, startup and operation of nuclear power facilities. Development of new maintenance equipment and techniques to provide improved safety, reliability and availability. Home: 495 N Link Rd Alpharetta GA 30201 Office: Gen Electric Co 22 Technology Park Norcross GA 30092

VEST, CHARLES MARSTILLER, mechanical engineer, educator, college dean; b. Morgantown, W.Va., Sept. 9, 1941; s. Marvin Lewis and Winifred Louise (Buzzerd) V.; m. Rebecca McCue, June 8, 1963; children: Ann Kemper, John Andrew. B.S.M.E., W.Va. U., 1963; M.S.E., U. Mich., 1964, Ph.D., 1967. Asst. prof. mech. engring. U. Mich., 1968-72, assoc. prof., 1972-77, prof., 1977—, head interferometric holography group, 1970-73, assoc. dean for acad. affairs, 1981—; vis. assoc. prof. Stanford U., 1974-75; honored guest Universidad Nacional de La Plata, Argentina, 1979. Author: Holographic Interferometry, 1979; contbr. articles on applied optics, fluid mechanics and heat transfer to profl. jours. Recipient Class of '38E Service award U. Mich., 1972. Fellow Optical Soc. Am. (assoc. editor jour. 1982—); mem. ASME, Sigma Xi, Tau Beta Pi, Pi Tau Sigma. Subspecialties: Holography; Mechanical engineering. Current work: Applied optics, holographic interferometry, computer tomography, holographic nondestructive testing, heat transfer, fluid mechanics. Home: 910 Kuebler Dr Ann Arbor MI 48103 Office: Dept Mech Engring U Mich Ann Arbor MI 48109

VEUM, TRYGVE LAURITZ, animal scientist, educator; b. Viroqua, Wis., Mar. 16, 1940; s. Kermit N. and Alice E. (Solberg) V.; m. Marjorie S., Dec. 27, 1967; children: Eric L., Kristen S. B.S. in Animal Sci, U. Wis., Madison, 1962; M.S. in Animal Nutrition, Cornell U., 1965, Ph.D., 1968. Asst. prof. animal scis. U.Mo., Columbia, 1967-74, assoc. prof., 1974-80, prof., 1980—. Contbr. numerous articles, primarily on monogastric nutrition, to profl. jours. Active ARC, PTA. Mem. Am. Inst. Nutrition, Am. Soc. Animal Sci. (Midwest sect. Research award 1980), U. Wis. Alumni Assn. Methodist. Subspecialties: Animal nutrition; Nutrition (biology). Current work: Monogastric nutrition with swine. Home: 916 Lathrop Rd Columbia MO 65201 Office: 110 Animal Sci Research Ctr U Mo Columbia MO 65211

VEZIROGLU, TURHAN NEJAT, mechanical engineer, educator, researcher, cons.; b. Istanbul, Turkey, Jan. 24, 1924; came to U.S., 1962, naturalized, 1983; s. Abdul Kadir and Ferruh (Burun) V.; m. Bengi Isikli, Mar. 17, 1960; children: Emre Alp, Oya Sureyya. A.C.G.I.

in Mech. Engring, City and Guilds Coll., London, 1946; B.Sc. with honors, U. London, 1947; Ph.D. in Heat Transfer, U. London, 1951; D.I.C., Imperial Coll., London, 1948. Engring, apprentice Alfed Herbert Ltd., Coventry, Eng., 1945; project engr. Office of Soil Products, Ankara, Turkey, 1954-56; tech. dir. M.K.V. Constrn. Co., Istanbul, Turkey, 1957-61; assoc. prof. mech. engring. U.Miami, Coral Gables, Fla., 1962-65, prof., 1966—, dir. grad. studies in mech. engring., 1965-71, chmn. dept. mech. engring., 1971-75, assoc. dean for research, 1975-79, dir. Clean Energy Research Inst., 1974—; UNESCO cons. on energy; invited lectr. in heat transfer and energy. Contbr. numerous articles on thermal contact conductance, two-phase flow instabilities, solar energy and hydrogen energy systems to profl. jours; editor 20 conf. procs. Pres. Learning Disabilities Found., Miami, Fla., 1972-73, adv., 1974—. Served to 1st lt. Ordnance Service Turkish Army, 1952-53. Recipient Turkish Presdl. Sci. award, 1975. Fellow Instn. Mech. Engrs., ASME, AAAS; mem. Am. Soc. Engring. Edn., AAUP, Am. Nuclear Soc., Internat. Solar Energy Soc., Soc. Engring. Sci., AIAA, Internat. Assn. Hydrogen Energy (editor Internat. Jour. Hydrogen Energy 1976—, pres. 1975—), Sigma Xi. Subspecialties: Mechanical engineering; Fuels and sources. Current work: Two-phase flow instabilities; thermal contact conductance; solar energy applications; hydrogen energy system. U. London chess champion, 1948. Home: 800 Paradiso Ave Coral Gables FL 33146 Office: U Miami Coral Gables FL 33124

VICENTE, PETER JAMES, psychologist, educator; b. Phila., May 12, 1947; s. James and Ana Rosa (Mendoza) V.; m. Margaret Ann Duddy, Dec. 27, 1969; children: Sean, Brian, Kevin. B.A., LaSalle Coll., 1969; M.A., Ohio State U., 1971, Ph.D., 1975. Diplomate: Diplomate Am. Bd. Profl. Neuropsychology, 1983. Dir. dept. psychol. services Ohio State U. Hosps., Columbus, 1976-80; dir. div. rehab. psychology Ohio State U. Coll. Medicine, 1976-80, clin. asst. prof., 1980—; clin. assoc. prof. Wright State U Sch. Profl. Psychology, Dayton, Ohio, 1981—; dir. dept. health psychology indsl. Commn. of Ohio, Columbus, 1980—; mem. profl. adv. bd. Nat. Head Injury Found. (Ohio chpt.), 1983-85. Co-editor: Foundations of Clinical Neuropsychology. Recipient Outstanding Profl. Contbns. award Nat. Acad Neuropsychologists, 1981. Fellow Nat. Acad. Neuropsychologists (v.p. 1982-84); mem. Ohio Acad. Neuropsychologists (pres. 1979-83), Internat. Neuropsychology Soc., Am. Psychol. Assn., Ohio Psychol. Assn. Subspecialty: Neuropsychology. Current work: Comprehensive bio-psycho-social assessment and treatment of industrially injured or diseased workers and their return to productivity. Home: 208 Erie Rd Columbus OH 43214 Office: Indsl Commn Ohio 107 N High St Columbus OH 43215

VICKERS, STANLEY, biochemical pharmacologist, researcher; b. Blackpool, Lancashire, Eng., Sept. 27, 1939; came to U.S., 1962, naturalized, 1979; s. Norman Stanley and Hannah (Snape) V. B.Sc., London U., 1962; Ph.D., SUNY-Buffalo, 1967. Fellow U. Kans., 1966-69; sr. research pharmacologist Merck & Co., West Point, Pa., 1969-71, research fellow, 1971-81, sr. research fellow, 1981—. Contbr. articles to profl. jours. Mem. Am. Soc. Pharmacology and Exptl. Therapeutics, Am. Chem. Soc., N.Y. Acad. Sci. Current work: Absorption, distribution and metabolism of drugs, assays of therapeutic concentration of drugs. Home: Box 243 RD 2 Slotter Rd Perkasie PA Office: Merck Institute for Therapeutic Research Bldg 44A West Point PA 19486

VICKERY, BRIAN CAMPBELL, librarian, educator; b. Sept. 11, 1918; s. Adam Cairns McCay and Violet (Watson) V.; m. Manuletta McMenamin, 1945; 2 children; m. Alina Gralewska, 1970. M.A., Oxford (Eng.) U. Chemist Royal Ordnance Factory, Somerset, Eng., 1941-45; librarian ICI Ltd., Welwyn, Eng., 1946-60; prin. sci. officer Nat. Lending Library for Sci. and Tech., 1960-64; librarian UMIST, 1964-66; head research and devel. Aslib, 1966-74; prof. library studies, dir. Sch. Library Archive and Info. Studies, Univ. Coll., London, Eng., 1973—. Author: Classification and Indexing in Science, 1958, 3d edit. 1975, On Retrieval System Theory, 1961, 2d edit., 1965, Techniques for Information Retrieval, 1970, Information Systems, 1973; contbr. articles to profl. jours. Home: 138 Midhurst Rd London England W13 9TP

VICKERY, ROBERT KINGSTON, JR., biologist, educator; b. Saratoga, Calif., Sept. 18, 1922; s. Robert Kingston and Ruth (Bacon) V.; m. Marcia Agnes Hoak, July 7, 1951; children: David Kingston, Peter Hoak. Student, U. Calif., Berkeley, summer 1942, Yale U., 1943; B.A., Stanford U., 1944, M.A., 1948, Ph.D., 1952. Instr. botany Pomona Coll., 1950-51; research fellow Calif. Inst. Tech., 1955; instr. U. Utah, 1952, asst. prof. genetics, 1952-57, assoc. prof. biology, 1957-64, prof. biology, 1964—, chmn. dept. genetics, 1962-65; panelist NIH Cell Biology and Genetics Fellowship, 1964-69. Author: Manual of the Common Native Plants of the Salt Lake Area, 1961, Case Studies in the Evolution of Species Complexes in Mimulus, 1978; contbr. articles to profl. jours. Served to 1st lt. USAAF, 1943-46; PTO. Recipient Disting. Teaching award U. Utah, 1972. Mem. Genetics Soc. Am., Am. Genetics Assn., AAAS, Calif. Bot. Soc., Bot. Soc. Am., Am. Soc. Plant Taxonomists, Internat., Soc. Plant Taxonomists, Soc. Plant Biosystematists, Evolution Soc., Sigma Xi. Democrat. Episcopalian. Subspecialties: Evolutionary biology; Ecology. Current work: Analysis by various techniques, for example experimental hybridizations, electrophoretic separations of enzymes, DNA/DNA hybridization studies, computer simulations, field trials, etc., of mechanisms of evolution. Home: 3376 Louise Ave Salt Lake City UT 84109 Office: Dept Biology U Utah Salt Lake City UT 84112

VICTOR, GEORGE A., physicist, educator; b. Ridgway, Pa., Nov. 15, 1936; s. Joseph A. and Mae R. (Segerstrom) V.; m. Sally Konopka, Mar. 10, 1963; children: Mae, Nancy. B.S. in Physics, Rensselaer Poly. Inst., 1958; Ph.D. in Applied Math. and Theoretical Physics, Queen's U. Belfast, Northern Ireland, 1966. Staff scientist GCA Corp., Burlington, Mass., 1961-71; physicist Smithsonian Astrophys. Obs.; lectr. in astronomy Harvard U., Cambridge, Mass.; sr. research fellow Queen's U. of Belfast, 1966-67; vis. fellow Joint Inst. Lab. Astrophysics, U. Colo. and Nat. Bur. Standards, Boulder, 1978-79. Contbr. papers to sci. jours. Fellow Am. Phys. Soc.; mem. Am. Assn. Physics Tchrs., Am. Geophys. Union. Subspecialties: Atomic and molecular physics; Aeronomy. Current work: Theoretical studies atomic and molecular processes and their role in atmospheric science, astrophysics, laboratory and fusion plasmas and lasers. Home: 23 Peter Rd North Reading MA 01864 Office: 60 Garden St Cambridge MA 02138

VIEHLAND, LARRY ALAN, chemist, educator, researcher; b. St. Louis, Apr. 30, 1947; s. Harold Henry and Beulah Fay (Allensworth) V.; m. Claudia Kimberling Winters, Aug.28, 1969; children: Jeremy Scott, Brian Daniel. B.S. in Chemistry, M.I.T., 1969, Ph.D., U. Wis., Madison, 1973. Research assoc. Brown U., 1973-76, asst. prof., research, 1976-77; asst. prof. chemistry Parks Coll. of St. Louis U., Cahokia, Ill., 1977-79, assoc. prof., 1979-83, prof., 1983—. Contbr. numerous articles to profl. jours. Research Corp. grantee, 1980; NSF grantee, 1980-84. Mem. Am. Phys. Soc., Sigma xi. Methodist. Subspecialties: Theoretical chemistry; Atomic and molecular physics. Current work: Theory of ion motion in gases; kinetic theory; stat. mechanics. Home: 2959 Arlmont Dr Saint Louis MO 63121 Office: Parks Coll of Saint Louis U Cahokia IL 62206

VIEST, IVAN M(IROSLAV), consulting structural engineer; b. Bratislava, Czechoslovakia, Oct. 10, 1922; came to U.S., 1947, naturalized, 1955; s. Ivan and Maria (Zacharova) V.; m. Barbara K. Stevenson, May 23, 1953. Ing., Slovak Tech. U., Bratislava, 1946; M.S., Ga. Inst. Tech., 1948; Ph.D., U. Ill., 1951. Registered profl. engr., Pa. Research asst. U. Ill., Urbana, 1948-50, research asso., 1950-51, research asst. prof., 1951-55, research asso. prof., 1955-57; bridge research engr. Am. Assn. State Hwy. Ofcls.; rd. test Nat. Acad. Scis., Ottawa, Ill., 1957-61; structural engr. Bethlehem Steel Corp., Pa., 1961-67, sr. structural cons., 1967-70, asst. mgr. sales engring. div., 1970-83; cons. structural engr., 1983—; lectr. in field. Author: Composite Construction, 1958. Recipient Constrn. award Engring. News Record, 1972. Fellow Am. Concrete Inst. (Wason Research medal 1956); mem. ASCE (hon., Research prize 1958, v.p. 1973-75), Am. Iron and Steel Inst., Internat. Assn. Bridge and Structural Engring., Nat. Acad. Engring., Transp. Research Bd., Soc. Automotive Engrs., AAAS. Club: Saucon Valley Country (Bethlehem). Subspecialties: Civil engineering; Theoretical and applied mechanics. Current work: Structural aspects of bridges; buildings and other civil engineering structures. Research, numerous publs. on various steel and concrete structures, especially bridges and bldgs., to profl. jours. Office: PO Box 1428 Bethlehem PA 18016

VIETS, HERMANN, university dean, mechanical engineering educator, researcher; b. Quedlinburg, Germany, Jan. 28, 1943; came to U.S., 1949, naturalized, 1954; s. Hans and Herta Betty (Heik) V.; m. Pamela Deane, June 30, 1968; children: Danielle, Deane, Hans, Hillar. B.S.A.E., Poly. Inst. Bklyn., 1965, M.S. in Astronautics, 1966, Ph.D., 1970. Research asst. Poly. Inst. Bklyn., 1968-69; USAF postdoctoral fellow von Karman Inst., Brussels, 1969-70; group leader Aerospace Research Labs., U.S. Air Force, Dayton, Ohio, 1970-76; assoc. prof. mech. engring. Wright State U., 1976-80, prof., 1980-81; assoc. dean., prof. engring. W.Va. U., Morgantown, 1981—; chmn. bd. Precision Stamping Inc., Beaumont, Calif.; cons. in field. Contbr. numerous articles to profl. jours. Recipient tech. achievement award USAF, 1974, sci. achievement award., 1975; NASA predoctoral trainee, 1965; NATO grantee, 1970. Fellow AIAA (assoc.); mem. Am. Soc. Engring. Edn., Sigma Xi, Sigma Gamma Tau, Tau Beta Pi. Subspecialties: Aeronautical engineering; Fluid mechanics. Current work: Unsteady flows, basic and applied; coherent structures; applied acoustics. Patentee in field, including for vortex ring generator, thrust augmentation system with oscillating jet nozzles and high frequency gust tunnel. Home: 144 Scenery Dr Morgantown WV 26505 Office: W Va U 151 Engring Scis Bldg Morgantown WV 26506

VIG, MADAN MOHAN, veterinarian, educator; b. Gujranwala, India, June 12, 1944; came to U.S., 1973, naturalized, 1980; s. Krishan Lal and Ved Kumari (Bedi) V.; m. Shashi Bajaj, Aug. 1, 1973; children: Chhavi, Pooja. B.Vet. Sci. and Animal Husbandry, Panjab Agrl. U., 1966; M.S., Haryana Agrl. U., 1968, Ph.D., 1972. Asst. prof. surgry Haryana Agrl. U., 1972-73; postdoctoral research assoc. U. Notre Dame, 1973-75; asst. prof. surgery, medicine and clinic Tuskegee Inst., Ala., 1975-80, assoc. prof., 1980—. Contbr. articles to profl. jours. U.S. Dept. Agr. grantee; Sr. Indian Council of Agrl. Research fellow; Govt. of India scholar. Mem. AVMA, Am. Animal Hosp. Assn., Indian Assn. Vet. Surgeons, Am. Assn. Vet. Clinicians, Sigma Xi, Phi Zeta. Subspecialties: Surgery (veterinary medicine); Internal medicine (veterinary medicine). Current work: Veterinary surgery and anesthesia; internal medicine (dermatology). Home: 326 Cricket Ln Auburn AL 36830 Office: Vet Teaching Hosp Tuskegee Institute AL 36088

VIGIL, MANUEL GILBERT, mechanical engineer; b. Medanales, N.Mex., Oct. 1, 1941; s. Manuel Juan and Georgia S. (Sisneros) V.; m. Barbara Josephine Vigil, Aug. 10, 1963; children: Mark Matthew, Gilbert Emmanuel. B.S.M.E., N.Mex. State U., Las Cruces, 1966; M.S.M.E., U. N.Mex., Albuquerque, 1968. Task leader Sandia Nat. Labs., Albuquerque, 1966—. Contbr. articles to profl. jours. Mem. ASME, Am. Nuclear Soc. Democrat. Roman Catholic. Club: Pajarito Classic T-Bird. Subspecialties: Fluid mechanics; Mechanical engineering. Home: 3325 June St NE Albuquerque NM 87111 Office: PO Box 5800 KAFB Albuquerque NM 87185

VIJH, ASHOK KUMAR, chemistry educator, researcher; b. Multan, India (now Pakistan), Mar. 15, 1938; s. Bishamber Nath and Prem Lata (Bahl) V.; m. Danielle Blais; 1 son, Aldous Ian. B.Sc. with honors, Panjab U., India, 1960, M.Sc., 1961; Ph.D., Ottawa U., 1966. Group leader Inst. Research Hydro-Quebec, 1969-74, program leader, 1975-81, maitre-de-research, 1973—; vis. prof., thesis dir. INRS-Energie, U. Que., 1970—. Editorial bd.: Solar Energy Materials; Contbr. over 170 articles to profl. jours.; author: Electrochemistry of Metals and Semiconductors, 1973; editor: Oxides and Oxide Films. Fellow Royal Soc. Chemistry, Chem. Inst. Can. (Noranda lectr. 1979), IEEE; mem. Electrochem. Soc. (Lash Miller award 1973). Subspecialties: Physical chemistry; Surface chemistry. Current work: Electrochemistry; maeterials science; surface chemistry. Office: 1800 Montee St Julie Varennes PQ Canada J0G 2P0

VILCEK, JAN TOMAS, medical educator, researcher; b. Bratislava, Czechoslovakia, June 17, 1933; s. Julius and Friderika (Fischer) V.; m. Marica Gerhath, July 28, 1962. M.D., Comenius U. Med. Sch., Bratislava, 1957; C.Sc., Inst. Virology Czechoslovak Acad. Sci., Bratislava, 1962. Asst. dept. microbiology Comenius U., 1953-57; research assoc. Inst. Virology Czechoslovak Acad. Sci., 1957-59, fellow, 1959-62, head lab., 1962-64; asst. prof. microbiology NYU, 1965-68, assoc. prof., 1968-73, prof., 1973—; chmn. Am. Cancer Soc. Adv. Com. on Microbiology and Virology. Author: Interferon, 1969; editor: (with T.C. Merigan and I Gresser) Regulatory Functions of Interferon, 1980, (with R. Kono) The Clinical Potential of Interferons, 1982; editor in chief: Archives of Viology, 1975—; mem. editorial bd.: Infection and Immunity, 1983—, Applied Biochemistry and Biotechnology, 1981—, Jour. of Interferon Research, 1981—, Interferon, 1979—, Virology, 1979-81; contbr. numerous articles to profl. jours. USPHS grantee, 1965—, 76—; recipient USPHS Career Devel. award, 1968-73. Mem. Soc. Gen. Microbiology London, Am. Soc. Microbiology, AAAS, N.Y. Acad. Sci., Am. Assn. Immunologists, Council Biology Editors, Am. Soc. Virology, Internat. Soc. Immunopharmacology. Subspecialties: Microbiology; Immunobiology and immunology. Current work: Mechanisms of interferon synthesis and action; properties of interferons and their functions; soluble mediators of the immune responses; anti-viral compounds. Office: Dept Microbiology NYU Sch Medicine 550 1st Ave New York NY 10016

VILKKI, ERKKI UUNO, astronomer; b. Soanlahti, Finland, July 5, 1922; came to U.S., 1967; s. Uuno and Bertta (Takkinen) V.; m. Josette Boulanger, June 20, 1952; 1 dau., Ann. Student, Nav. Sch. Rauma, Finland, 1947-48, ;51-52; Sea Capt. Cert., Nav. Sch. Kotka, Finland, 1959. With Scandianvian Mcht. Marine, 1945-62; collaborator Pic du Midi Obs., Hautes Pyŕenées, France, 1964-66; observer-research technician Yerkes Obs., Williams Bay, Wis., 1967-73, asst. astronomer, 1973-78, assoc. astronomer, 1979—. Contbr. articles to profl. jours. Served with Finnish Army, 1941-44. Mem. Am. Astron. Soc. Subspecialty: Optical astronomy. Current work: Trigonometric stellar parallaxes; photographic observations of double stars. Office: Yerkes Obs Williams Bay WI 53191

VILLA, JUAN FRANCISCO, educator; b. Matanzas, Cuba, Sept. 23, 1941; s. Urbvano and Eulalia M. (Graciaa) V.; m. Elena M. Baez, Feb. 4, 1967; children: John F., Ellen M., Paul A., Irene L. B.S., U. Miami, 1965, M.S., 1967, Ph.D., 1969. Postdoctoral assoc. U. N.C., Chapel Hill, 1969-71; asst. prof. Lehman Coll., CUNY, Bronx, 1971-74, assoc. prof., 1974-77, acting dean natural and social scis., 1980-81, prof. chemistry, 1977—. Contbr. articles to chem. jours. Bd. dirs. Hal Block Soccer League, Ramapo, N.Y., 1981—, Spring Valley (N.Y.) Little League, 1980-82. CUNY grantee, 1972, 74-75, 79-81. Fellow Am. Inst. Chemists, London Chem. Soc., N.Y. Acad. Scis.; mem. Am. Chem. Soc. (research grantee 1971-73), AAAS, Sigma Xi. Republican. Roman Catholic. Subspecialty: Inorganic chemistry. Current work: Synthesis of coordination compounds of biol. activity and their instrumental study, electron paramagnetic resonance, infrared spectroscopy, magnetic susceptibility, electronic spectroscopy, synthesis, coordination compounds. Home: 14 Clark Dr Spring Valley NY 10977 Office: Chemistry Dept H H Lehman Coll CUN Y Bronx NY 14068

VILLABLANCA, JAIME ROLANDO, medical scientist, educator; b. Chillan, Chile, Feb. 29, 1929; came to U.S., 1971; s. Ernesto and Teresa (Hernandez) V.; m. Guillermina Nieto, Dec. 3, 1955; children: Amparo C., Jaime G., Pablo J., Francis X, Claudio I. B. Biology, Nat. Inst. Chile, Santiago, 1946; licentiate in medicine, U. Chile, Santiago, 1953, M.D., 1954; cert. in neurophysiology, UCLA, 1968. Resident in neurology Neurol. Clinic, U. Chile Sch. Medicine, 1955-58; mem. faculty U. Chile Sch. Medicine, 1954-71, prof. exptl. medicine, 1970-71; assoc. research anatomist and psychiatrist Sch. Medicine, UCLA, 1971-72, assoc. prof. psychiatry, 1972-76, prof. psychiatry, 1976—, prof. anatomy, 1977—; postdoctoral fellow in physiology Johns Hopkins U. and Harvard U. med. schs., 1959-61; internat. research fellow in anatomy UCLA, 1966-68; mem. Mental Retardation Research Ctr., UCLA, Brain Research Inst., UCLA; mem. sci. council Internat. Inst. Research and Advice in Mental Deficiency, Madrid; cons. in field. Contbr. to profl. jours. numerous articles, chpts., abstracts on neurol., behavorial and electrophysiol. effects of lesions upon mature and developing brain; drugs and functional recovery and brain reorganization after brain injury. Rockefeller Found. fellow, 1959-61; NIH fellow, 1966-68; USAF Office Sci. research grantee, 1962-65; Found. Fund Research Psychiatry grantee, 1969-72; USPHS Nat. Inst. Child Health and Human Devel. grantee, 1972-84; USPHS Nat. Inst. Drug Abuse grantee, 1981-84. Mem. Internat. Brain Research Orgn., Am. Physiol. Soc., Soc. Neurosci., Brit. Brain Research Assn. (hon.), European Brain and Behavior Soc. (hon.), Sigma Xi. Subspecialties: Neurophysiology; Neuropharmacology. Current work: Recovery of function and anatomical reorganization following lesions of mature and of developing brain; role of basal ganglia on effects of opiates. Office: Dept Psychiatry UCLA 760 Westwood Plaza Los Angeles CA 90024

VILLAFANA, THEODORE, radiol. physicist, educator, researcher; b. N.Y.C., Sept. 23, 1936; s. Leovigildo and Rose Maria (Solivan) V.; m. Abigail Rodriguez, Feb. 2, 1942. M.Sc. (USPHS fellow), U. Pitts., 1965, Ph.D., Johns Hopkins U., 1969. Diplomate: Am. Bd. Radiology, 1978. Radiol. physicist Columbia-Presbyterian Med. Center, N.Y.C., 1959-63, Montefiore Hosp., Pitts., 1963-65; radiol. physicist, asst. prof. U. P.R. Sch. Medicine, 1969-72, George Washington U., 1972-75; radiol. physicist, asst. to assoc. prof. Sch. Medicine, Temple U., 1975-82, prof., 1982—. Contbr. numerous articles on computerized tomography and radiol. physics of diagnostic imaging to profl. jours. Served with USAFR, 1955-63. Mem. Am. Assn. Physicists in Medicine, Health Physics Soc., IEEE. Baptist. Subspecialty: Imaging technology. Current work: Research in computerized tomography, physics of diagnostic x-ray imaging. Office: Dept Diagnostic X-ra Temple U Hos Philadelphia PA 19140

VILLA-KOMAROFF, LYDIA, molecular biology educator; b. Las Vegas, Aug. 7, 1947; d. John and Drucilla (Jaramillo) Villa; m. Anthony L. Komaroff, June 18, 1970. Student, U. Wash., 1965-68; B.A. cum laude, Goucher Coll., 1970; Ph.D., M.I.T., 1975. Postdoctoral fellow Harvard U., Cambridge, Mass., 1975-78; vis. postdoctoral fellow, Cold Spring Harbor, N.Y., 1976-77; asst. prof. molecular genetics/gene structure U. Mass. Med. Sch., Worcester, 1978-82, assoc. prof., 1982—; mem. mammalian genetics study sect. NIH, 1982—. Contbr. articles to profl. jours. Helen Hay Whitney fellow, 1975-78; Basil O'Conner Starter grantee Nat. Found. March of Dimes, 1981-83. Mem. Am. Soc. Cell Biology, AAAS, N.Y. Acad. Scis., Fedn. Am. Scientists, Assn. Women in Sci., Soc. Advancement Chicanos and Native Ams., Sigma Xi. Subspecialties: Genetics and genetic engineering (agriculture); Genome organization. Office: Dept Molecular Genetics and Microbiology 55 Lake Ave North Worcester MA 01605

VILLAR-PALASI, CARLOS, pharmacology educator, researcher; b. Valencia, Spain, Mar. 3, 1928; came to U.S., 1957; s. Vicente Villar-Bolinga and Teresa (Palasi-Pinazo.). M.S., U. Valencia, 1951; Ph.D. in Biochemistry, U. Madrid, 1955; M.S. in Pharmacy, U. Barcelona, Spain, 1962. Research fellow Eppendorf, Hamburg, W.Ger., 1953-54; postdoctoral research fellow Western Res. U., Cleve., 1957-58; research assoc. in pharmacology, 1963-64; research assoc. in biochemistry U. Minn., 1964-65, asst. prof. biochemistry, 1965-69; assoc. prof. pharmacology U.Va., 1969-73, prof., 1973—. Research numerous publs. on control of carbohydrate metabolism, protein phosphorylation and related subjects. Mem. Western Albemarle Rescue Squad, Crozet, Va., 1979—. Served to 2d lt. arty. Spanish Army, 1948-53. Recipient Research Career award NIH, 1967-70; NIH grantee, 1965—; Diabetes Assn. grantee, 1969-71; NSF grantee, 1967-77; Abbott Labs. grantee, 1980-81. Mem. Am. Soc. Pharmacology, Am. Biochem. Soc., Brit. Biochem. Soc., AAAS. Roman Catholic. Subspecialties: Molecular pharmacology; Biochemistry (medicine). Current work: Control of metabolism by protein phosphorylation; study of protein kinases and phosphatases. Office: U Va 520 Jorday Hall Charlottesville VA 22908

VILLERE, KAREN R., astrophysicist; b. Teaneck, N.J., Apr. 9, 1944; d. Peter N. and Alice E. (Hall) Heere; m. Gary L. Villere, Aug. 28, 1967. B.A. summa cum laude, U. Pa., 1965; M.A., U. Calif.-Berkeley, 1968; Ph.D., U. Calif.-Santa Cruz, 1976. Research scientist Sci. Applications Inc., Palo Alto, Calif., 1974-76; Nat. Acad. Scis., NRC resident research assoc. NASA Ames Research Center, Moffett Field, Calif., 1977-79, asst. research astronomer, 1979—, U. Calif.-Santa Cruz, 1979—. Woodrow Wilson fellow, 1965. Mem. Am. Astron. Soc., Astron. Soc. Pacific. Subspecialty: Theoretical astrophysics. Current work: Stellar evolution, star formation. Home: 226 Flynn Ave Mountain View CA 94043 Office: NASA Ames Research Center MS 245-3 Moffett Field CA 94035

VINCENZI, FRANK FOSTER, pharmacologist, educator; b. Seattle, Mar. 14, 1938; s. Frank and Thelma Charlotte (McAllister) V.; m. Judith I. Heimbigner, Aug. 27, 1960; children: Ann, Franklin, Joseph. B.S. magna cum laude in Pharmacy, U. Wash., 1960, M.S. in Pharmacology, 1962, Ph.D., 1965. Research asst. dept. pharmacology U. Wash., 1963-65, USPHS trainee, 1965, asst. prof., 1967-72, assoc. prof., 1972-80, prof., 1980—, acting chmn., 1975-77, vice chmn., 1977—. Author revs. and articles in field. Mem. exec. com., bd. dirs. Wash./Alaska chpt. Cystic Fibrosis Found., 1977-80, chmn. med. adv. com., 1977-80. Mem. AAAS, Am. Soc. Pharmacology and Exptl. Therapeutics, Biophys. Soc., Cardiac Muscle Soc., Red Cell Club, Soc.

Exptl. Biology and Medicine, Sigma Xi, Phi Beta Kappa, Kappa Psi. Subspecialties: Molecular pharmacology; Pharmacology. Home: 3205 109th St SE Bellevue WA 98004 Office: U Washington Pharmacology SJ-30 Seattle WA 98195

VINEYARD, GEORGE HOAGLAND, physicist; b. St. Joseph, Mo., Apr. 28, 1920; s. George Hoagl and Mildred M. (Barkley) V.; m. Phyllis Ainsworth Smith, Feb. 3, 1945; children: John H., Barbara Gale. B.S., M.I.T., 1941, Ph.D., 1943. Mem. staff Radiation Lab. M.I.T., 1943-45; mem. faculty U. Mo., Columbia, 1946-54; prof. U. Md., 1952-54; mem. staff Brookhaven Nat. Lab., Upton, N.Y., 1954—, sr. physicist, 1960—, chmn. dept. physics, 1961-66, dep. dir. lab., 1966-72, Dir. lab., 1973-81; cons. to govt. and industry; fellow Poly. Inst. N.Y., 1978—; vis. com. materials Sci. Center, Cornell U., 1964-67; dept. physics M.I.T., 1969-73; mem. sci. and ednl. adv. com. Lawrence-Berkeley Lab., U. Calif., 1977-81; adv. com. Nat. Magnet Lab., 1963-67; adv. com. math. and phys. scis. NSF, 1966-71, chmn., 1971; mem. sci. policy bd. Stanford Synchotron Radiation Project, 1977-81; mem. solid state scis. panel NRC, 1954—, chmn., 1965-71, chmn. solid state scis. com., 1970-72; chmn. panel condensed matter physics survey com. Nat. Acad. Scis., 1969-72; mem. materials research council Def. Advanced Research Projects Agy., 1967—; mem. ad hoc com. U.S. participation Internat. Atomic Energy Agy., 1970-72. Co-editor: series Documents in Modern Physics, 1964-78; bd. assoc. editors: Am. Jour. Physics, 1948-50, Phys. Rev, 1959-61; bd. editors: Physics, 1964-68, Jour. Computational Physics, 1966-79, Physics and Chemistry of Liquids, 1968—; editor Phys. Rev. Letters, 1983—. Bd. dirs. L.I. Action Com., 1979-82; mem. adv. com. on sci. and tech. N.Y. State Legis. Commn., 1980—. Recipient award for disting. contbns. to higher edn. Stony Brook Found., 1975. Fellow Am. Acad. Arts and Scis., Am. Phys. Soc. (chmn. div. solid state physics 1972-73, councillor-at-large 1974-79), AAAS; mem. L.I. Assn. Commerce and Industry (dir. 1975—), Sigma Xi. Subspecialties: Condensed matter physics; Theoretical physics. Current work: Research in Physics of condensed matter. Home: 10 Brewster Ln Bellport NY 11713 Office: Brookhaven Nat Lab Upton NY 11973

VINOGRADOV, ALEKSANDRA M., civil engineering educator; b. Chernovtsy, Ukraine, USSR, Sept. 23, 1940; emigrated to Can., 1977; d. Mark and Malya (Litvin) Endelshtein; m. Oleg Vinogradov, Apr. 7, 1963; 1 son, Mark. M.Sc., Lvov (USSR) State U., 1962; Ph.D., Leningrad State U., 1972. Registered prof. engr. Engr. Design Inst. Coal Industry, Lugansk, USSR, 1962-66; research engr. Nat. Research Inst. Hydraulic Engring., Leningrad, USSR, 1971-77; research assoc. dept. civil engring. U. Calgary, Alta., Can., 1977-80, asst. prof., 1980—. Research fellow Natural Scis and Engring. Research Council, Can., 1980. Mam. Canadian Soc. Civil Engring., Am. Soc. Engring. Sci., Canadian Assn. Univ. Tchrs. Subspecialties: Civil engineering; Solid mechanics. Current work: Creep of structures, stability of structures, interaction problems, ice mechanics. Office: U Calgary 2500 University Dr NW Calgary AB Canada T2N 1N4 Home: 180 Ranch Estates Rd NW Calgary AB Canada T3G 2A9

VINTI, JOHN PASCAL, physicist, cons.; b. Newport, R.I., Jan. 16, 1907; s. John Joseph and Anna Catherine (Sild) V.; m. Ella Keen Johnson, Feb. 17, 1951. Sc.D. in Physics, M.I.T., 1932. Postdoctoral research fellow U. Pa., 1932-34; research asst. M.I.T., Cambridge, 1934-36; instr. Brown U., Providence, 1936-37; asst. prof. The Citadel, 1937-38; instr. Worcester (Mass.) Poly. U., 1939-41; research physicist Aberdeen (Md.) Proving Ground, 1941-57, Nat. Bur. Standards, Washington, 1957-65; prof. math. N.C. State U., 1966-67; cons. M.I.T. Measurement Systems Lab., 1967-73, lectr. celestial mechanics, 1967—; lectr. U. Del., 1948-49, U. Md., 1950-52; professorial lectr. Georgetown U., 1963-64; adj. prof. Cath. U. Am., 1964-65. Author papers on atomic physics, electromagnetic wave propogation, mathematics, satellite orbits, and celestial mechanics. Recipient award Nat. Bur. Standards, 1961. Fellow Am. Phys. Soc., Royal Astron. Soc., Brit. Interplanetary Soc., Washington Acad. Scis., AIAA (assoc.); mem. Astron. Soc., Am. Geophys. Union, Philos. Soc. Washington, Sigma Xi. Unitarian. Club: Cosmos (Washington). Subspecialties: Celestial mechanics; Astrodynamics. Current work: Celestial mechanics. Home: 44 Quint Ave Apt 8 Allston MA 02134 Office: Measurement Systems Lab MIT 59-216 Cambridge MA 02139

VIOLA, ALFRED, educator; b. Vienna, Austria, July 8, 1928; came to U.S., 1940, naturalized, 1945; s. Isidore and Greta (Broch) V.; m. Joy Darlene Winkie, Oct. 19, 1963. B.A., Johns Hopkins U., 1949, M.A., 1950; Ph.D., U. Md., 1955. Research assoc., vis. instr. biochemistry Boston U., 1955-57; asst. prof. Northeastern U., Boston, 1957-62, assoc. prof., 1962-68, prof. chemistry, 1968—; vis. prof. U. Munich, 1977. Contbr. articles to profl. jours. Mem. Am. Chem. Soc., Sigma Xi. Subspecialty: Organic chemistry. Current work: Reaction mechanisms, thermal rearrangements of organic compounds, stereochemistry of transition states, thermal reactions of acetylenic and allenic compounds, unimolecular kinetics. Home: 14 Glover Rd Wayland MA 01778 Office: Dept Chemistry Northeastern U Boston MA 02115

VIOLATO, CLAUDIO, psychology educator; b. Valdagno, Vicenza, Italy, May 23, 1952; emigrated to Can., 1958; s. Efrem and Marianna (Battistin) V. B.Sc., U. B.C., 1976, M.A., 1978; Ph.D., U. Alta., 1982. Instr. Kwantlen Coll., Surrey, B.C., 1981-82; asst. prof. psychology U. Victoria, B.C., 1982—; cons. sch. dists. Contbr. articles to profl. jours. Mem. Can. Psychol. Assn., Am. Psychol. Assn., Can. Soc. Study of Edn., Can. Assn. Ednl. Psychologists. Roman Catholic. Subspecialties: Developmental psychology; Social psychology. Current work: Personality organization and stability, IQ testing and assessment, creativity and giftedness, memory processes. Office: Dept Psychol Founds U Victoria Victoria BC Canada V8W 2Y2 Home: 1576 Midgard Ave Apt 308 Victoria BC Canada V8P 2Y1

VIRGO, JULIE ANNE CARROLL, library association executive, educator; b. Adaelaide, South Australia, Australia, June 14, 1944; came to U.S., 1966; d. Archibald Henry and Norma Mae (Gillett) Noolan; m. Daniel Thuering Carroll, Aug. 20, 1977. M.A., U. Chgo., 1968, Ph.D., 1974. Library fellow U. Chgo., 1967-68, mem. faculty, 1968—, dir.; with State Research Libraries, Chgo., 1977—. Author: Continuing Education: Needs Assessment and Model Programs; contbr. articles to profl. jours. Nat. Library of Medicine trainee, 1968-69; Higher Edn. Act. fellow, 1969-72. Mem. ALA, Am. Soc. Assn. Execs., Am. Mgmt. Assn., Assn. Am. Library Schs., Spl. Libraries Assn., AAAS, Am. Soc. Info. Scis. (doctoral award), Beta Phi Mu. Home: 900 Private Rd Winnetka IL 60093 Office: Assn Coll and Research Libraries 50 E Huron St Chicago IL 60611

VISHER, GLENN SHILLINGTON, oil properties corporation executive, lecturer; b. Evansville, Ind., May 20, 1930; s. John William and Marguerite Ruth (Miller) V.; m. Bettye Ruth Gentry, Oct. 17, 1953; children: Christine Ann, Lynn Ellen, Sara Catherine. B.S., U. Cin., 1952; M.S., Northwestern U., 1958, Ph.D., 1960. Cert. profl. geol. scientist. Exploration geologist Shell Oil Co., Casper, Wyo., 1958-60; research geologist Sinclair Oil Co., Tulsa, 1960-66; prof. U. Tulsa, 1966-80; pres. Geol. Services and Ventures, Inc., Tulsa, 1980—, BNJ Oil Properties, Inc., 1981—; lectr. continuing edn. Oil & Gas Cons. Internat., Tulsa, 1969—; convenor, rapporteur symposia, profl. confs. Contbr. writings to profl. publs. Pres. Tulsa chpt. ACLU, 1976; v.p. Okla. chpt. ACLU, 1964-66; mem. budget exec. com. United Way, Tulsa, 1974-78. Served as 1st lt. USAF, 1952-54. Fellow AAAS, Geol. Soc. Am.; mem. Soc. Econ. Paleontologists and Mineralogists, Am. Assn. Petroleum Geologists (lectr. Tulsa 1973-75), Internat. Assn. Sedimentologists. Democrat. Unitarian. Subspecialties: Sedimentology; Oceanography. Current work: Developing scientific and stratigraphic bases for petroleum exploration; use of tectonic, geomorphic, geochemical, sedimentologis processes to predict hydrocarbon occurrences. Home: 2920 E 73rd St Tulsa OK 74136

VISHNIAC, HELEN SIMPSON, microbiology educator, researcher; b. New Haven, Dec. 22, 1923; d. George Gaylord and Anne (Roe) Simpson; m. Wolf Vladimir Vishniac, Aug. 18, 1951 (dec. Dec. 1973); children: Obadiah (dec.), Ethan Tecumseh, Ephraim Meriwether. B.A., U. Mich., 1945; M.A., Radcliffe Coll., 1947; Ph.D., Columbia U., 1950. Instr. CUNY, 1948-52; lectr. Yale U., New Haven, 1953-61; research assoc. U. Rochester, N.Y., 1974-78; vis. assoc. prof. SUNY-Brockport, 1976-78; asst. prof. microbiology Okla. State U., Stillwater, 1978—. Mem. Am. Soc. Microbiology, Mycol. Soc. Am., Soc. Gen. Microbiology, AAAS, Phi Beta Kappa, Sigma Xi. Subspecialties: Ecology; Evolutionary biology. Current work: Microbial ecology of extreme environments; Antarctic yeasts and substrate limited habitats; molecular evolution; phylogeny and DNA, RNA homology. Office: Okla State U Stillwater OK 74078

VISKANTA, RAYMOND, mechanical engineer, educator, consultant; b. Lithuania, July 16, 1931; came to U.S., 1949, naturalized, 1956; s. Vincas and Genovaite (Vinickas) V.; m. Birute Barbara, Oct. 13, 1956; children: Renata, Vitas, Tadas. B.S.M.E., U. Ill., 1955; M.S.M.E., Purdue U., 1956, Ph.D., 1960. Registered profl. engr., Ill., 1961. Asst. mech. engr. Argonne Nat. Lab., Ill., 1956-61, assoc. mech. engr., 1961-62; mem. faculty Purdue U., West Lafayette, Ind., 1962—, assoc. prof., 1962-66, prof., 1966—; Springer prof. mech. engring. U. Calif., Berkeley, 1968, vis. prof., 1969; guest prof. Tech. U. Munich, W.Ger., 1976-77; cons. to numerous corps. Contbr. numerous articles on heat transfer, radiative transfer and applied thermodynamics to tech., sci. jours., U.S., Europe. Recipient Sr. U.S. Scientist award Alexander von Humboldt Found., Bonn,. W.Ger., 1975; Japan Soc. Promotion of Sci. fellow, 1982. Fellow ASME (Heat Transfer Meml. award 1976); mem. AIAA (Thermophysics award 1979), Combustion Inst., AAUP. Roman Catholic. Subspecialties: Mechanical engineering; Fluid mechanics. Current work: Heat transfer in combustion systems; heat and mass transfer in buoyancy driven flows; melting and solidification. Home: 123 Pawnee Dr West Lafayette IN 47906 Office: Sch Mech Engring Purdue U West Lafayette IN 47907

VISOTSKY, HAROLD MERYLE, educator, physician; b. Chgo., May 25, 1924; s. Joseph and Rose (Steinberg) V.; m. Gladys Mavrich, Dec. 18, 1955; children: Jeffrey, Robin. Student, Herzl Coll., Chgo., 1943-44, Baylor U., 1944-45, Sorbonne, 1945-46; B.S., U. Ill., 1947; M.D. magna cum laude, U. Ill., 1951. Intern Cin. Gen. Hosp., 1951-52; resident U. Ill., Ill. Research and Ednl. Hosp., also Neuropsychiat. Inst., Chgo., 1952-55; asst. prof. U. Ill. Coll. Medicine, 1957-61, asso. prof. psychiatry, 1965-69, dir. psychiat. residency tng. and edn., 1955-59; prof., chmn. dept. psychiatry and behavioral scis. Northwestern U. Med. Sch., Chgo., 1969—; dir. Psychiat. Inst., chmn. dept. psychiatry Northwestern Meml. Hosp.; sr. attending Evanston Hosp.; Polio respiratory center psychiat. cons. Nat. Found. Infantile Paralysis, U. Ill., 1955-59; dir. mental health Chgo. Bd. Health div. mental health services, 1959-63; dir. Ill. Dept. Mental Health, 1962-69; examiner Am. Bd. Psychiatry and Neurology, 1964—; dir. Center Mental Health and Psychiat. Services, Am. Hosp. Assn., 1979—; mem. 1st U.S. mission on mental health to USSR State Dept. Mission, 1967; chmn. task force V Joint Commn. on Mental Health of Children, 1967—; mem. adv. com. on community mental health service Nat. Inst. Mental Health, 1965—; mem. profl. adv. com. Jerusalem Mental Health Center; profl. adv. group Am. Health Services, Inc.; adv. com. Joint Commn. Accreditation Hosps., Council Psychiat. Facilities; mem. spl. panel mental illness for bd. dirs. ACLU; rector Lincoln Acad. of Ill. Faculty Social Service; mem. select com. psychiat. care and evaluation HEW; mem. faculty Practising Law Inst.; bd. overseers Spertus Coll. Judaica, Chgo., 1981—. Contbr. articles to psychiat. jours., chpts. psychiat. textbooks. Trustee Erikson Inst. Early Edn., Ill. Hosp. Assn., Mental Health Law Project, Washington. Served with AUS, 1942-46. Decorated D.S.C., Purple Heart, Bronze Star; recipient Edward A. Strecker award Inst. Pa. Hosp., 1969; Med. Alumnus of Year award U. Ill., 1976; Disting. Service award Chgo. chpt. Anti-Defamation League, B'nai B'rith, 1978. Fellow Am. Orthopsychiat. Assn. (dir. v.p. 1970-71, pres. 1976), Am. Psychiat. Assn. (chmn. council on mental health services 1967, v.p. 1973-74, sec. 1981—, chmn. council nat. affairs 1975-78, com. on abuse of psychiatry and psychiatrists), AAAS, Am. Coll. Psychiatrists (charter, bd. regents 1976-79, v.p. 1980, pres. 1983-84, Bowis award 1981), Chgo. Inst. Medicine; mem. Am. Assn. Chmn. Dept. Psychiatry, Am. Assn. Social Psychiatry (v.p. 1976), Council Med. Splty. Socs., AMA, Ill. Psychiat. Soc. (pres. 1965-66), Am. Coll. Psychoanalysts. Subspecialty: Psychiatry. Current work: Health care planning; stess research. Home: 1128 Ridge Ave Evanston IL 60202 Office: 320 E Huron St Chicago IL 60611

VISTE, ARLEN ELLARD, chemistry educator; b. Austin, Minn., Aug. 13, 1936; s. Arthur E. and Edith L. (Kehret) V.; m. Elizabeth Ann Lindbeck, June 14, 1959; children: Solveig, David, Mark. B.A., St. Olaf Coll., 1958; Ph.D., U. Chgo., 1962. Asst. prof. chemistry St. Olaf Coll., Northfield, Minn., 1962-63; NSF postdoctoral fellow Columbia U., N.Y.C., 1963-64; asst. prof. chemistry Augustana Coll., Sioux Falls, S.D., 1964-68, assoc. prof., 1968-73, prof., 1973—; faculty research participant Argonne (Ill.) Nat. Lab., 1970-71; vis. research assoc. prof. chemistry U. Waterloo, Ont., Can., summer 1973; vis. scientist Åbo Akademi, Turku, Finland, 1981-82. Contbr. articles to profl. jours. Mem. Am. Chem. Soc., Royal Soc. Chemistry (London), S.D. Acad. Sci., Midwest Assn. Chemistry Tchrs. in Liberal Arts Colls., Phi Beta Kappa, Sigma Xi. Lutheran. Subspecialties: Inorganic chemistry; Physical chemistry. Current work: Inorganic reaction mechanisms and electronic structure, relativistic effects in chemistry. Office: Dept Chemistry Augustana Coll Sioux Falls SD 57197

VISWANATHAN, CHAND RAM, electrical engineering educator, consultant; b. India, Oct. 23, 1929; came to U.S., 1957, naturalized, 1971; d. C. J. and S. N. (Pattammal) Ramachandran; m. Sree W. Viswanathan, Dec. 20, 1942; children: Usha, Meera, Asha. M.S., UCLA, 1959, Ph.D., 1962. Asst. prof. engring. UCLA, 1962-68, assoc. prof., 1968-74, prof., 1974—, asst. dean, 1974-77, chmn. dept. elec. engring., 1979—; cons. to semicondr. and integrated cir. industries. Contbr. numerous articles, abstracts to profl. publs. Recipient Disting. Teaching award Engring. Grad. Students Assn., 1972, Western Electric fund award Am. Soc. Engring. Educators, 1974, Disting. Teaching award Acad. Senate/UCLA Alumni Assn., 1976, Disting. Faculty award UCLA Engring. Alumni Assn., 1981. Fellow IEEE; mem. Am. Phys. Soc. Subspecialties: 3emiconductors; Microelectronics. Patentee in field. Home: 19573 Braewood Dr Tarzana CA 91356 Office: UCLA Sch Engring and Applied Sci Room 7732 Los Angeles CA 90024

VISWANATHAN, K., consulting engineer; b. Calicut, Kerala, India, Apr. 28, 1944; came to U.S., 1971; s. V. Thangam and V. Kirshnan. B.E., Birla Inst. Tech. and Sci., Pilani, India, 1966; M.S. in Indsl. Engring, Texas A&M U., 1972. Mech. engr. Brown & Root, Inc., Houston, 1972-74, sr. engr., 1980—; process engr. Wright Engr., Ltd., Vancouver, B.C., Can., 1974-76; design engr. Spraymould Ltd., Cambridge, Ont., Can., 1977-78, Shawinigan Energy Cons., Toronto, Ont., Can., 1978-80. Patrol leader Boy Scouts, 1960; active player Tamilnadu Cricket and Football Assn., 1966 active player Tamilnadu Cricket and Football Assn., all India; active United Way Campaigns, Houston, 1982. Sr. mem. Inst. Indstl. Engrs.; assoc. mem. Ops Research Soc. Am.; mem. Alpha Pi Mu, Phi Kappa Phi. Hindu. Clubs: Soccer (Missouri City, Tex.) (player coach 1982); Century (College Station, Tex.)). Subspecialties: Mechanical engineering; material handling. Current work: Planning and control; equipment and systems design. Home: 3126 Cherry Creek Dr Missouri City TX 77459

VITELLO, PETER ALFONSO JAMES, research physicist; b. Glendale, Calif., Sept. 15, 1950; s. Alfonso and Angelina (Petruzella) V.; m. Philomena Ann Cammuso, Aug. 28, 1976. B.S., U. So. Calif., 1972; Ph.D., Cornell U., 1977. Center for Astrophysics research fellow Harvard U., 1977-79; research scientist Sci. Applications, Inc., McLean, Va., 1979—. Mem. Am. Astron. Soc., Am. Phys. Soc., Sigma Xi. Roman Catholic. Subspecialties: Theoretical astrophysics; Plasma physics. Current work: Stellar winds, black holes, plasma physics, theoretical research in plasma physics and astrophysics. Office: 1710 Goodridge Dr PO Box 1303 McLean VA 22102

VIVEROS, OSVALDO HUMBERTO, neurobiologist, educator; b. Santiago, Chile, May 14, 1937; s. Humberto and Elena (Letelier) V.; m. Yolanda Faune, Mar. 24, 1964; children: Cristian. Claudia, Cristobal. M.D., U. Chile, 1962. Research fellow dept. physiology and biophysics U. Chile Med. Sch., 1962-65; research assoc. dept. physiology and pharmacology Duke U. Med. Center, Durham, N.C., 1966-67, USPHS postdoctoral fellow dept. biochemistry, 1977-79, Fulbright-Hays scholar dept. biochemistry, 1971; assoc. prof. dept. physiology and biophysics U. Chile Med. Sch., 1970-71; prof. dept. neurobiology Cath. U. Chile, Santiago, 1971-74; vis. scientist Lab. Clin. Scis., NIMH, Bethesda, Md., 1974-77; group leader dept. medicinal biochemistry Wellcome Research Labs., Research Triangle Park, N.C., 1977—; adj. assoc. prof. dept. pharmacology Duke U. Med. Center; reviewer div. physiology, cellular and molecular biology NSF. Contbr. articles to profl. jours.; mem. editorial and adv. bd.: Molecular Pharmacology, 1978—. Mem. Colegio Medico de Chile, Sociedad de Biologia de Chile, Latinoamerican Pharmacological Soc., Internat. Brain Research Orgn., Soc. Neuroscience, Am. Soc. Pharmacology and Exptl. Therapeutics, Am. Soc. Neurochemistry. Club: Duke Faculty (Durham, N.C.). Subspecialties: Neurobiology; Neuroendocrinology. Current work: Catecholamine and opiod peptide biosynthesis, storage, secretion; co-transmission epinephrine, norepinephrine, enkephalins, sympathetic system, adrenal medulla, adrenal cortex, tetrahydrobiopterin, pterins, excitation-secretion coupling. Office: 3030 Cornwallis Rd Research Triangle Park NC 27709

VLADUTIU, ADRIAN OCTAVIAN, physician, researcher; b. Bucharest, Romania, Aug. 5, 1940; came to U.S., 1969, naturalized, 1974; s. Octavian A. and Veturia (Chirescu) V.; m. Georgirene D. Therrien, Sept. 4, 1971; children: Christina Lynn, Catherine Joy. B.S., Spiru Haret, Bucharest, Romania, 1956; M.D., Bucharest U., 1962; Ph.D., Jassy U., 1968. Diplomate: Am. Bd. Pathology. Intern G. Marinesco Hosp., Bucharest, 1962-65, Millard Fillmore Hosp., Balt., 1971-72; resident E.J. Meyer Hosp., Buffalo, 1973-74, Buffalo Gen. Hosp., 1973-74, dir. clin. labs., 1982—, dir. chemistry, 1981—, dir. immunopathology, 1974—; prof. microbiology SUNY-Buffalo, 1982—, prof. pathology, 1981—; cons. Niagara Falls Meml. Hosp., N.Y., 1976-82; mem. Ctr. for Immunology, Buffalo, 1981—. Fellow ACP, Assn. Clin. Scientists; mem. Am. Assn. Immunologists, Am. Assn. Pathologists, N.Y. Acad. Scis. Subspecialties: Pathology (medicine); Immunology (medicine). Current work: Study of pathogenesis of experimental autoimmune diseases especially genetic control; chemical composition of pleural effusions in relation to differential diagnosis, creatine kinase and lactate dehydrogenase isoenzymes. Home: 80 Oakview Dr Amherst NY 14221

VLAY, GEORGE JOHN, engring. mgmt. exec.; b. Buffalo, Dec. 1, 1927; s. John and Victoria (Mili) V.; m. Betty Jo Wayland, July 21, 1949; children: Vanessa Michele, Susan Victoria, George John. B.S.E.E., U. Buffalo, 1953, postgrad., 1954-56. Registered profl. engr., Okla. Project engr. R.B. Warman, Inc., Buffalo, 1954-56, Aero Comdr., Inc., Norman, Okla., 1956-61, GTE/Sylvania, Williamsville, N.Y., 1961-66; mgr. advanced communication systems Philco-Ford Corp., Palo Alto, Calif., 1966-77; with Ford Aerospace & Communications Corp., Palo Alto, 1977—, dir. tech. affairs, 1982—; Mem. tech. council Electronics Industry Assn., Washington, 1978—. Served to sgt. USAF, 1946-49. Mem. IEEE (sr.), AIAA (space Systems com. 1976-78), Armed Forced Communications and Electronics Assn., Air Force Assn., Navy League, Am. Def. Preparedness Assn., Nat. Security Indsl. Assn. Republican. Methodist. Subspecialties: Aerospace engineering and technology; Satellite studies. Current work: Advanced communication satellite systems. Computer aided engineering, design, test and manufacturing. Home: 32 Yerba Buena Ave Los Altos CA 94022 Office: Ford Aerospace & Communication Corp 3939 Fabian Way Palo Alto CA 94313

VODKIN, MICHAEL HAROLD, geneticist; b. Boston, Dec. 4, 1942; s. Hyman Roy and Eva (Weiner) V.; m. Lila Ott, June 28, 1975. B.S., Boston Coll., 1964, M.S., 1969; Ph.D., U. Ariz., Tucson, 1971. Postdoctoral fellow Cornell U., Ithaca, N.Y., 1972-74; asst. prof. U. S.C., Columbia, 1974-79; staff fellow NIH, Bethesda, Md., 1979-81; NRC sr. staff fellow U.S. Army Med. Research Inst. Infectious Diseases, Ft. Detrick, Md., 1981-83, research chemist, 1983—. NIH grantee, 1975-79. Mem. Genetics Soc. Am., Am. Soc. Microbiology, Sigma Xi. Jewish. Subspecialties: Genetics and genetic engineering (biology); Genome organization. Current work: Cloning of prokaryotic genomes. Home: 900 Laredo Rd Silver Spring MD 20901 Office: USAMRIID Fort Detrick MD 21701

VOELKER, ROBERT ALLEN, research geneticist; b. Palmer, Kans., Jan. 24, 1943; s. Elmer W. and Helen D. (Moddelmog) V.; m. Darlene May Worthy, Aug. 22, 1965; 1 dau., Cheri Renee. B.S. in Edn, Concordia Tchrs. Coll., Seward, Nebr., 1965; M.S., U. Nebr., 1967; Ph.D. in Zoology, U. Tex.-Austin, 1970. NSF postdoctoral fellow U. Oreg., Eugene, 1970-71; research assoc. N.C. State U., Raleigh, 1971-73, asst. prof. dept. genetics, 1973-76; sr. staff fellow Nat. Inst. Environ. Health Scis., Research Triangle Park, N.C., 1976-80, research geneticist, 1980—. NIH fellow, 1967-70. Mem. Genetics Soc. Am. Democrat. Lutheran. Subspecialties: Gene actions; Genetics and genetic engineering (biology). Current work: RNA polymerase II structure and function; molecular analysis of genetic suppressors. Home: 709 Barbara Dr Raleigh NC 27606 Office: Nat Inst Environ Health Scis Research Triangle Park NC 27709

VOGEL, CHARLES LEWIS, physician, oncology educator; b. Belle Harbor, N.Y., Nov. 6, 1938; s. Samuel and Sylvia (Love) V.; m.; children: Stacey Lynn, Brian Arnold. A.B., Princeton U., 1960; M.D., Yale U., 1964. Diplomate: Nat. Bd. Med. Examiners, Am. Bd. Internal Medicine. Intern, med. resident Grady Meml. Hosp., Atlanta, 1964-66; clin. assoc. medicine br., solid tumor service Nat. Cancer Inst., Bethesda, Md., 1966-69; clin. instr. Georgetown U. Sch. Medicine, Washington, 1967-68; sr. investigator Nat. Cancer Inst., 1969-73; assoc. prof. medicine Emroy U., Atlanta, 1973-75; assoc. prof.

oncology U. Miami, Fla., 1975-82; prof., 1982—; chief div. breast cancer, clin. dir. Comprehensive Cancer Ctr., State of Fla., 1980—; sci. advisor Solid Tumor Center of Uganda Cancer Inst., 1969-73; cons. in field. Contbr. articles to profl. jours. Served with USPHS, 1966-73. Mem. Am. Assn. Cancer Research, Am. Soc. Clin. Oncology, Fla. Soc. Clin. Oncologists (dir. 1978—, sec. 1980-81). Subspecialty: Chemotherapy. Current work: Breast cancer research/chemotherapy. Office: 1475 NW 12 Ave PO Box 016960 Miami FL 33101 Home: 1420 NE 101st St Miami Shores FL 33183

VOGEL, GLENN CHARLES, chemistry educator; b. Columbia, Pa., Mar. 7, 1943; m. Kathy Rauser, June 15, 1969. B.S., Pa. State U., 1965; M.S., U. Ill., 1967, Ph.D., 1970. Asst. prof. chemistry Ithaca (N.Y.) Coll., 1970-74, assoc. prof., 1974-82, prof., 1982—. Contbr. articles to profl. jours. Petroleum Research Fund grantee, 1973-75; NSF grantee, 1975; Research Corp. grantee, 1979-80. Mem. Am. Chem. Soc. Subspecialty: Inorganic chemistry. Current work: Lewis acid base interactions; copper catechol complexes. Office: Dept Chemistry Ithaca Coll Ithaca NY 14850

VOGEL, HERBERT DAVIS, civil engineer; b. Chelsea, Mich., Aug. 26, 1900; s. Lewis P. and Pearl M. (Davis) V.; m. Loreine Elliott, Dec. 23, 1925; children: Herbert D., Richard E. B.S., U.S. Mil. Acad., 1924; M.S. in Civil Engring, U. Calif., 1928; Dr.-Ing., Berlin Tech. U., 1929; C.E., U. Mich., 1933. Registered profl. engr., N.Y., Tex., Tenn., D.C. Commd. 2d lt. U.S. Army, 1924, advanced through grades to brig. gen.; founder, builder, dir. U.S. Army Engrs. Waterways Expt. Sta., Vicksburg, Miss., 1929-34; student Command and Gen. Staff Sch., 1934-36, with 3d Engrs., Hawaii, 1936-38, instr. Army Engrs. Sch., Ft. Belvoir, 1938-40, asst. to dist. engr., chief inspection div., Pitts., 1940-41, dist. engr., 1942-43, grad. Army-Navy Staff Coll., 1943, , S.W. Pacific Theatre Operations, 1944; chief of staff intermediate sect. (New Guinea), then base comdr. Base M, Philippines, 1944-45, with Army of Occupation, Yokohama, Japan, 1945, dist. engr. C.E., Buffalo, 1945-49; engr. maintenance Panama Canal, v.p., dir. Panama R.R. Co., 1949-50, lt. gov. C.Z. Govt., v.p. Panama Canal Co., 1950-52, div. engr. Southwestern div. C.E., 1952- 54; chmn. bd. TVA, 1954-62; engr. adviser World Bank, 1963-67, including Indus Basin project, 1964-67; resource devel. engring. cons., 1967—; 1st occupant George W. Goethals chair mil. constrn. U.S. Army Engr. Sch., 1973—; Dir. Internat. Gen. Industries, 1973-80, Planning and Devel. Collaborative, 1968-74; Mem. Beach Erosion Bd., 1946-49, Internat. Boundary Commn., 1946-49; mem. Miss. River Commn., 1952-54, Bd. Engrs. for Rivers and Harbors, 1952-54; chmn. Ark., White, Red Basin Inter-Agy. Com., 1952-54; mem. permanent internat. com. Permanent Internat. Assn. Navigation Congresses, 1957-64, hon. mem., 1967; mem. cons. panel Ludington pumped storage project, 1969-73; chmn. bd. consultants to Panama Canal, 1971-72; engring. cons. GAO, 1977; Mem. visitors com. Sch. Engring., Vanderbilt U., 1967-80; vice chmn. WHO/FAOUNEP Panel of Experts on Environ. Mgmt. for Vector Control, 1982-83. Writer numerous papers on hydraulic models and lab. procedures, 1930-40, articles and speeches on TVA, 1954-63, papers defining U.S. position on inland navigation. Decorated D.S.M., Legion of Merit, Philippine Liberation and Independence medals, knight Grand Cross, Thailand, Colon Alfaro medal; recipient Distinguished Alumnus award U. Mich., 1953; Meritorious Service award for service to engring. profession Cons. Engrs. Council, 1967; Benjamin Franklin fellow Royal Soc. Arts. Fellow Am. Cons. Engrs. Council; mem. U.S. Com. Large Dams, Am. Power Conf., ASCE (hon., mem. nat. water policy com. 1971-74, chmn. 1974-76, mem. nat. energy policy com. 1976—, pres.'s award 1979), Soc. Am. Mil. Engrs. (hon., nat. dir. 1956-62, 66, treas. 1974-76), Nat. Acad. Engring., Nat. Soc. Profl. Engrs., Pub. Works Hist. Soc. (hon.), Order of Carabao. Episcopalian. Clubs: Army-Navy, Cosmos (Washington). Subspecialties: Civil engineering; Hydrogeology. Current work: Development of natural resource facilities, especially as serving interests of navigation; power and flood control for developing countries. Home: 3033 Cleveland Ave NW Washington DC 20008 Office: 434 Washington Bldg Washington DC 20005

VOGEL, STEVEN, zoology educator; b. Beacon, N.Y., Apr. 7, 1940; s. Max and Jeanette Rachel (Zucker) V.; m. Jane Gregory, Dec. 13, 1974; 1 son, Roger Booth. B.S., Tufts, U., 1961; A.M., Harvard U., 1963, Ph.D., 1966. Mem. faculty Duke U., Durham, N.C., 1966—, prof. zoology, 1979—; instr. Marine Biol. Lab., Woods Hole, Mass., 1972; mem. vis. faculty Marine Lab., U. Wash., Friday Harbor, 1979-83; cons. in field. Author: A Functional Bestiary, 1969, A model Menagerie, 1972, Life in Moving Fluids, 1981; contbr. numerous articles to profl. publs. Soc. of Fellow Jr. fellow Harvard U., 1964-66. Mem. AAAS, Soc. Arthropod Psychiatry, AAUP, Sigma Xi. Subspecialties: Physiology (biology); Fluid mechanics. Current work: Adaptations of organisms to movement of air and water around them. Home: 1212 Woodburn Rd Durham NC 27705 Office: Zoology Dept Duke Univ Durham NC 27706

VOGEL, THOMAS TIMOTHY, surgeon, educator, physiologist; b. Columbus, Ohio, Feb. 1, 1934; s. Thomas A. and Charlotte (Hogan) V.; m. Darina Kelleher, May 29, 1965; children: Thomas Timothy, Catherine Darina, Mark Patrick, Nicola Marie. A.B., Coll. Holy Cross, 1955; M.S., Ohio State U., 1960, Ph.D., 1962; M.D., Georgetown U., 1965. Diplomate: Nat. Bd. Med. Examiners, Am. Bd. Surgery. Intern Georgetown and D.C. Gen. hosps., 1965-66; resident Ohio State Univ. Hosps., 1966-70; instr. physiology Ohio State U., Columbus, 1960-62, instr. surgery, 1969-70, clin. asst. prof. surgery, 1974—. Contbr. articles to profl. jours. Fellow ACS; mem. Am. Physiol. Soc. (assoc.), Fedn. Am. Socs. Exptl. Biology, Soc. Acad. Surgery, Am. Soc. Parenteral and Enteral Nutrition, Am. Trauma Soc., AMA, Ohio Med. Assn., Ohio Acad. Sci., Sigma Xi. Lodge: Rotary. Subspecialties: Surgery; Physiology (medicine). Current work: Cardiovascular, endocrine and gastrointestinal physiology and surgery, nutrition of the hospitalized patient. Office: 621 S Cassingham Rd Columbus OH 43209 Home: 247 S Ardmore Rd Columbus OH 43209

VOGL, OTTO, polymer science and engineering educator; b. Traiskirchen, Austria, Nov. 6, 1927; came to U.S., 1953, naturalized, 1959; s. Franz and Leopoldine (Scholz) V.; m. Jane Cunningham, June 10, 1955; children: Eric, Yvonne. Ph.D., U. Vienna, 1950; Dr. rer. nat. h.c., U. Jena, 1983. Instr. U. Vienna, 1948-55; research asso. U. Mich., 1953-55, Princeton U., 1955-56; scientist E.I. Du Pont de Nemours & Co., Wilmington, Del., 1956-70; prof. polymer sci. and engring. U. Mass., 1970-83, prof. emeritus, 1983—; Herman F. Mark porf. polymer sci. Poly. Inst. N.Y., 1983—; guest prof. Kyoto U., 1968, 80, Osaka U., 1968, Royal Inst. Stockholm, 1971, U. Freiburg, Germany, 1973, U. Berlin, 1977, Strasbourg U., 1976, Tech. U. Dresden, 1982; guest Soviet Acad. Sci., 1973, Polish Acad. Sci., 1973, 75, Acad. Sci. Rumania, 1974, 76; cons. in field. Chmn. com. on macromolecular chemistry Nat. Acad. Sci. Author: Polyaldehydes, 1967, (with Furukawa) Polymerization of Heterocyclics, 1973, Ionic Polymerization, 1976, (with Simionescu) Radical Co and Graftpolymerization, 1978, (with Donaruma) Polymeric Drugs, 1978, (with Donaruma and Ottenbrite) Polymers in Biology and Medicine, 1980, (with Goldberg and Donaruma) Targeted Drugs, 1983; contbr. articles to profl. jours. Recipient Fulbright award, 1976; Humboldt award, 1977; Japan Soc. Promotion of Sci. sr. fellow, 1980. Mem. Am. Chem. Soc. (chmn. div. polymer chemistry 1974, chmn. Connecticut Valley sect. 1974), Am. Inst. Chem., AAAS, Austrian Chem. Soc., Japanese Soc. Polymer Sci. Subspecialty: Polymer chemistry. Current work: Polymer science and engineering, polymer synthesis, polymeric drugs, polymeric stabilizers, optically active polymers, ring opening polymerization aldehyde polymerization. Home: 212 Aubinwood Rd Amherst MA 01002 Office: Poly Inst NY Brooklyn NY 11201

VOGLER, WILLIAM RALPH, physician, educator; b. Louisville, July 2, 1929; s. William Ralph and Clara Frances (Rogers) V.; m. Marilyn Lydia Rubio, Sept. 13, 1952; children: Katherine, Gregory, Brian. M.D., Northwestern U., 1954. USPHS fellow Nat. Cancer Inst., 1955-58; resident in medicine Emory U., Atlanta, 1958-61; mem. faculty dept. medicine Emory U. Sch. Medicine, 1961—, now prof. Contbr. articles to profl. jours. Mem. Am. Assn. Cancer Research, Am. Soc. Clin. Oncology, Am. Soc. Hematology, Cell Kinetics Soc., AAAS, Internat. Soc. Exptl. Hematology, AMA. Subspecialty: Cancer research (medicine). Current work: Clinical and laboratory studies in leukemia; clinical studies and bone marrow transplantation, treatment of acute leukemia, laboratory studies in hematopoiesis and leukemia. Home: 2057 Clairmont Rd Decatur GA 30033 Office: Emory U 718 Woodruff Meml Bldg Atlanta GA 30322

VOGT, ROCHUS EUGEN, physicist, educator; b. Neckarelz, Germany, Dec. 21, 1929; came to U.S., 1953; s. Heinrich and Paula (Schaefer) V.; m. Micheline Alice Yvonne Bauduin, Sept. 6, 1958; children: Michele, Nicole. Student, U. Karlsruhe, Germany, 1950-52, U. Heidelberg, Germany, 1952-53; S.M., U. Chgo., 1957, Ph.D., 1961. Mem. faculty dept. physics Calif. Inst. Tech., Pasadena, 1962—, asso. prof., 1965-70, prof., 1970-82, R. Stanton Avery Disting. Service prof., 1982—, chmn. faculty, 1975-77, chmn. div. physics, math. and astronomy, 1978-83, v.p. and provost, 1983—, chief scientist, 1977-78; acting dir. Owens Valley Radio Obs., 1980-81. Fellow Am. Phys. Soc.; mem. Am. Assn. Physics Tchrs., AAAS, Sigma Xi. Subspecialties: Cosmic ray high energy astrophysics; Gamma ray high energy astrophysics. Research in astrophysics. Office: Calif Inst Tech Physics 206-31 Pasadena CA 91125

VOGT, STEVEN SCOTT, astrophysicist; b. Rock Island, Ill., Dec. 20, 1949; s. Calvin Roy and Jeanne (Josephson) V.; m. Zarmina Dastagir, June 15, 1980; 1 dau., Crystal. A.B. in Physics and Astronomy, U. Calif.-Berkeley, 1972; M.A., U. Tex.-Austin, 1975, Ph.D. in Astronomy, 1978. Research asst. astronomy dept. U. Tex.-Austin, 1972-78; asst. astronomer, asst. prof. Lick Obs.,U. Calif.-Santa Cruz, 1978—; cons. in field. Contbr.: articles to publs. including Nature. Recipient acad. fellowships; grantee NSF, NASA. Mem. Am. Astron. Soc., Soc. Photo-Optical Instrumentation Engrs., Astron. Soc. of Pacific, Internat. Astron. Union, Optical Soc. Am. Subspecialty: Optical astronomy. Current work: Development of solid state array detectors for low light level astronomical research; spectroscopy and direct imaging. Office: Natural Sci II U Calif Santa Cruz CA 95064

VOIGHT, JESSE CARROLTON, psychology educator, researcher; b. Memphis, Jan. 7, 1933; s. Sheldon Wayne and Gladys Ruth (Carrolton) V.; m. Betty Jane Hearn, Aug. 26, 1963; children: Tina, Timothy, Wendy. B.S., U. Ark., 1956, M.S. in Psychology, 1959, Ph.D., U. Mich., 1963. Asst. prof. psychology Va. State U., Petersburg, 1963-66, assoc. prof., 1966-70; assoc. prof. psychology Wayne State U., Detroit, 1970-75, prof., 1975—. Grantee Ford Found., 1973-74, NIMH, 1981—. Mem. Am. Psychol. Assn., Midwestern Psychol. Assn. Subspecialties: Cognition; Neuropsychology. Office: Werik Ctr 1380 Penobscot Bldg Detroit MI 48226

VOIGT, HERBERT FREDERICK, III, neuroscientist, educator; b. N.Y.C., Oct. 27, 1952; s. Herbert Frederick and Simona Rita (Comunale) V.; m. Ronit Gunst, Apr. 5, 1975; 1 son, Justin David. B.E. in Elec. Engring, City U. N.Y., 1974; Ph.D. in Biomed. Engring, Johns Hopkins U., 1979. Postdoctoral fellow Neural Encoding Lab. Johns Hopkins U., 1979-80; asst. prof. Boston U., 1981—. Contbr. articles to profl. jours. Research grantee Boston U., 1981; NIH award, 1982—. Mem. IEEE, Acoustical Soc. Am., AAAS, Soc. Neurosci. Subspecialties: Neurophysiology; Biomedical engineering. Current work: Neuronal circuitry of subcortical auditory nervous system; neural encoding of speech by primary, auditory nerve fibers; acquisition and analysis of multi-unit activity. Office: 110 Cummington St Boston MA 02215

VOLANAKIS, JOHN EMMANUEL, immunology educator, researcher; b. Thessaloniki, Greece, Mar. 17, 1938; came to U.S., 1968; s. Emmanuel John and Cleo (Agathonos) V.; m. JoAnne Somerville, May 16, 1970; children: Manolis, Marina. M.D., U. Thessaloniki, 1962; D.Med., U. Athens, 1968. Fellow in rheumatology Metropolitan Gen. Hosp., Cleve., 1968-71; instr. medicine U. Ala. in Birmingham, 1971-73, asst. prof. medicine and pathology, 1973-77, assoc. prof. medicine, pathology and microbiology, 1977—; vis. scientist U. Utrecht, Netherlands, 1978; mem. study sect. NIH, Bethesda, Md., 1981—. Editor: C-Reactive Protein and the Plasma Protein Response to Tissue Injury, 1982; assoc. editor: Jour. Immunology, 1982—. NIH fellow, 1978. Mem. Am. Assn. Immunologists, Am. Rheumatism Assn., AAAS, Sigma Xi. Subspecialties: Immunology (medicine); Biochemistry (biology). Current work: Biochemistry and biology of complement and acutephase proteins. Home: 3432 Old Wood Ln Birmingham AL 35243 Office: Univ Ala in Birmingham Univ Sta LHR-407 Birmingham AL 35294

VOLCANI, BENJAMIN ELAZARI, microbiology educator; b. Ben-Shemen, Israel, Jan. 4, 1915; came to U.S., 1956, naturalized, 1963; s. Isaac and Sarah (Krieger) V.; m. Eleanor Toni Solomons, Mar. 14, 1948; 1 son, Yanon. M.Sc., Hebrew U., 1936, Ph.D., 1941. Vis. scientist Inst. Tech. Delft, Netherlands, 1937-38, U. Utrecht, 1938-39; staff Sieff Research Inst., Weizmann Inst. Sci., Israel, 1939-58, head sect. microbiology, 1948-58; research fellow plant nutrition U. Calif.-Berkeley, 1945-46, Hopkins Marine Sta., Stanford (Calif.) U., 1946-47, Calif. Inst. Tech., Pasadena, 1947-48, U. Wis.-Madison, 1948; research assoc. Inst. Pasteur, Paris, 1951: research assoc. biochemistry virus lab. U. Calif.-Berkeley, 1957-59; vis. prof. Tenovus Cancer Inst., Cardiff, Wales, 1972-73; prof. microbiology U. Calif. Scripps Inst. Oceanography, La Jolla, 1959—; NASA cons., 1966. Editorial bd.: Ann. Rev. Microbiology, 1962-67, 78; editorial bd.: Sci. Total Environ, 1975-79, Bioinorganic Chemistry, 1976-78; Contbr. articles to profl. jours. NSF fellow, 1960-70; NIH/USPHS fellow, 1960—; Elsa U. Pardee Found. grantee, 1971-75. Mem. Am. Soc. Microbiology, Am. Soc. Cell Biology, Western Soc. Naturalists, So. Calif. Soc. Electron Microscopy, Soc. Gen. Microbiology (U.K.). Subspecialties: Microbiology; Biochemistry (biology). Current work: Role of silicon in cellular metabolism and diseases. Office: Univ Calif Scripps Inst Oceanography La Jolla CA 92093 Home: 6708 Muirlands Dr La Jolla CA 92037

VOLD, BARBARA SCHNEIDER, research biochemist, consultant; b. Oakland, Calif., Jan. 3, 1942; d. Julius Mackey and Thea Elfrieda (Riedel) Schneider. B.A. in Zoology with distinction, U. Calif., Berkeley, 1963; M.S. (univ. fellow), U. Ill., Urbana, 1964; Ph.D. in Cell Biology (NIH predoctoral fellow), U. Ill., Urbana, 1967. Postdoctoral fellow M.I.T., Cambridge, 1967-68; postdoctoral fellow Scripps Clinic and Research Found., La Jolla, Calif., 1968-69, research assoc., 1969-70, staff assoc., 1970-76; sr. biochemist SRI Internat., Menlo Park, Calif., 1977—; cons. NIH, 1973-77. Contbr. articles to profl. publs. NIH postdoctoral fellow, 1967-69; research career devel. awardee, 1971-76; NIH research grantee, 1970—; research grantee NSF, 1977-79, 80-82. Mem. Am. Soc. Biol. Chemists, Am. Soc. Microbiology, Phi Beta Kappa. Subspecialties: Molecular biology; Biochemistry (biology). Current work: Structure and function of transfer ribonucleic acids; sequence of tRNA and tRNA genes; differentiation in Bacillus subtilis; function of modified nucleosides; immunoassays and monoclonal antibodies to nucleosides. Patentee in field. Office: Biomed Research SRI Interna 333 Ravenswood Ave Menlo Park CA 9402

VOLICER, LADISLAV, pharmacologist, physician; b. Prague, Czechoslovakia, May 21, 1935; s. Ladislav and Vilma (Molnarova) V.; m. Olga Holeckova, July 14, 1959; children: Irena, Katerina; m. Beverly J. Beers, May 20, 1972; children: Susan, Marika, Nadine. M.D. with honors, Charles U., Prague, 1959; Ph.D., Czechoslovak Acad. Scis., 1964. Resident Hosp. Jindrichuv Hradec, Czechoslovakia, 1959-61; vis. assoc. Nat. Heart Inst., NIH, Bethesda, Md., 1965-66; research assoc. Inst. Pharmacology, Czechoslovak Acad. Sci., Prague, 1966-68; research asst. prof. pharmacology U. Munich, W.Ger., 1968-69; from asst. prof. to prof. pharmacology Boston U., 1969—; clin. pharmacologist E. N. Rogers Meml. VA Hosp., Bedford, Mass., 1978—; mem. Mass. Drug Formulary Commn., 1977-83. Editor/co-editor books in field; contbr. articles to sci. jours. Merck faculty devel. awardee, 1971-72; Nat. Inst. Alcoholism grantee, 1972-78; Nat. Inst. Drug Abuse grantee, 1973-77; others. Mem. Am. Soc. Pharmacology and Exptl. Therapeutics, Research Soc. Alcoholism, Soc. Neurosci. Unitarian. Club: Masaryk (Boston). Subspecialties: Pharmacology; Gerontology. Current work: Research on pharmacology of alcoholism and aging; teaching. Home: 11 Beverly Rd Bedford MA 01730 Office: 200 Springs Rd Bedford MA 01730

VOLKMAN, ALVIN, educator, physician; b. Bklyn., June 10, 1926; s. Henry Phillip and Sarah Lucille (Silverstein) V.; m. Carol Ann Fishel, Jan. 27, 1973; children: Jeffrey C., Natalie F.; m.; children from previous marriage: Karl F., Nicholas J., Rebecca J. Evans, Margaret R. Werrell, Deborah A. Falls. B.S., Union Coll., Schenectady, 1947; M.D., SUNY, Buffalo, 1951; D.Phil., Oxford U., 1963. Diplomate: Am. Bd. Pathology, 1959. Intern Mt. Sinai Hosp., Cleve., 1951-52; research fellow anatomy Western Res. U. Sch. Medicine, Cleve., 1952-54; resident, sr. resident in pathology Peter Bent Brigham Hosp., Boston, 1956-59, asst. in pathology, 1959-60; teaching fellow pathology Harvard Med. Sch., 1956-69; asst. prof. pathology Columbia U. Coll. Physicians and Surgeons, 1960-66; asst. to assoc. mem. Trudeau Inst., Saranac Lake, N.Y., 1960-77; prof. pathology East Carolina U. Sch. Medicine, Greenville, N.C.; mem. immunol. scis. study sect. NIH, 1974-79, chmn., 1977-79; cons. NIH; pathologist Pitt County Meml. Hosp., Greenville. Contbr. chpts. to books, articles to profl. jours. Served to lt. USNR, 1954-56. Mem. Am. Assn. Immunologists, Am. Soc. Microbiology, Am. Soc. Hematology, Reticuloendothelial Soc., N.Y. Acad. Scis., AAAS. Jewish. Subspecialties: Immunobiology and immunology; Infectious diseases. Current work: Origin, population regulation and functions of mononuclear phagocytes; regulation of population renewal and origins of functional diversity among subpopulations of mononuclear phagocytes. Office: Dept Pathology East Carolina U Sch Medicine Greenville NC 27834

VOLKMAN, DAVID J., physician, research scientist; b. Bklyn., Jan. 11, 1945; s. Clarence and Ruth (Fox) V.; m. Pamela Marion Bickerman, Jan. 29, 1967; children: Eric Solomon, Aaron Jon. B.S., Union Coll., 1966; Ph.D., U. Rochester, 1971, M.D., 1976. Diplomate: Am. Bd. Internal Medicine, Am. Bd. Allergy and Immunology. Research assoc. Sloan-Kettering Inst., N.Y.C., 1971-72; intern U. Pitts., 1976-77, resident in internal medicine, 1977-78; clin. assoc. NIH, Bethesda, Md., 1978-80; med. officer AID, 1980-82, sr. investigator, 1983—, vice chmn, clin. research subpanel. Contbr. numerous articles to sci. publs. Fellow ACP; mem. Am. Fedn. Clin. Research, Am. Assn. Immunologists, Sigma Xi. Subspecialties: Immunology (medicine); Internal medicine. Current work: elucidation of the mechanisms of human cellular immunology through techniques of human T cell cloning and constructing hybrid molecules linking toxins to cell specific moieties; applying these techniques to clinical diseases. Home: 2804 Blaine Dr Chevy Chase MD 20815 Office: NIH Bldg 10 Room 11 B 09 Bethesda MD 20205

VOLLUM, HOWARD, corp. exec.; b. 1913; m. (married). B.A., Reed Coll., 1936. With Murdock Radio & Appliance Co., 1936-41; a founder, chmn. Tektronix, Inc., 1946—; dir. U.S. Nat. Bank Oreg., Pacific Power & Light Co. Served with U.S. Signal Corps, 1941-45. Subspecialty: Electronics. Office: Tektronix Inc PO Box 500 Beaverton OR 97005

VOLSKY, DAVID JULIAN, pathology educator; b. Wroclaw, Poland, Nov. 10, 1948; came to U.S., 1981; s. Izaak and Ilana (Forc) V.; m. Barbara Paszt, Nov. 11, 1972; children: Karin, Nelly. B.Sc., Ben-Gurion U., 1972; M.Sc., Hebrew U., 1975, Ph.D., 1979. Teaching asst. Hebrew U., Jerusalem, 1976-78, teaching instr., 1978-79; postdoctoral researcher Karolinska Inst., Stockholm, 1979-81; asst. prof. pathology and biochemistry U. Nebr. Med. Center, Omaha, 1981-83, asso. prof., 1983—, mem. peer rev. com., 1982. Recipient Landau Prize in biochemistry, 1978; European Molecular Biology Orgn. fellow, 1978, 79-81; NIH grantee, 1981—. Mem. Am. Soc. Biol. Chemistry, Internat. Assn. Comparative Research on Leukemia and Related Diseases, AAAS. Subspecialties: Biochemistry (biology); Molecular biology. Current work: Tumor biology and virology; molecular mechanisms of cell transformation; DNA microinjection; membrane biology; receptors, reconstruction of membranes. Home: 3363 S 115th St Omaha NE 68144 Office: Univ Nebr Med Center 42nd and Dewey Ave Omaha NE 68105

VOLZ, RICHARD, electrical and computer engineering educator, researcher, university administrator; b. Woodstock, Ill., July 10, 1937; m., 1961; 2 children. B.S., Northwestern U., 1960, M.S., 1961, Ph.D. in Elec. Engring, 1964. Assoc. prof. U. Mich., Ann Arbor, 1964-77, prof. elec. and computer engring., 1977—, assoc. chmn. elec. and computer engring. dept., 1978-79, assoc. dir. computer ctr., 1979-82, dir. Robotics Lab., 1981—; mem. program com. Nat. Electronics Conf., 1968. NSF grantee, 1965-68, 70-72. Mem. IEEE, Assn. Computing Machinery, Soc. Mfg. Engrs., Internat. Inst. Robotics. Subspecialty: Robotics. Office: Dept Elec and Computer Engring U Mich Ann Arbor MI 48109

VON BAEYER, HANS CHRISTIAN, theoretical physicist, educator, researcher; b. Berlin, Apr. 6, 1938; m.; 2 children. A.B., Columbia U., 1958; M.Sc., U. Miami, 1961; Ph.D. in Physics, Vanderbilt U., 1964. Research assoc. in physics McGill U., Montreal, Que., Can., 1964-65, asst. prof., 1965-68; from asst. prof. to assoc. prof. Coll. William and Mary, Williamsburg, Va., 1968-75, prof. physics, 1975—, chmn. dept. physics, 1972-78; dir. Va. Assoc. Research Campus, 1979—; vis. prof. Tri-Univ Meson Facility and Simon Fraser U., 1978-79. Fellow Am. Phys. Soc.; mem. Fedn. Am. Scientists, AAUP. Subspecialty: Theoretical physics. Office: Dept Physics Coll William and Mary Williamsburg VA 23185

VON BERNUTH, ROBERT DEAN, engineering educator, researcher; b. Del Norte, Colo., Apr. 14, 1946; s. John D. and Bernice H. (Dunlap) von B.; m. Judy W. Wehrman, Dec. 29, 1969; children: Jeanie D., Suzann S. B.S., Colo. State U., Fort Collins, 1968; M.S., U.

Idaho-Moscow, 1970; M.B.A., Claremont Grad. Sch., 1980; Ph.D., U. Nebr.-Lincoln, 1982. Registered profl. engr., Nebr., Calif. Nuclear power engr. Betis Atomic Power Labs., Idaho Falls, Idaho, 1973-74; product mgr. Rain Bird Sprinkler Mfg., Glendora, Calif., 1974-80; instr. U. Nebr., Lincoln, 1980-82; assoc. prof engring. U. Tenn., Knoxville, 1982—; prin. Von-Solo Cons., Knoxville, 1982-83. Served with USN, 1969-73. Decorated D.F.C. (2); named Hon. Engr. Colo. Engring. Council, 1968. Mem. Am. Soc. Agr. Engrs., Irrigation Asn., Am. Mktg. Assn., Internat. Commn. Irrigation and Drainage, Sigma Xi, Alpha Zeta, Omicron Delta Kappa, Tau Beta Pi, Kappa Mu Epsilon, Gamma Sigma Delta. Republican. Subspecialty: Agricultural engineering. Current work: Research on irrigation and erosion of agricultural soils, application of economics and statistics to analysis and synthesis of problems associated with irrigation and erosion; measurement of size and velocity of falling water deposits. Patentee in field. Office: Agr Engring Dept U Tenn Knoxville TN 37901-1071

VONDER HAAR, THOMAS HENRY, atmospheric science educator, consultant; b. Quincy, Ill., Dec. 28, 1942; s. Paul and Helen Vonder H.; m. Dee M. Clark, July 19, 1980; children: Kim, Kurt, Nick, Krista. B.S., Parks Coll., 1963; M.S., U. Wis., 1964, Ph.D., 1968. Research asst. dept. meteorology U. Wis., 1963-65, research project supr., 1965-67, asst. scientist Space Sci. and Engring. Center, 1968-69, assoc. scientist, 1969; asst. prof. dept. atmospheric sci. Colo. State U., Ft. Collins, 1970-72, assoc. prof., 1972-78, head dept. atmospheric sci., 1974—, prof., 1978—, acting dean Coll. Engring., 1982—; mem. sub panel on radiation Gate adv. panel Nat. Acad. Sci., 1973-76; mem. com. on radiation (Energy Council), 1976—; mem. NASA Sci. Team, 1975-82; mem. sci. and analysis team (Earth Radiation Budget Expt.), 1980—. Contbr. articles to profl. jours. Mem. Ft. Collins Mayor's Blue Ribbon Panel; active Girl Scouts U.S.A., Boy Scouts Am. Nat. merit scholar, 1956; Ill. State scholar, 1956-60. Mem. Am. Meteorol. Soc. (1980, 2d half-century award, councilor, exec. com.), Nat. Acad. Sci. (bd. atmospheric sci. and climate), World Meteorol. Soc. Club: Colo. State U. Flying (Ft. Collins). Subspecialties: Meteorology; Satellite studies. Current work: Design and use of measurements from meteorological satellites, radiation physics, global climate,remote sensing and image processing teaching and research in department of atmospheric science Colorado State University. Home: 3501 Shiloh Dr Fort Collins CO 80521 Office: Dept Atmospheric Sci Colo State U Fort Collins CO 80523

VON FRAUNHOFER, JOSEPH ANTHONY, biomaterials science educator; b. London, Eng., Nov. 9, 1940; came to U.S., 1978; s. Hans and Jessie Josephine (Schoen) von F.; m. Anne Marsom, Sept. 7, 1962 (div. 1979); children: Nicola Anne, Michael Anthony. B.S., U. London, Eng., 1963, M.S., 1967; Ph.D., Council Nat. Acad. Awards, London, 1969. Chartered chemist; chartered engr. Sci. officer Brit. Rail Research Div., London, 1963-64; scientist Harris Plating Ltd., London, 1964-65; sr. officer research div. Gas Council, London, 1965-70; sr. lectr., dept. chmn. Inst. Dental Surgery, U. London, 1970-78; prof. biomaterials sci. U. Louisville., 1978—. Author: Potentiostat and Its Applications, 1972, Concise Corrosion Science, 1974, Paint Formulation, 1981, Protective Paint Coatings for Metals, 1976, Concise Paint Technology, 1977, Instrumentation in Metal Finishing, 1975, Basic Metal Finishing, 1976, Statistics in Medical, Dental and Biological Studies, 1976, Scientific Aspects of Dental Materials, 1975; contbr. sci. articles to profl. publs. Fellow Royal Soc. Chemistry, Instn. Corrosion Sci. and Tech. (sec. 1977-78); mem. Instn. Metallurgists. Club: Chemical (London). Subspecialties: Biomaterials; Corrosion. Current work: Biomaterials, wear and degradation of materials in the biosystem. Chemistry and metallurgy of corrosion of metals and alloys in the biosystem and in engineering. Protective coatings for metals against corrosion. Home: 2032 Eastern Pkwy Louisville KY 40204 Office: Health Scis Center Dental Sch Univ Louisville Louisville KY 40204

VON GIERKE, HENNING EDGAR, government official, educator; b. Karlsruhe, Germany, May 22, 1917; came to U.S., 1947, naturalized, 1977; s. Edgar and Julie (Braun) Von G.; m. (married); 2 children. Dipl. Ing., Karlsruhe Tech., 1943, Dr. Engr., 1944. Asst. in acoustics Karlsruhe Tech., 1944-47, lectr., 1946; cons. Aerospace Med. Research Labs, Wright-Patterson AFB, Ohio, 1947-54, chief bioacoustics br., 1954-63, dir. biodynamics and bionics div., 1963—; asso. prof. Ohio State U., 1963—; clin. prof. Wright State U., 1980—; mem. com. hearing and bioacoustics Armed Forces NRC, 1953, bio-astronaut, com., 1959-61; mem. adv. com. flight medicine and biology NASA, 1960-61. Author numerous tech. publs., book chpts. Recipient Dept. Def. Disting. Civilian Service award, 1963; Hubertus Strughold medal, 1980. Fellow Acoustical Soc. Am. (pres. 1979-80), Aerospace Med. Assn. (v.p. 1966-67, E. Liljenkrantz award 1966, A.D. Tuttle award 1974), Inst. Environ. Scis. (hon.), Internat. Acad. Aviation and Space Medicine; mem. Inst. Noise Control Engring., Biomed. Engring. Soc., Nat. Acad. Engring. Subspecialties: Biomedical engineering; Human factors engineering. Current work: Biodynanics; effects of acceleration and weightlessness in safety; health and performance; injury protection; bioacoustics. Researcher in bioacoustics, acoustics, biomechanics and bioengring. Home: 1325 Meadow Ln Yellow Springs OH 45387 Office: Biodynamics and Bionics Div Air Force Aerospace Med Research Lab Wright Patterson AFB OH 45433

VON HIPPEL, PETER HANS, chemistry educator; b. Goettingen, Germany, Mar. 13, 1931; came to U.S., 1937, naturalized, 1942; s. Arthur Robert and Dagmar (franck) von H.; m. Josephine Baron Raskind, June 20, 1954; children: David F., James A., Benjamin J. B.S., MIT, 1952, M.S., 1953, Ph.D., 1955. Phys. biochemist Naval Med. Research Inst., Bethesda, Md., 1956-59; from asst. prof. to assoc. prof. biochemistry Dartmouth, 1959-67; prof. chemistry, mem. Inst. Moledular Biology U. Oreg., 1967—; dir. inst. Inst. Moledular Biology of U. Oreg., 1969-80; chmn. dept. chemistry U. Oreg., 1980—; mem. study sect. USPHS, 1963-67; chmn. biopolymers Gordon Conf., 1968; mem. trustees' vis. com. biology dept. MIT, 1973-76; mem. bd. sci. counsellors Nat. Inst. Arthritis, Metabolic and Digestive Diseases, NIH, 1974-78; mem. council Nat. Inst. Gen. Med. Scis., NIH, 1982—. Editorial bd.: Jour. Biol. Chemistry, 1967-73, 76-82, Biochem. Biophys. Acta, 1965-70, Physiological Reviews, 1972-77, Biochemistry, 1977-80; Contbr. articles to profl. jours. and books. Served as lt. M.S.C. USNR, 1956-59. NSF predoctoral fellow, 1953-55; NIH postdoctoral fellow, 1955-56; NIH sr. fellow, 1959-67; Guggenheim fellow, 1973-74. Fellow Am. Acad. Arts and Scis.; mem. AAAS, Am. Chem. Soc., Am. Soc. Biol. Chemists, Biophys. Soc. (mem. council 1970-73, pres. 1973-74), Nat. Acad. Scis., Soc. Gen. Physiology, Fedn. Am. Scientists, Sigma Xi. Subspecialties: Molecular biology; Biophysical chemistry. Current work: Protein-nucleic acid intreaction; regulation of gene expression; DNA replecation; MRNA transcription. Home: 1900 Crest Dr Eugene OR 97405

VON LEDEN, HANS, surgeon, laryngologist, educator; b. Breslau, Prussia, Nov. 20, 1918; s. Peter Paul and Elizabeth (Freter) von L.; m. Mary Louise Shine, Jan. 10, 1948; children: Jon Eric, Lisa Maria. M.D., Loyola U., Chgo., 1941. Diplomate: Am. Bd. Otolaryngology. Intern Mercy Hosp.-Loyola U. Clinics, 1941-42; resident Presbyn. Hosp.-Rush Med. Coll., Chgo., 1942-43; fellow in otolaryngology Mayo Found., Rochester, Minn., 1943-44; asst. Mayo Clinic, 1945; instr. Loyola U., 1947-51; asst. prof. to assoc. prof. Northwestern U. Med. Sch., Chgo., 1951-61; assoc. prof. UCLA, 1961-66; prof. U. So. Calif. Med. Sch., Los Angeles, 1966—, med. dir., 1966—; cons. U.S. Navy, 1947—, W.Ger., 1960—, Lenox Hill Hosp., N.Y.C., 1968—, Juilliard Sch. Music, 1969—; mem. adv. staff Gov. Ill., 1953-61. Author: (with Rand and Rinfret) Cryosurgery, 1968, (with Cahan) Cryogenics in Surgery, 1971; contbr. articles, chpts. to med. publs.; author sci. films; chief editor: ORL Digest; editorial bd. med. jours. Col., a.d.c. Gov. La., 1976—. Served to lt. USNR, 1945-46. Recipient numerous awards, including Minerva award Third Internat. Festival for Medico-Sci. Films, 1957; Bucranio award U. Padua, 1958; Gold medal Italian Red Cross, 1959; award of Honor Am. Acad. Ophthalmology and Otolaryngology, 1959; Casselberry award Am. Laryngol. Assn., 1962; Manuel Garcia prize Internat. Assn. Logopedics and Phoniatrics, 1968; Gutzmann medaille Deutsche Gesellschaft fur Otorhinolrynglogie, 1980. Fellow ACS, Internat. Coll. Surgeons (past pres. U.S. sect., internat. bd. govs., hon. fellow), Am. Acad. Otolaryngology (life), AAAS, Am. Soc. Head and Neck Surgery (sr.), Am. Acad. Facial Plastic and Reconstructive Surgery (dir. 1962-76), Am. Speech and Hearing Assn., Collegium Medicorum Theatri (sec.-coordinator), Am. Coll. Cryosurgery, Academia Peruana de Cirugia (hon.); mem. AMA (Hektoen medal 1960, Sci. Achievement award 1980), Calif. Med. Assn. (past del.), Los Angeles County Med. Assn. (past pres., past chmn., trustee), Am. Council Otolaryngology (dir. 1973-76), Am. Fedn. Clin. Research, Assn. Mil. Surgeons U.S. (life), Soc. Med. Cons. to Armed Forces, Pan Am. Assn. Otorhinolaryngology and Bronchoesophagology (sec. gen. 1966-80, pres. 1980-82, cons. 1982—), Pacific Coast Oto-Ophthal. Soc., Paracelsus Soc. (past pres.), Los Angeles Soc. Otolaryngology, Bay Dist. Surg. Soc., Westwood Acad. Medicine and Dentistry (past pres.), Mil. Order World Wars (comdr., comdr. in chief 1962-63), Sigma Xi; hon. mem. Soc. Mil. Otolaryngologists, Deutsche Gesellschaft fur Hals-Nasen-Ohren-Heilkunde, Kopf-und Hals-Chirurgie, Japan Soc. Oto-Rhino-Laryngology, Sociedad Mexicana de Otorrinolaringologia, Sociedad Colombiana de Otorrinolaringologia y Bronco-Esofagologia, Associacion Argentina de Logopedia, Foniatria y Audiologia, Sociedad Chilena de Otorrinolaringologia, Sociedad Peruana de Otorrinolaringologia y Broncoesofagologia, Sociedad de Otorrinolaringologia de El Salvador, Sociedad Panamena de Otorrinolaringologia, Dallas So. Clin. Soc., Greek Soc. Noise Pollution Research; corr. mem. Dansk Oto-Laryngologisk Selskab, Sociedad Cubana de Cancerologia. Republican. Roman Catholic. Clubs: Marines Meml. (San Francisco); Army and Navy (Washington). Subspecialty: Otorhinolaryngology. Current work: Physiology and pathology vocal system; development of new technics and instruments for the diagnosis and treatment of laryngeal diseases and voice disorders. 90024X1 Home: 259 Tilden Ave Los Angeles CA 90049

VON TERSCH, LAWRENCE WAYNE, educator; b. Waverly, Iowa, Mar. 17, 1923; s. Alfred and Martha (Emerson) Von T.; m. LaValle Sills, Dec. 17, 1948; 1 son, Richard George. B.S., Iowa State U., 1943, M.S., 1948, Ph.D., 1953. From instr. to prof. elec. engring. Iowa State U., 1946-56; dir. computer lab. Mich. State U., 1956—, prof. elec. engring., chmn. dept., 1958-65, asso. dean engring., 1965-68, dean, 1968—. Author: (with A. W. Swago) Recurrent Electrical Transients, 1953. Mem. IEEE, Sigma Xi, Tau Beta Pi, Eta Kappa Nu, Phi Kappa Phi, Pi Mu Epsilon. Subspecialties: Computer engineering; Electrical engineering. Home: 4282 Tacoma Blvd Okemos MI 48864 Office: Coll Engring Michigan State U East Lansing MI 48823

VONVOIGTLANDER, PHILIP FRIEDRICH, pharmacologist; b. Jackson, Mich., Feb. 3, 1946; s. Frederick and Elizabeth (Nation) VonV.; m. Barbara J. Armstrong, June 10, 1968; children: Erika A., Charlotte E., Karin C. B.S., Mich. State U., 1968, D.V.M., 1969, M.S., 1970, Ph.D., 1972. Research scientist Upjohn Co., Kalamazoo, 1972—, sr. scientist, 1982—. Contbr. articles and abstracts to profl. lit. Active in monitoring local environ. hazards. Mem. Am. Soc. Pharmacology and Exptl. Therapeutics, Soc. Neurosci., AAAS, Phi Zeta. Club: Macatawa Bay Yacht. Subspecialties: Neuropharmacology; Neurochemistry. Current work: Development of animal models of psychiatric and neurological disorders for discovery and evaluation of new therapeutic agents. Home: 1 S Lake Doster Plainwell MI 49080 Office: Upjohn Co Kalamazoo MI 49001

VOOK, RICHARD WERNER, materials sci. educator, researcher; b. Milw., Aug. 2, 1929; s. Fred Ludwig and Hedwig Anna (Werner) V.; m. Julia Deskins, Sept. 7, 1957; children: Katherine, Elizabeth, Richard, Frederick. B.A., Carleton Coll., 1951; M.S. in Physics, U. Ill., 1952, Ph.D., 1957. Staff physicist IBM Research Lab., Yorktown Heights, N.Y., 1957-61; sr. research physicist Franklin Inst. Research Labs., Phila., 1961-65; assoc. prof. Syracuse (N.Y.) U., 1965-70, prof. materials sci., 1970—. Contbr. articles, revs. to profl. publs. Mem. Am. Phys. Soc., Am. Vacuum Soc., Electron Microscope Soc. Am., Metall Soc. of AIME, ASTM, Phi Beta Kappa, Sigma Xi, Pi Mu Epsilon. Republican. Lutheran. Subspecialty: Thin Films and Surfaces. Current work: Epitaxy, catalysis, electrical contact phenomena, tribology, electron microscopy (scanning and transmission), electron diffraction (low and high energy), Auger electron spectroscopy, X-ray diffraction and spectroscopy. Home: 5592 Sentinel Heights Rd Nedrow NY 13120 Office: Syracuse U 409 Link Hall Syracuse NY 13210

VOURNAKIS, JOHN NICHOLAS, biochemistry educator, pharmaceutical company executive; b. Cambridge, Ohio, Dec. 1, 1939; s. Nicholas John and Panayiota (Andritsakis) V.; m. Karen Ann Munro, Sept. 9, 1961; 1 son, Christopher N. B.A., Albion Coll., 1961; Ph.D. in Phys. Chemistry, Cornell U., 1968. Research assoc. MIT, Cambridge, 1969-72, Harvard U., 1972-73; mem. faculty Syracuse (N.Y.) U., 1973—, prof. biochemistry, 1983—; cons. Bristol-Meyers Co., Syracuse, 1982-83, dir. genetic engring., 1983—. Contbr. over 50 articles to tech. publs. NIH, NSF, Bristol-Myers Co., Alton-Jones Found. grantee, 1964—. Mem. AAAS, Fedn. Am. Biochemists, Biophys. Soc. Subspecialties: Biochemistry (biology); Genetics and genetic engineering (biology). Current work: Secondary structure of Eukaryotic mRNA as it relates to the initiation of protein synthesis; DNA and RNA structure and function; DNA structural changes induced by the binding of anti-tumor drugs. Home: 816 Oakwood St Fayetteville NY 13066 Office: Syracuse Univ 130 College Pl Syracuse NY 13210

VOVIS, GERALD FRANCIS, geneticist; b. Chgo., Feb. 15, 1943; s. Frank Joseph and Elsie Mary (Mucha) V.; m. Carol Ann Klail, Aug. 21, 1965. B.A., Knox Coll., 1965; Ph.D., Case Western Res. U., 1971. Research assoc., dept. genetics Rockefeller U., N.Y.C., 1970-76, asst. prof., 1976-80; sr. research scientist, core tech. Collaborative Research Inc., Waltham, Mass., 1980—82, EMIA research and devel. program mgr., Lexington, Mass., 1982-83, dir. RFLP devel., Waltham, Mass., 1983—. Mem. Am. Soc. Biol. Chemists, AAAS, Am. Soc. Microbiology, Genetics Soc. Am., N.Y. Acad. Sci., Sigma Xi. Subspecialties: Molecular biology; Genetics and genetic engineering (biology). Current work: Research in structure and function of nucleic acids, control of gene expression, expression of foreign genes in Saccharomyces cerevisiae, development of DNA based diagnostic tests for inherited diseases. Office: 1365 Main St Waltham MA 02154

VRBA, FREDERICK JOHN, research astronomer; b. Cedar Rapids, Iowa, May 25, 1949; s. Fred and Emily Katherine (Dobry) V.; m. Sheryl Lynn Blunk, June 1, 1971; 1 dau., Marya Katherine. B.A. in Astronomy and Physics, 1971, Ph.D., U. Ariz., Tucson, 1976. Astronomer U.S. Naval Obs., Flagstaff, Ariz., 1976—. Mem. Am. Astron. Soc., Internat. Astron. Union, Sigma Xi. Subspecialty: Infrared optical astronomy. Current work: Optical and infrared wavelength investigations of star formation and the interstellar medium. Office: US Naval Observatory Flagstaff Station PO Box 1149 Flagstaff AZ 86002

WACHSPRESS, EUGENE LEON, mathematician; b. N.Y.C., Apr. 17, 1929; s. Sidney and Jean (Lichtenstein) W.; m. Natalie Janet Gross, Mar. 16, 1952; children: Amy Lynn, William Sidney, Daniel Avram. B.M.E., Cooper Union, 1950; B.S. in Math, Union Coll, Schenectady, N.Y., 1956, Ph.D., Rensselaer Poly. Inst., 1968. Nuclear engr., mathematician Knolls Atomic Power Lab., Schenectady, 1953-70, mathematician, 1971—; prof. math. U. Tenn.-Knoxville, 1983—; vis. fellow U. Dundee, Scotland, 1970-71. Author: Iterative Solution of Elliptic Systems, 1966, A Rational Finite Element Basis, 1975. Fellow Am. Nuclear Soc. (chmn. math. and comp. div. 1969); mem. Soc. Indsl. and Applied Math. Jewish. Subspecialties: Numerical analysis; Applied mathematics. Current work: Numerical analysis associated with numerical solution of partial differential equations, especially neutron diffusion, Navier-Stokes, and structural dynamics. Algebraic geometry foundations for finite element basis function construction.

WACHTER, WILLIAM JOHN, research and development director; b. Alliance, Ohio, July 22, 1923; s. Ferdin and Helen Alberta (Ormesher) W.; m. Barbara Lee Keep, Oct. 8, 1950; children: William John, Gretchen Lee, Heidi Ann, Frederica Jo, Annelies. B.S.M.E., U. Pitts., 1946; M.S.A.E., Case Western Res. U., 1955. Cert. profl. engr., Pa. Mem. staff Lord Mfg., Erie, Pa., 1947-50; staff engr., vibration cons. NASA, Cleve., 1950-53; mgr. applied physics projects Crush Beryllium Co., Cleve., 1953-55; engr. fellow Westinghouse Electric Corp., Bettis Atomic Power Lab., West Mifflin, Pa., 1955-67; pres., prin. cons. Wachter Assocs. Inc., Gibsonia, Pa., 1967-81; mgr. research and devel. U.S. Tool & Die, Gibsonia, 1981—. Fellow Am. Chem. Soc.; mem. ASME, Am. Nuclear Soc. Subspecialties: Nuclear fission; Nuclear fusion. Current work: Designing high burnup nuclear fuel rods; developing new concepts in nuclear spent fuel storage. Patentee in field (50). Office: US Tool & Die 5410 Rt 8 North Gibsonia PA 15090 Home: 411 English Rd Wexford PA 15090

WACKER, WALDON BURDETTER, immunologist, researcher; b. Garrison, N.D., Aug. 13, 1923; s. Edmund E. and Annette Eleanora (Walter) W.; m. Priscilla Jean Johnson, Feb. 26, 1955; children: Janet, Nancy, Steven. A.B., Washington U., St. Louis, 1949; M.S., U. Mich., 1951; Ph.D., Ohio State U., 1957. Chief bacteriologist VA Hosp., Dayton, Ohio, 1952-55; research assoc. Ohio State U., Columbus, 1958-59; asst. prof. U. Louisville, 1959-69, assoc. prof., 1969-80, prof., dept. ophthalmology, 1980—. Contbr. articles to profl. jours. Served with USN, 1942-46; PTO. NIH predoctoral fellow, 1956; recipient Career Devel. award NIH, 1962; NIH research grantee, 1962—. Mem. Assn. Research in Vision and Ophthalmology, Sigma Xi. Democrat. Lutheran. Subspecialties: Immunology (medicine); Microbiology (medicine). Current work: Development of animal models of autoimmune uveoretinitis, isolation and characterization of the active antigens, and investigation of the immunologic mechanisms of the disease process and relevance to human disease. Home: 7320 Keisler Way Louisville KY 40222 Office: Dept Ophthalmology 301 E Walnut St Louisville KY 40202

WACKERNAGEL, H(ANS) BEAT, astronomer; b. Basel, Switzerland, Aug. 31, 1931; came to U.S., 1958, naturalized, 1964; s. Karl H. and Georgine (Hagenback) W.; m. Irene E. Chavez, Dec. 5, 1974; children: Barbara, Jacob, William, Patricia. Dr. phil. II, U. Basel, 1958. Physicist Hdqrs. NORAD, Peterson AFB, Colo., 1979—; lectr. on celestial mechanics U. Colo., 1965-72. Fellow Brit. Interplanetary Soc.; mem. Internat. Astron. Union, Am. Astron. Soc., Swiss Astron. Soc., Astron. Soc. Basel. Republican. Methodist. Club: Colorado Springs Racquet. Lodge: North Colorado Springs Rotary. Home: 51 Broadmoor Hills Dr Colorado Springs CO 80906 Office: Hdqrs Space Command Code D06 Peterson AFB CO 80914

WACKERS, FRANS JOZEF THOMAS, cardiology educator, researcher; b. Echt, Netherlands, May 29, 1939; came to U.S., 1977; s. Thomas F. and Miep (Koopman) W.; m. Marjan A. Meyer, Sept. 29, 1972; children: Michiel, Paul. M.D., U. Amsterdam, 1970, Ph.D., 1970. Diplomate: Netherlands Bd. Internal Medicine, Netherlands Bd. Cardiology. Intern U. Amsterdam, 1965-66, 69-70; resident in medicine, 1972-74, fellow in cardiology, 1974-77; asst. prof. cardiology Yale U., 1977-81; assoc. prof. cardiology U. Vt.-Burlington, 1981-84, dir. nuclear cardiology, 1981-84; assoc. prof. cardiology and radiology Yale U., 1984—. Editor: Thallium-201 Myocardial Imaging, 1978, Myocardial Imaging in CCU, 1980. Fellow Am. Coll. Cardiology (editorial bd. jour.); mem. Am. Fedn. Clin. Research, Soc. Nuclear Medicine. Subspecialties: Cardiology; Nuclear medicine. Current work: Nuclear cardiology; noninvasive assessment of cardiac function and perfusion (Tl-201, Tc-99m, Au-195m). Office: University of Vermont Burlington VT 05401

WADDELL, THOMAS GROTH, chemistry educator; b. Madison, Wis., July 29, 1944; s. Harvey Anderson and Frieda (Groth) W. B.S., U. Wis., 1966; Ph.D., UCLA, 1969. Postdoctoral chemist UCLA, 1969; NIH postdoctoral fellow U. Calif.-Berkeley, 1970-71; asst. prof. U. Tenn., Chattanooga, 1971-76, assoc. prof., 1976-81; found. prof., 1976-81, alumni disting. prof., 1981—. Contbr. articles to profl. jours. Named Outstanding Tchr. U. Tenn. Alumni, 1975; Petroleum Research Fund research grantee, 1977-82. Mem. Am. Chem. Soc. (disting. service award 1978), AAAS, Tenn. Acad. Sci., Sigma Xi. Subspecialty: Organic chemistry. Current work: Chemistry of medicinal herbs, organic synthesis, mechanism of action of bio-active natural products, natural products chemistry. Home: 409 Cameron Circle Apt 601 Chattanooga TN 37403x Office: Dept Chemistry U Tenn Chattanooga TN 37402

WADDINGTON, CECIL JACOB, astrophysicist, educator; b. Cambridge, Eng., July 6, 1929; came to U.S., 1957, naturalized, 1973; s. Conrad H. and Cecil E. (Lascelles) W.; m. Jean C. Bassett Webb, Sept. 15, 1956. B.Sc. with honors, U. Bristol, Eng., 1952, Ph.D., 1955. Royal Soc. research student, 1955-59; research assoc. U. Minn., 1957-58; lectr. U. Bristol, 1959-61; vis. prof. Goddard Space Sci. Ctr., 1959; assoc. prof. U. Minn., 1961-69, prof. physics, 1969—; vis. prof. Imperial Coll., London, 1971-72. Contbr. articles to profl. jours. Served with Royal Elec. and Mech. Engrs., 1947-49. Recipient NASA Exceptional Sci. Achievement Medal, 1980. Fellow Am. Phys. Soc.; mem. Am. Astron. Soc., AAAS, Internat. Astron. Union. Subspecialties: Cosmic ray high energy astrophysics; Nuclear physics. Current work: Mass, charge, and energy spectra of cosmic ray nuclei observed from balloons and satellites. The characteristics of high energy nucleus-nucleus interactions using cosmic ray and artificially accelerated nuclei. Office: U Minn 116 Church St SE Minneapolis MN 55455

WADE, ADELBERT ELTON, pharmacologist, educator, researcher; b. Hilliard, Fla., Apr. 29, 1926; s. Adelbert Elton and Esther (Sundberg) W.; m. Mary Lucy Wade, Jan. 22, 1950; children: William Elton, James Howard. B.S. in Pharmacy, U. Fla., 1954, M.S. in Pharmacology, 1956, Ph.D., 1959. Instr. U. Fla., 1954-56; asst. prof. U. Ga., Athens, 1959-62, assoc. prof., 1962-67, prof., 1967—, head dept. pharmacology, 1968—; research in field. Contbr. articles to profl. jours. Served with USN, 1944-46. Recipient Creative Research award

U. Ga., 1981; NIH grantee, 1972—. Mem. Am. Soc. Pharmacology and Exptl. Therapeutics, Internat. Soc. Biochem. Pharmacology, Soc. Exptl. Biol. Medicine. Democrat. Baptist. Club: Torch (Athens). Subspecialties: Molecular pharmacology; Cancer research (medicine). Current work: Drug activity enzymes in the initiation and promotion of chemically induced cancer. Office: Univ of Ga Sch of Pharm GA 30802

WADE, CHARLES EDWIN, biomedical researcher, educator; b. Oakland, Calif., Nov. 23, 1949; s. Robert Henry and Jean (Pace) W.; m. Maryla Wilbur, Nov. 9, 1980; 1 son, Samuel Robert. B.A., Occidental Coll., 1970; M.A., U. Calif. Davis, 1973, Ph.D., U. Hawaii Sch. Medicine, 1979. Physiology intern Tripler Army Med. Ctr., Honolulu, 1979; vis. asst. prof. U. Calif.-Santa Cruz, 1980; postdoctoral trainee U. Calif.-San Francisco, 1979-80, postdoctoral fellow, 1980-82; research physiologist, dept. clin. investigation Presidio of San Francisco, 1982—; vis. lectr. U. Calif., 1982—; cons. Rockefeller U., N.Y.C., 1982—; researcher NASA-Ames Research Ctr., Moffett Field, Calif., 1982—. NIH fellow, 1980. Mem. Am. Physiology Soc., N.Y. Acad. Sci., Am. Coll. Sports Medicine, Sigma Xi. Subspecialties: Neuroendocrinology; Physiology (medicine). Current work: Regulation of vasopressin secretion; fluid and electrolyte homeostasis; renal handling of water and solutes; hormonal regulation of renal function. Office: Care: HSHH-Q Letterman Army Med Ctr Presdio of San Francisco San Francisco CA 94129 Home: 1355 Felder Rd Sonoma CA 95476

WADE, PATRICIA DIANE, neurobiologist, physiologist, researcher; b. Cedar Rapids, Iowa, Nov. 25, 1946; d. Oliver Wendell and Margaret Amalia (Grunewald) W. Student, U. Iowa; A.B. in Zoology, U. Calif.-Berkeley; Ph.D. in Physiology, U. Calif.-Berkeley, 1978. Postdoctoral fellow Rockefeller U., 1978-82, research assoc. in cell biology, 1983—; postdoctoral fellow Brandeis U., 1982. U. Iowa acad. achievement scholar, 1967; N.Y. State Health Program fellow, 1979; Muscular Dystrophy Assn. fellow, 1980, 81; U. Calif. grantee, 1976. Mem. Soc. Neurosci., N.Y. Acad. Sci., Assn. Women in Sci., Internat. Soc. Devel. Neurosci., Sigma Xi. Subspecialties: Neurobiology; Physiology (biology). Current work: Stability and change in organisms, learning. Home: 500 E 63d St New York NY 10021 Office: Rockefeller U 1230 York Ave New York NY 10021

WADESON, HARRIET CLAIRE, art therapy educator, consultant, psychotherapist; b. Washington, Jan. 9, 1931; d. Max Meyer and Sophie (Spectot) Weisman; m.; children: Elisa, Eric, Keith, Sinrod. B.A., Cornell U., 1952; M.A., Goddard Coll., 1975; M.S.W., Catholic U., 1976; Ph.D., Union Grad. Sch., 1978. Cert. Acad. Cert. Social Workers; lic. clin. social worker, Md., Ill. Research psychologist NIMH, Bethesda, Md., 1962-75; mem. faculty Grad. Sch. NIH, Bethesda, 1974-78; mem. faculty, coordinator art therapy Lindenwood Coll., Washington, 1976-78; pvt. practice psychotherapy, Washington, 1968-78; dir. art therapy U. Houston, 1978-80, U. Ill., Chgo., 1980—; lectr. in field. Author: Art Psychotherapy, 1980; contbr. chpts. to books, articles to profl. jours. Recipient Rush Bronze medal Am. Psychiat. Assn., 1971; 1st prize for art Smithsonian Inst., 1968. Mem. Nat. Assn. Social Workers, Am. Psychol. Assn., Am. Art Therapy Assn. (registered art therapist, 1st prize for research 1978), Assn. Humanistic Psychology. Subspecialties: Psychotherapy; Art psychotherapy. Current work: Education for psychotherapists in training; art psychotherapy. Office: University Illinois PO Box 4348 Chicago IL 60680

WADLINGER, ROBERT LOUIS, chemistry educator, researcher; b. Phila., Mar. 20, 1932; s. Ralph Louis and Anna (Lutzuch) W.; m. Anna Carolyn (Shride), June 25, 1955; children: Mark Robert, Gregory David, Lynne Christine. A.B., LaSalle Coll., 1953; Ph.D., Cath. U. Am., 1961. Research asst. Franklin Inst. Labs., Phila., 1953-55; Nat. Bur. Standards, Washington, summers 1955, 56, 57; sr. research chemist Mobil Oil Corp., Paulsboro, N.J., 1960-62; assoc. prof. Oneonta State Coll., N.Y., 1962-65, Niagara U. (N.Y.), 1965-76; asst. prof. Pa. State U., Middletown, 1981—; cons. environ. chemistry Hooker Chem. Corp., Grand Island, N.Y., 1980. Chmn. Air Pollution Control Assn. Western N.Y.; coordinator Air Pollution Symposium Series. Mem. Am. Chem. Soc., Holy Name Soc. Roman Catholic. Subspecialties: Photochemistry; Physical chemistry. Current work: Novel zeolite synthesis and characterization; quantum mechanics modifications. Patentee Zeolites Alpha and Beta, 1960-62. Home: 7 Lynewood Bldg Middletown PA 17057 Office: Pa State U Capitol Campus Middletown PA 17057

WADSWORTH, DALLAS FREMONT, plant pathologist, educator; b. Arcadia, Okla., Mar. 2, 1922; s. Charles M. and Myrtle (Dowell) W.; m. Claribel Lee, Nov. 21, 1950; 1 son, Mike. Ph.D., U. Calif.-Davis, 1966. Faculty dept. plant pathology Okla. State U., 1949—, prof., 1974—; cons. in field. Served with USN, 1942-46. Mem. Am. Phytopathol. Soc., Am. Peanut Research and Edn. Assn. Subspecialty: Plant pathology. Current work: Soil-borne plant diseases. Office: Okla State U Stillwater OK 74078

WADSWORTH, WILLIAM STEELE, JR., chemistry educator; b. Hartford, Conn., May 6, 1927; s. William Steele and Arlene W. (Graham) W.; m. Nancy Hegan, Aug. 11, 1956; children: John, William, Thomas, Timothy. B.S., Trinity Coll., 1950, M.S., 1952; Ph.D., Pa. State U., 1955. Research scientist Rohm & Haas Co., Phila., 1955-63; asst. prof. S.D. State U., Brookings, 1963-68, prof., 1968—; vis. prof. U. Notre Dame, 1978; vis. faculty fellow Princeton U., 1979-80; cons. in field. Contbr. numerous articles on chemistry to profl. jours. Mem. Brookings (S.D.) Sch. Bd., 1973-79; scoutmaster Boy Scouts Am.; mem. Service Club. Mem. Am. Chem. Soc., Sigma Xi, Phi Lambda Upsilon. Lodge: Kiwanis. Subspecialties: Organic chemistry; Catalysis chemistry. Current work: Organophosphorus chemistry, phosphates, mechanisms, stereochemistry, substitutions. Office: SD State U 117b Shepard Hall Brookings SD 57007

WAELSCH, SALOME GLUECKSOHN, genetics educator; b. Danzig, Germany, Oct. 6, 1907; came to U.S., 1933, naturalized, 1938; d. Ilyia and Nadia Gluecksohn; m. Heinrich B. Waelsch, Jan. 8, 1943; children—Naomi Barbara, Peter Benedict. Student, U. Konigsberg, Ger., U. Berlin; Ph.D., U. Freiburg, Ger., 1932. Research asso. in genetics Columbia U., 1936-55; asso. prof. anatomy Albert Einstein Coll. Medicine, 1955-58, prof., 1958-63, prof. genetics, 1963—, chmn. dept. genetics, 1963-76; mem. study sects. NIH. Fellow Am. Acad. Arts and Scis.; mem. Nat. Acad. Scis., Am. Soc. Zoologists, Am. Assn. Anatomists, Genetics Soc., Soc. Devel. Biology, Am. Soc. Naturalists, Am. Soc. Human Genetics, Sigma Xi. Subspecialties: Gene actions; Developmental biology. Research, numerous publs. on devel. genetics. Office: Dept Genetics Albert Einstein Coll Medicine 1300 Morris Park Ave Bronx NY 10461

WAFFLE, ELIZABETH L., parasitologist, educator; b. Marion, Iowa, Feb. 14, 1938; d. Norman J. and LeNora E. and Decker W. Ph.D., Iowa State U., 1967. Mem. faculty Armstrong State Coll., 1966-67, Iowa Wesleyan Coll., 1967-68; asst. prof. Eastern Mich. U., Ypsilanti, 1968—; cons. Ann Arbor (Mich.) Biol. Ctr., 1970-81. Mem. Mich. Acad. Arts and Scis., Am. Heartworm Soc. Clubs: Ann Arbor Kennel, Ann Arbor Dog Tng., German Shepherd Dog of Kensington, Half Arabian Assn. Mich. Subspecialties: Parasitology; Microbiology (veterinary medicine). Current work: Behavior of vectors of disease—mosquitoes. Office: Dept Biology Eastern Mich U Ypsilanti MI 48197

WAGGONER, PAUL EDWARD, agricultural scientist, agricultural experiment station administrator; b. Appanoose County, Iowa, Mar. 29, 1923; s. Walter Loyal and Kathryn (Maring) W.; m. Barbara Ann Lockerbie, Nov. 3, 1945; children: Von Lockerbie, Daniel Maring. S.B., U. Chgo., 1946; M.S., Iowa State Coll., 1949, Ph.D., 1951. With Conn. Agrl. Exptl. Sta., New Haven, 1951—, vice dir., 1969-72, dir., 1972—; lectr. Sch. Forestry Yale U., 1962; dir. N.E. Bancorp Inc., Union Trust Co. Contbr. articles to profl. jours. Served to 1st lt. USAAF, 1943-46. Guggenheim fellow, 1963. Fellow Am. Soc. Agronomy, AAAS, Am. Phytopath. Soc.; mem. Am. Acad. Arts and Sci., Am. Meterol. Soc. (award outstanding achievement biometeorology 1967), Am. Soc. Plant Physiology, Nat. Acad. Sci., Conn. Acad. Sci. and Engring. (v.p. 1976—). Club: Grad. (New Haven). Subspecialties: Plant pathology; Integrated systems modelling and engineering. Current work: Effect of weather on plants and insects; epidemiology of plant disease. Home: 314 Vineyard Rd Guilford CT 06437 Office: 123 Huntington St New Haven CT 06511

WAGMAN, ALTHEA M.I., research psychophysiologist, educator; b. Knoxville, Tenn., Dec. 22, 1933; d. David Gerard and Althea Devecmon (Wimbrough) Iliff; m. William D. Wagman, June 5, 1954; children: Althea Susan, David Wolfe, Ida Lee. B.S., Coll. William and Mary, 1954; M.S., Columbia U., 1958; Ph.D., So. Ill. U., 1966. Lic. psychologist, Md. Lectr. So. Ill. U., Carbondale, 1961-72; asst.prof. Towson (Md.) State U., 1967-69, assoc. prof., 1969-71; research scientist State of Md., 1971-77; research assoc. in psychiatry Md. Psychiat. Research Center U. Md. Med. Sch., Balt., 1977-83; prof. Loyola Coll., Balt., 1980; research assoc. prof. U. Md. Med. Sch., Balt., 1983—; chmn. Tng. Adv. Bd. Psychology Md. Dept. Mental Hygiene. Contbr. articles to profl. jours. USPHS fellow, 1964-66; grantee; NIH; NIMH. Mem. Am. Psychol. Assn., Soc. Neurosci., Soc. Psychophysiol. Research. Democrat. Subspecialties: Psychophysiology; Neuropsychology. Current work: Evoked potentials in schizophrenia. Home: 1533 Park Ave Baltimore MD 21217 Office: Box 3235 Baltimore MD 21228

WAGNER, AUBREY JOSEPH, energy consultant; b. Hillsboro, Wis., Jan. 12, 1912; s. Joseph Michael and Wilhelmina Johanna (Filter) W.; m. Dorothea Johanna Huber, Sept. 9, 1933; children: Audrey Wagner Elam Joseph M., James R., Karl E. B.S.C.E. magna cum laude, U. Wis., 1933; LL.D. (hon.), Newberry Coll., 1966; Ph.D. in Public Adminstrn. (hon.), Lenoir-Rhyne Coll., 1970. Registered profl. engr., Tenn. With TVA, 1934-78, asst. gen. mgr., Knoxville, 1951-54, gen. mgr., 1954-61, bd. dirs., 1961-62, chmn. bd., 1962-78; cons. energy and resource-use matters, Knoxville, Tenn., 1978—; seminar lectr., Pakistan, 1967, Salzburg, Austria, 1968; bd. dirs. U.S. Nat. Com. of World Energy Conf., 1975—. Contrbr. chpts. numerous articles to various nat., regional publs. Bd. dirs. Citizens for Home Rule, Knoxville, 1979—. Recipient Walter H. Zinn award Am. Nuclear Soc., 1978; named Engr. of Distinction Tenn. Technol. U., 1981. Mem. Tenn. Soc. Profl. engrs., Nat. Soc. Profl. Engrs., Nat. Acad. Engring., Explorers Club, Lambda Chi Alpha (Order of Achievement). Lutheran. Subspecialties: Civil engineering; Nuclear engineering. Current work: Working actively as proponent of nuclear energy and especially breeder reactor.

WAGNER, BERNARD M., pathologist; b. Phila., Jan. 7, 1928; s. John and Kathrine W.; m. Patricia, Mar. 1950; children: Cynthia Wagner Ginsberg, Nancy Wagner Albert, Robert. Student, U. Pa.; M.D., Hahnemann Med. Coll., 1949. Diplomate: Am. Bd. Pathology; Nat. Bd. Med. Examiners.; Lic. physician N.Y., 1953, Pa., 1955, Wash., 1958, N.J., 1977. Intern Phila. Gen. Hosp., 1949-50, resident in pathology, 1950; lectr. Phila. Coll. Pharmacy and Sci., 1955-58; asst. prof. pathology U. Pa., 1955-58; assoc. prof. pathology U. Wash., 1958-60, Robert L. King chair cardiovascular research, 1958-60; prof., chmn dept. pathology N.Y. Mkd. Coll., 1960-66; prof. pathology Coll. Physicians and Surgeons, Columbia U., 1967—; vis. prof. numerous univs.; mem. Wash. Bd. Med. Examiners, 1959-60; biomed. panel, sci. adv. bd. Chief of Staff, USAF, 1960-65; rep. Dept. of State and Dept. of Def. to Moscow for exchange of space research data, 1962-64; NIH rep. to Ministry of Health, Warsaw, 1962-66; cons. Office of Dir., Nat. Inst. Child Health and Human Devel., 1963-67, mem. program project study sect., 1965-68; cons. FDA, 1964-66; 70-74; neoplastic disease panel N.Y.C. Health Research Council, 1965-68; cons. N.Y. State Dept. Mental health, 1969—, VA, 1970; pathology subcom. Nat. Research Council Toxicology Adv. Ctr., 1976-78; environ. pathology registry com. Armed Forces Inst. Pathology, 1978—; com. on industry USAF Manned Space Program, 1960-62; cons. Merck Inst. Research, 1963, U.S. Vitamin and Pharm. Corp., 1963-66; v.p. Warner-Lambert Research Inst., 1966-67, dir., 1967-72; cons. Hoffman-LaRoche, Inc., 1967—, Office of Chief Scientist State of Israel, 1970—; dir. Electro-Nucleonics, Inc., 1973—, cons., 1972-73, G.D. Searle & Co., Calorie Control Council. Editor: Human Pathology, 1971—; assoc. editor: Connective tissue Research, 1977—, Jour. Environ. Pathology and Toxicology, 1978—; editorial bd.: Archives of Toxicology, 1976—, Investigative Microtechniques in Biology and Medicine, 1979—, Pathology Update Series, 1978—, Am. Jour. Surg. Pathology, 1977-81; cons. editor: Jour. Ultrastructural Pathology, 1977—; contbr. numerous articles to profl. jours., chpts. in books. Simon Guggenheim Foudn. scholar, 1944; fellow Dazian Found. Med. Research, 1953, Hosp. for Sick Children, London, 1959; recipient Charles W. Burr prize Phila. Gen. Hosp., 1950, Silver Medal U.S. Air Force Sci. Adv. Bd., 1965, Alumni award Hahnemann Med. Coll., 1966. Mem. Soc. Exptl. Medicine and Biology, Histochem. Soc., Soc. Pediatric Research, Am. Assn. pathologists, Internat. Acad. pathologists (council 1973-76, pres. U.S.-Can. div. 1982-83), AIAA, AAAS, Harvey Soc., Am. Soc. Clin. Pathology, Soc. Toxicology, Am. Coll. Toxicology, German Soc. Pharmacology (hon.), N.J. Soc. Pathology, Phi Lambda Kappa, Sigma Alpha Mu, Alpha Omega Alpha. Subspecialty: Pathology (medicine). Office: Dept Pathology Overlook Hosp Summit NJ 07901

WAGNER, BILL *See also* **WAGNER, FREDERICK WILLIAM**

WAGNER, DAVID HENRY, herbarium curator, botanist, educator; b. Detroit, Aug. 18, 1945. B.A., U. Puget Sound, 1968; M.S., Wash. State U., 1974, Ph.D., 1976. Vis. assst. prof. U. Wash., Seattle, summer 1976; curator herbarium U. Oreg., Eugene, 1976-78, dir., curator, 1978—; cons. Pres. Mount Pisgah Arboretum, 1980-82. Contbr. articles to sci. jours. Mem. Native Plant Soc. Oreg. (pres. 1981-82), Bot. Soc. Am., Internat. Assn. for Plant Taxonomy, Am. Soc. Plant Taxonomists, Am. Bryological and Lichenological Soc., Am. Fern Soc., others. Democrat. Subspecialties: Evolutionary biology; Taxonomy. Current work: Biosystematic, taxonomic, ecological, biogeographical, evolutionary studies of bryophytes and vascular plants of the Pacific Northwest. Office: U Oreg Dept Biology Eugene OR 97403

WAGNER, FREDERICK WILLIAM (BILL WAGNER), lawyer; b. Daytona Beach, Fla., Apr. 13, 1933; s. Adam A. and Nella (Schroeder) W.; children: Alan Frederick, Darryl William, Thomas Adam. B.A., U. Fla., 1955, LL.B. with honors, 1960. Bar: Fla. 1960, U.S. Supreme Ct. 1967; Cert. civil trial lawyer Nat. Bd. Trial Advocacy, Fla. bar. Pvt. practice law, Miami, Fla., 1960-63, pvt. practice law, Orlando, Fla., 1963-65, Tampa, Fla., 1965—; partner law firm Nichols, Gaither, Beckham, Colson, Spence & Hicks, Tampa, Fla., 1965-67, Wagner, Cunningham, Vaughan & McLaughlin (P.A. and predecessor names), 1967—, pres., 1967—; Mem. Gov.'s Judicial Nominations Commn., 1971-72, Constnl. Judicial Nominations Commn., 1972-75; mem. Fla. Bd. Bar Examiners, 1974-77; chmn. Civil Procedure Rules Com. Fla. Bar, 1977-78; bd. govs. Fla. Bar, 1978-83. Contbr. articles to profl. jours. Served to capt. USAF, 1955-57. Roscoe Pound fellow Am. Trial Lawyers Found. Fellow Am. Coll. Trial Lawyers, Internat. Coll. Trial Lawyers; mem. Assn. Trial Lawyers of Am. (bd. govs. 1973-80, exec. com. 1975-78, chmn. nat. membership com. 1975-76, dir. internal affairs 1976-77, treas. 1982—), Acad. Fla. Trial Lawyers (bd. govs. 1965—, pres. 1972-83), Bay Area Trial Lawyers Assn. (v.p. 1966-68), Lawyer-Pilots Bar Assn., N.Y. State, Calif. trial lawyers assns., Fla. Bar Found., U. Fla. Law Center Assn., U. Fla. Alumni Assn., Order of Coif, Fla. Blue Key, Beta Theta Pi. Democrat. Methodist. Subspecialties: Nuclear fusion; Advanced methods of power generation.. Home: 6090 River Trace Tampa FL 33617 Office: 708 Jackson St Tampa FL 33602

WAGNER, GERALD GALE, immunologist, educator; b. Plainview, Tex., June 3, 1941; s. Gerald Weeber and Rogene (Shepard) W.; m. Beverly Ann Hamilton, Aug. 16, 1962; children: Kristin Van, Lisa Deann. B.S., Tex. Tech U., 1963; M.A., U. Kans., 1965; Ph.D., 1968. Microbiologist Plum Island Animal Disease Center, Greenport, N.Y., 1968-70; microbiologist, ofcl.-in-charge coop. research div. East African Vet. Research Orgn., Muguga, Kenya, 1971-77; assoc. prof. dept. vet. microbiology and parasitology Coll. Vet. Medicine, Tex. A&M U., College Station, 1977—. Contbr. chpts. to books, articles to sci. jours. Nat. Acad. Scis.-Agrl. Research Service postdoctoral fellow, 1968-70. Mem. Am. Assn. Immunologists, Am. Soc. Microbiology, Am. Soc. Parasitologists, Am. Soc. Zoologists, Sigma Xi. Subspecialties: Cell and tissue culture; Immunology (agriculture). Current work: Cellular effector mechanisms in host defense to hemoprotozoa and other intracellular infections; epidemiology of hemoprotozoan infections. Office: Coll Vet Medicine Tex A&M U College Station TX 77843

WAGNER, JAMES BRUCE, JR., solid state science educator; b. Hampton, Va., July 28, 1927; s. James Bruce and Mary (Hudgins) W.; m. Phyllis M. Mountjoy, Sept. 5, 1951; children: James Bruce III, Ashley Stephen, Rebecca Bland. B.S. in Chemistry, U.Va., 1950, Ph.D., 1955. Research assoc. MIT, Cambridge, 1954-56; asst. prof. metallurgy Pa. State U., State College, 1956-58, Yale U., New Haven, 1958-62; assoc. prof. materials sci. Northwestern U., Evanston, Ill., 1962-65, prof., 1965-77, dir., 1972-76; mem. faculty Ariz. State U., Tempe, 1977—, prof. solid state sci., 1977—, dir., 1980—, affiliate prof. physics and mech. and aerospace engring., 1981—; Ford. Found. residency in engring. Motorola, Inc., Phoenix, 1968-69. Mem. Electrochem. Soc. (pres. 1983), Am. Phys. Soc., Metall. Soc. AIME. Episcopalian. Subspecialties: Ceramics; Solid state chemistry. Current work: Solid state electrolytes; diffusion in non-stoichiometric compounds; high temperature corrosion. X1 Office: Center for Solid State Sci Ariz State Univ Tempe AZ 85287

WAGNER, JEAMES ARTHUR, research physiologist; b. New Prague, Minn., Sept. 5, 1944; s. Stanley Francis and Lillian (Sladek) W.; m. Sandra Spiegel, Nov. 10, 1979; children: Sarah, Matthew, Chad. B.S., St. John's U., 1966; M.A., U. S.D., 1967; Ph.D., U. Western Ont., 1970. Grad. tutor U. S.D., Vermillion, 1966-67; undergrad. tchr. U. Western Ont., London, 1967-69; vis. asst. prof. Ind. U., Bloomington, 1969-71; asst. research physiologist U. Calif. Santa Barbara, 1971-80, assoc. research physiologist, 1980—; guest lectr. Westmont Coll., Santa Barbara, 1977. Author: Environmental Stress, 1978; Contbr. articles to sci. publs. Mem. Santa Barbara Lung Commn., 1974-75; judge Santa Barbara Heart Assn. Sci. Fair, 1980. NSF fellow, 1966-67; NIH grantee, 1979-82. Mem. AAAS, Am. Coll. Sports Medicine, Am. Physiol. Soc. (mem. orgn. com. 1969-70), Fedn. Am. Scientists, N.Y. Acad. Sci. Democrat. Roman Catholic. Subspecialties: Physiology (medicine); Gerontology. Current work: Research on environmental stresses as affecting physiological functions in young and older men and women. Home: 1334 Kenwood Rd Santa Barbara CA 93109 Office: Inst Environmental Stress U Calif Santa Barbara CA 93106

WAGNER, JOHN GEORGE, research mechanical engineer, consultant; b. Bowmansville, N.Y., July 9, 1942; s. Norman J. and Lucy Ann (Wiedenbeck) W.; m. Elizabeth Josefa Anthony, Mar. 10, 1944; children: Katrin Elizabeth, Carla Ann, John George. B.S.M.E., SUNY-Buffalo, 1964; M.S. in Applied Mechanics, U. Calif.-Berkeley, 1965; Ph.D. in Solid Mechanics, Brown U., 1969. Asst. prof. mech. engring. U. Pitts., 1969-76; research engr. U.S. Steel Corp., Monroeville, Pa., summer 1975; Research engr. Philips Labs., Briarcliff Manor, N.Y., 1976-80, Philips Natuurkundig Laboratorium, Eindhoven, Netherlands, 1980-81, Briarcliff Manor, 1981—; cons. Contbr. articles to profl. jours. Mem. Big Bros. Assn., 1974-76. Mem. ASME, Am. Soc. Metals, Soc. Exptl. Stress Analysis, Am. Powder Metallurgy Inst., Sigma Xi. Roman Catholic. Clubs: Continental Village (N.Y.); Sportsman's. Subspecialties: Powder metallurgy; Theoretical and applied mechanics. Current work: Modelling - via the principles of continuum mechanics - the flow, consolidation and sintering processes of powder metallurgy and ceramurgy. Home: Grandview Rd RR 2 Box 142 South Salem NY 10590 Office: 345 Scarborough Rd Briarcliff Manor NY 10510

WAGNER, LORRY YALE, energy co. exec., research engr.; b. Miami Beach, Fla., June 24, 1951; s. Morris Maxwell and Thelma (Portman) W.; m. Joanne Elaine Gold, Jan. 6, 1973 (div. Nov. 1975); m. Susan Lee Hogan, Oct. 6, 1979. B.S. in Engring, Purdue U., 1973, M.S. in Nuclear Engring, 1975, Ph.D., 1981. Research instr. Purdue U., West Lafayette, Ind., 1974-81; exec. v.p. Phillips Electric Co., Cleve., 1981—; pres. P.E. Energy Resources, Cleve., 1981—; exec. v.p. Redmond Waltz Electric Co., Cleve., 1981—; dir. Remote Sensors, Inc., Cleve., Bond Electric Co. Contbr. articles to profl. jours. Coach West Lafayette Swim Club, 1974-81, Lafayette Jefferson High Sch., 1978-81, Purdue U., 1976-78, Cleveland Heights (Ohio) Community Services, 1982—. Mem. Am. Nuclear Soc., Assn. Energy Engrs., Order of the Engr., Tau Beta Pi. Club: University (Cleve.). Subspecialties: Systems engineering; Energy resource management. Current work: Integrating/designing automation equipment for computer control of industrial facilities, design/implementation of computer controlled energy management systems. Home: 2977 Coleridge Rd Cleveland Heights OH 44118 Office: Phillips Electric Energy Resources 4126 St Clair Ave Cleveland OH 44103

WAGNER, MORRIS, microbiologist, educator; b. Chgo., Aug. 6, 1917; s. Isador Joel and Ida (Rovner) W.; m. Zelma Zonenberg, Aug. 24, 1947; children: Nana L., Robert D., Joel I., Judith B. B.Sc., Cornell U., 1941; M.S., Notre Dame U., 1946; Ph.D., Purdue U., 1966. Bacteriologist Lobund Labs., U. Notre Dame, South Bend, Ind., 1943-46, instr. microbiology, 1946-51, asst. prof., 1951-54, assoc. prof., 1954-69, prof., 1969—; adj. prof. microbiology Ind. U. Med. Sch., South Bend Ctr. Med. Edn., 1977—; cons. Contbr. numerous articles to profl. jours. NSF Faculty fellow, 1963; recipient Chgo. Dental Soc. award, 1955; Am. Soc. Dentistry Children award, 1981. Mem. Am. Soc. Microbiology, Assn. Gnotobiotics, Am. Assn. Lab. Animal Sci. Democrat. Jewish. Lodge: B'nai Brith. Subspecialties: Microbiology;

Cariology. Current work: Gnotobiology, normal microbial flora, dental caries. Home: 1940 Briar Way South Bend IN 46614 Office: Dept Microbiology Notre Dame U Notre Dame IN 46556

WAGNER, RAYMOND LEE, software specialist; b. Kansas City, Mo., Aug. 21, 1946; s. Albert Louis and Esther Pauline (Anderson) W.; m. Cheri Charlene Adams, Aug. 28, 1969; children: Richard Lamar, Frederek Prescot. A.B., Rice U., 1968; Ph.D., U. Tex.-Austin, 1972. Asst. prof. astronomy U. Wash., Seattle, 1972-74; asst. prof. astronomy and physics La. State U., Baton Rouge, 1974-78; sr. astrodynamics software engr. Ford Aerospace & Communications Corp., Colorado Springs, Colo., 1979-81, prin. software engr., 1981-83, Supr. Security Design sect., 1983—. Contbr. articles to profl. jours. Cabot scholar Rice U., 1964-68; NSF summer fellow, 1970; U. Tex. fellow, 1970-72. Mem. Am. Astron. Soc., Internat. Astron. Union, Sigma Xi, Phi Kappa Phi. Republican. Methodist. Subspecialties: Theoretical astrophysics; Systems engineering. Current work: Secure systems, formal specification and verification, astrodynamics, stellar astrophysics. Home: 2610 Black Diamond Terr Colorado Springs Co 80918 Office: 10440 Hwy 84 C-75 Colorado Springs CO 80908-3699

WAGNER, ROSELIN SEIDER, chemist, educator; b. Bklyn., Aug. 6, 1928; d. David Meyer and Anna (Prebluda) Seider; m. David Morris Wagner, Sept. 3, 1950; children: Marcia, Jonathan, Kenneth. A.B. magna cum laude (N.Y. State Regents Scholar), Barnard Coll., 1950; M.S., N.Y. U., 1952, Columbia U., 1956. Research assoc. dept. chemistry Emory U., 1955-58; lectr. in chemistry Barnard Coll., 1959-60; asst. prof. chemistry Hofstra U., 1969—. Author: (with others) Ideas, Investigation and Thought, A Laboratory Manual In Introductory Chemistry, 1980. Mem. Am. Chem. Soc., AAAS, Phi Beta Kappa, Sigma Xi, Iota Sigma Pi. Subspecialties: Analytical chemistry; Inorganic chemistry. Current work: Spectroscopy (microwave, visible and ultraviolet); complex ion chemistry; research on effect of non-aqueous solvents on stability of complex ions; lab. expts. in innovative forms being devised for use in teaching. Office: Dept Chemistry Hofstra U Hempstead NY 11550

WAGNER, STEPHEN GREGORY, nuclear reactor physicist; b. Chgo., Jan. 31, 1946; s. Charles Henry and Eunice Elizabeth (Prince) W.; m. Karen Ella Anderson, June 29, 1968; children David, Alicia. B.S., Cornell U., 1968; S.M., Harvard U., 1970. Sr. physicist Combustion Engring., Windsor, Conn., 1974-79, supr., 1979—. Served to lt. (j.g.) USNR, 1970-74. Mem. Am. Nuclear Soc., Soc. for Computer Simulation, Bloomfield Jaycees (v.p. 1976-81), Jr. Soccer Assn. Bloomfield (registrar 1980-82). Democrat. Club: Bloomfield Civitan. Subspecialties: Nuclear engineering; Numerical analysis. Current work: Nuclear and thermal hydraulic analysis of reactor transients. Home: 7 Kent Ln Bloomfield CT 06002 Office: Combustion Engring Dept 9492-2403 1000 Prospect Hill Rd Windsor CT 06095

WAGNER, WILLIAM CHARLES, veterinarian; b. Elma, N.Y., Nov. 12, 1932; s. Frederick George and Doris Edna (Newton) W.; m. Donna Ann McNeill, Aug. 14, 1954; children: William Charles, Elizabeth Ann, Victoria Mary, Kathryn Farrington. D.V.M., Cornell U., 1956, Ph.D., 1968. Gen. practice vet. medicine, Interlaken, N.Y., 1956-57; research veterinarian Cornell U., 1957-65, NIH postdoctoral fellow dept. animal sci., 1965-68; asst. prof. vet. medicine Vet. Med. Research Inst., Iowa State U., Ames, 1968-69, asso. prof., 1969-74, prof., 1974-77; prof. physiology, head dept. vet. bioscis. U. Ill., Urbana, 1977—; gen. sec. Internat. Congress on Animal Reprodn., Urbana, 1984. Pres. Ames Community Theater, 1972-73, 76-77. Recipient Alexander von Humboldt U.S. Scientist award Humboldt Stiftung, Freising-Weihenstephan, W. Ger., 1973-74. Mem. Am. Coll. Theriogenologists (diplomate), AVMA, Physiol. Soc., Am. Soc. Animal Sci., Soc. Study Reprodn., Soc. Study Fertility, N.Y. Acad. Scis., Sigma Xi, Phi Kappa Phi, Phi Zeta, Gamma Sigma Delta, Alpha Zeta. Lutheran. Subspecialties: Reproductive biology; Animal physiology. Current work: Effects of stress on reproduction; prolactin regulation; fetal endocrinology. Home: 306 W Florida Ave Urbana IL 61801 Office: U Ill Coll Vet Medicine Urbana IL 61801 I believe that it is important to be friendly to others, to try to understand the other person's position or feelings and deal with colleagues and subordinates in an impartial and fair manner. One should always remember that talents are a gift to be used wisely and to the fullest extent possible.

WAGNER-BARTAK, CLAUS GUNTHER JOHANN, advanced technology executive; b. Munich, Bavaria, Ger., Sept. 9, 1937; emigrated to Can., 1974; s. Friedrich and Johanna A. (Trinschek) Wagner-B.; m. Maria Helene Reich, Aug. 23, 1969; children: Natalie, Nicolaus, Nadine. B.S., Ludwig-Maximilian U., Munich, 1962, M.S., 1966, Ph.D., 1969. Cert. physics and engring, Ger., Ont. Research engr. Siemens Co., Munich, 1962; research asst. U. Munich, 1966-69; aerospace engr and project mgr. Junkers, Messerschmitt-Boelkow-Blohm Co., Munich, 1969-74; program mgr. and program dir. Spar Aerospace Ltd., Toronto, Ont., 1974-80, v.p. and gen. mgr., 1980-81, v.p. research and tech., 1982-83; pres. Energy Dynamics, Inc., Toronto, 1983—; lectr. U. Toronto, 1982; dir. Energy Dynamics Inc., Toronto, 1975. Recipient NASA Pub. Service medal, 1982; Group Achievement awards NASA-Kennedy Space Ctr. and Johnson Space Ctr., 1982; NASA Astronaut award, 1983. Sr. mem. Robotics Internat.; mem. AIAA, Assn. Profl. Engrs. Ont. (Engring. medal 1982), Am. Soc. Quality Control, Can. Nuclear Assn. Club: Bd. Trade (Toronto). Subspecialties: Aerospace engineering and technology; Robotics. Current work: Developments in advanced technology; systems for aerospace and terrestrial applications; developments of robotics systems; energy and environmental science; operations research; marine technologies; computer science. Home: 32 Woodgreen Dr Woodbridge ON Canada L4L 3B3 Office: Energy Dynamics Inc 6303 Airport Rd Toronto ON Canada L4V 1R8

WAGONER, ROBERT VERNON, astrophysicist; b. Teaneck, N.J., Aug. 6, 1938; s. Robert Vernon and Marie Theresa (Clifford) W.; m. Lynne Ray Moses, Sept. 2, 1963; children: Alexa Frances, Kim Stephanie. B.M.E., Cornell U., 1961; M.S., Stanford U., 1962, Ph.D., 1965. Research fellow in physics Calif. Inst. Tech., 1965-68, Sherman Fairchild Disting. scholar, 1976; asst. prof. astronomy Cornell U., 1968-71, asso. prof., 1971-73; asso. prof. physics Stanford U., 1973-77, prof., 1977—; George Ellery Hale Disting. vis. prof. U. Chgo., 1978; mem. Com. on Space Astronomy and Astrophysics, 1979-82. Contbr. articles on theoretical astrophysics and gravitation to profl. publs., mags.; co-author Cosmic Horizons. Sloan Found. research fellow, 1969-71; Guggenheim Meml. fellow, 1979; NSF grantee, 1973—. Fellow Am. Phys. Soc.; mem. Am. Astron. Soc., Internat. Astron. Union, Sigma Xi, Tau Beta Pi, Phi Kappa Phi. Subspecialties: Theoretical astrophysics; Relativity and gravitation. Current work: Supernova atmospheres, sources of gravitational radiation, early universe. Patentee in field. Home: 984 Wing Pl Stanford CA 94305

WAH, BENJAMIN WAN-SANG, engineering educator; b. Hong Kong, Sept. 7, 1952; came to U.S., 1970; s. Hsien-Feng and Chi-Ching (Wong) W.; m. Christine Hai-Ling Lee, Nov. 6, 1981. B.Sc. in Elec. Engring, Columbia U., 1974, M.Sc., 1975; M.Sc. in Computer Sci, U. Calif.-Berkeley, 1976; Ph.D. in Engring, U. Calif.-Berkeley, 1979. Asst. prof. Purdue U., West Lafayette, Ind., 1979—. Author: Data Management on Distributed Systems, 1981; contbr. articles to profl. jours. NSF grantee, 1981-83. Mem. IEEE (Best Paper award 1981), Assn. of Computing Machinery, Tau Beta Pi, Eta Kappa Nu. Subspecialties: Computer architecture; Database systems. Current work: High speed parallel architecture supporting combinatorial search problems. Office: School of Electrical Engineering Purdue U West Lafayette IN 47907

WAHL, PATRICIA WALKER, statistics educator, epidemiologist; b. LaGrande, Oreg., Dec. 6, 1938; d. Chauncey Albert and Florence Lee (Jackson) Walker; m. Stanley Neal Blacker, Feb. 1960 (dec. Oct. 1961); m. William Dudley Wahl, Aug. 25, 1962; 1 dau., Gretchen Dudley. B.A., San Jose State U., 1960; Ph.D., U. Wash., 1971. Asst. prof. biostats. U. Wash., Seattle, 1971-78, assoc. prof., 1978—; cons. N.W. Kidney Ctr., Seattle, 1975—. NIH Biometry Tng. grantee, 1966-70. Mem. Am. Statis. Assn., Biometrics Soc., Women's Caucus in Stats., Phi Kappa Phi. Republican. Episcopalian. Subspecialties: Statistics; Epidemiology. Current work: Multivariate statistical data analysis applied to fields of lipid metabolism, coronary heart disease, End-Stage Renal failure, diabetes. Home: 700 SE Shoreland Dr Bellevue WA 98004 Office: Dept of Biostatistics University of Washington Seattle WA 98195

WAHLBECK, PHILLIP GLENN, chemistry educator and researcher; b. Kankakee, Ill., Mar. 29, 1933; s. John Hugo and Florence Ada (Jefferson) W.; m. Donna G. Frost, Aug. 25, 1956; children: Debra, Paul, Beth. B.S., U. Ill., 1954, Ph.D., 1958. Research assoc. U. Kans., Lawrence, 1958-60; instr. Ill. Inst. Tech., Chgo., 1960-61, asst. prof., 1961-67, assoc. prof., 1967-72; prof. chemistry Wichita (Kans.) State U., 1972—. Contbr. research articles to sci. jours. Research grantee Petroleum Research Fund, 1981—, NSF, 1971-77, 81, AEC, 1961-71. Mem. Am. Chem. Soc. (offices local sect. 1980-82), AAAS, Am. Sci. Affiliation, Sigma Xi, Pi My Epsilon, Phi Lambda Upsilon. Subspecialties: High temperature chemistry; Surface chemistry. Current work: Surface adsorption - desorption kinetics, vapor pressure measurements, properties of gases. Home: 7413 Norfolk Wichita KS 67206 Office: Dept Chemistry Wichita State Univ Wichita KS 67208

WAHLBIN, LARS BERTIL, mathematics educator; b. Linkoping, Sweden, Mar. 25, 1945; s. Harry Axel and Margit (Andersson) W.; m. Anita Inga Alvin, Apr. 7, 1966; 1 son, Steffan. B.A., U. Goteborg, 1967, M.A., 1968, Ph.D., 1971. Instr. U. Chgo., 1972-74; asst. prof. math Cornell U., Ithaca, N.Y., 1974-77, assoc. prof., 1977—. Served as sgt. Swedish Army, 1964-65. Subspecialty: Numerical analysis. Current work: Numerical analysis of partial differential equations. Home: 123 W Miller Rd Ithaca NY 14850 Office: Dept Math Cornell U Ithaca NY 14853

WAHLSTROM, RICHARD CARL, animal science educator, researcher; b. Craig, Nebr., Feb. 13, 1923; s. Carl Hugo and Edna Evelyn (Erickson) W.; m. LaRayne Frances Steyer, Aug. 17, 1947; children: Richard K., Mark W., Ronald E. B.S., U. Nebr., 1948; M.S., U. Ill., 1950, Ph.D., 1952. Research assoc. Merck Inst. Therapeutic Research, 1951-52; assoc. prof. animal husbandry S.D. State U., 1952-59, prof. animal sci., 1960—, head dept. animal sci., 1960-67. Contbr. numerous articles to sci. and popular publs. Served to cpl. U.S. Army, 1943-46; ETO. Recipient Outstanding Tchr. award Gamma Sigma Delta, 1980. Mem. Am. Soc. Animal Sci. (Award in animal mgmt. 1976), Inst. Nutrition. Methodist. Lodge: Rotary. Subspecialty: Animal nutrition. Current work: Swine nutrition, amino acids, minerals, feed additives, water quality, teaching feed technology and pork prodn. Home: 819 9th Ave Brookings SD 57006 Office: SD State U 106 AS Complex Brookings SD 57007

WAHREN, DOUGLAS, research administrator, engineering educator; b. Norrkoping, Sweden, Mar. 12, 1934; came to U.S., 1979; s. Helge K. and Jane I. (Agrell) W.; m. Inger V. Weeleen, Feb. 9, 1957; children: Caroline, Johan. Civilingenior, Royal Inst. Tech., Stockholm, 1957, Lic. in Engring, 1961, D.Sc., 1964, Docent, 1965. With Swedish Forest Products Research Lab., Stockholm, 1957-73, dir. research, paper tech. div., 1968-73; prof. paper tech. Royal Inst Tech., Stockholm, 1970-73, hon. prof., 1973; v.p. research AB Karlstads Mekaniska Werkstad, Karlstad, Sweden, 1974-78; v.p. research, prof. engring. Inst. of Paper Chemistry, Appleton, Wis., 1979—; research scientist Beloit Corp. Research, Wis., 1964-65; cons. Contbr. articles to tech. jours. Fellow TAPPI. Subspecialty: Paper Technology. Current work: Pressing, drying, high consistency processing. Patentee. Office: Institute of Paper Chemistry PO Box 1039 Appleton WI 54912

WAHRHAFTIG, AUSTIN LEVY, chemistry educator; b. Sacramento, May 5, 1917; s. Moses Solomon and Irma R. (Levy) W.; m. Ruby Martha Dixon, Aug. 24, 1957. A.B., U. Calif.-Berkeley, 1938; Ph.D., Calif. Inst. Tech., 1941. Research fellow Calif. Inst. Tech., Pasadena, 1941-45; research scientist W.E. Williams, Pasadena, 1945-46; univ. fellow Ohio State U., Columbus, 1946-47; asst. prof. U. Utah, Salt Lake City, 1947-53, assoc. prof., 1953-59, prof. chemistry, 1959—; vis. prof. Latrobe U., Melbourne, Australia, 1972, 80. Mem. Am. Chem. Soc., Am. Phys. Soc., ASTM, Soc. for Mass Spectrometry (dir. 1970-72), AAAS, Sigma Xi. Subspecialties: Physical chemistry; Atomic and molecular physics. Current work: Dense (supercritical) gas chromatography/mass spectrometry, theory of mass spectrometry, molecular spectroscopy. Home: 2239 Logan Ave Salt Lake City UT 84108 Office: Dept Chemistry Univ Utah Salt Lake City 84112

WAINBERG, MARK ARNOLD, med. researcher, immunologist, virologist; b. Montreal, Que, Can., Apr. 21, 1945; s. Abe and Fay (Haffner) W.; m. Susan F. Hubschman, Mar. 1, 1969; children: Zev, Jonathan. B.Sc., McGill U., Montreal, 1966; Ph.D., Columbia U., 1972. Lectr. dept. immunology Hebrew U., Hadassah Med. Sch., Jerusalem, 1972-74; staff investigator Lady Davis Inst., Montreal, 1974—; assoc. prof. microbiology and immunology McGill U., Montreal, 1979—; vis. scientist NIH, Bethesda, Md., 1980-81. Contbr. articles to profl. jours. Conseil de la Recherche en Sante du Que. research scholar, 1975-81; research assoc. Fonds de la Recherche en Sante du Que., 1981—. Mem. Am. Assn. Immunologists, Am. Soc. Microbiology, Am. Soc. Cell Biology, Am. Assn. Cancer Research. Jewish. Subspecialties: Cell study oncology; Infectious diseases. Current work: Mechanisms of tumor induction by viruses in animal models; mechanisms of viral suppression of immune responsiveness. Office: 3755 Cote Ste Catherine Rd Montreal PQ Canada H3T 1E2

WAINES, JOHN GILES, educator; b. Branton, U.K., June 8, 1940; came to U.S., 1964; s. John Winspear and Ulla (Kirk) W. B.Sc. with honors, U. Reading, Eng., 1963; M.S., UCLA, 1966; Ph.D., U. Calif., Riverside, 1969. Fellow in genetics U. Mo., Columbia, 1968-70; deptl. demonstrator Botany Sch., Oxford U., Eng., 1970-72; asst. prof. biology Va. Poly. Inst. and State U., Blacksburg, 1972-74; assoc. prof. genetics U. Calif., Riverside, 1974—, dir., 1981—. Contbr. articles to profl. jours. Mem. Genetics Soc. Am., Am. Bot. Soc., Soc. Econ. Botany, Crop Sci. Soc., Agronomy Soc. Subspecialties: Plant genetics; Evolutionary biology. Current work: Research in plant genetics, cytogenetics, breeding, stress tolerance and evolution. Office: U Calif Dept Botany and Plant Sciences Riverside CA 92521

WAINWRIGHT, STANLEY DUNSTAN, biologist, educator; b. Cottingham, Eng., Apr. 15, 1927; s. Stanley and Hannah (Dunstan) W.; m. Lillian Karelltz, Mar. 10, 1952; children: David Stanley, Peter Francis. B.A. with 1st class honors, Cambridge (Eng.) U., 1947; Ph.D. in Biochemistry, London U., 1950. Exchange research scholar Pasteur Inst., Paris, 1950-51; research assoc. zoology dept. Columbia U., N.Y.C., 1951-52; NRC fellow in biology Atomic Energy of Can. Ltd., Chalk River, Ont., 1952-54; research assoc. microbiology dept. Yale U., New Haven, 1954-56; career research investigator, research prof. biochemistry Dalhousie U., Halifax, N.S., 1956—. Author: Control Mechanisms and Protein Synthesis, 1972. Mem. Can. Fedn. Biol. Socs. (bd. dirs. 1979-82), Can. Soc. Cell Biology (past pres.), Can. Biochem. Soc. (past councillor), Am. Soc. Cell Biology, Genetics Soc. Am., AAAS, Am. Inst. Biol. Scis., Sigma Xi. Subspecialties: Chronobiology; Neurobiology. Current work: Molecular biology of chick pineal 'clock'. Home: 21 Torrington Dr Halifax NS Canada B3M 1Y5 Office: Biochemistry Dept Dalhousie U Halifax NS Canada B3H 4H7

WAISMAN, JERRY, hospital administrator; b. Borger, Tex., Sept. 14, 1934; s. Sammie and Hillie (Novit) W.; m. Jane Barbara Atkins, June 15, 1958; children: Eric A., Nina A, John L. B.A., U. Tex., 1956; M.D., U. Tex. Med. Br., Galveston, 1960. Intern SUNY Downstate, Bklyn., 1960-61; asst. resident U. Utah, 1961-62, resident, 1964-68; from asst. prof. to prof. UCLA, 1968-81; prof. and lab. dir. NYU, 1981—; cons. VA hosp., Sepulveda, Calif., 1976-81, N.Y.C., 1981—. Contbd.: chpts. to Urogenital Cancer, 1978, Textbook of Surgical Pathology, 1979. Served to capt. USAF, 1962-64. Yamagiwa-Yoshida fellow Internat. Union Against Cancer, Stockholm, 1979. Mem. Am. Assn. Pathologists, Internat. Acad. Pathologists, Election Soc. Am. Subspecialty: Pathology (medicine). Current work: Diagnostic election microscopy, fine-needle aspiration, urogenital pathology. Home: 4 Washington Sq Village New York NY 10012 Office: NYU Sch Medicine 550 1st Ave New York NY 10016

WAITE, DANIEL ELMER, oral surgeon; b. Grand Rapids, Mich., Feb. 19, 1926; s. Charles Austin and Phoebe Isabel (Smith) W.; m. Alice Darlene Carlile, June 20, 1948; children—Christine Ann, Thomas Charles, Peter Daniel, Julie Marilyn, Stuart David. A.A., Graceland Coll., 1948; D.D.S., State U. Ia. Coll. Dentistry, 1953; M.S., Grad. Coll., 1955, certificate oral surgery, certificate residency univ. hosps., 1955. Diplomate: Am. Bd. Oral Surgery. Resident oral surgery State U. Ia. Hosps., 1953-55; instr. oral surgery State U. Ia. Coll. Dentistry, 1955-56, asst. prof., 1956-57, asso. prof., acting head dept. oral surgery, 1957-59, prof. head dept., 1959-63; mem. staff Mayo Clinic, Mayo Grad. Sch. Medicine, Rochester, Minn., 1963-68; prof., also chmn. div. oral surgery U. Minn., 1968—; vis. prof. U. Adelaide, Australia, 1980. Author: Textbook of Practical Oral Surgery, 1972, 2d edit., 1978; Contbr. numerous articles to profl. jours. Active People to People Found.; served with Project HOPE, Peru, 1962 served with Project HOPE, Sri Lanka, 1969 served with Project HOPE, Egypt, 1975; Trustee Graceland Coll., Lamoni, Iowa, 1970-78, Park Coll., Parkville, Mo., 1972-78; bd. dirs. Hennepin County unit Am. Cancer Soc., 1970-73. Served with USAAF, 1944-46; sr. dental surgeon USPHS(R). Recipient Novice award Internat. Assn. Dental Research, 1955; named Man of Yr. U. Minn. Sch. Dentistry Century Club, 1980. Fellow Am. Coll. Dentists, Am. Soc. Oral Surgeons; mem. Midwestern Soc. Oral Surgeons, Ia. Soc. Oral Surgeons (sec. 1958-61, pres. 1962), Minn. Soc. Oral Surgeons (pres. 1974), Am. Dental Assn., Internat. Assn. Dental Research (sec. Ia. sect. 1957-62), Sigma Xi, Omicron Kappa Upsilon. Mem. Reorganized Ch. of Jesus Christ of Latter Day Saints. (patriarch, mem. med. council). Subspecialty: Oral and maxillofacial surgery. Home: 1801 Pennsylvania Ave N Golden Valley MN 55427

WAITE, PAUL JUNIOR, climatologist; b. New Salem, Ill., June 21, 1918; s. Wesley Philip and Edna Viola (Bartlett) W.; m. Margaret Elizabeth Cresson, June 13, 1943; children: Carolyn, Lawrence. B.Ed., Western Ill. State U., 1940; M.S., U. Mich.-Ann Arbor, 1966. Sci. tchr., coach Ill. Schs., 1938-39, 40-42, 46-48; meteorologist Nat. Weather Service, Kansas City, Mo., 1948-51, 52-56, Chgo., 1948-51, 52-56, state climatologist, Des Moines, Madison, Wis., 1956-59, 59-74; dep. project mgr. NOAA, Johnson Space Center, Houston, 1974-76; state climatologist Iowa Dept. Agr., Des Moines, 1976—; adj. prof. Drake U., Des Moines, 1970—; collaborator Iowa State U., Ames, 1959-73; asst. dir. Iowa Weather Service, Des Moines, 1959-70, dir., 1970-73. Author: series Climate of Iowa, 1979—; contbr. articles in field to profl. jours. Served to 1st lt. USAF, 1942-46, 51-52. Recipient Group Achievement award NASA, 1976. Fellow Iowa Acad. Sci.; mem. Am. Assn. State Climatologists (pres. 1977-78), Am. Meteorol. Soc., AAAS, Nat. Weather Assn. Republican. Mem. Evangelical Free Ch. Club: Toastmasters (Des Moines) (pres. 1981-82). Lodge: Masons. Subspecialty: Climatology. Current work: Applied climatology. Home: 6657 NW Timberline Dr Des Moines IA 50321 Office: Iowa Dept Agr Room 10 Municipal Airport Des Moines IA 50321

WAITZMAN, DONALD ANTHONY, chemical engineer; b. New Orleans, May 17, 1924; s. Samuel D. and Mollie A. (Dietlein) W.; m. Mary Ann Stroble; children: Paul D., Donald Anthony, Charles S., Caroline D. Waitzman Beck, Marianne Waitzman Stanhope. B.S. in Chem. Engring, Auburn U., 1948. Registered profl. engr., Ala. Technician City Service Refining Corp., Lake Charles, La., 1948-51, Stone & Webster Engring. Corp., Boston, 1951-53; asst. project engr. Rust Engring. Co., Birmingham, Ala., 1953-56; v.p. D.B. Gooch Assocs., Inc., Birmingham, 1956-60; sales engr. Crandall Engring. Co., Birmingham, 1960-62, project engr. chem. devel. div. design br., 1962-75; mgr. ammonia from coal projects, chem. devel. div. TVA, Muscle Shoals, Ala., 1975—. Contbr. numerous articles to sci. publs. Served to 1st lt. USAAF, 1943-45. Decorated Air medal with two oak leaf clusters. Mem. Am. Chem. Soc., Am. Inst. Chem. Engrs., Nat. Mgmt. Assn., Sigma Alpha Epsilon. Roman Catholic. Club: Florence Golf and Country. Subspecialties: Fuels and sources; Agricultural engineering. Current work: In responsible charge of project to obtain technical and economic information from prototype plant that substitutes coal for natural gas for the production of ammonia, and synthetic fuel gas. Home: 515 Windsor St Florence AL 35630 Office: TVA Chem Devel Div Ammonia from Coal Projects Muscle Schoals AL 35660

WAITZMAN, MORTON BENJAMIN, physiology educator; b. Chgo., Nov. 8, 1923; s. Joseph and Anna (Glickman) W.; m. Aviva Shedroff, June 9, 1949; children: Sherri, Brad, Rhonda. B.S., U. Miami, 1948; M.S., U. Ill., 1950, Ph.D., 1953. Instr. dept. pharmacology Western Res. U., Cleve., 1954-59, asst. prof., 1959-62; asst. prof. dept. physiology Emory U., Atlanta, 1962-67, assoc. prof., 1962-68, prof., 1968—, dir., 1962—. Contbr. articles to profl. jours.; sect. editor: Metabolic, Pediatric and Systemic Ophthalmology. Past chmn. tech. adv. com. Citizens for Clean Air. NIH grantee, 1954—. Mem. Assn. for Research in Vision and Ophthalmology; mem. Am. Physiol Soc.; Mem. AAUP (past chpt. pres.), Ga. Soc. for Prevention of Blindness (dir.), Sigma Xi, Phi Sigma. Subspecialties: Physiology (medicine); Ophthalmology. Current work: Cyclic nucleotide and prostaglandin metabolism as related to physiopathologic events in diabetic vascular disease and glaucoma and other inflammatory processes. Home: 1137 Mason Woods Dr NE Atlanta GA 30329 Office: Lab for Ophthalmic Research Emory U Atlanta GA 30322

WAITZMAN, MORTON BENJAMIN, educator; b. Chgo., Nov. 8, 1923; s. Joseph and Anna (Glickman) W.; m. Aviva Shedroff, June 9, 1949; children: Sherri, Brad, Rhonda. B.S., U. Miami, Fla., 1948; M.S., U. Ill., 1950, Ph.D., 1953. Instr. dept. pharmacology Case Western Res. U., Cleve., 1954-59, asst. prof., 1959-62; asst. prof. dept.

physiology Emory U., Atlanta, 1962-67, assoc. prof., dir., 1962-68, prof., dir., 1968—. Editorial bd.: Metabolic, Pediatric and Systemic Ophthalmology, 1982—; contbr. articles to profl. jours. Chmn. tech. adv. com. Citizens for Clean Air of Atlanta, 1971. NIH grantee, 1962. Mem. Assn. for Research in Vision and Ophthalmology, AAAS, Internat. Soc. Metabolic Eye Diseases, Am. Physiol. Soc., AAUP, Internat. Physiol. Soc., Sigma Xi, Phi Sigma. Subspecialties: Physiology (biology); Ophthalmology. Current work: Metabolic aspects of blood-tissue transport and vascular permeability, especially involving diabetes, glaucoma and inflammation. Home: 1137 Mason Woods Dr Atlanta GA 30329 Office: Lab for Ophthalmic Research Emory Univ Atlanta GA 30322

WAKEFIELD, ERNEST HENRY, economist, manufacturing company executive; b. Vermilion, Ohio, Feb. 11, 1915; s. Frederick Wright and Mary (Poley) W.; m.; children: Ann, John. B.S., U. Mich., 1938, M.A., 1939, Ph.D. in Elec. Engring, 1952. With Westinghouse Electric Corp., 1939; instr. U. Tenn., 1939-42; assoc. scientist Atomic Bomb Project, U. Chgo., 1943-46; pres. RCL, 1946-62, Linear Alpha, Inc., Evanston, Ill., 1962—, Imported Interiors, Inc., Evanston, 1963-69; chmn. Third World Energy Inst., Evanston, 1973—, Internat. Inst. of Mgmt. and Appropriate Tech. for Developing Nations, 1978—. Mem. Evanston Bd. Edn., 1962-67, pres., 1967. Served with U.S. Army, 1942-46. Recipient Pub. Service award Civil Def., 1962; Community Service award Skokie C. of C., 1962. Subspecialties: Agricultural economics; Biomass (energy science and technology). Current work: Economics for developing nations; improving the agriculture and economics of 3d World; design of electric vehicles. Office: 1000 Grove St Evanston IL 60201

WALBA, DAVID MARK, chemistry educator; b. Oakland, Calif., June 29, 1949; s. Harold and Beatrice (Alpert) W.; m. Cassandra B. Geneson, Oct. 30, 1981; 1 son, Paul. B.S. in Chemistry, U. Calif.-Berkeley, 1971, Ph.D., Calif. Inst. Tech., 1975. Postdoctoral fellow UCLA, 1975-77; asst. prof. chemistry U. Colo., Boulder, 1977—. Contbr. articles to profl. jours. NIH fellow, 1976-77; A. P. Sloan fellow, 1982-84. Mem. Am. Chem. Soc., Sigma Xi. Subspecialty: Organic chemistry. Current work: Organic chemistry, natural products total synthesis, host-guest chemistry, and stereochemical topology. Office: U Colo Dept Chemistry PO Box 215 Boulder CO 80309

WALBERG, HERBERT JOHN, research educator; b. Chgo., Dec. 27, 1937; s. Herbert J. and Helen (Bauer) W.; m. Madoka Bessho, Aug. 20, 1965; 1 son, Herbert. Ph.D., U. Chgo., 1964. Asst. prof. Harvard U., Cambridge, Mass., 1966-69; research assoc. U. Wis., Madison, 1969-70; research prof. U. Ill., Chgo., 1970—. Editor: books including Evaluating Educational Performance, 1974, Research on Teaching, 1979, Improving Educational Standards, 1982. Fellow Am. Psychol. Assn., Royal Statis. Soc., AAAS. Subspecialties: Learning; Social psychology. Current work: Productivity of science education; research synthesis; science policy. Office: University of Illinois College of Education on 4348 Chicago IL 60680

WALBORN, NOLAN REVERE, astronomer, researcher; b. Bloomsburg, Pa., Sept. 30, 1944; s. George Mark and Evelyn Loretta (Miller) W.; m. Gladys Olivares, Oct. 15, 1975; 1 son, Francis Augustus. B.A., Gettysburg Coll., 1966; Ph.D., U. Chgo., 1970. Postdoctoral research assoc. Yerkes Obs., U. Chgo., Williams Bay, Wis., 1971, David Dunlap Obs., U. Toronto, Richmond Hill., Ont., Can., 1971-73; staff astronomer Cerro Tololo Inter-Am. Obs., La Serena, Chile, 1973-81; sr. research assoc. NASA/NRC, Goddard Space Flight Ctr., Greenbelt, Md., 1982—. Contbr.: articles to profl. publs. including Astron. Jour., Astrophys. Jour.; publs. of Astron. Soc. of Pacific. Mem. Am. Astron. Soc., Can. Astron. Soc., Internat. Astron. Union. Subspecialties: Optical astronomy; Ultraviolet high energy astrophysics. Current work: Spectroscopy, early-type stars, spectral classification, interstellar lines, Magellanic Clouds, Eta Carinae. Office: Goddard Space Flight Center Code 683 Greenbelt MD 20771

WALCOTT, BENJAMIN, cell biologist; b. Boston, May 31, 1941; s. Charles Fulsom and Susan Mary (Cabot) W.; m. (married), June 29, 1946. B.A., Harvard U., 1963; Ph.D., U. Oreg., Eugene, 1968. Fellow dept. neurobiology Australian Nat. U., 1969-72; asst. prof. anatomical sci. SUNY Sch. Medicine, Stony Brook, 1972-79, assoc. prof., 1979—. Mem. Soc. Neuroscis., Soc. Exptl. Biology, Soc. Gen. Physiologists, Biophys. Soc., Soc. Cell Biology. Subspecialties: Comparative neurobiology; Cell biology (medicine). Current work: Cell biology, ultrastructure, physiology and biochemistry of muscle, nerves, and sensory systems. Office: Dept Anatomical Sci Sch Medicine SUNY Stony Brook NY 11794

WALD, ARNOLD, gastroenterologist, educator; b. N.Y.C., June 10, 1942; s. Jack and Ruth (Fox) W.; m. Ellen Faith Rashkow, June 26, 1966; children: Elissa Karen, Eric Lawrence. A.B., Colgate U., 1964; M.D., SUNY-Bklyn., 1968. Diplomate: Am. Bd. Internal Medicine, Am. Bd. Gastroenterology. Intern Kings County Hosp., Bklyn., 1968-69, resident, 1969-71; Fellow in gastroenterology Johns Hopkins Hosp., Balt., 1973-75, instr. medicine, 1975-76, asst. prof. medicine, 1976-78, U. Pitts., 1978—; physician Montefiore Hosp., Pitts., 1978—. Cons. Spina Bifida Assn. Western Pa., 1980—. Served to maj. U.S. Army, 1971-73. Fellow ACP; mem. Physicians for Social Responsibility, Am. Gastroent. Assn., Am. Fedn. Clin. Research, Nat. Found. Ileitis and Colitis (dir.), Midwest Gut Club. Democrat. Jewish. Subspecialties: Gastroenterology; Internal medicine. Current work: Disorders of gastrointestinal motility; behavioral modification and treatment of bowel dysfunction; treatment gastrointestinal and liver diseases. Home: 1143 Shady Ave Pittsburgh PA 15232 Office: Univ Pitts Sch Medicine 3459 Fifth Ave Pittsburgh PA 15213

WALD, GEORGE, biochemist; b. N.Y.C., Nov. 18, 1906; s. Isaac and Ernestine (Rosenmann) W.; m. Frances Kingsley, May 15, 1931 (div.); children—Michael, David; m. Ruth Hubbard, 1958; children—Elijah, Deborah. B.S., N.Y. U., 1927; M.A., Columbia U., 1928, Ph.D., 1932; M.D. (hon.), U. Berne, 1957; D.Sc., Yale U., 1958, Wesleyan U., 1962, N.Y. U., 1965, McGill U., 1966, Amherst Coll., 1968, U. Rennes, 1970, U. Utah, 1971, Gustavus Adolphus U., 1972. NRC fellow at Kaiser Wilhelm Inst., Berlin and Heidelberg, U. Zurich, U. Chgo., 1932-34; tutor biochem. scis. Harvard U., 1934-35, instr. biology, 1935-39, faculty instr., 1939-44, asso. prof. biology, 1944-48, prof., 1948—, Higgins prof. biology, 1968-77, prof. emeritus, 1977—; vis. prof. biochemistry U. Calif., Berkeley, summer 1956; Nat. Sigma Xi lectr., 1952; chmn. divisional com. biology and med. scis. NSF, 1954-56; Guggenheim fellow, 1963-64; Overseas fellow Churchill Coll., Cambridge U., 1963-64; participant U.S.-Japan Eminent Scholar Exchange, 1973; guest China Assn. Friendship with Fgn. Peoples, 1972; v.p. Permanent Peoples' Tribunal, Rome, 1980—. Co-author: General Education in a Free Society, 1945, Twenty Six Afternoons of Biology, 1962, 66, also sci. papers on vision and biochem. evolution. Recipient Eli Lilly prize Am. Chem. Soc., 1939; Lasker award Am. Pub. Health Assn., 1953; Proctor medal Assn. Research in Ophthalmology, 1955; Rumford medal Am. Acad.; Arts and Scis., 1959; Ives medal Optical Soc. Am., 1966; Paul Karrer medal in chemistry U. Zurich, 1967; co-recipient Nobel prize for physiology, 1967; T. Duckett Jones award Helen Hay Whitney Found., 1967; Bradford Washburn medal Boston Mus. Sci., 1968; Max Berg award, 1969; Priestley medal Dickinson Coll., 1970. Fellow Nat. Acad. Sci., Am. Acad. Arts and Scis., Am. Philos. Soc. Subspecialties: Biochemistry (biology); Biophysics (biology). Home: 21 Lakeview Ave Cambridge MA 02138 "A scientist lives with all reality. There is nothing better. To know reality is to accept it, and eventually to love it. A scientist is in a sense a learned child. There is something of the scientist in every child. Others must outgrow it. Scientists can stay that way all their lives." (Remarks on receiving the Nobel Prize, Stockholm, 1967)

WALDEN, DAVID BURTON, geneticist, educator, researcher; b. New Haven, Mar. 29, 1932; emigrated to went to Can., 1961; s. David Conger and Fannie (McFarl) W.; m. Carol Ann Isherwood, June 9, 1956; children: David John, Karen Ruth. B.A., Wesleyan U., 1954; M.Sc., Cornell U., 1958, Ph.D., 1959. Postdoctoral felow in genetics Ind. U., 1959-61; asst. prof. plant scis. U. Western Ont. (Can.), London, 1961-64, assoc. prof., 1964-71, prof., 1971—; vis. prof. genetics U. Birmingham, Eng., 1973-74, 81, U. Ill.-Urbana, 1974; pres. Biol. Council Can., 1974-76. Contbr. numerous articles, notes, abstracts to profl. publs. Decorated Queen's Silver Jubilee medal, Eng.; recipient Excellence in Teaching medal U. Western Ont., 1982. Mem. Genetics Soc. Can. (pres. 1981-83), Genetics Soc. U.S.A., AAAS, Am. Genetics Soc., N.Y. Acad. Sci., Crop Sci. Soc. Am. Subspecialties: Genetics and genetic engineering (agriculture); Plant cell and tissue culture. Current work: Cytogenetics; pollen biology; gene regulation by stress factors; primary gene products in heterosis; fluorography; electrophoresis. Home: 87 Glenburnie Crescent London ON Canada N5X 2A1 Office: Dept Plant Scis U Western Ont London ON Canada N6A 5B7

WALDMAN, RONALD JAY, epidemiologist; b. N.Y.C., Mar. 4, 1946; s. Milton and Helen (Jaray) W. B.A., U. Rochester, 1967; M.D., U. Geneva, Switzerland, 1975; M.P.H., Johns Hopkins U., 1979. Diplomate: Deiplomate Am. Bd. Preventive Medicine. Med. epidemiologist Smallpox Eradication Project, WHO, Bangladesh, 1975-76; intern, resident Highland Hosp., Rochester, N.Y., 1976-78; med. epidemiologist Centers for Disease Control Epidemic Intelligence Service, Atlanta, 1979-81, Internat. Health Program Office, 1981—; cons. in field. Contbr. articles on preventive medicine. to profl. jours. Active Physicians for Social Responsibility. Served as med. officer USPHS, 1979—. Mem. Soc. for Epidemiol. Research. Subspecialties: Epidemiology; Preventive medicine. Current work: Reye's syndrome, vaccine-preventable diseases, oral rehydration, international health, epidemiology, immunization. Office: Internat Health Program Office Centers for Disease Control 1600 Clifton Rd Atlanta GA 30333

WALDMANN, THOMAS ALEXANDER, immunology researcher, physician; b. N.Y.C.; s. Charles and Elizabeth (Sipos) W.; m. Katharine Emory Spreng, Mar. 29, 1958; children: Richard Allen, Robert James, Carol Ann. A.B., U. Chgo., 1951; M.D., Harvard U., 1955. Diplomate: Am. Bd. Allergy and Immunology. Med. Intern Mass. Gen. Hosp., Boston, 1955-56; clin. assoc. Nat. Cancer Inst., NIH, Bethesda, Md., 1956-58, sr. investigator, 1958-68, head immunophysiology sect., 1968-73, chief metabolism br., 1972—; cons. WHO, FTC, FDA; William Dameseek vis. prof. Tufts Med. Sch., Boston, 1983. Author 1 book, over 320 sci. articles. Bd. dirs. Found. Advanced Edn. in the Scis., Bethesda, 1981—. Recipient Michael Heidelberger award Columbia U., 1976; Henry Stratton medal Am. Soc. Hematology, 1977; named Man of Yr. Leukemia Soc. Am., 1980; G. Burroughs Mider lectr. NIH, 1980. Fellow Am. Acad. Allergy-Immunology (Bela Schick award 1977); mem. Am. Soc. Clin. Investigation (editorial bd. 1975-80), Am. Assn. Physicians (Kroc lectr. 1980), Am. Assn. Immunology (assoc. editor jour. 1982, Honor lectr 1981), Internat. Plasma Protein Study Group (dir., sec. 1970-75). Democrat. Clubs: NIH Camera (pres. 1970), Silver Spring (Md.) Camera.). Subspecialties: Immunology (medicine); Cancer research (medicine). Current work: Analysis of the regulatory mechanisms that control the human immune response and the disorders in these mechanisms in immunodeficiency diseases and cancer; special emphasis on use of recombinant DNA technology to study arrangement of immuoglobulin genes and on studies of suppressor and helper T lymphocytes. Discoverer intestinal lymphanglectasia, 1961, allergic gastroenteropathy, 1963, human suppressor T cells in immunodeficiency (Henry Berton award), 1977. Home: 3910 Rickover Rd Silver Spring MD 20902 Office: 9000 Rockville Pike Bethesda MD 20205

WALDRON, ACIE CHANDLER, agricultural research educator, adminstrator; b. Malad, Idaho, Feb. 4, 1930; s. Nathaniel Acie and Leah Mary (Chandler) W.; m. Catherine Venonne Moore, Aug. 20, 1957; children: Deborah, Michael, Joel, Laurie, Susan. B.Sc., Brigham Young U., 1957; M.Sc., Ohio State U., 1959, Ph.D., 1961. Research assoc. Ohio Agrl. Expt. Sta., Columbus, 1959-61; pesticide residue method devel. chemist Am. Cyanamid Co., Princeton, N.J., 1961-66; extension specialist pesticide chems. Ohio Coop. Extension Service, Ohio State U., Columbus, 1966-77; asst., then assoc. prof. Ohio State U., Columbus, 1966-74, prof. entomology and agronomy, 1974—; pesticide coordinator Ohio Coop. Extension Service, Ohio State U., 1969-77; coordinator N. Central Region Pesticide Impact Assessment Program, Ohio Agrl. Research and Devel. Ctr., 1977—. Contbr. articles to profl. jours. Served to 1st lt. U.S. Army, 1953-56. Mem. Am. Chem. Soc., Am. Soc. Agronomy, Soil Sci. Soc. Am., Assn. Ofcl. Analytical Chemists, Council Agrl. Sci. and Tech., AAAS, N. Central Weed Control Conf., Benson Inst., Ohio Acad. Sci., Ohio Pesticide Edn. Assn. (trustee 1970-79, pres. elect 1979), Ohio Fertilizer and Pesticide Assn. (co-chrmn. 1979, chmn. scholarship and honors com. 1979—), Sigma Xi, Gamma Sigma Delta, Epsilon Sigma Phi. Democrat. Mormon. Current work: Pesticide residues, pesticide legislation and regulations, pesticide impact assessment, environmental pollution. Home: 4220 Lyon Dr Columbus OH 43220 Office: Ohio State U Dept Entomology 1735 Neil Ave Columbus OH 43210

WALDRON, JOSEPH ANTHONY, psychologist, educator; b. Batavia, N.Y., Oct. 3, 1943; s. Elsworth Thomas and Dolores Agnes (Kanaley) W.; m. Irene Montgomery, Oct. 31, 1966; children: Wendy June, Joelle Dolores, Elizabet Hannah. B.A., SUNY-Buffalo, 1972; M.A., Ohio State U., 1973, Ph.D., 1975. Lic. psychologist, Ohio; registered Nat Register Health Service Providers. Psychologist, dept. head Ohio Youth Commn., Columbus, 1975-77; chief psychologist Mahoning County Diagnostic and Evaluation Clinic, Youngstown, Ohio, 1977-78; asst. prof. dept. criminal justice Youngstown State U., 1978—; owner Towne Square Psychol. Services, Inc., Canfield, Ohio, 1971—; pres. Mahoning Valley Acad. Psychology, 1981-82; founder, dir. Integrated Profl. Systems Inc., Austintown, Ohio, 1981-82; cons. to local and state agencys. Author manuals and computer programs; contbr. articles to profl. jours. Mem. Youth Services Council, Youngstown, 1979; bd. dirs. Gateways to Better Living, Youngstown, 1980-81. State of Ohio grantee, 1980-82. Mem. Am. Psychol. Assn., Ohio Psychol. Assn., Acad. Criminal Justice Scis., AAAS, Sigma Xi. Current work: Construction of microcomputer programs to be used for psychometric assessments, test scoring, and test interpretation. Home: 266 Bradford Dr Canfield OH 44406 Office: Towne Square Psychol Services Inc PO Box 567 Youngstown OH 44501

WALI, MOHAN KISHEN, environmental science educator; b. Kashmir, India, Mar. 1, 1937; came to U.S., 1969, naturalized, 1975; s. Jagan Nath and Somavati (Wattal) W.; m. Sarla Safaya, Sept. 25, 1960; children: Pamela, Promod. B.Sc., U. Jammu and Kashmir, India, 1957; M.Sc., U. Allahabad, India, 1960; Ph.D., U. B.C., Can., 1970. Lectr. S.P. Coll., Srinagar, Kashmir, 1963-65; research fellow U. Copenhagen, 1965-66; grad. fellow U. B.C., Vancouver, 1967-69; asst. prof. biology U. N.D., Grand Forks, 1969-73, assoc. prof., 1973-79, prof., 1979-83, Hill research prof., summer 1973, dir., 1970-79, 1975-83, spl. asst. to pre. univ., 1977-83; prof., dir. grad. program environ. sci. SUNY-Syracuse, 1983—; staff ecologist Grand Forks Energy Research Lab., U.S. Dept. Interior, part-time 1974-75. Contbr. articles to profl. jours.; editor: Some Environmental Aspects of Strip-Mining in North Dakota, 1973, Prairie: A Multiple view, 1975, Practices and Problems of Land Reclamation in Western North America, 1975, Ecology and Coal Resource Development, 1979; sr. editor: Reclamation Rev, 1976-80; chief editor, 1980-81, Reclamation & Revegetation Research, 1982—. Vice-chmn. N.D. Air Pollution Adv. Council. Recipient Outstanding Research award Sigma Xi-U. N.D., 1975; B.C. Gamble Disting. Teaching and Service award, 1977. Mem. Ecol. Soc. Am. (chmn. sect. internat. activities 1980—), Brit. Ecol. Soc., Can. Bot. Assn. (dir. ecology sect. 1976-79, v.p. 1982-83), Torrey Bot. Club, AAAS, Am. Soc. Agronomy, Am. Inst. Biol. Sci. (gen. chmn. ann. meeting), Internat. Assn. Ecology, Internat. Soc. Soil Sic., N.D. Acad. Sci. (chmn. editorial com. 1979-81), Sigma Xi. Subspecialties: Ecology; Ecosystems analysis. Office: Coll Environ Sci & Forestry SUNY Syracuse NY 13210

WALIA, AMRIK SINGH, surgery educator, immunologist; b. Punjab, India, Aug. 6, 1947; came to U.S., 1970; s. Harkishan Singh and Harbhajan K. (Sumitra) Ahluwalia; m. Shammi A. Chadha, Jan. 6, 1974; children: Jasmeet Singh, Shalini R. B.Sc., Punjab U., 1965; M.Sc., Meerut U., India, 1968; Ph.D., Loyola U., New Orleans, 1975. Prin. investigator alcoholism research VA Med. Ctr., Birmingham, Ala., 1982—; research asst. prof. surgery U. Ala.-Birmingham, 1980—, assoc. scientist, 1980—. U. Ala.-Birmingham Cancer Ctr. Faculty Devel. grantee, 1980-81; research fellow Nat. Cancer Inst., 1977-80; grantee Internat. Union Against Cancer, Sweden, 1978. Mem. Am. Assn. Immunologists, N.Y. Acad. Scis., Sigma Xi. Subspecialties: Immunology (medicine); Cancer research (medicine). Current work: Immunology and effects of alcohols, retinoids, divalent cations, trace metals on immune response; T cells, B cells, immune complex receptors, complement receptors. Home: 500 Cedar St Birmingham AL 35206 Office: Dept of Surgery University of Alabama Birmingham AL 35294

WALKER, BRIAN KEITH, medical oncologist, hematologist, internist; b. Bellefonte, Pa., Mar. 25, 1947; s. Eric Arthur and Josephine (Schmeiser) W.; m. Mary Ann Lightner, Apr. 15, 1972; 1 dau., Laura. B.A., Princeton U., 1969; M.D., Cornell U., 1969-73. Diplomate: Am. Bd. Internal Medicine. Intern Temple U. Hosp., Phila., 1973-74, resident, 1974-76; fellow in hematology Jefferson U., Phila., 1976-78; asst. prof. W.Va. U., 1978-81, assoc. prof., 1981-82; staff physician J.C. Blair Hosp., Huntingdon, Pa., 1982—, Centre Community Hospital, State College, Pa., 1982—; mem. Cancer and Leukemia Group, N.Y.C., 1978-82, Central Pa. Oncology Group, Hershey, Pa., 1982—. Contbr. articles to profl. jours. Bd. dirs. Am. Cancer Soc., Huntingdon, 1982. Fellow ACP; mem. Am. Soc. Hematology, Am. Soc. Clin. Oncology. Presbyterian. Subspecialties: Chemotherapy; Hematology. Current work: Cancer chemotherapy research; investigational chemotherapy protocols. Home: 945 Oak Ridge Ave State College PA 16801 Office: Internal Medicine Associates 3901 S Atherton St State College PA 16801

WALKER, DAN BERNE, biology educator, researcher; b. Connersville, Ind., Apr. 18, 1945; s. Merrell C. and Cathryn R. (Burkhart) W.; m. Denise G. Garriott, June 15, 1969; m. Susan S. Schroeder, May 29, 1982. A.B., Ind. U., 1968; Ph.D., U. Calif., Berkeley, 1974. Lectr. botany dept. U. Calif., Berkeley, 1973-74; asst. prof. botany U. Ga., Athens, 1974-78, UCLA, 1978—; cons. Contbr. articles to profl. jours. Recipient Outstanding Instr. award U. Ga., 1977. Mem. AAAS, Bot. Soc. Am., Am. Soc. Plant Physiologists, Soc. Devel. Biology, Electron Microscopy Soc. Am. Subspecialties: Plant growth; Developmental biology. Current work: Epidermal morphogenesis and pattern formation in plants, differentiation and dedifferentiation in epidermal and subepidermal plant cells, pattern formation in plant tissues. Office: Biology Dept U Calif Los Angeles CA 90024

WALKER, DAVID CROSBY, chemistry educator; b. York, Eng., June 16, 1934; emigrated to Can., 1964, naturalized, 1967; s. John Clement and Frances Mary (Alton) W.; m. Valerie Rosemary Jackson, Apr. 21, 1961; m. Gale Roberts Young, Apr. 28, 1978; children: Shannon Louise, Jennifer Jane, Elizabeth Jean. B.Sc. with honors, St. Andrews U., Scotland, 1956, D.Sc., 1974; Ph.D., Leeds U., Eng., 1959. Postdoctoral fellow NRC, Ottawa, 1959-61; warden Sadler Hall, research lectr. U. Leeds, Eng., 1961-64; prof. chemistry U. B.C., Vancouver, 1964—. Author: Origins of Optical Activity in Nature, 1979, Muon and Muonium Chemistry, 1983; Contbr. articles to research jours. Fellow Chem. Inst. Can.; mem. Am. Contract Bridge League (life master). Subspecialties: Physical chemistry; Laser-induced chemistry. Current work: Radiation chemistry, including studies of solvated electrons, muonium atoms and positron annihilation.

WALKER, ERIC ARTHUR, consulting engineer; b. Long Eaton, Eng., Apr. 29, 1910; came to U.S., 1923, naturalized, 1937; s. Arthur and Violet Elizabeth (Haywood) W.; m. L. Josephine Schmeiser, Dec. 20, 1937; children: Gail (Mrs. Peter Hearn), Brian. B.S., Harvard, 1932, M.S., 1933, Sc.D., 1935; LL.D., Temple U., 1957, Lehigh U., 1957, Hofstra Coll., 1960, Lafayette Coll., 1960, U. Pa., 1960, U. R.I., 1962; L.H.D., Elizabethtown Coll., 1958; D.Litt., Jefferson Med. Coll., 1960; D.Sc., Wayne State U., 1965, Thiel Coll., 1966, U. Notre Dame, 1968, U. Pitts., 1970. Registered profl. engr., Pa. Instr. math. Tufts Coll., 1933-34, asst. prof., asso. prof. elec. engring., 1935-38, head elec. engring. dept., 1935-40, U. Conn., 1940-43; asso. dir. Harvard Underwater Sound Lab., 1942-45; dir. Ordnance Research Lab., Pa. State U., 1945-52, head elec. engring. dept., 1945-51, dean, 1951-56, v.p. univ., 1956, pres., 1956-70; v.p. sci. and tech. Aluminum Co. Am., 1970-76; dir. Armstrong Cork Co.; Salem Corp. Exec. sec. Research and Devel. Bd., 1950-51; cons. NRC, 1949-50; mem. and past chmn. com. on undersea warfare; chmn. Pres.'s Com. on Tech. and Distbn. Research for Benefit of Small Bus., 1957; chmn. nat. sci. bd. NSF, 1964-66; chmn. Naval Research Adv. Com., 1963-65, 71—, Army Sci. Adv. Panel, 1956-58; vice chmn. Pres.'s Com. Scientists and Engrs., 1956-58; adv. panel on engring. and tech. manpower Pres.'s Sci. Adv. Com.; mem. Gov.'s Com. of 100 for Better Edn., 1960-61; bd. dirs. Engring. Found.; chmn. bd. Inst. for Def. Analysis. Contbr. to tech. mags. United bd. visitors U.S. Naval Acad., 1958-60, U.S. Mil. Acad., 1962-64. Recipient Horatio Alger award, 1959, Tasker H. Bliss award Am. Soc. Mil. Engrs., 1959; Golden Omega award Am. Inst. E.E and Nat. Elec. Mfg. Assn., 1962; DoD Pub. Service medal, 1970; Presdl. citation, 1970. Fellow IEEE, Am. Acoustical Soc., Am. Inst. E.E., Am. Phys. Soc.; mem. Am. Inst. Physics, Am. Soc. Engring. Edn. (Lamme award 1965, pres. 1961-62), Pa. Assn. Colls. and Univs. (pres. 1950-60), Middle States Assn. Colls. and Secondary Schs. (commn. higher edn. 1958-61), Engrs. Joint Council (pres. 1962-63), Nat. Assn. State Univs. and Land-Grant Colls. (exec. com. 1958-62), Nat. Acad. Engring. (pres. 1966-70), Am. Acad. Arts and Scis., Newcomen Soc., Royal Soc. Arts, Sigma Xi, Tau Beta Pi, Phi Kappa Phi. Clubs: Duquesne, Cosmos. Subspecialties: Electrical engineering; Acoustical engineering. Current work: New processes and products. Home: Rock Spring Farm Pennsylvania Furnace PA 16865

WALKER, EUGENE HOFFMAN, hydrologist, pleistacene geologist; b. N.Y.C., Mar. 28, 1915; s. John Baldwin and Mai Elmendorf (Hackstaff) W.; m. Mary Morris, Aug. 27, 1947; children: Arthur, Cynthia, Pamela. B.A., M.A., Ph.D., Harvard U. Geologist Shell Oil Co., Midland, Tex., 1939-41; teaching fellow Harvard U., 1941-43; geologist Patino Mines, Bolivia, 1943-46; instr. U. Mich., Ann Arbor, 1946-49; hydrologist U.S. Geol. Survey, various locations, 1949-81; vol. hydrologist, geologist dept. natural resources Town of Concord, Mass., 1981—. Contbr. articles to profl. jours. Mem. Geol. Soc. Am., Torrey Bot. Club, New Eng. Bot. Club. Democrat. Episcopalian. Subspecialties: Ground water hydrology; Geology. Current work: Hydrology and quarternary geology; advisory activities to town of Concord, Mass. on hydrology, geology and environmental problems. Address: 14 Chestnut St Concord MA 01742

WALKER, HARLEY JESSE, geography educator, coastal researcher; b. Bushnell, Mich., July 4, 1921; s. Clair John and Cassiebelle (Brown) W.; m. Rita Haus, Mar. 14, 1953; children: Winona, Angela, Tina. B.A., U. Calif.-Berkeley, 1947, M.A., 1954; Ph.D., La. State U., 1960. Asst. to assoc. prof. geography Ga. State U., Atlanta, 1950-59, chmn. geography dept., 1954-59; mem. faculty La. State U., Baton Rouge, 1960—, asst. prof. to prof. geography, 1960-76, chmn. geography dept., 1962-70, Boyd prof. in geography, 1976—. Contbr. numerous sci. articles to profl. publs. Served with USMCR, 1942-45. Recipient Disting. Research Master award La. State U., 1974; USAF ADTIC Research grantee, 1957-58. Mem. AAAS, Am. Geophys. Union, Assn. Am. Geographers (Honors award 1977), Explorers Club, Coastal Soc. (dir. 1980-82), Sigma Xi. Subspecialties: Geomorphology; Physical Geography. Current work: Coastal morphology; deltaic morphology; coastal ecology. Home: 1250 Thibodeaux Ave Baton Rouge LA 70806 Office: La State U Geography Dept Baton Rouge LA 70803

WALKER, HARRELL LYNN, plant pathologist; b. Minden, La., May 14, 1945; s. George Harrell and Janice Ora (Nix) W.; m. Diane Reynolds, Sept. 6, 1975; children: Thomas, Alan Reynolds, Mark Thomas. B.S., La. Tech. U., 1966; M.S., U. Ky., 1969, Ph.D., 1970. Plant pathologist Ala. Dept. Agr., 1974-75, asst. dir. plant industry div., 1975-76; research plant pathologist So. Weed Sci. Lab., U.S. Dept. Agr., Stoneville, Miss., 1976—. Contbr. articles to profl. jours. Served with U.S. Army, 1970-72. U.S. Dept. Agr. grantee. Mem. Weed Sci. Soc. Am., Am. Phytopath. Soc., Miss. Assn. Plant Pathologists, Sigma Xi. Republican. Baptist. Subspecialties: Plant pathology; Integrated pest management. Current work: Biological control of weeds with plant pathogens. Office: So Weed Sci Lab PO Box 225 Stoneville MS 38776

WALKER, HOMER FRANKLIN, mathematician; b. Beaumont, Tex., Sept. 7, 1943; s. John Harold and Esther Orlou (Hooks) W.; m. Susan Dorothy Proctor, June 13, 1970 (div. 1975). B.A., Rice U., 1966; M.S., N.Y.U., 1968, Ph.D., 1970. Asst. prof. math. Tex. Tech. U., Lubbock, 1970-73, assoc. prof. math., 1973-74, U. Houston, 1974-80, prof. math., 1980—; vis. assoc. prof. U. Denver, 1973-74, Cornell U., Ithaca, N.Y., 1978-79; vis. prof. U. N.Mex., Albuquerque, 1981-82; cons. Lawrence Livermore (Calif.) Nat. Lab., 1979—. Mem. Soc. Indsl. and Applied Math., Am. Math. Soc., Math. Assn. Am. Subspecialties: Applied mathematics; Numerical analysis. Current work: Numerical analysis, pattern recognition, partial differential equations. Office: Dept Math Univ Houston Cullen Blvd Houston TX 77004

WALKER, JAMES ROY, microbiologist, educator; b. Chestnut, La., Nov. 8, 1937; s. Clint Cortez and Annie Mae (Holl) W.; m. Barbara Ann Fess, Aug. 9, 1959; children: James Bryan, Melinda Lee. B.S., Northwestern State U., La., 1960; Ph.D., U. Tex., Austin, 1963. Postdoctoral research assoc. Princeton U., 1965-67; asst. prof. microbiology U. Tex., 1967-71, assoc. prof., 1971-78, prof., 1978—, chmn. dept., 1981—; research assoc. in biochemistry and molecular biology Harvard U., 1972-73. Served with U.S. Army, 1963-65. NIH, NSF, Am. Cancer Soc. fellow; various grants. Mem. Am. Soc. for Microbiology, Genetics Soc. Am. Subspecialties: Genetics and genetic engineering (biology); Molecular biology. Current work: Biochemical and genetic studies on chromosome replication and regulation of cell division in Escherichia coli. Home: 8504 Greenflint Ln Austin TX 78759 Office: Dept Microbiology U Tex Austin TX 78712

WALKER, JAMES WILLARD, botany educator; b. Taylor, Tex., Mar. 23, 1943; s. James Willard and Anna (Fritz) W.; m. Audrey Galusha, Jan. 6, 1973; 1 son, Jonathan Willard. B.A. in Botany magna cum laude, U. Tex., 1964; Ph.D. in Biology, Harvard U., 1970. Mem. faculty U. Mass.-Amherst, 1969—, prof. botany, 1983—. Editor: The Bases of Angiosperm Phylogeny, Annals of Mo. Bot. Garden, Vol. 62, Number 3, 1975; contbr. numerous articles to sci. publs. Jr. fellow U. Tex.-Austin, 1964; Harvard U. fellow Harvard U., 1967-69; NDEA Fellow, 1964-67; NSF grantee, 1972—. Mem. Bot. Soc. Am., Am. Soc. Plant Taxonomists (Cooley award 1972), Sigma Xi, Phi Eta Sigma, Phi Beta Kappa. Subspecialties: Systematics; Evolutionary biology. Current work: Pollen morphology of primitive angiosperms; ultrastructure of Lower Cretaceous angiosperm pollen; phylogeny of the flowering plants; origin and early evolution of the angiosperms. Home: 31 Morgan Circle Amherst MA 01002 Office: Botany Dept Univ Mass Amherst MA 01003

WALKER, JERRY TYLER, plant pathologist; b. Cin., Sept. 7, 1930; s. Wallace Burch and Edith Tyler W.; m. Mary Bridges, Sept. 26, 1930; children: Robert William, Ann Elizabeth. A.B., Miami U., Oxford, Ohio, 1952; M.Sc., Ohio State U., Columbus, 1957, Ph.D., 1960. Research assoc. Ohio Agrl. Research and Devel. Ctr., Wooster, 1959-61; plant pathologist Bklyn. Bot. Garden, 1961-69; chmn. Kitchawan Field Sta., Ossining, N.Y., 1968-69; prof., dept. head U. Ga., Ga. Agrl. Expt. Sta., 1969—. Contbr. articles to profl. jours. Served with Chem. Corps U.S. Army, 1953-55. NSF grantee, 1963. Mem. Am. Soc. Phytopath., Soc. Nematologists, Internat. Soc. Arboriculture, Sigma Xi. Episcopalian. Lodges: Rotary; Elks. Subspecialty: Plant pathology. Current work: Air pollution effects on plants, nematology, diseases of ornamentals. Office: Ga Sta Experiment GA 30212

WALKER, JOHN SCOTT, theoretical and applied mechanics educator, consultant; b. Washington, May 25, 1944; s. Irving Scott and Elizabeth (Haslacker) W.; m. Mary Beatrice Schwab, Oct. 26, 1970. B.S., Webb Inst. Naval Architecture, Glen Cove, N.Y., 1966; Ph.D., Cornell U., 1970. Research assoc., instr. Cornell U., Ithaca, N.Y., 1970-71; asst. prof. U. Ill., Urbana, 1971-75, assoc. prof., 1975-78, prof. theoretical and applied mechanics, 1978—; cons. Oak Ridge Nat. Lab., 1977-81, Westinghouse Research Lab., Pitts., 1981, IBM Research Lab., San Jose, Calif., 1981; prin. investigator NSF, Washington, 1973—. Assoc. editor: Jour. Applied Mechanics, 1981—, Mechanics, 1975-78. Recipient award Am. Bur. Shipping, 1966. Mem. ASME (Pi Tau Sigma Gold medal 1976), Am. Nuclear Soc., Am. Soc. Engring. Edn., Univ. Fusion Assn., Sigma Xi. Subspecialties: Nuclear fusion; Fluid mechanics. Current work: Theoretical studies of liquid-metal flows in strong magnetic fields (magnetohydrodynamics). Applications: thermal hydraulic analyses of liquid lithium blankets for magnetic confinement fusion reactors; magnetic suppression of convection in Czochralski crystal growth; electromagnetic machines for metal foundries such as magnetic shapers and stirers for castings. Office: Univ Illinois 104 S Wright St Urbana IL 61801

WALKER, ROBERT D., clinical microbiologist; b. Salt Lake City, Sept. 16, 1944; s. Garth Talmage and Mildred Ruth (Smith) W.; m. Sue Ann Corradini, Aug. 26, 1967; children: Paula Sue, Travis Aaron. A.S., Mesa Coll., 1968; B.S., U. Utah, 1971; M.S., U. N.D., 1973; Ph.D., Okla. State U., 1978. Grad. research asst. dept. microbiology U. N.D., 1972-74; research assoc. dept. vet. parasitology, microbiology and pub. health Coll. Vet. Medicine, Okla. State U., Stillwater, 1974-77, grad. research asst. dept. vet. parasitology , microbiology and pub. health, 1977-78; postdoctoral research assoc. dept. microbiology U. Tenn., Knoxville, 1978-79, asst. prof. dept. pathobiology, 1978—. Contbr. articles to profl. jours. Served to sgt. USMC, 1963-66. Decorated Purple Heart. Mem. Am. Soc. Microbiologists, Tenn. Soc. Clin. Microbiologists, Appalachian Zool. Soc. Mormon. Subspecialty: Microbiology (veterinary medicine). Current work: Research on bovine and equine respiratory diseases; antimicrobial therapy in veterinary medicine and the isolation and identification of clinically important anaerobic bacteria; canine osteomyelitis.

WALKER, ROBERT MOWBRAY, educator, physicist; b. Phila., Feb. 6, 1929; s. Robert and Margaret (Seivwright) W.; m. Alice J. Agedal, Sept. 2, 1951 (div. 1973); children: Eric, Mark; m. Ghislaine Crozaz, Aug. 24, 1973. B.S. in Physics, Union Coll., 1950, D.Sc., 1967; M.S., Yale, 1951, Ph.D., 1954; D.h.c., Université de Clermont-Ferrand, 1975. Physicist Gen. Electric Research Lab., Schenectady, 1954-62, 63-66; McDonnell prof. physics Washington U., St. Louis, 1966—; dir. McDonnell Center for Space Scis., 1975—; Vis. prof. U. Paris, 1962-63; adj. prof. metallurgy Rensselaer Poly. Inst., 1958, adj. prof. physics, 1965-66; vis. prof. physics and geology Calif. Inst. Tech., 1972, Phys. Research Lab., Ahmedabad, India, 1981, Institut d'Astrophysique, Paris, 1981; Pres. Vols. for Internat. Tech. Assistance, 1960-62, 65-66, founder, 1960, bd. dirs., 1961—; mem. Lunar Sample Analysis Planning Team, 1968-70, Lunar Sample Rev. Bd., 1970-72; adv. com. Lunar Sci. Inst., 1972-75; mem. temporary nominating group in planetary scis. Nat. Acad. Scis., 1973-75, bd. on sci. and tech. for internat. devel., 1974-76, com. planetary and lunar exploration, 1978—; Bd. dirs. Univs. Space Research Assn., 1969-71. Recipient Distinguished Service award Am. Nuclear Soc., 1964, Yale Engring. Assn. award for contbn. to basic and applied sci., 1966, Indsl. Research awards, 1964, 65; Exceptional Sci. Achievement award NASA, 1970; E.O. Lawrence award AEC, 1971; NSF fellow, 1962-63. Fellow Am. Phys. Soc., Meteoritical Soc., AAAS; mem. Am. Geophys. Union, Am. Astron. Soc., Nat. Acad. Scis. Subspecialty: Geophysics. Research and publs. on cosmic rays, nuclear physics, geophysics, radiation effects in solids, particularly devel. solid state track detectors and their application to geophysics and nuclear physics problems; discovery of fossil particle tracks in terrestrial and extra-terrestrial materials and fission track method of dating; application of phys. scis. to art and archaeology; lab. studies of interplanetary dust. Home: 3 Romany Park Ln Olivette MO 63132

WALKER, RUSSELL GLENN, astronomer; b. Cin., May 3, 1931; s. Glenn Herbert and Hazel Lilian (Dalton) W.; m. Dorisanne Walker, Mar. 21, 1953; children: Sandra J., Sharon L. B.Sc., Ohio State U., 1953, M.Sc., 1954; Ph.D., Harvard U., 1967. With Block Assocs., Cambridge, Mass., 1960-61; sr. scientist Air Force Geophys. Lab., 1956-60, chief infrared physics lab., 1961-75; staff scientist NASA, Ames Research Ctr., 1976-81; assoc. scientist Jamieson Sci. and Engring., Inc., Palo Alto, Calif., 1981—. Contbr. articles to profl. jours. Served to capt. USAF, 1954-56. Recipient USAF Sci. Achievement award, 1970. Mem. Am. Astron. Soc., Optical Soc. Am., Astron. Soc. Pacific, Am. Soc. Enologists. Subspecialties: Infrared optical astronomy; Satellite studies. Current work: Research in infrared physics and astronomy, development of cryogenic optical systems for use in space. Home and Office: PO Box 11698A Palo Alto CA 94306

WALKER, SHARYN MARIE, immunologist; b. Ontario, Oreg., Apr. 21, 1945; d. John Ralph and Ruth Reine (Feller) W. B.A., Stanford U., 1967, Ph.D., 1974. Postdoctoral fellow Stanford U., 1974-75; postdoctoral fellow Scripps Clinic and Research Found., La Jolla, Calif., 1975-78, asst. mem., 1979-83; assoc. prof. Children's Hosp., Los Angeles, 1983—. Contbr. articles to profl. jours. Arthritis Found. fellow, 1976-79; Am. Cancer Soc. jr. faculty awardee, 1979-82; NIH grantee, 1979-82. Mem. Am. Assn. Immunologists, N.Y. Acad. Scis. Subspecialty: Immunobiology and immunology. Current work: Antibody response; immunological tolerance; immunological adjuvants. Home: 12317 Otsego St North Hollywood CA 91607 Office: Children's Hosp 4650 Sunset Blvd Los Angeles CA 90054

WALKUP, JOHN FRANK, electrical engineer, educator; b. Oakland, Calif, Feb. 7, 1941; s. Francis Milton and Mabel Doreen (Lishman) W.; m. Patricia Ann Hagbom, June 26, 1965; children: Mary Kathleen, Amy Christine, Rebecca Joy. B.A., Dartmouth Coll., 1962, B.E.E., 1963; M.S., Stanford U., 1965, Engr., 1969, Ph.D., 1971. Registered profl. engr., Tex. Research asst. Stanford Electronics Labs., Stanford U., 1963-71; asst. prof. elec. engring. Tex. Tech. U., Lubbock, 1971-76, assoc. prof., 1976-81, prof., 1981—, interim assoc. dean engring., 1982-83; cons. in field. Author articles. Recipient Goodrich prize Dartmouth Coll., 1963; Halliburton award for excellence in teaching Tex. Tech U., 1980; Pres.'s award for excellence in teaching, 1981. Fellow Optical Soc. Am.; mem. IEEE, Soc. Photo-Optical Instrumentation Engrs., Am. Soc. Engring. Edn., Sigma Xi. Subspecialties: Optical signal processing; Graphics, image processing, and pattern recognition. Current work: Optical information processing, image processing, communication theory, electrical engineering education.

WALL, CONRAD, III, biomedical researcher and biomedical engineer; b. Boston, June 13, 1939; s. Conrad and Neil (Kennedy) W.; m. Susan Vieth, Feb. 1, 1961; children: Conrad C., Richard A. B.S. in Physics, Tulane U., 1962, M.S., 1968; Ph.D. in Bioengring, Carnegie-Mellon U., 1975. Mem. tech. staff Boeing Co., 1965-70; research assoc. dept. otolaryngology Sch. Medicine, U. Pitts., 1975-76; sci. mgr. R.E. Jordan Vestibular Lab., U. Pitts., 1977-82; asst. prof. dept. otolaryngology U. Pitts., 1977-82; dir. R.E. Jordan Vestibular Lab., 1982—; assoc. prof. dept. otolaryngology U. Pitts., 1982—; cons. vestibular testing. Contbg. author papers to profl. publs. and conf. Served to 1st lt. U.S. Army, 1962-64. Named to Apollo/Saturn V Roll of Honor. Mem. AAAS, Soc. for Neurosci., IEEE, Assn. for Research in Otolaryngology, Barany Soc., Sigma Xi. Democrat. Episcopalian. Club: Wahnita Lodge (Ft. Littleton, Pa.). Subspecialty: Biomedical engineering. Current work: Research in human balance. Office: Eye and Ear Hosp 230 Lothrop St Room 1101 Pittsburgh PA 15213

WALL, FREDERICK THEODORE, chemistry educator; b. Chisholm, Minn., Dec. 14, 1912; s. Peter and Fanny Maria (Rauhala) W.; m. Clara Elizabeth Vivian, June 5, 1940; children: Elizabeth Wall Ralston, Jane Vivian Wall-Meinike. B.Chemistry, U. Minn., 1933, Ph.D., 1937. Instr. chemistry U. Ill., 1937-39, assoc., 1939-41, asst. prof., 1941-43, asso. prof., 1943-46, prof. chemistry, 1946-64, acting dean grad. coll., 1951-52, head div. phys. chemistry, 1953-56, dean grad. coll., 1955-63; prof., chmn. dept. chemistry U. Calif., Santa Barbara, 1964-66, vice chancellor research, 1965-66; vice chancellor grad. studies and research, prof. chemistry U. Calif. at San Diego, 1966-69; exec. dir. Am. Chem. Soc., Washington, 1969-72; prof. chemistry Rice U., Houston, 1972-78, San Diego State U., 1978-81, U. Calif., San Diego, 1982—; Pres. Assn. Grad. Schs., 1961; trustee Inst. Def. Analyses, 1962-64; Mem. governing bd. Nat. Acad. Scis.-NRC, 1963- 67; mem. adv. com. for math. and phys. scis. NSF, 1964-68, chmn., 1967; chmn. bd. trustees Univs. Space Research Assn., 1970-73, chmn. council instns., 1974; mem. sci. adv. com. U. Calif. concerning Los Alamos and Livermore Labs., 1972-80. Author: Chemical Thermodynamics, 1958; Editor: Jour. Phys. Chemistry, 1965-69; Contbr. articles on phys. chemistry of polymers and statis. mechanics to sci. jours. Mem. Am. Chem. Soc. (Pure Chemistry award 1945, dir. 1962-64), Finnish Chem. Soc. (corr.), Am. Acad. Arts and Scis., Nat. Acad. Scis., Am. Phys. Soc., AAAS, Sigma Xi, Phi Kappa Phi, Phi Lambda Upsilon (hon.). Subspecialties: Physical chemistry; Theoretical chemistry. Current work: Statistical mechanics of polymer systems, thermodynamics. Home: 2468 Via Viesta La Jolla CA 92037

WALL, ROBERT ECKI, science administrator; b. Aurora, Ill., Aug. 1, 1935; s. Clifford Nathan and Mildred (Ecki) W.; m. Carol Porta, May 29, 1963; children: Laura, Andrea, Jason. B.A., Carleton Coll., 1957; Ph.D., Columbia U., 1965. Research assoc. Columbia U., 1965-66; sci. officer Office Naval Research, Washington, 1966-70; program dir. NSF, Washington, 1970-75, sect. head, 1975—. Mem. Am. Geophys. Union, AAAS, Geol. Soc. Am. Subspecialties: Oceanography; Geophysics. Current work: Technical administration of $55 million grants program supporting research largely in academic institutions in physical, chemical and biological oceanography and submarine geology and geophysics. Office: NSF 1800 G St NW Washington DC 20550

WALLACE, DOUGLAS CECIL, geneticist, educator; b. Cumberland, Md., Nov. 6, 1946; s. David H. and Elizabeth M. W. B.S., Cornell U., 1968; M.Ph., Yale U., 1972, Ph.D., 1975. Research microbiologist USPHS, Gig Harbor, Wash., 1968-70; fellow in microbiology Yale U., 1970-75, postdoctoral fellow in human genetics, 1975-76; asst. prof. genetics Stanford (Calif.) U., 1976-83; prof. biochemistry, assoc. prof. pediatrics in med. genetics Emory U., Atlanta, 1983—. Mem. Am. Soc. Human Genetics, Am. Soc. Microbiology, AAAS, Sigma Xi. Subspecialties: Genetics and genetic engineering (biology); Molecular biology. Current work: Organelle genetics. Office: Emory U Sch Medicine Atlanta GA 30322

WALLACE, JAMES, JR., marine engineer, consultant; b. New Brunswick, N.J., Oct. 6, 1932; s. James and Mary (Devaney) W.; m. Nancy E. Vivian, June 14, 1958; children: Patrick, Kathleen, Megan, Anne, Michael. B.S. in Marine Engring, U.S. Mcht. Marine Acad., 1954; M.S. in Engring. Sci, U. Notre Dame, 1959; Ph.D. in Aero. Engring, Brown U., 1963. Research scientist Avco Everett Research Lab., Mass., 1963-73; pres. Far Field, Inc., Sudbury, Mass., 1973—. Contbr. articles to profl. jours. Served with USN, 1954-56. Mem. AIAA, IEEE, Optical Soc. Am., N.Y. Acad. Scis., Sigma Xi. Democrat. Roman Catholic. Subspecialties: Atmospheric optics; Aeronautical engineering. Current work: Laser propagation including thermal blooming, adaptive optics and atmospheric turbulence. Research in fluid mechanics that includes flow around airplanes and ships. Home and Office: 6 Thoreau Way Sudbury MA 01776

WALLACE, JOHN EDWIN, meteorology consulting firm executive; b. Holton, Kans., Dec. 22, 1913; s. Verne P. and Clara Elizabeth (Fencil) W.; m. Elizabeth Ann Johnson, June 24, 1944. B.S., Rice U., 1937; M.S. in Meteorology, Calif. Inst. Tech., 1942. Cert. cons. meteorologist. Salesman Gulf Oil Corp., Houston, 1937-40; founder, pres. Weather Services Corp., Bedford, Mass., 1947—; co-founder, v.p. Weather Services Internat., 1979—. Served with USAAF, 1940-46. Recipient award for Outstanding Contbn. Am. Meteorol. Soc., 1977. Mem. Am. Meteorol. Soc. (past sec. exec. com.), Nat. Council Indsl. Meteorologists (past pres., dir.). Club: Belmont (Mass.) Hill. Subspecialties: Meteorology; Synoptic meteorology. Office: 131A Great Rd Bedford MA 01730

WALLACE, JOHN FRANCIS, educator; b. Boston, Oct. 26, 1919; s. John Joseph and Ellen Gertrude (Organ) W.; m. Agnes Teresa Fitzgerald, July 17, 1943; children—Susan, Sally, Nancy. B.S., Mass. Inst. Tech., 1941, M.S., 1953. Metallurgist, head metal processing lab. div. Watertown Arsenal (now Army Materials and Mechanics Research Center), Watertown, Mass., 1941-54; metallurgist, chief heat treatment div., dir. Rodman Lab., Case Western Res. U., Cleve., 1954—, prof. metallurgy and materials sci., 1961—, Republic Steel prof. metallurgy, 1979—, chmn. dept., 1974—; cons. to industry; Hoyt Meml. lectr. Am. Foundrymen's Soc., 1975. Author 3 books; editor; contbr. articles to profl. jours. Served with AUS, 1942-46. Decorated Bronze Star medal.; Recipient Pangborn Gold medal Am. Foundrymen's Soc., 1962; award of honor Steel Founders Soc. Am., 1967; Nyselius award Am. Die Casting Inst., 1967; Gold medal Gray and Ductile Soc., 1970. Fellow British Foundrymen's Soc.; mem. Am. Foundrymen's Soc. (hon. life, pres. tech. council 1972-74, chmn. gray iron div. 1970-72, Howard Taylor award 1981), Am. Inst. Mining and Metal Engrs., Am. Soc. Metals, Soc. Die Casting Engrs. (hon. life), Ret. Officers Assn., Sigma Xi. Subspecialty: Metallurgy. Current work: Solidification of metals, metallurgical processes, particularly die casting, centrifugal casting and sand casting. Home: 3326 Braemar Rd Cleveland OH 44120

WALLACE, ORSON JOSEPH, scientific programmer; b. Durango, Colo., Oct. 23, 1931; s. Joseph W. and Zelma L. (Dalton) W.; m. Jean Delores Fitpold, Sept. 18, 1961; 1 son, Alan Joseph. B.S., Western State Coll., 1949; M.S., U. Pitts., 1962. Programmer Westinghouse Electric Corp., East Pittsburg, Pa., 1955-66; sr. sci. programmer Westinghouse Bettis Lab, West Mifflin, Pa., 1966-74, prin. scientist, 1974—. Contbr. articles to profl. jours. Mem. Assn. Computing Machinery, Am. Nuclear Soc. Subspecialties: Mathematical software; Software engineering. Current work: Development of reactor shielding calculation methods. Home: 423 Temona Dr Pleasant Hills PA 15236 Office: Westinghouse Bettis Lab PO Box 79 West Mifflin PA 15122

WALLACE, RICHARD KENT, astrophysicist; b. Washington, Jan. 29, 1954; s. John Cheatham and Mary Ellen (Hutchison) W.; m. Susan Lee Wallace, June 19, 1976; 1 dau., Linda Sheryl. B.S., La. State U., 1975; M.S., U. Calif.-Santa Cruz, 1977, Ph.D., 1981. Research asst. Lick Obs., U. Calif.-Santa Cruz, 1976-81; staff mem., physicist Los Alamos (N. Mex.) Nat. Lab., 1981—. Contbr. articles to profl. jours. Mem. Am. Astron. Soc., Phi Kappa Phi, Pi Sigma, Phi Eta Kappa, Pi Mu Epsilon. Democrat. Presbyterian. Subspecialties: Theoretical astrophysics; Nuclear fusion. Current work: Theoretical nuclear astrophysics, including x and gamma ray burst mechanism, stellar instabilities, hydrodynamics physics or nuclear fusion reactions. Home: 3708-B Arizona St Los Alamos NM 87544 Office: Los Alamos Nat Lab X2 MS B220 Los Alamos NM 87545

WALLACE, ROBERT BRUCE, neurobiologist, educator; b. Stoneham, Mass., Jan. 16, 1937; s. William Sheperd and Dorothy Constance (Gilbert) W. A.B., Boston U., 1960, A.M., 1961, Ph.D., 1966. Lectr. in psychology Boston U., 1967-8; research assoc. M.I.T., Cambridge, 1966-68, Inst. of Living, Hartford, Conn., 1971—; asst. prof. U. Hartford, Conn., 1968-72, assoc. prof., 1972-80, prof. psychology and biology, 1980—, research assoc. II, 1974—; cons. Purdue U., 1968-70, State of Conn., 1971, 74, Nat. Cancer Inst., 1978. Contbr. articles to sci. jours. Bd. dirs. Univ. Research Inst. Conn., Inc.,

1975—. NSF fellow, 1960. Mem. Am. Assn. Anatomists, AAAS, Eastern Psychol. Assn., European Soc. Comparative Endocrinology, Internat. Neuropsychol. Soc. Assn., N.Y. Acad. Scis., Psychonomic Soc., Soc. Neurosci., Sigma Xi, Psi Chi. Republican. Clubs: Golf (Avon, Conn.); City (Hartford). Subspecialty: Comparative neurobiology. Current work: Transplantation of embryonic brain tissue in an effort to determine potential for recovery of function; exploration of developmental plasticity of the mammalian nervous system. Home: 48 Avonwood Rd Avon CT 06001 Office: U Hartford West Hartford CT 06117

WALLACE, ROBERT BRUCE, JR., computer software consultant; b. Ft. Bragg, N.C., Dec. 4, 1955; s. Robert Bruce and Bertha Jane (Bell) W. B.S. in Math. and Computer Sci, Fla. State U., 1977. Cons. Leon County Schs., Tallahassee, 1977-78; systems analyst STAR Operating Systems Group, Sunnyvale, Calif., 1978; project mgr. MAXBASIC, Nat. Info. Systems, Cupertino, Calif., 1978-79; mgr. software devel. Personal Software Inc., Sunnyvale, 1979-80; cons. Aydin Energy Div., Palo Alto, Calif., 1980-81, GenRad, Milpitas, Calif., 1981, OMEX, Santa Clara, Calif., 1981-82, Mgmt. Blueprint Software, Los Gatos, Calif., 1982—; pres. PolyGlot, San Jose, Calif., 1981—. Contbr. articles to mag. Nat. Merit scholar. Mem. Assn. Computing Machinery, Profl. and Tech. Cons. Assn. Subspecialties: Software engineering; Programming languages. Current work: Increase ease of use and flexibility of computer systems/human interfaces with application of artificial intelligence techniques and high order languages. Created game Asteroids in Space, 1980. Office: PolyGlot 6030 Calle de Suerte San Jose CA 95124

WALLACE, ROBERT DEAN, biochemist, nutritionist, researcher; b. Watertown, Wis., July 7, 1939; s. Elden Robert and Dorothy (Hurd) W.; m. Patti Rae Plautz, June 18, 1966. Quality control chemist Armour Pharm. Corp., Kankakee, Ill., 1967-69; mgr. Nutritional Research Lab., Gerber Products Co., Fremont, Mich., 1971—. Active Boy Scouts Am. Served with Hosp. Corps USN, 1958-62. Mem. Am. Soc. Microbiology, AAAS, Vitamin Chemists Assn. Lutheran. Subspecialties: Food science and technology; Biochemistry (biology). Current work: Survey of infant dietary intakes, sugars and cholesterol in foods, phytic acid, nitrosamines, hydrolysis of starches, inversions of sucrose, analytical techniques for high pressure liquid chromatography. Home: 6326 Lakeview Dr Fremont MI 49412 Office: 445 State Fremont MI 49412

WALLACE, ROBERT EARL, geologist, government administrator; b. N.Y.C., July 16, 1916; m., 1945; 1 child. B.S., Northwestern U., 1938; M.S., Calif. Inst. Tech., 1940, Ph.D. in Structural Geology and Vertebrate Paleontology, 1946. Geologist U.S. Geol. Survey, 1942—, chief southwestern br., 1960-65; chief Nat. Center Earthquake Research, 1972-73, regional geologist, 1970-73; chief scientist Office Earthquake Studies, Menlo Park, Calif., 1973—; from asst. prof. to assoc. prof. Wash. State U., 1946-51; vis. lectr. Stanford U., 1960; mem. com. seismology Nat. Acad. Sci.-NRC.; chmn. U.S./USSR Environ Agreement, U.S. Working Group Earthquake Prediction; mem. engrng. criteria rev. bd. San Francisco Bay Conservation and Devel. Commn., 1978—, chmn., 1981—. Fellow Geol. Soc. Am.; mem. Soc. Econ. Geology, Seismol. Soc. Am., Earthquake Engrng. Research Inst. Subspecialties: Geology; Tectonics. Office: US Geol Survey Menlo Park CA 94025

WALLACE, WILLIAM JAMES, chemistry educator; b. Knoxville, July 27, 1935; s. Homer Houston and Bessie May (Hargis) W.; m. Julia Alayne Wehner, Dec. 31, 1958; children: Nora Michelle, Lisa Renée. B.S., Carson-Newman Coll., 1956; Ph.D., Purdue U., 1961. Asst. prof. chemistry U. Miss., 1960-63; asst. prof. chemistry Muskingum Coll., New Concord, Ohio, 1963-68, assoc. prof., 1968-72, prof., 1972—; research assoc. U. Glasgow, Scotland, 1970-71. NSF grantee; NASA grantee, 1979, 80. Mem. Am. Chem. Soc., Sigma Xi (club pres. 1980-82). Baptist. Lodge: Lions. Subspecialties: Inorganic chemistry; Physical chemistry. Current work: Electrocatalysis; computer assisted laboratory teaching. Office: Dept Chemistry Muskingum Coll New Concord OH 43762 Home: 17 N Liberty St New Concord OH 43762

WALLACH, DONALD PINNY, research scientist, free-lance author; b. N.Y.C., Sept. 16, 1927; s. Michael and Estelle (Jamburger) W.; m. Vera Mathilde Sprinz, Sept. 1951. B.S., Mich. State U., 1947, M.S., 1948; Ph.D., U. Wis., 1953. Fellow Wis. Alumni Research Found., U. Wis., 1953-54; USPHS fellow U. Kans. Med. Center, Kansas City, 1954-56; sr. research scientist Upjohn Co., Kalalazoo, 1956—; cons. to attys. in Mich.'s PBB intoxication episode (cattle poisoning). Contbr. articles to sci. publs. Mem. Am. Soc. Pharmacology and Exptl. Therapeutics, Am. Soc. Biol. Chemists, Sigma Xi. Subspecialties: Biochemistry (biology); Pharmacology. Current work: Inhibition of enzymes for potential therapeutic purposes; enzymes of arachidonic acid cascade, phospholipids, prostaglandins, leucotrienes. Patentee in field. Home: 9679 Sterling Rd Richland MI 49083 Office: 301 Henrietta St Kalamazoo MI 49001

WALLER, BRUCE FRANK, cardiologist, med. researcher and educator; b. Austin, Minn., Oct. 18, 1947; s. Frank Joseph and Marcella Marie (Greenlee) W.; m. Nancy Lynn Morton, Aug. 26, 1978. B.A. in Biology, , , Luther Coll., 1969; M.D., U. Minn., 1973, M.S., 1976. Diplomate: Am. Bd. Internal Medicine, Am. Bd. Cardiovascular Diseases. Intern in medicine Mayo Clinic, Rochester, Minn., 1973-74, resident in medicine, 1974-76; fellow in cardiology Georgetown U., Washington, 1976-78, asst. prof. medicine (cardiology), 1979-82; staff assoc. Nat. Heart, Lung and Blood Inst., NIH, Bethesda, Md., 1978-82; assoc. prof. medicine and pathology Ind. U., Indpls., 1983—; cons. cardiology Silver Spring (Md.) Ultrasound, 1978-82; cons. cardiac pathology D.C. VA Hosp., Washington, 1979-82, Riley Children's Hosp., Indpls., 1982—, St. Vincent Hosp., 1982—. Co-author: monograph Exercise and Sudden Death, 1982. Served to lt. comdr. USPHS, 1978-82. Recipient Edgar Van Allen award Am. Heart Assn., Mayo Clinic, 1973. Fellow Am. Coll. Cardiology, ACP, Am. Coll. Chest Physicians, Am. Heart Assn.-Clin. Council Cardiology; mem. Internat. Acad. Pathology, Am. Coll. Sports Medicine. Presbyterian. Subspecialties: Cardiology; Pathology (medicine). Current work: Clinical and morphologic correlation in congenital and acquired diseases of the cardiovascular system. Office: Ind U Sch Medicine 926 W Michigan Indianapolis IN 46223

WALLERSTEIN, GEORGE, astronomy educator; b. N.Y.C., Jan. 13, 1930; s. Leo and Dorothy (Calman) W. B.A., Brown U., 1951; M.S., Calif. Inst. Tech., 1954, Ph.D., 1958. Research asso. Calif. Inst. Tech., Pasadena, 1957-58; instr. U. Calif., Berkeley, 1958-60, asst. prof., 1960-64, asso. prof., 1964-65; prof., chmn. astronomy U. Wash., Seattle, 1965-80, prof. astronomy, 1980—. Trustee Brown U., Providence, 1975-80. Served with U.S. Navy, 1951-53. Mem. Am. Astron. Soc., Astron. Soc. Pacific, AAAS, Arctic Inst. N. Am. Subspecialty: Optical astronomy. Current work: Chemical composition of stars; stellar evolution; variable stars; emission line stars. Home: 538 NE 92d St Seattle WA 98115 Office: Astronomy FM-20 U Washington Seattle WA 98195 It is not sufficient "to follow knowledge like a sinking star, beyond the utmost bounds of human thought." One must endeavor to create knowledge, and beyond that to create understanding.

WALLERSTEIN, ROBERT SOLOMON, psychiatrist; b. Berlin, Germany, Jan. 28, 1921; s. Lazar and Sarah (Guensberg) W.; m. Judith Hannah Saretsky, Jan. 26, 1947; children—Michael Jonathan, Nina Beth, Amy Lisa. B.A., Columbia, 1941, M.D., 1944; postgrad., Topeka Inst. Psychoanalysis, 1951-58. Asso. dir., then dir. research Menninger Found., Topeka, 1954-66; chief psychiatry Mt. Zion Hosp., San Francisco, 1966-78; tng. and supervising analyst San Francisco Psychoanalytic Inst., 1966—; clin. prof. U. Calif. Sch. Medicine, Langley-Porter Neuropsychiat. Inst., 1967-75, prof., chmn. dept. psychiatry, also dir. inst., 1975—; vis. prof. psychiatry La. State U. Sch. Medicine, also New Orleans Psychoanalytic Inst., 1973-73, Pahlavi U., Shiraz, Iran, 1977, Fed. U. Rio Grande do Sul, Porto Alegre, Brasil, 1980; Mem., chmn. research scientist career devel. com. NIMH, 1966-70; Fellow Center Advanced Study Behavioral Scis., Stanford, Calif., 1961-62, 81-82. Author books and monographs., Mem. editorial bd. 8 profl. jours.; Contbr. articles to profl. jours. Served with AUS, 1946-48. Recipient Heinz Hartmann award N.Y. Psychoanalytic Inst., 1968; Distinguished Alumnus award Menninger Sch. Psychiatry, 1972; J. Elliott Royer award U. Calif. at San Francisco, 1973. Fellow Am. Psychiat. Assn., A.C.P., Am. Orthopsychiat. Assn.; mem. Am. Psychoanalytic Assn. (pres. 1971-72), Internat. Psychoanalytic Assn. (v.p. 1977—), Group Advancement Psychiatry, Phi Beta Kappa, Alpha Omega Alpha. Subspecialty: Psychiatry. Current work: Psychotherapy research into processes and outcomes of psychotherapy and psychoanalysis psychotherapy supervision; nature of supervision process. Home: 290 Beach Rd Belvedere CA 94920 Office: Langley-Porter Neuropsychiat Inst 401 Parnassus St San Francisco CA 94143

WALLICK, EARL TAYLOR, pharmacologist; b. Monticello, Ark., Jan. 11, 1938; s. Earl James Henry and LaFran (Hankins) W.; m. Lillie Ruth Fair, June 6, 1962; children: Karl Taylor, James Lee. B.S., Miss. State U., 1960, M.S., 1962; Ph.D., Rice U., 1966. Research chemist DuPont, Kinston, N.C., 1966-67; assoc, prof. chemistry King Coll., Bristol, Tenn., 1967-71; postdoctoral fellow Baylor Coll. Medicine, 1971-73, instr., 1973-77; asst. prof. U. Cin., 1977-78, assoc. prof., 1978—; cons. NIH, 1981-85. NIH grantee, 1978-88, 1980-85. Mem. Am. Chem. Soc., Am. Soc. Pharmacology and Exptl. Therapeutics. Subspecialties: Cellular pharmacology; Molecular pharmacology. Current work: Na, K, -ATPase, cardiac glyosides, regulation of ion transport. Home: 283 Compton Rd Cincinnati OH 45215 Office: Dept Pharmacology and Cell Biophysics U Cin Coll Medicine Cincinnati OH 45267

WALLIN, JACK ROBB, plant pathologist; b. Omaha, Nov. 21, 1915; s. Carl A. and Elizabeth Josephine (Smith) W.; m. Janet May Melhus, Sept. 25, 1937; children: Jack I.M., Robb M. B.S., Iowa State U., 1939, Ph.D., 1944. Research grad. asst. U. Mo., Columbia, 1939-40, prof. plant pathology, 1975—; research grad. asst. Iowa State., Ames, 1941-44, research asst. prof., 1944-47; research plant pathologist U.S. Dept. Agr., Agrl. Research Service Iowa State U., Ames, 1947—, U.S. Dept. Agr. Research Service U. Mo., Columbia, 1947—. Recipient William F. Petersen award Internat. Soc. Biometeorology, 1966; Iowa State U. Pres. grantee, 1966-67. Mem. Internat. Soc. Plant Pathology, Am. Phytopath. Soc., Mo. Acad. Sci., Sigma Xi, Gamma Sigma Delta. Republican. Presbyterian. Lodge: Rotary. Subspecialties: Plant pathology; Plant genetics. Current work: Host plant resistance to plant diseases in corn search for genetic control in corn to aspergillus flavus, corn viruses, low level of aflatoxin in corn, stewart's wilt, other pathogens. Patentee in field.

WALLIN, JOHN DAVID, physician, researcher; b. Pasadena, Calif., June 30, 1937; s. Nathaniel Charles and Florence (Wade) W.; m. Karen Elder, June 20, 1959; 1 son, John. B.S., Stanford U., 1958; M.D., Yale U., 1962. Rotating intern Naval Hosp., San Diego, 1962-63, resident in internal medicine, Oakland, Calif., 1963-66; clin. instr. medicine U. P.R. Sch. Medicine, 1967-70; research assoc. internal medicine U. Tex. Southwestern Med. Sch., Dallas, 1970-72; head nephrology unit Naval Hosp., Oakland, Calif., 1972-78, dir. clin. investigation ctr., 1973-78; chief sect. nephrology VA Hosp., New Orleans, 1978-81; prof. medicine, adj. prof. physiology, chief sect. nephrology Tulane Med. Sch., New Orleans, 1978—. Served to capt. USN, 1962-78. Recipient Merit Rev. award VA, 1978—. Fellow ACP; mem. Am. Soc. Nephrology, Am. Fedn. Clin. Research, AMA, Internat. Soc. Nephrology. Subspecialties: Internal medicine; Nephrology. Current work: Examination of mechanisms of vasopressin release from neurohypophysis and effects of heavy metals. Office: Tulane U Med Sch 1430 Tulane Ave New Orleans LA 70112

WALLNER, RICHARD ALAN, research scientist; b. Bklyn., Apr. 28, 1945; s. Mathias and Margaret Louise (Boss) W.; m. Barbara Ann Zurawel, July 27, 1969; children: Richard N., Mary Catherine T. B.A., NYU, 1966, M.S., 1968; postgrad., U. N.Mex., 1974-76. Weapons dir. U.S. Air Force, Syracuse, N.Y., 1970-71, Hofn, Iceland, 1971-72, Petersburg, Va., 1972-73; chief Wavefront Analysis Lab. Air Force Weapons Lab., Albuquerque, 1973-77; asst. prof. physics U.S. Air Force Acad., Colorado Springs, Colo., 1977-81; research scientist Kaman Scis. Corp., Colorado Springs, 1981—. Contbr. articles to profl. jours. Served to capt. USAF. Mem. Am. Phys. Soc., Optical Soc. Am. Republican. Roman Catholic. Subspecialty: Laser diagnostics. Current work: Laser system threat analysis; laser optics, laser pointing and tracking, laser power generation, infrared sensors. Office: PO Box 7463 Colorado Springs CO 80933

WALLS, BETTY L., psychologist, consultant; b. Kansas City, Mo., Oct. 26, 1932; d. Donald (stepfather) and Gladys O. (Gillespie) Webb Morrison; m. William C. Walls, Apr. 6, 1954 (div. Jan. 1961); 1 son, Paul Kevin. R.N. diploma, Kansas City (Mo.) Gen. Hosp., 1957; B.A., U. Mo.-Kansas City, 1967, M.A., 1971, Ph.D. with distinction, 1974. Lic. psychologist, Mo.; cert. psychologist, Kans. Lectr. U. Mo.-Kansas City, 1973-80; chmn. dept. psychology Park Coll., 1974-80; contract psychologist Catholic Charities, Kansas City, 1980—; vis. prof. psychology U. Mo.-Kansas City, 1980-83; lectr. Rockhurst Coll., 1980, Benedictine Coll., 1980. Author: test bank for Byrne Introduction to Psychology, 1975. Bd. dirs. Alcoholism Recovery, Kansas City, 1974—; v.p. bd. dirs. Operation Discovery, Kansas City, 1975-82; v.p. Sherwood Ctr., Kansas City, 1978-83; cons. Kansas City Regional Assn. Mental Retardation, 1978-82. Named Outstanding Tchr. U. Mo.-Kansas City, 1973, Park Coll., 1980. Mem. Am. Psychol. Assn., N.Y. Acad. Sci., Assn. Behavior Analysts, AAAS, Am. Assn. Tension Control, Mo. Psychol. Assn., Greater Kansas City Psychol. Assn., Psi Chi. Democrat. Methodist. Clubs: Cotton Eyed Joe, Sugar Shack (Kansas City). Subspecialties: Behavioral psychology; Learning. Current work: Child development, fetal development, acquisition of various curricular material, reinforcement theory. Home: 8019 Kenwood Kansas City MO 64131 Office: U Mo 51st and Rockhill Kansas City MO 64110

WALSER, MACKENZIE, physician, educator; b. N.Y.C., Sept. 19, 1924; s. Kenneth Eastwood and Jean (Mackenzie) W.; m. Lynne Margaret White, Aug. 8, 1965; children—Karen D., Jennifer McK., Cameron M., Eric H. Grad., Phillips Exeter Acad., 1941; A.B., Yale, 1944; M.D., Columbia, 1948. Diplomate: Am. Bd. Internal Medicine. Intern Mass. Gen. Hosp., Boston, 1948-49, asst. resident in medicine, 1949-50; resident Parkland Hosp., Dallas, 1950-52; staff mem. Johns Hopkins Hosp., Balt., 1957—; instr. U. Tex. at Dallas, 1950-51, asst. prof., 1951-52; investigator Nat. Heart Inst., Bethesda, Md., 1954-57; asst. prof. pharmacology Johns Hopkins Med. Sch., 1957-61, asso. prof., 1961-70, prof., 1970—, asst. prof. medicine, 1957-64, asso. prof., 1964-74, prof., 1974—; Med. dir. USPHS, 1970—, pharmacology study sect., 1968-72. Co-author: Mineral Metabolism, 2d edit, 1969, Handbook of Physiology, 1973, The Kidney, 1976, 2d edit., 1981, also articles; co-editor: Branched-Chain Amino and Ketoacids, 1981. Served with USNR, 1942-45; to lt. M.C. USNR, 1952-54. Recipient Research Career Devel. award USPHS, 1959-69. Mem. Am. Soc. Clin. Investigation, Assn. Am. Physicians, Am. Fedn. Clin. Research, AAAS, Am. Physiol. Soc., AAUP (pres. Johns Hopkins 1970), Biophys. Soc., Am. Soc. Pharmacology and Exptl. Therapeutics (Exptl. Therapeutics award 1975), Am. Soc. Nephrology, Council Renal Disease, Am. Heart Assn., Sigma Xi, Zeta Psi, Nu Sigma Nu. Club: Century Assn. Subspecialties: Nephrology; Nutrition (medicine). Current work: Amino acid and protein metabolism and disorders thereof. Home: 7513 Club Rd Ruxton MD 21204 Office: Johns Hopkins U Sch Medicine Baltimore MD 21205

WALSH, GERALD MICHAEL, pharmacologist; b. Portland, Oreg., Sept. 1, 1944; s. Alfred Charles and Myrtle Mary (Derbes) W.; m. Ellen Reiko Nakada, June 20, 1970; children: Joseph, Patrick, Shannon, Colleen. B.S., U. Santa Clara, 1966; Ph.D., Oreg. State U., 1971. Asst. prof. U. Ga., 1970-74, U. Okla., 1974-76; staff researcher Alton Ochsner Med. Found., 1976-78; mgr. Travenol Labs., Morton Grove, Ill., 1978-81; group leader antihypertension G.D. Searle & Co., Skokie, Ill., 1981—. Contbr. numerous articles to med. jours. NIH grantee, 1976-79. Mem. Am. Heart Assn., Soc. Pharmacology and Exptl. Therapeutics, Soc. Exptl. Biology and Medicine. Subspecialty: Pharmacology. Current work: Development of cardiovascular drugs. Home: 529 Northgate St Lindenhurst IL 60046 Office: G D Searle Box 5110 Chicago IL 60680

WALSH, JACQUELINE ANN, systems engineer, consultant; b. Denver, Sept. 9, 1951; d. John James and Shyla Darlene (Burke) W. Student, S.D. State U., 1969-70; B.S. in Computer Sci, Iowa State U., 1973; postgrad., U. Minn., 1973-74, U. Houston, 1976-77. Computer programmer Sperry Univac Def. Systems, Eagan, Minn., 1973-75, Lockheed Electronics, Houston, 1975-77; sr. systems analyst Sperry Univac, Blue Bell, Pa., 1977-80; sr. systems engr. Space Systems div. Gen. Electric Co., Valley Forge, Pa., 1980-83; data processing consultant Sperry Computer Systems, Blue Bell, Pa., adr3—. Recipient commendation Lockheed Electronics, 1976, NASA, 1976. Mem. Assn. for Computing Machinery. Subspecialties: Operating systems; Computer architecture. Current work: Distributed systems and network communications. Primary interest in operating systems internals and in inter computer communications. Interest in new micro processor technologies. Home: PO Box 306 Warrington PA 18976 Office: Sperry Computer Systems PO Box 500 Blue Bell PA 19424

WALSH, JOHN HARLEY, internist, educator; b. Jackson, Miss., Aug. 22, 1938; s. John Howard and Aimee (Shands) H.; m. Courtney K. McFadden, June 12, 1963; children: Courtney, John Harley. B.A., Vanderbilt U., 1959, M.D., 1963. Diplomate: Am. Bd. Internal Medicine, Am. Bd. Gastroenterology. Intern and resident N.Y. Hosp.-Cornell Med. Ctr., N.Y.C., 1963-67, NIH, 1967-69, Bronx VA Hosp., 1969-70, Wadsworth VA Hosp., 1970-71; asst. prof. medicine UCLA, 1970-74, assoc. prof., 1974-78, prof., 1978—, assoc. dir., 1977—; adv. council NIH, Nat. Inst. Arthritis, Diabetes, Digestive and Kidney Diseases, Bethesda, Md., 1982—. Assoc. editor: Gastroenterology, 1976—; contbr. articles to profl. jours. Served to lt. comdr. USPHS, 1967-69. Mem. Am. Soc. Clin. Investigation, Gastroenterology Research Group, Am. Gastroent. Assn., Endocrine Soc. Episcopalian. Subspecialties: Gastroenterology; Neuroendocrinology. Current work: Defining chemical structure and physiological roles of gut-brain peptides and gastrointestinal hormones. Home: 247 S Carmelina Ave Los Angeles CA 90049

WALSH, KENNETH ANDREW, biochemistry educator; b. Sherbrooke, Que., Can., Aug. 7, 1931; s. G. Stanley and Dorothy (Sangster) W.; m. Deirdre A. Clarke, Aug. 22, 1953; children: Andrew, Michael, Erin. B.Sc. in Agr, McGill U., 1951; M.S., Purdue U., 1953; Ph.D., U. Toronto, 1959. Faculty mem. U. Wash., Seattle, 1962-65, assoc. prof., 1965-69, prof. biochemistry, 1969—. Mem. Am. Soc. Biol. Chemistry. Subspecialties: Biochemistry (biology); Biochemistry (medicine). Current work: Structure of proteins and relation to function and regulation. Office: Dept Biochemistry U Wash Seattle WA 98195

WALSH, SCOTT WESLEY, endocrinologist, reproductive physiologist; b. Wauwatosa, Wis., July 23, 1947; s. Virgil C. and Harriet E. (Jacobson) W.; m. Cynthia Lea Sorenson, Oct. 10, 1981. B.S., U. Wis.-Milw., 1970; M.S., U. Wis. Madison, 1972, Ph.D., 1975. Asst. prof. U. N.D., Grand Forks, 1975-76; asst. scientist Oreg. Primate Ctr., Beaverton, 1976-79, assoc. scientist, 1980; asst. prof. Oreg. Health Scis. U., Portland, 1978-80; asst. prof. dept. physiology Mich. State U., East Lansing, 1980—. NSF grantee, 1970; USPHS grantee, 1970-75, 83—. Mem. Endocrine Soc., Soc. Gynecologic Investigation, Am. Physiol. Soc., Soc. Study of Reprodn., Sigma Xi, Phi Kappa Phi. Subspecialties: Maternal and fetal medicine; Endocrinology. Current work: Endocrine control and placental modulation of parturition and fetal growth; experiment model-chronically catheterized rhesus monkey fetus; intrauterine surgery; placental prostacyclin and thromboxane in toxemia of human pregnancy.

WALSH, WILLIAM MICHAEL, educator, psychologist; b. Chgo., Mar. 15, 1943; s. Patrick J. and Elizabeth (Hargraves) W.; m. Kathleen E. Moran, July 28, 1962; children: Karen E., Kristen M. B.S., Loyola U., Chgo., 1966, M.Ed., 1968; Ph.D., U. Wyo., 1971. Registered profl. psychologist, Ill. Faculty Northeastern Ill. U., Chgo., 1971—, prof. psychology, 1983—; cons. Northbrook Schs., 1971-79. Author: Counseling Children, 1976, Primer Family Therapy, 1980; Contbr. articles to profl. jours. Mem. Glenbrook Bd. Edn., Northbrook/Glenview, 1980-83. Recipient Presdl. merit award Northeastern Ill, U., 1982; Northeastern Ill. U. research grantee, 1975, 82. Mem. Am. Psychol. Assn., Am. Personnel and Guidance Assn., Ill. Guidance Assn., Phi Delta Kappa. Roman Catholic. Subspecialties: Family therapy; Developmental psychology. Current work: Validation of family measurement instrument: dynamics of family life. Home: 1301 Brookside Ln Northbrook IL 60062 Office: 5500 N St Louis Ave Chicago IL 60625

WALSH-REITZ, MARGARET MARY, clin. researcher; b. Chgo., Mar. 7, 1943; d. William Henry and Mary Josephine (Bauer) Walsh; m. David J. Platt, Aug. 17, 1974; m. Roger L. Reitz, Sept. 5, 1980. B.S.Ed., No. Ill. U., 1965, M.S., 1967; Ph.D., Ind. U., 1975. Cert. high sch. tchr., Ill. Biology tchr. Sullivan High Sch., Chgo., 1976-78; genetic counseling, 1977-80, coordinator biochem. genetics, 1978-80; research assoc. U. Chgo., 1976-80; asst. prof. dept. medicine Billings Hosp., U. Chgo., 1981—. Contbr. articles to profl. jours. Mem. sci. adv. com. Mus. Sci. and Industry, Chgo.; vol. Anti-Cruelty Soc., Chgo. AEC fellow, 1964-65; NIH fellow, 1965-67; USPHS trainee, 1968-74. Mem. AAAS, Genetics Soc. Am., N.Y. Acad. Sci., Am. Soc. Human Genetics, Am. Genetic Assn., Am. Inst. Biol.Scis., Sigma Xi. Roman Catholic. Subspecialties: Cell and tissue culture; Nephrology. Current work: Cellular growth control mechanisms; tissue culture; physiologcial mitogens; growth control; ion modulation of cellular growth response. Office: 950 E 59th St Box 453 Chicago IL 60637

WALSTAD, ALLAN M(ARTIN), physics educator; b. Trenton, Jan. 30, 1947; divorced. B.S. in Physics, Ursinus Coll., 1969; M.S., U. Mass., 1971, Ph.D. in Astronomy, 1975. Asst. prof. Broome Community Coll., Binghamton, N.Y., 1975-77; temporary teaching positions Pa. State U.-Scranton, 1977, Bates Coll., 1978; asst. prof. natural scis. and physics U. Pitts.-Johnstown, 1978—. Contbr. articles to profl. jours. Mem. Am. Astron. Soc., Am. Phys. Soc., Am. Assn. Physics Tchrs. Subspecialties: Cosmology; Theoretical astrophysics. Current work: QSO surveys and cosmology; nature of time. Office: Dept Physics U Pitts Johnstown PA 15904

WALT, MARTIN, physicist, aircraft company executive; b. West Plains, Mo., June 1, 1926; s. Martin and Dorothy (Mantz) W.; m. Mary Estelle Thompson, Aug. 16, 1950; children: Susan Mary, Stephen Martin, Anne Elizabeth, Patricia Ruth. B.S., Calif. Inst. Tech., 1950; M.S., U. Wis., 1951, Ph.D., 1953. Staff mem. Los Alamos Sci. Lab., 1953-56; research scientist, mgr. physics Lockheed Aircraft Corp., Palo Alto (Calif.) Research Lab., 1956-71, dir. phys. scis., 1971—. Contbr. articles in field to sci. jours. Served with USNR, 1944-46. Wis. Research Found. fellow, 1950-51; AEC fellow, 1951-53. Fellow Am. Geophys. Union (officer, com. mem.), Am. Phys. Soc.; mem. AIAA. Club: Golden Gate Yacht. Subspecialties: Magnetospheric Physics; Plasma physics. Current work: Theory of aurora and geomagnetically trapped radiation. Home: 12650 Viscaino Ct Los Altos Hills CA 94022 Office: 3251 Hanover St Palo Alto CA 94304

WALT, MARTIN, IV, aerospace executive; b. West Plains, Mo., June 1, 1926; s. Martin and Dorothy (Mantz) W.; m. Mary Thompson, Aug. 15, 1950; children: Susan, Stephen, Anne, Patricia. B.S., Calif. Inst. Tech., 1950; M.S., U. Wis., 1951, Ph.D., 1953. Staff Los Alamos (N.Mex.) Sci. Lab., 1953-56; scientist, mgr., dir. phys. sci. Lockheed Missiles and Space Co., Palo Alto, Calif., 1956—. Contbr. articles to sci. jours. Served with USNR, 1944-46. AEC predoctoral fellow, 1951-53. Fellow Am. Phys. Soc., Am. Geophys. Union. Club: Golden Gate Yacht. Subspecialty: Space Plasma physics. Current work: Particles and fields in planetary magnetospheres; aurora, upper atmospheric physics. Home: 12650 Viscaino Ct Los Altos Hills CA 94022 Office: 3251 Hanover St Palo Alto CA 94304

WALTENBAUGH, CARL RALPH, immunologist, educator; b. Canton, Ohio, July 17, 1948; s. Carl Ralph and Vivian Olga (Rankin) W.; m. Karen Sue Jenkins, Jan. 20, 1950; children: Gretchen Lisa, Heidi Kirsten, Karl Dietrich. Student, Mt. Union Coll., Alliance, Ohio, 1966-68; B.S. in Zoology, Baldwin-Wallace Coll., 1970; M.S. in Anatomy, U. Ill., 1973; Ph.D. in Anatomy-Immunology, U. Ill., 1975. Research fellow dept. pathology Harvard Med. Sch., Boston, 1975-77, instr. pathology, 1977-79; asst. prof. dept. microbiology-immunology Northwestern U. Med. Sch., Chgo., 1979—. Contbr. articles to profl. jours. Recipient Career Devel. award Schweppe Found., 1981; Nat. Cancer Inst. grantee, 1978-79; 80—; Am. Cancer Soc. grantee, 1979-80; NIH grantee, 1981—. Mem. Am. Assn. Immunologists, Soc. for Devel. Biology, Reticuloendothelial Soc., Sigma Xi. Subspecialties: Immunogenetics; Immunobiology and immunology. Current work: Genetic regulation of immune response by suppressor T cells/factors; immunology, cellular immunology, immunogenetics, T cell subsets, immunoregulatory factors, major histocompatibility complex, I-J subregion. Office: Dept Microbiology-Immunology Northwestern U Med Sch 303 E Chicago Chicago IL 60611

WALTER, THOMAS HARRY, podiatric surgeon, educator; b. Allentown, Pa., Feb. 16, 1950; s. John Harold and Mae (Grammes) W. B.A., Pa. State U., 1971; D.P.M., Pa. Coll. Podiatric Medicine, 1978. Intern Parkview Hosp., Phila., 1978-79, resident, 1979-80; research fellow, instr. surgery Pa. Coll. Podiatric Medicine, Phila., 1980-81; fellow U. Md. Hosp. and Trauma Ctr., 1981-82; research cons. Sutter Biomech. Corp., San Diego, 1981—; sole practice, Phila. and New London, Conn., 1982—; dir. Broad St. Hosp. Pediatric Edn., Phila., 1983—, asst. dir. research tng. program, 1983—. Fellow Am. Podiatry Assn.; mem. Conn. podiatry Assn., Hartford County Podiatry Assn., Phila. Bone Club, Bioelec. Repair and Growth Stimulation Soc. Republican. Subspecialties: Orthopedics; Bioelectricity. Current work: Advances in podiatric surgery; research in electrical bone stimulation. Home: 12 Water St Apt B-4 Mystic CT 06355 Office: Dept Podiatric Surgery Pa Coll Podiatric Medicine 8th & Race Sts Philadelphia PA 19107

WALTER, WILLIAM TRUMP, physicist, educator; b. Jamaica, N.Y., Dec. 28, 1931; s. William Olcott and Elizabeth (Trump) W.; m. Susan Gail Tallman, Sept. 24, 1960; children: William Russell, Todd Frederick, Bruce Jonathan, Elizabeth Susan. A.B., Middlebury Coll., 1953; Ph.D., M.I.T., 1962. Mem. staff Sandia Corp., Albuquerque, 1956; research asst. Los Alamos Sci. Lab., 1958, Research Lab. of Electronics, M.I.T., Cambridge, 1959-62; sr. scientist TRG div. Control Data, Melville, N.Y., 1962-67; research scientist Poly. Inst. N.Y., Farmingdale, 1967-79; research assoc. prof. electrophysics, 1979—; pres. Laser Cons., Inc., 1968—. Pres. Walt Whitman Birthplace Assn., 1980-83; chmn. Huntington (N.Y.) Beautification Council, 1972—; trustee South Huntington Pub. Library, 1965-66; vice-chmn. Fedn. West Huntington Civic Assns., 1970—. Served with Signal Corps U.S. Army, 1953-55. Mem. Am. Phys. Soc., Optical Soc. Am., N.Y. Acad. Scis., Soc. Photo-Optical Instrumentation Engrs., Metall. Soc. of AIME, Am. Assn. Physics Tchrs., AAAS, Sigma Xi. Subspecialties: Optical engineering; Spectroscopy. Current work: Quantum electronics, lasers, development and applications, wave-matter interactions, microwaves, high power sources and propagation, atomic and molecular physics. Office: Poly Inst Route 110 Farmingdale NY 11735

WALTERS, JUDITH RICHMOND, neuropharmacology researcher; b. Concord, N.H., June 20, 1944; d. Samuel Smith and Hazel A. (Stewart) Richmond; m. James Wilson Walters, II, Aug. 21, 1969; children: James Richmond, Gregory Stewart. B.A., Mt. Holyoke Coll., 1966; Ph.D., Yale U. Med. Sch., 1972. Postdoctoral dept. psychiatry Yale U., New Haven, 1972-74, asst. prof., 1974-75; staff fellow Nat. Inst. Neurol. and Communicative Disorders and Stroke, NIH, Bethesa, Md., 1975-78, unit head exptl. therapeutics br., 1978-80, sect. chief, 1980—. Subspecialties: Neuropharmacology; Neurophysiology. Current work: Neuropharmacology and basic function of the basal ganglia and substantia nigra. Office: Exptl Therapeutics Br NIH-NINCDS Bldg 10 Rm 5C106 Bethesda MD 20205

WALTERS, THOMAS RICHARD, medical educator; b. Milw., May 9, 1929; s. Edwin C. and Eleanore R. (Koeppen) W.; m. Mary Anne Pearson, July 27, 1957 (div. May 1975); children: Benjamin, David, Sarah Anne, Tobin. Student, Milw. State Coll., 1947-48; M.D., Marquette U., 1954. Diplomate: Am. Bd. Pediatrics, Am. Bd. Pediatric Hematology-Oncology. Intern San Francisco Gen. Hosp., 1954-55; resident in pediatrics Stanford U., 1957-59, fellow in hematology, 1959-62; asst. prof. pediatrics U. Kans. Med. Sch., Kansas City, 1962-67; assoc. staff in hematology St. Jude Children's Research Hosp., Memphis, 1967-71; assoc. prof. pediatrics U. Medicine and Dentistry N.J., Newark, 1971-76, prof., 1976—, dir. div., 1976—. Author: Pediatric Hematology, rev. edit, 1975, Pediatric Hematology and Oncology, 1975. Served to capt. U.S. Army, 1955-57; Korea. Fellow Am. Acad. Pediatrics; mem. Am. Soc. Hematology, Am. Soc. Pediatric Hematology and Oncology, Am. Assn. Cancer Research, Am. Soc. Clin. Oncology., Am. Soc. Cancer Educators. Subspecialties: Pediatrics; Oncology. Current work: Immunologic function, hematologic-oncologic disorders. Office: New Jersey Medical School 100 Bergen St Newark NJ 07103

WALTMAN, PAUL ELVIS, mathematics educator; b. St. Louis, Oct. 17, 1931; s. Elvis P. and Imogene (Pate) W.; m. Ruth Major, Dec. 29, 1953; children: Fred, Dennis, Robert. B.A., St. Louis U., 1952; M.A., Baylor U., 1954, U. Mo.-Columbia, 1960, Ph.D., 1962. Mem. staff Mitre Corp., Bedford, Mass., 1962, Sandia Corp., Albuquerque, 1963-65; asst. prof. math. U. Iowa, 1965-66, assoc. prof., 1966-68, prof., 1968-83; prof., chmn. dept. math. and computer sci. Emory U., 1983—. Author: (with Bailey and Shampine) Two Point Nonlinear Boundary Value Problems, 1968, Deterministic Threshold Models in Theory of Epidemics, 1974. Mem. Soc. Indsl. and Applied Math., Am. Math. Soc. Subspecialties: Applied mathematics; Theoretical ecology. Current work: Mathematical modeling of ecological problems. Home: 2356 Sherbrooke Dr NE Altanta GA 30345 Office: U Iowa Dept Math Iowa City IA 52240

WALTMANN, WILLIAM LEE, mathematics educator; b. Cedar Falls, Iowa, July 5, 1934; s. Leo Herman and Alma Louise (Engel) W.; m. Carol Ann Johnson, July 12, 1958; children: Karen Jean, Ronald Dale, Diane Lynn. B.A., Warburg Coll., 1956; M.S., Iowa State U., 1958, Ph.D., 1964. Instr. math. Wartburg Coll., Waverly, Iowa, 1958-61, asst. prof., 1964-67, assoc. prof., 1967-72, prof., 1972—, chmn. dept. math. and computer sci., 1971—; instr. math. Iowa State U., Ames, 1963-64; participant various NSF-sponsored summer programs. Contbr. articles to profl. jours.; test item writer preliminary actuarial exam, 1978—. Pres. Waverly-Shell Rock PTA, 1969-70; bd. dirs. Waverly Day Care Ctr., 1973-77; trustee, sec. Waverly Mcpl. Hosp., 1976—. Hill Found. fellow in computer sci., 1975. Mem. Soc. Indsl. and Applied Math., Am. Math. Soc., Math. Assn. Am. (chmn. Iowa sect. 1966-67, bd. govs. 1980-83), Iowa Acad. Sci. (chmn. math. sect. 1966-67). Republican. Lutheran. Lodge: Kiwanis. Subspecialties: Numerical analysis; Applied mathematics. Current work: Computer use in linear algebra, microcomputer use in classroom teaching. Office: Dept Math Wartburg Coll Waverly IA 50677 Home: 1904 3d Ave NW Waverly IA 50677

WALTON, C. MICHAEL, civil engineering educator, transportation consultant; b. Hickory, N.C., July 28, 1941; s. Charles O. and Virginia R. (Hart) W.; m. Betty Grey Hughes; children: Susan, Camila, Michael, Gantt. B.S., Va. Mil. Inst., 1963; M.C.E., N.C. State U., 1969, Ph.D., 1971. Registered profl. engr., Tex. Transp. economist Office of Sec., Dept. Transp., Washington, 1969; transp. planning engr. N.C. Hwy. Comm., Raleigh, 1970; asst. prof. U. Tex., Austin, 1971-76, assoc. prof., 1976-83, prof. civil engring., 1983—, assoc. dir., 1980—; cons. Assn. Am. R.R.s, 1980—, also to bus. Contbr. numerous articles to profl. jours. Mem. and chair Urban Transp. Commn., Autin, 1977—; mem. Tex. Energy Efficiency Comm., 1981-82, Gov.'s Interagy. Transp. Council, 1973-79. Served to capt. U.S. Army, 1963-67; Ger. Recipient cert. of appreciation Fed. Hwy. Adminstrn., 1982. Mem. ASCE (chair urban transp. div. 1983—), Soc. Am. Mil. Engrs., Inst. Transp. Engrs., Transp. Research Bd. (chair tech. com. 1978—), Ops. Research Soc. Am. Democrat. Methodist. Clubs: Tarrytown Boat (sec.-tres. 1976—), West Austin Neighborhood Group (Austin) (sec.-treas. 1976—). Subspecialties: Civil engineering; Operations research (engineering). Current work: Transportation programming and policy-making, transportation management, transportation planning, cost effectiveness of transportation systems, and transportation energy planning and analysis. Home: 3404 River Rd Austin TX 78703 Office: U Tex Austin TX 78712

WALTON, HAROLD VINCENT, agricultural engineering educator; b. Christiana, Pa., June 17, 1921; s. Howard King and Alice Loretta (Kirk) W.; m. Velma Purvis Braun, June 24, 1946; children: H. Richard, Marilyn J. Walton Friedersdorf, Carol A. Walton Smith. B.S. in Agrl. Engring, Pa. State U., 1942, M.S., 1950, Ph.D., Purdue U., 1961. Registered profl. engr. Test engr. Gen. Electric Co., Schenectady, N.Y., 1943-45; instr. Pa. State U., University Park, from 1947, prof. agrl. engring., 1961—, head. agrl. engring. dept., 1976—; profl. agrl. engring. U. Mo., Columbia, 1962-69, 71-76, chmn. agrl. engring. dept., 1962-69, chief of party, Bhubaneswar, India, 1969-71. Fellow Am. Soc. Agrl. Engrs.; mem. Am. Soc. Engring. Edn. Republican. Subspecialty: Agricultural engineering. Current work: Food engineering. Home: 291 E McCormick Ave State College PA 16801 Office: Pa State U 250 Agrl Engring Bldg University Park PA 16802

WALTON, LEWIS ANTHONY, nuclear engineer; b. Wilmington, Del., Nov. 22, 1945; s. Warren and Helen (Harvilchuck) W.; m. Barbara Rae Brown, Oct. 22, 1969; children: Thomas Anthony, David Richard. B.S., Rensselaer Poly. Inst., 1967. Registered profl. engr., Va. Jr. engr. Boeing Aircraft, Seattle, 1967-71; engr. Getty Oil Co., Rockville, Md., 1971-73, Babcock & Wilcox, Lynchburg, Va., 1973-74, sr. engr., 1974-79, supervisory engr., 1979—; cons. engr., Forest, Va., 1978—. Contbr. articles to tech. jours. Treas. Forest PTA, 1982, pres., 1983. N.Y. State Regents scholar, 1963. Mem. Am. Nuclear Soc. (newsletter com. 1981—, program com. 1982—, pub. info. com. Va. sect. 1981), AIAA (sect. vice-chmn. 1964-66, chmn. 1966-67). Subspecialties: Mechanical engineering; Materials (engineering). Current work: Performance of materials and design in light water reactor cores (fuel elements) - advanced designs and performance models; high level waste disposal studies. Patentee in field. Office: PO Box 1260 Lynchburg VA 24505 Home: 102 Lake Ridge Dr Forest VA 24551

WALTON, PAUL TALMAGE, geologist; b. Salt Lake City, Feb. 4, 1914; s. Paul and Margaret (Watts) W.; m. Helen E. Baer, July 3, 1944; children: Holly, Paul, Ann. B.S. in Geol. Enging, U. Utah., 1935, M.S. in Geology, 1940; Ph.D., M.I.T., 1942. Cert. Am. Inst. Profl. Geologists. With U.S. Dept. Agr., Utah, 1935-38; geologist and geophysicist Standard Oil Calif., Saudi Arabia, 1938-39; exploration geologist Rocky Mountain area Tex. Co., 1942-44; div. geologist Pacific Western Oil Corp. div. Getty Oil Co., Rocky Mountains and Saudi Arabia, 1944-49; geologist, partner Morgan-Walton Oils, Rocky Mountains, 1949-55, Walton-Kearns, 1955-57; partner Paul T. Walton & Assocs., Salt Lake City, 1957—; cons. bd. dirs. Jackson Hole Land Trust. Contbr. to profl. bulls. and guidebooks. Mem. Am. Assn. Petroleum Geologists, Am. Inst. Profl. Geologists, Geol. Soc. Am., Am. Geol. Inst., Ind. Petroleum Assn. Am., others. Subspecialty: Geology. Current work: Exploration for oil and gas. Home: 2591 Brentwood Dr Salt Lake City UT 84121 Office: 1102 Walker Bldg Salt Lake City UT 84111

WALTON, THOMAS EDWARD, JR., veterinarian, researcher; b. McKeesport, Pa., Dec. 2, 1940; s. Thomas Edward and Matilda Lucy W.; m.; children: Anne Louise, Leigh Ellen. D.V.M., Purdue U., 1964; Ph.D., Cornell U, 1968. NIH trainee in microbiology Cornell U., 1964-68; research microbiologist Middle Am. research unit HEW-USPHS-NIH-Nat. Inst. Arthritis and Infectious Diseases, Ancon, C.Z., 1968-69, research vet. med. officer, 1969-72; vet. med. officer western region Agrl. Research Service, U.S. Dept. Agr., Arthropod-borne Animal Diseases Research Lab., Denver, 1972-74, research leader, 1974—. Contbr. articles to profl. jours. Recipient cert. merit Dept. Agr., 1974, 77, 82, Sci. and Edn. Adminstr. Dir.'s award, 1981. Mem. Am. Com. Arthropod-borne Viruses, Am. Soc. Tropical Medicine and Hygiene, AVMA, Am. Soc. Virology, Am. Soc. Microbiology, Nat. Assn. Fed. Veterinarians, U.S. Animal Health Assn., AAAS, N.Y. Acad. Scis. Club: Optimists (Lakewood, Colo.). Subspecialties: Microbiology (veterinary medicine); Virology (veterinary medicine). Current work: Vet. medicine, arbovirology, research mgmt. Office: PO Box 25327 Denver CO 80225

WALTZ, DAVID LEIGH, electrical engineering educator; b. Boston, May 28, 1943; s. Maynard Carlton and Lubov Clara (Leonovich) W.; m. Bonnie Elaine Freedson, Feb. 21, 1970; children: Vanessa Leigh, Jeremy Benjamin. S.B., MIT, 1965, S.M., 1968, Ph.D., 1972. Teaching asst. elec. engring. dept. MIT, Cambridge, Mass., 1966-67; research asst. Artificial Intelligence Lab., 1967-72, postdoctoral researcher, 1972-73; asst. prof. elec. engring. dept. U. Ill.-Urbana, 1973-79, assoc. prof., 1979-82, prof., 1982—; cons. Bolt Beranek & Newman, Inc., Machine Intelligence Corp., Symantec, Hughes Aircraft Co., Comtex Sci. Corp., NSF. Contbr. articles in field to profl. jours. Mem. Assn. Computing Machinery, Assn. Computational Linguistics, Am. Assn. Artificial Intelligence, Cognitive Sci. Soc., SIGART, Sigma Xi, Tau Beta Pi, Eta Kappa Nu. Subspecialties: Artificial intelligence; Automated language processing. Current work: Relationships between language and perception, understanding and generating scene and event descriptions, cognitive universals, and representational primitives for sensory/motor meaning, metaphor comprehension models. Home: 410 Sherwin Dr Urbana IL 61801 Office: Coordinated Sci Lab 1101 W Springfield Ave Urbana IL 61801

WANG, AN-CHUAN, immunogenetics educator; b. Tsing-Tao City, China, Dec. 28, 1936; came to U.S., 1961, naturalized, 1972; s. Chih-Chin and Ching-Chen Wu W.; m. Irene Y. F. Wang, Jan. 12, 1940; children: Jack B., J. Wang, Henry B. H. B.S., Nat. Taiwan U., 1959; Ph.D., U. Tex.-Austin, 1965. Research assoc. U. Tex.-Austin, 1966-67; postdoctoral fellow U. Calif.-San Francisco, 1967-70, asst. prof., 1970-72, assoc. prof., 1972-75; assoc. prof. immunology Med. U. S.C., Charleston, 1975-76, prof., 1976—. Contbr. articles to profl. jours. Recipient NIH Career Devel. award, 1974, Faculty Research award Am. Cancer Soc., 1974-79; NIH research grantee, 1970-78; NSF research grantee, 1970-85. Mem. Am. Assn. Immunologists, Am. Soc. Human Genetics, AAUP, AAAS, Genetics Soc. Am., Sigma Xi. Subspecialties: Immunology (agriculture); Animal genetics. Current work: Structure, genetics and evolution of immunoglobulins and several other serum proteins. Home: 855 Robert E Lee Blvd Charleston SC 29412 Office: Dept Basic and Clin Immunology and Microbiolog Med U SC 171 Ashley Av Charleston SC 29425

WANG, BOSCO SHANG, immunologist; b. Shanghai, China, Aug. 30, 1947; s. Tai-Yung and Sun-Sen Ho W.; m. Helen Ku; children: Burkon, Tammy. M.S., Mich. State U., 1973; Ph.D., Boston U., 1976. Asst. prof. Harvard Med. Sch., Boston, 1976-81; sr. scientist Lederle Labs., Am. Cyanamid Co., Pearl River, N.Y., 1981—. Contbr. articles to profl. jours. Recipient Wilson S. Stone Meml. award Tumor Inst. of Tex. System Cancer Ctr., 1977. Mem. Am. Assn. Cancer Research, Am. Assn. Immunologists, Am. Soc. Microbiology, N.Y. Acad. Scis., Internat. Soc. Immunopharmacology. Subspecialties: Immunopharmacology; Cancer research (medicine). Current work: Immunopharmacology, cancer research, cellular immunology. Office: Lederle Labs Bldg 60B Rm 209 Pearl River NY 10965

WANG, CHARLES P., aerospace scientist, educator; b. Shanghai, China, Apr. 25, 1937; s. Kuan Ying and Pinglu (Ming) W.; m. Lilly L. Lee, June 29, 1963. B.S., Taiwan U., 1959; M.S., Tshinghua U., 1961; Ph.D., Calif. Inst. Tech., 1967. Lectr. Taiwan U., Taipei, 1961-62; Inst. scholar Calif. Inst. Tech., Pasadena, 1963-67; mem. tech. staff Bellcomm, Washington, 1967-69; research engr. U. Calif., San Diego, 1969-74; adj. prof., 1979—; sr. scientist Aerospace Corp., Los Angeles, 1976—. Assoc. editor AIAA Jour., 1981—; Contbr. articles to profl. jours. Mem. AIAA (assoc. fellow), Am. Phys. Soc., Am. Optical Soc., IEEE, Chinese Engrs. and Scientist Assn. So. Calif. (pres., chmn. bd.), Sigma Psi. Democrat. Club: Manhattan Beach (Calif.) Badminton. Subspecialties: Aeronautical engineering; Laser research. Current work: Laser development and applications, gasdynamics, plasmadynamics, gas lasers, chemical laser, quantum electronics, excimer lasers, laser remote sensing, laser wavefront sensor. Home: 28509 Seamont Dr Palos Verdes CA 90274 Office: 2350 E El Segundo Blvd El Segundo CA 90045

WANG, CHEN-SHOW, physicist; b. Taiwan, Oct. 10, 1936; s. Po-Chen and Ming-Shia (Lin) W.; m. Nimia L. Leong, Aug. 28, 1963; children: James, Robert, Linda. B.S., Nat. Taiwan U., 1959; M.S., U. Iowa, 1964; Ph.D., U. Calif.-San Diego, 1968. Postdoctoral research physicist U. Calif.-San Diego, 1968-69; research fellow Harvard U., 1969-72; asst. prof. physics Bartol Research Found., U. Del., 1972-77, assoc. prof., 1977-79; mgr. research and devel. Gen. Optronics Corp., South Plainfield, N.J., 1979-82, dir., 1982—. Co-author: Nonlinear Optics, 1975; contbr. articles to profl. jours. Mem. Am. Phys. Soc., Sigma Xi. Subspecialties: Fiber optics; Semiconductor lasers. Current work: Fiber optics communication, semiconductor lasers, light emitting diode, detectors, telecommunications components. Home: 58 Mount Horeb Rd Warren NJ 07060 Office: 2 Olsen Ave Edison NJ 08820

WANG, CHIA PING, research physicist, educator; b. Philippines, Sept. 1; came to U.S., 1963, naturalized, 1972; s. Guan Can and Tah (Lin) W. B.Sc., U. London, 1950; M.Sc., U. Malaya, 1951; Ph.D. in Physics, Univs. of Malaya and Cambridge, 1953, D.Sc., U. Singapore, 1972. Asst. lectr. U. Malaya, 1951-53; assoc. prof. Nankai U., Tientsin, 1954-56, prof. physics, 1956-58, head electron physics div., 1955-58, mem. steering com. nuclear physics div., 1958; head electron physics div. Lanchow Atomic Project, 1958; sr. lectr., prof. physics, acting head dept. physics and math. U. Hong Kong and Chinese U. of Hong Kong, 1958-63; research assoc. Lab. Nuclear Studies, Cornell U., Ithaca, N.Y., 1963-64; assoc. prof. space sci. and applied physics Cath. U. Am., 1964-66; assoc. prof. physics Case Inst. Tech. and Case-Western Res. U., Cleve., 1966-70; vis. scientist, vis. prof. U. Cambridge, U. Leuven (Belgium), U.S. Naval Research Lab., U. Md., and M.I.T., 1970-75; research physicist U.S. Army Natick (Mass.) Research and Devel. Labs., 1975—. Contbr. articles to sci. jours., chpts. in books. Recipient Outstanding Performance award Dept. of Army, 1980, Quality Increase award, 1980. Mem. Am. Phys. Soc., Inst. Physics London, AAAS (life), N.Y. Acad. Scis., Sigma Xi. Subspecialties: Particle physics; Thermal physics. Current work: Particle interactions, radiation, heat transfer, laser interferometry, scientific applications of computers; director, planner, conductor research and development. First to convert picosecond time interval into pulse height, 1963; deduced from more than 50 expts. the many-subunit structure of the nucleon and other hadrons in 1968; formulated gen. integral survival fraction of bacteria during heat sterilization, 1978. Home: 28 Hallet Hill Rd Weston MA 02193 Office: US Army Natick Research and Development Laboratories Natick MA 01760

WANG, CHI-RONG, government scientist, aerospace engineer; b. Taipei, Taiwan, Aug. 3, 1940; came to U.S., 1966, naturalized, 1978; s. Kin-Luh and Sheue (Lin) W.; m. Lin-Lan Chin, Nov. 5, 1973; children: Susan C., Judith C. B.S.M.E., Nat. Taiwan U., Taipei, 1964; M.S.M.E., SUNY-Buffalo, 1968; Ph.D., NYU, 1973. Teaching asst. Nat. Taiwan U., 1965-66; research scientist NYU, 1973-79; aerospace engr. NASA Lewis Research Ctr., Cleve., 1979—; cons. engr. Gen. Applied Scis. Labs., Westbury, N.Y., 1978, 79. Contbr. articles to profl. publs. Served to lt. ROTC, Chinese Air Force Flying Sch., 1964-65. Recipient NYU Founder's Meml. award, 1974, NASA Spl. award Lewis Research Ctr., 1981. Mem. AIAA, ASME, North Olmstead League of Edn. for Gifted Children. Subspecialties: Aeronautical engineering; Applied mathematics. Current work: Turbulence transport modelling in surface heat transfer analysis of propulsion systems. Office: NASA Lewis Research Center 21000 Brookpark Rd Cleveland OH 44070

WANG, CHUN-JUAN K., mycology educator; b. Mukden, China, Jan. 10, 1928; d. Si-Ping and C. H. (Kao); m. Stephen S. Wang, July 30, 1955; children: Effie, Vivian, Andrew. B.S., Nat. Taiwan U., 1950; M.S., Vassar Coll., 1952; Ph.D., U. Iowa, 1955. Research assoc. Jewish Hosp. Clin. Lab., Cin., 1955-58; instr. med. mycology Cin. Sch. Medicine, 1956-58; research assoc. SUNY Coll. Environ. Sci. and Forestry, Syracuse, 1959-61, instr., 1961-64, asst. prof., 1964-67, assoc. prof., 1967-72, prof., 1972—. Author: Pulp and Paper Fungi of New York, 1965; contbr. articles to profl. jours. NSF, U.S. Forest Service, Electric Power Research Inst. grantee. Mem. Mycol. Soc. Am. (chmn. ann. lectr. com. 1981-82), Brit. Mycol. Soc., Sigma Xi. Subspecialties: Microbiology; Ecology. Current work: Ultrastructure, ecology and taxonomy of deuteromycetes. Office: SUNY Coll Environ Sci and Forestry Syracuse NY 13210

WANG, DEANE, ecologist; b. N.Y.C., Mar. 8, 1951; s. Derek H.T. and Dee (Kwok) W.; m. Carolynne Rebecca Norman, July 4, 1975; 1 dau., Carolynne. B.A., Harvard U., 1973; M.S., Cornell U., 1977; M.Phil., Yale U., 1979. Ecologist Roy F. Weston Inc., West Chester, Pa., 1975-78; assoc. in research Yale U. Sch. Forestry and Environ. Studies, New Haven, 1982—. Mem. Ecol. Soc. Am., Soil Sci. Soc. Am., Nat. Assn. Environ. Profls., Am. Inst. Biol. Scis. Subspecialties: Ecosystems analysis; Ecology. Current work: Research on the effect of air pollutants on the structure and function of natural ecosystems. Office: Yale U Sch Forestry and Environ Studies 370 Prospect St New Haven CT 06511

WANG, GEORGE SHIN-CHANG, industrial construction corporation executive; b. Anking, China, May 4, 1933; m. Ann Chen, Dec. 21, 1957; children: David, Julia, Gregory, Tracy. B.S., U. Mich., 1958, M.S. in Civil Engring, 1960, 1960; Ph.D., UCLA, 1970. Registered profl. engr., Calif. Various engring. positions Bechtel Power Corp., Los Angeles, 1962-78, mgr. engring., Norwalk, Calif., 1978-80; mgr. coal programs Bechtel Group, Inc., San Francisco, 1980-82, mgr. bus. devel. and comml. programs, 1982, dep. mgr. research and engring., 1982, mgr. research and engring., 1982—; asst. research engr. U. Mich. Engring. Research Inst., Ann Arbor, 1961-62. Contbr. sci. papers to profl. lit. NSF grantee, 1961-62. Mem. Am. Nuclear Soc., ASCE, Chi Epsilon. Club: World Trade (San Francisco). Subspecialties: Civil engineering; Environmental engineering. Current work: Management biotechnology, fission, fusion, metals. Office: Research and Engring Bechtel Group Inc 50 Beale St San Francisco CA 94105

WANG, HAO, philosopher, mathematician; b. Tsinan, Shantung, China, May 20, 1921; s. Chuchen and Tsecheng (Liu) W.; m. Yenking Kan, June 22, 1948 (div. Mar. 1977); children: Sanyu, Yiming, Jane Hsiaoching; m. Yu-Shih Chen. B.Sc., Southwestern Asso. U., China, 1943; M.A., Tsing Hua U., China, 1945, Balliol Coll., Oxford, 1956; Ph.D., Harvard, 1948; postgrad., Zurich, 1950-51. Tchr. math., China, 1943-46; research engr. Burroughs Corp., 1953-54; asst. prof. philosophy Harvard, 1951-56, Gordon McKay prof. math. logic, applied math., 1961-67; John Locke lectr. philosophy U. Oxford, 1954-55, reader philosophy of math., 1956-61; prof. Rockefeller U., 1967—; cons. U. Mich., summers 1956, 57, IBM, summer, 1957, 58, Mass. Inst. Tech., summer 1960, Bell Telephone Labs., 1962-64, mem. tech. staff, 1959-60; vis. prof. Rockefeller U., 1966-67; vis. scientist IBM Research Labs., 1973-74; visitor Inst. for Advanced Study, 1975-76; lectr. on math. logic Academia Sinica, 1979. Author: A Survey of Mathematical Logic, 1962, Reflections on China (in Chinese), 1973, From Mathematics to Philosophy, 1974, Popular Lectures on Mathematical Logic, English and Chinese edits, 1981; Contbr. numerous articles to math. philos. jours. Recipient 1st Mileston prize Joint Internat. Conf. on Artificial Intelligence, 1983; Jr. fellow Soc. of Fellows, Harvard, 1948-51; fellow Rockefeller Found., 1954-55. Mem. Assn. Symbolic Logic, Am. Acad. Arts and Sci., Brit. Acad. (fgn.). Subspecialties: Mathematical logic; Theoretical computer science. Home: 1230 York Ave New York NY 10021

WANG, HSIN-PANG, mechanical engineer; b. Nanking, China, Apr. 11, 1946; came to U.S., 1970, naturalized, 1978; s. I-Shen and Chin-Hsien (Wu) W.; m. Ting-Ting Wang; children: Kevin, Michelle. B.M.E., Nat. Cheng-Kung U., Tainan, Taiwan, 1969; M.M.E., U. Fla., 1972; Ph.D., U. R.I., 1976. Registered profl. engr., N.Y. Mem. research staff corp. research and devel. Gen. Electric Co., Schenectady, 1976. Contbr. articles to profl. jours. Chmn. Schenectady chpt. Friends of Free China, 1981. Mem. ASME, Am. Phys. Soc. Subspecialties: Mechanical engineering; Polymer engineering. Current work: Process modeling; computer aided engineering; finite element; manufacturing processes. Co-patentee measurement of residual stresses. Office: Corp Research and Devel Gen Electric Corp PO Box 43 Schenectady NY 12345

WANG, JON YI, aero. engr., researcher; b. Taiwan, June 22, 1943; came to U.S., 1966, naturalized, 1975; s. Ko-Ming and Ming-Hue (Lai) W.; m. Barbara C. Huang, Aug. 21, 1971; 1 son, Eric. B.S. in Mech. Engring, Nat. Taiwan U., 1965; M.S. in Aeros. and Astronautics, M.I.T., 1968; Ph.D. in Aero., Astronautical and Engring. Scis, Purdue U., 1971. Sr. research engr. Gen. Dynamics/Convair, San Diego, 1971-73, engring. staff scientist, 1975—; scientist Sci. Applications, Inc., La Jolla, Calif., 1973-75. Contbr. articles to profl. jours. Mem. Optical Soc. Am. Democrat. Subspecialties: Optical engineering; Applied optics. Current work: Applications of coherent laser radar; infrared sensor system performance analysis; applications of infrared spectroscopy; remote sensing of atmospheric parameters. Home: 2665 San Clemente Terr San Diego CA 92122 Office: PO Box 85357 MZ 42-6210 San Diego CA 92138

WANG, PAO-KUAN, meteorology educator; b. Tainan, Taiwan, Dec. 1, 1949; came to U.S., 1973; s. Shou and Luan-Chao (Chiu) W.; m. Li-Bi C. Wang, Aug. 28, 1976; children: Lawrence Chang-Yung, Victor Chang-Mien. B.S., Nat. Taiwan U., 1971; M.S., UCLA, 1975, Ph.D., 1978. Research atmospheric physicist UCLA, 1978-80, adj. asst. prof., 1980; asst. prof. meteorology U. Wis., Madison, 1980—. Contbr. articles to profl. jours. EPA grantee, 1982; NSF grantee, 1982. Fellow Royal Meteorol. Soc. (Eng.); mem. Am. Meteorol. Soc., AAAS. Subspecialties: Meteorology; Atmospheric chemistry. Current work: Microphysics of clouds and precipitation, cloud dynamics, aerosol physics, atmospheric chemistry, atmospheric electricity, historical climatology and climatic change. Office: Dept Meteorology U Wis 1225 W Dayton St Madison WI 53706

WANG, PATRICK SHEN-PEI, computer specialist; b. Shanghai, China, Dec. 27, 1946; came to U.S., 1972; s. Yung-hsi and Su-jen (Lo) W.; m. Yuan-fang Tung, July 30, 1972; children: Da-yuan, Da-wen. B.S.E.E., Nat. Chiao Tung U., 1968; M.S.E.E., Nat. Taiwan U., 1971; M.S.I.C.S., Ga. Inst. Tech., 1974; Ph.D., Oreg. State U., 1978. Asst. prof. U. Oreg, Eugene, 1976-80; sr. mem. tech. staff GTE Labs., Waltham, Mass., 1980-82; adj. assoc. prof. Boston U., 1982—; software engring. specialist Wang Labs., Lowell, Mass., 1982—; prof. Northwestern U., 1983. Editor: Artificial Intelligence, 1983. NSF grantee, 1979, 80, 81; Steward grantee, 1979; Chung Shan scholar award, 1971. Sr. mem. IEEE Computer Soc.; Mem. Assn. for Computing Machinery, Pattern Recognition Soc., Chinese Computer Soc. Subspecialties: Software engineering; Graphics, image processing, and pattern recognition. Patentee in field. Home: 7 Ledgelawn Ave Lexington MA 02173 Office: Wang Labs Lowell MA 01851

WANG, PAUL P., electrical engineer, educator; b. China, 1936; m. (married); 2 children. B.Sc., Nat. Taiwan U., 1958; M.Sc., U. N.B., Can., 1963; Ph.D. in Elec. Engring, Ohio State U., 1965. Grad. and research asst. elec. engring. dept. Ohio State U., 1962-65; mem. tech. staff Bell Labs., Inc., Holmdel, N.J., 1965-68; asst. prof. elec. engring. Duke U., 1968-70, assoc. prof., 1970-75, prof., 1975—; vis. prof. U. Mass., 1974-75; cons. Western Electric Co., LORD Corp.; mem. adv. bd. Knowledge System Inst., 1980—. Editor: Jour. Fuzzy Math, 1981—; editor: Internat. Jour. Public Policy Analysis and Info. Systems, 1977—; corr.: Bul. pour les Sous Ensembles Flour et Leurs Applications, 1981—; Author, editor works in field. NASA grantee. Mem. IEEE, Pattern Recognition Soc., Chinese Lang. Computer Soc. (founding mem.), Sigma Xi. Subspecialty: Electrical engineering. Office: Engring Dept Duke U Durham NC 27706

WANG, PONG-SHENG, computer research scientist; b. Taipei, Taiwan, Sept. 13, 1950; s. Man-po and Chin-Fan (Yin) W.; m. Ai-chu Yeh, Feb. 20, 1950; children: Annie An-Li, Charles Li-cheng. B.S., Nat. Taiwan U., Taipei, 1972; M.S., Ind. U., Bloomington, 1976; Ph. D., Ohio State U., Columbus, 1980. Assoc. instr. Ind. U., Bloomington, 1974-76; programmer/analyst Shoe Corp. Am., Columbus, Ohio, 1976-77; grad. research assoc. Ohio State U., Columbus, 1976-80; prin. research scientist Honeywell, Mpls., 1980-83; staff programmer IBM, San Jose, Calif., 1983—. Mem. IEEE, Assn. Computing Machinery. Subspecialties: Distributed systems and networks; Computer architecture. Current work: Storage Management systems. Home: 741 Mairwood Ct San Jose CA 95120 Office: 555 Bailey Ave D28-D20 San Jose CA 95150

WANG, REX YUE, pharmacology educator; b. Shanghai, China, July 17, 1947; s. John Ling and Chi (Wha) W.; m. Sharon S. Yun-Yen Shan, June 24, 1972; children: Audrey Yating, Shijay David. B.S. in Psychology, Nat. Taiwan U., 1969; Ph.D. in Physiol. Psychology, U. Del., 1975. Postdoctoral tng. in neuropharmacology Yale U., New Haven, 1974-77, research staff dept. psychiatry, 1974-76, research assoc., 1976-77, asst. prof., 1977-79; asst. prof. pharmacology St. Louis U. Sch. Medicine, 1979-83, assoc. prof., 1983—; participant numerous seminars. Contbr. articles to sci. jours. NSF fellow, 1970-71; NIMH grantee, 1981—. Mem. AAAS, Soc. Neurosci., Am. Soc. Pharmacology and Exptl. Therapeutics, N.Y. Acad. Sci., Sigma Xi. Subspecialties: Neuropharmacology; Psychopharmacology. Current work: To determine interactions among neurotransmitters or modulators and mode of action of psychoactive drugs (especially antipsychotic and anti-depressant drugs) in the CNS. Office: 1402 S Grand Blvd Saint Louis MO 63104

WANG, RONG, materials scientist; b. Szechwan, China, Aug. 3, 1939; came to U.S., 1963, naturalized, 1967; s. Sun Hsu and Min Lang (Wang) W.; m. Dora Fan, Apr. 12, 1941; 1 son, Wilson. B.S. in Chem. Engring, Cheng Kung U., Tainan, Taiwan, 1961, M.S., U. Tex., 1964, Ph.D. in Materials, 1967. Research fellow MIT, Cambridge, 1967-68; asst. prof. materials and chem. engring. U. So. Calif., Los Angeles, 1968-73; sr. research scientist Battelle Meml. Inst. Pacific N.W. Lab, Richland, Wash., 1973—; cons. in field. Contbr. over 50 tech. articles to profl. publs. Mem. Am. Crystallographic Assn., Am. Soc. Engring. Edn., Am. Ceramic Soc., Materials Research Soc., Electrochem. Soc., Nat. Assn. Corrosion Engrs. Subspecialties: Amorphous metals; Corrosion. Current work: Glassy stainless steel coatings; alloy phase stability; non-equilibrium phases; electrochemical and photoelectrochemical properties of materials, corrosion and corrosion prevention techniques. Home: 2453 Catalina St Richland WA 99352

WANG, SHING CHUNG, electrical engineer; b. Hsinchu, Taiwan, Nov. 20, 1934; came to U.S., 1967, naturalized, 1978; s. Wan Jen and Yu (Chen) W.; m. Chu Mei Kuo, Jan. 21, 1958; children: Jean, Philip, Fanny, Charlie. B.S., Taiwan U., 1957; M.S., Tohoku U., 1965; Ph.D. (scholar), Stanford U., 1971. Assoc. prof. elec. engring. Chiao Tung U., Hsinchu, Taiwan, 1965-67; research assoc. Microwave Lab., Stanford U., 1971-74; sr. research scientist Xerox Corp., Pasadena, Calif., 1974—. Contbr. articles in field to profl. jours. Japanese Govt. scholar, 1963-65. Mem. So. Calif. Taiwanese Assn. (chmn. bd. 1979-80), Optical Soc. Am., IEEE, Quantum Electronics and Application Soc., Sigma Xi. Buddhist. Subspecialties: Laser data storage and reproduction; 3emiconductors. Current work: Laser technology and applications; research and development of hollow cathode lasers for applications in reprographics and information processing. Patentee in field. Office: 300 N Halstead St Pasadena CA 91107

WANG, SHYH, elec. engr., educator; b. Wusih, Jiangsu, China, June 15, 1925; s. Chung Chuan and Pao Chen (Wong) W.; m. Dila Ben, Nov. 23, 1962. B.S., Chiao-tung U., Shanghai, China, 1945; M.A., Harvard U., 1949, Ph.D., 1951. Research fellow Harvard U., 1951-53; sr. engr., engring. specialist Sylvania Electric Products, GTE, 1953-58; assoc. prof. elec. engring. and computer scis. U. Calif., Berkeley, 1958-64, prof., 1964—; cons. in field. Author: Solid State Electronics, 1965; contbr. articles to profl. jours. Guggenheim Found. fellow, 1964-65; NSF grantee; Army Research Office grantee; Office Naval Research grantee; Air Force Office Sci. Research grantee. Fellow IEEE, Optical Soc. Am.; mem. Am. Phys. Soc. Subspecialties: Optical signal processing; 3emiconductors. Current work: Research in integrated and guided-wave optics, semiconductor lasers, electrical and optical properties of III-V Compound semiconductors, microelectronic devices of III-V Compounds; development of semiconductor lasers and integrated optic devices for optical signal processing. Patentee in field. Home: 8636 Thors Bay Rd El Cerrito CA 94530 Office: U Calif 467 Cory Hall Berkeley CA 94720

WANG, TEEN-MEEI THOMAS, dental educator; b. Shia-Man, Fu-Chieng, China, Sept. 10, 1945; s. Jit-Liang and Yu-Ying (Lin) W.; m. Chun-Meei Su, Jan. 29, 1975; children: Yie-Shuan, Yie-Ding. B. Dental Surgery, Nat. Def. Med. Center, Taipei, Taiwan, 1969; Ph.D., U. Utah, 1974. Teaching asst. Nat. Def. Med. Center, Taipei, 1969-72, asso. prof., 1974-79, prof., 1973-80; teaching asst. U. Utah, Salt Lake City, 1972-74; vis. research prof. Office of Continuing Edn., Baylor U. Med. Center, Dallas, 1980-81; prof. dept. dentistry, dean Sch. Dentistry China Med. Coll., Taichung, Taiwan, 1980—; cons. dept. dentistry Tri-Service Gen. Hosp., Taipei. Author: Basic Neuroanatomy, 1976, Clinical Neuroanatomy and Neurophysiology, 1979; editor-in-chief: Oral Sci. Bimonthly, 1981; adv. editor: Chinese Dental Jour, 1982—, China Dental Mag, 1982—. Served to lt. col. Chinese Army, 1963-79. Recipient Excellent Tchr. award Nat. Def. Med. Center, Taipei, 1976; Nat. Sci. Council Republic of China scholar, 1972-74; China Med. Coll. fellow, 1980; R. Jackson Found. fellow, 1981. Fellow Internat. Coll. Dentists; mem. Internat. Assn. Dental Research, Taipei Dental Assn. Republic of China, Dental Assn. Republic of China (trustee and editor-in-chief 1977-81), Chinese Med. Assn. Club: Fgn. Affairs (Taipei). Subspecialties: Dental growth and development; Anatomy and embryology. Current work: Cell Kinetics and Morphometrics studies of the effects of mitogenic agents upon hard tissue and periodontal ligament following orthodontic tooth movement; pulp response to dental materials. Office: School of Dentistry China Medical College 91 Sheh-Shih Rd Taichung Taiwan 400 Home: 2/F 34 Alley 5 Lane 626 Ding-chow Rd Taipei Taiwan 107

WANG, THEODORE SHENG-TAO, nuclear medicine educator, researcher; b. Mukden, Liaoning, China, Oct. 18, 1930; s. Teh-Wu and Shih (Loung) W.; m. Aileen M. P. Chao, Jan. 14, 1967; children: Angela A. M., Christopher S. C. B.M., Nat. Mukden Med. Coll., 1950; M.S., U. Nebr., 1958; Ph.D., U. Md., 1964. Postdoctoral research fellow U. Md. Sch. Medicine, Balt., 1964-67; sr. radio-organic chemist New Eng. Nuclear Corp., Boston, 1967-69; sr. research scientist Lexington (Mass.) research lab. Kendall Co., 1969-74; asst. prof. nuclear medicine U. Conn. Health Ctr., Farmington, 1974-76; sr. staff assoc. Columbia-Presbyn. Med. Ctr., N.Y.C., 1976-77; asst. prof. clin. radiology Columbia U. Coll. Phys. and Surgs., N.Y.C., 1977—; research collaborator Brookhaven Nat. La., Upton, N.Y., 1982—; adj. asst. prof. Manhattan Coll., Bronx, N.Y., 1979—. Contbr. chpts. to books, articles to profl. jours. Mem. Soc. Nuclear Medicine, Am. Chem. Soc., N.Y. Acad. Scis., AAAS, Sigma Xi, Rho Chi. Mem. Christian Ch. Subspecialties: Pharmacology; Medicinal chemistry. Current work: Radiopharmacology (quantitative structure activity relationship); radionuclide-cells labelling for diagnostic imaging; radionuclide-antibodies labelling for diagnostic and therapeutic nuclear medicine. Office: 630 W 168th St New York NY 10032

WANG, TSUEY TANG, research scientist; b. Tainan, Taiwan, Nov. 12, 1932; came to U.S., 1958, naturalized, 1971; s. Shih Neng and Tsun (Chen) W.; m. Margaret Mei-Tieh Lin, June 12, 1965; children: David, Marjorie, Vanessa. B.Sc., Cheng-Kung U., Tainan, 1955; M.Sc., Brown U., 1961, Ph.D., 1965. Research group leader Poly. Inst. N.Y., 1965-66, asst. prof. applied mechanics, 1966-67; mem. tech. staff AT&T Bell Labs., Murray Hill, N.J., 1967—; mem. Ph.D. thesis rev. com., dept. mechanics and materials Rutgers U., Piscatway, N.J., 1982-83. Contbr. numerous articles to profl. pubs. Fellow Am. Phys. Soc.; mem. Am. Acad. Mechanics, N.Y. Acad. Scis., Am. Chem. Soc., Soc. Rheology. Subspecialties: Polymer physics; Theoretical and applied mechanics. Current work: Piezoelectricity in polymers; relaxation and mechanical properties of polymers; structure-properties of polymer thin films. Patentee in field. Office: 600 Mountain Ave Murray Hill NJ 07974

WANG, WEI-YEH, geneticist; b. Sian, China, Oct. 10, 1944; m. Wenan L. Wang, Jan. 29, 1969; children: Conrad, Oliver. B.S., Nat. Taiwan U., 1966; Ph.D., U. Mo.-Columbia, 1972. Postdoctoral fellow Duke U., 1972-75; asst. prof. U. Iowa, 1975-80; vis. prof. Carlsberg Lab., Copenhagen, 1980-81; assoc. prof. U. Iowa, Iowa City, 1980—. NIH fellow, 1973-75; Haigitt fellow, 1972-73; NSF grantee, 1979, 80. Mem. Am. Soc. Cell Biology, Am. Soc. Plant Physiologists, Genetics Soc. Am. Subspecialty: Gene actions. Current work: Researcher in genetic regulation of chlorophyll biosynthesis. Office: Dept Botany U Iowa Iowa City IA 52242

WANN, ELBERT VAN, geneticist, researcher; b. Grange, Ark., Dec. 29, 1930; s. Everett Ray and Cora Sebrina (Mason) W.; m. Joyce Nadine Sawyer, Sept. 23, 1950; 1 dau., Vivian Leigh. B.S. in Agr, U. Ark., 1959; M.S. in Horticulture, 1960; Ph.D. in Genetics, Purdue U., 1962. Research assoc. U. Ill.-Urbana, 1962-63; research geneticist U.S. Dept. Agr., Charleston, S.C., 1963—, dir. vegetable lab., 1972—. Contbr. numerous articles to profl. jours. Served with U.S. Army, 1952-54. Recipient Asgrow award, 1972. Fellow Am. Soc. Hort. Sci.; mem. Agronomy Soc. Am., Crop Sci. Soc. Am. Methodist. Club: Toast Masters. Subspecialties: Plant genetics; Plant pathology. Current work: Plant breeding. Office: 2875 Savannah Hwy Charleston SC 29407

WANN, LEE SAMUEL, cardiologist, educator, echocardiographic researcher; b. Crawfordsville, Ind., July 28, 1946; s. Raymond Woodrow and Ora Lee (Riepie) W.; m. Mary Alice Fifer, June 8, 1968; children: Randie Leigh, Carrie Lynn. A.b., Ind. U.-Bloomington, 1968; M.D., Ind. U.-Indpls., 1971. Diplomate: Am. Bd. Internal Medicine. Intern Ind. U.-Indpsl., Hosps., 1971-72, resident, 1972-75, fellow in cardiology, 1975-77; asst. prof. medicine Ind. U.-Indpls., 1977-79; asst. prof. Med. Coll. Wis., 1979-81, assoc. prof., 1981—; research assoc. Krannert Inst. Cardiology, Indpls., 1977-79; dir. echocardiographic labs. Milw. County Med. Complex and Wood VA Med. Ctr., Milw., 1979—; mem. adv. com. Clin. Research Ctr., Milw., 1979-81. Contbr. chpts. to books. Am. Heart Assn. fellow, 1976; VA grantee, 1980; NIH grantee, 1981. Fellow Am. Coll. Cardiology, Council Clin. Cardiology of Am. Heart Assn.; mem. ACP, Am. Inst. Ultrasound in Medicine, Am. Fedn. Clin. Research, Am. Soc. Echocardiography (dir. 1982—). Subspecialties: Cardiology; Internal medicine. Current work: Clinical research in field of echocardiograph, including digital image processing, development of ultrasonic contrast agents, and hemodynamic modeling. Home: 4751 N Cumberland Blvd Whitefish Bay WI 53211 Office: Cardiology Div Med Coll Wis 8700 W Wisconsin Ave Milwaukee WI 53226

WARD, CALVIN HERBERT, biologist, educator; b. Strawberry, Ark., Mar. 1, 1933; s. Floyd E. and Ava (Roberts) W.; m. Barbara Ann Nunn, Aug. 15, 1954; children: Kirk A., Joel M., Peter T. B.S., N.Mex. State U., 1955; M.S., Cornell U., 1958, Ph.D., 1960; M.P.H., U. Tex., 1978. Research biologist USAF Sch. Aerospace Medicine, San Antonio, 1960-63, physiologist, dir. bioregenerative research program, 1963-66; assoc. prof. biology and environ. Sci. Rice U., Houston, 1966-70, prof., 1970—, chmn. dept. environ. sci. and engring., 1970-81; vis. prof. environ. health Sch. Pub. Health, U. Tex., Houston, 1973-74; mem. sci. adv. com. Hazardous Waste Research Ctr., La. State U., 1982—. Contbr. articles to profl. jours. Served with USAF, 1960-63. Recipient Bausch and Lomb award in sci., 1951; Group Achievement award NASA, 1981; Shell fellow, 1957-59. Mem. Am. Phytopathol. Soc., Soc. Nematologists, Soc. for Indsl. Microbiology (pres. 1983-84), Am. Inst. Biol. Scis. (governing bd., exec. com., pres. elect 1983-84), Phycological Soc. Am., AAAS, Am. Soc. Plant Physiologists, Internat. Assn. on Water Pollution Research, Internat. Water Resources Assn. (v.p. U.S. nat. com.), Sigma Xi, Alpha Zeta. Subspecialties: Species interaction; Microbiology. Current work: Aquatic toxicity of hazardous materials; biology and chemistry of ground water, aquatic and microbial physiology. Home: 10954 Beinhorn Rd Houston TX 77024 Office: Dept Environ Sci and Engrin Rice U PO Box 1892 Houston TX 77251

WARD, DAVID ALOYSIUS, research mgr.; b. Joliet, Ill., Nov. 13, 1930; s. Aloysius and Flora May (Picton) W.; m. Diana Louise Nainis, July 4, 1953; children: Rebecca, David Aloysius, Jr. Thomas, Kathleen, Eileen, Mary, Christopher, Matthew, Jonathan. B.S. in Mech. Engring, U. Ill.-Urbana, 1953. Research engr. E.I. DuPont De Nemours & Co. Savannah River Lab., Aiken, S.C., 1953-64; process engr. Savannah River Plant, 1964-67, engring. supr., 1967-72, chief engr. supr., 1972-75, dept. supt., 1975-80; research mgr. Savannah River Lab., 1980—; mem. adv. com. for reactor safeguards NRC, Washington, 1980—; mem. com. on advanced nuclear systems Nat. Acad. Sci., Washington, 1982—. Chmn. North Augusta (S.C.)

Republican Party, 1956-61; del. Republican Nat. Conv., Chgo., 1960; pres. Our Lady of Peace Sch. Bd., North Augusta, 1972-74. Mem. Am. Nuclear Soc. (chmn. nat. membership com. 1978-81). Roman Catholic. Subspecialties: Nuclear engineering; Fluid mechanics. Current work: Nuclear reactor safety, operations, human factors, experimental heat transfer and fluid flow. Home: 2108 Pisgah Rd North Augusta SC 29841 Office: E I DuPont De Nemours & Co Savanhnah River Laboratory Aiken SC 29808

WARD, H. BLAIR, JR., manufacturing engineer; b. Cranston, R.I., Sept 7, 1935; s. H. Blair and Edwinna Mary (Wilson) W.; m. Marilyn Eileen Hughes, Feb. 15, 1974; children: Michelle, Kristi, Kelly. Tool engrng. diploma (Imperial Oil scholar), Ryerson Poly. Inst., Toronto, Ont., Can., 1958. Registered profl. engr., Calif., Pa. Tool engr. Automatic Elec., Brockville, Ont., 1958-60; mfg. engr. Moog, East Aurora, N.Y., 1960-66; project engr. A. Smith, Erie, Pa., 1966-71; mfg. engr. Talon, Meadville, Pa., 1971-81; mgr. mfg. engrng. Leech Tool & Die, Meadville, 1981—. Mem. Soc. Mfg. Engrs. (cert. mfg. engr.), Nat. Soc. Profl. Engrs., ASME, Soc. Automotive Engrs. Subspecialties: Materials processing; Polymers. Current work: Developed manufacturing processes for consumer products in high-volume miniature sizes—in both polymers and metals. Patentee zipper mfg.; developer high speed automatic molding processes, high speed automatic stamping and forming processes and equipment. Home: 1 Box 291 Guys Mills PA 16327 Office: Leech Tool & Die RD 7 Meadville PA 16355

WARD, JERRY MACK, research physicist; b. Tallahassee, May, 23, 1943; s. James Holloman and Jewell Naomi (Maige) W.; m. Janice Marie Felesky, Aug. 31, 1968; children: Stephen Christopher, Emily Elizabeth. B.S., U. Fla., 1966, Ph.D., 1971. Postdoctoral research assoc. Naval Ordnance Lab., Silver Spring, Md., 1972-74, research physicist, 1974—. Mem. AIAA, Fed. Exec. and Profl. Assn. Democrat. Presbyterian. Subspecialties: Explosion effects; Aerospace engineering and technology. Current work: Fragments, airblast, explosion effects, explosion hazards, explosion safety, weapon safety, hydrodynamic trajectory calculations, debris effects, hazards. Office: Naval Surface Weapons Ctr White Oak Lab Silver Spring MD 20910

WARD, JOHN WESLEY, pharmacologist; b. Martin, Tenn., Apr. 8, 1925; s. Charles Wesley and Sara Elizabeth (Little) W.; m. Martha Isabelle Hendley, Dec. 7, 1947; children: Henry Russell, Judith Carol, Charles Wesley, Richard Little. A.A., George Washington U., 1948, B.S., 1950, M.S., 1955; Ph.D., Georgetown U., 1959. Research asso. in pharmacology Hazleton Labs., Falls Church, Va., 1950-55, head dept. pharmacology, 1955-58, chief depts. biochemistry and pharmacology, 1958-59; with A. H. Robins Co., Richmond, Va., 1959—, dir. biol. research, 1978-80, dir. research, 1980—, v.p. research, 1982—; lectr. in pharmacology Med. Coll. Va., 1960-64, adj. assoc. prof. pharmacology, 1982—; guest lectr. Seminar on Good Lab. Practices, FDA, Washington, 1979, Chgo., 1979, San Francisco, 1979. Contbr. articles on pharmacology, toxicology and medicinal chemistry to profl. publs. Served with USMC, 1943; Served with USN, 1944-46; Served with U.S. Army, 1944. Mem. AAAS, N.Y. Acad. Sci., Va. Acad. Sci., Am. Chem. Soc., Soc. Toxicology (charter), Am. Soc. Pharmacology and Exptl. Therapeutics, Pharm. Mfrs. Assn. (chmn. color additive toxicology com. 1978—), Am. Assn. for Accreditation Lab. Animal Care (chmn. bd. trustees 1976-80), Sigma Xi. Clubs: Willow Oaks (Richmond); Cosmos, Masons (Washington). Subspecialties: Pharmacology; Toxicology (medicine). Current work: Cardiovascular, central nervous system, gastrointestinal, inflammation, allergy; directs research and development in medicinal chemistry, pharmacologym molecular biology, drug metabolism, toxicology and pathology. Patentee in field. Home: 10275 Cherokee Rd Richmond VA 23235 Office: 1211 Sherwood Ave Richmond VA 23220 An appreciation of the responsibility we have to society has set the standards by which I live. These responsibilities are as important as the rights to be gained from society. Those who are unwilling to assume responsibility should have no rights.

WARD, LOUIS EMMERSON, retired physician; b. Mt. Vernon, Ill., Jan. 19, 1918; s. Henry Ben Pope and Aline (Emmerson) W.; m. Nan Talbot, June 5, 1942; children—Nancy, Louis, Robert, Mark. A.B., U. Ill., 1939; M.D., Harvard, 1943; M.S. in Medicine, U. Minn., 1949. Intern Ill. Research and Ednl. Hosp., Chgo., 1943; fellow medicine Mayo Found., 1946-49; cons. medicine, rheumatology Mayo Clinic, 1950-83, chmn. bd. govs., 1964-75; dir. Northwestern Bell Telephone Co., Bankers Life, Des Moines. Contbr. articles to profl. jours. Vice chmn. bd. trustees Mayo Found., 1964-76; past bd. dirs. Fund for Republic, bd. dirs. Center for Study Democratic Instns., past bd. dirs. Arthritis Found.; mem. Nat. Council Health Planning and Devel., 1976—. Recipient U. Ill. Alumni Achievement award, 1968; recipient disting. alumnus award Mayo Found., 1983. Mem. Inst. Medicine (Nat. Acad. Scis.), AMA, Am. Rheumatism Assn. (pres. 1969-70), Nat. Soc. Clin. Rheumatologists (pres. 1967-69), Central Soc. Clin. Research, Minn., Zumbro Valley med. socs., So. Minn. Med. Assn., Phi Beta Kappa, Sigma Xi, Alpha Omega Alpha, Phi Delta Theta. Subspecialties: Internal medicine; Rheumatology. Current work: Clinical investigation in rheumatic diseases. Home: 9 Raeburn Ct Port Ludlow WA 98365

WARD, O. BYRON, JR., psychology educator; b. Amelia, Va., Sept. 19, 1937; s. Otis Byron and Hazel (Carter) W.; m. Ingeborg Annemarie Lehmann, Jan. 13, 1963; children: Richard Byron, Michael Patrick. B.A., Duke U., 1958; M.S., U. Richmond, 1961; Ph.D., Tulane U., 1968. Research assoc. Tulane Med. Sch., New Orleans, 1964-66; asst. prof. psychology Villanova (Pa.) U., 1966-72, assoc. prof., 1972—. Contbr. chpts. to books, articles to profl. jours. Nat. Inst. Drug Abuse grantee, 1978. Mem. Soc. for Neurosci., Sigma Xi. Subspecialties: Physiological psychology; Neuropharmacology. Current work: Fetal exposure to opiates; neuroendocrine and behavioral effects. Office: Villanova U Villanova PA 19085

WARD, OSCAR G., JR., cytogeneticist, researcher; b. Denver, Feb. 16, 1932; s. Oscar G. and Elizabeth M. W.; m. Lea E. Ramirez, July 16, 1955; children: Oscar G., Anne E. B.S., U. Ariz., Tucson, 1958, M.S., 1960; Ph.D., Purdue U., 1966. Instr. biol. scis. Purdue U., 1960-64, David Ross research fellow, 1964-66; asst. prof. biology U. Ariz., Tucson, 1966-74, lectr., 1974-77, assoc. prof. dept. ecology and evolutionary biology, 1977—; sr. internat. fellow PHS Universidad Nacional Autonoma de Mexico, 1980. Served with USMC, 1951-54. NSF grantee, 1976-77; Fogarty Internat. Center and NIH sr. internat. fellow grantee to, Mex., 1980. Mem. AAAS, Am. Soc. Human Genetics, Genetics Soc. Am., Am. Genetics Assn., Am. Soc. Mammalogists, Sigma Xi, Tau Beta Pi. Subspecialties: Genome organization; Evolutionary biology. Current work: Plant and animal cytogenetics, emphasis on mammalian systems; Karyotype evolution in rodents and canids. Office: Dept Ecology/Evolutionary Biology Univ Ariz Tucson AZ 85721

WARD, PATRICK E., pharmacologist; b. Warren, Ohio, Mar. 2, 1947; s. Donal L. and Virginia (Kania) W.; m. (div.); 1 son, Christopher M. B.A., Kent State U., 1969; M.S., Miami U., Oxford, Ohio, 1971; Ph.D., Cambridge (Eng.) U., 1974. NIH postdoctoral fellow dept. pharmacology U. Tex. Health Sci. Center, Dallas, 1974-77, asst. prof., 1977-80; assoc. prof. dept. pharmacology N.Y. Med. Coll., Valhalla, 1980—, dir. grad. program pharmacology, 1980—; established investigator Am. Heart Assn., 1978-83. Contbr. articles to profl. jours. Mem. Am. Soc. Pharmacology and Exptl. Therapeutics, Am. Heart Assn. (council for high blood pressure research). Subspecialties: Cellular pharmacology; Molecular pharmacology. Current work: Metabolism of vasocative peptides and their relationship to blood pressure regulation and hypertension. Office: Dept Pharmacology New York Med Coll Valhalla NY 10595

WARD, PETER ALLAN, pathologist, educator; b. Winsted, Conn., Nov. 1, 1934; s. Parker J. and Mary Alice (McEvoy) W. B.S., U. Mich., Ann Arbor, 1958, M.D., 1960. Diplomate: Am. Bd. Anat. Pathology. Intern Bellevue Hosp., 1960-61; resident U. Mich. Hosp., Ann Arbor, 1961-63; chief immunobiology br. Armed Forces Inst. Patholgy, Washington, 1967-71; prof. dept. pathology, chmn. dept. U. Conn. Health Center, Farmington, 1971-80; prof., chmn. dept. pathology U. Mich., Ann Arbor, 1980—, interim dean, 1982—; cons. VA Hosp., 1980—; mem. research rev. com. Nat. Heart, Lund, Blood Inst., NIH, Bethesda, Md., 1978—; bd. dirs. Univ. Associated for Research and Edn. in Pathology, Inc., 1978—; chmn., bd. mem. sci. adv. bd. Armed Forces Inst. Pathology, Washington, 1981-83; mem. patholgy A study sect. NIH, Bethesda, Md., 1972-76, chmn., 1976-78. Served to capt., M.C. U.S. Army, 1965-67. Recipient Borden Research award U. Mich. Med. Sch., Ann Arbor, 1980, Research and Devel. award U.S. Army, 1969; Meritorious Civilian Service award Dept. Army, 1970; Park-Davis award Am. Soc. Exptl. Pathology, 1971. Mem. Am. Assn. Pathologists (pres. 1978-79), Am. Soc. Clin. Investigation, Am. Assn. Immunologists, Assn. Pathology Chmn., Mich. Soc. Pathologists. Subspecialty: Pathology (medicine). Current work: Basic clinical biomedical research. Home: 2815 Washenaw Ave Ann Arbor MI 48104 Office: Med Sci I 1315 Catherine Rd M5240 Ann Arbor MI 48109

WARD, SAMUEL, cell biologist, biology educator; b. Los Angeles; s. Morgan and Sigrid (Von Toll) W.; m. Anne F., Sept. 12, 1966; children: Timothy, Geoffrey. A.B., Princeton U., 1965; Ph.D., Calif. Inst. Tech., 1971. Postdoctoral fellow MRC Lab. Molecular Biology, Cambridge, Eng., 1971-73; asst.prof. dept. biol. chemistry Harvard Med. Sch., 1973-77; staff mem. dept. embryology Carnegie Instn., Washington, 1978—; assoc. prof. biology Johns Hopkins U., Balt., 1978-81, prof., 1981—. Mem. AAAS, Soc. Nematology, Am. Soc. Cell Biology, Genetics Soc. Am. Subspecialties: Developmental biology; Gene actions. Current work: Cell morphology, genetics, nematology, developmental biology. Office: 115 W University Pkwy Baltimore MD 21210

WARD, WILLIAM CORNELIUS, research psychologist; b. Balt., Nov. 21, 1939; s. William C. and Mary B. (Crisp) W. A.B., Johns Hopkins U., Balt., 1961; Ph.D., Duke U., 1966. Asst. prof. Stanford U., Palo Alto, Calif., 1965-68; research psychologist Ednl. Testing Service, Princeton, N.J., 1968-76, sr. research psychologist, 1976—. Mem. Am. Psychol. Assn., Soc. for Research in Child Devel., Am. Ednl. Research Assn., Nat. Council Measurement in Edn. Subspecialty: Cognition. Current work: Problem solving, creativity, computerized testing, innovative cognitive measurement. Office: Educational Testing Service Princeton NJ 08541

WARDELL, WILLIAM MICHAEL, clinical pharmacologist, educator; b. Christchurch, N.Z., Nov. 15, 1938; came to U.S., 1971; s. Thomas William Rae and Phyllis Ruth Margaret (Robinson) W.; m. Dorothy Muriel Rile, July 24, 1965; children: Steven, Michael. D.Phil., U. Oxford, 1964, B.M., B.Ch., 1967; D.M., U. Rochester/Oxford U., 1973. Assoc. physician Strong Meml. Hosp., U. Rochester, N.Y., 1973—, dir., 1975—; assoc. prof. pharmacology and toxicology in pharmacology, asst. prof. medicine U. Rochester Sch. Medicine and Dentistry, 1976—. Contbr. numerous articles on pharmacology to profl. jours.; author: Regulation and Drug Development, 1975, Controlling the Use of Therapeutic Drugs: An International Comparison, 1978, Drug Development, Regulatory Assessment, and Postmarketing Surveillance, 1981. Christopher Welch scholar, 1961; recipient Radcliffe Prize for Research in Medicine U. Oxford, 1967; Merck Internat. fellow, 1970. Mem. Nat. Council on Drugs, Am. Soc. clin. Pharmacology and Therapeutics, AMA. Subspecialties: Pharmacology; Clinical pharmacology. Current work: Clinical trials on analgesics and on prevention of myocardial infarction; drug development and regulation. Office: Dept Pharmacology U Rochester Med Ctr Rochester NY 14642

WARDEN, JOSEPH TALLMAN, JR., chemistry educator, researcher; b. Huntington, W.Va., Aug. 7, 1946; s. Joseph Tallman and Ruth Onleta (Dunbar) W.; m. Carolyn Edith Lee, June 1, 1968. B.S., Furman U., 1968; Ph.D., U. Minn., 1972. Vis. scientist biophysics dept. U. Leiden, 1972-73; postdoctoral assoc. chemistry dept. U. Calif.-Berkeley, 1973-75; asst. prof. Rensselaer Poly. Inst., Troy, N.Y., 1975-79, assoc. prof., 1980—; vis. prof. dept. botany and microbiology Univ. Coll., London, 1981; scientist in residence Argonne Nat. Lab., 1982. Contbr. articles on chemistry to profl. jours. Mem. Am. Chem. Soc., Am. Soc. for Photobiology. Subspecialties: Biophysical chemistry; Photosynthesis. Current work: Physical chemistry of light-induced electron transfers in vivo and in vitro, membranes, eletron spin resonance, photochemical processes on surfaces, laboratory automation. Office: Dept Chemistr Rensselaer Poly Inst Troy NY 12181

WARFIELD, JOHN NELSON, electrical engineering educator, management researcher; b. Sullivan, Mo., Nov. 21, 1925; s. John D. and Flora A. (L) W.; m. Rosamond Arline Howe, Feb. 2, 1948; children: Daniel, Nancy, Thomas. A.B., U. Mo.-Columbia, 1948, B.S.E.E., 1948, M.S.E.E., 1949; Ph.D., Purdue U., 1952. Mem. Faculty U. Mo.-Columbia, 1948, Pa State U., University Park, 1949-55, U. Ill.-Urbana, 1955-57, Purdue U., Lafayette, Ind., 1957-58, U. Kans., Lawrence, 1958-65; dir. research Wilcox Electric Co., Kansas City, Mo., 1965-66; sr. research leader Battelle Meml. Inst., Columbus, Ohio, 1966-74; prof. U. Va., Charlottesville, 1975—, Forsythprof. elec. engrng., 1975—; dir. Center Interactive Mgmt., 1981—; cons. consensus methodologies, orgn. design, long-range planning. Author 4 books, numerous articles. Served in U.S. Army, 1944-46. RCA fellow NRC, 1951-52; recipient Western Electric Fund award Midwest sect. Am. Soc. Engrng. Edn., 1966. Fellow IEEE (cert. for outstanding service 1973); mem. Soc. Gen. Systems Research (pres. 1982-83), Acad. Polit. Sci., Am. Mgmt. Assns., World Future Soc. Subspecialties: Systems engineering; Foundations of computer science. Current work: Research on interactive management, including developing and testing consensus methodologies, and creating productive environments for holding meetings. Patentee (2). Home: 96 Wild Flower Dr Charlottesville VA 22901 Office: U Va Thornton Hall Charlottesville VA 22901

WARGO, PHILIP MATTHEW, research forest pathologist; b. Danville, Pa., Mar. 1, 1940; s. Augustine A. and Eva C. (Marhefka) W.; m. Jane W. Fillmore, Aug. 3, 1963; 1 son, David Scott. B.A., Gettysburg Coll., 1962; M.S., Iowa State U., 1964, Ph.D., 1966. Research forest pathologist N.E. Forest Insect and Disease Lab. (now Ctr., for Biol. Control of Forest Insects and Disease), U.S. Forest Service), Handen, Conn., 1968—. Served to capt. U.S. Army, 1966-68. Mem. Am. Phytopath. Soc., Internat. Soc. Arboriculture. Subspecialties: Plant pathology; Plant physiology (agriculture). Current work: Stress physiology, changes induced by stress that induce disease; secondary organisms; root diseases; disease resistance. Home: 3 Spring Brook Rd Wallingford CT 06492 Office: 51 Mill Pond Rd Hamden CT 06514

WARING, GEORGE HOUSTOUN, zoology educator; b. Denver, July 15, 1939; s. G. Houstoun and Irene L. (Fender) W.; m. Ann-Meredith Kenney, Dec. 30, 1962; children: Sari, Houstoun, Heidi. B.S., Colo. State U., 1962; Ph.D., 1966; M.A., U. Colo., 1964. Teaching asst., research fellow Colo. State U., Ft. Collins, 1964-66; asst. prof. zoology So. Ill. U., Carbondale, 1966-72, assoc. prof., 1972—; vis. prof. U. Munich, W.Ger., 1972-73; research program dir. U.S. Marine Mammal Comm., 1974-75, cons., 1975—. Author: Survey of Federally Funded Marine Mammal Research, 1981, Horse Behavior, 1983. Served with USCGR, 1957-65. NSF fellow, 1965; NIH fellow, 1965-66; Colo-Wyo. Acad. Sci. awardee, 1965. Mem. Animal Behavior Soc., Am. Soc. Mammalogists (life), Am. Soc. Vet. Ethology, Am. Ornithologists Union, Nat. Assn. Environ. Profls., AAAS, Ecol. Soc. Am., Wildlife Soc., Sigma Xi. Subspecialties: Ethology; Behavioral ecology. Current work: Applied ethology of domesticated and wildlife species. Office: Dept Zoology So Ill U Carbondale IL 62901

WARME, JOHN EDWARD, geology educator; b. Los Angeles, Jan. 16, 1937; s. Clarence Herbert and Edna (Peterson) W.; m. (div.); children: Susan Lynn, Jane Kathleen. B.A., Augustana Coll., 1959; Ph.D., UCLA, 1966. Cert. petroleum geologist. Postdoctoral fellow U. Edinburgh, Scotland, 1966-67; from asst. prof. to assoc. prof. geology Rice U., Houston, 1967-76, W. Maurice Ewing prof. oceanography, 1976-79; prof. geology Colo. Sch. Mines, Golden, 1979—; cons. In Situ, Inc., Laramie, Wyo., 1981—, NSF, 1979-82. Co-editor: The Deep Sea Drilling Project: Decade of Progress, 1981; assoc. editor: Jour. Sedimentary Petrology, 1977-80, Palaeoclimatology, Palaeogeography, Palaeoecology, 1977—. Fulbright scholar, 1966-67; NSF grantee, 1970, 78, 81, 82, 83. Fellow Geol. Soc. Am., AAAS; mem. Soc. Econ. Paleontologists and Mineralogists (pres. 1983-84), Am. Assn. Petroleum Geologists, Paleonto. Soc., Internat. Assn. Sedimentologists, Internat. Palaeonto. Assn., Colo. Sci. Soc. Subspecialties: Sedimentology; Paleoecology. Current work: Basin analysis (stratigraphic, sedimentologic, paleoecologic) applied to petroleum exploration, submarine bioerosion, deep sea sedimentation, coral reef ecology. Home: 1420 Genesee Ridge Rd Golden CO 80401 Office: Colo Sch Mines Dept Geology Golden CO 80401

WARME, PAUL KENNETH, biochemist, research administrator; b. Westbrook, Minn., Jan. 23, 1942; s. Kenneth B. and Thelma M. (Johnson) W.; m. Beverly J. Ring, Dec. 22, 1962; 1 dau., Rebecca. B.Chem., U. Minn., 1964; Ph.D., U. Ill., 1969. Postdoctoral fellow Cornell U., 1969-72; asst. prof. biochemistry Pa. State U., 1972-79; pres. research dir. Interactive Microware, Inc., State College, Pa., 1979—. Mem. Am. Soc. Biol. Chemists, IEEE. Subspecialties: Software engineering; Biochemistry (biology). Current work: Laboratory data acquisition. Home: 120 S Patterson St State College PA 1680 Office: Interactive Microware Inc 135 N Gill St State College PA 16801

WARNAT, WINIFRED IRENE, educational researcher, social psychologist; b. Grosse Pointe, Mich., Feb. 8, 1943; d. Rudolf P. W. and Frieda (Lupp) W.; m. Henry Godfrey Scharles, Nov. 29, 1968 (div. 1976). B.A., Fla. Atlantic U., 1965, M.Ed., 1967; Ph.D., Am. U., 1971. Spl. edn. tchr. Deerfield Beach (Fla.) Jr. High Sch.; vocat. rehab. counselor, Ft. Lauderdale, Fla.; dir. placement Gallaudet Coll., Washington, 1963-68; asst. dean advanced studies and research Grad. Sch. Arts and Scis., clin. prof. dept. pediatrics and child health Coll. Medicine, prof./chair dept. curriculum and instrn. Sch. Edn., Howard U., Washington, 1969-77; research prof. Sch. Edn. Am. U., Washington, 1977-81, dir. Adult Learning Potential Inst., 1977-81; dir. Nat. Center Teaching and Learning, Eastern Mich. U., Ypsilanti, 1981—; lectr., speaker. Contbr. chpts. to books, articles to profl. jours. Mem. Montgomery County Commn. for Women, Md., 1974-76. Grantee Charles Stewart Mott Found., 1981-83, U.S. Office Edn., 1975-77, 77-80, 75-76, 74-76; others. Mem. Am. Ednl. Research Assn., Am.Psychol. Assn., Robotics Internat./Soc. Mfg. Engrs., N.Am. Soc. Corp. Planners. Subspecialties: Robotics; Learning. Current work: High technology and changing character of works. Robotics education and human factors; training and employment of youth displaced workers, women, minorities, and older workers; gender and technology; adult learning and development. Office: Eastern Mich U 111 King Hall Ypsilanti MI 48197

WARNER, JONATHAN ROBERT, cell and molecular biologist, researcher, educator, admininstrator; b. N.Y.C., Feb. 19, 1937; s. Robert and Anne Marie (Homer) W.; m. Nancy Elizabeth Heers, June 23, 1958; children: Deborah Pearson, Anne Homer. B.S. in Physics, Yale U., 1958; Ph.D. in Biophysics, M.I.T., 1963. Research assoc. M.I.T., 1963-64; research fellow Albert Einstein Coll. Medicine, Bronx, N.Y., 1964-65, asst. prof. biochemistry, 1965-68, assoc. prof. biochemistry and cell biology, 1968-74, prof. biochemistry and cell biology, 1974-83, prof., chmn. cell biology, 1983—, dir. Sue Golding Grad. Sch., 1973-83; vis. fellow Imperial Cancer Research Fund, Eng., 1971-72; sci. adv. com. Am. Cancer Soc. Editorial bd.: Jour. Biol. Chemistry. Guggenheim fellow, 1971. Mem. Am. Soc. Biol. Chemistry, Am. Soc. Microbiology, Am. Soc. Cell Biology. Subspecialties: Cell biology; Gene actions. Current work: Regulation of synthesis and localization of ribosomal components in eukaryotes: a problem of molecular inventory control. Office: 1300 Morris Park Ave Bronx NY 10461

WARNER, NOEL LAWRENCE, sci. adminstr.; b. Melbourne, Australia, Dec. 17, 1939; s. Lawrence E. and Elsa M. (Smith) W.; m. Pamela A. Tibbles, Mar. 9, 1962; children: Grant, Scott, Galen. Ph.D., U. Melbourne, 1963. Postdoctoral fellow Stanford U. and N.Y.U., 1964-67; head genetics unit Walter and Elza Hall Inst., Melbourne, 1967-77; prof. depts. pathology and medicine U. N.Mex., 1977-81; sci. dir. Becton Dickinson Monoclonal Center, Mountain View, Calif., 1981—. Contbr. 220 publs. to profl. jours. Fulbright fellow, 1964-66. Mem. Am. Assn. Immunologists, Am. Assn. Cancer Research. Subspecialties: Immunology (medicine); Cancer research (medicine). Current work: Immunology and cancer research. Office: 2375 Garcia Ave Mountain View CA 94043

WARNER, RONALD RAY, research physiologist; b. Atlanta, Feb. 1, 1944; s. Donald C. and Mary A. (Ray) W.; m. Virginia Anne Bilz, Mar. 29, 1969; children: Brian, Britten. B.S. in Physics, U. Tex.-Austin, 1967, B.A. in Math, 1967; Ph.D. in Biophysics, U. Rochester, 1972. Postdoctoral researcher dept. physiology Yale U., 1973-74; research assoc. Harvard Med. Sch., Boston, 1974-76, instr. physiology, 1976-79; research scientist Procter & Gamble Co., Cin., 1980—. Editor: Microbeam Analysis in Biology, 1979. Mem. Am. Physiol. Soc., Microbeam Analysis Soc., Soc. Gen. Physiologists. Subspecialties: Physiology (medicine); Microscopy. Current work: Application of microscopy and microprobe analysis to study cellular physiology, particularly in intestine, kidney and skin. Home: 4658 Celadon Ave Fairfield OH 45014 Office: Procter & Gamble Co Miami Valley Labs PO Box 39175 Cincinnati OH 45247

WARNKE, ROGER ALLEN, pathologist, educator; b. Peoria, Ill., Feb. 22, 1945; s. Delmar Carl and Ruth Armanelle (Peard) W.; m.

Joan Marie Gebhart, Nov. 18, 1972; children: Kirsten, Lisa. B.S., U. Ill., 1967; M.D., Washington U., St. Louis, 1971. Diplomate: Am. Bd. Pathology. Intern Stanford (Calif.) Med. Ctr., 1971-72, resident in pathology, 1972-73, fellow in pathology, 1973-75, immunology research fellow, 1975-76, asst. prof. pathology, 1976-82, assoc. prof., 1983—; cons. Becton Dickinson Monoclonal Antibodies, Mountain View, Calif., 1982—. Contbr. numerous sci. articles to profl. publs. Recipient Castleman award Mass. Gen. Hosp., 1981. Mem. Southbay Pathology Soc., Calif. Soc. Pathologists, Am. Soc. Pathologists, Internat. Acad. Pathologists. Subspecialty: Pathology (medicine). Current work: Hematopathology, immunopathology of lymphomas and leukemia. Office: Stanford Med Center Pathology Dept Stanford CA 94305

WARREN, DONALD WILLIAM, dental educator, researcher; b. Bklyn., Mar. 22, 1935; s. Sol B. and Frances (Plotkin) W.; m. Priscilla G. Girardi, June 10, 1956; children: Donald William, Michael C. B.S., U. N.C., 1956, D.D.S., 1959; M.S., U. Pa., 1961, Ph.D., 1963. Asst. prof. prosthodontics U. N.C., Chapel Hill, 1963-64, assoc. prof. dental ecology, 1964-69, prof., 1969-80, Kenan prof., 1980—; cons. in field. Contbr. chpts., numerous articles on cleft palate and speech to profl. publs. NIH tng. grantee, 1959-63; research grantee, 1963—; Robert Wood Johnson Found. grantee, 1979-83. Fellow Am. Speech and Hearing Assn.; mem. Am. Cleft Palate Assn. (pres. 1981-82, pres. ednl. found. 1976-77), ADA, N.Y. Acad. Sci., Acoustical Soc. Am., Internat. Assn. Dental Research. Subspecialties: Dental growth and development; Oral biology. Current work: Cleft palate, oral morphology and breathing, speech aerodynamics. Home: PO Box 2352 Chapel Hill NC 27514 Office: U NC Sch Dentistry Chapel Hill NC 27514

WARREN, GUYLYN REA, research molecular biologist, rancher; b. Butte, Mont., Aug. 16, 1941; s. Guy Lorimer and Evelyn (Barter) W. B.S., Mont. State U., 1963, Ph.D., 1967. Research assoc. Palo Alto (Calif.) Med. Research Found., 1970-72; asst. prof. dept. chemistry Mont. State U., 1973-80, adj. assoc. prof., 1980—, chmn. biosafety com., 1980—, mem. research adv. com., 1981—. Rev. author: Annual Rev. Phytopathology, 1979. Mem. research adv. com. Air Quality Bur., Helena, Mont., 1975-79. NSF fellow, 1963; NDEA fellow, 1963-67; NIH fellow, 1968-70; NIH grantee, 1979, 82; Proctor & Gamble grantee, 1979. Mem. AAAS, Environ. Mutagen Soc., Am. Soc. Microbiology, Mt. Columbia Sheep Breeders Assn. (dir. 1980—), Agrl. Preservation Assn. Lodge: Order Eastern Star. Subspecialties: Environmental toxicology; Cancer research (medicine). Current work: s: Molecular mode of action of inorganic mutagens/carcinogens; rapid bioassays applied to assessment of human risk for cancer. Home: PO Box 66 Willow Creek MT 59760 Office: Montana State University Biochemistry Johnson Hall Bozeman MT 59717

WARREN, HERMAN LECIL, plant pathology educator; b. Tyler, Tex., Nov. 13, 1932; s. Cicero and Leola W.; m. Mary Kingsberry, May 26, 1933; children: Michael James, Christopher Lecil, Mark Herman. B.S., Prairie View A&M U., 1953; M.S., Mich. State U., 1960; Ph.D., U. Minn., 1970. Research scientist Olin Chem. Corp., New Haven, 1962-67; plant pathologist U.S. Dept. Agr., Beltsville, Md., 1967-70; prof. plant pathology Purdue U., West Lafayette, Ind., 1971—; cons. Egyptian Major Cereal Improvement Program, Cairo, 1979—. Contbr. articles to profl. jours. Served with AUS, 1953-56. Mem. Am. Phytopathol. Soc., Mycol. Soc. Am., Agronomy Soc. Am. Methodist. Club: Optimist. Subspecialty: Plant pathology. Current work: Improvement methods and techniques to elucidate physiology of host-parasite relationships and gentics of host-parasite-relationships; fluorescence microscopy studies in relations to fungal viability. Home: 308 W Stadium Ave West Lafayette IN 47906 Office: Dept Botany and Plant Pathology Purdue U West Lafayette IN 47906

WARREN, HOLLAND DOUGLAS, research physicist; b. Wilkes County, N.C., July 31, 1932; s. Henry Harrison and Nannie (Shaver) W.; m. Nancy Wall, May 21, 1955; children: Douglas Alan, Jill Jeneen, Karen Kay. B.S. in Math, Wake Forest U., 1959; M.S. in Physics, U. Va., 1961, Ph.D., 1963. Research assoc. U. Va., summers 1960-62; devel. physicist Celanese Corp., Charlotte, N.C., 1963-64; sr. physicist Babcock & Wilcox Co., Lynchburg, Va., 1964-69, research specialist, 1969—. Contbr. numerous articles to profl. jours. Served with USN, 1951-55. Mem. Am. Phys. Soc., Am. Nuclear Soc., Phi Beta Kappa, Kappa Mu Epsilon. Republican. Baptist. Subspecialties: Nuclear engineering; Nuclear physics. Current work: Theoretical and experimental development of self-powered instrumentation for application inside the cores of nuclear power plants (incore instrumentation). Patentee in field. Office: Babcock & Wilcox Co PO Box 239 Lynchburg VA 24505

WARREN, JAMES VAUGHN, physician, educator; b. Columbus, Ohio, July 1, 1915; s. James Halford and Lucile (Vaughn) W.; m. Gloria Kicklighter, May 27, 1954. B.A., Ohio State U., 1935; M.D., Harvard, 1939; D.Sc., Emory U., 1974. Diplomate: Am. Bd. Internal Medicine (sec.-treas. 1970-71, vice chmn. 1971-72). Med. house officer Peter Bent Brigham Hosp., 1939-41, asst. resident medicine, 1941-42; research fellow medicine Harvard, 1941-42; med. investigator problems of shock and vascular injuries OSRD, 1942-46; instr. medicine Emory U. Med. Sch., 1942-46, asso. prof. medicine, prof. physiology, chmn. dept. physiology, 1947-51, prof. medicine, 1951-52; asst. prof. medicine Yale Med. Sch., 1946-47; prof. medicine Duke Sch. Medicine, 1952-58; prof. medicine, chmn. dept. internal medicine U. Tex. Sch. Medicine, 1958-61; prof. medicine Ohio State U. Med. Sch., 1961—, chmn. dept. medicine, 1961-79; Mem. cardiovascular sect. USPHS, 1952-56; mem. tng. grant com. Nat. Heart Inst., 1961-64; mem. Nat. Acad. Medicine-Inst. Medicine, 1971. Author: Cardiovascular Physiology, 1975; Editorial bd.: jour. Med. Opinion, 1971; Contbg. author: Pre-Eclamptic and Eclamptic Toxemia of Pregnancy, 1941, Methods in Medical Research, vol. VII; Contbr. numerous articles to profl. publs. Recipient Distinguished Alumnus award Duke U. Med. Center, 1970; Columbus Mayor's award for vol. service, 1978. Master A.C.P.; mem. AMA (chmn. sect. internal medicine 1964-65), Assn. Profs. Medicine (pres. 1967-68, Williams disting. chmn. medicine award 1979), Am. Clin. and Climatol. Assn., Soc. Univ. Cardiologists, Am. Heart Assn. (pres. 1962-63, Gold Heart award, James B. Herrick award 1976), Am. Soc. Clin. Investigation, Am. Fedn. Clin. Research (nat. pres. 1952-53), Am. Physiol. Soc., Osler Soc., Soc. Exptl. Biology and Medicine, Assn. Am. Physicians, So., Central socs. clin. research, Sigma Xi, Alpha Omega Alpha. Conglist. Clubs: Cosmos, Explorers, Harvard. Subspecialty: Internal medicine. Home: 5526 Ashford Rd Dublin OH 43017 Office: Ohio State U Hosp Columbus OH 43210

WARREN, JOHN EDWARD, JR., agricultural science educator; b. Lewisburg, W.Va., June 28, 1943; s. John Edward and Virginia (Neely) W.; m. Carol Ann, Aug. 6, 1966; children: Kelly Ann, Mark Edward. B.S. in Agr, W.Va. U., 1964, M.S. in Animal Physiology, 1966; Ph.D. in Reproductive Physiology, U. Md., 1970. Research physiologist Agrl. Research Service, U.S. Dept. Agr., Beltsville, Md., 1970-71; assoc. prof. animal sci. Lincoln U., Jefferson City, Mo., 1971-81, prof., 1981—, leader animal sci. research program. Contbr. articles to profl. publs. Bd. dirs. Meml. Community Hosp. Mem. Am. Soc. Animal Sci., Soc. Study of Reproduction. Club: Jefferson City Kennel. Subspecialty: Animal physiology. Current work: Postpartum female infertility and sperm transport. Office: Lincoln Univ Agr Dept Jefferson City MO 65101

WARREN, KENNETH S., physician; b. N.Y.C., June 11, 1929; m. Sylvia Marjorie Rothwell, Feb. 14, 1959; children: Christopher Harwood, Erica Marjorie. A.B., Harvard U., 1951, M.D., 1955. Intern, Harvard service Boston City Hosp., 1955-56; research asso. Lab. Tropical Diseases, NIH, Bethesda, Md., 1956-62; asst. prof. medicine Case Western Res. U., 1963-68, asso. prof., 1968-75, prof., 1975-77, prof. library sci., 1974-77; dir. health scis. Rockefeller Found., N.Y.C., 1977—; cons. WHO; mem. Inst. Medicine, Nat. Acad. Scis. Author: Schistosomiasis: The Evolution of a Medical Literature. Selected Abstracts and Citations, 1852-1972, 1973, Geographic Medicine for the Practitioner, 1978, Scientific Information Systems and the Principle of Selectivity, 1980, Coping with the Biomedical Literature, 1981, Immunology of Parasitic Diseases, 1983, Tropical and Geographical Medicine, 1983; contbr. numerous articles to profl. jours. Recipient Career Devel. award NIH, 1966-71. Fellow ACP; mem. Am. Soc. Clin. Investigation, Assn. Am. Physicians, Am. Assn. Immunologists, Am. Assn. Study Liver Diseases, Am. Soc. Tropical Medicine and Hygiene (Bailey K. Ashford award 1974), Infectious Diseases Soc. Am. (Squibb award 1975), Royal Soc. Tropical Medicine and Hygiene. Subspecialties: Immunology (medicine); Tropical medicine. Patentee diagnostic methods, drugs. Office: 1133 Ave of Americas New York NY 10036 As a young physician, I chose tropical medicine as a career for two reasons: to travel and to enjoy the freedom of a sparsely populated field of endeavor. Having worked in Brazil, St. Lucia, Kenya, Egypt, Thailand, the Philippines and China, I fulfilled my first goal. My second was also fulfilled, as unfortunately, tropical medicine still remains a frontier field. Based on firsthand experience of the vast amount of human misery caused by the great neglected diseases of mankind, I quickly gained a third goal: to do everything in my power to improve the health of our fellow human beings in the developing world.

WARREN, MARGUERITE QUEEN, psychology educator, researcher; b. Marion, Ohio, Jan. 30, 1920; d. Asa and Hazel Mae (Zieg) Queen; m. Martin Warren, July 31, 1963; children: Laurie Jane, Lisa Grace, Lesley Elaine. B.A., Western Res. U., 1942; M.A., U. Calif.-Berkeley, 1946, Ph.D., 1961. Prin. investigator Community Treatment Project, Sacramento and San Francisco, 1961-67; program dir. Am. Justice Inst., Sacramento, 1967-73; prof. psychology SUNY-Albany, 1972—; pres. Hindelang Criminal Justice Research Ctr., Albany, 1982. Assoc. editor: Jour. Research in Crime and Delinquency, Albany, 1977—; contbr. numerous articles to profl. jours. Mem. Am. Psychol. Assn., Am. Soc. Criminology (v.p. 1982, exec. counselor 1979-82), Am. Assn. Correctional Psychologists (Psychologist of Yr. 1975). Subspecialties: Cognition; Social psychology. Current work: Longitudinal studies of female offenders from adolescence through adulthood; restitution of offenders to victims of their crimes. Home: 32 Willow Dr Delmar NY

WARREN, RICHARD LLOYD, microbiologist; b. Jackson, Miss., June 6, 1947; s. Richard Lloyd and Nellie (Puckett) W.; m. Connie Jo., July 20, 1968; children: Richard, Keith, Shawn Christopher, Elizabeth Anne. B.S., Wright State U., 1969, M.S., 1972; Ph.D., U. Utah, 1974. Lab. technician Wright State U., Dayton, Ohio, 1969-70, teaching asst., 1970-71, research asst., 1971-72, asst. prof. dept. microbiology and immunology, 1979—; teaching asst. U. Utah, Salt Lake City, 1972-73; NIH postdoctoral U. Wis., Madison, 1974-77, research assoc., 1979; sr. research scientist Union Carbide Corp., Tarrytown, N.Y., 1977-79. Reviewer: Infection and Immunity. Asst. scout leader Tecumseh council Boy Scots Am. Served with USAF, 1969-70. Grantee Eli Lilly, 1981-83, Smith, Klein and French, 1981-82. Mem. Am. Soc. Microbiology. Subspecialties: Genetics and genetic engineering (biology); Microbiology (medicine). Current work: Genetics and regulation of proteins secreted by pseudomonas aeruginosa explored by constructing insertion mutation with transposons and recombinant DNA techniques. Home: 1447 Parkman Pl Beavercreek OH 45385 Office: Wright State Univ 409 Oleman Hall Dayton OH 45435

WARREN, WALTER RAYMOND, JR., research company executive; b. N.Y.C., Nov. 25, 1929; s. Walter Raymond and Helen Veronica (McNally) W.; m. Austine Rose Warren, Apr. 24, 1954; children: Michael, Christopher, Walter, Austine, Susan, Richard, Jennifer, John, David, James. B. Aeor. Engring., NYU, 1950; M.S.A.E., Princeton U., 1952, Ph.D. in Aero. Engring, 1957. Grad. asst. Princeton U., 1950-55; lab. leader Lockheed Missile & Space Co., Van Nuys, Calif., 1955-56; mgr. exptl. fluid physics Missile and Space div. Gen. Electric Co., Valley Forge, Pa., 1956-68, mgr., 1968; dir. Aerophysics Lab., Aerospace Corp., El Segundo, Calif., 1968-81; pres. Pacific Applied Research, Palos Verdes, Calif., 1981—; grad. techr. hypersonics U. Pa., 1962-68; mem. fluid dynamics adv. com. NASA, 1970; mem. laser subcom. USAF Sci. Adv. Bd., 1970-75; mem. device com. High Energy Laser Rev. Group, 1975-78. Contbr. articles to profl. jours. Convair scholar NYU, 1949-50; Guggenheim fellow Princeton U., 1952-53; recipient Sci. award Aerospace Corp., 1979. Fellow AIAA (chmn. pasmadynamics tech. com., assoc. editor). Democrat. Roman Catholic. Subspecialties: Fluid mechanics; High energy laser systems. Current work: To conduct and direct applied research studies in the support of the development by the U.S. Department of Defense primarily of high energy laser systems and also, of other aerospace systems. Patentee in field. Home and Office: 6 Crestwind Dr Rancho Palos Verdes CA 90274

WARREN, WAYNE HUTCHINSON, JR., astronomer; b. Newark, Dec. 11, 1940; s. Wayne H. and Grace E. (Klesick) W.; m. Martha Hope, May 7, 1967; children: Kenneth R., Sandra K., Katherine G.E. B.A. in Physics, Fairleigh Dickinson U., 1968; M.A. in Astronomy, Ind. U., 1970, Ph.D., 1975. Nat. Acad. Scis. NRC research assoc. NASA Goddard Space Flight Center, Greenbelt, Md., 1976-77; astronomer Nat. Space Sci. Data Center/World Data Center, 1978—. Editor: Astron. Data Center Bull, 1980—; contbr. articles to profl. jours. Mem. Internat. Astron. Union, Am. Astron. Soc., Astron. Soc. Pacific, Royal Astron. Soc. Subspecialties: Computerized astronomical data; Optical astronomy. Current work: Creation and distribution of astronomical data in machine-readable form. Office: NASA Goddard Space Flight Center Code 601 Greenbelt MD 20771

WARREN, WAYNE LAWRENCE, computer software engineer; b. Jacksonville, Fla., May 30, 1954; s. Lawrence Charles and Elizabeth Jane (Moore) W.; m. Jennifer Anne Littlepage, Dec. 15, 1973; children: Micah Wayne, Jesse David. B.S. in Computer Sci, Mesa Coll., 1976, M.S., Colo. State U., 1978. Software engr. Boeing Computer Services, Seattle, 1978-82; software engr., sec.-treas. SysJen, Inc., Renton, Wash., 1982—. Contbr. articles to profl. jours. Mem. Assn. Computing Machinery. Mem. Christian Ch. Subspecialties: Software engineering; Database systems. Current work: Microprocessor applications, data description and communications, programmer productivity, UNIX, software tools, inter-computer data communications, data dictionary. Home: 26871 172d Pl SE Kent WA 98031 Office: 11717 Rainier Ave S Seattle WA 98178

WARREN-MELTZER, STEPHANIE, psychologist; b. Los Angeles, Apr. 12, 1937; d. Paul and Frances B. (Koltnow) Warren-M. B.A., Duke U., 1957; M.S., U. Pitts., 1967, Ph.D., 1971. Cert. psychoanalyst Am. Inst. Psychoanalysis; registered clin. psychologist, N.J., Pa.; registered psychoanalyst, N.Y. Devel. psychologist Children's Hosp., Pitts., 1969-70; instr. psychology U. Pitts., 1969-70; staff psychologist Mercy Hosp., Pitts., 1971-74; chief psychology St. Joseph's Hosp. and Med. Ctr., Paterson, N.J., 1975-79; sect. chief psychology Inst. Child Devel., Hackensack (N.J.) Hosp., 1982—; adj. psychoanalyst Am. Inst. Psychoanalysis, 1980—; adj. asst. prof. U. Medicine and Dentistry N.J., Newark, 1975—. Bd. dirs., treas. Carnegie-Mellon U. Child Care Ctr., Pitts., 1971-72. NIMH fellow, 1966-67; Scottish Rite Found. grantee, 1970, 71. Mem. Am. Psychol. Assn., Eastern Psychol. Assn., Am. Inst. Psychoanalysis Student Orgn. (pres. 1977-78), Duke U. Alumni Assn., Phi Lambda Theta. Subspecialties: Developmental psychology; Cognition. Current work: Neonatology; sleep. Psychoanalytic—distress management. Ultrasound analysis of human fetal behaviors related to post-natal development and sleep patterns. Application of analytic theory to business stress-distress problems. Office: 49 E 78th St New York NY 10021

WARSHAW, RHODA, psychologist; b. N.Y.C., Nov. 17, 1930; d. Irving and Celia (Edlin) Adelstein; m. Seymour Warshaw, 1950 (div. 1972); children: Lynne, Sheryl, Michael. B.A., Hunter Coll., 1951; M.S., Queens Coll., 1971; Ph.D., Fordham U., 1978. Tchr. Long Beach (N.Y.) Pub. Schs., 1956-61 62-71; sch. psychologist Yonkers (N.Y.) Pub. Sch., 1971-72, Smithtown (N.Y.) Pub. Schs., 1974—; psychotherapist Commack (N.Y.) Cons. Ctr., 1979—. Mem. Nassau County Psychol. Assn., Am. Psychol. Assn., N.Y. State United Tchrs., Nat. Assn. Sch. Psychologists. Subspecialties: Developmental psychology; Learning. Current work: Counseling, also diagnosis, and remediation of learning disorders. Home: 22 Linden St Great Neck NY 11021 Office: Smithtown Public Schs Smithtown NY 11787 Commack Cons Center 154 Commack Rd Commack NY 11725

WARTELL, SUE ANN, biomedical researcher; b. Pottstown, Pa., Oct. 4, 1950; d. Harold Howard and Pearl Marea (Fronheiser) Hartman. B.A. with highest honors and distinction, U. Del., 1972; Ph.D., U. Conn., 1978. Postdoctoral fellow M.S. Hershey (Pa.) Med. Center, 1978-81, postdoctoral scholar, 1981-83, research assoc., 1983—. Mem. Am. Soc. Microbiology., Am. Physiol. Soc. Subspecialties: Cell and tissue culture; Biochemistry (biology). Current work: Regulation of protein metabolism; control of synthesis of cell-specific proteins in various cell types in lung. Home: 102 W Chocolate Ave Hershey PA 17033 Office: Dept Physiolog M S Hershey Med Center Hershey PA 17033

WARTIK, THOMAS, chemist, coll. dean; b. Cin., Oct. 1, 1921; s. Abraham and Lena (Monnes) W.; m. Louise Dreifus, Apr. 8, 1952; children—Nancy, Steven Philip. A.B., U. Cin., 1943; Ph.D., U. Chgo., 1949. Chemist, Manhattan project U. Chgo., 1944-46; faculty Pa. State U., University Park, 1950—, prof. chemistry, 1960—, chmn. dept., 1960-71, dean, 1971—; Vis. scientist U. Calif. Radiation Labs., 1957, 59, 61; cons. in field. Mem. adv. bd. Petroleum Research Fund, Am. Chem. Soc., 1968-71; mem. Fulbright Scholar Selection Com. in Chemistry, 1966-72, chmn., 1969-72; mem. N.Y. State Doctoral Evaluation Com., 1974; chmn. chemistry vis. com. Tex. A. and M. U., 1979—. Contbr. profl. articles to jours.; Editor: Borax to Boranes, 1961. Fellow AAAS; mem. Am. Chem. Soc. (councillor 1967-69), Phi Beta Kappa, Sigma Xi. Club: Lake Glendale Sailing (dir.). Subspeciality: Inorganic chemistry. Home: 939 Ringneck Rd State College PA 16801 Office: 211 Whitmore Lab University Park PA 16802

WARTZOK, DOUGLAS, biological sciences educator; b. Lansing, Mich., May 10, 1942; s. Leonard Gustive and Violette June (Cady) W.; m. Susan Jane Gibson, Aug. 29, 1966. B.A., Andrews U., 1963; M.S., U. Ill., 1965; Ph.D., Johns Hopkins U., 1971. Research physicist Dow Chem. Co., Midland, Mich., summer 1963; teaching and research asst., physics dept. U. Ill., Urbana, 1963-65; postdoctoral fellow Johns Hopkins U., Balt., 1971-72, asst. prof. pathobiology, 1972-78, assoc. prof., 1978-83, dir. div. ecology and behavior, 1979-82; prof., chmn. dept. biol. scis. Purdue U., Ft. Wayne, (Ind.), 1983—; vis. scientist Inst. Marine Sci., U. Alaska, Fairbanks, 1983. Mem. Delta Omega. Subspecialties: Comparative physiology; Psychophysics. Current work: Physiological ecology and behavior of marine mammals emphasizing sensory system and metabolic factors; development of remote sensing techniques for walrus; radio tracking whales. Home: 2210 Springmill Rd Fort Wayne IN 46825 Office: Purdue University Fort Wayne IN 46805

WARWICK, JAMES WALTER, astronomer, educator, consultant; b. Toledo, Ohio, May 22, 1924; s. Walter and Mary Alice (Mauk) W.; m. Constance Bragdon Sawyer, Sept. 6, 1947; m. June Deette Pellillo, Dec. 9, 1966; children: Sarah H., David I., Rachel J., Joel H., J. Walter, Julia D. A.B., Harvard U., 1947, A.M., 1948, Ph.D., 1951. Asst. prof. astronomy Wellesley Coll., 1950-52; research assoc. Harvard U., 1952-55; prof. dept. astrogeophysics U. Colo., Boulder, 1955—. Contbr. articles to profl. jours. Served with U.S. Army, 1943-46. Mem. Am. Astron. Soc., Am. Geophys. Union, math. Assn. Am., AAAS, Internat. Astron. Union, Internat. Sci. Radio Union. Subspecialties: Theoretical astrophysics; Planetary science. Current work: Voyager missions planetary radio astronomy program, application of radio astronomical techniques in geophysics, atmospheric physics, medical imaging.

WARWICK, WARREN JAMES, pediatrics educator, physician; b. Racine, Wis., Jan. 27, 1928; s. Edmund H. and Adeline C. (Holthusen) W.; m. Henrietta Holm, Sept. 13, 1952; children: Marion, Anne. B.A. magna cum laude, St. Olaf Coll., Minn., 1950; M.D., U. Minn., 1954. Intern White Cross Hosp., Columbus, Ohio, 1954-55; resident U. Minn. Hosp., Mpls, 1955-57, mem. staff, 1959—; practice medicine, specializing in pediatrics; dir. pediatric chest clinic U. Minn., Mpls., 1962—, dir., 1962—, prof. pediatrics, 1978—, now head pediatric lung disease, 1962—; mem. Cystic Fibrosis Found. Registry; med. cons. Minn. chpt. Nat. Cystic Fibrosis Found. Mem. ad hoc cm. metabolic newborn screening Minn. Health Dept., 1976—. Served to col. USAR, 1959—. NIH grantee, 1982-87. Mem. Internat. Cystic Fibrosis Assn., Found. Health Care Eval. Subspecialties: Pediatrics; Pulmonary medicine. Current work: Cystic fibrosis care, teaching and research; pediatric pulmonary care and research. Office: Box 184 Mayo Meml Bldg 420 Delaware St SE Minneapolis MN 55455

WARZECHA, LADISLAUS WILLIAM (LAD WARZECHA), electrical aerospace equipment manufacturing corporation executive; b. Cuero, Tex., Jan. 23, 1929; s. Vincent William and Suzie (Dreymala) W.; m. Elinor Ryan, Oct. 9, 1954; children: Janet, Carol, Gary. B.S., U. Tex., 1948. With Gen. Electric Co., since 1948—, mgr. systems engring., Phila., 1956-62, gen. mgr. NASA Houston programs, 1962-73, ground systems dept., 1973-77, space systems ops., 1977-79, gen. mgr. re-entry systems, Phila., 1980—. Contbr. articles to profl. jours. Bd. dirs. United Way, 1975-77. Mem. AIAA, IEEE, Am. Astronautical Soc., Am. Mgmt. Assn. Roman Catholic. Clubs: Waynesborough Country, Pelican Bay Golf and Country. Subspecialties: Systems engineering; Electrical engineering. Home: 1056 Beaumont Rd Berwyn PA 19312 Office: Gen Electric Co 3198 Chestnut St Philadelphia PA 19101

WASHBURN, JOHN GARRETT, medical and dental equipment consulting designer, dental researcher; b. Los Angeles, June 3, 1924; s. John Garrett and Rachel Gwendolyn (Struble) W.; m. Lavina S. Roloson, 1945 (div. 1954). Student, Pasadena (Calif.) City Coll., 1941-42, Coll. St. Thomas, St. Paul, 1942-43, U. So. Calif., 1943-45, UCLA, 1949-51. Pvt. practice med. and dental equipment consulting designer, Alhambra, Calif., 1969-82; project engr. Unitek Corp., Monrovia, Calif., 1968-69; cons. The Birtcher Corp., El Monte, Calif., 1971—. Founder, chmn. Commn. for the Reestablishment of Mountain Lion Hunting in Calif., 1977-82. Served with USNR, 1942-45. Mem. Internat. Assn. Dental Research, Am. Assn. Dental Research. Subspecialties: Biomedical engineering; Dental Materials. Current work: Design electromedical equipment (cardiology, intensive care, electro-surgery); design dental equipment, instruments and materials; dental materials research. Home: 207 S Cordova St Alhambra CA 91801

WASHBURNE, STEPHEN SHEPARD, chemistry educator, consultant; b. Hartford, Conn., Sept. 6, 1942; s. Francis Courtenay and Julia (Hopkins) W.; m. J. Denise Barbour, Dec., 1970 (div. 1977); son: Matthew Courtenay. B.S., Trinity Coll., 1963; Ph.D., MIT, 1968. Asst. prof. chemistry Temple U., 1967-72, assoc. prof., 1973-81, prof., 1982—; cons. chem. cos., 1973—; mem. area com., for Mediterranean countries Council for Internat. Exchange of Scholars, Washington, 1982—. Contbr. articles to sci., tech. jours.; Editor: Portu-Info Jour, 1981—. Fulbright grantee, 1980; NSF, NIH, U.S. Army grantee, 1967-77. Democrat. Episcopalian. Subspecialties: Synthetic chemistry; Organic chemistry. Current work: Organic synthesis with organometallic reagents; new techniques use of organosilicon compounds and reagents in organic synthesis. Home: 3116 W Penn St Philadelphia PA 19129 Office: Dept Chemistry Temple U Philadelphia PA 19129

WASHINGTON, WARREN MORTON, meteorologist; b. Portland, Oreg., Aug. 28, 1936; s. Edwin and Dorothy Grace (Morton) W.; m. LaRae Herring, July 30, 1959 (div. Aug. 1975); children: Teri, Kim, Marc, Tracy; m. Joan Ann, July 3, 1978. B.S. in Physics, Ore. State U., 1958, M.S. in Meteorology, 1960, Ph.D., Pa. State U., 1964. Sr. scientist Nat. Center Atmospheric Research, Boulder, Colo.; affiliate prof. meteorology oceanography U. Mich. at Ann Arbor, 1968-71; mem. Nat. Adv. Com. for Oceans and Atmospheres, 1978—. Contbr. articles to meteorol. jours. Mem. Boulder Human Relations Commn., 1969-71; mem. Gov.'s Sci. Adv. Com., 1975-78. Fellow Am. Meteorol. Soc., AAAS; mem. Am. Geog. Union. Subspecialty: Meteorology. Current work: Computer modeling of earth's climate. Home: 1480 Landis Ct Boulder CO 80303 Office: PO Box 3000 Boulder CO 80307

WASHOM, BYRON JOHN, energy research and development executive; b. Bethesda, Md., Mar. 24, 1949; s. Paul S. and Doris I. (Lee) W. B.S., U. So. Calif., 1971, M.B.A., 1972; postgrad., MIT, 1976. Coordinator Sea Grant Program, U. So. Calif., 1972-75, sr. research assoc., 1975-77; mgr. tech. and policy Fairchild-Stratos, Manhattan Beach, Calif., 1977-80; pres. Advance Corp., El Segundo, Calif., 1980—; cons. UN, 1980-81, U.S. Congress, 1976-77, MIT, 1976-78, Fairchild Industries, 1976-77; testified before Congress, 1977-83; mem. solar research and devel. panel Energy Research Adv. Bd., Washington. Mem. Renewable Energy Inst. (dir.), Solar Energy Industry Assn., Marine Tech. Soc., Soc. Naval Architects and Marine Engrs. (assoc.), Tech. Mktg. Soc. Am. Presbyterian. Subspecialties: Solar energy; Systems engineering. Current work: Parabolic dish Stirling solar electric systems sponsored by Dept. of Energy; established record conversion efficiencies of sunlight to electricity of 29%, 1982. Home: 3616 Manhattan Ave Manhattan Beach CA 90266 Office: Advanco Corp 999 N Sepulveda Blvd Suite 314 El Segundo CA 90245

WASLENCHUK, DENNIS GRANT, geochemical oceanography educator, environmental chemist; b. Calgary, Alta., Can., Aug. 18, 1951; s. Nick and Minnie Ellen (Gardner) W.; m. Lorraine Rose Letourneau, Aug. 26, 1972. B.Sc., U. B.C., 1973; M.Sc., U. Ottawa, 1975; Ph.D., Ga. Inst. Tech., 1977. Research asst. Skidaway Inst. Oceanography, Savannah, Ga., 1975-77; lectr., research asst. Ga. Inst. Tech., Atlanta, 1975-76; asst. prof. U. Conn., Groton, 1977—; dir., chief scientist Aquademia (Cons. Oceanography), New London, Conn., 1979—. Advisor, Marine Commerce and Devel. Com., New London, 1982—. NSF grantee, 1981-83. Mem. Am. Geophys. Union, Geochem. Soc., AAAS. Subspecialties: Oceanography; Geochemistry. Current work: The natural processes and human activities that exert controls on seawater chemistry, especially the influences of industrial waste, aquatic organisms and hydrodynamics on the minor elements of the coastal ocean, lakes and rivers. Home: 50 Center St New London CT 06320 Office: U Conn Avery Point Groton CT 06340

WASS, HANNELORE LINA, psychology educator; b. Heidelberg, Ger., Sept. 12, 1926; came to U.S., 1957; d. Hermann and Mina (Lasch) Kraft; m. Harry H. Sisler, Apr. 1, 1978; 1 son, Brian C. B.A., Tchrs. Coll., Heidelberg, 1951; M.A., U. Mich., 1960, Ph.D., 1968. Supervising tchr. U. Chgo. Lab. Sch., 1960-61, U. Mich. Lab. Sch., Ann Arbor, 1958-60, 63; asst. prof. Eastern Mich. U., Ypsilanti, 1964-69; prof. U. Fla., Gainesville, 1969—. Author: (with others) Death Education - An Annotated Resource Guide, Vol. I, 1980, Helping Children Cope with Death, 1982; editor, co-author: Dying-Facing the Facts, 1975; editor: Death Edn, 1977—. Mem. adv. bd. Compassionate Friends, Gainesville, 1982—, Meml. Soc., Gainesville, 1978—, Hospice, Gainesville, 1982—. EDPA fellow, 1968; Research award U. Fla., 1972; Fla. Dept. Health and Rehab. Services grantee, 1978. Mem. Am. Psychol. Assn., Fourm Death Edn. and Counseling (dir. 1977-80, program dir. 1979), Internat. Workgroup on Death, Dying and Bereavement, Gerontol. Soc., So. Gerontol. Soc. Current work: Children's death concepts and fears, death education for parents, teachers, children, health professionals, evaluation of educational gerontology programs effectiveness, bereavement, hospice work. Home: 6014 NW 54th Way Gainesville FL 32606 Office: U Fla 1418 Norman Hall Gainesville FL 32611

WASS, WALLACE MILTON, veterinarian, educator; b. Lake Park, Iowa, Nov. 19, 1929; s. Arthur C. and Esther M. (Moberg) W.; m. Doreen L., June 31, 1953; children: Karen, Kimberly, Christopher, Kirby. B.S., U. Minn., 1951, D.V.M., 1953, Ph.D., 1961. Gen. practice vet. medicine, Medford, Wis., 1955-58; mem. faculty U. Minn., 1958-64, Iowa State U., Ames, 1964—, now head dept. vet. clin. scis., dir. clin. programs. Contbr. articles to profl. jours. Served to 1st lt. Vet. Corps USAF, 1953-55. Mem. AVMA, Iowa Vet. med. Assn., Bovine Practitioners Assn., Am. Assn. Vet. Clinicians, Am., Coll. Vet. Internal Medicine, Phi Zeta, Alpha Zeta, Gamma Sigma Delta, Phi Kappa Phi. Methodist. Subspecialty: Internal medicine (veterinary medicine). Current work: Metabolic diseases and nutrition. Home: 2166 Ashmore St Ames IA 50010 Office: 1832 VM Iowa State U Ames LA 50010

WASSER, LARRY PAUL, internist, oncologist; b. Moline, Ill., July 27, 1943; s. Paul Earnest and Harriet Lenore (Swan) W.; m. Suzan Jean Wilson, Oct. 18, 1969 (div. 1974). B.A. magna cum laude, Baylor U., 1965; M.D. summa cum laude, Baylor Coll. Medicine, 1969. Diplomate: Am. Bd. Internal Medicine. Intern Presbyterian-St. Luke's Hosp., Chgo., 1969-70, resident in internal medicine, 1970-71; fellow in hematology Baylor Med. Sch., Houston, 1971-72; fellow in oncology U. Va. Hosp., Charlottesville, 1974-75; staff physician hematology and oncology dept. VA Med. Ctr., North Chicago, Ill., 1977—; assoc. prof. medicine U. Health Scis., Chgo. Med. Sch., North Chicago, 1977—. Contbr. sci. articles to profl. publs. Served to lt. comdr. USNR, 1972-74, 75-77. Mem. Am. Soc. Clin. Oncology, Eastern Coop. Oncology Group. Subspecialties: Internal medicine; Chemotherapy. Current work: Study chairman Eastern Cooperative Group Mesothelioma Trials. Office: North Chicago VA Hosp North Chicago IL 60064

WASSERMAN, GAIL A., psychologist, researcher; b. Boston, Feb. 16, 1946; d. Albert David and Janet (Koopetz) Abramson; m. Jeffry H. Gallet, Jan. 17, 1982; 1 dau. by previous marriage, Janet. Student, U. Mich., 1963-65; B.A., U. Pa., 1965-67; M.A., U. Ill., Chgo., 1970; Ph.D., CUNY, 1977. Lic. psychologist, N.Y. Teaching and research fellow U. Ill., 1968-69; instr. Manhattanville Coll., Purchase, N.Y., 1974-77; research assoc., instr. Cornell U. Med. Coll., N.Y.C., 1976-80; research assoc. in pediatrics Coll. Phys. and Surgs., Columbia U., N.Y.C., 1977-81, asst. clin. prof. psychiatry, 1981—; asst. attending psychologist Columbia-Presbyn. Hosp., 1981—; research psychologist N.Y. State Psychiat. Inst., N.Y.C., 1979—; Child fellow Inst. for Behavior Therapy, N.Y.C., 1982; research cons. Ctr. for Psychoanalytic Tng., Columbia U., 1982—; sr. child clin. cons. Inst. for Behavior Therapy, N.Y.C., 1982—; cons. psychologist, bd. dirs. Med. Ctr. Nursery Sch., N.Y.C., 1981—. Contbr. articles to profl. jours. Acting chmn. task force on mental health priorities Am. Jewish Congress, N.Y.C., 1982—. NIMH grantee, 1980—; March of Dimes Found. grantee, 1980—; Easter Seal Research Found. grantee, 1982—. Mem. Soc. for Research in Child Devel., Am. Psychol. Assn., N.Y. State Psychol. Assn., Soc. Developmental and Behavioral Pediatrics. Subspecialty: Developmental psychology. Current work: Research on mother-child interaction, especially potentially deviant groups, physically handicapped, victims of child abuse, etc. Development of assessment procedures to evaluate early child competence, parental style, and parenting adequacy. Office: NY State Psychiat Inst 722 W 168th St New York NY 10032

WASSERMAN, GERALD STEWARD, psychological sciences educator; b. Bklyn., Nov. 22, 1937; s. Julius and Bessie (Weissman) W.; m. Louise J. Mund, June 17, 1962; children: Mark D., Rachel L. B.A., NYU, 1961; Ph.D., MIT, 1965. Predoctoral fellow MIT, Cambridge, 1963-65; postdoctoral fellow NIH, Bethesda, Md., 1965-67; asst. prof. U. Wis., Madison, 1967-70, assoc. prof., 1970-75; prof. dept. psychol. scis. Purdue U., West Lafayette, Ind., 1975—, exec. com. neuroscis. program, 1980-83. Author: Color Vision: An Historical Introduction, 1978; adv. editor: Contemporary Psychology, 1981—; editorial bd.: Color Research and Application, 1977-80; editorial commentator: Brain and Behavioral Scis, 1977—; contbr. articles to profl. jours. 2d v.p. Temple Israel, West Lafayette, Ind., 1980-82, v.p., 1982—. Recipient Grants NIH, 1977—, 1975-78, NSF, 1969-73, 78-80. Fellow Optical Soc. Am; mem. Psychonomic Soc., Soc. Neurosci., Assn. Research in Vision and Ophthalmology. Jewish. Subspecialties: Psychobiology; Sensory processes. Current work: The psychobiology of sensory coding with special attention to coding in natural and artificial sensory receptors. Home: 3512 Capilano Dr West Lafayette IN 47906 Office: Dept Psychol Scis Purdue U West Lafayette IN 47907

WASSERMAN, MARTIN ALLAN, pharmacologist, pharmacist; b. Newark, N.J., Nov. 20, 1941; s. Charles and Betty (Schneider) W.; m. Cheryl Elyse Price, June 22, 1966; children: Dana Beth, Rick Darrin. B.S., Rutgers U., 1963; M.A. in Pharmacology, U. Tex. Med. Br., Galveston, 1971, Ph.D., 1972. Registered pharmacist, N.J., Tex., Mich. Pharmacist Shor's Med. Service Ctr., Elizabeth, N.J., 1964-68; research scientist Upjohn Co., Kalamazoo, Mich., 1972-81; asst. dir. pharmacology SK&F Labs., Phila., 1981-82; assoc. dir. pharmacology Smith Kline & French Labs., Phila., 1982—, 1982—; instr. pharmacology Kalamazoo Valley Community Coll., 1978-81. McLaughlin fellow, 1969-72; recipient Mead-Johnson excellence of research award, 1971. Mem. Am. Soc. Pharmacology and Exptl. Therapeutics, Am. Lung Assn., Sigma Xi. Jewish. Subspecialties: Pharmacology; Physiology (biology). Current work: Research in asthma and other lung diseases, protaglandins, thromboxanes, leukotrienes, calcium antagonists.

WASSMUNDT, FREDERICK WILLIAM, chemistry educator, researcher; b. Oak Park, Ill., Aug. 6, 1932; s. Ferdinand William and Shirley C. (Reiser) W.; m. Elizabeth Ann Tobin, May 27, 1964 (div. 1977); 1 son, Frederick William. B.A., DePauw U., 1953; Ph.D., U. Ill., 1956. Instr. U. Calif.-Berkeley, 1956-58; instr. U. Conn., Storrs, 1958-60, asst. prof., 1960-69, assoc. prof. chemistry, 1969—; cons. New Eng. Research Application Ctr., Storrs, 1975—; treas. 8th Biennial Conf. Chem. Edn., 1982 . Author: Exercises and Preparations in Organic Chemistry, 1977; also articles. Mem. Mansfield (Conn.) Republican Town Com., 1975—, treas., 1978—. Mem. Am. Chem. Soc., Royal Soc. Chemistry, N.Y. Acad. Scis., Phi Lambda Upsilon. Subspecialties: Organic chemistry; Synthetic chemistry. Current work: Studies of reactions of organic nitrogen compounds and their synthetic utility. Office: U Conn Storrs CT 06268

WATANABE, KOUICHI, pharmacologist, educator; b. Manchuria, Japan, Aug. 26, 1942; s. Tetsuya and Mine W.; m.; children: Toshikazu, Yoshihiro, Motohiro; m. Sumiko Abe, Aug. 18, 1977. B.S., Tokyo Coll. Pharmacy, 1966; M.S., Osaka U., 1968; Ph.D., 1971. Vis. fellow reprodn. research br. Nat. Inst. Child Health and Devel., NIH, Bethesda, Md., 1971-73; vis. scientist dept. pharmacology Coll. Medicine, Howard U., Washington, 1973-75, asst. prof., 1975-83; asst. prof. pharmacology U. Hawaii, 1983—. Contbr. articles to sci. jours. Am. Cancer Soc. grantee, 1980-81. Mem. Am. Soc. Pharmacology and Exptl. Therapeutics, N.Y. Acad. Scis. Subspecialties: Chemotherapy; Molecular pharmacology. Current work: Mechanism of action of various antineoplastic agents on calmodulin. Vinca alkaloids found to be calmodulin inhibitors. Suggested that amounts of calmodulin its binding proteins may be endogenous regulators of antineoplastic action or transport of these drugs. Home: 35-4 Sakaecho Kitaku Tokyo Japan Office: 1960 East West Rd Honolulu HI 96822

WATANABE, KYOICHI ALOISIUS, Medical researcher and educator; b. Amagasaki City, Japan, Feb. 28, 1935; s. Yujiro P. and Yoshiko F. W.; m. Kiyoko Agatha (maiden name), Nov. 22, 1962; children: Kanna, Kay, Kenneth, Kim, Kelly, Katherine. B.A., Hokkaido U., Japan, 1958, M.A., 1960, Ph.D. in Organic Chemistry, 1963. Lectr. Sophia U., Tokyo, 1963; research assoc. U. Alta., Can., 1966-68, Sloan Kettering Inst. for Cancer Research, Cornell U., Rye, N.Y., 1963-66, assoc., 1968-72, assoc. mem., 1972-80, mem., 1981--, assoc. prof., 1972-80, prof. pharmacology, 1981--; Mem. medicinal chemistry study sect. NIH Washington, 1981-84. Contbr. over 100 sci. articles to profl. publs. Mem. Am. Chem. Soc., Internat. Heterocyclic Chemistry Congress, Pharm. Soc. Japan. Roman Catholic. Subspecialties: Organic chemistry; Medicinal chemistry. Current work: Chemistry of carbohydrates, heterocycles, nucleosides and related natural products. Patentee in field.

WATANABE, TSUNEO, otolaryngology educator, head and neck surgeon, researcher; b. Tokyo, Mar. 20, 1950; U.S., 1975; s. Morio and Kayo W.; m. Marie-Ann McMahone, Apr. 19, 1980; children: Jennifer, Emily. M.D., Keio U., Tokyo, 1974. Diplomate: Am. Bd. Otolaryngology. Intern St. Louis U., 1975-76; resident in otolaryngology Duke U., 1976-80, asst. prof. otolaryngology, 1981—. Fellow Am. Acad. Otolaryngology. Subspecialties: Otorhinolaryngology; Microsurgery. Current work: Cancer immunology, microsurgery. Office: Cary Ear Nose and Throat Clinic 1151 Kildaire Farm Rd Cary NC 27511

WATERS, DAVID JOHN, molecular virologist; b. San Diego, Nov. 24, 1942; s. John David and Jane (Bates) W.; m. Marjorie, Mar. 17, 1962; children: Jeanne Louise, Marie Annette. B.S., San Diego State U., 1965,66; Ph.D., U. Kans., 1972. NIH trainee, dept. microbiology U. Kans., Lawrence, 1969-72; NIH multiple sclerosis research fellow Wistar Inst., Phila., 1973-74, asst. prof., 1974-78; asst. dir. sci. Mass. Biol. Labs., Boston, 1978—; asst. prof. medicine Tufts U., Boston, 1979—; lectr. dept. med. microbiology and immunology Harvard U. Med. Sch., Boston, 1981—. Contbr. articles to profl. jours. Chmn. Norfolk (Mass.) Bd. Health, 1980—. Mem. Am. Soc. Microbiology, Soc. Gen. Microbiology. Subspecialties: Virology (biology); Molecular biology. Current work: Research on the genetic and molecular biology of varicellazoster, protein fractionation of coagulant proteins from human plasma. Office: Massachusetts Biological Labs 375 South St Boston MA 02130

WATERS, JAMES HERMAN, psychologist; b. New Orleans, Feb. 9, 1946; s. Robert M. and Marcelle R. W. B.A. in Psychology, Swarthmore Coll., 1968, M.A., U. Colo., 1970, Ph.D., 1976. Staff psychologist Mental Health Ctr., Boulder, Colo., 1973-78, dir. intensive treatment services, 1978-80, psychologist research dept. and child, adolescent and family services, 1980—; pvt. practice psychology, 1981—; cons. on services to chronically mentally ill adults on criminal and civil legal issues on patient population, and on rehab. of brain injured patients. Contbr. papers to profl. confs. U. Colo. fellow, 1969-70. Mem. Am. Psychol. Assn., Fedn. Fly Fishermen. Democrat. Subspecialties: Behavioral psychology; Neuropsychology. Current work: Family therapy and individual therapy of children and adults; initiating program to retrain and rehabilitate brain functions in brain-injured population. Home: 1085 Grant Pl Boulder CO 80302 Office: 1333 Iris Ave Boulder CO 80302

WATERS, LARRY CHARLES, biochemical researcher and educator; b. Glennville, Ga., July 1, 1939; s. Joseph Charlie and Mary Mildred (Durrence) W.; m. Marian Yvonne Mullis, Sept. 10, 1960; 1 child, Laurie. B.S., Valdosta State Coll., 1961; M.S., U. Ga., 1964, Ph.D., 1965. Research scientist Oak Ridge Nat. Lab., 1967—; lectr. U. Tenn.-Knoxville, 1967—. Am. Cancer Soc. postdoctoral fellow, 1965-67. Mem. AAAS, Am. Soc. Biol. Chemists. Baptist. Club: Oak Ridge Country. Subspecialties: Biochemistry (medicine); Molecular biology. Current work: Biochemical mechanisms of mutagenesis. toxicology, cancer. Home: 20 Windhaven Ln Oak Ridge TN 37830 Office: Oak Ridge Nat Lab Biology Div PO Box Y Oak Ridge TN 37830

WATERWORTH, HOWARD EUGENE, plant pathology and nematology research administrator; b. Randolph, Wis., Sept. 3, 1936; s. Merlin I. and Dorothy A. (Wrede) W.; m. Pamela Dick, July 16, 1978; children: Mary E., Allison R., Rebeccah A. B.S. in Agrl. Edn, U. Wis.-Madison, 1958, Ph.D. in Plant Pathology, 1962. Research pathologist in virus diseases of fruits and ornamentals Dept. Agr., Glenn Dale, Md., 1964-82; chief plant germplasm resources lab. Agrl. Research Ctr., Beltsville, Md., 1980-82; nat. program leader for plant pathology-nematology Agrl. Research Service, Beltsville, 1982—. Contbr. numerous articles on viruses and virus diseases of plants to profl. jours., articles on plant germplasm and quarantines to popular mags., chpts. to books; editor: Virus Diseases of Pome Fruits. Pres. Craft Orgns., Washington, 1976-78, treas., 1974-76; pres. Musical Choral Groups, Bowie, Md., 1972-82, treas., 1972-75; v.p. Civic Assn., Seabrook, Md., 1970-71. Served with Mil. Police Corps U.S. Army, 1962-64; Berlin. Mem. Am. Phytopathol. Soc., Council Agrl. Sci. and Tech., Soc. Nematology, Hist. Auto Club, Alpha Zeta. Subspecialties: Plant pathology; Nematology. Current work: Administration of research on plant pathology and nematology. Office: Dept Agr Research Ctr Room 230 Bldg 005 Beltsville MD 20705

WATKINS, DAVID HYDER, surgeon; b. Denver, Nov. 26, 1917; s. David Milroy and Mary Rose (Hyder) W.; m. Lucile Maxine Pingel, Sept. 27, 1941; children: John David Hyder, Bryan David Pingel. A.B., U. Colo., 1937, M.D., 1940; M.S. in Surgery, U. Minn., 1947, Ph.D., 1949. Diplomate: Am. Bd. Surgery, Am. Bd. Thoracic Surgery. Intern U. Iowa Hosps., Iowa City, 1940-41; resident Mayo Clinic, Rochester, Minn., 1942-44, asst. surg. staff, 1945-49; instr. surgery Ohio State U., Columbus, 1949-50; asso. prof. surgery U. Colo., Denver, 1951-56, prof., 1956-67; practice medicine specializing in surgery, Des Moines, 1967—; mem. staff Iowa Meth Hosp., Broadlawns Hosp., VA Hosp.; clin. prof. surgery U. Iowa, Iowa City, 1967—. Contbr. articles in field to med. jours. Fellow A.C.S.; mem. Am. Assn. for Thoracic Surgery, Central, Western surg. assns., S.W. Surg. Congress, Soc. Univ. surgeons, Societe Internationale de Chirurgie, Am. Heart Assn., Am. Coll. Cardiology, Am. Coll. Chest Physicians, Am. Fedn. for Clin. Research, Am. Geriatrics Soc., Am. Soc. for Artificial Internal Organs, Priestley Soc., S.R., Phi Beta Kappa, Sigma Xi, Alpha Omega Alpha. Club: Univ. (Denver). Subspecialties: Surgery; Artificial organs. Current work: Development of acute and subactue devices for assisted circulation. Home: 6039 N Waterbury Rd Des Moines IA 50312 Office: 1200 Pleasant St Des Moines IA 50308 The historical imperative is as valid in science as in politics. Knowledge of the past is essential to predict the future.

WATKINS, ELTON, JR., biomedical research administrator, surgeon; b. Portland, Oreg., Aug. 16, 1921; s. Elton and Daniela Ruth (Sturges) W.; m.; children: Elton III, Sturges Benjamin. B.A., Reed Coll., 1941; M.D., U. Oreg., 1944. Diplomate: Am. Bd. Surgery, Am. Bd. Thoracic Surgery. Rotating intern U. Oreg. Med. Sch. Hosps. and Clinics, 1944-45, asst. resident in thoracic surgery, 1945-47, chief resident in thoracic surgery, 1947-48, dir., 1948-49, asst. resident in gen. surgery, 1949-50, chief resident in gen. surgery, 1950-51; instr. in physiology U. Oreg., 1948-49; research fellow Children's Hosp. Med. Ctr., Boston, 1951-53, sr. staff surgeon, 1956-75, Lahey Clinic Found., Burlington, Mass., 1957—, chmn. div. research, 1964—; asst. in surgery Harvard U. Med. Sch., 1953-54, instr. in surgery, 1954-57, asst. clin. prof., 1957, lectr., 1979—; mem. staff New Eng. Deaconess Hosp., Boston, New Eng. Baptist Hosp.; lectr. First Brit. Acad. Conf. in Otolaryngology, Royal Coll. Surgeons, London, 1963; ann. orator Danish Surg. Soc., Copenhagen, 1967. Contbr. numerous articles to profl. jours. Chmn. bd. dirs. Pub. Responsibility in Medicine and Research, Inc., Boston, 1977-79. Served to comdr. M.C. USNR, 1954-56. Nat. Cancer Inst. grantee, 1975-81; Nat. Inst. Heart, Lung and Blood grantee, 1957-75. Fellow ACS, Am. Heart Assn. Council Cardiovascular Surgery; mem. Am. Assn. Thoracic Surgery, Am. Assn. Cancer Research, New Eng. Surg. Soc., Soc. Vascular Surgery, Mass. Med. Soc., AMA, AAAS, Am. Fedn. Clin. Research, Am. Soc. Clin. Oncology, Soc. Surg. Oncology, Transplantation Soc. (charter), Tissue Culture Assn., Am. Assn. Med. Instrumentation, Tumor Registrars Assn. New Eng. (hon.), Societe Internationale de Chirurgie, Brit. Assn. Surg. Oncology (fgn. corr.), Brit. Assn. Cancer Research, Societa' Italiana di Terapie dei Tumori (fgn. corr.), Sociedad de Cirugia de Uruguay (fgn. corr.), Sociedad de Cirujanos del Chile (hon.), Sociedad Chilena de Cancerologia (hon.), Sigma Xi. Democrat. Episcopalian. Clubs: Harvard (Boston); Cosmos (Washington). Current work: Cancer immunology research, surgical research, human research risks.

Home: 16 Sagamore Way Waltham MA 02154 Office: Div Research Lahey Clinic Med Center 41 Mall Rd Box 541 Burlington MA 01805

WATKINS, LINDA MAY, neurophysiologist, educator; b. Norfolk, Va., June 29, 1954; d. Martin and Marion (May) Rothblum; m. Jeffrey Walter Watkins, Apr. 16, 1976. B.S. in Psychology and Biology, Va. Poly. Inst. and State Univ., 1976; postgrad, UCLA, 1976-77; Ph.D. in Physiology, Med. Coll. of Va., 1980. Postdoctoral fellow and instr. dept. physiology Med. Coll. Va., Richmond, 1981-82, asst. prof., 1982-83; asst. research neurophysiologist U. Calif.-Davis, 1983—. Contbr. articles on neurophysiology to profl. jours. NSF fellow, 1976, 78, 79; United Nuclear fellow, 1980; recipient Riese Neuroanatomy award 1979. Mem. Internat. Assn. for Study of Pain, Am. Pain Soc., Soc. for Neuroscis., Phi Kappa Phi. Subspecialties: Neurophysiology; Sensory processes. Current work: Neuroanatomical, neurochemical and behavioral investigations of endogenous pain inhibitory systems, studies of analgesia produced by morphine, brain stimulation, and environmental stimuli. Office: Animal Physiology U Calif Davis CA 95616

WATKINS, PAUL ROGER, accounting educator; b. Cedar City, Utah, Jan. 14, 1944; s. David Crockett and Ella (Sagers) W.; m. Sheila Victoria Cunningham, June 22, 1972; children: Jennifer, Kimberly, Michael, Jeremy. B.S., Ariz. State U., 1974, M.B.A., 1975, Ph.D., 1980. Systems cons. A-M Internat., Phoenix, 1968-73; v.p., controller O. Sullivan Woodside, Phoenix, 1973-76; research asst. Ariz. State U., Tempe, 1976-77, faculty assoc., 1977-78; asst. prof. dept. Sch. of Acctg. U. So. Calif., Los Angeles, 1978—. Mem. Am. Acctg. Assn., Am. Inst. Decision Scis., Inst. Internat. Forecasters, Inst. Mgmt. Scientists, Inst. Mgmt. Accts., Beta Gamma Sigma, Theta Tau. Republican. Mormon. Subspecialties: Cognition; Information systems, storage, and retrieval (computer science). Current work: Research in human judgement and decision making also info. systems, expert systems and artificial intelligence. Home: 31871 Via Puntero San Juan Capistrano CA 92675 Office: University of Southern California Grad School Business Los Angeles CA 90089

WATKINS, SALLIE ANN, physicist, educator; b. Jacksonville, Fla., June 27, 1922; d. Howard Rice and Mary (O'Brien) W. B.S. with honors (honors scholar), Notre Dame Coll., Cleve., 1945; M.S., Cath. U. Am., Washington, 1954, Ph.D., 1957. High sch. tchr., Cleve., 1945-50; prof. physics Notre Dame Coll., Cleve., 1950-66, U. So. Colo., Pueblo, 1966—; trustee Colo. Associated Univ. Press; cons. Harper & Row, Burgess, Ednl. Testing Service. Contbr. articles profl. jours. Mem. Am. Phys. Soc., Am. Assn. Physics Tchrs., Am. Women in Sci., AAUW, Mensa, Pueblo LWV, Sigma Xi. Democrat. Roman Catholic. Subspecialties: Acoustics; Solar physics. Current work: Development and testing of courses in acoustics for speech pathology students, solar physics for non-science students; research in history of physics. Home: 1081 S Lynx Dr PuebloWest CO 81007 Office: U So Colo Pueblo CO 81001

WATRAS, RONALD EDWARD, chemistry educator; b. Bayonne, N.J., Feb. 3, 1943; s. Edward T. and Mae T. (Bonczek) W.; m. Margaret M. Korneluk, July 9, 1966; children: Mary Lynn, Sandra Dawn, Rhonda. B.S., No. Mich. U., 1968; M.S. in Chemistry, U. Ariz., 1972, M.Ed., 1973; D.A. in Chemistry, U. No. Colo., 1979. Cancer researcher E.R. Squibbs, New Brunswick, N.J., 1963-64; commd. 2d lt. U.S. Air Force, 1968, advanced through grades lt. col., 1983; missile combat crew comdr., 1968-74; asst. prof. chemistry U.S. Air Force Acad., Colorado Springs, Colo., 1974-79, assoc. prof., 1979-80; assoc. prof. chemistry U.S. Naval Acad., Annapolis, 1981—; dir. State of Colo. Dept. Energy, Citizens Workshops on Energy, 1978-80. Mem. Am. Chem. Soc., Nat. Sci. Tchrs. Assn., Md. Sci. Tchrs. Assn., Soc. Coll. Sci. Tchrs., Internat. Union Pure and Applied Chemistry (Md. rep. chem. edn.). Roman Catholic. Subspecialties: Chemical education; Energy education. Current work: Chemical and energy education, chemical demonstrations. Office: Dept Chemistry US Naval Academy Annapolis MD 21402

WATSON, CHARLES SCHOFF, research scientist; b. Chgo., Aug. 16, 1932; s. Charles Burton Piatt and Anna Mary (Schoff) Frazer; m. Betty Unger; children: Ann, Katharine, Mary, Elizabeth. A.B., Ind. U., 1958, Ph.D., 1963. Research asst. Ind. U., 1959-61; asst. prof. exptl. psychology U. Tex., 1962-65; prof. Washington U., St. Louis, 1966-76; research dir. Boys Town Nat. Inst., Omaha, 1977-83; prof., chmn. dept. speech and hearing scis. Ind. U., Bloomington, 1983—. Contbr. articles to profl. jours. Served with U.S. Navy, 1951-54. NIH grantee, 1977, 78, 81, 83. Fellow Am. Psychol. Assn., Acoustical Soc. Am. (chmn. tech. com. on psycho. acoustics); mem. Am. Speech, Hearing and Lang. Assn., Soc. Math. Psychology, Psychonomic Soc., Cognitive Sci. Soc. Subspecialties: Sensory processes; Psychophysics. Current work: Research on hearing and deafness; perception of complex sounds; psychophysical methods. Home: 2384 Winding Brook Circle Bloomington IN 47401 Office: Ind U Speech and Hearing Ctr Bloomington IN 47405

WATSON, DARRELL GENE, chemistry educator; b. Quannah, Tex., Aug. 29, 1947; s. Ralph Walter and Bessie Fay (Jones) W.; m. Patricia Ann McAlister, Aug. 30, 1968; children: Michael, Ronald, Clyde Allen. B.S., Sul Ross State U., 1969; Ph.D., Tex. A&M U., 1975. Assoc. prof. chemistry U. Mary Hardin Baylor, Belton, Tex., 1976—, Jeffy and Agnes McBryde Ellis prof., 1982. Robert A. Welch Found. grantee, 1979—. Mem. Tex. Acad. Sci., Tex. Assn. Health Professions Advisors, Am. Chem. Soc. (reporter Heart O'Tex. sect. 1977, sec.-treas. 1978—). Subspecialties: Photochemistry; Organic chemistry. Home: 801 N Pearl St Belton TX 76513 Office: Dept Chemistry U Mary Hardin Baylor Belton TX 76513

WATSON, DEBORAH KAY, physics educator; b. Mt. Vernon, Ohio, Nov. 27, 1950; d. Miles Bert and Ruth Marie (Garber) W.; m. John Bartolett Frick, May 19, 1979. Ph.D. in Chemistry, Harvard U., 1977. Postdoctoral fellow Calif. Inst. Tech., Pasadena, 1977-80; research scientist Aerospace Corp., Los Angeles, 1980-81; asst. prof. physics U. Okla., Norman, 1981—; cons. Livermore Nat. Lab. Contbr. articles to profl. jours. Research Corp. grantee, 1982. Mem. Am. Phys. Soc. Subspecialty: Atomic and molecular physics. Current work: Interaction of electrons with atoms, ions and molecules. Home: 409 NW 19th St Oklahoma City OK 73103 Office: U Okla 440 W Brooks Norman OK 73019

WATSON, DONALD CHARLES, cardiovascular surgeon; b. Fairfield, Ohio, Mar. 15, 1945; s. Donald Charles and Priscilla (Hirons) W.; m. Susan Prince., June 23, 1973; children: Moira Huntington, Katherine Anne, Kirsten Prince. B.S., B.A., Lehigh U., 1968; M.S., Stanford U., 1969; M.D., Duke U., 1972. Diplomate: Am. Bd. Surgery, Am. Bd. Thoracic Surgery, Nat. Bd. Med. Examiners. Resident Stanford U. Med. Ctr., 1972-1974, 76-78, chief resident, 1978-80; clin. assoc., acting attending Nat. Heart and Lung Inst., NIH, Bethesda, Md., 1974-76; cardiovascular surgeon Children's Hosp., Washington, 1980—; asst. prof. surgery George Washington U., 1980—. Served to lt. comdr. USPHS, 1974-76. NSF fellow, 1968-69; Smith Kline & French fellow, 1968. Fellow Am. Coll. Chest Physicians, Am. Coll. Cardiology; mem. A.C.S., Assn. for Acad. Surgery, Am. Fedn. Clin. Research. Republican. Presbyterian. Subspecialties: Cardiac surgery; Transplant surgery. Current work: Surgical treatment of pediatric cardiovascular diseases. Preservation and transplantation of the mammalian heart. Home: 10908 Roundtable Ct Rockville MD 20852 Office: Children's Nat Med Center 111 Michigan Ave NW Washington DC 20010

WATSON, DONALD RALPH, architect, educator, writer; b. Providence, Sept. 27, 1937; s. Ralph Giles and Ethel Mae (Fletcher) W.; m. Marja Helena Watson, Sept. 9, 1966 (div. 1984); children: John [illegible] A.B., Yale U., [illegible] M.E.D., [illegible]. Architect Peace Corps, 1962-64; cons. Govt. of Tunisia, 1964-65; prin., pvt. practice as architect, Branford, Conn., 1967—; vis. prof. Yale Sch. Architecture; cons. UN, World Bank, other internat. orgns. Author Designing and Building a Solar House, 1977, Energy Conservation Through Building Design, 1979, Climatic Design, 1983; editor: Advances in Solar Energy, 1982—. Bd. dirs. Save the Children Fedn., 1979-82. AMAX/ACSA research fellow, 1967-69; Rockefeller Found. fellow, 1978; recipient various prizes in architecture. Fellow AIA; mem. Am. Solar Energy Soc. Subspecialty: Solar energy. Current work: Building energy conservation, solar building design. Office: PO Box 401 Guilford CT 06437

WATSON, EILEEN LORRAINE, pharmacologist, educator; b. Newark, Jan. 24, 1942; d. Joseph C. and Victoria (Kaminsky) W. B.S. in Pharmacy, Rutgers, U., 1963; Ph.D. in Pharmacology, U. Utah, 1970. Registered pharmacist, N.J., Utah, Wash. NIH postdoctoral fellow U. Wash., Seattle, 1970-72, research assoc., 1972-74, research asst. prof. pharmacology, 1974—; pharm. cons. to dental firm, 1982—. Contbr. articles to profl. jours., also monographs. Speaker community poison control program. Wash. State Heart Assn. grantee, 1971-75; Center for Research in Oral Biology grantee, 1978-80; NIH grantee, 1980—; Cystic Fibrosis Found. grantee, 1980-82. Mem. Internat. Assn. for Dental Research, Western Pharmacology Soc., Am. Soc. Pharmacology and Exptl. Therapeutics, N.Y. Acad. Scis., Sigma Xi, Rho Chi. Subspecialties: Pharmacology; Cellular pharmacology. Current work: Cellular mechanisms involved in the regulation of secretory events; development of animal models for the study of cellular events in cystic fibrosis. Home: 2411 161st St SE Bellevue WA 98008 Office: U Wash Pacific Ave Seattle WA 98195

WATSON, JAMES DEWEY, educator, molecular biologist; b. Chgo., Apr. 6, 1928; s. James Dewey and Jean (Mitchell) W.; m. Elizabeth Lewis, 1968; children: Rufus Robert, Duncan James. B.S., U. Chgo., 1947, D.Sc., 1961; Ph.D., Ind. U., 1950; D.Sc., 1963; LL.D., U. Notre Dame, 1965; D.Sc., L.I. U., 1970, Adelphi U., 1972, Brandeis U., 1973, Albert Einstein Coll. Medicine, 1974, Hofstra U., 1976, Harvard U., 1978, Rockefeller U., 1980, Clarkson Coll., 1981, SUNY, 1983. Research fellow NRC, U. Copenhagen, 1950-51; Nat. Found. Infantile Paralysis fellow Cavendish Lab., Cambridge U., 1951-53; sr. research fellow biology Calif. Inst. Tech., 1953-55; asst. prof. biology Harvard, 1955-58, asso. prof., 1958-61, prof., 1961-76; dir. Cold Spring Harbor Lab., 1968—. Author: Molecular Biology of the Gene, 1965, 2d edit., 1970, 3d edit., 1976, The Double Helix, 1968, (with John Tooze) The DNA Story, 1981, (with others) The Molecular Biology of the Cell, 1983. Hon. fellow Clare Coll., Cambridge U.; Recipient (with F. H. C. Crick) John Collins Warren prize Mass. Gen. Hosp., 1959; Eli Lilly award in biochemistry Am. Chem. Soc., 1959; Albert Lasker prize Am. Pub. Health Assn., 1960; with F.H.C. Crick Research Corp. prize, 1962; with F.H.C. Crick and M.H.F. Wilkins Nobel prize in medicine, 1962; Presdl. medal of freedom, 1977. Mem. Royal Soc. (London), Nat. Acad. Scis. (Carty medal 1971), Am. Philos. Soc., Danish Acad. Arts and Scis., Am. Assn. Cancer Research, Am. Acad. Arts and Sci., Am. Soc. Biol. Chemists. Subspecialty: Molecular biology. Home: Bungtown Rd Cold Spring Harbor NY 11724

WATSON, JAMES FREDERIC, energy research and development co. materials and chemistry administrator; b. Port Huron, Mich., Aug. 26, 1931; s. Norman Frederic and Christina (Carr) W.; m. Frances Joanne Brown, Jan. 28, 1952 (div. Apr. 1971); children: Russell, Timothy, Daniel; m. Linda Mae Johns, May 21, 1977. B.S., U. Mich., 1953, M.S., 1956, Ph.D., 1958. Registered profl. metall. engr., Calif. Staff scientist Gen. Dynamics, San Diego, 1958-62; dir. materials and chemistry div. GA Technols. Inc., San Diego, 1962—. Contbr. articles to profl. publs. Served as sgt. U.S. Army., 1953-55. Fellow Am. Soc. Metals; mem. Am. Nuclear Soc. Republican. Presbyterian. Subspecialties: High-temperature materials; Clad metals and coating technology. Current work: Management of research and development in: nuclear fuels; structural materials; radiochemistry; thermochemical watersplitting; nuclear waste disposal; welding; corrosion; creep-fatigue; friction and wear; chemical vapor deposition of pyrocarbon and SiC; Sol-Gel; stress-rupture. Patentee nuclear fuels. Office: GA Technologies Inc 10955 John Jay Hopkins Dr San Diego CA 92121

WATSON, KENNETH FREDRICK, biochemist, educator; b. Pasco, Wash., Feb. 17, 1942; s. Walter I. and Isabel D. (Frost) W.; m. Janice P. Wilson, June 6, 1964; children: Heidi M., Julie M. B.A., Northwest Nazarene Coll., 1964; Ph.D., Oreg. State U., 1969. NIH postdoctoral fellow Columbia U., N.Y.C., 1969-71; instr. dept. human genetics and devel. (Coll. Physicians and Surgeons), 1971-72; Internat. Agy. Research on Cancer Research fellow Robert Koch Inst., Berlin, W. Germ., 1972-73; asst. prof. Chemistry U. Mont., Missoula, 1973-77, assoc. prof., 1977-82, prof., 1982—; cons. recombinant DNA tech. Author research papers. Am. Cancer Soc. faculty research award and grantee, 1976—; NIH research grantee, 1974—. Mem. Am. Soc. Microbiology, Northwest Nazarene Coll. Alumni Assn. (bd. dirs. 1974-80, bd. regents, pres. 1980-83), Sigma Xi. Mem. Ch. of Nazarene. Subspecialties: Molecular biology; Virology (biology). Current work: Retrovirus replication; gene synthesis by reverse transcription; role of protein phosphorylation in retrovirus life cycle; mechanism of retrovirus RNA-directed DNA synthesis. Office: Chemistry Dept U Mont Missoula MT 59812

WATSON, WINSOR HAYS, III, neurobiologist, educator; b. New Rochelle, N.Y., Oct. 10, 1950; s. Winsor Hays and Janet Loretta (Burns) W.; m. Suzanne P. Lucas, July 19, 1980. B.A., Wesleyan U., 1972; Ph.D., U. Mass., Amherst, 1978. Research asst. Children's Hosp., Boston, 1972-74; teaching asst. U. Mass., 1974-75, research assoc., 1975-78; asst. prof. dept. zoology U. N.H., Durham, 1978—; Grass fellow Marine Biol. Lab., Woods Hole, 1978, assoc. dir., summer 1979. Contbr. articles to profl. jours. NSF grantee, 1979-82. Mem. Soc. Neuroscience, Am. Soc. Zoologists. Subspecialty: Neurobiology. Current work: Neuropeptides and biogenic amines; neural basis of behavior; invertebrate neurobiology. Office: Dept Zoology U New Hampshire Durham NH 03824

WATTENDORF, FRANK LESLIE, engineer; b. Boston, May 23, 1906; s. Frank Michael and Helen Ruth (Hurley) W.; m. Glenn Rogers, Jan. 11, 1941; 1 son, Roger Frank. A.B., Harvard U., 1926; M.S., M.I.T., 1928; Ph.D., Calif. Inst. Tech., 1933. Cert. profl. engr., Ohio. Guggenheim research fellow Calif. Inst. Tech., Pasadena, 1930-34; chief research engr. Metropolitan Water Dist., Hydraulic Research Labs., 1934-36; prof. aero. engring. Peiping (China) U., 1936-38; dir. wind tunnels Wright Field, Dayton, Ohio, 1939-45; civilian chmn. planning group Air Engring. Devel. Center (now Arnold Engring. Devel. Center), Tullahoma, Tenn., 1945-50, dep. chief sci. adv., 1950-52; dir. adv. group Aero. Research & Devel., NATO, Paris, 1952-63, vice chmn., 1963-68, hon. vice chmn., 1968—; cons. in field. Fellow AIAA, Royal Aero. Soc.(Eng.); mem. Internat. Council Aero. Scis. (founder), ASME, Am. Phys. Soc., Air Force Assn., Assn. Francais Technique de l'Aeronautique et de l'Espace(France), Deutsche Gesellschaft fur Luft und Raumfahrt(Ger.), Gen. Assembly Von Karman Inst.(Belgium). Subspecialty: Aerospace engineering and technology. Current work: Aerospace engineering and technology. Co-inventor Multi-component Gas Compressor, 1982; designer China's first large wind tunnel. Home: 3005 P St NW Washington DC 20007

WATTERS, KENNETH LYNN, chemistry educator, University official; b. Iowa City, Iowa, Jan. 21, 1939; s. Robert L. and Elizabeth (Chapman) W.; m. Mary-Lynn Robinson, Aug. 18, 1964; children: Matthew S., Geoffrey B., Robert S. B.S., U. Ill., 1962; Ph.D., Brown U., 1970. Postdoctoral research in chemistry SUNY-Buffalo, 1969-70; asst. prof., then assoc. prof. chemistry U. Wis.-Milw., 1970-82, assoc. dean letters and scis., 1977-82, asst. to vice chancellor, 1982—. Contbr. articles on profl. publs. Grantee Petroleum Research Fund/Am. Chem. Soc., 1972-78, NSF, 1978-81, NIH, 1974-80. Mem. Am. Chem. Soc. Unitarian/Universalist. Subspecialties: Catalysis chemistry; Surface chemistry. Current work: Inorganic surface chemistry and catalysis; infrared and laser raman spectroscopy; metal carbonyl chemistry. Home: 5059 N Hollywood Ave Milwaukee WI 53217 Office: Dept Chemistry U Wis Milwaukee WI 53217

WATTS, MALCOLM S.M., physician, medical educator; b. N.Y.C., Apr. 30, 1915; s. Malcolm S.M. and Elizabeth (Forbes) W.; m. Genevieve Moffitt, July 12, 1947; children: Pauline, Elizabeth, Malcolm, James. A.B., Harvard U., 1937; M.D., 1941. Diplomate: Pan Am. Med. Assn. Group practice internal medicine, San Francisco, 1948-76; clin. prof. medicine U. Calif. Sch. Medicine, 1972—, assoc. dean, 1966—, dir. Extended Programs in Med. Edn., 1973-82; dir. Calif. Statewide Area Health Edn. System, 1979—; chmn. bd. trustees San Francisco Consortium, 1968-74, trustee, 1974-80, exec. dir., 1981—; dir. Soc. Med. Coll. Dirs. Continuing Med. Edn., 1975-82, pres., 1980-81; trustee Hospice of San Francisco, v.p., 1979—; pres. Alliance Continuing Med. Edn., 1979-81. Editor: Western Jour. Medicine. Served to capt. M.C. AUS, 1942-46. Recipient Outstanding Community Leadership citation United Community Funds and Councils Am., 1964, U. Calif. San Francisco medal, 1983. Fellow A.C.P., Am. Coll. Hosp. Adminstrs. (hon.); mem. AMA, Calif. Acad. Scis., Calif. Acad. Medicine, AAAS, Am. Med. Writers Assn., San Francisco Med. Soc. (pres. 1961), Am. Soc. Internal Medicine (pres. 1964-65), Calif. Med. Assn. (dir. 1962—), Nat. Inst. Medicine, Soc. Med. Friends Wine, Council Biology Editors, Am. Acad. Polit. and Social Sci., Am. Acad. Polit. Sci., Academia Mexicana de Ciencias Mexicano de Cultura (corr.). Subspecialties: Internal medicine; Information systems, storage, and retrieval (computer science). Current work: Use of computer technology to assess physician practice performance in comparison with peers. Home: 270 Sea Cliff Ave San Francisco CA 94121 Office: U Calif Sch Medicine San Francisco CA 94143

WATZLAWICK, PAUL, psychotherapist, researcher; b. Villach, Austria, July 25, 1921; came to U.S., 1960; s. Paul and Emy (Casari) W. Ph.D., U. Venice, Italy, 1949; Diploma, C. G. Jung-Inst. for Analytical Psychology, Zurich, Switzerland, 1954. Lic. clin. psychologist, Calif. Prof. U. El Salvador, 1957-59; research assoc. Temple U. Med. Ctr., 1960, Mental Research Inst., Palo Alto, Calif., 1960—; clin. assoc. prof. psychiatry Stanford U. Med. Ctr., 1976—; numerous guest lectureships, N. and S. Am., Europe, 1964—. Author: (with Beavin and Jackson) books, including Pragmatics of Human Communication, 1967, (with Weakland and Fisch) Change, 1974, How Real is Real, 1976, The Language of Change, 1978; contbr.: numerous articles to profl. jours. The Language of Change; contbg. editor: books The Language of Change; adv. editor: profl. jours. The Language of Change, U.S., Germany, Gt. Britain, France, Switzerland. Mem. Am. Psychol. Assn., Internat. Soc. Hypnosis, PEN Club Austria, PEN Club Liechtenstein, Authors League Am. Subspecialties: Psychiatry; Social psychology. Current work: Brief psychotherapy; pathologies of large social systems; processes of reality construction. Office: Mental Research Inst 555 Middlefield Rd Palo Alto CA 94301

WAUGH, JOHN STEWART, chemist, educator; b. Willimantic, Conn., Apr. 25, 1929; s. Albert E. and Edith (Stewart) W.; m.; children: Alice Collier, Frederick Pierce. A.B., Dartmouth, 1949; Ph.D., Cal. Inst. Tech., 1953. Research fellow physics Cal. Inst. Tech., 1952-53; mem. faculty Mass. Inst. Tech., 1953—, prof. chemistry, 1962—, Albert Amos Noyes prof., 1973—; Vis. prof. U. Cal. at Berkeley, 1963-64; lectr. Robert Welch Found., 1968; Falk-Plaut lectr. Columbia, 1973; DuPont lectr. U. S.C., 1974; Lucy Pickett lectr. Mt. Holyoke Coll., 1978; Reilly lectr. U. Notre Dame, 1978; Spedding lectr. Iowa State U., 1979; McElvain lectr. U. Wis., 1981; Vaughan lectr. Rocky Mountain Conf., 1981; G.N. Lewis meml. lectr. U. Calif., 1982; sr. fellow Alexander von Humboldt-Stiftung; also vis. prof. Max Planck Inst., Heidelberg, 1972; mem. chemistry adv. panel NSF, 1966-69, vice chmn., 1968-69; mem. rev. com. Argonne Nat. Lab., 1970-74; mem. sci. and edn. adv. com. Lawrence Berkeley Lab., 1980—; exchange visitor USSR Acad. Scis., 1962, 75; mem. vis. com. Tufts U., 1966-69, Princeton, 1973-78; mem. fellowship com. Alfred P. Sloan Found., 1977-82. Author: New NMR Methods in Solid State Physics, 1978; Editor: Advances in Magnetic Resonance, 1965—; assoc. editor: Jour. Chem. Physics, 1965-67, Spectrochimica Acta, 1964-78; editorial bd.: Chem. Revs., 1978-82. Recipient Irving Langmuir award, 1976; Pitts. award Spectroscopic Soc. Pitts., 1979; Sloan fellow, 1958-62; Guggenheim fellow, 1963-64, 72. Fellow Am. Acad. Arts and Scis., Am. Phys. Soc.; mem. Nat. Acad. Scis., AAAS, Phi Beta Kappa, Sigma Xi. Subspecialties: Physical chemistry; Magnetic physics. Current work: Nuclear magnetic resonance, especially in solids and at low temperatures. Home: Conant Rd Lincoln MA 01773 Office: 77 Massachusetts Ave Cambridge MA 02139

WAXMAN, HERBERT SUMNER, internist, hematologist, medical educator; b. Boston, Sept. 1, 1936; s. Samuel David and Martha (Jacobs) W.; m. Paula M. Waxman, June 26, 1960; children: Matthew, Marcy, Eric. B.S., MIT, 1958; M.D., Harvard U., 1962. Diplomate: Am. Bd. Internal Medicine. Intern. Mass. Gen. Hosp., Boston, 1962-64, resident, 1966-67; research assoc. NIH-Nat. Cancer Inst., Bethesda, Md., 1964-66; fellow in hematology Washington U. Sch. Medicine, St. Louis, 1967-68; asst. prof. to prof. medicine Temple U. Sch. Medicine, Phila., 1968-77, prof., dep. chmn. dept. medicine, 1979—; chmn. dept. medicine Baystate Med. Ctr., Springfield, Mass., 1977-79, Albert Einstein Med. Ctr., Phila., 1979—. Author jour. articles and book chpt.; editor: Computer-Assisted Med. Diagnosis, Meditel, Inc., 1973—. Served with USPHS, 1964-66. Recipient Lindback award Temple U., Phila., 1974. Fellow ACP, Coll.

Physicians Phila.; mem. Am. Fedn. Clin. Research, Am. Soc. Hematology, Assn. Program Dirs. in Internal Medicine, Sigma Xi, Alpha Omega Alpha, Phi Lambda Upsilon. Subspecialties: Internal medicine; Hematology. Current work: Control of hemoglobin synthesis; computer-assisted medical diagnosis; cancer chemotherapy. Office: Albert Einstein Med Ctr York and Tabor Rds Philadelphia PA 19141

WAXMAN, MICHAEL FREDERICK, brewing company microbiologist; b. N.Y.C., Sept. 17, 1942; s. Oscar Henry and Helen (Bornstein) W.; m. Barbara Hurowitz, Sept. 30, 1946. B.A., Hunter Coll., 1965; Ph.D., CUNY, 1974. Lectr. Bklyn. Coll., 1970-74; research assoc. U. Chgo., 1974-76; asst. prof. Ohio State U., 1976-79; sr. microbiologist Molson Breweries of Can., Ltd., Montreal, 1979-81; mgr. brewing research and devel. and corp. mocrobiology Pabst Brewing Co., Milw., 1981—. Sr. research scholar CUNY, 1973-74; Cold Spring Harbor Labs. grantee, 1970, 71. Mem. Am. Soc. Brewing Chemists, Master Brewers of the Americas, Soc. Microbiology, Soc. Indsl. Microbiology, Can. Soc. Microbiology, Sigma Xi. Subspecialties: Gene actions; Microbiology. Current work: Industrial fermentations and effects of aberrant conditions, yeast genetics and natural selection and gene inheritance. Office: Pabst Brewing Co 1000 N Market St Milwaukee WI 53201

WAXMAN, STEPHEN G(EORGE), physician, scientist, neurologist, educator; b. Newark, Aug. 17, 1945; s. Morris and Beatrice (Levitch) W.; m. Merle Applebaum, June 25, 1968; children: Matthew Curtis, David Mitchell. A.B., Harvard U., 1967; Ph.D., Albert Einstein Coll. Medicine, 1971, M.D., 1972. Diplomate: Am. Bd. Psychiatry and Neurology. Resident Boston City Hosp., 1972-75; clin. fellow in neurology Harvard Med. Sch., Boston, 1972-75, asst. prof. neurology, 1975-77, assoc. prof. neurology, 1977-78; vis. asst. prof. biology MIT, Cambridge, 1975-77, vis. assoc. prof. biology, 1977-78; prof. neurology Stanford (Calif.) U. Med. Sch., 1978-81, prof., vice chmn. dept. neurology, 1981—; sci. adv. bd. Paralyzed Vets. Am., 1981—; nat. adv. bd. Nat. Multiple Sclerosis Soc., 1980-83; nat. sci. adv. bd. Spinal Cord Injury Assn., 1982—; cons. VA Central Office, Washington, 1979—. Editor: Physiology of Axons, 1978, Demyelinating Diseases, 1981; assoc. editor: jours. including Muscle and Nerve. Recipient NIH Trygve Tuve award, 1973, Career Devel. award, 1975, Am. Epilepsy Assn. Chauveau award, 1969. Mem. Am. Soc. Cell Biology, Am. Acad. Neurology, Am. Neurol. Assn., World Fedn. Neurology. Subspecialties: Neurology; Neurobiology. Current work: Clinical and experimental neurology; diseases of the brain, spinal cord, and peripheral nerves; mechanisms of recovery from neurological disease. Office: Dept Neurology 127 Stanford U Sch Medicine VA Med Center 3801 Miranda Ave Palo Alto CA 94304

WAY, ANTHONY B., physician, educator; b. Boston, Mar. 12, 1940; s. Milton Tremont and Rhoda Elizabeth (Biden) W.; m. Barbara Haight Shoemaker, June 10, 1967; children: Matthew Shoemaker, Sarah Shoemaker. B.A., Williams Coll, 1962; M.D., U. Pa., 1967; Ph.D., U. Wis., 1972. Diplomate: Am. Bd. Preventive Medicine. Assoc. prof. preventive medicine and community health Health Scis. Ctr., Tex. Tech U., Lubbock, 1972—, assoc. prof. anthropology, 1972—; Adviser grad. med. edn. nat. adv. com. HEW, 1980. Contbr. chpts. to books, articles to profl. publs. NIH trainee, 1964-66; Nat. Inst. Gen. Med. Scis. fellow, 1968-72; Can. Nat. Health Research and Devel. Program grantee, 1978; Sid W. Richardson Found. grantee, 1980. Fellow Am. Coll. Preventive Medicine (membership task force 1980—); mem. Assn. Tchrs. Preventive Medicine (bd. dirs. 1981—), Am. Assn. Phys. Anthropologists, Human Biology Council, Am. Pub. Health Assn. Subspecialty: Preventive medicine. Current work: Clinical preventive medicine; epidemiology; human adaptability; health care delivery. Home: 3311 41st St Lubbock TX 79413 Office: Tex Tech Univ Health Scis Center Preventive Medicine and Community Health Dept Lubbock TX 79430

WAY, JAMES LEONG, educator, pharmacologist/toxicologist; b. Watsonville, Calif., Mar. 21, 1927; s. Wong Bung Whee and Shew Lay Har; m. Helen Wong, Mar. 21, 1932; children: Lani, Jon, Lori. B.S. in Chemistry, U. Calif., Berkeley, 1951; Ph.D. in Pharmacology, George Washington U., 1955. Instr. pharmacology U. Wis., Madison, 1958-59, asst. prof., 1959-62; assoc. prof. Med. Coll. Wis., Milw., 1962-67; prof. Wash. State U., Pullman, 1957-82; Sheiton prof., Tex. A&M U., College Station, 1982—; vis. scientist Nat. Inst. Med. Research, London, 1973-75; vis. prof. Nat. Def. Med. Ctr., Taipei, Taiwan, 1981-82; mem. toxicology data bank com. Nat. Library Medicine, 1977—; mem. sci. adv. bd. Nat. Ctr. Toxicology and Research, 1979-82, mem. exec. council, 1980-82; mem. pesticide div. U.S. Dept. Agr., 1973-76; mem. toxicology study sect. NIH, 1974-78, mem. com. effect chems. in health Dept. fisheries, Washington, 1978—. Contbr. articles to profl. jours., chpts. to books. Greenwalde scholar, 1949-50; Baxter N.Am. fellow, 1951-52; Nat. Cancer Inst. fellow, 1952-1955, 55-57; NAt. Inst. Gen. Med. Sci. fellow, 1957-58; NIH spl. research fellow, 1974-75; NSF vis. scientist, 1981-82. Mem. Am. Soc. Pharmacology and Exptl. Therapeutics, Soc. Toxicology, Western Pharmacology Soc. (pres. 1978), Am. Chem. Soc., AAAS, N.Y. Acad. Scis., Soc. Biology and Exptl. Medicine, Am. Whitewater Assn., Nat. Ski Assn. Subspecialties: Toxicology (medicine); Pharmacology. Current work: Pharmacology and toxicology of cyanide, alkylphosphate antagonists, selective fish toxicants. Home: 18 Forest Dr College Station TX 77840 Office: Dept Pharmacology Tex A&M U Coll Medicine College Station TX 77843

WAY, RICHARD A(LVORD), nuclear engineer; b. Erie, Pa., Mar. 10, 1956; s. Robert Bailey and Audrey (Booth) Chase W. B.S., Rensselaer Poly. Inst., 1978. Field engr. Gen. Electric Co., Oak Brook, Ill., 1978-79, constrn. engr., Monroe, Mich., 1979, Perry, Ohio, 1979-80; instr. Boiling Water Reactor Tng. Ctr., Tulsa, 1980-81, startup engr. startup/testing ops., Marseilles, Ill., 1981—. Vol. Joliet Area Operation Snowball, 1982. Mem. Am. Nuclear Soc. Republican. Lutheran. Clubs: Rugby (Cleve.); (Tulsa). Subspecialties: Nuclear engineering; Nuclear fission. Current work: Providing supervision and technical direction to utility personnel on the startup and testing of LaSalle I, the first domestic operating BWR/5. Office: Gen Electric Co DA & ESO 8157 Cass Ave Darien IL 60559

WAYMAN, COOPER HARRY, lawyer, chemist, educator; b. Trenton, N.J., Jan. 29, 1927; s. Cooper Ott and Helen V. W.; m. Ruth T. Treier, Mar. 17, 1930; children: Carol B., Andrea L. B.S., Rutgers U., 1951; M.S., U. Pitts., 1954; Ph.D., Mich. State U., 1959; J.D., U. Denver, 1967. Bar: Colo 1969, Tex 1971; registered engr., real estate broker, Colo. Prof. chemistry Colo. Sch. Mines, Golden, 1965-71, prof. environ. sci., 1978-80; gen. counsel EPA, Dallas, 1971-74, dir. energy office, Denver, 1974-78; exec. asst. Mayor of Denver, 1980-83; corp. atty. Denver Water Bd., 1983—; pres. G.W. Squared Inc., Arvada, Colo., 1969-83; sec. Colo. Air Commn., Denver, 1969-71; asst. dir. Indsl. Ecology Inst., Golden, 1978-80, Mineral Research Inst., 1980-81; pres. Wayman Land Devel. Co., Arvada, 1980-83. Author: Permits Handbook for Coal Development, 1980. Served with USN, 1945-46; PTO. Republican. Methodist. Subspecialties: Environmental toxicology; Water supply and wastewater treatment. Current work: Carbon monoxide exposure; lake rehabilitation nutrients; consulti. Home: 10360 W 74th Pl Arvada CO 80005 Office: Office of Mayor 14th at Bannock Denver CO 80204

WAYNANT, RONALD WILLIAM, electronics engineer, educator; b. Gettysburg, Pa., Oct. 4, 1940; s. Vaughn William and Naomi Grace (Martin) W.; m. Louise Fisher, Dec. 22, 1962; m. Priscilla Jane Pilson, Aug. 24, 1977; children: Marcia Louise, William Ronald, Kristopher Vaughn. B.E.S., Johns Hopkins U., 1962; M.S.E.E., Catholic U. Am., 1966, Ph.D., 1971. Engr. Westinghouse Electric Corp., Balt., 1962-69; electronics engr. Naval Research Lab., Washington, 1969—; mem. faculty Catholic U.Am. Contbr. articles on quantum electronics to profl. jours. Recipient Publ. award Naval Research Lab., 1970. Sr. mem. IEEE; Mem. Optical Soc. Am., Am. Phys. Soc., Washington Acad. Scis., Soc. Photog. Instrumentation Engrs. Subspecialty: Laser research. Current work: Basic research in lasers; development of new laser materials, novel methods of producing lasing; development damage resistant components. Patentee in field. Home: 13101 Claxton Dr Laurel MD 20708 Office: Naval Research Lab Code 6541 Washington DC 20375

WAYNE, BURTON HOWARD, engineering educator; b. Acton, Mass., Nov. 18, 1924; s. James Arthur and Margaret (Quimby) W.; m. Joan Saynor, Nov. 29, 1945; children: Lawrence, Kenneth, Terrance. B.S.E.E., Mich. State U., 1951, M.S.E., 1954, Ph.D., 1960. Instr. Mich. State U., East Lansing, 1954-60, asst. prof., 1960-64; assoc. prof. dept. engrng. analysis U. N.C., Charlotte, 1964-70, prof., and chmn., 1970—; mem. steering com. Southeastern Symposium on Systems Theory, 1969—. Contbr. articles to profl. jours. Served with U.S. Army, 1941-47. Mem. IEEE (chmn. 1976, sec. 1979-81, vice chmn. N.C. council 1981—), Nat. Soc. Profl. Engrs. Subspecialties: Electrical engineering; Computer-aided design. Home: 5163 Grafton Dr Charlotte NC 28215 Office: NC University Station Dept Engrng Analysis and Design Charlotte NC 28223

WAYNE, STEVEN FALKO, materials engr., researcher; b. Manchester, N.H., Aug. 2, 1951; s. Lacon McCauley and Lillian Adrienne (Provencher) W.; m. Elizabeth Biase, June 29, 1979; 1 son, David Thomas. B.S.E. in Mech. Engrng, U. Conn., 1977, M.S. in Material Sci, 1980. Engrng. aide Naval Underwater Systems Center, New London, Conn., 1972-73; research asst., dept. mech. engrng. and Inst. Materials Sci. U. Conn., Storrs, 1976—. Contbr. articles to profl. jours., confs. Recipient Soc. Mfg. Engrs. award, 1972, Soc. Nondestructive Testing award, 1972. Mem. ASME, Am. Soc. Metals, AIME, AAUP. Democrat. Roman Catholic. Subspecialties: Metallurgical engineering; Allergy. Current work: Tribology research,. alloy (carbide and nitride) devel., conservation of strategic materials. Home: Box 416 Long Hill Rd Andover CT 06232 Office: U Conn Inst Materials Sci U-136 Storrs CT 06268

WAYNE, WILLIAM JOHN, geology educator; b. So. Porter Twp., Cass County, Mich., Apr. 23, 1922; s. Norris Wagner and May (Clark) W.; m. Naomi Luella Liebl, July 9, 1946; children: Nancy, John W., Annette E. A.B., Ind. U., 1943; A.M., 1950; Ph.D., 1952. Head glacial geologist Ind. Geol. Survey, Bloomington, 1952-68; vis. prof. U. Wis., Madison, 1966-67; assoc. prof. U. Nebr., Lincoln, 1968-71, prof. geology, 1971—. Mem., sec. Monroe County Plan Commn., Bloomington, Ind., 1957-64. Served with U.S. Army, 1943-46. Recipient Disting. Teaching award U. Nebr., 1974; NSF grantee, 1980, 82-83. Fellow Geol. Soc. Am. (sect. chmn. 1971), Ind. Acad. Sci. (pres. 1968); mem. Nat. Assn. Geology Tchrs. (sect. pres. 1971), Am. Quaternary Assn., Asociación Geológica Argentina. Congregationalist. Subspecialty: Geology. Current work: Pleistocene geology of U.S. Midwest, eastern Nevada and central Andes; alpine geomorphology and glacial geology; alpine permafrost; Pleistocene land and fresh-water mollusks, urban geology. Office: Dept Geolog U Nebr Lincoln NE 68588-0340

WEATHERFORD, CHARLES ALBERT, physicist, educator; b. Mobile, Ala., June 23, 1947; s. Charles William and Mildred Laverne (Wells) W.; m. Virginia Sara Weatherford, Aug. 16, 1982. B.S. in Physics, La. State U., 1969, Ph.D., 1974. Prof. physics Fla. A&M U., Tallahassee, 1974—. Author: ETO Multicenter Molecular Integrals; contbr. articles to profl. jours. NSF fellow, 1980. Mem. Am. Phys. Soc., N.Y. Acad. Scis., AAAS. Democrat. Subspecialties: Atomic and molecular physics; Physical chemistry. Current work: Electron-molecule scattering; applied math teaching, research in electron-molecule scattering, applied math, computer modeling.

WEATHERSBY, A(UGUSTUS) BURNS, entomologist-parasitologist; b. Pinola, Miss., May 19, 1913; s. Augustus Benton and Louis Jane (Burns) W.; m. Olive Pearl Hammons, Apr. 8, 1945; children: Richard Michael, Robert Benton. B.S., La. State U., 1938, M.S., 1940, Ph.D., 1954; postgrad., George Washington U., 1950-53. Lab. instr. La. State U., Baton Rouge, 1938-40; asst. entomologist at large La. Dept. Agr., Baton Rouge, 1940-42; commd. ensign U.S. Navy, 1942, advanced through grades to comdr., 1957, ret., 1962; faculty U. Ga., Athens, 1962—, prof. entomology, 1964—. Contbr. articles to profl. jours. on susceptibility of mosquitoes. Fellow Royal Soc. Tropical Medicine and Hygiene; mem. Am. Soc. Parasitologist, Am. Soc. Tropical Medicine and Hygiene, Entomol. Soc. Am., Southeastern Soc. Parasitologists. Baptist. Subspecialty: Entomology. Current work: Teaching entomology and research on susceptibility of mosquitoes to malaria, mosquito immunity, cryobiology. life cycles of plasmodium and time lapse photography. Home: 210 Bishop Dr Athens GA 30606 Office: U Ga Athens GA 30602

WEAVER, CHRISTOPHER SCOT, research scientist; b. N.Y.C., Feb. 6, 1951; s. Richard B. and Mildred J. (Stier) W. B.A., Hobart Coll., 1973; M.A., M.S., Wesleyan U., 1975, C.A.S., 1976; postgrad., Columbia U., 1977-79. Mgr. tech. research ABC, N.Y.C., 1977-79; v.p. sci. and tech. Nat. Cable TV Assn., Washington, 1979-81; prin. Media Tech. Assocs., Ltd., Bethesda, Md., 1981—; Instr. Aikido, Middletown, Conn., 1974; mem. Internat. Electrotech. Commn., Geneva, 1979-81; vis. scholar MIT, 1981—; mem. tech. advisor Subcom. on Communications, U.S. Congress, 1981—. Author: Aikidosho, 1975; mem. editorial bd.: Internat. Videotex/Teletext News, Washington, 1981—. Mem. IEEE (cable subcom. 1979-82), Soc. Motion Picture and TV Engrs. (new tech. com. 1979—), Soc. Info. Display, Am. Nat. Standards Inst. (com. V98 1979-81), Soc. Photo Optical Instrumentation Engrs. Jewish. Subspecialties: Information systems, storage, and retrieval (computer science); Laser data storage and reproduction. Current work: Engaged in the development of advanced videodisc, broadband, and computer applications. Office: 9208 Burning Tree Rd Bethesda MD 20817

WEAVER, JAMES BODE, JR., agronomy educator; b. b Hartwell, Ga., Jan. 28, 1926; s. James B. and Lula (Hilley) W.; m. Betty Dove, Dec. 18, 1949; 1 son, James B. B.S. in Agronomy, U. Ga., 1950; M.S., N.C. State U., 1952, Ph.D., 1955. Asst. prof. dept. agronomy U. Ga., Athens, 1967-72, assoc. prof., 1972-79, prof., 1979—; Part-time cons. FAO-UNDP, Rome, 1977—. Served with AUS, 1945-46; ETO. Recipient Nat. Cotton Council Am. Cotton genetics award, 1983. Mem. Am. Soc. Agronomy, Crop Sci. Soc. Am., Entomol. Soc. Am. Baptist. Subspecialty: Plant genetics. Current work: Hybrid cotton, breeding for insect resistance and non-preference. Home: 155 Harden Dr Athens GA 30605 Office: Dept Agronomy U Ga Athens GA 30602

WEAVER, MICHAEL JOHN, plant pathologist, pesticide coordinator; b. Greenville, Pa., July 30, 1952; s. Wayne Alvin and Virginia Grace (Malizia) W.; m. Nancy Jean Schiefferle, July 26, 1975; 1 dau., Jennifer Christine. B.S., Edinboro (Pa.) State Coll., 1974; M.S., W.Va. U., 1977; Ph.D., Va. Poly. Inst. and State U., 1982. Grad. asst., pesticide and chem. office W.Va. U., W.Va. Coop. Ext. Service, Morgantown, 1975-77; grad. asst. chem., drug and pesticide unit Va. Poly Inst. and State U., 1977-80, extension coordinator chem., drug and pesticide unit, 1980—, research assoc. in plant pathology, 1980-83, asst. prof., 1983—; state pesticide applicator tng. coordinator, state liaison rep. Inter-regional Research Project 4, Pesticide Impact Assessment Program. Mem. Am. Phytopathological Soc. Subspecialties: Plant pathology; Information systems (information science). Current work: Shade tree pathology, plant stress pathology, plant disease survey, pesticide information system, grape pathology. Office: Chem Drug and Pesticide Uni Va Poly Inst and State U 139 Smyth Hall Blacksburg VA 24061

WEBB, ALISTAIR IAN, veterinary anesthesiologist, researcher; b.; b. Datchet, Bucks, U.K., May 3, 1943; came to U.S., 1978; s. Frederick Constantine and Myrtle Iris (Higgins) W.; m. Jennifer Jill Wolstenholme, July 16, 1966; children: Jonathan, Catherine. B.V.Sc., U. Queensland, Brisbane, Australia, 1966; D.V.A., Royal Coll. Vet. Surgeons, London, 1977; Ph.D., U. Bristol, 1978. Diplomate: Am. Coll. Vet. Anesthesiologists. Lectr. U. Queensland, 1970-74; research assoc. U. Bristol, U.K., 1974-78; asst. prof. U. Fla., Gainesville, 1978-83, assoc. prof. vet. anesthesiology, 1983—. Mem. AVMA, Australian Vet. Assn. (fed. councillor 1972-74). Subspecialties: Anesthesiology; Comparative physiology. Current work: Clinical anesthesiology and the scientific evaluation of current and new proposed techniques, pharmacokinetics and pharmacodynamics of anesthetic drugs. Home: 4710 NW 18th Pl Gainesville FL 32605 Office: U Fla Box J-126 JHMHC Gainesville 32610

WEBB, BYRON KENNETH, agricultural engineering educator; b. Cross Anchor, S.C., Feb. 2, 1934; s. John Walker and Cora (Smith) W.; m. Sybil Boatwright, Apr. 7, 1961; children: Charles Kenneth. B.S., Clemson U., 1955, M.S., 1962; Ph.D., N.C. State U., 1966. Asst. agrl. engr. U.S. Dept. Agr., Agrl. Research Service, Agrl. Exptl. Sta., Clemson (S.C.) U., 1955-63; prof. agrl. engrng. Clemson U., 1965-76, head dept. agrl. engrng., 1976—. Contbr. articles to profl. jours. Mem. Am. Soc. Agrl. Engrs., Am. Soc. Engrng. Edn., Sigma Xi, Gamma Sigma Delta. Republican. Baptist. Subspecialty: Agricultural engineering. Current work: Design and development of fruit and vegetable harvesting and handling equipment, aquacultural engineering. Patentee in field. Office: Agrl Engrng Dept McAdams Hall Clemson SC 29631 Home: 217 Hunter Ave Clemson SC 29631

WEBB, KENNETH LOUIS, oceanographer, educator; b. Old Frot, Ohio, July 18, 1930; s. John Louis and Vera Marie (Schiffer) W.; m. Susan R. Stevick, June 16, 1973; m. Ilse Bloch, Mar. 30, 1952 (div. 1970). B.A., Antioch Coll., 1953; M.Sc., Ohio State U., 1954, Ph.D., 1959. Research assoc. U. Ga. Marine Inst., Sapelo Island, Ga., 1960-65; asst. prof. oceanography Coll. William and Mary, 1965-75, assoc. prof., 1975-79, prof., 1979—; mem. adv. com. Ocean Sci. Biol. Oceanography program NSF, Washington, 1978-81. Contbr. articles to profl. jours. Muellhaupt scholar, 1959-60; grantee in field. Mem. AAAS, Am. Soc. Limnology and Oceanography, Physol. Soc. Am., Am. Soc. Zoologists, Japanese Soc. Plant Physiologists, Atlantic Estuarine Research Soc., Western Soc. Naturalists, Va. Acad. Sci., Sigma Xi. Subspecialties: Oceanography; Ecology. Current work: Interdisciplinary research related to energy flow and nutrient cycling in marine environments; physiology of marine organisms; image analysis. Home: PO Box 220 Gloucester Point VA 23062 Office: College of William and Mary Virginia Institute of Marine Science Gloucester Point VA 23062

WEBB, ROBERT BRADLEY, microbial geneticist; b. Conway, Ark., Nov. 20, 1927; s. Raymond Horace and Byrtie Lou (Bradley) W.; m. Mary Jane Rose, Nov. 23, 1930; children: Rebecca Jane, Colleen Louise. B.S., Harding Coll., 1947; M.S., U. Okla., 1950, Ph.D., 1956. Asst. prof. U. Tenn., Martin, 1950-52; NRC postdoctoral fellow Argonne (Ill.) Nat. Lab., 1956-58, assoc. scientist, 1958-73, scientist, 1973—; adj. assoc. prof. No. Ill. Ky., DeKalb, 1973—. Contbr. articles to profl. jours. Recipient U. Chgo. medal for disting. performance, 1980. Mem. AAAS, Am. Soc. Microbiology, Environ. Mutagen Soc., Am. Soc. Photobiology, Radiation Research Soc., Sigma Xi. Subspecialties: Genetics and genetic engineering (biology); Microbiology. Current work: Research in molecular basis of mutagenesis; role of DNA repair processes in mutagenesis; genetic effects of mid-UV and near UV radiation. Office: Argonne Nat Lab 9700 S Cass Argonne IL 60439

WEBB, THOMPSON, III, geology educator; b. Los Angeles, Jan. 13, 1944; S. Thompson and Diana (Stimson) W.; m. Joan Moscovitch, Aug. 10, 1969; children: Rosanna V., Sarah M. B.A., Swarthmore Coll., 1966; Ph.D., U. Wis.-Madison, 1971. Research assoc. U. Mich., Ann Arbor, 1970-72; asst. research prof. Brown U., Providence, 1972-75, assoc. geology, 1975—; vis. assoc. prof. U. Wis., Madison, 1976; vis. fellow U. Cambridge, Eng., 1977-78. Mem. Am. Quaternary Assn., AAAS, Wis. Acad. Arts, Letters and Sci., Am. Meteorol. Soc., Quebec Quaternary Assn. Subspecialties: Paleoecology; Climatology. Current work: Mapping and modeling climates and climatic changes of past 10,000 years; Quaternary palynology and reconstruction of past vegetation patterns; concerned with possible climate effects of increased atmospheric CO_2. Home: 111 Lorimer Ave Providence RI 02906 Office: Dept Geol Scis Brown U Providence RI 02912

WEBB, WILLIS LEE, meteorologist, physicist, researcher; b. Nevada, Tex., July 9, 1923; s. Roy and Rintha Eldysse (Simmons) W.; m. Inda Lanice Bryant, Dec. 20, 1942; 1 son, Michael Floyd. B.S. in Math. and Physics, So. Methodist U., 1952; M.S. in Meteorology, U. Okla., 1970; Ph.D. in Atmospheric Scis, Colo. State U., 1972. Weather observer U.S. Weather Bur. Stas., Tex., 1942-52, research physicist, Washington, 1952-55; research meteorologist White Sands Missile Range, N. Mex., 1955-78; adj. prof. physics U. Tex.-El Paso, 1961—. Author: Structure of Stratopheric Mesosphere, 1966, Geoelectricity, 1980; editor: Stratospheric Circulation, 1969. Mem. Am. Meterol. Soc., Am. Rocketry Soc., Am. Geophys. Union, AIAA. Subspecialties: Meteorology; Planetary atmospheres. Current work: Blue Ridge Circle rocket network; satellite meteorology. Home: 4929 Blue Ridge Circle El Paso TX 79904 Office: Schellenger Research Lab U Tex El Paso TX 79968

WEBBE, FRANK MICHAEL, psychology educator; b. Vero Beach, Fla., Oct. 13, 1947; s. Richard St. Clare and Peggy (Buckles) W.; m. Ellen Kane, Sept. 6, 1969; children: Elizabeth St. Clare, Tristan Kane. B.A., U. Fla., 1969, M.S., 1971, Ph.D., 1974. Interim asst. prof. U. Fla., 1974-75; research asst. prof. U. Miss., 1975-78; asst. prof. psychology Fla. Inst. Tech., 1978-80, assoc. prof., 1980—; chmn. dept. psychology, 1979—; cons. Brevard County (Fla.) Mental Health Center, 1980—; mem. faculty diplomate in preventive medicine program Internat.

Preventive Medicine Found., Houston, 1981—. Author: Student Workbook for Psychology, 1982; contbr. numerous articles to profl. jours. Soccer and softball coach Palm Bay (Fla.) Youth Athletic Assn., 1980—; chmn. Cub Scout Pack 732 Central Fla. council Boy Scouts Am., 1982. Named Outstanding Young Man Am. U.S. Jaycees, 1979; NSF research trainee, 1970-74. Mem. Am. Psychol. Assn., Southeastern Psychol. Assn., Psychol. Assn. Brevard County (pres. 1980-82), N.Y. Acad. Scis., Phi Beta Kappa, Sigma Xi (pres. club 1983-84), Phi Kappa Phi. Democrat. Subspecialties: Behavioral psychology; Psychobiology. Current work: Examination of effects of continuously present stimulation on behavior. Home: 1380 SW Billiar Ave Palm Bay FL 32905 Office: Sch Psychology Fla Inst Tech Melbourne FL 32901

WEBBER, ANDREW, chemist; b. Williton, U.K., Oct. 19, 1954; came to U.S., 1980; s. Kenneth Henry John and Joan Vernon (West) W.; m. Barbro Carlsdotter Englund, July 11, 1981; 1 son, Michael Mark. B.Sc. with honors, U. London, 1975, Ph.D., 1979. Postdoctoral research chemist U. Lund, Sweden, 1979-80, SUNY-Buffalo, 1980—. Mem. Am. Chem. Soc. Subspecialties: Analytical chemistry; Organic chemistry. Current work: Research in electrochemistry of organic and biological compounds; development of new analytical methods in electrochemistry using computers. Office: SUNY Acheson Hall Buffalo NY 14214

WEBBER, JOHN CLINTON, astronomer; b. Shreveport, La., Apr. 2, 1943; s. Jack Calvin and Dorothy Elizabeth (Kennemer) W.; m. Fredda Frances Davidson, June 7, 1964; children: David, Michael. B.S., Calif. Inst. Tech., 1964, Ph.D., 1970. Research assoc. U. Ill., Urbana-Champaign, 1969-71, asst. prof., 1971-77, sr. research scientist, 1977-80; research staff NEROC Haystack Obs., MIT, Westford, 1980-83, asst. dir., 1983—. Contbr.: articles to profl. jours. including Radio Science. Mem. Am. Astron. Soc., Internat. Astron. Union, Internat. Union Radio Sci. (comm. J). Subspecialties: Radio and microwave astronomy; Tectonics. Current work: Structure of compact radio sources, very-long-baseline interferometry, quasars, geodesy, astrometry, process control. Home: 118 Concord Rd Westford MA 01886 Office: Haystack Obs Westford MA 01886

WEBBER, ROBERT PATRICK, Mathematics educator; b. Washington, May 12, 1945; s. Robert Franklin and Marjory (Horner) W.; m. Linda Joan Ferry, Aug. 11, 1973; 1 dau., Caroline. B.A., U. Richmond, 1966; M.S., Stephen F. Austin U., 1967; Ph.D., U. Tenn., 1972. Mathematician Naval Weapons Lab., Dahlgren, Va., 1966; tchr. math. Upward Bound, Knoxville, 1967-70; assoc. prof. math. Longwood Coll., Farmville, Va., 1972—; Reviewer Brooks/Cole, 1976—, Holt, Rinehart, 1978—. Author: Precalculus, 1980, Business Math, 1976; Contbr. articles to profl. jours. Actor Waterworks Players, Farmville, Va., 1976—, Longwood Players, Farmville, 1972—. Williams scholar, 1966; Nat. Merit scholar, 1962. Mem. Am. Math. Soc., Math. Assn. Am., AAUP, Pi Mu Epsilon, Alpha Psi Omega. Home: Route 4 PO Box 432 Farmville VA 23901 Office: Longwood Coll Farmville VA 23901

WEBER, CHARLES WALTER, nutritionist, biochemist; b. Harold, S.D., Nov. 30, 1931; s. Walter Eral and Vera Jean (Scott) W.; m. Marylou Adam, Feb. 3, 1961; children: Matthew Charles, Scott Adam. B.S., Colo. State U., 1956, M.S., 1958; Ph.D., U. Ariz., 1966. Research chemist U. Colo., Boulder, 1960-63; research asst. U. Ariz., Tucson, 1963-66, asst. prof., 1966-68, assoc. prof., 1969-72, prof. dept. nutrition and food sci., 1972—; pres. Nutrition Council Ariz., 1983—; cons. U. Sonora, Mex., 1981-82. Contbr. chpts. to books in field. Pres. Randolph Soccer Club, Tucson, 1978-80, Central Region Pima County Soccer, Tucson, 1981; trustee United Ch. of Christ, Tucson, 1981—. Served with U.S. Army, 1952-54. Mem. Am. Soc. Animal Sci., Am. Inst. Nutrition, Am. Soc. Clin. Nutrition, N.Y. Acad. Scis., Poultry Sci. Assn. Subspecialties: Nutrition (medicine); Biochemistry (medicine). Current work: Mineral metabolism, arid plant usage. Home: 4031 Calle de Jardin Tucson AZ 85711 Office: Dept Nutrition and Food Sci Univ Ariz 309 Agricultural Sci Bldg Tucson AZ 85721

WEBER, DARRELL JACK, plant biochemist, pathologist, researcher; b. Thornton, Idaho, Nov. 16, 1933; s. John and LaNorma Anna (Severson) W.; m. Carolyn Foremaster, Aug. 24, 1962; children: Brad, Becky, Brian, Todd, Kelly, Jason, Trent. B.S., U. Idaho, 1958, M.S., 1959; Ph.D., U. Calif.-Davis, 1963. Postdoctoral fellow U. Wis.-Madison, 1963-65; asst. prof., then assoc. prof. U. Houston, 1965-69; prof. botany Brigham Young U., Provo, Utah, 1969—; postdoctoral fellow Mich. State U., 1975-76. Contbr. articles to sci. jours.; author of 3 books. Active Ch. Jesus Christ of Latter-day Saints. Recipient Karl G. Masear award, 1973; Utah Acad. Sci. fellow, 1974; grantee in field. Mem. Am.Inst. Biol. Scis., Am. Phytopath Soc., Am. Microbiology Soc., Am. Mycol. Soc., Sigma Xi. Republican. Subspecialties: Biochemistry (biology); Plant pathology. Current work: Physiology of host parasite interaction; physiology of salt tolerance; fungal physiology. Office: Brigham Young U Provo UT 84602 Home: 560 E Robin Orem UT 84057

WEBER, DAVID ALEXANDER, med. physicist; b. Lockport, N.Y., Mar. 6, 1939; s. Fred Leonard and Gladys Catherine (Woodcock) W.; m. Sandra Watson, Aug. 26, 1961; children: Sarah D., David A. B.S., St. Lawrence U., 1960; Ph.D., U. Rochester, 1971. Grad. teaching asst. physics U. Buffalo, 1960-61; sr. research aide, research asst. biophysics div. Sloan-Kettering Research Inst., N.Y.C., 1961-68; asst. attending physicist, asst. physicist div. med. physics Meml. Hosp. Cancer and Allied Disease, N.Y.C., 1967-68, lab. chief radioactive isotopes sect. div. med. physics, 1967-68; AEC grad. lab. fellow dept. radiation biology and biophysics U. Rochester, N.Y., 1968-70, asst. prof. radiology, 1970-75, asst. prof. radiation biology and biophysics, 1970-81, acting chief div. nuclear medicine, 1974-75, assoc. prof. radiation biology and biophysics, 1981—, assoc. prof. radiology, 1975—; clin. faculty dept. clin. sci. Sch. Health Related Professions, Rochester (N.Y.) Inst. Tech., 1976—; sr. internat. fellow Fogarty Internat. Center, NIH, Bethesda, Md., 1978-79; sr. internat. fellow dept. orthopedic surgery Lundu. Hosp., Sweden, 1978-79. Recipient Vis. Scientist fellowship award Swedish Med. Research Council, Stockholm, Lund U. Hosp., 1978-79. Mem. Soc.Nuclear Medicine, Am. Assn. Physicists in Medicine, Am. Coll. Nuclear Medicine, Health Physics Soc., Assn. Univ. Radiologists. Subspecialties: Nuclear medicine; Imaging technology. Current work: Design and evaluation of image storage, processing and display systems for nuclear medicine imaging; tracer kinetics, especially in the study of benign and malignant bone disease. Office: U Rochester Div Nuclear Medicine 601 Elmwood Ave Rochester NY 14642

WEBER, EVELYN JOYCE, biochemistry educator, research chemist; b. Pana, Ill., Nov. 9, 1928; d. John Henry and Emma Caroline (Schoch) W. B.S., U. Ill., 1953; Ph.D., Iowa State U., 1961. Research assoc. biochemistry dept. U. Ill., Urbana, 1961-64, asst. prof. agronomy dept., 1965-72, assoc. prof., 1972-82, prof. plant biochemistry, 1982—, research chemist, 1965—. Contbg. author on biochemistry to profl. jours. USPHS fellow, 1960; Am. Inst. Chemists fellow, 1965. Mem. Am. Chem. Soc., Am. Oil Chemists Soc., Am. Soc. Plant Physiologists, AAAS, Iota Sigma Pi (pres. chpt. 1960, 62). Mem. United Ch. of Christ. Subspecialties: Biochemistry (biology); Plant physiology (agriculture). Current work: Biochemistry and genetics of plant lipids, fatty acids and fat-soluble vitamins. Office: S-320 Turner Hall USDA Univ Ill 1102 S Goodwin St Urbana IL 61801

WEBER, GEORGE, biochemist, cancer researcher; b. Budapest, Hungary, Mar. 29, 1922; came to U.S., 1959, naturalized, 1965; s. Salamon and Hajnalka (Arvai) W.; m. Catherine E. Forrest, June 30, 1958; children: Dolly, Julie, Jefferson. B.A., Queen's U. Kingston, Ont., Can., 1950; M.D., 1952; M.D. (hon.), Chieti (Italy) Med. Sch., 1979, U. Budapest, 1982. Lic. Med. Council Can. Postdoctorate fellow U. B.C., 1952-53; research assoc. Notre Dame Hosp.-Montreal Cancer Insti., U. Montreal, Que., Can., 1953-56, head sect. path. chemistry, 1956-59; vis. scientist Harvard U. Med. Sch., summer 1957; mem. faculty Ind. U. Med. Sch., 1959—, prof. pharmacology, 1961—, prof. dir. exptl. oncology, 1974—. Author numerous articles, book chpts. in field.; Editor: Advances in Enzyme Regulation, vols. 1-21, 1962—. Recipient Alecce prize cancer research, Rome, 1971, SAMA award Student AMA, 1966, 68; hon. mem. All-Union Biochem. Soc., Nat. Acad. Sci., USSR, 1981—. Fellow Royal Soc. Medicine; Mem. Am. Assn. Cancer Research (G.H.A. Clowes award 1982), Am. Soc. Pharmacology and Exptl. Therapeutics, Am. Physiol. Soc. Subspecialties: Cancer research (medicine); Molecular pharmacology. Current work: Biochemistry and pharmacology of cancer, chemotherapy and enzymology. Home: 7307 Lakeside Dr Indianapolis IN 46278 Office: Room 337 Riley Cancer Wing Ind U Med Sch 1100 W Michigan St Indianapolis IN 46223

WEBER, GREGORIO, biochemist, educator; b. Buenos Aires, Argentina, July 4, 1926; s. Leon and Rosa (Gerchunoff) W. M.D., U. Buenos Aires, 1942; Ph.D., U. Cambridge, Eng., 1947. Lectr. in biochemistry U. Sheffield, Eng., 1952—, reader, to 1962; prof. biochemistry U. Ill.-Urbana, 1962—. Recipient Rumford medal Am. Acad. Arts and Scis., 1982. Mem. Nat. Acad. Scis. Subspecialty: Biochemistry (biology). Current work: Spectroscopy and thermodynamics of biological interest. Office: U Ill 1209 W California St Urbana IL 61807

WEBER, JAMES ALAN, physicological plant ecologist; b. Santa Monica, Calif., Mar. 16, 1944; s. Leo Andreas and Wilma Kathleen (Lewis) W.; m. Nancy Jane Smith, Apr. 18, 1970. Student, El Camino Coll., Torrance, Calif., 1962-65; A.B. in Botany, U. Calif.-Berkeley, 1966, A.M., U. Mich., 1967, Ph.D., 1973. Postdoctoral scholar Biol. Sta., U. Mich., Ann Arbor, 1973-79, asst. research scientist, 1979—. Contbr. articles to profl. jours. Mem. exec. com. Mackinac chpt. Sierra Club, 1976-80. Mem. AAAS, Am. Inst. Biol. Scis., Am. Soc. Plant Physiologists, Am. Soc. Limnology and Oceanography, Bot. Soc. Am., Ecol. Soc. Am., Internat. Assn. Aquatic Vascular Plant Biologists, Sigma Xi. Subspecialties: Plant physiology (biology); Ecology. Current work: Control of carbon dioxide exchange in leaves as a function of environmental variables; the relationships between net carbon gain and plant growth. Office: Biological Station U Mich Ann Arbor MI 48109

WEBER, JOSEPH, physicist; b. Paterson, N.J., May 17, 1919; s. Jacob and Lena (Stein) W.; m. Anita Meinhardt Straus, Oct. 18, 1942; m. Virginia Louise Trimble, Mar. 16, 1972; children: Jonathan, Paul, James, David. B.S., U.S. Naval Acad., 1940; Ph.D. in Physics, Cath. U. Am., 1951. Physicist U.S. Navy Bur. Ships, 1945-48; U. Md., College Park, 1948—, U. Calif.-Irvine, 1973—; staff Inst. Advanced Study, Princeton, N.J., 1955-56, 62-63, 69-70. Author: Quantum Electronics, 1951, Gravitational Radiation Antennna, 1958. Served with USN, World War II. Guggenheim fellow, 1955, 62; NRC fellow, 1955-56; recipient First prize Gravity Research Found., 1959, Boris Pregel award N.Y. Acad. Scis., 1973. Fellow IEEE, Am. Phys. Soc. Subspecialties: Relativity and gravitation; Statistical physics. Current work: Gravitaton, relativity, statistical physics, weak interactions, quantum electronics. Office: Physics Dept U Md College Park MD 20742

WEBER, KENNETH CHARLES, physiologist, educator; b. Cold Springs, Minn., June 30, 1937; s. Christian M. and Laura (Stein) W.; m. Bertha Mae Williams (div. Nov. 1983); children: Charles Arthur, Gerald Michael, Terry Lynn, Patricia Diane. A.A., DeVry Tech. Inst., 1957; B.S. E.E., U. Minn., 1963, Ph.D., 1968. Asst. prof. W. Va. U., 1968-72, assoc. prof., 1972-76, prof., 1976—; chief physiology sect. Nat. Inst. Occupational Safety and Health, Ctr. Disease Control, Morgantown, W. Va., 1968—; v.p. Am. Heart Assn., Dallas, 1981-82; organizer, chmn. Gordon Research Conf., 1975. Mem. Am. Physiol. Soc., Am. Heart Assn., Biophysics Soc., N.Y. Acad. Sci., Sigma Xi. Roman Catholic. Subspecialties: Physiology (medicine); Pharmacology. Current work: Physiology and pharmacology of occupational respiratory diseases; animal lung physiology; lung mechanics; cardiopulmonary physiology. Home: 1375 Hunter Ln Morgantown WV 26505 Office: Physiology Sect NIOSH 944 Chestnut Ridge Rd Morgantown WV 26505

WEBER, MARVIN JOHN, physicist; b. Fresno, Calif., Feb. 26, 1932; s. John William and Louise (Grill) W.; m. Pauline Margaret Sikes, Feb. 2, 1957; children: Ann Hilary, Eve Kimberley. A.B., U. Calif., Berkeley, 1954, M.A., 1956, Ph.D., 1959. Research assoc. dept. physics U. Calif., Berkeley, 1959; prin. scientist Research Div., Raytheon Co., Waltham, Mass., 1960-73; vis. research asso. dept. physics Stanford U., 1966; group leader Lawrence Livermore Nat. Lab., 1973-80, asst. assoc. program leader, 1980-83; with Office Basic Energy Scis. Dept Energy, 1983—; editor-in-chief CRC Handbook Series of Laser Sci. and Tech., 1978—; adv. editorial bd. Jour. Non-Crystalline Solids, 1981—; cons. NSF, 1973-76. Contbr. articles to profl. jours. Recipient IR 100 award Indsl. Research Mag., 1979, George W. Money award Am. Ceramics Soc., 1983. Mem. Am. Phys. Soc. (fellow), Optical Soc. Am., Am. Ceramics Soc., Fedn. Am. Scientists, Sigma Xi, Phi Beta Kappa. Subspecialties: Condensed matter physics; Spectroscopy. Current work: Lasers, quantum electronics, luminescence, optical spectroscopy, optical materials. Home: 221 Loch Lomond Way Danville CA 94526 Office: PO Box 808 Livermore CA 94550

WEBER, STEPHEN GREGORY, chemistry educator, consultant; b. Boston, May 10, 1949; s. Frederick Reuel and Virginia Ann (Fair) W.; m. Carol Teasley, June 21, 1971; children: Jessica Lynn, Emily Anne. B.A., Case-Western Res. U., 1970; Ph.D., McGill U., 1979. Asst. prof. chemistry U. Pitts., 1979—; cons. Energy Conversion Devices, Troy, Mich., 1978-80, Scopas Tech. Corp., N.Y.C., 1983—. Contbr. articles to profl. jours. Served with USN, 1970-74. Research grantee Research Corp., 1981; Petroleum Research Fund, 1980; NIH, 1980. Mem. Am. Chem. Soc., Am. Assn. Clin. Chemists, Sigma Xi. Subspecialties: Analytical chemistry; Physical chemistry. Current work: To improve our ability to measure extremely low concentrations of important molecules existing in complex matrices. Patentee in field. Office: Dept Chemistry U Pitts Pittsburgh PA 15260 Home: 18 Sylvan Rd Forest Hills PA 15221

WEBER, THOMAS ANDREW, research scientist; b. Tiffin, Ohio, June 8, 1944; s. Stanley Walter and Anne Louise (Decker) W.; m. Stephanie Anthony, Aug. 17, 1968; children: Thomas, Lisa, Stefan, Alice, Andrea. B.S., U. Notre Dame, 1966; Ph.D., Johns Hopkins U., 1970. Mem. staff Bell Labs., Murray Hill, N.J., 1970—. Fellow Am. Phys. Soc.; mem. Phi Beta Kappa, Sigma Xi. Roman Catholic. Subspecialties: Theoretical chemistry; Polymer chemistry. Current work: Monte Carlo and molecular dynamics simulation of water, simple fluids and alkanes, simulation of polymers and glasses. Office: Bell Laboratories 600 Mountain Ave Murray Hill NJ 07974

WEBER, THOMAS RICHARD, pediatric surgery educator, researcher; b. Cleve., Feb. 19, 1945; s. Robert E. and Lois L. (Payne) W.; m. Suzanne M. Muehring, June 14, 1969; children: Amy, Jill, Patrick. B.A., Eastern Mich. U., 1967; M.D., Ohio State U., 1971. Diplomate: Am. Bd. Surgery. Intern U. Mich. Hosp., Ann Arbor, 1971-72, resident in surgery, 1972-77; resident in pediatric surgery Washington Children's Hosp., 1977-79; asst. prof. pediatric surgery Ind. U. Med. Ctr., Indpls., 1979—; attending surgeon Riley Children's Hosp., Indpls., 1979—. Fellow ACS; mem. Assn. Acad. Surgery (exec. council 1979-81), Am. Pediatric Surg. Assn. Subspecialties: Surgery; Cancer research (medicine). Current work: Neonatal bowel ischemia, pediatric solid tumors, pediatric septic shock—pharmacologic support, pediatric liver disease. Office: Riley Childrens Hospital 702 Barnhill Dr Indianapolis IN 46223

WEBER, WALTER JACOB, JR., engineering educator; b. Pitts., June 16, 1934; s. Walter Jacob and Anne Mae (Chando) W.; m. Ruth L. Stryker, Dec. 17, 1955 (div. Jan. 1975); children: Wendilyn Ruth, Elizabeth Anne, Pamela Jean, Linda Lorraine; m. Patricia L. Nagel Braden, July 20, 1981. Sc.B., Brown U., 1956; M.S.E., Rutgers U., 1959; A.M., Harvard, 1961, Ph.D., 1962. Registered profl. engr.; Diplomate Am. Acad. Environ. Engrs. Engr. Caterpillar Tractor Co., Peoria, Ill., 1956-57; instr. Rutgers U., 1957-59; engr. Soil Conservation Service, New Brunswick, N.J., 1957-59; research, teaching asso. Harvard, 1959-63; faculty U. Mich., Ann Arbor, 1963—, prof., chmn. water resources program, 1968—, Disting. prof., 1978—; Internat. cons. to industry, govt. Author: (with K.H. Mancy) Analysis of Industrial Wastewaters, 1971, Physicochemical Processes for Water Quality Control, 1972; editor-author: (with E. Matijevic) Adsorption from Aqueous Solution, 1968; Contbr. numerous articles and chpts. to tech., profl. jours. and books. Recipient Distinguished Faculty awards U. Mich., 1967, 78, Assn. Environ. Engring. Profs., 1968; Faraday lectr. U. Mich., 1970; Engr. of Distinction Engrs. Joint Council, 1973; Assn. Environ. Engring. Profs.-NALCO research award, 1979; Research Excellence award U. Mich., 1980. Mem. Am. Chem. Soc. (cert. of merit 1962, F. J. Zimmerman award 1982), Am. Inst. Chem. Engrs., ASCE (Rudolph Hering medal 1980, Thomas R. Camp award 1982, Simon W. Freese award 1984), Am. Water Works Assn. (Acad. Achievement award 1981), Assn. Environ. Engring. Profs., Water Pollution Control Fedn. (John R. Rumsey Meml. award 1975, Willard F. Shephard award 1980), Tau Beta Pi, Sigma Xi, Chi Epsilon, Delta Omega. Subspecialty: Water supply and wastewater treatment. Home: 1700 South Grove Rd Ypsilanti MI 48197 Office: Water Resources Program Coll Engring U Mich Ann Arbor MI 48109

WEBER-LEVINE, MARGARET LOUISE, neuroscientist; b. Bronx, Sept. 23, 1941; s. Irving and Anne (Shapiro) Weber; m. Stephen Mark Levine, May 30, 1969. B.A., Antioch Coll., 1964; postgrad. (NIMH trainee), Queens Coll., 1964-67; Ph.D., Stony Brook SUNY, 1972. Asst. prof. psychology Morehouse Coll., Atlanta, 1972-79, assoc. prof., 1979—. Instr. CPR, Midfulton County, 1978-81; judge Sci. Fair Atlanta Pub. Schs., 1980-82. NIH grantee, 1977-78; 1979-82; NIMH grantee, 1980—. Mem. AAAS, Am. Psychol. Assn., Soc. Neuroscience, Southeastern Psychol. Assn., N.Y. Acad. Sci., Gerontol. Soc. Subspecialties: Neurophysiology; Physiological psychology. Current work: Neurophysiological changes in aging; nutrition and brain function. Home: 373 Sargent Dr SE Atlanta GA 30315 Office: Dept Psychology Morehouse Coll Atlanta GA 30314

WEBSTER, EDWARD WILLIAM, med. physicist; b. London, Apr. 12, 1922; s. Edward and Bertha Louisa (Cornish) W.; m. Dorothea Anne Wood; children: John, Peter, Anne, Edward, Mark, Susan. B.Sc., U. London, 1943, Ph.D., 1946. Diplomate: Am. Bd. Radiology, Am. Bd. Health Physics. Research engr. English Electric Co., Stafford, Eng., 1945-49; research physicist div. indsl. cooperation M.I.T., 1950-52; lectr. nuclear energy and elec. engring. Queen Mary Coll., U. London, 1952-53; physicist, dir. radiol. scis. div. dept. radiology Mass. Gen. Hosp., Boston, 1953—, radiation safety officer, 1962-80; prof. radiology Harvard U., 1975—, (Harvard-M.I.T. Health Scis. and Tech. Div.), 1978—; examiner in physics Am. Bd. Radiology, 1958-78; dir. Nat. Council Radiation Protection, 1981—; mem. adv. com. on med. uses of isotopes U.S. Nuclear Regulatory Commn., 1971—; mem. com. on biol. effects of ionizing radiation U.S. Nat. Acad. Scis., 1977-80; cons. IAEA, 1960-61, WHO, 1964-67. Contbr. chpts. to books, articles to profl. jours. Robert Blair traveling fellow London County Council, 1949-50; USPHS fellow, 1965-66; NIH grantee. Mem. Radiol. Soc. N.Am. (v.p. 1977-78), Am. Assn. Physicists in Medicine (pres. 1963-64), Soc. Nuclear Medicine (trustee 1970-71, 73-77), Am. Coll. Radiology, New Eng. Radiologic Physics Orgn. (chmn. 1970-73), Radiation Research Soc., Health Physics Soc. Subspecialties: Imaging technology; Medical physics. Current work: Low dose radiation imaging technology, radiation dosimetry, biological effects of radiation, radiation protection and shielding design, medical uses of radiosotopes. Patentee in field of low energy x-ray shielding. Office: Dept Radiology Mass Gen Hosp Boston MA 02114

WEBSTER, GARY DEAN, geology educator, consultant geologist; b. Hutchinson, Kans., Feb. 15, 1934; s. John Raymond and Mable Fae (Randles) W.; m. Beverly Eileen Wilson, Aug. 29, 1964; children: Dean Wilson, Karissa Delight. Student, Hutchinson Jr. Coll., 1951-53; B.S. in Geol. Engring, U. Okla., 1953-56; M.S. in Geology, U. Kans., 1960, Ph.D., UCLA, 1961. Jr. geologist Amerada Petroleum, Williston, N.D., 1956-57; geologist Belco Petroleum, Big Piney, Wyo., 1960; asst. prof. geology San Diego State U., 1965-68; prof. geology, chmn. dept. geology Wash. State U., 1968—; cons. to oil cos., cons. firms, state surveys, 1968—. Author monographs, map; contbr. articles to profl. pubs. Fellow Geol. Soc. Am.; mem. Paleontol. Soc. (regional chmn. 1978-79), Soc. Econ. Paleontologists and Mineraologists, Palaeontol. Assn. Subspecialties: Paleobiology; Sedimentology. Current work: Research in Paleozoic stratigraphy of western U.S.; paleontologic research in corodonts and crinois. Home: NE 1155 Orchard Dr Pullman WA 99163 Office: Dept Geolog Wash State U Pullman WA 99164

WEBSTER, GEORGE CALVIN, biological sciences educator, researcher; b. South Haven, Mich., July 17, 1924; s. Eugene Homer Webster and Hazel Edna (Empson) Davis; m. Sandra Lee Whitman, Jan. 23, 1960; children: Jeffrey C., Kimberley Ann. B.S., Western Mich. U., 1948; M.S., U. Minn.-Mpls., 1949, Ph.D., 1952. Sr. research fellow Calif. Inst. Tech., 1952-55; assoc. prof. Ohio State U., 1955-61; vis. prof. U. Wis., Madison, 1961-64; chief environ. health lab., Cape Kennedy, Fla., 1964-70; prof. biol. scis. Fla. Inst. Tech., Melbourne, 1971—, head dept., 1971—. Author: Nitrogen Metabolism in Plants, 1959; Contbr. numerous articles to profl. jours. Served to 1st lt. USAF, 1942-45. Am. Heart Assn. investigator, 1963. Fellow AAAS; mem. Am. Soc. Biol. Chemists, Biochem. Soc. Gt. Britain, Am. Soc. Cell Biology, Gerontol. Soc. Democrat. Subspecialties: Molecular biology; Genetics and genetic engineering (biology). Current work: Research on the molecular basis of the aging process in animals, using recombinant DNA and related techniques. Home: 389 Chester Dr Cocoa FL 32922 Office: Department of Biological Sciences Florida Institute Technology 150 W University Blvd Melbourne FL 32901

WEBSTER, JOHN GOODWIN, biomedical engineering educator; b. Plainfield, N.J., May 27, 1932; s. Franklin F. and Emily (Boody) W.; m. Nancy Egan, Dec. 27, 1954; children: Paul, Robin, Mark, Amylark. B.E.E., Cornell U., 1953; M.S.E.E., U. Rochester, 1965, Ph.D., 1967. Registered profl. engr., Wis. Engr. N.Am. Aviation, 1954-55, Boeing Airplane Co., 1955-59, Radiation Inc., 1959-61; staff engr. Mitre Corp., 1961-62, IBM Corp., 1962-63; NIH predoctoral fellow U. Rochester, N.Y., 1963-67; asst. prof. elec. engring. U. Wis., Madison, 1967-70, assoc. prof., 1970-73, prof. elec. and computer engring., 1973—; cons. in field. Contbr. articles to profl. jours.; Editor: Medical Instrumentation: Application and Design, 1978, (with A.M. Cook) Clinical Engineering: Principles and Practices, 1979, (with W.J. Tompkins) Design of Microcomputer-based Medical Instrumentation, 1981, (with A.M. Cook) Therapeutic Medical Devices: Application and Design, 1982; Author: (with B. Jacobson) Medicine and Clinical Engineering, 1977. Recipient NIH Research Career Devel. award, 1971-76, Instrument Soc. Am. Donald P. Eckman Edn. award, 1974, Am. Soc. Engring. Edn. Western Electric Fund award, 1970. Fellow Instrument Soc. Am.; mem. IEEE, Biomed. Engring. Soc., Am. Soc. Engring. Edn., Assn. for Advancement Med. Instrumentation. Subspecialties: Biomedical engineering; Bioinstrumentation. Current work: Research in medical instrumentation and transducers; monitoring and stimulating electrodes, biopotential amplifiers, ambulatory monitors, bioimpedance measurements, electrogastrogram, human energy expenditure. Home: 1710 Hoyt St Madison WI 53705 Office: 1415 Johnson Dr Madison WI 53706

WEBSTER, LESLIE TILLOTSON, JR., pharmacologist, educator; b. N.Y.C., Mar. 31, 1926; s. Leslie Tillotson and Emily (de Forest) W.; m. Alice Katharine Holland, June 24, 1955; children—Katharine White, Susan Holland, Leslie Tillotson III, Romi Anne. B.A., Amherst Coll., 1947, Sc.D. (hon.), 1982; student, Union Coll., 1944; M.D., Harvard U., 1948. Diplomate: Am. Bd. Internal Medicine. Rotating intern Cleve. City Hosp., 1948-49, jr. asst. resident, 1949-50; asst. resident medicine Bellevue Hosp., N.Y.C., 1952-53; research fellow medicine Harvard and Boston City Hosp. Thorndike Meml. Lab., 1953-55; demonstrator Case Western Res. U. Sch. Medicine, 1955-60, research assoc. to sr. instr. biochemistry, 1959-60, asst. prof. medicine, 1960-70, asst. prof. biochemistry, 1960-65, asst. prof. pharmacology, 1965-67, asso. prof., 1967-70, prof. pharmacology, 1976—, chmn. pharmacology dept., 1976—, prof. medicine, 1980—; prof., chmn. pharmacology dept. Northwestern U. Med. and Dental Sch., 1970-76; cons. NIH, WHO, Rockefeller Found. Contbr. articles to sci. and med. jours. Served to lt. USNR, 1950-52. Russell M. Wilder fellow Nat. Vitamin Found., 1956-59; Sr. USPHS Research fellow, 1959-61; Research Career Devel. awardee, 1961-69; Macy faculty scholar, 1980-81. Mem. Am. Assn. Study Liver Diseases, ACP (life), Central Soc. Clin. Research (emeritus), Am. Soc. Clin. Investigation (emeritus), Am. Soc. Biol. Chemists, Assn. Med. Sch. Pharmacology, Am. Soc. Pharmacology and Exptl. Therapeutics. Subspecialties: Molecular pharmacology; Immunopharmacology. Current work: Molecular and immunopharma-cology metabolism of anti-parasitic drugs. Home: 2728 Leighton Rd Shaker Heights OH 44120 Office: Dept Pharmacology Case-Western Res U Sch Med 2119 Abington Rd Cleveland OH 44106

WEBSTER, MURRAY ALEXANDER, JR., sociologist; b. Manila, Philippines, Dec. 10, 1941; s. Murray A. and Patricia (Morse) W. A.B., Stanford U., 1963, M.A., 1966, Ph.D., 1968. Instr. Coll. San Mateo, Calif., 1967; asst. prof. Johns Hopkins U., Balt., 1968-74; assoc. prof. U. S.C., Columbia, 1974-76, prof. sociology, 1976—, adj. prof. psychology, 1979—; vis. prof. Stanford U., 1981-82; session organizer Am. Sociol. Assn. and So. Sociol. Soc., 1974-80. Contbr. articles to profl. jours.; editorial bd.: Am. Jour. Sociology, 1977-80, Social Psychology Quar, 1976-79; assoc. editor: Social Sci. Research, 1976—; author: Sources of Self-Evaluation, 1974, Actions and Actors, 1975. NIH trainee, 1965; Wilson fellow, 1964; NIH grantee, 1966-68; Nat. Inst. Edn. grantee, 1968-74; NSF grantee, 1976—. Republican. Presbyterian. Subspecialty: Social psychology. Current work: Status cues and effects on behavior; physical attractiveness; interrelations of social processes (liking, control, status, love/committment). Home: 1829 Senate St Apt 9A Columbia SC 29201 Office: Dept Sociology Univ SC Columbia SC 29208

WEBSTER, ROBERT OWEN, medicine and microbiology educator; b. Mitchel AFB, N.Y., Dec. 19, 1943; s. Daniel P. and Doris M. (Owens) W.; m. Carol R. Schulzetenberg, June 14, 1969; children: Sarah C., Bryan M. B.S., Purdue U., 1965; M.S., U. Minn., 1968; Ph.D., Albany Med. Coll., 1976. Research scientist N.Y. State Kidney Disease Inst., Albany, 1971-77; research fellow Nat. Jewish Hosp., Denver, 1977-81; asst. research rheumatologist U. Calif.-San Francisco, 1981-82; asst. prof. medicine and microbiology St. Louis U. Med. Sch., 1982—. Contbr. articles to profl. jours. Served to capt. U.S. Army, 1969-70. USPHS Grad. Sch. fellow, 1965-67; Albany Med. Coll. scholar, 1972-77; recipient Nat. Research Service award USPHS, 1979-81. Mem. Internat. Soc. Immunology and Pharmacology, Am. Assn. Immunologists, Am. Soc. Microbiology, AAAS. Subspecialties: Immunopharmacology; Pulmonary medicine. Current work: Rule of complement neurophils and platelets in acute lung injury, mechanisms of neurophil activation. Home: 825 Carman Woods Dr Manchester MO 63104 Office: Pulmonary Div Dept Medicine St Louis U Sch Medicine 1325 S Grand Blvd St Louis MO 63104

WEBSTER, TERRY RICHARD, biology educator; b. Hamilton, Ohio, Feb. 10, 1938; s. Richard Stanley and Florence Lucille (Moore) W.; m. Orra Joan, June 27, 1964; 1 dau., Jennifer Leigh. B.A., Miami U., Ohio, 1960; M.A., U. Sakatchewan, Ph.D., 1965. Prof. biology U. Conn., Storrs, 1965—. Contbr. articles to profl. jours. NSF grantee, 1967-69, 1981-82. Mem. Am. Fern Soc., Bot. Soc. Am. Congregationalist. Subspecialties: Morphology; Plant genetics. Current work: Morphology of lower vascular plants, reproductive biology of the genus Selaginella. Office: Biol Scis Group U Conn Storrs CT 06268

WEBSTER, WILLIAM MERLE, JR., electronics company executive; b. Warsaw, N.Y., June 13, 1925; s. William Merle and Carrie Melinda (Luce) W.; m. Mary Lambert Tourison, May 3, 1947; children: Melissa, Cecelia. B.S. in Physics, Union Coll., Schenectady, 1945; Ph.D. in Elec. Engring, Princeton, 1954. With RCA Corp., 1946—, v.p. labs., Princeton, N.J., 1969—; dir. Horizon Bancorp., Princeton Bank & Trust Co. Trustee Med. Center Princeton. Served to lt. (j.g.) USNR, 1942-46. Fellow IEEE; mem. Nat. Acad. Engring., Sigma Xi. Subspecialties: Electronics; Microelectronics. Current work: Technical Management. Home: 11 Morven Pl Princeton NJ 08540 Office: RCA Laboratories Princeton NJ 08540

WECHSLER, JAMES ALAN, molecular geneticist; b. Pitts., Oct. 26, 1940; s. Maurice B. and Pearl Janice (Wilner) W.; m. Josephine Ann Puddington, Feb. 2, 1963; children: Samantha Leigh, Jeremy Dylan. B.S., Yale U., 1962, Ph.D., 1968. Postdoctoral fellow U. Edinburgh, Scotland, 1968-70; asst. prof. Columbia U., 1970-75, assoc. prof., 1975-77; research assoc. prof. U. Utah, Salt Lake City, 1977-80; sr. scientist Pub. Health Research Inst. City N.Y., 1981—. Contbr. articles to profl. jours. USPHS Grad. fellow, 1964-68; Am. Cancer Soc. postdoctoral fellow, 1968-70; USPHS research grantee, 1971—. Mem. Am. Soc. Microbiology, AAAS, N.Y. Acad. Scis., Union of Concerned Scientists, Sierra Club, Fedn. Am. Scientists. Democrat. Club: Yale of Utah. Subspecialties: Gene actions; Molecular biology. Current work: Analysis of chromosome replication in Escherichia coli. Regulation of genes and interaction of gene products involved in DNA replication. Home: 2475 Emerson Ave Salt Lake City UT 84108 Office: 455 1st Ave 11th Floor New York NY 10016

WECHTER, WILLIAM JULIUS, Pharmacetical company research management executive; b. Louisville, Feb. 13, 1932; s. Louis and Elsa (Strauss) W.; m. Roselyn Ann Greenman, Aug. 24, 1952; children: Laurie Jo, Diane Joy, Julie Lynn; m. Kathryn Elaine Edwards, Apr. 16, 1982. A.B., U. Ill., Urbana, 1953, M.S., 1953; Ph.D., UCLA, 1957. Research scientist Upjohn Co., Kalamazoo, Mich., 1957-68, research head, 1968-79, research mgr. hypersensitivity diseases, 1979—; vis. scholar Stanford Med. Sch., 1967-68; vis. lectr. pathology Harvard Med. Sch., 1977-78; adj. lectr. biol. chemistry Kalamazoo Coll., 1974—. Contbr. articles in field to profl. jours. State rep. to conv. Democratic party, 1966; mem. Mayor's Com. on Housing, 1969; chmn. eval. com. Mich. State Dept. Health, 1979. Wright scholar U. Ill., 1954; U.S. Royal fellow UCLA, 1956-57; recipient Upjohn prize, 1975. Mem. ACLU, AAAS, Am. Chem. Soc., Am. Assn. Cancer Research, Transplantation Soc., Am. Assn. Immunologists. Democrat. Jewish. Subspecialties: Immunopharmacology; Immunobiology and immunology. Current work: Immunopharmacology, nucleic acid chem. and biochemistry research direction and adminstrn., respiratory and immunoinflammatory diseases. Home: 810 W South St Kalamazoo MI 49007 Office: Upjohn Co Kalamazoo MI 49001

WECKER, LYNN, pharmacologist, educator; b. N.Y.C., Sept. 27, 1947; d. Frank L. and Sue (Levin) W.; m. Roger E. Freeman, Jan. 10, 1981. Ph.D. in Pharmacology, U. Fla., 1972. Postdoctoral fellow in pharmacology Vanderbilt U. Sch. Medicine, Nashville, 1973-75; asst. prof. med. chemistry and pharmacology Northeastern U., Boston, 1975-76; asst. prof. pharmacology Vanderbilt U. Sch. Medicine, 1976-78, La. State U. Med. Center, New Orleans, 1978-80, assoc. prof., 1980—, assoc. prof. psychiatry, 1980—. Contbr. articles to profl. jours. Recipient Andrew Mellon Tchr.-Scientist award, 1975; Outstanding Young Scientist award SK&F Labs., 1979; NIMH grantee. Mem. Am. Soc. Neurochemistry, Am. Soc. Pharmacology and Exptl. Therapeutics, Soc. Neuroscience, Internat. Soc. Neurochemistry. Subspecialties: Neuropharmacology; Neurochemistry. Current work: Mechanisms regulating the metabolism of acetylcholine on the nervous system. Office: Dept Pharmacology Louisiana State Univ Med Center 1901 Perdido St New Orleans LA 70112

WEDEEN, RICHARD PETER, physician, researcher; b. N.Y.C., Jan. 19, 1934; s. Marcus D. and Dorothy (Feldman) Mason; m. Roberta Rubien, Oct. 27, 1957; 1 son, Timothy. A.B., Harvard U., 1955; M.D., NYU, 1959. Diplomate: Am. Bd. Internatl Medicine. Intern Beth Israel Hosp., N.Y.C., 1959-60, 1960-61; resident Mt. Sinai Hosp., N.Y.C., 1961-62, 63-64; research assoc. Mt. Sinai Med. Sch., N.Y.C., 1965-66, instr. medicine, 1966-68, asst. prof., 1968-72; assoc. prof. medicine N.J. Med. Sch., Newark, 1972-76, prof., 1976—; assoc. chief staff for research VA Med. Ctr., East Orange, N.J., 1978—; vis. lectr. Harvard U. Med. Sch., Boston, 1968. Contbr. numerous articles to med. jours. Polechek Found. fellow, 1967-68. Fellow ACP; mem. Am. Soc. Clin. Investigation, Am. Physiol. Soc., Med. History Soc. of N.J. (sec. 1982—), Harvey Soc. Subspecialties: Nephrology; Toxicology (medicine). Current work: Renal diseases and renal physiology, lead nephropathy and organic acid transport mechanisms in the renal tubule. Home: 574 S Forest Dr Teaneck NJ 07666 Office: VA Med Center East Orange NJ 07019

WEDGWOOD, RALPH JOSIAH PATRICK, pediatrician, educator; b. London, May 25, 1924; U.S., 1940, naturalized, 1951; s. Josiah and Dorothy Mary (Winser) W.; m. Virginia Lloyd Hunt, Oct. 25, 1943; children: Josiah Francis, James Cecil (dec.), Jeffrey Galton, John Christopher Ralph. M.D., Harvard U., 1947. Intern in pediatrics Bellevue Hosp., N.Y.C., 1974-48, resident, 1948-49; research fellow in pediatrics Harvard U., 1949-51; sr. instr. in pediatrics and biochemistry Western Res. U. Med. Sch., 1953-57, asst. prof. pediatrics and preventive medicine, 1957-62; assoc. prof. pediatrics U. Wash., 1962-63, prof., 1963—, chmn. dept. pediatrics, 1963-72; spl. cons., mem. gen. clin. research ctr. com. NIH, 1962-66; mem. nat. Adv. Research Resources Council, 1966-70; pres. Assn. Med. Sch. Pediatric Chmn., 1966-68; chmn. Joint council Nat. Pediatric Soc., 1967-69; mem. several research adv. coms. Birth Defects Found., 1969—; mem. sci. adv. bd. St. Jude Hosp., Memphis, 1970-76. Contbr. numerous articles to profl. jours.; editor several books in field; editorial bd.: Pediatrics, 1966-73. Served to capt. M.C. AUS, 1944-46, 51-53. Spl. USPHS research fellow Microbiol. Research Establishment, Parton, Eng., 1960-61, John and Mary Markle scholar in med. scis., 1960-65; vis. fellow St. John's Coll., Cambridge (Eng.) U., 1969-70. Mem. Assn. Am. Med. Colls. (Disting. Service Mem.), and numerous other profl. socs. Subspecialties: Pediatrics; Preventive medicine. Current work: 0540, 2250. Office: Sch Medicine U Wash Dept Pediatrics RD 20 Seattle WA 98195

WEDIG, JOHN H., toxicologist, pharmacologist; b. St. Louis, Dec. 25, 1941; s. John H. and Verla L. (Lampert) W.; m.; children: Sarah, Alison, Meredith, Robert. A.B in Zoology, Washington U., St. Louis, 1964; M.S. in Wildlife Mgmt, U. Mich., 1967; Ph.D. in Toxicology, U. Mich., 1971. Diplomate: Am. Bd. Toxicology. With Olin Corp., New Haven, Conn., 1971—, sr. toxicologist, 1976—; environ. toxicology cons. Nilo Earms Shooting Preserve, div. Winchester-Western, 1971—; non-resident lectr. U. Mich. Contbr. chpts. to books, articles to profl. jours. Mem. Am. Coll. Vet. Toxicologists, Soc. Toxicology, Am. Soc. Pharmacology and Exptl. Therapeutics. Subspecialties: Toxicology (medicine); Environmental toxicology. Current work: Neurotoxicology, reproductive toxicology, ecological toxicology, dermato-toxicology. Patentee in field. Office: Olin Corp 91 Shelton Ave New Haven CT 06511

WEDNER, H. JAMES, physician, educator; b. Pitts., May 12, 1941; s. Benjamin and Lucille (Jacobs) W.; m. Maureen Wedner, June 18, 1978. M.D., Cornell U., 1967. USPHS trainee in immunology Washington U. Sch. Medicine, St. Louis, 1971-75, instr. medicine, 1973-74, asst. prof., 1974-80, assoc. prof., 1980—; asst. physician Barnes Hosp., St. Louis, 1974-81, assoc. physician, 1982—; mem. spl. rev. com. Nat. Inst. Allergy and Immunologic Diseases, 1982. Served as sr. asst. surgeon USPHS, 1968-70. Recipient Clarence Coryell prize in Medicine, 1967; Bordon prize, 1967. Mem. Am. Assn. Immunologists, Am. Acad. Allergy, Internat. Soc. Immunopharmacology, Alpha Omega Alpha. Subspecialties: Allergy; Immunology (medicine). Current work: Cell growth and function, polymorphonuclear leukocyte function. Office: 660 Euclid Saint Louis MO 63110

WEED, HERMAN ROSCOE, biomedical engineering educator, consultant, researcher; b. Union City, Pa., Aug. 5, 1922; s. Roscoe Conklin and Leta Venettie (Bryner) W.; m. Sylvia Kathryn Yearick, Apr. 20, 1946; children: David Herman, Douglas Leonard, Kathryn Marie. B.S.E.E., Pa. State U., 1945; M.Sc., Ohio State U., 1948. Registered profl. engr., Ohio. Instr. Pa. State U., 1945-46; instr. Ohio State U., Columbus, 1946-49, asst. prof., 1949-55, assoc. prof., 1955-59, prof., elec. engring., 1959—, dir., 1971—; cons. HOPE, 1980—, others; vis. prof. Inst. Biocybernetics, U. Karlsruhe, W.Ger., 1979; UNESCO adv. Punjab (India) Agrl. U., 1978, 81, Ford Found. cons., 1966, 72, 78; guest Polish Acad. Sci. Biomed. Engring. Inst., Oct. 1975, French Ministry Industry and Research, Sept. 1975; vis. prof. U. Cairo, 1977. Contbr. numerous articles to sci. and tech. jours. Recipient Disting. Teaching award Eta Kappa Nu, 1965; cert. of commendation Ohio Ho. of Reps., 1977; Robert Critchfield award, 1968, 73. Mem. Internat. Fedn. Automatic Control (chmn. tech. com. systems bio med. engring.), Ohio Acad. Sci., Am. Soc. Engring. Edn. (chmn. COBECC), Alliance for Engring. in Medicine and Biology, Am. Automatic Control Council. Clubs: Ohio State U. Faculty, Photography. Subspecialty: Biomedical engineering. Current work: Biomedical engineering with emphasis on muscle stimulation, physiological controls, heartsound analysis, developing technology. Office: Room 257 2015 Neil Ave Columbus OH 43210

WEEDMAN, DANIEL WILSON, astronomer; b. Nashville, Oct. 19, 1942; s. Roy Lee and Luetta (Lutz) W.; m. Suzanne Dallas, Oct. 11, 1968; children: Diana, Sylvia. B.A., Vanderbilt U., Nashville, 1964; Ph.D., U. Wis., 1967. Faculty assoc. U. Tex., 1967-69; asst. prof., 1969-70, Vanderbilt U., 1970-73, assoc. prof., 1975-79; vis. assoc. prof. U. Minn., 1974-75; prof. astronomy Pa. State U., 1979—. Contbr. articles to profl. jours. NSF grantee; NASA grantee. Mem. Am. Astron. Soc. (mem. council 1980-82), Internat., Astron. Union, Assn. Univs. for Research in Astronomy (dir. 1978-81). Subspecialties: Optical astronomy; Radio and microwave astronomy. Current work: Observational astrophysics: galaxies and quasars. Office: Pa State Univ 525 Davey Lab University Park PA 16802

WEEKS, ALBERT WILLIAM, geologist, consultant; b. Chilton, Wis., Nov. 26, 1901; s. George Thomas and Katherine (Schneider) W.; m. Alice M. Dowse, May 20, 1950. B.A., U. Wis., 1923, M.A., 1924; Ph.D., U. Tex., 1941. Geologist Midwest Refining Co., Denver, 1924-25; geologist, dist. geologist Shell Oil Co., Houston, 1925-38; cons., Austin, Tex., 1938-41; Sr. petroleum devel. analyst Petroleum Adminstrn. for War, Washington and Denver, 1941-44; staff geologist Sun Co., Phila., 1944-66; cons., lectr., Wynnewood, Pa., 1966—; guest lectr. Temple U., 1978—. Contbr. articles to profl. jours. Fellow Geol. Soc. Am. (housing chmn. 1967); mem. Am. Assn. Petroleum Geologists (cert., dist. rep. Eastern sect. 1952-54, 65-67), AAAS (rgn. geology program chmn. 1951), N. Tex. Geol. Soc. (pres. 1937-38), Rocky Mountain Geol. Soc. (sec.-treas. 1942-44), Can. Soc. Petroleum Geologists, Phila. Geol. Soc. (treas. 1968-79). Republican. Episcopalian. Clubs: Adirondack Mountain, Green Mountain, Appalachian Mountain, Philadelphia Trail (excursions chmn. 1948-52), Philadelphia Trail (pres. 1952-55). Subspecialty: Geology. Current work: Petroleum exploration and lecturing at Temple U. Address: 1029 Nicholson Rd Wynnewood PA 19106

WEEKS, JAMES ROBERT, pharmacologist; b. Des Moines, Aug. 13, 1920; s. Leo and Roberta Elizabeth (Likes) W.; m. Mary Ellen, July 12, 1943; children: Diane Louise, Weeks Rinehart, James Robert, Linda Susan Weeks Kreft. B.Sc. in Pharmacy, U. Nebr., 1941, M.S. in Pharmacology, 1946, Ph.D., U. Mich., 1952. Lic. pharmacist, Nebr. Instr. Coll. Pharmacy, Drake U., Des Moines, 1946, prof., 1950-57; research assoc. Upjohn Co., Kalamazoo, 1957—. Contbr. articles to sci. publs. Served to capt., Chem. Warfare Service U.S. Army. Mem. Am. Soc. Pharmacology and Exptl. Therapeutics, Soc. Exptl. Biology and Medicine, AAAS, Sigma Xi, Phi Lambda Upsilon. Subspecialties: Pharmacology; Psychopharmacology. Current work: General pharmacology of prostaglandins; cardiovascular pharmacology especially methods for small animals; experimental addiction; introduced method of intravenous self-administration by animals. Office: Cardiovascular Diseases Research Upjohn Co Kalamazoo MI 49001

WEEKS, JOHN RANDEL, IV, metallurgist, administrator; b. Orange, N.J., Oct. 30, 1927; s. John Randel and Marion Roberta (Heberton) W.; m. Barbara Ann Brewster, July 16, 1951; children: Ann, John. Met.E., Colo. Sch. Mines, 1949; M.S., U. Utah, 1950, Ph.D., 1953. With Brookhaven Nat. Lab., Upton, N.Y., 1953—, sr. metallurgist, leader corrosion sci. group, 1979—; on assignment to AEC, Washington, 1972-74; adj. prof. Poly. Inst. N.Y., Bklyn., 1978—. Mem. Am. Soc. Metals (chmn. L.I. chpt. 1962-63, chmn. Am. Inst. Metal. Engring, nuclear metallurgy com. 1971-73), Am. Nuclear Soc. (treas. L.I. sect. 1980-82), Electrochem. Soc., Nat. Assn. Corrosion Engrs. Subspecialties: Corrosion; Metallurgical engineering. Current work: Materials degradation in water and liquid metal coded nuclear reactors, corrosion and stress corrosion cracking mechanisms. Home: 25 Acorn Ln Stony Brook NY 11790 Office: Brookhaven Nat Lab Bldg 703 Upton NY 11973

WEEKS, STEPHAN JOHN, research analytical chemist; b. Mpls., Apr. 13, 1950; s. John Norman and Betty Jo Ann (Hemingway) W.; m. Jean Carol Seifert, Nov. 24, 1977; children: Erik, Mark. B.A., St. Olaf Coll., 1972; Ph.D., U. Fla., 1977. Chem. technician Minn. Dept. Health, Mpls., 1971; grad. research asst. U. Fla., Gainesville, 1972-77; NRC postdoctoral research fellow Nat. Bur. Standards, Washington, 1977-79, research chemist, 1979—. Contbr. articles to profl. jours. Mem. Am. Chem. Soc. (analytical and petroleum divs.), Soc. Applied Spectroscopy, ASTM. Lutheran. Subspecialties: Analytical chemistry; Spectroscopy. Current work: Trace atomic and molecular analysis; petroleum analysis and characterization; characterize petroleum products through physical, chemical and performance testing and computer correlation and data analysis. Home: 67 Timber Rock Rd Gaithersburg MD 20878 Office: Nat Bureau of Standards Washington DC 20234

WEEKS, WILFORD FRANK, glaciologist; b. Champaign, Ill., Jan. 8, 1929; s. Frank Cook and B. Caroline (Pool) W.; m. Beverly Jean Weeks, June 8, 1952; children—Ellen Jean, Paul Russell. B.S., U. Ill., 1951, M.S., 1953; Ph.D. in Geochemistry, U. Chgo., 1956. Glaciologist Air Force Cambridge Research Center, 1955-56; asst. prof. geology Washington U., 1957-62; glaciologist U.S. Army Cold Regions Research and Engring. Lab., Hanover, N.H., 1962—; adj. prof. earth scis. Dartmouth Coll., 1972—; Japan Soc. for Promotion Sci. prof. Inst. Low Temperature Sci., Hokkaido U., Japan, 1973; Office of Naval Research chair of arctic marine sci. Naval Postgrad. Sch., 1978-79; mem. numerous govt. and industry panels. Bassist, Dartmouth Symphony, Monterey Symphony, Hanover Chamber Orch.; Contbr. numerous articles to profl. jours. Served with USAF, 1955-57. Recipient research and devel. achievement award U.S. Army, 1967, 81. Fellow Arctic Inst. N.Am.; mem. Internat. Glaciological Soc. (pres. 1972-75), Am. Geophys. Union, Am. Polar Soc., Nat. Acad. Engring., Phi Beta Kappa, Phi Kappa Phi, Sigma Xi. Subspecialties: Geophysics; Oceanography. Current work: Geophysical characteristics of ice in the sea as they relate to offshore engineering; remote sensing and material sciences; general glaciology. Office: Cold Regions Research and Engring Lab 72 Lyme Rd Hanover NH 03755

WEERTMAN, JOHANNES, materials science educator; b. Fairfield, Ala., May 11, 1925; s. Roelof and Christina (van Vlaardingen) W.; m. Julia Ann Randall, Feb. 10, 1950; children: Julia Ann, Bruce Randall. Student, Pa. State Coll., 1943-44; B.S., Carnegie Inst. Tech., (now Carnegie Mellon U.), 1948, D.Sc., 1951; postgrad., Ecole Normale Superieure, Paris, France, 1951-52. Solid State physicist U.S. Naval Research Lab., Washington, 1952-58, cons., 1960-67; sci. liaison officer U.S. Office Naval Research, Am. Embassy, London, Eng., 1958-59; faculty Northwestern U., Evanston, Ill., 1959—, prof. materials sci.

dept., 1961-68, chmn. dept., 1964-68, prof. geol. scis. dept., 1963—, Walter P. Murphy prof. materials sci., 1968—; vis. prof. geophysics Calif. Inst. Tech., 1964, Scott Polar Research Inst., Cambridge (Eng.) U., 1970-71; Cons. U.S. Army Cold Regions Research and Engring. Lab., 1960-75, Oak Ridge Nat. Lab., 1963-67, Los Alamos Sci. Lab., 1967—; co-editor materials sci. books MacMillan Co., 1962-76. Author: (with wife) Elementary Dislocation Theory, 1964; Editorial bd.: Metal. Trans, 1967-75, Jour. Glaciology, 1972—; asso. editor: Jour. Geophys. Research, 1973-75; Contbr. articles to profl. jours. Served with USMC, 1943-46. Honored with naming of Weertman Island in Antarctica.; Fulbright fellow, 1951-52; recipient Acta Metallurgica gold medal, 1980; Guggenheim fellow, 1970-71. Fellow Am. Soc. Metals, Am. Phys. Soc., Geol. Soc. Am., Am. Geophys. Union (Horton award 1962); mem. Nat. Acad. Engring., Am. Geophys. Union, Am. Inst. Mining, Metall. and Petroleum Engrs. (Mathewson gold medal 1977), Am. Inst. Physics, Internat. Glaciological Soc. (Seligman Crystal award 1983), AAAS, Arctic Inst., Am. Quaternary Assn., Explorers Club, Sigma Xi, Tau Beta Pi, Phi Kappa Phi, Alpha Sigma Mu, Pi Mu Epsilon. Club: Evanston Running. Current work: Fatigue and fracture of metals; creep of crystalline solids; dislocation theory; physics of glaciers, earthquake dislocations; ice ages theory. Home: 834 Lincoln St Evanston IL 60201

WEERTMAN, JULIA RANDALL, materials science educator, researcher; b. Muskegon, Mich., Feb. 10, 1926; d. Winslow Henry and Louise Elizabeth (Neumeister) Randall; m. Johannes Weertman, Feb.10, 1950; children: Julia Ann, Bruce Randall. B.S. in Physics, Carnegie Inst. Tech., 1946, M.S., 1947, D.Sc., 1951. Postdoctoral fellow Ecole Normale Superieure, Paris, 1951-52; solid state physicist Naval Research Lab., Washington, 1952-58; asst. prof. materials sci. Northwestern U., 1973-78, assoc. prof., 1978-82, prof., 1982—, asst to dean for research and grad. studies, 1973-76. Author: (with J. Weertman) Elementary Dislocation Theory, 1964, 68, 69, 70; contbr. chpts., articles, to profl. publs.; editor: (with M. E. Fine, J. Weertman) Macmillan Series on Materials Science, 1963-76. Recipient Environ. award City of Evanston, Ill., 1979; Creativity in Research award NSF, 1982; Rotary Internat. fellow, Paris, 1951-52. Mem. AIME, Am. Phys. Soc., Am. Inst.Physics, Am. Soc. Metals, Soc. Engring. Sci., Am. Crystallographic Assn. Methodist. Subspecialties: High-temperature materials; Metallurgy. Current work: Behavior of metals and alloys at high temperature; study of grain boundary cavitation; use of small angle neutron scattering to investigate cavitation and other microstructural changes; high temperature fatigue and creep; dislocation structures in fatigue. Home: 834 Lincoln St Evanston IL 60201 Office: Dept Materials Sci and Engring Northwestern U 2145 Sheridan Rd Evanston IL 60201

WEFEL, JAMES STERN, dentistry educator, researcher; b. Cleve., Mar. 10, 1947; s. Paul S. and Hilda A. (Unger) W.; m. Janice I. Meehan, Aug. 24, 1968; children: Jeffrey S., Jay S. Student, Am. U. Beirut, 1964-65; B.S., Valparaiso U., 1968; Ph.D., SUNY-Buffalo, 1972. Asst. research scientist U. Iowa, 1973-75, assoc. research scientist, 1975-79, adj. asst. prof. dentistry, 1977-79, asst. prof. dentistry, 1979-82, assoc. prof., 1982—; invited discussion leader Gordon Research Conf., N.H., 1979, 81; reviewer jours. pediatric dentistry, 1980—; project site visitor NIH Nat. Inst. Dental Research, 1981-82. Contbr. chpts., articles, numerous abstracts to profl. publs. Coach Iowa City Kickers, soccer team, 1979-82; asst. coach Iowa City Little League Baseball, 1981-82; cubmaster Hawkeye Area council Boy Scouts Am., 1981-83. Recipient Young Investigator award 3d Internat. Symposium on Tooth Enamel, 1978; NIH Nat. Inst. Dental Research grantee, 1978—. Mem. Internat. Assn. Dental Research, Am. Assn. Dental Research, Am. Chem. Soc., N.Y. Acad. Scis. Republican. Lutheran. Subspecialties: Cariology; Preventive dentistry. Current work: Use of fluorides, mechanisms and kinetics to prevent dental caries or root caries; remineralization—repair of carious lesions. Home: 4 Regal Ln Iowa City IA 52240 Office: U Iowa Coll Dentistry Dental Sci Bldg Iowa City IA 52242

WEG, JOHN GERARD, physician; b. N.Y.C., Feb. 16, 1934; s. Leonard and Pauline M. (Kanzleiter) W.; m. Mary Loretta Flynn, June 2, 1956; children—Diane Marie, Kathryn Mary, Carol Ann, Loretta Louise, Veronica Susanne, Michelle Celeste. B.A. cum laude, Coll. Holy Cross, Worcester, Mass., 1955; M.D., N.Y. Med. Coll., 1959. Diplomate: Am. Bd. Internal Medicine. Commd. 2nd lt. USAF, 1958, advanced through grades to capt., 1967; intern Walter Reed Gen. Hosp., Washington, 1959-60; resident, then chief resident in internal medicine Wilford Hall USAF Hosp., Lackland AFB, Tex., 1960-64, chief pulmonary sect., 1964-66, chief inhalation sect., 1964-66, chief pulmonary and infectious disease service, 1966-67; resigned, 1967; clin. dir. pulmonary disease div. Jefferson Davis Hosp., Houston, 1967-71; from asst. prof. to assoc. prof. medicine Baylor U. Coll. Medicine, Houston, 1967-71; assoc. prof. medicine U. Mich. Med. Sch. Univ. Hosp., Ann Arbor, 1971-74, prof., 1974—; physician-in-charge pulmonary div., 1971-81, physician-in-charge pulmonary and critical care med. div., 1981—; cons. Ann Arbor VA, Wayne County Gen. hosps.; advisory bd. Washtenaw County Health Dept., 1973—. Contbr. med. jours., reviewer, mem. editorial bds. Decorated Air Force Commendation medal; travelling fellow Nat. Tb and Respiratory Disease Assn., 1971; recipient Aesculpaius award Tex. Med. Assn., 1971. Fellow Am. Coll. Chest Physicians (chmn. bd. govs. 1976-79, gov. Mich. 1975-79, chmn. membership com. 1976-79, prof.-in-residence 1972—, chmn. critical care council 1982—), Am. Coll. Chest Physicians and Internat. Acad. Chest Physicians (exec. council 1976—), Internat. Acad. Chest Physicians unit of Am. Coll. Chest Physicians (pres. elect 1979-80, pres. 1980-81), ACP (chmn. Mich. program com. 1974); mem. AAAS, Am. Fedn. Clin. Research, AMA, Am. Thoracic Soc. (sec.-treas. 1974-76), Am. Assn. Inhalation Therapy, Air Force Soc. Internists and Allied Specialists, Soc. Med. Consultants to Armed Forces, Internat. Union Against Tb, Mich. Thoracic Soc. (pres. 1976-78), Mich. Lung Assn. (dir.), Am. Lung Assn., Research Club U. Mich., Assn. Advancement Med. Instrumentation, Central Soc. Clin. Research, Am. Bd. Internal Medicine (asso., subsplty. com. on pulmonary disease 1980), Alpha Omega Alpha. Subspecialties: Pulmonary medicine; Internal medicine. Current work: Pulmonary pathophysiology, critical care medicine, education. Home: 3060 Exmoor St Ann Arbor MI 48104 Office: 1405 E Ann St Ann Arbor MI 48109

WEGMAN, DAVID HOWE, physician, medical educator; b. Balt., Mar. 13, 1940; s. Myron Ezra and Isabel (Howe) W.; m. Cynthia Heynen, June 18, 1962; m. Margaret Nelson, June 7, 1969; children: Jesse Howe, Marya Nelson. B.A., Swarthmore Coll., 1963; M.D., Harvard U., 1966, M.S., 1972. Diplomate: Am. Bd. Preventive Medicine. Intern Met. Gen. Hosp., Cleve., 1966-67; med. epidemiologist Nat. Communicable Disease Ctr., N.Y.C., 1967-69; dir. Indsl. Health and Safety Project, Urban Planning Aid, Inc., Cambridge, Mass., 1969-71; occupational hygiene physician Div. Occupational Hygiene, Mass. Dept. Labor and Industries, Boston, 1972-77; asst. prof. occupational health Harvard U. Sch. Pub. Health, Boston, 1972-77, assoc. prof., 1977-83; prof., dir. div. environ. and computational health scis. UCLA Sch. Pub. Health, 1983—; Mem. occupational health and safety study sect. NIH, Washington, 1979; mem. Permanent Commn. Internat. Assn. Occupational Health. Author and co-editor: Occupational Health, 1983; mem. editorial bd.: Am. Jour. Pub. Health, 1976-79; adv. bd.: Harvard U. Med. Sch. Health Letter, 1980—. Served to lt. comdr. USPHS, 1967-69. Fellow Am. Coll. Preventive Medicine; mem. Mass. Pub. Health Assn. (Alfred L. Frechette award 1979), Am. Pub. Health Assn., Soc. Epidemiologic Research, Internat. Epidemiologic Assn. Subspecialties: Preventive medicine; Epidemiology. Current work: Epidemiology of occupational pulmonary disease; epidemiology of cancer associated with work through use of vital records; development of occupational disease surveillance schemes. Home: 10551 Rochester Ave Los Angeles CA 90024 Office: Harvard U Sch Public Health 665 Huntington Ave Boston MA 02115

WEHAUSEN, JOHN VROOMAN, mathematician, educator; b. Duluth, Sept. 23, 1913; s. George W. and Elizabeth (Vroman) W.; m. Mary Katherine Wertime, Aug. 19, 1938; children—Sarah, Peter Vrooman, Julia, John David. B.S., U. Mich., 1934, M.S., 1935, Ph.D., 1938. Instr. math. Brown U., 1937-38, Columbia, 1938-40, U. Mo., 1940-44; mathematician David Taylor Model Basin, Carderock, Md., 1946-49; acting head mechanics br. Office Naval Research, 1949-50; exec. editor Math. Revs., 1950-56; asso. research mathematician Inst. Engring. Research, U. Calif. at Berkeley, 1956-57, research mathematician, 1957—, asso. prof. engring. sci., 1958-59, prof., 1959—; Fulbright lectr. U. Hamburg, 1960-61; Consultant Operations Research Group USN, 1944-46. Mem. Am. Math. Soc., Schiffbautechnische Gesellschaft, Math. Assn. Am., Soc. Naval Architects and Marine Engrs., Soc. Naval Architects of Japan, Nat. Acad. Engring. Subspecialties: Fluid mechanics; Probability. Current work: Ship hydrodynamics; water-wave theory. Home: 15 Hillside Ct Berkeley CA 94704

WEHINGER, PETER AUGUSTUS, astronomer, educator; b. Goshen, N.Y., Feb. 18, 1938; s. George Edward and Elizabeth Marie (Goode) W.; m. Susan Wyckoff, July 29, 1967. B.S. in Physics, Union Coll., Schenectady, N.Y., 1960; M.A. in Astronomy, Ind. U., Bloomington, 1962; Ph.D. in Astronomy (NASA predoctoral fellow), Case Inst. Tech., 1966. Instr. astronomy U. Mich., 1965-67, asst. prof., 1967-70; assoc. prof. physics and astronomy U. Kans., 1970-72; vis. assoc. prof., Smithsonian research fellow in physics and astronomy Tel Aviv U., 1972-75; U.K. Sci. Research Council prin. research fellow Royal Greenwich Obs., Eng., 1975-78; adj. prof. astronomy Sussex U., 1975-78; Max Planck Gesellschaft vis. sr. scientist Max Planck Institut fur Astronomie, W. Ger., 1978-80; vis. sr. research assoc. astronomy Ohio State U., 1978-79; prof. physics Ariz. State U., Tempe, 1981—; vis. research prof. physics No Ariz. U., 1981-82, discipline scientist for spectroscopy and spectrophotometry Internat. Halley Watch, 1981-82. Contbr. numerous articles profl. jours. Fellow Royal Astron. Soc. London; mem. Am. Astron. Soc., Astron. Soc. Pacific, Internat. Astron. Union, Sigma Xi (research prize 1960). Subspecialty: Optical astronomy. Current work: Detection of extended sodium torus associated with Io and Jupiter; imaging and surface photometry of quasar galaxies; cometary spectroscopy; imagery and spectroscopy of quasars and active galaxies. Co-discoverer of ionized water in comets. Home: 1606 E West Wind Way Tempe AZ 85283 Office: Astronomy Group Physics Dept Arizona State University Tempe AZ 85287

WEHRY, EARL LUTHER, JR., chemist, educator, cons.; b. Reading, Pa., Feb. 13, 1941; s. Earl Luther and Florence E. (Webber) W. B.S., Juniata Coll., 1962; Ph.D. (NSF fellow), Purdue U., 1965. Asst. prof. chemistry Ind. U., Bloomington, 1965-70; asst. prof. chemistry U. Tenn., Knoxville, 1970-72, assoc. prof., 1972-77, prof., 1977—. Editor: Modern Fluorescence Spectroscopy, Vols 1 and 2, 1976, Vols 3 and 4, 1981; contbr. numerous articles and revs. on analytical spectroscopy to profl. jours. Research grantee NSF, Electric Power Research Inst., Dept. Energy, NIH, Petroleum Research Fund. Mem. Am. Chem. Soc., Optical Soc. Am., Soc. Applied Spectroscopy, AAAS, Inter-Am. Photochem. Soc., Sigma Xi, Phi Lambda Upsilon. Subspecialties: Analytical chemistry; Spectroscopy. Current work: Molecular fluorescence and phosphorescence spectroscopy, photochemistry analytical applications of laser spectroscopy. Home: 3636 Taliluna Ave Apt 113 Knoxville TN 37919 Office: Dept Chemistry U Tenn Knoxville TN 37996

WEI, JAMES, chemical engineering educator; b. Macao, China, Aug. 14, 1930; came to U.S., 1949, naturalized, 1960; s. Hsiang-chen and Nuen (Kwok) W.; m. Virginia Hong, Nov. 4, 1956; children: Alexander, Christina, Natasha, Randolph. B.S. in Chem. Engring, Ga. Inst. Tech., 1952; M.S., Mass. Inst. Tech., 1954, Sc.D., 1955; grad., Advanced Mgmt. Program Harvard, 1969. Research engr. to research asso. Mobil Oil, Paulsboro, N.J., 1956-62; sr. scientist, Princeton U., N.J., 1963-68, mgr. analysis, N.Y.C., 1969-70; Allan P. Colburn prof. U. Del., Newark, 1971-77; Sherman Fairchild distinguished scholar Calif. Inst. Tech., 1977; Warren K. Lewis prof., head. dept. chem. engring. Mass. Inst. Tech., Cambridge, 1977—; Vis. prof. Princeton, 1962-63, Calif. Inst. Tech., 1965; cons. Mobil Oil Corp., Milliken Corp.; cons. com. on motor vehicle emissions Nat. Acad. Sci., 1972-74, 79-80; mem. sci. adv. bd. EPA, 1976-79; mem. Presdl. Pvt. Sector Survey Task Force on Dept. Energy, 1982-83. Bd. editors: Chem. Tech, 1971—, Chem. Engring. Communications, 1972—; cons. editor chem. engring. series, McGraw-Hill, 1964—; editor-in-chief: Advances in Chemical Engineering, 1980; Contbr. papers, monographs to profl. lit., The Structure of Chemical Processing Industries, 1979. Recipient Am. Acad. Achievement Golden Plate award, 1966. Mem. Am. Inst. Chem. Engrs. (dir. 1970-72, Inst. lectr. 1968, Profl. Progress award 1970, Walker award 1980), Am. Chem. Soc. (award in petroleum chemistry 1966), Nat. Acad. Engring. (nominating com. 1981, peer com. 1980-82, membership com. 1983—), AAAS, Am. Acad. Arts and Scis., Academica Sinica of Taiwan, Sigma Xi. Subspecialties: Chemical engineering; Catalysis chemistry. Home: 420 Waverley Ave Newton MA 02158 Office: Chem Engring Dept Mass Inst Tech Cambridge MA 02139

WEI, JIM P(IAU), mechanical engineer; b. Hsin-Chu, Taiwan, Feb. 25, 1940; came to U.S., 1968, naturalized, 1974; s. Bing-Lang and Mang-Mei (Chen) W.; m. Rose L. Tang, June 15, 1968; children: Tracy, Sherry. B.S., Nat. Cheng-Kung U., Tainan (Taiwan), 1963; M.A.Sc., U. B.C. (Can.), Vancouver, 1967; Ph.D., U.Md., 1972. Sr. Engr. Bechtel Inc., San Francisco, 1972-74; Sr. engr. advanced reactor systems dept. Gen. electric Co., Sunnyvale, Calif., 1974—. Mem. Am. Nuclear Soc. Democrat. Subspecialties: Nuclear fission; Nuclear engineering. Current work: Liquid metal fast breeder reactor thermal-hydraulics and safety, especially in reactor core fluid flow and heat transfer. Office: Gen Electric Co/ARS 310 De Guigne Dr PO Box 508 Sunnyvale CA 94086

WEI, MILLET LUNCHIN, consulting engineer; b. Taiwan, July 1, 1937; came to U.S., 1962, naturalized, 1973; s. Yu Tzu and Mei (Cheng) W.; m. Betty Teresa Leung, June 3, 1967; children: Natalie Vennesa, Terence. B.S., Nat. Taiwan U., Taipei, 1960; M.S., U. R.I., Kingston, 1964; Ph.D., Carnegie-Mellon U., Pitts., 1967. Registered profl. engr., Pa. Research engr. Bethlehem Steel Co., Pa., 1967-73, project engr., 1973-74, cons. engr. tech. services, 1974—. Contbr. articles to profl. jours. Served to 2d lt. Chinese Marine Corps, 1960-61. Mem. Assn. Iron and Steel Engrs. (Kelly award 1981), AIME, ASME, ASCE, Chinese Inst. Engrs. U.S.A. (dir.). Republican. Roman Catholic. Club: Shepherd Hills Country (Wescosville, Pa.). Subspecialties: Civil engineering; Mechanical engineering. Current work: Hardware technology for steelmaking in basic oxygen furnaces and electric furnaces, cooling technology for blast furnaces, modern technology for tall coke oven batteries, equipment technology for rolling mills, ladles, slag pots, heat exchangers, pressure vessels, reheat furnaces and continuous casters. Patentee in field. Home: 29 Fairway Ln Wescosville PA 18106 Office: 1709 Martin Tower 8th Ave Bethlehem PA 18016

WEI, ROBERT, clinical biochemist; b. Hawaii, Apr. 16, 1939; m. Soraya Naghshineh. Ph.D., George Washington U., 1972. Diplomate: Am. Bd. Clin. Chemists. Sr. scientist Electro-Nucleonics Lab., Inc., Bethesda, Md., 1975-78; assoc. prof. clin. biochemistry Cleve. State U., 1978—. Served with U.S. Army, 1962-64. Mem. Am. Assn. Immunologists, Am. Assn. Clin. Chemistry. Democrat. Episcopalian. Subspecialties: Clinical chemistry; Immunobiology and immunology. Current work: Structure and function of biomolecules involved in immune reactions. Home: 3382 Hollister Rd Cleveland Heights OH 44118 Office: Dept Chemistry Cleve State U Cleveland OH 44115

WEI, WEI-ZEN, tumor immunologist, researcher; b. Taiwan, May 1, 1951; d. Cheng-Hwa and Shen-Hung Lin; m. Kuang-Chung Wei, Aug. 12, 1973; children: John Brown, Benjamin. Ph.D., Brown U., 1978. Postdoctoral fellow pathology U. Conn. Health Center, Farmington, 1978-79; immunopathology lab. supr. dept. pathology Wright State U., Dayton, Ohio, 1979-80; postdoctoral research assoc. Ohio State U., Columbus, 1980-81, instr., 1981—. Contbr. articles to sic. jours. Am. Cancer Soc. grantee, 1980-81, 81-82; Ohio State U. grantee, 1982-83. Mem. Am. Assn. Immunologists. Mem. Chinese Christian Ch. Subspecialties: Cancer research (medicine); Immunology (medicine). Current work: Tumor immunology and immunotherapy; using anti-tumor monoclonal antibodies to study cancer biology and therapy. Home: 1407 Meadow Moor Dr Xenia OH 45385 Office: 333 W 10th Ave Columbus OH 43210

WEIBLEN, WILLIAM ACHORN, mech. engr.; b. Whitefield, Maine, June 2, 1938; s. Ervin Charles and Ellen Achorn (Smith) W.; m. Eileen Brown; children: Peter Michael, Elizabeth Ellen. B.S.M.E., U.Maine, 1961; M.S.M.E., U. Conn., 1967, M.B.A., 1972. Registered profl. engr., Conn. With comml. engring. div. Pratt & Whitney Aircraft, East Hartford, Conn., 1961—, asst. design project engr., 1968-81, staff project engr., 1981—; lectr. to engring. groups. Cubmaster Pack 122 Long Rivers council Boy Scouts Am., 1975-78; scoutmaster Troop 122, 1981-82. Mem. ASME (nat. agenda bd., region 1 operating bd.), Soc. Am. Value Engrs. Subspecialty: Product cost reduction. Current work: Mgmt. of product cost reduction budget assigned to engring. dept. Office: 400 Main St East Hartford CT 06108

WEIDE, BRUCE WARREN, computer scientist; b. Toledo, Dec. 13, 1952; s. Harley Warren and Betty Jo Ann (Habig) W. B.S.E.E., U. Toledo, 1974; Ph.D., Carnegie-Mellon U., Pitts., 1978. Asst. prof. computer and info. sci. Ohio State U., Columbus, 1978-83, assoc. prof., 1983—. NSF grad. fellow, 1974-77; IBM grad. fellow, 1977-78; NSF grantee, 1979-81. Mem. Assn. Computing Machinery, IEEE. Subspecialties: Algorithms; Software engineering. Current work: Data structures, algorithms, process control systems design aids. Home: 5425 Rockport St Columbus OH 43220 Office: 2036 Neil Ave Mall Columbus OH 43210

WEIDLINGER, PAUL, consulting civil engineer; b. Budapest, Hungary, Dec. 22, 1914; came to U.S., 1944, naturalized, 1948; s. Andrew and Juliette W.; m. Solveig Hojberg, Dec. 26, 1963; children: Thomas Evarist, Pauline Dana, Jonathan Niels. M.S., Swiss Poly. Inst., Zurich, 1937. Dir. engring. Sociedad Constructora Nacional, La Paz, Bolivia, 1942; prof. structural engring. San Andres U., La Paz, 1942-43; chief engr. Bur. Reclamation, La Paz, 1943-46, Atlas Aircraft, N.Y.C., 1944-46; sr. ptnr. Weidlinger Assocs., N.Y.C., 1948-; vis. lectr. Harvard U., 1946, MIT, 1964. Contbr. numerous articles to profl. jours. Recipient Engring. News Record award, 1966. Mem. Nat. Acad. Engring., N.Y. Acad. Scis., ASCE (James R. Croes medal 1963, Moisseiff award 1975), Engring. Earthquake Research Inst., AIAA. Subspecialties: Civil engineering; Solid mechanics. Current work: Structural engineering, defense systems, engineering seismology. Home: 301 E 47th St New York NY 10017

WEIFFENBACH, JAMES MILTON, phychologist; b. St. Louis, Aug. 30, 1935; s. Milton Wesley and Gladys Henrietta W.; m. Antoinette D.; children: Jay Wesley, Margaret Emma. B.A., Brown U., 1957; Ph.D., McGill U., 1964, M.A., 1962. Research assoc., research asst. prof. U. Ill., 1964-66; research fellow Harvard U., 1966-67; staff fellow Nat. Inst. Child Health and Human Devel., 1967-69; sr. staff fellow, research psychologist Nat. Inst. Dental Research, Bethesda, Md., 1970—; instr. Found. Advanced Edn. in Scis., Bethesda. Editor: Taste and Development—The Genesis of Sweet Preference; contbr. articles to profl. jours. Served to lt. (j.g.) USN, 1957-60. Mem. Soc. Neurosci., Assn. Chemo Reception Scis., Sigma Xi. Democrat. Unitarian. Subspecialties: Sensory processes; Behavioral psychology. Current work: Psychophysical measurement of human oral sensory phenomena, taste. Office: NIH Bldg 10 Rm 1-A-05 Bethesda MD 20205

WEIGEL, PAUL HENRY, biochemist, educator; b. N.Y.C., Aug. 11, 1946; s. Helmut and Jeanne B. (Wakeman) W.; m. Nancy L. Shulman, June 15, 1968; 1 dau., Dana Jeanne. B.A., Cornell U., 1968; M.A., Johns Hopkins U., 1969, Ph.D., 1975. Asst. prof. biochemistry U. Tex. Med. Br., Galveston, 1978-82, assoc, prof., 1982—; reviewer grants and manuscripts Am. Cancer Soc., NSF and profl. jours., 1978—; cons. in field. Contbr. chpts. to books and articles to profl. jours. Served with U.S. Army, 1969-71. Nat. Cancer Inst. fellow, 1975-78; NIH grantee, 1979—, 81—; Robert A. Welch Found. grantee, 1979—. Mem. Am. Chem. Soc., Am. Soc. Cell Biology, Am. Soc. Biol. Chemists, AAAS, Sigma Xi. Unitarian. Subspecialties: Biochemistry (medicine); Cell biology (medicine). Current work: Receptor mediated endocytosis; mechanism of function and recycling of the hepatic asialoglycoprotein receptor; cell surface carbohydrate receptors. Home: 5100 Ash Ct Dickinson TX 77539 Office: University of Texas Medical Branch Division of Biochemistry Basic Science Bldg Room 630 Galveston TX 77550

WEIGEL, RICHARD GEORGE, psychologist; b. St. Louis, Feb. 23, 1937; s. George Dwane and Irene Katheryn (Bretz) W.; m. Virginia Ann Morris, Aug. 29, 1964; children—Paul Karl, Laura Katherine. B.A., DePauw U., 1959; M.A., U. Mo., 1962, Ph.D., 1968. Diplomate: Am. Bd. Profl. Psychology; lic. psychologist, Colo. Instr. U. Mo., Columbia, 1962-63; asst. prof. Oreg. State U., Corvallis, 1964-67; asst. prof., then assoc. prof. Colo. State U., Fort Collins, 1967-75, prof., 1975-78; cons. psychologist Rohrer, Hibler & Replogle, Inc., Denver, 1978-81, mgr., 1981—; pvt. cons. practice, Fort Collins, 1970-78; mem. profl. bd. Denver U. Sch. Profl. Psychology, 1976-78. Bd. dirs. Benton County Mental Health Assn., Corvallis, 1965-67. NIMH tng. grantee, 1977-82. Fellow Am. Psychol. Assn.; mem. Rocky Mountain Psychol. Assn. (life; pres. 1973-74), Colo. State Bd. Psychologist Examiners (v.p. 1974-76), Am. Bd. Profl. Psychology (chmn. intermountain regional bd. 1981), Colo. Psychol. Assn. (bd. dirs. 1973-74), Am. Bd. Profl. Psychology (dir. 1981—), Sigma Xi, Psi Chi, Phi Mu Alpha., Phi Gamma Delta. Current work: Research methodology pertinent to emerging field of consulting psychology. Home: 6104 W Leawood Dr Littleton CO 80123 Office: Rohrer Hibler and Replogle Inc 441 Wadsworth Blvd Suite 200 Denver CO 80226

WEIGHT, FORREST F., pharmacology researcher; b. Waynesboro, Pa., Apr. 17, 1936; s. Forrest F. and Nina K. (Beaver) W.; m. Yvonne F. (Tinklenberg), Dec. 27, 1961; children: Christopher, Eric, Elizabeth. A.B., Princeton U., 1958; M.D., Columbia U., 1962. Intern medicine and pediatrics U. N.C. Hosp., 1962-63; resident in medicine Mary Imogene Bassett Hosp., 1963-64; research assoc. NIHM, 1964-68, chief sect. synaptic pharmacology, 1969-78; chief Lab. Preclin. Studies, Nat. Inst. Alcohol Abuse and Alcoholism, 1978—; vis. scientist U. Goteberg, Sweden, 1968-69. Served with USPHS, 1976—. Mem. Am. Physiol. Soc., Am. Soc. Pharmacology and Exptl. Therapeutics, Soc. Neurosci. (pres. Potomac chpt. 1979-80), NIHM Assembly Scientist (sec.-treas. 1978-79), Soc. Biol. Psychiatry. Subspecialties: Neurophysiology; Neuropharmacology. Current work: Cellular mechanisms of communication and information processing in the nervous system. Office: Lab Preclin Studies NIAAA Rockville MD 20852

WEIKEL, MAURICE MARCEL, dental manufacturing company executive; b. Winnemucca, Nev., May 1, 1921; s. Charles Elmer and Julie (Sairde) W.; m. Lorraine H. Hansen, Sept. 1, 1944; children: Larry, Gary, Dennis, Kristi, Joni. B.S., U. Pacific, 1945; D.D.S., 1945. Owner Dental Clinic, San Diego, 1945-67; pres. CIA Minera Tecate, Tecate, Mex., 1955-58, N.K. Metals, San Diego, 1953-58, Am. Silver & Mercury, 1966-73, Dispersalloy Inc., El Cajon, Calif., 1973-74, Koberly Inc., El Cajon, 1974—; dir. U. Pacific Dental Sch., San Francisco, 1969-72; cons. in field. Served to lt. USN, 1945-46. Mem. ADA, Internat. Assn. Dental Researchers, Am. Powder Mettallurgy Inst., PSI-Omega. Republican. Lodge: Kiwanis. Current work: Dental filling material for posterior teeth. Patentee: Dental Anchor Prothesis, 1976; Disposable Dental All Container, 1978; Dental Tablet and Mercury Dispenser, 1978; Sliding Dental Dispenser, 1982. Office: Koberly Inc 1050 Greenfield Dr PO Drawer D El Cajon CA 92022 Home: 3537 S Buena Vista Dr Las Vegas NC 89132

WEIL, MARVIN L(EE), pediatrician, neurologist, educator, researcher; b. Gainesville, Fla., Sept. 28, 1924; s. Joseph and Anna (Abrams) W.; m. Joyce Sari Zimmerman, Mar. 21, 1928; children: Daniel, Clifford, Meredith. B.S., U. Fla., 1942; M.D., Johns Hopkins U., 1946. Diplomate: Am. Bd. Pediatrics, Am. Bd. Psychiatry and Neurology. Intern Duke Hosp., Durham, N.C., 1946-47, resident, 1947-48, Cin. Children's Hosp., 1950-52; clin. instr. in pediatrics U. Cin., 1952-53; practice medicine specializing in pediatrics, Miami Beach, Fla., 1953-65, North Miami Beach, Fla., 1953-65; Nat. Inst. Neurol. Diseases and Blindness spl. fellow in pediatric neurology Johns Hopkins Hosp., Balt., 1965-66, UCLA Ctr., Health Scis., 1966-68; asst. prof. pediatrics and neurology UCLA, 1968-72, assoc. prof., 1972-78, prof., 1978—; chief div. pediatric neurology Harbor-UCLA Med. Ctr., Torrance, Calif., 1968—; Fogarty Internat. Ctr. fellow Karolinska Inst., Stockholm, 1967-77. Contbr. sci. and tech. articles to profl. jours. Served to capt. M.C. AUS, 1948-50. Fellow Am. Acad. Pediatrics, Am. Assn. Immunology, Am. Assn. Mental Deficiency (exec. bd. 1971-76, chmn. Region II 1973-74), N.Y. Acad. Sci., Los Angeles Soc. Neurology and Psychiatry (sec.-treas. 1978-80, pres. 1981, dir. 1982-84). Democrat. Jewish. Subspecialties: Pediatrics; Neuroimmunology. Current work: Infections of central nervous system; child neurology; neuro-immunology. Home: 7030 Starstone Dr Rancho Palos Verdes CA 90274 Office: Div Pediatric Neurology Harbor-UCLA Med Center 1000 W Carson St Torrance CA 90509

WEIL, RICHARD, III, surgery educator; b. N.Y.C., Feb. 22, 1936; s. Richard and Allene (Hall) W.; m. Polly Wood Edgar, Aug. 22, 1959; children: Wendy Hollings, Richard MacCoy. A.B., Princeton U., 1957; M.D., Columbia U., 1961. Intern Columbia Presbyn. Med. Ctr., N.Y.C., 1961-62, resident, 1962-63, 65-69; instr. surgery Columbia U., N.Y.C., 1969, asst. prof. surgery, 1970-74; fellow transplant surgery U. Minn., Mpls., 1970; assoc. prof. surgery U. Colo., Denver, 1974-79, prof., 1979—; dir. organ transplantation Health Scis. Ctr., 1980—. Contbr. articles to profl. jours. Served to capt. U.S. Army, 1963-65. Fellow ACS; mem. Am. Soc. Transplant Surgeons, Am. Soc. Artificial Internal Organs, Am. Surg. Assn., Assn. for Acad. Surgery, Societe Internationale de Chirurgie, Soc. Univ. Surgeons, Transplantation Soc., Central Surg. Assn., Harvey Soc., Internat. Cardiovascular Soc. Subspecialties: Surgery; Transplant surgery. Current work: Transplantation surgery and transplantation biology. Home: 2012 E 4th Ave Denver CO 80206 Office: U Colo Health Sci Center Box C-305 4200 E 9th Ave Denver CO 80262

WEILER, JOHN MAYER, physician, researcher; b. Erie, Pa., Mar. 19, 1945; s. Ad R. and Ruth (Schlosser) W. B.S., U. Mich, 1967; M.D., Temple U., 1971. Diplomate: Am. Bd. Internal Medicine. Intern Ind. U. Hosp., 1971-72; resident Kans. U. Med. Ctr., 1974-75; research fellow Harvard U., 1975-77; asst. prof. U. Iowa, Iowa City, 1977-83, assoc. prof., 1983—. Served with USPHS, 1972-74. Recipient Research Assoc. Career award VA, 1979-82; recipient Clin. Investigation Career award VA, 1983-85; NIH research career devel. awardee, 1983—. Fellow ACP; mem. Am. Acad. Allergy, Am. Assn. Immunologists, Am. Rheumatism Assn. Subspecialties: Immunology (medicine); Allergy. Office: SW34E GH U Iowa Iowa City IA 52242

WEILER, KURT WALTER, science administrator; b. Phoenix, Mar. 16, 1943; s. Henry C. and Dorothy Marie (Esser) W.; m. Geertje Weiler-Stoelwinder, June 8, 1979. B.S., U. Ariz., 1964; Ph.D., Calif. Inst. Tech., 1970. Sr. sci. officer Westerbork Radio Obs., Dwingeloo, Netherlands, 1970-72; fgn. guest collaborator U. Groningen, Netherlands, 1972-74; sci. collaborator Laboratorio di Radioastronomia, Bologna, Italy, 1975-76; research assoc. Max Planck Inst. Radioastronomy, Bonn, W. Ger., 1976-79; program dir. NSF, Washington, 1979—. Contbr. numerous articles to profl. jours. Mem. Am. Astron. Soc., Royal Astron. Soc., Internat. Astron. Union, Nederlandse Astronomen Club. Subspecialties: High energy astrophysics; Radio and microwave astronomy. Current work: Astronomical research on supernovae, supernova remants. Patentee in field. Office: Div Astron Scis NSF Washington DC 20550

WEILER, MARGARET HORTON, physicist; b. Sewickley, Pa., Apr. 30, 1941; d. Clarence Reuben and Louise Carolyn (Gardner) Horton; m. William McCoy Weiler, June 16, 1962; children: Christopher William, Theodore Reuben. A.B., Radcliffe Coll., 1962; M.S., U. Maine, 1964; Ph.D., MIT, 1977. Instr. physics U. Maine, 1964-65; mem. staff Francis Bitter Nat. Magnet Lab., MIT, 1965-74, asst. prof. dept. physics, 1977-83; sr. research scientist, research div. Raytheon Corp., Lexington, Mass., 1983—. Contbr. articles to profl. jours. Mem. AAAS, IEEE, Am. Phys. Soc., Assn. Women in Sci. Subspecialty: Condensed matter physics. Current work: Research in monolithic microwave semiconductor devices. Home: 52 Willow St Belmont MA 02178 Office: Raytheon Research Div 131 Spring St Lexington MA 02173

WEILL, HANS, physician, educator; b. Berlin, Germany, Aug. 31, 1933; came to U.S., 1939; s. Kurt and Gerda (Philipp) W.; m. Kathleen Burton, Apr. 3, 1958; children: Judith, Leslie, David. B.S., Tulane U., 1955, M.D., 1958. Diplomate: Am. Bd. Internal Medicine. Intern Mt. Sinai Hosp., N.Y.C., 1958-59; resident Tulane Med. Unit, Charity Hosp. La., New Orleans, 1959-60, chief resident, 1961-62, sr. vis. physician, 1972—; NIH research fellow dept. medicine and pulmonary lab. Sch. Medicine Tulane U., New Orleans, 1960-61, instr. in medicine, 1962-64, asst. prof. medicine, 1964-67, asso. prof., 1967-71, prof. medicine, 1971—; chief pulmonary diseases sect. Tulane Med. Center, 1980—; dir. interdisciplinary research group in occupational lung diseases Nat. Heart, Lung and Blood Inst., 1972—, chmn. pulmonary disease adv. com., 1982-84; active staff Tulane Med. Center Hosp., 1976—; attending physician VA Hosp., New Orleans, 1963—; cons. pulmonary diseases and medicine USPHS Hosp., New Orleans, 1964—; cons. pulmonary diseases Touro Infirmary, New Orleans, 1962—; cons. NIH, Nat. Inst. Occupational Safety and Health, Occupational Safety and Health Adminstrn., USN, Nat. Acad. Scis., U.S. EPA. Editorial bd.: American Review of Respiratory Disease, 1980; editor: Respiratory Diseases Digest, 1981; guest editor: Byssinosis Conference Supplement, Chest, 1981; letr. participant workshops and confs. profl. groups in field, U.S., France, Can., U.K. Fellow Am. Acad. Allergy, Royal Soc. Medicine, A.C.P.; mem. Am. Thoracic Soc. (pres. 1976), Am. Lung Assn. (bd. dirs. 1975-78), New Orleans Acad. Internal Medicine (sec., treas. 1973-75), Am. Coll. Chest Physicians (gov. for La. 1970-75), Am. Fedn. Clin. Research, So. Soc. Clin. Investigation, N.Y. Acad. Scis., Brit. Thoracic Assn., Internat. Epidemiol. Assn., Am. Heart Assn. (task force on environment and cardiovascular system 1978), Phi Beta Kappa, Alpha Omega Alpha. Subspecialties: Pulmonary medicine; Epidemiology. Current work: Occupational lung dieases. Home: 333 Friedrichs Ave Metairie LA 70005 Office: 1700 Perdido St New Orleans LA 70012

WEIMBERG, RALPH, biochemist; b. San Diego, Dec. 22, 1924; s. Morris and Rose (Goodman) W.; m. Marilyn Rene Simon, June 20, 1952; children: Marcie Ann, Gary Richard. B.A., U. Calif.-Berkeley, 1949, M.A., 1951, Ph.D., 1955. Research asst. U. Calif.-Berkeley, 1951-55; research assoc. Oak Ridge (Tenn.) Nat. Lab., 1955-56; instr. Western Res. U., Cleve., 1956-58; biochemist No. Utility Research and Devel. Div., Peoria, Ill., 1958-65, U.S. Salinity Lab., Riverside, Calif., 1965—. Served with U.S. Army, 1942-45; ETO. Mem. Am. Soc. Biol. Chemists, AAAS, Am. Soc. Microbiology, Am. Soc. Plant Physiologist, N.Y. Acad. Sci. Jewish. Subspecialties: Plant physiology (biology); Biochemistry (biology). Current work: Effects of salinity and other harmful environments on plant physiology and metabolism. Home: 5444 Quince St Riverside CA 92506 Office: US Salinity Lab 4500 Glenwood Dr Riverside CA 92501

WEIN, ALAN JEROME, urologist, educator, researcher; b. Newark, Dec. 15, 1941; s. Isadore R. and Jeannette Francis (Abrams) W.; m. Frances Ann Martin, Dec. 30, 1975; children: Allison, Rebecca. A.B., Princeton U., 1962; M.D., U. Pa., 1966. Diplomate: Am. Bd. Urology. Instr. urology U. Pa., 1971-72, asst. prof. urology, 1974-76, assoc. prof., 1976—, chmn. div. urology, 1980—. Contbr. articles to profl. jours. Served to maj., M.C. U.S. Army, 1972-74. NIH; VA grantee. Mem. ACS, Soc. Univ. Urologists, Soc. Univ. Surgeons, Am. Assn. Surgery Trauma, Am. Soc. Pharmacology and Exptl. Therapeutics, Am. Soc. Clin. Oncology, Am. Urologic Assn., Sigma Xi, Alpha Omega Alpha. Republican. Jewish. Lodge: Rotary. Subspecialties: Urology; Neuropharmacology. Current work: Urology, neurophysiology and neuropharmacology of bladder function. Home: 502 Addison Ct Philadelphia PA 19147 Office: 5 Silverstein U Pa 3400 Spruce St Philadelphia PA 19104

WEINBERG, JERRY LLOYD, astronomer; b. Detroit, Dec. 2, 1931; s. Max and Ethel R. (Cherniak) W.; m. Marcia Hawver, July 2, 1961; children: Mark T., Scott T. B.S., St. Lawrence U., 1958; Ph.D., U. Colo., 1962. Asst. prof. astrogeophysics U. Hawaii, 1963-65; assoc. astronomer Haleakala Obs., Kula, Hawaii, 1965-68, supr., 1966-67; astronomer Dudley Obs., Albany, N.Y., 1968-70, astronomer, assoc. dir., 1970-73; research prof. astronomy and space sci. SUNY-Albany, 1973-80, dir., 1973-80; research scientist dept. astronomy, dir. Space Astronomy Lab., U. Fla., 1980—. Contbr. articles to sci. jours. Served with USAF, 1951-55. Recipient Exceptional Sci. Achievement medal NASA, 1975. Mem. Am. Astron. Soc., Astron. Soc. Pacific, AAAS, AIAA, Internat. Astron. Union (pres. Commn. 21 1973-76), Internat. Assn. Geomagnetism and Aeronomy, Com. Space Research, Optical Soc. Am., Am. Geophys. Union, Phi Beta Kappa, Sigma Xi, Pi Mu Epsilon, Sigma Pi Sigma. Subspecialties: Optical astronomy; Space astronomy. Office: Space Astronomy Lab U Fla 1810 NW 6th St Gainesville FL 32601

WEINBERG, KENNETH STEVEN, researcher, biology educator; b. Brookline, Mass., June 5, 1947; s. Joseph and Dora (Rubin) W.; m. Natalie Jane Edgers, July 4, 1971; children: Jonathan, Matthew. B.A., Boston U., 1969, Ph.D., 1979; M.Sc. in Hygiene, U. Pitts., 1970. Postdoctoral fellow New Eng. Med. Ctr., Boston, 1978-81, Parker B. Francis Found. fellow, 1981-82, mem. spl. and sci. research staff, 1982—, asst. prof. medicine, 1982—; lectr. Boston U., 1981-82. Instr. CPR, Am. Heart Assn./ARC, 1982—. Mem. AAAS, Am. Fedn. Clin. Research. Jewish. Club: Randolph Jaycees (pres. 1982-83). Subspecialties: Cell biology (medicine); Pathology (medicine). Current work: Pulmonary cellular biology, role of the endothelium in pulmonary injury and repair, morphologic and biochemical correlates of cellular injury and repair, pulmonary ultrastructure. Office: New England Medical Center 171 Harrison Ave Boston MA 02111

WEINBERG, RICHARD ALAN, computer scientist; b. Mpls., Sept. 7, 1951; s. Edward Maurice and June Adele (Lieberman) W. B.A., Cornell U., 1973; M.S., U. Minn.-Mpls., 1974-80. Electronics engr. NASA Johnson Space Ctr., Houston, 1977; engr. Lockheed Electronics, Houston, 1977-79; computer scientist Cray Research, Inc., Mpls., 1980—. Contbr. articles to Spl. Interest Group Computer Graphcs publs.; dir.: computer-animated film Euclidean Illusions, 1978 (Cine Golden Eagle 1978). Mem. Assn. Computing Machinery, Spl. Interest Group Computer Graphics, IEEE. Subspecialties: Graphics, image processing, and pattern recognition; Computer architecture. Current work: Research and development of high-speed computer architectures and algorithms for synthesis of complex computer graphics images; Very large scale integrated circuit architectures; computer-synthesized animation. Office: Cray Research Inc 3416 S La Cienega Blvd Los Angeles CA 90016

WEINBERG, RICHARD ALAN, psychologist, educator; b. Chgo., Jan. 28, 1943; s. Meyer and Mollie Idell (Soell) W.; m. Gail Ellen Blumberg, Aug. 25, 1964; children: Eric, Brett. B.S., U. Wis., 1964; M.A.T., Northwestern U., 1965; Ph.D., U. Minn., 1968. Asst. prof. Columbia Tchrs. Coll., N.Y.C., 1968-70; from asst. to assoc. prof. ednl. psychology, child psychology and psychology U. Minn., Mpls., 1970-77, prof., co-dir., 1977—, dir. psychology in the schs., 1975-80. Author: books, including Classroom Observer, 1977; Contbr. articles to profl. jours. Fellow Am. Psychol. Assn. (mem. accreditation com. 1977-81); mem. Soc. for Research in Child Devel., Am. Ednl. Research Assn. Subspecialties: Developmental psychology; School psychology. Current work: Psychoeducational assessment of early childhood development, individual differences; behavior genetics; adolescent involvement in sport. Office: Univ Minn N548 Elliott Hall 75 E River Rd Minneapolis MN 55455

WEINBERG, ROBERT ALLEN, biology educator; b. Pitts., Nov. 11, 1942; s. Fritz E. and Lore R. W.; m. Amy Shulman, Aug. 29, 1976; children: Aron, Leah Rosa. B.A., MIT, 1964, Ph.D., 1969. Postdoctoral fellow Weizmann Inst., Rehovoth, Israel, 1969-70, Salk Inst., La Jolla, Calif., 1970-72; research assoc. fellow MIT, 1972-73, asst. prof. biology dept. biology and, 1973-76, assoc. prof., 1976-79, prof., 1979—; research scholar Mass. div. Am. Cancer Soc., 1974-77. Named Scientist of Yr. Discover Mag., 1982; recipient nat. divisional award Mass. div. Am. Cancer Soc., 1982; Rita Allen Found. scholar, 1976-80. Subspecialties: Molecular biology; Cell and tissue culture. Current work: Molecular basis of carcinogenesis; isolation and characterization of tumor oncogenes. Office: Center for Cancer Research MIT 77 Massachusetts Ave Cambridge MA 02139

WEINBERG, ROBERT STEPHEN, sport psychologist; b. N.Y.C., June 15, 1948; s. Harold and Jeanette (Stone) W.; m. Heather Kay Hardy, Jan. 8, 1978 (div. Apr. 1981). B.S., Bklyn. Coll., 1970; M.S., UCLA, 1972, M.A., 1975, Ph.D., 1977. Postdoctoral fellow UCLA, 1978; tchr. N.Y.C. pub. schs., Bklyn., 1971-72; instr. Bklyn. Coll., 1972-74; teaching fellow UCLA, 1974-77, lectr., 1977-78; asst. prof. phys. edn. North Tex. State U., Denton, 1978—; cons. Dallas/Ft. Worth Sch. Dists., 1981—. Author: Health Related Fitness, 1982; Editor: Psychological Foundations in Sport, 1983; Contbr. articles to profl. jours. Fellow AAHPER; mem. Am. Psychol. Assn., N.Am. Soc. for Sport Psychology, Tex. Assn. for Health and Phys. Edn. (chmn. research div. 1982-83). Democrat. Jewish. Subspecialty: Social psychology. Current work: Mental preparation strategies for competition, confidence building, intrinsic motivation, stress management in sport. Office: North Tex State U Dept Phys Edn Denton TX 76203

WEINBERG, STEVEN, physicist, educator; b. N.Y.C., May 3, 1933; s. Fred and Eva (Israel) W.; m. Louise Goldwasser, July 6, 1954; 1 dau., Elizabeth. B.A., Cornell U., 1954; postgrad., Copenhagen Inst. Theoretical Physics, 1954-55; Ph.D., Princeton U., 1957; A.M. (hon.), Harvard U., 1973; Sc.D. (hon.), Knox Coll., 1978, U. Chgo., 1978, U. Rochester, 1979, Yale U., 1979, CUNY, 1980, Clark U., 1982. Research asso., instr. Columbia U., 1957-59; research physicist Lawrence Radiation Lab., Berkeley, Calif., 1959-60; mem. faculty U. Calif.-Berkeley, 1960-69, prof. physics, 1964-69; vis. prof. MIT, 1967-69, prof. physics, 1969-73; Higgins Prof. physics Harvard U., 1973-83; sr. scientist Smithsonian Astrophys. lab., 1973—; Josey prof. sci. U. Tex.-Austin, 1982—; cons. Inst. Def. Analyses, Washington, 1960-73, ACDA, 1970-73; mem. Pres.'s Com. on Nat. Medal of Sci., 1979-82, Council of Scholars, Library of Congress, 1983—; sr. adv. La Jolla Inst.; chair in physics Collège de France, 1971; sr. adv. NRC Com. on Internat. Security and Arms Control, 1981; vis. prof. Stanford U., 1976-77, U. Tex., 1981; dir. Jerusalem Winter Sch. Theoretical Physics, 1983—; Silliman lectr. Yale U., 1977; Richtmeyer lectr., 1974; Scott lectr. Cavendish Lab., 1975; Lauritsen Meml. lectr. Calif. Inst. Tech., 1979; Bethe lectr. Cornell U., 1979; de Shalit lectr. Weizman Inst., 1979; Schild lectr. U. Tex., 1979; Sloan fellow, 1961-65; Loeb lectr. in physics Harvard U., 1966-67; Cherwell-Simon lectr. Oxford U., 1983; Bampton lectr. Columbia U., 1983; Morris Loeb vis. prof. physics Harvard U., 1983-84. Author: Gravitation and Cosmology: Principles and Applications of the General Theory of Relativity, 1972, The First Three Minutes: A Modern View of the Origin of the Universe, 1977, The Discovery of Subatomic Particles, 1982; research and publs. on elementary particles, quantum field theory, cosmology.; co-editor, Cambridge Univ. Press; Co-editor monographs on math. physics; editorial bd., Progress in Sci. Culture, Advances in Applied Math., U. Chgo. Press, series on theoretical astrophysics. Recipient J. Robert Oppenheimer meml. prize, 1973, Dannie Heineman prize in math. physics, 1977; Am. Inst. Physics-U.S. Steel Found. sci. writing award, 1977; Nobel prize in physics, 1979; Elliott Cresson medal Franklin Inst., 1979. Mem. Am. Acad. Arts and Scis. (council), Am. Phys. Soc. (past councilor at large), Nat. Acad. Sci., Am. Astron. Soc., Council Fgn. Relations, Am. Philos. Soc., Royal Soc. London (fgn. mem.), Am. Mediaeval Acad. Subspecialties: Theoretical physics; Theoretical astrophysics. Office: Dept Physics U Tex Austin TX 78712

WEINBERG, UZI, endocrinology and biology clin investigator; b. Tel Aviv, Israel, Apr. 19, 1939; U.S., 1972; s. Elimelech and Hava (Listek) W.; m. Nira Yaron, Apr. 4, 1967; children: Odealia, Maya. M.D., U. Paris, 1968. Lic. physician, N.Y. Extern Hospitaux de Paris, 1964; asst. prof. N.Y. U., 1976-77; asst. prof. medicine and neurology Albert Einstein Coll. Medicine, 1979-82, assoc. prof., 1982—; cons. Eli Lilly Co., 1980—. Contbr. articles profl. jours.; reviewer profl. jours. Served to capt. Israeli Def. Force, 1957-60. Ford Found. grantee, 1977-78; NIH grantee, 1979-82. Mem. Am. Endocrine Soc., Am. Fedn. Clin. Research, Internat. Soc. Internat., Soc. Sleep Research, Internat. Soc. Endocrinology. Subspecialties: Neuroendocrinology; Biochemistry (medicine). Current work: Developmental neurobiology and endocrinology. Inventor RIA for melatonin, discoverer novel biochemical mechanism concering puberty.

WEINBERGER, MYRON HILMAR, physician; b. Cin., Sept. 21, 1937; s. Samuel and Helen Eleanor (Price) W.; m. Myrna Minnie Rosenberg, June 12, 1960; children: Howard David, Steven Neal, Debra Ellen. B.S., Ind. U., 1959; M.D., Ind. U.-Indpls., 1963. Intern Ind. U., Indpls., 1963-64, resident, 1964-66; fellow Ind. Heart Assn., 1962; USPHS trainee Stanford U. Med. Ctr., 1966-67, USPHS spl. fellow, 1968-69; asst. prof. medicine Ind. U.-Indpls., 1969-73, assoc. prof., 1973-76, prof., 1976—, dir., 1980—, 1976-80, investigator, 1970-76. Contbr. articles to profl. jours. Fellow ACP, Am. Heart Assn. (council on high blood pressure research), Am. Coll. Nutrition, Am. Coll. Cardiology; mem. Endocrine Soc., AAAS, Am. Heart Assn., Am. Soc. Clin. Pharmacology Therapeutics, Am. Soc. Nephrology, Internat. Soc. Nephrology, Internat. Soc. Hypertension, Soc. Exptl. Biology and Medicine, Central Soc. Clin. Research, Midwest Salt and Water Club. Subspecialties: Internal medicine; Endocrinology. Current work: Causes and treatment of hypertension. Home: 8015 Dartmouth St Indianapolis IN 46260 Office: Ind U Med Center 541 Clinical Dr Indianapolis IN 46223

WEINDRUCH, RICHARD HOWARD, biogerontologist, researcher, educator; b. Rock Island, Ill., Mar. 18, 1950; s. Jacob and Alice W. B.S. in Biology, U. Ill.-Urbana, 1972, M.S., 1973; postgrad., Northwestern U. Dental Sch., 1973-74; Ph.D. in Exptl. Pathology, UCLA, 1978. Postdoctoral fellow UCLA, 1978-80, adj. asst. prof. I pathology, 1980-82, adj. asst. prof. III, 1982—; lectr. on research on dietary restriction as modifier of aging process. Contbr. articles on dietary restriction and longevity to profl. publs. USPHS grantee, 1981-86. Mem. Gerontol. Soc., Am. Aging Assn., AAAS. Subspecialties: Gerontology; Immunobiology and immunology. Current work: Effects of nutrition on aging process, in particular, effects of dietary restriction on aging in mice.

WEINER, HENRY, biochemistry educator; b. Cleve., May 18, 1937; s. Philip and Hilda (Dapeer) W.; m. Esther Riza Blankfeld, June 11, 1960; children: Suzanna Lynn, Alexander James. B.S. in Chem. Engring., Case Inst. Tech., 1959; Ph.D., Purdue U., 1963. Research assoc. Brookhaven Nat. Lab., Upton, N.Y., 1963-65; NIH fellow Karolinska Inst., Stockholm, 1965-66; asst. prof. biochemistry Purdue U., West Lafayette, Ind., 1966-69, assoc. prof., 1969-76, prof., 1976—; mem. rev. bd. Ind. Heart Assn., 1974-78, Nat. Inst. Alcohol Abuse and Alcoholism, 1977, 82, 83—. Editor: Enzymology of Carbonyl Metabolism, 1982; contbr. articles to sci. jours. Pres. Temple Israel, West Lafayette, 1980-82; Democratic precinct chmn., 1970-72. NSF grantee, 1967-82; Nat. Inst. Alcohol Abuse grantee, 1978-83; also career devel. award. Mem. Am. Soc. Biol. Chemists, Am. Chem. Soc., Research Soc. on Alcoholism. Subspecialty: Biochemistry (biology). Current work: Enzymology; aldehyde and alcohol metabolism; enzymology of aldehyde oxidizing systems; alcohol and acetaldehyde metabolism. Home: 2253 Indiana Trail West Lafayette IN 47906 Office: Biochemistry Dept Purdue U West Lafayette IN 47907

WEINER, IRWIN M., pharmacology educator, researcher; b. N.Y.C., Nov. 5, 1930; s. Samuel and Pearl L. (Levine) W.; m. Lieselotte R. Roth, June 20, 1981; m. Lois M. Fuxman, Mar. 17, 1961 (div. 1980); children: Stephanie F., Jeffrey N. A.B., Syracuse U., 1952; M.D., SUNY-Upstate Medical Ctr.-Syracuse, 1956. Fellow Johns Hopkins U., 1956-58, instr., 1958-60, asst. prof., 1960-66; vis. assoc. prof. Albert Einstein Coll., 1964-65; assoc. prof. pharmacology SUNY Upstate Med. Ctr., 1966-68, prof., chmn., 1968—; mem. pharmacology com. Nat. Bd. Med. Examiners, 1977-82; cons. Sterling-Winthrop, Rensselaer, N.Y. Mem. editorial com.: Ann. Rev. of Pharmacology, Palo Alto, Calif., 1982—; editor: Uric Acid, 1978; mem. editorial bd.: Jour. Pharmacology and Exptl. Therapeutics, 1981—; contbr. articles to profl. jours. NSF fellow, 1958-60; NIH grantee, 1962—. Mem. Am. Soc. Pharmacology and Exptl. Therapeutics, AAAS, N.Y. Acad. Sci., Am. Soc. Nephrology, Internat. Soc. Biochem. Pharmacology. Jewish. Subspecialties: Pharmacology; Physiology (medicine). Current work: Renal pharmacology; effects of drugs on kidney, role of kidney in drug elimination. Home: 221 Deforest Rd Syracuse NY 13214 Office: State University New York Upstate Medical Center 766 Irving Ave Syracuse NY 13210

WEINER, MURRAY, physician, educator, cons. to pharm. industry; b. N.Y.C., Apr. 18, 1919; s. Samuel O. and Gussie (Begun) W.; m. Marilyn Rose, Jan. 14, 1951 (dec.); m. Helen Jane McNeely, June 15, 1976; children: Eve Gail Weiner Schauer, George Jay, Joan Sally. B.S., CCNY, 1939; M.S. in Biochemistry, N.Y.U., 1943. Diplomate: Am. Bd. Internal Medicine. Intern Sinai Hosp., Balt., 1943-44; resident N.Y.U. Research Service, Goldwater Meml. Hosp., 1946-49; practice medicine specializing in internal medicine, N.Y.C., 1950-71; mem. faculty N.Y.U. Coll. Medicine, N.Y.C., 1946-71; v.p. biologic research, also other positions Geigy Pharms., N.Y.C., 1957-71; v.p. research and sci. affairs Merrell Research Center, Cin., 1971-81; clin. prof. medicine, dir. clin. pharmacology Coll. Medicine, U. Cin., 1981—; pres. Weiner Cons., Inc., Cin., 1981—. Author: Coagulation. Thrombosis, and Dicumarol, 1949, Nicotinic Acid: Nutrient/Cofactor/Drug, 1983; novel The Medicine Makers, 1979; contrb. over 170 articles to sci. publs. Chmn. Cin. chpt. Project HOPE, 1979-81. Served to capt. M.C. U.S. Army, 1944-46; PTO. NIH, Am. Heart Assn., Nutrition Found. grantee; also others. Fellow ACP, N.Y. Acad. Scis., Am. Soc. for Pharmacology and Exptl. Therapeutics; mem. numerous other orgns. Subspecialties: Pharmacology; Internal medicine. Current work: Clinical pharmacology; drug disposition, safety, utility.

WEINER, WILLIAM JERROLD, neurology educator, neuropharmacology researcher; b. Chgo., June 28, 1945; s. Leonard and Maxine (Rappaport) W.; m. Susan Rosenband, Sept. 10, 1967; children: Monica, Miriam. B.S. in Zoology, U. Ill.-Urbana, 1966; M.D., U. Ill.-Chgo., 1969. Diplomate: Am. Bd. Psychiatry and Neurology. Intern Presbyn.-St. Luke's Hosp., Chgo., 1969-70, resident in neurology, 1971-73, asst. attending neurologist, 1973-75, assoc. attending neurologist, 1977-83; resident in neurology U. Minn., Mpls., 1970-71; asst. prof. medicine U. Chgo., 1975-77; assoc. prof. neurology and pharmacology Rush U., Chgo., 1977—83; assoc. attending neurologist Michael Reese Hosp., 1975-77; prof. neurology U. Miami Med. Sch. (Fla.), 1983—. Author: Textbook of Clinical Neuropharmacology, 1981; editor: Respiratory Dysfunction in Neurologic Disease, 1980, Neurology for Non-Neurologist, 1981. Served to lt. comdr. USNR, 1973-75. Edmund J. James scholar U. Ill., 1963-66. Fellow Am. Acad. Neurology; mem. Am. Neurol. Assn., Am. Geriatrics Soc., Soc. Neurosci., Internat. Soc. Neurochemistry, Phi Beta Kappa, Alpha Omega Alpha. Subspecialties: Neurology; Neuropharmacology. Current work: Clinical and pre-clinical interests in neuropharmacology of movement disorders (Parkinson's disease, Huntington's disease, dystonia, tremor, Tourette's syndrome). Home: 425 Bianca Coral Gables FL 33146 Office: Dept U Miami Med Sch 1501 NW 9th Ave Miami FL 33101 Home: 425 Bianca Coral Gables FL 33146

WEINFURTER, ERICH BRIAN, nuclear engineer; b. Appleton, Wis., Nov. 27, 1955; s. Robert Wayne and Cathryn Janice (Masterson) W. B.S. in Engring, Iowa State U., Ames, 1981. Tech. staff engr. Commonwealth Edison Co., Cordova, Ill., 1981—. Mem. Am. Nuclear Soc., Profl. Reactor Operator Soc. Subspecialty: Nuclear engineering. Current work: Design and development of safety related systems for a nuclear power station. Home: 3441 60th St Apt 6C Moline IL 61252 Office: Commonwealth Edison Co 22710 206th Ave N Cordova IL 61242

WEINHOLD, ALBERT RAYMOND, plant pathologist, educator; b. Evans, Colo., Feb. 14,1931; s. Albert Raymond and Ruth Evelyn (Stocks) W.; m. Connie Marie Seastrand, Mar. 15, 1952; children: Albert Raymond, Kathryn Beth. B.S., Colo. State U., 1953, M.S., 1955; Ph.D., U. Calif.-Davis, 1958. Mem. faculty U. Calif., Berkeley, 1960—, prof. plant pathology, 1972—, chmn, plant pathology dept., 1976—. Served to 1st lt. USAF, 1958-60. Fellow Am. Phytopath. Soc.; mem. AAAS, Sigma Xi. Republican. Presbyterian. Club: Mira Vista Golf and Country (Calif.). Subspecialties: Plant pathology; Microbiology. Current work: Biology and physiology of soilborne fungal plant pathogens. Home: 213 Arlington Ave Kensington CA 94707 Office: Univ Calif 147 Hilgard Hall Berkeley CA 94720

WEINHOLD, PAUL ALLEN, biological chemist, educator; b. Evans, Colo., Sept. 23, 1935; s. Albert Raymond and Ruth Evelyn (Stocks) W.; m. Yvonne Rose Livingstone, Sept. 1, 1956; children: Julie, Lisa, Scott, Michael. B.S., Colo. State U., 1957; Ph.D., U. Wis., 1961. Research fellow Harvard Med. Sch., Boston, 1963-65; research biochemist VA Med. Ctr., 1965—; from asst. to assoc. prof. dept. biol. chemistry U Mich., Ann Arbor, 1965—. Contbr. numerous articles to sci. publs. Trustee Ann Arbor Bd. Edn. Mem. Am. Soc. Biol. Chemistry. Republican. Subspecialties: Biochemistry (medicine); Pulmonary medicine. Current work: Phospholipid metabolism in lung; pulmonary surfactant formation; membrane biochemistry; enzyme regulation. Home: 3007 Lexington Ann Arbor MI 48105 Office: VA Hospital U Mich Ann Arbor MI 48105

WEINHOUS, MARTIN S., physicist; b. Bklyn., July 30, 1944; s. Samuel A. and Ceil W.; m. Janet S. Varney, Dec. 28, 1975. B.S., Rensselaer Poly. Inst., 1966; M.S., U. N.H., 1970, Ph.D., 1973. Mem. faculty North Adams State Coll., 1973-75, Norwich U., 1975-76, U. N.H., 1976-77, Keene State Coll., 1977-79; research assoc. dept. therapeutic radiology Yale U., New Haven, 1980—. Mem. Am. Assn. Physicists in Medicine. Subspecialties: Cancer research (medicine); Biophysics (physics). Current work: Magnetic enhancement of electron beam radiation therapy for cancer. Home: 104 Whittier Rd New Haven CT 06515 Office: Therapeutic Radiology Yale U PO Box 3333 New Haven CT 06510

WEINHOUSE, SIDNEY, biochemist, educator; b. Chgo., May 21, 1909; s. Harry and Dora (Cutler) W.; m. Sylvia Krawitz, Sept. 15, 1935 (dec. Aug. 1957); children: Doris Joan, James Lester, Barbara May; m. Adele Klein, Dec. 27, 1969. B.S., U. Chgo., 1933, Ph.D., 1936; D.M.S. (hon.), Med. Coll. Pa., 1973, D.Sc., Temple U., 1976. Eli Lilly fellow U. Chgo., 1936-38, Coman fellow, 1939-41; staff OSRD, 1941-44; with Houdry Process Corp., 1944-47; biochem. research dir. Temple U. Research Inst., 1947-50, prof. chemistry, 1952-77; emeritus prof. biochemistry Temple U. Med. Sch., 1977—; head dept. metabolic chemistry Lankenau Hosp. Research Inst. and Inst. Cancer Research, 1950-57; chmn. div. biochemistry Inst. Cancer Research, 1957-61; asso. dir. Fels Research Inst., Temple U. Med. Sch., Phila., 1961-64, dir., 1964-74; mem. bd. sci. advisers Inst. Environ. Health, NIH. Contbr. articles on original research to sci. jours.; editor: Jour. Cancer Research, 1969-79. Bd. dirs. Am. Cancer Soc. Mem. Am. Chem. Soc., Am. Soc. Biol. Chemists, Am. Assn. Cancer Research, Nat. Acad. Sci. Subspecialties: Biochemistry (medicine); Cancer research (medicine). Current work: Enzyme studies in cancer as manifestations of abnormalities of general regulation. Home: 1919 Chestnut St Philadelphia PA 19103 Office: Fels Research Inst Temple U Sch Medicine Philadelphia PA 19140

WEINKAM, ROBERT JOSEPH, medicinal chemist; b. Cin., Dec. 27, 1942; s. Robert Charles and Clare Alma (Pauley) W.; m. Mona Kwan, July 8, 1980; children: Patrick, Sara, Jennifer. B.S., Xavier U., Cin., 1964; Ph.D., Duquesne U., Pitts., 1967. Postdoctoral fellow Syva Research Inst., Palo Alto, Calif., 1968; postdoctoral fellow, then asst. prof. pharm. chemistry U. Calif.-San Francisco, 1969-79; assoc. prof. medicinal chemistry Purdue U., 1980-83, dir. mass. spectrometry lab., 1980-83; dir. drug metabolism and pharmokinetics Allergan Pharms, Irvine, Calif., 1983—. Author papers in field. Recipient Career Devel. award NIH, 1975-80; NASA predoctoral fellow, 1964-67. Mem. Am. Chem. Soc., Am. Soc. Cancer Research, Am. Soc. Mass Spectrometry. Subspecialties: Medicinal chemistry; Analytical chemistry. Current work: Metabolism and pharmacokinetics of drugs in the eye and skin; drug design based on biodispositive properties; drug action in cell culture, drug metabolism, new development mass spectrometry in bio medicine. Address: Allergan Pharms 2525 Dupont Dr Irvine CA 92713

WEINRICH, BRIAN ERWIN, educator; b. Passaic, N.J., Jan. 8, 1952; s. Erwin H. and Ann E. (Gall) W. B.S., Pa. State U., 1974, M.A., 1978; M.S., Shippensburg State U., 1983. Mathematician U.S.Dept. Agr., SEA-ARS, N.E. Watershed Research Ctr., University Park, Pa., 1974-80; instr. math and computer sci. Shippensburg (Pa.) State Coll., 1980—; cons. in field. Author: (with A. S. Rogowski) Water Movement and Quality on Strip-Mined Lands: A Compilation of Computer Programs, 1983; contbg. author (Surface Mining) Edit, II, 1983; contbr. articles to profl. jours. Mem. Missions bd. Calvary Bapt. Ch., State College, Pa., 1975-80; visitation team Prince St. United Brethren Ch., Shippensburg, 1982—. U.S. Dept. Age. grantee, 1982—. Mem. Soc. for Indsl. and Applied Math., Math. Assn. Am., Am. Math. Soc. Republican. Subspecialties: Mathematical software; Hydrology. Current work: Development of mathematical, statistical and numerical models and simulation techniques in hydrology, especially as applied to strip mining/erosion. Home: 116 S Prince St Shippensburg PA 17257 Office: Shippensburg State Coll Dept Math and Computer Sci Box F314 Shippensburg PA 17257

WEINSTEIN, CLAIRE ELLEN, psychologist, educator, researcher; b. Bklyn., Nov. 8, 1946; d. Sidney and Fannie (Silverstein) W. B.S., Bklyn. Coll., 1967; Ph.D., U. Tex.-Austin, 1975. Research dir. Learning Concepts, Austin, 1975-79; asst. prof. U. Tex.-Austin, 1977-82, assoc. prof., 1982—; advisor Nat. Inst. Edn., Washington, 1979—, Army Research Inst., 1979—, Hillel Found., Austin, 1977—, several pub. sch. dists., Tex., 1975—. Contbr. articles to profl. jours. Spencer fellow Nat. Acad. Edn., 1972-83; grantee Army Research Inst., 1979—; Spencer Found., 1978-79. Mem. Am. Psychol. Assn., Am. Ednl. Research Assn. (asst. chmn. program com. div.), S.W. Ednl. Research Assn. (mem. exec. com. 1981-83), AAAS, Internat. Assn. Applied Psychology (sec.-treas. div. ednl., instructional and sch. psychology 1982—), Phi Kappa Phi. Democrat. Subspecialties: Learning; Cognition. Current work: Human learning, cognition, and instruction. In particular, learning-to-learn phenomena including cognitive strategies and study skills. College teaching. Office: Univ Texas Dept Ednl Psychology EDB 504 Austin TX 78712 Home: 6108A Shadow Valley Dr Austin TX 78731

WEINSTEIN, DAVID, microbiologist; b. N.Y.C., June 24, 1928; s. Isidor J. and Helen (Taub) W.; m. Marcia Adria (Weiner), Oct. 24, 1954; children: Miriam Sara, Judith Ann. B.S., CCNY, 1950; M.A., Bklyn. Coll., 1954; Ph.D., Purdue U., 1959. Research asso. Duke U., 1959-62; research asso. Merck Inst., West Point, Pa., 1962-63; asso. The Wistar Inst., Phila., 1963-68; cythogeneticist Hoffmann La Roche, Nutley, N.J., 1968-78, cytogeneticist, 1968-76; research group chief Hoffman La Roche, 1978—. Contbr. articles to profl. jours. Mem. Am. Soc. Microbiology, AAAS, Am. Soc. Cell Biology, Environ. Mutagen Soc., Tissue Culture Assn., N.Y. Acad. Sci., Genetic Toxicology Assn., Sigma Xi. Subspecialties: Cell and tissue culture; Microbiology. Current work: Genotoxicology; short term carcinogenicity and mutagenicity testing. Home: 78 Luddington Rd West Orange NJ 07052 Office: Hoffmann La Roche Bldg 10 Nutley NJ 07110

WEINSTEIN, I. BERNARD, physician; b. Madison, Wis., Sept. 9, 1930; s. Max and Frieda (Blackman) W.; m. Joan Anker, Dec. 21, 1952; children: Tamara, Claudia, Matthew. B.S., U. Wis., 1952, M.D., 1955. Intern, then resident in medicine Montefiore Hosp., N.Y.C., 1955-57; clin. asso. NIH, 1957-59; research asso. Harvard U. Med. Sch., 1959-60, MIT, 1960-61; mem. faculty Columbia U. Coll. Phys. and Surg., 1961—, prof. medicine, 1973—; also dir. div. environ. sci. Sch. Pub. Health and dep. dir. carcinogenesis research Cancer Research Center; attending physician Presbyn. Hosp., N.Y.C.; cons. Nat. Cancer Inst., Roswell Park Meml. Inst., Brookhaven Nat. Lab., Internat. Agy. Research Cancer, Chem. Industry's Inst of Toxicology. Author research papers biochemistry, genetics, cancer biology, chem. carcinogenesis. Served with USPHS, 1957-59. Recipient Meltzer medal Soc. Exptl. Biology and Medicine; European Molecular Biology Orgn. fellow Imperial Cancer Research Fund, London, 1970-71. Mem. Am. Soc. Microbiology, Am. Soc. Clin. Investigation, Am. Soc. Biol. Chemists, Am. Soc. Cell Biology, Internat. Soc. Quantum Biology, Am. Assn. Cancer Research, Inst. Medicine of U.S. Nat. Acad. Scis., Phi Beta Kappa, Sigma Xi, Alpha Omega Alpha. Jewish. Subspecialties: Cancer research (medicine); Cell biology (medicine). Home: 249 Chestnut St Englewood NJ 07631 Office: 701 W 168th St New York NY 10032

WEINSTEIN, IRWIN MARSHALL, internist, hematologist; b. Denver, Mar. 5, 1926; m. Judith Braun, 1951. Student, Dartmouth Coll., 1943-44, Williams Coll., 1944-45; M.D., U. Colo., Denver, 1949. Diplomate: Am. Bd. Internal Medicine (assoc. bd. govs. hematology subcom.). Intern Montefiore Hosp., N.Y.C., 1949-50, jr. asst. resident in medicine, 1950-51; sr. asst. resident in medicine U. Chgo., 1951-52, resident in medicine, 1952-53, instr. in medicine, 1953-54, asst. prof. medicine, 1954-55; vis. asso. prof. medicine U. Calif. Center for Health Scis., Los Angeles, 1955-56, asso. clin. prof., 1957-60, clin. prof., 1970—; sect. chief in medicine, hematology sect. Wadsworth Gen. Hosp., VA Center, Los Angeles, 1956-59; pvt. practice medicine specializing in hematology and internal medicine, Los Angeles, 1959—; mem. staff Cedars-Sinai Med. Center, Los Angeles, 1959—, chief of med. staff, 1972-74; mem. staff U. Calif. Center Health Scis., Wadsworth Gen. Hosp., VA Center; vis. prof. Hadassah Med. Center, Jerusalem, 1967; adv. for health affairs to Hon. Alan Cranston, 1971—; mem. com. on space biology and medicine Space Sci. Bd., Nat. Acad. Scis., 1979—. Contbr. articles to profl. publs.; editor: (with Ernest Beutler) Mechanisms of Anemia, 1962. Fellow Israel Med. Assn. (hon.), A.C.P.; mem. AAAS, Am. Fedn. Clin. Research, Am. Soc. Hematology (exec. com. 1974-78, chmn. ad hoc com. on practice 1978—, mem. council 1974-78), Am. Soc. Internal Medicine, Assn. Am. Med. Colls., Internat. Soc. Hematology, Internat. Soc. Internal Medicine, Los Angeles Acad. Medicine, Los Angeles Soc. Nuclear Medicine, Inst. of Medicine of Nat. Acad. Scis., N.Y. Acad. Sci., Reticulo-Endothelial Soc., Royal Soc. Medicine, Western Soc. Clin. Research, UCLA Comprehensive Cancer Center, Alpha Omega Alpha. Subspecialties: Internal medicine; Hematology. Office: 8635 W 3d St Suite 1165 Los Angeles CA 90048

WEINSTEIN, MALCOLM SAMUEL, psychologist; b. Edmonton, Alta., Can., Feb. 11, 1942; s. Philip and Helen Bella (Laub) W.; m. Judith Marcia Kline, Aug. 10, 1965; children: Jason, Todd. B.A., U. B.C., 1960, M.A., 1964; Ph.D., U. Oreg., 1968. Registered psychologist, B.C. Research assoc. Oreg. Research Inst., Eugene, 1966-68; instr. U. Mo., Kansas City, 1968-69; assoc. prof. York U., Toronto, Ont., 1969-75; cons. Dellcrest Children's Ctr., Toronto, 1970-75; dir. Health Planning, Vancouver Health Dept., 1975—; clin. instr. U. B.C., 1976—; dir. Western Ctr. for Preventive and Behavioral Medicine, 1978—. Author: Impact Supervision, 1976, Health in the City, 1980, Stress Management Series for Clinicians, 1982. Can. Council fellow, 1965-68, 75; WHO travel fellow, 1979. Mem. Canadian Pub. Health Assn., Am. Pub. Health Assn., Am. Psychol. Assn., Canadian Psychol. Assn., B.C. Psychol. Assn. (dir.), Soc. Behavioral Medicine, N. Am. Soc. Corporate Planning, Vancouver Soc. for Evaluation and Research in Community Health (pres. 1978—). Current work: Research, teaching, consultation in relation to behavioral causes and consequences of health and illness in individuals, groups and organizations; stress and health management; community health services planning and evaluation. Home: 5850 Hudson St Vancouver BC Canada V6M 2Z3 Office: Vancouver Health Dept 1060 W 8th Ave Vancouver BC Canada V6H 1C4

WEINSTEIN, MILTON CHARLES, educator; b. Brookline, Mass., July 14, 1949; s. William and Ethel (Rosenbloom) W.; m. Rhonda Kruger, June 14, 1970; children: Jeffrey William, Daniel Jay. A.B., Harvard U., 1970; A.M., M.P.P., 1972, Ph.D., 1973. Asst. prof. John F. Kennedy Sch. Govt., Harvard U., Cambridge, Mass., 1973-76, assoc. prof., 1976-80; prof. policy and decision scis. Harvard Sch. Pub. Health, Boston, 1980—; vis. prof. community and family medicine Dartmouth Med. Sch., Hanover, N.H., 1981—; cons. Battelle Human Affairs Research Ctr., Seattle, 1981—, Mass. Health Data Consortium, Waltham, Mass., 1983. Author: Clinical Decision Analysis, 1980, Hypertension: A Policy Perspective, 1976; editorial bd.: Med. Decision Making, 1981—. NSF fellow, 1972. Mem. Soc. Med. Decision Making (v.p. 1982-83, trustee 1980-82), Soc. Risk Analysis, AAAS, Ops. Research Soc. Am., Phi Beta Kappa. Subspecialties: Health services research; Operations research (mathematics). Current work: Evaluation medical technology; cost-benefit and cost-effectiveness analysis; medical decision making; health care policy; risk analysis. Office: 677 Huntington Ave Boston MA 02115

WEINSTEIN, NORMAN JACOB, consulting corporation executive, chemical engineer; b. Rochester, N.Y., Dec. 31, 1929; s. Sol and Anne (Trapunsky) W.; m. Ann F. Keiles, June 30, 1957; children: Maury S., Aaron S., Kenneth B. B.Ch.E., Syracuse U., 1951, M.Ch.E., 1953; Ph.D., Oreg. State U., 1956. Registered profl. engr., N.J., N.Y., Pa., Fla. Chem. engr. Exxon Research and Engring. Co., Florham Park, N.J., 1956-60, sr. engr., 1960-65, engring. assoc., 1965-66; asst. dir. to dir. engring. and devel. Princeton (N.J.) Chem. Research, 1966-69; adj. prof. petroleum refining N.J. Inst. Tech., Newark, 1963-66; founder Recon Systems, Inc., Three Bridges, N.J., 1969, pres., 1969—. Contbr. sci. articles to profl. publs. Mem. Somerville (N.J.) Borough Council, 1977-79, pres., 1979; mem. Somerville Bd. Edn., 1980—, pres., 1981—. Mem. Am. Chem. Soc., N.Y. Acad. Sci., ASTM, Am. Inst. Chem. Engrs. Democrat. Jewish. Subspecialties: Chemical engineering; Environmental engineering. Current work: Used oil recycling; solid waste-to-energy; hazardous waste disposal; design of environmental control systems; testing of environmental control systems; chemical process design. Home: 105 Reimer St Somerville NJ 08876 Office: Route 202 N Box 460 Three Bridges NJ 08887

WEINSTEIN, SAM, orthodontics educator, researcher; b. Omaha, May 24, 1916; s. Abraham and Soney (Edelman) W.; m. Helen Margaret Davidson, Mar. 22, 1946; children: David, Laurie. D.D.S., Creighton U., 1941; M.S.D., Northwestern U., 1948. Diplomate: Am. Bd. Orthodontics. Asst. prof. U. Nebr. Dental Coll., Lincoln, 1954-57, assoc. prof., chmn. dept. orthodontics, 1957-65, prof., chmn., 1965-71; prof., dir. grad. orthodontics U. Conn. Health Ctr., Farmington, 1971-81, prof., acting head dept. orthodontics, 1982—; examiner in orthodontics Council on Edn., Can. Dental Assn., 1974-79; cons. study sect. Nat. Inst. Dental Research, Washington, 1969-73, ADA Council Dental Edn., Chgo., 1969-76. Co-author: Equilibrium Theory of Tooth Position, 1963-82. Served to lt. comdr. USN, 1942-46. Recipient awards U. Nebr. Coll. Dentistry, 1974, U. Nebr. Med. Ctr., Omaha, 1982. Fellow AAAS, Omicron Kappa Upsilon; mem. Am. Assn. Orthodontists, Internat. Soc. Craniofacial Biology (pres. 1969), Angle Soc. Midwest Component (pres. 1963). Democrat. Jewish. Subspecialties: Dental growth and development; Biomedical engineering. Current work: Physical properties of cheek and lips, cranio-facial growth. Home: 23 Mallard Dr Farmington CT 06032 Office: Univ Conn Health Center Farmington CT 06032

WEINSTEIN, STEPHEN H., xenobiologist; b. Bronx, N.Y., Apr. 14, 1937; s. Isidor and Helen Lucille (Taub) W.; m. Felice Doris Helman, Nov. 12, 1966; 1 dau., Jodi Lynn. B.S., Queens Coll., 1958; M.S., Adelphi Coll., 1961, Ph.D., 1967. Scientist Warner-Lambert, Morris Plains, NJ, 1967-68; sr. scientist Endo Labs., Garden City, N.Y., 1968-73, group leader, 1973-76; sr. research scientist DuPont Co., Wilmington, Del., 1976-80; sect. head Squibb Inst Med. Research, E.R. Squibb & Sons, Princeton, N.J., 1980—. Served with U.S. Army, 1961-63. Mem. Am. Soc. Pharmacology and Exptl. Therapeutics, Acad. Pharm. Scis., AAAS, Drug Metabolism Group. Current work: Drug metabolism, pharmacokinetics, biopharmaceutics. Home: 21 Pine Knoll Dr Lawrenceville NJ 08648 Office: PO Box 4000 Princeton NJ 08540

WEIR, BRUCE SPENCER, educator; b. Christchurch, N.Z., Dec. 31, 1943; came to U.S., 1976; s. Gordon Ralph and Margaret Annie (Hodder) W.; m. Elliazbetrh Anna Swainson, Aug. 7, 1971; children: Claudia Beth, Henry Bruce. B.Sc. (hons.), U. Canterbury, 1964; Ph.D., N.C. State U., 1968. Reader Massey U., N.Z., 1970-76; prof. statso/genetics N.C. State U., Raleigh, 1976—. Contbr. articles to profl. jours. Mem. Am. Statis. Assn., Biometric Soc, Genetic Soc. Am., Soc. Study of Evolution. Subspecialties: Genetics and genetic engineering (biology); Statistics. Current work: Statistical analysis of genetic data. Office: NC State Univ PO Box 5457 Raleigh NC 27650 Home: 3368 Boulder Ct Raleigh NC 27067

WEIR, MAURICE DEAN, mathematics educator; b. Seattle, June 6, 1939; s. William Deo and Flora Ann (Beaudin) W.; m. Gale Hempstead Weir, Dec. 26, 1961; children: Maia Deborah, Renee Elizabeth. B.A., Whitman Coll., 1961; M.S., Carnegie Mellon U., 1963, D.A., 1970. Instr. Whitman Coll., Walla Walla, Wash., 1963-66; asst. prof. math. Naval Postgrad. Sch., Monterey, Calif., 1969-73, assoc.

prof., 1973—. Author: Hewitt-Nachbin Spaces, 1975, Calculus Self-Study, 1979, Calculator Clout, 1981, Calculus by Calculator, 1982. Mem. Am. Math. Soc., Math. Assn. Am., Sigma Xi. Democrat. Unitarian. Subspecialties: Applied mathematics; Mathematical software. Current work: Textbook author working on mathematical modeling and its applications; differential equations emphasizing modeling. Office: Dept Math Naval Postgrad Sch Monterey CA 93940

WEIR, WILLIAM CARL, animal science educator; b. Lakeview, Oreg., Aug. 24, 1919; s. Robert Lawson and Fannie (Tonningsen) W.; m. Elizabeth Galvin Riley, Dec. 27, 1946; children: Robert William, Timothy Miles. B.S., Oreg. State U., 1940, M.S., 1941; Ph.D., U. Wis., 1948; postgrad., U. Ill., 1957, U. Western Australia, 1965-66. Research asst. Wis. Alumni Research, 1940-41, 45-46; instr. animal husbandry U. Wis., 1946-47; asso. prof. Oreg. State U., 1948; asst. prof. U. Calif.-Davis, 1948-54, asso. prof., 1954-60—, chmn. dept., 1973-81, dean of students, 1958-65, acting asso. dean agr. and environ. sci., 1979-80, asso. program dir. Small Ruminant Collaborative Research Support program, 1981—; with U.S. Dept. Agr., 1979. Contbr. to profl. jours. Served to maj. U.S. Army, 1941-45. Decorated Silver Star, Bronze Star, Purple Heart.; recipient Golden Fleece award Calif. Woolgrowers, 1973. Mem. Am. Inst. Nutition, Am. Soc. Animal Sci., Soc. Range Mgmt. Roman Catholic. Subspecialties: Animal nutrition; Nutrition (biology). Current work: Nutrition of the grazing animal; techn. transfer through research with sheep and goats; program funded AID with work sites in Brazil, Indonesia, Peru, Kenya and Morocco.

WEISBERG, AARON, internist, gastroenterologist, consultant, research, educator; b. Bklyn., July 21, 1915; s. Joseph and Yetta (Weisberg) W.; m. Ruth Hannah Mintz, Feb. 7, 1949; children: Harlene Edith, Sharon Esta Weisberg Shapiro. B.A. in Chemistry, NYU, 1935; M.D., Cin. Eclectic Med. Coll., 1939. Diplomate: Am. Bd. Internal Medicine. Intern Coney Island Hosp., Bklyn., 1939-41, resident, 1941-42, attending physician, 1946-74, attending physician emeritus, 1974—; dir. medicine Carson C. Peck Meml. Hosp., Bklyn., 1954-70; chief of medicine Meth. Hosp. Bklyn., 1970-74, cons. in medicine and gastroenterology, 1974—; attending staff mem. in medicine and gastroenterology Tampa VA Hosp., 1974 —, St. Petersburg (Fla.) Gen. Hosp., 1974—, Palms Pasadena Hosp., St. Petersburg, 1974—; med. dir. Spery Gyroscope, Clearwater, Fla., 1975—; clin. asst. prof. internal medicine and gastroenterology and clin. asst. prof. comprehensive medicine U. South Fla., 1975—. Contbr. numerous articles on cardiology, gastroenterology and cancer to profl. jours.; editorial staff: Colon and Rectal Surgery, 1980—. Served to capt. M.C. U.S. Army, 1942-46. Recipient Gold Medal award Coney Island Hosp., 1972, Silver cert. Meth. Hosp., 1974. Fellow ACP, Am. Coll. Gastroenterology, Internat. Acad. Proctology, Royal Soc. Medicine, Am. Coll Nutrition; mem. Am. Fedn. Clin. Research (sr.), Am. Soc. Gastrointestinal Endoscopy, AMA, Pan Am. Soc., Occupational Med. Assn., Am. Chem. Soc., Am. Heart Assn., Phi Lambda Kappa. Clubs: NYU (N.Y.C.); Seminole Country and Golf. Subspecialties: Internal medicine; Gastroenterology. Current work: Cancer research; thymus gland lymphocytes immunity. Office: 6499 Ave N Saint Petersburg FL 33710

WEISBERG, JOSEPH SIMPSON, geography and earth science educator, consultant; b. Jersey City, June 7, 1937; s. Samuel and Augusta (Biel) W.; m. Gloria Helen Kobren, June 21, 1964; children: Debra Susan, David Jeffrey. B.A., Jersey City State Coll, 1960; M.S., Montclair State Coll., 1964; Ed.D., Columbia U., 1969. Cert. tchr., N.J. Instr. Wayne Twp., N.J., 1960-64; prof. geosci. Jersey City State Coll, 1964—; dir. Joseph S. Weisberg Ednl. Assocs., 1976—, North Jersey Computer Acad., 1982—. Author: Meteorology, 1976, 2d edit., 1981; sr. author: Oceanography, 1974; co-author: Generation to Generation, 1969, InvestiGuides, 1967. Vice pres. Parsippany (N.J.) Bd. Edn., 1979—; 1st V.P. N.J. State Alliance for Environ Edn., 1980-81; pres. Lake Hiawatha (N.J.) Jewish Ctr., 1981-83. Recipient Disting. Alumnus award Jersey City State Coll., 1982, Merit award U.S. EPA, 1980, Educator of the Year award Jaycees, 1982. Mem. Geol. Soc. Am., Nat. Assn., Research in Sci. Teaching, Nat. Coll. Sci. Tchrs. Assn. Jewish. Subspecialties: Computer applications in learning; Learning. Current work: Computer applications in learning languages for microcomputers and the micro in the learning environment. Home: 4 Camelot Way Parsippany NJ 07054 Office: Jersey City State Coll 2039 Kennedy Blvd Jersey City NJ 07305

WEISBORD, NORMAN EDWARD, invertebrate paleontology educator; b. Jersey City, Oct. 1, 1901; s. Edward and Clara (Mirsky) W.; m. Nettie S. Schein, Dec. 19, 1939. B.A., Cornell U., 1923, M.S., 1926. Geologist Atlantic Refining Co., Latin Am., 1923-33; sr. field geologist Standard Oil Co. N.J., Argentina and Bolivia, 1933-35; geologist to asst. chief geologist Standard Vacuum Oil Co., Java, Sumatra, Borneo and New Guinea, 1935-42; chief geologist Mobil Oil Corp., Venezuela, 1942-57; research assoc. in geology Fla. State U., Tallahassee, 1957-64, prof. geology, 1964—; occasional cons. oil cos. Author 38 articles and books on geology and paleontology, 1926-81. Mem. Paleontol. Research Instn. of Ithaca (charter and life mem., v.p. 1956-59, pres. 1959-61), Alpha Epsilon Pi. Subspecialties: Geology; Paleobiology. Current work: Adviser to students; publishing on matters relating to geology and invertebrate paleontology. Home: 1910 Gibbs Dr Tallahassee FL 32303 Office: Dept Geology Fla State U Tallahassee FL 32306

WEISBROT, DAVID ROBERT, geneticist, educator; b. Bklyn., Dec. 29, 1931; s. Seymour and Celia (Zarobchick) W.; m. Phoebe Barbara (Goldberg), June 12, 1960; children: Joshua, Miriam, Ari. B.S., Bklyn. Coll., 1953, M.A., 1958; Ph.D., Columbia U., 1963. Postdoctoral fellow U. Calif., Berkeley, 1963-64; asst. prof. Tufts U., Boston, 1964-69; assoc. prof. SUNY, Binghamton, 1969-72; vis. lectr. Columbia U., 1975-81; prof. dept. biology William Paterson Coll., Wayne, N.J., 1973—. Served with U.S. Army, 1953-55. Mem. Genetics Soc. Am., Soc. Am. Naturalists, Sigma Xi. Democrat. Jewish. Subspecialty: Genetics and genetic engineering (agriculture). Current work: Gene action and cytogenetics. Office: Dept Biology William Paterson Coll Wayne NJ 07470

WEISBURGER, ELIZABETH KREISER, science administrator; b. Greenlane, Pa., Apr. 9, 1924; d. Raymond Samuel and Amy Elizabeth (Snavely) Kreiser; m. John Weisburger, Apr. 7, 1947 (div.); children: William R., Diane S., Andrew J. B.S., Lebanon Valley Coll., Annville, Pa., 1944; Ph.D., U. Cin., 1947, D.Sc. (hon.), 1981. Research assoc. U. Cin., 1947-49; postgrad. research Nat. Cancer Inst., Bethesda, Md., 1949-51, research scientist, 1951-81, asst. dir. chem. carcinogenesis div. of, 1981—; trustee Lebanon Valley Coll., 1970—; tchr. grad. sch. NIH, 1980—, Am. U., 1982—; chairperson Interagy. Testing Com., Washington, 1981—. Contbr. numerous articles and chpts. to sci. Jours. and books.; asst. editor: Jour. Nat. Cancer Inst, 1971—. Served to col. USPHS, 1951—. Recipient Meritorious Service awards USPHS, Hillebrand award Chem. Soc., Washington. Mem. AAAS, Am. Chem. Soc. (Garvin medal), Am. Soc. Biol. Chems., Am. Assn. Cancer Research, Internat. Soc. Study Xenobioltics, Soc. Toxicology, Royal Soc. Chemistry, Sigma Delta Epsilon, Iota Sigma Pi. Subspecialties: Cancer research (medicine); Toxicology (medicine). Current work: Metabolism of chemical carcinogens. Office: Nat Cancer Inst Nat Insts Health 9000 Rockville Pike Bethesda MD 20205

WEISER, MILTON MOSES, medicine and biochemistry educator, researcher, physician; b. Detroit, Mar. 17, 1930; s. Samuel and Bertha (Fettman) W.; m. Helen Laurie Freedman, June 14, 1959; children: Julia D., Carl S., Daniel E. B.S., Wayne State U., 1955; M.D., U. Mich., 1959. Intern So. Pacific Gen. Hosp., San Francisco, 1959-60; resident U. Mich. Med. Ctr., Ann Arbor, 1960-63; NIH research fellow in gastroenterology Albert Einstein Coll. Medicine, 1965-67; prof. in charge gastroenterology Harvard U.-MIT Program, 1973-78; assoc. physician Mass. Gen. Hosp., Boston, 1975-78; assoc. prof. medicine Harvard U., 1976-78; prof. medicine, chief gastroenterology SUNY-Buffalo, 1978—, prof. biochemistry, 1978—; chief gastroenterology Buffalo Gen. Hosp., 1982—; dir. gastroenterology Erie County Med. Ctr., Buffalo, 1978—. Am. Cancer Soc. grantee, 1974-80; Nat. Cancer Inst. grantee, 1974—; NIH research grantee, 1976—; NIH research tng. grantee, 1979—. Fellow ACP; mem. Am., Gastroent. Assn., Gastroenterology Research Group, Am. Assn. Physicians. Democrat. Jewish. Subspecialties: Gastroenterology; Cancer research (medicine). Current work: Differentiation; plasmalemma of glycoprotein synthesis; growth control through plasmalemmal receptors; galactosyltransferases; calcium absorption and vitamin D metabolism; inflammatory bowel disease. Office: SUNY-Buffalo Buffalo Gen Hosp Kimberly Bldg 100 High St Buffalo NY 14203

WEISMAN, JOEL, nuclear engineer, educator; b. N.Y.C., July 15, 1928; s. Abraham and Ethel (Marcus) W.; m. Bernice Newman, Feb. 6, 1955; 1 son, Jay (dec.). B.Ch.E., CCNY, 1948; M.S., Columbia U., 1949; Ph.D., U. Pitts., 1968. Registered profl. engr., Ohio, N.Y. Plant engr. Etched Products, N.Y.C., 1950-51; engr. Brookhaven Nat. Lab., Upton, N.Y., 1951-54; sr. engr. to supervisory engr. Atomic Power div. Westinghouse Electric Corp., Pitts., 1954-60; fellow engr. to mgr. Westinghouse Electric Corp., 1960-68; sr. engr. Nuclear Devel. Assocs., White Plains, N.Y., 1960; assoc. prof. to prof. nuclear engring. U. Cin., 1968—; cons. in field. Author: (with others) Thermal Analysis of PWR's, 1970, 2d edit., 1979, Introduction to Optimization Theory, 1973; Editor: Elements of Nuclear Reactor Design, 1977, 2d edit., 1983. Pres. Cin. Asian Art Soc., Cin., 1982-84. Sr. NATO fellow U.K. Atomic Energy Authority, 1972. Fellow Am. Nuclear Soc. (v.p. Pitts. sect. 1966); mem. Am. Inst. Chem. Engrs., Ops. Research Soc. Am., Sigma Xi. Subspecialties: Nuclear engineering; Nuclear fission. Current work: nuclear reactor core design and analysis, nuclear reactor safety, boiling heat transfer and two-phase flow. Office: Dept Chem and Nuclear Engring Univ Cin Cincinnati OH 45221

WEISS, DANIEL LEIGH, pathologist, government official; b. Long Branch, N.J., July 27, 1923; s. Harry and Liberty (Moisseiff) W.; m. Mary Caudill, May 26, 1951; children: Peter Caudill, Leah Weatherly, Harry Moiseiff. B.A., Columbia U., 1943, M.D., 1946. Diplomate: Am. Bd. Pathology. Intern Hosp. for Joint Diseases, N.Y.C., 1946-47; resident in medicine Mt. Sinai Hosp., N.Y.C., 1949, resident in pathology, 1950, Dazian Found. fellow in pathology, 1951-52, Life Ins. Research Found. fellow in pathology, 1952-53; fellow in pathology and exptl. medicine Beth Israel Hosp., N.Y.C., 1951-52; dir. dept. pathology and lab. medicine D.C. Gen. Hosp., 1953-63; prof. pathology U. Ky., 1963-77; exec. sec. div. med. scis. NRC, Washington, 1977-82; dep. dir. affiliated edn. programs service, office acad. affairs VA, Washington, 1983—; clin. prof. pathology Georgetown U., 1953-63, Howard U., 1959-63; George Washington U., 1953-63; 80—; guest research prof. U. Copenhagen, 1969-70; U.S. liaison rep. COGENE, WHO, Geneva, 1978-82; mem. biomechanics adv. com. Dept. Transp., Washington, 1979-82; presentations in field. Research numerous publs. in field. Pres. Chamber Music Soc. Central Ky., Lexington, 1970-75; vice chmn. Lexington Council of Arts, 1974. Served to capt. U.S. Army, 1947-49. Recipient Outstanding Preclin. Tchr. award Student AMA, Lexington, 1965. Fellow Coll. Am. Pathologists (chmn. sect. 1964-66, 70-72), Am. Soc. Clin. Pathologists (adv. council 1960-62), Internat. Acad. Pathology, Am. Assn. Pathologists, Assn. Clin. Studies (Sunderman Clin. Sci. award 1981, pres. 1982-83). Club: Cosmos (Washington). Subspecialties: Pathology (medicine); Health services research. Current work: Pathogenesis of cancer; environmental pathology; paleopathology; research manpower in medical sciences; medical education; medical and allied health manpower and training. Home: 7201 Park Terr Dr Alexandria VA 22307 Office: VA 810 Vermont Ave NW Washington DC 20420

WEISS, EMILIO, microbiologist; b. Pakrac, Yugoslavia, Oct. 4, 1918; s. Edoardo and Vanda (Schrenger) W.; m. Hilda Damick, June 23, 1943; children: Natalie A. W. Holzwarth, Elizabeth R. A.B., U. Kans., 1941; M.S., U. Chgo., 1942, Ph.D., 1948. Diplomate: Am. Acad. Microbiology. Research assoc. U. Chgo., 1948-50; asst. prof. Ind. U., 1950-53; br. chief Ft. Detrick Biol. Labs., 1953-54; with dept. microbiology Naval Med. Research Inst., Bethesda, Md., 1954—, chair sci., 1979—. Contbr. chpts. to books, articles to profl. jours. Served to capt. AUS, 1943-46. Mem. Am. Soc. Microbiology, Am. Assn. Immunologists, AAAS, Soc. Exptl. Biology and Medicine, Sigma Xi. Subspecialties: Microbiology (medicine); Microbiology. Current work: Microbial physiology with emphasis on obligate and facultative intracellular bacteria of medical importance. Home: 3612 Raymond St Chevy Chase MD 20815 Office: Naval Med Research Inst Bethesda MD 20814

WEISS, HAROLD SAMUEL, physiologist, researcher; b. N.Y.C., Sept. 10, 1922; s. Seymour and Evelyn (Vormund) W.; m. Ann Socolow, Aug. 29, 1948; children; Ronald, Karen, Pamela, Seymour. B.Sc., Rutgers U., 1946, M.Sc., 1949, Ph.D., 1950. Instr. Rutgers U., New Brunswick, N.J., 1950-51, asst. prof., 1954-56, assoc. prof. physiology, 1956-63, Ohio State U., Columbus, 1963-67, Prof., 1967—; cons. NSF, 1969, Nat. Inst. Occupational Safety and Health, 1980. Assoc. editor: Poultry Sci. Assn, 1961-63; contbr. articles to profl. jours. Mem. Franklin County Democratic Com., N.J., 1960. Served to capt. USAAF, 1942-46; Served to capt. USAF, 1951-54. Recipient Scholarship State N.J., 1941; HEW research grantee, 1955-79; NASA research grantee, 1963-77; EPA grantee, 1980-83. Fellow Aviation Med. Assn.; mem. AAAS, Am. Physiol. Soc., AAUP (treas. Ohio State U Chapt. 1982—), Soc. Exptl. Biology and Medicine, Phi Beta Kappa, Sigma Xi. Subspecialties: Physiology (medicine); Environmental toxicology. Current work: Environment physiology, temperature, altitude, acceleration, pressure, diving, pollutants; environment and disease, oncology, cardiopulmonary. Home: 1174 W 1st Ave Columbus OH 43212 Office: Dept Physiology Ohio State U 1645 Neil Ave Columbus OH 43210

WEISS, HARVEY JEROME, physician, medical educator; b. N.Y.C., June 30, 1929; s. Sidney and Henrietta (Horowitz) W.; m. Thirell M. Lipsey, Apr. 28, 1957; children: Deborah, Adrienne. A.B., Harvard Coll., 1951; M.D., Harvard U., 1955. Intern Bellevue Hosp., N.Y.C., 1955-56; resident Manhattan VA Hosp., N.Y.C., 1956-58, Mt. Sinai Hosp., 1958-59; instr. medicine NYU Sch. Medicine, 1962-64; asst. prof. medicine Mt. Sinai Sch. Medicine, 1966-69; assoc. prof. medicine Columbia U., 1972-75, prof., 1975—; dir. hematology-oncology Roosevelt Hosp., N.Y.C., 1969—; mem. internat. com. on thrombosis and hemostasis, Chapel Hill, N.C., 1979—. Author: Platelets: Pathophysiology and Anti-Platelet Drug Therapy, 1982; Editor: Platelets and Their Role in Hemostasis, 1971; Contbr. to books. Served to capt. MC AUS, 1959-62. NIH Research grantee, 1971—; recipient N.Y.C. Health Research Council Career Scientist award, 1969-72. Fellow ACP; mem. Am. Heart Assn. (exec. com. on thrombosis 1982—), Am. Soc. Hematology (subcom. on hemostasis 1981—), Assn. Am. Physicians, Am. Soc. Clin. Investigation, Am. Physiol. Soc. Subspecialties: Internal medicine; Hematology. Current work: Physiology and biochemistry of bleeding disorders; coagulation mechanisms: platelet physiology; thrombosis. Home: 520 St Nicholas Ave Haworth NJ 07641 Office: St Lukes-Roosevelt Hosp 428 W 59th St New York NY 10019

WEISS, HOWARD JACOB, operations research educator; b. Phila., Sept. 18, 1950; s. Ernest D. and Charlotte M. (Silverman) W.; m. Lucia Beck, Aug. 17, 1975; children: Lisa, Ernest. B.Sc., Washington U., St. Louis, 1972; M.S., Northwestern U., 1973, Ph.D., 1975. Asst. prof. Western Ill. U., Macomb, 1975-76; asst. prof. Temple U., Phila., 1976-81, assoc. prof. dept. mgmt., 1981—; cons. USI, Inc., N.Y.C., 1982-83, Matlack, Inc., Lansdowne, Pa., 1982-83. Author: (with Ben Lev) Introduction to Mathematical Programming, 1982. Mem. Ops. Research Soc. Am./Inst. Mgmt. Sci., Ops. Research Soc. Phila. (pres. 1977-78). Subspecialties: Operations research (engineering); Industrial engineering. Current work: Research interests in operations management, inventory, bulk service queues, job shop scheduling. Office: Dept Mgmt Temple U Sch Bus Adminstrn Philadelphia PA 19122 Home: 2157 Woodlawn Ave Glenside PA 19038

WEISS, IRA PAUL, neurophysiologist, educator; b. N.Y.C., Feb. 27, 1942; s. Edward A. and Helen (Friedman) W.; m. Susan J. Weiss, Aug. 27, 1942; children: Elaine R., Joanna G. Student, MIT, 1959-61; B.S., CCNY, 1965; Ph.D., Syracuse U., 1969. USPHS postdoctoral fellow Callier Ctr., Dallas, 1969-71; assoc. dir. Evoked Response Lab., Children's Hosp. Nat. Med. Ctr., Washington, 1972—; assoc. prof. child health and devel. George Washington U. Med. Sch., Washington, 1980—; also cons. Contbr. articles to profl. jours. Mem. Soc. Neurosci., AAAS. Subspecialties: Neurophysiology; Sensory processes. Current work: Application of auditory, visual and comatosensory evoked potentials to diagnosis of sensory, neurological and developmental disorders. Home: 11 Lily Pond Ct Rockville MD 20852 Office: 111 Michigan Ave NW Washington DC 20010

WEISS, JAMES MOSES AARON, educator, psychiatrist; b. St. Paul, Oct. 22, 1921; s. Louis Robert and Gertrude (Simon) W.; m. Bette Shapera, Apr. 7, 1946; children: Jenny Anne Weiss Ford, Jonathan James. A.B. summa cum laude, U. Minn., 1941, Sc.B., 1947, M.B., 1949, M.D., 1950; M.P.H. with honors, Yale U., 1951. Diplomate: Am. Bd. Psychiatry and Neurology (examiner 1963-83). Teaching asst. psychology St. Thomas Coll., St. Paul, 1941-42; intern USPHS Hosp., Seattle, 1949-50; resident, fellow psychiatry Yale Med. Sch., 1950-53; from instr. to asst. prof. psychiatry Washington U., St. Louis, 1954-60; mem. faculty U. Mo., 1959—, prof. psychiatry, 1961—, founding chmn. dept., 1960—, prof. community medicine, 1971—; vis. prof. Inst. Criminology, Cambridge (Eng.) U., 1968-69; internat. cons., 1958—. Author numerous articles in field; editor, co-author: Nurses, Patients, and Social Systems, 1968; corr. editor: Jour. Geriatric Psychiatry, 1967—; founding editor, chmn. bd.: Jour. Operational Psychiatry, 1970—; editorial adv.: Community Mental Health Jour., 1979—; trustee: Mo. Rev., 1982-83. Served with Med. Adminstrv. Corps, AUS, 1942-46; PTO; capt. M.C., AUS, 1953-54. Decorated Philippine Liberation medal, 1945; faculty fellow Inter-Univ. Council, 1958; recipient Sir Henry Wellcome award in mil. medicine, 1955; Israeli bronze medal for leadership, 1963; Basic Books award, 1974, Disting. Service commendation Nat. Council Community Mental Health Ctrs., 1982; named Chancellor's Emissary U. Mo., 1979; recipient Disting. Service commendation Nat. Council Community Mental Health Ctrs., 1982-83; sr. research fellow Am. Council Edn. and NSF, 1984. Found. fellow Royal Coll. Psychiatrists; fellow Royal Soc. Medicine, Am. Psychiat. Assn., Am. Pub. Health Assn., Royal Soc. Health, AAAS, Am. Coll. Psychiatrists, Am. Assn. Psychoanalytic Physicians (hon.); mem. Assn. Mil. Surgeons U.S. (hon. life), Assn. Acad. Psychiatry, Assn. Western Profs. Psychiatry (chmn. 1970-71), Mo. Acad. Psychiatry (1st pres. 1966-67), Mil. Order World Wars, Phi Beta Kappa, Sigma Xi, Psi Chi, Alpha Omega Alpha, Gamma Alpha. Clubs: Scholars (Cantab.); Wine Label (London); Yale (St. Louis). Subspecialties: Psychiatry; Epidemiology. Research in areas of suicide, homicide, antisocial behavior, aging, social psychiatry. Home: Crow Wing Farm Route 2 Columbia MO 65201 Only this endures: creativity, the pursuit of excellence, and continuing concern for human civilization.

WEISS, JONATHAN HYMAN, psychologist, consultant; b. Bklyn., May 19, 1936; s. Morris and Celia (Saltzberg) W.; m. Leah Goldberg, 1958 (dec. 1962); m. Jutta Volck, Dec. 19, 1967; children: Leslie, Danielle, Lori. B.A., Bklyn. Coll., 1957; Ph.D., U. Rochester, 1961. Asst. prof. psychology U. Rochester, N.Y., 1961-62; research clin. psychologist Children's Asthma Research Inst., Denver, 1962-63, chief clin. psychologist, 1963-68, dir. behavior sci. div., 1968-70; assoc. prof. psychology Yeshiva U., N.Y.C., 1970-76; pvt. practice psychology, N.Y.C., 1977—; cons. VA, 1971-72, St. Vincent's Med. Ctr., N.Y.C., 1971-76, Am. Lung Assn., 1978—; lectr., condr. workshops. Contbr. chpts. to books, articles, papers to profl. lit. Research grantee NIH, 1963, 67, 68, Sigma Xi, 1964, 68, Cornell Med. Ctr., 1978, Am. Lung Assn., 1978. Mem. Am. Psychol. Assn., Am. Psychosomatic Soc., N.Y. Acad. Sci., Sigma Xi, Psi Chi. Democrat. Jewish. Subspecialty: Behavioral psychology. Current work: Psychosomatic and behavioral medicine; research in self-care, psychotherapy, teaching, training. Office: 635 Park Ave New York NY 10021

WEISS, KAY, educator, astronomy; b. Elizabeth, N.J., June 28, 1950; d. Max and Bess (Levitte) Stollman; m. Steven Jay Weiss, June 18, 1972; 1 dau: Deborah Ann. B.A. in Astronomy, UCLA, 1972; M.S. in Physics, U. Mo., Kansas City, 1976. Teaching asst. UCLA, 1969-72; research asst. Kitt Peak Nat. Obs., 1970; mem. asst. tech. staff The Aerospace Copr., El Segundo, Calif., 1972-73; research asst. in physics U. Mich., Ann Arbor, 1973-74; teaching asst. U. Mo., Kansas City, Mo., 1974-76; instr. math., astronomy Kansas City (Kans.) Community Coll., 1976—; lectr. in field. Contbr. in field. Recipient Most Outstanding Undergraduate in Astronomy award UCLA, 1970. Mem. Am. Astron. Soc., Soc. Pacifice, Sigma Pi Sigma. Subspecialties: 1-ray high energy astrophysics; Optical astronomy. Current work: Herbig-Haro objects and stellar evolution, effective teaching math., astronomy and physics; editing math. Texts, devel. of TV courses in astronomy. Office: 7250 State Ave: Kansas City KS 66112

WEISS, L. LEONARD, cancer researcher, pathologist; b. London, June 15, 1928; U.S., 1964; m. Maureen Ann Jones, Feb. 23, 1951; children: Gregory, Simon, Emma. B.A., U. Cambridge, Eng., 1950, M.A., B.Chir., M.B., 1953, M.D., 1958, Ph.D., 1963, Sc.D., 1971. Jr. pathologist Westminster Med. Sch., London, 1954-58; scientist Med. Research Council, Mill Hill, U.K., 1958-60; cell physiologist Strangeways Research Labs., Cambridge, U.K., 1960-64; dir. exptl. pathology Roswell Park Meml. Inst., Buffalo, 1964—, chief cancer research clinician, 1979—; research prof. biophysics SUNY-Buffalo, 1969—. Author books and papers on cancer metastasis; also editor; author: Watchmaking, 1760-1820, 1982. Served to maj. Brit. Army, 1960-64. Fellow Royal Coll. Pathologists, Coll. Am. Pathologists, Inst. Biology (U.K.), AAAS. Club: Youngstown Yacht (dir. 1982—). Subspecialties: Oncology; Biofeedback. Current work: Metastasis of cancer; cell interaction biophysics: diagnostic ultrasound. Office: NY State Health Dept Roswell Park Meml Inst Buffalo NY 14263

WEISS, OLIN ERIC, mechanical engineer; b. Fredericksburg, Tex., Mar. 17, 1939; s. Benno F.K. and Bertha (Krause) W.; children: Deric Olin, Jason Karl. B.A., Tex. Christian U., 1962; B.S. in M.E, U. Tex., 1962, M.S., 1964, Ph.D., 1967. Registered profl. engr., Tex. Engr., dept. nuclear physics U. Tex.-Austin, 1962; sr. engr. Gen. Dynamics, Ft. Worth, 1967-76, project mf. tech. engr., 1976-78, supt. splty. fabrication, 1978-80, mgr. facility planning, 1980-81, project engr., structures and design dept., 1981—. Author articles. Alderman City of Annetta North, Tex., 1979—. Recipient various fellowships. Mem. ASME, ASTM, Am. Soc. Metals, Nat. Mgmt. Assn. Democrat. Methodist. Subspecialty: Mechanical engineering. Current work: Application of advanced composite materials. Patentee in field. Home: 700 Quail Ridge Aledo TX 76008

WEISS, RICHARD LOUIS, chemistry educator; b. Evanston, Ill., June 24, 1944; s. Louis Christian and Patty Jean (Campbell) W. B.S., Stanford U., 1966; Ph.D., U. Wash., 1971. Postdoctoral fellow U. Mich., Ann Arbor, 1971-73; asst. prof. UCLA, 1974-80, assoc. prof. dept. chemistry, 1980—; mem. study sect. NIH, Bethesda, Md., 1981—. NIH fellow, 1972; Am. Cancer Soc. fellow, 1973. Mem. Am. Chem. Soc., Am. Soc. Microbiology, Genetics Soc. Am., Am. Soc. Biol. Chemists. Subspecialties: Biochemistry (biology); Molecular biology. Current work: Compartmentation of enzymes and metabolites: organelle biogenesis; application of nuclear magnetic resonance spectroscopy to biology. Office: Dept Chemistry and Biochemistry U Calif 405 Hilgard Ave Los Angeles CA 90024

WEISS, STEFAN ADAM, microbiologist; b. Jablonica, Poland, Dec. 25, 1936; came to U.S., 1966, naturalized, 1971; s. Antony Jan and Mary Helen (Malyszczuk) W.; m.; children: Carl, Catherine, Tanya. M.S., Poly U. Lwow, Poland, 1955, 1957; postgrad., U. Toronto, 1962. Research assoc. Connaught Med. Research Labs., U. Toronto, 1959-62; virologist Food and Drug Directorate of Can. Dept. Nat. Health and Welfare, Ottawa, 1962-66; research virologist Bristol Labs., Syracuse, N.Y., 1966-69; virologist Alcon Labs., Inc., Ft. Worth, 1969-70; research virologist Monsanto Co., St. Louis, 1970-72; head cell culture and virology prodn. Inst. Molecular Virology, St. Louis U., 1972-73; sr. research virologist-microbiologist Dow Lepetit Ltd. of Dow Chem. Co., Indpls., 1973-75; mgr. tissue culture prodn. Litton Bionetics, Inc., Kensington, Md., 1975-77; head cell culture labs. microbiology and infectious diseases Southwest Found. Research and Edn., San Antonio, 1977—; cons. in field. Contbr. articles to profl. jours. Grantee in field. Mem. Am. Soc. Microbiology, Am. Tissue Culture Assn. Democrat. Roman Catholic. Lodge: KC. Subspecialties: Tissue culture; Virology (biology). Current work: Virology, human, animal, insect, chemotherapy; viral insecticides, cell culture, large scale, bioengineering, genetic toxicology. Patentee in field. Office: W Loop 410 at Military Dr San Antonio TX 78284

WEISS, THOMAS FISCHER, neuroscience educator; b. Prague, Czechoslovakia, Oct. 17, 1934; came to U.S., 1941, naturalized, 1951; s. Eugene and Erna (Frenkel) W.; m. Aurice Vernon, June 9, 1928; children: Max Philip, Elisa Lane, Eric Radford. B.E.E., CCNY, 1956; S.M. in Elec. Engring, M.I.T., 1959, Ph.D., 1963. Asst. prof., elec. engring. M.I.T., 1963-68, assoc. prof., 1968-78, prof., 1978—; research assoc. Mass. Eye and Ear Infirmary, 1964—, Harvard U., 1964-69, instr., 1969-74; mem. faculty div. health scis. and tech. Harvard-M.I.T., 1980—; mem. communicative disorders rev. com. NIH. Mem. AAAS, Am. Phys. Soc., Neuroscis. Soc., Sigma Xi, Eta Kappa Nu, Tau Beta Pi. Subspecialties: Electrical engineering; Neurophysiology. Current work: Research in sensory physiology, hearing, membrane biophysics, neurobiology.

WEISSBACH, ARTHUR, cell biologist, researcher, educator; b. N.Y.C., Aug. 27, 1927; s. Louis and Vivian W.; m. Joyce, Nov. 1, 1958; children: Lyle, Claudia. B.S., CCNY, 1947; Ph.D. (Life Ins. Med. Research Fund fellow), Columbia U., 1953. Nat. Found. Infantile Paralysis postdoctoral fellow NIH, 1953-55; NSF sr. postdoctoral fellow Institut Pasteur, Paris, 1959-60; research chemist in biochemistry Nat. Inst. Arthritis and Metabolic Diseases, NIH, 1960-68; head dept. cell biology Roche Inst. Molecular Biology, Nutley, N.J., 1968-83; adj. prof. human genetics and devel. Coll. Physicians and Surgeons, Columbia U., 1969-83; adj. prof. microbiology Coll. Medicine and Dentistry N.J., 1981—; lectr. in chemistry Georgetown U. Grad. Sch., 1957-58; vis. scientist Institut de Biologie Physico-chimique, Paris, 1968-69. Bd. dirs. Rapkine French Sci. Fund. Subspecialties: Cell and tissue culture; Biochemistry (biology). Current work: Enzymology and mechanism of DNA synthesis in mammalian and plant cells; control of viral and host DNA replication after infection with DNA containing viruses. Office: Roche Inst Molecular Biology Nutley NJ 07110

WEISSBACH, HERBERT, biochemist; b. N.Y.C., Mar. 16, 1932; s. Louis and Vivan (Ruhalter) W.; m. Renee Kohl, Dec. 27, 1953; children: Lawrence, Marjorie, Nancy, Robert. B.S., CCNY, 1953; M.S. in Biochemistry, George Washington U., 1955, Ph.D., 1957. Chemist Lab. Clin. Biochemistry, NIH, Bethesda, Md., 1953-68, acting chief, 1968-69; asso. dir. Roche Inst. Molecular Biology, Nutley, N.J., 1969—; adj. prof. dept. human genetics and devel. Columbia U., 1969—; adj. prof. dept. microbiology N.J. Coll. Medicine and Dentistry, 1981—. Editor: (with S. Pestka) Molecular Mechanisms of Protein Biosynthesis, 1977, (with R. Kunz) Health Research: Search for the Medicines of Tomorrow, 1978, (with M.A.Q. Siddiqui and M. Krauskopf) Molecular Approaches to Gene Expression and Protein Synthesis. Recipient Superior Service award HEW, 1968. Mem. Nat. Acad. Scis., Am. Soc. Biol. Chemistry, Am. Chem. Soc. (Enzyme award 1970), AAAS, Am. Soc. Microbiology. Subspecialties: Biochemistry (biology); Molecular biology. Current work: Regulation of gene expression. Home: 333 Crestmont Rd Cedar Grove NJ 07009 Office: Roche Inst Molecular Biology Nutley NJ 07110

WEISSBLUTH, MITCHEL, applied physics educator, researcher; b. Yampol, Russia, Jan. 7, 1915; came to U.S., 1922, naturalized, 1928; s. Elias and Miriam (Saltzman) W.; m. Margaret Hochhauser, Feb. 25, 1940; children: Stephen, Marc, Thomas. B.A., Bklyn. Coll., 1936; M.S., George Washington U., 1941; Ph.D., U. Calif.-Berkeley, 1950. Radio engr. Crosley Radio Corp., Cin., 1941-42; sr. research engr. Jet Propulsion Lab., Pasadena, Calif., 1942-46; instr. physics Stanford U., 1951-54, assoc. prof., 1967-76, prof., 1976—; dir. Biophysics Lab., 1964-68; liaison scientist Office Naval Research, London, 1967-68, Tokyo, 1978-79. Editor: Quantum Aspects of Polypeptides and Polynucleotides, 1964; author: Molecular Biophysics, 1965, Hemoglobin, 1974, Atoms and Molecules, 1978. Fulbright scholar, 1960-61. Mem. Am. Phys. Soc. (mem. exec. com. div. biology physics 1973-76), Internat. Soc. Quantum Biology (pres. 1973-75), AAUP. Republican. Jewish. Subspecialties: Atomic and molecular physics; Biophysics (physics). Current work: Interactions of electromagnetic waves with matter, quantum theory, laser physics. Home: 820 Pine Hill Rd Stanford CA 94305 Office: Department Applied Physics Stanford University Stanford CA 94305

WEISSE, ALLEN B., cardiologist, medical educator; b. N.Y.C., Dec. 6, 1929; s. Charles and Frieda (Lewitt) W.; m. Laura Van Raalte, Aug. 5, 1967; children; Danielle, Charles. B.A., NYU, 1950; M.D., SUNY Downstate Med. Ctr., Bklyn., 1958. Diplomate: Am. Bd. Internal Medicine. Intern Mt. Zion Hosp., San Francisco; resident Fort Miley VA Hosp., San Francisco Gen. Hosp.; fellow in cardiovascular research U. Utah Sch. Medicine, Salt Lake City; instr. medicine Seton Hall Coll. Medicine, Jersey City, 1963-65, N.J. Med. Sch., 1965-67; asst. prof. medicine Coll. Medicine and Dentistry of N.J.-N.J. Med. Sch., Jersey City, 1967-69; assoc. prof. medicine U. Medicine and Dentistry of N.J., Newark-N.J. Med. Sch., 1969-74, prof., 1974—; dir. cardiac ICU Coll. Hosp., Newark, 1972-77, pres. med.-dental staff, 1979-81. Author numerous cardiovascular research papers. Served to lt. USAF, 1952-54. Recipient Charles L. Brown award N.J. Med. Sch. Alumni Assn., 1977. Fellow ACP, Am. Coll. Cardiology; mem. Am. Heart Assn., Am. Fedn. Clin. Research, Am. Physiol. Soc., AAAS, Am. Soc. Echocardiology. Democrat. Jewish. Subspecialties: Internal medicine, Cardiology. Current work: Coronary artery disease; cardiac ultrasound; medical history and ethics. Home: 164 Hillside Ave Springfield NJ 07081 Office: U Medicine and Dentistry of NJ-NJ Med Sch 100 Bergen St Newark NJ 07103 x1

WEISSKOPF, MARTIN CHARLES, astronomer; b. Omaha, Apr. 21, 1942; s. Walter A. and Gertrude F. (Rosenfeld) W.; m. Vera Joan Hanfmann, Sept. 14, 1962; children: Antonia, Alexander. B.S., Oberlin Coll., 1964; Ph.D. in Physics, Brandeis U., 1969. Asst. prof. physics Columbia U., N.Y.C., 1972-77; sr. x-ray astronomer George C. Marshall Space Flight Center, NASA, Huntsville, Ala., 1977—. Author, editor numerous publs. Woodrow Wilson fellow; Revson fellow. Mem. Am. Phys. Soc., Am. Astron. Soc., Phi Beta Kappa, Sigma Xi. Subspecialties: 1-ray high energy astrophysics; High energy astrophysics. Current work: X-ray astronomy; analysis and interpretation of astronomical data; x-ray astronomer; instrument developer. Home: 3601 Georgetta Dr Huntsville AL 35801 Office: ES62 Marshall Space Flight Center AL 35812

WEISSKOPF, VICTOR FREDERICK, physicist; b. Vienna, Austria, Sept. 19, 1908; came to U.S., 1937, naturalized, 1942; s. Emil and Martha (Gut) W.; m. Ellen Tvede, Sept. 5, 1934; children: Thomas Emil, Karen Louise. Ph.D., U. Goettingen, Germany, 1931. Research asso. U. Copenhagen, Denmark, 1932-34, Inst. of Tech., Zûrich, Switzerland, 1934-37; asst. prof. physics U. Rochester, N.Y., 1937-43; with Manhattan Project, Los Alamos, N.M., 1943-46; prof. physics Mass. Inst. Tech., 1946-60; dir. gen. European Orgn. for Nuclear Research, Geneva, Switzerland, 1961-65; Inst. prof. Mass. Inst. Tech., 1965—; chmn. high energy physics adv. panel AEC, 1967-73. Author: (with J. Blatt) Theoretical Nuclear Physics, 1952, Knowledge and Wonder, 1962, Physics in the Twentieth Century, 1972; articles on nuclear physics, quantum theory, radiation theory, etc. in science jours. Recipient Max Planck medal, Germany, 1956, Hi Majorana award, 1970, G. Gamov award, 1971, Boris Pregel award, 1971, Prix Mondial Cino del Duca, France, 1972, L. Boltzmann prize, Austria, 1977; Nat. Medal of Sci., U.S., 1980; Wolf prize, Jerusalem, 1982. Fellow Am. Phys. Soc. (pres. 1960); mem. Nat. Acad. Scis., Am. Acad. Arts and Scis. (pres. 1975-79), French Academie des Scis. (corr.), Austrian Acad. Sci. (corr.), Danish Acad. Sci. (corr.), Bavarian Acad. Sci. (corr.), Scottish Acad. Sci. (corr.), Soviet Acad. Sci. (corr., Pontifical Acad. Sci.). Subspecialty: Nuclear physics. Home: 36 Arlington St Cambridge MA 02140

WEISSLER, GERHARD LUDWIG, physicist, educator; b. Eilenburg, Germany, Feb. 20, 1918; came to U.S., 1939, naturalized, 1944; s. Otto and Margret Louise (Wendt) W.; m. Claire Betty Weissler, Aug. 15, 1953; children: Roderick A., Robert Eric, Mark Gregory (dec.). B.S., Tech. U. Berlin, 1938; M.A., U. Calif.-Berkeley, 1941, Ph.D., 1942. Diplomate: Am. Bd. Radiology. Instr. radiol. physics U. Calif. Med. Sch.-San Francisco, 1942-44; asst. prof. physics U. So. Calif., Los Angeles, 1944-48, assoc. prof. physics, 1948-52, prof. physics, 1952—, chmn. dept. physics, 1951-56, dir. nuclear acceleration, 1955-65. Contbr. articles to sci. jours. Fellow Am. Phys. Soc. (chmn., mem. exec. com. div. electron and atomic physics), Optiocal Soc. Am., AAUP, Sigma Xi. Subspecialties: Atomic and molecular physics; Plasma physics. Current work: Vacuum ultraviolet and x-radiation interaction with gases and surfaces. Office: University Park U So Calif Los Angeles CA 90089

WEISSMAN, PAUL ROBERT, astronomer; b. Bklyn., Sept. 28, 1947; s. Jack H. and Edith (Lipshitz) W. A.B. in Physics, Cornell U., 1969; M.S. in Astronomy, U. Mass., 1971; Ph.D. in Planetary Physics, UCLA, 1978. Research asst. U. Mass., Amherst, 1969-71; UCLA, 1971-74; mission analyst Jet Propulsion Lab., Calif. Inst. Tech., Pasadena, 1974-77; sr. scientist Jet Propulsion Lab., Calif. Inst. Tech., 1977-80, mem. tech. staff, 1980—. Contbr. articles to profl. jours. Mem. Am. Astron. Soc., Internat. Astron. Union, AAAS. Subspecialties: Celestial mechanics; Planetary science. Current work: Physical and dynamical studies of small bodies in solar system, in particular, comets. Spacecraft studies of solar system bodies, trajectory design. Office: Mail stop 183-301 Jet Propulsion Lab 4800 Oak Grove Dr Pasadena CA 91109

WEITKAMP, WILLIAM GEORGE, nuclear physicist; b. Fremont, Nebr., June 22, 1934; s. Alvin Herman and Georgia Ann (Fuhrmeister) W.; m. Audrey Ann Jensen, June 2, 1956; children—Erick, Gretchen, Laurie. B.A., St. Olaf Coll., 1956; M.S., U. Wis., 1961, Ph.D., 1965. Research asst. prof. U. Wash., Seattle, 1965-67; asst. prof. U. Pitts., 1967-68; tech. dir., research prof. Nuclear Physics Lab., U. Wash., Seattle, 1968—. Served with USAF, 1956-59. Acad. guest Eidgenossische Technische Hochschule, Zurich, Switzerland, 1974-75. Mem. Am. Phys. Soc. Subspecialties: Nuclear physics; Electrical engineering. Current work: Polarization phenomena in nuclear reactions; Van de Graaff accelerator design and development; superconducting linac design and development. Home: 2019 E Louisa St Seattle WA 98112 Office: Nuclear Physics Lab GL-10 U Wash Seattle WA 98195

WELCH, BRUCE LYNN, biological researcher, educator; b. Atlanta, Nov. 18, 1931; s. Leonard Enoch and Eloise Jewel (Linn) W.; m. Ann Marie Stephenson, Aug. 23, 1959. Ph.D., Duke U., 1962. Postdoctoral researcher depts. biochemistry and exptl. surgery Duke U., Durham, N.C., 1962; asst. prof. dept. biology, dir. Lab. Population Ecology, Coll. William and Mary, Williamsburg, Va., 1962-66; sr. scientist, dir. Lab. Environ. Neurobiology, U. Tenn. Meml. Research Ctr. and Hosp., 1966-69; assoc. prof. depts. zoology and psychology U. Tenn., Knoxville, 1966-69; chief psychophysiology research Md. Psychiat. Research Ctr., Balt., 1969-70; dir. environ. neurobiology research Friends Med. Sci. Research Ctr., Balt., 1970-74; assoc. prof. behavioral biology Johns Hopkins U. Sch.Medicine, Balt., 1969-78; vis. lectr. dept. psychiatry Yale U. Sch Medicine, New Haven, 1975-78; sr. ptnr. Welch Assocs., Woodridge, Conn., 1975—; cons. in field. Author: (with Annemarie S. Welch) Physiological Effects of Noise, 1969; contbr. articles to profl. jours. Bd. dirs. various civic groups concerned with environ. quality. Mem. Am. Soc. Pharmacology and Exptl. Therapeutics, Am. Soc. Neurochemistry, Am. Physiol. Soc., Am. Psychosomatic Soc., Am. Soc. Biol. Chemists, Am. Pub. Health Assn., Soc. for Neurosci., Soc. Biolo. Psychiatry, Soc. Toxicology, Soc. Exptl. Biology and Medicine, Internat. Soc. for Research on Aggression. Democrat. Methodist. Subspecialties: Environmental neurobiology; Neuropharmacology. Current work: Environmental neurobiology; health effects noise; science in public policy. Home and Office: 1113 High Rd Kensington CT 06037

WELCH, ROSS MAYNARD, plant physiologist, research scientist; b. Lancaster, Calif., May 8, 1943; s. Lloyd Charles and Theda Wynonna (Slane) W.; m. Jill Suzanne Varley, Aug. 22, 1965; children: Renell Cherie, Brent Ross. B.S., Calif. State Poly. U., 1966; M.S., U. Calif.-Davis, 1969, Ph.D., 1971. Research assoc. agronomy dept. Cornell U., Ithaca, N.Y., 1971-72, asst. prof. agronomy dept., 1975-81, assoc. prof., 1981—; plant physiologist Agr. Research Service, U.S. Dept. Agr., Ithaca, 1972—; acting lab. dir. U.S. Plant, Soil and Nutrition Lab., Ithaca, 1977-78; disting. vis. scientist Murdoch (Western Australia) U., 1980-81. Editor: Crops as Sources of Nutrients for Humans, 1983; Contbr. sci. articles to publs. Pres. Ellis Hollow Community Assn., Ithaca, 1982-83. U.S. Dept. Agr., Agr. Research Service grantee, 1980-81, 81-82. Mem. Am. Soc. Plant Physiologists, Am. Soc. Agronomy, Soil Sci. Soc. Am., N.Y. Acad. Scis., Sigma Xi (v.p. Cornell Chpt. 1982-83). Republican. Lodge: Masons (jr. warden 1983—). Subspecialties: Plant physiology (agriculture); Nutrition (biology). Current work: Trace element function in plants; bioavailability of mineral nutrient to humans; physiology of ion transport; mechanisms of nutrient transport; establishment of new essential elements for plants. Home: 24 Hickory Circle Ithaca NY 14850 Office: U S Plant Soil and Nutrition Lab Tower Rd Ithaca NY 14853

WELCH, WILLIAM HENRY, biochemistry educator; b. Los Angeles, Dec. 13, 1940; s. William Henry and Lola Lucille (Ellsworth) W.; m. Marcia Delaney, Aug. 14, 1965; children: William, Deborah, Emily, Gregory. B.A., U. Calif.-Berkeley, 1963; Ph.D., U. Kans., 1969. Postdoctoral fellow Brandeis U., 1969-70; asst. prof. biochemistry U. Nev.-Reno, 1970-76, assoc. prof., 1976—. Contbr. articles profl. jours. Mem. AAAS, Am. Chem. Soc., Sigma Xi. Subspecialties: Biochemistry (biology); Biophysical chemistry. Current work: Biochemical mechanisms in adaptation to thermal stress, protein assembly, role of monovalent cations in enzyme catalysis. Office: Dept Biochemistry University of Nevada Reno NV 89557

WELCH, WILLIAM JOHN, astronomer; b. Chester, Pa., Jan. 17, 1934; s. William Taylor and Ruth (van Leuven) W.; m. Jill C. Tartar, July 4, 1980; children by previous marriage—Eric, Leslie, Jeanette. B.S., Stanford U., 1955; M.S., U. Calif., Berkeley, 1958, Ph.D., 1960; Hon. Dr., Universite de Bordeaux, 1979. Asst. prof. elec. engring. U. Calif., Berkeley, 1960-65, asso. prof., 1965-69, prof. astronomy and elec. engring., 1969—; dir. Radio Astronomy Lab., 1972. Contbr. numerous articles on radio astronomy and related fields to profl. jours. Bd. trustees Asso. Univs., Inc.; mem. Arecibo Adv. Bd. Grantee in radio astronomy NSF, NASA. Mem. Internat. Astron. Union, Internat. Union Radio Sci., AAAS, Am. Astron. Soc. Subspecialties: Radio and microwave astronomy; Planetary science. Current work: Radio astronomy instrumentation; interstellar medium; star formation. Home: 2727 Shasta Rd Berkeley CA 94708 Office: Radio Astronomy Laboratory University of California Berkeley CA 94720

WELDON, VIRGINIA V., pediatric endocrinologist, educator, medical administrator; b. Toronto, Ont., Can., Sept. 8, 1935; m., 1963; 2 children. A.B., Smith Coll., 1957; M.D., SUNY-Buffalo, 1962. Intern, then resident, then fellow Sch. Medicine, Johns Hopkins U. and Hosp., 1962-67; instr. pediatrics Johns Hopkins U., 1967-68; from instr. to assoc. prof. Washington U., St. Louis, 1968-79, prof. pediatrics, 1979—, asst. vice chancellor med. affairs, 1975-81, assoc. vice chancellor med. affairs, 1981—; co-dir. div. metabolism and endocrinology St. Louis Children's Hosp., 1973-77; cons. adv. com. endocrinology and metabolism FDA, 1973-76; mem. Mo. Health Manpower Planning Task Force, 1976—; mem. gen. clin. research ctr. adv. com. NIH, 1976-80, mem. nat. adv. research resources council, 1980-84. Mem. AAAS, Endocrine Soc., Soc. Pediatric Research, Am. Pediatric Soc., Inst. Medicine, Sigma Xi. Subspecialty: Pediatrics. Office: Washington U Sch Medicine PO Box 8106 660 S Euclid Ave Saint Louis MO 63110

WELDON, WILLIAM FORREST, mechanical engineer, researcher, cons.; b. San Marcos, Tex., Jan. 12, 1945; s. Forrest Jackson and Rubie Mae (Wilson) W.; m. Morey Sheppard McGonigle, July 27, 1968; children: William Embree, Seth Forrest. B.S. in Engring. Sci, Trinity U., 1967; M.S. in Mech. Engring, U. Tex., 1970. Registered profl. engr., Tex. Engr. Cameron Iron Works, Houston, 1967-68; project engr. Glastron Boat Co., Austin, Tex., 1970-71; chief project engr. Nalle Plastics, Austin, 1971-73; chief engr. Energy Storage Group, U. Tex., Austin, 1973-76; tech. dir. Center for Electromechanics, 1976—; cons. to industry, govt. Contbr. numerous articles on pulsed power to profl. jours. Mem. ASME. Subspecialties: Mechanical engineering; Fusion. Current work: Electromechanical machine design, pulsed power, homopolar generator, compulsator. Patentee homopolar generators and compulsators. Home: 4707 Peace Pipe Path Austin TX 78746 Office: U Tex 227 Taylor Hall Austin TX 78712

WELKER, J. REED, chemical engineering consultant; b. Rexburg, Idaho, Dec. 1, 1936; s. Jesse R. and Mary V. (Kauer) W.; m. Juanita M. Jensen, June 20, 1958; children: J. Ronald, Sheri L., Michele A. B.S., U. Idaho, 1959, M.S., 1961; Ph.D., U. Okla., 1965. Registered profl. engr., Okla. Group leader Oil Recovery Corp., Norman, Okla., 1961-63; project dir. U. Okla. Research Inst., 1963-73; v.p. Univ. Engrs., Inc., Norman, 1971-77; pres. Applied Tech. Corp., Norman, 1977—; prof. chem. engring. U. Ark., Fayetteville, 1983—; cons. engring., 1965—. Contbr. numerous articles to profl. publs.; assoc. editor: Jour. Fire and Flammability, 1970—, Jour. Consumer Products Flammability, 1974—, Jour. Energy Resources Tech, 1979-82. Mem. Am. Inst. Chem. Engrs., Am. Chem. Soc., Combustion Inst., Am. Gas Assn., Sigma Xi. Subspecialties: Chemical engineering; Fuels. Current work: Fundamental fire research, fire protection research, safety for liquefied gas systems. Home: 2124 Loren Circle Fayetteville AR 72701 Office: Dept Chem Engring U Ark Fayetteville AR 72701

WELLER, MILTON WEBSTER, animal ecologist, educator, researcher; b. St. Louis, May 23, 1929; m., 1947; 1 child. A.B., U. Mo., 1951, M.A., 1954, Ph.D. in Zoology, 1956. Instr. in zoology U. Mo., 1956-57; from asst. prof. to prof. Iowa State U., 1957-74, chmn. sect. fisheries and wildlife, 1967-74; prof. and head dept. entomology, fisheries and wildlife U. Minn.-St. Paul, 1974-83; prof., Kleberg chair in wildlife ecology Tex. A&M U., College Station, 1983—. NSF grantee, 1964-65, 70-74, 76-77. Fellow AAAS, Am. Ornithol. Union; mem. Ecol. Soc. Am., Cooper Ornithol. Soc., Wildlife Soc. Subspecialty: Animal ecology. Office: Wildlife and Fisheries Scineces Tex A&M Univ College Station TX 77843

WELLER, PETER FAHEY, physician, researcher; b. Boston, May 5, 1946; s. Thomas Huckle and Kathleen (Fahey) W.; m. Anne Nicholson, May 26, 1979; children: Susan R., Nathaniel N. A.B., Harvard U., 1968, M.D., 1972. Diplomate: Am. Bd. Internal Medicine, Am. Bd. Allergy and Immunology. Intern Peter Bent Brigham Hosp., Boston, 1972-73, resident, 1973-74; clin. and research fellow in medicine, infectious diseases Mass. Gen. Hosp., 1976-77; asst. in medicine Brigham and Women's Hosp., Boston, 1977—; asst. prof. medicine Harvard U., 1979—; asst. physician Beth Israel Hosp., 1981—; cons. infectious diseases New Eng. Deaconess Hosp., Boston, 1980—. Contbr. articles to profl. jours. Served with USPHS, 1974-76. Fellow ACP; mem. Am. Soc. Tropical Medicine and Hygiene (Janssen award 1979), Am. Soc. Microbiology, Royal Soc. Tropical Medicine and Hygiene, Am. Thoracic Soc., Am. Assn. Immunologists, Am. Acad. Allergy, Am. Fedn. Clin. Research, Phi Beta Kappa. Subspecialties: Infectious diseases; Parasitology. Current work:

Immunology of eosinophils and host defense against parasitic infections.

WELLER, SOL WILLIAM, engineering educator; b. Detroit, July 27, 1918; s. Ira and Bessie W.; m. Miriam D. Damick, June 11, 1943; children: Judith, Susan, Robert, Ira. B.S., Wayne U., 1938; Ph.D., U. Chgo., 1941. Mem. faculty SUNY, Buffalo, 1965—, prof. dept. chem. engring., 1965—, C.C. Furnas Meml. prof., 1983. Contbr. articles to profl. jours. Mem. Am. Inst. Chem. Engrs., Am. Chem. Soc. (H.H. Storch award 1981, E.V. Murphee award 1982), AAAS. Subspecialties: Coal; Chemical engineering. Current work: Coal liquefaction; catalyst characterization; reaction kinetics. Patentee. Home: 96 Carriage Circle Williamsville NY 14221 Office: SUNY 305 Furnas Hall Buffalo NY 14260

WELLINGS, SEFTON ROBERT, pathologist; b. Tacoma, Oct. 2, 1927; s. Donald and Norah Evelyn (Atkins) W.; m. Marjorie Plumb Wellings, June 15, 1951; children: Anne Katherine, Elizabeth Norah, Julie Virginia, Mary Martha, James Sefton; m. Carol Christine Wellings, Oct. 13, 1974. B.S. cum laude, U. Wash., 1951, M.D., 1953; Ph.D., U. Calif., Berkeley, 1961. Diplomate: Am. Bd. Pathology. Intern Highland Alameda County Hosp., Oakland, Calif., 1954-55; resident in pathology U. Calif. Hosps. and Clinics, San Francisco, 1955-59; asst. prof. pathology U. Oreg. Med. Sch., Portland, 1961-62, prof., chmn. dept., 1962-69, U. Calif., Davis, 1969-75, prof., 1975—; dir. Labs. Anatomical and Clin. Pathology, Sacramento Med. Ctr., 1969-75, surg. pathologist, 1975—. Contbr. articles to profl. jours. Served with USN, 1944-47. Mem. Am. Soc. Clin. Pathology, Internat. Acad. Pathology, Am. Fisheries Soc., Am. Ornithologists Union, Sigma Xi, Alpha Omega Alpha. Republican. Subspecialties: Pathology (medicine); Pathology (veterinary medicine). Current work: Mammary biology and pathology, cancer, comparative pathology, crustacean pathology, coelenterate pathology, fish neoplasia, mollusc pathology. Home: 1419 11th Ave Sacramento CA 95818 Office: Dept Pathology Med Sch Univ Calif Davis CA 95616

WELLNER, CHARLES AUGUST, research forester, forest ecologist; b. Enid, Okla., Jan. 3, 1911; s. August and Adelia Geneva (Anderson) W.; m. Ethel Dorothy Wolf, June 10, 1939 (dec. 1969); children: Jon, Chris, Sandra, Kent. B.S., U. Idaho, 1933; M.F., Yale U., 1938; cert. aero. engring., U.S. Naval Acad. Postgrad. Sch., 1944. Research silviculturist U.S. Dept. Agr. Forest Service, Missoula, Mont., 1933-48; leader Inland Empire Research Ctr., Spokane, 1948-58; div. chief Intermountain Sta., Ogden, Utah, 1958-65, asst. dir., 1965-71, program leader, 1971-73; affiliate prof. Idaho U, 1973—; chmn. Idaho Natural Areas Coordinating Com., 1974—. Contbr. articles to profl. jours. Served to lt. comdr. USNR, 1942-46. Superior Service award U.S. Dept. Agr., 1962, 72. Fellow Soc. Am. Foresters; mem. Ecol. Soc. Am., Northwest Sci. Assn. Subspecialties: Resource conservation; Resource management. Current work: Classification of natural diversity and the selection and establishment of research natural areas to preserve examples of natural diversity. Home: 439 Styner Ave Moscow ID 83843 Office: Dept Forest Resources U Idaho Moscow ID 83843

WELLS, ALAN HARVEY, nuclear data corporation executive, computer science educator; b. Ancon, C.Z., July 5, 1948; m. Kathy Jane Melanson, Aug. 8, 1970; children: Chandra, Krista, Michael. B.S. in Physics, Stevens Inst. Tech., 1970; M.S., Tex. A&M U., 1975, Ph.D., 1978. Data processing mgr. Nuclear Assurance Corp., Atlanta, 1979-81, sr. engr., 1981-82, cons., 1982—; engring. devel. mgr. Nuclear Data, Inc., Smyrna, Ga., 1982—; cons. Los Alamos Sci. Lab., 1978-79; adj. prof. computer sci. Ga. Inst. Tech., Atlanta, 1982—. Contbr. articles to profl. jours. Served to capt. USAF, 1970-74. Mem. Am. Nuclear Soc. (program com. 1974—). Subspecialties: Nuclear engineering; Artificial intelligence. Current work: Nuclear criticality safety for power reactor fuel, nuclear reactor operator training using knowledge-based computer aided instruction, control room diagnostic computers. Office: Nuclear Data Inc 2734 S Cobb Industrial Blvd Smyrna GA 30080

WELLS, EDWARD CURTIS, consultant, retired engineering executive; b. Boise, Idaho, Aug. 26, 1910; s. Edward Lansing and Laura Alice (Long) W.; m. Dorothy Evangeline Ostlund, Aug. 25, 1934; children: Laurie Jo (Mrs. William Tull), Edward Elliott. Student, Willamette U., 1927-29, D.Sc. (hon.), 1963; B.A. with gt. distinction, Stanford, 1931; LL.D., U. Portland, 1946. Draftsman Boeing Co., Seattle, summer 1930, draftsman, engr., 1931-33, group engr., 1933-34, asst. project engr., 1934-37, chief preliminary desgin engr., 1937-38, chief project engr., 1938-39, asst. chief engr., 1939-43, chief engr., 1943-47, v.p., chief engr., 1947-48, v.p. engring., 1948-58, 59-61, v.p., gen. mgr. mil. aircraft systems div., 1961-63, v.p. product devel., 1963-65, 66-67, group v.p. airplanes, 1965-66, sr. v.p., 1967-71, ret., 1972; v.p., gen. mgr. Systems Mgmt. Office, 1958-59, dir. in charge preliminary design, 1934; cons., 1972—; vis. prof. Stanford U., 1969-70; State Adv. Council Atomic Energy, 1958-60. Life trustee Willamette U.; mem. adv. bd. Wash. State Inst. Tech., 1957-61; mem. adv. council Stanford Engring. Sch.; pres. bd. Ryther Child Center, 1961-67, dir., 1961-80; mem. Def. Sci. Bd., 1969-72. Recipient Lawrence Sperry award Inst. Aero. Scis. for outstanding contbns. to art airplane design with spl. reference to 4-engined aircraft, 1942; Fawcett Aviation award, 1944; Young Man of Year, Seattle, 1943; Elder Statesman of Aviation award, 1978; Daniel Guggenheim Medal award, 1980. Fellow AIAA (hon.), Soc. Automotive Engrs.; mem. AAAS, Nat. Acad. Engring., Phi Beta Kappa Assos., Phi Delta Theta, Tau Beta Pi. Club: Tennis (Seattle). Subspecialty: Aeronautical engineering. Home: PO Box 2031 Bellevue WA 98009

WELLS, GARY LEROY, psychology educator; b. Hutchinson, Kan., Dec. 11, 1950; s. Richard J. and Bonnie (Wilson) W.; m. Teresa Diane Wilson, Nov. 2, 1968; children: Gary Jonathan, Kristopher Aaron. B.Sc., Kan. State U., 1973; M.Sc., Ohio State U., 1975, Ph.D., 1977. Instr. Ohio State U., Columbus, 1974-76; asst. prof. U. Alta., Edmonton, Can., 1977-80, assoc. prof., 1980—, assoc. chmn. dept. psychology, 1982—; spl. cons. Law Reform Commn. of Can., Ottawa, 1982—; reviewer NSF, Washington, 1978—; cons. on memory Can. and U.S. cts., 1979—; cons. The Investigators, Edmonton, 1982—. Editor, author: Eyewitness Test: Psychological Perspectives, 1983; contbr. articles to profl. jours.; guest editor: Law and Human Behavior, 1980. Eloise Worthy scholar, 1971-73; Ohio State U. fellow, 1974-76; Social Sci. and Humanities Research Council grantee, 1981—. Mem. Soc. Exptl. Social Psychology, Am. Psychol. Assn. Club: Social Sci. (pres.). Subspecialties: Cognition; Social psychology. Current work: Experimental research on eyewitness testimony, attitude change and propaganda; experimental research on rumor transmission. Office: Dept Psychology Univ Alta Edmonton AB Canada T6G 2E9

WELLS, HERBERT ARTHUR, mechanical engineer; b. Jersey City, Aug. 4, 1921; s. Herbert C. and Minnie E. (Banefuer) W.; m. June Korwan, Jan. 12, 1946; children: Carolyn Hotaling, Barbara Hotaling, Richard A. B.M.E., Cooper Union, 1947; M.S. in Mech. Engring, Newark Coll. Engrng., 1952. Registered profl. engr., N.Y. With Bell Telephone Labs., Inc., Whippany, N.J., 1947—, supr., 1956—, supr. installation and ship modification, 1965—. Served to capt. USAAF, 1942-46. Mem. ASME, Nat. Security Indsl. Assn. Subspecialty: Mechanical engineering. Current work: Cable ships, cable machinery, cable installation. Patentee in multiple drop wire clamp, cable handling machine. Home: 772 Norman Pl Westfield NJ 07090 Office: Whippany Rd Rm 3B-201 Whippany NJ 07981

WELLS, JACK NULK, pharmacologist; b. McLouth, Kans., May 17, 1937; s. Russell and Mabel Maria (Nulk) W.; m. Marjorie Elaine Wells, June 14, 1938; children: Daniel N., Douglas C. B.A., Park Coll., 1959; Ph.D., U. Mich., 1963. Asst. prof. medicinal chemistry Purdue U., West Lafayette, Ind., 1963-67, assoc. prof., 1967-72; asst. prof. physiology Vanderbilt U., Nashville, 1973-75, asst. prof. pharmacology, 1975-77, assoc. prof., 1977—. Mem. Fedn. Am. Socs. Explt. Biology, Am. Chem. Soc., AAAS, Am. Heart Assn., Phi Beta Kappa, Sigma Xi. Subspecialties: Molecular pharmacology; Medicinal chemistry. Current work: Molecular pharmacology of vascular smooth muscle; cyclic nucleotide metabolism; mechanism of action of the xanthines. Home: 3604 Saratoga Dr Nashville TN 37205 Office: Dept Pharmacology Med Sch Vanderbilt U Nashville TN 37232

WELLS, JAMES OPIE, JR., nephrologist, educator; b. Asheville, N.C., Oct. 5, 1937; s. James Opie and Carol (McDevitt) W.; m. Sybil Jean Beasley, Dec. 27, 1964. B.S., Wake Forest U., 1960, M.D., 1963. Diplomate: Am. Bd. Internal Medicine, Am. Bd. Nephrology. Intern Emory U. Affiliated Hosps., Atlanta, 1963-64, resident, 1964-67; NIH research fellow Mayo Clinic, Rochester, Minn., 1969-70; from asst. to assoc. prof. Emory U., Atlanta, 1971—, dir. nephrology/immunoassay lab., 1971—; med. dir. Dialysis Clinic, Inc., Atlanta, 1979-82. Served as capt. USAF, 1967-69. Fellow ACP, Am. Coll. Clin. Pharmacology; mem. Mayo Nephrology Soc. (pres. 1976-77), Am. Fedn. Clin. Research, Am. Soc. Nephrology, Internat. Soc. Nephrology, Am. Soc. for Artificial Internal Organs. Presbyterian. Subspecialties: Nephrology; Internal medicine. Current work: Immunoassay of hormones; hypertensive research and therapy; dialysis and renal transplantation. Home: 2249 Greencrest Dr NE Atlanta GA 30345 Office: Emory U Sch Medicine 69 Butler St SE Atlanta GA 30303

WELLS, MARION ROBERT, biology educator; b. Jackson, Miss., Feb. 9, 1927; s. Onous John and Virginia (Chancellor) W.; m. Tommie Sue Butts, Dec. 31, 1959; children: Cynthia Sue, Sherry Beth, Amy Paige, Marion Robert. B.S., Memphis State U., 1959, M.A., 1963; Ph.D., Miss. State U., 1971. Prof. dept. biology Troy State U., Ala., 1963-64, Middle Tenn. State U., Murfreesboro, Tenn., 1964—. Author: Food Composition and Analysis, 1983. Named Outstanding Tchr. Middle Tenn. State U., 1974. Fellow Tenn. Acad. Sci.; mem. Assn. Southeastern Biologists. Baptist. Clubs: Exchange (pres. 1978, dist. dir. 1979. Subspecialties: Cell biology; Molecular biology. Current work: Pesticides and non-target organisms; working on pesticides and scanning and transmitting electron microscopy cells. Home: Route 2 PO Box 215A Lascassas TN 37085 Office: Middle Tenn State U Murfressboro TN 37132

WELLS, ROBERT DALE, biochemist, educator; b. Uniontown, Pa., Oct. 2, 1938; s. Charles O. and Margaret Elizabeth (Sturm) W.; m. Dorothy Jackson Smart, June 25, 1960; children: Robert Kevin, Cynthia Gail. B.A., Ohio Wesleyan U., 1960; Ph.D., Sch. Medicine, U. Pitts., 1964. Postdoctoral fellow Inst. for Enzyme Research, U. Wis.-Madison, 1964-66; prof. dept. biochemistry Inst. Enzyme Research, U. Wis.-Madison, 1966-81; prof., chmn. dept. biochemistry U. Ala., Birmingham, 1982—; cons. in field. Contbr. chpts. to books, articles to profl. jours.; assoc. editor: Jour. Biol. Chemistry, 1978—. Guggenheim fellow, 1976-77. Mem. Am. Soc. Biol. Chemists, AAAS, Biophys. Soc., Am. Assn. Microbiology, Sigma Xi, Phi Lambda Upsilon, Alpha Sigma Phi. Subspecialties: Biochemistry (biology); Gene actions. Current work: DNA structure and gene regulation. Home: 2922 Westmoreland Dr Birmingham AL 35223 Office: Univ Alabama 201 Volker Hall Birmingham AL 35294

WELLS, RONALD ALLEN, planetary astronomer, cons. Ancient Egyptian Astronomy; b. Norton, Va., Sept. 12, 1942; s. Ray Washington and Maria Dorene (Nard) W.; m. Dorothy Gwynne Tompkinson, Sept. 17, 1966; children: Kimberleigh Erin, Christopher Cedric. Student, Va. Poly. Inst., 1960-62; A.B. in Astronomy, U. Calif., Berkeley, 1964; Diploma in Space Sci. (European Space Research Orgn. fellow, 1965-67), Univ. Coll., U. London, 1966; Ph.D. in Astronomy, Univ. Coll., U. London, 1967. Asso. research astronomer Space Scis. Lab. U. Calif., Berkeley, 1967-72; research asso., 1973-79, 1981—, research investigator; European Space Research Orgn. scholar, summer, 1965; project dir. NASA grant, 1971-72. Author: Geophysics of Mars, 1979; contbr. papers to profl. publs., profl. confs. Fellow Royal Astron. Soc. Gt. Brit.; mem. Am. Astron. Soc. (div. planetary sci.). Subspecialties: Planetary science; Planetology. Current work: Investigation origins and applications of astronomy in Ancient Egypt from written texts, temple inscriptions, heiroglyphs, precise orientations, stellar rise/set positions. Office: Dept Near Eastern Studies Univ Calif Berkeley CA 94720

WELLS, WILLIAM TERRY, fireplace manufacturer, inventor; b. Elfrida, Ariz., Dec. 11, 1919; s. Reubin Garret and Delma Montez (Terry) W.; m. Ruth Crum, Sept. 5, 1964; children: Berna and Sheila (twins), Mary and Michelle (twins), Rhonda, Marsha and Rebecca (twins). Student, U. Ariz., 1938-42. Quality engr. Boeing Aircraft, Seattle, 1946-59, Hughes Aircraft, Tucson, 1946-59; inventor, 1960--; devel. and mktg. energy-efficient fireplaces Wells Fireplaces, Tucson, 1976—; invented gearless automatic transmission and rotary engine. Served in USAF, 1942-45. Decorated Air medal (4). Mem. So. Ariz. Home Builders Assn., Tucson C. of C. Republican. Mormon. Subspecialties: Mechanical engineering; Theoretical and applied mechanics. Current work: Research and development of fireplaces, transmission and rotary engines. Patentee internal combusion engine, infinitely variable transmission, fireplaces. Office: 1221 W Monte Vista PO Box 7097 Tucson AZ 85725

WELSCH, CLIFFORD WILLIAM, JR., Oncologist, educator; b. St. Louis, Sept. 10, 1935; s. Clifford William and Lorraine Ann W.; m. Margaret Ann Goodrich, Mar. 19, 1980; children: Jeffrey Stephen, David Brian, Richard William. Ph.D. in Physiol. Chemistry, U. Mo., 1965. Instr. dept. biochemistry U. Mo., 1964-65; research assoc. dept. physiology Mich. State U., 1965-68; asst. prof., 1968-71, assoc. prof. dept. anatomy, 1971-76, prof., 1976—. Contbr. articles to profl. jours. Recipient Tchr.-Scholar of Yr. Mich. State U., 1971; Nat. Cancer Inst. spl. research fellow, 1966-68; research career devel. awardee, 1971-76. Mem. Am. Assn. Cancer Research, AAAS, Am. Physiol. Soc., Soc. Exptl. Biology and Medicine. Congregationalist. Subspecialties: Cancer research (medicine); Oncology. Current work: Factors controlling breast cancerigenesis. Office: Dept Anatomy Mich State U East Lansing MI 48824

WELSCH, FRANK, pharmacologist, toxicologist; b. Berlin, Germany, Apr. 14, 1941; s. Rudolf K. and Ilse (Hohn) W.; m. Melissa G. Hutchinson, Aug. 30, 1968; 1 son, Derek. Dr. med. vet., Free U. Berlin, 1964. Diplomate: Am. Bd. Toxicology. Research assoc. vet. pharmacology Free U. Berlin, 1964-67; research assoc. in neurochemistry Columbia U., 1967-68; research assoc. in pharmacology Vanderbilt U., 1968-71; mem. faculty Mich. State U., East Lansing, 1971-82, prof., 1981-82; sr. scientist Chem. Industry Inst. Toxicology, Research Triangle Park, N.C., 1982—. Contbr. articles to profl. jours. NIH grantee, 1972-82; March of Dimes grantee, 1975-78; Alexander von Humboldt Found. research fellow, 1977-78. Mem. Am. Soc. Pharmacology and Exptl. Therapeutics, AAAS, Teratology Soc., Soc. Toxicology. Subspecialties: Teratology; Toxicology (medicine). Current work: Effects of chemicals on reproduction, research in prenatal toxicology.

WELSEY, WAYNE CECIL, chemistry educator; b. Battle Creek, Mich., Nov. 12, 1936; s. Henry Cecil and Frieda Maxine (Samms) W.; m. Mary Lou Morris, July 31, 1965; children: Carole, Roger. B.S., Mich. State U., 1958; Ph.D., U. Kans., 1962. Teaching asst. U. Kans., 1958-59, NSF fellow, 1959-62; sr. research chemist P.P.G. Chems., Barberton, Ohio, 1962-65; mem. faculty dept. chemistry Macalester Coll., St. Paul, 1965—, prof., 1980—; vis. asst. prof. Ariz. State U., Tempe, 1971-72; Sr. research fellow U. Bristol, Eng., 1978-79. Co-author: Chemical Principles in the Laboratory, 1982. Mem. Minn. Acad. Sci. (pres. 1981-82), Am. Chem. Soc., AAAS, AAUP, Phi Beta Kappa, Phi Kappa Phi, Sigma Pi Sigma, Phi Lambda Upsilon. Subspecialties: Inorganic chemistry; Analytical chemistry. Current work: Organo metallic chemistry; computer interfacing. Home: 2197 Berkeley Ave Saint Paul MN 55105 Office: Dept Chemistry Macalester Coll Saint Paul MN 55105

WEN, WEN-YANG, chemistry educator, researcher; b. Taiwan, China, Mar. 7, 1931; came to U.S., 1953, naturalized, 1973; s. Chi and Yueh-Er (Yu) W.; m. Sue Liu, Aug. 1, 1959; children: Lilian, Alvin. B.S., Nat. Taiwan U., 1953; Ph.D., U. Pitts., 1957. Research assoc. U. Pitts., 1957-58; postdoctoral fellow Northwestern U., 1958-60; asst. prof. chemistry DePaul U., 1960-62; asst. prof. Clark U., 1962-66, assoc. prof., 1966-73, prof., 1973—; Alexander von Humboldt fellow Universitat Karlsruhe, W. Ger., 1970-71, summer 1973; vis. prof. Drittes Physikalisches Institut, Universitat Gottingen, Ger., summer 1976; faculty research participant Morgantown Energy Tech. Ctr., Dept. Energy, 1978-79. Contbr. articles to profl. publs. Li Found. fellow, 1953-55. Mem. AAUP, Am. Chem. Soc., AAAS. Democrat. Baptist. Subspecialties: Coal; Physical chemistry. Current work: Coal gasification; tar cracking. Home: 6 Old Brook Dr Worcester MA 01609 Office: 950 Main St Worcester MA 01610

WENDELBERGER, JAMES GEORGE, statistician, consultant; b. Milw., Mar. 13, 1953; s. Joseph Martin and Elizabeth (Neimon) W. B.S. with distinction in Math. and Physics, U. Wis.-Madison, 1976, M.S. in Stats, 1978, Ph.D., 1982. Research assoc. Space Sci. and Engring. Ctr., U Wis. - adison, 1982-83. Mem. Inst. Math. Stats., Soc. Indsl. and Applied Math., Am. Statis. Assn., AAAS, Sigma Xi. Subspecialties: Statistics; Mathematical software. Current work: Consulting research statistician; multiple time series analysis, multidimentional spline smoothing. Office: Dept Stats U Wis 1210 W Dayton St Madison WI 53706

WENE, EDWARD G., plant pathologist; b. Watseka, Ill., Nov. 8, 1946; s. Donald M. and Gladys L. (Lobdell) W.; m. Bonnie Lee Boyle, May 2, 1947. Ph.D., U. Ill., 1979. Plant pathologist Argonne Nat. Lab., 1980—. Served with U.S. Army, 1968-70. Mem. Am. Phytopathol. Soc. Subspecialty: Plant pathology. Current work: Microbial degradation of biomass. Office: 9700 S Cass Ave Argonne IL 60439

WENG, GEORGE JUENG-CIOUS, engineering educator; b. Taiwan, Oct. 8, 1944; s. Wan-Chung and Kuan-chia (Hsieh) W.; m. Shu-yu Huang, Oct. 26, 1949; children: Bruce, Joyce. B.S., Taiwan U., 1967; M.Phil., Yale U., 1971, Ph.D., 1974. Research fellow Delft (Netherlands) U. Tech., 1973-74; postdoctoral fellow Yale U., UCLA, 1974-76; sr. research engr. Gen. Motors Research Lab., Warren, Mich., 1976-77; asst. prof. mechanics and materials sci. Rutgers U., Piscataway, N.J., 1977-80, assoc. prof., 1980—. Contbr. articles to profl. jours. NSF grantee, 1978, 80; U.S. Dept. Enrgy grantee, 1980. Mem. Am. Acad. Mechanics, ASME, AIME, Soc. Rheology, Sigma Xi. Subspecialties: Solid mechanics; Metallurgical engineering. Current work: Micromechanics of plastic deformation of metals, creep and fracture at elevated temperature, interfacial problems in continuum plasticity and mechanical metallurgy, mechanics of composite materials, theory of inclusions. Home: 9 Langley Rd Kendall Park NJ 08824 Office: Coll Engring Rutgers U Piscataway NJ 08854

WENGER, GALEN ROSENBERGER, pharmacologist, educator, researcher; b. Sellersville, Pa., May 16, 1946; s. Warren Martin and Ethel Gargas (Rosenberger) W.; m. Carolyn Jean Liechty, Nov. 24, 1972; children: Alyssa Nicole, Aaron Joseph. B.A. in Biology, Goshen (Ind.) Coll., 1968; Ph.D., W.Va. U., 1971. Postdoctoral fellow dept. pharmacology U. Colo. Med. Center, Denver, 1972-73; research fellow Lab. Psychobiology dept. psychiatry Harvard U. Med. Sch., Boston, 1973-75, instr. pharmacology, 1975-78; asst. prof. pharmacology and interdisciplinary toxicology U. Ark. for Med. Sci., Little Rock, 1978-81, assoc. prof., 1981—; grant reviewer NSF, VA. Contbr. articles and abstracts to sci. publs., also monographs; reviewer: Jour. Pharmacology and Exptl. therapeutics, Psychopharmacology, Pharmacology Biochemistry and Behavior. Mem. Am. Soc. Pharmacology and Exptl. Therapeutics, Behavioral Pharmacology Soc., Neurobehavioral Toxicology Soc., Sigma Xi. Democrat. Mennonite. Subspecialties: Pharmacology; Behavioral psychology. Current work: Behavioral pharmacology of drug abuse, behavioral toxicology, operant conditioning. Home: 18 Hayfield Dr Little Rock AR 72207 Office: 4301 W Markham St Mail Slot 638 Little Rock AR 72205

WENK, EDWARD, JR., civil engineer, policy analyst, educator; b. Balt., Jan. 24, 1920; s. Edward and Lillie (Heller) W.; m. Carolyn Frances Lyford, Dec. 21, 1941; children: Lawrence Shelley, Robin Edward Alexander, Terry Allan. B.E., Johns Hopkins U., 1940, D.Eng., 1950; M.Sc., Harvard U., 1947; D.Sc. (hon.), U. R.I., 1968. Head structures div. USN David Taylor Model Basin, Washington, 1942-56; chmn. dept. engring. mechanics S.W. Research Inst., San Antonio, 1956-59; sr. specialist sci. and tech. Legis. Reference Service, Library of Congress, Washington, 1959-61, chief sci. policy research div., 1964-66; tech. asst. to U.S. President's sci. adviser and exec. sec. Fed. Council for Sci. and Tech., White House, Washington, 1961-64; exec. sec. Nat. Council on Marine Resources and Engring. Devel., Exec. Office of Pres., Washington, 1966-70; prof. engring. and pub. affairs U. Wash., Seattle, 1970—, dir. program in social mgmt. tech., 1973-79; dir. URS Corp.; lectr. U. Md.; Cons. in pub. policy for environ. and tech. affairs, risk assessments, ocean engring. Nat. Adv. Com. on Oceans and Atmosphere, 1972-73; vice chmn. U.S. Congress Tech. Assessment Adv. Council, 1973-79; adviser Congress, GAO, U.N. Secretariat, Wash. State, U.K., Sweden, Philippines, pub. interest groups; vis. scholar Woodrow Wilson Internat. Center for Scholars, 1970-72, Harvard, 1976, Woods Hole Oceanographic Instn., 1976, U. Sussex, 1977. Author: The Politics of the Ocean, 1972, Margins for Survival, 1979; Editor: Engring. Mechanics Jour, 1958-60, Exptl. Mechanics Jour, 1954-56; mem. editorial bd.: Tech. Forecasting; contbr. articles to profl. jours. Bd. dirs. Human Interaction Research Inst., Smithsonian Sci. Info. Exchange., 1977-82. Served as ensign USNR, 1944-45. Recipient Navy Meritorious Civilian Service award, 1946; named Disting. Alumnus Johns Hopkins U., 1979; Tchr. of Yr. Wash. State Engrs., 1980; Ford Found. grantee, 1970; Rockefeller Found. fellow, 1976. Fellow ASME; mem. Soc. Exptl. Stress Analysis (past pres. William M. Murray lectr.), Internat. Assn. Impact Assessment (pres. 1981-82), Nat. Acad. Engring. (chmn. com. on pub. policy 1970-75), Nat. Acad. Pub. Adminstrn., Am. Soc. for Pub. Adminstrn. (chmn. com. on sci. and tech. in govt. 1974-78), Assembly Engring.-NRC, ASCE, Nat. Soc. Profl. Engrs., Nat. Oceanography

Assn. (v.p. pub affairs 1970-72), Cousteau Soc. (chmn. adv. bd.), AAAS, Explorers Club, Sigma Xi (nat. lectr.), Tau Beta Pi, Chi Epsilon. Club: Cosmos (Washington). Subspecialty: Technology assessment. Current work: Technology and industrial productivity and economic vitality. Designer Aluminaut submarine. Home: 15142 Beach Dr NE Seattle WA 98155 Each of us has the opportunity, indeed responsibility, to contribute to the human experience and to enrich the lives of future generations. In a world of change, cultural diversity and uncertainty, we must be ourselves and not merely slaves of conventional thought. We must act on the basis of what we believe to be right rather than only in the long run from the desire to be loved.

WENNBERG, JEFFREY NORMAN, solar heating system designer; b. Milton, Mass., Feb. 11, 1953; s. Norman A. and Lynn B. W.; m. Nancy Bevins, Oct. 1, 1977. B.S. in Physics, Clarkson Coll., 1975, M.S. in Mgmt, 1977. Cons. Bienvenu Assocs., Rutland, Vt., 1977-79; pres. Delta G Solar Systems, Inc., Rutland, 1979—. Mem. Gov.'s Commn. Energy Independence, Gov.'s Commn. for Preservation of Vt.'s Heritage, Rutland City Sch. Bd., 1980—, pres., 1982-83. Tomorrow's Scientists and Engrs. scholar. Mem. Eastern N.Y. Solar Energy Soc. Subspecialty: Solar energy. Current work: Development of low cost commercial solar space heating systems. Office: 97 Center St Rutland VT 05701

WENTINK, TUNIS, JR., physical chemist, physics educator, researcher; b. Paterson, N.J., Feb. 3, 1920; m., 1946, 68. B.S., Rutgers U., 1941; Ph.D. in Chemistry, Cornell U., 1954. Research chemist Photoproducts div. E. I. du Pont de Nemours & Co., 1941-43; from research assoc. to mem. staff div. indsl. co-op. MIT, 1947-48; from research assoc. to specialist microwave spectroscopy Brookhaven Nat. Lab., 1948-50; asst. Cornell U., 1950-54; physicist Gen. Electric Co., 1953-55; prin. research scientist and supr. chem. lab. Avco-Everett Research Lab., Avco Corp., 1955-59; from prin. scientist to sr. cons. scientist Adv. Research and Devel. div., 1959-67; prin. scientist GCA Corp., 1967-68; head exptl. physics dept. Panametrics, Inc., 1968-70; assoc. dir., then dir. Inst. Arctic Environ. Engring. U. Alaska, Fairbanks, 1972-73, 73—; NSF vis. prof. Geophys. Inst., 1968, prof. physics, 1970—; cons. in field. Hon. mem. Sigma Xi. Subspecialty: Physical chemistry. Office: Geophys Inst U Alaska Fairbanks AK 99701

WENTZEL, DONAT GOTTHARD, astrophysicist, educator; b. Zurich, Switzerland, June 25, 1934; came to U.S., 1948, naturalized, 1954; s. Gregor and Anna (Wielich) W.; m. Maria Mayer, Mar. 21, 1959; 1 child, Tania. B.A., U. Chgo., 1954, B.S., 1955, M.S., 1956, Ph.D., 1960. Instr. to assoc. prof. astronomy U. Mich., Ann Arbor, 1960-66; mem. faculty U. Md.-College Park, 1967—, prof. astronomy, 1974—; vis. lectr. Princeton (N.J.) U., 1974; vis. prof. Tata Inst. Fundamental Research, Bombay, India, 1973; acad. guest. Fed. Inst. Tech., Zurich, Switzerland, 1978. Contbr. astrophys. and solar physics articles to profl. publs. Recipient Teaching of Sci. award Wash. Acad. Sci., 1975, Disting. Scholar-Tchr. award U. Md., 1983-84; Alfred P. Sloan fellow, 1962-66. Fellow AAAS (sec. astronomy sect. 1978-85); mem. Am. Astron. Soc., Internat. Astron. Union (pres. comm. on edn. in astronomy 1979-82). Subspecialties: Solar physics; Theoretical astrophysics. Current work: Plasma physics and magnetohydrodynamics related to solar corona and its radio emission as well as to cosmic rays; astronomy educator. Office: Astronomy Dept Univ Md College Park MD 20742

WENZ, MICHAEL FRANK, JR., nuclear engineer; b. Arlington, Va., Dec. 9, 1953; Michael Frank and Theo (Lambert) W. B.S. in Nuclear Engring, U. Va., 1975, 1975, M.E. in Nuclear Engring, 1977. Nuclear engr. Naval Reactors, Naval Sea Systems Command, Washington, 1977—. Mem. Am. Nuclear Soc. Subspecialties: Nuclear engineering; Nuclear fission. Current work: Design and fabrication of naval nuclear reactors. Home: 107 E Del Ray Ave Alexandria VA 22301 Office: Dept Navy Naval Sea Systems Command Nuclear Propulsion Directorate Washington DC 20362

WENZEL, JAMES GOTTLIEB, ocean engineering and marine systems company executive; b. Springfield, Minn., Oct. 16, 1926; s. Henry Gottlieb and Elvira (Runck) W.; m. Elaine Joyce Abrahamson, June 17, 1950; children: Lori Lynn Wenzel Taylor, Jodi Ann Wenzel Bjurman, Sheri Lee, James Gottlieb, II. B.Aero. Engring. with distinction, U. Minn., 1948, M.S., 1950; postgrad. in advanced mgmt, U. Hawaii, 1969, in advanced mgmt, Lockheed Exec. Inst., 1980. Aero. engring. Convair, San Diego, 1948-55; project mgr. anti-submarine warfare and ocean systems. Gen. Dynamics, San Diego, 1956-57, asst. to v.p. engring., 1958-59, asst. to sr. v.p. engring., 1960-61; mgr. govt. planning U.S. Navy, 1961-62; mgr. cruise missiles systems Lockheed, Sunnyvale, Calif., 1962-63, mgr. ocean systems, 1963-70, asst. gen. mgr. research and devel. div., dir. ocean systems, 1970-73, v.p. ocean systems, 1973—; v.p. Lockheed Petroleum Services Ltd., Vancouver, B.C., Can., 1970-76; chmn., pres. Ocean Minerals Co., Mountain View, Calif., 1977—; instr. U. Minn., 1948-50, UCLA Grad. Sch. Aerodyn., 1950-57; mem. panel Naval Research Adv. Com.; mem. Sea-Space Symposium; Theodore von Karman guest lectr. in oceanology. Contbr. in field. Vice chmn. bd. dirs. Jr. Achievement Santa Clara County, Calif., 1978; bd. dirs. Jr. Achievement Found., 1980-82; bd. regents Calif. Lutheran Coll. Served with USN, 1944-46; with USNR, 1948-59. Consol.-Vultee fellow, 1948; recipient City of San Diego Outstanding Achievement award, 1961, City of Los Angeles award of appreciation, 1967; Silver Knight of Mgmt. award Nat. Mgmt. Assn., 1977. Fellow Marine Tech. Soc.; mem. Nat. Acad. Engrs., Royal Aero. Soc., Inst. Aero. Scis. (award of appreciation 1971), Soc. Naval Architects and Engrs. Republican. Lutheran. Club: Saratoga (Calif.) Men's Cosmos. Subspecialties: Ocean engineering; Ocean thermal energy conversion. Current work: Deep ocean technology and system development, ocean mining and ocean thermal energy conversion research and engineering, ocean science and instrumentation, oceanography and exploration, systems engineering, program and general management.

WERBELOW, LAWRENCE GLEN, chemistry educator, researcher; b. Ross, Calif., Dec. 19, 1948; s. Arnold Glen and Helen Corrine (Freeburg) W.; m. Catherine Elizabeth Fouques, Dec. 28, 1979; 1 dau. Prisca. B.Sc., Humboldt State U., 1970; Ph.D., U. B.C., 1974; D.Sc., U. Provence, Marseille, France, 1979. Research assoc. U. Utah, Salt Lake City, 1974-78; vis. prof. U. Provence, Marseille, 1978-79, Mont. State U., Bozeman, 1979-80; assoc. prof. chemistry N.Mex. Inst. Mining and Tech., Socorro, 1980—; vis.scientist Los Alamose Nat. Lab., 1980—. Contbr. chpts. to books, articles to profl. jours. Nat. Research Council Can. fellow, 1974; Am. Chem. Soc. grantee, 1980; NSF grantee, 1980; NATO grantee, 1982; Research Corp. grantee, 1983. Mem. Internat. Soc. Magnetic Resonance. Subspecialties: Nuclear magnetic resonance (chemistry); Nuclear magnetic resonance (biotechnology). Current work: Time-dependent aspects of nuclear paramagnetism; creation and dissipation of transient multipolar spin order; quantum theory angular momentum. Home: 907 Michigan St Socorro NM 87801 Office: Dept Chemistry N Mex Inst Mining and Tech Socorro NM 87801

WERGIN, WILLIAM P., research scientist; b. Manitowoc, Wis., Apr. 20, 1942; s. Eugene A. and Catherine A. (Virnoche) W.; m. Mary Ester Guse, Aug. 25, 1962; children: Anne Michele, W. Peter. B.S. in Genetics, U. Wis., 1964, Ph.D. in Botany, 1970. With Agrl. Research Service, U.S. Dept. Agr., 1970—, assigned to a, 1970-72, 1972-74, 1974-78, 1974-79, 1979, lab. chief, Beltsville, Md., 1979—. Contbr. articles to profl. jours. H.S. Dept. Agr. grantee, 1977-79; Binat. Agrl. Research and Devel. Fund grantee, 1980-83. Mem. Am. Soc. Cell Biologists, AAAS, Am. Soc. Plant Physiology, Soc. Nematologists, Electron Microscopy Soc. Am., Helminthological Soc., Sigma Xi. Subspecialties: Plant physiology (agriculture); Cell and tissue culture. Current work: Ultrastructural alterations in plants that are subjected to environmental or biological stress. Home: 10108 Towhee Ave Adelphi MD 20783 Office: US Dept Agr Plant Stress Lab Bldg 001 Room 206 BARC-W Beltsville MD 20783

WERNER, BARBARA GRAHAM, microbiologist, research administrator, educator; b. Wilmington, Del., Oct. 17, 1942; d. Arthur Horace and Janet Elizabeth (Graham) W. A.B. (Nat. Presbyn. Coll. scholar), Wilson Coll., 1964; M.A., Case Western Res. U., 1971, Ph.D., 1973. Research asst. Inst. Cancer Research, Phila., 1964-67, postdoctoral researcher, 1973-74, research assoc., 1975-79; grad. research asst., predoctoral fellow Case Western Res. U., 1967-72; chief Hepatitis Lab., Mass. State Lab. Inst., Boston, 1979—, dir. Clin. Investigations Lab., 1982—; asst. prof. medicine Tufts U. Sch. Medicine, Boston, 1979—; vis. lectr. dept. microbiology Harvard U. Sch. Pub. Health, 1981—. Contbr. articles, primarily on immune response to hepatitis B virus, to profl. jours. Mem . Am. Soc. Microbiology. Subspecialties: Health services research; Infectious diseases. Current work: Hepatitis B and other viruses; seroepidemiology; passive/active immunization; research and public health service on hepatitis B, cytomegalovirus and other infectious diseases. Office: 305 South St Boston MA 02130

WERNER, CHRISTIAN THOR, mechanical engineer; b. Chgo., Mar. 25, 1916; s. Thor Christian and Anna Hedvig (Engstrom) Rothstein; m. Barbara Ruth Schneck, July 20, 1957 (div. 1972); 1 dau., Diane Lynn Werner Zink. B.S. in Aero. Engring, Aero. U., Chgo., 1937. Aero. engr. Boeing Aircraft Co., Seattle, 1938-43; aerodynamicist Republic Aviation Corp., Farmingdale, L.I., N.Y., 1944-46, sr. aerodynamicist, 1946-48; contract aerodynamics cons. Naval Air Devel. Ct., Johnsville, Pa., 1949-50; systems engr. Bendix Missile System Div., Mishawaka, Ind., 1951-57, sr. systems engr., 1958-67; sr. mech. engr. Sparton Electronics Div., Jackson, Mich., 1967-68, prin. mech. engr., 1968-69, lab. mgr., 1969-71, staff engr., 1972—; instr. Swedish Jackson Community Coll., 1976-80. Fellow AIAA (assoc.); mem. Marine Tech. Soc., Am. Def. Preparedness Assn. Club: Engineers of St. Joseph Valley (South Bend, Ind.). Subspecialties: Fluid mechanics; Theoretical and applied mechanics. Current work: Preliminary mechanical design and systems analysis of underwater anti-submarine and anti-surface vessel defense systems. Home: 313 Tecumseh St PO Box 608 Brooklyn MI 49230 Office: ASW Tech Ct Sparton Electronics Div 2400 E Ganson St Jackson M 49202

WERNICK, JACK HARRY, metallurgical engineer; b. St. Paul, May 19, 1923; s. Joseph and Eva (Legan) W.; m.; children: Phyllis Roberta, Rosanne Pauline. B. in Metall. Engring., U. Minn., 1947, M.S., 1948; Ph.D., Pa. State U., 1954. Metallurgist Manhattan Project, Los Alamos, 1944-46; instr. Pa. State U., State College, 1949-54; mem. tech. staff Bell Labs., Murray Hill, N.J., 1954-64, head phys, metallurgy research dept., 1964-73, head solid state chemistry research dept., 1973-81, head device materials research dept., 1981—. Contbr. articles on metall. engring. to profl. jours. Served with C.E. AUS, 1944-46. Fellow Metall. Soc. of AIME, Am. Phys. Soc., N.Y. Acad. Scis.; mem. Nat. Acad. Engring., IEEE, AAAS, Am. Soc. for Metals, Electrochem. Soc. Subspecialties: Electronic materials; Metallurgy. Current work: synthesis and study of new superconducting, semiconducting, and magnetic materials for possible use in optoelectronic, superconducting and magnetic devices. Patentee in field. Office: 600 Mountain Ave Murray Hill NJ 07974

WERT, CHARLES ALLEN, engineering educator; b. Battle Creek, Iowa, Dec. 31, 1919; s. John Henry and Anna (Spotts) W.; m. Lucille Vivian Mathena, Sept. 5, 1943; children: John Arthur, Sara Ann. B.A., Morningside Coll., Sioux City, 1941; M.S., State U. Iowa, 1943, Ph.D., 1948. Mem. staff Radiation Lab., Mass. Inst. Tech., 1943-45; instr. physics U. Chgo., 1948-50; mem. faculty U. Ill. at Urbana, 1950—, prof., 1955, head dept. metall. and mining engring., 1967—; cons. to industry. Author: Physics of Metals, 1970, Opportunities in Materials Science and Engineering, 1977; also articles.; Cons. editor, McGraw Hill Book Co. Recipient sr. scientist award von Humboldt-Stiftung, W. Ger. Fellow Am. Phys. Soc., Am. Soc. Metals, AAAS, AIME; mem. Sigma Xi. Subspecialties: Metallurgy; Condensed matter physics. Current work: Description of crystalline and chemical nature of fine-scale inclusions in metals and alloys; chemistry and structure of coal, especially as related to combustion and chemical conversion processes. Home: 1708 W Green St Champaign IL 61820 Office: Metallurgy and Mining Bldg Univ Ill Urbana IL 61801

WERTH, JEAN MARIE, biologist, educator; b. Rochester, N.Y., Jan. 21, 1943; d. Henry Richard and Marjorie Frances (Vrooman) W. Ph.D., Syracuse U., 1973, M.S., 1969; B.A., Nazareth Coll., Rochester, 1964. Asst. prof. biology William Paterson Coll., Wayne, N.J., 1972-75, assoc. prof., 1975-82, prof., 1982—. Contbr. articles to profl. jours. Pres. Lake Reality Homeowner's Assn., 1981. Mem. Am. Soc. Microbiology, N.Y. Acad. Sci., AAAS, Am. Women in Sci., Sigma Xi. Subspecialties: Molecular biology; Biochemistry (biology). Current work: Enzyme which reduces oxidized methionine residues. Office: Dept Biology William Patterson Coll 300 Pompton Rd Wayne NJ 07470

WERTZ, HARVEY JOE, computer scientist; b. Muskogee, Okla., May 1, 1936; s. Bradley Leo and Beulah (Snider) W.; m. Joan Margaret Orson, July 20, 1968. B.S. in Elec. Engring, U. Kans., 1958, M.S., 1959; Ph.D., U. Wis., 1962. Asst. prof. elec. engring. U. Wis., Madison, 1962-66; mem. tech. staff Aerospace Corp., Los Angeles, 1966-68; assoc. prof. U. Wis., Madison, 1968-69; head dept. Aerospace Corp., Los Angeles, 1969-79, prin. dir., 1979—. NSF fellow, 1959-60; U. Wis. fellow, 1960-62. Mem. IEEE, Assn. Computing Machinery, Soc. Indsl. and Applied Math, Sigma Xi. Subspecialties: Algorithms; Mathematical software. Current work: Application of mathematics and computing to the analysis and synthesis of physical systems; development of user-oriented software, especially library quality subroutines and precompilers. Home: 1004 Centinela Ave Santa Monica CA 90403 Office: PO Box 92957 Los Angeles CA 90009

WESSELINK, ADRIAAN JAN, emeritus research astronomer; b. Hellevoetsluis, Holland, Apr. 7, 1909; s. Jan Hendrik and Adriane Marine Nicolette (Stok) W.; m. Jeanette van Gogh, July 11, 1919; children: Josephine, Jan, Henriette. D.Sc., Leiden U., Holland, 1938. Observer Leiden Obs., 1945-50; chief asst., dep. dir. Radcliffe Obs., 1950-64; research assoc., sr. research assoc., sr. research astronomer Yale U., New Haven, 1964-77, emeritus, 1977—, sci. dir. So. Obs., 1965-74; referee sci. jours. and NSF. Author sci. publs. in field. Mem. Am. Astron. Soc., Internat. Astron Union, S. African Astron Soc. (Pres. 1962), Royal Astron. Soc. Subspecialties: Optical astronomy; Theoretical astrophysics. Current work: Study of variable stars. Home: 143 Falls Rd Bethany CT 06525

WESSON, PAUL STEPHEN, physics educator; b. Nottingham, Eng., Sept. 11, 1949; emigrated to Can., 1976; s. Stephen and Betty (Butler); m. Ellen Stauborg, Sept. 12, 1980. B.Sc. in Physics with honors, U. London, 1971; M.Math. with honors, Cambridge (Eng.) U., 1972; Ph.D. in Astronomy, Cambridge (Eng.) U., 1979. Mem. faculty dept. physics Queen's U., Kingston, Ont., Can., 1975-76; with dept. geophysics and astronomy U. B.C. (Can.), Vancouver, 1979-80; prof. dept. physics U. Alta. (Can.), Edmonton, 1980—. Author: Cosmology and Geophysics, 1978, Gravity, Particles and Astrophysics, 1980; author articles on astrophysics and geophysics. Sci. Research Council (Eng.) grantee, 1972-75; NRC (Can.) grantee, 1975-76; Brit. Council grantee, 1977; Royal Soc. London and NATO grantee, 1977-78; Natural Scis. and Engring. Research Council (Can.) grantee, 1980 . Fellow Royal Astron. Soc.; mem. Internat. Astron. Union. Subspecialties: Geophysics; Planetology. Current work: Rotation of earth, plate tectonics, planetary formation, interstellar dust, galaxies, clusters of galaxies, cosmology, general relativity. Office: Dept Physics U Alta Edmonton AB Canada T6G 2J1

WEST, ANTHONY ROBERT, computer engineer; b. Hannover, W.Ger., Aug. 1, 1953; s. Arthur John and Monica Waltraud (Dunkel) W.; m. Frances Barbara, Sept. 7, 1979; children: Barbara, Vanessa. B.Sc. with honors, U. Kent, g., 1974; Ph.D. in Computer Sci., U. London, 1981. Hardware engr. IBM, Hursley, Eng., 1974; research fellow Queen Mary Coll., London U., 1977-79; mem. research staff IBM Research Lab., Zurich, 1979-81, Xerox Palo Alto (Calif.) Research Ctr., 1981-83; mgr. systems engring. Sun Microsystems Inc., Palo Alto, Calif., 1983—. Contbr. articles to profl. jours. Mem. ACM, Internat. Fedn. Info. Processing. Subspecialties: Distributed systems and networks; Computer architecture. Current work: Local-area networks, distributed computing systems, personal information systems, computer architecture, operating systems, computer systems engineering, cryptography and data security, digital communications. Office: Sun Microsystems 2550 Garcia Ave Mountain View CA 94043

WEST, BOB, drug company executive; b. Ellenville, N.Y., Mar. 7, 1931; s. Harry and Elsie May Wicentowsky; m. Betty Parker, May 9, 1957 (div.); children: Debra Ellen, Elizabeth Ann, Sharon Lynn; m. Jacqueline Cutler, Mar. 3, 1982. B.S., Union U., 1952; M.S., Purdue U., 1954, Ph.D., 1956; postgrad. grad. mgmt. seminar, U. Chgo., 1972. Research pharmacologist Am. Cyanamid Co., Stamford, Conn., 1958-60; v.p. Rosner-Hixson Labs., Chgo., 1960-68; dir. sci. and regulatory affairs Vick Chem. Co., Mt. Vernon, N.Y., 1968-75; pres. Bob West Assocs., Inc., Stamford, Conn., 1975—. Contbr. articles to profl. jours. Mem. Am. Soc. Pharmacology and Exptl. Therapeutics, Soc. Toxicology, Acad. Pharm. Scis. Lodge: Rotary. Subspecialties: Pharmacology; Toxicology (agriculture). Current work: Scientific and regulatory consultant in health care and chemical diagnostics and devices.

WEST, CHARLES HUTCHISON KEESOR, neurophysiologist; b. Wheeling, W.Va., Aug. 9, 1948; s. Carl Hollas and Margaret Ellen (Keesor) W.; m. Sharon Lynn Harris, Aug. 12, 1972; children: Daniel, Laura. B.S., Ohio U.-Athens, 1970, M.S., 1972; Ph.D., Mich. State U., 1977. Postdoctoral fellow U. Wis.-Madison, 1977-80; research scientist Ga. Mental Health Inst., Atlanta, 1980—; asst. prof. dept. psychiatry Sch. Medicine Emory U., Atlanta, 1981—. Contbr. writings to profl. publs. in field. NIH fellow, 1977. Mem. AAAS, Am. Soc. Zoologists, Soc. for Neurosci. Subspecialties: Neurophysiology; Neuropsychology. Current work: Neural basis of reinforcement and attentional processes as they may be related to etiology of mental disorders. Office: Ga Mental Health Inst 1256 Briarcliff Rd Atlanta GA 30306

WEST, DAVID ARMSTRONG, educator; b. Beirut, Apr. 9, 1933; s. William Armstrong and Dorothy (Allen) W.; m. Lindsay Lattimore Butte, July 26, 1958; children: Peter A., Roger L., Susan T. B.A., Cornell U., 1955, Ph.D., 1959. Asst. prof. Cornell U., Ithaca, N.Y., 1959-60; postdoctoral fellow dept. zoology Liverpool U., Eng., 1960-62; asst. prof. Va. Poly. Inst. and State U., Blacksburg, 1962-68, assoc. prof., 1968—. Contbr. articles to profl. jours. NSF grantee, 1978-80. Mem. AAAS, Genetic Soc. Am., Lepidopterists Soc., Am. Soc. Naturalist Ornithologists Union, Brit. Ornithologists Union. Subspecialty: Evolutionary biology. Current work: Ecological genetics of threshold traits; genetics of mimicry. Office: Dept Biology Va Poly Inst and State Univ Blacksburg VA 24061 Home: 607 Giles Rd Blacksburg VA 24060

WEST, E. DALE, physicist; b. Ravenwood, Mo., Oct. 6, 1918; s. Athol Winslow and Bertha Vivian (Lewis) W.; m. Doris Todd., Sept. 22, 1940; children: Jon Todd, Janet Sue. B.S. in Chemistry, Calif. Inst. Tech., Pasadena, 1949; Ph.D., U. Md., 1968. Physicist Nat. Bur. Standards, Washington, 1950-69, Boulder, Colo., 1969-74; pres. Calorimetrics, Inc., Boulder, 1973—; cons. Argonne Nat. Lab., 1957-58. Contbr. articles to profl. jours. Served to lt. AUS, 1940-45. Recipient Silver Metal award Dept. Commerce, 1969. Mem. Calorimetry Conf. Republican. Unitarian. Lodge: Elks. Subspecialties: Thermodynamics; Physical chemistry. Current work: Design, building and testing calorimeters for laser measurements. Address: PO Box 4146 Boulder CO 80306

WEST, JOHN MERLE, nuclear engineer; b. Stilwell, Okla., Jan. 18, 1920; s. James and Maude (Bacon) W.; m. Navlion Farmer, Oct. 5, 1945; children: James Cornel, Leonard Clark. B.S., Northeastern Okla. U., 1939; M.S. in Physics, U. Iowa, 1941. Physicist DuPont Co., N.J. and Okla., 1941-43; physicist Manhattan Project, U. Chgo., 1943-44; supr. Hanford Works, DuPont and Gen. Electric Co., Richland, Wash., 1944-49; assoc. dir. engring. Argonne Nat. Lab., Ill., 1949-57; v.p., them exec. v.p. Gen. Nuclear Engring. Corp., Dunedin, Fla., 1957-65; with Combustion Engring., Inc., Windsor, Conn., 1965—, v.p. nuclear power systems, 1974—; chmn. bd. subs. Electro-Mechanics, Inc., 1981—. Recipient Charles A. Coffin award Gen. Electric Co., 1949. Fellow Am. Nuclear Soc. (charter mem.); mem. Nat. Acad. Engring. Republican. Presbyterian. Subspecialties: Nuclear fission; Nuclear engineering. Current work: Development, design and operation of power plants utilizing fission as their heat sources. Patentee nuclear reactor design and operation (15). Home: 154 Stoner Dr West Hartford CT 06107 Office: 1000 Prospect Hill Rd Windsor CT 06095

WEST, ROBERT ALAN, researcher; b. Valparaiso, Ind., June 14, 1951; s. Richard W. and Anna M. (Engwer) W.; m. Karen J. Reinhard, June 23, 1979. B.S. in Astronomy, Calif. Inst. Tech., 1973; Ph.D. in Planetary Sci, U. Ariz., 1977. Research assoc. Lab. for Atmospheric Space Physics, U. Colo., Boulder, 1978—, lectr. astro-geophysics dept., 1979—. Contbr. articles to profl. publs. Mem. Am. Astron. Soc. (div. planetary scis.), Am. Geophys. Union, Sierra Club. Subspecialties: Planetary atmospheres; Planetary science. Current work: Radiative transfer in planetary atmospheres, particularly scattering and polarization of sunlight. Office: U Colo Lab Atmospheric Space Physics Campus Box 392 Boulder CO 80309

WEST, ROBERT CULBERTSON, chemistry educator; b. Glen Ridge, N.J., Mar. 18, 1928; s. Robert C. and Constance (MacKinnon) W.; m.; children: David Russell, Arthur Scott. B.A., Cornell U., 1950; A.M., Harvard U., 1952, Ph.D., 1954. Asst. prof. Lehigh U., 1954-56; mem. faculty U. Wis.-Madison, 1956—, prof. chemistry, 1963—, Eugene G. Rochow prof., 1980; indsl. and govt. cons., 1961—; Abbott lectr. U. N.D., 1964; Fulbright lectr. Kyoto U., 1964-65; vis. prof. U. Würzburg, 1968-69, Haile Selassie I U., 1972, U. Calif., Santa Cruz, 1977, U. Utah, 1981; Jean Day Meml. lectr. Rutgers U., 1973; Japan

Soc. for Promotion Sci. vis. prof. Tohoku U., 1976; Lady Davis vis. prof. Hebrew U., 1979; Cecil and Ida Green honors prof. Tex. Christian U., 1983. Co-editor: Advances in Organometallic Chemistry, Vols. I-XXIII, 1964-83, Organometallic Chemistry—A Monograph Series, 1968—; contbr. articles to profl. jours. Pres. Madison Community Sch., 1970-71; bd. dirs. Women's Med. Fund, 1971—, Zero Population Growth, 1980—; bd. dirs., v.p. Protect Abortion Rights Inc., 1980; lay minister Prairie Unitarian Universalist Soc., 1982. Recipient F.S. Kipping award, 1970; Amoco Disting. Teaching award, 1974. Mem. Am. Chem. Soc., Chem. Soc. (London), Japan Chem. Soc., AAAS, Wis. Acad. Sci. Subspecialties: Inorganic chemistry; Organic chemistry. Current work: Discovered first compound containing a silicon-silicon double bond. Home: 305 Nautilus Dr Madison WI 53705

WEST, THEODORE LEE, periodontist, educator; b. N.Y.C., July 26, 1934; s. Jacob Martin and Belle (Lasky) w.; m. Amy Mae Freedman, June 29, 1958; children: Miles Kirby, Andrew David, Sharyn Rebecca. Student, Brown U., Providence, R.I., 1952-54; D.D.S., U. Pa., 1958, M.S.D., 1961; certificate, Boston U., 1960. Instr. U. Pa., Phila., 1960-64; assoc. prof. Fairleigh Dickinson U., Hackensack, N.J., 1972—; pres. Periodontal Assocs., Englewood, N.J., 1963—. Contbr. research reports, clin. articles to profl. jours. Fellow Internat. Coll. Dentistry; mem. Am. Acad. Periodontology, Northeastern Acad. Periodontology, Phila. Soc. Periodontics, Internat. Assn. Dental Research, Omicron Kappa Upsilon. Republican. Subspecialty: Periodontics. Current work: Wound healing; periodontal disease microbiology; genetic susceptibility to periodontal disease; bone transplants into periodontal defects; surgical and non-surgical periodontal therapy development. Home: 1 McCain Ct Closter NJ 07624 Office: 1 Periodontal Assocs 97 N Dean St Englewood NJ 07631

WEST, WILLIAM PHILIP, physicist; b. Dallas, June 18, 1948; s. Hoy E. and Dorthy (Harkins) W.; m. Marlys Larson, July 10, 1971; 1 son, Joseph Leo. B.S., Tex. Tech. U., 1971; M.A., Rice U., 1974, Ph.D., 1976. Postdoctoral fellow Joint Inst. Lab. Astrophysics, U. Colo., 1975-77; postdoctoral fell dept. physics U. Calif., Santa Barbara, 1977-79; physicist Gen Fusion div. Atomic Co., San Diego, 1979—. Contbr. articles to profl. jours. Mem. Am. Phys. Soc. Subspecialties: Atomic and molecular physics; Plasma physics. Current work: High temperature plasma diagnostics, spectroscopy and photochemistry in solids; development of novel high temperature plasma diagnostics; spectroscopy of molecules in solids. Office: L509 Gen Atomic Co P O Box 81608 San Diego CA 92138

WESTBY, CARL ANDREW, microbiologist, educator; b. Los Angeles, Feb. 8, 1936; s. Irving B. and Genevieve Clara (Gabel) W.; m. Merilyn Jean Pirtle, Aug. 6, 1958; children: John, Carolyn, Steven, Ann, Theresa, Carl. B.S., U. Calif.-Riverside, 1958; Ph.D., U. Calif.-Davis, 1964. Asst. prof. microbiology Utah State U., Logan, 1967-72; asst. prof. microbiology S.D. State U., Brookings, 1972-76, assoc. prof., 1976-80, prof., 1980—; cons. 3M, Brookings; head fuel alcohol plant S.D. State U., 1980—. Contbr. articles to profl. jours. Scoutmaster Boy Scouts Am., 1975—. U.S. Dept. Agr. grantee, 1980—. Mem. Am. Soc. Microbiology, Sigma Xi. Democrat. Roman Catholic. Lodge: K.C. Subspecialties: Microbiology; Biochemistry (biology). Current work: Role of purines in myxobacterial development, carbon metabolism in nitrogen fixing bacteria, farm-scale fuel alcohol production, biological indicators in sterilization. Home: 1702 Victory St Brookings SD 57006 Office: S D State Univ Dept Microbiology Brookings SD 57007

WESTCOTT, WILLIAM WARREN, utilities analyst, electrical engineer; b. Cleve., July 8, 1930; s. William Warren and Ann (Maus) W.; m. Cynthia Lall, May 30, 1967; 1 son, William Warren. B.S. in Econs, U. Pa., 1957; P.M.D., Harvard U., 1971. Indsl. economist Govt. of P.R., San Juan, 1957-59; owner, mgr. P.R. Electric Constrn. Co., 1959-70; utilities analyst, sr. mem. utilities Analysis Cons. Group, Cleve., 1973—. Active Cleve. and Cuyahoga County Election Com., Ohio Republican Election Com. Mem. Alpha Sigma Phi. Club: Harvard Bus. Sch. (Cleve.). Subspecialties: Electrical engineering. Current work: Reduction of utility costs for generation, transmission and consumption of electricity, natural gas, water, water treatment and steam through systems analyses.

WESTERBERG, ARTHUR WILLIAM, chemical engineering educator; b. St. Paul, Oct. 9, 1938; s. Kenneth Waldorf and Marjorie Clair (Darling) W.; m. Barbara Ann Dyson, July 14, 1963; children: Kenneth William, Karl Michael. B.S. in Chem. Engring, U. Minn., 1960, M.S., Princeton U., 1961; Ph.D., U. London, 1964. Pres. Farm Engring., Inc., 1964-65; sr. analyst Control Data Corp., La Jolla, Calif., 1965-67; asst. prof. U. Fla., Gainesville, 1967-71, assoc. prof., 1971-76, prof., 1976; prof. dept. chem. engring. Carnegie-Mellon U., Pitts., 1976—, head dept., 1980—, dir., 1978-80. Author: (with others) Process Flowsheeting, 1979; Contbr. articles on design research to profl. jours. Mem. Am. Inst. Chem. Engrs., Am. Soc. Engring. Edn., Sigma Xi. Subspecialty: Chemical engineering. Current work: Computer-aided design; process synthesis, analysis, optimization; application of computers to the design of chemical processes.

WESTERMAN, DAVID SCOTT, geologist, educator; b. Ann Arbor, Mich., July 12, 1946; s. Harold Scott and Shirley Martha (Mackey) W.; m. Jenny Anne Swanson, Sept. 13, 1968 (div. 1980); children: Lisa Anne, Matthew Evan. B.S. in Geology, Allegheny Coll., 1969, M.S., Lehigh U., 1971, Ph.D., 1972. Lic. geologist Maine. Asst. prof. earth scis. Northeastern U., Boston, Mass., 1972-78; vis. asst. prof. geology U. So. Maine, Portland, 1978-79; asst. prof. geology U. Maine, Orono, 1980, Colby Coll., Waterville, Maine, 1980-82; assoc. prof. earth scis. Norwich U., Northfield, Vt., 1982—; field geologist Maine Geol. Survey. Editor: Maine Geology. Recipient research grants Lehigh U., 1971, Northeastern U., 1976-77. Mem. Geol. Soc. Am., Geol. Soc. Maine (pres. 1978-80), Maine Minerals Resources Assn., Vt. Geol. Soc., Planetary Soc., Zero Population Growth (charter), Sigma Xi (grant-in-aid 1971). Unitarian. Subspecialties: Tectonics; Petrology. Current work: analysis of rock structures to understand current seismic activity in regions of earthquake concentration; resolution of ancient stress fields; the persistence of large-scale tectonic parameters. Home: PO Box 457 Moretown VT 05660 Office: Norwich U Dept Physical Sciences Northfield VT 05663

WESTERVELT, PETER JOCELYN, physics educator; b. Albany, N.Y., Dec. 16, 1919; s. William Irving and Dorothy (Jocelyn) W.; m. Alice Francis Brown, June 2, 1956; children: Dirck Edgell, Abby Brown. B.S., MIT, 1947, M.S., 1949, Ph.D., 1951. Mem. staff radiation lab. MIT, 1940-41, underwater sound lab., 1941-45, asst. in physics, 1946-47, research asso., 1948-50; asst. prof. physics Brown U., 1951-58, asso. prof., 1958-63, prof., 1963—; mem. subcom. aircraft noise NASA, 1954-59; mem. com. hearing and bio-acoustics NRC, 1957-83, exec. council, 1960-61, 78-83, chmn., 1967-68, 80-83; mem. sonic boom com. Nat. Acad. Sci., 1968-71; mem. R.I. Atomic Energy Commn., 1968-73; cons. applied research lab. U. Tex., Austin, 1973—. Fellow Am. Physics Soc., Acoustical Soc. Am. Club: Hope (Providence). Subspecialties: Acoustics; Relativity and gravitation. Current work: Theory of nonlinear waves; acoustic and gravitational. Home: 16 John St Providence RI 02906 Office: Brown U Dept Physics Providence RI 02912

WESTFALL, DAVID PATRICK, pharmacologist, educator; b. Harrisville, W.Va., June 9, 1942; s. Creed Simpson and Cecilia Rita (McKay) W.; m. Shirley Ann Spencer, June 13, 1942; children: Timothy David, Alison Spencer. A.B., Brown U., 1964; M.S., W. Va. U., 1966, Ph.D., 1968. Demonstrator in pharmacology Oxford (Eng.) U., 1968-70; asst. prof. pharmacology W.Va. U., Morgantown, 1970-73, assoc. prof., 1973-77, prof., 1977-82; prof., chmn. dept. pharmacology U. Nev. Sch. Medicine, Reno, 1982—. Contbr. chpts. to books, articles to profl. jours. Mem. Am. Soc. for Pharmacology and Exptl. Therapeutics, Soc. for Neurosci., Sigma Xi. Subspecialties: Pharmacology; Neuropharmacology. Current work: Autonomic and cardiovascular pharmacology; smooth muscle physiology. Neurotransmission; co-transmission; supersensitivity; anti-hypertensive drugs; smooth muscle electrophysiology. Office: Dept Pharmacology U Nev Sch Medicine Reno NV 89557

WESTHEAD, EDWARD WILLIAM, biochemist, educator; b. Phila., June 19, 1930; s. Edward William and Eleanore M. (Ritchie) W.; m.; children: Victoria, Edward G. B.S., Haverford (Pa.) Coll., 1951, M.S., 1952; Ph.D., Bklyn. Poly. Inst., 1955. NSF fellow, Uppsala, Sweden, 1955-57; research assoc. U. Minn. Med. Sch., Mpls., 1957-60; asst. prof. Dartmouth Med. Sch., 1960-65; assoc. prof. and head dept. U. Mass., Amherst, 1966-69, prof. biochemistry, 1971—; vis. scholar Calif. Inst. Tech., 1971; vis. prof. pharmacology U. Innsbruck, 1979-80. Contbr. articles to profl. jours. Recipient 15 yr. award Haverford Coll.; NIH fellow, 1972-73. Mem. Am. Soc. Biol. Chemists, Am. Soc. Neurochemistry, Soc. Neurosci., Am. Soc. Cell Biology, Phi Beta Kappa. Subspecialty: Biochemistry (biology). Current work: Metal-ion activation of enzymes; control of red cell metabolism; secretion and metabolism of catecholaminergic cells. Office: Biochemistry Dept U Mass Amherst MA 01003

WESTHEIMER, FRANK H(ENRY), chemistry educator; b. Balt., Jan. 15, 1912; s. Henry Ferdin and Carrie (Burgunder) W.; m. Jeanne Friedmann, Aug. 30, 1937; children: Ruth, Ellen. A.B., Dartmouth Coll., 1932, D.Sc. (hon.), 1962; M.A., Harvard U., 1933, Ph.D., 1935; NRC fellow, Columbia U., 1935-36; Sc.D. (hon.), U. Chgo., 1973, U. Cin., 1976, Tufts U., 1978, U. N.C., 1983, Bard Coll., 1983. Instr. chemistry U. Chgo., 1936-41, asst. prof., 1941-44, assoc. prof., 1946-48, prof. chemistry, 1948-53; vis. prof. Harvard U., Cambridge, Mass., 1953-54, prof. chemistry, from 1954, now Morris Loeb prof. emeritus, chmn. dept., 1959-62; Overseas fellow Churchill Coll., Cambridge (Eng.) U., 1962-63; Harrison Howe lectr. U. Rochester, 1954; Stieglitz lectr. U. Chgo., 1956; Morrell lectr. Cambridge U., 1962; Alexander Todd lectr., 1976, Centenary lectr. Chemistry Soc., 1963; Folkers lectr. U. Wis., 1963; Baker lectr. Cornell U., 1964; Priestly lectr. Pa. State U., 1968; Morris S. Kharasch lectr. U. Chgo., 1969; David Rivett Meml. lectr. U. Canberra, Australia, 1969; DuPont lectr. U. Tenn., 1973; Bachmann lectr. U. Mich., 1974; Werner lectr. U. Kans., 1975; Kolthoff lectr. U. Minn., 1980; Disting. Scientist lectr. Bard Coll., 1982. Assoc. editor: Jour. Chem. Physics; editorial bd.: Jour. Am. Chem. Soc, 1960-69; contbr. articles to sci. jours. Mem. Pres.'s Sci. Adv. Com., 1967-70; research supr. Explosives Research Lab., Nat. Def. Research Com., 1944-45; chmn. com. survey chemistry Nat. Acad. Scis., 1964-65, mem. council, 1971-75, 76-78. Recipient Naval Ordnance Devel. award, 1946; Army-Navy cert. of appreciation, 1948; James Flack Norris award, 1970; Willard Gibbs medal, 1970; Theodore Richards medal, 1976; award in chem. scis. Nat. Acad. Sci., 1980; Richard Kokes, award, 1980; Charles Frederick Chandler medal Columbia U., 1980; Rosenstiel award Brandeis U., 1981; Guggenheim fellow, 1962-63; Fulbright-Hays fellow, 1974; Welch award Robert A. Welch Found., 1982; Nichols medal Am. Chem. Soc., 1982; Arthur C. Cope award, 1982. Fellow Royal Soc.; mem. Nat. Acad. Sci., Am. Philos. Soc., Am. Chem. Soc., Am. Acad. Arts and Scis., Phi Beta Kappa. Subspecialties: Organic chemistry; Physical chemistry. Current work: Calculations of electrostatic effects and of steric effects in organic chemistry; determination of mechanisms of chromic acid oxidation; enzymic and metal-ion promoted decarboxylation, biochemical oxidation-reduction reacitons which require nicotine adenine dinucleotide as coenzyme; the mechanisms of the hydrolysis of phosphate estersl and photoaffinity labeling. Office: Harvard U 12 Oxford St Cambridge MA 02138

WESTHEIMER, FRANK HENRY, chemist, educator; b. Balt., Jan. 15, 1912; s. Henry Ferdinand and Carrie (Burgunder) W.; m. Jeanne Friedmann, Aug. 31, 1937; children: Ruth Susan, Ellen. A.B., Dartmouth Coll., 1932, Sc.D. (hon.), 1961; M.A., Harvard U., 1933, Ph.D., 1935; NRC fellow, Columbia, 1935-36; Sc.D. (hon.), U. Chgo., 1973, U. Cin., 1976, Tufts U., 1978, U. N.C., 1983, Bard Coll., 1983. Instr. chemistry U. Chgo., 1936-41, asst. prof., 1941-44, asso. prof., 1946-48, prof. chemistry, 1948-53, vis. prof. Harvard, 1953-54, prof. chemistry, 1954—, chmn. dept., 1959-62; Overseas fellow Churchill Coll., U. Cambridge, Eng., 1962-63; Mem. President's Sci. Advisory Com., 1967-70; research supr. Explosives Research Lab., Nat. Def. Research Com., 1944-45; chmn. com. survey chemistry Nat. Acad. Scis., 1964-65. Asso. editor: Jour. Chem. Physics, 1942-44, 52-54; editorial bd.: Jour. Am. Chem. Soc, 1960-69, Procs. Nat. Acad. Scis., 1983—; Contbr. articles to profl. jours. Recipient Naval Ordnance Development award, 1946; Army-Navy Certificate of Appreciation, 1948; James Flack Norris award phys.-organic chemistry, 1970; Willard Gibbs medal, 1970; Theodore Richards medal, 1976; award in chem. scis. Nat. Acad. Sci., 1980; Richard Kokes award, 1980; Charles Frederick Chandler medal, 1980; Rosenstiel award, 1981; Nichols medal, 1982; Robert A. Welch award, 1982; Ingold medal, 1983; Guggenheim fellow, 1962-63; Fulbright-Hays fellow, 1974. Mem. Nat. Acad. Sci. (council 1971-75, 76-79), Am. Philos. Soc. (council 1981-84), Am. Chem. Soc., Am. Acad. Arts and Scis., Phi Beta Kappa. Subspecialties: Organic chemistry; Biochemistry (biology). Current work: Enzyme mechanisms; phosphate ester chemistry. Home: 3 Berkeley St Cambridge MA 02138

WESTRUM, EDGAR FRANCIS, JR., chemistry educator, researcher; b. Albert Lea, Minn., Mar. 16, 1919; s. Edgar Francis and Nora Dorothy (Kipp) W.; m. Florence Emily Barr, June 13, 1943; children: Ronald Mark, James Scott, Michael Lauren, Margaret Kristin. Student, Hamline U., 1936-38; B.Chemistry, U. Minn., 1940; Ph.D., U. Calif.-Berkeley, 1944. Scientist Metall. Lab., U. Chgo., 1944-45; scientist Radiation Lab., U. Calif., 1945; asst. prof. chemistry U. Mich., Ann Arbor, 1947-56, assoc. prof., 1956-62, prof., 1963—; sec. gen. com. on data for sci. and tech. Internat. Council Sci. Unions, 1973-82. Editor: Jour. of Chem. Thermodynamics, 1970-80, Bull. Chem. Thermodynamics, 1955-77, Codata Directory, 1982—; contbr. articles to profl. jours. Recipient Bausch and Lomb hon. sci. award, 1936. Fellow AAAS, Am. Inst. Chem. Engrs., Am. Phys. Soc.; mem. Am. Chem. Soc., Netherlands Phys. Soc. Presbyterian. Subspecialties: Thermodynamics; Condensed matter physics. Current work: Cryogenic calorimetry, phase, ordering and electronic transitions, resolution of electronic and magnetic contributions. Home: 2019 Delaware Dr Ann Arbor MI 48103 Office: Department Chemistry University Michigan Ann Arbor MI 48109

WESTWATER, JAMES WILLIAM, chemical engineering educator; b. Danville, Ill., Nov. 24, 1919; s. John and Lois (Maxwell) W.; m. Elizabeth Jean Keener, June 9, 1942; children: Barbara, Judith, David, Beverly. B.S., U. Ill., 1941; Ph.D., U. Del., 1948. Mem. faculty U. Ill., Urbana, 1948—, prof. chem. engring., 1962—, head dept., 1962-80; papers chmn. 5th Nat. Heat Transfer Conf., Buffalo, 1960; chmn. 3d Internat. Heat Transfer Conf., Chgo., 1966; Reilly lectr. Notre Dame U., 1958; Donald L. Katz lectr. U. Mich., 1978. Contbr. articles profl. jours. Recipient Conf. award 8th Nat. Heat Transfer Conf., 1965; William H. Walker award Am. Inst. Chem. Engrs., 1966; Max Jakob award Am. Inst. Chem. Engrs.-ASME, 1972. Mem. Am. Inst. Chem. Engrs. (dir., past div. chmn., Inst. lectr. 1964, named Eminent Chem. Engr. 1983), Am. Chem. Soc., ASME, Am. Soc. Engring. Edn. (Vincent Bendix award 1974), Nat. Acad. Engring. Subspecialty: Chemical engineering. Current work: Heat transfer during boiling or condensation. Home: 116 W Iowa St Urbana IL 61801

WETHEY, DAVID SUNDERLAND, biology educator; b. Ann Arbor, Mich., Sept. 15, 1950; s. Harold Edwin and Alice Luella (Sunderl) W.; m. Sarah Ann Woodin, Jan. 5, 1980. B.A., Yale U., 1973; M.S., U. Mich., 1976, Ph.D., 1979. NSF postdoctoral fellow U. Coll. North Wales, Menai Bridge, 1979-80, U. Leeds, Robin Hood's Bay, Yorkshire, Eng., 1980; asst. prof. biology U. S.C., Columbia, 1980—. NSF fellow, 1974-77; NSF grantee, 1978-79, 82—; Office Naval Research grantee, 1982—. Mem. Ecol. Soc. Am., Brit. Ecol. Soc., Am. Soc. Zoologists, Am. Soc. Limnology and Oceanography, Am. Soc. for Study of Evolution, AAAS, Sigma Xi. Subspecialties: Ecology; Oceanography. Current work: Population dynamics, biogeography, rocky intertidal, community ecology, fouling organisms. Office: Dept Biology U SC Columbia SC 29208

WETMUR, JAMES GERARD, microbiologist; b. New Castle, Pa., July 1, 1941; s. Leon Gerard and Wilma Aileen (Lostetter) W.; m. Brigid M. Long, Sept. 4, 1965; children: Katherine, John, Tara. B.S., Yale U., 1963; Ph.D., Calif. Inst. Tech., 1967. Chief biochemistry U.S. Army Aeromed. Research Lab., Ft. Rucker, Ala., 1967-69; asst. prof. chemistry and biochemistry U. Ill., Urbana, 1969-74; assoc. prof. microbiology Mt. Sinai Sch. Medicine, CUNY, 1974-82, prof., 1982—. Contbr. articles to sci. jours. Served as capt. AUS, 1967-69. NIH grantee, 1969—. Mem. Am. Soc. Biol. Chemists, Am. Soc. Microbiology, Am. Chem. Soc., N.Y. Acad. Sci., Sigma Xi. Republican. Roman Catholic. Subspecialties: Microbiology; Biophysical chemistry. Current work: DNA reassociation kinetics; DNA protein interactions, equilibra and kinetics; recombination of DNAs. Home: 994 Post Rd Scarsdale NY 10583 Office: Dept Microbiolog Mt Sinai Sch Medicine 1 Gustave Levy Pl New York NY 10029

WETSTEIN, LEWIS, cardio-thoracic surgeon, cardiac researcher, electro-physiologist; b. N.Y.C., June 23, 1947; s. Benjamin and Rose (Finkilstein) W.; m.; children: Jennifer Sandra. B.A., Queens Coll, 1968; M.D., Autonoumous U., Barcelona, Spain, 1973. Diplomate: Am. Bd. Surgery, Am. Bd. Thoracic Surgery. Intern L.F. Jewish Hosp.-Hillside Med. Center, 1973-75; resident in surgery Kings County Hosp.-Downstate Med. Center, 1975-80; asst. instr. surgery Downstate Med. Ctr., Bklyn., 1975-80; instr. surgery U. Pa., Phila., 1980-82, research assoc., 1980-82; asst. prof. surgery Med. Coll. Pa., Phila., 1982—; cons. VA Hosp., Phila., 1982—. Contbr. articles to profl. jours. Served as maj. USAFR, 1976—. Recipient postdoctoral research service award NIH, 1980-82, spl. investigatorship award Am. Heart Assn., 1983-85. Fellow Assn. Acad. Surgery, ACS, Am. Coll. Cardiology, Am. Coll. Chest Physicians., Soc. Thoracic Surgeons. Jewish. Subspecialties: Cardiac surgery; Surgery. Current work: Electrophysiology: development of the surgical therapy for cardiac arrhythnias. Home: 296 Uxbridge Dr Cherry Hill NJ 08034 Office: Med Coll Pa 3300 Henry Ave Philadelphia PA 19129

WETTERHAHN, KAREN E., chemist, educator; b. Plattsburgh, N.Y., Oct. 16, 1948; d. Gustave G. and Mary (Thibault) W.; m. Leon H. Webb, June 19, 1982. B.S., St. Lawrence U., 1970; Ph.D., Columbia U., 1975. Chemist The Mearl Corp., Ossining, N.Y., 1970-71; research fellow Columbia U., 1971-75; postdoctoral fellow Inst. Cancer Research Columbia U., 1975-76; asst. prof. chemistry Dartmouth Coll., 1976-82, asso. prof. chemistry, 1982—. Contbr. articles to sci. jours. A.P. Sloan fellow, 1981—. Mem. Am. Chem. Soc., Am. Assn. Cancer Research, AAAS, N.Y. Acad. Sci. Subspecialties: Inorganic chemistry; Biophysical chemistry. Current work: Mechanisms of chemical carcinogenesis; carcinogen metabolism; interaction of inorganic carcinogens with nucleic acids and proteins. Office: Dept Chemistry Steele Hall Dartmouth Coll Hanover NH 03755

WEVER, ERNEST GLEN, psychology educator; b. Benton, Ill., Oct. 16, 1902; s. Ernest Sylvester and Mary Jane (Shirtz) W.; m. Suzanne Rinehart. A.B., Ill. Coll., 1922; A.M., Harvard U., 1924, Ph.D., 1926; D.Sc. (hon.), U. Mich., 1981. Instr. U. Calif.-Berkeley, 1926-27; instr. Princeton U., 1927-29, asst. prof. psychology, 1929-31, assoc. prof., 1931-41, prof, 1941-71, prof. emeritus, sr. research psychologist, 1971—. Author: books The Reptile Ear, 1978, The Amphibian Ear; contbr. articles to profl. jours. Mem. Am. Psychol. Assn., Nat. Acad. Scis., Soc. Exptl. Psychology, Acoustical Soc. Am. (Silver medal 1981), Am. Otolaryngology Soc. (assoc.), Assn. Research in Otolaryngology (award of Merit 1983, hon. mem.). Subspecialties: Physiological psychology; Sensory processes. Current work: The ear and hearing processes; evolution of vertebrate ear. Office: Princeton U 055 Green Hall Princeton NJ 08544

WHALEN, RICHARD EDWARD, psychologist; b. Holyoke, Mass., Mar. 29, 1934; s. John J. and Katheryn (O'Neil) W. A.B., Brown U., 1956; M.S., Yale U., 1957, Ph.D., 1960. Faculty UCLA, 1961-65; mem. faculty U. Calif., Irvine, 1965-79; prof. dept. psychiatry and behavioral sci. SUNY-Stony Brook, 1979-82; prof. psychology U. Calif., Riverside, 1982—. Contbr. articles to profl. jours. NIH, NIMH grantee, 1962—. Mem. Soc. Neurosci., Internat. Brain Research Orgn., Internat. Acad. Sex Research, Internat. Soc. Devel. Neurosci., Sigma Xi. Subspecialty: Psychobiology. Current work: Hormone brain interactions in development and regulation of neuroendocrine function and behavior.

WHALEN, RICHARD VINCENT, software engr.; b. Waterbury, Conn., May 4, 1958; s. Richard James and Louise Josephine (Laurelli) W. B.S., Worcester Poly. Inst., 1980. Software engr. Digital Equipment Corp., Worcester, Mass., 1980—. Mem. Assn. Computing Machinery. Subspecialties: Software engineering; Operating systems. Current work: Theoretical computer science, remote terminal emulator development, advanced systems engineering. Home: 83 1/2 Elm St Worcester MA 01609 Office: 77 Reed Rd Hudson MA 01749

WHALEY, JULIAN WENDELL, plant pathologist; b. Parkersburg, W.Va., Aug. 12, 1937; s. James William and Lucille Evelyn (Malone) W.; m. Jeanine Leann Fortner; children: Julie, Jay, Jill, Jennifer. B.S., West Liberty State coll., 1959; M.S., W.Va. U., 1961; Ph.D., U. Ariz., 1964. Sr. plant pathologist Eli Lilly & Co., Greenfield, Ind., 1964-70, Fresno, Calif., 1970; prof. plant pathology Calif. State U., Fresno, 1970—; cons. in crop injury investigations. Mem. Am. Phytopath. Soc. (dir.), Council Agrl. Sci. and Tech. Subspecialty: Plant pathology. Current work: Crop injury symptoms due to pesticides; plant disease control. Office: Dept Plant Sci Calif State U Fresno CA 93740

WHANGER, PHILIP DANIEL, animal nutrition researcher, educator; b. Lewisburg, W.Va., Aug. 30, 1936; s. Jesse J. and Madge (Coffman) W.; m. Lois R. Jones, Aug. 29, 1964; children: Darren Massey, Lanita Dawn. B.S., Berry Coll., 1959; M.S., W.Va. U., 1961; Ph.D., N.C. State U., 1965. Research assoc. Mich. State U., East

Lansing, 1965-66; asst. prof. Oreg. State U., Corvallis, 1965-72, assoc. prof., 1973-80, prof. dept. agrl. chemistry, 1981—; assoc. staff Harvard Med. Sch., Boston, 1972-73; vis. scientist Commonwealth Sci. and Indsl. Research Orgn., Wembley, Western Australia, 1980-81. Contbr. sci. articles to publs. NIH fellow; NSF exchange fellow. Mem. Am. Inst. Nutrition, Am. Soc. Animal Sci., Nutrition Today, Internat. Assn. Bioinorganic Scientists. Subspecialties: Animal nutrition; Biochemistry (biology). Current work: Selenium requirements and metabolism in animals including man, and influence and interacting element on zinc and metallothione in metabolism. Office: Dept Agrl Chemistry Oreg State Univ Corvallis OR 97331

WHAUN, JUNE M., research hematologist; b. North Vancouver, C., Can., June 26, 1935; d. Thomas Moore and Diamond (Kwan) W. M.D., U. B.C., 1960. Diplomate: Am. Bd. Pediatrics. Intern U. Md. Hosp., Balt., 1960-61; resident in pediatrics U. Mich. Hosp., Ann Arbor, 1961-62, U. Toronto Hosp. for Sick Children, 1962-64; resident in internal medicine Shaughnessy Hosp., U.B.C., Vancouver, 1964-65; clin. research fellow in hematology U. Toronto Hosp. for Sick Children, 1965-66; fellow King County Central Blood Bank, U. Wash., Seattle, 1966-67; research hematology fellow Hosp. of U. Pa., Phila., 1967-69; asst. hematologist and dir. coagulation lab. Children's Hosp. Phila., 1969-71, Foothills U. Hosp., Calgary, Alta., Can., 1971-72; from asst. to assoc. prof. pediatrics U. Calgary Faculty Medicine, 1971-78; assoc. prof. pediatrics Uniformed Services U. Health Scis., Bethesda, Md., 1981—; research hematologist Walter Reed Army Inst. Research, Washington, 1978—; vis. guest scientist Chinese Acad. Med. Scis., Beijing, 1982; Founder, dir. South Alta. Pediatric Oncology Program, Calgary, 1975-78, Alta. Children's Hosp. Regional Hemophilia Program, 1976-78; cons., vis. prof. pediatrics Tex. Tech Sch. Medicine, Amarillo, 1981—. Med. Research Council Can. Research grantee, 1971-78. Fellow Royal Coll. Physicians and Surgeons Can.; mem. Soc. Pediatric Research, Am. Soc. Hematology, Biophys. Soc., Am. Soc. Clin. Oncology, Soc. Protozoologists, Internat. Soc. Thrombosis and Haemostasis, Am. Fedn. Clin. Research. Club: Linden Hill Tennis (Bethesda). Subspecialties: Hematology; Cell biology (medicine). Current work: Investigative hematology; characterizing differences between host and parasite/tumor for potential chemotherapy targets. Home: 336 New Mark Esplanade Rockville MD 20850 Office: Walter Reed Army Inst Researc Hematology Dept Washington DC 20307

WHEALTON, JOHN H., physicist; b. Bklyn., Apr. 27, 1943. Ph.D., U. Del., 1971. Faculty Brown U., 1971, U. Colo., 1973; physicist Oak Ridge Nat. Lab., 1975—. Contbr. articles to profl. jours. Mem. Am. Phys. Soc. Subspecialties: Plasma physics; Nuclear fusion. Current work: Ion accelerators, optics algorithms. Office: Oak Ridge Nat Lab PO Box Y Oak Ridge TN 37830

WHEAT, ROBERT WAYNE, biochemist, immunologist, educator; b. Springfield, Mo., Nov. 10, 1926; s. Earl Franklin and Retha Louise (Wilson) W.; m. Johnnie Maxine Simmons, Dec. 24, 1948; children: Gregory, Jane, Anne. B.A., Southwest Mo. State U., 1948; M.S., U. N. Mex., 1950; Ph.D., Washington U., St. Louis, 1954. NIH postdoctoral fellow, Bethesda, Md., 1955-56; asst. prof. biochemistry Duke U. Med. Ctr., Durham, N.C., 1958—, assoc. prof. microbiology, 1966-74, prof. microbiology, 1974—. Served with USN, 1944-46. Mem. Am. Soc. Biol. Chemists, Am. Soc. Microbiology, Mycol. Soc. Am., Sigma Xi. Subspecialties: Infectious diseases; Biochemistry (medicine). Current work: polysaccharides and immunobiology of pathogenic bacteria and fungi. Office: Duke U Med Center Durham NC 27710 Home: 2720 Montgomery St Durham NC 27705

WHEDON, GEORGE DONALD, medical administrator, researcher; b. Geneva, N.Y., July 4, 1915; s. George Dunton and Elizabeth (Crockett) W.; m. Margaret Brunssen, May 12, 1942 (div. Sept., 1982); children: Karen Anne, David Marshall. A.B., Hobart Coll., 1936, Sc.D. (hon.), 1967; M.D., U. Rochester, 1941, Sc.D. (hon.), 1978. Diplomate: Am. Bd. Internal Medicine, Am. Bd. Nutrition. Intern in medicine Mary Imogene Bassett Hosp., Cooperstown, N.Y., 1941-42; asst. in medicine U. Rochester Sch. Medicine; also asst. resident physician medicine Strong Meml. Hosp., Rochester, 1942-44; instr. medicine Cornell U. Med. Coll., 1944-50, asst. prof. medicine, 1950-52; chief metabolic diseases br. Nat. Inst. Arthritis, Diabetes, Digestive and Kidney Diseases, NIH, Bethesda, Md., 1952-65, asst. dir., 1956-62, dir., 1962-81, sr. sci. adv., 1981-82; sr. assoc. dir. conf. program Kroc Found., Santa Ynez, Calif., 1982—; adj. prof. medicine (endocrinology) UCLA Sch. Medicine, 1982—; Mem. subcom. on calcium, com. dietary allowances Food and Nutrition Bd., NRC, 1959-64; cons. to office manned space flight NASA, 1963—, chmn. Am. Inst. Biol. Scis. med. program adv. panel to, 1971-75, chmn. life scis. com., 1974-78, mem. space program adv. council, 1974-78; cons. on endocrinology and metabolism adv. com. Bur. Drugs, FDA, 1977—; mem. subcommn. on gravitational biology Com. on Space Research, 1979—; mem. research adv. bd. Shriners Hosps., 1981—; cons. in medicine Wadsworth Gen. Hosp. VA Center, Los Angeles, 1982—. Editorial bd.: Jour. Clin. Endocrinology and Metabolism, 1960-67; adv. editor: Calcified Tissue Research, 1967-76; Contbr. articles to profl. publs. Mem. med. alumni council U. Rochester Sch. Medicine, 1971-76; mem. trustees' council U. Rochester, 1971-76, vice chmn., 1973-74, chmn., 1974-75; trustee Dermatology Found., 1978-82. Recipient Superior Service award USPHS, 1967, Alumni citation U. Rochester, 1971; Ayerst award Endocrine Soc., 1974; Exceptional Sci. Achievement medal NASA, 1974. Mem. Am. Fedn. Clin. Research, Assn. Am. Physicians, Pan Am. Med. Assn., Aerospace Med. Assn. (Arnold D. Tuttle meml. award 1978), Am. Rheumatism Assn., Md. Acad. Scis. (mem. sci. council 1964-70, 81-82), Endocrine Soc. (Robert H. Williams Disting. Leadership award in endocrinology 1982), N.Y. Acad. Scis., AAAS, Am. Physiol. Soc., Am. Diabetes Assn., Am. Gasteroenterol. Assn., Gerontol. Soc., Am. Inst. Nutrition, Acad. Orthopaedic Surgeons (hon.), Am. Soc. for Bone and Mineral Research (public affairs and devel. com.), Orthopaedic Research Soc., Am. Astronautical Soc., Dermatology Found., Theta Delta Chi. Episcopalian. Subspecialty: Nutrition (medicine). Home: 100 Oceano Ave Apt 15 Santa Barbara CA 93109 Office: Kroc Found PO Box 547 Santa Ynez CA 93460

WHEELER, ALFRED GEORGE, JR., entomologist; b. Nebraska City, Nebr., Apr. 11, 1944; s. Alfred George and Frances (Rudisill) W. B.A., Grinnell Coll., 1966; Ph.D., Cornell U., 1971. Research, teaching asst. dept. entomology Cornell U., 1966-70, instr., 1971; entomologist Pa. Dept. Agr., Harrisburg, 1971—; adj. asst. prof. Pa. State U., 1978, adj. assoc. prof., 1981—; cons. Dames & Moore, Cranford, N.J., 1974-75. Editor: Entomological Jour., Melsheimer Entomological Series, 1978—, Growers' mag., Regulatory Horticulture, 1975—. Mem. Entomol. Soc. Am., Entomol. Soc. Washington, Entomol. Soc. Pa., Entomol. Soc. N.Y., Pa. Acad. Sci., Hershey Jaycees (v.p. 1977). Subspecialties: Ecology; Entomology. Current work: Biology of Hemiptera-Heteroptera, especially Berytidae and Miridae; insect-plant associations. Office: Bur Plant Industry Pa Dept Agr Harrisburg PA 17110

WHEELER, GLYNN PEARCE, biochemist; b. Milan, Tenn., Oct. 13, 1919; s. Hollis E. and Vera M. (Pearce) W.; m. Annie Ford Lester, Jan. 20, 1943; children: William H., Anne P. B.A., Vanderbilt U., 1941, Ph.D., 1950; M.S., U. Akron, 1947. Analytical chemist Tenn. Coal Iron & R.R. Co., Birmingham, Ala., 1941, Govt. Lab. Ala. Ordnance Works, Childersburg, 1942; research chemist B.F. Goodrich Co., Akron, Ohio, 1942-46; head cancer biochemistry div. So. Research Inst., Birmingham, 1946-48, 50—; mem. ad hoc coms. Nat. Cancer Inst. Contbr. numerous articles in field to profl. jours. Trustee Homewood Pub. Library, Gorgas Scholarship Found. Inc. USPHS fellow, 1948-50. Mem. Am. Soc. Biol. Chemists, Am. Assn. Cancer Research, AAAS, Am. Chem. Soc., Ala. Acad. Sci., Cell Kinetics Soc., Southeastern Cancer Research Assn., Phi Beta Kappa, Sigma Xi. Methodist. Club: The Club (Birmingham). Subspecialties: Cancer research (medicine); Chemotherapy. Current work: Biochemistry related to cancer chemotherapy; mechanims of action of anticancer agents. Office: PO Box 66306 Birmingham AL 35266

WHEELER, JOHN ARCHIBALD, scientist; b. Jacksonville, Fla., July 9, 1911; s. Joseph Lewis and Mabel (Archibald) W.; m. Janette Hegner, June 10, 1935; children: Isabel Letitia Wheeler Ufford, James English, Alison Christie Wheeler Ruml. Ph.D., Johns Hopkins U., 1933, LL.D., 1977; Sc.D., Western Res. U., 1958, U. N.C., 1959, U. Pa., 1968, Middlebury Coll., 1969, Rutgers U., 1969, Yeshiva U., 1973, Yale U., 1974, U. Uppsala, 1975, U. Md., 1977, Gustavus Adolphus U., 1981, Cath. U. Am., 1982, U. Newcastle-upon-Tyne, 1983. NRC fellow, N.Y., Copenhagen, 1933-35; asst. prof. physics U. N.C., 1935-38, Princeton U., 1938-42, asso. prof., 1945-47, prof., 1947-76, Joseph Henry prof. physics, 1966-76, Joseph Henry prof. physics emeritus, 1976—; prof. physics U. Tex., Austin, 1976—, dir. Ctr. for Theoretical Physics, 1976—, Ashbel Smith prof., 1979—, Blumberg prof., 1981—; cons. and physicist on atomic energy projects Princeton U., 1939-42, U. Chgo., 1942, E.I. duPont de Nemours & Co., Wilmington, Del., and Richland, Wash., 1943-45, Los Alamos, 1950-53; dir. project Matterhorn Princeton U., 1951-53; Guggenheim fellow, Paris and Copenhagen, 1949-50; summer lectr. U. Mich., U. Chgo., Columbia U.; Lorentz prof. U. Leiden, 1956; Fulbright prof. Kyoto U., 1962; 1st vis. fellow Clare Coll. Cambridge U., 1964; Ritchie lectr. Edinburgh, 1958; vis. prof. U. Calif.-Berkeley, 1960; Battelle prof. U. Wash., 1975; sci. adviser U.S. Senate del. to 3d ann. conf. NATO Parliamentarians, Paris, 1957; mem. adv. com. Oak Ridge Nat. Lab., 1957-65, U. Calif., Los Alamos and Livermore, 1972-77; v.p. Internat. Union Physics, 1951-54; chmn. joint com. Am. Phys. Soc. and Am. Philos. Soc. on history theoretical physics in 20th Century, 1960-72; sci. adv. bd. USAF, 1961, 62; chmn. Dept. Def. Advanced Research Projects Agy. Project, 1958; mem. U.S. Gen. Adv. Com. Arms Control and Disarmament, 1969-72, 74-77. Author: Geometrodynamics, 1962, Gravitation Theory and Gravitational Collapse, 1965, Spacetime Physics, 1966, Einstein's Vision, 1968, Gravitation, 1973, Black Holes, Gravitational Radiation and Cosmology, 1974, Frontiers of Time, 1979, (with W. Zurek) Quantum Theory and Measurement, 1983; contbr. articles to profl. jours. Trustee Battelle Meml. Inst., 1959—. Recipient A. Cressy Morrison prize N.Y. Acad. Sci. for work on nuclear physics, 1947; Albert Einstein prize Strauss Found., 1965; Enrico Fermi award AEC, 1968; Franklin medal Franklin Inst., 1969; Nat. medal of Sci., 1971; Herzfeld award, 1975; Outstanding Grad. Teaching award U. Tex., 1981; Niels Bohr Internat. Gold medal, 1982; Oersted medal Am. Assn. Physics Tchrs., 1983. Fellow AAAS (dir. 1965-68), Am. Phys. Soc. (pres. 1966); mem. Am. Math Soc., Internat. Astron. Union, Am. Acad. Arts and Scis., Nat. Acad. Sci., Am. Philos. Soc. (councillor 1963-66, 76—, v.p. 1971-73), Royal Danish Acad. Scis., Royal Acad. Sci. (Uppsala, Sweden), l'Académie Internationale de Philosophie des Sciences, Internat. Union Physics (v.p. 1951-54), Phi Beta Kappa, Sigma Xi. Unitarian (trustee 1965). Subspecialties: Relativity and gravitation; Theoretical physics. Home: 1410 Wildcat Hollow Austin TX 78746 Office: Physics Dept U Tex Austin TX 78712 We will first understand how simple the universe is when we recognize how strange it is.

WHEELER, J(OHN) CRAIG, astrophysicist; b. Glendale, Calif., Apr. 5, 1943; s. George Lafayette and Margaret Ann (Pratt) W.; m. Hsueh Lie, Oct. 29, 1967; children: Diek Winters, J. Robinson. B.S. in physics, M.I.T., 1965; Ph.D. in Physics, U. Colo., 1969. Research fellow Calif. Inst. Tech., 1969-71; asst. prof. astronomy Harvard U., 1971-74; assoc. prof. U. Tex., Austin, 1974-79, prof., 1979—. Contbr. articles in field to profl. jours. Recipient Griffith Obs. Popular Sci. essay 1st prize, 1975. Mem. Am. Astron. Soc., Internat. Astron Union, Sigma Xi. Subspecialties: High energy astrophysics; Theoretical astrophysics. Current work: Supernova evolution, dynamics, binary stars; active galaxies research, teaching, book writing on astrophysics. Office: Department of Astronomy University of Texas Austin TX 78712

WHEELER, KENNETH THEODORE, JR., radiation biologist, educator; b. Dover, N.H., Sept. 11, 1940; s. Kenneth Theodore and Emma Belle (Yeaton) W.; m. C. Anne Wallen, 1980. B.A., Harvard U., 1962; M.A.T., Wesleyan U., Middletown, Conn., 1963; Ph.D. in Radiation biophysics, Kans. U., Lawrence, 1970. Asst. prof. radiology and radiation biology Colo. State U., 1972; asst, prof., then assoc. prof. neurol. surgery and radiology U. Calif., San Francisco, 1972-76; assoc. prof. radiation oncology, dir. Cell Separation Facility U. Rochester Cancer Center (N.Y.), 1976-81; assoc. prof. radiation medicine Brown U., 1981-83; biophysicist, radiation biologist R.I. Hosp., Providence, 1981-83; scientist, prof. radiation biophysics Kans. U., Lawrence, 1983—; mem. grant rev. coms. NSF, NIH, Dept. Energy, cons. in field. Author papers in field. Grantee NIH, Am. Cancer Soc. Mem. Radiation Research Soc. (chmn, edn. and tng. com. 1981-83), Am. Assn. Cancer Research, Biophysics Soc., Sigma Xi. Subspecialty: Biophysics (biology). Current work: DNA damage and repair, brain tumor therapy. Home: 3718 Westland Pl Lawrence KS 66044 Office: Dept Radiation Biophysics Nuclear Ractor Ctr. W 15th St Lawrence KS 66045

WHEELER, WALTER HALL, geology educator; b. Syracuse, N.Y., Dec. 21, 1923; s. George Carlos and Esther (Hall) W.; m. Eula Virginia Krueger, Jan. 16, 1945; children: Diana, Roger. B.S., U. Mich., 1945, M.S., 1948; Ph.D., Yale U., 1951. Faculty U. N.C., Chapel Hill, 1951—, prof. geology, 1968—. Served to 1st lt. USAAF, 1943-46. Fellow Geol. Soc. Am.; mem. Paleontol. Soc., Am. Inst. Profl. Geologists, Am. Assn. Petroleum Geologists, Soc. Vertebrate Paleontology. Democrat. Current work: Stratigraphy of Atlantic coastal plain, Triassic-Jurassic basins, large Cenozoic mammals. Home: 28 Mount Bolus Rd Chapel Hill NC 27514 Office: U NC Dept Geology 029A Chapel Hill NC 27514

WHEELESS, LEON LUM, JR., pathology educator, electrical engineering educator; b. Jackson, Miss., Nov. 6, 1935; s. Leon Lum and Frances (King) W.; m. Waldine Marie Jones, Aug. 24, 1957; children: Susan, Diane, Linda. S.B., MIT, 1958; M.S., U. Rochester, 1962, Ph.D., 1965. Research scientist Bausch & Lomb, Rochester, N.Y., 1958-67, sr. scientist, 1967-69, dir. biomed. research, 1969-72; assoc. prof. pathology U. Rochester, 1972-81, prof. pathology and elec. engring., 1981—, dir. analytical cytology div., 1975—. Contbr. chpts. to books. Grantee Nat. Cancer Inst., 1973-82, 81-84; others. Mem. Soc. Analytical Cytology (charter, pres. 1982—), Engring. in Medicine and Biology Soc. of IEEE (sr., pres. 1974-75), Am. Soc. Cytology (profl.), Internat. Acad. Cytology (profl.). Subspecialties: Biomedical engineering; Computer engineering. Current work: Biomedical instrumentation; instrumentation for recognition of abnormal cells; slit-scan static and flow instrumentation. Patentee in field. Home: 30 Maryvale Dr Webster NY 14580 Office: U Rochester Med Ctr Rochester NY 14642

WHEELON, ALBERT DEWELL, physicist; b. Moline, Ill., Jan. 18, 1929; s. Orville Albert and Alice Geltz (Dewell) W.; m. Nancy Helen Hermanson, Feb. 28, 1953 (dec. May 1980); children—Elizabeth Anne, Cynthia Helen. B.Sc., Stanford U., 1949; Ph.D., Mass. Inst. Tech., 1952. Teaching fellow, then research asso. physics Mass. Inst. Tech., 1949-52; with Douglas Aircraft Co., 1952-53, Ramo-Wooldridge Corp., 1953-62; dep. dir. sci. and tech. CIA, 1962-66; with Hughes Aircraft Co., 1966—; sr. v.p. and group pres. Space and Communications Group, El Segundo, Calif., 1970—; mem. Def. Sci. Bd., 1967-77; cons. President's Sci. Adv. Council, 1961-74, NSC, 1974—. Author 30 papers on radiowaves propagation and guidance systems. Recipient Distinguished Intelligence medal CIA, 1966. Fellow IEEE; mem. Nat. Acad. Engring., Am. Phys. Soc., Internat. Union Radio Sci., Sigma Chi. Republican. Episcopalian. Club: Cosmos (Washington). Subspecialties: Applied mathematics; Theoretical physics. Office: PO Box 92919 Los Angeles CA 90009

WHELAN, ELIZABETH M., epidemiology and public health administrator; b. N.Y.C., Dec. 4, 1943; d. Joseph F. and Marion (Barrett) Murphy; m. Stephen T. Whelan, Apr. 3, 1971; 1 dau., Christine. B.A., Conn. Coll., 1965; M.P.H., Yale U., 1967; M.S., Harvard U., 1968, D. Sc., 1971. Research assoc. Harvard Sch. Pub. Health, Boston, 1976-80; with Am. Council Sci. and Health, N.Y.C., 1977—; mem. adv. Com. EPA; dir. Nat. Agrl. Legal Fund, Washington, 1982—, Food and Drug Law Inst. Author: Preventing Cancer, 1978, Nutrition During Pregnancy, 1982, The Nutrition Hoax, 1983. Recipient Am. Pub. Health Assn. early career award, 1982. Mem. Am. Cancer. Soc., Media Inst. (bd. dirs.). Subspecialty: Epidemiology. Home: 3109 Ocean Blvd Brant Beach NJ 08008 Office: Am Council Sci and Health 1995 Broadway New York NY 10023

WHETSELL, WILLIAM OTTO, JR., neuropathologist; b. Orangeburg, S.C., Sept. 25, 1940; s. William Otto and Margaret Elizabeth (Daniel) W.; m. Anne Elizabeth Rodgers, Oct. 14, 1967; children: William Otto III, Helen Fern Elizabeth. B.S., Wofford Coll., Spartanburg, S.C., 1961; M.S., Med. U. S.C., 1964, M.D., 1966. Diplomate: Am. Bd. Pathology, Sub-Bd. Neuropathology. Fellow in neurobiology, instr. Columbia U. Coll. Physicians and Surgeons, 1968-69; resident in neurology, pathology and neuropathology N.Y. Neurol. Inst. and Columbia U., 1971-75; asst. prof. pathology and neurology Mt. Sinai Sch. Medicine, N.Y.C., 1975-78, asso. prof., 1978-79, cons. dept. neurology, 1979—; adj. asst. prof. Rockefeller U., N.Y.C., 1975-78; prof. pathology, dir. neuropathology U. Tenn. Center Health Sci., Memphis, 1979—; chief neuropathology VA Hosp., Memphis, 1979—. Contbr. articles to profl. jours. Served to lt. comdr. USN, 1969-71. USPHS grantee, 1975—. Mem. Am. Assn. Anatomists, Am. Assn. Neuropathologists, Am. Acad. Neurology, Assn. Research in Nervous and Mental Disorders, Soc. Neuroscience, AMA, N.Y. Acad. Scis. Republican. Current work: Development of exploitation of experimental models in organotypic nerve cell tissue culture for study of porphyrin-heme metabolism in nervous system, study of cellular mechanisms involved in neurodegenerative disorders of neostriatum. Home: 100 E Parkway N Memphis TN 38104 Office: 858 Madison Ave Memphis TN 38163

WHETSTONE, STANLEY LEROY, government energy program administrator; b. Newark, Aug. 30, 1925; s. Stanley L. and Mary C. (Rhoads) W.; m. Jeanne A. Clark, June 29, 1952; children: Karin, Kirsten, Sara, Steven. B.A., Williams Coll., 1949; Ph.D., U. Calif.-Berkeley, 1955. Staff mem. Los Alamos Sci. Lab., 1955-70, 75-76; vis. lectr. U. Wash., 1967-68; physics sect. head IAEA, Vienna, Austria, 1970-75; physicist U.S. Dept. Energy, Germantown, Md., 1976—, now program mgr. low energy nuclear physics program. Served with AUS, 1944-46. Subspecialty: Nuclear physics. Office: Dept Energy ER23/GTN Washington DC 20545

WHIKEHART, DAVID RALPH, physiol. optics and biochemistry educator; b. Pitts., Aug. 21, 1939. B.S., Duquesne U., 1965; Ph.D., W.Va. U., 1969. Postdoctoral fellow McLean Hosp., Belmont, Mass., 1969-71, asst. biochemist, 1971-72; spl. staff fellow NIH, Bethesda, Md., 1972-75, sr. staff fellow, 1975-78; asst. prof. physiol. optics, biochemistry U. Ala., 1978-81, assoc. prof., 1981—, dir. vis. scholars program, 1978—, co-chmn. faculty senate, 1980—. Reviewer: Current Eye Research, 1978—; Investigative: Ophthalmology and Visual Science, 1978—; Co-author: Clinical Ocular Pharmacology, 1984. NIH grantee, 1978—. Mem. Assn. Research in Vision and Ophthalmology. Club: Centennial Sertoma (v.p. 1982-83). Subspecialties: Biochemistry (medicine); Biochemistry of the eye. Current work: Transport and metabolic systems in the cornea of the eye; Meibomian gland functions; the effects of alkali burns on the cornea. Office: U Ala-Birmingham Sch Optometry Dept Physiol Optics Birmingham AL 35294

WHINNERY, JOHN ROY, electrical engineering educator; b. Read, Colo., July 26, 1916; s. Ralph V. and Edith Mable (Bent) W.; m. Patricia Barry, Sept. 17, 1944; children—Carol Joanne, Catherine, Barbara. B.S. in Elec. Engring, U. Calif. at Berkeley, 1937, Ph.D., 1948. Student engr. Gen. Electric Co., 1937-40, supr. high frequency course advanced engring. program, 1940-42, research engr., 1942-46; part-time lectr. Union Coll., Schenectady, 1945-46; asso. prof. elec. engring. U. Calif.-Berkeley, 1946-52, prof., vice chmn. div. elec. engring., 1952-56, chmn., 1956-59; dean Coll. Engring. U. Calif-Berkeley, 1959-63, prof. elec. engring., 1963-80, Univ. prof. Coll. Engring., 1980—; vis. mem. tech. staff. Bell Telephone Labs., 1963-64; research sci. electron tubes Hughes Aircraft Co., Culver City, 1951-52; bd. editors I.R.E., 1956. Author: (with Simon Ramo) Fields and Waves in Modern Radio, 1944, (with D. O. Pederson and J. J. Studer) Introduction to Electronic Systems, Circuits and Devices; also tech. articles. Chmn. Commn. Engring. Edn., 1966-68; mem. sci. and tech. com. Manned Space Flight, NASA, 1963-69; mem. Pres.'s Com. on Nat. Sci. Medal, 1970-73, 79-80; standing com. controlled thermonuclear research AEC, 1970-73. Recipient Edn. medal IEEE, 1967; Lamme medal Am. Soc. Engring. Edn., 1975; Microwave Career award IEEE Microwave Theory and Techniques Soc., 1977; Engring Alumni award U. Calif. at Berkeley, 1980; named to Hall of Fame Modesto High Sch. (Calif.), 1983; Guggenheim fellow, 1959. Fellow I.R.E. (dir. 1956-59), Optical Soc. Am., Am. Acad. Arts and Scis.; mem. Nat. Acad. Engring., Nat. Acad. Scis., IEEE (dir. 1969-71, sec. 1971), Phi Beta Kappa, Sigma Xi, Tau Beta Pi, Eta Kappa Nu. Conglist. Subspecialties: Electrical engineering; Fiber optics. Current work: Generation and use of short optical pulses. Home: One Daphne Ct Orinda CA 94563 Office: U Calif Berkeley CA 94720

WHIPPLE, FRED LAWRENCE, astronomer; b. Red Oak, Iowa, Nov. 5, 1906; s. Harry Lawrence and Celestia (MacFarl) W.; m. Dorothy Woods, 1928 (div. 1935); 1 son, Earle Raymond; m. Babette F. Samelson, Aug. 20, 1946; children—Dorothy Sandra, Laura. Student, Occidental Coll., 1923-24; A.B., UCLA, 1927, Ph.D., at Berkeley, 1931; A.M. (hon.), Harvard, 1945; Sc.D., Am. Internat. Coll., 1958; D.Litt. (hon.), Northeastern U., 1961, D.Sc., Temple U., 1961, U. Ariz., 1979, LL.D., C.W. Post Coll., L.I. U., 1962. Teaching fellow U. Calif. at Berkeley, 1927-29; Lick Obs. fellow, 1930-31; instr. Stanford, summer 1929, U. Calif., summer 1931; staff mem. Harvard Obs., 1931—; instr. Harvard, 1932-38, lectr., 1938-45; research asso. Radio Research Lab., 1942-45, asso. prof. astronomy, 1945-50, prof. astronomy, 1950—, chmn. dept., 1949-56, Phillips prof. astronomy, 1968-77; dir. Smithsonian Astrophys. Obs., 1955-73, sr. scientist,

1973—; mem. Rocket Research Panel U.S., 1946—; U.S. subcom NASA, 1946-52, U.S. Research and Devel. Bd. Panel, 1947-52; chmn. Tech. Panel on Rocketry; mem. Tech. Panel on Earth Satellite Program; other coms. Internat. Geophys. Year, 1955-59; mem., past officer Internat. Astron. Union; cons. missions to U.K. and MTO, 1944; del. Inter-Am. Astrophys. Congress, Mexico, 1942; active leader project on Upper-Atmospheric Research via Meteor Photog. sponsored by Bur. Ordnance, U.S. Navy, 1946-51, by, 1951-57, USAF, 1948-62; mem. com. meteorology, space sci. bd., com. on atmospheric sics. Nat. Acad. Scis.-NRC, 1958-65; project dir. Harvard Radio Meteor Project, 1958—; adviser Sci. Adv. Bd., USAF, 1963-67; spl. cons. com. Sci. and Astronautics U.S. Ho. Reps., 1960-73; chmn. Gordon Research Confs., 1963; dir. Optical Satellite Tracking Project, NASA, 1958-73; project dir. Orbiting Astron. Obs., 1958-72; dir. Meteorite Photography and Recovery Program, 1962-73, cons. planetary atmospheres, 1962-69; mem. space scis. working group on Orbiting Astron. Observatories, 1959-70; chmn. sci. council geodetic uses artifical satellites Com. Space Research, 1965—. Author: Earth, Moon and Planets, rev. edit, 1968, Orbiting The Sun: Planets and Satellites of The Solar System; co-author: Survey of the Universe; Contbr.: sci. papers on astron. and upper atmosphere to Ency. Brit; mags., other publs.; Asso. editor: Astronomical Jour, 1954-56, 64—; editor: Smithsonian Contributions to Astrophysics, 1956-73, Planetary and Space Science, 1958—, Science Revs, 1961—; editorial bd.: Earth and Planetary Sci. Letters; inventor tanometer, meteor bumper; a developer window as radar countermeasure, 1944. Decorated comdr. Order of Merit for research and invention, Esnault-Pelterie award, France; recipient Donohue medals for ind. discovery of 6 new comets; Presdl. Cert. of Merit for sci. work during World War II; J. Lawrence Smith medal Nat. Acad. Scis. for research on meteors, 1949; medal for astron. research U. Liege, 1960; Space Flight award Am. Astronautical Soc., 1961; Disting. Fed. Civilian Service award, 1963; Space Pioneers medallion for contbns. to fed. space program, 1968; Public Service award for contbns. to OAO2 devel. NASA, 1969; Leonard medal Meteoritical Soc., 1970; Kepler medal AAAS, 1971; Career Service award Nat. Civil Service League, 1972; Henry medal Smithsonian Instn., 1973; Alumnus of Yr. Achievement award UCLA, 1976; Golden Plate award Am. Acad. Achievement, 1981; Benjamin Franklin fellow Royal Soc. Arts, London, 1968—. Fellow Am. Astron. Soc. (v.p. 1962-64), Am. Rocket Soc., Am. Geophys. Union; asso. Royal Astron. Soc.; mem. Nat. Acad. Scis., AIAA Astronautics (aerospace tech. panel space physics 1960-63), Astronautical Soc. Pacific, Solar Assos., Internat. Sci. Radio Union (U.S.A. nat. com. 1949-61), Am. Meteoritical Soc., Am. Standards Assn., Am. Acad. Arts and Scis., Am. Philos. Soc. (councillor sect. astronomy and earth scis. 1966—), Royal Soc. Scis. Belgium (corr.), Internat. Acad. Astronautics (sci. advisory com. 1962—), Internat. Astronautical Fedn., AAAS, Am. Meteorol. Soc., Royal Astron. Soc. (asso.), Phi Beta Kappa, Sigma Xi, Pi Mu Epsilon. Clubs: Examiner (Boston); Cosmos (Washington). Subspecialty: Optical astronomy. Home: 35 Elizabeth Rd Belmont MA 02178 Office: 60 Garden St Cambridge MA 02138

WHISONANT, ROBERT CLYDE, geologist; b. Columbia, S.C., Apr. 20, 1941; s. Clyde and Mary (Lanford) Whisonant D.; m. Brenda Dale Lark, June 7, 1963; children: Dell Raye, Robert Dowling. B.S., Clemson U., 1963; M.S., Fla. State U., 1965, Ph.D., 1967. Petroleum geologist Humble Oil and Refining Co., Houston and Kingsville, Tex., 1967-71, Cons., 1971-72; prof. geology Radford (Va.) Coll., 1971—; cons. Nat. Geog. Soc., Washington, 1980-81. Contbr. articles to profl. jours. NASA trainee, 1963-66; Penrose grantee Geol. Soc. Am., 1971-72; Jeffress grantee Jeffress Meml. Trust, 1982-84. Fellow Geol. Soc. Am.; mem. Am. Assn. Petroleum Geologists, Soc. Econ. Paleontologists and Mineralologists, Va. Acad. Sci., Carolina Geol. Soc., Sigma Xi. Methodist. Lodge: Lions. Subspecialty: Sedimentology. Current work: Clastic sequences of late Precambrian-Paleozoic age, Southern Appalachians; intraclastic carbonates of early paleozoic age, southeastern Virginia. Home: 29 Round Hill Dr Radford VA 24141 Office: Dept Geology Radford U Radford VA 24142

WHITAKER, EWEN ADAIR, astronomer; b. London, June 22, 1922; U.S., 1958; s. George Frederick and Gladys Emily (Johnstone) W.; m. Beryl Joyce Horswell, June 22, 1946; children: Malcolm John, Graham David, Fiona Carolyn. Higher nat. cert. in mech. engring, Woolwich Poly., London, 1944. Spectographer, analytical chemist Siemens Bros. Co., Ltd., London, 1940-49; exptl. officer Royal Greenwich Obs., London, 1949-56, Hertmonceux Castle, Sussex, 1956-58; research assoc. Yerkes Obs., Williams Bay, Wis., 1958-60; assoc. research scientist Lunar & Planetary Lab., U. Ariz., Tucson, 1960—. Author: Photographic Lunar Atlas, 1960, Orthographic Atlas of the Moon, 1961, Rectified Lunar Atlas, 1963, Consolidated Lunar Atlas, 1967; author: NASA Catalogue of Lunar Nomenclature, 1982; contbr. articles to profl. jours. Recipient Walter Goodacre medal and gift Brit. Astron. Assn., 1982. Fellow Royal Astron. Soc.; mem. Am. Astron. Soc., Internat. Astron. Union. Subspecialty: Planetary science. Current work: Lunar surface topography, physics/chemistry, evolution. Home: 4332 E 6 St Tucson AZ 85711 Office: Lunar & Planetary Lab U Ariz Tucson AZ 85721

WHITE, CHARLES OLDS, aerospace engineer; b. Beirut, Lebanon, Apr. 2, 1931; came to U.S., 1945; s. Frank Laurence and Dorothy Alice (Olds) W.; m. Mary Carolyn Liechty, Sept. 3, 1955; children: Charles Cameron, Bruce Blair. B.S., MIT, 1953, M.S., 1954. Aero. engr. Douglas Aircraft Co., Long Beach, Calif., 1954-60; sr. engring. specialist Ford Aerospace & Communications Corp., Newport Beach, Calif., 1969-79, staff office gen. mgr., 1979-80, tech. mgr., 1981-82; supr. design and analysis DIVAD div., 1982—. Prin. collaborator: handbook USAF Stability and Control Methods; contbr. articles to profl. jours. Recipient Sigma Gamma Tau award MIT, 1953. Mem. AIAA, Nat. Mgmt. Assn., Am. Aviation Hist. Soc. Republican. Presbyterian. Clubs: Newport Beach and Tennis, Masters Swimming. Subspecialties: Fluid mechanics; Aerospace engineering and technology. Current work: High temperature fluid mechanics, aerodynamic heating, propellants, materials, system dynamics. Home: 2857 Alta Vista Dr Newport Beach CA 92660 Office: Ford Aerospace and Communications Corp Ford Rd Newport Beach CA 92660

WHITE, DAVID, chemistry educator; b. Ukraine, USSR, Jan. 14, 1925; s. Maurice and Mary White; m. Birdye White, Aug. 7, 1945; children: Sharon, Jacqueline, Edward. B.Sc., McGill U., 1944; Ph.D. in Chemistry, U. Toronto, 1947. Fellow Ohio State U., 1948-50, asst. dir. Cryogenic Lab., 1950-53, asst. prof. chemistry, 1954-57, assoc. prof., 1957-61, prof., 1961-66; prof. chemistry U. Pa., Phila., 1966—, chmn. chemistry dept., 1966-79, dir. Lab. for Research on Stucture of Matter, 1981—. Contbr. over 150 articles to chem. publs. Mem. Am. Chem. Soc., Am. Phys. Soc., Sigma Xi. Subspecialties: Physical chemistry; Nuclear magnetic resonance (chemistry). Current work: Low temperature thermodynamics and solid state nmr; molecular structure of inorganic vapor species; infrared and Raman spectroscopy. Patentee cryogenic expansion engine, Can. and U.S. Office: Chemistry Dept Univ Pa Philadelphia PA 19104

WHITE, DONALD EDWARD, geologist; b. Dinuba, Calif., May 7, 1914; s. Arthur Thomas and Alice Louise (Hughson) W.; m. Helen Beth Severance, Sept. 16, 1941; children: Margaret Anne, Eleanor Louise, Catherine Marie. A.B., Stanford U., 1936; Ph.D., Princeton U., 1939. Mem. staff Nfld. Geol. Survey, 1937-39; with U.S. Geol. Survey, 1939—; asst. chief Mineral Deposits Br., Washington, 1958-60, research geologist, Menlo Park, Calif., 1960-81; adviser Geothermal Energy Program, 1971—; adviser geothermal energy N.Z., Iceland, Japan, Italy, Am. industry; cons. prof. geophysics Stanford U., 1974; Davidson Meml. lectr. St. Andrews U., Scotland, 1975. Recipient Superior Performance award U.S. Geol. Survey, 1969; Disting. Service award Dept. Interior, 1971. Fellow Geol. Soc. Am., Soc. Econ. Geologists (Distinguished Lectr. 1967, 74, v.p. 1977, pres. 1982), AAAS; mem. Nat. Acad. Sci., Mineral. Soc. Am., Am. Geophys. Union, Geochem. Soc., AAAS. Subspecialties: Geochemistry; Economic geology. Current work: Mineral deposits and geothermal energy. Developer prins. for recognition and evaluation natural geothermal systems for utilizing geothermal energy, and for applying principles of active systems to fossil geothermal systems (hydrothermal mineral deposits). Home: 222 Blackburn Ave Menlo Park CA 94025

WHITE, DOUGLAS RECTOR, hematology and oncology educator; b. Washington, Nov. 24, 1941; s. John Carpenter and Margaret (Keister) W.; m. Mary Louise Cavan, Feb. 22, 1970; children: Sara Carpenter, Margaret Cavan. B.S., U. Chgo., 1963, M.D., 1967. Diplomate: Am. Bd. Internal Medicine, Am. Bd. Hematology, Am. Bd. Oncology. Intern Emory U., Atlanta, 1967-68, resident in internal medicine, 1968-69, U. Tex.-San Antonio, 1971-72; clin. fellow in hematology, 1972-73, research fellow, 1973-74; asst. prof. medicine Bowman Gray Sch. of Medicine, Winston-Salem, N.C., 1974-78, assoc. prof., 1978—. Contbr. articles profl. jours. Served to maj. M.C. U.S. Army, 1969-71; Vietnam. Fellow ACP; mem. Am. Soc. Hematology, Am. Soc. Clin. Oncology, AMA, Southeastern Cancer Research Assn. Subspecialties: Oncology; Hematology. Current work: Clinical cancer research. Office: Dept of Medicin Bowman Gray School of Medicine Winston Salem NC 27103

WHITE, HELEN LYNG, biochemist, pharmacologist; b. Oceanside, N.Y., Oct. 25, 1930; d. James and Irene Genevieve (Dilzer) Lyng; m. James Rushton White, Jan. 15, 1955; children: Jennifer, John Nelson. B.A., Russell Sage Coll., 1952; M.S., U. Del., 1963; Ph.D. (NASA fellow), U. N.C., 1967. Chemist E.I. duPont de Nemours & Co., Wilmington, Del., 1952-56; research assoc. dept. medicinal chemistry U. N.C., Chapel Hill, 1967-70; sr. research pharmacologist Wellcome Research Labs., Research Triangle Park, N.C., 1970—; ad hoc reviewer for several sci. jours. and NSF. Contbr. over 100 articles and abstracts to sci. publs. Recipient Crockett Alumnae award Russell Sage Coll., 1972. Mem. Am. Soc. for Pharmacology and Exptl. Therapeutics, Am. Soc. Biol. Scis., Am. Chem. Soc. (medicinal chemistry div.), Sigma Xi. Subspecialties: Biochemistry (biology); Pharmacology. Current work: Enzyme inhibitors; neurochemistry; prostaglandins; monoamine oxidase; drug development for therapeutic purposes. Patentee in field. Office: Wellcome Research Labs Research Triangle Park NC 27709

WHITE, IRVIN LINWOOD, JR., state official; b. Hertford, N.C., Mar 15, 1932; s. Irvin Linwood and Katherine Margarite (Winslow) W.; m. Patricia Ann Hathaway, Sept. 12, 1954 (div. 1978); children: Jonathan Randolph, David Hathaway; m. Mary Ruth Hamilton, May 6, 1978. B.A., Pa. State U., 1954; Ph.D., U. Ariz., 1967. Asst. prof. polit. sci. Purdue U.-West Lafayette, Ind., 1968-70; asst. dir. sci. and pub. policy program, prof. polit. sci. U. Okla.-Norman, 1970-80; spl. asst. and acting dir. Office of Strategic Assessment, EPA, Washington, 1978-80; asst. dir. Bur. Land Mgmt., Energy Research and Mineral Resources, U.S. Dept. Interior, Washington, 1980-81; pres. N.Y. State Energy Research and Devel. Authority, Albany, 1981—; mem. com. on producibility of oil and gas Nat. Acad. Scis., 1981-82. Author: (with C.E. Wilson and J.A. Voshburgh) Law and Politics in Outer Space: A Bibliography, 1974, (with others) Energy Under the Oceans: A Technology Assessment of Continental Shelf Oil and Gas Operations, 1973; also articles. Served with USN, 1954-63. NASA trainee, 1964-67. Mem. Am. Polit. Sci. Assn., AAAS, Policy Studies Orgn. Democrat. Current work: Chief operating officer of broad-gauged state energy (conservation, renewables, coal, electrical systems) organization; also active in resource management. Office: Two Rockefeller Plaza 10th Floor Albany NY 12223

WHITE, JOHN EVANS, aerospace engineering researcher, educator; b. Pampa, Tex., Apr. 7, 1954; s. Lonnie Joe and Nancy Louella (Evans) W. B.S. in Aerospace Engring, U. Okla., Norman, 1976, M.S., U. Tex., Austin, 1980. Engr. Halliburton Services, Duncan, Okla., summer 1975; engr. McDonnell Douglas Astronautics Co.-East, St. Louis, 1976-77; engr., remote site NASA White Sands Test Facility, Las Cruces, N. Mex., 1978; research asst. U. Tex., Austin, 1979—; engr. McDonnell Douglas Tech. Services Co., Houston, summer 1983; pvt. pilot, 1974—. U. Tex. Engring. Found. fellow, 1979. Mem. AIAA, Sigma Gamma Tau, Tau Beta Pi. Subspecialty: Aerospace engineering and technology. Current work: Modern control theory; fault tolerant flight control systems. Home: 3001 Duval Apt 203 S Austin Tx 78705 Office: U Tex Dept Aerospace Engring Austin TX 78712

WHITE, JOHN SPENCER, industrial engineer, educator; b. River Falls, Wis., Mar. 7, 1926; s. Kenneth Sidney and Helen (Kyle) W.; m. (widower); children: Jane Fritz, Jill Downey. B.A., U. Minn., 1947, M.A., 1951, Ph.D., 1955. Cert. reliability engr. Asst. prof. U. Man., Winnipeg, 1953-56; staff quality engr. Honeywell Co., Mpls., 1956-60; research mathematician Gen. Motors Co., Warren, Mich., 1960-68; prof. mech. engring. U. Minn., Mpls., 1968—; cons. Teledyne Corp., 1968-72, Medtronics Corp., 1972-73, Viking Sprinkler, 1981-83, Rosemount Engring., 1982-83; all Mpls. Served with USNR, 1942-45; ETO. Mem. Am. Soc. Quality Control (sr, Craig award 1979), IEEE (sr.), Inst. Math. Stats. Republican. Episcopalian. Subspecialties: Industrial engineering; Statistics. Current work: Quality control, reliability, applied statistics time series analysis, stochastic processes. Home: 5135 Ranier Pass Minneapolis MN 55421 Office: U Minn 111 Church St SE Minneapolis MN 55455

WHITE, MARY-ALICE, psychologist, educator; b. Washington, Mar. 18, 1920; d. Charles Stanley and Blanche M. (Strong) W.; m. Edward N. Kimball, Mar. 27, 1949 (div. 1968); children: Christopher, Katharine. B.A., Vassar Coll., 1941; Ph.D., Columbia U., 1948. Head dept. psychology Westchester div. N.Y. Hosp., White Plains, 1948-60; psychol. cons. Pelham (N.Y.) Sch. System, 1953-62; assoc. prof. Tchrs. Coll., Columbia U., N.Y.C., 1966-67, prof. psychology, 1967—; dir. Electronic Learning Lab., 1977—. Author: (with others) Parents Guide to School Testing, 1982; editor: Future of Electronic Learning, 1983; Contbr. articles to profl. jours. NIMH grantee, 1966-82. Fellow Am. Psychol. Assn. (pres. div. 1967, Disting. Service award 1971). Democrat. Subspecialties: Psychology of electronic learning; Psychology of computer learning. Current work: Development of electronic learning psychology—how humans learn from electronic communications and information technology. Home: Box 426 Salisbury CT 06068 Office: Tchrs Coll Columbia New York NY 10027

WHITE, MAURICE EDWARD, veterinarian, educator; b. Bklyn., Dec. 1, 1946; s. Maurice S. and Grace (O'Halloran) Twomey; m. Steffi White, Jan. 9, 1972; children: Katherine, Michael, Linda. D.V.M., Cornell U., 1975. Intern, resident U. Guelph, Ont., Can., 1975-77; pvt. practice vet. medicine, Springfield, Vt., 1977-78; ambulatory clinician, asst. prof. medicine N.Y. State Coll. Vet. Medicine, Cornell U., 1978—. Contbr. articles sci. jours. Served to sgt. USAF, 1965-68. Subspecialties: Preventive medicine (veterinary medicine); Internal medicine (veterinary medicine). Current work: Medical decision making, computer-assisted decision making. Developer Cons., computer-assisted differential diagnostic program. Home: 101 Eastern Heights Dr Ithaca NY 14850 Office: Dept Clin Scis NY State Coll Vet Medicine Cornell U Ithaca NY 14853

WHITE, NATHANIEL ALDRICH, II, veterinarian; b. Phila., May 13, 1946; s. Nathaniel E. and Anna (Paul) W.; m. Susan Lararine, June 23, 1973. B.S., Cornell U., 1969, D.V.M., 1971; M.S., Kans. State U., 1976. Diplomate: Am. Coll. Veterinary Surgeons. Intern U. Calif.-Davis, 1971-72, resident in surgery, 1972-73; practice veterinary medicine, Arcata, Calif., 1973-74; instr. equine medicine and surgery Kans. State U., 1975-76; asst. prof. equine surgery Coll. Veterinary Medicine, U. Ga., 1976-81, asso. prof., 1981—. Contbr. articles to profl. jours. Mem. Am. Veterinary Med. Assn., Am. Assn. Equine Practitioners, Acad. Veterinary Cardiology, Am. Coll. Veterinary Surgeons, Alpha Zeta, Omega Tau Sigma, Phi Zeta. Subspecialties: Surgery (veterinary medicine); Pathology (veterinary medicine). Current work: Equine surgery, large animal cardiology, equine actue abdomen and shock. Office: Coll Vet Medicine U Ga Athens GA 30602

WHITE, NEIL HARRIS, physician, researcher; b. N.Y.C., June 25, 1949; s. Alan Maurice and Edith Jean (Berman) W.; m. Ann Teresa Davida, June 11, 1975; children: Michael Steven, Justin Alan. B.S., SUNY-Albany, 1971; M.D., Albert Einstein Coll. Medicine, 1975. Diplomate: Nat. Bd. Med. Examiners, Am. Bd. Pediatrics. Resident in pediatrics St. Louis Children's Hosp., 1975-77, fellow in pediatric endocrinology, 1977-79, staff physician, 1979—; asst. prof. pediatrics Washington U., St. Louis, 1979—. Contbr. articles to profl. jours. Mem. Am Diabetes Assn., Am. Fedn. Clin. Research. Jewish. Subspecialties: Pediatrics; Endocrinology. Current work: Pediatric endocrinology with primary interest in diabetes mellitus; care of diabetic patients using mechanical devices of insulin delivery, evaluation of these devices. Home: 1420 Weatherby Dr Saint Louis MO 63141 Office: St Louis Children's Hosp PO Box 14871 Saint Louis MO 63178

WHITE, RAYMOND EDWIN, JR., astronomy educator, researcher; b. Freeport, Ill., May 6, 1933; s. Raymond Edwin and Beatrice Ellen (Rahn) W.; m. Ruby Elaine Fisk, Oct. 16, 1956; children: Raymond E. III, Kathleen M., Kevin D. B.S. in Math, U. Ill., 1955, Ph.D. in Astronomy, 1967. Instr. U. Ariz., 1964-65, asst. prof., 1965-71, lectr., 1971-81, assoc. prof., 1981—, asst. astronomer, 1965-71, assoc. astronomer, 1981—; program officer NSF, 1971-71. Contbr. articles in field to profl. jours. Vice chmn. Pima County (Ariz.) Air Quality Adv. Council, 1979—. Served with USAR, 1955-58. NSF grantee, 1968, 72,76,81; Earthwatch grantee, 1980, 82. Mem. Am. Astron. Soc., AAAS, Internat. Astron. Union, Royal Astron. Soc., Sigma Xi. Republican. Subspecialty: Optical astronomy. Current work: Observational astronomy; photometry and spectroscopy of population II subsystems; archaeoastronomy definition of inkaic astron. interests. Home: 2500 E 8th St Tucson AZ 85716 Office: Steward Observatory University of Arizona Tucson AZ 85721

WHITE, RAYMOND PETRIE, JR., dentist; b. N.Y.C., Feb. 13, 1937; s. Raymond Petrie and Mabel Sarah (Shutze) W.; m. Betty Pritchett, Dec. 27, 1961; children—Karen Elizabeth, Michael Wood. Student, Washington and Lee U., 1955-58; D.D.S., Med. Coll. Va., 1962, Ph.D., 1967. Diplomate: Am. Bd. Oral and Maxillofacial Surgery. Postdoctoral fellow anatomy Med. Coll. Va., Richmond, 1962-67, resident in oral surgery, 1964-67; asst. prof. U. Ky., Lexington, 1967-70, asso. prof., 1970-71, chmn. dept. oral surgery, 1969-71; prof., asst. dean adminstrn. Va. Commonwealth U., Richmond, 1971-74; prof. Sch. Dentistry, U. N.C., Chapel Hill, 1974—, dean, 1974-81, asso. dean, 1981—; cons. Portsmouth Naval Hosp., Fayetteville VA; mem. staff N.C. Meml. Hosp., mem. exec. com., 1974—, sec., 1977-78, asso. chief of staff, 1981—; mem. rev. com. health manpower br. USPHS, 1974-76; mem. rev. com. VA Health Manpower Tng. Assistance, 1976-79. Author: (with E.R. Costich) Fundamentals of Oral Surgery, 1971, (with Bell and Proffit) Surgical Correction of Dentofacial Deformities, 1980; contbr. sci. articles to profl. jours. Recipient A.D. Williams award for highest class standing Med. Coll. Va., 1962, Outstanding Tchr. award U. Ky., 1971. Mem. Am. Dental Assn., N.C. Dental Soc., Am. Acad. Oral Pathology, Atwood Wash Oral Surgery Soc., AAAS, Internat. Assn. for Dental Research (pres. Ky. sect. 1970), Inst. of Medicine of Nat. Acad. Scis., N.Y. Acad. Scis., Chalmers J. Lyons Acad. Oral Surgery, Am. Assn. Oral and Maxillofacial Surgeons (gen. chmn. sci. sessions com. 1974-76, Outstanding Service award as committeeman 1976), N.C. Soc. Oral and Maxillofacial Surgeons, Sigma Xi, Psi Omega, Delta Tau Delta, Alpha Sigma Chi, Sigma Zeta, Psi Omega (Scholarship award 1962), Omicron Kappa Upsilon. Roman Catholic. Subspecialties: Oral and maxillofacial surgery; Health services research. Current work: Facial growth and development; effects of treatment of dentofacial deformity; health services research and policy analysis in dentistry. Home: 1506 Velma Rd Chapel Hill NC 27514 Office: Dept Oral and Maxillofacial Surgery Univ NC Sch Dentistry Chapel Hill NC 27514

WHITE, RICHARD EDWARD, astronomer, educator; b. Chgo., Jan. 18, 1944; s. Edward Joseph and Eileen Dorothy (Prendergast) W.; m. Penny Conrad, Sept. 5, 1970. B.S. summa cum laude, St. Joseph's Coll., 1965; Ph.D., Columbia U., 1971. Carnegie fellow Hale Obs. Pasadena, Calif., 1970-72; resident research assoc. NASA Goddard Inst. for Space Studies, N.Y.C., 1972-74; adj. asst. prof. Columbia U., 1974; asst. prof. astronomy Smith Coll., Northampton, Mass., 1974-75, 76-82, lectr., 1975-76, assoc. prof., 1982—; vis. astronomer F.L. Whipple Obs., Smithsonian Instn., Amado, Ariz., 1980, 82, 83; mem. users adv. com. Kitt Peak Nat. Obs., 1977-79. Contbr. articles to profl. jours. NSF grantee, 1978-80; Am. Astron. Soc. grantee, 1981. Mem. Am. Astron. Soc., Astron. Soc.Pacific, AAAS, Sigma Xi. Democrat. Subspecialty: Optical astronomy. Current work: Optiocal studies of interstellar clouds.

WHITE, RICHARD HAMILTON, biology educator, biologist; b. Rochester, N.Y., Feb. 16, 1934; s. George Lynn and Edith Rosalie (Barager) W.; m. Sally Jean Martin, Feb. 18, 1939; children: Diane Elizabeth, James Barager, Erica Mary. B.A., U. Rochester, 1956; Ph.D., Washington U., St. Louis, 1961. Postdoctoral fellow U. Va., Charlottesville, 1961-62; asst. prof. Purdue U., West Lafayette, Ind., 1962-69; research fellow Harvard U., Cambridge, Mass., 1969-70; assoc. prof. U. Mass., Boston, 1970-77, prof. biology, 1977—; grants reviewer NSF, Washington, 1965—. Contbr. articles to profl. jours. NIH postdoctoral fellow, 1961, 69; NSF research grantee, 1962, 65, 73; NIH research grantee, 1977. Mem. AAAS, Soc. Devel. Biology, Assn. Research in Vision and Ophthalmology. Democrat. Subspecialties: Cell biology; Neurobiology. Current work: Ultrastructure, function and development of photoreceptor cells, formation and turnover of photoreceptor membranes. Office: Biology Dept U Mass Harbor Campus Boston MA 02125

WHITE, RICHARD MAHAFFEY, management technology, company executive, consultant; b. N.Y.C., June 14, 1944; s. Richard Thornton and Anna M. (Mahaffey) W. B.S. in Systems Analysis, Miami U., Oxford, Ohio, 1972, M.S. in Stats. cum laude, 1972. Cons. Ops. Analysis and Research, Washington, 1978—; dir. The Oar Corp., Washington, 1979—; cons. Naval Electronics Systems Command, 1978—. Served to lt. comdr. USN, 1965-78; Served to lt. comdr.

USNR, 1978—. Mem. Soc. Indsl. and Applied Math., Math Assn. Am., U.S. Naval Inst., Soc. Old Crows, Smithsonian Assocs., Cultural Alliance Greater Washington, Corcoran Assocs., Phi Beta Kappa. Subspecialties: Distributed systems and networks; Cryptography and data security. Current work: Underwater acoustics research, space technology, Navy command and control systems, world wide communications architectures. Office: Ops Analysis and Research 6005 Milo Dr Washington DC 20816

WHITE, ROBERT J., educator, neurosurgeon, neuroscientist; b. Duluth, Minn., Jan. 21, 1926; , married 1950; children: Robert, Christopher, Patricia, Michael, Daniel, Pamela, James, Richard, Marguerite, Ruth. B.S., Coll. St. Thomas, U. Minn., 1951, Ph.D. in Neurosurg. Physiology, 1962; M.D., Harvard, 1953; D.Sc. (hon.), John Carroll U., 1979, Cleve. State U., 1981. Intern surgery Peter Bent Brigham Hosp., Boston, 1953-54; resident Boston Children's Hosp. and Peter Bent Brigham Hosp., 1954-55; fellow neurosurgery Mayo Clinic, 1955-58; asst. to staff, 1958-59, research asso. neuro physiol., 1959-61; mem. faculty Case Western Res. U. Sch. Medicine, 1961—, prof. neurosurgery, 1966—, co-dir. neurosurgery, 1972, co-chmn. neurosurgery, 1973—; dir. neurosurg. and brain research lab. Cleve. Met. Gen. Hosp., 1961—; neurosurgeon Univ. Hosps.; also sr. attending neurosurgeon VA Hosp., 1961—; lectr. USSR, 1966, 68, 70, 72, 73, 78, 79, People's Republic China, 1977, 81. Gen. editor: Internat. Soc. Angiol. Jour, 1966; editor: Western Hemisphere Jour. Resuscitation, 1971—, Surg. Neurology, Resuscitation, Jour. of Trauma; co-editor: Surg. Neurology, 1973—; Contbr. numerous articles to profl. jours. Served with AUS, 1944-46. Recipient Mayo Clinic Research award, 1958; Med. Mut. Honor award, 1975; L.W. Freeman award PF, 1977. Mem. ACS (com. on trauma task force), Harvey Cushing Soc., Soc. Univ. Surgeons, Am. Physiol. Soc., Soc. Univ. Neurosurgeons (pres. 1978-79), Ohio Neurosurg. Soc. (pres. 1975), Northeast Ohio Neurosurg. Soc. (pres. 1971), Soc. Exptl. Biology, Internat. Soc. Cybernetic Medicine (mem. internat. bd. 1971—), Internat. Soc. Surgery, Soc. Neurol. Surgeons, Neurosurg. Soc. Am., A.C.S., AMA, Acad. Medicine Cleve. (dir., pres. 1978-79). Subspecialties: Neurochemistry; Neurophysiology. Current work: Operative brain surgery; brain transplantation; neurochemistry; neuroimmuology; head injury; spinal cord injury. Developed 1st isolated brain models; pioneered brain and cephalon transplant in animals, deep hypothermia techniques in brain and spinal surgery. Address: 3395 Scranton Rd Cleveland OH 44109 My professional work deals with the human brain, and I must treat its maladies through surgical and pharmacological intervention. These intimate experiences have convinced me of both the Divinity and the naturalness of man. For me, the very essence of what makes us what we are is centered in the tissues of the human brain. Operating and experimenting on this, the most complex organ of the body, has only intensified my respect for man and the acknowledgment that all of his (her) accomplishments have resulted through the processes of the single biological entity, the human brain.

WHITE, ROLFE DOWNING, medical educator; b. Annapolis, Md., Sept. 9, 1949; s. John and Jane (Angus) W.; m. Sarah Elizabeth Haughton, June 4, 1978; 1 dau., Kathryn Diane. B.S., Va. Mil. Inst., 1971; postgrad., George Washington U. Sch. Medicine, 1970; M.D., Med. Coll. Va., 1975. Diplomate: Nat. Bd. Med. Examiners, Am. Bd. Ob-Gyn. Resident in ob-gyn Naval Regional Med. Ctr., Portsmouth, Va., 1975-79; attending physician in gynecol. urology Nat. Naval Med. Ctr., Bethesda, Md., 1980-81; cons. in gynecol. urology Walter Reed Army Med. Ctr., Washington, 1980-81; chmn. dept. ob-gyn U.S. Naval Hosp., Patuxent River, Md., 1980-81; pres. Va. Urodynamics Labs., Virginia Beach, Va., 1982—; asst. prof. Eastern Va. Med. Sch., Norfolk, 1981—. Contbr. articles to profl. jours. Served to lt. comdr. USN, 1975-81. Recipient Physicians Recognition award AMA, 1983; Cert. of Achievement NASA, 1970. Fellow Am. Coll. Ob-Gyn, Am. Soc. Abdominal Surgeons, Southeastern Surg. Conf.; mem. Am. Fertility Soc., Am. Assn. Mil. Surgeons U.S., Am. Fedn. for Clin. Research. Republican. Episcopalian. Subspecialties: Obstetrics and gynecology; Gynecologic urology. Current work: Utilization of real-time ultrasonography for the evaluation of the lower urinary tract successfully adapted commonly used obstetric technology (fetal monitor systems) for electronic cystourethrometrics (bladder testing). Office: Va Urodynamics Labs 3386 Holland Rd Suite 205 Virginia Beach VA 23452

WHITE, STEPHEN HALLEY, biophysicist, educator; b. Wewoka, Okla., May 14, 1940; s. James Halley and Gertrude June (Wyatt) W.; m. Buff Ertl, Aug. 20, 1961; children: Saill, Shell, Storn, Sharr, Skye, Sunde. B.S., U. Colo., 1963; M.S., U. Wash., 1965, Ph.D., 1969; grad., Mgmt. Inst., U. Calif., 1981. Asst. prof. dept. physiology and biophysics U. Calif.-Irvine, 1972-75, assoc. prof., 1975-78, prof., 1978—, vice chmn., 1974-75, chmn., 1977—; guest assoc. physiologist Brookhaven Nat. Lab., Upton, N.Y., 1977—. Contbr. articles to profl. jours. Mem. adv. panel on molecular biology NSF. Served to capt. U.S. Army, 1969-71. Recipient Excellence in Teaching award Kaiser Permanente Found., 1975; Research Career Devel. award NIH, 1976; USPHS fellow, 1971. Mem. Biophys. Soc. (council 1981, exec. bd. 1981-83), Am. Physiol. Soc. (editorial bd. 1981—). Democrat. Subspecialties: Biophysics (biology); Physical chemistry. Current work: Structure and physical chemistry of cell membranes. Office: Dept Physiology U Calif Irvine CA 92717

WHITE, TIMOTHY PETER, physical education educator; b. Buenos Aires, Argentina, July 9, 1949; came to U.S., 1957, naturalized, 1969; s. Anthony Robert and Mary (Weston) W.; m. Nina Marie Kasper, Oct. 11, 1981. B.A. magna cum laude, Fresno State U., 1970; M.S., Calif. State U.-Hayward, 1972; Ph.D., U. Calif.-Berkeley, 1977; postgrad., U. Mich., 1976-78. Asst. prof. dept. phys. edn. U. Mich., Ann Arbor, 1978-82, assoc. prof., 1982-83; assoc. prof. dept. phys. edn. U. Calif.-Berkeley, 1983—. Fellow Am. Coll. Sports Medicine (New Investigator award 1981); mem. Am. Physiol. Soc., Phi Kappa Phi. Subspecialties: Physiology (biology). Current work: Research on the mechanisms by which exercise influences morphological, biochemical and physiological variables of normal and regenerating skeletal muscle. Home: 112 Patterson Blvd Pleasant Hill CA 94523 Office: Dept Physical Education University of California 103 Harmon Gym Berkeley CA 94720

WHITECAR, JOHN P., JR., medical oncologist; b. Phila., July 17, 1939; s. John P. and Patience (Daven) W.; m. Kathleen L. Hemelt, Dec. 3, 1966; children: Linnane Rene, John P. III, Michael Anthony, Colleen Jeanette. B.A. magna cum laude, LaSalle Coll., 1960; M.D., Jefferson Med. Coll., 1964. Diplomate: Am. Bd. Internal Medicine. Intern U. Minn. Hosp., Mpls., 1964-65, resident in medicine, 1965-66, 67-68; USPHS trainee in hematology 1966-67; clin. fellow Am. Cancer Soc., 1967-68; faculty assoc., chief leukemia and hematology sect., dept. devel. therapeutics U. Tex.-M.D. Anderson Hosp. and Tumor Inst., Houston, 1970-71; prin. investigator S.W. Oncology Group, San Antonio, 1971-74; clin. assoc. prof./dept. medicine U. Tex. Health Sci. Center, San Antonio, 1972-79, clin. prof. medicine (oncology), 1979—; prin. investigator clin. oncology S.W. Tex. Gen. Hosp., San Antonio, 1977-79; practice med. oncology S.W. Oncology Assocs. (P.A.), San Antonio, 1974—. Contbr. articles to profl. jours. Served with M.C., U.S. Army, 1971-74. Recipient Anatomy prize Jefferson Med. Coll., 1961, Henry M. Phillips prizes in medicine and surgery, 1964, Edward J. Moore prize, 1964. Fellow ACP; mem. N.Y. Acad. Scis., AAAS, Tex. Med. Assn., Bexar County Med. Soc., AMA, Am. Fedn. Clin. Research, Am. Soc. Hematology, Am. Assn. Cancer Research, Am. Soc. Clin. Oncology, Am. Soc. Internal Medicine, So. Med. Assn., San Antonio Club Internal Medicine (sec.-treas. 1980—), Alpha Omega Alpha. Subspecialties: Chemotherapy; Hematology. Current work: Clinical research in cancer therapy. Home: Route 1 Box 24 San Antonio TX 78023 Office: 1303 McCullough 348 San Antonio TX 78212

WHITEHEAD, FRANK ROGER, physicist, researcher; b. Biloxi, Miss., May 19, 1944; s. Thomas Frank and Willa (Robertson) W.; m. Ann Marie Tram, Apr. 25, 1971; children: Pamela, Brian. B.S. in Physics, Hamline U., 1966; M.S. in Optical Scis, U. Ariz., 1975, Ph.D., 1976. Prin. research physicist Searle Diagnostics, Des Plaines, Ill., 1973-79; systems engr. Gen. Electric Ultrasound, Rancho Cordova, Calif., 1979-81; staff scientist Siemens Gammasonics, Des Plaines, 1981-83; research scientist Sound Imaging, Folsom, Calif., 1983—; lectr. in field. Contbr. chpt., articles to profl. publs. Mem. Soc. Nuclear Medicine. Republican. Methodist. Subspecialties: Imaging technology; Electronics. Current work: Development of medical imaging systems; special interest in analysis and testing of imaging systems, performance and optimization of image display systems. Home: 8610 Gaines Ave Orangeville CA 95662 Office: 715 Sutter St Folsom CA 95630

WHITEHEAD, GEORGE WILLIAM, mathematician; b. Bloomington, Ill., Aug. 2, 1918; s. George W. and Mary (Gutschlag) W.; m. Kathleen E. Butcher, June 7, 1947. B.S., U. Chgo., 1937, M.S., 1938, Ph.D., 1941. Instr. Purdue U., 1941-45; mathematician Aberdeen Proving Ground, 1945; instr. Princeton U., 1945-47, vis. prof., 1958-59; asst. prof. Brown U., 1947-48, asso. prof., 1948-49; asst. prof. Mass Inst. Tech., 1949-51, asso. prof., 1951-57, prof. math., 1957—; vis. research fellow Birkbeck Coll., U. London, 1973, U. London, Oxford U., 1981. Author: Homotopy Theory, 1965, Recent Advances in Homotopy Theory, 1970, Elements of Homotopy Theory, 1978. Guggenheim fellow, Fulbright research scholar, 1955-56; NSF sr. postdoctoral fellow, 1965-66. Fellow Am. Acad. Arts and Scis.; mem. Am. Math. Soc. (v.p. 1978-79), Math. Assn. Am., London Math. Soc., Nat. Acad. Scis., Phi Beta Kappa, Sigma Xi. Subspecialty: Topology. Current work: Homotopy theory; particularly stable homotopy and generalized homology theories. Home: 25 Bellevue Rd Arlington MA 02174 Office: Room 2-284 Mass Inst Tech Cambridge MA 02139

WHITEHURST, DARRELL DUAYNE, organic chemist research and development company executive; b. Vernon, Ill., July 8, 1938; m., 1967; 2 children. A.B., Bradley U., 1960; M.S., U. Iowa, 1963; Ph.D. in Organic Chemistry, 1964. Research chemist Mobil Research & Devel. Corp., Princeton, N.J., 1964-65, sr. research chemist, 1965-68, group leader catalysis, 1968-73, research assoc., 1974-75, prin. investigator EPRI contracts, 1975—, group mgr. coal and heavy liquids research, 1980-83, research scientist, 1983—; mem. com. task force motor fuel and photochemistry smog Am. Petroleum Inst.; research assoc. BPRI, 1973, prin. investigator fundamental coal chemistry study, 1975. Mem. Am. Chem. Co. (Richard A. Glen award, Henry H. Storch award). Subspecialty: Organic chemistry. Office: Mobil Research & Devel Corp Princeton NJ 08540

WHITESELL, JAMES JUDD, science educator, researcher; b. Phila., Oct. 14, 1939; s. Melvin Winfield and Dorothy Percival (Jackson) W.; m. Tallulah Long, June 26, 1965; 1 son, William Long. B.S., Dickinson Coll., 1962; M.Ed., U. Fla., 1967, M.S., 1969, Ph.D., 1974. Tchr. Rickards Jr. High Sch., Ft. Lauderdale, Fla., 1963-67; research assoc. U. Fla., Gainesville, 1972-74; instr. Snead State Jr. Coll., Boaz, Ala., 1974-76; assoc. prof. sci. edn. Valdosta (Ga.) State Coll., 1976—, cross country coach, 1980—; cons. Tactical Air Command, Langley AFB, Va., 1982—. Contbr. articles to profl. jours. Office Naval Research grantee, 1981-82. Mem. Entomol. Soc. Am., Fla. Entomol. Soc., Sigma Xi, Phi Sigma, Gamma Sigma Delta. Republican. Methodist. Club: Valdosta Track. Lodge: Elks. Subspecialties: Behavioral ecology; Evolutionary biology. Current work: Using insects as sensors of infiltrators, elimination of cattle egret bird/airstrike hazard. Home: 2123 Pinecliff Dr Valdosta GA 31601 Office: Science Edn Valdosta State Coll N Patterson St Valdosta GA 31698

WHITESIDE, JACK OLIVER, plant pathologist; b. Barnstaple, Eng., June 5, 1928; s. Albert John and Edith Fredrika W.; m. Nora Irene Larsen, Aug. 4, 1951; children: Susan Erica, Lynne Judy. B.S., U. London, 1948, Ph.D., 1953. Plant pathologist Ministry Agr., Zimbabwe, Africa, 1953-67; plant pathologist Citrus Agrl. Research and Edn. Ctr., U. Fla., Lake Alfred, 1967—. Mem. Am. Phytopathological Soc., Internat. Soc. Citriculture, Fla. Hort. Soc. Subspecialty: Plant pathology. Current work: Epidemiology and control of fungal diseases of citrus fruits and trees. Office: 700 Experiment Station Rd Lake Alfred Fl 33850

WHITESIDES, GEORGE MCCLELLAND, chemistry educator; b. Louisville, Aug. 3, 1939; s. George Walter and Sarah Elizabeth (McClell) W.; m. Barbara Breasted, June 28, 1970; children: George Thomas, Benjamin Hull. B.A., Harvard U., 1960; Ph.D., Calif. Inst. Tech., 1964. With Mass. Inst. Tech., Cambridge, 1963-82, prof. chemistry, 1971-82, A.C. Cope prof., 1977-82; prof. chemistry Harvard U., 1982—; Alfred P. Sloan fellow, 1968. Contbr. articles to profl. jours. Mem. Nat. Acad. Scis., Am. Acad. Arts and Scis., Am. Chem. Soc. (award in pure chemistry 1975), AAAS. Subspecialty: Organic chemistry. Home: 124 Grasmere St Newton MA 02158 Office: Dept Chemistry Harvard U 12 Oxford St Cambridge MA 02138

WHITFIELD, JACK DUANE, technology company executive, consultant; b. Paoli, Okla., May 16, 1928; s. Lloyd H. and Ethel (Wigley) W.; m. Marcheta Rae Steward, Sept. 11, 1949; children: Donna, Jeffrey, Karen. B.S.A.E., U. Okla., 1951; M.S.M.E., U. Tenn., 1960; D.Sc., Royal Inst. Tech., Stockholm, 1972. Registered profl. engr., Tenn. Dir. von Karman Gas Dynamics Facility, Arnold Engring. Devel. Center, 1968-75, dir., 1975-76; exec. v.p. Sverdrup/ARO, Inc. and successor firm, Sverdrup Tech., Inc., Tullahoma, Tenn., 1974-81, pres., dir., 1981—; dir. Sverdrup & Parcel & Assocs., Sverdrup & Parcel Constrn. Mgmt., Sverdrup Corp., Nine Ten North Eleventh, Hotel Equity, Ltd., Conv. Plaza, W., Sverdrup & Parcel, Inc. Real Estate; v.p., corp. prin. advanced tech. Reviewer: Heat Transfer and Fluid mechanics. Recipient Gen. H.H. Arnold award, 1968. Assoc. fellow AIAA (Simulation and Ground Testing award 1979); mem. Sigma Xi, Tau Beta Pi, Tau Omega. Subspecialties: Fluid mechanics; Aeronautical engineering. Current work: Design and development of environmental ground test facilities, executive corporate management, project direction on major design or developement projects, advanced planning for corporate advanced technology programs. Home: Route 2 Box 150A Wartrace TN 37183 Office: 600 William Northern Blvd Tullahoma TN 37388

WHITMAN, PATRICK GENE, physicist, educator; b. Beaumont, Tex., Sept. 30, 1944; s. Norman L. and Roberta M. (Wright) W.; m. Darla J. Mason, Aug. 25, 1973; children: Jeremy A., Leslie A. B.Sc., Lamar Inst. Tech., 1970; M.Sc. in Math., Lamar U., 1971; postgrad., U. Tex.-Dallas, 1971-74; Ph.D., North Tex. State U., 1978. Asst. prof. physics Benedictine Coll., Atchison, Kans., 1978-79, U. Southwestern La., Lafayette, 1979—, mem. grad. council, 1981—. Contbr. articles to profl. jours. Active local programs to rehabilitate handicapped. Welch Found. predoctoral fellow, 1976-77. Mem. Am. Inst. Physics, Am. Phys. Soc., Am. Astron. Soc. Democrat. Subspecialties: Relativity and gravitation; Theoretical astrophysics. Current work: Exact solutions in relativity/close binary star systems; equations of state in ultra-high density regime; exact solutions to field equations for reasonable equations of state; mass flow in binaries.

WHITMAN, ROBERT LESLIE, research elec. engr.; b. Kansas City, Mo., Aug. 8, 1933; s. Leslie R. and Laura W.; m. Marge Lassen, May 30, 1958; 1 dau., Julie. B.S.E.E., U. Ill., 1961, postgrad. in physics, Ill. Inst. Tech., 1954-68. Design engr. Motorola Co., 1954-55, design engr., 1957-59; microwave design engr. Raytheon Co., 1960-61; research engr. Zenith Radio Corp., Northbrook, Ill., 1961-78; mgr. advanced devel. Extel Corp., Northbrook, 1978—. Contbr. articles on optics and acousto-optics to profl. jours. Served with U.S. Army, 1955-57. Mem. IEEE, Optical Soc. Am., Soc. Photo-Optical Instrumentation Engrs., Soc. Photographic Scientists and Engrs. Subspecialties: Optical signal processing; Acoustics. Current work: Optics, acousto-optics, acoustic imaging engineering, electrical, applied magnetics. Patentee optics, acousto-optics, video disc fields. Office: 4000 Commercial St Northbrook IL 60062

WHITMAN, ROBERT VAN DUYNE, civil engr.; b. Pitts., Feb. 2, 1928; s. Edwin A. and Elsie (Van Duyne) W.; m. Elizabeth Cushman, June 19, 1954; children—Jill Martyne, Martha Allerton (dec.), Gweneth Giles. B.S., Swarthmore Coll., 1948; S.M., Mass. Inst. Tech., 1949, Sc.D., 1951. Mem. faculty Mass. Inst. Tech., 1953—, prof. civil engring., 1963—, head structural engring., 1970-74, head soil mechanics div., 1970-72; vis. scholar U. Cambridge, Eng., 1976-77; cons. to govt. and industry, 1953—. Author: (with T. W. Lambe) Soil Mechanics. Mem. Town Meeting Lexington, Mass., 1962-76, mem. permanent bldg. com., 1968-75, mem. bd. appeals, 1979-81. Served to lt. (j.g.) USNR, 1954-56. Mem. Am. Soc. Civil Engrs. (Research award, 1962, Terzaghi Lecture 1981), Boston Soc. Civil Engrs. (Structural sect. prize 1963, Desmond Fitzgerald medal 1973, Ralph W. Horne Fund award 1977), Internat. Soc. Soil Mechanics and Found. Engrs., Seismol. Soc. Am., Nat. Acad. Engring., Earthquake Engring. Research Inst. (dir. 1978-81, v.p. 1979-81). Subspecialty: Civil engineering. Current work: Earthquake engineering; permanent deformations of soil; seismic hazard analysis. Research in soil mechanics, soil dynamics and earthquake engring. Home: 9 Demar Rd Lexington MA 02173 Office: Mass Inst Tech Cambridge MA 02139

WHITMER, JEFFREY THOMAS, pediatric cardiologist, researcher; b. Butler, Pa., Mar. 22, 1948; s. Lewis Lloyd and Helen Genieve (Shannon) W.; m. Kyra Marie Riegle, June 28, 1975. B.S., Pa. State U., 1970, Ph.D., 1974, M.D., 1977. Intern, then resident Upstate Med. Ctr., Syracuse, N.Y.; grad. asst. in physiology Pa. State U., Hershey, 1970-74; asst. instr. pediatrics SUNY-Syracuse, 1977-79; fellow in pediatrics cardiology div. cardiology Children's Hosp., Cin., 1979-82, cardiologist, asst. prof. pediatrics, 1982—. USPHS fellow, 1982; recipient Clin. Scientist award Am. Heart Assn., 1983, Clin. Investigator award NIH, 1983. Fellow Am. Acad. Pediatrics; mem. AAAS, Am. Heart Assn., Am. Fedn. Clin. Research. Republican. Presbyterian. Subspecialties: Cardiology; Biochemistry (medicine). Current work: Characterization of glucose and fatty acid metabolism in cardiomyopathic hearts; with this information available, attempt to develop new clinical treatment regimens. Office: Children's Hospital Division of Cardiology Elland Ave and Bethesda Ave Cincinnati OH 45229

WHITNEY, HASSLER, mathematics educator; b. N.Y.C., Mar. 23, 1907; m., 1930 and 1955; 2 children. Ph.B., Yale U., 1928, Mus.B., 1929; Ph.D., Harvard U., 1932; Sc.D. (hon.), Yale U., 1947. Instr. math. Harvard U., 1930-31, NRC lectr. and fellow, 1931-33, from instr. to prof., 1933-52, prof., 1952-77; Emer prof. math. Inst. Advanced Study, Princeton, N.J., 1977—. Mem. math. panel Nat. Def. Research Com., 1943-45. Recipient Wolf prize, 1982; Nat. medal of Sci., 1976. Mem. Nat. Acad. Sci., Am. Math. Soc. (v.p. 1948-50), Am. Philos. Soc. Subspecialty: Topology. Office: Sch Math Inst Advanced Study Princeton NJ 08540

WHITNEY, JAMES MARTIN, materials engineer, researcher; b. Owosso, Mich., Sept. 6, 1936; s. Karl Marcy and Mary Elizabeth (Hines) W.; m. Phyllis Ann, May 4, 1963; children: Thomas J., Sharon J., Jennifer A., Douglas A., Laura E. B.A., Ill. Coll., 1959; B.T.E., Ga. Inst. Tech., 1959, M.T.E., 1961; M.S., Ohio State U., 1964, Ph.D. in Engring. Mechs, 1968. With U.S. Air Force, 1961—, materials research engr., Wright-Patterson AFB, Ohio, 1961—; lectr. Ohio State U., 1969-70; adj. prof. Air Force Inst. Tech., 1971-73; Wright State U., 1968-69; cons. in field. Sr. author: Experimental Mechanics of Fiber Reinforced Composite Materials, 1982; contbr. numerous articles on mechanics of fiber reinforced composite materials to profl. jours. Cubmaster Pioneer council Boy Scouts Am., 1978-80; pres. edn. commn. St. Helen Sch., Dayton, Ohio, 1981—. Fellow AIAA (asso.), ASTM (chmn. com. on high modulus fibers and their composites 1976-81, award of appreciation 1982, award of merit 1983); mem. ASME, Soc. Exptl. Stress Analysis, Am. Acad. Mechanics (founder mem.), Soc. Engring. Sci. Roman Catholic. Subspecialties: Composite materials; Solid mechanics. Current work: Interested in mechanics of composite materials with emphasis on plates and shells, test methods, fracture, and fatigue. Home: 4371 Round Tree Dr Beavercreek OH 45432 Office: Air Force Wright Aero Labs MLB Wright-Patterson AFB OH 45433

WHITTEN, CHARLES ALEXANDER, JR., educator; b. Harrisburg, Pa., Jan. 20, 1940; s. Charles Alexander and Helen (Shoop) W.; m. Joan Emann, Nov. 20, 1965; 1 son, Charles Alexander. B.S. summa cum laude, Yale U., 1961; Ph.D. in Physics, Princeton U., 1966. Research asso. A.W. Wright Nuclear Structure Lab., Yale U., 1966-68; asst. prof. physics UCLA, 1968-74, asso. prof., 1974-80, prof., 1980—; vis. scientist Centre d'Etudes Nucléaires de Saclay-Moyenne Energie, 1980—. Contbr. articles to profl. jours. Mem. Am. Phys. Soc., Sigma Pi Sigma, Phi Beta Kappa. Subspecialties: Nuclear physics; Particle physics. Current work: Intermediate energy physics which uses high energy particles as probes of nuclear structure; polarized proton-nucleus scattering; small angle nucleon-nucleon scattering. Home: 9844 Vicar St Los Angeles CA 90034 Office: Dept Physics U Calif 405 Hilgard Ave Los Angeles CA 90024

WHITTEN, ERIC HAROLD TIMOTHY, geologist, university official; b. Ilford, Essex, Eng., July 26, 1927; came to U.S., 1958; s. Charles Alexander and Muriel Gladys (Smith) W.; m. Mary Cleopha Staciva, Feb. 28, 1976; children: Catherine, Peter, Jennifer, Adam, Joshua. B.Sc. with gen. honors, U. London, 1948, 1948, Ph.D., 1952, D.Sc., 1968. Clk. Rex Thomas (Ins.) Ltd., London, 1943-45; lectr. in geology Queen Mary Coll., London U., 1948-58; asso. prof. geology Northwestern U., 1958-63, prof., 1963-81, chmn. dept. geol. scis., 1977-81; v.p. for acad. affairs Mich. Technol. U., Houghton, 1981—. Author: Structural Geology of Folded Rocks, 1966, Quantitative Studies in the Geological Sciences, 1975. Geol. Soc. London Daniel Pidgeon Fund grantee, 1954. Fellow Geol. Soc. London, Geol. Soc. Am., AAAS; mem. Internat. Assn. Math. Geology (pres.), Geologists' Assn. (London), Queen Mary Coll. Student Union Soc. (hon. life), English Speaking Union (life). Clubs: Hope Town Sailing (Abaco, Bahama Islands); Onigaming Yacht (Houghton). Current work: Quantitative variability and petrogenetic evolution of the granitoids of the Lachlan Fold Belt, S.E. Australia. Office: Mich Technol U Houghton MI 49931

WHITTIER, DEAN PAGE, biology educator; b. Worcester, Mass., July 2, 1935; s. George Albert and Marian Louise (Johnson) W.; m. Virginia Elaine Johnson, June 7, 1958; children: Karen Susan, Ethan Dean. B.S., U. Mass., 1957; M.A., Harvard U., 1959, Ph.D., 1961. Asst. prof. biology Va. Poly. Inst. and State U., 1961-64; research fellow Harvard U., 1964-65; asst. prof. biology Vanderbilt U., 1965-68, assoc. prof. biology, 1968-77, prof. biology dept. gen. biology, 1977—. Mem. Bot. Soc. Am., Am. Fern Soc., Internat. Soc. Plant Morphologists, Assn. Southeastern Biologists. Subspecialties: Plant growth; Plant cell and tissue culture. Current work: Morphogenesis of pteridophytes. Office: Dept Gen Biology Vanderbilt U Nashville TN 37235

WHITWORTH, GARY WILLIAM, psychologist, consultant; b. Corpus Christi, Tex., Sept. 7, 1951; s. William Edmond and Doris Jean (Ware) W.; m. Patricia Maureen Gray, Aug. 18, 1973 (div. Feb. 1976); m. Francisca Longoria, July 21, 1978; children: Kevin Scott, Chelsea Marie. B.A., Tex. A&I U., 1974, M.S., 1975; Ph.D., U. Tex.-Austin, 1981. Lic. psychologist, Tex. Unit psychologist Corpus Christi (Tex.) State Sch., 1973-78; teaching asst. U. Tex.-Austin, 1979-80; legis. aide Tex. Ho. of Reps., Austin, 1979; residential dir. South Tex. Ind. Sch. Dist., Edinburg, 1981-82; pvt. practice psychology, McAllen, Tex., 1982—. Author: An Information Processing Analysis of Anxiety, 1981; author: copyrighted technique The Good Behavior Ladder, 1976; editor: Sch. Psychology Newsletter, 1982—. Hammond Found. fellow, 1973; NIMH trainee, 1979-80; U. Tex.-Austin grantee, 1981. Mem. Am. Psychol. Assn., Tex. Psychol. Assn., Assn. Advancement Behavior Therapy, Rio Grande Valley Psychol. Assn. (charter). Democrat. Subspecialties: Behavioral psychology; Learning. Current work: Information-processing models of psychology; cognitive behavior therapy; child psychology. Home: 332 Hollywood Dr Edinburg TX 78539 Office: Gary W Whitworth PhD 1111 N Tenth St McAllen TX 78501

WIBERG, JOHN SAMUEL, molecular geneticist; b. Plaistow, N.H., Dec. 4, 1930; s. Hugo William and Mary Josephine (Loeffler) W.; m. Elsie Nelson, Nov. 22, 1952; children: Kristina, Karl, Derek. B.S., Trinity Coll., 1952; Ph.D., U. Rochester, 1956. USPHS postdoctoral fellow M.I.T., 1959-60, research asso., 1961-63; asst. prof. radiation biology and biophysics U. Rochester, 1963-70, asso. prof., 1970—. Contbr. articles to sci. jours.; Editorial bd.: Jour. of Virology, 1970-79. Served with USAF, 1957-58. Mem. Am. Chem. Soc., AAAS, Am. Soc. Biol. Chemists, Genetics Soc. Am., Am. Soc. Microbiology, Fedn. Am. Scientists. Subspecialties: Molecular biology; Virology (biology). Current work: Regulation of protein synthesis by bacteriophage; nucleic acid metabolism. Home: 8 Woodside Ln Pittsford NY 14534 Office: RBB Dept U Rochester Med Sch Rochester NY 14642

WIBERG, KENNETH BERLE, educator, chemist; b. Bklyn., Sept. 22, 1927; s. Dan and Solveig Berle W.; m. Marguerite Koch, Mar. 18, 1951; children—Patricia, Robert, William. B.S., Mass. Inst. Tech., 1948; Ph.D., Columbia, 1950. Mem. faculty U. Wash., 1950-62, prof. chemistry, 1958-62, Yale, 1962—, Whitehead prof., 1968—, chmn. chem. dept., 1968-71; Boomer lectr. U. Alta., Can., 1959; Mem. chemistry advisory panel Air Force Office Sci. Research, 1960-66, NSF, 1965-68, chmn., 1967-68. Author: Laboratory Technique in Organic Chemistry, 1960, Interpretation of NMR Spectra, 1962, Physical Organic Chemistry, 1964, Computer Programming for Chemists, 1966, also articles.; Editor: Oxidation in Organic Chemistry, 1966, Sigma M.O. Theory, 1970; mem. editorial bd.: Organic Syntheses, 1963-71, Jour. Organic Chemistry, 1968-72; bd. editors: Jour. Am. Chem. Soc, 1969-72. Recipient James Flack Norris award, 1973; Sloan Found. fellow, 1958-62; Guggenheim fellow, 1961-62. Mem. Nat. Acad. Scis., Am. Acad. Arts and Scis., Am. Chem. Soc. (exec. com. organic div. 1961-63, Calif. sect. award 1963). Subspecialties: Organic chemistry; Theoretical chemistry. Current work: Small ring chemistry; thermochemistry; spectroscopy and theoretical calculations related to the above. Home: 160 Carmalt Rd Hamden CT 06517 Office: Chemistry Dept Yale U New Haven CT 06520

WIBERLEY, STEPHEN EDWARD, chemistry educator, consultant; b. Troy, N.Y., May 31, 1919; s. Irving Charles and Ruth (Stanley) W.; m. Mary Elizabeth Bartle, Feb. 21, 1942; children: Stephen Edward, Sharon Elizabeth. B.A., Williams Coll., 1941; M.S., Rensselaer Poly. Inst., 1948, Ph.D., 1950. Sr. chemist Congoleum Nairn, Inc., Kearny, N.J., 1941-44; prof. chemistry Rensselaer Poly. Inst., Troy, N.Y., 1946-63, 79—, acting chmn. chemistry, 1983—, dean, 1964-79, vice provost, 1969-79; vis. sr. physicist Brookhaven Nat. Lab., Upton, N.Y., 1952; cons. Imperial Paper & Color Corp., Glens Falls, N.Y., 1957—, Socony Mobil Oil Co., Bklyn., 1957—, Huyck Felt Co., Rensselaer, N.Y., 1958—, Schenectady Chems. Inc., 1961—, U.S. Gypsum Co., Buffalo, 1962—. Author: Instrumental Analysis, 1954, Laboratory Manual for General Chemistry, 1963, Introduction to Infrared and Raman Spectroscopy, 1964. Served with AUS, 1944-46. Mem. Am. Chem. Soc., AAUP, Sigma Xi, Phi Lambda Upsilon. Subspecialties: Analytical chemistry; Oil shale. Current work: Synthetic fuels from coal and oil shale; reduction of iron oxides. Home: 1676 Tibbits Ave Troy NY 12180

WICKELGREN, WAYNE ALLEN, psychology educator, researcher; b. Hammond, Ind., June 4, 1938; s. Herman and Alma Emelia (Larson) W.; m. Barbara Sue Gordon, June 10, 1962 (div. mar. 1972); children: Ingrid, Abe, Peter Kirsten. A.B., Harvard U., 1960; Ph.D., U. Calif.-Berkeley, 1962. Asst. prof. MIT, Cambridge, 1962-67, assoc. prof., 1967-69, prof., 1969; prof. psychology U. Oreg., Eugene, 1969—. Author: How to Solve Problems, 1974, Learning and Memory, 1977, Cognitive Psychology, 1979. NIMH grantee, 1963-72; NIH grantee, 1972-80; NSF grantee, 1972-82. Fellow AAAS; mem. Psychonomic Soc., Soc. Math. Psychology, IEEE Computer Soc., Assn. Computing Machinery. Subspecialties: Programming languages; Artificial intelligence. Current work: Programming languages, artificial intelligence, cognition, learning. Office: U Oreg Psychology Dept Eugene OR 97403

WICKER, ED FRANKLIN, research plant pathologist; b. Upper Tygart, Ky., Aug. 21, 1930; s. Leslie and Bessie Mae (Hamilton) W.; m. Veneta Carol Law, Dec. 20, 1953; children: Cynthia, Sonja. B.S. in Forestry, Wash. State U., Pullman, 1959, Ph.D. in Plant Pathology, 1965. Research forester Intermountain Forest and Range Expt. Sta., USDA Forest Service, Spokane, Wash., 1959-63, plant physiologist, 1963-78; staff research plant pathologist USDA Forest Service, Washington, 1978-82; asst. dir. Rocky Mountain Forest and Range Expt. Sta., Fort Collins, Colo., 1982—; vis. scientist sch. botany Cambridge (Eng.) U., 1970-71. Contbr. articles to profl. jours. Served with USAF, 1950-54. Recipient Research award Govt. of Japan, 1974. Mem. Am. Phytopathol. Soc., Soc. Am. Foresters, Mycol. Soc. Am. Subspecialties: Plant pathology; Resource conservation. Current work: Biology, ecology, and control of dwarf mistletoes; biology of conifer stem rusts; biological control of forest tree diseases. Home: 4118 Attleboro Ct Fort Collins CO 80525 Office: 240 W Prospect St Fort Collins CO 80526

WICKERSHEIM, KENNETH ALAN, solid state physicist, corporate executive; b. Fullerton, Calif., Mar. 4, 1928; m., 1952, 67; 2 children. A.B., UCLA, 1950, M.A., 1956, Ph.D. in Physics, 1959. Mem. staff Los Alamos Sci. Lab., 1953-55; asst. in physics UCLA, 1955-58; mem. staff research labs. Hughes Aircraft Co., 1958-61; research physicist Palo Alto Labs. Gen. Tel. & Electronics Lab., Inc., 1961-63; assoc. prof. materials sci. Stanford U., 1963-64; staff scientist in solid state physics Lockheed Palo Alto Research Labs., 1964-65, sr. mem., 1965-66, head advanced electronics, 1966-70; pres. Spectrotherm Corp., 1970-75; v.p. research and devel. UTI Corp., 1975-77; v.p., gen. mgr. Quantex Corp., 1977-78; pres. Luxtron Corp., 1978—; cons. in field. Fellow Am. Phys. Soc.; mem. AAAS. Home: 3895 Middlefield Rd Palo Alto CA 94303

WICKES, WILLIAM CASTLES, research and devel. project mgr., physicist, writer; b. Lynwood, Calif., Nov. 25, 1946; s. William Hopkins and Nancy Rose (Castles) W.; m. Susan Jane Monroe, Feb. 13, 1971; children: Kenneth William, Lara Kathleen. B.S., UCLA, 1967; M.A., Princeton U., 1969, Ph.D. in Physics, 1972. Asst. prof. physics Princeton U., 1972-77; asst. prof. U. Md., 1977-81; research and devel. project mgr. Hewlett-Packard Corp., Corvallis, Oreg., 1981—. Author: Synthetic Programming on the HP-41C, 1980. Mem. Am. Astron. Soc. Subspecialties: Mathematical software; Software engineering. Current work: Directing sci. software devel. for portable computers. Home: 4517 NW Queens Ave Corvallis OR 97330 Office: 1000 NE Circle Blvd Corvallis OR 97330

WICKRAMASEKERA (IAN EDWARD WICKRAM), psychiatry and behavioral sciences educator, clinical psychologist and psychophysiologist; b. Colombo, Ceylon, Oct. 23, 1938; came to U.S., 1959, naturalized, 64; s. Harry Stanley and Maude M. (Robinson) W.; m.; children: Melissa, Edward. B.A., Friend's U., 1961; student, London U., U.K., 1956-58; M.A., Roosevelt U., 1965; Ph.D., U. Ill., 1969. Diplomate: Am. Bd. Clin. Psychology, Am. Bd. Psychol. Hypnosis. Intern East Moline (Ill.) State Hosp., 1964-65, staff psychologist, 1965-66; research asst. Children's Research Center, Urbana, Ill., 1966-67; staff psychologist Peoria (Ill.) Mental Health Clinic, 1969-75; assoc. prof. psychiatry and behavioral medicine U. Ill. Med. Sch., Peoria, 1971-81; prof. psychiatry Eastern Va. Med. Sch., Norfolk, 1981—; cons. VA, Iowa City, 1974-80, Chgo., 1978-79. Author, editor: Biofeedback Behavior Therapy and Hypnosis, 1976; also articles; mem. editorial bd.: Behavior Therapy, 1975-78. Clin. fellow Behavior Therapy and Research Soc.; fellow Am. Psychol. Assn.; mem. AAAS, Ill. Psychol. Assn. (pres. 1980), Biofeedback Soc. Am. (dir. 1978-80). Roman Catholic. Subspecialties: Psychophysiology; Psychobiology. Current work: Risk factors for psychophsiological disorders placebo effect, hypnosis. Office: Eastern VA Med Sch 700 Olney Rd PO Box 1980 Norfolk VA 23501 Home: 922 W Ocean View Norfolk VA 23503

WICKSTROM, ERIC, biophysical chemist, biochemistry educator; b. Chgo., Dec. 21, 1946; s. Lester Eric and Lillian (Partnoy) W.; m. Lois June Sinsheimer, July 1, 1967; children: Erica Lorraine, Eileen Anitra. B.S. in Biology with honors, Calif. Inst. Tech., 1968; Ph.D. in Chemistry, U. Calif.-Berkeley, 1972. Research asst. U. Calif.-Berkeley, 1968-72; research assoc. U. Colo., Boulder, 1973-74; asst. prof. U. Denver, 1974-81; sr. research scientist So. Biotech, Inc., Tampa, Fla., 1981-82; vis. assoc. prof. chemistry U. South Fla., Tampa, 1982-83, asst. prof., 1983—. Contbr. articles to profl. jours. NSF research fellow, 1968-71; NIH grantee, 1976-81,83—. Mem. Am. Soc. Biol. Chemists, Biophys. Soc., Am. Chem. Soc., AAAS. Democrat. Jewish. Subspecialties: Biophysical chemistry; Molecular biology. Current work: Control of gene expression through messenger RNA sequence and secondary structure. Home: 10612 Altman St Tampa FL 33612 Office: Dept Chemistry Univ South Fla Tampa FL 33620

WIDDOES, LAWRENCE CURTIS, oil service company executive, consultant; b. Spokane, Wash., Nov. 10, 1919; s. Curtis E. and Hazel A. (North) W.; m. Marcella Elizabeth Maize, Feb. 28, 1942; children: Bonni, Patricia, Lawrence. B.S., Calif. Inst. Tech., 1941; M.S.E., U. Mich., 1947. Registered profl. engr., Calif., N.Y., Mo., Tex. Reseach engr. Nat. Dairy Co., Idlewild, N.Y., 1947-50; project mgr. Monsanto Co., St. Louis, 1950-56; pres. Internuclear Co., Clayton, Mo., 1956-60; dir. research Petrolite Corp., Webseter Groves, Mo., 1960-66; pres. Conresco Corp., Stamford, Conn., 1966-72; v.p. devel. Magna Corp., Houston, 1972—. Served to lt. comdr. USN, 1942-46; PTO. Fellow Am. Inst. Chem. Engrs. Republican. Club: Petroleum (Houston). Subspecialties: Fuels; Petroleum engineering. Current work: Development of enhanced oil recovery techniques and application thereof. Home: 5877 Sugar Hill Houston TX 77057 Office: PO Box 33387 Houston TX 77233

WIDERA, G.E.O., mechanical engineering educator, consultant; b. Dortmund, Germany, Feb. 16, 1938; came to U.S., 1950; s. Otto and Gertrude (Yzermann) W.; m. Kristel Kornas, June 21, 1974; 1 dau., Erika. B.S., U. Wis., 1960, M.S., 1962, Ph.D., 1965. Asst. prof., then prof. materials engring. dept. U. Ill.-Chgo., 1965-82, prof. mech. engring., 1982—, head dept., 1983—; gastdozent U. Stuttgart, W.Ger., 1968; vis. prof. U. Wis.-Milw., 1973-74, Marquette U., Milw., 1978-79; cons. Ladish Co., Cudahy, Wis., 1967-76, Howmedica, Inc., Chgo., 1972-75, Sargent & Lundy, 1970—, Nat. Bur. Standards, 1980; vis. scientist Argonne Nat. Lab., Ill., 1968. Editor: Procs. Innovations in Structural Engring, 1974; assoc. editor: Pressure Vessel Tech, 1977-81; editorial bd.: Pressure Vessels and Piping Design Technology, 1982; tech. editor: Pressure Vessel Technology, 1983—. Standard Oil Co. Calif. fellow, 1961-63; NASA fellow, 1966; Nat. Acad. Scis. travel grantee, Russia, 1972; von Humboldt fellow, W.Ger., 1968-69. Mem. ASME (chmn. (research com. pressure vessels 1982—, chmn. design and analysis com. pressure vessel and piping div. 1980-83, chmn. jr. awards com. applied mechanics div. 1973-76, chmn. machine design div. of Chgo. sect. 1967-68, editor newsletter 1971-73, exec. com. Chgo. sect. 1970-73), ASCE (sec.-treas. structural div. of Ill. sect. 1972-73, chmn. div. 1976-77), Internat. Assn. Dental Research, Gesellschaft für Angewandte Mathematik und Mechanik, Am. Acad. Mechanics. Subspecialties: Solid mechanics; Composite materials. Current work: Mechanics of composite materials, plates and shells, asymptotic methods in elasticity, pressure vessels and piping, mechanics of deformation processing. Home: 345 Greenleaf Wilmette IL 60091 Office: PO Box 4348 Chicago IL 60680

WIDEY, ROBERT LEROY, astrophysicist, educator, cons.; b. Los Angeles, Aug. 22, 1934; s. Charles D. and Lillian F. (Houts) W.; m. Diana H. Skolfield, June 27, 1959; children: Robert B., Wendy C., Herbert C. B.S., Calif. Inst. Tech., 1957, M.S., 1958, Ph.D., 1962. Research engr. Jet Propulsion Lab., Calif. Inst. Tech., 1959-60; research fellow in astronomy and geology Mt. Wilson and Palomar Obs. and Calif. Inst. Tech., 1962-65; astrophysicist U.S. Geol. Survey, 1965—; faculty No. Ariz. U., Flagstaff, 1971—, prof. astrophysics and astronomy, 1979—; vis. prof. U. Calif., Berkeley; cons. Nat. Acad. Scis. Contbr. articles to profl. jours. Phi Kappa Phi scholar, 1978; Cert. of Appreciation NASA. Fellow Am. Geophys. Union, Geol. Soc. Am., Royal Astron. Soc.; mem. Internat. Astron. Union, Am. Astron. Soc., Nat. Rifle Assn., C.A.P. Zen Buddhist. Subspecialties: Infrared optical astronomy; Remote sensing (geoscience). Current work: Extragalactic, stellar, planetary and geodetic astronomy, derivation of planetary surface topography from radar back-scatter imagery. Home: 414 E Cherry Ave Flagstaff AZ 86001 Office: Physics Dep No Ariz U Flagstaff AZ 86011

WIDNALL, SHEILA EVANS, aeronautics and astronautics educator, consultant; b. Tacoma, July 13, 1938; d. Rolland John and Genievieve (Krause) Evans; m. William Soule Widnall, June 11, 1960; children: Bill, Ann. B.S., MIT, 1960, M.S., 1961, Sc.D., 1964; Ph.D. (hon.), New Eng. Coll., 1975. Asst. prof. aeros. and astronautics MIT, 1964-70, assoc. prof., 1970-74, prof., 1974—; dir. univ. research Dept. Transp., Washington, 1974-75; cons. in field; mem. engring. adv. com. NSF, 1981—. Contbr. numerous articles to profl. publs.; assoc. editor: Jour. Aircraft, 1972-75, Physics of Fluids, 1981—, ASME Jour. Applied Mechanics, 1983—. Bd. visitors Air Force Acad., 1978—. Recipient Outstanding Achievement award Soc. Women Engrs., 1975. Fellow Am. Phys. Soc. (exec. com. div. fluid dynamics 1979-82), AAAS (dir. 1982—), AIAA (assoc., Lawrence Sperry Achievement award 1972, dir. 1975-77). Club: Seattle Mountaineers. Subspecialties: Aeronautical engineering; Fluid mechanics. Current work: Aerodynamics, fluid mechanics, acoustics, industrial aerodynamics, noise and vibration, turbulence and flow stability. Office: MIT 77 Massachusetts Ave Cambridge MA 02139

WIDOM, BENJAMIN, physical chemist, educator, researcher; b. Newark, Oct. 13, 1927; m., 1953; 3 children. A.B., Columbia U., 1949; Ph.D. in Chemistry, Cornell U., 1953. Research assoc. in chemistry U. N.C., 1952-54; instr. Cornell U., 1954-55, from asst. prof. to assoc. prof., 1955-63, prof. chemistry, 1963—; van der Waals prof. U. Amsterdam, 1972; IBM prof. Oxford U., 1978. Guggenheim and Fulbright fellow, 1961-62; NSF sr. fellow, 1965; Guggenheim fellow, 1969. Mem. Nat. Acad. Sci., Am. Phys. Soc., Am. Chem. Soc. Subspecialties: Physical chemistry; Statistical mechanics. Office: Dept Chemistry Cornell U Ithaca NY 14853

WIEBE, HERMAN HENRY, biology educator; b. Newton, Kans., May 30, 1921; s. J.E. and K. (Busenitz) W.; m. Melva Goering, June 8, 1951; children: William, Sara. A.B., Goshen Coll., 1947; M.S., U. Iowa, 1949; Ph.D., Duke U., 1953. Instr. N.C. State Coll., 1953; from asst. prof. to prof. dept biology Utah State U., Logan, 1954—; research participant Oak Ridge Inst. Nuclear Studies, 1953-54; NSF Sci. Faculty fellow Agrl. U., Stuttgart, W. Ger., 1964-65; Fulbright fellow Trinity Coll., Dublin, Ireland, 1973-74; research fellow U. Natal, Pietermaritzburg, S.Africa, 1981-82. Subspecialty: Plant physiology (biology). Current work: Plant water relations. Home: 710 N 7 St E Logan UT 84321 Office: Biology4 Utah State U Logan UT 84322

WIEBE, MICHAEL EUGENE, microbiologist, educator; b. Newton, Kans., Oct. 1, 1942; s. Austin Roy and Ruth Fern (Stucky) W.; m. Rebecca Ann Wiebe, June 12, 1965. B.S., Sterling Coll., 1965; Ph.D. in Microbiology, U. Kans., 1971. Research assoc. Duke U., 1971-73; asst. prof. microbiology Cornell U., 1973-81, assoc. prof., 1980-83, dir. leukocyte products, 1983—; assoc. dir. research and devel. N.Y. Blood Center, N.Y.C., 1980—. Contbr. articles to profl. jours. NIH fellow, 1971-73; recipient Alumni award Sterling Coll., 1979. Mem. Am. Soc. Microbiology, Am. Soc. Virology, Am. Soc. Tropical Medicine and Hygiene, Soc. Exptl. Biology and Medicine, N.Y. Acad. Sci. Democrat. Presbyterian. Subspecialties: Virology (medicine); Cell biology (medicine). Current work: Human interferon and lymphokine induction, synthesis and regulation, molecular virology. Home: 83 Haviland Ct Stamford CT 06903 Office: 310 E 67th St New York NY 10021

WIECH, NORBERT LEONARD, biochemical pharmacologist, educator; b. Chgo., Mar. 13, 1939; s. Chester Joseph and Stella (Ryzner) W.; m. Christine Marie Brown, Aug. 5, 1961; children: Stephanie M., David L., Christopher M. B.S., U. Notre Dame, 1960, M.S., 1963; Ph.D., Tulane U., 1966. Research assoc. Harvard U. Sch. Pub. Health, 1966-67; sect. head biochem. pharmacology Merrell Research Ctr., Dow Pharms., Cin., 1967—; assoc. prof. exptl. medicine U. Cin., 1970—, lectr. in chemistry, 1967—. Contbr. articles to profl. jours. Mem. Neurosci., Am. Oil Chemists' Soc., Am. Diabetes Assn., N.Y. Acad. Scis. Subspecialties: Neuropharmacology; Membrane biology. Current work: Receptor biochemistry, neuropharmacology, metabolic diseases. Patentee in field. Office: Merrell Research Ctr Dow Pharms 2110 E Galbraith Rd Cincinnati OH 45215

WIECH, RALPH BENJAMIN, industrial hygienist, environmental consultant; b. Bridgeport, Conn., Aug. 24, 1941; s. Benjamin John and Lillian Ann (Pavoni) W.; m. Toni Marie Esposito, July 28, 1965; children: Glenn David, Heather Melissa. B.A., U. Cin., 1968. Quality control chemist Glenbrook Labs., Stamford, Conn., 1968-70; environ. chemist York Research Corp., Stamford, 1970-77; pres., chmn. bd. Environ. Assocs., Fairfield, Conn., 1977—. Mem. Am. Indsl. Hygiene Assn., ASTM. Club: Fayerweather Yacht (Bridgeport). Subspecialties: Analytical chemistry; Fuels and sources. Current work: Resource recovery, asbestos monitoring and control, industrial toxicology. Office: 703 Post Rd Fairfield CT 06430

WIEGEL, ROBERT LOUIS, cons. engr.; b. San Francisco, Oct. 17, 1922; s. Louis Henry and Antionette L. (Decker) W.; m. Anne Pearce, Dec. 10, 1948; children—John M., Carol E., Diana L. B.S., U. Calif. at Berkeley, 1943, M.S., 1949. Mem. faculty U. Calif. at Berkeley, 1946—, prof. civil engring., 1963—, asst. dean, 1963-72, acting dean, 1972-73; dir. state tech. services program for Calif. U. Calif., 1965-68; vis. prof. Nat. U. Mexico, summer 1965, Polish Acad. Sci., 1976, U. Cairo, 1978; sr. Queen's fellow in marine sci., Australia, 1977, cons. to govt. and industry, 1946—; chmn. U.S. com. for internat. com. oceanic resources, mem. marine bd. Nat. Acad. Engring., 1975-81; pres. Internat. Engring. Com. on Oceanic Resources, 1972-75; mem. coastal engring. research bd. Dept. Army; mem. IDOE adv. panel NSF, 1974-77, Gov. Calif. Adv. Commn. Ocean Resources, 1967, Calif. Adv. Commn. on Marine and Coastal Resources, 1967-73, Tsunami Tech. Adv. Council, Hawaii, 1964-66; U.S. del. U.S.-Japan coop. sci. programs, 1964, 65. Author. Served to 1st lt. AUS, 1943-46. Fellow ASCE (chmn. exec. com. waterways, harbors, coastal engring. div. 1974-75, vice chmn. coastal engring. research council 1964-78, chmn. 1978—, chmn. task com. wave forces on structures 1960-67, chmn. com. on coastal engring. 1970-71, Research prize 1962, Moff-att-Nichol Coastal Engring. award 1978), AAAS; mem. Nat. Acad. Engring., Permanent Internat. Assn. Nav. Congresses, Internat. Assn. Hydraulic Research, Sigma Xi. Subspecialties: Civil engineering; Ocean engineering. Patentee in field. Home: 1030 Keeler Ave Berkeley CA 94708

WIELAND, STEVEN JOSEPH, neurobiologist; b. Lakewood, Ohio, Dec 5, 1948; s. Steven Joseph and Elvira Ann (Almassy) W.; m. Susan Elizabeth Barr, Oct. 2, 1977; 1 son, Benjamin. B.S. in Physics, U. Notre Dame, 1970; Ph.D., Harvard U., 1979. Research fellow dept. neruosci. Children's Hosp. Med. Center, Boston, 1979-80; research assoc. dept. biology Princeton U., 1980—. World Book Yearbook Nat. Merit scholar, 1966-70. Mem. Soc. Neurosci., AAAS, Phi Beta Kappa. Subspecialties: Neurobiology; Neurophysiology. Current work: Neuronal plasticity; hormone interactions with the nervous system;; neuromodulators, learning and memory. Office: Dept Biology Princeton U Princeton NJ 08544

WIERNIK, PETER HARRIS, physician; b. Crocket, Tex., June 16, 1939; s. Harris and Molly (Emmerman) W.; m. Roberta Joan Fuller, Sept. 6, 1961; children: Julie Anne, Lisa Britt, Peter Harrison. B.A. with distinction, U. Va., 1961, M.D., 1965; Dr. h.c., U. of the Republic, Montevideo, Uruguay, 1982. Diplomate: Am. Bd. Internal Medicine, Sub-Bd. Med. Oncology. Intern Cleve. Met. Gen. Hosp., 1965-66, resident, 1969-70, Osler Service Johns Hopkins Hosp., Balt., 1970-71; sr. asst. surgeon USPHS, 1966, advanced through grades to med. dir., 1976; sr. staff asso. Balt. Cancer Research Center, 1966-71, chief sect. med. oncology, 1971-76, chief clin. oncology br., 1976-81, dir., 1976-82; asso. dir. div. cancer treatment, 1976-81; asst. prof. medicine U. Md. Hosp., Balt., 1971-74, assoc. prof., 1974-76, prof., 1976-82; Gutman prof., chmn. dept. oncology Montefiore Med. Ctr., 1982—; head div. med. oncology Albert Einstein Coll., 1982—; assoc. dir. Albert Einstein Cancer Ctr., 1982—; cons. hematology and med. oncology Union Meml. Hosp., Greater Balt. Med. Center, Franklin Sq. Hosp; bd. dirs. Balt. City unit Am. Cancer Soc., 1971-78, chmn. patient care com., 1972-75; mem. med. adv. com. Nat. Leukemia Assn., 1976—; chmn. adult leukemia com. Cancer and Leukemia Group B, 1976-82; prin. investigator Eastern Corp. Oncology Group, 1982—. Editor: Controversies in Oncology, 1982; editor: Supportive Care of the Cancer Patient, 1983; Mem. editorial bd.: Cancer Treatment Reports, 1972-76, Leukemia Research, 1976—, Cancer Clin. Trials, 1977—, Hosp. Practice, 1979—; co-editor: Am. Jour. Med. Scis, 1976—; also articles, chpts. in books. Recipient Z Soc. award U. Va., 1961, Byrd S. Leavell Hematology award U. Va. Sch. Medicine, 1965. Fellow AAAS, Am. Coll. Clin. Pharmacology, Internat. Soc. Hematology, Royal Soc. Medicine (London), ACP; mem. Am. Soc. Clin. Investigation, Am. Soc. Clin. Oncology (chmn. edn. and tng. com. 1976-79, subcom. on clin. investigation 1980—), Am. Assn. Cancer Research, Am. Soc. Hematology, Am. Fedn. Clin. Research, Am. Acad. Clin. Toxicology, Internat. Soc. Experimental Hematology, N.Y. Acad. Sci., Am. Soc. Hosp. Pharmacy, Am. Soc. Clin. Pharmacology and Therapeutics, Am. Radium Soc., Phi Beta Kappa, Sigma Xi, Alpha Omega Alpha, Phi Sigma (award 1961). Subspecialties: Cancer research (medicine); Hematology. Current work: New treatments for cancer, especially leukemia and lymphoma. Home: 43 Longview Ln Chappaqua NY 10514 Office: Montefiore Med Ctr 111 E 210th St New York NY 10467 Always remember why you entered a profession in the first place. Leave the politics to those who have forgotten.

WIESE, MAURICE VICTOR, plant pathologist, educator, research administrator; b. Columbus, Nebr., Sept. 22, 1940; s. Frank J. and Helen M. (Pojar) W.; m. Suzanne J. Glenn, June 15, 1963; children: Patrick, Kristina, Steven. B.S. in Agr., U. Nebr., 1963, M.S. in Botany and Biochemistry, 1965; Ph.D., U. Calif., Davis, 1969. Asst. prof. wheat pathology research Mich. State U., East Lansing, 1969-74, assoc. prof., 1974-78; prof. crop loss assessment research U. Idaho, Moscow, 1978—, asst. head plant and soil sci. dept., 1981—; asst. dir. Idaho Agrl. Expt. Sta., 1983—. Author: Compendium of Wheat Diseases, 1977. Mem. Am. Phytopath. Soc., Am. Soc. Agronomy, Crop Sci. Soc. Am., Sigma Xi, Alpha Zeta, Gamma Sigma Delta. Subspecialties: Integrated systems modelling and engineering; Plant pathology. Current work: Crop Survey, yield loss assessments, integrated pest and crop management. Home: 711 Park Dr Moscow ID 83843 Office: Dept Plant Soil and Entomological Sci U Idaho Moscow ID 83843

WIESE, WOLFGANG LOTHAR, physicist; b. Tilsit, Germany, Apr. 21, 1931; came to U.S., 1957, naturalized, 1965; s. Werner Max and Charlotte (Donath) W.; m. Gesa Ladehoff, Oct. 12, 1957; children: Margrit, Cosima. B.S., U. Kiel, W.Ger., 1954, Ph.D., 1957. Research assoc. U. Md., 1958-59; research physicist Nat. Bur. Standards, Washington, 1960-62, chief plasma spectroscopy sect., 1962-77, chief atomic and plasma radiation div., 1977-; lectr. UCLA, summers 1963, 64; Guggenheim fellow Max-Planck Inst., Munich, W.Ger., 1966-67. Prin. author: Atomic Transition Probabilities, Vol. I, 1966, Vol. II, 1969; asso. ed.: Jour. of Quantitative Spectroscopy and Radiative Transfer, 1971—; contbr. numerous articles to profl. jours. Recipient Dept. Commerce Silver medal, 1962, Gold medal, 1971. Fellow Am. Phys. Soc., Optical Soc. Am.; mem. Internat. Astron. Union, Fusion Power Assos., Sigma Xi. Subspecialties: Atomic and molecular physics; Plasma physics. Current work: Atomic processes in high temperature plasmas; atomic transition probabilities; spectral line broadening by plasmas. Home: 8229 Stone Trail Dr Bethesda MD 20817 Office: Nat Bur Standards Room A267 Bldg 221 Washington DC 20234

WIESEL, TORSTEN NILS, neurobiologist, educator; b. Upsala, Sweden, June 3, 1924; came to U.S., 1955; s. Fritz Samuel and Anna-Lisa Elisabet (Bentzer) W.; 1 dau., Sara Elisabet. M.D., Karolinska Inst., Stockholm, 1954; A.M. (hon.), Harvard U., 1967. Instr. physiology Karolinska Inst., 1954-55; asst. dept. child psychiatry Karolinska Hosp., 1954-55; fellow in ophthalmology Johns Hopkins U., 1955-58, asst. prof. ophthalmic physiology, 1958-59; asso. in neurophysiology and neuropharmacology Harvard U. Med. Sch., Boston, 1959-60, asst. prof. neurophysiology and neuropharmacology, 1960-64, asst. prof. neurophysiology, dept. psychiatry, 1964-67, prof. physiology, 1967-68, prof. neurobiology, 1968-74, Robert Winthrop prof. neurobiology, 1974-83, chmn. dept. neurobiology, 1973-82; prof. neurobiology Rockefeller U., N.Y.C., 1983—; Ferrier lectr. Royal Soc. London, 1972; NIH lectr., 1975; Grass lectr. Soc. Neurosci., 1976; lectr. Coll. de France, 1977; Hitchcock prof. U. Calif.-Berkeley, 1980; Sharpey-Schafer lectr. Phys. Soc. London; George Cotzias lectr. Am. Acad. Neurology, 1983. Contbr. numerous articles to profl. jours. Recipient Jules Stein award Trustees for Prevention Blindness, 1971, Lewis S. Rosenstiel prize Brandeis U., 1972, Friedenwald award Trustees of Assn. for Research in Vision and Ophthalmology, 1975, Karl Spencer Lashley prize Am. Philos. Soc., 1977, Louisa Gross Horwitz prize Columbia U., 1978, Dickson prize U. Pitts., 1979, Nobel prize in Physiology/Medicine, 1981. Mem. Am. Physiol. Soc., Am. Philos. Soc., AAAS, Am. Acad. Arts and Scis., Nat. Acad. Arts and Scis., Swedish Physiol. Soc., Soc. Neurosci. (pres. 1978-79), Royal Soc. (fgn. mem.), Physiol. Soc. (Eng.) (hon. mem.). Subspecialty: Neurobiology. Office: Rockefeller U 1230 York Ave New York NY 10021

WIESENFELD, JAY MARTIN, physicist; b. New Brunswick, N.J., Sept. 24, 1950; s. Joel and Paula (Brenner) W.; m. Kay Ruth Granstrom, Dec. 24, 1979; 1 son, David Jason. A.B., Harvard U., 1972, A.M. in Physics, 1972; Ph.D. in Phys. Chemistry, U. Calif.-Berkeley, 1978. Postdoctoral fellow Bell Labs., Holmdel, N.J., 1977-80; mem. tech. staff Guided Wave Research Lab., Crawford Hill Lab., 1980—. Contbr. articles to profl. jours. NSF fellow, 1973-76. Mem. Am. Chem. Soc., Am. Phys. Soc., Optical Soc. Am., Sigma Xi. Subspecialties: Picosecond laser optics; Fiber optics. Current work: urrent work: Picosecond laser optics; development of ultrashortpulse sources for optical communication; measurement of rapid material relaxation processes. Office: Bell Labs Crawford Hill Lab Holmdel NJ 07733

WIESNER, JEROME BERT, engineering educator and researcher, former university president; b. Detroit, May 30, 1915; s. Joseph and Ida (Friedman) W.; m. Laya Wainger, Sept. 1, 1940; children: Stephen Jay, Zachary Kurt, Elizabeth Ann, Joshua A. B.S., U. Mich., 1937, M.S., 1938, Ph.D., 1950. Assoc. dir. U. Mich. Broadcasting Service, 1937-40; chief engr. Acoustical Record Lab., Library of Congress, 1940-42; staff Mass. Inst. Tech. Radiation Lab., 1942-45, U. of Calif. Los Alamos Lab., 1945-46; mem. faculty Mass. Inst. Tech., 1946-71, dir. research lab. of electronics, 1952-61, head dept. elec. engring., 1959-60, dean of sci., 1964-66, provost, 1966-71, pres., 1971-80, Inst. researcher and prof., 1980—; mem. Army sci. adv. com., 1956-61, spl. asst. to Pres. on sci. and tech., 1961-64; chmn. Pres.'s Sci. Adv. Com., 1961-64; chmn. tech. assessment adv. council Office Tech. Assessment, U.S. Congress, 1976-79; Dir. Automatix, Damon Engring., Dudley Sta. Corp., New Eng. TV Corp., Schlumberger Ltd., Raychem Corp. Author: Where Science and Politics Meet, 1965, ABM—An Evaluation, 1969. Bd. govs. Weizmann Inst. Sci., MacArthur Found.; trustee Am. Found. for Blind, Woods Hole Oceanographic Inst., Kennedy Meml. Trust; mem. corp. Mus. of Sci. in Boston. Fellow IEEE, Am. Acad. Arts and Scis.; mem. Am. Philos. Soc., AAUP, Am. Geophys. Union, Acoustical Soc. Am., Nat. Acad. Engring., Nat. Acad. Scis., Sigma Xi, Phi Kappa Phi, Eta Kappa Nu, Tau Beta Pi. Subspecialty: Electrical engineering. Home: 61 Shattuck Rd Watertown MA 02172 Office: Mass Inst Tech Cambridge MA 02139

WIEST, WILLIAM MARVIN, psychology educator; b. Loveland, Colo., May 8, 1933; s. William Walter and Katherine Elizabeth (Buxman) W.; m. Thelma Lee Bartel, Mar. 30, 1936; children: William Albert, Suzanne Kay, Cynthia May. B.A., Tabor Coll., 1955; M.A., U. Kans., 1957; Ph.D., U. Calif.-Berkeley, 1962. Asst. prof. psychology Reed Coll., Portland, Oreg., 1961-66, assoc. prof., 1967-78, prof., 1979—; vis. investigator Health Services Research Ctr., Portland, 1976-80; project coordinator WHO, Geneva, 1976-81; research cons. Bonneville Power Adminstrn., Portland, 1980—; vis. scientist Oceanic Inst., Waimanalo, Hawaii, 1967-68. NSF faculty sci. fellow, 1975-76. Mem. Am. Psychol. Assn., Am. Pub. Health Assn., Population Assn. Am., Sigma Xi, Phi Beta Kappa. Subspecialties: Behavioral psychology; Social psychology. Current work: Research on factors determining acceptability of new contraceptive methods including quantification of possible effects on sexual behavior in men within various cultural settings. Home: 5009 SE 46th Ave Portland OR 97206 Office: Psychology Dept Reed Coll Portland OR 97202

WIFF, DONALD RAY, research polymer physicist, polymer consultant; b. Youngstown, Ohio, Feb. 19, 1936; s. Ernest and Mildred Bietta (Kreps) W.; m. Carol June Skipper, Aug. 25, 1962; children: David Skipper, Devin Drew, Daniel Don. B.S. in Physics, Capital U., Columbus, Ohio, 1958; M.A., Kent State U., 1960; Ph.D., Tex. A&M U., 1967; M.B.A., U. Dayton, 1980. Instr. physics Tex. A&M U., 1960-67; polymer physicist U. Dayton Research Inst., 1967—. Contbr. articles to profl. jours. Mem. Am. Phys. Soc., Brit. Phys. Soc., European Phys. Soc., Am. Chem. Soc., Soc. Plastics Engrs., Am. Mgmt. Assn., Sigma Pi Sigma. Lutheran. Subspecialties: Polymer physics; Composite materials. Current work: Morphology-molecular structure-physical property correlation of new polymer concepts; molecular composites; conducting polymers; research and development management; mathematically ill-posed problems; advanced structural composites. Patentee in field. Home: 3764 Woodbrook Way Beavercreek OH 45430 Office: University Dayton Research Institute 300 College Park Dayton OH 45469

WIGGINS, RALPHE, geophysicist, researcher; b. Broadwater, Nebr., Apr. 4, 1940; s. Ralph Otis and Celia Elvina (Davis) W.; m. Rosemary Jackson, Dec. 31, 1973; 1 dau., Suzyn. B.S. in Geophys. Engring, Colo. Sch. Mines, Golden, 1957-61; Ph.D., MIT, 1965. Sr. scientist Geosci., Inc., Cambridge, Mass., 1965-66; research assoc. MIT, Cambridge, 1966-70; assoc. prof. U. Toronto, 1970-73, U. B.C., Vancouver, 1973-75; sr. research geophysicist Western Geophys. Co., Houston, 1975-77; prin. geophysicist DelMar Tech. Assocs., Calif., 1977-78; sr. research assoc. Mobil Research and Devel. Co., Dallas, 1978-82; dir. dept. geosci. Schlumberger-Doll Research, Ridgefield, Conn., 1982—. Contbr. articles to profl. jours. Mem. Am. Geophys. Union, European Assn. Exploration Geophysicists, Seismol. Soc. Am. (dir. 1980—), Soc. Exploration Geophysicists. Subspecialties: Geophysics; Algorithms. Current work: Exploration geophysics, information extraction from geophysical sensors, seismic exploration, borehole seismology, well logging, propagation of eleastic waves. Office: PO Box 307 Ridgefield CT 06877

WIGHTMAN, ARTHUR STRONG, educator, physicist; b. Rochester, N.Y., Mar. 30, 1922; s. Eugene Pinckney and Edith Victoria (Stephenson) W.; m. Anna-Greta Larsson, Apr. 28, 1945 (dec. Feb. 11, 1976); 1 dau., Robin Letitia; m. Ludmila Popova, Jan. 14, 1977. B.A., Yale, 1942; Ph.D., Princeton, 1949; D.Sc., Swiss Fed. Inst. Tech., Zurich, 1968. Instr. physics Yale, 1943-44; from instr. to asso. prof. physics Princeton, 1949-60, prof. math. physics, 1960—, Thomas D. Jones prof. math. physics, 1971—; vis. prof. Sorbonne, 1957. Served to lt. (j.g.) USNR, 1944-46. NRC postdoctoral fellow Inst. Teoretisk Fysik, Copenhagen, Denmark, 1951-52; NSF sr. postdoctoral fellow, 1956-57; recipient Dannie Heineman prize math. physics, 1969. Fellow Am. Acad. Arts and Scis., Royal Acad. Arts; mem. Nat. Acad. Scis., Am. Math. Soc., Am. Phys. Soc., AAAS, Fedn. Am. Scientists. Subspecialty: Theoretical physics. Current work: Quantum field theory. Home: 30 The Western Way Princeton NJ 08540

WIGNER, EUGENE PAUL, retired educator; b. Budapest, Hungary, Nov. 17, 1902; came to U.S., 1930, naturalized, 1937; s. Anthony and Elisabeth (Einhorn) W.; m. Amelia Z. Frank, Dec. 23, 1936 (dec. 1937); m. Mary Annette Wheeler, June 4, 1941 (dec. Nov. 1977); children: David Wheeler, Martha Faith; m. Eileen C.P. Hamilton, 1979. Chem. Engr. and Dr. Engring., Technische Hochschule, Berlin, 1925; hon. D.Sc., U. Wis., 1949, Washington U., 1950, Case Inst. Tech., 1956, U. Chgo., 1957, Colby Coll., 1959, U. Pa., 1961, Thiel Coll., 1964, U. Notre Dame, 1965, Technische Universität Berlin, 1966, Swarthmore Coll., 1966, Université de Louvain, Belgium, 1967; Dr.Jr., U. Alta., 1957; L.H.D., Yeshiva U., 1963. Lectr. Princeton U., 1930, halftime prof. math. physics, 1931-36; prof. physics U. Wis., 1936-38; Thomas D. Jones prof. theoretical physics Princeton U., 1938-71; on leave of absence, 1942-45; at Metall. Lab., U. Chgo., 1946-47; as dir. research and devel. Clinton Labs.; dir. CD Research Project, Oak Ridge, 1964-65; Lorentz lectr. Inst. Lorentz, Leiden, 1957; cons. prof. La. State U., 1971—; mem. gen. adv. com. AEC, 1952-57, 59-64; mem. math. panel NRC, 1952-54; physics panel NSF, 1953-56; vis. com. Nat. Bur. Standards, 1947-51. Decorated medal of Merit, 1946; recipient Franklin medal Franklin Inst., 1950, citation N.J. Tchrs. Assn., 1951, Enrico Fermi award AEC, 1958, Atoms for Peace award, 1960, Max Planck medal German Phys. Soc., 1961, Nobel prize for physics, 1963, George Washington award Am. Hungarian Studies Found., 1964, Semmelweiss medal Am. Hungarian Med. Assn., 1965, Nat. Sci. medal, 1969, Pfizer award, 1971, Albert Einstein award, 1972, Golden Plate medal Am. Acad. Achievement, 1974, Disting. Achievement award La. State U., 1977, Wigner medal, 1978, Founders medal Internat. Cultural Found., 1982, Medal of the Hungarian Central Research Inst., Medal of the Autonomous Univ. Barcelona; named Nuclear Pioneer Soc. Nuclear Medicine, 1977; recipient Colonel Gov. of La., 1983. Mem. Royal Soc. Eng. (fgn.), Royal Netherlands Acad. Sci. and Letters, Am. Nuclear Soc., Am. Phys. Soc. (v.p. 1955, pres. 1956), Am. Math. Soc., Am. Assn. Physics Tchrs., Am. Acad. Arts and Scis., Am. Philos. Soc., Nat. Acad. Scis., N.Y. Acad. Scis. (hon. life mem.), Austrian Acad. Scis., German Phys. Soc., Franklin Inst., AAAS, Sigma Xi, Acad. Sci., Gottingen, Germany (corr.), Hungarian Acad. Sci. (hon.), Austrian Acad. Scis. (hon.), Hungarian L. Eötuös Phys. Soc. (hon.). Subspecialties: Theoretical physics; Statistical physics. As to success, one achieves it best if one does not strive for it too vigorously, has a good deal of luck, does one's duty, follows his inclinations, tries to be understanding and useful. It sounds sanctimonious, and perhaps even boastful, but it is true nevertheless, that being considerate, mindful of the sensitivities of others, gives one more peace of mind, more satisfaction, than success or anything else.

WIITA, PAUL JOSEPH, astrophysicist, educator; b. Bronx, N.Y., Feb. 18, 1953; s. Paul Elias and Martha (Knippenberg) W.; m. Brinda Umberkoman, May 31, 1978; 1 son, Arun Paul. B.S., Cooper Union, 1972; M.A., Princeton U., 1974, Ph.D., 1976. Research assoc. Enrico Fermi Inst., U. Chgo., 1976-79; NSF-NATO postdoctoral fellow U. Cambridge, Eng., 1977-78; asst. prof. astronomy and astrophysics U. Pa., Phila., 1979—. Contbr. articles to profl. jours. NSF fellow, 1972-75; Compton lectr., 1977; vis. fellow Copernicus Astron. Center, Warsaw, Poland, 1978, Tata Inst., Bombay, India, 1981; U. Pa. faculty research grantee, 1980. Mem. Internat. Astron. Union, Am. Astron. Soc., Am. Phys. Soc., Royal Astron. Soc., Sigma Xi, Sigma Pi Sigma. Democrat. Subspecialties: Theoretical astrophysics; General relativity. Current work: Theoretical studies of radio galaxies, active galactic nuclei, relativistic astrophysics, rotating stars, accretion disks. Home: 431 S 45th St Philadelphia PA 19104 Office: Dept Astronomy and Astrophysics E1 U Pa Philadelphia PA 19104

WIJSMAN, ELLEN MARIE, research scientist; b. Berkeley, Calif., Apr. 19, 1954; d. Robert Arthur and Gertrud (Zierau) W.; m. Ethan Allen Merrit, Nov. 20, 1980. B.S., Mich. State U., 1975; Ph.D., U. Wis.-Madison, 1981. Postdoctoral fellow Stanford U., 1981—. NSF predoctoral fellow, 1975-78; NIH predoctoral fellow, 1978-81. Mem. Genetics Soc. Am. Subspecialties: Evolutionary biology; Genetics and genetic engineering (medicine). Current work: Population genetics of interacting populations, theory and analysis of migration in human populations by use of gene frequency data, evolution of the genome. Office: Dept Genetics Stanford U Stanford CA 94305

WILBANKS, JOHN RANDALL, geology educator; b. Foreman, Ark., June 10, 1938; s. Hubert Harrison and Nolia Antoinette (Blakely) W.; m. Dorothy Louise Ansley, Dec. 22, 1962; children: Holly Ann, Randall Wade. B.S., N.Mex. Inst. Mining and Tech., 1960; M.S., Tex. Tech U., 1963, Ph.D., 1969. Water rights technician N.Mex. State Engr., Santa Fe, 1960-61; geologist N.Mex. Hwy. Dept Geology Research Lab., Santa Fe, 1961-63; research assoc. Antarctica program Tex. Tech. U., Lubbock, 1966-71, asst. prof. geology dept. geoscis., 1969-71; assoc. prof. U. Nev.-Las Vegas, 1973—; cons. field trips various oil cos., 1978-81. Pres. Las Vegas Friends of Jung, 1982-83. Recipient U.S. Antarctic Service medal NSF Office Polar Programs, 1966. Mem. Geol. Soc. Am., Nat. Assn. Geology Tchrs., Antarctican Soc., Sigma Xi, Sigma Gamma Epsilon. Subspecialties: Geology; Tectonics. Current work: Geology, structural geology. Home: 6953 Erin Circle Las Vegas NV 89128 Office: Dept Geosci U Nev Las Vegas NV 89154

WILBER, CHARLES GRADY, forensic science educator, consultant; b. Waukesha, Wis., June 18, 1916; s. Charles Bernard and Charlotte Agnes (Grady) W.; m. Ruth Mary Bodden, July 12, 1944 (dec. 1950); childen: Maureen, Charles Bodden, Michael; m. Clare Marie O'Keefe, June 14, 1952; children: Thomas Grady (dec.), Kathleen, Aileen, John Joseph, Maureen, Charles, Michael. B.Sc., Marquette U., 1938; M.A., Johns Hopkins U., 1941, Ph.D., 1942. Asst. prof. physiology Fordham U., 1945-49; assoc. prof. physiology, dir. biol. labs. St. Louis U., 1949-52; leader Arctic expdns., 1943-44, 48, 50, 51; physiologist Chem. Corps, U.S. Army, 1952-61; assoc. physiology and pharmacology U. Pa., 1953-61, chief comparative physiology, 1956-61; profl. lectr. biol. scis. Loyola Coll., Balt., 1957-61, dir., 1958-61; prof. biol. scis., univ. research coordinator, dean Grad. Sch., Kent State U., 1961-64; dir. marine laboratories U. Del., 1964-67; chmn. prof. dept. zoology Colo. State U., 1967-73, prof., 1973—, also dir. forensic sci. lab.; dep. coroner, Larimer County, Colo.; mem. Center for Human Identification; expert witness fed. and state cts. on poisons, firearms, others. mem. Marine Biol. Lab., Woods Hole, Mass., 1947—; mem. U.S. Army Panel Environ. Physiology, 1952-61; mem. study group Nat. Acad. Scis.-USAF, 1958-61. Author: Biological Aspects of Water Pollution, 2d edit, 1971, Japanese edit., 1970; author: Forensic Biology for the Law Enforcement Officer, 1975, Contemporary Violence, 1975, Ballistic Science for the Law Enforcement Officer, 1977, Medicolegal Investigation of The President John F. Kennedy Murder, 1978, Chemical Trauma from Pesticides, 1979, Forensic Toxicology, 1980, Beryllium, 1980, Agent Orange, 1980; contbr. articles to profl. jours.; exec. editor: Adaption to the Environment, vol. in series, 1962; editor: Am. Lecture Series in Environ. Studies; mem. editorial bd.: Am. Jour. Forensic Medicine and Pathology; contbr.: Harper Ency. Nat; vis. lectr.: Am. Inst. Biol. Scis, 1957—. Served to capt. USAAF, 1942-46; col., ret. USAF. Fellow N.Y. Acad. Scis., Am. Acad. Forensic Sci.; mem. Am. Physiol. Soc., Phi Beta Kappa, Sigma Xi, Phi Sigma, Gamma Alpha. Republican. Catholic. Club: Cosmos (Washington). Subspecialties: Physiology (biology); Toxicology (medicine). Current work: Energy effect relations in wound ballistics, chemical trauma from pesticides. Home: 900 Edwards St Fort Collins CO 80524 Office: Dept Zoology Colo State U Fort Collins CO 80523 My most precious possession has been the right of freedom, and the responsibility for the consequences of my actions. This right of freedom, even to be wrong or stupid, is the essence of mankind. Any abridgement of that freedom, for whatever stated reason is subversive of mankind and dehumanizes the person. Any government that encroaches on that radical freedom in order "to protect me from myself" must be destroyed.

WILBUR, HENRY MILES, zoology educator; b. Bridgeport, Conn., Jan. 25, 1944; s. Robert Leonard and Martha (Miles) W.; m. Dorothy Ann Spales, Jan. 27, 1967 (div. 1980); 1 dau., Sarah Dustin; m. Rebecca Bea Burchell, May 22, 1981; 1 dau., Helen Margaret. B.S., Duke U., 1966; Ph.D., U. Mich., 1972. Asst. prof. Duke U., Durham, N.C., 1973-77, assoc. prof., 1977-82, prof. zoology, 1982—; NSF grad. fellow, 1967-70, U. Mich. Soc. Fellows jr. fellow, 1971-73. Mem. Ecol. Soc. Am. (assoc. editor jour. 1979-82), Am. Soc. Naturalists, Soc. Study Evolution (assoc. editor jour. 1979-80), Am. Soc. Ichthyologists and Herpetologists, AAAS. Democrat. Subspecialties: Population biology; Evolutionary biology. Current work: Evolution of species interactions, complex life cycle ecology. Office: Dept zoology Duke U Durham NC 27706

WILBUR, JAMES M(YERS), JR., chemist, educator, researcher; b. Phila., Oct. 31, 1929; s. James M. and Mary E. (Scherer) W.; m. Ann J.; children: Kirsten, Karen, Eric. B.S., Muhlenberg Coll., 1951; Ph.D. in Chemistry, U. Pa., Phila., 1959. Chemist J. T. Baker Chem. Co., Phillipsburg, N.J., 1951-53; postdoctoral fellow U. Minn., Mpls., 1958-60; research chemist E. I. duPont de Nemours & Co., Wilmington, Del., 1960-62; postdoctoral fellow U. Ariz., Tucson, 1962-63; prof. chemistry S.W. Mo. State U., 1963—. Served with Signal Corps U.S. Army, 1954-56. Mem. Am. Chem. Soc., Chem. Soc. (London), AAAS. Subspecialties: Organic chemistry; Polymer chemistry. Current work: Polymer synthesis; chemotherapy. Home: 1715 S National Springfield MO 65804 Office: 905 S National Springfield MO 65802

WILBUR, LYMAN DWIGHT, civil engineer; b. Los Angeles, Apr. 27, 1900; s. Curtis Dwight and Olive (Doolittle) W.; m. Henrietta Shattuck, July 6, 1925; children: Olive Wilbur. A.B. in Civil Engring, Stanford U., 1921; LL.D. (hon.), Coll. Idaho, 1962, Dr.Sci., U. Idaho, 1967. Registered civil engr., Calif., Idaho, Wash., Ariz. Field engr. City of San Francisco, 1921-24; designer Merced Irrigation Dist., Calif., 1924-26; design engr. East Bay Mcpl. Utility Dist., Oakland, Calif., 1926-29; asst. to chief consulting engr. Middle Asia Water Ecology Service, USSR, 1929-31; engr. to v.p. and dir. Morrison-Knudsen Co., Inc., Los Angeles and Boise, Idaho, 1932-70; concurrently exec. v.p. to chmn. Internat. Engring. Co., Inc., San Francisco, constrn. mgr. and resident ptnr. joint ventures; self employed cons. engr., Boise, 1971—. Contbr. articles to profl. jours. Pres. Good Samaritan League, 1982—; bd. dirs. Bench Sewer Dist., 1978—, St. Alphonsus Hosp. Found., Blue Cross of Idaho Health Services, Univ/Community Health Scis. Assn.; trustee Coll. Idaho. Served with U.S. Army, 1918. Recipient Golden Beaver award The Beavers, 1962; Constrn. Man of Yr. Engring. News Record, 1966; award for disting. humanitarian accomplishments through engring. Idaho Soc. Profl. Engrs., 1967; John Fritz medal Five Founder Engring. Socs., 1973; Moles award for outstanding achievement in constrn., 1974; ann. award Nat. Soc. Profl. Engrs., 1974; others. Mem. Nat. Acad. Engring., ASCE, Idaho Soc. Profl. Engrs., Nat. Soc. Profl. Engrs., Soc. Am. Mil. Engrs. Republican. Presbyterian. Clubs: Hillcrest, Arid. Subspecialty: Civil engineering. Current work: Consulting and helping with civic and non-profit orgns. Patentee in field. Address: 4502 Hillcrest Dr Boise ID 83705

WILCOX, JOHN MARSH, solar physics educator; b. Iowa City, Iowa, Jan. 31, 1925; s. Myron Jefferson and Marguerite (Marsh) W.; m. (div.); children: Sharon Ann, David Allen. B.S., Iowa State Coll., 1948; Ph.D., U. Calif.-Berkeley, 1954. Asst. physicist Ames (Iowa) AEC Lab., 1947-49; physicist Los Alamos Sci. Lab., summers 1949, 50; vis. physicist Royal Inst. Tech., Stockholm, 1961-62; physicist U. Calif. Lawrence Radiation Lab., 1951-61; research physicist U. Calif. Space Scis. Lab., 1964-71; adj. prof. solar physics Stanford U., 1971—. Contbr. articles to profl. jours. Served with USAAF, 1943-45. Fellow AAAS, Am. Phys. Soc.; mem. Am. Geophys. Union, Am. Astron. Soc., Astron. Soc. Pacific, Royal Astron Soc. (London), Internat. Astron Union, Am. Meteorol. Soc. Subspecialty: Solar physics. Home: 350 Sharon Park Dr Apt J-3 Menlo Park CA 94025 Office: Inst for Plasma Research Stanford U Via Crespi Stanford CA 94305

WILCOX, PATTI MARIE, nurse, clinician and clinical researcher; b. River Falls, Wis., Jan. 24, 1946; d. David Arthur Sr. and Patti Marie (Gambrell) W.; m. dau., Amanda. Diploma, Johns Hopkins Hosp. Sch. Nursing, 1967; cert., Johns Hopkins Sch. Health Services, 1976. Cert. adult nurse practitioner. Head nurse Johns Hopkins Hosp., Balt., 1964-70, Man Alive Research Inst., 1970-71, Johns Hopkins Oncology Ctr., 1971-75, adult nurse practitioner, 1976—; health care coordinator Threshold Inc., Balt., 1978-81; dir., developer NEED, Balt., 1979—; coordinator breast cancer program Am. Cancer Soc., Balt., 1978-80. Author, editor: NEED Co-leader Handbook, 1982. Active Big sister Big Bros./Big Sisters; bd. dirs. YWCA Greater Balt. Area, Inc. Recipient cert. merit Am. Cancer Soc., 1981; volo. of yr. award YWCA, 1982. Mem. Oncology Nursing Soc., Md. Oncology Nursing Soc. (pres. 1980-81), Am. Cancer Soc.; affiliate mem. Am. Soc. Clin. Oncology. Democrat. Methodist. Subspecialties: Oncology. Current work: Breast cancer-primary, metastic, reconstruction after, adjuvant chemotherapy, nausea and vomiting-anticipatory; alopecia-2 chemotherapy; psychosocial issues following mastectomy; body image after mastectomy; support groups; patient education. Office: Johns Hopkins Oncology Ctr B-111 600 N Wolfe St Baltimore MD 21205 Home: 2710 Southern Ave Baltimore MD 21214

WILCOX, W(EBSTER) WAYNE, forest products pathologist; b. Berkeley, Calif., Oct. 28, 1938; s. Webster Williamson, Jr. and Edith Jeanette (LaBelle) W.; m. Margaret Ruth Starkweather, Aug. 7, 1960; children: Melissa Margaret, Wynn William. B.S. in Forestry, U. Calif., Berkeley, 1960; M.S., U. Wis., 1962, Ph.D. in Plant Pathology, 1965. Plant pathologist U.S. Forest Products Lab., Madison, Wis., 1960-64; mem. faculty U. Calif., Berkeley, also U. Calif. Forest Product Lab., 1964—, forest products pathologist, prof. forestry, 1977—; cons. in field; mem. Calif. Structural Pest Control Bd., 1979-81. Fulbright-Hays sr. postdoctoral fellow, 1973-74. Fellow Internat. Acad. Wood Sci.; mem. Forest Products Research Soc. (Wood award 1965), Soc. Wood Sci. and Tech., Am. Inst. Biol. Scis., Sigma Xi, Xi Sigma Pi. Subspecialty: Plant pathology. Office: Forest Products Lab U Calif 47th and Hoffman Blvd Richmond CA 94804

WILCOXSON, ROY DELL, plant pathologist; b. Columbia, Utah, Jan. 12, 1926; s. Roy E. and Bertha M. (Karren) W.; m. Iva Wall, Apr. 15, 1949; children: Bonnie, Paul, Karren, John. B.S., Utah State U., 1953; M.S., U. Minn., 1955, Ph.D., 1957. Prof. dept. plant path. U. Minn., St. Paul, 1957—; spl. staff mem. Rockefeller Found., 1969; vis. prof. Indian Agrl. Research Inst., 1969, 80. Served with USN, 1944-47; Served with USAF, 1950-64. Mem. Am. Phytopath. Soc., Indian Phytopath. Soc., Sigma Xi. Mem. Ch. of Jesus Christ of Latter-day Saints. Subspecialties: Plant pathology; Integrated pest management. Current work: Breeding disease resistant cultivars.

WILCZEK, FRANK ANTHONY, physicist, educator; b. Mineola, N.Y., May 15, 1951; s. Frank John and Mary Rose (Cona) W.; m. Elizabeth Jordan Devine, July 3, 1973; children: Amity Michelina, Mira. B.S. in Math, U. Chgo., 1970, M.A., Princeton U., 1972, Ph.D. in Physics, 1973. Mem. staff Inst. for Advanced Studies, Princeton, N.J., 1976-77; asst. prof. to prof. Princeton U., 1974-81; prof. physics, mem. staff Inst. for Theoretical Physics, U. Calif.-Santa Barbara, 1981—; mem. high energy adv. com. Brookhaven Nat. Lab., Upton, N.Y., 1977—; Loeb lectr. Harvard U., Apr., 1981. Editor: Zeit F. Physik C, 1981—. A.P. Sloan fellow, 1975-77; John and Catherine MacArthur fellow, 1981-86. Subspecialties: Particle physics; Cosmology. Current work: Quantum field theory, symmetries, cosmology. Office: Inst for Theoretical Physics U Calif Santa Barbara CA 93106

WILCZYNSKI, WALTER, psychology educator; b. Trenton, N.J., Sept. 18, 1952; s. Alexander Walter and Eugenia Mary (DiGuiseppi) W. B.S. in Psychology, Lehigh U., 1974; Ph.D. in Neuroscis, U. Mich., 1978. Lectr. biology Cornell U., Ithaca, N.Y., 1979, postdoctoral fellow, 1979-83; asst. prof. psychology U. Tex., Austin, 1983—. Contbr. articles to profl. jours. NSF fellow, 1979-80; NIH fellow, 1980—83. Mem Soc. Neurosci., Am. Soc. Zoologists, Am. Assn. Anatomists, AAAS. Subspecialty: Comparative neurobiology. Current work: Research in nervous system evolution; neuroanatomy in nonmammalian vertebrates, sensory processing (mainly auditory system). Office: Dept Psychology U Tex Austin TX 78712

WILD, JAMES ROBERT, research scientist; b. Sedalia, Mo., Nov. 24, 1945; s. Robert Lee and Frances Elleta (Wheeler) W.; m. Ann Lynn Brenner, Aug. 1, 1973. B.S., U. Calif.-Davis, 1967; Ph.D., U. Calif.-Riverside, 1971. Teaching asst. dept. biology U. Calif.-Riverside, 1967-71; asst. prof. genetics dept. plant scis. Tex. A&M U., 1975-80, assoc. prof. biochemistry and genetics, dept. biochemistry and biophysics, 1980—. Contbr. numerous articles to profl. jours. Served with USNR, 1972-75. NIH grantee, 1981—; NSF grante, 1981—; Dept. Agr. grantee, 1980-82; Robert A. Welch Found. grantee, 1982—. Mem. Am. Soc. Biol. Chemists, Am. Soc. Microbiology, AAAS, Genetics Soc. Am., Sigma Xi, Phi Sigma. Subspecialties: Genetics and genetic engineering (agriculture); Plant cell and tissue culture. Current work: Analysis of gene structure-function; regulation of gene expression enzymology, molecular genetics, plant tissue culture, development metabolism, biochemical analogues, gene amplification and restructuring. Home: 1606 Todd Trail College Station TX 77840 Office: Dept Biochemistry and Biophysics Tex A&M U College Station TX 77843

WILDENTHAL, C(LAUD) KERN, university dean; b. San Marcos, Tex., July 1, 1941; s. Bryan and Doris (Kellam) W.; m. Margaret Dehlinger, Oct. 15, 1964; children—Pamela, Catharine. B.A., Sul Ross Coll., 1960; M.D., U. Tex., Dallas, 1964; Ph.D., U. Cambridge, Eng., 1970. Intern Bellevue Hosp., N.Y.C., 1964-65; resident in medicine, fellow cardiology Parkland Hosp., Dallas, 1965-67; research fellow Nat. Heart Inst., Bethesda, Md., 1967-68; vis. research fellow Strangeways Research Lab., Cambridge, 1968-70; asst. prof. to prof. internal medicine and physiology U. Tex., Dallas, 1970-76, prof., dean Grad. Sch., 1976-80; prof., dean Southwestern Med. Sch., 1980—; sci. cons. Strangeways Research Lab.; chmn. research rev. com. Nat. Heart, Lung and Blood Inst. Author: Regulation of Cardiac Metabolism, 1976, Degradative Processes in Heart and Skeletal Muscle, 1980; contbr. articles to profl. jours. Bd. dirs. Tex. br. Am. Heart Assn., 1978—. USPHS spl. research fellow, 1968-70; recipient Research Career Devel. award NIH, 1972, Sybil Eastwood research award, 1976; John Simon Guggenheim Meml. fellow, 1975-76. Mem. Am. Soc. Clin. Investigation, Am. Coll. Cardiology, Royal Soc. Medicine Gt. Britain, Am. Physiol. Soc., Internat. Soc. Heart Research, Am. Fedn. Clin. Research, Assn. Am. Med. Colls., AMA. Subspecialties: Physiology (medicine); Cell biology (medicine). Current work: Cardiac cellular physiology and metabolism; lysosomes; protein turnover. Home: 4128 Southwestern Blvd Dallas TX 75225 Office: 5323 Harry Hines Blvd Dallas TX 75235

WILDER, DAVID RANDOLPH, engineering educator; b. Lorimor, Iowa, June 11, 1929; s. Rex Marshall and Ethel Marie (Busch) W.; m. Donna Moore, June 17, 1951; children: Susan, Michael, Margaret, Bruce. B.S., Iowa State U., 1951, M.S., 1952, Ph.D., 1958. Registered profl. engr., Iowa. Mem. faculty Iowa State U., Ames, 1955—, prof. engring., 1961—, chmn. materials sci. and engring. dept., 1975—. Contbr. numerous engring. articles to profl. publs. Subspecialties: Ceramic engineering; Materials (engineering). Current work: Refractory oxides; sintering; phase equilibria. Patentee in field. Office: Iowa State U 110 Engring Annex Material Sci and Engring Dept Ames IA 50010

WILDUNG, RAYMOND EARL, soil scientist, environmental chemist; b. Van Nuys, Calif., Feb. 24, 1941; m., 1961; 2 children. B.S., Calif. state Poly. Coll., 1962; M.S., U. Wis., 1964, Ph.D. in Soil Sci, 1966. NIH fellow U. Wis., Madison, 1966-67; sr. research scientist Soil-Sediment Sci. Battelle PAC NW Labs., Richland, Wash., 1967-71, program leader, 1971-75, mgr. environ. chemistry, 1975—; affiliate prof. Wash. State U. and Calif. State U.; mem. com. accessory elements, chmn. oil shale panel, mem. com. soil as mineral resource Nat. Acad. Sci.; mem. exec. com. coordination solid waste mgmt. Dept. Energy Oil Shale Task Force. Dept. Agr. grantee, 1968-70; EPA grantee, 1968-71; Dept. Energy grantee, 1968—; nat. Inst. Environ. Health Sci. grantee, 1971—. Mem. AAAS, Am. Chem. Soc., Am. Soc. Agronomy, Internat. Soc. Soil Sci., Soil Sci. Soc. Am. Subspecialties: Soil chemistry; Environmental chemistry. Office: Battelle Pac NW Labs PO Box 999 Richland WA 99352

WILES, DAVID MCKEEN, chemist; b. Springhill, N.S., Can., Dec. 28, 1932; s. Roy McKeen and Olwen Gertrude (Jones) W.; m. Valerie Joan Rowlands, June 8, 1957; children: Gordon Stuart, Sandra Lorraine. B.Sc. with honors, McMaster U., 1954, M.Sc., 1955; Ph.D. in Chemistry, McGill U., 1957. Research officer chemistry div. Nat. Research Council of Can., Ottawa, 1959-66, head textile chemistry sect. chemistry div., 1966-75, dir. chemistry div., 1975—; chmn. Can. High Polymer Forum, 1967-69; v.p. N.Am. Chem. Congress, Mexico City, 1975. Contbr. articles to profl. jours.; mem. editorial adv. bd. numerous profl. jours. Can. Ramsay Meml. fellow, 1957-59. Fellow Chem. Inst. Can. (chmn. bd. dirs. 1972-74, pres. 1975-76, Dunlop Lectr. award 1981), Textile Inst., Royal Soc. Chem. London, Royal Soc. Can.; mem. Am. Chem. Soc., Fiber Soc. Subspecialties: Polymer chemistry; Polymers. Current work: Polymer photochemistry; fiber physics; UV stabilzation; composites, thermal analysis. Patentee in field. Home: 1927 Fairmeadow Crescent Ottawa ON Canada K1H 7B8 Office: Montreal Rd Ottawa ON Canada K1A 0R6

WILEY, ALBERT LEE, JR., physician, radiology and oncology educator; b. Forest City, N.C., June 9, 1936; s. Albert Lee and Mary Louise (Davis) W.; m. Janet Lee Pratt, June 18, 1960; children: Allison, Susan, Mary, Heather. B. Nuclear Engring., N.C. State U.-Raleigh, 1958; M.D., U. Rochester, N.Y., 1963; Ph.D., U. Wis.-Madison, 1972. Cert. Am. Bd. Radiology, Am. Bd. Nuclear Medicine. Engr. Lockheed Nuclear, Marietta, Ga., 1958; intern U. Va. Hosp., Charlottesville, 1963-64; Nat. Cancer Inst. fellow Stanford U. Med. Ctr., Palo Alto, Calif., 1964-65, U. Wis.-Madison, 1965-68; med. dir. U.S. Navy Radiol. Def. Lab., San Francisco, 1968-69; radiation therapist U.S. Navy Hosp., San Diego, 1969-70; asst. prof. U. Tex.; M.D. Anderson Hosp., Houston, 1972-73; prof. radiology and oncology U. Wis. Hosp., Madison, 1973—; assoc. dir. Radiation Oncology, 1979-82; cons. U.S. Nuclear Regulatory Commn., Washington, 1981-82. Contbr. chpts. to books, articles to profl. jours. Served to lt. comdr. USNR, 1959-76. Oak Ridge Inst. Nuclear Studies fellow, 1958-59. Fellow Am. Coll. Preventive Medicine; mem. IEEE, AMA, Soc. Nuclear Medicine, Am. Coll. Radiology, Am. Coll. Nuclear Physicists, Health Physics Soc., Sigma Xi. Republican. Lutheran. Subspecialties: Imaging technology; Nuclear medicine. Current work: Use of computerized tomography and nuclear medicine techniques in radiation therapy planning; combined intro-arterial chemotherapy and radiation therapy. Home: Route 3 Hwy 138 Stoughton WI 53589 Office: U Wis K4B100 CSC 600 Highland Ave Madison WI

WILEY, ALBERT LEE, JR., therapeutic radiologist, educator; b. Forest City, N.C., June 9, 1936; s. Albert Lee and Mary Louise (Davis) W.; m. Janet Lee Pratt, June 18, 1960; children: Allison, Susan, Mary Catherine, Heather. B.Nuclear Engring., N.C. State U., 1958; M.D., U. Rochester, 1963; Ph.D., U. Wis.-Madison, 1972. Diplomate: Am. Bd. Radiology, Am. Bd. Nuclear Medicine. Nuclear engr. Lockheed Nuclear Products, Lockheed Aircraft Corp., Marietta, Ga., 1958; intern in medicine and surgery U. Va. Hosp., Charlottesville, 1963-64; postdoctoral fellow Stanford U. Hosp., Palo Alto, Calif., 1964-65, U. Wis. Hosps., Madison, 1965-68; med. dir. U.S. Naval Radiol. Def. Lab., San Francisco, 1968-69; radiation therapist U.S. Naval Hosp., San Diego, 1969-70, M.D. Anderson Hosp., Houston, 1971-72; prof. human oncology and radiology, 1979—; assoc. dir., clin. dir. radiation oncology U. Wis. Hosps., Madison, 1979-82; cons. U.S. Nuclear Regulatory Commn., 1981-82; tech. advisor Wis. Dept. Health and Social Services; mem. exec. com. U Wis Biomed. Engring. Ctr. Contbr. chpts. to books, articles to profl. jours. Mem. Gov.'s Commn. on UN. Served to lt. comdr. USNR, 1959-76. Oak Ridge Inst. Nuclear Studies fellow, 1958-59. Fellow Am. Coll. Preventive Medicine; mem. Sigma Xi, Tau Beta Pi. Republican. Lutheran. Subspecialties: Radiology; Nuclear engineering. Current work: Medical research, teaching and administration; interfacing of engineering technologies into clinical practice of medicine. Office: U Wis Med Ctr K4 B100 CSC Madison WI 53792

WILEY, MICHAEL DAVID, educator; b. Long Beach, Calif., Nov. 28, 1939; s. David Michael and Elsie Louise (Magnuson) W.; m. Mary Alice Kuehne, Dec. 16, 1961; children: David Michael, Heather Jane. B.S., U. So. Calif., 1961; Ph.D., U. Wash., Seattle, 1969. Teaching asst. U. Wash., Seattle, 1961-63; asst. prof. chemistry Calif. Lutheran Coll., Thousand Oaks, Calif., 1968-74, assoc. prof., 1974—; research assoc. dept. organic chemistry U. Liverpool, Eng., 1981. Calif. State scholar, 1957-61; NSF coop. grad. fellow, 1963-64; NSF summer fellow, 1963, 65. Mem. Royal Soc. Chemistry, Am. Chem. Soc., AAAS, Calif. Assn. Chemistry Tchrs. Subspecialty: Organic chemistry. Office: Dept Chemistry Calif Luth Coll Thousand Oaks CA 91360

WILEY, RONALD GORDON, neurologist, neurobiologist; b. Akron, Ohio, Mar. 21, 1947; s. H.J. and Sara T. (Moore) W.; m. Karen Sue Steffy, July 24, 1970; children: Elizabeth Ann, Kathleen Sara, Allison Christine. B.S., Northwestern U., 1972, Ph.D., 1975, M.D. with distinction, 1975. Diplomate: Am. Bd. Psychiatry and Neurology, Am. Bd. Internal Medicine. Intern, then jr. asst. resident in medicine Peter Bent Brigham Hosp., Boston, 1975-77; resident in neurology N.Y. Hosp., N.Y.C., 1977-80; fellow Lab. Neurobiology, Cornell U. Med. Coll., 1980-82; assoc. attending neurologist La Guardia Hosp., N.Y.C., 1980-82; asst. prof. neurology, instr. pharmacology Vanderbilt U. Med. Sch., Nashville, 1982—; attending neurologist VA Med. Center, Nashville, 1982—. Contbr. articles to sci. publs. Recipient Roche awards in neurosci., G.D. Searle award for research; Tchr. of Yr. award Ill. Coll. Optometry, 1972. Mem. AAAS, Am. Acad. Neurology, Soc. Neurosci., N.Y. Acad. Scis., Sigma Xi (research award), Alpha Omega Alpha. Subspecialties: Neurology; Neurobiology. Current work: Development of suicide transport agents; axonal transport; mechanisms of neurotransmitter secretion, particularly role of synaptic vesicles. Office: 1310 24th Ave S Nashville TN 37203

WILEY, WILLIAM R., research dir.; b. Oxford, Miss., Sept. 5, 1931; s. William Russell and Edna Alberta (Threlkeld) W.; m. Myrtle Louise Smith, Nov. 10, 1952; 1 child, Johari. B.S., Tougaloo Coll., 1954; M.S. (Rockefeller Found. fellow), U. Ill., Urbana, 1960; Ph.D., Wash. State U., 1965. Mgr. cellular and molecular biology sect. Battelle-Northwest, Richmond, Wash., 1969-72, mgr. biology dept., 1974-79, dir. research, 1979—; coordinator Battelle Inst. Life Scis. Program, Richland, 1972-74; George Washington Carver Lectr. Tuskegee Inst., 1967; lectr. Black Exec. Exchange Program. Co-author 2 books; editor: Methods in Enzymology, 1974; contbr. articles to sci. publs. Trustee Gonzaga U., 1982—, bd. regents, 1968-81; bd. dirs. Richland City Library, chmn. bd., 1970-71; trustee Kadlec Hosp., 1972-74. Served with U.S. Army, 1954-56. Mem. Am. Soc. Microbiology, AAAS, Soc. Exptl. Biology and Medicine. Lodge: Kiwanis. Subspecialties: Cell biology; Biochemistry (biology). Current work: Mgmt. research and devel. environ., health, geoscis., space scis. and engring.

WILFRET, GARY JOE, plant breeder and geneticist; b. Sacramento, Oct. 13, 1943; s. Joseph Andrew and Della Catherine (McCurry) W.; m. Janice Eileen Ross, Dec. 6, 1944; children: Catherine, David. A.A., Sacramento City Coll., 1963; B.S., U. Hawaii, 1965, Ph.D., 1968. Asst. prof. Ga. So. Coll., 1968-69; asst. prof. plant genetics U. Fla., 1969-74, assoc. prof., 1974-80, prof., 1980—. Mem. Am. Soc. Hort. Sci., Bot. Soc. Am., Am. Hort. Soc., N. Am. Gladiolus Council. Republican. Lutheran. Subspecialties: Plant genetics; Plant cell and tissue culture. Current work: Breeding and genetics of horticultural crops and tissue culture of these for rapid propagation and production of disease-free explants. Home: 208-65th St Ct NW Bradenton FL 33529 Office: 5007 60th St E Bradenton FL 33508

WILHELM, STEPHEN, plant pathologist; b. Imperial, Calif., Apr. 19, 1919; s. John Henry and Martha (Burgacher) W.; m. Elizabeth Ruth Wilson, July 22, 1944; children: Stephen, Paul, George, Nicholas. B.S., UCLA, 1942; Ph.D., U. Calif.-Berkeley, 1948. Pest control specialist Paul J. Howards Hort. Establishment, Los Angeles, 1937-42; mem. faculty U. Calif.-Berkeley, 1948—, assoc. prof. plant pathology, 1958-60, prof., 1960—; cons. Niklor Chem. Co., Inc., Long Beach, Calif. Author: (with J.E. Sagen) A History of the Strawberry From Ancient Gardens to Modern Markets, 1974. Guggenheim fellow, 1958-59. Fellow Am. Phytopath. Soc.; mem. Am. Hort. Soc., Internat. Hort. Soc. Subspecialty: Plant pathology. Current work: Vascular wilts, root infecting fungi, soil fumigation, resistance breeding. Patentee verticillum wilt resistant olive rootstock. Home: 12770 Skyline Blvd Oakland CA 94619 Office: Dept Plant Pathology U Calif Berkeley CA 94720

WILHM, JERRY L., zoology educator; b. Kansas City, Kans., Apr. 27, 1930; s. Jerome and Ada (Wallace) W.; m. Nona E. Wilhm Reed, May 27, 1955; children: Jerry L., Jacqueline Lynn. B.S., Emporia State U., 1952, M.S., 1955; Ph.D., Okla. State U., 1966. Postdoctoral fellow Oak Ridge Nat. Lab., 1965-66; from asst. prof. to prof. dept zoology Okla. State U., Stillwater, 1966—, head dept. zoology, 1971-74, 81—; cons. Aquatic Life Cons., Stillwater, 1970—. Contbr. articles to profl. jours. Served with U.S. Army, 1952-54. Fulbright lectr., N.Z., 1979. Mem. Am. Soc. Limnology and Oceanography, Ecol. Assn. Am., Fulbright Assn., Okla. Acad. Sci. Current work: Lake restoration; biological effects of organic chemicals. Home: 1123 N Lincoln St Stillwater OK 74074 Office: Dept Zoology Okla State U Stillwater OK 74074

WILKE, CHARLES ROBERT, chemical engineer, educator, investment adviser; b. Dayton, Ohio, Feb. 4, 1917; s. Otto Alexander and Stella M. (Dodge) W.; m. Bernice Lucille Arnett, June 19, 1946. B. Chem. Engring., U. Dayton, 1940; M.S. in Chemistry, State Coll. Wash., 1942; Ph.D. in Chem. Engring, U. Wis., 1944. Asso. engr. Union Oil Co. of Calif., 1944-45, cons., 1952—; instr. chem. engring. Wash. State Coll., 1945-46, U. Calif. at Berkeley, 1946-47, asst. prof., 1947-51, asso. prof., 1951-53, prof., 1953—, chmn. dept. chem. engring., 1953-63, asst. to chancellor acad. affairs, 1967-69, prin. investigator Lawrence Berkeley Lab., 1950—; indsl. cons., 1952—; chmn. bd. C.R. Wilke Internat. Corp., 1970—; cons. editor Reinhold Pub. Co., 1958-64; mem. Calif. Bd. Registration Civil and Profl. Engrs., 1964-72, pres., 1967-69. Contbr. articles to profl. jours. Recipient Walker award Am. Inst. Chem. Engrs., 1965, Colburn award, 1951. Fellow Am. Inst. Chem. Engrs. (past chmn. No. Calif. sect.); mem. Soc. Applied Bacteriology, Am. Chem. Soc., Electrochem. Soc., Nat. Acad. Engring., Am. Soc. Engring. Edn. (Chem. Engring. lecture award 1964), Am. Soc. Microbiology, Sigma Xi, Tau Beta Pi. Clubs: Commonwealth of California, World Trade of San Francisco. Subspecialties: Chemical engineering; Enzyme technology. Current work: Production of chemicals by microbial processes, mircobial and enzyne reactor design, enzymatic conversion processes. Home: 1327 Contra Costa Dr El Cerrito CA 94530

WILKENING, LAUREL LYNN, planetary science educator, researcher; b. Richland, Wash., Nov. 23, 1944; d. Marvin H. and Ruby A. (Barks) W.; m. Godfrey T. Sill, June 18, 1974. B.A., Reed Coll., 1966; Ph.D., U. Calif.-San Diego, 1970. Research assoc. Enrico Fermi Inst., 1972-73; research assoc. chem. dept; U. Chgo., 1972-73; asst. prof. planetary sci. U. Ariz., Tucson, 1973-78, assoc. prof., 1978—, dept. head, 1981—, acting dean scis., 1982—, dir., 1981—; div. scientist planetary div. NASA Hdqrs., Washington, 1980. Editor:

Comets, 1982; contbr. sci. articles to profl. publs. Recipient Ninenger Meteorite award, 1971. Mem. Am. Geophys. Union, Am. Astron. Soc. (div. planetary sci.), Internat. Astron. Union, Meteoritical Soc., AAAS, Phi Beta Kappa. Subspecialties: Planetary science; Space chemistry. Current work: Meteorites, asteroids, comets. Office: U Ariz Space Sci Bldg Tucson AZ 85721

WILKERSON, GARY WARD, optical design engineer; b. Denver, Aug. 30, 1940; s. Dewey Lee and Hazel Blanche (Young) W.; m. Barbara Ann Barrett, Dec. 1, 1973; children: Gary Samuel, Hannah Barbara, Jonathan Dewey, Hazel Sarah, Helen Ruth. Student, Colo State U., 1958-60; B.S. in Math, U. Ariz., 1962; M.S. in Optical Scis, 1970. Research assoc. Steward Obs., U. Ariz., Tucson, 1964-66; research assoc. Optical Scis. Ctr., 1967-70; staff engr. Kollmorgen Corp., Northampton, Mass., 1966-67; mem. tech. staff Aerospace Corp., Los Angeles, 1970-72; with MTS-V Rockwell Internat. Space Div., Seal Beach, Calif., 1972-77; staff engr. Martin Marietta, Orlando, Fla., 1977-79; cons. Wilkerson Assocs., Kissimmee, Fla., 1979-80; asst. project engr. United Technologies Research Ctr., West Palm Beach, Fla., 1980—; lectr. on astronomy and NASA projects. Contbr. articles to profl. jours. Mem. Am. Astron. Soc. (assoc.), Soc. Photo-Optical Instrumentation Engrs., Optical Soc. Am., Am. Security Council.; mem. Am. Def Preparedness Assn. Baptist. Subspecialty: Optical engineering. Current work: Composite materials science, optical engineering, and systems engineering applied to structures and mirrors of remote sensing satellite and ground-based telescopes. Optical design engineering for laser microsurgery and general electro-optical infrared systems. Office: United Technologies Research Ctr PO Drawer 4181 West Palm Beach FL 33402

WILKERSON, M. SUSAN, air force officer, astronomer; b. San Diego, Nov. 8, 1953; d. Elmer Davis and Dyxie Dathyne (Canaday) W. B.A., U. Calif., San Diego, 1975; Ph.D., U. Ariz., 1979. Grad. research asst Steward Obs., U. Ariz., Tucson, 1975-77; research asst. Kitt Peak Nat. Obs., Tucson, 1977-79; commd. 2d lt. U.S. Air Force, 1980, advanced through grades to capt., 1984; sci. sr. staff Sacramento Peak Obs., Sunspot, N.Mex., 1980-81; chief, laser recorder systems br. Hdqrs. Space Div./YBE, Los Angeles, 1981—. Contbr. articles to profl. jours. Mem. Am. Astron. Soc., Astron. Soc. of Pacific. Subspecialties: Optical astronomy; Optical image processing. Current work: Astronomer in solar type stars, astronomical image reconstruction at near diffraction and surface photometry of galaxies. Home: 1616 W 256th St Harbor City CA 90710 Office: Hdqrs Space Division/YBE Box 92960 World Way Postal Center Los Angeles AFS CA 90009

WILKERSON, ROBERT DOUGLAS, pharmacologist; b. Wilson, N.C., Aug. 5, 1944; s. Rainey Bryan and Blanche (Boyette) W.; m. Dottie Bullock, July 31, 1965; children: Robert Douglas, Julie Diane. B.S., U. N.C., 1967; M.S., Med. U. S.C., 1969, Ph.D., 1972. Postdoctoral fellow Tulane U. Sch.Medicine, New Orleans, 1971-73; asst. prof. U. South Ala. Coll. Medicine, Mobile, 1973-76, assoc. prof., 1976-79; assoc. prof. pharmacology Med. Coll. Ohio, Toledo, 1979—. Editor: Cardiac Pharmacology, 1981; author articles. Mem. Am. Soc. Pharmacology and Exptl. Therapeutics, Am. Coll. Clin. Pharmacology. Presbyterian. Subspecialty: Pharmacology. Current work: Cardiovascular pharmacology; ischemic heart disease, cardiac drugs. Home: 7226 Winsford Ln Sylvania OH 43560 Office: Dept Pharmacology Med Coll Ohio Toledo OH 43699

WILKES, DONALD FANCHER, inventor, designer, mechanical engineer; b. Portland, Oreg., July 20, 1931; s. Gordon Buell and Catherine Amey (Fancher) W.; m. Joan Adell Wilkes, June 27, 1954; children: Martin Carey, Norma Jean, Roger Allen. B.S.M.E., Wash. State U., 1954; M.S.M.E., U. N. Mex., 1962. Staff research mech. engr. Sandia Corp., Albuquerque, 1964-67; sr. v.p., corp. dir., inventor, engr., dir. research and devel. Rolamite, Inc. (and successor firm Foothill Lab. div. Foothill Group, Inc.), Albuquerque, 1968-78; mgr., prin. tech. staff mem. Arco Solar Industries (subs. Atlantic Richfield Co.), Albuquerque, 1978-81; mgr., prin. tech. person Albuquerque Mech. Lab., 1981—; tech. cons. Elder Presbyterian Ch. Contbr. articles to profl. jours. Served with AUS, 1955-57. Mem. ASME, Phi Kappa Phi, Tau Beta Pi, Sigma Tau, Alpha Phi Omega, Pi Tau Sigma. Subspecialty: Theoretical and applied mechanics. Current work: Improved ways of accomplishing old and new objectives for human beings which are conservative of energy and basic resources. Holder 37 patents.

WILKES, H(ILBERT) GARRISON, biologist, educator, plant breeder; b. Los Angeles, Oct. 2, 1937; s. Hilbert Garrison and Margret Lee (Boggs) W.; m. Susan Kreps Redwood, May 29, 1965; m. Marie Dalton Gibby, Apr. 9, 1978; children: Nathan, Jennifer, Andrew, Katharan. B.A., Pomona Coll., 1959; Ph.D., Harvard U., 1966. Asst. prof. Tulane U., 1966-70; assoc. prof. biology U. Mass., Boston, 1970—; cons. tropical biology, plant genetic resources, maize germplasm; Fulbright fellow, India, 1959-60, Harvard travelling fellow, C. Am., 1963-64, Indo-Am. fellow, India, 1978-79; mem. exec. com. Assembly Life Sci., Nat. Acad. Sci., 1973-77. Contbr. numerous articles on maize evolution, econ. botany, plant genetic resources to profl. jours. Woodrow Wilson fellow, 1960-61; NSF grantee, 1968-70, 72, 74-78. Mem. Am. Bot. Soc., Soc. Econ. Botany, Soc. Study Evolution, Linnean Soc. Subspecialties: Plant genetics; Evolutionary biology. Current work: Evolution under domesticization; genetic resources; econ. botany; evolution crop plants, especially maize and its wild relatives; conservation of plant genetic resources; gene banks.

WILKINS, J. ERNEST, JR., mathematician, company executive; b. Chgo., Nov. 27, 1923; m. (widower); 2 children. S.B., U. Chgo., 1940, S.M., 1941, Ph.D. in Math, 1942; B.M.E., N.Y. U., 1957, M.M.E., 1960. Instr. in math. Tuskegee Inst., 1943-44; from assoc. physicist to physicist Manhattan Project Metall. Lab., U. Chgo., 1944-46; mathematician Am. Optical Co., 1946-50; sr. mathematician Nuclear Devel. Corp. Am., 1950-55, mgr. dept. physics and math., 1955-57, asst. mgr. research and devel., 1958-59, mgr., 1959-60; asst. chmn. theoretical physics dept. Gen. Atomic div. Gen. Dynamics Corp., 1960-65, asst. dir. lab., 1965-70; Disting. prof. applied math. physics Howard U., 1970-77; assoc. gen. mgr. EG&G Idaho, Inc., Idaho Falls, 1977-80, dep. gen. mgr., 1980—. Mem. AAAS, Am. Math. Soc., Optical Soc. Am., Am. Nuclear Soc., Soc. Indsl. and Applied Math. Subspecialty: Applied mathematics. Office: EG&G Idaho Inc PO Box 1625 Idaho Falls ID 83415

WILKINSON, DAVID STANLEY, pathologist, consultant, researcher; b. Richmond, Va., Feb. 2, 1945; s. Herbert Carroll and Hattie Mae (Vaughan) W.; m. Judith Farish Pace, June 16, 1967; children: Jill Marie, Julie Lynne, Virginia Ann. B.S. in Chemistry, Va. Mil. Inst., Lexington, 1967; Ph.D. in Exptl. Oncology and Pathology, U. Wis.-Madison, 1971; M.D., U. Miami, 1978. Diplomate: Am. Bd. Pathology. Commd. 2d lt. U.S. Army, 1967, advanced through grades to maj., 1982; fellow McArdle Lab. Cancer Research U. Wis., 1967-71; asst. prof. biochemistry U. South Fla., Tampa, 1972-76; resident in pathology Walter Reed Army Med. Ctr., Washington, 1978-82; instr. pathology Uniformed Services U. of Health Scis., Bethesda, 1979-82; chief clin. pathology Eisenhower Army Med. Ctr., Ft. Gordon, Ga., 1982—; mem. clin. faculty Med. Coll. Ga. Augusta; lectr. in field. Contbr. articles to profl. jours. Damon Runyon-Walter Winchell Cancer Fund grantee, 1973; Am. Cancer Soc. grantee, 1973; Nat. Cancer Inst. grantee, 1975. Mem. Am. Assn. Cancer Research, Soc. Exptl. Biology and Medicine, Am. Soc. Clin. Pathologists. Republican. Club: VMI Keydet (Lexington, Va.). Subspecialties: Pathology (medicine); Cancer research (medicine). Current work: Biochemical pathology, experimental cancer chemotherapy, mechanisms of action of antitumor agents, therapeutic drug monitoring. Office: Department Pathology Dwight D Eisenhower Army Medical Center Fort Gordon GA 30905

WILKINSON, GRANT ROBERT, pharmacologist, educator; b. Derby, Eng., Aug. 27, 1941; s. Arthur Henry and Gwendoline Mary (Fox) W.; m. Margaret Kay Fletcher, Aug. 8, 1964 (div. 1978); children: Grant R., Nicole E.; m. June Zoe Dass, July 12, 1978; children: Tracey A., Erika L. B.Sc., U. Manchester, Eng., 1963; Ph.D., U. London, 1966. Postdoctoral fellow dept. pharmacology U. Calif., San Francisco, 1966-68; asst. prof. Coll. Pharmacy, U. Ky., Lexington, 1968-71; asst. prof. dept. pharmacology Vanderbilt U., Nashville, 1971-73, assoc. prof., 1973-78, prof., 1978—; assoc. dir. Ctr. for Clin. Pharmacology, 1980—; cons. NIH; indsl. cons. Fellow AAAS; mem. Am. Soc. for Pharmacology and Exptl. Therapeutics, Am. Soc. Clin. Pharmacology and Therapeutics. Subspecialties: Pharmacology; Pharmacokinetics. Current work: Drug disposition in man; factors affecting this such as genetics, disease-states, aging, drug interactions. Pharmacokinetics. Bioanalysis. Office: Dept Pharmacology Vanderbilt U Nashville TN 37232

WILKINSON, ROBERT EUGENE, researcher, agronomy educator; b. Oilton, Okla., Oct. 24, 1926; s. Olney S. and Grace Elma (Curry) W.; m. Evelyn Dolores Smith, Oct. 31, 1951; children: Olney Thomas, Randall David. B.S., U. Ill., 1950; M.S., U. Okla., 1952; Ph.D., U. Calif.-Davis, 1956. Research plant physiologist U.S. Dept. Agr., Clarkedale, 1957-62, Los Lunas, N.Mex., 1962-65; prof. dept agronomy U. Ga. Agrl. Expt. Sta., Experiment, 1965—; cons. U. Piracicabe, Brazil, 1978; sr. Fulbright-Hayes lectr. U. Turku, Finland, 1974-75, U. Nova Sad, Yugoslavia, 1975. Author: How to Know the Weeds, 1972; Editor: Research Methods in Weed Science, 1972. Served with USN, 1944-45. Am. Soybean Research Found. grantee, 1980-81. Mem. Am. Soc. Plant Physiology, Weed Sci. Soc. Am., So. Weed Sci. Soc., Pesticide Biochemistry and Physiology. Methodist. Lodge: Kiwanis. Subspecialties: Plant physiology (agriculture); Plant growth. Current work: Herbicide mechanism of action, herbicide biochemistry, metabolism and degradation, plant response to environment, plant growth regulators. Office: Dept Agronomy GA Sta Experiment GA 30212 Home: 655 Laura Dr Griffin Ga 30223

WILKINSON, ROY MIELE, computer systems designer; b. Ithaca, N.Y., Apr. 23, 1949; s. Robert E. and Antoinette (Miele) W. A.B., Cornell U., 1970; M.S. in Physics, U. Wash.-Seattle, 1973, U. Wash.-Seattle, 1975. Programmer Burroughs Corp., Goleta, Calif., 1975-79; sr. software engr. Intel Corp., Hillsboro, Oreg., 1979—. Mem. Assn. for Computing Machinery, Am. Guild Organists. Subspecialties: Operating systems; Software engineering. Current work: Operating system design and system architecture, design and development of an advanced multiprocessing operating system for a state-of-the-art object-oriented microprocessor. Office: JF1-2-091 Intel Corp 2111 NE 25th Ave Hillsboro OR 97123

WILKNISS, PETER EBERHARD, oceanographer, radiochemist, government research administrator; b. Berlin, Sept. 28, 1934; m., 1963; 2 children. M.S., Munich (W.Ger.) Tech. U., 1959, Ph.D. in Radiochemistry, 1961. Program mgr. Nat. Ctr. Atmospheric Research Program NSF, 1975-76, program mgr., 1976-80, team mgr., 1980, dir., 1980-81, sr. sci. assoc., office dir., Washington, 1981-82, dep. asst. dir. sci., tech. and internat. affairs, 1982—; liaison mem. NSF-NRC Marine Bd., 1978-81. Mem. AAAS, Am. Geophys. Union, Sigma Xi. Subspecialty: Oceanography. Office: Sci Tech and Internat Affairs 1800 G St NW Washington DC 20550

WILL, CLIFFORD MARTIN, physicist, educator; b. Hamilton, Ont., Can., Nov. 13, 1946; s. Frank Earl and Marjorie Winnifred (Dunk) W.; m. Leslie Saxe, June 26, 1970; children: Elizabeth, Rosalie. B.S., McMaster U., 1968; Ph,.D., Calif. Inst. Tech., 1971. Instr. in physics Calif. Inst. Tech., Pasadena, 1971-72; Enrico Fermi Fellow U. Chgo., 1972-74; asst. prof. physics Stanford (Calif.) U., 1974-81; asso. prof. physics Washington U., St. Louis, 1981—; researcher. Author: Theory and Experiment in Gravitational Physics, 1981; contbr. articles to profl. publs. in field. Research grantee NASA, 1975-81, NSF, 1982—; Alfred P. Sloan Found. research fellow, 1975-79. Mem. Am. Phys. Soc., Am. Astron. Soc., Internat. Astron. Union., Internat. Soc. Gen. Relativity and Gravitation, Sigma Xi. Jewish. Subspecialties: Relativity and gravitation; Theoretical astrophysics. Current work: Relativity and observable consequences of gen. relativity, black holes, gravitational waves, cosmology, empirical founds. of gen. relativity. Office: Dept Physics Washington Univ Saint Louis MO 63130

WILL, LOREN AUGUST, veterinarian, epidemiologist, researcher; b. Mpls., Dec. 31, 1942; s. Vernon Eugene and Lillian Dorothy (LaFore) W.; m. Joan Heather Allibone, June 13, 1964; children: Heather, Michelle, Christina, Davina, John, Geoffrey. B.A., St. Mary's Coll., Winona, Minn., 1965; B.S., U. Minn., St. Paul, 1967, D.V.M., 1969, M.P.H., 1975. Tchr. St. Josephs Sch., Vanderhoof, B.C., 1963-64; vet. clinician, Prince George, B.C., 1969-70; environ. health and cancer research scientist U. Iowa, Iowa City, 1974-78; asst. prof. pub. health Coll. Vet. Medicine, Iowa State U., Ames, 1978—. Pack chmn. Cub Scouts. Mem. AAAS, Assn. Animal Allergic Veterinarians (founder , dir., newsletter pub.), Assn. Tchrs. of Vet. Pub. Health and Preventive Medicine, Assn. Am. Vet. Med. Colls., Acad. Vet. Allergy. Roman Catholic. Subspecialties: Allergy; Epidemiology. Current work: Allergy to animals in occupational groups, nature of the problem, its extent, its medical, physiologic, and social consequences. Home: 1910 George Allen Dr Ames IA 50010 Office: 2116 Veterinary Medicine Iowa State University Ames IA 50011

WILL, PETER MILNE, computer scientist; b. Peterhead, Scotland, Nov. 2, 1935; came to U.S., 1964; s. James and Margaret (Milne) W.; m. Angela Hay Giulianotti, Mar. 21, 1959; children: Christopher D.P., Jonathan R. G., Gabrielle H.C. B.Sc., U. Aberdeen, 1958, Ph.D., 1960. With AMF, Stamford, Conn., 1961-65, with, U.K., 1961-65; computer scientist IBM Research, Yorktown Heights, N.Y., 1965-78, IBM Advanced Systems, Boca Raton, Fla., 1978-80, Schlumberger Well Services, Houston, 1980-83; computer scientist, dir. systems sci. Schlumberger-Doll Research, Ridgfield, Conn., 1983—; cons. in field. Contbr. articles to profl. jours. Mem. IEEE, ACM, Nat. Research Council (mem. computer sci. & tech. bd.). Subspecialties: Artificial intelligence; Robotics. Current work: Artificial intelligence, interactive systems, work station based computing environments and CAD/CAM. Developer of vision devices. Office: Schlumberger-Doll Research Ctr Old Quarry Rd Ridgefield CT 06877

WILLARD-GALLO, KAREN ELIZABETH, molecular biologist; b. Oak Ridge, July 8, 1953; d. Harvey Bradford and Isabella Victoria (Rallis) Willard; m. James Paul Gallo, July 31, 1982. Student, U. Reading, Eng., 1973-74; A.B. in Biology, Randolph-Macon Woman's Coll., 1975; M.S. in Immunology, Va. Poly. Inst., 1978; Ph.D. in Molecular Biology, Va. Poly. Inst., 1981. Grad. teaching asst. immunology Va. Poly. Inst., Blacksburg, 1976-78; resident student assoc. Argonne Nat. Lab., 1978-81, postdoctoral fellow, 1981-82; research scientist Ludwig Inst. Cancer Research, Brussels, 1982—; Cons. in field. Contbr. chpts. to books and articles to profl. jours. Recipient Teaching Excellence award Va. Poly. Inst., 1977, 78; Research Inst. Cell Biology fellow, 1977. Mem. Am. Soc. Cell Biology, Electrophoresis Soc. Subspecialty: Molecular biology. Current work: Development of tests for more accurate and earlier diagnosis of various forms of human leukemia. Patentee method for early detection infectious mononucleosis. Home: Rue de Pinchart 16 1340 Ottignie Belgium Office: Ludwig Inst Cancer Research Ave H Hippocrate UCL 74.59 B-1200 Brussels Belgium

WILLCOX, PHILLIP JAMES, project engineer, scientific consultant; b. Winterset, Iowa, June 28, 1935; s. Irl F. and Margaret J. (Perry) W.; m. Marlene N. Clark, June 18, 1955 (dec. Feb. 1982); children: James M., John A. B.S. in Math, Ind. Inst. Tech., Ft. Wayne, 1961, 1962, M.S., Akron U., 1965; postgrad. in astrophysics and philosophy, UCLA and U. Calif.-Berkeley, 1972-78. Staff scientist Goodyear Aerospace Co., Akron, Ohio, 1962-65; sr. staff scientist Heliodyne Corp., Norton AFB, Calif., 1965-66; project engr., sect. mgr. TRW DSSG, Redondo Beach, Calif., 1966-79; ind. cons. Magnavox, TRW, 1979-82; project engr. Magnavox GIEC, Ft. Wayne, 1982—; cons. TRW DSSG, 1979—; instr. physics and math. Huntingdon Cath. High Sch., Ind. Inst. Tech., 1979—; Assoc. fellow AIAA; guest lectr. Author: The UFO Question, 1976, Modern Cosmology, 1982, High Energy Astronomy, in preparation; contbr. articles tech. jours. Served with USMC, 1952-55. Mem. AAAS, Planetary Soc. (charter). Democrat. Subspecialties: Electrical engineering; High energy astrophysics. Current work: Electronic warfare development, ballistic missile development, high energy astrophysics, scientific education. Home: 4585N 615W Huntington IN 46750 Office: Magnavox GIEC 1313 Production Rd Fort Wayne IN 46808

WILLENBROCK, FREDERICK KARL, engr., educator; b. N.Y.C., July 19, 1920; s. Berthold Daniel and Anna Marie (Koniger) W.; m. Mildred Grace White, Dec. 20, 1944. Sc.B., Brown U., 1942; M.A., Harvard U., 1947, Ph.D., 1950. Research fellow, lectr. and asso. dean Harvard U., Cambridge, Mass., 1950-67; provost, prof. faculty engring. and applied sci. SUNY, Buffalo, 1967-70; dir. Inst. Applied Tech., Nat. Bur. Standards, Washington, 1970-76; dean Sch. Engring. and Applied Sci., So. Meth. U., Dallas, 1976-81, Cecil H. Green prof. engring., 1976—. Contbr. articles to profl. jours. Served with USN, 1943-46. Recipient Disting. Engring. award Brown U., 1962; Gold medal U.S. Dept. Commerce, 1975. Fellow IEEE, AAAS; mem. Nat. Acad. Engring., Am. Phys. Soc., Am. Soc. for Engring. Edn., ASTM, Sigma Psi, Tau Beta Phi. Office: Sch of Engring and Applied Sci So Meth Univ Dallas TX 75275

WILLETT, ROGER DUWAYNE, chemist, educator; b. Northfield, Minn., July 13, 1936; s. Leon Clifford and Inga (Overby) W.; m. Thelma Clarice, June 7, 1957; children: Juanita Willett Hansen, DuWayne, Kathryn, Andrea, Eric, Dallas. B.S., St. Olaf Coll., Northfield, 1958; Ph.D., Iowa State U., 1962. Instr. chemistry Wash. State U., Pullman, 1962-64, asst. prof., 1964-67, assoc. prof., 1967-72, prof., 1972—, chmn. dept. chemistry, 1973-77. Contbr. articles to profl. jours. NATO sr. postdoctoral fellow, 1978; Fulbright-Hays vis. scholar, Netherlands, 1981. Mem. Am. Chem. Soc., Am. Crystallographic Assn., Sigma Xi. Subspecialties: Solid state chemistry; Magnetic physics. Current work: Design and synthesis of new magnetic materials; use of magneto-structural correlations to design new magnetic materials, characterization of magnetic materials; study of phase transitions. Office: Dept Chemistry Wash State U Pullman WA 99164

WILLIAMS, ALAN KEISER, nuclear fuel cycle executive; b. Harrisburg, Pa., Dec. 19, 1928; s. Paul Rupp and Margaret Helen (Keiser) W.; m. Barbara Elaine Martin Hanson, Aug. 8, 1952 (div. 1975); children: Margaret Vivian Westfall, Bryn Barbara Stuart, Andrew Hanson Williams; m. Carolyn Boatner, Aug. 15, 1975. B.A., U. No. Colo., Greeley, 1952. Sr. research mgr. Dow Chem. Co., Rocky Flats, Colo., 1952-74; v.p. Allied Gen. Nuclear Services, Barnwell, S.C., 1974—; cons. Los Alamos Nat. Lab., 1982—; co-chmn. tech. Internat. Am. Nuclear Soc.-European Nuclear Soc. Conf. Fuel Cycles, Brussels, Belgium, 1982. Served with AUS, 1946-47. Mem. Am. Nuclear Soc. (dir. 1982—), Am. Chem. Soc., AAAS. Subspecialties: Nuclear fission; Nuclear engineering. Current work: Separations chemistry with respect to the actinide elements, spent fuel reprocessing, nuclear waste management. Home: 715 Winged Foot Dr Aiken SC 29801 Office: Allied Gen Nuclear Services PO Box 847 Barnwell SC 29812

WILLIAMS, CAROL ANN, astronomer, mathematician, educator, consultant, researcher; b. Stratford, N.J., Oct. 3, 1940; d. Thomas Ambrose and Evelyn Benton (Faulhaber) W. B.A. in Math, Conn. Coll., 1962; Ph.D. in Astronomy, Yale U., 1967. Instr. Conn. Coll., 1966-67; research astronomer Yale U., 1967-68; assoc. prof. math. U. South Fla., Tampa, 1968—; assoc. prof. astronomy U. Fla., Gainesville, 1973—; assoc. research engr. Jet Propulsion Lab., 1964-65, cons., 1965—, U.S. Naval Obs., 1980—, Space Telescope Sci. Inst., 1983—. Editor: Celestial Mechanics; contbr. articles to astron. jours. NSF grantee, 1972, 75. Mem. Am. Astron. Soc. (div. on dynamical astronomy), Internat. Astron. Union, Royal Astron. Soc. Subspecialties: Ocean thermal energy conversion; Satellite studies. Current work: Celestial mechanics, specializing in planetary theory, three body problems; astrometry—star positions. Office: U South Fla PHY 356 Math Dept Tampa FL 33620

WILLIAMS, CARROLL MILTON, biologist, entymolaigist, educator; b. Richmond, Va., Dec. 2, 1916; s. George Leslie and Jessie Ann (Henricks) W.; m. Muriel Anne Voter, June 26, 1944; children: John Leslie (dec.), Wesley Conant, Peter Glenn (dec.), Roger Lee. B.S., U. Richmond, 1937, D.Sc. (hon.), 1960; M.A., Harvard U., 1938, Ph.D., 1941, M.D., 1946. Jr. prize fellow Harvard U., Cambridge, Mass., 1941-46, asst. prof. biology, 1946-48, assoc. prof. biology, 1943-53, prof. biology, 1953-65, Benjamin Bussey prof. biology, 1965—, chmn. dept. biology, 1959-62, chmn., 1975-79; cons. in field. Author, co-author numerous papers in sci. jours. Mem. ACLA, Planned Parenthood, Zero Population Growth, Citizens for a Livable World. Cambridge U. Guggenheim fellow, 1955-56; recipient Harvard U. Med. Sch. Boylston medal, 1961, Harvard U. George Ledlie award, 1967, U. Chgo. Howard Taylor Ticketts award, 1969. Fellow AAAS (council 1952-55, 74-77, Newcomb Cleveland prize 1950); mem. Nat. Acad. Scis. (chmn. zoology and anatomy sect. 1970-73), Am. Philos. Soc., Inst. Medicine, Phi Beta Kappa, Sigma Xi. Democrat. Clubs: Harvard Faculty, Examiner, Cambridge Scientific, Signet Soc. Subspecialties: Developmental biology. Current work: Endocrinology and developmental biology of insects. Office: Harvard Biological Laboratories 16 Divinity Ave Cambridge MA 02138 Home: 27 Eliot Rd Lexington MA 02173

WILLIAMS, CHRISTOPHER JOHN, biometrician, consultant, researcher; b. Alor Star, Malaysia, Aug. 3, 1926; s. Fabian Sebastian Nicholas and Joyce Angahard (Jenkins) W.; m. Catherine Anne, Oct. 7, 1961; children: Christopher James, David Cameron, Russell Fabian. B.Sc. A., U. B.C., Can., 1967, M.Sc., 1971, Ph.D., 1973. Dairy farmer, Duncan, B.C., 1947-66; researcher scientist Research br. Agr. Can., Ottawa, Ont., 1973—. Contbr. numerous articles to various publs. Served with airborne inf. Can. Army. Mem. Geneics Soc. Am., Statis. Soc. Can., Genetic Soc. Can., Can. Soc. Animal Sci., Agr. Inst. Can. Subspecialties: Statistics; Animal genetics. Current work: Kinetics of uptake and disposition of vitamins and minerals in farm ruminants; analysis of covariance and regression. Offic: Engring and Stat Research Ins Research Br Agr Can Bldg 54 CE Ottawa ON Canada K1A OC

WILLIAMS, CURTIS ALVIN, JR., educator, researcher, editor; b. Moorestown, N.J., June 26, 1927; s. Curtis A. and Nola Allen (Johnson) W.; m. Marjorie R. King, Jan. 20, 1960; children: Jennifer, Scott, Elisabeth. Student, Swarthmore Coll., 1946-49; B.S., Pa. State U., 1950; Ph.D., Rutgers U., 1954. Fellow Pasteur Inst., Paris, 1952-54, Carlsberg Lab., Copenhagan, 1954-56; research assoc. Rockefeller U., 1955-60, asst. prof., 1960-65, assoc. prof., 1965-70, adj. prof., 1970-78; prof. biology SUNY, Purchase, 1969—, dean, 1969-80. Editor: Methods in Immunology and Immunochemistry, Volumes I-V; contbr. numerous articles in field. Grantee in field. Mem. Am. Assn. Immunology, Soc. Neurosci., Am. Soc. Microbiology, Electrophoresis Soc. (recipient Founders award 1982). Subspecialties: Neuroimmunology; Neurochemistry. Current work: Research on animal models and mechanism for neuropsychiat. disorders associated with autoimmune diseases. Home: 436 Manor Ridge Rd Pelham NY 10803 Office: Division of Natural Science State University of New York Purchase NY 10577

WILLIAMS, DANIEL FRANK, zoology educator; b. Redmond, Oreg., Nov. 20, 1942; s. John Frank and Margaret Lucille (Zehner) W.; m. Susan Diane Waltman, Nov. 28, 1981; children by previous marriage: Matthew, Amy. B.A., Central Wash. State Coll., 1966; M.S., U. N.Mex., 1968, Ph.D., 1971. Lab. technician Shell Oil Co., Seattle, 1962-63; grad. teaching asst. U. N.Mex., Albuquerque, 1966-71; asst. prof. biology Calif. State U.-Stanislaus, 1971-75, assoc. prof., 1975-80, prof. zoology, 1980—; postdoctoral fellow Carnegie Inst., Pitts., 1977-78, research assoc., 1978—; dir. Inst. Ecology, Calif. State Coll., 1977-82; cons. Forest Service, U.S. Dept. Agr., Berkeley, Calif., 1977—. Author: Systematics and Ecogeographic Variation of the Apache Pocket Mouse, 1978, Mammalian Species of Special Concern in California, California Wildlife and Their Habitats, 1980; Editor: (with Sydney Anderson) Mammalian Species, 1978-82. NDEA Title IV fellow HEW, 1966-69; Ford Found. fellow, 1964-66; U.S. Dept. Agr. grantee, 1977—. Mem. Am. Soc. Mammalogists, Soc. Study of Evolution, Soc. Systematic Zoology, Ecol. Soc. Am., Wildlife Soc., Phi Sigma. Subspecialties: Resource management; Evolutionary biology. Current work: Management and conservation of nongame wildlife species; ecology, evolution and systematics of mammals; habitat inventory of small mammals; cytoegenetics. Office: Dept Biology Calif State Coll Turlock CA 95380

WILLIAMS, DAVID JOHN, III, veterinarian, educator, consultant; b. Cordele, Ga., Sept. 22, 1927; s. David John and Mary Boisclair (Kiker) W.; m. Mary Carolyne Mace, Feb. 20, 1948; children: Carole, David, Mace. D.V.M., U. Ga., 1953, B.S., 1961; M.S., Auburn U., 1963. Diplomate: Am. Coll. Theriogenologists. Pvt. practice vet. medicine, Cordele, 1953-60; instr. U. Ga., 1960-61; instr. dept. medicine and surgery Auburn (Ala.) U., 1961-63, asst. prof., 1963-66, assoc. prof., 1966, U. Ga. Sch. Vet. Medicine, Athens, 1966-73, prof. dept. large animal medicine, 1973—, dir. satellite herd health unit, 1976-80; research assoc. Royal Vet. Coll., Stockholm, 1964-65; cons. Palma Ranch, Tupa, Sao Paulo, Brazil, 1975, Colegio Taylor, Bahia, Brazil, 1979, Drs. Club Internat. Contbr. articles to profl. jours. Served with Hosp. Corps. USN, 1945-47. Recipient A.M. Mills award, 1981. Mem. AVMA, Ga. Vet. Medicine Assn., Soc. Theriogenology, Sigma Xi, Phi Kappa Phi, Phi Zeta. Baptist. Club: Doctors Internat. (N.Y.C.). Subspecialty: Theriogenology. Current work: Animal reproduction; theriogenology.

WILLIAMS, DONALD ELMER, chemistry educator; b. Kansas City, Mo., Mar. 7, 1930; s. Elmer Edwin and Esther Mary (Stephens) W.; m. Audrey Parker; children: Bruce David, Joanna Alaine. A.B., William Jewell Coll., 1950; Ph.D., Iowa State U., 1964. Asst. chemist Ames (Iowa) Lab., AEC, 1962-67, assoc. chemist, 1967; assoc. prof. U. Louisville, 1967-72, prof. chemistry, 1972—. Contbr. articles to profl. jours. NIH Research grantee, 1983. Mem. Am. Chem. Soc., Am. Crystallographic Assn. Subspecialties: Physical chemistry; Crystallography. Current work: Intermolecular forces in crystals. Office: Dept Chemistry U Louisville Louisville KY 40292

WILLIAMS, DONALD JOHN, research physicist; b. Fitchburg, Mass., Dec. 25, 1933; s. Toivo John and Ina (Kokkinen) W.; m. Priscilla Mary Gagnon, July 4, 1953; children—Steven John, Craig Mitchell, Eino Stenroos. B.S., Yale U., 1955, M.S., 1958, Ph.D., 1962. Sr. staff physicist Applied Physics Lab., Johns Hopkins U., 1961-65; head particle physics br. Goddard Space Flight Center, NASA, 1965-70; dir. Space Environ. Lab., NOAA, Boulder, Colo., 1970—; prin. investigator Energetic Particles expt. NASA Galileo Mission, 1977—; research physicist Johns Hopkins U. Applied Physics Lab., 1982—; Mem. nat. and internat. sci. planning coms. Author: (with L. R. Lyons) Quantitative Aspects of Magnetospheric Physics, 1983; Assoc. editor: Jour. Geophys. Research, 1967-69; editor: (with G.D. Mead) Physics of the Magnetosphere, 1969, Physics of Solar-Planetary Environments, 1976; mem. editorial bd.: Space Sci. Revs, 1975—; contbr. (with G.D. Mead) articles to profl. jours. Served to lt. USAF, 1955-57. Sci. Research award, 1974; Disting. Authorship award, 1976. Mem. AAAS, Am. Geophys. Soc., Am. Phys. Soc., Explorers Club. Subspecialties: Satellite studies; Planetary science. Current work: Research in space plasma physics applied to solar system physics, solar-terrestrial physics and planetary environments including the earth. Home: 5611 Suffield Court Columbia MD 21044

WILLIAMS, GARY MURRAY, biomedical researcher; b. Regina, Sask., Can., May 7, 1940; s. Murray Austin and Selma Ruby (Domstad) W.; m. Christine Lundberg, Nov. 26, 1966; children: Walter, Jeffrey, Ingrid. B.A., Washington and Jefferson Coll., 1963; M.D., U. Pitts., 1967. Diplomate: Am. Bd. Pathology, Am. Bd. Toxicology. Intern, then resident Mass. Gen. Hosp., Boston, 1967-69; staff assoc. Nat. Cancer Inst., Bethesda, Md., 1969-71; asst. prof. Temple U., 1971-75; chief div. pathology and toxicology Naylor Dana Inst. Disease Prevention, Valhalla, N.Y., 1979-80, chief div. pathology and toxicology and assoc. dir. inst., 1980—; research prof. N.Y. Med. Coll. Contbr. articles to profl. jours. Served with USPHS, 1969-71. Recipient Sandford award Am. Soc. Clin. Pathologists, 1967; Arnold J. Lehman award Soc. Toxicology, 1982. Mem. Am. Assn. Cancer Research, Am. Assn. Pathologists, Internat. Acad. Pathology, Environ. Mutagen Soc.; mem. Assn. Univ. Pathologists; Mem. Soc. Toxicologic Pathologists; mem. Toxicology Soc.; Mem. Phi Beta Kappa, Alpha Omega Alpha. Lutheran. Club: Vasa Order. Subspecialties: Pathology (medicine); Cell study oncology. Current work: Mechanism of action of chemical carcinogens. Home: 8 Elm Rd Scarsdale NY 10583 Office: Naylor Dana Inst Am Health Found 1 Dana Rd Valahalla NY 10595

WILLIAMS, GORDON LEE, research molecular geneticist, genetic engr., educator; b. Mechanicsburg, Pa., Sept. 9, 1947; s. Frank B. and Eva M. (Bowman) W.; m. Phyllis Jaye Reynolds, June 14, 1969; children: Eric Alan, Paul Andrew, Jennifer Lynn. B.A. with honors in Biology and Psychology, Lehigh U., 1969; Ph.D. in Genetics (Alumni research fellow), U. N.H., 1972. Asst. prof. biology Wilmington (Ohio) Coll., 1972-76; asst. prof. biology, coordinator biomed. scis. SUNY, Fredonia, 1976-79, asso. prof., chmn. biology dept., dir. biomed. scis., 1979-80; sr. research scientist, project mgr. molecular genetics Battelle Pacific N.W. Labs., Richland, Wash., 1980-83; research and devel. mgr. Northwest Biotech., Inc., 1983—; cons. Jamestown (N.Y.) Gen. Hosp., 1977-80; adj. prof. grad. genetics Wash. State U., Pullman, 1982—; mem. oil shale task force U.S. Dept. Energy, 1980-81. Contbr. articles to profl. jours.; patentee in field; discovered Quasmid DNA in bacteria, 1983. Bd. dirs. S.W. N.Y. Chpt. Am. Cancer Soc., 1978-80; mem. research bd. Western N.Y. chpt. Am. Heart Assn., 1979-80. NIMH fellow, 1968-69; grantee N.Y. Dept. Public Health, SUNY Research Corp., NSF, Sigma Xi; recipient citation for program excellence Am. Soc. Allied Health Professions, 1978, Chancellor's award SUNY, 1979. Mem. N.Y. Acad. Scis., Ohio Acad. Scis., Genetics Soc. Am., Am. Soc. Microbiology, Sigma Xi. Subspecialties: Genetics and genetic engineering (biology); Molecular biology. Current work: Applied and basic recombinant DNA research; genetic engineering of microorganisms and eucaroyotes; genetic transformation; proprietary gene exchange systems. Home: 741 Grosscup Blvd West Richland WA 99352 Office: Battelle Pacific NW Lab Richland WA 99352

WILLIAMS, JAMES GERARD, astronomer, researcher; b. New Kensington, Pa., Apr. 12, 1941; s. James Emerson and Mary Frances (Bolick) W. B.S., Calif. Inst. Tech., 1963; Ph.D., UCLA, Westwood, 1969. Mem. tech. staff N.Am. Rockwell, Downey, Calif., 1962-68; research scientist Jet Propulsion Lab., Pasadena, Calif., 1969—. Contbr. articles to profl. jours. Recipient Exceptional Sci. Achievement medal NASA, 1976. Mem. Am. Astron. Soc., Am. Geophys. Union, Internat. Astron. Union. Subspecialties: Astronautics; Geophysics. Current work: Analysis of lunar laser range data; orbit of moon; rotations of Earth and moon; dynamical evolution of asteroid orbits; planet crossing asteroids; asteroid families and main belt morphology. Office: Jet Propulsion Lab 264/781 4800 Oak Grove Dr Pasadena CA 91109

WILLIAMS, JAMES HENRY, JR., mechanical engineer, educator, cons.; b. Newport News, Va., Apr. 4, 1941; s. James H. Williams and Margaret L. (Holt) Mithcell; 1 son, James Henry III. Mech. designer (Homer L. Ferguson scholar), Newport News Apprentice Sch., 1965; S.B., M.I.T., 1967, S.M., 1968; Ph.D., Trinity Coll., Cambridge U., 1970. Sr. design engr. Newport News Shipyard, 1960-70; asst. prof. mech. engring. M.I.T., 1970-74, assoc. prof., 1974-81, prof., 1981—, duPont prof., 1973, Edgerton prof., 1974-76, head, 1974—; cons. engring. to numerous cos. Contbr. numerous articles on stress analysis, vibration, fracture mechanics, composite materials and nondestructive testing to profl. jours. Recipient Charles F. Bailey Bronze medal, 1961, Charles F. Bailey Silver medal, 1962, Charles F. Bailey Gold medal, 1963, Baker award M.I.T., 1973, Den Hartog Disting. Educator award, 1981. Mem. ASME, Am. Soc. Nondestructive Testing, Nat. Tech. Assn. Subspecialties: Theoretical and applied mechanics; Composite materials. Office: 77 Massachusetts Ave Room 3-360 Cambridge MA 02139

WILLIAMS, JAMES MCSPADDEN, anatomist; b. Pauls Valley, Okla., May 12, 1952; s. Clymer Vinston and Marika (Chaharyn) W.; m. Sammy Carol Hearne, Dec. 31, 1973; children: James Tebe, Cecelia Jane. B.S. in Zoology, U. Okla., 1975; Ph.D., Ind. U., 1980. Lab. technician Wishard Hosp., Indpls., 1978-79; assoc. instr. dept. anatomy Ind. U. Sch. Medicine, Indpls., 1976-80, Arthritis Found. post-doctoral fellow dept. medicine/rheumatology div., 1981—; adj. faculty biology dept. Ind. U.-Purdue U., Indpls., 1981-82. Contbr. articles to med. jours. Deacon Faith Missionary Ch., Indpls., 1981-82. Assn. for Mental Health Research and Edn. grantee, 1978-80. Fellow Arthritis Found.; mem. Am. Fedn. for Clin. Research, Christian Med. Soc. (assoc.). Democrat. Subspecialties: Morphology; Anatomy and embryology. Current work: Connective tissue research, synovial joint morphology, physiology and repair and arthritis related research. Office: Dept Medicine Rheumatology Div Ind U 541 Clinical Dr Long 492 Indianapolis IN 46223

WILLIAMS, JERRY ALBERT, consulting meteorologist; b. St. Petersburg, Fla., Dec. 1, 1933; s. Arthur William and Beulah Gladys (Atherton) W.; m. Adele Mae Cooper, Sept. 11, 1954; children: Jerry Martin, Alan James, Reed Edward (dec.). B.G.S., U. Nebr.-Omaha, 1974. Weather observer/forecaster USAF, Air Weather Service, 1953-74; meteorologist Bechtel, Inc., Fairbanks, Alaska, 1974-75, Oceanroutes, Inc., Palo Alto, Calif., 1976—. Leader Boy Scouts Am., Big Spring, Tex., 1964-67, Eielson AFB, Alaska, 1967-70; commr. Offutt AFB, Omaha, 1970-76; commr., dist. chmn. Eielson AFB, San Jose, Calif., 1976-82. Served with USAF, 1953-74. Mem. Am. Meteorol. Soc. (v.p. farthest north chpt. 1967-70). Republican. Congregationalist. Subspecialties: Meteorology; Climatology. Current work: Meteorology/oceanography consulting for offshore construction and petroleum industry, operational offshore forecasting. Office: Oceanroutes Inc 3260 Hillview Ave Palo Alto CA 95117

WILLIAMS, KEITH ALAN, dental researcher; b. Indpls., June 7, 1951; s. Herbert Otto and Gladys A. (Ruff) W.; m. Kimberly S. Anderson, May 6, 1978; 1 dau.: Kassandra Ann. B.S. in Chemistry, Purdue U., 1976. Research technician Oral Health Research Inst., Indpls., 1977-79, sr. research technologist, 1979—. Contbr. articles in field to profl. jours. Mem. Internat. Assn. Dental Research, Am. Assn. Dental Research. Subspecialties: Analytical chemistry; Cariology. Current work: The investigation of smokeless tobaccos and their effect on dental caries; anti-plaque /anticaries substances of plant origin. Home: 1326 N Ewing Indianapolis IN 46201 Office: Oral Health Research Inst 415 N Lansing Indianapolis IN 46202

WILLIAMS, KENNETH LEE, biology educator; b. Saybrook, Ill., Sept. 4, 1934; s. Louis Lee and Dorthy (Rowley) W.; m. Viola K. Stoehr, Aug. 23, 1956; children: Michele, Christine, Kimberly. B.S., U. Ill., 1959, M.S., 1961; Ph.D., La. State U., 1970. Instr. Tulane U., New Orleans, summer 1964; instr. Millikin U., Decatur, Ill., 1962-64; asst. prof. zoology Northwestern State U., Natchitoches, La., 1964-70, assoc. prof., 1970-79, prof., 1979—. Author: Systematics and Natural History Milksnakes, 1978; contbr. articles to profl. jours. Served with U.S. Army, 1954-56. Mem. Am. Soc. Ichthyologists and Herpetologists, Southwestern Assn. Naturalists (treas.), Herpetologists League, Soc. Study of Amphibians and Reptiles, Societas Europaea Herpetologica. Subspecialties: Systematics; Morphology. Current work: Reptiles and amphibians, especially snakes; taxonomy and morphology. Home: Route 1 PO Box 643 Natchitoches LA 71457 Office: Dept Biol Sc Northwestern State U Natchitoches LA 71457

WILLIAMS, LAWRENCE ERNEST, imaging physicist; b. Youngstown, Ohio, Nov. 29, 1937; s. William Karapanza and Dorothy Anne (Randulovich) W.; m. Sonia Bell Bredmeyer, Nov. 12, 1966; children: Erica, Beverley. B.Sc. in Physics, Carnegie-Mellon U., 1959; M.Sc., U. Minn., 1962, Ph.D., 1965. Cert, radiol. physicist Am. Bd. Radiology. NIH spl. fellow in nuclear medicine U. Minn., Mpls., 1971-73, research asso. radiology, 1973, asst. prof. radiology, 1973-78, asso. prof. radiology, 1978-80; imaging physicist City of Hope, Duarte, Calif., 1980—; cons. Jet Propulsion Lab., Calif. Inst. Tech., 1981—; adj. asso. prof. med. physics UCLA, 1982—. Contbr. articles to sci. jours.; Author: Biophysical Science, 2d edit, 1979. Recipient Raymond J. Wean scholarship, 1957. Mem. Soc. Nuclear Medicine, Calif. Radiol. Soc., Am. Coll. Radiology, Am. Assn. Physicists in Medicine, N.Y. Acad. Scis., Sigma Xi. Democrat. Methodist. Subspecialties: Imaging technology; Nuclear medicine. Current work: Image enhancement; computer modeling of physiological systems; computed tomography; tumor imaging with phospholipid vesicles; monoclonal antibody imaging. Office: 1500 E Duarte Rd Duarte CA 91010

WILLIAMS, LESLEY LATTIN, educator; b. New Bedford, Mass., Aug. 10, 1939; d. Bruce Wallace and Lesley (Olcott) W.; m. Felix Schreiner, July 12, 1969. A.B., Hollins Coll., 1961; Ph.D., U. Wis., 1968. Asst. prof. dept. phys. scis. Chgo. State U., 1968-73, assoc. prof. chemistry, 1973-78, prof., 1978—; lectr. chemistry U. Md. Munich Campus, 1971-72; faculty research participant Argonne Nat. Lab., 1978. Mem. Am. Phys. Soc., AAAS, Am. Fedn. Tchrs., Sigma Xi, Sigma Delta Epsilon. Subspecialty: Physical chemistry. Current work: Teacher physical chemistry, instrumental analysis, research interests in radiochemistry, NMR relaxation mechanisms in liquids, physics of music. Office: Chicago State U Dept Phys Scis Chicago IL 60628

WILLIAMS, MARJORIE JOAN, physician, consultant; b. Calcutta, India, Nov. 22, 1919; came to U.S., 1945; d. Alfred Norman and Catherine Mary (Bagshawe) Dix; m. Bill H. Williams, Feb. 4, 1943; 1 son, Robert J. M.B. Ch.B., U. Bristol, Eng., 1944. Diplomate: Am. Bd. Pathology. Intern Queen Elizabeth Hosp., Birmingham, Eng., 1944; resident U. Bristol, Eng., Tulane Med. Sch., New Orleans, 1945-47; chief lab. service VA Ctr., Temple, Tex., 1949-62; dep. dir. pathology service VA, Washington, 1962-63, dir. pathology service, 1962-80; Nat. Com. for Clin. Lab. Standards, 1981—; cons. in field. Contbr. articles to med. jours. Bd. dirs. Coll. Am. Pathologists Found., 1976—. Recipient Fed. Womens award, 1967; Career Service award Nat. Civil Service League, 1967; Stitt award Assn. Mil. Surgeons, 1972; Disting. Career award VA, 1980. Fellow Coll. Am. Pathologists (bd. govs. 1971-77, pathologist of yr. 1978), Am. Soc. Clin. Pathologists; mem. Am. Assn. Pathologists, AMA, Royal Soc. Medicine. Subspecialties: Pathology (medicine); Health services research. Current work: Pathology practice and education; research in health services; international health; consultant in pathology management, practice and education. Home: 1225 4th St SW Washington DC 20024

WILLIAMS, MARSHALL HENRY, JR., physician, educator; b. New Haven, July 15, 1924; s. Marshall Henry and Henrietta (English) W.; m. Mary Butler, Aug. 27, 1948; children: Stuart, Patricia, Marshall, Frances, Richard. Grad., Pomfret Sch., 1942; B.S., Yale, 1945, M.D., 1947. Diplomate: Nat. Bd. Med. Examiners, Am. Bd. Internal Medicine. Intern Presbyn. Hosp., N.Y.C., 1947-48, asst. resident medicine, 1948-49, New Haven Hosp., 1949-50, asst. in medicine, 1950; trainee Nat. Heart Inst., 1950; practice medicine, specializing in internal medicine, Bronx, N.Y.; chief respiratory sect., dept. cardiorespiratory diseases Army Med. Service Grad. Sch., Walter Reed Army Med. Center, 1953-55; dir. cardiorespiratory lab. Grasslands Hosp., Valhalla, N.Y., 1955-59; dir. chest service Bronx Municipal Hosp. Center, 1959; vis. asst. prof. physiology Albert Einstein Coll. Medicine, Bronx, N.Y., 1955-59, assoc. prof. medicine and physiology, 1959-66, prof. medicine, 1966—; dir. pulmonary div. Albert Einstein Coll. Medicine—MontefioreHosp. and Med. Ctr., 1966—. Author: Clinical Applications of Cardiopulmonary Physiology, 1960; Contbr. articles to profl. jours. Served from 1st lt. to capt. U.S. Army, 1950-52. Mem. Am. Physiol. Soc., A.A.A.S., Am., Westchester (past pres.) heart assns., Am. Thoracic Soc., Am. Fedn. Clin. Research, N.Y. Acad. Sci., N.Y. Trudeau Soc. (past pres.), Am. Soc. Clin. Investigation, Soc. Urban Physicians (past pres.), N.Y. Tb. and Health Assn. (past dir.), Alpha Omega Alpha. Subspecialty: Pulmonary medicine. Home: 13 Colvin Rd Scarsdale NY 10583 Office: Albert Einstein Coll Medicine Bronx NY 10461

WILLIAMS, MARSHALL VANCE, JR., microbiologist, cons.; b. Memphis, Mar. 22, 1948; s. Marshall Vance and Edna Marie (Tate) W.; m. Marilyn L., Aug. 15, 1970; children: Laurie Michele, Jennifer Leigh. B.S., Memphis State U., 1970, M.S., 1973; Ph.D., U. Ga., Athens, 1976. Postdoctoral fellow dept. exptl. therapeutics Roswell Park Meml. Inst., Buffalo, 1976-78; asst. prof. dept. microbiology and immunology Kirksvills (Mo.) Coll. Osteo. Medicine, 1978-82; asst. prof. dept. med. microbiology and immunology Ohio State U., Columbus, 1982—; cons. Bio Diesel Fuels Inc. Contbr. articles to profl. jours. Mem. Am. Soc. Microbiology, Soc. Indsl. Microbiology. Methodist. Subspecialties: Microbiology; Biochemistry (biology). Current work: Deoxyuridine metabolism in eucaryotic cells and in cells infected with herpes simplex virus. Office: 333 W 10th St Columbus OH 43081

WILLIAMS, MARTHA ETHELYN, chemistry educator; b. Chgo., Sept. 21, 1934; d. Harold Milton and Alice Rosemond (Fox) W. B.A., Barat Coll., 1955; M.A., Loyola U., 1957. With IIT Research Inst., Chgo., 1957-72, mgr. info. scis., 1962-72, mgr. computer search center, 1968-72; adj. asso. prof. sci. info. Ill. Inst. Tech., Chgo., 1965-73, lectr. chemistry dept., 1968-70; research prof. info. sci., coordinated sci. lab. (Coll. Engring.); also dir. info. retrieval research lab., affiliate computer sci. dept. U. Ill., Urbana, 1972—; chmn. large data base conf. Nat. Acad. Sci./NRC, 1974, mem. ad hoc panel on info. storage and retrieval, 1977, numerical data adv. bd., 1979-82; mem. task force on sci. info. activities NSF, 1977; U.S. rep. review com. for project on broad system of ordering, UNESCO, Hague, Netherlands, 1974; vice chmn. Gordon Research Conf. on Sci. Info. Problems in Research, 1978, chmn., 1980; cons. to numerous cos. and research founds. Contbr. numerous articles to profl. jours.; editor-in-chief: Computer-Readable Databases—A Directory and Data Sourcebook, 1976; editor, Ann. Rev. Info. Sci. and Tech., 1976—, Online Rev., 1979—; contbg. editor: column on Databases to Bull. Am. Soc. Info. Sci., 1974-78. Trustee Engring. Index, 1974-83, bd. dirs., 1976—, chmn. bd. dirs., 1982-84, v.p., 1978-79, pres., 1980-81; regent Nat. Library Medicine, 1978-82, chmn. bd. regents, 1981. Recipient best paper of year award H. W. Wilson Co., 1975; NSF travel grantee, Luxembourg, 1972, Honolulu, 1973, Tokyo, 1973, Mexico City, 1975, Scotland, 1976.

Fellow AAAS (computers, info. and communication mem.-at-large 1978-81, nominating com. 1983); Mem. Am. Chem. Soc., Am. Soc. Info. Sci. (councilor 1971-72, chmn. networks com. 1973-74, chmn. spl. interest group on SDI 1974-75), Assn. for Computing Machinery (pub. bd. 1972-76), Assn. Sci. Info. Dissemination Centers (v.p. 1971-73, pres. 1975-77), Nat. Acad. Sci. (joint com. with NRC on chem. info. 1971-73), U.S. Nat. Com. for Internat. Fedn. for Documentation. Subspecialties: Information systems (information science); Information systems, storage, and retrieval (computer science). Current work: Information Science, storage and retrieval databases; online systems; systems analysis and design. Home: Route 1 Monticello IL 61856 Office: Univ of Ill CSL 5 135 1101 W Springfield St Urbana IL 61801

WILLIAMS, MARVIN WRIGHT, clinical psychologist, educator, researcher; b. Houston, May 10, 1949; s. Marvin Wright and Mary Katherine (Lacey) W. B.A., U.Tex.-Austin, 1971; M.S., Fla. State U., 1976, Ph.D., 1978. Lic. and cert. in psychology, Tex. Staff psychologist VA Med. Ctr., Houston, 1979—; adj. asst. prof. Baylor Coll. Medicine, Houston, 1980—; adj. asst. prof. psychology U. Houston, 1982—. USPHS fellow NIMH, 1973-76; recipient cert. of recognition DAV, 1980. Mem. Am. Psychol. Assn. (cert. of recognition for pub. service 1981), Southeastern Psychol. Assn., Am. Group Psychotherapy Assn., Am. Assn. Correctional Psychologists. Current work: Dangerous behavior, the law/psychology interface, neuropsychology. Currently involved in clinical work, research, teaching and training in a large VA Medical Center. Office: VA Medical Ctr Psychology Service 2002 Holcombe Blvd Houston TX 77211

WILLIAMS, PAUL TENNYSON, family medicine educator, physician; b. Hillsboro, Ohio, Jan. 11, 1925; s. Harry Dana and Kathryn (Sanders) W.; m. Marianna Hamilton, Sept. 14, 1947; children: Tracey, Shellaine, Dayna, Matthew Tennyson. B.A., Western Res. U., 1948, M.D., 1951. Diplomate: Am. Bd. Family Physicians. Intern Miami Valley Hosp., Dayton, Ohio, 1951-52; practice medicine specializing in family practice, Delaware, Ohio, 1952-74; clin. instr. preventive medicine Ohio State U., Columbus, 1969-74, prof., chmn. dept. family medicine, 1974—. Contbr. articles to profl. jours. Mem. data standing com. Consortium of Cancer Control Ohio, 1980-82. Served with USN, 1945-48; with U.S. Army, 1943-44. Named Outstanding Local Pres. Ohio Jaycees, 1958-59, Outstanding Team Physician Ohio Med. Assn., Ohio High Sch. Athletic Assn., Joint Adv. Com. Sports Medicine, 1982. Mem. Am. Acad. Family Physicians, Soc. Tchrs. Family Medicine, Assn. Depts. Family Medicine, Ohio Med. Assn., Ohio Acad. Family Physicians, Central Ohio Acad. Family Physicians, Columbus and Franklin County Acad. Medicine. Lutheran. Subspecialty: Family practice. Current work: Manpower survey of Ohio Family Physicians, projection of future needs of family physicians, streptococcal pharyngitis. Home: 6300 Riverside Dr Dublin OH 43017 Office: 456 Clinic Dr 1114 UHC Columbus OH 43210

WILLIAMS, PETER JOHN, geotechnical science educator; b. Croydon, Eng., Sept. 27, 1932; s. John George and Kathleen Emily (White) W.; m. Kari Fuglesang, June 1, 1957; children: Dag, Beatrice, Inger. B.A., Cambridge U., 1956, M.A., 1958; Fil. lic., Fil. dr., U. Stockholm, 1969. Research officer NRC Can., 1957-69; research fellow Norwegian Geotech. Inst., 1962-65; assoc. prof. geography Carleton U., Ottawa, Ont., Can., 1969-71, prof., 1971—, dir., 1979—; cons. Peter J. Williams & Assocs., Ltd., Ottawa, 1979—. Author: Pipelines and Permafrost, 1979, The Surface of the Earth; An Introduction to Geotechnical Science, 1982. Mem. Canadian Geotech. Soc., Norwegian Geotech. Soc., Canadian Soc. Soil Sci., Internat. Glaciological Soc. Subspecialties: Freezing of soils; Geotechnical Science. Current work: Interaction of pipelines with freezing soils, properties of earth materials in relation to engineering behavior, significance of natural phenomena to geotechnical design. Office: Geotech Sci Labs Carleton U Ottawa ON Canada K1S 5B6

WILLIAMS, PETER MACLELLAN, nuclear engineer, project manager; b. N.Y.C., Aug. 30, 1931; s. Gilbert Harris and Evelyn (Buss) W.; m. Lois Crane, Oct. 9, 1956; children: Jane, Gilbert, Katherine, Anne, Louise, Robert. B. Chem. Engring., Cornell U., 1954; M.S., MIT, 1957; Ph.D., U. Md., 1971. Engr. Dupont Savannah River, Aiken, S.C., 1954-55; task engr. AGN, San Ramon, Calif., 1957-60; project mgr. Am. Machine & Foundry, Greenwich, Conn., 1960-62; research staff mem. Princeton U., 1962-67; sr. project mgr. U.S. Nuclear Regulatory Commn., Washington, 1967—. Scoutmaster Boy Scouts Am., Potomac, Md., 1972; pres. Winston Churchill High Sch. PTA, Potomac, 1981. Recipient U.S. Nuclear Regulatory Commn. High Quality Service award, 1982. Mem. ASME, Am. Nuclear Soc., Sigma Xi. Democrat. Unitarian. Subspecialties: Nuclear fission; Fluid mechanics. Current work: High temperature gas cooled ractors for commercial power, safety and licensing requirements. Patentee liquid core nuclear rocket. Home: 9418 Thrush Ln Potomac MD 20854 Office: US Nuclear Regulatory Commn Washington DC 20555

WILLIAMS, RAYMOND WARREN, chemical process engineer; b. Kansas City, Mo., Sept. 4, 1951; s. Raymond and Jean Marie (Kiechle) W.; m. Kathleen Erna King, July 31, 1982. B.S. in Chem. Engring, U. Mo.-Columbia, 1973, M.S. in Nuclear Engring, 1976. Registered profl. engr., Mo. Chem. engr. Black & Veatch (Cons. Engrs.), Kansas City, Mo., 1975-82; process engr. Syntex Agribusiness, Inc., Springfield, Mo., 1982—. Author profl. reports. Recipient Alt. Energy award Dept. Energy, 1980. Mem. Nat. Soc. Profl. Engrs., Am. Nuclear Soc. (vice chmn. chpt. 1982-83), Am. Inst. Chem. Engrs. (assoc. sec. 1981, treas. 1982), AAAS, Alpha Chi Sigma, Tau Beta Pi. Club: Springfield Corvette. Subspecialties: Chemical engineering; Nuclear engineering. Current work: Fusion energy, alternative energy sources, energy conservation, process development and optimization, computerized process control. Home: 3227 W Village Ln Springfield MO 65807 Office: Syntex Agribusiness Inc PO Box 1246 SSS Springfield MO 65805

WILLIAMS, ROBERT JOSEPH, psychologist, educator; b. Durango, Colo., Feb. 14, 1948; s. Owen Clement and Florence Kathryn (Fairchild) W.; m. Rebecca Sue Webb Gott, Sept. 30, 1967 (div. 1972); 1 dau., Robin Lu; m. Lay Lynn Noda, Mar. 24, 1973; children: Matthew Owen, Nicholas Robert. B.A., U. Colo., 1970; M.A., U. No. Colo., 1976; Ph.D., U. Minn., 1979. Lic. psychologist, Colo. Tchr. math. Jefferson County Schs., Lakewood, Colo., 1970-76; intern in psychology VA Med. Ctr., Mpls., 1977-79; clin. dir. Pikes Mental Health Ctr., Colorado Springs, Colo., 1979—; asst. prof. U. Denver, 1981—; mem. Placement Alternatives Commn., Colorado Springs, 1981—; cons. CETA, Colorado Springs, 1979-81, Dept. Social Service, Colorado Springs and Canon City, 1979—. Author manuals in field of computerized psychol. testing; Contbr. articles to profl. jours. Mem. El Paso County Democratic Com. Served with USMCR, 1970-76. Mem. Am. Psychol. Assn., Colo. Psychol. Assn. (ad hoc com. on computer testing 1981), El Paso County Psychol. Soc. Presbyterian. Lodge: Masons. Current work: Computerized psychological testing and assessment; assessment and treatment of dysfunctional families. Home: 5125 Escapardo Way Colorado Springs CO 80917 Office: Inst Family and Personal Devel 3707 Parkmoor Village Dr Suite 107 Colorado Springs CO 80907

WILLIAMS, ROBERT REID, computer scientist; b. Ottawa, Ont., Can., Aug. 30, 1956; s. Alan James and Marion Alma (Potts) W. B.Math., U. Waterloo, Ont., 1979, M.Math., 1982. Contract research specialist Computer Communications Networks group U. Waterloo, 1981-82; software specialist Electrohome Ltd., Waterloo, 1982; software designer Mitel Corp., Kanata, Ont., 1982—. Descartes fellow, 1975. Mem. Assn. Computing Machinery. Subspecialties: Distributed systems and networks; Graphics, image processing, and pattern recognition. Current work: Videotex systems, distributed systems, performance eval. Home: 592 Seyton Dr Nepean ON Canada K2H 7K5 Office: Mitel Corp Kanata ON Canada

WILLIAMS, ROGER JOHN, retired chemist, educator; b. Ootacumund, India, Aug. 14, 1893; s. Robert Runnels and Alice Evelyn (Mills) W.; m. Hazel Elizabeth Wood, Aug. 1, 1916 (dec. 1952); children—Roger John, Janet, Arnold; m. Mabel Phyllis Hobson, May 9, 1953; 1 stepson, John W. Hobson. Student, U. Calif., 1914-15; B.S., U. Redlands, Calif., 1914, D.Sc. (hon.), 1934; M.S., U. Chgo., 1918, Ph.D., 1919; D.Sc. (hon.), Columbia, 1942, Oreg. State Coll., 1956. Research chemist Fleischmann Co., Chgo., 1919-20; asst. prof. chemistry U. Oreg., Eugene, 1920-21, asso. prof., 1921-28, prof., 1928-32; prof. chemistry Oreg. State Coll., Corvallis, 1932-39, U. Tex., Austin, 1939-71, emeritus, 1971—; research scientist Clayton Found. Biochem. Inst. Tex., Austin, 1971—, dir., 1941-63; Mem. Pres.'s Adv. Panel on Heart Disease, 1972. Author or co-author: books relating to chemistry and nutrition; latest The Biochemistry of B Vitamins, 1950, (with R.E. Eakin, E. Beerstecher, W. Shive) Nutrition and Alcoholism, 1951, Free and Unequal, 1953, Biochemical Individuality, 1956, Alcoholism: The Nutritional Approach, 1959, Nutrition in a Nutshell, 1962, You Are Extraordinary, 1967, Nutrition Against Disease: Environmental Prevention, 1971, Physicians' Handbook of Nutritional Science, 1975, The Wonderful World Within You: Your Inner Nutritional Environment, 1977, The Prevention of Alcoholism Through Nutrition, 1981, The Advancement of Nutrition, 1982. Recipient Mead-Johnson award Am. Inst. Nutrition, 1941; Chandler medal Columbia, 1942; S.W. Region award Am. Chem. Soc., 1950. Fellow AAAS; mem. Am. Chem. Soc. (pres. 1957), Am. Soc. Biol. Chemists, Am. Assn. Cancer Research, Soc. Exptl. Biology and Medicine, Nat. Acad. Scis., N.Y. Acad. Sci., Internat. Acad. Preventive Medicine, Am. Inst. Nutrition, Biochem. Soc. London, Phi Beta Kappa, Sigma Xi, Phi Kappa Phi, Phi Sigma, Pi Kappa Delta, Phi Lambda Upsilon. Methodist. Subspecialties: Preventive medicine; Nutrition (medicine). Current work: Biochemical individuality; unified education; nurtition against alchoholism; human nutrition; inborn individuality as it relates to human problems. Discoverer pantothenic acid; concentrated and named folic acid; also made microbiol. study of vitamins, individual metabolic patterns. Home: 1604 Gaston Ave Austin TX 78703 Office: Dept Chemistry U Tex Austin TX 78712

WILLIAMS, RONALD OSCAR, systems engineer; b. Denver, May 10, 1940; s. Oscar H. and Evelyn V. (Johnson) W. B.S. in Applied Math, U. Colo.-Boulder, 1964; postgrad., 1968-74; postgrad., U. Colo., 1975-82. Computer programmer Apollo systems dept. Missile & Space div. Gen. Electric Co., Kennedy Space Center, Fla., 1965-67, Manned Spacecraft Ctr., Houston, 1967-68; computer programmer U. Colo., Boulder, 1968-73; computer programmer analyst Def. Systems div. System Devel. Corp., N.Am. Aerospace Def. Command, Colorado Springs, Colo., 1974-75; engr. Def. Systems div. and Command and Info. Systems div. Martin Marietta Aerospace, Denver, 1976-80; systems engr. Space and Communications group Def. Systems div. Hughes Aircraft Co., Aurora, Colo., 1980—. Vol. fireman Clear Lake City Fire Dept., Tex., 1968; Officer, rescue squadman Boulder (Colo.) Emergency Squad, Boulder, Colo., 1969-76; spl. police officer Boulder Police Dept., 1970-75; spl. dep. sheriff Boulder County Sheriff's Dept., 1970-71; mem. nat. adv. bd. Am. Security Council, Coalition of Peace through Strength; mem. nat. com. Republican party, mem. nat. senatorial com. Served with USMCR, 1958-66. Recipient awards Hughes Aircraft Co., 1982. Mem. AAAS, Math. Assn. Am., Am. Math. Soc., Soc. Indsl. and Applied Math., AIAA, Armed Forces Communications and Electronics Assn., Assn. Old Crows, Am. Def. Preparedness Assn., Marine Corps Assn., Air Force Assn., Colo. Hist. Soc., Historic Denver Inc., Mensa, Historic Boulder Inc. Republican. Lutheran. Club: Eagles. Subspecialty: Aerospace engineering and technology. Home: 7504 W Quarto Ave Littleton CO 80123 Office: Hughes Aircraft Co PO Box 31979 Aurora CO 80041

WILLIAMS, T. H. LEE, geography educator; b. Deganwy, Wales, U.K., May 31, 1951; came to U.S., 1976; m. Trevor Glynne and Grace Myfanwy (Owen) W.; m. D.M. Naila Mendes, Apr. 14, 1973; 1 son, Owen-John Roderiques. B.Sc. with honors, U. Bristol, Eng., 1972; Ph.D. in Geography, 1977. Tchr. sci. Antigua (W.I.) Grammar Sch., 1972-73; vis. research assoc. U. Okla., Norman, 1976-77; asst. prof. geography U. Kans., Lawrence, 1977-81, assoc. prof., 1981—; assoc. fellow Inst. Math. and Its Applications, Eng., 1978; cons. UNESCO, 1978-79, Nat. Park Services, 1979. Mem. Am. Soc. Photogrammetry, Assn. Am. Geographers, Inst. Brit. Geographers. Subspecialties: Remote sensing (geoscience); Graphics, image processing, and pattern recognition. Current work: Applications of visible and microwave imaging systems and development of geographic information analysis systems for natural resource management. Home: 1022 Wellington St Lawrence KS 66044 Office: Dept Geography-Meteorology U Kans Lawrence KS 66045

WILLIAMS, THEODORE JOSEPH, engineering educator; b. Black Lick, Pa., Sept. 2, 1923; s. Theodore Finley and Mary Ellen (Shields) W.; m. Isabel Annette McAnulty, July 18, 1946; children: Theodore Joseph, Mary Margaret, Charles Augustus, Elizabeth Ann. B.S.Ch.E., Pa. State U., 1949, M.S.Ch.E., 1950, Ph.D., 1955; M.S. in Elec. Engring., Ohio State U., 1956. Research fellow Pa. State U., University Park, 1947-51; asst. prof. Air Force Inst. Tech., 1953-56; technologist Monsanto Co., 1956-57, sr. engring. supr., 1957-65; vis. prof. Washington U., St. Louis, 1962-65; prof engring., dir. control and information systems lab. Purdue U., Lafayette, Ind., 1965—; dir. Purdue Lab. Applied Indsl. Control, 1966—. Author: Systems Engineering for the Process Industries, 1961, Automatic Control of Chemical and Petroleum Processes, 1961, Progress in Direct Digital Control, 1969, Interfaces with the Process Control Computer, 1971, Modeling and Control of Kraft Production Systems, 1975, Modelling, Estimation and Control of the Soaking Pit, 1983, The Use of Digital Computers in Process Control, 1983; editor: Computer Applications in Shipping and Shipbuilding, 6 vols., 1973-79, Proceedings Advanced Control Confs., 9 vols., 1974-83. Served to 1st lt. USAAF, 1942-45; to capt. USAF, 1951-56. Decorated Air medal with 3 oak leaf clusters. Fellow Instrument Soc. America (pres. 1968-69), Am. Inst. Chemists, Am. Inst. Chem. Engrs., Inst. Measurement and Control (London), AAAS; sr. mem. IEEE; mem. Am. Chem. Soc., Am. Automatic Control Council (pres. 1965-67), Am. Fedn. Info. Processing Socs. (pres. 1976-78), Sigma Xi, Tau Beta Pi, Phi Kappa Phi, Phi Lambda Upsilon. Subspecialties: Distributed systems and networks; Systems engineering. Current work: Industrial automation; hierarchical computer control systems; mathematical modeling of industrial processes for automatic control. Home: 208 Chippewa St West Lafayette IN 47906 Office: Purdue Lab Applied Indsl Control Potter Research Ctr Purdue Univ Lafayette IN 47907

WILLIAMS, WALTER CHARLES, JR., aero. engr., govt. ofcl.; b. New Orleans, July 30, 1919; s. Walter C. and Emilia A. (Brunnier) W.; m. Helen Manning, Dec. 16, 1939; children—Charles Manning, Howard Lee, Elizabeth Anne. B.S. in Aero. Engring, La. State U., 1939; D.Eng. (hon.), 1963. Asst. head stability and control br. flight div. NACA, Langley Field, Va., 1940-46, flight research project engr., 1944-46; dir. flight test activities NACA-NASA, Muroc AFB; asso. dir. Manned Spacecraft Center; v.p. Aerospace Corp., 1964-75; chief engr. NASA, Downey, Calif., 1975—; mem. bd. advs. Missile, Space and Range Pioneers, from 1966. Contbr. articles to profl. publs. Recipient Sylvanus Albert Reed award Inst. Aerospace Scis., 1963; Medal of Honor City of New York, 1962; Disting. Service medal NASA, 1962; named Fed. Engr. of Year, 1981. Fellow AIAA (Astronautics award 1964), Am. Astronautical Soc., Am. Rocket Soc.; mem. Nat. Soc. Aerospace Profls. Club: Masons. Subspecialty: Aerospace engineering and technology. Office: Chief Engr care Rockwell International 12214 Lakewood Blvd Downey CA 90241

WILLIAMSON, ARTHUR ELDRIDGE, JR., research physicist; b. Montgomery, Ala., July 6, 1926; s. Arthur E. and May Ellen (Bray) W.; m. Jean Phillips, Oct. 13, 1951; children: Arthur Mark, Paul Steven. B. Engring. Physics, Auburn U., 1950, M.S., 1951; postgrad., Ga. Inst. Tech., 1955-59. Asst. prof. physics U. Richmond, 1952-53; research physicist So. Research Inst., 1953-55, head electro-optics sect., 1973—; asst. prof., project dir. Ga. Inst. Tech., 1955-59; mktg. mgr. Martin Marietta, Orlando, Fla., 1959-73. Served with USNR, 1944-46. Mem. Am. Phys. Soc., Optical Soc. Am., Sigma Phi Sigma, Phi Kappa Phi. Home: 501 Benbow Dr Birmingham AL 35226 Office: 2000 9th Ave S Birmingham AL 35255

WILLIAMSON, HAROLD EMANUEL, pharmacologist; b. Racine, Wis., Aug. 8, 1930; s. Harold E. and Grace Mae (McIntyre) W.; m. Joan Louise Chase, Apr. 28, 1957; children: Timothy, Julie, Eric. B.S., U. Wis., 1953, Ph.D., 1959. Project assoc. pharmacology U. Wis.-Madison, 1959-60; instr. to prof. U. Iowa, Iowa City, 1960—. Contbr. articles on pharmacology to profl. jours. Fellow Am. Coll. Clin. Pharmacology; mem. Am. Soc. for Pharmacology and Exptl. Therapeutics, Am. Soc. Nephrology, Soc. for Exptl. Biology and Medicine, Internat. Soc. Nephrology, Am. Heart Assn., Am. Fedn. for Clin. Research, AAAS. Subspecialty: Pharmacology. Current work: Research in renal pharmacology and physiology, mechanisms of action and toxicity of diuretic agents, factor influencing renal blood flow. Home: 131 S Mt Vernon Dr Iowa City IA 52240 Office: 2-456 BSB U Iowa Iowa City IA 52242

WILLIAMSON, PETER GEORGE, geology educator, paleontology curator; b. Hemel Hempstead, Hertfordshire, U.K., May 5, 1952; s. George Bardsley and Jean (Goodwill) W. B.S., Bristol U., U.K., 1974; Ph.D., 1980. NATO research fellow Mus. Comparative Zoology; Harvard U., Cambridge, Mass., 1980-81, asst. curator paleontology, asst. prof. geology, 1982—. Contbr. articles on geology to profl. jours. Brit. Govt. grantee, 1974-77; NATO fellow, 1980-81. Mem. Geologists Assn. (U.K.), Conchological Soc. (U.K.), Malacological Assn. (U.K.), Am. Soc. Systematic Zoologists. Subspecialties: Paleobiology; Evolutionary biology. Current work: Morphometric evaluation of fossil sequences in order to document patterns of evolutionary mode and tempo, geometric analysis of shell coiling, biostratigraphy of East African Hominid deposits. Office: Harvard U Room 153 Mus Comparative Zoology Cambridge MA 02139

WILLIAMSON, THOMAS GARNETT, nuclear engineering and engineering physics educator; b. Quincy, Mass., Jan. 27, 1934; s. Robert Burwell and Elizabeth B. (McNeer) W.; m. Kay Darlan Love, Aug. 16, 1961; children: Allen, Sarah, David. B.S. in Physics, Va. Mil. Inst., 1955, M.S., Rensselaer Poly. Inst., 1957, Ph.D., U. Va., 1960. Asst. prof. U. Va.-Charlottesville, 1960-62, assoc. prof., 1962-69, prof., 1969—, chmn. dept. nuclear engring. and engring. physics, 1977—; physicist Gen. Atomic, Calif., 1965, Combustion Engring., Windsor, Conn., 1970-71, Los Alamos Sci. Lab., 1969; cons. Philippine Atomic Energy Commn., summer 1963, 1982, Va. Electric and Power Co., 1975—, Babcock & Wilcox, Lynchburg, Va., 1975—. Vestryman Ch. of Our Savior, Charlottesville. Fellow Am. Nuclear Soc.; Mem. AAAS, Am. Soc. for Engring. Edn., Sigma Xi. Episcopalian. Subspecialties: Nuclear engineering; Nuclear fission. Current work: Nuclear engineering education, radioisotope production and utilization, radiation shielding and dosimetry. Office: U Va Reactor Facility Charlottesville VA 22901

WILLIARD, PAUL G., organic chemistry educator; b. Mt. Carmel, Pa., Dec. 18, 1950; s. Ray Elmer and Victoria (Greco) W.; m. Faith Joan Hutchison, June 16, 1979. B.S., Bucknell U., 1972, M.S., 1972; M.Phil., Columbia U., 1974, Ph.D., 1976. NIH trainee MIT, Cambridge, Mass., 1976-78, NIH fellow, 1978-79; asst. prof. organic chemistry Brown U., Providence, 1979—. Mem. Am. Chem. Soc., Brit. Chem. Soc., Phi Beta Kappa. Subspecialties: Organic chemistry; X-ray crystallography. Current work: Marine natural products chemistry, synthesis and x-ray structure determination. Office: Dept Chemistr Brown U Box H Providence RI 02912

WILLIG, MICHAEL ROBERT, biology educator; b. Pitts., June 7, 1952; s. Robert Michael and Mary Ruth (Kronenberger) W. B.S., U. Pitts., 1974, Ph.D., 1982. Teaching asst. U. Pitts., 1974-76, Mellon fellow, 1980-81; research fellow Brazilian Acad. Sci., Rio de Janeiro, 1976-78; vis. prof. LaRoche Coll., Pitts., 1979; asst. prof. Loyola U., New Orleans, 1981—; research assoc. Audubon Zoo, New Orleans, 1982—; Cons. Dept. Energy, 1983. Contbr. articles to profl. jours. Mem. research adv. bd. Audubon Park and Zool. Garden, 1982—. Dept. Energy grantee, 1982. Mem. AAAS, Ecol. Soc. Am., Soc. for Study of Evolution, Am. Soc. Mammalogists, Am. Soc. Naturalists. Democrat. Roman Catholic. Subspecialties: Evolutionary biology; Theoretical ecology. Current work: Application of quantitative-statistical techniques to problems in animal ecology; experimental design and analysis. Office: Dept Biology Loyola U Box 28 New Orleans LA 70118

WILLIS, ISAAC, dermatologist; b. Albany, Ga., July 13, 1940; s. R.L. and Susie M. (Miller) W.; m. Alliene Horne, June 12, 1965; children: Isaac Horne, Alliric Isaac. B.S., Morehouse Coll., 1961; M.D., Howard U., 1965. Diplomate: Am. Bd. Dermatology. Intern Phila. Gen. Hosp., 1965-66; fellow Howard U., Washington, 1966-67; resident, fellow U. Pa., Phila., 1967-69, assoc. in dermatology, 1969-70; instr. dept. dermatology U. Calif.-San Francisco, 1970-72; asst. prof. Johns Hopkins U. and Johns Hopkins Hosp., Balt., 1972-73, Emory U., Atlanta, 1973-77, assoc. prof., 1977-82; prof. Morehouse Sch. Medicine, Atlanta, 1982—; attending staff Phila. Gen. Hosp., 1969-70, Moffit Hosp., U. Calif., 1970-72, Johns Hopkins Hosp., Balt. City Hosp., Good Samaritan Hosp., 1972-74, Crawford W. Long Meml. Hosp., Atlanta, 1974—, W. Paces Ferry Hosp., 1974—, others.; cons. in field. Author: Textbook of Dermatology, 1971—; Contbr. articles to profl. jours. Chmn. bd. med. dirs. Lupus Erythematrosus Found., Atlanta, 1975-83; bd. dirs. Jacquelyn McClure Lupus Erythematrosus Clinic, 1982—; bd. med. dirs. Skin Cancer Found., 1980—; trustee Friendship Bapt. Ch., Atlanta, 1980—. Nat. Cancer Inst. grantee, 1974-77, 78—; EPA grantee, 1980—, 82—. Fellow Am. Acad. Dermatolgy, Am. Dermtol. Assn.; mem. Soc. Investigative Dermatology, Am. Fedn. Clin. Research, AAAS, Am. Soc. Photobiology, Am. Med. Assn., Nat. Med. Assn., Internat. Soc. Tropical Dermatology, Pan Am. Med. Assn., Phi Beta Kappa. Clubs: Frontiers Internat., Sportsman Internat. Subspecialties: Dermatology;

Cancer research (medicine). Current work: Clinical therapy and research in dermatology and photomedicine including cancer research. Office: NW Med Center Suite 342 3280 Howell Mill Rd NW Atlanta GA 30327 Home: 1141 Regency Rd NW Atlanta GA 30327

WILLMAN, MICHAEL KAREL, radiologist; b. Detroit, Feb. 6, 1941; s. Michael Joseph and Minnie Ann (Simerka) W.; m. Janet Faye Burchett, May 20, 1965; children: Michael, Paul, Amy. D.O., Kansas City Coll. Osteopathic Medicine, 1965. Family practice osteopathic medicine, Rochester, Mich., 1966-67; resident in radiology Kirksville (Mo.) Coll. Osteopathic Medicine, 1969-72, radiologist, 1972—. Bd. dirs. YMCA, 1972-81, pres., 1979. Served to maj. AUS, 1967-69; Vietnam. Decorated Bronze Star. Mem. Am. Osteopathic Coll. Radiology (sec. 1975-82, v.p. 1982-83), Am. Osteopathic Assn., Mo. Assn. Osteopathic Physicians and Surgeons, Northeast Assn. Osteopathic Physicians and Surgeons, Radiol. Soc. N. Am. Club: Kirksville Country (pres. 1980-81). Subspecialties: Diagnostic radiology; Nuclear medicine. Home: East Hwy 6 Kirksville MO 63501 Office: Kirksville Osteopathic Health Center 800 W Jefferson St Kirksville MO 63501

WILLMOTT, ANDREW JOHN, oceanographer, applied mathematician, educator; b. Bushey, Hertfordshire, Eng., Aug. 1, 1954; came to U.S., 1981; s. David Edward and Margret Ethel (Punt) W.; m. Sasithorn Aranuvachapan, Sept. 9, 1950. B.Sc. with honors, U. Bristol, Eng., 1975; M.Sc., U. East Anglia, Norwich, Eng., 1976, Ph.D., 1978. Postdoctoral fellow U. B.C. (Can.), Vancouver, 1978-81; asst. prof. oceanography Naval Postgrad. Sch., Monterey, Calif., 1981—. Contbr.: articles on deep-sea research, dynamics of atmospheres and oceans, geophys. and astrophys. fluid dynamics to Jour. Phys. Oceanography. Mem. Am. Meteorol. Soc., Am. Geophys. Union, Can. Meterol. and Oceanographic Soc. Subspecialties: Oceanography; Fluid mechanics. Current work: Rossby wave dynamics; topographic Rossby wave dynamics low frequency oceanic waves; dynamics of eastern boundary currents; geophysical fluid dynamics. Home: 69 Work Ave Del Ray Oaks CA 93940 Office: Dept Oceanography Naval Postgrad Sch Monterey CA 93940

WILLNER, STEVEN PAUL, astronomy researcher, consultant; b. Louisville, Mar. 19, 1950; s. Alan and Eugenia (Moser) W. A.B., Harvard U., 1971; Ph.D., Calif. Inst. Tech., 1976. Postgrad. research physicist U. Calif., San Diego, 1976-78, asst. research physicist, 1978-81; astronomer Smithsonian Astrophys. Obs., Cambridge, Mass., 1981—. Contbr. articles to profl. jours. Mem. Am. Astron. Soc., Internat. Astron. Union, Astron. Soc. Pacific. Subspecialties: Infrared optical astronomy; Infrared spectroscopy. Current work: Active galactic nuclei and QSOs, galactic centers, ionized hydrogen regions, detector devel., infrared spectroscopy.

WILLOUGHBY, NANCY BHARUCHA, nuclear engineer, consultant; b. Aug. 6, 1945; U.S., 1959; d. Pesi N. and Tehmi (Mistry) Bharucha; m. James Willoughby, 1962 (div. 1963); 1 son, Scot Cleon. B.S., Royal Inst. Sci., 1958; A.T.C.L., Trinity Coll. Music, London; S.M., Harvard U., 1961; postgrad., N.C. State Coll.-Raleigh, 1961-62, 63-65. Registered profl. engr., N.Y., Md. Nuclear engr. Babcock & Wilcox Co., Lynchburg, Va., 1962-63; research/teaching asst. N.C. State U., Raleigh, 1963-65; physicist Picatinny Arsenal, Dover, N.J., 1965-68; nuclear/environ. supr. Consol. Edison, N.Y.C., 1968-70; asst. dir. nuclear tech. N.Y. State Atomic and Space Devel., N.Y.C., 1970-71; project engr. Bechtel Corp., Gaithersburg, Md., 1972—; chmn. standard com. Am. Nat. Standards Inst., 1979—. Contbr. articles to profl. jours. N.C. State U. fellow, 1961, 63, 64; recipient award of merit Bechtel Corp., 1975, 80; Pres. of Year award Exptl. Aircraft Assn., 1972. Mem. Am. Nuclear Soc. (sec. 1977-79, exec. com. 1977-82), ASME, AAAS, IEEE, Exptl. Aircraft Assn. (pres. 1970-80). Zoroastrian. Club: Harvard Engring. (N.Y.C.). Subspecialties: Nuclear engineering; Environmental engineering. Current work: Reactor safety, radiation protection, aircraft construction. Home: 18700 Walker's Choice Rd Gaithersburg MD 20879 Office: Bechtel Corp 19760 Shady Grove Rd Gaithersburg MD 20877

WILLSON, LEE ANNE MORDY, astrophysics educator, researcher; b. Honolulu, Mar. 14, 1947; m., 1969; 2 children. A.B., Harvard U., 1968; M.S., U. Mich., 1970, Ph.D. in Astronomy, 1973. Instr. Iowa State U., Ames, 1973-75, asst. prof., 1975-79, assoc. prof. astrophysics, 1979—. Mem. Internat. Astron. Union, Am. Astron. Soc. (Annie J. Cannon award 1980-81). Subspecialty: Theoretical astrophysics. Office: Dept Physics Iowa State U Ames IA 50011

WILLSON, MARK JOSEPH, computer systems designer; b. Ft. Worth, Nov. 27, 1945; s. Robert and Virginia (Lambert) W.; m. Michelle Kindt, Mar. 1, 1979; children: Christopher, Clifton. B.S.E.E., Princeton U., 1967; Ph.D., U. Pa., 1979. Pres. Interactive Structures, Bala Cynwyd, Pa., 1974—. Office: 146 Montgomery Ave Suite 201 Bala Cynwyd PA 19004

WILLSON, RICHARD CLAYTON, solar radiation scientist; b. Austin, May 18, 1937; s. Richard James and Evelyn Ann W.; m. Marie Elizabeth, Aug. 20, 1977. B.S., U. Chgo., 1957, M.S., 1963; Ph.D., UCLA, 1975. Research engr. Jet Propulsion Lab., Calif. Inst. Tech., Pasadena, 1963-65, sr. engr., 1966-75, sr. scientist, 1975-81, mem. tech. staff., 1981—; prin. investigator solar irradiance monitoring NASA. Contbr. numerous sci. articles to profl. publs. Recipient Exceptional Sci. Achievement medal NASA, 1982, others. Mem. Am. Geophys. Union, Am. Meteorol. Soc. Club: Pasadena Athletic. Subspecialties: Solar-terrestrial relationships; Solar physics. Current work: The link between solar irradiance variability and the earth's climate; solar-terrestrial relationships; solar total irradiance variability; measurements of solar irradiance in space flight experiments; weather and climate response to solar variability. Developed active cavity radiometer instrumentation approach. Office: 4800 Oak Grove Dr 171-400 Pasadena CA 91109

WILNER, GEORGE DUBAR, clinical pathologist, investigator; b. N.Y.C., Dec. 7, 1940; s. Joseph Walter and Marie (Spring) W.; m. Caroline T. Hauptmann, Nov. 1, 1963; children: Michael, Jessica, Marissa. A.A., George Washington U., 1960; B.S., Northwestern U., 1962, M.D., 1965. Diplomate: Am. Bd. Pathology. Intern Presbyn. Hosp., N.Y.C., 1965-66, resident, 1966-67, Montefiore Hosp., N.Y.C., 1967-69; from asst. prof. to assoc. prof. Columbia U., N.Y.C., 1969-78; assoc. prof. Washington U., St. Louis, 1978—; dir. lab. medicine Jewish Hosp., St. Louis, 1978—. Contbr. sci. articles to profl. publs. Served to maj. U.S. Army, 1971-73. Research career devel. awardee NIH, Nat. Heart Lung Inst., 1973-78; various research grants. Mem. Am. Soc. Hematology, Am. Assn. Pathologists, Am. Fedn. Clin. Research, Am. Heart Assn., Internat. Soc. on Thrombosis and Haemostasis. Subspecialties: Hematology; Pathology (medicine). Current work: Fibrinogen immunology and immunochemistry; cell interactions with thrombin. Office: Dept Pathology Jewish Hosp St Louis 216 S Kingshighway Saint Louis MO 63110

WILSON, ALEXANDER THOMAS, chemist, researcher; b. b Wellington, N.Z., Feb. 8, 1930; s. Percy Frederick and Dorothy (Evans) W. B.S., Victoria U. Coll., N.Z., 1949; M.S. with 1st class honors, 1950; Ph.D., U. Calif., Berkeley, 1954; D.Sc. (hon.), U. Waikato, N.Z., 1981. With Standard Oil Co. Ind., 1954-56; div. nuclear sci. N.Z. Dept. Sci. and Indsl. Research, 1957-59; sr. lectr., assoc. prof. U. Wellington, 1960-69; dean of Sci., prof. chemistry U. Waikato, 1969-79; dir. research Duval Corp. (subs. Pennzoil Co.), Tucson, 1979—. Contbr. numerous articles to profl. jours. Recipient Easterfield medal N.Z. Inst. Chemistry, 1963; Nuffield Dominion fellow in natural scis., 1965; decorated Polar medal Queen Elizabeth II, 1978. Subspecialty: Physical chemistry.

WILSON, ALLAN BYRON, interactive graphics co. exec.; b. Jackson, Miss., Aug. 19, 1948; s. Allen Bernice and Mary Pickering (Levereault) W.; m. Ines Ghinato, May 19, 1975; 1 son, Lucas Ghinato. B.S., Rice U., 1970, M.S. in Elec. Engring, 1971. Systems adminstr. Max Planck Institut für, Kohlenforschung, Mulheim, W. Ger., 1971; systems programmer Digital Equipment Corp., Maynard, Mass., 1972-74, mktg. specialist, 1974-75, mktg. mgr., 1976-79; internat. ops. dir. Intergraph Corp., Huntsville, Ala., 1980-82, v.p. corp. and internat. ops., 1982—. Contbr. articles to profl. jours. Mem. Assn. Computing Machinery, IEEE. Subspecialties: Graphics, image processing, and pattern recognition; Operating systems. Current work: International operations/mktg. text processing; operating systems. Home: 576 Rainbow Dr Madison AL 35758 Office: 1 Madison Industrial Park Huntsville AL 35807

WILSON, ALLAN CHARLES, biochemist; b. Ngaruawahia, N.Z., Oct. 18, 1934; m., 1958; 2 children. B.Sc., U. Otago, N.Z., 1955; M.S., Wash. State U., 1957; Ph.D. in Biochemistry, U. Calif.-Berkeley, 1961. Fellow in biochemistry Brandeis U., 1961-64; from asst. prof. to assoc. prof. U. Calif.-Berkeley, 1964-72, prof. biochemistry, 1972—; mem. Alpha Helix Expdn., New Guinea, 1969. Assoc. editor: Biochemical Genetics, 1975—, Jour. Molecular Evolution, 1978—. Recipient Guggenheim Meml. Found. award Weizmann Inst. Sci., U. Nairobi, 1972-73; grantee NSF, 1965—, NIH, 1974—. Mem. Am. Soc. Biol. Chemistry, Soc. Systematic Zoology. Subspecialty: Genetics and genetic engineering (biology). Office: U Calif 401 Biochemistry Bldg Berkeley CA 94720

WILSON, ANDREW STEPHEN, astronomer; b. Doncaster, Eng., Mar. 26, 1947; s. Norman and Mary Alice (Beckett) W.; m. Kaija Annikki, Oct. 4, 1975; children: Daniel Marcus, Caroline Johanna. B.A. with 1st class honors, U. Cambridge, 1969, M.A., Ph.D., 1973. Royal Soc. European Program fellow Leiden (Netherlands) Obs., 1973-75; research fellow U. Sussex, Eng., 1975-78; asst. prof. astronomy U. Md., College Park, 1978-81, assoc. prof., 1981—. Contbr. articles to profl. jours. Grantee NASA, NSF, 1979—. Fellow Royal Astron. Soc.; mem. Am. Astron. Soc. Subspecialties: High energy astrophysics; Optical astronomy. Current work: Active galactic nuclei, supernova remnants. Office: Astronomy Program University of Maryland College Park MD 20742

WILSON, ARTHUR JESS, clinical psychologist, consultant; b. Yonkers, N.Y., Oct. 25, 1910; s. Samuel Louis and Anna Lee (Gilbert) W.; m. Lillian Moss, Sept. 16, 1941; children: Warren David, Anton Francis. B.S., NYU, 1935, M.A., 1949, Ph.D., 1961; LL.B., St. Lawrence U., 1940; J.D., Bklyn. Law Sch., 1967. Supr. rehab. N.Y. State Edn. Dept., N.Y.C., 1942-44; rehab. field sec. N.Y. Health Assn., N.Y.C., 1946-48; dir. rehab. Westchester County Med. Ctr., Valhalla, N.Y., 1948-67; dir. N.Y. State Drug Abuse Rehab. Ctr, N.Y.C., 1967-68; clin. psychologist F. D Roosevelt VA Hosp., Montrose, N.Y., 1968-72; pvt. practice clin. psychology, Yonkers, 1972—; instr. in psychology Westchester Community Coll., 1966-67; founder Manhattan Narcotic Rehab. Ctr., 1967-68; cons. in field; participant Clin. Study Tour of mental health facilities in Mainland China, 1980. Author: Emotional Life of the Ill and Injured, 1950. Served with USN, 1944-46. Recipient Founders Day award NYU, 1961; honored as Westchester author Westchester County Hist. Soc. at Washington Irving Celebration, 1957; named Pioneer N.Y. State Narcotic Addiction Control Commn. Mem. N.Y. Acad. Scis., Am. Psychol. ASsn., N.Y. State Psychol. Assn., Internat. Mark Twain Soc. (hon.), Kappa Delta Pi, Phi Dela Kappa, Epsilon Pi Tau. Current work: Intensive research in psychotherapeutic techniques involving eidetic imagery; application of research to psychotherapy in private practice. Home: 487 Park Ave Yonkers NY 10703

WILSON, BENJAMIN JAMES, biochemistry educator, consultant; b. Pennsboro, W.Va., Jan. 7, 1923; s. Benjamin and Gertrude (Doak) W.; m. Christina Ara Hufford, Apr. 6, 1944 (dec. June 1974); children: Suzanne Hadley, Barbara Bryant, James Christopher, Rebecca E.; m. Nancy Wynell Ligon, Dec. 23, 1978. A.B., W.Va. U., 1943, M.S., 1947; Ph.D., George Washington U., 1955. Dir. clin. labs. Hopemount Sanitorium, W.Va., 1947-49; assoc. prof. biology David Lipscomb Coll., Nashville, 1959-63; from asst. prof. to prof. biochemistry Sch. Medicine, Vanderbilt U., Nashville, 1963—; adj. prof. Coll. Vet. Medicine, U. Tenn., Knoxville, 1979—; cons. food cos. FDA; mem. coms. Nat. Acad. Scis., 1968—. Contbr. chpts. to books. Served in USN, 1943-45. Mem. N.Y. Acad. Scis., Am. Soc. Microbiology, Soc. Toxicology, Am. Chem. Soc. Republican. Mem. Ch. of Christ. Subspecialties: Food science and technology; Toxicology (agriculture). Current work: Naturally occurring toxicants of fungi and higher plants. Home: Route 1 Box 241-C White Bluff TN 37187 Office: Vanderbilt U Nashville TN 37232

WILSON, CHARLES LINDSAY, plant pathologist, educator; b. Bristol, Va., Apr. 9, 1932; s. Crawford Lindsay and Nannie Virginia (Rollins) W.; m. Mirian Janet Williams, Dec. 14, 1973; children: Cathy, Nancy, Barrie, Rebecca, Patrick, Charles. B.A., U. Va., 1953; M.S., U.Va. U., 1956, Ph.D., 1958. Prof. forest pathology U. Ark., Fayetteville, 1958-68; research leader Fruit Pathology and Nematology Unit, Applachian Fruit Research Sta., U.S. Dept. Agr., Kearneysville, W.Va., 1980—; adj. prof. plant pathology W.Ua. U., Morgantown, 1980—; cons. in field. Author: The World of Terrarivas, 1975, The Gardener's Hint Book, 1979; editor: Exotic Pests and North American Agriculture, 1983; contbr. over 150 sci. articles to profl. publs., chpts. to books. Mem. Am. Phytopath. Soc., Ark. Alumni Assn. (Disting. Research award 1968). Subspecialties: Plant pathology; Membrane biology. Current work: Biological control-nature of resistance in plants; research leader-plant pathology and nematology unit conducting research on biological control and nature of plant resistance. Home: Box 1194 Shepherdstown WV 25443 Office: US Dept Agr Appalachian Fruit Research Sta Kearneysville WV 25430

WILSON, DAVID BUCKINGHAM, biochemist, educator; b. Cambridge, Mass., Jan. 15, 1940; s. E. Bright and Emily (Buckingham) W.; m. Nancy Jane Heffelfinger, Dec. 23, 1962; children: Allison K., Ashley L., Laurie E. B.A., Harvard U., 1961; Ph.D., Stanford U., 1966. Postdoctoral fellow Johns Hopkins U. Med. Sch., Balt., 1966-67; asst. prof. biochemistry Cornell U., Ithaca, N.Y., 1967-73, assoc. prof., 1973—. Bd. mgrs. Tompkins Community Hosp. Mem. Am. Soc. Biol. Chemists, Am. Soc. Microbiology, AAAS. Democrat. Subspecialties: Biochemistry (biology); Biomass (energy science and technology). Current work: Ecoli inner membrane; enzymology of cellulose degradation and genetic engineering of cellulase genes. Home: 232 Troy Rd Ithaca NY 14850 Office: Dept Biochemistry Cornell U Ithaca NY 14853

WILSON, DAVID MERL, JR., plant pathologist; b. Mosca, Colo., May 25, 1941; s. David Merl and Beulah Katherine (Hunt) W.; m. Susan Margaret Hay, Aug. 27, 1966; children: Keri Rebecca, Mark Jeremy. Student, Adams State Coll., 1959-61; B.S., Colo. State U., 1964, M.S., 1966, Ph.D., 1968. Research assoc. N.C. State U., Raleigh, 1968-69; asst. prof. U. Vt., Burlington, 1969-73; asst. prof. plant pathology Coastal Plain Sta., U. Ga., Tifton, 1973-77, assoc. prof., 1977-82, prof., 1982—; founder Mycotoxin Analysis Research Ctr., 1978. Contbr. articles to profl. jours. Mem. Am. Chem. Soc., Am. Phytopath. Soc., Assn. Ofcl. Analytical Chemists. Lutheran. Subspecialties: Plant pathology; Analytical chemistry. Current work: Mycotoxin contamination of foods and feeds; research on control of toxic fungal metabolites that may contaminate animal or human foods or feeds. Home: 702 W 12th St Tifton GA 31794 Office: Plant Pathology Coastal Plain Station Tifton GA 31793

WILSON, DWIGHT ELLIOTT, biologist; b. Greensburg, Pa., June 7, 1932; s. Dwight E. and Thelma C. (Chadwick) W.; m.; children: Charles, Robert, David, Todd. B.S., Yale U., 1953, Ph.D., 1957. Asst. prof. biology Rensselaer Poly. Inst., 1956-61, assoc. prof., 1961-68, prof., 1968—. Contbr. articles in field to profl. jours. Trustee Bender Lab. Recipient Rensselaer Poly. Inst. Disting. faculty award, 1971. Mem. AAAS, Am. Soc. Human Genetics. Subspecialty: Genome organization. Current work: Organization and evolution of the mammmalion genome. Office: Rensselaer Polytechnic Institute 122 Cogswell Laboratory Troy NY 12181

WILSON, EDGAR BRIGHT, chemistry educator; b. Gallatin, Tenn., Dec. 18, 1908; s. Edgar Bright and Alma (Lackey) W.; m. Emily Buckingham, June 15, 1935 (dec. 1954); children: Kenneth, David, Nina (Nina W. Cornell); m. Therese Bremer, July 25, 1955; children: Anne, Paul, Steven. B.S., Princeton U., 1930, A.M., 1931; Ph.D., Calif. Inst. Tech., 1933; A.M., Harvard, 1942; D. honoris causa, U. Brussels, 1975; D.Sc. (hon.), Dickinson Coll., 1976, Columbia U., 1979, Princeton U., 1981, Clarkson Coll., 1983, Harvard U., 1983; Dr. Chem., U. Bologna, 1976. Research fellow Calif. Inst. Tech., 1933-34; jr. fellow Soc. of Fellows Harvard U., 1934-36, asst. prof. chemistry, 1936-39, asso. prof., 1939-46, prof., 1946-79, Theodore William Richards prof. chemistry, 1947-79, prof. emeritus, 1979—; research dir. Underwater Explosives Research Lab., 1942-44; chief div. 2 Nat. Def. Research Com., 1944-46; research dir., weapons systems evaluation group Dept. Def., 1952-53. Author: (with Linus Pauling) Introduction to Quantum Mechanics, 1935, Introduction to Scientific Research, 1952, (with P.C. Cross, J.C. Decius) Molecular Vibrations, 1955. Hon. trustee Woods Hole Oceanographic Instn., 1979—. Recipient Am. Chem. Soc. award, 1937, medal for Merit U.S. govt., 1948, Debye award, 1962; Distinguished Service award Calif. Inst. Tech., 1966; Pauling award, 1972; Rumford medal, 1973, Nat. Medal Sci., 1976; Feltrinelli award, 1976; Ferst award Sigma Xi, 1977; Pitts. Spectroscopy award, 1978; Robert A. Welch award, 1978; Willard Gibbs award, 1979; Lippincott medal, 1979; Guggenheim fellow, 1949-50, 70-71; Fulbright grantee Queen's Coll., Oxford, Eng., 1949-50. Mem. Am. Chem. Soc. (Norris award N.E. sect. 1966, Lewis award Calif. sect. 1969, T.W. Richards medal N.E. sect. 1978), Am. Phys. Soc. (Plyler award 1978), Am. Philos. Soc., Am. Acad. Arts and Scis., Internat. Acad., Quantum Molecular Scis., Nat. Acad. Scis., Phi Beta Kappa. Subspecialties: Physical chemistry. Current work: Molecular spectroscopy and structure. Office: 12 Oxford St Cambridge MA 02138

WILSON, EDWARD OSBORNE, entomology educator; b. Birmingham, Ala., June 10, 1929; s. Edward Osborne and Inez (Freeman) W.; m. Irene Kelley, Oct. 30, 1955; 1 dau., Catherine Irene. B.S., U. Ala., 1949, M.S., 1950; Ph.D., Harvard U., 1955; D.Sc. (hon.), Duke U., Grinnell Coll., Lawrence U., U. West Fla.; L.H.D., U. Ala.; LL.D., Simon Fraser U. Jr. fellow Soc. Fellows, Harvard, 1953-56, mem. faculty, 1956—, Baird prof. sci., 1976—, curator entomology, 1971—. Author: The Insect Societies, 1971, Sociobiology: The New Synthesis, 1975, On Human Nature, 1978, Promethean Fire, 1983. Recipient Nat. Medal Sci., 1976; Pulitzer prize for nonfiction, 1979; Leidy medal Acad. Natural Sci., Phila., 1979; Disting. Service award Am. Inst. Biol. Scis., 1976; Mercer award Ecol. Soc. Am., 1971; Founders Meml. award Entomol. Soc. Am., 1972; Archie Carr medal U. Fla., 1978; Disting. Service award Am. Humanist Soc., 1982; others. Fellow Am. Acad. Arts and Scis., Am. Phil. Soc., Deutsche Akad. Naturforsch.; mem. Nat. Acad. Sci. Subspecialties: Sociobiology; Evolutionary biology. Home: 9 Foster Rd Lexington MA 02173 Office: Mus Comparative Zoology Harvard Univ Cambridge MA 02138

WILSON, FREDERICK ALLEN, medical educator and administrator; b. Winchester, Mass., Aug. 22, 1937; s. Warren Archibald and Alice Jane (Springall) W.; m. Lynne Stewart Cantley, Feb. 24, 1962; children: Douglas, Victoria. B.A., Colgate U., 1959; M.D., Albany Med. Coll., 1963. Intern and resident Hartford (Conn.) Hosp., 1963-66; postdoctoral fellow in gastroenterology Albany (N.Y.) Med. Coll., 1966-67; USPHS postdoctoral fellow in gastroenterology U. Tex. Southwestern Med. Sch., Dallas, 1969-72; asst. prof. medicine Vanderbilt U. Sch. Medicine, Nashville, 1972-76, assoc. prof. medicine, 1976-82; attending physician Vanderbilt U. Hosp., 1972-82; prof. medicine, chief div. gastroenterology Milton S. Hershey Med. Ctr., Pa. State U., Hershey, 1982—; attending physician Milton S. Hershey Med. Ctr. Hosp., 1982—; staff physician Met. Nashville Gen. Hosp., 1979-82; cons. VA Med. Ctr., Nashville, 1976-82, St. Thomas Hosp., 1980-82, VA Med. Ctr., Lebanon, Pa., 1982. Contbr. chpts. to books, writings to publs. in field; reviewer: sci. jours. including Diabetes. Served to maj. U.S. Army, 1967-69. Clin. investigator VA, 1972-75; investigator Howard Hughes Med. Inst., 1975-78; sr. internat. fellow NIH Fogarty Internat. Ctr., Vanderbilt U. Research Council; Fellow Max-Planck Inst. for Biophysics, Frankfurt, W.Ger., 1979-80; Nat. Inst. Arthritis, Diabetes and Digestive and Kidney Diseases grantee, 1981—. Mem. Am. Fedn. Clin. Research, Am. Gastroenterology Assn., Central Soc. Clin. Research, Am. Assn. for Study Liver Diseases, Am. Soc. Clin. Investigation, N.Y. Acad. Scis. Subspecialty: Gastroenterology. Current work: Teaching medical students and house staff, patient care and research in area of intestinal transport, especially bile salts. Office: Milton S Hershey Med Ctr Pa State U Div Gastroenterology Box 850 Hershey PA 17033

WILSON, GABRIEL HENRY, radiologist; b. Caruthersville, Mo., Jan. 30, 1929; s. Michael Earl and Alma Alberta (Cecil) W. Student, St. Mary's Coll., 1946-48, U. Calif. at Los Angeles, 1952-53; B.S., Loyola U., Chgo., 1955; M.D., Creighton U., 1959. Diplomate: Nat. Bd. Med. Examiners, Am. Bd. Radiology. Rotating intern Los Angeles County Harbor Gen. Hosp., Torrance, Calif., 1959-60, resident in radiology, 1960-63, staff radiologist, 1963-64, 65-66, vis. physician in neuroradiology, 1964-67, instr. in radiology, 1963-64, asst. prof., 1964-66; asst. prof. radiology UCLA, 1966-70, asso. prof., 1970-74, prof., 1974—, chmn. dept. radiol. scis., 1973—; med. staff sec. UCLA Hosp. and Clinics, 1978-80, chief of staff, 1980-82; cons. neuroradiology VA Hosp., Long Beach, Calif., 1965-66, cons., Los Angeles, 1965—. Served with USNR, 1948-52. Fellow Am. Coll. Radiology (councilor 1976-81); mem. AMA, Calif. Med. Assn. (asst. sec. sect. on radiology 1970-71, sec. 1971-72, chmn. 1972-74, chmn. adv. panel radiology 1972-73, 76-82), Calif. Radiol. Soc. (exec. com. 1976—, pres. 1980-81), Los Angeles Radiol. Soc. (exec. com. 1971—, treas. 1971-72, sec. 1972-73, pres. 1974-75), Japan Radiol. Soc., Radiol. Soc. So. Calif. (bd. dirs. 1976-83, sec.-treas. 1979-80, pres.-elect 1980-81, pres. 1981-82), Radiol. Soc. N.Am. (councilor 1976-81), Am. Soc. Neuroradiology (v.p. 1979-80, pres.-elect 1980-81, pres. 1981-82), Western Neuroradiol. Soc. (pres. 1968-70, chmn. nominating com. 1971-72, chmn. membership com. 1974), Fedn. Western Socs. Neurol. Sci. (bd. dirs. 1968-76, chmn. nominating com. 1968-69, chmn. auditing com. 1969-70, sec.-treas. 1971-73, chmn. 1973-75), Inter-Am. Fedn. Neuroradiology, Los Angeles Soc. Neurology and Psychiatry, Soc. Chairmen of Acad. Radiology Depts., Am. Roentgen Ray Soc., Alpha Omega Alpha. Subspecialties: Diagnostic radiology; Imaging technology. Current work: Nuclear magnetic resonance. Home: 4009 Ocean Front Walk Venice CA 90292 Office: U California Dept of Radiological Sciences Center for Health Sciences Los Angeles CA 90024

WILSON, GEORGE PORTER, III, veterinary surgeon; b. Flint, Mich., Nov. 10, 1927; s. George Porter and Eva Harris (Spencer) W.; m.; children: George A., Todd S., Amy P., Laurie B. B.S., U. Ill., 1951; V.M.D., U. Pa., 1955; M.Sc., Ohio State U., 1959. Intern Angell Meml. Animal Hosp., Boston, 1955-56; prof. dept. vet. clin. sci. Ohio State U., 1956—, prof. vet. anatomy, 1983—; vis. scientist Armed Forces Inst. Pathology, 1979; vis scientist Nat. Cancer Inst., 1978. Served with U.S. Navy, 1945-46. Mem. N.Y. Acad. Sci., AAAS, AVMA, AM. Coll. Vet. Surgeons, Sigma Xi, Phi Zeta. Subspecialties: Surgery (veterinary medicine); Epidemiology. Current work: Developmental anatomy of mammalian reporductive tract, neoplasms hormone dependency. Home: 378 Fenway Rd Columbus OH 43214 Office: 1900 Coffey Rd Columbus OH 43210

WILSON, GLENN FRANCIS, research psychologist, psychology educator; b. Hawthorne, Calif., Aug. 15, 1942; s. Alfred R. and Kathryn (Powelson) W.; m. Marietta R. Chaves, July 28, 1963; children: Todd F., Kimerly A. B.A., Idaho State U., 1965; M.A., U. Ariz., 1967, Ph.D., 1968. Postgrad. research psychologist UCLA, 1968-70, asst. research psychologist, 1970-72; asst. prof. psychology Wittenberg U., 1972-78, assoc. prof., 1978—; research psychologist Wright-Patterson AFB, Ohio, 1981-83; cons. Clark County Mental Health Unit. Contbr. articles to profl. jours. NDEA predoctoral fellow, 1965-68. Mem. Soc. Psychophysiol. Research, Soc. Neurosci., Human Factors Soc., AAAS, Sigma Xi. Subspecialties: Physiological psychology; Neuropsychology. Current work: Psychophysiological correlates of human information processing and workload, evoked potentials, cognition, psychophysiological correlates of personality. Office: Dept Psychology Wittenberg U PO Box 720 Springfield OH 45501

WILSON, GREGORY BRUCE, medical educator, consultant; b. Columbus, Ohio, Oct. 15, 1948; s. Bruce N. and Miriam J. (Allen) W.; m. Nancy Lee Maddux, Apr. 13, 1975 (div. June 1978); m. Judy Hunter Jennings, Dec. 4, 1982. B.A. in Zoology, UCLA, 1971, Ph.D. in Biology, 1974. Research asst. UCLA, 1972-74, postdoctoral research fellow dept. biology, 1974; postdoctoral research fellow dept. medicine U. Calif.-San Francisco, 1974-75; assoc. dept. basic and clin. immunology and microbiology Med. U. S.C., Charleston, 1975-76, asst. prof., 1976-79, assoc. prof., 1979—, assoc. prof. dept. pediatrics, 1982—; cons. Intron, Inc., Charleston, 1982—. Co-editor: Immunological Aspects of Cystic Fibrosis, 1983; contbr. numerous articles and revs. to sci. jours. Pres. regional council Cystic Fibrosis Found., 1982—, state bd. dirs., 1979—, numerous other coms.; county mem. med. adv. com. Nat. Found. March of Dimes, 1977-81. Nat. Cystic Fibrosis Found. research fellow, 1974-76; Nat. Found. March of Dimes Basil O'Connor grantee, 1976-79; Nat. Cystic Fibrosis Found. grantee, 1980—. Mem. AAAS, Soc. Exptl. Biology and Medicine, Reticuloendothelial Soc., N.Y. Acad. Scis., Am. Fedn. Clin. Research, Electrophoresis Soc., Soc. Am. Inventors, Sigma Xi. Subspecialties: Genetics and genetic engineering (medicine); Immunology (medicine). Current work: Immunoregulation - development of immunotherapeutic and immunoprophylactic agents and regimens for treatment of immunodeficiency diseases; development of immunoassays for detection of subjects with genetic diseases and probes for elucidation of their basic genetic defect(s). Patentee method of diagnosing cystic fibrosis patients and asymptomatic carriers of the CF gene, 1982. Home: 2706 Cameron Blvd PO Box 13 Isle of Palms SC 29451 Office: 171 Ashley Ave Charleston SC 29425

WILSON, G(USTAVUS) EDWIN, JR., chemistry educator; b. Phila., Oct. 6, 1939; s. G(ustavus) Edwin and Alberta (Seuffert) W.; m. Marcia Virginia Paullin, Aug. 19, 1961; children: Kristine Susan, Karin Elaine, Jennifer Lynne, Kiersten Beth. S.B., MIT, 1961; Ph.D., U. Ill., 1964. Instr. chemistry Poly. Inst. N.Y., Bklyn., 1964-65, asst. prof., 1965-69, assoc. prof., 1969-75, prof., 1975-80; prof. chemistry Clarkson Coll., Potsdam, N.Y., 1980—. Fellow N.Y. Acad. Scis. (chmn. chem. sci. 1971-72); mem. Am. Chem. Soc., The Chem. soc., Sigma Xi. Subspecialties: Organic chemistry; Biophysical chemistry. Current work: Application of organic chemistry and NMR to the solution of biological problems, biochemistry of protein biosynthesis. Home: 41 Hitchner Ave Bridgeton NJ 08302 Office: Clarkson Coll Potsdam NY 13676

WILSON, JEAN DONALD, physician; b. Wellington, Tex., Aug. 26, 1932; s. J. D. and Maggie E. (Hill) W. B.A. in Chemistry, U. Tex., 1951, M.D., 1955. Diplomate: Am. Bd. Internal Medicine. Intern, then resident in internal medicine Parkland Meml. Hosp., Dallas, 1955-58; clin. assoc. Nat. Heart Inst., Bethesda, Md., 1958-60; instr. internal medicine U. Tex. Southwestern Med. Sch., Dallas, 1960-61, prof., 1968—. Editor: Jour. Clin. Investigation, 1972-77. Served as sr. asst. surgeon USPHS, 1958-60. Recipient Oppenheimer award Endocrine Soc., 1972; Amory prize Am. Acad. Arts and Scis., 1977; Fuller prize Am. Urol. Assn., 1983. Mem. Am. Fedn. Clin. Research, Am. Soc. Clin. Investigation, Soc. Exptl. Biology and Medicine, Am. Soc. Biol. Chemists, Endocrine Soc., Assn. Am. Physicians., Am. Acad. Arts and Scis., Nat. Acad. Scis. Subspecialty: Receptors. Current work: The molecular mechanisms of normal androgen action in the developing male embryo and in postembryonic life and inrestigation of clinical disorders that impair androgen action and cause abnormal sexual development. Home: 4517 Watauga Rd Dallas TX 75209 Office: Dept Internal Medicine U Tex Health Sci Center Dallas TX 75235

WILSON, JOHN SHERIDAN, chem. engr.; b. Morgantown, W.Va., June 17, 1944; s. Jack Belmont and Vivian Jean (Fike) W.; m. Diane Barbara Wilson, July 14, 1944; children: David Bailey, Kevin Mark, Matthew Joseph, Adam Jason. B.S. in Chem. Engring, W.Va. U., 1966, M.S. in Nuclear Engring, 1968, Ph.D. in Chem. Engring, 1975. Project leader and project engr. Bur. Mines, U.S. Dept. Interior, Morgantown, W.Va., 1968-76; project mgr. combustion research and devel. br. Morgantown Energy Research Center, ERDA, 1976, research supr., 1976-77; asst. dir. energy conversion and utilization div. Morgantown (W.Va.) Energy Tech. Center, U.S. Dept. Energy, 1977-79, dir. coal projects mgmt. div., 1979—. Contbr. articles to profl. jours. Bd. dirs. Morgantown Hockey Assn. Mem. Am. Chem. Soc., ASME, Sigma Xi. Baptist. Subspecialties: Coal; Chemical engineering. Current work: Coal combustion and gasification processes along with environmental aspects of coal use; utilization of coal and coal-derived fuels in gas turbines, fuel cells and boilers. Office: PO Box 880 Morgantown WV 26505

WILSON, JOHN TUZO, museum executive, geophysicist; b. Ottawa, Ont., Can., Oct. 24, 1908; m., 1938; G62 children. B.A., U. Toronto, Ont., 1930, D.Sc. (hon.), 1977; M.A., U. Cambridge, Eng., 1932, Sc.D., 1958; Ph.D. in Geology, Princeton U., 1936; LL.D. (hon.), Carleton U., 1958, Simon Fraser U., 1978, D.Sc., U. Western Ont., 1958, Acadia U., Meml. U. Nfld., 1968, McGill U., 1974, Laurentian U., 1978, Middlebury Coll., 1981, Sc.D., Franklin and Marshall Coll., 1969, D. Univ., U. Calgary, 1974. Asst. geologist Geol. Survey Can., 1936-39; prof. geophysics U. Toronto, 1946-74, Disting. lectr., 1974-77; prin. Erindale Coll., 1967-74; dir.-gen. Ont. Sci. Centre, Don Mills, 1974—; vis. prof. Australian Nat. U., 1950, 65; pres. Internat. Union Geologists and Geophysicists, 1957-60; mem. NRC Can., 1958-64, Def. Research Bd. Can., 1960-66. Recipient Blaylock medal Can. Inst. Mining and Metallurgy, 1959; Bucher medal Am. Geophys. Union, 1968; M. Ewing medal, 1980; Companion award Order Can., 1974; Wollaston medal Geol. Soc. London, 1978; Soc. Exploration Geophysics award, 1981; A. G. Huntsman award Bedford Inst. Oceanography, 1981. Fellow Geol. Soc. Am. (Penrose medal 1969), Royal Soc. Can. (Miller medal 1956, pres. 1972-73); mem. Nat. Acad. Sci. (Fgn. assoc., J. J. Carty medal 1975), Am. Philos. Soc. (fgn. mem.). Subspecialty: Geophysics. Office: Ont Sci Centre 770 Don Mills Rd Don Mills ON Canada M3C 1R7

WILSON, KENNETH GEDDES, physical science educator; b. Waltham, Mass., June 8, 1936; s. Edgar Bright and Emily Fisher (Buckingham) W.; m. Alison Brown, Oct. 1982. A.B., Harvard U., 1956, Ph.D. (hon.), 1981, Calif. Tech. U., 1961, U. Chgo., 1976. Asst. prof. to prof. phys. sci. Cornell U., Ithaca, N.Y., 1963—. Contbr. articles to profl. jours., conf. presentations on theoretical physics. Recipient Nobel Prize, 1982, Franklin medal, 1983. Subspecialty: Theoretical physics. Current work: Elementary particles theory; statistical mechanics; large scale scientific computing. Office: Cornell U 316 Newman Lab Ithaca NY 14853

WILSON, KENNETH GLADE, botanist, educator; b. Payson, Utah, May 18, 1940; s. Boyd Carson and Jennie V (Tanner) W.; m. Kathryn Jermain, Nov. 14, 1959; children: Cynthia, Carla, Kenneth Boyd. B.S., U. Utah, 1962, Ph.D., 1968. Reliability engr. Hercules Corp., Salt Lake City, 1962-63; prof. botany Miami U., Oxford, Ohio, 1967—. Mem. Am. Soc. Plant Physiologists, Am. Soc. Microbiology, Am. Genetics Assn., Internat. Soc. for Plant Molecular Biology, Sigma Xi. Democrat. Mormon. Club: 1809 (Oxford) (pres. 1972-73). Subspecialties: Genetics and genetic engineering (agriculture); Tissue culture. Current work: Plant molecular biology, motochondria, plastids, tissue culture, mass culture, gene engineering. Office: Dept Botany Miami U Oxford OH 45056 Home: 209 Beechpoint Dr Oxford OH 45056

WILSON, KENNETH SHERIDAN, biologist, educator, consl; b. Waterloo, Iowa; s. Paul W. and Dorothy J. (Dunsworth) W.; m. Regina Virginia Vilutis, June 2, 1962; 1 dau., Jennifer Ann. B.S., Colo. Coll., 1949; M.S., U. Wyo., 1950; Ph.D., Purdue U., 1954. Instr. Purdue U., Hammond, Ind., 1954-57, asst. prof., 1957-59, asso. prof., 1959-64, prof., 1964—, head dept. biology, 1959-74. Author sci. articles. Served in USMC, 1942-45. Decorated Purple Heart with two gold stars; recipient AMOCO Outstanding Teaching award, 1976. Mem. AAAS, Bot. Soc. Am., Mycol. Soc. Am., Soc. Indsl. Microbiology, Am. Soc. Microbiology. Subspecialties: Morphology; Ecology. Current work: Morphology and taxonomy of the Orchidaceae. Home: 189 W 150 N Valparaiso IN 46383 Office: Purdue U-Calumet Hammond IN 46323

WILSON, MARJORIE PRICE, physician, university dean; b. Pitts., Sept. 25, 1924; d. Robert John and Grace Alma (McMillen) Price; m. Lynn Minford Wilson, Sept. 15, 1951; children: Lynn Deyo, Liza Price. Student, Bryn Mawr Coll., 1942-45; M.D., U. Pitts., 1949. Diplomate: Nat. Bd. Med. Examiners. Intern U. Pitts. Med. Center Hosps., 1949-50; resident Children's Hosp. U. Pitts., 1950-51, Jackson Meml. Hosp., U. Miami Sch. Medicine, 1954-56; chief residency and internship div. edn. service Office of Research and Edn., VA, Washington, 1956, chief profl. tng. div., 1956-60, asst. dir. edn. service, 1960; chief tng. br. Nat. Inst. Arthritis and Metabolic Disease, NIH, 1960-63; asst. to asso. dir. for tng. Office of Dir. NIH, 1963-64; asso. dir. extramural programs Nat. Library Medicine, 1964-67; asso. dir. program devel. OPPD NIH, 1967-69, asst. dir. program planning and evaluation, Bethesda, Md., 1969-70; dir. dept. instl. devel. Assn. Am. Med. Colls., Washington, 1970-81; sr. asso. dean U. Md. Sch. Medicine, Balt., 1981—; mem. Inst. Medicine, Nat. Acad. Scis., 1974—; bd. visitors U. Pitts. Sch. Medicine, 1974—; mem. adv. com. Md. Cancer Registry; mem. Gov.'s Commn. on Toxic Wastes. Contbr. articles to profl. jours. Mem. governing bd. Robert Wood Johnson Health Policy Fellowships, 1975—; trustee Analytic Services, Inc., Falls Church, Va., 1976—. Mem. Assn. Am. Med. Colls., Am. Fedn. Clin. Research, AAAS, IEEE. Episcopalian. Subspecialties: Preventive medicine; Management, health/medical systems. Current work: Research and teaching in management of medical systems, medical center organization and management; medical education; information sciences. Office: Univ Maryland Sch Medicine Baltimore MD 21201

WILSON, MAX KEARNS, mechanical engineer; b. Durham, N.C., Mar. 20, 1947; s. Max and Emagene (Kearns) W.; m. Susan Gail Whattam, Aug. 12, 1978; 1 dau., Karen Leigh. B.S.M.E., N.C. State U., 1969. Engr. Western Electric Co., Balt., 1969, devel. engr., 1970-78, sr. devel. engr., 1978—. Contbr. articles to profl. jours. Mem. ASME, Soc. Plastics Engrs., ASTM. Democrat. Methodist. Subspecialty: Mechanical engineering. Current work: High speed wire processing for communications industry (extrusion, wire drawing, cooling, high speed take-ups). Patentee in field. Office: Dept 3710 2500 Broening Hwy Baltimore MD 21224

WILSON, OLIN C(HADDOCK), astronomer; b. San Francisco, Jan. 13, 1909; s. Olin C. and Sophie (Clary) W.; m. Katherine E. Johnson, Sept. 3, 1943; children: Nicole Wilson McMillin, Randall S. A.B., U. Calif.-Berkeley, 1930; Ph.D., Calif. Inst. Tech., 1934. Asst. Mt. Wilson Obs., Pasadena, Calif., 1931-41; astronomer Mt. Wilson Obs., Palomar Obs., Pasadena, 1946-74, emeritus astronomer, 1975—. Mem. Nat. Acad. Sci., Am. Astron. Soc., AAAS. Democrat. Subspecialty: Optical astronomy. Current work: Stellar chromospheres, activity cycles in stars. Office: 813 Santa Barbara St Pasadena CA 91101

WILSON, RANDALL JOE, nuclear engineer, researcher; b. Wayne, Nebr., June 15, 1952; s. Loren Grove and Helen Jane (Bearinger) W.; m. Janae Marie Lerum, June 23, 1973; 1 son, Nicholas Lee. B.A., Midland Lutheran Coll., Fremont, Nebr., 1974; M.S., U. Ill., 1975, Ph.D., 1979; M.B.A. candidate, U. Chgo., 1980—. Asst. nuclear engr. Argonne Nat. Lab., Ill., 1979—. Contbr. articles to profl. jours. Mem. Am. Nuclear Soc., Scientists and Engrs. for Secure Energy. Subspecialties: Nuclear fission; Fluid mechanics. Current work: Liquid metal fast breeder reactor safety analysis, fuel motion, basic fluid mechanics. Home: 613 Buttonwood Circle Naperville IL 60540 Office: Argonne Nat Lab 9700 S Cass Ave Argonne IL 60439

WILSON, RAYMOND HIRAM, JR., astronomer, mathematician; b. Gap, Pa., Feb. 14, 1911; s. Raymond Hiram and Agnes (Wright) W.; m. Irene Gladys Louise Hansing, Aug. 21, 1940; 1 dau., Kristin Marie Wilson Young. A.B., Swarthmore Coll., 1931; A.M., U. Pa., 1933, Ph.D., 1935, postgrad., 1950-51; postgrad., Harvard U., 1937. Research asso., instr. math. and astronomy various colls., 1929-40; astronomer Naval Obs., 1940-42; prin. investigator contracts Office Naval Research, 1949-52; asst. prof. math. and astronomy, cons. research inst. Temple U., 1946-51; asst. prof. math. U. Louisville, 1951-54; physicist Naval Research Lab. and Project Vanguard, 1954-58; applied mathematician Goddard Space Flight Center, 1958-71; chief applied math. br. NASA, Washington, 1958-71; profl. lectr. astronomy Georgetown U. Grad. Sch., Washington, 1962-68; prof. astronomy and applied math., dir. obs. U. Aegean, Turkey, 1971-74, 77-80; astronomer Armagh Obs., No. Ireland, 1974-77; self-employed, 1980—; liaison com. mem. div. math. Nat. Acad. Sci., 1963-68. Contbr. articles in field to profl. jours. Served to comdr. USNR, 1942-46. Recipient Incentive award NASA, 1963. Fellow AAAS; mem. Internat. Astron. Union, Am. Astron. Soc. (grant 1981), Math. Assn. Am., Rittenhouse Astron. Soc. (pres. 1949), Sigma Xi, Sigma Pi Sigma. Democrat. Episcopalian. Clubs: Statesman's (Washington); Kings Ridge Swim (Fairfax, Va.). Subspecialties: Optical astronomy; Applied mathematics. Current work: Measurement of visual double stars and computation of their orbits for publication. Patentee in field. Home: 5325 Gainsborough Dr Fairfax VA 22032

WILSON, RICHARD CAMERON, physiologist, educator; b. Warrenton, Va., Jan. 22, 1953; s. Ray Cameron and Virginia Dean (Shreckhise) W.; m. Linda Louise Pascale, Nov. 27, 1977. B.S., Coll. William and Mary; Ph.D. in Physiology, U. Va. Postdoctoral fellow U. Pitts., 1980-82; asst. prof. dept. neurobiology and anatomy U. Tex. Med. Sch., Houston, 1982—. Contbr. articles to profl. jours. Mem. Soc. Neurosci. Subspecialties: Neuroendocrinology; Neurophysiology. Current work: Neural regulation of female primate reproductive cycle. Role of temporal signal patterns in regulation of reproductive function. Office: Dept Neurobiology and Anatom U Tex Med Sch 6431 Fannin Suite 7264 Houston TX 77030

WILSON, ROBERT R., physicist; b. Frontier, Wyo., Mar. 4, 1914; s. Platt E. and Edith W. (Rathbun) W.; m. Jane Scheyer, Aug. 20, 1940; children: Daniel, Jonathan, Rand. A.B., U. Calif., Berkeley, 1936, Ph.D., 1940; hon. degrees, U. Bonn., W.Ger., U. Notre Dame. Instr., asst. prof. Princeton U., 1940-43; head cyclotron group and research div. Los Alamos Nat. Lab., 1943-46; assoc. prof. physics Harvard U., 1946-47; dir. Lab. Nuclear Studies, Cornell U., Ithaca, N.Y., 1947-67, Fermi Lab., Batavia, Ill., 1967-78; prof. U. Chgo., 1967-80; prof. physics Columbia U., 1980—. Recipient Elliott Cresson medal Franklin Inst.; Nat. medal of Sci., 1973. Mem. Am. Phys. Soc., Nat. Acad., Scis., Am. Acad. Scis., Philos. Soc. Subspecialty: Nuclear physics. Sculpture exhibitied Inst. Advanced Studies, Princeton, N.J., Harvard Sci. Ctr., Fermi Lab. Home: 916 Stewart Ave Ithaca NY 14850 Office: Newman Lab Cornell U Ithaca NY 14853

WILSON, ROBERT THOMPSON, publisher; b. Chgo., June 19, 1918; s. Robert T. and Anna M. (Schalk) W.; m. Doris Jane Quant, Feb. 19, 1943; children: Robin Jane, Scott Alan. B.A. in Econs, Lawrence Coll., Appleton, Wis., 1940. Profl. musician with dance orchs., 1934-42, life ins. underwriter, 1940-42; advt. space salesman, then gen. sales mgr. Gillette Publishing Co., Chgo., 946-60; sales mgr. Dun-Donnelley Pub. Corp., Chgo., 1960-70, publisher, 1970—; dir. The Road Info. Program. Served with AUS, 1942-46. Mem. Bus.-Profl. Advertisers Assn., Chgo. Bus. Pubs. Assn., Am. Bus. Press, Constrn. Industries Mfg. Assn. (dir.), Asso. Gen. Contractors Am., Am. Rd. and Transp. Builders Assn. (dir.), Asso. Equipment Distbrs., Nat. Sand and Gravel Assn. (dir.). Office: 1301 Grove Ave Barrington IL 60010

WILSON, ROBERT WOODROW, radio astronomer; b. Houston, Jan. 10, 1936; s. Ralph Woodrow and Fannie May (Willis) W.; m. Elizabeth Rhoads Sawin, Sept. 4, 1958; children—Philip Garrett, Suzanne Katherine, Randal Woodrow. B.A. with honors in Physics, Rice U., 1957; Ph.D., Calif. Inst. Tech., 1962. Research fellow Calif. Inst. Tech., Pasadena, 1962-63; mem. tech. staff Bell Labs., Holmdel, N.J., 1963-76, head radio physics research dept., 1976—. Recipient Henry Draper medal Royal Astron. Soc., London, 1977, Herschel medal Nat. Acad. Scis., 1977; Nobel prize in physics, 1978; NSF fellow, 1958-61; Cole fellow, 1957-58. Mem. Am. Astron. Soc., Internat. Astron. Union, Am. Phys. Soc., Internat. Sci. Radio Union, Nat. Acad. Scis., Phi Beta Kappa, Sigma Xi. Subspecialty: Radio and microwave astronomy. Current work: Structure of interstellar clouds and galaxies using the radiation from simple molecules. Discoverer of 3 deg. k Microwave Backgraound Radiation, 1965; discoverer of CO and other molecules in interstellar space using their millimeter wavelength radiation. Home: 9 Valley Point Dr Holmdel NJ 07733 Office: Bell Labs HOH L239 Holmdel NJ 07733

WILSON, SAMUEL H., biochemist; b. Washington, Aug. 5, 1939; s. Samuel H. and Sue (Whatley) W.; m. Dorothea Cowart, June 29, 1957; children: Samuel, Katherine. A.B., U. Denver, 1961, postgrad., 1961-62, 62-64; M.D., Harvard U., 1968. Postdoctoral fellow dept. biochemistry Dartmouth Med. Sch., 1967-68; research assoc. Lab. Biochem. Genetics, Nat. Heart and Lung Inst., NIH, Bethesda, Md., 1968-70; sr. staff fellow Lab. Biochemistry, Nat. Cancer Inst., NIH, Bethesda, 1970-73, sr. investigator, 1973—. Contbr. articles to profl. jours. Served with USPHS, 1968-70, 77—. Mem. Am. Assn. For Cancer Research, AAAS, Am. Chem. Soc., Am. Soc. for Cell Biology, Am. Soc. Biol. Chemists. Subspecialties: Biochemistry (biology); Molecular biology. Current work: Biochemistry of DNA replication. Home: 7803 Fulbright Ct Bethesda MD 20817 Office: Nat Cancer Inst Bldg 37 Room 4D-23 Bethesda MD 20205

WILSON, STEPHEN THOMAS, chemist; b. Eden, N.C., Oct. 20, 1946; s. Warren Burton and Louise (Hodnett) W.; m. Laura Stringfellow, July 16, 1949; 1 son, Benjamin. B.S., Wake Forest U., 1968; Ph.D., Harvard U., 1975; postgrad., U. N.C., 1973-76. Sr. staff chemist Union Carbide Corp., Tarrytown, N.Y., 1976-81, research scientist, 1981—. Danforth fellow, 1968-72. Mem. Am. Chem. Soc., Internat. Zeolite Assn., N.Y. Catalysis Soc. Subspecialties: Inorganic chemistry; Solid state chemistry. Current work: Molecular sieve synthesis, catalysis synthesis of novel molecular sieve materials. Patentee in field. Home: 1024 E Main St Shrub Oak NY 10588 Office: Union Carbide Corp Tarrytown NY 10591

WILSON, THOMAS LEON, physicist; b. Alpine, Tex., May 21, 1942; s. Homer Marvin and Ogarita Maude (Bailey) W.; m. Joyce Ann Krevosky, May 7, 1978; 1 son, Kenneth Edward Byron. B.A., Rice U., 1964, B.S., 1965, M.A., 1974, Ph.D., 1976. With NASA, Houston, 1965—, astronaut instr., 1965-74, high-energy theoretical physicist, 1969—. Contbr. articles in field to profl. jours. Recipient Hugo Gernsback award IEEE, 1964; NASA fellow, 1969-76. Mem. Am. Phys. Soc., AAAS, N.Y. Acad. Scis., Am. Assn. Physicists in Medicine. Subspecialties: Relativity and gravitation; High energy astrophysics. Current work: Theoretical work in grand unified field theory, relativistic quantum field theory, quantum chromodynamics, supergravity, cosmology, astrophysics, deep inelastic scattering, authority on neutrino tomography. Patentee in field. Home: 206 Woodcombe Dr Houston TX 77062 Office: NASA Johnson Space Center Houston TX 77058

WILSON, THORNTON ARNOLD, airplane company executive; b. Sikeston, Mo., Feb. 8, 1921; s. Thornton Arnold and Daffodil (Allen) W.; m. Grace Miller, Aug. 5, 1944; children: Thornton Arnold III, Daniel Allen, Sarah Louise Wilson Anderson. Student, Jefferson City (Mo.) Jr. Coll., 1938-40; B.S., Iowa State Coll., 1943; M.S., Calif. Inst. Tech., 1948, MIT, 1952-53. With Boeing Co., Seattle, 1943—, asst. chief tech. staff, project engring. mgr., 1957-58, v.p., mgr., 1962-64, v.p. ops. and planning, 1964-66, exec. v.p., dir., 1966-68, pres., 1968—, chief exec. officer, chmn. bd., 1972—; dir. PACCAR, Inc., Seattle-First

Nat. Bank, Weyerhaeuser Co., U.S. Steel Co. Bd. govs. Iowa State U. Found.; mem. Bus. Council., Trilateral Commn. Mem. Aerospace Industries Assn. Subspecialty: Aeronautical engineering. Office: The Boeing Co PO Box 3707 Seattle WA 98124

WILSON, VICTOR JOSEPH, educator; b. Berlin, Germany, Dec. 24, 1928; s. Andrew A. and Lydia (Yampolsky) W.; m. Isa Hermer, June 7, 1953; children—Janet, Denise. B.S., Tufts Coll., 1948, M.S., 1949; research student, U. Cambridge, Eng., 1949-50; Ph.D., U. Ill., 1953. Research asso. Rockefeller U., N.Y.C., 1953-58, asst. prof., 1958-62, asso. prof., 1962-69, prof. physiology, 1969—. Served to 1st lt. U.S. Army, 1954-56. Mem. Am. Physiol. Soc., Soc. Neurosci., Harvey Soc., Sigma Xi, Phi Beta Kappa. Subspecialties: Neurobiology; Physiology (biology). Current work: Neurophysiological analysis of postoral reflexes; study of vestibular system and spinal cord. Research physiology central nervous system. Home: 4525 Henry Hudson Pkwy New York NY 10471

WILSON, WALTER ERVIN, physicist; b. Salem, Oreg., Apr. 1, 1934; s. Ralph A. and Erma L. (Simmons) W.; m. Connie Sue Harding, May 2, 1939; children: Bruce, Douglas, Gregory, DeeAnn. B.A., Willamette U., 1956; M.S., U. Wis., 1958, Ph.D., 1961. Sr. scientist Battelle N.W., Richland, Wash., 1965—; adj. asst. prof. radiol. physics Joint Center Grad. Studies, Richland, 1977—. Mem. Am. Phys. Soc., Radiation Research Soc., Am. Assn. Physicists in Medicine, Health Physics Soc. Subspecialties: Atomic and molecular physics; Numerical analysis. Current work: Research in effects of radiation. Office: PO Box 999 Richland WA 99352

WILSON, WILLIAM JOHN, microwave research engineer; b. Spokane, Wash., Dec. 16, 1939; s. William Edward and Margret Avis (Drury) W.; m. Julia Ann Haselton, Aug. 7, 1982; children by previous marriage: Elizabeth Joan, Amy Kathleen. B.S.E.E., U. Wash., 1961; M.S.E.E., M.I.T., 1963, Ph.D., 1970. Research asst. M.I.T., Cambridge, 1967-70; sect. mgr. Aerospace Corp., El Segundo, Calif., 1970-76, 77-80; asst. prof. elec. engring. U. Tex., Austin, 1976-77; group supr. microwave advanced systems Jet Propulsion Lab., Pasadena, 1980—. Contbr. articles to profl. jours. Served to capt. USAF, 1964-67. Decorated Air Force Commendation medal.; Whitney fellow, 1961. Mem. IEEE, Am. Astron. Soc., Internat. Union Radio Sci., Internat. Astron. Union. Subspecialties: Electronics; Radio and microwave astronomy. Current work: Development of low-noise millimeter-wave radiometers and systems, including mixers, solid state sources and quasi-optical components. Office: 4800 Oak Grove 168-327 Pasadena CA 91109

WILSON, WILLIAM STANLEY, oceanographer; b. Alexander City, Ala., June 5, 1938. B.S., Coll. William and Mary, 1959, M.A., 1964; Ph.D., Johns Hopkins U., 1972. Program mgr. phys. oceanography Office Naval Research, 1972-79; chief oceanic processes br. NASA Hdqrs., Washington, 1979—. Recipient Superior Civilian Service award Dept. Navy, 1979; Exceptional Sci. Achievement medal NASA, 1981. Mem. Am. Geophys. Union, Am. Meteorol. Soc. Subspecialties: Oceanography; Satellite studies. Current work: Oceanography from space.

WIMMER, ECKARD, microbiologist, educator; b. Berlin, Germany, May 22, 1936; came to U.S., 1966; m. Astrid Brose, Sept. 4, 1965; children: Thomas, Susanne. Vordiplom, U. Gottingen, Germany, 1956, Diploma in Chemistry, 1959, Doctor rerum naturalis in Organic Chemistry, 1962. Research asso. dept. biochemistry U. B.C. (Can.), Vancouver, 1964-66; research asso. dept. botany U. Ill., Urbana, 1966-68; asst. prof. dept. microbiology St. Louis U. Sch. Medicine, 1968-73; asso. prof. dept. microbiology SUNY-Stony Brook, 1974-79, prof., 1979—. Josiah Macy Jr. faculty scholar, 1981-82. Mem. AAAS, Am. Soc. for Microbiology, Am. Soc. for Virology. Subspecialty: Virology (biology). Current work: Structure and function of viruses, particularly of poliovirus; devel. of new vaccines.

WINARSKI, DANIEL JAMES, mechanical engineer; b. Toledo, Dec. 16, 1948; s. Daniel Edward and Marguerite (Pietersien) W.; m. Donna Ilene Robinson, Oct. 10, 1970; 1 son, Tyson York. B.S.M.E., U. Mich., 1970; M.S.M.E., U. Colo., 1973; Ph.D., U. Mich., 1976. Registered profl. engr., Ariz., Colo. Research engr. Exxon Prodn. Research, Houston, 1976-77; staff engr. gen. products div. IBM, Boulder, Colo., 1977-78, Tucson, 1978—, staff engr. computer peripherals devel., 1978—; instr. U.S. Mil. Acad., West Point, 1980—; instr. Computer Camp for Minorities, No. Ariz. U., Flagstaff, 1983—. Co-v.p. Gale Sch. PTA, 1979-80, co-pres., 1981-82. Served to capt. C.E., U.S. Army, 1970-72. Recipient Invention Achievement awards IBM, 1981, 82; NSF fellow, 1973-76; Nat. Student Dissertation grantee U. Mich., 1975-76. Mem. ASME, Res. Officers Assn., Soc. Lunar and Planetary Observers. Republican. Methodist. Subspecialties: Information systems, storage, and retrieval (computer science); Biomedical engineering. Current work: Computer data archive and retrival systems, design and development, design artificial leg, study of knee stresses. Patentee in field. Home: 647 S Woodstock Dr Tucson AZ 85710 Office: IBM 64L/071-2 Tucson AZ 85744

WINBURY, MARTIN MAURICE, pharmacologist, pharm. co. exec.; b. N.Y.C., Aug. 4, 1918; s. Ervin and Helen (Stein) WINBURY.; m. Blanche M. Simons, July 11, 1942; children: Nancy E. Winbury Griffith, Gail E. B.S., L.I.U., 1940; M.S., U. Md., 1942; Ph.D., N.Y.U., 1951. Research fellow Merck Dohme Co., 1944-47; pharmacologist Searle Co., 1947-55; dir. pharmacology Schering Co., 1955-61, Warner-Lambert Co., Ann Arbor, Mich., 1961-79, dir. sci. devel., 1979—; organizer symposia and workshops. Contbr. articles to profl. jours., chpts. to books. Mem. Am. Soc. Pharmacology and Exptl. Therapeutics, AAAS, Soc. Exptl. Biology and Medicine, Am. Coll. Cardiology, Am. Heart Assn., Sigma Xi. Subspecialties: Pharmacology; Cardiology. Current work: Action of drugs for coronary disease; locate, identify and evaluate new therapeutic agents and technologies. Home: 3600 Windemere Dr Ann Arbor MI 48105 Office: Warner Lambert 2800 Plymouth Rd Ann Arbor MI 48105

WINCHESTER, JAMES FRANK, physician, medical educator; b. Glasgow, Lanarkshire, Scotland, Mar. 24, 1944; s. Alexander and Elizabeth (McKillop) W.; m. Patricia Jane Gray, May 16, 1968; children: J. Craig, Jane E. M.B.Ch.B., Glasgow U., 1969, M.D., 1980. Sr. registrar Royal Infirmary, Glasgow, Scotland, 1974-76; assoc. in medicine Georgetown U. Med. Ctr., Washington, 1976-78, asst. prof. medicine, 1978-82, assoc. prof., 1982—; dir. self-care dialysis program Georgetown U. Hosp., 1977—, assoc. dir. hemodialysis, hemoperfusion and transplant program, 1977—. Author, editor: Clinical management of Poisoning and Overdose, 1983. Chmn. profl. adv. bd. Nat. Capital Area chpt. Nat. Kidney Found., Washington, 1982-83. Fellow Royal Coll. Physicians; mem. Am. Soc. for Artificial Internal Organs, Am. Fedn. Clin. Research, Am. Soc. Nephrology, Internat. Soc. Nephrology. Subspecialties: Internal medicine; Nephrology. Current work: Clinical nephrology, poisoning, sorbent hemoperfusion, artificial organ development, hypertension. Home: 6502 Machodoc Ct Falls Church VA 22043 Office: Georgetown U Med Ctr 3800 Reservoir Rd NW Washington DC 20007

WINDELS, CAROL ELIZABETH, plant pathologist, research scientist; b. Long Prairie, Minn., July 12, 1948; d. Jerome Joseph and Genevieve Anna Marie (Clasemann) Schrenk; m. Mark Bernard Windels, Apr. 4, 1970. B.A. in Biology, St. Cloud State U., 1970; M.S. in Plant Pathology, U. Minn., 1972, Ph.D., 1980. Asst. scientist dept. plant pathology U. Minn.-St. Paul, 1974-77, assoc. scientist, 1977-80, scientist, 1980—. Contbr. chpts. to books, articles to profl. jours. Mem. Am. Phytopathol. Soc., Sigma Xi, Gamma Sigma Delta, Gamma Sigma Epsilon. Roman Catholic. Subspecialties: Plant pathology; Biological control. Current work: Soil microbiology; biological seed treatment to control soil-borne pathogens; relationship of biological seed treatments to other spermosphere and rhizosphere microorganisms. Home: 7380 Melody Dr NE Fridley MN 55432 Office: U Minn Dept Plant Pathology 209 Stakman Hall Saint Paul MN 55108

WINER, JANE LOUISE, psychology educator; b. Albany, N.Y., Nov. 1, 1947; d. Harold and Elizabeth Gertrude (Jensen) W.; m. Monty Joseph Strauss, Nov. 4, 1978. B.A., SUNY-Albany, 1969, M.L.S., 1970; M.A., Ohio State U., 1971, Ph.D., 1975. Lic. psychologist, Tax.; registered psychologist Nat. Register Health Service Providers in Psychology. Asst. prof. psychology Tex. Tech U., Lubbock, 1975-81, assoc. prof., 1981—. Contbr. articles to profl. jours. NSF grad. trainee Ohio State U., 1970-73. Mem. Am. Psychol. Assn., Southwestern Psychol. Assn., Tex. Psychol. Assn., Am. Personnel and Guidance Assn., Assn. Women in Sci. Democrat. Jewish. Current work: Vocational psychology, including vocational choice, person-environment congruence, computer literacy and applications. Home: 7010 Nashville Dr Lubbock TX 79413 Office: Dept Psychology Tex Tech U PO Box 4100 Lubbock TX 79409

WING, EDWARD JOSEPH, physician, educator; b. Mineola, N.Y., June 19, 1945; s. Maurice J. and Frances (Elliott) W.; m. Rena Rimsky, Aug. 27, 1967; children: Jonathan, Kenneth. B.A., Williams Coll., 1967; M.D., Harvard U., 1971. Diplomate: Am. Bd. Internal Medicine, Am. Bd. Infectious Diseases, Nat. Bd. Med. Examiners. Med. intern, asst. resident Peter Bent Brigham Hosp., Boston, 1971-73; fellow in infectious diseases Stanford U. and Palo Alto (Calif.) Med. Research Found.; clin. instr. medicine Harvard U., 1971-73, U. Pitts., 1973-75, asst. prof., 1977-82, assoc. prof., 1982—. Contbr. numerous articles to profl. jours. Served with USPHS, 1973-75. Mem. Am. Soc. Microbiology, Am. Assn. Immunologists, Am. Fedn. Clin. Research, Infectious Diseases Soc. Am., Reticuloendothelial Soc., Central Soc. Clin. Research, Phi Beta Kappa. Subspecialties: Infectious diseases; Immunology (medicine). Current work: Immunology of infectious dieseases, macrophage physiology, nutrition. Office: Montefiore Hosp Pittsburgh PA 15213

WING, ELIZABETH SCHWARZ, mus. curator in zooarcheology; b. Cambridge, Mass., Mar. 5, 1932; d. Henry Frederick and Maria Lisa (Gutherz) Schwarz; m. James Edward Wing, Apr. 18, 1957; children: Mary Elizabeth, Stephen Richard. B.A., Mt. Holyoke Coll., 1955; M.S., U. Fla., 1957, Ph.D., 1962. Interum asst. curator Fla. State Mus., U. Fla., Gainesville, 1961-69, asst. curator, 1969-73, assoc. curator, 1974-78, curator, 1978—. Author: (with A.B. Brown) Paleonutrition, 1979. U.S. rep. Internat. Council for Archaeozoology, 1980—. NSF grantee. Mem. Am. Soc. Mammalogists, Soc. Am. Archaeology, AAAS. Subspecialty: Zooarcheology. Current work: Research on human use of animal resources particularly prehistoric uses based upon study of animal remains excavated from archeological sites. Office: Fla State Mu U Fla Gainesville FL 32611

WING, ROBERT FARQUHAR, astronomer, educator; b. New Haven, Oct. 31, 1939; s. Donald Goddard and Charlotte (Farquhar) W.; m. Ingrid McCowen, July 27, 1963; children: Sylvia, Roger Vincent, James Donald. B.S., Yale U., 1961; postgrad., Cambridge (Eng.) U., 1961-62; Ph.D., U. Calif.- Berkeley, 1967. Asst. prof. Dept. Astronomy, Ohio State U., Columbus, 1967-71, assoc. prof., 1971-76, prof., 1976—. Contbr. articles to profl. jours. Smithsonian Inst. grantee, 1977-78; NSF grantee, 1968—; NASA grantee, 1978—. Mem. Internat. Astron. Union, Am. Astron. Soc., Royal Astron. Soc., Astron. Soc. Pacific. Subspecialties: Infrared optical astronomy; Variable stars/stellar classification. Current work: Stellar astronomy, determination of temperatures, luninosities, chem. compositions, molecular spectroscopy, galactic structure, others. Address: Astronomy Dept Ohio State U 174 W 18th Ave Columbus OH 43210

WINGARD, LEMUEL BELL, JR., pharmacology and anesthesiology educator; b. Pitts., July 10, 1930; s. Lemuel B. and Emily (King) W.; m. Cynthia Bowden, Sept. 1965; children: Amy, Diane. B.Ch.E., Cornell U., 1953, Ph.D. in Biochem. Engring. and Biochemistry, 1965. Asst. prof. chem. engring. Cornell U., Ithaca, N.Y., 1965-66; assoc. prof. chem. engring. U. Denver, 1966-67; mem faculty U. Pitts. Sch. Medicine, 1967—, prof. pharmacology and anesthesiology, 1980—. Editor: Enzyme Engineering, Vol. 1, 1972, Vol. 2, 1974, Vol. 4, 1978, Vol. 6, 1982, Applied Biochemistry and Bioengineering, Vol. 1, 1976, Vol. 2, 1979, Vol. 3, 1981, Vol. 4, 1983, Enzyme Engineering: Future Directions, 1980; Contbr. numerous sci. articles to profl. publs. NIH spl. fellow, 1970-72. Mem. Am. Soc. Pharmacology and Exptl. Therapeutics, AAAS, N.Y. Acad. Sci., Am. Inst. Chem. Engrs., Am. Soc. Artificial Internal Organs, Sigma Xi. Subspecialties: Pharmacology; Cancer research (medicine). Current work: Immobilized enzymes; enzyme electrodes; anticancer drugs; artificial pancreas; pharmacokinetics. Home: 4444 Gateway Dr Monroeville PA 15146 Office: U Pitts Sch Medicine 518 Scaife Hall Pharmacology Dept Pittsburgh PA 15261

WINGATE, CATHARINE LOUISE, radiol. physicist, sci. adminstr., rev. editor; b. Boston. B.S., Simmons Coll.; M.A., Radcliffe Coll.-Harvard U.; Ph.D., Columbia U., 1961. Research scientist Columbia U., 1954-63; radiol. physicist Naval Radiol. Def. Lab., San Francisco, 1963-66; assoc. scientist Brookhaven Nat. Lab., Upton, N.Y., 1967-70; research asst. prof. radiology SUNY, Stony Brook, 1970-75, research assoc. prof., 1975-78; health sci. adminstr. NIH, Bethesda, Md., 1978—. Rev. editor: Med. Physics, 1979—; Contbr. numerous articles on radiation research, radiology, sci. to profl. jours. Mem. Radiation Research Soc., Am. Assn. Physicists in Medicine, AAAS, Sigma Xi. Subspecialties: Imaging technology; Biophysics (physics). Office: Westwood Bldg NIH 5333 Westbard Ave Bethesda MD 20205

WINGET, CHARLES MERLIN, physiologist, educator; b. Garden City, Kans., Dec. 26, 1925; s. Charles Ansel and Ruth May (Coburn) W.; m. Katherine Barkas; children: Jean Ann, Jo Anne. B.A., San Francisco State U., 1951; Ph.D., U. Calif.-Davis, 1957. Assoc. prof. U. Guelph, Ont., 1959-63; prof. U. Calif.-Davis, 1964—; clin. prof. Wright State U. Sch. Medicine, Dayton, Ohio, 1975—; prof. Fla. A&M U., Tallahassee, 1975—; research scientist NASA-Ames Research Center, 1963—. Contbr. numerous articles to sci. jours., chpts. to books. Served with USNR, 1944-46. Postdoctoral fellow Nat. Inst. Neurol. Disease and Blindness, 1957; recipient Paul Bert award Aerospace Physiologists, 1977; Fla. Bd. Regents Disting. Vis. Scholar, 1978, Arthur D. Tuttle award in aerospace medicine, 1982. Mem. Am. Physiol. Soc., Aerospace Med. Assn., Endocrine Soc., Biophys. Soc. Subspecialties: Physiology (medicine); Space medicine. Current work: Jet lag, shift work schedules, space medicine. Home: 10459 Mary Ave Cupertino CA 95014 Office: Biomedical Research Div 239 7 NASA Moffett Field CA 94035

WINICOV, ILGA BUTELIS, biochemist, educator; b. Riga, Latvia, May 16, 1935; d. Arturs and Zenta (Gutmanis) Butelis; m. Herbert B. Winicov, Aug. 30, 1958; children: Eric, Mark; m. Rodney E. Harrington, Jan. 26, 1979. M.S., U. Wis-Madison, 1958; A.B., U. Pa., 1956, Ph.D., 1971. Postdoctoral fellow Inst. for Cancer Research, Phila., 1972-74, research assoc., 1974-76; research asst. prof. biochemistry Fels Research Inst., Temple U. Sch. Medicine, Phila., 1976-78; asst. prof. biochemistry U. Nev. Sch. Medicine, Reno, 1979—; cons. in field. Contbr. numerous sci. articles to profl. publs. NSF grantee, 1978-83; NIH grantee, 1980-83. Mem. Am. Soc. Biol. Chemists, Am. Soc. Microbiology, Am. Assn. Cancer Research. Subspecialties: Biochemistry (medicine); Molecular biology. Current work: Research in gene expression at the transcriptional and RNA processing level of eucaryotic cells using genetic engineering technology; also the effect of chemical carcinogens on this process. Office: U Nev Biochemistry Dept Reno NV 89557

WINKELHAKE, JEFFREY LEE, immunologist, educator, researcher; b. Champaign, Ill., Oct. 5, 1945; s. Claude Arthur and Marjorie Elsie (Seigwart) W.; m. Lynn Loise Winkelhake, June 5, 1982. B.S. in Zoology and Chemistry, U. Ill., 1967, M.S. in Microbiology, 1, 969, Ph.D. in Immunochemistry and Microbiology, 1974. Research assoc. The Salk Inst., San Diego, Calif., 1974-76; asst. prof. microbiology Med. Coll. Wis., Milw., 1976-79, assoc. prof., 1980—; assoc. dir. Freshwater/Marine Biomed. Center, 1978—; cons. Blood Center Southeastern Wis. Contbr. articles to profl. jours. Served with AUS, 1969-72. Fellow Jane Coffin Childs Meml. Fund for Med. Research, 1974-76; grantee NIH, NSF, Am. Cancer Soc., 1976—; Fulbright-Hays research scholar, 1982. Mem. Am. Assn. Immunologists, Am. Assn. Cancer Research, Am. Chem. Soc., Am. Soc. Biol. Chemists, Biochem. Soc. London, Internat. Soc. Developmental and Comparative immunology, Sigma Xi. Club: South Shore Yacht (Milw.). Subspecialties: Immunobiology and immunology; Cellular engineering. Current work: Structure: function relationships of immunoglobulins, monoclonal antibodies, evolution of antibody structure and function. Office: PO Box 26509 Milwaukee WI 53226

WINKELSTEIN, JERRY ALLEN, pediatrician; b. Syracuse, N.Y., Sept. 5, 1940; s. Warren W. and Lillian (Sirkin) W.; m. Marilyn A. Link, June 21, 1969; children: Beth, Amy. B.S., Syracuse U., 1961; M.D., Albert Einstein Coll. Medicine, 1965. Diplomate: Am. Bd. Pediatrics. Intern John Hopkins Hosp., Balt., 1965-66, resident, 1966-68; asst. prof. pediatrics and microbiology Johns Hopkins U., Balt., 1973-77, assoc. prof. pediatrics and immunology, 1978-83, prof. pediatrics, 1983—. Served with USPHS, 1968-70. Recipient Stettler award Johns Hopkins Hosp., 1968, Mead Johnson award Am. Acad. Pediatrics, 1982. Mem. Am. Soc. Clin. Investigation, Am. Assn. Immunologists, Am. Soc. Microbiology, Soc. Pediatric Research, Infectious Disease Soc. Subspecialties: Pediatrics; Infectious diseases. Current work: Complement research. Office: Johns Hopkins Hosp Baltimore MD 21205

WINKELSTEIN, WARREN, JR., educator, physician; b. Syracuse, N.Y., July 1, 1922; s. Warren and Evelyn (Neiman) W.; m. Veva Kerrigan, Feb. 14, 1976; children by previous marriage: Rebecca Winkelstein Yamin, Joshua, Shoshana. B.A., U. N.C., 1942; M.D. cum laude, Syracuse U., 1947; M.P.H., Columbia U., 1950. Diplomate: Am. Bd. Preventive Medicine. Intern Charity Hosp., New Orleans, 1947-48; with I.C.A., Vietnam, 1951-53; dir. div. communicable disease control to 1st dep. commr. local and environmental health services Erie County Health Dept., 1953-62; asso. prof. to prof. SUNY-Buffalo, 1962-68; prof. epidemiology, dean pub. health U. Calif., 1972—. Author: Basic Readings in Epidemiology, 1972; Contbr. articles profl. jours. Served with AUS, 1944-46. Mem. Am. Pub. Health Assn., AAAS, Internat., Am. epidemiol. socs., Am. Heart Assn. Subspecialty: Epidemiology. Home: 560 Washington Ave Point Richmond CA 94801

WINKER, DAVID M(ICHAEL), physicist; b. Mpls., Aug. 28, 1954; s. James A. and Marlene J. (Modjeske) W. B.Physics, U. Minn., Mpls., 1977; M.S., U. Ariz., 1982, Ph.D., 1984. Research asst. Optical Sci. Ctr., U. Ariz., 1977-79; research assoc. Inst. Atmospheric Physics, 1979—. Mem. Optical Soc. Am., Am. Meterol. Soc., Tau Beta Pi. Subspecialty: Remote sensing (atmospheric science). Current work: Research into methods of using lasers to measure atmospheric parameters. Office: Inst for Atmospheric Physics U Ariz Tucson AZ 85721

WINKLER, PAUL FRANK, JR., astrophysicist, educator; b. Nashville, Nov. 10, 1942; s. Paul Frank and Estelle (Pye) W.; m.; children: Katherine, Johanna. B.S., Calif. Inst. Tech., 1964; A.M., Harvard U., 1965, Ph.D. Instr. Middlebury (Vt.) Coll., 1969-70, asst. prof. physics, 1970-77, assoc. prof., 1977-82, prof., 1982—, chmn. dept. physics, 1980—; vis. scientist Center for Space Research, M.I.T., 1973-74, 78-80. Contbr. numerous articles to profl. jours. Alfred P. Sloan research fellow, 1976-80. Mem. Am. Astron. Soc., Am. Phys. Soc., Internat. Astron. Union. Subspecialties: High energy astrophysics; Optical astronomy. Current work: Observational astrophysics; x-ray and optical astronomy; physics and astronomy education. Office: Dept Physics Middlebury Coll Middlebury VT 05753

WINKLER, PETER, physicist; b. Zwickau, Germany, Feb. 2, 1937; s. Robert Gotthard and Martha Ida (Franz) W.; m. Erika Caecilie Bock, Feb. 21, 1962; children: Michael, Ulrike. Diploma in Physics, Universitaet Frankfurt am Main, W.Ger., 1966; Dr. rer. nat., U. Erlangen-Nuernberg, W.Ger., 1969, Dr. rer. nat. habil, 1977. Asst. U. Erlangen-Nuernberg, 1966-72; Max-Kade fellow SUNY, Stony Brook, 1972-74; asst. U. Erlangen-Nurnberg, 1974-77, sr. research assoc., 1977-79; assoc. prof. physics U. Nev., Reno, 1979—. Contbr. articles to profl. jours. Mem. Am. Phys. Soc., Deutsche Physikalische Gesellschaft, Sigma Pi Sigma. Subspecialties: Theoretical physics; Atomic and molecular physics. Current work: Theory of resonances; new mathematical methods in physics; theoretical atomic, molecular and chemical physics. Home: 205 Bisby St Reno NV 89512 Office: Dept Physics U Nev Reno NV 89557

WINN, RICHARD EARL, physician, educator; b. Houston, Aug. 21, 1950; s. Earl Hardy and Elvera Elayne (Balas) W.; m. Sandra Lee Sutterfield, Sept. 14, 1974; children: Sara Sutterfield, Alice Lee Sutterfield. B.S., U.S. Air Force Acad., 1972; M.D., U. Ariz., 1975. Diplomate: Am. Bd. Internal. Am. Bd. Medicine, Infectious Diseases, Am. Bd. Pathology, Am. Bd. Microbiology. Resident in internal medicine David Grant Med. Ctr., Fairfield, Calif., 1975-78; fellow in infectious diseases U. Oreg., Portland, 1978-80; staff infectious diseases Wilford Hall Med. Ctr., San Antonio, 1980—, asst. chief infectious diseases, 1980—, dir. infectious diseases research lab., 1980-82; asst. prof. Uniformed Services U. Health Scis., Washington, 1981—; clin. asst. prof. U. Tex.-San Antonio, 1981—. Served to maj. USAF, 1972—. Fellow ACP; mem. Am. Soc. Microbiology, Soc. Air Force Physicians, Tex. Soc. Infectious Diseases. Subspecialties: Internal medicine; Infectious diseases. Current work: Mechanisms of human fungal infections, mucormycosis, staphylococcal bacterial diseases, osteomyelitis. Home: Route 1 Box 1661 Boerne TX 78006 Office: Wilford Hall Med Ctr Lackland AFB TX 78236

WINOGRAD, NICHOLAS, chemistry educator; b. New London, Conn., Dec. 27, 1945; s. Arthur Selig and Winifred (Schaefer) W.; m. Barbara J. Garrison, Apr. 15, 1978. B.S., Rensselaer Poly. Inst., 1967; Ph.D., Case Western Res. U., 1970. Asst. prof. Purdue U., Lafayette,

Ind., 1970-74, assoc. prof., 1975-79; prof. chemistry Pa. State U.-University Park, 1979—; cons. Shell Oil Co., Houston. Contbr. over 90 sci. articles to profl. publs. Sloan fellow, 1972-75; Guggenheim fellow, 1977-78; NSF grantee, 1970-83; Air Force Office of Sci. Research grantee, 1971-83; Navy Office of Naval Research grantee, 1979-83. Mem. Am. Chem. Soc., AAAS, Sigma Xi. Subspecialties: Analytical chemistry; Surface chemistry. Current work: Surface science. Home: 415 Nimitz Ave State College PA 16801 Office: 152 Davey Lab University Park PA 16802

WINOGRAD, SHMUEL, mathematician; b. Tel Aviv, Jan. 4, 1936; U.S., 1956, naturalized, 1965; s. Pinchas Mordechai and Rachel W.; m. Elaine Ruth Yates, Jan. 5, 1958; children—Danny H., Sharon A. B.S. in Elec. Engring, M.I.T., M.S.; Ph.D. in Math, N.Y. U., 1968. Mem. research staff IBM, Yorktown Heights, N.Y., 1961-70, dir. math. sci. dept., 1970-74, IBM fellow, 1972—; permanent vis. prof. Technion, Israel. Author: (with J.D. Cowan) Reliable Computations in the Presence of Noise. Fellow IEEE (W. Wallace McDowell award 1974); mem. Am. Math. Soc., Math. Assn. Am., Assn. Computing Machines, Nat. Acad. Scis., Soc. Indsl. and Applied Math. Subspecialty: Applied mathematics. Current work: Directory Mathematical Sciences department. Research on complexity of computations. Home: 235 Glendale Rd Scarsdale NY 10483 Office: IBM Research PO Box 218 Yorktown Heights NY 10598

WINOGRAD, TERRY ALLEN, computer science educator, consultant; b. Takoma Park, Md., Feb. 24, 1946; s. Harold S. and Florence L. (Winograd); m. Carol Hutner, Aug. 24, 1968; children: Shoshana, Avra. B.A. in Math, Colo. Coll., 1966; postgrad., Univ. Coll., London, 1967; Ph.D. in Applied Math, MIT, 1970. Research programmer Penrose Hosp., Colorado Springs, Colo., 1965-66; instr. math. MIT, Cambridge, 1970-71, asst. prof. elec. engring., 1971-74; vis. asst. prof. computer sci. and linguistics Stanford (Calif.) U., 1973-74, asst. prof., 1974-79, assoc. prof., 1979—; cons. Xerox Palo Alto Research Ctr., 1972—, Hermenet, Inc., San Francisco, 1981—; spl. cons. to French Govt., 1981, Fuji Xerox, Japan, 1982; mem. govt. panels and other panels in field; advisor Cognitive Sci. Program U. Calif.-San Francisco, 1979—; dir. Live Oak Inst., Berkeley, Calif., 1980—. Author: Language as a Cognitive Process: Vol. I—Syntax, 1983, Understanding Natural Language, 1972; contbr. numerous articles on liguistics and computer sci. to profl. jours.; mem. editoral bd.: Artificial Intelligence, 1973—, Am. Jour. Computational Linguistics, 1974-77, Cognitive Sci, 1977-80, Discourse Processes, 1978—, Behaviorial and Brain Scis. (commentator), 1978—. Recipient Dean's award Stanford U., 1977; Danforth fellow, 1967-70; NSF (hon.) fellow, 1966; Woodrow Wilson fellow (hon.), 1966; Fulbright fellow, 1966-67; Boettcher scholar, 1962-66; NSF grantee, 1975-77, 1982-83; ARPA grantee, 1969-73, 73-75; Xerox grantee, 1975-80; Sloan Found. grantee, 1978—; System Devel. Found. grantee, 1982-83; Mellon Jr. Faculty fellow, 1977. Mem. Assn. for Computational Linguistics, Linguistics Soc. Am., Am. Assn. Artificial Intelligence (nat. bd. dirs. 1979-81), Union Concerned Scientists, Computer Profls. for Social Responsibility. Subspecialties: Artificial intelligence; Software engineering. Current work: The development of system description languages. Office: Stanford U Stanford CA 94305

WINOKUR, GEORGE, psychiatrist, educator; b. Phila., Feb. 10, 1925; s. Louis and Vera P. W.; m. Betty Stricklin, Sept. 15, 1951; children: Thomas, Kenneth, Patricia. A.B., Johns Hopkins U., 1944; M.D., U. Md., 1947. Intern Church Home and Hosp., Balt., 1947-48; asst. resident Seton Inst., Balt., 1948-50; asso. Washington U., St. Louis, 1950-51; resident in neuropsychiatry Barnes Hosp., St. Louis, 1950-51; asst. prof. psychiatry Washington U., St. Louis, 1955-59, asso. prof., 1959-66, prof., 1966-71; asso. psychiatrist Barnes Hosp., 1963-71; cons. in psychiatry Homer G. Phillips Hosp., 1954-64; instr. psychiatry Meharry Med. Coll., Nashville, 1954-55; prof., head dept. psychiatry U. Iowa, 1971—; dir. Iowa Psychiat. Hosp. Author: Manic Depressive Illness, 1969, Depression: The Facts, 1981; chief Am. editor: Jour. Affective Disorders, 1979—; editorial bd. 8 profl. jours.; contbr. numerous articles on clin. genetics of affective disorders, alcoholism and schizophrenia to profl. jours. Served to capt. M.C. USAF, 1952-54. Recipient Anna-Monika 1st prize award, 1973, Hofheimer prize, 1972, Samuel W. Hamilton award, 1977, Leonard Crammer Meml. award, 1980, Paul Hoch award, 1981; Vol. Service award Nat. Council Alcoholism, 1974. Fellow Am. Psychiat. Assn.; Mem. Am. Psychopath. Assn. (pres. 1975-77), Am. Soc. Human Genetics, Internat. Group Study of Affective Disorders, Psychiat. Research Soc., Am. Fedn. Clin. Research, Assn. Research in Nervous and Mental Disorders, Sigma Xi. Club: Tudor and Stewart (Balt.). Subspecialties: Psychiatry; Genetics and genetic engineering (medicine). Current work: Epidemiology, genetics and classification of affective disorders, alcoholism and schizophrenia; treatment studies in depression. Office: 500 Newton Rd Iowa City IA 52242

WINSKILL, ROBERT WALLACE, fuel systems manufacturing company executive; b. Tacoma, Oct. 30, 1925; s. Edward F. W. and Margaret Eyre (Myers) W. B.A., Coll. Puget Sound, 1947. Field mgr. Ray Burner Co., San Francisco, 1954-57; div. sales mgr., v.p. Western Boiler Co., Los Angeles, 1957-60; 1960-69, nat. sales mgr., San Francisco, 1973—; v.p., gen. mgr. Orr & Sembower, Inc., Middletown, Pa., 1969-73. Contbr. articles on biomass combustion to tech. jours. Served with U.S. Army, 1943-44. Mem. ASME, ASHRAE. Club: Olympic (San Francisco). Subspecialties: Biomass (energy science and technology); Combustion processes. Current work: Design and marketing of medium-size solid fuel systems to replace oil and gas fired boiler systems. Office: 1301 San Jose Ave San Francisco CA 94112

WINSTEAD, JOE EVERETT, plant ecology educator; b. Wichita Falls, Tex., Mar. 17, 1938; s. Leonard Elbert and Helen Nell (Reaser) W.; m. Sara Sidney Carlisle, Sept. 3, 1960 (div. 1976); 1 dau., Cynthia Jo; m. Catherine Elaine Poole, Aug. 10, 1980; 1 dau., Julia Courtney. B.S., Midwestern U., 1960; M.S., Ohio U., 1962; Ph.D., U. Tex., 1968. Instr. Delta Coll., Saginaw, Mich., 1962-63; asst. prof. biology Western Ky. U., Bowling Green, 1968-72, assoc. prof., 1972-78, prof., 1978—. Served to capt. U.S. Army, 1963-65. Mem. Ecol. Soc. Am., Bot. Soc. Am., Assn. S.E. Biologists, Ky. Acad. Sci. (pres. elect 1984), Audubon Soc. (pres. S. Central Ky. chpt. 1974, editor 1975—). Democrat. Methodist. Subspecialties: Ecosystems analysis; Evolutionary biology. Current work: Ecosystems analysis; populational adaptations of hard wood tree species and mechanisms of adaptations of native plants to strip mined habitats. Office: Dept Biology Western Ky U Bowling Green KY 42101

WINSTON, PATRICK HENRY, computer science educator; b. Peoria, Ill., Feb. 5, 1943; s. Robert Watson and Dorothy Ann (Zeis) W. B.S., M.I.T., 1965, M.S., 1967, Ph.D., 1970. Asst. prof. computer sci. M.I.T., Cambridge, 1970-72, assoc. prof., 1972-82, prof., 1982—, dir. Artificial Intelligence Lab., 1973—. Author: Artificial Intelligence, 1977, LISP, 1981. Subspecialty: Artificial intelligence. Current work: Learning, analogy, common-sense reasoning. Office: 545 Technology Square Cambridge MA 02139

WINTER, ARTHUR, neurosurgeon; b. Newark, Sept. 7, 1922; s. Benjamin and Rose W.; m. Ruth N. Grosman, June 16, 1957; children: Robin, Craig, Grant. B.A., Drew U., 1947; P.C.B. cum laude, U. Montreal, 1948, M.D., 1953. Cert. physician, N.Y.; Cert. physician, N.J. Intern U. Montreal Hosps., 1952-53; resident Newark Beth Israel Hosp., 1953-55; resident in neurol. surgery Baylor Med. Coll.-Tex. Med. Center, Houston, 1955-56, Albert Einstein Med. Center, 1956-59; practice medicine specializing in neurol. surgery, Livingston, N.J.; attending staff St. Barnabas Med. Center, Hosp. Center at Orange (N.J.), N.J. Orthopedic Hosp., VA Hosp.; cons. numerous hosps. and rehab. insts.; clin. instr. N.J. Coll. Medicine and Dentistry. Author: The Moment of Death, 1969; author: Surgical Control of Behavior, 1971, Life and Death Decisions; also numerous articles; developer Winter head dressing, microsug. brain retractor, pupicon. Served with U.S. Army, 1943-46; ETO. Decorated Purple Heart, Battle Stars (3); grantee Multiple Sclerosis Research Fund, 1974, N.J. Dept. Health Beth Israel Hosp. Fellow Am. Acad. Neurol. Sand Orthopedic Surgeons, Am. Coll. Emergency Physicians, N.J. Acad. Medicine, Royal Soc. Health; mem. AMA, EEG Soc. (assoc.), AAAS, Congress of Neurol. Surgeons, Essex County Med. Assn., Soc. Neurosci., Am. Soc. Stereotactic and Functional Neurosurgery, others. Subspecialties: Neurology; Neurophysiology. Current work: Research in cancer (microwaves), multiple sclerosis (amino acids) and brain and spinal cord injury' central nervous system functions and dysfunctions and applied research. Office: 22 Old Short Hills Rd Suite 110 Livingston NJ 07039

WINTER, CHARLES GORDON, biochemist, educator; b. Hanover, Pa., Dec. 28, 1936; s. Charles and Emma May (Crouse) W.; m. Betty Ann Fuhrman, June 8, 1958; children: C. David, Douglas A., John K. B.S. in Chemistry, Juniata Coll., 1958; M.S. in Biol. Chemistry, U. Mich., 1963, Ph.D., 1964. Fellow physiol. chemistry Johns Hopkins U. Coll. Medicine, Balt., 1964-66; asst. prof. biochemistry U. Ark. Coll. Medicine, Little Rock, 1966-73, assoc. prof. biochemistry, 1973—; hon. research assoc. Harvard U., Cambridge, 1978-79. Mem. Am. Soc. Biol. Chemists, Am. Chem. Soc., Biophys. Soc., AAAS, Sigma Xi. Presbyterian. Subspecialties: Biochemistry (biology); Biochemistry (medicine). Current work: Biological transport mechanisms; enzymology; membrane structure and function. Home: 7 Pyeatt Circle Little Rock AR 72205 Office: Dept Biochemistry Coll Medicine 4301 W Markham St Little Rock AR 72205

WINTER, J. RONALD, engineering mechanicist, failure analyst; b. Welch, W. Va., Sept. 3, 1942; s. William J. and Lillian J. W.; m. Sherry Carter, Mar. 11, 1947; children: R. Christopher, Lorie Ann. B.S. in Engring. Mechanics with honors, Va. Poly. Inst. and State U., 1964. Process engr. Chem. Propulsion div. Hercules RAAP, Radford, Va., 1965-68; structural dynamicist, launch systems br. Boeing Co., Huntsville, Ala., 1968-69; sr. engring. mechanicist Tenn. Eastman Co., Eastman Chems. div. Eastman Kodak, Kingsport, 1969—; tchr. engring. mechanics Va. Poly. Inst., 1965—; mem. pressure vessel research com. Welding Research Council. Contbr. articles to profl. jours. Mem. ASME, Am. Acad. Mechanics, Soc. Exptl. Stress Analysis, Tau Beta Pi, Phi Kappa Phi. Methodist. Club: Camera. Subspecialties: Theoretical and applied mechanics; Materials. Current work: Finite element analysis, advanced flange design and structural dynamics. Home: Route 11 Box 241 Gray TN 37615 Office: Tenn Eastman Co B-150B Kingsport TN 37662

WINTER, PETER MICHAEL, physician, anesthesiologist, educator; b. Sverdlovsk, Russia, Aug. 5, 1934; came to U.S., 1938, naturalized, 1944; s. George and Anne W.; m. Madge Sato, Aug. 22, 1964; children: Karin Anne, Christopher George. B.A., Cornell U., 1958; M.D., U. Rochester, 1962. Intern U. Utah, Salt Lake City, 1962-63; resident in anesthesiology Mass. Gen. Hosp., Boston, 1963-65; USPHS fellow Harvard U. Med. Sch., 1964-66; Buswell fellow dept. physiology, asst. prof. SUNY, Buffalo, 1966-69; asso. prof. dept. anesthesiology Sch. Medicine, U. Wash., Seattle, 1969-74, prof., 1974-79; prof., chmn. dept. anesthesiology U. Pitts. Sch. Medicine, 1979—; cons. Union Carbide Corp., NIH; med. officer Tektite II (underwater habitation project); anesthesiologist in chief Univ. Health Center Hosps., Pitts. Mem. editorial bd.: Jour. Critical Care Medicine; editorial cons.: Anesthesiology; contbr. chpts. to books, papers and abstracts on anesthesia, environ. phys. pharmacology and med. edn. to publs. Served with U.S. Army, 1953-56. Recipient NIH career devel. award, 1971. Mem. AMA, Am. Coll. Chest Physicians, Am. Soc. Anesthesiologists, N.Y. Acad. Scis., Undersea Med. Soc. Internat. Anesthesia Research Soc., Soc. Acad. Anesthesia Chairmen, Assn. Univ. Anesthetists. Subspecialties: Anesthesiology; Physiology (medicine). Current work: Anesthesiology; mechanism of action of central nervous system depressants; environmental physiology. Office: 1385 E Scaife Hall Univ Pittsburgh School Medicine Pittsburgh PA 15260

WINTER, WILLIAM THOMAS, chemistry educator, researcher; b. N.Y.C., Nov. 14, 1944; s. Garrett Henry and Dorothea (Babcock) W.; m. Helen Dorothy Dalmaso, Apr. 12, 1969. B.S., SUNY, 1966, Ph.D., 1974. Vis. prof. sci. Purdue U., West Lafayette, Ind., 1973-77; asst. head, assoc. prof. chemistry Poly. Inst. N.Y., Bklyn., 1977—; physics tchr. Peace Corps, Kota Bharu, Kelantan, Malaysia, 1966-68; cons. Dart Kraft, Paramus, N.J., 1982. NIH-USPHS grantee, 1979; NSF trainee, 1969-72; Arthritis Found. postdoctoral fellow, 1974-77. Mem. Am. Chem. Soc. (grantee 1977, 82), Soc. Complex Carbohydrates, N.Y. Acad. Scis., Sigma Xi. Democrat. Roman Catholic. Subspecialties: Polymer chemistry; Biophysical chemistry. Current work: X-ray fiber diffraction, conformational energy calculations and NMR studies of polysaccharides, filamentous viruses and novel synthetic polymers. Home: 65 Middagh St Brooklyn NY 11201 Office: Poly Inst NY 333 Jay St Brooklyn NY 11201

WINTERLIN, WRAY LAVERNE, environmental chemist, educator; b. Siox City, Iowa, July 20, 1930; s. William and Nettie Priscilla (Larson) W.; m. Arlene F. Harper, June 15, 1956; children: Larry, Jerry, Dwight. Student, Morningside Coll., Sioux City, 1948-50, Iowa State U., Ames, 1950-51; B.S., U. Nebr., 1955, M.S., 1956; postgrad., Fuller Theol. Sem., Pasadena, Calif., 1956-57, Berkeley Bapt. Sem., 1957-58. Jr. chemist Dept. Water Resources State of Calif., Bryte, 1958-59; staff research assoc. U. Calif.-Davis, 1959-65, exptl. sta. specialist, 1965-79, lectr., 1979—, environ. chemist, 1981—, chairperson dept. environ. toxicology, 1972, dir. pesticide, 1965—. Contbr. articles to profl. jours. Served with U.S. Army, 1951-53. Grantee in field. Mem. Am. Chem. Soc., Am. Soc. Agronomy, Soc. Environ. Toxicology and Chemistry, Sigma Xi. Republican. Mem. Covenant Ch. Lodge: Kiwanis. Subspecialties: Environmental toxicology; Analytical chemistry. Current work: Characterization and development of analytical methods for trace quantities of environmental agents in the environment; isolation and confirmation of trace organic agents in these samples and metabolism and transformation of biologically active environmental agents. Home: 1931 Amador Ave Davis CA 95616 Office: Department of Environmental Toxicology University of California Davis CA 95616

WINTROBE, MAXWELL MYER, medical educator; b. Halifax, N.S., Can., Oct. 27, 1901; came to U.S., 1927, naturalized, 1933; s. Herman and Ethel (Swerling) W.; m. Becky Zanphir, Jan. 1, 1928; children: Susan Hope, Paul William H. (dec.). B.A., U. Man., Can.), Winnipeg, 1921, M.D., 1926, B.Sc. in Medicine, 1927, D.Sc. (hon.), 1958; Ph.D., Tulane U., 1929; D.Sc. (hon.), U. Utah, 1967, Med. Coll. Wis., Milw., 1974, M.D., U. Athens, 1981. Diplomate Am. Bd. Internal Medicine. Gordon Bell fellow, Manitoba, 1926-27; instr. in medicine Tulane U., 1927-30; also asst. vis. physician Charity Hosp., New Orleans; instr. medicine Johns Hopkins, 1930-35, asso. in medicine, 1935-43, also asso. physician, 1935-43; physician in charge, clinic for nutritional, gastro-intestinal and hemopoietic disorders Johns Hopkins Hosp., 1941-43; prof. medicine U. Utah, 1943-70, Disting. prof. internal medicine, 1970—, head dept. medicine, 1943-67, dir. cardiovascular research and tng. inst., 1969-73; physician-in-chief Salt Lake Gen. Hosp., 1943-65, U. Utah Med. Center, 1965-67; cons. to surgeon-gen. U.S. Army; chief cons. Va Hosp., Salt Lake City; dir. lab. for study of hereditary and metabolic disorders U. Utah, 1945-73; spl. cons. nutritional anemias WHO; UN. Council mem. Nat. Adv. Arthritis and Metabolic Diseases Council, USPHS, 1950-54, chmn. hematology study sect., 1950-59; council mem. Nat. Allergy and Infectious Disease Council, 1967-70; mem. com. research in life scis. Nat. Acad. Sci., 1966-69; chmn. sci. adv. bd. Scripps Clinic and Research Found., 1964-74; med. cons. AEC; mem., then chmn. adv. council Life Ins. Med. Research Fund, 1949-53; dir. Am. Soc. Human Genetics, 1948; chmn. hematology com. Research and Devel. Command Surg. Gen.'s U.S. Army; cons. FDA, Dept. Health, Edn. and Welfare; vis. prof. many univs. including Rochester, Vanderbilt, Marquette, N.Y., Tufts, Johns Hopkins, Tulane, U. Calif. at San Diego, Brown U., N.C., Emory, U. Fla., Gainesville, Ala., Southwestern at Dallas, Harvard, UCLA, U. Toronto, Ottawa, McGill U., Dalhousie U. Mng. editor: Bull. Johns Hopkins Hosp., 1942-43; asso. editor: Nutrition Revs, Boston, 1943-49; adv. editor: Tice Practice of Medicine, 1943-76; asso. editor: Internat. Med. Digest, 1944- 58, Blood, 1945-75, Am. Jour. Medicine, 1946-56, Medicine, Cancer, Jour. Clin. Pathology; editorial bd.: Jour. Clin. Investigation, 1948-49, Gen. Practitioner, 1949-59, Jour. Clin. Nutrition, 1952-65, Jour. Chronic Diseases, 1955-75; Author: Clinical Hematology, 1942, 8th edit., 1981, Blood, Pure and Eloquent, 1980, also numerous sci. articles.; Editor: Harrison's Principles of Internal Medicine, 1950; editor-in-chief, 6th edit., 1970, 7th edit., 1974. Bd. dirs., v.p. Salt Lake Chamber Music Soc.; nat. adv. bd. Utah Symphony Orch.; bd. dirs. Pro-Utah.; Mem. Anti-Anemia Preparations adv. bd., U.S. Pharmacopeia, 1941-49, com. revision, 1950-60. Recipient gold medals in polit. econs. and French, 1921, Isbister prizes, 1920-25; all U. Man.)., Physician of Excellence award Med. Times, 1979. Fellow A.C.P. (John Philips Meml. award 1967, master 1973); corr. mem. Italian, Swiss, Brit. Assn. Clin. Pathology; mem. Am. Soc. Hematology (pres. 1971-72), AMA (vice chmn. council on drugs 1964-68, chmn. sect. on adverse reactions), Assn. Am. Physicians (Kober medal 1974, councillor 1957-63, pres. 1964-65), Assn. Profs. Medicine (Robert H. Williams award 1973, councillor 1962-63, pres. 1965-66), Western Assn. Physicians (pres. 1956-57), Am. Soc. Exptl. Pathology, AAAS, Nat. Acad. Scis. (com. on sci. and public policy, chmn. sect. on human genetics, hematology and oncology 1976-79), Soc. Exptl. Biology and Medicine, Leukemia Soc. (chmn. nat. med. adv. bd.), Pacific Interurban Clin. Club, Western Soc. Clin. Research (Mayo Soley award 1970), European Soc. Hematology (corr.), Internat. Soc. Hematology (councillor-at-large 1972-74, v.p. 1976-78, pres. 1978-80, Ferrata award Rome 1958), Harvey Soc. (hon.), Assn. Clin. Pathologists (Eng.) (corr.), Am. Fedn. Clin. Research, Phi Beta Kappa (hon.), Sigma Xi, Alpha Omega Alpha, Sigma Alpha Mu. Subspecialty: Internal medicine. Current work: Internal medicine; hematology; medical education; history of medicine; especially hematology. Home: 5882 Brentwood Dr Salt Lake City UT 84121 Office: U Utah Med Center Salt Lake City UT 84132 Never guess, secure the facts; measure rather than estimate; differentiate fact from hypothesis. Whatever one undertakes should receive one's best efforts, without consideration of reward other than the satisfaction of a job well done.

WIPKE, WILL TODD, chemistry educator, researcher; b. St. Charles, Mo., Dec. 16, 1940. B.S., U. Mo., 1962; Ph.D., U. Calif., 1965. Research fellow in chemistry Harvard U., Cambridge, Mass., 1967-69; asst. prof. chemistry Princeton (N.J.) U., 1969-75; mem. faculty U. Calif.-Santa Cruz, 1975—, prof. chemistry, 1981—; cons. Merck, Sharp and Dohme, Rahway, N.J., 1970-80, Squibb, Princeton, 1976-81, BASF, Ludwigshafen, W. Ger., 1974-78, Molecular Design, Ltd., Hayward, Calif., 1978—, Giba-Geigy, Basle, Switzerland., 1978-82; mem. adv. bd. Chem. Abstracts Service, Columbus, 1970-73; dir. Advanced Study Inst., NATO, Noordwijker-hout, Netherlands, 1973; mem. com. to determine policy for Nat. Computing Lab., Nat. Acad. Scis., 1974-77; mem. program com. Nat. Resource for Computation in Chemistry, Lawrence Berkeley Labs., 1978-80. Editor: Computer Representation and Manipulation of Chemical Information, 1973, Computer-Assisted Organic Synthesis, 1977; mem. editorial bd.: Chem. Substructure Indes, 1975—, Computers in Chemistry, 1975—, Jour. Chem. Info. and Computer Sci, 1975-77. Served with U.S. Army, 1966-67. Recipient Eastman Kodak Research award, 1964, Texaco Outstanding Sr. Research award, 1964; NIH fellow, 1964-65; Merck Career Devel. grantee, 1970. Mem. Am. Chem. Soc. (role of computers in chem. edn. com. 1970-76, chem. lit. div. program com. 1974-76), Sigma Xi. Subspecialties: Organic chemistry; Artificial intelligence. Current work: Organic chemistry, structure and synthesis; computer applications to creative chemical thought processes. Office: Univ Calif Natural Sci Div Thimann Labs Santa Cruz CA 95064

WIREN, ROBERT CRAIG, electronics firm nuclear engineer-designer; b. Seattle, Apr. 23, 1954; s. Robert and Agnes (Selle) W. B.S. Engring., U. Wash., 1976, M.S., 1979. Engr. Gen. Electric Co., San Jose, Calif., 1979-81, Electronic Assocs., West Long Branch, N.J., 1981—. Mem. Nat. Republican Congl. Com., 1981-82. Mem. Am. Nuclear Soc. Mem. Ch. of Scientology. Subspecialties: Nuclear engineering; Nuclear fission. Current work: Increasing the ability of reactor operators, exteriorization in the control room and operator processing to clear and operating thetan, BWR ECCS logic systems simulation, reactor protection system simulation for boiling water reactors. Home: 1382 Ocean Ave Apt B-15 Sea Bright NJ 07760

WISE, DAVID STEPHEN, computer science educator, researcher; b. Findlay, Ohio, Aug. 10, 1945; s. Jerome and Ruth (Rosenfeld) W.; m. Rita Kathryn, Nov. 25, 1971; children: Jeremy, Kathryn. B.S. in Math, Carnegie Inst. Tech., 1967; M.S., U. Wis., 1969, Ph.D. in Computer Sci, 1971. Lectr. U. Edinburgh, 1971-72; asst. prof. computer sci. Ind. U., 1972-77, assoc. prof., 1977—. Mem. Assn. Computing Machinery, IEEE Computer Soc., European Assn. for Theoretical Computer Sci. Subspecialties: Programming languages; Foundations of computer science. Current work: Applicative programming languages and systems, snytax and semantics for expressing computer programs as mathematical functions, especially as applied to parallel/multiprocessing systems. Office: 101 Lindley Hall Bloomington IN 47405

WISE, GARY LAMAR, electrical engineering and mathematics educator, researcher; b. Texas City, Tex., July 29, 1945; s. Calder Lamar and Ruby Lavon (Strom) W.; m. Mary Estella Warren, Dec. 28, 1974; 1 dau., Tanna Estella. B.A., Rice U., 1971; M.S.E., Princeton U., 1973, M.A., 1973, Ph.D., 1974. Postdoctoral research assoc. Princeton (N.J.) U., 1974; asst. prof. Tex. Tech U., Lubbock, 1975-76, U. Tex.-Austin, 1976-80, assoc. prof. elec. engring. and math., 1980—; tech. reviewer Army Research Office, Durham, N.C., 1976, Air Force Office Sci. Research, Washington, 1980, 83, Harper and Row, N.Y.C., 1982-83; speaker tech. confs. Contbr. chpts., numerous articles to profl. publs. Air Force Office Sci. Research grantee, 1976—; Carroll D. Simmons Centennial teaching fellow in engring. U. Tex.-Austin, 1982—. Mem. IEEE, Soc. Indsl. and Applied Math., Phi Beta Kappa, Tau Beta Pi. Methodist. Subspecialties: Electrical engineering; Probability. Current work: Statistical communication theory, random processes, signal processing, signal detection and estimation, quantization theory, and applied mathematics. Home: 8705 Collingwood Dr Austin TX 78745 Office: Dept Elec Engring U Tex Austin TX 78712

WISE, HENRY, physical chemist; b. Cienchanow, Poland, Jan. 14, 1919; m., 1943, 60; 6 children. S.B., U. Chgo., 1941, S.M., 1944, Ph.D. in Phys. Chemistry, 1947. Research assoc. U. Chgo., 1941-46; dir. field lab. N.Y. U., 1946-47; scientist Nat. Adv. Com. Aeros., Ohio, 1947-49; phys. chemist Calif. Inst. Tech., 1949-55; chmn. dept. chem. dynamics SRI Internat., Menlo Park, Calif., 1955-71, sci. fellow, 1971—; lectr. Sch. Engring., Stanford U., 1960-68, adj. prof. materials sci., 1983—; vis. prof. Israel Inst. Tech., 1965; lectr. in chem. engring. U. Calif.-Berkeley, 1977-80; mem. com. motor vehicle emission Nat. Acad. Sci. Recipient Fullbright award, 1965, McBean-SRI award, 1983. Mem. Am. Chem. Soc., Am. Phys. Soc., Catalysis Soc., The Chem. Soc. Subspecialty: Physical chemistry. Office: Material Sci SRI Internat Menlo Park CA 94025

WISE, JOHN J., oil company executive; b. Cambridge, Mass., Feb. 28, 1932; s. Daniel and Alice W.; m. Rosemary B.; children. B.S., Tufts U., 1953; Ph.D., MIT, 1965. Researcher Mobil Oil Corp., N.Y.C., 1953-62, research mgr. refining and petrochems., 1965-77, v.p. planning, 1977-83, Mobil Research & Devel. Corp., N.Y.C., 1983—. Contbr. articles to profl. jours. Mem. Am. Chem. Soc., Am. Inst. Chem. Engrs., AAAS, Sigma Upsilon. Subspecialties: Chemical engineering; Catalysis chemistry. Current work: Research and engineering management. Patentee in field. Office: 150 E 42d St New York NY 10017

WISEMAN, SANDOR ELLIOT, psychologist, educator; b. Montreal, Ont., Can., Aug. 21, 1947. B.Sc., McGill U., 1968; Ph.D., Cornell U., 1973. Registered psychologist, Ont. Asst. prof. psychology U. Toronto, 1975—; dir. Can. Found. for Children and the Law, Toronto, 1982—. Contbr. articles to profl. jours. Ford Found. fellow, 1968. Fellow Psychonomic Soc.; mem. Can. Psychol. Assn., Am. Psychol. Assn., Ont. Psychol. Assn., Ont. Soc. for Clin. and Exptl. Hypnosis. Subspecialties: Cognition. Current work: Forensic psychology, cognitive psychology. Office: Dept Psychology U Toronto Toronto ON Canada M5S 1A1

WISHNER, LAWRENCE ARNDT, chemistry educator, biochemistry and ethology researcher; b. N.Y.C., Sept. 7, 1932; s. Samuel J. and Kathryn Arndt (Thomas) W.; m. Janet L. Fraser, Mar. 6, 1982; children: Nancy T., Catherine M. B.S., U. Md., 1954, M.S., 1961, Ph.D., 1964. Grad. asst. U. Md., 1957-61; asst. prof. chemistry Mary Washington Coll., 1961-64, assoc. prof., 1964-68, prof., 1968—, chmn. dept. chemistry, 1967-71, asst. dean coll., 1971-77. Author: Eastern Chipmunks, 1982; contbr. popular and research articles to various publs. Served to 1st lt. USAF, 1954-57. Mem. Am. Oil Chemists Soc., Am. Chem. Soc., Am. Inst. Chemists, AAAS, N.Y. Acad. Sci., Sigma Xi. Subspecialties: Biochemistry (biology); Ethology. Current work: Autoxidation of lipids; metabolism and vitamin E; behavior and life history of squirrels.

WISLOCKI, PETER GREGORY, research scientist; b. Derby, Conn., Jan. 21, 1947; s. Peter Daniel and Eva (Lechus) W.; m. Mary Kay Anderson, Aug. 5, 1972; children: Daniel, Andrew. B.S. in Chemistry, Fairfield U., 1968; Ph.D. in Oncology, U. Wis.-Madison, 1974. Postdoctoral fellow Hoffman-LaRoche, Nutley, N.J., 1974-76, vis. scientist, 1976-77; asst. prof. oncology Eppley Cancer Inst., U. Nebr., Omaha, 1977-78; research fellow Merck, Sharp Dohme, Rahway, N.J., 1978—. Contbr. numerous sci. articles to profl. publs. Mem. Am. Assn. for Cancer Research, Internat. Soc. for Study of Xenobiotics. Subspecialties: Toxicology (medicine); Cancer research (medicine). Current work: Metabolic activation of carcinogens, toxicological assessment of animal drug residues. Office: MSDRL-80nlA-10 Rahway NJ 07068

WITCOMB, STANLEY ERNEST, physics educator; b. Denver, Jan. 23, 1951; s. Ernest Albert and Ida Mae Whitcomb; m. Laurie Ann Whitcomb, Sept. 24, 1977. B.S., Calif. Inst. Tech., 1973; postgrad., Cambridge U., 1973-74; Ph.D., U. Chgo., 1980. Nat. Needs fellow U. Chgo., 1980; asst. prof. physics Calif. Inst. Tech., Pasadena, 1980—. Mem. Am. Phys. Soc., Am. Astron. Soc., Optical Soc. Am. Subspecialty: Relativity and gravitation. Current work: Experiments in relativity and gravitation; development of high precision laser interferometers for gravitational wave detection. Office: California Institute Technology 130-33 Pasadena CA 91125

WITHERELL, EGILDA DEAMICIS, radiological physicist; b. Fall River, Mass., Nov. 1, 1922; d. Berardino and Gina (Arcesi) DeAmicis; m. Dana Grover Witherell, May 12, 1956. S.B., M.I.T., 1944. Cert. health physicist Am. Bd. Health Physics, 1960; cert. radiol. physicist Am. Bd. Radiology, 1953. Physicist, radiation lab. M.I.T., Cambridge, 1944-45; chemistry instr. Northeastern U., Boston, 1945-46; mathematician dynamicanalysis and control lab. MIT, Cambridge, 1946-47; radiol. physicist New Eng. Deaconess Hosp., Boston, 1947-66, Peter Bent Brigham Hosp., 1966-67, Newton Wellesley Hosp., Newton, Mass., 1967—; radiation safety officer, dosimetry, calibration and radiation surveys, educator nuclear medicine and radiologic tech. Contbr.: writings to profl. publs. in field. including Science, Jour. AMA, New Eng. Jour. Medicine, Nucleonics. Mem. Am. Assn. Physicist in Medicine, Health Physics Soc., Soc. Nuclear Medicine, Am. Coll. Radiology. Subspecialty: Radiological physics. Current work: Radiation dosimetry; safety and education. Office: 2014 Washington St Newton MA 02162

WITHERS, HUBERT RODNEY, radiobiologist; b. Stanthorpe, Queensland, Australia, Sept. 21, 1932; came to U.S., 1966; s. Hubert and Gertrude Ethel (Tremayne) W.; m. (div.); 1 dau: Genevieve. M.B.B.S., U. Queensland, 1956; Ph.D., U. London, 1965. Diplomate: Am. Bd. Radiology, Royal Australasian Coll. Radiology. Radiotherapist Prince of Wales Hosp., Randwick, Sydney, Australia, 1966; vis. research scientist Lab. Physiology, Nat. Cancer Inst., Bethesda, Md., 1966-68; assoc. radiotherapist, assoc. prof. radiotherapy sect. exptl. radiotherapy U. Tex. System Cancer Center, M.D. Anderson Hosp and Tumor Inst., Houston, 1968-71, radiotherapist, prof. radiotherapy, chief sect., 1971-80; assoc. grad. faculty U. Tex. Grad. Sch. Biomed. Scis., Houston, 1969-73, mem. grad. faculty, 1973-80; prof. dept. radiotherapy U. Tex. Health Sci. Center, Houston, 1975-80; prof., dir. exptl. radiation oncology, dept. radiation oncology Center for Health Scis., UCLA, 1980—; cons. ad hoc revs. div. research grants Nat. Cancer Inst., 1969—, cons. clin. cancer centers program project rev. com., 1976-79, Nat. Surg. Adjuvant Breast and Bowel Program, 1976-81; mem. radiation study sect. NIH, 1970-74; mem. Internat. Com. on Radiation Protection Task Force on Non-Stochastic Effects of Radiation, 1980—, Com. on Radiation Oncology Studies, Subcom. on Particle Therapy, 1979-81. Assoc. editor: Radiation Research, 1973-76, Internat. Jour. Radiation Oncology Biology Physics, 1977—; contbr. articles in field to profl. jours. Recipient Finzi Bequest prize Brit. Inst. Radiology, 1974; Commemorative medal of achievement U. Tex. System Cancer Center, 1981; Gaggin fellow cancer research U. Queensland, 1963-66. Fellow Royal Australasian Coll. Radiologists; mem. Radiation Research Soc. (pres. 1982-83), Am. Soc. Therapeutic Radiologists (mem. com. on long-range planning, sci. program com.), Am. Coll. Radiology (mem. com. on radiotherapeutic devel., radiotherapy written exams. com.), Radiation Therapy Oncology Group (mem. radiobiology com.), Am. Assn. Cancer Research, Am. Radium Soc., Brit. Inst. Radiology. Subspecialties: Cancer research (medicine); Radiology. Current work: Radiation biology in cancer treatment. Office: Dept Radiation Oncology UCLA Center Health Scis Los Angeles CA 90024

WITHERSPOON, JOHN PINKNEY, JR., industrial scientist; b. Hamlet, N.C., Feb. 28, 1931; s. John Pinkney and Beatrice Lula (Turbeville) W.; m. Ulilla Ann Treon, Dec. 26, 1952; children: Michael, Sharon, Susan, Patricia, Robert. B.A. in Biology, Emory U., 1953, M.S., 1954; Ph.D. in Botany, U. Tenn.-Knoxville, 1962. With Oak Ridge Nat. Lab., 1962—, sr. scientist health and safety div., 1977—, assessments applications group leader, 1978—; adj. prof. ecology U. Tenn.-Knoxville, 1976—; cons. NRC, Nat. Acad. Scis. Contbr. over 70 articles to sci. publs. Mem. Oak Ridge Environ. Quality Bd., 1977-80. Mem. Soc. Tech. Communication (award of merit 1980), Ecol. Soc. Am., Health Physics Soc., Assn. Southeastern Biologists. Methodist. Subspecialties: Environmental toxicology; Nuclear fission. Current work: Assessment of radiological impacts of nuclear technologies. Home: 100 Wade Ln Oak Ridge TN 37830

WITHRINGTON, ROGER JOHN, optical engr.; b. London, Sept. 1, 1942; U.S., 1970; s. John W. and Nine B. (Mann) W.; m. Anita M. Andrews, Jan. 2, 1968; children: David T., Nina C. Ph.D., Imperial Coll., London, 1969. Optical engr. Bendix Aerospace Systems DIv. Bendix Corp., Ann Arbor, Mich., 1970-72; optical system engr. Hughes Aircraft Co., Culver City, Calif., 1972—, sr. scientist, 1982—. Contbr. articles on holographic and laser optical system design to profl. publs. Mem. Optical Soc. Am. Subspecialties: Optical engineering; Holography. Current work: Optical system design and analysis of space, laser, and adaptive optical systems; design of heads up and head mounted displays utilizing holographic optical elements. Office: Hughes Aircraft Co Culver City CA 90230

WITIAK, DONALD T., medicinal chemistry educator, university pharmaceutical and toxicological research institute director; b. Milw., Nov. 16, 1935; s. Theodore and Elvi W.; m. Deanne Beth, Nov. 11, 1956; children: Mark, Elizabeth. B.S. in Pharmacy, U. Wis.-Madison, 1958, Ph.D. in Medicinal Chemistry, 1961. Asst. prof. medicinal chemistry U. Iowa, 1961-66, assoc. prof., 1966-67; mem. faculty Ohio State U., Columbus, 1967—, prof. medicinal chemistry and pharmacognosy, 1971—, chmn. medicinal chemistry dept. pharmacognosy, 1973-83; dir. Pharm. & Toxicol. Research Inst., 1982—; cons. Adria Labs., Inc.; lectr. in field. Contbr. numerous sci. articles to profl. publs., chpts. to books. NIH fellow, 1959-61; Acad. Pharm. Scis. fellow. Mem. Am. Pharm. Assn., Am. Assn. Colls. of Pharmacy, Am. Chem. Soc., Columbus C. of C. (mem. tech. roundtable 1983-84). Subspecialties: Organic chemistry; Medicinal chemistry. Current work: Antilipidemic drugs; drug synthesis; pharmacology and carcinogenesis; central and peripheral nervous system drugs; antineoplastic agents; carcinogenesis. Patentee in field. Office: Ohio State Univ 200G Medicine Adminstrn Center 370 W Ninth Ave Columbus OH 43210

WITKIN, STEVEN S., immunologist; b. Bklyn., Oct. 19, 1943; s. Bernard A. and Ruth Sarah (Hymowitz) W.; m. Lynne J. Lipset, Dec. 22, 1966; children: Jolene, Keren. A.B., Hunter Coll., 1965; M.S., U. Conn., 1967; Ph.D., UCLA, 1970. Research assoc. Coll. Physicians and Surgeons Columbia U., 1972-74; assoc. Meml. Sloan-Kettering Cancer Ctr., N.Y.C., 1974-81; assoc. prof. dept. Ob-Gyn Cornell U. Med. Coll., N.Y.C., 1981—. NIH fellow. Mem. Am. Soc. Microbiology, Am. Assn. Immunologists, Am. Fertility Soc., Soc. Study of Reproduction. Subspecialties: Immunology (medicine); Obstetrics and gynecology. Current work: Reproductive immunology, basic and clinical research in immunology related to reproductive and gynecological disorders. Office: 515 E 71st St New york NY 10021

WITKOP, BERNHARD, chemist; b. Freiburg, Baden, Germany, May 9, 1917; came to U.S., 1947, naturalized, 1953; s. Philipp W. and Hedwig M. (Hirschhorn) W.; m. Marlene Prinz, Aug. 8, 1945; children—Cornelia Johanna, Phyllis, Thomas. Diploma, U. Munich, 1938, Ph.D., 1940; Sc.D., Privat-Dozent, 1947. Matthew T. Mellon research fellow Harvard, 1947-48, mem. faculty, 1948-50; spl. USPHS fellow NIH Nat. Heart Inst., 1950-52; vis. sci. Nat. Inst. Arthritis and Metabolic Diseases, 1953, chemist, 1954-55, chief sect. metabolities, 1956—, chief lab. chemistry, 1957—; vis. prof. U. Kyoto, Japan, 1961, U. Freiburg, Germany, 1962; adj. prof. U. Md. Med. Sch., Balt.; Nobel symposium lectr. Stockholm-Karlskoga, 1981; Mem. bd. Internat. Sci. Exchange, 1974; mem. exec. Com. NRC, 1975; mem. Com. Internat. Exchange, 1977, Paul Ehrlich Award Com., Frankfurt, 1980. Editor: Fedn. European Biochem. Soc. Letters, 1979—. Recipient Superior Service award USPHS, 1967; Paul Karrer gold medal U. Zurich, 1971; Kun-ni-to (medal of sci. and culture 2d class) Emperor of Japan, 1975; Alexander von Humboldt award for sr. U.S. scientists, 1978. Mem. Am. Chem. Soc. (Hillebrand award 1958), Nat. Acad. Scis., Acad. Leopoldina (fgn.), Pharm. Soc. Japan (hon.), Chem. Soc. Japan (hon.). Subspecialties: Organic chemistry; Biochemistry (medicine). Current work: Naturaltoxins, receptors, drugs, neurotransmitters. Office: Nat Insts Health Bethesda MD 20205 A career between two worlds and two wars, spanning 45 years of research aims changing from structural to dynamic aspects, may be considered epigonal in the sense that my teacher H. Wieland (Nobel Prize 1928) always considered biochemistry as a neglected area of organic chemistry. In a small way I tried to follow his example and interests, such as oxidation mechanisms, natural products and highly active toxins.

WITMAN, GARY B., internist, oncology laboratory executive, educator; b. N.Y.C., Aug. 3, 1949; m. Dianne S. Zarum, Dec. 24, 1950; children: Samantha, Zachary. B.A., Rutgers U., 1971; M.D., SUNY Downstate Med. Center, Bklyn., 1975. Diplomate: Am. Bd. Internal Medicine. Intern Miriam Hosp., Providence, 1975-76, resident in internal medicine, 1976-78; program dir. clin. oncology Nat. Cancer Inst., Bethesda, Md., 1980-82, reviewer autologous bone marrow transplant working task force, 1981-82; dir. clin. research Microbiol. Scis., Inc., Fiskville, R.I., 1982—, v.p. med. affairs, 1982—; pres. Oncology Labs., Inc., West Warwick, R.I., 1982—; clin. instr. Brown U., Providence, 1977-78, Yale U., New Haven, 1978-80; cons. Nat. Cancer Inst. Can., Kingston, Ont., 1980-82, Scientist Inst. Pub. Info., N.Y.C., 1981—. Am. Cancer soc. fellow Yale U. Comprehensive Cancer Ctr., 1978-79. Mem. AMA (Physicians Recognition award 1981), ACP, Tissue Culture Assn., Am. Fedn. Clin. Research, AAAS, N.Y. Acad. Scis. Democrat. Jewish. Club: Brown Univ. (Providence). Subspecialties: Cancer research (medicine); Oncology. Home: 64 Hazard Ave Providence RI 02906 Office: Scott Labs Inc 771 Main St West Warwick RI 02923

WITT, FOUNTAIN JOEL, mechanical engineer; b. Los Angeles, Dec. 8, 1929; s. Samuel Newton and Mary Mable (White) W.; m. Margaret Elizabeth Brock, Aug. 11, 1951; children: David Joel, Mary Elizabeth. B.S., Tenn. Tech. U., 1950; M.A., Duke U., 1951. Asst. prof. math. Tenn. Tech. U., 1952-55; instr. math. U. Tenn.-Knoxville, 1955-56; research engr. Oak Ridge Nat. Lab., 1956-74; fellow engr. Westinghouse Electric Corp. (Water Reactor Div.), 1974—. Contbr. articles on stress analysis and fracture mechanics to tech. jours. Mem. ASTM, ASME. Republican. Methodist. Subspecialties: Fracture mechanics; Probability. Current work: Elastic plastic fracture mechanics, probabilistic fracture mechanics, fracture toughness testing and evaluation.

WITTE, KURT ALLEN, electronics engr.; b. Mpls., May 27, 1949; s. Elmer Reinhardt and Thelma Anna (Tessmer) W.; m. Randi Elaine Johnson, June 27, 1969; m. Mary Lynne Bundy, Dec. 16, 1978; 1 son, David Francis. B.A. in Math, U. Minn., 1972, 1974. Instrumentation engr. Metallurgical, Inc., Mpls., 1977; supr. spl. products dept. Mobile Radio Engring., Inc., Mpls., 1978; design engr. research dept. CPT Corp., Mpls., 1978-82; electronics design engr. Solid Controls, Inc., Mpls., 1982—. Mem. IEEE, Assn. Computing Machinery, Mensa. Subspecialties: Distributed systems and networks; Computer engineering. Current work: Multimicroprocessor systems architecture and software design. Home: 225 Monroe Ave N Hopkins MN 55343 Office: 6925 Washington Ave S Minneapolis MN 55435

WITTE, MICHAEL, chemist, cons., solar energy researcher; b. Poland, Mar. 15, 1911; s. Jake and Rose (Schlesinger) W.; m. Louise, Dec. 21, 1940; children: Michael S., Janet L., Lois J., John C. B.S., Loyola U., Chgo., 1937; M.S., U. Ill., 1938, Ph.D., 1941. Asst. chemistry dept. U. Ill., 1938-41; research chemist Nat. Aniline div. Allied Chem. Corp., Buffalo, 1941-47; product supr. Gen. Aniline & Film Corp., Rensselaer, N.Y., 1947-53, product mgr., Linden, N.J., 1953-56; pres. Simpson Labs. Inc., Newark, 1957-59, Carnegies Fine Chems. div. Rexall Drug & Chem. Corp., Kearny, N.J., 1959-60, M. Witte Assocs., Chatham, N.J., 1960—; cons., dir. Tennant Devel. Corp., N.Y.C., 1961-65. Explorer chmn. Boy Scouts Am., Berkeley Heights, N.J., 1959-60. Dept. Energy grantee, 1980-81. Mem. Am. Chem. Soc., AAAS, Sigma Xi. Subspecialties: Organic chemistry; Solar energy. Current work: Organic syntheses of polarizing crystals; solar energy absorption; sulfur dioxide absorption; basic and applied chemical process and research development; research in dyestuff mfg., pharm. intermediates, organic chems., sulfur dioxide absorption, solar energy collection, polarizing crystals. Patenteein field (U.S. and fgn). Home and Office: 420 River Rd C-11 Chatham NJ 07928

WITTEBORN, FRED CARL, astronomer, physicist; b. St. Louis, Dec. 27, 1934; s. Fred and Erna Sophia Wilhelmina (Giese) W.; m. Nancy Durham, June 15, 1957; 1 son, Fred James. B.S., Calif. Inst. Tech., 1956; M.S., Stanford U., 1958, Ph.D., 1965. Research scientist physics br. NASA, Ames Research Center, Moffett, Calif., 1958-62, research scientist, 1968-71, chief astrophysics br., 1971-75, research scientist, 1975—; research asst. physics dept. Stanford (Calif.) U., 1962-65, research assoc., 1965-68. Served to capt. USAF, 1958-62. Mem. Am. Astron. Soc., Am. Phys. Soc., AAAS, Planetary Soc., Sigma Xi. Subspecialties: Infrared optical astronomy; Planetary science. Current work: Infrared astronomy, planet formation, infrared instruments. Office: 245-6 NASA Ames Research Center Moffett Field CA 94035

WITTEN, LOUIS, physics educator; b. Balt., Apr. 13, 1921; s. Abraham and Bessie (Perman) W.; m. Lorraine Wollach, Mar. 27, 1949; children: Edward, Celia, Matthew, Jesse. B.E., Johns Hopkins U., 1941, Ph.D. in Physics, 1951; B.S. in Meteorology, NYU, 1944. Research assoc. Princeton U., 1951-53, U. Md., College Park, 1953-54; staff scientist Lincoln Lab., MIT, 1954-55; assoc. dir. Martin-Marietta Research Lab., Balt., 1955-68; prof. physics U. Cin., 1968—; v.p. Gravity Research Found., 1967—. Editor: Gravitation: An Introduction to Current Research, 1962, Relativity, 1964, Asymptotic Structure of Space-Time, 1977; contbr. articles to profl. jours.; patentee in field. Served to 1st lt. USAAF, 1942-46. Fulbright lectr. Weizmann Inst. Sci., Rehovot, Israel, 1963-64; NSF grantee; Dept. Def. grantee. Fellow Am. Phys. Soc.; mem. Internat. Astron. Union, Am. Astron. Soc., Am. Math. Soc., Am. Assn. Physics Tchrs., AAAS. Jewish. Subspecialties: Relativity and gravitation; Particle physics. Current work: Theory of supergravity—an attempt to unite general theory of relativity with theory of elementary particles. Patentee in field. Home: 7920 Rollingknolls Dr Cincinnati OH 45237 Office: Dept Physics U Cin Cincinnati OH 45221

WITTENBORN, JOHN RICHARD, psychology educator, clinical researcher, consultant; b. Ft. Gage, Ill., May 22, 1915; s. Richard Edward and Mabel (Mulholl) W.; m. Sarah Elizabeth Alwood, Apr. 29, 1938; children: Sarah Elizabeth, Gretchen Ann, Richard, Christopher Dirk. Ed.B., So. Ill. U., 1937; M.S., U. Ill., 1939, Ph.D., 1942. Instr. Yale U., New Haven, 1942-45, asst. prof., 1945-50, assoc. prof., 1950-54; clin. psychologist dept. univ. health, 1942-46; Univ. prof. psychology and edn. Rutgers U., New Brunswick, N.J., 1954—, dir, interdisciplinary research ctr., 1958-80; adv. com. psychopharmacology FDA, Washington, 1970 73; cons. pharm industries. Author: Wittenborn Psychiatric Rating Scale, 1955, rev., 1964, Clinical Pharmacology of Anxiety, 1966, Placement of Adoptive Children, 1957, Guidelines for Clinical Trials of Psychotropic Drugs, 1977. USPHS fellow, 1948-50; recipient disting. teaching and research award Rutgers U., 1961; Devel. Sch. Psychology award N.J. Assn. Sch. Psychologists, 1973. Fellow Am. Coll. Neuropsychopharmacology (sec.-treas. 1964-72, pres. 1973, Paul Hoch Disting. Service award 1968), Collegium International Neuropsychopharmacologicum (v.p. 1982—), Am. Psychol. Assn. Club: Cosmos (Washington). Subspecialties: Psychopharmacology; Psychopathology. Current work: Design of clinical investigations of behavior disorders and their evaluation and treatment, age-related behavior changes. Office: Rutgers U 13 Senior St New Brunswick NJ 07979

WITTIE, LARRY DAWSON, computer scientist, educator; b. Bay City, Tex., Mar. 9, 1943; s. L.D. and Mildred (Brown) W.; m. Diane M. Fischer, July 15, 1972; children: Lea D.F., Loren D.F. B.S., Calif. Inst. Tech., 1966; M.S., U. Wis., 1967, Ph.D., 1973. Systems programmer Calif. Inst. Tech., Pasadena, 1963-66, IBM, Sunnyvale, Calif., 1966; NASA trainee in computer sci. U. Wis., Madison, 1966-69, research asst., 1969-72; vis. asst. prof. computer sci. Purdue U., Lafayette, Ind., 1972-73; asst. prof. computer sci. SUNY, Buffalo, 1973-79, assoc. prof., 1979-82, assoc. prof. computer sci., Stony Brook, 1982—; tech. adviser Boeing, M. Wile & Co., Raytheon, Calspan, U.S. Army, NIH. Contbr. articles to profl. publs. Research grantee NSF, USAF, NASA, U.S. Army. Mem. Assn. Computing Machinery, IEEE, Soc. Neurosci., Sigma Xi. Subspecialties: Distributed systems and networks; Operating systems. Current work: Distributed information processing in large networks: portable distributed operating systems for network computers and efficient interconnection topologies for networks of millions of computers. Designer MICRONET computer and MICROS distributed operating system. Home: 7 Woodhull Rd East Setauket NY 11733 Office: Computer Sci Dept Room 1426 Lab Office Bldg SUNY Stony Brook NY 11794

WITTING, LLOYD ALLEN, technical director; b. Chgo., May 18, 1930; s. Theodore Allen and Elsie Martha (Korinek) W.; m. Lucille Ruth Gerches, Aug. 9, 1956; children: Sandra, Cynthia, Michael. B.S., U. Ill., 1952, M.S., 1953, Ph.D., 1956. Project assoc. U. Wis.-Madison, 1957-59; research scientist Ill. Dept. Mental Health, Elgin, 1959-72; asst. prof. U. Ill. Coll. Medicine, Chgo., 1962-73; assoc. prof. Tex. Woman's U., Denton, 1972-74; cons. North Tex. Ednl. and Tng. Coop, Inc., Gainesville, 1975-76; tech. dir. Supelco, Inc., Bellefonte, Pa., 1976—, cons., 1975-76. Editor: Glycolipid Methodology, 1976; co-editor: Modification of Lipid Metabolism, 1975. NIH research grantee, 1966-72. Mem. Am. Soc. Biol. Chemists, Am. Inst. Nutrition,

Am. Oil Chemists Soc. (assoc. editor 1972-81), Am. Chem. Soc., AAAS, Sigma Xi, Gamma Sigma Delta. Lutheran. Subspecialty: Analytical chemistry. Current work: Research and development in high performance liquid chromatography, thin layer chromatography, and chemical standards. Home: 249 Oakley Dr State College PA 16801 Office: Supelco Inc Supelco Park Bellefonte PA 16823

WITTKOWER, ANDREW B., manufacturing company executive; b. London, Nov. 7, 1934; s. Eric and Clair (Weil) W.; m. Mary Nora Shotter, Aug. 17, 1957; children: David A., Elizabeth A. B.Sc., McGill U., 1955; M.Sc., U. Cambridge, 1959; Ph.D., Univ. Coll. London, 1967. Scientist High Voltage Engring., Burlington, Mass., 1959-71; v.p., gen. mgr. Extrion (now Varian/Extrion), Gloucester, Mass., 1971-78; v.p., gen. mgr. Nova (now Eaton/Nova), Beverly, Mass., 1978—; pres. Zymet Inc., 1983—. Contbr. numerous articles to profl. jours. Recipient McPherson prize McGill U., 1954. Fellow Am. Phys. Soc., Inst. Physics U.K. Subspecialties: 3emiconductors; Atomic and molecular physics. Current work: Conceptualizing, designing and building ion implants for use in semiconductorsand metals. Patentee in field. Home: Folly Cove Rockport MA 01966 Office: 33 Cherry Hill Dr Danvers MA 01923

WITTWER, SYLVAN HAROLD, research administrator, horticulture educator; b. Hurricane, Utah, Jan. 17, 1917; s. Joseph and Mary Ellen (Stucki) W.; m. Maurine Cottle, July 27, 1938; children: La Ree Wittwer Farrar, Alice Wittwer Sowards, Arthur John, Carl Thomas. B.S., Utah State Coll., 1939; Ph.D., U. Mo., 1943; D.Sc. (hon.), Utah State U., 1982. Instr. horticulture U. Mo., 1943-46; asst. prof. horticulture Mich. State U., 1964-68, East Lansing assoc. prof., 1948-50, prof., 1950—, dir., 1965-83, assoc. deanColl. Agr. and Natural Resources, 1982—. Contbr. numerous articles on horticulture, plant physiology, agrl. tech., agrl. communications, priorities in agrl. research, food policy and agrl. policy, and sci. and tech. for food prodn. in 21st century to profl. publs. Recipient Disting. Faculty award Mich. State U., 1965, Benjamin Duggar award in plant physiology Auburn U., 1967, Disting. Service to Agr. award Mich. Farm Bur., 1976, James E. Talmage Sci. Achievement award Brigham Young U., 1977, citation of merit Coll. Agr., U. Mo. Mem. Am. Soc. Hort. Sci., AAAS, Am. Soc. Plant Physiologists, Bot. Soc. Am., Soc. Devel. Biology, V.I. Lenin All Union Acad. Agrl. Scis. (USSR). Mormon. Subspecialties: Plant physiology (agriculture); Plant physiology (biology). Current work: Biological limits in crop and livestock productivity; telecommunication and electronic communications; genetic engineering, research priorities in food and agriculture, science and technology-the further frontiers in food production for the 21st century. Office: Mich State U 108 Agr Hall East Lansing MI 48824

WIZENBERG, MORRIS JOSEPH, radiation oncologist; b. Toronto, Ont., Can., Apr. 9, 1929; came to U.S., 1953, naturalized, 1961; s. Eli and Dora (Flanceman) W.; m. Evelyn R. Mills, May 20, 1955; m. Toni N. Mack, Dec. 24, 1976; children: Elizabeth, Adam, David, Margaret. M.D., U. Toronto, 1953. Diplomate: Am. Bd. Radiology. Intern Sinai Hosp., Balt., 1953-54; resident, 1954-58, Univ. Hosp., Balt., 1959-61; prof. radiology U. Md., 1967-74, U. Okla., 1974-81, clin. prof., 1981—; staff attending radiation therapist Mercy Health Center, Oklahoma City, 1981—, South Community Hosp., 1981—; cons. in field. Contbr. articles in field to profl. jours. Nat. Cancer Inst. fellow, 1961-63. Mem. AMA, Radiol. Soc. N.Am., Am. Soc. Therapeutic Radiologists, Am. Radium Soc., Am. Coll. Radiology. Democrat. Jewish. Subspecialties: Radiology; Oncology. Current work: Clinical radiation oncology, clinical cancer research; consultant National Cancer Institute. Office: 4300 W Memorial Rd Oklahoma City OK 73120

WNEK, GARY EDMUND, science educator; b. Amsterdam, N.Y., Sept. 9, 1955; s. Edmund J. and Joan (Malicki) W.; m. Maria Dufresne, June 17, 1978; 1 dau., Janice. B.S.Ch.E., Worcester Poly. Inst., 1977; Ph.D. in Polymer Sci., U. Mass., 1980. Asst. prof. dept. materials sci. MIT, Cambridge, 1980—. NSF undergrad. research fellow SUNY-Albany, 1975. Mem. Am. Chem. Soc. (polymer div. speakers bur. chmn. 1983), Am. Phys. Soc., AAAS. Roman Catholic. Subspecialties: Polymer chemistry; Electronic materials. Current work: Synthesis modification and characterization of electrically conductive polymers, sythesis of new polymers, ion implantation of polymers. Home: 45 Shore Rd Ashland MA 01721 Office: MIT 77 Massachusettes Ave Rm 13-5094 Cambridge MA 02139

WNUK, MICHAEL PETER, mechanics engineering educator; b. Katowice, Poland, Sept. 12, 1936; came to U.S., 1966, naturalized, 1972; s. Marian and Helena (Zaluska) W.; m. Renata Budzilowicz, June 7, 1964; 1 dau., Jennifer. M.S. in Mechanics, Tech. U. Krakow, Poland, 1959, Ph.D. in Theoretical and Applied Mechanics, 1962; M.S. in Physics, Jagiellonian U.-Krakow, 1965. Disting. vis. scholar U. Cambridge, Eng., 1970; sr. research fellow Calif. Inst. Tech., Pasadena, 1967-68; asst. prof. mech. engring. S.D. State U., Brookings, 1966-67, assoc. prof., 1968-76, prof., 1976-82; prof. engring. mechanics U. Wis.-Milw., 1982—. Author: Introduction to Fracture Mechanics, 1977; editor: Proceedings 1st Yugoslav Summer Sch. on Fracture Mechanics, 1980. Mem. Cambridge Philos. Soc., N.Y. Acad. Scis. (life), Sigma Xi. Roman Catholic. Subspecialties: Mechanical engineering; Applied mathematics. Current work: Stability problems in elastic-plastic fracture; numerical methods in continum mechanics. Home: 3436 N Dousman St Milwaukee WI 53201 Office: U Wis Coll of Engring and Applied Scis 1530 Cramer St Milwaukee WI 53201

WODARSKI, JOHN STANLEY, research center adminstr.; b. Phila., Feb. 27, 1943; s. John S. and Estelle (Sapieha) W.; m. Lois Ann Moon, Aug. 15, 1964; 1 dau., Ann Christine. B.S., Fla. State U., 1965; M.S.S.W., U. Tenn., 1967; Ph.D., Washington U., 1970. Instr. Sam Houston State U., Huntsville, Tex., 1967-68; asst. prof. research assoc. Washington U., St. Louis, 1970-74; grant devel. cons. John Hopkins U., Balt., 1975-77; assoc. prof. U. Md., Balt., 1975-78; dir. research ctr. U. Ga., Athens, 1978—; project cons. Kettering Found., Dayton, Ohio, 1980-81; prin. investigator Ford Found., Athens, 1982—, Ga. Dept. Human Resources, 1982—, NIMH, 1979—. Author: (with D. Bagarozzi) Behavioral Social Work, 1979, (with S. Lenhart) Curriculum Guide, Alcohol Edn. by the T-G-T Method, 1982; author: Role of Research in Clinical Practice, 1981, Rural Community Mental Health Practice, 1983. Children's Bur. fellow, 1968-70. Mem. Nat. Assn. Social Workers, Am. Psychol. Assn., Council on Social Work Edn., Am. Sociol. Assn. Democrat. Subspecialties: Behavioral psychology; Social psychology. Current work: Research administration; teenage alcoholism. Home: 150 Green Hills Rd Athens GA 30605 Office: Research Ctr Sch of Social Work Tucker Hall U Ga Athens GA 30602

WODEHOUSE, EDMUND BERKELEY, botany and biology educator; b. Nyack, N.Y., Mar. 6, 1946; s. Roger Philip and Ellys (Butler) W; m. Artis Stiffey, June 5, 1971. B.A., U. Vt., 1968, M.S., 1970; Ph.D., Va. Poly. Inst. and State U., 1973. Instr. botany/biology Skyline Coll., San Bruno, Calif., 1973—. Recipient Antarctic Service medal NSF, 1980. Mem. Nat. Geog. Soc., Am. Inst. Biol. Scis., Audubon Soc., Smithsonian Assocs. Subspecialty: Ecology. Current work: Aquatic ecology, primary productivity, nutrient cycling in aquatic ecosystems. Home: 3792 Park Blvd Palo Alto CA 94306 Office: Skyline Coll 3300 College Dr San Bruno CA 94066

WOERTH, JANICE KAY, dentistry educator; b. Santa Ana, Calif., May 25, 1948; d. J. Daniel and Irene Elizabeth (Dennis) Detwiler; m. Duane Edward Woerth, June 18, 1971. B.S., U. Nebr., 1971, B.A., 1973; M.P.H., U. Okla., 1977; Ph.D., U. Mo.-Kansas City, 1983; B.S. (hon.), U. Pueblo, Mex., 1978. Dental hygienist, Omaha, 1972-76; intern in health planning Okla. Health Systems Agy., Oklahoma City, 1977; asst. prof. dentistry U. Mo.-Kansas City Sch. Dentistry, 1977—; mem. patient edn. planning com. St. Mary's Hosp., Kansas City, Mo., 1982. Recipient grad. stipend U. Okla., 1977; U. Mo.-Kansas City Grad. Women J. Coffey scholar, 1981. Mem. Mo. Dental Hygienist Assn. (v.p. 1981-82, pres.-elect 1982-83), Am. Dental Hygienist Assn., Internat. Assn. Dental Researchers, Am. Assn. Dental Schs. Methodist. Subspecialties: Preventive dentistry; Health services research. Current work: Occupational orientation and demographic characteristics of women in health occupations. Home: 8029 NW Waukomis Dr Kansas City MO 64151 Office: U Mo 650 E 25th St Kansas City MO 64108

WOGAN, GERALD NORMAN, educator; b. Altoona, Pa., Jan. 11, 1930; s. Thomas B. and Florence E. (Corl) W.; m. Henrietta E. Hoenicke, Aug. 24, 1957; children—Christine F., Eugene E. B.S., Juniata Coll., 1951; M.S., U. Ill., 1953, Ph.D., 1957. Asst. prof. physiology Rutgers U., New Brunswick, N.J., 1957-61; asst. prof. toxicology Mass. Inst. Tech., Cambridge, 1962-65, asso. prof., 1965-69, prof., 1969—, head dept. nutrition and food sci., 1979—; cons. to nat. and internat. govt. agys., industries. NIH grantee, 1963—. Mem. editorial bd.: Cancer Research, 1971—, Applied Microbiology, 1971—, Chem.-Biol. Interactions, 1975—, Toxicology, Environ. Health, 1974—; Contbr. articles and revs. to profl. jours. NSF grantee, 1965-68. Mem. Nat. Acad. Scis. U.S., Am. Assn. Cancer Research, Am. Soc. Pharmacology and Exptl. Therapeutics, Am. Soc. Microbiology, Soc. Toxicology, Am. Inst. Nutrition, AAAS, Sigma Xi. Subspecialties: Cancer research (medicine); Biochemistry (medicine). Current work: Mechanisms of chemical carcinogens; identification of carcinogens. Home: 125 Claflin St Belmont MA 02178 Office: Dept Nutrition Food Sci Mass Inst Tech Cambridge MA 02139

WOHL, ROBERT CHAIM, biochemist; b. Budapest, Hungary, Mar. 27, 1938; came to U.S., 1962, naturalized, 1971; s. Nandor W. B.S., Columbia U., 1967; Ph.D., SUNY-Buffalo, 1972. Sr. research assoc. Michael Reese Research Found., Chgo., 1975—. NIH grantee, 1980, 82. Mem. Am. Chem. Soc., Am. Soc. Biol. Chemists, N.Y. Acad. Scis., Am. Heart Assn. Subspecialties: Biochemistry (biology); Biochemistry (medicine). Current work: Research on the causes and solutions of dissolution of clots in thrombosis. Office: Michael Reese Research Found 530 E 31st St Chicago IL 60616

WOHLFEIL, PAUL FREDERICK, health maintenance organization counselor; b. Saginaw, Mich., Aug. 28, 1934; s. Herman Frederick and Rose (Kueffner) W.; m. Shirli Jean Setzer, July 20, 1976; children: Paul John, Ondria Rose. B.A., Mich. State U., 1962; M.A., Eastern Mich. U., 1966; Ph.D., Independence (Mo.) U., 1973; postgrad., Rutgers U., 1974. Lic. psychologist, marriage counselor, Mich., cert. social worker, lic. marriage and family therapist, cert. clin. mental health counselor, Fla. Psychologist Boys Tng. Sch., Whitmore Lake, Mich., 1962-65; psychologist, assoc. dir. Whaley Home for Disturbed Children, Flint, Mich., 1963-68; psychologist Adult Mental Health Clinic, Bay City, Mich., 1968-73, Salman Psychiat. Clinic, Bay City, 1973-79; dir. counseling Group Health Services of Mich., Saginaw, 1978—; cons. Saginaw Steering Gear, 1973—, Gen. Electric Corp., Pittsfield, Mass., 1978-79, Gratiot Community Mental Health, Alma, Mich., 1969-71; dir. Personal Growth, Inc., Saginaw, 1975—. Author: (1977.) Awareness Counseling for Nurses. Served with AUS, 1954-56. Recipient award Bay City Bd. Commrs., 1971, Flint (Mich.) Bd. Dirs., 1963. Mem. Am. Assn. Marriage and Family Therapists, Mich. Assn. Marriage and Family Therapists, Mich. Psychol. Assn., Am. Psychol. Assn. (assoc.), Mich. Assn. Group Psychotherapy. Lutheran. Club: Riverview Rod and Gun (Grayling, Mich.). Subspecialties: Health psychology; Cognition. Current work: Psychological and physiological aspects in the health care field, using computer data-based information for counseling development. Home: 3102 Sharon Rd Midland MI 48640 Office: Group Health Service Mich 4200 Fashion Square Blvd Saginaw MI 48603

WOJCIECHOWSKI, WITOLD STANISLAW, computer scientist, hardware and software consultant; b. Poznan, Poland, July 11, 1949; s. Edward Jozef and Maria (Noyszewska) W. M.S., A. Mickiewicz U., 1971, Poznan (Poland) U., 1972; Ph.D., Ill. Inst. Tech., 1980. Researcher, lectr. Poznan Technical U., 1972-76; prin. cons. SEI Info. Tech., Chgo., 1980—; part-time instr. computer sci. dept. Ill. Inst. Tech. Mem. Assn. Computing Machinery. Subspecialties: Distributed systems and networks; Computer engineering. Current work: Hardware and software; microprocessors, operating systems, specialized hardware and application software; VLSI and integrated hardware-software systems; distributed systems; data communications, data bases. Home: 4430 N Malden St Apt 1S Chicago IL 60640 Office: 450 E Ohio St Chicago IL 60611

WOLBARSHT, MYRON LEE, biophysicist, educator, researcher; b. Balt., Sept. 18, 1924; s. Samuel and Rose (Levine) W.; m.; children: Seth, Selah, Jeremy. A.B., St. John's Coll., 1950; Ph.D., Johns Hopkins U., 1958. Research assoc. Psychiat. Inst., U. Md., 1954-60; research cons. neurology U. Md. Med. Sch., 1960-63; research fellow Johns Hopkins U., Balt., 1959-68; research cons. neurology York (Pa.) Hosp., 1963-68; physicist Naval Med. Research Inst., Bethesda, Md., 1958-68, head biophysics div., 1966-68; prof. ophthalmology and biomed. engring. Duke U., Durham, N.C., 1968—, assoc. prof. physiology, lectr. psychology; cons. in field of laser safety. Author: (with D.H. Sliney) Safety with Lasers and Opther Optical Sources, 1980; assoc. editor: Quar. Rev. Biology, 1963-68, IRE-IEEE Transactions on Biomed. Electronics, 1961-69; contbr. chpts. to books, articles to profl. jours. Served with USAAF, 1943-46. Mem. Am. Nat. Standards Inst. (chmn. eye hazards subcom. 1969—), U.S. Nat. Com. Photobiology, Laser Inst. Am. (pres.-elect 1982), Am. Soc. Laser Medicine and Surgery (bd. dirs.), Armed Forces-RC Com. on Vision, Am. Physiol. Soc., Am. Soc. Photobiology, Assn. Research in Vision and Ophthalmology, Biophys. Soc., IEEE, Soc. Gen. Physiologists, N.Y. Acad. Scis., Optical Soc. Am., Soc. Photo-Optical Instrumentation Engrs., Royal Soc. Medicine, Sigma Xi. Club: Cosmos (Washington). Subspecialties: Neurophysiology; Ophthalmology. Current work: Research in laser tissue interaction, vascular complications of diabetes, laser safety, automated visual function testing. Patentee in field. Office: Dept Ophthalmology Duke U Med Ctr Durham NC 27710

WOLBERG, GERALD, immunologist; b. N.Y.C., Aug. 18, 1937; s. Sam and Fannie (Lipshitz) W.; m. Marilynn Harriet Goldstein, July 1, 1967; children: Alisa Sue, Lori Melanie. B.A., N.Y.U., 1958; M.S., U. Ky., 1963; Ph.D., Tulane U., 1967. Postdoctoral tng. N.Y. Pub. Health Research Labs., Research Triangle Park, N.C., 1970—. Mem. Am. Soc. for Microbiology, Am. Assn. Immunologists, Sigma Xi. Subspecialties: Immunobiology and immunology; Immunopharmacology. Current work: Chemical control of the immune response. Home: 1109 Troon Ct Gary NC 27511 Office: 3030 Cornwallis Rd Research Triangle Park NC 27709

WOLEN, ROBERT LAWRENCE, pharmacologist, educator; b. N.Y.C., May 20, 1928; s. Albert and Yetta W.; m. Marion Jacobs, Apr. 9, 1953; children: Sonya M., Y. Rosalind. B.S., West Chester State U., 1950; M.S., U. Del., 1951, Ph.D., 1961. Lectr. Oak Ridge Inst. Nuclear Studies, 1961-62; research scientist St. Joseph Hosp., Lancaster, Pa., 1962-63; sr. biochemist Eli Lilly & Co., Indpls., 1963-66, sr. clin. biochemist, 1966-68, research scientist, 1968-69, chief clin. chemistry and drug metabolism dept., 1969-75, research assoc., 1975-79, research adv., 1979—. Served to lt. USN, 1951-55. Mem. Am. Chem. Soc., Am. Soc. Pharmacology and Exptl. Therapeutics, Am. Soc. Clin. Pharmacology and Therapeutics, Am. Assn. Clin. Chemists. Jewish. Subspecialties: Pharmacology; Biochemistry (medicine). Current work: Drug metabolism and pharmacokinetics. Office: 307 E McCarty St Indianapolis IN 46285

WOLF, BARBARA ANNE, research administrator, educator; b. N.Y.C., July 24, 1947; d. Boris and Molly (Gruberg) W.; m. Robert, Aug. 25, 1977. B.A. in Chemistry, Queens Coll., CUNY, 1968; Ph.D. in Biology, M.I.T., 1973. Research asst. Sloan-Kettering Inst. Cancer Research, Rye, N.Y., 1967; undergrad. researcher Queens Coll., 1968; teaching asst. M.I.T., 1969-70, 72-73; research asst. in virology Rockefeller U., 1973-75; fellow Nat. Cancer Inst., 1975-77, research assoc., 1977; mgr. biol. services Revlon Research Ctr., Bronx, N.Y., 1977-80, dir. biol. scis., 1981—; attendee U. Tex., 1979; assoc. prof. St. John's U., 1981—. Contbr. articles to profl. jours. Recipient Stanley Koncal award, 1968; N.Y. State Regents scholar, 1964-68. Mem. Environ. Mutagen Soc., Genetic Toxicology Assn., Am. Coll. Toxicology, Soc. Toxicology, N.Y. Acad. Scis., AAAS, Am. Soc. Microbiologists, Soc. Cosmetic Chemists, N.Y . Soc. Electron Microscopists, Phi Beta Kappa, Sigma Xi, Beta Delta Chi. Subspecialties: Toxicology (medicine); Cell biology (medicine). Current work: Animal toxicology, genotoxicity (mutagenicity), clinical testing, antimicrobial testing.

WOLF, EDWARD DEAN, electrical engineering educator; b. Quinter, Kans., May 30, 1935; s. Ezra Lawrence and Zora Blanche (Jamison) W.; m. Marlene Kay Simpson, Aug. 12, 1955; children: Julie Christine Wolf Saline, LeAnn Cynthia, Shelly Diane. Student, Kans. State U., 1953-54; B.S., McPherson Coll., 1957; Ph.D., Iowa State U., 1961; postgrad., Princeton U., 1962, U. Calif.-Berkeley, 1968. Mem. tech. staff Rockwell Internat. Sci. Ctr., Thousand Oaks, Calif., 1963-65, Hughes Research Labs., Malibu, Calif., 1965-67, sr. mem. tech. staff, 1967-72, sect. head, 1972-78, sr. scientist, 1974-78; dir. nat. research and resource facility for submicron structures, prof. elec. engring. Cornell U., Ithaca, N.Y., 1978—; cons. to industry and govt.; mem. steering coms. various tech. confs.; co-organizer 1st Gordon Research Conf. on Chemistry and Physics of Microstructure Fabrication, 1976. Contbr. articles to profl. jours. Fellow Am. Inst. Chemists, IEEE (guest editor spl. issues on electron devices and Jour. Vacuum Sci. and Tech.); mem. Bohmische Phys. Soc., Am. Phys. Soc., Am. Vacuum Soc., Electron Microscopy Soc. Am. Republican. Subspecialties: Microchip technology (engineering); Surface chemistry. Current work: Advanced electron and ion beam lithography, reactive ion beam etching and microminiaturization research. Patentee in field. Home: 8 Highgate Circle Ithaca NY 14850 Office: Cornell U M105 Knight Lab NRRFSS Ithaca NY 14853

WOLF, GERALD LEE, internist, radiologist, educator; b. Sidney, Nebr., Apr. 2, 1938; m. Lynn Beckenhauer, Mar. 14, 1964; children: David, Darin, Kristin, Jody; m. Karen Christiansen, Aug. 3, 1979. Student, Hastings Coll., 1957-60; B.S., U. Nebr., 1962, M.S., 1963, Ph.D., 1965; M.D., Harvard Med. Sch., 1968. Diplomate: Am. Bd. Radiology. Vis. staff Bishop Clarkson Hosp., 1968-78; asst. prof. physiology U. Nebr. Coll. Medicine, 1968-71, asst. prof. dept. internal medicine, 1969-71, asst. prof. pharmacology and radiology, 1971-74, dir. clin. pharmacology, 1973-75, dir. radiology research, 1973-80, assoc. prof. pharmacology and radiology, 1974-80, dir. cardiovascular radiology, 1974-80; staff physician, acting chief, then chief radiology Omaha VA Med. Center, 1976-80; prof. radiology U. Pa. Med. Center, Phila., 1981—, chief radiology service and assoc. chief staff for research, 1982—; cardiovascular radiology cons. NIH Clin. Center. Contbr. articles, chpts. and books to profl. lit. Tenor soloist, mem. fin. commn. Lima Meth. Ch. Morseman fellow, 1962-63; Life Ins. Med. Research fellow, 1964-68; research career devel. awardee NIH, 1973-75; Premier Investigation award Omaha Mid-West Clin. Soc., 1974-76; Marshak award Am. Jour. Gastroenterology; Winthrop scholar U. Pa., 1981-82. Fellow Am. Coll. Clin. Pharmacology; mem. Assn. Univ. Radiologists, Soc. Cardiovascular Radiologists, Radiol. Soc. N.Am., Alpha Chi. Republican. Subspecialties: Radiology; Pharmacology. Current work: Vascular radiology, nuclear magnetic resonance, clinical pharmacology. Home: 441 S Feathering Rd Media PA 19063 Office: University and Woodland Aves Research Philadelphia PA 19104

WOLF, GREGORY THOMAS, otolaryngologist; b. Racine, Wis., Apr. 28, 1948; s. Lee J. and Margaret Rose (Friedlg) W.; m. Suzanne M. Vojtisek, Feb. 14, 1970; children: Michael, Melissa; Jenifer. B.S., U. Notre Dame, 1970; M.D., U. Mich., 1973. Diplomate: Am. Bd. Otolaryngology (head and neck surgery), Nat. Bd. Med. Examiners. Intern, resident in surgery Georgetown U., 1973-75; resident in otolaryngology SUNY Upstate Med. Center, Syracuse, 1975-78; project officer Nat. Cancer Inst., 1978-80; asst. prof. otolaryngology U. Mich. Med. Sch., 1980—; chief otolaryngology VA Med. Center, Ann Arbor, 1980—. Mem. Am. Acad. Ooloryngology, Soc. Univ. Otalrygnologists. Subspecialties: Otorhinolaryngology; Cancer research (medicine). Current work: Head and neck cancer immunology. Office: Box 61 C-6170 Out-Patient Dept Univ Hosp Ann Arbor MI 48109

WOLF, JOSEPH ALLEN, JR., mechanical engineer; b. Tacoma, Nov. 26, 1933; s. Joseph Allen and Wilma Audrey (Murphy) W.; m. Sally Blackwood Doyle, June 25, 1960; 1 son, Joseph Allen. M.E. with honor, Stevens Inst. Tech., 1955; M.S. in Applied Physics, UCLA, 1957; Sc.D. in Mech. Engring. (NSF fellow), M.I.T., 1967. Registered profl. engr., Calif. Mem. tech. staff Hughes Aircraft Co., Culver City, Calif., 1955-62; asst. prof. engring. and applied sci. UCLA, 1966-71; staff research engr. dept. engring. mechanics Gen. Motors Research Labs., Warren, Mich., 1971—. Co-author and co-editor: Modern Automotive Structural Analysis, 1982; contbr. articles, chpts. to profl. publs. Mem. ASME, Sigma Xi, Tau Beta Pi. Republican. Episcopalian. Subspecialties: Theoretical and applied mechanics; Acoustical engineering. Current work: Dynamics, mechanical and structural vibrations, and structural analysis of automotive components and structures. Home: 438 S Glenhurst Birmingham MI 48009 Office: Engring Mechanics Dept Gen Motors Research Labs Warren MI 48090

WOLF, KENNETH EDWARD, research microbiologist; b. Chgo., Oct. 22, 1921; s. Frank A. and Marguerite (Ziegler) W.; m. Betty C. Wolf, 1948; children: Mark, Gregory, Anthony. B.S., Utah State U., 1951, M.S., 1952; Ph.D., 1956. Bacteriologist Eastern Fish Disease Lab. (name changed to Nat. Fish Health Research Lab.), U.S. Fish and Wildlife Service), Ithaca, N.Y., 1954-59, microbiologist, 1959-61, research microbiologist, 1961-72, dir., 1972-77, sr. research scientist, 1977—. Subspecialties: Microbiology; Virology (biology). Office: Nat Fish Health Research Lab Cornell U Ithaca NY 14853

WOLF, LARRY LOUIS, biology educator, researcher; b. Madison, Wis., Oct. 21, 1938; s. Mark A. and Dorothy (Test) W.; m. Janet N. Sorlien, Sept. 5, 1965; children: Alan, Frederick. B.S., U. Mich., 1961; Ph.D., U. Calif.-Berkeley, 1966. Asst. prof. biology Syracuse U., 1967-71, assoc. prof., 1971-76, prof., 1976—; mem. ecology rev. panel NSF, Washington, 1977-79. Author: General Ecology, 1973, 2d edit., 1978, Species Relationships in Aimophila, 1979. NSF grantee, 1968—. Fellow Am. Ornithologists Union; mem. Ecol. Soc. Am., Brit. Ecol. Soc., Am. Soc. Naturalists, Animal Behavior Soc. Subspecialties: Behavioral ecology; Ecology. Current work: Population and community ecology; ecological factors influencing behavior. Office: Dept Biology Syracuse U Syracuse NY 13104

WOLF, RICHARD CLARENCE, educator; b. Lancaster, Pa., Nov. 28, 1926; s. Clarence Lester and Bertha Mae (Felker) W.; m. Marilyn Jean Miller, Aug. 23, 1952; children—Mark, Eric. B.S., Franklin and Marshall Coll., 1950; Ph.D., Rutgers U., 1954. Faculty dept. physiology U. Wis.-Madison, 1957—, prof., 1966—, chmn., 1971—, mem. endocrinology-reproductive physiology program, 1963—, co-dir., 1968-70, dir., 1970—; Cons. NIH, Ford Found.; mem. sci. bd. Yerkes Regional Primate Center, Emory U., 1972—. Contbr. articles to profl. jours. Served with AUS, 1945-46. Waksman-Merck fellow Rutgers U., 1954-55; Milton fellow Harvard, 1955-56; USPHS fellow, 1956-57. Fellow N.Y. Acad. Scis.; Mem. Endocrine Soc., Am. Soc. Zoologists, Am. Physiol. Soc., Soc. for Endocrinology (Gt. Britain), Soc. Exptl. Biology and Medicine (sec. Wis.), Soc. Study Fertility, Internat. Soc. for Research in Biology of Reproduction, Soc. for Study Reproduction, Sigma Xi., Phi Beta Pi. Lutheran. Club: Mason. Subspecialty: Endocrinology. Current work: Reproductive biology; follicle growth and implantation. Home: 4205 Manitou Way Madison WI 53711

WOLF, THOMAS MARK, psychology educator; b. Cin., Dec. 25, 1944; s. Herbert and Ursula (Wachtel) W.; m. Valerie Barbara Winchester, Sept. 20, 1969; 1 son, Mark Benjamin. B.A., U. Cin., 1966; M.A., Miami (Ohio) U., 1967; Ph.D., U. Waterloo, Ont, 1971; postgrad., St. Louis U., 1974-76. Lic. psychologist, La. Asst. prof. SUNY-Cortland, 1970-74, assoc. prof., 1974-75; assoc. prof. psychology La. State U. Med. Ctr., New Orleans, 1975-82, prof., 1982—; cons. psychologist Youth Study Ctr., New Orleans, 1977, New Orleans Pub. Schs., 1979-80, St. Bernard Group Home, Chalmette, La., 1979—, Central City Mental Health Clinic, 1980—. Contbr. articles to profl. jours. Mem. early childhood com. Jewish Community Ctr., New Orleans, 1981—; mem. children's com. Mental Health Assn. New Orleans, 1982—. La. Heart Assn. grantee, 1977-79, 80-81; others. Mem. Am. Psychol. Assn., Southeastern Psychol. Assn., Soc. for Research in Child Devel., Soc. Behavioral Medicine, Assn. Am. Med. Colls. Subspecialty: Developmental psychology. Current work: Child development, lifestyle characteristics, stress and disease of children and medical students; family intervention. Home: 7046 Camp St New Orleans LA 70118 Office: Dept Psychiatry La State U Med Sch 1542 Tulane Ave New Orleans LA 70112

WOLF, WALTER ALAN, chemistry educator, researcher; b. N.Y.C., Mar. 9, 1942; s. Theodore B. and Tasia (Richmond) W.; m. Doris I. Buckley, June 13, 1964; children: Terri-Lynn, Patricia. B.A., Wesleyan U., 1962; M.A., Brandeis U., 1964, Ph.D., 1967. Research assoc. MIT, Cambridge, Mass., 1967-70; asst. prof. Colgate U., Hamilton, N.Y., 1970-77; assoc. prof. Eisenhower Coll., Seneca Falls, N.Y., 1977-82, Rochester (N.Y.) Inst. Tech., 1982—. Editor: Jour. Chem. Edn, Flagstaff, 1974—. Mem. Am. Chem. Soc., AAAS, Chem. Soc. London, Sigma Xi. Subspecialties: Biochemistry (biology); Computer modelling of insect systems. Current work: Research on the enzymes involved in the biosynthesis of sex hormones in moths, identification of reaction and control mechanisms. Home: RD 3 Box 76 Waterloo NY 13165 Office: Rochester Inst Tech Rochester NY 14623

WOLF, WERNER PAUL, educator, physicist; b. Vienna, Austria, Apr. 22, 1930; came to U.S., 1963, naturalized, 1977; s. Paul and Wilhelmina (Wagner) W.; m. Elizabeth Eliot, Sept. 23, 1954; children: Peter Paul, Mary-Anne Githa. B.A., Oxford (Eng.) U., 1951, D.Phil, 1954, M.A., 1954; M.A. (hon.), Yale U., 1965. Research fellow Harvard U., 1956-57; Fulbright travelling fellow, 1956-57; Imperial Chem. Industries research fellow Oxford U., 1957-59, univ. demonstrator, lectr., 1959-63; lectr. New Coll., 1957-63; faculty Yale U., 1963—, prof. physics and applied sci., 1965—, dir. grad. studies dept. engring. and applied sci., 1973-76, Becton prof., 1976—, chmn. dept. engring. and applied sci., 1976-81, chmn. council engring., 1981—; cons. to industry, 1957—; Sigma Xi vis. prof. Technische Hochschule, Munich, Germany, 1969; Sci. Research Council sr. vis. fellow Oxford U., 1980; mem. program com. Conf. Magnetism and Magnetic Materials, 1963, 65, chmn., 1968, mem. adv. com., 1964-65, 70-76, chmn., 1972, steering com., 1970-71, conf. gen. chmn., 1971; mem. organizing, program coms. Internat. Congress on Magnetism, 1967, internat. program com., 1978-79, planning com., 1979; vis. physicist Brookhaven Nat. Lab., 1966, 68, vis. sr. physicist, 1970, research collaborator, 1972, 74, 75, 77, 80; mem. vis. com. dept. physics U. Del., 1980—. Contbr. papers on magnetic materials and low temperature physics. Fellow Am. Phys. Soc. (edn. com. 1977-80, program dir. Indsl. Grad. Intern Program 1978—, chmn. fellowship com. 1981, mem. exec. com. Div. Condensed Material Physics 1980-83); sr. mem. IEEE; mem. Am. Assn. Crystal Growers, Conn. Acad. Sci. and Engring., Yale U. Sci. and Engring. Assn. (hon. v.p. 1976). Subspecialties: Magnetic physics. Current work: Properties of magnetic materials; especially rare earth compounds, at low temperatures; phase transitions and critical phenomena. Home: Apple Tree ln Woodbridge CT 06525 Office: Becton Center Yale U New Haven CT 06520

WOLFE, ARTHUR MICHAEL, astrophysicist, educator; b. Bklyn., Apr. 29, 1939; s. Benjamin and Sara (Jacobson) W.; m. Constance Eve Wolfe, Oct. 15, 1965; children: David B., Diana S. B.S., Queens Coll., 1961; M.S., Stevens Inst. Tech., 1963; Ph.D., U. Tex., 1967. Postdoctoral fellow U. Calif.-San Diego, 1967-70, Inst. Astronomy, Cambridge, 1970-72; NAS-Exchange Fellow, Moscow, 1972; research fellow U. Manchester, Eng., 1972; asst. prof. U. Pitts., 1973-77, assoc. prof., 1977-81, prof. physics and astronomy, 1981—; mem. astronomy adv. com. NSF, 1979-82; mem. adv. com. Arecibo Obs. Contbr. articles to profl. jours. NSF grantee, 1977—. Mem. Internat. Astron. Union, Am. Astron. Soc. Democrat. Jewish. Subspecialties: Cosmology; Theoretical astrophysics. Current work: Investigation of matter in early universe, using radiation emitted by distant quasars. Office: Dept Physics and Astronomy U Pitts Pittsburgh PA 15260 Home: 6537 Darlington Rd Pittsburgh PA 15217

WOLFE, BARRY B., neuropharmacologist, med. educator; b. Denver, June 16, 1945; s. Russell and Doris (Barlow) W.; m. Claire Cameron, Sept. 12, 1971. B.S., UCLA, 1967; M.S., Calif. State U.,-Northridge, 1969; Ph.D., U. Calif.-Santa Barbara, 1973. Postdoctoral fellow U. Colo. Health Sci. Center., Denver, 1973-76, instr., 1976-78, asst. prof., 1978-81, U. Pa. Sch. Medicine, Phila., 1981—. Contbr. articles to profl. jours. NIH fellow, 1975-76; grantee Colo. Heart Assn., 1978, 81, NIH, 1979-82, NIMH, 1980. Mem. Am. Soc. Pharmacology and Exptl. Therapeutics (Fgn. Travel award 1978), Soc. Neurosci. Subspecialties: Neuropharmacology; Neurobiology. Current work: Neurotransmitter receptors.

WOLFE, BRADLEY ALLEN, electrical engineer; b. Seattle, Nov. 1, 1935; s. Adolph Edward and Bernice Eleanor (Younglowe) W.; m. Louis Mae Weber, Feb. 14, 1974; 1 stepdau., Kimberly Mae; m. Zelma Jean Harvey, Apr. 2, 1955 (div. Nov. 1973); children: Scott Bradley, Colleen Jean, Cynthia Joann. B.S.E.E., Heald Engring Coll., San Francisco, 1959. Registered profl. engr., Calif. Lead design engr. Boeing Co., Seattle, 1959-61, research and devel. mgr., Oak Ridge, 1965-78; project engr. Collins Radio Co., Cedar Rapids, Iowa, 1961-63; systems engr. Sperry Rand Corp., St. Paul, 1963-65; div. mgr. Sci. Applications, Inc., Oak Ridge, 1978-81; group mgr., projects mgr. Rockwell Internat., Richland, Wash., 1981—. Contbr. articles to profl. jours. Chmn. citizens adv. council Granite Falls (Wash.) Sch. Dist., 1972. Served with USN, 1953-56. Recipient Apollo Medallion NASA/Boeing, 1967; Outstanding Achievement award U.S. Dept. Energy, 1976. Mem. System Safety Soc. (chmn. N.W. chpt. 1967-68), IEEE (vice chmn, engring. mgmt. 1971-72), N.Y. Acad Scis., Am. Nuclear Soc., Nat. Mgmt. Assn. Subspecialties: Systems engineering; Operating systems. Current work: Application of step-by-step planning methods for conduct of complex projects in efficient and effective manner. Effective application of computers to petro-chemical process monitor and control. Home: 7015 W 8th St Kennewick WA 99336 Office: Rockwell Internat PO Box 800 Richland WA 99352

WOLFE, GENE HENRY, physicist; b. Calumet City, Ill., May 18, 1936; s. Henry Leo and Sophie S. (Goreski) Wojciechowski; m. Elaine Marie Mateja, Aug. 27, 1960; children: Lynne Marie, Lisa Ann, Eugene Michael, Karen Sue, Kimberly Ann, Amy Lynn. B.S., U. Ill., 1958, M.S. in Physics, 1959. Asst. physicist Argonne (Ill.) Nat. Lab., 1959; assoc. physicist Ill. Inst. Tech. Research Inst., Chgo., 1960-64; research physicist R.R. Donnelley & Sons, Chgo., 1964—. Contbr. articles to profl. jours. Trustee Thornton Twp., 1981—; del. Republican Conv., 1980; mem. Chgo. Community Schs. Study Commn., 1982. Mem. Optical Soc. Am., Am. Numismatic Assn. Roman Catholic. Lodge: KC. Current work: Electrostatics and electrostatic printing research; laser-optical image processing. Home: 1309 Buffalo Calumet City IL 60409 Office: 2223 S King Dr Chicago IL 60616

WOLFE, HOWARD FRANCIS, mechanical and aeronautical engineer, research administrator; b. Dayton, Ohio, Feb. 17, 1938; s. Marion Francis and Ethel Louella (Dempsey) W.; m. Phyllis Ann, June 17, 1961; children: Timothy, Douglas, Brian, Gregory, Elizabeth, Matthew. B.M.E., U. Dayton, 1961; M.S.M.E., Ohio State U., 1970. Registered profl. engr., Ohio. Aerospace engr. Air Force Wright-Aero. Lab., Wright-Patterson AFB, Ohio, 1964-77; tech. mgr. Acoustics and Sonic Fatigue Group, 1977—. Mem. Beavercreek (Ohio) Music Parents Assn., 1979—, chmn. band camp, 1981. Served to 1st lt. Ordnance Corps U.S. Army, 1961-64. Decorated Army Commendation medal. Mem. ASME (Dayton sect. Outstanding Achievement award 1980), Nat. Mgmt. Assn. Roman Catholic. Subspecialties: Aerospace engineering and technology; Mechanical engineering. Current work: Acoustic fatigue of structures, acoustics, structural dynamics, random vibration, modal analysis, fatigue behavior. Patentee wide band siren system. Home: 2834 Southfield Dr Xenia OH 45385 Office: Air Force Wright-Aero Lab/FIBED Wright-Patterson AFB OH 45433

WOLFE, MICHAEL DAVID, mathematician; b. Houston, Feb. 16, 1950; s. Alfred Sigmund and Raquel (Azcarraga) W. B.S., U. Tex.-Austin, 1972, Ph.D., 1978; postgrad., Cambridge (Eng.) U., 1974. Asst. instr. U. Tex.-Austin, 1972-77; lectr. U. Tex.-San Antonio, 1978; staff mem. BDM Corp., Albuquerque, 1978-80; scientist Mission Research Corp., Albuquerque, 1980-83; mem. tech. staff Rocketdyne div. Rockwell Internat. Corp., Albuquerque, 1983—. Author: Long Range Potential Scattering, 1978. Mem. Am. Math. Soc., Soc. Indsl. and Applied Math., IEEE, Inst. Mgmt. Sci., Phi Kappa Phi. Clubs: N.Mex. Orchid Soc. (state fair chmn. 1981), Sierra.). Subspecialties: Mathematical software; Operations research (mathematics). Current work: Mathematics of nuclear effects, electromagnetic pulse theory, measurement of efficiency, scheduling algorithms, scattering theory. Home: 4016 Douglas Macarthur NE Albuquerque NM 87110 Office: Rocketdyne Div Rockwell Internat Corp PO Box 5670 Kirkland AFB NM 87185

WOLFE, RALPH STONER, microbiologist, educator; b. New Windsor, Md., July 18, 1921; s. Marshall Richard and Jennie Naomi (Weybright) W.; m. Gretka Margaret Young, Sept. 9, 1950; children—Daniel Binns, Jon Marshall, Sylvia Suzanne. B.S., Bridgewater (Va.) Coll., 1942; M.S., U. Pa., 1949, Ph.D., 1953. Mem. faculty U. Ill., 1953—, prof. microbiology, 1961—; cons. USPHS, Nat. Inst. Gen. Med. Scis. Author research papers microbial physiology, biochemistry. Guggenheim fellow, 1961, 75; USPHS spl. postdoctoral fellow, 1967; recipient Pasteur award Ill. Soc. for Microbiology, 1974. Mem. Nat. Acad. Scis., Am. Acad. Arts and Scis., Am. Soc. Microbiology (Carski Disting. Teaching award 1971), Am. Soc. Biol. Chemists. Subspecialty: Microbiology. Address: 131 Burrill Hall Univ Ill Urbana IL 61801

WOLFENSTEIN, LINCOLN, physicist, educator; b. Cleve., Feb. 10, 1923; s. Leo and Anna (Koppel) W.; m. Wilma Caplin, Feb. 3, 1957; children: Frances, Leonard, Miriam. S.B., U. Chgo., 1943, S.M., 1944, Ph.D., 1949. Physicist Nat. Adv. Com. Aeros., 1944-46; mem. faculty dept. physics Carnegie-Mellon U., Pitts., 1948—, asso. prof., 1957-60, prof., 1960-78, Univ. prof., 1978—. Contbr. articles to profl. jours. Guggenheim fellow, 1973, 83. Mem. Am. Phys. Soc., Fedn. Am. Scientists, AAAS, Nat. Acad. Sci. Subspecialty: Particle physics. Current work: Theory of weak interactions; neutrino theory. Office: Carnegie Mellon U Pittsburgh PA 15213

WOLFF, DAVID ALWIN, microbiologist, govt. adminstr.; b. Cleve., Nov. 2, 1934; s. Alwin Emil and Marion Eloise (Perkins) W.; m. Carol Ann Kish, June 14, 1958; m. Linda Diane Heding, July 24, 1976; children: Kurt, Christopher, Andrew, Lauren. A.B., Coll. Wooster, 1956; M.S., U. Cin., 1960, Ph.D., 1965. Asst. prof. microbiology Ohio State U., 1964-67, assoc. prof., 1967-76, prof., 1976-78; health scientist adminstr. NIH, Bethesda, Md., 1978—. Contbr. articles to profl. jours. Served with U.S. Army, 1956-59. Recipient Alumni Disting. Teaching award Ohio State U., 1974; USPHS-NIH grantee, 1966-71. Mem. Am. Soc. Microbiology, AAAS. Subspecialties: Virology (biology); Virology (medicine). Current work: Management of grants and contracts on basic medical research. Home: 2212 Lomond Ct Vienna VA 22180 Office: NIH Nat Inst General Med Scis Bethesda MD 20205

WOLFF, FREDERICK WILLIAM, medical scientist, educator, researcher, consultant; b. Berlin, Aug. 21, 1920; U.S., 1959, naturalized, 1967; s. Bruno and Elizabeth (Landau) W.; m. Catherine M. Chura, Feb. 14, 1967; children: Susan, Peter, Catherine. M.B., B.S., U. Durham, Eng., 1947, M.D., 1957. Resident, chief resident teaching hosps. of Newcastle and London, 1946-58; asst. prof. Johns Hopkins Hosp., Balt., 1959-64; prof. medicine Sch. Medicine, George Washington U., Washington, 1965—; pres. Inst. for Drug Devel., Washington, 1979—; vis. scientist Nat. Heart Inst., NIH, 1982—; cons. in field. Contbr. numerous articles to sci. and med. jours. Served with RAF, 1950-52. Named hon. diplomate Georgian Acad. Scis., 1981. Mem. Am. Diabetes Assn., Am. Heart Assn., Soc. Pharmacology and Exptl. Therapeutics, Am. Soc. for Clin. Pharmacology, Brit. Pharm. Soc., Brit. Med. Assn., Am. Fedn. for Clin. Research, AAAS, AAUP, UN Assn. (bd. dirs. Nat. Capital area 1982—). Club: Cosmos (Washington). Subspecialties: Pharmacology; Endocrinology. Current work: New drug development; organization of interdisciplinary pharmaceutical research. Home and Office: 800 Notley Rd Silver Spring MD 20904

WOLFF, GEORGE LOUIS, biologist, researcher; b. Hamburg, Germany, Aug. 24, 1928; came to U.S., 1940, naturalized, 1946; s. Adolf and Eva (Nathan) W.; m. Eleanor Herstein, Aug. 30, 1953; children: David B., Adrienne A. B.S cum laude, Ohio State U., 1950; Ph.D., U. Chgo., 1954. Postdoctoral fellow Nat. Cancer Inst., 1954-56, biologist, 1956-58; research assoc., supr. animal colony Inst. Cancer Research, Phila., 1958-63, asst. mem., supr. animal colony, 1963-68, asst. mem., geneticist, 1968-72; chief mammalian genetics br. Nat. Center Toxicol Research, 1972-73, chief div. mutagenic research, 1973-79; sr. sci. coordinator genetics, research biologist Nat. Center Toxicol. Research, 1979—; coordinator Interdisciplinary Toxicology Program, 1982—; adj. asst. prof. dept. biochemistry U. Ark. for Med. Scis., 1973-81, adj. assoc. prof. dept. biochemistry and div. toxicology dept. pharmacology, 1981—. Contbr. articles to sci. publs. Mem. Genetics Soc., Am. Genetic Assn., Soc. Toxicology, Am. Assn. Cancer Research, Soc. Exptl. Biology and Medicine, Environ. Mutagen Soc. Subspecialties: Animal genetics; Toxicology (medicine). Current work: Definition of metabolic bases for differential susceptibility of inbred strains, F-1 hybrids and mutant experimental animals to induction of toxicologic endpoints. Characterization of inbred strains, etc. for metabolic characteristics of importance in in vivo toxicologic assays; design and testing of improved toxicologic assays. Office: Nat Center Toxicological Research Jefferson AR 72079

WOLFF, JAMES ALEXANDER, physician, emeritus educator; b. N.Y.C., June 19, 1914; s. William F. and Blanch R. W.; m. Janet Wolff, June 24, 1946; children: James A., John, Barbara, Timothy. B.A., Harvard U., 1935; M.D., N.Y.U., 1940. Diplomate: Am. Bd. Pediatrics. Prof. pediatrics Coll. Physicians and Surgeons, Columbia U., N.Y.C., 1972-82, prof. emeritus, 1982—; dir. Valerie Fund Children's Ctr., N.Y.C., 1981—. Contbr. articles to profl. jours. Served to capt., M.C. U.S. Army, 1942-45. Decorated Silver Star. Mem. Soc. Pediatric Research, Am. Pediatric Soc., Am. Soc. Pediatric Hematology and Oncology, Internat. Soc. Pediatric Oncology. Subspecialties: Cancer research (medicine); Hematology. Current work: Research on childhood cancer, disorders of the blood. Office: Valerie Childrens Center Overlook Hosp Summit NJ 07901

WOLFF, PETER ADALBERT, physicist, educator; b. Oakland, Calif., Nov. 15, 1923; s. Adalbert and Ruth Margaret W.; m. Catherine C. Carroll, Sept. 11, 1948; children: Catherine Mia, Peter Whitney. A.B. in Physics, U. Calif., Berkeley, 1945, Ph.D., 1951. Research scientist Lawrence Radiation Lab., 1951-52; staff scientist Bell Telephone Lab., Murray Hill, N.J., 1952-63, dept. head, dir. electronic research lab., 1964-70; prof. physics U. Calif., San Diego, 1963-64; prof. physics, head solid state and atomic physics div., asso. dir. material sci. center M.I.T., Cambridge, 1970-76, prof. physics, 1976—, dir. research lab. of electronics, 1976-81; dir. Francis Bitter Nat. Magnet Lab., 1981—; dir. Draper Lab. Contbr. articles to profl. jours. Served with C.E. U.S. Army, 1945-46. Mem. Am. Phys. Soc., Am. Acad. Arts and Scis. Subspecialty: Theoretical physics. Office: MIT 77 Massachusetts St Cambridge MA 02139

WOLFF, ROGER GLEN, hydrologist; b. Eureka, S.D., Sept. 7, 1932; s. Otto and Bertha Thurn W.; m. Barbara A. Mutschler, Sept. 23, 1951 (div. 1955); m. Mary R. Varga, Aug. 1, 1959; children: Steven C., Mark R. Student, No. State Tchrs. Coll., Aberdeen, S.D., 1950-51, 56; B.S., S.D. Sch. Mines, 1958; M.S., U. Ill.-Urbana, 1960, Ph.D., 1961. Research project chief U.S. Geol. Survey, Washington, 1961-75, regional research hydrologist, Reston, Va., 1975-78, research project chief, 1979-80, dep. asst. chief hydrologist for research, 1980—. Treas. Walnut Woods Citizens Assn., Rockville, Md., 1972, pres., 1973. Served with U.S. Army, 1953-55. Calif. Co. fellow, 1958-59; Nat. Lead Co. fellow, 1959-61. Fellow Geol. Soc. Am., Pinedale Research Inst.; mem. Am. Geophys. Union, N.Am. Clay Mineral Soc., Geol. Soc. Washington (councillor 1980-82), Sigma Tau. Subspecialties: Hydrogeology; Ground water hydrology. Current work: Hydraulic conductivity of confining layers; subsurface waste disposal, research program management. Home: 11909 Hitching Post Ln Rockville MD 20852 Office: US Geol Survey MS 413 Reston VA 22092

WOLFF, SHELDON, radiobiologist, educator; b. Peabody, Mass., Sept. 22, 1928; s. Henry Herman and Goldie (Lipchitz) W.; m. Frances Faye Farbstein, Oct. 23, 1954; children: Victor Charles, Roger Kenneth, Jessica Raye. B.S. magna cum laude, Tufts U., 1950; M.A., Harvard U., 1951, Ph.D., 1953. Teaching fellow Harvard U., 1951-52; sr. research staff biology div. Oak Ridge Nat. Lab., 1953-66; prof. cytogenetics U. Calif., San Francisco, 1966—, dir. Lab. Radiobiology and Environ. Health, 1983—; Vis. prof. radiation biology U. Tenn., 1962, lectr., 1953-65; cons. several fed. sci. agys. Editor: Chromosoma, 1983—; assoc. editor: Cancer Research, 1983—; Editorial bd.: Radiation Research, 1968-72, Photochemistry and Photobiology, 1962-72, Radiation Botany, 1964—, Mutation Research, 1964—, Caryologia, 1967—, Radiation Effects, 1969-81, Genetics, 1972—; Contbr. articles to sci. jours. Recipient E.O. Lawrence meml. award U.S. AEC, 1973. Mem. Genetics Soc. Am., Radiation Research Soc. (counselor for biology 1968-72), Am. Soc. Naturalists, Am. Soc. Cell Biology, Environmental Mutagen Soc. (council 1972—, pres.-elect 1979, pres. 1980-81), Internat. Assn. Environ. Mutagen Socs. (treas. 1978—), Sigma Xi. Democrat. Subspecialties: Genome organization; Cell and tissue culture. Current work: Radiation mutagenesis, enviornmental mutagenesis, cytogenetics. Home: 41 Eugene St Mill Valley CA 94941 Office: Lab Radiobiology U Calif San Francisco CA 94143

WOLFF, SIDNEY CARNE, astronomer; b. Sioux City, Iowa, June 6, 1941; d. George Albert and Ethel Larose (Smith) Carne; m. Richard James Wolff, Oct. 7, 1940. B.A., Carleton Coll., 1962; Ph.D., U. Calif-Berkeley, 1966. Astronomer Inst. Astronomy, U. Hawaii, Honolulu, 1967-82, assoc. dir., 1976—. Mem. Am. Astron. Soc., Astron. Soc. Pacific, Internat. Astron. Union. Subspecialty: Optical astronomy. Office: 2680 Woodlawn Dr Honolulu HI 96822

WOLFSON, PAUL MARTIN, cardiologist, educator; b. Atlantic City, Dec. 20, 1943; s. Jack and Sadie (Finkelstein) W.; m. Marlene I. Rosenberg, July 7, 1968; children: Jack Michael, Eric Marc, Dana Joy. B.S., Muhlenberg Coll., 1965; D.O., Chgo. Coll. Osteo. Medicine, 1970. Diplomate: Am. Bd. Internal Medicine. Intern Cleve. Clinic, 1970-71, resident in internal medicine, 1971-72; fellow in cardiology U. Iowa Hosps., 1972-74; asst. prof. medicine Chgo. Coll. Osteo. Medicine, 1974-77, assoc. prof., 1977-83, prof., 1983—; med. dir. Health Care and Rehab. Ctr., Chgo., 1981-82; mem. com. Chgo. Heart Assn., 1979—; mem. adminstrv. group Chgo. Cardiology Group, 1981—. Contbr. articles to profl. jours.; author, editor: self-instructional videotapes Doppler Ultrasound, 1977; author sci. exhibit, 1976 (cert. merit). Chgo. Coll. Osteo. Medicine grantee, 1975-83; pharm. cos. grantee, 1975-83; Am. Heart Assn. grantee, 1982-83. Fellow Am. Coll. Cardiology, ACP, Am. Coll. Angiology, Council Clin. Cardiology, Inst. Medicine Chgo. Subspecialty: Cardiology. Current work: NMR evaluation of K^+ and Mg^+ cardioplegia with Ca blockers; noninvasive diagnosis of cardiovascular disease; noninvasive diagnosis of extra-cranial cerebral vascular disease; thrombolytic therapy; percutaneous transluminal angioplasty; newer pharmacologic agents. Office: Chgo Osteo Med Ctr 5200 S Ellis Ave Chicago IL 60615

WOLGIN, DAVID L., psychology educator; b. Elizabeth, N.J., Oct. 17, 1945; s. Richard and Alma (Cohen) W.; m. Maxine Dale, July 3, 1968. B.A., Rutgers U., 1967, Ph.D., 1973; M.A., Vanderbilt U., 1968. Postdoctoral fellow U. Pa., 1973-74; research assoc. U. Ill., 1974-75; asst. prof. psychology Fla. Atlantic U., Boca Raton, 1975-80, assoc. prof., 1981—. Contbr. articles to profl. jours. Mem. Soc. Neurosci., Eastern Psychol. Assn., Sigma Xi. Subspecialties: Physiological psychology; Psychopharmacology. Current work: Recovery from brain damage, behavioral tolerance to drugs. Office: Dept Psychology Fla Atlantic U Boca Raton FL 33431

WOLINSKY, JERRY SAUL, neurology educator, neurovirologist; b. Balt., Nov. 26, 1943; s. Morris and Anne (Smith) W.; m. Gerlind Stahler, Jan. 18, 1969; children: Anja Kerstin, Jean-Paul. Student, Ill. inst. Tech., 1962-65; M.D., U. Ill., 1969. Diplomate: Nat. Bd. Med. Examiners, Am. Bd. Psychiatry and Neurology. Resident in neurology U. Calif.-San Francisco, 1970-73; research assoc. VA Hosp., San Francisco, 1975-78; asst. prof. neurology U. Calif.-San Francisco, 1975-78; assoc. prof. neurology Johns Hopkins U. Sch. Medicine, Balt., 1978-83, assoc. prof. immunology and infectious diseases, 1981-83; prof. neurology U. Tex. Health Sci. Center at Houston, 1983—. Contbr. over 100 articles to sci. jours. Served to capt. USAR, 1970-77. David M. Olkon scholar, 1968; Basil O'Connor Research fellow Nat. Found.-March of Dimes, 1975-78; NIH Research Career Devel. awardee, 1979-84. Fellow Am. Acad. Neurology; mem. Am. Neurol. Assn., Am. Assn. Neuropathologists, Am. Soc. Microbiology, Alpha Omega Alpha. Jewish. Subspecialties: Neurology; Virology (medicine). Current work: Acute and chronic viral infection of central nervous system; mechanisms of virulence of viral infections of brain; role of immune complexes in neural damage in chronic rubella virus infections of man; treatment of viral encephalitis. Home: 3311 Rice Blvd Houston TX 77005 Office: PO Box 20708 Room 7044 Houston TX 77025

WOLIVER, ROBERT EDWARD, university president, clinical psychologist; b. Hanalei, Kauai, Hawaii, May 29, 1947; s. Edward C. and Esther C. (Cottingham) W.; m.; 1 child, Keiko. B.S., Georgetown U., 1969; M.A., U. Hawaii, 1976, Ph.D., 1979. Lic. psychologist, Hawaii. Prof. Chaminade U., Honolulu, 1979-80, U. Hawaii, 1980-81; pres. Hawaii Sch. Profl. Psychology, Honolulu, 1980—, pvt. practice clin. psychology, 1980—; cons. psychologist Queen's Med. Ctr., Mental Health, Honolulu, 1978-83. Contbg. author: Functional Psychological Test, 1983; contbr. articles to profl. jours. Mem. Am. Psychol. Assn., Western Psychol. Assn., Hawaii Psychol. Assn. Club: Hawaii Yacht (Honolulu). Subspecialties: Social psychology; Clinical psychology. Current work: Research in area of mate selection and intercultural relations. Office: Hawaii Sch Profl Psychology 2424 Pali Hwy Honolulu HI 96817

WOLLIN, MYRON, radiation physicist; b. N.Y.C., May 6, 1940; s. Sol and Lillian (Mendick) W.; m. Carol Miriam Goodsten, June 19, 1961; children: Jennifer Eva, Gordon Sol; m. Cynthia Sue Gordon, May 2, 1982. M.S., Columbia U., 1964. Diplomate: Am. Bd. Radiology. Researcher Sloan Kettering Inst., N.Y.C., 1964-65; radiation physicist Univ. Hosp., Ann Arbor, Mich., 1965-66, Cedars of Lebanon Hosp., Los Angeles, 1966-72; med. radiation physicist So. Calif. Permanente Med. Group, Los Angeles, 1972—. Contbr. articles to profl. jours. Mem. Am. Assn. Physicists in Medicine, Health Physics Soc. Los Angeles, Am. Soc. Therapeutic Radiologists, Am. Coll. Radiology, So. Calif. Fedn. Scientists. Subspecialty: Medical physics. Current work: Radiation therapy physics, clinical radiobiology, computers in medicine. Office: 1510 N Edgemont St Los Angeles CA 90027 Home: 4448 Keever Ave Long Beach CA 90807

WOLLMAN, HARRY, medical educator; b. Bklyn., Sept. 26, 1932; s. Jacob and Florence Roslyn (Hoffman) W.; m. Anne Carolyn Hamel, Feb. 16, 1957; children—Julie Ellen, Emily Jane, Diana Leigh. A.B. summa cum laude (hon. John Harvard scholar 1950-53, hon. Harvard Coll. scholar 1953-54, Detur award 1951), Harvard, 1954, M.D., 1958. Diplomate: Am. Bd. Anesthesiology. Intern U. Chgo. Clinics, 1958-59; resident U. Pa., 1959-63, asso. in anesthesia, 1963-65, mem. faculty, 1965—, prof. anesthesia, 1970—, prof. pharmacology, 1971—, Robert Dunning Dripps prof., chmn. dept. anesthesia, 1972—; prin. investigator Anesthesia Research Center, 1972-78; program dir. Anesthesia Research Tng. Grant, 1972—; Mem. anesthesia drug panel, drug efficacy study, com. on anesthesia Nat. Acad. Scis.-NRC, 1970-71, com. on adverse reactions to anesthesia drugs, 1971-72; mem. pharm. and toxicology tng. grants com. NIH, 1966-68, anesthesia tng. grants com., 1971-73, surgery, anesthesia and trauma study sect., 1974-78; chmn. com. on studies involving human beings U. Pa., 1972-76, chmn. clin. practice exec. com., 1976-80. Asso. editor for revs.: Anesthesiology, 1970-75; Contbr. and editor books. NIH research traineeship fellow, 1959-63; Pharm. Mfg. Assn. fellow, 1960-61. Mem. Pa. Soc. Anesthesiologists (pres. 1972-73), Am. Physiol. Soc., Assn. U. Anesthetists (exec. council 1971-74, chmn. sci. adv. bd. 1975-77), Soc. Acad. Anesthesia Chairmen (chmn. com. on financial resources 1973-77, pres.-elect 1976-77, pres. 1977-78), Am. Soc. Anesthesiologists, Phila. Soc. Anesthesiologists, AMA, Pa. Med. Soc., Phila. County Med. Soc., John Morgan Soc., Coll. Physicians Phila., Phi Beta Kappa, Sigma Xi. Republican. Unitarian. Subspecialties: Anesthesiology; Critical care. Current work: Anesthesiology and critical care. Home: 2203 Delaney Pl Philadelphia PA 19103 Office: Dept Anesthesia Hosp University Pa 3400 Spruce St Philadelphia PA 19104

WOLMAN, M. GORDON, geography educator; b. Balt., Aug. 16, 1924; s. Abel and Anna (Gordon) W.; m. Elaine Mielke, June 20, 1951; children: Elsa Anne, Abel Gordon, Abby Lucille, Fredericka Jeannette. Student, Haverford Coll.; A.B. in Geology, Johns Hopkins U., 1949, M.A., Harvard U., 1951, Ph.D., 1953. Geologist U.S. Geol. Survey, 1951-58, part-time, 1958—; assoc. prof. geography Johns Hopkins U., Balt., 1958-62; prof. Johns Hopkins, 1962—; chmn. dept. geography and environ. engring., 1958—; Mem. adv. com. geography U.S. Office Naval Research; mem. exec. com. Div. Earth Sci., NRC; mem. com. internat. environ. programs, mem. environ. studies bd.; mem. com. water Nat. Acad. Sci.; mem. exec. com. Earth Sci. div., Nat. Acad. Scis., mem. com. mineral resources and environment, chmn. nat. commn. water quality policy; cons. in field to, City of Balt., Balt. County, State of Md.; Mem. Balt. City Charter Revision Commn.; mem. Community Action Com., Balt. Author: Fluvial Processes in Geomorphology, 1964; Editorial bd.: Science mag. Trustee Park Sch., Balt., Md. Acad. Scis.; v.p. bd. trustees Sinai Hosp., Balt.; pres. bd. dirs. Resources for Future. Served with USNR, 1943-46. Recipient Meritorious Contribution award Assn. Am. Geographers, 1972. Fellow Am. Acad. Arts and Scis.; mem. Am. Geophys. Union (chmn. subcom. sedimentation, pres. hydrol. sect.), Geol. Soc. Am. (v.p. 1983), Washington Geol. Soc., ASCE, Agrl. Hist. Soc., Am. Geog. Soc. (councilor 1965-70), Assn. Am. Geographers, Phi Beta Kappa, Sigma Xi, Md. Acad. Scis. (mem. exec. com.). Subspecialties: Geology; Hydrology. Current work: Geomorphology, allurial river channel processes, hydrology, quaternary geology, physical geography, environmental change and natural processes, energy and environmental policy. Home: 2104 W Rogers Ave Baltimore MD 21209

WOLMAN, SANDRA R., pathologist, geneticist, educator, researcher; b. N.Y.C., Nov. 23, 1933; d. Alexander J. and Sophie (Raffel) Rosman; m. Alan J. Bell, June 10, 1956; m. Eric Wolman, July 27, 1963; children: Karin, Alec. B.A. cum laude, Radcliffe Coll., 1955; M.D., N.Y.U., 1959. Diplomate: Am. Bd. Med. Genetics. Intern Bellevue Hosp., N.Y.C., 1959-60, resident, 1960-63; USPHS fellow; instr. N.Y.U. Med. Ctr., 1962-64; asst. pathologist Morristown (N.J.) Meml. Hosp., 1964-66, Monmouth (N.J.) Med. Ctr., 1966-67; asst. prof. pathology N.Y.U. Sch. Medicine, 1967-76, assoc. prof., 1976—; dir. cytogenetics VA, 1967—; mem. staff Bellevue Hosp., Univ. Hosp.; cons. NIH, mem. pathology B study sect., 1976-80; disting. lectr. dept. community medicine Rutgers U., 1980. Research numerous publs. in field; editorial reviewer: for various jours., including Cancer Research. Mem. adv. bd. Women's Center, Brookdale Community Coll., 1975—; mem. bd. Planned Parenthood Monmouth County, 1980—, chmn. nominating com., 1981, 82. Recipient Merit award Monmouth County Bd. Chosen Freeholders, 1976. Mem. Am. Assn. Cancer Research, Am. Assn. Pathologists, Am. Soc. Clin. Pathologists, Am. Soc. Human Genetics, AAAS, Environ. Mutagen Soc., Tissue Culture Assn., Harvey Soc., Somatic Cell Genetics, Internat. Soc. Exptl. Hematology. Clubs: Radcliffe (N.Y.C.); Sea Bright Beach. Subspecialties: Cancer research (medicine); Genetics and genetic engineering (medicine). Current work: Cytogenetics of cancer cells; differentiation of leukemic stem cells; mutation testing. Office: 550 1st Ave New York NY 10016

WOLTER, KIRK MARCUS, mathematical statistician, educator; b. Evanston, Ill., Nov. 28, 1948; s. Alfred E. and Joyce M. (Neslow) W.; m. Mary Jane Vanderford, Mar. 28, 1981; children: Nicole, Alexander, Asilinn. B.A., St. Olaf Coll., Minn., 1970; M.S., Iowa State U., 1972, Ph.D., 1974. Math. statistician Research Ctr. Measurement Methods, U.S. Bur. Census, Washington, 1974-78, asst. div. chief statis. research, 1978-81, sr. math. statistician, 1981-83; chief statis. research div., 1983—; assoc. professorial lectr. George Washington U., Washington, 1975-81, professorial lectr., 1981—; cons. Author: Introduction to Variance Estimation, 1983, also articles. Recipient George W. Snedecor award Iowa State U., 1972, Outstanding Young Alumnus award, 1980; Bronze medal U.S. Dept. Commerce, 1982. Fellow Am. Statis. Assn.; mem. inst. Math. Stats., Internat. Assn. Survey Statisticians. Lutheran. Subspecialties: Statistics; Mathematical software. Current work: Variance estimation for complex sample surveys; dual system estimation or capture-recapture methodology; estimation and adjustment for undercount error in the U.S. decennial census; measurement error in economic data. Home: PO Box 184 Owings MD 20736 Office: US Bur Census Room 3065 FB #3 Washington DC 20233

WOLTHUIS, ROGER ALLEN, physiologist; b. Champaign, Ill., Mar. 30, 1937; s. Enno and Jessie (Zylstra) W.; m. Donna Jean Beutema, May 13, 1965; children: Susan, Janet. B.A., U. Mich., 1963; M.S., Mich. State U., 1965, Ph.D., 1968; postgrad., U. Tex.-San Antonio. Prin. scientist Tech., Inc., Houston, 1968-74; researcher U.S. Air Force Sch. Medicine, San Antonio, 1974-79; sr. scientist Medtronic, Inc., Mpls., 1979-81; dir. research and devel. Squibb Med. Systems, Bellevue, Wash., 1981-82; v.p. Tech. Dynamics, Inc., Woodinville, Wash., 1982—. Contbr. articles to profl. jours.; patentee in field. Served with USAF, 1956-60. Recipient Skylab Achievement award NASA, 1974, Cert. of Recognition, 1973, 76; USAF Incentive award, 1977. Fellow Am. Coll. Cardiology, Aerospace Med. Assn.; mem. IEEE., Am. Physiol. Soc., Am. Heart Assn. Subspecialties: Bioinstrumentation; Biomedical engineering. Address: 15430 NE 162 St Woodinville WA 98072

WOMACK, JAMES E., geneticist; b. Anson, Tex., Mar. 30, 1941; s. C.E. and Eva M. (Hollums) W.; m. Raby J. Beakley; children: Wendy Anne, James Michael. B.S., Abilene Christian U., 1963; Ph.D., Oreg. State U., 1968. Faculty Abilene Christian U., Tex., 1968-73; staff Jackson Lab., Bar Harbor, Maine, 1973-77; assoc. prof. vet. pathology Tex. A&M U., College Station, 1977—. Contbr. articles to profl. jours. Mem. AAAS, Am. Soc. Human Genetics, Am. Genetics Assn., Genetics Soc. Am. Subspecialties: Genetics and genetic engineering (biology); Gene actions. Current work: Genetic engineering of farm animals. Office: Texas A&M U Dept Veterinary Pathology College Station TX 77843

WOMACK, JOSEPH DONALD, SR., manufacturing company executive, consultant; b. Ft. Worth, Aug. 15, 1926; s. Walter William and Hortense (McDonald) W.; m. Teresa Rose Parrette, June 1951; 1 son, Joseph Donald; m. Lewis Brennan O'Brien, Nov. 29, 1959; 1 dau., Courtney Cralle; m. Marie Annette Thomas, Jan. 12, 1974. Student, U. Tex., 1943, Sophia U., Tokyo, 1952-53, Am. U., 1954-55. Fgn. service officer, Seoul, Tokyo, Beirut, 1951-63; exec. IBM World Trade, 1964; pres. Fed. Mktg. Services, Washington, 1965-69, AVMedia, Inc., Alexandria, Va., 1969-77; pres., chief exec. officer Solactor Corp., Alexandria, 1977—; cons. OEO, 1964, Army Chief Staff, 1969. Served with U.S. Maritime Service and U.S. Mcht. Marines, 1944-48; PTO. Mem. Internat. Solar Energy Soc., U.S. Solar Energy Soc., Solar Energy Industries Assn. Subspecialties: Solar energy; Electrical engineering. Current work: Development of high performance, tracking solar radiation concentrators for coupling to advanced external combustion engine/generators and/or furnaces for use in distillation processes, industrial process heat, remote electrical power, etc. Home: 5104 Remington Dr Alexandria VA 22309 Office: PO Box 6026 Alexandria VA 22306

WONG, CLEMENT PO-CHING, nuclear engineer, researcher; b. Hong Kong, Mar. 20, 1944; U.S., 1963; s. Kai Ng and Kum (Sin) Shum; m. Siu-Yin Lee, June 24, 1972; 1 son, Dong Ping. B.S., U. Wis., 1967, M.S., 1969; Ph.D., U. Tex., 1977. Sr. engr. GA Technologies, Inc., San Diego, 1977—. Contbr. numerous articles to profl. jours. Mem. Am. Nuclear Soc., Am. Phys. Soc., AAAS. Subspecialties: Nuclear engineering; Nuclear fusion. Current work: Fusion reactor blanket design, thermal-hydraulics, tritium considerations, hybride blanket, power conversion. Home: 7116 Cather Ct San Diego CA 92122 Office: GA Technologies Inc PO Box 81608 San Diego CA 92138

WONG, DAVID T., biochemist; b. Hong Kong, Nov. 6, 1935; s. Chi-Keung and Pui-King W.; m. Christina lee, Dec. 28, 1963; children: Conrad, Melvin, Vincent. Student, Nat. U. Taiwan, 1955-56; B.S., Seattle Pacific Coll., 1960; M.S., Oreg. State U., 1964; Ph.D., U. Oreg., 1966. Postdoctoral fellow U. Pa., Phila., 1966-68; sr. biochemist Lilly Research Labs., Indpls., 1968-72, research biochemist, 1973-77, research asso., 1978—. Contbr. numerous articles to sci. jours. Charter mem. Chinese Christian Fellowship, Indpls., 1969, Chinese Community Ch., Indpls., 1973; charter mem. Indpls. Assn. Chinese Ams., 1974; deacon 1st Presby. Ch. of Soutport, Ind., 1981—. Mem. Am. Soc. Pharmacology and Exptl. Therapeutics, Internat. Soc. Neurochemistry, Am. Soc. Neurochemistry, Biophys. Soc., Soc. Neurosci., N.Y. Acad. Scis., Sigma Xi. Republican. Subspecialties: Neurochemistry; Neuropharmacology. Current work: Biochemistry and pharmacology of neurotransmission; discovery of the first selective inhibitor of serotonin uptake and development of new type of antidepressive drug, Fluoxetine and a selective inhibitor of norepinephrine uptake, Tomoxetine; studies of potentially useful substances which activate transmission of norepinephrine, dopamine, serotonin, acetylcholine and GABA-neurons; studies of natural products led to the discovery of carboxylic ionophores: Narasin, A28695 and A204, which increase transport of cations across biomembranes. Office: Lilly Research Labs Indianapolis IN 46285

WONG, EDWARD CHOR-CHEUNG, computer engineer; b. Hong Kong, Jan. 16, 1952; s. Kan-Wen and Wai-Ying (Ip) W.; m. Rosaline Kowk-Chun, Aug. 4, 1973; children: Duane, Fun-wah, Edward Chiu-Wah, Elliott Chun-Wah. B.A., Fordham U., 1973; B.S., Columbia U., 1974, M.S., 1974, Computer Systems Engr., 1980. With IBM, Poughkeepsie, N.Y., 1974—, staff engr., 1979-82, devel. engring. mgr., 1982—; Adj. lectr. Columbia U., 1982; adj. prof. Marist Coll., 1983. Mem. IEEE, Assn. Computing Machinery, Soc. Indsl. and Applied Math. Democrat. Subspecialties: Computer architecture; Logic design. Current work: Error detection fauly isolation technique, with incorporation into design automation; algorithms, cyrptology and switching circuit theory application on designs. Patentee cache control for concurrent access. Home: 109 E Cedar St Poughkeepsie NY 12601 Office: IBM PO Box 390 Poughkeepsie NY 12602

WONG, JULIUS PAN, engineering educator, consultant; b. Shanghai, China, May 8, 1937; came to U.S., 1960, naturalized, 1973; s. Kai S. and Po L. (Hsu) W. B.S., Hong Kong Bap. Coll., 1960; M.S., La. Tech. U., 1962; Ph.D., Okla. State U.-Stillwater, 1966. Analytical specialist Bendix Corp., South Bend, Ind., 1966-69; assoc. prof. engring. U. Louisville, 1970-80, prof., 1980—. Contbr. articles to profl. jours. Recipient Outstanding Faculty award Tau Beta Pi, 1976; Outstanding Tchr. award, 1982. Mem. ASME, Am. Soc. Engring. Edn., Am. Acad. Mechanics, Soc. Indsl. and Applied Math., AAUP. Subspecialties: Theoretical and applied mechanics; Mechanical engineering. Current work: Stress analysis, optimal design, computer-aided engineering. Office: U Louisville Louisville KY 40292

WONG, MARTIN R., psychologist; b. Trenton, N.J., May 26, 1936; s. Ivan Yau Sun and Janet (Middagh) W.; m. Joyce M. Bredahl, June 12, 1967 (div. 1972); 1 son, Thaddeus J.R.; m. Gretchen S. Barbatsis, Aug. 11, 1974. A.A., Los Angeles City Coll., 1956; B.A., San Jose State Coll., 1960; M.A., Mich. State U., 1967, Ph.D., 1969. Lic. psychologist, Minn., Nebr., Mich. Asst. prof. psychology U. South Fla., Tampa, 1969-72, U. Minn., Mpls., 1972-75; assoc. prof. psychology U. Nebr., Omaha, 1975-77, Peabody Coll., Vanderbilt U., Nashville, 1977-79; postdoctoral fellow Mich. State U., East Lansing, 1979-81; psychologist VA Med. Center, Battle Creek, Mich., 1981—; Mem. adv. bd. Asian Am. Psychol. Assn., Bloomington, Ind., 1982—. Author: Systematic Instructional Design, 1974, Why Drugs: The Psychology of Drug Abuse, 1974; co-editor: The Psychology of School Learning, 1974; editor: Asian Am. Psychol. Assn. Jour, 1982—. Served with U.S. Army, 1956-59. Recipient various grants. Mem. Am. Psychol. Assn., Am. Ednl. Research Assn. (chmn. spl. interest groups). Democrat. Unitarian. Subspecialties: Cognition; Learning. Current work: Research in male psychology, androgyny, sex-role development. Home: 1623 Colorado Dr East Lansing MI 48823 Office: Psychology 116B VA Med Center Battle Creek MI 48823

WONG, MING MING, parasitology educator and researcher, consultant; b. Singapore, Jan. 3, 1928; d. Siong-Cie and Mary S.S. (Hsu) W. B.S., Wilmington Coll, 1952; M.S., Ohio State U., 1953; Ph.D., Tulane U., 1963. Med. technician Am. Soc. Clin. Pathologists. High sch. biology tchr., 1955-56; instr. microbiology, faculty medicine Hong Kong U., 1956-59; research fellow Tulane U., 1964-65; lectr. in parasitology, faculty medicine, Malaya, 1965-67; research parasitologist Calif. Primate Research Ctr., U. Calif., Davis, 1967-78; pres. Paradiagnostics, Inc., Davis, 1977-81; assoc. prof. parasitology Sch. Vet. Medicine, U. Calif., Davis, 1980-82, prof., 1982—; cons. canine heartworm serodiagnosis. Contbr. articles to publs. Recipient dissertation prize Sigma Xi, 1963; grantee NIH, WHO, others. Mem. Royal Soc. Tropical Medicine and Hygiene, Am. Soc. Parasitology, Am. Soc. Tropical Medicine and Hygiene, Am. Heartworm Soc. (charter). Subspecialties: Parasitology; Infectious diseases. Current work: Canine heartworm immunology; serodiagnosis of bovine cysticerosis, primate parasites. Developer IFS-mf slide test for occult dirofilariasis. Office: Dept Vet Medicine and Immunology Univ Calif Davis CA 95616

WONG, PATRICK YUI-KWONG, biochemical and pharmacology educator; b. Kiangsai, China, Nov. 25, 1944; came to U.S., 1969, naturalized, 1983; s. Kuan Kuen and Bai Chei (Cheng) W.; m. Patsy H-W Chao, Sept. 26, 1969. B.Sc., Nat. Taiwan Normal U., 1967; Ph.D., U. Vt., 1974. Postdoctoral fellow Med. Coll. Wis., Milw., 1974-75; instr. pharmacology U. Tenn. Med. Center, Memphis, 1975-78, asst. prof., 1978-79; assoc. prof. pharmacology N.Y. Med. Coll., Valhalla, 1979—. Contbr. chpts. to books, articles to profl. jours. Recipient Young Investigator award NIH, 1978-80, Research Career Devel. award, 1981—; spl. dental award Nat. Inst. Dental Research, 1978-81; NIH grantee, 1980—. Mem. Am. Soc. Biol. Chemists, Am. Soc. Pharmacology and Exptl. Therapeutics, Am. Heart Assn. (high blood pressure council, thrombosis council). Subspecialties: Biochemistry (medicine); Cellular pharmacology. Current work: Relationship of prostaglandins in hypertension; prostaglandins, leukotrienes; calcium and calmodulin; hypertension and thrombosis. Home: 2307 William Ct Yorktown Hwights NY 10598 Office: Dept Pharmacology NY Med Coll Valhalla NY 10595

WONG, PO KEE, mathematician; b. Canton City, China, May 5, 1934; came to U.S., 1959, naturalized, 1971; s. K. F. and W. C. (Lam) W.; m. Ruby Ching, Aug. 18, 1965; children: Adam, Anita. B.Sc., Cheng-Kung U., 1956; Engr., U. Utah, 1961; M.M.E., Calif. Inst. Tech., 1966; Ph.D., Stanford U., 1970. Registered profl. engr., Taiwan.; cert. tchr., Mass. Teaching asst. Cheng-Kung U., 1958-59; tchr. math. and sci. Hong Kong YMCA Coll., 1959; research/teaching asst. U. Utah, 1959-61, Calif. Inst. Tech., Pasadena, 1961-65; sr. scientist Lockheed Missiles & Space Co., Palo Alto, Calif., 1966-68; research asst. Stanford U., 1968-70; instr., researcher U. Santa Clara, 1970-71; staff Moffet Field, Calif. Ames Center, NASA, 1970; engr. I, dept. breeder reactors Gen. Electric Co., Sunnyvale, Calif., 1972-73; specialist engr. Nuclear Service Co., Campbell, Calif., 1973; engr. Stone & Webster Engring. Co., Boston, 1974; tchr. math. and sci. Boston Pub. Schs., 1979—; pres., chief exec. officer Systems Research Co., Brookline, Mass., 1976—. Reviewer: Applied Mechanics Rev, 1972—; contbr. articles to profl. jours. Mem. ASME, Am. Acad. Mechanics, Sigma Xi. Subspecialties: Theoretical physics; Applied mathematics. Current work: Initiated trajectory solid angle to solve the P2 problem of statis. mechanics, three dimensional unsteady stream functions, magnetovisco-elastodynamics; formulated phys. econ. model via solution of indeterminate structure problem; provided formulation and solution of multireservoir transient problem. Address: 238 Cypress St S3 Brookline MA 02146

WONG, ROBERT KING-SUEN, neuroscientist; b. Chungsha, Hunan, China., May 25, 1946; came to U.S., 1975; s. Kai and Edith (Wu) Wang; m. Ruth Lo-Nor Wong, Mar. 30, 1974; 1 son, Andrew. B.S., McGill U., 1970; Ph.D., U. Alta., 1975; postdoctoral student, Stanford U., 1975-79. Asst. prof. La. State U., Shreveport, 1979-81; asst. prof. U. Tex. Med. Br., Galveston, 1981-83, assoc. prof. dept. physiology, 1983—. Contbr. chpt. to book, articles to profl. jours. Research grantee NIH, 1980—, Am. Heart Assn., 1981—; neurosci. fellow Ester and John Klingenstein Found., 1981—. Mem. Soc. for Neurosci., Am. Physiol. Soc. Subspecialties: Physiology (medicine); Neurophysiology. Current work: Basic mechanism of epilepsy using electrophysiological studies on In Vitro brain slices; biophysics of mammalian brain cells using individual neurons isolated from the cerebral cortex. Home: 16451 Parksley Houston TX 77059 Office: Univ Tex Med Br Galveston TX 77550

WONG, STEWART, pharmacologist; b. Toronto, Ont., Can., Jan. 2, 1930; came to U.S., 1960; s. Sai Moy and Mabel (Mar) W.; m.; children: John, Clifford. B.A., U. Toronto, 1958; M.A. in Physiology, 1960; Ph.D. in Biochem. Pharmacognosy, Purdue U., 1963. Asso. prof. pharmacology N.D. State U., Fargo, 1963-65; instr., postdoctoral trainee dept. pharamcology U. Iowa, Iowa City, 1965-67; head pharmacology sect. Applied Sci. div. Litton Industries, Inc., Bethesda, Md., 1968; head cardiovascular sect. dept. pharmacology Union Carbide Co., 1968-69; group leader exptl. diseases dept. pharmacology McNeil Labs., Inc., Ft. Washington, Pa., 1969-78, research fellow in immunopharmacology, 1978; group leader dept. pharmacology Boehringer-Ingelheim Ltd., Ridgefield, Conn., 1979—; mem. adv. bd. Lupus Research Inst., 1980—. Contbr. articles to sci. jours., chpt. to book. Recipient medal Rotary Club, 1952. Mem. Inflammation Research Assn. (treas. 1975-80, pres. 1980), Am. Soc. Pharmacology and Exptl. Therapeutics, Am. Chem. Soc. (medicinal chemistry div.), AAAS, N.Y. Acad. Scis., Am. Coll. Toxicology. Subspecialties: Immunopharmacology; Immunotoxicology. Current work: Immunomodulation for anti-rheumatic diseases; immunopharmacology and immunotoxicology; etiology of rheumatic diseases. Home: 6 Elbow Hill Rd Brookfield CT 06804 Office: Boehringer Ingelheim Ltd PO Box 368 Ridgefield CT 06877

WONG-MCCARTHY, WILLIAM JAMES, research psychologist; b. Paris, May 20, 1951; s. John Robert and Helen Ruth (House) McC.; m. Angela Wong, Mar. 23, 1974. B.A., Columbia U., 1973; M.A., U. Ill., 1976; Ph.D., Yale U., 1980. Vis. asst. prof. Hampshire Coll., Amherst, Mass., 1978-79; lectr. Pepperdine U., Los Angeles, 1979, Calif. Sch. Profl. Psychology, 1979; research assoc. UCLA, 1980, asst. research psychologist, 1980—; cons. Am. Heart Assn., 1981-82, Am. Cancer Soc., Am. Lung Assn. Cons. editor: Psychology of Women Quar. Mem. Coalition of So. Calif. Psychologists, 1980—, Concerned Faculty of UCLA, 1981—, Coalition for Clean Air, 1981—. NIH grantee, 1980—; Am. Lung Assn., 1982. Mem. Am. Psychol. Assn., Fedn. Am. Scientists, Am. Sociol. Assn., AAAS, Internat. Communication Assn.; mem. Am. Pub. Health Assn. Democrat. Subspecialties: Social psychology; Preventive medicine. Current work: Longitudinal research into the psychosocial antecedents of cigarette smoking; smoking cessation in the workplace; research on sex difference in language behavior. Home: 1323 Carmelina Ave Unit 208 Los Angeles CA 90025 Office: Dept Psychology 1282 Franz Hall UCLA Los Angeles CA 90024 Home: 1323 Carmelina Ave Unit 208 Los Angeles CA 90025

WONHAM, WALTER MURRAY, electrical engineering educator, system theorist; b. Montreal, Que., Can., Nov. 1, 1934. B. Engring., McGill U., Montreal, 1956; Ph.D., U. Cambridge, Eng., 1961. Assoc. prof. div. applied maths. Brown U., 1964-68; prof. elec. engring. U. Toronto, 1970—. Author: Linear Multivariable Control: A Geometric Approach, 1974, 2d edit., 1979. Nat. Acad. Sci. fellow, 1967-69. Fellow IEEE; mem. Soc. Indsl. and Applied Maths. Subspecialty: Systems engineering. Current work: Control theory. Office: Dept Elec Engring U Toronto Toronto ON Canada M5S 1A4

WOO, RICHARD, radio scientist; b. Portland, Oreg., June 24, 1941; s. Robert P.Y. and Lilliam H. (Ho) W.; m. Bobbie Wong, Dec. 20, 1964; 1 son, Dennis. B.S.E.E., U. Wash., 1962, M.S.E.E., 1964. Engr. Jet Propulsion Lab., Pasadena, Calif., 1964-68, sr. engr., 1968-73, mem. tech. staff, 1973-79, research scientist engr., 1979—; guest lectr. Calif. Inst. Tech., 1982. Contbr. articles to profl. jours. Recipient Exceptional Sci. Achievement medal NASA, 1979. Mem. Am. Astron. Soc., Am. Geophys. Union, Internat. Radio Sci. Union. Subspecialties: Planetary atmospheres; Solar wind. Current work: Wave propagation theory; remote sensing of planetary atmospheres and the solar wind with radio wave; radio scintillations. Patentee in field.

WOO, SAVIO LAU CHING, molecular medical geneticist; b. Shanghai, China, Dec. 20, 1944; came to U.S., 1966; s. Kwok-Cheung and Fun-sin (Yu) W.; m. Emily H. Chang, July 14, 1973; children: Audrey C. C., Brian Y.Y. B.Sc., Loyola Coll., Montreal, Can., 1966; Ph.D., U. Wash., 1971. Asst. prof. cell biology Baylor Coll. Medicine, Houston, 1975-78, assoc. prof., 1979-83; assoc. investigator Howard Hughes Med. Inst., Coconut Grove, Fla., 1977-79, investigator, 1979—; cons. Cooper Lab., Palo Alto, Calif., 1982—, Zymos Corp., Seattle, 1982—. Contbr. sci. articles to prof. publs. Mem. bd. dirs. March of Dimes Birth Defects Found., Met. Houston chpt., 1979—. Mem. Am. Soc. Human Genetics, Am. Soc. Biol. Chemists, Am. Soc. Cell Biology. Subspecialties: Genetics and genetic engineering (medicine); Gene actions. Current work: Development of analytical methodologies using recombinant DNA technology for prenatal diagnosis of human genetic disorders.

WOO, SAVIO LAU-YUEN, educator, bioengineer; b. Shanghai, China, June 3, 1942; came to U.S., 1961; s. Kwok Chong and Fung Sing (Yu) W.; m. Patricia Tak-kit Cheong, Sept. 6, 1969; children: Kirstin Wei-Chi, Jonathan I-Huei. B.S. in Mech. Engring, Chico (Calif.) State U., 1965, M.S., U. Wash., Seattle, 1966, Ph.D., 1971. Research assoc. U. Wash., 1968-70; asst. research prof. U. Calif.-San Diego, 1970-74, Assoc. research prof., 1974-75, assoc. prof., 1975-80, prof. surgery and bioengring., 1980—; prin. investigator VA Med. Ctr., San Diego, 1972—; cons. bioengr. Childrens Hosp., San Diego, 1973—; cons. med. implant cos., 1978—; vis. prof. biomechanics Kobe (Japan) U., 1981-82. Assoc. editor: Jour. Biochem. Engring, 1979—, Jour. Biomechanics, 1978—, Jour. Orthopedic Research, 1983—; contbr. articles to profl. jours. Vice pres. Bone and Joint Disease Found., San Diego, 1977. Recipient Elizabeth Winston Lanier award Kappa Delta, 1983, award for excellence in basic sci. research Am. Orthopedic Soc. Sports Medicine, 1983; Japan Soc. for Promotion of Sci., fellow, 1981; Research Career, Devel. award NIH, 1977-82. Mem. ASME (sec., chmn. biomechanics com., chmn. honors com. bioengring. div.), Am. Soc. Biomechanics (sec.-treas.), Orthopedic Research Soc. (chmn. program com.), Western Orthopaedic Assn. (hon.), others. Subspecialties: Orthopedics; Biomedical engineering. Current work: Experimental and theoretical biomechanics; analyses of nonlinear and viscoelastic properties of musculoskeletal tissues; stress and motion dependent homeostasis of ligaments, tendons, and bone remodeling, healing and repair. Inventor internal fixation plates. Home: 442 Jolina Way Encinitas CA 92024 Office: Dept Orthopedic Surgery M-004 Univ Calif San Diego CA 92093

WOOD, ALLEN JOHN, electrical engineer; b. Milw., Oct. 1, 1925; m. Barbara Ann Cook, Oct. 29, 1949; children: John, Susan. B.E.E., Marquette U., 1949; M.E.E., Ill. Inst. Tech., 1952; Ph.D. Rensselear Poly. Inst, 1959. Registered profl. engr., N.Y. Engr. Gen. Electric Co., Lynn, Mass., 1952-53, sr. engr., Schenectady, 1953-59, 61-69; mem. tech. staff Hughes Aircraft Co., Culver City, Calif., 1959-60; prin. engr. and dir. Power Technologies, Inc., Schenectady, 1969—; adj. prof. elec. engring. Rensslear Poly. Inst., Troy, N.Y., 1966—. Author: Power System Reliability Calculations, 1973, Power Generation, Operation and Control, 1983; contbr. numerous tech. articles to profl. publs. Served with U.S. Army, 1943-46. Fellow IEEE; mem. Am. Nuclear Soc. Club: Mohawk (Schenectady.). Subspecialties: Electrical engineering; Operations research (engineering). Current work: Electric power and energy; reliability; optimum operations and applications; development and application of computers. Home: 901 Vrooman Ave Schenectady NY 12309 Office: Power Technologies Inc PO Box 1058 Schenectady NY 12301

WOOD, CHARLES CRESSON, management systems consultant, foundation administrator; b. Phila., Feb. 22, 1955; s. Charles Wistar and Margaret Davis (Ansley) W. B.S.E. with honors in Acctg, U. Pa., 1976, M.S.E. in Computer and Info. Sci, 1979, M.B.A. in Fin, 1979. C.P.A., Calif. Teaching fellow computer sci. U. Pa., Phila., 1976-79; system performance engr. Booz-Allen & Hamilton, Washington, 1976; systems designer Am. Mgmt. Systems, Washington, 1977; acct. Richard Eisner & Co., N.Y.C., 1978; cons., analyst, designer specializing in fin. info. systems, applications involving computer security and privacy, cryptography; cons. in acctg. procedures and computer systems Stanford Research Inst., Menlo Park, Calif., 1979—. Contbr. articles on computer security to profl. jours. Founder and pres. Found. for Alternative Research; bd. dirs. Mid-Peninsula Peace Ctr. EDP Auditors' Assn. Quaker. Subspecialties: Cryptography and data security; Information systems (information science). Current work: Electronic funds transfer system security, computer crime, information resource management, identified and analyzed policy options available to the Federal Government for the restriction of non-classified computer related research that threatens national security. Office: 333 Ravenswood Ave Suite BN 145 Menlo Park CA 94025

WOOD, DAVID ALLEN, computer system designer; b. Santa Monica, Calif., Aug. 24, 1959; s. Roger Charles and Ann Elizabeth (Wilson) W. Student, U. Calif., Santa Barbara, 1977-78, B.S. with highest honors, 1981. Sr. engring. aide U. Calif. Berkeley Electronics Research Lab., 1980-81; mem. tech. staff, database Synapse Computer Corp., Milpitas, Calif., 1981—. Mem. IEEE, Assn. for Computing Machinery. Subspecialties: Database systems; Distributed systems and networks. Current work: Design and implementation of a distributed relational database system. Research interests are concurrency control control in a distributed environment. Office: 801 Buckeye Ct Milpitas CA 95035

WOOD, DAVID DUDLEY, medical research immunologist; b. Wilmington, Del., May 3, 1943; s. William Herman and Barbara Emma (Rose) W.; m. Carole Lee Putnam, Sept. 5, 1964; children: Kristen Elizabeth, Whitney Lynn. B.A., Harvard U., 1965; Ph.D., Rockefeller U., 1970. Helen Hay Whitney fellow Harvard U. Med. Sch., Boston, 1970-72; sr. research immunologist Merck, Sharp and Dohme Research Labs., Rahway, N.J., 1972-76, research fellow, 1976-82, sr. research fellow, assoc. dir., 1982-83; dir. immunology Ayerst Research Labs., Princeton, N.J., 1983—. Contbr. articles to profl. jours. Mem. Summit Bd. Edn., 1979-81. Mem. N.Y. Acad. Scis., Am. Assn. Immunology. Subspecialties: Immunobiology and immunology; Immunopharmacology. Current work: Biochemistry and biology of interleukin-1; immunoregulation; natural resistance. Home: 280 Woodland Ave Summit NJ 07901 Office: Ayerst Research Lab CN 8000 Princeton NJ 08540

WOOD, GORDON HARRY, JR., coal geologist; b. Poteau, Okla., Jan. 30, 1919; s. Gordon Harry and Lyde Jane (Lobdell) W.; m. Eleanor Elizabeth Suydam, Apr. 20, 1946; 1 dau., Ellen Louise Wood Revel. B.S., U. N.Mex., 1942. Topographic engr. U.S. Coast and Geodetic Survey, 1942-44; geologist U.S. Geol. Survey, Washington and Reston, Va., 1944—; chmn. exec. com. World Coal Resource and Res. Data Service, Internat. Energy Agy., London, 1979—. Author over 100 books, articles, maps in field. Recipient meritorious service award Dept. Interior, U.S. Geol. Survey, 1982. Fellow Geol. Soc. Am. (editor guide books); mem. Am. Assn. Petroleum Geologists (pres. Eastern sect. 1970), Geol. Soc. Washington (del. Am. Assn. Petroleum Geologists), AAAS, Geol. Soc. N.Mex. (dir. 1953). Republican. Club: Cosmos (Washington). Lodge: Elks. Subspecialties: Geology; Tectonics. Current work: U.S. coal resources, world coal resources; coal fields of the world; coal tectonics; editor coal reports, books, maps, resource data; coal resource terminology, methodology, research. Office: US Geological Survey National Center Reston VA 22092

WOOD, HARLAND GOFF, educator, biochemist; b. Delavan, Minn., Sept. 2, 1907; s. William C. and Inez (Goff) W.; m. Mildred Lenora Davis, Sept. 14, 1929; children—Donna, Beverly, Louise. A.B., Macalester Coll., 1931, D.Sc. (hon.), 1946; Ph.D., Iowa State Coll., 1935; D.Sc., Northwestern U., 1972. Fellow NRC, biochem. dept. U. Wis., 1935-36; asst. prof. bacteriology dept. Iowa State Coll., 1936-43; asso. prof. physiol. chemistry dept. U. Minn., 1943-46; prof., dir. biochemistry dept. Case Western Res. U., Cleve., 1946-65, prof. biochemistry, 1965—, dean scis., 1967-69, Univ. prof., 1970—; chmn. isotope panel NRC (com. on growth), 1947-48; adv. council Life Ins. Med. Research Fund, 1957-62; mem. tng. grant com. NIH, 1965-69; advisory bd. Case Inst. Tech., 1971-78; advisory com. div. biology and medicine U.S. AEC, 1957-62; mem. Presdl. Sci. Advisory Com., Washington, 1968-72, mem. physiol. chemistry study sect., 1973-77. Recipient Eli Lilly research award in bacteriology, 1942, Carl Neuberg award, 1952, Nat. Acad. Scis., 1953, Glycerine award, 1954, Modern Med. award Distinguished Achievement, 1968; Lynen lectr. and medal, 1972; Humbolt prize sr. U.S. scientist, W. Ger., 1979; Fulbright fellow, 1955; Guggenheim fellow, 1962. Mem. Am. Acad. Arts and Scis., Am. Heart Assn. (research com. 1949), Am. Chem. Soc., Am. Soc. Biol. Chemists (pres.), Biochem. Soc. Gt. Britain, N.Y. Acad. Scis., Am. Cancer Soc. (advisory bd. 1965-69), Internat. Union Biochemistry (mem. council 1967-76, sec. gen. 1970-73, pres. elect 1979), Soc. Am. Bacteriologists, Sigma Xi; corr. mem. Der Bayerischen Akademie der Wissenschaften. Methodist. Subspecialties: Microbiology; Biochemistry (medicine). Current work: Mechanism of enzyme reactions; synthesis of acetate by bacteria from CO or CO2 and H2. Office: 2109 Adelbert Rd Cleveland OH 44106

WOOD, HARRY ALAN, virologist; b. Albany, N.Y., Apr. 24, 1941; s. Frank Clifton and Mary Adair (Jones) W.; children: Kirsten Maria, Tiffany. B.A., Middlebury Coll., 1063; M.S., Purdue U., 1965, Ph.D., 1968. Asst. virologist Boyce Thompson Inst., Ithaca, N.Y., 1968-72, assoc. virologist, 1972-83, prof. virology, 1983—; vis. fellow U. Calif.-Davis, 1974-75. Contbr. articles in field to profl. jours. NIH fellow, 1966-68; grantee in field. Mem. Am. Soc. Virology, Am. Soc. Microbiology, Soc. Invertebrate Pathology, Sigma Xi. Subspecialties: Animal virology; Genetics and genetic engineering (agriculture). Current work: Insect virology is the general research interest with special emphasis of viral genetics and modes of virus infection and replication. Office: Boyce Thompson Inst Tower Rd Ithaca NY 14853

WOOD, HOUSTON GILLEYLEN, III, engineering educator; b. Tupelo, Miss., Oct. 4, 1944; s. Houston G. and Jane (Poe) W.; m. Grace Andres Maier, Aug. 28, 1965; children: Andrea Maier, Heather Gay. B.A., Miss. State U., 1965, M.S., 1967; Ph.D., U. Va., 1978. Devel. engr. Union Carbide Corp., Oak Ridge, 1967-73; mgr. centrifuge physics, 1977-81, cons., 1981—; research engr. U. Va., Charlottesville, 1973-77, assoc. prof., 1981—. Contbr. articles to profl. jours. Mem. Am. Phys. Soc., Soc. for Indsl. and Applied Math., Math. Assn. Am., Sigma Xi. Subspecialties: Applied mathematics; Nuclear fission. Current work: Research in dynamics of rotating fluids and isotope separation by gas centrifuge process. Applications of classical and numerical analysis. Home: 2303 Whippoorwill Rd Charlottesville VA 22901 Office: Dept Mech and Aerospace Engring Thornton Hall Va Charlottesville VA 22901

WOOD, JACKIE DALE, physiology educator; b. Picher, Okla., Feb. 16, 1937; s. Aubrey T. Wood and Wilma (Coleman) Wood P. B.S., Kans. State U.-Pittsburg, 1964, M.S., 1966; Ph.D., U. Ill.-Urbana, 1969. Asst. prof. biology Williams Coll., Williamstown, Mass., 1969-71; mem. faculty U. Kans. Med. Ctr., Kansas City, 1971-79, prof. physiology, 1978-79, U. Nev. Sch. Medicine, Reno, 1979—, chmn. dept., 1979—. Recipient Teaching Excellence award U. Kans., Lawrence, 1975; Prof. of Yr. award Med. Class. U. Nev., 1980; NIH grantee, 1974—; Humboldt fellow, 1976. Mem. Am. Physiol. Soc., Soc. Neurosci., Am. Gastroent. Assn., Am. Soc. Zoologists, Am. Fedn. Clin. Research. Democrat. Subspecialties: Physiology (medicine); Gastroenterology. Current work: Neurobiology of the gastrointestinal tract. Office: Univ Nev Sch Medicine Physiology Dept Reno NV 89557

WOOD, OBERT REEVES, II, laser researcher; b. Sacramento, Jan. 18, 1943; s. Obert Reeves and Marietta (Korff) W.; m. Nancy Jean Stanger, Apr. 21, 1973; 1 son, Obert Reeves III. B.S., U. Calif.-Berkeley, 1964, M.S., 1965, Ph.D., 1969. Research asst. Electronics Research Lab., U. Calif.-Berkeley, 1965-69; mem. tech. staff Bell Labs., Holmdel, N.J., 1969—; Cons. on carbon dioxide laser eye surgery N.Y. Hosp.-Cornell U. Med. Ctr., N.Y.C., 1978—. Contbr. articles to profl. jours. Mem. Am. Phys. Soc., Optical Soc. Am., IEEE, AAAS, N.Y. Acad. Scis., Sigma Xi, Tau Beta Pi, Eta Kappa Nu. Subspecialties: Atomic and molecular physics; Laser research. Current work: Research on short-wave length lasers and laser-produced plasmas; involved in experimental tests of quantum electrodynamics and use of lasers in medicine. Patentee quantum electronics, optics, atomic physics and laser fields. Home: 19 Fox Hill Dr Little Silver NJ 07739 Office: Bell Labs Room 4C434 Holmdel NJ 07747

WOOD, OWEN LESLIE, virologist, microbiologist; b. Westbrook, Maine, Aug. 20, 1936; s. Owen Ellis and Dora Adell (Rines) W.; m. Nancy Jean Belden, June 22, 1962; children: Jeane Leslie, Betsy Caroline, Andrew Owen. B.A. magna cum laude, Bates Coll., 1958; M. Phil., Yale U., 1968, Ph.D., 1972. Diplomate: Am. Bd. Med. Microbiology. Commd. ensign U.S. Navy, 1960, advanced through grades to comdr., 1981; asst. prof. dept. microbiology U. Nebr., Lincoln, 1970-73; head virology div. NAMRU-3, Addis Ababa, Ethiopia, 1973-77; research assoc. Yale Arbovirus Research Unit, Yale Med. Sch., 1977-79; head virology div. U.S. Naval Med. Research Unit 3, Cairo, Egypt, 1979-83; mem. staff Nat. Naval Med. Ctr., Bethesda, Md., 1983—; lectr. Quinnipiac Coll., Hamden, Conn., 1977-79. Contbr. articles to profl. jours. Active Boy Scouts, Cairo, 1981-82; vestryman Anglican Ch., Addis Ababa, 1976-77. USPHS trainee, 1966-70; NSF grantee, 1972. Mem. Am. Soc. for Microbiology, Am. Soc. Tropical Medicine and Hygiene, Am. Soc. for Electron Microscopy, Phi Beta Kappa, Sigma Psi. Subspecialties: Virology (veterinary medicine); Virology (medicine). Current work: Arboviruses and hemorrhagic fevers of viral origin; Rift Valley fever vaccine development and pathogenesis; Congo Crimean hemorrhagic fever epidemiology; rapid diagnosis of viral CNS disease. Office: NMRI Nat Naval Med Ctr Bethesda MD 20014

WOOD, RANDALL DUDLEY, scientist; b. Palmer, Ky., Aug. 3, 1936; s. Ocie B. and Tena (Harris) W.; m. Arlene Baker, Sept. 6, 1961; 1 dau., Marjorie Arlene. B.S., U. Ky., 1959, M.S., 1961; Ph.D., Tex. A&M U., 1965. Research scientist Oak Ridge Assoc. Univs., 1965-70; assoc. prof. Loyola U. Stritch Med. Sch., 1970-71, U. Mo., 1971-76; prof. Tex. A&M U., 1976—. Contbr. articles to profl. jours. Recipient MacGee award Am. Oil Chemists Soc., 1963; Bond award, 1967; grantee in field. Mem. AAAS, Am. Assn. Cancer Research, Am. Chem. Soc., Am. Oil Chemists Soc., Am. Inst. Nutrition, Am. Soc. Biol. Chemists, Soc. Complex Carbohydrates, Tissue Culture Assn., Sigma Xi. Subspecialties: Biochemistry (medicine); Biochemistry (biology). Current work: Lipid biochemistry of embryonic, normal and cancer cells. Office: Dept Biochemistry & Biophysics Tex A&M U College Station TX 77843

WOOD, WENDY, psychology educator; b. Amersham, Eng., June 17, 1954; came to U.S., 1956, naturalized, 1979; d. Charles and Pamela (Cole) W. B.S., U. Ill., 1975; M.S., U. Mass.-Amherst, 1978, Ph.D., 1980. Asst. prof. psychology U. Wis.-Milw., 1980-82; asst. prof. psychology Tex. A&M U., 1982—. Contbr. articles to profl. jours. U. Wis.-Milw. Grad. Sch. Research awardee, 1980-81. Mem. Am. Psychol. Assn., Assn. Women in Psychology, Soc. Advancement Social Psychology. Subspecialty: Social psychology. Current work: Research on attitude formation and change; sex differences in social interaction. Home: 2613B Pecan Knoll Bryan TX 77801 Office: Tex A&M U College Station TX 77843

WOOD, WILLIAM BARRY, III, biologist, educator; b. Balt., Feb. 19, 1938; s. William Barry, Jr. and Mary Lee (Hutchins) W.; m. Marie-Elisabeth Renate Hartisch, June 30, 1961; children: Oliver Hartisch, Christopher Barry. A.B., Harvard U., 1959; Ph.D., Stanford U., 1963. Asst. prof. biology Calif. Inst. Tech., Pasadena, 1965-68, assoc. prof., 1968-69, prof. biology, 1970-77; prof. molecular, cellular and developmental biology U. Colo., Boulder, 1977—; chmn. dept., 1978-83; mem. panel for developmental biology NSF, 1970-72; physiol. chemistry study sect. NIH, 1974-78; mem. com. on sci. and public policy Nat. Acad. Scis., 1979-80. Author: (with J.H. Wilson, R.M. Benbow, L.E. Hood) Biochemistry: A Problems Approach, 2d edit, 1981, (with L.E. Hood and J.H. Wilson) Molecular Biology of Eucaryotic Cells, 1975, (with L.E. Hood and I.L. Weissman) Immunology, 1978, (with L.E. Hood, I.L. Weissman, and J.H. Wilson)

Immunology, 2d edit., 1984, (with L.E. Hood and I.L. Weissman) Concepts in Immunology, 1978; Contbr. articles to profl. jours. Recipient U.S. Steel Molecular Biology award, 1969; NIH research grantee, 1965—; Guggenheim fellow, 1975-76. Mem. Nat. Acad. Scis., Am. Acad. Arts and Scis., Am. Soc. Biol. Chemists, Soc. for Developmental Biology, Soc. Nematology, Am. Soc. Virology, AAAS. Subspecialties: Gene actions; Developmental biology. Current work: Genetic control of invertebrate development; genetic control of virus assembly. Office: Dept MCD Biology Box 347 U Colorado Boulder CO 80309

WOODBURN, WILTON ALLEN, process engineer; b. Pitts., Nov. 2, 1926; s. Wilton Allen and Georgia (Guepner) W.; m. Joan Berry, Nov. 3, 1956; children: Susan, Gail, Robert. Student, U. Louisville, 1944-46; B.S.M.E., Carnegie Inst. Tech., 1948, M.S.M.E., 1956. Research engr. engring. design div. Alcoa Research Labs., Alcoa Center, Pa., 1947-65, research engr. fabricating metallurgy div., 1965-71, engring. assoc., 1971-80, sr. tech. specialist, 1980—. Served with USN, 1950-52. Mem. ASME (assoc.). Subspecialties: Materials processing; Theoretical and applied mechanics. Current work: Research and development in primary fabricating processes, particularly hot and cold rolling; computerization of existing equipment, developing/analyzing new techniques, training courses for plant engineers. Patentee in field. Home: 144 Claremont Dr Lower Burrell PA 15068 Office: Alcoa Labs Alcoa Tech Ctr Alcoa Center PA 15069

WOODBURNE, MICHAEL OSGOOD, geology educator, vertebrate paleontologist; b. Ann Arbor, Mich., Mar. 8, 1937; s. Russell Thomas and Elizabeth Florence (Osgood) W.; m. Janice Mary, Mar. 20, 1979; children: Andrew, Paul, Marilyn, Christine, Meg, Emily. B.S., U. Mich., M.S. in Geology; Ph.D., U. Calif.-Berkeley. Asst. prof. geology U. Calif.-Riverside, 1966-72, assoc. prof., 1972-76, prof., 1976—. Fellow Geol. Soc. Am.; mem. Paleontol. Soc., Soc. Vertebrate Paleontology, Soc. Study of Evolution. Subspecialties: Paleontology; Geology. Current work: Evolution of Australian marsupials, geology and vertebrate paleontology of Mojave Desert (Calif.), systematics and evolution of Neogene Equidae. Office: Dept Earth Scis Univ Calif Riverside CA 92521

WOODBURY, GEORGE W., JR., chemist educator, researcher; b. Moscow, Idaho, Oct. 13, 1937; s. George W. and Kathryn Cox (Schanley) W.; m. Carolyn Ruth Anderson, June 11, 1960; children: Joan, David. B.S., U. Idaho, 1959; Ph.D., U. Minn., 1964. Research assoc. U. Minn., 1964-65; research assoc. Cornell U., 1965-66, vis. assoc. prof., 1973; asst. prof. chemistry to prof. U. Mont., 1966—. Contbr. articles on surface properties of fluids and adsorption prperties of fluids to sci. jour. NSF grantee, 1968-70, 70-72, 74-76. Mem. Am. Phys. Soc., Am. Chem. Soc. Subspecialties: Statistical mechanics; Surface chemistry. Current work: Statistical mechanical theory of physical adsorption of fluids on solid surfaces. Office: U Mont Dept Chemistry Missoula MT 59812

WOODE, GERALD NOTTIDGE, veterinarian, educator; b. Hartlepool, Eng., July 16, 1934; came to U.S., 1978; s. Meredith Marshall and Evelyn Grace (Baker) W.; m. Shirley Wilkin, July 4, 1959; children: Clair, Lindsay, Mary. B.Vet. Med., U. London, 1960, D.Vet. Med., 1978. Gen practive veterinary medicine, Ashford, Eng., 1961-62; virologist Glaxo, Ltd., Eng., 1962-66; lectr. Royal Sch. Veterinary Studies, Edinburgh, 1966-71; prin. veterinary research officer Inst. Research on Animal Diseases, Eng., 1971-77; prof. veterinary medicine Iowa State U., 1978—; cons. in field. Contbr. to profl. jours. Served with Brit. Army, 1953-55. U.S. Dept. Agr. grantee, 1979-82; Nat. Pork Producers Council grantee, 1979-81; State of Iowa grantee, 1978-82. Mem. Am. Soc. Virology, WHO, Royal Coll. Veterinary Surgeons. Subspecialties: Animal virology; Virology (veterinary medicine). Current work: Virus infections of the intestinal tract causing diarrhea and death in children: calves, piglets, horses, lambs, dogs. Office: Department Veterinary Mircobiology College of Veterinary Medicine Iowa State University Ames IA 50011

WOODFORD, DAVID A., research metallurgist; b. Cleethorpes, Lincolnshire, Eng., Sept. 17, 1937; s. Aubrey Stanley and Gertrude Anne (Norris) W.; m. Catherine Yvonne Edwards, Dec. 30, 1961; children: Karen J., Paul D., Allison R. B.Sc. with honors, Birmingham (Eng.) U., 1960, Ph.D., 1963, Sc.D., 1981. Research metallurgist Materials and Processes Lab., Gen. Electric Co., Schenectady, 1964-74, staff metallurgist corp. research and devel., 1974—. Assoc. editor: Jour. Engring. Materials and Tech; mem. editorial bd.: Metal Sci. Jour; contbr. articles to profl. jours. Recipient Alfred H. Giesler award, 1972. Mem. Am. Soc. Metals, ASME. Subspecialties: High-temperature materials; Metallurgy. Current work: High temperature fatigue. High temperature creep. Effect of gaseous environments on mechanical properties. Patentee in field. Home: 1166 Avon Rd Schenectady NY 12308 Office: Gen Electric Co Bldg K-1 Room 231M Schenectady NY 12301

WOODHOUR, ALLEN FRANCIS, microbiologist; b. Newark, Feb. 21, 1930; s. Frank E. and Alice (Rawa) W.; m. Rosamond C. Woodhour, Nov. 5, 1955 (dec. Sept. 1982); 1 dau., Rosamond M. A.B., St. Vincent Coll., Latrobe, Pa., 1952; M.S., Cath. Univ. Am., 1954; Ph.D., Cath. U. Am., 1956. Microbiologist Walter Reed Army Inst. Research, Washington, 1956-57; Microbiologist, Chas. Pfizer & Co., Inc., Terre Haute, Ind., 1957-60; with Merck & Co., Inc., West Point, Pa., 1960—; exec. dir. bacterial vaccines/adminstrv. affairs, virus and cell biology research Merck Sharp & Dohme Research Labs., 1978—. Contbr. articles to sci. jours. Mem. Am. Soc. Microbiology, AAAS, Internat. Assn. Biol. Stantardization, Am. Soc. Immunologists, Sigma Xi. Subspecialties: Microbiology (medicine); Virology (medicine). Current work: Direct research on human bacterial vaccines. Patentee in field (5). Home: PO Box 187 Brandon Dr Lederach PA 19450 Office: Merck Labs West Point PA 19406

WOODIN, SARAH ANN, educator; b. N.Y.C., Dec. 27, 1945; d. William Graves and Barbara (Mason) W.; m. David Sunderland Wethey, Jan. 5, 1980. B.A., Goucher Coll., 1967; Ph.D., U. Wash., 1972. Asst. prof. U. Md., College Park, 1972-75; asst. prof. Johns Hopkins U., Balt., 1975-80; research assoc. prof. ecology U. S.C., Columbia, 1980—; marine editor Ecol. Soc. Am., 1978-82; Mem. editorial bd. Am. Zoologist, 1983—. NSF grantee; recipient Recognition Award for Young Scholars AAUW, 1977. Fellow AAAS; mem. Ecol. Soc. Am., Am. Soc. Zoologists, Brit. Ecol. Soc., Sigma Xi (nat. com. on grants-in-aid of research), Phi Beta Kappa. Subspecialties: Ecology; Species interaction. Current work: Population dynamics and community structure of organisms in marine soft-sediment environments. Address: Dept Biology U SC Columbia SC 29208

WOODLAN, DONALD RAY, consulting engineer; b. Painesville, Ohio, Aug. 9, 1946; s. Donald A. and June E. (Ward) W.; m. Judith Anne Zuber, Dec. 28, 1967; children: Melissa Sharee, Donald Scott. B.S. U.S. Naval Acad, 1968; M.S.E.E., Mich. State U., 1969. Registered profl. engr., Ohio. Ops. engr. Cleve. Electric Illuminating, 1975-79; licensing engr. Tex. Utilities Services, Inc., Dallas, 1979-80, sr. licensing engr., 1980—; mem. adv. group equipment qualification Electric Power Research Inst., 1979—; mem. ad hoc cm. equipment qualification 1980; mem. subcom. equipment qualification Atomic Indsl. Forum, 1979—; chmn. Westinghouse Users Group on 323-74, 1981—. Served to lt. USN, 1968-75; MTO. Mem. IEEE, Am. Nuclear Soc. Subspecialties: Nuclear engineering; Materials (engineering). Current work: Establishment of adequate/realistic safety goals for uranium power plants; develop criteria for adequate environmental qualification of equipment. Home: 925 Canyon Ridge DeSoto TX 75115 Office: Tex Utilities Services Inc 2001 Bryan Tower Dallas TX 75201

WOODLEY, DAVID TIMOTHY, dermatology educator; b. St. Louis, Aug. 1, 1946; s. Raoul Ramos-Mimosa and Marian (Schlueter) W.; m. Christina Paschall Prentice, May 4, 1974; 1 son, David Thatcher. A.B., Washington U., St. Louis, 1968; M.D., U. Mo., 1973. Diplomate: Am. Bd. Internal Medicine, Am. Bd. Dermatology, Nat. Bd. Internal Medicine. Intern Beth Israel Med. Center, Mt. Sinai Sch. Medicine, N.Y. Hosp., Cornell U. Sch. Medicine, N.Y.C., 1973-74; resident in internal medicine U. Nebr., Omaha, 1974-76; resident in dermatology U.N.C., Chapel Hill, N.C., 1976-78, asst. prof. dept. dermatology, 1983—; research fellow U. Paris, 1978-80; expert NIH, Bethesda, Md., 1980-82; cons. VA Med. Ctr., AHEC Med. Ctr., Fayetteville, N.C., 1983—. Contbr. chpts. to books and articles in field to profl. jours. Mem. Clean Water Action Project, Washington, 1982-83; mem. Potomac Albicore Fleet, Washington, 1982-83, Friends of the Art Sch., Chapel Hill, 1983—, Jungian Soc. Triangle Area, Chapel Hill, 1983—. Fellow Am. Acad. Dermatology; mem. Dermatology Found., Soc. Investigative Dermtology, ACP (assoc.), Assn. Physican Poets. Subspecialties: Dermatology; Cell biology (medicine). Current work: The biochemistry of basement membrane molecules in skin and the influence of these molecules on epidermal cells. Home: 46 Laurel Ridge Apts Chapel Hill NC 27514 Office: U NC Med Sch Dept Dermatology Rm 137 NC Meml Hosp Chapel Hill NC 27514

WOODRING, JOHN HOWELL, diagnostic radiology educator, physician; b. Louisville, Sept. 10, 1951; s. Franklyn Howell and Dorothy Moore (McInteer) W.; m. Catherine Anne Martin, Aug. 27, 1977; 1 son, Paul Martin. B.S., U. Louisville, 1973, M.D., 1976. Diplomate: Am. Bd. Radiology. Intern Louisville Gen. Hosp., 1976-77; resident in radiology U. Ky. Med. Center, Lexington, 1977-80; asst. prof. diagnostic radiology U. Ky. Coll. Medicine, Lexington, 1980—; chest radiologist, dept. diagnostic radiology, 1980—. Fellow U. Ky. Fellows; mem. AMA, Am. Coll. Radiology, Radiol. Soc. N.Am., Bluegrass Radiol. Soc., Phi Kappa Phi. Methodist. Subspecialty: Diagnostic radiology. Current work: Research and teaching chest trauma, carcinoma of the lung, adult respiratory distress syndrome, computed tomography of the chest, mycobacterial and fungal diseases of the lung. Office: Dept Diagnostic Radiology HX315B U Ky Med Ctr 800 Rose St Lexington KY 40536 Home: 162 N Arcadia Park Lexington KY 40503

WOODROOFE, MICHAEL BARRETT, statistician; b. Corvallis, Oreg., Mar. 17, 1940; s. Robin Russell and Helen Lucille (Barrett) W.; m. Frances Smock, July 6, 1974; children—Russell, Carolyn, Blake. B.S. in Math, Stanford U., 1962, M.A., U. Oreg., 1964, Sc., 1965. Research asso. Stanford U., 1965-66; asst. prof. stats. Carnegie Mellon U., 1966-68; mem. faculty U. Mich., 1968—, prof. math. and stats., 1973-83, chmn. dept. stats., 1977-83; prof. stats. Rutgers U., 1983—; vis. assoc. prof. Columbia U., 1970-71; vis. prof. MIT, 1976-77. Author: Probability with Application, 1974, Non Linear Renewal Theory in Sequential Analysis, 1982. Grantee NIH, NSF, U.S. Army. Mem. Am. Math. Soc., Inst. Math. Stats., Phi Beta Kappa. Subspecialties: Probability; Statistics. Current work: Sequential analysis; repeated significance tests; non linear renewal theory; optimal stopping; sequential estimation. Home: 707 Abbott St Highland Park NJ 80903 Office: Stats Dept Rutgers U New Brunswick NJ

WOODRUFF, CALVIN WATTS, child health educator, consultant; b. New Haven, July 16, 1920; s. William Watts and Myra Cannon (Kilborn) W.; m. Betty P. Perry, June 17, 1950; children: Virginia, Carl, Bill. B.A., Yale Coll., 1941, M.D., 1944. Diplomate: Am. Bd. Pediatrics, Am. Bd. Nutrition. Resident Vanderbilt U., Nashville, 1947, from asst. prof. to assoc. prof., 1947-60; prof., chmn. pediatrics Am. U., Beirut, Lebanon, 1960-63; prof. nutrition Sch. Pub. Health, U. Mich., Ann Arbor, 1963-65; prof. child health U Mo., Columbia, 1965—; sec. Edn. Com., Columbia, 1970-72; chmn. Perspectives in Medicine, Columbia, 1971-73; chmn. Adv. Com. on PKU and Related Disorders, Mo. Div. Health. Contbr. articles to profl. jours. Served to lt. M.C. USNR, 1952-53. Markle scholar, 1952. Fellow Am. Acad. Pediatrics; mem. Am. Pediatric Soc., Soc. Pediatric Research, Am. Inst. Nutrition, Am. Soc. Clin. Nutrition, Lebanese Pediatric Soc. (hon.). Democrat. Subspecialties: Nutrition (medicine); Gastroenterology. Current work: Pediatric nutrition with emphasis on fat soluble essential substances and trace elements, especially in low birth weight infants and children with malabsorption syndromes. Home: 910 Wayne Rd Columbia MO 65201 Office: Dept Child Health U Mo Columbia 1 Hospital Dr Columbia MO 65212

WOODRUFF, DIANA STENEN, psychologist, educator; b. Pasadena, Calif., Apr. 10, 1946; d. Norman Hanson and Ruth Lee (Helsel) Stenen; m. Hyung Woong Pak, Aug. 5, 1975; children: Jonathan Tong-Hee, Michelle Hyun-Mi. A.B. magna cum laude, UCLA, 1968; A.M., U. So. Calif., 1970, Ph.D., 1972. Research asst. dept. psychology UCLA, 1972-74; asst. prof. psychology U. So. Calif., Los Angeles, 1974-75, Temple U., Phila., 1975-77, assoc. prof., 1977-83, prof., 1983—; vis. prof. psychology Stanford U., 1983-85. Author: Can You Live to Be 100?, 1977, (with others) Developmental Psychology; A Life-Span Approach, 1981; editor: (with James E. Birren) Aging: Scientific Perspectives and Social Issues, 1st edit, 1975, 2d edit., 1983; Contbr. articles to profl. jours. Nat. Inst. Aging research awardee, 1979-81; sr. fellow, 1983-85. Fellow Am. Psychol. Assn., Gerontol. Soc. Am.; Mem. Soc. for Psychophysiol. Research (chmn. ethics com. 1980—), Soc. for Neurosci., Phi Beta Kappa. Subspecialties: Developmental psychology; Physiological psychology. Current work: Psychophysiology of aging; infant perceptual development. Home: 80 Peter Coutts Circle Stanford CA 94305 Office: Dept Psychology Stanford U Stanford CA 94305

WOODRUFF, MICHAEL LESTER, neurobiologist, med. psychologist, educator; b. Abingdon, Va., June 29, 1947; s. Galen Sargent and Dorothy (Scent) W.; m. Judity Dale Rosling, Sept. 23, 1969. A.B., U. Mich., 1969; M.S., U. Fla., 1971, Ph.D., 1973. Asst. prof. psychology U. Fla., 1973-74, Middlebury (Bt.) Coll., 1974-79; adj. asst. prof. psychology SUNY-Albany, 1978-79; asst. prof. anatomy Coll. Medicine, East Tenn. State U., Johnson City, 1979-81, assoc. prof. anatomy and med. psychology, 1981—; research cons. in surgery Mountain Home VA Center, 1981—; extramural reviewer NSF. Manuscript referee: Am. Jour. Psychology; asso.: Behavioral and Brain Scis; contbr. articles and book revs. to sci. jours., chpts. to books. NIH grantee, 1976-77; Am. Heart Assn. grantee, 1981-83. Mem. Am. Assn. Anatomists (abstract referee), Soc. for Neurosci., Psychonomic Soc., N.Y. Acad. Scis., Tenn. Acad. Sci., Sigma Xi. Club: State of Franklin Track (Johnson City) (treas. 1981). Subspecialties: Neurobiology; Physiological psychology. Current work: Neural control of emotional behavior; neuroanatomic and neurochemical basis of cardiovascular modulation. Home: 516 Lamont St Johnson City TN 37601 Office: Dept Anatomy Quiller-Dishner Coll Medicine East Tenn State U Johnson City TN 37614

WOODRUFF, RONNY CLIFFORD, geneticist, educaor; b. Greenville, Tex., Mar. 12, 1943; s. Clifford W. and Julia Vinita (Swann) W.; m. Brenda Ann Woodruff, July 10, 1945; 1 dau., Marya. B.S., E. Tex. State U., 1966, M.S., 1967; Ph.D., Utah State U., 1971. NIH postdoctoral fellow dept. zoology U. Tex., Austin, 1972-74; sr. asst. in research, dept. genetics Cambridge (Eng.) U., 1974-76; vis. research assoc. dept. zoology U. Okla., Norman, 1976-77; asst. prof. dept. biol. sci. Bowling Green (Ohio) State U., 1977-80, assoc. prof., 1980—; teaching, research dir. Mid-Am. Drosophila Stock Center. Contbr. articles to profl. jours. Recipient NIH Research Career Devel. award, 1980—; Sigma Xi Outstanding Young Scientist award, 1981; NSF grantee, 1977—; NIH grantee, 1978—. Mem. Genetics Soc. Am., Environ. Mutagen Soc., Am. Soc. Naturalists, Soc. Study of Evolution, Sigma Xi. Subspecialty: Gene actions. Current work: Genetic control of mutation and chemical mutagenesis.

WOODS, CHARLES ARTHUR, zoology educator, curator; b. Sherman, Tex., Dec. 23, 1940; s. Hicks Arthur and Janice (Gertrude) (McCallum) W.; m. Ellen Stott Woods, Mar. 23 1963; children: C. Stott, Patricia, Bryan. Student, Middlebury Coll., 1959-62; B.A., U. Denver, 1964; Ph.D., U. Mass., 1970. Asst. prof. biology U. Denver, 1970-71; asst. prof. zoology U. Vt., 1971-76, assoc. prof., 1976-79; prof. zoology U. Fla., 1979—, assoc. curator, 1979-82, curator, 1982—; chmn., 1979—. Contbr. articles to profl. jours. Served with USMC Res., 1962-68. NSF fellow, 1977-79. Mem. Am. Soc. Mammalogists, Am. Soc. Naturalists, Soc. Study Evolution, Soc. Vertebrate Paleontologists, Sigma Xi. Subspecialties: Evolutionary biology; Systematics. Current work: Functional biology, evolution and systematics of mammals with an emphasis on rodents; conservation of island mammals. Home: PO Box 218 Micanopy FL 32667 Office: Florida State Museum Gainesville FL 32611

WOODS, DELMA MARIA, psychologist; b. N.Y.C., June 11; d. Edmund and Maria Williams (Tucker) Prioleau; m. Roy Woods, Mar. 17, 1974; children: Roy, Edmund, Delma. M.S., Howard U., 1971. Tchr. emotionally disturbed children D.C. Pub. Schs., Washington, 1964-66, psychologist, 1967-74, 76-78; clin. psychologist Head Start, Washington, 1969, psychol. cons., 1973; 76-78; psychologist Dade County (Fla.) Schs., Miami, 1974-76; cons. Roy Woods, Charleston, S.C., 1979—; guardian ad-litem for neglected and abused children, Charleston County, S.C., 1979-82; mem. chmn. Charleston County Multidisciplinary Commn., 1978-79; coordinator The Health Line, Sta. WPAL, Charleston, S.C., 1979—. Chmn. Jack and Jill Am., Inc., Charleston, 1979—; co-chmn, youth leadership com. Alpha Kappa Alpha, Charleston, 1979-80, co-chmn. health edn. com., 1979-80; founder Morris Brown Players, Charleston, 1982; bd. dirs. Pam Robinson Contemporary Sch. Performing Arts, Charleston, 1982—. Recipient Achievement awards Washington Sch. Psychiatry, 1972, Children's Hosp. Nat. Ctr., 1973. Mem. Am. Psychol. Assn., Assn. Black Psychologists, AAAS, S.C. Assn. Black Psychologists, Charleston County Med. Soc. Aux. (sec.), Palmetto Med. Soc. Aux. (exec. bd.). Subspecialty: Behavioral psychology. Current work: The interrelationships between physical and psychological disorders, behavior modification in young children. Home: 1106 Woodhaven Dr Charleston SC 29407 Office: 19 Hagood Ave Suite 901 Charleston SC 29407

WOODS, EUGENE FRANCIS, health care products corporation executive; b. Richmond, Va., June 29, 1931; s. Edward Joseph and Wilhelmena (Walker) W.; m. E. Ruth Butler, Oct. 30, 1951 (div. Nov. 1980); children: Janet, Ralph, Randolph; m. Janet M. Scolari, Apr. 3, 1982. Faculty mem. Med. U. S.C., Charleston, 1957-75, prof. pharmacology, 1975; dir. pharmacology Baxter–Travenol Inc., Morton Grove, Ill, 1975-79, dir. safety assessment, 1979—. Served to maj. U.S. Army Res., 1950-67. Recipient USPHS (NIH) Research Career Devel. award, 1960-70; NIH, Am. Heart Assn., S.C. Heart Assn. research grantee, 1960-75. Mem. Am. Soc. Pharmacology and Exptl. Therapeutics, Am. Physiol. Soc., Am. Soc. for Artificial Internal Organs. Subspecialties: Pharmacology; Toxicology (medicine). Current work: Research and administration in broad areas of medical devices and drug agents. Office: Baxter–Travenol Inc 6301 Lincoln Ave Morton Grove IL 60053

WOODS, JAMES STERRETT, toxicologist; b. Lewistown, Pa., Feb. 26, 1940; s. James Sterrett Jr. and Jane S. (Parker) W.; m. Nancy Fugate, Dec. 20, 1969; 1 dau., Erin E. A.B., Princeton U., 1962; M.S., U. Wash., 1968, Ph.D., 1970; M.P.H., U. N.C., 1978. Research asso. Yale U. Sch. Medicine, New Haven, 1970-72; staff fellow Environ. Toxicology br. Nat. Inst. Environ. Health Scis., NIH, Research Triangle Park, N.C., 1972-75, head biochem. toxicology sect., 1975-78; program leader epidemiology and environ. health research program Battelle Research Ctr., Seattle, 1978—; research assoc. prof. environ. health Sch. Pub. Health, U. Wash., Seattle. Contbr. articles to profl. publs. Served as officer USN, 1962-66; Vietnam. USPHS scholar; Am. Cancer Soc. research assoc. Mem. AAAS, Am. Assn. Cancer Research, Am. Pub. Health Assn., Am. Soc. Pharmacology and Exptl. Therapeutics, N.Y. Acad. Scis., Soc. Epidemiology Research, Soc. Exptl. Biology and Medicine, Soc. Occupational and Environ. Health, Soc. Toxicology. Subspecialties: Toxicology (medicine); Epidemiology. Current work: Biochemical toxicology of environmental chemicals with specific reference to effects of trace metals and other toxic substances on regulation of heme synthesis and heme-dependent processes in mammalian tissues. Office: 4000 NE 41st St Seattle WA 98105

WOODS, JOHN GALLOWAY, mechanical engineer, high technology manufacturing company executive; b. N.Y.C., July 21, 1926; s. John Burrows Collings and Elizabeth (Galloway) W.; m. Merilyn Baron, Sept. 15,1948; children: Anne Helen, Elizabeth Ruth. B.M.E., Cornell U., 1949; M.S.M.E., Drexel U., 1958. Registered profl. engr., Pa. Test engr. Babcock & Wilcox Co., Alliance, Ohio, 1949-51; sr. engr. Philco Corp., Phila., 1951-54; sr. devel. engr. IRC Inc., Phila., 1954-69, TRW Inc. (merger IRC Inc. with TRW Inc.), 1969-79, sr. project engr., 1979-81; program mgr. Electronic Components Group, Research and Devel. Labs., 1981-82; chief engr. Fiber Optic Products Electronic Components Group, Research and Devel. Labs., 1983—. Asst. treas. Awbury Arboretum Neighbors Assn., Phila., 1979—. Served with USN, 1945—46. Mem. Nat. Soc. Profl. Engrs. (vice chmn. Profl. Engrs. in Industry 1977-79, Outstanding Service award 1979), Pa. Soc. Profl. Engrs. (pres. Phila. chpt. 1979-80, v.p. S.E. Pa. region 1980—, mem., state pres.-elect 1983-84, Disting. Service award Profl. in Industry 1979, Mem., Outstanding Service award Phila. chpt. 1980), ASME, IEEE. Subspecialties: Fiber optics; Electronic materials. Current work: Fiber optic interconnections and sensors. Patentee electronic components. Home: 5928 Devon Pl Philadelphia PA 19138 Office: TRW 401 N Broad St Philadelphia PA 19108

WOODS, MARY, chemistry educator, nun; b. Webster Groves, Mo., Dec. 22, 1923; d. John Charles and Mary Catherine (Langdon) W. B.A., Rosary Coll., River Forest, Ill., 1945; M.A., U. Ill.-Urbana, 1947; Ph.D., U. Wis.-Madison, 1961. Joined Sinsinawa Dominicans Roman Catholic Ch., 1947; mem. faculty dept. chemistry Rosary Coll., 1953—; research assoc. Argonne Nat. Lab., Ill., 1970-83. Mem. Am. Chem. Soc., AAAS, Sigma Xi. Subspecialties: Inorganic chemistry; Kinetics. Current work: Kinetics of actinide elements in solution. Address: 7900 W Division River Forest IL 60305

WOODS, RAYMOND FRANCIS, statistician, mfg. co. exec.; b. Danville, N.Y., Jan. 11, 1927; s. John Francis and Frances Amelia (Nies) W.; m. Patricia O'Brien, June 21, 1952; children: Kathleen, Carol, Debra, Lori, Tim, Mary. B.S. in Math, Canisius Coll., 1953; M.A., Bowling Green State U., 1954. Cert. reliability and quality engr. With Eastman Kodak Co., Rochester, N.Y., 1954—, asst. to mgr. product quality, in charge quality assurance and stats., 1981—; tchr. math. U. Rochester Evening Div., 1954-77; tchr. stats. Rochester Inst. Tech. Evening Div., 1975-83. Contbr. articles to profl. jours. Served with USMC, 1945-46; to sgt., 1950-51. Fellow Am. Soc. Quality Control; mem. Am. Statis. Assn., Am. Nat. Standards Inst. Republican. Roman Catholic. Subspecialties: Quality assurance engineering; Statistics. Current work: Quality systems.

WOODS, ROY, internist, immunologist; b. Birmingham, Ala., Mar. 25, 1938; s. Abraham Lincoln and Maggie Rosalee (Wallace) W.; m. Delma Maria Prioleau, Mar. 17, 1974; children: Roy, Edmund Prioleau, Delma Maria. A.B., Miles Coll., 1958; M.S., U. Wis.-Madison, 1960, Ph.D., 1963; M.D., U. Miami, Fla., 1976. Diplomate: Am. Bd. Medical Examiners. Assoc. prof. biology Miles Coll., Birmingham, 1963-64; postdoctoral research fellow U. Uppsala, Sweden, 1964-66, Stanford (Calif.) U., 1966-67; sr. scientist, dir. immunoglogulin reference Ctr. Nat. Cancer Inst., Springfield, Va., 1967-74; practice medicine specializing in internal medicine and immunology, Charleston, S.C., 1979—; mem. staff Roper Hosp., St. Francis-Xavier Hosp., Charleston.; cons., prin. investigator WHO. Author: Qualitation and Quantitation of the Immunoglobulins, 1972; columnist: Charleston Chronicle; contbr. numerous articles top profl. jours.; panel host Health Line, Sta. WPAL. NIH grantee. Mem. Nat. Med. Assn., AMA, Am. Assn. Immunologists, AAAS, N.Y. Acad. Scis., S.C. Med. Assn., Charleston County Med. Soc. Baptist. Club: Whist (Charleston). Subspecialties: Internal medicine; Immunology (medicine). Current work: Private practice of internal medicine and the investigation of immunodeficiency syndromes. Home: 1106 Woodhaven Dr Charleston SC 29407 Office: 19 Hagood Ave Suite 901 Charleston SC 29403

WOODS, WALTER ABNER, marketing executive, educator, consumer behavior researcher; b. Lingle, Wyo., Jan. 16, 1915; s. James Abner and Mazeppa (Israel) W.; m. Margaret C. Edmiston, June 15, 1955 (div. 1974); 1 dau., Dana Jeanne. A.B., U. Wyo., 1937; M.A., Syracuse U., 1942; Ph.D., Columbia U., 1952; student, Art Students League, N.Y.C., 1946-47. Research psychologist Art Sch., Pratt Inst., Bklyn., 1948-51; assoc. prof. psychology Richmond (Va.) Profl. Inst., 1952-55; v.p., sr. dir. research Nowland & Co., Greenwich, Ct., 1955-61; pres., sr. researcher Products & Concepts Research Inc., Spaata, (N.J.), Brussels, Sydney, 1961—; prof. mktg. West Ga. Coll., Carrollton, 1971—; dir. Prognosis S.A., Brussels, 1963—; cons. in field. Author: Consumer Behavior, 1980. County chmn. Ford for Pres., Carroll County, Ga., 1976; coordinator Anderson for Pres. campaign, Ga., 1980. Served to lt. USNR, 1942-46. Recipient disting. mktg. service award Sales Execs. Club of N.Y., 1968; named Outstanding Educator of Am., 1973. Fellow Acad. Mktg. Sci.; mem. Am. Psychol. Assn., Am. Mktg. Assn., Internat. Assn. for Empirical Aesthetics, AAAS. Subspecialties: Behavioral psychology; Cognition. Current work: Supra level purposes and motives in consuming behavior motivation and perception in art experiencing. Developer: personality test Polychrome Index, 1954; color aptitude test, 1951. Home: 389 Smyrna Church Rd Carrollton GA 30117 Office: West Ga Coll Carrollton GA 30118

WOODS, WALTER RALPH, animal scientist; b. Grant, Va., Dec. 2, 1931; s. John Wythe and Hazel Gladys (Hash) W.; m. Jacqulyn Rose Miller, Sept. 14, 1953; children: Neal Ralph, Diana Lyn. B.S., Murray (Ky.) State U., 1954; M.S., U. Ky., 1955; Ph.D., Okla. State U., 1957. Instr. animal sci. Okla. State U., 1956-57; asst. prof., then asso. prof. Iowa State U., 1957-62; asso. prof., then prof. U. Nebr., 1962-71; prof. animal sci., head dept. Purdue U., 1971—. Author papers, articles in field. Bd. dirs. Ind. 4-H Found., 1979-81. Recipient Disting. Agrl. Alumni award Murray State U., 1969, Meritorious Service award Ind. Pork Producers Assn., 1975. Mem. Am. Soc. Animal Sci. (sec.-treas. Midwest sect. 1979-81, pres. Midwest sect. 1983-84), Am. Inst. Nutrition, Am. Dairy Sci. Assn., Poultry Sci. Assn., Ind. Beef Cattle Assn. (dir.), Sigma Xi, Gamma Sigma Delta. Baptist. Club: Lafayette Rotary (dir. 1976-79). Subspecialty: Animal nutrition. Current work: Nutrition and feeding programs for food producing animals. Home: 201 N Sharon Chapel West Lafayette IN 47906 Office: Dept Animal Scis Purdue U West Lafayette IN 47907

WOODS, WALTER THOMAS, cardiovascular physiologist, educator; b. Nashville, Mar. 13, 1947; s. Walter Thomas and Evelyn Eugenia (Cooper) W.; m. Kathleen Gage, Frye, Aug. 23, 1969; children: Thomas Cooper, Kathleen Gage, Helen Frye. B.A., U. of the South, Sewanee, Tenn., 1969; M.A., Appalachian State U., 1971; Ph.D., Bowman Gray Sch. Medicine Wake Forest U., 1975. With dept. physiology and biophysics U. Ala., Birmingham, 1975—, asst. prof., 1971—; outside expert in cardiac electrophysiology Med. Research Council, Can., 1983—. Contbr. articles to profl. jours. NSF trainee, 1973-74; Am. Heart Assn. grantee, 1981; Lilly Research Labs. grantee, 1982; NIH grantee, 1983. Fellow Council Circulation Am. Heart Assn.; mem. Am. Physiol. Soc. Republican. Episcopalian. Club: Mt. Brook (Birmingham). Subspecialties: Cell biology; Physiology (biology). Current work: Research investigates mechanisms by which spontaneous electrical activity is generated by pacemaker cells in the heart; how heart rate is controlled is ultimate goal. Home: PO Box 7662A Birmingham AL 35253 Office: U Ala Sch Medicine University Station Birmingham AL 35294

WOODS, WILLIAM GUARD, pediatric oncologist, educator; b. Balt., Oct. 11, 1946; s. William G. and Claire (Weldon) W.; m. Kathleen Grimes, Nov. 26, 1971; 1 son, Andrew Keane. B.S. in Biology, Bucknell U., 1968; M.D., U. Pa., 1972. Diplomate: Am. Bd. Pediatrics, 1977, Am. Bd. Pediatric Hematology-Oncology, 1978. Intern. U. Minn. Hosps., Mpls., 1972-73, resident in pediatrics, 1973-75, asst. prof. dept. pediatrics, 1979-83; assoc. prof. pediatrics U. Minn. Hosp., 1983—; sci. coordinator Late Effects Study Group. Contbr. articles to profl. jours. Daymon Runyon-Walter Winchell cancer fellow, 1977. Mem. Am. Assn. Cancer Research, Am. Soc. Hematology, Am. Soc. Clin. Oncology, Am. Soc. Pediatric Hematology-Oncology; mem. Soc. Pediatric Research; Mem. N.W. Pediatric Soc., Physicians for Social Responsibility. Unitarian. Subspecialties: Oncology; Cell study oncology. Current work: DNA repair mechanisms; genetics of childhood cancer; clinical pediatric oncology. Office: Box 454 Mayo U Minn Hosps Minneapolis MN 55455

WOODSIDE, JEFFREY ROBERT, physician, urology educator, researcher; b. Pasadena, Calif., Nov. 24, 1942; s. Albion Belmont, Jr. and Marianne (Starbuck) W.; m. Marilyn Elizabeth Duke, Aug. 29, 1964; children: Pamela Michelle, Shauna Janelle. B.S., Oreg. State U., 1964; M.D., U. Oreg., 1968. Diplomate: Am. Bd. Urology. Intern U. Oreg. Med. Sch., Portland, 1968-69; resident in surgery U. N.Mex. Sch. Medicine, Albuquerque, 1973-74, resident in urology, 1974-77, asst. prof. urology, 1977-81, assoc. prof. urology, 1981—; staff urologist VA Hosp., Albuquerque, 1977—; cons. Kirtland AFB Hosp., Albuquerque, 1977—, Presbyterian Hosp., 1977—, St. Joseph Hosp., 1977—. Contbr. articles to profl. jours. Bd. dirs. Resolve of N.Mex., Inc., 1980—. Served to lt. comdr. USN, 1969-73. Research grantee N.Mex. Kidney Found., 1976; U. N.Mex. Found., 1982. Mem. Am. Urol. Assn., Internat Continence Soc., Urodynamic Soc., Am. Fertility Soc., Spina Bifida Assn. Am. Republican. Baptist. Subspecialties: Urology; Microsurgery. Current work: Clinical research in treatment of neurogenic bladder dysfunction in children, basic research in neurogenic vesical dysfunction in the Manx cat. Home: 7501 Arroyo del Oso NE Albuquerque NM 87109 Office: Div Urology Univ N Mex Sch Medicine 2211 Lomas NE Albuquerque NM 87131

WOODSIDE, KENNETH HALL, biochemist; b. Northampton, Mass., June 18, 1938; s. Gilbert Llewellyn and Mary Calhoun (Livingston) W.; m. Laura Elizabeth Adams, June 18, 1960; children: Mary Ellen, Elizabeth Ann. A.B., Oberlin Coll., 1959; Ph.D., U. Rochester, 1969. Asst. prof. Pa. State U., Hershey, 1970-76; asst. prof. U. Miami, Fla., 1976-79; research biochemist Mt. Sinai Med. Ctr., Miami Beach, Fla., 1979-81; prof., chmn. biochemistry Southeastern Coll. Osteo. Medicine, North Miami Beach, Fla., 1981—. Contbr. articles in field to profl. jours. Pres. Greater Miami Youth Symphony, 1982-83. NIH research grantee, 1978-81. Mem. Am. Physiol. Soc., AAAS, Sigma Xi. Mem. Soc. of Friends. Subspecialty: Biochemistry (medicine). Current work: Lung macrophages, protein metabolism, surfactant composition and metabolism in the lung, dynamics of lung cell populations, phagocytosis by macrophages, teaching of biochemistry to medical students. Home: 7531 SW 137th St Miami FL 33158 Office: Southeastern Coll Osteo Medicine 1750 NE 168th St North Miami Beach FL 33162

WOODSON, HERBERT HORACE, electrical engineering educator; b. Stamford, Tex., Apr. 5, 1925; s. Herbert Viven and Floy (Tunnell) W.; m. Blanche Elizabeth Sears, Aug. 17, 1951; children: William Sears, Robert Sears, Bradford Sears. S.B., S.M., M.I.T., 1952, Sc.D. in Elec. Engring, 1956. Registered profl. engr., Tex., Mass. Instr. elec. engring., also project leader magnetics div. Naval Ordnance Lab., 1952-54; mem. faculty M.I.T., 1956-71, prof. elec. engring., 1965-71, Philip Sporn prof. energy processing, 1967-71; prof. elec. engring., chmn. dept. U. Tex. at Austin, 1971-81, Alcoa Found. prof., 1972-75, Ernest H. Cockrell Centennial prof. engring., 1982—, dir. Center for Energy Studies, 1973—; prof. engring. Tex. Atomic Energy Research Found., 1980-82; staff engr. elec. engring. div. AEP Service Corp., N.Y.C., 1965-66; cons. to industry, 1956—. Author: (with others) Electromechanical Dynamics, parts I, II, III. Served with USNR, 1943-46. Fellow IEEE (pres. Power Engring. Soc. 1978-80); mem. Am. Soc. Engring. Edn., Nat. Acad. Engring., AAAS. Subspecialties: Electrical engineering; Fusion. Current work: Electromechanical pulse power supplies and their application to industrial processes; military systems and fusion research. Patentee in field. Home: 7603 Rustling Rd Austin TX 78731

WOODWARD, EDITH J., astronomer, educator; b. Waldron, Ind., Aug. 15, 1914; d. James R. Jones; m.; children: James, Barbara, Paul. B.S., Purdue U., 1935; Ph.D., Radcliffe Coll., 1941. Mem. faculty Mt. Holyoke Coll., Mass., 1938-40, Hunter Coll., N.Y.C., 1950-52; prof. astronomy William Paterson Coll., Wayne, N.J., 1959-83. Contbr. articles to profl. jours. NASA grantee, 1979, 80. Mem. Am. Astron. Soc., LWV, Phi Beta Kappa, Sigma Xi. Subspecialty: Optical astronomy. Current work: Eclipsing binary stars.

WOODWARD, JAMES FRANKLIN, history educator; b. Boston, Dec. 22, 1941; s. William Redin and Edith (Jones) W.; m. Paulette Grignon, Oct. 16, 1966. A. B., Middlebury Coll., 1964; M.S., N.Y.U., 1969; Ph.D., U. Denver, 1972. Asst. prof. history Calif. State U., Fullerton, 1972-76, assoc. prof., 1976-80, prof., 1980—, chmn. dept., 1982—, adj. prof. physics, 1980—. Contbr. articles to profl. jours. Mem. Internat. Soc. on Gen. Relativity and Gravitation, N.Y. Acad. Scis., AAAS, Phi Alpha Theta. Subspecialties: Relativity and gravitation; Theoretical astrophysics. Current work: Experimental work on the coupling of the electromagnetic and gravitational fields. Theoretical work on the evolution of pulsars and quasars. Office: Calif State U Fullerton CA 92634

WOODWARD, LEE ALBERT, geology educator, consulting geologist; b. Omaha, Apr. 22, 1931; s. Luman Albert and Ellen (Hutchison) W.; m. Kathleen McKenna, Dec. 27, 1952; children: Leslie, Ann, Joseph, Madeline. B.Sc., U. Mont., 1958, M.Sc., 1959; Ph.D., U. Wash., 1962. Geologist U.S. Bur. Reclamation, Missoula, 1958, Pan Am. Petroleum, Billings, Mont., 1962-63; instr. Olympic Coll., Bremerton, Wash., 1963-65; asst. prof. geology U. N.Mex., Albuquerque, 1965-68; assoc. prof. 1968-73, prof., 1973—. Served to 1st lt. U.S. Army, 1953-56. NSF grad. fellow, 1959-61; NATO postdoctoral fellow, Edinburgh, Scotland, 1973. Fellow Geol. Soc. Am. (assoc. editor 1976-82); mem. Am. Assn. Petroleum Geologists (v.p. Rocky Mountain sect. 1976). Subspecialty: Tectonics. Current work: Tectonics of overthrust belt and Rocky Mountains foreland of U.S. Office: Dept Geology U NMex Albuquerque NM 87131

WOODWARD, PAUL RALPH, computational physicist, astrophysicist; b. Rockville Centre, N.Y., Aug. 25, 1946; s. William Redin and Edith (Jones) W.; m. Judith Hansburg, Jan. 5, 1973; 1 son, Thomas Christopher. B.A., Cornell U., Ithaca, N.Y., 1967; Ph.D. in Physics (Woodrow Wilson fellow), U. Calif.-Berkeley, 1973. Research assoc. Nat. Radio Astronomy Obs., Charlottesville, Va., 1974-75, Leiden U. Obs., Netherlands, 1975-78; computational physicist Lawrence Livermore (Calif.) Nat. Lab., 1978—. Mem. Internat. Astron. Union, Am. Astron. Soc. Subspecialties: Theoretical astrophysics; Algorithms. Current work: Simulation of astrophysical systems on computers for study of evolution of spiral galaxies and formation of stars. Co-developer Piecewise-Parabolic Method, technique for calculating fluid flow with computers. Office: Lawrence Livermore Nat Lab PO Box 808 Livermore CA 94550

WOODY, CHARLES DILLON, neurobiology educator, researcher; b. Bklyn., Feb. 6, 1937; s. Dillon and Margaret (Deree) W.; m. Patricia Ann Crawford, Sept. 12, 1959; children: Jan Dillon, Mila Mir. A.B., Princeton U., 1957; M.D., Harvard U., 1962. Cert. medicine, Calif. Research asst. communications biophysics group MIT, 1959; intern Strong Meml. Hosp.-U. Rochester, N.Y., 1962-63; research fellow neurology Harvard Med Sch., 1963-64; resident in neurology Boston City Hosp., 1963-64; research assoc. Lab. Neurophysiology, NIMH, NIH, Bethesda, Md., 1964-67; Harvard Mosely fellow Inst. Physiology, Czechoslovakian Acad. Sci., Prague, 1967-68; research officer Lab. Neural Control, NIH, Bethesda, 1968-71; assoc. prof. in residence anatomy, physiology and psychiatry UCLA, 1971-76, assoc. prof. in residence anatomy, psychiatry, 1976-77, prof. in residence anatomy, psychiatry and biobehavioral sci., 1977—. Author: Memory, Learning and Higher Function, 1982; editor: Conditioning: Representation of Involved, 1982; editorial bd.: Physiol. Revs, 1975-80. Recipient Resnick Research prize Harvard Med. Sch., 1962, Nightingale prize Biol. Engring. Soc./Internat. Fedn. Med. Biol. Engring., 1969. Mem. Am. Physiol. Soc., Soc. Neuroscience, IBRO. Subspecialties: Neuropsychology; Neurophysiology. Current work: Cellular basis of memory and learning in mammals as studied in model system of conditioned behavior. Office: UCLA Medical Center 760 Westwood Plaza Los Angeles CA 90024

WOOLDRIDGE, DEAN EVERETT, scientist, exec.; b. Chickasha, Okla., May 30, 1913; s. Auttie Noonan and Irene Amanda (Kerr) W.; m. Helene Detweiler, Sept. 1936; children—Dean Edgar, Anna Lou, James Allan. A.B., U. Okla., 1932, M.S., 1933; Ph.D., Calif. Inst. Tech., 1936. Mem. tech. staff Bell Telephone Labs., N.Y.C., 1936-46; co-dir. research and devel. labs Hughes Aircraft Co., Culver City, Calif., 1946-51, dir., 1951-52, v.p. research and devel., 1952-53; pres., dir. Ramo-Wooldridge Corp., Los Angeles, 1953-58, Thompson Ramo Wooldridge, Inc., Los Angeles, also Cleve., 1958-62; research asso. Calif. Inst. Tech., 1962—. Author: The Machinery of the Brain, 1963, The Machinery of Life, 1966, Mechanical Man, 1968, Sensory Processing in the Brain, 1979, also articles. Recipient citation of Honor Air Force Assn., 1950, Raymond E. Hackett award, 1955, Westinghouse Sci. Writing award AAAS, 1963. Fellow Am. Acad. Arts and Sci., Am. Phys. Soc., IEEE, Am. Inst. Aeros. and Astronautics; mem. Nat. Acad. Scis., Nat. Acad. Engring., Calif. Inst. Assos., Am. Inst. Physics, AAAS, Phi Beta Kappa, Sigma Xi, Tau Beta Pi, Phi Eta Sigma, Eta Kappa Nu. Subspecialty: Electrical engineering. Address: 4545 Via Esperanza Santa Barbara CA 93110

WOOLF, NEVILLE JOHN, astronomer, educator; b. London, Sept. 15, 1932; U.S., 1959, naturalized, 1970; s. Henry and Lily (Diminsky) W.; m. Patricia Martin, Sept. 2, 1972; children: by previous marriage: David, Martin. B.Sc. with honors in Physics, Manchester (Eng.) U., 1956, Ph.D., 1959. Research assoc. Lick Obs., 1959-61, Princeton (N.J.) U. Obs., 1962-65; assoc. prof. U. Tex.-Austin., 1965-67; prof., dir. obs. U. Minn., Mpls., 1968-74; astronomer, prof. Steward Obs., U. Ariz., Tucson, 1974—. Served with U.K. Army, 1951-53. Mem. Am. Astron. Soc., Internat. Astron. Union. Subspecialties: Infrared optical astronomy; Optical imaging. Current work: Giant telescopes for ground and space, astronomical seeing. Home: 4318 E Whitman St Tucson AZ 85711 Office: D212 Steward Obs Univ Ariz Tucson AZ 85721

WOOLFORD, THOMAS LANEY, mathematical physicist, aerospace engineer, researcher; b. Potosi, Mo., Mar. 22, 1937; s. William Tyler and Currie Roberta (Laney) W.; m. Barbara Joanne Biggs, June 29, 1968. B.A. in Chemistry, N. Mex. Highlands U., 1958; M.S. in Physics, Ariz. State U., Tempe, 1969, postgrad., 1969-71, U. Houston, 1978-83. Research aerospace engr. U.S. Army Missile Command, Redstone Arsenal, Ala., 1962-66; research physicist U.S. Dept. Agr. Forest Service, Tempe, 1972-73; prin. engr. Link div. Singer Corp., Houston, 1974-78; sr. engr. Computer Sci. Corp., Houston, 1978-81; prin. scientist Lockheed-EMSCO, Houston, 1981—. Mem. Houston Mus. Find Arts, 1980—. Served with U.S. Army, 1960-62. Recipient NASA ALT Award, 1978. Mem. AIAA, AAAS, Am. Phys. Soc., Soc. Indsl. and Applied Math., Am. Soc. Photogrammetry. Current work: Research remote sensing of agricultural crop cover and condition: specializing in visible and infrared interactions, plant growth models and radiative transfer theorgy; work in fluid and flight dynamics, aerospace systems analysis, guidance and navigation, microwave-plasma interactions, and combustion. Designer Shuttle Mission Simulator astrodynamics and rigid body motion simulation software, 1976-78. Home: 734 Ramada Dr Houston TX 77062 Office: Lockheed EMSCO 1830 NASA Rd 1 Houston TX 77058

WOOLMAN, MYRON, psychologist; b. Phila., Jan. 18, 1917; s. Irvin and Rose (Goldberg) W.; m. Eugenia Marr Johnston, Nov. 1, 1941; children—Marcia Isabel, Bruce Douglas, Diana Eugenia. B.S. magna cum laude, Columbia, 1950, M.A. (Pres.'s scholar), 1951, Ph.D. in Psychology, 1955. Chief proficiency measurements, McConnell AFB, Kans., 1952-56; sr. scientist Human Resources Office, George Washington U., 1956-61; dir., prin. investigator Inst. Ednl. Research, 1961—; cons., prin. investigator for major projects U.S. Dept. Labor, NIMH, OEO, Office Child Devel., govts. Nigeria and Ghana, 1961—; pres. Myron Woolman, Inc., 1969—; William H. Robinson lectr. Hampton (Va.) Coll., 1978; systems cons. Nat. Urban League, 1979, U.S. Dept. Agr./AID, 1979; adviser to mayor of Detroit in riot prevention, 1967, prin. investigator, designer manpower devel. and learning systems for edn., govt. and industry in, U.S. and abroad, 1967—. Author numerous papers in field; patentee in field. Bd. dirs. James Farmer Center Community Action, 1964-65; mem. Consortium Developmental Continuity Coll. Human Ecology, Cornell U. Served with USAAF, 1942-46. Menninger Found. fellow, 1980. Mem. Am. Psychol. Assn., AAAS, Am. Assn. Mental Deficiency, AAUP, Council Exceptional Children, Kappa Delta Pi, Phi Delta Kappa. Subspecialties: Learning; Information systems, storage, and retrieval (computer science). Current work: Computerized accelered learning environments, management information systems, accelerated learning/ training methods for international development. Home: 10649 Montrose Ave Bethesda MD 20814 My work rests on the firm belief that truly civilized societies focus on the development of human potential; the great issue is whether we possess the wit and will to direct some of the resources now devoted to destructive technologies to techniques for upgrading human skills and symbolic capabilities.

WOOLSEY, CLINTON NATHAN, neurophysiology educator, consultant; b. Bklyn., Nov. 30., 1904; s. Joseph Woodhull and Matilda (Aichholz) W.; m. Harriet Runion, May 24, 1942; children: Thomas Allen, John David, Edward Alexander. A.B., Union Coll., 1928, Sc.D. (hon.), 1968; M.D., Johns Hopkins U., 1933. Asst. in physiology, assoc. prof. physiology Johns Hopkins U. Sch. Medicine, 1933-48; Charles Sumner Slichter research prof. neurophysiology, dir. Lab. Neurophysiology, U. Wis., Madison, 1948-75, coordinator biomed. research unit, 1973-80, Slichter prof. emeritus, 1975—; trainer grad. students and postdoctoral fellows in neurosci. NIH Exchange Mission to Russia, 1958. NIH grantee, 1948-75; recipient Ralph Gerard. Mem. Am. Physiol. Soc., AAAS, Soc. for Neurosci. (Ralph Gerard award 1982), Internat. Brain Research Orgn. Democrat. Subspecialties: Animal physiology; Neurobiology. Current work: Research on animals and man on cerebral cortical localization; present efforts devoted to preparing for publication of several studies not previously published. Home: 106 Virginia Terr Madison WI 53705 Office: U Wis 627 Waisman Center Madison WI 53706

WOOLSON, WILLIAM ANDREW, engineering executive, researcher; b. Norristown, Calif., May 12, 1944; s. William A. and Elsie (Emery) W.; m. Carol A. Clark, Dec. 8, 1965 (div. Feb. 1978); children: Shelly Lynn, Laura Lee, Lisa Ann. B.S., Pa. State U., 1966, Ph.D, 1970. Scientist Aerojet Nuclear Systems, Sacramento, Calif., 1970-71; scientist Sci. Applications, Inc., La Jolla, Calif., 1971-78, div. mgr., 1978—, asst. v.p., 1980—. NDEA fellow, 1966; AEC fellow, 1969. Mem. Am. Nuclear Soc., Health Physics Soc., Sigma Xi, Phi Kappa Phi, Tau Beta Pi. Subspecialties: Nuclear engineering; Mathematical software. Current work: The Development and application of Monte Carlo methods for the solution of radiation transport problems in nuclear fission and fusion systems and geophysical exploration instruments. Office: Sci Applications Inc 1200 Prospect St La Jolla CA 92038

WOOSLEY, STANFORD EARL, astrophysicist, educator; b. Texarkana, Tex., Dec. 8, 1944; s. Homer Earl and Wanda Faye (Fisher) W.; m. Annegret Irene Beuse, Dec. 5, 1972. B.S., Rice U., 1966, Ph.D. in Astrophysics, 1971. Research assoc. Kellogg Radiation Lab., Calif. Inst. Tech., Pasadena, 1973-75; asst. prof. astronomy and astrophysics U. Calif.-Santa Cruz, 1975-78, assoc. prof., 1978-83, prof., 1983—; cons., part-time employee Lawrence Livermore (Calif.) Nat. Lab. Contbr. writings in field to profl. publs. NSF and NASA grantee. Mem. Am. Astron. Soc., Am. Phys. Soc., Internat. Astron. Union.

Subspecialties: High energy astrophysics; Nuclear physics. Current work: Nucleosynthesis (origin of the elements); supernova models, models for X-ray and gamma-ray bursts; nuclear astrophysics. Home: 115 Auburn Ave Santa Cruz CA 95060 Office: Lick Obs U Calif Santa Cruz CA 95064

WOOSTER, WARREN S(CRIVER), marine educator; b. Westfield, Mass., Feb. 20, 1921; s. Harold Abbott and Violet (Scriver) W.; m. Clarissa Pickles, Sept. 13, 1948; children: Susan Wooster Allen, Daniel, Dana Wooster Pawka. Sc.B., Brown U., 1943; M.S., Calif. Inst. Tech., 1947; Ph.D., UCLA, 1953. From research asst. to prof. Scripps Instn. Oceanography, U. Calif., 1948-73; dir. UNESCO Office Oceanography, 1961-63; dean Rosenstiel Sch. Marine Atmospheric Sci., U. Miami, 1973-76; prof. marine studies and fisheries U. Wash., Seattle, 1976—. Contbr. to books, profl. jours. Served with USNR, 1943-46. Fellow Am. Geophys. Union, Am. Meterol Soc.; mem. Sci. Com. Oceanic Research, Sigma Xi. Subspecialty: Oceanography. Current work: Establishing abiotic variability effects on fish stock abundance in ocean. Office: U Wash Inst for Marine Studies Seattle WA 98105

WOOTEN, HENRY ALWYN, astronomer, educator; b. Salisbury, Md., May 3, 1948; s. Alwyn Henry and Elizabeth Caroline (Bonnar) W.; m. Ida Lee Wooten, Jan. 5, 1980; 1 son, Nathaniel Alwyn. B.S., U. Md., 1970; M.S., U. Tex., 1976, Ph.D., 1978. Robert A. Welch fellow U. Tex.- Austin, 1976-78; jr. research fellow Calif. Inst. Tech., 1979-82; research assoc. Rensselaer Poly. Inst., 1982—. Contbr. articles in field to profl. jours. Served with USNG, 1970-76. NATO grantee, 1982—; recipient Alumni Assn. Award U. Md. Alumni Assn., 1970, Dudley award The Dudley Obs., 1982. Mem. Am. Astron. Soc., Internat. Astron. Union. Methodist. Club: Am. Orchid Soc. Subspecialties: Radio and microwave astronomy; Infrared optical astronomy. Current work: Development of millimeter and submillimeter spectroscopy to determine the physics of the processes of star formation in our own and other galaxies. Office: Dept Physics Rensselaer Poly Inst Troy NY 12181

WORDEN, FREDERIC GARFIELD, neuroscientist; b. Syracuse, N.Y., Mar. 22, 1918; s. Vivien S. and Alice Garfield (Davis) W.; m. Katharine Cole, Jan. 8, 1944; children: Frederic, Dwight, Philip, Barbara, Katharine. A.B., Dartmouth Coll., 1939; M.D., U. Chgo., 1942. Diplomate: Am. Bd. Psychiatry and Neurology, Nat. Bd. Med. Examiners. Intern Osler Med. Clinic, Johns Hopkins Hosp., Balt., 1942-43; house officer Henry Phipps Psychiat. Clinic, 1943, house officer to chief resident, 1946-50; asst. in psychiatry, Commonwealth Found. fellow Johns Hopkins Med. Sch., 1946-48; instr. psychiatry, 1949-52; tng. Balt. Psychoanalytic Inst., 1947-53; clin. dir. Sheppard and Enoch Pratt Hosp., Towson, Md., 1950-52, supr. therapy, 1952-53; research psychiatrist, prof. psychiatry Med. Sch., UCLA, 1953-69, head div. adult psychiatry, 1968-69; prof. psychiatry, dir. neuroscis. research program MIT, 1969-83, prof. emeritus, 1983—; mem. research scientist devel. rev. com. NIMH, 1971-74, mem. bd. sci. counselors, 1975-78; mem. com. on brain scis. NRC, 1971-74; Bd. dirs. Founds. Fund for Research in Psychiatry, 1973-76; bd. overseers Dartmouth Med. Sch., 1979—; mem. Nat. Adv. Mental Health Council, 1980-83. Author: (with R. Galambos) Auditory Processing of Biologically Significant Sounds, 1972, (with F.O. Schmitt) The Neurosciences: Third Study Program, 1974, (with J.P. Swazey and G. Adelman) The Neurosciences: Paths of Discovery, 1975; also research pubs. in psychiatry, neurophysiology. Served to maj. AUS, 1943-46. Fellow Am. Acad. Arts and Scis.; mem. Am. Psychiat. Assn. (chmn. task force on research tng. council on med. edn. and devel.), Soc. for Neurosci. (chmn. program com. 1972), Acoustical Soc. Am., AAAS, Am. Psychoanalytic Assn., UCLA Brain Research Inst., Psychiat. Research Soc., N.Y. Acad. Scis. Club: St. Botolph (Boston). Subspecialty: Neurophysiology. Home: Rural Route 2 Box 127AA Jamestown RI 02835 Office: 165 Allandale St Jamaica Plain Station Boston MA 02130

WORLEY, RAY EDWARD, research horticulturist; b. Robbinsville, N.C., May 4, 1932; s. Ambrose Nolan and Mary Ethel (Hyde) W.; m. Billie Jean Adams, Jan. 30, 1955; children: David Ray, Diane Lea, Miriam Eve. B.S., N.C. State U., 1954, M.S., 1958; Ph.D., Va. Poly. Inst. and State U., 1961. Research horticulturist Coastal Plain Expt. Sta., U. Ga., Tifton, 1961—. Served to 1st Lt. U.S. Army, 1954-56. Mem. Am. Soc. Hort. Sci., So. Sect. Am. Soc. Hort. Sci. (L.M. Ware Research award 1981), Southeastern Pecan Growers Assn., Ga. Pecan Growers Assn. Club: Tifton Exchange (pres. 1961-81). Current work: Pecan physiology and nutrition. Office: Coastal Plain Experiment Station University of Georgia Moore Hwy Tifton GA 31793 Home: 2601 Camellia Dr Tifton GA 31794

WORLEY, S. D., chemistry educator; b. Russellville, Ala., Jan. 31, 1942; s. Shelby L. and Betty (Davis) W.; m. Karen H., June 20, 1964; children: Christopher G., Brian S. B.S., Auburn U., 1964; Ph.D., U. Tex.-Austin, 1969. Research chemist NASA Manned Spacecraft Center, Houston, 1969-72; asst. prof. chemistry Cleve. State U., 1972-73; sci. officer Office of Naval Research, Arlington, Va., 1973-74; asst. prof. chemistry Auburn (Ala.) U., 1974-78, assoc. prof., 1978—, Alumni prof. chemistry, 1982—. Contbr. numerous articles on chemistry to profl. jours. Served to capt. U. S. Army, 1969-72. Research Corp. grantee, 1975, 81; NSF grantee, 1981. Mem. Am. Chem. Soc. Subspecialties: Catalysis chemistry; Physical chemistry. Current work: Infrared and photoelecton spectroscopy, water disinfection, catalysis, theoretical chemistry. Office: Chemistry Dept Auburn U Auburn AL 36849

WORTH, NORMAN PAUL, systems analyst; b. Oakland, Calif., Jan. 18, 1938; s. Russell Girard and Marguerite May (Porter) W. BA., San Jose State U., 1964; M.E.A., U. Utah, 1974. Programmer, analyst State of Utah, Salt Lake City, 1974, U. Utah Research Inst., 1975-78; systems analyst Becton Dickinson, Salt Lake City, 1979-83; mgmt. info. scis. mgr. Controls div. Bourns Inc., Ogden, Utah, 1983—. Contbr. articles to profl. jours. Served to capt. USAF, 1965-73. Mem. AAAS, Am. Chem. Soc., Assn. for Computing Machinery, IEEE, Am. Prodn. and Inventory Control Soc. Subspecialties: Industrial engineering; Software engineering. Current work: Design, integration, and use of large systems, particularly for manufacturing, including modularization, user interaction, and control of these systems. Office: Bourns Inc 2533N 1500 W Ogden UT Home: 2486 Taylor Ogden UT 84401

WORTHINGTON, ELLIOTT ROBERT, management educator, consultant, researcher; b. New Milford, Conn., May 5, 1937; s. Elmer Harry and Mildred (Knight) W.; m. Anita Elliott, Sept. 3, 1959; children: Susan Bontley, Julie, Karen. B.A., Dartmouth Coll., 1961; M.A., No. Ariz. U, 1970; Ph.D., U. Utah, 1973; M.A., Webster Coll., 1978; student, U.S. Army Spl. Warfare Sch., 1965, Def. Lang. Inst., 1965, U.S. Army Command and Gen. Staff Coll., 1974. Commd. 2d lt. U.S. Army, 1961, advanced through grades to lt. col., 1977; served in, Vietnam, 1966-67, 68-69; psychologist/researcher in doctoral/postdoctoral tng. U.S. Army, Salt Lake City, 1971-73, El Paso, 1973-74; cons. psychologist U.S. Army Hosp., Ft. Polk, La., 1974-75; psychology cons. U.S. Army Health Service Command, Ft. Sam Houston, Tex., 1975-81; chief psychology service Brooke Med. Ctr., Ft. Sam Houston, 1976-78, dir. community mental health service, 1978-81; ret., 1981; prof. mgmt. Sch. Bus., West Tex. State U., 1981—; mgmt. cons. Worthington & Worthington, 1974—. Author: research paper Post Service Adjustment and Vietnam Era Veterans, 1975 (John Hopkins U. Research award 1975); editor: (with A.D. Mangelsdorff) Current Trends in Army Medical Department Psychology, 1979; sr. editor: Veterans Career Guidelines For Army Psychologists, 1982. Decorated Legion of Merit; recipient Pres.'s award St. Mary's U., San Antonio, 1979; U.S. Army Med. Dept. scholar, 1971-73; postdoctoral fellow, 1973-74. Mem. Acad. Mgmt., Consortium on Vet. Studies, Phi Kappa Phi. Subspecialties: Social psychology; Clinical psychology. Current work: Researching small to medium businesses as cases for college text on management strategies. Home: 23 Eagle Pass Canyon TX 79015 Office: Dept Bus Administrn Sch Bus West Tex State U Canyon TX 79016

WOSILAIT, WALTER DANIEL, pharmacologist; b. Racine, Wis., Feb. 4, 1924; s. Julius George and Louise (Kalweit) W.; m. Marilyn Anne, Aug. 14, 1948; 1 dau., Karen Anne. Ph.D., Johns Hopkins U., 1953. Research assoc. Western Res. U., 1953-56; asst. prof. and assoc. prof. SUNY Downstate Med. Center, Bklyn., 1956-65; prof. pharmacology U. Mo. Sch. Medicine–Columbia, 1965—. Pres. Columbia Art League. Served with USN, 1943-46. Mem. Am. Soc. Biol. Chemists, Am. Soc. Pharmacology and Exptl. Therapeutics, AAAS. Subspecialties: Pharmacology; Pharmacokinetics. Current work: Drug interactions affecting anticoagulants, pharmacokinetics.

WOSKOBOINIKOW, PAUL PETER, physicist; b. Watkins Glen, N.Y., Apr. 25, 1950; s. Tichon and Julia (Marchenko) W.; m. Constance Ann Golba, Aug. 29, 1976. B.S., Rensselaer Poly. Inst., 1972, M.S. in Elec. Engring, 1974, Ph.D. in Electrophysics, 1976. Postdoctoral fellow Francis Bitter Nat. Magnet Lab., M.I.T., Cambridge, 1976-77, staff scientist, 1977-80; project leader Plasma Fusion Ctr., 1980—. Mem. Am. Phys. Soc., IEEE. Subspecialties: Laser research; Plasma physics. Current work: Infrared, far infrared, millimeter high power source development including lasers; application of far-infrared technology to fusion plasma diagnostics. Office: Mass Inst Tech NW16-258 Cambridge MA 02139

WOSNICK, MICHAEL ALAN, research scientist; b. Toronto, Ont., Can., Nov. 3, 1951; s. Hyman L. and Bella (Levy) W.; m. Katherine J. Wosnick, May 30, 1973; 1 dau., Allison. B.Sc., U. Toronto, 1974; Ph.D., Queens U., Kingston, Ont., 1978. Fellow U. Calgary, Alta., 1978-81; research scientist Connaught Research Inst., Toronto, 1981—. Contbr. numerous articles in field to profl. jours. Med. Research Council Can. fellow, 1978-81; Govt. Ont. scholar, 1976-77; Nat. Research Council Can. scholar, 1977. Mem. AAAS, Canadian Soc. Cell Biology, Genetics Soc. Am. Subspecialties: Genetics and genetic engineering (biology); Molecular biology. Current work: Genetic engineering, recombinant DNA technology, expression of genes in foreign hosts, regulation of gene expression. Home: 28 Darby Way Thornhill ON Canada L3T 5V1 Office: 1755 Steeles Ave W Willowdale ON Canada M2R 3T4

WOTIZ, HERBERT HENRY, biochemist, educator; b. Vienna, Austria, Oct. 8, 1922; came to U.S., 1938, naturalized, 1944; s. Edward and Irene (Politzer) Wottitz; m. Miriam S. Rose, June 15, 1947; children: Sue W. Goldstein, Robert P., Richard A. B.Sc., Providence Coll., 1944; Ph.D., Yale U., 1951. Asst. prof. biochemistry Boston U. Sch. Medicine, 1951-55, assoc. prof., 1955-63, research prof. urology, 1978—, prof. biochemistry, 1963—; dep. dir. H. H. Humphrey Cancer Research Center, Boston U., 1981—; cons. and dir. Seragen, Inc., Boston, 1979—; Mem. Milton (Mass.) Sch. Com., 1959-65, 66-69. Mem. editorial bd.: Chem. Abstracts, 1969—, Steroids, 1956—, Jour. Chromatographic Sci, 1963-83. Served with U.S. Army, 1944-45. NIH sr. research fellow, 1960-65; career devel. fellow, 1965-70. Mem. Am. Chem. Soc., Endocrine Soc., Am. Soc. Biol. Chemists, Soc. Exptl. Biology and Medicine, AAAS, Am. Assn. Cancer Research. Subspecialties: Receptors; Cancer research (medicine). Current work: The role of hormones in cancer, mechanism of estrogen action, analysis of hormone receptors, role of estriol in carcinogenesis. Home: 9 Cape Cod Ln Milton MA 02187 Office: 80 E Concord St Boston MA 02118

WOYCZYSKI, WOJBOR ANDRZEJ, mathematician, research, educator, administrator; b. Czestochowa, Poland, Oct. 24, 1943; came to U.S., 1970; s. Eugeniusz and Otylia Sabina (Borkiewicz) W.; m. Aleksandra Henryka Krasna, May 25, 1976; 1 son, Martin Wojbor. M.Sc. in Elec. Engring, Wroclaw (Poland) U., 1966, Ph.D. in Math, 1968. Teaching asst. Wroclaw U., 1966-68, asst. prof., 1968-72, assoc. prof., 1972-77; postdoctoral fellow Carnegie-Mellon U., 1970-72; prof. math. and stats. Cleve. State U., 1977-82; prof., chmn, dept. math. and stats. Case Western Res. U., 1982—; vis. prof. U. Paris, 1974, U. Wis.-Madison, 1976; Northwestern U., 1976-77, U.S.C., 1979. Author: Geometry and Martingales in Banach Spaces, part I, 1975, Part II, 1979; editor: Martingale theory in harmonic analysis and Banach spaces, 1982. NSF grantee, 1970, 71, 76, 77, 81; Polish Acad. Scis. grantee, 1972-76. Mem. Polish Math. Soc. (v.p. br. 1973-75, Gt. prize 1973), Inst. Math. Stats., Am. Math. Soc., Wroclaw Sci. Soc. Roman Catholic. Clubs: Park East Racquet (Beechwood, Ohio); Acad. Sports (Wroclaw). Subspecialty: Probability. Current work: Probability theory on infinite dimensional spaces, stochastic processes, harmonic analysis, functional analysis, theory of martingales. Home: 18417 Scottsdale Blvd Shaker Heights OH 44122 Office: Dept Math and Stats Case Western Res U University Circle Cleveland OH 44106

WRAY, JOHN LAWRENCE, mechanical engineer; b. Maryville, Mo., June 17, 1935; s. Lawrence Paul and Roberta Inez (Cook) W.; m. Sally Blair Gerdes, Dec. 27, 1958; children: Mary, Nancy, Carolyn. B.S., U. Mo., 1957; M.S., Stanford U., 1958; M.B.A., U. Santa Clara, 1966. Registered profl. engr., N.Y., Ill., Mo., Pa., others. Instr. George Mason U., Arlington, Va., 1961-62; with Gen. Electric Co., San Jose, Calif., 1962-78, mgr. product planning and market research, 1972-76, mgr. market research and planning, 1977-78; v.p. engring. Quadrex Corp., Campbell, Calif., 1978-82, v.p. computer systems and ops., 1982—; dir., treas. Mich. Nuclear Services Corp., Detroit, 1982—; dir. Quadrex FLIC Computer Systems, Inc., Princeton, N.J. Mem. fin. com. Elem. Sch. Dist., Saratoga, Calif., 1974-78. Served to capt. USAF, 1958-62. AEC fellow, 1957-58. Mem. Am. Nuclear Soc., ASME, Stanford U. Alumni Club (pres. 1970-71), Beta Theta Phi. Republican. Club: Brookside Swim and Tennis (pres., dir. 1974-75). Lodge: Masons. Subspecialties: Information systems, storage, and retrieval (computer science); Systems engineering. Current work: Computer systems applied to process control industry. Office: Quadrex Corp 1700 Dell Ave Campbell CA 95008 Home: 14961 Haun Ct Saratoga CA 95070

WRAY, JOHN ROBERT, health physics engineer; b. Passaic, N.J., June 24, 1951; s. Arthur and Mary (DeLeo) W.; m. Marybeth Glasheen, July 17, 1976. B.S., Rensselaer Poly. Inst., 1973; M.S., Northeastern U., 1978. Registered profl. engr., Mass.; cert. health physicist. Radiation protection engr. Stone & Webster Engring. Co., Boston, 1973-78; health physics inspr. U.S. Nuclear Regulatory Commn., Atlanta, 1979—. Recipient Cert. of Appreciation U.S. Nuclear Regulatory Commn., 1980, 82, 83, Cert. of Recognition, 1981. Mem. Am. Nuclear Soc., Health Physics Soc. Subspecialty: Nuclear fission. Current work: Inspection and evaluation of radiation protection and radioactive waste management programs at nuclear power plants, research and test reactors, and fuel facilities. Respond to nuclear power reactor plant emergencies. Home: 1740 Charmeth Rd Lithonia GA 30058 Office: US Nuclear Regulatory Commn 101 Marietta St Suite 2900 Atlanta GA 30303

WRENN, SIMEON MAYO, molecular pharmacology educator, consultant; b. Goldsboro, N.C., Nov. 17, 1944; s. Simeon Mayo and Ann (Christensen) W.; m. Ann Zelinka, Oct. 4, 1969; children: Katerine E., Simeon Mayo. B.S., Fla. Atlantic U., 1967; M.S., Emory U., 1972, Ph.D., 1973. Research fellow Harvard Med. Sch. also Mass. Gen. Hosp., 1973-77; asst. prof. dept. cell biology, sect. cardiovascular scis. dept. medicine Baylor Coll. Medicine, 1977-80, assoc. prof. dept. pathology U. Pa., 1980—; cons. Centocor Inc. NIH grantee, 1977—; Am. Heart Assn. grantee, 1977—. Mem. Am. Chem. Soc., Am. Soc. Cell Biology, Am. Soc. Pharmacology and Exptl. Therapeutics. Episcopalian. Club: Conestoga. Subspecialties: Molecular pharmacology; Biochemistry (medicine). Current work: Molecular pharmacology of B-ardrenergic and adenosine receptors, and interaction with adenylate cyclas, cardiovascular pharmacology and ischemia, cancer antigens and immunology and immunopharmacology. Office: 4001 Spruce St Philadelphia PA 19104

WRIGHT, ALDEN HALBERT, mathematics educator; b. Missoula, Mont., Apr. 23, 1942; s. Philip Lincoln and Margaret (Halbert) W.; m. Sally Fant, Mar. 23, 1967; children: Eric, Kevin. A.B., Dartmouth Coll., 1964; Ph.D., U. Wis., 1969. Asst. prof. U. Utah, Salt Lake City, 1969-70; asst. prof. dept. math. Western Mich. U., Kalamazoo, 1970-74, assoc. prof., 1974-81, prof., 1981—. Mem. Soc. for Indsl. and Applied Math., Inst. Mgmt. Sci., Math. Programming Soc., Assn. for Computing Machinery, Phi Beta Kappa. Subspecialties: Operations research (mathematics); Numerical analysis. Current work: Homotopy methods for solution of systems of nonlinear equations, linear programming, artificial intelligence. Home: 6062 Litchfield St Kalamazoo MI 49009 Office: Dept Math Western Mich Univ Kalamazoo MI 49009

WRIGHT, CHRISTOPHER PEARCE, mechanical engineer; b. Buffalo, May 19, 1939; s. Sidney Melton and Martha Dunn (Pearce) W.; m. Suzanne DeLaine, Aug. 19, 1962; children: Angela, Timothy Noel, Benjamin James. B.S., Va. Poly. Inst., 1962. Registered engr., Minn., Fla. With IBM, Endicott, N.Y., 1962-64; with Pratt & Whitney Aircraft, West Palm Beach, Fla., 1964-66; v.p. Underseas Engring. Inc., Riviera Beach, Fla., 1966-73; with Fluidyne Engring., Mpls., 1973-78, United Computing Systems, 1978-80; cons. engr., Minnetonka, Minn., 1980—. Mem. ASME (safety standard com.), Tau Beta Pi. Episcopalian. Subspecialties: Mechanical engineering; Materials. Current work: Computer applications in engineering productivity improvement. Patentee jet fluid amplifier. Address: 4903 Royal Oaks Dr Minnetonka MN 55343

WRIGHT, CLARENCE PAUL, biologist, educator, researcher; b. Cliffside, N.C., Apr. 15, 1939; s. Gordon Clifford and Cora May (Gettys) W.; m. Janice Fay Monteith, June 23, 1973. B.S. in Biology, Lenoir Rhyne Coll., 1962; M.S. in Genetics, U. Utah, 1965, Ph.D., 1968. Asso. prof. biology Western Carolina U., Cullowhee, N.C., 1968—; researcher in field. Mem. Genetics Soc. Am. Subspecialty: Gene actions. Current work: Developmental genetics of Drosophila melanogaster. Office: Dept Biology Western Carolina Univ Cullowhee NC 28723

WRIGHT, CREIGHTON BOLTER, medical educator, cardiovascular surgeon; b. Washington, Jan. 29, 1939; s. Benjamin W. and Catherine B. (Zeller) W.; m. Carolyn Eleanor Craver, Jan. 29, 1966; children: Creighton, Benson, Kathryn, Elizabeth. B.A. in Chemistry, Duke U., 1961, M.D., 1965. Diplomate: Am. Bd. Surgery, Am. Bd. Thoracic Surgery. Intern Duke U., 1965; resident U. Va., 1966-71; chief cardiovascular physiology Walter Reed Inst. Research, Washington, 1971-74; assoc. prof. George Washington U. and chief cardiovascular surgery Washington VA Med. Ctr., both Washington, 1974-76; prof. surgery U. Iowa, Iowa City, 1976-81; also chief surgery VA Med. Ctr., 1979-81; prof. surgery U. Cin. and staff Christ and Jewish hosps., Cin., 1982—; cons. VA Med. Ctr., Cin., 1982—. Editor: Vascular Grafting: Clinical Application and Techniques, 1983; co-editor: Venous Trauma: Pathophysiology Diagnosis and Surgical Management, 1983, Vascular Occlusive Disorders: Medical and Surgical Management, 1981; contbr. chpts. to books, articles to profl. jours. Served to col., M.C. USAR, 1966—. Recipient Kindred award U. Va., 1967; Meritorious Service medal U.S. Army, 1974; Golden Apple award Georgetown U. Students Assn., 1975. Fellow ACS; mem. Assn. Acad. Surgery (pres. and council 1979-80), Muller Surg. Soc. (v.p. 1981-83), Soc. Vascular Surgery (program chmn. 1980), Alpha Omega Alpha. Subspecialties: Surgery; Cardiac surgery. Current work: Clinical and experimental cardiovascular research; teaching and active thoracic and cardiovascular surgeon. Home: 1242 Edwards Rd Cincinnati OH 45208 Office: 2139 Auburn Ave Cincinnati OH 45219

WRIGHT, DANIEL GODWIN, physician, biomedical researcher, army officer; b. St. Louis, July 27, 1945; s. Whitbeck and Jane Montgomery (Godwin) W.; m. Elizabeth Chalmers, Sept. 4, 1967; 1 son, Christopher Talcott. B.A., Yale U., 1967, M.D., 1971. Diplomate: Am. Bd. Internal Medicine. Intern, resident in medicine Yale-New Haven Hosp., 1971-73; clin. assoc., med. officer Nat. Inst. Allergy and Infectious Diseases, NIH, Bethesda, Md., 1973-77; fellow dept. medicine Johns Hopkins U. Sch. Medicine, Balt., 1977-78; sr. investigator Nat. Cancer Inst., NIH, Bethesda, 1978-80; commd. maj. M.C. U.S. Army, 1979, advanced through grades to lt. col., 1982; chief dept. hematology Walter Reed Army Inst. Research, Washington, 1979—; mem. hematology study sect. NIH, 1982—. Served to lt. comdr. USPHS, 1973-77. Mem. Am. Soc. Hematology, Am. Fedn. Clin. Research, Am. Assn. Immunology, N.Y. Acad. Scis. Episcopalian. Subspecialties: Hematology; Infectious diseases. Current work: Neutrophil physiology, neutropenias, regulation of myelopoiesis. Office: Dept Hematology Walter Reed Army Inst Research Washington DC 20307

WRIGHT, GEORGE G(REEN), microbiologist; b. Ann Arbor, Mich., Aug. 17, 1916; s. George Green and May (Bradbeer) W.; m. Marjory Gray Hawley, 1941; 1 son, George; m. Mary Griffith West, 1957; children: Laurence, Mary. B.A., Olivet Coll., Mich., 1936; Ph.D. in Microbiology, U. Chgo., 1941. Instr. immunology U. Chgo., 1941-42; research fellow Calif. Inst. Tech., Pasadena, 1942-46; with CIA, 1946-48; chief immunology br. Dept. Army, Ft. Detrick, Frederick, Md., 1948-71; asst. dir. Mass. Pub. Health Biologic Labs., Boston, 1971-77, dir., 1977-82; research asst. prof. dept. internal medicine U. Va. Sch. Medicine, 1983—; lectr. medicine Tufts U., 1972-83; vis. lectr. applied microbiology Harvard U. Sch. Pub. Health, 1972-83. Contbr. chpts. to books, articles to sci. jours. Recipient Ricketts prize U. Chgo., 1941; Exceptional Civilian Service award Dept. Army, 1954; Sec. Army fellow Oxford U., 1957-58. Mem. Am. Assn. Immunologists, Soc. Exptl. Biology and Medicine, Am. Soc. Microbiology. Methodist. Subspecialties: Immunology (medicine); Microbiology (medicine). Current work: Research on mode of action of bacterial Toxins on mammalian cells. Office: Div Infectious Disease Box 385 U Va Sch Medicine Charlottesville VA 22908

WRIGHT, GEORGE LEONARD, JR., immunologist, cons.; b. Ludington, Mich., Feb. 8, 1937; s. George Leonard and Alma Edel (Schorder) W.; m. Yolonda Mary Jackson, Aug. 2, 1964; children:

Juliana, Christopher George. B.A., Albion Coll., 1959; M.S., Mich. State U., 1962, Ph.D., 1966; postdoctoral, George Washington U., 1966-67. Asst. prof. microbiology George Washington U., 1967-73, assoc. research prof., 1973; assoc. prof. Eastern Va. Med. Sch., Norfolk, 1973-76, prof., 1976—, dir. immunology program, 1975—; cons. in field. Contbr. articles to profl. jours. Fellow Am. Thoracic Soc., 1966, 67; NIH career devel.award, 1975-80. Mem. Am. Microbiology Soc., Am. Assn. Cancer Research, Am. Assn. Immunologists, AAAS, Sigma Xi. Subspecialties: Immunology (medicine); Infectious diseases. Current work: Tumor immunology. Home: 829 Moultrie Ct Virginia Beach VA 23455 Office: 700 Olney Rd Norflk VA 23451

WRIGHT, GEORGE NELSON, psychologist, educator; b. Earlington, Ky., Dec. 10, 1921; s. James Nelson and Jennie (Carr) W.; m. Patricia Gilmer, June 8, 1952; children: John Nelson, Elizabeth, Robert Carr, James Nelson. B.S., Ind. U., 1947; M.S., Purdue U., 1954, Ph.D., 1958. Employment interviewer Mo. Valley Bridge Co., Evansville, Ind., 1941-45; rehab. counselor Ind. Div. Vocat. Rehab., Lafayette, Ind., 1947-58; research fellow Purdue U., West Lafayette, Ind., 1958-59; nat. program dir. Nat. Epilepsy League, Chgo., 1959-62; prof. rehab. counselor edn., dir. Rehab. Research Inst., U. Wis.-Madison, 1962—. Author: Epilepsy Rehabilitation, 1975, Total Rehabilitation, 1980, (with H.H. Remmers) Handicap Problems Inventory, 1960, (with F.A. Gibbs) Total Rehabilitation of Epileptics, 1962, (with A.J. Butler) Rehabilitation Counselor Functions, 1968, (with A.B. Trotter) Rehabilitation Research, 1968, (with K.W. Reagles and S. Katz) Rehabilitation in Israel, 1974; contbr. articles to profl. jours. Mem. Am. Psychol. Assn. (fellow divs. counseling and rehab., pres. 1983), Am. Personnel and Guidance Assn., Am. Rehab. Counseling Assn. (pres. 1975), Nat. Council Rehab. Counselor Educators, Nat. Rehab. Assn., Nat. Vocat. Guidance Assn., Sigma Xi. Subspecialty: Rehabilitation psychology. Home: 492 Presidential Lane Madison WI 53711 Office: Rehab Research Inst U Wis Madison WI 53706

WRIGHT, HARLAN TONIE, biochemistry educator; b. Hackensack, N.J., Feb. 5, 1941; s. Harlan Dalzell and Dorothy (Cullingford) W.; m. Christine Gerda Schubert, Aug. 31, 1965; children: Colin Eliot, Angelique Colleen. A.B., Princeton U., 1963; Ph.D., U. Calif.-San Diego, 1968. Fellow U. Calif.-San Diego, La Jolla, 1968-69, MRC Lab., Cambridge, Eng., 1969-71, Princeton U., 1971-73, instr. biochemistry, 1973-78; asst. prof. biochemistry Med. Coll. Va./Va. Commonwealth U., Richmond, 1980—; non-resident fellow Linus Pauling Inst., Menlo Park, Calif., 1978-80, Oreg. Inst. Sci. and Medicine, Cave Junction, 1981—. Contbr. chpts. to books, articles to profl. jours. Committeeman Boy Scouts Am., Richmond, 1981—; investigator Am. Heart Assn., 1973-78. Mem. AAAS, Am. Crystallographic Assn. Republican. Episcopalian. Subspecialties: Biochemistry (biology); X-ray crystallography. Current work: Structure of biological macromolecules; properties and functions of enzymes, nucleic acids and membrane associated proteins. Home: 105 N Wilton Rd Richmond VA 23226 Office: Dept Biochemistry Box 614 MCV Station Med Coll Va Richmond VA 23298

WRIGHT, JACKSON T., JR., pharmacology educator; b. Pitts., Apr. 28, 1944; s. Jackson T. and Lillian (Doak) W.; m. Molly L. Richardson, Sept. 2, 1967; 1 dau., Adina Marie. B.A., Ohio Wesleyan U., 1967; M.D., U. Pitts., 1976, Ph.D., 1977. Diplomate: Am. Bd. Internal Medicine. Med. intern U. Mich.-Ann Arbor, 1971-78, med. resident, 1978-80; asst. prof. pharmacology and medicine Med. Coll. Va.-Va. Commonwealth U., Richmond, 1980—. Served to capt. USAF, 1967-71. Woodrow Wilson-Martin Luther King fellow, 1971-73. Mem. Am. Soc. Clin. Pharmacology and Therapeutics, Am. Fedn. Clin. Research, ACP, Nat. Med. Assn., Sigma Xi. Subspecialties: Internal medicine; Pharmacology. Current work: Renal prostaglandins and hypertension; cardiovascular pharmacology. Office: Med Coll Va 426 McGuire Hall MCV Box 613 Richmond VA 23298 Home: 538 Rossmore Rd Richmond VA 23225

WRIGHT, JOHN CURTIS, chemistry educator; b. Lubbock, Tex., Sept. 17, 1943; s. John Edward and Jean (Love) W.; m. Carol Swanson Wright, Aug. 17, 1968; children: Dawna Lynn, John David. B.S. in Physics, Union Coll. Schenectady, 1965, Ph.D., Johns Hopkins U., 1970. Postdoctoral fellow Purdue U., West Lafayette, Ind., 1970-72; asst. prof. chemistry U. Wis.–Madison, 1972-78, assoc. prof., 1979-83, prof., 1980—. Contbr. articles in field to profl. jours. Recipient William F. Meggers award, 1981. Mem. Am. Chem. Soc., Am. Phys. Soc. Presbyterian. Subspecialties: Analytical chemistry; Solid state chemistry. Current work: Laser spectroscopy, analytical chemistry, defects. Office: Dept Chemistry U Wis Madison WI 53706

WRIGHT, JOHN RICKEN, chemist, educator, cons.; b. Batesville, Ark., Jan. 3, 1939; s. John Adam and Alice Lanelle (Burge) W.; m. (married), Feb. 1, 1964; 1 dau.: Karen Elizabeth. Student, Ark. Coll., Batesville, 1956-57; B.S., Ark. State U., Jonesboro, 1960; M.S., U. Miss., 1967, Ph.D., 1971. High sch. sci. tchr., Wynne, Ark., 1960-61; research asst. Washington U., St. Louis, 1967-68; postdoctoral research assoc. in chemistry Fla. State U., Tallahassee, 1972-73; assoc. prof. chemistry Southeastern Okla. State U., Durant, 1973—; cons. Dept. Energy, NASA. Contbr. chem. articles to profl. jours. Served to capt. USAF, 1961-65. NIH grantee, 1974—. Mem. Sigma Xi. Lodge: Kiwanis. Subspecialties: Inorganic chemistry; Immunobiology and immunology. Current work: Development of anticancer immunotoxins which double as staining agents for light and electron microscopy. Copper cluster chemistry. Office: Southeastern Okla State U Box 4181 Sta A Durant OK 74701

WRIGHT, LARRY LYLE, immunologist, researcher; b. Rochester, N.Y., July 15, 1941; s. Raymond A. and Eleanor N. W. B.A., Andrews U., 1965; M.S., U. Ga., 1973, Ph.D., 1975. Tchr. Greece Olympia High Sch., Rochester, 1966-70; instr. Med. Coll. Va., 1975-76; med. research assoc. Duke U. Med. Center, 1976-83; research fellow Nat. Inst. Environ. Health Sci., Research Triangle Park, N.C., 1981—. Contbr. articles to profl. jours. Mem. Am. Soc. Microbiology. Subspecialties: Immunobiology and immunology; Immunotoxicology. Current work: Immunotoxicology; generating monoclonal antibodies; using enzyme immunoassays in determining change in antibody activity against marker molecules from mice treated with suspected mutagens. Office: NIEHS MD 406 Research Triangle Park NC 27231

WRIGHT, MELVYN CHARLES HARMAN, astronomer; b. Oadby, Eng., Sept. 1, 1944; came to U.S., 1970; s. Charles William and Florence Nellie (Harman) W. B.A., St. John's Coll., Cambridge (Eng.) U., 1966, M.A., 1969, Ph.D., 1970. Research assoc. Nat. Radio Astronomy Obs., 1970-72; research fellow Calif. Inst. Tech., Pasadena, 1972, vis. assoc., 1972-76; asst. research astronomer Radio Astronomy Lab., U. Calif.-Berkeley, 1972-80, assoc. research astronomer, 1980—; staff mem. Owens Valley Radio Obs., 1972-76. Mem. Internat. Astron. Union, Royal Astron. Soc., Am. Astron. Soc., Internat. Union Radio Sci., Common Cause, Union Concerned Scientists. Subspecialty: Radio and microwave astronomy. Current work: Millimeter Radio Astronomy, Aperture synthesis techniques, spiral and radio galaxies intergalactic medium, interstellar medium in regions of star formation. Home: 1136 Euclid Ave 4 Berkeley CA 94708 Office: Radio Astronomy Lab U Calif Berkeley CA 94720

WRIGHT, PAUL KENNETH, mechanical engineer, educator; b. Watford, Herts, Eng., Aug. 24, 1947; came to U.S., 1979; s. Kenneth Browitt and Violet Annie (Woodland) W.; m. Frances June Ody, Oct. 24, 1970; children: Samuel, Joseph, Thomas. B.Sc., U. Birmingham, Eng., 1968, Ph.D. in Indsl. Metallurgy, 1971. Cons. to N.Z. Govt., 1972-74; sr. lectr. U. Auckland, N.Z., 1975-79; research assoc. Cavendish Lab., Cambridge, Engr., 1978-79; asso. prof. dept. mech. engring. and robotics Carnegie Mellon U., Pitts., 1979—. Contbr. articles to profl. jours. Recipient George Tallmann Ladd award Carnegie Mellon U., 1981; Royal Soc. Commonwealth Bursary grantee, 1978. Mem. Robotics Inst. Am., ASME, Soc. Mfg. Engrs. (Outstanding Young Mfg. Engr. of Yr. award 1980). Subspecialties: Materials processing; Robotics. Current work: Computer aided manufacturing and materials processing. Home: 5604 Fair Oaks St Pittsburgh PA 15217 Office: Dept Mech Engring Carnegie Mellon U Pittsburgh PA 15213

WRIGHT, RICHARD KENNETH, comparative immunologist; b. Richmond, Ind., Sept. 22, 1939; s. Willard Kenneth and Dorothy Janelle (Moon) W. B.S., San Diego State U., 1967, M.S., 1970; Ph.D., U. Calif., Santa Barbara, 1973. Teaching asst. San Diego State U., 1967-69; assoc. biol. scis. U. Calif., Santa Barbara, 1971-73; USPHS postdoctoral fellow UCLA, 1973-75, asst. research anatomist, 1975-81, assoc. research anatomist, 1981—. Author: Phylogeny of Thumus and Bone Marrow-Bursa Cells, 1976; contbr. articles to profl. jours. Served with USN, 1960-64. Mem. Am. Assn. Immunologists, Internat. Soc. Devel. and Comparative Immunology, Am. Soc. Zoologists, AAAS, Sigma Xi. Republican. Subspecialties: Immunobiology and immunology; Immunology (medicine). Current work: Immunobiology of invertebrates (tunicates), amphibians and reptiles; evolution of immune responses; immunobiology of tunicates, amphibians and reptiles; invertebrate hemocyte membrane receptors for foreignness. Home: 1315 Stanford St #1 Santa Monica CA 90404 Office: Dept Anatomy Sch Medicine UCLA Los Angeles CA 90024

WRIGHT, SEWALL, geneticist, educator; b. Melrose, Mass., Dec. 21, 1889; s. Philip and Elizabeth (Sewall) W.; m. Louise Lane Williams, Sept. 10, 1921; children: Richard, Robert, Elizabeth Quincy (Mrs. John Rose). B.S., Lombard Coll., Galesburg, Ill., 1911; M.S., U. Ill., 1912, Sc.D. (hon.), 1961, Harvard U., 1915; hon. Sc.D., U. Rochester, 1942, Yale U., 1949, Knox Coll., 1957, Western Res. U., 1958, U. Chgo., 1959, U. Wis., 1965; LL.D. (hon.), Mich. State U., 1955. Sr. animal husbandman U.S. Dept. Agr., Washington, 1915-25; assoc. prof. zoology U. Chgo., 1926-29, prof., 1930-37, Ernest D. Burton Disting. Service prof., 1938-54; Leon J. Cole prof. genetics U. Wis., Madison, 1955-60, prof. emeritus, 1960—; Hitchcock prof. U. Calif., Berkeley, spring 1943; Fulbright prof. U. Edinburgh, Scotland, 1949-50; pres. 10th Internat. Congress Genetics, 1958. Author: Evolution and the Genetics of Populations, 4 vols, 1968-78; over 200 articles. Recipient Weldon Meml. medal Oxford (Eng.) U., 1947; Nat. medal of Sci., 1966; Darwin medal Royal Soc. London, 1980; Thomas Hunt Morgan medal Genetics Soc. Am., 1982. Mem. Nat. Acad. Sci. (Girard Elliott medal 1945, Kimber Genetics award 1956), Am. Soc. Zoologists (pres. 1934), Genetics Soc. Am. (pres. 1944), Am. Philos. Soc. (Lewis prize 1950), Am. Soc. Naturalists (pres. 1952), AAAS, Soc. Study of Evolution (pres. 1955), Royal Soc. Edinburgh (hon.), Royal Soc. London, Royal Danish Acad. Arts and Sci., others. Democrat. Unitarian. Club: University (Madison). Subspecialties: Animal genetics; Evolutionary biology. Current work: Theory of evolution. Office: U Wis Madison WI 53706

WRIGHT, WILLIAM THOMAS, clinical psychologist, educator; b. Winfield, Kans., Dec. 27, 1923; s. William Thomas and Gladys Sara (Hatfield) W.; m. Betty Lou Weekley, June 2, 1947; children: Claudia Ann, Thomas Brian, Allison Lynn. B.S., Southwestern Coll., Winfield, 1947; M.S., Kans. State U., 1949; Ph.D., U. Denver, 1958. Cert. clin. psychologist, Kans. Grad. asst. instr. Kans. State Coll., 1949; clin. psychologist Winfield State Hosp., 1950-51, Larned State Hosp., 1951-53; chief clin. psychologist Hertzler Clinic, Halstead, Kans., 1953-56; cons. psychologist in pvt. practice, Denver, 1956-58; chief clin. psychologist Hertzler Clinic, 1958-61, Prairie View, Inc., Newton, Kans., 1961—; also asst. prof. psychology U. Kans. Med. Sch.-Wichita, 1980—. Served with M.C. AUS, 1942-46. Mem. Am. Psychol. Assn., Am. Group Therapy Assn., Kans. Psychol. Assn. Presbyterian. Lodges: Elks; Lions; Masons. Current work: Clinical general practice, psychotherapy, neuropsychology. Office: PO Box 467 1902 E 1st St Newton KS 67114 Home: 1205 Parkwood Ln Newton KS 67114

WRIGHTON, MARK STEPHEN, chemistry educator; b. Jacksonville, Fla., June 11, 1949; s. Robert D. and Doris (Cutler) W.; m. Deborah Ann Wiseman, Aug. 10, 1968; children: James Joseph, Rebecca Ann. B.S., Fla. State U., 1969; Ph.D., Calif. Inst. Tech., 1972; D.Sc. (hon.), U. West Fla., 1983. Asst. prof. chemistry M.I.T., 1972-76, assoc. prof., 1976-77, prof., 1977—, Frederick G. Keyes prof. chemistry, 1981—; cons. Gen. Electric; Alfred P. Sloan fellow, 1974-76. Author: Organometallic Photochemistry, 1979; editor books in field; cons. editor, Houghton-Mifflin. Recipient Herbert Newby McCoy award Calif. Inst. Tech., 1972; Dreyfus Tchr.-Scholar, 1975-80; MacArthur fellow, 1983—. Mem. Am. Chem. Soc. (award in pure chemistry 1981), Electrochem. Soc., AAAS. Subspecialties: Inorganic chemistry; Photochemistry. Current work: Energy conversion, catalysis, photochemistry. Office: Dept Chemistry MIT Cambridge MA 02139

WROBLEY, ARTHUR RAY, resource co. exec., cons.; b. Denver, Dec. 6, 1938; s. Matthew B. and Helen H. (Lejon) W.; m. Darlene M. Hines, Nov. 7, 1978. B.S. in Forestry, U. Mo., 1966; postgrad., No. Ariz. U., 1975-77. Registered forester, Okla., Ark. Forester U.S. Forest Service, Mo., Wis. and Ariz., 1966-74; adminstrv. asst. Tupper Tree Farm, Ariz., 1974; pres. SEC, Inc., Sedona, Ariz., 1974—. Served with USCG, 1958-62. Mem. Soc. Am. Foresters, Assn. Cons. Foresters, Am. Forestry Assn. Subspecialties: Resource conservation; Resource management. Current work: Actively engaged in the management of our natural resources with an emphasis on the forest resource. Office: PO Box 1471 Rojo Vista Bldg Sedona AZ 86336

WU, ANDREW, medical physicist; b. Shanghai, China, Jan. 23, 1938; came to U.S., 1963, naturalized, 1970; s. Cheng-Hsiang and Ming-Tsu (Ma) W.; m. Angela Fu, Jan. 13, 1940; 1 dau., Christine. Ph.D., Temple U., 1971. Diplomate: Am. Bd. Radiology. Asst. physicist Cooper Med. Center, Camden, N.J., 1972-74; chief physicist Divine Providence Hosp., Williamsport, Pa., 1974-81; head clin. physicist Tufts-New Eng. Med. Ctr., Boston, 1981—, clin. asst. prof., 1981—. Contbr. articles to sci. jours. NIH fellow, 1971. Mem. Am. Inst. Physics, Am. Assn. Physicists in Medicine, Am. Soc. Therapeutic Radiology. Subspecialty: Medical physics. Current work: Medical physics and hyperthermia. Home: 14 Regis Rd Wellesley MA 02181 Office: Dept Therapeutic Radiology Tufts New Eng Med Ctr 171 Harrison Ave Boston MA 02111

WU, CHENG-WEN, biochemistry educator; b. Taipei, Taiwan, June 19, 1938; s. Hai-Chu and Fan-Po (Chen) W.; m. Felicia Y. H. Chen, Nov. 10, 1963; children: David, Faith, Albert. M.D., Nat. Taiwan U., 1964; Ph.D., Case Western Res. U., 1969. Postdoctoral assoc. Cornell U., Ithaca, N.Y., 1969-70; NIH spl. fellow Yale U., New Haven, 1970-72; from asst. prof. to prof. biophysics Albert Einstein Coll. Medicine, Bronx, N.Y., 1972-77, prof. biochemistry, 1978-79; vis. prof. Institut Pasteur, Paris, 1979-80; prof. pharm. sci. SUNY-Stony Brook, 1980—; mem. spl. rev. sect. NIH, Bethesda, Md., 1982—, physiol. chemistry study sect., 1978-82; lectr. NATO Sch. Molecular Biology, 1976. Editor: Archives of Biochemistry and Biophysics, 1978—. Catacosinos prof. for cancer research Stony Brook Found., 1980; recipient Irma T. Hirschl Sci. award, 1977; Philippe Found. Sci. award, 1978; NIH career devel. awardee, 1972. Mem. Am. Soc. Biol. Chemists, Am. Chem. Soc., Biophys. Soc., N.Y. Acad. Scis., Sigma Xi. Subspecialties: Biochemistry (medicine); Genetics and genetic engineering (medicine). Current work: Gene regulation and carcinogenesis, nucleic acid-protein interactions, fast reactions in biological systems, spectroscopic studies of macromolecules. Home: 7 W Meadow Rd Setauket NY 11733 Office: Dept Pharm Scis SUNY Stony Brook NY 11794

WU, CHIEN HENG, mechanics educator, academic administrator; b. Szechuan, China, June 15, 1935; came to U.S., 1960, naturalized, 1972; s. Kao Tien and Yin (Chen) W.; m. Sylvia Pan, Dec. 18, 1965; children: Helen, Anita, Patricia. B.S., Nat. Taiwan U., Taiwan, 1957; M.S., U. Kans.-Lawrence, 1962; Ph.D., U. Minn.-Mpls., 1965. Asst. prof. mechanics U. Ill.-Chgo., 1966-70, assoc. prof., 1970-74, prof., 1974—, head dept. civil engring., mechanics and metallurgy, 1982—. Research publs. in field. Bd. dirs. Orgn. Chinese Ams.-Chgo. North Chinese Lang. Sch. NSF grantee, 1979-82; Army Research Office grantee, 1970-78. Mem. ASME, Soc. Indsl. and Applied Math., Orgn. Chinese Ams. Subspecialties: Theoretical and applied mechanics; Fracture mechanics. Current work: Linear and nonlinear elasticity, fracture mechanics, nonlinear oscillations, asymptotic methods. Home: 3716 Davis St Skokie IL 60076 Office: Dept Civil Engring Mechanics and Metallurgy U Ill-Chgo Box 4348 Chicago IL 60680

WU, CHIEN-SHIUNG, physicist; b. Shanghai, China, 1912; d. Zong-Ye W. and F.H. Fan; ; d. Zong-Ye W. and; m. Dr. Luke Chia-liu Yuan, May 30, 1942; 1 son, Vincent W.C. Yuan. B.S., Nat. Central U., China, 1934; Ph.D., U. Calif. at Berkeley, 1940; D.Sc., Princeton, 1958, Smith Coll., 1959, Goucher Coll., 1960, Yale, 1967, Russel Sage Coll., 1971, Harvard, 1974; LL.D., Chinese U. Hong Kong, 1969, others. Physicist, specializing nuclear physics; Pupin prof. physics Columbia, 1957—. Recipient research award Research Coop. Am., 1958; Woman of Year award AAUW, 1962; Comstock award Nat. Acad. of Sci., 1964; Chia Hsin Achievement award, 1965; Nat. Medal of Scis., 1975; 1st Wolf prize, Israel, 1978; named Scientist of Yr. Indsl. Research, 1974. Fellow Royal Soc. Edinburgh (hon.); mem. Nat. Acad. Sci., Am. Acad. Arts and Sci., Am. Phys. Soc. (pres. 1975), Acadamie Sinica of China. Subspecialty: Nuclear physics. Current work: Nuclear beta decays and weak interactions; exotic atoms; ultra-low temperature nuclear physics. Experimentally established non-conservation of parity in beta-decay, 1957. Home: 15 Claremont Ave New York NY 10027

WU, CHIH, mech. engr., educator, researcher; b. Changsha, Hunan, China, Apr. 13, 1936; came to U.S., 1961, naturalized, 1973; s. K.T. and D.R. W.; m. Holly H.Y., Jan. 27, 1966; children: Anna, Joy, Sheree, Patricia. B.S., Cheng Kung U., Taiwan, 1957; Ph.D. in Mech. Engring, U. Ill.-Urbana, 1966. Asst. prof. to prof. mech. engring. U.S. Naval Acad., Annapolis, Md., 1966—; prof. evening coll. Johns Hopkins U., Balt., 1969—. Contbr. articles profl. jours. Mem. ASME, Am. Soc. Engring. Edn. Subspecialty: Mechanical engineering. Current work: Thermodynamics, fluid dynamics, heat transfer, energy conversion, statistics, computer application, education. Home: 1705 Tarleton Way Crofton MD 21114 Office: Dept Mech Engring US Naval Acad Annapolis MD 21402

WU, CHUN-FANG, biology educator; b. Fujian, China, Feb. 4, 1947; came to U.S., 1970; s. Chung-Chuen and Yun (Chen) W.; m. Mei-Lien Lin, Sept. 25, 1971; children: Daw-An, Yusing, Tonying. B.S., Tunghai U., Taiwan, 1969; Ph.D., Purdue U., 1976. Research asst. Purdue U., West Lafayette, Ind., 1971-76, research assoc., 1976; research fellow Calif. Inst. Tech., Pasadena, 1976-79; asst. prof. U. Iowa, Iowa City, 1979-83, assoc. prof., 1983—. Contbr. articles to profl. jours. Searle scholar, 1981-84; Spencer research fellow, 1976-78; recipient Research Career Devel. award NIH, 1982-87. Mem. Soc. Neurosci., Biophys. Soc., AAAS, Phi Kappa Phi. Subspecialties: Neurobiology; Neurophysiology. Current work: Genetic dissection of nerve and muscle membrane excitability in Drosophila. Office: Dept Zoology U Iowa Iowa City IA 52242

WU, FELICIA YING-HSIUEH, biochemical educator, researcher; b. Taipei, Taiwan, China, Feb. 27, 1939; came to U.S., 1965, naturalized, 1976; d. I-Sung and Ti (Yen) Chen; m. Cheng-Wen Wu, Nov. 10, 1963; children: David, Faith, Albert. B.S., Nat. Taiwan U., 1961; M.S., U. Minn., 1963; Ph.D., Case Western Res. U., 1969. Med. technician U.S. Naval Med. Research Unit #2, Taipei, 1963-65; research assoc. Cornell U. Sect. Biochemistry and Molecular Biology, 1969-71; research assoc. dept. pharmacology Yale U., 1971; assoc. dept. biophysics Albert Einstein Coll. Medicine, 1972-73, instr., 1973-78, asst. prof. dept. biochemistry, 1978-79; assoc. prof. dept. pharm. scis. SUNY-Stony Brook, 1979—; vis. prof. Pasteur Inst., Paris, 1979-80, Inst. Gustave-Roussy, Villejuif, France, 1980. Contbr. numerous articles to profl. jours. NIH grantee, 1972-84; Am. Cancer Soc. grantee, 1972-84; NSF grantee, 1980-83; recipient Catacosinos Cancer Research award, 1980; Chinese Govt. Model Youth award, 1967. Mem. Am. Chem. Soc., Biophys. Soc., Am. Soc. Biol. Chemists. Subspecialties: Biochemistry (biology); Cancer research (medicine). Current work: Role of metal ions in genetic information transfer; mechanism of action of a new antitumor drug; chemical carcinogenesis; DNA-protein interactions. Home: 7 W Meadow Rd Setauket NY 11733 Office: Dept Pharmacol Scis SUNY Basic Health Science Center 7T-182 Stony Brook NY 11794

WU, GUANG-JER, microbiology educator, researcher; b. Shin-Chu, Taiwan, Republic of China, Jan. 15, 1943; s. Tsai-Wang and Tsai-Chin W.; m. Mei-Whey H., Mar. 26, 1971; children: Jeanette Shin-Jin, Felise Shin-Shaw, Jonathan Chung-Ying. B.S. in Agrl. Chemistry, Nat. Taiwan U., Taipei, 1965; Ph.D. in Biochemistry and Biophysics, U. Calif.-Davis, 1970. Teaching asst. U. Calif.-Davis, 1966-70; postdoctoral fellow dept. embrology Carnegie Inst., Washington, Balt., 1970-72; research assoc. dept. biol. scis. Columbia U., 1972-76; asst. prof. microbiology Emory U. Sch. Medicine, 1976-81, assoc. prof. microbiology and immunology, 1981—. Contbr. articles to sci. jours. Nat. Cancer Inst. grantee, 1979-82. Mem. Am. Soc. Microbiology, Am. Soc. Virology, AAAS, Am. Assn. Cancer Research, Sigma Xi. Subspecialties: Molecular biology; Gene actions. Current work: Regulation of transcription in normal and tumor cells. Office: Dept Microbiology and Immunology Emory U Sch Medicine Atlanta GA 30322

WU, JAIN-MING (JAMES WU), aerospace engineering educator, university administrator, consultant; b. Nanking, China, Aug. 13, 1933; came to U.S., 1957, naturalized, 1969; m. Ying-Chu (Susan) Lin, June 13, 1959; children: Ernest H., Albert H., Karen H. B.S. in Mech. Engring, Nat. Taiwan U., Taiwan, 1955; M.S. in Aero. Engring, Calif. Inst. Tech., 1965; Ph.D. in Aeros, Calif. Inst. Tech., 1965. Mem. tech. staff Nat. Engr. Sci. Co., Pasadena, Calif., 1960-63; asst. prof. aerospace engring. U. Tenn. Space Inst., 1965-67, assoc. prof., 1967-72, prof., 1972—, dir. gas dynamics div., 1975—; vis. prof. Von Karman Inst., Rhode-St-Genese, Belgium, 1970; cons. in field. Contbr. numerous articles to profl. publs. Vice pres. Tullahoma (Tenn.)

Unitarian Fellowship, 1969. Anthony scholar, 1965; grantee USAF, Def. Advanced Projects Research Agy., NASA. Assoc. fellow AIAA (Gen. H. H. Arnold award Tenn. sect. 1970); mem. Am. Rocket Soc., Southeastern Conf. Theoretical and Applied Mechanics, Order of Engrs., Sigma Xi. Club: Tullahoma Country. Subspecialties: Aeronautical engineering; Fluid mechanics. Current work: Fluid mechanics and aerothermal dynamics in aerospace engineering, both theoretical and experimental; jet propulsion, external and internal flows. Office: U Tenn Space Inst Tullahoma TN 37388 Home: 111 Lakewood Tullahoma TN 37388

WU, JAMES *See also* **WU, JAIN-MING**

WU, JIA-HSI, biologist, educator; b. Taiwan, July 6, 1926; m.; children: Su-Ming, Su-Lin. B.S., Taiwan U., 1950; M.S., Cornell U., 1952; Ph.D., Washington U., St. Louis, 1958. Research assoc. U. Wis., Madison, 1958-59; asst. botanist UCLA, 1959-63; asst. prof. Tex. Technol. U., Lubbock, 1963-66; prof. dept. biol. sci. Calif. State Poly. U., Pomona, 1966—. Contbr. articles to profl. jours. Mem. AAAS, Am. Soc. Plant Physiologists, Am. Soc. Photobiologists. Subspecialties: Plant virology; Plant pathology. Current work: Mechanism of host cell resistance to spread of viruses. Office: 3801 W Temple St Pomona CA 91768

WU, JOHN NAICHI, elec. co. exec.; b. Soochow, China, Sept. 10, 1932; m. Mary C. Chan; children: Winthrop J., Jarvis C. Ph.D., U. Fla., 1965. Sr. research specialist Alliance Research Center, Babcock & Wilcox Co., Alliance, Ohio, 1962-66, group supr., 1967-77; mngr. Materials and Processes Lab., Transp. System Bus. Operation, Gen. Electric Co., Erie, Pa., 1977—, chmn. research and tech. papers com., 1980—, engring. design coordinator, 1981—. Contbr. articles to profl. jours. Bd. dirs. Erie Internat. Inst., 1981—; pres. Erie Chinese Assn., 1982. Mem. Am. Acad. Mechanics, ASME, Acoustic Soc. Am., Am. Metals Engring., Phi Kappa Phi. Subspecialties: Theoretical and applied mechanics; Mathematical software. Current work: Applied mechanics. Home: 139 Putnam Dr Erie PA 16511 Office: 2901 E Lake Rd Erie PA 16531

WU, JOSEPH MAN-HAY, biochemist, researcher; b. Shanghai, China, Aug. 1, 1947; came to U.S., 1970, naturalized, 1983; s. Eisen Yen-Sun and Yui-Wei (Cheung) W.; m. Susan H. Chou, Nov. 24, 1974; children: Amy H., Mary H. B.S., McGill U., Montreal, Que., Can., 1970; M.S., Fla. State U., Tallahassee, 1972, Ph.D., 1975. Postdoctoral research assoc. Temple U., Phila., 1976-77, research instr., 1977-78; asst. prof. N.Y. Med. Coll., Valhalla, 1978-82, assoc. prof., 1982—. NIH grantee, 1980-83; Nat. Inst. Alcoholism research grantee, 1982-83. Mem. Am. Soc. Biol. Chemists. Subspecialties: Biochemistry (biology); Developmental biology. Current work: Research on interferon action, gene regulation, development and differentiation. Home: 6 Hollywood St Mohegan Lake NY 10547 Office: Dept Biochemistry NY Med Coll Valhalla NY 10595

WU, KENNETH KUN-YU, physician, educator; b. Koahsiung, Taiwan; s. Chuan W. and Chin-Piau W.; m. Lung-chin Shih, Mar. 29, 1969; children: Stanley, David. M.D., Nat. Taiwan U., 1966; M.S., Yale U., 1968. Diplomate: Am. Bd. Internal Medicine. Resident and fellow in internal medicine and hematology U. Iowa, Iowa City, 1969-72, asst. prof. medicine, 1974-76; assoc. prof. medicine Rush U. Med. Coll., Chgo., 1976-81, prof. medicine, 1981—; dir. coagulation and thrombosis unit Rush-Presbyn.-St. Luke's Med. Ctr., Chgo., 1976—; mem. program project rev. com. Nat. Inst. Neurol. Disease and Stroke, NIH, 1977-81; mem. research com. Chgo. Heart Assn., 1981—. Contbr. articles to profl. jours. NIH, Am. Heart Assn. grantee. Fellow ACP; mem. Am. Soc. Hematology, Am. Assn. Immunologists, Am. Heart Assn., Central Soc. Clin. Research, Am. Fedn. Clin. Research. Subspecialties: Hematology; Biochemistry (medicine). Current work: Biochemical mechanisms of thrombosis and hemostasis; particularly interested in prostaglandins, platelet physiology and biochemistry. Home: 1642 Robin Ln Glenview IL 60025 Office: 1753 W Congress Pkwy Chicago IL 60612

WU, RAYMOND KEE-KIN, medical radiation physicist, educator; b. Canton, China, Feb. 4, 1948; came to U.S., 1969, naturalized, 1983; s. Leung and Wai-Tak (Chow) W.; m. Dulcie W. Wu, Apr. 20, 1975; 1 son, Jeffrey Kai. B.Sc. in Physics, Chinese U. Hong Kong, 1969; Ph.D., Dartmouth Coll., 1973. Diplomate: Am. Bd. Radiology. Med. physicist Dept. Radiation Therapy, Thomas Jefferson U. Hosp., Phila., 1973-75; dir. Div. Med. Physics, St. Joseph Hosp., Milw., 1975-77; asst. prof., dir. radiation therapy physics Temple U. Hosp., Phila., 1977-78, assoc. prof., 1978—, dep. dir. dept. radiation oncology, 1982—; cons. physicist dept. radiation oncology Chang Gung Meml. Hosp., Taipei, Taiwan, 1978; mem. adv. com. AEC, Republic of China, 1981—. Contbr. articles to profl. jours. Mem. Am. Assn. Physicists in Medicine (pres. Delaware Valley chpt.), Am. Soc. Therapeutic Radiologists, Am. Coll. Radiology. Republican. Club: Chinese Community Center of South Jersey. Subspecialties: Radiology; Biophysics (physics). Current work: Radiation therapy physics, radiation dosimetry in therapeutic radiology and nuclear medicine. Office: 3401 N Broad St Philadelphia PA 19140

WU, SING-YUNG, physician, clinical researcher, educator; b. Chengtu, China, July 5, 1939; s. Samuel Sung-Ching and Shih Lin W.; m. Yvonne Y.C. Yu, Aug. 2, 1982. M.B.D., Nat. Taiwan U., Taipei, 1963; Ph.D., U. Wash., 1969; M.D., Johns Hopkins U., 1972. Research asst. Washington U., Balt., 1964-65; fellow U. Wash., Seattle, 1965-69, NIH fellow in medicine, 1973-75; med. intern U. Chgo., 1971-72; med. resident U. Calif.-Irvine, 1972-73, asst. prof., 1977—; NIH fellow UCLA, 1975-77; staff physician, prin. investigator VA Med. Ctr., Long Beach, Calif., 1977—. Author: Microsomal Electron Transport and Drug Oxidation, 1969, Atlas of Nuclear Medicine, 1982; contbr. articles to med. jours. VA grantee, 1977, 79, 82. Mem. Am. Fedn. Clin. Research, Am. Thyroid Assn., Nuclear Medicine Soc. Subspecialties: Endocrinology; Nuclear medicine. Current work: Thyroid physiology, intermediate thyroid hormone metabolism, monodeiodinating enzymes in thyroid and in peripheral tissues; clinical endocrinology and nuclear medicine. Office: VA Med Ctr 5901 E 7th St Long Beach CA 90822

WU, SOUHENG, physical chemist; b. Tainan, Taiwan, Jan. 16, 1936; came to U.S., 1961, naturalized, 1972; s. Pei-Song and Long-Yu (Hsu) W.; m. Tung Ching Wang, Apr. 3, 1965; 1 son, Lawren Chialun. B.S., Nat. Cheng Kung U., Taiwan, 1958; Ph.D., U. Kans., 1965. Researcher Taiwan Sugar Corp., 1958-61; research chemist DuPont Co., Wilmington, Del., 1965-68, staff chemist, 1968-70, research assoc., 1970-76, mem. research staff, 1976—. Author: Polymer Interface and Adhesion, 1982; Contbr. numerous articles to sci. jours. Mem. Am. Chem. Soc., Am. Phys. Soc., Soc. Rheology. Subspecialties: Polymer physics; Polymers. Current work: Polymer rheology, polymer interface and adhesion; polymer blends and composites; polymer coatings; molecular dynamics. Patentee in field. Home: 711 Taunton Rd Wilmington DE 19803 Office: EI duPont de Nemours & Co Exptl Sta 356/207 Wilmington DE 19898

WU, STEVE, mechanical engineer; b. Macau, Nov. 11, 1923; U.S., 1959, naturalized, 1968; s. Pak Tin and Shio Chi (Soo) W.; m.; 1 dau., Theresa. B.S.M.S., Aero. Inst., Chung King, China, 1947; M.S.M.E., U. Portland, 1976. Registered profl. engr., Oreg. Mem. tech. staff Civil Air Transport, China, 1953-60; engr. Cascade Corp., Portland, Oreg., 1960-79; asst. prof. mech. engring. Portland State U., 1979-80; pres. Wemco Crang Co., Portland, 1980—; adj. prof. mech. engring. Portland State U. and U. Portland, 1970-79; cons. in field. Mem. ASME. Subspecialties: Fluid mechanics; Theoretical and applied mechanics. Current work: Hydraulic machinery; mechanic robot. Patentee hydraulic crane, hydraulic seal, rotating mechanism; developed formulas for critical load of n-sects. column, 180 degree phase error in direction finding for aircraft. Home: 11740 SE Salmon St Portland OR 97216

WU, SUSAN *See also* **WU, YING-CHU LIN**

WU, WEN-LI, polymer physicist, researcher; b. Chentu, Szuchuan, China, Nov. 13, 1945; came to U.S., 1968, naturalized, 1981; s. Chinting and Shih-Chieh (Chung) W.; m. Katie K. Young, June 18, 1971; children: Ju-ru, Yu-jing. B.S., Nat. Taiwan U., Taipei, 1967; M.S., MIT., 1969, Ph.D., 1972. Sr. research engr. Monsanto Co., St. Louis, 1973-75, research specialist, 1975-77, sr. research specialist, 1977-79; material engr. Nat. Bur. Standards, Washington, 1979—. Contbr. articles to profl. jours. Mem. Am. Physics Soc., Am. Chem. Soc., Am. Assn. Dental Research. Subspecialties: Polymers; Polymer physics. Current work: Theoretical and experimental studies of deformation mechanisms and structure-properties relation of polymers and composites. Patentee in field. Home: 8 Manette St Gaithersburg MD 20878 Office: Nat Bur Standards Room 143A Bldg 224 Washington DC 20234

WU, WILLIAM GAY, microbiology educator; b. Portland, Oreg., Feb. 5, 1931; s. George Pon and Ida Lillian (Moy) W.; m. Gladys Kam Min, Jan. 5, 1957; children: Jeffrey S., Randall K., Mitchel G. B.S., Oreg. State U., 1954, M.S., 1959; Ph.D., U. Utah, 1962. Lab. asst. microbiology Oreg. State U., Corvallis, 1957-58, research asst. vet. microbiology, 1958-59; research asst., USPHS fellow U. Utah, Salt Lake City, 1959-62; faculty San Francisco State U., 1962—, prof. biology, 1969—, chmn. dept. microbiology, 1967-72, chmn. dept. biology, 1976-81. Contbr. articles to profl. jours. Grantee Research Corp., 1964-65, NSF, 1965-67, USPHS, 1967-71. Mem. AAAS, Am. Assn. Immunologists, Am. Soc. Microbiologists. Subspecialties: Microbiology; Immunobiology and immunology. Current work: Virulence mechanisms of pathogenic bacteria; immune mechanisms to bacterial capsular antigens; antigenic constitution of T. pallidum. Office: Dept Biol Scis San Francisco State U San Francisco CA 94132

WU, YING VICTOR, chemist, researcher; b. Peking, China, Nov. 1, 1931; came to U.S., 1949, naturalized, 1960; s. Hsien and Daisy (Yen) W.; m. Mildred Ling, June 18, 1960; 1 dau., Julia. B.S., U. Ala., 1953; Ph.D. in Phys. Chemistry, MIT, 1958. Research asst. MIT, 1953-57, fellow, 1957-58; research assoc. Cornell U., 1958-61; research chemist No. Regional Research Ctr., Peoria, Ill., 1961—. Contbr. numerous articles to profl. jours. Mem. Am. Chem. Soc., Am. Assn. Cereal Chemists, Inst. Food Technologists, AAAS, Am. Soc. Biol. Chemists, Sigma Xi, Sigma Pi Sigma, Pu Mu Epsilon, Gamma Sigma Epsilon, Alpha Chi Sigma. Subspecialties: Biophysical chemistry; Food science and technology. Current work: Physical chemistry of protein, protein conformation, cereal protein concentrate, reducing costs and improving byproducts from production of alcohol. Office: Northern Regional Research Center 1815 N University St Peoria IL 61604

WU, YING-CHU LIN (SUSAN WU), university administrator, educator, aeronautical engineer; b. Peking, China, June 23, 1932; came to U.S., 1957, naturalized, 1969; d. Chi-yu and Kuo-chun (Kung) Lin; m. Jain-Ming James Wu, June 13, 1959; children: Ernest H., Albert H., Karen H. B.S. in Mech. Engring, Nat. Taiwan U., Taipei Taiwan, 1955; M.S. in Aero. Engring, Ohio State U., 1959; Ph.D. in Aeros, Calif. Inst. Tech., 1963. Sr. engr. Electro-Optics Systems, Inc., Pasadena, Calif., 1963-65; asst. prof. aerospace engring. U. Tenn. Space Inst., Tullahoma, 1965-67, assoc. prof., 1967-73, prof., 1973—, mgr. research and devel. lab., 1977-81, adminstr., 1981—. Contbr. numerous articles to profl. jours.; editor: Procs. 14th Symposium on Engring. Aspects of Magneto Hydro Dynamics, 1974; guest co-editor: Jour. Energy, 1982. Recipient Best Scholastic award Calif. Inst. Tech. chpt. Inst. Aerospace Scis., 1962, Chancellor's Research Scholar award U. Tenn., 1978; Amelia Earhart fellow, 1958, 59, 62. Fellow AIAA (assoc.; sect. chmn. 1978-79, conf. chmn. 1976); mem. ASME, Sigma Xi. Subspecialties: Plasma; Combustion processes. Current work: Magnetohydrodynamics energy conversion, fluid dynamics, high-temperature gasdynamics and non-equilibrium flows. Office: U Tenn Space Inst Tullahoma TN 37388 Home: 111 Lakewood Dr Tullahoma TN 37388

WUBBELS, GENE GERALD, chemistry educator, researcher; b. Preston, Minn., Sept. 21, 1942; s. Victor and Genevieve (Sikkink) W.; m. Joyce Ruth Honebrink, Aug. 26, 1967; children: Kristen, Benjamin, John. B.S., Hamline U., 1964; Ph.D., Northwestern U., 1968. From asst. prof. to prof. Grinnell (Iowa) Coll., 1968—, prof. chemistry, 1979—; councillor Council Undergrad. Research, Mpls., 1978—. Editor: Survey of Progress in Chemistry, 1980—; mem. editorial adv. bd.: Accounts of Chem. Research, 1978-83. Moderator United Ch. of Christ Congregationalist, Grinnell, 1980-82. NIH grad. fellow, 1965-68; research grantee Petroleum Research Found. Am. Chem. Soc., 1971-83; NSF research ednl. grantee, 1971-83. Mem. Am. Chem. Soc., Iowa Acad. Sci., Sigma Xi. Republican. Club: Fortnightly (Grinnell). Subspecialties: Organic chemistry; Photochemistry. Current work: Defining and exploring catalytic photochemical phenomena; photo-Smiles rearrangement; chemical reactions of exciplexes. Home: 1403 Summer St Grinnell IA 50112 Office: Grinnell Coll Grinnell IA 50112

WUEBBLES, DONALD JAMES, atmospheric scientist; b. Breese, Ill., Jan. 28, 1948; s. James E. and Helen M. (Isaak) W.; m. Barbara J. Yaley, June 13, 1970; children: Ryan David, Kevin Kyle. B.S., U. Ill., 1970, M.S., 1972; Ph.D., U. Calif.-Davis, 1983. Research atmospheric scientist NOAA, Boulder, Colo., 1972-73, Lawrence Livermore Nat. Lab., Calif., 1973—. Contbr. numerous sci. articles to profl. publs. Recipient Spl. Achievement award NOAA, 1973, Group Achievement award NASA, 1982. Mem. Am. Geophys. Union, Am. Meteorol. Soc., AAAS, Eta Kappa Nu, Sigma Tau, Phi Eta Sigma, Tau Beta Pi. Democrat. Roman Catholic. Subspecialties: Aeronomy; Atmospheric chemistry. Current work: Computational modeling of the physics and chemistry of the atmosphere; atmospheric modeling; atmospheric photochemistry; ozone; stratosphere; climate. Office: Lawrence Livermore Nat Lab PO Box 808 L-262 Livermore CA

WUENSCH, BERNHARDT JOHN, engineering and ceramics educator; b. Paterson, N.J., Sept. 17, 1933; s. Bernhardt and Ruth Hannah (Slack) W.; m. Mary Jane Harriman, June 4, 1960; children: Stefan Raymond, Katrina Ruth. S.B. in Physics, MIT, 1955, S.M., 1957, Ph.D. in Crystallography, 1963. Research fellow U. Bern, Switzerland, 1963-64; Ford Found. fellow in engring. MIT, Cambridge, 1964-66, asst. prof. ceramics, 1964-69, assoc. prof., 1969-74, prof., 1974—, acting head materials sci. and engring. dept., 1980; vis. prof. crystallography U. Saarland, Saarbruecken, W. Ger., 1973; physicist Max Planck Inst. for Solid State Research, Stuttgart, W. Ger., 1981; Mem. U.S. nat. com. for crystallography Nat. Acad. Scis., 1980-83; mem. N.E. Regional Com. for Selection of Marshall Scholars, 1970-73, chmn., 1974-80. Assoc. editor: Can. Mineralogist Jour, 1978-80; adv. editor: Physics and Chemistry of Minerals, 1976—; editor: Zeitschrift fuer Kristallographie, 1981—; contbr. numerous articles to sci. publs. Recipient Outstanding Grad. Teaching award MIT, 1975, 79. Fellow Am. Ceramic Soc., Mineral Soc. Am.; mem. Am Crystallographic Assn., Mineral Assn. Can., Sigma Xi. Episcopalian. Subspecialties: Ceramics; Crystallography. Current work: X-ray and neutron diffraction; crystal chemistry of oxides, sulfides and fast-ion conductors; diffusion and point defects. Home: 190 Southfield Rd Concord MA 01742 Office: Mass Inst Tech Room 13-4037 Cambridge MA 02139

WUEST, PAUL J., plant pathologist; b. Phila., Feb. 20, 1937; s. Frank C. and Catherine R. (McLaughlin) W.; m. Janet A. Smith, Sept. 16, 1961; children: Paula, Greta, Philip, Becky. B.S., Pa. State U., 1958, Ph.D. in Plant Pathology, 1963. Grad. asst. Pa. State U., 1958-63, asst. prof. dept plant pathology, 1964-69, assoc. prof., 1969-75, prof., 1975—; mushroom ext. specialist Pa. State U. Coop. Ext. Service, 1964—; v.p. research Gro-Plus Farms, Inc., 1981—; pres. 20th Century Agr., Inc.; dir. Champignon Slack Ltee. Contbr. articles to profl. jours. Served to capt. U.S. Army, 1958-59. Mem. Am. Phytopathological Soc., Mycological Soc. Am., Soc. Nematologists, Am. Mushroom Inst., Can. Mushroom Growers Assn. Subspecialties: Plant pathology; Integrated pest management. Current work: Homeopathic crop culture; diseases and control of mushroom diseases; impact of pesticides; phytobacteriology. Office: 211 Buckhout Lab University Park Pa 16802

WUJEK, DANIEL EVERETT, botanist, educator; b. Bay City, Mich., Oct. 26, 1939; s. Edwin and Norma (Smith) W.; m. Mildred G. Wujek, May 23, 1935; children: Cynthia, Kassia. B.S., Central Mich. U., 1961, M.A., 1962; Ph.D., U. Kans., 1966. Mem. faculty Central Mich. U., Mt. Pleasant, 1968—, prof. botany, 1973—; cons. govt. and industry. Contbr. articles to sci. jours. Mem. AAAS, Bot. Soc. Am., Electron Micorscopy Soc. Am., Brit. Phycol. Soc., Am. Phycol. Soc., Internat. Phycol. Soc., N.Am. Benthological Soc., Aquatic Plant Management Soc. Methodist. Subspecialties: Systematics; Taxonomy. Current work: Electron microscopy of algae. Office: Central Mich U Dept Biology Mount Pleasant MI 48859

WULFF, CLAUS ADOLF, chemist, educator; b. July 20, 1938; s. Edgar E. and Grete (Kiefer) W.; m. Suzanne Singer, Aug. 28, 1960; children: Wendie A., Warren E.A. A.B., Cornell U., 1959; Ph.D., M.I.T., 1962. Inst. Sci. and Tech. fellow U. Mich., 1962-63; asst. prof. Carnegie-Mellon U., Pitts., 1963-65; asst. prof. to assoc. prof. U. Vt., 1965-73, prof. chemistry, 1973—; Berquist fellow U. Lund, Sweden, 1971-72, vis. prof. and external examiner, 1977. Contbr. articles to profl. jours. Du Pont fellow, 1961-62. Mem. Am. Chem. Soc., Am. Phys. Soc., Chem. Soc. (London), Calorimetry Conf., Sigma Xi. Subspecialties: Thermodynamics; Physical chemistry. Current work: Chemical thermodynamics and statistical mechanics. Office: Dept Chemistry Cook Hall U Vt Burlington VT 05405

WUNSCH, CARL ISAAC, oceanographer, educator; b. Bklyn., May 5, 1941; s. Harry and Helen (Gellis) W.; m. Marjory Markel, June 6, 1970; children—Jared, Hannah. S.B., M.I.T., 1962, Ph.D., 1967. Asst. prof. phys. oceanography M.I.T., 1967-70, asso. prof., 1970-75, prof., 1975-76, Cecil and Ida Green prof., 1976—, head dept. earth and planetary scis., 1977-81; sr. vis. fellow U. Cambridge, Eng., 1969, 74-75, 81-82; cons. NASA, NSF, Nat. Acad. Scis. Asso. editor: Jour. Phys. Oceanography, 1977-80, Revs. of Geophysics and Space Physics, 1981—; co-editor: Evolution of Physical Oceanography, 1981; contbr. articles to profl. jours. Recipient Tex. Instruments Found. Founders prize, 1975; Fulbright sr. scholar, 1981-82; Guggenheim fellow, 1981-82. Fellow Am. Geophys. Union (James R. Macelwane award 1971), Royal Astron. Soc., Am. Acad. Arts and Scis.; mem. Nat. Acad. Scis. Club: Cosmos. Subspecialty: Oceanography. Current work: Study of ocean circulation using satellite altimetry and scatterometry, ocean acoustic tomography and inverse methods. Home: 16 Crescent St Cambridge MA 02138 Office: Department of Earth and Planetary Sciences Massachusetts Institute Technology Cambridge MA 02139

WURST, GLEN GILBERT, biology educator, developmental geneticist; b. Mount Holly, N.J., Apr. 17, 1945; s. Melvin N. and Miriam E.S. (Garwood) W.; m. Gloria Zettle, Aug. 14, 1966; m. Paula Coyle, July 10, 1982. B.S., Juniata Coll., 1967; Ph.D., U. Pitts., 1975. Asst. prof. biology Allegheny Coll., Meadville, Pa., 1975—; vis. assoc. research scientist Johns Hopkins U., Balt., 1980-81. Mem. Genetics Soc. Am., Soc. Developmental Biology, Nat. Sci. Tchrs. Assn., Sigma Xi. Democrat. Subspecialty: Gene actions. Current work: Mechanism of transdetermination in Drosophila Melanogaster. Home: 125 Glenwood Ave Meadville PA 16335 Office: Dept Biology Allegheny Coll 210 Carnegie Hall Meadville PA 16335

WURTMAN, RICHARD JAY, physician, educator; b. Phila., Mar. 9, 1936; s. Samuel Richard and Hilda (Schreiber) W.; m. Judith Joy Hirschhorn, Nov. 15, 1959; children: Rachael Elisabeth, David Franklin. A.B., U. Pa., 1956; M.D., Harvard U., 1960. Intern, Mass. Gen. Hosp., 1960-61, resident, 1961-62, fellow medicine, 1965-66; research asso., med. research officer NIMH, 1962-67; mem. faculty MIT, 1967—, prof. endocrinology and metabolism, 1970-80, prof. neuroendocrine regulation, 1980—; lectr. medicine Harvard Med. Sch., 1969—; prof. Harvard-MIT Div. Health Scis. and Tech., 1978—; invited prof. U. Geneva, 1981; Sterling vis. prof. Boston U., 1981; mem. small grants study sect. NIMH, 1967-69, preclin. psychopharmacolgy study sect., 1971-75; behavioral biology adv. panel NASA, 1969-72; council basic sci. Am. Heart Assn., 1969-74; research adv. bd. Parkinson's Disease Found., 1972-80, Am. Parkinson's Disease Assn., 1978—; com. phototherapy in newborns NRC-Nat. Acad. Scis., 1972-74, com. nutrition, brain devel. and behavior, 1976, mem. space applications bd., 1976—; mem. task force on drug devel. Muscular Dystrophy Assn., 1980—; chmn. life scis. adv. com. NASA, 1979—; mem. adv. bd. Alzheimer's Disease Assn., 1981—; asso. neuroscis. research program MIT, 1974-82; Bennett lectr. Am. Neurol. Assn., 1974; Flexner lectr. U. Pa., 1975. Author: Catecholamines, 1966, (with others) The Pineal, 1968; editor: (with Judith Wurtman) Nutrition and the Brain, Vols. I and II, 1977, Vols. III, IV, V, 1979, Vol. VI, 1983; also articles.; editorial bd.: Endocrinology, 1967-73, Jour. Pharmacology and Exptl. Therapeutics, 1968-75, Jour. Neural Transmission, 1969—, Neuroendocrinology, 1969-72, Metabolism, 1970-80, Circulation Research, 1972-77, Jour. Neurochemistry, 1973-82, Life Scis., 1973-81, Brain Research, 1977—. Recipient Alvarenga prize and lectureship Phila. Coll. Physicians, 1970. Mem. Am. Soc. Clin. Investigation, Endocrine Soc. (Ernst Oppenheim award 1972), Am. Physiol. Soc., Am. Soc. Biol. Chemists, Am. Soc. Pharmacology and Exptl. Therapeutics (John Jacob Abel award 1968), Am. Soc. Neurochemistry, Soc. Neuroscis., Am. Soc. Clin. Nutrition. Club: Harvard (Boston). Subspecialties: Neurochemistry; Neuroendocrinology. Current work: Amino acid metabolism; neurotransmitters; drug development; appetite; memory disorders. Home: 193 Marlborough St Boston MA 02116 Office: Mass Inst Tech Cambridge MA 02139

WURTZ, ROBERT HENRY, neuroscientist; b. St. Louis, Mar. 28, 1936; s. Robert Henry and Alice Edith (Popplewell) W.; m. Sally Smith, Dec. 20, 1958 (div.); children: William, Erica. A.B. in Chemistry, Oberlin (Ohio) Coll., 1958; Ph.D. in Physiol. Psychology, U. Mich., Ann Arbor, 1962. Research assoc. Com. Nuclear Info., St. Louis, 1962-63; postdoctoral fellow Washington U., St. Louis, 1962-

65; scientist NIH-NIMH, Bethesda, Md., 1966-78; chief Lab. Sensorimotor Research, Nat. Eye Inst., NIH, Bethesda, 1978—. Contbr. articles profl. jours. Grass Found. research fellow, 1961. Mem. Soc. Neurosci., Am. Physiol. Soc., Internat. Brain Research Orgn., Phi Beta Kappa. Subspecialties: Neurophysiology; Neurobiology. Current work: Behavioral and neurophysiological mechanisms of vision and movement. Home: 4907 Cumberland Ave Chevy Chase MD 20815 Office: National Eye Institute NIH Bldg 10 Rm 60420 Bethesda MD 20205

WUST, CARL JOHN, microbiologist, immunologist; b. Providence, July 2, 1928; s. Louis Antoine and Ida A. (Jauernig) W.; m. Barbara Marion Russin, Sept. 5, 1951; children: Carl John, Stephen Louis, Catherine, Gregory, Elizabeth. B.S., Providence Coll., 1950; M.S., Brown U., 1953; Ph.D., Ind. U., 1967. NIH-Nat. Cancer Inst. postdoctoral fellow Yale U., 1957-59; biologist, chemist Oak Ridge Nat. Lab., 1959-70; now prof. U. Tenn., Knoxville. Contbr. numerous articles to profl. jours. Nat. Cancer Inst. grantee, 1975-81. Mem. Am. Assn. Immunologists, Am. Soc. Microbiology, Soc. Exptl. Biology and Medicine, Soc. Tropical Medicine and Hygiene, Sigma Xi. Roman Catholic. Lodge: KC. Subspecialties: Immunobiology and immunology; Microbiology (medicine). Current work: Immune protection to viruses, leukemia antigens. Home: 132 Iroquois Rd Oak Ridge TN 37830 Office: M409 Dept Microbiology U Tenn Knoxville TN 37996

WYATT, HARRY JOEL, vision scientist, educator; b. Chgo., Apr. 13, 1942; s. Saul Henry and Sophia (Edelman) W. B.A. in Physics, Pomona Coll., 1964, M.A., U. Calif.-Berkeley, 1968, Ph.D. in Biophysics, 1971. NIH trainee, research fellow Washington U. Med. Sch., St. Louis, 1971-75; asst. prof. biol. scis. SUNY Coll. Optometry, N.Y.C., 1975-78, assoc. prof., 1978—. Contbr. articles to profl. jours. Mem. Assn. for Research in Vision and Ophthalmology, Soc. Neurosci., N.Y. Acad. Scis. Subspecialties: Neurobiology; Physiological psychology. Current work: Oculomotor research: visual information processing and its use in motor behavior. Office: SUNY Coll Optometry 100 E 24th St New York NY 10010

WYATT, ROBERT EDWARD, botanist, educator; b. Charleston, S.C., July 15, 1950; s. Lloyd Edward and Helen Jean (Holder) W.; m. Ann Hudson Stoneburner, Mar. 8, 1978. A.B. in Botany, U. N.C., 1972; Ph.D. in Botany (NSF fellow and grantee), Duke U., 1977. Instr. botany Duke U., Durham, N.C., 1976-77; asst. prof. biology Tex. A&M U., College Station, 1977-79; asst. prof. botany U. Ga., Athens, 1979-83, assoc. prof. botany, 1983—, dir. botany plant growth facilities, 1981—; vis. asst. prof. botany Mountain Lake Biol. Sta., U. Va., summer 1980. Contbr. articles to sci. jours. Mem. Bot. Soc. Am., Am. Soc. Plant Taxonomists, Soc. for Study Evolution, Am. Soc. Naturalists, Ecol. Soc. Am., Torrey Bot. Club, Am. Bryological and Lichenological Soc., Phi Beta Kappa, Sigma Xi, Phi Eta Sigma. Subspecialties: Population biology; Evolutionary biology. Current work: Plant reproductive biology, evolution of breeding systems, pollination ecology. Home: 220 Trailwood Dr Watkinsville GA 30677 Office: Dept Botany U Ga Athens GA 30602

WYCKOFF, JOHN WYNN, geography educator; b. Des Moines, Oct. 29, 1949; s. Paul Clifford and Velma (Price) W.; m. Ann Marie Kidston, Sept. 19, 1970; children: Kristine Ann, Jennifer Marie. Asst. curator vertebrate research collections U. Utah, Salt Lake City, 1969-79, NASA research asst., 1976-78, research asst., 1976-79; asst. prof. geography U. N.D., Grand Forks, 1979—, research assoc., archeol. research, 1981—, exec. sec., 1979—; environ. cons. Sigma Xi grantee 1973; U. Utah environ studies grantee, 1974; U. N.D. faculty research grantee, 1980. Mem. Assn. Am. Geographers, Am. Soc. Photogrammetry, Wildlife Soc., N.D., Natural Sci. Soc., Assn. N.D. Geographers, Sigma Xi, Gamma Theta Upsilon. Subspecialties: Remote sensing (geoscience); Ecology. Current work: Application of remote sensing to wildlife management and geoarcheology. Office: Dept Geography Univ North Dakota Grand Forks ND 58201

WYER, ROBERT SELDEN, JR., psychologist, educator; b. Delhi, N.Y., Apr. 18, 1935; s. Robert Selden and Wilhelmina (Sebesta) W.; m. Joan Marie Winters, Dec. 16, 1960 (div. 1978); children: Kathryn Lynn, Natalie Ann. B.E.E., Rensselaer Poly. Inst., Troy, N.Y., 1957; M.S. in Elec. Engring, NYU, 1959; Ph.D., U. Colo., 1962. Mem. staff Bell Telephone Labs., Whippany, N.J., 1957-59; asst. prof. U. Iowa, Iowa City, 1963-65; asst. prof. to prof. psychology U. Ill., Chgo., 1965-73, Urbana, 1973—; guest research prof. U. Mannheim, Germany, 1977, 78, 80; vis. prof. Katholieke U. Leuven, Belgium, 1982. Author: Cognitive Organization and Change, 1974, (with D. Carlston) Social Cognition, Inference and Attribution, 1979; Editor: (with others) Person Memory, 1980, (with T. Snull) Handbook of Social Cognition, 1983. Fellow Am. Psychol. Assn.; mem. Soc. Exptl. Social Psychology. Subspecialties: Cognition; Social psychology. Current work: Social information processing, person memory, attitude and belief formation and change. Home: 406 Holmes St Urbana IL 61801 Office: Dept Psychology Univ Illinois 603 E Daniel St Champaign IL 61820

WYMAN, BOSTWICK FRAMPTON, mathematics educator; b. Aiken, S.C., Aug. 22, 1941; s. Bostwick F. and Myra (Faust) W.; m. Lockhart Moore, Oct. 28, 1967 (dec. Nov. 7, 1973); m. Linda Curtis, Nov. 29, 1975; children: Tracy, John. B.S., MIT, 1962; M.A., U. Calif.-Berkeley, 1964, Ph.D., 1966. Lectr., math. dept. Princeton U., 1966-68; asst. prof. dept. math. Stanford U., 1968-72; vis. asst. prof. U. Oslo, 1971; mem. faculty dept. math. Ohio State U., Columbus, 1972—, prof. math., 1982—. Mem. Am. Math. Soc., Math. Assn. Am., Soc. Indsl. and Applied Math., IEEE. Democrat. Episcopalian. Current work: Applications of abstract algebra to system and control theory. Address: 242 W New England Ave Worthington OH 43085

WYMAN, JEFFREY ALAN, entomology educator; b. Cirencester, Eng., June 10, 1945; s. Sydney Arthur and Freda (Tompkins) W.; m. Patricia Ann Dussling, June 21, 1969; children: Andrew, Megan, Benjamin. B.Sc. with honors, U. Manchester, Eng., 1966; M.S., U. Wis., 1969, Ph.D., 1971. Research fellow U. Bath, Eng., 1971-73; project assoc. U. Wis., 1973-75; asst. prof. entomology U. Calif.-Riverside, 1976-79, U. Wis., 1979-81, assoc. prof., 1981—. Contbr. articles to profl. jours. Fulbright scholar, 1966-71. Mem. Entomol. Soc. Am. (chmn. sect. 1974-75). Subspecialty: Integrated pest management. Current work: Integrated pest management systems; insect bionomics; insect plant interactions. Office: Dept Entomology U Wis Madison WI 53706

WYMAN, ROBERT J., neurophysiologist; b. Syracuse, N.Y., June 8, 1940; s. Ralph and Selma (Franklin) W. A.B., Harvard U., 1960; M.S., U. Calif., Berkeley, 1963, Ph.D., 1965; M.A. (hon.), Yale U., 1980. Math. analyst Tech. Research Group, N.Y.C., 1959; NSF research fellow Calif. Inst. Tech., 1966; asst. prof. Yale U., 1966-70, assoc. prof., 1970-80, prof., 1980—; vis. scientist Nobel Inst., Stockholm, 1970-71, Med. Research Council, Cambridge, Eng., 1974, U. Basel, Switzerland, 1977. Bd. dirs. Urban League; bd. sponsors Nat. Com. for an Effective Congress. Mem. Soc. Neurosci., Internat. Brain Research Orgn., Soc. Exptl. Biology, Sigma Xi. Subspecialties: Comparative neurobiology; Genetics and genetic engineering (biology). Current work: Genes which specify development of the nervous system. Home: 11J Cedar Ct East Haven CT 06513 Office: Dept Biology 646 KBT Yale U New Haven CT 06511

WYMER, RAYMOND GEORGE, research and development engineer; b. Colton, Ohio, Oct. 1, 1927; s. George F. and and Elsie R. (Fry) Ridgway. B.S., Memphis State Coll., 1950; M.A., Vanderbilt U., 1953. Chemist Oak Ridge Nat. Lab., 1953-59, group leader, 1959-65, sect. chief, 1965-73, assoc. div. dir., 1973-82, div. dir., 1983—. Author: Chemistry in Nuclear Technology, 1963; editor: Chemistry of the Light Water Reactor Fuel Cycle, 1981, Jour. Radiochimica Acta, 1978—. Served with USN, 1945-46. Fellow Am. Nuclear Soc.; mem. Am. Chem. Soc., Am. Inst. Chem. Engrs. Subspecialty: Nuclear fission. Home: 188-A Outer Dr Oak Ridge TN 37830 Office: Oak Ridge Nat Lab Oak Ridge TN 37830

WYNGAARDEN, JAMES BARNES, physician; b. East Grand Rapids, Mich., Oct. 19, 1924; s. Martin Jacob and Johanna (Kempers) W.; m. Ethel Vredevoogd, June 20, 1946 (div. 1977); children: Patricia (Mrs. Michael Fitzpatrick), Joanna (Mrs. William Gandy), Martha (Mrs. Richard Krauss), Lisa, James Barnes. Student, Calvin Coll., 1942-43, Western Mich. Coll. Edn., 1943-44; M.D., U. Mich., 1948. Diplomate: Am. Bd. Internal Medicine. Intern Mass. Gen. Hosp., Boston, 1948-49, resident, 1949-51; vis. investigator Pub. Health Research Inst., N.Y.C., 1952-53; investigator NIH, USPHS, Bethesda, Md., 1953-56; asso. prof. medicine and biochemistry Duke Med. Sch., 1956-61, prof., 1961-65; prof., chmn. U. Pa. Med. Sch., 1965-67; Frederic M. Hanes prof., chmn. dept. medicine Duke Med. Sch., 1967-82; dir. NIH, Bethesda, MD, 1982—; mem. staff Duke, VA, Durham County hosps.; cons. Office Sci. and Tech., Exec. Office of President, 1966-72; Mem. President's Sci. Adv. Com., 1972-73; mem. Pres.'s Com. for Nat. Medal of Sci., 1977-80; mem. adv. com. biology and medicine AEC, 1966-68; mem. bd. sci. counselors NIH, 1971-74; mem. adv. bd. Howard Hughes Med. Inst., 1969-82; mem. adv. council Life Ins. Med. Research Fund, 1967-70; adv. bd. Sci. Yr., 1977-81; vice chmn. Com. on Study Nat. Needs for Biomed. and Behavioral Research Personnel, NRC, 1977-81. Author: (with W.N. Kelley) Gout and Hyperuricemia, 1976; Mem. editorial bd.: Jour. Biol. Chemistry, 1971-74, Arthritis and Rheumatism, 1959-66, Jour. Clinical Investigation, 1962-66, Ann. Internal Medicine, 1964-74, Medicine, 1963—; editor: (with J.B. Stanbury, D.S. Fredrickson) The Metabolic Basis of Inherited Disease, 1960, 66, 72, 78, (with O. Sperling and A. DeVries) Purine Metabolism in Man, 1974, (with L.H. Smith, Jr.) Cecil Textbook of Medicine, 16th edit, 1981, (with L.H. Smith Jr.) Rev. of Internal Medicine: A Self-Assessment Guide, 1979. Bd. dirs. Royal Soc. Medicine Found., 1971-76, The Robert Wood Johnson Found. Clin. Scholar Program. Served with USNR, 1943-46; sr. surgeon USPHS, 1951-56. Recipient Borden Undergrad. Research award U. Mich., 1948; Dalton scholar in medicine Mass. Gen. Hosp., 1950; vis. scientist Inst. de Biologie-Physicochemique, Paris, France, 1963-64; Recipient N.C. Gov.'s award for sci., 1974. Mem. Am. Rheumatism Assn., Am. Fedn. Clin. Research, So. Soc. Clin. Investigation (pres. 1974, founder's medal 1978), A.C.P. (Phillips medal 1980), Am. Soc. Clin. Investigation, AAAS, Am. Soc. Biol. Chemistry, Assn. Am. Physicians (councillor 1973-77, pres. 1978), Endocrine Soc., Nat. Acad. Sci., Am. Acad. Arts and Sci., Inst. Medicine, Sigma Xi. Club: Interurban Clinical (Balt.). Subspecialties: Biochemistry (medicine); Medical research administration. Office: Nat Inst Health 9000 Rockville Pike Bethesda MD 20205

WYRICK, PRISCILLA BLAKENEY, microbiologist, educator; b. Greensboro, N.C., Apr. 28, 1940; d. Carnie Lee and Prestine (Blakeney) W. B.S., U. N.C., 1962, M.S., 1967, Ph.D., 1971. Supr. clin. microbiol. lab. N.C. Meml. Hosp., Chapel Hill, 1962-66; Med. Research Council fellow Nat. Inst. Med. Research, London, 1971-73; asst. prof. dept. microbiology U. N.C., Chapel Hill, 1973-79, assoc. prof., 1979—. Contbr. articles to profl. jours. Recipient faculty teaching award U. N.C. Sch. Nursing, 1979; NIH grantee. Mem. Am. Soc. for Microbiology (pres. N.C. br. 1982-83), Am. Acad. Microbiology, Sigma Xi. Subspecialties: Microbiology; Molecular biology. Current work: The molecular basis of bacterial pathogenesis. Chlamydial infectious diseases. Office: Dept Microbiology U NC Sch Medicine 804 FLOB 231H Chapel Hill NC 27514

WYSKIDA, RICHARD MARTIN, indsl. engring. educator; b. Perrysburg, N.Y., Sept. 2, 1935; s. Martin Joseph and Mary (Mirek) W.; m. Betty Jo Long, Sept. 8, 1962; children: Alan, Carol. B.S.E.E., Tri-State U., Angola, Ind., 1960; M.S.I.E., U. Ala., 1964; Ph.D., Okla. State U., 1968. Elec. engr. Philco Corp., Huntsville, Ala., 1960-62; in aerospace tech. NASA-MSFC, Huntsville, 1962-68; prof. indsl. engring. U. Ala., Huntsville, 1968—; cons. Revere Copper & Brass, Scottsboro, Ala., 1974, Research Triangle Inst., Durham, N.C., 1980, Battelle, Redstone Arsenal, Ala., 1982. Author: Modeling of Cushioning Systems, 1980. Mem. Inst. Indsl. Engrs. (sr.; named Outstanding Indsl. Engr. 1981), Ops. Research Soc. Am., Am. Soc. Engring. Edn. Subspecialties: Operations research (engineering); Industrial engineering. Current work: Cushioning systems (thermosensitive), cost modeling. Home: 225 Spring Valley Ct Huntsville AL 35802 Office: U Ala University Dr Huntsville AL 35800

WYSS, JAMES MICHAEL, anatomist; b. Ft. Wayne, Ind., Mar. 11, 1948; s. Alen George and Anne (Winicker) W.; m. Gloria F. Wyss, Apr. 25, 1973; children: Dana Ann, William Alen. A.A., Concordia Coll., Ann Arbor, Mich., 1968, B.A., 1970; M.Div., Luth. Sch. Theology, Chgo., 1974; Ph.D., Washington U., St. Louis, 1976. Sloan postdoctoral fellow Washington U., St. Louis, 1976-79, instr. anatomy, 1976-79; asst. prof. psychology U. Ala., Birmingham, 1979-83, assoc. prof., 1983—; mem. Neurosci. Research Ctr., 1979—, Cardiovascular Research and Tng. Ctr., 1980—. Contbr. articles to profl. jours. Ordained to ministry Lutheran Ch., 1976; dir. alcholism and youth services Project Promised Land, St. Louis, 1974-79; Chmn. Law Enforcement Assistance Adminstrn. Regional Program, North St. Louis, 1974-77. Washington U. fellow, 1975; NIH fellow, 1977-79. Mem. AAAS, Soc. Neurosci., Am. Assn. Anatomists. Subspecialties: Neurobiology; Cardiology. Current work: Determination of structure and function of various cerebral cortex regions, especially limbic cortex regions and neurogenic control of blood pressure and heart rate. Home: 1925 Old Creek Trail Vestavia Hills AL 35216 Office: Dept Anatomy Univ Ala Birmingham AL 35294

WYSS, JERRY C., physicist; b. Palo Alto, Calif., Mar. 10, 1952; s. John A. and Janet (Campbell) W. B.S., U. Calif.-Santa Barbara, 1974, M.A., 1976, Ph.D., 1978; student, U. Edinburgh, Scotland, 1972-73. Postdoctoral research assoc. in physics U. Iowa, Iowa City, 1979-80; postdoctoral research assoc. U. Colo., Boulder, 1980-81; physicist, electronics engr. Nat. Bur. Standards, Boulder, Colo., 1981—. Contbr. articles to profl. jours. Mem. Am. Phys. Soc., Optical Soc. Am., AAAS, Sigma Xi. Democrat. Methodist. Subspecialties: Fiber optics; Atomic and molecular physics. Current work: Design of fiber optical, electro-magnetic sensors. Characterize, study and measure electromagnetic fields using fiber optical technology. Home: 302 27th St Boulder CO 80303 Office: 325 Broadway Boulder CO 80303

WYSSBROD, HERMAN ROBERT, JR., physiology, biophysics educator; b. Louisville, Oct. 17, 1941; s. Herman Robert and Julia Annamarie (Hoerni) W.; m. Kay Elaine Davis, June 15, 1963; children: Robert Lloyd, Mark Lawrence, Karen Marie. B.E.E., U. Louisville, 1963, Ph.D., 1968, M.Engring., 1973. Asst. prof. biophys. chemistry Mt. Sinai Sch. Medicine, N.Y.C., 1968-74, asst. prof. physiology, 1971-73, assoc. prof. physiology, 1974—, prof., 1982—; vis. asst. prof. Rockefeller U., N.Y.C., 1971-73, vis. assoc. prof. phys. biochemistry, 1974-78. NIH Research Career Devel. awardee, 1972-77; named sr. investigator N.Y. Heart Assn., 1977-81. Fellow Am. Chem. Soc., N.Y. Acad. Medicine (assoc.); mem. Am. Physiol. Soc., Biophys. Soc., Internat. Soc. Magnetic Resonance. Subspecialties: Physiology (medicine); Nuclear magnetic resonance (chemistry). Current work: Study of the conformation of peptides by nuclear magnetic resonance spectroscopy, establishment of conformation-function relationships for biologically active peptides. Home: 571 Rockland St Westbury NY 11590 Office: Ctr for Polypeptides and Membrane Research Mt Sinai Sch Medicine 1 Gustave L Levy Pl New York NY 10029

YABLONOVITCH, ELI, physicist; b. Puch, Austria, Dec. 15, 1946; s. Hercz and Mina (Birnberg) Jablonowicz; m. Karen Sue Freede, Sept. 20, 1979. B.Sc., McGill U., 1967; Ph.D., Harvard U., 1972. Mem. profl. staff Bell Telephone Labs., Holmdel, N.J., 1972-74; assoc. prof. Harvard U., Cambridge, Mass., 1974-79; research assoc. Exxon Research Center, Linden, N.J., 1979—. Recipient Adolph Lomb medal Optical Soc. Am., 1978. Mem. Am. Phys. Soc., IEEE, Optical Soc. Am. Subspecialty: Optics research.

YACOBI, AVRAHAM, pharmacologist, researcher; b. Iran, Nov. 3, 1945; came to U.S., 1971; s. Yacov and Fahima (Attarzadeh) Y.; m. Diana Buckler, Sept. 14, 1971; children: Oran, Lily-Yonat, Eva. M.Pharmacy summa cum laude, Hebrew U., Jerusalem, 1970; Ph.D. in Pharmaceutics, SUNY-Buffalo, 1975. Registered pharmacist, Israel. Assoc. dir. clin. pharmacology Am. Critical Care div. Am. Hosp. Supply Corp., McGaw Park, Ill., 1976-78, sect. head clin. pharmacokinetics and drug metabolism, 1978-82; dept. head pharmacodynamics Med. Research div. Am. Cyanamid, Pearl River, N.Y., 1983—; lectr. Northwestern U. Med. Sch., 1978-82. Contbr. numerous articles to profl. publs. Recipient award for sci. and tech. excellence Am. Critical Care, 1979. Fellow Acad. Pharm Scis. (various offices); Mem. Am. Pharm. Assn., Acad. Pharm. Scis. (various offices), Am. Soc. Clin. Pharmacology and Therapeutics, Am. Soc. Pharmacology and Exptl. Therapeutics. Jewish. Subspecialty: Pharmacology. Current work: Pharmacokinetics, drug disposition and bioanalytical manager in research and development. Supervise a department involved in research in drug disposition and pharmacokinetics. Office: Am Cyanamid Pearl River NY 10965

YAFFE, LEO, chemistry educator; b. Devil's Lake, N.D., July 6, 1916; s. Samuel and Mary (Cohen) Y.; m. Betty Workman, Mar. 18, 1945; children: Carla Joy Yaffe Krasnick, Mark John. B.Sc. with honors, U. Man., 1940, M.Sc., 1941, D.Sc. (hon.), 1981; Ph.D., McGill U., Montreal, Que., Can., 1943; D. Lett. (hon.), U. Trent, Peterborough, Ont., Can., 1980. Nuclear chemistry and tracer research project leader Atomic Energy Can., Ltd., Chalk River, Ont., 1943-52; Macdonald prof. chemistry McGill U., Montreal, 1952—, vice prin. administrn., 1974-82; dir. research and labs. IAEA, Vienna, Austria, 1963-65. Contbr. over 150 nuclear chemistry articles to profl. publs. Fellow Royal Soc. Can., AAAS, Am. Phys. Soc., Chem. Inst. Can. (Montreal medal 1979); mem. Am. Chem. Soc. (Nuclear Chemistry award 1982). Subspecialty: Nuclear Chemistry. Current work: Nuclear chemistry; nuclear fission; application of nuclear chemistry to medicine and to archaeology. Home: 5777 McAlear Ave Montreal PQ Canada H4W 2H2 Office: Chemistry Dept McGill Univ Montreal PQ H3A 2K6 Canada

YAFFE, SUMNER JASON, pediatrician, devel. and pediatric pharmacologist; b. Boston, May 9, 1923; s. Henry H. and Ida E. (Fisher) Y.; m.; children: Steven, Kris, Jason, Noah. A.B., Harvard U., 1945; M.A., 1950; M.D., U. Vt., 1954. Diplomate: Am. Bd. Pediatrics. Intern Children's Hosp., Boston, 1954-55; resident 1955-56; exchange resident St. Mary's Hosp., London, 1956-57; asst. prof. pediatrics Stanford U., Palo Alto, Calif., 1960-63; assoc. prof. pediatrics SUNY, Buffalo, 1963-66, prof., 1966-75; dir. Poison Control Center, Children's Hosp., Buffalo, 1967-75; prof. pediatrics and pharmacology U. Pa., Phila., 1975-81; dir. Ctr. for Research for Mothers and Children, Nat. Inst. Child Health and Human Devel., NIH, Bethesda, Md., 1981—; vis. prof. pharmacology Karolinska Inst., Stockholm, 1969-70; Wall Meml. lectr. Children's Hosp., Washington, 1968; Dr. W.E. Upjohn lectr. Can. Med. Assn., 1974; William N. Creasy vis. prof. clin. pharmacology SUNY, Buffalo, 1976; mem. expert adv. panel on maternal and child health WHO; cons. Am. Found. for Maternal and Child Health, Inc.; dir. div. clin. pharmacology Children's Hosp., Phila., 1975-81; mem. adv. panel in pediatrics U.S. Pharmacopeia. Editor-in-chief: Pediatric Pharmacology; mem. editorial bd.: Pharmacology, Developmental Pharmacology and Therapeutics; contbr. articles to profl. jours. Served with U.S. Army, 1943-44. Recipient Lederle Med. Faculty award, 1962; Fulbright scholar, Eng., 1956-57. Mem. Am. Pharm. Assn., Am. Soc. Clin. Pharmacology and Therapeutics (dir.), Am. Soc. Pharmacology and Exptl. Therapeutics, Acad. Pharm. Scis., Am. Acad. Pediatrics, AAAS, Am. Assn. Poison Control Centers, AAUP, Am. Pediatric Soc., Soc. Pediatric Research, Wilderness Soc. Subspecialties: Pediatrics; Pharmacology. Current work: Direct program of research in biomedical and behavioral sciences with special emphasis on pregnancy, perinatal biology ad human biological and behavioral development from conception through adolescence to maturity. Home: 8144 Inverness Ridge Rd Potomac MD 20854 Office: NIH Room 7C03 Landow Bldg 7910 Woodmont Ave Bethesda MD 20205

YAGIELA, JOHN ALLEN, dental educator, researcher; b. Washington, July 23, 1947; s. Stanley and Kathryn (Gilkeson) Y.; m. Dolores Jean Mitchell, Mar. 21, 1970; children: Gregory Mitchell, Leanne Elizabeth. D.D.S., UCLA, 1971; Ph.D. in Pharmacology, U. Utah, 1975. Asst. prof. dept. oral biology Emory U. Sch. Dentistry, Atlanta, 1975-78, assoc. prof., 1978-82, UCLA Sch. Dentistry, 1982—, coordinator anesthesia and pain control, 1982—; cons. Astra Pharm. Co., Worcester, Mass., 1981—, C.V. Mosby Co., St. Louis, 1981—; outside reviewer U.S. Pharmacopeia Dispensing Info., 1981—. Co-author: Regional Anesthesia of the Oral Cavity, 1981; co-editor: Pharmacology and Therapeutics for Dentistry, 1980. Recipient award of achievement Am. Coll. Dentists, 1971; regents scholar UCLA, 1969-71. Fellow Am. Dental Soc. Anesthesiology; Mem. Internat. Assn. Dental Research (sec. Atlanta sect. 1979-81, pres. so. Calif. sect. 1983—), Am. Assn. Dental Schs. (chmn. sect. pharmacology and therapeutics 1982), Dental Research Inst. (UCLA), Omicron Kappa Upsilon, Alpha Omega. Methodist. Subspecialties: Oral biology; Pharmacology. Current work: Research concerning local and systemic toxicity of local anesthetic drugs; application of therapeutic agents for pain control in dentistry. Home: 7956 Glade Ave Canoga Park CA 91304 Office: UCLA Sch Dentistry Center for Health Scis Los Angeles CA 90024

YAKIN, MUSTAFA ZAFER, management science operations research educator, researcher; b. Isparta, Turkey, May 8, 1952; s. Suleyman and Zubeyde (Mutlukul) Y.; m. Hikmet Buyukoz, Sept. 14, 1979. B.S. magna cum laude, Tech. U. Istanbul, Turkey, 1973; M.S.E., U. Mich., 1975, Ph.D., 1980. Instr. U. Mich.-Ann Arbor, 1978-79; lectr. Eastern Mich. U., Ypsilanti, 1979-80; asst. prof. mgmt. sci. U. Houston, 1980—. Sci. and Tech. Research Inst. Turkey scholar, 1969-73; grantee in field. Mem. Ops. Research Soc. Am. (assoc.), Inst. Mgmt. Scis. (assoc.), Math. Programming Soc. Subspecialties: Industrial engineering; Operations research (engineering). Current work: Development of nonlinear optimization algorithms, modeling and solution of optimization problems in energy production. Home:

6550 Hillcroft #374 Houston TX 77081 Office: U Houston 258 McElhinney Bldg Houston TX 77004

YALOW, ROSALYN SUSSMAN, medical physicist; b. N.Y.C., July 19, 1921; d. Simon and Clara (Zipper) Sussman; m. A. Aaron Yalow, June 6, 1943; children: Benjamin, Elanna. A.B., Hunter Coll., 1941; M.S., U. Ill., Urbana, 1942, Ph.D., 1945, D.Sc. (hon.), 1974, Phila. Coll. Pharmacy and Sci., 1976, N.Y. Med. Coll., 1976, Med. Coll. Wis., Milw., 1977, Yeshiva U., 1977, Southampton (N.Y.) Coll., 1978, Bucknell U., 1978, Princeton U., 1978, Jersey City State Coll., 1979, Med. Coll. Pa., 1979, Manhattan Coll., 1979, U. Vt., 1980, U. Hartford, 1980, Rutgers U., 1980, Rensselaer Poly. Inst., 1980, Colgate U., 1981, U. So. Calif., 1981, Clarkson Coll., 1982, U. Miami, 1983, Washington U., St. Louis, 1983, Adelphi U., 1983, U. Alta. (Can.), 1983, L.H.D., Hunter Coll., 1978, Sacred Heart U., Conn., 1978, St. Michael's Coll., Winooski Park, Vt., 1979, Johns Hopkins U., 1979; D. honoris causa, U. Rosario, Argentina, 1980. Diplomate: Am. Bd. Scis. Lectr., asst. prof. physics Hunter Coll., 1946-50; physicist, asst. chief radiosotope service VA Hosp., Bronx, N.Y., 1950-70, chief nuclear medicine, 1970-80, acting chief radiosotope service, 1968-70; research prof. Mt. Sinai Sch. Medicine, City U. N.Y., 1968-74, Disting. Service prof., 1974-79; Disting. prof.-at-large Albert Einstein Coll. Medicine, Yeshiva U., 1979—; chmn. dept. clin. Scis. Albert Einstein Coll. Medicine, Montefiore Med. Center, Bronx, 1980—; cons. Lenox Hill Hosp., N.Y.C., 1956-62, WHO, Bombay, 1978; sec. U.S. Nat. Com. on Med. Physics, 1963-67; mem. nat. com. Radiation Protection, Subcom. 13, 1957—; mem. Pres.'s Study Group on Careers for Women, 1966—; sr. med. investigator VA, 1972; dir. Solomon A. Berson Research Lab., VA Hosp., Bronx, N.Y., 1973. Co-editor: Hormone and Metabolic Research, 1973-79; editorial adv. council: Acta Diabetologica Latina, 1975-77, Ency. Universalis, 1978—; editorial bd.: Mt. Sinai Jour. Medicine, 1976-79, Diabetes, 1976—, Endocrinology, 1967-72; contbr. numerous articles to profl. jours. Bd. dirs. N.Y. Diabetes Assn., 1974. Recipient VA William S. Middleton Med. Research award, 1960; Eli Lilly award Am. Diabetes Assn., 1961; Van Slyke award N.Y. met. sect. Am. Assn. Clin. Chemists, 1968; award A.C.P., 1971; Dickson prize U. Pitts., 1971; Howard Taylor Ricketts award U. Chgo., 1971; Gairdner Found. Internat. award, 1971; Commemorative medallion Am. Diabetes Assn., 1972; Bernstein award Med. Soc. State N.Y., 1974; Boehringer-Mannheim Corp. award Am. Assn. Clin. Chemists, 1975; Sci. Achievement award AMA, 1975; Exceptional Service award VA, 1975; A. Cressy Morrison award N.Y. Acad. Scis., 1975; sustaining membership award Assn. Mil. Surgeons, 1975; Distinguished Achievement award Modern Medicine, 1976; Albert Lasker Basic Med. Research award, 1976; La Madonnina Internat. prize, Milan, 1977; Golden Plate award Am. Acad. Achievement, 1977; Nobel prize for medicine and physiology, 1977; citation of esteem St. John's U., 1979; G. von Hevesy medal, 1978; Rosalyn S. Yalow Research and Devel. award established Am. Diabetes Assn., 1978; Banting medal, 1978; Torch of Learning award Am. Friends Hebrew U., 1978; Virchow gold medal Virchow-Pirquet Med. Soc., 1978; Gratum Genus Humanum gold medal World Fedn. Nuclear Medicine or Biology, 1978; Jacobi medallion Asso. Alumni Mt. Sinai Sch. Medicine, 1978; Jubilee medal Coll. of New Rochelle, 1978; VA Exceptional Service award, 1978; Fed. Woman's award, 1961; Harvey lectr., 1966; Am. Gastroenterol. Assn. Meml. lectr., 1972; Joslyn lectr. New Eng. Diabetes Assn., 1972; Franklin I. Harris Meml. lectr., 1973; 1st Hagedorn Meml. lectr. Acta Endocrinologica Congress, 1973; Sarasota Med. award for achievement and excellence, 1979; gold medal Phi Lambda Kappa, 1980; Achievement in Life award Ency. Brit., 1980; Theobald Smith award, 1982; Pres.'s Cabinet award U. Detroit, 1982; John and Samuel Bard award in medicine and sci. Bard Coll., 1982; Disting. Research award Dallas Assn. Retarded Citizens, 1982; numerous others. Fellow N.Y. Acad. Scis. (chmn. biophysics div. 1964-65), Am. Coll. Radiology (asso. in physics), Clin. Soc. N.Y. Diabetes Assn.; mem. Nat. Acad. Scis., Am. Acad. Arts and Scis., Am. Phys. Soc., Radiation Research Soc., Am. Assn. Physicists in Medicine, Biophys. Soc., Soc. Nuclear Medicine, Endocrine Soc. (Koch award 1972, pres. 1978), Am. Physiol. Soc., Phi Beta Kappa, Sigma Xi, Sigma Pi Sigma, Pi Mu Epsilon, Sigma Delta Epsilon. Subspecialties: Neuroendocrinology; Gastroenterology. Current work: Radioimmunoassay. Office: 130 W Kingsbridge Rd Bronx NY 10468

YAMAGUCHI, MAMORU, electron microscopist; b. Taiki, Hokkaido, Japan, Aug. 26, 1943; s. Jiro and Miyo (Tanabe) Y.; m. Carol Ellen Cochrane, Nov. 16, 1948; children: Ken, Katarina. B.S., Obihiro U., 1966; M.S., Hokkaido U., 1968; Ph.D., U. Wis., 1972. Postdoctoral fellow Muscle Biology Lab., U. Wis., 1973-74; asst. prof. Okayama U. Med. Sci., 1974-75; postdoctoral fellow Iowa State U., 1975-78; sr. electron microscopist, instr. asst. prof. Ohio State U., 1978—. Contbr. articles to profl. jours. Grantee Am. Heart Assn., 1980-83; Muscular Dystrophy Assn. fellow, 1976-78. Mem. Japanese Biochem. Soc., Am. Soc. Cell Biology, Japanese Cancer Soc., Biophys. Soc., Am. Heart Assn., Japanese Cell Biology Soc., Sigma Xi. Subspecialties: Cell biology; Morphology. Current work: Muscle Diseases. Office: 1900 Coffey Rd Columbus OH 43210 Home: 4860 Larwell Dr Columbus OH 43220

YAMAGUCHI, SHOGO, botanist, former educator; b. Gresham, Oreg., Mar. 29, 1916; s. Tokuji and Shina (Watanabe) Y.; m. Elizabeth May Hughes, May 30, 1928; children: Nichola, Matthew, Theodore. Ph.D. in Bot. Scis, UCLA, 1954. Asst. to plant anatomist U. Calif.-Riverside, 1953; herbicide researcher U. Calif.-Davis, 1954-65; assoc. prof. Tuskegee Inst., 1966-81. Author: (with A.S. Crafts) Autoradiography of Plant Materials, 1964. Served to sgt. U.S. Army, 1941-45. Decorated Bronze Star.; Recipient W. Kelly Mosley Environ. award Ala. Coop. Extension Service, Auburn U., 1980; Agrl. Research Service grantee, 1967-71. Democrat. Roman Catholic. Subspecialty: Plant physiology (biology). Home: 1031 Hughes Ln Fallbrook CA 92028

YAMAMOTO, RICHARD SUSUMU, biochemist, cancer researcher; b. Honolulu, May 15, 1920; s. Brown Kumanosuke and Yoshi Y.; m. Fumie Fuke, Mar. 30, 1946; 1 dau., Joyce Akiko Casso. A.B., U. Hawaii, 1946; M.A., George Washington U., 1949; D.Sc., Johns Hopkins U., 1954. Research assoc. Johns Hopkins U., Balt., 1954-55; research chemist Nat. Inst. Arthritis and Metabolic Disease NIH, Bethesda, Md., 1955-61; biochemist Nat. Cancer Inst. NIH, 1961—. Contbr. articles in field to profl. jours. Served with U.S. Army, 1943-46. Decorated Bronze Star.; Japanese Sci. and Tech. Agy. vis. scientist grantee, 1972-73. Mem. Am. Assn. Cancer Research, Am. Inst. Nutrition, Am. Chem. Soc., Soc. Exptl. Biology and Medicine, Am. Soc. Exptl. Biology, AAAS, Soc. Nutrition Edn., Soc. Toxicology, Japanese Cancer Assn., Sigma Xi. Subspecialties: Cancer research (medicine); Nutrition (medicine). Current work: Chemical carcinogenesis, mutagenesis, toxicology; carcinogen metabolism; development of cancer chemically, inhibition and promotion; role of nutrition and endocrinology in cancer; biochemistry and physiology of obesity. Home: 11109 Jolly Way Kensington MD 20895 Office: National Cancer Institute NIH Bldg 37 Bethesda MD 20205

YANAGIHARA, TAKEHIKO, neurologist; b. Kobe, Japan, Feb. 10, 1936; came to U.S., 1971. M.D., Osaka (Japan) U., 1960. Diplomate: Am. Bd. Neurology. Fellow Mayo Grad. Sch. Medicine, Rochester, Minn., 1962-66, Inst. Neurology, U. London, 1966-67, Inst. Neurobiology, U. Goteborg, Sweden, 1968-71; staff neurology Mayo Clinic, Rochester, 1971—, asst. prof. neurology, 1973-75, assoc. prof., 1975-78, prof., 1978—. Mem. Am. Neurol. Assn., Am. Acad. Neurology, Am. Soc. Neurochemistry, Internat. Soc. Neurochemistry, Soc. for Neurosci. Subspecialties: Neurology; Neurochemistry. Current work: Molecular mechanism of cerebral ischemia and its reversibility. Office: Mayo Clinic 200 1st St SW Rochester MN 55905

YANCEY, ASA GREENWOOD, physician; b. Atlanta, Aug. 19, 1916; s. Arthur H. and Daisy L. (Sherard) Y.; m. Carolyn E. Dunbar, Dec. 28, 1944; children—Arthur H. II, Carolyn L., Caren L., Asa Greenwood. B.S., Morehouse Coll., 1937; M.D., U. Mich., 1941. Diplomate: Am. Bd. Surgery. Intern City Hosp., Cleve., 1941-42; resident Freedmen's Hosp., Washington, 1942-45, U.S. Marine Hosp., Boston, 1945; instr. surgery Meharry Med. Coll., 1946-48; chief surgery VA Hosp., Tuskegee, Ala., 1948-58; practice medicine, specializing in surgery, Atlanta, 1958—; med. dir. Grady Meml. Hosp., Atlanta, 1972—; mem. staff Hughes Spalding, St. Joseph, hosps.; asst. prof. surgery Emory U., 1958-72, prof., 1975—, asso. dean, asso. prof., 1972-75. Contbr. articles to profl. jours. Mem. Atlanta Bd. Edn., 1967-77; Trustee Ga. chpt. Am. Cancer Soc. Served to 1st lt. M.C. AUS, 1942. Fellow A.C.S., Am. Surg. Assn.; mem. Nat. Med. Assn. (trustee 1960-66, editorial bd. jour. 1964-80), Inst. Medicine of Nat. Acad. Scis. Baptist. Subspecialty: Surgery. Home: 2845 Engle Rd NW Atlanta GA 30318 Office: Grady Memorial Hosp Atlanta GA 30303

YANG, CHUNG SHU, biochemist, educator; b. Beijing, China, Aug. 8, 1941; came to U.S., 1963, naturalized, 1970; s. Su Chuan and Sue Fen (Li) Y.; m. Sue Pai, June 25, 1966; children: Arlene, Jenny. B.S., Nat. Taiwan U., 1962; M.N.S., Cornell U., 1965, Ph.D., 1967. Mem. faculty U. Medicine and Dentistry N.J.-N.J. Med. Sch., 1971—, prof. biochemistry, 1979—; mem. pathology study sect. NIH, 1979; mem. spl. rev. com. (site visit) Nat. Cancer Inst., 1979, 80. Author papers in field. Recipient Faculty Research award Am. Cancer Soc., 1971; Future Leaders award Nutrition Found., 1973; Internat. Union Against Cancer award to visit Cancer Inst. of Chinese Acad. Med. Scis., Peking, 1979; research scholar (China) Nat. Acad. Scis., 1980. Mem. Am. Soc. Biol. Chemists, Soc. Pharmacology and Exptl. Therapeutics, Am. Inst. Nutrition, Am. Assn. Cancer Research. Subspecialties: Biochemistry (medicine); Cancer research (medicine). Current work: Biochemistry of cytochrome P-450 and carcinogen activation, cancer etiology and modification of carcinogenesis. Office: Dept Biochemistry NJ Med Sch Newark NJ 07103

YANG, DAVID CHIH-HSIN, chemistry educator; b. Tia-Hua, Hesin-Chiang, China, Jan. 8, 1947; s. Warren M.S. and Su-si (Chang) Y.; m. Linda Li-Ta, June 12, 1971; children: Stephen, Alan. B.S., Nat. Taiwan U., 1968; Ph.D., Yale U., 1973. Research assoc. Rockefeller U., N.Y.C., 1973-75; asst. prof. Georgetown U., Washington, 1975-81, assoc. prof. chemistry, 1981—. Grantee NSF, NIH. Mem. Am. Soc. Biol. Chemists, Am. Chem. Soc., AAAS, Sigma Xi. Subspecialty: Biochemistry (biology). Current work: Conformation of transfer RNA, amino Acyl-t RNA synthetase complex, animal lactins. Home: 1604 Woodmoor Ln McLean VA 22101 Office: Georgetown U 37th and Q Sts Washington DC 20057

YANG, JEN TSI, biochemistry educator; b. Shanghai, China, Mar. 18, 1922; came to U.S., 1947; s. Dao-Kai and Ho-Ching (Yu) Y.; m. Yee-Mui Lee, Aug. 8, 1949; children: Janet, Frances. B.S., Nat. Central U., Nanking, China, 1944; Ph.D., Iowa State U., 1952, postdoctoral, 1952-54; postdoctoral, Harvard U., 1954-56. Research chemist Am. Viscose Corp., Marcus Hook, Pa., 1956-59; assoc. prof. Dartmouth Med. Sch., Hanover, N.H., 1959-60, U. Calif.-San Francisco, 1960-64, prof. biochemistry, 1964—; Vis. prof. Japan Soc. for Promotion of Sci., 1975; Guggenheim fellow, 1959, Commonwealth Fund fellow, 1967. Mem. Am. Chem. Soc., Am. Soc. Biol. Chemists, Biophys. Soc., AAAS. Subspecialty: Biophysical chemistry. Current work: Chiroptical properties of biopolymers; conformation of macromolecules; structure-function relationship of proteins. Home: 1375 20th Ave San Francisco CA 94122 Office: Dept Biochemistry U Calif San Francisco San Francisco CA 94143

YANG, SHEN KWEI, chemist, toxicologist, educator, researcher; b. China, May 4, 1941; came to U.S., 1966, naturalized, 1976; s. Zhi Gian and Jong Farn (Ju) Y.; m. Sandra Chung, Sept. 25, 1945; 1 son, Michael. B.S., Nat. Taiwan U., 1964; M.A., Wesleyan U., Middletown, Conn., 1969; M.Ph., Yale U., 1970, Ph.D., 1972. Postdoctoral assoc. Yale U., 1971-73; research fellow Calif. Inst. Tech., 1973-75; sr. staff fellow Nat. Cancer Inst., 1975-77; assoc. prof. dept. pharmacology Uniformed Services U. Health Scis., Bethesda, Md., 1977-81, prof., 1981—. Contbr. numerous articles on nucleic acid and chem. carcinogenesis to profl. jours. Recipient Sci. Achievement award Chinese Med. and Health Assn., 1978; Nat. Cancer Inst. grantee, 1981—. Mem. Am. Soc. Pharmacology and Exptl. Therapeutics, Am. Assn. for Cancer Research, Am. Soc. Biol. Chemists, Internat. Soc. Study Xenobiotics. Subspecialties: Environmental toxicology; Cancer research (medicine). Current work: Chemical carcinogenesis, drug metabolism; teacher/researcher. Home: 5 Spruce Tree Ct Bethesda MD 20814 Office: Dept Pharmacology, Uniformed Services U Health Scis 4301 Jones Bridge Rd Bethesda MD 20814

YANG, TSU-JU (THOMAS YANG), veterinary immunologist, pathologist; b. Taiwan, Aug. 14, 1932; s. Fu-Su and Mon-Tau Y.; m. Sue N. Yang, July 2, 1961; children: Kai H., Andrew T., Michael B. B.V.M., Nat. Taiwan U., 1955; D.V.M., Ministry of Exam., Taiwan, 1959; Ph.D. in Immunology, McGill U., Montreal, 1971. Research asso. U. Pa., Phila., 1964-66; research fellow U. Minn., Mpls., 1966-67; demonstrator McGill U., Montreal, Que., Can., 1968-71; from asst. prof. to assoc. prof. immunology and hematology U. Tenn., Knoxville, 1971-75; assoc. prof. pathobiology and immunobiology U. Conn., Storrs, 1975-78, prof., 1978—; vis. fellow Walter and Eliza Hall Inst. Med. Research, Melbourne, Australia, 1983. Contbr. articles to profl. jours. Served with Chinese Veterinary Corps, 1955-56. Recipient Chancellor's award Nat. Taiwan U., 1952-55; ICA fellow, 1959-60; Med. Research Council Can. grantee, 1968-71; NCI grantee, 1974—; U.S. Dept. Agr. grantee, 1981—. Mem. Am. Assn. Immunologists, Am. Assn. for Cancer Research, Am. Soc. for Microbiology, AAAS. Subspecialties: Immunology (agriculture); Cancer research (veterinary medicine). Current work: Canine immunology; canine hematology; tumor immunology; serologic and immunologic techniques; tumor-host interaction; tumor rejection mechanism. Home: 129 Davis Rd Storrs CT 06268 Office: Dept Pathobiology U Conn Storrs CT 06268

YANG, YEE-HONG, electrical engineer, computer scientist; b. Hong Kong, Feb. 29, 1952; came to U.S., 1977; s. Yuan-Yung and Wai-King (Lau) Y.; m. Evelyn Lai-Ngor Hui, May 31, 1975. B.Sc. with honors, U. Hong Kong, 1974; M.Sc., Simon Fraser U., 1977; M.S.E.E., U. Pitts., 1980, Ph.D., 1982. Teaching asst. Simon Fraser U., Burnaby, Vancouver, Can., 1974-77; staff engr. Pattern Recognition Lab., U. Pitts., 1977-80; scientist Mellon Inst. Computer Engring. Ctr., Pitts., 1980—. Contbr. articles to profl. jours. Mem. IEEE, Assn. Computing Machinery, Am. Phys. Soc. Subspecialties: Computer engineering; Graphics, image processing, and pattern recognition. Current work: Image processing, hardware descriptive languages, simulation, language design, very high speed computation and computer architecture. Office: Mellon Inst Computer Engring Ctr 4616 Henry St Pittsburgh PA 15213

YANKAUER, ALFRED, pediatrician, educator; b. N.Y.C., Oct. 12, 1913; s. Alfred and Teresa (Loewy) Y.; m. Marian Wynn May, May 21, 1948; children: Douglas, Kenneth. A.B. with honors, Dartmouth Coll., 1934; M.D., Harvard U., 1938; M.P.H., Columbia U., 1947. Diplomate: Am. Bd. Pediatrics, Am. Bd. Preventive Medicine. Intern Albany (N.Y.) Hosp., 1938-39; resident in pediatrics Children's Hosp., Boston, 1939-40, Mass. Gen. Hosp., 1940-41; dist. health officer, N.Y.C., 1947-50; asst. commr. health City of Rochester, N.Y., 1950-52; dir. bur. Maternal and Child Health, N.Y. State Dept. Health, Albany, 1952-60; regional adv. maternal and child health Pan-Am Health Organ. WHO, 1961-66; WHO prof. child health Madras (India) Med. Coll., 1958-60; sr. research assoc., lectr. Harvard U. Sch. Pub. Health, Boston, 1966—; prof. family and community medicine, prof. pediatrics U. Mass. Med. Sch., Worcester, 1968—. Editor: Am. Jour. Pub. Health, 1975—; contbr. over 150 articles to med. publs., also chpts. to books. Served to maj. M.C. U.S. Army, 1941-46. Mem. Am. Acad. Pediatrics (Job Lewis Smith award 1979), Ambulatory Pediatric Assn., Am. Pub. Health Assn., Phi Beta Kappa, Alpha Omega Alpha. Subspecialty: Preventive medicine. Current work: Editor of American Journal of Public Health. Home: Ashley Rd North Brookfield MA 01535

YANKEE, RONALD A., physician, educator, researcher; b. Franklin, Mass., May 24, 1934; s. Theodore H. and Georgia C. (Webb) Y. B.S., Tufts Coll., 1956; M.D., Yale U., 1960. Intern U. Va., 1961; resident in medicine U. Mich., 1962-63; spl. fellow NIH, 1963-65; sr. investigator Nat. Cancer Int., Bethesda, Md., 1963-73; dir. Blood Bank, Sidney Farber Cancer Inst., 1973-79; assoc. prof. medicine Harvard Med. Sch., 1973-79; prof. medicine Brown U., Providence, 1980—; dir. R.I. Blood Center, Providence, 1979—. Contbr. articles to profl. jours. Served with USPHS, 1963-66. Mem. AMA, Am. Soc. Hematology, Am. Assn. Blood Banks. Subspecialty: Internal medicine. Current work: Blood and blood products; platelets and marrow transplantation. Office: 551 W Main St Providence RI 02940

YANKELL, SAMUEL LEON, dental researcher, educator, consultant; b. Bridgeton, N.J., July 4, 1935; s. Morris Harry and Francis (Fishman) Yankelowitz; m. Kuna Bea Rickler, June 22, 1958; children: Morris, Stuart, Elaine. B.S., Ursinus Coll., 1956; M.S., Rutgers U., 1957, Ph.D., 1960. Diplomate: Registered dental hygienist U. Pa. Sch. Dental Medicine. Research fellow Rutgers U., 1957-60; sr. research biochemist Colgate-Palmolive, Piscataway, N.J., 1960-63; head dept. biochem. pharmacology Smith Miller & Patch, New Brunswick, N.J., 1963-65; head dept. biol. scis. Menley & James Labs., Phila., 1965-74; research assoc. prof. Sch. Dental Medicine, U. Pa., Phila., 1975—; cons. Dental Research Ltd., Spinnerstown, Pa. Co-author: Comprehensive Dental Hygiene Care, 1980; contbr. articles to profl. jours. Mem. AAAS, Am. Assn. for Lab. Animal Sci., Am. Chem. Soc., Am. Soc. Pharmacol. Exptl. Therapeutics, Internat. Assn. for Dental Research, N.Y. Acad. Scis., Soc. Toxicology, Sigma Xi. Subspecialties: Dental research; Pharmacology. Current work: Plaque, gingivitis, hypersensitivity, analgesics, irritation and sensitization testing; investigational new drug and new drug applications. Patentee. Home: 405 E 2d Moorestown NJ 08057 Office: University of Pennsylvania School of Dental Medicine 4001 Spruce St Philadelphia PA 19104

YANNAS, IOANNIS VASSILIOS, engineering educator; b. Athens, Greece, Apr. 14, 1935; s. Vassilios Pavlos and Thalia eleni (Sarafoglou) Y.; children: Tania, Alexis. A.B., Harvard U., 1957; S.M., MIT, 1959; M.A., Princeton U., 1965, Ph.D., 1966. Research chemist W.R. Grace and Co., Cambridge, Mass., 1959-63; asst. prof. fibers and polymers div., dept. mech. engring. MIT, Cambridge, 1966-68, DuPont asst. prof., 1968-69, assoc. prof., 1969-78, prof. polymer sci. and engring., 1978—; cons. to industry. Contbr. articles to profl. jours. Recipient Fred O. Conley award Soc. Plastics Engrs., 1982; Cutty Sark/Sci Digest award, 1982; Esso Standard Oil fellow, 1958; USPHS fellow, 1963. Mem. N.Y. Acad. Scis., Biomed. Engring. Soc. (charter), Am. Chem. Soc., AAAS, Assn. Harvard Chemists, Soc. for Biomaterials (founders award 1982), ASME, Am. Soc. for Artificial Internal Organs, Sigma Xi. Subspecialties: Polymers; Biomaterials. Current work: Design of Polymers for use in tissue and organ repair. Patentee in field. Home: 149 Baldpate Rd Newton Center MA 02159 Office: Dept Mech Engring MIT Room 3-334 77 Massachusetts Ave Cambridge MA 02139

YANOFSKY, CHARLES, educator; b. N.Y.C., Apr. 17, 1925; s. Frank and Jennie (Kopatz) Y.; m. Carol Cohen, June 19, 1949; children—Stephen David, Robert Howard, Martin Fred. B.S., Coll. City N.Y., 1948; M.S., Yale, 1950, Ph.D., 1951, D.Sc. (hon.), 1981, U. Chgo., 1980. Research Asst. Yale, 1951-54; asst. prof. microbiology Western Res. U. Med. Sch., 1954-57; mem. faculty Stanford, 1958—, prof. biology, 1961—. Career investigator Am. Heart Assn., 1969—. Served with AUS, 1944-46. Recipient Lederle Med. Faculty award, 1957, Eli Lilly award bacteriology, 1959, U.S. Steel Co. award molecular biology, 1964, Howard Taylor Ricketts award U. Chgo., 1966, Albert and Mary Lasker award, 1971; Townsend Harris medal Coll. City N.Y., 1973; Louisa Gross Horwitz prize in biology and biochemistry Columbia U., 1976; V.D. Mattia award Roche Inst., 1982; medal Genetics Soc. Am., 1983. Mem. Nat. Acad. Scis. (Selman A. Waksman award in microbiology 1972), Am. Acad. Arts and Scis., Genetics Soc. Am. (pres. 1969). Subspecialty: Molecular biology. Current work: Regulation of gene expression; evolution of genes; proteins and regulatory regions. Home: 725 Mayfield Ave Stanford CA 94305

YAO, ANDREW CHI-CHIH, computer science educator; b. Shanghai, China, Dec. 24, 1946; came to U.S., 1968; s. Joseph and Ann Y.; m. Foong Frances, Aug. 15, 1971. B.S., Nat. Taiwan U., Taipei, 1967; Ph.D., Harvard U., 1972, U. Ill., 1975. Asst. prof. MIT, 1975-76; asst. prof. Stanford (Calif.) U., 1976-81, prof. computer sci., 1982—; prof. U. Calif., Berkeley, 1981-82. Editor: SIAM Jour. on Computing, 1981—, Jour. Algorithms, 1980—; area editor: Jour ACM, 1983—. Mem. Am. Math. Soc., Soc. Indsl. and Applied Math., Assn. Computing Machinery. Subspecialties: Algorithms; Theoretical computer science. Current work: Analysis of algorithms, computational complexity, combinatorics. Office: Dept Computer Scis Stanford U Stanford CA 94305

YAO, SHANG JEONG, chemist; b. Canton, China, June 6, 1934; s. Wan Wien and Ly Yiu (Kong) Y.; m. Huei-Ying Sun, Sept. 2, 1966; 1 son, Gene J. Diploma Chem. Engring., Taipei Inst. Tech., 1955; M.A., U. Oreg., 1961; Ph.D., U. Minn., 1966. Robert A. Welch Found. postdoctoral fellow Tex. A&M U., 1966-67; postdoctoral fellow chemistry Northwestern U., 1967-68; asst. prof. chemistry Wilbur Wright Coll., Chgo., 1968-69; instr. U. Chgo. Med. Sch., 1969-71; research assoc. Michael Reese Hosp. and Med. Ctr., Chgo., 1969-71;

asst. prof. U. Pitts. Med. Sch., 1971-78, research assoc. prof., 1978—; sr. scientist Montefiore Hosp., Pitts., 1971—; co-chmn., co-editor biomed. session 7th Intersociety Energy Conversion Engring. Conf., San Diego, 1972; chmn. session on implantable energy sources 15th Ann. Meeting Assn. Advanced of Med. Instrumentation, 1980; co-prin. investigator NIH grants, 1969-78, prin. investigator John A. Hartford Found. grant, 1977-80. Patentee in field. Recipient Achievement award Montefiore Hosp., 1982; NIH grantee, 1979—. Mem. Am. Phys. Soc., Am. Soc. Artificial Internal Organs, Soc. Neurosci., Brit. Brain Research Assn. (hon.); mem. Sigma Xi; Mem. Phi Lambda Upsilon. Roman Catholic. Subspecialty: Biophysics (biology). Current work: Bioenergetics, quantum theory of enzyme specificity, theory and catalyst facilitated molecular tunnelling; artificial organs and neuroscience. Office: Research Lab U Pitts 3459 5th Ave Pittsburgh PA 15213

YARBRO, JOHN WILLIAMSON, oncologist, educator; b. Chattanooga, Sept. 15, 1931; s. Joseph Jarvis and Francys Ann (Williamson) Y.; m. Connie Henke, July 7, 1979. B.A., U. Louisville, 1952, M.D., 1956; Ph.D., U. Minn., 1965. Diplomate: Am. Bd. Internal Medicine, 1963. Intern Tripler Army Hosp., Honolulu, 1956-57; resident in medicine U. Minn. Hosp., Mpls., 1959-63; instr. U. Minn., 1963-65, asst. prof., 1965-68; assoc. prof., dir. hematology U. Ky., Lexington, 1968-70; assoc. prof. U. Pa., Phila., 1970-72; assoc. dir. Nat. Cancer Inst., NIH, Washington, 1972-75; prof. medicine U. Mo., Columbia, 1975--. Author: (with Grune and Stratton) Oncologic Emergencies, 1981, Toxicity Chemotherapeutic Agents, 1983; editor-in-chief: Seminars in Oncology, 1974--; contbr. numerous sci. articles to profl. publs. Served with U.S. Army, 1956-59. Fellow ACP; mem. Am. Soc. Clin. Oncology, Am. Assn. Cancer Research. Subspecialties: Oncology; Internal medicine. Current work: Investigation and consultation in oncology. Home: 2604 Luan Ct Columbia MO 65201 Office: U Mo Med Center Columbia MO 65201

YARCHOAN, ROBERT, immunologist, researcher; b. N.Y.C., July 21, 1950; s. Zachary and Anne Mae (Veneroso) Y.; m. Giovanna Tosato, Nov. 7, 1981. B.A., Amherst Coll., 1971; M.D., U. Pa., 1975. Diplomate: Nat. Bd. Internal Medicine, Am. Bd. Allergy/Immunology. Intern U. Minn. Hosps., Mpls., 1975-76, resident in internal medicine, 1976-78; clin. assoc. Metabolism Br., NCI, NIH, Bethesda, Md., 1978-80, expert, 1980, investigator, 1980—. Contbr. articles to profl. jours. Served with USPHS, 1978—. Mem. Am. Fedn. Clin. Research, Sigma Xi, Phi Beta Kappa. Club: Audubon Soc. Subspecialties: Immunology (medicine); Internal medicine. Current work: Research in clinical immunology, particularly the immunoregulation of specific antibody production by human lymphocytes; care of patients with immunodeficient diseases. Office: NIH Bldg 10 Rm 4N113 Bethesda MD 20205

YARGER, FREDERICK LYNN, physics educator; b. Lindsey, Ohio, Mar. 8, 1925; s. Sherman W. and Burgena (Lynn) Y.; m. Mary Jean Petersen, Aug. 21, 1948; children: Frederick Daniel, Peter David. B.S., Capital U., 1950; M.S., Ohio State U., 1953, Ph.D., 1960. Research asst. Los Alamos Sci. Lab., 1952, mem. staff, 1953-55; sr. engr. N.Am. Avaition Co., Columbus, Ohio, 1956-58; scientist Nat. Bur. Standards, Boulder, Colo., 1960-63; sci. specialist EGG, Inc., Las Vegas, Nev., 1964-65; sr. engr. Falcon Research & Devel., Denver, 1965-66; prof. physics N.Mex. Highlands U., Las Vegas, 1966—; vis. staff Los Alamos Nat. Lab., 1974—; vis. prof. Nat. U. Mex., 1968. Contbr. articles to sci. jours. Vol. firefighter, 1971—, fire chief, 1982-83. Served in USN, 1943-47. Mem. Optical Soc. Am., Am. Phys. Soc., Sigma Xi, Sigma Pi Sigma. Lutheran. Subspecialties: Condensed matter physics; Infrared spectroscopy. Current work: High pressure physics of solids, liquids and gases. Home: PO Box 875 Las Vegas NM 87701 Office: NMex Highlands U Las Vegas NM 87701

YARIV, AMNON, scientist, educator; b. Tel Aviv, Israel, Apr. 13, 1930; came to U.S., 1951, naturalized, 1964; s. Shraga and Henya (Davidson) Y.; m. Frances Pokras, Apr. 10, 1972; children: Elizabeth, Dana, Gabriela. B.S., U. Calif., Berkeley, 1954, M.S., 1956, Ph.D., 1958. Mem. tech. staff Bell Telephone Labs., 1959-63; dir. laser research Watkins-Johnson Co., 1963-64; mem. faculty Calif. Inst. Tech., 1964—, Thomas G. Myers prof. elec. engring. and applied physics, 1966—; chmn. bd. ORTEL Inc.; cons. in field. Author: Quantum Electronics, 1967, 75, Introduction to Optical Electronics, 1971, 77, Theory and Applications of Quantum Mechanics. Served with Israeli Army, 1948-50. Fellow IEEE, Am. Optical Soc., Am. Acad. Arts and Scis.; mem. Am. Phys. Soc., Nat. Acad. Engring. Subspecialties: Fiber optics; 3emiconductors. Current work: Integrated optoelectronics (integrating transistors and lasers); real-time holography. Office: 1201 California Ave Pasadena CA 91125

YARMOLINSKY, MICHAEL BEZALEL, molecular geneticist, researcher; b. N.Y.C., Jan. 18, 1929; s. Avrahm and Babette Estelle (Deutsch) Y.; m. Sirpa Helvi, June 22, 1962; 1 dau., Miriam Laura. A.B., Harvard U., 1950; Ph.D., Johns Hopkins U., 1954. Instr. pharmacology NYU Coll. Medicine, 1954-55; research assoc., asst. prof. dept. biology McCollum Pratt Inst., Johns Hopkins U., Balt., 1958-63; NSF fellow Pasteur Inst., Paris, 1963-64; research chemist Lab. Molecular Biology, NIH, Bethesda, Md., 1964-71; dir. research CNRS Inst. Molecular Biology, U. Paris, 1971-76; head molecular genetics sect. Cancer Biology Program, NCI-Frederick (Md.) Cancer Research Facility, 1976—. Contbr. articles to profl. jours. Served with USPHS, 1955-58. Mem. Am. Soc. Biol. Chemists. Subspecialties: Molecular biology; Genetics and genetic engineering (biology). Current work: Area of present interest microbial molecular genetics, particularly as it illuminates plasmid maintenance. Home: 4711 Waverly Ave Garrett Park MD 20896 Office: Frederick Cancer Research Facility PO Box B Frederick MD 21701

YASBIN, RONALD ELIOTT, microbiologist, educator; b. Bklyn., Apr. 27, 1947; s. Ben and Olga (Ritchek) Y.; m. Sherrill Carla Elkin, July 23, 1972; children: Lorne, Iric, Todd Ian. B.S. in Zoology, Pa. State U., 1968; M.S. in Genetics, Cornell U., 1970; Ph.D. in Microbiology, U. Rochester, 1974. Postdoctoral fellow NIH, 1975; research assoc. Brookhaven Nat. Lab., 1976; asst. prof. microbiology and molecular genetics Pa. State U., University Park, 1976-81; asst. prof. microbiology U. Rochester (N.Y.) Sch. Medicine, 1981—82, assoc. prof., 1982—, assoc. dir., 1982—. Contbr. articles to profl. jours. Mem. State Coll. Bd. Health, 1980-81. Am. Soc. for Microbiology fellow, 1973; recipient Wallace O. Fenn award U. Rochester. Mem. Genetics Soc. Am., Am. Soc. for Microbiology, Environ. Mutagen Soc. Subspecialty: Genetics and genetic engineering (biology). Current work: Genetics of DNA repair systems and genetic engineering in Bacillus subtilis and molecular biology virulence in Neisseria genus. Office: University of Rochester Box 672 Rochester NY 14642

YASUDA, NAOKI, endocrinologist, medical educator; b. Tokyo, Mar. 12, 1945; U.S., 1972; m. Yuko Hoshino, May 1, 1971; children: Hajime, Makiko, Tsutomu. M.D., Tokyo Med. and Dental U., 1969. Intern Mercy Hosp. and Med. Ctr., Chgo., 1972-73; resident in medicine Toranomon Hosp., Tokyo, 1970-72; fellow in endocrinology Oreg. Health Scis. U., Portland, 1973-76, asst. prof. medicine, 1976-82, assoc. prof., 1982—. Tartar Research fellow Med. Research Found. Oreg., Portland, 1980; NIH grantee, 1982. Mem. Am. Endocrine Soc., Am. Fedn. Clin. Research, Western Soc. Clin. Research. Subspecialty: Neuroendocrinology. Current work: Neuroendocrine mechanism regulating AC TH secretion; elucidation of biochemical nature of corticotropin-releasing hormones. Home: 2634 SW Boundary St Portland OR 97201 Office: Oreg Health Scis U Endocrinology Div 3181 SW Sam Jackson Park Rd Portland OR 97201

YATES, ALLAN JAMES, neuropathologist, neurochemist; b. Calgary, Can., May 23, 1943; s. James and Evelyn May (Scorah) Y.; m. Charlotte Ann, June 15, 1968; 1 son, Robert Bruce. Student, U. Calgary, 1961-63; M.D., U. Alta., 1967; Ph.D., U. Toronto, 1972. Diplomate: Am. Bd. Pathology. Faculty McGill U., 1972-73, U. Toronto, 1973-75; asst. prof. neuropathology Ohio State U., Columbus, 1975-79, assoc. prof., 1979—, dir. div. neuropathology, 1982—; cons. neuropathologist Columbus Children's Hosp.; NIH. Contbr. articles to profl. jours. Am. Cancer Soc. grantee. Fellow Royal Coll. Physicians and Surgeons Can.; mem. Am. Assn. Neuropathology, Am. Soc. Neurochemistry, Am. Assn. Pathologists, Soc. Neurosci., AAAS. Subspecialties: Neurochemistry; Neuropathology. Current work: Neurochemistry, glycolipids and neurological disorders. Office: Ohio State U 105 Upham Hall Columbus OH 43210

YATES, JOHN HARRY, research chemist, educator; b. Atlantic City, N.J., Oct. 12, 1948; s. Charles S. and Lillian I. (Sears) Y. B.S., Case Western Res. U., 1970; M.S. in X-Ray Crystallography, Ohio State U., 1973; Ph.D. in Theoretical Chemistry, Ohio State U., 1976. Research assoc. dept. crystallography U. Pitts., 1977-79, research asst. prof. chemistry and crystallography, 1982—; research assoc. dept. chemistry and chem. engring. Stevens Inst. Tech., Hoboken, N.J., 1980-82. Contbr. articles to profl. jours. Mem. Am. Chem. Soc., Am. Inst. Physics, Am. Crystallographic Assn., AAAS, Am. Phys. Soc. Subspecialties: Theoretical chemistry; X-ray crystallography. Current work: Accurate molecular structure determination using state-of-the-art methods, both theoretical and crystallographic, with extensive use of computers; fundamental research in theoretical chemistry and x-ray crystallography. Home: 321 Denniston Ave Pittsburgh PA 15206 Office: Dept Crystallography U Pitts Pittsburgh PA 15260

YATSU, FRANK MICHIO, neurologist, educator; b. Los Angeles, Nov. 28, 1932; s. Frank K. and Helen A. Y.; m. Michiko Yamane, Sept. 10, 1955; 1 dau., Libby. A.B., Brown U., 1955; M.D., Case-Western Res. U., 1959. Diplomate: Am.Bd. Psychiatry and Neurology, 1967. Intern U. Hosp., Cleve., 1959-60, resident, 1960-63; resident in neurology Neurol. Inst. N.Y., 1961-63; fellow in neurochemistry Albert Einstein Med. Coll., N.Y.C., 1964-65; from asst. prof. to assoc. prof. neurology U. Calif.-San Francisco Sch. Medicine, 1967-75; prof., chmn. dept. neurology Oreg. Health Scis. U., Portland, 1975-82, U. Tex. Med. Sch., Houston, 1982—. Trustee Brown U., 1969-74. Served to lt. comdr. USN, 1965-67. John and Mary Markle scholar in academic medicine, 1969-74. Mem. AMA, Am. Acad. Neurology, Am. Neurol. Assn. Democrat. Subspecialty: Dermatology. Current work: Strokes. Office: University of Texas Medical School Dept Neurology Houston TX 77025

YATSU, LAWRENCE YONEO, research chemist; b. Pasadena, Calif., Aug. 2, 1925; s. Frank K. and Helen I. Y.; m. Teruko Yeya, Aug. 14, 1954; children: Helen Mariko, Elizabeth Y. B.S., Mich. State U., 1949; M.S., U. Calif., Davis, 1950; Ph.D., Cornell U., 1960. Analytical chemist Strong-Cobb, Cleve., 1954-55; research assoc. Cornell U., Ithaca, N.Y., 1955-60; plant physiologist Dept. Agr., Bogalusa, La., 1960-62, research chemist, New Orleans, 1962—; adj. assoc. prof. Tulane U. Served with AUS, 1944-45. Decorated Purple Heart, Bronze Star.; Nat. Cottonseed Products Assn. postdoctoral fellow, 1963. Mem. Am. Soc. Plant Physiologists, AAAS, Am. Inst. Biol. Scis., Bot. Soc. Am., Japan Soc. of New Orleans, Sigma Xi, Phi Kappa Phi. Subspecialties: Plant physiology (agriculture); Cell biology. Current work: Cytochemistry. Home: 7611 Dalewood Rd New Orleans LA 70126 Office: PO Box 19687 New Orleans LA 70179

YAU, SHING-TUNG, mathematician; b. Swatow, Kwantung, China, Apr. 4, 1949; s. Chan-Ying and Yeuk-Lam (Leung) Chiou; m. Yu-Yun Kuo, Sept. 4, 1976. Ph.D., U. Calif., Berkeley, 1971; D.Sc., Chinese U. Hong Kong, 1980. Mem. Inst. Advanced Study, 1971-72; asst. prof. SUNY, Stony Brook, 1972-73; vis. asst. prof. Stanford (Calif.) U., 1973-74, asso. prof., 1975-77, prof. math., 1977-80; prof. Inst. for Advanced Study, Princeton U., 1980—. Recipient Carty prize Nat. Acad. Sci., 1981; named Calif. Scientist of, 1979. Mem. Am. Math. Soc. (Veblen prize 1981). Home: 7 Locust Ln Princeton NJ 08540

YAU, STEPHEN SIK-SANG, electrical engineering educator, computer scientist; b. Wusei, Kiangsu, China, Aug. 6, 1935; came to U.S., 1958, naturalized, 1968; s. Pen-Chi and Wen-Chum (Shum) Y.; m. Vickie Liu, June 14, 1964; children: Andrew, Philip. B.S. in Elec. Engring, Nat. Taiwan U., China, 1958, M.S., U. Ill., Urbana, 1959, Ph.D., 1961. Research asst. elec. engring. lab. U. Ill., Urbana, 1959-61; asst. prof. elec. engring. Northwestern U., Evanston, Ill., 1961-64, asso. prof., 1964-68, prof., 1968—, prof. computer scis., 1970—, also chmn. dept. computer scis., 1972-77, chmn. dept. elec. engring. and computer sci., 1977—; Conf. chmn. IEEE Computer Conf., Chgo., 1967; symposium chmn. Symposium on feature extraction and selection in pattern recognition Argonne Nat. Lab., 1970; gen. chmn. Nat. Computer Conf., Chgo., 1974, First Internat. Computer Software and Applications Conf., 1977; Trustee Nat. Electronics Conf., Inc., 1965-68. Editor-in-chief Computer mag., 1981—; Contbr. numerous articles on computer sci., elec. engring. and related fields to profl. publs. Recipient Louis E. Levy medal Franklin Inst., 1963, Golden Plate award Am. Acad. of Achievement, 1964. Fellow IEEE (mem. governing bd. Computer Soc. 1967-76, pres. 1974-75, dir. Inst. 1976-77, Richard E. Merwin award Computer Soc. 1981), AAAS, Franklin Inst.; mem. Assn. for Computing Machinery, Soc. for Indsl. and Applied Math., Am. Soc. for Engring. Edn., Am. Fedn. Info.-Processing Socs. (mem. exec. com. 1974-76, 79-82, dir. 1972-82, chmn. award com. 1979-82, v.p. 1982—, Chmn. Nat. Computer Conf. Bd. 1982-83), Sigma Xi. Subspecialties: Software engineering; Distributed systems and networks. Current work: Computer systems reliability and maintainability, software engineering, distributed computer systems. Patentee in field. Home: 2609 Noyes St Evanston IL 60201 Office: Dept of Elec Engring and Computer Sciences Northwestern Univ 2145 Sheridan Rd Evanston IL 60201

YAZULLA, STEPHEN, neuroscientist, educator; b. Jersey City, Sept. 3, 1945; s. Stephen and Elsie Alvina (Smith) Y.; m. Margarett Ann Stanley; children: Lisa, Debra, stepchildren: Caroline, Marie. B.S., U. Scranton, Pa., 1967; M.A., U. Del., 1969, Ph.D., 1971. Asst. prof. biology SUNY-Stony Brook, 1974-79, assoc. prof. neurobiology, 1979—. Contbr. articles in field to profl. jours. NASA fellow, 1967-70; NIH fellow, 1972, 73-74; NIH grantee, 1976-79, 1979—. Mem. AAAS, Assn. Research in Vision and Ophthalmology, Soc. Neuroscience, Sigma Xi. Subspecialties: Neurobiology; Neurophysiology. Current work: Vertebrate retina, synaptic transmission, electrophysiology, electron microscopical autoradiography and immunohistochemistry, neurochemistry. Office: Dept Neurobiology and Behavior SUNY Stony Brook NY 11794

YEAGER, ANDREW MICHAEL, hematology/oncology educator, pediatrician; b. Newark, July 10, 1950; s. Andrew Arthur and Winifred Thora (Webster) Y.; m. Helen Elizabeth Blumberg, June 16, 1973; children: Jonathan W., Elizabeth B. B.A., Johns Hopkins U., 1972, M.D., 1975. Diplomate: Am. Bd. Pediatrics (sub-bd. Pediatric Hematology-Oncology). Intern, resident in pediatrics Johns Hopkins U. Hosp., Balt., 1975-78, pediatric hematology-oncology fellow, 1978-80, chief resident pediatrics, 1980-82, asst. prof. oncology, pediatrics, 1982—. Editor: The Harriet Lane Handbook, 1981. Evaluator, Westinghouse Corp. Nat. Sci. Talent Search, Washington, 1979—. So. Med. Assn. research project grantee, 1980. Fellow Am. Acad. Pediatrics; mem. Am. Soc. Pediatric Hematology-Oncology, So. Med. Assn., Am. Fedn. for Clin. Research, Phi Beta Kappa. Subspecialties: Marrow transplant; Pediatrics. Current work: Bone marrow transplantation, megakaryocytopoiesis. Office: Oncology Ct 3-127 Johns Hopkins Hos 600 N Wolfe St Baltimore MD 21205

YEAGER, VERNON LEROY, anatomy educator; b. Williston, N.D., Nov. 20, 1926; s. Walter O. and Elva Leona (Parkhurst) Y.; m. Grethe Elsie Spoklie, Oct. 4, 1947; children: Thomas Albert, Donna Joy, Susan Elaine, Robin Vernon. B.S., Minot State Tchrs. Coll., 1949; Ph.D., U. N.D., 1955. Tchr. Garrison (N.D.) High Sch., 1949-51; faculty mem. U. N.D., Grand Forks, 1955-67, prof., 1966-67; assoc. prof. St. Louis U., 1967-68, 71-75, prof. anatomy, 1975—; vis. prof. Mahidol U., Bangkok, Thailand, 1968-71. Contbr. articles in field to profl. sci. jours. Served with inf. AUS, 1944-46. Recipient Hektoen Gold medal AMA, 1978, Alumni Golden award Minot State Coll., 1980. Mem. Am. Assn. Anatomists, Sigma Xi. Lutheran. Subspecialty: Anatomy and embryology. Current work: Research in cancer of larynx, pathology of connective tissue, computed tomography. Office: Dept Anatomy Saint Louis U 1402 S Grand Blvd Saint Louis MO 63104

YEARGAN, KENNETH VERNON, entomology educator; b. Clanton, Ala., Feb. 12, 1947; s. R. V. and Loraine (Wright) Y.; m. Michelle Renée Barbeau, June 15, 1972; 1 son: Bret Vincent. B.S., Auburn U., 1969; Ph.D., U. Calif.-Davis, 1974. Asst. prof. entomology U. Ky., 1974-79, assoc. prof., 1979—. Contbr. articles to profl. jours., chpts. to books. U. Ky. Research Found. Outstanding Research awardee, 1982; Gamma Sigma Delta Master Tchr. award, 1983. Mem. Entomol. Soc. Am., Entomol. Soc. Can., Ecol. Soc. Am., Internat. Orgn. Biol. Control of Pests, Am. Arachnol. Soc. Democrat. Subspecialties: Integrated pest management; Ecology. Current work: Biological control of insect pests of agricultural crops; integration of new chemical and non-chemical methods for suppressing insect pest populations. Office: Dept Entomology U Ky Lexington KY 40546

YEATMAN, HARRY CLAY, biology educator, researcher; b. Ashwood, Tenn., June 22, 1916; s. Trezevant Player and Mary Eastin (Wharton) Y.; m. Jean Hansford Anderson, Nov. 24, 1949; children: Henry Clay, Jean Hansford. A.B., U. N.C.-Chapel Hill, 1939, M.A., 1942, Ph.D., 1953. Cert. biologist. Asst. prof. biology U. of South, Sewanee, Tenn., 1950-54, assoc. prof., 1954-60, prof., 1960-82, Kenan prof. emeritus, 1982—, chmn. dept. biology, 1972-76; vis. prof. marine biology Va. Inst. Marine Sci., 1967; cons. Smithsonian Instn., Washington, 1939—. Contbr.: Encyclopedia of Science & Technology, 1982; contbr. articles to profl. jours. Served with U.S. Army, 1942-46. Fellow AAAS; mem. Soc. Systematic Zoology, Soc. Limnology and Oceanography, Soc. Ichthyology and Herpetology, Am. Microscopy Soc., Phi Beta Kappa, Sigma Xi. Republican. Episcopalian. Subspecialties: Zooplankton limnology; Deep-sea biology. Current work: Preparing identification guides for freshwater and marine copepods and being a consultant of museums and ecology groups; also identifying copepods as a consultant. Office: Biology Dept U of South University Ave Sewanee TN 37375 Home: PO Box 356 Sewanee TN 37375

YECIES, LEWIS DAVID, med. educator, physician; b. Newark, Mar. 21, 1944. A. B., U. Pa., 1966, M.D., 1971. Diplomate: Am. Bd. Allergy and Immunology, Am. Bd. Internal Medicine. Medicine. Intern U. Tex., Dallas, 1971-72, resident in internal medicine, 1972-73; asst. prof. medicine SUNY, Stony Brook, 1979—. Served to lt. comdr. USPHS, 1973-76. Nat. Cancer Inst. fellow, 1976-79. Mem. Am. Acad. Allergy, Am. Assn. Immunologists, N.Y. Acad. Scis., AAAS. Subspecialties: Allergy; Immunology (medicine). Current work: Lipids of inflammation. Office: HSC Tower 16-040 Stony Brook NY 11794

YEE, ALFRED ALPHONSE, structural engineer, consultant; b. Honolulu, Aug. 5, 1925; s. Yun Sau and Kam (Ngo) Y.; m. Elizabeth Wong, June 24, 1975; children: Lailan Yee Fell, Mark, Eric, Malcolm, Ian, Suling, Trevor, I'Ling. B.S.C.E., Rose Poly. Inst., 1948; M.Eng., Yale U., 1949; D.Eng. (hon.), Rose-Hulman Inst. Tech., 1976. Registered profl. engr., Hawaii. Civil engr. Ter. Hawaii Dept. Pub. Works, 1949-50; structural engr. 14th Naval Dist., Pearl Harbor, Hawaii, 1951-54; prin. Alfred A. Yee (Structural Engr.), Honolulu, 1954-55; cons. engr. Park & Lee Ltd., Honolulu, 1955-60; pres. Alfred A. Yee & Assocs., Inc., Honolulu, 1960-81; v.p./tech. dir. Alfred A. Yee div. Leo A. Daly, Honolulu, 1982—; dir. Kaiser Cement Corp.; spl. cons. Singapore Housing and Devel. Bd. Contbr. articles to profl. jours. Mem. spl. com. offshore installations Am. Bur. Shipping. Served in U.S. Army, 1946-47. Fellow ASCE, Yale Engring. Assocs., Am. Concrete Inst.; mem. Nat. Acad. Engring., Hawaii Soc. Profl. Engrs. (Engr. of Yr. 1969), Prestressed Concrete Inst. (Martin P. Korn award 1965), Nat. Soc. Profl. Engrs., Internat. Assn. Bridge and Structural Engrs., Structural Engrs. Assn. Hawaii. Subspecialties: Structural engineering; Civil engineering. Current work: Development of engineering devices to enhance economy and speed of construction; development of ocean-going offshore concrete vessels and platforms for marine environment. Patentee. Office: 1441 Kapiolani Blvd Suite 810 Honolulu HI 96814

YEH, BILLY KUO-JIUN, cardiologist, pharmacologist; b. Foochow, China, Aug. 28, 1937; came to U.S., 1962, naturalized, 1973; s. Shin-Hwa and Que-Lu (Mao) Y.; m. Lydia Lo-Pi Ou, Aug. 21, 1965; children: Elizabeth Shih-I, Biran Shih-Heng, William Shih-Tseng. M.D., Nat. Taiwan U., 1961; M.S., U. Okla., 1963; Ph.D., Columbia U., 1967. Diplomate: Am. Bd. Internal Medicine, Sub-Bd. Cardiovascular Disease. Intern Nat. Taiwan U., Taipei, 1960-61; resident Emory U. Affiliated Hosps., Atlanta, 1967-69; sect. chief clin. pharmacology, also chief hypertension clinic, attending cardiologist Mt. Sinai Med. Center of Greater Miami, 1969-71; acting co-dir. Heart Sta., Jackson Meml. Hosp., Miami, 1972-73; assoc. dir. div. clin. investigation Miami Heart Inst., 1973-76; asst. prof. medicine U. Miami Sch. Medicine, 1970-76, assoc. clin. prof. medicine, 1976—; practice medicine specializing in internal medicine and cardiology, Coral Gables, Fla., 1976—; cons. clin. cardiologist and pharmacologist to several maj. med. centers and pharm. firms. Contbr. articles to profl. publs.; mem. editorial bd.: Miami Medicine, 1978—. Fellow ACP, Am. Coll. Cardiology; mem. Am. Heart Assn. (fellow council on clin. cardiology, past mem. bd. dirs. Miami chpt.), Am. Physiol. Soc., Am. Soc. Pharmacology and Exptl. Therapeutics, Fla. Heart Assn. (profl.

edn. com.). Presbyterian. Subspecialties: Cardiology; Pharmacology. Current work: Mechanisms and therapies of cardiac arrhythmias; ambulatory monitor of blood pressures; efficacy and safety of new antihypertensive and antiarrhythmic drugs. Home: 7110 SW 148 Terr Miami FL 33158 Office: 315 Palermo Ave Coral Gables FL 33134

YEH, CHENG SHIN, nuclear engineer; b. Chu-Chi, Chekiang, China, Nov. 20, 1918; came to U.S., 1957; s. Chin Chong and (Maiyah) Kok; m. Nancy F. Huang, Jan. 15, 1951; children: Ta-Chien, Ruby. Student, Nat. Sun Yat-Sen U., 1938-40; B.S., Huachung U., Wuchang, China, 1942; M.S., N.C. State U.-Raleigh, 1959. Electronic engr. Curtiss-Wright Corp., Princeton (N.J.) div., 1953-60, physicist Atomic Power Devel. Assocs., Detroit, 1960-73; sr. engr., engring. specialist Burns & Roe, Inc., Oradell, N.J., 1973—. Contbr. articles to profl. jours. Mem. Am. Nuclear Soc. Subspecialties: Nuclear fission; Fast breeder reactor physics. Current work: Radiation protection and shielding, nuclear plant safety evaluation and design, radioactive waste management. Home: 1 Rambling Woods Dr Morristown NJ 07960 Office: Burns and Roe Inc 800 Kinderkamack St Oradell NJ 07649

YELLIN, ABSALOM M(OSES), psychiatry and psychology educator, researcher, administrator; b. Tel-Aviv, July 25, 1936; U.S., 1960; s. N. Isaac and Ann (Trachtenberg) Y.; m. Judith Ann Himell, Aug. 29, 1965; children: Elana, Talia. B.A., U. Del., 1965, M.A., 1968, Ph.D., 1970. Postdoctoral scholar Neuropsychiat. Inst., UCLA, 1969-71, research scientist, 1972-71; asst. prof. and dir. research child study unit U. Calif.-Davis, 1972-74; asst. prof. U. Minn., Mpls., 1974-80, assoc. prof. dept. psychiatry, 1980—, dir research, dir. lab. neurosics, div. child/adolescent psychiatry. Contbr. writings to publs. in field. Mem. Am. Psychol. Assn., N.Y. Acad. Scis., Soc. Psychophysiol. Research, Internat. Soc. Devel. Psychobiology, Sigma Xi. Jewish. Subspecialties: Psychophysiology; Psychopharmacology. Current work: Attention, information processing,control of autonomic functions, biorhythms, psychopharmacology, sleep, psychophysiology, electrophysiology, attention deficits. Office: Dept Psychiatry Univ Minn Box 95 Mayo Bldg Minneapolis MN 55455

YELLIN, ALBERT ELLIOT, medical educator, surgeon; b. Lawrence, Mass., Mar. 21, 1937; s. Isaac and Dorothy (Hochman) Y.; m. Elissa Lifland, June 24, 1959; children: Ian Geoffrey, Mia Rachel. A.B., Harvard U., 1958; M.D., U. So. Calif., 1962. Diplomate: Am. Bd. Surgery. Instr. surgery U. So. Calif. Sch. Medicine, Los Angeles, 1969-70, asst. prof., 1970-75, assoc. prof., 1975—; sr. research scientist Ctr. Laser Studies, Los Angeles, 1975-76; assoc. med. dir. Los Angeles County-USC Med. Ctr., 1977-81, chief profl. services, 1977-81, dir. vascular surgery, 1982. Contbr. articles to sci. jours. Chmn. So. Calif. Com. on Trauma, 1982-83; cons. Calif. Emergency Med. Services Authority, 1982-83. Served to capt. U.S. Army, 1963-65. Fellow ACS (chpt. recorder 1979-80, chpt. sec. 1981-82); mem. Western Surg. Soc., Trauma Soc. (chmn. 1983), Pacific Coast Surg. Assn., Internat. Cardiovascular Soc. Democrat. Jewish. Subspecialty: Surgery. Current work: Vascular trauma, atherosclerosis, portal hypertension. Office: USC Sch Medicine 1200 N State St Rm 9420 Los Angeles CA 90033

YELLIN, EDWARD LEON, cardiac physiologist, educator; b. N.Y.C., July 2, 1927; s. Alex and Sarah (Mendelowitz) Y.; m. Jean Fagan, Dec. 17, 1948; children: Peter Fagan, Elizabeth Mitchell, Michael Fagan. B.S., Colo. State U., 1959; Ph.D., U. Ill., 1964. Postdoctoral fellow U. Wash., Seattle, 1964-65; assoc. in surgery Albert Einstein Coll. Medicine, Bronx, 1965-68, asst. prof., 1968-73, assoc. prof., 1973-79, prof. surgery and assoc. prof. physiology, 1979—. Editor: (with Baan and Artzenius) Cardiac Dynamics, 1981. Pres. Glenwood Lake Assn., New Rochelle, N.Y., 1973-75. Served with USN, 1945-46. Ford Found. teaching intern, 1959-64; research grantee NIH, 1974-76, 76-82, 79—. Mem. Am. Physiol. Soc., ASME, Cardiovascular System Dynamics Soc. (dir. 1978-82, pres.-elect 82-84). Subspecialties: Cardiology; Physiology (medicine). Current work: Cardiac dynamics; left ventricular filling dynamics; normal and abnormal mitral valve function. Home: 38 Lakeside Dr New Rochelle NY 10801 Office: Albert Einstein Coll Medicine 1300 Morris Park Ave Bronx NY 10461

YELLIN, JOEL, environ. scientist; b. Los Angeles, Dec. 12, 1940; s. Martin J. and Lisa R. Y.; m. Nicole P. Peter, Sept. 23, 1969; children: Kimon A., Aram M. B.S., Calif. Inst. Tech., 1962; S.M., U. Chgo., 1964, Ph.D., 1965. Formerly staff Brookhaven Nat. Lab., Upton, L.I., N.Y., U. Tel-Aviv, Israel; physicist U. Calif. Radiation Lab., Berkeley, 1967-69; mem. Sch. Natural Scis., Inst. Advanced Study, Princeton, 1969-72; assoc. prof. environ. sci. and law, visitor program in Social Scis. M.I.T., Cambridge, Mass., 1972-82, sr. research scientist, 1982—. Contbr. articles to profl. lit. Grantee Sloan Found., 1978-80, NSF, 1977-81, Ford Found., 1975-78, others. Subspecialties: Resource management; Population biology. Current work: Optimal management depletable resources; theory and practice of environmental decisions; combined genetic and ecological dynamics. Office: MIT E51-201A Cambridge MA 02139

YEMANSKI, MICHAEL RALPH, clinical chemist; b. Augusta, Ga., Apr. 21, 1943; s. Irving Charles and Doris Gayle (Merton) Y.; m. Barbara Hunt Lockwood, Feb. 26, 1969; children: Charles, Gayle, Eric. B.S. in Chemistry, U. Ga., 1964, M.S., 1966; Ph.D. in Biochemistry, U. Pitts., 1970. Postdoctoral fellow U. Kans., Kansas City, 1970-72, assoc. dir. clin. chemistry, 1972-76; assoc. prof. clin. chemistry Baylor U. Med. Ctr., Dallas, 1976-82, prof., 1982—; cons. Baxter Travenol Labs., Inc., Deerfield, Ill., 1983—. Bd. advisors: Chemical Abstracts, 1982—. Served with U.S. Army, 1966-67. Fellow Am. Soc. Clin. Pathologists; mem. Am. Assn. Clin. Chemistry, Assn. Clin. Scientists, Am. Fedn. Clin. Research. Subspecialty: Clinical chemistry. Office: Werik Lab 5635 Yale Blvd 1st Fl Dallas TX 75206

YEN, ANDREW, med. sci. researcher and educator; b. N.Y.C., Mar. 4, 1948; s. Andrew Hsu-tung and Josephine Chiu (Liu) Y.; m. Kathleen Mary Betit, Feb. 1, 1975; children: Jennifer, Christopher. B.A., Haverford Coll., 1969; M.S., U. Wash., 1970; Ph.D., Cornell U., 1976. Research fellow Harvard U., 1976-78; profl. staff assoc. Sloan-Kettering Inst., N.Y.C., 1976-78; asst. prof. depts. internal medicine, physiology, and biophysics, dir. flow cytometry facility U. Iowa Sch. Medicine, Iowa City, 1981—. Contbr. articles to profl. jours. Woodrow Wilson fellow, 1969; grantee NIH, Nat. Cancer Inst., Leukemia Soc. Am. Mem. Am. Soc. Cell Biology, Biophys. Soc., Am. Assn. Cancer Research, Cell Kinetics Soc. Subspecialties: Cell biology; Biophysics (biology). Current work: Research on regulation of cell growth and differentiation. Office: Dept Internal Medicine U Iowa Sch Medicine Iowa City IA 52242

YEN, TERENCE TSINTSU, research scientist; b. Shanghai, China, May 2, 1937; came to U.S., 1961, naturalized, 1972. B.S., Nat. Taiwan U., Taipei, 1958; Ph.D., U. N.C.-Chapel Hill, 1966. Sr. scientist Lilly Research Labs., Indpls., 1965-70, research scientist, 1971—. Mem. Am. Soc. Biol. Chemists, AAAS, Am. Chem. Soc., N.Am. Assn. Study Obesity, Sigma Xi. Subspecialty: Biochemistry (medicine). Current work: Biochemistry and treatment of metabolic diseases. Office: Lilly Research Lab Eli Lilly and Co 307 E McCarty St Indianapolis IN 46285

YENSEN, RICHARD, psychologist, computer cons.; b. Washington, Aug. 28, 1949; s. Elwood and Maria Amalia (Perez-Venero) Y.; m. Joyce Marie Compton, May 1972; 1 son, John Alexander. B.A., U. Calif.-Irvine, 1971, 1972, M.A. in Social Scis, 1975, Ph.D. in Psychology, 1975. Lic. psychologist, Md. Research fellow clin. scis. div. Md. Psychiat. Research Ctr., Balt., 1972-76 mem. core faculty in applied behavioral scis. Johns Hopkins U., Balt., 1976—; pvt. practice psychology, Balt., 1976—; dir. Inst. for Human Devel., Balt., 1980—, adj. prof. Union Grad. Sch., Yellow Springs, Ohio, 1980—; cons.-clinician Ctr. for Vaccine Devel., U. Md., Balt., 1978—, Grassroots Crisis Ctr., Columbia, Md., 1976—, Oldfields Sch. for Girls, Sparks, Md., 1979—. Creator, producer multimedia prodns. and video tapes. Bd. dirs., pres. Brotherhood of Man Counseling Ctr., Towson, Md., 1976-77; bd. dirs. Grassroots Crisis Ctr., 1976-77. Mem. Assn. for Computing Machinery, Am. Psychol. Assn., Md. Psychol. Assn., World Future Soc., Internat. Transpersonal Assn., N.Y. Acad. Scis., AAAS. Subspecialties: Psychotherapy/personality/consciousness; Information systems (information science). Current work: Psychedelic drugs as adjuncts to psychotherapy. Use of computer technology to manage audio-visual database generated environments. Use of audiovisual environments for consciousness alteration and psychotherapy. Office: Inst for Human Devel 2403 Talbot Rd Baltimore MD 21216

YEOMAN, LYNN CHALMERS, cancer researcher, educator; b. Evanston, Ill., May 17, 1943; s. Kenneth Chalmers and Lillian (Worner) Y.; m. Ann Craven Yeoman, Aug. 20, 1966; children: Caroline, Christopher, Sarah. B.A., DePauw U., Greencastle, Ind., 1965; Ph.D., U. Ill., 1970. Postdoctoral fellow Baylon Coll. Medicine, Houston, 1970-72; mem. faculty Baylor Coll. Medicine, 1972—, assoc. prof. pharmacology, 1976; lectr. pharmacology, drug metabolism, biochemistry cancer chemotherapy. Author papers in field.; Co-editor: Methods in Cancer Research, vols. 19, 20. Mem. adminstrv. bd. St. Luke's United Meth. Ch., Houston, 1976-78, 80-82; active local Cub Scouts. Grantee Nat. Cancer Inst., 1975—. Mem. Am. Chem. Soc., Am. Assn. Cancer Research, Cell Biology Soc., Am. Soc. Bio. Chemists, Tissue Culture Assn., Sigma Xi. Subspecialties: Cancer research (medicine); Biochemistry (medicine). Current work: Biology and chemistry of cancer cells, nuclear nonhistone proteins, histones, protein chemistry, micro-injection, DNA-binding proteins, tumor markers, tumor antigens, monoclonal antibodies. Office: Dept Pharmacology Baylor Coll Medicine 1200 Moursund Ave Houston TX 77030

YEOMANS, DONALD KEITH, astronomer; b. Rochester, N.Y., May 3, 1942; s. George E. and Jessie (Sutherl) Y.; m. Laurie Ernst, May 20, 1970; children Sarah K., Keith A. B.S., Middlebury Coll., 1964; M.S., U. Md., 1967, Ph.D., 1970. Tech. supr. Computer Scis. Corp., Silver Spring, Md., 1972-76; mem. tech. staff Jet Propulsion Lab., Pasadena, Calif., 1976—; mem. NASA sci. working groups to study space missions to comets and asteroids, discipline specialist for astrometry within Internat. Halley Watch orgn., 1982. Contbr. articles to profl. jours. Mem. Internat. Astron. Union, Am. Astron. Soc., History of Sci. Soc., Explorer's Club. Democrat. Subspecialty: Planetary science. Current work: Research on long-term motion and behavior of comet Halley. Support of European, Japanese and Russian space missions to comet Halley. Home: 833 Chehalem Rd La Canada CA 91011 Office: Jet Propulsion Lab 264-664 Pasadena CA 91109

YERKES, WILLIAM DILWORTH, JR., environmental health educator; b. Wilkes-Barre, Pa., May 29, 1922; s. William D. and Carol Virginia (Walker) Y.; m. Marion Smith, June 24, 1947; children: Christine, Todd, Emily. B.S., Wash. State U., 1948, Ph.D., 1952. Research scientist Rockefeller Found., Mexico City, 1952-60; research scientist Kimberley-Clark Corp., Neenah, Wis., 1960-72; mem. faculty dept. environ. scis. Grand Valley State Coll., Allendale, 1972—, prof., 1972—; cons. Gov.'s Hazardous Waste Task Force, 1978-80, Dept. Health, Grand Rapids, Mich., 1982-83. Served with AUS, 1942-45. Mem. Water Pollution Control Fedn., ASCE, Nat. Environ. Health Assn., Sierra Club (regional v.p. 1978-80), Phi Beta Kappa, Sigma Xi, Phi Kappa Phi. Episcopalian. Subspecialties: Environmental engineering; Water supply and wastewater treatment. Current work: Hazardous wastes, on-site wastewater treatment, solid waste management. Home: 3679 Blackfoot Ct Grandville MI 49418 Office: Grand Valley State Coll Allendale MI 49401

YESKE, LANNY ALAN, marine engineer; b. Aberdeen, S.D., June 26, 1938; s. Walter Albert and Delores Arlene (Wilke) Peck Y.; m. Jacqueline Lee Janecek, Dec. 27, 1960 (div. Aug. 1977); children; Jay, Troy; m. Molly M. McGaughy, Sept. 3, 1978. B.S., U. Nebr.-Lincoln, 1960; M.S., Naval Postgrad. Sch., 1968; Ph.D., U. Wis.-Madison, 1970-73. Commd. ensign USN, 1960, advanced through grades to comdr., 1981; communications and supply officer (USS Sea Poacher), Key West, Fla., 1961-63, navigator, ops. officer, Holylock, Scotland, 1963-66, exec. officer, engr., San Diego, Calif., 1968-70, internat/interagy. affairs/environmental quality, Washington, 1974-78, dir. air-ocean programs, Monterey, Calif., 1978-81; mgr. marine surveys Tracor Marine, Fort Lauderdale, Fla., 1981—. Contbr. articles to profl. jours. Sr. mil. rep. U.S. Senate Youth Program, Washington, 1975-77; cubmaster Boy Scouts Am., Springfield, Va., 1975-77; v.p. Newington Civic Assn., Va., 1976-77; baseball and soccer coach Springfield (Va.) teams, 1975-77. Served to comdr. USN, 1960-81. Recipient Meritorious Service medal Pres. of U.S., 1974-78. Mem. Marine Tech. Soc., Hydrographic Soc., AAAS, Sigma Xi, Mu Epsilon Nu. Republican. Subspecialties: Offshore technology; Resource management. Current work: Oceanographic, hydrographic and geophysical surveying and project management. Home: 1487 NE 63d Ct Fort Lauderdale FL 33334 Office: Tracor Marine Inc PO Box 13114 Port Everglades FL 33316

YESSIOS, CHRIS IOANNIS, architecture and computer-aided design educator, architect, planner, consultant; b. Edessa, Greece, Aug. 10, 1938; came to U.S., 1968, naturalized, 1979; s. Ioannis C. and Aikaterini (Papachristou) Y.; m. Alexandra Varsamis, Sept. 1, 1971; children: Yiannis, Katerina, Dorina, Christina. Diploma in Law, Aristotelian U., Thesaloniki, Greece, 1962, diploma in Architecture, 1967; Ph.D., Carnegie-Mellon U., 1973. Tech. dir., ptnr. Gorgo: Workshop for Interior Designs and Popular Crafts, Athens and Thessaloniki, Greece, 1960-62; with various archtl. and planning firms, 1962-73, pvt. practice architecture specializing in single family houses and apt. bldgs., Thessaloniki and Columbus, Ohio, 1974—; teaching asst. dept. urban affairs U. Pitts., 1968; teaching and research assoc. dept. architecture Carnegie-Mellon U., 1969-71, lectr., 1971-73; asst. prof. architecture Ohio State U., 1973-79, assoc. prof. architecture and computer-aided design, 1979-83, prof., 1983—; cons. in computer-aided design and computer graphics techiques for practice architecture and planning, 1975—. Research publs. and presentations in field; author user manuals of implemented systems. Recipient Best Research Paper Contest citation 15th Design Automation Conf., Las Vegas, Nev., 1978, Mayor's prize for Vernacular Theme Mayor Thessaloniki, 1962; IBM grantee, 1982—. Mem. Greek Inst. Architects, Tech. Chamber Greece, Environ, Design and Research Assn., Design Methods Group, Assn. Computer Aided Design in Architecture, Assn. Computing Machinery, Spl. Interest Group for Computer Graphics, Spl. Interest Group for Design Automation, Nat. Computer Graphics Assn., Anatolia Coll. Alumni Assn., Carnegie-Mellon U. Alumni Assn. Greek Orthodox. Subspecialties: Graphics, image processing, and pattern recognition; Computer-aided design. Current work: Development of innovative techniques for use of computer (primarily computer graphics) in environmental design (computer-aided architectural design). Home: 4367 Mumford Dr Columbus OH 43220 Office: Dept Architecture Ohio State U 190 W 17th Ave Columbus OH 43210

YETERIAN, EDWARD HARRY, psychology educator, consultlant; b. New Britain, Conn., Mar. 5, 1948; s. Arthur and Mary T. (Harutunian) Y.; m. Margaret Emily Wellock, Sept. 13, 1975; 1 son, Robert. B.S., Trinity Coll., Conn., 1970; M.A., U. Conn., 1974, Ph.D., 1975. Postdoctoral fellow and research fellow in neurology and neuroanatomy Harvard U. Med. Sch., Boston, 1975-78; asst. prof. psychology Colby Coll., Waterville, Maine, 1978—; cons. Maine State Bur. Mental Retardation. Contbr. articles to profl. jours. NIMH fellow, 1974-75, 77-78. Mem. AAAS, Eastern Psychol. Assn., Soc. Neurosci. Democrat. Subspecialties: Physiological psychology; Neuropsychology. Current work: Anatomical and functional investigations of cortical-subcortical relationships in primate brains. Home: RFD 1 Box 360 Ohio Hill Rd Fairfield ME 04937 Office: Colby College Waterville ME 04901

YETIV, JACK ZEEV, pediatrician, writer, consultant, editor; b. Haifa, Israel, July 27, 1956; came to U.S., 1967; s. Isaac and Susan (Reuveni) Y. B.A., Brandeis U., 1976; M.D., Ohio State U., 1979, Ph.D., 1981. Lic. physician, Ohio, Calif. Intern Children's Hosp. of Los Angeles, 1981-82; editorial cons. Grune & Stratton, San Diego, Calif., 1982—; physician Comphealth, Salt Lake City, 1983. Editor: Recent Advances in Clinical Therapeutics, Vol 1, 1981, Vol. 2, 1983, Vol. 3, 1983. Mem. Am. Fedn. Clin. Research. Jewish. Subspecialties: Pharmacology; Endocrinology. Current work: Prostaglandins, especially prostacyclin and its relationship to various chronic diseases, especially cardiovascular and diabetes. Home: 5312 Cole St San Diego CA 92117

YEUNG, KATHERINE LU, pharmacologist, researcher; b. Shanghai, China, July 28, 1943; d. Kuang Sheng and Lin (Chang) Lu; m. Joseph C. Yeung, Aug. 24, 1968; children: Wilfred, Malcolm. B.Sc., U. Houston, 1965, M.Sc., 1968. Research asst. U. Tex. M.D. Anderson Hosp., Houston, 1968-75, research assoc., 1975-80, asst. pharmacologist, 1980—. Mem. N.Y. Acad. Sci., Am. Assn. Cancer Research, Am. Soc. Pharmacology and Exptl. Therapeutics, Sigma Xi. Subspecialties: Epidemiology; Chemotherapy. Current work: Clinical pharmacology of chemotherapeutic agents. Home: 1102 Stoney Hill Houston TX 77077 Office: U Tex Cancer Center MD Anderson Hosp Room BF6006 Houston TX 77030

YIELDING, LERENA WADE, biochemist, educator, researcher; b. Wilmington, Del., Sept. 16, 1943; d. Erling Bjornstjerne and Louise Mills (Faircloth) Hauge; m. Gary Edwin Blodgett, Aug. 21, 1965; children: Elaine Louise, Laura Carlen; m. K. Lemone, Dec. 8, 1973; 1 dau., Katrina Elizabeth. B.S., U. Ala.-Tuscaloosa, 1965; Ph.D., U. Ala.-Birmingham, 1970. Postdoctoral trainee Inst. dental Research, Birmingham, Ala., 1970-71; research assoc. biochemistry U. Ala., Birmingham, 1972-76, instr., 1976-78, asst. prof., 1978-80; asst. prof. biochemistry dept. biochemistry dept. U. South Ala., Mobile, 1980—, grad. student faculty adv., 1981—. Fund raising chamn. Children's Service League, Birmingham, Ala., 1971-80, E. R. Dickson Elem. Sch., Mobile, 1982-83. NIH grantee, 1978-84; NSF undergrad. fellow, 1964-65. Mem. Am. Soc. Biol. Chemists (travel grantee 1982), Am. Chem. Soc., AAAS, Sigma Xi (First prize 1969). Subspecialties: Molecular pharmacology; Biochemistry (medicine). Current work: The design and synthesis of azido analogs to be used as photoaffinity probes for identifying and characterizing receptor sites of biologically important compounds. Office: U South Ala Coll Medicine 307 University Blvd Mobile AL 36688

YILMA, TILAHUN, veterinary microbiology educator; b. Ethiopia, Dec. 15, 1943; s. Yilma and Getenesh (Negewo) Wolde Ab. B.S., U. Calif.-Davis, 1968, D.V.M., 1970, Ph.D., 1976. Lic. Veterinarian, Calif. Head Vet. Service for Eastern Ethiopia, 1970-71; lectr. Sch. Animal Health FAO, Ethiopia, 1971-72; asst.prof. microbiology Sch. Vet. Medicine, Wash. State U., Pullman, 1980—. Contbr. articles to profl. jours. Mem. Am. Soc. Microbiology, AAAS, Calif. Vet. Med. Assn., Am. Coll. Vet. Tchrs. Subspecialty: Microbiology (veterinary medicine). Current work: Animal Interferons, subumit and synthetic vaccines. Office: Coll Vet Medicine Wash State U Pullman WA 99163

YIN, FRANK, CHI-PONG, cardiovascular physician, educator; b. Kunming, Yunnan, China, June 21, 1943; came to U.S., 1948, naturalized, 1961; s. Peter Yi-Ming and Hua-Nien (Chien) Y.; m. Grace Lu-Chi Chen, Apr. 19, 1975; children: Gregory, Jeffrey. B.S., MIT, 1965, M.S., 1967; Ph.D., U. Calif.-San Diego, 1970, M.D., 1973. Diplomate: Nat. Bd. Med. Examiners. Intern U. Calif. Hosps., San Diego, 1973-74, asst. resident, 1974-75; clin. assoc. Nat. Inst. Aging, Balt., 1975-77; fellow cardiology Johns Hopkins Med. Inst., Balt., 1977-78; asst. prof. Johns Hopkins U., Balt., 1978—; mem. study sect. NIH Cardiovascular and Pulmonary Sect. A, Bethesda, 1983—. Reviewer: Circulation Research, 1978—, Circultion, 1978—, Biophys. Soc, 1980—; contbr. articles to profl. jours. Chmn. bd. dirs. Chinese Lang. Sch. Balt., 1983—. Frank T. McClure fellow, 1980-82; Phi Tau Phi fellow, 1973. Mem. Biophys. Soc., Am. Physiol. Soc., Am. Heart Assn., Am. Fedn. Clin. Research, Rho Psi. Subspecialties: Biomedical engineering; Cardiology. Current work: Biomedical engineer, materials science, cardiology, physiology, biomaterials, solid mechanics, applied mechanics, gerontology, rheology. Home: 529 Saint Francis Rd Towson MD 21204 Office: Johns Hopkins Med inst 600 N Wolfe St Baltimore MD 21205

YING-SIN, LI, chemistry educator; b. Kwangtung, China, July 26, 1936; came to U.S., 1963, naturalized, 1977; s. Mu-Sun and Tzu-Chun (Lin) L.; m. Jackie T.L. Tu, Dec. 30, 1968; children: Ming-Po, Ming-Lin, Ming-Way, Ming-Yen Jason. B.S., Cheng Kung U., Taiwan, 1968; Ph.D., U. Kans., 1961. Research asst. Taiwan Sugar Research Inst., 1961-63; research assoc. U. Kans., Lawrence, 1963-66, Princeton U., 1968-70, U. S.C. Columbia, 1970-82; asst. prof. chemistry Memphis State U., 1982—; vis. prof. Cheng Kung U., 1975-76. Contbr. articles to profl. jours. Mem. Am. Chem. Soc., Sigma Xi, Phi Lambda Upsilon. Roman Catholic. Subspecialties: Physical chemistry; Infrared spectroscopy. Current work: Molecular spectroscopy and structural chemistry. Office: Dept Chemistry Memphis State U Memphis TN 38152

YOCUM, CHARLES FREDRICK, biology educator; b. Storm Lake, Iowa, Oct. 31, 1941; s. Vincent Gary and Olive Lucille (Cammack) Y.; m. Patricia Bury, Jan. 1, 1982; 1 son, Erik Charles. B.S., Iowa State U., 1963; M.S., Ill. Inst. Tech., 1968; Ph.D., Ind. U., 1971. Assoc. research biochemist Research Inst., Ill. Inst. Tech., 1963-68; fellow Ind. U., 1968-71; postdoctoral fellow Cornell U., Ithaca, N.Y., 1971-73; asst. prof. biol. scis. and chemistry U. Mich., Ann Arbor, 1973-77, assoc.

prof., 1978-82, prof., 1983—; mem. metabolic biology adv. panel NSF. Contbr. articles to profl. jours. Recipient Henry Russel award U. Mich., 1978. Mem. Am. Chem. Soc., AAAS, Am. Soc. Plant Physiologists, Biophys. Soc., Am. Soc. Photobiology. Subspecialties: Biochemistry (biology); Biophysical chemistry. Current work: Research on mechanisms of oxygen evolution and electron transport in photosynthesis. Office: Div of Biol Scis Univ Mich Ann Arbor MI 48109

YODER, HATTEN SCHUYLER, JR., petrologist; b. Cleve., Mar. 20, 1921; s. Hatten Schuyler and Elizabeth Katherine (Knieling) Y.; m. Elizabeth Marie Bruffey, Aug. 1, 1959; children: Hatten Schuyler III, Karen. A.A., U. Chgo., 1940, S.B., 1941; student, U. Minn., summer 1941; Ph.D., Mass. Inst. Tech., 1948; Dr. h.c., U. Paris VI, 1981. Petrologist Geophys. Lab., Carnegie Instn., Washington, 1948-71, dir., 1971—. Author: Generation of Basaltic Magma, 1976; Editor: The Evolution of the Igneous Rocks: Fiftieth Anniversary Perspectives; co-editor: Jour. of Petrology, 1959-69; asso. editor: Am. Jour. Sci, 1972—; Contbr. articles to sci. jours. Served to lt. comdr. USNR, 1942-46. Recipient award Mineral. Soc. Am., 1954, Bicentennial medal Columbia, 1954; Arthur L. Day medal Geol. Soc. Am., 1962; A.L. Day prize and lectureship Nat. Acad. Sci., 1972; A.G. Werner medal German Mineral. Soc., 1972; Golden Plate award Am. Acad. Achievement, 1973; Wollaston medal Geol. Soc. London, 1979; mineral, Yoderite, named in his honor. Fellow Geol. Soc. Am. (mem. council 1966-68), Mineral. Soc. Am. (council 1962-64, 69-73, pres. 1971-72), Am. Geophys. Union (pres. volcanology, geochemistry and petrology sect. 1962-64); mem. Mineral Soc. London (hon. 1983—), Geol. Soc. Edinburgh (corr.), Geol. Soc. Finland, All-Union Mineral. Soc. USSR (hon.), Geochem. Soc. (council 1956-58), Am. Chem. Soc., Mineral. Assn. Can., Nat. Acad. Sci. (chmn. geology sect. 1973-76), Washington Acads. Sci., Geol. Soc. Washington, Chem. Soc. Washington, Am. Philos. Soc. (council 1983—), Am. Acad. Arts and Scis., Explorers Club, Sigma Xi, Phi Delta Theta. Subspecialties: Petrology; Geophysics. Current work: Experimental petrology; mineral stability; high-pressure synthesis; heat and mass transfer; phase equilibria. Address: 2801 Upton St NW Washington DC 20008

YODER, JANICE DANA, psychology educator; b. Reading, Pa., Dec. 31, 1952; d. Daniel White and Helen (Petzar) Y.; m. John F. Zipp, Jr., June 6, 1981. B.A., Gettysburg (Pa.) Coll., 1974; M.A., SUNY-Buffalo, 1977, Ph.D., 1979. Vis. asst. prof. Washington U., St. Louis, 1979-80; Disting. vis. prof. U.S. Mil. Acad., West Point, N.Y., 1980; asst. prof. psychology Webster Coll., St. Louis, 1981—, coordinator women's studies program, 1981—, adminstrv. intern, 1981-82. Author: (with others) Leadership Effectiveness, 1983; contbr. articles to profl. jours. Coll. rep. women's program council Higher Edn. Ctr., St. Louis, 1981—. Mem. Am. Psychol. Assn., Midwestern Psychol. Assn., Assn. Women in Psychology, Soc. Psychol. Study of Social Issues, Soc. Personality and Social Psychology, NOW, Coalition for Environ. Democrat. Subspecialties: Social psychology; Psychology of women. Current work: Women's issues, particularly tokenism, sex-role stereotypes, and leadership. Office: Webster Coll 470 E Lockwood Saint Louis MO 63119

YODER, LARRY RICHARD, environ. ctr. dir., botany educator; b. Ft. Wayne, Ind., June 7, 1942; s. Loyal R. and Ruby Mae Y.; m. Ilse Ruth Hodel, Aug. 12, 1967; children: Erika Elizabeth, Laura Anne. B.A., Manchester (Ind.) Coll., 1964; M.A., Ind. U., 1966, M.A.T., 1970, Ph.D., 1972. Asst. prof. botany Ohio State U., Marion, 1972-79, assoc. prof., 1979-81; assoc. prof., dir. Merry Lea Environ. Learning Center, Goshen (Ind.) Coll., 1981—. Fellow Ohio Acad. Sci., Ind. Acad. Sci.; mem. Bot. Soc. Am. Mennonite. Subspecialties: Resource management; Plant anatomy. Current work: Preservation of ecotypes in Northern Indiana lake region. Office: Goshen Coll Goshen IN 46526

YODER, OLEN C., plant pathology educator; b. Fairview, Mich., Jan. 26, 1942; s. Curtis O. and Mabel (Beachy) Y.; m. (div.); 1 son: Brandon P.Y. Kneen. B.A., Goshen Coll., 1964; M.S., Mich. State U., 1968, Ph.D., 1971. Asst. prof. plant pathology Cornell U., 1971-77, assoc. prof., 1977—; vis. Stanford U., 1978-79. Contbr. articles to tech. jours., chpts. to books. Mem. Am. Phytopath. Soc., Am. Soc. Microbiology, AAAS. Subspecialties: Molecular biology; Genetics and genetic engineering (biology). Current work: Molecular biology of pathogenicity of microbes to plants.

YOERGER, ROGER RAYMOND, agrl. engr., educator; b. LeMars, Iowa, Feb. 17, 1929; s. Raymond Herman and Crystal Victoria (Ward) Y.; m. Barbara M. Ellison, Feb. 14, 1953; 1 dau., Karen Lynne; m. Laura M. Summitt, Dec. 23, 1971; stepchildren—Daniel L. Summitt, Linda Summitt Canull, Anita Summitt Smith. B.S., Iowa State U., 1949, M.S., 1951, Ph.D., 1957. Registered profl. engr., Ill., Pa., Iowa. Instr., asst. prof. agrl. engring. Iowa State U., 1949-56; asso. prof. agrl. engring. Pa. State U., 1956-58; prof. agrl. engring. U. Ill., Urbana, 1959—, head agrl. engring. dept., 1978—. Contbr. articles to profl. jours. Mem. Ill. Noise Task Force, 1974—. Fellow Am. Soc. Agrl. Engrs.; mem. Am. Soc. Engring. Edn., AAAS, Phi Kappa Phi (dir. fellowships, dir. 1971—). Roman Catholic. Clubs: Rotary, Moose. Subspecialties: Agricultural engineering; Acoustical engineering. Current work: Administration of an educational and research program in agricultural engineering. Patentee in field. Home: 107 W Holmes Urbana IL 61801 Office: 1208 W Peabody Dr Urbana IL 61801

YOKEL, ROBERT ALLEN, pharmacologist, researcher, educator; b. Rockford, Ill., June 22, 1945; s. Edward Clarence and Lilyann Lucille (Ehlert) Y.; m. Susan Jeanne Brown, Dec. 27, 1969; children: Erich Matthew, Kimberly Allison. B.S. in Pharmacy, U. Wis., 1968; Ph.D. in Pharmacology, U. Minn., 1973. Registered pharmacist, Wis., Ohio. Postdoctoral research assoc. Corcordia U., Montreal, Que., Can., 1973-75; asst. prof. pharmacology U. Cin., 1975-79; asst. prof. Coll. Pharmacy, U. Ky., Lexington, 1979—. Contbr. articles to sci. publs., chpts. to books. Mem. Soc. Neurosci., Behavioral Pharmacology Soc., Am. Soc. Pharmacology and Exptl. Therapeutics, AAAS. Subspecialties: Psychopharmacology; Toxicology (medicine). Current work: Neurobehavioral toxicology; aluminum toxicology; developmental toxicology; aluminum analysis.

YOKELL, MICHAEL DAVID, economist, cons.; b. Plattsburgh, N.Y., Nov. 21, 1946; s. Stanley and Edith Helen (Gersen) Y.; m. Jane Bunin, Apr. 13, 1946. B.Sc. in Physics, MIT, 1968; M.A. in Econs, U. Colo., 1973, Ph.D., 1975. Instr. math. econs. U. Colo., 1975; asst. prof. money and banking Wash. State U., 1976; vis. asst. prof., NSF research fellow U. Calif.-Berkeley, 1977; sr. economist Solar Energy Research Inst., 1977-79; prin., econs., chief adminstrv. officer Energy & Resource Cons., Inc., Boulder, Colo., 1979—; cons., lectr., speaker profl. meetings. Author: (with Geoff Sanders) The Economics of Minerals Reclamation, 1981, The Environmental Benefits and Costs of Solar Energy, 1980, Yellowcake: The International Uranium Cartel, (with June Taylor), 1979. Mem. Am. Econ. Assn., AAAS, Nat. Assn. Bus. Economists, Fedn. Am. Scientists, Internat. Assn. Energy Economists. Club: Am. Alpine. Subspecialties: Resource management; Fuels. Current work: Research and cons. in mgmt. of energy and natural resources.

YOKOYAMA, SHOZO, population geneticist; b. Miyazaki, Japan, Jan. 15, 1946; s. Iwanori and Masu Y.; m. Ruth Weaver. B.S., Miyazaki U., 1968; M.S., Kyushu U., 1971; Ph.D., U. Wash., 1977. Research assoc. U. Tex., Houston, 1977-78; instr. depts. psychiatry and genetics Washington U., St. Louis, 1978-79, asst. prof. dept. psychiatry, 1980—, asst. prof. dept. genetics, 1981—, asst. prof. dept. biology, 1982—. Contbr. articles to sci. jours. NIH grantee, 1981—. Mem. Soc. Study of Evolution, Genetics Soc. Japan, Genetics Soc. Am., AAAS, Am. Soc. Human Genetics, Am. Soc. Naturalists. Subspecialties: Population biology; Genome organization. Current work: Population genetics; evolution; human genetics; molecular biology; mathematical models; deleterious genes; mutation; recombinant DNA. Home: 7405 Melrose Saint Louis MO 63130 Office: Washington U Saint Louis MO 63110

YOO, DAL, physician, medical researcher; b. Kwang-Ju, Korea, Nov. 20, 1943; came to U.S., 1967, naturalized, 1972; s. Bong Soo and Yang Hee (Kim) Y.; m. Charlotte M. Nordanlycke, Apr. 14, 1974; children: Derek Torgny, Nora Ottilia. B.S., Seoul (Korea) Nat. U., 1963, M.D., 1967. Diplomate: Am. Bd. Internal Medicine, Am. Bd. Hematology, Am. Bd. Med. Oncology. Intern St. Luke's Hosp., Newburgh, N.Y., 1967-68, asst. resident in internal medicine, 1968, resident in anatomic pathology, 1968-69; resident Thomas Jefferson U. Hosp., 1969-70, George Washington U. Hosp., 1970-71, fellow in hematology and oncology, 1971-72, 74-75, assoc. clin. prof. medicine, 1975—; fellow in blood banking and immunochematology ARC, Washington, 1975; practice medicine specializing in internal medicine, hematology and med. oncology, Washington, 1975—; sr. attending physician Washington Hosp. Ctr., 1975—; mem. active staff Providence Hosp., Washington, Capitol Hill Hosp.; mem. courtesy staff Suburban Hosp., Washington Adventist Hosp., Holy Cross Hosp. Contbr. articles to profl. jours. Served to maj. U.S. Army, 1972-74. Fellow ACP; mem. Med. Soc. D.C., Washington Blood Club, Washington Soc. Oncology, Am. Soc. Hematology, Am. Soc. Clin. Oncology. Presbyterian. Subspecialties: Hematology; Oncology. Current work: Clincial and laboratory research of hematology and medical oncology. Home: 5827 Magic Mountain Dr Rockville MD 20852 Office: Providence Hosp 1150 Varnum St NE Washington DC 20017

YOO, TAI JUNE, allergist, immunologist, educator, researcher; b. Seoul, Korea, Mar. 7, 1935; came to U.S., 1959, naturalized, 1966; s. Heung Jin and Cho Soon (Rhee) Y.; m. Marie Anne Kwanghee Han, Sept. 21, 1963; children: Stephanie Misook, Christine Mia, Katherine Deborah. Diploma, M.D., Seoul Nat. U., 1959; Ph.D. in Biophysics, U. Calif.-Berkeley, 1963. Intern, resident in medicine, fellow in immunology Barnes Hosp./Washington U., St. Louis, 1963-66; sr. research fellow in immunology Roswell Park Meml. Inst., Buffalo, 1966-68; sr. resident dept. medicine N.Y.U. Med. Center, 1968-69, asst. prof. medicine, 1972-75; assoc. prof. medicine U. Iowa, 1975-80, adj. prof. bioengring., 1975-80; prof. medicine, assoc. prof. microbiology and immunology, dir. alergy-immunology div. U. Tenn. Center Health Scis., Memphis, 1980—; chief allergy and clin. immunology, chief allergy research program VA Med. Center, Memphis, 1980—; asst. chief allergy and immunology service Letterman Army Med. Center, 1971-72. Served with AUS, 1969-72; col. Res. Mem. Biophys. Soc., AAAS, Am. Assn. Immunology, N.Y. Acad. Scis., Am. Fedn. Clin. Research, Am. Acad. Allergy, Reticuloendothelial Soc., Central Soc., Am. Assn. Cancer Research, Am. Soc. Exptl. Pathology. Presbyterian. Subspecialties: Allergy; Internal medicine. Current work: Immunologic mechanisms in the pathogenesis of hearing loss, otosclerosis and Meniere's Disease, mechanism of immunotherapy, structure and function of allergens. Home: 6760 Slash Pine Cove Memphis TN 38163 Office: Allergy Immunology Div Room 3H05 956 Court Ave Memphis TN 38163

YOON, HYO SUB, biomedical engineer; b. Bian, Kyungbook, Korea, Apr. 17, 1935; came to U.S., 1962, naturalized, 1977; s. Bok Young and Mahl Soon (Kim) Y.; m. Chong Young Kim, June 3, 1972. B.S., Seoul Nat. U., 1959; M.S., U. Cin., 1965; Ph.D., Pa. State U., 1971. Research fellow U. Cin., 1962-66; postdoctoral assoc. Rensselaer Poly. Inst., Troy, N.Y., 1971-74, instr., research assoc., 1974—. Mem. Am. Soc. Metals, Am. Ceramic Soc., Am. Phys. Soc., N.Y. Acad. Sci., IEEE, Am. Crystallographic Assn., AAAS, Sigma Xi. Democrat. Subspecialties: Biomedical engineering; Bioinstrumentation. Current work: Biomedical ultrasonics; biomechanics; calcified tissues; clinical applications of acoustic emission; animal sound. Home: 6 Tower Heights Loudonville NY 12211 Office: Dept Biomedical Engring Rensselaer Poly Inst Troy NY 12181

YOON, JI-WON, microbiologist; b. Korea, Mar. 28, 1939; came to U.S., 1969, naturalized, 1977; s. Bakin and Ducksoon Y.; m. Chungja Rhim, Aug. 17, 1968; children: John, James. M.S., U. Conn., 1971, Ph.D., 1973. Asst. prof. microbiology Chosun U., Kwangju, Korea, 1965-67, assoc. prof. microbiology, 1967-69; staff fellow NIH, Bethesda, Md., 1974-74, sr. staff fellow, 1976-78, sr. investigator, 1978—, mem. faculty Grad. Sch., 1983—. Mem. Am. Soc. Microbiology, N.Y. Acad. Sci., Am. Tissue Culture Assn., AAAS. Baptist. Subspecialties: Virology (biology); Virology (medicine). Current work: Role of viruses and autoimmunity in pathogenesis of diseases in man and animal; virus-induced diabetes mellitus. Office: NIH Room 232 Bldg 30 Bethesda MD 20205

YOON, JONG SIK, geneticist; b. Suwon, Korea, Jan. 25, 1937; s. Ki and Pil (Kang) Y.; m. Kyung-soon A. Yoon, Sept. 10, 1962; children: Edward, Mimi, Sunny. B.S., Yonsei U., Seoul, 1961; M.A., U. Tex., Austin, 1964, Ph.D., 1965. Research scientist U. Tex., Austin, 1965-68; research scientist, faculty mem., 1971-78; asst. prof. Yonsei U., 1968-71; assoc. prof. Bowling Green (Ohio) State U., 1978-83, prof., 1983—. Contbr. articles to profl. jours. Served with Korean Army, 1958-60. China Med. Bd. N.Y. fellow, 1968-70; Baylor Coll. Medicine research fellow, 1970; recipient Young Scientist award Internat. Union Against Cancer, 1970; NIH grantee, 1976-77; NSF grantee, 1976—; Ohio Bd. Regents grantee, 1980-82. Fellow Tex. Acad. Sci.; mem. Genetics Soc. Am., Am. Genetic Assn., Soc. Study Evolution, Am. Soc. Naturalists, AAAS, Environ. Mutagen Soc., Tex. Assn. Radiation Research, Ohio Acad. Sci., Sigma Xi. Subspecialties: Genetics and genetic engineering (biology); Genome organization. Current work: Research in cytogenetics, mutations, oncogenetics, genome organization and genetic engineering. Home: 4 Picardie Ct Bowling Green OH 43402 Office: Dept Biol Scis Bowling Green State U Bowling Green OH 43403

YORK, DONALD GILBERT, astronomer, astrophysical sciences educator; b. Shelbyville, Ill., Oct. 28, 1944; s. Maurice Alfred and Virginia Maxine (Huntwork) Y.; m. Anna Sue Hinds, June 12, 1966; children: Sean, Maurice, Chandler, Jeremy. B.S., MIT, 1966; Ph.D., U. Chgo., 1970. Research asst. Princeton (N.J.) U., 1970-71, research assoc., 1971-73, research staff, 1973-75, research astronomer, 1975-78, sr. research astronomer, 1978-82; assoc. prof. dept. astrophys. scis. U. Chgo., 1982—. Contbr. articles to Astrophys. Jour. Recipient Disting. Service award NASA, 1975. Mem. Internat. Astron. Union, Am. Astron. Soc. Subspecialties: Ultraviolet high energy astrophysics; Optical astronomy. Current work: Abundances of elements, origin of elements, nature of interstellar matter, charge coupled device cameras, two dimensionalphoton counting cameras, echelle spectrographs, ultraviolet optics, earth-orbiting satellites. Office: 5640 S Ellis Ave Chicago IL 60637

YORK, DONALD HAROLD, physiology and neurosurgery educator; b. Moose Jaw, Sask., Can., Jan. 30, 1944; came to U.S., 1975; s. Henry Harold and Dorothy Marguerite (Sparrow) Y.; m. Catherine Louise Robbie, Oct. 1, 1966; children: Andrew, Deborah, James. B.Sc. with honors, U. B.C., Vancouver, 1965, M.Sc., 1966; Ph.D., Monash U., 1969. Asst. prof. Queen's U., Kingston, Ont., Can., 1968-73, assoc. prof., 1973-75; assoc. prof. physiology and neurosurgery U. Mo., Columbia, 1975-82, prof., 1982—. Contbr. in field. Coach Columbia Soccer Assn., 1973-83; asst. scoutmaster Great Rivers council Boy Scouts Am., Columbia, 1981-83. Med. Research Council Can. scholar, 1970-75. Mem. Soc. Neurosci. (chpt. pres. 1977-78), Can. Physiol. Soc., Am. Physiol. Soc., IEEE, Pharmacolo. Soc. Can., Sigma Xi. Subspecialties: Neurophysiology; Biomedical engineering. Office: Dept Physiology U Mo M412 Medical Sciences Bldg Columbia MO 65212

YORK, JAMES LESTER, behavioral pharmacologist, educator; b. Peoria, Ill., Nov. 12, 1942; s. Wayne and Lucy (Aupperle) Y.; m. Patricia Stanton, Aug. 22, 1970; children: Benjamin, Nora. A.B. in Psychology, Bradley U., 1965; Ph.D. in Pharmacology, U. Ill., Chgo., 1972. Postdoctoral trainee dept. pharmacology SUNY, Buffalo, 1972-74; research scientist Research Inst. on Alcoholism, Buffalo, 1975—; research assoc. prof. dept. psychology SUNY, Buffalo, 1981—. Contbr. articles to profl. jours. Nat. Inst. Alcohol Abuse and Alcoholism grantee, 1976-78; N.Y. State Health Research Council grantee, 1978-79, 81-82. Mem. Am. Soc. Pharmacology and Exptl. Therapeutics, Research Soc. on Alcoholism, Soc. Stimulus Properties of Drugs. Subspecialties: Pharmacology; Developmental biology. Current work: Etiology of addictive behaviors; discriminative stimulus properties of psychoactive agents, age-related determinants of responsiveness to drugs. Home: 2228 Blakeley Rd East Aurora NY 14052 Office: 1021 Main St Buffalo NY 14203

YORKE, JAMES ALAN, mathematics educator; b. Tokyo, Aug. 3, 1941; U.S., 1945; s. Edward Thomas and Margaret (Tiernan) Y. A.B., Columbia U., 1963; Ph.D., U. Md.-College Park, 1966. Mem. faculty dept math. and Inst. Phys. Sci. and Tech., U. Md., College Park, 1966—, prof., 1973—. Guggenheim Found. fellow, 1980. Mem. Am. Math. Soc., Soc. Indsl. and Applied Math. Current work: Study irregular oscillatiors of deterministic processes, and the role of topology in applied mathematics. Home: 5465 Mystic Ct Columbia MD 21044 Office: U Md IPST Bldg College Park MD 20742

YORKS, PAMELA FLORENCE, biologist, cons.; b. Syracuse, N.Y., Mar. 30, 1949; d. Kenneth Preston and Mabel Lois (Andes) Y. A.A., Graceland Coll., 1969; B.S., SUNY Coll. Forestry, Syracuse, 1971; Ph.D. (Regents fellow), U. Calif.-Berkeley, 1980. Research asst. U. Calif. Herbarium, Berkeley, 1971-72, teaching asst. botany dept., 1972-74, 1975-76; asst. prof. biology U. Puget Sound, Tacoma, Wash., 1976-82. Mem. AAAS, Bot. Soc. Am., Ecol. Soc. Am., Calif. Bot. Soc., Sierra Club, Wash. Native Plant Soc., Audubon Soc. Subspecialties: Ecology; Evolutionary biology. Current work: Pollination biology, community ecology.

YOSHIDA, TAKESHI, immunology researcher and educator; b. Fukuoka, Japan, July 24, 1938; came to U.S., 1967; s. Kiyoshi and Yae Y.; m. Tomoko Tasaka, Nov. 7, 1964; children: Atsushi, Keiko. M.D., U. Tokyo, 1963, D.Med.Sci., 1970. Research officer NIH, Tokyo, 1964-71; research assoc. NYU, N.Y.C., 1967-68; vis. scientist Nat. Inst. Allergy and Infectious Diseases, NIH, Bethesda, Md., 1968-69, 70; research asst. prof. SUNY-Buffalo, 1971-74; prof. pathology U. Conn. Health Ctr., Farmington, 1974—; cons. immunology sci. study sect. NIH, Bethesda, Md., 1979-83. Co-editor: Basic and Clinical Aspects of Granulom Diseases, 1980, Immunobiology of Eosinophils, 1982, Basic and Clinical Aspects of Cancer Metastasis, 1983; mem. editorial bd.: Clin. Immunology and Immunopathology, 1977—, Jour. Reticuloendothelial Systems, 1978-80, Immunological Communications, 1981—. Recipient Research Career Devel. award NIH, 1975-80, research grantee Nat. Heart, Lung, and Blood Inst., Nat. Inst. Allergy and Infectious Diseases, 1976—; Research grantee Nat. Cancer Inst., 1977-80. Mem. Am. Assn. Immunologists (travel awards 1974, 77), Am. Assn. Pathologists, Reticuloendothelial Soc., Am. Soc. Microbiology, N.Y. Acad. Scis. Club: Pluto. Subspecialties: Cellular engineering; Infectious diseases. Current work: Mechanisms of delayed type hypersensitivity and cell-mediated immunity, studies on their soluble mediators, mechanisms of granulomatous inflammation and its regulation. Office: Dept Pathology U Conn Health Ctr Farmington Ave Farmington CT 06032

YOSHIKAWA, HERBERT HIROSHI, nuclear engineer; b. South Dos Palos, Calif., May 13, 1929; s. Ichiji and Chiyo (Morita) Y.; m. Helen Hisako Sadataki, Sept. 24, 1960. Ph.B., U. Chgo., 1948, M.S., 1951; Ph.D., U. Pa., 1958. Sr. engr. Gen. Electric, Richland, Wash., 1958-65; mgr. Battelle-N.W., Richland, 1965-70; mgr. tech. Westinghouse Hanford Co., Richland, 1970—. Mem. Am. Phys. Soc., Am. Nuclear Soc., AAAS, Sigma Xi. Lodge: Kiwanis. Subspecialties: Nuclear engineering; Materials. Current work: Fast breeder and fusion reactor technology; materials irradiations, liquid metals, fission and fusion related neutron dosimetry and materials development. Office: Westinghouse Hanford Co PO Box 1970 W/C 44 Richland WA 99352 Home: 2712 W Klamath Ave Kennewick WA 99336

YOUNG, A. THOMAS, govt. ofcl.; b. Nassawadox, Va., Apr. 19, 1938; s. William Thomas and Margaret Sara (Colonna) Y.; m. Page Hayden Young, June 24, 1961; children—Anne Blair, Thomas Carer. B.A.E., B.M.E., U. Va., 1961; M.S., M.I.T., 1972. Design engr. Newport News Shipbldg. and Dry Dock Co., Va., 1961; with NASA, 1961—, mission ops. mgr., Hampton, Va., 1974-75, Viking mission dir., Pasadena, Calif., 1975, dir. planetary program, Washington, 1976-79, dep. dir., Moffett Field, Calif., 1979-80, dir., Greenbelt, Md., 1980—. Fellow Am. Astron. Soc., AIAA; mem. Washington Acad. Scis. Methodist. Subspecialty: Aerospace engineering and technology. Home: 947 Placid Ct Arnold MD 21012 Office: Goddard Space Flight Center NASA Greenbelt MD 20771

YOUNG, ANDREW TIPTON, astronomer; b. Canton, Ohio, Apr. 4, 1935; s. Clarence Willard and Margaret Ethel (Tipton) Y.; m. Jeannette Noel Eiseman, June 19, 1954; m. Louise Dillon Gray, Dec. 14, 1968; children: Susan Margaret Young Kreider, Alan Samuel. B.A., Oberlin (Ohio) Coll., 1955; M.A., Harvard U., 1957, Ph.D., 1962. Asst. prof. astronomy U. Tex., Austin, 1965-67; mem. tech. staff Jet Propulsion Lab., Pasadena, Calif., 1968-73; research scientist Tex. A&M U., College Station, 1974-80; adj. assoc. prof. San Diego State U., 1981—; cons. to NASA. Mem. Am. Astron. Soc., Internat. Astron. Union, Royal Astron. Soc., Optical Soc. Am. Subspecialties: Optical astronomy; Planetary science. Current work: High accuracy photometry and data analysis research on clouds and atmosphere of Venus; optical properties of cloud pigments, teaching. Office: Astronomy Dept San Diego State U San Diego CA 92182

YOUNG, CHARLES GILBERT, engineer; b. Fawn Grove, Pa., Feb. 25, 1930; s. Charles C. and Mabel (Bomberger) Y.; m. Louise Dickey MacKeen, July 10, 1959; children: Douglas Harold, Steven Charles,

David Andrew. B.A. in Math. and Physics, Elizabethtown (Pa.) Coll., 1952; M.S. in Physics, U. Conn., 1956, Ph.D., 1961. Instr. U. Conn., Storrs, 1960-62; sr. physicist, dept. head, gen. mgr. Am. Optical Co., Southbridge, Mass., 1962-77, dir. product devel., 1973-77; dir. engring. Kollmorgen, Northampton, Mass., 1977-79; asst. tech. mgr. NERAC, Storrs, 1979-80; dir. tech. devel. Combustion Engring. Co., Stamford, Conn., 1980—; cons. Contbr. numerous articles to profl. jours. Recipient Indsl. Design award Indsl. Design Mag., 1972; Indsl. Research, Inc. grantee, 1969, 71, 72. Mem. IEEE (sr.), Optical Soc. Am., Am. Phys. Soc., Soc. Photog. and Instrumentation Engrs., Am. Nuclear Soc., ASME. Current work: Corporate direction of new technology, research and development in energy, metallic and non-metallic and composite materials, automation, fiber optics, multi-decade technology futures. Patentee in field. Office: Combustion Engring 900 Long Ridge Rd PO Box 9308 Stamford CT 06904

YOUNG, DAVID MONAGHAN, JR., mathematics and computer science educator, university administrator, researcher; b. Boston, Oct. 20, 1923; m., 1949; 3 children. B.S., Webb Inst. Naval Architecture, 1944; M.A., Harvard U., 1947, Ph.D. in Math, 1950. Instr., research assoc. in math. Harvard U., 1950-51; mathematician, Aberdeen Proving Ground, 1951-52; assoc. prof. math. U. Md., 1952-55; mgr. math. analysis dept. Ramo-Wooldridge Corp., 1955-58; prof. math., dir. Computer Center U. Tex.-Austin, 1958-70, prof. math. and computer sci., dir., 1970—. Mem. Am. Math. Soc., Soc. Indsl. and Applied Math., Math., Assn. Am., Assn. Computing Machinery. Subspecialty: Numerical analysis. Office: Ctr Numerical Analysi U Te Austin TX 78712

YOUNG, DELANO VICTOR, chemistry educator, researcher; b. Honolulu, Nov. 17, 1945; s. Lum Fai and Gladys Sau Pung (Wong) Y.; m. Chin-Yi Yang, Jan. 31, 1970; 1 dau., Heather Tien. B.S., Stanford U., 1967; Ph.D., Columbia U., 1973. Postdoctoral fellow Salk Inst., San Diego, 1973-75; asst. prof. Boston U., 1975—; vis. scholar Harvard U., Cambridge, Mass., 1982-83; textbook cons. D. Van Nostrand, N.Y.C., Willard Grant Press, Boston, Prentice Hall, N.Y.C. Contbr. articles in field to profl. jours. Nat. Cancer Inst. research grantee Boston U., 1981-84, 76-80; Elsa U. Pardee Found. research grantee Boston U., 1979-81; Jane Coffin Childs Fund postdoctoral fellow Salk Inst., 1973-75; Eugene Higgins fellow Columbia U., 1967-68; Gen. Motors scholar, 1963-67. Mem. Am. Soc. Microbiology, AAAS, Am. Soc. Biol. Chemists, Phi Beta Kappa, Sigma Xi, Phi Lambda Upsilon. Democrat. Roman Catholic. Subspecialties: Biochemistry (biology); Molecular biology. Current work: Growth control of normal and neoplastic cells in tissue culture, regulation of gene expression, mechanism of growth factors and growth-regulating hormones. Home: 476 Old Town Way Hanover MA 02339 Office: Boston U Dept Chemistry 685 Commonwealth Ave Boston MA 02215

YOUNG, DONALD FRANCIS, mathematician, educator; b. Washington, Oct. 26, 1944; s. Martin Greene and Alberta Elizabeth (Francis) Y. B.S., Duke U., 1966; M.S., U. Va., 1972, Ph.D., 1975. Asst. prof. U S.C., Columbia, 1975-78; asst. prof. math. Agnes Scott Coll., Decatur, Ga., 1978-83; asst. prof. Oxford (Ga.) Coll., 1983—. Served to lt. USN, 1966-71. Mem. Am. Math. Soc., Math. Assn. Am., Soc. Indsl. and Applied Math., Phi Beta Kappa. Unitarian Universalist. Subspecialty: Applied mathematics. Current work: Controllability and optimal control of systems governed by linear and non-linear functional equations. Home: PO Box 226 Oxford GA 30267 Office: Dept Math Oxford Coll Oxford GA 30267

YOUNG, DOUGLAS WILFORD, operations research analyst; b. Salt Lake City, July 9, 1944; s. Wilford T. and Ada J. (Asper) Y.; m. Crystal Jean Burrup, Dec. 2, 1968; children: Gregory, Kimberly, Wendy, Brenda, Linda. B.S., U. Utah, 1967, M.B.A., 1970. Cert. cost analyst. Ops. research analyst U.S. Army Communications Command, Ft. Huachuca, Ariz., 1979-80, 80-81, 82—, budget analyst, 1971-79, U.S. Air Traffic Control Ctr., Ft. Huachuca, 1980; program analyst Dugway Proving Ground, Utah, 1981-82. Served with U.S. Army, 1968-70; Vietnam. Recipient profl. awards. Mem. Ops. Research Soc. Am. (assoc.), Am. Soc. Mil. Comptrollers (award 1975). Subspecialties: Operations research (mathematics). Office: US Army Communications Command Fort Huachuca AZ 85613

YOUNG, EDWARD JOSEPH, geologist; b. Roselle, N.J., Feb. 18, 1923; s. Michael Joseph and Anne Antoinette (Brodowska) Y.; m. Amelia Frances Painter, Feb. 1, 1955; children: Michael, Ann. B.S., Rutgers U., 1948; M.S., MIT, 1950, Ph.D., 1954. Geologist U.S. Geol. Survey, Denver, 1952—. Contbr. numerous articles on geology and mineralogy to profl. jours., 1951—. Served with U.S. Army, 1943-45; ETO. Fellow Geol. Soc. Am; mem. Mineral. Soc. Am., Mineral, Assn. Can., Phi Beta Kappa, Sigma Xi. Club: Colo, Mountain (Denver). Subspecialties: Mineralogy; Petrology. Current work: Topics in geology, mineralogy, petrology and geochemistry. Home: 3000 Union St Lakewood CO 80215 Office: US Geol Survey Mail Stop 916 Box 25046 Fed Center Denver CO 80225

YOUNG, FRANCIS ALLAN, psychologist; b. Utica, N.Y., Dec. 29, 1918; s. Frank Allan and Julia Mae (McOwen) Y.; m. Judith Wadsworth Wright, Dec. 21, 1945; children—Francis Allan, Thomas Robert. B.S., U. Tampa, 1941; M.A., Western Res. U., 1945; Ph.D., Ohio State U., 1949. Instr. Wash. State U., Pullman, 1948-50, asst. prof., 1950-56, assoc. prof., 1956-61, prof. psychology, 1961—, dir. primate research center, 1957—; vis. prof. ophthalmology U. Oreg., Portland, 1964; vis. prof. pharmacology U. Uppsala (Sweden) Med. Sch., 1971; vis. prof. optometry U. Houston, 1979-80. Editor: (with Donald B. Lindsley) Early Experience and Visual Information Processing in Perceptual and Reading Disorders, 1970. Named Disting. Psychologist State of Wash., Wash. Psychol. Assn., 1973; recipient Paul Yarwood Meml. award Calif. Optometric Assn., 1978; Apollo award Am. Optometric Assn., 1980; Nat. Acad. Sci.-NRC sr. postdoctoral fellow in physiol. psychology U. Wash., 1956-57; research grantee NSF, 1950-53, USAF, 1965-72, NIH, 1960-78. Fellow Am. Acad. Optometry, Am. Pychol. Assn. (pres. Div. 31 1974-75); mem. Common Cause, Ams. Dem. Action, Assn. Research in Vision and Ophthalmology, Internat. Soc. Myopia Research (sec.-treas. 1978—), AAAS, Psychonomic Soc., Wash. State Psychol. Assn. (exec. sec. 1965-77), Western Psychol. Assn.; Mem. N.Y. Acad. Scis.; mem. Sigma Xi, Psi Chi (nat. pres. 1968-70). Subspecialties: Sensory processes; Physiological psychology. Current work: Role of behavior and genetics in development of myopia including influence of accomodation and convergence; its morbidity and prevention and epidemiological character. Home: NW 344 Webb St Pullman WA 99163 Office: Wash State U Pullman WA 99164

YOUNG, FRANKLIN ALDEN, JR., biological and physical sciences educator, consulting company executive; b. Harrisburg, Pa., Mar. 14, 1938; s. Franklin Alden and Elizabeth (Catterton) Y.; m. Ann Marie Ruttkay, June 13, 1959 (div. Dec. 1973); children: Kathleen A., Mary-Elizabeth, Thomas Franklin; m. Carolyn Joyce Herron, Dec. 14, 1973; stepchildren: Philip B. Watson, Suzanne C. Watson, Warren J. Watson. B.I.E., U. Fla., 1960, M.S.E., 1963; D.Sc., U. Va., 1968. Mfg. engr. Gen. Electric Co., N.Y.C., 1960-61; instr. Clemson U., 1963-65; asst. prof., 1968-70; assoc. prof. biol. and phys. scis. Med. U. S.C., 1970-75, prof., chmn. dept. biol. and phys. scis., 1975—; pres. Young and Assocs.; cons. NIH, 1973-82, mem. study sect., 1975-79. Editor: Bioceramics, Engineering in Medicine, 1972, Ceramics and Surgical Implants, 1979. USPHS grantee, 1972—, 77-82. Mem. Am. Soc. Metals (chpt. chmn. 1974-75), Internat. Assn. Dental Research (councilor 1979-80), Soc. Biomaterials, Sigma Xi, Omicron Kappa Upsilon. Subspecialties: Biomaterials; Implantology. Current work: Interactions between materials and biosystems. Office: Med U SC 171 Ashley Ave Charleston SC 29425

YOUNG, GARRY GEAN, nuclear/licensing engr.; b. Ft. Smith, Ark., Sept. 7, 1951; s. Lee Leonard and Lola Belle (Bartlett) Y.; m. Patricia Ann Farmer, June 24, 1977; 1 dau., Kathryn Elizabeth. B.S. in M.E, U. Ark., 1974, M.S., 1975. Registered profl. engr. Ark. D.C. Prodn. engr. Ark. Power & Light Co., Little Rock, 1975-79; fellow Adv. Com. on Reactor Safeguards, Washington, 1979-80, reactor engr., 1980-81; supervising engr. United Energy Services Corp., Atlanta, 1981—. Author tech. papers. Youth leader, Sunday Sch. tchr. Knollwood Bapt. Ch., Burke, Va., 1980-81; co-dir. children's ch. First Bapt. Ch. of Pelham, Ala., 1982. Mem. ASME, Am. Nuclear Soc., Health Physics Soc., Pi Tau Sigma. Subspecialties: Nuclear fission; Mechanical engineering. Current work: Commercial nuclear power plant technical reviews and evaluation of engineered safety features and reactor protection systems. Office: United Energy Services Corp 8235 Dunwoody Pl Altanta GA 30338

YOUNG, JAMES FORREST, elec. engr., educator; b. Meadville, Pa., June 22, 1943; s. David George and Carolyn Hope (Spinney) Y.; m. Cecily Sweet, June 10, 1971. B.S., M.I.T., 1965, M.S., 1966; Ph.D. in Elec. Engring, Stanford U., 1970. Research assoc. Stanford U., 1970-75, prof. elec. engring., 1975—, asst. dir. tech. ops., 1977—; cons. Bell Telephone Labs, Spectra-Physics, Quanta-Ray, Coherent. Contbr. articles to profl. jours. Fellow Optical Soc. Am.; mem. IEEE, Fedn. Am. Scientists, Tau Beta Pi. Subspecialties: Optical engineering; Spectroscopy. Current work: Research and development of new lasers, short wave length sources and new laser applications. Office: Ginzton Lab Stanford CA 94305 Home: 940 Cottrell Way Stanford CA 94305

YOUNG, JERRY WESLEY, nutrition educator; b. Mulberry, Tenn., Aug. 19, 1934; s. Rufus William and Annie Jewell (Sweeney) Y.; m. Charlotte Sullenger, July 8, 1959; children: David Wesley, Jeretha Lynn. B.S., Berry Coll., 1957; M.S., N.C. State U.-Raleigh, 1959, Ph.D., 1963. Research asst. N.C. State U., Raleigh, 1957-63; NIH fellow U. Wis., Madison, 1963-65; asst. prof. animal sci Iowa State U., Ames, 1965-71, assoc. prof., 1971-74, prof., 1974—. Contbr. numerous articles to profl. publs. NIH grantee, 1966-69, 71-78; U.S. Dept. Agr. grantee, 1979-82. Mem. Am. Inst. Nutrition, Am. Dairy Sci. Assn., Am. Soc. Animal Sci., Sigma Xi, Phi Kappa Phi. Democrat. Baptist. Subspecialties: Animal nutrition; Nutrition (biology). Current work: Metabolism and kinetics of glucose in cattle; ketosis in lactating cows; digestive physiology and metabolism of ruminants. Home: 1515 20th St Ames IA 50010 Office: Iowa State U 313 Kildee Hall Ames LA 50011

YOUNG, JOHN ARTHUR, meteorology educator; b. Washington, July 4, 1939; s. Arthur Eugene and Helen Betty (Marsau) Y.; m.; children: David, Melissa. B.A., Miami U., Oxford, Ohio, 1961; Ph.D., MIT, 1966. Mem. faculty U. Wis., Madison, 1966—, now prof. meteorology; vis. assoc. prof. MIT, 1973-74. Contbr. articles to meteorol. jours. NSF postdoctoral fellow, 1966. Mem. Am. Meteorol. Soc. Subspecialties: Meteorology; Synoptic meteorology. Current work: Dynamic meteorology and oceanography; monsoons. Office: 1503 Meteorology U of Wis Madison WI 53706

YOUNG, LAURENCE RETMAN, biomedical engineer, educator; b. N.Y.C., Dec. 19, 1935; s. Benjamin and Bess (Retman) Y.; m. Joan Marie Fisher, June 12, 1960; children—Eliot Fisher, Leslie Ann, Robert Retman. A.B., Amherst Coll., 1957; S.B., MIT, 1957; S.M., Mass. Inst. Tech., 1959, Sc.D., 1962; Certificat de License (French Govt. fellow), Faculty of Sci. U. of Paris, France, 1958. Registered profl. engr., Mass. Engr. Sperry Gyroscope Co., Great Neck, N.Y., 1957; engr. Instrumentation Lab., Mass. Inst. Tech., 1958-60, asst. prof. aero. and astronautics, 1962-67, asso. prof., 1967-70, prof., 1970—; summer lectr. U. Ala., Huntsville, 1966-68; lectr. Med. Sch. Harvard U., 1970-78; mem. tng. com. biomed. engring. NIH, 1971-73; mem. com. space medicine and biology Space Sci. Bd., Nat. Acad. Scis., 1974—, chmn. vestibular panel summer study of life scis. in space, 1977; mem. com. engring. and clin. care Nat. Acad. Engring., 1970; mem. Air Force Sci. Adv. Bd., 1979—; vis. prof. Swiss Fed. Inst. Tech., Zurich, 1972-73, Conservatoire Nationale des Arts and Metiers, Paris, 1972-73; vis. scientist Kantonsspital Zurich, 1972-73; prin. investigator vestibular expts. on Spacelabs—1, 4 and D-1, 1977—; cons. Arthur D. Little, NASA, Gulf & Western, Link div. Singer Co., Boeing Corp. Contbg. author: chpt. on vestibular system Medical Physiology, 1974; Editorial bd.: Internat. Jour. Man-Machine Studies, 1966-75, Neurosci, 1976—; Contbr. numerous articles to profl. jours. Fellow IEEE (Franklin V. Taylor award 1963); mem. Nat. Acad. Engring., Biomed. Engring. Soc. (founding/charter mem., dir. 1972-75, pres. 1979-80), Aerospace Med. Assn., ASTM (com. sports safety, subcom. skiing safety, chmn. skiing statistics subcom. 1975—), Internat. Fedn. Automatic Control (tech. com. biomed. engring. 1975—), AIAA (working group for simulator facilities 1976—, Dryden lectr. in research 1982), Internat. Soc. Skiing Safety (dir. 1977—), Barany Soc., Tau Beta Pi. Subspecialties: Biomedical engineering; Space medicine. Current work: Human adaptation to weightlessness; human factors. Inventor eye movement monitor. Home: 8 Devon Rd Newton Centre MA 02159 Office: Mass Inst Tech Dept Aeros and Astronautics Cambridge MA 02139

YOUNG, MATT, physicist; b. Bklyn., Jan. 30, 1941; s. Arthur and Florence (Turner) Y.; m. Deanna Clair, May 22, 1964; children: David, Rachel. B.S., U. Rochester, 1962, Ph.D., 1967. Research assoc. U. Rochester, 1967; asst. prof. physics U. Waterloo, Ont., Can., 1967-70; asst. prof. electrophysics Rensselaer Poly. Inst., Troy, N.Y., 1970-74; assoc. prof. natural sci. Verrazzano Coll., Saratoga Springs, N.Y., 1974-75; physicist Nat. Bur. Standards, Boulder, Colo., 1976—; cons. in field. Author: Optics and Lasers, 1977, 2d edit., 1984; assoc. editor: Jour. Optical Soc. AM., 1972-78; contbr. articles to profl. jours. Sec. Com. for Open Edn., 1973-74. Mem. Optical Soc. Am. (pres. Rocky Mountain sect. 1982-83), Am. Assn. Physics Tchrs., AAAS, Fedn. Am. Scientists. Jewish. Subspecialties: Optics research; Laser research. Current work: Optical fiber measurements, precision measurement related to optics and laser technology. Home: 3145 Fremont Boulder CO 80302 Office: Electromagnetic Tech Div Nat Bur Standards Boulder CO 80303

YOUNG, MICHAEL WARREN, geneticist; b. Miami, Fla., Mar. 28, 1949; s. Lloyd G. and Mildred L. (Tillery) Y.; m. Laurel Ann Eckhardt, Dec. 27, 1978. B.A., U. Tex., 1971, Ph.D., 1975. Postdoctoral fellow Stanford U., 1975-77; asst. prof. genetics Rockefeller U., N.Y.C., 1978—. Contbr. articles to profl. jours. Meyer Found. fellow, 1978—. Mem. Genetics Soc. Am., AAAS. Subspecialties: Gene actions; Genome organization. Current work: Eukaryote chromosome organization, transposable elements, genes controlling development and behavior. Office: Rockefeller U 1230 York Ave New York NY 10021

YOUNG, PEGGY SANBORN, psychologist; b. Painesville, Ohio, Aug. 25, 1926; d. Philip Harold and Josephine Diana (Masters) Sanborn; m. Philip Percy Young, Nov. 14, 1947 (div. Sept. 1968); children: Philip Harold, Timothy Mark, Don Sanborn. B.S. Ed., Baldwin Wallace Coll., 1956; M.A., Case Western Res. U., 1963; Ph.D., Kent State U., 1977. Tchr. spl. edn. Willoughby (Ohio) East Schs., 1960-61; vocat. counseling psychologist Salvation Army Hosp., Cleve., 1961-62; psychologist Willoughby Eastlake Schs.(Ohio), 1962-63, Mentor (Ohio) Pub. Schs., 1963-65, Tuslaw and Fairless Local Schs., Stark County, Ohio, 1965-67; chief psychologist Mentor (Ohio) Pub. Schs., 1967-75; coordinator spl. edn. and related services, 1975—, dir. ops., 1981—; mem. adv. bd. Comprehensive Program for Hearing Impaired, Mayfield, Ohio, 1975; mem. Lake County Welfare Bd. Childrens Services, Painesville, 1979-82. Mem. Am. Psychol. Assn., Nat. Assn. Sch. Psychologists, Ohio Psychol. Assn., Ohio Assn. Sch. Psychologists, Cleve. Area Sch. Psychologists. Republican. Episcopalian. Subspecialties: Learning; Behavioral psychology. Current work: Handicapped children, personnel psychology, personnel selection, body image and self concept; obesity. Home: 9956 Johnnycake Ridge C-5 Painesville OH 44077 Office: Mentor Exempted Village Schs 6451 Center St Mentor OH 44060

YOUNG, ROBERT CLARINGBOLD, physician; b. Binghamton, N.Y., May 22, 1947; s. Claude and Suzanne Viola (Conant) Y.; m. Ellice Chase, June 12, 1975; 1 son, Ian James. B.A., Williams Coll., 1969; M.D., Cornell U., 1974. Chief resident dept. psychiatry Washington U. Sch. Medicine, St. Louis, 1978-79; instr. dept. psychiatry Cornell U. Med. Coll. (and clin. affiliate N.Y. Hosp.-Westchester Div.), 1979-80, asst. prof. psychiatry, 1981—; asst. attending psychiatrist N.Y. Hosp., White Plains, 1981—; cons. in field. Contbr. articles to profl. jours. Recipient Harold Wolff Research prize Cornell U., 1973, 74. Mem. Am. Psychiat. Assn., Assn. Research in Nervous and Mental Disease, Soc. Neurosci., Sigma Xi, Phi Beta Kappa. Subspecialties: Psychopharmacology; Gerontology. Current work: Research in psychopharmacology, geriatrics. Office: 21 Bloomingdale Rd White Plains NY 10605 Home: 6 Winding Ct Mohegan Lake NY 10547

YOUNG, ROBERT CLELAND, materials scientist; b. Beloit, Wis., June 18, 1931; s. Miles Martin and Alice Isabel (Cleland) Y.; m. Nancy F. Smith, Aug. 14, 1954; children: Robert M., Patricia S. B.S., Beloit Coll., 1953; postgrad., U. Wis., 1953-54. Registered profl. engr., Ohio. Research engr. Babcock & Wilcox, Alliance, Ohio, 1954-71, materials engr., 1971-76; project engr. Nevada Engring. & Tech. Inc., Long Beach, Calif., 1976-78; research specialist Lockheed-Calif. Co., Saugus, 1978—. Contbr. articles to profl. jours. Fellow ASTM (Award of Merit 1982); mem. ASME, Soc. Advancement of Material and Process Engring. Republican. Presbyterian. Subspecialties: Composite materials; Ceramic engineering. Current work: Development and preliminary application of composite materials. Patentee in field. Home: 24031 Via Aranda Valencia CA 91355 Office: PO Box 551 D74-71 B211 P2 Burbank CA 91520

YOUNG, ROBERT JOHN, educator; b. Calgary, Alta., Canada, Feb. 10, 1923; came to U.S., 1956, naturalized, 1965; s. Harold P. and Kate A. (Thomson) Y.; m. Greta G. Milne, June 16, 1950; children—Kenneth W., Donna E. B.S.A. with honors, U. B.C., 1950; Ph.D., Cornell U., 1953. Research asso. dept. med. research Barting & Best, Toronto, Ont., Can., 1953-56; research chemist Internat. Minerals and Chem. Corp., Skokie, Ill., 1956-58, Proctor and Gamble Co., Cin., 1958-60; prof. animal nutrition Cornell U., 1960—, chmn. dept. poultry sci., 1965-76, chmn. dept. animal sci., 1976—. Author: (with M.L. Scott, M.C. Nesheim) Nutrition of the Chicken, 1976; Contbr. articles in field to profl. jours. Served with RCAF, 1942-45. Mem. Am. Inst. Nutrition, Poultry Sci. Assn., Am. Soc. Animal Sci., Am. Soc. Dairy Sci. Subspecialty: Animal nutrition. Current work: Poultry nutrition. Office: Morrison Hall Cornell Univ Ithaca NY 14853

YOUNG, ROBERT RICE, clinical neurophysiologist-neurologist, reseacher, educator; b. Washington, Pa., Aug. 26, 1934; s. Robert Sterling and Elizabeth Linnaea (Jackson) Y.; m. Katharine M. Stehle, June 26, 1959; children: Alexander S., Geoffrey S., Nicholas B. S.B. summa cum laude, Yale Coll., 1956; postgrad., Australian Nat. U., Canberra, 1956-57; M.D. cum laude, Harvard U., Boston, 1961, Oxford (Eng.) U., 1965-67. Diplomate: Am. Bd. Psychiatry and Neurology. Med. house officer Peter Bent Brigham Hosp., Boston, 1961-62; resident in neurology Mass. Gen. Hosp., Boston, 1962-65; NIH spl.fellow Oxford U., 1965-67; mem. faculty dept. neurology Harvard U. Med. Sch., 1968—, asst. prof. neurology, 1970-73, assoc. prof., 1973-83, prof., 1983—; mem. staff Mass. Gen. Hosp., 1968—, dir. clin. Neurophysiology Lab., 1968—, dir. Movement Disorder Clinic, 1975—; trustee Grass. Found.; praelector U. Dundee, Scotland; vis. scientist Swedish Med. Research Council, 1979; Sydney Watson Smith lectr. Royal Coll. Physicians Edinburgh, Scotland, 1981. Author: Spasticity: Disordered Motor Control, 1980, Clinical Neurophysiology, 1981; contbr. numerous articles to profl. jours.; editorial bd.: Annals of Neurology, 1983—, Muscle and Nerve, 1980—, Internat. Med. Revs.-Neurology, 1979—. Recipient Faculty Scholar award Josiah Macy Jr. Found., 1979; Fulbright scholar, 1956-57. Mem. Am. Neurol. Assn., Am. Acad. Neurology, Am. EEG Soc., Am. Assn. Electromyography and Electrodiagnosis, Internat. Microneurography Soc. (founding pres. 1981-83), Soc. Neurosci., Mass. Med. Soc., Phi Beta Kappa, Sigma Xi. Clubs: Longwood Cricket (Chestnut Hill, Mass.); Badminton and Tennis (Boston). Subspecialties: Neurology; Neurophysiology. Current work: Clinical neurophysiology as quantitative foundation for restorative neurology (rehabilitation); fundamental mechanisms responsible for tremor; research and development of quantitative neurological techniques; research into pathophysiology of tremor and other common movement disorders; basic science in rehabilitation. Home: 397 Newton St Chestnut Hill MA 02167 Office: Mass Gen Hosp Boston MA 02114

YOUNG, ROY ALTON, biological science educator; b. McAlister, N.Mex., Mar. 1, 1921; s. John Arthur and Etta Julia (Sprinkle) Y.; m. Marilyn Ruth Sandman, May 22, 1950; children: Janet Elizabeth, Randall Owen. B.S., N.Mex. A&M Coll., 1941; M.S., Iowa State U., 1942, Ph.D., 1948; LL.D. (hon.), N.Mex. State U., 1978. Teaching fellow Iowa State U., 1941-42, instr., 1946-47, Indsl. fellow, 1947-48; asst. prof. Oreg. State U., 1948-50, asso. prof., 1950-53, prof., 1953—, head dept. botany and plant pathology, 1958-66, dean research, 1966-70, acting pres., 1969-70, v.p. for research and grad. studies, 1970-76; chancellor U. Nebr., Lincoln, 1976-80; mng. dir. Boyce Thompson Inst. Plant Research, Cornell U., Ithaca, N.Y., 1980—; Dir. Pacific Power & Light Co., First Bank of Ithaca.; Mem. Commn. on Undergrad. Edn. in Biol. Scis., 1963-68; cons. State Expt. Stas. div. U.S. Dept. Agr.; chmn. subcom. plant pathogens, agr. bd. Nat. Acad. Scis.-NRC, 1965-68, mem. exec. com. study on problems of pest control, 1972-75; mem. exec. com. Nat. Govs.' Council on Sci. and Tech., 1970-74; mem. U.S. com. man and biosphere UNESCO, 1973—; mem. com. to rev. U.S. component Internat. Biol. Program, Nat. Acad. Scis., 1974-76; mem. adv. panel on post-doctoral fellowships in environ. sci. Rockefeller Found., 1974-78; bd. dirs. Boyce Thompson Inst. Plant Research, 1975—, Boyce Thompson Southwestern Arboretum, 1981—; mem. adv. com. Directorate for Engring. and Applied Sci., NSF, 1977-81, mem. sea grant adv. panel, 1978-80. Trustee Ithaca Coll., 1982—. Served to lt. USNR, 1943-46. Fellow AAAS (exec. com. Pacific div. 1963-67, pres. div. 1971), Am. Phytopath Soc. (pres. Pacific div. 1957); mem. Oreg. Acad. Sci., Nat. Assn. State Univs. and Land Grant Colls. (chmn. council for research policy and adminstrn. 1970, chmn. standing com. on environment and energy 1974-82), Sigma Xi, Phi Kappa Phi, Phi Sigma, Sigma Alpha Epsilon. Subspecialty: Plant pathology. Home: 17-B Strawberry Hill Rd Eastwood Commons Ithaca NY 14850 Office: Boyce Thompson Inst Tower Rd Cornell U Ithaca NY 14853

YOUNG, WEI, biophysicist; b. Loting, Hopei, China, Feb. 10, 1919; came to U.S., 1949, naturalized, 1961; s. Shu-Tong Yang and Hai-Lan Chang; m. Ho Lee, Dec. 28, 1949; 1 dau., Linda. B.S., Cath. U. Peking, 1943; Ph.D., U. Calif.-Berkeley, 1957. Instr. Med. Coll., Shanghai, China, 1945-49; biophysicist U. Calif.-Berkeley, 1957-63, sr. biophysicist, 1979—; biophysicist-8 U. Calif. Livermore Lab., 1963-70; sr. fellow Nat. Research Council (Moffett, Calif.), 1971-73; adj. research prof. San Jose State U., Calif., 1975-79. Author: Biological Effects of Magnetic Fields, 1969; Contbr. articles to profl. publs. Ames/NASA NRC Grantee, 1971-73, 1975-78, 1976—. Mem. Am. Physiol. Soc., AAAS, N.Y. Acad. Sci., Cryobiology Assn. (charter). Democrat. Subspecialties: Biophysics (biology); Comparative neurobiology. Current work: Interaction of microwave/laser on biomembrane energy barrier, zero magnetic field on Ga-As Microcuits; membrane bound enzyme kinetics; microwave reflectometry for arteriosclerosis. Home: 5978 Greenridge Rd Castro Valley CA 94546 Office: U Calif Berkeley CA 94720

YOUNG, WILLIAM ROBERT, air force officer, meteorologist; b. Foxboro, Mass., Dec. 27, 1947; s. Robert Benjamin and Margaret Louise (Moore) Y.; m. Deanna Louise McConnell Lusk, June 13, 1982. B.S. in Social Sci, Colo. State U., 1969; M. Meteorology, Tex. A&M U., 1977. Cert. meteorologist. Commd. 2d lt. U.S. Air Force, 1969, advanced through grades to maj., 1982; weather forecaster (Detachment 31 15 Weather SQ, Dobbins AFB), Marietta, Ga., 1971-73, asst. team chief,global team, Offutt AFB, Nebr., 1973-76, staff meteorologist space shuttle, El Segundo, Calif., 1978-81, comdr., Giebelstadt AAF, W. Ger., 1981—. Decorated Air Force Commendation Medal, 1973, 76; Meritorious Service Medal, 1981. Mem. Air Force Assn., AIAA. Republican. Methodist. Subspecialty: Synoptic meteorology. Current work: Forecasting of freezing precipitation types in attempt to develop a decision graph. Home: 12002 Dunklee Ln Garden Grove CA 92640 Office: Detachment 10 7th Weather SQ APO New York NY 09182

YOUNGER, MELANIE MOORE, veterinarian, educator; b. Cookeville, Tenn., Feb. 6, 1950; d. William F. and Betty (Fickel) Moore; m. George Winston Younger, Mar. 15, 1975; children: Katherine Moore, Elizabeth Sims. B.A., Ohio Wesleyan U., 1972; D.V.M., Auburn U., 1975. Lic. veterinarian, Tenn., Ala., Ohio, Ky., La. Instr. Auburn U., 1975-76; staff veterinarian Grady Vet. Hosp., Cin., 1976-77, Brentwood Vet. Hosp., Tenn., 1979-80; asst. prof. dept. agr. Northwestern State U., Natchitoches, La., 1980—; pvt. practice vet. medicine, 1981—. Vol. Natchitoches Animal Shelter. Named to Athletic Hall of Fame Ohio Wesleyan U., 1979. Mem. Central La. Vet. Medicine, La. Vet. Medicine Assn., Am. Assn. Animal Tech. Educators, AAUW, Bus. and Profl. Women. (local pres.). Methodist. Subspecialties: Preventive medicine (veterinary medicine); Internal medicine (veterinary medicine). Home: 1405 Barclay Narchitoches LA 71457 Office: Small Animal Clinic 126 Touline St Natchitoches LA 71457 Agr Dept Vet Tech Program Northwestern State U 126 Touline St Natchitoches LA 71457

YOUSSEF, MAHMOUD Z. HASSAN, research engineer; b. Cairo, Egypt, Oct. 9, 1945; came to U.S., 1974; s. Hassan M. and S. (Youssef) Y.; m. Mona A. Ismail, Jan. 23, 1974; children: Amr M., Susan. B.Sc., U. Alexandria, Egypt, 1967, M.Sc., 1973; M.Sc., U. Wis., 1976, Ph.D. in Nuclear Engring, 1980. Teaching asst. Atomic Energy Authority, Cairo, Egypt, 1971-74; research asst. Nuclear Study Ctr. of Casaccia, Rome, Italy, 1974; research asst. dept. nuclear engring. U. Wis., Madison, 1974-80; research engr. Sch. Engring. and Applied Sci., UCLA, 1980—; cons. TRW, Redondo Beach, Calif., 1981—. Contbr. articles to profl. jours. IAEA fellow, 1974. Mem. Am. Nuclear Soc. Subspecialties: Nuclear fusion; Nuclear fission. Current work: Fusion and fusion-fission engineering; cross-section processing; sensitivity analysis; perturbation theory. Office: UCLA Sch Engring & Applied Sci Los Angeles CA 90024

YOZAWITZ, ALLAN, neuropsychologist; b. Bklyn., Jan. 8, 1949; s. Louis and Sylvia Claire Y.; m. Arlene Susan Greenfield, Jan. 20, 1973; children: Elissa Gayle, Justin Mark. B.S., Poly. Inst. Bklyn., 1970; M.A., Queens Coll. CUNY, 1973; Ph.D., CUNY, 1977. Asst. research scientist biometrics N.Y. State Dept. Mental Hygiene, N.Y.C., 1970-77; trainee clin. neuropsychology Montefiore Hosp. and Med. Center, Bronx, N.Y., 1974-75; cons. gerontology sect. N.Y. State Psychiat. Inst., N.Y.C., 1975-76; dir. neuropsychology lab. Hutchings Psychiat. Center, N.Y. State Office Mental Health, Syracuse, 1977—; asst. prof. Med. Coll. SUNY Upstate Med. Center, Syracuse, 1979—; adj. asst. prof. psychology Syracuse U., 1979—; cons. Syracuse Devel. Center, 1979—, Benjamin Rush Center, Syracuse, 1980—; pvt. practice, Syracuse, 1979—. Cons. editor: Jour. Clin. Neuropsychology, 1983—; Contbr. articles to profl. jours. NIMH grantee, 1974-77, 79-82. Mem. AAAS, Am. Psychol. Assn. (charter mem. div. clin. neuropsychology 1979—), Internat. Neuropsychol. Soc. (task force on edn., accreditation and credentialing 1979), N.Y. State Psychol. Assn., N.Y. Acad. Scis., Soc. Neuroscience. Subspecialties: Neuropsychology; Psychiatry. Current work: Cognitive rehabilitation of psychiatric patients based on neuropsychological diagnosis, computer software design for cognitive rehabilitation, theories of neuropsychological basis of psychiatric disorder. Home: 150 Brookside Ln Fayetteville NY 13066 Office: Hutchings Psychiat Cente Neuropsychology Lab Syracuse NY 13210

YU, CHEN-CHENG WILLIAM, electrical engineer; b. Hanyang, Hupei, China, Aug. 31, 1946; came to U.S., 1969, naturalized, 1981; s. Kwang Yuen and Shaw Tze Y.; m. Mei-Ning Ho, July 7, 1979; 1 son, James Jacob. B.S., Nat. Taiwan U., 1968; Ph.M., Yale U., 1971, M.S., 1972, Ph.D., 1974. Teaching fellow physics dept. Yale U., New Haven, 1971-73; sr. assoc. engr. IBM, Poughkeepsie, N.Y., 1974-78, staff engr., 1978-81, adv. engr., 1981—. Contbr. articles to profl. jours. Served to 2d lt. Chinese Air Force, 1968-69. Mem. Soc. Indsl. and Applied Math. (session chmn. 1978, nat. meeting), Optical Soc. Am. Subspecialties: Microelectronics; Numerical analysis. Current work: Very large scale integration device reliability; numerical simulation of semiconductor devices; reliability statistics. Office: IBM Box 390 Dept All Bldg 052 Poughkeepsie NY 12602

YU, DAVID TAK YAN, physician, researcher; b. Hong Kong, Feb. 20, 1943; s. Peter Shiu On and Doris Yuet King (Kaan) Y.; m. Christina Chau Wan, Jan. 11, 1967; 1 child, Hong Sze. M.B., B.S., U. Hong Kong, 1966. Diplomate: Am. Bd. Internal Medicine. Intern U. Hong Kong, Queen Elizabeth Hosp., Hong Kong, 1966-67; pathologist Hong Kong Govt. Inst. Pathology, 1967-68; intern L.I. Coll. Hosp., Bklyn., 1968-69; resident in medicine Montefiore Hosp. and Med. Ctr., Bronx, N.Y., 1969-71; rheumatology fellow UCLA Sch. Medicine, 1971-74, asst. prof. medicine dept. medicine div. rheumatology, 1974-80; guest investigator, postdoctoral fellow Rockefeller U., N.Y.C., 1978-79, asst. prof. medicine, 1979-80; assoc. prof. medicine div. rheumatology UCLA Sch. Medicine, 1980—. Contbr. articles to profl. jours. Arthritis Found. fellow, 1975-78; sr. fellow, 1981—; NIH grantee, 1981—; Treadwell Found. grantee, 1982—. Mem. Am. Rheumatism Assn., Am. Fedn. Clin. Research, Am. Assn. Immunologists. Subspecialties: Internal medicine; Immunogenetics. Current work: Application of basic aspects of immunology especially immunogenetics to the study of arthritis. Office: 1000 Veteran Ave Los Angeles CA 90024

YU, FU LI, biochemist, educator, researcher; b. Peking, China, May 2, 1934; came to U.S, 1958, naturalized, 1972; s. Ling Ko and Dian Ying (Chang) Y.; m. Jie Feng, Apr. 20, 1980; children: Chan Ching, Chan Mei. B.S., Chung-Shing U., Taiwan, 1956; M.S., U. Ala., Birmingham, 1962; Ph.D., U. Calif., San Francisco, 1965. Instr. biochemistry U. N.Mex., 1965-66; research assoc. Inst. Cancer Research, Columbia U., 1966-73; asst. prof. biochemistry Jefferson Med. Coll., 1973-79, U. Ill., Rockford, 1979-80, assoc. prof., 1980—. Contbr. articles to profl. jours. Grantee NIH, Am. Cancer Soc. Mem. Am. Soc. Biol. Chemists, Am. Assn. Cancer Research, Am. Chem. Soc., N.Y. Acad. Scis., Harvey Soc., AAAS, Chinese Biochem. Soc., Sigma Xi. Subspecialties: Biochemistry (medicine); Cancer research (medicine). Current work: Chemical carcinogenesis, hormone action, nucleic acid metabolism, RNA polymerase and gene regulation in mammalian cells. Home: 1810 Hollyhock Dr Rockford IL 61107 Office: Dept Biomed Scis Univ Ill Coll Medicine 1601 Parkview Ave Rockford IL 61107-1897

YU, JIA-HUEY, pharmacologist, researcher; b. Taiwan, China, May 16, 1941; came to U.S., 1969, naturalized, 1976; d. Te-Fang and Chin-Len (Chang) Lin; m. Henry Hongjen Ye, June 22, 1968; children: Deborah, Tyson. D.D.S., Nat. Taiwan U., Taipei, 1966; M.S., U. Alta., Can.), Edmonton, 1968; Ph.D., U. Mich., 1973. Research assoc. Food and Drug Directorate, Ottawa, Ont., Can., 1968-69; lectr. U. Mich., 1973-74; research asst. prof. Boston U., 1974-77; research asst. prof. physiology U. Ala.-Birmingham, 1977—. Recipient Spl. Dental Research award Nat. Inst. Dental Research, 1980—; Nat. Inst. Dental Research postdoctoral fellow, 1971-73. Mem. Internat. Assn. Dental Reseearch, AAAS, Sigma xi. Subspecialties: Pharmacology; Physiology (medicine). Current work: Modulating roles of prostaglandins in regulating secretory activities of salivary glands during stimulation of autonomic innvervations. Home: 3471 Loch Ridge Trail Birmingham AL 35216 Office: U Ala 1600 University Blvd Birmingham AL 35294

YU, SIMON SHYI-JIAN, insect toxicologist, researcher, educator; b. Ilan, Taiwan, Sept. 11, 1935; came to Can., 1963; came to U.S., 1968; s. Song-Wei and Ah-So (Liaw) Y.; m. Rachel Ruey-Chih Yeh, Sept. 16, 1967; children:Robert, Edmund P. B.S., Nat. Taiwan U., 1959; M.S., McGill U., 1965, Ph.D., 1968. Research entomologist Taiwan Sugar Co., Kaohsiung, 1961-62; research asst. McGill U., Montreal, Que., Can., 1963-68; postdoctoral fellow Cornell U., Ithaca, N.Y., 1968-69; research assoc. Oreg. State U., Corvallis, 1969-74, asst. prof., 1974-79; assoc. prof. dept. entomology and nematology U. Fla., Gainesville, 1980—, prin. investigator, 1980—. Grantee NIH, 1979—, EPA, 1981, U.S. Dept. Agr., 1982. Mem. Entomol. Soc. Am., AAAS, Am. Chem. Soc., Fla. Entomol. Soc., Sigma Xi. Subspecialties: Toxicology (agriculture); Environmental toxicology. Current work: Insecticide toxicology, biochemical toxicology, detoxication mechanisms, insecticide resistance, pest management. Home: 3560 NW 30th Blvd Gainesville FL 32605 Office: Dept Entomology and Nematology Univ Fla Gainesville FL 32611

YU, YI-YUAN, academic dean, mechanical and aerospace engineer; b. Tientsin, China, Jan. 29, 1923; came to U.S., 1947, naturalized, 1962; s. Tsi-Chi and Hsiao-Kung (Wang) Y.; m. Eileen Wu, June 14, 1952; children: Yolanda, Lisa. B.S., Tientsin U., 1944; M.S., Northwestern U., 1950, Ph.D., 1951. Asst. prof. applied mechanics Washington U., St. Louis, 1951-54; assoc. prof. mech. engring. Syracuse U., 1954-57; prof. mech. engring. Poly. Inst. Bklyn., 1957-66; cons. engr. Space div. Gen. Electric Co., Valley Forge, Pa., 1966-71; Disting. prof. aero. engring. Wichita State U., 1972-75; mgr. components and analysis Rocketdyne div. Rockwell Internat., Canoga Park, Calif., 1975-79, exec. engr., 1979-81; dean engring. Newark Coll. Engring., N.J. Inst. Tech., 1981—; vis. prof. Cambridge (Eng.) U., 1960; lectr. Gen. Electric Modern Engring. Course, 1963-73; adv. Middle East Tech. U., Ankara, Turkey, 1966; cons. Chinese U. Devel. Project., 1983; cons. to govt. agys., industry; prin. investigator research grants Air Force Office Sci. Research, 1956-66, NASA Marshall Space Flight Center, 1967-69, NASA Langeley Research Center, 1974-75. Contbr.: chpt. to Handbook of Engineering Mechanics; articles to profl. publs.; reviewer tech., sci. jours. Guggenheim fellow, 1959-60. Fellow AIAA (assoc.); mem. ASME, Am. Soc. Engring Edn., N.Y. Acad. Scis., Sigma Xi, Phi Kappa Phi, Pi Tau Sigma, Tau Beta Pi. Subspecialties: Theoretical and applied mechanics; Aerospace engineering and technology. Current work: Dynamics, structural mechanics, wind turbine. Home: 22 Linden Aven West Orange NJ 07052 Office: 323 High Street Newark NJ 07102

YUAN, JIAN-MIN, physics educator; b. Chungking, China, Aug. 31, 1944; came to U.S., 1969, naturalized, 1982; s. Wen-Kai and Wen-Ming (Liao) Y.; m. Barbara O'Ching Yuan, Dec. 19, 1971; children: Jean, Weicon. B.S., Nat. Taiwan U., Taipei, 1966, M.S., 1968; Ph.D., U. Chgo., 1973. Postdoctoral fellow U. Fla., Gainesville, 1973-75; research assoc., instr. U. Rochester, 1975-78; asst. prof. dept. physics and atmospheric sci. Drexel U., Phila., 1978—. Contbr. articles to profl. jours. Recipient Univ. Research award, 1981. Mem. Am. Phys. Soc., Am. Chem. Soc., Sigma Xi. Subspecialties: Atomic and molecular physics; Laser-induced chemistry. Current work: Laser-induced rate processes, multiphoton dissociations, nonlinear dynamics, molecular reaction dynamics and photophysics. Office: Dept Physics and Atmospheric Sci Drexel U Philadelphia PA 19104

YUE, MIKE YUAN, nuclear engineer; b. Chungking, Szechuan, China, June 6, 1941; came to U.S., 1964, naturalized, 1976; s. Chien-nan and Fong-shan (Young) L.; m. Anna Mo-Chee Fung, June 5, 1976; children: Vincent John, Joey John. B.S. in Mech. Engring, Nat. Cheng Kung U., Taiwan, 1963, M.S., N.C. State U., 1968, 1972, postgrad., 1969-72. Engr. The Babcock & Wilcox Inc., Lynchburg, Va., 1966-69; sr. nuclear steam supply system engr. Combustion Engring. Inc., Windsor, Conn., 1972-76; advanced engr. Westinghouse Hanford Co., Richland, Wash., 1976-82; sr. core analysis engr., supr. Duquesne Light Co., Pitts., 1982—. Contbr. articles to profl. jours. Mem. Am. Nuclear Soc. Baptist. Subspecialties: Nuclear fission; Nuclear engineering. Current work: Senior core analysis engineering supervisor; core analysis and safety licensing group supervisor; supervise core physics personnel and safety analysis personnel to perform analysis and research to support Beaver Valley nuclear reactors safety analysis activities and fast breeder reactor core physics research. Home: 802 Carriage Rd Pittsburgh PA 15220 Office: Duquesne Light Company 435 6th Ave Pittsburgh PA 15219

YUND, MARY ALICE, research scientist, endocrinologist; b. Xenia, Ohio, Feb. 12, 1943; d. John Edward and Ethel Louise (Jemison) Stallard; m. E. William Yund, June 11, 1966. B.A., Knox Coll., 1965; M.A., Harvard U., 1967, Ph.D., 1970. Postdoctoral fellow dept. genetics U. Calif.-Berkeley, 1970-73, asst. research geneticist, 1975—; asst. prof. biology Wayne State U., Detroit, 1973-75. Danforth grad. fellow, 1965-70; Woodrow Wilson grad. fellow, 1965-66; NIH fellow, 1970-73; research grantee. Mem. Soc. for Devel. Biology, Genetics Soc. Am., Am. Soc. Zoologists, AAAS, Phi Beta Kappa, Sigma Xi. Subspecialties: Developmental biology; Gene actions. Current work: Steroid hormone action, hormone receptors, gene regulation, developmental biology, ecdysteroids, drosophila. Home: 723 Woodhaven Rd Berkeley CA 94708 Office: Dept Genetics U Calif Berkeley CA 94720

YUNG, SHU-CHIEN, nuclear energy corporation engineer, engineering educator; b. Ching-tu, Szechwan, China, Nov. 13, 1936; came to U.S., 1963, naturalized, 1972; s. Fu-Min and Ching-Dir (Ho) Y.; m. Shu-Shih Chu, Mar. 25, 1967; children: Jane, Delphine, Irene. B.S., Nat. Taiwan U., 1959; M.S., U. Mo.-Rolla, 1966; Ph.D., U. Ill.-Chgo., 1973. Registered profl. engr., N.J., N.Y. Mgr. quality control China Textile Ind. Corp., Nayli, Taiwan, 1961-63; assoc. engr. Allis-Chalmers Mfg. Co., Harvey, Ill., 1965-66; sr. engr. Curtiss-Wright Corp., Wood-Ridge, N.J., 1973-75, Westinghouse Hanford Co., Richland, Wash., 1975—; research asst. U. Ill.-Chgo., 1966-73; lectr. Joint Ctr. for Grad. Study, Richland, 1977—. Contbr. papers in field to profl. jours. Pres. Tri-Cities Chinese Christian Fellowship, Richland, 1976-79. Mem. Am. Nuclear Soc. Subspecialties: Nuclear fission; Nuclear engineering. Current work: Liquid metal fast breeder reactor safety analysis: including intrasubassembly incoherencies studies on liquid metal fast breeder reactor unprotected transient overpower accidents, sodium fire modelling, nuclear reactor core debris and substrates interaction modelling, containment code development. Home: 2133 Cascade Ct Richland WA 99352 Office: Westinghouse Hanford Co PO Box 1970b Richland WA 99352

YUNGBLUTH, THOMAS ALAN, biology educator; b. Warren, Ill., Dec. 12, 1934; s. Thomas Adam and Esther Mae (Trude) Y. B.S., U. Ill., 1956; Ph.D., U. Minn., 1966. Research asst. U. Minn.-Mpls., 1956-66; faculty biology Western Ky. U., Bowling Green, 1966—, prof., 1980—. Mem. Am. Soc. Agronomy, Crop Sci. Soc. Am., Ky. Acad. Sci. Republican. Methodist. Subspecialties: Plant genetics; Genetics and genetic engineering (agriculture). Office: Western Ky U Bowling Green KY 42101

YUNGER, LIBBY MARIE, pharmacologist; b. East Cleveland, Ohio, Feb. 20, 1944; d. Ladimer and Eleanore Wilma (Svasek) Y.; m. Richard D. Cramer III, May 22, 1979. B.A., Earlham Coll., 1966; postgrad., U. Chgo., 1967-68; M.A., U. Iowa, 1971, Ph.D., 1974. Nat. Inst. Neurol. and Communicable Diseases and Stroke postdoctoral fellow U. Pitts., 1974-75; research biochemist Lederle Labs., Pearl River, N.Y., 1975-78; assoc. sr. investigator Smith Kline & French Labs., Phila., 1978-83; mgr. bioanalytical scis. Internat. Minerals & Chem. Corp., 1983—. Contbr. articles to profl. jours. Mem. AAAS, Soc. Neuroscience, N.Y. Acad. Scis., Internat. Soc. Immunopharmacology, Sigma Xi, Phi Beta Kappa. Subspecialties: Immunopharmacology; Neurochemistry. Current work: Production of polyclonal and monoclonal antibodies to, and development of immunoassays and receptor binding assays for neurotransmitters and other chemical mediators of cellular functions; using antibodies as tools to determine physiological mechanism of action of these mediators. Office: Dept Life Sci Internat Minerals and Chem Corp Box 207 Terre IN 47808

YUSKA, HENRY B., chemistry educator; b. Bklyn., Nov. 7, 1914; s. John and Margaret (Renkevichius) Y.; m. Lillian Z., July 25, 1941; children: Kenneth, Reynold. B.S., CCNY, 1935; M.S., Bklyn. Poly. Inst., 1939; Ph.D., U. Ill.-Urbana, 1942. Chemist Jewish Hosp., Bklyn., 1935-39; research chemist Allied Chem. and Dye Co., Edgewater, N.J., 1942-43; sr. chemist, research dir. Interchem. Corp., N.Y.C., 1943-62; research dir. Sun Chem. Corp., N.Y.C., 1962-66; prof. chemistry Bklyn. Coll., 1966—. Mem. Am. Chem. Soc., Sigma Xi. Subspecialties: Polymer chemistry; Synthetic chemistry. Current work: Organic polymers, organic syntheses. Patentee in field. Home: 113-09 107th Ave Richmond Hill NY 11419 Office: Bklyn Coll Dept Chemistry Bedford Ave and Ave H Brooklyn NY 11210

YUSPA, STUART HOWARD, cancer research scientist, physician; b. Balt., July 19, 1941; s. Michael and Rose Y.; m. Eleanor M. Hecht, Aug. 1, 1965; children: Catharine, Margaret. B.S., Johns Hopkins U., 1962; M.D., U. Md., 1966. Diplomate: Am. Bd. Internal Medicine. Intern and resident in medicine Hosp. of U. Pa., Phila., 1966-67, 1970-72; commd. med. officer USPHS, HEW, 1972-75; sr. scientist Nat. Cancer Inst., Bethesda, Md., 1972-75, sect.chief, 1975-80, lab. chief, 1980—. Contbr. articles to profl. jours. Served with USPHS, 1967-70. Recipient Balder prize U. Md. Med. Sch., 1966; recipient Commendation medal USPHS, 1978. Mem. Am. Assn. for Cancer Research, AAAS, Am. Soc. Cell Biology. Subspecialties: Cancer research (medicine); Cell biology. Current work: Chemical carcinogenesis, cellular differentiation. Office: Nat Cancer Inst Bethesda MD 20205

ZABARA, JACOB, neurophysiologist, biophysicist; b. Phila., May 8, 1932; s. Joseph and Manya (Cohen) Z.; m. Ezliabeth Louise Omand; children: Joseph, Daniel. B.A. in Physics, Johns Hopkins U., 1953; M.S. in Physiology, U. Pa., 1958, Ph.D., 1959. Inst. Neurol. Scis. fellow U. Pa., 1954-59, USPHS fellow, 1959-60; instr. dept. pharmacology Dartmouth Coll., N.H., 1960-61; instr. Inst. Neurol. Sci., U. Pa., Phila., 1961-63, dept. pharmacology, 1961-63, spl. fellow biomath., 1963-64, instr. dept. physiology, 1964-67, asso. dept. pharmacology, 1965-67; asso. prof. dept. physiology/biophysics Temple U., Phila., 1967—; vis. prof. Hebrew U. Hadassah Med. Sch., Jerusalem, 1974. Contbr. articles to sci. jours. Mem. exec. bd. Soviet Jewry Council, 1975—; mem. Cardinal's Interfaith Commn. Phila., 1977-80. NIH grantee, 1974-79; Lily Found. fellow, 1980-82. Mem. Aerospace Med. Soc., Undersea Med. Soc., Am. Soc. Cybernetics, N.Y. Acad. Scis., Internat. Assn. Cybernetics, Am. Assn. Anatomists, Soc. Neurosci., Biopys. Soc., Am. Physiol. Soc., Internat. Center Cybernetics and Systems, World Orgn. Gen. Systems and Cybernetics. Subspecialties: Neurophysiology; Biophysics (physics). Current work: Autorhythmic structure of the brain; neurocybernetic and biophysical analysis of synaptic transformations; control of seizures; neurotropic viral infections; receptors. Office: 3223 N Broad St Philadelphia PA 19140

ZABORSKY, OSKAR RUDOLF, biological chemist, government administrator; b. Neuwaldorf, Czechoslovakia, Oct. 6, 1941; m., 1968; G62 children. B.Sc., Phila. Coll. Pharmacy and Sci., 1964; Ph.D. in Chemistry, U. Chgo., 1968. NIH fellow Harvard U., 1968-69; sr. research chemist Corp. Research Labs., Esso Research & Engring. Co., Linden, N.J., 1969-74; program mgr. enzyme tech. renewable resources program NSF, Washington, 1974—. Mem. AAAS, Am. Chem. Soc. Subspecialties: Enzyme technology; Biomass (energy science and technology). Office: Renewable Resources Program NSF Washington DC 20550

ZACHARIAH, GERALD LEROY, univ. inst. dean, agrl. engr., educator; b. McLouth, Kans., June 12, 1933; s. Kenneth Martin and Flornece Irene (Baker) Z.; m. Rita, Aug. 23, 1953; children: Stephen, Michael, Mark (dec.). B.S. in Agrl. Engring, Kans. State U., 1955, M.S., 1959, Ph.D., Purdue U., 1963. Registered profl. engr., Kans. Instr. in agrl. engring. Kans. State U., 1955-60; instr. Purdue U., 1960-63, asst. prof. agrl. engring., 1965-67, assoc. prof., 1967-71, prof., 1971-75; asst. prof. U. Calif., Davis, 1963-65; prof. U. Fla., 1975—, chmn. dept. agrl. engring., 1975-80, dean for resident instruction, 1980—; cons. to industry and govt. Author: (with B.A. McKenzie) Understanding and Using Electricity, 1975, 2d edit., 1982; contbr. chpts., numerous articles to profl. publs. Served to capt. Signal Corps U.S. Army, 1956. Recipient Teaching award Purdue U., 1972, also 4 paper awards, 11 grants. Mem. Soc. Agrl. Engrs., Am. Soc. Engring.

Edn., Inst. Food Technologists, Nat. Soc. Profl. Engrs., Fla. Engring. Soc. (Engr. of Yr. 1978). Methodist. Subspecialties: Agricultural engineering; Systems engineering. Current work: Food engineering, automatic control, processing agricultural products. Patentee grain drying. Home: 3121 NW 9th Pl Gainesville FL 32605 Office: U Fla 1001 McCarty Hall Gainesville FL 32611

ZACHERT, VIRGINIA, psychologist, educator; b. Jacksonville, Ala., Mar. 1, 1920; d. R.E. and Cora H. (Massee) Z. Student, Norman Jr. Coll., 1937; A.B., Ga. State Woman's Coll., 1940; M.A., Emory U., 1947; Ph.D., Purdue U., 1949. Diplomate: Am. Bd. Profl. Psychologists. Statistician [illegible], 1941-44, research psychologist Mil. Contracts, Auburn Research Found., Ala. Poly. Inst.; indsl. and research psychologist Sturm & O'Brien (cons. engrs.), 1958-59; research project dir. Western Design, Biloxi, Miss., 1960-61; self-employed cons. psychologist, Norman Park, Ga., 1961-71, Good Hope, Ga., 1971—; research assoc. med. edn. Med. Coll. Ga., Augusta, 1963-65, assoc. prof., 1965-70, research prof., 1970—, chief learning materials div., 1973—, mem. faculty senate, 1976—, mem. acad. council, 1976-82, pres. acad. council, 1983, sec., 1978; mem. Ga. Bd. Examiners of Psychologists, 1974-79, v.p., 1977, pres., 1978; Mem. adv. bd. Comdr. Gen. ATC USAF, 1967-70; cons. Ga. Legislature, 1980. Author: (with P.L. Wilds) Essentials of Gynecology-Oncology, 1967, Applications of Gynecology-Oncology, 1967. Del. White House Conf. on Aging, 1981. Served as aerologist USN, 1944-46; aviation psychologist USAF, 1949-54. Fellow Am. Psychol. Assn.; mem. Am. Statis. Assn., AAUP (chpt. pres. 1977-80), Sigma Xi. (chpt. pres. 1980-81). Baptist. Subspecialties: Learning; Obstetrics and gynecology. Current work: Use of multi-media techniques to instruct students in cognative aspects of subject matter; self-teaching or training; all ages, especially elderly. Home: 1126 Highland Ave Augusta GA 30904 Office: Dept Obstetrics and Gynecology Med Coll Ga Augusta GA 30912 It's really quite simple—I find that, if I wish to be understood or heard, that simplicity is necessary but not ever easy. Simplicity is basic, essential and always the major factor in my search for truth.

ZACKS, JAMES LEE, psychology educator, researcher; b. Iron Mountain, Mich., Mar. 23, 1941; s. Maurice and Naomi (Chudacoff) Z.; m. Rose Toby Greenbloom, June 19, 1966; children: Jeffrey Martin, Rebecca Leah. B.A., Harvard U., 1963; M.S., U. Calif.-Berkeley, Ph.D., 1967. Asst. prof. psychology U. Pa., Phila., 1967-71; assoc. prof. psychology Mich. State U., East Lansing, 1972-77, prof. psychology, 1977—. Fellow Optical Soc. Am., Am. Acad. Optometry; mem. Psychonomic Soc., Assn. Research in Vision and Ophthalmology, Sigma Xi. Jewish. Subspecialties: Sensory processes; Psychophysics. Current work: Visual system, comparative color vision, visual capacities of the white-tailed deer, visual-perceptual training and athletics; visual factors in reading X-rays. Home: 4446 Calgary Blvd Okemos MI 48864 Office: Mich State U Dept Psychology East Lansing MI 48824

ZAFFARONI, ALEJANDRO CESAR, pharmaceutical company executive; b. Montevideo, Uruguay, Feb. 27, 1923; s. Carlos and Luise (Alfaro) Z.; m. Lyda Russomanno, July 5, 1946; children: Alejandro C., Elisa. B., U. Montevideo, 1943; Ph.D., U. Rochester, 1949, D.Sc. (hon.). Assoc. dir. biol. research Syntex S.A., Mexico City, 1951-52, dir. biochemistry, 1952-54, dir. research and devel., 1954-56; exec. v.p. Syntex, Palo Alto, Calif., 1956-68; pres. Syntex Research div. Syntex Labs., Inc., 1962-68; pres. bd. govs. U. Tel-Aviv, 1972-73; pres., dir. research, chmn. exec. com. ALZA Corp., Palo Alto, Calif., 1968—; chmn. bd. DNAX; fin. com. Inst. Medicine; cons. prof. pharmacology Stanford U.; chmn. bd. Internat. Psoriasis Research Found.; bd. govs. Weizmann Inst. Sci., Rehovot, Israel; pharm. panel com. on tech. and internat. econ. and trade issues Nat. Acad. Engring.'s Office Fgn. Sec. and Assembly of Engring., Washington. Contbr. articles profl. jours. Trustee Linus Pauling Inst. Sci. and Medicine, Menlo Park, Calif.; founding life mem. MIT Sustaining Fellows; incorporator Neuroscis. Research Found. of MIT; bd. trustees U. Rochester. Recipient Barren medal Barren Found., 1974; decorated Caballero Order Vasco Nunez de Balboa, Panama, 1967; recipient Chem. Pioneer award, 1979. Fellow Am. Acad. Arts and Scis., Am. Pharm. Assn.; mem. Nat. Acad. Medicine Mex. (hon.), Am. Chem. Soc., Am. Soc. Biol. Chemists, Am. Soc. Pharmacology and Exptl. Therapeutics, Biomed. Engring. Soc., Internat. Soc. Chronobiology, Internat. Pharm. Fedn., N.Y. Acad. Scis. Subspecialties: Enzyme technology. Office: 950 Page Mill Rd Palo Alto CA 94304

ZAGER, PHILIP GEORGE, medical educator; b. N.Y.C., Oct. 1, 1941; s. Max Coleman and Ruth Elaine (Katz) Z. B.A., Northwestern U., 1963; M.D., Tulane U., 1967. Diplomate: Am. Bd. Internal Medicine (Endocrinology and Metabolism, Nephrology), Am. Bd. Nuclear Medicine. Intern Genesee Hosp., Rochester, N.Y., 1967-68; resident Univ. Hosp., Ann Arbor, Mich., 1968-69, St. Joseph-Mercy Hosp., Ann Arbor, 1969-70, Northwestern U. Med. Ctr., Chgo., 1970-71; assoc. div. diabetes and endocrinology Scripps Clinic and Research Found., La Jolla, Calif., 1975-76; research fellow dept. medicine Stanford (Calif.) U. Sch. Medicine, 1977-78; assoc. prof. U.N.Mex. Sch. Medicine, Albuquerque, 1979—; med. dir. hemodialysis unit U.N.Mex. Hosp., 1981—, co-dir. hypertension clinic, 1979—. Mem. Am. Fedn. Clin. Research, Endocrine Soc., Am. Soc. Nephrology, Internat. Soc. Nephrology, Am. Soc. Clin. Pharmacology and Therpeutics, ACP. Subspecialties: Nephrology; Neuroendocrinology. Current work: Clinical research on hormonal modulation of blood pressure; secretion, distribution, binding of adrenal steroids in normals and patients with hypertension and renal disease. Home: 10048 Menaul NE #K-28 Albuquerque NM 87112 Office: 2211 Lomas Blvd NE Albuquerque NM 87131

ZAGON, IAN STUART, anatomist, educator, researcher; b. N.Y.C., Mar. 28, 1943; s. Bert and Beatrice (Shaffer) Z.; m. Eileen Kostel, Nov. 26, 1964. B.S., U. Wis.-Madison, 1965; M.S., U. Ill.-Urbana, 1969; Ph.D. in Anatomy, U. Colo, 1972. Asst. prof. biol. structure U. Miami (Fla.) Med. Sch., 1972-74; asst. prof. anatomy M. S. Hershey Med. Center, Pa. State U., 1974-78, assoc. prof., 1978—; cons., researcher. Editorial bd.: Brain Research Bill; Contbr. over 100 articles to sci. jours. Biol. Stain Commn. fellow, 1968, 71; grantee Am. Cancer Soc., NIH, Nat. Inst. Drug Abuse. Mem. Am. Assn. Anatomists, Soc. Neurosci., Soc. Developmental Neurosci., AAAS. Subspecialties: Anatomy and embryology; Developmental neuroscience. Current work: Developmental neurobiology focusing on normal and abnormal brain devel. relationship of opiates and endorphins to brain devel. biological influences of opiates and endorphins in cancer. Office: MS Hershey Med Center Hershey PA 17033

ZAHLER, STANLEY ARNOLD, geneticist; b. N.Y.C., May 28, 1926; s. Irving and Clara (Heimowitz) Z.; m. Eleanor Janette Haugness, Nov. 1, 1952; children: Kathy Ann, Diane Louise, Peter Irving. A.B., N.Y.U., 1948; M.S., U. Chgo., 1950, Ph.D., 1952. Postdoctoral fellow U. Ill., Urbana, 1952-54; asst. prof. U. Wash., Seattle, 1954-59, W.Va. U., Morgantown, 1959, Cornell U., Ithaca, N.Y., 1959-64, assoc. prof., 1964-80, assoc. dir. div. biol. sci., 1976-79, prof. microbiology sect. genetics and devel., 1980—; cons. in field. Assoc. editor: Jour. Bacteriology, 1968-74, Applied & Environ. Microbiology, 1980—. Served with USNR, 1944-46. Mem. AAAS, Am. Soc. Microbiology, Genetics Soc. Am., Soc. Gen. Microbiology. Subspecialties: Genetics and genetic engineering (biology); Molecular biology. Current work: Genetics and genetic engineering of Bacillus and its bacteriophages. Office: Cornell Univ 317 Bradfield Hall Ithaca NY 14853

ZAIDER, MARCO A., radiol. biophysicist; b. Bacau, Romania, Jan. 3, 1946; came to U.S., 976; s. Abraham and Sarah (Granach) Z.; m. Edith F. Borkovi, Oct. 12, 1968; children: Arik, Talia-Irit, Ian-Hagai. M.Sc., Bucharest (Romania) U., 1968; Ph.D., Tel Aviv U., 1976. Research fellow Inst. Atomic Physics, Bucharest, 1968-70; teaching and research asst. Tel Aviv (Israel) U., 1971-76; vis. scientist Centre d'Etudes Nucleaire, Saclay, France, summers, 1972, 73, 74; postdoctoral fellow Los Alamos (N.Mex.) Nat. Lab., 1976-78, staff mem., 1979; research assoc. Radiol. Physics Lab., Columbia U., N.Y.C., 1980—. Contbr. articles to profl. publs. Mem. Radiation Research Soc., Am. Assn. Physicists in Medicine, ACLU. Subspecialty: Radiation biophysics. Current work: Effects of ionizing radiation on living systems. Dosimetry, microdosimetry, radiobiology, radiation biophysics. Office: 630 W 168th St Suite VC 11-241 New York NY 10032

ZAIS, ELLIOT JACOB, petroleum geothermal engr.; b. San Francisco, Nov. 27, 1943; s. Henry Moses and Deborah Isaevna (Maremant) Z.; m. Kathleen Adele Kerr, May 21, 1977. Student, Calif. Inst. Tech., 1962-64; B.S. in Petroleum Engring, U. Calif.-Berkeley, 1966; M.S., U. Tex.; Ph.D. in Petroleum Engring, U. Tex., 1972. Registered profl. engr., Tex., Calif., Oreg. Engr. Tex. Air Control Bd., Austin, 1972-74; research assoc. Getty Oil, Houston, 1974-76, reservoir engr., Bakersfield, Calif., 1976-77; pvt. cons., Corvallis, Oreg., 1977—; prin. Northwest Geophys. Assocs., Inc., Corvallis, 1981—; pres. E. Zais & Assocs., Inc., 1977—. Author: Analysis of Production Decline in Geothermal Reservoirs, 1980. Mem. Soc. Petroleum Engrs., Geothermal Resources Council. Lodge: Elks. Subspecialties: Petroleum engineering; Geothermal power. Current work: Well testing and reservoir engineering, fractured reservoirs testing, petroleum geothermal. Home: 7915 NW Siskin Corvallis OR 97330 Office: PO Box 1063 Corvallis OR 97339

ZAITLIN, MILTON, plant pathology educator; b. Mt. Vernon, N.Y., Apr. 2, 1927; s. Isadore and Martha (Padis) Z.; m. Marjorie Atkins, Sept. 5, 1951; children: David, Michael, Deborah, Paul. B.S. in Plant Pathology, U. Calif.-Berkeley, 1949; Ph.D. in Bot. Sci., UCLA, 1954. Technician, Calif. Inst. Tech., 1949-50; research officer sect. microbiology Commonwealth Sci. and Indsl. Research Orgn., Canberra, Australia, 1954-58; asst. prof. horticulture U. Mo., Columbia, 1958-60; asst. prof. agrl. biochemistry U. Ariz., Tucson, 1960-61, assoc. prof., 1962-66, prof., 1966-73; prof. plant pathology Cornell U., 1973—, assoc. dir. biotech. program, 1983—; vis. research scientist biochemistry U. Calif., Davis, 1979-80; mem. virology task force NIH, 1976-77; mem. sci. adv. bd. Plant Genetics Inc., Davis, 1981—; mem. com. basic research in plant biology McKnight Found., Mpls., 1982—. Assoc. editor: Virology, 1966-71, 82—; editor, 1972-81; contbr. articles to profl. jours. Served with USNR, 1945-46. Fulbright fellow, 1966-67; Guggenheim fellow, 1966-67. Fellow Am. Phytopath. Soc.; mem. Am. Soc. Virology (counselor 1982—), Am. Soc. Plant Physiologists, Soc. Gen Microbiology, Sigma Xi. Subspecialties: Virology (biology); Plant pathology. Current work: Replication of plant viruses and viroids. Office: Dept Plant Pathology Cornell U Ithaca NY 14853

ZAKIAN, VIRGINIA ARAXIE, Molecular biologist; b. Phila., Feb. 10, 1948; d. Aram and Charlotte Arpine (Boyajian) Z.; m. Robert Neil Sandberg, Sept. 8, 1973; 1 dau., Megan. A.B., Cornell U., 1970; Ph.D. (NSF fellow), Yale U., 1975. Postdoctoral fellow Princeton (N.J.) U., 1975-76, U. Wash., 1976-79; faculty Hutchinson Cancer Research Ctr., Seattle, 1979—; faculty Grad. Sch., U. Wash. Contbr. articles to profl. jours. NIH fellow, 1975-78. Mem. Am. Soc. Cell Biology (program com. 1980, membership com. 1981-83), Genetics Soc. Am., Am. Women in Sci. Subspecialties: Genome organization; Molecular biology. Current work: Structural and functional organization of eukaryotic genome, initiation of DNA replication. Office: Hutchinson Cancer Research Ctr 124 Columbia St Seattle WA 98104

ZAKIN, JACQUES LOUIS, chemical engineering educator; b. N.Y.C., Jan. 28, 1927; s. Mordecai and Ada Davies (Fishbein) Z.; m. Laura Pienkny, June 11, 1950; children: Richard Joseph, David Fredric, Barbara Ellen, Emily Anne, Susan Beth. B.Chem. Engring., Cornell U., 1949; M.S. in Chem. Engring, Columbia U., 1950; D.Engring. Sci. (Socony-Mobil Employee Incentive fellow), N.Y. U., 1959. Chem. engr. Flintkote Research Labs., Whippany, N.J., 1950-51; research technologist, research dept. Socony-Mobil, Bklyn., 1951-53, sr. research technologist, 1953-56, supervising technologist, 1959-62; asso. prof. chem. engring. U. Mo., Rolla, 1962-65, prof., 1965-77, dir. minority engring. program, 1974-77, dir. women in engring. program, 1975-77; chmn. dept. chem. engring. Ohio State U., Columbus, 1977—. Co-editor: Proc. Turbulence Symposium, 1969, 71, 73, 75, 77, 79, 81; contbr. articles to profl. jours. Bd. dirs. Rolla Community Concert Assn., 1966-77, 2d v.p., 1975-77; bd. dirs. Ozark Mental Health Assn., 1976-77; trustee Ohio State Hillel Found., 1981—, Congregation Beth Likoch, 1983; co-chmn. Concerned Academics and Scientists for Soviet Refuseniks. Served with USNR, 1945-46. Recipient Outstanding Research award U. Mo., Rolla, 1970; Am. Chem. Soc. Petroleum Research Fund Internat. fellow, 1968-69. Mem. Am. Inst. Chem. Engrs., Am. Chem. Soc., Soc. of Rheology, Am. Soc. Engring. Edn., Sigma Xi, Phi Lambda Upsilon, Phi Eta Sigma, Alpha Chi Sigma, Tau Beta Pi. Jewish. Subspecialty: Chemical engineering. Current work: Turbulent drag reduction; rheology of dilute polymer solutions; rtransport of viscous crude oil as oil-in-water emulsions. Patentee in field. Office: 140 W 19th Ave Ohio State U Columbus OH 43210

ZAKKAY, VICTOR H., aeronautical engineering educator; b. Baghdad, Iraq, Sept. 8, 1927; came to U.S., 1946, naturalized, 1955; s. Haron and Massouda Isac (David) Z. B.Ae.E., Poly. Inst. Bklyn., 1952, M.S. Ae.E., 1953, Ph.D., 1959. Research asst. prof. Poly. Inst. Bklyn., 1959-62, research assoc. prof., 1962-64; assoc. prof. aeronaut. engring. N.Y.U., N.Y.C., 1964-65, prof., 1965—; asst. dir. Antonio Ferri Labs., 1970-76, dir., 1976—, chmn. dept. applied sci., 1977—; cons. to various orgns. Contbr. articles to profl. jours. NSF grantee, India, 1982. Mem. AIAA (fluid dynamics tech. adv. panel 1972-75), Sigma Xi, Tau Beta Pi, Sigma Gamma Tau. Subspecialties: Fluid mechanics; Combustion processes. Current work: Research in fluid mechanics, turbulence, oil combustion, and fluidized bed coal combustion. Office: 425 Merrick Ave Westbury NY 11590

ZAKOWSKI, JACK J., clinical chemist; b. Salzburg, Austria, Aug. 8, 1950; s. Leon and Tosia Z.; m. Ruth, July 14, 1973; children: Aaron, Rebecca. B.A., UCLA, 1973; Ph.D., U. Calif.-Davis, 1978. Postdoctoral fellow microbiology U. Va., 1978-80, postdoctoral fellow in clin. chemistry dept. pathology, 1981—. Contbr. articles profl. jours. Nat. Cancer Inst. fellow, 1981-82. Mem. Am. Assn. Clin. Chemistry, N.Y. Acad. Scis., AAAS, Sigma Xi. Jewish. Lodge: B'nai B'rith. Subspecialties: Clinical chemistry; Biochemistry (medicine). Current work: Clinical chemistry and biochemistry, tumor markers. Home: 711 Lexington Ave Charlottesville VA 22901 Office: Dept Pathology U Va Hosp Charlottesville VA 22908

ZALESKI, MAREK BOHDAN, immunologist; b. Krzemieniec, Poland, Oct. 18, 1936; came to U.S., 1969, naturalized, 1977; s. Stanislaw and Jadwiga (Zienkowicz) Z. M.D., Sch. Medicine, Warsaw, 1960, Dr. Med. Sci., 1963. Instr. dept. histology Sch. Medicine, Warsaw, 1955-60, asst. prof., 1960-69; research asst. prof. (Henry C. and Bertha H. Buswell fellow) dept. microbiology SUNY, Buffalo, 1969-72, asso. prof., 1976-78, prof., 1978—; vis. scientist Inst. Exptl. Biology and Genetics, Czechoslovac Acad. Sci., Prague, 1965; Brit. Council's scholar, research lab. Queen Victoria Hosp., East Grinstead, Eng., 1966-67; asst. prof. dept. anatomy Mich. State U., East Lansing, 1972-75, assoc. prof., 1975-76. Contbg. author: Transplantation and Preservation of Tissues in Human Clinic, 1966, [illegible], Cytophysiology, 1970, Principles of Immunology, 1978, Medical Microbiology, 1982; Co-author: Immunogenetics, 1981, Molecular Immunology, 1982; co-editor: Immunobiology of Major Histocompatibility Complex, 1981; co-transl.: (J. Tischner) Ethics of Solidarity; editorial com.: Immunol. Communications; contbr. articles in field to med. jours. NIH grantee, 1976—. Mem. Polish Anatomical Soc., Transplantation Soc., Internat. Soc. Exptl. Hematology, Ernest Witebski Center Immunology, Am. Assn. Immunologists, Buffalo Collegium of Immunology, N.Y. Acad. Scis., Solidarity and Human Rights Assn. Roman Catholic. Subspecialties: Immunogenetics; Transplantation. Current work: Genetic regulation of the immune response to normal and neoplastic cell-surface alloantigens. Office: Dept Microbiology SUNY Buffalo NY 14214

ZAMENHOF, ROBERT GEORGE, med. physicist; b. Tanganyika, East Africa, Oct. 24, 1946; s. Julian and Olga Eugenia (Nietupska) Z.; m. Ruth Lilian Dlugi, Apr. 19, 1949. B.Sc., Poly. N. London, 1970; M.Sc., U. Strathclyde, Glasgow, Scotland, 1972; Ph.D., M.I.T., 1977. Research fellow in radiology Harvard Med. Sch., 1977-79; asst. prof. engring. Boston U., 1978-80; asst. prof. therapeutic and diagnostic radiology Tufts U. Sch. Medicine, Boston, 1979-82, assoc. prof., 1982—. Contbr. articles to profl. jours. Brit. Sci. Research Council scholar, 1971-72; Harvard-M.I.T. Health Scis. and Tech. fellow, 1972-74; M.I.T. Health Scis. Fund fellow, 1975-77; Nat. Cancer Inst. tng. fellow, 1978-79. Mem. Am. Assn. Physicists in Medicine (pres. New Eng. chpt. 1982-83), Biomedical Engring. Soc., Roentgen Ray Soc., Inst. Electronic and Radio Engrs., Nat. Commn. Radiation Protection. Subspecialty: Medical physics. Current work: Digital imaging, activation analysis, clinical physics. Home: 129 Clinton Rd Brooklin MA 02146 Office: 171 Harrison Ave Boston MA 02111

ZAMENHOF, STEPHEN, biochemist, educator; b. Warsaw, Poland, June 12, 1911; s. Henry Gregory and Sabina (Szpinak) Z.; m. Patrice J. Driskell, May 2, 1961. Ph.D. in Biochemistry, Columbia U., 1949. Asst. prof., assoc. prof. biochemistry UCLA, 1951-64, prof. microbial genetics and biol. chemistry, 1964—, prof. emeritus, 1978, recalled for active duty. Author: The Chemistry of Heredity, 1959; contbr. 210 articles to jours. Guggenheim fellow, 1958-59. Mem. Am. Soc. Biol. Chemists, Am. Soc. Neurochemistry, Am. Inst. Nutrition, Am. Soc. Anatomists, Soc. Neurosci., Internat. Soc. Devel. Psychobiology, Internat. Soc. Neurochemistry, Sigma Xi. Subspecialties: Biochemistry (biology); Neurochemistry. Current work: Neuroscience; factors affecting prenatal brain development. Home: 333 Medio Dr Los Angeles CA 90049 Office: UCLA Sch Medicine Los Angeles CA 90024

ZANDI, IRAJ, educator; b. Tehran, Iran, June 30, 1931; came to U.S., 1962; s. Housain and Ahtram (Batmaughelidj) Z.; m. Annette M. Grantham, June 20, 1958; children—Mark M., Richard H., Meriam R., Karl D., Peter P. B.S. in Electro-Mech. Engring, U. Tehran, 1952; M.S. in Civil Engring, U. Okla., 1957; Ph.D., Ga. Inst. Tech., 1959; M.A. (hon.), U. Pa., 1971. Dir. Dept. Environ. Engring., Ministry of Health, Tehran, 1961; asso. prof. Abadan Inst. Tech., 1962; asst. prof. U. Del., Newark, 1962-66; asso. prof. U. Pa., Phila., 1966-69, prof. dept. civil engring., 1969—; chmn. Nat. Center Energy Mgmt. and Power, 1971-72; partner Resoumetric U.S.A., Inc., Radnor, Pa., 1975—; sci. cons. U.S. Congressman Lawrence R. Coughlin, 1973. Contbr. articles to various publs. Recipient M.A. Ferst award Sigma Xi, Ga. Inst. Tech., 1960. Mem. ASCE, Am. Inst. Chem. Engrs. Subspecialty: Civil engineering. Home: 260 Highview Dr Radnor PA 19087 Office: Dept Civil and Urban Engring U Pa Philadelphia PA 19104

ZAR, JERROLD H(OWARD), biologist, statistician; b. Chgo., June 28, 1941; s. Max and Sarah (Brody) Z.; m. Carol B., Jan.15, 1967; children: David Michael, Adam Joseph. B.S., No. Ill. U., 1962; M.S., U. Ill., Urbana, 1964, Ph.D., 1967. NSF fellow in marine sci. Duke U. Marine Lab., Beaufort, N.C., 1965; research assoc. dept. zoology U. Ill., Urbana, 1967-68; asst. prof. biol. scis. No. Ill. U., 1968-71, assoc. prof., 1971-78, prof., chmn. dept. biol. scis., 1978—; vis. scientist Argonne (Ill.) Nat. Lab., 1974; cons. to govt. agys., industry. Author: Biostatistical Analysis, 1974, (with J. E. Brower) Field and Laboratory Methods for General Ecology, 1977; contbr. numerous chpts., articles to profl. publs. Founder, dir. ENCAP, Inc., 1974—. NIH fellow, 1965-67; NSF marine sci. fellow, 1965. Mem. AAAS, Am. Inst. Biol. Scis., Am. Ornithologists' Union, Am. Physiol. Soc., Am. Stat. Assn., Biometric Soc., Cooper Ornithol. Soc., Am. Soc. Zoologists, Ecol. Soc. Am., Internat. Assn. Ecology, Nat. Assn. Biology Tchrs., Nat. Assn. Environ. Profls., Wilson Ornithol. Soc. Subspecialties: Ecology; Comparative physiology. Current work: Physiological adaptations of animals to environmental stress; statistical analysis of biological data. Home: 918 Sunnymeade DeKalb IL 60115 Office: Dept Biol Scis No Ill U DeKalb IL 60115

ZARATZIAN, VIRGINIA LOUISE, pharmacologist; b. Highland Park, Mich., Nov. 15, 1918; d. Vahan Oskihan and Makrouhie (Kevorkian) Z. B.S., U. Mich., 1942, 1946, M.S., 1949; Ph.D., Wayne State U., 1956. Research assoc. U. Ill. Coll. Medicine, Chgo., 1956-59; pharmacologist, acting chief, pesticide sect. FDA, HEW, Washington, 1959-61; research pharmacologist, acting chief, pharmacology sect., div. air pollution USPHS, HEW, Cin., 1961-63; pharmacologist, chief pharmacology br. U.S. Environ. Hygiene Agy., Edgewood Arsenal, Md., 1963-66; pharmacologist pesticides regulation U.S. Dept. Agr., Washington, 1966-68, toxicologist, food safety inspection service, 1978—; pharmacologist, psychopharmacology research br. div. NIMH, Rockville, MD., 1968-78. Contbr. articles to profl. jours. Mem. Am. Chem. Soc., AAAS, Internat. Soc. Study Zenobiotics, Soc. Toxicology, Pan Am. Med. Soc., Iota Sigma Pi. Subspecialties: Toxicology (agriculture); Neurobiology. Current work: Researcher in toxicology. Home: PO Box 2217 Gaitherburg MD 20879 Office: US Dept Agriculture Cotton Annex 300 12th St SW Washington DC 20250

ZARCO, ROMEO M., immunologist; b. Caloocan, Rizal, Philippines, Oct. 7, 1920; s. Pablo V. and Marciana (Morales) Z.; m. Soledad Arcenas, Jan. 31, 1948; children: Cynthia, David, Sylvia. M.D., U. Philippines, 1943; M.P.H., Johns Hopkins U., 1954. Prof. microbiology U. Philippines, Manila, 1946-66; asst. dir. Howard Hughes Med. Inst., Miami, Fla., 1964-67; asst. prof. U. Miami, Fla., 1964—; with Cordis Labs., Inc., Miami, 1967—, pres., 1980—. WHO fellow, 1953; USPHS fellow, 1959. Mem. Am. Assn. Immunologists, Filipino-Am. Assn. Fla. (pres. 1982), Philippine-Am. Acad. Sci. and Engring. (chmn. 1982-83), Phi Sigma. Republican. Subspecialty: Immunobiology and immunology. Office: 2140 N Miami Ave Miami FL 33127

ZARE, RICHARD NEIL, chemistry educator; b. Cleve., Nov. 19, 1939; s. Milton and Dorothy (Amdur) Z.; m. Susan Leigh Shively,

Apr. 20, 1963; children—Bethany Jean, Bonnie Sue, Rachel Amdur. B.A., Harvard, 1961; postgrad., U. Calif. at Berkeley, 1961-63; Ph.D. (NSF predoctoral fellow), Harvard, 1964. Postdoctoral fellow Harvard, 1964; postdoctoral research asso. Joint Inst. for Lab. Astrophysics, 1964-65; asst. prof. chemistry Mass. Inst. Tech., 1965-66; asst. prof. dept. physics and astrophysics U. Colo., 1966-68, assoc. prof. physics and astrophysics, asso. prof. chemistry, 1968-69; prof. chemistry Columbia, 1969-77, Higgins prof. natural sci., 1975-77; prof. Stanford U., 1977—, Shell Disting. prof. chemistry, 1980—; Cons. Aeronomy Lab., NOAA, 1966-77, radio standards physics div. Nat. Bur. Standards, 1968-77, Lawrence Livermore Lab., U. Calif., 1974—, Stanford Research Inst., 1974—, Los Alamos Sci. Lab., U. Calif., 1975—; mem. IBM Sci. Advisory Com., 1977—. Recipient Fresenius award Phi Lambda Upsilon, 1974; Michael Polanyi medal, 1979; award Spectroscopy Soc. Pitts., 1983; Nonresident fellow Joint Inst. for Lab. Astrophysics, 1970—; Alfred P. Sloan fellow, 1967-69; Christensen fellow St. Catherine's Coll., Oxford U., 1982. Mem. Nat. Acad. Sci., Am. Acad. Arts and Scis., AAAS, Am. Phys. Soc. (Earle K. Plyler prize 1981), Am. Chem. Soc., Chem. Soc. London, Phi Beta Kappa. Subspecialties: Physical chemistry; Laser-induced chemistry. Current work: Reaction dynamics. Research and publs. on molecular luminescence, photochemistry and chem. physics. Office: Dept Chemistry Stanford U Stanford CA 94305

ZARIT, STEVEN HOWARD, psychology educator; b. Chgo., Nov. 1, 1945; s. Albert and Sara Rose (Maslov) Z.; m. Leora Aline Berns, Sept. 6, 1969 (div. 1977); children: Benjamin, Megan, Matthew; m. Judy Maes, Sept. 5, 1980; stepchildren: Michael Weston, Thomas Weston. B.A., U. Mich.-Ann Arbor, 1963-67; M.A., U. Pa., 1968; Ph.D., U. Chgo., 1972. Lic. clin. psychologist, Calif. Assoc. project dir. dept. psychiatry U. Chgo., 1971-73; teaching assoc. dept. psychology City Coll., CUNY, 1973-75; research assoc. Jewish Home and Hosp. for the Aged, N.Y.C., 1973-75; asst. prof. Andrus Gerontology Ctr., U. So. Calif., 1975-81, assoc. prof., 1981—; cons. Ctr. for Partially Sighted, Santa Monica, Calif., 1978—. Author: Aging and Mental Disorders, 1980; editor: Readings in Aging and Death, 1976, 82. Mem. adv. bd. Am. Found. for Blind, N.Y.C., 1980—. Mem. Am. Psychol. Assn., Western Psychol. Assn., Gerontol. Soc. Am. Subspecialties: Behavioral psychology; Developmental psychology. Current work: Senile dementia, depression, training in mental health. Office: Andrus Gerontology Ctr U So Calif Los Angeles CA 90089-0191

ZAROMB, SOLOMON, environmental scientist; b. Belchatow, Poland, Aug. 15, 1928; came to U.S., 1947, naturalized, 1955; s. Psakhya and Frayda (Dzialowska) Z.; m. Esther Almer, May 28, 1976; children: Franklin, Mendel, Earl Isaac. B.Ch.E., Cooper Union, 1950; Ph.D., Poly. Inst. N.Y., 1954. Research assoc. M.I.T., Cambridge, 1953-55; assoc. chemist IBM, Poughkeepsie, N.Y., 1956-58; project physicist, research specialist Philco Corp., Phila. and Blue Bell, Pa., 1958-61; scientist research and devel. Martin-Marietta Corp., Balt., 1962-63; pres. Zaromb Research Corp., Passaic and Newark, N.J., 1963-81; environ. scientist Argonne (Ill.) Nat. Lab., 1981—; cons. electrochemistry, product devel., patents. Contbr. articles to profl. jours. NIH grantee, 1969-72. Mem. Am. Astron. Soc., Am. Chem. Soc., Electrochem. Soc., Sigma Xi, Phi Lambda Upsilon. Subspecialties: Physical chemistry; Analytical chemistry. Current work: Developing instrumentation for toxic gas detection and monitoring; interests in electrochemical energy conversion devices and especially in aluminum-consuming power sources. Patentee in field. Office: 9700 S Cass Ave Argonne IL 60439

ZARTMAN, DAVID LESTER, geneticist; b. Albuquerque, July 6, 1940; s. Lester Grant and Mary Elizabeth (Kitchel) Z.; m. Micheal Aline Zartman, July 6, 1963; children: Kami Renee, Dalan Lee. B.Sc., N.Mex. State U., 1962; M.Sc., Ohio State U., 1966, Ph.D., 1968. Jr. partner Marlea Guernsey Farm, Albuquerque, 1962-64; research asst. Ohio State U., Columbus, 1964-68; asst. prof. N.Mex. State U., Las Cruces, 1968-71, assoc. prof., 1971-78, prof., 1978—; pres. Mary K. Zartman, Inc., Albuquerque, 1977—; lectr. in field; Fulbright Hayes lectr., Malaysia, 1976. Contbr. articles to profl. jours. Supt. dairy div. So. N.Mex. State Fair, 1978—; judge 4-H. NIH fellow, 1973; Fulbright grantee, 1976. Fellow AAAS; mem. Am. Inst. Biol. Sci., Am. Dairy Sci. Assn., Am. Soc. Animal Sci., N.Mex. Farm and Livestock Bur., Sigma Xi, Phi Kappa Phi. Republican. Subspecialties: Biomedical engineering; Genetics and genetic engineering (agriculture). Current work: Bio-engineering of farm animals, genetic improvement of livestock. Patentee in field. Office: PO Box 3-I New Mexico State Univ Las Cruces NM 88003

ZAVADA, MICHAEL STEPHAN, botanist, educator; b. Bridgeport, Conn., Aug. 25, 1952; s. Michael Joseph and Helen (Kokoruda) Z.; m. Maria Francisca Chavez, June 3, 1972; children: Yolanda, Rebecca, Sarah. B.S., Ariz. State U., 1974, M.S., 1976; B.A., U. Conn., 1982, Ph.D., 1982. Teaching asst. Ariz. State U., 1974-76; tchr. biology Fairfield (Conn.) Coll. Prep. Sch., 1977-78; teaching asst. U. Conn., Storrs, 1978-82; postdoctoral research fellow Ind. U., Bloomington, 1982—. Contbr. articles to profl. jours. Fulbright-Hayes grantee, Skopje, Yugoslavia, 1976-77; Bulgarian Govt. grantee, 1981; NSF grantee, 1981. Mem. Am. Assn. Stratigraphic Palynologists, Bot. Soc. Am., Internat. Orgn. Paleobotanists, Internat. Assn. Angiosperm Paleobotanists, Am. Assn. Tchrs. Slavic and East European Langs., Sigma Xi, Alpha Mu Gamma, Beta Beta Beta. Roman Catholic. Subspecialties: Evolutionary biology; Paleobiology. Current work: Angiosperm paleobotany and evolution, stratigraphic palynology (Mesozoic), pollen morphology and ultrastructure (fossil and extant). Office: Biology Dep Ind U Bloomington IN 47405

ZAVITSAS, ANDREAS ATHANASIOS, chemist, educator, cons.; b. Athens, Greece, July 14, 1937; came to U.S., 1954, naturalized, 1963; s. Athanasios A. and Catherine C. (Calliakoudas) Z.; m. Lourdes R.; 1 son, Athanasios. B.S. magna cum laude, CCNY, 1959; M.A., Columbia U., 1961, Ph.D., 1962. Research assoc. Brookhaven Nat. Labs., Upton, N.Y., 1962-64; research chemist Monsanto Co., Springfield, Mass., 1964-67; asst. prof. chemistry L.I. U., 1967-70, assoc. prof., 1970-73, prof., 1973—, grad. dean, 1975-79. Contbr. articles to profl. jours. Chmn. Sch. bd. Holy Cross Sch., Bklyn. Recipient Trustees award L.I. U., 1980. Fellow Am. Inst. Chemists; mem. Am. Chem. Soc., N.Y. Acad. Scis. Subspecialties: Organic chemistry; Kinetics. Current work: Organic free radicals, electrochemical sensors, physical-organic chemistry, semi-empirical calculations. Office: Dept Chemistry LI U Brooklyn NY 11201

ZAVOS, PANAYIOTIS MICHAEL, reproductive physiologist, educator, researcher, consultant; b. Tricomo, Famagusta, Cyprus, Feb. 23, 1944; came to U.S., 1966, naturalized, 1977; s. Michael Demetrios and Theodora (Vasiliou) Z.; m. Mary Susan Zavos, Feb. 22,1969; children: Michael John, Charles David, Christina Theodora. B.A., Emporia State U., 1970, M.S., 1972, Ed.S., 1976; Ph.D., U. Minn., 1978. Agrl. research asst. Agrl. Research Inst. of Cyprus, 1963-65; teaching asst. Emporia (Kans.) State U., 1970-73; instr. Johnson County Community Coll., 1973-74; research specialist U. Minn., St. Paul, 1974-78; vis. scholar Northwestern U., Chgo., 1978-79; asst. prof. animal sci. U. Ky., Lexington, 1979—; dir. labs. Fertility Inst. Chgo., 1978-79; chief reproductive physiologist Environ. Health Research and Testing Labs., Lexington, 1981—; lectr., condr. seminars and workshops in reproductive physiology. Contbr. articles to profl. jours. Served with N.G. Cyprus, 1965-66. Recipient Scholarship award Agronomique Institute de Montpellier, France, 1966, Outstanding Grad. Asst. award Emporia State U., 1971, Student Leadership and Service Recognition award U. Minn., 1974-75, U. Ky. Research Found. award, 1979-80; Govt. Cyprus Fulbright Found. Award scholar, 1966. Mem. Am. Fertility Soc., Am. Physiol. Soc., Internat. Soc. Cryobiology, Am. Soc. Animal Sci. Greek Orthodox. Clubs: Minn. Friends of Cyprus (treas. 1977-78); Tates Creek Country (Lexington); Optimists (v.p. 1981—). Subspecialties: Reproductive biology; Cryogenics. Current work: Investigation of male infertility; methods of assessing and practicing male fertility; separation of x- and y-bearing spermatozoa; embryo transfer; cryopreservation of male-female gametes.

ZAWADA, EDWARD THADDEUS, JR., physician; b. Chgo., Oct. 3, 1947; s. Edward Thaddeus and Evelyn Mary (Kovarek) Z.; m. Nancy Ann Stephen, Mar. 26, 1976; 1 dau., Elizabeth Ann. B.S., Loyola U., 1969, M.D., 1973. Intern UCLA Hosps. and Clinics, 1973-74, resident, 1974-78; asst. prof. medicine UCLA, 1978-79; chief nephrology SLC VA Hosp., Salt Lake City, 1980-81; asst. prof. medicine U. Utah, Salt Lake City, 1979-81; asst. chief medicine McGuire VA Hosp., Richmond, Va., 1981—; asso. prof. medicine Med. Coll. Va., Richmond, 1981—; cons. in field. Contbr. articles to profl. jours. Recipient VA Merit Rev., 1982—. Fellow ACP; mem. Alpha Omega Alpha. Roman Catholic. Subspecialties: Internal medicine; Nephrology. Current work: Role of calcium in blood pressure regulation; prostaglandins in renal physiology. Home: 1903 Grove Ave Richmond VA 23220 Office: McGuire VA Hosp 1201 Broad Rock Richmond VA 23249

ZAWADZKI, ZBIGNIEW APOLINARY, physician; b. Sosnowiec, Poland, July 23, 1921; s. Stanislaw and Sabina (Paliga) Z.; m. Danuta Irena Nowotczynska, Oct. 15, 1947; children: Barbara E., Joanna K. M.D., Sch. Medicine U. Warsaw, Poland, 1947, Dr.Sci., 1951; A.M. (hon.), Brown U., 1975. Asst. prof. Inst. Hematology, Warsaw, 1951-60; fellow in hematology New Eng. Ctr. Hosp., 1957-59; research assoc. VA Hosp., Pitts., 1961-67, chief hematology and oncology sect., 1961-67; asst. prof. U. Pitts., 1967-72, assoc. prof. medicine, 1972-74, Brown U., 1974—; dir. div. clin. immunology and oncology Meml. Hosp., Pawtucket, R.I., 1974—. Contbr. articles to profl. jours. Mem. AMA, Am. Assn. Immunologists, Am. Soc. Clin. Oncology, Am. Soc. Hematology, Am. Rheumatism Assn., R.I. Med. Soc., Sigma Xi. Roman Catholic. Club: Brown U. Faculty. Subspecialties: Oncology; Immunology (medicine). Current work: Study of paraproteinemias and the immunologic aspects of cancer disease. Home: 21 Wingate Rd Providence RI 02906 Office: Meml Hosp Pawtucket RI 02860

ZDYBEL, FRANK, JR., computer scientist; b. Sacramento, Jan. 16, 1950; s. Frank and Lillian Louise (Schuetz) Z. Student Calif. Inst. Tech., 1967-72; U. Calif.-Irvine, 1973-74. Cons. System Devel. Corp., Santa Monica, Calif., 1971; sr. coder dept. info. sci. U. Calif.-Irvine, 1972-74; systems analyst Perceptronics, Inc., Woodland Hills, Calif., 1974-76; computer specialist Naval Personnel Research & Devel. Ctr., Port Loma, Calif., 1976-79; scientist Bolt Beranek & Newman, Inc., Cambridge, Mass., 1979-82; mem. research staff Xerox PARC, Palo Alto, Calif., 1982—. Mem. Assn. Computing Machinery, Am. Assn. Artificial Intelligence. Subspecialties: Artificial intelligence; Programming languages. Current work: Knowledge based systems; programming language design; interactive graphics; simulation. Office: Software Concepts Group Xerox Research Center 3333 Coyote Hill Rd Palo Alto CA 94304

ZEBOVITZ, EUGENE, health sci. adminstr.; b. Chgo., Feb. 24, 1926; s. Harry and Anna (Chap) Z.; m. Marion Mogilewsky; children: Thomas C., Steven, Margaret, Shirley, Edward, Samuel H., Seth. B.S., Roosevelt U., 1950; M.S., U. Chgo., 1952, Ph.D., 1955. Med. microbiologist Ft. Detrick, Frederick, Md., 1955-58, med. virologist, 1962-70; research microbiologist, chief product control lab. Universal Foods Corp., Milw., 1958-62; sect. chief virology U.S. Naval Med. Research Inst., Bethesda, Md., 1970-74; exec. sec., exptl. virology study sect., div. research grants NIH, Bethesda, 1974—. Vice pres., treas. Beth Sholom Congregation, Frederick, 1970-73. Served with U.S. Army, 1944-46. Mem. Am. Soc. for Microbiology, AAAS, Sigma Xi. Subspecialties: Virology (medicine); Health sciences administration. Current work: Medical virology, genetics, immunology, infectious diseases, molecular biology. Office: NIH Exptl Virology Sect Div Research Grants Room 206 Westwood Bldg Bethesda MD 20205

ZEBROSKI, EDWIN LEOPOLD, nuclear safety engineer; b. Chgo., Apr. 1, 1921; s. P. Paul and Sophi Z.; m. Gisela Karin, Sept. 6, 1969; children: Lars, Zoe, Susan, Peggy. B.S., U. Chgo., 1941; Ph.D., U. Calif.-Berkeley, 1947. Profl. engr. Calif. Research assoc. Radiation lab. U. Calif.-Berkeley, 1947, Gen. Electric Research lab., 1947-54; project engr. Iriton submarine Standanford (Calif.) Research Inst., 1954-58; advance engr., nuclear engr., sect. mgr. Physics div. Gen. Electrics Nuclear Div., 1958-74; devel. engr. program mgmt. fuels Electric Power Research Inst., 1974-79; dir. Nuclear Safety Analysis Center, 1979-81; v.p. engring. Inst. Nuclear Power Ops., Palo Alto, Calif., 1981-83, v.p., chief scientist, 1983—; cons. in field. Contbr. numerous articles to profl. publs. Pres. bd. dirs. Palo Alto Unitarian Ch., 1967-70. Recipient Gen. Electric Charles A. Coffin award, 1954. Fellow Am. Nuclear Soc., Am. Inst. Chemists; mem. AAAS, Am. Phys. Soc., Nat. Acad. Engring., Phi Beta Kappa, Sigma Xi. Club: Los Altos Hill Country. Current work: Risk analysis, reactor systems safety analysis; power systems materials. Patentee in field. Office: 3412 Hillview Ave Palo Alto CA 94303

ZEE, DAVID SAMUEL, neurologist, neuroscientist, educator; b. Chgo., Aug. 14, 1944; s. Harry and Pearl (Taube) Z.; m.; children: Nathaniel, Landon. B.A., Northwestern U., 1965; M.D., Johns Hopkins U., 1969. Diplomate: Am. Bd. Neurology and Psychiatry. Intern N.Y. Hosp., N.Y.C., 1969-70; resident in neurology Johns Hopkins Hosp., Balt., 1970-73, asst. prof., 1975-78, assoc. prof. ophthalmology and neurology, 1978—; vis. scientist NIH, Bethesda, Md., 1982; cons. Nat. Eye Inst., Bethesda, 1978—. Co-author: Neurology of Eye Movements, 1983; co-editor: Functional Basis of Ocular Motility Disorders, 1982; mem. editorial bd.: Investigative Ophthalmology, 1977-80, Jour. Clin. Neuroophthalmology, 1982—, Annals of Neurology, 1983—. Served to lt. comdr. USPHS, 1973-75. Research grantee NIH/Nat. Eye Inst., 1976—; Career Devel. awardee, 1980; tchr.-investigator awardee NIH/Nat. Inst. Neurol. and Communicative Disorders and Stroke, 1975. Fellow Am. Acad. Neurology; mem. Assn. for Research in Vision and Ophthalmology, Barany Soc., Soc. for Neurosci., Am. Assn. Neurology. Subspecialties: Neurology; Neurophysiology. Current work: Disorders of ocular motility in patients using quantitative analysis and computer simulations; develop experimental animal models of human eye disorders. Home: 9718 Basket Ring Rd Columbia MD 21045 Office: Johns Hopkins Hosp 600 N Wolfe St Baltimore MD 21205

ZEHR, ELDON IRVIN, plant pathology, educator; b. Manson, Iowa, June 25, 1935; s. Clarence Dorance and Clara (Horsch) Z.; m. Rosa Rebecca, Aug. 31, 1957; children: Jeffrey, Darrell, Russell. B.A., Goshen Coll., 1960; M.S., Cornell U., 1965, Ph.D., 1969. Asst. prof. plant pathology Clemson (S.C.) U., 1969-73, assoc. prof., 1973-78, prof., 1978—. Contbr. articles to profl. jours. Mem. AAAS, Am. Phytopathol. Soc., Soc. Nemetologists. Subspecialty: Plant pathology. Current work: Teacher and researcher in control of fruit diseases and pesticide development, biological and chemical control of nematodes in peach orchards. Home: PO Box 245 Sandy Springs SC 29677 Office: Clemson University Dept Plant Pathology and Physiology Clemson SC 29631

ZEIGER, ALLEN RICHARD, biochemist, educator; b. N.Y.C., July 29, 1941; s. Harry and Ann Esther (Hochheiser) Z.; m. Naomi Ruth, Dec. 16, 1967; children: Mordechai, Yael, Shlomit, Amiram, Elisha, Malka. B.A., Brandeis U., 1962; Ph.D., Johns Hopkins U., 1967. NIH postdoctoral fellow, 1967-69; asst. prof. biochemistry Thomas Jefferson U., 1969-76, assoc. prof., 1976—. Contbr. articles, abstracts to profl. jours. Bd. dirs. Torah Acad. Elem. Sch., Phila. Mem. Am. Assn. Immunologists, Am. Soc. Microbiology. Jewish. Subspecialties: Biochemistry (biology); Immunobiology and immunology. Current work: Chemical and immunological properties of bacterial peptidoglycans; the role of peptidoglycans from pathogenic and indigenous flora in infectious diseases. Office: 1020 Locust St Philadelphia PA 19107

ZEISSIG, GUSTAVE ALEXANDER, systems engineer; b. Ithaca, N.Y., June 2, 1940; s. Alexander and Edith M. (Cuervo) Z.; m. Olga I. Rodriguez, May 3, 1969; children: Gustavo A., Walesca, Eric G., Karl A. B.E.E., Cornell U., 1964, Ph.D., 1971. Assoc. prof. physics dept. U. P.R., Rio Piedras, 1971-78; tech. staff M.I.T./Lincoln Lab., Lexington, Mass., 1978—; asst. leader Kiernan Reentry Measurements Site, Kwajalein, Marshall Islands, 1982-83. Contbr. articles to profl. jours. Mem. Kwajalein Parents Orgn.; pres. Kwajalein Swim Team. Mem. AAAS, Am. Astron. Soc., N.Y. Acad. Scis., Sigma Xi. Clubs: Kwajalein Scuba, Kwajalein Yacht. Subspecialties: Systems engineering; Radio and microwave astronomy. Current work: Radar systems, radar imaging, multi-static radar systems, pattern recognition (as applied to imaging techniques), simulation techniques, coherent radar processing techniques. Home: 216 Harwood Ave Littleton MA 01460 Office: MIT/Lincoln Lab (KREMS) PO Box 73 Lexington MA 02173

ZELAZO, PHILIP ROMAN, developmental psychologist, researcher, clinician; b. Ludlow, Mass., Oct. 3, 1940; s. Wanda and Szydlik (Zelazo); m. Nancy Ann Burl, Sept. 3, 1962; children: Philip David, Kirsten Marie, Suzanne Beth, Seana Ann. B.A., Am. Internat. Coll., 1962; M.S., N.C. State U., 1965; Ph.D., U. Waterloo, Ont., Can., 1967. Lic. psychologist, Mass. Asst. prof. Queen's U., Kingston, Ont., 1967-69; postdoctoral fellow Harvard U., 1969-70, research assoc., lectr., 1970-75; assoc. prof. psychology, co-dir. Tufts U. and New Eng. Med. Ctr. Hosp., Boston, 1974-82; assoc. prof. Tufts U., Medford, Mass., 1982—; dir. New Eng. Med. Ctr. Hosp. and St. Elizabeth's Hosp., Brighton, Mass., 1982—; co-dir., editor Ctr. for Behavioral and Pediatrics and Infant Devel., Boston, 1974-81; dir. Ctr. for Infant-Toddler Devel., Brighton, 1981—; panelist NSF/Social and Devel. Psychology, Washington, 1981—. Author: (with Kagan and Kearsley) Infancy: Its Place in Human Development, 1978, (with Kearsley and Ungerer) Learning to Speak: A Manual for Parents, 1981; editorial bd.: Infant Behavior and Devel, 1981-83, Jour. Applied Devel. Psychology, 1983—. Office Spl. Edn. grantee, 1976-80; Nat. Found. March of Dimes, 1978-82; Carnegie Corp. N.Y. grantee, 1982-83. Mem. AAAS, Am. Psychol. Assn., Soc. Research Child Devel. Democrat. Subspecialties: Developmental psychology; Cognition. Current work: Developing information processing procedures to assess intelligence in infants and toddlers and treatment strategies to reverse mild mental retardation in young children, including a program to teach speech-delayed children to talk. Inventor Zelaz-Keersley Test of Central Processing, 1976. Office: Center Infant-Toddler Devel Brighton Marine Public Health Center 77 Warren St Brighton MA 02135

ZELDIN, MARTEL, chemistry educator; b. N.Y.C., Aug. 11, 1937; s. Jacob and Sarah (Vogel) Z.; m. Cynthia Rutstein, June 15, 1958; children: Darryl C., Todd L., Tayna, James H. B.S., CUNY, Bklyn., 1962; Ph.D., Pa. State U., 1966. Chemist Interchem. Corp., N.Y.C., 1959-62; research scientist Union Carbide Corp., Tarrytown, N.Y., 1966-68; assoc. prof. Poly. Inst. N.Y., Bklyn., 1968-81; prof. chemistry, chmn. dept. Ind. U./Purdue U., Indpls., 1981—; research vis. scholar IBM, Yorktown Hts., summer 1977. Contbr. articles to profl. jours. Recipient Pa. State U. DuPont teaching award, 1965; Monsanto fellow, 1963; NIH fellow, 1964. Mem. Am. Chem. Soc., Sigma Xi, Phi Lambda Upsilon. Subspecialties: Inorganic chemistry; Polymer chemistry. Home: 8738 Lancaster Rd Indianapolis IN 46260 Office: Ind Univ Purdue Univ 1125 E 38th St PO Box 647 Indianapolis IN 46223

ZELDIS, JEROME BERNARD, physician, educator; b. Waterbury, Conn., Apr. 6, 1950; s. Miltonand Norma and (Gratz) Z.; m. Sharon W. Stamm, May 28, 1978. A.B., M.S., Brown U., 1972; M.Phil., Yale U., 1977, M.D., 1978, Ph.D., 1978. Intern in medicine UCLA Med. Ctr., 1978-79, jr. asst. resident, 1979-80, resident in medicine, 1980-81; research fellow in medicine Harvard Med. Sch., Boston, 1981—; clin. and research fellow in medicine Mass. Gen. Hosp., Boston, 1981—. Mem. editorial bd.: Jour. History of Medicine and Allied Scis, 1973-74, Yale Jour. Biology and Medicine, 1974-78; contbr. articles to profl. jours. Mem. ACP, Am. Assn. Immunologists, Phi Beta Kappa, Sigma Xi. Subspecialties: Gastroenterology; Immunogenetics. Current work: Gastroenterology, immunochemistry. Home: 115 Elinor Rd Newton Highlands MA 02161 Office: Mass Gen Hosp Fruit St Boston MA 02114

ZELLER, EDWARD JACOB, geology and physics educator, researcher; b. Peoria, Ill., Nov. 6, 1925; s. John George and Mabel Gertrude (Singer) Z.; m. Anke M. Neuman, June 18, 1966 (div. Mar. 1968). A.B., U. Ill.-Urbana, 1946; M.A., U. Kans.-Lawrence, 1948; Ph.D., U. Wis.-Madison, 1951. Project assoc. U. Wis., Madison, 1951-56; asst. prof. geology U. Kans.- Lawrence, 1956-59, assoc. prof., 1959-63, prof., 1963—, prof. physics, 1969—, dir. radiation physics lab., 1971—; disting. lectr. Am. Assn. Petroleum Geologist, Tulsa, Okla., 1971. NSF sr. postdoctoral fellow, Bern, Switzerland, 1961-62; recipient Antarctica Service medal Nat. Acad. Sci., 1966. Fellow Geol. Soc. Am.; mem. AAAS, Am. Geophys. Union, Antarctican Soc., Sigma Xi. Subspecialties: Geophysics; Geochemistry. Current work: Primary activity is polar geophysics and ice chemistry, radioactive resource surveys in Antarctica and studies of the relationship between solar activity and variations in polar ice chemistry. Patentee in field. Home: 2908 W 19th St Lawrence KS 66044 Office: Space Tech Center U Kans 2291 Irving Hill Rd Lawrence KS 66045

ZEN, E-AN, research geologist; b. Peking, China, May 31, 1928; came to U.S., 1946, naturalized, 1963; s. Hung-chun and Heng-chi'h (Chen) Z. A.B., Cornell U., 1951; M.A., Harvard U., 1952, Ph.D., 1955. Research fellow Woods Hole Oceanographic Inst., 1955-56, research asso., 1956-58; asst. prof. U. N.C., 1958-59; geologist U.S. Geol. Survey, 1959-80, research geologist, 1981—; vis. asso. prof. Calif. Inst. Tech., 1962; Crosby vis. prof. M.I.T., 1973; Harry H. Hess sr. vis. fellow Princeton U., 1981. Contbr. articles to profl. jours. Fellow Geol. Soc. Am., AAAS, Am. Acad. Arts and Scis., Mineral. Soc. Am. (council 1975-77, pres. 1975-76); mem. Geol. Soc. Washington (pres. 1973), Nat. Acad. Scis., Mineral. Assn. Can. Subspecialty: Petrology. Office: Mail Stop 959 US Geol Survey Nat Center Reston VA 22092

ZENSER, TERRY VERNON, research biochemist, geriatric and gerontology research coordinator; b. Port Clinton, Ohio, Aug. 1, 1945; s. Vernon S. and Hazel Z.; m. Barbara Jean Morrison, Aug. 10, 1968; children: Nathan, Jason. B.S. in Biol. Scis, Ohio State U., 1967; Ph.D. in Biochemistry, U. Mo., 1971. Lectr. Hood Coll., 1974-75; renal research VA Med. Center, Pitts., 1975-76; adj. asst. prof. biochemistry U. Pitts., 1975-76; asst. prof. biochemistry and medicine St. Louis U., 1976-80, assoc. prof., 1980—; core coordinator Geriatric Research, Edn. and Clin. Center, VA Med. Center, St. Louis, 1976—. Contbr. numerous articles to profl. jours., chpts. to books. Served to capt., Med. Service Corps U.S. Army, 1971-75. VA grantee, 1976—; Am. Cancer Soc. grantee, 1981—; Nat. Cancer Inst. grantee, 1980—, EPA grantee, 1980-81. Mem. Am. Chem. Soc., Am. Fedn. Clin. Research, Gerontol. Soc., Am. Soc. Biol. Chemists, Am. Soc. Pharmacology and Exptl. Therapeutics, Sigma Xi. Subspecialties: Biochemistry (medicine); Cancer research (medicine). Current work: Biology of aging and drug metabolism; investigating mechanism of initiation of toxic and carcinogenic effects of certain chemicals and renal metabolism. Home: 1200 Dunloe Rd Manchester MO 63011 Office: VA Med Center GRECC 111G-JB Saint Louis MO 63125

ZENTALL, THOMAS ROBERT, psychology educator; b. Beziers, France, Sept. 29, 1940; came to U.S., 1942, naturalized, 1948; s. Robert and Elizabeth (Aigner) Z.; m. Sydney Snider, Aug. 29, 1965; children: Gabriel, Shannon. B.S., Union Coll., 1963, B.E.E., 1963; Ph.D., U. Calif.-Berkeley, 1969. Assoc. mem. tech. staff Bell Labs., Murray Hill, N.J., 1965-66; asst. prof. U. Pitts., 1969-75; from asst. prof. to prof. psychology U. Ky., Lexington, 1976—, acting chmn. dept. psychology, 1982. Contbr. articles to profl. jours. NIMH grantee, 1972, 74, 80. Mem. Am. Psychol. Assn., Psychonomic Soc., AAAS. Subspecialties: Behavioral psychology; Learning. Current work: Animal intelligence - cognitive behavior in animals, concept learning, memory, imitation. Home: 252 Market St Lexington KY 40508 Office: Dept Psychology Univ Ky Lexington KY 40506

ZENTMYER, GEORGE AUBREY, JR., plant pathology educator; b. North Platte, Nebr., Aug. 9, 1913; s. George Aubrey and Mary Elizabeth (Strahorn) Z.; m. Dorothy Anne Dudley, May 24, 1941; children: Elizabeth Zentmyer Dossa, Jane Zentmyer Fernald, Susan Dudley. A.B., UCLA, 1935; M.S., U. Calif., 1936, Ph.D., 1938. Asst. forest pathologist U.S. Dept. Agr., San Francisco, 1937-40; asst. pathologist Conn. Agrl. Expt. Sta., New Haven, 1940-44; asst. plant pathologist to plant pathologist U. Calif. at Riverside, 1944-62, prof. plant pathology, 1962—, chmn. dept., 1968-73; Cons. NSF, Trust Ty. of Pacific Islands, 1964, 66, Commonwealth of Australia Forest and Timber Bur., 1968, AID, Ghana and Nigeria, 1969, Govt. of South Africa, 1980. Author: Plant Disease Development and Control, 1968, Recent Advances in Pest Control, 1957, Plant Pathology, an Advanced Treatise, 1977, The Soil-Root Interface, 1979, Phytophthora: Its Biology, Taxonomy, Ecology and Pathology, 1983; assoc. editor: Ann. Rev. of Phytopathology, 1971—; Asso. editor: Jour. Phytopathology, 1951-54; contbr. articles to profl. jours. Bd. dirs. Riverside YMCA, 1949-58; pres. Town and Gown Orgn., Riverside, 1962; mem. NRC, 1968-73. Guggenheim fellow, Australia, 1964-65; NATO sr. sci. fellow, Eng., 1971; recipient award of honor Calif. Avocado Soc., 1954; NSF research grantee, 1963, 68, 71, 74, 78. Fellow Am. Phytopath. Soc. (pres. 1966, pres. Pacific div. 1955), AAAS (pres. Pacific div. 1974-75); mem. Nat. Acad. Scis., Mycol. Soc. Am., Am. Inst. Biol. Scis., Bot. Soc. Am., Brit. Mycol. Soc., Australian Plant Pathology Soc., Philippine Phytopath. Soc., Indian Phytopath. Soc., Assn. Tropical Biology, Internat. Soc. Plant Pathology (councilor 1973-78), Explorers Club, Sigma Xi. Subspecialty: Plant pathology. Current work: The genus phytophthora-biology; physiology, control, ecology; diseases of subtropical and tropical plants; especially avocado and cacao; soil-borne pathogens; fungicides. Home: 3892 Chapman Pl Riverside CA 92506

ZEPPETELLA, ANTHONY JOHN, actuary; b. Bklyn., Dec. 26, 1949; s. Michael Lucio and Filemena (Peluso) Z.; m. Barbara Bahna, July 9, 1972; children: Peter, Thomas. B.S., Bklyn. Coll., CUNY, 1971; M.S., Courant Inst., N.Y.C., 1973, Ph.D., 1976. Asst. prof. math. NYU, N.Y.C., 1975-76, Clarkson Coll., Potsdam, N.Y., 1976-81; asst. actuarial dir. Mut. Life Ins. Co. N.Y., N.Y.C., 1980-82; actuary McKay-Barlow Co., Butler, N.J., 1982-83; assoc. actuary Mut. Life Ins. Co. N.Y., N.Y.C., 1983—. Research asst. Courant Inst., 1973-76; research grantee NIH, NSF, 1979-81. Fellow Soc. Actuaries (assoc., instr. risk theory 1982); Mem. Soc. for Indsl. and Applied Math., Pi Mu Epsilon. Club: N.Y. Actuaries (instr. risk theory 1982—). Current work: Mathematical risk thoery. Home: 540 75th St Brooklyn NY 11209 Office: McKay-Barlow Co 10 Park Pl Butler NJ 07405

ZERBE, JOHN IRWIN, wood technologist, researcher; b. Hegins, Pa., June 4, 1926; s. Allen and Rosa Jane (Miller) Z.; m. Ruby June Deitrich, Sept. 1, 1951; children: Lynne Diane Zerbe Durst, Eric Alan, Donna Lee Zerbe Blaser. B.S., Pa. State U., 1951; M.S., N.Y. State Coll. Environ. Sci. and Forestry, 1953, Ph.D., 1956. Research asst. prof. U. Ill., Urbana, 1956-58; mgr. govt. specifications Nat. Forest Products Assn., Washington, 1958-59, asst. to dir. tech. services, 1959-65, asst. v.p., 1965-70; dir. forest products engring. research U.S. Forest Service, Dept. Agr., Washington, 1970-76; mgr. energy research devel. and application Forest Products Lab., Madison, Wis., 1976—; adj. prof. U. Wis. Served with USNR, 1944-46. Timber Engring. Co. scholar, 1951. Mem. ASTM, Forest Products Research Soc., Soc. Wood Sci. and Tech., Soc. Am. Foresters, Xi Sigma Pi. Lutheran. Subspecialties: Biomass (energy science and technology); Fuels and sources. Current work: Research on ethanol and related fuels from wood, wood combustion, gasification of wood, energy from municipal waste. Home: 3310 Heatherdell Ln Madison WI 53713 Office: Gifford Pinchot Dr PO Box 5130 Madison WI 53705

ZERNER, MICHAEL CHARLES, chemistry educator; b. Boston, Jan. 31, 1940; s. Maurice Bernard and Blanche (Deutsch) Z.; m. Anna Gunilla Fojerstam, May 15, 1966; children: Erik Mark, Emma Danielle. B.Sc., Carnegie-Mellon U., 1961; M.Sc., Harvard U., 1962, Ph.D., 1966. Asst. prof. U. Guelph, Ont., Can., 1970-74, assoc. prof., 1974-79, prof., 1979-82, adj. prof., 1982—; prof. U. Fla., Gainesville, 1981—; vis. prof. Pontificia Universidade Catolica do Rio de Janeiro, 1980, U. Uppsala, Sweden, 1978-79; cons. Eastman Kodak Co., Rochester, N.Y., 1978—. Assoc. editor: Internat. Jour. Quantum Chemistry, 1981—, Applied Quantum Theory, 1981—; contbr. articles to profl. jours. Trail coordinator Guelph Trail Assn., 1972-82. Served to capt. U.S. Army, 1966-68. NIH grantee, 1968-70; grantee Natural Scis. and Engring. Research Council Can., 1970—. Mem. Am. Inst. Physics, Am. Chem. Soc., Quantum Chemistry Program Exchange (bd. adv.), Internat. Soc. Quantum Biology (mem. exec.), AAAS. Subspecialties: Theoretical chemistry; Physical chemistry. Current work: Theoretical chemistry, especially molecular electronic structure, bonding, reaction and spectroscopy. Home: 3505 NW 31st St Gainesville FL 32605 Office: Dept Chemistry U Florida Gainesville FL 32611

ZEROKA, DANIEL, chemistry educator; b. Plymouth, Pa., June 22, 1941; s. Michael and Mary (Klimchak) Z.; m. Alexandra S. Kotulak, May 27, 1967; children: Daniel M., Andrea M. B.S. in Chemistry, Wilkes Coll., 1963, Ph.D., U. Pa., 1966. Asst. prof. chemistry Lehigh U., Bethlehem, Pa., 1967-74, assoc. prof., 1974—. Author research publs. in field. NSF postdoctoral fellow, 1966-67. Mem. Am. Chem. Soc., Am. Phys. Soc., Combustion Inst., Sigma Xi. Subspecialties: Physical chemistry; Theoretical chemistry. Current work: Quantum chemistry; statistical thermodynamics; effect of magnetic and electric fields on matter. Office: Dept Chemistry Lehigh U Bethlehem PA 18015

ZETTLER, FRANCIS WILLIAM, plant pathology educator; b. Easton, Pa., Aug. 13, 1938; s. Francis Joseph and Frances Ellen (Ivey) Z.; m. Carol Rose Schurz, Jan. 1, 1961; children: Lawrence William, Jennifer Ann. B.S., Pa. State U., 1961; M.S., Cornell U., 1964, Ph.D., 1966. Asst. prof. plant pathology U. Fla., Gainesville, 1966-72, assoc. prof., 1972-76, prof., 1976—. Contbr. numerous articles to profl. publs. U.S. Army C.E. grantee, 1970-73; Fla. Dept. Natural Resources grantee, 1970-73; Office of Water Resources research grantee, 1970-73; U.S. Dept. Agr. grantee, 1974-78, 82-85; Binat. Agrl. Research and Devel. Fund, grantee, 1982-85. Mem. Am. Phytopath. Soc. Subspecialties: Plant virology; Plant pathology. Current work: Virus diseases and control of viruses of ornamental plants and certain food staples of the tropics. Home: 31 Grassy Lake Rd Archer FL 32618 Office: Plant Pathology Dept U Fla HSPP Bldg Gainesville FL 32611

ZEVELOFF, SAMUEL IRA, zoologist, ecologist, educator; b. N.Y.C., May 2, 1950; s. Harold Michael and Muriel (Rubins) Z.; m. Linda B., Aug. 19, 1973; 1 dau., Abigail, Debra. B.A., SUNY-Binghamton, 1972; M.S.Ed., CCNY, 1973, N.C. State U., 1976; Ph.D., U. Wyo., 1982. Vis. instr. N.C. State U., Raleigh, 1976-78, vis. asst. prof., 1982-83; supply instr. U. Wyo., Laramie, 1982. Contbr. articles to profl. jours. Bd. dirs. Wake County Audubon Soc., 1977-78; pres. N.C. State U. Wildlife Soc., 1982—. Mem. Am. Soc. Mammalogists, Am. Soc. Naturalists, Sigma Xi. Subspecialties: Ecology; Population biology. Current work: Reproductive patterns in mammals; life history of evolution, community ecology. Office: Dept Zoology NC State Univ Raleigh NC 27650

ZHU, NAI JUE, chemist; b. Zun Yi, Guizhou, China, Jan. 24, 1942; s. Shi Kai and Boruo (Pan) Z.; m. Li Liang, Dec. 21, 1967; children: Nan Zhu, Hua Zhu. B.S., Peking U., Beijing, 1965. Research asst. Inst. Chemistry, Beijing, China, 1965-78, research assoc., group leader, 1979—; cons. Beijing Grad. Sch., Wuhan Geol. Coll., 1980; vis. scholar SUNY-Buffalo, 1980-82. Recipient Inst. Chemistry, Academia Sinica invention award, 1979. Mem. Chinese Chem. Soc., Am. Crystallographic Soc. Subspecialties: Crystallography; X-ray crystallography. Current work: Electron density distributions and chemical bond, using x-ray diffractometer at low temperatures, relationship between molecular structure and properties for novel materials and organometallic compounds, metallic clusters structures. Office: Inst Chemistry Academia Sinica Zhong Guan Dun Beijing People's Republic of China

ZIEGLER, MICHAEL GREGORY, nephrologist, educator, researcher; b. Chgo., May 12, 1946; s. William J. and Agnes (Finlayson) Z.; m. Carole Lowrey, Dec. 12, 1971; children: Barbara, Matthew. B.S., Loyola U., Chgo., 1967; M.D., U. Chgo., 1971. Intern Kans. Med. sch., 1971-72, resident, 1972-73; attending physician NIH Clin. Ctr., Bethesda, Md., 1974-76; cons. internal medicine NIH Sect. Exptl. Therapeutics, Bethesda, 1974-76; pharmacologist NIMH, Bethesda, 1975-76; asst. prof., clin. pharmacologist U. Tex. Med. Br., 1976-80; asst. prof. medicine, div. nephrology U. Calif.-San Diego, 1980—. Editor: Norepinephrine, 1984, Psychiatric Disease and the Calecholamines, 1984; editorial bd.: Hypertension, 1981—. Served to lt. comdr. USPHS, 1973-75. Mem. Am. Heart Assn., Am. Fedn. Clin. Research, Am. Soc. Nephrology. Subspecialties: Neuropharmacology; Internal medicine. Current work: Elucidate the role of the nervous system in control of blood pressure. Office: U Calif Med Ctr H-781-B 225 Dickinson St San Diego CA 92103 Home: 342 W Lewis St San Diego CA 92103

ZIELINSKI, PAUL BERNARD, civil engineering educator; b. West Allis, Wis., Sept. 9, 1932; s. Stanley Charles and Lottie Caroline (Pliszkiewicz) Z.; m. Monica Theresa Beres, July 13, 1957; children: Daniel Paul, Gregory John, Robert Mathias, Sarah Ann. B.S.C.E., Marquette U., 1956; M.S., U. Wis-Madison, 1961, Ph.D., 1965. Registered profl. engr., Wis., S.C. Asst. instr. Marquette U., Milw., asst. prof., 1964-67; instr. U. Wis.-Madison, 1959-64; from asst. prof. to prof. civil engring. dept. Clemson (S.C.) U., 1967—; dir. Water Resources Research Inst., Clemson, 1978—; cons. Am. Pub. Works Assn., Chgo., 1974-78. Mem. S.C. Water Resources Commn., Columbia, 1978—; chmn. Clemson City Planning Commn., 1971-74. Mem. ASCE, Am. Soc. Engring. Edn., Sigma Xi. Roman Catholic. Lodge: Sertoma (pres. 1979-80). Subspecialties: Water supply and wastewater treatment; Fluid mechanics. Current work: consultant on several hydrauliclaboratory and field demonstration projects of swirl chamber as a storm water separator, a primary sedimentation chamber and a soil erosion control device. Home: 506 W Shorecrest Dr Clemson SC 29631 Office: Civil Engring Dept Clemson U Clemson SC 29631

ZIMM, BRUNO HASBROUCK, physical chemistry educator; b. Woodstock, N.Y., Oct. 31, 1920; s. Bruno L. and Louise S. (Hasbrouck) Z.; m. Georgianna S. Grevatt, June 17, 1944; children: Louis H., Carl B. Grad., Kent (Conn.) Sch., 1938; A.B., Columbia, 1941, M.S., 1943, Ph.D., 1944. Research asso. Columbia, 1944; research asso., instr. Polytech. Inst. Bklyn., 1944-46; instr. chemistry U. Calif. at Berkeley, 1946-47, asst. prof., 1947-50, asso. prof., 1950-51; vis. lectr. Harvard, 1950-51; research asso. research lab. Gen. Electric Co., 1951-60; prof. chemistry U. Calif., La Jolla, 1960—. Asso. editor: Jour. Chem. Physics, 1947-49; adv. bd.: Jour. Polymer Sci, 1953-62, Jour. Bio-Rheology, 1962-73, Jour. Biopolymers, 1963—, Jour. Phys. Chemistry, 1963-68, Jour. Biophys. Chemistry, 1973—. Recipient Bingham Medal Soc. Rheology, 1960—, also the High Polymer Physics prize from Am. Phys. Soc., 1963. Mem. Biophys. Soc., Am. Soc. Biol. Chemists, Am. Chem. Soc. (Baekeland award 1957), Nat. Acad. Scis. (award in Chem. Scis. 1981), Am. Acad. Arts and Scis., Am. Phys. Soc. Subspecialties: Biophysical chemistry; Theoretical chemistry. Current work: Theory of macromolecular solutions; properties and structure of biopolymers. Home: 2605 Ellentown Rd La Jolla CA 92037

ZIMMACK, HAROLD LINCOLN, zoology educator, insect pathologist; b. Chgo., Feb. 12, 1925; s. Harold Nicholas and Wanda T. (Dobbs) Z.; m. Barbara Jean Keen, Sept. 2, 1951; children: Cinda Lou, John Mark, Lissa Carol. B.S., Eastern Ill. U., 1951; M.S., Iowa State U., 1953, Ph.D., 1956. Insect pathology collaborator U.S. Dept. Agr., Ankeny, Iowa, 1952-56; asst. prof. zoology Eastern Ky. U., Richmond, 1956-63; prof. zoology Ball State U., Muncie, Ind., 1963—; collaborator Office for Agr., Renewable Resources, Washington, 1980—. Mem. governing bd. Hazelwood Christian Ch., Muncie, 1983. Fellow Ind. Acad. Sci.; mem. AAAS, Am. Inst. Biol. Scis., Entomol. Soc. Am. (chmn. entomology sect. 1973), Soc. for Invertebrate Zoology. Republican. Subspecialties: Animal physiology; Physiology (biology). Current work: Rapid screening technique to identify potential insect microbial pathogens. Home: RR 9 Box 495 Muncie In 47302 Office: Dept Biology Ball State U Muncie IN 47306

ZIMMANCK, FRANK ROBERT, JR., nuclear reactor operator; b. St. Petersburg, Fla., Feb. 5, 1950; s. Frank Robert and Helen E. (Warns) Z.; m. Naoma Lee Kinsman, May 17, 1974; children: Leesa, Patricia, Frank. Cert. sr. reactoroperator. Aux. operator Fla. Power Corp., Crystal River, 1975-77, reactor operator, 1977-80, sr. reactor operator, 1980-81, nuclear tech. support analyst, 1981—. Mem. PowerPAC, St. Petersburg, 1982—; bd. dirs. Fla. Power Employees Fed. Credit Union, 1982—. Mem. Am. Nuclear Soc. Democrat. Mem. Ch. of God/Ind.. Subspecialty: Nuclear fission. Current work: Supporting a PWR nuclear power generating facility with technical and operating data. Home: Rt 5 Box 800B Homosassa FL 32646 Office: PO Box 1240 Crystal River FL 32629

ZIMMER, JAMES FRANCIS, veterinarian, educator; b. Rochester, N.Y., Nov. 16, 1943; s. Herbert Joseph and Gladys M. (Casey) Z.; m.; children: Matthew, Michael, Mark, Peter. D.V.M., N.Y. State Coll. Vet. Medicine, 1968; Ph.D., Cornell U., 1979. Asst. prof. vet. medicine Cornell U., 1979—. Mem. AVMA, Am. Animal Hosp. Assn. Subspecialties: Internal medicine (veterinary medicine); Animal physiology. Current work: Clinical gastroenterology, gastrointestinal physiology. Office: NY State Coll Vet Medicine Ithaca NY 14853

ZIMMER, JAMES GRIFFITH, preventive medicine educator; b. Lynbrook, N.Y., Apr. 10, 1932; s. James Henry and Orpha Marie (Spicer) Z.; m. Mary Jane Scriggins, June 27, 1953 (div. Apr. 1970); children: Heidi, Eric, Roger; m. Anne Wilder, May 15, 1971. B.A., Cornell U., 1953; M.D., Yale U., 1957; D.T.P.H., London Sch. Hygiene and Tropical Medicine, 1966. Diplomate: Am. Bd. Internal Medicine. Intern Grace-New Haven Hosp., 1957-58; resident in internal medicine Strong Meml. Hosp., Rochester, N.Y., 1958-60; assoc. prof., dir. grad. program dept. preventive, family and rehab. medicine U. Rochester, 1963—; med. dir. Regional Utilization and Med. Rev. Project, Genesee Valley Med. Found., 1971—. Contbr. articles to profl. jours., chpts. to books. Chmn. profl. adv. com. Vis. Nurse Service, 1979—, Monroe County Long-Term Care, Inc., 1982—; chmn. regional utilization com. Genesee Valley Med. Found., 1971—. Served to capt. U.S. Army, 1960-63. Milbank Faculty fellow, 1964-71; recipient Okeke and Simpson prizes London Sch. Hygiene and Tropical Medicine, 1966. Fellow Am. Coll. Preventive Medicine; assoc. Am. Coll. Epidemiology; mem. Internat. Epidemiol. Assn., Am. Pub. Health Assn., Royal Soc. Tropical Medicine and Hygiene. Clubs: Photog. Hist. Soc., Genesee Valley Scuba (sec.). Subspecialties: Health services research; Epidemiology. Current work: Long-term care, care of elderly, quality of care, health care program evaluation, grad. education in preventive medicine and community health. Home: 145 Penn Ln Rochester NY 14625 Office: Dept of Preventive Family and Rehabilitation Medicine University of Rochester Rochester NY 14642

ZIMMERMAN, ARTHUR MAURICE, zoology educator, researcher; b. N.Y.C., May 24, 1929; m., 0953; 3 children. B.A., N.Y. U., 1950, M.S., 1954, Ph.D in Cell Physiology, 1956. Technician N.Y. U., 1951-52; research assoc. N.Y. U. and Marine Biol. Lab., Woods Hole, Mass., 1955-56; Lalor research fellow Marine Biol. Lab., 1956; Nat. Cancer Inst. research fellow U. Calif., 1956-58; from instr. to asst. prof. pharmacology Coll. Medicine SUNY Downstate Med. Center, 1958-64; prof. zoology U. Toronto, Ont., Can., 1964—, sec. dept. zoology, 1970-75, assoc. chmn. grad. affairs, 1975-78, assoc. dean Div. IV Sch. Grad. Studies, 1978-81; acting dir. Inst. Immunology, 1980-81; mem. corp. Marine Biol. Lab.; vis. prof. anatomy U. Tex. Health Sci. Center, San Antonio.; vis. scientist Weizmann Inst. Sci, Israel; vis. prof. U. Fla. Assoc. editor: Can. Jour. Biochemistryand Cell Biology., 1980—; editor: Cell Biology Series; editorial bd. Expt. Cell Research. Subspecialty: Cell biology. Office: Dept Zoolog U Toronto Toronto ON Canada M5S 1A1

ZIMMERMAN, DANIEL HILL, immunologist, biochemist; b. Los Angeles, June 3, 1941; s. Clinton Ballard and Alma Jane (Hill) Z.; m. Scott Wood Delius, June 8, 1963; children: David Anders, Malissa Wood, Emily Shannon. B.S., Emory and Henry Coll., 1963; M.S., U. Fla., 1966, Ph.D. in Biochemistry, 1969. Staff fellow NIAMDD, NIH, Bethesda, Md., 1969-71, fellow, 1971-73, scientist, 1973-77, sr. scientist, 1978-81, program mgr. hybridoma, 1981-82; tech. dir. Cell Sci. Lab., Electronuceonics Inc., Silver Springs, Md., 1982—. Contbr. articles to profl. jours. Pres. Ayrlawn Elem. Sch. PTA, 1977-78; Treas. North Bethesda Congress Civic Assn., 1981—; pres. Ayrlawn Civic Assn., 1978-79. Mem. Am. Assn. Immunologists, AAAS, Am. Assn. Clin. Chemistry. Republican. Episcopalian. Subspecialties: Immunobiology and immunology; Virology (biology). Current work: Monoclonal antibodies, hybridoma, cellular immunology, virology. Home: 5527 Oakmont Ave Bethesda MD 20817 Office: 12050 Tech Rd Silver Spring MD 20950

ZIMMERMAN, EARL ABRAM, neuroscientist, neurologist; b. Harrisburg, Pa., May 5, 1937; s. Earl B. and Hazel M. (Myers) Z. B.S. in Chemistry, Franklin and Marshall Coll., 1959; M.D., U. Pa., 1963. Diplomate: Am. Bd. Internal Medicine, Am. Bd. Psychiatry and Neurology. Intern in medicine Presbyterian Hosp., N.Y.C., 1963-64, resident in medicine, 1964-65, fellow in endocrinology, 1970-72; resident in neurology Neurol. Inst., N.Y.C., 1965-68; asst. prof. neurology Columbia U., 1972-77, assoc. prof., 1977-80, prof., 1980—. Contbr. numerous articles, chpts., abstract to profl. jours. Served to maj. USAF, 1968-70. Recipient 1907 Meml. prize for neurology research U. Pa., 1963, Lucy B. Moses prize Columbia U., 1974, Tchr.-Investigator award NIH Nat. Inst. Neurol. and Communicative Diseases and Strokes, 1972-77. Mem. Am. Acad. Neurology, Am. Neurol. Assn., Endocrine Soc., Assn. Research in Nervous and Mental Diseases, Soc. Neurosci., AAAS, Histochem. Soc. Democrat. Subspecialties: Neurobiology; Endocrinology. Current work: Neuroendocrinology, neuropeptide research; localization of brain neuropeptide networks by immunohistochemistry. Office: 630 W 168th St 328 Black Bldg New York NY 10023

ZIMMERMAN, HOWARD ELLIOT, chemist, educator; b. N.Y.C., July 5, 1926; s. Charles and May (Cohen) Z.; m. Jane Kirschenheiter, June 3, 1950 (dec. Jan. 1975); children: Robert, Steven, James; m. Martha L. Bailey Kaufman, Nov. 7, 1975; stepchildren: Peter and Tanya Kaufman. B.S., Yale U., 1950, Ph.D., 1953. NRC fellow Harvard U., 1953-54; faculty Northwestern U., 1954-60, asst. prof., 1955-60; assoc. prof. U. Wis., Madison, 1960-61, prof. chemistry, 1961—, Arthur C. Cope prof. chemistry, 1975—; Chmn. 4th Internat. Union Pure and Applied Chemistry Symposium on Photochemistry, 1972. Author: Quantum Mechanics for Organic Chemists, 1975; Mem editorial bd.: Jour. Organic Chemistry, 1967-71, Molecular Photochemistry, 1969—, Jour. Am. Chem. Soc., 1982—; Contbr. articles to profl. jours. Recipient Halpern award for photochemistry N.Y. Acad. Scis., 1979. Mem. Am. Chem. Soc. (James Flack Norris award 1976), Chem. Soc. London, German Chem. Soc., Inter-Am. Photochemistry Assn. (co-chmn. organic div. 1977-79, exec. com. 1979—), Nat. Acad. Scis., Phi Beta Kappa, Sigma Xi. Subspecialties: Organic chemistry; Photochemistry. Current work: Organic photochemical research to explore reactivity of electronically excited organic molecules; search for new organic photochemical reactions and theory; study of unusual reactions and mechanisms; synthesis of unusual organic molecules. Home: 1 Oconto Ct Madison WI 53705

ZIMMERMAN, HYMAN JOSEPH, physician, educator; b. Rochester, N.Y., July 14, 1914; s. Philip and Rachel (Marine) Z.; m. Kathrin J. Jones, Feb. 28, 1943; children: Philip M., David J., Robert L., Diane E. A.B., U. Rochester, 1936; M.A., Stanford, 1938, M.D., 1942. Diplomate: Am. Bd. Internal Medicine. Intern Stanford U. Hosp., 1942-43; resident George Washington U. div. Gallinger

Municipal Hosp., 1946-48, clin. instr. medicine, 1948-51; practice medicine, specializing in internal medicine, Washington, 1948-49; asst. chief med. service VA Hosp., Washington, 1949-51; dir. Liver and Metabolic Research Lab., 1965-68, chief med. service, 1971-78, sr. clinician, 1978—; asst. prof. medicine Coll. Medicine U. Nebr.; also chief med. service VA Hosp., Omaha, 1951-53; chief med. service West Side VA Hosp., Chgo.; also clin. asso. prof. medicine Coll. Medicine U. Ill., 1953-57; prof., chmn. dept. medicine Chgo. Med. Sch.; also chmn. dept. medicine Mt. Sinai Hosp., Chgo., 1957-65; prof. medicine George Washington Sch. of Medicine, Washington, 1965-68, 71—; chief med. service Boston VA Hosp., 1968-71; prof. medicine Boston U. Sch. Medicine, 1968-71; lectr. medicine Tufts U. Sch. Medicine, 1968-71; chief med. service VA Hosp., Washington, 1971-78, sr. clinician, 1978—; prof. medicine George Washington Sch. Medicine, 1971—; clin. prof. medicine Georgetown U., 1971—, Howard U., 1971—, Uniformed Services U. Health Scis., 1978—. Contbr. numerous articles to med. jours. Served to maj. AUS, 1943-46. Fellow ACP; mem. AMA (council on drugs, gastroenterology panel 1968-70), AAAS, Am. Fedn. for Clin. Research, Am. Diabetes Assn., Endocrine Soc., Assn. for Study Liver Diseases, Am. Soc. Clin. Investigation, N.Y. Acad. Scis., Central Soc. Clin. Research, Soc. for Exptl. Biology and Medicine, Am. Soc. for Pharmacology and Therapeutics, Am. Gastroenterol. Assn., Sigma Xi, Alpha Omega Alpha. Subspecialty: Gastroenterology. Home: 7913 Charleston Ct Bethesda MD 20817 Office: 2150 Pennsylvania Ave NW Washington DC 20037

ZIMMERMAN, JOHN FREDERICK, chemist; b. Monticello, Iowa, June 22, 1937; s. Herman F. and Anna (Lubben) Z.; m. (married), June 7, 1959; children: Kathryn, Christopher, John, Michael. B.S. in Chemistry, Iowa State U., 1959; Ph.D., U. Kans., 1964. Chemist E.I. duPont Co., Clinton, Iowa, summer 1958, Am. Cyanamid Co., Stamford, Conn., summer 1959; mem. faculty Wabash Coll., Crawfordsville, Ind., 1963—, prof. chemistry, 1979—. Named Tchr. of Yr. Wabash Coll., 1976; NSF fellow, 1960-63. Mem. Am. Chem. Soc., Midwestern Assn. Chemistry Tchrs. in Liberal Arts Colls., Sigma Xi. Lutheran. Subspecialties: Analytical chemistry; Instrumentation. Current work: Information/instrument/computer interfaces. Computer on-line instrumentation. Office: Wabash Coll Crawfordsville IN 47933

ZIMMERMAN, MICHAEL RAYMOND, pathologist, anthropologist, educator; b. Newark, Dec. 26, 1937; s. Edward Louis and Bessie (Herman) Z.; m. Barbara Elaine Hoffman, Dec. 25, 1960; children: Jill, Wendy. B.A., Washington and Jefferson Coll., 1959; M.D., NYU, 1963; Ph.D. in Anthropology, U. Pa., 1976. Diplomate: Nat. Bd. Med. Examiners, Am. Bd. Pathology. Intern Kings County Hosp. Ctr., Bklyn., 1963-64; resident, fellow in pathology NYU-Bellevue Med. Ctr., N.Y.C., 1964-68; asst. pathology Lankenau Hosp., Phila., 1970-72; asst. prof. pathology U. Pa., Phila., 1972-77, assoc. prof. anthropology, 1980—; pathologist Wayne County Gen. Hosp., Westland, Mich., 1977-80; assoc. prof. pathology and anthropology U. Mich., Ann Arbor, 1977-80; assoc. prof. pathology Hahnemann U., Phila., 1980-82, clin. assoc. prof. pathology, 1982—; chief anatomic pathology Jeanes Hosp., Phila., 1982—. Author: Foundations of Medical Anthropology, 1980, (with M.A. Kelley) Atlas of Human Paleopathology, 1982. Served to maj. U.S. Army, 1968-70. Rackham faculty research grantee, 1978-80. Mem. Internat. Acad. Pathology, Am. Soc. Clin. Pathology, Paleopathology Assn., AAAS, Am. Assn. Phys. Anthropology, Soc. Med. Anthropology, Am. Anthrop. Assn. Subspecialties: Pathology (medicine); Paleopathology. Current work: Anatomic pathology; research - paleopathology (evidence of disease in ancient remains). Office: Jeanes Hosp Dept Pathology 7600 Central Ave Philadelphia PA 19111

ZIMMERMAN, PATRICK ROBERT, atmospheric scientist; b. Spokane, Wash., May 23, 1950; s. Clarence Robert and Patricia Jean (Rule) Z.; m. Marilyn Eileen Thorvig, May 17, 1971; children: Robert Scott, Melissa Lynn. B.S. in Environ. Scis, Wash. State U., 1973, M.S., 1975; postgrad. range sci., Colo. State U. Jr. environ. scientist, mem. faculty Wash. State U., 1973-76, asst. environ. scientist air pollution research sect., 1976-79; scientist (Nat. Ctr. for Atmospheric Research), 1979—; affiliate assoc. prof. Wash. State U. Contbr. articles on atmospheric sci. to profl. jours. NATO grantee, 1981. Subspecialties: Atmospheric chemistry. Current work: Identification of biological processes which may affect atmospheric chemistry, atmospheric science, biogenic emissions, organic fluxes termite emissions, vegetation emssions, gas chromatography. Office: Nat Center for Atmospheric Research PO Box 3000 Boulder CO 80307

ZIMMERMAN, SARAH E., immunologist; b. Indpls., Oct. 29, 1937; s. Abraham and Mary (Caplan) Z. A.B., Ind. U., 1959, M.A., 1961; Ph.D., Wayne State U., 1969. Postdoctoral research assoc. dept. chemistry Ind. U., Bloomington, 1969-71; postdoctoral research assoc. dept. microbiology Ind. U. Sch. Medicine, 1971-73, research scientist dept. pathology, 1973—. Mem. Am. Chem. Soc., Am. Assn. Immunologists, Am. Soc. Microbiology, Sigma Xi. Subspecialties: Immunobiology and immunology; Infectious diseases. Current work: Development of immunological diagnostic tests for infectious diseases (example: enzyme immunoassays). Office: Dept Pathology Riley Hosp A-20 Ind U Sch Medicine Indianapolis IN 46223

ZIMMERMAN, STUART ODELL, biomathematician, university department chairman; b. Chgo., July 27, 1935; s. O. C. and Matilda (Kribs) Z.; m. Mary Joan Spiegel, Aug. 8, 1959; 1 son, Kurt Michael. B.A., U. Chgo., 1954, Ph.D., 1964. Research assoc., asst. prof. biomath. U. Chgo., 1963-67, assoc. biomathematician, acting head dept. biomath., 1967-68; assoc. prof., head dept. biomath. U. Tex. M.D. Anderson Hosp. and Tumor Inst., Houston, 1968-72, prof., chmn. dept., 1972—, head div. biomed. info. resources, 1981—; cons. statis. group Am. Dental Assn., Chgo., 1966—; cons. Task Force on Design and Analysis of Dental Clin. Trials, N.J. Dept. Health, 1970—; site visitor NIH, Bethesda, Md., 1969—; chmn. exec. bd. U. Tex.-Houston Edn. and Research Computer Ctr., 1973—. Nat. Cancer Inst.-NIH Research grantee, 1968-84; U. Chgo. gen. univ. scholar, 1955-59; Zoller fellow, 1960-64. Mem. Assn. Computing Machinery (pres. Tex. chpt. 1975-76), Soc. Math. Biology (mem. council 1980-82), Assn. Am. Cancer Insts. (chmn. task 2 com. on data processing requirements 1974-76). Roman Catholic. Current work: Biomathematics, biomedical computing, image processing, computer Karyotyping, cell kinetics. Home: 9906 Bob White St Houston TX 77096 Office: U Tex System Cancer Center M D Anderson Hosp and Tumor Inst Houston TX 77030

ZIMMERMAN, THOMAS PAUL, research biochemist; b. Plainfield, N.J., Sept. 3, 1942; s. Carl Paul and Anna Marie (Deere) Z.; m. Barbara Ann Ouimet, June 18, 1966; children: Michael Thomas, Catherine Ann. B.S. in Chemistry, Providence Coll., 1964; Ph.D. in Biochemistry, Brown U., 1969. USPHS postdoctoral fellow Brown U., Providence, R.I., 1969-71; sr. research biochemist Burroughs Wellcome Co., Research Triangle Park, N.C., 1971—. Contbr. articles to sci. publs. Mem. Am. Soc. Pharmacology and Exptl. Therapeutics, Internat. Soc. Immunopharmacology. Roman Catholic. Subspecialties: Biochemistry (biology); Molecular pharmacology. Current work: Metabolic studies of purine antimetabolites. Determination of mechanisms by which pharmacological agents modulate immunological function of lymphocytes. Patentee in field. Home: 1415 Brunson Ct Cary NC 27511 Office: 3030 Cornwallis Rd Research Triangle Park NC 27709

ZIMMERMANN, F(RANCIS) J(OHN), mechanical engineering educator; b. Jersey City, Apr. 21, 1924; s. William Henry and Philomena Clare (Krysa) Z.; m. Margaret Malinda Stephens, Apr. 14, 1950; children: Stephen Robert, Marcy Jean. B.Engring., Yale U., 1948; S.M., MIT, 1950, Mech. Engr., 1951, Sc.D., 1953. Staff engr. Arthur D. Little, Inc., Cambridge, Mass., 1952-55; asst. prof. Yale U., 1955-61, assoc. prof., 1961-62; prof. mech. engring. Lafayette Coll., Easton, Pa., 1962—, dept. head, 1978-82; asst. program dir. engring. sect. NSF, 1961-62; cons. Arthur D. Little, 1955-62, Air Products & Chems., Allentown, Pa., 1963—. Bd. dirs. Pa. Sinfonia Orch., Allentown, 1982—. Served with AUS, 1943-45. Recipient Lindback award Lafayette Coll., 1977. Mem. ASME, AIAA, Am. Soc. for Engring. Edn., Cryogenic Soc. Am. Republican. Subspecialties: Cryogenics; Mechanical engineering. Current work: Teaching thermodynamics, heat transfer, and cryogenic engineering. Patentee container for liquefied gas, miniature refrigeration device, liquefaction low boiling gases, cryogenic refrigeration system. Home: 325 Reeder St Easton PA 18042 Office: Dept Mech Engring Lafayette Coll Easton PA 18042

ZINDER, NORTON DAVID, geneticist, educator; b. N.Y.C., Nov. 7, 1928; s. Harry Jean and (Gottesman) Z.; m. Marilyn Estreicher, Dec. 24, 1949; children—Stephen, Michael. A.B., Columbia U., 1947; M.S., U. Wis., 1949, Ph.D., 1952. Asst. Rockefeller U., N.Y.C., 1952-56, asso., 1956-58, asso. prof. genetics, 1958-64, prof., 1964—, John D. Rockefeller Jr. prof., 1977—; Cons. genetic-biology NSF, 1962—; chmn. ad hoc com. to rev. viral cancer program Nat. Cancer Inst., 1973-74; Mem. vis. com. dept. biology Harvard U., 1975—; sect. virology Yale U., 1975—; dept. biochemistry Princeton U., 1975—. Asso. editor: Virology. Recipient Eli Lilly award in microbiology and immunology, 1962; U.S. Steel Found. award in molecular biology, 1966; medal of excellence Columbia U., 1969. Fellow Am. Acad. Arts and Scis.; mem. Nat. Acad. Scis. (exec. com. Assembly of Life Scis. 1975—), Soc. Am. Biol. Chemists, Genetics Soc. Am., Am. Soc. for Microbiology, Harvey Soc., Sigma Xi. Subspecialty: Genetics and genetic engineering (biology). Spl. research in microbial genetics. Home: 450 E 63d St New York NY 10021 Office: Rockefeller U 66th St and York Ave New York NY 10021

ZINN, GENE MARVIN, veterinarian, educator; b. Kellerton, Iowa, Apr. 9, 1932; s. Raymond Bryon and Pearl Lavon (Watts) Z.; m. Muriel Lue McLain, June 16, 1957; children: Karl Gene, Kurt Ray, Sandra Lee, Eric Marvin. D.V.M., Iowa State U., 1956; Ph.D., U. Mo., 1970. Gen. practice vet. medicine, Bethany, Mo., 1956-65; 70-78, Gallatin, Mo., 1966-67; instr. large animal surgery Kans. State U., Manhattan, 1965-66; research assoc. dept. vet. physiology and pharmacology U. Mo., Columbia, 1967-70, asst. prof. food animal medicine and surgery, 1979—. Contbr. articles to profl. jours. County chmn. Harrison County Republican Central Com., 1972-76. Recipient various awards. Mem. AVMA, Am. Assn. Bovine Practitioners, Am. Assn. Swine Practitioners, Mo. Vet. Med. Assn., Am. Acad. Vet. Nutrition, Am. Assn. Vet. Clinicians, Am. Soc. Animal Sci., Council for Agr. Sci. and Tech. Methodist. Subspecialties: Food Animal Medicine and Surgery; Animal nutrition. Current work: Swine disease, swine nutrition. Home: 1204 Longwell Dr Columbia MO 65201 Office: 1600 E Rollins Rd Columbia MO 65211

ZIPES, DOUGLAS PETER, physician, researcher, educator; b. White Plains, N.Y., Feb. 27, 1939; s. Robert Samuel and Josephine Helen (Weber) Z.; m. Marilyn Joan Jacobus, Feb. 18, 1961; children: Debra, Jeffrey, David. B.A., Dartmouth Coll., 1961, B.Med. Sci., 1962; M.D., Harvard U., 1964. Diplomate: Am. Bd. Internal Medicine, Am. Bd. Cardiovascular Disease. Intern, resident, fellow Duke U. Med. Ctr., Durham, N.C., 1964-68; scientist Masonic Med. Research Lab., Utica, N.Y., 1970-71; asst. prof. medicine Ind. U. Sch. Medicine, Indpls., 1970-73, assoc. prof., 1973-76, prof., 1976—; cons. Medtronic Inc., Mpls., 1975—. Author: Comprehensive Cardiac Care, 1968, 71, 75, 79, 83; editor: Slow Inward Current, 1980; editor-in-chief: Cardiac Impulse; mem. editorial bd.: Am. Heart Jour; assoc. editor: Am. Jour. Medicine. Vice-pres. Indpls. Opera Co., Indpls., 1981-83. Served to lt. comdr. USN, 1968-70. NIH research grantee, 1977—. Fellow ACP, Am. Coll. Cardiology, Council Clin. Cardiology Am. Heart Assn.; mem. Am. Soc. Clin. Investigation, Assn. Univ. Cardiologists, Cardiac Electrophysiology Soc. (sec./treas.), Am. Heart Assn. (chmn. com. electrocardiography and clin. electrophysiology), Am. Coll. Cardiology. Current work: Clinical and animal electrophysiological research into mechanisms responsible for and therapy of cardiac arhythmias. Inventor transvenous cardioverter, 1982. Home: 2113 Brewster Rd Indianapolis IN 46260 Office: Ind U Sch Medicine 1100 W Michigan St Indianapolis IN 46223

ZIRIN, HAROLD, educator, astronomer; b. Boston, Oct. 7, 1929; s. Jack and Anna (Buchwalter) Z.; m. Mary Noble Fleming, Apr. 20, 1957; children: Daniel Meyer, Dana Mary. A.B., Harvard U., 1950, A.M., 1951, Ph.D., 1952. Asst. phys. scientist RAND Corp., 1952-53; lectr. Harvard, 1953-55; research staff High Altitude Obs., Boulder, Colo., 1955-64; prof. astrophysics Calif. Inst. Tech., 1964—; staff mem. Hale Obs., 1964-80; chief astronomer Big Bear Solar Obs., 1969-80; dir. (Big Bear Solar Obs.), 1980—; U.S.- USSR exchange scientist, 1960-61. Author: The Solar Atmosphere, 1966; Adv. editor: Soviet Astronomy, 1965-69. Trustee Polique Canyon Assn., 1977—. Agassiz fellow, 1951-52; Sloan fellow, 1958-60; Guggenheim fellow, 1960-61. Mem. Am. Astron. Soc., Internat. Astron. Union, AURA (dir. 1977-83). Current work: Study of the sun, primarily observational, optical, radio and space techniques. Home: 1178 Sonoma Dr Altadena CA 91001 Office: California Inst Tech Pasadena CA 91125

ZIRKER, JACK BERNARD, solar astronomer; b. Bklyn., July 19, 1927; s. Joseph and Rose Z.; m. Lorette Zuckerman, Jan. 21, 1951; children—Robin, Alizon, Pamela. B.M.E., CCNY, 1948; M.S., N.Y. U., 1953; Ph.D., Harvard U., 1956. Scientist Sacramento Peak Obs., Sunspot, N.Mex., 1956-64, dir. obs., 1976—; solar astronomer U. Hawaii, 1964-76, prof., 1965-76; cons. NASA. Served with U.S. Army, 45-47. Mem. Am. Astron. Soc., Internat. Astron. Union. Subspecialty: Solar physics.

ZISK, STANLEY HARRIS, research scientist; b. Boston, July 11, 1931; s. Morris and Edith (Lewenberg) Z.; m.; children: Jonathan Lee, Stephen Robert, Matthew Bruce. B.S., M.I.T., 1953, M.S., 1953; Ph.D., Stanford U., 1965. Faculty dept. elec. engring. M.I.T., Cambridge, 1965-68; staff scientist, head cryogenics sect. Haystack Obs., Westford, 1968—; cons. in field; lectr. in field. Contbr. articles to profl. jours. Served with USNR, 1956-59. NASA grantee, 1968—. Mem. Am. Astron. Soc., Am. Geophys. Union, Union Radio Sci. Internat., IEEE. Subspecialties: Planetary science; Cryogenics. Current work: Remote sensing of planetary surfaces; maser amplifier development. Address: MIT Haystack Obs Westford MA 01886

ZITTER, THOMAS ANDREW, plant pathologist; b. Saginaw, Mich., Dec. 30, 1941; s. Andrew and Helen Martha (Meulenbeck) Z.; m. Judith M. Zitter, Aug. 13, 1966; children: Timothy, Daniel, Julie. B.S., Mich. State U., 1963, Ph.D., 1968. Asst. plant pathologist U. Fla., Belle Glade, 1968-74, assoc. prof., assoc. plant pathologist, 1974-79; assoc. prof., plant pathologist Cornell U., Ithaca, N.Y., 1979—. Contbr. articles to profl. jours. Mem. Am. Phytopathol. Soc., Fla. State Hort. Soc., Assn. Applied Biologists. Lutheran. Lodge: Lions. Subspecialties: Plant virology. Current work: Epidemiology and control of plant viruses; insect transmission of virus diseases; plant resistance. Home: 4 Bean Hill Ln Ithaca NY 14850 Office: Dept Plant Pathology Cornell U Ithaca NY 14853

ZITZEWITZ, PAUL W., physicist; b. Chgo., June 5, 1942; s. Walter William and Barbara Lois (Ohl) Z.; m. Barbara Shaw, June 11, 1966; children: Eric, Karin. A.B., Carleton Coll., 1964; M.A., Harvard U., 1965, Ph.D., 1970. Postdoctoral fellow U. Western Ont., London, 1970-72; sr. physicist Corning Glass Works, N.Y., 1972-73; asst. prof. physics U. Mich.-Dearborn, 1973-78, assoc. prof., 1978-83, prof., 1983—; vis. prof. Bielefeld (W.Ger.) U., 1979-80. Contbr. articles to profl. jours. Alexander von Humboldt fellow, 1979-80. Mem. Am. Phys. Soc., Am. Assn. Physics Tchrs., Internat. Soc. for Study of Origins of Life. Subspecialty: Atomic and molecular physics. Current work: Study of fundamental physical theories with positrons and positronium. Testing quantum electrodynamics with positronium. Low energy polarized positron beams. Office: 4901 Evergreen Dearborn MI 48128

ZIVIN, JUSTIN ALLEN, neurologist, neurochemistry researcher, educator; b. Chgo., Aug. 17, 1946; s. Simon and Mabel Edith (Libert) Z.; m. Reni-Zoe Mandell, June 23, 1968; children: Kara, Leslie. B.S., Northwestern U., 1967, M.S., 1970, Ph.D., 1971, M.D., 1972. Diplomate: Am. Bd. Psychiatry and Neurology. Intern U. Mich., Ann Arbor, 1972-73; research assoc. Lab. Clin. Sci., NIH, 1973-75; resident in neurology U. Calif., San Francisco, 1975-78; assoc. prof. neurology U. Mass. Med. Center, Worcester, 1978—, assoc. prof. pharmacology, 1980—. Contbr. articles to sci. jours. Served as surgeon USPHS, 1973-75. Recipient Tchr. Investigator Devel. award Nat. Inst. Neurol. and Communicative Disease and Stroke, 1979—. Mem. Am. Acad. Neurology, Soc. for Neurosci., Am. Soc. for Neurochemistry, Am. Heart Assn. Subspecialties: Neurology; Neurochemistry. Current work: Studies of biogenic amines, energy metabolism and blood flow in central nervous system ischemia, trauma and metabolic disorders. Office: 55 Lake Ave N Worcester MA 01605

ZMOLEK, WILLIAM GEORGE, extension livestock specialist, farm manager; b. Toledo, Iowa, July 3, 1921; s. George A. and Agnes M. (Kubic) Z.; m. Jean A. McCormick; children: Steven, Gary, Gloria, John, Paul. B.S., Iowa State U., 1944, M.S., 1951. Extension dir. Counties of Marshall and Jasper, 1944-47; farm rep. Newton Nat. Bank, Iowa, 1948; extension livestock specialist Iowa State U., Ames, 1948—. Contbr. in field. Mem. Am. Soc. Animal Sci., Council for Agrl. Sci. and Tech., Am. Grassland Council. Republican. Roman Catholic. Lodge: Moose. Subspecialty: Animal nutrition. Current work: Animal management and nutrition.

ZOEBISCH, OSCAR C., chem. co. scientist; b. Albany, Minn., Dec. 16. 1919; s. Philip and Minne (Johnson) Z.; m. Virginia Paulson, Aug. 18, 1945; children: Ann, Philip, Amy, Eve. B.S., U. Minn., 1946, M.S., 1950, Ph.D., 1950. Faculty U. Hawaii, 1950-53; with Libb Food Co., 1953-65, E.I. DuPont Co., 1965—. Served with USN, 1940-46. Republican. Lutheran. Home: 121 Edgewood Ardomore PA 19003 Office: DuPont E 268 Wilmington DE 19898

ZOLLER, BROC GERALD, agricultural consultant, consulting firm executive; b. Oakland, Calif., Apr. 17, 1944; s. Dudley Francis and Edcil Fitzgerald (Doerr) Z.; m. Sharron Ann Higgins, July 31, 1971; children: Molly Elizabeth, Zachary Joseph, Analiese Maria. Student, Georg August U., Goettingen, W. Ger., 1965-66; B.S. in Plant Pathology with highest honors, U. Calif.-Davis, 1966, Ph.D., 1972. Agrl. cons. Agrl. Advs., Inc., Yuba City, Calif., 1972-79; agrl . cons., pres. Pear Doctor, Inc., Yuba City, 1979—. Contbr. articles, abstracts to profl. publs. Served with U.S. Army, 1967-69. Recipient departmental citation, dept. plant pathology U. Calif.-Davis, 1966; Achievement award Lake County Farm Bur., 1982; Title IV fellow, 1969. Mem. Am. Phytopathol. Soc. (vice chmn. com. on cons.), Assn. Applied Insect Ecologists (dir. 1978-82, program chmn. 1980, pres. 1981), Nat. Alliance Ind. Crop Cons., Sierra Club, Phi Beta Kappa, Sigma xi. Subspecialties: Integrated pest management; Plant pathology. Current work: Economics and application of integrated pest management in deciduous tree fruit crops; agricultural loss investigation and expert testimony; disease control in deciduous tree crops; production of deciduous tree crops. Office: PO Box 952 1441 Garden Hwy Yuba City CA 95992

ZOLLWEG, ROBERT JOHN, physicist; b. Medina, N.Y., Aug. 1, 1924; s. Louis August and Martha Mildred (Lagasse) Z.; m. Aileen Boules, Sept. 7, 1946; children: Lynn. Vicky. B.S., Northwestern U., 1949, M.S., 1950; Ph.D., Cornell U., 1955. Research physicist Westinghouse Research & Devel. Center, Pitts., 1954-67, fellow scientist, 1967-68, adv. scientist, 1968-80, cons. scientist, 1980—. Contbr. articles to profl. jours. Served with USN, 1943-46. Mem. Am. Phys. Soc. Subspecialties: Atomic and molecular physics; High temperature chemistry. Current work: Research in high temperature plasmas, arc modeling, arc lamps, plasma chemistry, electrodes. Patentee in field. Home: 4560 Bulltown Rd Murrysville PA 15668 Office: Westinghouse Research and Development Center 1310 Beulah Rd Pittsburgh PA 15235

ZOMBECK, MARTIN VINCENT, physicist; b. Peekskill, N.Y., Aug. 14, 1936; s. Martin Francis and Sophie Blanche (Ceglecki) Z.; m. Monique Jeanne Lagouthe, July 14, 1963; children: Richard, Yann. B.S., M.I.T., 1957, Ph.D., 1969; postgrad., U. Calif., Berkeley, 1958, Freie Universitat, West Berlin, 1957-58, U. Munich, W.Ger., 1962-63. Research asst. MIT, 1963-69; satellite tracker Smithsonian Astrophys. Obs., Cambridge, Mass., 1960-62, physicist, 1973—; staff scientist Am. Sci. & Engring., Cambridge, 1969-72. Author: Handbook of Space Astronomy and Astrophysics, 1982; also articles. Alfred P. Sloan Nat. scholar, 1954-57; Fulbright scholar, 1957-58; NASA Skylab Achievement award, 1974. Mem. Am. Astron. Soc., Internat. Astron. Union. Subspecialty: 1-ray high energy astrophysics. Current work: X-ray astronomy from rockets and satellites; x-ray instrumentation development. Home: 42 Fletcher St Winchester MA 01890 Office: 60 Garden St Cambridge MA 02138

ZOOK, BERNARD CHARLES, research pathologist; b. Beach, N.D., Nov. 1, 1935; s. Frank Nicholas and Elizabeth Fern (Kramer) Z.; m. Elinore Ann Schillo, Oct. 1, 1955; children: Bernita Elin, Melinda Susan, Andrew Kramer. B.S., Colo. State U., 1962, D.V.M., 1963. Diplomate: Am. Coll. Vet. Pathologists. Trainee, tchr., researcher Harvard U. Med. Sch. and Angell Meml. Animal Hosp., Boston, 1963-66, 67-69; NIH spl. fellow Harvard U. Sch. Public Health, 1966-67; asst. prof. pathology George Washington U., 1969-73, asso. prof., 1974—, dir., 1972—; research assoc. Smithsonian Instn., 1969—; cons. in pathology Litton Bionetics, Kensington, Md. Contbr. chpts. to books, articles to profl. jours. Bd. dirs. Nat. Soc. for Med. Research; mem. Stonegate Citizens Assn. Served with U.S. Army, 1955-56. Recipient Mary Mitchell award Mass. Soc. for Protection of Animals, 1966; grantee NIH, 1974, 77, Nat. Cancer Inst., 1975—, Murray Corp., 1981, Nat. Bur. Standards, 1980, Rockefeller U., 1982. Mem. AVMA, Am. Assn. Lab. Animal Scis., Am. Coll. Vet. Pathologists, Radiation Research Soc., AAAS. Republican. Roman Catholic. Club: Duplicate Bridge. Lodge: K.C. Subspecialties: Pathology (veterinary medicine);

Environmental toxicology. Current work: Radiation biology, developmental anomalies, toxicologic pathology. Office: 2300 I St NW Washington DC 20037

ZORN, GUS TOM, physics educator; b. Ada, Okla., June 18, 1924; s. Charles August and Nell (McLachlan) Z.; m. Bice Sechi, Nov. 9, 1955. B.S. in E.E, Okla. A&M Coll., 1948; M.S. in Physics, U. N.Mex., 1951; D.Physics, U. Padua, Italy, 1954. Assoc. physicist Brookhaven Nat. Lab., Upton, N.Y., 1954-62; assoc. prof. physics U. Md., College Park, 1962-72, prof., 1972—. Contbr. articles to profl. jours. Served with U.S. Army, 1942-45; ETO. Mem. Am. Phys. Soc., Italian Phys. Soc., N.Y. Acad. Sci. Subspecialties: Particle physics; Plasma physics. Current work: Study of particle physics using high energy electron-positron collision. Home: 8722 23d Ave Adelphi MD 20783 Office: Dept Physics U Md College Park MD 20783

ZOUROS, ELEFTHERIOS, population geneticist; b. Lesbos, Greece, Aug. 31, 1939; s. George and Efstratia (Orphanos) Z.; m. Efstratia Amorianos, Aug. 15, 1968; children: Alexander George, Irene Adamantia. M.Sc. (govt. scholar), Agrl. Coll. Athens, 1963; Ph.D., U. Chgo., 1972. Lectr., Agrl. Coll. Athens, 1965-68; postdoctoral fellow U. Chgo., 1969-73; prof. biology Dalhousie U., Halifax, N.S., 1973—; prof. biology U. Crete, Greece. Author: Introduction to Biology; also articles. Grantee Ford Found., 1969-73, NATO, 1980, 82, Natural Sci. and Engrng. Research Council Can., 1973—. Mem. Genetics Soc. Am., Am. Soc. Naturalists, Soc. Study Evolution Genetics Soc. Can. Subspecialties: Evolutionary biology; Population biology. Current work: Evolutionary genetics; genetic basis of speciation in drosophila; genetics of growth rate in marine molluscs. Home: 2361 Macdonald St Halifax NS Canada B3L 3H3 Office: Dept Biology Dalhousie U Halifax NS Canada B3H 4J1

ZUCCARELLI, ANTHONY JOSEPH, microbiologist, educator, researcher; b. N.Y.C., Aug. 11, 1944; s. Anthony D. and Rose Marie (Lisanti) Z.; m. Sharron Adele Ames, Dec. 23, 1968; children: Cara Nicole, Anthony Alexander. B.S. (Roberts scholar), Cornell U., 1966; M.A., Loma Linda U., 1968; Ph.D., Calif. Inst. Tech., 1973. Am. Cancer Soc. postdoctoral fellow U. Konstanz, W.Ger., 1974-76; asst. prof. biology Loma Linda U., 1976-80, assoc. prof. microbiology, 1980—. Research publs. on genetics. NSF grantees, 1977, 78-79; Loma Linda U. Research Adv. Com. grantee, 1978-79. Mem. Am. Soc. Microbiology, AAAS, Sigma Xi. Adventist. Subspecialties: Genetics and genetic engineering (biology); Molecular biology. Current work: Genetic engrng. of small DNA viruses; protein-DNA interactions; site-specific mutagenesis; enzymology of DNA replication in E. coli. Home: 22255 Mavis St Grand Terrace CA 92324 Office: Dept Microbiology Loma Linda U Loma Linda CA 92350

ZUDECK, STEVEN, laboratory executive; b. N.Y.C., Nov. 4, 1932; s. Maurice H. and Sylvia (Berg) Z.; m. Sylvia G. Zudeck, Sept. 6, 1959; children: Jeffrey, Melissa. B.S. in Biology, Hofstra U., 1955. With Center Labs., Port Washington, N.Y., 1957-81, nat. sales mgr., 1968-71, dir. quality assurance, 1975-81; v.p. Barry Labs., Inc., Pompano Beach, Fla., 1981—; cons. in field. Dir. L.I. Sci. Congress, 1980-81. Served with U.S. Army, 1955-57. Mem. Parenteral Drug Assn, Pharm. Mfrs. Assn., Am. Mgmt. Assn., AAAS, Long Island Computer Assn. Subspecialties: Pharmaceutical manufacturing; Allergenic extract preparations. Current work: Research and development and manufacture of allergenic extracts. Home: 7108 NW 43d St Coral Springs FL 33065 Office: PO Box 1967 Pompano Beach FL 33061

ZUNDE, PRANAS, information science educator, researcher; b. Kaunas, Lithuania, Nov. 26, 1923; U.S., 1960, naturalized, 1964; s. Pranas and Elzbieta (Lisajevic) Z.; m. Alge R. Bizauskas, May 29, 1945; children: Alge R., Audronis K., Aurelia R., Aidis L., Gytis J. Dipl. Ing., Hanover Inst. Tech., 1947; M.S., George Washington U., 1965; Ph.D., Ga. Inst. Tech., 1968. Project dir. Documentation, Inc., Bethesda, Md., 1961-64, mgmt. info. system mgr., 1964-65; sr. research scientist Ga. Inst. Tech., Atlanta, 1965-68, assoc. prof., 1968-72, prof. dept. computer sci., 1973—; cons. UNESCO, Caracas, Venezuela, 1970-72, Esquela Polit. Nacional, Quito, Ecuador, 1974-75, State of Ga., Atlanta, 1976-78; Fulbright prof. Nat. Acad. Sci., 1975; vis. prof. Simon Bolivar U., Caracas, 1976, J. Kepler U., Austria, 1981. Author: Agriculture in Soviet Lithuania, 1962; Editor: Procs. Information Utilities, 1974, Procs. Foundations of Information and Software Science, 1983. Mem. Am. Soc. Info. Sci., Semiatic Soc. Am. Roman Catholic. Subspecialties: Software engineering. Current work: Theoretical foundations of information and software sciences, system theory, information systems design, mathematical modeling, pattern recognition. Home: 1808 Timothy Dr NE Atlanta GA 30329 Office: Ga Inst Tech North Ave Atlanta GA 30332

ZURAWSKI, VINCENT RICHARD, JR., biotechnology company executive, researcher; b. Irvington, N.J., June 10, 1946; s. Vincent Richard and Norma Mary (Alliston) Z.; m. Mary Rita Stanziola, Aug. 18, 1968; children: Daniel Vincent, John Alliston. B.A., Montclair State Coll., 1968; Ph.D., Purdue U., 1973. Research fellow dept. medicine Harvard U.- Mass. Gen. Hosp., 1975-78, instr., 1978-79; sr. v.p., tech. dir. Centocor, Malvern, Pa., 1979-83, exec. v.p, tech. dir., 1983—. Contbr. articles to profl. jours. Served with USAR, 1969-79. NIH fellow, 1976-78; Med. Found. research fellow, 1978-79. Mem. AAAS, Am. Chem. Soc., Am. Soc. Microbiologists. Subspecialties: Biochemistry (biology); Immunobiology and immunology. Patentee in field. Office: 244 Great Valley Pkwy Malvern PA 19355

ZUSMAN, DAVID ROBERT, microbiology educator, researcher; b. Bklyn., Dec. 31, 1943; s. Morris and Evelyn (Cohen) Z.; m. Ami Saperstein, Oct. 12, 1969. B.A., Bklyn. Coll., CUNY, 1965; M.A., U. Calif., 1967; Ph.D., UCLA, 1970. Postdoctoral fellow, dept. biochem. sci. Princeton (N.J.) U., 1970-72; asst. prof. microbiology U. Calif., Berkeley, 1973-79, assoc. prof. microbiology, 1979—. Contbr. chpts. to monographs, articles to sci. jours. in field. Mem. Am. Soc. Microbiology, AAAS. Subspecialties: Microbiology; Genetics and genetic engineering (biology). Current work: Regulation of gene expression and cell-cell interactions during development of fruiting bacterium, Myxococcus xanthus; developmental bioloy, gene expression, regulation, cloning. Office: Dept Microbiology Univ Calif Berkeley CA 94720

ZUSPAN, FREDERICK PAUL, physician, educator; b. Richwood, Ohio, Jan. 20, 1922; s. Irl Goff and Kathryn (Speyer) Z.; m. Mary Jane Cox, Nov. 23, 1943; children: Mark Frederick, Kathryn Jane, Bethany Anne. B.A., Ohio State U., 1947, M.D., 1951. Intern Univ. Hosps., Columbus, Ohio, 1951-52, resident, 1952-54, Western Res. U., Cleve., 1954-56, Oblebay fellow, 1958-60, asst. prof., 1958-60; chmn. dept. ob-gyn McDowell (Ky.) Meml. Hosp., 1956-58, chief clin. services, 1957-58; prof., chmn. dept. ob-gyn Med. Coll. Ga., Augusta, 1960-66; Joseph Boliver DeLee prof. ob-gyn, chmn. dept. U. Chgo., 1966-75; obstetrician, gynecologist in chief Chgo. Lying-In Hosp., 1966-75; prof., chmn. dept. ob-gyn Ohio State U. Sch. Medicine, Columbus, 1975—. Founding editor: Lying In, Jour. Reproductive Medicine; editor: (with Lindheimer and Katz) Hypertension in Pregnancy, 1976, Current Developments in Perinatology, 1977, (with Quilligan) Operative Obstetrics, 1981, Manual of Practical Obstetrics, 1981, Am. Jour. Ob-Gyn, Clin. and ExpH. Hypertension in Pregnancy, (with Rayburn) Drug Therapy in Ob-Gyn, 1981; Mem. editorial bd.: Exerpta Medica; editor: (with Christian) Obstetrics and Gynecology; Contbr. (with Rayburn) articles to med. jours., chpts. to books. Pres. Barren Found., 1974-76. Served with USNR, 1942-43; to 1st lt. USMCR, 1943-45. Decorated DFC, Air medal wth 10 oak leaf clusters. Mem. Soc. Gynecol. Investigation, Chgo. Gynecol. Soc., Am. Assn. Ob-Gyn, Am. Acad. Reproductive Medicine (pres.), Am. Coll. Obstetricians and Gynecologists, Assn. Profs. Gynecology and Obstetrics (pres. 1972), South Atlantic Assn. Obstetricians and Gynecologists (Found. prize for research 1962), Central Assn. Ob-Gyn (cert. of merit, research prize 1970), Am. Soc. Clin. Exptl. Hypnosis (exec. sec. 1968, v.p. 1970), So. Gynecol. Soc., Internat. Soc. Study Hypertension in Pregnancy (pres. 1981—), Sigma Xi, Alpha Omega, Alpha Kappa Kappa. Subspecialties: Obstetrics and gynecology; Maternal and fetal medicine. Current work: Catecholamine research on hypertension in pregnancy and fetal neuroendocrine regulation. Home: 2400 Coventry Rd Upper Arlington Columbus OH 43211 The strength of our nation rests in the quality of our offspring. Every fetus has the privilege of being wellborn.

ZUZOLO, RALPH CARMINE, biologist, educator, researcher; b. Dente Cane, Avellino, Italy, Sept. 5, 1929; came to U.S., 1930; s. Antonio and Assunto (Nardone) Z.; m. Betty Ann Fong, July 22, 1972. B.A. in Biology, N.Y.U., 1956, MS. in Cell Physiology, 1960, Ph.D. in cell physiology and Micro-Surgery, 1965. Asst. prof. CCNY of CUNY, 1965, assoc. prof., 1975—; co-dir. Robert Chambers Lab., 1978—, Sch. Gen. Studies supr. dept. biology, 1968—; adj. research sci. N.Y.U., 1965-78. Served to cpl. USMC, 1951-52; Korea. Recipient NYU Founders Day award, 1965; U. Tex. research fellow, 1966-68. Mem. AAAS, N.Y. Acad. Sci., Soc. Applied Spectroscopy, AAUP, Sigma Xi. Current work: Continue developing Robert Chambers Lab for cellular microsurgery; produce simplified instrumentations and methods for mechanical microinjection of macromolecule into single cells; serve as a center for other interested scientist. Office: CCNY of CUNY Convent Ave at 138th St New York NY 10031

ZWEIG, GEORGE, educator, physicist; b. Moscow, USSR, May 20, 1937; came to U.S., 1938, naturalized, 1944; s. Alfred and Rachael (Froehlich) Z.; m.; children—Geoffrey, Christopher. B.S. in Math, U. Mich., 1959; Ph.D. in Physics, Calif. Inst. Tech., 1963. With Orgn. Europeenne poar la Recherche Nucleaire, 1963-64; asst. prof. physics Calif. Inst. Tech., 1964-66, asso. prof., 1966-67, prof., 1967—. Woodrow Wilson fellow, 1959; Nat. Acad. Sci. NRC fellow, 1963; Sloan Found. fellow in physics, 1966; Sloan Found. fellow in neurophysiology, 1975. Subspecialty: Particle physics. Research on particle classification, originator of the quark model; research in hearing. Office: Lauritsen Lab Calif Inst Tech Pasadena CA 91125

ZWEIG, GILBERT, mech. engr.; b. N.Y.C., Apr. 5, 1938; s. Aaron and Esther (Tashbook) Z.; m.; children: David A., Steven A. B.M.E., B.A., N.Y.U., 1960, M.S., 1965; postgrad., Columbia U., 1965-68. Research staff IBM, Yorktown Heights, N.Y., 1961-63; supr. advanced research Pitney-Bowes, Stamford, Conn., 1963-73; founder, v.p. Imtex Inc., Milford, Conn., 1973-74; dir. tech. bus. devel. Arkwright Inc., Fiskeville, R.I., 1974-81; v.p. MCI Optonix, Inc., Cedar Knolls, N.J., 1981—. Contbr. articles to profl. jours. Mem. Am. Assn. Physicists in Medicine, Soc. Photog. Scientists and Engrs., Soc. Photo Optical Instrument Engrs., Sierra Club, Sigma Xi. Subspecialties: Photo-optical imaging; Reprographics. Current work: Research in radioluminescent imaging systems as applied to fluoroscopic and x-ray intensifying screens. Patentee in field. Home: 24 Stiles Ave Morris Plains NJ 07950 Office: PO Box I Cedar Knolls NJ 07927

ZWEIG, JOSEPH B., psychologist, consultant; b. N.Y.C., Mar. 29, 1949; s. Milton and Rose (Engelson) Z.; m. Michelle Bonnie Bloch, Dec. 19, 1970; children: Jeremy, Samantha. B.A., SUNY-Stony Brook, 1969; M.A., Fairleigh Dickinson U., 1970; Ph.D., Hofstra U., 1980. Diplomate: Am. Acad. Behavioral Medicine; cert. psychologist, biofeedback therapist, rehab. counselor. Psychologist Bergen Pines County Hosp., Paramus, N.J., 1970-72, Jewish Vocat. Services, East Orange, N.J., 1972-74; program dir. East Plains Mental Health Services, Hicksville, N.Y., 1974-80; psychologist Rockland Ctr. for Physically Handicapped, New City, N.Y., 1980—; dir./ptnr. Rockland Psychol. Assocs., Pomona, N.Y., 1981—; adj. faculty Hofstra U., Hempstead, N.Y., 1982—; steering com. Mothers' Ctr., Hicksville, 1975-78. Bd. Edn. citizens adv. com. mem., Clarkstown, N.Y., 1980. Mem. Am. Psychol. Assn., Biofeedback Soc. Am., Soc. for Psychosomatic Obs/Gyn. Jewish. Subspecialties: Behavioral psychology. Current work: Stress management; biofeedback: first study of use of biofeedback iun preparation for childbirth. Office: Rockland Psychol Assocs Route 45 Pomona NY 10970

ZWEIG, RICHARD LEE, publisher, psychologist; b. St. Louis, July 15, 1927; s. Julius and Clara Z.; m. Phoebe A. Johnson, Sept. 7, 1957 (div. 1976); children: Julie, Lisa, Steven; m. Nancy M. Sweatt, July 7, 1977. B.S., U. Ill., 1949; M.S., Western Ill. U., 1950; Ph.D., San Gabriel U., 1971. Lic. psychologist, Calif. Psychologist, counselor, tchr. pub. schs., Calif., Ill., 1951-55, cons. sch. psychologist, Orange County, Calif., 1957-59; exec. dir., pres., chmn. bd. Reading Guidance Ctr., Inc., Huntington Beach, Calif., 1955-69; psychologist Abbott Labs., Los Angeles, 1969-70, Orange County Child Guidance Ctr., 1969-70; pres. Richard L. Zweig Assocs., Huntington Beach, 1970-80; pres. Zweig Assocs. div. Skillcorp Pubs., Inc. sub. Highlights for Children, Irvine, Calif., 1980—; asst. prof. (part-time) Calif. State Coll.-Long Beach, 1957-62; lectr. U. Calif.-Santa Barbara, summer 1964. Contbr. articles to profl. jours. Mem. Am. Psychol. Assn., AAAS, Am. Personnel and Guidance Assn., Western Psychol. Assn., Southeastern Psychol. Assn., Phi Delta Kappa, Psi Chi. Lodge: Masons. Subspecialties: Learning; Cognition. Current work: Neuropsychology, programming and retraining. Home: 1217 Palm Ave Huntington Beach CA 92648 Office: Zweig Assocs 20800 Beach Blvd Huntington Beach CA 92648

ZWEIMAN, BURTON, physician, scientist, educator; b. N.Y.C., June 7, 1931; s. Charles and Gertrude (Levine) Z.; m. Claire Traig, Dec. 30, 1962; children: Amy Beth, Diane Susan. A.B., U. Pa., 1952, M.D., 1956. Intern Mt. Sinai Hosp., N.Y.C.; resident in medicine Hosp. U. Pa., Bellevue Hosp. Center, 1957-60; fellow N.Y. U. Sch. Medicine, 1960-61; mem. faculty dept. medicine U. Pa. Sch. Medicine, Phila., 1963—, prof. medicine, chief allergy and immunology sect., 1969—; cons. U.S. Army, NIH; co-chmn. Am. Bd. Allergy and Immunology, 1978-81. Contbr. articles to med. jours. Served with M.C. USNR, 1961-63. Allergy Found. Am. fellow, 1959-61. Fellow Am. Acad. Allergy, A.C.P.; mem. Am. Assn. Immunologists, Am. Fedn. Clin. Research, Phi Beta Kappa, Alpha Omega Alpha. Subspecialties: Allergy; Immunology (medicine). Current work: Immune and inflammatory mechanisms in allergic connective tissue and neurologic disorder. Office: 512 Johnson Pavilion 36th and Hamilton Walk Philadelphia PA 19104

ZWEMER, THOMAS JOHN, dental educator, dentist, orthodontist; b. Mishawaka, Ind., Mar. 23, 1925; s. John Dewy and Ruth C. (Brooks) Z.; m. Betty Johnson, June 30, 1949; children: John Thomas, Stephen James, Carol Ann. Student, Emmanuel Missionary Coll., 1939-43, Atlantic Union Coll., 1946; D.D.S., U. Ill.-Chgo., 1950; M.S.D., Northwestern U., 1954. Diplomate: Am. Bd. Orthodontists. Assoc. prof., dir. Marquette U., Milw., 1950-58; prof., chmn dentistry Loma Linda (Calif.) U., 1958-66; prof. assoc. dean dentistry Med. Coll. Ga., Augusta, 1966—. Guest editor: The Dental Clinics of North America, 1976; editor: Boucher's Clinical Dental Terminology, 3d edit, 1982; sect. editor: Am. Jour Orthodontics, 1982. Trustee Loma Linda U., Calif., 1967-76, So. Coll., Collegedale, Tenn., 1969-82; bd. dirs. United Way, Augusta, Ga., 1978, Health Systems Agy., Augusta, 1979. Served with U.S. Army, 1943-46. Fellow Am. Coll. Dentists (grad. scholar 1952-54, research fellow 1965), Internat. Coll. Dentists; mem. Am. Dental Assn., Eastern Dist. Dental Soc., Ga. Dental Assn. Subspecialties: Health services research; Preventive medicine. Current work: Academic administration; consultant on programmatic accreditation; teaching postdoctoral level cranio-facial growth and development; biomechanics. Office: Med Coll Ga Sch Dentistry Augusta GA 30912

ZWISLOCKI, JOZEF JOHN, communication science educator, researcher; b. Lwow, Poland, Mar. 19, 1922; came to U.S., 1952, naturalized, 1960; s. Tadeus and Helena (Moscicki) Z.; m. Sylvia Goldman, July 11, 1954. Diploma in elec. engring, Fed. Inst. Tech., Zurich, 1944, Sc.D., 1948. Research asst., head Electroacoustic Lab. dept. otolaryngology U. Basel, Switzerland, 1945-51; research fellow Psychoacoustics Lab., Harvard U., 1951-57; mem. faculty Syracuse U. (N.Y.), 1957—, prof. elec. engring., 1962-64, dir. Lab. Sensory Communication, 1963-74, prof. sensory sci., 1974—, profl. spl. edn. communicative disorders, 1982—, dir. Inst. Sensory Research, 1974—; mem. research faculty SUNY Upstate Med. Center, Syracuse, 1961—, research prof., 1967—. Contbr. chpts. to books and articles to profl. jours. NIH grantee, 1962—; recipient Internat. Centro Ricerche e Studi Amplifon Prize, Milan, Italy, 1976. Fellow Acoustical Soc. Am., Am. Speech and Hearing Assn.; mem. Psychonomic Soc., IEEE, AAAS, Internat. Soc. Audiology, Soc. Neurosci., Polish Inst. Arts and Scis., Collegium Oto-Rhino-Laryngologicum Amicitiae Sacrum, Eastern Psychol. Assn., N.Y. Acad. Scis., Assn. Research in Otolaryngology. Subspecialties: Acoustics; Psychophysics. Current work: Mathematical theory and experiments on acoustics and mechanics of the outer, middle and inner ear; auditory instrumentation; phychophysics including signal detection and scaling. Patentee in field. Office: Institute for Sensory Research Syracuse University Syracuse NY 13210 Home: RD 1 Pompey Hollow Rd Cazenovia NY 13035

ZYSKIND, JUDITH WEAVER, biologist, educator, researcher; b. Cin., July 2, 1939; d. Max Correy and Mary Catherine (Landis) Weaver; m. George Zyskind, May 12, 1964 (dec. 1974); children: Aviva Deborah, Joy Esther; m. Douglas Wemp Smith, Aug. 16, 1975. B.S., U. Dayton, 1961; M.S., Iowa State U., 1964, Ph.D. (USPHS predoctoral fellow, 1965-67), 1968. Postdoctoral research assoc. dept. genetics Iowa State U., Ames, 1970-71, vis. asst. prof. dept. genetics, 1971-73, postdoctoral research assoc. dept. biochemistry, 1973-74; NIH postdoctoral research fellow dept., biology U. Calif., San Diego, 1974-77, asst. research biologist, 1977-82; assoc. prof. biology San Diego State U., 1982—; prin. investigator research NIH. Contbr. articles to sci. jours. Mem. Am. Soc. Microbiology, AAAS, Genetics Soc. Am., Sigma Xi. Subspecialties: Genetics and genetic engineering (biology); Molecular biology. Current work: Initiation of DNA replication in bacteria: components and control.

Index

Fields and Subspecialties

Pearson, Lorentz C.
Pellett, Harold Melvin
Peterson, Peter Andrew
Pfahler, Paul Leighton
Phillips, Gregory Conrad
Raju, Namboori Bhaskara
Rick, Charles Madeira, Jr.
Rines, Howard Wayne
Robbins, Marion LeRon
Rockwood, Donald Lee
Rogers, Owen Maurice
Sandhu, Shahbeg Singh
Schoenike, Roland Ernest
Sears, Ernest Robert
Shands, Henry Lee
Shaner, Gregory Ellis
Sinclair, John Henry
Soost, Robert Kenneth
Sprague, George Frederick
Stavely, Joseph Rennie
Steele, Oliver Leon
Stewart, Donald Martin
Theurer, J. Clair
Tisserat, Brent Howard
Townsend, Alden Miller
Waines, John Giles
Wallin, Jack Robb
Wann, Elbert Van
Weaver, James Bode, Jr.
Webster, Terry Richard
Wilfret, Gary Joe
Wilkes, H(ilbert) Garrison
Yungbluth, Thomas Alan

Immunology
Anderson, Robert Simpers
Antczak, Douglas Francis
Bankert, Richard Burton
Barta, Ota
Castello, John Donald
Cinader, Bernhard
Costea, Nicolas Vincent
Farah, Fuad Salim
Gallagher, Michael Terrence
Grogan, James Bigbee
Hill, John Hemmingson
Kasel, Julius Albert
Kleinschuster, Stephen John
Klesius, Phillip Harry
Kochan, Ivan
Ligler, Frances Smith
Medzihradsky, Joseph Ladislas
Melvold, Roger Wayne
Munster, Andrew Michael
Piraino, Frank Francis
Quilligan, James Joseph, Jr.
Quimby, Fred William
Sanford, Barbara Hendrick
Saravis, Calvin Albert
Silverman, Paul Hyman
Solorzano, Robert F.
Stimpfling, Jack Herman
Stoddard, Patricia Ann
Stone, William Harold
Sundsmo, John Sievert
Thomas, Francis T.
Ting, Chou-Chik
Varga, Janos M.
Wagner, Gerald Gale
Wang, An-Chuan
Yang, Tsu-Ju

Integrated pest management
Adkisson, Perry Lee
Aldrich, Richard John
Alexander, Samuel Adam
Amador, Jose Manuel
Barnes, George Lewis
Becker, William Nicholas
Bega, Robert V.
Bergstrom, Gary Carlton
Burnside, Orvin Charles
Carroll, Robert Buck
Drye, Charles Edwin
Engelhard, Arthur William
Froyd, James Donald
Geyer, Lynette Arnason
Gillaspy, James Edward
Glaze, Norman Cline
Gottwald, Tim R.
Graham, Harry Morgan
Hart, Elwood Roy
Hart, Richard Allan
Hartwig, Nathan LeRoy
Hoffmann, James Allen
Hopen, Herbert J.
Hoy, Marjorie Ann
Hulst, David Clark
Jones, Alan Lee
Jones, John Paul
Kempe, Robert Aron
Keppelmann, Frank Alfred
Kliejunas, John Thomas
Koch, Henry George
Kuntz, James Eugene
Lavigne, Robert James
Leigh, Thomas Francis
Lewellen, Robert Thomas
Lewis, Gwynne David
Long, William Henry, III
Manley, Donald Gene
Marion, Daniel Francis
Martinson, Charlie Anton
Mason, Curtis L.
McDaniel, Kirk Cole
McDonald, Geral Irving
McIntyre, Gary Allen
Merrill, William, Jr.
Miller, Terry Dee
Nelson, Dennis Raymond
Nordgaard, John Thomas
Ogawa, Joseph Minoru
Overman, Amegda Jack
Pass, Bobby Clifton
Pienkowski, Robert L(ouis)
Poinar, George Orlo, Jr.
Roberts, Donald Wilson
Robertson, Robert Lafon
Rosenberger, David A.
Ryan, Roger Baker
Saettler, Alfred William
Sayre, Richard Martin
Sill, Webster Harrison, Jr.
Skilling, Darroll Dean
Smith, Richard Stanley, Jr.
Thomas, Claude Earle
Tidwell, Timothy Eugene
Varner, Reed William
Vaughn, James Lloyd
Walker, Harrell Lynn
Wilcoxson, Roy Dell
Wuest, Paul J.
Wyman, Jeffrey Alan
Yeargan, Kenneth Vernon
Zoller, Broc Gerald

Integrated systems modelling and engineering. *See also* **ENVIRONMENTAL SCIENCE, Ecosystems analysis.**
Becking, Rudolf Willem
Bensen, David Warren
Brennan, Robert William
Butcher, John Edward
Erb, Hollis Nancy
Gardner, John C.
Hillel, Daniel
Holst, Robert Weigel
Irvine, Eileen M.
Jordan, Darryl Allyn
Knievel, Daniel Paul
Kydd, George Herman
Landolph, Joseph Richard, Jr.
Lawrence, Gary Wright
Lee, Van Ming
Levine, Gilbert
Medina, Miguel Angel, Jr.
Miller, Wilbur Charles
Nesbitt, Patricia Marie
Rogers, Hugo Homer, Jr.
Scudder, Harvey Israel
Self, Hazzle Lafayette
Spisak, John Francis
Stoner, Martin Franklin
Venette, James Raymond
Waggoner, Paul Edward
Wiese, Maurice Victor

Microbiology. *See* **BIOLOGY.**

Nitrogen fixation
Bishop, Paul Edward
Blevins, Dale Glenn
Brill, Winston J.
Burris, Robert Harza
Craig, Burton MacKay
Currier, William Wesley
Davis, Lawrence Clark
Evans, Harold J.
Hardy, Ralph W.F.
Hirsch, Ann Mary
Humphrey, Ronald De Vere
Imsande, John D.
Pueppke, Steven G.
Rebeiz, Constantin Anis
Shanmugam, Keelnatham Thirunavukkarasu
Tiffney, Wesley Newell, Jr.
Urban, James Edward

Photosynthesis. *See* **BIOLOGY.**

Plant cell and tissue culture. *See also* **BIOLOGY, Cell biology, cell and tissue culture.**
Agrios, George N.
Anderson, James Otto
Bare, Charles Edgar
Barr, Richard Arthur
Carman, John G(riffith)
Chaleff, Roy Scott
Chen, Chen-Ho
Conger, Bob Vernon
Craig, Burton MacKay
Davis, David Gerhardt
Epp, Melvin David
Gamborg, Oluf Lind
Goldfarb, Norman Marc
Gudmestad, Neil Carlton
Hartman, Robert Dale
Hiatt, Thomas Andrew
Hodges, Thomas Kent
Holcomb, Gordon Ernest
Horst, Ralph Kenneth
Huang, Jeng-sheng
Israel, Herbert William
Janick, Jules
Jauhar, Prem Prakash
Jensen, Roy A.
Johnson, John Morris
Joppa, Leonard Robert
Karnosky, David Frank
Kasha, Kenneth John
Krikorian, Abraham Der
Loescher, Wayne Harold
Lurquin, Paul Francis
Magill, Clint William
Mahlberg, Paul Gordon
McCown, Brent Howard
McKenzie, Robert James
Mizicko, John Richard
Obendorf, Ralph Louis
Peng, Ying-shin Christine
Phillips, Gregory Conrad
Preece, John Earl
Radin, David N.
Rines, Howard Wayne
Rogers, Owen Maurice
Ruesink, Albert William
Salters, Grace Heyward
Schaeffer, Gideon W.
Shafi, Muhammad Iqbal
Shuler, Michael Louis
Sjolund, Richard David
Staba, Emil John
Stewart, Donald Martin
Sun, Mike
Tisserat, Brent Howard
Van't Hof, Jack
Vasil, Indra Kumar
Walden, David Burton
Whittier, Dean Page
Wild, James Robert
Wilfret, Gary Joe

Plant pathology
Aist, James Robert
Alcorn, Stanley Marcus
Alexander, Samuel Adam
Allen, Thomas Cort, Jr.
Alves, Leo Manuel
Amador, Jose Manuel
Anagnostakis, Sandra Lee
Anderson, Neil Albert
Arny, Deane Cedric
Baker, Con Jacyn
Baker, Kenneth Frank
Baker, R. Ralph
Barnes, George Lewis
Barrat, Joseph George
Bateman, Durward Franklin
Becker, William Nicholas
Beer, Steven Vincent
Bega, Robert V.
Benson, David Michael
Bergstrom, Gary Carlton
Berry, Charles Richard
Beute, Marvin Kenneth
Bhattacharya, Pradeep K.
Blanchard, Robert Osborn
Bloom, James Richard
Brennan, Eileen G.
Burr, Thomas James
Bush, David Lynn
Bushnell, William Rodgers
Byler, James W.
Cameron, H. Ronald
Carroll, Robert Buck
Caruso, Frank Lawrence
Castello, John Donald
Charudattan, Raghavan
Cheo, Pen Ching
Chou, Liu-Gei
Clark, Raymond Loyd
Coakley, Stella Melugin
Cobb, Fields White, Jr.
Cobb, William Thompson
Conway, Kenneth Edward
Conway, William Scott
Crall, James Monroe
Crane, Joseph Leland
Csinos, Alexander Stephen
Currier, William Wesley
Daniels-Hetrick, Barbara Ann
Davis, David
Davis, Michael Jay
Davis, Robert Gene
Dean, Jack Lemuel
DeVay, James Edson
Diener, Theodor Otto
Diener, Urban Lowell
Dooley, Harrison LeRoy
Douglas, Dexter Richard
Drye, Charles Edwin
Duniway, John M.
Edmonds, Robert Leslie
Emmatty, Davy Alleasu
Engelhard, Arthur William
Eslyn, Wallace Eugene
Essenberg, Margaret Kottke
Feldman, Albert William
Fett, William Frederick
Ford, Richard Earl
Fravel, Deborah R.
French, Alexander Murdoch
Froyd, James Donald
Fulbright, Dennis Wayne
Gardner, Wayne Scott
Garnsey, Stephen Michael
Gillaspie, Athey Graves, Jr.
Goode, Monroe Jack
Gottwald, Tim R.
Gough, Francis Jacob
Gowen, Patricia Elizabeth
Gregory, Garold Fay
Grichar, William James
Hagedorn, Donald James
Haglund, William Arthur
Hale, Maynard George
Halloin, John McDonell
Hampton, Richard Owen
Hancock, Joseph Griscom, Jr.
Hanlin, Richard Thomas
Harman, Gary Elvan
Harrison, Martin Bernard
Hart, John Henderson
Heath, Michele Christine
Heggestad, Howard Edwin
Heidrick, Lee Edward
Helton, Audus Winzle
Herr, Leonard Jay
Hesseltine, Clifford William
Hibben, Craig Rittenhouse
Highley, Terry Leonard
Hilty, James Willard
Himelick, Eugene Bryson
Hinds, Thomas Edward
Ho, Hon Hing
Hoefert, Lynn Lucretia
Hoffmann, James Allen
Holcomb, Gordon Ernest
Hollis, John Percy, Jr.
Holtzmann, Oliver Vincent
Hooker, Arthur Lee
Hooper, Gary Ray
Houck, Laurie Gerald
Houseworth, Lloyd Douglas
Houston, David Royce
Huang, Jeng-sheng
Huber, Don Morgan
Hulst, David Clark
Humaydan, Hasib Shaheen
Hunter, Richard Edmund
Hussey, Richard Sommers
Israel, Herbert William
Jedlinski, Henryk
Johnston, Robert Howard
Jones, Alan Lee
Jones, John Paul
Joshi, Madan Mohan
Keen, Noel Thomas
Kelman, Arthur
Keppelmann, Frank Alfred
Khan, Rudolph A.
Kiesling, Richard Lorin
Kimble, Kenneth Alan
Kingsland, Graydon Chapman
Kliejunas, John Thomas
Knott, Douglas Ronald
Komm, Dean Albert
Kommedahl, Thor
Krausz, Joseph Philip
Kress, Lance Whitaker
Krigsvold, Dale Thomas
Krusberg, Lorin Ronald
Kuc, Joseph
Kulik, Martin Michael
Kuntz, James Eugene
Lacy, George Holcombe
Lai, Carl Mingtan
Lamey, Howard Arthur
Lawrence, Ernest Grey, Jr.
Lawrence, Gary Wright
Leach, Jeanette
Leath, Kenneth Thomas
Leathers, Chester Ray
Lee, Richard Frank
Lee, Thomas A
Lewis, Gwynne David
Lewis, Robert, Jr.
Line, Roland F.
Lister, Richard Malcolm.
Litton, Columbus C.
Livingston, Clark H.
Lockwood, John LeBaron
Lowe, Sunny Ken
Lukezic, Felix Lee
Lumsden, Robert Douglas
Lyda, Stuart Davisson
Mace, Marshall Ellis
MacSwan, Iain Christie
Mai, William Frederick
Marion, Daniel Francis
Marlatt, Robert Bruce
Martinson, Charlie Anton
Mason, Curtis L.
Mathre, Donald Eugene
Matthysse, Ann Gale
Mayol, Pete Syting
McCarter, States Marion
McCool, Patrick Michael
McCoy, Randolph Edward
McCracken, Francis Irvin
McIntyre, Gary Allen
McLaughlin, Michael Ray
McNew, George Lee
Melouk, Hassan Aly
Merrill, William, Jr.
Michell, Richard Edward
Miller, Raymond Michael
Miller, Terry Dee
Millikan, Daniel Franklin, Jr.
Mircetich, Srecko Mirko
Mixson, Wayne Clark
Mizicko, John Richard
Moline, Harold E.
Moody, Arnold Ralph
Moore, John Duain
Moore, Larry Wallace
Moore, Laurence D.
Moore, William Fred
Moorman, Gary Williams
Moseman, John Gustav
Mount, Mark Samuel
Muchovej, James John
Nelson, Earl Edward
Nordgaard, John Thomas
Norton, Don Carlos
Ogawa, Joseph Minoru
Ostazeski, Stanley Anthony
Overman, Amegda Jack
Owens, Charles Wesley
Partyka, Robert Edward
Paxton, Jack Dunmire
Percich, James Angelo
Phipps, Patrick Michael
Politis, Demetrios John
Pueppke, Steven G.
Quinn, James Allen
Raghunathan, Rengachari
Rebois, Raymond Victor
Rich, Avery Edmund
Rich, Jimmy Ray
Riggs, Robert Dale
Roberts, Daniel Altman
Roelfs, Alan Paul
Rogers, Jack David
Rogers, Marlin Norbert
Roistacher, Chester N.
Rosenberger, David A.
Rowe, Randall Charles
Ruppel, Earl George
Saad, Sami Michel
Saettler, Alfred William
Sayre, Richard Martin
Schadler, Daniel Leo
Schafer, John Francis
Scheffer, Robert Paul
Schipper, Arthur Louis, Jr.
Schlegel, David Edward
Schmitt, Donald Peter
Sequeira, Luis
Seravalli, Egilde Paola
Shands, Henry Lee
Shaner, Gregory Ellis
Shea, Keith Raymond
Sheen, Shuh Ji
Shoemaker, Paul Beck, III
Skilling, Darroll Dean
Slack, Derald Allen
Sleesman, John Paul
Smidt, Mary Louise
Smith, Richard Stanley, Jr.
Smith, Samuel Howard
Sommer, Noel Frederick
Spalding, Donald Hood
Spencer, James Alphus
Staley, John Merrill
Stall, Robert Eugene
Staples, Richard Cromwell
Stavely, Joseph Rennie
Steadman, James Robert
Stoner, Martin Franklin
Stuteville, Donald Lee
Sylvia, David Martin
Tattar, Terry Alan
Tessier, Bruce Joseph
Thomas, Claude Earle
Thompson, Samuel Stanley, Jr.
Tidwell, Timothy Eugene
Timian, Roland Gustav
Tsai, James Hsi-cho
Tweedy, Billy G.
Ullrich, Robert Carl
Van Alfen, Neal K.
Van Dyke, Cecil Gerald
Van Etten, Hans D.
Varner, Reed William
Venette, James Raymond
Wadsworth, Dallas Fremont
Waggoner, Paul Edward
Walker, Harrell Lynn
Walker, Jerry Tyler
Wallin, Jack Robb
Wann, Elbert Van
Wargo, Philip Matthew
Warren, Herman Lecil
Waterworth, Howard Eugene
Weaver, Michael John
Weber, Darrell Jack
Weinhold, Albert Raymond
Wene, Edward G.
Whaley, Julian Wendell
Whiteside, Jack Oliver
Wicker, Ed Franklin
Wiese, Maurice Victor
Wilcox, W(ebster) Wayne
Wilcoxson, Roy Dell
Wilhelm, Stephen
Wilson, Charles Lindsay
Wilson, David Merl, Jr.
Windels, Carol Elizabeth
Wu, Jia-Hsi
Wuest, Paul J.
Young, Roy Alton
Zaitlin, Milton
Zehr, Eldon Irvin
Zentmyer, George Aubrey, Jr.
Zettler, Francis William
Zoller, Broc Gerald

Plant physiology. *See also* **BIOLOGY.**
Akers, Stuart William
Arntzen, Charles Joel
Baker, Con Jacyn
Bare, Charles Edgar
Barr, Richard Arthur
Bateman, Durward Franklin
Bechtel, Donald Bruce
Bhattacharya, Pradeep K.
Bhella, Harbans Singh
Blevins, Dale Glenn
Bogorad, Lawrence
Burris, Joseph Stephen
Burton, Glenn Willard
Bushnell, William Rodgers
Caruso, Frank Lawrence
Castelfranco, Paul Alexander
Chaney, William Reynolds
Clayton, Roderick Keener
DeVay, James Edson
Downs, Robert Jack
Duniway, John M.
Ellenson, James L.
Emino, Everett Raymond
Epstein, Emanuel
Evans, Lance Saylor
Feldman, Albert William
Fett, William Frederick
Finkle, Bernard J.
Fisher, Donald Boyd
French, Charles Stacy
Funt, Richard Clair
Gardner, John C.
Gatherum, Gordon Elwood
Glaze, Norman Cline
Gorske, Stanley Francis
Graham, Terrence Lee
Green, Victor Eugene, Jr.
Greenwood, Michael Sargent
Gull, Dwain D.
Hageman, Richard Harry
Hale, Maynard George
Halloin, John McDonell
Hanson, Kenneth Warren
Hardy, Ralph W.F.
Hatzios, Kriton Kleanthis
Heck, Walter Webb
Heidrick, Lee Edward
Helton, Audus Winzle
Hess, Charles E.
Hodges, Thomas Kent
Hodgson, Richard Holmes
Holst, Robert Weigel
Holt, Donald Alexander
Homann, Peter Hinrich Fritz
Houck, Laurie Gerald
Hrazdina, Geza
Huber, Donald John
Hurkman, William James, II
Ingle, L. Morris
Jensen, Richard Grant
Jensen, Roy A.
Johnson, Virgil Allen
Kawase, Makoto

Knievel, Daniel Paul
Koontz, Harold Vivien
Kozlowski, Theodore Thomas
Kramer, Paul Jackson
Krikorian, Abraham Der
Kurtzman, Ralph Harold, Jr.
Laing, Frederick Mitchell
Lambeth, Victor Neal
Larson, Philip Rodney
Lee, Thomas A
Leone, Ida A.
Lewis, Archie Jefferson, III
Li, Paul Hsiang
McCown, Brent Howard
McNew, George Lee
Moore, Laurence D.
Obendorf, Ralph Louis
Ohki, Kenneth
Partyka, Robert Edward
Pellett, Harold Melvin
Peterson, David Maurice
Proebsting, Edward Louis
Roberts, Bruce Roger
Rogers, Marlin Norbert
Rouse, Roy Dennis
Schipper, Arthur Louis, Jr.
Schonbeck, Mark Walter
Sequeira, Luis
Shortle, Walter Charles
Simpson, Frederick James
Sjolund, Richard David
Stebbins, George Ledyard
Sung, Zinmay Renee
Synder, Freeman Woodrow
Tessier, Bruce Joseph
Vanderhoef, Larry Neil
Wargo, Philip Matthew
Weber, Evelyn Joyce
Welch, Ross Maynard
Wergin, William P.
Wilkinson, Robert Eugene
Wittwer, Sylvan Harold
Yatsu, Lawrence Yoneo

Plant virology
Agrios, George N.
Allen, Thomas Cort, Jr.
Barrat, Joseph George
Brakke, Myron Kendall
Cameron, H. Ronald
Cheo, Pen Ching
Christie, Stephen Rolland
Clark, Raymond Loyd
Diener, Theodor Otto
Edwardson, John Richard
Ford, Richard Earl
Gardner, Wayne Scott
Garnsey, Stephen Michael
Gillaspie, Athey Graves, Jr.
Goodman, Robert Merwin
Gudmestad, Neil Carlton
Hadidi, Ahmed Fahmy
Hampton, Richard Owen
Hartman, Robert Dale
Hill, John Hemmingson
Hoefert, Lynn Lucretia
Hooper, Gary Ray
Horst, Ralph Kenneth
Howarth, Alan Jack
Hsu, Hei-ti
Jedlinski, Henryk
Kimble, Kenneth Alan
Lee, Richard Frank
Lister, Richard Malcolm.
Livingston, Clark H.
McLaughlin, Michael Ray
Millikan, Daniel Franklin, Jr.
Moline, Harold E.
Nuss, Donald Lee
Roberts, Daniel Altman
Roistacher, Chester N.
Ruppel, Earl George
Schlegel, David Edward
Shepherd, Robert James
Siegel, Albert
Sill, Webster Harrison, Jr.
Smith, Samuel Howard
Sun, Mike
Timian, Roland Gustav
Tolin, Sue Ann
Tsai, James Hsi-cho
Wu, Jia-Hsi
Zettler, Francis William
Zitter, Thomas Andrew

Resource conservation
Becker, Donald August
Berry, Charles Richard
Bradley, Katharine Tryon
Coate, Barrie Douglas
Dahl, Billie Eugene
Davis, John Rowland
Dwinell, Lew David
Enk, Gordon A.
Finkle, Bernard J.
Flikke, Arnold Maurice
Granger, Clark Allen
Grover, John Harris
Hellmann, Robert A.
Johnson, Carl Maurice
Kenney, Dennis Raymond
Lee, Harry William
Ligon, James Teddie
Linn, D. Wayne
Nickerson, Norton Hart
Palmer, Melville Louis
Parr, James Floyd, Jr.
Ramig, Robert Ernest
Ross, Henry A.
Triplett, Glover Brown, Jr.
Wellner, Charles August
Wicker, Ed Franklin
Wrobley, Arthur Ray

Soil chemistry
Bennett, William F., Sr.
Bhella, Harbans Singh
Bloom, Paul Ronald
Cheng, H(wei)-H (sien)
Grunes, David Leon
Gupta, Gian Chand
Harter, Robert Duane
Himes, Frank Lawrence
Holmes, Neal J.
Kenney, Dennis Raymond
Logan, Terry James
Lunt, Owen Raynal
Lyle, Everett Samuel, Jr.
McFee, William Warren
Olsen, Ralph A.
Parr, James Floyd, Jr.
Perkins, Henry Frank
Rouse, Roy Dennis
Rue, Rolland Ray
Sanchez, Pedro Antonio
Wildung, Raymond Earl

Solar energy. ***See*** **ENERGY SCIENCE.**

Toxicology
Ahmed, Farid El Mamoun
Baskin, Steven Ivan
Bass, Eugene Lawrence
Bell, Marvin Carl
Bennett, Jesse Harland
Cutkomp, Laurence Kremer
Dahl, Alan Richard
Davis, Brian Kent
Falk, Hans Ludwig
Floyd, Robert A.
Freed, Virgil Haven
Giri, Shri N.
Grosch, Daniel Swartwood
Guthrie, Frank Edwin
Hsia, Mong-Tseng S.
Kilgore, Wendell Warren
Klicka, John Kenneth
Lue, Louis Ping-Sion
Megel, Herbert
Mehendale, Harihara Mahadeva
Moldenke, Alison Feerick
Nigg, Herbert Nicholas
Polin, Donald
Pollock, Gerald Arthur
Salem, Harry
Schiefer, Hans Bruno
Schneider, Norman Richard
Singh, Harpal
Stahr, Henry Michael
Stone, William Lawrence
Taylor, Steve Lloyd
Tu, Anthony Tsuchien
West, Bob
Wilson, Benjamin James
Yu, Simon Shyi-Jian
Zaratzian, Virginia Louise

Other
Adams, John Lester, Jr., *Animal husbandry*
Aldrich, Jeffrey Richard, *Entomology*
Alm, Alvin Arthur, *Forestry*
Andrews, Luther David, *Poultry management*
Armistead, Willis William, *Administration, agricultural research*
Bentley, Orville George, *Agricultural research administration.*
Bloom, James Richard, *Nematology*
Carlquist, Sherwin, *Plant anatomy*
Charudattan, Raghavan, *Weed science*
Chitwood, David Joseph, *Plant nematology*
Cobb, William Thompson, *Weed science*
Cooper, Arthur Wells, *Forestry*
Daniels, Raymond Bryant, *Soil science*
Dregne, Harold Ernest, *Soil science*
Eisen, Eugene J., *Animal breeding*
Erickson, Eric Herman, Jr., *Apiculture, Crop Pollination*
Everhart, Watson Harry, *Fishery Biology*
Faust, Miklos, *Pomology*
Fravel, Deborah R., *Biocontrol*
Gardner, Charles Olda, *Plant breeding*
Gorske, Stanley Francis, *Weed science*
Green, Victor Eugene, Jr., *Sunflower crop production*
Grichar, William James, *Weed control*
Hess, Charles E., *Horticulture*
Hills, Frederick Jackson, *Agronomy*
Hole, Francis Doan, *Soil science*
Honma, Shigemi
Howell, Robert Wayne, *Agronomy*
Janski, Alvin Michael, *Animal products*
Johnson, George Robert, *Animal Science Administration*
Johnson, Jack Donald, *Arid lands agriculture*
Jordan, Darryl Allyn, *Appropriate agriculture*
Kendrick, James Blair, Jr., *Agricultural research administration*
Kingsland, Graydon Chapman, *Tropical agriculture*
Kinney, Terry B., Jr., *Animal research administration*
Kommedahl, Thor, *Biological control*
Koontz, Harold Vivien
Kramer, Paul Jackson, *Physiology of plant stress*
Laben, Robert Cochrane, *Animal production*
Leiser, Andrew Twohy, *Ornamental horticulture*
Lucey, Robert Francis, *Agronomy*
McCool, Patrick Michael, *Air pollution effects on plants*
Michell, Richard Edward, *Plant nematology*
Minvielle, Francis Paul Georges, *Quantitative genetics*
Mixson, Wayne Clark, *Turf management*
Muniappan, Rangaswamy Naicker, *Biological Control*
Niedbalski, Joseph Francis, *Agricultural chemicals*
Preece, John Earl
Proebsting, Edward Louis, *Pomology*
Putnam, Paul A(din), *Agricultural research adminstration*
Rempel, William Ewert, *animal breeding*
Rich, Jimmy Ray, *Nematology*
Ross, Mary Harvey, *Entomology*
Rothschild, Brian James, *Fishery management*
Ryan, James Bernard, *Agricultural chemicals development, Weed science*
Ryder, Edward Jonas, *Plant breeding*
Saladini, John Louis, *Agrichemicals*
Schmidt, Berlie Louis, *Agronomy, Soil science*
Smith, Thomas Lloyd, *Urban forest resource management*
Sommer, Noel Frederick, *Postharvest pathology*
Swakon, Doreen H.D., *Agronomy-Forage Crops*
Thomson, George Willis, *Forestry*
Triplett, Glover Brown, Jr., *Agronomy*
VanDemark, Noland Leroy, *Research administration*
Waterworth, Howard Eugene, *Nematology*
Weller, Milton Webster, *Animal ecology*
Wheeler, Alfred George, Jr., *Entomology*

ASTRONOMY. *See also* SPACE SCIENCE.

Cosmology
Aaronson, Marc
Bahcall, Neta Assaf
Blumenthal, George Ray
Burbidge, Geoffrey
Chaisson, Eric Joseph
Chincarini, Guido Ludovico
Emslie, A. Gordon
Faber, Sandra Moore
Fox, Kenneth
Fry, James N.
Gaposchkin, Peter John Arthur
Gatrousis, Christopher
Gott, John Richard, III
Groth, Edward John, III
Gudehus, Donald Henry
Gunn, James Edward
Guth, Alan Harvey
Hegyi, Dennis Jerome
Helou, George
Hickson, Paul
Hively, Ray Michael
Huchra, John Peter
Hut, Piet
Kaplan, Irving Eugene
Kierein, John
Kirshner, Robert Paul
Kristian, Jerome
Langacker, Paul George
Lecar, Myron
Markus, Lawrence
Mather, John Cromwell
McGraw, John Thomas
Moody, Elizabeth Anne
Morabito, David Dominic
Nanos, George Peter, Jr.
Noonan, Thomas Wyatt
Pagels, Heinz Rudolf
Partridge, Robert Bruce
Penzias, Arno Allan
Press, William Henry
Ramond, Pierre M.
Richstone, Douglas Orange
Roeder, Robert Charles
Sagan, Carl Edward
Sandage, Allan Rex
Sargent, Wallace Leslie William
Schild, Rudolph Ernest
Schramm, David Norman
Segal, Irving Ezra
Seldner, Michael
Smith, Harding Eugene
Soneira, Raymond M.
Stecker, Floyd William
Steigman, Gary
Thiemens, Mark H.
Thuan, Trinh Xuan
Tifft, William Grant
Tipler, Frank Jennings, III
Tryon, Edward Polk
Tully, Richard Brent
Turner, Edwin Lewis
Walstad, Allan M(artin)
Wilczek, Frank Anthony
Wolfe, Arthur Michael

General relativity
Brecher, Kenneth
Chandrasekhar, Subrahmanyan
Detweiler, Steven Lawrence
Gott, John Richard, III
Hacyan, Shahen
Hill, Henry Allen
Laubscher, Roy Edward
Page, Don Nelson
Roeder, Robert Charles
Teukolsky, Saul Arno
Tipler, Frank Jennings, III
Wiita, Paul Joseph

High energy astrophysics
Basinska-Lewin, Ewa Maria
Baum, Peter Joseph
Baumert, John Henry
Brecher, Kenneth
Bridge, Herbert Sage
Bunner, Alan Newton
Burbidge, Geoffrey
Burns, Jack O'Neal, jr.
Conner, Jerry Power
Cordes, James Martin
Dennison, Brian Kenneth
Fichtel, Carl Edwin
Frank, Louis Albert
Friedman, Herbert
Garmire, Gordon Paul
Hegyi, Dennis Jerome
Henriksen, Richard Norman
Holt, Stephen S.
Katz, Jonathan Isaac
Kraushaar, William Lester
Kronberg, Philipp Paul
Lin, Robert Peichung
Meekins, John Fred
Morrison, Philip
Murray, Stephen S.
Ormes, Jonathan F.
Osterbrock, Donald Edward
Ostriker, Jeremiah Paul
Phillips, Robert Boone
Rose, William Kenneth
Sarazin, Craig Leigh
Shapiro, Maurice Mandel
Starrfield, Sumner Grosby
Straka, William Charles
Terrell, James, (Jr.)
Terzian, Yervant
Tucker, Wallace Hampton
Weiler, Kurt Walter
Weisskopf, Martin Charles
Wheeler, J(ohn) Craig
Willcox, Phillip James
Wilson, Andrew Stephen
Wilson, Thomas Leon
Winkler, Paul Frank, Jr.
Woosley, Stanford Earl

High energy astrophysics, cosmic ray
Adams, James Hall, Jr.
Ahluwalia, Harjit Singh
Balasubrahmanyan, Vriddhachalam Krishnaswamy
Chasson, Robert Lee
Doolittle, Robert F., II
Fireman, Edward Leonard
Greisen, Kenneth Ingvard
Hadler, Herbert Isaac
Israel, Martin Henry
Jokipii, Jack Randolph
Kobetich, Edward John
Leighton, Robert Benjamin
Linsley, John (David)
Marshak, Marvin Lloyd
Meyer, Peter
Ormes, Jonathan F.
Perez-Peraza, Jorge A.
Price, Paul Buford
Rossi, Bruno
Shapiro, Maurice Mandel
Silberberg, Rein
Stone, Edward Carroll, Jr.
Van Allen, James Alfred
Vogt, Rochus Eugen
Waddington, Cecil Jacob

High energy astrophysics, gamma ray
Cline, Thomas Lytton
Doolittle, Robert F., II
Fichtel, Carl Edwin
Fishman, Gerald Jay
Gehrels, Neil
Greisen, Kenneth Ingvard
Gruber, Duane Edward
Jacobson, Allan Stanley
Klebesadel, Ray William
Leventhal, Marvin
MacCallum, Crawford John
Metzger, Albert Emanuel
Thompson, David John
Ulmer, Melville Paul
Vogt, Rochus Eugen

High energy astrophysics, ultraviolet
Augensen, Harry John
Bolton, Charles Thomas
Brugel, Edward William
Cassinelli, Joseph Patrick
Cordova, France Anne-Dominic
Davidsen, Arthur Falnes
Davis, Robert James
Eaton, Joel A.
Ferland, Gary Joseph
Green, Richard Frederick
Gull, Theodore Raymond
Harrington, J(ames) Patrick
Hathaway, William Howard
Hobbs, Lewis Mankin
Holberg, Jay Brian
Hutchings, John Barrie
Johnson, Hollis Ralph
Kondo, Yoji
Kuhi, Leonard Vello
Linsky, Jeffrey Lawrence
Maran, Stephen Paul
Mariska, John Thomas
McIlwain, Carl Edwin
Michalitsianos, Andrew Gerasimos
Moos, Henry Warren
Olson, Gordon Lee
Paresce, Francesco
Plavec, Mirek Josef
Shipman, Harry Longfellow
Smith, Peter Lloyd
Walborn, Nolan Revere
York, Donald Gilbert

High energy astrophysics, X-ray
Ayasli, Serpil
Bahcall, John Norris
Basinska-Lewin, Ewa Maria
Becker, Robert Howard
Blake, Richard Lee
Bleach, Richard David
Boldt, Elihu, A.
Bradt, Hale Van Dorn
Buff, James Steve
Canizares, Claude Roger
Clark, George Whipple
Cordova, France Anne-Dominic
Davis, John Moulton
Dolan, Joseph Francis
Feigelson, Eric Dennis
Fried, Peter Marc
Garmire, Gordon Paul
Giacconi, Riccardo
Gorenstein, Paul
Griffiths, Richard Edwin
Gruber, Duane Edward
Haisch, Bernhard Michael
Helfand, David John
Henry, J. Patrick
Holt, Stephen S.
Jacobson, Allan Stanley
Kalata, Kenneth
Kylafis, Nikolaos Dimitriou
Leahy, Denis Alan
Lewin, Walter H(endrik) G(ustav)
Liller, William
Malina, Roger Frank
Margon, Bruce H.
Mastronardi, Richard
Meekins, John Fred
Murray, Stephen S.
Pradhan, Anil Kumar
Pravdo, Steven H.
Priedhorsky, William Charles
Sanders, Wilton Turner, III
Serlemitsos, Peter J.
Tananbaum, Harvey Dale
Tucker, Wallace Hampton
Ulmer, Melville Paul
Vanspeybroeck, Leon Paul
Weiss, Kay
Weisskopf, Martin Charles
Zombeck, Martin Vincent

Optical astronomy
Adelman, Saul Joseph
Africano, John Louis
Aikman, George Christopher Lawrence
Aller, Lawrence Hugh
Anderson, Kurt Steven Jarl
Anderson, Per Holme
Augensen, Harry John
Babcock, Hope Madeline
Bahcall, Neta Assaf
Bahng, John Deuck Ryong
Barker, Timothy
Baumert, John Henry
Beckers, Jacques Maurice
Benedict, George Frederick
Blanco, Victor Manuel
Boatright, Tony James
Boeshaar, Patricia Chikotas
Bolton, Charles Thomas
Bradt, Hale Van Dorn
Breakiron, Lee Allen
Brugel, Edward William
Burbidge, E. Margaret
Butcher, Harvey Raymond, III
Canizares, Claude Roger
Cardona, Octavio, *Optical astronomy*
Carney, Bruce William
Chincarini, Guido Ludovico
Christensen, Clark Gardner
Christian, Carol Ann
Clarke, John Terrel
Clements, Gregory Leland
Cohen, Judith Gamora
Condal, Alfonso Ramon
Connolly, Leo Paul
Corwin, Harold Glenn, Jr.
Cowie, Lennox Lauchlan
Coyne, George Vincent
Crawford, David Livingston
Cudworth, Kyle McCabe
Daehler, Mark
Davis, Robert James
Dick, Steven James
Doherty, Lowell Ralph
Dolan, Joseph Francis
Dorren, John David
Dressler, Alan
Dube, Roger Raymond
Dufour, Reginald James
DuPuy, David Lorraine
Eaton, Joel A.
Elliot, James Ludlow
Elmegreen, Debra Anne Meloy
Evans, David Stanley
Evans, Nancy Remage
Faber, Sandra Moore
Federman, Steven Robert
Fisher, Richard Royal
Fleischer, Robert
Fliegel, Henry Frederick
Flower, Phillip John
Fontaine, Gilles Joseph
Fredrick, Laurence William
Frogel, Jay Albert
Garfinkel, Boris
Garrison, Robert Frederick
Genet, Russell Merle
Giovanelli, Riccardo
Goodell, John Boyden
Green, Richard Frederick
Greenstein, Jesse Leonard
Groth, Edward John, III

Gudehus, Donald Henry
Guetter, Harry Hendrik
Gull, Theodore Raymond
Hardorp, Johannes C.
Harris, Alan William
Havlen, Robert James
Heckman, Timothy Martin
Heiser, Arnold Melvin
Hemenway, Mary Kay
Henry, J. Patrick
Herr, Richard Baessler
Hesser, James E(dward)
Hewitt, Anthony Victor
Hickson, Paul
Hildner, Ernest Gotthold, III
Hiltner, William Albert
Hoag, Arthur Allen
Hobbs, Lewis Mankin
Hoffleit, Ellen Dorrit
Howard, Sethanne
Hube, Douglas Peter
Humphreys, Roberta Marie
Hutchings, John Barrie
Ianna, Philip A.
Irvine, Cynthia Emberson
Jenkins, Edward Beynon
Jones, Dayton Loren
Kaler, James B.
King, Ivan Robert
Kirshner, Robert Paul
Kormendy, John
Kowal, Charles Thomas
Kreidl, Tobias Joachim
Kristian, Jerome
Kuhi, Leonard Vello
Landstreet, John Darlington
Lanning, Howard Hugh
Lester, John Bernard
Liller, Martha H(azen)
Liller, William
Lindenblad, Irving Werner
Lippincott, Sarah Lee
Livingston, William Charles
Lutz, Thomas Edward
Mallama, Anthony
Maran, Stephen Paul
Margon, Bruce H.
Martins, Donald Henry
Matsushima, Satoshi
Mattei, Janet Akyuz
McCollough, Michael Leon
McElroy, Douglas Boyden
McGraw, John Thomas
Mc Millan, Robert Scott
Meinel, Aden Baker
Michalitsianos, Andrew Gerasimos
Millikan, Allan Grosvenor
Mink, Douglas John
Moffat, Anthony Frederick John
Moffett, Thomas Joseph
Mohler, Orren Cuthbert
Moore, Elliott Paul
Newburn, Ray Leon, Jr.
Newkirk, Gordon Allen, Jr.
Newsom, Gerald Higley
Odell, Andrew Paul
Oliver, John Parker
Olson, Edward Cooper
Osborn, Wayne Henry
Osterbrock, Donald Edward
Papp, Kim Alexander
Percy, John Rees
Perry, Charles Lewis
Pesch, Peter
Peterson, Bradley Michael
Peterson, Charles John
Poland, Arthur Ira
Pravdo, Steven H.
Prince, Helen Dodson
Sandage, Allan Rex
Sanders, Walter L.
Sargent, Wallace Leslie William
Schild, Rudolph Ernest
Schmidt, Edward George
Schoening, William Edward
Schweizer, Francois
Seidelmann, P. Kenneth
Shao, Cheng-Yuan
Shields, Gregory Alan
Shipman, Harry Longfellow
Siegel, Michael Jason
Smith, Harlan James
Smith, Horace Alden
Smith, Myron Arthur
Soderblom, David Robert
Spitzer, Lyman, Jr.
Stachnik, Robert Victor
Stoeckley, Thomas Robert
Sumners, Carolyn Taylor
Tandberg-Hanssen, Einar Andreas
Tedesco, Edward Francis
Teske, Richard Glenn
Thomas, Norman Gene
Thonnard, Norbert
Tifft, William Grant
Toller, Gary Neil
Trimble, Virginia Louise
Tully, Richard Brent
Turner, Edwin Lewis
Underhill, Anne Barbara
Upgren, Arthur Reinhold, Jr.
van den Bergh, Sidney
Vilkki, Erkki Uuno
Vogt, Steven Scott
Walborn, Nolan Revere
Wallerstein, George
Warren, Wayne Hutchinson, Jr.
Weedman, Daniel Wilson
Wehinger, Peter Augustus
Weinberg, Jerry Lloyd
Weiss, Kay
Wesselink, Adriaan Jan
Whipple, Fred Lawrence
White, Raymond Edwin, Jr.
White, Richard Edward
Wilkerson, M. Susan
Wilson, Andrew Stephen
Wilson, Olin C(haddock)
Wilson, Raymond Hiram, Jr.
Winkler, Paul Frank, Jr.
Wolff, Sidney Carne
Woodward, Edith J.
York, Donald Gilbert
Young, Andrew Tipton

Optical astronomy, infrared

Aaronson, Marc
Bartel, Norbert Harald
Bell, Roger Alistair
Briotta, Daniel A., Jr.
Carney, Bruce William
Caroff, Lawrence John
Clark, Thomas Alan
Cohen, Judith Gamora
Cohen, Martin
Craine, Eric Richard
Cruikshank, Dale Paul
Cudaback, David Dill
Daehler, Mark
Eddy, John Allen
Edwards, Suzan
Evans, Neal John, II
Fay, Theodore Denis, Jr.
Frogel, Jay Albert
Geballe, Thomas Ronald
Gehrz, Robert Douglas
Gezari, Daniel Ysa
Goorvitch, David
Grasdalen, Gary Lars
Guetter, Harry Hendrik
Harper, Doyal Alexander, Jr.
Harvey, Paul M.
Huchra, John Peter
Humphreys, Roberta Marie
Johnston, Kenneth J.
Knacke, Roger Fritz
Kostiuk, Theodor
Krassner, Jerry
Krisciunas, Kevin
Lada, Charles Joseph
Lebofsky, Larry Allen
Leighton, Robert Benjamin
Lo, Kwok-Yung
Low, Frank James
Lutz, Barry Lafean
Martin, Terry Zachry
Mather, John Cromwell
Miner, Ellis Devere, Jr.
Morrison, David
Neugebauer, Gerry
Ney, Edward Purdy
Price, R(ichard) Marcus
Russell, Ray William
Sargent, Anneila Isabel
Shivanandan, Kandiah
Shoore, Joseph David
Smith, Harding Eugene
Smith, Howard Alan
Telesco, Charles Michael
Thompson, Rodger Irwin
Thuan, Trinh Xuan
Tokunaga, Alan Takashi
Townes, Charles Hard
Traub, Wesley Arthur
Vrba, Frederick John
Walker, Russell Glenn
Widey, Robert LeRoy
Willner, Steven Paul
Wing, Robert Farquhar
Witteborn, Fred Carl
Woolf, Neville John
Wooten, Henry Alwyn

Planetary science

Anders, Edward
Appleby, John Frederick
Apt, Jerome, III
Arvidson, Raymond Ernst
Backus, George Edward
Benner, Drayton Chris
Bering, Edgar Andrew, III
Black, David Charles
Blanford, George Emmanuel, Jr.
Boss, Alan Paul
Brownlee, Donald Eugene, II
Brunk, William Edward
Burgess, Eric
Burke, Bernard Flood
Burns, Joseph Arthur
Cameron, Alastair Graham Walter
Clarke, John Terrel
Cohen, Marshall Harris
Cordell, Bruce Monteith
Counselman, Charles Claude, III
Cowan, John James
Cruikshank, Dale Paul
Davies, Merton Edward
Donahue, Thomas Michael
Edberg, Stephen J.
Elachi, Charles
El-Baz, Farouk
Elliot, James Ludlow
Eshleman, Von Russel
Esposito, Larry Wayne
Flasar, F. Michael
Gehrels, Tom
Goldreich, Peter Martin
Gradie, Jonathan Carey
Gustafson, Bo Ake Sture
Hapke, Bruce William
Harris, Alan William
Hartung, Jack Burdair
Heeschen, David Sutphin
Herbert, Floyd Leigh
Holberg, Jay Brian
Hubbard, William Bogel, Jr.
Irving, Donald J.
James, Philip Benjamin
Khare, Bishun Narain
Klein, Michael John
Kliore, Arvydas Joseph
Knacke, Roger Fritz
Krimigis, Stamatios Mike
Lambert, John Vincent
Lebofsky, Larry Allen
Lewis, John Simpson, Jr.
Lutz, Barry Lafean
Martin, Leonard James
Martin, Terry Zachry
McConnell, John Charles
Mc Cord, Thomas Bard
McDonough, Thomas Redmond
McIlwain, Carl Edwin
Metzger, Albert Emanuel
Miner, Ellis Devere, Jr.
Mink, Douglas John
Misconi, Nebil Yousif
Morrison, David
Nelson, Robert M.
Newburn, Ray Leon, Jr.
Owen, Tobias Chant
Peale, Stanton Jerrold
Penzias, Arno Allan
Pettengill, Gordon H(emenway)
Pleskot, Larry Kenneth
Pollack, James Barney
Reasenberg, Robert David
Reiff, Patricia Hofer
Reitsema, Harold James
Reynolds, John Hamilton
Reynolds, Ray Thomas
Russell, Christopher Thomas
Sagan, Carl Edward
Sandel, Bill Roy
Seidelmann, P. Kenneth
Sekanina, Zdenek
Shao, Cheng-Yuan
Shoemaker, Eugene Merle
Simpson, Richard Allan
Smith, Harlan James
Standish, Erland Myles, Jr.
Stevenson, David John
Stone, Edward Carroll, Jr.
Strom, Robert Gregson
Swenson, George Warner, Jr.
Tedesco, Edward Francis
Thomas, Norman Gene
Tokunaga, Alan Takashi
Warwick, James Walter
Weissman, Paul Robert
Welch, William John
Wells, Ronald Allen
West, Robert Alan
Whitaker, Ewen Adair
Wilkening, Laurel Lynn
Williams, Donald John
Witteborn, Fred Carl
Yeomans, Donald Keith
Young, Andrew Tipton
Zisk, Stanley Harris

Radio and microwave astronomy

Backer, Donald Charles
Ball, John Allen
Bally, John
Bartel, Norbert Harald
Bash, Frank Ness
Becker, Robert Howard
Blitz, Leo
Bowers, Phillip Frederick
Broderick, John Joseph
Burns, Jack O'Neal, jr.
Burton, William Butler
Carter, William Eugene
Chaisson, Eric Joseph
Clark, Frank Oliver
Cohen, Martin
Cohen, Richard S.
Condon, James Justin
Cordes, James Martin
Counselman, Charles Claude, III
Cudaback, David Dill
Damashek, Marc
Dennison, Brian Kenneth
Dickey, John Miller
Dickman, Robert Laurence
Donivan, Frank Forbes, Jr.
Drake, Frank Donald
Edwards, Suzan
Ekers, Ronald David
Elitzur, Moshe
Elmegreen, Bruce Gordon
Elmegreen, Debra Anne Meloy
Evans, Neal John, II
Ewing, Martin Sipple
Feigelson, Eric Dennis
Fleischer, Robert
Galt, John Alexander
Giovanelli, Riccardo
Gold, Thomas
Gordon, Mark Aitken
Gottesman, Stephen Thancy
Haddock, Frederick Theodore, Jr.
Hankins, Timothy Hamilton
Hansen, Stanley Severin, II
Harvey, Paul M.
Havlen, Robert James
Haynes, Martha Patricia
Heckman, Timothy Martin
Helfand, David John
Herbst, Eric
Higgs, Lloyd Albert
Hjellming, Robert Michael
Hogg, David Edward
Hollis, Jan Michael
Howard, William Eager, III
Hughes, Victor Augustine
Hurford, Gordon James
Irving, Donald J.
Jackson, Bernard Vernon
Johnson, Donald Rex
Johnson, Douglas William
Jones, Dayton Loren
Kellermann, Kenneth Irwin
Kierein, John
Klein, Michael John
Kliore, Arvydas Joseph
Knowles, Stephen Howard
Kronberg, Philipp Paul
Kuiper, Thomas Bernardus Henricus
Kundu, Mukul R.
Kutner, Marc Leslie
Kwok, Sun
Kylafis, Nikolaos Dimitriou
Lada, Charles Joseph
Langer, William David
Leung, Chun Ming
Lewis, Brian Murray
Linfield, Roger Paul
Lo, Kwok-Yung
Marsh, Kenneth Albert
Matsakis, Demetrios Nicholas
Meeks, Marion Littleton
Mirabel, Igor Felix
Morabito, David Dominic
Morris, Mark Root
Moseley, Gerard Franklin
Myers, Philip Cherdak
Newell, Robert Terry
Oliver, Bernard More
Papagiannis, Michael D.
Parrish, Alan deSchweinitz
Partridge, Robert Bruce
Phillips, Robert Boone
Price, R(ichard) Marcus
Rankin, Joanna Marie
Roberts, Morton Spitz
Rodriguez, Luis Felipe
Rubin, Robert Howard
Salpeter, Edwin Ernest
Sanders, Walter L.
Schmidt, Maarten
Schupler, Bruce Ralph
Schwartz, Philip Raymond
Seling, Theodore Victor
Shapiro, Irwin Ira
Shuter, William Leslie Hazlewood
Sullivan, Woodruff Turner, III
Swanson, Paul Norman
Terzian, Yervant
Thieman, James Richard
Thompson, Anthony Richard
Thonnard, Norbert
Tyler, George Leonard
Unwin, Stephen Charles
Webber, John Clinton
Weedman, Daniel Wilson
Weiler, Kurt Walter
Welch, William John
Wilson, Robert Woodrow
Wilson, William John
Wooten, Henry Alwyn
Wright, Melvyn Charles Harman
Zeissig, Gustave Alexander

Theoretical astrophysics

Anderson, Kurt Steven Jarl
Anderson, Per Holme
Arons, Jonathan
Ayasli, Serpil
Barker, Timothy
Barnes, Aaron
Bash, Frank Ness
Becker, Stephen Allan
Bell, Roger Alistair
Bieniek, Ronald James
Black, David Charles
Blandford, Roger David
Blumenthal, George Ray
Bodenheimer, Peter Herman
Boss, Alan Paul
Bregman, Joel Norman
Brunish, Wendee M.
Cameron, Alastair Graham Walter
Cardona, Octavio
Caroff, Lawrence John
Cassinelli, Joseph Patrick
Chandrasekhar, Subrahmanyan
Chanmugam, Ganesar
Cheng, Andrew Francis
Comins, Neil Francis
Cowan, John James
Cowie, Lennox Lauchlan
Cox, Arthur Nelson
Dalgarno, Alexander
Deupree, Robert Gaston
DeYoung, David Spencer
Dickey, John Miller
Edwards, Terry Winslow
Elitzur, Moshe
Elmegreen, Bruce Gordon
Federman, Steven Robert
Ferland, Gary Joseph
Flannery, Brian Paul
Flower, Phillip John
Fontaine, Gilles Joseph
Fry, James N.
Gelman, Donald
Gerola, Humberto C.
Gilliland, Ronald Lynn
Goldberg, Leo
Goode, Philip Ranson
Greenstein, Jesse Leonard
Haisch, Bernhard Michael
Hansen, Carl John
Harrington, J(ames) Patrick
Hathaway, David Henry
Held, Ronald Dennis
Henriksen, Richard Norman
Hills, Jack Gilbert
Hinata, Satoshi
Hjellming, Robert Michael
Holman, Gordon Dean
Holt, Alan Craig
Hut, Piet
Iben, Icko, Jr.
Johnson, Hollis Ralph
Jokipii, Jack Randolph
Jones, Eric Manning
Katz, Jonathan Isaac
Kaufman, Michele
Kirkpatrick, Ronald Crecelius
Krolik, Julian Henry
Kwok, Sun
Kyrala, Ali
Lamb, Donald Quincy
Larson, Richard Bondo
Leahy, Denis Alan
Lecar, Myron
Lee, Martin Alan
Leung, Chun Ming
Levy, Eugene Howard
Lewis, Brian Murray
Linsky, Jeffrey Lawrence
Littleton, John Edward
Low, Boon-Chye
Marks, Dennis William
Matsushima, Satoshi
Max, Claire Ellen
Mazurek, Thaddeus John
McCluskey, George Eadon, Jr.
McCollough, Michael Leon
McCray, Richard Alan
McDonald, Keith Leon
McKee, Christopher Fulton
Merritt, David Roy
Mihalas, Dimitri Manuel
Miller, Glenn Edward
Mitalas, Romas
Morris, Mark Root
Newell, Robert Terry
Nickas, George Demosthenes
Odell, Andrew Paul
Olson, Gordon Lee
Ostriker, Jeremiah Paul
Papp, Kim Alexander
Parker, Eugene Newman
Petschek, Albert George
Plavec, Mirek Josef
Press, William Henry
Price, Clifton William
Purcell, Edward Mills
Regev, Oded
Richstone, Douglas Orange
Rose, William Kenneth
Rubin, Robert Howard
Sackmann, I-Juliana
Salpeter, Edwin Ernest
Sarazin, Craig Leigh
Schramm, David Norman
Schwarzschild, Martin
Shields, Gregory Alan
Smith, Dean F.
Soneira, Raymond M.
Spitzer, Lyman, Jr.
Sreenivasan, S. Ranga
Starrfield, Sumner Grosby
Stecker, Floyd William
Steigman, Gary
Stein, Robert Foster
Stellingwerf, Robert F.
Straka, William Charles
Tarter, C. Bruce
Tassoul, Jean-Louis
Terrell, James, (Jr.)
Teukolsky, Saul Arno
Thompson, Rodger Irwin
Thorne, Kip Stephen
Toomre, Alar
Tremaine, Scott Duncan
Trimble, Virginia Louise
Tubbs, David Lee
Underhill, Anne Barbara
Vandervoort, Peter Oliver
Van Horn, Hugh Moody
Van Hoven, Gerard
Villere, Karen R.
Vitello, Peter Alfonso James
Wagner, Raymond Lee
Wagoner, Robert Vernon
Wallace, Richard Kent
Walstad, Allan M(artin)
Warwick, James Walter
Weinberg, Steven
Wentzel, Donat Gotthard
Wesselink, Adriaan Jan
Wheeler, J(ohn) Craig
Whitman, Patrick Gene
Wiita, Paul Joseph
Will, Clifford Martin
Willson, Lee Anne Mordy
Wolfe, Arthur Michael
Woodward, James Franklin
Woodward, Paul Ralph

Other

Austin, Sam M., *Nuclear Astrophysics*
Aveni, Anthony Francis, *Archaeoastronomy, History of Astronomy*
Blanco, Victor Manuel, *Observational optical astronomy*
Branham, Richard Lacy, Jr., *Astrometry*
Carlson, John B., *Archaeoastronomy, Extragalactic astronomy*
Cowley, Anne Pyne, *Spectroscopy*
Dick, Steven James, *astrometry*
Doggett, LeRoy Elsworth, *Celestial Mechanics*
Duley, Walter Winston, *Laboratory astrophysics*
Eddy, John Allen, *Archaeoastronomy*
Evans, David Stanley, *History of Astronomy*
Garfinkel, Boris, *Celestial mechanics*
Gatewood, George David, *Astrometry*
Guenter, Joseph Martin, *Astronomy education*
Harrington, Robert Sutton, *Astrometry, Celestial mechanics*
Helou, George, *Galaxies*
Hively, Ray Michael, *Archaeoastronomy*
Hughes, Victor Augustine, *Star formation*
Ianna, Philip A., *Astrometry*
Jordan, Stuart Davis, *Solar, stellar astrophysics*
Krisciunas, Kevin, *History of astronomy*

Lester, John Bernard, *Stellar composition*
Lindenblad, Irving Werner, *Geodetic astronomy*
Mallama, Anthony, *Binary stars*
Mathews, Grant James, *Nuclear astrophysics*
Mattei, Janet Akyuz, *Variable stars*
McCarthy, Dennis Dean, *Astrometry*
McDonough, Thomas Redmond, *The search for extraterrestrial intelligence*
Miller, Freeman Devold, *Cometary physics*
Mitalas, Romas, *Stellar structure and evolution*
Oesterwinter, Claus, *Celestial mechanics*
Oliver, John Parker, *Astronomical instruments*
Olson, Edward Cooper, *Close binary stars*
Osmer, Patrick Stewart, *Astrophysics*
Pellerin, Charles James, Jr., *Astrophysics*
Preston, George W., III, *Observatory administration*
Ramsey, Lawrence William, *Astronomical instrumentation*
Rich, John Charles, *Astrophysics*
Sargent, Anneila Isabel, *Millimeter/submillimeter wave astrophysics*
Vandervoort, Peter Oliver, *Dynamical astronomy*
Vinti, John Pascal, *Celestial mechanics*
Warren, Wayne Hutchinson, Jr., *Computerized astronomical data*
Weissman, Paul Robert, *Celestial mechanics*
Wing, Robert Farquhar, *Variable stars/stellar classification*

ASTROPHYSICS. *See* ASTRONOMY.

ATMOSPHERIC SCIENCE. *See also* ENVIRONMENTAL SCIENCE.

Aeronomy
Akasofu, Syun-Ichi
Benner, Drayton Chris
Bowhill, Sidney Allan
Brown, Neal Boyd
Caledonia, George Ernest
Cartwright, David Chapman
Donahue, Thomas Michael
Evans, John Vaughan
Ferguson, Eldon Earl
Fremouw, Edward Joseph
Goldberg, Richard Aran
Grams, Gerald William
Hanson, William Bert
Johnson, Francis Severin
Kuhn, William Richard
McConnell, John Charles
Nisbet, John Stirling
Paresce, Francesco
Schunk, Robert Walter
Sheridan, John Roger
Smith, Henry John Peter
Torr, Marsha Russell
Traub, Wesley Arthur
Victor, George A.
Wuebbles, Donald James

Atmospheric chemistry
Bradshaw, John David
Cahill, Thomas Andrew
de Pena, Rosa G.
Dingle, Albert Nelson
Farwell, Sherry Owen
Ferguson, Ronald Max
Fishman, Jack
Friend, James Philip
Hobbs, Peter Victor
Holloway, Thomas Thornton
Johnston, Harold Sledge
Kolb, Charles Eugene
Lenschow, Donald Henry
Lewis, David Kenneth
Martin, L(aurence) Robbin
Mauersberger, Konrad
Pitter, Richard Leon
Saxena, Vinod Kumar
Seinfeld, John Hersh
Wang, Pao-Kuan
Wuebbles, Donald James
Zimmerman, Patrick Robert

Climatology
Barry, Roger Graham
Bryson, Reid Allen
Bushnell, Robert Hempstead
Coakley, Stella Melugin
Dando, William Arthur
Davis, Joel Stephen
Decker, Wayne Leroy
DeLuisi, John James
Derr, Vernon Ellsworth
Donn, William L.
Essenwanger, Oskar Maximilian Karl
Estrada, Galindo, Ignacio
Gates, William Lawrence
Gerstl, Siegfried Adolf Wilhelm
Havens, A. Vaughn
Hoyt, Douglas Vincent
Ives, Philip Truman
James, Philip Benjamin
Kellogg, William Welch
Kuhn, William Richard
La Marche, Valmore Charles, Jr.
Landsberg, Helmut E(rich)
Langway, Chester Charles, Jr.
Marcus, Melvin Gerald
Mitchell, John Murray, Jr.
Namias, Jerome
Pollack, James Barney
Ramanathan, Veerabhadran
Rosenberg, Norman Jack
Schneider, Stephen Henry
Smagorinsky, Joseph
Suess, Hans Eduard
Toon, Owen Brian
Trapasso, Louis Michael
Untersteiner, Norbert
Waite, Paul Junior
Webb, Thompson, III
Williams, Jerry Albert

Meteorology
Atlas, David
Austin, Geoffrey Leonard
Badgley, Franklin Ilsley
Baer, Ferdinand
Bannon, Peter Richard
Boudreau, Robert Donald
Boyd, John Philip
Brown, Neal Boyd
Cerni, Todd Andrew
Cox, Stephen Kent
Deepak, Adarsh
Dingle, Albert Nelson
Djuric, Dusan
Dutton, John Altnow
Epstein, Edward S.
Frank, Neil LaVerne
Fultz, Dave
Gates, William Lawrence
Ghil, Michael
Gilman, Donald Lawrence
Goldman, Joseph L.
Hallett, John
Hasler, Arthur Frederick
Hess, Wilmot Norton
Hobbs, Peter Victor
Holton, James R.
Hosler, Charles Luther, Jr.
Houghton, David Drew
Jayaweera, Kolf
Kasahara, Akira
Kassner, James Lyle, Jr.
Kellogg, William Welch
Kessler, Edwin
Krishnamurti, Tiruvalan N.
Landis, Robert Clarence
Liou, Kuo-Nan
Long, Robert Radcliffe
Lorenz, Edward Norton
Maddox, Robert Alan
Mc Carthy, John
Miyakoda, Kikuro
Murino, Clifford John
Norment, Hillyer Gavin
Pandolfo, Joseph Peter
Peterson, Vern LeRoy
Pfeffer, Richard Lawrence
Pitter, Richard Leon
Ramanathan, Veerabhadran
Robert, Andre
Shapiro, Ralph
Shuman, Frederick Gale
Smagorinsky, Joseph
Smith, Phillip Joseph
Stearns, Charles Richard
Theon, John Speridon
Vaughan, William Walton
Vonder Haar, Thomas Henry
Wallace, John Edwin
Wang, Pao-Kuan
Washington, Warren Morton
Webb, Willis Lee
Williams, Jerry Albert
Young, John Arthur

Meteorology, meteorologic instrumentation
Berry, Edwin X
Cerni, Todd Andrew
Gill, Gerald Clifford
Grams, Gerald William
Greenfield, Stanley Marshall
Pallmann, Albert Josef
Palmer, Thomas Yealy
Payne, Richard Earl
Porch, William Morgan
Saxena, Vinod Kumar
Stearns, Charles Richard

Meteorology, micrometeorology
Blackadar, Alfred Kimball
Braham, Roscoe Riley
Disraeli, Donald Jay
Hatfield, Jerry Lee
Havens, A. Vaughn
Kelley, Neil Davis
Landsberg, Helmut E(rich)
Lenschow, Donald Henry
Lumley, John Leask
Rosenberg, Norman Jack

Meteorology, synoptic meteorology
Blackadar, Alfred Kimball
Djuric, Dusan
Frank, Sidney Raymond
Gaffney, Paul Golden, II
Gannon, Patrick Thomas
Maddox, Robert Alan
Namias, Jerome
Smith, Phillip Joseph
Wallace, John Edwin
Young, John Arthur
Young, William Robert

Planetary atmospheres
Allen, Mark Andrew
Appleby, John Frederick
Barth, Charles Adolph
Bufton, Jack Lytle
Capen, Charles Franklin
Chamberlain, Joseph Wyan
Elson, Lee Stephen
Flasar, F. Michael
Hanson, William Bert
Jastrow, Robert
Leite, Richard Joseph
Lewis, John Simpson, Jr.
Lo, Aloysius Kou-fang
Mauersberger, Konrad
Nier, Alfred Otto Carl
Nisbet, John Stirling
Owen, Tobias Chant
Pallmann, Albert Josef
Pomilla, Frank R.
Sandel, Bill Roy
Sato, Makiko
Scattergood, Thomas W.
Schunk, Robert Walter
Toon, Owen Brian
Trafton, Laurence Munro
Travis, Larry Dean
Webb, Willis Lee
West, Robert Alan
Woo, Richard

Remote sensing
Atlas, David
Bevan, Thomas Edward
Booker, Henry George
Bowhill, Sidney Allan
Brook, Marx
Bufton, Jack Lytle
Calio, Anthony John
Clark, Frank Oliver
Condal, Alfonso Ramon
Cox, Stephen Kent
Dando, William Arthur
Davidson, Gilbert
DeLuisi, John James
Derr, Vernon Ellsworth
Elliott, Richard Amos
Elson, Lee Stephen
Evans, John Vaughan
Hasler, Arthur Frederick
Hatfield, Jerry Lee
Herget, William Frederick
Hilgeman, Theodore William
Hinkley, Everett David, Jr.
Hiser, Homer Wendell
Jayaweera, Kolf
Jobst, Joel Edward
Johnson, Douglas William
Jory, Virginia Vickery
LaPorte, Daniel D'Arcy
Liou, Kuo-Nan
McCormick, Norman Joseph
Meeks, Marion Littleton
Murino, Clifford John
Palmquist, John Charles
Parrish, Alan deSchweinitz
Pleskot, Larry Kenneth
Post, Madison John
Sassen, Kenneth
Sawyer, Constance Bragdon
Smith, William Leo
Sweet, Haven Colby
Theon, John Speridon
Trapasso, Louis Michael
Winker, David M(ichael)

Other
Albers, Mark Alan, *Heat Transfer*
Baer, Ferdinand, *Atmospheric modeling*
Benninghoff, Anne Stevenson, *Aerobiology*
Bierly, Eugene Wendell, *Atmospheric science research management*
DeMeo, V(incent) James, Jr., *Desertification studies*
Donn, William L., *Atmospheric Infrasound*
Fleming, Rex James, *Government program administration*
Goldberg, Richard Aran, *Atmospheric Electrodynamics*
Haurwitz, Bernhard, *Dynamic meteorology*
Layton, Richard Gary, *Cloud physics*
Porch, William Morgan, *Atmospheric optics*
Sagalyn, Rita C., *Ionospheric physics*
Smith, William Leo, *Satellite meteorology*

BIOLOGY

Behaviorism
Arave, Clive Wendell
Casseday, John Herbert
Day, Stacey Biswas
Gabel, J. Russel
Gara, Robert Imre
Greene, Elias Louis
Jacobs, Merle E(mmor)
Lardy, Henry Arnold
Novak, John Allen
Potter, Rosario H. Yap
Russell-Hunter, W(illiam) D(evigne)
Seiger, Marvin Barr
Stollnitz, Fred

Biochemistry. *See also* MEDICINE.
Abbott, Mitchel Theodore
Adler, Julius
Ames, Bruce N(athan)
Ames, Giovanna Ferro-Luzzi
Anderson, Louise Eleanor
Anderson, Thomas Richard
Anfinsen, Christian Boehmer
Apirion, David
Armstrong, Donald
Arnon, Daniel I(srael)
Arnott, Struther
Asakura, Toshio
Aull, John Louis
Baggett, Billy
Bailin, Gary
Ball, Laurence Andrew
Bambara, Robert Anthony
Bamburg, James Robert
Banerjee, Sipra
Banik, Narendra Lal
Barden, Roland Eugene
Barnes, Larry Dean
Barrett, Edward Joseph
Bates, Margaret Westbrook
Baumgold, Jesse
Beck, William Samson
Behrisch, Hans Werner
Bell, Richard
Belman, Sidney
Bender, Myron Lee
Benson, Andrew Alm
Berg, Paul
Berg, Richard Alan
Bernfeld, Peter Harry William
Bernlohr, Robert William
Berns, Donald Sheldon
Bernstein, Eugene Harold
Birnbaum, Linda Silber
Blincoe, Clifton Robert
Bloch, Konrad
Boyer, Herbert Wayne
Boyer, Paul D.
Braatz, James Anthony
Brakke, Myron Kendall
Braun, Andrew George
Brooker, Gary
Brown, John Clifford
Brown, Kenneth Lawrence
Brownie, Alexander C.
Brutlag, Douglas Lee
Bundy, Hallie Flowers
Burleigh, Bruce Daniel
Burris, Robert Harza
Butler, John Edward
Butler, Larry Gene
Byerrum, Richard Uglow
Callewaert, Denis Marc
Calvo, Joseph Marle
Carraway, Kermit Lee
Carroll, Robert Byers
Carson, Dennis Anthony
Carson, Steven Douglas
Carter, Herbert Edmund
Caruthers, Marvin Harry
Castellino, Francis Joseph
Cathou, Renata Egone
Cederbaum, Arthur I.
Cerami, Anthony
Chalmers, John H., Jr.
Chan, Phillip C.
Chang, Kwen-Jen
Chase, John William
Chatterjee, Sunil Kumar
Cherry, John Paul
Chitwood, David Joseph
Choi, Yong Sung
Christman, Judith Kershaw
Chuang, Ronald Y(an-li)
Chytil, Frank
Clark, John Magruder, Jr.
Clarke, Steven Gerard
Cohen, Philip Pacy
Cohen, Robert Jay
Coleman, Joseph Emory
Coligan, John Ernest
Collins, Jimmy Harold
Coniglio, John Giglio
Connett, Richard James
Cook, Paul Fabyan
Cooper, Alan Douglas
Cossins, Edwin Albert
Costa, Max
Counselman, Clarence James
Cox, G. Stanley
Cross, Richard Lester
Crouch, Rosalie Kelsey
Cunningham, Glenn Norman
Curnow, Randall Thomas
Cushman, David Wayne
D'Angelo, Gaetano
Daron, Harlow H.
Dashman, Theodore
Davidson, Eugene Abraham
Davis, George Kelso
DeBari, Vincent (Anthony)
de Duve, Christian Rene
De Haas, Herman
De Luca, Luigi Maria
DeLuca, Marlene A(nderegg)
De Moss, Ralph Dean
De Vries, George Henry
Dilley, Richard A.
Dixon, Jack E.
Doolittle, Russell F.
Douglas, Michael Gilbert
Downey, Ronald Joseph
Drickamer, Kurt
Duker, Nahum Johanan
Dunlap, R. Bruce
Dus, Karl M(aria)
Eagon, Robert Garfield
Eden, Francine Claire
Edmonds, Mary Patricia
Edwards, Lois Adele
Ehrenfeld, Elvera
Ehrlich, Kenneth C.
Eilert, Jeffries Harvey
Elbein, Alan David
Ellsworth, Robet King
Esser, Alfred F.
Etzler, Marilynn Edith
Farley, John Randolph
Ferone, Robert
Fink, Anthony Lawrence
Fischer, Edmond H.
Fisher, Ronald Richard
Fishman, William H(arold)
Flory, William Evans Sherlock
Floyd, Robert A.
Forman, Henry Jay
Forrest, Hugh Sommerville
Franson, Richard Carl
Fresco, Jacques Robert
Fujimura, Robert Kanji
Fukuda, Minoru
Ganesan, Adayapalam Tyagarajan
Garrett, Reginald Hooker
Gassman, Merrill Loren
Gelboin, Harry Victor
Gennis, Robert Bennett
Geren, Collis Ross
Gerwin, Brenda Isen
Gibian, Morton J.
Gilbertson, John Robert
Gill, D. Michael
Gingery, Roy Evans
Glasel, Jay Arthur
Godfrey, Henry Philip
Goldfine, Howard
Goldstein, Menek
Goll, Darrel Eugene
Goodridge, Alan Gardner
Graham, Terrence Lee
Greaser, Marion Lewis
Green, David Ezra
Greenaway, Frederick Thomas
Greengard, Paul
Gregory, John Delafield
Griffin, Charles Campbell
Gross, Paul Randolph
Grossman, Lawrence
Guillory, Richard John
Gumport, Richard I.
Gupta, Sohan Lal
Guterman, Sonia Kosow
Hadler, Herbert Isaac
Hager, Lowell Paul
Hahn, Fred Ernst
Hall, Madeline Molnar Hall
Hammes, Gordon G.
Hansford, Richard Geoffrey
Hare, James Frederic
Harrington, Rodney Elbert
Harris, Don Navarro
Hatch, Frederick Tasker
Hatcher, Victor Bernard
Hay, Donald Ian
Hayes, Dora Kruse
Heck, Henry d'Arcy
Hector, Mina L.
Hefferren, John James
Hegeman, George Downing
Heimer, Ralph
Hendler, Richard Wallace
Herr, Earl Binkley, Jr.
Herrmann, Robert Lawrence
Hershberger, Charles Lee
Herting, David Clair
Herz, Fritz
Hierholzer, John Charles
Higgins, Edwin Stanley
Hirschberg, Carlos Benjamin
Hitchings, George Herbert
Ho, Chien
Hoagland, Mahlon Bush
Hoch, George Edward
Holden, Joseph Thaddeus
Horecker, Bernard Leonard
Horowitz, Jack
Horwitt, Max Kenneth
Houston, L.L.
Huang, Ching-hsien
Huber, Donald John
Idler, David Richard
Inman, John K.
Jacobson, Karl Bruce
Jagendorf, Andre Tridon
Janski, Alvin Michael
Jenkins, W(inborne) Terry
Johnson, Eric Foster
Johnson, Kenneth Allen
Johnson, Ronald Roy
Johnston, James Bennett
Kaback, H. Ronald
Kaiser, Armin Dale
Kamath, Savitri Krishna
Kamin, Henry
Kaplan, Stanley Albert
Karavolas, Harry J(ohn)
Karnovsky, Manfred L.
Karpel, Richard Leslie
Kasvinsky, Peter John
Keen, Noel Thomas
Keller, Patricia J.
Kennedy, Eugene Patrick
Kenney, Francis Thomas
Kern, Harold Lloyd
Khorana, Har Gobind
Kiechle, Frederick Leonard
Kimball, Edward Saul
Kimmich, George Arthur
Kindt, Thomas James
King, Jonathan Alan
King, Jonathan Stanton
Kisliuk, Roy Louis
Klibanov, Alexander Maxim
Kline, Edward Samuel
Klinman, Judith Pollock
Knoche, Herman William
Knowles, Jeremy Randall
Konrad, Michael
Korn, Edward David
Kritchevsky, David
Krogmann, David William
Kuby, Stephen Allen
Kuc, Joseph
Kupke, Donald Walter
Kushner, Sidney Ralph
LaBelle, Edward Francis
Lacko, Andras Gyorgy
Lacks, Sanford Abraham
Landon, Leroy James
Larson, Bruce L.
Lata, Gene Frederick
Lazzari, Eugene Paul
Leach, Franklin Rollin
Lebherz, Herbert G(rover)
Lee, James Ching

Lee, Ted Choong Kil
Lefkowitz, Robert Joseph
Lehninger, Albert Lester
Leive, Loretta
Levitz, Mortimer
Li, Choh Hao
Liao, Shutsung
Liao, Ta-Hsiu
Libby, Paul Robert
Lin, Ying-Ming
Lipmann, Fritz (Albert)
Litov, Richard Emil
Little, Henry Nelson
Liu, Edwin H.
Liu, Teh-Yung
Liu, Yung-Pin
Lockridge, Oksana
Lodish, Harvey Franklin
Loomis, Stephen Henry
Lorance, Elmer Donald
Lovell, Richard Arlington
Low, Philip Stewart
Luck, Dennis Noel
Lukin, Marvin
Lumeng, Lawrence
Lundblad, Roger Lauren
Lygree, David Gerald
Macdonald, Timothy Lee
Mace, Marshall Ellis
Mangano, Richard Michael
Marcoullis, George Panayiotis
Margoliash, Emanuel
Marks, Neville
Marsh, Benjamin Bruce
Martin, David Lee
Martin, Joseph Patrick, Jr.
Martinez-Carrion, Marino
Massaro, Edward Joseph
Matthews, Harry Roy
Matthews, Kathleen Shive
Max, Stephen Richard
McClure, William Robert
Mc Elroy, William David
McGroarty, Estelle Josephine
McHenry, Charles Steven
McIntire, Floyd Cottam
McLaughlin, Calvin S.
Melius, Paul
Melton, Russell Paul
Menzies, Robert Allen
Merrifield, Robert Bruce
Meselson, Matthew Stanley
Mihalyi, Elemer
Miller, Edward Joseph
Model, Peter
Modrich, Paul L.
Moody, Terry William
Morre, D. James
Morrison, Martin
Mortlock, Robert Paul
Mosbach, Erwin Heinz
Moscatelli, Ezio Anthony
Moudgil, Virinder Kumar
Müller, Miklós
Müller-Eberhard, Hans Joachim
Natelson, Samuel
Nes, William Robert
Neufeld, Harold Alex
Nirenberg, Marshall Warren
Nowak, Thomas L.
Oeltmann, Thomas Napier
Oldfield, James Edmund
Orme-Johnson, William H.
Oro, Juan
Osborn, Mary Jane Merten
Paech, Christian Gerolf
Pall, Martin Lawrence
Palmer, Sushma Mahyera
Pardee, Arthur Beck
Parker, Frank S.
Paselk, Richard Alan
Patterson, Manford K., Jr.
Penniston, John Thomas
Perejda, Andrea Jeanne
Petsko, Gregory Anthony
Pettit, Flora Hunter
Pierce, John Grissim
Pincus, Jack Howard
Pogo, A, Oscar
Pollack, Ralph Martin
Pollack, Robert Leon
Pomeranz, Yeshajahu
Ponnamperuma, Cyril Andrew
Posky, Leon
Potter, Neil Harrison
Powell, Gary Lee
Powers, Daniel Duffy
Powers, Dennis Alpha
Preiss, Jack
Price, Alan Roger
Prigge, Edward Christian, Jr.
Prough, Russell Allen
Ptashne, Mark Stephen
Quinlan, Dennis Charles
Rabinowitz, Jesse Charles
Racker, Efraim
Ralston, Douglas Edmund
Reazin, George Harvey, Jr.
Rebeiz, Constantin Anis
Reddy, Chinthamani Channa
Reed, Lester James
Reichenbecher, Vernon Edgar, Jr.
Richardson, John Paul
Rizack, Martin Arthur
Robbins, Phillips Wesley
Roberts, John D.
Robinson, Joseph Douglass
Robson, Richard Morris
Robyt, John Francis
Roelofs, Wendell Lee
Rogan, Eleanor Groeniger
Rosenthal, Harold Leslie
Ross, Elliott M(orton)
Roufa, Donald Jay
Rutman, Robert J.
Saier, Milton Herman, Jr.
Salsbury, Robert Lawrence
Samuels, Robert
Samy, Anantha T. S.
Sanders, Robert Burnett
Sani, Brahma P.
Santella, Regina Maria
Sarich, Vincent M.
Satyanarayana, T.
Savage, Carl Richard, Jr.
Scher, William
Scherberg, Neal Harvey
Schnoor, Ronald Lee
Schwartz, Ira S.
Scora, Rainer Walter
Searls, Robert Louarn
Segal, Joseph
Sevall, Jack Sanders
Shapiro, Irving Meyer
Shatkin, Aaron Jeffrey
Shemin, David
Shichi, Hitoshi
Shih, Jason Chia-Hsing
Shortle, Walter Charles
Siedler, Arthur James
Silver, Simon David
Silverman, Richard Bruce
Simmons, William Howard
Simpson, Robert Todd
Sinha, Navin Kumar
Siuta-Mangano, Patricia
Skipski, Vladimir P(avlovich)
Slavkin, Harold Charles
Smith, Emil L.
Smith, James Cecil, Jr.
Smith, Leonard Charles
Smith, Robert Charles
Snyder, Fred Leonard
Somerville, Ronald Lamont
Spitsberg, Vitaly Lev
Spivey, Howard Olin
Spremulli, Linda Lucy
Stadtman, Earl Reece
Stadtman, Thressa Campbell
Steck, Theodore Lyle
Steiner, Robert Frank
Stekoll, Michael Steven
Stenesh, Jochanan
Stephenson, Mary Louise
Stoner, Clinton Dale
Stowe, Bruce Bernot
Stromer, Marvin Henry
Sullivan, Ann Clare
Suzuki, Tsuneo
Terner, Charles
Thomas, Paul Elbert
Thomas, William Eric
Thompson, David Jerome
Thornber, James Philip
Tischler, Marc Eliot
Tőkés, Zoltán András
Trimble, Robert Bogue
Turco, Salvatore Joseph
Tweedy, Billy G.
Udenfriend, Sidney
Underwood, Arthur Louis, Jr.
Vanaman, Thomas C.
Van Kley, Harold James
Van Wart, Harold Edgar
Veis, Arthur
Volanakis, John Emmanuel
Volcani, Benjamin Elazari
Vold, Barbara Schneider
Volsky, David Julian
Vournakis, John Nicholas
Wald, George
Wallace, Robert Dean
Wallach, Donald Pinny
Walsh, Kenneth Andrew
Warme, Paul Kenneth
Wartell, Sue Ann
Weber, Darrell Jack
Weber, Evelyn Joyce
Weber, Gregorio
Weimberg, Ralph
Weiner, Henry
Weiss, Richard Louis
Weissbach, Arthur
Weissbach, Herbert
Welch, William Henry
Wells, Robert Dale
Werth, Jean Marie
Westby, Carl Andrew
Westhead, Edward William
Westheimer, Frank Henry
Whanger, Philip Daniel
White, Helen Lyng
Wiley, William R.
Williams, Marshall Vance, Jr.
Wilson, David Buckingham
Wilson, Samuel H.
Winter, Charles Gordon
Wishner, Lawrence Arndt
Wohl, Robert Chaim
Wolf, Walter Alan
Wood, Randall Dudley
Wright, Harlan Tonie
Wu, Felicia Ying-Hsiueh
Wu, Joseph Man-Hay
Yang, David Chih-Hsin
Yocum, Charles Fredrick
Young, Delano Victor
Zamenhof, Stephen
Zeiger, Allen Richard
Zimmerman, Thomas Paul
Zurawski, Vincent Richard, Jr.

Biophysics

Absolom, Darryl Robin
Arnott, Struther
Bean, Charles Palmer
Bendet, Irwin Jacob
Berg, Howard Curtis
Bisson, Mary Aldwin
Blanchard, Robert Osborn
Bremermann, Hans J.
Brink, Peter Richards
Bronk, Burt V.
Burke, Patricia Virginia
Carpenter, David Orlo
Chance, Britton
Cheung, Herbert Chiu-Ching
Cohen, Robert Jay
Coleman, Joseph Emory
Cranefield, Paul Frederic
D'Arrigo, Joseph Salvatore
Davis, Brian Keith
DeLisi, Charles
DeVault, Don Charles
Dewey, Thomas Gregory
Dunham, Philip Bigelow
Dusenbery, David Brock
Edgerton, Mary Elizabeth
Fabiato, Alexandre
Fernández, Salvador M.
Gardner, Daniel
Gibbons, Walter Ray
Goldman, Lawrence
Goldstein, Dora Benedict
Goll, Darrel Eugene
Goode, Melvyn Dennis
Gross, Leo
Gueft, Boris
Haak, Richard Arlen
Haynes, Robert Hall
Hill, Terrell Leslie
Hobbie, Russell Klyver
Hollander, Philip Ben
Hui, Chiu Shuen
James, Thomas L.
Jankelson, Bernard
Johnson, Walter Curtis, Jr.
Jung, Chan Yong
Justesen, Don Robert
Karle, Isabella Lugoski
Kessel, David
Krishna, Nepalli Rama
Krohmer, Jack Stewart
Kupke, Donald Walter
La Celle, Paul Louis
Lange, Christopher Stephen
Lasater, Eric Martin
Leblanc, Roger Maurice
Lee, Chi Ok
Leibovic, K. Nicholas
Lieberman, Edward Marvin
Lindley, Barry Drew
Llinas, Rodolfo Riascos
Loewenstein, Werner Randolph
Lontz, John Frank
Lucas, William John
Makinen, Marvin William
Martinez-Carrion, Marino
Mauzerall, David Charles
McGroarty, Estelle Josephine
Merilan, Charles Preston
Murphy, Richard Alan
O'Leary, Dennis Patrick
Owen, Charles Scott
Pallotta, Barry S
Patrick, Michael Heath
Peachey, Lee DeBorde
Petty, Howard Raymond
Piette, Lawrence Hector
Plocke, Donald Joseph
Pohl, Herbert Ackland
Purcell, Edward Mills
Ritchie, Joseph Murdoch
Robinson, Joseph Douglass
Ross, William Noel
Salzberg, Brian Matthew
Seaman, Ronald Leon
Setlow, Richard Burton
Silvidi, Anthony Alfred
Song, Chang Won
Stevens, Charles F.
Suddath, Fred Leroy, Jr.
Takashima, Shiro
Tanford, Charles
Thomas, John Alva
Tien, Ho Ti
Ts'o, Paul On-Pong
Wald, George
Wheeler, Kenneth Theodore, Jr.
White, Stephen Halley
Yao, Shang Jeong
Yen, Andrew
Young, Wei

Cell biology. *See also* **MEDICINE.**

Aist, James Robert
Alberte, Randall Sheldon
Ananthaswamy, Honnavara
Narasimhamurthy
Anderson, Thomas Richard
Ault, Jeffrey George
Ball, Wilfred Randolph
Bamburg, James Robert
Barnes, Larry Dean
Benjamini, Eliezer
Billen, Daniel
Bleyman, Lea Kanner
Blystone, Robert Vernon
Bodenstein, Dietrich Hans Franz
Alexander
Bonner, Thomas Patrick
Bortenfreund, Ellen
Bouma, Hessel, III
Bozzola, John Joseph
Branton, Daniel
Braun, Andrew George
Bregman, Allyn Aaron
Brokaw, Charles Jacob
Bryan, John Henry Donald
Burke, Derek Clissold
Byers, Breck Edward
Carraway, Kermit Lee
Casciano, Daniel Anthony
Chabot, Jean Fincher
Chapman, Russell Leonard
Chobanian, Aram Van
Coleman, Annette Wilbois
Collins, Jimmy Harold
Cottrell, Stephen F.
Crick, Francis Harry Compton
Danielli, James Frederic
Darnell, James Edwin, Jr.
de Duve, Christian Rene
DeLisi, Charles
DiPaolo, Joseph Amedeo
Diwan, Joyce Johnson
Draznin, Boris
Dritschilo, Anatoly
Dute, Roland Roy
Eddy, Edward Mitchell
Edelson, Paul Jeffrey
Fallon, Ann Marie
Fisher, Steven Kay
Forman, David Sholem
Fox, C. Fred
Friend, Charlotte
Fussell, Catharine Pugh
Gale, Robert Peter
Gall, Joseph Grafton
Gealt, Michael Alan
Gelehrter, Thomas David
Goldstein, Byron Bernard
Goldstein, Lester
Goode, Melvyn Dennis
Gueft, Boris
Hackett, Charles Joseph
Hand, Arthur Ralph
Hatcher, Victor Bernard
Herbert, Edward
Herlyn, Meenhard Folkeus
Herman, William Sparkes
Higgins, Paul Joseph
Hilfer, Saul Robert
Hirschberg, Carlos Benjamin
Hitchcock-DeGregori, Sarah Ellen
Hofer, Kurt Gabriel
Holland, John Joseph
Hotta, Yasuo
Houser, Steven Robert
Howse, Harold Darrow
Jeon, Kwang Wu
Johnson, Kenneth Allen
Juliano, Rudolph L.
Kennedy, John Robert
Kramarsky, Bernhard
Kretsinger, Robert H.
LaChance, Leo Emery
La Claire, John Willard, II
Laffler, Thomas G.
Ledbetter, Myron Calvert
Libby, Paul Robert
Lindquist, Susan Lee
Lipton, Allan
Lodish, Harvey Franklin
Maguire, Marjorie Paquette
Marengo, Norman Playson
Martz, Eric
Marzella, Libero Louis
Mazia, Daniel
Metcalf, John Franklin
Meyer, Dale R.
Miller, Morton W(illiam)
Miller, Oscar Lee, Jr.
Moldenke, Alison Feerick
Nandi, Satyabrata
Nicklas, Robert Bruce
Nicolson, Garth L.
Olins, Donald Edward
Oross, John William
Painter, Richard Grant
Palade, George Emil
Pappas, Peter William
Pardee, Arthur Beck
Pardue, Mary Lou
Parthasarathy, Mandayam
Veerambhudi
Peachey, Lee DeBorde
Pfenninger, Karl Hans
Pickett-Heaps, Jeremy David
Pleshkewych, Alexander
Politis, Demetrios John
Porter, Keith Roberts
Powers, Edward Lawrence
Press, Newtol
Pritchard, Hayden Nelson
Raikow, Radmila Boruvka
Reynafarje, Baltazar Davila
Ris, Hans
Rivera, Ezequiel Ramirez
Robbins, Robert Raymond
Roseman, Saul
Ruddle, Francis Hugh
Salzman, Norman Post
Samuels, Robert
Sanadi, D. Rao
Scharff, Matthew Daniel
Scher, William
Schiff, Jerome Arnold
Schlegel, Robert Allen
Schvartzman, Jorge Bernardo
Shen, Wei-Chiang
Siekevitz, Philip
Sivak, Andrew
Smith, Gary Lee
Sreevalsan, Thazepadath
Stein, Gary Stephen
Stoner, Clinton Dale
Strike, Terry Lee
Stromer, Marvin Henry
Tamm, Igor
Thompson, Elizabeth Barnes
Tobey, Robert A.
Tomasz, Alexander
Trimble, Robert Bogue
Tytell, Michael
Warner, Jonathan Robert
Wells, Marion Robert
White, Richard Hamilton
Wiley, William R.
Woods, Walter Thomas
Yamaguchi, Mamoru
Yatsu, Lawrence Yoneo
Yen, Andrew
Yuspa, Stuart Howard
Zimmerman, Arthur Maurice

Cell biology, cell and tissue culture. ***See also*** **AGRICULTURE, Plant cell and tissue culture.**

Abid, Syed Hasan
Andrews, Peter Walter
Avner, Ellis David
Bantle, John Albert, II
Baserga, Renato Luigi
Bassham, James Alan
Baumann, [illegible]
Becker, Robert Otto
Bell, George Irving
Berg, Richard Alan
Berliner, Martha D.
Bernd, Paulette Sally
Birckbichler, Paul Joseph
Bortenfreund, Ellen
Bouck, Noel Patrick
Breitman, Theodore Ronald
Broadhurst, John Henry
Bronson, David Lee
Brostrom, Margaret Ann
Brown, Bruce Leonard
Brown, Donald David
Bucher, Nancy L.R.
Burleigh, Bruce Daniel
Came, Paul E.
Chou, Iih-Nan
Chu, Ernest Hsiao-Ying
Cooper, Norman Streich
Crain, Stanley M.
Crang, Richard Francis Earl
Cronshaw, James
Culp, Lloyd Anthony
Davis, Bill David
Diacumakos, Elaine G.
Dickinson, Winifred Ball
Dow, Lois Weyman
Drickamer, Kurt
Duff, Ronald George
Egan, Marianne Louise
Eil, Charles
Eppstein, Deborah Anne
Epstein, Joshua
Erickson, Leonard Charles
Etzler, Marilynn Edith
Farley, John Randolph
Fawcett, Don Wayne
Fotos, Peter George
Fox, Michael Henry
Fried, Jerrold
Gardner, Eldon John
Gentile, Arthur Christopher
Glick, J. Leslie
Godlfeder, Anna
Gurney, Elizabeth Tucker Guice
Hall, Stanton Harris
Hanley, Kevin Joseph
Hare, James Frederic
Hatfield, G. Wesley
Hawrot, Edward
Hayflick, Leonard
Heinz, Don J
Helson, Lawrence
Henney, Christopher Scot
Heppel, Leon Alma
Hill, Ray Allen
Hoagland, Mahlon Bush
Hoegerman, Stanton Fred
Holley, Robert William
Horner, Harry Theodore
Hosick, Howard Lawrence
Huberman, Eliezer
Hyndman, Arnold Gene
Iannaccone, Philip Monroe
Jeon, Kwang Wu
Johnson, John Morris
Kaighn, Morris Edward
Kakunaga, Takeo
Kennedy, John Robert
Kollar, Edward James
Kubitschek, Herbert Ernest
Kucherlapati, Raju S.
Ledbetter, Mary Lee Stewart
Ledinko, Nada
Lembach, Kenneth James
Leung, Benjamin Shuet-kin
Lostroh, Ardis June
Macario, Alberto Juan Lorenzo
Mahlberg, Paul Gordon
Maslow, David E(zra)
Mayne, Richard
Moehring, Joan Marquart
Moldenhauer, Jeanne Elisabeth
Munro, Hamish Nisbet
Nandi, Satyabrata
Nelson, Phillip Gillard
Norin, Allen J.
Oberley, Terry De Wayne
Pariza, Michael Willard
Park, Chan Hyung
Park, Chung Suru
Patterson, Manford K., Jr.
Patterson, Rosalyn Victoria Mitchell
Paxton, Jack Dunmire
Pfeffer, Lawrence Marc
Pinero, Gerald Joseph
Pollack, Robert Elliot
Pollard, Harvey Bruce
Quinlan, Dennis Charles
Riser, Mary Elizabeth
Robbins, Phillips Wesley
Robson, Richard Morris
Rosenbusch, Ricardo Francisco
Rost, Thomas Lowell
Schlenker, Robert Alison
Seravalli, Egilde Paola
Shafi, Muhammad Iqbal
Shapiro, Burton Leonard
Shires, Thomas Kay
Silver, Robert Benjamin
Smith, Gary Lee
Stabler, Timothy Allen
Steuer, Anton Francis
Stevenson, Harlan Quinn
Takemoto, Kenneth K.
Tedeschi, Henry

Thimann, Kenneth Vivian
Thompson, Edward Ivins Bradbridge
Tihon, Claude
Ting, Yu-Chen
Topper, Yale J.
Toran-Allerand, C(laude) Dominique
Trowbridge, Richard Stuart
Unbehaun, Laraine Marie
Van De Water, Thomas Roger
Van't Hof, Jack
Wagner, Gerald Gale
Walsh-Reitz, Margaret Mary
Wartell, Sue Ann
Weinberg, Robert Allen
Weinstein, David
Weissbach, Arthur
Wergin, William P.
Wolff, Sheldon

Chronobiology
Beljan, John Richard
Cutkomp, Laurence Kremer
Evans, James Warren
Glenn, Loyd LeRoy
Guillaume, Germaine Gabrielle Cornelissen
Hayes, Dora Kruse
Hrushesky, William John Michael
Jacklet, Jon W.
Kafka, Marian Adele Stern
Lieb, Margaret
Morin, Lawrence Porter
Morris, Ralph William
Pauly, John Edward
Perlow, Mark Jacob
Repenning, Charles Albert
Rosenberg, Gary David
Smolensky, Michael Hale
Sturtevant, Frank Milton
Wainwright, Stanley Dunstan

Developmental biology
Adler, Paul Neil
Aduss, Howard
Amen, Ralph DuWayne
Andrews, Peter Walter
Bacon, Jonathan Peter
Baird, John Jeffers
Band, Henretta Trent
Bantle, John Albert, II
Benbow, Robert Michael
Bender, Welcome W.
Bentley, Michael Martin
Bernd, Paulette Sally
Bernstein, Sanford Irwin
Blystone, Robert Vernon
Bodenstein, Dietrich Hans Franz Alexander
Bohn, Martha D.
Bonner, John Tyler
Boylan, Elizabeth S.
Britten, Roy John
Bromley, Stephen C.
Brown, Donald David
Brunso-Bechtold, Judy Karen
Cancro, Michael Paul
Castelfranco, Paul Alexander
Chang, Ernest Sun-Mei
Cherbas, Peter Thomas
Costello, Walter James
Cress, Dean Ervin
Cunningham, Bruce Arthur
Davidson, Eric Harris
Dawid, Igor Bert
Denburg, Jeffrey Lewis
Dey, Sudhansu K.
Dickinson, Winifred Ball
Doane, Winifred Walsh
Douthit, Harry Anderson, Jr.
Downs, Robert Jack
Duncan, Ian William
Dute, Roland Roy
Edds, Louise Luckenbill
Edelman, Gerald Maurice
Edwards, John Stuart
Emmons, Scott Wilson
Engelhardt, Dean Lee
Ennis, Herbert L.
Erikson, Raymond Leo
Frankel, Joseph
Fritz, Irving Bamdas
Fulton, Chandler Montgomery
Fussell, Catharine Pugh
Gifford, Ernest Milton, Jr.
Goldberg, Stephen
Goldwasser, Eugene
Goodman, Corey Scott
Gordon, Richard
Goss, Richard Johnson
Green, Paul Barnett
Greene, Robert Morris
Gross, Jerome
Gross, Paul Randolph
Grunwald, Gerald Bruce
Hadden, Edward Leal, Jr.
Hall, Brian Keith
Hammill, Terrence Michael
Hauke, Richard Louis
Hemmendinger, Lisa M.
Hibbard, Emerson
Hilfer, Saul Robert
Hogness, David Swenson
Homyk, Theodore, Jr.
Horner, Harry Theodore
Hough-Evans, Barbara Raymond
House, Verl Lee
Howe, Chin C.
Huskey, Robert John
Imberski, Richard Bernard
Jeffery, William Richard
Jensh, Ronald Paul
Kafatos, Fotis C.
Kidwell, William Robert
Kleinschuster, Stephen John
Koch, Henry George
Kornberg, Thomas Bill
Kvist, Tage Nielsen
Lauder, Jean Miles
Leopold, Roger Allen
Lowe, Irene Posner
Macagno, Eduardo Roberto
Maderson, Paul F.A.
Mahowald, Anthony P.
Martin, Presley Frank
Mascarenhas, Joseph Peter
Mayne, Richard
McCorkle, George Maston
McElligott, Sandra G.
McGaughey, Robert William
McLoon, Linda Kirschen
Miksche, Jerome Phillip
Mogensen, Hans Lloyd
Monder, Harvey
Moscona, Aron Arthur
Mueller, Nancy Alice Schnieder
Nelson, Phillip Gillard
Nemer, Martin Joseph
O'Farrell, Patrick Henry
Olsen, George Duane
Opitz, John Marius
Palmiter, Richard DeForest
Parthasarathy, Mandayam Veerambhudi
Peters, Marvin Arthur
Petersen, Nancy Sue
Peterson, Curt Morris
Petri, William Hugh
Poccia, Dominic Louis
Postlethwait, Samuel Noel
Raghunathan, Rengachari
Rakic, Pasko
Raper, Carlene Allen
Riddle, Donald Lee
Schreckenberg, Gervasia Mary
Schuetze, Stephen Mark
Searls, Robert Louarn
Sheffield, Joel B.
Sherald, Allen Franklin, III
Shinnick, Thomas Michael
Simpson, Robert Todd
Slavkin, Harold Charles
Smith, Lewis Dennis
Spiess, Luretta D(avis)
Spitzer, Nicholas Canaday
Springer, Alan David
Stabler, Timothy Allen
Steph, Nick Charles
Swanson, Margaret MacMorris
Thonney, Michael Larry
Timberlake, William Edward
Topper, Yale J.
Udin, Susan Boymel
Van De Water, Thomas Roger
Van Etten, James Lee
Van Houten, Judith Lee
Waelsch, Salome Gluecksohn
Walker, Dan Berne
Ward, Samuel
Williams, Carroll Milton
Wood, William Barry, III
Wu, Joseph Man-Hay
York, James Lester
Yund, Mary Alice

Ecology. ***See*** **ENVIRONMENTAL SCIENCE.**

Ethology
Able, Kenneth Paul
Allen, Theresa Ohotnicky
Askins, Robert Arthur
Banks, Edwin Melvin
Barlow, George Webber
Barrows, Edward Myron
Caldwell, Willard E.
Carico, James Edwin
Clark, George Alfred, Jr.
Crews, David P.
DeGhett, Victor John
de Lanerolle, Nihal Chandra
Eisner, Thomas
Fernald, Russell Dawson
Fine, Michael Lawrence
Franklin, Robert Fraser, Jr.
Gorlick, Dennis Lester
Gray, Philip Howard
Greenberg, Neil
Griffin, Donald R(edfield)
Grove, Patricia Ann
Hailman, Jack Parker
Hayward, James Lloyd
Herrnkind, William Frank
Horn, Henry Stainken
Johnsgard, Paul Austin
Kovach, Joseph K.
Labov, Jay Brian
Lavigne, Robert James
Lishak, Robert Stephen
Lloyd, James Edward
Lutton, Lewis Montfort
Marler, Peter Robert
Miller, David Bennett
Newman, John Dennis
Partridge, Brian Lloyd
Perelle, Ira B.
Perrill, Stephen Arthur
Pietrewicz, Alexandra Theresa
Pitts, Charles William
Rebach, Steve
Rettenmeyer, Carl William
Robbins, Robert John
Root, Thomas Michael
Rowland, William Joseph
Saladin, Kenneth Stanley
Waring, George Houstoun
Wishner, Lawrence Arndt

Evolutionary biology
Abrahamson, Warren Gene, II
Adler, Kraig (Kerr)
Alexander, Richard Dale
Anderson, Gregory Joseph
Anderson, Wyatt Wheaton
Andrews, Henry Nathaniel, Jr.
Aquadro, Charles Frederick
Atchley, William Reid
Ayala, Francisco Jose
Baker, Herbert George
Barron, Sarah Kathryn Braswell
Bawa, Kamaljit Singh
Bazinet, Lester
Beason, Robert Curtis
Bell, Clyde Ritchie
Benseler, Rolf Wilhelm
Benzing, David Hill
Bernhardt, Peter
Bierzychudek, Paulette Francine
Blair, William Franklin
Boaz, Noel Thomas
Bonner, John Tyler
Bookman, Susan Stone
Brett, Betty Lou Hilton
Brush, Alan Howard
Burch, John Bayard
Burggren, Warren William
Campbell, Carlos Boyd Godfrey
Campbell, Russell Bruce
Carr, Gerald Dwayne
Carson, Hampton Lawrence
Case, Ted Joseph
Chambers, Kenton Lee
Chapman, Russell Leonard
Chitaley, Shyamala Dinkar
Choate, Jerry Ronald
Colbert, Edwin H.
Colinvaux, Paul Alfred
Commisso, Franklyn W.
Coney, Charles Clifton
Cornette, James Lawson
Cotter, David James
Cotter, William Bryan, Jr.
Cox, Paul Alan
Crews, David P.
Dancik, Bruce Paul
Davidson, Donald W.
Dilcher, David Leonard
Dolph, Gary Edward
Doolittle, Donald Preston
Druger, Marvin
Dykhuizen, Daniel Edward
Ellstrand, Norman Carl
Erpino, Michael James
Fitch, Walter Monroe
Frazzetta, Thomas H.
Fritze, Karen J.
Fryxell, Paul Arnold
Funk, Vicki Ann
Futuyma, Douglas Joel
Gans, Carl
Giddings, Luther Val
Gilbert, Donald George
Gilmartin, Amy Jean
Gingerich, Philip Dean
Glanz, William Edward
Gojobori, Takashi
Goodman, Major M.
Goodman, Morris
Gould, Stephen Jay
Grant, Verne Edwin
Hailman, Jack Parker
Hall, Barry Gordon
Hamrick, James Lewis, III
Haubrich, Robert Rice
Highton, Richard Taylor
Hilu, Khidir Wanni
Hsiao, Ting Huan
Hubbs, Clark
Hudson, William Donald, Jr.
Jain, Subodh Kumar
Johnston, Stephen Albert
Jones, Claris Eugene, Jr.
Kaneshiro, Kenneth Yoshimitsu
Kevan, Peter Graham
Kiang, Yun-Tzu
King, James Clement
Knoll, Andrew Herbert
Kreutzer, Richard David
Labanick, George Michael
Laipis, Philip James
Landing, Benjamin Harrison
Larson, (Leonard) Allan
Legler, John Marshall
Leigh, Egbert Giles, Jr.
Levinthal, Mark
Lewis, Edward B.
Lipps, Jere Henry
Lloyd, James Edward
Lombard, R. Eric
Luykx, Peter van Oosterzee
Lynch, Carol Becker
Martin, Joseph Patrick, Jr.
Matten, Lawrence Charles
Mayr, Ernst
McClendon, John Haddaway
McCommas, Steven Andrew
McInnis, Donald Owen
McKinney, Frank Kenneth
Mertz, David Byron
Michod, Richard Earl
Millay, Michael Alan
Mock, Douglas Wayne
Moore, John Alexander
Moore, Richard Harlan
Nei, Masatoshi
Nitecki, Matthew Henry
Nyberg, Dennis
Ohta, Alan Takashi
Olsen, John Stuart
Olson, Everett Claire
Pettus, David
Pianka, Eric Rodger
Pierce, Benjamin Allen
Powers, Dennis Alpha
Raven, Peter Hamilton
Reeder, John R(aymond)
Richmond, Rollin Charles
Rick, Charles Madeira, Jr.
Riley, Monica
Rogers, Claude Marvin
Rogers, Jack David
Rollins, Reed Clark
Rose, Michael Robertson
Roth, Jesse
Rozen, Jerome George, Jr.
Rubinoff, Ira
Ruibal, Rodolfo
Sarich, Vincent M.
Satyanarayana, T.
Schanfield, Moses Samuel
Scheckler, Stephen Edward
Schilling, Edward E.
Schneider, Edward Lee
Schopf, Thomas Joseph Morton
Schwartz, Karlene V.
Scora, Rainer Walter
Sears, Ernest Robert
Seavey, Steven R.
Selander, Robert Keith
Shetler, Stanwyn Gerald
Simpson, George Gaylord
Smiley, Charles Jack
Smith, Albert Charles
Sokolowski, Marla Berger
Spiess, Eliot Bruce
Spolsky, Christina Maria
Stanley, Steven Mitchell
Stearns, Stephen Curtis
Stebbins, George Ledyard
Stern, Kingsley Rowland
Stidd, Benton Maurice
Stini, William Arthur
Sulzbach, Daniel Scott
Tabachnick, Walter Jay
Tamarin, Robert Harvey
Tattersall, Ian Michael
Taylor, Charles Ellett
Taylor, Ronald J.
Thaeler, Charles Schropp, Jr.
Thompson, Vinton
Tiffney, Bruce Haynes
Tseng, Charles C(hiao)
Turner, Bruce Jay
Uhl, Charles H.
Utech, Frederick Herbert
Van Devender, Robert Wayne
Vasek, Frank C(harles)
Vickery, Robert Kingston, Jr.
Vishniac, Helen Simpson
Wagner, David Henry
Waines, John Giles
Walker, James Willard
Ward, Oscar G., Jr.
West, David Armstrong
Whitesell, James Judd
Wijsman, Ellen Marie
Wilbur, Henry Miles
Wilkes, H(ilbert) Garrison
Williams, Daniel Frank
Williamson, Peter George
Willig, Michael Robert
Wilson, Edward Osborne
Winstead, Joe Everett
Woods, Charles Arthur
Wright, Sewall
Wyatt, Robert Edward
Yorks, Pamela Florence
Zavada, Michael Stephan
Zouros, Eleftherios

Genetics and genetic engineering. ***See also*** **AGRICULTURE, MEDICINE.**
Ahmed, Farid El Mamoun
Allard, Robert Wayne
Ames, Bruce N(athan)
Anderson, Gary Bruce
Apirion, David
Ayala, Francisco Jose
Ayles, G. Burton
Baker, William Kaufman
Banerjee, Ranjit
Barban, Stanley
Bartlett, Alan Claymore
Bassham, James Alan
Baxter, John Darling
Beaty, Terri Hagan
Bender, Harvey Alan
Bender, Welcome W.
Bennett, Joan Wennstrom
Benzer, Seymour
Berg, Claire M.
Berg, Douglas E.
Berg, Patricia E.
Berg, Raissa Lvovna
Berger, Edward Michael
Bernstein, Harris
Bertani, Giuseppe
Beutler, Ernest
Bhattacharjee, Jnanendra K.
Bollon, Arthur Peter
Brenchley, Jean Elnora
Brown, Gregory Gaynor
Bullas, Leonard Raymond
Byers, Breck Edward
Byrne, Barbara Jean
Calvo, Joseph Marle
Came, Paul E.
Campbell, Judith Lynn
Carlson, Marian Bille
Carroll, Dana
Carson, Hampton Lawrence
Case, Mary Elizabeth
Caspari, Ernest W(olfgang)
Cavalli-Sforza, Luigi Luca
Chakrabarty, Ananda Mohan
Chakravarti, Aravinda
Chesney, Robert Harold
Christman, Judith Kershaw
Claxton, Larry Davis
Clough, David William
Coffino, Philip
Copeland, James Clinton
Croce, Carlo Maria
Crow, James Franklin
Cummings, Michael Richard
Curiale, Michael Steven
Curtiss, Roy, III
Davidson, Richard Laurence
Davis, Brian Kent
Dean, Donald Harry
Deering, Reginald Atwell
Dennis, Anthony Joseph
Dickson, Robert Carl
Dixon, Gordon Henry
Douglas, Michael Gilbert
Drexler, Henry
Dubes, George Richard
Duckworth, Donna Hardy
Duncan, Ian William
Dutta, Sisir Kamal
Eberle, Helen I.
Eckroat, Larry Raymond
Edenberg, Howard Joseph
Elander, Richard Paul
El-Gewely, Mohamed Raafat
Engelhardt, Dean Lee
Enquist, Lynn William
Epstein, Wolfgang
Erickson, John (Elmer)
Esen, Asim
Ferl, Robert Joseph
Fink, Gerald Ralph
Finnerty, William Robert
Flagg, Raymond Osbourn
Fraser, Thomas Hunter
Friedman, Orrie Max
Friedman, Selwyn Marvin
Galas, David John
Gamborg, Oluf Lind
Gardner, Eldon John
Gartland, William Joseph
Gates, Frederick Taylor, III
Gautsch, James Willard
Gealt, Michael Alan
Geiger, Jon Ross
Gensler, Helen Lynch
Gentile, James Michael
Giddings, Luther Val
Giles, Norman Henry
Glick, J. Leslie
Goldberg, Edward Bleier
Goldberg, Ivan D.
Goldschmidt, Raul Max
Goldstein, Steven B.
Gowans, Charles Shields
Graham, John Borden
Grant, David Miller
Grosch, Josephine Cataldo
Grossman, Lawrence
Guild, Greg
Guild, Walter Rufus
Gumport, Richard I.
Gurney, Elizabeth Tucker Guice
Gutman, George Andre
Hall, Banjamin Downs
Hall, Barry Gordon
Hall, Linda M.
Hamerton, John Laurence
Harding, John Delano
Harriman, Philip Darling
Hartman, Philip Emil
Hatfield, G. Wesley
Haynes, Robert Hall
Hecht, Norman Bernard
Hector, Mina L.
Hershberger, Charles Lee
Herzenberg, Leonard Arthur
Hill, Walter Ernest
Hoch, James Alfred
Hoegerman, Stanton Fred
Hoffman, George Robert
Hoggan, M. David
Hogness, David Swenson
Homyk, Theodore, Jr.
Hood, Leroy Edward
Horiuchi, Kensuke
Hughes, Karen Woodbury
Hundley, Louis Reams
Huskey, Robert John
Ihler, Garret Martin
Infanger, Ann
Ingraham, John Lyman
Ingram, Lonnie O'Neal
Ippen-Ihler, Karin Ann
Jackson, David Archer
Jackson, Ethel Noland
Jacobson, Gunnard Kenneth
Jervis, Herbert Hunter
Johnston, James Bennett
Joslyn, Dennis Joseph
Kaback, David B.
Kaiser, Armin Dale
Kao, Fa-Ten
Kedes, Laurence Herbert
Kennedy, Samuel Ian T.
Kiang, Yun-Tzu
Kirsch, Donald, R.
Kline, Ellis Lee
Knudson, Gregory Blair
Kogoma, Tokio
Krause, Eliot
Kung, Shain-dow
Kushner, Sidney Ralph
Lacks, Sanford Abraham
Laffler, Thomas G.
Lago, Barbara Drake
Laipis, Philip James
Larimer, Frank William
Larkin, Edward Charles
Larson, (Leonard) Allan
Lauer, Florian I.
Lawrence, Christopher William
Lederberg, Joshua
Lee, Lihsyng Stanford
Leibowitz, Michael Jonathan
Levin, Barbara Chernov
Levner, Mark Henry
Ley, Timothy James
Long, George L.
Lovett, Paul Scott
Lundblad, Roger Lauren
Lurquin, Paul Francis
MacQuillan, Anthony Mullens
Magee, Paul Terry

Malamy, Michael Howard
Malone, Robert Edward
Markert, Clement Lawrence
Marrs, Barry Lee
Martin, Presley Frank
Mayeda, Kazutoshi
Mayor, Heather Donald
McFall, Elizabeth
McInnis, Donald Owen
McLaughlin, Calvin S.
Mendelson, Neil Harland
Mertens, Thomas Robert
Meyerowitz, Elliot Martin
Mickelson, Claudia A(nn)
Mielenz, Jonathan Richard
Miller, Robert Verne
Miller, Walter L.
Mittler, Sidney
Mohney, Leone Laura
Montelone, Beth Ann
Moore, Martha May
Mosmann, Timothy Richard
Nei, Masatoshi
Nicklas, Janice Ann
Nightingale, Elena Ottolenghi
Novick, Richard Paul
Novotny, Jiri
Overbye, Karen Marie
Owen, Ray David
Ozer, Harvey Leon
Paule, Marvin R.
Perkins, David D(exter)
Pero, Janice Gay
Peter, Thomas Douglas
Peterson, Peter Andrew
Piatak, Michael, Jr.
Pierce, Benjamin Allen
Prakash, Louise
Prakash, Satya
Press, Newtol
Puck, Theodore Thomas
Pukkila, Patricia Jean
Pun, Pattle Pak Toe
Rabussay, Dietmar Paul
Rathmann, George Blatz
Rawson, James Rulon Young
Ream, Lloyd Walter, Jr.
Redei, Gyorgy Pal
Reed, Albert Paul
Riley, Monica
Riser, Mary Elizabeth
Robertson, Donald Sage
Roman, Herschel Lewis
Sager, Ruth
Salkoff, Lawrence Benjamin
Sarachek, Alvin
Scandalios, John George
Schoeny, Rita Sue
Schrier, Bruce Kenneth
Schwartz, Drew
Sclair, Morton H.
Segal, Shoshana
Severn, David Jones
Shapiro, James Alan
Shepherd, Robert James
Shew, Harry Wayne
Shub, David A.
Simmon, Vincent Fowler
Sinha, Navin Kumar
Sinibaldi, Ralph Michael, sr.
Smith, Cassandra L.
Smith, Hamilton Othanel
Smith, Kendric Charles
Smith, Oliver Hugh
Somerville, Ronald Lamont
Spreitzer, Robert Joseph
Steiner, William Wallace Mokahi
Stevenson, Harlan Quinn
Strelkauskas, Anthony James
Strike, Terry Lee
Stuart, William Dorsey, Jr.
Swanson, Robert A.
Taber, Harry Warren
Temin, Howard Martin
Tye, Bik-Kwoon
Van Alfen, Neal K.
VandenBerg, John Lee
van der Meer, John Peter
Vodkin, Michael Harold
Voelker, Robert Allen
Vournakis, John Nicholas
Vovis, Gerald Francis
Walker, James Roy
Wallace, Douglas Cecil
Warren, Richard Lloyd
Webb, Robert Bradley
Webster, George Calvin
Weir, Bruce Spencer
Williams, Gordon Lee
Wilson, Allan Charles
Womack, James E.
Wosnick, Michael Alan
Wyman, Robert J.
Yarmolinsky, Michael Bezalel
Yasbin, Ronald Eliott
Yoder, Olen C.
Yoon, Jong Sik
Zahler, Stanley Arnold
Zinder, Norton David
Zuccarelli, Anthony Joseph
Zusman, David Robert
Zyskind, Judith Weaver

Genetics, gene actions
Adair, Gerald Michael
Adelberg, Edward Allen
Adler, Paul Neil
Alexander, Mary Louise
Ames, Giovanna Ferro-Luzzi
Anderson, Neil Albert
Aquadro, Charles Frederick
Atchley, William Reid
Atkins, Charles Gilmore
Balbinder, Elias
Barron, Sarah Kathryn Braswell
Bell, A. Earl
Bell, Robin Graham
Bennett, Cecil Jackson
Bentley, Michael Martin
Berger, Edward Michael
Billen, Daniel
Birchler, James Arthur
Bleyman, Lea Kanner
Bonner, James Jose
Bouma, Hessel, III
Brewer, George J.
Bruns, Peter J.
Bryan, John Henry Donald
Bukovsan, Laura Ann
Busch, Robert Henry
Byrne, Barbara Jean
Caspari, Ernest W(olfgang)
Champe, Sewell Preston
Childs, Barton
Chiu, Jen-Fu
Chovnick, Arthur
Christensen, J(ames) Roger
Chu, Ernest Hsiao-Ying
Chung, Young Sup
Clough, David William
Cohen, Stanley Norman
Cole, Charles Norman, Jr.
Cori, Carl Ferdinand
Costa, Max
Couse, Nancy Lee
Crandall, Marjorie Ann
Daniel, William Louis
Dickinson, William Joseph
Doane, Winifred Walsh
Doniger, Jay
Elgin, Sarah Carlisle Roberts
Ely, Berten E., III
Esposito, Rochelle Easton
Eves, Eva Mae
Farmer, James Lee
Figurski, David Henry
Forest, Charlene Lynn
Forrest, Hugh Sommerville
Fournier, Raymond Emile Keith
Fowler, Elizabeth
Frankel, Joseph
Frelinger, Jeffrey Allen
Game, John Charles
Gilboa, Eli
Giles, Norman Henry
Godson, G. Nigel
Goldstein, Elliott S.
Gottesman, Susan
Gross, Robert H.
Hartl, Daniel Lee
Hattman, Stanley Martin
Hickey, Donal Aloysius
Hochman, Benjamin
House, Verl Lee
Howe, Henry Branch, Jr.
Hung, Paul P(erwen)
Imberski, Richard Bernard
Jackson, Ethel Noland
Jacobs, Merle E(mmor)
Johnson, Edward Michael
Johnson, Thomas Eugene
Karn, Robert Cameron
Kenney, Francis Thomas
King, Jonathan Alan
Kohel, Russell James
Kovach, Joseph K.
Kuemmerle, Nancy Benton Stevens
Leach, Jeanette
Levinthal, Mark
Lewin, Benjamin
Lewis, James Alexander
Lieb, Margaret
Lindahl, Lasse Allan
Liu, Leroy Fong
Loper, John Carey
Low, Kenneth Brooks, Jr.
Lucas, Myron Cran
Maguire, Marjorie Paquette
Mahowald, Anthony P.
Margolin, Paul
Maroni, Gustavo Primo
Matthews, Harry Roy
Maxson, Stephen Clark
Maxwell, Joyce Bennett
McClintock, Barbara
McFall, Elizabeth
McGuire, Terry Russell
Mc Kusick, Victor Almon
McMillin, David Edwin
McMorris, F(rederick) Arthur
Melton, Russell Paul
Michels, Corrine Anthony
Modak, Mukund Janardan
Moore, Jay W.
Nebert, Daniel Walter
Nelson, Jr. Oliver Evans
Nichols, Warren Wesley
Nicholson-Guthrie, Catherine Shirley
O'Farrell, Patrick Henry
Ohta, Alan Takashi
Oiler, Larry Wayne
Overbye, Karen Marie
Pall, Martin Lawrence
Palmiter, Richard DeForest
Pandey, Janardan Prasad
Patterson, David
Pelzer, Charles Francis
Petersen, Nancy Sue
Petri, William Hugh
Polacek, Laurie Ann
Postlethwait, John Harvey
Prakash, Satya
Procunier, James Douglas
Radin, David N.
Raper, Carlene Allen
Redei, Gyorgy Pal
Richardson, John Paul
Riddle, Donald Lee
Rogolsky, Marvin
Rossi, John Joseph
Scandalios, John George
Schwartz, Drew
Segal, Shoshana
Seiger, Marvin Barr
Seyfried, Thomas Neil
Shank, Peter R.
Sherald, Allen Franklin, III
Sherman, Fred
Shreffler, Donald Cecil
Shub, David A.
Siciliano, Michael J.
Smith, George Franklin
Smith, Gerald Ralph
Sollner-Webb, Barbara Thea
Steitz, Joan Argetsinger
Stock, David Allen
Stonehill, Elliott H.
Sutton, Harry Eldon
Taylor, Milton William
Tereba, Allan Michael
Timberlake, William Edward
Tompkins, Laurie
Vaidya, Akhil Babubhai
Voelker, Robert Allen
Waelsch, Salome Gluecksohn
Wang, Wei-yeh
Ward, Samuel
Warner, Jonathan Robert
Waxman, Michael Frederick
Wechsler, James Alan
Wells, Robert Dale
Womack, James E.
Woo, Savio Lau Ching
Wood, William Barry, III
Woodruff, Ronny Clifford
Wright, Clarence Paul
Wu, Guang-jer
Wurst, Glen Gilbert
Young, Michael Warren
Yund, Mary Alice

Genetics, genome organization
Adair, Gerald Michael
Asato, Yukio
Basilico, Claudio
Bernstein, Sanford Irwin
Betterley, Donald Alan
Bradley, Walter G.
Bregman, Allyn Aaron
Britten, Roy John
Bruns, Peter J.
Bukhari, Ahmad Iqbal
Bullas, Leonard Raymond
Campbell, Allan McCulloch
Carrano, Anthony Vito
Catlin, B. Wesley
Chovnick, Arthur
Coffin, John Miller
Cohen, Stanley Norman
Cummings, Michael Richard
Daniel, William Louis
DeGiovanni-Donnelly, Rosalie F.
Deininger, Prescott Leonard
Dickinson, William Joseph
Dingle, Richard Douglas Hugh
Douglas, Tommy C.
Dutta, Sisir Kamal
Elgin, Sarah Carlisle Roberts
Emmons, Scott Wilson
Endrizzi, John E.
Engels, William Robert
Erickson, John (Elmer)
Esposito, Rochelle Easton
Farmer, James Lee
Galinat, Walton Clarence
Gerwin, Brenda Isen
Goldstein, Paul
Gray, Joe William
Gustafson, John Perry
Hallick, Richard Bruce
Harding, John Delano
Harford, Agnes Gayler
Henderson, Ann Shirley
Hickey, Donal Aloysius
Hochman, Benjamin
Holmquist, Gerald Peter
Hyde, Beal Baker
Hyman, Richard Walter
Jackson, Raymond Carl
Johnson, Edward Michael
Kaback, David B.
Kao, Fa-Ten
King, James Clement
Lange, Christopher Stephen
Lee, Robert Wingate
Lewin, Benjamin
Lin, Yue Jee
Luykx, Peter van Oosterzee
Martin, Scott McClung
Mathis, Philip Monroe
McCommas, Steven Andrew
Mc Kusick, Victor Almon
Mickey, George Henry
Miksche, Jerome Phillip
Patterson, Rosalyn Victoria Mitchell
Peter, Thomas Douglas
Rao, Dabereu Chandrasekhara
Robbins, Leonard G(ilbert)
Robbins, Robert John
Robertson, Donald Sage
Rose, Ann Marie
Rose, Michael Robertson
Rosner, Judah Leon
Rossi, John Joseph
Rubin, Gerald M.
Sauerbier, Walter
Schvartzman, Jorge Bernardo
Scriver, Charles Robert
Sheehy, Ronald J.
Sheffer, Richard Douglas
Shepherd, Hurley Sidney
Siciliano, Michael J.
Sinclair, John Henry
Singer, Maxine Frank
Sokolowski, Marla Berger
Stein, Diana B.
Swanson, Margaret MacMorris
Thompson, Vinton
Tye, Bik-Kwoon
Villa-Komaroff, Lydia
Vodkin, Michael Harold
Ward, Oscar G., Jr.
Wilson, Dwight Elliott
Wolff, Sheldon
Yokoyama, Shozo
Yoon, Jong Sik
Young, Michael Warren
Zakian, Virginia Araxie

Gravitational biology. *See* **SPACE SCIENCE.**

Immunobiology and immunology
Alevy, Yael Gris
Ansari, Aftab A.
Arthur, Larry Ottis
Austen, K(arl) Frank
Baltimore, David
Barisas, Bernard George, Jr.
Barrett, James Thomas
Beck, Lee Randolph
Bell, George Irving
Bell, Robin Graham
Beller, David I.
Benacerraf, Baruj
Benjamini, Eliezer
Berczi, Istvan
Beutner, Ernst Herman
Bice, David E.
Bloom, Eda Terri
Bona, Constantin A(tanasie)
Bonavida, Benjamin
Borsos, Tibor
Braciale, Thomas Joseph
Braciale, Vivian Lam
Brooks, Colin G.
Brunda, Michael John
Butler, John Edward
Callewaert, Denis Marc
Cancro, Michael Paul
Caren, Linda Ann Davis
Carlo, Jaime Rafael
Carrier, E. Bernard
Carson, Dennis Anthony
Castro, Alberto
Cathou, Renata Egone
Cerini, Costantino Peter
Chi, David S.
Choi, Yong Sung
Chused, Thomas M.
Clark, William R.
Coligan, John Ernest
Conway, Thomas Patrick
Conway de Macario, Everly
David, Gary Samuel
Davies, Philip
Day, Eugene Davis
DeLustro, Frank Anthony
DeMeio, Joseph Louis
Dennis, Anthony Joseph
Douglas, Tommy C.
Dreesman, Gordon Ronald
Dusanic, Donald Gabriel
Dyminski, John W(ladyslaw)
Egan, Marianne Louise
Elgert, Klaus Dieter
Ettinger, Anna Marie
Fernandez-Cruz, Eduardo P.
Fidler, John Michael
Finkelstein, Richard Alan
Fitch, Frank Wesley
Flaherty, Lorraine Amelia
Frasch, Carl Edward
Freed, John Howard
Fritz, Robert Bartlett
Fuccillo, David Anthony
Gately, Maurice Kent
Gates, Frederick Taylor, III
Genco, Robert Joseph
Ghaffar, Abdul
Gleicher, Norbert
Glovsky, M. Michael
Goldenberg, Marvin M.
Goodman, Joel Warren
Goust, Jean-Michel Christian
Granlund, David John
Graziano, Kenneth Donald
Green, William Robert
Greene, Elias Louis
Gregory, Francis J.
Gregory, Richard Lee
Gutman, George Andre
Hackett, Charles Joseph
Halonen, Marilyn Jean
Hargis, Betty Jean
Hashim, George A.
Havas, Helga Francis
Henney, Christopher Scot
Heppner, Gloria Hill
Herscowitz, Herbert Bernard
Herzenberg, Leonard Arthur
Hillis, William Daniel, Sr.
Hogan, Yvonne Holland
Hokama, Yoshitsugi
Holowka, David Allan
Hood, Leroy Edward
Horng, Wayne J(ing-Wei)
Hsu, Clement C.S.
Hsu, Hei-ti
Hsu, Shu Ying
Huard, Thomas King
Ichiki, Albert T.
Inman, John K.
Iverson, Gilbert Michael
Jacobs, Diane Margaret
Kaplan, Alan Marc.
Kapral, Frank Albert
Karp, Richard Dale
Kaufman, Leo
Kemp, Walter Michael
Kim, Byung Suk
Kim, Yoon Berm
Kimball, Edward Saul
Kobilinsky, Lawrence
Koo, Peter Hung-Kwan
Koprowski, Hilary
Koshland, Marian Elliott
Kuo, Cho-Chou
Kushner, Irving
Ladisch, Stephan
Lafuse, William Perry
Lahita, Robert George
Lala, Peeyush Kanti
Lamm, Michael Emanuel
Lamster, Ira Barry
Laux, David Charles
Lawrence, David A.
Lee, Sang He
Lehrer, Samuel B.
Leon, Myron A.
Leong, Stanley Pui-Lock
Levy, Elinor M.
Liao, Shuen-Kuei
Linscott, William Dean
Loughman, Barbara Ellen Evers
Low, Teresa Lingchun Kao
Lubaroff, David Martin
Mage, Michael Gordon
Makinodan, Takashi
Martz, Eric
Mathieson, Bonnie Jean
McNamara, T(homas) F(rancis)
Medzihradsky, Joseph Ladislas
Merluzzi, Vincent James
Meyers, Paul
Michael, Jacob Gabriel
Mickelson, Claudia A(nn)
Mickley, Harold Somers
Montgomery, Paul Charles
Moody, Charles Edward, Jr.
Morgan, Donald O'Quinn
Mortensen, Richard F.
Mosmann, Timothy Richard
Mueller, Nancy Alice Schnieder
Muscoplat, Charles Craig
Nakamura, Ichiro
Nash, Donald Robert
Nielsen, Klaus H.B.
Nisonoff, Alfred
Norin, Allen J.
Olsen, Richard George
Panagides, John
Paque, Ronald Edward
Paul, William Erwin
Pierce, Carl William
Plate, Janet Margaret-Dieterle
Platsoucas, Chris Dimitrios
Pollack, Sylvia Byrne
Ragland, William Lauman, III
Raikow, Radmila Boruvka
Ramsey, Robert Bruce
Ray, Prasanta Kumar
Reiss, Carol Shoshkes
Rodkey, Leo Scott
Rossen, Roger Downey
Rothberg, Richard Martin
Roux, Kenneth Henry
Scharff, Matthew Daniel
Schwaber, Jerrold
Schwartz, Ronald Harris
Segre, Diego
Segre, Mariangela
Sher, Franklin Alan
Shreffler, Donald Cecil
Sidky, Younan Abdel-Malik
Sindelar, William Francis
Smith, Hamilton Othanel
Solotorovsky, Morris
Sonnenfeld, Gerald
Specter, Steven Carl
Stamper, Hugh Blair, Jr.
Stanton, Glennon John
Storb, Ursula
Strelkauskas, Anthony James
Strober, Samuel
Talmage, David Wilson
Till, Gerd Oskar
Trown, Patrick Willoughby
Tse, Harley Y.
Tseng, Jeenan
Vilcek, Jan Tomas
Volkman, Alvin
Walker, Sharyn Marie
Waltenbaugh, Carl Ralph
Wechter, William Julius
Wei, Robert
Weindruch, Richard Howard
Winkelhake, Jeffrey Lee
Wolberg, Gerald
Wood, David Dudley
Wright, John Ricken
Wright, Larry Lyle
Wright, Richard Kenneth
Wu, William Gay
Wust, Carl John
Zarco, Romeo M.
Zeiger, Allen Richard
Zimmerman, Daniel Hill
Zimmerman, Sarah E.
Zurawski, Vincent Richard, Jr.

Immunocytochemistry
Benno, Robert Howard
Carlo, Jaime Rafael
Crim, Joe William
Eldred, William Doughty, III
Everhart, Donald Lee
Gaffar, Abdul
Goebel, Walther Frederick
Hawthorne, Marion Frederick
Ho, Raymond How-Chee
Hoffmann, Louis Gerhard
Iverson, Gilbert Michael
King, Joan Caluda
Lafuse, William Perry
Lowe, Irene Posner
Mage, Michael Gordon
McNeill, Thomas Hugh
Nilaver, Gajanan
Pratt, Lee Herbert
Reiner, Anton John
Reiss, Errol
Sibley, Carol Hopkins

Stuart, William Dorsey, Jr.
Vacca, Linda L.

Membrane biology
Aberg, Gunnar A.K.
Adelberg, Edward Allen
Al-Bazzaz, Faiq Jaber
Andrew, Clifford George
Banschbach, Martin Wayne
Baumann, Heinz
Birckbichler, Paul Joseph
Bonavida, Benjamin
Braun, Phyllis Cellini
Burke, Patricia Virginia
Candia, Oscar A.
Carson, Steven Douglas
Chacko, George Kutty
Cone, Robert Edward
Conway, Thomas Patrick
Cori, Carl Ferdinand
Diwan, Joyce Johnson
Earhart, Charles Franklin, Jr.
Eaton, Barbara Ruth
Epstein, Wolfgang
Esser, Alfred F.
Fischbarg, Jorge
Forest, Charlene Lynn
Fox, C. Fred
Frazier, Loy William, Jr.
Gilbert, Daniel Lee
Gilbertson, John Robert
Godlfeder, Anna
Goldfine, Howard
Goldin, Stanley Michael
Goodman, David Barry Poliakoff
Goodman, Steven Richard
Haywood, Anne Mowbray
Hazelbauer, Gerald Lee
Hendler, Richard Wallace
Heppel, Leon Alma
Hester, Richard Kelly
Holden, Joseph Thaddeus
Jacquez, John Alfred
Janis, Ronald Allen
Kaback, H. Ronald
Kamin, Henry
Katzung, Bertram George
Kempson, Stephen Allan
Kennedy, Eugene Patrick
Kern, Harold Lloyd
Kimelberg, Harold Keith
Kimmich, George Arthur
Koblick, Daniel Cecil
Kung, Ching
LaBelle, Edward Francis
Ledbetter, Mary Lee Stewart
Low, Philip Stewart
Lyles, Douglas Scott
Markwell, Mary Ann K.
Mayer, Manfred Martin
Mendelsohn, Richard
Morre, D. James
Morrison, Martin
Moscatelli, Ezio Anthony
Painter, Richard Grant
Penniston, John Thomas
Perrone, Ronald David
Petty, Howard Raymond
Powell, Gary Lee
Ralston, Douglas Edmund
Randall, Linda Lea
Ray, Tushar Kanti
Reddy, Neelupalli Bojji
Ritter, Carl Alan
Roberts, Robert Michael
Robinson, Richard Bruce
Roseman, Saul
Rosenthal, Kenneth Steven
Schlegel, Robert Allen
Schroeder, Friedhelm
Segrest, Jere Palmer
Steck, Theodore Lyle
Stephens, Cathy Lamar
Stokes, John Bispham, III
Stroud, Robert Michael
Sweadner, Kathleen J.
Tanford, Charles
Tedeschi, Henry
TenEick, Robert Edwin
Tosteson, Daniel Charles
Troy, Frederic Arthur, II
Van Houten, Judith Lee
Wiech, Norbert Leonard
Wilson, Charles Lindsay

Microbiology
Anderson, Mauritz Gunnar
Atkins, Charles Gilmore
Baker, Kenneth Frank
Baker, Thomas Irving
Bambara, Robert Anthony
Banwart, George J.
Baross, John Allen
Barrett, James Thomas
Beck, Doris Jean
Beer, Steven Vincent
Belikoff, Steven William
Bennett, Joan Wennstrom
Berg, Howard Curtis
Berliner, Martha D.
Bernlohr, Robert William
Bernstein, Carol
Bertani, Giuseppe
Betterley, Donald Alan
Billhimer, Ward Loren, Jr.
Bishop, Paul Edward
Bopp, Lawrence Howard
Borgatti, Alfred Lawrence
Bozzola, John Joseph
Braun, Phyllis Cellini
Brenchley, Jean Elnora
Brierley, Corale Louise
Brock, Thomas Dale
Broker, Thomas Richard
Bruce, Barbara Jean
Bukhari, Ahmad Iqbal
Bukovsan, Laura Ann
Bulla, Lee Austin, Jr.
Bullerman, Lloyd Bernard
Carrier, E. Bernard
Catlin, B. Wesley
Chadha, Kailash Chandra
Chakrabarty, Ananda Mohan
Champe, Sewell Preston
Chang, Te-Wen
Chesney, Robert Harold
Chou, Liu-Gei
Cohen, Joel Ralph
Cohen, Paul S.
Coleman, Annette Wilbois
Collins, Arlene Rycombel
Colwell, Rita Rossi
Conway, Kenneth Edward
Crang, Richard Francis Earl
Cronholm, Lois S.
Curtiss, Roy, III
Daniels-Hetrick, Barbara Ann
Daoust, Donald Roger
deFiebre, Conrad William
Dehority, Burk Allyn
Della-Latta, Phyllis
De Moss, Ralph Dean
Diener, Urban Lowell
Docherty, John Joseph
Doi, Roy Hiroshi
Domer, Judith Elaine
Dougherty, Robert Malvin
Douthit, Harry Anderson, Jr.
Dow, Martha Anne
Downey, Ronald Joseph
Drexler, Henry
Durbin, Richard Duane
Eagon, Robert Garfield
Earhart, Charles Franklin, Jr.
Edmonds, Robert Leslie
Eiserling, Frederick Allen
Elander, Richard Paul
Elbein, Alan David
Elliott, Larry Phillip
Ely, Berten E., III
Eslyn, Wallace Eugene
Essex, Myron Elmer
Everhart, Donald Lee
Fabricant, Catherine Grenci
Ferchau, Hugo Alfred
Ferone, Robert
Finkelstein, Richard Alan
Finnerty, William Robert
Fitzgerald, Edward Aloysius
Flournoy, Dayl Jean
Foster, Edwin Michael
Founds, Henry William, Jr.
Fox, Eugene Noah
Frasch, Carl Edward
Friedman, Selwyn Marvin
Friedmann, E(merich) Imre
Frisell, Wilhelm Richard
Fryer, John L.
Fulbright, Dennis Wayne
Gaffar, Abdul
Garfinkle, Barry David
Geiger, Jon Ross
Gerhardt, Philipp
Gill, D. Michael
Gleicher, Norbert
Goebel, Edwin Mark
Goldberg, Ivan D.
Gorzynski, Eugene Arthur
Gowans, Charles Shields
Gregory, Francis J.
Gregory, Richard Lee
Grogan, James Bigbee
Guterman, Sonia Kosow
Hadden, Edward Leal, Jr.
Hagen, Charles Alfred
Hamilton, Velda Marie
Hammill, Terrence Michael
Hammond, Benjamin Franklin
Harman, Gary Elvan
Harris, Denny Olan
Hartman, Paul Arthur
Hegeman, George Downing
Hemmingsen, Barbara Bruff
Hesseltine, Clifford William
Highley, Terry Leonard
Ho, Hon Hing
Hopkins, Donald Lee
Horowitz, Bruce Richard
Howe, Henry Branch, Jr.
Huber, Don Morgan
Humphrey, Ronald De Vere
Hurwitz, Jerard
Ingraham, John Lyman
Ippen-Ihler, Karin Ann
Jacobson, Gunnard Kenneth
Jacobson, Richard Otto
Jannasch, Holger Windekilde
Johnson, F. Brent
Johnson, Roy Melvin
Johnston, Eugene Benedict
Joklik, Wolfgang Karl
Jong, Shung Chang
Joshi, Madan Mohan
Kadis, Solomon
Kapral, Frank Albert
Karp, Richard Dale
Kasai, George Joji
Kelleher, William Joseph
Keller, John Randall
Kelman, Arthur
Kline, Ellis Lee
Krieg, Noel Roger
Krigsvold, Dale Thomas
Kuo, Cho-Chou
Kuserk, Frank Thomas
Lago, Barbara Drake
Larkin, John Montague
Laux, David Charles
Leach, Franklin Rollin
Leath, Kenneth Thomas
Leive, Loretta
Lerner, Stephen Alexander
Leslie, John Franklin
Lewis, Donald Howard
Lipmann, Fritz (Albert)
Lippincott, James Andrew
Litchfield, John Hyland
Lockwood, John LeBaron
Lukezic, Felix Lee
Lumsden, Robert Douglas
Luria, Salvador Edward
Lyda, Stuart Davisson
MacQuillan, Anthony Mullens
Maramorosch, Karl
Marchin, George L.
Marquis, Robert Edward
Marshall, Vincent dePaul
Martin, Scott Elmore
Martin, Scott McClung
Masker, Warren Edward
Mayberry, William Roy
Mayol, Pete Syting
McCoy, Randolph Edward
Mc Elroy, William David
McNamara, T(homas) F(rancis)
McNicol, Lore Anne
Mehta, Bipin Mohanlal
Mendelson, Neil Harland
Meyer, Harry M., Jr.
Miller, Chris H.
Miller, Glendon Richard
Moehring, Joan Marquart
Mohney, Leone Laura
Montgomery, Paul Charles
Moody, Arnold Ralph
Moody, Charles Edward, Jr.
Moore, Larry Wallace
Moorman, Gary Williams
Morris, J(oseph) Anthony
Mortlock, Robert Paul
Najarian, Haig Hacop
Nathans, Daniel
Neidhardt, Frederick Carl
Ng, Thomas K.
Nielsen, Klaus H.B.
Orland, Frank Jay
Ostazeski, Stanley Anthony
Otero, Raymond B.
Pattee, Peter Arthur
Phillips, John Howell, Jr.
Phipps, Patrick Michael
Pomper, Seymour
Poston, John Michael
Pratt, Charles Walter
Pun, Pattle Pak Toe
Raju, Namboori Bhaskara
Reeves, Fontaine Brent
Reiss, Errol
Reissig, Jose Luis
Richards, Gary Paul
Rogolsky, Marvin
Rosan, Burton
Saad, Sami Michel
Salsbury, Robert Lawrence
Samuelson, Don Arthur
Sarachek, Alvin
Savageau, Michael Antonio
Schadler, Daniel Leo
Schaechter, Moselio
Schoknecht, Jean Donze
Setliff, Edson Carmack
Shan, Hsin-Tsan Grace
Shanmugam, Keelnatham Thirunavukkarasu
Shaw, James Elwood
Sherris, John Charles
Shew, Harry Wayne
Shih, Jason Chia-Hsing
Silver, Simon David
Simonson, Lloyd Grant
Simpson, Frederick James
Singh, Shiva Pujan
Six, Howard Ronald
Smidt, Mary Louise
Smith, Oliver Hugh
Sonenshein, Abraham Lincoln
Sreevalsan, Thazepadath
Stadtman, Earl Reece
Stadtman, Thressa Campbell
Staubus, John Reginald
Stevens, Roy Harris
Stock, David Allen
Stotzky, Guenther
Taylor, Bernard Franklin
Thacore, Harshad Rai
Thomas, Donald Charles
Torriani, Annamaria Gorini
Tzianabos, Theodore
Underdahl, Norman Russell
Vilcek, Jan Tomas
Volcani, Benjamin Elazari
Wagner, Morris
Wang, Chun-Juan K.
Ward, Calvin Herbert
Waxman, Michael Frederick
Webb, Robert Bradley
Weinhold, Albert Raymond
Weinstein, David
Weiss, Emilio
Westby, Carl Andrew
Wetmur, James Gerard
Williams, Marshall Vance, Jr.
Wolf, Kenneth Edward
Wolfe, Ralph Stoner
Wood, Harland Goff
Wu, William Gay
Wyrick, Priscilla Blakeney
Zusman, David Robert

Molecular biology
Aaslestad, Halvor Gunerius
Adams, Steven Paul
Adler, Julius
Alberts, Bruce Michael
Amer, M. Samir
Anderson, Carl William
Anderson, Thomas Foxen
Anderson, W. French
Asato, Yukio
Baker, Robert Frank
Baker, Thomas Irving
Baldwin, Robert Lesh
Ballal, Raghu Veer
Baltimore, David
Banerjee, Amiya Kumar
Banerjee, Ranjit
Baserga, Renato Luigi
Beachy, Roger Neil
Bean, Charles Palmer
Bean, William Joseph, Jr.
Bello, Jake
Benade, Leonard Edward
Benbow, Robert Michael
Benham, Craig John
Berg, Douglas E.
Berg, Patricia E.
Bernstein, Carol
Bernstein, Harris
Blair, Carol Dean
Blake, Richard Douglas
Bond, Clifford Walter
Bonner, James
Bonner, James Jose
Brockman, William Warner
Brown, Gregory Gaynor
Bruce, Barbara Jean
Brutlag, Douglas Lee
Burger, Richard M.
Camiolo, Sarah May
Campbell, Judith Lynn
Carlson, John Edward
Carlson, Marian Bille
Carroll, Dana
Carter, Timothy Howard
Cech, Thomas Robert
Chalmers, John H., Jr.
Chen, Chong-Maw
Cherbas, Peter Thomas
Chin, William Waiman
Chow, Louise Tsi
Chung, Young Sup
Clark, John Magruder, Jr.
Coffino, Philip
Cohen, Paul S.
Cohen, Philip Pacy
Cole, Charles Norman, Jr.
Collett, Marc Stephen
Conway, Thomas William
Cottrell, Stephen F.
Cox, G. Stanley
Croce, Carlo Maria
Crooke, Stanley Thomas
Cunningham, Bruce Arthur
Cunningham, Glenn Norman
Danna, Kathleen Janet
Davidson, Eric Harris
Dawid, Igor Bert
Dean, Donald Harry
Deering, Reginald Atwell
DeGiovanni-Donnelly, Rosalie F.
Dickson, Robert Carl
Dixon, Gordon Henry
Dixon, Jack E.
Doi, Roy Hiroshi
Doniger, Jay
Duker, Nahum Johanan
Dutko, Francis Joseph, Jr.
Eastman, Alan Richard
Eberle, Helen I.
Eckhart, Walter
Edelman, Gerald Maurice
Eden, Francine Claire
Edenberg, Howard Joseph
Edmonds, Mary Patricia
Eichhorn, Gunther Louis
Eiserling, Frederick Allen
Engelking, Henry Mark
Ennis, Herbert L.
Enquist, Lynn William
Evans, Mary Jo
Eves, Eva Mae
Fallon, Ann Marie
Fan, Hung Y
Fanning, George Richard
Felsenfeld, Gary
Ferl, Robert Joseph
Fernandes, Daniel James
Figurski, David Henry
Flanegan, James Bert
Fluck, Michele Marguerite
Fournier, Raymond Emile Keith
Fowler, Elizabeth
Fox, George Edward
Fraenkel-Conrat, Heinz
Fresco, Jacques Robert
Friedberg, Errol Clive
Fujimura, Robert Kanji
Fulton, Chandler Montgomery
Furth, John Jacob
Galas, David John
Game, John Charles
Ganesan, Adayapalam Tyagarajan
Garrett, Reginald Hooker
Gartland, William Joseph
Gautsch, James Willard
Geiduschek, E(rnest) Peter
Gelfand, David H.
Ghangas, Gurdev S.
Gilbert, Walter
Gilboa, Eli
Gillham, Nicholas Wright
Ginsberg, Harold Samuel
Glitz, Dohn George
Goeddel, David Van Norman
Goldschmidt, Raul Max
Goldstein, Elliott S.
Goldstein, Steven B.
Goodman, Howard M.
Goodman, Joel Warren
Goodridge, Alan Gardner
Goorha, Rakesh Mohan
Gottesman, Susan
Grant, David Miller
Green, Harry
Green, Maurice
Greenlee, Lorance Lisle
Grosch, Josephine Cataldo
Gross, Robert H.
Grubman, Marvin
Guild, Greg
Guild, Walter Rufus
Hadidi, Ahmed Fahmy
Hahn, Fred Ernst
Hamerton, John Laurence
Hamilton, Leonard Derwent
Hanafusa, Hidesaburo
Hand, Roger
Harford, Agnes Gayler
Harriman, Philip Darling
Harrington, Rodney Elbert
Harrington, William Fields
Haseltine, William Alan
Hashimoto, Shuichi
Hattman, Stanley Martin
Hays, John Bruce
Hayward, William S.
Hazelbauer, Gerald Lee
Hecht, Norman Bernard
Henderson, Ann Shirley
Herbert, Edward
Hill, Walter Ernest
Ho, Chien
Horecker, Bernard Leonard
Horiuchi, Kensuke
Horowitz, Jack
Hough-Evans, Barbara Raymond
Howe, Chin C.
Hung, Paul P(erwen)
Hunter, Eric
Hurkman, William James, II
Hurwitz, Jerard
Iglewski, Wallace Joseph
Jackson, David Archer
Jacobson, Karl Bruce
Jacobson, Richard Otto
Jaworski, Ernest George
Jernigan, Robert Lee
Johnston, Stephen Albert
Kalimi, Mohammed Yahya
Kaplan, Barry Bernard
Kashmiri, Syed V.S.
Kessler, Robert Evans
Kettlewell, Neil MacKewan
Khoury, George
Kingsbury, David Wilson
Kirsch, Donald, R.
Knudson, Gregory Blair
Kopecko, Dennis J.
Kornberg, Thomas Bill
Kretsinger, Robert H.
Kuemmerle, Nancy Benton Stevens
Kung, Shain-dow
Langridge, Robert
Larimer, Frank William
Lawrence, Christopher William
Leamnson, Robert Neal
Ledinko, Nada
Lehman, William Jeffrey
Lerner, Richard Alan
Lindahl, Lasse Allan
Liu, Leroy Fong
Liu, Teh-Yung
Livingston, David Morse
Long, George L.
Lovett, Paul Scott
Lu, Ponzy
Luck, Dennis Noel
Magee, Paul Terry
Mahler, Inga R.
Major, Eugene Oliver
Malamy, Michael Howard
Malone, Robert Edward
Marchin, George L.
Margoliash, Emanuel
Margolin, Paul
Marrs, Barry Lee
Martin, Malcolm Alan
Marzluf, George Austin
Masker, Warren Edward
Matthews, Brian Wesley
Matthysse, Ann Gale
McClure, William Robert
McCorkle, George Maston
McHenry, Charles Steven
McKelvy, Jeffrey Forrester
McNicol, Lore Anne
Meehan, Thomas
Meyerowitz, Elliot Martin
Michels, Corrine Anthony
Miller, Robert Verne
Modak, Mukund Janardan
Model, Peter
Montelone, Beth Ann
Mulligan, Richard C.
Nayak, Debi Prosad
Neidhardt, Frederick Carl
Nemer, Martin Joseph
Novick, Richard Paul
Novotny, Jiri
Ochoa, Severo
Oeltmann, Thomas Napier
Oiler, Larry Wayne
Pardue, Mary Lou
Pato, Martin Leon
Patrick, Michael Heath
Pero, Janice Gay
Perrault, Jacques
Piatak, Michael, Jr.
Plocke, Donald Joseph
Poccia, Dominic Louis
Pogo, A, Oscar
Polatnick, Jerome
Pons, Marcel William
Prakash, Louise
Pratt, Charles Walter
Preiss, Jack
Procunier, James Douglas
Ptashne, Mark Stephen
Pukkila, Patricia Jean
Rabussay, Dietmar Paul
Raj, Harkisan Dunichand
Randall, Linda Lea
Ream, Lloyd Walter, Jr.
Reed, Albert Paul
Reed, Lester James

Reichenbecher, Vernon Edgar, Jr.
Reichmann, Manfred Eliezer
Reilly, Joseph Garrett
Reissig, Jose Luis
Reynafarje, Baltazar Davila
Riggin, Charles Henry, Jr.
Robin, Eugene Debs
Roman, Ann
Rose, Ann Marie
Rosenkranz, Herbert S.
Rosner, Judah Leon
Rottman, Fritz, M.
Roufa, Donald Jay
Rownd, Robert Harvey
Rubin, Gerald M.
Rueckert, Roland Rudyard
Rutman, Robert J.
Samuel, Charles E.
Sarkar, Nurul Haque
Schaechter, Moselio
Schlesinger, Milton Joseph
Schuster, Todd Mervyn
Schwartz, Ira S.
Sclair, Morton H.
Scott, June Rothman
Sehgal, Pravinkumar Bhagatram
Sen, Ganes C(handra)
Shatkin, Aaron Jeffrey
Shepherd, Hurley Sidney
Sherman, Fred
Sherman, Louis Allen
Shinnick, Thomas Michael
Simons, Samuel Stoney, Jr.
Singer, Maxine Frank
Singh, Arjun
Smith, Cassandra L.
Smith, Emil L.
Smith, Gerald Ralph
Smith, Lewis Dennis
Smith, Robert Charles
Sollner-Webb, Barbara Thea
Sonenshein, Abraham Lincoln
Spolsky, Christina Maria
Spremulli, Linda Lucy
Stein, Diana B.
Stein, Gary Stephen
Steinberg, Martin H.
Steitz, Joan Argetsinger
Stenesh, Jochanan
Stent, Gunther Siegmund
Stephenson, Mary Louise
Storb, Ursula
Strauss, Bernard S.
Taber, Harry Warren
Tershak, Daniel Richard
Tomizawa, Junichi
Torriani, Annamaria Gorini
Udenfriend, Sidney
Vaidya, Akhil Babubhai
Vande Woude, George F., Jr.
Vold, Barbara Schneider
Volsky, David Julian
von Hippel, Peter Hans
Vovis, Gerald Francis
Walker, James Roy
Wallace, Douglas Cecil
Waters, David John
Waters, Larry Charles
Watson, James Dewey
Watson, Kenneth Fredrick
Webster, George Calvin
Wechsler, James Alan
Weinberg, Robert Allen
Weiss, Richard Louis
Weissbach, Herbert
Wells, Marion Robert
Werth, Jean Marie
Wiberg, John Samuel
Wickstrom, Eric
Willard-Gallo, Karen Elizabeth
Williams, Gordon Lee
Wilson, Samuel H.
Winicov, Ilga Butelis
Wosnick, Michael Alan
Wu, Guang-jer
Wyrick, Priscilla Blakeney
Yanofsky, Charles
Yarmolinsky, Michael Bezalel
Yoder, Olen C.
Young, Delano Victor
Zahler, Stanley Arnold
Zakian, Virginia Araxie
Zuccarelli, Anthony Joseph
Zyskind, Judith Weaver

Morphology
Adler, Irving
Archibald, Patricia A.
Bayer, Shirley Ann
Berrios-Ortiz, Angel
Bhatnagar, Kunwar P.
Boyne, Alan Frederick
Brotzman, Harold George
Bruck, David Kenneth
Brush, Alan Howard
Byers, George William
Byrd, Kenneth Elburn
Carothers, Zane Bland
Chiasson, Robert Breton
Clark, George Alfred, Jr.
Commisso, Franklyn W.
Daily, Fay Kenoyer
Edgar, Arlan Lee
Emino, Everett Raymond
Frazzetta, Thomas H.
Gabel, J. Russel
Gans, Carl
Ghoshal, Nani Gopal
Gona, Amos Gnanaprakasham
Grant, Richard Evans
Harris, Denny Olan
Hickey, Leo Joseph
Jee, Webster Shew Shun
Juncosa, Adrian Martin
Koutnik, Daryl Lee
Lane, Roger Lee
Lersten, Nels Ronald
Lombard, R. Eric
Maderson, Paul F.A.
McConnell, Dennis Brooks
Mikesell, Jan Erwin
Morey, Elsie D.
Moseley, Maynard Fowle, Jr.
Olive, Lindsay Shepherd
Oliver, Douglas Lamar
Puglisi, Susan Grace
Rembert, David Hopkins, Jr.
Rennels, Marshall Leigh
Rogers, Donald Philip
Scalia, Frank Richard
Schaefer, Carl Walter, II
Sherman, Robert George
Shively, James Nelson
Snider, Jerry Allen
Spurr, Arthur Richard
Sybers, Harley Duane
Tseng, Charles C(hiao)
Varkey, Alexander
Webster, Terry Richard
Williams, James McSpadden
Williams, Kenneth Lee
Wilson, Kenneth Sheridan
Yamaguchi, Mamoru

Neurobiology
Agrawal, Harish Chandra
Allen, Theresa Ohotnicky
Andrew, Clifford George
Asanuma, Hiroshi
Bacon, Jonathan Peter
Baizer, Joan Susan
Baldino, Frank, Jr.
Barker, Jeffery Lange
Barlow, Robert Brown, Jr.
Barnes, David Edward
Barnes, Karen Louise
Bayer, Shirley Ann
Beattie, Michael Stephen
Beck, Mary McLean
Beitz, Alvin James
Bennett, Michael Vander Laan
Benno, Robert Howard
Benzer, Seymour
Berger, Theodore William
Berman, Nancy E.
Bernstein, Jerald J(ack)
Biber, Michael Peter
Bishop, Beverly P(etterson)
Bizzi, Emilio
Bloom, Floyd Elliott
Bodian, David
Bohn, Martha D.
Boyd, Eleanor H.
Boyne, Alan Frederick
Bradley, Walter G.
Brimijoin, William Stephen
Broadwell, Richard Dow
Brown, Paul Burton
Browning, Michael Douglas
Brunjes, Peter Crawford
Brunso-Bechtold, Judy Karen
Burne, Richard Allen
Busis, Neil Amdur
Calvin, William Howard
Campbell, Carlos Boyd Godfrey
Carr, Daniel Barry
Casey, Michael Allen
Caviness, Verne Strudwick, Jr.
Chang, Kwen-Jen
Cheng, Mei-Fang Hsieh
Church, Allen Charles
Clemens, James Allen
Cohen, David Harris
Coleman, Paul D.
Costello, Walter James
Cowan, William Maxwell
Crain, Stanley M.
Crick, Francis Harry Compton
Dafny, Nachum Frenkel
Dasheiff, Richard Mitchell
Davis, Leonard George
de Lanerolle, Nihal Chandra
de la Torre, Jack Carlos
Demeter, Steven
Denburg, Jeffrey Lewis
DeSantis, Mark Edward
De Vries, George Henry
Dibner, Mark Douglas
Dietrich, W. Dalton, III
Donovick, Peter J.
Dowling, John Elliott
Dubner, Ronald
Dusenbery, David Brock
Easter, Stephen Sherman, Jr.
Edds, Louise Luckenbill
Edwards, John Stuart
Egger, Maurice David
Ehrenpreis, Seymour
Eldred, William Doughty, III
Ellman, George
Erulkar, Solomon David
Etgen, Anne Marie
Faber, Donald Stuart
Faden, Alan Ira
Feldman, Susan C.
Fernald, Russell Dawson
Fisher, Steven Kay
Fite, Katherine Virginia
Flexner, Louis Barkhouse
Flood, Dorothy Garnett
Forman, David Sholem
Friedhoff, Arnold
Friesen, Wolfgang Otto
Frost, Douglas Owen
Funch, Paul Gerard
Furshpan, Edwin Jean
Gash, Don Marshall
George, Stephen Anthony
Gerber, Joseph Charles, III
Gershon, Michael David
Ghetti, Bernardino
Gilbert, Charles D.
Glaser, Donald A(rthur)
Glendenning, Karen Kircher
Glusman, Silvio
Gobel, Stephen
Goldberg, Alan Marvin
Goldin, Stanley Michael
Gona, Amos Gnanaprakasham
Goodman, Corey Scott
Gorlick, Dennis Lester
Grenell, Robert Gordon
Grisell, Ronald David
Gross, Paul Munn
Grossfeld, Robert Michael
Grunwald, Gerald Bruce
Guillery, Rainer Walter
Haigler, Henry James
Hall, Linda M.
Halperin, John Jacob
Hamill, Robert Wallace
Hand, Peter James
Harris, Charles Leon
Hawkins, Robert Drake
Hawrot, Edward
Helke, Cinda Jane
Heller, Allen Harvey
Hemmendinger, Lisa M.
Hildebrand, John G(rant), III
Hirsch, Helmut Villard Buntenbroich
Ho, Raymond How-Chee
Hoffman, Paul Ned
Hornung, David Eugene
Hubel, David Hunter
Humbertson, Albert O., Jr.
Hyndman, Arnold Gene
Hyson, Michael Terry
Irwin, Louis Neal
Ishii, Douglas Nobuo
Jacklet, Jon W.
Jen, Philip Hung Sun
Kaczmarek, Leonard Konrad
Kafka, Marian Adele Stern
Kandel, Eric Richard
Kaplan, Barry Bernard
Kendig, Joan Johnston
Kennedy, Linda Mann
Kimelberg, Harold Keith
Kimes, Alane Susan
King, Frederick Alexander
Kirby, Margaret Loewy
Koelle, George Brampton
Kornguth, Steven Edward
Krebs, Helmut Waldemar Graf von Thorn
Kung, Ching
Kvist, Tage Nielsen
Lauder, Jean Miles
Leibovic, K. Nicholas
Lennard, Paul Ross
Levitan, Herbert
Levy, Nelson Louis
Lewis, James Alexander
Lewis-Pinke, Ellen Ruth
Light, Alan Ray
Loewenstein, Werner Randolph
Loh, Yoke Peng
Lowndes, Herbert Edward
Lukas, Scott Edward
Macagno, Eduardo Roberto
Macrides, Foteos
Malmgren, Leslie Theodore, Jr.
Markelonis, George Joseph, Jr.
Marrocco, Richard Thomas
McBride, William Joseph, Jr.
McCrea, Robert Alan
McElligott, Sandra G.
McKelvy, Jeffrey Forrester
McLoon, Linda Kirschen
McNamara, James O'Connell
McNeill, Thomas Hugh
Mendelson, Thea
Messing, Rita Bailey
Miselis, Richard R.
Monjan, Andrew Arthur
Morest, Donald Kent
Mountcastle, Vernon Benjamin, Jr.
Neary, Joseph Thomas
Newman, John Dennis
Nicklas, William John
Norton, William Thompson
Oliver, Douglas Lamar
Olivo, Margaret Anderson
O'Steen, Wendall Keith
Pallotta, Barry S
Park, Dong Hwa
Patrick, George Walter
Pazoles, Christopher James
Pellmar, Terry C.
Perlow, Mark Jacob
Peusner, Kenna Dale
Pfaff, Donald W.
Pfenninger, Karl Hans
Phillips, Christine Elaine
Pitts, Nathaniel Gilbert
Powers, Maureen Kennedy
Prendergast, Jocelyn
Price, Steven
Proenza, Luis Mariano
Provine, Robert Raymond
Purpura, Dominick Paul
Ray, Radharaman
Reese, Thomas Sargent
Reiner, Anton John
Rennels, Marshall Leigh
Rockland, Kathleen Skiba
Roper, Stephen David
Ross, William Noel
Routtenberg, Aryeh
Salcman, Michael
Salkoff, Lawrence Benjamin
Salzberg, Brian Matthew
Saporta, Samuel
Saunders, James Charles
Scalia, Frank Richard
Scheff, Stephen William
Schein, Stanley Jay
Schneider, Richard Joel
Schreckenberg, Gervasia Mary
Schuetze, Stephen Mark
Scott, John W(atts)
Setler, Paulette Elizabeth
Sheffield, Joel B.
Sherman, S. Murray
Shinowara, Nancy Lee
Silberberg, Donald H.
Singer, Marcus Joseph
Soller, R. William
Sparrow, Janet Ruthe
Spencer, Robert Frederick
Spitzer, Nicholas Canaday
Sprague, James Mather
Stach, Robert William
Stanford, Laurence Ralph
Stefano, George B.
Stein, Barry Edward
Stent, Gunther Siegmund
Stern-Tomlinson, Wendy Barbara
Steward, Oswald
Streicher, Eugene
Stuart, Ann Elizabeth
Swarz, Jeffrey Robert
Sweadner, Kathleen J.
Talamo, Barbara Ruth
Teyler, Timothy James
Thompson, Elizabeth Barnes
Tieman, Suzannah Bliss
Tompkins, Laurie
Toran-Allerand, C(laude) Dominique
Trojanowski, John Quinn
Trubatch, Janett
Trune, Dennis Royal
Turner, James Eldridge
Tuttle, Jeremy Ballou
Tytell, Michael
Udin, Susan Boymel
Ueda, Tetsufumi
Vacca, Linda L.
Vernadakis, Antonia
Viveros, Osvaldo Humberto
Wade, Patricia Diane
Wainwright, Stanley Dunstan
Watson, Winsor Hays, III
Waxman, Stephen G(eorge)
White, Richard Hamilton
Wieland, Steven Joseph
Wiesel, Torsten Nils
Wiley, Ronald Gordon
Wilson, Victor Joseph
Wolfe, Barry B.
Woodruff, Michael Lester
Woolsey, Clinton Nathan
Wu, Chun-Fang
Wurtz, Robert Henry
Wyatt, Harry Joel
Wyss, James Michael
Yazulla, Stephen
Zaratzian, Virginia Louise
Zimmerman, Earl Abram

Neurobiology, comparative
Adams, David Bachrach
Alley, Keith Edward
Allman, John Morgan
Baird, John Jeffers
Butler, Ann Benedict
Casseday, John Herbert
Delcomyn, Fred
Eaton, Robert Charles
Emerson, Victor F.
Fine, Michael Lawrence
Fite, Katherine Virginia
Friedman, David Paul
Goldberg, Stephen
Greenberg, Neil
Griffin, Donald R(edfield)
Heath, James Edward
Howland, Howard Chase
Jerison, Harry Jacob
Jones, Stephen Wallace
Kennedy, Donald
Kennedy, Michael Craig
Kicliter, Ernest Earl, Jr.
Koestner, Adalbert
Macrides, Foteos
Martin, George Franklin
Nelson, Margaret Christina
Page, Charles Henry
Partridge, Brian Lloyd
Puglisi, Susan Grace
Rakic, Pasko
Roberts, John Lewis
Root, Thomas Michael
Scharrer, Berta Vogel
Sharma, Sansar C.
Sherman, Robert George
Skeen, Leslie Carlisle
Sze, Paul Yi Ling
Walcott, Benjamin
Wallace, Robert Bruce
Wilczynski, Walter
Wyman, Robert J.
Young, Wei

Nutrition
Abernathy, Richard Paul
Alfano, Michael Charles
Anderson, Jay Oscar
Aurand, Leonard William
Bates, Margaret Westbrook
Bhattacharyya, Ashim K.
Briley, Margaret Elizabeth
Brockman, Ronald Paul
Caldwell, Elwood Fleming
Calloway, Doris Howes
Campbell, Thomas Colin
Chen, Tung-Shan
Davis, Robert Harry
de la Noue, Joel Jean-Louis
Del Valle, Francisco Rafael
Dewart, Dorothy Boardman
Dreizen, Samuel
Easter, Robert Arnold
Evans, Edwin Victor
Ferguson, John Carruthers
Flodin, Nestor Winston
Freedland, Richard A.
Garrett, William N.
Grivetti, Louis Evan
Guthrie, Helen A.
Harper, Alfred Edwin
Harrell, Ruth Flinn
Hegsted, David Mark
Higgins, Edwin Stanley
Hunter, James Edward
Jansen, Gustav Richard
Kamath, Savitri Krishna
Ketola, Henry George
Kritchevsky, David
Kuftinec, Mladen Matija
Labuza, Theodore Peter
Levine, Gary M.
Lichton, Ira Jay
Litov, Richard Emil
Mahan, Donald Clarence
Mann, Robert Arthur
McLaughlin, Carol Lynn
Mertz, Walter
Mitchell, George Ernest, Jr.
Morris, James Grant
Mosbach, Erwin Heinz
Munro, Hamish Nisbet
Padberg, Daniel Ivan
Palmer, Sushma Mahyera
Phillips, Lawrence S(tone)
Pollack, Robert Leon
Poston, Hugh Arthur
Prothro, Johnnie Watts
Reeves, Robert Donald
Rogers, Quinton Ray
Ross, Morris H.
Ruberg, Robert Lionel
Shamberger, Raymond Joseph, Jr.
Smith, Leonard Charles
Solomons, Noel W.
Stini, William Arthur
Stockland, Wayne Luvern
Thomas, John William
Thurman, Lloy Duane
Tumbleson, M.E.
Van Campen, Darrell R.
Veum, Trygve Lauritz
Weir, William Carl
Welch, Ross Maynard
Young, Jerry Wesley

Parasitology. *See* **MEDICINE.**

Photosynthesis
Alberte, Randall Sheldon
Anderson, Louise Eleanor
Arnon, Daniel I(srael)
Arntzen, Charles Joel
Bjorkman, Olle Erik
Bowes, George Ernest
Boyer, John Strickland
DeVault, Don Charles
Dilley, Richard A.
Duke, Stephen Oscar
Ellsworth, Robet King
French, Charles Stacy
Hatzios, Kriton Kleanthis
Henderson, Robert Edward
Hoch, George Edward
Homann, Peter Hinrich Fritz
Izawa, Seikichi
Jensen, Richard Grant
Kelly, Jeffrey John
Krogmann, David William
Malkin, Richard
Mooney, Harold Alfred
Paech, Christian Gerolf
Prezelin, Barbara Berntsen
Rawson, James Rulon Young
Sharp, Robert Richard
Sherman, Louis Allen
Smith, Bruce Nephi
Spreitzer, Robert Joseph
Thornber, James Philip
Warden, Joseph Tallman, Jr.

Physiology. *See also* **MEDICINE.**
Abrams, Robert Marlow
Anderson, Donald Keith
Asterita, Mary Frances
Axen, Kenneth
Ball, Wilfred Randolph
Bealer, Steven Lee
Beames, Calvin G., Jr.
Behrisch, Hans Werner
Berry, Christine Albachten
Beyenbach, Klaus Werner
Bishop, Beverly P(etterson)
Boulpaep, Emile Louis J.B.
Bourdeau, James Edward
Brenner, Barry Morton
Brown, Paul Burton
Burke, Robert Emmett
Burnett, Louis Elwood, Jr.
Butler, James Preston
Campbell, James Wayne
Capen, Ronald Leroy
Carson, Virginia Rosalie Gottschall
Chowdhury, Parimal
Connett, Richard James
Corradino, Robert A.
Coulson, Patricia Bunker
Cronin, Michael John
Davis, James (Othello)
Diamond, Jared Mason
Dobson, Alan
Dunham, Philip Bigelow
Fabiato, Alexandre
Feir, Dorothy Jean
Ferguson, John Carruthers
Fingerman, Milton
Forman, Henry Jay
Francis, Kennon Thompson
Fried, George Herbert
Gallagher, Kim Patrick
Gatto, Louis Albert
George, Stephen Anthony
Gilbert, Daniel Lee

Grim, Eugene
Hageman, Gilbert Robert
Harland, Barbara Ferguson
Hazelwood, Robert L(eonard)
Heath, Hunter, III
Hoffman, Eric Alfred
Honig, Carl R.
Horwitz, Barbara Ann
Hossner, Kim Lee
House, Edwin Wesley
Howland, Howard Chase
Imig, Charles Joseph
Jankelson, Bernard
Kalimi, Mohammed Yahya
Karagueuzian, Hrayr Sevag
Katz, Murray Alan
Kendig, Joan Johnston
Ketola, Henry George
Knobil, Ernst
Koblick, Daniel Cecil
Krausz, Stephen
LaBarbera, Andrew Richard
Lai, Yih-Loong
Larkin, John Montague
Lawson, Edward Earle
Lee, Chung
Leffler, Charles William
Lehman, William Jeffrey
Levinsky, Norman George
Loomis, Stephen Henry
Lutton, Lewis Montfort
Lutz, Peter Louis
Maack, Thomas Michael
Mason, Charles Perry
Metcalf, John Franklin
Mines, Allan Howard
Moore-Ede, Martin Christopher
Murphy, Richard Alan
Oatley, David Herbert
Ohata, Carl Andrews
Olivo, Margaret Anderson
Pashley, David Henry
Petrofsky, Jerrold Scott
Pfeffer, Janice Marie
Pharriss, Bruce Bailey
Pincus, Irwin J.
Pitts, Charles William
Platzer, Edward George
Poston, Hugh Arthur
Price, Steven
Rayford, Phillip Leon
reid, Donald House
Rhodes, James Benjamin
Roales, Robert R.
Roelofs, Wendell Lee
Russell-Hunter, W(illiam) D(evigne)
Samaras, George Michael
Schwartz, Karlene V.
Shepherd, A.P., Jr.
Shirley, Barbara Anne
Sillman, Arnold Joel
Simmons, William Howard
Smallridge, Robert Christian
Smolensky, Michael Hale
Sokoloff, Louis
Spaziani, Eugene
Tenney, Stephen Marsh
Tobe, Stephen Solomon
Tonna, Edgar Anthony
Vogel, Steven
Wade, Patricia Diane
Waitzman, Morton Benjamin
Wasserman, Martin Allan
White, Timothy Peter
Wilber, Charles Grady
Wilson, Victor Joseph
Woods, Walter Thomas
Zimmack, Harold Lincoln

Plant growth
Bennett, William F., Sr.
Bowers, Maynard Claire
Bruck, David Kenneth
Burris, Joseph Stephen
Chen, Chong-Maw
Cleland, Robert Erskine
Coate, Barrie Douglas
Davies, Peter John
Davis, Bill David
Funt, Richard Clair
Green, Paul Barnett
Greenfield, Sydney Stanley
Gregory, Robert Aaron
Grunes, David Leon
Hill, Jane Foster
Jacobs, Mark
Kawase, Makoto
Larson, Philip Rodney
Lauer, Florian I.
Ledbetter, Myron Calvert
Lewis, Archie Jefferson, III
Lyle, Everett Samuel, Jr.
Maravolo, Nicholas C.
Marx, Donald Henry
McConnell, Dennis Brooks
Miller, Carlos Oakley
Nooden, Larry D.
Ohki, Kenneth
Olsen, Ralph A.
Philip, A G Davis
Postlethwait, Samuel Noel
Pritchard, Hayden Nelson
Relf, Paula Diane
Rost, Thomas Lowell
Ruddat, Manfred
Salters, Grace Heyward
Schonbeck, Mark Walter
Schwartzkopf, Steven Henry
Seago, James Lynn, Jr.
Singh, Adya Prasad
Smith, Thomas Lloyd
Spiess, Luretta D(avis)
Strader, Herman Lee
Switzer, George Lester
Theurer, J. Clair
Thomas, Robert James
Vanderhoef, Larry Neil
van der Meer, John Peter
Walker, Dan Berne
Whittier, Dean Page
Wilkinson, Robert Eugene

Plant physiology. ***See also*** **AGRICULTURE.**
Aboulela, Mohamed
Alves, Leo Manuel
Amen, Ralph DuWayne
Bennett, Jesse Harland
Benzing, David Hill
Bisson, Mary Aldwin
Boyer, John Strickland
Briggs, Winslow Russell
Byerrum, Richard Uglow
Cleland, Robert Erskine
Cossins, Edwin Albert
Cowles, Joe Richard
Davies, Peter John
Davis, David Gerhardt
Dekker, Eugene Earl
Doss, Robert Paul
Duke, Stanley Housten
Duke, Stephen Oscar
Durbin, Richard Duane
Einhellig, Frank Arnold
Evans, Lance Saylor
Feinleib, Mary Ella (Harman)
Gassman, Merrill Loren
Gentile, Arthur Christopher
Gibbs, Martin
Goodin, Joe Ray
Greenfield, Sydney Stanley
Gregory, Garold Fay
Henderson, James Henry Meriwether
Hendrix, John Edwin
Hill, Ray Allen
Hirsch, Ann Mary
Hopkins, Donald Lee
Horwitt, Max Kenneth
Jacobs, Mark
Jagendorf, Andre Tridon
Jensen, William August
Kaufman, Peter Bishop
Keeley, Jon Edward
Kennedy, Robert Alan
Kerstetter, Rex Eugene
Kim, Chong-Kyun
Li, Paul Hsiang
Lincoln, David Erwin
Lippincott, James Andrew
Lockhart, James Arthur
Loescher, Wayne Harold
Lucas, William John
Lunt, Owen Raynal
McClendon, John Haddaway
Miller, Carlos Oakley
Nobel, Park S.
Nooden, Larry D.
Oross, John William
Pallardy, Stephen Gerard
Peterson, Curt Morris
Pratt, Lee Herbert
Prien, Samuel David
Reazin, George Harvey, Jr.
Ruddat, Manfred
Ruesink, Albert William
Scheffer, Robert Paul
Schiff, Jerome Arnold
Smith, Bruce Nephi
Spurr, Arthur Richard
Stafford, Helen Adele
Staples, Richard Cromwell
Stekoll, Michael Steven
Stowe, Bruce Bernot
Sweet, Haven Colby
Synder, Freeman Woodrow
Tatina, Robert Edward
Thimann, Kenneth Vivian
Thomas, Robert James
Unbehaun, Laraine Marie
Vance, B(enjamin) Dwain
Weber, James Alan
Weimberg, Ralph
Wiebe, Herman Henry
Wittwer, Sylvan Harold
Yamaguchi, Shogo

Population biology
Abrahamson, Warren Gene, II
Anderson, Wyatt Wheaton
Arnold, Jonathan
Band, Henretta Trent
Bawa, Kamaljit Singh
Berry, James Frederick
Bookman, Susan Stone
Brynjolfsson, Ari
Bucklin, Ann Cone
Case, Ronald Mark
Case, Ted Joseph
Caswell, Hal
Christian, John Jermyn
Cox, Paul Alan
Craighead, John J.
Cuellar, Orlando
Davidson, Donald W.
Dykhuizen, Daniel Edward
Ehrlich, Paul Ralph
Ellstrand, Norman Carl
Gilbert, Donald George
Ginzburg, Lev
Gojobori, Takashi
Grant, Verne Edwin
Gullion, Gordon Wright
Hamrick, James Lewis, III
Hartl, Daniel Lee
Horn, Henry Stainken
Howe, Henry Franklin
Jain, Subodh Kumar
Johnson, Eric Van
Kahler, Alex LeRoy
Kaneshiro, Kenneth Yoshimitsu
Keeley, Jon Edward
Kinney, Terry B., Jr.
Kozloff, Lloyd M.
Labanick, George Michael
Lewis, Allen Rogers
Lockhart, James Arthur
Mason, John Montgomery, Jr.
Matzinger, Dale Frederick
May, Robert McCredie
Mead, Albert Raymond
Parker, William Skinker
Pearson, William Dean
Pettus, David
Pfahler, Paul Leighton
Pianka, Eric Rodger
Pitelka, Louis Frank
Quinn, James Amos, Jr.
Rothschild, Brian James
Runkle, James Reade
Sassaman, Clay A.
Schuder, Donald Lloyd
Sokoloff, Alexander
Sturges, Franklin Wright
Tamarin, Robert Harvey
Taylor, Charles Ellett
Wilbur, Henry Miles
Wyatt, Robert Edward
Yellin, Joel
Yokoyama, Shozo
Zeveloff, Samuel Ira
Zouros, Eleftherios

Psychobiology. ***See*** **PSYCHOLOGY.**

Reproductive biology
Albert, Mary Day
Amoss, Max St. Clair, Jr.
Baggett, Billy
Baker, Irene
Bartke, Andrzej
Bell, Clyde Ritchie
Benseler, Rolf Wilhelm
Boylan, Elizabeth S.
Brinster, Ralph Lawrence
Bronson, Franklin Herbert
Bryant-Greenwood, Gillian Doreen
Buss, Edward George
Carothers, Zane Bland
Chung, Melvin Chung-Hing
Coleman, Marilyn Ruth Adams
Coulson, Patricia Bunker
Davis, Brian Keith
Dey, Sudhansu K.
Edgren, Richard Arthur
Erpino, Michael James
Evans, James Warren
Farner, Donald Sankey
Ford, Stephen Paul
Freeman, Marc Edward
Fritz, Irving Bamdas
Fujimoto, George I.
Geula, Gibori
Grogan, William McLean
Hagen, Daniel Russell
Hannan, Gary Louis
Harper, Michael John Kennedy
Hawk, Harold William
Herr, John Mervin, Jr.
Hotta, Yasuo
Howards, Stuart S.
Inskeep, Emmett Keith
Kalra, Satya Paul
Kesler, Darrel Joe
Lala, Peeyush Kanti
Leopold, Roger Allen
Lerner, Leonard Joseph
Leto, Salvatore
Lu, John Kuew-Hsiung
Markert, Clement Lawrence
Martin, David Edward
McGaughey, Robert William
Menino, Alfred Rodrigues, Jr.
Moudgil, Virinder Kumar
Niklas, Karl J.
Nutting, Ehard Forrest
Parrish, John Wesley, Jr.
Pekary, Albert Eugene
Pharriss, Bruce Bailey
Phemister, Robert David
Pineda, Mauricio Hernan
Pippen, Richard Wayne
Pitkow, Howard Spencer
Plotka, Edward Dennis
Postlethwait, John Harvey
Reel, Jerry Royce
Riehl, Robert Michael
Rivier, Catherine Laure
Robbins, Robert Raymond
Roberts, Robert Michael
Shirley, Barbara Anne
Spies, Harold Glen
Stover, Janet
Terner, Charles
Tobe, Stephen Solomon
van Tienhoven, Ari
Verma, Om Prakash
Wagner, William Charles
Zavos, Panayiotis Michael

Sociobiology
DeGhett, Victor John
Haubrich, Robert Rice
Howard, Lauren Davis
Hrdy, Sarah Blaffer
Jaco, E. Gartly
Kistner, David Harold
Labov, Jay Brian
Michener, Charles Duncan
Michod, Richard Earl
Mock, Douglas Wayne
Rowland, William Joseph
Uetz, George William
Wilson, Edward Osborne

Systematics
Anderson, Gregory Joseph
Armstrong, David Michael
Ashlock, Peter Dunning
Beadles, John Kenneth
Boss, Kenneth Jay
Bowers, Maynard Claire
Breckon, Gary John
Burch, John Bayard
Burr, Brooks Milo
Carico, James Edwin
Carr, Gerald Dwayne
Chambers, Kenton Lee
Choate, Jerry Ronald
Chuey, Carl Francis
Crane, Joseph Leland
Croat, Thomas Bernard
Dent, James Norman
Erwin, Terry Lee
Estes, James Russell
Flagg, Raymond Osbourn
Fox, George Edward
Fritze, Karen J.
Funk, Vicki Ann
Hannan, Gary Louis
Hatch, Stephan LaVor
Haynes, Robert Ralph
Henrickson, James Solberg
Herr, John Mervin, Jr.
Highton, Richard Taylor
Hill, Steven Richard
Ho, Ju-shey
Hubbs, Clark
Jackson, Raymond Carl
Jensen, Richard Jorg
Jones, Claris Eugene, Jr.
Juncosa, Adrian Martin
Keeley, Sterling Carter
Keil, David John
Kohn, Alan Jacobs
Koutnik, Daryl Lee
Lersten, Nels Ronald
Levin, Michael Howard
Lillegraven, Jason Arthur
Martin, Larry Dean
Mason, Charles T., Jr.
McCafferty, William Patrick
Michener, Charles Duncan
Moore, W(alter) E(dward) C(ladek)
Nakamura, Ichiro
Norden, Carroll Raymond
Novak, John Allen
Olsen, John Stuart
Phillips, Raymond Bruce
Pippen, Richard Wayne
Raven, Peter Hamilton
Redfearn, Paul Leslie, Jr.
Rollins, Reed Clark
Rozen, Jerome George, Jr.
Schaefer, Carl Walter, II
Schaeffer, Robert L., Jr.
Schultes, Richard Evans
Seavey, Steven R.
Seigler, David Stanley
Sheffer, Richard Douglas
Shetler, Stanwyn Gerald
Smith, Albert Charles
Smith, Dale Metz
Sohmer, Seymour Hans
Spencer, Lorraine Barney
Steiner, William Wallace Mokahi
Stuessy, Tod Falor
Taylor, Ronald J.
Thaeler, Charles Schropp, Jr.
Thomas, William Wayt
Turner, Bruce Jay
Utech, Frederick Herbert
Varkey, Alexander
Vasek, Frank C(harles)
Walker, James Willard
Williams, Kenneth Lee
Woods, Charles Arthur
Wujek, Daniel Everett

Taxonomy
Babero, Bert Bell
Balick, Michael Jeffrey
Baranov, Andrey I(ppolitovich)
Bazinet, Lester
Beadles, John Kenneth
Byers, George William
Chuey, Carl Francis
Crosby, Marshall Robert
Curry, Mary Grace
Daily, Fay Kenoyer
Day, William Hartwell Eveleth
Dehgan, Bijan
deLaubenfels, David John
Drapalik, Donald Joseph
Duke, James Alan
Eleuterius, Lionel Numa
Ellison, Marlon Louis
Estes, James Russell
Fryxell, Paul Arnold
Griffin, Dana Gove, III
Hanlin, Richard Thomas
Hatch, Stephan LaVor
Haynes, Robert Ralph
Hellquist, Carl Barre
Hill, Steven Richard
Hilu, Khidir Wanni
Holland, Richard Darlan
Howard, Lauren Davis
Hudson, William Donald, Jr.
Huft, Michael John
Huttleston, Donald Grunert
Jensen, Richard Jorg
Johnsgard, Paul Austin
Johnson, Carl Maurice
Kaplan, Eugene Herbert
Keil, David John
Khalaf, Kamel Toma
Korf, Richard Paul
Lago, Paul Keith
Leiser, Andrew Twohy
Lellinger, David Bruce
Little, Elbert Luther, Jr.
Main, Stephen Paul
Manley, Donald Gene
Mason, Charles T., Jr.
Mead, Albert Raymond
Meijer, Willem
Miller, James Frederick
Miller, Regis Bolden
Minton, Sherman Anthony
Olive, Lindsay Shepherd
Phillips, Raymond Bruce
Reeder, John R(aymond)
Reveal, James Lauritz
Rogers, Claude Marvin
Rogers, Donald Philip
St.Clair, Larry Lee
Salamun, Peter J(oseph)
Schaeffer, Robert L., Jr.
Schilling, Edward E.
Schuder, Donald Lloyd
Schultes, Richard Evans
Sharp, Aaron John
Sibley, Charles Gald
Siddiqi, Shaukat Mahmood
Smith, Clifford Winston
Snider, Jerry Allen
Sohmer, Seymour Hans
Spencer, Lorraine Barney
Staley, John Merrill
Stern, Kingsley Rowland
Thomas, William Wayt
Tucker, Arthur Oliver
Wagner, David Henry
Wujek, Daniel Everett

Tissue culture
Avampato, James Erwin
Behbehani, Abbas M.
Bissell, Michael Gilbert
Giaever, Ivar
Goodin, Joe Ray
Hamilton, Velda Marie
Henderson, James Henry Meriwether
Hughes, Karen Woodbury
Lamar, Carlton Hine
Oberley, Terry De Wayne
Pieczynski, William John
Punnett, Hope Suzanne
Rose, George Gibson
Sapatino, Bruno Vasile
Vaughn, James Lloyd
Weiss, Stefan Adam
Wilson, Kenneth Glade

Virology
Anderson, Carl William
Anderson, Thomas Foxen
Arthur, Larry Ottis
Ball, Laurence Andrew
Barban, Stanley
Basilico, Claudio
Bean, William Joseph, Jr.
Bernstein, Eugene Harold
Biswal, Nilambar
Blair, Carol Dean
Bond, Clifford Walter
Brockman, William Warner
Broker, Thomas Richard
Brown, Bruce Leonard
Brown, Hannah R(eeva)
Burke, Carroll Nutile
Burke, Derek Clissold
Campbell, Allan McCulloch
Carroll, Robert Byers
Carter, Timothy Howard
Cerini, Costantino Peter
Choppin, Purnell Whittington
Chow, Louise Tsi
Christensen, J(ames) Roger
Coffin, John Miller
Cox, Donald Cody
Danna, Kathleen Janet
Deininger, Prescott Leonard
DeMeio, Joseph Louis
Docherty, John Joseph
Dougherty, Robert Malvin
Dubes, George Richard
Duff, James Thomas
Duff, Ronald George
Dutko, Francis Joseph, Jr.
East, James Lindsay
Eckhart, Walter
Ehrenfeld, Elvera
Engelking, Henry Mark
Eppstein, Deborah Anne
Fabricant, Catherine Grenci
Fan, Hung Y
Fitzgerald, Edward Aloysius
Flanegan, James Bert
Founds, Henry William, Jr.
Fuccillo, David Anthony
Garfinkle, Barry David
Geiduschek, E(rnest) Peter
Gingery, Roy Evans
Ginsberg, Harold Samuel
Goldberg, Edward Bleier
Goorha, Rakesh Mohan
Granlund, David John
Green, Maurice
Gregoriades, Anastasia
Grubman, Marvin
Haywood, Anne Mowbray
Hillis, William Daniel, Sr.
Hoggan, M. David
Hsiung, Gueh-Djen
Hunter, Eric
Johnson, F. Brent
Kasel, Julius Albert
Keller, John Randall
Kennedy, Samuel Ian T.
Kingsbury, David Wilson
Knight, Vernon
Koprowski, Hilary
Kozloff, Lloyd M.
Kramarsky, Bernhard
Leamnson, Robert Neal
Leibowitz, Michael Jonathan
Lewis, Andrew Morris, Jr.
Li, Joseph K(wok) K(wong)
Lyles, Douglas Scott
Major, Eugene Oliver
Markwell, Mary Ann K.
Martignoni, Mauro Emilio

Martin, Malcolm Alan
Melnick, Joseph L.
Meyers, Paul
Morris, J(oseph) Anthony
Nayak, Debi Prosad
Ozer, Harvey Leon
Panem, Sandra
Pato, Martin Leon
Person, Donald Ames
Petersen, Lawrence John
Pfeffer, Lawrence Marc
Polatnick, Jerome
Pollack, Robert Elliot
Pons, Marcel William
Potash, Louis
Price, Alan Roger
Quilligan, James Joseph, Jr.
Rapp, Ulf Fuediger
Reichmann, Manfred Eliezer
Reiss, Carol Shoshkes
Richards, Gary Paul
Roane, Philip Ransom, Jr.
Roman, Ann
Rosenthal, Kenneth Steven
Rueckert, Roland Rudyard
Samuel, Charles E.
Sapatino, Bruno Vasile
Schwabel, Mary Jane
Sen, Ganes C(handra)
Shapshak, Paul
Shaw, James Elwood
Sidwell, Robert William
Singh, Shiva Pujan
Soike, Kenneth Fieroe, Jr.
Specter, Steven Carl
Stanton, Glennon John
Takemoto, Kenneth K.
Tamm, Igor
Taylor, Bernard Franklin
Taylor, Milton William
Thacore, Harshad Rai
Thomas, Donald Charles
Thormar, Halldor
Trowbridge, Richard Stuart
Trown, Patrick Willoughby
Tyler, Kenneth Laurence
Tzianabos, Theodore
Van Etten, James Lee
Varnos, Harold Eliot
Vasington, Paul John
Waters, David John
Watson, Kenneth Fredrick
Weiss, Stefan Adam
Wiberg, John Samuel
Wimmer, Eckard
Wolf, Kenneth Edward
Wolff, David Alwin
Yoon, Ji-Won
Zaitlin, Milton
Zimmerman, Daniel Hill

Other
Aaslestad, Halvor Gunerius, *Science Administration*
Archibald, Patricia A., *Phycology (Algology)*
Balick, Michael Jeffrey, *Economic botany*
Baranov, Andrey I(ppolitovich), *Ethnobotany*
Beeton, Alfred Merle, *Limnology*
Benson, Andrew Alm, *Marine metabolism*
Berends, Lawrence Keith, *marine biology*
Billingham, John, *Exobiology*
Bleckmann, Charles Allen, *Biological aspects of petroleum production*
Borgatti, Alfred Lawrence, *Insect pathology*
Briggs, Winslow Russell, *Photobiology*
Brotzman, Harold George, *Mycology*
Brown, Hannah R(eeva), *Immunoelectron microscopy*
Bundy, Hallie Flowers, *Enzymology*
Butcher, James Walter, *Entomology*
Carlquist, Sherwin, *Insular biology*
Collins, Margaret Strickland, *Entomology*
Colwell, Rita Rossi, *Bacteriology*
Craighead, John J., *Wildlife biology*
Cranefield, Paul Frederic, *Electrphysiology*
Cronkite, Eugene Pitcher, *Radiation Biology*
Deibel, Robert Howard, *Bacteriology*
Delevoryas, Theodore, *Botany*
Edmondson, W(allace) Thomas
Feng, Sung Yen, *Marine Biology*
Foulke, Judith Diane, *Radiobiology*
Gifford, Ernest Milton, Jr., *Plant Anatomy*
Gollub, Edith Goldberg, *gene regulation, genetics*
Grahn, Douglas, *Radiobiology*
Griffin, Dana Gove, III, *Floristics*
Gruber, Samuel Harvey, *Marine Biology*
Grundy, Scott Montgomery, *Metabolism*
Guthrie, Frank Edwin, *Entomology*
Hahn, Eric Walter, *Radiation biology*
Halstead, Bruce Walter, *Biotoxicology*
Hanebrink, Earl Lee
Harris, Charles Leon, *History and philosophy of biology*
Harshbarger, John Carl, Jr., *Pathobiology*
Hart, Ronald Wilson, *Radiation biology*
Ho, Ju-shey, *Biogeography*
Honig, Carl R., *Microcirculation*
Hudecki, Michael Stephen, *Muscle pathology*
Jackson, Kenneth Lee, *Radiation biology*
Jeffries, Harry Perry
Jensh, Ronald Paul, *Teratology*
Jong, Shung Chang, *Culture collections*
Kaiser, Hans Elmar, *Comparative pathology*
Kallman, Robert Friend, *Radiation biology*
Kaufman, Leo, *Medical mycology*
Kocol, Henry, *Radiation biological effects*
Korf, Richard Paul, *Mycology*
Kurtzman, Ralph Harold, Jr., *Mycology*
La Claire, John Willard, II, *Marine Phycology*
Lane, Roger Lee, *Histology and Histochemistry of Invertebrates*
Leopold, Estella Bergere, *Botany*
Lestrel, Pete Ernest, *Morphometrics*
Lewis, Edward B., *Radiation biology*
Lichtwardt, Robert William, *Mycology*
Lustick, Sheldon, *Phyaiological ecology*
Manglitz, George Rudolph, *Entomology*
McGuire, Terry Russell, *Behavior genetics*
Migdalof, Bruce Howard, *Xenobiology*
Miller, Morton W(illiam), *Radiation Biology*
Mittler, Sidney, *Radiation biology*
Moore, Richard Harlan, *Marine Zoology*
Pittendrigh, Colin Stephenson, *Biological research management*
Plonsey, Robert, *Cardiac electrophysiology*
Polacek, Laurie Ann, *Human genetics*
Porter, Keith Roberts, *Electron miroscopy*
Potter, Frank Walter, Jr., *Paleobotany*
Powers, Edward Lawrence, *Radiation biology*
Ragotzkie, Robert Austin, *Aquatic ecology*
Rahn, Joan Elma, *General biology*
Rembert, David Hopkins, Jr., *Botanical history*
Reveal, James Lauritz, *Botanical history*
Roberts, Donald Wilson, *Insect pathology*
Sarett, Lewis Hastings, *Life sciences research and administration*
Smith, Dale Metz, *Phytochemistry*
Sparling, Donald Wesley, *Biostatistics*
Spitsberg, Vitaly Lev, *Biochemical evolution*
Steingraeber, David Allen, *Plant morphology*
Stevens, Reggie Harrison, *Radiation biology*
Tattersall, Ian Michael, *Primatology*
Tiffany, Lois Hattery, *Mycology*
Vagelos, Pindaros Roy, *Biomedical research management*
Van Dyke, Cecil Gerald, *Biological control*
Weathersby, A(ugustus) Burns, *Entomology*
Whikehart, David Ralph, *Biochemistry of the eye*
Windels, Carol Elizabeth, *Biological control*
Wing, Elizabeth Schwarz, *Zooarcheology*
Yeatman, Harry Clay
Yoder, Larry Richard, *Plant anatomy*

BIOTECHNOLOGY

Artificial organs. *See also* **MEDICINE, Surgery.**
Ackerman, Roy Alan
Belikoff, Steven William
Brodhagen, Thomas Warren
Chang, Thomas Ming Swi
Daniel, Michael Andrew
DeVore, Dale Paul
Dorson, William John, Jr.
Galletti, Pierre Marie
Hambrecht, Frederick Terry
Jaron, Dov
Loeb, Gerald Eli
Miller, Irving Franklin
Pierce, William Schuler
Popovich, Robert Peter
Soeldner, John Stuart
Vasington, Paul John

Bioinstrumentation
Barth, Daniel Stephen
Black, William Carter, Jr.
Bradley, A. Freeman, Jr.
Chang, Wei
Combs, Claud Steve
Del Guercio, Louis Richard Maurice
Ellis, Donald Griffith
Fatt, Irving
Friesen, Wolfgang Otto
Gable, Ralph Kirkland
Gallant, Nanette
Haselhorst, Donald Duane
Horrocks, Donald Leonard
Jarzembski, William Bernard
Jaszczak, Ronald Jack
Jurmain, Jacob Harry
Kaplan, Joel Howard
Kim, Myunghwan
Kinzel, Augustus Braun
Knoll, Glenn Frederick
Kramer, Stephen Leonard
Mitra, Jyoti
Mok, Machteld Cornelia
Morris, Lucien Ellis
Mummert, Thomas Allen
Murray, Rodney Brent
Patti, Robert Dale
Prohl, A(medeus) Edward
Pykett, Ian Lewis
Royds, Robert B.
Rugh, John Douglas
Schultz, Jerome Samson
Scott, Norman Roy
Shamos, Morris Herbert
Shapiro, Howard Maurice
Spergel, Philip
Stavinoha, William B.
Webster, John Goodwin
Wolthuis, Roger Allen
Yoon, Hyo Sub

Bioinstrumentation, CAT scan
George, Ajax Elis
Gould, Robert George
Laning, David Bruce
Peters, Terence Malcolm
Pollack, Richard Stuart
Rugge, Henry Ferdinand

Bioinstrumentation, mass spectrometry
Andresen, Brian Dean
Boutton, Thomas William, Jr.
Brown, George Barremore
Djerassi, Carl
Green, Donald Eugene
Hattox, Susan Ellen
Knapp, Daniel Roger
Libbey, Leonard Morton
Smith, Ronald Gene

Bioinstrumentation, PET scan
Gerber, Joseph Charles, III
Goldman, Stephen Shepard
Gur, Ruben C.
Mullani, Nizar Abdul
Tedeschi, David Henry

Biomaterials. *See* **MATERIALS SCIENCE.**

Biomedical engineering. *See also* **ENGINEERING.**
Albisser, Anthony Michael
Altschuler, Martin David
Attinger, Ernst Otto
Babb, Albert Leslie
Bak, Martin Joseph
Becker, Robert Otto
Bischoff, Kenneth Bruce
Boston, John Robert
Bradley, A. Freeman, Jr.
Buchwald, Henry
Cetas, Thomas Charles
Chaffin, Don B.
Chen, Ching-Nien
Childress, Dudley Stephen
Corrsin, Stanley
Cox, Jerome Rockhold, Jr.
Dell, Ralph B.
DiBianca, Frank Anthony
Duffy, Frank Hopkins
Dunn, Floyd
Eisenberg, Lawrence
Finney, Roy Pelham
Forney, LeRoy S.
Garfinkel, David
Goldstein, Steven Alan
Graham, James Alexander
Graupe, Daniel
Green, Daniel G.
Hamilton, Robert William
Hellums, Jesse David
Homsy, Charles Albert
Hung, Tin-Kan
Jarvik, Robert K.
Johns, Richard James
Katona, Peter Geza
Knowles, Lloyd George
Koeneman, James Bryant
Kristol, David Sol
Kuc, Roman Basil
Kydd, George Herman
Langer, Robert Samuel
Latham, Allen, Jr.
Lee, Wylie In-Wei
Levin, Stephen Michael
Levinthal, Elliott Charles
Liebig, William John
Lih, Marshall Min-Shing.
Lin, Heh-Sen
Lyman, John
Machalski, Robert Chester
McInerney, Joseph John
Michels, Lester David
Murphy, Eugene Francis
Myers, Philip Cherdak
Nunan, Craig Spencer
Oman, Charles McMaster
Petrofsky, Jerrold Scott
Popovich, Robert Peter
Quigley, Michael J.
Remmel, Ronald Sylvester
Reynolds, David Burkman
Saha, Subrata
Sandhu, Taljit Singh
Tancrell, Roger Henry
Von Gierke, Henning Edgar
Wall, Conrad, III
Weed, Herman Roscoe
Weinstein, Sam
Wolthuis, Roger Allen
Woo, Savio Lau-Yuen
Yin, Frank, Chi-Pong
York, Donald Harold
Young, Laurence Retman

Enzyme technology
Ackerman, Roy Alan
Bassi, Sukh Dev
Bollon, Arthur Peter
Butler, Larry Gene
Chang, Thomas Ming Swi
Chase, John William
Cheryan, Munir
Chuang, Ronald Y(an-li)
Cooney, Charles Leland
Copeland, James Clinton
Daron, Harlow H.
Evans, Harold J.
Fink, Anthony Lawrence
Fraser, Thomas Hunter
Glover, George I.
Greenstein, Teddy
Hamilton, Bruce King
Hulcher, Frank Hope
Humphrey, Arthur Earl
Kempe, Lloyd Lute
Klibanov, Alexander Maxim
Knodel, Elinor Livingston
Koshland, Daniel Edward, Jr.
Krenitsky, Thomas Anthony
Loftfield, Robert Bernard
Mann, Robert Arthur
Melius, Paul
Mielenz, Jonathan Richard
Ng, Thomas K.
Ortiz-Suarez, Humberto Jose
Pettit, Flora Hunter
Pomper, Seymour
Quinn, John Albert
Reilly, Peter John
Roberts, Joseph
Royer, Garfield Paul
Schultz, Jerome Samson
Shafer, Jules Alan
Shemin, David
Simonson, Lloyd Grant
Tedeschi, David Henry
Wilke, Charles Robert
Zaborsky, Oskar Rudolf
Zaffaroni, Alejandro Cesar

Genetics and genetic engineering. *See* **BIOLOGY, MEDICINE.**

Nuclear magnetic resonance. *See also* **CHEMISTRY.**
Aull, John Louis
Berliner, Lawrence Jules
Budinger, Thomas Francis
Chance, Britton
Cowburn, David Alan
Dunlap, R. Bruce
Greiner, Jack Volker
Gutowsky, Herbert Sander
Keeler, Elaine Kathleen
Kopp, Stephen James
Kressel, Herbert Y.
Lange, Robert Carl
Lauterbur, Paul C(hristian)
Markley, John Lute
Pykett, Ian Lewis
Silvidi, Anthony Alfred
Springer, Charles Sinclair, Jr.
Werbelow, Lawrence Glen

Other
Bone, Donald Robert, *Hybridoma Technology*
Devore, Paul Warren, *Regenerative technical systems*
Fry, Francis J(ohn), *Ultrasound*
Jervis, Herbert Hunter, *Biotechnology patents*
Kazarian, Leon Edward, *Biomechanics*
Moshy, Raymond Joseph, *Agricultural Biotechnology Management.*

BOTANY. *See* BIOLOGY.

CHEMISTRY

Analytical chemistry
Anderson, Charles Thomas
Armstrong, Andrew Thurman
Asher, Sanford Abraham
Asleson, Gary Lee
Aue, Walter Alois
Bard, Allen Joseph
Benson, Royal Henry
Blotcky, Alan Jay
Boggess, Robert Keith
Bornmann, John Arthur
Bower, Nathan Wayne
Bradshaw, John David
Braun, Robert Denton
Burnett, John Nicholas
Bursey, Maurice M.
Bushaw, Thomas Henry
Chakrabarti, Chuni Lal
Chan, Kwan Ming
Christian, Gary Dale
Ciaccio, Leonard Louis
Cone, Edward J.
Crippen, Raymond C.
Currah, Walter Everett
Danielson, Neil David
Davis, Thomas Paul
Dea, Phoebe K.
Dessy, Raymond Edwin
Dus, Karl M(aria)
Edelson, Martin Charles
Elkin, Robert Glenn
Ellman, George
Elzerman, Alan William
Evans, Frederick E.
Fales, Henry Marshall
Farwell, Sherry Owen
Fateley, William Gene
Fawcett, Newton Creig
Fendler, Janos Hugo
Ferguson, Ronald Max
Finston, Harmon Leo
Fleck, George Morrison
Fried, Bernard
Frye, Herschel Gordon
Ganapathy, Ramachandran
Gavini, Mural B.
Giddings, John Calvin
Glascock, Michael Dean
Grob, Robert Lee
Groth, Richard Henry
Guilbault, George Gerald
Guth, Joseph Henry
Harrison, Willard Wayne
Hausmann, Werner Karl
Heininger, Clarence George, Jr.
Heydegger, H(elmut) Roland
Hinthorne, James Roscoe
Hinze, Willie Lee
Hoffman, Norman Edwin
Hofstetter, Kenneth John
Hubbard, Harold Mead
Irgolic, Kurt Johann
Ito, Yoichiro
Jespersen, Neil David
Johnson, Phyllis Elaine
Johnson, Robert Chandler
Juvet, Richard Spalding, Jr.
Kan, Alla
Karger, Barry Lloyd
Karle, Jean Marianne
Katz, Hyman Bernard
Kegel, Gunter Heinrich Reinhard
Kimlin, Mary Jayne
Kohn, Erwin
Krull, Ira Stanley
Landon, Leroy James
Lindauer, Maurice William
Loach, Kenneth William
Lynch, John August
Lyon, William Southern, Jr.
Macdonald, James Ross
MacDonald, John Chisholm
Malbica, Joseph Orazio
Malinowski, Edmund Robert
Margolis, Sam Aaron
Marks, Peter Jacob
Marquart, John Robert
Marshall, Alan George
Mayberry, William Roy
McDowell, William Jackson
McLafferty, Fred Warren
Meal, Larie
Meyer, John Austin
Michaels, Adlai Eldon
Middleditch, Brian Stanley
Miller, Sidney Israel
Mojavarian, Parviz
Mosier, Benjamin
Mottola, Horacio A.
Murray, Royce Wilton
Nieman, Timothy Alan
Nyssen, Gerard Allan
O'Reilly, James Emil
Pacey, Gilbert Ellery
Pankow, James Frederick
Parker, Lloyd Robinson, Jr.
Parsons, Michael Loewen
Pease, David Nathaniel
Perry, Mildred Elizabeth
Pietri, Charles Edward
Preiss, Ivor Louis
Purdy, William Crossley
Reiss, Howard
Rekers, Robert George
Riley, John Thomas
Rosen, Joseph David
Rumpel, Max Leonard
Saalfeld, Fred Eric
Scheeline, Alexander
Schmid, Gerhard Martin
Shalvoy, Richard Barry
Sherren, Anne Terry
Shore, Karen Fay
Sivertson, John Neilos
Smith, John Elvans
Stahr, Henry Michael
Stookey, George K.
Swartz, William Edward, Jr.
Taylor, Larry Thomas
Tivin, Fred
Turnquist, Truman Dale
Underwood, Arthur Louis, Jr.
Wagner, Roselin Seider
Webber, Andrew
Weber, Stephen Gregory
Weeks, Stephan John
Wehry, Earl Luther, Jr.
Weinkam, Robert Joseph
Welsey, Wayne Cecil
Wiberley, Stephen Edward
Wiech, Ralph Benjamin
Williams, Keith Alan
Wilson, David Merl, Jr.
Winograd, Nicholas
Winterlin, Wray LaVerne
Witting, Lloyd Allen
Wright, John Curtis
Zaromb, Solomon
Zimmerman, John Frederick

Biophysical chemistry
Adler, Alan David
Amey, Ralph Leonard
Asher, Sanford Abraham
Baldwin, Robert Lesh
Barrett, Terence William
Bello, Jake
Berg, Paul

Berliner, Lawrence Jules
Berns, Donald Sheldon
Bettelheim, Frederick Abraham
Bier, Milan
Bigelow, Charles Cross
Birnbaum, Edward R.
Blake, Richard Douglas
Bloomfield, Victor Alfred
Blout, Elkan Rogers
Boskey, Adele Ludin
Boxer, Steven George
Breslow, Ronald Charles
Bush, C. Allen
Castellino, Francis Joseph
Cheung, Herbert Chiu-Ching
Chiang, Joseph Fei
Clementi, Enrico
Cohn, Mildred
Cook, Paul Fabyan
Cooper, Alan Douglas
Crothers, Donald Morris
Darnall, Dennis Wayne
Dea, Phoebe K.
Dewey, Thomas Gregory
Dill, Kenneth Austin
Dixon, Earl, Jr.
Edmondson, Dale Edward
Ehrlich, Robert Stark
Epstein, Irving Robert
Felsenfeld, Gary
Fisher, Ronald Richard
Frankel, Richard Barry
Fried, Josef
Fu, Shou-Cheng Joseph
Fung, Leslie Wo-Mei
Gennis, Robert Bennett
Glaser, Michael
Glusker, Jenny Pickworth
Greenaway, Frederick Thomas
Guillory, Richard John
Hammes, Gordon G.
Harrison, John Henry, IV
Haurowitz, Felix
Hill, Terrell Leslie
Hill, Walter Ensign
Holowka, David Allan
Holtzer, Marilyn Emerson
Holzwarth, George Michael
Huang, Ching-hsien
Jain, Mahendra Kumar
Johnson, Walter Curtis, Jr.
Karpel, Richard Leslie
Kelly, Jeffrey John
Khalifah, Raja Gabriel
Klinman, Judith Pollock
Klotz, Irving Myron
Knox, James Russell, Jr.
Kollman, Peter Andrew
Koshland, Daniel Edward, Jr.
Kraus, Marjorie Patt
Krause, Sonja
Kurtin, William Eugene
Kushick, Joseph N.
Kwiram, Alvin L.
Lakowicz, Joseph Raymond
Lancaster, Jack R., Jr.
Lee, James Ching
Lee, Ted Choong Kil
Lehninger, Albert Lester
Lerman, Sidney
Lever, Alfred Beverley Philip
Live, David Harris
Lu, Ponzy
Markley, John Lute
Marshall, Alan George
Matthews, Kathleen Shive
McCammon, James Andrew
McConnell, Harden Marsden
Mendelsohn, Richard
Mihalyi, Elemer
Miller, Kenneth John
Minch, Michael Joseph
Myers, William Graydon
Nowak, Thomas L.
Olson, Wilma King
Ottenbrite, Raphael Martin
Parker, Frank S.
Parr, Gary Raymond
Patel, Ramesh Chandra
Pauling, Linus Carl
Petsko, Gregory Anthony
Pimentel, George Claude
Richards, Frederic Middlebrook
Richmond, Jonas E.
Roberts, David Craig
Rose, George David
Ross, J(ohn) B(randon) Alexander
Schachman, Howard Kapnek
Scheraga, Harold Abraham
Schroeder, Friedhelm
Schuh, Merlyn Duane
Schultz, Richard Michael
Schuster, Todd Mervyn
Schwartz, A(lbert) Truman
Scovell, William Martin
Shafer, Jules Alan
Silverman, David Norman
Spiro, Thomas George
Spivey, Howard Olin
Springer, Charles Sinclair, Jr.
Starzak, Michael Edward
Steiner, Robert Frank
Stroud, Robert Michael
Sweet, Robert Mahlon
Thomas, George Joseph, Jr.
Tu, Anthony Tsuchien
Vallee, Bert Lester
Van Wart, Harold Edgar
Veis, Arthur
von Hippel, Peter Hans
Warden, Joseph Tallman, Jr.
Welch, William Henry
Wetmur, James Gerard
Wetterhahn, Karen E.
Wickstrom, Eric
Wilson, G(ustavus) Edwin, Jr.
Winter, William Thomas
Wu, Ying Victor
Yang, Jen Tsi
Yocum, Charles Fredrick
Zimm, Bruno Hasbrouck

Catalysis chemistry
Bloch, Herman Samuel
Brumberger, Harry
Cheh, Huk Yuk
Chen, Hoffman Hor-Fu
Collman, James Paddock
Crabtree, Robert Howard
Cratty, Leland Earl, Jr.
Cusumano, James Anthony
Davies, Geoffrey
Forster, Denis
Gassman, Paul George
Good, Mary Lowe
Goodman, David Wayne
Haensel, Vladimir
Halpern, Jack
Hemstock, Glen Alton
Henry, Patrick M.
Hettinger, William Peter, Jr.
Hightower, Joe Walter
Jumper, Eric John
Katz, Thomas Joseph
Klemm, LeRoy Henry
Klotz, Irving Myron
Markham, Claire Agnes
Meek, Devon Walter
Orchin, Milton
Parshall, George William
Roth, James Frank
Schmitz, Roger Anthony
Shapley, John Roger
Sinfelt, John Henry
Somorjai, Gabor A(rpad)
Staley, Ralph Horton
Storm, David Anthony
Wadsworth, William Steele, Jr.
Watters, Kenneth Lynn
Wei, James
Wise, John J.
Worley, S. D.

Clinical chemistry
Auerbach, Victor Hugo
Bhattacharya, Syamal Kanti
Bruns, David Eugene
Byers, James Martin
Christian, Gary Dale
DeBari, Vincent (Anthony)
Gershbein, Leon Lee
Gress-Gordon, Jean Anne
Guilbault, George Gerald
Hankes, Lawrence Valentine
Hinze, Willie Lee
Hoffman, Norman Edwin
Jurmain, Jacob Harry
Kachmar, John Frederick
King, Jonathan Stanton
Knodel, Elinor Livingston
Ladenson, Jack Herman
LeBlanc, Robert Bruce
Mojavarian, Parviz
Natelson, Samuel
Paselk, Richard Alan
Porter, William Hudson
Purdy, William Crossley
Richardson, Keith Erwin
Rosner, Anthony Leopold
Saffran, Judith
Shamberger, Raymond Joseph, Jr.
Simon, Marcia
Van Kley, Harold James
Wei, Robert
Yemanski, Michael Ralph
Zakowski, Jack J.

Crystallography
Azaroff, Leonid Vladimirovitch
Balascio, Joseph Francis
Camerman, Arthur
Duchamp, David James
Freeman, Wade Austin
Jeffrey, George Alan
Karle, Jerome
McEnally, Terence Ernest, Jr.
Sands, Donald Edgar
Schneer, Cecil J.
Veblen, David Rodli
Williams, Donald Elmer
Wuensch, Bernhardt John
Zhu, Nai Jue

Crystallography, X-ray
Beasley, Wayne Machon
Bednowitz, Allan Lloyd
Cantrell, Joseph Sires, Jr.
Carpenter, Gene Blakely
Cotton, Frank Albert
Duax, William Leo
Freeman, Wade Austin
Frueh, Alfred Joseph
Glusker, Jenny Pickworth
Karle, Isabella Lugoski
Knox, James Russell, Jr.
Kopp, Otto Charles
Macintyre, Walter MacNeil
Pavkovic, Stephen Frank
Potenza, Joseph Anthony
Radonovich, Lewis Joseph
Robinson, William Robert
Sayre, David
Scheidt, W. Robert
Schwartz, Lyle Howard
Shoemaker, David Powell
Sweet, Robert Mahlon
Szalda, David Joseph
Williard, Paul G.
Wright, Harlan Tonie
Yates, John Harry
Zhu, Nai Jue

High temperature chemistry
Ames, Lynford Lenhart
Barnes, Hubert Lloyd
Crerar, David Alexander
Cubicciotti, Daniel
Dalal, Nar Singh
Hastie, John William
Howald, Reed Anderson
Keneshea, Francis Joseph
Krikorian, Oscar Harold
Leary, Joseph Aloysius I
Lipschutz, Michael Elazar
Margrave, John Lee
Muan, Arnulf
Osterheld, R(obert) Keith
Shafizadeh, Fred
Vala, Martin Thorvald, Jr.
Wahlbeck, Phillip Glenn
Zollweg, Robert John

Immunocytochemistry. *See* **BIOLOGY.**

Inorganic chemistry
Abrahamson, Harmon Bruce
Adamson, Arthur Wilson
Alper, Howard
Anderson, Charles Thomas
Armendarez, Peter Xavier
Atwood, Jim D.
Barnes, James Alford
Bartlett, Neil
Basolo, Fred
Bauman, John E., Jr.
Beran, Jo Allan
Birk, James Peter
Boggess, Robert Keith
Bond, Walter Dayton
Brill, Thomas Barton
Burmeister, John Luther
Burow, Duane Frueh
Campbell, Donald Leroy
Caslavska, Vera Barbara
Chang, James C.
Chisholm, Malcolm Harold
Choppin, Gregory Robert
Clemens, Donald Faull
Clifford, Alan Frank
Cohen, Paul Shea
Collman, James Paddock
Coops, Melvin (Sterline)
Corey, Joyce Yagla
Cotton, Frank Albert
Crabtree, Robert Howard
Cratty, Leland Earl, Jr.
Cummings, Sue Carol
Dahl, Alan Richard
Davies, Geoffrey
Deiters, Joan Adele
Dodson, Charles Leon, Jr.
Earley, Joseph Emmet
Eaton, Gareth Richard
El-Awady, Abbas Abbas
Eliezer, Isaac
Emerson, Kenneth
Finston, Harmon Leo
Ford, Peter Campbell
Forster, Denis
George, Arnold
Gray, Harry Barkus
Hagen, Arnulf Peder
Halpern, Jack
Hanson, John Elbert
Harrington, Roy Victor
Hawthorne, Marion Frederick
Hill, Sister Ann Gertrude
Hill, Orville Farrow
Hinckley, Conrad Cutler
Hunt, Harold Russell
Hutchison, James Robert
Hyde, Kenneth Edwin
Ibers, James Arthur
Irgolic, Kurt Johann
Johnson, David Alfred
Kirschenbaum, Louis Jean
Klingen, Theordore James
Kochi, Jay Kazuo
Kriegsman, William Edwin
Kuznesof, Paul Martin
Lamberger, Paul Henry
Lane, George Ashel
Langley, Richard Howard
Lanoux, Sigred
Lawton, Emil Abraham
Lees, Alistair John
Leuhrs, Dean Carl
Logan, Terry James
Long, Gary John
MacInnes, David Fenton, Jr.
Margerum, Dale William
Margrave, John Lee
Martin, James Cullen
Mason, W. Roy
McCarley, Robert Eugene
McDowell, William Jackson
McElhattan, Glenn Richard
Meek, Devon Walter
Milburn, Ronald McRae
Muetterties, Earl Leonard
Nechamkin, Howard
Nyssen, Gerard Allan
Osterheld, R(obert) Keith
Patel, Kantilal Chaturbhai
Patel, Ramesh Chandra
Pearson, Ralph Gottfrid
Petersen, John David
Pillay, Sivasankara K.K.
Porterfield, William Wendell
Potts, Richard Allen
Pribush, Robert Allen
Purcell, Keith Frederick
Radonovich, Lewis Joseph
Randall, Roger Ellis
Reed, James Leslie
Reis, Arthur Henry, Jr.
Rossman, George Robert
Rumpel, Max Leonard
Sager, Ray Stuart
Schaeffer, Charles David
Scheidt, W. Robert
Schug, Kenneth Robert
Scovell, William Martin
Segal, Alvin
Shapley, John Roger
Sherren, Anne Terry
Sisler, Harry Hall
Sodd, Vincent Joseph
Spencer, John Brockett
Spiro, Thomas George
Stephens, James Francis
Stucky, Gary Lee
Szalda, David Joseph
Tarr, Donald Arthur
Taube, Henry
Taylor, Larry Thomas
Telkes, Maria
Thompson, Larry Clark
Thompson, Mary Eileen
Thompson, Richard Claude
Villa, Juan Francisco
Viste, Arlen Ellard
Vogel, Glenn Charles
Wagner, Roselin Seider
Wallace, William James
Wartik, Thomas
Welsey, Wayne Cecil
West, Robert Culbertson
Wetterhahn, Karen E.
Wilson, Stephen Thomas
Woods, Mary
Wright, John Ricken
Wrighton, Mark Stephen
Zeldin, Martel

Kinetics
Alekman, Stanley Lawrence
Baughcum, Steven Lee
Benson, Sidney William
Birely, John Horton
Bopp, Charles Dan
Bowman, Joel Mark
Crim, Forrest Fleming, Jr.
Datz, Sheldon
Dorko, Ernest A(lexander)
Earley, Joseph Emmet
El-Awady, Abbas Abbas
Epstein, Irving Robert
Field, Richard Jeffrey
Firestone, Richard Francis
Gill, Piara Singh
Greenstock, Clive Lewis
Hadley, Fred Judson
Hill, E(lgin) Alexander
Hunt, Harold Russell
Kirkien-Rzeszotarski, Alicja Maria
Kirschenbaum, Louis Jean
Knight, David Bates
Kurz, Joseph L.
Kwok, Munson Arthur
Lampe, Frederick Walter
Lee, Yuan Tseh
Lepley, Arthur Ray
Lewis, David Kenneth
Lin, Ming-Chang
Lippmann, David Zangwill
Lynch, John August
Marcus, Rudolph Arthur
Martin, L(aurence) Robbin
Meisels, Gerhard George
Moore, William Marshall
Moran, Thomas Francis
Muschlitz, Earle Eugene, Jr.
Noyes, Richard Macy
Parker, Richard C.
Parr, Gary Raymond
Peplinski, Daniel Raymond
Perrin, Charles Lee
Raw, Cecil John Gough
Ridge, Douglas Poll
Slichter, William Pence
Stanbro, William David
Sturm, James Edward
Tarr, Donald Arthur
Taube, Henry
Thompson, Richard Claude
Woods, Mary
Zavitsas, Andreas Athanasios

Laser-induced chemistry. *See* **LASER.**

Neurochemistry. *See* **NEUROSCIENCE.**

Nuclear magnetic resonance. *See also* **BIOTECHNOLOGY.**
Asleson, Gary Lee
Bailey, William F.
Baldeschwieler, John Dickson
Blout, Elkan Rogers
Bovey, Frank Alden
Brame, Edward Grant, Jr.
Brey, Wallace Siegfried
Bush, C. Allen
Chen, Ching-Nien
Cohn, Mildred
Cowburn, David Alan
Cunningham, Clarence Marion
Dailey, Benjamin Peter
Dalal, Nar Singh
Dybowski, Cecil Ray
Eliel, Ernest Ludwig
Evans, Frederick E.
Froemsdorf, Donald H.
Gutowsky, Herbert Sander
Haselhorst, Donald Duane
Hurd, Ralph Eugene
James, Thomas L.
Jonas, Jiri
Jones, Alan Anthony
Karabatsos, Gerasimos John
Khalifah, Raja Gabriel
Krishna, Nepalli Rama
LaPlanche, Laurine Anna
Lauterbur, Paul C(hristian)
Lepley, Arthur Ray
McCall, David Warren
Mc Dowell, Charles Alexander
Minch, Michael Joseph
Noggle, Joseph Henry
Ostercamp, Daryl Lee
Paudler, William Wolfgang
Reynolds, G(arth) Fredric
Schaeffer, Charles David
Sharp, Robert Richard
Slichter, William Pence
Stary, Frank Edward
Steinmetz, Wayne Edward
Stephens, James Francis
Swinehart, James Stephen
Werbelow, Lawrence Glen
White, David
Wyssbrod, Herman Robert, Jr.

Organic chemistry
Adams, Steven Paul
Agosta, William Carleton
Aguiar, Adam Martin
Aldrich, Jeffrey Richard
Alekman, Stanley Lawrence
Alper, Howard
Armbruster, Charles William
Ault, Addison
Bahner, Carl Tabb
Bailey, William F.
Bak, David Arthur
Barborak, James Carl
Barrett, Edward Joseph
Bartlett, Paul Doughty
Battiste, Merle Andrew
Baumgarten, Reuben Lawrence
Baumstark, Alfons Leopold
Bean, Gerritt Post
Becker, Ernest I.
Bender, Myron Lee
Benson, Royal Henry
Berson, Jerome Abraham
Blossey, Erich Carl
Bodor, Nicholas Stephen
Boekelheide, Virgil Carl
Borowitz, Grace Burchman
Bradshaw, Jerald Sherwin
Brauman, John I.
Breslow, Ronald Charles
Brown, Herbert Charles
Buchi, George Hermann
Bunnett, Joseph Frederick
Burke, Luke Anthony
Cairns, Theodore LeSueur
Carlson, Kenneth T.
Carpenter, Barry Keith
Carroll, Felix Alvin, Jr.
Caswell, Lyman Ray
Chan, Tak Hang
Chang, Chi Kwong
Chapman, Orville Lamar
Chen, Hoffman Hor-Fu
Christensen, Larry Wayne
Clark, Ronald Duane
Closs, Gerhard Ludwig
Cohen, Irwin
Cook, Clarence Edgar
Counts, Wayne Boyd
Covington, Edward Royals
Cram, Donald James
Crandall, Elbert Williams
Cristol, Stanley Jerome
Curran, Dennis Patrick
Curtin, David Yarrow
Dannenberg, Joseph J.
Dauben, William Garfield
Daves, Glenn Doyle, Jr.
de Mayo, Paul
Djerassi, Carl
Durst, Tony
Ehrlich, Kenneth C.
Eilert, Jeffries Harvey
Eliel, Ernest Ludwig
Elsenbaumer, Ronald Lee
Evans, Latimer Richard
Fales, Henry Marshall
Fatiadi, Alexander Johann
Filipescu, Nicolae
Fisher, Charles Harold
Flory, William Evans Sherlock
Fodor, Gabor Bela
Ford, Warren Thomas
Freeman, Peter Kent
Fried, Josef
Froemsdorf, Donald H.
Fry, James Leslie
Gandour, Richard David
Gassman, Paul George
Gates, Marshall DeMotte, Jr.
Gibian, Morton J.
Gillespie, Jesse Samuel, Jr.
Gilliom, Richard D.
Glover, George I.
Graham, Donald Lee
Graybill, Bruce Myron
Gribble, Gordon Wayne
Groth, Richard Henry
Grunwald, Ernest Max
Guida, Wayne Charles
Haberfield, Paul
Hammond, George Simms
Hansch, Corwin Herman
Harvey, Ronald Gilbert
Hausmann, Werner Karl
Helsley, Grover Cleveland
Hill, E(lgin) Alexander
Hirschberg, Albert Irwin
Hornback, Joseph Michael
Horning, Evan Charles
Hudgin, Donald Edward
Hudrlik, Paul Frederick
Huffman, John William
Huyser, Earl Stanley
Jacobson, Barry Martin
Jeffrey, George Alan
Jencks, William Platt
Jones, Donald George
Karabatsos, Gerasimos John
Katz, Thomas Joseph
Keehn, Philip Moses

Kennedy, Carol Tyler
Khorana, Har Gobind
Kingston, David George Ian
Klein, Jack S.
Klemm, LeRoy Henry
Klimstra, Paul Dale
Knapp, Daniel Roger
Knight, David Bates
Knowles, Jeremy Randall
Kochi, Jay Kazuo
Kovac, Pavol
Kowalski, Conrad John
Kristol, David Sol
Kristy, Thomas Wayne
Krull, Ira Stanley
Kurtz, David Williams
Langler, Richard Francis
Lazarus, Allan K.
Lazzari, Eugene Paul
Lee, James Travis, Jr.
Leonard, Nelson Jordan
Lerner, Leon Maurice
Lichter, Robert Louis
Liebman, Joel Fredric
Lipkowitz, Kenneth Barry
Lokensgard, Jerrold Paul
Longroy, Allan Leroy
Lorance, Elmer Donald
Lown, James William
Lukin, Marvin
Lwowski, Walter W.
Macdonald, Timothy Lee
Magid, Ronald Michael
Mallory, Clelia Wood
Martin, James Cullen
McMurry, John Edward
Meal, Larie
Meinwald, Jerrold
Mendenhall, George David
Michejda, Christopher Jan
Miljkovic, Momcilo
Miller, Bernard
Miller, Sidney Israel
Millich, Frank
Mitscher, Lester Allen
Moffatt, John Gilbert
Montgomery, John Atterbury
Moore, Leonard Oro
Mosher, Melvyn Wayne
Moshy, Raymond Joseph
Mowery, Dwight Fay
Nagasawa, Herbert Taukasa
Nakanishi, Koji
Nechamkin, Howard
Negishi, Ei-ichi
Nes, William Robert
Neumeyer, John Leopold
Newcomb, Martin Eugene
Newman, Melvin Spencer
Nicolae, Gheorghe
Nordlander, J(ohn) Eric
Olah, George Andrew
Oro, Juan
Ostercamp, Daryl Lee
Overberger, Charles Gilbert
Pacey, Gilbert Ellery
Pappas, Socrates Peter
Paudler, William Wolfgang
Perrin, Charles Lee
Pettit, George Robert
Pirrung, Michael Craig
Pollack, Ralph Martin
Pomerantz, Martin
Popp, Frank Donald
Potter, Neil Harrison
Powers, Daniel Duffy
Pyle, James Lawrence
Quirk, Roderic Paul
Rabjohn, Norman
Ramirez, Fausto
Reich, Hans Jurgen
Reich, Ieva L.
Rice, Kenner Cralle
Rickborn, Bruce F.
Rislove, David Joel
Roberts, David Craig
Roberts, John D.
Robins, Morris Joseph
Roush, William Richard
Russell, Glen Allan
Schenck, Hans Uwe
Schroeder, Leland Roy
Schwartz, Herbert
Searles, Arthur Langley
Seigler, David Stanley
Shafizadeh, Fred
Simmons, Howard Ensign, Jr.
Sipe, Herbert James, Jr.
Skolnik, Herman
Snider, Barry Bernard
Sohn, John Edwin
Spessard, Gary Oliver
Stang, Peter John
Stille, John Kenneth
Stolberg, Marvin A.
Stork, Gilbert (Josse)
Sund, Eldon H(arold)
Swinehart, James Stephen
Swisher, Joseph Vincent
Sytsma, Louis F.
Taber, Douglass Fleming
Tarbell, Dean Stanley
Taylor, Stephen Keith
Terrell, Ross Clark
Tishler, Max
Trachtenberg, Edward Norman
Trost, Barry Martin
van Tamelen, Eugene Earle
Varughese, Pothen
Veber, Daniel Frank
Vercellotti, John Raymond
Viola, Alfred
Waddell, Thomas Groth
Wadsworth, William Steele, Jr.
Walba, David Mark
Washburne, Stephen Shepard
Wassmundt, Frederick William
Watanabe, Kyoichi Aloisius
Watson, Darrell Gene
Webber, Andrew
West, Robert Culbertson
Westheimer, Frank Henry
Whitehurst, Darrell Duayne
Whitesides, George McClelland
Wiberg, Kenneth Berle
Wilbur, James M(yers), Jr.
Wiley, Michael David
Williard, Paul G.
Wilson, G(ustavus) Edwin, Jr.
Wipke, Will Todd
Witiak, Donald T.
Witkop, Bernhard
Witte, Michael
Wubbels, Gene Gerald
Zavitsas, Andreas Athanasios
Zimmerman, Howard Elliot

Photochemistry
Abrahamson, Harmon Bruce
Agosta, William Carleton
Calvin, Melvin
Carroll, Felix Alvin, Jr.
Cech, Thomas Robert
Creed, David
Cristol, Stanley Jerome
Dauben, William Garfield
de Mayo, Paul
Dorko, Ernest A(lexander)
Ford, Peter Campbell
Gaylord, Norman Grant
Gray, Harry Barkus
Haberfield, Paul
Halpern, Arthur M(errill)
Hornback, Joseph Michael
Juvet, Richard Spalding, Jr.
Kuntz, Robert Roy
Kurtin, William Eugene
Kurtz, David Williams
Lerman, Sidney
Lever, Alfred Beverley Philip
Lin, Sheng Hsien
Lwowski, Walter W.
Makinen, Marvin William
Mallory, Clelia Wood
Markham, Claire Agnes
Mc Clure, Donald Stuart
Mendenhall, George David
Moore, William Marshall
Nauman, Robert Vincent
Pappas, Socrates Peter
Petersen, John David
Pribush, Robert Allen
Reed, James Leslie
Rennert, Joseph
Schwerzel, Robert Edward
SMith, Allan Laslett
Smith, Gerald Duane
Strickler, Stewart Jeffery
Sturm, James Edward
Trozzolo, Anthony Marion
Turro, Nicholas John
Wadlinger, Robert Louis
Watson, Darrell Gene
Wrighton, Mark Stephen
Wubbels, Gene Gerald
Zimmerman, Howard Elliot

Photochemistry, laser
Baughcum, Steven Lee
Berry, Michael James
Butler, James Ehrich
Delap, James Harve
Earl, Boyd Lorel
Fendler, Janos Hugo
Flynn, George William
Grossweiner, Leonard Irwin
Jackson, William Morgan
Johnston, Harold Sledge
Kliger, David Saul
Lampe, Frederick Walter
Lees, Alistair John
Mauzerall, David Charles
McLaughlin, Donald Reed
Piltch, Martin Stanley
Rettner, Charles Thomas
Robinson, Dean Wentworth
Schuh, Merlyn Duane
Schwerzel, Robert Edward

Physical chemistry
Acrivos, Juana Vivo
Addy, John Keith
Adler, Alan David
Alberty, Robert Arnold
Albright, John Grover
Allara, David Lawrence
Allen, Martin
Ames, Lynford Lenhart
Amey, Ralph Leonard
Armendarez, Peter Xavier
Baer, Eric
Bahe, Lowell Warren
Baldeschwieler, John Dickson
Bancroft, George Michael
Bard, Allen Joseph
Barnes, James Alford
Bauer, Simon Harvey
Benson, Sidney William
Berkowitz, Joseph
Bernheim, Robert Allan
Bernstein, Elliot Roy
Bernstein, Richard Barry
Bigeleisen, Jacob
Birely, John Horton
Black, James Francis
Boggs, James Ernest
Bohme, Diethard Kurt
Bond, Walter Dayton
Bornmann, John Arthur
Bowen, Lawrence Hoffman
Boxer, Steven George
Bramwell, Fitzgerald Burton
Brauman, John I.
Braun, Charles Louis
Brewer, Leo
Brill, Thomas Barton
Brooks, Philip Russell
Broughton, Donald Beddoes
Brumberger, Harry
Buettner, Garry Richard
Burow, Duane Frueh
Butler, James Ehrich
Callis, Patrik Robert
Campbell, David Owen
Cantrell, Joseph Sires, Jr.
Carpenter, Gene Blakely
Chance, Ronald Richard
Chase, Grafton D.
Chiang, Joseph Fei
Chu, Benjamin
Clark, Donald Eldon
Clark, Hugh Kidder
Clark, Roy White
Clarke, George Alton
Clifton, David Geyer
Closs, Gerhard Ludwig
Colson, Steven Douglas
Companion, Audrey Lee
Conan, Robert James, Jr.
Coursey, Bert Marcel
Cox, Hollace Lawton, Jr.
Crawford, Bryce Low, Jr.
Criss, Cecil M.
Crosley, David Risdon
Cross, Richard James, Jr.
Cunningham, Clarence Marion
Curl, Robert Floyd, Jr.
Dailey, Benjamin Peter
Davis, Howard Ted
Debye, Nordulf W. G.
Delap, James Harve
Denio, Allen A(lbert)
Devlin, Frank Joseph
Dodson, Charles Leon, Jr.
Donoghue, Edward Sylvester
Drickamer, Harry George
Duchamp, David James
Dunbar, Robert Copeland
Dunn, Thomas M.
Dybowski, Cecil Ray
Dye, James Louis
Dykstra, Clifford Elliot
Eastman, Michael Paul
Eaton, Gareth Richard
Ebdon, David William
Eisenberg, Morris
Eliason, Morton Albert
Eliezer, Isaac
Engelking, Paul Craig
Eyler, John Robert
Farrar, James Martin
Fateley, William Gene
Field, Richard Jeffrey
Firestone, Richard Francis
Fleck, George Morrison
Flynn, George William
Freeman, Gordon Russel
Freeman, Mark Phillips
Freund, Robert Stanley
Fried, Vojtech
Friedlander, Gerhart
Friedrich, Donald Martin
Fuller, Milton Eugene
Fung, Bing-Man
Glogovsky, Robert Louis
Goates, James Rex
Goldstein, Mark Kingston Levin
Goodfriend, Paul Louis
Goodisman, Jerry
Gordon, Roy Gerald
Gouterman, Martin Paul
Graybeal, Jack Daniel
Guillet, James Edwin
Gulari, Erdogan
Hach, Edwin Ellison, Jr.
Hadley, Fred Judson
Halpern, Arthur M(errill)
Hamby, Drannan C.
Hammond, George Simms
Han, Charles Chih-Chao
Harmony, Marlin Dale
Harris, Harold Hart
Hedberg, Kenneth Wayne
Heinemann, Heinz
Heininger, Clarence George, Jr.
Hernandez, Samuel P.
Herschbach, Dudley Robert
Higuchi, Takeru
Hodge, Ian Moir
Hollingsworth, Charles Alvin
Holly, Frank Joseph
Holtzer, Marilyn Emerson
Hornig, Donald Frederick
Hsia, Yu-Ping
Hutchison, James Robert
Isganitis
Johnson, David Alfred
Johnson, Donald Rex
Johnson, Robert Chandler
Jonas, Jiri
Jones, Donald George
Jones, Francis Thomas
Jones, M(arvin) Thomas
Kan, Alla
Kapral, Raymond Edward
Katz, Joseph Jacob
Kaufman, Frederick
Kaufman, Joyce Jacobson
Kauzmann, Walter Joseph
Kay, Robert Leo
Kemp, Marwin King
Kimlin, Mary Jayne
King, Frederick Warren
Kirkien-Rzeszotarski, Alicja Maria
Kolb, Charles Eugene
Krishnan, Pallassana Narayanier
Kuffner, Roy Joseph
Kuist, Charles Howard
Kuntz, Robert Roy
Kunzler, John Eugene
Kwiram, Alvin L.
Laane, Jaan
Ladanyi, Branka Maria
Lagunas-Solar, Manuel Claudio
Lakowicz, Joseph Raymond
Lane, George Ashel
Lang, Conrad Marvin
LaPietra, Joseph Richard
Lee, Yuan Tseh
Leung, Wing Hai
Lin, Sheng Hsien
Lindauer, Maurice William
Lippmann, David Zangwill
Lipscomb, William Nunn, Jr.
Litovchenko, Vladimir Alexei
Long, Franklin A.
Lonsdale, Harold Kenneth
Lyons, John Winship
MacKay, John Kelvin
Mackay, Raymond Arthur
Maclay, William Nevin
Malinowski, Edmund Robert
Mantei, Kenneth Alan
Mark, James Edward
Marquart, John Robert
Marshall, Walter Lincoln
Martin, Richard Lee
Mason, W. Roy
Mathews, C(ollis) Weldon
McAtee, James Lee, Jr.
McCall, David Warren
Mc Dowell, Charles Alexander
Meckstroth, Wilma Koenig
Meisels, Gerhard George
Meiser, John Henry
Melton, Charles Estel
Mettee, Howard Dawson
Michaels, Adlai Eldon
Miller, William Hughes
Millero, Frank Joseph, Jr.
Miner, Bryant Albert
Mohilner, David Morris
Moore, John Hays
Moran, Thomas Francis
Moreno, Edgard Camacho
Morris, George Vincent
Mortensen, Earl Miller
Mortimer, Robert George
Muschlitz, Earle Eugene, Jr.
Nauman, Robert Vincent
Navangul, Himanshoo Vishnu
Nelson, Robert Norton
Noggle, Joseph Henry
Oblad, Alexander Golden
Orwoll, Robert Arvid
Osthoff, Robert C.
Owens, Charles Wesley
Pan, Chai-Fu
Parker, Richard C.
Patterson, Gary David
Pauling, Linus Carl
Pecora, Robert
Pedersen, Lee Grant
Perkins, Richard Scott
Person, Willis Bagley
Pitzer, Kenneth S.
Plazek, Donald John
Polanyi, John Charles
Polavarapu, Prasad Leela
Potenza, Joseph Anthony
Pratt, David Wixon
Prausnitz, John Michael
Pye, Earl Louis
Raw, Cecil John Gough
Reiss, Howard
Rennert, Joseph
Rentzepis, Peter M.
Reynolds, G(arth) Fredric
Rice, Stuart Alan
Rider, Paul Edward
Ridge, Douglas Poll
Roeder, Stephen Bernhard Walter
Rossington, David Ralph
Rossini, Frederick Dominic
Ruch, Richard Julius
Rue, Rolland Ray
Russell, Allen Stevenson
Saalfeld, Fred Eric
Salzman, William Ronald
Samuel, Aryeh Hermann Albert
Sands, Donald Edgar
Schellman, John Anthony
Scheraga, Harold Abraham
Schoonmaker, Richard Clinton
Schufle, Joseph Albert
Schuler, Robert Hugo
Schuman, Robert Paul
Schwartz, A(lbert) Truman
Selmanowitz, Victor Joel
Shirley, David Arthur
Sipe, Herbert James, Jr.
Smalley, Richard Errett
Spencer, John Brockett
Spindel, William
Staley, Ralph Horton
Starkweather, Howard Warner, Jr.
Steinmetz, Wayne Edward
Stevens, John Gehret
Stockmayer, Walter Hugo
Strickler, Stewart Jeffery
Stucky, Gary Lee
Summitt, (William) Robert
Telkes, Maria
Thompson, H. Bradford
Thompson, Mary Eileen
Trajmar, Sandor
Tuan, Debbie Fu-tai
Tyrrell, James
Vijh, Ashok Kumar
Viste, Arlen Ellard
Wadlinger, Robert Louis
Wahrhaftig, Austin Levy
Walker, David Crosby
Wall, Frederick Theodore
Wallace, William James
Waugh, John Stewart
Weatherford, Charles Albert
Weber, Stephen Gregory
Wen, Wen-Yang
Wentink, Tunis, Jr.
West, E. Dale
Westheimer, Frank H(enry)
White, David
White, Stephen Halley
Widom, Benjamin
Williams, Donald Elmer
Williams, Lesley Lattin
Wilson, Alexander Thomas
Wilson, Edgar Bright
Wise, Henry
Worley, S. D.
Wulff, Claus Adolf
Ying-Sin, Li
Zare, Richard Neil
Zaromb, Solomon
Zerner, Michael Charles
Zeroka, Daniel

Polymer chemistry
Addy, John Keith
Baer, Eric
Bailey, Frederick Eugene, Jr.
Berry, Guy Curtis
Blossey, Erich Carl
Bovey, Frank Alden
Bowen, Rafael Lee
Coleman, Michael Murray
Counselman, Clarence James
Covington, Edward Royals
Crandall, Elbert Williams
Danielson, Neil David
Darsey, Jerome Anthony
Denio, Allen A(lbert)
Elias, Hans Georg
Fawcett, Newton Creig
Ferretti, James Alfred
Ferry, John Douglass
Fisher, Charles Harold
Flory, Paul John
Ford, Warren Thomas
Gaylord, Norman Grant
Han, Charles Chih-Chao
Harris, Milton
Harvey, Leonard A.
Hawkins, Walter Lincoln
Hoeg, Donald Francis
Hoffman, Dennis Mark
Holzwarth, George Michael
Horowitz, Carl
Huang, Sun-Yi
Hudgin, Donald Edward
Jernigan, Robert Lee
Jones, Alan Anthony
Kambour, Roger Peabody
Koenig, Jack Leonard
Kohn, Erwin
Kong, Eric Siu-Wai
Kratzer, Reinhold Hermann
Krause, Sonja
Kumar, Ganesh Narayanan
Lambuth, Alan Letcher
Lawton, Emil Abraham
LeBlanc, Robert Bruce
Lee, Lieng-Huang
Live, David Harris
MacInnes, David Fenton, Jr.
MacKnight, William John
Maclay, William Nevin
Manson, John Alexander
Mark, James Edward
Massa, Dennis Jon
Meyer, John Austin
Millich, Frank
Mittal, Kashmiri Lal
Moore, Leonard Oro
Morgan, Paul Winthrop
Mosier, Benjamin
Olson, Wilma King
Orwoll, Robert Arvid
Ottenbrite, Raphael Martin
Overberger, Charles Gilbert
Pecora, Robert
Pohl, Herbert Ackland
Quirk, Roderic Paul
Restaino, Alfred Joseph
Reynard, Kennard Anthony
Ryan, Charles Luce, Jr.
Salamone, Joseph Charles
Schaefer, Henry Frederick, III
Schenck, Hans Uwe
Senich, George A.
Silverman, Joseph
Simha, Robert
Skeist, Irving
Sohn, John Edwin
Stahl, Joel S.
Starkweather, Howard Warner, Jr.
Stary, Frank Edward
Stille, John Kenneth
Stockmayer, Walter Hugo
Turner, Derek Terence
Van De Mark, Michael Roy
Vandenberg, Edwin James
Vogl, Otto
Weber, Thomas Andrew
Wilbur, James M(yers), Jr.
Wiles, David McKeen
Winter, William Thomas
Wnek, Gary Edmund
Yuska, Henry B.
Zeldin, Martel

Solid state chemistry
Acrivos, Juana Vivo
Baker, William Oliver
Balascio, Joseph Francis
Bartlett, Neil
Baughman, Ray Henry
Borg, Richard John
Bowen, Lawrence Hoffman
Bramwell, Fitzgerald Burton
Braun, Charles Louis
Carter, Forrest Lee
Curtin, David Yarrow
Emerson, Kenneth

Evans, Billy Joe
Giessen, Bill Cormann
Herber, Rolfe H.
Hochmann, Petr Tomás
Ibers, James Arthur
Jones, M(arvin) Thomas
Klingen, Theordore James
Kuznesof, Paul Martin
Langley, Richard Howard
Laudise, Robert Alfred
Long, Gary John
Mac Diarmid, Alan Graham
McCarroll, William Henry
Meiser, John Henry
Metzger, Robert Melville
Oesterreicher, Hans Karl
Ovshinsky, Stanford Robert
Phillips, James Charles
Reis, Arthur Henry, Jr.
Risen, William Maurice, Jr.
Robin, Eugene Debs
Robinson, William Robert
Sarma, Abul Chandra
Sato, Hiroshi
Shoemaker, David Powell
Sienko, Michell J.
Stafford, Fred E.
Stevens, John Gehret
Trozzolo, Anthony Marion
Wagner, James Bruce, Jr.
Willett, Roger DuWayne
Wilson, Stephen Thomas
Wright, John Curtis

Space chemistry. ***See*** **SPACE SCIENCE.**

Statistical mechanics
Adelman, Steven A.
Andrews, Frank Clinton
Emptage, Michael R(ollins)
Fisher, Michael Ellis
Fixman, Marshall
Fried, Vojtech
Helfand, Eugene
Ladanyi, Branka Maria
Lee, John Francis
Mayer, Klaus
Mazo, Robert Marc
Mortimer, Robert George
Paul, Edward
Pratt, Lawrence R.
Prigogine, Ilya
Stratt, Richard Mark
Strieder, William Christian
Widom, Benjamin
Woodbury, George W., Jr.

Surface chemistry
Adamson, Arthur Wilson
Allara, David Lawrence
Allen, Martin
Andres, Ronald Paul
Boudart, Michel
Brey, Wallace Siegfried
Conan, Robert James, Jr.
Cusumano, James Anthony
D'Arrigo, Joseph Salvatore
Davis, Rodney James
Davison, Sydney George
Dresser, Miles Joel
Dunlap, Brett Irving
Ebdon, David William
Ehrlich, Gert
Engel, Thomas Walter
Fogler, Hugh Scott
Freeman, Mark Phillips
Fuerstenau, Douglas Winston
Gomer, Robert
Good, Mary Lowe
Goodisman, Jerry
Green, Michael Enoch
Hagstrum, Homer Dupre
Hance, Robert Lee
Handy, Lyman Lee
Hemstock, Glen Alton
Holly, Frank Joseph
Hubert, Jay Marvin
Kaldor, Andrew Peter
Kassner, James Lyle, Jr.
Katz, William
Kelsh, Dennis J.
Koenig, Jack Leonard
Krauss, Alan Robert
Kuffner, Roy Joseph
Layton, Richard Gary
Leblanc, Roger Maurice
Lee, Lieng-Huang
Leidheiser, Henry, Jr.
Leung, Wing Hai
Matijevic, Egon
McAtee, James Lee, Jr.
Mittal, Kashmiri Lal
Moudgil, Brij Mohan
Muetterties, Earl Leonard
Powell, Cedric John
Ramaker, David Ellis
Rhodin, Thor Nathaniel
Rosenberg, Richard Allan
Ruch, Richard Julius
Schechter, Robert Samuel
Schmid, Gerhard Martin
Schoonmaker, Richard Clinton
Shirley, David Arthur
Somorjai, Gabor A(rpad)
Stafford, Fred E.
Swartz, William Edward, Jr.
Thomas, H. Ronald
Tuul, Johannes
Van De Mark, Michael Roy
Vijh, Ashok Kumar
Wahlbeck, Phillip Glenn
Watters, Kenneth Lynn
Winograd, Nicholas
Wolf, Edward Dean
Woodbury, George W., Jr.

Synthetic chemistry
Battiste, Merle Andrew
Becker, Ernest I.
Brown, Herbert Charles
Buchi, George Hermann
Burmeister, John Luther
Chan, Tak Hang
Cheng, Chia-Chung
Christensen, Larry Wayne
Corey, Joyce Yagla
Counts, Wayne Boyd
Crouch, Rosalie Kelsey
Cummings, Sue Carol
Curran, Dennis Patrick
Daves, Glenn Doyle, Jr.
Evans, Latimer Richard
Fatiadi, Alexander Johann
Fodor, Gabor Bela
Gates, Marshall DeMotte, Jr.
Ginos, James Zissis
Guida, Wayne Charles
Huang, Sun-Yi
Hudrlik, Paul Frederick
Huffman, John William
Jerina, Donald Michael
Kabalka, George Walter
Kovac, Pavol
Kowalski, Conrad John
Langler, Richard Francis
Lokensgard, Jerrold Paul
Miljkovic, Momcilo
Nair, Madhavan
Negishi, Ei-ichi
Newcomb, Martin Eugene
Newman, Melvin Spencer
Nicolae, Gheorghe
Nordlander, J(ohn) Eric
Patel, Kantilal Chaturbhai
Ramirez, Fausto
Reich, Hans Jurgen
Rice, Kenner Cralle
Rislove, David Joel
Rosen, William Michael
Roth, James Frank
Schug, Kenneth Robert
Searles, Arthur Langley
Shen, Tsung Ying
Sisler, Harry Hall
Snider, Barry Bernard
Spessard, Gary Oliver
Sund, Eldon H(arold)
Taber, Douglass Fleming
Terrell, Ross Clark
Trost, Barry Martin
van Tamelen, Eugene Earle
Washburne, Stephen Shepard
Wassmundt, Frederick William
Yuska, Henry B.

Theoretical chemistry
Adams, William Henry
Adelman, Steven A.
Alder, Berni Julian
Alexander, Millard Henry
Benham, Craig John
Boggs, James Ernest
Bowman, Joel Mark
Bunge, Carlos Federico
Burke, Luke Anthony
Carpenter, Barry Keith
Clarke, George Alton
Clementi, Enrico
Cohen, Irwin
Combs, Leon Lamar
Companion, Audrey Lee
Cook, Robert C.
Dannenberg, Joseph J.
Del Bene, Janet Elaine
Delos, John Bernard
Dunlap, Brett Irving
Dykstra, Clifford Elliot
Eaker, Charles William
Eliason, Morton Albert
Fixman, Marshall
Freed, Karl Frederick
Gilliom, Richard D.
Gimarc, Benjamin Maurice
Gislason, Eric Arni
Goodfriend, Paul Louis
Gouterman, Martin Paul
Guberman, Steven Lawrence, *Theoretical chemistry*
Gund, Peter Herman
Hirschfelder, Joseph Oakland
Hochmann, Petr Tomás
Hoffmann, Roald
Hollingsworth, Charles Alvin
Hopfield, John Joseph
Hopper, Darrel Gene
Howard, Robert Ernest
Joshi, Bhairav Datt
Julienne, Paul S.
Kaufman, Joyce Jacobson
Kay, Kenneth George
Keller, Jaime
King, Frederick Warren
Klein, Douglas Jay
Kollman, Peter Andrew
Konowalow, Daniel Dimitri
Kushick, Joseph N.
Langhoff, Peter Wolfgang
Levin, George Benjamin
Liebman, Joel Fredric
Lipkowitz, Kenneth Barry
Lykos, Peter George
Malik, David Joseph
Marcus, Rudolph Arthur
Martin, Richard Lee
McCammon, James Andrew
McLaughlin, Donald Reed
Metzger, Robert Melville
Miller, William Hughes
Mortensen, Earl Miller
Pan, Yuh Kang
Parr, Robert Ghormley
Payne, Philip Warren
Pearson, Ralph Gottfrid
Pedersen, Lee Grant
Pitzer, Kenneth S.
Pitzer, Russell Mosher
Pratt, Lawrence R.
Ramaker, David Ellis
Rhodes, William Clifford
Rice, Stuart Alan
Richardson, James Wyman
Rider, Paul Edward
Salzman, William Ronald
Schaefer, Henry Frederick, III
Schneider, Barry Irwin
Sloane, Christine Scheid
Sparks, Morgan
Stover, Betsy Jones
Stratt, Richard Mark
Thompson, H. Bradford
Tuan, Debbie Fu-tai
Tyrrell, James
Van Uitert, LeGrand Gerard
Viehland, Larry Alan
Wall, Frederick Theodore
Weber, Thomas Andrew
Wiberg, Kenneth Berle
Yates, John Harry
Zerner, Michael Charles
Zeroka, Daniel
Zimm, Bruno Hasbrouck

Thermodynamics
Alberty, Robert Arnold
Albright, John Grover
Andrews, Frank Clinton
Bahe, Lowell Warren
Barron, Saul
Bauman, John E., Jr.
Clifton, David Geyer
Criss, Cecil M.
Debye, Nordulf W. G.
Fletcher, Edward Abraham
Germano, Don Joseph
Goates, James Rex
Gorges, Heinz A.
Gupta, Vijay Kumar
Gyftopoulos, Elias Panayiotis
Haberman, William Lawrence
Hatsopoulos, George Nicholas
Helgeson, Harold Charles
Higuchi, Takeru
Howald, Reed Anderson
Kellogg, Herbert Humphrey
Krikorian, Oscar Harold
Kurz, Joseph L.
Lee, John Francis
Lindenmeyer, Paul Henry
Lyon, Richard Evan
Mayer, Klaus
Millero, Frank Joseph, Jr.
Miner, Bryant Albert
Mohilner, David Morris
O'Connell, John Paul
Pan, Chai-Fu
Paul, Edward
Prigogine, Ilya
Readnour, Jerry Michael
Serrin, James Burton
Van Hecke, Gerald Raymond
West, E. Dale
Westrum, Edgar Francis, Jr.
Wulff, Claus Adolf

Other
Angus, John Cotton, *Electrochemistry*
Armbruster, Charles William, *Chemical education*
Barborak, James Carl, *Oraganometallic chemistry*
Bartlett, Paul Doughty, *Reaction mechanisms*
Basolo, Fred, *Nuclear chemistry*
Batzel, Roger Elwood, *Nuclear chemistry*
Beran, Jo Allan, *Environmental chemistry*
Brame, Edward Grant, Jr., *Spectroscopy*
Brown, Glenn Halstead, *Liquid crystals*
Brown, Kenneth Lawrence, *Organometallic chemistry*
Cerny, Joseph, III, *Nuclear chemistry*
Chang, Ching-jer, *Phytochemistry*
Chisholm, Malcolm Harold, *Organometallic chemistry*
Cleland, W(illiam) Wallace, *Biochemistry*
Crawford, Bryce Low, Jr., *Molecular spectroscopy*
Crippen, Raymond C., *Biochemistry*
Daniel, Samuel Henderson, III, *Nuclear power plant chemistry*
Douglas, Bryce, *Medicinal chemistry*
Field, Frank Henry, *Mass spectrometry*
Filby, Royston Herbert, *Nuclear chemistry*
Finn, Ronald Dennet, *Radiochemistry*
Folkers, Karl August, *Bromedical research administration*
Forster, William Owen, *Chemical oceanography*
Friedlander, Gerhart, *Nuclear chemistry*
Frye, Cecil Leonard, *Environmental chemistry, Silicon chemistry*
Frye, Herschel Gordon, *Forensic chemistry*
Gardner, Marjorie Hyer, *Chemical Education*
Gibbs, Ann, *Nuclear chemistry*
Gilman, Henry, *Organometallic chemistry.*
Given, Peter Harvey, *Coal chemistry*
Goldstein, Irving Solomon, *Wood chemistry*
Hamby, Drannan C., *Electrochemistry*
Harman, Denham, *Free radical chemistry*
Harris, Milton, *Natural and synthetic fibers.*
Haustein, Peter Eugene, *Nuclear chemistry*
Hedberg, Kenneth Wayne, *Molecular structure*
Hinckley, Conrad Cutler, *Bioinorganic chemistry*
Hoeg, Donald Francis, *Organometallic Chemistry*
Hoffman, Darleane Christian, *Nuclear chemistry*
Katz, Frances R., *Carbohydrate chemistry*
Kay, Robert Leo, *Solution chemistry*
Keil, Klaus, *Cosmochemistry*
Klehr, Edwin Henry, *Nuclear analytical chemistry*
Lang, Conrad Marvin, *Electron spin resonance*
Lazarus, Allan K., *Synthetic lubricants*
Matijevic, Egon, *Colloid chemistry*
McDonald, Charles Cameron, *Research management*
Mercer, Kermit Ray, *Electron Spin resonance*
Norman, Jack C., *Nuclear chemistry*
Olhoeft, Gary Roy, *Electrochemistry*
Orgel, Leslie Eleazer, *Prebiotic chemistry*
Pavkovic, Stephen Frank, *Chemical education*
Perkins, Richard Scott, *Electrochemistry*
Pillay, Sivasankara K.K., *Nuclear and Radiochemistry*
Poskanzer, Arthur M., *Nuclear chemistry*
Puchtler, Holde, *Histochemistry*
Rayudu, Garimella V. S., *Nuclear and radiochemistry*
Riley, John Thomas, *Coal Chemistry*
Rowland, Frank Sherwood, *Atmospheric chemistry*
Sackett, William Malcolm, *Marine Chemistry*
Schroeder, Leland Roy, *Carbohydrate chemistry*
Schuler, Robert Hugo, *Radiation chemistry*
Schweiker, George Christian, *Chemical research administration*
Scotti, Vincent Guy, *Radiochemistry*
Seaborg, Glenn Theodore, *Nuclear Chemistry*
Soloway, Albert Herman, *Medicinal chemistry*
Song, Leila Shia, *Environmental Chemistry*
Spindel, William, *Science policy, chemical science administration*
Stolberg, Marvin A., *Radiochemistry*
Throdahl, Monte Corden, *Chemical research management*
Turkevich, Anthony Leonid, *Nuclear Chemistry*
Vandenbosch, Robert, *Nuclear chemistry*
Watras, Ronald Edward, *Chemical education*
Wildung, Raymond Earl, *Environmental chemistry*
Yaffe, Leo, *Nuclear Chemistry*
Zimmerman, John Frederick, *Instrumentation*

COGNITIVE SCIENCE. ***See*** **PSYCHOLOGY, Cognition.**

COMPUTER SCIENCE. ***See also*** **ENGINEERING, Electrical.**

Algorithms
Amsbury, Wayne
Barzilai, Jonathan
Bauer, Michael Anthony
Baybars, Ilker
Bent, Samuel Watkins
Brown, R(oy) Leonard
Burkhart, Richard Henry
Burstein, Joseph
Chandra, Ashok Kumar
Childs, Joseph Edwin
Chin, Francis Yuk-Lun
Corder, Michael Paul
Day, William Hartwell Eveleth
Dobkin, David Paul
Fienup, James Ray
Ford, Lester R., Jr.
Franco, John Vincent
Frank, Ellen Ryan
Gajski, Daniel Danko
Garey, Michael Randolph
Garzia, Ricardo Francisco
Gupta, Udaiprakash Induprakash
Hale, William Kent
Hamacher, Horst Wilhelm
Hopcroft, John Edward
Itoga, Stephen Yukio
Jasiulek, Joachim Norbert
Kennedy, William Jo
Kleiman, Howard
Knuth, Donald Ervin
Kobler, Virginia Ponds
Koskelo, Markku Juhani
Krisnamoorthy, Mukkai Subramaniam
Lagarias, Jeffrey Clark
Langston, Michael Allen
Laskowski, Sharon J.
Ledbetter, Carl Scotius, Jr.
Lee, E. Bruce
Leifman, Lev Jacob
Lekoudis, Spiro G(eorge)
Lieberherr, Karl Josua
Lin, Shen
Lincoln, Walter Butler, Jr.
Marquardt, Donald Wesley
Mendelsohn, Nathan Saul
Monier, Louis Marcel
Morgan, Stephen Lyle
Newman, Morris
Pan, Victor Yakovlevich
Pollnow, Gilbert Frederick
Rabin, Michael O.
Rosen, Judah Ben
Savage, John Edmund
Schwartz, Jacob Theodore
Stanton, Ralph
Tarjan, Robert Endre
Urry, Vern William
Weide, Bruce Warren
Wertz, Harvey Joe
Wiggins, Ralphe
Woodward, Paul Ralph
Yao, Andrew Chi-Chih

Artificial intelligence
Adams, Richard L.
Apple, Martin Allen
Ball, William Ernest
Barstow, David Robbins
Berliner, Hans Jack
Boyle, Brian John
Brady, J(ohn) Michael
Breland, Hunter Mansfield
Briggs, Arthur Brailsford, Jr.
Conway, Lynn Ann
Corff, Nicholas J.
Daly, John Anthony
Decker, Bruce Michael
Dickey, Thomas Edgar
Feigenbaum, Edward Albert
Firestone, Roger Morris
Fu, King-sun
Galambos, James Andrew
Green, Cordell
Greenfeld, Norton Robert
Grisell, Ronald David
Hansen, James Vernon
Hecht, Lee Martin
Heilmeier, George Harry
Hirschman, Lynette
Hoffman, Frederick
Hofstadter, Douglas Richard
Holtzman, Samuel
Hunt, Bobby Ray
Joshi, Aravind Krishna
Kalet, Ira Joseph
Kanal, Laveen Nanik
Kedes, Laurence Herbert
Klabosh, Charles
Kobler, Virginia Ponds
Kochen, Manfred
Kopstein, Felix Friedrich
Krug, Harry Everistus Peter, Jr.
Lachman, Roy
Lederberg, Joshua
Lieberman, Philip
Lind, Henrik Olav
Loatman, Robert Bruce
Marder, Barry M(ichael)
McCarthy, John
McDermott, John Donovan
Mendelsohn, Nathan Saul
Meystel, Alexander Michael
Millen, Jonathan Kaye
Miller, Lance Arnold
Minsky, Marvin Lee
Moore, Michael Hart
Moravec, Hans Peter
Moses, Joel
Mumford, David Bryant
Newell, Allen
Park, Ok-Choon
Parsons, Michael Loewen
Plastock, Roy A.
Popov, Dan
Powell, Edward Gordon
Reggia, James Allen
Reilly, Kevin Denis
Roberts, Steven Kurt
Sage, Andrew Patrick, Jr.
Schank, Roger Carl
Schwartz, William Benjamin
Shapiro, Linda Gail
Shortliffe, Edward Hance
Sidlinger, Bruce Douglas
Siegel, Melvin Walter
Triffet, Terry
Turchin, Valentin Fedorovich
Utting, Kenneth
Waltz, David Leigh
Wells, Alan Harvey
Wickelgren, Wayne Allen
Will, Peter Milne
Winograd, Terry Allen
Winston, Patrick Henry
Wipke, Will Todd
Zdybel, Frank, Jr.

Computer architecture
Amdahl, Gene Myron
Amsbury, Wayne
Amundsen, Keith Byron
Arvind
Aupperle, Kenneth Robert
Banerjee, Utpal
Bartlett, Peter Greenough
Bell, Chester Gordon
Bertrand, John
Bic, Lubomir
Black, James Emmett, Jr.
Bouhana, James P.
Brooks, Frederick Phillips, Jr.
Burkhardt, Kenneth John, Jr.

Cheung, John Yan-Poon
Childs, Jeffrey John
Chu, Wesley Wei-chin
Chuang, Henry Ying Huang
Conery, John Simpson
Conway, Lynn Ann
Cox, Jerome Rockhold, Jr.
Crabb, Charles Frederick
Cray, Seymour R.
Davidson, Scott
DeGroot, Richard Douglas
Dickey, Thomas Edgar
Dietz, William Bruce
Donnelley, James Ellis
Egan, John Thomas
Ekanadham, Kattamuri
Evans, David C.
Feigenbaum, Edward Albert
Fernández, Eduardo Buglioni
Fleck, David Charles
Gajski, Daniel Danko
Gordon, Edward Barry
Gottlieb, Allan Joseph
Gray, Festus Gail
Greco, Richard James
Greenberg, Kenneth Freeman
Hammer, Carl
Heard, Harry Gordon
Heuft, Richard William
Holub, Richard Anthony
Hopkins, William Christopher
Howe, John Edward
Irwin, Mary Jane
Jacobs, Ronald Michael
Kain, Richard Y(erkes)
Kalos, Malvin Howard
Kaman, Charles Henry
Karshmer, Arthur Israel
Kavipurapu, Krishna M.
Kirsch, Lawrence Edward
Leuze, Michael Rex
Liptay, John Stephen
Merritt, Charles Randall
Monier, Louis Marcel
Mooney, James Donald
Morris, Robert
Myers, Wade Hampton, Jr.
Oh, Richard Young
Olson, Robert Allen
Pahwa, Ashok
Parce, Donald Lewis
Patterson, David Andrew
Patton, Peter Clyde
Protopapas, Dimitrios
Quek, Swee-Meng
Rafalko, Edward Dennis
Ramamoorthy, Chittoor V.
Rigg, Carl Wilson
Roberts, Steven Kurt
Schwartz, Jacob Theodore
Shaw, David Elliot
Shriver, Bruce Douglas
Siewiorek, Daniel Paul
Stewart, Robert Murray, Jr.
Stock, Rodney Dennis
Strasen, Stephen M.
Surmacz, Joseph George
Svarrer, Robert W.
Thammavaram, N. Rao
Wah, Benjamin Wan-Sang
Walsh, Jacqueline Ann
Wang, Pong-Sheng
Weinberg, Richard Alan
West, Anthony Robert
Wong, Edward Chor-Cheung

Cryptography and data security
Berkovits, Shimshon
Blakley, George Robert, Jr.
Blum, Manuel
Davida, George I.
DeMillo, Richard A.
Denning, Dorothy Elizabeth
Figueres, Maurice Christian
Hammer, Carl
Hellman, Martin Edward
Kriegsman, William Edwin
Linn, John
Makar, Boshra Halim
Millen, Jonathan Kaye
Miller, James Edward
Morris, Robert
Myers, Eugene Dolan
Rabin, Michael O.
Richards, Roger Thomas
Simmons, Gustavus James
White, Richard Mahaffey
Wood, Charles Cresson

Database systems
Alavian, Farid
Aubin, William M.
Ball, William Ernest
Bhargava, Bharat Kumar
Bic, Lubomir
Blasgen, Michael William
Boylan, Stephen P.
Breitbart, Yuri Jacob
Butler, Shahla
Carino, Felipe, Jr.
Cattell, Roderic Geoffrey Galton
Chiang, Tung Ching
Chin, Francis Yuk-Lun
Clark, Jon D(ennis)
Conway, Richard Walter
Coppoc, Gordon Lloyd
Courtheoux, Richard James
De Pree, Robert Wilson
De Smith, Donald Albert
Dougherty, George John
Du, David Hung-Chang
Duncan, Doris Gottschalk
Fagin, Ronald
Fedorowicz, Jane
Fernández, Eduardo Buglioni
Flores, Ivan
Freeman, Harvey Allen
Graham, Tad Laury
Gupta, Udaiprakash Induprakash
Hanson, Trevor Russell
Heller, Jack
Hevner, Alan Raymond
Hoplin, Herman Peter
Howard, Jay Lloyd
Hull, Richard Baxter
Itoga, Stephen Yukio
Jennings, Larry Eugene
Kellner, Richard George
Koenig, Michael Edward Davison
Krishnamurthy, Ravindran
Lee, Edward Yue Shing
Levy, David Edward
Li, Victor On-Kwok
Lipka, Stephen Erik
Lohman, Guy Maring
Loomis, Mary Elizabeth Snuggs
Machalski, Robert Chester
Marashi, Musa S.
Martin-Robinson, Kera Gayle
McGarvey, John James
Mohan, C.
Nigam, Rajendra C.
Pahwa, Ashok
Rose, Glenn Robert
Ryder, Benjamin Mills
Scheuermann, Peter I.
Selinger, Patricia Griffiths
Sokol, Robert James
Stewart, David Harry
Taborek, Jerry
Ullman, Jeffrey David
Vernazza, Jorge Enrique
Wah, Benjamin Wan-Sang
Warren, Wayne Lawrence
Wood, David Allen

Distributed systems and networks
Alavian, Farid
Amar, Amar-Dev
Amundsen, Keith Byron
Appelbe, William Frederick
Audet, John James, Jr.
Aupperle, Eric Max
Avadian, John Mark
Beck, Robert Donald
Bedrosian, Samuel Der
Berkovits, Shimshon
Bernstein, Herbert Jacob
Bielawski, W. Bart
Black, James Emmett, Jr.
Blasgen, Michael William
Blum, Howard Stanley
Borochoff, Robert M.
Breitbart, Yuri Jacob
Bridges, Alan Lynn
Carson, George Stephen
Carter, Jeff Crossett
Chintapalli, Hemantha Kumar
Christ, Duane Marland
Chu, Wesley Wei-chin
Curtis, Ronald Sanger
Dessy, Raymond Edwin
Doherty, Mark Fitzgerald
Donahue, Michael James
Donnelley, James Ellis
Dougherty, George John
Du, David Hung-Chang
Farah, Badie Naiem
Ferrari, Domenico
Freeman, Harvey Allen
Fry, James Palmer
Gillman, Clifford Brian
Gobbel, John Randall
Gottlieb, Allan Joseph
Grand, Diana Leigh
Green, Paul Eliot, Jr.
Gurwitz, Robert Frey
Hale, William Kent
Hall, Nancy Rose
Hampson, Bradford Ellsworth
Hevner, Alan Raymond
Hickson, Edward Lilliott
Irwin, Mary Jane
Jacobs, Ronald Michael
Jennings, Steven Fletcher
Johnson, Brian Weatherred
Kain, Richard Y(erkes)
Kellner, Richard George
Krishnamurthy, Ravindran
Kueishiong, Tu
Landweber, Lawrence Hugh
Lantz, Keith Allen
Laub, Leonard Joseph
Lazowska, Edward Delano
Lee, E. Bruce
Lee, Edward Yue Shing
Lee, Hikyu
Lewis, Ralph Jay, III
Li, Victor On-Kwok
Lientz, Bennet Price
Lind, Henrik Olav
Linn, John
Lloyd, Evan Elliott Morgan
Lohman, Guy Maring
Loomis, Mary Elizabeth Snuggs
Lu, Chun Chian
Maier, John Joseph
Manthey, Michael John
Masloff, Jacqueline Jo
Meditch, James Stephen
Miller, Richard Hans
Mohan, C.
Moorhead, Deborah Kay
Moss, John Eliot Blakeslee
Murata, Tadao
Myers, Eugene Dolan
Myers, Wade Hampton, Jr.
Nikora, Allen Peter
Oh, Richard Young
Oran, David Robert
Polonsky, Ivan Paul
Robinson, James Robert
Scheuermann, Peter I.
Schwab, Jeffrey Richard
Shen, Yi-Shang
Shreve, Gregory Monroe
Shriver, Bruce Douglas
Skinner, Thomas Paul
Springer, Donald Harold
Stein, Scott Allen
Stewart, Robert Murray, Jr.
Sweeney, Patrick J.
Tripathi, Anand Vardhan
  Raghunandan
Wang, Pong-Sheng
West, Anthony Robert
White, Richard Mahaffey
Williams, Robert Reid
Williams, Theodore Joseph
Witte, Kurt Allen
Wittie, Larry Dawson
Wojciechowski, Witold Stanislaw
Wood, David Allen
Yau, Stephen Sik-sang

Foundations of computer science
Backus, John
Dymond, Patrick William
Fagin, Ronald
Fendrich, John William
Fuller, Milton Eugene
Hartmanis, Juris
Hill, Sister Ann Gertrude
Hull, Richard Baxter
Krisnamoorthy, Mukkai
  Subramaniam
Manthey, Michael John
Metropolis, Nicholas C.
O'Donnell, Michael James
Warfield, John Nelson
Wise, David Stephen

Graphics, image processing, and pattern recognition
Abrams, Robert Jay
Anderson, David Carleton
Aydelotte, Lee C.
Ball, George William
Barsky, Brian Andrew
Bartels, Richard Harold
Battista, Jerry Joseph
Bednowitz, Allan Lloyd
Bedrosian, Samuel Der
Benedict, George Frederick
Bernstein, Ralph
Bitzer, Donald Lester
Bono, Peter Richard
Bowyer, Allen Frank
Boyse, John Wesley
Brady, J(ohn) Michael
Branham, Richard Lacy, Jr.
Bridges, Alan Lynn
Brooks, Frederick Phillips, Jr.
Burkhart, Richard Henry
Carberry, James Joseph
Carino, Felipe, Jr.
Carson, George Stephen
Chakravarty, Indranil
Chappell, Gary Alan
Chasen, Sylvan Herbert
Christian, Carol Ann
Clark, Alan Lee
Coblitz, David Barry
Cohen, Edgar Allan, Jr.
Cohen, Norman Edward
Cooper, (Howard) Gordon
Corff, Nicholas J.
Craig, Donald M.
Croft, Thomas A(rthur)
Cross, Kenneth James
Dean, Edwin Becton
De Fanti, Thomas Albert
De Pree, Robert Wilson
Dertouzos, Michael Leonidas
Dobkin, David Paul
Doherty, Mark Fitzgerald
Dudley, Alden Woodbury, Jr.
Dudnik, Elliott Eliasaf
Dufour, Reginald James
Dusko, Harold George
Eaker, Charles William
Egan, John Thomas
Eigen, Daryl Jay
Epstein, Sheldon Lee
Evans, David C.
Fakharzadeh, Ali M.
Feeser, Larry James
Fischell, David Ross
Fougere, Paul Francis
Franzblau, Daniel Eric
Fu, King-sun
Garfield, Alan J.
Ghanta, Babu Madhur
Gibson, John Egan
Gil, Joan
Glackin, David Langdon
Gnanadesikan, Ramanathan
Grace, Thomas Peter
Greco, Richard James
Hansche, Bruce David
Harrington, Steven Jay
Hasegawa, Tony Seisuke
Haynes, Mack W., Jr.
Hertel, Richard James
Hewitt, Anthony Victor
Hillman, Gilbert Rothschild
Holloway, Thomas Thornton
Hughett, Paul William
Hunt, Bobby Ray
Hutchinson, Frank David, III
Jansson, Peter Allan
Jenkins, Arnold Milton
Jordan, Byron Dale
Jordan, Scott Wilson
Joseph, John Louis
Kanal, Laveen Nanik
Kelley, Michael Stephen
Kiefhaber, Nikolaus Josef
Knowles, Richard James Robert
Kratz, Lawrence John
Kreidl, Tobias Joachim
Krell, Mitchell
Kruse, David Harold
Kuei, Chih-Chung
Landgrebe, David Allen
Langridge, Robert
Latta, John Neal
Lee, Roy Yuewing
Leonard, Myer Samuel
Levin, Joshua Zev
Lieberman, Philip
Ligler, George Todd
Lipkin, Bernice Sacks
MacDonald, John Chisholm
Maiman, Theodore Harold
Marsh, Kenneth Albert
Marshall, Garland Ross
Martin-Robinson, Kera Gayle
Martins, Donald Henry
Meagher, Donald Joseph
Merrill, Marshall Leigh
Mitchell, Nancy Brown
Mittelman, Phillip Sidney
Moore, Elliott Paul
Moravec, Hans Peter
Morgan, Stephen Lyle
Morrison, John B.
Olins, Donald Edward
Owens, James Carl
Pollack, Bary William
Reiffel, Leonard
Ritter, Gerhard X.
Roger, David Freeman
Rose, George David
Roth, Mitchell Godfrey
Schoech, William Joseph
Schweizer, Francois
Shani, Uri
Shapiro, Linda Gail
Siler, William MacDowell
Sproull, Robert Fletcher
Stock, Rodney Dennis
Surmacz, Joseph George
Tou, Julius T.
Trower, William Peter
Utting, Kenneth
Vesel, Richard Warren
Walkup, John Frank
Wang, Patrick Shen-pei
Weinberg, Richard Alan
Williams, Robert Reid
Williams, T. H. Lee
Wilson, Allan Byron
Yang, Yee-Hong
Yessios, Chris Ioannis

Information systems, storage, and retrieval. *See also* **INFORMATION SCIENCE.**
Ackerman, Allan Douglas
Beiser, Leo
Blakley, George Robert, Jr.
Bookstein, Abraham
Borochoff, Robert M.
Branscomb, Lewis McAdory
Braun, Stephen Hughes
Brinson, Donald Edward
Broome, Douglas Ralph, Jr.
Burford, Hugh Jonathan
Cazes, Albert N.
Clark, Jon D(ennis)
Cohen, Marvin Sidney
Cummings, Martin Marc
Demmerle, Alan Michael
Diedrichsen, Loren Dale
Doszkocs, Tamas Endre
Dube, Rajesh
Dumas, Neil Stephen
Du Wors, Robert Jerome
Dwinell, Lew David
Everett, Robert Rivers
Farhi, Leon Elie
Fisher, H. Leonard
Fortna, John David Edward
Fried, Jerrold
Fry, James Palmer
Gardin, T. Hershel
Goncher, Richard Sidney
Gordon, James Arthur
Graham, Tad Laury
Hagel, Andrew Richard
Hagemark, Kjell Ingvar
Haueisen, William David
Heard, Harry Gordon
Hickson, Edward Lilliott
Howard, John Hayes
Hoye, Robert Earl
Humi, Mayer
Hyman, Edward Jay
Jacobs, Keith William
Jaffe, Norman J.
Konzak, Calvin Francis
Kricka, Hanna Halyna
Krynicki, Victor Edward
Lesko, Robert Joseph
Lipkin, Bernice Sacks
Livingston, John David
Loach, Kenneth William
Macintyre, Walter MacNeil
Mansur, Ovad Mordecai
Masloff, Jacqueline Jo
Maxwell, Donald Lee
Miller, James Edward
Mohler, Ronald Rutt
Newell, Allen
Packer, Katherine Helen
Parker, Robert B.
Pearlstein, Sol
Pesch, William Allan
Peterson, Darwin Wilson
Powell, John Edwin
Pratt, Arnold W.
Pribor, Hugo Casimer
Quirke, Terence Thomas, Jr.
Rine, David C.
Rush, Richard William
Ryder, Benjamin Mills
Sager, Naomi
Schniederjans, Marc James
Schroer, Bernard Jon
Smith, Linda Cheryl
Smith, Robert David
Stewart, David Harry
Summit, Roger Kent
Sweeney, Urban Joseph
Taaffe, Gordon
Tucker, Marc Stephen
Watkins, Paul Roger
Watts, Malcolm S.M.
Weaver, Christopher Scot
Williams, Martha Ethelyn
Winarski, Daniel James
Woolman, Myron
Wray, John Lawrence

Mathematical software
Ablow, Clarence Maurice
Alfeld, Peter Wilhelm
Altschuler, Martin David
Anderson, James Bryan
Barker, William Hamblin
Benaroya, Haym
Bickart, Theodore Albert
Boisvert, Ronald Fernand
Boley, Daniel Lucius
Bownds, John Marvin
Brown, Richard Don
Brown, R(oy) Leonard
Bunch, Phillip Carter
Burr, Baldwin Gwynne
Chang, Albert Fuwu
Chang, Peter Hon-You
Clay, Charles George, Jr.
Cody, William James, Jr.
Craig, Richard G.
Dediu, Mihai Virgil
Denny, William Francis, II
Dixon, Wilfrid Joseph
Dongarra, Jack Joseph
Driessel, Kenneth R.
Engel, Lars Norlick
Farr, William Rogers
Ferry, Jason Hughes
Forbes, George Franklin
Ford, Byron Milton
Franke, Richard Homer
Garzia, Ricardo Francisco
Gautschi, Walter
Gear, Charles William
Grace, Thomas Peter
Gray, George Amelung
Hageman, Louis Alfred
Hall, Charles Allan
Harbaugh, John Warvelle
Harrington, Steven Jay
Held, Ronald Dennis
Hendricks, John Stanley
Hough, David Granville
Humphrey, John P.
Johnson, Robert Shepard
LaBudde, Robert Arthur
Larson, John Leonard
Larson, Richard Gustavus
Lee, Sang Moon
Levin, Joshua Zev
Ley, Richard Wayne
Lin, Char-Lung Charles
Lin, Shen
Ling, Robert F.
Lunde, Peter J.
Lupash, Lawrence O(Vidiu)
Lyness, James N.
Lyons, William Kimbel
Markatos, Nicolas-Christos Gregory
Matthews, John Brian
McGrath, Joseph Fay
McKinney, Stanley Joe
Meyer, Carl Dean, Jr.
Mochizuki, Leslie Yasuko
Moore, Polly
Morgan, Alexander Payne
Moses, Joel
Mowery, Dwight Fay
Nash, John Christopher
Nguyen, Chinh Trung
Niccolai, Marino John
Pitts, Thomas Griffin
Rall, Louis B(aker)
Romboski, Lawrence David
Sheridan, Robert Emmett, Jr.
Silverman, Barry George
Sincovec, Richard F(rank)
Smith, Henry John Peter
Sorensen, Danny Chris
Stahl, Raymond Earl
Thompson, Sylvester
Tretter, Marietta Joan
Tsao, Nai-Kuan
Vandergraft, James Saul
Wallace, Orson Joseph
Weinrich, Brian Erwin
Weir, Maurice Dean
Wendelberger, James George
Wertz, Harvey Joe
Wickes, William Castles
Wolfe, Michael David
Wolter, Kirk Marcus
Woolson, William Andrew
Wu, John Naichi

Numerical analysis
Alfeld, Peter Wilhelm
Allgower, Eugene Leo
Altiero, Nicholas James
Ancona, Antonio
Anderson, Dale Arden
Anderson, James Bryan
Anderson, William Judson
Andrews, George Harold
Andrushkiw, Roman Ihor
Aulick, Charles Mark
Bartels, Richard Harold
Barzilai, Jonathan

Basehore, Kerry Lee
Beny, Neta
Berger, Alan Eric
Bernstein, Herbert Jacob
Bingham, Billy Elias
Blackburn, Jacob Floyd
Boal, Jan List
Boisvert, Ronald Fernand
Boley, Daniel Lucius
Bollinger, Richard Coleman
Bownds, John Marvin
Bradley, James Henry
Bramble, James Henry
Branco, Maria dos Milagres
Burton, Howard Alan
Carasso, Alfred Sam
Carter, Leland LaVelle
Chang, Albert Fuwu
Ciment, Melvyn
Cody, William James, Jr.
Concordia, Charles
Concus, Paul
Cornyn, John Joseph
Cott, Donald W(ing)
Cullum, Jane Kehoe
Dantzig, George Bernard
Di Donato, Armido Richard
Dietenberger, Mark Anthony
Doggett, LeRoy Elsworth
Dongarra, Jack Joseph
Ehrlich, Louis William
Epperson, James Felts
Faber, Vance
Field, David Anthony
Fink, Joanne Krupey
Finlayson, Bruce Alan
Fix, George Joseph
Fletcher, John Edward
Franke, Richard Homer
Gallie, Thomas Muir, Jr.
Gautschi, Walter
Gear, Charles William
Geer, James Francis
Gilmartin, Amy Jean
Glaz, Harland Mitchell
Glimm, James Gilbert
Golub, Gene H.
Goodman, Richard H.
Grimm, Louis J.
Grosse, Eric H.
Gunzburger, Max Donald
Haber, Seymour
Hach, Edwin Ellison, Jr.
Hageman, Louis Alfred
Hakala, Reino William
Halsey, Norman Douglas
Hansen, Richard Olaf
Hanson, Floyd Bliss
Harpavat, Ganesh Lal
Hicks, Darrell Lee
Hlavacek, Vladimir
Hoagland, Gordon Wood
Hough, David Granville
Howard, Sethanne
Howe, Timothy Max
Ingram, Glenn R.
Jasiulek, Joachim Norbert
Jedruch, Jacek
Johnson, Carol William
Johnson, Olin Glynn
Kadi, Kamal Sif-el
Kammler, David William
Kascic, Michael Joseph, Jr.
Kaufman, Linda
Kaul, Maharaj Krishen
Kikuchi, Noboru
Kjaer-Pedersen, Niels
Kratz, Lawrence John
Kriegsmann, Gregory Anthony
Kunisch, Karl
Larsen, Kenneth Martin
Larson, John Leonard
Latoza, Kenneth Charles
Lax, Peter David
Leon, Steven Joel
Leuze, Michael Rex
Leventhal, Stephen Henry
Lu, Allen An-hua
Lyness, James N.
Mansfield, Lois
Markham, Thomas Lowell
McAuley, Van Alfon
McBroom, Robert Chism
McCall, Edward Huffaker
McRae, D(avid) Scott
Meyer, Carl Dean, Jr.
Mills, Wendell Holmes, Jr.
Monash, Ellis Alan
Morita, Toshio
Neumann, Herschel
Nguyen, Vietchau
Orszag, Steven Alan
Painter, Jeffrey Farrar
Pan, Victor Yakovlevich
Pang, Jong-Shi
Peaceman, Donald William
Pence, Dennis Dale
Pierce, Sam
Poole, George Douglas
Post, Douglass Edmund, Jr.
Rall, Louis B(aker)
Rheinboldt, Werner Carl
Rigterink, Paul Vernon
Rosen, Judah Ben
Roth, Mitchell Godfrey
Schneider, Gerald Elmore
Schoenstadt, Arthur Loring
Sereny, Aron
Sorensen, Danny Chris
Steihaug, Trond
Tsao, Nai-Kuan
Vandergraft, James Saul
Vasilakis, John Dimitri
Wachspress, Eugene Leon
Wagner, Stephen Gregory
Wahlbin, Lars Bertil
Walker, Homer Franklin
Waltmann, William Lee
Wilson, Walter Ervin
Wright, Alden Halbert
Young, David Monaghan, Jr.
Yu, Chen-Cheng William

Operating systems
Beck, Robert Donald
Belady, Laszlo Antal
Boyd, Donald Loren
Brinch Hansen, Per
Brinson, Donald Edward
Brower, Joseph Gilbert
Browne, James Clayton
Carter, Jeff Crossett
Chintapalli, Hemantha Kumar
Corbató, Fernando José
Craig, Donald M.
Darby, Clifton Floyd
DeGroot, Richard Douglas
De Smith, Donald Albert
Doria, Victor
Du Wors, Robert Jerome
Ekanadham, Kattamuri
Ferrari, Domenico
Flores, Ivan
Friedberg, Carl E.
Gobbel, John Randall
Gurwitz, Robert Frey
Hampson, Bradford Ellsworth
Hansen, Stanley Severin, II
Hanson, Trevor Russell
Haynes, Mack W., Jr.
Hollis, Jan Michael
Howard, John Hayes
Irons, Edgar T(owar)
Jennings, Steven Fletcher
Jones, Anita Katherine
Joseph, John Louis
Karshmer, Arthur Israel
Kelley, Michael Stephen
Lantz, Keith Allen
Lazowska, Edward Delano
Lee, Hikyu
Mayes, Leslie William
Merrill, Marshall Leigh
Merritt, Charles Randall
Moldenhauer, Jeanne Elisabeth
Mooney, James Donald
Olson, Robert Allen
Oran, David Robert
Parce, Donald Lewis
Paris, Steven Mark
Rapsey, Laurie Adele, *Operating systems*
Rigg, Carl Wilson
Schwab, Jeffrey Richard
Schwetman, Herbert Dewitt
Sherman, Gordon Rae
Skinner, Thomas Paul
Stein, Scott Allen
Tripathi, Anand Vardhan Raghunandan
Walsh, Jacqueline Ann
Whalen, Richard Vincent
Wilkinson, Roy Miele
Wilson, Allan Byron
Wittie, Larry Dawson
Wolfe, Bradley Allen

Programming languages
Appelbe, William Frederick Arvind
Backus, John
Bergquist, James William
Bondy, Jonathan
Boylan, Stephen P.
Brinch Hansen, Per
Cattell, Roderic Geoffrey Galton
Conery, John Simpson
Council, Edward Latimer
Curtis, Ronald Sanger
Dediu, Mihai Virgil
De Fanti, Thomas Albert
Firestone, Roger Morris
Goguen, Joseph Amadee
Graham, Robert Montrose
Graham, Susan Lois
Habermann, Arie Nicolaas
Irons, Edgar T(owar)
Jaffe, Norman J.
Joshi, Bhairav Datt
Kaufman, Raymond
Kinney, John James
Knuth, Donald Ervin
Krell, Mitchell
Lee, Van Ming
Mantei, Kenneth Alan
Mayes, Leslie William
McCormick, Ferris Ellsworth
Meseguer, Jose Guaita
Meyers, Richard Anthony
Moss, John Eliot Blakeslee
O'Donnell, Michael James
Peelle, Howard Arthur
Perlis, Alan J.
Polonsky, Ivan Paul
Poole, George Douglas
Richards, Howell Alan
Rine, David C.
Sowers, Joseph Louis
Teitelbaum, Ray
Turchin, Valentin Fedorovich
Wallace, Robert Bruce, Jr.
Wickelgren, Wayne Allen
Wise, David Stephen
Zdybel, Frank, Jr.

Software engineering
Adams, Richard L.
Adrion, William Richard
Afshar, Siroos K.
Aydelotte, Lee C.
Barstow, David Robbins
Bauer, Michael Anthony
Bayer, Jesse Abraham
Belady, Laszlo Antal
Bhargava, Bharat Kumar
Bondy, Jonathan
Bono, Peter Richard
Boyd, Donald Loren
Bradley, James Henry
Brakefield, James Charles
Briggs, Arthur Brailsford, Jr.
Brody, Steven
Brower, Joseph Gilbert
Burgess, Eric
Burns, Robert Donald, III
Butler, Shahla
Campoy, Leonel Perez
Chang, Peter Hon-You
Chiang, Tung Ching
Christian, John Thomas
Chuang, Henry Ying Huang
Clay, Charles George, Jr.
Clements, Gregory Leland
Cohen, Norman Edward
Conway, Richard Walter
Corbató, Fernando José
Council, Edward Latimer
Cross, Kenneth James
Curtis, Bill
Daniels, Carole Angela
Danner, David Lee
Darby, Clifton Floyd
DeMillo, Richard A.
Dick, Daniel Eggleston
Dietz, William Bruce
Dlhopolsky, Joseph Gerald
Doria, Victor
Ebeling, Dolph George
Fakharzadeh, Ali M.
Falk, James Robert
Farah, Badie Naiem
Fendrich, John William
Fenves, Steven Joseph
Figueres, Maurice Christian
Foss, Donald John
Franzblau, Daniel Eric
Friesen, Larry Jay
Galbiati, Louis Joseph
Gardner, Willard Hale
Ghanta, Babu Madhur
Giannetti, Ronald A.
Goguen, Joseph Amadee
Goldman, Lee William
Goncher, Richard Sidney
Goodhue, William Lehr
Gordon, Edward Barry
Grafton, Robert Bruce
Graham, Robert Montrose
Grand, Diana Leigh
Graves, Harvey Wilbur, Jr.
Greenberg, Kenneth Freeman
Hacken, George
Hagel, Andrew Richard
Hall, Nancy Rose
Harrison, Michael Alexander
Heller, Jack
Hertel, Richard James
Hopkins, William Christopher
Howe, John Edward
Hummel, Myron Floyd
Hyde, Kenneth Edwin
Jacob, Robert Joseph Kassel
Kavipurapu, Krishna M.
Kelsoe, Lynda Carol
Khalil, Hatem Mohamed
Kiefhaber, Nikolaus Josef
Kruse, David Harold
Kutlik, Roy Lester
Lee, Bert Gentry
Lee, Roy Yuewing
Ligler, George Todd
Lipka, Stephen Erik
Maier, John Joseph
Manly, Philip James
Mansur, Ovad Mordecai
Marashi, Musa S.
Martin, Paul Joseph
McCormick, Ferris Ellsworth
McGill, Scott Douglas
Meseguer, Jose Guaita
Miller, Joseph S.
Miller, Robert Alan
Miller, Thomas Paul
Mochizuki, Leslie Yasuko
Moorhead, Deborah Kay
Morganstein, Stanley
Niccolai, Marino John
Nieh, Bill
Nikora, Allen Peter
Paris, Steven Mark
Patton, Peter Clyde
Perlis, Alan J.
Perrenod, Stephen Charles
Pohl, Jens Gerhard
Pollack, Bary William
Puchyr, Peter Joseph
Ramamoorthy, Chittoor V.
Reilly, Kevin Denis
Richards, Howell Alan
Robinson, Earl James
Rothrock, Ray Alan
Sammet, Jean E.
Seldner, Michael
Shani, Uri
Shen, Yi-Shang
Shih, Charles Chien
Sidlinger, Bruce Douglas
Sincovec, Richard F(rank)
Spital, Robin David
Springer, Donald Harold
Strasen, Stephen M.
Sweeney, Patrick J.
Teitelbaum, Ray
Town, Donald Earl
Wallace, Orson Joseph
Wallace, Robert Bruce, Jr.
Wang, Patrick Shen-pei
Warme, Paul Kenneth
Warren, Wayne Lawrence
Weide, Bruce Warren
Whalen, Richard Vincent
Wickes, William Castles
Wilkinson, Roy Miele
Winograd, Terry Allen
Worth, Norman Paul
Yau, Stephen Sik-sang
Zunde, Pranas

Theoretical computer science
Afshar, Siroos K.
Baker, Brenda S.
Bedinger, Joseph Arnold
Bent, Samuel Watkins
Berman, Joel D.
Blum, Manuel
Carlyle, Jack Webster
Chandra, Ashok Kumar
Denning, Peter James
Dennis, Jack Bonnell
Dix, Rollin Cumming
Dymond, Patrick William
Faber, Vance
Franco, John Vincent
Goldstine, Herman Heine
Grafton, Robert Bruce
Guida, Peter Matthew
Harrison, Michael Alexander
Hartmanis, Juris
Killian, Barbara Germain
Landweber, Lawrence Hugh
Langston, Michael Allen
Larson, Richard Gustavus
Laskowski, Sharon J.
Lewis, Philip M.
Lloyd, Evan Elliott Morgan
Loui, Michael Conrad
Purdom, Paul Walton, Jr.
Savage, John Edmund
Sherman, Gordon Rae
Ullman, Jeffrey David
Wang, Hao
Yao, Andrew Chi-Chih

Other
Atwood, Charles LeRoy, *Computer applications in engineering*
Avadian, John Mark, *Software engineering*
Batson, Alan Percy, *Computer systems*
Bennett, John Roscoe, *Computer company management*
Birk, James Peter, *Educational software*
Bouhana, James P., *Computer Performance Evaluation*
Briotta, Daniel A., Jr., *Laboratory microcomputing*
Brown, Stephen Woody, *Computer applications in education and psychology*
Burkhardt, Kenneth John, Jr., *System design*
Chase, Robert Arthur, *Computers in Medicine*
Clark, Alan Lee, *Geometric modelling of solids*
Curtis, Kent Krueger, *Computer science research administration*
Damashek, Marc, *Digital signal processing*
Danaher, Brian Grayson, *Computer-based instruction*
Dertouzos, Michael Leonidas, *Personal computers*
Dixon, Paul Nichols, *Microcomputers in education*
Dupree, Samuel Hardy, Jr., *Scientific Computing*
Estrin, Gerald, *Digital computer systems*
Everett, Robert Rivers, *Digital computer design*
Fano, Robert Mario, *Computer science education*
Fletcher, John Dexter, *Computer-assisted instruction*
Gatewood, George David, *Computer interfacing*
Gatewood, Lael Cranmer, *Biomedical computation*
Hacken, George, *Real-time, concurrent systems*
Hurley, Thomas John, Jr., *Nuclear reactor process control*
Inman, Bobby Ray, *Computer technology research and development management*
Jobs, Steven Paul, *Computer company management*
Johnson, Olin Glynn, *Vector processing/processors*
Kaman, Charles Henry, *Applications software*
Komanduri, Ayyangar M., *Scientific Computer Programming*
Laxer, Cary, *Education biomedical applications*
LeMay, Moira Kathleen, *Human-computer interaction*
Levine, Randolph Herbert, *Computer Education*
Levy, Stephen Raymond, *Computer company administration.*
Meyers, Richard Anthony, *Computer science education*
Nadin, Mihai, *Semiotics of computer use and applied artificial intelligence*
Navangul, Himanshoo Vishnu, *Use of computers in education*
Oettinger, Anthony Gervin, *Information resources policy*
Perkins, Glenn Richard, *Data management*
Pinney, Frank Batchelder, *Computer-aided manufacturing*
Purdom, Paul Walton, Jr., *Compiler design*
Rafalko, Edward Dennis, *Computervision*
Ramm, Dietolf, *Medical applications*
Rapsey, Laurie Adele, *Real time work*
Rawlinson, Stephen John, *Microprogramming*
Selinger, Patricia Griffiths, *Office systems*
Siegel, Michael Jason, *Data analysis programming*
Smith, Robert David, *Human resource information systems*
Snyder, James Newton, *Administration*
Sowers, Joseph Louis, *Simulation*
Tarjan, Robert Endre, *Group theory, Data structures*
Weisberg, Joseph Simpson, *Computer applications in learning*
White, Mary-Alice, *Psychology of computer learning*
Wolf, Walter Alan, *Computer modelling of insect systems*
Wong, Edward Chor-Cheung, *Logic design*

DENTISTRY AND ODONTOLOGY

Cariology
Besic, Frank Charles
Bowen, William Henry
Call-Smith, Kathy Meredith
Clark, George Eugene
Dawes, Colin
Gantt, David Graham
Harris, Norman Oliver
Hein, John William
Kleber, Carl Joseph
Leverett, Dennis H.
Levy, John Stuart
Mallatt, Mark Edward
Mellberg, James Richard
Moreno, Edgard Camacho
Newbrun, Ernest
Pape, Harry Rudolph, Jr.
Putt, Mark Stuart
Rawls, Henry Ralph
Shrestha, Buddhi Man
Stookey, George K.
Wagner, Morris
Wefel, James Stern
Williams, Keith Alan

Endodontics
Block, Robert Michael
Cappuccino, Carleton C.
DePalma, Robert Anthony
Heuer, Michael Alexander
Lin, Louis Min-Tsu
Taintor, Jerry Frank
Torneck, Calvin David

Growth and development
Chow, Michael Hung-Chun
Doyle, Walter Arnett
Duperon, Donald Francis
Garner, LaForrest Dean
Goldsmith, Douglas Howard
Green, Larry Joy
Hall, Stanton Harris
Harris, Suzanne Straight
Kohn, Donald William
Lestrel, Pete Ernest
Malhotra, Shyam Kumar
Nahoum, Henry Isaac
Nakamoto, Tetsuo
Owen, David Gray
Peterson, Thomas Mark
Porter, Chastain Kendall
Potter, Rosario H. Yap
Tonna, Edgar Anthony
Wang, Teen-Meei Thomas
Warren, Donald William
Weinstein, Sam

Implantology
Chappell, Robert Paul
Cranin, Abraham Norman
Hodosh, Milton
McKinney, Ralph Vincent, Jr.
Sendax, Victor Irven
Stallard, Richard Elgin
Taylor, Ross Lawton
Young, Franklin Alden, Jr.

Oral and maxillofacial surgery
Beirne, Owen Ross
Bertolami, Charles Nicholas
Clayman, Lewis
Cranin, Abraham Norman
DeVore, Duane Thomas
Goldberg, Allen Fred
Goldsmith, Douglas Howard
Gotcher, Jack Everett, Jr.
Gross, Bob Dean
Joy, Edwin Douglas, Jr.
Keith, David Alexander
Leake, Donald Lewis
Leonard, Myer Samuel
Machado, Lester
Mc Leran, James Herbert
Mercier, Paul
Petri, William Henry, III
Steelman, Robert Joe
Waite, Daniel Elmer
White, Raymond Petrie, Jr.

Oral biology
Alley, Keith Edward
Avery, James Knuckey
Bahn, Arthur Nathaniel
Bertolami, Charles Nicholas
Boyan, Barbara Dale

Buchanan, William
Burich, Raymond Lucas
Cappuccino, Carleton C.
Ciarlone, Alfred Edward
Clark, Rudolph Ernest
Davenport, William Daniel, Jr.
Dawes, Colin
Dirksen, Thomas Reed, II
Dubner, Ronald
Feldman, Roy Samuel
Fotos, Peter George
Gangaros, Louis Paul, Sr.
Gantt, David Graham
Gaynor, Harold Marvin
Goepp, Robert August
Goldberg, Louis J.
Golub, Lorne Malcolm
Gonzalez, Luis Francisco
Gossling, Jennifer
Hall, Brian Keith
Hancock, Everett Brady
Harn, Stanton Douglas
Hay, Donald Ian
Hefferren, John James
Heys, Ronald Jay
Keith, David Alexander
Keller, Patricia J.
Kiely, Michael Lawrence
Lechner, Joseph Hadrian
Mashimo, Paul Akira
Mayhall, John Tarkington
McCulloch, Christopher Allan
McIntire, Floyd Cottam
Menaker, Lewis
Miller, Chris H.
Newbrun, Ernest
Orland, Frank Jay
Pashley, David Henry
Pollock, Jerry Joseph
Rao, Gopal Subba
Rifkin, Barry Richard
Rosan, Burton
Rose, George Gibson
Scott, David Bytovetzski
Siegel, Ivens Aaron
Slomiany, Bronislaw Leszek
Stevens, Roy Harris
Torneck, Calvin David
Warren, Donald William
Yagiela, John Allen

Oral pathology
Alexander, William Nebel, *Oral Pathology*
Archard, Howell Osborne
Farman, Allan George
Fischman, Stuart Lee
Fullmer, Harold Milton
Gorlin, Robert James
Lin, Louis Min-Tsu
Lynch, Denis Patrick
Main, James Hamilton Prentice
McKinney, Ralph Vincent, Jr.
Nelson, John Franklin
Porter, Chastain Kendall

Orthodontics
Aduss, Howard
Chow, Michael Hung-Chun
Chumbley, Alvin Brent
Di Salvo, Nicholas Armand
Doyle, Walter Arnett
Garner, LaForrest Dean
Green, Larry Joy
Hanley, Kevin Joseph
Kuftinec, Mladen Matija
Legan, Harry Lewis
Litton, Stephen Frederick
Malhotra, Shyam Kumar
Nahoum, Henry Isaac
Nikolai, Robert Joseph
Peterson, Thomas Mark
Riedel, Richard Anthony
Roberts, Wilbur Eugene
Storey, Arthur Thomas

Periodontics
Alfano, Michael Charles
Alvares, Olav Filomeno
Bandt, Carl Lee
Blank, Lawrence William
Buchanan, William
Chilton, Neal W(arwick)
Ciancio, Sebastian Gene
Cohen, Ronald Alex
Cox, Donald Stephen
De Marco, Thomas Joseph
Eggert, Frank Michael
Evian, Cyril Ian
Feldman, Roy Samuel
Gaynor, Harold Marvin
Genco, Robert Joseph
Gher, Marlin Eugene, Jr.
Gold, Steven Ira
Golub, Lorne Malcolm
Gonzalez, Luis Francisco
Hancock, Everett Brady
Hinrichs, James Edward
Horton, John Edward
Jeffcoat, Marjorie
Kamen, Paul Raphael
Kaslick, Ralph Sidney
Kiel, Robert Allen
Lamster, Ira Barry
Lopatin, Dennis Edward
Mashimo, Paul Akira
McCulloch, Christopher Allan
Mehta, Noshir Rustom
Mellonig, James Thomas
Morris, Melvin Lewis
Pihlstrom, Bruce Lee
Simring, Marvin
Suzuki, Jon Byron
West, Theodore Lee

Preventive dentistry
Besic, Frank Charles
Bowen, William Henry
Call-Smith, Kathy Meredith
Caslavska, Vera Barbara
Christen, Arden Gale
Davila, Jorge M.
Faunce, Frank Roland
Greene, John Clifford
Grover, Pushpinder Singh
Harris, Norman Oliver
Hill, Iden Naylor
Joos, Richard William
Kandelman, Daniel
Kaslick, Ralph Sidney
Kleber, Carl Joseph
Leverett, Dennis H.
Mallatt, Mark Edward
Mellberg, James Richard
Menaker, Lewis
Pape, Harry Rudolph, Jr.
Putt, Mark Stuart
Ross, Norton Morris
Simring, Marvin
Stallard, Richard Elgin
Wefel, James Stern
Woerth, Janice Kay

Prosthodontics
Anusavice, Kenneth John
Asawa, George Nobuo
Barolet, Ralph Yvon
Baxter, Joann Crystal
Bertolotti, Raymond Lee
Bolender, Charles L.
Churgin, Lawrence S.
Cohen, Ronald Alex
Gavelis, Jonas Rimvydas
Javid, Nikzad Sabet
Lenchner, Nathaniel Herbert
Lorton, Lewis
Mehra, Rita Virmani
Sakaguchi, Ronald Louis
Schwartzman, Boris
Sendax, Victor Irven
Taylor, Ross Lawton

Other
Alexander, William Nebel, *Oral Medicine*
Barkmeier, Wayne Walter, *Dental materials, Restorative dentistry*
Bastawi, Aly Eloui, *Pedodontics*
Blank, Lawrence William, *General dentistry*
Bowen, Rafael Lee, *Biomaterials*
Bush, Francis Marion, *Dental education*
Campbell-Smith, Rosemary Gilles, *Dental practice consulting*
Ciancio, Sebastian Gene, *Dental materials*
Civjan, Simon, *Dental materials*
Crim, Gary Allen, *Operative dentistry*
Douglas, William Hugh, *Restorative dentistry*
Duperon, Donald Francis, *Pediatric dentistry*
Falkler, William Alexander, Jr., *Immunology of periodontal diseases*
Feasby, Wilfrid Harold, *Pedodontics*
Fischman, Stuart Lee, *Forensic dentistry*
Gavelis, Jonas Rimvydas, *Dental materials*
Gershen, Jay Alan, *Behavioral sciences in dentistry*
Goldstein, Ronald E(rwin), *Cosmetic dentistry*
Greene, John Clifford, *Epidemiology*
Grodberg, Marcus Gordon, *Pharmaceutical products research and development management*
Heyde, John Bradley, *Dental materials*
Howley, Thomas Patrick, *Experimental design in dentistry*
Johnson, Lewis Benjamin, Jr., *Dental materials*
Joos, Richard William, *Dental materials*
Kohn, Donald William, *Pediatric dentistry*
Kress, Gerard Clayton, Jr., *Dental health policy and education research*
Lipton, James Abbott, *Facial pain*
Lund, Melvin Robert, *Dental materials, operative dentistry*
Malhotra, Manohar Lal, *Dental materials*
Mealiea, Wallace Laird, Jr., *Behavioral sciences in dentistry*
Meckel, Alfred Hans, *Dental research consulting*
Mehra, Rita Virmani, *Dental materials*
Mehta, Noshir Rustom, *Cranio-mandibular cervical pain and dysfunction*
Mellonig, James Thomas, *Oral Medicine*
Mueller, Herbert Joseph, *Dental materials*
Nelson, John Franklin, *Oral medicine*
Olsen, Norman Harry, *Dental education/admininstration, Pediatric dentistry*
Reiskin, Allan Burt, *Radiology*
Riviere, George Robert, *Pediatric dentistry*
Schoen, Max Howard, *Public health*
Washburn, John Garrett, *Dental Materials*
Yankell, Samuel Leon, *Dental research*

ENERGY SCIENCE AND TECHNOLOGY

Biomass. *See also* **AGRICULTURE.**
Abeles, Tom Peter
Antal, Michael Jerry, Jr.
Archer, Richard Earl
Bagnall, Larry Owen
Barnett, Stockton Gordon
Basic, John Nicholas, Sr.
Bassi, Sukh Dev
Benson, Peter Howard
Bertuglia, Lynn Ellen
Boubel, Richard William
Bungay, Henry Robert, III
Busche, Robert Marion
Cork, Douglas James
Dawson, David Henry
Dolan, Linda Sutliff
Flikke, Arnold Maurice
Goldstein, Irving Solomon
Hamrick, Joseph Thomas
Henry, Patrick M.
Hiler, Edward Allan
Ingram, Lonnie O'Neal
Irgon, Joseph
Johnson, Jack Donald
Kalvinskas, John Joseph
Parker, Harry William
Pate, John Ray
Peters, Max Stone
Ramsay, William Charles
Rockwood, Donald Lee
Roy, Dipak
Wakefield, Ernest Henry
Wilson, David Buckingham
Winskill, Robert Wallace
Zaborsky, Oskar Rudolf
Zerbe, John Irwin

Combustion processes
Agnew, William George
Allen, John Malone
Audette, Louis Girard, II
Basic, John Nicholas, Sr.
Berlad, Abraham Leon
Berman, Herbert L(awrence)
Broughton, Paul Leonard
Busch, Christopher William
Carrier, George Francis
Cashdollar, Kenneth Leroy
Cashin, Kenneth Delbert
Chen, Ching Jen
Churchill, Stuart Winston
Cranberg, Lawrence
Elghobashi, Said (Elsayed)
Gelinas, Robert Joseph
Gilmour, William Alexander
Glassman, Irvin
Gollahalli, Subramanyam Ramappa
Gronhovd, Gordon Harlan
Hamrick, Joseph Thomas
Harsha, Philip Thomas
Hjertager, Bjorn Helge
Hochwalt, Carroll Alonzo, Jr.
Johnson, Terry R.
Jones, Charles
Kratz, Richard L., Jr.
Krishnamurthy, Lakshminarayanan
Kruger, Charles Herman, Jr.
Kukin, Ira
Laurendeau, Normand Maurice
Lefebvre, Arthur Henry
Lightman, Allan Joel
Longwell, John Ploeger
Lortie, John William
Loth, John Lodewyk
Massey, Evan Morgan
Matalon, Moshe
McGowan, Jon Gerald
Myers, Phillip Samuel
Palmer, Thomas Yealy
Penner, Stanford Solomon
Proctor, Charles Lafayette, II
Reuther, James Joseph
Rosenberg, Robert Brinkmann
Schmieder, Robert William
Sirignano, William Alfonso
Strain, John Willard
Suzuki, Tateyuki
Winskill, Robert Wallace
Wu, Ying-chu Lin
Zakkay, Victor H.

Fuels
Adams, William Eugene
Astley, Eugene Roy
Backhus, DeWayne Allan
Batay-Csorba, Peter Andrew
Bell, Jimmy Todd
Berty, Jozsef Mihaly
Binstock, Martin H.
Bloch, Herman Samuel
Bodily, David Martin
Borg, Iris Yvonne
Bowman, Clement Willis
Bowsher, Arthur LeRoy
Broughton, Paul Leonard
Cassidy, Martin Macdermott
Collipp, Bruce Garfield
Cordell, Robert James
Creagan, Robert Joseph
Epperly, William Robert
Everhart, Donald Lough
Glassman, Irvin
Goldberg, Ivan
Gollahalli, Subramanyam Ramappa
Grant, Philip Robert, Jr.
Gueron, Henri Maximilien
Harbay, Edward William
Harmon, David E., Jr.
Hoffman, Edward Jack
Hoover, L. John
Jawetz, Pincas
Johanson, Jerry Ray
Kessler, George William
Kratzer, Reinhold Hermann
Kydd, Paul Harriman
Lefebvre, Arthur Henry
Leonard, Ellen Marie
Linden, Henry Robert
Longwell, John Ploeger
Looney, Paul Bryan
Lowe, Phillip A(rnold)
Marsden, Sullivan Samuel, Jr.
Mc Afee, Jerry
Mc Ketta, John J., Jr.
Merriam, Daniel F(rancis)
Miernyk, William Henry
Myers, Phillip Samuel
Pfeiffer, Heinz Gerhard
Potter, Paul Edwin
Rosenberg, Robert Brinkmann
Rossini, Frederick Dominic
Sheriff, Robert Edward
Shinnar, Reuel
Tenison, Robert Blake
Welker, J. Reed
Widdoes, Lawrence Curtis
Yokell, Michael David

Fuels, coal
Aldridge, Melvin Dayne
Bodily, David Martin
Carlton, Donald Morrill
Chun, Sun Woong
Clark, Ronald Duane
Cohen, Arthur David
Fisher, Ray W.
Frölicher, Franz
Gorbaty, Martin Leo
Gronhovd, Gordon Harlan
Gueron, Henri Maximilien
Hagen, Arnulf Peder
Hochwalt, Carroll Alonzo, Jr.
Hsia, Yu-Ping
Jansen, George James
Kelsh, Dennis J.
Kessler, Robert
Kottlowski, Frank Edward
Massey, Evan Morgan
Mathewson, Christopher Colville
McNeese, Leonard Eugene
Meyer, Edmond Gerald
Meyer, Howard Stuart
Miernyk, William Henry
Skeist, Irving
Taylor, David John
Tenison, Robert Blake
Weller, Sol William
Wen, Wen-Yang
Wilson, John Sheridan

Fuels, oil shale
Alexander, Peter
DuBow, Joel Barry
Gorbaty, Martin Leo
Gratt, Lawrence Barry
Miknis, Francis Paul
Pei, Richard Yusien
Pforzheimer, Harry, Jr.
Wiberley, Stephen Edward

Fuels, other fuels and sources
Acheson, Willard P(hillips)
Audette, Louis Girard, II
Barrett, Richard John
Benson, Peter Howard
Bowden, Bryant Baird
Bowman, Clement Willis
Clemens, Donald Faull
Craig, Robert Bruce
Fan, Liang-tseng
Freiwald, David Allen
Gupta, Vijay Kumar
Hartz, Kenneth E.
Kratz, Richard L., Jr.
Leber, Ralph Eric
Lortie, John William
Lumb, Ralph Francis
Mandra, York T.
Othmer, Donald Frederick
Peerenboom, James Peter
Penner, Stanford Solomon
Popper, Steven Herbert
Starr, Chauncey
Stergakos, Elias P(anagiotes)
Stetler, Dwight L(awrence)
Veziroglu, Turhan Nejat
Waitzman, Donald Anthony
Wiech, Ralph Benjamin
Zerbe, John Irwin

Fusion. *See also* **LASER.**
Ashworth, Clinton Paul
Auer, Peter Louis
Baker, Charles Clayton
Berwald, David H.
Bohachevsky, Ihor Orest
Burrell, Charles Frederick
Bussard, Robert William
Chang, Joseph Yung
Channon, Stephen R.
Dean, Stephen Odell
Ellis, William R.
Finn, Patricia Ann
Fleischmann, Hans Hermann
Freeman, Marsha Gail
Furth, Harold Paul
George, Thycodam Varkkey
Gilleland, John Rogers
Gross, Robert Alfred
Henderson, Dale B(arlow)
Jackson, David Phillip
Jesser, William Augustus
Johnson, John Lowell
Kiang, Robert L.
Kirkpatrick, Ronald Crecelius
Kyrala, George Amine
Leonard, Ellen Marie
Levine, Jerry David
Miley, George Hunter
Miller, Robert Alan
Moir, Ralph Wayne
Norem, James H.
Perry, Erik David
Post, Douglass Edmund, Jr.
Priedhorsky, William Charles
Schlachter, Alfred Simon
Sheffield, John
Shkarofsky, Issie Peter
Turchan, Otto Charles
Weldon, William Forrest
Woodson, Herbert Horace

Geothermal power
Goldsberry, Fred Lynn
Gough, Denis Ian
Grant, Philip Robert, Jr.
Iceman, Larry
Koenig, James Bennett
Loxley, Thomas Edward
Lund, John William
Philbrick, David Alan
Rinehart, John Sargent
Russell, Eugene A.
Zais, Elliot Jacob

Nuclear fission
Abdou, Mohamed Aziz
Adamantiades, Achilles G.
Agee, Lance James
Alapour, Adel
Alcouffe, Raymond Edmond
Allman, Norris C.
Alter, H. Ward
Andrews, James Barclay, II
Apley, Walter Julius, Jr.
Ascher, Michael Charles
Ashworth, Clinton Paul
Asprey, Margaret Williams
Austin, Edward Marvin
Babb, Albert Leslie
Baer, Robert Lloyd
Barbehenn, Craig E(dwin)
Barrett, Richard John
Barschall, Henry Herman
Barsky, Arnold M(ilton)
Bartine, David Elliott
Barton, Ronald Winston
Bauer, Richard Carlton
Beck, William Nelson
Beckjord, Eric Stephen
Bell, Barbara Jean
Bell, Gregory Lee
Bell, Jimmy Todd
Benedict, Bruce John
Benedict, Manson
Berté, Frank Joseph
Besmann, Theodore Martin
Bettenhausen, Lee H.
Bezella, Winfred August
BIANCHERIA, Amilcare
Bickel, John Henry
Bingham, Billy Elias
Block, Robert Charles
Blomeke, John Otis
Boller, Bruce Raymond
Boltax, Alvin
Booth, James Albert
Boyer, Vincent Saull
Bridges, Donald Norris
Bullock, Scott Verne
Bunch, Wilbur Lyle
Burge, Charles Arthur
Burley, Elliott Lenhard
Burris, Leslie
Cacuci, Dan Gabriel
Campbell, David Owen
Cantrell, Weldon Kermit
Carbon, Max William
Carter, Robert LeRoy
Cella, Alexander
Chan, Shih Hung
Chandler, John Christopher
Chao, Jiatsong Jason
Chen, Shih-Yew
Chewning, June Spangler
Chipman, Gordon Leigh, Jr.
Cho, Joon Ho
Choi, Seung Hoon
Choppin, Gregory Robert
Chulick, Eugene Thomas
Chung, Chien
Chung, Hee Mok
Clare, George Hadley
Clark, David Delano
Cohen, Karl Paley
Cole, Thomas Earle
Conte, Richard Joseph
Coops, Melvin (Sterline)
Courtney, John Charles
Crawley, Paul F.
Critoph, Eugene
Croff, Allen Gerald
Crouthamel, Carl Eugene
Cullingford, Hatice S(adan)
Cunningham, John Edward
Dahlberg, Richard Craig
Davis, Rodney James
Dean, Richard Anthony
Defilippo, Felix Carlos
Demas, Nicholas George
Detrick, Carl Anthony
Detterman, Robert Linwood
De Volpi, Alexander
Dickens, Justin Kirk
Dickerman, Charles Edward
Dickson, Paul Wesley, Jr.
Dimmick, David Michael
Dobson, Wayne Lawrence
Dolan, Linda Capano
Doster, Joseph Michael
Dube, Donald Arthur
Duda, Richard Frank
Duke, Winston Lavelle
Du Temple, Octave Joseph
Edlund, Milton Carl
Einziger, Robert Emanuel

Evans, Albert E.
Evans, Ersel Arthur
Fainberg, Anthony
Fast, Edwin
Feemster, John Ronald
Feinberg, Robert Jacob
Ferguson, James Malcolm
Flanagan, Charles Allen
Fleishman, Morton Robert
Fluegge, Ronald Marvin
Flynn, Kevin Francis
Frantz, Frederick Strassner, Jr.
Frederickson, Robert
Freeman, Louis Barton
Fritsch, Arnold Rudolph
Fulford, Phillip James
Garrick, B. John
Garritson, Grant Richard
Gat, Uri
George, Thomas Carl
Gibbs, Ann
Gilligan, John Gerard
Goldberg, Ivan
Goldman, Marvin
Goldsmith, William Alee
Goldstein, Mark Kingston Levin
Goode, Glenn Amos, Jr.
Gowda, Byre Venkataramana
Gradin, Lawrence Paul
Green, Lawrence
Greenman, Gregory Michael
Groenier, William Samuel
Grotenhuis, Marshall
Haas, Paul Arnold
Hagmann, Dean Berry
Haire, Marvin Jonathan
Haler, Lawrence Eugene
Halverson, Thomas George
Hancox, William Thomas
Hannum, William Hamilton
Hard, James Ellsworth
Hardy, Judson, Jr.
Harkness, Samuel Dacke, III
Harms, Archie Arkadius
Heckman, Richard Ainsworth
Hepworth, Harry Kent
Hickman, Jack William
Highberger, Paul Feightner
Hildebrandt, Thomas Owen
Hill, Orville Farrow
Hill, Thomas Johnathan
Hofmann, Peter Ludwig
Hofstetter, Kenneth John
Holder, Nadine Duguid
Holzer, Joseph Mano
Huang, Hai Chow
Hurd, Edward Nelson, III
Hurley, Thomas John, Jr.
Ibrahim, Shawki Amin
Ireland, John Richard
Jacoby, William Richard
Jain, Parveen Kumar
Jaquess, James Fletcher
Jeffers, John Bryant
Jensen, Betty Klainminc
Joksimovich, Vojin
Kai, Michael S.
Kalinauskas, Gediminas Leonardas
Kaplan, Maureen Flynn
Karam, Ratib Abraham
Kempe, Robert Aron
Keneshea, Francis Joseph
Kerr, William
Kiang, Robert L.
Kikuchi, Chihiro
Kim, Yong Su
Kittel, John Howard
Kjaer-Pedersen, Niels
Klein, Keith Allen
Kniazewycz, Bohdan George
Koussa, Harold Alan
Kouts, Herbert John Cecil
Kramer, Rex Williard, Jr.
Kulcinski, Gerald LaVern
Laning, David Bruce
Larrimore, James Abbott
Larsen, Edward William
Laubenthal, Neil David
Lawrence, Leo Albert
Lee, John Chaeseung
Lee, Robert Bongkyu
Leech, J(ames) Nathan
Lentsch, Jack Wayne
Lester, David Brent
Leuze, Rex Ernest
Levin, Alan Edward
Levine, Melvin Mordecai
Levine, Samuel Harold
Liu, Yung Yuan
Lorenzini, Paul Gilbert
Ma, Benjamin Minglee
MacPherson, Herbert Grenfell
Mahaffey, Michael Kent
Malin, Douglas Harwell
Maxwell, Donald Lee
McArthur, David Alexander
McArthur, Wilson Cooper
McBroom, Robert Chism
McCormack, Michael G.
McFarlane, Harold Finley
McKnight, Richard D.
McSherry, Arthur James
Meem, James Lawrence, Jr.
Melese d'Hospital, Gilbert Bernard
Menlove, Howard Olsen
Misra, Balabhadra
Mittelman, Phillip Sidney
Morewitz, Harry Alan
Murphy, Edward Thomas
Newman, Darrell Francis
Nicolosi, Stephen Louis
Oatley, David Herbert
Okrent, David
Pagel, Deborah Joanne
Parry, John O.
Pasqua, Pietro, F(ernando)
Paustian, Harold Herman
Paxton, Hugh Campbell
Peddicord, Kenneth Lee
Pelletier, Charles A.
Pense, Alan Wiggins
Phelps, James Parkhurst
Pietri, Charles Edward
Pizzica, Philip Andrew
Raab, Harry Frederick, Jr.
Rao, Surendar Purushothay
Rasmussen, Norman Carl
Redmond, Robert Francis
Richman, Jack William
Rogers, Vern Child
Rose, Ronald Palmer
Ruby, Lawrence
Saluja, Jagdish Kumar
Sanders, Arthur
Sarram, Mehdi
Schaeffer, Norman M.
Schaub, John Robert
Seale, Robert Lewis
Serdula, Kenneth James
Shaffer, Clinton John
Shum, Raymond Hing-Yan
Sideris, Antonios George
Snyder, Thoma Mees van't Hoff
Stewart, James Edmund, III
Stiefel, John T.
Storrs, Charles L.
Thomas, Frank Joseph
Thompson, Dudley
Tomlinson, Richard Lee
Tomonto, James Robert
Turinsky, Paul Josef
Ugelow, Albert Jay
Umek, Anthony M.
Varley, Ronald Arthur
Vest, Anthony Leon
Wachter, William John
Way, Richard A(lvord)
Wei, Jim P(iau)
Weisman, Joel
Wenz, Michael Frank, Jr.
West, John Merle
Williams, Alan Keiser
Williams, Peter MacLellan
Williamson, Thomas Garnett
Wilson, Randall Joe
Wiren, Robert Craig
Witherspoon, John Pinkney, Jr.
Wood, Houston Gilleylen, III
Wray, John robert
Wymer, Raymond George
Yeh, Cheng Shin
Young, Garry Gean
Youssef, Mahmoud Z. Hassan
Yue, Mike Yuan
Yung, Shu-Chien
Zimmanck, Frank Robert, Jr.

Nuclear fusion
Abdou, Mohamed Aziz
Amherd, Noel A.
Attaya, Hosny M(oustafa)
Block, Robert Charles
Briggs, William Benajah
Carter, Robert LeRoy
Chao, Jiatsong Jason
Cheng, Edward Teh-Chang
Cooper, Ralph Sherman
Cullingford, Hatice S(adan)
Demas, Nicholas George
Draper, Ernest Linn, Jr.
Ehst, David Alan
Flanagan, Charles Allen
Fonck, Raymond John
Fuller, James Leslie
Furth, Harold Paul
Garner, James Kirkland
Ghoniem, Nasr Mostafa
Gilligan, John Gerard
Gohar, Mohamed Yousry Ahmed
Gould, Roy Walter
Green, Lawrence
Grossbeck, Martin Lester
Grove, Don J.
Hamasaki, Seishi
Harms, Archie Arkadius
Hoffman, Myron Arnold
Holdren, John Paul
Horton, Wendell Claude, Jr.
Hovingh, Jack
Jensen, Betty Klainminc
Kaye, Stanley Martin
Kazimi, Mujid S.
Kim, Kyekyoon
Kozman, Theodore Albert
Krauss, Alan Robert
Kugel, Henry W.
Kulcinski, Gerald LaVern
Lazareth, Otto William
Liu, Yung Yuan
Maglich, Bogdan C.
Mattas, Richard Francis
McCormack, Michael G.
McNally, James Rand, Jr.
Mense, Allan Tate
Miley, George Hunter
Misra, Balabhadra
Phillips, James Alfred
Piet, Steven James
Ragheb, Magdi
Richman, Jack William
Rose, Ronald Palmer
Roth, J(ohn) Reece
Ruby, Lawrence
Scott, Franklin Robert
Shaffer, Clinton John
Shieh, Paulinus Shee-Shan
Simon, Albert
Starr, Chauncey
Sudan, Ravindra Nath
Sweeney, Mary Ann
Taylor, John Joseph
Wachter, William John
Wagner, Frederick William
Walker, John Scott
Wallace, Richard Kent
Whealton, John H.
Wong, Clement Po-Ching
Youssef, Mahmoud Z. Hassan

Ocean thermal energy conversion
Berry, William Benjamin Newell
Daniel, Thomas Henry
Kreith, Frank
Monney, Neil Thomas
Reid, Allen Francis
Venkataramiah, Amaraneni
Wenzel, James Gottlieb
Williams, Carol Ann

Plasma. *See also* ENGINEERING.
Chen, Hollis Ching
Cordero, Julio
Cott, Donald W(ing)
Ehst, David Alan
Gilleland, John Rogers
Gross, Robert Alfred
Hamasaki, Seishi
Mihas, Faquir Ullah
Phillips, James Alfred
Seshadri, Sengadu Rangaswamy
Trivelpiece, Alvin William
Wu, Ying-chu Lin

Solar energy
Aadland, Donald Ingvald
Abeles, Tom Peter
Alexander, Charles Kenneth, Jr.
Allen, William Hand
Antal, Michael Jerry, Jr.
Archer, Richard Earl
Arnas, Ozer Ali
Backhus, DeWayne Allan
Backus, Charles Edward
Bauch, Tamil Daniel
Beard, James Taylor
Becker, Albert Walter
Berry, William Bernard
Bier, Charles James
Blue, Todd Irwin
Boehm, Robert Foty
Boer, Karl Wolfgang
Bowman, Thomas Eugene
Bull, Stanley Raymond
Burr, Baldwin Gwynne
Bush, George Edward
Bushnell, Robert Hempstead
Byrd, James William, Sr.
Casperson, Richard L.
Chiang, Chao-Wang
Chiou, Jiunn Perng
Clausing, Arthur Marvin
Coates, Gary Joseph
Cohen, Marshall Jay
Craine, Eric Richard
Creed, David
Dean, Thomas Scott
DeBlasio, Richard
Doud, Eric Leo
Dudnik, Elliott Eliasaf
Duffie, John Atwater
Ebenezer, Job Selvarayan
Eltimsahy, Adel H.
Emery, Keith Allen
Farber, Joseph
Fletcher, Edward Abraham
Gay, Charles Francis
Glaser, Peter Edward
Glogovsky, Robert Louis
Gould, Thurman Woodrow, Jr.
Gross, Eric Taras Benjamin
Haberman, William Lawrence
Heerwagen, Dean Reese
Henderson, Robert Edward
Hodges, Carl Norris
Holloway, Dennis Robert
Holmes, John Thomas
Houlihan, John Frank
Howell, John Reid
Hubbard, Harold Mead
Hummel, Myron Floyd
Icerman, Larry
Irgon, Joseph
Jennings, Burgess Hill
Johnson, Elwin Leroy
Jones, Robert Edwin, Jr.
Jones, Walter Allan
Jordan, Richard Charles
Kelley, Mark Elbridge, III
Kim, Rhyn Hyun
Kokoropoulos, Panos
Kreith, Frank
Kremers, Jack Alan
Kristy, Thomas Wayne
Lampert, Seymour
Leboeuf, Cecile Marie
Lechner, Norbert Manfred
Lillywhite, Malcolm Alden
Little, Roger George
Loeb, William A.
Lorenzen, Robert T.
Loxley, Thomas Edward
Lunde, Peter J.
Machovec, George Stephen
MacKay, John Kelvin
Makofske, William Joseph
Maloney, John Patrick
Manalis, Melvyn Samuel
Martin, Jose Ginoris
Mc Cord, Thomas Bard
McDaniels, David Keith
McGarity, Arthur Edwin
McGowan, Jon Gerald
McKee, James Stanley Colton
McKee, Robert Bruce, Jr.
McKown, Cora F.
Meilleur, Steven Grant
Meredith, David Bruce
Mettee, Howard Dawson
Meyer, John (Hans) Forrest
Miluschewa, Sima
Morse, Frederick H.
Nash, Jonathon Michael
Nelson, David Torrison
Newman, Jerry Okey
Norman, Jack C.
Olah, Stephen
Pate, John Ray
Philbrick, David Alan
Pohl, Jens Gerhard
Prieto, Robert
Ramakumar, Ramachandra Gupta
Rapp, Donald
Reinert, Charles Peter
Remo, John Lucien
Reynolds, John Spencer
Rosenberg, Paul
Russell, Eugene A.
Salomon, Robert Ephraim
Saluja, Jagdish Kumar
Schneiderwent, Myron Otto
Schwartz, David Michael
Shahryar, Ishaq M.
Shalleck, Alan Bennett
Shing, Yuh-Han
Shkedi, Zvi
Sletten, Carlyle Joseph
Smith, Otto J.M.
Smith, Phyllis Sterling
Stange, James Henry
Stetler, Dwight L(awrence)
Sumners, Carolyn Taylor
Sutton, George Walter
Szalai, Imre A.
Szego, George Charles
Theisen, Jeffrey A.
Thomas, Carlton Eugene
Tien, Ho Ti
Vant-Hull, Lorin Lee
Washom, Byron John
Watson, Donald Ralph
Wennberg, Jeffrey Norman
Witte, Michael
Womack, Joseph Donald, Sr.

Wind power
Berry, Edwin X
Clews, Henry Madsion
Dick, Daniel Eggleston
Foreman, Kenneth M(artin)
Gross, Eric Taras Benjamin
Keith, Theo Gordon, Jr.
Kelley, Neil Davis
Kempf, Gary William
Kennell, E. Edison, III
Liu, Chang Yu
Manalis, Melvyn Samuel
Mathur, Radhey Mohan
Maydew, Randall Clinton
Meyer, John (Hans) Forrest
Miller, Rene Harcourt
Pfister, Philip Carl
Reitan, Daniel Kinseth
Riegler, Gerold Ernst
Sheff, James Robert
Smith, Otto J.M.
Smith, Phyllis Sterling

Other
Ackerman, Allan Douglas, *energy conservation*
Averbuch, Martin Philip, *Heat Recovery*
Balzhiser, Richard Earl, *Energy research management*
Bauch, Tamil Daniel, *Energy conservation*
Bonnet, Juan Amedee, *Energy research administration*
Cawley, Charles Nash, *Environmental effects of energy technologies*
Ceperley, Peter Hutson, *Energy conversion devices*
Chan, Sek Kwan, *Explosives*
Chewning, June Spangler, *Employment aspects of energy production*
Chuang, Henry Ning, *Energy management*
Coleman, James Stafford, *Energy Research Management*
Culler, Floyd LeRoy, Jr., *Energy science and engineering*
Curtis, George Darwin, *Alternate energy sources*
Dean, Thomas Scott, *Energy conservation*
Decora, Andrew W., *Energy consulting*
Dilmore, Roger H., *Utility operations*
Ebenezer, Job Selvarayan, *Pedal power*
Firester, Arthur Herbert, *Photovoltaics*
Gradin, Lawrence Paul, *Nuclear power*
Gyftopoulos, Elias Panayiotis, *Energy conservation*
Hamilton, Leonard Derwent, *Environmental effects of energy technologies*
Hansborough, Lash Devous, *Accelerator Engineering*
Harned, Joseph William, *Policy formulation*
Hassoun, Hussein Ali, *Energy planning*
Hatcher, S(tanley) Ronald, *Nuclear energy research administration*
Hester, Jarrett Charles, *Energy systems*
Hibshman, Henry Jacob, *Energy consulting*
Hoffman, Kenneth Charles, *Resource policy*
Holdren, John Paul, *Energy technology assessment*
Inhaber, Herbert, *Energy Science and Technology*
Jaquess, James Fletcher, *Nuclear Power generation-quality engineering*
Kendall, Ernest Terry, *Energy-risk analysis*
Kennell, E. Edison, III, *Micro-hydro electric power*
Kerr, Donald MacLean, Jr., *Energy research and development administration*
Kincaide, William Charles, *Hydrogen production systems*
Klock, Peter Illitch, Jr., *Energy conservation*
Konigsberg, Jan, *Other energy policy research*
Kovach, Paul Joseph, *Applied health physics*
Krachman, Howard Ellis, *Energy management*
Lederer, Charles Michael, *Energy use in buildings*
Lee, Bernard Shing-Shu, *Energy research management*
Lee, Thomas Henry, *Energy systems*
Leech, J(ames) Nathan, *Nuclear power plant quality assurance*
Meredith, David Bruce, *Energy conservation*
Morse, Richard Stetson, *Batteries and industrial instruments, Management of new enterprises*
Murray, Peter, *Nuclear energy research and development administration*
Parker, Harry William, *Comparative energy costs*
Picazo, Esteban David, *Radioactive waste management*
Ramsay, William Charles, *Energy policy research*
Reiss, Howard R., *Energy science*
Roddis, Louis Harry, Jr., *Energy research and development*
Russell, B. Don, *Power and energy transmission*
Salomon, Robert Ephraim
Saxon, George Edward, *Energy conservation*
Schultz, Frederick H. C., *energy sources and uses*
Shanker, Roy James, *Regulated utilities*
Siem, Laurie Helen, *Energy conservation*
Smith, John Elvans, *Energy conservation*
Spencer, Dwain Frank, *Energy Sciences Technology*
Staszesky, Francis Myron, *Energy consulting*
Sterling, Raymond Leslie, *Earth-sheltered buildings*
Swallom, Daniel Warren, *Magnetohydrodynamic energy conversion*
Thompson, Benny Louis, *Gasification*
Tien, Chang Lin, *Heat transfer*
Tyler, Loraine Lyon, *Energy Education*
Vallery, Stafford Jean, II, *Energy conservation*
Wagner, Frederick William, *Advanced methods of power generation.*
Wagner, Lorry Yale, *Energy resource management*
Watras, Ronald Edward, *Energy education*

ENGINEERING. *See also* BIOTECHNOLOGY, COMPUTER SCIENCE, ENERGY SCIENCE, ENVIRONMENTAL SCIENCE, GEOSCIENCE, INFORMATION SCIENCE, LASER, MATERIALS SCIENCE, OPTICS, PHYSICS, SPACE SCIENCE.

Acoustical
Beranek, Leo Leroy
Cyr, Reginald John
Davidson, James Blaine
Duerinckx, Andre Jozef
Eberhardt, Allen Craig
Eldred, Kenneth McKechnie
Ellison, William Theodore
Farnell, Gerald William
Fink, Martin Ronald
Fischell, David Ross
Goldstein, Marvin E(manuel)
Harris, Cyril Manton
Harris, Wesley Leroy
Hellman, Rhona Phyllis
Heymann, Frank Joseph
Johnson, Glen Eric
Koch, Werner
Kremkau, Frederick W.
Lang, William Warner
Leed, Peter Lewis
Malosh, James Boyd
Mathews, Max Vernon
Moller, Aage Richard
Munson, John Christian
Olson, Harry F.
Packman, Allan B.
Schlegel, Ronald Gene
Walker, Eric Arthur
Wolf, Joseph Allen, Jr.
Yoerger, Roger Raymond

Aeronautical
Adamczak, Robert Leonard

Adamson, Thomas Charles, Jr.
Allen, Jerry Michael
Ashley, Holt
Ballhaus, William Francis
Blackwelder, Ron
Bluestein, Theodore
Bodonyi, Richard James
Bogdonoff, Seymour Moses
Brilliant, Howard Michael
Busch, Christopher William
Carlson, Richard Merrill
Caughey, David Alan
Chow, Chuen-Yen
Colborn, Joseph Nelson
Conly, John Franklin
Cosner, Raymond Robert
Covert, Eugene Edzards
Dolling, David Stanley
Donaldson, Coleman duPont
Duffy, Robert Aloysius
Elfstrom, Gary Macdonald
Fearn, Richard Lee
Fernández, Fernando Lawrence
Fink, Martin Ronald
Flax, Alexander Henry
Fleisig, Ross
Francis, Michael Skok
Gavin, Joseph Gleason, Jr.
Gessow, Alfred
Goland, Martin
Gregoriou, Gregor Georg
Habashi, Wagdi George
Haberman, Charles Morris
Hall, Charles Frederick
Halsey, Norman Douglas
Harris, Wesley Leroy
Harvey, William Donald
Hawkins, Willis Moore, Jr.
Hayes, Wallace D(ean)
Hazen, David Comstock
Heinemann, Edward H.
Hoff, Nicholas John
Hsia, Henry Tao-sze
Hsieh, Tsuying Carl
Hua, Hsichun Mike
Hughett, Paul William
Jones, Robert Thomas
Jumper, Eric John
Killian, Barbara Germain
Lau, Jark C(hong)
Lekoudis, Spiro G(eorge)
Lewis, David Sloan, Jr.
Liebowitz, Harold
Liu, Chen-Huei
Lo, Aloysius Kou-fang
Loewy, Robert Gustav
Long, Lyle Norman
Lund, Charles Edward
Malvern, Donald
Marvin, Joseph George
Maydew, Randall Clinton
Mc Carthy, John Francis, Jr.
McCormick, Barnes Warnock
Mikulla, Volker
Nomura, Yasumasa
Olstad, Walter Ballard
Oman, Charles McMaster
Orlik-Rückemann, Kazimierz Jerzy
Oved, Yoel
Owczarek, Jerzy Antoni
Pai, Shih I.
Panton, Ronald Lee
Payne, Fred Ray
Perkins, Courtland Davis
Perkins, Porter J., Jr.
Plotkin, Allen
Plotkin, Kenneth Jay
Puckett, Allen Emerson
Puskas, Elek
Reshotko, Eli
Roger, David Freeman
Roshko, Anatol
Russell, David Allison
Rutan, Burt
Savage, William Frederick
Schmit, Lucien André, Jr.
Schwanhausser, Robert Rowland
Sears, William Rees
Seginer, Arnan
Shank, Maurice Edwin
Shannon, Jack
Shea, John R., III
Siuru, William Dennis, Jr.
Speas, Robert Dixon
Sun, John
Van Dyke, Milton Denman
Viets, Hermann
Wallace, James, Jr.
Wang, Charles P.
Wang, Chi-Rong
Wells, Edward Curtis
Whitfield, Jack Duane
Widnall, Sheila Evans
Wilson, Thornton Arnold
Wu, Jain-Ming

Aerospace. *See* **SPACE SCIENCE.**

Agricultural
Albright, Louis DeMont
Anderson, James Henry
Bagnall, Larry Owen
Bockhop, Clarence William
Buelow, Frederick Henry
Creagan, Robert Joseph
Davis, John Rowland
Edwards, Donald Mervin
Haan, Charles Thomas
Hahn, George LeRoy
Haugh, Clarence Gene
Hendrick, James G., III
Hiler, Edward Allan
Jacobson, E. Paul
Jones, James W.
Krishnan, Palaniappa
Larson, Dennis L.
Levine, Gilbert
Ligon, James Teddie
Lorenzen, Robert T.
McLendon, Bennie Derrell
Millar, Gordon Halstead
Nelson, Gordon Leon
Newman, Jerry Okey
Palmer, Melville Louis
Paulsen, Marvin Russell
Peterson, Dean Freeman, Jr.
Pratt, George L.
Privette, Charles Victor, Jr.
Rehkugler, Gerald Edwin
Rotz, C. Alan
Scott, Norman Roy
Scura, Lawrence Thomas, Jr.
Shoup, William David
Splinter, William Eldon
Turnquist, Paul Kenneth
Von Bernuth, Robert Dean
Waitzman, Donald Anthony
Walton, Harold Vincent
Webb, Byron Kenneth
Yoerger, Roger Raymond
Zachariah, Gerald LeRoy

Applied Mathematics. *See* **MATHEMATICS.**

Biomedical. *See also* **BIOTECHNOLOGY; MEDICINE, Physical medicine, Radiology.**
Abbrecht, Peter Herman
Adkisson, William Milton
Alexander, Harold
Anderson, David Walter
Axen, Kenneth
Balster, Frederick Werden
Bertolotti, Raymond Lee
Bier, Milan
Bilotto, Gerardo
Bowman, H(arry) Frederick
Boyle, Brian John
Brighton, John Austin
Brown, Jack Harold Upton
Chambers, Lawrence Paul
Chen, Michael Ming
Cholvin, Neal Robert
Chu, James Chien Hua
Cokelet, Giles Roy
Colahan, Patrick Timothy
Cowin, Stephen Corteen
Cox, Mary E.
Craik, Rebecca Lynn
Curran, Bruce Howlett
Davis, Philip Keith
Dean, Robert Charles, Jr.
Del Guercio, Louis Richard Maurice
Dell-Osso, Louis Frank
De Luca, Carlo John
Dennis, Melvin Best, Jr.
Dickson, James Francis, III
Dorson, William John, Jr.
Doubek, Clifford James
Ellis, Donald Griffith
Erickson, Howard Hugh
Fearnot, Neal Edward
Fletcher, Ronald Darling
Francis, John Elbert
Frazer, David George, Jr.
Geddes, Leslie Alexander
Gezari, Daniel Ysa
Ghoshal, Nani Gopal
Goodenough, Samuel Henry
Graham, James Alexander
Gross, Joseph Francis
Hambrecht, Frederick Terry
Haque, Promod
Hefzy, Mohamed Samir W. M.
Horváth, Csaba Gyula
Jacobs, John Edward
Jaron, Dov
Jarzembski, William Bernard
Johnston, Daniel
Kallok, Michael John
Katz, J. Lawrence
Kreifeldt, John Gene
Lakes, Roderic Stephen
Laxer, Cary
Leavitt, Robert LaVerne
Lee, Jen-shih
Lenchner, Nathaniel Herbert
Lin, Pei-Jan Paul
Little, Robert Colby
Llaurado, Josep G.
Lopez, Hector
Luttges, Marvin Wayen
Macovski, Albert
Mann, Robert Wellesley
McDonagh, Paul Francis
McElhaney, James Harry
Meyer, Andrew U(lrich)
Meyer, Richard Arthur
Miller, Irving Franklin
Mummert, Thomas Allen
Nunan, Craig Spencer
Nyquist, Gerald Warren
Pierson, Richard Norris, Jr.
Plonsey, Robert
Rain, Robert L.
Rasor, Ned Shaurer
Reswick, James Bigelow
Reynolds, David Burkman
Reynolds, Larry Owen
Rhodine, Charles Norman
Riso, Ronald Raymond
Rosenberg, Gerson
Rosenblith, Walter Alter
Rubal, Bernard J.
Rudy, Yoram
Rugge, Henry Ferdinand
Rushmer, Robert Frazer
Saha, Subrata
Sakaguchi, Richard Louis
Samaras, George Michael
Schultz, Albert Barry
Schwan, Herman Paul
Seaman, Ronald Leon
Seireg, Ali A(bdel Hay)
Shapiro, Ascher Herman
Shepherd, A.P., Jr.
Sheppard, Louis Clarke
Siler, William MacDowell
Skelskie, Stanley Irvin
Smiley, Parker Clark
Soutas-Little, Robert William
Sprawls, Perry, Jr.
Takashima, Shiro
Tarr, Richard Robert
Voigt, Herbert Frederick, III
Washburn, John Garrett
Webster, John Goodwin
Wheeless, Leon Lum, Jr.
Winarski, Daniel James
Yoon, Hyo Sub
Zartman, David Lester

Ceramic
Agnihotri, Krishna Venktesh
Bergeron, Clifton George
Friedberg, Arthur Leroy
Johnson, Elwin Leroy
Johnson, James Robert
Kingery, William David
Ott, Walter Richard
Pentecost, Joseph Luther
Rossington, David Ralph
Smiley, Parker Clark
Wilder, David Randolph
Young, Robert Cleland

Chemical
Acrivos, Andreas
Anderson, Donald Keith
Andres, Ronald Paul
Angus, John Cotton
Aris, Rutherford
Ayers, Arnold Leslie, Sr.
Barron, Saul
Beck, Theodore Richard
Benedict, Manson
Bennett, George Alan
Berman, Herbert L(awrence)
Berty, Jozsef Mihaly
Bhagat, Phiroz Maneck
Bigelow, John Ealy
Bischoff, Kenneth Bruce
Blakely, John Paul, *publications management*
Boundy, Ray Harold
Boylan, David Ray
Brenner, Howard
Broughton, Donald Beddoes
Buckley, Page Scott
Bungay, Henry Robert, III
Burris, Leslie
Busche, Robert Marion
Canon, Ronald Martin
Caplan, John DAvid
Carr, William Hoge, Jr.
Cashin, Kenneth Delbert
Characklis, William Gregory
Cheh, Huk Yuk
Chesworth, Robert Hadden
Chin, Jin H.
Chun, Sun Woong
Churchill, Stuart Winston
Cohen, Robert Edward
Cooney, Charles Leland
Cooper, Stuart Leonard
Cota, Harold Maurice
Couper, James Riley
Croff, Allen Gerald
Crooke, Philip Schuyler, III
Dahlstrom, Donald Albert
Dang, Vi Duong
Davis, Howard Ted
Dickson, Philip F.
Doubek, Clifford James
Duda, Richard Frank
Duffey, Dick
Duffie, John Atwater
Dunigan, Paul Francis Xavier
Eagleton, Lee Chandler
Eckert, Charles Alan
Epperly, William Robert
Erdogan, Haydar
Fair, James Rutherford, Jr.
Fan, Liang-tseng
Finlayson, Bruce Alan
Fisher, Edward Richard
Fisher, Harold Wallace
Fleck, Robert Davis
Fogler, Hugh Scott
Foroulis, Z. Andrew A.
Franke, Richard Herbert
Friedlander, Sheldon Kay
Frisch, Norman W(illiam)
Frumerman, Robert
Gautreaux, Marcelian Francis, Jr.
Ghavamikia, Hamid
Goldstein, Paul
Graham, Donald Lee
Greenkorn, Robert Albert
Greenstein, Teddy
Groenier, William Samuel
Gulari, Erdogan
Haas, Paul Arnold
Haensel, Vladimir
Haller, Gary Lee
Halligan, James Edmund
Hanratty, Thomas Joseph
Hauser, Ray Louis
Heath, Colin Arthur
Heckman, Richard Ainsworth
Hellums, Jesse David
Hendrickson, Waldemar Forrsel
Hightower, Joe Walter
Hiller, Judith Irene
Hlavacek, Vladimir
Hoffman, Edward Jack
Holder, Nadine Duguid
Holmes, John Thomas
Hopkins, Harvey Childs, Jr.
Horváth, Csaba Gyula
Hudson, John Lester
Humphrey, Arthur Earl
Humphrey, John P.
Isakoff, Sheldon Erwin
Ito, Yoichiro
Johnson, Benajmin Martineau
Johnson, Karl Otto, Jr.
Johnson, Terry R.
Jones, Francis Thomas
Kadlec, Robert Henry
Kalvinskas, John Joseph
Katz, Donald L.
Kampe, Lloyd Lute
King, Cary Judson, III
Koenig, Louis
Koerner, Ernest Lee
Kydd, Paul Harriman
Lamberger, Paul Henry
Laster, Richard
Leuze, Rex Ernest
Levenspiel, Octave
Lih, Marshall Min-Shing.
Linden, Henry Robert
Loven, Andrew Witherspoon
Luus, Rein
Lyons, William Kimbel
Macosko, Christopher Ward
Marchaterre, John Frederick
Markatos, Nicolas-Christos Gregory
Marshall, William Robert, Jr.
May, Walter Grant
Mc Afee, Jerry
McHenry, Keith Welles, Jr.
Mc Ketta, John J., Jr.
McNeese, Leonard Eugene
Mesler, Russell Bernard
Metil, Ignatius
Metzner, Arthur Berthold
Meyer, Walter
Mickley, Harold Somers
Mills, Patrick Leo
Nicolosi, Stephen Louis
Nobe, Ken
Oblad, Alexander Golden
O'Connell, John Paul
Ojalvo, Morris S(olomon)
Oldshue, James Y.
Othmer, Donald Frederick
Overcash, Michael Ray
Paul, Donald Ross
Peppas, Nikolaos Athanassiou
Perlmutter, Daniel D.
Peters, Max Stone
Pickard, David Kenneth
Pike, Ralph Webster
Poettmann, Fred Heinz
Prausnitz, John Michael
Quinn, John Albert
Reid, Allen Francis
Reid, Robert Clark
Reid, Robert Lelon
Reilly, Peter John
Robb, Walter Lee
Robins, Norman Alan
Rudd, Dale Frederick
Schechter, Robert Samuel
Schmitz, Roger Anthony
Schowalter, William Raymond
Seaver, Philip Henry
Seinfeld, John Hersh
Shinnar, Reuel
Shuler, Michael Louis
Smith, Joe Mauk
Song, Leila Shia
Spencer, Dwain Frank
Springer, Allan Matthew
Staats, William Richard
Stockinger, Siegfried Ludwig
Storm, David Anthony
Strieder, William Christian
Swanson, William Mason
Szego, George Charles
Taborek, Jerry
Thomas, Leo John
Timmerhaus, Klaus Dieter
Tsao, George T.
Turner, Howard Sinclair
Varma, Arvind
Wei, James
Weinstein, Norman Jacob
Welker, J. Reed
Weller, Sol William
Westerberg, Arthur William
Westwater, James William
Wilke, Charles Robert
Williams, Raymond Warren
Wilson, John Sheridan
Wise, John J.
Zakin, Jacques Louis

Civil
Amirtharajah, Appiah
Anand, Subhash C.
Arnold, Orville Edward
Bacon, Vinton Walker
Baron, Melvin Leon
Beard, Leo Roy
Bechtel, Stephen Davison, Jr.
Beedle, Lynn Simpson
Bellport, Bernard Philip
Blue, Todd Irwin
Blume, John August
Bovay, Harry Elmo, Jr.
Brannon, H(ezzie) Raymond, Jr.
Breen, John Edward
Cady, Philip Dale
Chen, Wai-Fah
Chesson, Eugene, Jr.
Christian, John Thomas
Clough, Ray William, Jr.
Cohen, Edward
Craven, John Pinna
D'Appolonia, Elio
Deere, Don Uel
Degenkolb, Henry John
Dobrovolny, Jerry Stanley
Drnevich, Vincent Paul
Durgun, Kanat
Dym, Clive Lionel
Elder, Rex Alfred
Ellingwood, Bruce Russell
Eng, Norman
Ewing, Benjamin Baugh
Fadum, Ralph Eigil
Fairhurst, Charles
Feeser, Larry James
Fenves, Steven Joseph
Fujishiro, Katakazu Kenneth
Gaither, William Samuel
Gallagher, Richard Hugo
Garbarini, Edgar Joseph
Gould, Phillip Louis
Graff, William John
Graham, Frederick Mitchell
Gunaji, Narendra Nagesh
Hall, William Joel
Halpin, Daniel William
Hanson, Robert D(uane)
Harris, Lee Errol
Hatheway, Allen Wayne
Hawkins, Neil Middleton
Hoel, Lester A.
Hognestad, Eivind
Hough, James Emerson
Hughes, Thomas Joseph
Jawad, Maan H(amid)
Jennings, Aaraon Austin
Johnson, James Allen
Johnston, Bruce Gilbert
Kane, Harrison
Kerri, Kenneth Donald
Kesler, Clyde Ervin
Kimmons, George Harvey
Kino, Gordon Stanley
Ko, Hon-Kim
Krishnamurthy, Muthusamy
Kuesel, Thomas Robert
Kulhawy, Fred Howard
Lamarre, Pierre
Lambe, Thomas William
Lean, Eric G.
Lear, George Emory
Lee, Griff Calicutt
Lee, Harry William
Leidersdorf, Craig B.
Lund, John William
Lundgren, James Reinhold
Lynn, Walter Royal
Medina, Miguel Angel, Jr.
Michel, Bernard
Mitchell, James Kenneth
Monismith, Carl Leroy
Moorhouse, Douglas Cecil
Noble, Charles Carmin
Paulson, Boyd Colton, Jr.
Pearson, Samuel Dibble, III
Peck, Ralph Brazelton
Peltier, Eugene Joseph
Penzien, Joseph
Peterson, Dean Freeman, Jr.
Pfrang, Edward Oscar
Pister, Karl Stark
Popov, Egor Paul
Popper, Steven Herbert
Potter, Miles Buttles
Price, Bobby Earl
Rajchman, Jan Aleksander
Ramanuja, Teralandur Krishnaswamy
Rao, Adiseshappa Ramachandra
Reese, Lymon Clifton
Richart, Frank Edwin, Jr.
Robeck, Gordon G.
Rogers, Wilbur Frank
Rolfe, Stanley Theodore
Rosenblueth, Emilio
Saada, Adel Selim
Saul, William Edward
Schmit, Lucien André, Jr.
Scisson, Sidney E.
Scordelis, Alexander Costicas
Seed, Harry Bolton
Siess, Chester Paul
Sozen, Mete Avni
Spiegel, Zane
Turner, Earl James
Vanoni, Vito August
Vaseen, Vesper Albert
Viest, Ivan M(iroslav)
Vinogradov, Aleksandra M.
Vogel, Herbert Davis
Wagner, Aubrey Joseph
Walton, C. Michael
Wang, George Shin-Chang
Wei, Millet Lunchin
Weidlinger, Paul
Whitman, Robert Van Duyne
Wiegel, Robert Louis
Wilbur, Lyman Dwight
Yee, Alfred Alphonse
Zandi, Iraj

Civil, water supply and wastewater treatment. *See* **ENGINEERING, Environmental.**

Computer Engineering. *See* **COMPUTER SCIENCE; ENGINEERING, Electrical.**

Corrosion
Anderson, Robert Clark
Ansuini, Frank Joseph
Beck, Theodore Richard
Cubicciotti, Daniel
Daniel, Samuel Henderson, III
Foster, William Samuel
French, David Nichols
Heidersbach, Robert Henry, Jr.
Jeffreys, James Victor
Johnson, Lewis Benjamin, Jr.
Leidheiser, Henry, Jr.
Munn, Raymond Shattuck
Nobe, Ken

Nowak, Welville Berenson
Pettit, Frederick Sidney
Pickerling, Howard William
Pye, Earl Louis
Summitt, (William) Robert
Verink, Ellis Daniel, Jr.
von Fraunhofer, Joseph Anthony
Wang, Rong
Weeks, John Randel, IV

Cryogenics
Adams, Ludwig
Barron, Randall Franklin
Birmingham, Bascom Wayne
Coombe, John Raymond
Cravalho, Ernest George
Geist, Jacob Myer
Hartnett, George Joseph, Jr.
Klockzien, Vernon George
Singhal, Anil Kumar
Timmerhaus, Klaus Dieter
Zavos, Panayiotis Michael
Zimmermann, F(rancis) J(ohn)
Zisk, Stanley Harris

Electrical
Aadland, Donald Ingvald
Aldridge, Melvin Dayne
Allen, William Hand
Arzbaecher, Robert
Atwood, Donald Jesse, Jr.
Bailey, Jake S.
Balanis, George Nick
Balwanz, William Walter
Becker, Albert Walter
Beckjord, Eric Stephen
Behnke, Wallace Blanchard, Jr.
Benedict, Anthony Gorman
Blanck, A.R.
Bloor, W(illiam) Spencer
Boerner, Wolfgang-Martin
Brown, George Harold
Brown, John Lawrence, Jr.
Burdick, Glenn Arthur
Cachat, John F.
Carroll, Lee Francis
Casperson, Lee Wendel
Cermak, Ivan Anthony
Chang, William Shen Chie
Chen, Wai-Kai
Chen, Wayne H.
Chien, Robert Tienwen
Clark, Melville, Jr.
Claus, Richard O.
Cohn, Nathan
Collin, Robert Emanuel
Cookson, Albert Ernest
Corry, Andrew Francis
Crain, Cullen Malone
Crowley, Joseph Michael
Cruz, Jose Bejar
Darlington, Sidney
David, Edward Emil, Jr.
DeBlasio, Richard
DeLucia, Frank Charles
de Wolf, David Alter
Dinneen, Gerald Paul
Edgerton, Harold Eugene
Eltimsahy, Adel H.
Fano, Robert Mario
Farnell, Gerald William
Fazio, Michael Vincent
Flanagan, James Loton
Friberg, Emil Edwards
Fry, Francis J(ohn)
Gandhi, Shirish Manilal
Gaylord, Thomas Keith
George, Thycodam Varkkey
Gerhardt, Lester A.
Gordon, Eugene Irving
Gray, Paul Edward
Greenwood, Allan Nunns
Haas, Violet Bushwick
Hansen, Robert C(linton)
Harris, Stephen Ernest
Helliwell, Robert Arthur
Holley, Charles H.
Holmes, Dyer Brainerd
Huggins, William Herbert
Ishimaru, Akira
Jackson, William David
Jacobs, John Edward
Joel, Amos Edward, Jr.
Kazek, Gregory Joseph
Kincaide, William Charles
Klimas, Edward John
Kong, Jin Au
Kopplin, Julius Otto
Kornhauser, Edward Theodore
Kuc, Roman Basil
Kuh, Ernest Shin-Jen
Kumar, Panganamala Ramana
Lafferty, James Martin
Lamm, (August) Uno
Lanzkron, Rolf Wolfgang
Lee, Thomas Henry
Li, Tingye
Liechti, Charles A.
Lischer, Ludwig Frederick
Lonngren, Karl Erik
Marcus, Steve Irl
Marcuse, Dietrich
Marcuvitz, Nathan
Mathews, Max Vernon
Mathur, Radhey Mohan
Mayo, John Sullivan
McMillan, Robert Walker
Melcher, James Russell
Melsa, James Louis
Meystel, Alexander Michael
Miller, Stewart Edward
Mohanty, Mirode Chandra
Mohler, Ronald Rutt
Mueller, George E.
Newley, Patrick Foster
Novotny, Donald Wayne
Nunn, Walter Melrose, Jr.
Olsen, Kenneth Harry
Packard, David
Paulk, Charles Jasper, Jr.
Paxton, Kenneth Bradley
Pease, Robert Louis
Pierre, Donald Arthur
Powers, Edward Joseph, Jr.
Pravda, Milton Frank
Quate, Calvin Forrest
Rabinow, Jacob
Ramakumar, Ramachandra Gupta
Rechtin, Eberhardt
Reed, Irving Stoy
Riseberg, Leslie Allen
Roe, Kenneth Andrew
Ross, Hugh Courtney
Ross, Ian Munro
Rowe, Joseph Everett
Rudolph, Luther Day
Russell, B. Don
Sandberg, Irwin Walter
Scott, Larry Donald
Sebo, Stephen Andrew
Seshadri, Sengadu Rangaswamy
Shank, Charles Vernon
Shepherd, Mark, Jr.
Shofner, Frederick Michael
Shum, Raymond Hing-Yan
Slemon, Gordon Richard
Slepian, David
Smith, John Barber
South, Hugh Miles
Staszesky, Francis Myron
Straiton, Archie Waugh
Suciu, Spiridon N.
Thomas, Frank Joseph
Thompson, James Elton
Thouret, Wolfgang Emery
Ulaby, Fawwaz Tayssir
Vanderweil, Raimund Gerhard, Jr.
Von Tersch, Lawrence Wayne
Walker, Eric Arthur
Wang, Paul P.
Warzecha, Ladislaus William
Wayne, Burton Howard
Weiss, Thomas Fischer
Weitkamp, William George
Westcott, William Warren
Whinnery, John Roy
Wiesner, Jerome Bert
Willcox, Phillip James
Wise, Gary Lamar
Womack, Joseph Donald, Sr.
Wood, Allen John
Woodson, Herbert Horace
Wooldridge, Dean Everett

Electrical, applied magnetics
Avery, Robert Tolman
Bainbridge, Kenneth Tompkins
Boerner, Wolfgang-Martin
Camras, Marvin
Hawke, Ronald Samuel
Humphrey, Floyd Bernard
Kryder, Mark Howard
McIntosh, Robert E., Jr.
Mullett, Charles Edwin
Reynolds, Larry Owen
Straiton, Archie Waugh
Sugai, Iwao

Electrical, computer-aided design
Bedinger, Joseph Arnold
Bhattacharyya, Shankar
Bollini, Raghupathy
Boyse, John Wesley
Chace, Milton
Chasen, Sylvan Herbert
Clayton, David Lawrence
Davidson, Scott
Donahue, Michael James
Frisch, Joseph
Goldstein, Lawrence Howard
Hancock, John Coulter
Krause, Irvin
Lieberherr, Karl Josua
Lupash, Lawrence O(Vidiu)
Meagher, Donald Joseph
Newley, Patrick Foster
Orthwein, William Coe
Parker, Alice Cline
Rhodine, Charles Norman
Salsburg, Kevyn Anne
Siewiorek, Daniel Paul
Smith, Kevin Richard
Speciale, Ross Aldo
Szilagyi, Mike Nicholas
Temes, Gabor Charles
Thiel, Leo Albert
Van Valkenburg, Mac Elwyn
Wayne, Burton Howard
Yessios, Chris Ioannis

Electrical, computer engineering
Abraham, George
Adams, Karyl Ann
Albisser, Anthony Michael
Alexander, Charles Kenneth, Jr.
Amdahl, Gene Myron
Anderson, Richard Cooper
Aupperle, Eric Max
Aupperle, Kenneth Robert
Bell, Chester Gordon
Benedict, Bruce John
Berlekamp, Elwyn Ralph
Bernstein, Ralph
Bertrand, John
Bickart, Theodore Albert
Bitzer, Donald Lester
Bloch, Erich
Brakefield, James Charles
Branscomb, Lewis McAdory
Bush, George Edward
Butler, Thomas Warwick, Jr.
Cermak, Ivan Anthony
Chakravarty, Indranil
Chen, Hollis Ching
Chen, Hsiang Tsun
Cheung, John Yan-Poon
Childs, Jeffrey John
Coates, Clarence Leroy, Jr.
Cooper, (Howard) Gordon
Crabb, Charles Frederick
Curtis, Kent Krueger
Dabkowski, Krzysztof Jerzy
Decker, Bruce Michael
Drozd, Andrew Louis Stephan
Epstein, Sheldon Lee
Evans, Bob Overton
Ewing, Martin Sipple
Fearnot, Neal Edward
Fleck, David Charles
Genet, Russell Merle
Gray, Festus Gail
Hankins, Timothy Hamilton
Hedges, Harry George
Hellman, Martin Edward
Heuft, Richard William
Jaluria, Rajiv
Jump, J. Robert
Keller, Robert Marion
Keyes, Robert William
Kim, Myunghwan
Kopplin, Julius Otto
Landauer, Rolf William
Levitin, Lev Berovich
Liptay, John Stephen
Lucky, Robert Wendell
Meditch, James Stephen
Mergler, Harry Winston
Miller, Thomas Paul
Minich, Card Edward
Mondal, Kalyan
Murata, Tadao
Nguyen, Chinh Trung
Osborne, Adam
Parker, Alice Cline
Protopapas, Dimitrios
Quek, Swee-Meng
Rabiner, Lawrence Richard
Rawlinson, Stephen John
Reed, Irving Stoy
Rumsey, Victor Henry
Silvern, Leonard Charles
Sproull, Robert Fletcher
Svarrer, Robert W.
Thammavaram, N. Rao
Van Valkenburg, Mac Elwyn
Von Tersch, Lawrence Wayne
Wheeless, Leon Lum, Jr.
Witte, Kurt Allen
Wojciechowski, Witold Stanislaw
Yang, Yee-Hong

Electrical, microchip technology. *See also* MATERIALS SCIENCE.
Armstrong, John Allan
Baertsch, Richard Dudley
Detterman, Robert Linwood
Early, James Michael
Economou, Nicholas Philip
Gillespie, Sherry Jacqueline
Holmes, Neal J.
Kestenbaum, Ami
Kilby, Jack St. Clair
Kressel, Henry
Moore, Gordon E.
Mutter, Walter Edward
Shibib, Muhammed Ayman
Spicer, William Edward, III
Sumney, Larry W.
Thompson, Eric Douglas
Wolf, Edward Dean

Electrical, semiconductors
Barnes, Frank Stephenson
Bube, Richard Howard
Burrus, Charles Andrew, Jr.
Champlin, Keith S(chaffner)
Channin, Donald Jones
Chu, Ting L.
Conwell, Esther Marly
Dacey, George Clement
Das, Pankaj K.
Dereniak, Eustace Leonard
Dow, Daniel Gould
DuBow, Joel Barry
Emery, Keith Allen
Erickson, Robert Arlen
Gillespie, Sherry Jacqueline
Haddad, George Ilyas
Heilmeier, George Harry
Holonyak, Nick, Jr.
Holton, William Coffeen
Jordan, Angel Goni
Larson, Donald Clayton
Lee, Sanboh
Lee, Tzuo-chang
Liu, Hua-Kuang
Martin, Thomas Lyle, Jr.
McCarroll, William Henry
Moore, Gordon E.
Mutter, Walter Edward
Noyce, Robert Norton
O'Clock, George Daniel, Jr.
Owen, S. John T.
Pearson, Gerald Leondus
Phelan, Robert Joseph, Jr.
Rosi, Fred David
Rozenbergs, Janis
Sangrey, Dwight A.
Shibib, Muhammed Ayman
Shockley, William Bradford
Sollner, Traugott Carl Ludwig Gerhard
Soukup, Rodney Joseph
Sparks, Morgan
Tauc, Jan
Van der Ziel, Aldert
Van Vliet, Carolyn Marina
Viswanathan, Chand Ram
Wang, Shing Chung
Wang, Shyh
Wittkower, Andrew B.
Yariv, Amnon

Electrical, superconductors
Anderson, Carl John
Black, William Carter, Jr.
Clark, Alan Fred
Kressel, Henry
Lefkowitz, Issai
Piore, Emanuel Ruben
Tinkham, Michael

Electronics
Adler, Robert
Allen, Jonathan
Atwood, Donald Jesse, Jr.
Bailey, Jake S.
Bak, Martin Joseph
Ball, John Allen
Barna, Arpad Alex
Bartlett, Peter Greenough
Blade, Richard Allen
Borten, William H.
Brown, Burton Primrose
Brown, George Harold
Brumm, Douglas Bruce
Buck, Richard Forde
Bussgang, Julian Jakob
Butler, Thomas Warwick, Jr.
Button, Kenneth J(ohn)
Camras, Marvin
Clark, Melville, Jr.
Collin, Robert Emanuel
Crain, Cullen Malone
Croft, Thomas A(rthur)
Cyr, Reginald John
Dow, Daniel Gould
Doyle, Robert Joseph
Dufilho, Harold Louis
Dunn, James
Fowler, Charles Albert
Freehill, Peter Eugene
Friedrich, Otto Martin, Jr.
Gablehouse, Reuben Harold
Galbiati, Louis Joseph
Geddes, Leslie Alexander
Gerken, Louis Charles
Ginzton, Edward Leonard
Gordon, Eugene Irving
Gray, Eoin Wedderburn
Greenberg, Harold Paul
Groeber, Edward Otto, Jr.
Gross, Leo
Grove, Andrew S.
Grow, Richard W.
Gunderson, Leslie Charles
Hammond, Donald L.
Hannay, N(orman) Bruce
Hanover, Paul Norden
Haque, Promod
Hartig, Elmer Otto
Haus, Hermann Anton
Headley, R. Paul
Herwald, Seymour W(illis)
Hines, Robin Hinton
Hogan, Clarence Lester
Hull, Harvard Leslie
Jobs, Steven Paul
Johnson, Brian Weatherred
Jordan, Arthur Kent
Kang, Min Ho
Kaufman, Raymond
Keicher, William Eugene
Keiser, Bernhard Edward
Kiser, Donald Owen
Klimas, Edward John
Kravitz, David William
Lafferty, James Martin
Larson, Larry Gale
Leavitt, Robert LaVerne
Lytle, James Mark
Marcatili, Enrique Alfredo José
Martin, Thomas Lyle, Jr.
McIntosh, Robert E., Jr.
Morris, Charles Reginald
Morrow, Walter Edwin, Jr.
Mullett, Charles Edwin
Nunn, Walter Melrose, Jr.
Owen, S. John T.
Patti, Robert Dale
Rajchman, Jan Aleksander
Rhodes, Donald Frederick
Rose, Albert
Ross, Hugh Courtney
Salvatori, Vincent Louis
Seling, Theodore Victor
Shah, Harshad
Sheth, Chandrakant Hakmichand
Sletten, Carlyle Joseph
Smith, Wayne D.
Speciale, Ross Aldo
Spellman, Donald Jerome
Stone, Gregory Michael
Stone, William Ross
Sturdevant, Eugene J.
Sweet, Ray Douglas
Swenson, George Warner, Jr.
Taylor, George William
Townsend, Marjorie Rhodes
Vollum, Howard
Webster, William Merle, Jr.
Whitehead, Frank Roger
Wilson, William John

Electronics, integrated circuits
Abraham, George
Anderson, Richard Cooper
Baertsch, Richard Dudley
Chen, Wai-Kai
Das, Pankaj K.
de Wit, Michiel
Goldstein, Lawrence Howard
Grove, Andrew S.
Guttag, Karl Marion
Krinsky, Jeffrey Alan
Libby, Vibeke
Linvill, John Grimes
Manor, Robert Edward
Meindl, James Donald
Moll, John Lewis
Mondal, Kalyan
Schober, Robert Charles
Temes, Gabor Charles
Varghese, Sankoorikal Lonappan

Electronics, microelectronics
Bucy, J. Fred
Connelly, Will Arthur
Dirks, Leslie Chant
Early, James Michael
Economou, Nicholas Philip
Eisenberg, Lawrence
Feinstein, Joseph
Fetterman, Harold Ralph
Greenhouse, Harold Mitchell
Haddad, George Ilyas
Heck, Daniel Curtis, Jr.
Holonyak, Nick, Jr.
Liu, Yung Sheng
Marcatili, Enrique Alfredo José
Martin, Frederick Wight
Michener, John Russell
Morrison, Roderick Gordon
O'Clock, George Daniel, Jr.
Ramm, Dietolf
Raschke, Curt Robert
Soukup, Rodney Joseph
Tien, Ping King
Viswanathan, Chand Ram
Webster, William Merle, Jr.
Yu, Chen-Cheng William

Environmental
Allen, John Malone
Armstrong, Neal Earl
Auer, Martin Tucker
Aulenbach, Donald Bruce
Baer, Thomas Strickland
Behrens, William Wohlsen, Jr.
Betts, Austin Wortham
Bolch, William Emmett, Jr.
Boubel, Richard William
Bregman, Jacob Israel
Cota, Harold Maurice
Daniels, Raphael Sanford
Edwards, Ray Conway
Eldred, Kenneth McKechnie
Elzerman, Alan William
Emrich, Grover Harry
Fay, James Alan
Friedlander, Sheldon Kay
Fujishiro, Katakazu Kenneth
Golay, Michael Warren
Goldsmith, William Alee
Goldstein, Paul
Happ, Stafford Coleman
Harleman, Donald Robert Fergusson
Harris, Sigmund Paul
Hart, Fred Clinton
Hartz, Kenneth E.
Hatheway, Allen Wayne
Hilbert, Morton Shelly
Holliday, George Hayes
Huang, Ju-Chang
Hull, Clark Ramsey
Jennings, Aaraon Austin
Jennings, Burgess Hill
Kathren, Ronald Lawrence
Kim, Geung-Ho
Kokoropoulos, Panos
Kremers, Jack Alan
Kuehn, Thomas Howard
LaGrega, Michael Denny
Lane, Dennis Del
Lerman, Abraham
Liebman, Jon Charles
Ling, Joseph Tso-Ti
Loven, Andrew Witherspoon
Ludwig, John Howard
Lugar, Robert Myers
Mar, Brian Wayne
Marks, Peter Jacob
Maugham, James Henry-Eamon, III
Metzler, Dwight Fox
Middleditch, Brian Stanley
Moeller, Dade William
Moore, William Walter
Morey, Philip Richard
Nelson, Gordon Leon
O'Shaughnessy, James Colin
Pankow, James Frederick
Perdue, Philip Taw
Perrine, Richard Leroy
Picazo, Esteban David
Plunkett, Robert Dale
Rogers, Vern Child
Rubin, Edward Stephen
Schwartz, David Michael
Siem, Laurie Helen
Sloane, Christine Scheid
Stange, James Henry
Stockinger, Siegfried Ludwig
Tatom, Frank Buck
Vaseen, Vesper Albert
Wang, George Shin-Chang
Weinstein, Norman Jacob
Willoughby, Nancy Bharucha
Yerkes, William Dilworth, Jr.

Environmental, water supply and wastewater treatment
Amirtharajah, Appiah
Aulenbach, Donald Bruce
Bacon, Vinton Walker
Bleckmann, Charles Allen
Bouwer, Herman
Bowen, Paul Tyner
Bregman, Jacob Israel
Brierley, Corale Louise
Brooks, Norman Herrick
Chambers, James Vernon
Characklis, William Gregory
Chulick, Eugene Thomas

Cooper, Robert Chauncey
Cronholm, Lois S.
Danner, David Lee
de la Noue, Joel Jean-Louis
Dornfeld, Richard Louis
Dow, Martha Anne
Drum, Donald A.
Eaton, David J.
Engelbrecht, Richard Stevens
Erdogan, Haydar
Ewing, Benjamin Baugh
Gloyna, Earnest Frederick
Gupta, Gian Chand
Hamilton, Joseph Hants, Jr.
Henley, Melvin Brent
Higgins, Irwin Raymond
Huang, Chin Pao
Huang, Ju-Chang
Johnson, James Allen
Kadlec, Robert Henry
Kendall, Ernest Terry
Kerfott, William Buchanan, Jr.
Kerri, Kenneth Donald
Kim, Kyo Sool
Koenig, Louis
Koerner, Ernest Lee
Kraus, Marjorie Patt
La Motta, Enruqie Jaime
Loh, Philip Choo-Seng
Long, David Ainsworth
Ludwig, John Howard
Malina, Joseph Francis, Jr.
Mc Kinney, Ross Erwin
Metzler, Dwight Fox
Myrick, Henry Nugent
Nesbitt, Patricia Marie
Okun, Daniel Alexander
Olesen, Douglas Eugene
O'Shaughnessy, James Colin
Potter, Miles Buttles
Probstein, Ronald Filmore
Robeck, Gordon G.
Rohlich, Gerard Addison
Roy, Dipak
Springer, Allan Matthew
Teuscher, Michael Cook
Wayman, Cooper Harry
Weber, Walter Jacob, Jr.
Yerkes, William Dilworth, Jr.
Zielinski, Paul Bernard

Fusion. *See* **ENERGY SCIENCE, LASER.**

Human factors
Bauer, Robert William
Bell, Barbara Jean
Bevan, Thomas Edward
Boff, Kenneth Richard
Boggs, George Johnson
Burgess, John Henry
Carley, John Wesley, III
Chaffin, Don B.
Cogbill, Charles Lipscomb, III
Cohen, Edwin
Collins, Paul Francis
Davies, Ivor Kevin
Dutton, James William
Eigen, Daryl Jay
Englund, Carl Ernest
Eschenbach, Arthur Edwin
Fergenson, P. Everett
Garvin, Everett Arthur
Goldman, Alexander
Hill, Ernest Elwood
Hill, Percy Holmes
Hodge, David Charles
Hoeller, Louise
Jacob, Robert Joseph Kassel
Johnston, Dorothy Mae
Jones, Daniel Todd
Kantowitz, Barry Howard
Kreifeldt, John Gene
LeMay, Moira Kathleen
Lincoln, Charles Ebenezer
Lindquist, Oiva Herbert
Lodge, George Townsend
Lyman, John
Malone, Thomas Becker
Miller, James Woodell
Mitchell, Nancy Brown
Onorato, Howard Louis
Peterson, James Robert
Rohles, Frederick Henry, Jr.
Salzenstein, Marvin A.
Shea, Daniel Joseph
Siegel, Joel Marvin
Singer, Timothy James
Smith, Wayne D.
Valfer, Ernst Siegmar
Von Gierke, Henning Edgar

Industrial
Amar, Amar-Dev
Bishop, Albert Bentley, III
Bovay, Harry Elmo, Jr.
Burge, Charles Arthur
Burnham, Donald Clemens
Cate, Martin Edward
Dizer, John Thomas, Jr.
ElGomayel, Joseph Ibrahim
Elzinga, Donald Jack
Frey, Henry Edwin, Jr.
Hedstrom, Joseph Charles
Huang, Chin Pao
Johnson, Harold Arthur
Kapur, Kailash Chander
Katz, Hyman Bernard
Landers, Jack Maxam
Litwhiler, Daniel W.
Litzenberg, David P.
Lowe, Timothy Joe
Majewski, Frank Thomas
Marshall, Harry Dwight
McKee, Keith Earl
Meier, Wilbur Leroy, Jr.
Oliver, Larry Ray
Raafat, Feraidoon
Sargent, Robert George
Schowalter, William Raymond
Schroer, Bernard Jon
Schwarz, Emil Arthur
Weiss, Howard Jacob
White, John Spencer
Worth, Norman Paul
Wyskida, Richard Martin
Yakin, Mustafa Zafer

Materials. *See also* **MATERIALS SCIENCE.**
Adler, William Fred
Aller, John Earl
Bailey, Stuart Lohr
Baker, William Oliver
Bauman, Thomas Charles
BIANCHERIA, Amilcare
Cady, Philip Dale
Carlson, Roy Washington
Chin, Gilbert Yukyu
Chubb, Walston
Cunningham, John Edward
Dai-Shu-Ho
Davis, Grayum Lloyd
De Vries, Kenneth Lawrence
Dresselhaus, Mildred Spiewak
Drucker, Daniel Charles
Eiber, Robert James
Finkin, Eugene Felix
Finn, Patricia Ann
Fiore, Nicholas Francis
Flemings, Merton Corson
Forbes, Judith
Friedberg, Arthur Leroy
Gallagher, Joseph Patrick
Goins, William (Doris), III
Goldenberg, Herbert Jay
Grace, Richard Edward
Green, Robert Edward, Jr.
Greenfield, Irwin Gilbert
Haggag, Fahmy Mahmoud
Hemmings, Robert Leslie
Irwin, George Rankin
Jacoby, William Richard
Johnson, James Robert
Jordan, Charles Ralph
Kapp, Joseph Alexander
Kenig, M(arvin) Jerry
Kesler, Clyde Ervin
Kinzel, Augustus Braun
Kiser, Donald Owen
Ko, Hon-Kim
Kramer, Bruce Michael
Kusy, Robert P
LaCourse, William Carl
Loss, Frank J.
Lyons, John Winship
Marshall, Walter Lincoln
Mielenz, Richard Childs
Mordfin, Leonard
Munn, Raymond Shattuck
Ott, Walter Richard
Payne, Robert William
Pellicane, Patrick Joseph
Polonis, Douglas Hugh
Promisel, Nathan E.
Restaino, Alfred Joseph
Roehrs, Robert Jesse
Salerni, John Vincent
Sanderson, Ian Scott
Schneberger, Gerald Leo
Shack, William John
Shahinpoor, Mohsen
Shirkey, William Dan
Smith, John Henry
Steinberg, Morris Albert
Stregowski, Thomas John
Teller, Cecil Martin, II
Thornton, J(oseph) Scott, Jr.
Verink, Ellis Daniel, Jr.
Walton, Lewis Anthony
Wilder, David Randolph
Woodlan, Donald Ray

Mechanical
Adams, William Eugene
Agnew, William George
Agrawal, Ram Kumar
Alexander, Dennis Jay
Aller, John Earl
Alley, Thomas Leroy
Anderson, conrad Victor
Anderson, David Carleton
Anderson, Edward Everett
Anderson, James Hilbert
Arnas, Ozer Ali
Atkinson, Steven Albert
Avery, Robert Tolman
Bachman, Walter Crawford
Baer, Robert Lloyd
Bailey, Stuart Lohr
Bair, Scott Slaybaugh, III
Balsley, Lawrence Edward
Balster, Frederick Werden
Ban, Stephen Dennis
Barkan, Philip
Baron, Seymour
Barron, Randall Franklin
Bateman, Alfred Chandler
Beachley, Norman Henry
Beadle, Charles Wilson
Beale, William Taylor
Beard, James Taylor
Beggs, James Montgomery
Bell, Cornelius
Benedict, Anthony Gorman
Bennett, Bruce Alan
Berkof, Richard Stanley
Bhagat, Phiroz Maneck
Bigler, William C.
Birmingham, Bascom Wayne
Bisplinghoff, Raymond Lewis
Blackwelder, Ron
Boehm, Robert Foty
Bohr, Bruce A.
Bosy, Brian Joseph
Boulger, Francis William
Bowden, Bryant Baird
Bowman, H(arry) Frederick
Bowman, Thomas Eugene
Boyer, Charles Benjamin
Brighton, John Austin
Brown, Wayne Samuel
Bujtas, Mark Steven
Cahill, William Joseph, Jr.
Calder, Clarence Andrew
Caplan, John DAvid
Carroll, Dyer Edmund
Chao, Bei Tse
Chattopadhyay, Somnath
Chen, Chih Ping
Chen, Michael Ming
Chiang, Chao-Wang
Chiou, Jiunn Perng
Cho, Joon Ho
Cho, Soung Moo
Chuang, Henry Ning
Clausing, Arthur Marvin
Coburn, Herbert Dightman, Jr.
Cole, Kenneth Dean
Conlon, Bartholomew Frederick, Jr.
Cook, William John
Copes, John Carson, III
Cornell, Robert Witherspoon
Cross, Ralph Emerson
Cugini, Edward Thomas
Cummings, Garth Ellis
Cwycyshyn, Walter
Czernik, Daniel Edward
Dabkowski, Krzysztof Jerzy
Daily, James Wallace
Dai-Shu-Ho
Damianov, Vladimir Blagoi
Dao, Kim C.
Dasgupta, Aaron
Davis, Grayum Lloyd
Dean, Richard Anthony
Dean, Robert Charles, Jr.
De Santo, Daniel Frank
DesChamps, Nicholas Howard
De Vries, Kenneth Lawrence
Dharan, C.K. Hari
DiOrio, Mark Lewis
Dix, Rollin Cumming
Dizer, John Thomas, Jr.
Donis, Jose Maria
Dornfeld, Richard Louis
Drake, Michael Lee
Eberhardt, Allen Craig
Eck, Bernard John
Eckert, Ernst R. G.
Eiss, Norman Smith, Jr.
Elias, Thomas Ittan
Engel, Peter Andras
Eyden, Bernard
Faccini, Ernest Carlo
Finkin, Eugene Felix
Fischer, Traugott Erwin
Flack, Ronald Dumont, Jr.
Fletcher, Leroy S(tevenson)
Foley, Cray Lyman
Forkel, Curt Emil
Fourney, M(ichael) E(ugene)
Francis, John Elbert
Frederickson, Robert
Freudenheim, Milton B.
Frey, Henry Edwin, Jr.
Friberg, Emil Edwards
Frisch, Joseph
Gaines, Albert Lowery
Gajewski, Walter Michael
Gakenheimer, David Charles
Gandhi, Shirish Manilal
Garg, Devendra Prakash
Garner, Andrew Morris
Garner, James Kirkland
Gat, Uri
Gill, Daniel Emmett
Glenn, Alan Holton
Glower, Donald Duane
Goldstein, Richard Jay
Goldstein, Steven Alan
Goodenough, Samuel Henry
Gorges, Heinz A.
Gouse, S. William, Jr.
Graff, William John
Grant, Donald Andrew
Grosh, Richard Joseph
Gunn, Michael Richard
Haan, David Charles
Haberman, Charles Morris
Hagmann, Dean Berry
Hart, David Anderson
Hartman, Patrick James
Hartnett, George Joseph, Jr.
Harvey, Robert Darnell
Hatsopoulos, George Nicholas
Haupt, H. James
Hawkins, Robert C(leo)
Hayes, Edward J(ames)
Hayner, Denis Robert
Heck, Daniel Curtis, Jr.
Heckman, Thomas Paul
Hester, Jarrett Charles
Heymann, Frank Joseph
Hill, Percy Holmes
Hills, Willard Andrew
Hoffman, John Robert, Jr.
Homer, Percy Albert
Hopkins, Stephen William
Hovingh, Jack
Howe, Timothy Max
Howell, John Reid
Hrones, John Anthony
Hrycak, Peter
Hughes, Thomas Joseph
Hughes, William Perry
Husain, Zakiud Din
Hussain, A.K.M. Fazle
Huston, Ronald Lee
Hutchings, William Frank
Jacobs, David
Jain, Sulekh Chand
Janzow, Edward Frank
Jawad, Maan H(amid)
Johnson, Benajmin Martineau
Johnson, Glen Eric
Johnson, Harold Arthur
Johnson, Stephen Thomas
Jones, Charles
Jordan, Richard Charles
Juvinall, Robert Charles
Kautz, David Johnathan
Keith, Theo Gordon, Jr.
Kelley, Mark Elbridge, III
Kelly, Catherine Louise
Kemelhor, Robert Elias
Kempf, Gary William
Kessler, George William
Kestin, Joseph
Kim, Rhyn Hyun
Kinney, Robert Bruce
Kinra, Vikram K.
Kirchner, Larry A.
Kirik, Michael John
Kirk, James Allen
Kline, Stephen Jay
Knapp, Ronald Harrison
Koch, Leonard John
Koeneman, James Bryant
Korwin, Paul
Kovacs, Gyula
Kozman, Theodore Albert
Krause, Irvin
Kuehn, Thomas Howard
Kumar, Nirmal
Kwik, Robert Julius
Lahoti, Goverdhan Das
Lai, Ying-San
Langner, Carl Gottlieb
Latham, Allen, Jr.
Laurenson, Robert Mark
Leboeuf, Cecile Marie
Lednicky, Raymond Anthony
Leed, Peter Lewis
Leitmann, George
Lewis, James Bryson
Lewis, Michael Dolan
Lilje, Karl David
Lin, Tung-Hua
Ling, Frederick Fongsun
Litzenberg, David P.
Long, Leonard Michael
Look, Dwight Chester, Jr.
Lorusso, Paul David
Lowe, Phillip A(rnold)
Lucas, Robert Alan
Lunchick, Myron Edwin
Lund, Charles Edward
Lundgren, James Reinhold
Macke, Harry Jerry
Maier, Leo Robert, Jr.
Majewski, Frank Thomas
Malik, Mujeeb Rehman
Marshall, Harry Dwight
McClure, Carl Kenneth
McFadden, Peter William
McGean, Thomas James
McLerran, Archie Ralph
McMenamin, Edward William
Melcher, James Russell
Merchant, Howard Carl
Merilo, Mati
Merritt, Joseph Claude
Milestone, Wayne Donald
Miller, Harry F.
Miller, Paul Leroy, Jr.
Mitson, Herbert Henry, Jr.
Morris, John William
Morris, Robert Howard
Mraz, George Jaroslav
Mumma, Albert G.
Nahm, Alexander Hong
Narayana, Anand Deo
Netzel, James Phillip
Nichols, Paul Arthur
O'Brien, Kenneth Stanley
Ojalvo, Morris S(olomon)
Olah, Stephen
Oliver, Larry Ray
Olson, Donald Richard
Oppenheim, Antoni Kazimierz
Orthwein, William Coe
Ovens, William George
Park, U. Young
Parker, Frank Wayne
Patel, Bhikhubhai L.
Patterson, Dwight Robert
Peach, Robert Westly
Perry, Erik David
Peters, Alexander Robert
Pfister, Philip Carl
Pflederer, Fred Raymond
Philips, Gerald John
Phillips, James Woodward
Phillips, William Evans
Phillips, Winfred Marshall
Pletcher, Richard Harold
Powe, Ralph Elward
Prasad, Birendra
Psaros, George Emanuel
Rabinow, Jacob
Rath, Robert Michael
Reid, Robert Lelon
Reswick, James Bigelow
Reynolds, William Craig
Richardson, Herbert Heath
Richardson, Rick Lee
Richter, George Brownell
Ricks, Stephen Andrew
Roberts, Albert Sidney, Jr.
Roe, Kenneth Keith
Rohsenow, Warren Max
Romero, Alejandro Francisco
Romesberg, Laverne Eugene
Rosa, David
Rosenberg, Gerson
Rowlands, Robert Edward
Rubin, Edward Stephen
Salerni, John Vincent
Salzenstein, Marvin A.
Sanderson, Ian Scott
San Juan, Eduardo Carrion
Scher, Robert Sander
Schilmoeller, Neil Herman
Schlegel, Ronald Gene
Schneider, Robert William
Schoen, George Janssen
Schoerer, Frank
Schorry, Robert Elmer
Schultz, Albert Barry
Schwarz, Emil Arthur
Scisson, Sidney E.
Scura, Lawrence Thomas, Jr.
Seely, John Henry
Seireg, Ali A(bdel Hay)
Shahinpoor, Mohsen
Shank, Maurice Edwin
Shaw, Milton Clayton
Sheth, Chandrakant Hakmichand
Sienicki, James Joseph
Silsby, Graham Forbes
Silvestri, George Joseph, Jr.
Simon, Richard L.
Simonetti, Joseph Lawrence
Simonson, Simon Christian, III
Singhal, Anil Kumar
Siuru, William Dennis, Jr.
Smith, John Barber
Smith, John Douglas
Smith, Kevin Richard
Smith, Richard James
Snyder, Glenn Jacob
Solberg, Ruell Floyd, Jr.
Spielvogel, Lawrence George
Springer, George Stephen
Stalker, Kenneth Walter
Starkey, Walter Leroy
Steinitz, Louis Joseph
Stumpe, Warren Robert
Suchora, Daniel Henry
Suciu, Spiridon N.
Suh, Nam Pyo
Swallom, Daniel Warren
Sweet, Ray Douglas
Swenson, Donald Otis
Synder, James Nevin
Szewczyk, Albin Anthony
Taraman, Khalil Showky
Taylor, Malcolm Ernest
Teller, Cecil Martin, II
Teuscher, Michael Cook
Thompson, Jack Mansfield, Jr.
Throner, Guy Charles
Timmerman, Robert Wilson
Timo, Dominic Peter
Tome, Richard Earle
Tomita, Nobuya
Traiforos, Spyros Anthony
Treder, John David
Trumpler, William E.
Ugelow, Albert Jay
Umek, Anthony M.
Vallery, Stafford Jean, II
Vanderweil, Raimund Gerhard, Jr.
Vaughan, David Sherwood
Vest, Charles Marstiller
Veziroglu, Turhan Nejat
Vigil, Manuel Gilbert
Viskanta, Raymond
Viswanathan, K.
Walton, Lewis Anthony
Wang, Hsin-Pang
Wei, Millet Lunchin
Weiss, Olin Eric
Weldon, William Forrest
Wells, Herbert Arthur
Wells, William Terry
Wilson, Max Kearns
Wnuk, Michael Peter
Wolfe, Howard Francis
Wong, Julius Pan
Wright, Christopher Pearce
Wu, Chih
Young, Garry Gean
Zimmermann, F(rancis) J(ohn)

Mechanics, fluid
Abramson, Hyman Norman
Ahmadi, Goodarz
Anderson, Edward Everett
Aref, Hassan
Bair, Scott Slaybaugh, III
Bandyopadhyay, Promode Ranjan
Bernard, Peter Simon
Bird, Robert Byron
Brilliant, Howard Michael
Brooks, Norman Herrick
Callens, E(arl) Eugene, Jr.
Carelli, Mario Domenico
Carrier, George Francis
Carrigan, Charles Roger
Cartlidge, Edward Sutterley
Caughey, David Alan
Caulk, David Allen
Chan, Shih Hung
Chao, Bei Tse
Chen, Ching Jen
Chin, Jin H.
Collins, Frank Gibson
Conly, John Franklin
Corrsin, Stanley
Cosner, Raymond Robert
Covert, Eugene Edzards
Crowley, Joseph Michael
Cumberbatch, Ellis
Daily, James Wallace
Dallman, John Clay
Damianov, Vladimir Blagoi
Darden, Christine Mann
Dash, Sanford Mark
Davis, Philip Keith
Deissler, Robert George
Denysyk, Bohdan
DesChamps, Nicholas Howard
Donnelly, Russell James

Doster, Joseph Michael
Dukler, Abraham Emanuel
Eaton, John Kelly
Eisenberg, Phillip
Elder, Rex Alfred
Elghobashi, Said (Elsayed)
Elias, Thomas Ittan
Fay, James Alan
Fearn, Richard Lee
Flack, Ronald Dumont, Jr.
Flippen, Luther Daniel, Jr.
Francis, Michael Skok
Freiwald, David Allen
Friedlander, Susan Jean
Goldschmied, Fabio Renzo
Goldstein, Marvin E(manuel)
Grosch, Chester Enright
Habashi, Wagdi George
Hancox, William Thomas
Handler, Robert Alphonse
Hanratty, Thomas Joseph
Harpavat, Ganesh Lal
Harsha, Philip Thomas
Hart, David Anderson
Hepworth, Harry Kent
Herring, H(ugh) James
Hjelmfelt, Allen Talbert, Jr.
Hjertager, Bjorn Helge
Ho, Chih-Ming
Hoffman, Myron Arnold
Hrycak, Peter
Hsu, Ming-Teh
Hung, Tin-Kan
Husain, Zakiud Din
Jawad, Sarim Naji
Kantrowitz, Arthur
Katsanis, Theodore
Kennedy, John Fisher
Kinney, Robert Bruce
Kitchens, Clarence Wesley, Jr.
Klebanoff, Philip Samuel
Kline, Stephen Jay
Koch, Werner
Krishnamurthy, Lakshminarayanan
Lake, Bruce Meno
Lampert, Seymour
Landweber, Louis
Lau, Jark C(hong)
Leonard, Anthony
Lewis, John Allen
Liu, Chang Yu
Liu, Chen-Huei
Liu, Paul Chi
Liu, Wing Kam
Long, Lyle Norman
Long, Robert Radcliffe
Lumley, John Leask
Lunchick, Myron Edwin
Malik, Mujeeb Rehman
Marvin, Joseph George
Massier, Paul Ferdinand
Matalon, Moshe
McRae, D(avid) Scott
Merilo, Mati
Mikulla, Volker
Miller, Harvey Philip
Morse, Theodore Frederick
Muirhead, Vincent Uriel
Norment, Hillyer Gavin
O'Brien, Morrough Parker
Oldshue, James Y.
Orlik-Rückemann, Kazimierz Jerzy
Orszag, Steven Alan, *Fluid mechanics*
Ostendorf, David William
Ostrach, Simon
Oved, Yoel
Owczarek, Jerzy Antoni
Panton, Ronald Lee
Patel, Vithal A.
Paul, John Francis
Peters, Alexander Robert
Phillips, Winfred Marshall
Phinney, Ralph E(dward)
Pletcher, Richard Harold
Plotkin, Allen
Powe, Ralph Elward
Probstein, Ronald Filmore
Proctor, Charles Lafayette, II
Reshotko, Eli
Reynolds, William Craig
Riegler, Gerold Ernst
Ritter, Alfred
Roshko, Anatol
Rouhani, Sayd Zia
Rovick, Allan Asher
Rudavsky, Alexander Bohdan
Schneider, Gerald Elmore
Sears, William Rees
Seebold, Otto Paul, Jr.
Seginer, Arnan
Sereny, Aron
Shapiro, Ascher Herman
Sheridan, Robert Emmett, Jr.
Sieracki, Leonard Mark
Siginer, Aydeniz
Sirignano, William Alfonso
Soo, Shao Lee
Stahl, Charles Drew
Street, Robert Lynnwood
Suzuki, Tateyuki
Szewczyk, Albin Anthony
Tai, Tsze Cheng
Tatom, Frank Buck
Thomas, John Howard
Turinsky, Paul Josef
Van Atta, Charles William
Van Dyke, Milton Denman
Viets, Hermann
Vigil, Manuel Gilbert
Viskanta, Raymond
Vogel, Steven
Walker, John Scott
Ward, David Aloysius
Warren, Walter Raymond, Jr.
Wehausen, John Vrooman
Werner, Christian Thor
White, Charles Olds
Whitfield, Jack Duane
Widnall, Sheila Evans
Williams, Peter MacLellan
Willmott, Andrew John
Wilson, Randall Joe
Wu, Jain-Ming
Wu, Steve
Zakkay, Victor H.
Zielinski, Paul Bernard

Mechanics, fracture
Adler, William Fred
Apostal, Michael Christopher
Burger, Christian Pieter
Caddell, Robert Macormac
Chen, Er-Ping
Chuang, Tze-jer
Coffin, Louis Fussell, Jr.
Dempsey, John Patrick
de Wit, Roland
Dunayevsky, Victor Arkady
Eiber, Robert James
Gallagher, Joseph Patrick
Hopkins, Stephen William
Hussain, Moayyed A.
Irwin, George Rankin
Jolles, Mitchell Ira
Jones, Douglas Linwood
Kanninen, Melvin Fred
Kapp, Joseph Alexander
Kobayashi, Albert S.
Landes, John David
Lange, Eugene Albert
Lewis, James Bryson
Liebowitz, Harold
Loss, Frank J.
May, Rodney Alan
Michener, John Russell
Nagar, Arvind Kumar
Needleman, Alan
Norris, Douglas Monroe
Pindera, Jerzy Tadeusz
Rolfe, Stanley Theodore
Saada, Adel Selim
Salama, Mamdouh M.
Sciammarella, Cesar Augusto
Sha, George Tzeng-Tsun
Smith, C. William
Smith, John Henry
Socie, Darrell Frederick
Stout, Robert Daniel
Sun, Chang-Tsen
Swedlow, Jerold Lindsay
Trumpler, William E.
Tseng, Ampere An-Pei
Witt, Fountain Joel
Wu, Chien Heng

Mechanics, solid
Adams, Donald Frederick
Adams, George G.
Altan, Taylan
Altiero, Nicholas James
Balsley, Lawrence Edward
Batra, Romesh C(hander)
Beadle, Charles Wilson
Bhushan, Bharat
Bogner, Fred Karl
Budiansky, Bernard
Burger, Christian Pieter
Caddell, Robert Macormac
Calder, Clarence Andrew
Chen, Er-Ping
Chen, Wai-Fah
Chiang, Fu-pen
Clifton, Rodney James
Conrad, Nicholas
Cozzarelli, Francis Anthony
de Richemond, Albert Leo
Dunayevsky, Victor Arkady
Durelli, August Joseph
Fabrikant, Valery Isaac
Foral, Ralph Francis
Freund, Lambert Ben
Gallagher, Richard Hugo
Gould, Phillip Louis
Hettche, Leroy R.
Hodge, Philip Gibson, Jr.
Hudson, Donald Ellis
Hutchings, William Frank
Jolles, Mitchell Ira
Jones, Orval Elmer
Kanninen, Melvin Fred
Kikuchi, Noboru
Knapp, Ronald Harrison
Kunin, Isaak A.
Lahoti, Goverdhan Das
Lakes, Roderic Stephen
Landes, John David
Langner, Carl Gottlieb
Laymon, Stephen Alan
Lee, Lawrence Hwa-Ni
Levinson, Mark
Lin, Tung-Hua
Liu, Wing Kam
Maier, Leo Robert, Jr.
Malvern, Lawrence Earl
Marscher, William Donnelly
McClure, Carl Kenneth
Mehrabadi, Morteza M(irzaie)
Min, Byung Kon
Mitson, Herbert Henry, Jr.
Nagar, Arvind Kumar
Naghdi, Paul Mansour
Narayana, Anand Deo
Needleman, Alan
Nikolai, Robert Joseph
Noor, Ahmed Khairy
Nordgren, Ronald Paul
Phillips, James Woodward
Popov, Egor Paul
Reiss, Robert
Rice, James Robert
Rinehart, John Sargent
Romesberg, Laverne Eugene
Rowlands, Robert Edward
Roylance, David Kaye
Sancaktar, Erol
Schey, John Anthony
Schile, Richard Douglas
Seksaria, Dinesh Chand
Senseny, Paul Edward
Shack, William John
Shaw, Milton Clayton
Smith, C. William
Starkey, Walter Leroy
Stoffer, Donald Carl
Subudhi, Manomohan
Sun, Chang-Tsen
Swedlow, Jerold Lindsay
Tandanand, Sathit
Tao, Li-Chung
Thakkar, Bharatkumar S.
Tseng, Ampere An-Pei
Vargas, John David
Vinogradov, Aleksandra M.
Weidlinger, Paul
Weng, George Jueng-cious
Whitney, James Martin
Widera, G.E.O.

Mechanics, theoretical and applied
Abramson, Hyman Norman
Achenbach, Jan Drewes
Adams, George G.
Ahmadi, Goodarz
Alexander, Leckie Frederick
Anand, Subhash C.
Apostal, Michael Christopher
Arkilic, Galip Mehmet
Barfield, Walter David
Baron, Melvin Leon
Batra, Romesh C(hander)
Benaroya, Haym
Benjamin, Roland John
Bogner, Fred Karl
Boley, Bruno Adrian
Bryson, Arthur Earl, Jr.
Budiansky, Bernard
Burns, Joseph Arthur
Carlson, Richard Merrill
Carson, James Matthew
Casperson, Richard L.
Caulk, David Allen
Chace, Milton
Chambliss, Joe Preston
Chattopadhyay, Somnath
Chow, Pao-Liu
Clifton, Rodney James
Cohen, Harley
Cowin, Stephen Corteen
Czernik, Daniel Edward
Czyzewski, Harry
Dasgupta, Aaron
Datta, Subhendu K(umar)
Dempsey, John Patrick
de Richemond, Albert Leo
Dick, William Allen
Dowell, Earl Hugh
Drucker, Daniel Charles
Dym, Clive Lionel
Engel, Peter Andras
Ericksen, Jerald Laverne
Fistedis, Stanley H.
Fitzgerald, Edwin Roger
Fleisig, Ross
Gerstle, Frank P., Jr.
Gill, Daniel Emmett
Gipson, Gary Steven
Goland, Martin
Graham, Frederick Mitchell
Grant, Donald Andrew
Haines, Charles Wills
Hall, William Joel
Handler, Robert Alphonse
Hefzy, Mohamed Samir W. M.
Hemp, Gene W(illard)
Herrmann, George
Ho, Chih-Ming
Hodge, Philip Gibson, Jr.
Howland, Robert Alden, Jr.
Huston, Ronald Lee
Johanson, Jerry Ray
Johnson, Carol William
Kaul, Maharaj Krishen
Kenig, M(arvin) Jerry
Kirik, Michael John
Kobayashi, Albert S.
Laheru, Ken Liem
Landweber, Louis
Latoza, Kenneth Charles
Leadon, Bernard Matthew
Lee, Jen-shih
Lee, Lawrence Hwa-Ni
Leipholz, Horst Hermann Eduard
Little, Robert Colby
Maatuk, Josef
Macke, Harry Jerry
Malosh, James Boyd
Man, Chi-Sing
Marscher, William Donnelly
McMeeking, Robert Maxwell
McNitt, Richard Paul
Mehrabadi, Morteza M(irzaie)
Merchant, Howard Carl
Miller, Richard Keith
Mindlin, Raymond David
Mitler, Henri Emmanuel
Mordfin, Leonard
Nachman, Arje
Naghdi, Paul Mansour
Nguyen, Vietchau
Norris, Douglas Monroe
Nyquist, Gerald Warren
Parker, Frank Wayne
Pindera, Jerzy Tadeusz
Pister, Karl Stark
Plunkett, Robert
Prasad, Birendra
Reiss, Robert
Reissner, Max Erich
Rosen, B. Walter
Rosenblueth, Emilio
Rubin, Sheldon
Saul, William Edward
Schorry, Robert Elmer
Sciammarella, Cesar Augusto
Seksaria, Dinesh Chand
Sethna, Patarasp Rustomji
Sidorowicz, Kenneth Joseph
Siginer, Aydeniz
Sikarskie, David Lawrence
Simitses, George John
Solberg, Ruell Floyd, Jr.
Soutas-Little, Robert William
Stevens, Karl Kent
Stoffer, Donald Carl
Subudhi, Manomohan
Suchora, Daniel Henry
Talapatra, Dipak Chandra
Tang, Ruen C.
Tomita, Nobuya
Triffet, Terry
Vasilakis, John Dimitri
Vaughan, David Sherwood
Viest, Ivan M(iroslav)
Wagner, John George
Wang, Tsuey Tang
Wells, William Terry
Werner, Christian Thor
Wilkes, Donald Fancher
Williams, James Henry, Jr.
Winter, J. Ronald
Wolf, Joseph Allen, Jr.
Wong, Julius Pan
Woodburn, Wilton Allen
Wu, Chien Heng
Wu, John Naichi
Wu, Steve
Yu, Yi-Yuan

Metallurgical. ***See also*** **MATERIALS SCIENCE.**
Agnihotri, Krishna Venktesh
Anderson, Stephen Clark
Aukrust, Egil
Bauman, Thomas Charles
Berardi, Matteo P.
Berg, Bengt Henrik
Binstock, Martin H.
Birkle, A(dolph) John
Bush, Spencer Harrison
Canonico, Domenic Andrew
Carroll, Dyer Edmund
Chang, Ji Young
Czyzewski, Harry
Dahlman, Geoffrey Edwin
Dahlstrom, Donald Albert
Dalgarno, Alexander
Dodd, Richard Arthur
Duerr, J. Stephen
Ebeling, Dolph George
Eck, Bernard John
Einziger, Robert Emanuel
Elliott, John Frank
Fiore, Nicholas Francis
Fischer, George J.
Foroulis, Z. Andrew A.
Fuerstenau, Douglas Winston
German, Randall M(ichael)
Gibson, George William
Glodowski, Robert John
Goins, William (Doris), III
Grace, Richard Edward
Heidersbach, Robert Henry, Jr.
Henderer, Willard E(verett), III
Higgins, Irwin Raymond
Kellogg, Herbert Humphrey
Khare, Ashok K.
Kirchner, Larry A.
Kittel, John Howard
Lane, Joseph Robert
Lange, Eugene Albert
Lautzenheiser, Clarence Eric
Lewis, Michael Dolan
Luerssen, Frank Wonson
McNitt, Richard Paul
Moudgil, Brij Mohan
Nahm, Alexander Hong
Niederkorn, Ioan Stefan
Patel, Bhikhubhai L.
Payne, Robert William
Perrin, James Stuart
Pettit, Frederick Sidney
Queneau, Paul Etienne
Robins, Norman Alan
Schwer, Roger Edwin
Shewmaker, Russell Newton
Shewmon, Paul Griffith
Spisak, John Francis
Stalker, Kenneth Walter
Stout, Robert Daniel
Wayne, Steven Falko
Weeks, John Randel, IV
Weng, George Jueng-cious

Nuclear. ***See also*** **ENERGY SCIENCE, Nuclear.**
Abrams, Richard Francis
Adamantiades, Achilles G.
Agee, Lance James
Alapour, Adel
Alcouffe, Raymond Edmond
Alexander, Dennis Jay
Amer, Ahmad (El Sayed)
Andrews, James Barclay, II
Apley, Walter Julius, Jr.
Arnold, William Howard
Asprey, Margaret Williams
Astley, Eugene Roy
Atkinson, Steven Albert
Attaya, Hosny M(oustafa)
Ayers, Arnold Leslie, Sr.
Baer, Thomas Strickland
Baker, Charles Clayton
Ballard, Charles Henry
Baratta, Anthony John
Barbehenn, Craig E(dwin)
Baron, Seymour
Barsky, Arnold M(ilton)
Bartine, David Elliott
Barton, Ronald Winston
Basehore, Kerry Lee
Bauer, Richard Carlton
Beakes, John Herbert
Beaton, Roy Howard
Behnke, Wallace Blanchard, Jr.
Benedetto, Anthony Richard
Bennett, George Alan
Berwald, David H.
Bettenhausen, Lee H.
Bevelacqua, Joseph John
Bezella, Winfred August
Bickel, John Henry
Bigeleisen, Jacob
Bigelow, John Ealy
Blomeke, John Otis
Bogert, Gary Michael
Bolch, William Emmett, Jr.
Booth, James Albert
Bournia, Anthony
Branco, Maria dos Milagres
Bray, Arthur Philip
Bridges, Donald Norris
Bridgman, Charles James
Brown, Steven Harry
Bullock, Scott Verne
Bunch, Wilbur Lyle
Burley, Elliott Lenhard
Burns, Robert Donald, III
Bush, Spencer Harrison
Cahill, William Joseph, Jr.
Callihan, Dixon
Campoy, Leonel Perez
Carbon, Max William
Card, Darrell Holder
Carelli, Mario Domenico
Carr, William Hoge, Jr.
Carter, Benjamin Dudley
Carter, Leland LaVelle
Chandler, John Christopher
Chen, Chih Ping
Chen, Shih-Yew
Cheng, Edward Teh-Chang
Chesworth, Robert Hadden
Chipman, Gordon Leigh, Jr.
Cho, Soung Moo
Choi, Seung Hoon
Chung, Chien
Clare, George Hadley
Clinard, Frank Welch, Jr.
Cogbill, Charles Lipscomb, III
Cohen, Karl Paley
Colby, Nathaniel Fred
Cole, Thomas Earle
Collins, Paul Francis
Conte, Richard Joseph
Coombe, John Raymond
Courtney, John Charles
Cranberg, Lawrence
Crawley, Paul F.
Crouthamel, Carl Eugene
Cummings, Garth Ellis
Cuttler, Jerry Milton
Dahlberg, Richard Craig
Dallman, John Clay
Dang, Vi Duong
Darby, John Littrell
Deen, James Robert
Defilippo, Felix Carlos
DeMott, Diana L(ynn)
Diamond, David Joseph
Dickerman, Charles Edward
Dickson, Paul Wesley, Jr.
Dimmick, David Michael
Dobson, Wayne Lawrence
Doerner, Robert Carl
Dolan, Linda Capano
Donis, Jose Maria
Draper, Ernest Linn, Jr.
Duffey, Dick
Duke, Winston Lavelle
Dunigan, Paul Francis Xavier
Dunn, Michael James
Du Temple, Octave Joseph
Dutton, James William
Ebert, Marlin J.
Eichholz, Geoffrey G.
Elleman, Thomas Smith
Eng, Norman
Fast, Edwin
Fellows, W(alter) Scott, Jr.
Fields, Carl Clarence
Fistedis, Stanley H.
Fleishman, Morton Robert
Flippen, Luther Daniel, Jr.
Fluegge, Ronald Marvin
Forkel, Curt Emil
Freehill, Peter Eugene
Freeman, Louis Barton
Friedland, Aaron J.
Fritsch, Arnold Rudolph
Fulford, Phillip James
Gaines, Albert Lowery
Gajewski, Walter Michael
Galbreth, Timothy Michael
Ganapol, Barry Douglas
Garrick, B. John
Garritson, Grant Richard
George, Thomas Carl
Ghoniem, Nasr Mostafa
Gibson, George William
Glower, Donald Duane
Gohar, Mohamed Yousry Ahmed
Golay, Michael Warren
Goode, Glenn Amos, Jr.
Goodman, Julius
Goodwin, Richard Clarke
Graves, Harvey Wilbur, Jr.
Greenman, Gregory Michael
Groeber, Edward Otto, Jr.
Grotenhuis, Marshall
Haake, Eugene Vincent
Haan, David Charles
Hager, Eugene Randolph
Haggag, Fahmy Mahmoud
Haire, Marvin Jonathan

Halverson, Thomas George
Hannum, William Hamilton
Hansen, Kent Forrest
Hanson, John Edward
Harbay, Edward William
Hard, James Ellsworth
Hatcher, S(tanley) Ronald
Haupt, H. James
Heath, Colin Arthur
Hemmings, Robert Leslie
Hendricks, John Stanley
Hendrickson, Waldemar Forrsel
Hendrie, Joseph Mallam
Hendron, John Alden
Hill, Ernest Elwood
Hill, Thomas Johnathan
Hilley, James Roger, Jr.
Hills, Willard Andrew
Hoffman, John Robert, Jr.
Hofmann, Peter Ludwig
Holzer, Joseph Mano
Hootman, Harry Edward
Hopkins, Harvey Childs, Jr.
Howerton, Robert James
Hsia, Henry Tao-sze
Hsu, Ming-Teh
Huang, Hai Chow
Hurd, Edward Nelson, III
Huston, Norman Earl
Hutchinson, Frank David, III
Ireland, John Richard
Jackson, Peter Sterling
Jacovitch, John
Jain, Parveen Kumar
Janzow, Edward Frank
Jedruch, Jacek
Jeffers, John Bryant
Johnson, Melvin Lawrence
Joksimovich, Vojin
Kadambi, Narasimha Prasad
Kai, Michael S.
Kalinauskas, Gediminas Leonardas
Karam, Ratib Abraham
Kazi, Abdul Halim
Kazimi, Mujid S.
Kelly, Kevin Anthony
Kennedy, Michael Francis
Kerr, William
Kim, Kyo Sool
Kim, Yong Su
Kimmons, George Harvey
Klein, Keith Allen
Klema, Ernest Donald
Kniazewycz, Bohdan George
Knoll, Glenn Frederick
Koch, Leonard John
Kolar, Oscar Clinton
Koskelo, Markku Juhani
Koussa, Harold Alan
Kouts, Herbert John Cecil
Kovach, Julius Louis
Kramer, Rex Williard, Jr.
Krug, Harry Everistus Peter, Jr.
Kwik, Robert Julius
Lake, James Alan
Larrimore, James Abbott
Lear, George Emory
Leary, Joseph Aloysius I
Lee, John Chaeseung
Lee, Robert Bongkyu
Lentsch, Jack Wayne
Lester, David Brent
Lester, Richard Keith
Levenson, Milton
Levin, George Benjamin
Levine, Jerry David
Levine, Melvin Mordecai
Levine, Samuel Harold
Lischer, Ludwig Frederick
Lorenzini, Paul Gilbert
Loyalka, Sudarshan K.
Lucas, Glenn Eugene
Lucier, Ronald David
Lumb, Ralph Francis
Lundin, Bruce Theodore
Lykoudis, Paul S.
Ma, Benjamin Minglee
MacPherson, Herbert Grenfell
Mahaffey, Michael Kent
Malin, Douglas Harwell
Mallay, James Francis
Malliakos, Asimios
Marchaterre, John Frederick
Martens, Frederick Hilbert
Martin, Jose Ginoris
Martin, Lawrence Ronald
Mason, Edward Archibald
McArthur, David Alexander
McArthur, Wilson Cooper
McCormick, Norman Joseph
McFarlane, Harold Finley
McGuire, Stephen Craig
McKlveen, John William
McKnight, Richard D.
Meem, James Lawrence, Jr.
Melese d'Hospital, Gilbert Bernard
Menlove, Howard Olsen
Mesler, Russell Bernard
Meyer, Walter
Michael, Eugene Joseph
Miller, Joseph S.
Monard, Joyce Anne
Morewitz, Harry Alan
Morita, Toshio
Morris, Charles Reginald
Morris, Robert Howard
Morrison, Roderick Gordon
Morton, Randall Eugene
Newman, Darrell Francis
Nichols, Paul Arthur
Niederkorn, Ioan Stefan
Okrent, David
Onorato, Howard Louis
Pagel, Deborah Joanne
Palladino, Nunzio Joseph
Pandey, Sudhakar
Park, U. Young
Parry, John O.
Parsly, Lewis Fuller
Pasqua, Pietro, F(ernando)
Pastorelle, Peter John
Patterson, Dwight Robert
Paustian, Harold Herman
Paxton, Hugh Campbell
Pearce, William Richard
Pearson, Samuel Dibble, III
Peddicord, Kenneth Lee
Perrin, James Stuart
Perry, Nelson Allen
Phelps, James Parkhurst
Piet, Steven James
Pigford, Thomas Harrington
Pizzica, Philip Andrew
Platt, Allison Michael
Plebuch, Richard Karl
Podowski, Michael Zbigniew
Pravda, Milton Frank
Prelas, Mark Anthony
Prieto, Robert
Profant, Richard Thomas, Jr.
Profio, A(medeus) Edward
Ragheb, Magdi
Ramanuja, Jayalakshmi
Krishnadesikachar
Rao, Surendar Purushothay
Redmond, Robert Francis
Reid, Robert William
Roberds, Richard Mack
Roberts, Albert Sidney, Jr.
Roe, Kenneth Andrew
Roe, Kenneth Keith
Rothrock, Ray Alan
Rouhani, Sayd Zia
Rust, James Harold
Sanders, Arthur
Sarram, Mehdi
Savage, William Frederick
Schaeffer, Norman M.
Schaub, John Robert
Schilmoeller, Neil Herman
Schoerer, Frank
Schuman, Robert Paul
Scotti, Vincent Guy
Seale, Robert Lewis
Seebold, Otto Paul, Jr.
Serdula, Kenneth James
Sheff, James Robert
Shieh, Paulinus Shee-Shan
Shih, Charles Chien
Sholtis, Joseph Arnold, Jr.
Sideris, Antonios George
Siegel, Joel Marvin
Sienicki, James Joseph
Siess, Chester Paul
Silverman, Joseph
Simon, Albert
Simonson, Simon Christian, III
Snyder, Thoma Mees van't Hoff
Somsel, Joseph Kent
Speis, Themis P.
Spellman, Donald Jerome
Stergakos, Elias P(anagiotes)
Stewart, James Edmund, III
Stiefel, John T.
Storrs, Charles L.
Strasser, Alfred Anthony
Teed, Douglas Earle
Thompson, Dudley
Thompson, Loren B.
Tomlinson, Richard Lee
Traiforos, Spyros Anthony
Uhrig, Robert Eugene
Varley, Ronald Arthur
Vaurio, Jussi Kalervo
Wagner, Aubrey Joseph
Wagner, Stephen Gregory
Ward, David Aloysius
Warren, Holland Douglas
Way, Richard A(lvord)
Wei, Jim P(iau)
Weinfurter, Erich Brian
Weisman, Joel
Wells, Alan Harvey
Wenz, Michael Frank, Jr.
West, John Merle
Wiley, Albert Lee, Jr.
Williams, Alan Keiser
Williams, Raymond Warren
Williamson, Thomas Garnett
Willoughby, Nancy Bharucha
Wiren, Robert Craig
Wong, Clement Po-Ching
Woodlan, Donald Ray
Woolson, William Andrew
Yoshikawa, Herbert Hiroshi
Yue, Mike Yuan
Yung, Shu-Chien

Operations research. *See also* **MATHEMATICS.**
Aleshire, Merle J.
Boodman, David Morris
Burgess, John Henry
Carberry, James Joseph
Cate, Martin Edward
Cullen, Daniel Edward
Daly, John Anthony
Datz, I. Mortimer
Daubert, Raymond Leo
Davis, Joel Stephen
Dube, Rajesh
Eaton, David J.
Eaves, Burchet Curtis
Fang, Shu-Cherng
Farr, William Rogers
Ferchek, Gary Randall
Frank, Ellen Ryan
Gibson, John Egan
Graff, Samuel M.
Gross, Donald
Hall, William Franklin, Jr.
Hedstrom, Joseph Charles
Holtzman, Samuel
Horvath, William John
Hwang, William G.
Jewell, William Sylvester
Jones, James Thomas, Jr.
Kapur, Kailash Chander
Karroll, Joseph E.
Klein, Cerry M.
Kleinrock, Leonard
Kocaoglu, Dundar F.
Kushner, Harvey David
LeDoux, Chris Bob
Lee, Sang Moon
Lee, William Wai Lim
Levary, Reuven Robert
Liberatore, Matthew John
Liebman, Jon Charles
Lowe, Timothy Joe
Luus, Rein
Lynn, Walter Royal
Magnanti, Thomas Lee
McGarity, Arthur Edwin
Meier, Wilbur Leroy, Jr.
Merdith, Orsell Montgomery
Miller, Bernard Paul
Mulvey, John Michael
Nieh, Bill
Pang, Jong-Shi
Pesch, William Allan
Raafat, Feraidoon
Ruckle, William Henry
Saigal, Romesh
Samuel, Aryeh Hermann Albert
Sargent, Robert George
Shea, Daniel Joseph
Shea, John R., III
Stanford, Robert Ernest
Steinberg, Frederick
Verry, William Robert
Walton, C. Michael
Weiss, Howard Jacob
Wood, Allen John
Wyskida, Richard Martin
Yakin, Mustafa Zafer

Optical. *See also* **OPTICS.**
Abbott, Fred
Axelrod, Norman Nathan
Brown, Gordon M(arshall)
Church, Eugene Lent
Crawford, David Livingston
Devereux, William Patrick
Dickson, LeRoy David
Dube, George
Dumas, Herbert Monroe, (Jr.)
Dunn, Anne Roberts
Durelli, August Joseph
Fein, Michael E.
Feldman, Martin
Gallo, Charles Francis, Jr.
Gelles, Rubin
Gill, Dennis Howard
Gould, Gordon
Gundersen, Martin Adolph
Hammond, Thomas Joseph
Hansche, Bruce David
Hilgeman, Theodore William
Hines, Robin Hinton
Johnson, Peter Dexter
Kinzly, Robert Edward
Kornstein, Edward
Lamberts, Robert Lewis
Liu, Hua-Kuang
Looft, Donald John
Marshall, Gerald Francis
Miles, Richard Bryant
Nutter, Gene Douglas
Post, Madison John
Silverstein, Elliot Morton
Skelton, Dennis Lee
Sollid, Jon Erik
Spoelhof, Charles P.
Thompson, Brian John
Thompson, James Elton
Walter, William Trump
Wang, Jon Yi
Wilkerson, Gary Ward
Withrington, Roger John
Young, James Forrest

Petroleum. *See also* **GEOSCIENCE, Oceanography; ENERGY SCIENCE, Fuels.**
Acheson, Willard P(hillips)
Ayoub, George Tanios
Brannon, H(ezzie) Raymond, Jr.
Cadden, James Monroe
Chilingarian, George Varos
Clark, Brian Oliver
Coats, Douglas A.
Coburn, Herbert Dightman, Jr.
Conrad, Nicholas
Coonts, Harvey Lee
Curtis, Christopher Michael
Dickson, Philip F.
Dorfman, Myron Herbert
Ewing, Richard Edward
Ferer, Kenneth Michael
Fisher, James Harold
Frey, Elmer J(acob)
Garb, Forrest Allan
Geer, Ronald Lamar
Glimm, James Gilbert
Goldsberry, Fred Lynn
Gray, Kenneth Eugene
Greenkorn, Robert Albert
Halbouty, Michel Thomas
Handy, Lyman Lee
Hettinger, William Peter, Jr.
Huff, Kenneth O.
Jahns, Hans O(tto)
Kadi, Kamal Sif-el
Leventhal, Stephen Henry
Mancini, Ernest Anthony
Marsden, Sullivan Samuel, Jr.
Naftchi, Nosrat Eric
Nordgren, Ronald Paul
Peaceman, Donald William
Perrenod, Stephen Charles
Poettmann, Fred Heinz
Puchyr, Peter Joseph
Russell, Donald Glenn
Sammon, Peter
Schubel, Jerry Robert
Smith, Joe Mauk
Stahl, Charles Drew
Widdoes, Lawrence Curtis
Zais, Elliot Jacob

Plasma. *See also* **ENERGY SCIENCE.**
Balwanz, William Walter
Fortna, John David Edward
Kazek, Gregory Joseph
Kessler, Robert
Kim, Kyekyoon
Kruger, Charles Herman, Jr.
Mense, Allan Tate
Seidl, Milos

Polymer
Cohen, Robert Edward
Dao, Kim C.
Elias, Hans Georg
Hoffman, Dennis Mark
Kumar, Ganesh Narayanan
Lednicky, Raymond Anthony
Lyon, Richard Evan
Macosko, Christopher Ward
Paul, Donald Ross
Peppas, Nikolaos Athanassiou
Plazek, Donald John
Pochan, John Michael
Wang, Hsin-Pang

Robotics
Angel, Thomas Michael
Bollinger, John Gustave
Bosy, Brian Joseph
Brockett, Roger Ware
Cwycyshyn, Walter
Daubert, Raymond Leo
ElGomayel, Joseph Ibrahim
Engelberger, Joseph Frederick
Fodor, Magda Maria
Franklin, Robert Fraser, Jr.
Gelles, Rubin
Hanifin, Leo Eugene
Hawkes, Graham Sidney
Hopcroft, John Edward
Hull, Harvard Leslie
Jawad, Sarim Naji
Johnson, (Charles) Bruce
Kalman, Rudolf Emil
Kraus, Julian (Joe) David
Lytle, James Mark
McClamroch, N. Harris
McKee, Keith Earl
Mergler, Harry Winston
Miluschewa, Sima
Moore, Michael Hart
Polcyn, Stanley Joseph
Rechnitzer, Andreas Buchwald
Rice, Stephen Landon
Robinson, James Robert
Sangrey, Dwight A.
Siegel, Melvin Walter
Simon, Richard L.
Stone, Gregory Michael
Thiel, Leo Albert
Thompson, Jack Mansfield, Jr.
Tiras, Herbert Gerald
Vesel, Richard Warren
Volz, Richard
Wagner-Bartak, Claus Gunther
Johann
Warnat, Winifred Irene
Will, Peter Milne
Wright, Paul Kenneth

Systems engineering
Adkisson, William Milton
Andrews, Mary Lou
Attinger, Ernst Otto
Austin, Edward Marvin
Baillieul, John Brouard
Baird, Samuel Dempsey
Baumann, Robert Coile
Bayer, Jesse Abraham
Bennett, Bruce Alan
Betts, Austin Wortham
Beutler, Frederick Joseph
Bhattacharyya, Shankar
Bigler, William C.
Bishop, Albert Bentley, III
Bloor, W(illiam) Spencer
Bollinger, John Gustave
Brown, Burton Primrose
Burnham, Donald Clemens
Bussgang, Julian Jakob
Cheng, David
Chestnut, Harold
Christ, Duane Marland
Clark, Richard LeFors
Cohen, Edwin
Cohn, Nathan
Concordia, Charles
Conley, Carolynn Lee
Cordero, Julio
Cruz, Jose Bejar
Cullen, Donald Lee
Curtis, George Darwin
Darlington, Sidney
Datz, I. Mortimer
Diedrichsen, Loren Dale
Drozd, Andrew Louis Stephan
Dunn, Michael James
Farber, Joseph
Farinola, Anthony Larry
Fish, Andrew Joseph, Jr.
Fletcher, James Chipman
Forbes, Judith
Forsee, Aylesa
Frey, Elmer J(acob)
Gaither, William Samuel
Garg, Devendra Prakash
Garzia, Mario Ricardo
Gault, Charles S.
Gerhardt, Lester A.
Gillette, Dean
Goel, Narendra Swarup
Goldschmied, Fabio Renzo
Gollobin, Leonard Paul
Goodhue, William Lehr
Goodson, Raymond Eugene
Gouse, S. William, Jr.
Graupe, Daniel
Gravitz, Sidney I.
Guastaferro, Angelo
Halpin, Daniel William
Hawkes, Graham Sidney
Hoag, David Garratt
Hoffman, Kenneth Charles
Hoover, L. John
Huang, Richard Shih-Chiu
Hughes, William Perry
Jaluria, Rajiv
Jameson, William James, Jr.
Jones, Daniel Todd
Jones, James W.
Jones, Orval Elmer
Kahne, Stephen James
Karroll, Joseph E.
Keehn, Neil Francis
Keil, Alfred Adolf Heinrich
Kelley, Albert Joseph
Kinzly, Robert Edward
Knowles, Lloyd George
Kraus, Julian (Joe) David
Kueishiong, Tu
Kushner, Harvey David
Kutlik, Roy Lester
Lamm, (August) Uno
Landgren, John Jeffrey
Lautzenheiser, Clarence Eric
Leber, Ralph Eric
LeDoux, Chris Bob
Lincoln, Walter Butler, Jr.
Loeb, William A.
Lucier, Ronald David
Marcus, Steve Irl
Markus, Lawrence
Martin, James Arthur
McClamroch, N. Harris
McGean, Thomas James
Mertens, Lawrence Edwin
Meyer, Andrew U(lrich)
Miller, Bernard Paul
Morton, Randall Eugene
Mumma, Albert G.
Nash, Jonathon Michael
Olson, Donald Richard
O'Neill, Russell Richard
Parker, Norman Francis
Peerenboom, James Peter
Plebuch, Richard Karl
Porter, Alan Leslie
Purser, Paul Emil
Rath, Robert Michael
Richardson, Herbert Heath
Roberts, Edward Baer
Ross, Ian Munro
Rotz, C. Alan
Sage, Andrew Patrick, Jr.
Salvatori, Vincent Louis
Sargent, Ernest Douglas
Schaefer, Jacob Wernli
Schuetz, Cary Edward
Seifer, Arnold David
Shalleck, Alan Bennett
Shea, Joseph Francis
Shelley, Edwin Freeman
Shoup, William David
Silverman, Barry George
Silvern, Leonard Charles
Silvestri, George Joseph, Jr.
Skalafuris, Angelo James
Snyder, Glenn Jacob
Sorenson, Harold Wayne
Spoelhof, Charles P.
Staehle, Charles Michael
Steinberg, Frederick
Stumpe, Warren Robert
Summers, George Donald
Sumney, Larry W.
Taylor, Malcolm Ernest
Todd, Deborah Ann
Verry, William Robert
Wagner, Lorry Yale
Wagner, Raymond Lee
Warfield, John Nelson
Warzecha, Ladislaus William
Washom, Byron John
Williams, Theodore Joseph
Wolfe, Bradley Allen
Wonham, Walter Murray
Wray, John Lawrence
Zachariah, Gerald LeRoy
Zeissig, Gustave Alexander

Other
Atwood, Charles LeRoy, *Architectural computer-aided design*
Baciocco, Albert Joseph, Jr., *Research administration*
Baird, Samuel Dempsey, *Sensor research and design*
Beale, William Taylor, *Stirling engines*
Blasingame, Benjamin Paul, *Aeronutical instrumentation*
Blevis, Bertram Charles, *Communications*
Boundy, Ray Harold, *International research management*
Bull, Stanley Raymond, *Engineering physics*
Bullock, Robert Morton, III, *Audio engineering*
Burke, Doyle, *Forestry engineering*
Cain, Charles Alan, *bioengineering*
Carcaterra, Thomas, *Structural engineering*

Cartner, John Aubrey, *Naval architecture and marine engineering*
Champlin, Keith S(chaffner), *Microwave electronics*
Charyk, Joseph Vincent, *Telecommunications*
Cook, Charles J., *Research Management, Technology transfer*
Cowpland, Michael Christopher John, *Telecommunications management*
Crombie, Douglass Darnill, *Telecommunications Systems*
Davenport, Alan Garnett, *Structural engineering*
Deissler, Robert George, *Heat Transfer*
De Lauer, Richard D., *Research engineering administration*
De Volpi, Alexander, *Instrumentation*
Donahue, Francis Martin, *Electrochemical engineering*
Duffy, Robert Aloysius, *Research Administration*
Edwards, Ray Conway, *Heat transfer*
Ellingwood, Bruce Russell, *Structural reliability*
Eyden, Bernard, *Hydroelectric Power*
Faccini, Ernest Carlo, *Explosives phenomena*
Fairhurst, Charles, *Mining engineering*
Fenton, Noel John, *Electronics company management*
Fisher, Ray W., *Mining engineering*
Flaschen, Steward Samuel, *Research and development managent*
Friedman, Edward Alan, *Technological Education*
Fung, Yuan-Cheng Bertram, *Biomechanics*
German, Richard Barry, *Mining engineering*
Goodson, Raymond Eugene, *Transportation engineering*
Gould, Lawrence, *Telecommunications Management*
Grace, Donald J., *Engineering research administration*
Gratt, Lawrence Barry, *Risk analysis*
Greenberg, Harold Paul, *Quality engineering*
Grenier, Edward Joseph, *Quality control*
Hager, Eugene Randolph, *Remote handling*
Hamilton, Angus Cameron, *Surveying engineering*
Hamilton, Bruce King, *Biochemical engineering*
Hanifin, Leo Eugene, *Manufacturing productivity*
Heerwagen, Dean Reese, *Heat transfer*
Helliwell, Robert Arthur, *Radioscience*
Herbich, John Bronislaw, *Coastal Engineering, Ocean Engineering*
Hilbertz, Wolf Hartmut, *Marine Architecture*
Hirschhorn, Joel S(tephen), *Science and technology policy*
Hoff, Marcian Edward, Jr., *Electronics industry research and development management.*
Holloway, John Thomas, *Radioactive Waste Management*
Hudson, Donald Ellis, *Earthquake Engineering*
Hwang, William G., *Communications systems engineering*
Jacobs, Joseph Donovan, *Construction Engineering*
Jarrett, Noel, *Electrochemical Engineering*
Johnson, (Charles) Bruce, *Photoelectronics*
Johnston, Bruce Gilbert, *Structural Engineering*
Jordan, Angel Goni, *Engineering education administration*
Jordan, Arthur Kent, *Inversion methods*
Jordan, Charles Ralph, *Naval engineering*
Kazi, Abdul Halim, *Radiation effects*
Keiser, Bernhard Edward, *Telecommunications*
Killian, James R., Jr., *Engineering administration*
Kirk, James Allen, *Automobile accident reconstruction*
Kocaoglu, Dundar F., *Engineering management*
Lamarre, Pierre, *Hydroelectric engineering*
Leidersdorf, Craig B., *Coastal engineering*
Levin, Alan Edward, *Heat transfer and fluid dynamics*
Lupo, Michael Vincent, *Reliability and maintainability*
Lyons, John W(inship), *Engineering research administration*
Manly, Philip James, *Radiological Engineering*
Marash, Stanley Albert, *Quality assurance*
May, Rodney Alan, *Applied mechanics*
Mc Cune, William James, Jr., *Manufacturing, engineering administration*
McKay, Kenneth G(ardiner), *Telecommunications*
McKlveen, John William, *Radiation Measurement*
Mc Lucas, John Luther, *Telecommunications company management*
Mettler, Ruben Frederick, *Electronics, engineering management*
Morgan, Jack Brandon, *Nondestructive Testing*
Nesbit, Richard Allison, *Scientific instruments research and development management*
Nuzzo, Salvatore Joseph, *Electronics company management*
O'Brien, Kenneth Stanley, *Quality assurance*
O'Neill, Eugene Francis, *Communication engineering*
Parkola, Walter R(obert), *Ink jet printing*
Paulson, Boyd Colton, Jr., *Construction engineering*
Peach, Robert Westly, *Marine engineering*
Perkins, Glenn Richard, *Non-destructive testing*
Porter, Alan Leslie, *Technology forecastingand assessment*
Ramo, Simon, *Engineering management*
Rasor, Ned Shaurer, *Engineering physics*
Ratliff, Thomas A., Jr., *Quality control*
Roberts, Edward Baer, *Management of technology*
Robertson, Leslie Earl, *Structural engineering*
Roehrs, Robert Jesse, *Nondestructive testing*
Rosa, David, *Plastics engineering*
Ross, Douglas Taylor, *Software engineering*
Rubin, Sheldon, *Shock and vibration*
Saperstein, Lee Waldo, *Mining engineering*
Saville, Thorndike, Jr., *Coastal, port, harbor engineering*
Schmitt, Roland Walter, *Research management*
Schoech, William Joseph, *Manufacturing engineering*
Schuetz, Cary Edward, *Thermophysics engineering*
Sebo, Stephen Andrew, *High voltage*
Shock, D'Arcy Adriance, *Solution Mining*
Slemon, Gordon Richard, *Electrical Propulsion*
Smith, Wilbur Stevenson, *Transportation Engineering*
Sorenson, Harold Wayne, *Control/ Feedback Systems*
Speis, Themis P., *Nuclear power plant risk assessment.*
Spergel, Philip, *Quality assurance*
Sterling, Raymond Leslie, *Underground engineering*
Summers, David Archibold, *Mining engineering*
Swanson, David Henry, *Technology transfer administration*
Throner, Guy Charles, *Ordnance technology*
Todd, Deborah Ann, *Power Engineering*
Utlaut, William Frederick, *Radio spectrum, Telecommunication systems*
Vanderslice, Thomas Aquinas, *Electrical research; manufacturing management*
Viswanathan, K., *material handling*
Wahren, Douglas, *Paper Technology*
Weiblen, William Achorn, *Product cost reduction*
Wenk, Edward, Jr., *Technology assessment*
Woods, Raymond Francis, *Quality assurance engineering*
Yee, Alfred Alphonse, *Structural engineering*

ENVIRONMENTAL SCIENCE. *See also* ATMOSPHERIC SCIENCE.

Ecology
Ahlgren, Clifford Elmer
Aldrich, Richard John
Almodovar, Luis Raul
Anderson, Roger Clark
Armstrong, Neal Earl
Askins, Robert Arthur
Baker, Herbert George
Bamforth, Stuart Shoosmith
Barbour, Michael George
Barrows, Edward Myron
Becker, Donald August
Becking, Rudolf Willem
Beeton, Alfred Merle
Bellis, Edward David
Benninghoff, Anne Stevenson
Bentley, Barbara Lee
Berry, James Frederick
Bierzychudek, Paulette Francine
Bigford, Thomas Edward
Bjorkman, Olle Erik
Blair, William Franklin
Bliss, Lawrence Carroll
Breckon, Gary John
Brock, Thomas Dale
Bucklin, Ann Cone
Burr, Brooks Milo
Case, Ronald Mark
Cates, Rex Gordon
Chittenden, Mark Eustace, Jr.
Cobb, Fields White, Jr.
Colinvaux, Paul Alfred
Coney, Charles Clifton
Cooper, Arthur Wells
Cotter, David James
Coull, Bruce Charles
Craig, Robert Bruce
Crites, John Lee
Crossley, D.A., Jr.
Cunningham, Harry Norman, Jr.
Curry, Mary Grace
Cutler, Irving
Dame, Richard Franklin
Day, Frank P., Jr.
Deacon, James Everett
Deevey, Edward Smith, Jr.
DeSelm, Henry (Hal) Rawie
Diamond, Jared Mason
Disraeli, Donald Jay
Drapalik, Donald Joseph
Earle, Sylvia Alice
Edgar, Arlan Lee
Edwards, William Charles
Ehrlich, Paul Ralph
Eisner, Thomas
Eleuterius, Lionel Numa
Erwin, Terry Lee
Faaborg, John Raynor
Ferchau, Hugo Alfred
Fisher, Stuart Gordon
Franz, Eldon Henry
Friedmann, E(merich) Imre
Frost, Melvin Jesse
Futuyma, Douglas Joel
Gammon, James Robert
Gara, Robert Imre
Gates, David Murray
Gavini, Mural B.
Glanz, William Edward
Golley, Frank Benjamin
Graham, Harry Morgan
Grant, Michael Clarence
Hanebrink, Earl Lee
Hanson, Joe Allan
Harper, Kimball Taylor
Harris, William Franklin, III
Hasler, Arthur Davis
Hilbert, Morton Shelly
Hill, Jane Foster
Hobbie, John Eyres
Holland, Richard Darlan
Holloway, Harry Lee, Jr.
Houston, David Royce
Howe, Henry Franklin
Hulbert, Lloyd Clair
Hull, Clark Ramsey
Hurt, Valina Kay
Ibrahim, Shawki Amin
Ilg, Ronald Jon
Inouye, David William
Iverson, Louis Robert
Johnson, Howard Ernest
Johnson, Philip Lewis
Jones, Walter Allan
Kaplan, Eugene Herbert
Karlson, Ronald Henry
Keeley, Sterling Carter
Kerstetter, Rex Eugene
Kevan, Peter Graham
Kistner, David Harold
Kiviat, Erik
Kohn, Alan Jacobs
Koplin, James Ray
Kunz, Thomas Henry
Kuserk, Frank Thomas
Lago, Paul Keith
Landis, Wayne G.
Lees, Lester
Leffler, John Warren
Legler, John Marshall
Leigh, Thomas Francis
Leopold, Aldo Starker
Leopold, Estella Bergere
Levin, Michael Howard
Likens, Gene Elden
Linn, D. Wayne
Little, Elbert Luther, Jr.
Liu, Edwin H.
Lutz, Richard Arthur
Main, Stephen Paul
Marion, Wayne Richard
Marten, Gordon Cornelius
Mason, Charles Perry
Mathieson, Arthur Curtis
Maughan, O. Eugene
McCrain, Gerald Ray
McIntosh, Robert Patrick
McNaughton, Samuel Joseph
Medve, Richard John
Meijer, Willem
Mertz, David Byron
Miller, Lee W.
Moll, Don L.
Mooney, Harold Alfred
Moore, Allen Murdoch
Mullin, Michael Mahlon
Nickerson, Norton Hart
Nixon, Scott West
Nobel, Park S.
Norden, Carroll Raymond
Norton, Don Carlos
Oliver, Kelly Hoyet, Jr.
Pamatmat, Mario Macalalag
Parker, William Skinker
Pearson, William Dean
Perry, David Anthony
Peterle, Tony John
Phoel, William C.
Pienkowski, Robert L(ouis)
Pieper, Rex Delane
Pimentel, David
Pitelka, Louis Frank
Pomeroy, Lawrence Richards
Prezelin, Barbara Berntsen
Quinn, James Amos, Jr.
Redfearn, Paul Leslie, Jr.
Reeves, Fontaine Brent
Ruibal, Rodolfo
Runkle, James Reade
Ryan, Roger Baker
St.Clair, Larry Lee
Salamun, Peter J(oseph)
Sanchez, Pedro Antonio
Sanderson, H. Reed
Seliger, Howard H.
Sharp, Aaron John
Sheldon, Joseph Kenneth
Singer, Robert
Smith, Clifford Winston
Sparling, Donald Wesley
Stalter, Richard
Steingraeber, David Allen
Stern, Daniel Henry
Strader, Herman Lee
Sudia, Theodore William
Summerfelt, Robert Clar
Thomson, Donald Arthur
Tiffney, Wesley Newell, Jr.
Ungar, Irwin A.
Vance, B(enjamin) Dwain
Vickery, Robert Kingston, Jr.
Vishniac, Helen Simpson
Wali, Mohan Kishen
Wang, Chun-Juan K.
Wang, Deane
Webb, Kenneth Louis
Weber, James Alan
Wethey, David Sunderland
Wheeler, Alfred George, Jr.
Wilson, Kenneth Sheridan
Wodehouse, Edmund Berkeley
Wolf, Larry Louis
Woodin, Sarah Ann
Wyckoff, John Wynn
Yeargan, Kenneth Vernon
Yorks, Pamela Florence
Zar, Jerrold H(oward)
Zeveloff, Samuel Ira

Ecology, behavioral
Adler, Kraig (Kerr)
Alexander, James L.
Atema, Jelle
Barlow, George Webber
Beason, Robert Curtis
Christian, John Jermyn
Collins, Margaret Strickland
Colson, Elizabeth Florence
Crowder, Larry B.
Dingle, Richard Douglas Hugh
Ellison, William Theodore
Engel, John William
Faaborg, John Raynor
Finck, Elmer John
Grove, Patricia Ann
Gruber, Samuel Harvey
Herrnkind, William Frank
Jaco, E. Gartly
Jaeger, Robert Gordon
Khalaf, Kamel Toma
Kunz, Thomas Henry
Lewis, Allen Rogers
Lindquist, David Gregory
McKown, Cora F.
Rebach, Steve
Rettenmeyer, Carl William
Rubinoff, Ira
Sheehy, Ronald J.
Sturges, Franklin Wright
Thomson, Donald Arthur
Uetz, George William
Waring, George Houstoun
Whitesell, James Judd
Wolf, Larry Louis

Ecology, theoretical
Caswell, Hal
Cushing, Jim Michael
Dolph, Gary Edward
Elliott, Dana Ray
Evenson, William Edwin
Ginzburg, Lev
Golley, Frank Benjamin
Jeffries, Harry Perry
Karr, James Richard
Lees, Lester
Leigh, Egbert Giles, Jr.
Maughan, O. Eugene
Miller, Lee W.
Stearns, Stephen Curtis
Van Devender, Robert Wayne
Waltman, Paul Elvis
Willig, Michael Robert

Ecosystems analysis. *See also* AGRICULTURE, Integrated systems.
Abbott, Robinson Shewell
Agnew, Douglas Craig
Alder, Guy Michael
Auer, Martin Tucker
Bensen, David Warren
Bliss, Lawrence Carroll
Carpenter, Richard Amon
Crossley, D.A., Jr.
Dame, Richard Franklin
Day, Frank P., Jr.
DeMeo, V(incent) James, Jr.
Duke, James Alan
Elliott, Dana Ray
Fisher, Stuart Gordon
Gammon, James Robert
Giddings, John Calvin
Grob, Robert Lee
Hanson, Joe Allan
Harris, William Franklin, III
Hart, Elwood Roy
Irvine, Eileen M.
Jespersen, Neil David
Jordan, Carl Frederick
Kennedy, Joseph Lane
Lane, Dennis Del
Lauenroth, William Karl
Likens, Gene Elden
Lim, Daniel V.
Ling, Joseph Tso-Ti
McCafferty, William Patrick
McFee, William Warren
McNaughton, Samuel Joseph
Miller, Raymond Michael
Moore, Allen Murdoch
Nixon, Scott West
Parker, Robert Hallett
Pearson, Lorentz C.
Perry, David Anthony
Pieper, Rex Delane
Riker, Joseph Thaddeus, III
Rogers, Hugo Homer, Jr.
Simons, Roy Kenneth
Switzer, George Lester
Wali, Mohan Kishen
Wang, Deane
Winstead, Joe Everett

Environmental toxicology
Abou-Donia, Mohamed Bahie
Amirkhanian, John David
Anderson, Robert Simpers
Ansari, Aftab A.
Beck, Barbara Doris
Bradford, Mark Lee
Brown, Steven Harry
Cawley, Charles Nash
Chanana, Arjun Dev
Chandler, Jerry LeRoy
Chang, Louis Wai-Wah
Cheng, H(wei)-H (sien)
Chowdhury, Parimal
Christian, John Edward
Claxton, Larry Davis
Coleman, Ronald L.
Couse, Nancy Lee
Dalvi, Ramesh R.
DeLucia, Anthony John
Edwards, Gordon Stuart
Eisenbud, Merril
Eisinger, Josef
Fouts, James Ralph
Freed, Virgil Haven
Fromm, Paul Oliver
Goldberg, Alan Marvin
Goldman, Marvin
Greenberg, Stephen Robert
Gurtoo, Hira L.
Hatch, Frederick Tasker
Heck, Walter Webb
Heggestad, Howard Edwin
Higginson, John
Hook, Jerry Bruce
Hornig, Donald Frederick
Jeffrey, Alan Miles
Johnson, Howard Ernest
Jones, Maurice (Mo), Jr.
Karnosky, David Frank
Kilburn, Kaye Hatch
Koschier, Francis Joseph, III
Kress, Lance Whitaker
Kuschner, Marvin
Landis, Wayne G.
Lasley, Stephen Michael
Leffler, John Warren
Leone, Ida A.
Lesko, Stephen Albert
Levinskas, George Joseph
Loper, John Carey
Mactutus, Charles Francis
Malins, Donald Clive
McClellan, Roger Orville
Mehendale, Harihara Mahadeva
Moore, Martha May
Neel, James Van Gundia
Nigg, Herbert Nicholas
Oberdorster, Gunter
Oliver, Kelly Hoyet, Jr.
Owen, David Gray
Paladino, Frank Vincent
Paul, Robert William, Jr.
Perrin, Eugene Victor Debs
Peterle, Tony John
Phelps, Harriette Longacre
Pilson, Michael Edward Quinton
Pitkow, Howard Spencer
Rattner, Barnett Alvin
Rech, Richard Howard
Reddy, Chinthamani Channa
Reilly, Christopher Aloysius, Jr.
Roales, Robert R.
Roggli, Victor Louis
Roth, Robert Andrew, Jr.
Sabri, Mohammad Ibrahim
Schneiderman, Marvin Arthur
Schoeny, Rita Sue
Shaikh, Zahir Ahmad
Shapiro, Irving Meyer
Simmon, Vincent Fowler
Singh, Harpal
Singh, Jarnall
Stasiak, Roger Stanley
Stern, Daniel Henry
Stevens, Reggie Harrison
Stotzky, Guenther
Sutton, Harry Eldon
Warren, Guylyn Rea
Wayman, Cooper Harry
Wedig, John H.
Weiss, Harold Samuel
Winterlin, Wray LaVerne
Witherspoon, John Pinkney, Jr.
Yang, Shen Kwei
Yu, Simon Shyi-Jian
Zook, Bernard Charles

Gas cleaning systems
Abrams, Richard Francis
Brook, Marx
Burchsted, Clifford Arnold
Carlton, Donald Morrill
Hall, Herbert Joseph
Kovach, Julius Louis
Meyer, Howard Stuart
Moeller, Dade William

Moore, William Walter
Swenson, Donald Otis
Thompson, Benny Louis

Resource management
Abbott, Robinson Shewell
Agnew, Douglas Craig
Beller, William Stern
Bertuglia, Lynn Ellen
Bier, Charles James
Bigford, Thomas Edward
Bisson, Robert Anthony
Black, James Francis
Bowersox, Todd Wilson
Boyd, James
Bradley, Katharine Tryon
Bradley, Michael Douglas
Brining, Dennis Lee
Burroughs, Richard H., III
Carpenter, Richard Amon
Childs, Joseph Edwin
Dahl, Billie Eugene
Dawson, James Clifford
Deacon, James Everett
Dilmore, Roger H.
Ekberg, Donald Roy
Enk, Gordon A.
Everhart, Watson Harry
Franz, Eldon Henry
Frost, Melvin Jesse
Gatherum, Gordon Elwood
Gullion, Gordon Wright
Hamon, Danny Joe
Harper, Kimball Taylor
Hart, John Henderson
Hellmann, Robert A.
Hodgson, Gordon Wesley
Hurt, Valina Kay
Ilg, Ronald Jon
Jacoby, Henry Donnan
Jayne, Benjamin Anderson
Johnson, Howard A(rthur)
Karr, James Richard
Kennedy, Joseph Lane
Kiviat, Erik
Kukin, Ira
Lauenroth, William Karl
Lee, William Wai Lim
Leopold, Aldo Starker
Levary, Reuven Robert
Marion, Wayne Richard
Marsh, William Michael
Marx, Donald Henry
Mason, John Montgomery, Jr.
McCrain, Gerald Ray
McDaniel, Kirk Cole
Miller, Jacquelin Neva
Monash, Ellis Alan
Myrick, Henry Nugent
Parker, Garald Gordon
Pease, David Nathaniel
Pfeiffer, Heinz Gerhard
Psuty, Norbert Phillip
Ranney, J. Warren
Renard, Kenneth George
Resler, Steven Charles
Riker, Joseph Thaddeus, III
Rohlich, Gerard Addison
Schoenike, Roland Ernest
Schubel, Jerry Robert
Schuh, G(eorge) Edward
Siegel, Gilbert Byron
Snyder, J. Herbert
Springer, Joseph Tucker
Sudia, Theodore William
Summerfelt, Robert Clar
Trzyna, Thaddeus Charles
Ullman, Joseph James
Wellner, Charles August
Williams, Daniel Frank
Wrobley, Arthur Ray
Yellin, Joel
Yeske, Lanny Alan
Yoder, Larry Richard
Yokell, Michael David

Species interaction
Armstrong, David Michael
Barbour, Michael George
Bentley, Barbara Lee
Berkland, James Omer
Bernhardt, Peter
Brett, Betty Lou Hilton
Bridgman, John Francis
Cates, Rex Gordon
Crowder, Larry B.
Cunningham, Harry Norman, Jr.
DeSelm, Henry (Hal) Rawie
Hamon, Danny Joe
Jaeger, Robert Gordon
Karlson, Ronald Henry
Koplin, James Ray
Li, Joseph K(wok) K(wong)
Lincoln, David Erwin
Lindquist, David Gregory
Mathis, Philip Monroe
McCormick, Michael Edward
Meinwald, Jerrold
Moll, Don L.
Morrison, Donald Michael
Nickol, Brent Bonner
O'Neill, Russell Richard
Siddiqi, Shaukat Mahmood
Springer, Joseph Tucker
Ward, Calvin Herbert
Woodin, Sarah Ann

Wastewater treatment systems. *See* **ENGINEERING, Environmental.**

Other
Aubert, Eugene James, *Environmental research management*
Boland, J. Robert, *Radioactive waste disposal*
Boyd, James, *Nuclear waste*
Bradford, Mark Lee, *Evnrionmental pollution contingency planning*
Brining, Dennis Lee, *Environmental monitoring and assessment*
Cain, William S., *Environmental health*
Ciaccio, Leonard Louis, *Environmental analysis of air and water*
deLaubenfels, David John, *Biogeography*
El-Ashry, Mohamed Taha, *Environmental management*
Ellison, Alfred Harris, *Environmental Science Research Administration*
Flynn, Kevin Francis, *Environmental effects of nuclear technology*
Folk, George Edgar, Jr., *Mammalian environmental physiology*
Greenfield, Stanley Marshall, *Air pollution dispersion*
Hall, William Franklin, Jr., *Environmental consulting*
Hellquist, Carl Barre, *Aquatic freshwater ecology*
Hiatt, Howard H., *Environmental health science*
Hibshman, Henry Jacob, *Environmental control systems*
Inhaber, Herbert, *Environmental Science*
Jacobson, E. Paul, *Flood Control*
Johnson, Eric Van, *Endangered species conservation and management*
Johnson, Roy Melvin, *Bacterial ecology*
Klehr, Edwin Henry, *Environmental chemistry*
Mayfield, Donald Lewis, *Radiological protection*
Overcash, Michael Ray, *Terrestrial systems*
Perhac, Ralph Matthew, *Environmental assessment*
Pilson, Michael Edward Quinton, *Marine ecosystems*
Porter, Warren Paul, *Biophysical ecology, Physiological ecology*
Reid, John Reynolds, *Environmental geology*
Reitan, Daniel Kinseth, *Energy Management*
Resler, Steven Charles, *Marine environmental sciences*
Roy, Peter Alan, *Contaminant exhaust systems*
Sanderson, H. Reed, *Range and wildlife habitat*
Schneider, Stephen Henry, *Environmental Policy*
Shock, D'Arcy Adriance, *Radioactive Waste Disposal*

GENETICS. *See* **AGRICULTURE, BIOLOGY, MEDICINE, VETERINARY MEDICINE.**

GEOSCIENCE

Geochemistry
Alter, H. Ward
Anderson, Duwayne Marlo
Angino, Ernest Edward
Baker, Donald Roy
Barden, Roland Eugene
Barker, Colin G.
Barker, Daniel Stephen
Barnes, Hubert Lloyd
Berner, Robert Arbuckle
Bikerman, Michael
Bishop, Richard Stearns
Brimhall, George H.
Brown, Harrison Scott
Brownlee, Donald Eugene, II
Burkart, Burke
Burtner, Roger Lee
Buseck, Peter R.
Chave, Keith Ernest
Craig, Harmon
Crawford, William Arthur
Crerar, David Alexander
Dachille, Frank
Eastman, Michael Paul
Edmond, John Marmion
Eggler, David H(ewitt)
Ehlers, Ernest George
Ericksen, George Edward
Ernst, Wallace Gary
Everett, Ardell Gordon
Filby, Royston Herbert
Foland, Kenneth Austin
Frost, John Elliott
Ganapathy, Ramachandran
Garrels, Robert Minard
Geidel, Gwendelyn
Glassley, William Edward
Goldberg, Edward David
Goodell, Horace Grant
Gregor, Clunie Bryan
Grossman, Lawrence
Halpern, Martin
Hanson, Gilbert Nikolai
Helgeson, Harold Charles
Heydegger, H(elmut) Roland
Hodgson, Gordon Wesley
Holland, Heinrich Dieter
Horvitz, Leo
Howd, Frank Hawver
Jensen, Mead LeRoy
Kauzmann, Walter Joseph
Kay, Robert Woodbury
Keeling, Charles David
Kemp, Marwin King
Kirkpatrick, R. James
Kutina, Jan
Kvenvolden, Keith Arthur
Lanphere, Marvin Alder
Lepp, Henry
Lerman, Abraham
Liou, John G.
Lofgren, Gary Ernest
Luth, William Clair
McKenzie, William Frank
Meyers, Philip Alan
Miknis, Francis Paul
Miller, Donald Spencer
Moore, Carleton Bryant
Mopper, Kenneth
Mueller, Paul Allen
Murthy, Varanasi Rama
Osborn, Elburt Franklin
Perdue, Philip Taw
Perhac, Ralph Matthew
Pickering, Ranard Jackson
Pigott, John Dowling
Prewitt, Charles Thompson
Quinn, James Gerard
Reynolds, John Hamilton
Rice, Donald Lester
Richardson, Catherine Kessler
Ruckmick, John Christian
Savin, Samuel Marvin
Schrader, Edward Leon
Schufle, Joseph Albert
Shieh, Yuch-Ning
Siever, Raymond
Stange, Morton Douglas
Ting, Francis Ta-Chuan
Turekian, Karl Karekin
Van Schmus, William Randall
Waslenchuk, Dennis Grant
White, Donald Edward
Zeller, Edward Jacob

Geology
Addicott, Warren Oliver
Allard, Gilles Olivier
Anderson, James Arthur
Andrews, John Thomas
Angino, Ernest Edward
Ash, Sidney Roy
Bagwell, Joyce Marie Burris
Bartholomew, Mervin Jerome
Berg, Henry Clay
Berkland, James Omer
Bhatt, Jagdish Jeyshanker
Bikerman, Michael
Bishop, Richard Stearns
Bonham, Harold Florian, Jr.
Boyer, Robert Ernst
Bradley, John Samuel
Briskin, Madeleine
Brooks, James Elwood
Brown, Jim McCaslin
Burchfiel, Burrell Clark
Burnett, John Laurence
Burton, Robert Clyde
Cady, Wallace Martin
Calkin, Parker Emerson
Callahan, John Edward
Campbell, Catherine Chase
Carew, James L.
Carlson, Carl Edward
Carter, William Douglas
Caruccio, Frank Thomas
Cassidy, Martin Macdermott
Chenoweth, Philip Andrew
Chilingarian, George Varos
Chronic, John
Churnet, Habte Giorgis
Clark, Sandra Helen Becker
Clopine, Gordon Alan
Cohen, Arthur David
Conley, James Franklin
Cordell, Robert James
Coveney, Raymond Martin, Jr.
Craddock, Campbell
Craig, Richard G.
Curtis, Graham Ray
Damuth, John Erwin
Dana, Stephen Winchester
Daviess, Steven Norman
Donovan, Terrence John
Doyle, Frank Lawrence
Draper, Grenville
DuBar, Jules Ramon
DuMontelle, Paul Bertrand
Dunne, Thomas
El-Ashry, Mohamed Taha
Elder, Curtis Harold
Ellwood, Brooks Beresford
Evenson, Edward Bernard
Everett, Ardell Gordon
Everhart, Donald Lough
Fassett, James Ernest
Fawcett, James Jeffrey
Ferrians, Oscar John, Jr.
Fisk, Lanny Herbert
Folger, David Winslow
Forsyth, Jane Louise
Friedman, Jules Daniel
Gabelman, John Warren
Galey, John Taylor
Garb, Forrest Allan
Garrels, Robert Minard
Genes, Andrew Nicholas
Goldthwaite, Duncan
Grantz, Arthur
Gregor, Clunie Bryan
Gryc, George
Gustafson, Lewis Brigham
Haeberle, Frederick Roland
Halbouty, Michel Thomas
Hall, Robert Dean
Hall, William Bartlett
Halpern, Martin
Haq, Bilal U.
Hargraves, Robert Bero
Harmon, David E., Jr.
Hartung, Jack Burdair
Hein, James Rodney
Heinrichs, Walter Emil, Jr.
Herrmann, Raymond
Herz, Norman
Higgins, Charles Graham
High, Lee Rawdon, Jr.
Hobbs, Carl Heywood, III
Holden, Frederick Thompson
Holzer, Thomas Lequear
Hook, John William
Hough, James Emerson
Howard, Keith Arthur
Howd, Frank Hawver
Hubbert, Marion King
Huff, Kenneth O.
Jamison, Harrison Clyde
Jensen, Mead LeRoy
Judson, Sheldon
Keigwin, Lloyd Denslow
Klasner, John Samuel
Koenig, James Bennett
Kottlowski, Frank Edward
Kvenvolden, Keith Arthur
Lackey, Laurence E(van)
Langway, Chester Charles, Jr.
Lanphere, Marvin Alder
Lemish, John
Lepp, Henry
Lowry, Wallace Dean
Luther, Edward Turner
Lynts, George Willard
Mandra, York T.
Mangus, Marvin Dale
Massell, Wulf Friedrich
Masursky, Harold
Mathewson, Christopher Colville
McCauley, John Francis
McCoy, Scott, Jr.
McGookey, Donald Paul
McGrain, Preston
McKenzie, William Frank
Mc Laren, Digby Johns
Melton, William Grover, Jr.
Menard, Henry William
Merriam, Daniel F(rancis)
Merrill, Robert David
Metsger, Robert William
Milling, Marcus Eugene
Mintz, Leigh Wayne
Montagne, John
Moore, George William
Moore, James Robert
Moran, William Rodes
Morrison, Roger Barron
Muller, Ernest Hathaway
Murphy, Michael Arthur
Myers, Donald Arthur
Peck, Dallas Lynn
Pees, Samuel Thomas
Perry, William James, IV
Peters, William Callier
Peterson, Frank Lynn
Péwé, Troy Lewis
Pierce, William Gamewell
Ponder, Herman
Porter, Stephen Cummings
Reid, John Reynolds
Riggs, Karl A., Jr.
Robinson, Joseph Edward
Rodgers, John
Ruckmick, John Christian
Rutford, Robert Hoxie
Schmidt, Ruth A.M.
Schwartz, Maurice Leo
Sharp, Robert Phillip
Shipman, Ross Lovelace
Shoemaker, Eugene Merle
Simpson, Eugene Sidney
Stott, Donald Franklin
Thierstein, Hans Rudolf
Thorson, Robert Mark
Ting, Francis Ta-Chuan
Twiss, Page Charles
Van Schmus, William Randall
Walker, Eugene Hoffman
Wallace, Robert Earl
Walton, Paul Talmage
Wayne, William John
Weeks, Albert William
Weisbord, Norman Edward
Wilbanks, John Randall
Wolman, M. Gordon
Wood, Gordon Harry, Jr.
Woodburne, Michael Osgood

Geology, mineralogy
Allen, John Christopher, Jr.
Berry, Richard Warren
Borg, Iris Yvonne
Buseck, Peter R.
Davies, David Keith
Ericksen, George Edward
Ewing, Rodney Charles
Finney, Joseph J.
Frueh, Alfred Joseph
Hagni, Richard Davis
Hamilton, Charles Leroy
Hinthorne, James Roscoe
Jansen, George James
Kidwell, Albert Laws
Kopp, Otto Charles
Kutina, Jan
Leonard, Benjamin Franklin, III
Milton, Charles
Phillips, William Revell
Prewitt, Charles Thompson
Rossman, George Robert
Schneer, Cecil J.
Thompson, James Burleigh, Jr.
Veblen, David Rodli
Young, Edward Joseph

Geology, petrology
Abbott, Richard Newton, Jr.
Allard, Gilles Olivier
Allen, John Christopher, Jr.
Amos, Dewey Harold
Baker, Donald Roy
Barker, Daniel Stephen
Bird, John Malcolm
Brimhall, George H.
Burtner, Roger Lee
Cain, J(ames) Allan
Churnet, Habte Giorgis
Crawford, William Arthur
Darby, Dennis Arnold
Dick, Henry Jonathan Biddle
Dixon, Helen Roberta
Eggler, David H(ewitt)
Ehlers, Ernest George
Fawcett, James Jeffrey
Fisher, George Wescott
Foland, Kenneth Austin
Glassley, William Edward
Goodwin, Bruce Kesseli
Hagni, Richard Davis
Hamilton, Charles Leroy
Hanson, Gilbert Nikolai
Hargraves, Robert Bero
Hayes, John Bernard
Herz, Norman
Joesten, Raymond
Kamb, Walter Barclay
Kay, Robert Woodbury
Kirkpatrick, R. James
Liou, John G.
Lipman, Peter Waldman
Lofgren, Gary Ernest
Lukert, Michael Thomas
Luth, William Clair
McBride, Earle Francis
McHone, James Gregory
Mielenz, Richard Childs
Milton, Charles
Miyashiro, Akiho
Mueller, Paul Allen
Murthy, Varanasi Rama
Phillips, William Revell
Shieh, Yuch-Ning
Thompson, James Burleigh, Jr.
Treves, Samuel Blain
Westerman, David Scott
Yoder, Hatten Schuyler, Jr.
Young, Edward Joseph
Zen, E-an

Geology, sedimentology
Anderson, Franz Elmer
Bachman, Richart T.
Berner, Robert Arbuckle
Berry, Richard Warren
Bloomer, Richard Rodier
Bowsher, Arthur LeRoy
Boyer, Paul Slayton
Brande, Scott
Brooks, James Elwood
Brown, Leonard Franklin, Jr.
Carlson, Carl Edward
Carozzi, Albert Victor
Chamberlain, Charles Franklin
Chitaley, Shyamala Dinkar
Clopine, Gordon Alan
Creager, Joe Scott
Curtis, Christopher Michael
Darby, Dennis Arnold
Davies, David Keith
Dawson, James Clifford
Dorfman, Myron Herbert
Dugolinsky, Brent Kerns
Ekdale, Allan Anton
Fagerstrom, John A.
Fisher, James Harold
Friedman, Gerald Manfred
Garrison, Robert, Edward
Gryc, George
Hand, Bryce Moyer
Happ, Stafford Coleman
Hay, William Winn
Hayes, John Bernard
Hein, James Rodney
Hobbs, Carl Heywood, III
Hoyt, William Henry
Kidwell, Albert Laws
Klein, George deVries
Kopf, Rudolph William
Kravitz, Joseph Henry
Larue, David Knight
Lindemann, Richard Henry
Lowright, Richard Henry
Ludwick, John Calvin, Jr.
Luther, Edward Turner
Mangus, Marvin Dale
Mann, Christian John
McBride, Earle Francis
McGookey, Donald Paul
Merrill, Robert David
Milling, Marcus Eugene
O'Kelley, Joseph Charles
Pigott, John Dowling
Potter, Paul Edwin
Psuty, Norbert Phillip
Rea, David Kenerson
Reinhardt, Juergen
Sadler, Peter Michael
Siever, Raymond
Twiss, Page Charles
Vail, Peter Robbins
Visher, Glenn Shillington
Warme, John Edward
Webster, Gary Dean
Whisonant, Robert Clyde

Geology, tectonics
Aki, Keiiti
Allen, Clarence Roderic
Alvarez, Walter
Amos, Dewey Harold
Arden, Daniel Douglas
Bartholomew, Mervin Jerome
Berg, Henry Clay
Bird, John Malcolm
Bloomer, Richard Rodier
Burchfiel, Burrell Clark
Burkart, Burke

Geophysics

Hydrology

Oceanography. ***See also*** **ENERGY SCIENCE, Ocean thermal energy conversion; ENGINEERING, Petroleum.**

Oceanography, deep-sea biology

Oceanography, ocean engineering

Oceanography, offshore technology

Oceanography, sea floor spreading

Paleontology

Paleontology, paleobiology

Paleontology, paleoecology

Planetology

Remote sensing

Other

Berg, Edward, *Seismology*
Biggs, Maurice Earl, *Seismology*
Bogard, Donald Dale, *Meteorites*
Bradley, Michael Douglas, *Water resources policy*
Bukry, J(ohn) David, *Oceanography, biochronology*
Bull, Colin Bruce Bradley, *Glaciology*
Burk, Creighton, *Marine geology*
Burnett, John Laurence, *Exploration geoglogy*
Cain, J(ames) Allan, *Mineral resources*
Carter, William Eugene, *Geodesy*
Catacosinos, Paul Anthony, *Stratigraphy*
Chapin, Charles Edward, *Volcanology*
Cisne, John Luther, *Stratigraphy*
Clark, John Desmond, *Paleoanthropology*
Elston, Wolfgang Eugene, *Economic geology, Volcanology*
Ettensohn, Frank Robert, *Geology of gas shales*
Fassett, James Ernest, *Stratigraphy*
Fields, Patrick F., *Paleobotany*
Frost, John Elliott, *Geoscience management*
Gaposchkin, Edward Michael, *Geodesy*
Hanson, Roy Eugene, *Geoscience Program Administration*
Harbaugh, John Warvelle, *Mathematical geology*
Harris, DeVerle Porter, *Mineral and energy resources*
Heinrichs, Donald Frederick, *Marine geophysics*
Helsley, Charles Everett, *Marine geophysics*
High, Lee Rawdon, Jr., *Petroleum geology*
Hudson, Robert Frank, *Petroleum geology*
Judson, Sheldon, *Geomorphology*
Keil, Klaus, *Meteoritics*
Kirk, John Gallatin, *Geodesy*
Koucky, Frank Louis, *Archaeogeology*
Kulhawy, Fred Howard, *Engineering geology*
La Marche, Valmore Charles, Jr., *Dendrochronology*
Leonard, Benjamin Franklin, III, *Ore or mineral deposits*
Mackay, John Ross, *Arctic studies*
Mandra, York T., *Micropaleontology*
Mann, Christian John, *Mathematical geology*
Marcus, Melvin Gerald, *Glaciology*
Marsh, William Michael, *Geomorphology*
McBryde, Felix Webster, *Geography, space relationship analysis, Thematic cartography, cartographic design*
McCarthy, Dennis Dean, *Geodesy*
McGrain, Preston, *Economic geology*
Myers, Donald Arthur, *Biostratigraphy*
Odishaw, Hugh, *Earth sciences education*
O'Kelley, Joseph Charles, *Precambrian geology*
Olsson, Richard Keith, *Biostratigraphy*
Perry, Mary Jane, *Biological oceanography*
Richards, Paul Granston, *Seismology*
Rutford, Robert Hoxie, *Geomorphology*
Saville, Thorndike, Jr., *Nearshore oceanography*
Scheid, Vernon Edward, *Economic geology*
Schrader, Edward Leon, *Ore deposit origins, distributions*
Schupler, Bruce Ralph, *Geodesy*
Sharp, Robert Phillip, *Geomorphology*
Thorson, Robert Mark, *Geological hazards*
Walker, Harley Jesse, *Geomorphology, Physical Geography*
White, Donald Edward, *Economic geology*
Williams, Peter John, *Freezing of soils, Geotechnical Science*

INFORMATION SCIENCE

Automated language processing
Borko, Harold
Cooper, Franklin Seaney
Doszkocs, Tamas Endre
Hirschman, Lynette
Miller, Lance Arnold
Oettinger, Anthony Gervin
Peters, Paul Stanley, Jr.
Sager, Naomi
Shreve, Gregory Monroe
Waltz, David Leigh

Information systems. *See also* COMPUTER SCIENCE, Information systems.
Aborn, Murray
Audet, John James, Jr.
Bailey, William James
Barbieri, Richard Charles
Beakes, John Herbert
Borko, Harold
Cummings, Martin Marc
Day, Stacey Biswas
Duncan, Doris Gottschalk
Edelman, Ann Lynn
Edwards, Carl Normand
Fedorowicz, Jane
Fiene, Richard John
Gardner, Willard Hale
Gillette, Dean
Gordon, James Arthur
Haeberle, Frederick Roland
Hamilton, Angus Cameron
Hansen, Grant Lewis
Hansen, James Vernon
Hoplin, Herman Peter
Howard, Jay Lloyd
Hoye, Robert Earl
Jenkins, Arnold Milton
Johnson, Karl Otto, Jr.
Joseph, Earl Clark
Koenig, Michael Edward Davison
Kricka, Hanna Halyna
Landgrebe, David Allen
Larson, Larry Gale
Leventhal, Gerald Seymour
Lide, David Reynolds, Jr.
Lientz, Bennet Price
Livingston, John David
Lukasik, Stephen Joseph
Lynch, Robert Michael
Machovec, George Stephen
McGarvey, John James
McGill, Scott Douglas
Meilleur, Steven Grant
Meyer, Fred Lewis
Mitra, Saibal K(umar)
Mohanty, Mirode Chandra
Nadin, Mihai
Packer, Katherine Helen
Powell, James Charles
Reiffel, Leonard
Ritt, Paul Edward
Schmandt-Besserat, Denise
Severn, David Jones
Shelley, Edwin Freeman
Skolnik, Herman, *Chemical information science*
Smith, Linda Cheryl
Summit, Roger Kent
Sweeney, Urban Joseph
Teddlie, Charles Benton
Tou, Julius T.
Tucker, Marc Stephen
Tummala, V. M. Rao
Turek, Jeffery Lee
Weaver, Michael John
Williams, Martha Ethelyn
Wood, Charles Cresson
Yensen, Richard

Other
Adams, John Lester, Jr., *Communications research*
Blakely, John Paul, *Information science*
Davis, Charles Hargis, *Library and information science education administration*
Fisher, H. Leonard, *Societal effects of information technology*
Fujimura, Osamu, *Linguistics and speech science*
Gallo, Charles Francis, Jr.
Griffith, Belver Callis, *Information science policy studies*
Harris, Zellig Sabbettai, *Information representation*
Joseph, Earl Clark, *Futures*
Joshi, Aravind Krishna, *Natural language analysis and processing*
Kirschenbaum, Donald Monroe, *Data compilations*
Kochen, Manfred, *Learning systems*
Patterson, Robert Logan, *Library systems automation*
Ross, Douglas Taylor, *Systems analysis*
Schroeder, James Ernest, *Computer-assisted tranining and simulation*
Zweig, Gilbert, *Reprographics*

LASER. *See also* OPTICS.

Data storage and reproduction
Axelrod, Norman Nathan
Barnes, Frank Stephenson
Beiser, Leo
Blazey, Richard Nelson
Cheng, David
Derderian, George
Gupta, Mool Chand
LaBudde, Robert Arthur
Laub, Leonard Joseph
Maiman, Theodore Harold
Mossberg, Thomas William
Owens, James Carl
Tomlinson, W. John
Urbach, John Charles
Wang, Shing Chung
Weaver, Christopher Scot

Fusion. *See also* ENERGY.
Ahearne, Daniel Paul
Dean, Stephen Odell
Devaney, Joseph James
Dube, George
Hertz, Richard Cornell
Kopp, Roger Alan
Kyrala, George Amine
Liu, Kwok-On Elisha
Manor, Robert Edward
McGuire, Eugene Joseph
Sanger, Gregory M.
Scott, Franklin Robert
Sollid, Jon Erik
Stellingwerf, Robert F.

Laser-induced chemistry
Bauer, Simon Harvey
Bel Bruno, Joseph J(ames)
Beri, Avinash Chandra
Berry, Michael James
Bhatnagar, Ravi
Bogert, Gary Michael
Brooks, Philip Russell
Caird, John Allyn
Cantrell, Cyrus Duncan, III
Clark, John Hamilton
Cooper, Ralph Sherman
Cox, Hollace Lawton, Jr.
Crim, Forrest Fleming, Jr.
Crosley, David Risdon
Davis, James Ivey
Diels, Jean-Claude Marcel
Dunbar, Robert Copeland
Earl, Boyd Lorel
Eyler, John Robert
Gill, Dennis Howard
Grunwald, Ernest Max
Hackett, Peter Andrew
Harmon, Gary R.
Hootman, Harry Edward
Jackson, William Morgan
Kaldor, Andrew Peter
Keehn, Philip Moses
Lin, Ming-Chang
Moseley, John Travis
Moskowitz, Paul A.
Paisner, Jeffrey Alan
Rettner, Charles Thomas
Robinson, Dean Wentworth
Siegman, Anthony Edward
Smalley, Richard Errett
Stanbro, William David
Sutton, George Walter
Taylor, David John
Walker, David Crosby
Yuan, Jian-Min
Zare, Richard Neil

Medicine
Bass, Michael
Bellina, Joseph Henry
Bennett, William Ralph, Jr.
Bourgelais, Donna Belle Chamberlain
Clayman, Lewis
Cosman, Bard
Dougherty, Thomas John
Evans, James Thomas
Fisher, John Courtney
Gelb, Arthur Franklin
Goldbaum, Michael Henry
Goldman, Leon
Ichiye, Takashi
Lee, Garrett
Lee, Wylie In-Wei
Newell, Frank William
Pepine, Carl John
Robinson, WalkerLee
Sanborn, Timothy Allen
Schwartz, William Benjamin
Sliney, David Hammond
Solon, Leonard R(aymond)

Spectroscopy
Armstrong, John Allan
Barisas, Bernard George, Jr.
Bass, Michael
Bayfield, James Edward
Bernstein, Elliot Roy
Birnbaum, Edward R.
Bloembergen, Nicolaas
Bloom, Arnold Lapin
Brink, Gilbert O.
Callis, Patrik Robert
Cantrell, Cyrus Duncan, III
Colson, Steven Douglas
Compaan, Alvin Dell
Cone, Rufus Lester
Cooper, Charles Dewey
Curl, Robert Floyd, Jr.
Devlin, Frank Joseph
Dugan, Charles Hammond
Duley, Walter Winston
Edelson, Martin Charles
Engelking, Paul Craig
Fairbank, William Martin, Jr.
Farley, John William
Farrar, James Martin
Fernández, Salvador M.
Fortson, Edward Norval
Gelbwachs, Jerry Avron
Hackel, Lloyd Anthony
Harney, Robert Charles
Harris, Harold Hart
Hilborn, Robert Clarence
Hinkley, Everett David, Jr.
Holt, Richard A.
Isganitis
Kelley, Paul Leon
Kliger, David Saul
Kocher, Carl Alvin
Kramer, Steven David
Krishnan, Pallassana Narayanier
Laane, Jaan
Lam, Leo K.
Lapatovich, Walter Peter
Laurendeau, Normand Maurice
Lee, Long C.
Liao, Paul F(oo-Hung)
Littman, Michael Geist
Lombardi, Gabriel G.
Mack, Michael Edward
Mathews, C(ollis) Weldon
Mc Clure, Donald Stuart
Mc Ilrath, Thomas James
Miles, Richard Bryant
Nelson, Robert Norton
Oettinger, Peter Ernest
Oka, Takeshi
O'Shea, Donald Charles
Paisner, Jeffrey Alan
Pappalardo, Romano G.
Patel, Chandra Kumar Naranbhai
Perry, Mildred Elizabeth
Person, Willis Bagley
Piltch, Martin Stanley
Powell, Richard Conger
Radziemski, Leon Joseph
Rand, Stephen Colby
Roessler, David Martyn
Ross, J(ohn) B(randon) Alexander
Roy, Rajarshi
Schawlow, Arthur Leonard
Scheeline, Alexander
Siegman, Anthony Edward
Siomos, Konstadinos
SMith, Allan Laslett
Smyth, Kermit Campbell
Sorokin, Peter Pitirimovich
Stone, Julian
Stwalley, William Calvin
Thomas, George Joseph, Jr.
Tolles, William Marshall
Vala, Martin Thorvald, Jr.
Walter, William Trump
Weber, Marvin John
Weeks, Stephan John
Wehry, Earl Luther, Jr.
Young, James Forrest

Other
Altschuler, Bruce Robert
Ammann, Eugene Otto, *Laser research*
Baldwin, Gary Dale, *Laser Research*
Bederson, Benjamin, *Laser research*
Birkitt, John Clair, *High energy chemical lasers*
Bjorkholm, John Ernst, *Nonlinear optics*
Bridges, William Bruce, *Physics research*
Brown, Lorin W., *Laser physics*
Carter, William Harold, *Laser research*
Casperson, Lee Wendel, *Laser research*
Chao, Shui Lin, *High energy laser systems*
Chen, Ying-Chih, *Semiconductor lasers*
Crisp, Michael Dennis, *Laser physics*
De Maria, Anthony J., *Laser physics*
Dickson, LeRoy David, *Laser research*
Diels, Jean-Claude Marcel, *Laser Optics*
Dowley, Mark William, *Laser research*
Faxvog, Frederick R., *Laser research*
Fein, Michael E., *Laser design*
Feinstein, Joseph, *Laser power generation*
Firester, Arthur Herbert, *Laser applications*
Fisher, Robert Alan, *Nonlinear laser optics, Optical phase conjugation*
Flom, Terrence Edsel, *Laser communications, Laser radar*
Friedrich, Otto Martin, Jr., *Laser research*
Geslicki, Mark Louis, *Laser scanning systems*
Goodwin, Richard Clarke, *Laser physics*
Gregson, Victor Gregory, *Industrial laser processing*
Hansler, Richard Lowell, *Laser processing*
Hauck, James Pierre, *Laser gyroscopes and radars*
Haun, Robert Dee, Jr., *Laser research*
Hill, Alan Eugene, *Laser Research and Development*
Hirleman, Edwin Daniel, Jr., *Laser instrumentation*
Holmes, Dale Arthur, *high energy laser systems*
Hunter, Robert Olin, Jr., *Laser physics*
Hutchinson, Donald Patrick, *Laser design*
Hyman, Howard Allan, *Laser physics*
Jackson, John Edwin, *Laser research*
Jacobs, Stephen Frank, *Laser metrology*
Kantrowitz, Arthur, *High energy laser*
Kelly, Patrick Joseph, *Laser surgery*
Kestenbaum, Ami, *Laser applications*
Krinsky, Jeffrey Alan, *Laser efficiency*
Kwok, Munson Arthur, *Chemical lasers*
Lee, Ching Tsung, *X-ray lasers*
Lee, Kotik Kai, *Laser physics*
Lightman, Allan Joel, *Laser-optical diagnostics*
Lin, Chinlon, *Laser technology*
Linder, Solomon Leon, *Guidance systems*
Linz, Arthur, *Laser research*
Litynski, Daniel Mitchell, *Laser research*
Liu, Yung Sheng, *Laser Optics*
Maccabee, Bruce Sargent, *Laser physics*
Mandel, Leonard, *Laser optics*
Marshall, Gerald Francis, *Laser alignment*
Martin, Lawrence Ronald, *Laser applications*
Martin, William Eugene, *Laser propagation physics, free electron lasers*
Massey, Gail Austin, *Laser instrumentation*
Melville, Richard Devern Samuels, Jr., *Laser technology*
Morgan, Lucian Lloyd, *Laser weapons*
Nanos, George Peter, Jr., *Laser physics*
Nelson, Donald Frederick, *Laser research*
Pernick, Benjamin, *Laser research*
Remo, John Lucien, *Lasers/ resonator theory*
Rich, John Charles, *High-energy lasers*
Sentman, Lee Hanley, *High energy lasers*
Sidorowicz, Kenneth Joseph, *Laser welding stress analysis*
Smith, Peter William, *Laser research*
Snavely, Benjamin Breneman, *Tunable lasers*
Steinbruegge, Kenneth Brian, *Laser materials, systems, applications*
Stotts, Larry Bruce, *Laser communication*
Tomren, Douglas Roy, *Laser research*
Vander Sluis, Kenneth L., *Laser systems*
Wallner, Richard Alan, *Laser diagnostics*
Wang, Charles P., *Laser research*
Wang, Chen-Show, *Semiconductor lasers*
Warren, Walter Raymond, Jr., *High energy laser systems*
Waynant, Ronald William, *Laser research*
Wiesenfeld, Jay Martin, *Picosecond laser optics*
Wood, Obert Reeves, II, *Laser research*
Woskoboinikow, Paul Peter, *Laser research*
Young, Matt, *Laser research*

MATERIALS SCIENCE. *See also* ENGINEERING, Materials.

Biomaterials
Absolom, Darryl Robin
Anusavice, Kenneth John
Apostolou, Spyridon F.
Bapna, Mahendra Singh
Barenberg, Sumner A(rnold)
Barolet, Ralph Yvon
Brodhagen, Thomas Warren
Chappell, Robert Paul
Chuang, Hanson Yii-Kuan
Civjan, Simon
Cooper, Stuart Leonard
Crim, Gary Allen
Doane, William McKee
Douglas, William Hugh
Faunce, Frank Roland
Forney, LeRoy S.
Galil, Khadry Ahmed
Galletti, Pierre Marie
Greener, Evan H.
Grossman, David Gary
Heuer, Michael Alexander
Ifju, Geza
Jenkins, W(inborne) Terry
Keller, John Charles
Lambuth, Alan Letcher
Langer, Robert Samuel
Leake, Donald Lewis
Lorton, Lewis
Mueller, Herbert Joseph
Northup, Sharon Joan
Polay, Janet Skinner
Rawls, Henry Ralph
Stannard, Jan Gregory
Stith, William Joseph
von Fraunhofer, Joseph Anthony
Yannas, Ioannis Vassilios
Young, Franklin Alden, Jr.

Ceramics
Bergeron, Clifton George
Bowen, Harvey Kent
Bradt, Richard Carl
Burke, Joseph Eldrid
Chen, Ho Sou
Gonczy, Stephen Thomas
Grossman, David Gary
Gulden, Terry D.
Kalyoncu, Rustu Sumer
Kingery, William David
LaCourse, William Carl
MacKenzie, John Douglas
Osborn, Elburt Franklin
Pask, Joseph Adam
Reynard, Kennard Anthony
Rowse, Robert Alfred
Roy, Rustum
Thornton, H. Richard
Wagner, James Bruce, Jr.
Wuensch, Bernhardt John

Materials
Balluffi, Robert Weierter
Bapna, Mahendra Singh
Barrett, Charles Sanborn
Beasley, Wayne Machon
Bendow, Bernard
Bever, Berliner Michael
Blakely, John McDonald
Blanck, A.R.
Boltax, Alvin
Bowen, Harvey Kent
Bujtas, Mark Steven
Burke, Joseph Eldrid
Cahn, John Werner
Carlson, Roy Washington
Carson, James Matthew
Clark, Donald Eldon

Clinard, Frank Welch, Jr.
Compton, W. Dale
Crandall, William Brooks
Daniels, Carole Angela
Dawson, George Eugene, Jr.
de Wit, Roland
Duwez, Pol Edgard
Ewing, Rodney Charles
Fontana, Mars Guy
Freund, Lambert Ben
Fromhold, Albert Thomas, Jr.
Galeener, Frank Lee
Giamei, Anthony Francis
Glicksman, Martin Eden
Gonczy, Stephen Thomas
Green, Robert Edward, Jr.
Greenfield, Irwin Gilbert
Gulden, Terry D.
Gunn, Michael Richard
Gupta, Mool Chand
Hannay, N(orman) Bruce
Harkness, Samuel Dacke, III
Harrington, Roy Victor
Harris, William James, Jr.
Hirth, John Price
Jensen, Barbara Lynne
Jette, Archelle Norman
Johnson, William Lewis
Kelly, Kevin Anthony
Knollman, Gil Carl
Laymon, Stephen Alan
Lazareth, Otto William
Liedl, Gerald LeRoy
Mansur, Louis Kenneth
Maurer, Robert Distler
McMeeking, Robert Maxwell
Meshii, Masahiro
Morris, George Vincent
Osthoff, Robert C.
Pehlke, Robert Donald
Plunkett, Robert
Rice, James Thomas
Richardson, Rick Lee
Risen, William Maurice, Jr.
Ritt, Paul Edward
Rowse, Robert Alfred
Sannella, Joseph Lee
Seely, John Henry
Seitz, Frederick
Sellers, Gregory Jude
Sha, George Tzeng-Tsun
Shepard, Marion LaVerne
Strasser, Alfred Anthony
Testardi, Louis Richard
Thornton, J(oseph) Scott, Jr.
Turnbull, David
Turner, Derek Terence
Tyndall, Bruce Mapes
Van Uitert, LeGrand Gerard
Versnyder, Francis Louis
Winter, J. Ronald
Wright, Christopher Pearce
Yoshikawa, Herbert Hiroshi

Materials, composite
Adams, Donald Frederick
Alper, Allen Myron
Bares, Jan
Baum, Gary Allen
Bradt, Richard Carl
Chow, Tsu Sen
Davis, LeRoy Wellington
Dharan, C.K. Hari
Dick, William Allen
Ewing, Richard Edward
Fleck, Robert Davis
Foral, Ralph Francis
Gerstle, Frank P., Jr.
Gujrati, Bitthal Das
Hong, Su-Don
Hoover, William Leichliter
Ifju, Geza
Jayne, Benjamin Anderson
Jones, Douglas Linwood
Kong, Eric Siu-Wai
Kuist, Charles Howard
Lindenmeyer, Paul Henry
Mallick, Pankaj Kumar
McKee, Robert Bruce, Jr.
McKinney, John Edward
Min, Byung Kon
Pearce, Malcolm Bulkeley, Jr.
Pflederer, Fred Raymond
Pipes, Robert Byron
Promisel, Nathan E.
Richardson, George Campbell
Ricks, Stephen Andrew
Rosen, B. Walter
Schile, Richard Douglas
Slykhous, Stewart James
Tang, Ruen C.
Tao, Li-Chung
Thornton, H. Richard
Whitney, James Martin
Widera, G.E.O.
Wiff, Donald Ray
Williams, James Henry, Jr.
Young, Robert Cleland

Materials, electronic
Bagley, Brian G.
Berry, William Bernard
Boring, Arthur Michael
Bube, Richard Howard
Cho, Alfred Y.
Chu, Ting L.
Dickens, Elmer Douglas, Jr.
Evans, Billy Joe
Goldner, Ronald B.
Gordon, Roy Gerald
Greenhouse, Harold Mitchell
Harpster, Joseph W. C.
Johnson, Robert E.
Katz, William
Laudise, Robert Alfred
Lee, Sanboh
Mendelsohn, Lawrence Barry
Mitchell, Dean Lewis
Nowak, Welville Berenson
Parker, Michael Andrew
Raschke, Curt Robert
Rosi, Fred David
Shing, Yuh-Han
Taylor, George William
Wernick, Jack Harry
Wnek, Gary Edmund
Woods, John Galloway

Materials, high-temperature
Alexander, Leckie Frederick
Alper, Allen Myron
Ansell, George Stephen
Bassford, Thomas Harvey
Besmann, Theodore Martin
Boone, Donald H(erbert)
Bopp, Charles Dan
Bragg, Robert Henry
Cezairliyan, Ared
Chang, Ji Young
Chuang, Tze-jer
Chubb, Walston
Coffin, Louis Fussell, Jr.
Elliott, John Frank
Fink, Joanne Krupey
Friedman, Melvin
Gowda, Byre Venkataramana
Green, Harry Western, II
Hanson, John Edward
Hastie, John William
Jarrett, Noel
Korwin, Paul
Muan, Arnulf
Perkins, Roger Allan
Pettit, Frederick Sidney
Phillips, William Evans
Ray, William Edward
Rohsenow, Warren Max
Roy, Rustum
Sarma, Abul Chandra
Schwer, Roger Edwin
Smith, Ronald William
Timo, Dominic Peter
Versnyder, Francis Louis
Watson, James Frederic
Weertman, Julia Randall
Woodford, David A.

Materials processing
Altan, Taylan
Amer, Ahmad (El Sayed)
Bailey, John Albert
Boulger, Francis William
Boyer, Charles Benjamin
Brondyke, Kenneth James
Canon, Ronald Martin
Cullen, Donald Lee
Flinn, Richard Aloysius
Foster, William Samuel
German, Randall M(ichael)
Gregson, Victor Gregory
Gujrati, Bitthal Das
Henderer, Willard E(verett), III
Hoover, William Leichliter
Jain, Sulekh Chand
Kramer, Bruce Michael
Kuhn, Howard Arthur
MacDougall, John Archibald
Morris, John William
Olson, David LeRoy
Ovens, William George
Pask, Joseph Adam
Pinney, Frank Batchelder
Ponder, Herman
Ray, William Edward
Rivkin, Maxcy Calvin
Rosenberger, Franz Ernst
Rush, John Edwin, Jr
Schey, John Anthony
Shirkey, William Dan
Smith, Ronald William
Stregowski, Thomas John
Taraman, Khalil Showky
Thakkar, Bharatkumar S.
Valyi, Emery I.
Ward, H. Blair, Jr.
Woodburn, Wilton Allen
Wright, Paul Kenneth

Metallurgy
Anderson, Stephen Clark
Ansell, George Stephen
Ansuini, Frank Joseph
Azaroff, Leonid Vladimirovitch
Bailey, John Albert
Balluffi, Robert Weierter
Bassford, Thomas Harvey
Baty, David Lee
Beck, Paul Adams
Berardi, Matteo P.
Berg, Bengt Henrik
Bever, Berliner Michael
Boone, Donald H(erbert)
Breyer, Norman Nathan
Brondyke, Kenneth James
Bruner, Ralph Clayburn
Chen, Ho Sou
Chin, Gilbert Yukyu
Chung, Hee Mok
Cohen, Jerome Bernard
Conrad, Hans
Copeland, William D.
Crandall, William Brooks
Dahlman, Geoffrey Edwin
Davis, LeRoy Wellington
Dawson, George Eugene, Jr.
DiOrio, Mark Lewis
Driscoll, Timothy John
Duerr, J. Stephen
Duwez, Pol Edgard
Fine, Morris Eugene
Flinn, Richard Aloysius
French, David Nichols
Gagnebin, Albert Paul
Giamei, Anthony Francis
Giessen, Bill Cormann
Glicksman, Martin Eden
Glodowski, Robert John
Goldenberg, Herbert Jay
Gordon, Robert Boyd
Grant, Nicholas John
Harris, William James, Jr.
Hart, Raymond Kenneth
Hasegawa, Ryusuke
Hirth, John Price
Hochman, Robert F(rancis)
Hollomon, John Herbert
Jaffee, Robert Isaac
Jesser, William Augustus
Johnson, William Lewis
Johnston, Mary Helen
Kautz, David Johnathan
Khare, Ashok K.
Koucky, Frank Louis
Kudryk, Val
Lane, Joseph Robert
Liedl, Gerald LeRoy
Lucas, Glenn Eugene
Lucas, William Ray
MacDougall, John Archibald
Maddin, Robert
Malhotra, Manohar Lal
Malozemoff, Plato
Mattas, Richard Francis
Mc Mahon, Charles Joseph, Jr.
McMenamin, Edward William
McSherry, Arthur James
Meshii, Masahiro
Morgan, Jack Brandon
Olson, David LeRoy
Opie, William Robert
Owen, Walter Shepherd
Parker, Earl Randall
Paxton, Harold William
Pehlke, Robert Donald
Pense, Alan Wiggins
Perkins, Roger Allan
Pettit, Frederick Sidney
Pfann, William Gardner
Pickerling, Howard William
Polonis, Douglas Hugh
Queneau, Paul Etienne
Roberts, George Adam
Salama, Mamdouh M.
Savage, Warren Fairbank
Schwartz, Lyle Howard
Sellmyer, David Julian
Shepard, Marion LaVerne
Socie, Darrell Frederick
Steinberg, Morris Albert
Swalin, Richard Arthur
Swanson, William Mason
Taggart, George Bruce
Teleshak, Stephen
Tundermann, John Hayes
Wallace, John Francis
Wang, Rong
Watson, James Frederic
Weertman, Julia Randall
Wernick, Jack Harry
Wert, Charles Allen
Woodford, David A.

Microchip technology. ***See also*** **ENGINEERING, Electrical.**
De Hodgins, Ofelia Canales
Liu, Kwok-On Elisha
Parker, Michael Andrew
Salsburg, Kevyn Anne
Venkatesan, T.

Polymers
Barenberg, Sumner A(rnold)
Bhushan, Bharat
Boyer, Raymond Foster
Brostow, Witold Konrad
Chance, Ronald Richard
Cohen-Addad, Jean-Pierre
Couchman, Peter Robert
Doane, William McKee
Ferry, John Douglass
Fornes, Raymond Earl, Sr.
Germano, Don Joseph
Ghavamikia, Hamid
Hartmann, Bruce
Hauser, Ray Louis
Hawkins, Walter Lincoln
Hsu, Shaw Ling
Koberstein, Jeffrey Thomas
Kusy, Robert P
Lando, Jerome B.
Lunn, Anthony C.
Mallick, Pankaj Kumar
Manson, John Alexander
Morgan, Paul Winthrop
Pochan, John Michael
Rice, James Thomas
Richardson, George Campbell
Roylance, David Kaye
Sancaktar, Erol
Sannella, Joseph Lee
Schneberger, Gerald Leo
Sellers, Gregory Jude
Senich, George A.
Skotheim, Terje Asbjorn
Skutnik, Bolesh S. Joseph
Stahl, Joel S.
Uralil, Francis Stephen
Valyi, Emery I.
Ward, H. Blair, Jr.
Wiles, David McKeen
Wu, Souheng
Wu, Wen-li
Yannas, Ioannis Vassilios

Other
Califano, Joseph Michael, *Filtration membranes*
Ericksen, Jerald Laverne, *Theories of crystals*
Grossbeck, Martin Lester, *Fusion reactor materials*
Harwood, Julius J., *Materials science research and development administration.*
Howden, David Gordon, *Welding technology*
Jacobs, David, *Elastomerics*
Johnson, Herbert Harrison, *Materials science research administration*
Kuhn, Howard Arthur, *powder metallurgy*
Lawrence, Leo Albert, *Nuclear reactor fuels*
Nosanow, Lewis Harold, *Materials research administration*
Rice, Stephen Landon, *Wear of Materials*
Schweiker, George Christian, *Materials science research administration*
Teleshak, Stephen, *Failure anaylsis*
Vook, Richard Werner, *Thin Films and Surfaces*
Wagner, John George, *Powder metallurgy*

MATHEMATICS. ***See also*** **COMPUTER SCIENCE.**

Applied
Ablow, Clarence Maurice
Ablowitz, Mark Jay
Achenbach, Jan Drewes
Adler, Irving
Allgower, Eugene Leo
Andrus, Jan Frederick
Andrushkiw, Roman Ihor
Aris, Rutherford
Arkilic, Galip Mehmet
Arnold, Leslie Kingsland
Baillieul, John Brouard
Baker, Lawrence John
Balanis, George Nick
Baxter, Judith Lee
Bellman, Richard Ernest
Berger, Alan Eric
Berkey, Dennis Dale
Berkovitz, Leonard David
Berkowitz, Jerome
Berman, Robert Hiram
Bernard, Peter Simon
Birkhoff, Garrett
Birnbaum, Zygmunt William
Blackburn, Jacob Floyd
Boal, Jan List
Bodonyi, Richard James
Bohachevsky, Ihor Orest
Bojadziev, George Nikolov
Boley, Bruno Adrian
Bollinger, Richard Coleman
Bramble, James Henry
Bremermann, Hans J.
Brockett, Roger Ware
Brown, John Lawrence, Jr.
Bullock, Robert Morton, III
Burstein, Joseph
Cacuci, Dan Gabriel
Caflisch, Russel Edward
Campbell, Russell Bruce
Campbell, Stephen LaVern
Carasso, Alfred Sam
Carr, Ralph W.
Caster, William Oviatt
Causer, Gary Lee
Cazes, Albert N.
Censor, Yair
Cerceo, John Michael
Charnes, Abraham
Chen, Wayne H.
Chow, Pao-Liu
Chudnovsky, David Volf
Ciment, Melvyn
Clark, Rudolph Ernest
Cohen, Harley
Cohen, Michael Paul
Colvin, Burton Houston
Concus, Paul
Conrad, Bruce Phillips
Cook-Ioannidous, Leslie Pamela
Corduneanu, Constantin C.
Cornette, James Lawson
Criminale, William Oliver, Jr.
Crooke, Philip Schuyler, III
Cumberbatch, Ellis
Cushing, Jim Michael
Dana, Martin P.
Danes, Zdenko Frankenberger
Darden, Christine Mann
Datta, Subhendu K(umar)
Davison, Mark Edward
Deliyannis, Platon Constantine
Dendy, Joel Eugene, Jr.
Denny, William Francis, II
Dhaliwal, Ranjit Singh
Di Donato, Armido Richard
di Franco, Roland Bartholomew
Ditto, Frank Haselwood
Dobbins, James Gregory Hall
Donoghue, Edward Sylvester
Driessel, Kenneth R.
Durgun, Kanat
Dutton, John Altnow
Ehrlich, Louis William
Epperson, James Felts
Evernden, Jack Foord
Fabrikant, Valery Isaac
Ferry, Jason Hughes
Field, David Anthony
Fish, Andrew Joseph, Jr.
Fitzgibbon, William Edward, III
Fix, George Joseph
Fletcher, Harvey
Fletcher, John Edward
Follingstad, Henry George
Forbes, George Franklin
Fowler, Howland Auchincloss
Freiberger, Walter Frederick
Friedlander, Susan Jean
Friedrichs, Kurt Otto
Ganapol, Barry Douglas
Garabedian, Paul Roesel
Garzia, Mario Ricardo
Geer, James Francis
Ghil, Michael
Gilbert, Robert Pertsch
Gipson, Gary Steven
Glaz, Harland Mitchell
Goldstein, Charles Irwin
Goldstein, Jerome Arthur
Goldstine, Herman Heine
Gomory, Ralph Edward
Grad, Harold
Graff, Samuel M.
Gray, George Amelung
Gray, Samuel Hutchison
Grenier, Edward Joseph
Grimm, Louis J.
Grosch, Chester Enright
Gross, Kenneth Irwin
Grotte, Jeffrey Harlow
Guckenheimer, John
Gunzburger, Max Donald
Haas, Violet Bushwick
Haber, Seymour
Haberman, Richard
Haines, Charles Wills
Hall, Charles Allan
Hamilton, Eugene Phillip
Hanson, Floyd Bliss
Hanson, Marvin Harold
Hartfiel, Darald Joe
Harvey, Charles Arthur
Hemp, Gene W(illard)
Herrmann, Robert Arthur
Hicks, Darrell Lee
Hoagland, Gordon Wood
Hobart, Robert H.B.W.S., Jr.
Hodes, Louis
Hoffman, Frederick
Hoffman, William Charles
Holland, Robert Louis
Holt, William R.
Homer, Percy Albert
Howland, Robert Alden, Jr.
Hubbell, John Howard
Huddleston, Philip Lee
Humi, Mayer
Huneycutt, James Ernest, Jr.
Hunter, Christopher
Hussain, Moayyed A.
Ismail, Mourad E. H.
Jameson, William James, Jr.
Johnsen, Eugene Carlyle
Johnson, Robert Shepard
Jory, Virginia Vickery
Kalia, Ravindra Nath
Kalman, Rudolf Emil
Kammler, David William
Kamp, William Paul
Kanwal, Ram Prakash
Kaper, Hans Gerard
Keener, James Paul
Keesling, James Edgar
Kelley, Henry Joseph
Kemp, L(ouis) Franklin, Jr.
Keyfitz, Barbara Lee
Khan, Winston
Kimme, Ernest Godfrey
Kitchens, Clarence Wesley, Jr.
Klein, John Sharpless
Kockinos, Constantin Neophytos
Kravitz, David William
Kriegsmann, Gregory Anthony
Kroll, John Ernest
Krumhansl, James Arthur
Kruskal, Martin David
Kumar, Panganamala Ramana
Kunin, Isaak A.
Kunisch, Karl
Kuttler, James Robert
Landgren, John Jeffrey
Langford, William Finlay
Larsen, Edward William
Larsen, Kenneth Martin
Lax, Melvin David
Lax, Peter David
Leipholz, Horst Hermann Eduard
Leon, Steven Joel
Leonard, Anthony
Levinson, Mark
Levitin, Lev Berovich
Lewis, John Allen
Lieberman, Gerald Jacob
Lindquist, Anders Gunnar
Lomen, David Orlando
Loomis, Harold George
Looney, Paul Bryan
Lowengrub, Morton
Lu, Kau U.
Lucas, Richard John
Luce, R(obert) Duncan
Malkus, Willem VanRensseler
Maloney, John Patrick
Malvern, Lawrence Earl
Marder, Barry M(ichael)
Matkowsky, Bernard Judah
May, Robert McCredie
McGrath, Joseph Fay
McKinney, Stanley Joe
McLenithan, Kelly Daniel
McLinden, Lynn
McMahon, Maribeth
Menaldi, José Luis
Miles, John Wilder
Miller, Harvey Philip
Miller, Wilbur Charles
Miller, Willard, Jr.
Mills, Patrick Leo
Mills, Wendell Holmes, Jr.
Mityagin, Boris Samuel
Miura, Robert Mitsuru
Montroll, Elliott Waters
Moore, Polly

MEDICINE

Cooper, Herbert Asel
Creighton, Donald John
Cross, Richard Lester
Cryer, Dennis Robert
Cunningham, Earlene Brown
Dakshinamurti, Krishnamurti
Darnall, Dennis Wayne
Davie, Earl W.
Dawson, Earl Bliss
Dawson, Glyn
De Haas, Herman
Deibel, Martin Robert, Jr.
Dekker, Eugene Earl
DeLuca, Marlene A(nderegg)
de Miranda, Paulo
DeVore, Dale Paul
Di Pasquale, Gene
Dirksen, Thomas Reed, II
Distler, Jack Jounior
Donnelly, Thomas Edward, Jr.
Dupont, Jacqueline (Louise)
Eaton, Barbara Ruth
Edelman, Isidore Samuel
El Kouni, Mahmoud Hamdi
Ellenbogen, Leon
Erecinska, Maria
Essenberg, Margaret Kottke
Estabrook, Ronald Winfield
Finkelstein, James David
Fowler, Bruce Andrew
Fox, Irving Harvey
Franson, Richard Carl
Free, Charles Alfred
Freedland, Richard A.
Fridovich, Irwin
Friedenson, Bernard Allen
Friedmann, Naomi Kraus
Frisell, Wilhelm Richard
Fu, Shou-Cheng Joseph
Furie, Bruce
Gabriel, Othmar
Garfinkel, David
Geren, Collis Ross
Gershbein, Leon Lee
Ghebrehiwet, Berhane
Giles, Ralph Edson
Gilmore, Richard Allen
Glaser, Michael
Glenner, George Geiger
Glitz, Dohn George
Goh, Edward Hua Seng
Goldstein, Jack
Goldyne, Marc Ellis
Gormus, Bobby Joe
Gray, Robert Dee
Griffith, Owen Wendell
Grogan, William McLean
Gurd, Ruth Sights
Guth, Joseph Henry
Hanson, Douglas M.
Harland, Barbara Ferguson
Harris, Robert Allison
Harrison, John Henry, IV
Harrison, Yvonne Elois
Hartzell, Charles Ross, III
Hattox, Susan Ellen
Heinrich, Milton Rollin
Heinz, Erich
Henshaw, Edgar Cummings
Hitchcock-DeGregori, Sarah Ellen
Hofmann, Klaus Heinrich
Hollenberg, Paul Frederick
Holloway, Caroline Tobia
Holman, Ralph Theodore
Hommes, Fritz Aukustinus
Hopfer, Ulrich
Horning, Evan Charles
Horrocks, Lloyd Allen
Hostetler, Karl Yoder
Hougland, Arthur Eldon
Hsu, Ih-Chang
Huang, Cheng-Chun
Huang, Leaf
Hubbard, Walter Clyde
Hulcher, Frank Hope
Ihler, Garret Martin
Inagami, Tadashi
Iqbal, Zafar
Iqbal, Zafar Mohd
Jacobsohn, Gert Max
Jainchill, Jerome
Jamdar, Subhash Chandrashekhar
James, Margaret O.
Jencks, William Platt
Johnson, David Andrew
Kachmar, John Frederick
Kadis, Barney Morris
Kaistha, Krishan, Kumar
Kaltenbach, John Paul
Kampschmidt, Ralph Fred
Kao, Winston Whei-Yang
Kappas, Attalllah
Karp, Warren Bill
Kastl, Peter Robert
Kasvinsky, Peter John
Keeler, Elaine Kathleen
Kefalides, Nicholas Alexander
Kelleher, William Joseph
Kennedy, Frank Scott
Kim, Jin Kyung
Kirschenbaum, Donald Monroe
Kishimoto, Yasuo
Kitto, George Barrie
Klingman, Jack Dennis
Koide, Samuel Saburo
Korn, Edward David
Kornberg, Arthur
Krasny, Harvey Charles
Kuby, Stephen Allen
Kumar, Vijaya Buddhiraju
Kuntzman, Ronald Grover
Kuo, Jyh-Fa
Kurosky, Alexander
Kushwaha, Rampratap Singh
Lacko, Andras Gyorgy
Lancaster, Jack R., Jr.
Lane, Daniel McNeel
Lane, Stanley Earl
Last, Jerold Alan
Layman, Donald K(eith)
Lebherz, Herbert G(rover)
Lechner, Joseph Hadrian
Lefkowitz, Robert Joseph
Leis, Jonathan Peter
Lembach, Kenneth James
Lenard, John
Lerner, Leon Maurice
Lesko, Stephen Albert
Lester, David
Levy, Hilton Bertram
Lewis, Randolph Vance
Lewis, Urban James
Li, Li-Hsieng
Liao, Ta-Hsiu
Liburdy, Robert Peter
Lichtenwalner, Diane Marie
Lieberman, Seymour
Lin, Ying-Ming
Liu, Maw-Shung
Loftfield, Robert Bernard
Longmore, William Joseph
Loo, Yen-Hoong
Luduena, Richard Froilan
Lui, May So-Ying
MacKenzie, Robert Douglas
Makman, Maynard Harlan
Malbica, Joseph Orazio
Malins, Donald Clive
Malkinson, Alvin Maynard
Mansour, Tag Eldin
Marchalonis, John Jacob
Margolis, Frank Leonard
Margolis, Sam Aaron
Markland, Francis Swaby, Jr.
Marquis, Norman Ronald
Mathias, Melvin Merle
Matthews, Richard Hugh
Matthews, Rowena Green
McGilvery, Robert Warren
Meister, Alton
Mercer, L. Preston
Meyer, Ralph Roger
Miller, Francis Peter
Miller, James Edward
Miller, Richard Lee
Mittal, Chandra Kant
Monder, Carl
Montgomery, Rex
Mulligan, Richard C.
Murayama, Makio
Murthy, Vadiraja Venkatesa
Mushinski, J. Frederic
Neary, Joseph Thomas
Nelson, James Arly
Neufeld, Harold Alex
Noble, Nancy Lee
Nordlie, Robert Conrad
Oka, Takami
O'Leary, Gerard Paul, Jr.
Olson, James Allen
Pangburn, Michael Kent
Patsch, Josef Rudolf
Patsch, Wolfgang
Paul, Thomas Daniel
Pearlmutter, A. Frances
Pegg, Anthony Edward
Peterson, Rudolph Nicholas
Pezzuto, John Michael
Pfefferkorn, Elmer Roy
Phan, Sem Hin
Pilgeram, Laurence Oscar
Pizzo, Salvatore Vincent
Pollock, Jerry Joseph
Pories, Walter J.
Poston, John Michael
Potts, John Thomas, Jr.
Purdy, Ralph Earl
Quay, Steven Carl
Rall, Joseph Edward
Rao, Kalipatnapu Narasimha
Rawat, Arun Kumar
Ray, Richard Schell
Reddy, Neelupalli Bojji
Reichle, Frederick Adolph
Reitz, Ronald Charles
Reyes, Philip
Rhoads, Allen Roy
Richardson, Carol Lynn
Richardson, Keith Erwin
Riley, Michael Verity
Rose, Kathleen Mary
Rosen, Martin Howard
Rosenkrantz, Harris
Rosenthal, Harold Leslie
Roskoski, Robert, Jr.
Rovetto, Michael Julien
Sacks, William
Sahu, Saura Chandra
Samuel, Paul
Sanders, Robert Burnett
Schachman, Howard Kapnek
Schafer, Andrew Imre
Schechter, Alan Neil
Schenkman, John Boris
Schlender, Keith Kendall
Schlesinger, David Harvey
Schneider, Donald Leonard
Schor, Joseph Martin
Schrohenloher, Ralph Edward
Schultz, Richard Michael
Seetharam, Bellur
Segrest, Jere Palmer
Seidler, Frederic John
Shapira, Raymond
Sharpless, Nansie Sue
Shichi, Hitoshi
Siakotos, Aristotle Nicholas
Siegel, Abraham Lazarus
Silverberg, Michael
Sky-Peck, Howard H.
Slakey, Linda L.
Smeby, Robert Rudolph
Smith, Roberts Angus
Squire, Phil G.
Stacpoole, Peter Wallace
Stollar, Bernard David
Subbiah, Ravi Mandepanda Thimmiah
Suddath, Fred Leroy, Jr.
Sun, Grace Yan Chi
Sundsmo, John Sievert
Sussman, Ira Israel
Tabor, Herbert
Thomas, John Alva
Thompson, William Joseph
Tischler, Marc Eliot
Tolman, Edward Lauria
Torrence, Paul Frederick
Tritsch, George Leopold
Tumbleson, M.E.
Turco, Salvatore Joseph
Vanaman, Thomas C.
Van Dyke, Russell Austin
Venter, J. Craig
Vercellotti, John Raymond
Villar-Palasi, Carlos
Walsh, Kenneth Andrew
Waters, Larry Charles
Weber, Charles Walter
Weigel, Paul Henry
Weinberg, Uzi
Weinhold, Paul Allen
Weinhouse, Sidney
Wheat, Robert Wayne
Whikehart, David Ralph
Whitmer, Jeffrey Thomas
Winicov, Ilga Butelis
Winter, Charles Gordon
Witkop, Bernhard
Wogan, Gerald Norman
Wohl, Robert Chaim
Wolen, Robert Lawrence
Wong, Patrick Yui-Kwong
Wood, Harland Goff
Wood, Randall Dudley
Woodside, Kenneth Hall
Wrenn, Simeon Mayo
Wu, Cheng-Wen
Wu, Kenneth Kun-yu
Wyngaarden, James Barnes
Yang, Chung Shu
Yen, Terence Tsintsu
Yeoman, Lynn Chalmers
Yielding, Lerena Wade
Yu, Fu Li
Zakowski, Jack J.
Zenser, Terry Vernon

Biofeedback
Asterita, Mary Frances
Baskin, Steven Marc
Blumenthal, James Alan
Boudewyns, Patrick Alan
Burish, Thomas Gerard
Credidio, Steven George
Crider, Andrew Blake
Doerr, Robert Douglas
Doros, Maria Heczey
Engel, Bernard Theodore
Ford, Lincoln Edmond
Freedman, Robert Russell
Grimsley, Douglas Lee
King, Dennis R.
Krasner, Paul R.
Largen, John William, Jr.
Lefebvre, Richard Craig
McKee, Michael Geoffrey
Palladino, Joseph James
Perryman, James Harvey
Pope, Alan Thomas
Ruegsegger, Paul Melchior
Staats, Joan
Tansey, Michael Anselme
Verma, Ram Sagar
Weiss, L. Leonard

Cancer research. ***See also*** **Oncology.**
Ablashi, Dharam Vir
Ablin, Richard Joel
Agrawal, Krishna Chandra
Ahmed, Nahed K.
Aisenberg, Alan C.
Aisner, Joseph
Albano, William A.
Allaudeen, Hameedsulthan Sheik
Al-Sarraf, Muhyi
Ames, Matthew Martin
Ananthaswamy, Honnavara Narasimhamurthy
Anderson, Wayne Keith
Archer, Michael Christopher
Arseneau, James Charles
Auerbach, Oscar
Baker, Laurence Howard
Balbinder, Elias
Balinsky, Doris
Ballal, Raghu Veer
Banerjee, Sipra
Baron, Jeffrey
Bast, Robert Clinton, Jr.
Bean, Michael Arthur
Beattie, Craig Warren
Beirne, Owen Ross
Belman, Sidney
Benade, Leonard Edward
Berkelhammer, Jane
Berlin, Nathaniel Isaac
Bernacki, Ralph James
Berry, David Lester
Bichsel, Hans
Biswal, Nilambar
Blair, Donald George Ralph
Blitzer, Andrew
Bloom, Eda Terri
Borden, Ernest Carleton
Borsos, Tibor
Bouck, Noel Patrick
Boyd, Ann Lewis
Bradner, William Turnbull
Brady, Luther W., Jr.
Brattain, Michael G.
Broder, Samuel
Bronson, David Lee
Brown, Jay Clark
Broxmeyer, Hal Edward
Bruce, William Robert
Brysk, Miriam Mason
Bubbers, John Eric
Buehler, Robert John
Bull, Frances Eleanor
Bunn, Paul Axtell, Jr.
Burdette, Walter James
Burk, Martyn William
Burt, Michael Edward
Burzynski, Stanislaw Rajmund
Cameron, Deborah Jane
Camiolo, Sarah May
Campbell, Thomas Colin
Cardiff, Robert Darrell
Carrano, Anthony Vito
Castro, Joseph Ronald
Cave, William Thompson, Jr.
Cavins, John Alexander
Chaganti, Raju Sreerama Kamalasana
Chakravarti, Aravinda
Chan, Po Chuen
Chandra, Satish
Chang, Kenneth Shueh-Shen
Chang, William Wei-Lien
Chavin, Walter
Chechik, Boris E.
Cheng, Yung-chi
Chernicoff, David Paul
Chin, Hong W.
Chiu, Jen-Fu
Chiuten, Delia Fung She
Chou, Iih-Nan
Chou, Ting-Chao
Chu, Ann Maria
Cohen, Samuel M.
Coleman, C. Norman
Coltman, Charles Arthur, Jr.
Conney, Allan Howard
Cooney, David Anthony
Cooper, Robert Arthur, Jr.
Cox, Donald Cody
Creech, Richard Hearne
Currie, Violante Earlscort
Dalla-Favera, Riccardo
Daniel, Muthiah D.
Datta, Surjit Kumar
Davidson, Eugene Abraham
Dawson, Jeffrey Robert
Day, Calvin Lee, Jr.
Dean, Judith Carol Hickman
DelVillano, Bert C., Jr.
DeVita, Vincent Theodore, Jr.
DeWys, William Dale
De Young, Lawrence Mark
Diamandopoulos, George Theodore
Dicello, John Francis, Jr.
DiPaolo, Joseph Amedeo
Douglas, J. Fielding
Dow, Lois Weyman
Drewinko, Benjamin
Duff, James Thomas
Dunham, Wolcott Balestier
East, James Lindsay
Eastman, Alan Richard
Ebert, Paul Stoudt
Economou, Steven George
Edwards, Gordon Stuart
Egorin, Merrill Jon
Elgert, Klaus Dieter
El Kouni, Mahmoud Hamdi
Engler, Mark J
Epstein, Alan Lee
Epstein, Lois Barth
Erickson, Leonard Charles
Essex, Myron Elmer
Evans, Charles Hawes, *Cancer research*
Evans, James Thomas
Evans, Mary Jo
Evans, Robert
Evans, William Edward
Ewing, David Leon
Fairchild, Ralph Grandison
Falk, Hans Ludwig
Fay, Joseph Wayne
Feinerman, Burton
Felton, James Steven
Fernandes, Daniel James
Fine, Donald Lee
Fishman, William H(arold)
Fluck, Michele Marguerite
Frank, Irwin Norman
Freedman, Herbert Allen
Fu, Karen King-Wah
Fuks, Joachim Zbigniew
Fukuda, Minoru
Gale, Robert Peter
Gallager, Harry Stephen
Gallo, Robert Charles
Garnick, Marc Bennett
Garvin, Paul Joseph, Jr.
Gass, George Hiram
Gentile, James Michael
Geran, Ruth I.
Gifford, George Edwin
Glasgow, Glenn Patrick
Glazer, Robert Irwin
Golberg, Leon
Gordon, Robert Thomas
Gotay, Carolyn Cook
Gottlieb, Arlan Jay
Gray, Joe William
Green, William Robert
Greenstock, Clive Lewis
Greenwald, Edward S.
Griffin, Thomas William
Grimm, Elizabeth Ann
Gross, Ludwik
Grossweiner, Leonard Irwin
Gullino, Pietro Michele
Gurpide, Erlio
Hacker, Miles Paul
Hager-Rich, Jean Carol
Hall, Stephen William
Hall, Thomas Christopher
Hampar, Berge
Hampton, James Wilburn
Hanafusa, Hidesaburo
Hand, Roger
Hartman, Philip Emil
Harvey, Ronald Gilbert
Hawrylko, Eugenia Anna
Haymond, Herman Ralph
Hedgcoth, Charles
Heine, Ursula Ingrid
Henderson, Isaac Craig
Henderson, James Stuart
Henshaw, Edgar Cummings
Heppner, Gloria Hill
Herlyn, Meenhard Folkeus
Herz, Fritz
Higgins, Paul Joseph
Higginson, John
Hill, Donald Lynch
Hillman, Elizabeth Ann
Hofer, Kurt Gabriel
Hokama, Yoshitsugi
Horton, John
Hosick, Howard Lawrence
Houghton, Alan Hourse, Jr.
Houghton, Janet Anne
Houghton, Peter James
Howell, Stephen Barnard
Hrushesky, William John Michael
Hsu, Ih-Chang
Huberman, Eliezer
Hueser, James Nicholas
Huggins, Charles
Iannaccone, Philip Monroe
Iqbal, Zafar Mohd
Ishii, Douglas Nobuo
Jackson, Don Vernon, Jr.
Jacobs, Jerome Barry
Jeffrey, Alan Miles
Jensen, Elwood Vernon
Jensen, Keith Edwin
Jerina, Donald Michael
Jerry, Laurence Martin
Jessup, John Milburn
Johnson, Randall K.
Johnson, Terry Charles
Jones, Russell Allen
Jordan, V. Craig
Kaighn, Morris Edward
Kakunaga, Takeo
Kallman, Robert Friend
Kaminskas, Edvardas
Kamiyama, Mikio
Kaplan, Richard Stephen
Kapp, John Paul
Karle, Jean Marianne
Katzenellenbogen, Benita S(chulman)
Kellen, John Andrew
Kelsen, David Paul
Kennedy, Byrl James
Kennett, Roger Howard
Kessel, David
Kessinger, Margaret Anne
Khan, Amanullah
Kiang, David Teh-ming
Kim, Jae Ho
Kim, Yoon Berm
Kim, Yung Dai
King, David Kyle
King, (Mary) Margaret
Kirkman, Hadley
Kisliuk, Roy Louis
Kithier, Karel
Klicka, John Kenneth
Knudson, Alfred George, Jr.
Koo, Peter Hung-Kwan
Kori, Shashidhar Halappa
Kovacs, Charles Jeffrey
Koven, Bernard J.
Krakoff, Irwin Harold
Kramer, Barnett Sheldon
Krause, Charles Joseph
Krishan, Awtar
Krown, Susan Ellen
Kumar, Prasanna K.
Ladisch, Stephan
Lamon, Eddie William
Larrick, James William
Laurie, John Andrew
Lazo, John Stephen
Leavitt, Richard Delano
Lee, Ching-Li
Lee, Lihsyng Stanford
Leon, Shalom A.
Leong, Stanley Pui-Lock
Lerch, Irving Abram
LeVeen, Harry
Leventhal, Brigid Gray
Levin, Robert David
Li, Jonathan J.
Li, Li-Hsieng
Liao, Shuen-Kuei
Ligler, Frances Smith
Lint, Thomas Franklin
Lipkin, George
Livingston, David Morse
London, William Thomas
Longnecker, Daniel Sidney
Look, A. Thomas
Lotlikar, Prabhakar Dattaram
Lown, James William
Lubaroff, David Martin
Luderer, Albert August
Ludlum, David Blodgett
Luhby, A. Leonard
Lui, May So-Ying
Lupulescu, Aurel Peter
Lvovsky, Edward Abraham
Magee, Peter Noel
Maher, Veronica Mary
Maheshwar, Prem Narain
Main, James Hamilton Prentice
Mainigi, Daivender Kumar

Cardiology

Cell biology. ***See also*** **BIOLOGY.**

Cytology and histology

Critical care

Dermatology

Endocrinology

Duax, William Leo
Edgren, Richard Arthur
Eil, Charles
Fajans, Stefan Stanislaus
Fortier, Claude
Freinkel, Norbert
Friedmann, Naomi Kraus
Galton, Valerie Anne
Ganda, Om P.
Gass, George Hiram
Gaut, Zane Noel
Genest, Jacques
Gershberg, Herbert
Geula, Gibori
Ginsberg-Fellner, Fredda Vita
Goldberg, Lee Dresden
Goodman, David Barry Poliakoff
Gray, Timothy Kenney
Gregerman, Robert Isaac
Gruber, Helen Elizabeth
Gueriguian, John Leo, *Endocrinology*
Güllner, Hans-Georg
Hagen, Arthur Ainsworth
Hamilton, Carlos Robert, Jr.
Harrison, Robert Walker, III
Heath, Hunter, III
Hertz, Roy
Hirsch, Philip Francis
Hofeldt, Fred Dan
Hogness, John Rusten
Holtkamp, Dorsey E(mil)
Horwitz, David Larry
Hossner, Kim Lee
Hostetler, Karl Yoder
Hotes, Lawrence Steven
Jewelewicz, Raphael
Johnson, Joseph Alan
Johnsonbaugh, Roger Earl
Jordan, V. Craig
Kabadi, Udaya Manohar
Kadis, Barney Morris
Kahn, C. Ronald
Kalu, Dike Ndukwe
Kenny, Alexander Donovan
Komanicky, Pavel
Kostyo, Jack Lawrence
Landsberg, Lewis
Lang, Robert
Lerner, Leonard Joseph
Levitz, Mortimer
Lewis, Urban James
Li, Choh Hao
Liao, Shutsung
Liddle, Grant Winder
Lieberman, Seymour
Liechty, Richard Dale
Lin, Tu
Lis, Martin
Lostroh, Ardis June
Lupulescu, Aurel Peter
Lymangrover, John Robert
Matz, Robert
Mendel, Verne Edward
Miller, William W., III
Mohla, Suresh
Monder, Carl
Nocenti, Mero Raymond
Nutting, Ehard Forrest
Odell, William Douglas
Pak, Charles Y.C.
Parker, Lawrence Neil
Patel, Dhanooprasad Gordhanbhai
Penhos, Juan Carlos Jacobo
Pento, J. Thomas
Permutt, Marshall Alan
Phillips, Lawrence S(tone)
Pisu (Pisunyer), F. Xavier
Plymate, Stephen Rex
Potts, John Thomas, Jr.
Raizada, Mohan Kishore
Rall, Joseph Edward
Ramos-Gabatin, Angelita
Reel, Jerry Royce
Rees, Earl Douglas
Rezvani, Iraj
Rice, Bernard Francis
Rich, Clayton
Riggs, Byron Lawrence, Jr.
Roberts, Hyman Jacob
Rodriquez-Rigau, Luis Jose
Rosenfeld, Ron Gershon
Roth, Jesse
Rubenstein, Arthur Harold
Sandberg, Avery Aba
Sandberg, Hershel
Scherberg, Neal Harvey
Schrott, Helmut Gunther
Schultz, Alvin Leroy
Segal, Barry M.
Segaloff, Albert
Shimaoka, Katsutaro
Siegel, Abraham Lazarus
Silva, J. Enrique
Smallridge, Robert Christian
Soeldner, John Stuart
Solomon, David Harris
Spaziani, Eugene
Stacpoole, Peter Wallace
Stoffer, Sheldon Saul
Theoharides, Theoharis Constantin
Threlfall, Walter Ronald
Umminger, Bruce Lynn
Walsh, Scott Wesley
Weinberger, Myron Hilmar
White, Neil Harris
Wolf, Richard Clarence
Wolff, Frederick William
Wu, Sing-Yung
Yetiv, Jack Zeev
Zimmerman, Earl Abram

Endocrinology, neuroendocrinology
Albers, Henry Elliott
Allen, John Paul
Amoss, Max St. Clair, Jr.
Anderson, Lloyd Lee
Antunes, João Lobo
Aronin, Neil
Baksi, Samarendra Nath
Berelowitz, Michael
Berlind, Allan
Beyer-Mears, Annette
Blackman, Marc Roy
Brito, Gilberto Ottoni
Brosnihan, K. Bridget
Brush, F. Robert
Bryson, George Gardner
Buckman, Maire Tults
Bukovsan, William
Carlson, Joseph Ralph
Carr, Daniel Barry
Carr, Laurence A.
Chappel, Scott Carlton
Chen, Hsien-Jen James
Cheng, Mei-Fang Hsieh
Chin, William Waiman
Chung, Melvin Chung-Hing
Conn, P. Michael
Cronin, Michael John
Dailey, Robert Arthur
Dana, Richard Charles
Debons, Albert Frank
Della-Fera, Mary Anne
Desjardins, Claude
Dunn, Adrian John
Einhorn, Daniel
Eisenbarth, George Stephen
Eldridge, John Charles
Elkind-Hirsch, Karen Elizabeth
Ellingboe, James
Etgen, Anne Marie
Feldman, Susan C.
Fernstrom, Madelyn Hirsch
Fernstrrom, John Dickson
Fischette, Christine Theresa
Ford, Donald Herbert
Foreman, Mark Mortensen
Fortier, Claude
Freeman, Marc Edward
Fujimoto, George I.
Gambert, Steven Ross
Gash, Don Marshall
Gorbman, Aubrey
Grandison, Lindsey James
Grota, Lee James
Guillemin, Roger
Güllner, Hans-Georg
Halbreich, Uriel
Hatton, Glenn Irwin
Herman, William Sparkes
Holaday, John Waldron
Izzo, Joseph L., Jr.
Jacoby, Jacob Herman
Kalra, Satya Paul
Kastin, Abba Jeremiah
Keefer, Donald Ashby
King, Joan Caluda
Knobil, Ernst
Leeman, Susan Epstein
Levine, Seymour
Lewis, Randolph Vance
Loh, Yoke Peng
Loosen, Peter Thomas
Lu, John Kuew-Hsiung
Luine, Victoria Nall
Manger, William Muir
Martin, James Tillison
McGuire, John L.
McIntosh, Tracy Kahl
McMillin, John Michael
Mellin, Theodore Nelson
Menninger, Richard Price
Meyer, Donald Charles
Nilaver, Gajanan
Pekary, Albert Eugene
Phillips, Michael Ian
Plotka, Edward Dennis
Plotsky, Paul Mitchell
Ragavan, Vanaja Vijaya
Reiter, Russel Joseph
Rivier, Catherine Laure
Rockhold, Robin William
Ryan, Kenneth John
Ryder, Steven William
Samaan, Naguib Abdelmalik
Sawyer, Charles Henry
Schally, Andrew Victor
Scharrer, Berta Vogel
Schlesinger, David Harvey
Schwartz, William Joseph
Severs, Walter Bruce
Shambaugh, George Elmer, III
Sladek, Celia Davis
Soliman, Karam Farag Attia
Spies, Harold Glen
Sridaran, Rajagopala
Steger, Richard Warren
Stumpf, Walter Erich
Timiras, Paola Silvestri
Tischler, Arthur S.
Uno, Hideo
Van Loon, Glen Richard
Velardo, Joseph Thomas
Viveros, Osvaldo Humberto
Wade, Charles Edwin
Walsh, John Harley
Weinberg, Uzi
Williams, Carroll Milton
Wilson, Richard Cameron
Wurtman, Richard Jay
Yalow, Rosalyn Sussman
Yasuda, Naoki
Zager, Philip George

Endocrinology, receptors
Archer, Juanita Almetta Hinnant
Aronow, Lewis
Bardin, Clyde Wayne
Baumann, Gerhard
Beattie, Craig Warren
Bhalla, Vinod K.
Bhathena, Sam Jehangirji
Ciaraldi, Theodore Paul
Clark, James Henry
Conn, P. Michael
Davies, Albert Owen
Galton, Valerie Anne
Gurd, Ruth Sights
Harrison, Robert Walker, III
Hornbrook, Kent Roger
Jacobs, Steven Jay
Jensen, Elwood Vernon
Johnson, Leonard Roy
Katzenellenbogen, Benita S(chulman)
Keefer, Donald Ashby
LaBarbera, Andrew Richard
Lata, Gene Frederick
Lee, Chung
Leung, Benjamin Shuet-kin
Li, Jonathan J.
Mehta, Rajendra G.
Molteni, Agostino
Peach, Michael Joe
Raam, Shanthi
Read, George Wesley
Reitz, Richard Elmer
Riehl, Robert Michael
Rosner, Anthony Leopold
Saffran, Judith
Savage, Carl Richard, Jr.
Segal, Joseph
Silva, J. Enrique
Simons, Samuel Stoney, Jr.
Singhakowinta, Amnuay
Sridaran, Rajagopala
Varga, Janos M.
Venter, J. Craig
Wilson, Jean Donald
Wotiz, Herbert Henry

Epidemiology
Alderman, Michael Harris
Alexander, Edward Russell
Baine, William Brennan
Band, Jeffrey David
Baskin, David
Brandt, Carl David
Breslow, Lester
Caldwell, Glyn Gordon
Christenfeld, Roger Michael
Craven, Donald Edward
Denny, Floyd Wolfe, Jr.
Densen, Paul Maximillian
Diehl, Andrew Kemper
Eisenberg, Leon
Fagan, Raymond
Feinstein, Alvan Richard
Finkel, Madelon Lubin
Fletcher, Robert H.
Fox, John Perrigo
Friedman, Gary David
Gardner, John Willard
Goldstein, Murray
Grayston, J. Thomas
Griffiss, J(ohn) McLeod
Gross, Peter Alan
Haynes, Suzanne G.
Henderson, Maureen McGrath
Hennekens, Charles H.
Hochberg, Marc Craig
Holliman, Rhodes Burns
Horstmann, Dorothy Millicent
Hulka, Barbara Sorenson
Hutchison, George Barkley
James, Sherman Athonia
Kandelman, Daniel
Kannel, William Bernard
Kaplan, Robert Malcolm
Kass, Edward Harold
Keehn, Robert John
Kendal, Alan Philips
Kilbourne, Edwin Michael
Kurland, Leonard Terry
Lawrence, Dale Nolan
Lebowitz, Michael David
Locke, Ben Z(ion)
London, William Thomas
Lyman, Gary Herbert
Mahboubi, Ezzat Ollah
Marienfeld, Carl Joseph
Mattson, Margaret Ellen
Neuhauser, Duncan von Briesen
Nickol, Brent Bonner
Nomura, Abraham Michael Yozaburo
Ostrow, David G.
Parsons, James Eugene
Purcell, Robert Harry
Robins, Lee Nelken
Schnurrenberger, Paul Robert
Schrott, Helmut Gunther
Schuman, Leonard Michael
Scrimshaw, Nevin Stewart
Shelokov, Alexis
Sheps, Cecil George
Soller, R. William
Spence, Mary Anne
Starfield, Barbara Helen
Van Peenen, Peter F(ranz) D(irk)
Wahl, Patricia Walker
Waldman, Ronald Jay
Wegman, David Howe
Weill, Hans
Weiss, James Moses Aaron
Whelan, Elizabeth M.
Will, Loren August
Wilson, George Porter, III
Winkelstein, Warren, Jr.
Woods, James Sterrett
Yeung, Katherine Lu
Zimmer, James Griffith

Family practice
Antonucci, Toni Claudette
Bennett, Donald Raymond
Burket, George Edward, Jr.
Fischer, (Albert) Alan
Horenstien, David
Lopez, Larry Markell
Marzella, Libero Louis
Posso, Manuel
Price, James Gordon
Shahady, Edward John
TePoorten, Bernard A.
Williams, Paul Tennyson

Gastroenterology
Achord, James Lee
Alphin, Reevis Stancil
Barone, Frank Carmen
Boland, C. Richard
Bozymski, Eugene Michael
Burk, Raymond Franklin, Jr.
Chalmers, Thomas Clark
Christensen, James
Clay, George A
Cooke, Allan Roy
Dajani, Esam Zapher
Dearing, William Hill
Della-Fera, Mary Anne
Deschner, Eleanor Elizabeth
Dietschy, John Maurice
Dubois, Andre
El-Ackad, Tarek M.
Euler, Arthur Ray
Gardner, Jerry David
Gelb, Alvin Meyer
Gidda, Jaswant Singh
Goetsch, Carl Allen
Goldman, Ira Steven
Gottlieb, Leonard Solomon
Holzbach, R. Thomas
Hoskins, Johnny Durr
Hoyumpa, Anastacio Maningo
Humphries, Thomas Joel
Isselbacher, Kurt Julius
Javitt, Norman B.
Katz, Julian
Kirsner, Joseph Barnett
Klein, Gordon Leslie
Land, Ivan Marshall
Lebenthal, Emanuel
Levine, Gary M.
Lewis, Terence David
Liebow, Charles
Long, Billy Wayne
Lumeng, Lawrence
Meyer, George Wilbur
Middleton, Henry Moore, III
Moody, Frank Gordon
Morrissey, John Fielding
Mortillaro, Nicholas A.
Olsen, Ward Alan
Pincus, Irwin J.
Pitchumoni, Capecomorin S.
Plaut, Andrew George
Popper, David Henry
Powell, Don Watson
Rajan, Kannan Ramalingam
Ram, Madhira Dasaradhi
Rayford, Phillip Leon
Reichen, Juerg
Reitemeier, Richard Joseph
Rhodes, James Benjamin
Ritchie, Wallace Parks, Jr.
Rosenberg, Irwin Harold
Roth, James Luther Aumont
Russell, Robert Mitchell
Salom, Ira Louis
Sarva, Rajendra P.
Schade, Robert Richard
Schmidt, Moshe
Schuster, Marvin Meier
Scott, Gerald William
Seetharam, Bellur
Shafritz, David Andrew
Shearin, Nancy Louise
Shriver, David Allen
Slomiany, Bronislaw Leszek
Smith, Philip Lawrence
Snape, William John, Jr.
Solomons, Noel W.
Stein, George Nathan
Stern, Robert Morris
Summers, Robert Wendell
Szabo, Sandor
Tache, Yvette France
Wald, Arnold
Walsh, John Harley
Weisberg, Aaron
Weiser, Milton Moses
Wilson, Frederick Allen
Wood, Jackie Dale
Woodruff, Calvin Watts
Yalow, Rosalyn Sussman
Zeldis, Jerome Bernard
Zimmerman, Hyman Joseph

Genetics and genetic engineering. *See also* **BIOLOGY.**
Allen, Robert Carter
Al Saadi, Abdul Amir
Anderson, W. French
Axel, Richard
Baker, Robert Frank
Baron, Miron
Baumiller, Robert Cahill
Bearn, Alexander Gordon
Bender, Harvey Alan
Benz, Edward John, Jr.
Bopp, Lawrence Howard
Bowman, James Edward
Brady, Roscoe O.
Brewer, George J.
Brown, Michael Stuart
Brown, William Ted
Carson, Paul Elbert
Chaganti, Raju Sreerama Kamalasana
Chan, Wai-Yee
Chandler, Jerry LeRoy
Chassy, Bruce Matthew
Clewell, Don Bert
Crusberg, Theodore Clifford
Cryer, Dennis Robert
Cummins, Joseph Edward
D'Ambrosio, Steven Mario
Diacumakos, Elaine G.
Dorn, Gordon Lee
Drees, Thomas Clayton
Duckworth, Donna Hardy
Epstein, Henry Fredric
Felton, James Steven
Ferretti, Joseph Jerome
Frias, Jaime Luis
Friedman, Orrie Max
Gartler, Stanley Michael
Gelehrter, Thomas David
German, James Lafayette, III
Giblett, Eloise Rosalie
Gilmore, Richard Allen
Goldstein, David Joel
Goldstein, Joseph Leonard
Goldstein, Paul
Gorlin, Robert James
Gottesman, Irving Isadore
Graham, John Borden
Hansen, Carl Tams
Harris, Harry
Hecht, Frederick
Holmquist, Gerald Peter
Holtzman, Neil Anton
Hommes, Fritz Aukustinus
Howard-Peebles, Patricia Nell
Huang, Eng-Shang
Jackson, Charles Eugene
Kan, Yuet Wai
Karn, Robert Cameron
Kennett, Roger Howard
Knudson, Alfred George, Jr.
Konrad, Michael
Kopecko, Dennis J.
Krieger, Ingeborg
Kucherlapati, Raju S.
Kuramitsu, Howard Kikuo
Kurosky, Alexander
Law, Peter Koi
Lebo, Roger Van
Lee, Ching Y.
Lindquist, Susan Lee
Littlefield, John Walley
Low, Kenneth Brooks, Jr.
Lowy, Douglas Ronald
Lubiniecki, Anthony Stanley
Manley, Audrey Forbes
Marinus, Martin Gerard
Mayeda, Kazutoshi
Merkel, Christian Gottfried
Middleton, Richard Burton
Morse, Melvin Laurance
Motulsky, Arno Gunther
Mukherjee, Anil Baran
Murray, Robert Fulton, Jr.
Nebert, Daniel Walter
Neel, James Van Gundia
Nitowsky, Harold Martin
Omenn, Gilbert Stanley
Opitz, John Marius
Orkin, Stuart H.
Parisi, Joseph Thomas
Paul, Thomas Daniel
Pelzer, Charles Francis
Permutt, Marshall Alan
Pincus, Jack Howard
Punnett, Hope Suzanne
Pyeritz, Reed Edwin
Rao, Dabereu Chandrasekhara
Reilly, Joseph Garrett
Rifkind, Richard Allen
Riggin, Charles Henry, Jr.
Robbins, Leonard G(ilbert)
Rogers, Edgar Stanfield
Rose, Kathleen Mary
Rossman, Toby G.
Rothschild, Henry
Rotter, Jerome Israel
Rottman, Fritz M.
Rowley, Peter Templeton
Schechter, Alan Neil
Schlessinger, David
Schroeder, Alice Louise
Schwaber, Jerrold
Scriver, Charles Robert
Shafritz, David Andrew
Shapiro, Burton Leonard
Shaw, Margery Wayne Schlamp
Spence, Mary Anne
Taggart, Robert Thomas
Thirion, Jean-Paul Joseph
Thompson, Edward Ivins Bradbridge
Trent, Dennis Wayne
Trosko, James Edward
Vesell, Elliot Saul
Wijsman, Ellen Marie
Wilson, Gregory Bruce
Winokur, George
Wolman, Sandra R.
Woo, Savio Lau Ching
Wu, Cheng-Wen

Gerontology
Abrass, Itamar B.
Al-Bagdadi, Fakhri Abdul Kareem
Albright, Julia Wan
Alikhan, Mahmood
Antonucci, Toni Claudette
Aronson, Miriam Klausner
Bartus, Raymond T.
Baxter, Joann Crystal
Bender, A(llan) Douglas
Bosmann, Harold Bruce
Branch, Laurence George
Brown, William Ted
Busse, Ewald William
Butler, Robert Neil
Calkins, Evan
Chen, Linda Huang
Cutler, Neal R.
Dean, Reginald Langworthy, III
Dimant, Jacob
Finley, Gordon Ellis
Gambert, Steven Ross
Gregerman, Robert Isaac

Greulich, Richard Curtice
Hansford, Richard Geoffrey
Harman, Denham
Hayflick, Leonard
Hirsch, Henry Richard
Inglis, James
Johnson, Thomas Eugene
Kay, Marguerite M. Boyle
Klein, Lawrence Elliot
Lodge, George Townsend
Makinodan, Takashi
Martinson, Ida Marie
Masters, Robert Edward Lee
Mensh, Ivan Norman
Meyer, John Stirling
Miller, Arnold Reed
Moore, Francis Daniels
Morrison, Mary Dyer
Mosteller, Frederick
Petrie, William Marshall
Rapoport, Stanley I.
Richmond, Rollin Charles
Rossman, Isadore
Rothschild, Henry
Schwartz, Arthur Gerald
Solomon, David Harris
Steger, Richard Warren
Uno, Hideo
Volicer, Ladislav
Wagner, Jeames Arthur
Weindruch, Richard Howard
Young, Robert Claringbold

Gynecology. ***See* Obstetrics.**

Health services research
Appel, Antoinette Ruth
Beck, John Robert
Becker, Joan Alaine
Boostrom, Eugene Richard
Branch, Laurence George
Brown, Jack Harold Upton
Connelly, John Peter
Densen, Paul Maximillian
Donabedian, Avedis
Edwards, Carl Normand
Finkel, Madelon Lubin
Frazier, Howard Stanley
Gardin, T. Hershel
Gerety, Robert John
Greenlick, Merwyn Ronald
Hamburg, David A.
Hatoff, Alexander
Hazdra, James J.
Hiatt, Howard H.
Hoeller, Louise
Horvath, William John
Hubbell, Floyd Allan
Hulka, Barbara Sorenson
Jacobs, Durand Frank
Kennedy, Clive Dale
Kohn, Robert M.
Lambird, Perry Albert
Lev, Maurice
Manohar, Murli Vaid
Mishra, Shri Kant
Monson, Roberta Ann Mills
Morey, Philip Richard
Neuhauser, Duncan von Briesen
Pauker, Stephen Gary
Pruss, Thaddeus Paul
Reichgott, Michael J.
Rich, Clayton
Rogers, David Elliott
Rushmer, Robert Frazer
Sheppard, Louis Clarke
Shimosato, Shiro
Starfield, Barbara Helen
Stark, Jack Alan
Taylor, Roger Norris
Thomas, Charles Samuel
Weinstein, Milton Charles
Weiss, Daniel Leigh
Werner, Barbara Graham
White, Raymond Petrie, Jr.
Williams, Marjorie Joan
Woerth, Janice Kay
Zimmer, James Griffith
Zwemer, Thomas John

Hematology
Akbar, Huzoor
Alcena, Valiere
Amir, Jacob
Amjad, Hassan
Anderson, Tom
Andreeff, Michael
Asakura, Toshio
Ascensão, João L.
Awad, William Michel, Jr.
Bainton, Dorothy Dee Ford
Balcerzak, Stanley Paul
Barger, James Daniel
Beck, William Samson
Benz, Edward John, Jr.
Berlin, Nathaniel Isaac
Bernstein, Shelly Corey
Bessman, Joel David
Beutler, Ernest
Bond, William Holmes
Bottomley, Sylvia Stakle
Boxer, Laurence Alan
Brinkhous, Kenneth Merle
Brown, Loren Dennis
Broxmeyer, Hal Edward
Budzynski, Andrei Zygmunt
Bull, Brian Stanley
Burd, Robert M.
Burns, Edward R.
Butler, John J.
Butler, Thomas Parke
Butler, William Manion
Carroll, David Stewart
Case, Delvyn Caedren, Jr.
Cassileth, Peter Anthony
Cavins, John Alexander
Chanana, Arjun Dev
Chang, Jae Chan
Chernoff, Amoz Immanuel
Chou, Albert Chung-Ho
Coleman, Morton
Colman, Robert Wolf
Comp, Philip Cinnamon
Cooper, Herbert Asel
Cooper, Richard Alan
Costea, Nicolas Vincent
Cronkite, Eugene Pitcher
Dalla-Favera, Riccardo
Davila, Enrique
DeVita, Vincent Theodore, Jr.
Dollinger, Malin Roy
Drees, Thomas Clayton
Durie, Brian George Martin
Fadulu, Sunday O.
Finch, Stuart Cecil
Fisher, James W.
Fleming, James Stuart, Jr.
Flessa, Herbert Christian
Franks, John Julian
Frenkel, Eugene Phillip
Furie, Bruce
Giblett, Eloise Rosalie
Goldberg, Jack
Golde, David William
Goldstein, Jack
Gordon, David Hugh
Gottlieb, Arlan Jay
Greenberg, Lowell Herbert
Hall, Charles A.
Hampton, James Wilburn
Haraf, Frank Joseph
Hauer, Jerome Maurice
Hays, Esther Fincher
Heyssel, Robert Morris
Hocking, William Gray
Hoovis, Marvin Lorin
Hoyer, Leon William
Hyman, Carol Brach
Iatridis, Panayotis George
Ichiki, Albert T.
Itano, Harvey Akio
Jacobson, Leon Orris
Jaffe, Ernst Richard
Jain, Sushil Kumar
Joishy, Suresh K.
Joist, Johann Heinrich
Juneja, Harinder Singh
Kan, Yuet Wai
Kasimis, Basil Spiros
Kass, Lawrence
Kennealey, Gerard
Klock, John Charles
Kochwa, Shaul
Kornfeld, Stuart Arthur
Kosow, David Phillip
Kough, Robert Hamilton
Kowalyshyn, Theodore Jacob
Kyle, Robert Arthur
La Celle, Paul Louis
Lane, Frank Benjamin
Lanzkowsky, Philip
Laufman, Leslie Rodgers
Lawrason, F. Douglas
Lazarchick, John
Lazerson, Jack
Levin, William Cohn
Ley, Timothy James
Lichtman, Marshall A.
Liu, Yung-Pin
Loeb, Virgil, Jr.
London, Irving Myer
Longenecker, Gesina (Louise) Lizana
Louis, John
Lynch, Sean Roborg
MacKenzie, Robert Douglas
Marcoullis, George Panayiotis
Masouredis, Serafeim P.
McCredie, Kenneth Blair
McIntyre, Oswald Ross
Minwada, Jun
Mirand, Edwin Albert
Morrison, Francis Secrest
Morse, Bernard S.
Nakai, George S.
Nelson, Eric Charles
Oken, Mertin M.
Orkin, Stuart H.
Papayannopoulou, Thalia
Patten, Ethel Doudine
Perry, David John
Petursson, Sigurdur Ragnar
Pirofsky, Bernard
Rahman, Fazlur
Ramanan, Sundaram Venkata
Rassiga, Anne Louise
Ratnoff, Oscar Davis
Rausen, Aaron Reuben
Richard, Joseph
Richards, Frederick, II
Rifkind, Richard Allen
Robinson, Stephen Howard
Rowley, Janet Davison
Rozen, Simon
Sandler, Sumner Gerald
Schafer, Andrew Imre
Schilling, Robert Frederick
Schwartz, Allen David
Scott, Robert Bradley, Sr.
Scott, Roland Boyd
Shapiro, Donald Michael
Shaw, Michael Trevor
Sherry, Sol
Silverberg, Michael
Smith, John Bryan
Sprague, Charles Cameron
Stein, Ira David
Steinberg, Martin H.
Steinherz, Peter Gustav
Straus, David Jeremy
Sussman, Ira Israel
Swisher, Scott Neil
Sydorak, Jaroslava Kuzmycz
Taetle, Raymond
Tavassoli, Mehdi
Walker, Brian Keith
Waxman, Herbert Sumner
Weinstein, Irwin Marshall
Weiss, Harvey Jerome
Whaun, June M.
White, Douglas Rector
Whitecar, John P., Jr.
Wiernik, Peter Harris
Wilner, George Dubar
Wolff, James Alexander
Wright, Daniel Godwin
Wu, Kenneth Kun-yu
Yoo, Dal

Hematology, marrow transplant
Armitage, James Olen
Bealmear, Patricia Maria
Bozdech, Marek Jiri
Fay, Joseph Wayne
Gee, Adrian Philip
Heim, Lyle Raymond
Herzig, Geoffrey Peter
Meredith, Ruby Frances
Patenaude, Andrea Farkas
Singer, Jack Wolfe
Strong, Douglas Michael
Stuart, Robert Kenneth
Thomas, Edward Donnall
Trigg, Michael Edward
Yeager, Andrew Michael

Immunology
Abdou, Nabih I.
Ablin, Richard Joel
Adkinson, N. Franklin, Jr.
Agnello, Vincent
Albini, Boris
Albright, Joseph Finley
Albright, Julia Wan
Alspaugh, Margaret Ann
Altman, Leonard
Amos, Dennis Bernard
Anderson, Arthur Osmund
Ariano, Marjorie A.
Aune, Thomas Martin
Bankhurst, Arthur Dale
Bardana, Emil John, Jr.
Barnett, John Brian
Barriga, Omar Oscar
Barth, Rolf Frederick
Bealmear, Patricia Maria
Bean, Michael Arthur
Beck, Lee Randolph
Beller, David I.
Bergstresser, Paul Richard
Berkelhammer, Jane
Bernstein, I. Leonard
Bianco, Celso
Birgit, Hertel-Wulff
Blaese, (Robert) Michael
Boyle, Michael Dermot
Broder, Samuel
Brown, Eric Reeder
Brown, John Clifford
Buehler, Robert John
Bulloch, Karen
Busse, William Walter
Bystryn, Jean-Claude
Callahan, Hugh James
Carsons, Steven Eric
Catalano, Michael Alfred
Cavallero, Joseph John
Chechik, Boris E.
Chen, Yi-Hsiang (Alan)
Chen, Yu Min
Chi, David S.
Chiorazzi, Nicholas
Chused, Thomas M.
Clagett, James Albert
Clinton, James Michael
Coe, John Emmons
Cohen, Philip Lawrence
Collison, Betty Christine
Conway de Macario, Everly
Cooper, Norman Streich
Cooper, Sheldon Mark
Cox, Donald Stephen
Crowell, Richard Lane
Crowle, Alfred John
Curd, John Gary
Dau, Peter Caine
Davis, John Mihran
Dawson, Jeffrey Robert
Dean, Jack Hugh
DeLustro, Frank Anthony
DelVillano, Bert C., Jr.
Dienstag, Jules Leonard
Dixon, Frank James
Dreesman, Gordon Ronald
Dyminski, John W(ladyslaw)
Eby, William Clifford
Eggert, Frank Michael
Eisenbarth, George Stephen
Engleman, Edgar George
Ennis, Francis A.
Epstein, Lois Barth
Espinoza, Luis Rolan
Evans, Robert
Fahey, John Leslie
Fierer, Joshua A(llan)
Fink, Mary Alexander
Finland, Maxwell
Foster, Charles Stephen
Fremount, Henry Neil
Friedenson, Bernard Allen
Fritz, Robert Bartlett
Fu, Shuman
Fudenberg, H. Hugh
Gallagher, Joan Shodder
Gee, Adrian Philip
Gerblich, Adi Abraham
Ghebrehiwet, Berhane
Gifford, George Edwin
Goebel, Walther Frederick
Goldman, Armond Samuel
Goodman, Michael Gordon
Gough, Patricia Marie
Grebenau, Mark David
Grimm, Elizabeth Ann
Gusdon, John Paul, Jr.
Hadden, John Winthrop
Hager-Rich, Jean Carol
Halme, Jouko Kalervo
Hamburger, Max I.
Hargis, Betty Jean
Havas, Helga Francis
Hawrylko, Eugenia Anna
Heim, Lyle Raymond
Heimer, Ralph
Heiner, Douglas Cragun
Henley, Walter L.
Herberman, Ronald Bruce
Herscowitz, Herbert Bernard
Hildreth, Eugene A.
Hillyer, George Van Zandt
Hoffman, Donald Richard
Hoffmann, Louis Gerhard
Horton, John Edward
Horwitz, David Allen
Hoyer, Leon William
Hunter, Robert L.
Huston, David Paul
Hyde, Richard Moorehead
Jacoby, Robert Ottinger
Jain, Naresh Kumar
Jarrett, Mark Paul
Jessup, John Milburn
Johnston, Marilyn F. M.
Kabat, Elvin Abraham
Kamiyama, Mikio
Kaplan, Joel Howard
Katz, David Harvey
Kaufman, Donald Barry
Kellen, John Andrew
Khan, Amanullah
Kier-Schroeder, Ann B.
Kierszenbaum, Felipe
Kim, Yung Dai
Kimberly, Robert P.
Kind, Phyllis Dawn
Kite, Joseph Hiram, Jr.
Klapper, David Gary
Kobayashi, Roger Hideo
Kobilinsky, Lawrence
Kochwa, Shaul
Kohler, Peter Francis
Korn, Joseph H.
Krown, Susan Ellen
Kulczycki, Anthony, Jr.
Kung, Patrick Chung-Shu
Kunkel, Henry George
Kyle, Robert Arthur
Lakin, James Dennis
Lamon, Eddie William
Lawrence, Ernest Clinton
Leavitt, Richard Delano
Lee, Ching-Li
Leon, Shalom A.
Leung, Christopher Chung-Kit
Levy, Elinor M.
Levy, Nelson Louis
Lint, Thomas Franklin
Linthicum, Darwin Scott
Lischner, Harold Will
Lopatin, Dennis Edward
Luderer, Albert August
Lum, Lawrence George
Luskin, Allan Tessler
Malley, Arthur
Marchalonis, John Jacob
Marx, James John, Jr.
Masouredis, Serafeim P.
Mathur, Subbi
Mayer, Manfred Martin
McCalmon, Robert Thomas, Jr.
McConahey, Patricia Jane
McDevitt, Hugh O'Neill
Mellors, Robert Charles
Miller, Glenn Allan
Miller, John Johnston, III
Mills, John Alexander
Mitchell, Malcolm Stuart
Morse, Bernard S.
Morse, Stephen Scott
Mudgett-Hunter, Meredith
Müller-Eberhard, Hans Joachim
Nahm, Moon Hea
Nardella, Francis Anthony
Nash, Donald Robert, immunologist/microbiologist
Nowell, Peter Carey
Nussenzweig, Victor
Pangburn, Michael Kent
Panush, Richard Sheldon
Parker, Charles Ward
Patterson, Loyd Thomas
Pinnas, Jacob Louis
Pirofsky, Bernard
Plotz, Paul Hunter
Pollack, William
Ratajczak, Helen Vosskuhler
Reaman, Gregory Harold
Reif, Arnold Eugene
Riviere, George Robert
Rothberg, Richard Martin
Rowley, Donald A.
Rudofsky, Ulrich Hubert Waldemar
Schlager, Seymour Irving
Schrohenloher, Ralph Edward
Schwartz, Anthony
Schwartz, Stanley Allen
Sell, Kenneth Walter
Senterfit, Laurence Benfred
Sery, Theodore Wilson
Shultz, Leonard Donald
Smith, Jackson Bruce
Smith, Kendall Owen
Sonnenfeld, Gerald
Soter, Nicholas Arthur
South, Mary Ann
Spilberg, Isaias
Spitzer, Roger Earl
Steblay, Raymond William
Steinberg, Alfred David
Stollar, Bernard David
Strober, Samuel
Stromberg, Bert Edwin, Jr.
Suzuki, Jon Byron
Suzuki, Tusneo
Tan, Eng Meng
Taylor, Roger Norris
Tennenbaum, James Irving
Terman, David Stephen
Thompson, James Jarrard
Tripodi, Daniel
Tsang, Alfred Kwong-Y
Valentine, Martin Douglas
Vandenbark, Arthur Allen
Vaughan, John Heath
Veltri, Robert William
Vessella, Robert Louis, Jr.
Vladutiu, Adrian Octavian
Volanakis, John Emmanuel
Volkman, David J.
Wacker, Waldon Burdetter
Waldmann, Thomas Alexander
Walia, Amrik Singh
Warner, Noel Lawrence
Warren, Kenneth S.
Wedner, H. James
Wei, Wei-Zen
Weiler, John Mayer
Wilson, Gregory Bruce
Wing, Edward Joseph
Witkin, Steven S.
Woods, Roy
Wright, George G(reen)
Wright, George Leonard, Jr.
Wright, Richard Kenneth
Yarchoan, Robert
Yecies, Lewis David
Zawadzki, Zbigniew Apolinary
Zweiman, Burton

Immunology, cellular engineering
Attallah, Abdelfattah
Bone, Donald Robert
Collins, Arlene Rycombel
Corley, Ronald Bruce
Duncan, William R.
Eckels, David Dean
Folds, James Donald
Gillis, Steven
Graziano, Kenneth Donald
Kim, Byung Suk
Larrick, James William
Macario, Alberto Juan Lorenzo
Malley, Arthur
Milani, Cyrus Saeed
Miller, Stephen Douglas
Moticka, Edward J.
Neefe, John Robert, Jr.
Nigam, Vijai Nandan
Nisfeldt, Michael Lee
Oldham, Robert Kenneth
Owen, Charles Scott
Paque, Ronald Edward
Reddy, Mohan M.
Ritts, Roy Ellot, Jr.
Roth, Jack Alan
Schreiber, Hans
Shevach, Ethan Menahem
Steele, Glenn Daniel, Jr.
Strom, Terry Barton
Tom, Baldwin H.
Tondreau, Suk-Pan, (Sue)
Tsoi, Mang-So
Winkelhake, Jeffrey Lee
Yoshida, Takeshi

Immunology, immunogenetics
Bach, Marilyn Lee
Benacerraf, Baruj
Birgit, Hertel-Wulff
Birshtein, Barbara Kathryl
Bubbers, John Eric
Carpenter, Charles Bernard
Cederqvist, Lars Lennart
Clark, Edward Alan
Cohen, Nicholas
Cramer, Donald Vernon
Dixon, Frank James
Duncan, William R.
Eckels, David Dean
Fathman, Charles Garrison
Flaherty, Lorraine Amelia
Freed, John Howard
Freedman, Herbert Allen
Frelinger, Jeffrey Allen
Fudenberg, H. Hugh
Garovoy, Marvin R.
Gill, Thomas James, III
Horng, Wayne J(ing-Wei)
Kagan, Irving George
Kaplan, Alan Marc.
Kenny, George Edward
Kindt, Thomas James
Knowles, Barbara B.
Krause, Richard Michael
Lewis, Alan James
Mathieson, Bonnie Jean
McCombs, Candace Cragen
Miller, Stephen Douglas
Mittal, Kamal Kant
Mudgett-Hunter, Meredith
Mushinski, J. Frederic
Nathenson, Stanley Gail
Nicklas, Janice Ann
Norman, Douglas James
Pandey, Janardan Prasad
Paul, William Erwin
Pierce, Carl William
Popp, Diana Marriott
Purtilo, David Theodore
Roux, Kenneth Henry
Schanfield, Moses Samuel
Schultz, Jane Schwartz
Sears, Duane William
Shevach, Ethan Menahem
Shultz, Leonard Donald

Snell, George Davis
Solinger, Alan Michael
Sondel, Paul M.
Stansfield, William Donald
Steinberg, Alfred David
Stormont, Clyde Junior
Taggart, Robert Thomas
Waltenbaugh, Carl Ralph
Yu, David Tak Yan
Zaleski, Marek Bohdan
Zeldis, Jerome Bernard

Immunology, immunopharmacology
Attallah, Abdelfattah
Bach, Michael Klaus
Bankhurst, Arthur Dale
Bast, Robert Clinton, Jr.
Bayer, Barbara Moore
Beaven, Michael Anthony
Borden, Ernest Carleton
Brunda, Michael John
Cabrera, Edelberto Jose
Cameron, Deborah Jane
Casey, Francis B.
Dianzani, Ferdinando
Evans, Charles Hawes
Feinerman, Burton
Fidler, John Michael
Gabourel, John Dustan
Gillespie, Elizabeth
Gillis, Steven
Glasky, Alvin Jerald
Godfrey, Henry Philip
Goldenberg, Marvin M.
Good, Robert Alan
Griswold, Don E.
Grunberg, Emanuel
Hadden, John Winthrop
Hahn, Bevra H(annahs)
Halonen, Marilyn Jean
Hastings, Robert Clyde
Herzig, David Jacob
Huard, Thomas King
Jensen, Keith Edwin
Kampschmidt, Ralph Fred
Kolodny, Abraham Lewis
Lahita, Robert George
Loughman, Barbara Ellen Evers
Lvovsky, Edward Abraham
Mathews, Kenneth Pine
Moolten, Frederick London
Norman, Philip Sidney
Panagides, John
Pisko, Edward John
Rosenberger, John Knox
Rothman, John M.
Schlager, Seymour Irving
Schultz, Richard Michael
Schwartz, Anthony
Spitler, Lynn E(llen)
Stewart, Patrick Brian
Stinnett, Jimmy Dwight
Tarr, Melinda Jean
Thueson, David Orel
Trigg, Michael Edward
Wang, Bosco Shang
Webster, Leslie Tillotson, Jr.
Webster, Robert Owen
Wechter, William Julius
Wolberg, Gerald
Wong, Stewart
Wood, David Dudley
Yunger, Libby Marie

Immunology, immunotoxicology
Bice, David E.
Caren, Linda Ann Davis
Dean, Jack Hugh
Faith, Robert Earl, Jr.
Gainer, Joseph Henry
Ghaffar, Abdul
Lawrence, David A.
Liburdy, Robert Peter
Megel, Herbert
Minton, Sherman Anthony
Pazdernik, Thomas Lowell
Schwartz, Lester William
Tarr, Melinda Jean
Wong, Stewart
Wright, Larry Lyle

Immunology, infectious diseases
Abrutyn, Elias
Albright, Joseph Finley
Allen, Robert Charles
Andersen, Burton Robert
Apicella, Michael Allen
Austrian, Robert
Bahn, Arthur Nathaniel
Band, Jeffrey David
Beam, Thomas Roger
Blacklow, Neil Richard
Braciale, Thomas Joseph
Buchholz, Donna Marie
Cabrera, Edelberto Jose
Carpenter, Charles Colcock Jones
Casali, Paolo
Cavallero, Joseph John
Chang, Te-Wen
Choppin, Purnell Whittington
Clark, Edward Alan
Coe, John Emmons
Cohn, Zanvil A.
Collins, Frank Miles
Cooney, Marion Kathleen
Corbeil, Lynette Bundy
Cramblett, Henry Gaylord
Craven, Donald Edward
Crowle, Alfred John
Cukor, George
Despommier, Dickson Donald
Diggs, Carter Lee
Domer, Judith Elaine
DuPont, Herbert Lancashire
Durack, David Tulloch
Edelson, Paul Jeffrey
Ennis, Francis A.
Everett, Elwood Dale
Fadulu, Sunday O.
Folds, James Donald
Forghani, Bagher
Fox, Eugene Noah
Frank, Michael M.
Friedman, Harvey Michael
Funderburk, Noel Roger
Galasso, George John
Gallin, John I.
Gerety, Robert John
Glasky, Alvin Jerald
Gold, Jonathan W.M.
Gorzynski, Eugene Arthur
Gough, Patricia Marie
Grayston, J. Thomas
Greenberg, Richard Neil
Greene, Bruce McGehee
Griffiss, J(ohn) McLeod
Gross, Peter Alan
Hackett, Joseph Leo
Hamburger, Max I.
Hardin, Hilliard Frances
Harford, Carl Gayler
Hendley, Joseph Owen
Higashi, Gene Isao
Holmes, King Kennard
Hsu, Clement C.S.
Hyde, Richard Moorehead
Hyslop, Newton Everett, Jr.
Ito, Michio
Jacoby, George Alonzo, Jr.
Jain, Naresh Kumar
James, Stephanie Lynn
Jesen, James Burt
Jeska, Edward L(awrence)
Joshi, Jai Hind
Kass, Edward Harold
Kieff, Elliott
Kirk, Billy Edward
Kirkpatrick, Charles Harvey
Klebanoff, Seymour Joseph
Klein, Richard Joseph
Klimek, Joseph John
Kochan, Ivan
Kohl, Steve
Krugman, Saul
Larson, Vivian M.
Lauter, Carl Burton
Lawrence, Dale Nolan
Lehman, Thomas Joseph Ansorge
Lieberman, Michael Merril
Marchette, Nyven John
Marquardt, Warren William
Martin, Christopher Michael
Matsumoto, Masakazu
McCarty, Maclyn
McHenry, Martin Christopher
Merigan, Thomas Charles, Jr.
Monjan, Andrew Arthur
Morahan, Page Smith
Morse, Stephen Scott
Mortensen, Richard F.
Muchmore, Harold Gordon
Nash, Donald Robert
Neurath, Alexander Robert
Neva, Franklin Allen
Newman, Simon Louis
Nisfeldt, Michael Lee
Noble, Robert Cutler
O'Connor, G(eorge) Richard
Parker, John Clarence
Paterson, Philip Y.
Pelletier, Lawrence Lee, Jr.
Pesanti, Edward Louis
Peter, Georges
Petersdorf, Robert George
Pierce, Nathaniel Field
Plaut, Andrew George
Plotkin, Gary Robert
Prince, Alfred Mayer
Reddy, Mohan M.
Reed, William Patrick
Rissing, John Peter
Roberts, Glenn Dale
Sanford, Jay Philip
Sapico, Francisco Lejano
Schaeffer, Morris
Schwab, John Harris
Shope, Thomas Charles
Simon, Gary Leonard
Sinkovics, Joseph G.
Six, Howard Ronald
Solotorovsky, Morris
South, Mary Ann
Spencer, Mary Josephine Mason
Tengerdy, Robert Paul
Volkman, Alvin
Wainberg, Mark Arnold
Weller, Peter Fahey
Werner, Barbara Graham
Wheat, Robert Wayne
Wing, Edward Joseph
Winkelstein, Jerry Allen
Winn, Richard Earl
Wong, Ming Ming
Wright, Daniel Godwin
Wright, George Leonard, Jr.
Yoshida, Takeshi
Zimmerman, Sarah E.

Immunology, neuroimmunology
Arnason, Barry Gilbert Wyatt
Bulloch, Karen
Calabrese, Vincent Paul
Cohen, Nicholas
Cook, Stuart Donald
Coyle, Patricia K.
Dau, Peter Caine
Day, Eugene Davis
Decker, Walter Johns
del Cerro, Manuel
Dore-Duffy, Paula
Fujinami, Robert Shin
Gadberry, Joseph Lafayette
Grota, Lee James
Hashim, George A.
Karpiak, Stephen Edward, Jr.
Kori, Shashidhar Halappa
Laudenslager, Mark Leroy
Lees, Marjorie Berman
Lennon, Vanda Alice
Linthicum, Darwin Scott
Lisak, Robert Philip.
Lublin, Fred David
Lyons, Michael Joseph
McIntosh, Tracy Kahl
Merrill, Jean Elizabeth
Paterson, Philip Y.
Pelletier, Kenneth R.
Repine, John Edward
Resch, Joseph Anthony
Shearer, David Ross
Sydorak, Jaroslava Kuzmycz
Vandenbark, Arthur Allen
Weil, Marvin L(ee)
Williams, Curtis Alvin, Jr.

Immunology, transplantation
Amos, Dennis Bernard
Bach, Marilyn Lee
Carpenter, Charles Bernard
Chatterjee, Satya Narayan
Clark, William R.
Cone, Robert Edward
Datta, Surjit Kumar
Esher, Henry Jemil
Fawwaz, Rashid Adib
Garovoy, Marvin R.
Gill, Thomas James, III
Hardy, Mark Adam
Helderman, J. Harold
Jacobson, Leon Orris
Joison, Julio
Kung, Patrick Chung-Shu
Lobo, Peter Isaac
Lum, Lawrence George
Maki, Takashi
Mathur, Subbi
McCombs, Candace Cragen
Meredith, Ruby Frances
Mittal, Kamal Kant
Monaco, Anthony Peter
Montefusco, Cheryl Marie
Munda, Rino
Nathenson, Stanley Gail
Nghiem, Dai Dao
Oh, Jung Hee
Rapaport, Felix Theodosius
Schultz, Jane Schwartz
Sears, Duane William
Simonian, Simon John
Snell, George Davis
Stiller, Calvin Ralph
Strom, Terry Barton
Strong, Douglas Michael
Tsoi, Mang-So
Zaleski, Marek Bohdan

Laser. ***See*** **LASER, Medicine.**

Medicine, internal
Aagaard, George Nelson
Abrutyn, Elias
Achord, James Lee
Ahmann, David Lawrence
Alavi, Abass
Alderman, Michael Harris
Alexander, Jonathan
Ambrose, Audrey Belson
Antman, Elliott Marshall
Apicella, Michael Allen
Archer, Juanita Almetta Hinnant
Austen, K(arl) Frank
Azarnoff, Daniel Lester
Barbosa-Saldivar, Jose Luis
Beam, Thomas Roger
Becker, Ernest Lovell
Belzer, Folkert Oene
Bennett, Joe Claude
Berelowitz, Michael
Bessman, Alice Neuman
Biglieri, Edward George
Bilitch, Michael
Bloomfield, Saul S.
Blythe, William Brevard
Boland, C. Richard
Bottomley, Sylvia Stakle
Boudoulas, Harisios
Bozymski, Eugene Michael
Bransome, Edwin Dagobert, Jr.
Braunwald, Eugene
Breslau, Neil Art
Briggs, Arthur Harold
Bulger, Roger James
Burzynski, Stanislaw Rajmund
Caldwell, Glyn Gordon
Calesnick, Benjamin
Calkins, Evan
Callen, Jeffrey Phillip
Carbone, Paul Peter
Carpenter, Charles Colcock Jones
Castle, Charles Hilmon
Cha, Se Do
Challoner, David Reynolds
Charan, Nirmal Biswas
Chernoff, Amoz Immanuel
Chiorazzi, Nicholas
Christlieb, Albert Richard
Cintron, Guillermo Bo.
Clark, Mervin Leslie
Cluff, Leighton Eggertsen
Cohn, Major Lloyd
Conolly, Matthew Ellis
Cooper, Edward Sawyer
Cooper, Sheldon Mark
Costanza, Mary E.
Couser, William Griffith
Crawford, Michael Howard
Curd, John Gary
Cushman, Paul, Jr.
Davis, Donald Crawford
Dearing, William Hill
Deedwania, Prakash Chandra
Desser, Kenneth Barry
Diehl, Andrew Kemper
Dietschy, John Maurice
Dimant, Jacob
Dougherty, Joseph Charles
Drayer, Jan Ignatius
DuPont, Herbert Lancashire
Durack, David Tulloch
Einhorn, Daniel
Epstein, Murray
Estes, Edward Harvey, Jr.
Ettinger, David Seymour
Ettinger, Philip Owen
Everett, Elwood Dale
Farshidi, Ardeshir B.
Fathman, Charles Garrison
Feinstein, Alvan Richard
Felts, William Robert, Jr.
Finch, Clement A.
Finch, Stuart Cecil
Finkelstein, James David
Fletcher, Robert H.
Forsham, Peter Hugh
Fox, Irving Harvey
Franciosa, Joseph Anthony
Franks, John Julian
Frazier, Howard Stanley
Frei, Emil, III
Friedman, Gary David
Frohlich, Edward David
Frommer, J. Pedro
Fruend, Gerhard
Gallin, John I.
Ganda, Om P.
Ganguly, Sunilendu Narayan
Garcia, Julio M.
Gault, N.L., Jr.
Gelb, Alvin Meyer
Gerblich, Adi Abraham
Gershberg, Herbert
Gheorghiade, Mihai
Gilbert, Fred Ivan, Jr.
Gold, Jonathan W.M.
Goldberg, Leon Isadore
Goldman, Ira Steven
Goldman, Lee
Goldsmith, Stanley Joseph
Graham, David Tredway
Grant, Brydon John Bruce
Gray, Timothy Kenney
Grebenau, Mark David
Greenberg, Richard Neil
Greene, Bruce McGehee
Grenfell, Raymond Frederic
Grob, David
Hadler, Nortin M.
Hadlock, Daniel C.
Hamilton, Carlos Robert, Jr.
Haviland, James West
Hayden, Frederick Glenn
Hays, Esther Fincher
Heidenberg, William Jay
Heyssel, Robert Morris
Hildreth, Eugene A.
Hiller, Frederick Charles
Hochberg, Marc Craig
Hoffman, Neil Robert
Hogness, John Rusten
Hollister, Alan Scudder
Holmes, King Kennard
Holtzman, Jordan Loyal
Holzbach, R. Thomas
Hotes, Lawrence Steven
Hoyumpa, Anastacio Maningo
Hubbell, Floyd Allan
Humphries, Thomas Joel
Hunninghake, Donald Bernard
Hunter, Harry Laymond
Hyman, Edward Sidney
Iatridis, Panayotis George
Jackson, Charles Eugene
Jaffe, Ernst Richard
James, Thomas Naum
Johns, Dearing Ward
Kabadi, Udaya Manohar
Kahn, C. Ronald
Kahn, Thomas
Kaminskas, Edvardas
Kaplan, Kenneth Charles
Kassan, Stuart S.
Kastor, John Alfred
Kay, Marguerite M. Boyle
Kefalides, Nicholas Alexander
Kilbourne, Edwin Michael
Kirsner, Joseph Barnett
Klausner, Steven Charles
Klein, Lawrence Elliot
Klein, Milton Samuel
Klimek, Joseph John
Koide, Samuel Saburo
Kolodny, Abraham Lewis
Komanicky, Pavel
Kowalyshyn, Theodore Jacob
Kramer, Barnett Sheldon
Krishna, Gollapudi Gopal
Kronfol, Nouhad O.
Krothapalli, Radha Krishna
Krupp, Marcus Abraham
Kushner, Irving
Laddu, Atul Ramchandra
Lang, Robert
Larson, Steven Mark
Laskin, Oscar Larry
Lavietes, Marc Harry
Lawrason, F. Douglas
Lazarchick, John
Leaf, Alexander
Lee, David
Lee, Sang Hoon
Levin, Robert John
Levitan, Alexander Allen
Levy, Robert Isaac
Lewis, Terence David
Lipicky, Raymond John
Lohr, Kristine Marie
London, Irving Myer
Long, Billy Wayne
Lopez-Ovejero, Jorge Andres
Lourenco, Ruy Valentim
Lutas, Elizabeth M.
Lutz, Frank Brobson, Jr.
Lyons, Kenneth Paul
Macdonald, John Stephen
Mahapatra, Rajat Kanti
Mao, Chi Chian
Matz, Robert
McHenry, Martin Christopher
Mehta, Jawahar L.
Merrill, Joseph Melton
Metzger, Walter James
Meyer, George Wilbur
Middleton, Henry Moore, III
Mills, John Alexander
Mohiuddin, Syed Maqdoom
Monson, Roberta Ann Mills
Morrissey, John Fielding
Moser, Robert Harlan
Motz, Robin Owen
Muchmore, Harold Gordon
Mukherjee, Anil Baran
Murray, Robert Fulton, Jr.
Murthy, Vishnubhakta Shrinivas
Nakai, George S.
Nardella, Francis Anthony
Nelson, Harold Stanley
Novick, David Miles
Oberfield, Richard Alan
Odell, William Douglas
O'Duffy, John Desmond
Okada, Robert Dean
Olsen, Ward Alan
Orloff, Jack
Pagano, Joseph Stephen
Panush, Richard Sheldon
Parisi, Alfred Francis
Patsch, Josef Rudolf
Patsch, Wolfgang
Patten, Ethel Doudine
Pelletier, Lawrence Lee, Jr.
Petersdorf, Robert George
Pettinger, William A.
Phillips, John Hunter
Pierce, Nathaniel Field
Pinnas, Jacob Louis
Plotkin, Gary Robert
Plotz, Paul Hunter
Plymate, Stephen Rex
Poe, Robert Hilleary
Pollnow, Robert Edward
Powers, Thomas Allen
Probstfield, Jeffrey Lynn
Pyeritz, Reed Edwin
Rajan, Kannan Ramalingam
Rao, Venkateswara Koppanadham
Reed, William Patrick
Reidenberg, Marcus Milton
Relman, Arnold Seymour
Repine, John Edward
Rhodes, Mitchell Lee
Richardson, Joseph Hill
Riely, Caroline Armistead
Rissing, John Peter
Roberts, Hyman Jacob
Rosenberg, Michael J.
Rossing, Thomas Harry
Rozen, Simon
Rubenstein, Arthur Harold
Rubenstein, Edward
Russell, Robert Mitchell
Salom, Ira Louis
Sandberg, Hershel
Sandler, Sumner Gerald
Sandstead, Harold Hilton
Sanford, Jay Philip
Sapico, Francisco Lejano
Sarva, Rajendra P.
Schade, Robert Richard
Schilsky, Richard Lewis
Schmid, Rudi Rudolf
Schoenfeld, Myron Paul
Schonfeld, Gustav
Schrier, Robert William
Schultz, Alvin Leroy
Schumacher, Harry, Ralph
Schwartz, Mortimer Leonard
Schwarz, Anton J.
Segal, Bernard Louis
Seldin, Donald Wayne
Shambaugh, George Elmer, III
Shortliffe, Edward Hance
Simon, Gary Leonard
Slick, Gary Lee
Smith, Jackson Bruce
Smith, James Douglas
Smith, Lynwood Herbert
Smith, Thomas Woodward
Sohmer, Marcus Frank, Jr.
Somani, Pitambar
Soskel, Norman Terry
Spilker, Bert
Spitler, Lynn E(llen)
Sprung, Charles Leon
Stambaugh, John Edgar, Jr.
Stemmler, Edward Joseph
Stolleman, Gene Howard
Stone, Lawrence Allan
Summers, Robert Wendell
Summers, William Koopmans
Swisher, Scott Neil
Tan, Eng Meng
Theologides, Athanasios
Thier, Samuel Osiah
Thomas, Lewis
Thompson, Wilmer Leigh
Tuma, Samir Naif
Vaziri, Nostratola Dabir
Veray, Francisco X.
Volkman, David J.
Wald, Arnold
Wallin, John David
Wann, Lee Samuel
Ward, Louis Emmerson
Warren, James Vaughn

Wasser, Larry Paul
Watts, Malcolm S.M.
Waxman, Herbert Sumner
Weg, John Gerard
Weinberger, Myron Hilmar
Weiner, Murray
Weinstein, Irwin Marshall
Weisberg, Aaron
Weiss, Harvey Jerome
Weisse, Allen B.
Wells, James Opie, Jr.
Winchester, James Frank
Winn, Richard Earl
Wintrobe, Maxwell Myer
Woods, Roy
Wright, Jackson T., Jr.
Yankee, Ronald A.
Yarbro, John Williamson
Yarchoan, Robert
Yoo, Tai June
Yu, David Tak Yan
Zawada, Edward Thaddeus, Jr.
Ziegler, Michael Gregory

Microbiology
Albach, Richard Allen
Albrecht, Thomas Blair
Alexander, Edward Russell
Austrian, Robert
Baccanari, David Patrick
Baine, William Brennan
Baum, Stephen Graham
Bhattacharjee, Jnanendra K.
Boyle, Michael Dermot
Bradner, William Turnbull
Brown, Jay Clark
Buchholz, Donna Marie
Chambliss, Glenn Hilton
Chen, Kirk Ching Shyong
Clewell, Don Bert
Clinton, James Michael
Collins, Frank Miles
Collins, Michael Thomas
Consigli, Richard Albert
Cornett, James Bryce
Crandall, Marjorie Ann
Davis, Bernard David
Davis, Starkey Dee
Della-Latta, Phyllis
Diggs, Carter Lee
Donikian, Mary Adrienne
Dorn, Gordon Lee
Esher, Henry Jemil
Falkler, William Alexander, Jr.
Falkow, Stanley
Ferretti, Joseph Jerome
Fine, Donald Lee
Fletcher, Ronald Darling
Formal, Samuel Bernard
Francis, Robert Dorl
Funderburk, Noel Roger
Gadberry, Joseph Lafayette
Georgiades, Jerzy Alexander
Gerhardt, Philipp
Gerone, Peter John
Gormus, Bobby Joe
Gossling, Jennifer
Grunberg, Emanuel
Haak, Richard Arlen
Hackett, Joseph Leo
Hahon, Nicholas
Hardin, Hilliard Frances
Hoch, James Alfred
Horstmann, Dorothy Millicent
Hughes, John Henry
Jackson, George John
Jacoby, George Alonzo, Jr.
Jones, Wilbur Douglas, Jr.
Jorgensen, James Hartley
Kenny, George Edward
Kessler, Robert Evans
Kettrick-Marx, Mary Alice
Kilbourne, Edwin Dennis
Kim, Kwang-Shin
Kite, Joseph Hiram, Jr.
Krieger, John Newton
Kumar, Vijaya Buddhiraju
Kuramitsu, Howard Kikuo
Lavender, John Francis
Leathers, Chester Ray
Levner, Mark Henry
Lieberman, Michael Merril
Lim, Daniel V.
Macasaet, Francisco Friginal
Marchette, Nyven John
Martin, Christopher Michael
McCarty, Maclyn
Mc Sharry, James John
Middleton, Richard Burton
Miller, Glendon Richard
Miller, Glenn Allan
Miller, Richard Lynn
Moe, James Burton
Moore, W(alter) E(dward) C(ladek)
Morahan, Page Smith
Nash, Donald Robert, immunologist/microbiologist
Noble, Robert Cutler
Novick, William Joseph, Jr.
Parisi, Joseph Thomas
Parker, John Clarence
Parsons, James Eugene
Pirtle, Eugene Claude
Pollard, Morris
Puhvel, Sirje Madli
Quarles, John Monroe
Raj, Harkisan Dunichand
Ray, Usharanjan
Richardson, Carol Lynn
Roberts, Glenn Dale
Robinson, Harry John
Roesing, Timothy George
Rustigian, Robert Carning
Saier, Milton Herman, Jr.
Schlessinger, David
Schwab, John Harris
Scott, June Rothman
Senterfit, Laurence Benfred
Shaw, Eugene Douglas
Smith, Ralph Earl
Stephenson, Edward Hayes
Stewart, John Alvin
Stinski, Mark Francis
Stolleman, Gene Howard
Thompson, James Jarrard
Tripodi, Daniel
Tseng, Jeenan
Urban, James Edward
Wacker, Waldon Burdetter
Warren, Richard Lloyd
Weiss, Emilio
Woodhour, Allen Francis
Wright, George G(reen)
Wust, Carl John

Microscopy
Anderson, Mauritz Gunnar
Barnes, David Edward
Bertalanffy, Felix Dionysius
Breisch, Eric Alan
Burns, Edward R.
Drakontides, Anna B.
Fremount, Henry Neil
Hart, Raymond Kenneth
Hillman, Elizabeth Ann
Kim, Kwang-Shin
Krupp, Patricia Powers
Leblond, Charles Philipps
Lee, Tony Jer-Fu
Mc Gregor, Douglas Hugh
Melcher, Antony Henry
Murayama, Makio
Norton, William Nicholson, Jr.
Prien, Samuel David
Schoknecht, Jean Donze
Stenback, Wayne Albert
Warner, Ronald Ray

Nephrology
Agus, Zalman S.
Albini, Boris
Avasthi, Pratap Shanker
Aviv, Abraham
Avner, Ellis David
Beck, Nama
Becker, Ernest Lovell
Bello-Reuss, Elsa N.
Berlyne, Geoffrey Merton
Berndt, William Oscar
Bernstein, Jay
Blythe, William Brevard
Bonventre, Joseph Vincent
Bourdeau, James Edward
Brenner, Barry Morton
Carter, Mary Kathleen
Cavallo, Tito
Chevalier, Robert Louis
Chiu, Peter Jiunn-Shyong
Chou, Shyan-Yih
Coelho, Jaime Bernardino
Couser, William Griffith
Covit, Andrew B.
Davis, James (Othello)
Dougherty, Joseph Charles
Duarte, Cristobal G.
Epstein, Murray
Falk, Sandor A.
Freeman, Richard Benton
Frommer, J. Pedro
Gambertoglio, John Gino
Genest, Jacques
Gibson, Thomas Patrick
Goldfarb, Stanley
Grantham, Jared James
Helderman, J. Harold
Huang, Chia Ming
Husted, Russell Forest
Kahn, Thomas
Katz, Murray Alan
Kaufman, Donald Barry
Kempson, Stephen Allan
Keyl, Milton Jack
Kim, Jin Kyung
Klein, William, J., Jr.
Krishna, Gollapudi Gopal
Kronfol, Nouhad O.
Krothapalli, Radha Krishna
Leaf, Alexander
Lee, David
Levinsky, Norman George
Lobo, Peter Isaac
Maack, Thomas Michael
Mandal, Anil Kumar
Misra, Raghunath Prasad
Norman, Douglas James
Oh, Jung Hee
Orloff, Jack
Osborn, Jeffrey Lynn
Pak, Charles Y.C.
Pearson, James Eldon
Perrone, Ronald David
Rao, Venkateswara Koppanadham
Schrier, Robert William
Schwartz, George John
Seldin, Donald Wayne
Selkurt, Ewald Erdman
Slick, Gary Lee
Smith, James Douglas
Smith, Lynwood Herbert
Spitzer, Roger Earl
Steblay, Raymond William
Stiller, Calvin Ralph
Stokes, John Bispham, III
Thier, Samuel Osiah
Tuma, Samir Naif
Vaziri, Nostratola Dabir
Wallin, John David
Walser, Mackenzie
Walsh-Reitz, Margaret Mary
Wedeen, Richard Peter
Wells, James Opie, Jr.
Winchester, James Frank
Zager, Philip George
Zawada, Edward Thaddeus, Jr.

Neurology
ACKERMAN, Robert Harold
Allen, John Paul
Allocca, John Anthony
Arnason, Barry Gilbert Wyatt
Asbury, Arthur Knight
Barchi, Robert Lawrence
Bates, Stephen Roger Denis
Bergey, Gregory Keith
Biber, Michael Peter
Black, Perry
Booker, Harold Edward
Brooke, Michael Howard
Broughton, Roger James
Burns, John Joseph
Busis, Neil Amdur
Caviness, Verne Strudwick, Jr.
Chu, Nai-Shin
Conomy, John Paul
Cook, Stuart Donald
Cordingley, Gary Edward
Coyle, Patricia K.
Dasheiff, Richard Mitchell
Davis, James Norman
Demeter, Steven
Dentinger, Mark Peter
Deshmukh, Vinod Dhundiraj
Dichter, Marc Allen
Drayer, Burton Paul
Dreifuss, Fritz Emanuel
Duchowny, Michael Samuel
Dudley, Alden Woodbury, Jr.
Elizan, Teresita S.
Faden, Alan Ira
Fahn, Stanley
Ferrendelli, James Anthony
Ford, Donald Herbert
Fromm, Gerhard Hermann
Galaburda, Albert Mark
Gallagher, Brian Boru
Geschwind, Norman
Gilman, Sid
Glaser, Gilbert Herbert
Goldberg, Michael E.
Goldstein, Murray
Goust, Jean-Michel Christian
Hallett, Mark
Halperin, John Jacob
Hamill, Robert Wallace
Harter, Donald Harry
Haywood, H(erbert) Carl(ton)
Hekmatpanah, Javad
Heyer, Eric John
Hoffert, Marvin Jay
Housepian, Edgar Minas
Jacobs, Lawrence David
Johnson, Richard Tidball
Kasarskis, Edward Joseph
Kennedy, Charles
Kennedy, William Robert
Kraig, Richard Paul
Krall, Ronald Lee
Kruger, Lawrence
Kuncl, Ralph William
Kurland, Leonard Terry
Lee, Soo Ik
Leigh, Richard John
Lesser, Ronald Peter
Levy, David Edward
Levy, Deborah Louise
Lipton, Howard Lee
Lisak, Robert Philip.
Lublin, Fred David
Malone, Michael Joseph
Manyam, Bala Venktesha
Marcus, Elliot Meyer
Martinez, Augusto Julio
Marwaha, Jwaharlal
Masters, Colin Louis
Mayer, Richard Frederick
McLean, John Robert
McNamara, James O'Connell
Mesulam, M. Marsel
Meyer, John Stirling
Millichap, J(oseph) Gordon
Mitsumoto, Hiroshi
Moskowitz, Michael Arthur
Mueller, Shirley Maloney
Nathanson, Morton
Ng, Lorenz Keng Yong
Ohata, Carl Andrews
Pappas, Carol Lynn
Pickett, Jackson Brittain Elbridge
Porter, Roger John
Ransom, Bruce Robert
Reggia, James Allen
Resch, Joseph Anthony
Richmond, Barry Jay
Riggs, Jack Edward
Rimel, Rebecca Webster
Roel, Lawrence Edmund
Rogawski, Michael Andrew
Rosen, Arthur D.
Rowland, Lewis Phillip
Ruff, Robert Louis
Sachs, Howard
Sanders, Donald Benjamin
Sax, Daniel Saul
Schuelein, Marianne
Schwartz, William Joseph
Seltzer, Benjamin
Shahani, Bhagwan Topandas
Silberberg, Donald H.
Slevin, John Thomas
Spehlmann, Rainer
Swaiman, Kenneth Fred
Teasdall, Robert Douglas
Tourtellotte, Wallace William
Traub, Roger Dennis
Tyler, Kenneth Laurence
Waxman, Stephen G(eorge)
Weiner, William Jerrold
Wiley, Ronald Gordon
Winter, Arthur
Wolinsky, Jerry Saul
Yanagihara, Takehiko
Young, Robert Rice
Zee, David Samuel
Zivin, Justin Allen

Neuroscience. ***See*** **NEUROSCIENCE.**

Nuclear medicine. ***See also*** **Radiology.**
Alavi, Abass
Benedetto, Anthony Richard
Bergmann, Steven Robert
Biello, Daniel Robert
Chandra, Ramesh
Chaudhuri, Tapan Kumar
Clanton, Jeffrey Alan
Coursey, Bert Marcel
Devous, Michael David, Sr.
Dunn, William Lee
Ebert, Marlin J.
Evens, Ronald Gene
Fawwaz, Rashid Adib
Fitzgerald, Joseph James
Francis, Marion David
Freeman, Leonard Murray
Gilbert, Fred Ivan, Jr.
Ginos, James Zissis
Godwin, John Thomas
Goldsmith, Stanley Joseph
Grissom, Michael Phillip
Holman, B. Leonard
Hornof, William J.
Kabalka, George Walter
Klingensmith, William Claude, III
Lange, Robert Carl
Larson, Steven Mark
Llaurado, Josep G.
Loken, Merle Kenneth
López-Marjano, Vincent
Luckett, Larry Wayne
Lyons, Kenneth Paul
MacIntyre, William James
Malamud, Herbert
Mansfield, Carl Major
McCarthy, David Murray
McCombs, Rollin Koenig
Myers, William Graydon
Oster, Zvi Herman
Patton, Dennis David
Pierson, Richard Norris, Jr.
Pollack, Richard Stuart
Powers, Thomas Allen
Rao, Dandamudi Vishnuvardhana
Rayudu, Garimella V. S.
Robertson, James Sydnor
Robinson, Walter Lloyd
Schelbert, Heinrich Ruediger
Segal, Barry M.
Siegel, Barry Alan
Siegel, Michael Elliot
Smith, Anna Clanton
Smith, Charles Irvel
Smith, Gerald Duane
Sodd, Vincent Joseph
Solomon, Nathan A.
Wackers, Frans Jozef Thomas
Weber, David Alexander
Wiley, Albert Lee, Jr.
Wu, Sing-Yung

Nutrition
Abraira, Carlos
Ahrens, Richard August
Askew, Eldon Wayne
Asplund, John Malcolm
Bidlack, Wayne Ross
Bistrian, Bruce Ryan
Blumberg, Jeffrey Bernard
Boutton, Thomas William, Jr.
Bull, Leonard Seth
Caddell, Joan Louise
Caster, William Oviatt
Chan, Wai-Yee
Chen, Linda Huang
Chi, Myung Sun
Chinn, Kenneth Sai-Keung
Combs, Gerald Fuson
Dawson, Earl Bliss
Demetrakopoulos, George Evangelos
DeWys, William Dale
Dupont, Jacqueline (Louise)
Ellenbogen, Leon
Etzel, Kenneth Raymond
Flodin, Nestor Winston
Fomon, Samuel Joseph
Fried, George Herbert
Ganguli, Mukul Chandra
Gaut, Zane Noel
Geliebter, Allan
Glick, Zvi
Goetsch, Carl Allen
Grivetti, Louis Evan
Grundy, Scott Montgomery
Hall, Charles A.
Hankes, Lawrence Valentine
Harris, Suzanne Straight
Hazdra, James J.
Hegsted, David Mark
Hodges, Robert Edgar
Holloway, Caroline Tobia
Holman, Ralph Theodore
Horwitz, David Larry
Jacobsohn, Gert Max
Jain, Sushil Kumar
Jansen, Gustav Richard
Johnson, Phyllis Elaine
Kennedy, Frank Scott
King, Janet Carlson
Klavins, Janis Viliberts
Klein, Peter Douglas
Krieger, Ingeborg
Kushwaha, Rampratap Singh
Layman, Donald K(eith)
Lee, Kyu Taik
Levander, Orville Arvid
Levitsky, David Aaron
Liepa, George Uldis
Lynch, Sean Roborg
Madura, James Anthony
Mathias, Melvin Merle
McCarthy, Robert D.
McElhattan, Glenn Richard
Mendeloff, Albert Irwin
Mercer, L. Preston
Milner, John A.
Milner, John Austin
Morgan, Brian Leslie Gordon
Nakamoto, Tetsuo
Nichols, Buford Lee, Jr.
Nordlie, Robert Conrad
Olson, James Allen
Olson, Robert Eugene
Omaye, Stanley Teruo
Owen, George Murdock
Pinapaka, Murthy V.L.N.
Pisu (Pisunyer), F. Xavier
Pitchumoni, Capecomorin S.
Pond, Wilson Gideon
Poydock, Mary Eymard
Prasad, Ananda Shiva
Ranhotra, Gurbachan Singh
Rassin, David Keith
Reddy, Bandaru S.
Reeves, Robert Donald
Resnick, Oscar
Reynolds, Robert David
Rogers, Quinton Ray
Sandstead, Harold Hilton
Schingoethe, David John
Scrimshaw, Nevin Stewart
Shank, Robert Ely
Shanklin, Douglas Radford
Sisson, Thomas Randolph Clinton
Spiller, Gene Alan
Stinnett, Jimmy Dwight
Toma, Ramses Barsoum
Tsang, Reginald Chun-Nau
Turk, Donald Earle
Walser, Mackenzie
Weber, Charles Walter
Whedon, George Donald
Williams, Roger John
Woodruff, Calvin Watts
Yamamoto, Richard Susumu

Obstetrics and gynecology
Andresen, Brian Dean
Bellina, Joseph Henry
Burry, Kenneth Arnold
Ehrlich, Clarence Eugene
Filipescu, Nicolae
Foster, Henry Wendell
Gusdon, John Paul, Jr.
Jewelewicz, Raphael
Jones, Howard Wilbur, Jr.
Kasdon, S. Charles
Kirschbaum, Thomas Harry
Lee, Si Gaph
Little, Alan Brian
Mattison, Donald Roger
Mishell, Daniel Randolph, Jr.
Moore, John George, Jr.
Pitkin, Roy Macbeth
Prystowsky, Harry
Stenchever, Morton Albert
Toth, Attila
White, Rolfe Downing
Witkin, Steven S.
Zachert, Virginia
Zuspan, Frederick Paul

Obstetrics, gynecological oncology. ***See also*** **Oncology.**
Abdulhay, Gazi
Kasdon, S. Charles
Lurain, John Robert, III
Pattillo, Roland A.
Peterson, Rudolph Nicholas
Prem, Konald Arthur
Thigpen, James Tate
Twiggs, Leo Brookhart
van Nagell, John Rensselaer, Jr.

Obstetrics, maternal and fetal medicine. ***See also*** **Surgery, fetal.**
Berkowitz, Richard L.
Cederqvist, Lars Lennart
Cohen, Wayne Roy
Fabro, Sergio Edigio
Fox, Harold Edward
Gabbe, Steven Glenn
Hess, Orvan Walter
Louis, Thomas Michael
Maulik, Debabrata
Miller, Richard Kermit
Mueller-Heubach, Eberhard August
Petrie, Roy Howard
Pitkin, Roy Macbeth
Sokol, Robert James
Towbin, Abraham
Walsh, Scott Wesley
Zuspan, Frederick Paul

Obstetrics, perinatal diagnosis and therapy
Fox, Harold Edward
Golbus, Mitchell Sherwin
Lea, Jean Hedrick
Maulik, Debabrata
Petrie, Roy Howard
Stenchever, Morton Albert

Obstetrics, reproductive biology
Abrams, Robert Marlow
Baumiller, Robert Cahill
Brackett, Benjamin Gaylord
Chen, Hsien-Jen James
Cohen, Wayne Roy
Fabro, Sergio Edigio
Gustafsson, Borje Karl
Halme, Jouko Kalervo
Harper, Michael John Kennedy
Louis, Thomas Michael

Mueller-Heubach, Eberhard August
Naftolin, Frederick
Pattillo, Roland A.
Sturtevant, Frank Milton

Obstetrics, reproductive endocrinology
Blackwell, Richard Edgar
Burry, Kenneth Arnold
Eldridge, John Charles
Gurpide, Erlio
Hodgen, Gary Dean
Lee, Si Gaph
Mastroianni, Luigi, Jr.
Reiter, Russel Joseph
Rodriquez-Rigau, Luis Jose
Ryan, Kenneth John
Stoffer, Sheldon Saul

Oncology. *See also* **Cancer research.**
Abdulhay, Gazi
Abeloff, Martin David
Ahmann, David Lawrence
Aisenberg, Alan C.
Aisner, Joseph
Ames, Ira H.
Amols, Howard Ira
Armitage, James Olen
Ascensão, João L.
Awad, William Michel, Jr.
Banks, William Louis, Jr.
Bell, Richard
Berjian, Richard A.
Bitran, Jacob David
Bozdech, Marek Jiri
Brown, Eric Reeder
Brown, Loren Dennis
Bull, Frances Eleanor
Bunn, Paul Axtell, Jr.
Burd, Robert M.
Bush, Raymond Sydney
Butler, Thomas Parke
Carbone, Paul Peter
Case, Delvyn Caedren, Jr.
Cassileth, Peter Anthony
Castro, Joseph Ronald
Chang, Jae Chan
Chatterjee, Sunil Kumar
Chernicoff, David Paul
Chin, Hong W.
Chu, Ann Maria
Chung, Chung-Taik
Cole, Jack Westley
Coleman, C. Norman
Corder, Michael Paul
Costanza, Mary E.
Creech, Richard Hearne
Davis, Thomas Edward
Demetrakopoulos, George Evangelos
Dobelbower, Ralph Riddal, Jr.
Donegan, William Laurence
Drago, Joseph Rosario
Drapkin, Robert L.
Dreizen, Samuel
Durie, Brian George Martin
Dutcher, Janice Phillips
Economou, Steven George
Erturk, Erdogan
Ettinger, David Seymour
Fink, Louis Maier
Fink, Mary Alexander
Flessa, Herbert Christian
Frenkel, Eugene Phillip
Friedman, Nathan Baruch
Fu, Karen King-Wah
Gensler, Helen Lynch
Giuliano, Armando E.
Golde, David William
Goldsmith, Michael Allen
Goodman, Michael Gordon
Gormley, Paul Edward
Greene, Frederick Leslie
Greenwald, Edward S.
Griffin, Thomas William
Hadlock, Daniel C.
Harshbarger, John Carl, Jr.
Henderson, Isaac Craig
Hocking, William Gray
Hoffman, Neil Robert
Holder, Walter Dalton, Jr.
Hoovis, Marvin Lorin
Horton, John
Houghton, Alan Hourse, Jr.
Hunter, Harry Laymond
Hyman, Carol Brach
Ignoffo, Robert John
Jackson, Don Vernon, Jr.
Jerry, Laurence Martin
Joishy, Suresh K.
Joshi, Jai Hind
Kaiser, C. William
Kaiser, Hans Elmar
Kaplan, Richard Stephen
Kartha, Mukund Krishna
Kashmiri, Syed V.S.
Kasimis, Basil Spiros
Kennedy, Byrl James
Kiang, David Teh-ming
King, David Kyle
Klock, John Charles
Konits, Philip H.
Landolph, Joseph Richard, Jr.
Lane, Daniel McNeel
Lee, Yeu-Tsu N.
Levin, Robert David
Levin, William Cohn
Levitan, Alexander Allen
Lotlikar, Prabhakar Dattaram
Lowe, Nicholas James
Lurain, John Robert, III
Lyman, Gary Herbert
Mansour, Edward George
Marchetta, Frank Carmelo
Marks, Paul Alan
Maruyama, Yosh
Mc Credie, Kenneth Blair
McCulloch, John Hathorn
McIntyre, Oswald Ross
Miller, Elizabeth Cavert
Miller, James Alexander
Minwada, Jun
Mitchell, Malcolm Stuart
Moertel, Charles George
Nag, Subir K.
Nathanson, Larry
Oettgen, Herbert F.
Oken, Mertin M.
Olsson, Carl Alfred
Order, Stanley Elias
Oster, Martin William
Park, Chan Hyung
Pazdur, Richard
Petursson, Sigurdur Ragnar
Pitot, Henry Clement
Raaf, John Hart
Raam, Shanthi
Ramming, Kenneth Paul
Rassiga, Anne Louise
Rausen, Aaron Reuben
Reaman, Gregory Harold
Richards, Frederick, II
Salcman, Michael
Salinas, Fernando A.
Schaefer, Steven David
Schally, Andrew Victor
Schneider, Robert Jay
Schreiber, Hans
Schwartz, Allen David
Shimaoka, Katsutaro
Shnider, Bruce I.
Siegel, Stuart Elliott
Silver, Hulbert Keyes Belford
Sitarz, Anneliese Lotte
Sobin, Leslie Howard
Sondel, Paul M.
Steeves, Richard Allison
Sternhagen, Charles James
Straus, David Jeremy
Stuart, Robert Kenneth
Sugarbaker, Everett Van Dyke
Taetle, Raymond
Temple, Walley John
Theologides, Athanasios
Thomas, Edward Donnall
Tormey, Douglass C.
Trump, Donald Lynn
Twiggs, Leo Brookhart
Umsawasdi, Theera
Urano, Muneyasu
Walters, Thomas Richard
Weiss, L. Leonard
Welsch, Clifford William, Jr.
White, Douglas Rector
Wilcox, Patti Marie
Witman, Gary B.
Wizenberg, Morris Joseph
Woods, William Guard
Yarbro, John Williamson
Yoo, Dal
Zawadzki, Zbigniew Apolinary

Oncology, chemotherapy
Abeloff, Martin David
Ahmed, Nahed K.
Alcena, Valiere
Al-Sarraf, Muhyi
Amir, Jacob
Amjad, Hassan
Anderson, Tom
Arseneau, James Charles
Bahner, Carl Tabb
Baker, Laurence Howard
Bitran, Jacob David
Blumenreich, Martin Sigvart
Bond, William Holmes
Bosmann, Harold Bruce
Butler, John J.
Butler, William Manion
Carroll, David Stewart
Chiuten, Delia Fung She
Chu, Barbara C.F.
Coleman, Morton
Crawford, E. David
Currie, Violante Earlscort
Davila, Enrique
Davis, Thomas Edward
Dollinger, Malin Roy
Drapkin, Robert L.
Ehrlich, Clarence Eugene
Frei, Emil, III
Fuks, Joachim Zbigniew
Garcia, Julio M.
Garnick, Marc Bennett
Geran, Ruth I.
Giner-Sorolla, Alfred
Goldsmith, Michael Allen
Gordon, David Hugh
Green, Daniel Michael
Greenberg, Lowell Herbert
Greene, Robert Jay
Hall, Stephen William
Hall, Thomas Christopher
Hande, Kenneth Robert
Haraf, Frank Joseph
Helson, Lawrence
Herzig, Geoffrey Peter
Hodes, Louis
Houghton, Janet Anne
Houghton, Peter James
Houston, L.L.
Howell, Stephen Barnard
Ignoffo, Robert John
Johnson, Randall K.
Jones, Russell Allen
Karp, Judith Esther
Kass, Lawrence
Kelsen, David Paul
Kennealey, Gerard
Kessinger, Margaret Anne
Konits, Philip H.
Korytnyk, Wsewolod
Koven, Bernard J.
Krishan, Awtar
Lane, Frank Benjamin
Laufman, Leslie Rodgers
Laurie, John Andrew
Lee, Sang Hoon
LeVeen, Harry
Lichtenfeld, Karen Moss
Link, Michael Paul
Lipton, Allan
Litterst, Charles Lawrence
Loeb, Virgil, Jr.
Loo, Ti Li
Louis, John
Luce, James Kent
Macdonald, John Stephen
Magrath, Ian Trevor
Marx, James John, Jr.
McCredie, Kenneth Blair
Merrin, Claude Emile Andre
Morris, Don Melvin
Nathanson, Larry
Nelson, Eric Charles
Oberfield, Richard Alan
Ohnuma, Takao
Oster, Martin William
Patt, Yehuda Z.
Pazdur, Richard
Perry, David John
Perry, Michael Clinton
Peterson, Douglas Edward
Pollnow, Robert Edward
Presant, Cary A.
Rahman, Fazlur
Reitemeier, Richard Joseph
Richard, Joseph
Roberts, Joseph
Rose, William Carl
Rutman, Robert J.
Sakol, Marvin J(ay)
Samy, Anantha T. S.
Sartiano, George P.
Schilsky, Richard Lewis
Schwartz, Pauline Mary
Shapiro, Donald Michael
Shaw, Michael Trevor
Shnider, Bruce I.
Silver, Richard Tobias
Sinks, Lucius Frederick
Stefanini, Mario
Stolinsky, David C.
Stone, Lawrence Allan
Tannock, Ian Frederick
Tobey, Robert A.
Vaughn, Clarence Benjamin
Vogel, Charles Lewis
Walker, Brian Keith
Wasser, Larry Paul
Watanabe, Kouichi
Wheeler, Glynn Pearce
Whitecar, John P., Jr.
Yeung, Katherine Lu

Oncology, cell study
Aaronson, Stuart Alan
Allen, Patton Tolbert
Ames, Ira H.
Andreeff, Michael
Breitman, Theodore Ronald
Busch, Harris
Chen, Yi-Hsiang (Alan)
Civin, Curt Ingraham
Collett, Marc Stephen
Culp, Lloyd Anthony
Deschner, Eleanor Elizabeth
Dwivedi, Chandradhar
Ebert, Paul Stoudt
Epstein, Alan Lee
Epstein, Joshua
Erikson, Raymond Leo
Fenoglio, Cecilia Mettler
Fisher, Paul B.
Freeman, Aaron Eliot
Glaser, Ronald
Haseltine, William Alan
Hayward, William S.
Heine, Ursula Ingrid
Karp, Judith Esther
Klement, Vaclav
Kovacs, Charles Jeffrey
Lewis, Andrew Morris, Jr.
Magrath, Ian Trevor
McCann, Frances V.
Moyer, Mary Patricia
Pieczynski, William John
Plasse, Terry Freeman
Rapp, Ulf Fuediger
Scher, Charles David
Schiffer, Lewis Martin
Scott, Robert Bradley, Sr.
Sherman, Merry Rubin
Smith, Helene Sheila Carettnay
Tejada, Francisco
Tomei, L. David
Wainberg, Mark Arnold
Williams, Gary Murray
Woods, William Guard

Ophthalmology
Armstrong, Donald
Bettelheim, Frederick Abraham
Beyer-Mears, Annette
Blackwell, Harold Richard
Campbell, Charles John
Candia, Oscar A.
Colasanti, Brenda Karen
Collier, Linda Lee
del Cerro, Manuel
Doughman, Donald James
Drance, S. M.
Eifrig, David Eric
Foster, Charles Stephen
Goldbaum, Michael Henry
Goodman, Arden Patricia
Gragoudas, Evangelos Stelios
Greiner, Jack Volker
Havener, William Henry
Hoffman, Paul Ned
Kastl, Peter Robert
Kulkarni, Prasad Shrikrishna
Lorenzetti, Ole John
Lotti, Victor Joseph
Morse, Peter Hodges
Munger, Robert John
Newell, Frank William
O'Connor, G(eorge) Richard
Oh, Jang
Records, Raymond Edwin
Riley, Michael Verity
Sadun, Alfredo Arrigo Umberto
Samuelson, Don Arthur
Schwartz, Bernard
Shoch, David Eugene
Streeten, Barbara Anne Wiard
Tripathi, Ramesh Chandra
Waitzman, Morton Benjamin
Wolbarsht, Myron Lee

Optometry
Bailey, Ian Laurence
Enoch, Jay Martin
Fatt, Irving
Goldrich, STanley Gilbert
Lit, Alfred
Takahashi, Ellen Shizuko

Orthopedics
Boskey, Adele Ludin
Chan, Donald Pin Kwan
Compere, Clinton Lee
Cooper, Reginald Rudyard
Cruess, Richard Leigh
Frankel, Victor Hirsch
Herndon, James Henry
Ingram, Alvin John
Jay, Richard Martin
Levin, Stephen Michael
Lozman, Jeffrey
Mac Ewen, George Dean
Matthews, Leslie Scott
Opgrande, John Donald
Reider, Bruce
Stith, William Joseph
Tarr, Richard Robert
Walter, Thomas Harry
Woo, Savio Lau-Yuen

Otorhinolaryngology
Black, Franklin Owen
Blitzer, Andrew
Brandenburg, James H.
Cohen, Glenn Milton
Cummings, Charles William
Daigneault, Ernest Albert
Farmer, Joseph Clarence, Jr.
Gregg, John Bailey
Homburger, Freddy
Huang, Cheng-Chun
Krause, Charles Joseph
Ludlow, Christy Leslie
Malmgren, Leslie Theodore, Jr.
Mc Cabe, Brian Francis
Musiek, Frank Edward
Peterson, Ernest A.
Schaefer, Steven David
Sismanis, Aristides
Snow, James Byron, Jr.
Stockwell, Charles Warren
Sypert, George Walter
von Leden, Hans
Watanabe, Tsuneo
Wolf, Gregory Thomas

Pathology
Allen, Robert Carter
Alvares, Olav Filomeno
Anderson, Arthur Osmund
Appelman, Henry D.
Aronson, Stanley Maynard
Auerbach, Oscar
Barger, James Daniel
Barth, Rolf Frederick
Batsakis, John George
Beck, John Robert
Benditt, Earl Philip
Bernstein, Jay
Blackburn, Will R.
Bloom, Sherman
Bowman, James Edward
Brinkhous, Kenneth Merle
Brown, Arnold Lanehart, Jr.
Bruns, David Eugene
Bull, Brian Stanley
Byers, James Martin
Cardiff, Robert Darrell
Cavallo, Tito
Chang, Louis Wai-Wah
Chang, William Wei-Lien
Chedid, Antonio
Cho, Chaidong
Clagett, James Albert
Cohen, Samuel M.
Conn, Rex Boland, Jr.
Cooper, Robert Arthur, Jr.
Cuppage, Francis Edward
Damjanov, Ivan
Diamandopoulos, George Theodore
Eby, William Clifford
Factor, Stephen Michael
Feldman, Joseph David
Fenoglio, Cecilia Mettler
Fierer, Joshua A(llan)
Fink, Louis Maier
Fishbein, Michael Claude
Fitch, Frank Wesley
Flournoy, Dayl Jean
Freeman, Robert Glen
French, A. James
Friedman, Nathan Baruch
Fullmer, Harold Milton
Furth, John Jacob
Gallager, Harry Stephen
Garrie, Stuart Allen
Ghetti, Bernardino
Glenner, George Geiger
Godwin, John Thomas
Goltz, Robert William
Gottlieb, Leonard Solomon
Greenberg, Stephen Robert
Gullino, Pietro Michele
Handorf, Charles Russell
Heffner, Reid Russell, Jr.
Henderson, James Stuart
Ho, Kang-Jey
Homburger, Freddy
Hruban, Zdenek
Hu, Chung-Hong
Inhorn, Stanley Lee
Jeska, Edward L(awrence)
Johnson, Peter Charles
Johnston, Marilyn F. M.
Joist, Johann Heinrich
Jordan, Scott Wilson
Jorgensen, James Hartley
Kiechle, Frederick Leonard
Kim, Hun
Kim, James C(hin) S(oo)
Kithier, Karel
Klavins, Janis Viliberts
Kleinerman, Jerome
Koestner, Adalbert
Korenyi-Both, Andras Levente
Koss, Leopold G.
Krigman, Martin Ross
Kuschner, Marvin
Lambird, Perry Albert
Lamm, Michael Emanuel
Landing, Benjamin Harrison
Lane, Bernard Paul
Lanks, Karl William
Large, H. Lee, Jr.
Lawson, Alfred James
Lee, Kyu Taik
Lee, Shuishih Sage
Lev, Maurice
Libcke, John Hanson
LiVolsi, Virginia Anne
Longnecker, Daniel Sidney
Lynch, Denis Patrick
Macasaet, Francisco Friginal
Mandal, Anil Kumar
Manhold, John Henry
Martinez, Augusto Julio
Masters, Colin Louis
Mc Gregor, Douglas Hugh
Mellors, Robert Charles
Mihm, Martin Charles, Jr.
Mikat, Eileen Marie
Milani, Cyrus Saeed
Miller, Carol Ann
Min, Kyung-Whan
Mirkin, Lazaro David
Misra, Raghunath Prasad
Molteni, Agostino
Moody, David Edward
Mowry, Robert Wilbur
Naeye, Richard L.
Nahm, Moon Hea
Nakamura, Robert Motoharu
Normann, Sigurd Johns
Novikoff, Alex Benjamin
Nowell, Peter Carey
O'Donoghue, John (Lipomi)
Pavelic, Zlatko Paul
Perrin, Eugene Victor Debs
Pertschuk, Louis Philip
Phan, Sem Hin
Philpot, Van Buren, Jr.
Pitcock, James Allison
Pitot, Henry Clement
Pizzo, Salvatore Vincent
Popper, David Henry
Posso, Manuel
Pour, Parviz M.
Powers, James Matthew
Pretlow, Thomas Garrett
Pribor, Hugo Casimer
Puchtler, Holde
Quay, Steven Carl
Rabson, Alan Saul
Race, George Justice
Rao, Sambasiva M.
Raska, Karel Frantisek, Jr.
Rifkin, Barry Richard
Rogers, Edgar Stanfield
Roggli, Victor Louis
Ross, Russell
Roth, Daniel
Roth, Lawrence Max
Rowlands, David Thomas
Rudofsky, Ulrich Hubert Waldemar
Russo, Irma Haydee Alvarez
Shanklin, Douglas Radford
Shrestha, Buddhi Man
Siakotos, Aristotle Nicholas
Siegal, Gene Philip
Sirica, Alphonse Eugene
Sobin, Leslie Howard
Sogandares-Bernal, Franklin
Sommers, Sheldon Charles
Song, Joseph
Stefanini, Mario
Steinberg, Jacob Jonah
Stenback, Wayne Albert
Stowell, Robert Eugene
Straus, Francis Howe, II
Strayer, David Sheldon
Streeten, Barbara Anne Wiard
Sudilovsky, Oscar
Swarm, Richard Lee
Sybers, Harley Duane
Szabo, Sandor
Till, Gerd Oskar
Tischler, Arthur S.
Toth, Attila
Toth, Bela
Towbin, Abraham
Tripathi, Ramesh Chandra
Trojanowski, John Quinn
Upton, Arthur Canfield
Uzman, Betty Geren
Vladutiu, Adrian Octavian
Wagner, Bernard M.
Waisman, Jerry

Waller, Bruce Frank
Ward, Peter Allan
Warnke, Roger Allen
Weinberg, Kenneth Steven
Weiss, Daniel Leigh
Wellings, Sefton Robert
Wilkinson, David Stanley
Williams, Gary Murray
Williams, Marjorie Joan
Wilner, George Dubar
Zimmerman, Michael Raymond

Parasitology
Albach, Richard Allen
Arlian, Larry G.
Babero, Bert Bell
Barriga, Omar Oscar
Beaver, Paul Chester
Bennett, James Leroy
Bers, Lipman
Bonner, Thomas Patrick
Bridgman, John Francis
Brohn, Frederick Herman
Castro, Gilbert Anthony
Cerami, Anthony
Collard, Sneed Body
Crites, John Lee
Despommier, Dickson Donald
Dusanic, Donald Gabriel
Fawcett, Don Wayne
Ferris, Deam Hunter
Fried, Bernard
Gleason, Larry Neil
Higashi, Gene Isao
Hogan, Yvonne Holland
Holliman, Rhodes Burns
Holloway, Harry Lee, Jr.
Honigberg, Bronislaw Mark
Hsu, Shu Ying
James, Stephanie Lynn
Jensen, Emron Alfred
Jesen, James Burt
Kagan, Irving George
Kemp, Walter Michael
Kierszenbaum, Felipe
Kreutzer, Richard David
Langlands, Robert Phelan
Loomis, Edmond Charles
McCall, John W.
McDougald, Larry Robert
Miller, Lynne Cathy
Müller, Miklós
Najarian, Haig Hacop
Neva, Franklin Allen
Nussenzweig, Victor
Pappas, Peter William
Pesanti, Edward Louis
Pfefferkorn, Elmer Roy
Platzer, Edward George
Poinar, George Orlo, Jr.
Prestwood, Annie Katherine
Race, George Justice
Rivera, Ezequiel Ramirez
Rota, Gian-Carlo Carlo
Saladin, Kenneth Stanley
Sher, Franklin Alan
Silverman, Paul Hyman
Sogandares-Bernal, Franklin
Stromberg, Bert Edwin, Jr.
Trager, William
Waffle, Elizabeth L.
Weller, Peter Fahey
Wong, Ming Ming

Pediatrics
Alpert, Joel Jacobs
August, Gilbert Paul
Baird, Henry Welles, III
Bates, Stephen Roger Denis
Behrman, Richard Elliot
Berlin, Cheston Milton, Jr.
Bernstein, Shelly Corey
Blaese, (Robert) Michael
Boxer, Laurence Alan
Brent, Robert Leonard
Chevalier, Robert Louis
Connelly, John Peter
Cowett, Richard Michael
Cramblett, Henry Gaylord
Dean, Raymond S.
Deibel, Rudolf
Dell, Ralph B.
Denny, Floyd Wolfe, Jr.
D'Ercole, Augustine Joseph
Dhanireddy, Ramasubbareddy
Enzer, Norbert Beverly
Euler, Arthur Ray
Fomon, Samuel Joseph
Frias, Jaime Luis
Garrettson, Lorne Keith
Gartner, Lawrence Mitchel
Ginsberg-Fellner, Fredda Vita
Goldman, Armond Samuel
Goldstein, David Joel
Good, Robert Alan
Green, Daniel Michael
Haggerty, Robert Johns
Harrison, Gunyon M.
Hatoff, Alexander
Hendley, Joseph Owen
Henley, Walter L.
Hill, Reba Michels
Holtzman, Neil Anton
Jaffe, Norman
Jay, Richard Martin
Johnsonbaugh, Roger Earl
Karp, Warren Bill
Katz, Michael
Kempe, Charles Henry
Klein, Gordon Leslie
Kobayashi, Roger Hideo
Kohl, Steve
Krivit, William
Krugman, Saul
Lanzkowsky, Philip
Lawson, Edward Earle
Lazerson, Jack
Lebenthal, Emanuel
Lee, Kwang-Sun
Lehman, Thomas Joseph Ansorge
Leventhal, Brigid Gray
Link, Michael Paul
Littlefield, John Walley
Luhby, A. Leonard
Magen, Myron Shimin
Manley, Audrey Forbes
Marienfeld, Carl Joseph
Martinson, Ida Marie
Merenstein, Gerald B.
Miller, C. Arden
Miller, John Johnston, III
Miller, Walter L.
Mirkin, Lazaro David
Mueller, Shirley Maloney
Nadler, Henry Louis
Newth, Christopher John Lester
Nichols, Buford Lee, Jr.
Nitowsky, Harold Martin
Owen, George Murdock
Pachman, Lauren Merle
Park, Myung Kun
Percy, Alan Kenneth
Person, Donald Ames
Peter, Georges
Pickoff, Arthur Steven
Rezvani, Iraj
Richmond, Julius Benjamin
Riely, Caroline Armistead
Robbins, Frederick Chapman
Rosenfeld, Ron Gershon
Schubert, William Kuenneth
Schuelein, Marianne
Schwartz, George John
Schwartz, Stanley Allen
Shinefield, Henry Robert
Sidbury, James Buren
Siegel, Stuart Elliott
Silver, Henry K.
Sinks, Lucius Frederick
Sitarz, Anneliese Lotte
Soyka, Lester Frank
Spencer, Mary Josephine Mason
Spitzer, Adrian, *Pediatrics*
Steinherz, Peter Gustav
Strunk, Robert Charles
Walters, Thomas Richard
Warwick, Warren James
Wedgwood, Ralph Josiah Patrick
Weil, Marvin L(ee)
Weldon, Virginia V.
White, Neil Harris
Winkelstein, Jerry Allen
Yaffe, Sumner Jason
Yeager, Andrew Michael

Pediatrics, neonatology
Behrman, Richard Elliot
Clements, John Allen
Cowett, Richard Michael
Dhanireddy, Ramasubbareddy
Gartner, Lawrence Mitchel
Hess, Orvan Walter
Hill, Reba Michels
Hodgman, Joan Elizabeth
Kotas, Robert Vincent
Lee, Kwang-Sun
Lewis, Michael
Merenstein, Gerald B.
Paxson, Charles L., Jr.
Sisson, Thomas Randolph Clinton
Tsang, Reginald Chun-Nau

Pharmacology
Aagaard, George Nelson
Aberg, Gunnar A.K.
Adams, Max D.
Adler, Martin W.
Akbar, Huzoor
Akera, Tai
Alexander, John Charles
Altura, Burton Myron
Ambre, John Joseph
Amer, M. Samir
Ames, Matthew Martin
Antonaccio, Michael John
Aronson, Carl Edward
Aston, Roy
Azarnoff, Daniel Lester
Bacopoulos, Nicholas G.
Balazs, Tibor
Barnett, Allen
Baskin, Steven Ivan
Beaven, Michael Anthony
Beaver, William Thomas
Beck, Nathan
Bender, A(llan) Douglas
Bennett, Donald Raymond
Berkowitz, Barry A.
Berlin, Cheston Milton, Jr.
Berman, David Albert
Bernacki, Ralph James
Berndt, William Oscar
Bernstein, Jerrold
Besch, Henry Roland, Jr.
Beyer, Karl Henry, Jr.
Bito, Laszlo Z.
Blaine, Edward Homer
Bleiberg, Marvin Jay
Blumberg, Harold
Boctor, Amal Morgan,
Boisse, Norman Robert
Borgstedt, Harold Heinrich
Borzelleca, Joseph Francis
Bosin, Talmage Raymond
Boyd, Eleanor H.
Breese, George Richard
Brezenoff, Henry Evans
Brodie, Harlow Keith Hammond
Brody, Michael J.
Brody, Theodore Meyer
Brooker, Gary
Brooks, Harold Lloyd
Brown, Richard Don
Buccafusco, Jerry Joseph
Buckley, Joseph Paul
Bull, Richard James
Bullock, Francis Jeremiah
Burford, Hugh Jonathan
Burkman, Allan Maurice
Burks, Thomas Franklin, II
Busch, Harris
Buyniski, Joseph Paul
Byrne, Jeffrey Edward
Calesnick, Benjamin
Cameron, John Stanley
Cantor, Elinor H.
Carrano, Richard A
Carroll, Paul T.
Carson, Paul Elbert
Carson, Steven
Carter, Mary Kathleen
Castronovo, Frank Paul, Jr.
Chan, Arthur Wing Kay
Chan, W. Y.
Chang, Richard Li-chai
Chasin, Mark
Chaturvedi, Arvind Kumar
Chen, Chi-Po
Chiu, Peter Jiunn-Shyong, *Pharmacology*
Clark, Julia Berg
Cluff, Leighton Eggertsen
Cohen, Marlene Lois
Cohn, Victor Hugo
Collins, J.G.
Collins, John H.
Concannon, James Thomas
Cone, Edward J.
Conolly, Matthew Ellis
Cook, Clarence Edgar
Cooper, Theodore
Cosmides, George James
Cowan, Alan
Curro, Frederick Anthony
Cushman, Paul, Jr.
Dailey, John William
Dajani, Esam Zapher
Davis, Donald Crawford
Davis, Lloyd Edward
de Miranda, Paulo
Demis, D. Joseph
De Salva, Salvatore Joseph
DiMicco, Joseph Anthony
Di Pasquale, Gene
Dixit, Balwant Narayan
Dixon, Robert L.
Do, Hien Duc
Domer, Floyd Ray
Dorris, Roy Lee
Downey, James Merritt
Drayer, Jan Ignatius
DuCharme, Donald Walter
Duckles, Sue Piper
Dunham, Earl Wayne
Dvorchik, Barry Howard
Egorin, Merrill Jon
Ehrreich, Stewart Joel
El-Ackad, Tarek M.
Ellingboe, James
Ellis, Keith Osborne
Emele, Jane Frances
Engel, Toby Ross
Eshelman, Fred Neville
Fedan, Jeffrey Stephen
Feigen, Larry Philip
Ferko, Andrew Paul
Fielding, Stuart
Findlay, John William Addison
Flaim, Stephen Frederick
Fleisch, Jerome Herbert
Fleming, James Stuart, Jr.
Flynn, John Thomas
Fouts, James Ralph
Frankl, William Stewart
Franko, Bernard Vincent
Freed, Curt Richard
Frishman, William Howard
Fuller, Ray Ward
Gabourel, John Dustan
Gallardo-Carpentier, Adriana
Gangaros, Louis Paul, Sr.
Garg, Lal Chand
Garrettson, Lorne Keith
Gatz, Edward Erwin
Gautieri, Ronald Francis
Geber, William Frederick
Gibson, Thomas Patrick
Giles, Ralph Edson
Giri, Shri N.
Glassman, Jerome Martin
Goh, Edward Hua Seng
Goldberg, Leon Isadore
Goldstein, Dora Benedict
Gomoll, Allen Warren
Graeme, Mary Lee
Gray, Grace Warner
Greenberg, Roland
Greenberg, Stan Shimen
Gross, Garrett John
Guarino, Anthony Michael
Gueriguian, John Leo,
Hacker, Miles Paul
Hagen, Arthur Ainsworth
Hall, Edward Dallas
Hall, Madeline Molnar Hall
Hance, Anthony James
Hancock, John Charles
Hande, Kenneth Robert
Harakal, Concetta
Harris, Louis Selig
Harrison, Steadman Darnell, Jr.
Harrison, Yvonne Elois
Hastings, Robert Clyde
Hatch, Roger Conant
Hava, Milos
Heffner, Thomas Gary
Henningfield, Jack Edward
Herzig, David Jacob
Hess, Marilyn E.
Hietbrink, Bernard Edward
Hill, Donald Lynch
Hirsch, Philip Francis
Hirsh, Kenneth Roy
Hitchcock, Margaret
Hofmann, Lorenz Martin
Hollister, Alan Scudder
Holtkamp, Dorsey E(mil)
Hoover, Donald Barry
Horakova, Zdenka Zahutova
Horovitz, Zola Philip
Hosko, Michael J., Jr.
Hubbard, Walter Clyde
Huber, William George
Hudak, William John
Hudgins, Patricia Montague
Husain, Syed
Hyman, Albert Lewis
Hynes, Martin Dennis, III
Imondi, Anthony Rocco
Izzo, Joseph L., Jr.
Jacoby, Henry I.
Jandhyala, Bhagavan Srikrishna
Juliano, Rudolph L.
Kammerer, Richard Craig
Kappas, Attalllah
Karagueuzian, Hrayr Sevag
Katzung, Bertram George
Kau, Sen T.
Kauker, Michael Lajos
Kaul, Pushkar Nath
Kenny, Alexander Donovan
Keplinger, Moreno L.
Kimura, Eugene Tatsuru
King, (Mary) Margaret
King, Theodore Oscar
Klaassen, Curtis Dean
Knott, Peter Jeffrey
Kodama, Jiro Kenneth
Koelle, George Brampton
Kohli, Jai Dev
Kosersky, Donald S.
Koss, Michael Campbell
Krebs, Edwin Gerhard
Krell, Robert Donald
Krivoy, William Aaron
Kulkarni, Prasad Shrikrishna
Kuntzman, Ronald Grover
Kupersmith, Joel
Lai, Fong Mao
Lanman, Robert Charles
Laskin, Oscar Larry
Lazo, John Stephen
Lee, Cheng-Chun
Lee, Tony Jer-Fu
Leeling, Jerry L.
Lerner, Stephen Alexander
Lester, David
Levin, Robert John
Liang, Chang-seng
Lin, Tsung-Min
Lish, Paul Merrill
Liu, Ching-Tong
Longenecker, Gesina (Louise) Lizana
Loo, Ti Li
Loran, Muriel Rivian
Loux, Joseph J.
Lowensohn, Howard Stanley
Ludlum, David Blodgett
Lukas, Scott Edward
MacFarlane, Malcolm David
Mackerer, Carl Robert
Maickel, Roger Philip
Makman, Maynard Harlan
Malanga, Carl Joseph
Malick, Jeffrey B.
Mallov, Samuel
Malone, Marvin Herbert
Mandel, H(arold) George
Marshall, Franklin Nick
Martin, Billy R.
Mathur, Pershottam P(rasad)
Maxwell, Donald Robert
Mayhew, Eric George
McAllister, Russell Greenway, Jr.
McCarty, Leslie Paul
McColl, John Duncan
McConnell, William Ray
McDermott, Daniel Joseph
McGovren, James Patrick
McGuire, John L.
Melmon, Kenneth Lloyd
Miceli, Joseph N(icola)
Millard, Ronald Wesley
Miller, Francis Peter
Miller, Jack W.
Miller, James Edward
Miller, Lynne Cathy
Miller, Richard Kermit
Mir, Ghulam Nabi
Misra, Anand Lal
Mokler, Corwin Morris
Moore, Alan Frederic
Moore, Robert Haldane, III
Morris, Ralph William
Mullane, John Francis
Murthy, Vishnubhakta Shrinivas
Musgrave, Gary Eugene
Mustafa, Syed Jamal
Nelson, James Arly
Novick, William Joseph, Jr.
Nwangwu, Peter Uchenna
Ochillo, Richard Frederick
Ogilvie, Richard Ian
Olsen, George Duane
Omaye, Stanley Teruo
Osterberg, Arnold Curtis
Page, John Gardner
Palatino, Richard Duane
Pandya, Krishnakant Hariprasad
Pang, Kim-Ching Sandy
Park, Myung Kun
Parker, Robert B.
Patel, Dhanooprasad Gordhanbhai
Pavelic, Zlatko Paul
Pento, J. Thomas
Perel, James Maurice
Perhach, James Lawrence, Jr.
Peters, Marvin Arthur
Pettinger, William A.
Phillips, Barrie Maurice
Piercey, Montford Frederic
Pindell, Merle Herbert
Pisko, Edward John
Pool, Peter Edward
Porter, Roger John
Powis, Garth
Prabhu
Price, Henry Locher
Prichard, John B.
Privitera, Philip Joseph
Proakis, Anthony George
Probstfield, Jeffrey Lynn
Proctor, Kenneth Gordon
Pruss, Thaddeus Paul
Purdy, Ralph Earl
Quay, John Ferguson
Quebbemann, Aloysius John
Quirion, Remi
Rall, David Platt
Rao, Gopal Subba
Ray, Richard Schell
Reichen, Juerg
Reichgott, Michael J.
Reidenberg, Marcus Milton
Reilly, Joseph Francis
Reilly, Margaret Anne
Reinke, Lester Allen
Reynolds, Robert Donald
Ribeiro, Lair Geraldo) Theodoro
Rickert, Douglas Edward
Rifkind, Arleen B.
Robinson, Casey Perry
Rocci, Mario Louis, Jr.
Rockhold, Robin William
Rosen, Michael Robert
Rosenberg, Howard Charles
Rosenkranz, Roberto Pedro
Ross, Norton Morris
Rubin, Bernard
Ruffolo, Robert Richard, Jr.
Russell, Robert Lee
Rutledge, Charles Ozwin
Saini, Ravinder Kumar
Salem, Harry
Salvador, Richard A.
Sanner, John Harper
Sartorelli, Alan Clayton
Schanker, Lewis Stanley
Schentag, Jerome J.
Schneider, Frederick H.
Sethi, Vidya Sagar
Setler, Paulette Elizabeth
Shargel, Leon David
Shepherd, Alexander M.M.
Shriver, David Allen
Siegel, Ivens Aaron
Siemens, Albert John
Slikker, William, Jr.
Slotkin, Theodore Alan
Smith, Cedric Martin
Smith, Roger Powell
Sofia, Robert Duane
Solomon, Nathan A.
Somani, Pitambar
Soyka, Lester Frank
Spector, Reynold
Spilker, Bert
Spratt, James Leo
Stancel, George Michael
Steinfeis, George Francis
Stevens, James Thomas
Sticht, Frank Davis
Strosberg, Arthur Martin
Su, Che
Sullivan, Ann Clare
Swingle, Karl Frederick
Sybertz, Edmund J., Jr.
Taber, Robert Irving
Takemori, Akira Eddie
Takesue, Edward I.
Taves, Donald R.
Taylor, Samuel Edwin
Thomas, John Arlen
Thompson, Wilmer Leigh
Thurman, Ronald Glenn
Tobin, Thomas
Tseng, Leon
Turlapaty, Prasad Durga Mallikharjuna Vara
Tuttle, Ronald Ralph
Urquilla, Pedro Ramon
Velamakanni, Krishnamurty Seetarama
Verebey, Karl G.
Vesell, Elliot Saul
Vincenzi, Frank Foster
Volicer, Ladislav
Wallach, Donald Pinny
Walsh, Gerald Michael
Wang, Theodore Sheng-Tao
Ward, John Wesley
Wardell, William Michael
Wasserman, Martin Allan
Watson, Eileen Lorraine
Way, James Leong
Weber, Kenneth Charles
Weeks, James Robert
Weiner, Irwin M.
Weiner, Murray
Wenger, Galen Rosenberger
West, Bob
Westfall, David Patrick
White, Helen Lyng
Wilkerson, Robert Douglas
Wilkinson, Grant Robert
Williamson, Harold Emanuel
WINBURY, Martin Maurice
Wingard, Lemuel Bell, Jr.
Wolen, Robert Lawrence
Wolf, Gerald Lee
Wolff, Frederick William
Woods, Eugene Francis
Wosilait, Walter Daniel
Wright, Jackson T., Jr.

Yacobi, Avraham
Yaffe, Sumner Jason
Yagiela, John Allen
Yankell, Samuel Leon
Yeh, Billy Kuo-Jium
Yeh, Billy Kuo-Jiun
Yetiv, Jack Zeev
York, James Lester
Yu, Jia-Huey

Pharmacology, cellular
Acosta, Daniel
Allen, Donald Orrie
Bayer, Barbara Moore
Beck, Nathan
Bennett, James Leroy
Bosin, Talmage Raymond
Bowen, John Metcalf
Brendel, Klaus
Chen, Chi-Po
Chen, Theresa, Shang-tsing
Chiang, Peter K.
Chu, Barbara C.F.
Clark, Julia Berg
Courtney, Kenneth Randall
Davies, Philip
Drewinko, Benjamin
Ezrin, Alan Mark
Fisher, James W.
Greenberg, Stan Shimen
Greenspan, Kalman
Haigler, Henry James
Hava, Milos
Hester, Richard Kelly
Hollander, Philip Ben
Hollinger, Mannfred Alan
Hornbrook, Kent Roger
Lai, Fong Mao
Lee, Insu Peter
Mallov, Samuel
Margolius, Harry Stephen
Marsh, James Dalton
McIsaac, Robert J(ames)
Momparler, Richard Lewis
Nechay, Bohdan Roman
Peach, Michael Joe
Rhee, Hee Min
Ritter, Carl Alan
Rizack, Martin Arthur
Robinson, Richard Bruce
Robison, G(eorge) Alan
Rubin, Ronald Philip
Russell, Diane Haddock
Russell, Robert Lee
Salganicoff, Leon
Smith, John Bryan
Sperelakis, Nick
Stitzel, Robert Eli
TenEick, Robert Edwin
Theoharides, Theoharis Constantin
Tolman, Edward Lauria
Van Dyke, Knox
Wallick, Earl Taylor
Ward, Patrick E.
Watson, Eileen Lorraine
Wong, Patrick Yui-Kwong

Pharmacology, molecular
Agrawal, Krishna Chandra
Allaudeen, Hameedsulthan Sheik
Alphin, Reevis Stancil
Alvares, Alvito Peter
Aronow, Lewis
Baksi, Samarendra Nath
Bastian, James Winslow
Brendel, Klaus
Briggs, Arthur Harold
Brostrom, Margaret Ann
Bunag, Ruben David
Bylund, David B.
Camerman, Arthur
Cantor, Elliot II
Chaudhari, Anshumali
Cheney, Darwin LeRoy
Cheng, Yung-chi
Chignell, Colin Francis
Cho, Arthur Kenji
Chou, Ting-Chao
Consroe, Paul Francis
Cooper, Jack Ross
Corbascio, Nicola Aldo
Creveling, Cyrus Robbins
Cuatrecasas, Pedro Martin
Cushman, David Wayne
Dailey, John William
Dashman, Theodore
Davis, Oscar F.
Donnelly, Thomas Edward, Jr.
Eldefrawi, Amira Toppozada
Enna, Salvatore Joseph
Fedan, Jeffrey Stephen
Free, Charles Alfred
Glazer, Robert Irwin
Goldstein, Joyce Allene
Gormley, Paul Edward
Gurtoo, Hira L.
Harris, Don Navarro
Hess, Marilyn E.
Hiller, Jacob Moses
Holtzman, Jordan Loyal
Insel, Paul Anthony
James, Margaret O.
Janis, Ronald Allen
Jarboe, Charles Harry
Karliner, Joel Samuel
Kellar, Kenneth Jon
Khairallah, Philip Asad
Kohn, Kurt William
Kuo, Jyh-Fa
Lane, Stanley Earl
LaPlanche, Laurine Anna
Lasslo, Andrew
Leibman, Kenneth Charles
Leviit, Morton
Lichtenwalner, Diane Marie
Lockridge, Oksana
Long, John Paul
Magee, Donal F.
Mansour, Tag Eldin
Marshall, Garland Ross
Mieyal, John Joseph
Miller, Kenneth John
Miller, Richard Lee
Mittal, Chandra Kant
Murthy, Vadiraja Venkatesa
Newman, Robert Alwin
Pandya, Krishnakant Hariprasad
Poland, Alan Paul
Randerath, Kurt
Rao, Gundu Hirisave Rama Rao
Read, George Wesley
Robinson, Harry John
Robison, G(eorge) Alan
Roseke, William Robert
Ross, Elliott M(orton)
Salganicoff, Leon
Scheibel, Leonard William
Schenkman, John Boris
Schlender, Keith Kendall
Schwartz, Pauline Mary
Shires, Thomas Kay
Silverman, David Norman
Smith, Roberts Angus
Somberg, John Charin
Stefano, George B.
Stohs, Sidney John
Taylor, John Edward
Tephly, Thomas Robert
Tew, Kenneth David
Thomas, Paul Elbert
Thompson, William Joseph
Villar-Palasi, Carlos
Vincenzi, Frank Foster
Wade, Adelbert Elton
Wallick, Earl Taylor
Ward, Patrick E.
Watanabe, Kouichi
Weber, George
Webster, Leslie Tillotson, Jr.
Wells, Jack Nulk
Wrenn, Simeon Mayo
Yielding, Lerena Wade
Zimmerman, Thomas Paul

Pharmacology, neuropharmacology. *See* **NEUROSCIENCE.**

Physical medicine and rehabilitation
Alexander, James L.
Barnard, Roy James
Betts, Henry Brognard
Cohen, Glenn Milton
Compere, Clinton Lee
Cook, Daniel Walter
Goodman, Arden Patricia
Hamilton, Byron Bruce
Homsy, Charles Albert
Kewman, Donald Glenn
Kottke, Frederic James
Landsberg, Lewis
Lehmkuhl, L(loyd) Don
Malinow, Manuel Rene
Moffroid, Mary Thompson
Quigley, Michael J.
Sabbahi, Mohamed Ahmed
Sanders, Gloria Tolson
Shahani, Bhagwan Topandas
Spencer, William Albert

Physiology. *See also* **BIOLOGY.**
Abbrecht, Peter Herman
Agus, Zalman S.
Alexander, Harold
Altura, Burton Myron
Bagchi, Nandalal
Baker, John Patton, Jr.
Banerjee, Mukul Ranjan
Barker, Harold Grant
Barnard, Roy James
Belloni, Francis Louis
Bello-Reuss, Elsa N.
Bennett, Peter Brian
Berliner, Robert William
Berlyne, Geoffrey Merton
Berry, Christine Albachten
Bers, Donald Martin
Beyenbach, Klaus Werner
Bhattacharyya, Ashim K.
Blaine, Edward Homer
Blaustein, Mordecai P.
Bligh, John
Bloom, Sherman
Blum, Paul Solomon
Boerboom, Lawrence Edward
Boulpaep, Emile Louis J.B.
Brobeck, John Raymond
Brooks, Chandler McCuskey
Brosnihan, K. Bridget
Bulkley, Gregory Bartlett
Burich, Raymond Lucas
Byrne, Jeffrey Edward
Cahill, George Francis, Jr.
Cain, Stephen Malcolm
Cassin, Sidney
Castro, Gilbert Anthony
Chapman, Lloyd William
Chew, Catherine Strong
Clements, John Allen
Cohn, Stanton Harry
Comroe, Julius Hiram, Jr.
Cooper, George, IV
Cope, Michael Keith
Cordova-Salinas, Maria Asuncion
Cournand, Andre F.
Courtney, Kenneth Randall
DeLucia, Anthony John
Desjardins, Claude
Dobson, Alan
Douglas, Ben Harold
Downey, James Merritt
Dunham, Earl Wayne
Ehrlich, Walter
Eldridge, Frederic Louis
Ellis, Keith Osborne
Emmers, Raimond
Erickson, Howard Hugh
Falk, Sandor A.
Fater, Dennis Carroll
Feigen, Larry Philip
Fischbarg, Jorge
Flaim, Stephen Frederick
Flynn, John Thomas
Ford, Lincoln Edmond
Forrester, Thomas
Forster, Robert E., II
Foulkes, Ernest C(harles)
Francis, Kennon Thompson
Frazer, David George, Jr.
Frazier, Donald Tha
Frazier, Loy William, Jr.
Ganguli, Mukul Chandra
Gibbons, Walter Ray
Gladfelter, Wilbert Eugene
Glusman, Silvio
Goldfarb, Roy David
Goldfarb, Stanley
Gottschalk, Carl William
Grant, Brydon John Bruce
Grantham, Jared James
Greene, Murray A.
Greenspan, Kalman
Gross, Paul Munn
Gurll, Nelson Joseph
Hagan, Raymond Donald
Hamilton, Robert William
Hartline, Haldan Keffer
Hauer, Jerome Maurice
Hawkins, Rilchard Albert
Heinz, Erich
Helstad, Donald Dean
Hertz, Roy
Hirsch, Henry Richard
Holloszy, John Otto
Homer, Louis David
Hopfer, Ulrich
Hornung, David Eugene
Houk, James Charles
House, Edwin Wesley
Houser, Steven Robert
Hudgins, Patricia Montague
Hui, Chiu Shuen
Husted, Russell Forest
Imig, Charles Joseph
Jensen, Emron Alfred
Johannsen, Ulmer James
Johnson, Joseph Alan
Johnson, Leonard Roy
Kallok, Michael John
Kalu, Dike Ndukwe
Kao, Race L.
Kass, Robert S.
Kauker, Michael Lajos
Kellogg, Ralph Henderson
Keyl, Milton Jack
Kolbeck, Ralph Carl
Kopp, Stephen James
Kraman, Steve Seth
Lai, Yih-Loong
Land, Ivan Marshall
Laragh, John Henry
Laughlin, Maurice Harold
Lee, Chi Ok
Leffler, Charles William
Levitzky, Michael Gordon
Lichtman, Marshall A.
Lichton, Ira Jay
Liebow, Charles
Lin, Tsung-Min
Lindley, Barry Drew
Liu, Ching-Tong
Liu, Maw-Shung
Loeppky, Jack Albert
Lowensohn, Howard Stanley
Lymangrover, John Robert
Lynch, Peter Robin
Malindzak, George Steve, Jr.
Manning, John William
Maron, Michael Brent
Marquis, Norman Ronald
Martin, David Edward
Martin, Loren Gene
Martin, Paul Joseph
McDermott, Daniel Joseph
McDonagh, Paul Francis
Menninger, Richard Price
Miletich, David John
Milic-Emili, Joseph
Millard, Ronald Wesley
Miller, Michael James
Miller, Thomas Allen
Miller, Virginia May
Millhorn, David Eugene
Milnor, William Robert
Moe, Gordon Kenneth
Mokler, Corwin Morris
Montefusco, Cheryl Marie
Mortillaro, Nicholas A.
Musgrave, Gary Eugene
Nocenti, Mero Raymond
Osborn, Jeffrey Lynn
Paganelli, Charles Victor
Pappenheimer, John Richard
Partridge, L. Donald
Penhos, Juan Carlos Jacobo
Penney, David George
Peterson, Darwin Wilson
Petri, William Henry, III
Pfeffer, Janice Marie
Pindell, Merle Herbert
Pindok, Marie Theresa
Popovic, Vojin
Powell, Don Watson
Price, Alvin Audis
Prichard, John B.
Proctor, Kenneth Gordon
Quinton, Paul Marquis
Randall, David Clark
Recknagel, Richard Otto
Rennie, Ian Drummond
Reynolds, Robert Donald
Riley, David Joseph
Robertson, Howard Thomas, II
Robertson, James Sydnor
Rosen, Arthur L(eonard)
Rosen, Michael Robert
Rottenberg, David Allan
Rovetto, Michael Julien
Rovick, Allan Asher
Rubal, Bernard J.
Rubin, Lewis J.
Ruff, Robert Louis
Ruwe, William David
Schanker, Lewis Stanley
Schopp, Robert Thomas
Schultz, Stanley George
Schwartz, Marshall Zane
Selkurt, Ewald Erdman
Shah, Sudhir Amratlal
Shearin, Nancy Louise
Sherman, James Howe
Shvartz, Esar
Siegman, Marion Joyce
Silver, Donald
Skoryna, Stanley Constantine
Smeby, Robert Rudolph
Smith, Elvin Estus, Jr.
Smith, Philip Lawrence
Sparks, Harvey Vise, Jr.
Sperelakis, Nick
Spiller, Gene Alan
Stanek, Karen Ann
Stinson, Joseph McLester
Susskind, Herbert
Sybertz, Edmund J., Jr.
Tai, Yuan-Heng
Thurber, Robert Eugene
Trippodo, Nick Charles
Turlapaty, Prasad Durga
Mallikharjuna Vara
Vassalle, Mario
Vogel, Thomas Timothy
Wade, Charles Edwin
Wagner, Jeames Arthur
Waitzman, Morton Benjamin
Warner, Ronald Ray
Weber, Kenneth Charles
Weiner, Irwin M.
Weiss, Harold Samuel
Wildenthal, C(laud) Kern
Winget, Charles Merlin
Winter, Peter Michael
Wong, Robert King-Suen
Wood, Jackie Dale
Wyssbrod, Herman Robert, Jr.
Yellin, Edward Leon
Yu, Jia-Huey

Physiology, comparative
Bass, Eugene Lawrence
Bito, Laszlo Z.
Bunag, Ruben David
Burggren, Warren William
Burnett, Louis Elwood, Jr.
Cameron, John Stanley
Chou, Shyan-Yih
Cohn, Jay Binswanger
D'Angelo, Gaetano
Dantzler, William Hoyt
Dawson, William Ryan
Folk, George Edgar, Jr.
Fromm, Paul Oliver
Garner, Harold E.
Glick, Zvi
Gootman, Phyllis Myrna
Gur, David
Harmon, John Watson
Hazelwood, Robert L(eonard)
Heath, James Edward
Kayar, Susan R.
Koizumi, Kiyomi
Krausz, Stephen
Kumbaraci, Nuran Melek
Magee, Donal F.
Malanga, Carl Joseph
McCann, Frances V.
Miller, Virginia May
Noda, Hiroharu
Paladino, Frank Vincent
Parrish, John Wesley, Jr.
Rahn, Hermann
Rattner, Barnett Alvin
Roberts, John Lewis
Schmidt-Nielsen, Knut
Schneider, Richard Joel
Smith, Charles Roger
Teasdall, Robert Douglas
Timiras, Paola Silvestri
Umminger, Bruce Lynn
Urthaler, Ferdinand
Velamakanni, Krishnamurty
Seetarama
Venkataramiah, Amaraneni
Wartzok, Douglas
Webb, Alistair Ian
Zar, Jerrold H(oward)

Physiology, neurophysiology. *See* **NEUROSCIENCE.**

Physiology, psychophysiology
Abildskov, J.A.
Barratt, Ernest Stoelting
Bastawi, Aly Eloui
Burdick, James Alan
Cacioppo, John Terrance
Chen, Richard Yuan Zin
Cooke, Allan Roy
Corley, Karl Coates, Jr.
Credidio, Steven George
del Regato, Juan Angel
Dixen, Jean Marie
Eisenberg, M(yron) Michael
Engel, Bernard Theodore
Etzel, Kenneth Raymond
Farhi, Leon Elie
Feuerstein, Michael
Graham, David Tredway
Hofer, Myron Arms
Kety, Seymour S(olomon)
Lee, Richard M
London, Ray William
Manhold, John Henry
Michels, Lester David
Miller, Neal Elgar
Odom, James Vernon
Orlans, Flora Barbara
Pelletier, Kenneth R.
Pivik, Rudolph Terry
Randall, David Clark
Schwartz, Irving Leon
Silverman, Albert Jack
Wagman, Althea M.I.
Wickramasekera
Yellin, Absalom M(oses)

Preventive medicine. *See also* **Epidemiology.**
Adams, John David
Alexander, John Charles
Aronson, Stanley Maynard
Bailey, William James
Barker, Lewellys Franklin
Bloomfield, Daniel Kermit
Boostrom, Eugene Richard
Breslow, Lester
Brownell, Kelly David
Charlesworth, Edward Allison
Chesney, Margaret Ann
Deuschle, Kurt Walter
Estes, Edward Harvey, Jr.
Farquhar, John William
Flay, Brian Richard
Gardner, John Willard
Greenwald, Peter
Hagan, Raymond Donald
Halstead, Bruce Walter
Hennekens, Charles H.
Hinman, Edward John
Holloszy, John Otto
Hutchison, George Barkley
Kuo, Peter Te
Mahboubi, Ezzat Ollah
Margulies, Robert Allan
Miller, C. Arden
Ng, Lorenz Keng Yong
Nightingale, Elena Ottolenghi
Safer, Martin Allen
Schaeffer, Morris
Scheibel, Leonard William
Schneiderman, Marvin Arthur
Schonfeld, Gustav
Schwabel, Mary Jane
Shank, Robert Ely
Shearer, David Ross
Sheps, Cecil George
Stasiak, Roger Stanley
Stewart, John Alvin
Van Peenen, Peter F(ranz) D(irk)
Waldman, Ronald Jay
Way, Anthony B.
Wedgwood, Ralph Josiah Patrick
Wegman, David Howe
Williams, Roger John
Wilson, Marjorie Price
Wong-McCarthy, William James
Yankauer, Alfred
Zwemer, Thomas John

Psychiatry
Abuzzahab, Faruk Said, Sr.
Acosta, Gustavo
Arthur, Ransom James
Baron, Miron
Barone, Robert Michael
Beck, Aaron Temkin
Berger, Henry
Boyd, Jeffrey Lynn
Butler, Robert Neil
Clayton, Paula Jean
Crawshaw, Ralph
Davis, Oscar F.
DePaulo, Joseph Raymond, Jr.
Eisenberg, Leon
Engel, George Libman
English, Joseph T.
Enzer, Norbert Beverly
Erlenmeyer-Kimling, L.
Extein, Irl Lawrence
Fensterheim, Herbert
Figley, Charles Ray
Gillin, John Christian
Glass, Richard McLean
Gold, Judith Hammerling
Goode, David John
Guze, Samuel Barry
Halaris, Angelos
Hamburg, David A.
Heston, Leonard Lancaster
Hobson, J. Allan
Jandhyala, Bhagavan Srikrishna
Kandel, Eric Richard
Katz, Jay
Klein, Marjorie Hanson
Loosen, Peter Thomas
Maibaum, Matthew
Mallinger, Alan Gary
McCarley, Robert William
McCurdy, Layton
Mefferd, Roy Balfour, Jr.
Menninger, William Walter
Moe, Gordon Kenneth
Monroe, Russell Ronald
Newhouse, Paul Alfred
Pearlson, Godfrey David
Pettinati, Helen Marie
Rakoff, Vivian Morris
Richmond, Julius Benjamin
Robins, Lee Nelken
Sarma, P.S. Bala
Sattin, Albert
Schuster, Marvin Meier
Shamoian, Charles A.
Silverman, Albert Jack
Sternberg, David Edward
Van Kammen, Daniel Paul

Visotsky, Harold Meryle
Wallerstein, Robert Solomon
Watzlawick, Paul
Weiss, James Moses Aaron
Winokur, George
Yozawitz, Allan

Psychiatry, psychopharmacology
Abuzzahab, Faruk Said, Sr.
Balster, Robert Louis
Barchas, Jack David
Barrett, James Edward
Borison, Richard Lewis
Brown, Gerald LaVonne
Buckman, Maire Tults
Burns, John Joseph
Byck, Robert
Byrd, Larry Donald
Chan, Arthur Wing Kay
Ciaranello, Roland David
Clark, Mervin Leslie
Cohn, Victor Hugo
Consroe, Paul Francis
Cutler, Neal R.
Davidson, Arnold B.
Deneau, Gerald Antoine
DePaulo, Joseph Raymond, Jr.
Diamond, Bruce I.
Dornbush, Rhea L.
Dren, Anthony Thomas
Ellinwood, Everett Hews, Jr.
Extein, Irl Lawrence
Freedman, Daniel X.
Gamzu, Elkan Raphael
Gershon, Elliot Sheldon
Giannini, A. James
Gilbert, Jeffrey Morton
Gillin, John Christian
Glass, Richard McLean
Gold, Mark Stephen
Goldstein, Jeffrey Marc
Goode, David John
Halaris, Angelos
Halbreich, Uriel
Hartmann, Ernest Louis
Hingtgen, Joseph Nicholas
Jacoby, Jacob Herman
Javaid, Javaid Iqbal
Johnson, David Norseen
Kaul, Pushkar Nath
Kety, Seymour S(olomon)
Kim, S. Peter
Kline, Nathan Schellenberg
Lake, Charles Raymond
Leavitt, Fred I.
Lucot, James Bernard
Malitz, Sidney
Mallinger, Alan Gary
Meltzer, Herbert Yale
Mendels, Joseph
Menon, Madhavan Krishna
Messiha, Fathy Sabry
Meyerhoff, James Lester
Miczek, Klaus A.
Nemeroff, Charles Barnet
Newhouse, Paul Alfred
Orzack, Maressa Hecht
Peek, Leon Ashley
Petrie, William Marshall
Pihl, Robert Olander
Pollard, Gerald Tilman
Post, Robert Morton
Pottash, A.L.C.
Pozuelo, Jose
Pradhan, Sachin N.
Rech, Richard Howard
Reiser, Morton Francis
Robinson, Susan Estes
Roffman, Mark
Romano, James Anthony, Jr.
Sanberg, Paul Ronald
Schechter, Martin David
Scholes, Norman Wallace
Schoolar, Joseph Clayton
Shamoian, Charles A.
Siegel, Ronald Keith
Skolnick, Phil
Spohn, Herbert Emil
Stern, Warren Charles
Sternberg, David Edward
Stiller, Richard Louis
Szara, Stephen
Taylor, John Edward
Uyeno, Edward Teiso
Van Kammen, Daniel Paul
Verebey, Karl G.
Wang, Rex Yue
Weeks, James Robert
Wittenborn, John Richard
Wolgin, David L.
Yellin, Absalom M(oses)
Yokel, Robert Allen
Young, Robert Claringbold

Pulmonary medicine
Adams, Donald Robert
Al-Bazzaz, Faiq Jaber
Block, A. Jay
Cantanzaro, Antonino
Charan, Nirmal Biswas
Cohen, Allen Barry
Comroe, Julius Hiram, Jr.
Cudkowicz, Leon
Fishman, Alfred Paul
Forsham, Peter Hugh
Gelb, Arthur Franklin
Goldman, Allan Larry
Harrison, Gunyon M.
Hiller, Frederick Charles
Hooper, Robert George
Horton, Frank O., III
Kilburn, Kaye Hatch
Kotas, Robert Vincent
Kraman, Steve Seth
Last, Jerold Alan
Lavietes, Marc Harry
Lawrence, Ernest Clinton
Lebowitz, Michael David
Levitzky, Michael Gordon
López-Marjano, Vincent
Lourenco, Ruy Valentim
Mayock, Robert Lee
Milic-Emili, Joseph
Miller, Michael James
Passero, Michael Anthony
Poe, Robert Hilleary
Quan, Stuart Fun
Rhodes, Mitchell Lee
Riley, David Joseph
Robertson, Howard Thomas, II
Rossing, Thomas Harry
Rubin, Lewis J.
Ryan, Una Scully
Smith, Joseph Lorenzo
Soskel, Norman Terry
Stamper, Hugh Blair, Jr.
Stemmler, Edward Joseph
Stinson, Joseph McLester
Susskind, Herbert
Warwick, Warren James
Webster, Robert Owen
Weg, John Gerard
Weill, Hans
Weinhold, Paul Allen
Williams, Marshall Henry, Jr.

Radiology
Abrams, Herbert LeRoy
Amols, Howard Ira
Ascoli, Frank Anthony
Berger, Harvey James
Bloch, Peter H.
Brady, Luther W., Jr.
Bush, Raymond Sydney
Cano, Elmer Raul
Chen, Tao-seng
Chu, James Chien Hua
del Regato, Juan Angel
DeWerd, Larry A.
Dotter, Charles Theodore
Dritschilo, Anatoly
Epp, Edward Rudolph
Evens, Ronald Gene
Fivozinsky, Sherman Paul
Gibbs, Samuel Julian
Goepp, Robert August
Goodman, Robert L.
Hahn, Eric Walter
Hubbard, Lincoln Beals
Hungerford, Gordon Douglas
Isenburger, Herbert R.
Jacobson, Harold Gordon
Kaplan, Henry Seymour
Kartha, Mukund Krishna
Khan, Faiz Mohammad
Klement, Vaclav
Lawson, Alfred James
Lerch, Irving Abram
Lin, Hsiu-San
Littleton, Jesse Talbot, III
Mansfield, Carl Major
McCombs, Rollin Koenig
Mills, Michael David
Moyer, Robert Findley
Narayanan, C. S.
Nibhanupudy, Jagannadha Rao
Noriega, Brian Keith
Order, Stanley Elias
Orton, Colin George
Perry, Nelson Allen
Purdy, James Aaron
Rao, Ravindra P.
Reinstein, Lawrence Elliot
Rockwell, Sara Campbell
Rollo, F. David
Rosemark, Peter Jay
Sakover, Raymond Paul
Seaman, William Bernard
Seydel, Horst Gunter
Sharma, Subhash Chander
Siegel, Michael Elliot
Sontag, Marc Robert
Szal, Marcel Michael
Tengerdy, Catherine Elizabeth
Wiley, Albert Lee, Jr.
Withers, Hubert Rodney
Wizenberg, Morris Joseph
Wolf, Gerald Lee
Wu, Raymond Kee-Kin

Radiology, diagnostic
ACKERMAN, Robert Harold
Alfidi, Ralph Joseph
Axel, Leon
Baum, Stanley
Biello, Daniel Robert
Bluth, Edward Ira
Cacak, Robert Kent
Carroll, Barbara Anne
Chakraborty, Dev Prasad
Courtney, John Vincent
Doi, Kunio
Dotter, Charles Theodore
Drayer, Burton Paul
Dunnick, Nicholas Reed
Eisenberg, Ronald Lee
Fennessy, John James
Fischer, Harry W.
George, Ajax Elis
Graham, Leslie Stephen
Gratt, Barton Michael
Jacobson, Harold Gordon
Jereb, Marjan Josip
Joseph, Peter Maron
Klingensmith, William Claude, III
Kressel, Herbert Y.
Kugel, Jeffrey Auriel
Lieberman, Robert Perry
Lin, Dorothy Sung
Lodwick, Gwilym Savage
Meyers, Morton Allen
Moore, William E.
Moseley, Robert David, Jr.
Rapport, Robert
Robinson, Walter Lloyd
Rosenberg, Eric Ronald
Rossi, Raymond Paul
Rowlands, John Alan
Seaman, William Bernard
Steckel, Richard J.
Stein, George Nathan
Tortorici, Marianne Rita
Willman, Michael Karel
Wilson, Gabriel Henry
Woodring, John Howell

Radiology, imaging technology
Agard, Eugene Theodore
Alfidi, Ralph Joseph
Arnold, Ben Allen
Ascoli, Frank Anthony
Axel, Leon
Banjavic, Richard Alan
Becker, Joan Alaine
Bencomo, Jose Antonio
Bluth, Edward Ira
Bruels, Mark C.
Budinger, Thomas Francis
Bunch, Phillip Carter
Carroll, Barbara Anne
Carson, Paul Langford
Censor, Yair
Chakraborty, Dev Prasad
Chang, Wei
Clanton, Jeffrey Alan
Correia, John Arthur
Davison, Mark Edward
DiBianca, Frank Anthony
Dick, Charles Edward
Doi, Kunio
Doyle, Robert Joseph
Duerinckx, Andre Jozef
Dunn, William Lee
Goitein, Michael
Goldman, Lee William
Goldstein, Albert
Gordon, Richard
Gould, Robert George
Gratt, Barton Michael
Grissom, Michael Phillip
Gur, David
Haacke, Ewart Mark
Haymond, Herman Ralph
Hoffman, Eric Alfred
Holman, B. Leonard
Jaszczak, Ronald Jack
Knowles, Richard James Robert
Kremkau, Frederick W.
Krohmer, Jack Stewart
Kugel, Jeffrey Auriel
Lea, Jean Hedrick
Lin, Pei-Jan Paul
Lodwick, Gwilym Savage
Luckett, Larry Wayne
Lynch, Peter Robin
MacIntyre, William James
Macovski, Albert
Malamud, Herbert
McInerney, Joseph John
Mooney, Richard T.
Moore, William E.
Moseley, Robert David, Jr.
Nickoloff, Edward L.
Peschmann, Kristian Ralf
Peters, Terence Malcolm
Rao, Dandamudi Vishnuvardhana
Reiskin, Allan Burt
Renner, Wendel Dean
Revesz, George
Rollo, F. David
Rossi, Raymond Paul
Rottenberg, David Allan
Rowlands, John Alan
Rubenstein, Edward
Shaw, Chorng-Gang
Slutsky, Robert Allen
Strom, Joel Andrew
Tai, Douglas Leung-Tak
Villafana, Theodore
Weber, David Alexander
Webster, Edward William
Whitehead, Frank Roger
Wiley, Albert Lee, Jr.
Williams, Lawrence Ernest
Wilson, Gabriel Henry
Wingate, Catharine Louise

Radiology, nuclear medicine. ***See also*** **Nuclear medicine.**
Adams, Ralph Melvin
Castronovo, Frank Paul, Jr.
Chandra, Ramesh
Croft, Barbara Yoder
Freeman, Leonard Murray
Graham, Leslie Stephen
Harris, Gale Ion
Hiscock, Robert Russell
Hoory, Shlomo
Lin, Dorothy Sung
Linnemann, Roger Edward
Loken, Merle Kenneth
McVey, James Thomas
Mooney, Richard T.
Nickoloff, Edward L.
Pastorelle, Peter John
Patton, Dennis David
Ruegsegger, Donald Ray, Jr.
Sakover, Raymond Paul
Williams, Lawrence Ernest
Willman, Michael Karel

Space medicine. ***See*** **SPACE SCIENCE.**

Surgery
Albano, William A.
Allison, David Coulter
Badder, Elliott
Baker, Daniel Clifton
Barker, Clyde Frederick
Barker, Harold Grant
Baue, Arthur Edward
Beal, John M.
Buchwald, Henry
Bulkley, Gregory Bartlett
Burdette, Walter James
Burk, Martyn William
Burt, Michael Edward
Campbell, Gilbert Sadler
Chase, Robert Arthur
Clowes, Alexander Whitehill
Coran, Arnold Gerald
Cosman, Bard
Crowley, Lawrence G.
Davis, John Mihran
De Bakey, Michael Ellis
Donegan, William Laurence
Dos, Serge Jacques
Eisenberg, M(yron) Michael
Farmer, Joseph Clarence, Jr.
Fisher, John Courtney
Friedman, Harold Ira
Geelhoed, Glenn William
Giuliano, Armando E.
Greene, Frederick Leslie
Greene, Robert Jay
Guida, Peter Matthew
Gurll, Nelson Joseph
Harmon, John Watson
Hirose, Teruo Terry
Holder, Walter Dalton, Jr.
Horton, Charles E.
Housepian, Edgar Minas
Hutchinson, William Burke
Imbembo, Anthony Louis
Jamplis, Robert Warren
Joison, Julio
Lee, James Travis, Jr.
Lee, Yeu-Tsu N.
Lewis, Victor Lamar
Liechty, Richard Dale
Macon, William Linus, IV
Madura, James Anthony
Mansour, Edward George
Marchetta, Frank Carmelo
Mathes, Stephen John
McCarthy, Joseph Gerald
McCulloch, John Hathorn
McDougal, W(illiam) Scott
Miller, Thomas Allen
Mohs, Frederic Edward
Moody, Frank Gordon
Moore, Francis Daniels
Morris, Don Melvin
Munster, Andrew Michael
Nahrwold, David Lange
Nichols, Walter Kirt
Orloff, Marshall J.
Parsons, Robert Wilton
Peters, Thomas Guy
Pomerantz, Marc Abraham
Pories, Walter J.
Prout, George Russell, Jr.
Pruitt, Basil Arthur
Raaf, John Hart
Ram, Madhira Dasaradhi
Ramming, Kenneth Paul
Ranson, John Hugh Charles
Reckard, Craig Reginald
Reed, William Piper, Jr.
Reemtsma, Keith
Reichle, Frederick Adolph
Ritchie, Wallace Parks, Jr.
Rosenberg, Jerry C.
Ruberg, Robert Lionel
Schenck, Robert Roy
Schilling, John Albert
Schwartz, Marshall Zane
Sears, Henry Francis, II
Shires, George Thomas
Silver, Donald
Sindelar, William Francis
Singh, Rajendra Pratap
Sollinger, Hans Werner
Spellman, Mitchell Wright
Spratt, John Stricklin
Sugarbaker, Everett Van Dyke
Temple, Walley John
Tobin, Gordon Ross
Turcotte, Jeremiah George
Vogel, Thomas Timothy
Watkins, David Hyder
Weber, Thomas Richard
Weil, Richard, III
Wetstein, Lewis
Wright, Creighton Bolter
Yancey, Asa Greenwood
Yellin, Albert Elliot

Surgery, artificial organs. ***See also*** **BIOTECHNOLOGY.**
DeVries, William Castle
Imbembo, Anthony Louis
Jarvik, Robert K.
Kantrowitz, Adrian
Kolff, Willem Johan
Reed, William Piper, Jr.
Rosen, Arthur L(eonard)
Scott, Gerald William
Watkins, David Hyder

Surgery, cardiac
Baue, Arthur Edward
Boerboom, Lawrence Edward
De Bakey, Michael Ellis
DeVries, William Castle
Dickson, James Francis, III
Dos, Serge Jacques
Frazier, Oscar Howard
Furman, Seymour
Geha, Alexander Salim
Gordon, Robert Thomas
Gott, Vincent Lynn
Hirose, Teruo Terry
Kantrowitz, Adrian
Kao, Race L.
Kilman, James William
Lowe, James Edward
Mc Goon, Dwight Charles
Pierce, William Schuler
Reitz, Bruce Arnold
Watson, Donald Charles
Wetstein, Lewis
Wright, Creighton Bolter

Surgery, fetal. ***See also*** **Obstetrics.**
Brodner, Robert Albert
Golbus, Mitchell Sherwin
Hodgen, Gary Dean

Surgery, microsurgery
Baker, Daniel Clifton
Blackwell, Richard Edgar
DeVore, Duane Thomas
Gallant, Nanette
Hubschmann, Otakar Rudolf
Kleinert, Harold Earl
Lewis, Herschel Paul
Mathes, Stephen John
Mullan, John Francis
Opgrande, John Donald
Schenck, Robert Roy
Shaw, William Wei-Lien
Shons, Alan Rance
Watanabe, Tsuneo
Woodside, Jeffrey Robert

Surgery, neurosurgery
Becker, Donald Paul
Bering, Edgar Andrew, Jr.
Black, Peter McLaren
Brodner, Robert Albert
Campbell, James Norman
Chou, Shelley Nien-chun
Davis, Ross
Girvin, John Patterson
Hekmatpanah, Javad
Hoff, Julian Theodore
Hubschmann, Otakar Rudolf
Jane, John Anthony
Javid, Manucher J.
Kapp, John Paul
Kelly, Patrick Joseph
Lewis, Herschel Paul
McQuarrie, Irvine Gray
Miller, Carole Ann
Mullan, John Francis
Ommaya, Ayub K.
Ortiz-Suarez, Humberto Jose
Raimondi, Anthony John
Reynolds, Arden Faine, Jr.
Robertson, James Thomas
Robinson, WalkerLee
Rosenblum, Mark Lester
Schwartz, Henry Gerard
Simpson, Richard Kendall, Jr.
Smith, Barry Hamilton
Sypert, George Walter
Velasco-Suarez, Manual M.

Surgery, transplants
Barker, Clyde Frederick
Belzer, Folkert Oene
Chatterjee, Satya Narayan
Frazier, Oscar Howard
Geelhoed, Glenn William
Geha, Alexander Salim
Hardy, Mark Adam
Monaco, Anthony Peter
Munda, Rino
Nghiem, Dai Dao
Orloff, Marshall J.
Peters, Thomas Guy
Rapaport, Felix Theodosius
Reckard, Craig Reginald
Shaw, William Wei-Lien
Shons, Alan Rance
Simonian, Simon John
Sollinger, Hans Werner
Thomas, Francis T.
Turcotte, Jeremiah George
Veith, Frank James
Watson, Donald Charles
Weil, Richard, III

Teratology
Beck, Sidney Louis
Blackburn, Will R.
Brent, Robert Leonard
Damjanov, Ivan
Daniel, Michael Andrew
Gartner, Leslie Paul
Gautieri, Ronald Francis
Geber, William Frederick
Gibson, James Edwin
Greene, Robert Morris
Harbison, Raymond Dale
Hayes, A. Wallace
Hood, Ronald David
Joffe, Justin Manfred
Johnson, Elmer Marshall
Kelman, Bruce Jerry
Rao, Suryanarayana Koppal
Sikov, Melvin Richard
Singh, Jarnall
Slikker, William, Jr.
Stephens, Trent Dee
Welsch, Frank

Toxicology
Aaroe, William Henry
Acosta, Daniel
Adams, Max D.
Akera, Tai
Alvares, Alvito Peter
Ambre, John Joseph
Anderson, Rebecca J.
Archer, Michael Christopher
Aronson, Carl Edward
Aust, Steven Douglas
Baker, Thomas
Balazs, Tibor
Baron, Jeffrey
Beck, Barbara Doris
Bernfeld, Peter Harry William

Bernstein, Isadore Abraham
Berry, David Lester
Besch, Henry Roland, Jr.
Bidlack, Wayne Ross
Bleiberg, Marvin Jay
Blumberg, Jeffrey Bernard
Borgstedt, Harold Heinrich
Borzelleca, Joseph Francis
Bull, Richard James
Carrano, Richard A
Carson, Steven
Chaturvedi, Arvind Kumar
Chen, Theresa, Shang-tsing
Chhabra, Rajendra Singh
Cochran, Kenneth William
Coleman, Ronald L.
Cooper, David Young
Cosmides, George James
Cummins, Joseph Edward
Dalvi, Ramesh R.
D'Ambrosio, Steven Mario
Davis, Wilbur Marvin
Decker, Walter Johns
Desaiah, Durisala
De Salva, Salvatore Joseph
Deutsch, Dale George
Dixon, Robert L.
Do, Hien Duc
Dolenz, John Joseph
Douglas, Ben Harold
Douglas, J. Fielding
Emele, Jane Frances
Emmerson, John L.
Foulkes, Ernest C(harles)
Fowler, Bruce Andrew
Franko, Bernard Vincent
Freeman, Aaron Eliot
Gandolfi, Allen Jay
Garvin, Paul Joseph, Jr.
Gibson, James Edwin
Glassman, Jerome Martin
Golberg, Leon
Goldstein, Joyce Allene
Green, Donald Eugene
Gress-Gordon, Jean Anne
Guarino, Anthony Michael
Hanson, Douglas M.
Harbison, Raymond Dale
Harris, Jane Ellen
Harrison, Steadman Darnell, Jr.
Hart, Ronald Wilson
Hatch, Roger Conant
Hayes, A. Wallace
Heck, Henry d'Arcy
Hietbrink, Bernard Edward
Hill, Jim Tom
Hitchcock, Margaret
Ho, Ing K.
Hoffman, George Robert
Hollenberg, Paul Frederick
Hollinger, Mannfred Alan
Hood, Ronald David
Hook, Jerry Bruce
Horakova, Zdenka Zahutova
Horowitz, Bruce Richard
Hsia, Mong-Tseng S.
Huber, William George
Husain, Syed
Iturrian, William Ben
Jarboe, Charles Harry
Jee, Webster Shew Shun
Johnson, Elmer Marshall
Joy, Robert McKernon
Kaistha, Krishan, Kumar
Kammerer, Richard Craig
Keller, John Charles
Kelman, Bruce Jerry
Keplinger, Moreno L.
Keplinger, Moreno Lavon
Kilgore, Wendell Warren
Kimura, Eugene Tatsuru
King, Theodore Oscar
Klaassen, Curtis Dean
Kodama, Jiro Kenneth
Koschier, Francis Joseph, III
Kurtz, Perry James
Lanman, Robert Charles
Lee, Cheng-Chun
Lee, Insu Peter
Levin, Barbara Chernov
Levinskas, George Joseph
Lish, Paul Merrill
Litterst, Charles Lawrence
Loomis, Ted Albert
Loux, Joseph J.
Lowndes, Herbert Edward
Mackerer, Carl Robert
Maibach, Howard I.
Mainigi, Daivender Kumar
Malone, Marvin Herbert
Mattison, Donald Roger
Maurissen, Jacques Paul Jean
Mayer, Ramona Ann
Mays, Charles William
McClellan, Roger Orville
McColl, John Duncan
McConnell, William Ray
Messiha, Fathy Sabry
Meyer, Dale R.
Miceli, Joseph N(icola)
Miller, Matthew Steven
Milman, Harry Abraham
Mir, Ghulam Nabi
Mitchell, Clifford L.
Moody, David Edward
Nechay, Bohdan Roman
Newman, Robert Alwin
Northup, Sharon Joan
Nwangwu, Peter Uchenna
Oberdorster, Gunter
O'Callaghan, James Patrick
Ochillo, Richard Frederick
O'Donoghue, John (Lipomi)
Olajos, Eugene Julius
Page, John Gardner
Passero, Michael Anthony
Pearson, James Eldon
Pepelko, William Edward
Pereira, Michael Alan
Perhach, James Lawrence, Jr.
Phillips, Barrie Maurice
Pollock, Gerald Arthur
Porter, William Hudson
Rao, Suryanarayana Koppal
Reddy, Chada Sudershan
Reilly, Joseph Francis
Rickert, Douglas Edward
Rifkind, Arleen B.
Robertson, James Bragg
Robinson, Farrel Richard
Rosen, Joseph David
Rosenkrantz, Harris
Rossman, Toby G.
Roth, Robert Andrew, Jr.
Saslaw, Leonard David
Schach von Wittenau, Manfred Eberhard
Schechter, Martin David
Schwartz, Edward
Scribner, John David
Shaikh, Zahir Ahmad
Shargel, Leon David
Sharma, Raghubir Prasad
Shertzer, Howard Grant
Shirkey, Harry Cameron
Sikov, Melvin Richard
Sivak, Andrew
Smith, Frank Ackroyd
Smith, Roger Powell
Sofia, Robert Duane
Stavinoha, William B.
Steinberg, Marshall
Stevens, James Thomas
Stiller, Richard Louis
Stover, Betsy Jones
Swarm, Richard Lee
Tardiff, Robert George
Taves, Donald R.
Tephly, Thomas Robert
Tepper, Lloyd Barton
Tobin, Thomas
Tyler, Tipton Ransom
Vadlamudi, Sri Krishna
Valdes, James John
Van Dyke, Knox
Ward, John Wesley
Way, James Leong
Wedeen, Richard Peter
Wedig, John H.
Weisburger, Elizabeth Kreiser
Welsch, Frank
Wilber, Charles Grady
Wislocki, Peter Gregory
Wolf, Barbara Anne
Wolff, George Louis
Woods, Eugene Francis
Woods, James Sterrett
Yokel, Robert Allen

Urology
Ansell, Julian Samuel
Crawford, E. David
Drago, Joseph Rosario
Finney, Roy Pelham
Frank, Irwin Norman
Grayhack, John Thomas
Hampel, Nehemia
Howards, Stuart S.
Krieger, John Newton
McDougal, W(illiam) Scott
Merrin, Claude Emile Andre
Olsson, Carl Alfred
Pong, Raymond S.
Prout, George Russell, Jr.
Resnick, Martin I.
Senior, David Frank
Wein, Alan Jerome
Woodside, Jeffrey Robert

Virology
Aaronson, Stuart Alan
Ablashi, Dharam Vir
Allen, Patton Tolbert
Balfour, Henry H., Jr.
Behbehani, Abbas M.
Bishop, John Michael
Blacklow, Neil Richard
Boyd, Ann Lewis
Brandt, Carl David
Chang, Kenneth Shueh-Shen
Christensen, Mary Lucas
Cochran, Kenneth William
Cooney, Marion Kathleen
Crowell, Richard Lane
Cukor, George
Deibel, Rudolf
Dianzani, Ferdinando
Dienstag, Jules Leonard
Donikian, Mary Adrienne
Dunham, Wolcott Balestier
Falk, Lawrence Address, Jr.
Fisher, Paul B.
Forghani, Bagher
Fox, John Perrigo
Francis, Robert Dorl
Friedman, Harvey Michael
Fujinami, Robert Shin
Funk, Glenn Albert
Gajdusek, Daniel Carleton
Galasso, George John
Georgiades, Jerzy Alexander
Gerone, Peter John
Gibbs, Clarence Joseph, Jr.
Glaser, Ronald
Gross, Ludwik
Hahon, Nicholas
Hampar, Berge
Harford, Carl Gayler
Harter, Donald Harry
Hashimoto, Shuichi
Hayden, Frederick Glenn
Hierholzer, John Charles
Hougland, Arthur Eldon
Hsiung, Gueh-Djen
Huang, Eng-Shang
Hyman, Richard Walter
Iglewski, Wallace Joseph
Johnson, Richard Tidball
Kasai, George Joji
Kettrick-Marx, Mary Alice
Kieff, Elliott
Kilbourne, Edwin Dennis
Kirk, Billy Edward
Klein, Richard Joseph
Larson, Vivian M.
Lavender, John Francis
Leis, Jonathan Peter
Levy, Hilton Bertram
Lipton, Howard Lee
Loh, Philip Choo-Seng
Lowy, Douglas Ronald
Lyons, Michael Joseph
Mayor, Heather Donald
Mc Sharry, James John
Melnick, Joseph L.
Merigan, Thomas Charles, Jr.
Meyer, Harry M., Jr.
Michalski, Frank Joseph
Miller, Richard Lynn
Neurath, Alexander Robert
Oh, Jang
Pagano, Joseph Stephen
Parrott, Robert Harold
Perrault, Jacques
Potash, Louis
Prince, Alfred Mayer
Purcell, Robert Harry
Quarles, John Monroe
Rabson, Alan Saul
Raska, Karel Frantisek, Jr.
Ray, Usharanjan
Rhim, Johng Sik
Roane, Philip Ransom, Jr.
Roesing, Timothy George
Roizman, Bernard
Roy-Burman, Pradip
Rustigian, Robert Carning
Schlesinger, Milton Joseph
Schmidt, Nathalie Joan
Schwarz, Anton J.
Seto, Joseph Tobey
Shank, Peter R.
Shaw, Eugene Douglas
Shelokov, Alexis
Shope, Thomas Charles
Smith, Kendall Owen
Smith, Ralph Earl
Soike, Kenneth Fieroe, Jr.
Stinski, Mark Francis
Stollar, Victor
Stringfellow, Dale Alan
Tershak, Daniel Richard
Trent, Dennis Wayne
Wiebe, Michael Eugene
Wolff, David Alwin
Wolinsky, Jerry Saul
Wood, Owen Leslie
Woodhour, Allen Francis
Yoon, Ji-Won
Zebovitz, Eugene

Other
Alspaugh, Margaret Ann, *Rheumatology*
Arzbaecher, Robert, *Computers in medicine*
Attix, Frank Herbert, *Radiation Dosimetry*
Baldonado, Ardelina-Erika Albano, *Research methodology*
Beaulnes, Aurele, *Medical research administration*
Berjian, Richard A.
Berliner, Robert William, *Circulatory system*
Billhimer, Ward Loren, Jr., *Human safety and efficacy evalutions*
Bliznakov, Emile George, *Medical research administration*
Bloomfield, Saul S., *Clinical pharmacology*
Bois, Pierre, *Medical research administration*
Bole, Giles G., *Rheumatology*
Catalano, Michael Alfred, *Rheumatology*
Chedid, Antonio, *Hepatology*
Collen, Morris Frank, *Medical Research Administration*
Cooper, John Allen Dicks, *Medical education*
Cope, Michael Keith
Coriell, Lewis L., *Medical research administration*
Crowe, Dennis Timothy, Jr., *Traumatology-Critical Care*
Crowley, Lawrence G., *Medical administration*
David, Gary Samuel, *Hybridoma technology*
Dean, Richard H., *Vascular Surgery*
Dell-Osso, Louis Frank, *Neuro-ophthalmology*
Dishman, Rodney King, *Sports medicine*
Donabedian, Avedis, *Health Care Organization and Administration*
Eichholz, Geoffrey G., *Radiation Protection*
Espinoza, Luis Rolan, *Rheumatology*
Ewing, David Leon, *Radiation Biology*
Farber, Phillip Andrew, *Cytogenetics*
Foege, William Herbert, *Public health agency administration*
Fry, Richard Jeremy Michael, *Radiobiology*
Garg, Lal Chand, *Renal Pharmacology*
Gault, N.L., Jr., *Medical Education*
Gibson, Sam Thompson, *Regulatory medicine*
Goldstein, Norman, *Medical photography*
Gregg, John Bailey, *Medical anthropology and paleopathology*
Hadler, Nortin M., *Rheumatology*
Hahn, Bevra H(annahs), *Rheumatology*
Hecht, Frederick, *Biomedical research administration*
Henderson, Donald Ainslie, *Public health education administration*
Hudecki, Michael Stephen, *Muscular dystrophy*
Ingram, Alvin John, *Health Care Delivery*
Johnson, Armead Howard, *Histocompatibility*
Johnson, William K., *Radiology and Nuclear Medicine Safety Procedures*
Jones, Wilbur Douglas, Jr., *Mycobacteriophages*
Kassan, Stuart S., *Rheumatology*
Katz, Michael, *Tropical medicine*
Kau, Sen T., *renal physiology*
Kaufmann, Peter G., *Hyperbaric Medicine*
Kindwall, Eric Post, *Hyperbaric Medicine*
Kleinman, Arthur Michael, *Medical anthropology*
Kline, Nathan Schellenberg, *Transcultural psychiatry*
Kohli, Jai Dev, *Cardiovascular pharmacology*
Krakauer, Randall Sheldon, *Rheumatology*
Krevans, Julius Richard, *Educational administration*
Krishnan, Leela, *Radiation oncology*
Leevy, Carroll Moton, *Hepatology*
Leto, Salvatore, *Andrology*
Lipton, James Abbott, *Medical sociology*
Lohr, Kristine Marie, *Rheumatology*
Ludwig, James J., *Medical research administration*
Margulies, Robert Allan, *Emergency Medicine*
McGaghie, William Craig, *Medical education*
Mustard, James Fraser, *Arteriosclerosis, Research administration*
Nath, Ravinder, *Radiological Physics*
Nelson, Norton, *Environmental medicine*
Ottensmeyer, David Joseph, *Health care administration.*
Pachman, Lauren Merle, *Rheumatology*
Parthasarathy, Srisailam V., *Radiation therapy*
Popp, Diana Marriott, *Genetic Toxicology*
Pratt, Arnold W., *Computers in medicine.*
Pratt, Philip Chase, *Occupational and environmental lung disease, Pulmonary pathology*
Prystowsky, Harry, *Medical administration*
Purpura, Dominick Paul, *Medical education and research admnistration*
Rabkin, Mitchell Thornton, *Medical administration*
Rosemark, Peter Jay, *Medical physics*
Rosenberg, Eric Ronald, *Diagnostic ultrasound*
Royds, Robert B., *Clinical Pharmacology*
Ruegsegger, Donald Ray, Jr., *Radiation therapy*
Ruegsegger, Paul Melchior, *Thermography of diseases*
Sabbagha, Rudy E., *Diagnostic ultrasound*
Selikoff, Irving John, *Environmental medicine*
Shaw, Margery Wayne Schlamp, *Health law*
Shrivastava, Prakash Narayan, *Medical physics*
Slack, Stephen Thomas, *Radiation safety*
Solinger, Alan Michael, *Rheumatology*
Solomon, Richard Lester, *Experimental psychology*
Spitzer, Adrian, *Pediatric nephrology*
Sprague, Charles Cameron, *Medical administration*
Steele, Glenn Daniel, Jr., *Surgical oncology*
Stetten, DeWitt, Jr., *Medical research administration*
Sullivan, Carole A., *Radiation oncology*
Sullivan, Louis Wade, *Medical education administration.*
Swazey, Judith Pound, *Social and ethical aspects of biomedicine*
Takesue, Edward I., *Clinical research, pharmacology*
TePoorten, Bernard A.
Tepper, Lloyd Barton, *Environmental medicine*
Walter, Thomas Harry, *Bioelectricity*
Ward, Louis Emmerson, *Rheumatology*
Wardell, William Michael, *Clinical pharmacology*
Warren, Kenneth S., *Tropical medicine*
White, Rolfe Downing, *Gynecologic urology*
Wilson, Marjorie Price, *Management, health/medical systems*
Wyngaarden, James Barnes, *Medical research administration*
Zebovitz, Eugene, *Health sciences administration*
Zimmerman, Michael Raymond, *Paleopathology*
Zudeck, Steven, *Allergenic extract preparations*

NEUROSCIENCE

Neurobiology. *See* **BIOLOGY.**

Neurochemistry
Abou-Donia, Martha May
Abou-Donia, Mohamed Bahie
Agrawal, Harish Chandra
Ariano, Marjorie A.
Aronstam, Robert Steven
Azmitia, Efrain Charles
Bacopoulos, Nicholas G.
Banik, Narendra Lal
Barchi, Robert Lawrence
Baumgold, Jesse
Bautz, Gordon Thomas
Blumenthal, Kenneth Michael
Brecher, Arthur Seymour
Brimijoin, William Stephen
Broderick, Patricia Ann
Browning, Michael Douglas
Burt, David Reed
Calabrese, Vincent Paul
Callaghan, Owen Hugh
Carroll, Paul T.
Chan, Pak Hoo
Chang, Raymond Shen-Long
Ciaranello, Roland David
Clouet, Doris Helen
Cohn, Major Lloyd
Cooper, Jack Ross
Coscina, Donald Victor
Cox, Brian Martyn
Craig, Charles Robert
Crider, Andrew Blake
Dafny, Nachum Frenkel
Dakshinamurti, Krishnamurti
Davis, Leonard George
Dawson, Glyn
Dettbarn, Wolf-Dietrich
Deutsch, Dale George
Dietrich, W. Dalton, III
Domangue, Barbara B.
Dora, Eors Istvan
Duffy, Thomas Edward
Dunn, Adrian John
Dwivedi, Chandradhar
Edwards, David Joel
Egel, Lawrence
Ehrlich, Yigal H.
Erecinska, Maria
Fidone, Salvatore Joseph
Fischette, Christine Theresa
Freed, Curt Richard
Frey, William Howard II
Frye, Gerald Dalton
Gilbert, Jeffrey Morton
Glasel, Jay Arthur
Gold, Barry I(ra)
Goldman, Stephen Shepard
Goldstein, Menek
Green, Harry
Grossfeld, Robert Michael
Grunewald, Gary Lawrence
Haber, Bernard
Hanin, Israel
Harris, Jane Ellen
Harris-Warrick, Ronald Morgan
Hawkins, Rilchard Albert
Hendley, Edith D.
Hildebrand, John G(rant), III
Hingtgen, Joseph Nicholas
Hirschhorn, Ira Daniel
Horrocks, Lloyd Allen
Hoss, Wayne P.
Iqbal, Zafar
Irwin, Louis Neal
Karpiak, Stephen Edward, Jr.
Kasarskis, Edward Joseph
Kimes, Alane Susan
Kinnier, William James
Kishimoto, Yasuo
Klingman, Gerda Isolde
Klingman, Jack Dennis
Lajtha, Abel
Larrabee, Martin Glover
Lasley, Stephen Michael
Ledeen, Robert Wagner
Lees, Marjorie Berman
Leviit, Morton
Lin, Shih-chia Chen
London, Edythe Danick
Loo, Yen-Hoong
Lovell, Richard Arlington
Luine, Victoria Nall
Maher, Timothy John
Malone, Michael Joseph
Mangano, Richard Michael
Marangos, Paul Jerome
Margolis, Frank Leonard
Markelonis, George Joseph, Jr.
Marks, Neville
Marrazzi, Mary Ann
Martin, David Lee
Max, Stephen Richard
McBride, William Joseph, Jr.
McEwen, Bruce Sherman
McMorris, F(rederick) Arthur
Medina, Miguel Angel
Merkel, Christian Gottfried
Meyerhoff, James Lester
Meyerson, Laurence R.

Miller, Matthew Steven
Moody, Terry William
Morell, Pierre
Morton, Bruce Eldine
Moskowitz, Michael Arthur
Murrin, Leonard Charles, II
Narasimhachari, Nedathur
Nicklas, William John
Norton, William Thompson
O'Donohue, Thomas Leo
Olajos, Eugene Julius
Palmer, Gene Charles
Park, Dong Hwa
Patrick, Robert L.
Pearlmutter, A. Frances
Percy, Alan Kenneth
Poduslo, Shirley Ellen
Pohorecky, Larissa A.
Pozuelo, Jose
Prentky, Robert Alan
Rainbow, Thomas Charles
Rassin, David Keith
Ratcheson, Robert Allan
Ray, Radharaman
Rebec, George Vincent
Reis, Donald Jeffery
Reith, Maarten Eduard Anton
Resnick, Oscar
Retz, Konrad Charles
Riker, Donald Kay
Roberts, Eugene
Roel, Lawrence Edmund
Rosenberg, Philip
Roskoski, Robert, Jr.
Roth, Jerome Allan
Routtenberg, Aryeh
Rowell, Peter Putnam
Sabri, Mohammad Ibrahim
Sachs, Howard
Sacks, William
St. Omer, Vincent Edmund Victor
Schein, Stanley Jay
Schnaar, Ronald Lee
Schrieber, Robert Alan
Scott, Irena McCammon
Seyfried, Thomas Neil
Shapira, Raymond
Sharp, Charles William
Sharpless, Nansie Sue
Shellenberger, Melvin Kent
Shih, Tsung-Ming Anthony
Snow, Anne Evelyn
Soifer, David
Sourkes, Theodore Lionel
Spector, Reynold
Squires, Richard Felt
Stach, Robert William
Stone, Eric Andrew
Streicher, Eugene
Sun, Grace Yan Chi
Sze, Paul Yi Ling
Talamo, Barbara Ruth
Tengerdy, Catherine Elizabeth
Thomas, William Eric
Tjioe, Sarah Archambault
Tolbert, Lelland Clyde
Ueda, Tetsufumi
VonVoigtlander, Philip Friedrich
Wecker, Lynn
White, Robert J.
Williams, Curtis Alvin, Jr.
Wong, David T.
Wurtman, Richard Jay
Yanagihara, Takehiko
Yates, Allan James
Yunger, Libby Marie
Zamenhof, Stephen
Zivin, Justin Allen

Neuroendocrinology. *See* **MEDICINE, Endocrinology.**

Neuroimmunology. *See* **MEDICINE, Immunology.**

Neuropharmacology
Abou-Donia, Martha May
Adams, Perrie Milton
Adler, Martin W.
Aghajanian, George Kevork
Alford, Geary Simmons
Alkana, Ronald Lee
Amirkhanian, John David
Anderson, Rebecca J.
Anisman, Hymie
Antonaccio, Michael John
Aronstam, Robert Steven
Aston, Roy
Babington, Ronald Glenn
Baker, Thomas
Baldino, Frank, Jr.
Barnes, Charles Dec
Barnett, Allen
Barone, Frank Carmen
Bartus, Raymond T.
Bautz, Gordon Thomas
Berney, Stuart Alan
Bird, Stephanie Jean
Bloom, Floyd Elliott
Blum, Kenneth
Blumberg, Harold
Boisse, Norman Robert
Borison, Richard Lewis
Braitman, David Jeffrey
Brase, David Arthur
Breese, George Richard
Brezenoff, Henry Evans
Broderick, Patricia Ann
Brown, Gerald LaVonne
Brown, Lucy Leseur
Browne, Ronald Gregory
Buccafusco, Jerry Joseph
Burchfiel, James Lee
Burkman, Allan Maurice
Burks, Thomas Franklin, II
Burt, David Reed
Byck, Robert
Bylund, David B.
Cannon, J. Timothy
Carr, Laurence A.
Carson, Virginia Rosalie Gottschall
Chang, Raymond Shen-Long
Chu, Nai-Shin
Church, Allen Charles
Chute, Douglas Lawrence
Ciarlone, Alfred Edward
Clay, George A
Clemens, James Allen
Clouet, Doris Helen
Cohen, Marlene Lois
Colasanti, Brenda Karen
Concannon, James Thomas
Cordingley, Gary Edward
Cowan, Alan
Cox, Brian Martyn
Craig, Charles Robert
Crawley, Jacqueline N.
Creese, Ian Nigel Richard
Creveling, Cyrus Robbins
Crow, Lowell Thomas
Daigneault, Ernest Albert
Davidson, Arnold B.
Davis, Hasker Pat
Davis, Michael
Davis, Thomas Paul
Davis, Wilbur Marvin
Dean, Reginald Langworthy, III
de Groat, William Chesney
de Jong, Rudolph H.
Deneau, Gerald Antoine
Desaiah, Durisala
Dettbarn, Wolf-Dietrich
Diamond, Bruce I.
Dibner, Mark Douglas
DiMicco, Joseph Anthony
Dingledine, Raymond Joseph
Domer, Floyd Ray
Dorris, Roy Lee
Dray, Andre
Dreifuss, Fritz Emanuel
Dren, Anthony Thomas
Duckles, Sue Piper
Dudek, Bruce Craig
Dun, Nae J.
Edmonds, Harvey Lee, Jr.
Edwards, David Joel
Ehrenpreis, Seymour
Eison, Michael Steven
Eldefrawi, Amira Toppozada
Ellinwood, Everett Hews, Jr.
Enna, Salvatore Joseph
Esposito, Ralph Umberto
Ettenberg, Aaron
Fahn, Stanley
Faingold, Carl Lawrence
Felpel, Leslie P.
Ferko, Andrew Paul
Fernstrom, Madelyn Hirsch
Fernstrom, John Dickson
Ferrendelli, James Anthony
Ferris, Steven Howard
Foreman, Mark Mortensen
Franz, Donald Norbert
Frederickson, Robert C.A.
Freedman, Daniel X.
Friedhoff, Arnold
Fromm, Gerhard Hermann
Fruend, Gerhard
Frye, Gerald Dalton
Fuller, Ray Ward
Gallagher, Brian Boru
Gallardo-Carpentier, Adriana
Gebhart, Gerald Francis
Geller, Herbert Miles
Gentry, R., Thomas
Glisson, Silas Nease, III
Gold, Barry I(ra)
Goldstein, Jeffrey Marc
Grandison, Lindsey James
Greengard, Paul
Hackett, John Taylor
Hackman, John Clement
Hall, Edward Dallas
Hance, Anthony James
Hancock, John Charles
Hanin, Israel
Hansl, Nikolaus Rudolf
Harris, Louis Selig
Hartmann, Ernest Louis
Harvey, John Adriance
Haskins, John Thomas
Heavner, James Edward
Hecht, Elizbaeth Anne
Heffner, Thomas Gary
Helke, Cinda Jane
Heller, Allen Harvey
Henderson, Edward George
Hendley, Edith D.
Hernandez, Linda Louise
Hiller, Jacob Moses
Hillman, Gilbert Rothschild
Hirschhorn, Ira Daniel
Hirsh, Kenneth Roy
Ho, Ing K.
Hodge, Gordon Karl
Holaday, John Waldron
Hollister, Leo Edward
Hoover, Donald Barry
Hoss, Wayne P.
Huffman, Ronald Dean
Hynes, Martin Dennis, III
Iorio, Louis Carmen
Iturrian, William Ben
Jacquet, Yasuko Filby
Jasinski, Donald Robert
Javaid, Javaid Iqbal
Johnson, David Norseen
Jones, David Joseph
Joy, Robert McKernon
Kaczmarek, Leonard Konrad
Kadzielawa, Krzysztof
Kellar, Kenneth Jon
Kinnier, William James
Klemm, William Robert
Klingman, Gerda Isolde
Knott, Peter Jeffrey
Kornetsky, Conan
Kosersky, Donald S.
Koss, Michael Campbell
Krall, Ronald Lee
Krivoy, William Aaron
Ksir, Charles Joseph, Jr.
Kuhar, Michael J.
Lake, Charles Raymond
Lalley, Peter Michael
Levin, Harvey Steven
Levitan, Herbert
Levy, Deborah Louise
Liebman, Jeffrey Mark
Lin, Shih-chia Chen
London, Edythe Danick
Long, John Paul
Lotti, Victor Joseph
Lucot, James Bernard
Luttinger, Daniel Alan
Maher, Timothy John
Malick, Jeffrey B.
Manyam, Bala Venktesha
Mao, Chi Chian
Maran, Janice Wengerd
Marangos, Paul Jerome
Marczynski, Thaddeus John
Marder, Eve E.
Mark, Marvin Robert
Marrazzi, Mary Ann
Marwaha, Jwaharlal
Maxwell, Donald Robert
McCarty, Leslie Paul
Mc Elligott, James George
McIsaac, Robert J(ames)
McKearney, James William
McLean, John Robert
Meiss, Dennis Earl
Meltzer, Herbert Yale
Menon, Madhavan Krishna
Messing, Rita Bailey
Millichap, J(oseph) Gordon
Mitchell, Clifford L.
Mitler, Merrill Morris
Miyamoto, Michael Dwight
Moore, Robert Haldane, III
Murray, Rodney Brent
Murrin, Leonard Charles, II
Myslinski, Norbert Raymond
Naftchi, Nosrat Eric
Narahashi, Toshio
Narasimhachari, Nedathur
Nemeroff, Charles Barnet
Novack, Gary Dean
O'Callaghan, James Patrick
O'Donohue, Thomas Leo
Orkand, Richard Kenneth
Osterberg, Arnold Curtis
Ostrow, David G.
Overton, Donald A.
Palmer, Gene Charles
Patrick, Robert L.
Pazdernik, Thomas Lowell
Pazoles, Christopher James
Phillis, John Whitfield
Pohorecky, Larissa A.
Poschel, Bruno Paul Henry
Post, Robert Morton
Pottash, A.L.C.
Prabhu
Pradhan, Sachin N.
Proudfit, Herbert Kerr, III
Quirion, Remi
Raja, Srinivasa N.
Rawat, Arun Kumar
Rebec, George Vincent
Reilly, Margaret Anne
Reith, Maarten Eduard Anton
Reitz, Ronald Charles
Retz, Konrad Charles
Richter, Judith Anne
Riker, Donald Kay
Ritchie, Joseph Murdoch
Roberts, Eugene
Robinson, Susan Estes
Roffman, Mark
Rogawski, Michael Andrew
Rosenberg, Howard Charles
Rosenberg, Philip
Rosenkranz, Roberto Pedro
Roth, Jerome Allan
Rowell, Peter Putnam
Rutledge, Charles Ozwin
Ruwe, William David
Ryder, Steven William
Salzman, Steven Kerry
Sattin, Albert
Schlesinger, Edward Bruce
Scholes, Norman Wallace
Schwark, Wayne Stanley
Seidler, Frederic John
Severs, Walter Bruce
Sharp, Charles William
Sharpe, Lawrence Grady
Shellenberger, Melvin Kent
Shih, Tsung-Ming Anthony
Shrivastav, Brij Bhushan
Skolnick, Phil
Sladek, Celia Davis
Slevin, John Thomas
Slotkin, Theodore Alan
Smith, Cedric Martin
Smith, Dennis B.
Smith, Richard Petri
Snow, Anne Evelyn
Soliman, Karam Farag Attia
Sourkes, Theodore Lionel
Squires, Richard Felt
Steinfels, George Francis
Stern, Warren Charles
Stitzel, Robert Eli
Stone, Eric Andrew
Study, Robert Edward, Jr.
Summers, William Koopmans
Szara, Stephen
Takemori, Akira Eddie
Tjioe, Sarah Archambault
Tolbert, Lelland Clyde
Tseng, Leon
Turkanis, Stuart Allen
Valdes, James John
Van Loon, Glen Richard
Van Woert, Melvin Holmes
Vaupel, Donald Bruce
Villablanca, Jaime Rolando
VonVoigtlander, Philip Friedrich
Walters, Judith Richmond
Wang, Rex Yue
Ward, O. Byron, Jr.
Wecker, Lynn
Weight, Forrest F.
Wein, Alan Jerome
Weiner, William Jerrold
Welch, Bruce Lynn
Westfall, David Patrick
Wiech, Norbert Leonard
Wolfe, Barry B.
Wong, David T.
Ziegler, Michael Gregory

Neurophysiology
Albano, Joanne Edvige
Alberts, Walter Watson
Allen, Gary Irving
Allman, John Morgan
Anderson, James Alfred
Andresen, Michael Christian
Asanuma, Hiroshi
Atema, Jelle
Baird, Henry Welles, III
Baker, John Patton, Jr.
Baldo, George Jesse
Barbas, Helen
Barlow, Robert Brown, Jr.
Barmack, Neal Herbert
Barnes, Charles Dec
Barnes, Karen Louise
Barth, Daniel Stephen
Beattie, Michael Stephen
Beck, Mary McLean
Bennett, Michael Vander Laan
Berger, Theodore William
Bergey, Gregory Kent
Berkley, Mark A.
Berlind, Allan
Bernstein, Jerrold
Berry, Stephen Daniel
Bhattacharya, Syamal Kanti
Bilotto, Gerardo
Bird, Stephanie Jean
Bizzi, Emilio
Black, Franklin Owen
Blaustein, Mordecai P.
Bligh, John
Blum, Paul Solomon
Bolanowski, Stanley John, Jr.
Boston, John Robert
Bowen, John Metcalf
Bowman, Mary Bronwyn
Braitman, David Jeffrey
Brink, Peter Richards
Brobeck, John Raymond
Brooks, Chandler McCuskey
Brown, George Barremore
Burchfiel, James Lee
Burke, Robert Emmett
Burne, Richard Allen
Byrd, Kenneth Elburn
Calvin, William Howard
Campbell, James Norman
Carlson, Joseph Ralph
Carpenter, David Orlo
Chou, Shelley Nien-chun
Cohen, David Harris
Cohen, Morton I.
Coleman, James R.
Coppola, Richard
Corley, Karl Coates, Jr.
Costanzo, Richard Michael
Craik, Rebecca Lynn
Crampton, George Harris
Dana, Richard Charles
Davis, Ross
Debons, Albert Frank
de Groat, William Chesney
de la Torre, Jack Carlos
Delcomyn, Fred
De Luca, Carlo John
Deshmukh, Vinod Dhundiraj
Dichter, Marc Allen
Dingledine, Raymond Joseph
Dora, Eors Istvan
Doty, Robert William
Dow, Bruce MacGregor
Dray, Andre
Duffy, Frank Hopkins
Dun, Nae J.
Easton, Dexter Morgan
Eaton, Robert Charles
Ebner, Ford Francis
Egger, Maurice David
Eldred, Earl
Eldridge, Frederic Louis
Emerson, Robert Charles
Emmers, Raimond
Engelhardt, John Kerch
Erickson, Robert Porter
Erulkar, Solomon David
Evarts, Edward Vaughan
Ewing, June Swift
Faber, Donald Stuart
Faingold, Carl Lawrence
Fidone, Salvatore Joseph
Franz, Donald Norbert
Frazier, Donald Tha
Frederickson, Robert C.A.
Funch, Paul Gerard
Furshpan, Edwin Jean
Gardner, Daniel
Gebhart, Gerald Francis
Geller, Herbert Miles
Giannini, A. James
Gidda, Jaswant Singh
Gilman, Sid
Girvin, John Patterson
Gladfelter, Wilbert Eugene
Glenn, Loyd LeRoy
Goldberg, Michael E.
Gonzalez-Lima, Francisco M.
Gootman, Phyllis Myrna
Greenspan, Joel Daniel
Grenell, Robert Gordon
Grob, David
Guinan, John Joseph, Jr.
Hackett, John Taylor
Hackman, John Clement
Hagiwara, Susumu
Haier, Richard Jay
Hallett, Mark
Hamilton, Robert Bruce
Hand, Peter James
Harris-Warrick, Ronald Morgan
Haskins, John Thomas
Hatton, Glenn Irwin
Henderson, Edward George
Heyer, Eric John
Hirsch, Helmut Villard Buntenbroich
Hobson, J. Allan
Hoffert, Marvin Jay
Holland, Robert Campbell
Holub, Richard Anthony
Hosko, Michael J., Jr.
Huffman, Ronald Dean
Jahan-Parwar, Behrus
Jen, Philip Hung Sun
Johnston, Daniel
Jones, Stephen Wallace
Kadzielawa, Krzysztof
Kastin, Abba Jeremiah
Kaufmann, Peter G.
Kennedy, Duncan Tilly
Klemm, William Robert
Koizumi, Kiyomi
Kraig, Richard Paul
Krauthamer, George Michael
Kukulka, Carl George
Kumbaraci, Nuran Melek
Lalley, Peter Michael
Larrabee, Martin Glover
Larsen, Kenneth David
Lasater, Eric Martin
Law, Peter Koi
Lawrence, Merle
Leeman, Susan Epstein
Lehmkuhl, L(loyd) Don
Leigh, Richard John
Lennard, Paul Ross
Levine, Michael Steven
Lieberman, Edward Marvin
Lindsley, David Ford
Lindsley, Donald Benjamin
Lipicky, Raymond John
Llinas, Rodolfo Riascos
Loeb, Gerald Eli
Low, Walter Cheney
MacLean, Paul Donald
Maguire, William Michael
Maran, Janice Wengerd
Marcus, Elliot Meyer
Marczynski, Thaddeus John
Marder, Eve E.
Marks, William Byron
Marrocco, Richard Thomas
Massopust, Leo Carl
Mayer, David Jonathan
Mayer, Richard Frederick
Mc Cabe, Brian Francis
McCarley, Robert William
McCrea, Robert Alan
Meiss, Dennis Earl
Meyer, Richard Arthur
Miller, Alan Douglas
Millhorn, David Eugene
Mitra, Jyoti
Miura, Robert Mitsuru
Miyamoto, Michael Dwight
Moller, Aage Richard
Moss, Robert Louis
Mountcastle, Vernon Benjamin, Jr.
Munson, John Bacon
Musiek, Frank Edward
Myslinski, Norbert Raymond
Nakamura, Richard Ken
Narahashi, Toshio
Nathanson, Morton
Nelson, Jeremiah I.
Nelson, Randall Jay
Newton, Roberta Ann
Noda, Hiroharu
O'Leary, Dennis Patrick
Orkand, Richard Kenneth
Pagala, Murali Krishna
Page, Charles Henry
Pappas, Carol Lynn
Partridge, L. Donald
Patterson, Michael Milton
Patterson, Terence Edward
Pellmar, Terry C.
Perryman, James Harvey
Pfaff, Donald W.
Phillips, Christine Elaine
Phillips, Michael Ian
Phillis, John Whitfield
Piercey, Montford Frederic
Pitts, Nathaniel Gilbert
Pivik, Rudolph Terry
Plotsky, Paul Mitchell
Proenza, Luis Mariano
Puente, Antonio Enrique
Raja, Srinivasa N.
Rall, Wilfrid
Ransom, Bruce Robert
Ratliff, Floyd
Ray, Richard Hallett
Reis, Donald Jeffery
Remmel, Ronald Sylvester
Reynolds, Arden Faine, Jr.
Richmond, Barry Jay
Riso, Ronald Raymond

Robinson, David Lee
Rockland, Kathleen Skiba
Rosen, Arthur D.
Rosenblith, Walter Alter
Sabbahi, Mohamed Ahmed
Salzman, Steven Kerry
Sanders, Donald Benjamin
Sax, Daniel Saul
Schlesinger, Edward Bruce
Scott, Irena McCammon
Scott, Thomas Russell, Jr.
Scrima, Lawrence
Semba, Kazue
Sharma, Sansar C.
Shrivastav, Brij Bhushan
Siegel, Jerome
Sillman, Arnold Joel
Simpson, Richard Kendall, Jr.
Smith, Barry Decker
Smith, Dennis B.
Smith, Robert L.
Smith, Thomas Graves, Jr.
Snape, William John, Jr.
Solomon, Paul Robert
Sparks, David Lee
Spehlmann, Rainer
Stanford, Laurence Ralph
Stein, Barry Edward
Stephens, Philip John
Stern-Tomlinson, Wendy Barbara
Stevens, Charles F.
Stockwell, Charles Warren
Storey, Arthur Thomas
Stuart, Ann Elizabeth
Study, Robert Edward, Jr.
Taber-Pierce, Elizabeth
Tache, Yvette France
Taub, Edward
Teyler, Timothy James
Thompson, Richard Frederick
Tourtellotte, Wallace William
Traub, Roger Dennis
Trehub, Arnold
Trubatch, Janett
Turkanis, Stuart Allen
Tuttle, Jeremy Ballou
Villablanca, Jaime Rolando
Voigt, Herbert Frederick, III
Walters, Judith Richmond
Watkins, Linda May
Weber-Levine, Margaret Louise
Weight, Forrest F.
Weiss, Ira Paul
Weiss, Thomas Fischer
West, Charles Hutchison Keesor
White, Robert J.
Wieland, Steven Joseph
Wilson, Richard Cameron
Winter, Arthur
Wolbarsht, Myron Lee
Wong, Robert King-Suen
Woody, Charles Dillon
Worden, Frederic Garfield
Wu, Chun-Fang
Wurtz, Robert Henry
Yazulla, Stephen
York, Donald Harold
Young, Robert Rice
Zabara, Jacob
Zee, David Samuel

Neuropsychology
Albano, Joanne Edvige
Albers, Henry Elliott
Allen, Robert Arthur
Anisman, Hymie
Appel, Antoinette Ruth
Barratt, Ernest Stoelting
Bauer, Richard Henry
Becker, James Thompson
Berg, Richard Alan
Berger, Henry
Berman, Marlene Oscar
Bigler, Erin David
Bodnar, Richard Julius
Bolanowski, Stanley John, Jr.
Bornstein, Robert A.
Bowman, Mary Bronwyn
Brito, Gilberto Ottoni
Brodie, Harlow Keith Hammond
Brooker, Alan Edward
Broughton, Roger James
Bryden, Mark Philip
Bryson, George Gardner
Buck, Ross Workman
Burdick, Charles Kenneth
Burton, Howard Alan
Busse, Ewald William
Campbell-Smith, Rosemary Gilles
Cannon, J. Timothy
Chute, Douglas Lawrence
Cleeland, Charles Samuel
Corkin, Suzanne
Coutts, Robert LaRoy
Craig, Paul Lawrence
Crovitz, Herbert Floyd
Crown, Barry Michael
Dalby, (John) Thomas
Davis, Michael
Dean, Raymond S.
Deese, James Earle
Demarest, David Steele
Dolenz, John Joseph
Donchin, Emanuel
Donovick, Peter J.
Doty, Robert William
Dunlop, Terrence Ward
Eckardt, Michael Jon
Egel, Lawrence
Erickson, Robert Porter
Eslinger, Paul Joseph
Esposito, Ralph Umberto
Fanselow, Michael Scott
Farrer, Donald Nathanael
Fisher, Jerid Martin
Fogel, Max Leonard
Gallo, Mario Martin
Gardner, Howard Earl
Gasparrini, William Gerard
Geschwind, Norman
Gianutsos, Rosamond R.
Glendenning, Karen Kircher
Goldman, Alexander
Gordon, Wayne Alan
Grafman, Jordan Henry
Greenblatt, Samuel Harold
Gur, Ruben C.
Guy, James David, Jr.
Harvey, John Adriance
Hecht, Elizbaeth Anne
Helffenstein, Dennis Alan
Hellige, Joseph Bernard
Hill, A. Lewis
Hinrichs, James Victor
Hirst, William Charles
Hoffman, William Charles
Holloway, Frank Albert
Holmstrom, Valerie Louise
Howard, James Lawrence
Inglis, James
Jackson, Thomas Larry
Jahan-Parwar, Behrus
Jones, Ben Morgan
Kamback, Marvin Carl
Kellar, Lucia Ames
Kettlewell, Neil MacKewan
Kimura, Doreen
King, Frederick Alexander
Kodanaz, Hatice Altan
Kosslyn, Stephen Michael
Krasner, Paul R.
Krynicki, Victor Edward
Kurlychek, Robert Thomas
Lacher, Miriam Browner
Lansdell, Herbert Charles
Largen, John William, Jr.
Latham, Eleanor Earthrowl
Leli, Dano Anthony
Levin, Harvey Steven
Levinthal, Charles Frederick
Lewin, Mark Henry
Lewis, Lisa
Litle, Patrick Alan
Lloyd, John Tracy
Lolli, Peter Patrick
Lubar, Joel Fredric
Ludlow, Christy Leslie
MacDonald, G(eorge) Wayne
Mack, James Lewis
Mahan, Harry Clinton
Malatesta, Victor Julio
Malitz, Sidney
Malone, Daniel Richard
Mapou, Robert Lewis
Masters, Robert Edward Lee
Matarazzo, Joseph Dominic
McDonald, Stanford Laurel
Mc Elligott, James George
McEwen, Bruce Sherman
Miller, Nancy E(llen)
Miller, Patricia Lynn
Mitchell, James Curtis
Moffie, Robert Wayne
Morton, Bruce Eldine
Nakamura, Richard Ken
Naranjo, Jennings Neal
Neff, William Duwayne
Netley, Charles Thomas
Olton, David Stuart
Orzack, Maressa Hecht
Parker, Jerry Calvin
Patterson, Michael Milton
Pavlidis, George Theophilou
Pearlson, Godfrey David
Peniston, Eugene G.
Perez, Francisco Ignacio
Pitcher, Georgia Ann
Plakosh, Paul, Jr.
Poschel, Bruno Paul Henry
Proudfoot, Ruth Ellen
Puente, Antonio Enrique
Purisch, Arnold David
Rainbow, Thomas Charles
Reitan, Ralph Meldahl
Riscalla, Louise Beverly
Robinson, Daniel Nicholas
Robinson, David Lee
Rosenzweig, Mark Richard
Russell, Elbert Winslow
Sachs, Benjamin David
Sbordone, Robert Joseph
Schaeffer, Benson
Schiffman, Susan Stolte
Sharpe, Lawrence Grady
Silverman, Hirsch Lazaar
Skeen, Leslie Carlisle
Soper, Henry Victor
Squire, Larry Ryan
Stapleton, Leroy Earl
Stellar, Eliot
Steward, Oswald
Symmes, David
Tansey, Michael Anselme
Valenstein, Elliot Spiro
Vicente, Peter James
Voight, Jesse Carrolton
Wagman, Althea M.I.
Waters, James Herman
West, Charles Hutchison Keesor
Wilson, Glenn Francis
Woody, Charles Dillon
Yeterian, Edward Harry
Yozawitz, Allan

Regeneration
Bernstein, Jerald J(ack)
Clemente, Carmine Domenic
Creese, Ian Nigel Richard
Davis, James Norman
Dentinger, Mark Peter
DeSantis, Mark Edward
Drakontides, Anna B.
Easter, Stephen Sherman, Jr.
Fass, Barry
Goodrich, James Tait
Guth, Lloyd
Haber, Bernard
Hibbard, Emerson
Kostrzewa, Richard Michael
Krebs, Helmut Waldemar Graf von Thorn
Kromer, Lawrence Frederick
Lewis-Pinke, Ellen Ruth
McQuarrie, Irvine Gray
Mendelson, Thea
Munson, John Bacon
Nelson, Randall Jay
Ommaya, Ayub K.
Politis, Michael James
Roper, Stephen David
Scheff, Stephen William
Sparrow, Janet Ruthe
Springer, Alan David
Stein, Donald Gerald
Turner, James Eldridge
Uzman, Betty Geren

Other
Adelman, George, *Interdisciplinary neuroscience*
Barbas, Helen, *Anatomy*
Berg, Richard Alan, *Pediatric neuropsychology*
Brick, John, *neuroscience*
Brown, Lucy Leseur, *Neuroanatomy*
Coleman, James R., *Developmental neuroscience*
Elizan, Teresita S., *Neurovirology*
Emerson, Robert Charles, *Neuroanatomy*
Friedman, David Paul, *Neuroanatomy*
Frost, Douglas Owen, *Nervous system development and plasticity*
Goldstein, Joel William, *Neuroscience*
Hamilton, Robert Bruce, *Neuroanatomy*
Henkel, Craig Kenneth, *Neuroanatomy*
Hope, George Marion, *Visual neuroscience*
Hopfield, John Joseph, *Neural Modeling*
Isaacson, Robert Lee, *Neuroscience*
Kaplan, Harriet Gevirtz, *Epilepsy*
Kennedy, Charles, *Brain energy metabolism*
Korenyi-Both, Andras Levente, *Neuromuscular disease*
Krauthamer, George Michael, *Neuroanatomy*
Krigman, Martin Ross, *Neuropathology*
Kromer, Lawrence Frederick, *Neurodeveleopment*
Kurtz, Perry James, *Neurobehavioral toxicology*
Low, Walter Cheney, *Neural transplantation*
MacLean, Paul Donald, *Neuranatomy*
Mesulam, M. Marsel, *Neuroanatomy*
Miller, Carol Ann, *Neuropathology*
Mitsumoto, Hiroshi, *Neuropathology*
Oster-Granite, Mary Lou, *Developmental neurogenetics*
Politis, Michael James, *Neuropathology*
Reiter, Lawrence W., *Neurotoxicology*
Sadun, Alfredo Arrigo Umberto, *Neuroanatomy*
Seltzer, Benjamin, *Neuroanatomy*
Shapshak, Paul, *Neuro-viral-immunology*
Swaiman, Kenneth Fred, *Developmental neuroscience*
Welch, Bruce Lynn, *Environmental neurobiology*
Yates, Allan James, *Neuropathology*
Zagon, Ian Stuart, *Developmental neuroscience*

OPERATIONS RESEARCH. *See* **ENGINEERING, MATHEMATICS.**

OPTICS. *See also* **LASER.**

Fiber optics
Akers, Francis Irving
Bagley, Brian G.
Bendow, Bernard
Bielawski, W. Bart
Brown, Lorin W.
Brumm, Douglas Bruce
Buckler, Michael J.
Burrus, Charles Andrew, Jr.
Channin, Donald Jones
Chen, Ying-Chih
Columbia, Timothy Francis
Dakss, Mark Ludmer
Davis, Mark Hezekiah, Jr.
Fang, Shu-Cherng
Farinola, Anthony Larry
Gould, Gordon
Gunderson, Leslie Charles
Hammer, Jacob Meyer
Kang, Min Ho
Kao, Charles Kuen
Kino, Gordon Stanley
Kovacs, Gyula
Li, Tingye
Lin, Chinlon
Marcuse, Dietrich
Maurer, Robert Distler
McMahon, Donald Howland
Miller, Stewart Edward
Newton, Steven Arthur
Phelan, Robert Joseph, Jr.
Philips, Gerald John
Prisco, John Joseph
Rozenbergs, Janis
Skutnik, Bolesh S. Joseph
Steinbruegge, Kenneth Brian
Stone, Julian
Tien, Ping King
Wang, Chen-Show
Whinnery, John Roy
Wiesenfeld, Jay Martin
Woods, John Galloway
Wyss, Jerry C.
Yariv, Amnon

Holography
Brandt, Gerald Bennett
Brown, Gordon M(arshall)
Bryan, David A.
Chang, B(yung) Jin
Cox, Mary E.
Decker, Arthur John
Erf, Robert K.
George, Nicholas
Jaffe, Bernard Mordecai
Lamberts, Robert Lewis
Latta, John Neal
Leibhardt, Edward
Leith, Emmett Norman
Owen, Robert Barry
Reynolds, George Owen
Thompson, Brian John
Tomren, Douglas Roy
Upatnieks, Juris
Urbach, John Charles
Vest, Charles Marstiller
Withrington, Roger John

Infrared spectroscopy
Cashdollar, Kenneth Leroy
Curnutte, Basil, Jr.
Faxvog, Frederick R.
Garing, John Seymour
Gilmour, William Alexander
Heacock, E(arl) Larry
Herget, William Frederick
Herzberg, Gerhard
Kostiuk, Theodor
LaPorte, Daniel D'Arcy
McCubbin, Thomas King, Jr.
Mizushima, Masataka
Polavarapu, Prasad Leela
Shoore, Joseph David
Smith, Sheldon Magill
Willner, Steven Paul
Yarger, Frederick Lynn
Ying-Sin, Li

Optical image processing
Casasent, David Paul
Coltman, John Wesley
Derderian, George
Fay, Theodore Denis, Jr.
Fienup, James Ray
George, Nicholas
Jaffe, Bernard Mordecai
Kessler, Bernard V.
Kornstein, Edward
Leith, Emmett Norman
Looft, Donald John
McMahon, Donald Howland
Mc Millan, Robert Scott
Melville, Richard Devern Samuels, Jr.
Revesz, George
Reynolds, George Owen
Shaw, Chorng-Gang
Stachnik, Robert Victor
Tomei, L. David
Wilkerson, M. Susan

Optical signal processing
Altschuler, Bruce Robert
Brandt, Gerald Bennett
Bryan, David A.
Casasent, David Paul
Chang, B(yung) Jin
Chen, Hsiang Tsun
Gault, Charles S.
Gaylord, Thomas Keith
Gibbs, Hyatt McDonald
Haus, Hermann Anton
Lee, Tzuo-chang
Litynski, Daniel Mitchell
Newton, Steven Arthur
Pernick, Benjamin
Scarl, Donald
Schumer, Douglas Brian
Smith, Peter William
Soffer, Bernard H.
Upatnieks, Juris
Walkup, John Frank
Wang, Shyh
Whitman, Robert Leslie

Other
Ammann, Eugene Otto, *Electro-optics*
Benjamin, Roland John, *Optical aspheric surfaces*
Berreman, Dwight Winton, *Optic research*
Bloom, Arnold Lapin, *Thin films*
Bottema, Murk, *Optical design*
Carter, William Harold, *optics research*
Chiang, Fu-pen, *stress analysis*
Coffey, Brian Joseph, *Non-linear optics*
Delisle, Claude A(rmand), *Optics Research*
Dereniak, Eustace Leonard, *Infrared physics*
Dowley, Mark William, *Optical engineering*
Dunn, Anne Roberts, *Infared technology*
Elliott, Richard Amos, *Atmospheric Optics*
Feldman, Martin, *Optics research management*
Geslicki, Mark Louis, *Laser optics design*
Goldner, Ronald B., *Optoelectronics materials/devices/systems*
Green, Daniel G., *Physiological optics*
Harney, Robert Charles, *Optical radar*
Hirleman, Edwin Daniel, Jr., *Optical instrumentation*
Holmes, Dale Arthur, *Laser optics*
Hughes, Carroll Garvin, III, *Optical radiation measurements*
Ishimaru, Akira, *Optical scattering*
Jackel, Janet Lehr, *Guided-wave optics*
Jackson, John Edwin, *Optics research*
Jacobs, Stephen Frank, *Laser optics*
Jansson, Peter Allan, *Optics research*
Keicher, William Eugene, *Optical systems design*
Kessler, Bernard V., *Infrared surveillance*
Kuyatt, Chris E., *Electron and ion optics*
Leibhardt, Edward, *Diffraction Gratings*
Liaw, Haw-ming, *Electro-optical systems*
Lilje, Karl David, *Opto-mechanics consulting*
Linder, Solomon Leon, *Electro-optical systems*
Lindquist, Oiva Herbert, *Infrared sensor systems*
Maccabee, Bruce Sargent, *Atmospheric optics*
Massey, Gail Austin, *Nonlinear optics*
McMahon, Maribeth, *Optical properties of materials*
McManamon, Paul Francis, *Thermal imaging*
Meinel, Aden Baker, *Optical system engineering*
O'Shea, Donald Charles, *Optical design*
Owen, Robert Barry, *Optical Measurement Systems*
Pendleton, J. David, *Particle sizing instrument simulation*
Phillips, Richard Arlan, *Optics research*
Prisco, John Joseph, *Nonlinear optics*
Sanger, Gregory M., *Advanced optical manufacturing and testing*
Schaefer, Albert Russell, *Photo-optical instrumentation*
Schultz, Frederick H. C., *polarized infrared absorption and reflection*
Sedlacek, Blahoslav Jan, *Physical optics*
Spiller, Eberhard, *Optics of X-rays*
Stotts, Larry Bruce, *Optics research*
Tomlinson, W. John, *Optics research*
Wallace, James, Jr., *Atmospheric optics*
Wang, Jon Yi, *Applied optics*
Woolf, Neville John, *Optical imaging*
Yablonovitch, Eli, *Optics research*
Young, Matt, *Optics research*
Zweig, Gilbert, *Photo-optical imaging*

PHARMACEUTICS

Medicinal chemistry
Aguiar, Adam Martin
Anderson, Wayne Keith
Banker, Gilbert Stephen
Beamer, Robert Lewis
Bodor, Nicholas Stephen
Bullock, Francis Jeremiah
Burger, Richard M.
Cates, Lindley Addison, Jr.
Chang, Ching-jer
Cheng, Chia-Chung
Chignell, Colin Francis
Cho, Arthur Kenji
Francis, Marion David
Giner-Sorolla, Alfred
Gold, Mark Stephen
Griffith, Owen Wendell
Grunewald, Gary Lawrence
Gund, Peter Herman
Handorf, Charles Russell
Helsley, Grover Cleveland
Henkel, James G.
Iorio, Louis Carmen
Karim, Aziz
Korytnyk, Wsewolod
Krenitsky, Thomas Anthony
LaRocca, Joseph Paul
Lasslo, Andrew
Lee, Henry Joung
Misra, Anand Lal
Montgomery, John Atterbury
Nagasawa, Herbert Taukasa
Neumeyer, John Leopold
Paul, Ara Garo
Popp, Frank Donald
Prabhu, Vilas Anandrao
Rabjohn, Norman
Robins, Morris Joseph
Rodriguez, Eloy
Sarett, Lewis Hastings
Schach von Wittenau, Manfred Eberhard

Shen, Tsung Ying
Silverman, Richard Bruce
Smith, Charles Irvel
Stohs, Sidney John
Struck, Robert Frederick
Tishler, Max
Torrence, Paul Frederick
Varughese, Pothen
Veber, Daniel Frank
Wang, Theodore Sheng-Tao
Watanabe, Kyoichi Aloisius
Weinkam, Robert Joseph
Wells, Jack Nulk
Witiak, Donald T.

Pharmacokinetics
Berney, Stuart Alan
Birnbaum, Linda Silber
Block, Lawrence Howard
Dvorchik, Barry Howard
Eshelman, Fred Neville
Evans, William Edward
Findlay, John William Addison
Finland, Maxwell
Galinsky, Raymond Ethan
Gambertoglio, John Gino
Graeme, Mary Lee
Gross, Joseph Francis
Hill, Jim Tom
Hofmann, Lorenz Martin
Imondi, Anthony Rocco
Jainchill, Jerome
Jusko, William Joseph
Kaplan, Stanley Albert
Karim, Aziz
Krasny, Harvey Charles
Kwan, King Chiu
Leeling, Jerry L.
Lopez, Larry Markell
Loran, Muriel Rivian
Mayock, Robert Lee
McIlhenny, Hugh Meredith
Mehta, Bipin Mohanlal
Mendels, Joseph
Migdalof, Bruce Howard
Pang, Kim-Ching Sandy
Perel, James Maurice
Quay, John Ferguson
Rocci, Mario Louis, Jr.
Schentag, Jerome J.
Schor, Joseph Martin
Siemens, Albert John
Singhvi, Sampat Manakchand
Sisenwine, Samuel Fred
Smith, Ronald Gene
Stambaugh, John Edgar, Jr.
Sullivan, Hugh Richard, Jr.
Tocco, Dominick Joseph
Tyler, Tipton Ransom
Wilkinson, Grant Robert
Wosilait, Walter Daniel

Other
Brady, Lynn Robert
Chavkin, Leonard Theodore, *Pharmaceutical Research Management*
Huang, Leaf, *Drug carriers*
Leibman, Kenneth Charles, *Drug Metabolism*
McIlhenny, Hugh Meredith, *Drug research and regulation*
Paul, Ara Garo
Rain, Robert L., *Pharmaceutical quality control*
Rodriguez, Eloy
Sisenwine, Samuel Fred, *Drug metabolism*
Staba, Emil John
Tivin, Fred, *Pharmaceutical quality assurance*
Zudeck, Steven, *Pharmaceutical manufacturing*

PHYSICS

Acoustics
Adler, Robert
Banjavic, Richard Alan
Barger, James Edwin
Beranek, Leo Leroy
Blazey, Richard Nelson
Cerceo, John Michael
Cheung, Lim Hung
Claus, Richard O.
Coltman, John Wesley
Cornyn, John Joseph
DeSanto, John Anthony
Dufilho, Harold Louis
Erf, Robert K.
Etter, Paul Courtney
Feldman, Henry R.
Fried, Peter Marc
Fujimura, Osamu
Gibbons, Michael Francis, Jr.
Goldstein, Albert
Griffy, Thomas Alan
Hargrove, Logan Ezral
Harris, Cyril Manton
Hodge, David Charles
Ingard, Karl Uno
Knollman, Gil Carl
Lang, William Warner
Liaw, Haw-ming
Lopez, Hector
Massier, Paul Ferdinand
Mintzer, David
Morse, Philip McCord
Munson, John Christian
Mutschlecner, Joseph Paul
Packman, Allan B.
Plotkin, Kenneth Jay
Richards, Roger Thomas
Sauder, William Conrad
Shivanandan, Kandiah
South, Hugh Miles
Tancrell, Roger Henry
Watkins, Sallie Ann
Westervelt, Peter Jocelyn
Whitman, Robert Leslie
Zwislocki, Jozef John

Astrophysics. *See* **ASTRONOMY.**

Atomic and molecular physics
Adams, Gail D.
Alexander, Millard Henry
Anderson, Carl John
Azziz, Nestor Jalil
Baer, Thomas Michael
Bay, Zoltan Lajos
Bayfield, James Edward
Baylis, William Eric
Beaty, Earl Claude
Bederson, Benjamin
Bel Bruno, Joseph J(ames)
Bennett, William Ralph, Jr.
Berkowitz, Joseph
Best, Philip Ernest
Bichsel, Hans
Bieniek, Ronald James
Bloembergen, Nicolaas
Bourgelais, Donna Belle Chamberlain
Brewer, Richard George
Brink, Gilbert O.
Brody, Burton Alan
Brown, Ellen Ruth
Brown, George Stephen
Bryant, Howard Carnes
Bunge, Carlos Federico
Burke, Edward Aloysius
Burns, Jay, III
Burr, Alexander Fuller
Caird, John Allyn
Caledonia, George Ernest
Callaway, Joseph
Cardamone, Michael J.
Cartwright, David Chapman
Chang, Tu-Nan
Cheng, Kwok-Tsang
Childs, William Jeffries
Church, David Arthur
Cohen, James Samuel
Conway, John G.
Cooper, Charles Dewey
Coulter, Philip Wylie
Cowan, Robert Duane
Cox, Arthur Nelson
Crasemann, Bernd
Crisp, Michael Dennis
Curnutte, Basil, Jr.
Currah, Walter Everett
Curtis, Lorenzo Jan
Dang, Richard Kaoyu
Datz, Sheldon
Davidson, Gilbert
Day, Michael Hardy
Dehmelt, Hans Georg
Delos, John Bernard
DeLucia, Frank Charles
Deslattes, Richard Day, Jr.
de Wit, Michiel
Dick, Charles Edward
Dugan, Charles Hammond
Emmerich, Werner Sigmund
Fairbank, William Martin, Jr.
Fano, Ugo
Farley, John William
Feld, Michael Stephen
Feldman, Henry R.
Fetterman, Harold Ralph
Fonck, Raymond John
Fontana, Peter Robert
Ford, Albert Lewis, Jr.
Fortson, Edward Norval
Fox, Kenneth
Freed, Karl Frederick
Freund, Robert Stanley
Friedrich, Donald Martin
Fuhr, Jeffrey Robert
Fystrom, Dell Orren
Gallagher, Thomas Francis
Galt, John Alexander
Garing, John Seymour
Gay, Timothy James
George, James
Gien, Tran Trong
Gillespie, George Hubert
Ginter, Marshall Lloyd
Gislason, Eric Arni
Golden, David Edward
Goorvitch, David
Gordy, Walter
Greene, Arthur Edward
Greene, Chris H.
Gregory, Donald Clifford
Gupta, Rajendra
Hance, Robert Lee
Harpster, Joseph W. C.
Harter, William George
Haun, Robert Dee, Jr.
Hayden, Howard Corwin, Jr.
Heer, Clifford V.
Henneberger, Walter Carl
Herbst, Eric
Herzberg, Gerhard
Hilborn, Robert Clarence
Holt, Richard A.
Hopper, Darrel Gene
Hu, Patrick Hung-Sun
Huang, Keh-Ning
Hughes, Raymond Hargett
Hulse, Russell Alan
Hyman, Howard Allan
Inokuti, Mitio
Isler, Ralph Charles
Jette, Archelle Norman
Johnson, Brant Montgomery
Johnson, Peter Dexter
Johnson, Robert E.
Judd, Brian Raymond
Julienne, Paul S.
Kerwin, Larkin
Kim, Yong-ki
Kobe, Donald Holm
Kocher, Carl Alvin
Konowalow, Daniel Dimitri
Kramer, Steven David
Kusch, Polykarp
Kuyatt, Chris E.
Lam, Leo K.
Langhoff, Peter Wolfgang
Lapatovich, Walter Peter
Lapicki, Gregory
Larson, Daniel John
Lee, Long C.
Leventhal, Marvin
Lewis, Lindon L.
Li, Ming Chiang
Lide, David Reynolds, Jr.
Lin, Chun Chia
Lindau, Ingolf Evert
Littman, Michael Geist
Lombardi, Gabriel G.
Lu, Chun Chian
Lubell, Michael Stephen
Lucatorto, Thomas Benjamin
Mack, Michael Edward
Madison, Don Harvey
Magnuson, Gustav Donald
Malik, David Joseph
Malik, Fazley Bary
Mandel, Leonard
Manson, Steven Trent
Mark, Hans Michael
Martin, Frederick Wight
Matulic, Ljubomir Francisco
Mavroyannis, Constantine
McColm, Douglas Woodruff
McConkey, John William
McCubbin, Thomas King, Jr.
McGregor, Wheeler Kesey, Jr.
McGuire, Eugene Joseph
McGuire, James Horton
Mc Ilrath, Thomas James
Mendelsohn, Lawrence Barry
Menendez, Manuel Gaspar
Meyerhof, Walter Ernst
Miller, William Robert, Jr.
Mizushima, Masataka
Monce, Michael Nolen
Moore, John Hays
Moseley, John Travis
Mossberg, Thomas William
Narducci, Lorenzo M.
Neumann, Herschel
Newsom, Gerald Higley
Nitz, David Edwin
Norcross, David Warren
Oka, Takeshi
Orel, Ann Elizabeth
Painter, Linda Robinson
Park, John Thornton
Parr, Albert Clarence
Patel, Chandra Kumar Naranbhai
Paul, Derek A. L.
Payne, Philip Warren
Peek, James Mack
Peplinski, Daniel Raymond
Peschmann, Kristian Ralf
Piore, Emanuel Ruben
Pomilla, Frank R.
Postma, Herman
Pradhan, Anil Kumar
Pratt, David Wixon
Purbrick, Robert Lamburn
Radziemski, Leon Joseph
Ramsey, Norman
Randall, Russel R.
Rao, Penmaraju Venugopala
Rapp, Donald
Reading, John Frank
Rescigno, Thomas Nicola
Rich, Arthur
Richardson, James Wyman
Rinker, George Albert, Jr.
Robinson, Edward J.
Roy, Denis L.
Rush, John Edwin, Jr.
Ryan, Stewart Richard
Sabin, John Rogers
St. John, Robert Mahard
(Salk), Sung Ho
Sarachman, Theodore Nicholas
Sauder, William Conrad
Scarl, Donald
Schaefer, Albert Russell
Schawlow, Arthur Leonard
Schectman, Richard Milton
Schlachter, Alfred Simon
Schmieder, Robert William
Schneider, Barry Irwin
Sellin, Ivan Armand
Shah, Saiyid Masroor
Shahin, Michael M.
Sharpton, Francis Arthur
Sheridan, John Roger
Siomos, Konstadinos
Sitterly, Charlotte Moore
Smith, Howard Alan
Smith, Peter Lloyd
Smyth, Kermit Campbell
Snow, William Rosebrook
Sorokin, Peter Pitirimovich
Sprawls, Perry, Jr.
Steph, Nick Charles
Stockdale, John Alexander Douglas
Stwalley, William Calvin
Tai, Chen-Yu
Talman, James Davis
Thomsen, John Stearns
Tolles, William Marshall
Townes, Charles Hard
Trajmar, Sandor
Varghese, Sankoorikal Lonappan
Varlashkin, Paul Gregory
Venkatesan, T.
Victor, George A.
Viehland, Larry Alan
Wahrhaftig, Austin Levy
Watson, Deborah Kay
Weatherford, Charles Albert
Weissbluth, Mitchel
Weissler, Gerhard Ludwig
West, William Philip
Wiese, Wolfgang Lothar
Wilson, Walter Ervin
Winkler, Peter
Wittkower, Andrew B.
Wood, Obert Reeves, II
Wyss, Jerry C.
Yuan, Jian-Min
Zitzewitz, Paul W.
Zollweg, Robert John

Biophysics
Agard, Eugene Theodore
Alberts, Walter Watson
Alfano, Robert Richard
Allocca, John Anthony
Alpen, Edward Lewis
Anderson, David Walter
Banerjee, Krishnadas
Barrett, Terence William
Battista, Jerry Joseph
Bearden, Alan Joyce
Bencomo, Jose Antonio
Benedek, George Bernard
Bigler, Rodney E.
Bronk, Burt V.
Butler, James Preston
Chu, William Tongil
Clayton, Roderick Keener
Cohen, Morrel Herman
Cokelet, Giles Roy
Dicello, John Francis, Jr.
Dunn, Floyd
Eisinger, Josef
Elbaum, Charles
Engler, Mark J
Fairchild, Ralph Grandison
Feinberg, Robert Jacob
Ferretti, James Alfred
Ferrier, Jack Moreland
Frauenfelder, Hans
Goad, Walter Benson, Jr.
Goitein, Michael
Goldstein, Byron Bernard
Gordy, Walter
Green, Michael Enoch
Harth, Erich
Hill, Walter Ensign
Hiscock, Robert Russell
Ho, John Ting-Sum
Hobbie, Russell Klyver
Hollaender, Alexander
Huang, Huey-Wen
Huber, H. Stephen
Hyman, Edward Sidney
Ingber, Lester
Kaesberg, Paul Joseph
Kalata, Kenneth
Kass, Robert S.
Katz, J. Lawrence
Khan, Faiz Mohammad
Krimm, Samuel
Lipicky, Raymond John
Love, Warner Edwards
Malsky, Stanley Joseph
Marino, Andrew Anthony
Matthews, Rowena Green
McNulty, Peter J.
Mercer, Kermit Ray
Merdith, Orsell Montgomery
Mielczarek, Eugenie Vorburger
Mills, Michael David
Noordergraaf, Abraham
Panitz, John Andrew
Purdy, James Aaron
Rall, Wilfrid
Rao, Ravindra P.
Rhodes, William Clifford
Rich, Alexander
Ritter, Rogers Charles
Rotenberg, A. Daniel
Rothstein, Jerome
Rudy, Yoram
Sandhu, Taljit Singh
Schlenker, Robert Alison
Schneider, Marilyn Beth
Schwan, Herman Paul
Schwartz, Irving Leon
Scott, Walter Neil
Shamos, Morris Herbert
Shulman, Robert Gerson
Siegman, Marion Joyce
Simpson, Larry Dean
Sliney, David Hammond
Smith, Orville Lee
Smith, Rene Jose
Smith, Thomas Graves, Jr.
Somlyo, Avril Virginia
Starzak, Michael Edward
Tai, Yuan-Heng
Thurber, Robert Eugene
Weinhous, Martin S.
Weissbluth, Mitchel
Wingate, Catharine Louise
Wu, Raymond Kee-Kin
Zabara, Jacob

Condensed matter physics
Alfano, Robert Richard
Ambler, Ernest
Andrews, Hugh Robert
Ashcroft, Neil William
Banavar, Jayanth Rama Rao
Baratta, Anthony John
Bardeen, John
Baum, Gary Allen
Beri, Avinash Chandra
Berreman, Dwight Winton
Best, Philip Ernest
Bienenstock, Arthur Irwin
Blakely, John McDonald
Boer, Karl Wolfgang
Boring, Arthur Michael
Brewer, Richard George
Brown, George Stephen
Burns, Jay, III
Button, Kenneth J(ohn)
Cahn, John Werner
Callaway, Joseph
Cezairliyan, Ared
Chan, Iu-Yam
Chan, Sek Kwan
Chen, Tao-seng
Chiang, Chwan-Kang
Cho, Alfred Y.
Clarke, John
Clem, John R.
Cohen, Marshall Jay
Cohen, Morrel Herman
Compaan, Alvin Dell
Compton, W. Dale
Cone, Rufus Lester
Conway, John G.
Conwell, Esther Marly
Davison, Sydney George
DeLong, Lance Eric
Demarest, Harold Hunt, Jr.
Dresselhaus, Mildred Spiewak
Dresser, Miles Joel
Drickamer, Harry George
Dundon, Jeffrey Michael
Family, Fereydoon
Feldman, Albert
Fischer, Traugott Erwin
Fontanella, John Joseph
Fowler, Howland Auchincloss
Fritzsche, Hellmut
Fromhold, Albert Thomas, Jr.
Galeener, Frank Lee
Gardner, John Arvy, Jr.
Gersten, Joel Irwin
Giaever, Ivar
Gibbs, Hyatt McDonald
Goldman, Allen Marshall
Gollub, Jerry Paul
Gomer, Robert
Hagemark, Kjell Ingvar
Hagstrum, Homer Dupre
Hakala, Reino William
Hall, Herbert Joseph
Hallett, John
Halperin, Bertrand Israel
Herber, Rolfe H.
Herring, William Conyers
Ho, John Ting-Sum
Holstein, Theodore David
Holton, William Coffeen
Houlihan, John Frank
Hu, Patrick Hung-Sun
Hubbard, William Bogel, Jr.
Hubert, Jay Marvin
Hulm, John Kenneth
Hwang, Dah-Min David
Jackson, David Phillip
Jensen, Barbara Lynne
Jona, Franco Paul
Jones, Robert Edwin, Jr.
Jordan, Byron Dale
Kadanoff, Leo Philip
Keller, Jaime
Kelley, Paul Leon
Kepler, Raymond Glen
Kerr, William Clayton
Keyes, Robert William
Kim, Young Hwa
Kohn, Walter
Krumhansl, James Arthur
Larson, Donald Clayton
Leath, Paul Larry
Lee, Ching Tsung
Lefkowitz, Issai
Levy, Peter Michael
Liao, Paul F(oo-Hung)
Libby, Vibeke
Liebenberg, Donald Henry
Lin, Chun Chia
Lindau, Ingolf Evert
Litster, James David
Long, Jerome Rudisill
Macdonald, James Ross
Maher, James Vincent, Jr.
Martin, Paul Cecil
Massey, Walter Eugene
Mavroyannis, Constantine
McMillan, Robert Walker
Mielczarek, Eugenie Vorburger
Miller, Allen H.
Miller, William Robert, Jr.
Mills, Robert Laurence
Mitchell, Dean Lewis
Nagel, David Joseph
Nam, Sang Boo
Nathan, Marshall I.
Nelson, Donald Frederick
Ngai, Kia Ling
Nosanow, Lewis Harold
O'Connell, Robert Francis
Page, John Boyd
Painter, Linda Robinson
Panitz, John Andrew
Pappalardo, Romano G.
Pearson, Gerald Leondus
Perkowitz, Sidney
Phillips, James Charles
Pierce, Daniel Thornton
Pollak, Michael
Powell, Cedric John
Powell, Richard Conger
Reneker, Darrell Hyson
Rhodin, Thor Nathaniel
Rice, Michael John
Rinker, George Albert, Jr.
Roeder, Stephen Bernhard Walter
Roessler, David Martyn
Rose, Frank Edward
Rosenberg, Richard Allan
Rosenberger, Franz Ernst
Sato, Hiroshi
Schneider, Marilyn Beth

Schwensfeir, Robert James, Jr.
Seiler, David George
Sellmyer, David Julian
Shockley, William Bradford
Shull, Clifford G.
Simmons, Ralph Oliver
Simonis, George Jerome
Singwi, Kundan Singh
Skofronick, James G.
Slichter, Charles Pence
Snavely, Benjamin Breneman
Sollner, Traugott Carl Ludwig Gerhard
Spicer, William Edward, III
Stearns, Mary Beth
Surko, Clifford Michael
Swenson, Clayton A.
Taggart, George Bruce
Tauc, Jan
Testardi, Louis Richard
Thompson, Eric Douglas
Turnbull, David
Tuul, Johannes
Van Driel, Henry Martin
Van Hecke, Gerald Raymond
Varlashkin, Paul Gregory
Vineyard, George Hoagland
Weber, Marvin John
Weiler, Margaret Horton
Wert, Charles Allen
Westrum, Edgar Francis, Jr.
Yarger, Frederick Lynn

Low temperature physics
Ambler, Ernest
Bardeen, John
Bartlett, David Farnham
Brody, Burton Alan
Cabrera, Blas
Clarke, John
Clem, John R.
DeLong, Lance Eric
Donnelly, Russell James
Dundon, Jeffrey Michael
Fairbank, William Martin
Garland, James C.
Geballe, Theodore Henry
Gehrz, Robert Douglas
Hamilton, William Oliver
Honig, Arnold
Hulm, John Kenneth
Liebenberg, Donald Henry
Moskowitz, Paul A.
Moss, Frank Edward
Noltimier, Hallan Costello
Perkowitz, Sidney
Richards, Paul Linford
Rivier, Nicolas Yves
Rose, Frank Edward
Simmons, Ralph Oliver
Swenson, Clayton A.
Tinkham, Michael
Vant-Hull, Lorin Lee

Magnetic physics
Clark, Alan Fred
Donoho, Paul Leighton
Hasegawa, Ryusuke
Hawke, Ronald Samuel
Heller, Gerald S.
Hillyer, George Van Zandt
Holt, Alan Craig
Kikuchi, Chihiro
Levy, Peter Michael
Magnuson, Gustav Donald
McEnally, Terence Ernest, Jr.
Shull, Clifford G.
Slichter, Charles Pence
Stearns, Mary Beth
Waugh, John Stewart
Willett, Roger DuWayne
Wolf, Werner Paul

Medical physics
Adams, Gail D.
Arnold, Ben Allen
Attix, Frank Herbert
Basavatia, Ram A.
Cacak, Robert Kent
Cetas, Thomas Charles
Cohn, Stanton Harry
Cruty, Michael Robert
Curran, Bruce Howlett
DeWerd, Larry A.
Gibson, Jean M.
Gustafson, David Earl
Kalet, Ira Joseph
Komanduri, Ayyangar M.
Liu, James Chi-Wing
Mistry, Vitthalbhai Dahyabhai
Moyer, Robert Findley
Nalcioglu, Orhan
Nosil, Josip
Orton, Colin George
Renner, Wendel Dean
Shah, Saiyid Masroor
Sisterson, Janet Margot
Skofronick, James G.
Slack, Stephen Thomas
Smith, Rene Jose
Starkschall, George
Sullivan, Carole A.
Tolbert, Donald Dean
Webster, Edward William
Wollin, Myron
Wu, Andrew
Zamenhof, Robert George

Nuclear physics
Aaroe, William Henry
Adair, Robert Kemp
Agnew, Harold Melvin
Ancona, Antonio
Andrews, Hugh Robert
Austin, Sam M.
Azziz, Nestor Jalil
Bainbridge, Kenneth Tompkins
Barschall, Henry Herman
Batay-Csorba, Peter Andrew
Becchetti, Frederick Daniel, Jr.
Becker, Stephen Allan
Bevelacqua, Joseph John
Bigler, Rodney E.
Blosser, Henry Gabriel
Blotcky, Alan Jay
Bradbury, Norris Edwin
Braid, Thomas Hamilton
Braun-Munzinger, Peter
Broadhurst, John Henry
Brynjolffson, Ari
Cahill, Thomas Andrew
Caldwell, David Orville
Callihan, Dixon
Carlson, Richard Frederick
Chamberlain, Owen
Chase, Grafton D.
Christian, John Edward
Clark, David Delano
Clark, Hugh Kidder
Clarke, Robert Francis
Clendenin, James Edwin
Crane, Horace Richard
Crasemann, Bernd
Crewe, Albert Victor
Cuttler, Jerry Milton
Darby, John Littrell
Dickens, Justin Kirk
Doerner, Robert Carl
Ellis, Paul, John
Emmerich, Werner Sigmund
Engel, Lars Norlick
England, Talmadge Ray
Evans, Albert E.
Ferguson, James Malcolm
Feshbach, Herman
Fireman, Edward Leonard
Firk, Frank William Kenneth
Fivozinsky, Sherman Paul
Fowler, William Alfred
Franco, Victor
Frauenfelder, Hans
Gatrousis, Christopher
Ghorai, Susanta K.
Glascock, Michael Dean
Glasgow, Glenn Patrick
Goldhaber, Maurice
Gorenstein, Paul
Gove, Harry Edmund
Griffy, Thomas Alan
Gustafson, David Earl
Hamilton, Joseph Hants, Jr.
Hardy, Judson, Jr.
Harris, Gale Ion
Harris, Sigmund Paul
Hausman, Hershel Judah
Helmer, Richard Guy
Hendrie, Joseph Mallam
Henley, Ernest Mark
Herb, Raymond George
Holloway, John Thomas
Horrocks, Donald Leonard
Howerton, Robert James
Iachello, Francesco
Jacovitch, John
Janecke, Joachim Wilhelm
Jobst, Joel Edward
Jones, William Barclay
Kazaks, Peter A.
Kegel, Gunter Heinrich Reinhard
Keyworth, George A.
Klema, Ernest Donald
Kolar, Oscar Clinton
Komorek, Michael, Jr.
Krisst-Krisciokaitis, Raymond John
Kumar, Prasanna K.
Lake, James Alan
Lancman, Henry
Lederer, Charles Michael
Lederman, Leon Max
Levinthal, Elliott Charles
Lind, V. Gordon
Lindsay, William Francis
Makofske, William Joseph
Malik, Fazley Bary
Mallay, James Francis
Mann, Frederick Michael
Mathews, Grant James
Mays, Charles William
McDaniels, David Keith
McGuire, Stephen Craig
McKee, James Stanley Colton
Mc Millan, Edwin Mattison
McNulty, Peter J.
McVey, James Thomas
Michael, Eugene Joseph
Mistry, Vitthalbhai Dahyabhai
Monard, Joyce Anne
Neilson, George Croydon
Newman, David Edward
Nierenberg, William Aaron
Paul, Derek A. L.
Paulk, Charles Jasper, Jr.
Payne, Gerald Lew
Pearlstein, Sol
Perkins, Sterrett Theodore
Peters, Charles William
Peters, Gerald Joseph
Peterson, Roy Jerome
Pieper, George F(rancis)
Pitts, Thomas Griffin
Poskanzer, Arthur M.
Preiss, Ivor Louis
Price, Paul Buford
Profant, Richard Thomas, Jr.
Prosser, Francis W(are)
Purbrick, Robert Lamburn
Raab, Harry Frederick, Jr.
Rabi, Isidor Isaac
Ramanuja, Jayalakshmi Krishnadesikachar
Randall, Russel R.
Rao, Penmaraju Venugopala
Rapaport, Jacobo
Reading, John Frank
Redish, Edward Frederick
Reiss, Howard R.
Rhodes, Donald Frederick
Richardson, Clarence Robert
Roberds, Richard Mack
Rosen, Louis
Saperstein, Alvin Martin
Schima, Francis Joseph
Schwensfeir, Robert James, Jr.
Segrè, Emilio
Sholtis, Joseph Arnold, Jr.
Silberberg, Rein
Sisterson, Janet Margot
Smith, Anna Clanton
Smith, Orville Lee
Sorensen, Raymond Andrew
Sprung, Donald Whitfield Loyal
Szalai, Imre A.
Teller, Edward
Tomonto, James Robert
Vegors, Stanley Henry, Jr.
Waddington, Cecil Jacob
Warren, Holland Douglas
Weisskopf, Victor Frederick
Weitkamp, William George
Whetstone, Stanley Leroy
Whitten, Charles Alexander, Jr.
Wilson, Robert R.
Woosley, Stanford Earl
Wu, Chien-Shiung

Particle physics
Abolins, Maris Arvids
Abrams, Robert Jay
Adair, Robert Kemp
Adler, Stephen Louis
Auvil, Paul R.
Boehm, Felix Hans
Borenstein, Jeffrey Mark
Bryant, Howard Carnes
Caldwell, David Orville
Chu, William Tongil
Clendenin, James Edwin
Coleman, Sidney Richard
Cool, Rodney Lee
Courant, Ernest David
Cronin, James Watson
Dehmelt, Hans Georg
Demos, Peter Theodore
Dowd, John Peter
Ehlers, Kenneth Warren
Fainberg, Anthony
Feldman, Gary Jay
Fitch, Val Logsdon
Friedberg, Carl E.
Friedman, Jerome Isaac
Georgi, Howard Mason, III
Goldberger, Marvin L.
Goldman, Joseph Ilya
Good, Roland Hamilton, Jr.
Goulianos, Konstantin
Guth, Alan Harvey
Haacke, Ewart Mark
Han, Moo-Young
Hendry, Archibald Wagstaff
Heusch, Clemens August
Jackson, (John) David
Johnson, David Edwin
Johnson, James Russell
Jones, Lorella Margaret
Kamal, Abdul Naim
Kendall, Henry Way
Kirsch, Lawrence Edward
Kramer, Stephen Leonard
Kraushaar, Philip Frederick, Jr.
Krisst-Krisciokaitis, Raymond John
Kroll, Norman Myles K.
Langacker, Paul George
Leacock, Robert Arthur
Lind, V. Gordon
Lindsay, William Francis
Linsley, John (David)
Lubatti, Henry Joseph
Lubell, Michael Stephen
Mann, Alfred Kenneth
Marshak, Marvin Lloyd
Marshak, Robert Eugene
Martin, William Eugene
McDaniel, Boyce Dawkins
Moffat, John William
Orr, J. Richie
Pagels, Heinz Rudolf
Pais, Abraham
Panofsky, Wolfgank Kurt Hermann
Pati, Jogesh Chandra
Perl, Martin Lewis
Peters, Gerald Joseph
Poirier, John Anthony
Ramond, Pierre M.
Ramsey, Norman
Reines, Frederick
Richardson, Clarence Robert
Richter, Burton
Roe, Byron Paul
Rosen, Louis
Sachs, Robert Green
Samios, Nicholas Peter
Samuel, Mark Aaron
Sandweiss, Jack
Schwartz, Melvin
Schwitters, Roy Frederick
Shapero, Donald Campbell
Shochet, Melvyn Jay
Treiman, Sam Bard
Trower, William Peter
Tryon, Edward Polk
Wang, Chia Ping
Whitten, Charles Alexander, Jr.
Wilczek, Frank Anthony
Witten, Louis
Wolfenstein, Lincoln
Zorn, Gus Tom
Zweig, George

Plasma physics
Ahearne, Daniel Paul
Akasofu, Syun-Ichi
Alfven, Hannes Olof Gosta
Arons, Jonathan
Auer, Peter Louis
Bakshi, Pradip M.
Berman, Robert Hiram
Birmingham, Thomas Joseph
Bleach, Richard David
Boehm, Felix Hans
Booker, Henry George
Buchsbaum, Solomon Jan
Buff, James Steve
Burrell, Charles Frederick
Chanin, Loren Maxwell
Chappell, Charles Richard
Colgate, Stirling A.
Davidson, Ronald Crosby
Davis, James Ivey
Dawson, John Myrick
Dewar, Robert Leith
Drummond, William Eckel
Eames, David Robson
Ehlers, Kenneth Warren
Ellis, William R.
Fleischmann, Hans Hermann
Forslund, David Wallace
Gottlieb, Melvin Burt
Gould, Roy Walter
Grad, Harold
Gray, Eoin Wedderburn
Greene, Arthur Edward
Grove, Don J.
Hackel, Lloyd Anthony
Hammond, Thomas Joseph
Hansler, Richard Lowell
Harmon, Gary R.
Hauck, James Pierre
Henderson, Dale B(arlow)
Henson, Bob Londes
Hinnov, Einar
Hollweg, Joseph Vincent
Horton, Wendell Claude, Jr.
Hulse, Russell Alan
Hutchinson, Donald Patrick
Isler, Ralph Charles
Jahn, Robert George
Johnson, Brant Montgomery
Johnson, John Lowell
Kalman, Gabor Jeno
Kan, Joseph R.
Kaye, Stanley Martin
King, David Quimby
Kugel, Henry W.
Langer, William David
Littleton, John Edward
Lonngren, Karl Erik
Lucatorto, Thomas Benjamin
Marcuvitz, Nathan
Max, Claire Ellen
McCoy, James Ernest
McNally, James Rand, Jr.
Meyerand, Russell Gilbert, Jr.
Miller, Hillard Craig
Moir, Ralph Wayne
Moos, Henry Warren
Motz, Robin Owen
Newman, David Edward
Oettinger, Peter Ernest
Oleson, Norman Lee
Parker, Eugene Newman
Payne, Gerald Lew
Pe, Maung Hla
Perkins, Sterrett Theodore
Postma, Herman
Prelas, Mark Anthony
Radoski, Henry Robert
Rosenbluth, Marshall Nicholas
Roth, J(ohn) Reece
Seidl, Milos
Shahin, Michael M.
Sheffield, John
Shkarofsky, Issie Peter
Snow, William Rosebrook
Sreenivasan, S. Ranga
Sudan, Ravindra Nath
Surko, Clifford Michael
Sweeney, Mary Ann
Symon, Keith Randolph
Thouret, Wolfgang Emery
Trivelpiece, Alvin William
Van Hoven, Gerard
Vernazza, Jorge Enrique
Vitello, Peter Alfonso James
Walt, Martin
Weissler, Gerhard Ludwig
West, William Philip
Whealton, John H.
Wiese, Wolfgang Lothar
Woskoboinikow, Paul Peter
Zorn, Gus Tom

Polymer physics. *See also* **MATERIALS SCIENCE.**
Apostolou, Spyridon F.
Bailey, Frederick Eugene, Jr.
Bares, Jan
Baughman, Ray Henry
Berry, Guy Curtis
Boyer, Raymond Foster
Chow, Tsu Sen
Chu, Benjamin
Cohen-Addad, Jean-Pierre
Cook, Robert C.
Couchman, Peter Robert
Darsey, Jerome Anthony
Dickens, Elmer Douglas, Jr.
Dill, Kenneth Austin
Family, Fereydoon
Fitzgerald, Edwin Roger
Flory, Paul John
Fontanella, John Joseph
Fornes, Raymond Earl, Sr.
Gaylord, Richard J
Hartmann, Bruce
Helfand, Eugene
Hodge, Ian Moir
Hong, Su-Don
Hsu, Shaw Ling
Kambour, Roger Peabody
Kepler, Raymond Glen
Kim, Young Hwa
Koberstein, Jeffrey Thomas
Kobetich, Edward John
Lando, Jerome B.
Lohse, David John
Lontz, John Frank
Massa, Dennis Jon
McKinney, John Edward
Ngai, Kia Ling
Patterson, Gary David
Pollnow, Gilbert Frederick
Reneker, Darrell Hyson
(Salk), Sung Ho
Sedlacek, Blahoslav Jan
Simha, Robert
Skotheim, Terje Asbjorn
Thomas, H. Ronald
Tripathy, Sukant Kishore
Uralil, Francis Stephen
Wang, Tsuey Tang
Wiff, Donald Ray
Wu, Souheng
Wu, Wen-li

Psychophysics
Bailey, Ian Laurence
Berkley, Mark A.
Birnbaum, Michael Henry
Blackwell, Harold Richard
Boggs, George Johnson
Borresen, C. Robert
Boynton, Robert Merrill
Chastain, Garvin
Cooper, Franklin Seaney
de Haan, Henry John
Drum, Bruce Alan
Finke, Ronald Alan
Garner, Wendell Richard
Gescheider, George Albert
Glaser, Donald A(rthur)
Green, David Marvin
Greenspan, Joel Daniel
Hart, John Birdsall
Hirsh, Ira Jean
Howard, Ian Porteous
Jameson, Dorothea
Kicliter, Ernest Earl, Jr.
Lockhead, Gregory Roger
Luria, Saul Martin
Makous, Walter
Marks, Lawrence Edward
Maurissen, Jacques Paul Jean
Raslear, Thomas Gregory
Riggs, Lorrin Andrews
Robinson, Daniel Nicholas
Robinson, George Horine, Jr.
Roederer, Juan Gualterio
Staley, Frederick Joseph
Verrillo, Ronald Thomas
Wartzok, Douglas
Watson, Charles Schoff
Zacks, James Lee
Zwislocki, Jozef John

Relativity and gravitation
Bartlett, David Farnham
Bay, Zoltan Lajos
DeWitt, Bryce Seligman
Dicke, Robert Henry
Dube, Roger Raymond
Follingstad, Henry George
Gaposchkin, Peter John Arthur
Hamilton, William Oliver
Lieske, Jay Henry
Long, Daniel Russel
Marks, Dennis William
Moffat, John William
Noonan, Thomas Wyatt
Page, Don Nelson
Reasenberg, Robert David
Ritter, Rogers Charles
Ross, Dennis Kent
Smalley, Larry Lee
Snider, Joseph Lyons
Unruh, William George
Wagoner, Robert Vernon
Weber, Joseph
Westervelt, Peter Jocelyn
Wheeler, John Archibald
Whitman, Patrick Gene
Will, Clifford Martin
Wilson, Thomas Leon
Witcomb, Stanley Ernest
Witten, Louis
Woodward, James Franklin

Solar physics. *See* **SPACE SCIENCE.**

Statistical physics
Alder, Berni Julian
Banavar, Jayanth Rama Rao
Brostow, Witold Konrad
Edwards, Terry Winslow
Fisher, Michael Ellis
George, James
Heer, Clifford V.
Huang, Huey-Wen
Ingber, Lester
Kac, Mark
Kalman, Gabor Jeno
Kalos, Malvin Howard
Kerr, William Clayton
Lebowitz, Joel Louis
Lee, M. Howard
Litster, James David
Maher, James Vincent, Jr.
Malkus, Willem VanRensseler
Martin, Paul Cecil
Mazo, Robert Marc
Miller, Allen H.
Mintzer, David
Moss, Frank Edward
Pollak, Michael
Rivier, Nicolas Yves
Rothstein, Jerome

Tarter, C. Bruce
Van der Ziel, Aldert
Van Vliet, Carolyn Marina
Weber, Joseph
Wigner, Eugene Paul

Theoretical physics
Adler, Stephen Louis
Anderson, Philip Warren
Aref, Hassan
Ashcroft, Neil William
Auvil, Paul R.
Baker, Howard Crittenden
Bakshi, Pradip M.
Baylis, William Eric
Baym, Gordon Alan
Bethe, Hans Albrecht
Blade, Richard Allen
Brown, David Gordon
Brown, Ellen Ruth
Brueckner, Keith Allan
Burdick, Glenn Arthur
Chang, Tu-Nan
Charpie, Robert Alan
Cheng, Kwok-Tsang
Chow, Paul Chuan-Juin
Chudnovsky, David Volf
Clark, Richard LeFors
Coffey, Brian Joseph
Cohen, James Samuel
Coleman, Sidney Richard
Cooper, Leon N.
Coulter, Philip Wylie
Courant, Ernest David
Cowan, Robert Duane
Dashen, Roger Frederick
Day, Michael Hardy
DeSanto, John Anthony
Detweiler, Steven Lawrence
Devaney, Joseph James
de Wolf, David Alter
Drummond, William Eckel
Dyson, Freeman John
Ellis, Paul, John
England, Talmadge Ray
Evenson, William Edwin
Fano, Ugo
Feigenbaum, Mitchell Jay
Feinberg, Gerald
Feshbach, Herman
Feynman, Richard Philipps
Flannery, Brian Paul
Fontana, Peter Robert
Forslund, David Wallace
Fradkin, David Milton
Franco, Victor
Frieman, Edward Allan
Gell-Mann, Murray
Gelman, Donald
Georgi, Howard Mason, III
Gersten, Joel Irwin
Gerstl, Siegfried Adolf Wilhelm
Gien, Tran Trong
Gillespie, George Hubert
Glashow, Sheldon Lee
Goldberger, Marvin L.
Goldman, Joseph Ilya
Good, Roland Hamilton, Jr.
Goode, Philip Ranson
Goodman, Julius
Greene, Chris H.
Han, Moo-Young
Harter, William George
Hendry, Archibald Wagstaff
Henley, Ernest Mark
Henneberger, Walter Carl
Herman, Robert
Herring, William Conyers
Hobart, Robert H.B.W.S., Jr.
Holstein, Theodore David
Huang, Keh-Ning
Huber, H. Stephen
Huddleston, Philip Lee
Jackson, (John) David
Jones, Eric Manning
Judd, Brian Raymond
Kadanoff, Leo Philip
Kamal, Abdul Naim
Kazaks, Peter A.
Kelly, Hugh P.
Kerman, Arthur Kent
Klein, Douglas Jay
Kobe, Donald Holm
Kockinos, Constantin Neophytos
Kohn, Walter
Kroll, Norman Myles K.
Lamb, Donald Quincy
Lamb, Willis Eugene, Jr.
Lapicki, Gregory
Lauer, Dennis Errol
Leacock, Robert Arthur
Leath, Paul Larry
Lee, Kotik Kai
Lee, M. Howard
Li, Ming Chiang
Litovchenko, Vladimir Alexei
Low, Francis Eugene
Luttinger, Joaquin Mazdak
Manning, Irwin
Manson, Steven Trent
Marshak, Robert Eugene
Massey, Walter Eugene
Mazurek, Thaddeus John
McDonald, Keith Leon
McGuire, James Horton
McLenithan, Kelly Daniel
Merzbacher, Eugen
Metropolis, Nicholas C.
Mills, Robert Laurence
Nam, Sang Boo
Nambu, Yoichiro
Narducci, Lorenzo M.
Nickas, George Demosthenes
Overhauser, Albert Warner
Page, John Boyd
Pais, Abraham
Pati, Jogesh Chandra
Pease, Robert Louis
Peek, James Mack
Pines, David
Quinn, Helen Rhoda
Redish, Edward Frederick
Rice, Michael John
Robinson, Edward J.
Romero, Alejandro Francisco
Rosenkilde, Carl Edward
Ross, Dennis Kent
Sabin, John Rogers
Sachs, Robert Green
Samuel, Mark Aaron
Saperstein, Alvin Martin
Schneiderwent, Myron Otto
Schnitzer, Howard Joel
Schrieffer, John Robert
Schwinger, Julian
Seeley, William Glover
Segal, Irving Ezra
Seitz, Frederick
Seltzer, Stephen Michael
Shapero, Donald Campbell
Simon, Barry
Singwi, Kundan Singh
Sorensen, Raymond Andrew
Spital, Robin David
Sprung, Donald Whitfield Loyal
Stith, James Herman
Stone, Marshall Harvey
Stone, William Ross
Symon, Keith Randolph
Talman, James Davis
Teller, Edward
Thorne, Kip Stephen
Treiman, Sam Bard
Tu, Yih-O
Unruh, William George
Vineyard, George Hoagland
Von Baeyer, Hans Christian
Weinberg, Steven
Wheeler, John Archibald
Wheelon, Albert Dewell
Wightman, Arthur Strong
Wigner, Eugene Paul
Wilson, Kenneth Geddes
Winkler, Peter
Wolff, Peter Adalbert
Wong, Po Kee

Other
Ahluwalia, Bhagwat Datta, *Medical Physics*
Albers, Mark Alan, *Aerosoles and Particulates*
Alessandro, Daniel, *Medical physics*
Apel, John Ralph, *Physics laboratory administration*
Asche, David Robert, *Therapeutic radiological physics*
Banerjee, Krishnadas, *Radiological physics*
Barfield, Walter David, *Radiative transfer*
Baym, Gordon Alan, *Quantum mechanics*
Bergeson, Haven Eldred, *High energy physics*
Berman, Alan, *Research management*
Bhatnagar, Jagdish Prasad, *Radiological physics*
Bjorkholm, John Ernst, *Applied physics*
Bostrom, Carl Otto, *Space physics*
Braid, Thomas Hamilton, *Scientific instrumentation*
Bridges, William Bruce, *Microwaves*
Brown, Harold, *National security policy*
Brueckner, Keith Allan, *Fusion*
Bruels, Mark C., *Radiological physics*
Burstein, Elias, *Solid State Physics*
Burstein, Paul Harris, *Physics - X-Ray applications*
Byrd, James William, Sr., *Energy transport*
Ceperley, Peter Hutson, *Accelerator physics*
Cherry, Robert Newton, Jr., *Medical health physics*
Cipolla, Sam Joseph, *Radiation physics and radiation dosimetry*
Crombie, Douglass Darnill, *Electromagnetic Wave Propogation*
Dacey, George Clement, *Solid state physics*
Datta, Ratna, *Medical/health physicist*
Deupree, Robert Gaston, *Nuclear shock effects*
Driscoll, Timothy John, *Surface physics*
Eastman, Dean Eric, *Solid state physics*
Eden, William Murphey, *Health Physics*
Ehrlich, Gert, *Surface Physics*
Elbaum, Charles, *Solid State Physics*
Epp, Edward Rudolph, *Medical Physics*
Esaki, Leo, *Solid-state physics*
Ewing, Ronald Ira, *Lightning*
Feigenbaum, Mitchell Jay, *Chaotic Phenomena*
Fisher, Edward Richard, *Chemical physics*
Fortney, Lloyd Ray, *High energy physics*
Garwin, Richard Lawrence, *Applied Physics*
Gelinas, Robert Joseph, *Non-equilibrium fluid dynamics*
Gemmell, Donald Stewart, *Physics research administration*
Grow, Richard W., *Microwave physics*
Guenter, Joseph Martin, *Radiation physics*
Gundersen, Martin Adolph, *Quantum electronics*
Guttman, Lester, *Solid state physics*
Hall, Robert Noel, *Solid state physics*
Hammond, Donald L., *Physics research administrator*
Hargrove, Logan Ezral, *Optical physics*
Hart, John Birdsall, *Operational general physics*
Hausman, Hershel Judah, *Nuclear astrophysics*
Hayden, Howard Corwin, Jr., *Ion implantation*
Hefferlin, Ray, *Periodic systems of small molecules*
Heller, Gerald S., *Solid state physics*
Hellwarth, Robert Willis, *Optical physics*
Henson, Bob Londes, *Mathematical modeling*
Heydemann, Peter Ludwig Martin, *Chemical physics research administration*
Hogg, David Clarence, *Radiophysics*
Holeman, George Robert, *Health physics*
Hsieh, Jen-shu, *Medical/health physics*
Hubbell, John Howard, *Radiation physics*
Huston, Norman Earl, *Instrumentation*
Inokuti, Mitio, *Radiation physics*
Iona, Mario, *Cosmic ray physics*
Jaffe, Arthur Michael, *Mathematical physics*
Javan, Ali, *Quantum electronics*
Johnson, David Edwin, *Accelerator Physics*
Jones, Richard Victor, *Solid state physics*
Karle, Jerome, *Diffraction Physics*
Kathren, Ronald Lawrence, *Health physics*
Kerr, Donald MacLean, Jr., *Physics research and development administration*
Khan, Winston, *Physics of fluids*
Kincaid, John Franklin, *Applied physics*
Kocol, Henry, *Radiation safety*
Kutzscher, Edgar Walter, *Infrared detectors and technology*
Land, Edwin Herbert, *Photographic science*
Landauer, Rolf William, *Solid state physics*
Landers, Roy Eslyn, Jr., *Therapeutic Radiological Physics*
Lowy, David Charles, *Health physics*
Lukasik, Stephen Joseph, *Fluid Dynamics*
Mayfield, Donald Lewis, *Health physics*
McCombe, Bruce Douglas, *Solid state physics*
Mitler, Henri Emmanuel, *Mathematical modeling*
Narayanan, C. S., *Medical radiation physics*
Nibhanupudy, Jagannadha Rao, *Medical physics*
Norem, James H., *Accelerator physics*
Nutter, Gene Douglas, *Metrology*
Overhauser, Albert Warner, *Solid state physics*
Pake, George Edward, *Research management*
Parthasarathy, Srisailam V., *Radiological physics*
Pe, Maung Hla, *Medical physics*
Pendleton, J. David, *Electromagnetic wave scattering theory*
Petschek, Albert George, *Applied Physics*
Reinstein, Lawrence Elliot, *Medical physics*
Rez, Peter, *Electron interaction with materials*
Riseberg, Leslie Allen, *Solid state physics*
Romanowski, Thomas Andrew, *High Energy Physics*
Rosenkilde, Carl Edward, *Fluid dynamics*
Rotenberg, A. Daniel, *Physics consulting*
Ryan, Stewart Richard, *Applied physics*
Sayre, David, *X-ray microscopy*
Schumer, Douglas Brian, *Ultrasonics*
Shalvoy, Richard Barry, *Surface physics*
Simonis, George Jerome, *Wave research*
Smith, Horace Vernon, Jr., *Accelerator physics*
Solon, Leonard R(aymond), *Health physics*
Sparks, Ronald Wayne, *Enhanced heat transfer*
Stith, James Herman, *Physics education*
Sturdevant, Eugene J., *Electrohydrodynamics in heat transfer*
Tai, Douglas Leung-Tak, *Radiological physics*
Tubbs, David Lee, *Applied physics*
Wang, Chia Ping, *Thermal physics*
Ward, Jerry Mack, *Explosion effects*
Witherell, Egilda DeAmicis, *Radiological physics*
Yeh, Cheng Shin, *Fast breeder reactor physics*
Zaider, Marco A., *Radiation biophysics*

PSYCHOLOGY

Behavioral
Aborn, Murray
Abramson, Edward E(ric)
Alford, Geary Simmons
Andersen, Barbara Lee
Anderson, Eric Edward
Annis, Lawrence Vincent, Jr.
Axelrood, Helen Blau
Babb, Harold
Bacon-Prue, Ansley
Baldonado, Ardelina-Erika Albano
Barbach, Lonnie
Barton, Edward James
Becker, Joel Leonard
Becker, Robert Earl
Bedell, Ralph Clairon
Beickel, Sharon Lynne
Belding, Hiram Hurlburt, IV
Bennett, Lawrence Allen
Berger, Audrey Marilyn
Berman, Mark Laurence
Beutler, Larry Edward
Bloom, Richard Fredric
Boeringa, James Alexander
Boor, Myron Vernon
Bootzin, Richard Ronald
Borresen, C. Robert
Boudewyns, Patrick Alan
Boyd, Jeffrey Lynn
Bradshaw, Howard Holt
Bratter, Thomas Edward
Brooker, Alan Edward
Brownell, Kelly David
Bucher, Bradley Dean
Bufford, Rodger Keith
Burish, Thomas Gerard
Caddy, Glenn Ross
Carpenter, Patricia
Carr, Edward Gary
Casler, Lawrence
Cerreto, Mary Christine
Charlesworth, Edward Allison
Chermol, Brian Hamilton
Chesney, Margaret Ann
Christian, Barry Theodore
Cleeland, Charles Samuel
Clifford, Margaret Louise
Coche, Erich Henry Ernst
Cohen, Ira Larry
Collins, William Edward
Connors, Gerard Joseph
Cook, Daniel Walter
Cotler, Sherwin Barry
Craig, Paul Lawrence
Crown, Barry Michael
Cullari, Salvatore Santino
Dalby, (John) Thomas
Danaher, Brian Grayson
de Lorge, John Oldham
Demarest, David Steele
Denkowski, George Carl
DeNoble, Victor John
D'Errico, Albert Pasquale, Jr.
DeVries, David Lee
Dixen, Jean Marie
Dreyfus, Edward A.
Dunlop, Terrence Ward
Dunn, Charleta Jessie
Ebert-Flattau, Pamela
Elliott, Charles H.
Engen, Eugene Paul
Engstrom, David Ralph
Etu, Paul David
Eurich, Alvin Christian
Farrer, Donald Nathanael
Fensterheim, Herbert
Ferguson, Kinglsey George
Firestone, Philip
Flory, Randall Kean
Force, Ronald C(larence)
Ford, Mary Elizabeth
Foreyt, John Paul
Fox, Robert Alan
Foxx, Richard Michael
Frame, Roger Everett
Franke, Richard Herbert
Franken, Robert Earl
Freeberg, Carlin Henry
Freedman, Robert Russell
Frey, William Howard II
Gallatin, Judith Estelle
Garvin, Everett Arthur
Gasparrini, William Gerard
George, Clay Edwin
Ghiselli, William Barron
Gingold, William
Glazer-Waldman, Hilda Ruth
Goh, David Shuh Jen
Goldberg, Isadore
Goldfried, Marvin R.
Goldstein, Steven Edward
Gollub, Lewis Raphael
Gordon, Wayne Alan
Gray, Philip Howard
Gubar, George
Hakel, Milton Daniel
Hall, Howard Ralph
Hall, R. Vance
Hamad, Charles Dean
Hardy, Miles Willis
Harpin, Raoul Edward
Harris, Mary Bierman
Henningfield, Jack Edward
Hensley, John Higgins
Hester, Reid Kevin
Heward, William Lee
Heyer, Miriam Harriet
Heying, Robert Hilarius
Hickis, Charles Francis
Hirsch, Jerry
Holmstrom, Valerie Louise
Horn, Wade Fredrick
Hover, Gerald Robert
Humphries, Joan Ropes
Ince, Laurence Peter
Jackson, Thomas Larry
Jacobs, Durand Frank
Jacobs, Keith William
Johnson, Richard Frederick
Kaczkowski, Henry Ralph
Kaiser, Charles Frederick
Kamback, Marvin Carl
Kaplan, Harriet Gevirtz
Kaplan, Robert Malcolm
Kendall, Philip Charles
Kennedy, Clive Dale
Kewman, Donald Glenn
Kimmel, Ellen Bishop
Klarreich, Samuel Henry
Klinman, Cynthia Stone
Krass, Alvin
Kraus, William Arnold
Krauss, Herbert H.
Kress, Gerard Clayton, Jr.
Ksir, Charles Joseph, Jr.
Lachter, Gerald David
Lack, Dorothea Z.
Latham, Gary Phillip
Laws, D(onald) R(ichard)
Lazarus, Arnold Allan
LeBow, Michael David
Lent, Robert William
Levine, Marvin
Lewin, Mark Henry
Lichtman, Robert Mark
Linehan, Marsha M.
Lingren, Ronald Hal
Litle, Patrick Alan
Litrownik, Alan Jay
Locke, Bill J.
Lofquist, Lloyd Henry
Lundin, Robert William
Lupiani, Donald Anthony
Luria, Saul Martin
Luttinger, Daniel Alan
Lutzker, John R.
MacDonald, G(eorge) Wayne
Malatesta, Victor Julio
Maloney, Dennis Michael
Markley, Kenneth Alan
Martin, Donald Vincent
Marx, Melvin Herman
Mash, Eric Jay
Mathews, R. Mark
Matson, Johnny Lee
Mattson, Margaret Ellen
May, Eugene Pinkney
Mayo, Clyde Calvin
McAbee, Thomas Allen
McCraw, Ronald Kent
McDonald, Stanford Laurel
McKee, Michael Geoffrey
McWhirter, James Jeffries
Mealiea, Wallace Laird, Jr.
Mehearg, Lillian Erl
Mensh, Ivan Norman
Miller, Thomas William
Mills, Harry Lee, Jr.
Montgomery, Doil Dean
Montgomery, Robert Lew
Moore, Benjamin L.
Morra, Michael Anthony
Morris, Edward Knox, Jr.
Morris, Richard Jules
Morrison, Robert Floyd
Morrow, Gary Robert
Mulick, James Anton
Nezu, Arthur Maguth
O'Farrell, Timothy James
O'Keefe, Edward John
Ollendick, Thomas Hubert
Padilla, Geraldine V(aldes)
Painter, Genevieve
Pazzaglini, Mario Peter
Pekala, Ronald James
Pelc, Robert Edward
Penick, Elizabeth Carnel
Peniston, Eugene G.
Perelle, Ira B.
Perera, Thomas Biddle
Perez, Francisco Ignacio
Pihl, Robert Olander
Pitta, Patricia Joyce
Plakosh, Paul, Jr.
Pleck, Joseph Healy
Poling, Alan Dale
Pollard, Gerald Tilman
Powell, Jay Raymond
Raslear, Thomas Gregory
Rehm, Lynn Paul
Reisman, Scott
Rosella, John Daniel
Sacco, William P(atrick)
Seligson, M. Ross
Siegel, Ronald Keith
Silverstein, Charles I.
Simpson, Dennis Dwayne
Smith, Ira Austin
Smith, John Philip
Stallworth, Charles Dorothea, Jr.
Stapleton, Leroy Earl
Stark, Jack Alan
Suedfeld, Peter
Thompson, William Warren
Tolchinsky, Paul D.
Trzasko, Joseph Anthony
Tyre, Timothy Edward
Uyeno, Edward Teiso
Walls, Betty L., Webb
Waters, James Herman
Webbe, Frank Michael
Weiffenbach, James Milton
Weiss, Jonathan Hyman
Wenger, Galen Rosenberger

Whitworth, Gary William
Wiest, William Marvin
Wodarski, John Stanley
Woods, Delma Maria
Woods, Walter Abner
Young, Peggy Sanborn
Zarit, Steven Howard
Zentall, Thomas Robert
Zweig, Joseph B.

Clinical
Beck, Niels Christian, Jr.
Bernstein, Anne Carolyn
Carpenter, Patricia
Chambless, Dianne L.
Cleveland, Sidney Earl
Coche, Erich Henry Ernst
Connors, Gerard Joseph
Davis, Thompson Elder, Jr.
Domino, George
Dunn, William Lawrence
Feifel, Herman
Feuerstein, Michael
Fisher, Jerid Martin
Frank, Robert George
Gallo, Mario Martin
Gerber, Gwendolyn L.
Giannetti, Ronald A.
Gubar, George
Guiora, Alexander Zeev
Henning, James Scott
Hodges, William Fitzgerald
Horenstien, David
Johnson, James Harmon
Leli, Dano Anthony
Lewis, Lisa
Llorca, Arthur Lee
Mandel, Harvey Phillip
Mapou, Robert Lewis
Megargee, Edwin Inglee
Morris, Richard Jules
Passman, Richard Harris
Pazzaglini, Mario Peter
Pikunas, Justin
Pine, Charles Joseph
Plaut, Thomas Franz Alfred
Pollack, Stephen Lewis
Raymond, Beth
Rehm, Lynn Paul
Rubert, Mary Lou
Schlossberg, Harvey
Sladen, Bernard Jacob
Woliver, Robert Edward
Worthington, Elliott Robert

Cognition
Anderson, Daniel R(aymond)
Anderson, James Alfred
Arenberg, David Lee
Arnoult, Malcolm Douglas
Beach, Lee Roy
Beck, Aaron Temkin
Becker, Joel Leonard
Belmont, John Mark
Bergquist, James William
Bieger, George R.
Bigelow, Brian John
Blechner, Mark J.
Bourne, Lyle Eugene, Jr.
Bower, Gordon Howard
Breland, Hunter Mansfield
Brown, Robert Michael
Brown, Roger William
Bruhn, Arnold Rahn, Jr.
Bryan, Glenn LeVan
Bryden, Mark Philip
Caporael, Linnda Rose
Carroll, John Stephen
Chastain, Garvin
Clemmer, Edward Joseph
Cohen, Marvin Sidney
Cole, Michael
Cooper, George David
Corno, Lyn
Cotler, Sherwin Barry
Crovitz, Herbert Floyd
Curtis, Bill
Dannenbring, Gary Lee
Das, Jagannath Prasad
Davies, Ivor Kevin
Deese, James Earle
D'Errico, Albert Pasquale, Jr.
Domangue, Barbara B.
Dreyfus, Edward A.
Duncan, Starkey Davis, Jr.
Dunn, Charleta Jessie
Eliot, John
Ellis, Henry Carlton
Engel, Joanne Boyer
Engle, Randall Wayne
Engstrom, David Ralph
Erber, Joan T.
Evans, Nancy Jean
Federico, Pat-Anthony
Ferris, Steven Howard
Finke, Ronald Alan
Fletcher, John Dexter
Flowers, John Hawkins
Fogel, Max Leonard
Forisha-Kovach, Barbara Ellen
Foss, Donald John
Foulkes, David
France, Olin Kenneth, Jr.
Franken, Robert Earl
Freeberg, Carlin Henry
Freedle, Roy Omer
Fulkerson, Samuel Cole
Fuller, Renee Nuni
Gagne, Ellen Dalton
Galambos, James Andrew
Gallatin, Judith Estelle
Garner, Wendell Richard
Getzels, Jacob Warren
Giammatteo, Michael Charles
Gianutsos, Rosamond R.
Gibson, Eleanor Jack
Glenberg, Arthur Mitchell
Goh, David Shuh Jen
Goldfried, Marvin R.
Gooding, Charles Thomas
Gottwald, Richard Landolin
Grafman, Jordan Henry
Guskey, Thomas Robert
Gyr, John Walter
Hansl, Nikolaus Rudolf
Harkins, Stephen Wayne
Healy, Alice Fenvessy
Hellige, Joseph Bernard
Hemingway, Peter
Herrmann, Douglas J.
Hilgard, Ernest Ropiequet
Hilliard, Asa Grant, III
Hinrichs, James Victor
Hirst, William Charles
Hodge, Milton Holmes, Jr.
Holley, Charles DeWayen
Horn, John Leonard
Howard, Darlene Vaglia
Jaeger, Theodore B.
Jensen, Arthur Robert
Jones, Ben Morgan
Jordan, Theresa Joan
Kachuck, Beatrice
Kantowitz, Barry Howard
Karpicke, John Arthur
Kellar, Lucia Ames
Kendall, Philip Charles
Kessen, William
Kidd, Robert Fletcher
Kiesler, Charles Adolphus
Klarreich, Samuel Henry
Kopstein, Felix Friedrich
Kosslyn, Stephen Michael
Lachman, Roy
Laird, James Douglas
Lazarus, Arnold Allan
Levine, Marvin
Lichtman, Robert Mark
Lincoln, Charles Ebenezer
Litrownik, Alan Jay
Lloyd, John Tracy
Locher, Paul John
Lockhead, Gregory Roger
Loftus, Elizabeth F.
London, Ray William
Loveland, Katherine Anne
MacLachlan, James Morrill
Maguire, William Michael
Mallory-Barkley, Barbara Zommer
Mandler, George
Manning, Susan Karp
Mason, Susan Elizabeth
Masson, Michael Edward Joseph
Matlin, Margaret White
May, Richard Beard
McCabe, Aliyssa Kim
McCombs, Barbara Leona
McDowell, David Jamison
McShane, Damian Anthony
Metzl, Marilyn Newman
Meyer, Bonnie June Francis
Meyers, Lawrence Stanley
Monty, Richard A.
Mosberg, Ludwig
Moshman, David Stewart
Muller, John Paul
Muma, John Ronald
Muse, Mark Dana
Nezu, Arthur Maguth
Norman, Donald Arthur
Patterson, Terence Edward
Pekala, Ronald James
Peters, Paul Stanley, Jr.
Pettinati, Helen Marie
Pinker, Steven Arthur
Pitta, Patricia Joyce
Pollock, Steven Edward
Pomerantz, James R(obert)
Pope, Alan Thomas
Powell, James Charles
Preston, Joan Muriel
Proudfoot, Ruth Ellen
Rota, Gian-Carlo Carlo
Russell, Elbert Winslow
Saxe, Geoffrey B.
Sbordone, Robert Joseph
Schank, Roger Carl
Schmandt-Besserat, Denise
Schunk, Dale Hansen
Schwartz, Steven
Schwebel, Milton
Shade, Barbara Jean
Silverman, Hirsch Lazaar
Silverstein, Charles I.
Stallworth, Charles Dorothea, Jr.
Straussner, Joel Harvey
Suedfeld, Peter
Teevan, Richard Collier
Trehub, Arnold
Vandendorpe, Mary Moore
Voight, Jesse Carrolton
Ward, William Cornelius
Warren, Marguerite Queen
Warren-Meltzer, Stephanie
Watkins, Paul Roger
Weinstein, Claire Ellen
Wells, Gary Leroy
Wiseman, Sandor Elliot
Wohlfeil, Paul Frederick
Wong, Martin R.
Woods, Walter Abner
Wyer, Robert Selden, Jr.
Zelazo, Philip Roman
Zweig, Richard Lee

Developmental
Achenbach, Thomas M.
Ackerman, Ralph Emil
Anderson, Daniel R(aymond)
Anderson, Eric Edward
Asher, Steven Robert
Atwood, Joan Dolores
Aubrey, Roger Frederick
Barocas, Harvey A(aron)
Barry, John Reagan
Bearison, David J.
Beickel, Sharon Lynne
Belmont, John Mark
Bennett, Kelly Randolph
Berndt, Thomas Joseph
Berninger, Virginia Wise
Bernstein, Anne Carolyn
Bigelow, Brian John
Black, Kathryn Norcross
Bocknek, Gene
Brannigan, Gary G(eorge)
Brown, Robert Michael
Bucher, Bradley Dean
Burke, Ronald John
Carr, Edward Gary
Cerreto, Mary Christine
Chibucos, Thomas Robert
Childers, John Stephen
Cleland, Charles Carr
Clifford, Margaret Louise
Cohen, Ira Larry
Cox, Rachel Dunaway
Cummings, Edward Mark
Damon, William V.B.
Danner, David William
Das, Jagannath Prasad
Downs, Asa Chris
Draper, Thomas William
El Ghatit, Zeinab Mohammed
Engle, Randall Wayne
Enright, Robert David
Erber, Joan T.
Eron, Leonard David
Etu, Paul David
Feshbach, Seymour
Fiene, Richard John
Finley, Gordon Ellis
Firestone, Philip
Forisha-Kovach, Barbara Ellen
Fouts, Gregory Taylor
Fox, Robert Alan
Friedman, Sarah L(andau)
Froming, William John
Fuller, Renee Nuni
Gardner, Howard Earl
Gibson, Eleanor Jack
Golland, Jeffrey Harris
Gorman, Bernard Samuel
Gottlieb, Barbara Weintraub
Gounard, Beverly Elaine
Grallo, Richard Martin
Gutierrez, Fernando Jose
Hainline, Louise
Hall, R. Vance
Hall, William Sterling
Harris, Mary Bierman
Held, Richard Marx
Henning, James Scott
Hensley, John Higgins
Herman, Marc Steven
Hochhauser, Mark
Hodges, William Fitzgerald
Hogan, John Daniel
Hollenbeck, Albert Russell
Holtzman, Wayne Harold
Horn, John Leonard
Horn, Wade Fredrick
Howard, Darlene Vaglia
Hyde, Janet Shibley
Joffe, Justin Manfred
Johnson, James Harmon
Jones, Molly Modrall
Jordan, Theresa Joan
Katz, Phyllis Alberts
Kessen, William
Kirschenbaum, Neal
Korner, Anneliese Friede
Kurtz, Theodore Stephen
Lacher, Miriam Browner
Ladd, Gary Wayne
Landa, Beth Kim
Latham, Eleanor Earthrowl
Lewis, Michael
Lolli, Peter Patrick
Loveland, Katherine Anne
Lynch, Mervin Dean
Lyons-Ruth, Karlen
Maccoby, Eleanor Emmons
Machtiger, Harriet Gordon
Mandel, Harvey Phillip
Martin, John Ayres
Mash, Eric Jay
Mason, Susan Elizabeth
Mason, William A(lvin)
Massad, Carolyn Emrick
Matter, Darryl Edgar
Matteson, David Roy
May, Richard Beard
McBride, Angela Barron
McCabe, Aliyssa Kim
McCall, Robert Booth
Mc Gee, Mark Gregory
McKinney, John Paul
McShane, Damian Anthony
Metzl, Marilyn Newman
Meyer, Bonnie June Francis
Midlarsky, Elizabeth Ruth
Miller, Nancy E(llen)
Moore, Benjamin L.
Morris, Edward Knox, Jr.
Morrison, Mary Dyer
Morrison, Robert Floyd
Mosatche, Harriet Sandra
Moshman, David Stewart
Muma, John Ronald
Murray, John Patrick
Netley, Charles Thomas
Netzer, Carol
Pacheco-Maldonado, Angel Manuel
Painter, Genevieve
Palisin, Helen
Parke, Ross Duke
Passman, Richard Harris
Pavlidis, George Theophilou
Peck, Robert F(ranklin)
Peske, Patric O'Connell
Pikunas, Justin
Pitcher, Georgia Ann
Plaut, Simon Michael
Poresky, Robert Harold
Porges, Stephen William
Powell, Louisa Rose
Preston, Joan Muriel
Price, Gary Glen
Robin, Mitchell Wolfe
Ross, Hildy S(haron)
Rubert, Mary Lou
Safarjan, William Robert
Sameroff, Arnold Joshua
Saxe, Geoffrey B.
Schaeffer, Benson
Schwebel, Milton
Seaton, Craig Edward
Sechzer, Jeri Altneu
Shaffer, David Reed
Solomons, Hope Cowen
Spear, Norman Eberman
Suomi, Stephen John
Tough, Allen MacNeill
Tramill, James Louis
Vandendorpe, Mary Moore
Violato, Claudio
Walsh, William Michael
Warren-Meltzer, Stephanie
Warshaw, Rhoda
Wasserman, Gail A.
Weinberg, Richard Alan
Wolf, Thomas Mark
Woodruff, Diana Stenen
Zarit, Steven Howard
Zelazo, Philip Roman

Learning
Ackerman, Ralph Emil
Babb, Harold
Berg, Paul Conrad
Berninger, Virginia Wise
Bhushan, Vidya
Bieger, George R.
Bitterman, Morton Edward
Brannigan, Gary G(eorge)
Bratter, Thomas Edward
Bryan, Glenn LeVan
Carney, Richard Edward
Christian, Barry Theodore
Clemens, William Jenkins
Colotla, Victor Adolfo
Corno, Lyn
Cotton, Eileen Giuffre
Drummond, Robert John
Ellis, Henry Carlton
Engel, Joanne Boyer
Estes, William Kaye
Eurich, Alvin Christian
Fanselow, Michael Scott
Federico, Pat-Anthony
Fielding, Stuart
Flory, Randall Kean
Fouts, Gregory Taylor
Foxx, Richard Michael
Freedle, Roy Omer
Gagne, Ellen Dalton
Gamzu, Elkan Raphael
Gardner, Marjorie Hyer
Ghiselli, William Barron
Glazer-Waldman, Hilda Ruth
Glenberg, Arthur Mitchell
Gooding, Charles Thomas
Gottlieb, Barbara Weintraub
Gounard, Beverly Elaine
Grallo, Richard Martin
Green, Donald Ross
Guskey, Thomas Robert
Hall, Robert Dilwyn
Hamilton, Leonard W.
Hardy, Miles Willis
Harrell, Ruth Flinn
Hawkins, Robert Drake
Heinberg, Paul Julius
Hochhauser, Mark
Humphries, Joan Ropes
Jensen, Arthur Robert
Kachuck, Beatrice
Kaczkowski, Henry Ralph
Karpicke, John Arthur
Kirschenbaum, Neal
Klinman, Cynthia Stone
Krass, Alvin
Lachter, Gerald David
Lack, Dorothea Z.
Loftus, Elizabeth F.
Luchins, Edith Hirsch
Lundin, Robert William
Mallory-Barkley, Barbara Zommer
Maple, Terry Lee
Marx, Melvin Herman
Masson, Michael Edward Joseph
McAreavy, John Francis
McCombs, Barbara Leona
Means, Larry Williams
Medley, Donald Matthias
Miller, D. Merrily
Mills, Harry Lee, Jr.
Mosberg, Ludwig
Ollendick, Thomas Hubert
Overmier, James Bruce
Overton, Donald A.
Park, Ok-Choon
Parker, Linda Alice
Payne, Robert Bryan
Peelle, Howard Arthur
Peske, Patric O'Connell
Pines, Ariel Leon
Powell, Donald A.
Quartermain, David
Revusky, Sam H.
Romano, James Anthony, Jr.
Rosenzweig, Mark Richard
Safarjan, William Robert
Sanders, Gilbert Otis
Schroeder, James Ernest
Schunk, Dale Hansen
Spear, Norman Eberman
Steinacker, Antoinette
Straub, Richard Otto
Tough, Allen MacNeill
Walberg, Herbert John
Walls, Betty L., Webb
Warnat, Winifred Irene
Warshaw, Rhoda
Weinstein, Claire Ellen
Weisberg, Joseph Simpson
Whitworth, Gary William
Wong, Martin R.
Woolman, Myron
Young, Peggy Sanborn
Zachert, Virginia
Zentall, Thomas Robert
Zweig, Richard Lee

Neuropsychology. *See* **NEUROSCIENCE.**

Physiological
Adams, David Bachrach
Andreassi, John L(awrence)
Atwell, Constance Woodruff
Balster, Robert Louis
Bauer, Richard Henry
Becker, James Thompson
Berman, Marlene Oscar
Black, Perry
Bodnar, Richard Julius
Brown, William Samuel
Burdick, James Alan
Butler, Ann Benedict
Carney, Richard Edward
Cheal, MaryLou
Clemens, William Jenkins
Coscina, Donald Victor
Crow, Lowell Thomas
Doerr, Robert Douglas
Don, Norman Stanley
Eison, Michael Steven
Eschenbach, Arthur Edwin
Everly, George Stotelmyer, Jr.
Geliebter, Allan
Gentry, R., Thomas
Goldrich, STanley Gilbert
Goy, Robert William
Grimsley, Douglas Lee
Grosser, George Samuel
Hall, Robert Dilwyn
Hamilton, Leonard W.
Heffner, Henry Edward
Hernandez, Linda Louise
Hodge, Gordon Karl
Howard, James Lawrence
Ince, Laurence Peter
Jones, Molly Modrall
Justesen, Don Robert
Kornetsky, Conan
Lansdell, Herbert Charles
Laudenslager, Mark Leroy
Laws, D(onald) R(ichard)
Levinthal, Charles Frederick
Liebman, Jeffrey Mark
Lindsley, Donald Benjamin
Lubar, Joel Fredric
Lukas, Jeffrey Hilton
Martin, Donald Vincent
McLaughlin, Carol Lynn
Means, Larry Williams
Mitchell, James Curtis
Mitler, Merrill Morris
Montgomery, Doil Dean
Olton, David Stuart
Perera, Thomas Biddle
Pfaffmann, Carl
Powers, J. Bradley
Quartermain, David
Ratliff, Floyd
Rechtschaffen, Allan
Reitan, Ralph Meldahl
Rohles, Frederick Henry, Jr.
Rugh, John Douglas
Scrima, Lawrence
Sechzer, Jeri Altneu
Siegel, Jerome
Solomon, Paul Robert
Soper, Henry Victor
Sparks, David Lee
Stein, Donald Gerald
Stellar, Eliot
Stern, Robert Morris
Taub, Edward
Teitelbaum, Philip
Tramill, James Louis
Van Hartesveldt, Carol Jean
Ward, O. Byron, Jr.
Weber-Levine, Margaret Louise
Wever, Ernest Glen
Wilson, Glenn Francis
Wolgin, David L.
Woodruff, Diana Stenen
Woodruff, Michael Lester
Wyatt, Harry Joel
Yeterian, Edward Harry
Young, Francis Allan

Psychobiology
Adams, Perrie Milton
Alkana, Ronald Lee
Allen, Robert Arthur
Barrett, James Edward
Berry, Stephen Daniel
Bigler, Erin David
Bitterman, Morton Edward
Brick, John
Brunjes, Peter Crawford
Brush, F. Robert
Byrd, Larry Donald
Caldwell, Willard E.
Cheal, MaryLou
Chennault, Madelyn Joanne
Davis, Hasker Pat
de Lorge, John Oldham
DeNoble, Victor John
Dillon, Donald Joseph
Donchin, Emanuel

Doros, Maria Heczey
Ehrlich, Yigal H.
Ellison, Carol Rinkleib
Erickson, Carl John
Erlenmeyer-Kimling, L.
Ettenberg, Aaron
Everly, George Stotelmyer, Jr.
Fass, Barry
Fiss, Harry
Geyer, Lynette Arnason
Glazer, Howard Irwin
Gollub, Lewis Raphael
Gonzalez-Lima, Francisco M.
Groves, Philip Montgomery
Grunberg, Neil Everett
Harkins, Stephen Wayne
Hellman, Rhona Phyllis
Hickis, Charles Francis
Hofer, Myron Arms
Holloway, Frank Albert
Isaacson, Robert Lee
Jerison, Harry Jacob
Kaiser, Charles Frederick
Kimura, Doreen
Korner, Anneliese Friede
Lee, Richard M
Levine, Michael Steven
Levine, Seymour
Levitsky, David Aaron
Luce, R(obert) Duncan
Lynds-Cherry, Patricia Gail
Mactutus, Charles Francis
Maple, Terry Lee
Marler, Peter Robert
Martin, James Tillison
Mason, William A(lvin)
Matarazzo, Joseph Dominic
Maxson, Stephen Clark
Mayer, David Jonathan
Mc Gee, Mark Gregory
McKearney, James William
Miczek, Klaus A.
Miller, David Bennett
Monder, Harvey
Morin, Lawrence Porter
Overmier, James Bruce
Pietrewicz, Alexandra Theresa
Plaut, Simon Michael
Porges, Stephen William
Powell, Donald A.
Prentky, Robert Alan
Provine, Robert Raymond
Reiser, Morton Francis
Revusky, Sam H.
Riscalla, Louise Beverly
Sachs, Benjamin David
Sanberg, Paul Ronald
Schiffman, Susan Stolte
Schrieber, Robert Alan
Schwartz, Steven
Smith, Barry Decker
Smith, Richard Petri
Sperry, Roger Wolcott
Squire, Larry Ryan
Stollnitz, Fred
Straub, Richard Otto
Suomi, Stephen John
Thompson, Richard Frederick
Thurman, Lloy Duane
Valenstein, Elliot Spiro
Van Hartesveldt, Carol Jean
Wasserman, Gerald Steward
Webbe, Frank Michael
Whalen, Richard Edward
Wickramasekera

Psychophysics. *See* **PHYSICS.**

Sensory processes
Adams, Henry B(ethune)
Andreassi, John L(awrence)
Arnoult, Malcolm Douglas
Boff, Kenneth Richard
Boynton, Robert Merrill
Brown, William Samuel
Burdick, Charles Kenneth
Cain, William S.
Collins, William Edward
Cooper, George David
Coppola, Richard
Corso, John Fiermonte
Crampton, George Harris
Dillon, Donald Joseph
Drum, Bruce Alan
Dunn, William Lawrence
Emerson, Victor F.
Gescheider, George Albert
Giammatteo, Michael Charles
Gillman, Clifford Brian
Green, David Marvin
Hainline, Louise
Harth, Erich
Heffner, Henry Edward
Held, Richard Marx
Hirsh, Ira Jean
Hope, George Marion
Howard, Ian Porteous
Jaeger, Theodore B.
Jameson, Dorothea
Johnston, Dorothy Mae
Kamen, Joseph M.
Kennedy, Linda Mann
Lawrence, Merle
Lit, Alfred
Lukas, Jeffrey Hilton
Makous, Walter
Marks, Lawrence Edward
Mirenburg, Barry Leonard
Neff, William Duwayne
Nelson, Jeremiah I.
Odom, James Vernon
Parker, Donald Edward
Pfaffmann, Carl
Pomerantz, James R(obert)
Powers, Maureen Kennedy
Ray, Richard Hallett
Riggs, Lorrin Andrews
Robinson, George Horine, Jr.
Smith, Robert L.
Takahashi, Ellen Shizuko
Tieman, Suzannah Bliss
Verrillo, Ronald Thomas
Wasserman, Gerald Steward
Watkins, Linda May
Watson, Charles Schoff
Weiffenbach, James Milton
Weiss, Ira Paul
Wever, Ernest Glen
Young, Francis Allan
Zacks, James Lee

Social psychology
Adams, John David
Adler, Leonore Loeb
Allen, Vernon L(eslie)
Archer, Richard Lloyd
Atwood, Joan Dolores
Austin, William George
Banas, Paul Anthony
Baron, Robert Alan
Barron, William Loring, III
Barry, John Reagan
Battistich, Victor Anthony
Bazerman, Max H(al)
Beach, Lee Roy
Beck, Kenneth Harold
Bennett, Lawrence Allen
Berger, Seymour Maurice
Birnbaum, Michael Henry
Black, Percy
Blake, Brian Francis
Boor, Myron Vernon
Bronzaft, Arline Lillian
Brown, Roger William
Brunson, Bradford Ira
Buck, Ross Workman
Burke, Wyatt Warner
Bush, David Frederic
Cacioppo, John Terrance
Caple, Richard Basil
Caporael, Linnda Rose
Carducci, Bernardo Joseph
Carroll, John Stephen
Carsrud, Alan Lee
Casler, Lawrence
Chennault, Madelyn Joanne
Chermol, Brian Hamilton
Childers, John Stephen
Christenfeld, Roger Michael
Cohen, Sheldon Avery
Conner, Ross F.
Crocker, Jennifer
Damon, William V.B.
Danner, David William
Diener, Edward Francis
Dixon, Paul Nichols
Doehrman, Steven R(alph)
Draper, Thomas William
Drummond, Robert John
Edelman, Ann Lynn
Endler, Norman Solomon
Engel, John William
Feldman, Jack Michael
Fiedler, Fred Edward
Figley, Charles Ray
Fisher, Jeffrey David
Flay, Brian Richard
Foley, Daniel Patrick
Franklin, Paula Anne
Froming, William John
Funder, David Charles
Gable, Ralph Kirkland
Gabrenya, William Karl, Jr.
Garland, Howard
Gastorf, John Wayne
Geibel, Valerie Henken
George, Clay Edwin
Gerber, Gwendolyn L.
Getzels, Jacob Warren
Glidewell, John Calvin
Gollob, Harry Frank
Gotay, Carolyn Cook
Gottfredson, Gary Don
Green, Dorothy Eunice
Grosser, George Samuel
Grunberg, Neil Everett
Gutierrez, Fernando Jose
Hatcher, John Christopher
Hautaluoma, Jacob Edward
Heinberg, Paul Julius
Hemingway, Peter
Heyer, Miriam Harriet
Heying, Robert Hilarius
Hilton, Thomas Frederick
Hoffman, L. Richard
Hofmann, Julius Joseph
Holtzman, Wayne Harold
Hull, Diana
Hyman, Edward Jay
Insko, Chester Arthur, Jr.
James, Sherman Athonia
Johnson, Robert Dennis
Kahn, Arnold S.
Kalick, Sheldon Michael
Kanungo, Rabindra Nath
Kegan, Daniel L.
Keys, Christopher Bennett
Kidd, Robert Fletcher
Kiesler, Charles Adolphus
Kilmann, Ralph Herman
Korabik, Karen Sue
Kraus, William Arnold
Krauss, Herbert H.
Kravitz, David Albert
Laird, James Douglas
Lasry, Jean-Claude Maurice
Leary, Mark Richard
Leventhal, Gerald Seymour
Levine, John Myron
Lewis, Ralph Jay, III
Lindsay-Hartz, Janice
Lindzey, Gardner
Lynch, Mervin Dean
Maibaum, Matthew
Markley, Kenneth Alan
Marsh, Jeanne Cay
Martin, John Ayres
Matteson, David Roy
McAndrew, Francis Thomas
McAreavy, John Francis
McBride, Angela Barron
McCombs, Harriet G.
McWhirter, James Jeffries
Michener, H. Andrew
Midlarsky, Elizabeth Ruth
Miller, Fayneese Sheryl
Miller, Rowland Spence
Miller, Thomas William
Mirenburg, Barry Leonard
Montgomery, Robert Lew
Morell, Jonathan Alan
Mosatche, Harriet Sandra
Murray, John Patrick
Nelson, Carnot Edward
Pacheco-Maldonado, Angel Manuel
Padilla, Geraldine V(aldes)
Palmer, Edward Leo
Pennebaker, James Whiting
Perlman, Daniel
Peterson, Gerald Leonard
Plaut, Thomas Franz Alfred
Pleck, Joseph Healy
Posner, Barry Zane
Powell, Louisa Rose
Rosnow, Ralph L(eon)
Rubin, Zick
Safer, Martin Allen
Saleh, Shoukry Dawood
Schlenker, Barry Richard
Shaffer, David Reed
Shaffer, Howard Jeffrey
Sirgy, Magdy Joseph
Smith, John Philip
Sparacino, Jack Robert
Stevens, Gwendolyn Ruth
Teddlie, Charles Benton
Teevan, Richard Collier
Valfer, Ernst Siegmar
Violato, Claudio
Walberg, Herbert John
Warren, Marguerite Queen
Watzlawick, Paul
Webster, Murray Alexander, Jr.
Weinberg, Robert Stephen
Wells, Gary Leroy
Wiest, William Marvin
Wodarski, John Stanley
Woliver, Robert Edward
Wong-McCarthy, William James
Wood, Wendy
Worthington, Elliott Robert
Wyer, Robert Selden, Jr.
Yoder, Janice Dana

Other
Abbott, Robert Dean, *Quantitative psychology*
Adams, Henry B(ethune), *Alocholism and substance abuse*
Adler, Leonore Loeb, *Comparative psychology*
Andrysco, Robert Michael, *Human/animal bond*
Austin, William George, *Clinical and forensic psychology*
Bach, Deborah, *Clinical psychology*
Banas, Paul Anthony, *Organizational*
Barbach, Lonnie, *Clinical psychology*
Barcus, Robert A., *Clinical psychology*
Bardach, Joan L(ucile), *Clinical Psychology*
Barocas, Harvey A(aron), *Clinical-community psychology*
Barron, William Loring, III, *Psychology of humor*
Barton, Edward James, *Industrial/organizational psychology*
Baskin, David, *Psychological evaluation and measurement*
Baskin, Steven Marc, *Clinical psychology*
Battistich, Victor Anthony, *Personality*
Bauer, Robert William, *Industrial psychology*
Bauserman, Deborah Nylen, *Clinical psychology, Tests and measurement*
Bazerman, Max H(al), *Oranizational psychology*
Bedell, Ralph Clairon, *Counseling*
Bell, Charolette Renee, *Testing and evaluation*
Blanchard, Ray Milton, *Gender indentity disorders*
Bloom, Richard Fredric, *Psychotherapy*
Blumenthal, James Alan, *Clinical psychology*
Bocknek, Gene, *Psychotherapy and intervention theory*
Bodin, Arthur Michael, *Clinical psychology*
Boeringa, James Alexander, *Behavioral medicine*
Bradley, John Michael, *Reading-related cognition and perception*
Braun, Stephen Hughes, *Clinical psychology*
Bronzaft, Arline Lillian, *Environmental psychology*
Brown, Stephen Woody, *Psychometrics*
Bruhn, Arnold Rahn, Jr., *Clinical psychology*
Brunson, Bradford Ira, *Counseling psychology*
Bufford, Rodger Keith, *Psychology and religion*
Burke, Wyatt Warner, *Organizational psychology*
Bush, David Frederic, *Health psychology*
Cadogan, Donald Andrew, *Psychotherapy*
Capanzano, Charles Thomas, *Psychology clinic*
Caple, Richard Basil, *Counseling psychology*
Carducci, Bernardo Joseph, *Personality*
Carsrud, Alan Lee, *Applied social/personality psychology*
Chibucos, Thomas Robert, *Child development and family studies*
Clemmer, Edward Joseph, *Psycholinguistics*
Coan, Richard Welton, *Personality, Psychology of science*
Coles, Gary John, *Applied research psychology*
Colotla, Victor Adolfo, *Behavioral toxicology*
Conner, Ross F., *Evaluationresearch*
Corso, John Fiermonte, *Aging*
Cotton, Eileen Giuffre, *Educational psychology*
Cox, Rachel Dunaway, *Clinical*
Cupchik, William, *Psychotherapy*
Dannenbring, Gary Lee, *Human factors in computing*
Davis, Donald Dean, *Organizational and community psychology*
Dean, Judith Carol Hickman, *Onoclogy Counseling Psychology*
Dederick, Judith Garrettson, *Clinical-developmental psychology*
de Haan, Henry John, *Perception*
Dellario, Donald J(oseph), *Psychiatric rehabilitation*
Denkowski, George Carl, *Rehabilitation psychology*
deTorres, Cory Delgado, *Neurolinguistic programming*
Dewart, Dorothy Boardman, *Diagnosis and control of chronic pain*
Dishman, Rodney King, *Execise and sport psychology*
Distefano, Michael Kelly, Jr., *Industrial and organizational psychology*
Doehrman, Steven R(alph), *Health Psychology*
Don, Norman Stanley, *Consciousness*
Dosamantes-Alperson, Erma, *Imagory and movement, Psychotherapy*
Dreitlein, Raymond Paul, *Alcoholism rehabilitation counseling, Burnout alocholism treatment personnel*
Dudek, Bruce Craig, *Behavior genetics*
Dumas, Neil Stephen, *Industrial psychology*
Eckardt, Michael Jon, *Medical psychology*
Edwards, John Robert, *Sociolinguistics*
El Ghatit, Zeinab Mohammed, *clinical psychology*
Eliot, John, *Spatial ability*
Elliott, Charles H., *Pediatric psychology*
Ellison, Carol Rinkleib, *Psychology of human sexuality*
Emener, William George, *Counseling Psychology, Rehabilitation Psychology*
Endler, Norman Solomon, *Personality*
Englund, Carl Ernest, *Applied experimental psychology*
Epstein, Seymour, *Personality*
Evans, Gary William, *Environmental psychology*
Fagot, Robert Frederick, *Measurement in psychology*
Feher, Leslie, *Perinatal psychology*
Feifel, Herman, *Thanatology*
Fergenson, P. Everett, *Consumer behavior*
Ferguson, Kinglsey George, *Primary prevention in mental health*
Ferraio, Nicholas LaVerne, *Clinical psychology*
Ferraro, Douglas Peter, *Behavioral Medicine, Behavioral Pharmacology*
Feshbach, Seymour, *Personality*
Fiedler, Fred Edward, *Organizational psychology*
Fine, Bernard J., *Individual differences and environmental stress*
Finger, Dennis Robert, *Psychology*
Fischl, Myron Arthur, *Industrial and organizational psychology*
Fisher, Lawrence, *Clinical research*
Fiss, Harry, *Psychology education*
Fontenelle, Don Harris, *Child psychology*
Force, Ronald C(larence), *Corrections psychology*
Ford, Larry Howard, *Organizational and educational psychology*
Frame, Roger Everett, *School psychology*
Frank, Robert George, *Health psychology*
Funder, David Charles, *Personality*
Gabrenya, William Karl, Jr., *Psychological anthropology*
Garland, Howard, *Organizational psychology*
Gastorf, John Wayne, *Health psychology*
Girden, Ellen Robinson, *Research design*
Glazer, Howard Irwin, *Clinical/Organizational Psychology*
Glidewell, John Calvin, *Organizational psychology*
Goldberg, Isadore, *Human resource development*
Goldstein, Joel William, *Psychology research program planning*
Goldstein, Steven Edward, *Stress management*
Goldstein, Yonkel Noah, *Health psychology, Hypnosis*
Golland, Jeffrey Harris, *Psychoanalysis*
Gould, Richard Bruce, *Industrial/organizational psychology*
Grandon, Gary Michael, *Educational measurement*
Green, Donald Ross, *Psychometrics*
Greer, Joanne Marie G., *Psychoanalysis*
Greuling, Jacquelin Wren, *Health Psychology*
Gribbons, Warren David, *Counseling psychology*
Guiora, Alexander Zeev, *Psycholinguistics*
Guy, James David, Jr., *Psychotherapy*
Gyr, John Walter, *Self-organizing systems*
Hafer, Marilyn Durham, *Psychometrics, test construction, personnel selection, Rehabilitation*
Haier, Richard Jay, *Personality*
Harpin, Raoul Edward, *Behavior therapy*
Hassan, Mark David, *Psychoanalytic psychotherapy*
Hautaluoma, Jacob Edward, *Organizational behavior*
Hayek, Theodore Craig, *Human resources management and development, Organizational and industrial psychology*
Haywood, H(erbert) Carl(ton), *Mental retardation*
Herrans, Laura Leticia, *Educational psychology*
Heward, William Lee, *Special education*
Hill, A. Lewis, *Psychology*
Hilliard, Asa Grant, III, *Cultural bias and testing*
Hilton, Thomas Frederick, *Health psychology*
Hoffman, L. Richard, *Organizational psychology*
Howard, Kenneth Irwin, *Psychotherapy Research*
Huba, George John, *Psychology*
Huber, R. John, *Personality*
Hull, Diana, *Population and environment, media psychology*
Hyde, Janet Shibley, *Psychology of women*
Hyman, Irwin A., *Discipline, behavioral management of children, School psychology*
Jacobson, George Robert, *Addictive disorders*
Johnson, Richard Frederick, *Human Factors Psychology*
Johnson, Thomas Folsom, *Family therapy*
Jones, Lyle Vincent, *Psychometrics*
Jurjevich, Ratibor-Ray M., *Psychotherapy*
Kamen, Joseph M., *Consumer behavior*
Kanungo, Rabindra Nath, *Organizational psychology*
Kegan, Daniel L., *Organizational psychology*
Keys, Christopher Bennett, *Community psychology*
Khatena, Joe, *Creativity giftedness*
Kilmann, Ralph Herman, *Organizational sciences*
Kim, S. Peter, *Trans/cross-cultural medical psychology*
Kimmel, Ellen Bishop, *Status of women*
King, Dennis R., *Pain Management*
Kinghorn, Carol Ann, *Giftedness*
Kleinknecht, Ronald Arthur, *Health psychology*
Kodanaz, Hatice Altan, *Clinical-behavioral studies*
Komechak, Marilyn Gilbert, *Psychotherapy*
Kopelman, Richard Eric, *Industrial and organizational psychology*
Korabik, Karen Sue, *Program evaluation*
Krug, Samuel Edward, *Measurement/personality assessment*
Kupst, Mary Jo, *Psychological aspects of illness*
L'Abate, Luciano, *Family psychology*
Landa, Beth Kim, *Statistics in psychology*
Lasry, Jean-Claude Maurice, *Psychotherapy*
Latham, Gary Phillip, *Industrial-organizational psychology*
Leddick, George Russell, *Counseling Psychology*
Lefebvre, Richard Craig, *Health Psychology*
Levant, Ronald F., *Counseling psychology*
Lewine, Richard Ralph Jean, *Psychology research*
Lewis, Mary Anita, *Applied behavioral research*

Lindsay-Hartz, Janice, *Phenomenological pshchology*
Linville, Malcolm Eugene, Jr., *Counseling psychology*
Locascio, Joseph Jasper, *Evaluation research, Quantitative/statistical methodology*
Lofquist, Lloyd Henry, *vocational psychology*
Lowery, Carol Rotter, *Family therapy*
Lynds-Cherry, Patricia Gail, *School psychology*
Machtiger, Harriet Gordon, *Psychotherapy*
Malgady, Robert George, *Psychometrics/quantitative psychology*
Malone, Thomas Becker, *Experimental psychology*
Martindale, Colin Eugene, *Psychological Aesthetics, Psychological Hedonics*
Massad, Carolyn Emrick, *Psychometrics*
Matter, Darryl Edgar, *Educational psychology*
May, Eugene Pinkney, *health psychology*
Mayo, Clyde Calvin, *Industrial/organizational psychology*
McAbee, Thomas Allen, *Community psychology*
McGaghie, William Craig, *Medical psychology*
Medley, Donald Matthias, *Educational psychology*
Mefferd, Roy Balfour, Jr., *Human resources technology*
Merenda, Peter Francis, *Measurement and evaluation*
Miner, John Burnham, *Organizational psychology*
Mitchell, James Vincent, Jr., *Personality theory and assessment*
Moffie, Robert Wayne, *Clinical pathology*
Morganstein, Stanley, *Engineering psychology*
Morra, Michael Anthony, *Airport psychology*
Mulick, James Anton, *Mental retardation*
Muller, John Paul, *Psychoanalysis*
Myers, Ernest Ray, *Community psychology*
Nelson, Carnot Edward, *Organizational psychology*
Novick, Melvin Robert, *Psychometrics*
O'Keefe, Edward John, *Multimodal psychology*
Parker, Jerry Calvin, *Pain management*
Patenaude, Andrea Farkas, *The psychological aspects of critical care medicine*
Patterson, Robert Logan, *Psychotherapy*
Payne, Robert Walter, *Experimental psychopathology*
Peck, Robert F(ranklin), *Management psychology*
Pelc, Robert Edward, *Applied psychology*
Pennebaker, James Whiting, *Health psychology*
Phillips-Jones, Linda, *Psychological counseling*
Pine, Charles Joseph, *Health psychology*
Pinker, Steven Arthur, *Developmental psycholinguistics*
Pohlman, Edward Wendell, *Population psychology*
Poling, Alan Dale, *Behavioral pharmacology*
Polivy, Janet, *Personality/abnormal psychology*
Pollack, Stephen Lewis, *Systems theory*
Poltrock, Steven Edward, *Perception*
Popov, Dan, *Organizational/Clinical Psychology*
Rodensky, Robert L., *Child Abuse, Program Evaluation*
Rodgers, Joseph Lee, III, *Psychometrics*
Saleh, Shoukry Dawood, *Organizational behavior*
Schlenker, Barry Richard, *Personality*
Schlossberg, Harvey, *Forensic Psychology*
Shaffer, Howard Jeffrey, *Substance use and abuse*
Shepard, Roger Newland, *Experimental psychology*
Simpson, Dennis Dwayne, *Evaluation*
Sirgy, Magdy Joseph, *Consumer psychology*
Smith, Ira Austin, *Mental retardation/developmental disabilities*
Spohn, Herbert Emil, *Experimental psychopathology*
Stewart, Horace Floyd, *Transpersonal psychology*
Straussner, Joel Harvey, *Educational psychology*
Thompson, William Warren, *School psychology*
Turkat, David Mark, *Media psychology*
Tzeng, Oliver Chun Shun, *Cross-cultural Psychology, Psychosemantics.*
Urry, Vern William, *Test theory*
Wadeson, Harriet Claire, *Art psychotherapy, Psychotherapy*
Walsh, William Michael, *Family therapy*
Weinberg, Richard Alan, *School psychology*
White, Mary-Alice, *Psychology of electronic learning*
Wittenborn, John Richard, *Psychopathology*
Wohlfeil, Paul Frederick, *Health psychology*
Wright, George Nelson, *Rehabilitation psychology*
Yensen, Richard, *Psychotherapy/personality/consciousness*
Yoder, Janice Dana, *Psychology of women*

SPACE SCIENCE. *See also* ASTRONOMY, ENGINEERING, PHYSICS.

Aerospace engineering and technology
Adamczak, Robert Leonard
Adams, Karyl Ann
Adams, Ludwig
Adamson, Thomas Charles, Jr.
Aleshire, Merle J.
Allen, Jerry Michael
Amherd, Noel A.
Anderson, Dale Arden
Anderson, William Judson
Androulakis, John George
Ashley, Holt
Aubin, William M.
Bandyopadhyay, Promode Ranjan
Bareiss, Lyle Eugene
Bastedo, William Gardner
Battin, Richard Horace
Baumann, Robert Coile
Beggs, James Montgomery
Bell, Cornelius
Bertrando, Bertrand Robert
Birkitt, John Clair
Bisplinghoff, Raymond Lewis
Blasingame, Benjamin Paul
Bluestein, Theodore
Bodington, Sven Henry Marriot
Boileau, Oliver Clark
Borten, William H.
Bottema, Murk
Brennan, Donald Francis
Briggs, William Benajah
Brodsky, Robert Fox
Brody, Steven
Bromberg, Robert
Broome, Douglas Ralph, Jr.
Buddington, Patricia Arrington
Bunner, Alan Newton
Bussard, Robert William
Callens, E(arl) Eugene, Jr.
Chambliss, Joe Preston
Chuan, Raymond Lu-po
Cibosky, William
Clarke, Larry Denman
Cloud, James Douglas
Coblitz, David Barry
Cohen, Edward
Colborn, Joseph Nelson
Collins, Frank Gibson
Conley, Carolynn Lee
Cornell, Robert Witherspoon
Currie, Malcolm Roderick
Dana, Martin P.
Dash, Sanford Mark
Decker, Arthur John
Deloffre, Bernard Thierry
De Santo, Daniel Frank
Devereux, William Patrick
Dietenberger, Mark Anthony
Dirks, Leslie Chant
Dolling, David Stanley
Dowell, Earl Hugh
Drake, Michael Lee
Ducoffe, Arnold L.
Duggin, Michael John
Elliott, John Othniel
Faget, Maxime A(llan)
Fansler, Kevin Spain
Feemster, John Ronald
Fletcher, Leroy S(tevenson)
Fodor, Magda Maria
Foley, Cray Lyman
Foreman, Kenneth M(artin)
Fourney, M(ichael) E(ugene)
Friedman, Louis Dill
Friesen, Larry Jay
Garner, Andrew Morris
Gartrell, Charles Frederick
Gessow, Alfred
Gilruth, Robert Rowe
Glaser, Peter Edward
Goodell, John Boyden
Gould, Charles Laverne
Gravitz, Sidney I.
Griffin, Gerald Duane
Guastaferro, Angelo
Hall, Charles Frederick
Hanover, Paul Norden
Hansen, Grant Lewis
Hartig, Elmer Otto
Harvey, William Donald
Hasegawa, Tony Seisuke
Hawkins, Robert C(leo)
Heacock, E(arl) Larry
Headley, R. Paul
Hearth, Donald Payne
Hendrick, Ira Grant
Hendron, John Alden
Hertz, Richard Cornell
Hnatiuk, Bohdan Taras
Hoag, David Garratt
Hoff, Nicholas John
Holmes, Dyer Brainerd
Howard, William Eager, III
Hsieh, Tsuying Carl
Huang, Richard Shih-Chiu
Hurd, Walter Leroy, Jr.
Hussain, A.K.M. Fazle
Jahn, Robert George
Jeffreys, James Victor
Jeffs, George W.
Johnson, Dale Robert
Johnson, Robert Louis
Jones, Robert Thomas
Katsanis, Theodore
Kautzmann, William Elwood
Kelley, Albert Joseph
Kelley, Henry Joseph
Kellogg, Robert Lea
Kelsoe, Lynda Carol
Khadduri, Farid Majid
King, Kenneth Roy
Kinra, Vikram K.
Kirk, Robert L.
Klabosh, Charles
Klockzien, Vernon George
Krachman, Howard Ellis
Krassner, Jerry
Krejci, Robert Henry
Laheru, Ken Liem
Laurenson, Robert Mark
Leadon, Bernard Matthew
Lee, Bert Gentry
Leite, Richard Joseph
Loewy, Robert Gustav
Loth, John Lodewyk
Low, George Michael
Lucas, William Ray
Lundin, Bruce Theodore
Maatuk, Josef
Mager, Artur
Mark, Hans Michael
Martin, James Arthur
Mastronardi, Richard
McAuley, Van Alfon
Mc Carthy, John Francis, Jr.
McGregor, Wheeler Kesey, Jr.
McKay, Kenneth G(ardiner)
Mettler, Ruben Frederick
Mihas, Faquir Ullah
Miller, Rene Harcourt
Miller, Richard Keith
Milstead, Andrew Hammill
Mollicone, Richard Anthony
Moody, Elizabeth Anne
Morgan, Lucian Lloyd
Muirhead, Vincent Uriel
Nomura, Yasumasa
Noor, Ahmed Khairy
Oder, Frederic Carl Emil
Pai, Shih I.
Paine, Thomas Otten
Park, Chul
Paschall, Lee McQuerter
Pearce, Malcolm Bulkeley, Jr.
Pearson, Jerome
Perkins, Porter J., Jr.
Peterson, James Robert
Petrone, Rocco A.
Phinney, Ralph E(dward)
Pickering, William Hayward
Porter, Frederick Charles
Psaros, George Emanuel
Purser, Paul Emil
Puskas, Elek
Randolph, James Eugene
Reese, Bruce Alan
Reitsema, Harold James
Ritter, Alfred
Rogers, Thomas Francis
Rudiger, Carl Ernest, Jr.
Ryker, Norman Jenkins, Jr.
San Juan, Eduardo Carrion
Santi, Gino P.
Sargent, Ernest Douglas
Schairer, George Swift
Schwanhausser, Robert Rowland
Sentman, Lee Hanley
Sewell, Kenneth Glenn
Shea, Joseph Francis
Sherrill, Thomas Joseph
Shkedi, Zvi
Silverstein, Elliot Morton
Simitses, George John
Simonetti, Joseph Lawrence
Singer, Timothy James
Skurla, George Martin
Strain, John Willard
Swanson, Paul Norman
Tai, Tsze Cheng
Talapatra, Dipak Chandra
Tapley, Byron Dean
Thomas, Richard Eugene
Throner, Guy Charles
Townsend, John William, Jr.
Townsend, Marjorie Rhodes
Turchan, Otto Charles
Turner, Earl James
Vaughan, William Walton
Vlay, George John
Wagner-Bartak, Claus Gunther Johann
Ward, Jerry Mack
Wattendorf, Frank Leslie
White, Charles Olds
White, John Evans
Williams, Ronald Oscar
Williams, Walter Charles, Jr.
Wolfe, Howard Francis
Young, A. Thomas
Yu, Yi-Yuan

Astronautics
Armstrong, Neil A.
Boatright, Tony James
Ditto, Frank Haselwood
Donivan, Frank Forbes, Jr.
Fellows, W(alter) Scott, Jr.
Fletcher, James Chipman
Freeman, Marsha Gail
Friedman, Louis Dill
Gavin, Joseph Gleason, Jr.
Gilruth, Robert Rowe
Hearth, Donald Payne
Hnatiuk, Bohdan Taras
Holland, Robert Louis
Jastrow, Robert
Lanzano, Paolo
Moore, Jesse William
Nigam, Rajendra C.
Oesterwinter, Claus
Pearson, Jerome
Peters, Charles Frederick
Randolph, James Eugene
Schairer, George Swift
Williams, James Gerard

Gravitational biology
Baker, R. Ralph
Heinrich, Milton Rollin
Kaufman, Peter Bishop
Laughlin, Maurice Harold
Loeppky, Jack Albert
Miller, Alan Douglas
Parker, Donald Edward
reid, Donald House

Satellite studies
Akers, Stuart William
Androulakis, John George
Baker, Kile Barton
Bareiss, Lyle Eugene
Barkley, Linda Dorothy
Bering, Edgar Andrew, III
Bertrando, Bertrand Robert
Blevis, Bertram Charles
Boudreau, Robert Donald
Brennan, Donald Francis
Brodsky, Robert Fox
Buck, Richard Forde
Cahill, Laurence James, Jr.
Calio, Anthony John
Chappell, Charles Richard
Deloffre, Bernard Thierry
Duggin, Michael John
Dumas, Herbert Monroe, (Jr.)
El-Baz, Farouk
Freeman, John Wright, Jr.
Fremouw, Edward Joseph
Gaposchkin, Edward Michael
Gartrell, Charles Frederick
Goel, Narendra Swarup
Gould, Charles Laverne
Griffiths, Richard Edwin
Gustafson, Bo Ake Sture
Hernandez, Samuel P.
Hurd, Walter Leroy, Jr.
Hyson, Michael Terry
Johnson, Francis Severin
Johnson, Richard Dameran
Kautzmann, William Elwood
Kellogg, Robert Lea
King, Kenneth Roy
Knowles, Stephen Howard
Kondo, Yoji
Lambert, John Vincent
Landecker, Peter Bruce
Lanzerotti, Louis J.
Largman, Kenneth
Laubscher, Roy Edward
Lieske, Jay Henry
Lupo, Michael Vincent
Masursky, Harold
McCluskey, George Eadon, Jr.
Milstead, Andrew Hammill
Mollicone, Richard Anthony
Nelson, Robert M.
Papagiannis, Michael D.
Peters, Charles Frederick
Peterson, Vern LeRoy
Pieper, George F(rancis)
Pike, Charles P.
Porter, John Robert, Jr.
Pritchard, Wilbur Louis
Radoski, Henry Robert
Reiff, Patricia Hofer
Rigterink, Paul Vernon
Rudiger, Carl Ernest, Jr.
Schwartzkopf, Steven Henry
Shawhan, Stanley Dean
Sherrill, Thomas Joseph
Smith, Sheldon Magill
Spiegel, Stanley Lawrence
Tapley, Byron Dean
Thieman, James Richard
Torr, Marsha Russell
Tousey, Richard
Travis, Larry Dean
Van Allen, James Alfred
Vlay, George John
Vonder Haar, Thomas Henry
Walker, Russell Glenn
Williams, Carol Ann
Williams, Donald John
Wilson, William Stanley

Solar physics
Anderson, Kinsey A.
Bai, Taeil
Baker, Kile Barton
Bame, Samuel Jarvis, Jr.
Barnes, Aaron
Baum, Peter Joseph
Beckers, Jacques Maurice
Blake, Richard Lee
Bohlin, John David
Brunish, Wendee M.
Chasson, Robert Lee
Claflin, Robert Malden
Clark, Thomas Alan
Cook, John W.
Davis, John Moulton
Davis, Raymond, Jr.
Edberg, Stephen J.
Emslie, A. Gordon
Fisher, Richard Royal
Friedman, Herbert
Gilliland, Ronald Lynn
Glackin, David Langdon
Gosling, John Thomas
Gould, Thurman Woodrow, Jr.
Hathaway, David Henry
Hildner, Ernest Gotthold, III
Hill, Henry Allen
Hinata, Satoshi
Hollweg, Joseph Vincent
Holman, Gordon Dean
Hoyt, Douglas Vincent
Hurford, Gordon James
Jackson, Bernard Vernon
Jefferies, John Trevor
Jordan, Stuart Davis
Joselyn, Jo Ann
Kan, Joseph R.
Kopp, Roger Alan
Kundu, Mukul R.
Landecker, Peter Bruce
Lee, Martin Alan
Levine, Randolph Herbert
Lin, Robert Peichung
Livingston, William Charles
Low, Boon-Chye
Mac Queen, Robert Moffat
Mariska, John Thomas
Michels, Donald Joseph
Mohler, Orren Cuthbert
Mutschlecner, Joseph Paul
Newkirk, Gordon Allen, Jr.
Perez-Peraza, Jorge A.
Poland, Arthur Ira
Prince, Helen Dodson
Sawyer, Constance Bragdon
Sitterly, Charlotte Moore
Skelton, Dennis Lee
Smith, Dean F.
Snider, Joseph Lyons
Stein, Robert Foster
Strong, Keith Temple
Tandberg-Hanssen, Einar Andreas
Teske, Richard Glenn
Thomas, John Howard
Tousey, Richard
Vegors, Stanley Henry, Jr.
Watkins, Sallie Ann
Wentzel, Donat Gotthard
Wilcox, John Marsh
Willson, Richard Clayton
Zirker, Jack Bernard

Space chemistry
Allen, Mark Andrew
Anders, Edward
Bohme, Diethard Kurt
Borg, Richard John
Davis, Raymond, Jr.
Kaplan, Irving Eugene
Khare, Bishun Narain
Lipschutz, Michael Elazar
Moore, Carleton Bryant
Ponnamperuma, Cyril Andrew
Scattergood, Thomas W.
Suess, Hans Eduard
Thiemens, Mark H.
Turkevich, Anthony Leonid
Wilkening, Laurel Lynn

Space medicine
Beljan, John Richard
Billingham, John
Chambers, Lawrence Paul
Johnson, Richard Dameran
Luttges, Marvin Wayen
Moore-Ede, Martin Christopher
Popovic, Vojin
Shvartz, Esar
Winget, Charles Merlin
Young, Laurence Retman

Other
Adelman, Saul Joseph, *space science*
Allen, Joe Haskell, *Solar-Terrestrial Physics*
Allen, Lew, Jr., *Space science research management.*
Bame, Samuel Jarvis, Jr., *Space plasma physics*
Banks, Peter Morgan, *Space physics*
Barth, Charles Adolph, *Space physics*
Berlad, Abraham Leon, *Gravitational effects on Combustion*
Birmingham, Thomas Joseph, *Cosmic rays*
Boller, Bruce Raymond, *Magnetospheric physics and geomagnetism*
Bournia, Anthony, *Nuclear rocketry*
Cahill, Laurence James, Jr., *Space physics*
Conner, Jerry Power, *Space project management*
Deleeuw, J.H., *Aerospace research administration*
Dlhopolsky, Joseph Gerald, *Space psychology*
Elliott, John Othniel, *Unmanned vehicle launches*
Feinberg, Gerald, *Extraterrestrial life*
Golike, Ann Elizabeth, *Computers in satellite tracking systems*
Gosling, John Thomas, *Space plasma physics*
Grosch, Daniel Swartwood, *Space Biology*
Hannah, David, Jr., *Space Launch Service*
Hathaway, William Howard, *Space observatory operations and calibration*
Hesser, James E(dward), *Space astronomy*

Hlass, I. Jerry, *Space technology management*
Jenkins, Edward Beynon, *Space astronomy (ultraviolet)*
Joselyn, Jo Ann, *Solar-Terrestrial relationships*
Keehn, Neil Francis, *National security space systems*
Kessler, Donald Joe, *Orbital debris studies.*
Khadduri, Farid Majid, *Solid rocket propulsion*
King, David Quimby, *Electric propulsion*
Kirk, John Gallatin, *Orbital mechanics*
Klebesadel, Ray William, *Space instrumentation.*
Krimigis, Stamatios Mike, *Space plasma physics*
Largman, Kenneth, *Military space systems*
Martin, Leonard James, *Planetary imaging*
Michels, Donald Joseph, *Solar-terrestrial physics*
Neugebauer, Gerry, *Astronomy from space*
Roederer, Juan Gualterio, *Space physics*
Rogers, Thomas Francis, *Space communications*
Sugiura, Masahisa, *Space plasma physics*
Thomas, Richard Eugene, *Aircraft aerodynamics*
Tien, Chang Lin, *Thermophysics*
Vinti, John Pascal, *Astrodynamics*
Walt, Martin, *Magnetospheric Physics*
Walt, Martin, IV, *Space Plasma physics*
Weinberg, Jerry Lloyd, *Space astronomy*
Willson, Richard Clayton, *Solar-terrestrial relationships*
Woo, Richard, *Solar wind*

VETERINARY MEDICINE. *See also* AGRICULTURE.

Biomedical engineering. ***See*** **ENGINEERING, Biomedical.**

Cancer research
Bankert, Richard Burton
Benjamin, Stephen Alfred
Bloch, Peter H.
Chang, Richard Li-chai
Cho, Byung-Ryul
Coppoc, Gordon Lloyd
Koller, Loren D.
Lee, Lucy Fang
MacCoy, Douglas Maidlow
Mayer, Ramona Ann
Merkley, David Frederick
Pedersen, Niels Christian
Purchase, Harvey Graham
Szebenyi, Emil Steven
Yang, Tsu-Ju

Embryo transplants. ***See*** **AGRICULTURE, Animal breeding, embryo transplants.**

Internal medicine
Bayly, Warwick Michael
Breitschwerdt, Edward Bealmear
Byars, T. Douglas
Davis, Lloyd Edward
Detweiler, David Kenneth
DiBartola, Stephen Paul
Gompf, Rebecca Elaine
Hoskins, Johnny Durr
Hull, Bruce Lansing
Kirk, Robert Warren
Marshak, Robert Reuben
Mohanty, Sashi B.
Pearson, Erwin Gale
Senior, David Frank
Vig, Madan Mohan
Wass, Wallace Milton
White, Maurice Edward
Younger, Melanie Moore
Zimmer, James Francis

Microbiology
Burger, Dieter
Claflin, Robert Malden
Coles, Embert Harvey, Jr.
Collins, Michael Thomas
Corbeil, Lynette Bundy
Dardiri, Ahmed Hamed
Dierks, Richard Ernest
Ferris, Deam Hunter
Fryer, John L.
Klesius, Phillip Harry
Larsen, Austin Ellis
Loomis, Edmond Charles
Matsumoto, Masakazu
McLennan, Jean Glinn
Myers, Lyle Leslie
Rosenbusch, Ricardo Francisco
Shadduck, John Allen
Sharma, Jagdev Mittra
Staples, George Emmett
Stephenson, Edward Hayes
Stoddard, Patricia Ann
Waffle, Elizabeth L.
Walker, Robert D.
Walton, Thomas Edward, Jr.
Yilma, Tilahun

Pathology
Benjamin, Stephen Alfred
Bickford, Arthur Alton
Bicknell, Edward J.
Brobst, Duane Franklin
Buergelt, Claus Dietmar
Cera, Lee Marie
Cheville, Norman Frederick
Coles, Embert Harvey, Jr.
Collier, Linda Lee
Cork, Linda K. Collins
Cramer, Donald Vernon
Hall, LeRoy Brooks, Jr.
Kier-Schroeder, Ann B.
Kim, Hyun Young
Kim, James C(hin) S(oo)
Kintner, Loren Don
Koller, Loren D.
Melby, Edward Carlos, Jr.
Moe, James Burton
Nagode, Larry Allen
Phemister, Robert David
Prestwood, Annie Katherine
Ragland, William Lauman, III
Robinson, Farrel Richard
Schiefer, Hans Bruno
Schwartz, Lester William
Shadduck, John Allen
Shively, James Nelson
Squire, Phil G.
Stoloff, David Robert
Torres-Medina, Alfonso
Wellings, Sefton Robert
White, Nathaniel Aldrich, II
Zook, Bernard Charles

Preventive medicine
Abinanti, Francis Ralph
Cypess, Raymond Harold
Dorn, C. Richard
El Dareer, Salah Mohammed
Erb, Hollis Nancy
Genigeorgis, Constantin
Gillespie, James Howard
Houser, Ronald Edward
McDaniel, Hugh Thomas, Jr.
McDougald, Larry Robert
Myers, Lyle Leslie
Reddy, Chada Sudershan
Schnurrenberger, Paul Robert
Thawley, David Gordon
Thedford, Thomas Ray
Underdahl, Norman Russell
Vadlamudi, Sri Krishna
White, Maurice Edward
Younger, Melanie Moore

Surgery
Bramlage, Lawrence Robert
Cholvin, Neal Robert
Colahan, Patrick Timothy
Crowe, Dennis Timothy, Jr.
Dennis, Melvin Best, Jr.
Egger, Erick Lowell
Ettinger, Helen Clarice
Hull, Bruce Lansing
Krahwinkel, Delbert Jacob, Jr.
MacCoy, Douglas Maidlow
Merkley, David Frederick
Miller, Joan Ellen
Pavletic, Michael Mark
Pearson, Phillip Theodore
Runyon, Caroline Louise
Skoryna, Stanley Constantine
Stoloff, David Robert
Vig, Madan Mohan
White, Nathaniel Aldrich, II
Wilson, George Porter, III

Virology
Abinanti, Francis Ralph
Avampato, James Erwin
Banerjee, Amiya Kumar
Burger, Dieter
Cho, Byung-Ryul
Dierks, Richard Ernest
Easterday, Bernard Carlyle
Falk, Lawrence Address, Jr.
Gainer, Joseph Henry
Gillespie, James Howard
Houser, Ronald Edward
Joklik, Wolfgang Karl
Kelling, Clayton Lynn
Kendal, Alan Philips
Kim, Hyun Young
Lee, Lucy Fang
Lehmkuhl, Howard Duane
Marquardt, Warren William
Mengeling, William Lloyd
Mohanty, Sashi B.
Morgan, Donald O'Quinn
Olsen, Richard George
Purchase, Harvey Graham
Sidwell, Robert William
Thawley, David Gordon
Torres-Medina, Alfonso
Walton, Thomas Edward, Jr.
Wood, Owen Leslie
Woode, Gerald nottidge

Other
Armistead, Willis William, *Administration, veterinary medical research*
Bachrach, Howard L.
Barta, Ota, *Clinical Immunology*
Bratton, Gerald Roy, *Veterinary anatomy, Veterinary toxicology*
Clark, James Derrell, *Laboratory Animal Medicine, Toxicology*
Cypess, Raymond Harold, *Parasitology*
Czarnecki, Caroline Mary Anne, *Histology, Veterinary anatomy*
Diesem, Charles David, *Ophthalmology*
Dieterich, Robert Arthur, *Wildlife diseases*
Faith, Robert Earl, Jr., *Laboratory animal medicine*
Fox, James Gahan, *Laboratory animal medicine*
Gompf, Rebecca Elaine, *Veterinary cardiology*
Gustafsson, Borje Karl, *Theriogenology*
Heavner, James Edward, *Veterinary Neurology*
Hornof, William J., *Veterinary Medical Research*
Hsu, Walter Haw, *Veterinary Pharmacology*
Hullinger, Ronald Loral, *Veterinary anatomy*
Jacoby, Robert Ottinger, *laboratory animal medicine*
Kincaid, Steven Alan, *Veterinary anatomy*
Klide, Alan Marshall, *Veterinary Anesthesiology*
Krahwinkel, Delbert Jacob, Jr., *Veterinary anesthesiology*
Larsen, Austin Ellis, *Laboratory Animal Medicine*
McDaniel, Hugh Thomas, Jr., *Reproduction*
McEntee, Kenneth, *Comparative Reproductive Pathology*
Melby, Edward Carlos, Jr., *Laboratory animal medicine*
Mengeling, William Lloyd
Miselis, Richard R., *Anatomy, neuroanatomy*
Mishra, Shri Kant
Munger, Robert John, *Veterinary ophthalmology*
Myer, Carole Wendy, *Veterinary radiology*
Oliver, Jack (Wallace), *Veterinary pharmacology*
Pineda, Mauricio Hernan, *Animal contraception*
Saini, Ravinder Kumar, *Veterinary cardiology*
Schneider, Norman Richard, *Diagnostic veterinary toxicology*
Schultz, Ronald David, *Immunology*
Schwark, Wayne Stanley, *Veterinary pharmacology*
Skarda, Roman Thomas, *Vet. med. anesthesia*
Staples, George Emmett, *Dietary therapeutics for animals*
Thedford, Thomas Ray, *wildlife diseases-foreign animal diseases*
Trim, Cynthia Mary, *Veterinary Anesthesiology*
Williams, David John, III, *Theriogenology*
Zinn, Gene Marvin, *Food Animal Medicine and Surgery*

ZOOLOGY. *See* BIOLOGY.